# 2021 County and City Extra
## Annual Metro, City, and County Data Book
### *29th Edition*

# 2021 County and City Extra

## Annual Metro, City, and County Data Book

*29th Edition*

Edited by Deirdre A. Gaquin
and Mary Meghan Ryan

Lanham • Boulder • New York • London

Published by Bernan Press
An imprint of The Rowman & Littlefield Publishing Group, Inc.
4501 Forbes Boulevard, Suite 200, Lanham, Maryland 20706
www.rowman.com

86-90 Paul Street, London EC2A 4NE

ISBN 978-1-63671-000-6 (hardback)
ISBN 978-1-63671-001-3 (ebook)

♾™ The paper used in this publication meets the minimum requirements of American National Standard for Information Sciences—Permanence of Paper for Printed Library Materials, ANSI/NISO Z39.48-1992.

# Contents

Page

vii  **Introduction**
New and Updated Information for the 2020 Edition
Rankings
Subjects Covered and Volume Organization
Symbols
Sources

xi  **Subjects Covered, by Geography Type**

xiv  **National Data Maps**

1  **Part A. States**
3  State Highlights and Rankings
17  State Column Headings
24  Table A

51  **Part B. States and Counties**
53  County Highlights and Rankings
69  State and County Column Headings
73  Table B

773  **Part C. Metropolitan Areas**
775  Metropolitan Area Highlights and Rankings
793  Metropolitan Area Column Headings
797  Table C

895  **Part D. Cities of 25,000 or More**
897  City Highlights and Rankings
907  City Column Headings
910  Table D

1175  **Part E. Congressional Districts of the 116th Congress**
1177  Congressional District Highlights and Rankings
1184  Congressional District Column Headings
1186  Table E

**Appendices**
A-1  A. Geographic Concepts and Codes
B-1  B. Metropolitan Statistical Areas, Metropolitan Divisions, and Components (as defined September 2018)
C-1  C. Core Based Statistical Areas (Metropolitan and Micropolitan), Metropolitan Divisions, and Components (as defined March, 2020)
D-1  D. Changes to Metropolitan Areas in the Definitions of OMB Bulletin 18-04
E-1  E. Cities by County
F-1  F. Source Notes and Explanations

# INTRODUCTION

*County and City Extra* is an annual publication that provides the most up-to-date statistical information available for every state, county, metropolitan area, and congressional district, as well as all cities in the United States with a 2010 census population of 25,000 or more. Data for places, including towns and cities with populations of fewer than 25,000 people are published by Bernan Press in a separate companion volume, *Places, Towns and Townships,* now in its seventh edition. These two volumes are designed to meet the needs of libraries, businesses, and other organizations or individuals who desire convenient and timely sources of the most frequently sought information about geographic entities within the United States. The annual updating of *County and City Extra* for 29 years ensures its stature as a reliable and authoritative source for statistical information.

Bernan Press also publishes a companion volume, the *State and Metropolitan Area Data Book,* previously published by the Census Bureau. The recently published third edition provides an expanded collection of data about states and metropolitan areas, including micropolitan areas and their component counties. Another recent addition is the *County and City Extra: Special Historical Edition, 1790-2010* with data from the earliest days of the nation, and states, counties, and cities from their beginnings.

*County and City Extra, Places, Towns and Townships,* and *State and Metropolitan Area Data Book* are large volumes, but are not big enough to accommodate the wealth of information from the decennial census and the American Community Survey. Two additional volumes in the *County and City Extra* series include this information. *County and City Extra—Special Decennial Census Edition* provides detailed population and housing data from the 2010 census and was published by Bernan Press in December 2011. *The Who, What, and Where of America—Understanding the American Community Survey*, recently released in its sixth edition, includes social and economic details from the ongoing American Community Survey, and the *County and City Extra* Series includes additional books on special topics, such as the recently published *Education and the American Workforce.*

The American Community Survey (ACS) is a national survey that has replaced the census long form as the key source of detailed social and economic data. *County and City Extra* includes data from both the 2010 census and the ACS.

## New and Updated Information for the 2021 Edition

This edition includes state data from the 2020 election. Updated data include 2020 population estimates for states, counties, metropolitan areas, and cities. Also included are the latest available data for education, vital statistics, income and poverty, employment and unemployment, residential construction, production by industry, health resources, crime, land use, city government finances, and many other topics.

Table E (Congressional Districts) includes a wide selection of 2019 American Community Survey data, Business Patterns, and Social Security data for the 116th Congress, as well as data from the 2017 Census of Agriculture for the congressional districts of the 116th Congress, along with the 117th Congressional representatives.

In September 2018, the Office of Management and Budget released an updated list of Core Based Statistical Areas (metropolitan and micropolitan areas) based on the 2010 census and some changes in the way these areas are defined. These changes were substantial, resulting in many new names and component counties. The 2020 population estimates, and a few other key data sources for this book used the new list of metropolitan areas. For other sources that still used the earlier metropolitan definitions, we aggregated the county data so that all metropolitan area data use the new definitions. Appendixes B and C provide details about the component counties of these metropolitan and micropolitan areas and their 2010 census and 2020 estimated populations. Appendix D lists all the changes made since the earlier delineations, and provides a link to a map of the new metropolitan areas.

This edition includes data from the 2010 census, 2020 population estimates, and the ACS. Annual ACS data are available for all states and all metropolitan areas (all geographic areas with populations of 65,000 or more), but five years are needed to build a sample large enough for reliable estimates for all counties The Census Bureau no longer releases 3-year data which were previously available for areas with populations of 20,000 or more. These have now been replaced with 1-year Supplemental Estimates which are used in this book for Table D, cities with populations of 25,000 or more.

With the now annual release of 1-year and 5-year estimates, *County and City Extra* uses ACS data for all geographic areas. ACS 1-year data for 2019 are included in Table A (States), metropolitan areas (Table C), and Table E (Congressional Districts)—all areas with populations of 65,000 or more. Table D (cities) uses the 1-year supplemental estimates which are less detailed than the regular 1-year estimates. Table B (Counties) includes 5-year data (2015–2019). The release of 5-year data for even the smallest geographic areas means that annual social and economic characteristics are available for all counties and cities.

Although some of the state data are also included in Table B (States and Counties), the separate state data table offers several important features:

- Additional data not available at the county level are provided. Examples include population projections, health insurance coverage, number of immigrants, personal tax payments, information about health service firms not subject to federal tax, and exports by state of origin.

- Additional data that exceeds the space limitations for counties can be found for states. Examples are age of householder, more detailed information about employment in retail trade and services, and detailed government employment and finance data.

- State totals can be found more quickly and compared more readily.

Appendix F, **Source Notes and Explanations**, includes internet references for all data sources. This is especially helpful in today's environment where the data sources are updated at a faster pace. The sources referenced here can be used to track down additional information too cumbersome for this book. Some of the data can be directly found in data tables on the websites; some can be assembled through on-line access tools; others can be obtained by downloading files and processing them with statistical software; and some need to be ordered from the agencies.

## Rankings

The rankings present the geography types by various subjects, including population, land area, population density, population change, age, immigration, birth rate, housing characteristics, race, Hispanic origin, educational attainment, income, unemployment rate, per capita local taxes, poverty rate, defense contracts, value of agricultural products, and violent crime rate.

## Subjects Covered and Volume Organization

A summary of the **subjects covered** in each of the five tables appears on **page xi**. The **colored map** portfolio begins on **page xv**.

The main body of this volume contains five basic parts. Each part includes a table that is preceded by highlights and rankings, as well as the complete column headings for the table. **Part A**, which begins on **page 1**, contains data for states. **Part B**, beginning on **page XXX**, contains information for states and counties. The county geography codes include county typology codes from the Economic Research Service of the Department of Agriculture. These codes characterize counties by size of the largest place as well as by other criteria for nonmetropolitan counties. (See Appendix A for the definition of each code.) **Part C**, beginning on **page XXX**, contains information for metropolitan areas. Statistics for cities with a 2010 census population of 25,000 or more can be found in **Part D**, which begins on **page XXX**. **Part E**, beginning on **page XXX**, contains data for the congressional districts of the 116th Congress.

A contents page preceding tables B through E lists the page number on which the data for a given geographic area begin. Counties and cities are listed alphabetically by state. Metropolitan areas are listed alphabetically, except that metropolitan divisions are listed alphabetically within the metropolitan statistical area of which they are components. Congressional districts are listed in numeric order within states.

The appendixes include definitions of geographic concepts (**Appendix A**), sources and definitions of each data item included in this volume (**Appendix F**), an alphabetical listing of metropolitan areas with their component counties delineated as of September 2018, with 2010 census populations (**Appendix B**), a listing of metropolitan and micropolitan areas and their component counties as of March 2020, with 2010 census populations and 2020 estimated populations (**Appendix C**), a list of cities by county (**Appendix E**), and a list of the changes resulting in the new set of metropolitan areas (**Appendix D**).

## Symbols

**D**    Indicates that the number has been withheld to avoid disclosure of information pertaining to a specific organization or individual, or because the number does not meet statistical standards for publication.

**NA**    Indicates that data are not available.

**X**    Indicates that data are not applicable or are not meaningful for this geographic unit.

In this volume, a figure that is less than half the unit of measure shown will appear as zero.

## Sources

All of the data in this volume have been obtained from federal government sources. For a complete list of these sources, see **Appendix F**.

Data included in this volume meet the publication standards established by the U.S. Census Bureau and the other federal statistical agencies from which they were obtained. Every effort has been made to select data that are accurate, meaningful, and useful. All data from censuses, surveys, and administrative records are subject to errors arising from factors such as sampling variability, reporting errors, incomplete coverage, nonresponse, imputations, and processing error. Responsibility of the editors and publishers of this volume is limited to reasonable care in the reproduction and presentation of data obtained from sources believed to be reliable.

*County and City Extra: Annual Metro, City, and County Data Book* is part of Bernan Press's *County and City Extra* series. The editors of *County and City Extra* acknowledge the contributions of the late Courtenay Slater and George Hall, the originators of this publication. Their initial contributions continue to enrich the *County and City Extra* series. As always, we are especially grateful to the many federal agency personnel who assisted us in obtaining the data, provided excellent resources on their websites, and patiently answered questions.

**Deirdre A. Gaquin** has been a data use consultant to private organizations, government agencies, and universities for over 35 years. Prior to that, she was Director of Data Access Services at Data Use & Access Laboratories, a pioneer in private sector distribution of federal statistical data. A former President of the Association of Public Data Users, Ms. Gaquin has served on numerous boards, panels, and task forces concerned with federal statistical data and has worked on five decennial censuses. She holds a Master of Urban Planning (MUP) degree from Hunter College. Ms. Gaquin is also an editor of Bernan Press's *The Who, What, and Where of America: Understanding the American Community Survey*; *Places, Towns and Townships*; *The Congressional District Atlas*, *The Almanac of American Education, Race and Employment in America*, and *the State and Metropolitan Area Data Book*.

**Mary Meghan Ryan** is the senior research editor for Bernan Press. She is also the editor for the *Handbook of U.S. Labor Statistics*, *State Profiles*, and the associate editor for *Business Statistics of the United States*.

# SUBJECTS COVERED, BY GEOGRAPHY TYPE

State data begin on page 1
County data begin on page 51
Metropolitan area data begin on page 773
City data begin on page 895
Congressional district data begin on page 1175

| Subject | Table A. States | Table B. States and Counties | Table C. Metropolitan Areas | Table D. Cities | Table E. Congressional Districts |
|---|---|---|---|---|---|
| Land area | 1 | 1 | 1 | 1 | 1 |
| Population | | | | | |
| Total persons, 1990 | 31 | | | | |
| Total persons, 2000 | 32 | 20 | 20 | 23 | |
| Total persons, 2010 | 33 | 21 | 21 | 24 | |
| Total persons, 2019 | | | | | 2 |
| Total persons, 2020 | 2 | 2 | 2 | 2 | |
| Rank, 2020 | 3 | 3 | 3 | 3 | |
| Persons per square mile | 4 | 4 | 4 | 4 | |
| Race and Hispanic or Latino origin, 2010 | 45-50 | | | | |
| Race and Hispanic or Latino origin, 2019 | | | | 5-10 | 4-11 |
| Race and Hispanic or Latino origin, 2020 | 5–9 | 5-9 | 5-9 | | |
| Percent female | 21 | 19 | 19 | 22 | 12 |
| Foreign-born population | 22 | | | 11 | 13 |
| Percent born in state of residence | 23 | | | | 14 |
| Immigrants | 24 | | | | |
| Age distribution, 2010 | 52-61 | | | | |
| Age distribution, 2019 | | | | 12-20 | 15-23 |
| Age distribution, 2020 | 10-19 | 10-18 | 10-18 | | |
| Median age | 20, 62 | | | 21 | 24 |
| Percent population change, 1990–2000 | 34 | | | | |
| Percent population change, 2000–2010 | 35 | 22 | 22 | 25 | |
| Percent population change, 2010–2020 | 36 | 23 | 23 | 26 | |
| Components of population change | 37-41 | 24-26 | 24-26 | | |
| Daytime population | | 33-34 | 33-34 | | |
| Population projections | 42-44 | | | | |
| Households | | | | | |
| Total households, 2010 | 64 | | | | |
| Total households, 2019 | 25 | 27 | 27 | 27 | 28 |
| Total households, 2015–2019 | | 27-31 | | | |
| Percent change in number of households | 26, 65 | | | | |
| Household type | 28-30, 67-68 | 29-31 | 29-31 | 29-33 | 30-33 |
| Persons per household | 27, 66 | 28 | 28 | 28 | 29 |
| Persons in group quarters | | 32 | 32 | 34 | 34-39 |
| Housing | | | | | |
| Housing units in 2010 | 69-78 | | | 47-49 | |
| Housing units in 2019 | 79-92 | | 89-96 | 50-58 | 40-45 |
| Housing units in 2020 | | 87-88 | 87-88 | | |
| Housing units in 2015–2019 | | 89-96 | | | |
| Percent change in number of housing units | 70, 80 | 88 | 88 | 48 | |
| Housing costs | 73-77, 83-90 | 91-95 | 91-95 | 52 | 43-45 |
| Substandard housing units | 78, 91 | 96 | | | |
| Percent with computer and internet access | | | 96 | 57-58 | |
| Commuting patterns | | | | 55-56 | |
| Percent who lived in same house one year ago | 92 | | | 59 | |
| Percent who lived in different place one year ago | | | | 60 | |
| New residential construction | 93-95 | 169-170 | 169-170 | 69-71 | |
| Manufactured housing | 96 | | | | |

# SUBJECTS COVERED, BY GEOGRAPHY TYPE — Continued

State data begin on page 1
County data begin on page 51
Metropolitan area data begin on page 773
City data begin on page 895
Congressional district data begin on page 1175

| Subject | Column number | | | | |
| --- | --- | --- | --- | --- | --- |
| | Table A. States | Table B. States and Counties | Table C. Metropolitan Areas | Table D. Cities | Table E. Congressional Districts |
| **Vital statistics** | | | | | |
| Births | 97-98 | 35-36 | 35-36 | | |
| Deaths | 99-103 | 37-38 | 37-38 | | |
| **Health** | | | | | |
| Persons in nursing facilities | | | | 33 | 37 |
| Medicare enrollees | 106 | 41-43 | 41-43 | | |
| Persons lacking health insurance | 104-105 | 39-40 | 39-40 | | 59 |
| **Crime** | 107-110 | 44-47 | 44-47 | 35-38 | |
| **Education** | | | | | |
| School enrollment | 111-112 | 48-49 | 48-49 | | 25 |
| Educational attainment | 113-116 | 50-51 | 50-51 | 39-41 | 26-27 |
| Expenditures for education | 117-118 | 52-53 | 52-53 | | |
| **Income** | | | | | |
| Personal income | 134-149 | 62-71 | 62-71 | | |
| Per capita income | 122, 136, 149 | 54, 64 | 54, 64 | | 46 |
| Household income | 123-126 | 55-58 | 55-58 | 42-44 | 47-48 |
| Family and non-family income | | | | 45-46 | |
| Earnings by gender | | | | 47-49 | |
| Poverty | 127-133 | 59-61 | 59-61 | | 49-50 |
| Food stamps | | | | | 51 |
| Personal income by type | 138-140 | 66-70 | 66-70 | | |
| Earnings by industry | 150-158 | 72-83 | 72-83 | | |
| Transfer payments | 141-146 | 71 | 71 | | |
| Gross state product | 159 | | | | |
| Personal tax payments | 147 | | | | |
| Disposable personal income | 148-149 | | | | |
| Social Security | 160-162 | 84-86 | 84-86 | | 60-62 |
| Individual income taxes | | 197-199 | 197-199 | | |
| **Labor Force and Employment** | | | | | |
| Labor force and unemployment | 167-171 | 97-100 | 97-100 | 61-68 | 52-54 |
| Employment in selected occupations | 163-166 | 101-103 | 101-103 | | 55-58 |
| Employment by industry | 172-183, 207-216 | 104-112 | 104-112 | | 73-84 |
| **Exports of goods produced** | 119-121 | | | | |
| **Establishments, employment, sales, and payroll** | | | | | |
| Manufacturing | 207-216 | 151-154 | 151-154 | 88-91 | |
| Construction | 217-221 | | | | |
| Wholesale trade | 222-226 | 135-138 | 135-138 | 72-75 | |
| Retail trade | 227-235 | 139-142 | 139-142 | 76-79 | |
| Information | 236-246 | | | | |
| Utilities | 247-251 | | | | |
| Transportation and warehousing | 252-256 | | | | |
| Finance and insurance | 257-261 | | | | |
| Real estate and rental and leasing | 262-266 | 143-146 | 143-146 | 80-83 | |
| Professional, scientific, and technical services | 267-275 | 147-150 | 147-150 | 84-87 | |
| Health care and social assistance | 276-289 | 159-162 | 159-162 | 100-103 | |

# SUBJECTS COVERED, BY GEOGRAPHY TYPE — Continued

State data begin on page 1
County data begin on page 51
Metropolitan area data begin on page 773
City data begin on page 895
Congressional district data begin on page 1175

| Subject | Column number | | | | |
| --- | --- | --- | --- | --- | --- |
| | Table A. States | Table B. States and Counties | Table C. Metropolitan Areas | Table D. Cities | Table E. Congressional Districts |
| Arts, entertainment and recreation | 290-294 | | | 96-99 | |
| Accommodation and food services | 295-300 | 155-158 | 155-158 | 92-95 | |
| Other services, except public administration | 301-308 | 163-166 | 163-166 | 104-108 | |
| Nonemployer businesses | | 167-168 | 167-168 | | |
| Government employment | 309-314 | 171, 194-196 | 171, 194-196 | 108 | |
| Government payroll | 315-330 | 172-179 | 172-179 | 109-116 | |
| Government finances | 331-350 | 180-193 | 180-193 | 117-139 | |
| Agriculture | 184-202 | 113-132 | 113-132 | | 63-72 |
| Land and water | 203-206 | 133-134 | 133-134 | | |
| Voting and elections | 351-355 | | | | |
| Climate | | | | 140-146 | |

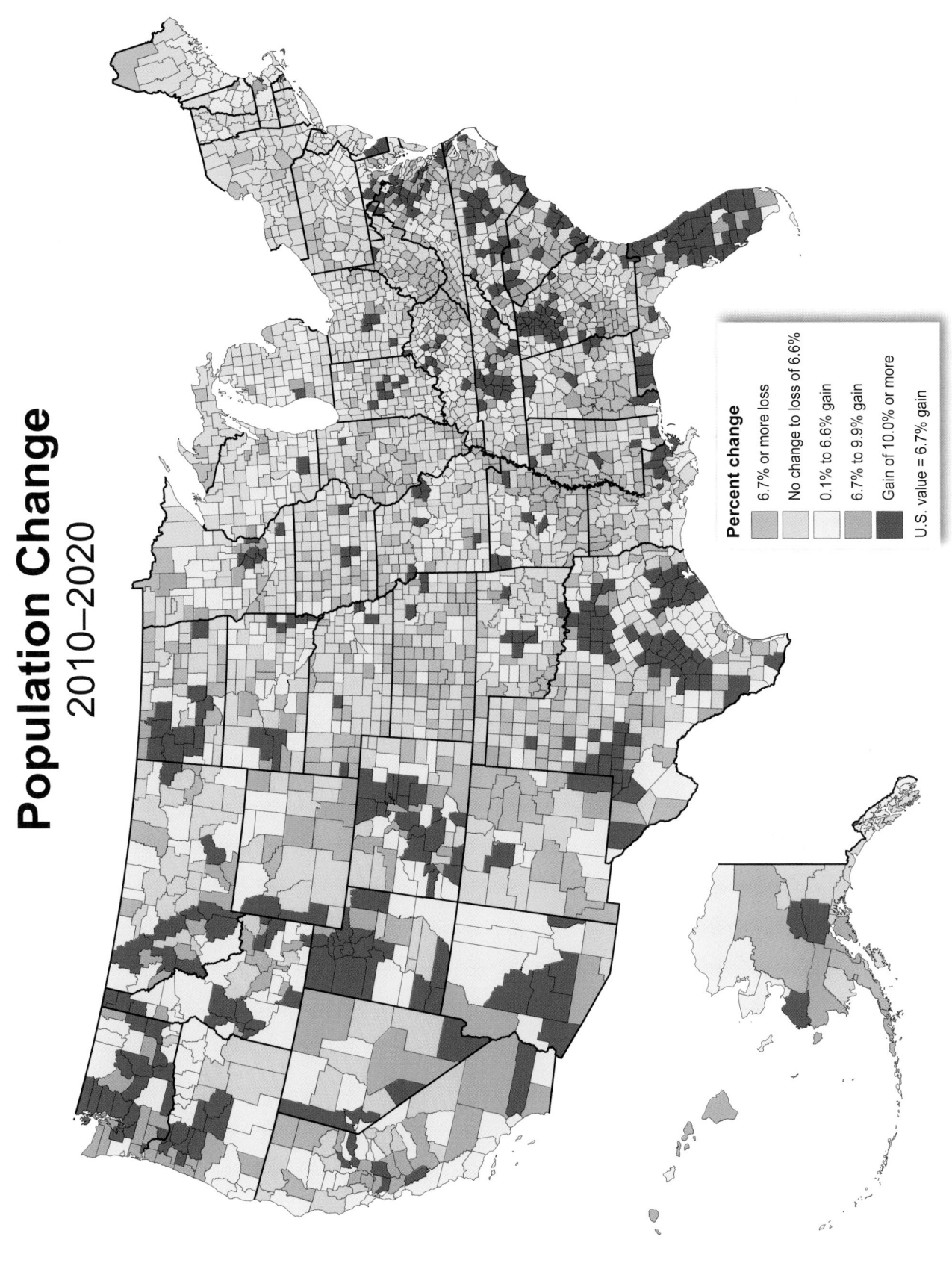

# Population Change
## 2010–2020

**Percent change**

6.7% or more loss

No change to loss of 6.6%

0.1% to 6.6% gain

6.7% to 9.9% gain

Gain of 10.0% or more

U.S. value = 6.7% gain

# Black, Not Hispanic or Latino, Population
## 2020

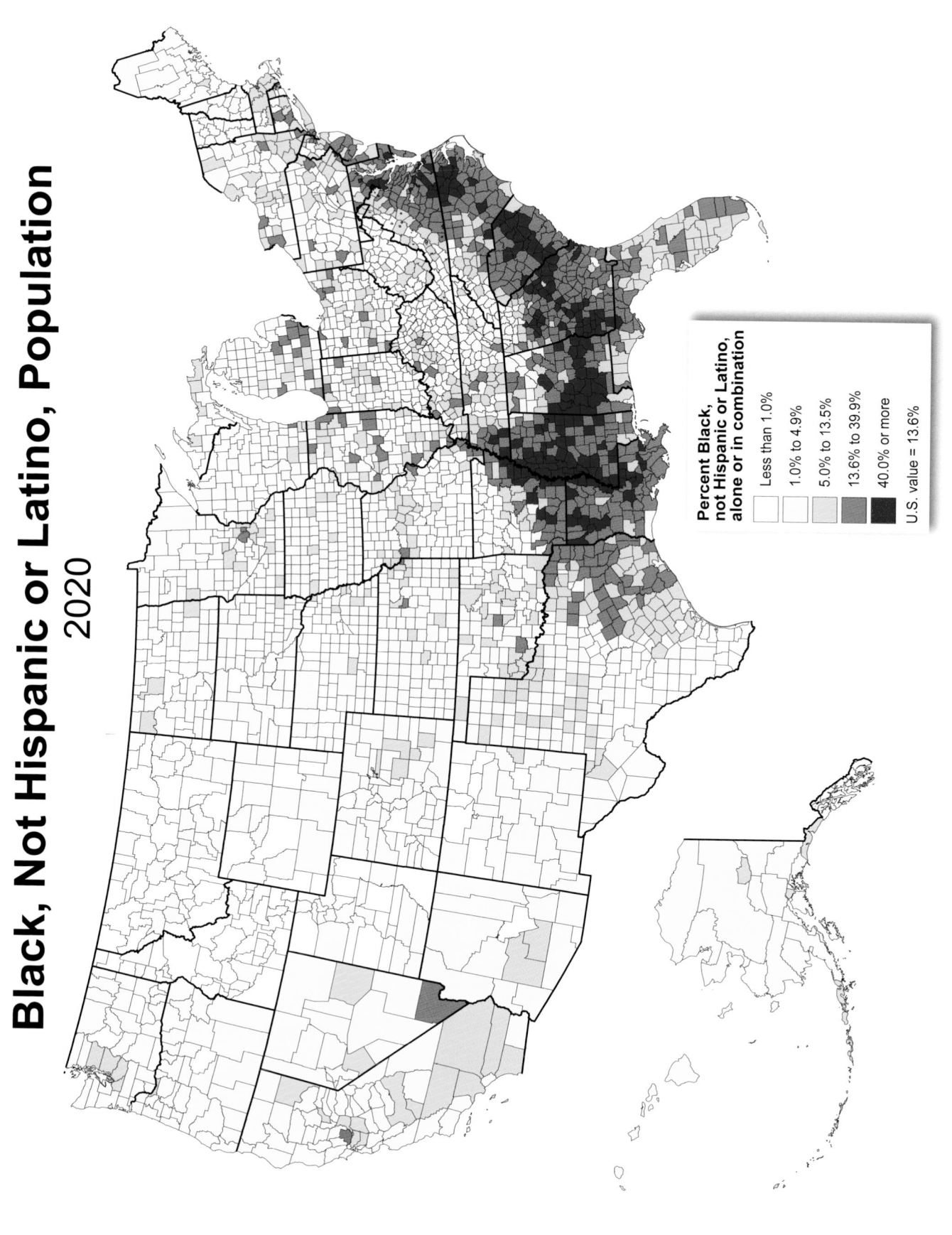

Percent Black,
not Hispanic or Latino,
alone or in combination

Less than 1.0%

1.0% to 4.9%

5.0% to 13.5%

13.6% to 39.9%

40.0% or more

U.S. value = 13.6%

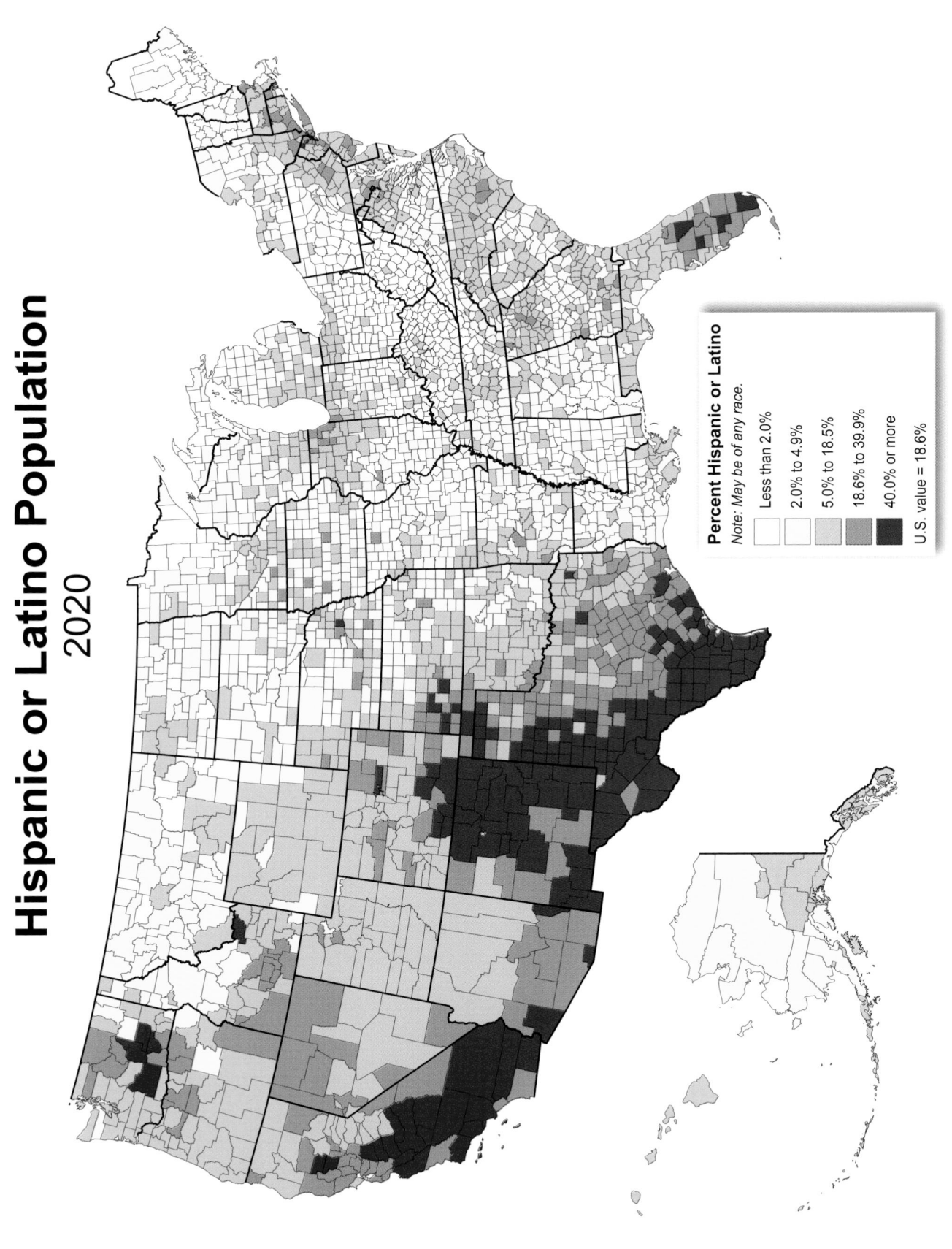

# Hispanic or Latino Population
## 2020

**Percent Hispanic or Latino**
*Note: May be of any race.*

Less than 2.0%
2.0% to 4.9%
5.0% to 18.5%
18.6% to 39.9%
40.0% or more

U.S. value = 18.6%

# Population Under 15 Years Old
## 2020

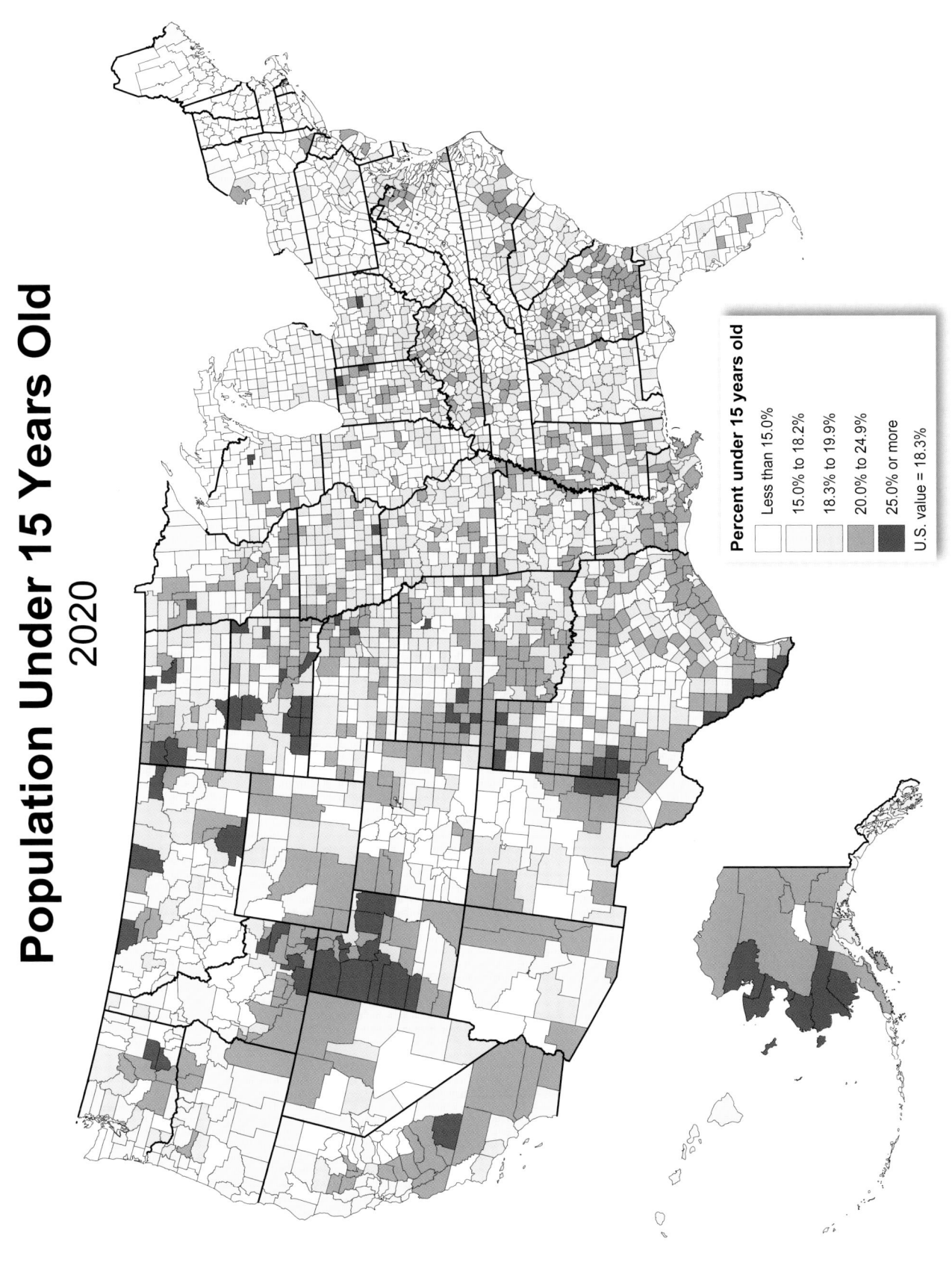

**Percent under 15 years old**

- Less than 15.0%
- 15.0% to 18.2%
- 18.3% to 19.9%
- 20.0% to 24.9%
- 25.0% or more

U.S. value = 18.3%

# Population 65 Years Old and Over
## 2020

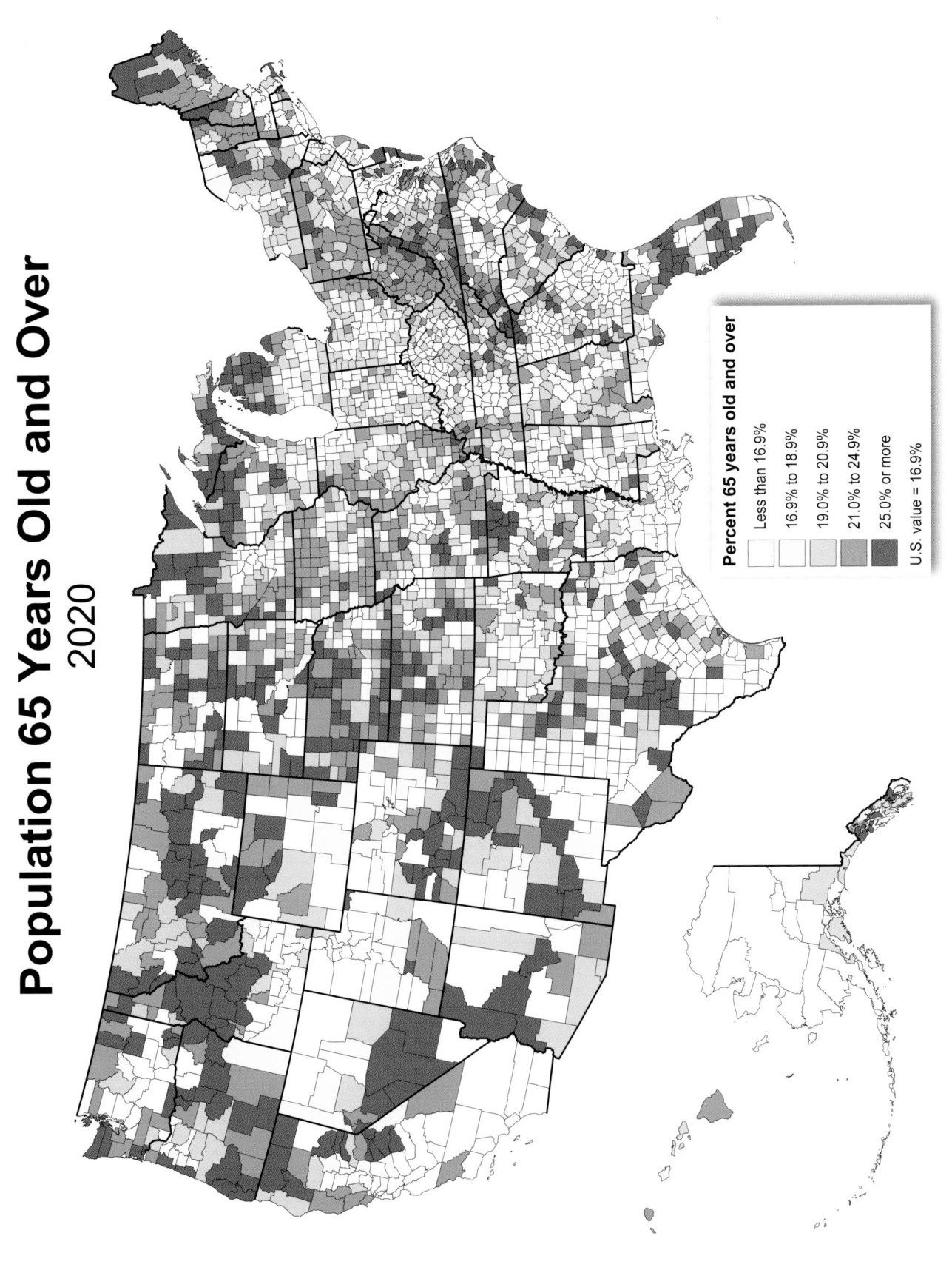

**Percent 65 years old and over**

Less than 16.9%

16.9% to 18.9%

19.0% to 20.9%

21.0% to 24.9%

25.0% or more

U.S. value = 16.9%

# Population Density

## 2020

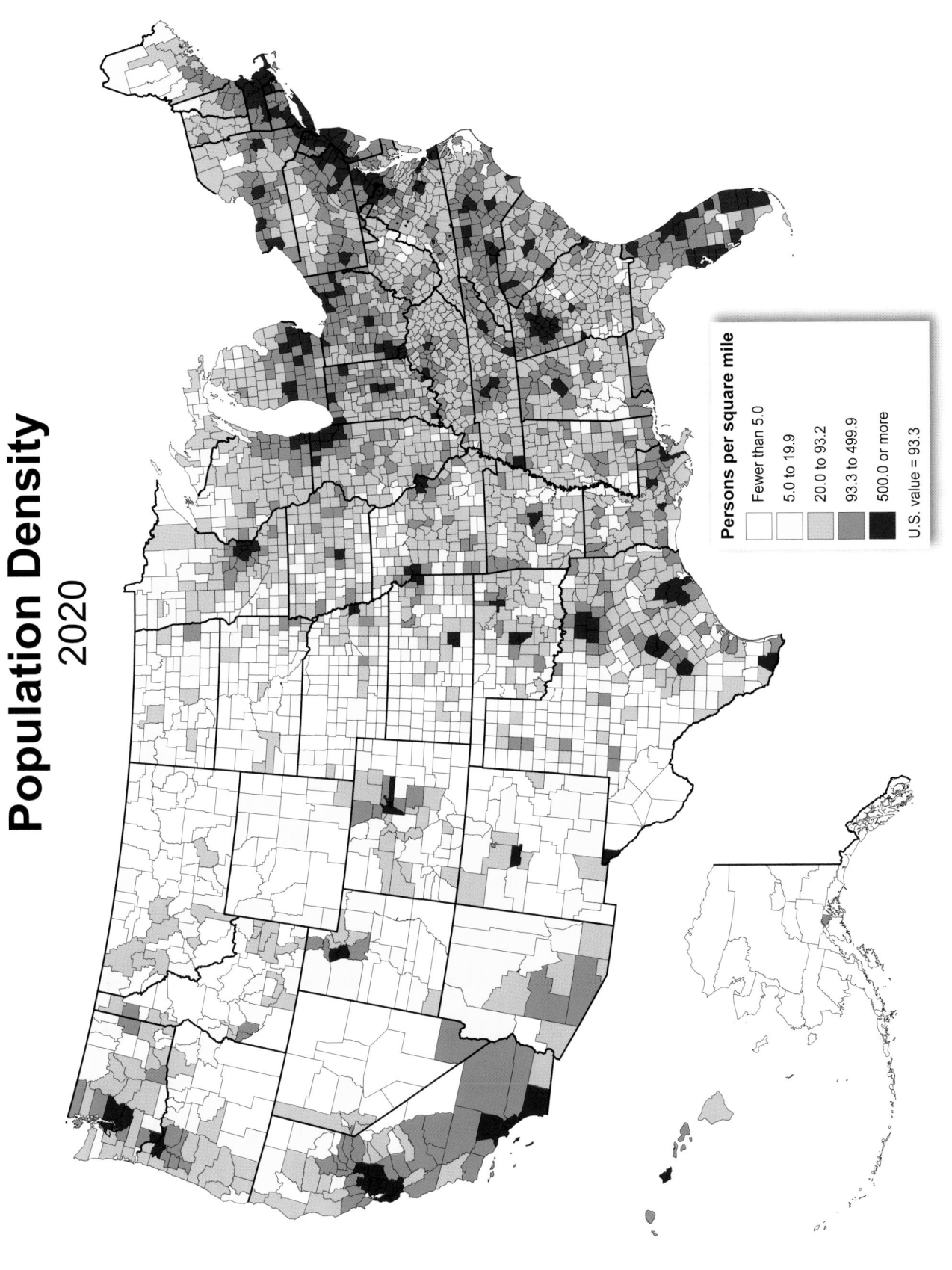

**Persons per square mile**

Fewer than 5.0
5.0 to 19.9
20.0 to 93.2
93.3 to 499.9
500.0 or more

U.S. value = 93.3

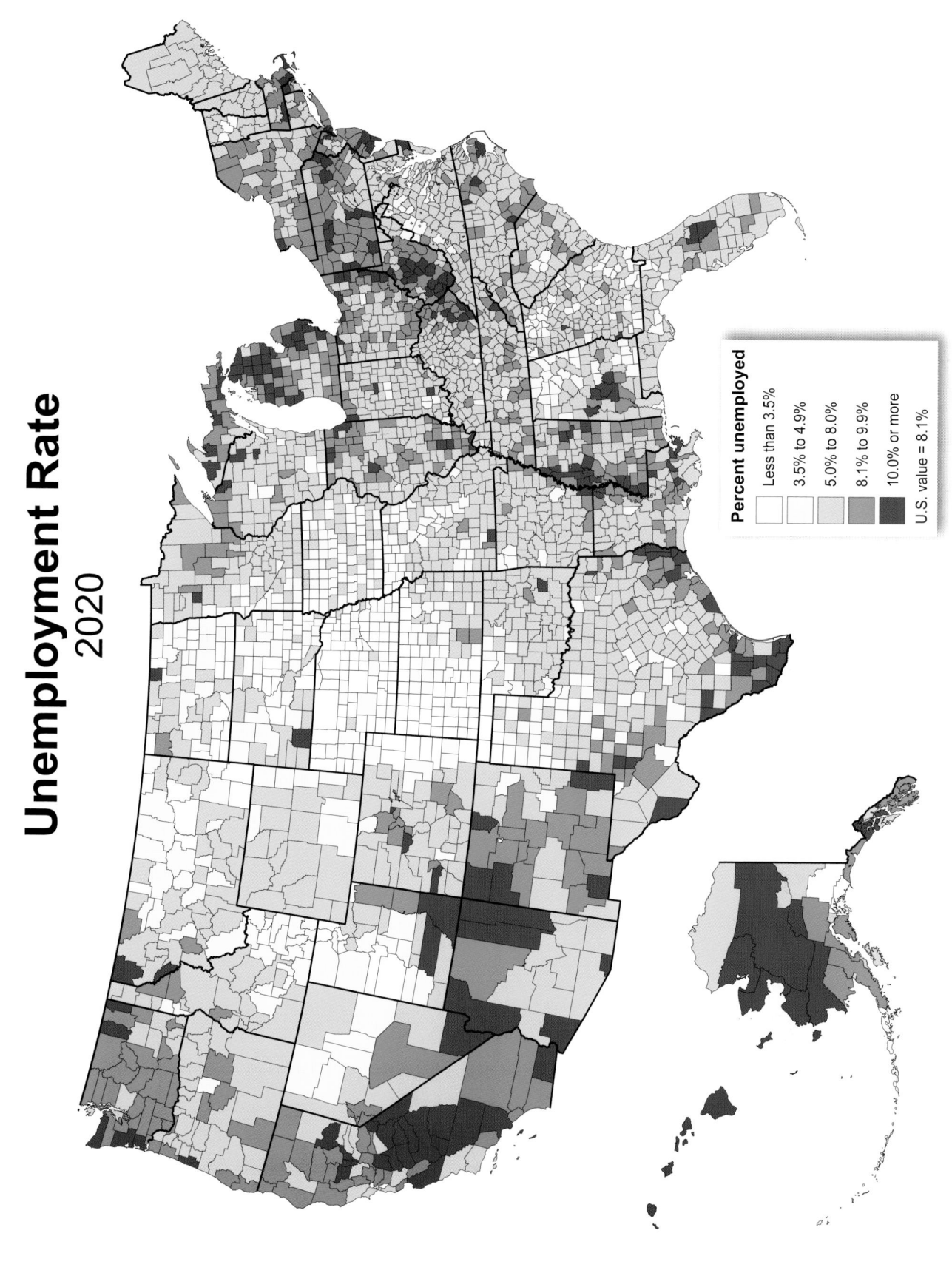

# Unemployment Rate
## 2020

**Percent unemployed**

Less than 3.5%
3.5% to 4.9%
5.0% to 8.0%
8.1% to 9.9%
10.0% or more

U.S. value = 8.1%

# Earnings from Manufacturing
## 2019

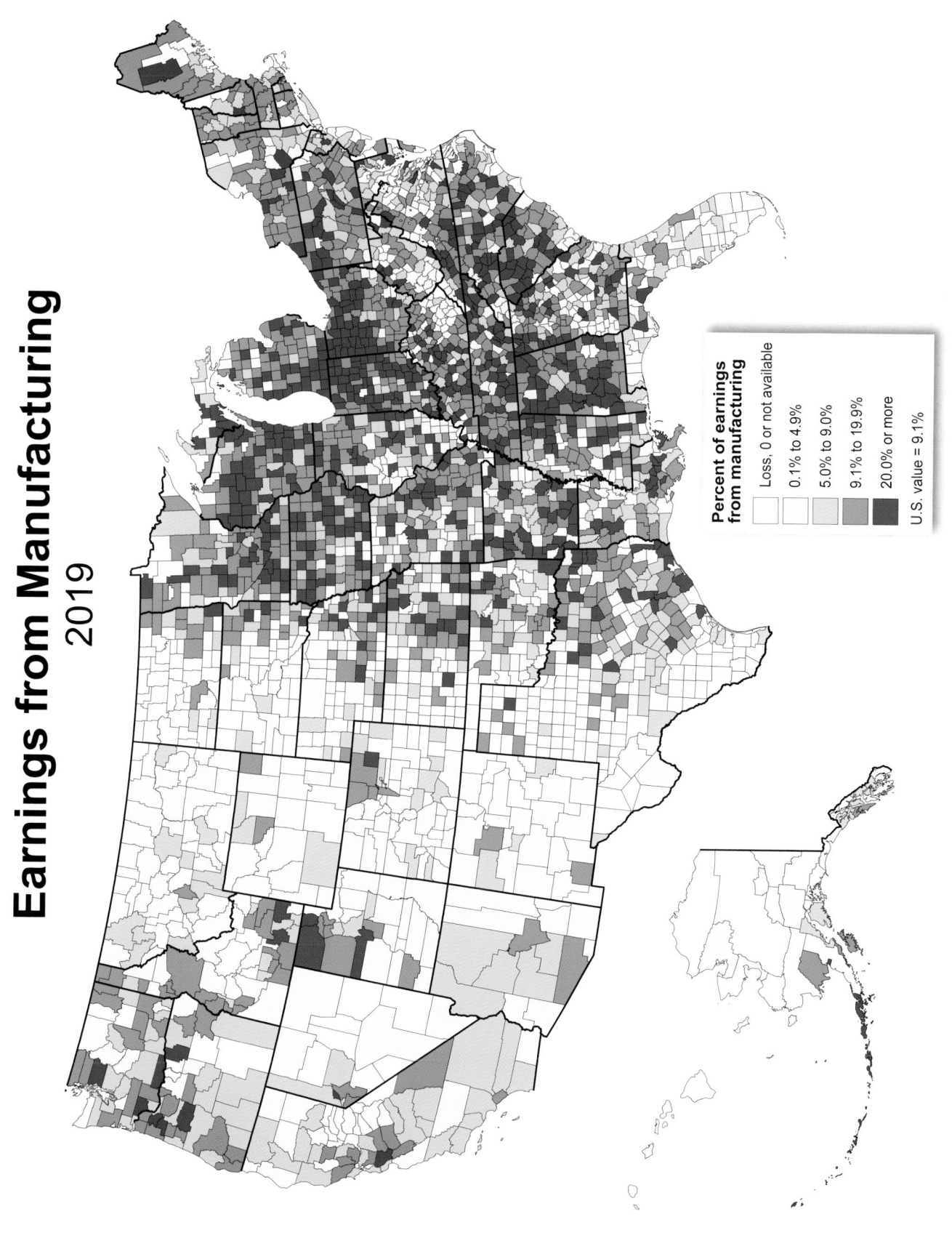

**Percent of earnings from manufacturing**

Loss, 0 or not available
0.1% to 4.9%
5.0% to 9.0%
9.1% to 19.9%
20.0% or more

U.S. value = 9.1%

# States

(For explanation of symbols, see page viii)

Page
| | |
|---|---|
| 3 | State Highlights and Rankings |
| 17 | State Column Headings |
| 24 | Table A |

**Part A—States**

1

# State Highlights and Rankings

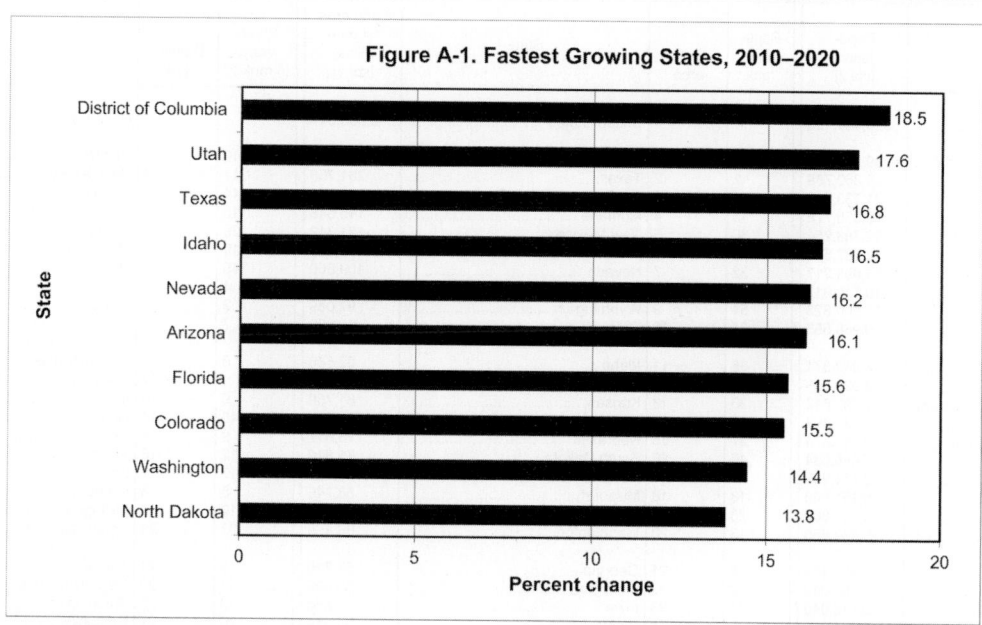

**Figure A-1. Fastest Growing States, 2010–2020**

| State | Percent change |
|---|---|
| District of Columbia | 18.5 |
| Utah | 17.6 |
| Texas | 16.8 |
| Idaho | 16.5 |
| Nevada | 16.2 |
| Arizona | 16.1 |
| Florida | 15.6 |
| Colorado | 15.5 |
| Washington | 14.4 |
| North Dakota | 13.8 |

There is no simple relationship between population size and land area for most of the geographic entities included in this publication. According to the Census Bureau's 2020 estimates, state populations ranged from a high of over 39 million in California to a low of 582,328 in Wyoming. (The median population for states—with half having a larger population and half having a smaller population—was over 4.4 million people.) California was also one of the largest states in land area (ranking third). Alaska was by far the largest state in area; it was more than twice the size of Texas, the second-largest state, even though its population rank was close to the bottom (ranked 48th). Texas was also the second-largest state in terms of total population with over 29.0 million residents. At the other end of the geographic size spectrum were many of the New England states (with Rhode Island ranking as the smallest), plus Delaware, Hawaii, and New Jersey. As a consequence of differing area size and population rank, New Jersey was the most densely settled state, with 1,207.7 persons per square mile while Alaska was the least densely settled, with about 1.3 persons per square mile. California, which had the largest population and third-largest land area, ranked 12th in terms of population density (252.6 persons per square mile). The 15 most populous states remained almost unchanged between 2010 and 2020—Arizona moved into the top 15 while Indiana dropped out—but there were changes within their ranks. Florida became the 3rd most populous state, pushing ahead of New York while Georgia became the 8th most populous state. North Carolina moved up to 9th place while Michigan dropped to 10th.

Not surprisingly, states with higher population density also had higher proportions of developed land. According to the Department of Agriculture's most recent National Resources Inventory, 35.7 percent of New Jersey's land was developed. Connecticut had the second highest proportion with 34.1 percent, followed by Massachusetts at 33.0 percent. Among the reporting states, Nevada had the lowest proportion of developed land, at just 0.8 percent, followed by Wyoming and Montana, with 1.1 percent and 1.2 percent of their land developed. Nearly 85 percent of Nevada's land was owned by the federal government, stemming from a provision in Nevada's original constitution of 1864. This was by far the highest percentage in the nation. Federal land accounted for almost 21 percent of the United States' total land area. (Estimates are not available for Alaska, and the District of Columbia. See Appendix F for definitions and additional information.)

The total population of the United States increased 6.3 percent between 2010 and 2020, with 20 states matching or exceeding this rate of growth and the remainder growing more slowly. The District of Columbia experienced a higher growth rate than any of the states (18.5 percent). Texas, Utah, Florida, Nevada, Idaho, Colorado, and Arizona all had growth rates above 15 percent. The District of Columbia ranked 49th by population size—more than either Vermont or Wyoming—and gained more than 110,000 residents in the 10-year period. Texas has the second largest population among all the states and gained more than 4.2 million residents. North Dakota grew nearly 14 percent between 2010 and 2020. Despite its growth, North Dakota ranked 47th for total population and 48th for density, with only 11.1 persons per square mile. Texas and Florida ranked among the top five states for total population and for population growth from 2010 to 2020. Florida, with the third-largest population, grew by 15.6 percent, increasing its population by over 2.9 million people. Rhode Island and Vermont both ranked among the 10 least populous states, as well as among the 10 states with the lowest population growth between 2010 and 2020. While most states have increased their populations in the 10 years, the populations of West Virginia, Vermont, Illinois, Connecticut, Mississippi, and New York all declined. Three other states experienced increases below one percent. Louisiana's population has rebounded from a loss of about 250,000 residents after Hurricane Katrina hit the state in August 2005. Its 2020 population of over 4.6 million is higher than its 2005 estimated population on July 1 of that year.

# States and the District of Columbia, Selected Rankings

| Population, 2020 | | | Land Area, 2020 | | | | Population density, 2020 | | | |
|---|---|---|---|---|---|---|---|---|---|---|
| Population rank | State | Population [col 2] | Population rank | Land area | State | Land area (square miles) [col 1] | Population rank | Density rank | State | Density (per square mile) [col 4] |
| | United States | 329,484,123 | | | United States | 3,533,044 | | | United States | 93.3 |
| 1 | California | 39,368,078 | 48 | 1 | Alaska | 571,017 | 49 | 1 | District of Columbia | 11,666.4 |
| 2 | Texas | 29,360,759 | 2 | 2 | Texas | 261,263 | 11 | 2 | New Jersey | 1,207.7 |
| 3 | Florida | 21,733,312 | 1 | 3 | California | 155,854 | 44 | 3 | Rhode Island | 1,022.5 |
| 4 | New York | 19,336,776 | 43 | 4 | Montana | 145,548 | 15 | 4 | Massachusetts | 883.7 |
| 5 | Pennsylvania | 12,783,254 | 36 | 5 | New Mexico | 121,312 | 29 | 5 | Connecticut | 734.5 |
| 6 | Illinois | 12,587,530 | 14 | 6 | Arizona | 113,653 | 19 | 6 | Maryland | 623.6 |
| 7 | Ohio | 11,693,217 | 32 | 7 | Nevada | 109,860 | 45 | 7 | Delaware | 506.4 |
| 8 | Georgia | 10,710,017 | 21 | 8 | Colorado | 103,638 | 4 | 8 | New York | 410.3 |
| 9 | North Carolina | 10,600,823 | 51 | 9 | Wyoming | 97,089 | 3 | 9 | Florida | 405.1 |
| 10 | Michigan | 9,966,555 | 27 | 10 | Oregon | 95,988 | 7 | 10 | Ohio | 286.2 |
| 11 | New Jersey | 8,882,371 | 38 | 11 | Idaho | 82,645 | 5 | 11 | Pennsylvania | 285.7 |
| 12 | Virginia | 8,590,563 | 30 | 12 | Utah | 82,377 | 1 | 12 | California | 252.6 |
| 13 | Washington | 7,693,612 | 35 | 13 | Kansas | 81,759 | 6 | 13 | Illinois | 226.7 |
| 14 | Arizona | 7,421,401 | 22 | 14 | Minnesota | 79,626 | 40 | 14 | Hawaii | 219.1 |
| 15 | Massachusetts | 6,893,574 | 37 | 15 | Nebraska | 76,817 | 9 | 15 | North Carolina | 218.0 |
| 16 | Tennessee | 6,886,834 | 46 | 16 | South Dakota | 75,810 | 12 | 16 | Virginia | 217.6 |
| 17 | Indiana | 6,754,953 | 47 | 17 | North Dakota | 68,995 | 17 | 17 | Indiana | 188.5 |
| 18 | Missouri | 6,151,548 | 18 | 18 | Missouri | 68,746 | 8 | 18 | Georgia | 185.6 |
| 19 | Maryland | 6,055,802 | 28 | 19 | Oklahoma | 68,596 | 10 | 19 | Michigan | 176.1 |
| 20 | Wisconsin | 5,832,655 | 13 | 20 | Washington | 66,455 | 23 | 20 | South Carolina | 173.6 |
| 21 | Colorado | 5,807,719 | 8 | 21 | Georgia | 57,716 | 16 | 21 | Tennessee | 167.0 |
| 22 | Minnesota | 5,657,342 | 10 | 22 | Michigan | 56,606 | 41 | 22 | New Hampshire | 152.6 |
| 23 | South Carolina | 5,218,040 | 31 | 23 | Iowa | 55,854 | 13 | 23 | Washington | 115.8 |
| 24 | Alabama | 4,921,532 | 6 | 24 | Illinois | 55,514 | 26 | 24 | Kentucky | 113.4 |
| 25 | Louisiana | 4,645,318 | 20 | 25 | Wisconsin | 54,167 | 2 | 25 | Texas | 112.4 |
| 26 | Kentucky | 4,477,251 | 3 | 26 | Florida | 53,648 | 20 | 26 | Wisconsin | 107.7 |
| 27 | Oregon | 4,241,507 | 33 | 27 | Arkansas | 52,038 | 25 | 27 | Louisiana | 107.5 |
| 28 | Oklahoma | 3,980,783 | 24 | 28 | Alabama | 50,647 | 24 | 28 | Alabama | 97.2 |
| 29 | Connecticut | 3,557,006 | 9 | 29 | North Carolina | 48,620 | 18 | 29 | Missouri | 89.5 |
| 30 | Utah | 3,249,879 | 4 | 30 | New York | 47,124 | 39 | 30 | West Virginia | 74.2 |
| 31 | Iowa | 3,163,561 | 34 | 31 | Mississippi | 46,926 | 22 | 31 | Minnesota | 71.0 |
| 32 | Nevada | 3,138,259 | 5 | 32 | Pennsylvania | 44,742 | 50 | 32 | Vermont | 67.6 |
| 33 | Arkansas | 3,030,522 | 25 | 33 | Louisiana | 43,205 | 14 | 33 | Arizona | 65.3 |
| 34 | Mississippi | 2,966,786 | 16 | 34 | Tennessee | 41,238 | 34 | 34 | Mississippi | 63.2 |
| 35 | Kansas | 2,913,805 | 7 | 35 | Ohio | 40,859 | 33 | 35 | Arkansas | 58.2 |
| 36 | New Mexico | 2,106,319 | 26 | 36 | Kentucky | 39,491 | 28 | 36 | Oklahoma | 58.0 |
| 37 | Nebraska | 1,937,552 | 12 | 37 | Virginia | 39,482 | 31 | 37 | Iowa | 56.6 |
| 38 | Idaho | 1,826,913 | 17 | 38 | Indiana | 35,826 | 21 | 38 | Colorado | 56.0 |
| 39 | West Virginia | 1,784,787 | 42 | 39 | Maine | 30,845 | 27 | 39 | Oregon | 44.2 |
| 40 | Hawaii | 1,407,006 | 23 | 40 | South Carolina | 30,064 | 42 | 40 | Maine | 43.8 |
| 41 | New Hampshire | 1,366,275 | 39 | 41 | West Virginia | 24,041 | 30 | 41 | Utah | 39.5 |
| 42 | Maine | 1,350,141 | 19 | 42 | Maryland | 9,711 | 35 | 42 | Kansas | 35.6 |
| 43 | Montana | 1,080,577 | 50 | 43 | Vermont | 9,218 | 32 | 43 | Nevada | 28.6 |
| 44 | Rhode Island | 1,057,125 | 41 | 44 | New Hampshire | 8,953 | 37 | 44 | Nebraska | 25.2 |
| 45 | Delaware | 986,809 | 15 | 45 | Massachusetts | 7,801 | 38 | 45 | Idaho | 22.1 |
| 46 | South Dakota | 892,717 | 11 | 46 | New Jersey | 7,355 | 36 | 46 | New Mexico | 17.4 |
| 47 | North Dakota | 765,309 | 40 | 47 | Hawaii | 6,422 | 46 | 47 | South Dakota | 11.8 |
| 48 | Alaska | 731,158 | 29 | 48 | Connecticut | 4,843 | 47 | 48 | North Dakota | 11.1 |
| 49 | District of Columbia | 712,816 | 45 | 49 | Delaware | 1,949 | 43 | 49 | Montana | 7.4 |
| 50 | Vermont | 623,347 | 44 | 50 | Rhode Island | 1,034 | 51 | 50 | Wyoming | 6.0 |
| 51 | Wyoming | 582,328 | 49 | 51 | District of Columbia | 61 | 48 | 51 | Alaska | 1.3 |

# States and the District of Columbia, Selected Rankings

| Percent population change, 2010–2020 | | | | Percent Under 15 years old, 2020 | | | | Percent 65 years old and over, 2020 | | | |
|---|---|---|---|---|---|---|---|---|---|---|---|
| Popu-lation rank | Percent change rank | State | Percent change [col 36] | Popu-lation rank | Under 15 years old rank | State | Percent under 15 years old [cols 10 + 11] | Popu-lation rank | 65 years old and over rank | State | Percent 65 years old and over [cols 17 + 18 + 19] |
| | | United States | 6.7 | | | United States | 18.3 | | | United States | 16.9 |
| 49 | 1 | District of Columbia | 18.5 | 30 | 1 | Utah | 23.6 | 42 | 1 | Maine | 21.8 |
| 30 | 2 | Utah | 17.6 | 2 | 2 | Texas | 21.0 | 3 | 2 | Florida | 21.3 |
| 2 | 3 | Texas | 16.8 | 48 | 3 | Alaska | 20.6 | 39 | 3 | West Virginia | 20.9 |
| 38 | 4 | Idaho | 16.5 | 46 | 4 | South Dakota | 20.5 | 50 | 4 | Vermont | 20.7 |
| 32 | 5 | Nevada | 16.2 | 37 | 5 | Nebraska | 20.4 | 45 | 5 | Delaware | 20.0 |
| 14 | 6 | Arizona | 16.1 | 38 | 6 | Idaho | 20.3 | 43 | 6 | Montana | 19.7 |
| 3 | 7 | Florida | 15.6 | 47 | 7 | North Dakota | 20.1 | 40 | 7 | Hawaii | 19.6 |
| 21 | 8 | Colorado | 15.5 | 28 | 8 | Oklahoma | 20.0 | 41 | 8 | New Hampshire | 19.2 |
| 13 | 9 | Washington | 14.4 | 35 | 9 | Kansas | 19.8 | 5 | 8 | Pennsylvania | 19.2 |
| 47 | 10 | North Dakota | 13.8 | 25 | 10 | Louisiana | 19.5 | 23 | 10 | South Carolina | 18.7 |
| 23 | 11 | South Carolina | 12.8 | 34 | 11 | Mississippi | 19.3 | 27 | 11 | Oregon | 18.6 |
| 9 | 12 | North Carolina | 11.2 | 8 | 12 | Georgia | 19.2 | 14 | 12 | Arizona | 18.5 |
| 27 | 13 | Oregon | 10.7 | 33 | 13 | Arkansas | 19.1 | 36 | 12 | New Mexico | 18.5 |
| 8 | 14 | Georgia | 10.6 | 17 | 13 | Indiana | 19.1 | 10 | 14 | Michigan | 18.2 |
| 45 | 15 | Delaware | 9.9 | 22 | 13 | Minnesota | 19.1 | 44 | 14 | Rhode Island | 18.2 |
| 46 | 16 | South Dakota | 9.6 | 31 | 16 | Iowa | 19.0 | 29 | 16 | Connecticut | 18.1 |
| 43 | 17 | Montana | 9.2 | 51 | 16 | Wyoming | 19.0 | 20 | 17 | Wisconsin | 18.0 |
| 16 | 18 | Tennessee | 8.5 | 1 | 18 | California | 18.5 | 31 | 18 | Iowa | 17.9 |
| 12 | 19 | Virginia | 7.4 | 26 | 18 | Kentucky | 18.5 | 7 | 18 | Ohio | 17.9 |
| 22 | 20 | Minnesota | 6.7 | 18 | 18 | Missouri | 18.5 | 51 | 18 | Wyoming | 17.9 |
| 37 | 21 | Nebraska | 6.1 | 32 | 18 | Nevada | 18.5 | 24 | 21 | Alabama | 17.8 |
| 28 | 21 | Oklahoma | 6.1 | 36 | 22 | New Mexico | 18.4 | 18 | 22 | Missouri | 17.7 |
| 1 | 23 | California | 5.7 | 24 | 23 | Alabama | 18.3 | 33 | 23 | Arkansas | 17.6 |
| 15 | 24 | Massachusetts | 5.3 | 14 | 23 | Arizona | 18.3 | 4 | 24 | New York | 17.5 |
| 19 | 25 | Maryland | 4.9 | 19 | 23 | Maryland | 18.3 | 46 | 24 | South Dakota | 17.5 |
| 17 | 26 | Indiana | 4.2 | 6 | 26 | Illinois | 18.2 | 15 | 26 | Massachusetts | 17.4 |
| 33 | 27 | Arkansas | 3.9 | 7 | 26 | Ohio | 18.2 | 26 | 27 | Kentucky | 17.1 |
| 31 | 28 | Iowa | 3.8 | 16 | 26 | Tennessee | 18.2 | 11 | 27 | New Jersey | 17.1 |
| 41 | 28 | New Hampshire | 3.8 | 12 | 29 | Virginia | 18.1 | 9 | 27 | North Carolina | 17.1 |
| 40 | 30 | Hawaii | 3.4 | 11 | 30 | New Jersey | 18.0 | 16 | 27 | Tennessee | 17.1 |
| 51 | 31 | Wyoming | 3.3 | 9 | 30 | North Carolina | 18.0 | 34 | 31 | Mississippi | 16.9 |
| 26 | 32 | Kentucky | 3.2 | 13 | 30 | Washington | 18.0 | 38 | 32 | Idaho | 16.8 |
| 24 | 33 | Alabama | 3.0 | 21 | 33 | Colorado | 17.8 | 22 | 32 | Minnesota | 16.8 |
| 48 | 34 | Alaska | 2.9 | 40 | 33 | Hawaii | 17.8 | 35 | 34 | Kansas | 16.7 |
| 18 | 35 | Missouri | 2.7 | 23 | 33 | South Carolina | 17.8 | 6 | 35 | Illinois | 16.6 |
| 20 | 36 | Wisconsin | 2.6 | 20 | 33 | Wisconsin | 17.8 | 32 | 35 | Nevada | 16.6 |
| 25 | 37 | Louisiana | 2.5 | 43 | 37 | Montana | 17.6 | 17 | 37 | Indiana | 16.5 |
| 36 | 38 | New Mexico | 2.3 | 10 | 38 | Michigan | 17.5 | 25 | 37 | Louisiana | 16.5 |
| 35 | 39 | Kansas | 2.1 | 45 | 39 | Delaware | 17.1 | 28 | 37 | Oklahoma | 16.5 |
| 42 | 40 | Maine | 1.6 | 4 | 39 | New York | 17.1 | 19 | 40 | Maryland | 16.4 |
| 7 | 41 | Ohio | 1.4 | 5 | 41 | Pennsylvania | 16.9 | 37 | 40 | Nebraska | 16.4 |
| 11 | 42 | New Jersey | 1.0 | 27 | 42 | Oregon | 16.8 | 12 | 42 | Virginia | 16.3 |
| 10 | 43 | Michigan | 0.8 | 29 | 43 | Connecticut | 16.4 | 13 | 42 | Washington | 16.3 |
| 5 | 44 | Pennsylvania | 0.6 | 39 | 43 | West Virginia | 16.4 | 47 | 44 | North Dakota | 16.1 |
| 44 | 45 | Rhode Island | 0.4 | 3 | 45 | Florida | 16.2 | 21 | 45 | Colorado | 15.2 |
| 34 | 46 | Mississippi | 0.0 | 49 | 46 | District of Columbia | 16.0 | 1 | 46 | California | 15.1 |
| 4 | 47 | New York | -0.2 | 15 | 47 | Massachusetts | 15.9 | 8 | 47 | Georgia | 14.8 |
| 50 | 48 | Vermont | -0.4 | 44 | 48 | Rhode Island | 15.7 | 2 | 48 | Texas | 13.2 |
| 29 | 49 | Connecticut | -0.5 | 42 | 49 | Maine | 15.0 | 48 | 49 | Alaska | 13.1 |
| 6 | 50 | Illinois | -1.9 | 41 | 49 | New Hampshire | 15.0 | 49 | 50 | District of Columbia | 12.5 |
| 39 | 51 | West Virginia | -3.7 | 50 | 51 | Vermont | 14.8 | 30 | 51 | Utah | 11.7 |

# States and the District of Columbia, Selected Rankings

| Percent born in state of residence, 2019 | | | | Number of immigrants, 2019 | | | | Birth rate, 2019 | | | |
|---|---|---|---|---|---|---|---|---|---|---|---|
| Population rank | Born in state of residence rank | State | Percent born in state of residence [col 23] | Population rank | Immigrant rank | State | Number of immigrants [col 24] | Population rank | Birth rate rank | State | Birth rate (per 1,000 population) [col 98] |
| | | United States | 58.0 | | | United States | 1,031,765 | | | United States | 11.4 |
| 25 | 1 | Louisiana | 77.6 | 1 | 1 | California | 193,093 | 30 | 1 | Utah | 14.6 |
| 10 | 2 | Michigan | 76.2 | 4 | 2 | New York | 124,026 | 47 | 2 | North Dakota | 13.7 |
| 7 | 3 | Ohio | 74.7 | 3 | 3 | Florida | 118,140 | 48 | 3 | Alaska | 13.4 |
| 34 | 4 | Mississippi | 71.5 | 2 | 4 | Texas | 107,955 | 2 | 4 | Texas | 13.0 |
| 5 | 5 | Pennsylvania | 71.4 | 11 | 5 | New Jersey | 48,754 | 49 | 5 | District of Columbia | 12.9 |
| 20 | 6 | Wisconsin | 70.9 | 6 | 6 | Illinois | 37,958 | 46 | 5 | South Dakota | 12.9 |
| 31 | 7 | Iowa | 69.7 | 15 | 7 | Massachusetts | 30,834 | 37 | 7 | Nebraska | 12.8 |
| 24 | 8 | Alabama | 69.2 | 8 | 8 | Georgia | 27,246 | 25 | 8 | Louisiana | 12.7 |
| 39 | 9 | West Virginia | 68.5 | 13 | 9 | Washington | 25,570 | 28 | 9 | Oklahoma | 12.4 |
| 26 | 10 | Kentucky | 68.3 | 5 | 10 | Pennsylvania | 25,329 | 38 | 10 | Idaho | 12.3 |
| 22 | 11 | Minnesota | 67.7 | 12 | 11 | Virginia | 24,784 | 34 | 10 | Mississippi | 12.3 |
| 17 | 12 | Indiana | 67.7 | 19 | 12 | Maryland | 22,304 | 33 | 12 | Arkansas | 12.1 |
| 6 | 13 | Illinois | 67.4 | 9 | 13 | North Carolina | 18,419 | 35 | 12 | Kansas | 12.1 |
| 18 | 14 | Missouri | 65.9 | 14 | 14 | Arizona | 18,181 | 24 | 14 | Alabama | 12.0 |
| 37 | 15 | Nebraska | 64.7 | 10 | 15 | Michigan | 17,414 | 17 | 14 | Indiana | 12.0 |
| 46 | 16 | South Dakota | 63.6 | 7 | 16 | Ohio | 16,467 | 8 | 16 | Georgia | 11.9 |
| 4 | 17 | New York | 63.1 | 22 | 17 | Minnesota | 13,833 | 40 | 16 | Hawaii | 11.9 |
| 47 | 18 | North Dakota | 61.8 | 21 | 18 | Colorado | 13,558 | 31 | 16 | Iowa | 11.9 |
| 30 | 19 | Utah | 61.7 | 32 | 19 | Nevada | 11,718 | 26 | 16 | Kentucky | 11.9 |
| 42 | 20 | Maine | 61.6 | 29 | 20 | Connecticut | 10,630 | 18 | 20 | Missouri | 11.8 |
| 28 | 21 | Oklahoma | 60.6 | 16 | 21 | Tennessee | 9,507 | 16 | 20 | Tennessee | 11.8 |
| 33 | 22 | Arkansas | 60.5 | 17 | 22 | Indiana | 8,527 | 22 | 22 | Minnesota | 11.7 |
| 2 | 23 | Texas | 59.5 | 27 | 23 | Oregon | 8,521 | 19 | 23 | Maryland | 11.6 |
| 15 | 24 | Massachusetts | 59.4 | 20 | 24 | Wisconsin | 7,138 | 7 | 24 | Ohio | 11.5 |
| 35 | 25 | Kansas | 59.2 | 18 | 25 | Missouri | 6,900 | 32 | 25 | Nevada | 11.4 |
| 16 | 26 | Tennessee | 59.1 | 30 | 26 | Utah | 6,838 | 4 | 25 | New York | 11.4 |
| 44 | 27 | Rhode Island | 56.6 | 26 | 27 | Kentucky | 6,006 | 12 | 25 | Virginia | 11.4 |
| 1 | 28 | California | 56.1 | 31 | 28 | Iowa | 5,534 | 1 | 28 | California | 11.3 |
| 9 | 29 | North Carolina | 56.0 | 35 | 29 | Kansas | 5,139 | 9 | 28 | North Carolina | 11.3 |
| 23 | 30 | South Carolina | 55.0 | 25 | 30 | Louisiana | 5,116 | 51 | 28 | Wyoming | 11.3 |
| 8 | 31 | Georgia | 54.4 | 40 | 31 | Hawaii | 4,917 | 11 | 31 | New Jersey | 11.2 |
| 36 | 32 | New Mexico | 53.8 | 23 | 32 | South Carolina | 4,675 | 6 | 32 | Illinois | 11.1 |
| 29 | 33 | Connecticut | 53.8 | 28 | 33 | Oklahoma | 4,642 | 23 | 32 | South Carolina | 11.1 |
| 43 | 34 | Montana | 53.4 | 37 | 34 | Nebraska | 4,279 | 13 | 32 | Washington | 11.1 |
| 40 | 35 | Hawaii | 52.3 | 36 | 35 | New Mexico | 3,736 | 14 | 35 | Arizona | 10.9 |
| 11 | 36 | New Jersey | 51.5 | 24 | 36 | Alabama | 3,669 | 21 | 35 | Colorado | 10.9 |
| 12 | 37 | Virginia | 49.5 | 44 | 37 | Rhode Island | 3,445 | 36 | 35 | New Mexico | 10.9 |
| 50 | 38 | Vermont | 48.8 | 33 | 38 | Arkansas | 2,915 | 20 | 35 | Wisconsin | 10.9 |
| 19 | 39 | Maryland | 47.4 | 49 | 39 | District of Columbia | 2,398 | 45 | 39 | Delaware | 10.8 |
| 13 | 40 | Washington | 46.4 | 38 | 40 | Idaho | 2,312 | 10 | 39 | Michigan | 10.8 |
| 38 | 41 | Idaho | 46.1 | 45 | 41 | Delaware | 2,084 | 5 | 41 | Pennsylvania | 10.5 |
| 27 | 42 | Oregon | 45.6 | 41 | 42 | New Hampshire | 2,046 | 43 | 42 | Montana | 10.4 |
| 45 | 43 | Delaware | 44.0 | 34 | 43 | Mississippi | 1,687 | 3 | 43 | Florida | 10.2 |
| 51 | 44 | Wyoming | 43.0 | 47 | 44 | North Dakota | 1,494 | 39 | 44 | West Virginia | 10.1 |
| 48 | 45 | Alaska | 42.9 | 48 | 45 | Alaska | 1,374 | 15 | 45 | Massachusetts | 10.0 |
| 21 | 46 | Colorado | 42.2 | 42 | 46 | Maine | 1,361 | 27 | 46 | Oregon | 9.9 |
| 41 | 47 | New Hampshire | 40.6 | 46 | 47 | South Dakota | 1,199 | 29 | 47 | Connecticut | 9.6 |
| 14 | 48 | Arizona | 39.9 | 39 | 48 | West Virginia | 817 | 44 | 47 | Rhode Island | 9.6 |
| 49 | 49 | District of Columbia | 37.2 | 50 | 49 | Vermont | 648 | 42 | 49 | Maine | 8.8 |
| 3 | 50 | Florida | 35.8 | 43 | 50 | Montana | 545 | 41 | 50 | New Hampshire | 8.7 |
| 32 | 51 | Nevada | 27.2 | 51 | 51 | Wyoming | 410 | 50 | 51 | Vermont | 8.6 |

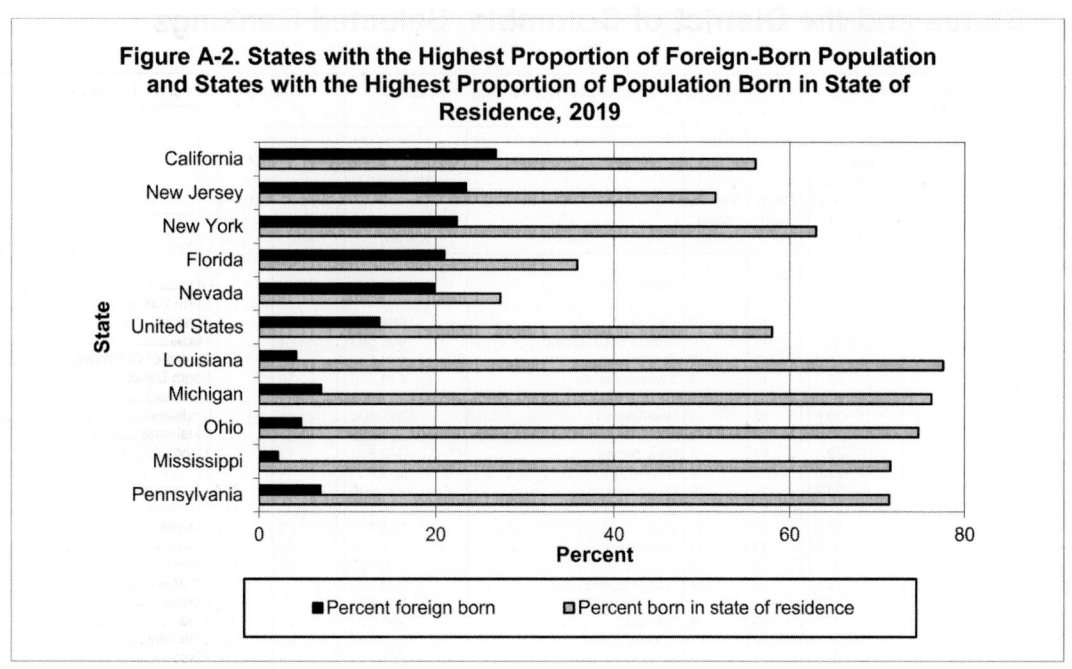

**Figure A-2. States with the Highest Proportion of Foreign-Born Population and States with the Highest Proportion of Population Born in State of Residence, 2019**

■ Percent foreign born    ▫ Percent born in state of residence

The U.S. median age increased slightly from 37.2 years in 2010 to 38.5 years in 2020, primarily caused by the aging Baby Boomer population. This increase was much less than the jump from 32.1 years to 35.3 years between 1990 and 2000. From 2010 to 2020, the population between 65 and 74 years showed the largest proportional increase, while the proportion between 5 and 17 years showed the largest decrease. The 25 to 34 Millennial group increased by 14 percent and they outnumber the peak Baby Boomers in the 55 to 64 age group. The median age by state ranged from 31.2 years in Utah to 45.1 years in Maine. Utah had the highest proportion of young residents; in 2020, 23.6 percent of the state's population was younger than 15 years old. The population age 65 years and over ranged from 11.7 percent in Utah to 21.8 percent in Maine. Alaska and North Dakota had the lowest proportions of female residents, and were among just eleven states in which men outnumbered women. The District of Columbia had the highest proportion of female residents with 52.6 percent, followed by Alabama and Delaware at 51.7 percent.

Natural growth is the difference between the number of births and the number of deaths. Vermont had the fewest births between 2010 and 2019. Maine and West Virginia were the only states to have more deaths than births. A net migration of nearly 37,000 people prevented Maine from having a population loss from 2010 to 2020 but West Virginia, with a net outmigration of 36,295, did experience a population decline. California, the largest state in the nation, had more than 2.3 million more births than deaths and gained 177,201 new residents through net migration. From 2010 to 2020, California gained 985,729 residents from foreign countries and lost 1,162,930 residents to other states. Florida had a net gain of over 2.6 million new residents during this period, with 55 percent from other states and 45 percent from other countries. Texas gained over 2.0 million new residents, 60.0 percent of them from other states. Eighteen states had a net loss of residents due to migration including New York which lost nearly 1.6 million residents to other states but gained 736,389 new residents from other countries.

In six states, 70 percent or more of the residents were born in that same state. Louisiana ranked highest with 77.6 percent. The ten highest rates were mainly in the Midwest and the South. Fourteen states and the District of Columbia had proportions less than 50 percent. Nevada had the lowest proportion by far, with just 27.2 percent of its residents having been born in the state. Nationally, 58.0 percent of Americans lived in the state of their birth.

The U.S. birth rate in 2019 was 11.4 per 1,000 population, a slight decrease from the previous three years. Utah had the highest birth rate in the nation, with 14.6 births per 1,000 population. North Dakota had the second highest birth rate at 13.7 followed by Alaska with a birth rate of 13.4. Connecticut, Maine, Vermont, New Hampshire, and Rhode Island had the lowest birth rates in the nation, all below 10 births per 1,000 population. Utah had the lowest crude death rate with 5.8 deaths per 1,000 population followed by Alaska with a crude death rate of 6.0. However, both states had relatively young populations (In Utah, 40.3 percent of the population was under 25 years old while 36.9 percent of the population was under 25 in Alaska). Once adjusted for age, Alaska's death rate increased to 7.0, and Utah's to 6.9, just under the U.S. rate of 7.2. West Virginia, Alabama, and Maine had the highest crude death rates in the nation. West Virginia had the highest age-adjusted death rate followed by Mississippi, both with a mix of younger and older residents. Maine had the highest proportion of senior citizens followed by Florida. However, Florida also had a high proportion of younger people, which helped give the state a crude death rate of 9.6 per 1,000 population, ranking it 17th, tied with Iowa and Vermont. When Florida's death rate was age-adjusted, it dropped to 6.6, which was well below the national age-adjusted rate of 7.2 and among the lowest ten states. Hawaii had the lowest age-adjusted death rate at 5.7, one of 14 states with rates below 7.0. Mississippi, Arkansas, and Louisiana had the highest infant mortality rates, while New Jersey and New Hampshire had the lowest infant death rates.

# States and the District of Columbia, Selected Rankings

| Percent of owners with a mortgage paying 30 percent or more of income for housing expenses, 2019 | | | | Median value of owner-occupied housing units, 2019 | | | | Median gross rent of renter-occupied housing units, 2019 | | | |
|---|---|---|---|---|---|---|---|---|---|---|---|
| Population rank | Percent of income for housing rank | State | Percent paying 30% or more of income for housing [col 83] | Population rank | Median value rank | State | Median value (dollars) [col 88] | Population rank | Median rent rank | State | Median rent (dollars) [col 90] |
| | | United States | 26.5 | | | United States | 240,500 | | | United States | 1,097 |
| 40 | 1 | Hawaii | 40.9 | 40 | 1 | Hawaii | 669,200 | 40 | 1 | Hawaii | 1,651 |
| 1 | 2 | California | 36.9 | 49 | 2 | District of Columbia | 646,500 | 1 | 2 | California | 1,614 |
| 11 | 3 | New Jersey | 32.5 | 1 | 3 | California | 568,500 | 49 | 3 | District of Columbia | 1,603 |
| 3 | 4 | Florida | 32.2 | 15 | 4 | Massachusetts | 418,600 | 19 | 4 | Maryland | 1,401 |
| 4 | 5 | New York | 31.3 | 21 | 5 | Colorado | 394,600 | 11 | 5 | New Jersey | 1,376 |
| 29 | 6 | Connecticut | 30.6 | 13 | 6 | Washington | 387,600 | 21 | 6 | Colorado | 1,369 |
| 41 | 7 | New Hampshire | 29.1 | 27 | 7 | Oregon | 354,600 | 15 | 7 | Massachusetts | 1,360 |
| 44 | 7 | Rhode Island | 29.1 | 11 | 8 | New Jersey | 348,800 | 13 | 8 | Washington | 1,359 |
| 15 | 9 | Massachusetts | 29.0 | 4 | 9 | New York | 338,700 | 4 | 9 | New York | 1,309 |
| 43 | 10 | Montana | 28.9 | 19 | 10 | Maryland | 332,500 | 12 | 10 | Virginia | 1,254 |
| 27 | 11 | Oregon | 28.8 | 30 | 11 | Utah | 330,300 | 3 | 11 | Florida | 1,238 |
| 13 | 12 | Washington | 28.6 | 32 | 12 | Nevada | 317,800 | 48 | 12 | Alaska | 1,201 |
| 32 | 13 | Nevada | 28.3 | 12 | 13 | Virginia | 288,800 | 27 | 13 | Oregon | 1,185 |
| 21 | 14 | Colorado | 27.9 | 44 | 14 | Rhode Island | 283,000 | 29 | 14 | Connecticut | 1,177 |
| 36 | 15 | New Mexico | 27.5 | 41 | 15 | New Hampshire | 281,400 | 32 | 15 | Nevada | 1,168 |
| 48 | 16 | Alaska | 27.1 | 48 | 16 | Alaska | 281,200 | 41 | 16 | New Hampshire | 1,147 |
| 2 | 17 | Texas | 26.5 | 29 | 17 | Connecticut | 280,700 | 45 | 17 | Delaware | 1,116 |
| 19 | 18 | Maryland | 26.0 | 45 | 18 | Delaware | 261,700 | 14 | 18 | Arizona | 1,101 |
| 50 | 19 | Vermont | 25.9 | 14 | 19 | Arizona | 255,900 | 30 | 19 | Utah | 1,098 |
| 42 | 20 | Maine | 25.8 | 38 | 20 | Idaho | 255,200 | 2 | 20 | Texas | 1,091 |
| 34 | 21 | Mississippi | 25.6 | 43 | 21 | Montana | 253,600 | 8 | 21 | Georgia | 1,049 |
| 14 | 22 | Arizona | 25.5 | 22 | 22 | Minnesota | 246,700 | 44 | 22 | Rhode Island | 1,043 |
| 6 | 22 | Illinois | 25.5 | 3 | 23 | Florida | 245,100 | 6 | 23 | Illinois | 1,020 |
| 49 | 24 | District of Columbia | 25.2 | 51 | 24 | Wyoming | 235,200 | 22 | 24 | Minnesota | 1,016 |
| 45 | 25 | Delaware | 25.1 | 50 | 25 | Vermont | 233,200 | 50 | 25 | Vermont | 980 |
| 25 | 26 | Louisiana | 24.6 | 6 | 26 | Illinois | 209,100 | 5 | 26 | Pennsylvania | 951 |
| 12 | 27 | Virginia | 24.5 | 47 | 27 | North Dakota | 205,400 | 9 | 27 | North Carolina | 931 |
| 30 | 28 | Utah | 24.3 | 8 | 28 | Georgia | 202,500 | 23 | 28 | South Carolina | 922 |
| 8 | 29 | Georgia | 24.2 | 42 | 29 | Maine | 200,500 | 16 | 29 | Tennessee | 904 |
| 38 | 30 | Idaho | 23.5 | 2 | 30 | Texas | 200,400 | 10 | 30 | Michigan | 888 |
| 5 | 31 | Pennsylvania | 23.4 | 20 | 31 | Wisconsin | 197,200 | 38 | 31 | Idaho | 880 |
| 23 | 31 | South Carolina | 23.4 | 9 | 32 | North Carolina | 193,200 | 42 | 32 | Maine | 870 |
| 9 | 33 | North Carolina | 23.2 | 5 | 33 | Pennsylvania | 192,600 | 20 | 33 | Wisconsin | 867 |
| 16 | 34 | Tennessee | 22.8 | 16 | 34 | Tennessee | 191,900 | 25 | 34 | Louisiana | 866 |
| 24 | 35 | Alabama | 22.6 | 46 | 35 | South Dakota | 185,000 | 35 | 35 | Kansas | 862 |
| 46 | 36 | South Dakota | 22.5 | 36 | 36 | New Mexico | 180,900 | 37 | 36 | Nebraska | 859 |
| 51 | 37 | Wyoming | 22.4 | 23 | 37 | South Carolina | 179,800 | 36 | 37 | New Mexico | 847 |
| 10 | 38 | Michigan | 22.1 | 37 | 38 | Nebraska | 172,700 | 17 | 38 | Indiana | 840 |
| 33 | 39 | Arkansas | 21.5 | 25 | 39 | Louisiana | 172,100 | 18 | 39 | Missouri | 834 |
| 20 | 40 | Wisconsin | 21.3 | 10 | 40 | Michigan | 169,600 | 43 | 40 | Montana | 831 |
| 26 | 41 | Kentucky | 21.2 | 18 | 41 | Missouri | 168,000 | 51 | 41 | Wyoming | 822 |
| 22 | 42 | Minnesota | 20.7 | 35 | 42 | Kansas | 163,200 | 28 | 42 | Oklahoma | 814 |
| 28 | 43 | Oklahoma | 20.5 | 31 | 43 | Iowa | 158,900 | 7 | 43 | Ohio | 813 |
| 18 | 44 | Missouri | 20.4 | 7 | 44 | Ohio | 157,200 | 31 | 44 | Iowa | 808 |
| 39 | 45 | West Virginia | 20.0 | 17 | 45 | Indiana | 156,000 | 24 | 45 | Alabama | 807 |
| 7 | 46 | Ohio | 19.7 | 24 | 46 | Alabama | 154,000 | 47 | 46 | North Dakota | 804 |
| 37 | 47 | Nebraska | 19.6 | 26 | 47 | Kentucky | 151,700 | 34 | 47 | Mississippi | 777 |
| 35 | 48 | Kansas | 19.3 | 28 | 48 | Oklahoma | 147,000 | 26 | 48 | Kentucky | 773 |
| 31 | 49 | Iowa | 18.8 | 33 | 49 | Arkansas | 136,200 | 46 | 49 | South Dakota | 769 |
| 17 | 50 | Indiana | 18.7 | 34 | 50 | Mississippi | 128,200 | 33 | 50 | Arkansas | 742 |
| 47 | 51 | North Dakota | 16.3 | 39 | 51 | West Virginia | 124,600 | 39 | 51 | West Virginia | 727 |

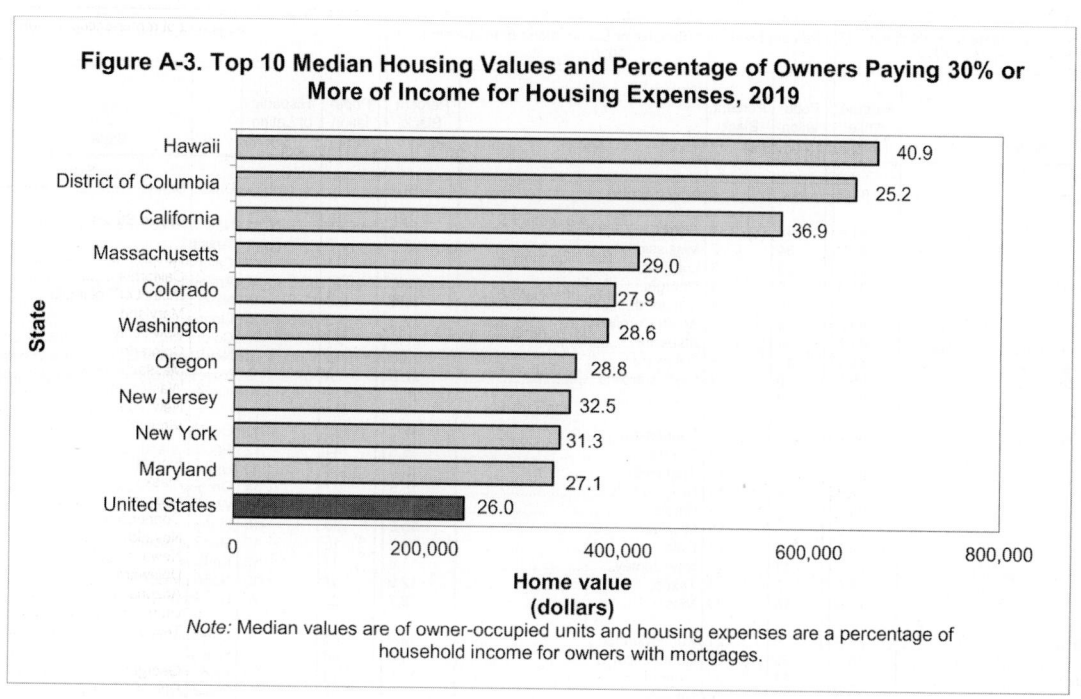

**Figure A-3. Top 10 Median Housing Values and Percentage of Owners Paying 30% or More of Income for Housing Expenses, 2019**

| State | Value |
|---|---|
| Hawaii | 40.9 |
| District of Columbia | 25.2 |
| California | 36.9 |
| Massachusetts | 29.0 |
| Colorado | 27.9 |
| Washington | 28.6 |
| Oregon | 28.8 |
| New Jersey | 32.5 |
| New York | 31.3 |
| Maryland | 27.1 |
| United States | 26.0 |

Home value
(dollars)

*Note:* Median values are of owner-occupied units and housing expenses are a percentage of household income for owners with mortgages.

In 2019, homeowners with mortgages paid a median of 20.8 percent of their incomes for monthly owner costs (mortgage, insurance, taxes, utilities, fuel, etc.). This ranged from a high of 26.4 percent in Hawaii to 17.8 percent in West Virginia. Nationally, 26.5 percent of owners with a mortgage paid 30 percent or more of income for housing expenses in 2019. Hawaii and California had the highest proportion of mortgage holders paying 30 percent or more of their income with 40.9 and 36.9 percent respectively, followed by New Jersey at 32.5 percent, Florida at 32.2 and New York at 31.3 percent. In 34 states and the District of Columbia, the proportion of mortgaged owners who paid more than 30 percent for housing was lower than the national average. North Dakota had the lowest proportion in the nation with 16.3 percent. Ten states and the District of Columbia had median home values exceeding $300,000 in 2019, led by Hawaii with a median home value of $669,200 Nationally, the median value of owner-occupied housing units was $240,500. Hawaii also had the highest median gross rent, at $1,651. The District of Columbia, California, and Washington all had median gross monthly rents exceeding $1,400.

Many minority groups had above average growth rates since 2000. Currently, in four states and the District of Columbia, the minority population outnumbers the non-Hispanic White population. Nationally, 61.8 percent of the U.S. population was non-Hispanic White alone or in combination, but the racial and ethnic compositions of the states varied widely. In Hawaii, the state with the highest proportion of minorities, Asian and Pacific Islander alone or in combination was the largest race group, representing 75.6 percent of the state's population. Hispanic or Latino residents made up over 49 percent of New Mexico's population and over 39 percent of residents in both California and Texas. The District of Columbia had the highest proportion of Black residents at 45.3 percent, down from 60.5 percent in 2000. Among the states, Mississippi and Louisiana ranked first and second, with Black populations of 38.3 and 33.3 percent, respectively. Alaska had the highest proportion of American Indians and Alaska Natives, who made up 19.2 percent of the population. Oklahoma, New Mexico, and South Dakota all had high proportions of American Indian populations. As might be expected, the states with the largest number of minorities were among the states with the highest total populations. New York was home to approximately 3.0 million Black residents, and California had the largest number of Hispanics, Asian and Pacific Islanders, and American Indians and Alaska Natives. California had over 390,000 non-Hispanic American Indian and Alaska Native residents, though they made up just 1.0 percent of the state's population. Despite having only about 85,000 Native American residents, South Dakota had the second highest proportion in the nation at 9.6 percent.

# States and the District of Columbia, Selected Rankings

| Population rank | Percent White rank | State | Percent White [col 5] | Population rank | Percent Black rank | State | Percent Black [col 6] | Population rank | Hispanic or Latino rank | State | Percent Hispanic or Latino [col 9] |
|---|---|---|---|---|---|---|---|---|---|---|---|
| | | United States | 61.8 | | | United States | 13.6 | | | United States | 18.6 |
| 42 | 1 | Maine | 94.4 | 49 | 1 | District of Columbia | 45.3 | 36 | 1 | New Mexico | 49.7 |
| 50 | 2 | Vermont | 94.2 | 34 | 2 | Mississippi | 38.3 | 2 | 2 | Texas | 39.8 |
| 39 | 3 | West Virginia | 93.5 | 25 | 3 | Louisiana | 33.3 | 1 | 3 | California | 39.5 |
| 41 | 4 | New Hampshire | 91.0 | 8 | 4 | Georgia | 33.0 | 14 | 4 | Arizona | 31.9 |
| 43 | 5 | Montana | 88.2 | 19 | 5 | Maryland | 31.6 | 32 | 5 | Nevada | 29.5 |
| 31 | 6 | Iowa | 86.2 | 23 | 6 | South Carolina | 27.4 | 3 | 6 | Florida | 26.5 |
| 26 | 7 | Kentucky | 85.7 | 24 | 7 | Alabama | 27.3 | 21 | 7 | Colorado | 21.9 |
| 51 | 8 | Wyoming | 85.4 | 45 | 8 | Delaware | 23.8 | 11 | 8 | New Jersey | 21.1 |
| 47 | 9 | North Dakota | 85.3 | 9 | 9 | North Carolina | 22.6 | 4 | 9 | New York | 19.3 |
| 38 | 10 | Idaho | 83.5 | 12 | 10 | Virginia | 20.6 | 6 | 10 | Illinois | 17.6 |
| 46 | 11 | South Dakota | 83.2 | 16 | 11 | Tennessee | 17.7 | 29 | 11 | Connecticut | 17.2 |
| 20 | 12 | Wisconsin | 82.2 | 3 | 12 | Florida | 16.5 | 44 | 12 | Rhode Island | 16.7 |
| 18 | 13 | Missouri | 81.1 | 33 | 13 | Arkansas | 16.3 | 30 | 13 | Utah | 14.5 |
| 22 | 14 | Minnesota | 80.8 | 4 | 14 | New York | 15.5 | 27 | 14 | Oregon | 13.7 |
| 7 | 15 | Ohio | 80.1 | 6 | 15 | Illinois | 15.0 | 13 | 15 | Washington | 13.4 |
| 17 | 16 | Indiana | 79.9 | 10 | 15 | Michigan | 15.0 | 38 | 16 | Idaho | 13.0 |
| 37 | 17 | Nebraska | 79.7 | 7 | 17 | Ohio | 14.2 | 15 | 17 | Massachusetts | 12.6 |
| 30 | 17 | Utah | 79.7 | 11 | 18 | New Jersey | 13.8 | 35 | 18 | Kansas | 12.4 |
| 27 | 19 | Oregon | 77.9 | 2 | 19 | Texas | 12.9 | 37 | 19 | Nebraska | 11.6 |
| 35 | 20 | Kansas | 77.6 | 18 | 20 | Missouri | 12.7 | 49 | 20 | District of Columbia | 11.4 |
| 5 | 21 | Pennsylvania | 76.9 | 5 | 21 | Pennsylvania | 12.0 | 28 | 20 | Oklahoma | 11.4 |
| 10 | 22 | Michigan | 76.7 | 29 | 22 | Connecticut | 11.6 | 40 | 22 | Hawaii | 10.9 |
| 16 | 23 | Tennessee | 75.1 | 17 | 23 | Indiana | 10.9 | 19 | 23 | Maryland | 10.8 |
| 33 | 24 | Arkansas | 73.6 | 32 | 24 | Nevada | 10.8 | 51 | 24 | Wyoming | 10.4 |
| 44 | 25 | Rhode Island | 72.9 | 26 | 25 | Kentucky | 9.5 | 8 | 25 | Georgia | 10.0 |
| 15 | 26 | Massachusetts | 72.3 | 28 | 26 | Oklahoma | 9.0 | 12 | 25 | Virginia | 10.0 |
| 13 | 27 | Washington | 70.7 | 15 | 27 | Massachusetts | 8.4 | 9 | 27 | North Carolina | 9.9 |
| 28 | 28 | Oklahoma | 69.8 | 22 | 28 | Minnesota | 8.2 | 45 | 28 | Delaware | 9.8 |
| 21 | 29 | Colorado | 69.7 | 44 | 29 | Rhode Island | 7.5 | 5 | 29 | Pennsylvania | 8.1 |
| 29 | 30 | Connecticut | 66.9 | 20 | 30 | Wisconsin | 7.4 | 33 | 30 | Arkansas | 8.0 |
| 24 | 31 | Alabama | 66.6 | 35 | 31 | Kansas | 7.1 | 17 | 31 | Indiana | 7.4 |
| 48 | 32 | Alaska | 65.9 | 1 | 32 | California | 6.5 | 48 | 32 | Alaska | 7.3 |
| 23 | 33 | South Carolina | 65.2 | 37 | 33 | Nebraska | 6.0 | 20 | 32 | Wisconsin | 7.3 |
| 9 | 34 | North Carolina | 64.1 | 14 | 34 | Arizona | 5.5 | 31 | 34 | Iowa | 6.5 |
| 12 | 35 | Virginia | 63.3 | 13 | 34 | Washington | 5.5 | 23 | 35 | South Carolina | 6.1 |
| 45 | 36 | Delaware | 63.2 | 21 | 36 | Colorado | 5.0 | 16 | 36 | Tennessee | 5.9 |
| 6 | 37 | Illinois | 62.0 | 31 | 36 | Iowa | 5.0 | 22 | 37 | Minnesota | 5.7 |
| 25 | 38 | Louisiana | 59.5 | 48 | 38 | Alaska | 4.7 | 25 | 38 | Louisiana | 5.4 |
| 34 | 39 | Mississippi | 57.3 | 39 | 39 | West Virginia | 4.6 | 10 | 38 | Michigan | 5.4 |
| 4 | 40 | New York | 56.6 | 47 | 40 | North Dakota | 4.0 | 24 | 40 | Alabama | 4.6 |
| 14 | 41 | Arizona | 55.8 | 40 | 41 | Hawaii | 3.0 | 18 | 41 | Missouri | 4.5 |
| 11 | 42 | New Jersey | 55.4 | 46 | 41 | South Dakota | 3.0 | 46 | 42 | South Dakota | 4.4 |
| 3 | 43 | Florida | 54.5 | 27 | 43 | Oregon | 2.9 | 47 | 43 | North Dakota | 4.3 |
| 8 | 44 | Georgia | 53.2 | 36 | 44 | New Mexico | 2.5 | 43 | 44 | Montana | 4.2 |
| 19 | 45 | Maryland | 51.7 | 42 | 45 | Maine | 2.2 | 41 | 44 | New Hampshire | 4.2 |
| 32 | 46 | Nevada | 50.6 | 41 | 46 | New Hampshire | 2.1 | 7 | 44 | Ohio | 4.2 |
| 2 | 47 | Texas | 42.2 | 50 | 47 | Vermont | 1.9 | 26 | 47 | Kentucky | 4.0 |
| 49 | 48 | District of Columbia | 39.6 | 30 | 48 | Utah | 1.7 | 34 | 48 | Mississippi | 3.4 |
| 1 | 49 | California | 38.5 | 51 | 49 | Wyoming | 1.6 | 50 | 49 | Vermont | 2.1 |
| 36 | 50 | New Mexico | 37.8 | 38 | 50 | Idaho | 1.2 | 42 | 50 | Maine | 1.9 |
| 40 | 51 | Hawaii | 36.2 | 43 | 51 | Montana | 1.0 | 39 | 51 | West Virginia | 1.8 |

1. May be of any race.

# States and the District of Columbia, Selected Rankings

| Percent high school graduates or more,[1] 2019 | | | | Percent college graduates (bachelor's degree or more),[1] 2019 | | | | Median household income, 2019 | | | |
|---|---|---|---|---|---|---|---|---|---|---|---|
| Popu-lation rank | Percent high school graduates rank | State | Percent high school graduates [col 115] | Popu-lation rank | Percent college graduates rank | State | Percent college graduates [col 116] | Popu-lation rank | Median income rank | State | Median income (dollars) [col 123] |
| | | United States | 88.6 | | | United States | 33.1 | | | United States | 65,712 |
| 51 | 1 | Wyoming | 94.5 | 49 | 1 | District of Columbia | 59.7 | 49 | 1 | District of Columbia | 92,266 |
| 43 | 2 | Montana | 94.2 | 15 | 2 | Massachusetts | 45.0 | 19 | 2 | Maryland | 86,738 |
| 48 | 3 | Alaska | 93.6 | 21 | 3 | Colorado | 42.7 | 15 | 3 | Massachusetts | 85,843 |
| 22 | 3 | Minnesota | 93.6 | 11 | 4 | New Jersey | 41.2 | 11 | 4 | New Jersey | 85,751 |
| 47 | 5 | North Dakota | 93.5 | 19 | 5 | Maryland | 40.9 | 40 | 5 | Hawaii | 83,102 |
| 41 | 6 | New Hampshire | 93.3 | 29 | 6 | Connecticut | 39.8 | 1 | 6 | California | 80,440 |
| 42 | 7 | Maine | 93.2 | 12 | 7 | Virginia | 39.6 | 29 | 7 | Connecticut | 78,833 |
| 50 | 8 | Vermont | 93.1 | 50 | 8 | Vermont | 38.7 | 13 | 8 | Washington | 78,687 |
| 30 | 9 | Utah | 93.0 | 4 | 9 | New York | 37.8 | 41 | 9 | New Hampshire | 77,933 |
| 20 | 10 | Wisconsin | 92.8 | 41 | 10 | New Hampshire | 37.6 | 21 | 10 | Colorado | 77,127 |
| 31 | 11 | Iowa | 92.6 | 22 | 11 | Minnesota | 37.3 | 12 | 11 | Virginia | 76,456 |
| 21 | 12 | Colorado | 92.4 | 13 | 12 | Washington | 37.0 | 30 | 12 | Utah | 75,780 |
| 40 | 12 | Hawaii | 92.4 | 6 | 13 | Illinois | 35.8 | 48 | 13 | Alaska | 75,463 |
| 46 | 14 | South Dakota | 92.1 | 1 | 14 | California | 35.0 | 22 | 14 | Minnesota | 74,593 |
| 37 | 15 | Nebraska | 92.0 | 44 | 15 | Rhode Island | 34.8 | 4 | 15 | New York | 72,108 |
| 49 | 16 | District of Columbia | 91.9 | 30 | 15 | Utah | 34.8 | 44 | 16 | Rhode Island | 71,169 |
| 35 | 17 | Kansas | 91.8 | 27 | 17 | Oregon | 34.5 | 45 | 17 | Delaware | 70,176 |
| 13 | 18 | Washington | 91.7 | 35 | 18 | Kansas | 34.0 | 6 | 18 | Illinois | 69,187 |
| 38 | 19 | Idaho | 91.5 | 40 | 19 | Hawaii | 33.6 | 27 | 19 | Oregon | 67,058 |
| 10 | 20 | Michigan | 91.4 | 43 | 19 | Montana | 33.6 | 51 | 20 | Wyoming | 65,003 |
| 27 | 20 | Oregon | 91.4 | 45 | 21 | Delaware | 33.2 | 47 | 21 | North Dakota | 64,577 |
| 15 | 22 | Massachusetts | 91.3 | 42 | 21 | Maine | 33.2 | 20 | 22 | Wisconsin | 64,168 |
| 5 | 23 | Pennsylvania | 91.0 | 37 | 21 | Nebraska | 33.2 | 2 | 23 | Texas | 64,034 |
| 7 | 24 | Ohio | 90.8 | 8 | 24 | Georgia | 32.5 | 5 | 24 | Pennsylvania | 63,463 |
| 29 | 25 | Connecticut | 90.7 | 9 | 25 | North Carolina | 32.3 | 32 | 25 | Nevada | 63,276 |
| 18 | 25 | Missouri | 90.7 | 5 | 25 | Pennsylvania | 32.3 | 37 | 26 | Nebraska | 63,229 |
| 19 | 27 | Maryland | 90.4 | 20 | 27 | Wisconsin | 31.3 | 50 | 27 | Vermont | 63,001 |
| 45 | 28 | Delaware | 90.3 | 2 | 28 | Texas | 30.8 | 35 | 28 | Kansas | 62,087 |
| 11 | 28 | New Jersey | 90.3 | 3 | 29 | Florida | 30.7 | 14 | 29 | Arizona | 62,055 |
| 12 | 30 | Virginia | 90.0 | 47 | 30 | North Dakota | 30.4 | 8 | 30 | Georgia | 61,980 |
| 6 | 31 | Illinois | 89.8 | 48 | 31 | Alaska | 30.2 | 31 | 31 | Iowa | 61,691 |
| 17 | 32 | Indiana | 89.6 | 14 | 31 | Arizona | 30.2 | 38 | 32 | Idaho | 60,999 |
| 44 | 33 | Rhode Island | 89.3 | 18 | 31 | Missouri | 30.2 | 10 | 33 | Michigan | 59,584 |
| 9 | 34 | North Carolina | 88.6 | 10 | 34 | Michigan | 30.0 | 46 | 34 | South Dakota | 59,533 |
| 3 | 35 | Florida | 88.4 | 46 | 35 | South Dakota | 29.7 | 3 | 35 | Florida | 59,227 |
| 28 | 35 | Oklahoma | 88.4 | 23 | 36 | South Carolina | 29.6 | 42 | 36 | Maine | 58,924 |
| 23 | 37 | South Carolina | 88.3 | 31 | 37 | Iowa | 29.3 | 7 | 37 | Ohio | 58,642 |
| 16 | 38 | Tennessee | 88.0 | 7 | 37 | Ohio | 29.3 | 17 | 38 | Indiana | 57,603 |
| 8 | 39 | Georgia | 87.9 | 51 | 39 | Wyoming | 29.1 | 18 | 39 | Missouri | 57,409 |
| 14 | 40 | Arizona | 87.6 | 38 | 40 | Idaho | 28.7 | 9 | 40 | North Carolina | 57,341 |
| 4 | 40 | New York | 87.6 | 16 | 40 | Tennessee | 28.7 | 43 | 41 | Montana | 57,153 |
| 33 | 42 | Arkansas | 87.5 | 36 | 42 | New Mexico | 27.7 | 23 | 42 | South Carolina | 56,227 |
| 26 | 43 | Kentucky | 87.2 | 17 | 43 | Indiana | 26.9 | 16 | 43 | Tennessee | 56,071 |
| 24 | 44 | Alabama | 87.1 | 24 | 44 | Alabama | 26.3 | 28 | 44 | Oklahoma | 54,449 |
| 39 | 44 | West Virginia | 87.1 | 28 | 45 | Oklahoma | 26.2 | 26 | 45 | Kentucky | 52,295 |
| 32 | 46 | Nevada | 86.9 | 32 | 46 | Nevada | 25.7 | 36 | 46 | New Mexico | 51,945 |
| 25 | 47 | Louisiana | 86.0 | 26 | 47 | Kentucky | 25.1 | 24 | 47 | Alabama | 51,734 |
| 36 | 48 | New Mexico | 85.9 | 25 | 48 | Louisiana | 25.0 | 25 | 48 | Louisiana | 51,073 |
| 34 | 49 | Mississippi | 85.3 | 33 | 49 | Arkansas | 23.3 | 33 | 49 | Arkansas | 48,952 |
| 2 | 50 | Texas | 84.6 | 34 | 50 | Mississippi | 22.3 | 39 | 50 | West Virginia | 48,850 |
| 1 | 51 | California | 84.0 | 39 | 51 | West Virginia | 21.1 | 34 | 51 | Mississippi | 45,792 |

1. Population 25 years and older.

# States and the District of Columbia, Selected Rankings

| Unemployment rate, 2020 | | | | Per capita state taxes, 2019 | | | | Exports of goods by state of origin, 2020 | | | |
|---|---|---|---|---|---|---|---|---|---|---|---|
| Population rank | Unemployment rate rank | State | Unemployment rate [col 171] | Population rank | State taxes rank | State | State taxes per capita (dollars) [col 337] | Population rank | Exports rank | State | Exports (millions of dollars) [col 119] |
| | | United States | 8.1 | | | United States | X | | | United States | 1,431,638 |
| 32 | 1 | Nevada | 12.8 | 47 | 1 | North Dakota | 6,521 | 2 | 1 | Texas | 279,294 |
| 40 | 2 | Hawaii | 11.6 | 40 | 2 | Hawaii | 5,797 | 1 | 2 | California | 156,112 |
| 1 | 3 | California | 10.1 | 50 | 3 | Vermont | 5,495 | 4 | 3 | New York | 61,888 |
| 4 | 4 | New York | 10.0 | 29 | 4 | Connecticut | 5,047 | 25 | 4 | Louisiana | 59,622 |
| 10 | 5 | Michigan | 9.9 | 22 | 5 | Minnesota | 4,996 | 6 | 5 | Illinois | 53,512 |
| 11 | 6 | New Jersey | 9.8 | 1 | 6 | California | 4,764 | 3 | 6 | Florida | 45,807 |
| 6 | 7 | Illinois | 9.5 | 45 | 7 | Delaware | 4,719 | 7 | 7 | Ohio | 45,039 |
| 44 | 8 | Rhode Island | 9.4 | 4 | 8 | New York | 4,710 | 10 | 8 | Michigan | 44,001 |
| 5 | 9 | Pennsylvania | 9.1 | 15 | 9 | Massachusetts | 4,614 | 13 | 9 | Washington | 41,281 |
| 15 | 10 | Massachusetts | 8.9 | 11 | 10 | New Jersey | 4,373 | 8 | 10 | Georgia | 38,784 |
| 36 | 11 | New Mexico | 8.4 | 19 | 11 | Maryland | 3,905 | 11 | 11 | New Jersey | 38,021 |
| 13 | 11 | Washington | 8.4 | 13 | 12 | Washington | 3,676 | 5 | 12 | Pennsylvania | 37,455 |
| 25 | 13 | Louisiana | 8.3 | 51 | 13 | Wyoming | 3,647 | 17 | 13 | Indiana | 35,283 |
| 39 | 13 | West Virginia | 8.3 | 36 | 14 | New Mexico | 3,542 | 23 | 14 | South Carolina | 30,295 |
| 34 | 15 | Mississippi | 8.1 | 44 | 15 | Rhode Island | 3,515 | 9 | 15 | North Carolina | 28,439 |
| 7 | 15 | Ohio | 8.1 | 42 | 16 | Maine | 3,477 | 16 | 16 | Tennessee | 28,134 |
| 49 | 17 | District of Columbia | 8.0 | 35 | 17 | Kansas | 3,443 | 27 | 17 | Oregon | 24,977 |
| 14 | 18 | Arizona | 7.9 | 20 | 18 | Wisconsin | 3,442 | 15 | 18 | Massachusetts | 24,893 |
| 29 | 18 | Connecticut | 7.9 | 33 | 19 | Arkansas | 3,386 | 26 | 19 | Kentucky | 24,750 |
| 48 | 20 | Alaska | 7.8 | 5 | 20 | Pennsylvania | 3,369 | 20 | 20 | Wisconsin | 20,504 |
| 45 | 20 | Delaware | 7.8 | 31 | 21 | Iowa | 3,355 | 22 | 21 | Minnesota | 20,077 |
| 3 | 22 | Florida | 7.7 | 6 | 22 | Illinois | 3,354 | 14 | 22 | Arizona | 19,718 |
| 27 | 23 | Oregon | 7.6 | 39 | 23 | West Virginia | 3,313 | 30 | 23 | Utah | 17,713 |
| 2 | 23 | Texas | 7.6 | 27 | 24 | Oregon | 3,310 | 24 | 24 | Alabama | 17,131 |
| 16 | 25 | Tennessee | 7.5 | 32 | 25 | Nevada | 3,164 | 12 | 25 | Virginia | 16,431 |
| 21 | 26 | Colorado | 7.3 | 30 | 26 | Utah | 3,109 | 29 | 26 | Connecticut | 13,802 |
| 9 | 26 | North Carolina | 7.3 | 12 | 27 | Virginia | 3,080 | 18 | 27 | Missouri | 12,770 |
| 17 | 28 | Indiana | 7.1 | 10 | 28 | Michigan | 3,031 | 19 | 28 | Maryland | 12,672 |
| 19 | 29 | Maryland | 6.8 | 17 | 29 | Indiana | 2,996 | 31 | 29 | Iowa | 12,635 |
| 41 | 30 | New Hampshire | 6.7 | 37 | 30 | Nebraska | 2,975 | 35 | 30 | Kansas | 10,399 |
| 26 | 31 | Kentucky | 6.6 | 43 | 31 | Montana | 2,965 | 32 | 31 | Nevada | 10,348 |
| 8 | 32 | Georgia | 6.5 | 26 | 32 | Kentucky | 2,886 | 34 | 32 | Mississippi | 10,272 |
| 20 | 33 | Wisconsin | 6.3 | 9 | 33 | North Carolina | 2,795 | 21 | 33 | Colorado | 8,283 |
| 22 | 34 | Minnesota | 6.2 | 34 | 34 | Mississippi | 2,785 | 37 | 34 | Nebraska | 6,984 |
| 23 | 34 | South Carolina | 6.2 | 21 | 35 | Colorado | 2,756 | 41 | 35 | New Hampshire | 5,453 |
| 12 | 34 | Virginia | 6.2 | 38 | 36 | Idaho | 2,733 | 28 | 36 | Oklahoma | 5,376 |
| 33 | 37 | Arkansas | 6.1 | 28 | 37 | Oklahoma | 2,712 | 33 | 37 | Arkansas | 5,197 |
| 18 | 37 | Missouri | 6.1 | 7 | 38 | Ohio | 2,579 | 47 | 38 | North Dakota | 5,189 |
| 28 | 37 | Oklahoma | 6.1 | 25 | 39 | Louisiana | 2,527 | 48 | 39 | Alaska | 4,612 |
| 24 | 40 | Alabama | 5.9 | 14 | 40 | Arizona | 2,495 | 39 | 40 | West Virginia | 4,568 |
| 35 | 40 | Kansas | 5.9 | 48 | 41 | Alaska | 2,434 | 45 | 41 | Delaware | 3,919 |
| 43 | 40 | Montana | 5.9 | 24 | 42 | Alabama | 2,361 | 36 | 42 | New Mexico | 3,691 |
| 51 | 43 | Wyoming | 5.8 | 8 | 43 | Georgia | 2,328 | 38 | 43 | Idaho | 3,403 |
| 50 | 44 | Vermont | 5.6 | 46 | 44 | South Dakota | 2,193 | 49 | 44 | District of Columbia | 2,776 |
| 38 | 45 | Idaho | 5.4 | 2 | 45 | Texas | 2,184 | 44 | 45 | Rhode Island | 2,402 |
| 42 | 45 | Maine | 5.4 | 41 | 46 | New Hampshire | 2,184 | 50 | 46 | Vermont | 2,399 |
| 31 | 47 | Iowa | 5.3 | 23 | 47 | South Carolina | 2,179 | 42 | 47 | Maine | 2,328 |
| 47 | 48 | North Dakota | 5.1 | 16 | 48 | Tennessee | 2,171 | 43 | 48 | Montana | 1,433 |
| 30 | 49 | Utah | 4.7 | 18 | 49 | Missouri | 2,148 | 46 | 49 | South Dakota | 1,377 |
| 46 | 50 | South Dakota | 4.6 | 3 | 50 | Florida | 2,086 | 51 | 50 | Wyoming | 1,165 |
| 37 | 51 | Nebraska | 4.2 | 49 | X | District of Columbia | X | 40 | 51 | Hawaii | 320 |

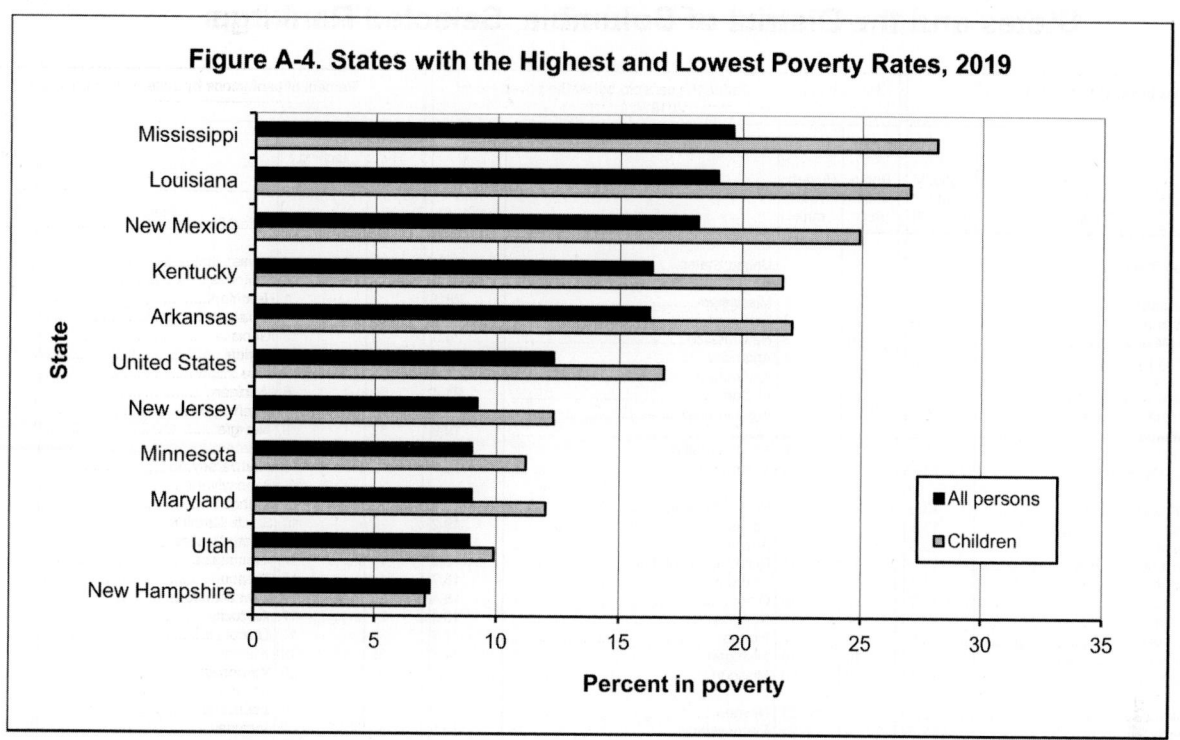

**Figure A-4. States with the Highest and Lowest Poverty Rates, 2019**

■ All persons
□ Children

Percent in poverty

State

Nationally, 88.6 percent of the population 25 years old and over had graduated from high school. Thirty states and the District of Columbia had high school attainment levels of 90 percent or more, led by Montana with 94.5 percent. States in the Midwest and the West tended to have above average high school attainment rates although California had the lowest rate at 84.0 percent, followed by Texas at 84.6 percent. States with above average high school attainment levels do not necessarily have high proportions of college graduates. Nationally, 33.1 percent of the population held bachelor's degrees. In the District of Columbia, 59.7 percent of the population had graduated from college. Even when compared with other large cities, the District of Columbia had among the 10 highest proportions of college graduates in the nation. Of the 50 states, Massachusetts, Colorado, Maryland, Connecticut, New Jersey, Virginia, New Hampshire, Vermont, Minnesota, New York, Washington, California, and Illinois each had 35 percent or more of their populations holding bachelor's degrees or more. States in the Northeast tended to have above average college attainment levels, while states in the South had below average rates.

Median household income ranged from $45,792 in Mississippi to $92,266 in the District of Columbia. Nationally, the median household income was $65,712. In Mississippi, 28.4 percent of households had incomes below $25,000. The District of Columbia had the highest proportion of households earning $100,000 or more, at 47.5 percent followed by New Jersey and Massachusetts at 41.6 percent each.

The poverty threshold for an individual was $13,011 in 2019. Mississippi had the highest poverty rate in the nation, with 19.6 percent of its population living in poverty. Louisiana was the only other state that had a poverty rate at 19 percent or above. The poverty threshold for a four-person family was $26,172. Among children under 18 years old, 16.8 percent were living in poverty. Over 28 percent of children in Mississippi lived in poverty. New Hampshire had the lowest proportion of children in poverty, at 7.1 percent. New Mexico had the highest proportion of residents 65 years and over living in poverty at 13.5 percent, followed by the District of Columbia at 13.3 percent.

The United States labor force decreased by 1.7 percent between 2019 and 2020. From 2000 to 2008, it grew about an average of 1 percent a year but then declined from 2009 to 2011 followed by small increases in recent years. Forty states experienced a decline in their labor force between 2019 and 2020 compared with only nine the previous year. COVID-19 had a major impact on the labor force between 2019 and 2020. Vermont experienced the largest decline, dropping -4.2 percent. Meanwhile, the labor force in Arizona and Washington grew by 1.4 percent in the same period. In 2020, the unemployment rate was 8.1 percent, more than double from 3.7 percent in 2019. Only three states had an unemployment rate below 5.0 percent . Nevada had the highest unemployment rate in the nation at 12.8 percent followed by Hawaii at 11.6 percent and California at 10.1 percent. Nebraska had the lowest unemployment rate in 2020, at 4.2 percent followed by South Dakota at 4.6 percent and Utah at 4.7 percent.

# States and the District of Columbia, Selected Rankings

| Population rank | Poverty rate rank | State | Poverty rate [col 127] | Population rank | Poverty rate rank | State | Poverty rate [col 128] | Population rank | Percent lacking health insurance rank | State | Percent lacking health insurance [col 104] |
|---|---|---|---|---|---|---|---|---|---|---|---|
| | | **Percent of persons below the poverty level, 2019** | | | | **Percent of children under 18 years old below the poverty level, 2019** | | | | **Percent of persons lacking health insurance, 2019** | |
| | | United States | 12.3 | | | United States | 16.8 | | | United States | 9.2 |
| 34 | 1 | Mississippi | 19.6 | 34 | 1 | Mississippi | 28.1 | 2 | 1 | Texas | 18.4 |
| 25 | 2 | Louisiana | 19.0 | 25 | 2 | Louisiana | 27.0 | 28 | 2 | Oklahoma | 14.3 |
| 36 | 3 | New Mexico | 18.2 | 36 | 3 | New Mexico | 24.9 | 8 | 3 | Georgia | 13.4 |
| 26 | 4 | Kentucky | 16.3 | 33 | 4 | Arkansas | 22.1 | 3 | 4 | Florida | 13.2 |
| 33 | 5 | Arkansas | 16.2 | 26 | 5 | Kentucky | 21.7 | 34 | 5 | Mississippi | 13.0 |
| 39 | 6 | West Virginia | 16.0 | 24 | 6 | Alabama | 21.4 | 51 | 6 | Wyoming | 12.3 |
| 24 | 7 | Alabama | 15.5 | 39 | 7 | West Virginia | 20.1 | 48 | 7 | Alaska | 12.2 |
| 28 | 8 | Oklahoma | 15.2 | 28 | 8 | Oklahoma | 19.9 | 32 | 8 | Nevada | 11.4 |
| 16 | 9 | Tennessee | 13.9 | 23 | 9 | South Carolina | 19.7 | 14 | 9 | Arizona | 11.3 |
| 23 | 10 | South Carolina | 13.8 | 16 | 9 | Tennessee | 19.7 | 9 | 9 | North Carolina | 11.3 |
| 9 | 11 | North Carolina | 13.6 | 9 | 11 | North Carolina | 19.5 | 38 | 11 | Idaho | 10.8 |
| 2 | 11 | Texas | 13.6 | 2 | 12 | Texas | 19.2 | 23 | 11 | South Carolina | 10.8 |
| 14 | 13 | Arizona | 13.5 | 14 | 13 | Arizona | 19.1 | 46 | 13 | South Dakota | 10.2 |
| 49 | 13 | District of Columbia | 13.5 | 49 | 14 | District of Columbia | 18.9 | 16 | 14 | Tennessee | 10.1 |
| 8 | 15 | Georgia | 13.3 | 8 | 15 | Georgia | 18.7 | 18 | 15 | Missouri | 10.0 |
| 7 | 16 | Ohio | 13.1 | 7 | 16 | Ohio | 18.4 | 36 | 15 | New Mexico | 10.0 |
| 10 | 17 | Michigan | 13.0 | 4 | 17 | New York | 18.1 | 24 | 17 | Alabama | 9.7 |
| 4 | 17 | New York | 13.0 | 3 | 18 | Florida | 17.7 | 30 | 17 | Utah | 9.7 |
| 18 | 19 | Missouri | 12.9 | 10 | 19 | Michigan | 17.6 | 35 | 19 | Kansas | 9.2 |
| 3 | 20 | Florida | 12.7 | 18 | 20 | Missouri | 17.1 | 33 | 20 | Arkansas | 9.1 |
| 43 | 21 | Montana | 12.6 | 32 | 21 | Nevada | 16.9 | 25 | 21 | Louisiana | 8.9 |
| 32 | 22 | Nevada | 12.5 | 5 | 21 | Pennsylvania | 16.9 | 17 | 22 | Indiana | 8.7 |
| 5 | 23 | Pennsylvania | 12.0 | 45 | 23 | Delaware | 16.4 | 43 | 23 | Montana | 8.3 |
| 17 | 24 | Indiana | 11.9 | 6 | 24 | Illinois | 15.7 | 37 | 23 | Nebraska | 8.3 |
| 46 | 24 | South Dakota | 11.9 | 1 | 25 | California | 15.6 | 21 | 25 | Colorado | 8.0 |
| 1 | 26 | California | 11.8 | 17 | 26 | Indiana | 15.2 | 42 | 25 | Maine | 8.0 |
| 6 | 27 | Illinois | 11.5 | 46 | 27 | South Dakota | 15.0 | 11 | 27 | New Jersey | 7.9 |
| 35 | 28 | Kansas | 11.4 | 43 | 28 | Montana | 14.9 | 12 | 27 | Virginia | 7.9 |
| 27 | 28 | Oregon | 11.4 | 35 | 29 | Kansas | 14.7 | 1 | 29 | California | 7.7 |
| 45 | 30 | Delaware | 11.3 | 29 | 30 | Connecticut | 14.1 | 6 | 30 | Illinois | 7.4 |
| 38 | 31 | Idaho | 11.2 | 44 | 31 | Rhode Island | 14.0 | 27 | 31 | Oregon | 7.2 |
| 31 | 31 | Iowa | 11.2 | 42 | 32 | Maine | 13.8 | 47 | 32 | North Dakota | 6.9 |
| 42 | 33 | Maine | 10.9 | 20 | 33 | Wisconsin | 13.5 | 39 | 33 | West Virginia | 6.7 |
| 44 | 34 | Rhode Island | 10.8 | 12 | 34 | Virginia | 13.4 | 45 | 34 | Delaware | 6.6 |
| 47 | 35 | North Dakota | 10.6 | 38 | 35 | Idaho | 13.2 | 7 | 34 | Ohio | 6.6 |
| 20 | 36 | Wisconsin | 10.4 | 27 | 36 | Oregon | 13.1 | 13 | 34 | Washington | 6.6 |
| 50 | 37 | Vermont | 10.2 | 48 | 37 | Alaska | 13.0 | 26 | 37 | Kentucky | 6.4 |
| 48 | 38 | Alaska | 10.1 | 31 | 37 | Iowa | 13.0 | 41 | 38 | New Hampshire | 6.3 |
| 51 | 38 | Wyoming | 10.1 | 40 | 39 | Hawaii | 12.4 | 19 | 39 | Maryland | 6.0 |
| 29 | 40 | Connecticut | 10.0 | 11 | 40 | New Jersey | 12.3 | 29 | 40 | Connecticut | 5.9 |
| 37 | 41 | Nebraska | 9.9 | 19 | 41 | Maryland | 12.0 | 10 | 41 | Michigan | 5.8 |
| 12 | 41 | Virginia | 9.9 | 13 | 41 | Washington | 12.0 | 5 | 41 | Pennsylvania | 5.8 |
| 13 | 43 | Washington | 9.8 | 15 | 43 | Massachusetts | 11.6 | 20 | 43 | Wisconsin | 5.7 |
| 15 | 44 | Massachusetts | 9.4 | 51 | 43 | Wyoming | 11.6 | 4 | 44 | New York | 5.2 |
| 21 | 45 | Colorado | 9.3 | 22 | 45 | Minnesota | 11.2 | 31 | 45 | Iowa | 5.0 |
| 40 | 45 | Hawaii | 9.3 | 37 | 46 | Nebraska | 11.0 | 22 | 46 | Minnesota | 4.9 |
| 11 | 47 | New Jersey | 9.2 | 21 | 47 | Colorado | 10.9 | 50 | 47 | Vermont | 4.5 |
| 19 | 48 | Maryland | 9.0 | 47 | 48 | North Dakota | 10.2 | 40 | 48 | Hawaii | 4.2 |
| 22 | 48 | Minnesota | 9.0 | 50 | 48 | Vermont | 10.2 | 44 | 49 | Rhode Island | 4.1 |
| 30 | 50 | Utah | 8.9 | 30 | 50 | Utah | 9.9 | 49 | 50 | District of Columbia | 3.5 |
| 41 | 51 | New Hampshire | 7.3 | 41 | 51 | New Hampshire | 7.1 | 15 | 51 | Massachusetts | 3.0 |

# States and the District of Columbia, Selected Rankings

| Popu-rank | State government employment rank | State | State government employment [col 312] | Popu-lation rank | Agricultural sales rank | State | Value of sales (millions of dollars) [col 197] | Popu-lation rank | Violent crime rate rank | State | Violent crime rate (per 100,000 population) [col 108] |
|---|---|---|---|---|---|---|---|---|---|---|---|
| | | United States | 4,459,084 | | | United States | 388,522.7 | | | United States | 379.4 |
| 1 | 1 | California | 439,267 | 1 | 1 | California | 45,154.4 | 49 | 1 | District of Columbia | 1049.0 |
| 2 | 2 | Texas | 314,952 | 31 | 2 | Iowa | 28,956.5 | 48 | 2 | Alaska | 867.1 |
| 4 | 3 | New York | 252,429 | 2 | 3 | Texas | 24,924.0 | 36 | 3 | New Mexico | 832.2 |
| 3 | 4 | Florida | 182,112 | 37 | 4 | Nebraska | 21,983.4 | 16 | 4 | Tennessee | 595.2 |
| 5 | 5 | Pennsylvania | 154,995 | 35 | 5 | Kansas | 18,782.7 | 33 | 5 | Arkansas | 584.6 |
| 10 | 6 | Michigan | 150,483 | 22 | 6 | Minnesota | 18,395.4 | 25 | 6 | Louisiana | 549.3 |
| 9 | 7 | North Carolina | 149,437 | 6 | 7 | Illinois | 17,010.0 | 23 | 7 | South Carolina | 511.3 |
| 11 | 8 | New Jersey | 145,736 | 9 | 8 | North Carolina | 12,900.7 | 24 | 8 | Alabama | 510.8 |
| 7 | 9 | Ohio | 135,613 | 20 | 9 | Wisconsin | 11,427.4 | 18 | 9 | Missouri | 495.0 |
| 8 | 10 | Georgia | 131,128 | 17 | 10 | Indiana | 11,107.3 | 32 | 10 | Nevada | 493.8 |
| 13 | 11 | Washington | 129,855 | 18 | 11 | Missouri | 10,525.9 | 14 | 11 | Arizona | 455.3 |
| 12 | 12 | Virginia | 126,239 | 46 | 12 | South Dakota | 9,721.5 | 19 | 12 | Maryland | 454.1 |
| 6 | 13 | Illinois | 123,673 | 33 | 13 | Arkansas | 9,651.2 | 1 | 13 | California | 441.2 |
| 15 | 14 | Massachusetts | 105,911 | 13 | 14 | Washington | 9,634.5 | 10 | 14 | Michigan | 437.4 |
| 17 | 15 | Indiana | 97,137 | 8 | 15 | Georgia | 9,573.3 | 28 | 15 | Oklahoma | 431.8 |
| 24 | 16 | Alabama | 94,644 | 7 | 16 | Ohio | 9,341.2 | 45 | 16 | Delaware | 422.6 |
| 21 | 17 | Colorado | 94,421 | 47 | 17 | North Dakota | 8,234.1 | 2 | 17 | Texas | 418.9 |
| 23 | 18 | South Carolina | 84,751 | 10 | 18 | Michigan | 8,220.9 | 35 | 18 | Kansas | 410.8 |
| 18 | 19 | Missouri | 84,702 | 5 | 19 | Pennsylvania | 7,758.9 | 6 | 19 | Illinois | 406.9 |
| 22 | 20 | Minnesota | 84,449 | 38 | 20 | Idaho | 7,567.4 | 43 | 20 | Montana | 404.9 |
| 19 | 21 | Maryland | 84,213 | 21 | 21 | Colorado | 7,491.7 | 46 | 21 | South Dakota | 399.0 |
| 26 | 22 | Kentucky | 81,509 | 28 | 22 | Oklahoma | 7,465.5 | 21 | 22 | Colorado | 381.0 |
| 16 | 23 | Tennessee | 79,800 | 3 | 23 | Florida | 7,357.3 | 3 | 23 | Florida | 378.4 |
| 25 | 24 | Louisiana | 76,689 | 34 | 24 | Mississippi | 6,196.0 | 9 | 24 | North Carolina | 371.8 |
| 27 | 25 | Oregon | 74,268 | 24 | 25 | Alabama | 5,980.6 | 17 | 25 | Indiana | 370.8 |
| 14 | 26 | Arizona | 74,127 | 26 | 26 | Kentucky | 5,737.9 | 4 | 26 | New York | 358.6 |
| 20 | 27 | Wisconsin | 71,640 | 4 | 27 | New York | 5,369.2 | 8 | 27 | Georgia | 340.7 |
| 28 | 28 | Oklahoma | 64,235 | 27 | 28 | Oregon | 5,006.8 | 15 | 28 | Massachusetts | 327.6 |
| 30 | 29 | Utah | 63,247 | 12 | 29 | Virginia | 3,960.5 | 39 | 29 | West Virginia | 316.6 |
| 33 | 30 | Arkansas | 62,409 | 14 | 30 | Arizona | 3,852.0 | 5 | 30 | Pennsylvania | 306.4 |
| 29 | 31 | Connecticut | 60,627 | 16 | 31 | Tennessee | 3,798.9 | 37 | 31 | Nebraska | 300.9 |
| 40 | 32 | Hawaii | 56,571 | 43 | 32 | Montana | 3,520.6 | 13 | 32 | Washington | 293.9 |
| 35 | 33 | Kansas | 54,456 | 25 | 33 | Louisiana | 3,173.0 | 7 | 33 | Ohio | 293.2 |
| 34 | 34 | Mississippi | 53,994 | 23 | 34 | South Carolina | 3,008.7 | 20 | 34 | Wisconsin | 293.2 |
| 31 | 35 | Iowa | 51,032 | 36 | 35 | New Mexico | 2,582.3 | 40 | 35 | Hawaii | 285.5 |
| 36 | 36 | New Mexico | 46,585 | 19 | 36 | Maryland | 2,472.8 | 47 | 36 | North Dakota | 284.6 |
| 39 | 37 | West Virginia | 38,877 | 30 | 37 | Utah | 1,838.6 | 27 | 37 | Oregon | 284.4 |
| 37 | 38 | Nebraska | 35,201 | 51 | 38 | Wyoming | 1,472.1 | 34 | 38 | Mississippi | 277.9 |
| 32 | 39 | Nevada | 30,167 | 45 | 39 | Delaware | 1,466.0 | 31 | 39 | Iowa | 266.6 |
| 45 | 40 | Delaware | 25,447 | 11 | 40 | New Jersey | 1,098.0 | 22 | 40 | Minnesota | 236.4 |
| 38 | 41 | Idaho | 24,893 | 50 | 41 | Vermont | 781.0 | 30 | 41 | Utah | 235.6 |
| 43 | 42 | Montana | 24,648 | 39 | 42 | West Virginia | 754.3 | 38 | 42 | Idaho | 223.8 |
| 48 | 43 | Alaska | 24,457 | 42 | 43 | Maine | 667.0 | 44 | 43 | Rhode Island | 221.1 |
| 42 | 44 | Maine | 20,776 | 32 | 44 | Nevada | 665.8 | 51 | 44 | Wyoming | 217.4 |
| 44 | 45 | Rhode Island | 19,068 | 29 | 45 | Connecticut | 580.1 | 26 | 45 | Kentucky | 217.1 |
| 41 | 46 | New Hampshire | 18,873 | 40 | 46 | Hawaii | 563.8 | 12 | 46 | Virginia | 208.0 |
| 47 | 47 | North Dakota | 18,259 | 15 | 47 | Massachusetts | 475.2 | 11 | 47 | New Jersey | 206.9 |
| 46 | 48 | South Dakota | 14,281 | 41 | 48 | New Hampshire | 187.8 | 50 | 48 | Vermont | 202.2 |
| 50 | 49 | Vermont | 13,932 | 48 | 49 | Alaska | 70.5 | 29 | 49 | Connecticut | 183.6 |
| 51 | 50 | Wyoming | 12,869 | 44 | 50 | Rhode Island | 58.0 | 41 | 50 | New Hampshire | 152.5 |
| 49 | X | District of Columbia | X | 49 | NA | District of Columbia | NA | 42 | 51 | Maine | 115.2 |

Table A. States — **Land Area and Population Characteristics**

| State code | STATE | Population, 2020 | | | | Race alone or in combination, not Hispanic or Latino (percent) | | | | Hispanic or Latino[2] (percent) | Age (percent) | | | | | |
|---|---|---|---|---|---|---|---|---|---|---|---|---|---|---|---|---|
| | | Land area,[1] 2020 (sq mi) | Total persons, 2020 | Rank | Per square mile | White | Black | American Indian, Alaska Native | Asian and Pacific Islander | | Under 5 years | 5 to 14 years | 15 to 24 years | 25 to 34 years | 35 to 44 years | 45 to 54 years |
| | | 1 | 2 | 3 | 4 | 5 | 6 | 7 | 8 | 9 | 10 | 11 | 12 | 13 | 14 | 15 |

1. Dry land or land partially or temporarily covered by water.    2. May be of any race.

Table A. States — **Population Characteristics, Immigration, and Households**

| STATE | Population characteristics, 2020 (cont.) | | | | | | | | | Households, 2019 | | | | | |
|---|---|---|---|---|---|---|---|---|---|---|---|---|---|---|---|
| | Age (percent) (cont.) | | | | Median age 2019 | Percent female 2020 | Percent foreign born 2019 | Percent born in state of residence 2019 | Immigrants admitted to legal status, 2019 | Number | Percent change, 2017–2018 | Persons per house-hold | Household type | | |
| | 55 to 64 years | 65 to 74 years | 75 to 84 years | 85 years and over | | | | | | | | | Married Couple family | Female house-holder family | House-holder living alone |
| | 16 | 17 | 18 | 19 | 20 | 21 | 22 | 23 | 24 | 25 | 26 | 27 | 28 | 29 | 30 |

Table A. States — **Population Change**

| STATE | Population, 1990–2010 | | | Population change, 1990–2020 | | | | | | | | Population, 2020–2030 | | |
|---|---|---|---|---|---|---|---|---|---|---|---|---|---|---|
| | Census counts | | | Percent change | | | Components of change, 2010–2020 | | | | | Projections | | |
| | 1990 | 2000 | 2010 | 1990–2000 | 2000–2010 | 2010–2020 | Births | Deaths | Migration | | | 2020 | 2025 | 2030 |
| | | | | | | | | | Net migration | Inter-national | Net internal | | | |
| | 31 | 32 | 33 | 34 | 35 | 36 | 37 | 38 | 39 | 40 | 41 | 42 | 43 | 44 |

Table A. States — **Population Characteristics**

| STATE | Population characteristics, 2010 | | | | | | | | | | | | | | | | | | |
|---|---|---|---|---|---|---|---|---|---|---|---|---|---|---|---|---|---|---|---|
| | Race (percent) | | | | | | | Age (percent) | | | | | | | | | | Median age | Percent female |
| | White alone | Black alone | American Indian, Alaska Native alone | Asian and Pacific Islander alone | Some other race or two or more races | Percent Hispanic or Latino[1] | Percent foreign born | Under 5 years | 5 to 17 years | 18 to 24 years | 25 to 34 years | 35 to 44 years | 45 to 54 years | 55 to 64 years | 65 to 74 years | 75 to 84 years | 85 years and over | | |
| | 45 | 46 | 47 | 48 | 49 | 50 | 51 | 52 | 53 | 54 | 55 | 56 | 57 | 58 | 59 | 60 | 61 | 62 | 63 |

1. May be of any race.

## Table A. States — Households and Housing Units

| STATE | Households, 2010 | | | | | Housing units, 2010 | | | | | | | | | |
|---|---|---|---|---|---|---|---|---|---|---|---|---|---|---|---|
| | | | | Percent | | | | | Occupied units | | | | | | |
| | | | | | | | | | Owner-occupied | | | | Renter-occupied | | |
| | | | | | | | | | | | Median owner cost as a percent of income | | | | |
| | Number | Percent change, 2000-2010 | Persons per house-hold | Female family house-holder[1] | One-person house-holds | Total | Percent change, 2000-2010 | Total | Percent | Median value[2] (dollars) | With a mort-gage | Without a mort-gage[3] | Median gross rent[4] (dollars) | Median rent as a percent of income | Sub-standard units[5] (percent) |
| | 64 | 65 | 66 | 67 | 68 | 69 | 70 | 71 | 72 | 73 | 74 | 75 | 76 | 77 | 78 |

1. No spouse present.    2. Specified owner-occupied units.    3. Median monthly costs is often in the minimum category—10.0 percent or less, which is indicated as 10.0 percent.
4. Specified renter-occupied units.    5. Overcrowded or lacking complete plumbing facilities.

## Table A. States — Housing Units

| STATE | Housing units, 2019 | | | | | | | | | | | | | |
|---|---|---|---|---|---|---|---|---|---|---|---|---|---|---|
| | | | Occupied units | | | | | | | | | | | |
| | | | | | Percent who pay 30 percent or more of income for housing expenses[1] | | Median owner cost as a percent of income | | | | | | | Percent living in a different house than 1 year ago |
| | Total | Percent change, 2018-2019 | Total | Percent owner-occupied | Owners with a mortgage | Renter | With a mort-gage | Without a mort-gage[2] | Median monthly housing costs (dollars) | Median value of units[3] (dollars) | Percent valued over $500,000 | Median gross rent[4] (dollars) | Sub-standard units[5] (percent) | |
| | 79 | 80 | 81 | 82 | 83 | 84 | 85 | 86 | 87 | 88 | 89 | 90 | 91 | 92 |

1. Excludes units where owner costs or gross rent as a percentage of houshold income cannot be calculated.    2. Median monthly costs is often in the minimum category—10.0 percent or less, which is indicated as 10.0 percent.    3. Specified owner-occupied units.    4. Specified renter-occupied units.    5. Overcrowded or lacking complete plumbing facilities.

## Table A. States — Residential Contruction, Vital Statistics, and Health

| STATE | Value of residential construction authorized by building permits, 2020 | | | New manufactured homes shipped, 2019 | Births, 2019 | | Deaths, 2018 | | | | | Percent lacking health insurance, 2019 | | Medicare beneficiaries, 65 years and older, 2019 |
|---|---|---|---|---|---|---|---|---|---|---|---|---|---|---|
| | | | | | | | Number | | Rate | | | | | |
| | | | | | | | | | Total | | | | | |
| | New construction ($1,000) | Number of housing units | Percent single family | | Total | Rate[1] | Total | Infant[2] | Crude[1] | Age-adjusted[1] | Infant[3] | All persons | Children under 19 years | |
| | 93 | 94 | 95 | 96 | 97 | 98 | 99 | 100 | 101 | 102 | 103 | 104 | 105 | 106 |

1. Per 1,000 resident population.    2. Deaths of infants under 1 year old.    3. Deaths of infants under 1 year old per 1,000 live births.

## Table A. States — Crime and Education

| STATE | Serious crime known to police,[1] 2019 | | | | Public elementary and secondary school enrollment, 2019–2020 | | Educational attainment[3] (percent) | | | | Local government expenditures for education, 2017-2018 | |
|---|---|---|---|---|---|---|---|---|---|---|---|---|
| | Violent crime | | Property crime | | | | 2010 | | 2019 | | | |
| | | | | | | Student/ teacher ratio | High school graduate or more | Bach-elor's degree or more | High school graduate or more | Bach-elor's degree or more | Total current expen-ditures (mil dol) | Current expen-ditures per student (dollars) |
| | Number | Rate[2] | Number | Rate[2] | Total | | | | | | | |
| | 107 | 108 | 109 | 110 | 111 | 112 | 113 | 114 | 115 | 116 | 117 | 118 |

1. Data for serious crimes have not been adjusted for underreporting; this may affect comparability between geographic areas and over time.    2. Per 100,000 population estimated by the FBI.
3. Persons 25 years old and over.

## Table A. States — Exports, Income, and Poverty

| STATE | Exports of goods by state of origin, 2020 (mil dol) | | | Income, 2019 | | | | | Percent below poverty level, 2019 | | | | | | |
| | | | | | Households | | | | | | | | Families with related children under 18 years | | |
| | Total | Manu-factured | Non-manu-factured | Per capita income (dollars) | Median income (dollars) | Percent with income of $24,999 or less | Percent with income of $100,000 or more | Median income of family of four | All persons | Children under 18 years | Persons 65 years and over | All families | Married-couple families | Male house-holder[1] families | Female house-holder[1] families |
| | 119 | 120 | 121 | 122 | 123 | 124 | 125 | 126 | 127 | 128 | 129 | 130 | 131 | 132 | 133 |

1. No spouse present.

## Table A. States — Personal Income

| STATE | Personal income | | | | | | | | | | | | | |
| | | | Per capita[1], 2020 | | Sources of personal income (mil dol) | | | | | | | | | |
| | | | | | | | | | | Transfer payments, 2019 | | | | |
| | | | | | | | | | | | Government payments to individuals | | | |
| | Total, 2020 (mil dol) | Percent change, 2019–2020 | Dollars | Rank | Wages and salaries[2], 2020 | Proprietors' income, 2020 | Dividends, interest, and rent, 2020 | Total | Total | Social Security | Medical payments | Income main-tenance | Unemploy-ment insurance |
| | 134 | 135 | 136 | 137 | 138 | 139 | 140 | 141 | 142 | 143 | 144 | 145 | 146 |

1. Based on the resident population estimated as of July 1 of the year shown.     2. Includes supplements to wages and salaries.

## Table A. States — Personal Income and Earnings

| STATE | | Disposable personal income, 2019 | | Earnings, 2020 | | | | | | | | | | Gross state product, 2020 (mil dol) |
| | | | | | | Percent by selected industries | | | | | | | | |
| | | | | | | Goods-related[2] | | Service-related and other[3] | | | | | | |
| | Personal tax payments, 2018 (mil dol) | Total (mil dol) | Per capita[1] (dollars) | Total (mil dol) | Farm | Total | Manu-facturing | Total | Retail trade | Finance, insurance, real estate, rental and leasing | Health care and social assistance | Govern-ment | |
| | 147 | 148 | 149 | 150 | 151 | 152 | 153 | 154 | 155 | 156 | 157 | 158 | 159 |

1. Based on the resident population estimated as of July 1 of the year shown.     2. Includes mining, construction, and manufacturing.     3. Includes private sector earnings in forestry, fishing, related activities, and other; utilities; wholesale trade; transportation and warehousing; and information.

## Table A. States — Social Security, Employment, and Labor Force

| STATE | Social Security beneficiaries, December 2019 | | Supple-mental Security Income recipients, December 2019 | Civilian employment and selected occupations,[2] 2019 | | | | Civilian labor force (annual average), 2020 | | | | |
| | | | | | Percent | | | | | | Unemployed | |
| | Number | Rate[1] | | Total | Management, business, science, and arts occupations | Services, sales, and office | Construction and production | Total (1,000) | Percent change, 2019–2020 | Employed (1,000) | Total (1,000) | Rate[3] |
| | 160 | 161 | 162 | 163 | 164 | 165 | 166 | 167 | 168 | 169 | 170 | 171 |

1. Per 1,000 resident population estimated as of July 1 of the year shown.     2. Persons 16 years old and over.     3. Percent of civilian labor force.

## Table A. States — **Nonfarm Employment and Earnings**

| STATE | Nonfarm employment and earnings, 2020 | | | | | | | | | | | |
|---|---|---|---|---|---|---|---|---|---|---|---|---|
| | Employed | | Manufacturing | | | Employment (1,000) | | | | | | |
| | Total (1,000) | Percent change, 2019–2020 | Employment (1,000) | Average earnings of production workers | | Construction | Transportation and public utilities | Whole-sale trade | Retail trade | Information | Financial activities | Services[1] |
| | | | | Hourly | Weekly | | | | | | | |
| | 172 | 173 | 174 | 175 | 176 | 177 | 178 | 179 | 180 | 181 | 182 | 183 |

1. Includes professional and business services, educational and health services, leisure and hospitality, and other services.

## Table A. States — **Agriculture**

| STATE | Agriculture, 2017 | | | | | | | | | | | |
|---|---|---|---|---|---|---|---|---|---|---|---|---|
| | Farms | | | Farm producers whose primary occupation is farming (percent) | Government payments, average per farm that recieves payments (dollars) | Land in farms | | | | | Value of land and buildings (dollars) | |
| | | Percent with: | | | | | | Acres | | | | |
| | Number | Fewer than 50 acres | 1,000 acres or more | | | Acreage (1,000) | Percent change, 2012–2017 | Average size of farm | Total irrigated (1,000) | Total cropland (1,000) | Average per farm | Average per acre |
| | 184 | 185 | 186 | 187 | 188 | 189 | 190 | 191 | 192 | 193 | 194 | 195 |

## Table A. States — **Agriculture, Land, and Water**

| STATE | Agriculture, 2017 (cont.) | | | | | | | | Land, 2015 | | | |
|---|---|---|---|---|---|---|---|---|---|---|---|---|
| | Value of machinery and equipment, average per farm (dollars) | Value of products sold | | | | | Organic Farms (number) | Farms with Internet access (percent) | Owned by the federal government (percent) | Developed land (percent) | Rural land (percent) | Public water supply withdrawn, 2015 (mil gal per day) |
| | | Total (mil dol) | Average per farm (dollars) | Percent from: | | | | | | | | |
| | | | | Crops | Livestock and poultry products | | | | | | | |
| | 196 | 197 | 198 | 199 | 200 | | 201 | 202 | 203 | 204 | 205 | 206 |

## Table A. States — **Manufactures and Construction**

| STATE | Manufactures, 2017 | | | | | | | | | | | Construction, 2017 | | | |
|---|---|---|---|---|---|---|---|---|---|---|---|---|---|---|---|
| | All employees | | | Production workers[1] | | | | | Value added by manu-facture (mil dol) | Sales, values of ship-ments or revenue (mil dol) | Total cost of materials (mil dol) | Employees | | | |
| | | | | | | Wages | | | | | | Number of estab-lishments | | | |
| | Number (1,000) | Percent change, 2012–2017 | Annual payroll (mil dol) | Number (1,000) | Work hours (millions) | Total (mil dol) | Average per worker (dollars) | | | | | | Number | Percent change, 2012–2017 | Sales, values of ship-ments or revenue (mil dol) | Annual payroll (mil dol) |
| | 207 | 208 | 209 | 210 | 211 | 212 | 213 | 214 | 215 | 216 | 217 | 218 | 219 | 220 | 221 |

1. Production and/or development and exploration workers

## Table A. States — **Wholesale Trade and Retail Trade**

| STATE | Wholesale trade, 2017 | | | | | Retail trade,[1] 2017 | | | | | | | | |
|---|---|---|---|---|---|---|---|---|---|---|---|---|---|---|
| | | Employees | | | | | | Employees | | | | | | |
| | Number of establishments | Number | Percent change, 2012–2017 | Sales (mil dol) | Annual payroll (mil dol) | Number of establishments | Total | Percent change, 2012–2017 | Motor vehicle and parts dealers | Food and beverage stores | Clothing and clothing accessory stores | General merchandise stores | Sales (mil dol) | Annual payroll (mil dol) |
| | 222 | 223 | 224 | 225 | 226 | 227 | 228 | 229 | 230 | 231 | 232 | 233 | 234 | 235 |

1. Establishments with payroll.

## Table A. States — **Information**

| STATE | Information, 2017 | | | | | | | | | | |
|---|---|---|---|---|---|---|---|---|---|---|---|
| | | Employees | | | | | | | | Sales, values of shipments or revenue (mil dol) | Annual payroll (mil dol) |
| | Number of establishments | Number | Percent change, 2012–2017 | Publishing, except Internet | Motion picture and sound recording | Broadcasting, except Internet | Internet publishing and broadcasting and web search portals | Tele-communications | Data processing, hosting, and related services | | |
| | 236 | 237 | 238 | 239 | 240 | 241 | 242 | 243 | 244 | 245 | 246 |

## Table A. States — **Utilities, Transportation and Warehousing, and Finance and Insurance**

| STATE | Utilities, 2017 | | | | | Transportation and warehousing, 2017 | | | | | Finance and insurance, 2017 | | | | |
|---|---|---|---|---|---|---|---|---|---|---|---|---|---|---|---|
| | | Employees | | Sales, values of shipments or revenue (mil dol) | Annual payroll (mil dol) | | Employees | | Sales, value of shipments, or revenue (mil dol) | Annual payroll (mil dol) | | Employees | | Sales, values of shipments or revenue (mil dol) | Annual payroll (mil dol) |
| | Number of establishments | Number | Percent change, 2012–2017 | | | Number of establishments | Number | Percent change, 2012–2017 | | | Number of establishments | Number | Percent change, 2012–2017 | | |
| | 247 | 248 | 249 | 250 | 251 | 252 | 253 | 254 | 255 | 256 | 257 | 258 | 259 | 260 | 261 |

## Table A. States — **Real Estate and Rental and Leasing and Professional, Scientific, and Technical Services**

| STATE | Real estate and rental and leasing, 2017 | | | | | Professional, scientific, and technical services, 2017 | | | | | | | | |
|---|---|---|---|---|---|---|---|---|---|---|---|---|---|---|
| | | Employees | | Sales, values of shipments or revenue (mil dol) | Annual payroll (mil dol) | | | Employees | | | | | Sales, values of shipments or revenue (mil dol) | Annual payroll (mil dol) |
| | Number of establishments | Number | Percent change, 2012–2017 | | | Number of establishments | Total | Percent change, 2012–2017 | Legal services | Accounting, tax preparation, bookkeeping, and payroll services | Architectural, engineering, and related services | Computer systems design and related services | | |
| | 262 | 263 | 264 | 265 | 266 | 267 | 268 | 269 | 270 | 271 | 272 | 273 | 274 | 275 |

## Table A. States — Health Care and Social Assistance

| STATE | Health care and social assistance, 2017 | | | | | | | | | | | | |
| | Subject to federal tax | | | | | | Tax-exempt | | | | | | |
| | | Employees | | | | Sales, value of shipments, or revenue (mil dol) | Annual payroll (mil dol) | | Employees | | | | Sales, values of shipments or revenue (mil dol) | Annual payroll (mil dol) |
| | Number of estab-lishments | Total | Percent change, 2012–2017 | Ambulatory health care services | Hospitals | | | Number of estab-lishments | Total | Percent change, 2012–2017 | Ambulatory health care services | Hospitals | | |
| | 276 | 277 | 278 | 279 | 280 | 281 | 282 | 283 | 284 | 285 | 286 | 287 | 288 | 289 |

## Table A. States — Arts, Entertainment, and Recreation and Accommodation and Food Services

| STATE | Arts, entertainment, and recreation, 2017 | | | | | Accommodation and food services, 2017 | | | | | |
| | | Employees | | | | | | Employees | | | |
| | Number of estab-lishments | Number | Percent change, 2012–2017 | Sales, value of shipments, or revenue (mil dol) | Annual payroll (mil dol) | Number of estab-lishments | Total | Percent change, 2012–2017 | Food services and drinking places | Sales, values of shipments or revenue (mil dol) | Annual payroll (mil dol) |
| | 290 | 291 | 292 | 293 | 294 | 295 | 296 | 297 | 298 | 299 | 300 |

## Table A. States — Other Services, Except Public Administration and Government Employment

| STATE | Other services, except public administration, 2017 | | | | | | | | Government employment, 2019 | | |
| | | Employees | | | | | Sales, values of shipments or revenue (mil dol) | Annual payroll (mil dol) | Federal civilian | Federal military | State and local |
| | Number of estab-lishments | Total | Percent change, 2012–2017 | Repair and mainte-nance | Personal and laundry services | Religious, civic, and similar services | | | | | |
| | 301 | 302 | 303 | 304 | 305 | 306 | 307 | 308 | 309 | 310 | 311 |

## Table A. States — Federal Funds

| STATE | State government employment and payroll, 2019 | | | | | | | | | | | |
| | State government employment, 2019 | | | State government payroll, 2019 | | Full-time equivalent payroll (1000 dollars) | | | | | | |
| | | | | | | | Percent of total for: | | | | | |
| | Full-time equivalent employees | Full-time employees | Part-time employees | Full-time March payroll (1000 dollars) | Part-time March payroll (1000 dollars) | Total March payroll (1000 dollars)[1] | Administration | Judicial and legal | Police | Corrections | Highways and transporta-tion | Welfare |
| | 312 | 313 | 314 | 315 | 316 | 317 | 318 | 319 | 320 | 321 | 322 | 323 |

1. Includes program categories not shown separately.

| STATE | State government employment and payroll, 2019 (cont.) | | | | | | | State government finances, 2019 | | | | | | | |
| | Full-time equivalent payroll (1000 dollars) (cont.) | | | | | | | General revenue (mil dol) | | | | | | | |
| | Percent of total for: | | | | | | | | From federal government | | | From own sources | | | |
| | | | | | | | | | | | | | Taxes | | Taxes per capita[1] (dollars) | |
| | Health | Hospitals | Social Insurance administration | Natural resources and parks | Utilities, sewerge, and waste management | Elementary and secondary education and libraries | Higher education | Total | Total | Per capita[1] (dollars) | Total | Total | Sales and gross receipts | Total | Sales and gross receipts |
| | 324 | 325 | 326 | 327 | 328 | 329 | 330 | 331 | 332 | 333 | 334 | 335 | 336 | 337 | 338 |

1. Based on resident population estimated as of July 1 of the year shown.

| STATE | State government finances, 2019 (cont.) | | | | | | | | | | | | Voting and registration, November 2020 | | Presidential election, 2020 (percent of vote cast) | | |
| | General expenditures (mil dol) (cont.) | | | | | | | | | | Debt outstanding | | | | | | |
| | | | Direct expenditures | | By selected function | | | | | | | | | | | | |
| | Total | To local govern-ments | Total | Per capita[1] (dollars) | Educa-tion | Health and hospitals | Highways | Public safety | Public welfare | Natural resources, parks, and recreation | Total (mil dol) | Per capita[1] | Percent registered | Percent voted | Demo-cratic | Repub-lican | All other |
| | 339 | 340 | 341 | 342 | 343 | 344 | 345 | 346 | 347 | 348 | 349 | 350 | 351 | 352 | 353 | 354 | 355 |

1. Based on resident population estimated as of July 1 of the year shown.

# Table A. States — Land Area and Population Characteristics

| State code | STATE | Land area,[1] 2,020 (sq mi) | Total persons, 2020 | Rank | Per square mile | White | Black | American Indian, Alaska Native | Asian and Pacific Islander | Hispanic or Latino[2] (percent) | Under 5 years | 5 to 14 years | 15 to 24 years | 25 to 34 years | 35 to 44 years | 45 to 54 years |
|---|---|---|---|---|---|---|---|---|---|---|---|---|---|---|---|---|
| | | 1 | 2 | 3 | 4 | 5 | 6 | 7 | 8 | 9 | 10 | 11 | 12 | 13 | 14 | 15 |
| 0 | United States | 3,533,044 | 329,484,123 | X | 93.3 | 61.8 | 13.6 | 1.3 | 7.2 | 18.6 | 5.9 | 12.4 | 12.9 | 14.0 | 12.8 | 12.3 |
| 1 | Alabama | 50,647 | 4,921,532 | 24 | 97.2 | 66.6 | 27.3 | 1.2 | 2.0 | 4.6 | 5.9 | 12.4 | 12.9 | 13.2 | 12.1 | 12.4 |
| 2 | Alaska | 571,017 | 731,158 | 48 | 1.3 | 65.9 | 4.7 | 19.2 | 10.6 | 7.3 | 6.8 | 13.8 | 13.0 | 15.9 | 13.4 | 11.4 |
| 4 | Arizona | 113,653 | 7,421,401 | 14 | 65.3 | 55.8 | 5.5 | 4.4 | 4.7 | 31.9 | 5.7 | 12.6 | 13.3 | 13.9 | 12.3 | 11.6 |
| 5 | Arkansas | 52,038 | 3,030,522 | 33 | 58.2 | 73.6 | 16.3 | 1.7 | 2.5 | 8.0 | 6.1 | 13.0 | 13.1 | 13.2 | 12.3 | 11.8 |
| 6 | California | 155,854 | 39,368,078 | 1 | 252.6 | 38.5 | 6.5 | 1.0 | 17.6 | 39.5 | 5.9 | 12.6 | 13.0 | 15.3 | 13.5 | 12.5 |
| 8 | Colorado | 103,638 | 5,807,719 | 21 | 56.0 | 66.9 | 11.6 | 1.3 | 4.7 | 21.9 | 5.6 | 12.2 | 12.9 | 15.9 | 14.1 | 12.1 |
| 9 | Connecticut | 4,843 | 3,557,006 | 29 | 734.5 | 66.9 | 11.6 | 0.7 | 5.6 | 17.2 | 5.0 | 11.4 | 13.3 | 12.6 | 12.1 | 13.0 |
| 10 | Delaware | 1,949 | 986,809 | 45 | 506.4 | 63.2 | 23.8 | 0.9 | 4.9 | 9.8 | 5.5 | 11.6 | 12.0 | 13.1 | 11.7 | 11.8 |
| 11 | District of Columbia | 61 | 712,816 | 49 | 11,666.4 | 39.6 | 45.3 | 0.8 | 5.6 | 11.4 | 6.2 | 9.8 | 12.3 | 23.1 | 15.9 | 10.4 |
| 12 | Florida | 53,648 | 21,733,312 | 3 | 405.1 | 54.5 | 16.5 | 0.6 | 3.7 | 26.5 | 5.2 | 11.0 | 11.4 | 13.0 | 12.2 | 12.4 |
| 13 | Georgia | 57,716 | 10,710,017 | 8 | 185.6 | 53.2 | 33.0 | 0.7 | 5.2 | 10.0 | 6.0 | 13.2 | 13.6 | 14.1 | 13.0 | 13.0 |
| 15 | Hawaii | 6,422 | 1,407,006 | 40 | 219.1 | 36.2 | 3.0 | 1.7 | 75.6 | 10.9 | 6.0 | 11.8 | 11.5 | 14.0 | 12.9 | 11.7 |
| 16 | Idaho | 82,645 | 1,826,913 | 38 | 22.1 | 83.5 | 1.2 | 1.9 | 2.7 | 13.0 | 6.2 | 14.1 | 13.5 | 13.2 | 12.8 | 11.2 |
| 17 | Illinois | 55,514 | 12,587,530 | 6 | 226.7 | 62.0 | 15.0 | 0.5 | 6.7 | 17.6 | 5.8 | 12.4 | 13.0 | 13.8 | 13.0 | 12.4 |
| 18 | Indiana | 35,826 | 6,754,953 | 17 | 188.5 | 79.9 | 10.9 | 0.7 | 3.2 | 7.4 | 6.1 | 13.0 | 13.7 | 13.3 | 12.4 | 12.0 |
| 19 | Iowa | 55,854 | 3,163,561 | 31 | 56.6 | 86.2 | 5.0 | 0.7 | 3.4 | 6.5 | 6.1 | 12.9 | 13.8 | 12.6 | 12.3 | 11.3 |
| 20 | Kansas | 81,759 | 2,913,805 | 35 | 35.6 | 77.6 | 7.1 | 1.8 | 4.0 | 12.4 | 6.3 | 13.5 | 14.1 | 13.1 | 12.5 | 11.1 |
| 21 | Kentucky | 39,491 | 4,477,251 | 26 | 113.4 | 85.7 | 9.5 | 0.7 | 2.2 | 4.0 | 6.0 | 12.5 | 13.0 | 13.2 | 12.3 | 12.4 |
| 22 | Louisiana | 43,205 | 4,645,318 | 25 | 107.5 | 59.5 | 33.3 | 1.2 | 2.3 | 5.4 | 6.4 | 13.1 | 12.8 | 14.0 | 12.9 | 11.6 |
| 23 | Maine | 30,845 | 1,350,141 | 42 | 43.8 | 94.4 | 2.2 | 1.4 | 1.9 | 1.9 | 4.7 | 10.3 | 11.2 | 12.2 | 11.6 | 12.6 |
| 24 | Maryland | 9,711 | 6,055,802 | 19 | 623.6 | 51.7 | 31.6 | 0.8 | 7.9 | 10.8 | 5.9 | 12.4 | 12.3 | 13.6 | 13.1 | 12.8 |
| 25 | Massachusetts | 7,801 | 6,893,574 | 15 | 883.7 | 72.3 | 8.4 | 0.6 | 8.2 | 12.6 | 5.1 | 10.8 | 13.4 | 14.5 | 12.5 | 12.6 |
| 26 | Michigan | 56,606 | 9,966,555 | 10 | 176.1 | 76.7 | 15.0 | 1.3 | 4.1 | 5.4 | 5.6 | 11.9 | 13.1 | 13.2 | 11.8 | 12.3 |
| 27 | Minnesota | 79,626 | 5,657,342 | 22 | 71.0 | 80.8 | 8.2 | 1.7 | 6.1 | 5.7 | 6.1 | 13.0 | 12.7 | 13.4 | 13.1 | 11.7 |
| 28 | Mississippi | 46,926 | 2,966,786 | 34 | 63.2 | 57.3 | 38.3 | 0.9 | 1.5 | 3.4 | 6.1 | 13.2 | 13.5 | 13.1 | 12.3 | 12.0 |
| 29 | Missouri | 68,746 | 6,151,548 | 18 | 89.5 | 81.1 | 12.7 | 1.2 | 3.0 | 4.5 | 6.0 | 12.5 | 12.9 | 13.4 | 12.4 | 11.7 |
| 30 | Montana | 145,548 | 1,080,577 | 43 | 7.4 | 88.2 | 1.0 | 7.6 | 1.7 | 4.2 | 5.5 | 12.1 | 12.7 | 13.0 | 12.4 | 10.8 |
| 31 | Nebraska | 76,817 | 1,937,552 | 37 | 25.2 | 79.7 | 6.0 | 1.4 | 3.4 | 11.6 | 6.6 | 13.8 | 13.8 | 13.1 | 12.7 | 11.0 |
| 32 | Nevada | 109,860 | 3,138,259 | 32 | 28.6 | 50.6 | 10.8 | 1.5 | 11.7 | 29.5 | 5.9 | 12.6 | 11.8 | 14.7 | 13.4 | 12.6 |
| 33 | New Hampshire | 8,953 | 1,366,275 | 41 | 152.6 | 91.0 | 2.1 | 0.8 | 3.7 | 4.2 | 4.6 | 10.4 | 12.4 | 12.8 | 11.7 | 13.0 |
| 34 | New Jersey | 7,355 | 8,882,371 | 11 | 1,207.7 | 55.4 | 13.8 | 0.5 | 10.9 | 21.1 | 5.8 | 12.2 | 12.2 | 13.0 | 12.8 | 13.2 |
| 35 | New Mexico | 121,312 | 2,106,319 | 36 | 17.4 | 37.8 | 2.5 | 9.4 | 2.3 | 49.7 | 5.6 | 12.8 | 13.3 | 13.5 | 12.3 | 11.1 |
| 36 | New York | 47,124 | 19,336,776 | 4 | 410.3 | 56.6 | 15.5 | 0.7 | 10.0 | 19.3 | 5.7 | 11.4 | 12.4 | 14.7 | 12.6 | 12.8 |
| 37 | North Carolina | 48,620 | 10,600,823 | 9 | 218.0 | 64.1 | 22.6 | 1.7 | 3.9 | 9.9 | 5.7 | 12.3 | 13.2 | 13.6 | 12.4 | 12.8 |
| 38 | North Dakota | 68,995 | 765,309 | 47 | 11.1 | 85.3 | 4.0 | 6.2 | 2.3 | 4.3 | 6.9 | 13.2 | 14.4 | 14.8 | 12.4 | 10.1 |
| 39 | Ohio | 40,859 | 11,693,217 | 7 | 286.2 | 79.8 | 14.2 | 0.7 | 3.2 | 4.2 | 5.9 | 12.3 | 12.8 | 13.4 | 12.1 | 12.1 |
| 40 | Oklahoma | 68,596 | 3,980,783 | 28 | 58.0 | 69.8 | 9.0 | 12.7 | 3.3 | 11.4 | 6.4 | 13.6 | 13.6 | 13.7 | 12.7 | 11.3 |
| 41 | Oregon | 95,988 | 4,241,507 | 27 | 44.2 | 77.9 | 2.9 | 2.4 | 6.9 | 13.7 | 5.2 | 11.6 | 12.0 | 14.3 | 13.6 | 12.1 |
| 42 | Pennsylvania | 44,742 | 12,783,254 | 5 | 285.7 | 76.9 | 12.0 | 0.5 | 4.4 | 8.1 | 5.4 | 11.5 | 12.4 | 13.4 | 12.0 | 12.4 |
| 44 | Rhode Island | 1,034 | 1,057,125 | 44 | 1,022.5 | 72.9 | 7.5 | 1.0 | 4.3 | 16.7 | 5.1 | 10.6 | 13.6 | 14.1 | 12.0 | 12.2 |
| 45 | South Carolina | 30,064 | 5,218,040 | 23 | 173.6 | 65.2 | 27.4 | 0.9 | 2.4 | 6.1 | 5.6 | 12.2 | 12.6 | 13.2 | 12.1 | 12.2 |
| 46 | South Dakota | 75,810 | 892,717 | 46 | 11.8 | 83.2 | 3.0 | 9.5 | 2.2 | 4.4 | 6.8 | 13.7 | 13.3 | 12.9 | 12.1 | 10.5 |
| 47 | Tennessee | 41,238 | 6,886,834 | 16 | 167.0 | 75.1 | 17.7 | 0.8 | 2.5 | 5.9 | 5.9 | 12.3 | 12.6 | 14.0 | 12.4 | 12.5 |
| 48 | Texas | 261,263 | 29,360,759 | 2 | 112.4 | 42.2 | 12.9 | 0.7 | 6.0 | 39.8 | 6.7 | 14.3 | 14.0 | 14.7 | 13.7 | 12.2 |
| 49 | Utah | 82,377 | 3,249,879 | 30 | 39.5 | 79.7 | 1.7 | 1.4 | 5.1 | 14.5 | 7.4 | 16.2 | 16.2 | 14.8 | 13.9 | 10.4 |
| 50 | Vermont | 9,218 | 623,347 | 50 | 67.6 | 94.2 | 1.9 | 1.2 | 2.6 | 2.1 | 4.5 | 10.3 | 13.7 | 12.1 | 11.6 | 12.2 |
| 51 | Virginia | 39,482 | 8,590,563 | 12 | 217.6 | 63.3 | 20.6 | 0.8 | 8.4 | 10.0 | 5.9 | 12.2 | 13.0 | 13.9 | 13.1 | 12.6 |
| 53 | Washington | 66,455 | 7,693,612 | 13 | 115.8 | 70.7 | 5.5 | 2.4 | 12.9 | 13.4 | 5.8 | 12.2 | 12.1 | 15.5 | 13.7 | 12.0 |
| 54 | West Virginia | 24,041 | 1,784,787 | 39 | 74.2 | 93.5 | 4.6 | 0.8 | 1.2 | 1.8 | 5.1 | 11.3 | 12.1 | 12.0 | 11.8 | 12.6 |
| 55 | Wisconsin | 54,167 | 5,832,655 | 20 | 107.7 | 82.2 | 7.4 | 1.4 | 3.7 | 7.3 | 5.6 | 12.2 | 13.1 | 12.7 | 12.4 | 12.0 |
| 56 | Wyoming | 97,089 | 582,328 | 51 | 6.0 | 85.4 | 1.6 | 2.9 | 1.7 | 10.4 | 5.8 | 13.2 | 12.9 | 13.1 | 13.0 | 11.0 |

1. Dry land or land partially or temporarily covered by water.   2. May be of any race.

# Table A. States — Population Characteristics, Immigration, and Households

| STATE | Population characteristics, 2020 (cont.) — Age (percent) (cont.) | | | | Median age 2019 | Percent female 2020 | Percent foreign born 2019 | Percent born in state of residence 2019 | Immigrants admitted to legal status, 2019 | Households, 2019 | | | Household type | | |
|---|---|---|---|---|---|---|---|---|---|---|---|---|---|---|---|
| | 55 to 64 years | 65 to 74 years | 75 to 84 years | 85 years and over | | | | | | Number | Percent change, 2018–2019 | Persons per house-hold | Married Couple family | Female house-holder family | House-holder living alone |
| | 16 | 17 | 18 | 19 | 20 | 21 | 22 | 23 | 24 | 25 | 26 | 27 | 28 | 29 | 30 |
| United States | 12.9 | 9.9 | 5.0 | 2.0 | 38.5 | 50.8 | 13.7 | 58.0 | 1,031,765 | 122,802,852 | 1.1 | 2.61 | 47.5 | 12.3 | 28.3 |
| Alabama | 13.4 | 10.5 | 5.4 | 1.9 | 39.4 | 51.7 | 3.6 | 69.2 | 3,669 | 1,897,576 | 2.3 | 2.52 | 47.0 | 14.0 | 29.8 |
| Alaska | 12.5 | 8.8 | 3.3 | 1.0 | 35.0 | 47.8 | 8.0 | 42.9 | 1,374 | 252,199 | -0.9 | 2.79 | 49.4 | 10.2 | 27.4 |
| Arizona | 12.1 | 10.6 | 5.9 | 2.0 | 38.3 | 50.3 | 13.4 | 39.9 | 18,181 | 2,670,441 | 2.1 | 2.67 | 47.6 | 12.0 | 27.3 |
| Arkansas | 12.7 | 10.2 | 5.4 | 2.0 | 38.8 | 50.9 | 5.1 | 60.5 | 2,915 | 1,163,647 | 0.6 | 2.52 | 47.5 | 12.9 | 29.5 |
| California | 12.1 | 8.8 | 4.4 | 1.9 | 37.0 | 50.3 | 26.7 | 56.1 | 193,093 | 13,157,873 | 0.7 | 2.94 | 49.3 | 12.8 | 24.1 |
| Colorado | 12.2 | 9.4 | 4.2 | 1.6 | 37.1 | 49.6 | 9.5 | 42.2 | 13,558 | 2,235,103 | 2.7 | 2.52 | 50.0 | 9.0 | 27.1 |
| Connecticut | 14.4 | 10.2 | 5.4 | 2.5 | 41.2 | 51.2 | 14.8 | 53.8 | 10,630 | 1,377,166 | -0.1 | 2.51 | 47.2 | 12.4 | 29.0 |
| Delaware | 14.2 | 12.0 | 5.9 | 2.1 | 41.4 | 51.7 | 10.0 | 44.0 | 2,084 | 376,239 | 2.3 | 2.52 | 47.3 | 11.9 | 28.8 |
| District of Columbia | 9.8 | 7.2 | 3.6 | 1.7 | 34.3 | 52.6 | 12.1 | 37.2 | 2,398 | 291,570 | 1.4 | 2.29 | 26.9 | 13.1 | 44.8 |
| Florida | 13.5 | 11.7 | 6.9 | 2.7 | 42.4 | 51.1 | 21.1 | 35.8 | 118,140 | 7,905,832 | 1.2 | 2.66 | 46.4 | 12.8 | 28.6 |
| Georgia | 12.4 | 9.0 | 4.3 | 1.5 | 37.2 | 51.4 | 10.3 | 54.4 | 27,246 | 3,852,714 | 1.3 | 2.69 | 46.8 | 14.7 | 28.0 |
| Hawaii | 12.6 | 11.0 | 5.6 | 3.0 | 39.6 | 49.9 | 19.3 | 52.3 | 4,917 | 465,299 | 2.2 | 2.95 | 49.9 | 12.5 | 25.2 |
| Idaho | 12.1 | 10.2 | 4.9 | 1.7 | 36.9 | 49.9 | 5.8 | 46.1 | 2,312 | 655,859 | 2.4 | 2.68 | 54.5 | 9.3 | 24.6 |
| Illinois | 13.0 | 9.6 | 4.9 | 2.1 | 38.6 | 50.9 | 13.9 | 67.4 | 37,958 | 4,866,006 | 0.0 | 2.54 | 46.3 | 11.9 | 30.8 |
| Indiana | 12.9 | 9.8 | 4.8 | 1.9 | 38.0 | 50.7 | 5.3 | 67.7 | 8,527 | 2,597,765 | -0.1 | 2.52 | 47.0 | 11.0 | 30.4 |
| Iowa | 13.0 | 10.2 | 5.2 | 2.5 | 38.5 | 50.2 | 5.6 | 69.7 | 5,534 | 1,287,221 | 1.5 | 2.38 | 48.9 | 9.6 | 29.6 |
| Kansas | 12.6 | 9.7 | 4.8 | 2.2 | 37.2 | 50.2 | 7.2 | 59.2 | 5,139 | 1,138,329 | 0.4 | 2.49 | 50.2 | 9.7 | 29.1 |
| Kentucky | 13.3 | 10.3 | 5.0 | 1.8 | 39.2 | 50.7 | 4.4 | 68.3 | 6,006 | 1,748,732 | 0.9 | 2.48 | 47.5 | 11.9 | 29.1 |
| Louisiana | 12.9 | 9.9 | 4.8 | 1.8 | 37.7 | 51.3 | 4.2 | 77.6 | 5,116 | 1,741,076 | 0.2 | 2.6 | 42.3 | 16.4 | 30.2 |
| Maine | 15.6 | 13.0 | 6.3 | 2.5 | 45.1 | 51.0 | 3.9 | 61.6 | 1,361 | 573,618 | 0.6 | 2.28 | 46.8 | 9.0 | 31.2 |
| Maryland | 13.5 | 9.6 | 4.8 | 2.0 | 39.0 | 51.6 | 15.4 | 47.4 | 22,304 | 2,226,767 | 0.5 | 2.65 | 47.2 | 13.6 | 28.2 |
| Massachusetts | 13.7 | 10.1 | 5.0 | 2.3 | 39.7 | 51.4 | 17.3 | 59.4 | 30,834 | 2,650,680 | 1.0 | 2.51 | 46.7 | 11.5 | 28.6 |
| Michigan | 13.9 | 10.8 | 5.3 | 2.1 | 39.8 | 50.8 | 7.0 | 76.2 | 17,414 | 3,969,880 | 0.3 | 2.46 | 46.0 | 11.8 | 30.5 |
| Minnesota | 13.3 | 9.8 | 4.8 | 2.2 | 38.4 | 50.2 | 8.4 | 67.7 | 13,833 | 2,222,568 | 1.3 | 2.48 | 49.5 | 8.8 | 29.4 |
| Mississippi | 12.9 | 10.1 | 5.0 | 1.8 | 38.3 | 51.6 | 2.1 | 71.5 | 1,687 | 1,100,229 | -0.8 | 2.62 | 43.8 | 16.8 | 30.3 |
| Missouri | 13.4 | 10.2 | 5.3 | 2.2 | 38.9 | 50.9 | 4.3 | 65.9 | 6,900 | 2,458,337 | 1.0 | 2.43 | 47.2 | 11.1 | 30.3 |
| Montana | 13.7 | 12.0 | 5.6 | 2.1 | 40.5 | 49.6 | 2.3 | 53.4 | 545 | 437,651 | 1.4 | 2.38 | 48.8 | 8.1 | 30.7 |
| Nebraska | 12.4 | 9.5 | 4.7 | 2.2 | 36.8 | 50.0 | 7.4 | 64.7 | 4,279 | 771,444 | 0.8 | 2.44 | 50.4 | 9.1 | 28.7 |
| Nevada | 12.4 | 10.1 | 5.0 | 1.5 | 38.4 | 49.9 | 19.8 | 27.2 | 11,718 | 1,143,557 | 1.2 | 2.66 | 44.3 | 12.8 | 28.5 |
| New Hampshire | 15.7 | 11.6 | 5.4 | 2.2 | 43.0 | 50.4 | 6.4 | 40.6 | 2,046 | 541,396 | 1.9 | 2.44 | 50.6 | 8.8 | 27.6 |
| New Jersey | 13.8 | 9.7 | 5.1 | 2.3 | 40.2 | 51.1 | 23.4 | 51.5 | 48,754 | 3,286,264 | 1.1 | 2.65 | 50.7 | 12.5 | 26.4 |
| New Mexico | 12.8 | 11.0 | 5.5 | 2.0 | 38.6 | 50.5 | 9.6 | 53.8 | 3,736 | 793,420 | -0.1 | 2.59 | 41.8 | 14.7 | 31.3 |
| New York | 13.4 | 9.9 | 5.2 | 2.4 | 39.2 | 51.4 | 22.4 | 63.1 | 124,026 | 7,446,812 | 1.1 | 2.54 | 43.3 | 13.9 | 30.5 |
| North Carolina | 13.0 | 10.2 | 5.1 | 1.8 | 39.1 | 51.4 | 8.4 | 56.0 | 18,419 | 4,046,348 | 0.9 | 2.52 | 47.6 | 12.9 | 28.6 |
| North Dakota | 12.0 | 9.1 | 4.6 | 2.4 | 35.5 | 48.8 | 4.1 | 61.8 | 1,494 | 323,519 | 1.3 | 2.28 | 47.0 | 7.5 | 33.7 |
| Ohio | 13.6 | 10.5 | 5.2 | 2.2 | 39.6 | 51.0 | 4.8 | 74.7 | 16,467 | 4,730,340 | 1.0 | 2.4 | 45.0 | 12.3 | 31.1 |
| Oklahoma | 12.3 | 9.6 | 5.0 | 1.9 | 37.0 | 50.5 | 6.1 | 60.6 | 4,642 | 1,495,151 | 0.7 | 2.57 | 47.7 | 12.2 | 28.6 |
| Oregon | 12.5 | 11.3 | 5.3 | 2.0 | 39.7 | 50.4 | 9.7 | 45.6 | 8,521 | 1,649,352 | 0.6 | 2.5 | 47.3 | 10.0 | 28.1 |
| Pennsylvania | 14.0 | 11.0 | 5.6 | 2.6 | 40.8 | 51.0 | 7.0 | 71.4 | 25,329 | 5,119,249 | 1.0 | 2.42 | 46.4 | 11.8 | 30.2 |
| Rhode Island | 14.1 | 10.4 | 5.2 | 2.6 | 40.1 | 51.3 | 13.7 | 56.6 | 3,445 | 407,174 | 0.1 | 2.5 | 44.2 | 11.9 | 31.0 |
| South Carolina | 13.4 | 11.4 | 5.5 | 1.8 | 39.9 | 51.6 | 5.6 | 55.0 | 4,675 | 1,975,915 | 2.5 | 2.54 | 47.0 | 13.7 | 29.0 |
| South Dakota | 13.1 | 10.5 | 4.7 | 2.3 | 37.7 | 49.5 | 4.1 | 63.6 | 1,199 | 353,799 | 2.4 | 2.4 | 50.2 | 8.7 | 29.9 |
| Tennessee | 13.1 | 10.2 | 5.1 | 1.8 | 39.0 | 51.2 | 5.5 | 59.1 | 9,507 | 2,654,737 | 2.0 | 2.51 | 48.0 | 12.3 | 28.7 |
| Texas | 11.2 | 8.0 | 3.8 | 1.4 | 35.1 | 50.3 | 17.1 | 59.5 | 107,955 | 9,985,126 | 2.1 | 2.84 | 49.6 | 13.7 | 25.6 |
| Utah | 9.4 | 7.1 | 3.4 | 1.2 | 31.2 | 49.6 | 8.6 | 61.7 | 6,838 | 1,023,855 | 2.5 | 3.08 | 60.5 | 9.1 | 19.2 |
| Vermont | 15.0 | 12.5 | 5.9 | 2.3 | 42.8 | 50.6 | 4.7 | 48.8 | 648 | 262,767 | 0.5 | 2.28 | 46.1 | 8.8 | 31.6 |
| Virginia | 13.0 | 9.6 | 4.9 | 1.8 | 38.5 | 50.8 | 12.7 | 49.5 | 24,784 | 3,191,847 | 0.5 | 2.6 | 49.8 | 11.4 | 27.7 |
| Washington | 12.4 | 9.9 | 4.6 | 1.8 | 37.9 | 49.9 | 14.9 | 46.4 | 25,570 | 2,932,477 | 1.3 | 2.55 | 49.4 | 9.9 | 26.6 |
| West Virginia | 14.0 | 12.4 | 6.2 | 2.3 | 42.9 | 50.5 | 1.6 | 68.5 | 817 | 728,175 | -0.9 | 2.4 | 48.4 | 10.9 | 29.6 |
| Wisconsin | 14.1 | 10.6 | 5.2 | 2.2 | 39.9 | 50.2 | 5.1 | 70.9 | 7,138 | 2,386,623 | 0.6 | 2.38 | 47.8 | 9.3 | 30.5 |
| Wyoming | 13.3 | 11.0 | 5.0 | 1.9 | 38.1 | 49.0 | 3.1 | 43.0 | 410 | 233,128 | 1.2 | 2.42 | 53.0 | 8.1 | 27.4 |

# Table A. States — Population Change

| STATE | Population, 1990–2010 Census counts | | | Percent change | | | Components of change, 2010–2020 | | Migration | | | Population, 2020–2030 Projections | | |
|---|---|---|---|---|---|---|---|---|---|---|---|---|---|---|
| | 1990 | 2000 | 2010 | 1990–2000 | 2000–2010 | 2010–2020 | Births | Deaths | Net migration | International | Net internal | 2020 | 2025 | 2030 |
| | 31 | 32 | 33 | 34 | 35 | 36 | 37 | 38 | 39 | 40 | 41 | 42 | 43 | 44 |
| United States | 248,709,873 | 281,421,906 | 308,745,538 | 13.2 | 9.7 | 6.7 | 40,009,421 | 27,751,753 | 8,468,350 | 8,468,350 | 0 | 335,804,546 | 349,439,199 | 363,584,435 |
| Alabama | 4,040,587 | 4,447,100 | 4,779,736 | 10.1 | 7.5 | 3.0 | 600,479 | 532,587 | 74,102 | 45,403 | 28,699 | 4,728,915 | 4,800,092 | 4,874,243 |
| Alaska | 550,043 | 626,932 | 710,231 | 14.0 | 13.3 | 2.9 | 111,891 | 44,413 | -47,260 | 19,445 | -66,705 | 774,421 | 820,881 | 867,674 |
| Arizona | 3,665,228 | 5,130,632 | 6,392,017 | 40.0 | 24.6 | 16.1 | 863,381 | 561,112 | 725,176 | 159,514 | 565,662 | 8,456,448 | 9,531,537 | 10,712,397 |
| Arkansas | 2,350,725 | 2,673,400 | 2,915,918 | 13.7 | 9.1 | 3.9 | 388,332 | 320,002 | 45,657 | 29,018 | 16,639 | 3,060,219 | 3,151,005 | 3,240,208 |
| California | 29,760,021 | 33,871,648 | 37,253,956 | 13.8 | 10.0 | 5.7 | 4,963,885 | 2,658,968 | -177,201 | 985,729 | -1,162,930 | 42,206,743 | 44,305,177 | 46,444,861 |
| Colorado | 3,294,394 | 4,301,261 | 5,029,196 | 30.6 | 16.9 | 15.5 | 666,380 | 370,913 | 477,921 | 97,975 | 379,946 | 5,278,867 | 5,522,803 | 5,792,357 |
| Connecticut | 3,287,116 | 3,405,565 | 3,574,097 | 3.6 | 4.9 | -0.5 | 367,838 | 312,319 | -72,975 | 146,353 | -219,328 | 3,675,650 | 3,691,016 | 3,688,630 |
| Delaware | 666,168 | 783,600 | 897,934 | 17.6 | 14.6 | 9.9 | 112,071 | 89,006 | 65,890 | 17,116 | 48,774 | 963,209 | 990,694 | 1,012,658 |
| District of Columbia | 606,900 | 572,059 | 601,723 | -5.7 | 5.2 | 18.5 | 96,247 | 51,168 | 64,816 | 38,285 | 26,531 | 480,540 | 455,108 | 433,414 |
| Florida | 12,937,926 | 15,982,378 | 18,801,310 | 23.5 | 17.6 | 15.6 | 2,244,645 | 1,985,189 | 2,661,420 | 1,199,099 | 1,462,321 | 23,406,525 | 25,912,458 | 28,685,769 |
| Georgia | 6,478,216 | 8,186,453 | 9,687,653 | 26.4 | 18.3 | 10.6 | 1,327,232 | 814,702 | 505,498 | 211,016 | 294,482 | 10,843,753 | 11,438,622 | 12,017,838 |
| Hawaii | 1,108,229 | 1,211,537 | 1,360,301 | 9.3 | 12.3 | 3.4 | 185,462 | 114,918 | -22,986 | 61,240 | -84,226 | 1,412,373 | 1,438,720 | 1,466,046 |
| Idaho | 1,006,749 | 1,293,953 | 1,567,582 | 28.5 | 21.1 | 16.5 | 229,433 | 133,664 | 162,950 | 18,017 | 144,933 | 1,741,333 | 1,852,627 | 1,969,624 |
| Illinois | 11,430,602 | 12,419,293 | 12,830,632 | 8.6 | 3.3 | -1.9 | 1,578,161 | 1,094,778 | -728,322 | 250,176 | -978,498 | 13,236,720 | 13,340,507 | 13,432,892 |
| Indiana | 5,544,159 | 6,080,485 | 6,483,802 | 9.7 | 6.6 | 4.2 | 849,083 | 637,936 | 61,821 | 109,031 | -47,210 | 6,627,008 | 6,721,322 | 6,810,108 |
| Iowa | 2,776,755 | 2,926,324 | 3,046,355 | 5.4 | 4.1 | 3.8 | 395,193 | 300,959 | 22,741 | 59,990 | -37,249 | 3,020,496 | 2,993,222 | 2,955,172 |
| Kansas | 2,477,574 | 2,688,418 | 2,853,118 | 8.5 | 6.1 | 2.1 | 391,375 | 267,583 | -63,016 | 55,070 | -118,086 | 2,890,566 | 2,919,002 | 2,940,084 |
| Kentucky | 3,685,296 | 4,041,769 | 4,339,367 | 9.7 | 7.4 | 3.2 | 563,501 | 471,069 | 46,257 | 58,419 | -12,162 | 4,424,431 | 4,489,662 | 4,554,998 |
| Louisiana | 4,219,973 | 4,468,976 | 4,533,372 | 5.9 | 1.4 | 2.5 | 633,464 | 453,494 | -68,962 | 56,624 | -125,586 | 4,719,160 | 4,762,398 | 4,802,633 |
| Maine | 1,227,928 | 1,274,923 | 1,328,361 | 3.8 | 4.2 | 1.6 | 129,051 | 143,362 | 36,945 | 15,673 | 21,272 | 1,408,665 | 1,414,402 | 1,411,097 |
| Maryland | 4,781,468 | 5,296,486 | 5,773,552 | 10.8 | 9.0 | 4.9 | 741,516 | 489,918 | 33,437 | 219,431 | -185,994 | 6,497,626 | 6,762,732 | 7,022,251 |
| Massachusetts | 6,016,425 | 6,349,097 | 6,547,629 | 5.5 | 3.1 | 5.3 | 729,892 | 581,114 | 201,059 | 391,402 | -190,343 | 6,855,546 | 6,938,636 | 7,012,009 |
| Michigan | 9,295,297 | 9,938,444 | 9,883,640 | 6.9 | -0.6 | 0.8 | 1,151,986 | 972,901 | -95,692 | 201,218 | -296,910 | 10,695,993 | 10,713,730 | 10,694,172 |
| Minnesota | 4,375,099 | 4,919,479 | 5,303,925 | 12.4 | 7.8 | 6.7 | 703,159 | 435,727 | 88,673 | 125,999 | -37,326 | 5,900,769 | 6,108,787 | 6,306,130 |
| Mississippi | 2,573,216 | 2,844,658 | 2,967,297 | 10.5 | 4.3 | 0.0 | 389,874 | 319,113 | -72,970 | 21,304 | -93,977 | 3,044,812 | 3,069,420 | 3,092,410 |
| Missouri | 5,117,073 | 5,595,211 | 5,988,927 | 9.3 | 7.0 | 2.7 | 763,481 | 608,997 | 10,057 | 74,450 | -64,393 | 6,199,882 | 6,315,366 | 6,430,173 |
| Montana | 799,065 | 902,195 | 989,415 | 12.9 | 9.7 | 9.2 | 123,123 | 99,738 | 67,564 | 9,848 | 57,716 | 1,022,735 | 1,037,387 | 1,044,898 |
| Nebraska | 1,578,385 | 1,711,263 | 1,826,341 | 8.4 | 6.7 | 6.1 | 266,099 | 165,518 | 11,190 | 37,238 | -26,048 | 1,802,670 | 1,812,787 | 1,820,247 |
| Nevada | 1,201,833 | 1,998,257 | 2,700,551 | 66.3 | 35.1 | 16.2 | 365,815 | 236,627 | 307,049 | 36,141 | 270,908 | 3,452,283 | 3,863,298 | 4,282,102 |
| New Hampshire | 1,109,252 | 1,235,786 | 1,316,470 | 11.4 | 6.5 | 3.8 | 126,460 | 120,351 | 44,426 | 31,374 | 13,052 | 1,524,751 | 1,586,348 | 1,646,471 |
| New Jersey | 7,730,188 | 8,414,350 | 8,791,894 | 8.9 | 4.5 | 1.0 | 1,054,140 | 751,791 | -212,010 | 324,296 | -536,306 | 9,461,635 | 9,636,644 | 9,802,440 |
| New Mexico | 1,515,069 | 1,819,046 | 2,059,179 | 20.1 | 13.2 | 2.3 | 258,988 | 182,459 | -29,366 | 32,437 | -61,803 | 2,084,341 | 2,106,584 | 2,099,708 |
| New York | 17,990,455 | 18,976,457 | 19,378,102 | 5.5 | 2.1 | -0.2 | 2,400,759 | 1,590,348 | -849,381 | 736,389 | -1,585,770 | 19,576,920 | 19,540,179 | 19,477,429 |
| North Carolina | 6,628,637 | 8,049,313 | 9,535,483 | 21.4 | 18.5 | 11.2 | 1,229,915 | 907,793 | 737,659 | 188,694 | 548,965 | 10,709,289 | 11,449,153 | 12,227,739 |
| North Dakota | 638,800 | 642,200 | 672,591 | 0.5 | 4.7 | 13.8 | 108,282 | 64,102 | 46,920 | 14,356 | 32,564 | 630,112 | 620,777 | 606,566 |
| Ohio | 10,847,115 | 11,353,140 | 11,536,504 | 4.7 | 1.6 | 1.4 | 1,408,338 | 1,204,444 | -43,576 | 197,444 | -241,020 | 11,644,058 | 11,605,738 | 11,550,528 |
| Oklahoma | 3,145,585 | 3,450,654 | 3,751,351 | 9.7 | 8.7 | 6.1 | 530,768 | 404,645 | 102,731 | 61,452 | 41,279 | 3,735,690 | 3,820,994 | 3,913,251 |
| Oregon | 2,842,321 | 3,421,399 | 3,831,074 | 20.4 | 12.0 | 10.7 | 455,763 | 360,602 | 314,917 | 62,857 | 252,060 | 4,260,393 | 4,536,418 | 4,833,918 |
| Pennsylvania | 11,881,643 | 12,281,054 | 12,702,379 | 3.4 | 3.4 | 0.6 | 1,431,212 | 1,352,118 | 9,078 | 292,565 | -283,487 | 12,787,354 | 12,801,945 | 12,768,184 |
| Rhode Island | 1,003,464 | 1,048,319 | 1,052,567 | 4.5 | 0.4 | 0.4 | 110,543 | 101,181 | -4,890 | 36,638 | -41,528 | 1,154,230 | 1,157,855 | 1,152,941 |
| South Carolina | 3,486,703 | 4,012,012 | 4,625,364 | 15.1 | 15.3 | 12.8 | 586,324 | 483,494 | 487,169 | 64,469 | 422,700 | 4,822,577 | 4,989,550 | 5,148,055 |
| South Dakota | 696,004 | 754,844 | 814,180 | 8.5 | 7.9 | 9.6 | 123,483 | 77,930 | 32,742 | 18,971 | 13,771 | 801,939 | 801,845 | 800,462 |
| Tennessee | 4,877,185 | 5,689,283 | 6,346,105 | 16.7 | 11.5 | 8.5 | 825,523 | 678,711 | 392,626 | 83,792 | 308,834 | 6,780,670 | 7,073,125 | 7,380,634 |
| Texas | 16,986,510 | 20,851,820 | 25,145,561 | 22.8 | 20.6 | 16.8 | 3,962,570 | 1,935,156 | 2,173,519 | 869,640 | 1,303,879 | 28,634,896 | 30,865,134 | 33,317,744 |
| Utah | 1,722,850 | 2,233,169 | 2,763,885 | 29.6 | 23.8 | 17.6 | 512,904 | 174,989 | 148,583 | 50,238 | 98,345 | 2,990,094 | 3,225,680 | 3,485,367 |
| Vermont | 562,758 | 608,827 | 625,741 | 8.2 | 2.8 | -0.4 | 59,672 | 59,104 | -2,579 | 9,219 | -11,798 | 690,686 | 703,288 | 711,867 |
| Virginia | 6,187,358 | 7,078,515 | 8,001,024 | 14.4 | 13.0 | 7.4 | 1,041,285 | 674,614 | 221,973 | 302,228 | -80,255 | 8,917,395 | 9,364,304 | 9,825,019 |
| Washington | 4,866,692 | 5,894,121 | 6,724,540 | 21.1 | 14.1 | 14.4 | 897,866 | 557,404 | 628,387 | 257,129 | 371,258 | 7,432,136 | 7,996,400 | 8,624,081 |
| West Virginia | 1,793,477 | 1,808,344 | 1,852,994 | 0.8 | 2.5 | -3.7 | 200,151 | 231,570 | -36,295 | 11,106 | -47,401 | 1,801,112 | 1,766,435 | 1,719,959 |
| Wisconsin | 4,891,769 | 5,363,675 | 5,686,986 | 9.6 | 6.0 | 2.6 | 678,202 | 522,217 | -8,205 | 68,382 | -76,587 | 6,004,954 | 6,088,374 | 6,150,764 |
| Wyoming | 453,588 | 493,782 | 563,626 | 8.9 | 14.1 | 3.3 | 74,224 | 48,935 | -7,214 | 4,450 | -11,664 | 530,948 | 529,031 | 522,979 |

# Table A. States — **Population Characteristics**

| STATE | Race (percent) | | | | | Percent Hispanic or Latino[1] | Percent foreign born | Age (percent) | | | | | | | | | | Median age | Percent female |
|---|---|---|---|---|---|---|---|---|---|---|---|---|---|---|---|---|---|---|---|
| | White alone | Black alone | American Indian, Alaska Native alone | Asian and Pacific Islander alone | Some other race or two or more races | | | Under 5 years | 5 to 17 years | 18 to 24 years | 25 to 34 years | 35 to 44 years | 45 to 54 years | 55 to 64 years | 65 to 74 years | 75 to 84 years | 85 years and over | | |
| | 45 | 46 | 47 | 48 | 49 | 50 | 51 | 52 | 53 | 54 | 55 | 56 | 57 | 58 | 59 | 60 | 61 | 62 | 63 |
| United States | 72.4 | 12.6 | 0.9 | 5.0 | 9.1 | 16.3 | 12.9 | 6.5 | 17.5 | 9.9 | 13.3 | 13.3 | 14.6 | 11.8 | 7.0 | 4.2 | 1.8 | 37.2 | 50.8 |
| Alabama | 68.5 | 26.2 | 0.6 | 1.2 | 3.5 | 3.9 | 3.5 | 6.4 | 17.3 | 10.0 | 12.6 | 13.0 | 14.5 | 12.3 | 7.8 | 4.4 | 1.6 | 37.9 | 51.5 |
| Alaska | 66.7 | 3.3 | 14.8 | 6.4 | 8.9 | 5.5 | 6.9 | 7.6 | 18.8 | 10.5 | 14.5 | 13.1 | 15.6 | 12.1 | 5.0 | 2.1 | 0.6 | 33.8 | 48.0 |
| Arizona | 73.0 | 4.1 | 4.6 | 3.0 | 15.3 | 29.6 | 13.4 | 7.1 | 18.4 | 9.9 | 13.4 | 12.9 | 13.2 | 11.4 | 7.8 | 4.4 | 1.6 | 35.9 | 50.3 |
| Arkansas | 77.0 | 15.4 | 0.8 | 1.4 | 5.4 | 6.4 | 4.5 | 6.8 | 17.6 | 9.7 | 12.6 | 12.6 | 14.0 | 12.0 | 8.0 | 4.6 | 1.8 | 37.4 | 50.9 |
| California | 57.6 | 6.2 | 1.0 | 13.4 | 21.9 | 37.6 | 27.2 | 6.8 | 18.2 | 10.5 | 14.2 | 13.9 | 14.1 | 10.8 | 6.1 | 3.7 | 1.6 | 35.2 | 50.3 |
| Colorado | 81.3 | 4.0 | 1.1 | 2.9 | 10.6 | 20.7 | 9.8 | 6.8 | 17.5 | 9.7 | 14.4 | 13.9 | 14.8 | 11.9 | 6.2 | 3.4 | 1.4 | 36.1 | 49.9 |
| Connecticut | 77.6 | 10.1 | 0.3 | 3.8 | 8.2 | 13.4 | 13.6 | 5.7 | 17.2 | 9.1 | 11.8 | 13.6 | 16.1 | 12.4 | 7.1 | 4.6 | 2.4 | 40.0 | 51.3 |
| Delaware | 68.9 | 21.4 | 0.5 | 3.2 | 6.1 | 8.2 | 8.0 | 6.2 | 16.7 | 10.1 | 12.3 | 12.9 | 14.9 | 12.4 | 8.1 | 4.5 | 1.8 | 38.8 | 51.6 |
| District of Columbia | 38.5 | 50.7 | 0.3 | 3.6 | 7.0 | 9.1 | 13.5 | 5.4 | 11.3 | 14.5 | 20.9 | 13.4 | 12.6 | 10.6 | 6.1 | 3.5 | 1.8 | 33.8 | 52.8 |
| Florida | 75.0 | 16.0 | 0.4 | 2.5 | 6.1 | 22.5 | 19.4 | 5.7 | 15.6 | 9.3 | 12.1 | 12.9 | 14.6 | 12.4 | 9.2 | 5.7 | 2.4 | 40.7 | 51.1 |
| Georgia | 59.7 | 30.5 | 0.3 | 3.3 | 6.1 | 8.8 | 9.7 | 7.1 | 18.6 | 10.0 | 13.6 | 14.4 | 14.4 | 11.0 | 6.3 | 3.1 | 1.2 | 35.3 | 51.2 |
| Hawaii | 24.7 | 1.6 | 0.3 | 48.6 | 24.8 | 8.9 | 18.2 | 6.4 | 15.9 | 9.6 | 13.3 | 13.0 | 14.2 | 12.9 | 7.4 | 4.7 | 2.3 | 38.6 | 49.9 |
| Idaho | 89.1 | 0.6 | 1.4 | 1.3 | 7.6 | 11.2 | 5.5 | 7.8 | 19.6 | 9.9 | 13.3 | 12.2 | 13.3 | 11.5 | 7.0 | 3.8 | 1.7 | 34.6 | 49.9 |
| Illinois | 71.5 | 14.5 | 0.3 | 4.6 | 9.0 | 15.8 | 13.7 | 6.5 | 17.9 | 9.7 | 13.9 | 13.5 | 14.6 | 11.5 | 6.6 | 4.0 | 1.9 | 36.6 | 51.0 |
| Indiana | 84.3 | 9.1 | 0.3 | 1.6 | 4.7 | 6.0 | 4.6 | 6.7 | 18.1 | 10.0 | 12.7 | 13.0 | 14.6 | 11.9 | 7.0 | 4.3 | 1.7 | 37.0 | 50.8 |
| Iowa | 91.3 | 2.9 | 0.4 | 1.8 | 3.6 | 5.0 | 4.6 | 6.6 | 17.3 | 10.0 | 12.6 | 12.0 | 14.4 | 12.2 | 7.4 | 5.0 | 2.5 | 38.1 | 50.5 |
| Kansas | 83.8 | 5.9 | 1.0 | 2.5 | 6.9 | 10.5 | 6.5 | 7.2 | 18.3 | 10.1 | 13.0 | 12.2 | 14.2 | 11.6 | 6.7 | 4.3 | 2.1 | 36.0 | 50.4 |
| Kentucky | 87.8 | 7.8 | 0.2 | 1.2 | 3.0 | 3.1 | 3.2 | 6.5 | 17.1 | 9.5 | 13.0 | 13.3 | 14.8 | 12.4 | 7.5 | 4.2 | 1.6 | 38.1 | 50.8 |
| Louisiana | 62.6 | 32.0 | 0.7 | 1.5 | 3.1 | 4.2 | 3.8 | 6.9 | 17.7 | 10.5 | 13.7 | 12.5 | 14.4 | 11.8 | 6.9 | 3.9 | 1.5 | 35.8 | 51.0 |
| Maine | 95.2 | 1.2 | 0.6 | 1.0 | 1.9 | 1.3 | 3.4 | 5.2 | 15.4 | 8.7 | 10.9 | 12.9 | 16.5 | 14.5 | 8.5 | 5.3 | 2.1 | 42.7 | 51.1 |
| Maryland | 58.2 | 29.4 | 0.4 | 5.6 | 6.5 | 8.2 | 13.9 | 6.3 | 17.1 | 9.7 | 13.2 | 13.8 | 15.6 | 12.1 | 6.7 | 3.9 | 1.7 | 38.0 | 51.6 |
| Massachusetts | 80.4 | 6.6 | 0.3 | 5.3 | 7.3 | 9.6 | 15.0 | 5.6 | 16.1 | 10.4 | 12.9 | 13.5 | 15.5 | 12.3 | 7.0 | 4.6 | 2.2 | 39.1 | 51.6 |
| Michigan | 78.9 | 14.2 | 0.6 | 2.4 | 3.8 | 4.4 | 6.0 | 6.0 | 17.7 | 9.9 | 11.7 | 12.9 | 15.3 | 12.7 | 7.3 | 4.5 | 1.9 | 38.9 | 50.9 |
| Minnesota | 85.3 | 5.2 | 1.1 | 4.0 | 4.3 | 4.7 | 7.1 | 6.7 | 17.6 | 9.5 | 13.5 | 12.8 | 15.2 | 11.9 | 6.7 | 4.2 | 2.0 | 37.4 | 50.4 |
| Mississippi | 59.1 | 37.0 | 0.5 | 0.9 | 2.4 | 2.7 | 2.1 | 7.1 | 18.4 | 10.3 | 12.6 | 12.6 | 14.1 | 11.7 | 7.2 | 4.1 | 1.5 | 36.0 | 51.4 |
| Missouri | 82.8 | 11.6 | 0.5 | 1.7 | 3.4 | 3.5 | 3.9 | 6.5 | 17.3 | 9.8 | 12.9 | 12.5 | 14.8 | 12.1 | 7.5 | 4.5 | 2.0 | 37.9 | 51.0 |
| Montana | 89.4 | 0.4 | 6.3 | 0.7 | 3.1 | 2.9 | 2.0 | 6.3 | 16.3 | 9.6 | 12.3 | 11.4 | 15.1 | 14.0 | 8.2 | 4.7 | 2.0 | 39.8 | 49.8 |
| Nebraska | 86.1 | 4.5 | 1.0 | 1.9 | 6.5 | 9.2 | 6.1 | 7.2 | 17.9 | 10.0 | 13.2 | 12.1 | 14.2 | 11.7 | 6.7 | 4.7 | 2.2 | 36.2 | 50.4 |
| Nevada | 66.2 | 8.1 | 1.2 | 7.8 | 16.7 | 26.5 | 18.8 | 6.9 | 17.7 | 9.2 | 14.3 | 14.2 | 13.9 | 11.7 | 7.3 | 3.6 | 1.1 | 36.3 | 49.5 |
| New Hampshire | 93.9 | 1.1 | 0.2 | 2.2 | 2.5 | 2.8 | 5.3 | 5.3 | 16.5 | 9.4 | 11.0 | 13.6 | 17.2 | 13.5 | 7.4 | 4.4 | 1.8 | 41.1 | 50.7 |
| New Jersey | 68.6 | 13.7 | 0.3 | 8.3 | 9.1 | 17.7 | 21.0 | 6.2 | 17.3 | 8.7 | 12.6 | 14.1 | 15.7 | 11.9 | 7.0 | 4.5 | 2.0 | 39.0 | 51.3 |
| New Mexico | 68.4 | 2.1 | 9.4 | 1.5 | 18.7 | 46.3 | 9.9 | 7.0 | 18.1 | 9.9 | 12.8 | 12.1 | 14.2 | 12.5 | 7.5 | 4.3 | 1.5 | 36.7 | 50.6 |
| New York | 65.7 | 15.9 | 0.6 | 7.3 | 10.4 | 17.6 | 22.2 | 6.0 | 16.4 | 10.2 | 13.7 | 13.5 | 14.9 | 11.9 | 7.0 | 4.5 | 2.0 | 38.0 | 51.6 |
| North Carolina | 68.5 | 21.5 | 1.3 | 2.3 | 6.5 | 8.4 | 7.5 | 6.6 | 17.3 | 9.8 | 13.0 | 13.9 | 14.4 | 11.9 | 7.3 | 4.1 | 1.6 | 37.4 | 51.3 |
| North Dakota | 90.0 | 1.2 | 5.4 | 1.0 | 2.3 | 2.0 | 2.5 | 6.6 | 15.7 | 12.0 | 13.1 | 11.2 | 14.1 | 12.2 | 7.0 | 5.1 | 2.5 | 37.0 | 49.5 |
| Ohio | 82.7 | 12.2 | 0.2 | 1.7 | 3.2 | 3.1 | 4.1 | 6.2 | 17.4 | 9.5 | 12.4 | 12.8 | 15.1 | 12.6 | 7.4 | 4.7 | 2.0 | 38.8 | 51.2 |
| Oklahoma | 72.2 | 7.4 | 8.6 | 1.8 | 10.0 | 8.9 | 5.5 | 7.0 | 17.7 | 10.2 | 13.5 | 12.3 | 14.0 | 11.7 | 7.5 | 4.4 | 1.6 | 36.2 | 50.5 |
| Oregon | 83.6 | 1.8 | 1.4 | 4.0 | 9.1 | 11.7 | 9.8 | 6.2 | 16.4 | 9.4 | 13.7 | 13.0 | 14.1 | 13.3 | 7.6 | 4.3 | 2.0 | 38.4 | 50.5 |
| Pennsylvania | 81.9 | 10.8 | 0.2 | 2.7 | 4.3 | 5.7 | 5.8 | 5.7 | 16.2 | 9.9 | 11.9 | 12.7 | 15.3 | 12.8 | 7.7 | 5.3 | 2.5 | 40.1 | 51.3 |
| Rhode Island | 81.4 | 5.7 | 0.6 | 3.0 | 9.3 | 12.4 | 12.8 | 5.5 | 15.8 | 11.4 | 12.0 | 13.0 | 15.4 | 12.4 | 7.0 | 4.9 | 2.5 | 39.4 | 51.7 |
| South Carolina | 66.2 | 27.9 | 0.4 | 1.4 | 4.2 | 5.1 | 4.7 | 6.5 | 16.8 | 10.3 | 12.7 | 13.0 | 14.3 | 12.6 | 8.0 | 4.1 | 1.6 | 37.9 | 51.4 |
| South Dakota | 85.9 | 1.3 | 8.8 | 0.9 | 3.0 | 2.7 | 2.7 | 7.3 | 17.6 | 10.0 | 12.8 | 11.4 | 14.4 | 12.0 | 7.1 | 4.8 | 2.4 | 36.9 | 50.0 |
| Tennessee | 77.6 | 16.7 | 0.3 | 1.5 | 3.9 | 4.6 | 4.5 | 6.4 | 17.1 | 9.6 | 12.8 | 13.5 | 14.6 | 12.4 | 7.7 | 4.2 | 1.6 | 38.0 | 51.3 |
| Texas | 70.4 | 11.8 | 0.7 | 3.9 | 13.2 | 37.6 | 16.4 | 7.7 | 19.6 | 10.2 | 14.3 | 13.8 | 13.7 | 10.3 | 5.9 | 3.3 | 1.2 | 33.6 | 50.4 |
| Utah | 86.1 | 1.1 | 1.2 | 2.9 | 8.7 | 13.0 | 8.0 | 9.5 | 22.0 | 11.5 | 16.2 | 12.0 | 11.1 | 8.7 | 5.0 | 2.9 | 1.1 | 29.2 | 49.8 |
| Vermont | 95.3 | 1.0 | 0.4 | 1.3 | 2.0 | 1.5 | 4.4 | 5.1 | 15.5 | 10.4 | 11.2 | 12.5 | 16.4 | 14.4 | 7.9 | 4.5 | 2.1 | 41.5 | 50.7 |
| Virginia | 68.6 | 19.4 | 0.4 | 5.6 | 6.1 | 7.9 | 11.4 | 6.4 | 16.8 | 10.0 | 13.6 | 13.9 | 15.2 | 11.9 | 6.9 | 3.8 | 1.5 | 37.5 | 50.9 |
| Washington | 77.3 | 3.6 | 1.5 | 7.8 | 9.9 | 11.2 | 13.1 | 6.5 | 17.0 | 9.7 | 13.9 | 13.5 | 14.7 | 12.4 | 6.8 | 3.7 | 1.8 | 37.3 | 50.2 |
| West Virginia | 93.9 | 3.4 | 0.2 | 0.7 | 1.8 | 1.2 | 1.2 | 5.6 | 15.3 | 9.1 | 11.7 | 12.8 | 14.9 | 14.3 | 8.8 | 5.2 | 2.0 | 41.3 | 50.7 |
| Wisconsin | 86.2 | 6.3 | 1.0 | 2.3 | 4.2 | 5.9 | 4.5 | 6.3 | 17.3 | 9.7 | 12.6 | 12.8 | 15.4 | 12.3 | 7.0 | 4.6 | 2.1 | 38.5 | 50.4 |
| Wyoming | 90.7 | 0.8 | 2.4 | 0.9 | 5.2 | 8.9 | 2.8 | 7.1 | 16.9 | 10.0 | 13.6 | 11.9 | 14.8 | 13.0 | 7.0 | 3.8 | 1.6 | 36.8 | 49.0 |

1. May be of any race.

| STATE | Households, 2010 Number | Percent change, 2000-2010 | Persons per house-hold | Percent Female family house-holder[1] | Percent One-person house-holds | Housing units, 2010 Total | Percent change, 2000-2010 | Occupied units Owner-occupied Total | Percent | Median value[2] (dollars) | Median owner cost as a percent of income With a mort-gage | Without a mort-gage[3] | Renter-occupied Median gross rent[4] (dollars) | Median rent as a percent of income | Sub-standard units[5] (percent) |
|---|---|---|---|---|---|---|---|---|---|---|---|---|---|---|---|
| | 64 | 65 | 66 | 67 | 68 | 69 | 70 | 71 | 72 | 73 | 74 | 75 | 76 | 77 | 78 |
| United States | 116,716,292 | 10.7 | 2.58 | 13.1 | 26.7 | 131,791,065 | 13.7 | 114,567,419 | 65.4 | 179,900 | 25.1 | 12.8 | 855 | 31.6 | 3.9 |
| Alabama | 1,883,791 | 8.4 | 2.48 | 15.3 | 27.4 | 2,174,428 | 10.7 | 1,815,152 | 70.1 | 123,900 | 23.0 | 12.3 | 667 | 32.2 | 2.4 |
| Alaska | 258,058 | 16.5 | 2.65 | 10.7 | 25.6 | 307,065 | 17.7 | 254,610 | 63.9 | 241,400 | 23.3 | 10.8 | 981 | 29 | 10.2 |
| Arizona | 2,380,990 | 25.2 | 2.63 | 12.4 | 26.1 | 2,846,738 | 30.0 | 2,334,050 | 65.2 | 168,800 | 26.5 | 11.3 | 844 | 31.6 | 5.1 |
| Arkansas | 1,147,084 | 10.0 | 2.47 | 13.4 | 27.1 | 1,317,818 | 12.3 | 1,114,902 | 67.4 | 106,300 | 21.5 | 10.9 | 638 | 29.9 | 3.2 |
| California | 12,577,498 | 9.3 | 2.90 | 13.3 | 23.3 | 13,682,976 | 12.0 | 12,406,475 | 55.6 | 370,900 | 30.6 | 11.4 | 1,163 | 33.8 | 9.1 |
| Colorado | 1,972,868 | 19.0 | 2.49 | 10.1 | 27.9 | 2,214,262 | 22.5 | 1,960,585 | 65.9 | 236,600 | 25.2 | 10.7 | 863 | 31.2 | 3.2 |
| Connecticut | 1,371,087 | 5.3 | 2.52 | 12.9 | 27.3 | 1,488,215 | 7.4 | 1,358,809 | 68.0 | 288,800 | 26.8 | 17.6 | 992 | 32.1 | 2.4 |
| Delaware | 342,297 | 14.6 | 2.55 | 14.2 | 25.6 | 406,489 | 18.5 | 328,765 | 73.0 | 243,600 | 24.8 | 11.8 | 952 | 32.3 | 2.9 |
| District of Columbia | 266,707 | 7.4 | 2.11 | 16.4 | 44.0 | 296,836 | 8.0 | 252,388 | 42.5 | 426,900 | 24.8 | 11.0 | 1,198 | 30.4 | 3.5 |
| Florida | 7,420,802 | 17.1 | 2.48 | 13.5 | 27.2 | 8,994,091 | 23.2 | 7,035,068 | 68.1 | 164,200 | 29.5 | 14.4 | 947 | 35.5 | 3.1 |
| Georgia | 3,585,584 | 19.3 | 2.63 | 15.8 | 25.4 | 4,091,482 | 24.7 | 3,482,420 | 66.2 | 156,200 | 25.2 | 12.3 | 819 | 32.4 | 3.2 |
| Hawaii | 455,338 | 12.9 | 2.89 | 12.6 | 23.3 | 519,992 | 12.9 | 445,812 | 58.0 | 525,400 | 30.1 | 10.1 | 1,291 | 33.5 | 9.1 |
| Idaho | 579,408 | 23.4 | 2.66 | 9.6 | 23.8 | 668,634 | 26.7 | 576,709 | 69.6 | 165,100 | 24.7 | 10.6 | 683 | 30.5 | 3.6 |
| Illinois | 4,836,972 | 5.3 | 2.59 | 12.9 | 27.8 | 5,297,077 | 8.4 | 4,752,857 | 67.7 | 191,800 | 25.9 | 13.8 | 848 | 31.5 | 3.1 |
| Indiana | 2,502,154 | 7.1 | 2.52 | 12.4 | 26.9 | 2,797,172 | 10.5 | 2,470,905 | 70.3 | 123,300 | 21.6 | 11.0 | 683 | 30.8 | 2.2 |
| Iowa | 1,221,576 | 6.3 | 2.41 | 9.3 | 28.4 | 1,337,563 | 8.5 | 1,223,439 | 72.4 | 123,400 | 21.3 | 11.5 | 629 | 28.1 | 1.7 |
| Kansas | 1,112,096 | 7.1 | 2.49 | 10.4 | 27.8 | 1,234,037 | 9.1 | 1,101,658 | 68.1 | 127,300 | 21.8 | 11.8 | 682 | 28.1 | 2.2 |
| Kentucky | 1,719,965 | 8.1 | 2.45 | 12.7 | 27.5 | 1,928,617 | 10.1 | 1,684,348 | 68.6 | 121,600 | 22.2 | 11.3 | 613 | 29.8 | 2.5 |
| Louisiana | 1,728,360 | 4.4 | 2.55 | 17.2 | 26.9 | 1,967,947 | 6.5 | 1,689,822 | 67.6 | 137,500 | 21.6 | 10.5 | 736 | 31.7 | 3.7 |
| Maine | 557,219 | 7.5 | 2.32 | 10.0 | 28.6 | 722,217 | 10.8 | 545,417 | 72.7 | 179,100 | 24.1 | 13.9 | 707 | 29.8 | 2.5 |
| Maryland | 2,156,411 | 8.9 | 2.61 | 14.6 | 26.1 | 2,380,605 | 11.0 | 2,127,439 | 67.0 | 301,400 | 25.4 | 12.9 | 1,131 | 30.8 | 2.4 |
| Massachusetts | 2,547,075 | 4.2 | 2.48 | 12.5 | 28.7 | 2,808,727 | 7.1 | 2,520,419 | 62.2 | 334,100 | 26.1 | 15.3 | 1,009 | 30.4 | 2.0 |
| Michigan | 3,872,508 | 2.3 | 2.49 | 13.2 | 27.9 | 4,531,231 | 7.0 | 3,806,621 | 72.8 | 123,300 | 24.6 | 13.9 | 730 | 33.3 | 2.1 |
| Minnesota | 2,087,227 | 10.1 | 2.48 | 9.5 | 28.0 | 2,348,242 | 13.7 | 2,091,548 | 73.0 | 194,300 | 24.1 | 11.9 | 764 | 30.2 | 2.3 |
| Mississippi | 1,115,768 | 6.6 | 2.58 | 18.5 | 26.3 | 1,276,441 | 9.9 | 1,079,999 | 69.8 | 100,100 | 23.5 | 12.0 | 672 | 33.2 | 4.0 |
| Missouri | 2,375,611 | 8.2 | 2.45 | 12.3 | 28.3 | 2,714,017 | 11.1 | 2,350,628 | 69.0 | 139,000 | 22.6 | 11.7 | 642 | 28.3 | 2.2 |
| Montana | 409,607 | 14.2 | 2.35 | 9.0 | 29.7 | 483,006 | 17.1 | 402,747 | 69.7 | 181,200 | 24.1 | 11.3 | 669 | 27.7 | 2.8 |
| Nebraska | 721,130 | 8.2 | 2.46 | 9.8 | 28.7 | 797,677 | 10.4 | 719,304 | 67.4 | 127,600 | 21.4 | 12.6 | 669 | 27.7 | 2.2 |
| Nevada | 1,006,250 | 34.0 | 2.65 | 12.7 | 25.7 | 1,175,070 | 42.0 | 989,811 | 57.2 | 174,800 | 28.1 | 12.2 | 952 | 31.6 | 5.0 |
| New Hampshire | 518,973 | 9.3 | 2.46 | 9.7 | 25.6 | 614,996 | 12.4 | 515,431 | 71.7 | 243,000 | 26.5 | 16.5 | 951 | 30.3 | 1.9 |
| New Jersey | 3,214,360 | 4.9 | 2.68 | 13.3 | 25.2 | 3,554,909 | 7.4 | 3,172,421 | 66.4 | 339,200 | 28.7 | 18.9 | 1,114 | 32.4 | 4.1 |
| New Mexico | 791,395 | 16.7 | 2.55 | 14.0 | 28.0 | 902,242 | 15.6 | 765,183 | 67.9 | 161,200 | 24.3 | 10.0 | 699 | 29.3 | 4.7 |
| New York | 7,317,755 | 3.7 | 2.57 | 14.9 | 29.1 | 8,108,211 | 5.6 | 7,196,427 | 54.3 | 296,500 | 26.3 | 15.5 | 1,020 | 31.7 | 5.5 |
| North Carolina | 3,745,155 | 19.6 | 2.48 | 13.7 | 27.0 | 4,333,479 | 23.0 | 3,670,859 | 67.2 | 154,200 | 24.0 | 12.5 | 731 | 31.3 | 2.8 |
| North Dakota | 281,192 | 9.3 | 2.30 | 8.2 | 31.5 | 318,099 | 9.8 | 280,412 | 66.9 | 123,000 | 19.6 | 10.0 | 583 | 25.8 | 1.3 |
| Ohio | 4,603,435 | 3.5 | 2.44 | 13.1 | 28.9 | 5,128,113 | 7.2 | 4,525,066 | 68.4 | 134,400 | 23.4 | 13.1 | 685 | 31.1 | 1.8 |
| Oklahoma | 1,460,450 | 8.8 | 2.49 | 12.3 | 27.5 | 1,666,205 | 10.0 | 1,432,959 | 67.8 | 111,400 | 21.9 | 11.2 | 659 | 28.9 | 3.0 |
| Oregon | 1,518,938 | 13.9 | 2.47 | 10.5 | 27.4 | 1,676,476 | 15.4 | 1,507,137 | 62.5 | 244,500 | 27.3 | 12.9 | 816 | 32.7 | 3.4 |
| Pennsylvania | 5,018,904 | 5.1 | 2.45 | 12.2 | 28.6 | 5,568,820 | 6.1 | 4,936,030 | 70.1 | 165,500 | 23.8 | 13.7 | 763 | 30.4 | 1.6 |
| Rhode Island | 413,600 | 1.3 | 2.44 | 13.5 | 29.6 | 463,416 | 5.4 | 402,295 | 60.8 | 254,500 | 27.7 | 15.4 | 868 | 30.9 | 2.6 |
| South Carolina | 1,801,181 | 17.4 | 2.49 | 15.6 | 26.5 | 2,140,337 | 22.0 | 1,761,393 | 68.7 | 138,100 | 23.5 | 12.0 | 728 | 32.2 | 2.7 |
| South Dakota | 322,282 | 11.0 | 2.42 | 9.7 | 29.4 | 364,031 | 12.6 | 318,955 | 68.0 | 129,700 | 21.9 | 10.9 | 591 | 26.9 | 2.7 |
| Tennessee | 2,493,552 | 11.7 | 2.48 | 13.9 | 26.9 | 2,815,087 | 15.4 | 2,440,663 | 68.1 | 139,000 | 23.7 | 11.4 | 697 | 31.4 | 2.5 |
| Texas | 8,922,933 | 20.7 | 2.75 | 14.1 | 24.2 | 9,996,209 | 22.5 | 8,738,664 | 63.6 | 128,100 | 23.4 | 12.5 | 801 | 30.2 | 5.8 |
| Utah | 877,692 | 25.2 | 3.10 | 9.7 | 18.7 | 981,821 | 27.7 | 880,025 | 69.9 | 217,200 | 24.8 | 10.0 | 796 | 29.5 | 4.6 |
| Vermont | 256,442 | 6.6 | 2.34 | 9.6 | 28.2 | 322,698 | 9.6 | 256,922 | 70.4 | 216,800 | 26.0 | 16.3 | 823 | 31.8 | 2.2 |
| Virginia | 3,056,058 | 13.2 | 2.54 | 12.4 | 26.0 | 3,368,674 | 16.0 | 2,992,732 | 67.7 | 249,100 | 24.7 | 11.4 | 1,019 | 30.2 | 2.6 |
| Washington | 2,620,076 | 15.4 | 2.51 | 10.5 | 27.2 | 2,888,594 | 17.9 | 2,606,863 | 63.1 | 271,800 | 26.7 | 12.1 | 908 | 30.6 | 3.4 |
| West Virginia | 763,831 | 3.7 | 2.36 | 11.2 | 28.4 | 882,213 | 4.5 | 741,940 | 74.6 | 95,100 | 20.1 | 10.0 | 571 | 29.7 | 1.8 |
| Wisconsin | 2,279,768 | 9.4 | 2.43 | 10.3 | 28.2 | 2,625,477 | 13.1 | 2,279,532 | 68.7 | 169,400 | 24.5 | 14.0 | 715 | 25.3 | 2.2 |
| Wyoming | 226,879 | 17.2 | 2.42 | 8.9 | 28.0 | 262,286 | 17.2 | 222,803 | 69.7 | 180,100 | 22.0 | 10.0 | 693 | 25.3 | 2.8 |

1. No spouse present.     2. Specified owner-occupied units.     3. Median monthly costs is often in the minimum category—10.0 percent or less, which is indicated as 10.0 percent.     4. Specified renter-occupied units.     5. Overcrowded or lacking complete plumbing facilities.

# Table A. States — **Housing Units**

| STATE | Housing units 2019 | | Occupied units | | | | | | | | | | | |
|---|---|---|---|---|---|---|---|---|---|---|---|---|---|---|
| | Total | Percent change, 2018-2019 | Total | Percent owner-occupied | Percent who pay 30 percent or more of income for housing expenses[1] | | Median owner cost as a percent of income | | Median monthly housing costs (dollars) | Median value of units[3] (dollars) | Percent valued over $500,000 | Median gross rent[4] (dollars) | Sub-standard units[5] (percent) | Percent living in a different house than 1 year ago |
| | | | | | Owners with a mortgage | Renter | With a mort-gage | Without a mort-gage[2] | | | | | | |
| | 79 | 80 | 81 | 82 | 83 | 84 | 85 | 86 | 87 | 88 | 89 | 90 | 91 | 92 |
| United States | 139,686,209 | 0.8 | 122,802,852 | 64.1 | 26.5 | 45.1 | 20.8 | 10.8 | 1,112 | 240,500 | 17.0 | 1,097 | 3.6 | 13.7 |
| Alabama | 2,284,922 | 0.4 | 1,897,576 | 68.8 | 22.6 | 40.9 | 18.9 | 10.0 | 778 | 154,000 | 5.0 | 807 | 1.9 | 13.4 |
| Alaska | 319,867 | 0.5 | 252,199 | 64.7 | 27.1 | 41.0 | 21.7 | 10.0 | 1,292 | 281,200 | 10.8 | 1,201 | 9.1 | 15.6 |
| Arizona | 3,076,048 | 1.3 | 2,670,441 | 65.3 | 25.5 | 43.4 | 20.6 | 10.0 | 1,069 | 255,900 | 12.3 | 1,101 | 4.9 | 16.1 |
| Arkansas | 1,389,159 | 0.6 | 1,163,647 | 65.5 | 21.5 | 38.6 | 18.7 | 10.0 | 709 | 136,200 | 3.6 | 742 | 3.4 | 14.4 |
| California | 14,367,012 | 0.6 | 13,157,873 | 54.9 | 36.9 | 50.7 | 24.7 | 10.5 | 1,695 | 568,500 | 56.9 | 1,614 | 8.5 | 12.0 |
| Colorado | 2,464,109 | 1.6 | 2,235,103 | 65.9 | 27.9 | 47.7 | 21.5 | 10.0 | 1,417 | 394,600 | 30.4 | 1,369 | 2.9 | 17.6 |
| Connecticut | 1,524,959 | 0.3 | 1,377,166 | 65.0 | 30.6 | 46.3 | 22.4 | 14.2 | 1,442 | 280,700 | 17.0 | 1,177 | 2.1 | 12.0 |
| Delaware | 443,764 | 1.2 | 376,239 | 70.3 | 25.1 | 46.6 | 20.2 | 10.0 | 1,149 | 261,700 | 9.2 | 1,116 | 2.0 | 12.6 |
| District of Columbia | 322,814 | 1.0 | 291,570 | 41.5 | 25.2 | 40.6 | 21.0 | 10.0 | 1,748 | 646,500 | 64.2 | 1,603 | 2.9 | 19.2 |
| Florida | 9,674,053 | 1.3 | 7,905,832 | 66.2 | 32.2 | 52.4 | 22.7 | 11.5 | 1,129 | 245,100 | 12.0 | 1,238 | 3.2 | 15.0 |
| Georgia | 4,378,350 | 1.2 | 3,852,714 | 64.1 | 24.2 | 44.8 | 19.9 | 10.0 | 1,054 | 202,500 | 9.5 | 1,049 | 2.4 | 14.0 |
| Hawaii | 550,328 | 0.7 | 465,299 | 60.2 | 40.9 | 49.7 | 26.4 | 10.0 | 1,665 | 669,200 | 68.7 | 1,651 | 9.1 | 12.9 |
| Idaho | 751,113 | 2.1 | 655,859 | 71.6 | 23.5 | 42.7 | 20.2 | 10.0 | 928 | 255,200 | 11.7 | 880 | 3.1 | 16.2 |
| Illinois | 5,388,210 | 0.2 | 4,866,006 | 66.0 | 25.5 | 42.4 | 20.6 | 12.2 | 1,126 | 209,100 | 9.4 | 1,020 | 2.5 | 12.1 |
| Indiana | 2,921,115 | 0.6 | 2,597,765 | 69.3 | 18.7 | 41.7 | 18.0 | 10.0 | 858 | 156,000 | 3.9 | 840 | 1.9 | 13.8 |
| Iowa | 1,418,600 | 0.6 | 1,287,221 | 70.5 | 18.8 | 38.6 | 18.7 | 11.1 | 870 | 158,900 | 4.0 | 808 | 2.1 | 13.9 |
| Kansas | 1,288,430 | 0.6 | 1,138,329 | 66.5 | 19.3 | 39.5 | 19.0 | 11.1 | 902 | 163,200 | 5.0 | 862 | 2.2 | 15.7 |
| Kentucky | 2,006,335 | 0.6 | 1,748,732 | 67.0 | 21.2 | 38.3 | 18.9 | 10.0 | 779 | 151,700 | 4.2 | 773 | 2.5 | 14.2 |
| Louisiana | 2,089,824 | 0.7 | 1,741,076 | 66.5 | 24.6 | 44.4 | 19.2 | 10.0 | 809 | 172,100 | 5.4 | 866 | 2.7 | 12.5 |
| Maine | 750,964 | 0.6 | 573,618 | 72.2 | 25.8 | 39.8 | 20.6 | 12.4 | 944 | 200,500 | 7.5 | 870 | 2.1 | 12.7 |
| Maryland | 2,470,307 | 0.5 | 2,226,767 | 66.8 | 26.0 | 47.4 | 21.0 | 10.0 | 1,521 | 332,500 | 22.1 | 1,401 | 2.6 | 12.6 |
| Massachusetts | 2,928,818 | 0.5 | 2,650,680 | 62.2 | 29.0 | 46.3 | 22.1 | 13.2 | 1,581 | 418,600 | 37.0 | 1,360 | 2.3 | 12.5 |
| Michigan | 4,629,605 | 0.3 | 3,969,880 | 71.6 | 22.1 | 44.3 | 19.2 | 11.3 | 902 | 169,600 | 5.4 | 888 | 1.9 | 12.9 |
| Minnesota | 2,477,515 | 0.9 | 2,222,568 | 71.9 | 20.7 | 41.8 | 19.3 | 10.4 | 1,125 | 246,700 | 9.1 | 1,016 | 2.4 | 13.2 |
| Mississippi | 1,339,047 | 0.5 | 1,100,229 | 67.3 | 25.6 | 40.2 | 19.7 | 10.0 | 709 | 128,200 | 3.0 | 777 | 3.0 | 12.2 |
| Missouri | 2,819,334 | 0.5 | 2,458,337 | 67.1 | 20.4 | 40.7 | 19.0 | 10.5 | 863 | 168,000 | 5.3 | 834 | 2.3 | 14.3 |
| Montana | 519,938 | 0.9 | 437,651 | 68.9 | 28.9 | 41.1 | 21.8 | 10.5 | 859 | 253,600 | 12.7 | 831 | 2.5 | 16.5 |
| Nebraska | 851,167 | 0.7 | 771,444 | 66.3 | 19.6 | 38.4 | 19.4 | 11.2 | 933 | 172,700 | 4.5 | 859 | 2.1 | 15.1 |
| Nevada | 1,285,681 | 1.3 | 1,143,557 | 56.6 | 28.3 | 48.2 | 21.8 | 10.0 | 1,192 | 317,800 | 16.3 | 1,168 | 4.8 | 17.6 |
| New Hampshire | 642,298 | 0.7 | 541,396 | 71.0 | 29.1 | 44.3 | 22.1 | 14.3 | 1,372 | 281,400 | 11.3 | 1,147 | 2.2 | 13.0 |
| New Jersey | 3,641,854 | 0.4 | 3,286,264 | 63.3 | 32.5 | 46.6 | 23.3 | 15.0 | 1,617 | 348,800 | 25.5 | 1,376 | 3.4 | 10.3 |
| New Mexico | 948,470 | 0.6 | 793,420 | 68.1 | 27.5 | 41.7 | 21.1 | 10.0 | 802 | 180,900 | 7.3 | 847 | 4.1 | 12.4 |
| New York | 8,404,205 | 0.5 | 7,446,812 | 53.5 | 31.3 | 47.3 | 22.1 | 12.8 | 1,362 | 338,700 | 33.0 | 1,309 | 5.2 | 10.5 |
| North Carolina | 4,748,148 | 1.3 | 4,046,348 | 65.3 | 23.2 | 42.6 | 19.3 | 10.0 | 941 | 193,200 | 8.1 | 931 | 2.6 | 14.8 |
| North Dakota | 379,974 | 0.6 | 323,519 | 61.3 | 16.3 | 35.5 | 18.4 | 10.0 | 822 | 205,400 | 6.5 | 804 | 2.2 | 18.0 |
| Ohio | 5,232,943 | 0.3 | 4,730,340 | 66.0 | 19.7 | 40.4 | 18.5 | 10.9 | 871 | 157,200 | 4.0 | 813 | 1.8 | 14.4 |
| Oklahoma | 1,749,520 | 0.4 | 1,495,151 | 65.5 | 20.5 | 39.5 | 18.9 | 10.0 | 805 | 147,200 | 4.2 | 814 | 3.0 | 16.0 |
| Oregon | 1,808,482 | 1.1 | 1,649,352 | 62.9 | 28.8 | 45.5 | 22.4 | 11.6 | 1,250 | 354,600 | 23.3 | 1,185 | 3.9 | 15.6 |
| Pennsylvania | 5,732,580 | 0.3 | 5,119,249 | 68.4 | 23.4 | 43.6 | 19.6 | 11.6 | 981 | 192,600 | 7.7 | 951 | 1.8 | 12.5 |
| Rhode Island | 470,177 | 0.2 | 407,174 | 61.7 | 29.1 | 45.9 | 22.4 | 12.8 | 1,248 | 283,000 | 13.1 | 1,043 | 2.2 | 11.4 |
| South Carolina | 2,351,364 | 1.4 | 1,975,915 | 70.3 | 23.4 | 42.0 | 19.2 | 10.0 | 880 | 179,800 | 8.3 | 922 | 2.6 | 13.3 |
| South Dakota | 401,749 | 1.1 | 353,799 | 67.8 | 22.5 | 35.9 | 19.7 | 10.5 | 840 | 185,000 | 5.2 | 769 | 2.2 | 15.3 |
| Tennessee | 3,028,437 | 1.2 | 2,654,737 | 66.5 | 22.8 | 42.1 | 19.4 | 10.0 | 877 | 191,900 | 8.6 | 904 | 2.1 | 14.4 |
| Texas | 11,283,892 | 1.6 | 9,985,126 | 61.9 | 26.5 | 44.8 | 20.9 | 11.0 | 1,096 | 200,400 | 8.9 | 1,091 | 5.1 | 15.0 |
| Utah | 1,133,543 | 2.2 | 1,023,855 | 70.6 | 24.3 | 41.8 | 20.7 | 10.0 | 1,214 | 330,300 | 18.3 | 1,098 | 3.7 | 15.8 |
| Vermont | 339,412 | 0.7 | 262,767 | 70.9 | 25.9 | 50.6 | 22.0 | 14.0 | 1,134 | 233,200 | 7.4 | 980 | 2.4 | 13.1 |
| Virginia | 3,562,258 | 0.7 | 3,191,847 | 66.1 | 24.5 | 42.9 | 20.5 | 10.0 | 1,286 | 288,800 | 22.0 | 1,254 | 2.3 | 14.7 |
| Washington | 3,195,098 | 1.5 | 2,932,477 | 63.1 | 28.6 | 45.3 | 22.4 | 10.6 | 1,434 | 387,600 | 33.0 | 1,359 | 3.9 | 16.9 |
| West Virginia | 894,983 | 0.1 | 728,175 | 73.4 | 20.0 | 36.3 | 17.8 | 10.0 | 607 | 124,600 | 2.3 | 727 | 1.7 | 11.8 |
| Wisconsin | 2,725,153 | 0.5 | 2,386,623 | 67.2 | 21.3 | 40.0 | 19.4 | 11.4 | 951 | 197,200 | 5.9 | 867 | 1.8 | 13.2 |
| Wyoming | 280,281 | 0.6 | 233,128 | 71.9 | 22.4 | 39.5 | 19.5 | 10.0 | 910 | 235,200 | 9.8 | 822 | 2.3 | 17.4 |

1. Excludes units where owner costs or gross rent as a percentage of household income cannot be calculated. 2. Median monthly costs is often in the minimum category—10.0 percent or less, which is indicated as 10.0 percent. 3. Specified owner-occupied units. 4. Specified renter-occupied units. 5. Overcrowded or lacking complete plumbing facilities.

# Table A. States — Residential Contruction, Vital Statistics, and Health

| STATE | Value of residential construction authorized by building permits, 2020 | | | New manufactured homes shipped, 2019 | Births, 2019 | | Deaths, 2018 | | | | | Percent lacking health insurance, 2019 | | Medicare beneficiaries, 65 years and older, 2019 |
| | | | | | | | Number | | Rate | | | | | |
| | | | | | | | | | Total | | | | | |
| | New construction ($1,000) | Number of housing units | Percent single family | | Total | Rate[1] | Total | Infant[2] | Crude[1] | Age-adjusted[1] | Infant[3] | All persons | Children under 19 years | |
| | 93 | 94 | 95 | 96 | 97 | 98 | 99 | 100 | 101 | 102 | 103 | 104 | 105 | 106 |
| United States | 307,209,904 | 1,471,141 | 66.6 | 94,390 | 3,747,540 | 11.4 | 2,839,205 | 21,467 | 8.7 | 7.2 | 5.7 | 9.2 | 5.7 | 60,242,615 |
| Alabama | 4,358,692 | 19,982 | 87.8 | 4,530 | 58,615 | 12.0 | 54,352 | 405 | 11.1 | 9.2 | 7.0 | 9.7 | 3.5 | 1,046,588 |
| Alaska | 381,356 | 1,420 | 71.8 | 64 | 9,822 | 13.4 | 4,453 | 60 | 6.0 | 7.0 | 6.0 | 12.2 | 9.4 | 100,297 |
| Arizona | 13,798,358 | 60,342 | 70.1 | 2,234 | 79,375 | 10.9 | 59,282 | 451 | 8.3 | 6.7 | 5.6 | 11.3 | 9.2 | 1,325,833 |
| Arkansas | 2,353,746 | 12,493 | 70.5 | 1,732 | 36,564 | 12.1 | 32,336 | 279 | 10.7 | 8.8 | 7.5 | 9.1 | 5.9 | 637,556 |
| California | 25,423,120 | 106,075 | 55.7 | 3,232 | 446,479 | 11.3 | 268,818 | 1,909 | 6.8 | 6.1 | 4.2 | 7.7 | 3.6 | 6,277,166 |
| Colorado | 10,266,541 | 40,469 | 65.8 | 917 | 62,869 | 10.9 | 38,526 | 298 | 6.8 | 6.5 | 4.7 | 8.0 | 5.5 | 911,545 |
| Connecticut | 1,061,756 | 5,471 | 45.9 | 99 | 34,258 | 9.6 | 31,230 | 147 | 8.7 | 6.4 | 4.2 | 5.9 | 3.5 | 680,357 |
| Delaware | 1,099,377 | 8,455 | 84.0 | 266 | 10,562 | 10.8 | 9,433 | 62 | 9.8 | 7.6 | 5.8 | 6.6 | 4.8 | 208,278 |
| District of Columbia | 886,594 | 7,370 | 1.9 | 0 | 9,079 | 12.9 | 5,008 | 64 | 7.1 | 7.2 | 7.0 | 3.5 | 2.0 | 94,058 |
| Florida | 36,884,176 | 164,074 | 70.2 | 6,681 | 220,002 | 10.2 | 205,426 | 1,332 | 9.6 | 6.6 | 6.0 | 13.2 | 7.6 | 4,556,611 |
| Georgia | 11,592,571 | 55,827 | 85.9 | 3,772 | 126,371 | 11.9 | 85,202 | 888 | 8.1 | 7.9 | 7.0 | 13.4 | 7.4 | 1,727,412 |
| Hawaii | 1,169,770 | 3,164 | 68.6 | 11 | 16,797 | 11.9 | 11,415 | 116 | 8.0 | 5.7 | 6.8 | 4.2 | 2.8 | 274,091 |
| Idaho | 3,948,590 | 19,130 | 78.2 | 416 | 22,063 | 12.3 | 14,261 | 109 | 8.1 | 7.3 | 5.1 | 10.8 | 5.0 | 334,445 |
| Illinois | 3,646,298 | 18,058 | 51.2 | 1,267 | 140,128 | 11.1 | 110,022 | 942 | 8.6 | 7.2 | 6.5 | 7.4 | 4.0 | 2,233,242 |
| Indiana | 5,872,298 | 24,919 | 76.2 | 1,935 | 80,859 | 12.0 | 65,693 | 556 | 9.8 | 8.3 | 6.8 | 8.7 | 7.1 | 1,260,564 |
| Iowa | 2,701,682 | 12,623 | 68.4 | 591 | 37,649 | 11.9 | 30,367 | 191 | 9.6 | 7.2 | 5.1 | 5.0 | 2.9 | 625,217 |
| Kansas | 1,991,033 | 8,211 | 72.8 | 762 | 35,395 | 12.1 | 27,537 | 236 | 9.5 | 7.7 | 6.5 | 9.2 | 5.8 | 534,927 |
| Kentucky | 2,142,572 | 11,281 | 78.7 | 3,455 | 53,069 | 11.9 | 48,707 | 315 | 10.9 | 9.2 | 5.8 | 6.4 | 4.3 | 928,872 |
| Louisiana | 3,427,774 | 17,283 | 89.4 | 4,264 | 58,941 | 12.7 | 46,048 | 454 | 9.9 | 8.7 | 7.6 | 8.9 | 4.4 | 869,795 |
| Maine | 1,194,138 | 5,304 | 78.8 | 632 | 11,779 | 8.8 | 14,715 | 67 | 11.0 | 7.5 | 5.4 | 8.0 | 5.6 | 339,400 |
| Maryland | 3,853,541 | 17,982 | 72.3 | 84 | 70,178 | 11.6 | 50,568 | 432 | 8.4 | 7.1 | 6.1 | 6.0 | 3.4 | 1,036,953 |
| Massachusetts | 4,761,486 | 17,025 | 39.7 | 114 | 69,117 | 10.0 | 59,152 | 290 | 8.6 | 6.7 | 4.2 | 3.0 | 1.5 | 1,331,279 |
| Michigan | 4,579,658 | 19,735 | 76.8 | 3,478 | 107,886 | 10.8 | 98,903 | 684 | 9.9 | 7.8 | 6.2 | 5.8 | 3.4 | 2,064,900 |
| Minnesota | 6,179,311 | 28,148 | 52.2 | 805 | 66,027 | 11.7 | 44,745 | 342 | 8.0 | 6.5 | 5.1 | 4.9 | 3.1 | 1,021,819 |
| Mississippi | 1,428,531 | 7,810 | 92.1 | 3,888 | 36,636 | 12.3 | 32,301 | 306 | 10.8 | 9.3 | 8.3 | 13.0 | 6.1 | 601,628 |
| Missouri | 4,021,896 | 19,839 | 66.1 | 1,464 | 72,127 | 11.8 | 63,117 | 460 | 10.3 | 8.2 | 6.3 | 10.0 | 6.5 | 1,227,462 |
| Montana | 1,051,886 | 5,980 | 55.5 | 244 | 11,079 | 10.4 | 9,992 | 55 | 9.4 | 7.2 | 4.8 | 8.3 | 6.2 | 230,309 |
| Nebraska | 1,643,788 | 9,483 | 60.6 | 213 | 24,755 | 12.8 | 16,904 | 149 | 8.8 | 7.2 | 5.9 | 8.3 | 5.7 | 346,150 |
| Nevada | 4,057,456 | 19,716 | 69.5 | 651 | 35,072 | 11.4 | 24,715 | 217 | 8.1 | 7.4 | 6.1 | 11.4 | 8.0 | 531,564 |
| New Hampshire | 1,059,391 | 4,320 | 70.2 | 307 | 11,839 | 8.7 | 12,774 | 43 | 9.4 | 7.1 | 3.6 | 6.3 | 3.7 | 298,735 |
| New Jersey | 4,796,438 | 36,146 | 34.0 | 479 | 99,585 | 11.2 | 75,765 | 390 | 8.5 | 6.7 | 3.9 | 7.9 | 4.3 | 1,615,169 |
| New Mexico | 1,225,981 | 5,219 | 84.7 | 1,351 | 22,960 | 10.9 | 19,007 | 131 | 9.1 | 7.5 | 5.7 | 10.0 | 5.7 | 421,972 |
| New York | 6,406,410 | 37,330 | 25.5 | 1,370 | 221,539 | 11.4 | 157,183 | 974 | 8.0 | 6.3 | 4.3 | 5.2 | 2.4 | 3,629,623 |
| North Carolina | 16,111,594 | 80,474 | 75.2 | 5,505 | 118,725 | 11.3 | 93,885 | 797 | 9.0 | 7.7 | 6.7 | 11.3 | 5.8 | 1,985,259 |
| North Dakota | 726,755 | 3,493 | 62.2 | 321 | 10,454 | 13.7 | 6,445 | 59 | 8.5 | 6.9 | 5.6 | 6.9 | 7.8 | 131,045 |
| Ohio | 6,418,479 | 29,686 | 61.8 | 1,927 | 134,461 | 11.5 | 124,264 | 938 | 10.6 | 8.4 | 6.9 | 6.6 | 4.8 | 2,340,027 |
| Oklahoma | 2,852,519 | 13,733 | 88.1 | 2,125 | 49,143 | 12.4 | 40,933 | 353 | 10.4 | 8.9 | 7.1 | 14.3 | 8.6 | 738,850 |
| Oregon | 4,067,001 | 18,665 | 61.6 | 1,399 | 41,858 | 9.9 | 36,187 | 176 | 8.6 | 6.9 | 4.2 | 7.2 | 4.1 | 863,650 |
| Pennsylvania | 5,077,342 | 25,706 | 61.6 | 1,648 | 134,230 | 10.5 | 134,702 | 805 | 10.5 | 7.6 | 5.9 | 5.8 | 4.6 | 2,732,824 |
| Rhode Island | 298,949 | 1,374 | 71.0 | 50 | 10,175 | 9.6 | 10,083 | 53 | 9.5 | 7.0 | 5.0 | 4.1 | 1.9 | 220,546 |
| South Carolina | 9,505,326 | 42,340 | 85.6 | 4,767 | 57,038 | 11.1 | 50,640 | 406 | 10.0 | 8.2 | 7.2 | 10.8 | 5.8 | 1,075,124 |
| South Dakota | 1,166,890 | 6,660 | 55.9 | 358 | 11,449 | 12.9 | 7,971 | 70 | 9.0 | 7.2 | 5.9 | 10.2 | 7.8 | 175,579 |
| Tennessee | 9,207,252 | 49,719 | 66.4 | 3,118 | 80,450 | 11.8 | 71,078 | 560 | 10.5 | 8.9 | 6.9 | 10.1 | 5.0 | 1,356,706 |
| Texas | 42,986,433 | 230,503 | 68.7 | 16,609 | 377,599 | 13.0 | 202,211 | 2,083 | 7.0 | 7.3 | 5.5 | 18.4 | 12.7 | 4,158,858 |
| Utah | 7,287,441 | 31,775 | 70.2 | 313 | 46,826 | 14.6 | 18,354 | 261 | 5.8 | 6.9 | 5.5 | 9.7 | 8.3 | 400,958 |
| Vermont | 396,203 | 2,077 | 55.6 | 122 | 5,361 | 8.6 | 6,027 | 35 | 9.6 | 7.1 | 6.4 | 4.5 | 2.1 | 147,222 |
| Virginia | 6,342,240 | 33,813 | 71.7 | 1,317 | 97,429 | 11.4 | 69,359 | 558 | 8.1 | 7.1 | 5.6 | 7.9 | 4.9 | 1,509,513 |
| Washington | 9,488,095 | 43,881 | 53.6 | 1,305 | 84,895 | 11.1 | 56,877 | 401 | 7.5 | 6.7 | 4.7 | 6.6 | 3.1 | 1,362,048 |
| West Virginia | 573,944 | 3,204 | 87.5 | 1,098 | 18,136 | 10.1 | 23,478 | 130 | 13.0 | 9.5 | 7.1 | 6.7 | 3.5 | 438,643 |
| Wisconsin | 4,733,365 | 21,226 | 57.3 | 760 | 63,270 | 10.9 | 53,684 | 393 | 9.2 | 7.2 | 6.1 | 5.7 | 3.8 | 1,171,957 |
| Wyoming | 703,859 | 2,128 | 83.1 | 99 | 6,565 | 11.3 | 5,070 | 35 | 8.8 | 7.5 | 5.3 | 12.3 | 10.6 | 109,989 |

1. Per 1,000 resident population.   2. Deaths of infants under 1 year old.   3. Deaths of infants under 1 year old per 1,000 live births.

# Table A. States — Crime and Education

| STATE | Serious crime known to police,[1] 2019 | | | | Public elementary and secondary school enrollment, 2019–2020 | | Educational attainment[3] (percent) | | | | Local government expenditures for education, 2017-2018 | |
|---|---|---|---|---|---|---|---|---|---|---|---|---|
| | Violent crime | | Property crime | | | | 2010 | | 2019 | | | |
| | Number | Rate[2] | Number | Rate[2] | Total | Student/ teacher ratio | High school graduate or more | Bachelor's degree or more | High school graduate or more | Bachelor's degree or more | Total current expenditures (mil dol) | Current expenditures per student (dollars) |
| | 107 | 108 | 109 | 110 | 111 | 112 | 113 | 114 | 115 | 116 | 117 | 118 |
| United States ............ | 1,245,410 | 379.4 | 6,925,677 | 2,109.9 | 50,720,122 | 15.9 | 85.6 | 28.2 | 88.6 | 33.1 | 639,951.9 | 12,654 |
| Alabama........................ | 25,046 | 510.8 | 131,133 | 2,674.4 | 744,235 | 17.7 | 82.1 | 21.9 | 87.1 | 26.3 | 7,214.1 | 9,717 |
| Alaska.......................... | 6,343 | 867.1 | 21,294 | 2,910.8 | 132,017 | 17.6 | 91.0 | 27.9 | 93.6 | 30.2 | 2,355.3 | 17,726 |
| Arizona......................... | 33,141 | 455.3 | 177,638 | 2,440.5 | 1,152,586 | 23.6 | 85.6 | 25.9 | 87.6 | 30.2 | 9,182.5 | 8,373 |
| Arkansas...................... | 17,643 | 584.6 | 86,250 | 2,858.0 | 496,927 | 12.9 | 82.9 | 19.5 | 87.5 | 23.3 | 5,044.1 | 10,168 |
| California...................... | 174,331 | 441.2 | 921,114 | 2,331.2 | 6,163,001 | 22.7 | 80.7 | 30.1 | 84.0 | 35.0 | 79,838.7 | 12,664 |
| Colorado....................... | 21,938 | 381.0 | 149,189 | 2,590.7 | 913,223 | 16.9 | 89.7 | 36.4 | 92.4 | 42.7 | 9,319.5 | 10,238 |
| Connecticut.................. | 6,546 | 183.6 | 50,862 | 1,426.6 | 523,690 | 12.4 | 88.6 | 35.5 | 90.7 | 39.8 | 10,703.9 | 20,147 |
| Delaware...................... | 4,115 | 422.6 | 21,931 | 2,252.2 | 139,930 | 14.4 | 87.7 | 27.8 | 90.3 | 33.2 | 2,082.8 | 15,282 |
| District of Columbia........... | 7,403 | 1,049.0 | 30,821 | 4,367.1 | 95,271 | 12.9 | 87.4 | 50.1 | 91.9 | 59.7 | 2,021.8 | 23,155 |
| Florida......................... | 81,270 | 378.4 | 460,846 | 2,145.7 | 2,858,461 | 17.2 | 85.5 | 25.8 | 88.4 | 30.7 | 27,371.0 | 9,663 |
| Georgia........................ | 36,170 | 340.7 | 252,249 | 2,375.8 | 1,769,657 | 15.0 | 84.3 | 27.3 | 87.9 | 32.5 | 19,031.0 | 10,760 |
| Hawaii.......................... | 4,042 | 285.5 | 40,228 | 2,841.2 | 181,088 | 14.8 | 89.9 | 29.5 | 92.4 | 33.6 | 2,756.3 | 15,242 |
| Idaho........................... | 4,000 | 223.8 | 21,793 | 1,219.5 | 311,096 | 18.1 | 88.3 | 24.4 | 91.5 | 28.7 | 2,363.0 | 7,846 |
| Illinois......................... | 51,561 | 406.9 | 233,984 | 1,846.5 | 1,943,117 | 14.6 | 86.9 | 30.8 | 89.8 | 35.8 | 31,848.9 | 15,912 |
| Indiana......................... | 24,966 | 370.8 | 132,694 | 1,971.0 | 1,051,411 | 17.0 | 87.0 | 22.7 | 89.6 | 26.9 | 10,576.8 | 10,033 |
| Iowa............................ | 8,410 | 266.6 | 54,699 | 1,733.7 | 517,324 | 14.5 | 90.6 | 24.9 | 92.6 | 29.3 | 6,000.9 | 11,724 |
| Kansas......................... | 11,968 | 410.8 | 67,428 | 2,314.5 | 497,963 | 13.6 | 89.2 | 29.8 | 91.8 | 34.0 | 5,515.1 | 11,095 |
| Kentucky...................... | 9,701 | 217.1 | 84,769 | 1,897.4 | 691,996 | 16.4 | 81.9 | 20.5 | 87.2 | 25.1 | 7,546.1 | 11,081 |
| Louisiana...................... | 25,537 | 549.3 | 146,993 | 3,162.0 | 710,439 | 18.4 | 81.9 | 21.4 | 86.0 | 25.0 | 8,321.4 | 11,636 |
| Maine.......................... | 1,548 | 115.2 | 16,743 | 1,245.6 | 180,291 | 12.2 | 90.3 | 26.8 | 93.2 | 33.2 | 2,719.6 | 15,069 |
| Maryland...................... | 27,456 | 454.1 | 117,901 | 1,950.2 | 909,404 | 14.8 | 88.1 | 36.1 | 90.4 | 40.9 | 13,543.6 | 15,155 |
| Massachusetts.............. | 22,578 | 327.6 | 81,317 | 1,179.8 | 959,394 | 12.8 | 89.1 | 39.0 | 91.3 | 45.0 | 17,682.7 | 18,328 |
| Michigan....................... | 43,686 | 437.4 | 158,296 | 1,585.0 | 1,495,925 | 17.6 | 88.7 | 25.2 | 91.4 | 30.0 | 17,723.9 | 11,688 |
| Minnesota .................... | 13,332 | 236.4 | 117,236 | 2,078.8 | 893,203 | 16.1 | 91.8 | 31.8 | 93.6 | 37.3 | 11,424.4 | 12,910 |
| Mississippi.................... | 8,272 | 277.9 | 70,707 | 2,375.8 | 466,002 | 14.8 | 81.0 | 19.5 | 85.3 | 22.3 | 4,261.4 | 8,909 |
| Missouri....................... | 30,380 | 495.0 | 161,946 | 2,638.7 | 910,466 | 13.2 | 86.9 | 25.6 | 90.7 | 30.2 | 10,101.3 | 11,034 |
| Montana....................... | 4,328 | 404.9 | 23,440 | 2,193.2 | 149,917 | 14.0 | 91.7 | 28.8 | 94.2 | 33.6 | 1,720.7 | 11,512 |
| Nebraska...................... | 5,821 | 300.9 | 39,449 | 2,039.3 | 330,018 | 13.7 | 90.4 | 28.6 | 92.0 | 33.2 | 4,148.4 | 12,813 |
| Nevada......................... | 15,210 | 493.8 | 71,525 | 2,322.1 | 500,855 | 19.6 | 84.7 | 21.7 | 86.9 | 25.7 | 4,391.7 | 9,040 |
| New Hampshire .............. | 2,074 | 152.5 | 16,442 | 1,209.2 | 177,351 | 12.1 | 91.5 | 32.8 | 93.3 | 37.6 | 2,976.5 | 16,588 |
| New Jersey ................... | 18,375 | 206.9 | 118,637 | 1,335.7 | 1,411,917 | 12.1 | 88.0 | 35.4 | 90.3 | 41.2 | 28,607.6 | 20,316 |
| New Mexico .................. | 17,450 | 832.2 | 65,269 | 3,112.7 | 331,206 | 15.2 | 83.3 | 25.0 | 85.9 | 27.7 | 3,331.0 | 9,963 |
| New York ..................... | 69,764 | 358.6 | 267,155 | 1,373.3 | 2,692,589 | 12.4 | 84.9 | 32.5 | 87.6 | 37.8 | 62,984.8 | 23,686 |
| North Carolina............... | 38,995 | 371.8 | 247,236 | 2,357.3 | 1,560,350 | 15.5 | 84.7 | 26.5 | 88.6 | 32.3 | 14,412.7 | 9,277 |
| North Dakota................. | 2,169 | 284.6 | 15,066 | 1,977.0 | 116,185 | 12.5 | 90.3 | 27.6 | 93.5 | 30.4 | 1,542.6 | 13,783 |
| Ohio............................ | 34,269 | 293.2 | 240,291 | 2,055.7 | 1,689,867 | 15.9 | 88.1 | 24.6 | 90.8 | 29.3 | 21,975.4 | 12,893 |
| Oklahoma..................... | 17,086 | 431.8 | 112,587 | 2,845.3 | 703,719 | 16.3 | 86.2 | 22.9 | 88.4 | 26.2 | 5,681.4 | 8,174 |
| Oregon......................... | 11,995 | 284.4 | 115,170 | 2,730.6 | 610,649 | 20.2 | 88.8 | 28.8 | 91.4 | 34.5 | 6,911.8 | 11,903 |
| Pennsylvania................. | 39,228 | 306.4 | 179,665 | 1,403.4 | 1,732,449 | 13.9 | 88.4 | 27.1 | 91.0 | 32.3 | 28,279.6 | 16,337 |
| Rhode Island................. | 2,342 | 221.1 | 16,259 | 1,534.8 | 143,557 | 13.4 | 83.5 | 30.2 | 89.3 | 34.8 | 2,423.5 | 16,954 |
| South Carolina.............. | 26,323 | 511.3 | 151,389 | 2,940.3 | 786,879 | 14.7 | 84.1 | 24.5 | 88.3 | 29.6 | 8,322.9 | 10,705 |
| South Dakota ............... | 3,530 | 399.0 | 15,667 | 1,771.0 | 139,949 | 14.1 | 89.6 | 26.3 | 92.1 | 29.7 | 1,414.5 | 10,263 |
| Tennessee.................... | 40,647 | 595.2 | 181,153 | 2,652.6 | 1,014,744 | 15.7 | 83.6 | 23.1 | 88.0 | 28.7 | 9,618.3 | 9,599 |
| Texas.......................... | 121,474 | 418.9 | 693,204 | 2,390.7 | 5,495,398 | 15.1 | 80.7 | 25.9 | 84.6 | 30.8 | 52,233.5 | 9,670 |
| Utah............................ | 7,553 | 235.6 | 69,546 | 2,169.3 | 684,694 | 22.6 | 90.6 | 29.3 | 93.0 | 34.8 | 5,063.0 | 7,576 |
| Vermont....................... | 1,262 | 202.2 | 8,888 | 1,424.4 | 87,125 | 10.8 | 91.0 | 33.6 | 93.1 | 38.7 | 1,773.7 | 20,149 |
| Virginia........................ | 17,753 | 208.0 | 140,213 | 1,642.7 | 1,297,012 | 14.9 | 86.5 | 34.2 | 90.0 | 39.6 | 15,786.3 | 12,224 |
| Washington................... | 22,377 | 293.9 | 204,224 | 2,681.9 | 1,142,073 | 18.4 | 89.8 | 31.1 | 91.7 | 37.0 | 14,418.1 | 12,985 |
| West Virginia................. | 5,674 | 316.6 | 28,376 | 1,583.4 | 263,486 | 14.0 | 83.2 | 17.5 | 87.1 | 21.1 | 3,150.6 | 11,572 |
| Wisconsin..................... | 17,070 | 293.2 | 85,672 | 1,471.4 | 855,400 | 14.3 | 90.1 | 26.3 | 92.8 | 31.3 | 10,712.5 | 12,446 |
| Wyoming...................... | 1,258 | 217.4 | 9,093 | 1,571.1 | 94,616 | 12.8 | 92.3 | 24.1 | 94.5 | 29.1 | 1,520.8 | 16,134 |

1. Data for serious crimes have not been adjusted for underreporting; this may affect comparability between geographic areas and over time.   2. Per 100,000 population estimated by the FBI.
3. Persons 25 years old and over.

## Table A. States — Exports, Income, and Poverty

| STATE | Exports of goods by state of origin, 2020 (mil dol) | | | Income, 2019 | | | | | Percent below poverty level, 2019 | | | | | | |
|---|---|---|---|---|---|---|---|---|---|---|---|---|---|---|---|
| | | | | Households | | | | | | | | | Families with related children under 18 years | | |
| | Total | Manu-factured | Non-manu-factured | Per capita income (dollars) | Median income (dollars) | Percent with income of $24,999 or less | Percent with income of $100,000 or more | Median income of family of four | All persons | Children under 18 years | Persons 65 years and over | All families | Married-couple families | Male house-holder[1] families | Female house-holder[1] families |
| | 119 | 120 | 121 | 122 | 123 | 124 | 125 | 126 | 127 | 128 | 129 | 130 | 131 | 132 | 133 |
| United States .................. | 1,431,638 | 956,306 | 226,689 | 35,672 | 65,712 | 18.1 | 31.4 | 99,048 | 12.3 | 16.8 | 9.4 | 8.6 | 5.7 | 16.2 | 33.5 |
| Alabama.......................... | 17,131 | 14,805 | 1,725 | 28,650 | 51,734 | 24.5 | 22.8 | 80,845 | 15.5 | 21.4 | 10.5 | 11.2 | 6.9 | 21.2 | 42.1 |
| Alaska............................ | 4,612 | 589 | 3,985 | 36,978 | 75,463 | 13.9 | 36.8 | 101,575 | 10.1 | 13.0 | 6.9 | 6.6 | 4.5 | 11.2 | 28.9 |
| Arizona........................... | 19,718 | 11,703 | 2,580 | 32,173 | 62,055 | 17.6 | 28.1 | 84,669 | 13.5 | 19.1 | 9.0 | 9.5 | 7.8 | 20.2 | 32.5 |
| Arkansas......................... | 5,197 | 3,451 | 473 | 27,274 | 48,952 | 25.2 | 19.4 | 67,349 | 16.2 | 22.1 | 10.5 | 11.7 | 10.3 | 22.9 | 38.9 |
| California......................... | 156,112 | 99,417 | 20,914 | 39,393 | 80,440 | 14.9 | 40.5 | 105,232 | 11.8 | 15.6 | 10.5 | 8.2 | 6.5 | 14.6 | 29.5 |
| Colorado ........................ | 8,283 | 7,186 | 323 | 41,053 | 77,127 | 13.7 | 37.7 | 114,066 | 9.3 | 10.9 | 7.2 | 5.8 | 4.4 | 10.9 | 26.8 |
| Connecticut..................... | 13,802 | 11,237 | 1,067 | 45,359 | 78,833 | 16.1 | 39.8 | 129,379 | 10.0 | 14.1 | 7.3 | 6.8 | 3.6 | 14.2 | 32.2 |
| Delaware......................... | 3,919 | 3,133 | 119 | 36,858 | 70,176 | 16.1 | 32.9 | 107,204 | 11.3 | 16.4 | 7.3 | 7.6 | 6.7 | 10.6 | 31.6 |
| District of Columbia........... | 2,776 | 2,310 | 392 | 59,808 | 92,266 | 16.3 | 47.5 | 171,779 | 13.5 | 18.9 | 13.3 | 8.8 | 1.3 | 15.4 | 29.3 |
| Florida............................ | 45,807 | 32,687 | 3,263 | 32,887 | 59,227 | 19.1 | 26.2 | 84,165 | 12.7 | 17.7 | 10.7 | 8.7 | 6.3 | 13.3 | 30.8 |
| Georgia........................... | 38,784 | 28,922 | 3,927 | 32,657 | 61,980 | 19.2 | 28.4 | 91,161 | 13.3 | 18.7 | 10.4 | 9.6 | 6.0 | 12.4 | 33.9 |
| Hawaii............................ | 320 | 134 | 137 | 36,989 | 83,102 | 13.5 | 41.1 | 118,223 | 9.3 | 12.4 | 8.7 | 6.4 | 4.9 | 15.9 | 26.9 |
| Idaho............................. | 3,403 | 2,223 | 601 | 29,606 | 60,999 | 17.0 | 25.2 | 89,661 | 11.2 | 13.2 | 6.9 | 7.4 | 6.9 | 19.0 | 27.6 |
| Illinois............................ | 53,512 | 41,127 | 2,832 | 37,728 | 69,187 | 17.6 | 33.7 | 107,226 | 11.5 | 15.7 | 8.6 | 7.9 | 4.9 | 14.6 | 31.3 |
| Indiana........................... | 35,283 | 31,237 | 513 | 30,988 | 57,603 | 19.3 | 24.8 | 90,654 | 11.9 | 15.2 | 7.7 | 7.8 | 4.6 | 14.7 | 32.8 |
| Iowa.............................. | 12,635 | 10,505 | 1,678 | 33,107 | 61,691 | 17.6 | 26.2 | 95,199 | 11.2 | 13.0 | 7.4 | 7.3 | 4.2 | 14.3 | 32.0 |
| Kansas........................... | 10,399 | 7,506 | 2,133 | 32,885 | 62,087 | 17.7 | 27.2 | 92,890 | 11.4 | 14.7 | 7.2 | 7.5 | 5.2 | 20.9 | 35.1 |
| Kentucky......................... | 24,750 | 18,837 | 319 | 29,029 | 52,295 | 24.2 | 21.6 | 81,619 | 16.3 | 21.7 | 11.6 | 11.8 | 7.9 | 26.7 | 42.9 |
| Louisiana........................ | 59,622 | 28,119 | 30,989 | 28,662 | 51,073 | 27.1 | 23.2 | 82,529 | 19.0 | 27.0 | 13.2 | 14.3 | 7.3 | 14.7 | 46.5 |
| Maine............................. | 2,328 | 1,375 | 803 | 34,078 | 58,924 | 19.2 | 25.3 | 91,651 | 10.9 | 13.8 | 8.5 | 6.5 | 4.3 | | 28.7 |
| Maryland ........................ | 12,672 | 8,602 | 2,311 | 43,325 | 86,738 | 13.0 | 43.3 | 130,252 | 9.0 | 12.0 | 7.8 | 5.8 | 3.6 | 12.0 | 22.8 |
| Massachusetts................. | 24,893 | 19,918 | 1,164 | 46,241 | 85,843 | 15.5 | 43.6 | 140,309 | 9.4 | 11.6 | 9.1 | 6.0 | 2.6 | 10.6 | 28.5 |
| Michigan......................... | 44,001 | 37,391 | 2,079 | 32,892 | 59,584 | 19.2 | 26.1 | 97,970 | 13.0 | 17.6 | 8.4 | 8.8 | 6.1 | 18.7 | 35.3 |
| Minnesota ....................... | 20,077 | 16,885 | 1,185 | 39,025 | 74,593 | 14.1 | 35.3 | 118,646 | 9.0 | 11.2 | 7.4 | 5.2 | 2.8 | 12.5 | 27.3 |
| Mississippi...................... | 10,272 | 7,081 | 1,262 | 25,301 | 45,792 | 28.4 | 18.1 | 70,656 | 19.6 | 28.1 | 13.2 | 14.8 | 8.0 | 29.5 | 46.7 |
| Missouri.......................... | 12,770 | 10,826 | 1,130 | 31,756 | 57,409 | 20.5 | 24.7 | 89,418 | 12.9 | 17.1 | 8.9 | 8.8 | 5.2 | 21.0 | 34.9 |
| Montana.......................... | 1,433 | 761 | 568 | 32,625 | 57,153 | 20.4 | 24.0 | 81,958 | 12.6 | 14.9 | 8.6 | 7.9 | 6.1 | 16.4 | 35.3 |
| Nebraska......................... | 6,984 | 5,411 | 1,282 | 33,272 | 63,229 | 16.8 | 27.4 | 96,749 | 9.9 | 11.0 | 8.1 | 6.2 | 3.2 | 13.8 | 29.5 |
| Nevada........................... | 10,348 | 5,507 | 707 | 33,575 | 63,276 | 17.9 | 28.4 | 83,731 | 12.5 | 16.9 | 9.5 | 8.7 | 6.7 | 13.8 | 30.2 |
| New Hampshire ................ | 5,453 | 4,134 | 226 | 41,241 | 77,933 | 13.5 | 37.7 | 128,157 | 7.3 | 7.1 | 6.2 | 4.2 | 2.0 | 11.5 | 22.2 |
| New Jersey...................... | 38,021 | 22,502 | 4,612 | 44,888 | 85,751 | 13.8 | 43.6 | 132,708 | 9.2 | 12.3 | 8.8 | 6.2 | 3.8 | 12.1 | 27.6 |
| New Mexico ..................... | 3,691 | 1,624 | 216 | 28,423 | 51,945 | 25.0 | 21.9 | 66,343 | 18.2 | 24.9 | 13.5 | 13.7 | 12.5 | 20.9 | 38.0 |
| New York......................... | 61,888 | 32,807 | 10,045 | 41,857 | 72,108 | 18.8 | 36.7 | 111,054 | 13.0 | 18.1 | 12.0 | 9.3 | 7.0 | 17.5 | 32.3 |
| North Carolina.................. | 28,439 | 23,403 | 1,467 | 32,021 | 57,341 | 20.4 | 25.3 | 88,942 | 13.6 | 19.5 | 9.1 | 9.6 | 6.0 | 20.3 | 37.0 |
| North Dakota.................... | 5,189 | 3,045 | 2,054 | 36,611 | 64,577 | 17.7 | 29.8 | 103,996 | 10.6 | 10.2 | 8.0 | 6.5 | 5.0 | 14.0 | 32.6 |
| Ohio.............................. | 45,039 | 35,461 | 4,448 | 32,780 | 58,642 | 20.1 | 25.5 | 95,003 | 13.1 | 18.4 | 8.3 | 9.2 | 4.9 | 17.9 | 38.4 |
| Oklahoma........................ | 5,376 | 4,220 | 314 | 29,666 | 54,449 | 21.8 | 22.9 | 78,458 | 15.2 | 19.9 | 9.7 | 10.8 | 7.8 | 19.0 | 38.9 |
| Oregon........................... | 24,977 | 19,117 | 2,353 | 35,531 | 67,058 | 16.6 | 31.1 | 100,533 | 11.4 | 13.1 | 8.1 | 6.6 | 5.1 | 13.5 | 27.9 |
| Pennsylvania.................... | 37,455 | 27,730 | 4,203 | 35,804 | 63,463 | 18.7 | 29.4 | 103,857 | 12.0 | 16.9 | 8.3 | 8.2 | 4.7 | 18.1 | 35.8 |
| Rhode Island.................... | 2,402 | 1,559 | 639 | 37,525 | 71,169 | 17.8 | 33.1 | 108,105 | 10.8 | 14.0 | 8.9 | 6.6 | 4.2 | 9.3 | 30.3 |
| South Carolina.................. | 30,295 | 27,681 | 677 | 31,295 | 56,227 | 21.1 | 24.7 | 85,227 | 13.8 | 19.7 | 10.1 | 9.8 | 4.9 | 22.3 | 38.3 |
| South Dakota ................... | 1,377 | 1,208 | 125 | 31,550 | 59,533 | 18.2 | 25.1 | 90,951 | 11.9 | 15.0 | 7.7 | 7.3 | 4.6 | 15.4 | 33.8 |
| Tennessee ...................... | 28,134 | 20,016 | 1,594 | 31,224 | 56,071 | 20.9 | 24.0 | 85,923 | 13.9 | 19.7 | 9.7 | 10.0 | 7.1 | 21.2 | 35.6 |
| Texas ............................ | 279,294 | 154,144 | 74,454 | 32,267 | 64,034 | 18.1 | 30.5 | 88,109 | 13.6 | 19.2 | 10.6 | 10.5 | 7.2 | 15.3 | 37.0 |
| Utah.............................. | 17,713 | 15,882 | 976 | 31,771 | 75,780 | 12.3 | 34.7 | 95,430 | 8.9 | 9.9 | 6.2 | 6.1 | 4.9 | 7.6 | 27.4 |
| Vermont.......................... | 2,399 | 2,088 | 50 | 35,702 | 63,001 | 17.4 | 28.4 | 111,095 | 10.2 | 10.2 | 6.1 | 5.8 | 2.8 | 10.9 | 29.2 |
| Virginia........................... | 16,431 | 11,704 | 3,285 | 40,635 | 76,456 | 15.0 | 37.9 | 114,910 | 9.9 | 13.4 | 7.1 | 6.5 | 4.0 | 11.6 | 31.0 |
| Washington...................... | 41,281 | 23,071 | 14,971 | 41,521 | 78,687 | 13.8 | 38.8 | 112,182 | 9.8 | 12.0 | 7.5 | 6.2 | 4.0 | 11.5 | 30.2 |
| West Virginia.................... | 4,568 | 2,818 | 1,561 | 27,446 | 48,850 | 25.8 | 19.1 | 73,600 | 16.0 | 20.1 | 9.3 | 11.1 | 7.9 | 23.8 | 42.1 |
| Wisconsin........................ | 20,504 | 17,479 | 965 | 34,568 | 64,168 | 16.8 | 28.1 | 103,708 | 10.4 | 13.5 | 7.4 | 6.2 | 3.7 | 12.1 | 30.1 |
| Wyoming......................... | 1,165 | 1,087 | 56 | 34,104 | 65,003 | 17.3 | 27.9 | 100,012 | 10.1 | 11.6 | 7.4 | 6.6 | 4.1 | 13.9 | 36.8 |

1. No spouse present.

# Table A. States — **Personal Income**

| STATE | Total, 2020 (mil dol) | Percent change, 2019–2020 | Per capita[1], 2020 Dollars | Rank | Wages and salaries[2], 2020 | Proprietors' income, 2019 | Dividends, interest, and rent, 2019 | Total | Total | Social Security | Medical payments | Income maintenance | Unemployment insurance |
|---|---|---|---|---|---|---|---|---|---|---|---|---|---|
| | 134 | 135 | 136 | 137 | 138 | 139 | 140 | 141 | 142 | 143 | 144 | 145 | 146 |
| United States .................. | 19,679,715 | 6.1 | 59,729 | X | 9,320,570 | 1,702,572 | 3,709,407 | 3,125,174 | 3,052,923 | 1,030,743 | 1,425,049 | 268,682 | 28,075 |
| Alabama............................ | 230,861 | 6.7 | 46,908 | 47 | 103,854 | 14,950 | 38,687 | 50,500 | 49,420 | 18,478 | 19,388 | 5,431 | 171 |
| Alaska .............................. | 47,365 | 3.1 | 64,780 | 10 | 21,274 | 3,731 | 8,680 | 7,939 | 7,780 | 1,633 | 3,367 | 751 | 78 |
| Arizona............................. | 363,274 | 8.4 | 48,950 | 43 | 173,292 | 24,639 | 65,521 | 66,453 | 64,847 | 23,277 | 29,925 | 5,006 | 256 |
| Arkansas.......................... | 142,765 | 6.0 | 47,109 | 46 | 61,412 | 7,516 | 32,022 | 31,559 | 30,896 | 10,857 | 14,344 | 2,590 | 125 |
| California.......................... | 2,814,011 | 6.9 | 71,480 | 6 | 1,354,270 | 256,549 | 570,562 | 373,130 | 364,399 | 97,810 | 183,951 | 38,035 | 5,018 |
| Colorado.......................... | 368,920 | 4.8 | 63,522 | 11 | 184,441 | 35,532 | 77,151 | 45,459 | 44,187 | 14,897 | 20,218 | 2,948 | 376 |
| Connecticut...................... | 283,747 | 3.0 | 79,771 | 2 | 118,398 | 30,194 | 60,161 | 36,204 | 35,424 | 12,480 | 17,717 | 2,461 | 593 |
| Delaware.......................... | 56,019 | 5.6 | 56,768 | 23 | 27,651 | 4,534 | 9,629 | 11,071 | 10,857 | 3,892 | 5,178 | 684 | 67 |
| District of Columbia........... | 62,061 | 5.4 | 87,064 | 1 | 79,532 | 7,184 | 10,478 | 7,062 | 6,911 | 1,317 | 4,218 | 889 | 86 |
| Florida............................. | 1,202,648 | 6.8 | 55,337 | 27 | 506,050 | 81,865 | 307,480 | 216,742 | 211,992 | 77,445 | 91,209 | 16,856 | 424 |
| Georgia............................ | 547,976 | 7.0 | 51,165 | 38 | 272,547 | 49,687 | 93,375 | 86,129 | 83,794 | 30,037 | 33,372 | 8,959 | 341 |
| Hawaii.............................. | 85,446 | 5.8 | 60,729 | 18 | 36,453 | 6,713 | 17,092 | 12,673 | 12,363 | 4,542 | 5,243 | 1,165 | 151 |
| Idaho............................... | 88,817 | 8.1 | 48,616 | 44 | 37,619 | 9,552 | 18,093 | 14,970 | 14,573 | 5,716 | 5,839 | 1,021 | 94 |
| Illinois............................. | 792,729 | 6.5 | 62,977 | 13 | 390,411 | 61,918 | 148,704 | 112,157 | 109,364 | 38,245 | 48,797 | 12,129 | 1,710 |
| Indiana............................. | 346,802 | 5.8 | 51,340 | 36 | 160,973 | 31,980 | 52,446 | 63,282 | 61,806 | 23,304 | 28,901 | 4,561 | 253 |
| Iowa................................ | 174,685 | 6.8 | 55,218 | 28 | 79,183 | 17,504 | 31,108 | 29,100 | 28,409 | 10,879 | 12,778 | 1,958 | 385 |
| Kansas............................ | 163,385 | 5.0 | 56,073 | 25 | 75,120 | 19,687 | 29,983 | 24,724 | 24,084 | 9,490 | 10,361 | 1,627 | 150 |
| Kentucky.......................... | 208,222 | 6.5 | 46,507 | 48 | 96,173 | 13,348 | 32,749 | 47,036 | 46,057 | 15,599 | 21,900 | 3,842 | 319 |
| Louisiana.......................... | 232,437 | 5.4 | 50,037 | 41 | 100,517 | 21,183 | 39,079 | 48,362 | 47,342 | 14,091 | 24,204 | 4,766 | 176 |
| Maine............................... | 73,212 | 7.6 | 54,225 | 30 | 31,915 | 5,288 | 12,515 | 15,069 | 14,774 | 5,383 | 6,763 | 958 | 93 |
| Maryland .......................... | 413,359 | 5.8 | 68,258 | 8 | 188,201 | 33,189 | 75,879 | 54,797 | 53,464 | 17,748 | 26,514 | 4,265 | 451 |
| Massachusetts.................. | 549,565 | 7.5 | 79,721 | 3 | 279,587 | 42,694 | 103,544 | 71,419 | 69,919 | 21,694 | 36,370 | 6,498 | 1,353 |
| Michigan........................... | 528,093 | 7.4 | 52,987 | 34 | 239,866 | 34,499 | 87,338 | 103,673 | 101,469 | 38,693 | 46,625 | 8,339 | 805 |
| Minnesota ........................ | 348,152 | 4.9 | 61,540 | 16 | 177,172 | 26,444 | 64,224 | 52,837 | 51,594 | 17,931 | 25,174 | 3,804 | 765 |
| Mississippi........................ | 123,850 | 6.9 | 41,745 | 51 | 51,551 | 8,604 | 18,452 | 30,683 | 30,032 | 10,325 | 13,648 | 3,060 | 90 |
| Missouri........................... | 314,818 | 5.4 | 51,177 | 37 | 155,548 | 21,716 | 56,240 | 58,950 | 57,605 | 21,176 | 26,324 | 4,169 | 277 |
| Montana .......................... | 57,626 | 8.4 | 53,329 | 32 | 23,222 | 5,473 | 13,214 | 10,411 | 10,176 | 3,750 | 4,327 | 613 | 109 |
| Nebraska.......................... | 112,266 | 6.5 | 57,942 | 21 | 53,600 | 12,917 | 21,048 | 16,383 | 15,958 | 5,802 | 6,423 | 1,196 | 74 |
| Nevada............................. | 168,319 | 6.8 | 53,635 | 31 | 73,315 | 11,479 | 38,022 | 26,263 | 25,577 | 8,939 | 11,284 | 2,177 | 303 |
| New Hampshire ................ | 90,745 | 5.1 | 66,418 | 9 | 40,814 | 8,555 | 14,707 | 12,968 | 12,671 | 5,409 | 5,388 | 607 | 63 |
| New Jersey ....................... | 668,354 | 6.8 | 75,245 | 5 | 282,591 | 62,812 | 111,354 | 87,021 | 85,058 | 29,919 | 40,128 | 6,867 | 1,935 |
| New Mexico ...................... | 96,476 | 6.2 | 45,803 | 49 | 42,220 | 5,869 | 17,268 | 21,597 | 21,133 | 6,834 | 9,940 | 2,187 | 134 |
| New York ......................... | 1,460,860 | 4.7 | 75,548 | 4 | 723,571 | 138,719 | 297,123 | 229,559 | 225,297 | 62,019 | 126,902 | 21,751 | 2,076 |
| North Carolina.................. | 530,956 | 6.0 | 50,086 | 40 | 263,016 | 35,554 | 92,508 | 97,751 | 95,447 | 35,069 | 39,496 | 8,186 | 216 |
| North Dakota.................... | 45,450 | 4.2 | 59,388 | 19 | 23,315 | 4,915 | 9,375 | 6,308 | 6,142 | 2,175 | 2,738 | 399 | 79 |
| Ohio................................. | 623,207 | 6.2 | 53,296 | 33 | 302,373 | 44,938 | 101,040 | 114,048 | 111,482 | 38,722 | 54,395 | 9,292 | 852 |
| Oklahoma......................... | 196,051 | 4.7 | 49,249 | 42 | 83,325 | 23,157 | 34,707 | 36,445 | 35,577 | 12,790 | 14,392 | 3,156 | 250 |
| Oregon............................. | 240,771 | 7.3 | 56,765 | 24 | 113,077 | 20,589 | 46,840 | 42,427 | 41,496 | 14,725 | 19,320 | 3,017 | 535 |
| Pennsylvania..................... | 795,093 | 7.0 | 62,198 | 15 | 349,135 | 71,132 | 129,968 | 143,936 | 141,142 | 48,688 | 69,867 | 11,460 | 1,834 |
| Rhode Island..................... | 64,313 | 7.7 | 60,837 | 17 | 28,550 | 4,266 | 10,610 | 11,768 | 11,538 | 3,798 | 5,632 | 973 | 148 |
| South Carolina.................. | 247,869 | 6.0 | 47,502 | 45 | 109,176 | 16,602 | 44,053 | 51,047 | 49,916 | 19,326 | 20,096 | 3,868 | 173 |
| South Dakota ................... | 51,128 | 7.1 | 57,273 | 22 | 21,475 | 7,272 | 11,222 | 7,504 | 7,312 | 2,875 | 2,933 | 597 | 30 |
| Tennessee ....................... | 348,109 | 4.7 | 50,547 | 39 | 165,155 | 43,857 | 48,772 | 66,256 | 64,751 | 23,833 | 28,318 | 5,658 | 218 |
| Texas .............................. | 1,610,182 | 5.1 | 54,841 | 29 | 785,853 | 197,544 | 266,992 | 224,585 | 218,177 | 68,941 | 99,226 | 22,324 | 2,205 |
| Utah ................................ | 169,810 | 8.2 | 52,251 | 35 | 88,087 | 13,343 | 34,431 | 19,880 | 19,167 | 6,970 | 7,424 | 1,691 | 150 |
| Vermont ........................... | 36,560 | 6.0 | 58,650 | 20 | 15,691 | 2,956 | 6,902 | 7,030 | 6,895 | 2,509 | 3,269 | 474 | 63 |
| Virginia............................ | 535,727 | 5.2 | 62,362 | 14 | 266,282 | 29,519 | 106,293 | 71,293 | 69,422 | 26,276 | 28,696 | 4,807 | 288 |
| Washington ...................... | 525,643 | 6.6 | 68,322 | 7 | 262,568 | 39,176 | 110,690 | 68,247 | 66,562 | 23,620 | 28,600 | 4,801 | 1,105 |
| West Virginia..................... | 80,510 | 6.2 | 45,109 | 50 | 32,062 | 4,936 | 11,461 | 21,561 | 21,167 | 7,704 | 9,326 | 1,712 | 183 |
| Wisconsin......................... | 323,635 | 4.4 | 55,487 | 26 | 154,693 | 21,985 | 59,221 | 53,864 | 52,582 | 21,177 | 22,803 | 3,854 | 394 |
| Wyoming.......................... | 36,840 | 2.4 | 63,263 | 12 | 14,397 | 4,308 | 10,397 | 5,043 | 4,916 | 1,935 | 1,796 | 241 | 47 |

1. Based on the resident population estimated as of July 1 of the year shown. 2. Includes supplements to wages and salaries.

# Table A. States — **Personal Income and Earnings**

| STATE | Personal tax payments, 2018 (mil dol) | Disposable personal income, 2019 Total (mil dol) | Per capita[1] (dollars) | Earnings, 2020 Total (mil dol) | Farm | Goods-related[2] Total | Manu-facturing | Service-related and other[3] Total | Retail trade | Finance, insurance, real estate, rental and leasing | Health care and social assistance | Govern-ment | Gross state product, 2020 (mil dol) |
|---|---|---|---|---|---|---|---|---|---|---|---|---|---|
| | 147 | 148 | 149 | 150 | 151 | 152 | 153 | 154 | 155 | 156 | 157 | 158 | 159 |
| United States | 2,199,985 | 16,342,277 | 49,774 | 13,132,608 | 0.8 | 16.3 | 8.8 | 67.4 | 5.7 | 9.8 | 11.2 | 15.6 | 20,936,558 |
| Alabama | 20,345 | 196,104 | 39,956 | 143,417 | 0.1 | 21.0 | 13.7 | 59.1 | 6.6 | 6.9 | 11.3 | 19.8 | 224,871 |
| Alaska | 3,562 | 42,383 | 57,774 | 32,205 | 0.1 | 17.2 | 3.0 | 51.1 | 5.5 | 4.1 | 12.9 | 31.6 | 50,247 |
| Arizona | 34,339 | 300,904 | 41,266 | 235,544 | 0.7 | 15.5 | 7.6 | 68.7 | 7.0 | 11.5 | 12.7 | 15.1 | 372,461 |
| Arkansas | 13,003 | 121,680 | 40,278 | 82,380 | 1.0 | 19.3 | 12.6 | 63.5 | 6.9 | 6.3 | 13.3 | 16.2 | 129,074 |
| California | 364,716 | 2,267,564 | 57,497 | 1,903,151 | 1.4 | 14.6 | 8.8 | 68.5 | 5.0 | 9.0 | 9.6 | 15.5 | 3,091,872 |
| Colorado | 42,930 | 309,255 | 53,704 | 256,718 | 0.6 | 18.2 | 5.7 | 66.0 | 5.2 | 9.0 | 8.9 | 15.2 | 390,099 |
| Connecticut | 40,534 | 235,023 | 65,906 | 174,555 | 0.2 | 16.3 | 10.6 | 71.2 | 5.4 | 17.1 | 12.1 | 12.3 | 280,900 |
| Delaware | 6,196 | 46,859 | 47,978 | 38,985 | 0.8 | 12.0 | 5.5 | 71.0 | 5.6 | 21.0 | 13.7 | 15.9 | 75,513 |
| District of Columbia | 8,523 | 50,341 | 71,078 | 107,612 | 0.0 | 1.7 | 0.1 | 59.9 | 1.0 | 4.5 | 5.8 | 38.5 | 143,533 |
| Florida | 130,060 | 995,924 | 46,339 | 695,289 | 0.6 | 12.4 | 4.8 | 73.6 | 7.3 | 10.4 | 12.5 | 13.4 | 1,095,888 |
| Georgia | 56,590 | 455,548 | 42,863 | 381,643 | 0.4 | 14.4 | 8.1 | 70.0 | 6.0 | 9.7 | 9.8 | 15.2 | 619,240 |
| Hawaii | 8,867 | 71,860 | 50,762 | 53,711 | 0.6 | 10.4 | 1.6 | 58.5 | 5.9 | 7.0 | 11.2 | 30.5 | 89,856 |
| Idaho | 7,666 | 74,483 | 41,632 | 56,425 | 4.8 | 19.7 | 10.7 | 60.5 | 8.2 | 6.7 | 11.8 | 15.0 | 84,032 |
| Illinois | 89,675 | 654,966 | 51,706 | 541,056 | 0.7 | 15.4 | 10.3 | 70.7 | 5.1 | 11.9 | 10.6 | 13.1 | 863,517 |
| Indiana | 31,816 | 295,896 | 43,960 | 230,939 | 0.8 | 25.5 | 18.6 | 61.4 | 5.8 | 9.6 | 13.1 | 12.3 | 372,637 |
| Iowa | 16,691 | 146,948 | 46,508 | 116,947 | 6.5 | 22.8 | 15.5 | 55.1 | 5.7 | 10.0 | 10.1 | 15.7 | 192,710 |
| Kansas | 16,107 | 139,541 | 47,909 | 112,354 | 3.7 | 18.5 | 11.6 | 62.0 | 5.2 | 9.8 | 11.3 | 15.7 | 173,298 |
| Kentucky | 20,063 | 175,486 | 39,238 | 133,609 | 0.9 | 21.0 | 14.3 | 60.9 | 6.1 | 7.5 | 13.1 | 17.1 | 210,024 |
| Louisiana | 19,591 | 201,039 | 43,157 | 145,692 | 0.6 | 22.2 | 9.6 | 60.1 | 6.5 | 6.7 | 12.6 | 17.1 | 241,991 |
| Maine | 6,891 | 61,172 | 45,455 | 45,118 | 0.6 | 16.8 | 9.1 | 65.6 | 7.7 | 7.5 | 16.5 | 17.0 | 66,196 |
| Maryland | 51,466 | 339,327 | 56,041 | 266,205 | 0.2 | 11.4 | 4.4 | 64.5 | 4.8 | 9.4 | 11.0 | 24.0 | 422,726 |
| Massachusetts | 77,354 | 433,980 | 62,942 | 380,501 | 0.0 | 13.5 | 7.4 | 75.0 | 4.5 | 11.5 | 12.9 | 11.5 | 584,039 |
| Michigan | 54,429 | 437,203 | 43,787 | 330,458 | 0.5 | 21.4 | 15.2 | 64.7 | 6.0 | 8.0 | 12.5 | 13.4 | 515,928 |
| Minnesota | 43,713 | 288,089 | 51,079 | 242,041 | 1.6 | 18.4 | 11.7 | 67.4 | 5.2 | 10.4 | 13.3 | 12.6 | 374,352 |
| Mississippi | 9,222 | 106,593 | 35,791 | 72,718 | 1.1 | 19.9 | 13.2 | 57.0 | 7.7 | 5.2 | 12.0 | 22.1 | 114,201 |
| Missouri | 31,328 | 267,292 | 43,529 | 214,837 | 0.7 | 17.5 | 10.8 | 67.3 | 6.1 | 9.0 | 12.9 | 14.5 | 321,709 |
| Montana | 5,594 | 47,574 | 44,457 | 34,560 | 3.4 | 16.3 | 4.4 | 61.9 | 8.2 | 7.0 | 14.5 | 18.4 | 51,489 |
| Nebraska | 10,601 | 94,854 | 49,082 | 79,933 | 5.1 | 15.7 | 9.3 | 63.8 | 5.3 | 9.7 | 11.3 | 15.5 | 128,809 |
| Nevada | 15,545 | 142,039 | 45,956 | 102,017 | 0.2 | 15.4 | 4.8 | 69.2 | 7.0 | 7.6 | 10.0 | 15.1 | 172,598 |
| New Hampshire | 8,811 | 77,534 | 56,978 | 58,359 | 0.1 | 19.3 | 11.1 | 68.8 | 7.9 | 9.0 | 12.8 | 11.9 | 85,109 |
| New Jersey | 83,364 | 542,573 | 61,023 | 408,975 | 0.1 | 13.6 | 7.8 | 72.7 | 5.9 | 10.8 | 11.5 | 13.6 | 619,061 |
| New Mexico | 8,036 | 82,811 | 39,441 | 58,309 | 1.9 | 14.3 | 3.7 | 57.8 | 6.4 | 5.5 | 12.2 | 26.0 | 100,310 |
| New York | 220,947 | 1,174,200 | 60,329 | 1,029,235 | 0.2 | 8.4 | 3.8 | 75.7 | 4.4 | 18.0 | 11.7 | 15.8 | 1,699,045 |
| North Carolina | 54,526 | 446,448 | 42,513 | 357,382 | 0.4 | 16.9 | 10.2 | 64.7 | 6.1 | 9.7 | 10.1 | 17.9 | 586,136 |
| North Dakota | 4,153 | 39,461 | 51,669 | 33,628 | 5.3 | 19.5 | 5.8 | 57.7 | 6.0 | 7.7 | 13.1 | 17.5 | 54,033 |
| Ohio | 62,334 | 524,450 | 44,838 | 418,549 | 0.3 | 20.0 | 13.5 | 65.3 | 5.8 | 8.1 | 13.4 | 14.4 | 675,037 |
| Oklahoma | 17,054 | 170,273 | 42,991 | 126,464 | 1.2 | 20.9 | 8.9 | 58.2 | 6.1 | 5.7 | 10.7 | 19.7 | 186,581 |
| Oregon | 28,776 | 195,570 | 46,386 | 160,909 | 1.1 | 18.7 | 10.9 | 64.3 | 6.4 | 7.6 | 12.5 | 15.9 | 250,459 |
| Pennsylvania | 84,127 | 658,797 | 51,473 | 506,415 | 0.3 | 16.1 | 9.7 | 71.1 | 5.3 | 8.0 | 14.7 | 12.5 | 780,176 |
| Rhode Island | 6,713 | 52,994 | 50,081 | 39,606 | 0.1 | 14.4 | 7.9 | 68.7 | 6.2 | 10.5 | 13.9 | 16.6 | 60,225 |
| South Carolina | 23,135 | 210,813 | 40,874 | 153,318 | 0.0 | 20.2 | 13.1 | 60.4 | 7.1 | 7.5 | 9.5 | 19.4 | 241,689 |
| South Dakota | 3,859 | 43,879 | 49,462 | 34,093 | 6.5 | 16.6 | 8.8 | 62.1 | 6.8 | 12.3 | 15.2 | 14.9 | 54,852 |
| Tennessee | 26,995 | 305,478 | 44,724 | 244,986 | 0.0 | 18.7 | 11.2 | 68.6 | 6.8 | 8.3 | 14.3 | 12.7 | 364,486 |
| Texas | 148,532 | 1,382,815 | 47,705 | 1,148,495 | 0.3 | 23.0 | 8.2 | 62.8 | 5.6 | 9.2 | 9.3 | 13.9 | 1,759,734 |
| Utah | 17,445 | 139,451 | 43,533 | 121,350 | 0.5 | 18.8 | 9.0 | 65.5 | 7.4 | 10.1 | 8.5 | 15.2 | 194,986 |
| Vermont | 3,608 | 30,895 | 49,507 | 22,469 | 1.2 | 17.3 | 9.9 | 63.0 | 7.2 | 6.1 | 15.6 | 18.5 | 32,797 |
| Virginia | 64,124 | 445,077 | 52,015 | 354,970 | 0.1 | 11.3 | 5.4 | 65.9 | 4.8 | 7.7 | 9.0 | 22.7 | 551,760 |
| Washington | 50,932 | 442,196 | 58,077 | 359,006 | 1.3 | 15.1 | 8.0 | 66.9 | 9.3 | 6.3 | 9.7 | 16.8 | 618,705 |
| West Virginia | 6,897 | 68,937 | 38,399 | 45,750 | -0.1 | 19.0 | 8.2 | 59.3 | 6.7 | 5.0 | 17.3 | 21.8 | 73,709 |
| Wisconsin | 34,614 | 275,295 | 47,264 | 216,979 | 1.4 | 24.0 | 17.1 | 60.7 | 5.8 | 8.0 | 12.7 | 13.9 | 338,678 |
| Wyoming | 3,589 | 32,404 | 55,858 | 22,744 | 1.4 | 22.5 | 4.3 | 51.9 | 5.5 | 5.3 | 7.5 | 24.1 | 36,242 |

1. Based on the resident population estimated as of July 1 of the year shown.  2. Includes mining, construction, and manufacturing.  3. Includes private sector earnings in forestry, fishing, related activities, and other; utilities; wholesale trade; transportation and warehousing; and information.

# Table A. States — Social Security, Employment, and Labor Force

| STATE | Social Security beneficiaries, December 2019 Number | Rate[1] | Supplemental Security Income recipients, December 2019 | Civilian employment and selected occupations,[2] 2019 Total | Percent Management, business, science, and arts occupations | Services, sales, and office | Construction and production | Civilian labor force (annual average), 2020 Total (1,000) | Percent change, 2019–2020 | Employed (1,000) | Unemployed Total (1,000) | Rate[3] |
|---|---|---|---|---|---|---|---|---|---|---|---|---|
| | 160 | 161 | 162 | 163 | 164 | 165 | 166 | 167 | 168 | 169 | 170 | 171 |
| United States | 62,502,630 | 190.4 | 8,075,821 | 158,758,794 | 39.9 | 38.1 | 22.1 | 160,742,000 | -1.7 | 147,795,000 | 12,947,000 | 8.1 |
| Alabama | 1,159,320 | 236.4 | 160,187 | 2,153,467 | 35.9 | 36.8 | 27.3 | 2,230,118 | -0.3 | 2,099,062 | 131,056 | 5.9 |
| Alaska | 104,568 | 142.9 | 12,471 | 338,011 | 38.3 | 36.8 | 24.8 | 347,414 | -1.8 | 320,219 | 27,195 | 7.8 |
| Arizona | 1,399,403 | 192.3 | 119,532 | 3,305,302 | 37.9 | 41.5 | 20.6 | 3,570,220 | 0.9 | 3,288,150 | 282,070 | 7.9 |
| Arkansas | 703,896 | 233.2 | 104,312 | 1,325,091 | 33.9 | 37.7 | 28.4 | 1,354,296 | -0.8 | 1,272,344 | 81,952 | 6.1 |
| California | 6,070,395 | 153.6 | 1,222,078 | 19,078,101 | 40.7 | 38.5 | 20.9 | 18,821,167 | -2.8 | 16,913,078 | 1,908,089 | 10.1 |
| Colorado | 897,886 | 155.9 | 72,761 | 3,032,173 | 45.1 | 36.3 | 18.6 | 3,122,237 | -0.1 | 2,895,473 | 226,764 | 7.3 |
| Connecticut | 689,996 | 193.5 | 66,783 | 1,823,915 | 44.7 | 37.9 | 17.4 | 1,872,631 | -2.3 | 1,724,621 | 148,010 | 7.9 |
| Delaware | 219,490 | 225.4 | 17,059 | 466,061 | 40.5 | 38.8 | 20.6 | 484,358 | -0.8 | 446,458 | 37,900 | 7.8 |
| District of Columbia | 83,899 | 118.9 | 25,521 | 387,826 | 67.8 | 26.2 | 6.0 | 409,734 | -1.3 | 376,839 | 32,895 | 8.0 |
| Florida | 4,747,364 | 221.0 | 576,861 | 9,958,518 | 36.5 | 43.4 | 20.1 | 10,114,329 | -2.1 | 9,332,838 | 781,491 | 7.7 |
| Georgia | 1,871,862 | 176.3 | 259,199 | 5,002,153 | 39.1 | 37.0 | 23.8 | 5,072,155 | -1.5 | 4,741,191 | 330,964 | 6.5 |
| Hawaii | 277,013 | 195.6 | 22,694 | 671,768 | 36.1 | 45.7 | 18.2 | 648,191 | -3.1 | 572,796 | 75,395 | 11.6 |
| Idaho | 358,633 | 200.7 | 31,007 | 848,223 | 35.5 | 39.4 | 25.1 | 892,151 | 1.4 | 844,365 | 47,786 | 5.4 |
| Illinois | 2,267,082 | 178.9 | 264,282 | 6,286,647 | 40.7 | 37.3 | 22.0 | 6,249,147 | -3.1 | 5,657,532 | 591,615 | 9.5 |
| Indiana | 1,370,231 | 203.5 | 127,800 | 3,272,728 | 35.1 | 36.5 | 28.4 | 3,319,010 | -1.9 | 3,082,982 | 236,028 | 7.1 |
| Iowa | 657,019 | 208.2 | 51,633 | 1,618,556 | 37.8 | 35.2 | 27.1 | 1,666,420 | -3.9 | 1,578,765 | 87,655 | 5.3 |
| Kansas | 561,634 | 192.8 | 47,365 | 1,455,746 | 40.0 | 36.0 | 24.0 | 1,497,003 | 0.2 | 1,408,995 | 88,008 | 5.9 |
| Kentucky | 1,001,700 | 224.2 | 171,487 | 1,997,773 | 36.2 | 36.4 | 27.4 | 2,019,887 | -2.4 | 1,885,645 | 134,242 | 6.6 |
| Louisiana | 922,223 | 198.4 | 173,485 | 2,040,325 | 35.4 | 40.7 | 23.9 | 2,076,643 | -2.0 | 1,905,238 | 171,405 | 8.3 |
| Maine | 349,962 | 260.3 | 36,599 | 684,413 | 40.1 | 37.8 | 22.2 | 676,547 | -2.8 | 639,759 | 36,788 | 5.4 |
| Maryland | 1,020,436 | 168.8 | 121,691 | 3,098,870 | 48.0 | 35.1 | 16.9 | 3,172,796 | -3.0 | 2,958,287 | 214,509 | 6.8 |
| Massachusetts | 1,287,830 | 186.8 | 182,701 | 3,700,243 | 48.4 | 35.5 | 16.1 | 3,658,321 | -3.3 | 3,334,126 | 324,195 | 8.9 |
| Michigan | 2,236,852 | 224.0 | 270,396 | 4,755,016 | 37.7 | 37.3 | 25.0 | 4,840,843 | -2.2 | 4,362,728 | 478,115 | 9.9 |
| Minnesota | 1,053,166 | 186.7 | 93,151 | 3,010,452 | 42.8 | 35.4 | 21.8 | 3,094,702 | 0.1 | 2,903,562 | 191,140 | 6.2 |
| Mississippi | 677,464 | 227.6 | 115,638 | 1,240,752 | 32.8 | 38.6 | 28.7 | 1,259,347 | -1.6 | 1,157,546 | 101,801 | 8.1 |
| Missouri | 1,312,639 | 213.9 | 136,094 | 2,942,459 | 38.4 | 37.9 | 23.7 | 3,052,700 | -0.9 | 2,867,162 | 185,538 | 6.1 |
| Montana | 239,410 | 224.0 | 17,677 | 520,261 | 39.1 | 38.3 | 22.6 | 539,883 | 0.5 | 508,095 | 31,788 | 5.9 |
| Nebraska | 352,880 | 182.4 | 28,692 | 1,008,957 | 40.0 | 35.5 | 24.4 | 1,035,175 | -0.5 | 991,388 | 43,787 | 4.2 |
| Nevada | 552,219 | 179.3 | 56,627 | 1,479,868 | 30.4 | 47.3 | 22.3 | 1,530,872 | -2.3 | 1,334,416 | 196,456 | 12.8 |
| New Hampshire | 312,043 | 229.5 | 18,328 | 735,493 | 42.5 | 35.6 | 21.9 | 761,732 | -1.5 | 710,817 | 50,915 | 6.7 |
| New Jersey | 1,646,792 | 185.4 | 178,009 | 4,496,699 | 44.9 | 36.3 | 18.8 | 4,495,166 | -0.6 | 4,055,260 | 439,906 | 9.8 |
| New Mexico | 445,742 | 212.6 | 62,064 | 907,775 | 37.6 | 41.0 | 21.3 | 943,287 | -1.8 | 863,874 | 79,413 | 8.4 |
| New York | 3,667,022 | 188.5 | 621,220 | 9,611,029 | 43.1 | 40.0 | 16.9 | 9,289,171 | -2.3 | 8,361,006 | 928,165 | 10.0 |
| North Carolina | 2,144,804 | 204.5 | 228,518 | 4,937,837 | 39.3 | 37.0 | 23.7 | 4,950,728 | -2.5 | 4,587,407 | 363,452 | 7.3 |
| North Dakota | 136,520 | 179.1 | 8,301 | 405,699 | 38.3 | 36.1 | 25.6 | 406,839 | -0.6 | 386,006 | 20,833 | 5.1 |
| Ohio | 2,386,362 | 204.2 | 307,783 | 5,692,943 | 38.1 | 37.4 | 24.5 | 5,754,286 | -1.5 | 5,285,484 | 468,802 | 8.1 |
| Oklahoma | 802,326 | 202.8 | 96,804 | 1,791,692 | 36.2 | 38.4 | 25.4 | 1,848,485 | 0.2 | 1,734,924 | 113,561 | 6.1 |
| Oregon | 891,726 | 211.4 | 88,912 | 2,045,338 | 40.4 | 38.2 | 21.4 | 2,104,657 | -0.1 | 1,945,212 | 159,445 | 7.6 |
| Pennsylvania | 2,861,155 | 223.5 | 354,037 | 6,285,109 | 40.3 | 37.2 | 22.5 | 6,387,869 | -1.7 | 5,807,942 | 579,927 | 9.1 |
| Rhode Island | 228,257 | 215.5 | 32,552 | 547,914 | 42.2 | 38.8 | 18.9 | 541,680 | -3.1 | 490,845 | 50,835 | 9.4 |
| South Carolina | 1,174,399 | 228.1 | 114,706 | 2,359,714 | 36.9 | 38.0 | 25.1 | 2,384,590 | 0.7 | 2,237,407 | 147,183 | 6.2 |
| South Dakota | 182,793 | 206.6 | 14,438 | 452,975 | 39.2 | 36.7 | 24.1 | 463,256 | 0.0 | 441,745 | 21,511 | 4.6 |
| Tennessee | 1,478,145 | 216.4 | 174,588 | 3,215,401 | 36.2 | 38.2 | 25.6 | 3,289,426 | -1.2 | 3,043,894 | 245,532 | 7.5 |
| Texas | 4,338,301 | 149.6 | 644,093 | 13,830,576 | 37.6 | 38.4 | 23.9 | 13,983,319 | -0.4 | 12,915,337 | 1,067,982 | 7.6 |
| Utah | 419,037 | 130.7 | 31,730 | 1,598,530 | 40.4 | 37.3 | 22.3 | 1,632,215 | 1.4 | 1,555,782 | 76,433 | 4.7 |
| Vermont | 153,124 | 245.4 | 15,009 | 331,247 | 42.9 | 36.1 | 20.9 | 330,058 | -4.2 | 311,645 | 18,413 | 5.6 |
| Virginia | 1,559,858 | 182.7 | 155,582 | 4,229,399 | 45.4 | 36.1 | 18.5 | 4,346,644 | -1.8 | 4,075,237 | 271,407 | 6.2 |
| Washington | 1,376,287 | 180.7 | 148,731 | 3,749,409 | 43.3 | 35.4 | 21.3 | 3,914,869 | 0.1 | 3,585,782 | 329,087 | 8.4 |
| West Virginia | 478,209 | 266.8 | 70,844 | 744,981 | 33.3 | 40.6 | 26.1 | 792,156 | -0.8 | 726,023 | 66,133 | 8.3 |
| Wisconsin | 1,257,850 | 216.0 | 116,794 | 3,002,074 | 37.9 | 35.7 | 26.4 | 3,065,402 | -0.9 | 2,872,609 | 192,793 | 6.3 |
| Wyoming | 115,406 | 199.4 | 6,995 | 283,750 | 36.3 | 36.6 | 27.1 | 296,801 | 0.2 | 279,462 | 17,339 | 5.8 |

1. Per 1,000 resident population estimated as of July 1 of the year shown.     2. Persons 16 years old and over.     3. Percent of civilian labor force.

# Table A. States — Nonfarm Employment and Earnings

Nonfarm employment and earnings, 2020

| STATE | Employed Total (1,000) | Employed Percent change, 2019–2020 | Manufacturing Employment (1,000) | Mfg. Avg. earnings — Hourly | Mfg. Avg. earnings — Weekly | Construction | Transportation and public utilities | Wholesale trade | Retail trade | Information | Financial activities | Services[1] |
|---|---|---|---|---|---|---|---|---|---|---|---|---|
| | 172 | 173 | 174 | 175 | 176 | 177 | 178 | 179 | 180 | 181 | 182 | 183 |
| United States | 142,185 | -5.8 | 12,179 | 22.79 | 928.28 | 7,269 | 6,097.0 | 5,639.9 | 14,853.1 | 2,694.0 | 8,724.0 | 62,202 |
| Alabama | 1,986 | -4.3 | 257.8 | 20.37 | 837.21 | 93.0 | 78.2 | 72.5 | 223.9 | 19.3 | 96.5 | 750.7 |
| Alaska | 301 | -8.7 | 11.6 | 25.62 | 950.50 | 16.1 | 19.0 | 6.2 | 33.1 | 4.9 | 10.8 | 111.6 |
| Arizona | 2,850 | -3.1 | 155.1 | 20.77 | 828.72 | 173.4 | 130.8 | 96.9 | 319.2 | 11.8 | 231.9 | 1,251.6 |
| Arkansas | 1,242 | -3.0 | 146.5 | 18.77 | 732.03 | 53.2 | 65.8 | 46.3 | 135.7 | | 63.7 | 496.9 |
| California | 16,141 | -7.4 | 1,261.7 | 24.06 | 964.81 | 855.1 | 727.8 | 643.4 | 1,523.6 | 74.6 | 815.3 | 7,277.6 |
| Colorado | 2,645 | -5.2 | 153.6 | 26.52 | 1,063.45 | 174.6 | 99.0 | 107.8 | 261.5 | 29.3 | 172.4 | 1,145.3 |
| Connecticut | 1,565 | -7.7 | 25.6 | 29.01 | 1,177.81 | 56.2 | 61.2 | 56.0 | 159.8 | 3.7 | 120.1 | 704.3 |
| Delaware | 439 | -6.0 | 4.8 | 21.24 | 902.70 | 22.4 | 19.1 | 10.9 | 47.7 | 0.4 | 47.2 | 197.0 |
| District of Columbia | 747 | -6.3 | 1.0 | – | – | 15.1 | 4.1 | 5.0 | 20.1 | 19.8 | 29.0 | 413.1 |
| Florida | 8,499 | -5.2 | 376.7 | 24.02 | 1,001.63 | 561.4 | 343.5 | 343.5 | 1,052.5 | 130.1 | 592.6 | 3,984.0 |
| Georgia | 4,407 | -4.6 | 385.3 | 20.99 | 879.48 | 201.2 | 235.7 | 208.6 | 478.3 | 109.6 | 250.7 | 1,847.4 |
| Hawaii | 557 | -15.4 | 11.9 | 23.48 | 899.28 | 36.0 | 27.2 | 16.5 | 60.8 | 7.2 | 27.3 | 249.6 |
| Idaho | 755 | -0.7 | 68.2 | 19.60 | 809.48 | 55.8 | 26.9 | 31.2 | 88.3 | 7.4 | 37.5 | 311.8 |
| Illinois | 5,690 | -7.1 | 554.2 | 22.11 | 921.99 | 216.5 | 322.6 | 281.1 | 553.7 | 87.4 | 406.2 | 2,476.1 |
| Indiana | 2,987 | -5.5 | 505.1 | 20.98 | 872.77 | 144.1 | 165.3 | 118.8 | 304.8 | 25.7 | 140.2 | 1,164.9 |
| Iowa | 1,506 | -5.1 | 216.6 | 20.94 | 841.79 | 76.8 | 67.7 | 64.8 | 168.3 | 19.1 | 109.4 | 529.2 |
| Kansas | 1,359 | -4.6 | 159.0 | 20.77 | 851.57 | 63.4 | 67.4 | 56.0 | 136.3 | 16.9 | 77.0 | 525.8 |
| Kentucky | 1,835 | -5.7 | 235.7 | 22.48 | 946.41 | 77.8 | 121.2 | 72.4 | 201.3 | 20.3 | 92.8 | 708.2 |
| Louisiana | 1,837 | -7.8 | 131.5 | 22.41 | 968.11 | 120.6 | 82.2 | 64.9 | 213.1 | 18.2 | 89.5 | 768.5 |
| Maine | 596 | -6.4 | 50.7 | 22.43 | 917.39 | 30.0 | 17.5 | 18.6 | 75.6 | 6.4 | 32.6 | 264.9 |
| Maryland | 2,581 | -6.8 | 108.4 | 22.15 | 963.53 | 160.2 | 109.5 | 80.7 | 255.8 | 32.8 | 137.1 | 1,197.0 |
| Massachusetts | 3,368 | -9.0 | 229.8 | 24.96 | 1,028.35 | 152.3 | 95.5 | 116.1 | 317.3 | 88.8 | 217.0 | 1,712.1 |
| Michigan | 4,033 | -9.2 | 557.1 | 22.51 | 916.16 | 164.8 | 157.6 | 161.2 | 427.7 | 50.4 | 223.3 | 1,702.4 |
| Minnesota | 2,776 | -6.9 | 308.7 | 23.69 | 933.39 | 123.3 | 103.5 | 124.4 | 273.9 | 42.7 | 193.6 | 1,193.7 |
| Mississippi | 1,109 | -4.3 | 139.4 | 21.27 | 838.04 | 44.0 | 61.7 | 33.7 | 132.4 | 9.6 | 43.0 | 402.9 |
| Missouri | 2,774 | -4.8 | 265.6 | 22.04 | 839.72 | 126.0 | 116.2 | 119.7 | 291.2 | 44.9 | 174.3 | 1,207.2 |
| Montana | 470 | -3.1 | 20.4 | 20.55 | 725.42 | 30.6 | 18.5 | 16.8 | 57.0 | 5.8 | 26.3 | 198.3 |
| Nebraska | 989 | -3.7 | 97.5 | 20.68 | 872.70 | 55.0 | 50.4 | 40.1 | 101.3 | 16.2 | 74.6 | 384.9 |
| Nevada | 1,274 | -10.4 | 56.1 | 20.10 | 884.40 | 93.7 | 78.5 | 36.5 | 141 | 13.5 | 66.7 | 613.9 |
| New Hampshire | 638 | -6.7 | 67.3 | 22.98 | 951.37 | 27.8 | 17.2 | 27.4 | 88.7 | 11.7 | 33.9 | 278.1 |
| New Jersey | 3,847 | -8.4 | 237.6 | 24.15 | 1,002.23 | 151.0 | 221.5 | 201.8 | 406.6 | 68.2 | 243.9 | 1,736.1 |
| New Mexico | 797 | -6.8 | 26.8 | 20.29 | 738.56 | 48.4 | 24.8 | 20.2 | 86.3 | 8.9 | 34.3 | 347.6 |
| New York | 8,778 | -10.3 | 400.8 | 23.06 | 910.87 | 362.2 | 275.7 | 294.5 | 800.9 | 267.1 | 707.6 | 4,220.1 |
| North Carolina | 4,381 | -4.3 | 453.0 | 19.03 | 791.65 | 229.1 | 171 | 181.6 | 486.5 | 73.4 | 256.3 | 1,810.8 |
| North Dakota | 412 | -6.6 | 25.2 | 21.33 | 799.88 | 26.2 | 22.1 | 23.2 | 43.8 | 5.8 | 24.5 | 146.4 |
| Ohio | 5,254 | -6.1 | 652.9 | 21.67 | 892.80 | 218.9 | 247.7 | 225.5 | 530.9 | 63.8 | 302.4 | 2,243.3 |
| Oklahoma | 1,621 | -4.9 | 131.3 | 20.08 | 845.37 | 78.8 | 70.3 | 55.0 | 174.6 | 18.3 | 76.7 | 640.3 |
| Oregon | 1,826 | -6.6 | 185.4 | 23.05 | 880.51 | 107.9 | 74.6 | 74.1 | 200.5 | 32.9 | 101.6 | 757.5 |
| Pennsylvania | 5,603 | -7.6 | 537.7 | 22.56 | 913.68 | 241.2 | 295.4 | 204.1 | 564.5 | 82.6 | 325.4 | 2,643.8 |
| Rhode Island | 460 | -8.8 | 37.4 | 20.29 | 746.67 | 18.9 | 12.2 | 14.9 | 44.0 | 5.2 | 34.4 | 229.2 |
| South Carolina | 2,080 | -5.0 | 244.2 | 21.87 | 901.04 | 103.8 | 82.6 | 72.2 | 245.5 | 24.7 | 105.7 | 830.1 |
| South Dakota | 425 | -3.4 | 43.2 | 19.58 | 791.03 | 24.2 | 13.3 | 21.0 | 49.6 | 5.0 | 28.4 | 163.1 |
| Tennessee | 2,999 | -4.0 | 335.1 | 20.33 | 803.04 | 129.1 | 186.3 | 117.6 | 323.4 | 43.2 | 171.0 | 1,256.2 |
| Texas | 12,266 | -4.3 | 868.7 | 23.40 | 952.38 | 737.7 | 600.2 | 590.1 | 1,278.1 | 198.5 | 804.8 | 5,030.4 |
| Utah | 1,532 | -1.6 | 135.9 | 20.92 | 817.97 | 115.4 | 66.4 | 52.8 | 169.8 | 38.2 | 93.3 | 606.4 |
| Vermont | 287 | -9.3 | 28.1 | 20.65 | 811.55 | 14.1 | 7.7 | 8.6 | 33.6 | 4.0 | 12.0 | 125.1 |
| Virginia | 3,856 | -5.0 | 233.7 | 20.19 | 799.52 | 202.4 | 143.4 | 105.9 | 387.1 | 65.4 | 209.4 | 1,789.7 |
| Washington | 3,285 | -5.3 | 272.0 | 28.32 | 1,183.78 | 213.9 | 112.4 | 130.0 | 386.4 | 149.0 | 158.3 | 1,295.8 |
| West Virginia | 674 | -6.6 | 44.5 | 21.79 | 906.46 | 29.8 | 24.7 | 19.2 | 76.4 | 7.3 | 29.3 | 276.1 |
| Wisconsin | 2,818 | -5.7 | 459.7 | 21.90 | 882.57 | 124.0 | 111.9 | 119.0 | 287.3 | 44.9 | 152.2 | 1,129.3 |
| Wyoming | 273 | -6.0 | 9.5 | 26.82 | 1,107.67 | 21.2 | 14.1 | 7.7 | 28.6 | 3.0 | 10.9 | 94.9 |

1. Includes professional and business services, educational and health services, leisure and hospitality, and other services.

# Table A. States — **Agriculture**

| STATE | Agriculture, 2017 | | | | | | | | | | | |
|---|---|---|---|---|---|---|---|---|---|---|---|---|
| | Farms | | | Farm producers whose primary occupation is farming (percent) | Government payments, average per farm that recieves payments (dollars) | Land in farms | | | | | Value of land and buildings (dollars) | |
| | | Percent with: | | | | | | Acres | | | | |
| | Number | Fewer than 50 acres | 1,000 acres or more | | | Acreage (1,000) | Percent change, 2012–2017 | Average size of farm | Total irrigated (1,000) | Total cropland (1,000) | Average per farm | Average per acre |
| | 184 | 185 | 186 | 187 | 188 | 189 | 190 | 191 | 192 | 193 | 194 | 195 |
| United States ................... | 2,042,220 | 41.9 | 8.5 | 41.1 | $13,906 | 900,218 | -1.6 | 441 | 58,013.9 | 396,433.8 | $1,311,808 | $2,976 |
| Alabama.............................. | 40,592 | 40.1 | 4.2 | 37.4 | $8,892 | 8,581 | -3.6 | 211 | 142.0 | 2,818.8 | $630,736 | $2,984 |
| Alaska................................. | 990 | 67.0 | 4.3 | 41.9 | $9,293 | 850 | 1.9 | 858 | 2.4 | 83.7 | $616,112 | $718 |
| Arizona............................... | 19,086 | 69.1 | 11.0 | 53.3 | $29,735 | 26,126 | -0.5 | 1,369 | 910.9 | 1,286.6 | $1,110,303 | $811 |
| Arkansas............................ | 42,625 | 30.3 | 7.2 | 41.7 | $38,624 | 13,889 | 0.6 | 326 | 4,855.1 | 7,825.9 | $1,030,741 | $3,163 |
| California............................ | 70,521 | 64.0 | 6.3 | 46.6 | $24,112 | 24,523 | -4.1 | 348 | 7,833.6 | 9,597.4 | $3,252,414 | $9,353 |
| Colorado............................ | 38,893 | 46.2 | 14.9 | 38.2 | $22,206 | 31,821 | -0.2 | 818 | 2,761.2 | 11,056.3 | $1,315,440 | $1,608 |
| Connecticut........................ | 5,521 | 70.9 | 0.7 | 39.0 | $7,551 | 382 | -12.6 | 69 | 7.4 | 148.6 | $862,636 | $12,483 |
| Delaware............................ | 2,302 | 55.7 | 6.9 | 52.8 | $18,604 | 525 | 3.3 | 228 | 163.3 | 452.2 | $1,920,109 | $8,414 |
| District of Columbia........... | NA | NA | NA | NA | NA | NA | NA | NA | NA | NA | NA | NA |
| Florida................................ | 47,590 | 71.0 | 3.2 | 41.3 | $14,795 | 9,732 | 1.9 | 204 | 1,519.4 | 2,825.8 | $1,206,788 | $5,901 |
| Georgia .............................. | 42,439 | 42.3 | 5.3 | 39.4 | $18,310 | 9,954 | 3.5 | 235 | 1,287.5 | 4,372.1 | $822,958 | $3,509 |
| Hawaii ................................ | 7,328 | 89.5 | 1.7 | 46.7 | $12,631 | 1,135 | 0.5 | 155 | 45.5 | 191.2 | $1,445,188 | $9,328 |
| Idaho.................................. | 24,996 | 56.0 | 9.7 | 40.8 | $21,306 | 11,692 | -0.6 | 468 | 3,398.3 | 5,894.7 | $1,340,738 | $2,866 |
| Illinois................................ | 72,651 | 35.6 | 10.8 | 43.4 | $10,727 | 27,006 | 0.3 | 372 | 612.5 | 24,003.1 | $2,705,291 | $7,278 |
| Indiana............................... | 56,649 | 46.4 | 7.1 | 38.1 | $12,628 | 14,970 | 1.7 | 264 | 555.4 | 12,909.7 | $1,737,741 | $6,576 |
| Iowa................................... | 86,104 | 31.7 | 9.8 | 45.0 | $11,146 | 30,564 | -0.2 | 355 | 222.0 | 26,546.0 | $2,506,812 | $7,062 |
| Kansas............................... | 58,569 | 21.8 | 20.2 | 41.9 | $14,089 | 45,759 | -0.8 | 781 | 2,503.4 | 29,125.5 | $1,443,891 | $1,848 |
| Kentucky............................ | 75,966 | 40.1 | 2.5 | 36.4 | $7,502 | 12,962 | -0.7 | 171 | 83.9 | 6,630.4 | $643,019 | $3,769 |
| Louisiana............................ | 27,386 | 46.5 | 7.6 | 37.7 | $22,822 | 7,998 | 1.2 | 292 | 1,235.8 | 4,345.8 | $889,146 | $3,045 |
| Maine.................................. | 7,600 | 47.2 | 2.4 | 43.2 | $10,806 | 1,308 | -10.1 | 172 | 32.3 | 472.5 | $446,614 | $2,596 |
| Maryland ............................ | 12,429 | 54.7 | 3.2 | 42.1 | $12,471 | 1,990 | -2.0 | 160 | 124.8 | 1,426.7 | $1,258,691 | $7,861 |
| Massachusetts................... | 7,241 | 67.8 | 0.3 | 42.8 | $7,583 | 492 | -6.1 | 68 | 23.9 | 171.5 | $739,711 | $10,894 |
| Michigan............................. | 47,641 | 46.3 | 4.5 | 43.1 | $10,892 | 9,764 | -1.9 | 205 | 670.2 | 7,924.5 | $1,015,631 | $4,955 |
| Minnesota.......................... | 68,822 | 28.8 | 9.3 | 45.5 | $9,568 | 25,517 | -2.0 | 371 | 611.6 | 21,786.8 | $1,799,201 | $4,853 |
| Mississippi......................... | 34,988 | 31.6 | 6.4 | 37.8 | $14,986 | 10,415 | -4.7 | 298 | 1,814.5 | 4,960.6 | $817,041 | $2,745 |
| Missouri.............................. | 95,320 | 29.6 | 6.2 | 38.8 | $10,366 | 27,782 | -1.7 | 291 | 1,529.2 | 15,599.4 | $986,481 | $3,385 |
| Montana.............................. | 27,048 | 30.9 | 31.7 | 48.5 | $27,014 | 58,123 | -2.7 | 2,149 | 2,061.2 | 16,406.3 | $1,968,381 | $916 |
| Nebraska............................ | 46,332 | 23.8 | 23.8 | 51.6 | $20,745 | 44,987 | -0.8 | 971 | 8,588.4 | 22,242.6 | $2,674,492 | $2,754 |
| Nevada............................... | 3,423 | 51.7 | 13.7 | 50.4 | $16,183 | 6,128 | 3.6 | 1,790 | 790.4 | 794.7 | $1,627,858 | $909 |
| New Hampshire ................. | 4,123 | 57.1 | 0.8 | 38.6 | $11,344 | 425 | -10.3 | 103 | 2.2 | 108.0 | $539,732 | $5,231 |
| New Jersey ........................ | 9,883 | 75.2 | 1.1 | 39.6 | $10,071 | 734 | 2.7 | 74 | 86.8 | 463.0 | $1,000,464 | $13,469 |
| New Mexico ....................... | 25,044 | 52.2 | 18.2 | 42.1 | $18,436 | 40,660 | -5.9 | 1,624 | 626.0 | 1,825.8 | $845,740 | $521 |
| New York ........................... | 33,438 | 36.7 | 3.3 | 48.1 | $9,162 | 6,866 | -4.4 | 205 | 53.3 | 4,291.4 | $663,082 | $3,229 |
| North Carolina.................... | 46,418 | 47.9 | 3.8 | 42.7 | $10,746 | 8,431 | 0.2 | 182 | 143.4 | 5,000.7 | $843,154 | $4,642 |
| North Dakota...................... | 26,364 | 11.7 | 40.0 | 54.3 | $22,770 | 39,342 | 0.2 | 1,492 | 263.9 | 27,951.7 | $2,546,783 | $1,707 |
| Ohio.................................... | 77,805 | 47.4 | 3.5 | 37.3 | $12,301 | 13,965 | 0.0 | 179 | 50.7 | 10,960.7 | $1,112,700 | $6,199 |
| Oklahoma........................... | 78,531 | 29.6 | 10.3 | 37.5 | $11,248 | 34,156 | -0.6 | 435 | 573.8 | 11,715.7 | $754,099 | $1,734 |
| Oregon............................... | 37,616 | 67.1 | 6.2 | 40.3 | $22,918 | 15,962 | -2.1 | 424 | 1,664.9 | 4,726.1 | $1,032,545 | $2,433 |
| Pennsylvania...................... | 53,157 | 42.1 | 1.4 | 45.7 | $6,823 | 7,279 | -5.5 | 137 | 32.1 | 4,651.2 | $897,125 | $6,552 |
| Rhode Island...................... | 1,043 | 72.5 | 0.4 | 38.6 | $14,205 | 57 | -18.3 | 55 | 3.0 | 17.7 | $897,835 | $16,468 |
| South Carolina ................... | 24,791 | 49.8 | 3.8 | 36.1 | $10,400 | 4,745 | -4.6 | 191 | 210.4 | 2,035.3 | $683,873 | $3,573 |
| South Dakota...................... | 29,968 | 19.3 | 32.0 | 52.4 | $19,416 | 43,244 | 0.0 | 1,443 | 492.5 | 19,813.5 | $2,984,426 | $2,068 |
| Tennessee ......................... | 69,983 | 45.2 | 2.3 | 35.8 | $6,254 | 10,874 | 0.1 | 155 | 184.9 | 5,286.3 | $608,739 | $3,918 |
| Texas.................................. | 248,416 | 44.1 | 8.3 | 35.8 | $20,984 | 127,036 | -2.4 | 511 | 4,363.3 | 29,360.2 | $980,409 | $1,917 |
| Utah.................................... | 18,409 | 62.1 | 6.9 | 32.1 | $12,633 | 10,812 | -1.5 | 587 | 1,097.2 | 1,654.4 | $1,067,323 | $1,817 |
| Vermont ............................. | 6,808 | 41.1 | 2.3 | 42.1 | $8,355 | 1,193 | -4.7 | 175 | 3.0 | 479.7 | $620,691 | $3,541 |
| Virginia .............................. | 43,225 | 42.2 | 3.1 | 40.1 | $10,122 | 7,798 | -6.1 | 180 | 63.4 | 3,084.1 | $834,254 | $4,624 |
| Washington ........................ | 35,793 | 66.6 | 7.2 | 39.9 | $30,692 | 14,680 | -0.5 | 410 | 1,689.4 | 7,488.6 | $1,143,889 | $2,789 |
| West Virginia...................... | 23,622 | 34.7 | 1.5 | 36.7 | $4,853 | 3,662 | 1.5 | 155 | 1.7 | 947.7 | $411,482 | $2,654 |
| Wisconsin........................... | 64,793 | 35.3 | 3.6 | 45.6 | $4,609 | 14,319 | -1.7 | 221 | 454.4 | 10,085.0 | $1,083,640 | $4,904 |
| Wyoming............................. | 11,938 | 32.7 | 25.0 | 43.0 | $14,410 | 29,005 | -4.5 | 2,430 | 1,567.6 | 2,587.5 | $1,892,340 | $779 |

# Table A. States — Agriculture, Land, and Water

| STATE | Value of machinery and equipment, average per farm (dollars) | Value of products sold Total (mil dol) | Average per farm (dollars) | Crops | Livestock and poultry products | Organic Farms (number) | Farms with Internet access (percent) | Owned by the federal government (percent) | Developed land (percent) | Rural land (percent) | Public water supply withdrawn, 2015 (mil gal per day) |
|---|---|---|---|---|---|---|---|---|---|---|---|
| | 196 | 197 | 198 | 199 | 200 | 201 | 202 | 203 | 204 | 205 | 206 |
| United States .................. | 133,363 | $388,522.7 | $190,245 | 49.8 | 50.2 | 20,806 | 75.4 | 20.8 | 5.9 | 70.6 | 38,595.83 |
| Alabama............................ | 88,528 | $5,980.6 | $147,334 | 20.3 | 79.7 | 57 | 72.6 | 2.8 | 8.7 | 84.5 | 761.53 |
| Alaska ............................... | 91,623 | $70.5 | $71,171 | 42.1 | 57.9 | 18 | 87.9 | NA | NA | NA | 99.18 |
| Arizona ............................. | 77,604 | $3,852.0 | $201,824 | 54.4 | 45.6 | 84 | 57.4 | 41.5 | 2.9 | 55.4 | 1,195.15 |
| Arkansas.......................... | 126,667 | $9,651.2 | $226,420 | 37.6 | 62.4 | 81 | 73.6 | 9.6 | 5.5 | 82.3 | 363.06 |
| California.......................... | 165,070 | $45,154.4 | $640,297 | 73.9 | 26.1 | 3,794 | 82.0 | 46.7 | 6.2 | 45.4 | 5,147.74 |
| Colorado.......................... | 117,337 | $7,491.7 | $192,623 | 29.9 | 70.1 | 323 | 81.4 | 36.1 | 3.0 | 60.4 | 843.95 |
| Connecticut...................... | 62,250 | $580.1 | $105,074 | 72.4 | 27.6 | 113 | 82.4 | 0.5 | 34.1 | 61.4 | 239.93 |
| Delaware.......................... | 198,096 | $1,466.0 | $636,826 | 22.2 | 77.8 | 13 | 78.7 | 1.6 | 19.4 | 59.2 | 86.35 |
| District of Columbia............ | NA | NA | NA | NA | NA | NA | NA | NA | NA | NA | 0.00 |
| Florida ............................. | 72,754 | $7,357.3 | $154,599 | 77.5 | 22.5 | 251 | 76.3 | 10.3 | 14.7 | 66.5 | 2,384.85 |
| Georgia ........................... | 115,773 | $9,573.3 | $225,577 | 34.2 | 65.8 | 139 | 76.0 | 5.5 | 12.3 | 79.3 | 1,069.91 |
| Hawaii .............................. | 50,701 | $563.8 | $76,938 | 74.0 | 26.0 | 167 | 76.1 | 14.8 | 6.0 | 78.6 | 266.92 |
| Idaho ............................... | 175,951 | $7,567.4 | $302,746 | 42.4 | 57.6 | 295 | 83.9 | 62.5 | 1.7 | 34.7 | 275.79 |
| Illinois ............................. | $220,485 | $17,010.0 | $234,133 | 81.4 | 18.6 | 328 | 76.9 | 1.4 | 9.6 | 87.0 | 1,475.66 |
| Indiana ............................ | $163,136 | $11,107.3 | $196,073 | 64.1 | 35.9 | 657 | 71.9 | 2.1 | 11.0 | 85.3 | 627.84 |
| Iowa ................................ | $230,716 | $28,956.5 | $336,296 | 47.8 | 52.2 | 785 | 79.6 | 0.6 | 5.4 | 92.6 | 390.38 |
| Kansas............................. | $180,725 | $18,782.7 | $320,694 | 34.4 | 65.6 | 117 | 76.5 | 0.9 | 4.0 | 94.0 | 351.15 |
| Kentucky .......................... | $82,740 | $5,737.9 | $75,533 | 44.3 | 55.7 | 227 | 72.4 | 4.9 | 8.2 | 84.4 | 552.83 |
| Louisiana......................... | $121,758 | $3,173.0 | $115,861 | 65.0 | 35.0 | 29 | 69.9 | 4.0 | 6.3 | 75.2 | 708.92 |
| Maine .............................. | $81,792 | $667.0 | $87,758 | 61.3 | 38.7 | 621 | 83.6 | 1.0 | 4.1 | 88.7 | 84.97 |
| Maryland .......................... | $124,871 | $2,472.8 | $198,955 | 38.3 | 61.7 | 134 | 76.9 | 2.1 | 19.4 | 57.1 | 749.52 |
| Massachusetts.................. | $65,382 | $475.2 | $65,624 | 76.5 | 23.5 | 208 | 84.1 | 1.4 | 33.0 | 58.6 | 648.06 |
| Michigan........................... | $154,740 | $8,220.9 | $172,560 | 56.5 | 43.5 | 764 | 77.2 | 8.6 | 11.2 | 77.1 | 1,030.44 |
| Minnesota ........................ | $223,666 | $18,395.4 | $267,289 | 55.4 | 44.6 | 735 | 79.0 | 6.4 | 4.5 | 83.2 | 515.24 |
| Mississippi....................... | $109,875 | $6,196.0 | $177,088 | 37.0 | 63.0 | 37 | 66.0 | 5.5 | 6.3 | 85.5 | 400.36 |
| Missouri........................... | $104,066 | $10,525.9 | $110,427 | 52.0 | 48.0 | 415 | 72.5 | 4.5 | 6.7 | 86.8 | 797.09 |
| Montana........................... | $164,524 | $3,520.6 | $130,162 | 45.0 | 55.0 | 221 | 81.4 | 28.7 | 1.2 | 69.0 | 153.19 |
| Nebraska.......................... | $268,968 | $21,983.4 | $474,476 | 42.4 | 57.6 | 292 | 81.3 | 1.2 | 2.6 | 95.3 | 275.18 |
| Nevada............................. | $155,033 | $665.8 | $194,495 | 41.5 | 58.5 | 51 | 82.9 | 84.1 | 0.8 | 14.6 | 558.26 |
| New Hampshire ................. | $68,629 | $187.8 | $45,548 | 57.4 | 42.6 | 156 | 87.2 | 13.5 | 12.3 | 70.2 | 95.52 |
| New Jersey ....................... | $86,532 | $1,098.0 | $111,095 | 89.7 | 10.3 | 122 | 80.9 | 3.4 | 35.7 | 50.1 | 1,175.42 |
| New Mexico ...................... | $63,619 | $2,582.3 | $103,112 | 25.2 | 74.8 | 183 | 60.5 | 33.9 | 1.7 | 64.2 | 254.10 |
| New York .......................... | $135,626 | $5,369.2 | $160,572 | 39.3 | 60.7 | 1,497 | 77.1 | 0.7 | 12.3 | 82.5 | 2,424.65 |
| North Carolina................... | $112,477 | $12,900.7 | $277,924 | 29.0 | 71.0 | 465 | 75.2 | 7.1 | 14.4 | 70.2 | 938.01 |
| North Dakota..................... | $375,872 | $8,234.1 | $312,324 | 81.1 | 18.9 | 129 | 78.9 | 3.9 | 2.3 | 91.0 | 84.18 |
| Ohio ................................ | $129,614 | $9,341.2 | $120,059 | 58.1 | 41.9 | 872 | 74.6 | 1.4 | 15.9 | 81.1 | 1,306.28 |
| Oklahoma......................... | $90,442 | $7,465.5 | $95,065 | 20.3 | 79.7 | 54 | 72.9 | 2.7 | 4.9 | 90.0 | 611.24 |
| Oregon............................. | $100,328 | $5,006.8 | $133,103 | 65.6 | 34.4 | 659 | 85.7 | 51.7 | 2.2 | 44.7 | 567.04 |
| Pennsylvania..................... | $109,024 | $7,758.9 | $145,962 | 35.8 | 64.2 | 1,142 | 69.3 | 2.3 | 15.4 | 80.6 | 1,391.70 |
| Rhode Island..................... | $62,786 | $58.0 | $55,607 | 70.5 | 29.5 | 22 | 84.1 | 0.5 | 28.7 | 52.1 | 97.46 |
| South Carolina .................. | $83,077 | $3,008.7 | $121,364 | 36.4 | 63.6 | 75 | 72.6 | 5.3 | 13.5 | 77.1 | 633.39 |
| South Dakota .................... | $282,162 | $9,721.5 | $324,397 | 53.1 | 46.9 | 87 | 81.0 | 5.6 | 2.0 | 90.6 | 71.95 |
| Tennessee ........................ | $80,447 | $3,798.9 | $54,284 | 57.4 | 42.6 | 122 | 72.7 | 5.2 | 11.6 | 80.3 | 849.69 |
| Texas ............................... | $83,627 | $24,924.0 | $100,332 | 27.7 | 72.3 | 466 | 72.6 | 1.8 | 5.4 | 90.3 | 2,885.33 |
| Utah ................................ | $97,789 | $1,838.6 | $99,876 | 30.5 | 69.5 | 95 | 77.7 | 64.1 | 1.6 | 31.4 | 785.91 |
| Vermont............................ | $100,672 | $781.0 | $114,713 | 24.0 | 76.0 | 679 | 86.4 | 7.4 | 6.5 | 81.7 | 42.66 |
| Virginia ............................ | $86,136 | $3,960.5 | $91,625 | 34.4 | 65.6 | 252 | 74.0 | 8.7 | 11.9 | 72.4 | 695.63 |
| Washington....................... | $121,662 | $9,634.5 | $269,172 | 72.5 | 27.5 | 933 | 84.1 | 28.5 | 5.7 | 62.2 | 866.53 |
| West Virginia ..................... | $56,120 | $754.3 | $31,931 | 20.3 | 79.7 | 63 | 70.0 | 8.2 | 7.5 | 83.2 | 184.96 |
| Wisconsin......................... | $156,689 | $11,427.4 | $176,368 | 35.6 | 64.4 | 1,708 | 76.1 | 5.1 | 7.8 | 83.6 | 479.38 |
| Wyoming........................... | $126,844 | $1,472.1 | $123,313 | 21.6 | 78.4 | 69 | 80.5 | 47.2 | 1.1 | 50.9 | 101.35 |

# Table A. States — Manufactures and Construction

| STATE | Manufactures, 2017 | | | | | | | | | | Construction, 2017 | | | | |
|---|---|---|---|---|---|---|---|---|---|---|---|---|---|---|---|
| | All employees | | | Production workers[1] | | | | Value added by manu-facture (mil dol) | Sales, values of ship-ments or revenue (mil dol) | Total cost of materials (mil dol) | Number of estab-lishments | Employees | | Sales, values of ship-ments or revenue (mil dol) | Annual payroll (mil dol) |
| | | | | | | Wages | | | | | | | | | |
| | Number (1,000) | Percent change, 2012–2017 | Annual payroll (mil dol) | Number (1,000) | Work hours (millions) | Total (mil dol) | Average per worker (dollars) | | | | | Number | Percent change, 2012–2017 | | |
| | 207 | 208 | 209 | 210 | 211 | 212 | 213 | 214 | 215 | 216 | 217 | 218 | 219 | 220 | 221 |
| United States | 11,522 | 2.7 | 670,678 | 8,192 | 16,111 | 388,508 | 47,425 | 2,463,183 | 5,548,797 | 3,098,297 | 715,364 | 6,647,047 | 17.2 | 1,994,166 | 398,816 |
| Alabama | 246 | 5.6 | 12,916 | 186 | 372 | 8,301 | 44,510 | 49,725 | 137,441 | 88,511 | 7,527 | 82,090 | 4.3 | 22,990 | 4,133 |
| Alaska | 13 | 0.5 | 585 | 11 | 21 | 443 | 41,161 | 2,279 | 8,617 | 6,305 | 2,456 | 17,370 | -28.2 | 5,841 | 1,323 |
| Arizona | 143 | 8.2 | 9,149 | 91 | 175 | 4,223 | 46,495 | 29,862 | 56,636 | 26,518 | 12,397 | 150,709 | 21.7 | 41,597 | 8,096 |
| Arkansas | 153 | -0.7 | 7,025 | 125 | 240 | 5,025 | 40,178 | 25,529 | 58,364 | 32,938 | 5,427 | 46,325 | 7.7 | 11,083 | 2,221 |
| California | 1,161 | -0.2 | 76,483 | 742 | 1,457 | 36,245 | 48,847 | 262,442 | 510,859 | 251,572 | 74,696 | 779,585 | 30.5 | 242,288 | 49,970 |
| Colorado | 121 | 5.9 | 7,357 | 82 | 158 | 4,078 | 49,976 | 23,132 | 50,809 | 27,789 | 18,531 | 162,864 | 31.9 | 54,261 | 9,699 |
| Connecticut | 157 | -4.3 | 11,022 | 97 | 189 | 5,330 | 54,984 | 31,091 | 57,882 | 27,284 | 7,898 | 58,224 | 4.1 | 18,655 | 3,932 |
| Delaware | 28 | 4.5 | 1,544 | 20 | 40 | 877 | 43,710 | 6,585 | 16,689 | 10,163 | 2,249 | 21,227 | 19.4 | 5,658 | 1,190 |
| District of Columbia | 1 | -19.8 | 50 | 1 | 1 | 31 | 43,659 | 157 | 277 | 118 | 452 | 10,386 | 15.2 | 3,163 | 679 |
| Florida | 296 | 7.0 | 16,576 | 203 | 388 | 8,930 | 44,006 | 56,986 | 106,342 | 49,852 | 53,316 | 424,041 | 43.9 | 118,374 | 20,555 |
| Georgia | 369 | 10.5 | 19,072 | 281 | 558 | 11,994 | 42,718 | 73,673 | 169,059 | 95,294 | 18,202 | 178,499 | 23.1 | 63,791 | 9,988 |
| Hawaii | 12 | 3.6 | 559 | 8 | 15 | 344 | 42,040 | 2,251 | 6,056 | 4,049 | 2,877 | 32,115 | 16.1 | 10,844 | 2,232 |
| Idaho | 59 | 12.8 | 3,486 | 45 | 88 | 2,219 | 49,831 | 8,739 | 20,263 | 11,611 | 6,929 | 40,671 | 26.9 | 10,059 | 1,870 |
| Illinois | 528 | -2.6 | 30,116 | 378 | 741 | 17,763 | 47,011 | 100,883 | 241,484 | 141,684 | 29,188 | 215,621 | 5.3 | 74,495 | 15,550 |
| Indiana | 496 | 9.6 | 28,131 | 379 | 761 | 18,509 | 48,798 | 104,043 | 246,672 | 143,610 | 13,439 | 132,101 | 6.6 | 36,549 | 7,818 |
| Iowa | 211 | 3.4 | 11,374 | 155 | 304 | 7,010 | 45,300 | 45,135 | 109,728 | 64,809 | 8,743 | 68,457 | 3.7 | 20,219 | 3,950 |
| Kansas | 156 | 2.3 | 9,032 | 113 | 226 | 5,490 | 48,601 | 32,593 | 83,418 | 51,155 | 7,031 | 66,965 | 13.8 | 18,339 | 3,597 |
| Kentucky | 234 | 9.6 | 12,769 | 183 | 363 | 8,846 | 48,399 | 46,460 | 133,415 | 87,255 | 7,327 | 73,032 | 14.1 | 19,305 | 3,786 |
| Louisiana | 118 | -13.5 | 8,257 | 85 | 177 | 5,180 | 60,769 | 51,488 | 187,440 | 136,659 | 8,121 | 138,421 | 1.5 | 29,264 | 7,788 |
| Maine | 49 | -1.3 | 2,599 | 37 | 71 | 1,703 | 46,636 | 8,045 | 15,089 | 7,097 | 5,041 | 26,421 | 4.1 | 6,188 | 1,414 |
| Maryland | 98 | -2.1 | 6,343 | 63 | 124 | 3,072 | 49,061 | 22,561 | 41,777 | 19,316 | 14,184 | 154,895 | 7.5 | 48,512 | 9,591 |
| Massachusetts | 232 | -1.1 | 15,749 | 140 | 272 | 7,058 | 50,264 | 45,306 | 82,308 | 36,573 | 19,310 | 140,470 | 16.3 | 51,184 | 10,416 |
| Michigan | 582 | 13.3 | 33,349 | 425 | 837 | 20,770 | 48,827 | 101,893 | 262,495 | 160,900 | 19,250 | 152,413 | 15.6 | 46,132 | 9,724 |
| Minnesota | 309 | 3.8 | 17,926 | 209 | 409 | 9,858 | 47,110 | 56,470 | 122,014 | 65,797 | 16,546 | 124,908 | 9.9 | 44,435 | 8,848 |
| Mississippi | 138 | 4.3 | 6,654 | 110 | 214 | 4,581 | 41,568 | 22,956 | 60,906 | 38,241 | 3,850 | 43,010 | 3.5 | 10,016 | 2,217 |
| Missouri | 259 | 6.7 | 14,217 | 190 | 366 | 8,764 | 46,070 | 51,822 | 118,634 | 67,448 | 13,505 | 130,055 | 17.8 | 37,872 | 7,667 |
| Montana | 18 | 14.1 | 936 | 13 | 24 | 576 | 45,722 | 4,101 | 10,944 | 6,928 | 5,466 | 25,767 | 7.5 | 7,041 | 1,398 |
| Nebraska | 94 | 1.2 | 4,728 | 74 | 153 | 3,246 | 43,975 | 18,363 | 53,129 | 34,845 | 6,472 | 47,140 | 16.3 | 12,497 | 2,463 |
| Nevada | 44 | 15.9 | 2,517 | 31 | 61 | 1,542 | 48,966 | 8,605 | 16,408 | 7,823 | 4,973 | 77,924 | 42.8 | 20,639 | 4,231 |
| New Hampshire | 65 | -2.1 | 3,875 | 43 | 83 | 1,955 | 45,642 | 10,046 | 20,304 | 10,173 | 4,166 | 28,351 | 16.0 | 7,713 | 1,818 |
| New Jersey | 220 | -4.7 | 13,909 | 147 | 287 | 7,136 | 48,384 | 46,344 | 95,483 | 49,168 | 21,605 | 163,968 | 11.5 | 51,338 | 11,359 |
| New Mexico | 23 | -13.1 | 1,287 | 16 | 31 | 754 | 47,410 | 5,113 | 13,724 | 8,636 | 4,338 | 40,228 | 6.5 | 8,698 | 1,912 |
| New York | 411 | -3.6 | 23,751 | 279 | 539 | 13,072 | 46,791 | 80,269 | 155,572 | 76,478 | 49,116 | 375,488 | 12.5 | 124,319 | 25,884 |
| North Carolina | 423 | 4.8 | 21,480 | 318 | 629 | 13,283 | 41,716 | 103,378 | 200,381 | 96,707 | 23,121 | 196,113 | 8.6 | 58,218 | 9,998 |
| North Dakota | 24 | 2.9 | 1,266 | 19 | 38 | 862 | 45,859 | 5,010 | 13,605 | 8,600 | 3,010 | 24,140 | -6.5 | 7,105 | 1,607 |
| Ohio | 652 | 4.0 | 36,257 | 479 | 942 | 22,623 | 47,241 | 129,535 | 306,222 | 178,309 | 19,908 | 197,082 | 8.8 | 58,063 | 10,345 |
| Oklahoma | 123 | -7.5 | 6,740 | 91 | 182 | 4,240 | 46,517 | 23,920 | 61,143 | 37,431 | 8,354 | 73,775 | 5.9 | 17,846 | 3,782 |
| Oregon | 172 | 13.3 | 10,668 | 120 | 236 | 5,882 | 49,020 | 30,739 | 62,411 | 31,782 | 12,870 | 95,607 | 37.0 | 26,206 | 5,516 |
| Pennsylvania | 537 | -1.2 | 30,393 | 383 | 749 | 18,191 | 47,529 | 112,379 | 227,812 | 115,356 | 26,672 | 242,502 | 1.8 | 70,345 | 15,230 |
| Rhode Island | 40 | 1.5 | 2,390 | 27 | 52 | 1,252 | 45,940 | 6,507 | 12,416 | 6,013 | 3,165 | 18,906 | 8.0 | 6,008 | 1,211 |
| South Carolina | 225 | 8.6 | 12,777 | 169 | 335 | 8,088 | 47,865 | 56,238 | 138,587 | 79,003 | 9,864 | 87,040 | 26.7 | 25,647 | 4,435 |
| South Dakota | 43 | 2.4 | 2,057 | 32 | 65 | 1,345 | 41,428 | 7,341 | 16,780 | 9,526 | 3,394 | 20,493 | 3.5 | 5,265 | 1,042 |
| Tennessee | 321 | 9.4 | 17,022 | 241 | 472 | 10,917 | 45,312 | 69,500 | 155,422 | 86,145 | 10,051 | 112,271 | 10.3 | 32,115 | 6,140 |
| Texas | 759 | -1.1 | 48,004 | 528 | 1,065 | 26,584 | 50,308 | 229,126 | 575,218 | 348,050 | 45,034 | 717,040 | 25.1 | 214,508 | 42,044 |
| Utah | 122 | 12.3 | 7,171 | 84 | 159 | 4,057 | 48,373 | 24,402 | 55,788 | 31,500 | 9,590 | 83,923 | 30.1 | 24,423 | 4,386 |
| Vermont | 29 | -9.1 | 1,610 | 20 | 39 | 909 | 45,393 | 4,112 | 8,652 | 4,529 | 2,694 | 13,757 | -13.0 | 3,182 | 756 |
| Virginia | 233 | 2.0 | 12,987 | 170 | 336 | 8,039 | 47,366 | 56,808 | 99,344 | 42,655 | 19,736 | 186,390 | 5.2 | 54,426 | 10,345 |
| Washington | 263 | 6.0 | 17,944 | 175 | 337 | 9,803 | 56,146 | 75,637 | 140,382 | 64,804 | 23,619 | 193,770 | 39.7 | 60,286 | 12,683 |
| West Virginia | 49 | -0.3 | 2,748 | 36 | 71 | 1,864 | 51,303 | 11,604 | 24,602 | 13,133 | 3,082 | 24,143 | -9.4 | 5,427 | 1,290 |
| Wisconsin | 456 | 4.3 | 25,181 | 331 | 647 | 15,205 | 45,969 | 79,112 | 171,896 | 93,118 | 13,851 | 112,462 | 9.7 | 37,628 | 7,941 |
| Wyoming | 9 | -7.3 | 643 | 7 | 13 | 440 | 63,791 | 2,891 | 7,898 | 5,033 | 2,721 | 17,892 | -16.0 | 4,112 | 994 |

1. Production and/or development and exploration workers

# Table A. States — Wholesale Trade and Retail Trade

| STATE | Wholesale trade, 2017 | | | | | Retail trade,[1] 2017 | | | | | | | | |
|---|---|---|---|---|---|---|---|---|---|---|---|---|---|---|
| | Number of establishments | Employees | | Sales (mil dol) | Annual payroll (mil dol) | Number of establishments | Employees | | Motor vehicle and parts dealers | Food and beverage stores | Clothing and clothing accessory stores | General merchandise stores | Sales (mil dol) | Annual payroll (mil dol) |
| | | Number | Percent change, 2012–2017 | | | | Total | Percent change, 2012–2017 | | | | | | |
| | 222 | 223 | 224 | 225 | 226 | 227 | 228 | 229 | 230 | 231 | 232 | 233 | 234 | 235 |
| United States | 408,551 | 6,527,427 | 11.0 | 8,517,446 | 431,791,587 | 1,064,210 | 15,960,369 | 8.5 | 2,009,860 | 3,186,678 | 1,796,234 | 2,803,586 | 4,976,024 | 442,635 |
| Alabama | 5,225 | 75,178 | 2.5 | 88,032 | 4,115,973 | 17,958 | 230,069 | 5.3 | 31,412 | 35,070 | 22,347 | 52,669 | 63,740 | 5,819 |
| Alaska | 748 | 8,680 | -0.7 | 9,162 | 537,383 | 2,480 | 34,498 | 2.3 | 4,268 | 7,373 | 2,460 | 7,520 | 10,385 | 1,095 |
| Arizona | 6,508 | 98,089 | 10.3 | 113,376 | 6,359,888 | 17,918 | 324,912 | 13.5 | 44,267 | 55,332 | 33,316 | 61,354 | 104,366 | 9,505 |
| Arkansas | X | X | X | X | X | X | X | X | X | X | X | X | X | X |
| California | 58,772 | 860,268 | 2.1 | 1,192,475 | 67,632,172 | 108,233 | 1,723,278 | 11.9 | 204,691 | 357,427 | 249,697 | 262,809 | 594,861 | 54,229 |
| Colorado | 7,179 | 102,499 | 10.6 | 135,159 | 7,453,581 | 19,056 | 279,982 | 14.0 | 36,214 | 56,152 | 27,107 | 44,919 | 84,931 | 8,420 |
| Connecticut | 4,061 | 76,120 | 0.9 | 149,834 | 6,361,890 | 12,391 | 186,297 | 2.1 | 21,461 | 44,728 | 23,917 | 23,526 | 55,404 | 5,561 |
| Delaware | 1,146 | 17,586 | 21.6 | 21,934 | 1,836,752 | 3,648 | 58,201 | 12.6 | 7,336 | 10,732 | 6,492 | 8,853 | 17,668 | 1,558 |
| District of Columbia | 395 | 4,540 | 6.6 | 5,167 | 357,039 | 1,743 | 23,133 | 17.0 | 175 | 7,152 | 4,704 | 2,252 | 5,534 | 673 |
| Florida | X | X | X | X | X | X | X | X | X | X | X | X | X | X |
| Georgia | X | X | X | X | X | X | X | X | X | X | X | X | X | X |
| Hawaii | 1,604 | 18,963 | 1.1 | 16,969 | 998,662 | 4,644 | 72,908 | 6.7 | 7,056 | 13,947 | 14,591 | 12,901 | 21,659 | 2,185 |
| Idaho | 2,141 | 29,855 | 13.9 | 37,786 | 1,632,433 | 6,133 | 82,312 | 12.8 | 13,707 | 12,965 | 5,326 | 16,134 | 24,936 | 2,312 |
| Illinois | 18,068 | 333,541 | 6.1 | 583,865 | 24,657,027 | 38,189 | 629,878 | 6.2 | 74,094 | 142,758 | 70,734 | 114,511 | 173,474 | 16,247 |
| Indiana | 7,585 | 121,772 | 9.9 | 143,451 | 7,426,911 | 21,327 | 336,615 | 8.7 | 43,510 | 52,350 | 26,526 | 70,245 | 102,106 | 8,660 |
| Iowa | 5,019 | 71,382 | 6.3 | 77,650 | 3,847,452 | 11,479 | 181,416 | 3.9 | 23,774 | 40,944 | 12,213 | 32,102 | 50,063 | 4,519 |
| Kansas | 4,416 | 63,614 | -2.3 | 96,904 | 3,780,377 | 10,095 | 149,845 | 3.0 | 19,895 | 28,887 | 12,847 | 28,386 | 39,338 | 3,785 |
| Kentucky | 4,259 | 71,042 | 5.1 | 125,944 | 4,022,946 | 15,021 | 225,127 | 11.1 | 28,368 | 39,150 | 17,918 | 46,827 | 64,294 | 5,630 |
| Louisiana | 5,330 | 74,158 | -3.4 | 77,677 | 4,155,627 | 16,564 | 233,385 | 6.0 | 30,595 | 39,555 | 23,920 | 50,131 | 65,001 | 5,988 |
| Maine | 1,517 | 18,360 | 9.6 | 18,858 | 1,015,627 | 6,250 | 81,733 | 2.0 | 11,068 | 18,269 | 5,632 | 11,641 | 23,879 | 2,250 |
| Maryland | 5,401 | 89,504 | 4.1 | 100,900 | 6,192,615 | 17,911 | 291,814 | 3.6 | 38,945 | 63,940 | 36,634 | 45,444 | 84,966 | 8,241 |
| Massachusetts | 7,521 | 153,303 | 9.9 | 214,395 | 13,078,823 | 23,928 | 364,204 | 3.6 | 37,828 | 98,789 | 46,773 | 42,020 | 110,195 | 10,911 |
| Michigan | X | X | X | X | X | X | X | X | X | X | X | X | X | X |
| Minnesota | 7,901 | 146,637 | 8.9 | 191,615 | 11,337,187 | 18,827 | 302,886 | 4.8 | 36,797 | 57,795 | 26,556 | 57,811 | 91,994 | 8,118 |
| Mississippi | 2,702 | 37,710 | 8.5 | 37,440 | 1,877,614 | 11,525 | 141,410 | 4.0 | 18,093 | 19,821 | 14,313 | 35,880 | 36,921 | 3,384 |
| Missouri | 7,565 | 126,395 | 0.1 | 154,926 | 7,064,082 | 20,694 | 312,616 | 3.3 | 41,346 | 51,652 | 27,218 | 62,336 | 100,394 | 8,148 |
| Montana | 1,597 | 14,875 | 3.0 | 17,276 | 767,973 | 4,754 | 59,032 | 6.5 | 8,828 | 10,613 | 3,234 | 9,598 | 16,936 | 1,623 |
| Nebraska | 3,192 | 41,471 | 1.6 | 63,229 | 2,442,473 | 7,154 | 109,729 | 3.6 | 13,622 | 20,908 | 8,307 | 19,779 | 31,215 | 2,853 |
| Nevada | 3,115 | 38,586 | 16.0 | 34,808 | 2,286,362 | 8,745 | 145,773 | 12.2 | 16,174 | 23,801 | 26,270 | 24,292 | 45,111 | 4,220 |
| New Hampshire | 1,826 | 26,417 | 7.6 | 26,679 | 1,928,099 | 6,032 | 96,591 | 1.0 | 12,381 | 21,789 | 9,016 | 13,857 | 30,039 | 2,805 |
| New Jersey | 13,844 | 288,082 | 9.4 | 463,247 | 26,201,731 | 31,200 | 469,615 | 7.6 | 48,577 | 111,658 | 61,699 | 63,997 | 149,171 | 13,453 |
| New Mexico | X | X | X | X | X | X | X | X | X | X | X | X | X | X |
| New York | 30,484 | 374,642 | 1.5 | 450,832 | 25,910,640 | 78,260 | 945,360 | 4.4 | 81,180 | 221,446 | 151,222 | 122,750 | 291,725 | 27,815 |
| North Carolina | 11,681 | 196,695 | 13.9 | 218,600 | 13,588,472 | 34,926 | 496,081 | 11.1 | 70,495 | 98,783 | 49,012 | 92,173 | 141,134 | 12,673 |
| North Dakota | 1,733 | 23,078 | 7.2 | 26,258 | 1,390,598 | 3,277 | 49,579 | 5.1 | 6,950 | 7,582 | 3,601 | 9,203 | 19,251 | 1,454 |
| Ohio | 13,630 | 238,916 | 4.2 | 276,251 | 14,577,328 | 35,500 | 588,060 | 7.1 | 76,594 | 123,389 | 49,953 | 106,112 | 174,300 | 14,861 |
| Oklahoma | X | X | X | X | X | X | X | X | X | X | X | X | X | X |
| Oregon | 5,355 | 77,990 | 7.7 | 80,531 | 5,041,622 | 14,318 | 211,222 | 12.7 | 27,054 | 41,835 | 19,924 | 40,269 | 61,699 | 6,067 |
| Pennsylvania | 14,246 | 259,620 | 7.1 | 310,515 | 18,332,703 | 42,514 | 662,560 | 2.9 | 84,065 | 148,289 | 65,071 | 104,995 | 234,836 | 17,354 |
| Rhode Island | 1,304 | 21,253 | 5.0 | 20,829 | 1,381,130 | 3,769 | 48,753 | 2.2 | 5,438 | 11,265 | 5,193 | 6,350 | 13,844 | 1,444 |
| South Carolina | 5,004 | 72,946 | 13.2 | 78,646 | 4,314,807 | 17,700 | 253,384 | 14.9 | 32,022 | 51,364 | 26,084 | 47,880 | 69,980 | 6,206 |
| South Dakota | 1,566 | 18,705 | 5.7 | 21,755 | 981,313 | 3,884 | 53,134 | 6.6 | 7,951 | 9,472 | 3,281 | 9,907 | 14,674 | 1,372 |
| Tennessee | 6,842 | 118,612 | 5.3 | 165,628 | 7,111,100 | 22,593 | 322,218.0 | 5.3 | 43,344 | 47,069 | 32,772 | 64,169 | 101,978 | 8,574 |
| Texas | X | X | X | X | X | X | X | X | X | X | X | X | X | X |
| Utah | 3,691 | 54,922 | 8.2 | 55,094 | 3,372,334 | 9,995 | 153,633 | 15.1 | 20,468 | 25,314 | 14,482 | 26,155 | 50,008 | 4,446 |
| Vermont | 777 | 11,048 | 2.8 | 14,273 | 633,048 | 3,219 | 38,390 | -1.3 | 5,241 | 9,871 | 2,545 | 3,157 | 10,811 | 1,103 |
| Virginia | 6,926 | 105,989 | -0.4 | 137,827 | 7,018,515 | 27,134 | 429,072 | 4.4 | 57,859 | 86,322 | 45,845 | 76,193 | 120,162 | 11,476 |
| Washington | 9,186 | 137,876 | 13.2 | 163,034 | 9,012,299 | 21,751 | 347,728 | 13.2 | 44,014 | 64,161 | 34,147 | 63,369 | 160,285 | 11,413 |
| West Virginia | 1,406 | 18,216 | -8.5 | 23,491 | 890,351 | 5,963 | 82,985 | -2.7 | 11,330 | 12,440 | 5,414 | 19,778 | 23,058 | 2,004 |
| Wisconsin | 6,944 | 122,143 | 8.0 | 117,299 | 7,335,932 | 18,908 | 317,668 | 7.0 | 41,782 | 55,212 | 22,692 | 59,776 | 91,764 | 8,274 |
| Wyoming | 813 | 6,961 | -16.0 | 7,664 | 406,997 | 2,583 | 29,786 | -1.0 | 4,522 | 5,228 | 1,663 | 5,812 | 9,124 | 853 |

1. Establishments with payroll.

| STATE | Number of estab-lishments | Employees Number | Percent change, 2012–2017 | Publishing, except Internet | Motion picture and sound recording | Broad-casting, except Internet | Internet publishing and broad-casting and web search portals | Tele-communi-cations | Data processing, hosting, and related services | Sales, values of ship-ments or revenue (mil dol) | Annual payroll (mil dol) |
|---|---|---|---|---|---|---|---|---|---|---|---|
| | 236 | 237 | 238 | 239 | 240 | 241 | 242 | 243 | 244 | 245 | 246 |
| United States | 154,096 | 3,719,890 | 11.4 | 1,042,420 | 342,048 | 275,797 | 265,848 | 1,143,243 | 603,236 | 175,519 | 59,956,856 |
| Alabama | 1,735 | 34,078 | 10.2 | 8,124 | 1,937 | D | 329 | 15,945 | 4,569 | 842 | 283,180 |
| Alaska | 416 | 6,553 | 4.3 | 615 | 707 | D | 21 | 4,186 | 286 | 47 | 14,114 |
| Arizona | 2,521 | 55,357 | 19.1 | 13,529 | 5,503 | 3,461 | 899 | 21,511 | 10,163 | 2,065 | 803,701 |
| Arkansas | X | X | X | X | X | X | X | X | X | X | X |
| California | 25,103 | 678,261 | 14.5 | 192,359 | 123,009 | 36,371 | 116,072 | 109,847 | 97,682 | 35,432 | 12,305,792 |
| Colorado | 3,610 | 91,856 | 19.4 | 22,548 | 5,185 | 4,433 | 3,707 | 39,292 | 16,318 | 4,240 | 1,211,517 |
| Connecticut | 1,796 | 42,615 | 7.2 | 9,348 | 2,650 | 7,650 | 3,003 | 12,283 | 6,229 | 1,915 | 676,358 |
| Delaware | 580 | 7,169 | 39.1 | 1,659 | 464 | 232 | e | 2,912 | 1,438 | 337 | 139,226 |
| District of Columbia | 778 | 23,787 | 5.0 | 8,742 | 964 | 5,151 | 3,274 | 2,232 | 2,331 | 497 | 190,120 |
| Florida | X | X | X | X | X | X | X | X | X | X | X |
| Georgia | X | X | X | X | X | X | X | X | X | X | X |
| Hawaii | 533 | 8,218 | -1.8 | 1,366 | 1,135 | D | 72 | D | 672 | 162 | 46,559 |
| Idaho | 793 | 13,410 | 21.3 | 3,408 | 847 | D | 108 | 5,147 | 2,846 | 699 | 160,421 |
| Illinois | 6,010 | 130,749 | 11.2 | 40,010 | 8,814 | 7,701 | 11,633 | 39,273 | 21,323 | 5,356 | 1,882,723 |
| Indiana | 2,440 | 47,015 | 11.8 | 13,327 | 3,419 | 3,748 | 2,133 | 14,919 | 9,252 | 1,538 | 504,267 |
| Iowa | 1,618 | 29,753 | 4.7 | 9,619 | 1,805 | D | e | 8,933 | 6,724 | 1,019 | 381,630 |
| Kansas | 1,445 | 30,009 | 2.8 | 6,507 | 1,976 | 1,925 | 859 | 15,352 | 3,212 | 662 | 271,784 |
| Kentucky | 1,801 | 29,021 | 16.0 | 5,111 | 1,978 | 2,713 | 625 | 11,653 | 6,767 | 879 | 340,971 |
| Louisiana | 1,537 | 24,091 | 7.8 | 3,096 | 2,541 | D | 192 | 12,482 | 2,988 | 518 | 157,705 |
| Maine | 817 | 11,461 | -3.3 | 3,110 | 846 | 1,043 | 112 | 4,286 | 1,465 | 256 | 84,844 |
| Maryland | 2,550 | 54,141 | 7.1 | 13,280 | 3,255 | 4,785 | 2,005 | 17,936 | 11,867 | 2,678 | 1,045,571 |
| Massachusetts | 3,879 | 127,108 | 5.6 | 57,523 | 4,716 | 5,170 | 12,750 | 25,659 | 19,614 | 5,962 | 2,469,086 |
| Michigan | X | X | X | X | X | X | X | X | X | X | X |
| Minnesota | 2,788 | 61,911 | 5.0 | 24,483 | 3,728 | 4,366 | 2,426 | 16,253 | 10,249 | 2,144 | 941,665 |
| Mississippi | 1,019 | 14,262 | 10.0 | 2,038 | 769 | D | 60 | 7,783 | 1,953 | 193 | 70,206 |
| Missouri | 2,701 | 60,546 | 11.5 | 15,118 | 3,519 | 4,482 | 1,145 | 26,202 | 9,677 | 2,034 | 696,126 |
| Montana | 693 | 8,637 | 12.7 | 2,436 | 654 | D | 57 | 3,656 | 807 | 129 | 55,272 |
| Nebraska | 983 | 20,967 | 3.0 | 6,515 | 1,179 | 1,593 | 1,874 | 5,913 | 3,851 | 1,041 | 334,034 |
| Nevada | 1,424 | 17,896 | 15.4 | 3,503 | g | 1,788 | f | 6,819 | D | X | 257,100 |
| New Hampshire | 773 | 16,782 | -3.6 | 6,667 | 957 | 537 | 223 | 5,274 | 2,896 | 491 | 201,331 |
| New Jersey | 3,885 | 96,288 | 4.9 | 23,089 | 5,650 | 4,336 | 3,718 | 40,914 | 17,510 | 5,194 | 1,742,653 |
| New Mexico | X | X | X | X | X | X | X | X | X | X | X |
| New York | 11,769 | 298,902 | 3.8 | 72,078 | 37,805 | 41,500 | 37,287 | 64,632 | 33,810 | 11,955 | 4,388,956 |
| North Carolina | 3,920 | 94,977 | 9.8 | 29,306 | 5,942 | 5,251 | 1,603 | 35,114 | 17,384 | 5,075 | 1,339,036 |
| North Dakota | 353 | 7,284 | -2.2 | 2,745 | e | D | b | 2,183 | 723 | 68 | 32,215 |
| Ohio | 4,295 | 102,069 | 8.6 | 35,950 | 5,154 | 6,213 | 7,517 | 32,531 | 14,177 | 3,054 | 1,346,784 |
| Oklahoma | X | X | X | X | X | X | X | X | X | X | X |
| Oregon | 2,388 | 39,690 | 18.8 | 15,175 | 3,348 | 2,492 | 1,612 | 11,282 | 5,625 | 2,133 | 453,027 |
| Pennsylvania | 5,562 | 113,792 | 8.9 | 31,835 | 6,681 | 7,653 | 4,841 | 38,621 | 18,633 | 5,454 | 1,783,508 |
| Rhode Island | 456 | 6,863 | 4.8 | 2,181 | 507 | 482 | b | 2,649 | 572 | 120 | 43,411 |
| South Carolina | 1,639 | 37,970 | 14.1 | 8,614 | 2,429 | 2,491 | 661 | 18,356 | 4,992 | 977 | 379,536 |
| South Dakota | 478 | 7,254 | 7.2 | 1,705 | 482 | 1,194 | e | D | 533 | 75 | 25,542 |
| Tennessee | 2,788 | 49,464 | 12.0 | 9,548 | 5,883 | 5,627 | 1,881 | 21,252 | 5,057 | 1,523 | 455,942 |
| Texas | X | X | X | X | X | X | X | X | X | X | X |
| Utah | 1,746 | 52,159 | 23.5 | 19,526 | 8,365 | 1,553 | 2,766 | 10,931 | 8,699 | 1,750 | 672,201 |
| Vermont | 513 | 8,052 | 1.2 | 1,705 | 416 | 644 | 337 | 1,551 | 3,088 | 1,088 | 260,644 |
| Virginia | 4,113 | 105,182 | 5.0 | 27,615 | 6,203 | 6,503 | 5,327 | 33,077 | 25,386 | 11,209 | 2,537,977 |
| Washington | 3,864 | 139,868 | 17.8 | 75,180 | 5,760 | 3,893 | 8,656 | 29,078 | 16,292 | 6,511 | 1,821,032 |
| West Virginia | 677 | 10,196 | 0.6 | 1,564 | 565 | 1,092 | 15 | 5,544 | 1,358 | 256 | 74,829 |
| Wisconsin | 2,634 | 59,134 | 14.8 | 25,848 | 3,577 | 4,214 | 1,141 | 16,382 | 7,806 | 1,208 | 452,176 |
| Wyoming | 401 | 4,268 | 21.1 | 889 | 427 | 387 | 90 | 1,983 | 485 | 83 | 35,768 |

| | Utilities, 2017 | | | | | Transportation and warehousing, 2017 | | | | | Finance and insurance, 2017 | | | | |
|---|---|---|---|---|---|---|---|---|---|---|---|---|---|---|---|
| | | Employees | | Sales, value of shipments, or revenue (mil dol) | Annual payroll (mil dol) | | Employees | | Sales, value of shipments, or revenue (mil dol) | Annual payroll (mil dol) | | Employees | | Sales, values of shipments or revenue (mil dol) | Annual payroll (mil dol) |
| STATE | Number of establishments | Number | Percent change, 2012–2017 | | | Number of establishments | Number | Percent change, 2012–2017 | | | Number of establishments | Number | Percent change, 2012–2017 | | |
| | 247 | 248 | 249 | 250 | 251 | 252 | 253 | 254 | 255 | 256 | 257 | 258 | 259 | 260 | 261 |
| United States ................ | 18,913 | 658,384 | 1.1 | 577,100 | 67,666,854 | 237,095 | 4,954,931 | 15.1 | 895,225,411 | 242,145,488 | 475,780 | 6,499,871 | 7.6 | 4,340,011 | 638,823 |
| Alabama............ | 344 | 13,298 | -14.0 | NA | 1,293,734 | 3,098 | 61,687 | 6.7 | 8,990,180 | 2,797,539 | 7,349 | 72,311 | -0.6 | NA | 4,949 |
| Alaska ................ | 104 | 2,200 | 8.6 | NA | 209,831 | 1,177 | 18,923 | -0.2 | 5,609,025 | 1,358,059 | 784 | 7,279 | 0.9 | NA | 515 |
| Arizona .............. | 290 | 11,882 | -2.5 | NA | 1,204,829 | 3,415 | 92,227 | 14.2 | 14,565,768 | 4,602,977 | 9,602 | 161,152 | 25.2 | NA | 11,297 |
| Arkansas ........... | 331 | 7,495 | 6.1 | NA | 663,572 | 2,456 | 51,508 | 1.4 | 8,011,461 | 2,352,309 | 4,448 | 38,275 | 7.8 | NA | 2,329 |
| California ........... | 1,237 | 62,686 | -6.2 | NA | 7,692,844 | 24,817 | 545,870 | 23.6 | 99,160,737 | 28,780,713 | 50,972 | 650,176 | 8.0 | NA | 73,582 |
| Colorado ............ | 401 | 9,358 | 10.1 | NA | 910,091 | 3,764 | 70,347 | 13.5 | 15,882,039 | 3,592,089 | 10,393 | 111,376 | 12.8 | NA | 9,690 |
| Connecticut........ | 131 | 8,573 | -18.7 | NA | 940,016 | 1,660 | 44,938 | 2.1 | 5,591,991 | 1,911,700 | 5,924 | 115,871 | -2.9 | NA | 17,038 |
| Delaware ............ | 52 | 2,527 | -3.6 | NA | 241,892 | 708 | 10,604 | -11.2 | 1,281,651 | 467,175 | 1,925 | 44,120 | 21.1 | NA | 4,030 |
| District of Columbia ....... | 55 | 2,143 | X | NA | 318,311 | 159 | 4,137 | -51.2 | 2,155,311 | 297,759 | 997 | 18,843 | 4.2 | NA | 2,948 |
| Florida ............... | 860 | 27,460 | 0.4 | NA | 2,795,166 | 15,619 | 265,209 | 26.7 | 68,145,959 | 13,283,748 | 32,200 | 372,525 | 9.9 | NA | 29,569 |
| Georgia .............. | 605 | 24,314 | 7.5 | NA | 2,294,017 | 6,961 | 194,654 | 26.2 | 35,564,898 | 9,505,164 | 14,991 | 187,894 | 14.3 | NA | 15,563 |
| Hawaii ................ | 68 | 3,596 | 6.4 | NA | 377,019 | 935 | 31,863 | 18.7 | 6,431,447 | 1,508,627 | 1,482 | 19,718 | 5.5 | NA | 1,434 |
| Idaho ................. | 211 | 3,873 | 4.4 | NA | 345,445 | 1,791 | 19,059 | 10.8 | 3,057,829 | 756,640 | 2,923 | 23,040 | 6.2 | NA | 1,319 |
| Illinois ............... | 512 | 30,800 | 2.8 | NA | 3,561,458 | 16,223 | 260,488 | 12.9 | 51,368,341 | 13,137,725 | 21,302 | 309,134 | 4.4 | NA | 34,782 |
| Indiana .............. | 638 | 16,501 | 6.1 | NA | 1,537,932 | 5,416 | 128,442 | 8.6 | 20,385,955 | 5,631,026 | 9,477 | 102,822 | 6.1 | NA | 7,012 |
| Iowa .................. | 271 | 7,716 | 0.8 | NA | 702,759 | 3,677 | 57,161 | 2.5 | 8,913,447 | 2,521,466 | 6,357 | 96,808 | 5.5 | NA | 7,069 |
| Kansas .............. | 246 | 7,042 | -3.0 | NA | 692,983 | 2,637 | 57,920 | 15.8 | 7,839,739 | 2,481,241 | 5,997 | 61,278 | 3.7 | NA | 4,226 |
| Kentucky ........... | 358 | 9,275 | 6.8 | NA | 821,512 | 2,951 | 93,363 | 10.2 | 14,378,607 | 4,884,055 | 6,355 | 78,417 | 15.5 | NA | 5,271 |
| Louisiana ........... | 481 | 10,653 | -4.3 | NA | 978,709 | 3,834 | 67,721 | -3.3 | 14,583,333 | 3,707,576 | 7,542 | 65,556 | 1.1 | NA | 4,374 |
| Maine ................ | 117 | 2,437 | 3.1 | NA | 215,414 | 1,238 | 17,202 | 15.4 | 1,846,035 | 713,950 | 1,900 | 27,150 | 5.7 | NA | 1,900 |
| Maryland ............ | 158 | 10,259 | 8.2 | NA | 1,293,025 | 3,504 | 73,564 | 13.3 | 11,314,342 | 3,595,588 | 7,422 | 102,326 | 2.1 | NA | 10,068 |
| Massachusetts ... | 292 | 13,284 | -0.2 | NA | 1,294,522 | 3,973 | 91,247 | 17.2 | 13,596,083 | 4,261,295 | 9,630 | 198,360 | -2.2 | NA | 27,287 |
| Michigan ............ | 414 | 24,691 | 7.2 | NA | 2,598,511 | 6,637 | 118,638 | 11.6 | 25,019,797 | 5,919,128 | 12,987 | 169,105 | 11.5 | NA | 12,531 |
| Minnesota .......... | 348 | 12,677 | -1.0 | NA | 1,271,159 | 4,835 | 88,812 | 7.9 | 18,342,371 | 4,086,380 | 9,584 | 166,181 | 5.5 | NA | 15,836 |
| Mississippi ........ | 594 | 9,469 | 7.6 | NA | 785,640 | 2,121 | 39,819 | 20.3 | 5,564,028 | 1,714,951 | 4,715 | 33,499 | -2.8 | NA | 1,931 |
| Missouri............. | 372 | 15,265 | -5.4 | NA | 1,520,155 | 4,884 | 88,413 | 7.4 | 15,071,280 | 3,854,874 | 10,747 | 141,494 | 6.8 | NA | 10,874 |
| Montana ............. | 210 | 2,882 | -1.5 | NA | 254,931 | 1,370 | 11,939 | 0.7 | 2,151,726 | 504,079 | 1,971 | 15,804 | -2.5 | NA | 971 |
| Nebraska ........... | 116 | 1,066 | 14.6 | NA | 99,896 | 2,497 | 55,100 | 10.6 | 7,483,576 | 1,355,937 | 4,420 | 68,117 | 9.1 | NA | 4,840 |
| Nevada .............. | 122 | 4,460 | -10.6 | NA | 473,186 | 1,647 | 55,100 | 26.0 | 7,654,009 | 2,437,670 | 4,192 | 37,906 | 10.2 | NA | 2,596 |
| New Hampshire ............... | 126 | 3,306 | -0.7 | NA | 325,590 | 814 | 13,901 | 0.8 | 1,583,610 | 572,593 | 2,035 | 29,386 | 21.2 | NA | 2,642 |
| New Jersey ......... | 365 | 22,029 | 8.5 | NA | 2,583,564 | 7,573 | 177,147 | 10.5 | 30,902,528 | 9,205,205 | 11,601 | 198,542 | -0.1 | NA | 23,977 |
| New Mexico ........ | D | D | X | NA | D | 1,404 | 18,165 | 3.7 | 2,816,277 | 806,118 | 2,715 | 25,035 | 10.2 | NA | 1,417 |
| New York ........... | 671 | 39,488 | -7.3 | NA | 4,070,431 | 13,083 | 259,694 | 7.9 | 50,015,058 | 12,357,227 | 26,296 | 564,569 | 4.6 | NA | 109,705 |
| North Carolina ... | 858 | 22,509 | 11.9 | NA | 2,232,231 | 6,110 | 125,437 | 15.3 | 16,823,524 | 5,683,607 | 13,543 | 190,259 | 13.1 | NA | 18,128 |
| North Dakota ...... | D | D | X | NA | D | 1,467 | 17,380 | -7.8 | 4,367,087 | 936,199 | 1,738 | 17,896 | 4.1 | NA | 1,072 |
| Ohio .................. | 703 | 24,526 | -6.5 | NA | 2,496,000 | 7,698 | 189,666 | 19.4 | 29,460,997 | 9,020,346 | 17,229 | 260,044 | 7.6 | NA | 19,921 |
| Oklahoma ........... | 368 | 10,505 | 28.1 | NA | 927,214 | 2,736 | 51,368 | 15.4 | 12,698,585 | 2,744,277 | 6,867 | 59,479 | 3.0 | NA | 3,696 |
| Oregon .............. | 279 | 7,818 | -3.1 | NA | 767,993 | 3,225 | 60,206 | 15.0 | 8,813,005 | 2,857,091 | 6,121 | 63,459 | 10.5 | NA | 4,747 |
| Pennsylvania...... | 847 | 31,639 | 3.1 | NA | 3,315,983 | 8,822 | 230,350 | 9.8 | 28,710,741 | 10,004,774 | 17,607 | 282,836 | 6.0 | NA | 25,349 |
| Rhode Island ...... | 38 | 1,300 | 6.8 | NA | 129,367 | 700 | 11,082 | -1.7 | 1,366,689 | 467,134 | 1,452 | 27,061 | 7.3 | NA | 2,419 |
| South Carolina ... | 417 | 12,951 | 8.3 | NA | 1,150,406 | 2,805 | 63,980 | 31.4 | 7,567,557 | 2,597,555 | 7,449 | 73,990 | 11.5 | NA | 4,790 |
| South Dakota ..... | 154 | 2,124 | -4.2 | NA | 186,178 | 1,259 | 10,207 | 6.9 | 1,686,473 | 421,105 | 1,981 | 26,270 | -0.8 | NA | 1,512 |
| Tennessee ......... | 150 | 3,263 | 0.2 | NA | 240,114 | 4,324 | 148,152 | 11.5 | 22,870,848 | 6,626,122 | 10,210 | 128,741 | 14.7 | NA | 9,257 |
| Texas ................ | 1,967 | 58,364 | 10.3 | NA | 5,838,442 | 19,943 | 485,957 | 25.6 | 101,947,033 | 26,437,929 | 40,589 | 551,228 | 14.4 | NA | 44,410 |
| Utah .................. | 222 | 3,546 | -17.9 | NA | 338,018 | 2,342 | 61,942 | 34.8 | 10,193,997 | 2,816,339 | 5,289 | 67,500 | 27.5 | NA | 4,628 |
| Vermont ............. | 70 | 1,187 | X | NA | 124,756 | 491 | 6,082 | 6.1 | 772,885 | 246,632 | 945 | 8,535 | -5.6 | NA | 646 |
| Virginia ............. | 335 | 16,574 | 10.0 | NA | 1,688,285 | 5,062 | 104,497 | 16.0 | 20,071,064 | 5,026,540 | 11,404 | 154,331 | 0.7 | NA | 13,462 |
| Washington ........ | 314 | 9,291 | 0.8 | NA | 1,003,750 | 5,474 | 100,012 | 13.8 | 21,845,661 | 5,560,125 | 9,992 | 104,294 | 7.2 | NA | 8,933 |
| West Virginia ..... | 229 | 6,775 | 15.8 | NA | 600,615 | 1,181 | 15,094 | -3.9 | 3,068,787 | 692,207 | 2,087 | 16,270 | -9.0 | NA | 856 |
| Wisconsin .......... | 344 | 12,182 | -13.3 | NA | 1,239,788 | 5,635 | 104,143 | 6.6 | 15,551,340 | 4,504,979 | 8,972 | 146,594 | 4.4 | NA | 11,019 |
| Wyoming ............ | 146 | 2,615 | 7.4 | NA | 233,272 | 943 | 9,892 | -2.4 | 2,130,700 | 539,944 | 1,070 | 7,015 | 4.6 | NA | 406 |

# Table A. States — Real Estate and Rental and Leasing and Professional, Scientific, and Technical Services

| STATE | Real estate and rental and leasing, 2017 | | | | | Professional, scientific, and technical services, 2017 | | | | | | | | |
|---|---|---|---|---|---|---|---|---|---|---|---|---|---|---|
| | Number of estab-lishments | Employees Number | Employees Percent change, 2012–2017 | Receipts (mil dol) | Annual payroll (mil dol) | Number of estab-lishments | Total | Employees Percent change, 2012–2017 | Legal services | Accounting, tax preparation, bookkeeping, and payroll services | Architectural, engineering, and related services | Computer systems design and related services | Sales, values of ship-ments or revenue (mil dol) | Annual payroll (mil dol) |
| | 262 | 263 | 264 | 265 | 266 | 267 | 268 | 269 | 270 | 271 | 272 | 273 | 274 | 275 |
| United States | 410,820 | 2,194,885 | 14.1 | 674,147 | 113,410 | 913,624 | 9,015,366 | 9.9 | 1,174,350 | 1,133,903 | 1,514,699 | 1,894,668 | 1,844,781 | 729,371 |
| Alabama | 4,362 | 22,778 | -0.3 | 5,785 | 968 | 9,496 | 100,857 | 12.1 | 13,320 | 11,627 | 27,913 | 22,558 | 20,681 | 7,188 |
| Alaska | 965 | 4,622 | 9.7 | 1,188 | 223 | 1,959 | 17,869 | 1.3 | 1,619 | 1,639 | 6,583 | 2,429 | 3,235 | 1,269 |
| Arizona | 9,937 | 48,305 | 19.3 | 12,992 | 2,267 | 18,100 | 146,970 | 21.1 | 17,326 | 21,944 | 27,167 | 28,543 | 24,119 | 10,563 |
| Arkansas | 3,115 | 12,776 | -0.7 | 2,588 | 473 | 5,914 | 38,088 | 18.2 | D | 7,251 | 7,567 | 5,094 | 5,565 | 2,071 |
| California | 57,434 | 314,273 | 14.9 | 110,822 | 18,275 | 127,023 | 1,241,452 | -4.7 | 150,205 | 139,959 | 186,928 | 273,580 | 300,626 | 118,664 |
| Colorado | 12,087 | 47,642 | 23.1 | 12,757 | 2,388 | 26,404 | 192,964 | 7.2 | 18,244 | 20,505 | 47,612 | 41,715 | 39,655 | 15,801 |
| Connecticut | 3,459 | 20,224 | 2.3 | 6,691 | 1,115 | 9,184 | 108,479 | 11.2 | 13,331 | 14,992 | 13,821 | 23,127 | 21,600 | 9,710 |
| Delaware | 1,286 | 6,096 | 12.8 | 5,164 | 288 | 2,991 | 29,716 | X | D | 2,989 | 3,623 | 7,163 | 7,003 | 2,942 |
| District of Columbia | 1,350 | 11,000 | 8.9 | 4,525 | 883 | 5,812 | 108,016 | 10.7 | 31,915 | 4,257 | 9,363 | 17,237 | 37,245 | 13,126 |
| Florida | 37,660 | 176,886 | 26.4 | 49,175 | 8,184 | 79,224 | 513,798 | 16.5 | 97,015 | 71,989 | 80,036 | 89,935 | 89,601 | 34,673 |
| Georgia | 12,426 | 63,935 | 15.1 | 23,009 | 3,795 | 29,737 | 266,523 | X | 32,158 | 41,891 | 38,896 | 73,357 | 55,149 | 20,189 |
| Hawaii | 2,069 | 13,101 | 15.2 | 4,409 | 675 | 3,380 | 22,668 | 4.8 | 3,354 | 3,306 | 5,345 | 3,659 | 3,799 | 1,466 |
| Idaho | 2,644 | 7,521 | 20.0 | 1,635 | 277 | 4,686 | 33,246 | 3.6 | D | 4,111 | 7,364 | 3,780 | 5,046 | 1,921 |
| Illinois | 13,589 | 82,763 | 7.8 | 32,355 | 4,824 | 38,805 | 404,450 | 11.0 | 58,309 | 55,152 | 52,957 | 80,580 | 87,119 | 33,951 |
| Indiana | 6,686 | 34,736 | 9.5 | 8,392 | 1,520 | 13,061 | 121,821 | 21.9 | 14,188 | 18,958 | 23,175 | 20,766 | 22,391 | 8,538 |
| Iowa | 3,130 | 13,999 | 16.4 | 2,890 | 576 | 6,460 | 52,607 | 8.4 | 7,363 | 9,299 | 7,185 | 10,164 | 7,937 | 3,172 |
| Kansas | 3,415 | 14,532 | 1.9 | 3,943 | 586 | 7,201 | 65,648 | 7.6 | 6,981 | 9,166 | 13,696 | 15,414 | 10,972 | 4,174 |
| Kentucky | 3,902 | 18,070 | -1.0 | 5,125 | 731 | 8,125 | 67,615 | 7.6 | 10,891 | 12,984 | 11,112 | 12,561 | 10,272 | 3,636 |
| Louisiana | 5,121 | 29,345 | -6.2 | 7,301 | 1,386 | 12,072 | 95,652 | 8.6 | 19,593 | 15,134 | 31,802 | 8,520 | 16,082 | 6,185 |
| Maine | 1,821 | 7,138 | 14.4 | 1,484 | 287 | 3,543 | 23,381 | 1.9 | 3,966 | 3,197 | 4,753 | 3,303 | 3,874 | 1,488 |
| Maryland | 6,811 | 49,157 | 14.8 | 18,087 | 2,852 | 20,974 | 283,999 | 16.1 | D | 19,575 | 47,076 | 89,511 | 55,405 | 23,645 |
| Massachusetts | 7,584 | 52,315 | 22.3 | 17,912 | 3,369 | 21,985 | 310,313 | 21.7 | 29,184 | 28,962 | 41,821 | 69,413 | 78,598 | 32,256 |
| Michigan | 8,467 | 54,808 | 12.5 | 17,783 | 2,374 | 21,832 | 282,246 | X | 26,529 | 34,657 | 69,425 | 42,499 | 39,436 | 19,670 |
| Minnesota | 7,218 | 38,077 | 10.4 | 10,433 | 1,915 | 16,689 | 186,597 | 32.4 | D | 19,948 | 23,440 | 32,043 | 34,697 | 14,712 |
| Mississippi | 2,403 | 9,683 | -5.4 | 1,997 | 345 | 4,746 | 30,477 | 0.9 | 7,154 | 6,993 | 5,184 | 3,731 | 4,598 | 1,618 |
| Missouri | 6,644 | 37,144 | 11.1 | 8,976 | 1,626 | 14,171 | 161,595 | 17.1 | 21,788 | 22,741 | 24,757 | 43,618 | 30,112 | 11,584 |
| Montana | 2,036 | 5,951 | 14.3 | 1,145 | 206 | 3,836 | 17,711 | 6.3 | 2,906 | 3,165 | 4,789 | 1,414 | 2,517 | 939 |
| Nebraska | 2,354 | 11,293 | 12.2 | 2,292 | 487 | 4,699 | 39,566 | -46.9 | D | 6,242 | 6,470 | 7,323 | 6,278 | 2,385 |
| Nevada | 4,684 | 30,562 | 36.4 | 7,359 | 1,312 | 9,018 | 60,168 | 25.5 | 9,457 | 7,337 | 11,600 | 8,421 | 10,816 | 3,951 |
| New Hampshire | 1,523 | 7,920 | 12.4 | 2,000 | 381 | 3,674 | 32,387 | 7.4 | 3,888 | 5,499 | 5,635 | 6,863 | 5,583 | 2,383 |
| New Jersey | 9,622 | 61,052 | 13.6 | 21,184 | 3,435 | 28,962 | 325,516 | 5.8 | 38,484 | 35,573 | 42,275 | 99,461 | 67,232 | 28,735 |
| New Mexico | 2,408 | 9,229 | -5.4 | 2,186 | 376 | 4,728 | 56,695 | 28.3 | D | 4,710 | 8,315 | 4,707 | 10,082 | 4,190 |
| New York | 34,076 | 193,442 | 16.3 | 70,693 | 11,357 | 61,744 | 668,196 | 13.5 | 128,077 | 97,737 | 67,837 | 94,294 | 172,936 | 60,227 |
| North Carolina | 12,450 | 56,360 | 19.5 | 14,647 | 2,639 | 24,766 | 221,438 | 12.8 | 24,381 | 29,133 | 34,200 | 44,672 | 40,878 | 16,048 |
| North Dakota | 1,100 | 5,440 | 5.5 | 1,280 | 242 | 1,821 | 14,880 | 8.5 | 2,038 | 2,071 | 3,899 | 2,450 | 2,425 | 947 |
| Ohio | 10,782 | 62,902 | 3.2 | 20,524 | 2,986 | 23,854 | 250,438 | 7.1 | 32,148 | 37,333 | 46,869 | 44,866 | 43,625 | 17,059 |
| Oklahoma | 4,461 | 22,135 | 4.1 | 4,696 | 942 | 9,736 | 70,920 | -1.5 | D | 14,752 | 13,173 | 9,779 | 11,247 | 4,456 |
| Oregon | 6,771 | 29,773 | 14.4 | 6,774 | 1,263 | 12,620 | 92,358 | 9.3 | 11,825 | 12,272 | 16,238 | 12,719 | 14,720 | 6,902 |
| Pennsylvania | 10,662 | 66,715 | 13.9 | 18,861 | 3,472 | 29,991 | 340,361 | 7.5 | 50,588 | 50,111 | 65,253 | 55,904 | 64,179 | 26,674 |
| Rhode Island | 1,110 | 5,287 | -5.8 | 1,351 | 248 | 3,026 | 23,910 | 13.0 | 4,787 | 2,899 | 4,089 | 4,717 | 4,043 | 1,566 |
| South Carolina | 5,890 | 26,764 | 15.4 | 6,510 | 1,128 | 10,859 | 100,377 | 25.7 | 14,550 | 11,436 | 27,015 | 13,704 | 17,038 | 6,498 |
| South Dakota | 1,143 | 4,330 | 22.8 | 829 | 155 | 1,967 | 13,063 | 17.2 | 1,871 | 2,668 | 2,490 | 1,761 | 1,763 | 655 |
| Tennessee | 6,048 | 36,212 | 18.4 | 10,679 | 1,760 | 11,411 | 116,205 | 11.1 | D | 20,277 | 19,988 | 17,191 | 18,739 | 7,661 |
| Texas | 32,290 | 198,712 | 16.9 | 56,443 | 10,560 | 70,033 | 735,744 | 15.0 | 87,848 | 90,118 | 156,673 | 165,740 | 150,903 | 59,488 |
| Utah | 5,577 | 19,457 | 20.1 | 5,296 | 877 | 10,873 | 92,444 | 21.1 | 9,359 | 12,820 | 13,900 | 18,315 | 15,305 | 5,610 |
| Vermont | 784 | 2,938 | -5.0 | 652 | 119 | 2,096 | 12,563 | -21.2 | 1,896 | 1,926 | 2,224 | 2,168 | 2,079 | 842 |
| Virginia | 10,051 | 55,778 | 2.8 | 17,208 | 2,949 | 31,431 | 470,265 | 9.4 | 27,806 | 40,367 | 67,638 | 186,672 | 101,433 | 41,553 |
| Washington | 11,826 | 53,706 | 18.8 | 15,627 | 2,757 | 22,144 | 214,628 | 28.1 | D | 21,192 | 49,424 | 45,588 | 42,429 | 17,333 |
| West Virginia | 1,432 | 6,061 | 0.8 | 1,369 | 229 | 2,813 | 22,856 | -7.9 | 5,530 | 3,671 | 4,304 | 2,725 | 3,295 | 1,218 |
| Wisconsin | 4,956 | 27,163 | 14.3 | 5,908 | 1,106 | 11,553 | 106,041 | 6.9 | 14,149 | 15,987 | 19,489 | 18,104 | 18,880 | 7,307 |
| Wyoming | 1,199 | 4,777 | 5.1 | 1,216 | 220 | 2,395 | 9,589 | 5.0 | D | 1,381 | 2,311 | 810 | 1,539 | 532 |

# Table A. States — Health Care and Social Assistance

| | Health care and social assistance, 2017 | | | | | | | | | | | |
| | Subject to federal tax | | | | | | Tax-exempt | | | | | |
| | | Employees | | | | | | | Employees | | | | |
| STATE | Number of estab-lishments | Total | Percent change, 2012–2017 | Ambulatory health care services | Hospitals | Sales, value of shipments, or revenue (mil dol) | Annual payroll (mil dol) | Number of estab-lishments | Total | Percent change, 2012–2017 | Ambulatory health care services | Hospitals | Sales, values of ship-ments or revenue (mil dol) | Annual payroll (mil dol) |
|---|---|---|---|---|---|---|---|---|---|---|---|---|---|---|
| | 276 | 277 | 278 | 279 | 280 | 281 | 282 | 283 | 284 | 285 | 286 | 287 | 288 | 289 |
| United States ................. | 744,628 | 11,103,994 | 16.4 | 6,669,762 | 751,064 | 1,258,937 | 514,343 | 147,617 | 9,402,508 | 6.0 | 893,376 | 5,423,087 | 1,268,966 | 475,713 |
| Alabama.......................... | 8,794 | 156,644 | 10.0 | 87,140 | 22,879 | 17,755 | 7,038 | 1,842 | 101,755 | 0.9 | 8,864 | 67,808 | 13,484 | 4,911 |
| Alaska ............................ | 2,009 | 22,324 | 4.4 | 13,480 | 2,202 | 3,381 | 1,316 | 670 | 29,801 | 9.1 | 6,093 | 13,540 | 4,475 | 1,732 |
| Arizona........................... | 16,871 | 233,019 | 26.5 | 131,519 | 21,262 | 26,801 | 10,435 | 1,945 | 145,996 | 11.5 | 16,655 | 84,708 | 21,329 | 7,974 |
| Arkansas........................ | 6,345 | 98,148 | 12.8 | 49,112 | 12,964 | 10,550 | 4,269 | 1,533 | 83,978 | 5.7 | 7,430 | 46,260 | 9,295 | 3,692 |
| California........................ | 99,272 | 1,198,086 | 19.2 | 740,129 | 76,748 | 159,332 | 63,328 | 14,118 | 845,031 | 9.6 | 88,774 | 499,331 | 151,980 | 56,173 |
| Colorado ........................ | 14,297 | 181,974 | 29.3 | 106,304 | 10,251 | 20,903 | 8,265 | 2,362 | 138,839 | 18.5 | 22,327 | 80,549 | 19,153 | 7,455 |
| Connecticut.................... | 8,563 | 158,995 | 20.7 | 84,973 | 4,028 | 16,249 | 7,493 | 2,502 | 136,088 | -2.5 | 15,603 | 64,530 | 19,054 | 6,576 |
| Delaware........................ | 2,094 | 34,764 | 14.4 | 21,953 | D | 3,887 | 1,746 | 554 | 33,584 | 6.6 | 2,600 | D | 4,968 | 1,975 |
| District of Columbia........... | 1,438 | 30,898 | 26.9 | 20,540 | X | 3,817.001 | 1,696 | 765 | 43,236 | -0.3 | 3,329 | D | 7,583 | 2,680 |
| Florida........................... | 55,996 | 718,280 | 17.0 | 444,187 | 82,648 | 97,268 | 35,551 | 5,558 | 408,875 | 8.4 | 39,531 | 241,661 | 58,016 | 19,921 |
| Georgia.......................... | 22,386 | 311,585 | 21.3 | 202,073 | 20,983 | 38,230 | 15,454 | 2,874 | 209,561 | 9.4 | 13,688 | 145,371 | 30,530 | 10,317 |
| Hawaii............................ | 2,844 | 29,918 | 11.7 | 22,564 | X | 3,771.589 | 1,585 | 833 | 43,633 | 9.1 | 6,514 | 23,312 | 6,014 | 2,414 |
| Idaho............................. | 4,725 | 59,441 | 19.8 | 30,208 | 5,271 | 5,729 | 2,234 | 585 | 38,659 | 14.1 | 2,306 | 29,506 | 4,740 | 2,058 |
| Illinois........................... | 28,608 | 422,680 | 10.1 | 251,137 | 15,018 | 46,173 | 19,337 | 5,627 | 395,053 | 2.2 | 29,205 | 232,126 | 51,453 | 18,752 |
| Indiana........................... | 13,698 | 249,114 | 20.6 | 150,472 | 15,742 | 26,772 | 11,332 | 2,970 | 187,502 | -1.5 | 16,506 | 112,181 | 25,065 | 8,309 |
| Iowa.............................. | 5,984 | 82,550 | 4.5 | 46,045 | X | 8,820.735 | 3,819 | 2,626 | 134,415 | 5.1 | 8,763 | D | 13,599.112 | 5,422 |
| Kansas........................... | 6,195 | 97,818 | 5.0 | 54,150 | 8,907 | 11,056 | 4,410 | 1,909 | 98,123 | -1.0 | 6,990 | 55,588 | 10,384 | 4,112 |
| Kentucky........................ | 9,600 | 137,974 | 8.3 | 79,800 | 9,299 | 14,980 | 6,411 | 1,997 | 130,737 | 5.1 | 12,887 | 88,082 | 17,389 | 6,001 |
| Louisiana....................... | 10,878 | 181,347 | 8.1 | 97,290 | 21,048 | 18,910 | 7,274 | 1,807 | 121,061 | 3.3 | 5,411 | 81,761 | 15,708 | 5,639 |
| Maine............................ | 3,107 | 44,753 | 7.3 | 24,807 | X | 3,954.996 | 1,767 | 1,664 | 67,841 | 0.5 | 9,032 | D | 7,822.29 | 3,233 |
| Maryland ........................ | 13,949 | 185,545 | 10.3 | 125,549 | X | 22,447.831 | 9,526 | 2,851 | 198,551 | 3.7 | 14,580 | D | 26,227.901 | 9,768 |
| Massachusetts................ | 13,717 | 271,634 | 14.3 | 164,867 | 21,831 | 31,795 | 14,422 | 5,632 | 363,378 | 3.9 | 36,356 | 170,639 | 42,229 | 17,616 |
| Michigan......................... | 22,148 | 317,165 | 17.2 | 202,891 | 16,752 | 33,917 | 14,583 | 4,829 | 310,643 | -1.3 | 29,288 | 190,508 | 40,278 | 14,727 |
| Minnesota....................... | 12,846 | 212,834 | 12.4 | 120,207 | X | 19,562.022 | 8,897 | 4,220 | 260,504 | 3.8 | 25,739 | D | 30,929.025 | 12,309 |
| Mississippi...................... | 5,356 | 87,759 | 3.9 | 47,881 | 8,931 | 9,520 | 3,654 | 1,035 | 81,251 | 11.0 | 3,278 | 60,548 | 9,232 | 3,879 |
| Missouri......................... | 15,917 | 213,163 | 15.3 | 116,761 | 12,850 | 22,077 | 9,086 | 3,180 | 209,894 | -2.4 | 20,948 | 124,164 | 26,116 | 9,596 |
| Montana......................... | 2,700 | 30,706 | 25.7 | 18,629 | D | 3,563 | 1,514 | 1,016 | 42,548 | 3.2 | 3,561 | D | 4,884.756 | 1,790 |
| Nebraska........................ | 4,606 | 62,134 | 8.6 | 34,657 | D | 6,875 | 2,831 | 1,211 | 73,557 | 7.8 | 6,402 | D | 9,186 | 3,286 |
| Nevada.......................... | 6,774 | 97,187 | 25.1 | 49,796 | 20,450 | 13,230 | 4,807 | 598 | 34,906 | 12.9 | 2,375 | 22,035 | 4,882 | 1,813 |
| New Hampshire ................ | 2,863 | 41,250 | 16.5 | 24,837 | D | 4,920 | 2,151 | 874 | 53,344 | 3.2 | 8,059 | D | 7,011 | 2,591 |
| New Jersey..................... | 24,064 | 352,009 | 18.2 | 231,014 | 17,104 | 42,098 | 16,936 | 3,941 | 261,397 | 7.6 | 21,211 | 149,483 | 32,625 | 13,198 |
| New Mexico..................... | 4,119 | 73,984 | 9.6 | 37,118 | 7,966 | 6,894 | 2,804 | 1,015 | 53,824 | 9.7 | 5,868 | 29,685 | 6,709 | 2,602 |
| New York ........................ | 44,830 | 677,837 | 25.5 | 480,628 | 2,593 | 69,277 | 28,220 | 14,072 | 976,756 | 5.1 | 97,152 | 483,633 | 124,231 | 52,468 |
| North Carolina................. | 20,223 | 329,678 | 12.6 | 195,220 | 9,644 | 35,214 | 14,481 | 3,857 | 272,766 | 15.2 | 19,344 | 185,174 | 37,518 | 13,146 |
| North Dakota................... | 1,368 | 16,368 | 8.8 | 11,030 | X | 2,324.254 | 801 | 689 | 46,087 | 10.8 | 4,260 | D | 4,974 | 2,112 |
| Ohio.............................. | 23,844 | 428,546 | 8.8 | 254,307 | 16,318 | 43,867 | 18,946 | 5,751 | 428,248 | 5.8 | 40,316 | 269,602 | 53,251 | 20,523 |
| Oklahoma....................... | 9,221 | 137,610 | 7.1 | 71,466 | 22,221 | 15,697 | 5,894 | 1,814 | 88,852 | 4.8 | 6,962 | 54,228 | 11,334 | 4,059 |
| Oregon........................... | 10,959 | 130,894 | 24.3 | 82,937 | D | 14,414 | 6,136 | 2,989 | 132,384 | 17.9 | 14,923 | D | 18,670 | 6,970 |
| Pennsylvania................... | 29,203 | 507,878 | 17.1 | 305,298 | 31,112 | 53,578 | 24,403 | 8,496 | 528,093 | 1.2 | 50,964 | 256,477 | 63,044 | 23,482 |
| Rhode Island................... | 2,458 | 43,682 | 22.7 | 23,589 | X | 4,493.564 | 2,028 | 719 | 43,864 | -9.5 | 4,870 | D | 5,080 | 2,064 |
| South Carolina ................ | 8,970 | 147,388 | 11.4 | 79,726 | 17,490 | 16,409 | 6,488 | 1,618 | 96,810 | 20.9 | 7,180 | 63,418 | 13,046 | 4,562 |
| South Dakota.................. | 1,623 | 21,521 | 4.9 | 12,287 | D | 2,486 | 968 | 797 | 50,300 | 17.0 | 2,921 | D | 6,229 | 2,490 |
| Tennessee...................... | 13,248 | 249,523 | 13.3 | 147,182 | 27,722 | 30,270 | 12,048 | 2,643 | 165,075 | 3.1 | 11,450 | 107,960 | 21,819 | 8,172 |
| Texas............................ | 63,412 | 1,127,958 | 19.3 | 654,109 | 132,686 | 119,276 | 46,467 | 6,540 | 453,619 | 13.4 | 37,542 | 305,689 | 66,833 | 23,432 |
| Utah.............................. | 7,490 | 92,983 | 20.7 | 52,728 | 8,909 | 10,675 | 3,970 | 764 | 53,382 | 8.6 | 5,754 | 37,205 | 8,033 | 2,733 |
| Vermont......................... | 1,345 | 16,268 | 8.6 | 9,600 | X | 1,590.873 | 706 | 757 | 32,017 | 9.6 | 6,800 | 14,913 | 3,860.753 | 1,524 |
| Virginia.......................... | 17,831 | 276,332 | 16.9 | 167,605 | 17,696 | 31,163.794 | 13,461 | 2,691 | 178,169 | 2.0 | 15,373 | 103,608 | 27,629 | 9,260 |
| Washington..................... | 18,081 | 229,068 | 18.0 | 137,208 | 2,813 | 24,847 | 11,014 | 3,183 | 209,767 | 16.5 | 28,963 | 123,489 | 31,594.945 | 11,836 |
| West Virginia................... | 3,607 | 59,426 | 3.5 | 29,948 | 5,355 | 6,403.191 | 2,460 | 1,262 | 73,201 | 2.1 | 8,890 | 45,140 | 8,834 | 3,384 |
| Wisconsin....................... | 12,596 | 200,596 | 14.5 | 113,937 | 2,096 | 20,053 | 9,188 | 3,387 | 216,769 | 2.7 | 28,527 | 114,852 | 28,425 | 10,011 |
| Wyoming........................ | 1,556 | 14,756 | 7.7 | 8,867 | D | 1,659 | 694 | 445 | 18,784 | 6.5 | 1,012 | D | 2,213 | 963 |

# Table A. States — Arts, Entertainment, and Recreation and Accommodation and Food Services

| STATE | Arts, entertainment, and recreation, 2017 | | | | | Accommodation and food services, 2017 | | | | | |
|---|---|---|---|---|---|---|---|---|---|---|---|
| | Number of establishments | Employees | | Sales, value of shipments, or revenue (mil dol) | Annual payroll (mil dol) | Number of establishments | Employees | | Food services and drinking places | Sales, values of shipments or revenue (mil dol) | Annual payroll (mil dol) |
| | | Number | Percent change, 2012–2017 | | | | Total | Percent change, 2012–2017 | | | |
| | 290 | 291 | 292 | 293 | 294 | 295 | 296 | 297 | 298 | 299 | 300 |
| United States | 142,938 | 2,390,279 | 14.8 | 265,620 | 82,256 | 726,081 | 14,002,624 | 16.6 | 11,881,174 | 678,148 | 202,539 |
| Alabama | 1,168 | 17,939 | 4.5 | 1,158 | 333 | 9,085 | 183,590 | 16.7 | 164,731 | 8,201 | 2,335 |
| Alaska | 575 | 5,433 | 7.5 | 449 | 104 | 2,214 | 28,853 | 7.5 | 21,693 | 1,557 | 505 |
| Arizona | 2,041 | 47,060 | 11.0 | 4,812 | 1,582 | 13,079 | 299,628 | 19.2 | 244,665 | 13,483 | 4,103 |
| Arkansas | 787 | 9,919 | 11.7 | 1,057 | 218 | 5,977 | 106,280 | 10.9 | 95,188 | 4,576 | 1,369 |
| California | 26,390 | 349,752 | 15.1 | 53,107 | 17,209 | 89,596 | 1,739,010 | 24.7 | 1,467,354 | 95,667 | 28,608 |
| Colorado | 3,028 | 58,229 | 13.7 | 5,681 | 1,935 | 14,121 | 290,915 | 21.0 | 235,601 | 13,366 | 4,324 |
| Connecticut | 1,744 | 30,320 | 14.5 | 2,406 | 785 | 8,762 | 146,456 | 8.9 | 122,550 | 7,676 | 2,322 |
| Delaware | 440 | 8,466 | 4.5 | 854 | 224 | 2,141 | 39,950 | 12.2 | 35,841 | 2,061 | 611 |
| District of Columbia | 380 | 9,817 | 30.7 | 1,736 | 830 | 2,733 | 72,890 | 20.7 | 58,276 | 4,153 | 1,370 |
| Florida | 8,883 | 208,733 | 22.9 | 23,435 | 6,639 | 42,071 | 955,006 | 21.5 | 772,959 | 45,113 | 13,288 |
| Georgia | 3,452 | 50,796 | 18.5 | 5,245 | 1,752 | 21,201 | 426,884 | 20.7 | 381,155 | 20,818 | 5,928 |
| Hawaii | 510 | 11,912 | 12.1 | 1,026 | 314 | 3,865 | 112,743 | 14.6 | 71,189 | 4,934 | 1,430 |
| Idaho | 804 | 9,644 | 7.8 | 602 | 174 | 3,856 | 65,463 | 20.7 | 54,616 | 2,636 | 786 |
| Illinois | 5,218 | 91,647 | 18.8 | 10,067 | 2,830 | 29,025 | 536,245 | 14.1 | 478,471 | 27,978 | 8,351 |
| Indiana | 2,320 | 37,937 | 12.5 | 3,999 | 1,119 | 13,647 | 280,340 | 9.8 | 250,491 | 11,856 | 3,586 |
| Iowa | 1,451 | 18,336 | -13.6 | 1,158 | 315 | 7,283 | 123,866 | 7.6 | 101,394 | 4,600 | 1,405 |
| Kansas | 1,092 | 17,313 | 20.7 | 1,231 | 323 | 6,253 | 118,905 | 11.3 | 106,572 | 4,860 | 1,487 |
| Kentucky | 1,406 | 19,605 | 12.9 | 1,965 | 441 | 8,228 | 174,910 | 11.4 | 159,803 | 7,781 | 2,329 |
| Louisiana | 1,526 | 25,649 | 9.7 | 2,789 | 845 | 9,877 | 215,048 | 10.9 | 173,404 | 9,431 | 2,740 |
| Maine | 887 | 8,027 | 9.9 | 677 | 188 | 4,257 | 55,746 | 12.2 | 44,641 | 2,610 | 831 |
| Maryland | 2,172 | 43,573 | 21.3 | 4,970 | 1,588 | 12,139 | 237,730 | 16.4 | 207,377 | 12,732 | 3,701 |
| Massachusetts | 3,477 | 66,160 | 19.0 | 7,037 | 2,360 | 17,773 | 311,058 | 13.9 | 278,738 | 18,222 | 5,700 |
| Michigan | 3,469 | 49,733 | 7.5 | 4,998 | 1,750 | 20,696 | 399,032 | 14.9 | 348,803 | 17,285 | 5,274 |
| Minnesota | 3,012 | 48,380 | 14.3 | 4,162 | 1,561 | 12,022 | 239,194 | 7.8 | 202,894 | 10,353 | 3,363 |
| Mississippi | 694 | 8,203 | -7.2 | 583 | 156 | 5,651 | 129,836 | 11.7 | 96,924 | 4,517 | 1,280 |
| Missouri | 2,274 | 40,484 | 6.0 | 3,953 | 1,554 | 12,896 | 263,644 | 10.2 | 229,465 | 11,044 | 3,420 |
| Montana | 1,234 | 10,492 | -3.8 | 877 | 180 | 3,568 | 52,415 | 13.3 | 40,479 | 2,158 | 632 |
| Nebraska | 913 | 14,879 | 13.7 | 891 | 249 | 4,621 | 76,386 | 8.9 | 68,262 | 3,253 | 979 |
| Nevada | 1,635 | 31,918 | 19.5 | 5,143 | 1,068 | 6,810 | 319,584 | 7.7 | 128,481 | 8,817 | 2,485 |
| New Hampshire | 815 | 12,442 | -3.9 | 978 | 291 | 3,784 | 59,531 | 10.1 | 50,102 | 2,886 | 905 |
| New Jersey | 3,842 | 65,529 | 16.1 | 5,967 | 1,934 | 21,495 | 318,734 | 9.2 | 272,795 | 18,267 | 4,993 |
| New Mexico | 700 | 12,062 | -1.6 | 1,120 | 272 | 4,392 | 91,601 | 10.9 | 72,054 | 3,610 | 1,115 |
| New York | 13,019 | 185,076 | 13.7 | 29,270 | 8,695 | 54,797 | 824,806 | 21.4 | 718,695 | 50,152 | 15,216 |
| North Carolina | 3,868 | 69,027 | 18.6 | 6,354 | 2,057 | 21,437 | 429,125 | 19.7 | 382,079 | 19,803 | 5,763 |
| North Dakota | 470 | 5,418 | 10.5 | 273 | 84 | 2,080 | 36,648 | 2.7 | 27,623 | 1,342 | 426 |
| Ohio | 3,999 | 76,914 | 26.7 | 8,596 | 2,832 | 24,346 | 474,616 | 8.5 | 439,944 | 21,350 | 6,390 |
| Oklahoma | 1,152 | 26,523 | 0.6 | 3,470 | 799 | 8,397 | 159,826 | 11.3 | 137,769 | 6,752 | 1,955 |
| Oregon | 1,992 | 29,222 | 21.6 | 2,256 | 776 | 11,708 | 182,613 | 21.4 | 155,042 | 8,988 | 2,837 |
| Pennsylvania | 4,862 | 101,095 | 1.5 | 10,655 | 3,265 | 28,843 | 481,682 | 9.7 | 424,046 | 22,462 | 6,493 |
| Rhode Island | 575 | 8,673 | -1.4 | 739 | 219 | 3,167 | 50,642 | 14.9 | 44,151 | 2,593 | 805 |
| South Carolina | 1,687 | 28,317 | 13.6 | 1,881 | 542 | 10,847 | 223,081 | 20.4 | 191,404 | 10,006 | 2,917 |
| South Dakota | 697 | 6,799 | 9.6 | 541 | 129 | 2,495 | 40,704 | 7.2 | 31,118 | 1,522 | 462 |
| Tennessee | 2,824 | 37,967 | 16.9 | 5,037 | 1,720 | 13,518 | 287,534 | 19.1 | 254,036 | 13,349 | 4,029 |
| Texas | 7,620 | 153,566 | 28.9 | 14,842 | 4,514 | 57,098 | 1,201,419 | 23.0 | 1,078,923 | 60,835 | 17,640 |
| Utah | 1,180 | 28,240 | 36.1 | 1,760 | 553 | 5,931 | 118,734 | 23.8 | 98,351 | 5,098 | 1,497 |
| Vermont | 457 | 7,739 | 8.3 | 517 | 151 | 1,979 | 32,891 | 4.9 | 20,555 | 1,128 | 370 |
| Virginia | 3,071 | 61,703 | 13.7 | 4,447 | 1,283 | 18,199 | 358,010 | 11.7 | 311,131 | 17,107 | 5,098 |
| Washington | 3,097 | 65,091 | 10.8 | 6,483 | 2,078 | 17,828 | 289,371 | 23.6 | 246,233 | 15,680 | 5,089 |
| West Virginia | 775 | 6,681 | -20.0 | 488 | 108 | 3,614 | 68,102 | 2.7 | 55,458 | 2,655 | 781 |
| Wisconsin | 2,809 | 47,048 | 8.0 | 3,494 | 1,218 | 14,840 | 244,099 | 10.2 | 208,576 | 9,811 | 2,900 |
| Wyoming | 445 | 4,791 | 20.6 | 384 | 107 | 1,839 | 27,248 | -1.2 | 19,072 | 1,014 | 315 |

# Other Services, Except Public Administration and Government Employment

| STATE | Other services, except public administration, 2017 | | | | | | | | Government employment, 2019 | | |
|---|---|---|---|---|---|---|---|---|---|---|---|
| | Number of establishments | Employees | | Repair and mainte-nance | Personal and laundry services | Religious, civic, and similar services | Sales, values of shipments or revenue (mil dol) | Annual payroll (mil dol) | Federal civilian | Federal military | State and local |
| | | Total | Percent change, 2012–2,017 | | | | | | | | |
| | 301 | 302 | 303 | 304 | 305 | 306 | 307 | 308 | 309 | 310 | 311 |
| United States | 560,845 | 3,696,831 | 7.8 | 1,285,621 | 1,473,881 | 937,329 | 544,128 | 133,751 | 2,879,000 | 1,946,000 | 19,911,000 |
| Alabama | 6,149 | 38,476 | 3.2 | 18,544 | 14,572 | 5,360 | 5,844 | 1,303 | 54,134 | 29,000 | 325,987 |
| Alaska | 1,357 | 7,271 | 0.5 | 2,649 | 2,267 | 2,355 | 1,351 | 270 | 14,866 | 25,806 | 61,699 |
| Arizona | 9,271 | 68,350 | 10.1 | 25,735 | 26,959 | 15,656 | 8,727 | 2,306 | 56,554 | 34,735 | 367,505 |
| Arkansas | 3,910 | 21,132 | -3.2 | 8,685 | 7,729 | 4,718 | 2,874 | 655 | 20,352 | 15,022 | 190,156 |
| California | 62,302 | 437,372 | 10.5 | 154,783 | 180,043 | 102,546 | 64,254 | 15,638 | 248,689 | 204,524 | 2,384,397 |
| Colorado | 11,667 | 74,317 | 15.4 | 25,318 | 28,209 | 20,790 | 10,974 | 2,785 | 53,124 | 53,846 | 403,856 |
| Connecticut | 7,507 | 46,490 | 7.2 | 13,655 | 22,178 | 10,657 | 6,055 | 1,646 | 18,251 | 13,638 | 220,785 |
| Delaware | 1,608 | 10,252 | 4.2 | 3,113 | 4,976 | 2,163 | 1,221 | 339 | 5,751 | 8,681 | 60,601 |
| District of Columbia | 3,502 | 73,756 | 25.0 | 579 | 6,979 | 66,198 | 24,226 | 5,459 | 191,495 | 14,158 | 42,658 |
| Florida | 38,932 | 222,512 | 14.0 | 75,815 | 94,573 | 52,124 | 29,617 | 6,946 | 143,148 | 92,838 | 968,609 |
| Georgia | 14,976 | 97,721 | 9.4 | 39,762 | 40,576 | 17,383 | 14,343 | 3,348 | 103,398 | 93,172 | 587,914 |
| Hawaii | 2,912 | 20,219 | 4.5 | 3,644 | 8,251 | 8,324 | 2,561 | 652 | 34,305 | 55,796 | 92,748 |
| Idaho | 2,749 | 14,016 | 15.0 | 6,719 | 4,691 | 2,606 | 1,574 | 435 | 13,234 | 9,033 | 114,765 |
| Illinois | 24,296 | 171,261 | 3.6 | 58,442 | 61,347 | 51,472 | 29,599 | 7,164 | 79,611 | 44,379 | 751,365 |
| Indiana | 10,777 | 75,446 | 3.4 | 31,173 | 25,609 | 18,664 | 11,121 | 2,562 | 39,033 | 20,441 | 388,680 |
| Iowa | 5,975 | 31,306 | 2.1 | 12,788 | 11,621 | 6,897 | 4,186 | 1,022 | 17,645 | 11,391 | 242,922 |
| Kansas | 5,069 | 27,987 | -5.2 | 10,906 | 10,652 | 6,429 | 3,986 | 928 | 25,536 | 32,232 | 241,779 |
| Kentucky | 5,900 | 38,040 | -0.8 | 16,045 | 15,291 | 6,704 | 4,755 | 1,198 | 34,918 | 45,355 | 273,713 |
| Louisiana | 6,531 | 43,472 | -0.1 | 20,396 | 15,461 | 7,615 | 5,702 | 1,572 | 31,692 | 33,774 | 293,551 |
| Maine | 2,921 | 14,477 | 5.2 | 5,355 | 4,523 | 4,599 | 1,809 | 475 | 16,090 | 6,807 | 85,870 |
| Maryland | 10,355 | 81,469 | 2.6 | 25,123 | 32,874 | 23,472 | 12,118 | 3,519 | 174,698 | 50,680 | 347,487 |
| Massachusetts | 14,810 | 100,088 | 7.1 | 26,498 | 47,503 | 26,087 | 13,094 | 3,606 | 46,226 | 19,363 | 402,690 |
| Michigan | 16,545 | 104,291 | 8.5 | 42,688 | 40,937 | 20,666 | 13,205 | 3,415 | 52,384 | 17,693 | 548,392 |
| Minnesota | 11,339 | 77,400 | 6.4 | 24,362 | 31,434 | 21,604 | 10,077 | 2,554 | 32,187 | 19,829 | 377,623 |
| Mississippi | 3,417 | 18,345 | -4.6 | 8,375 | 6,453 | 3,517 | 2,279 | 603 | 25,562 | 27,625 | 218,148 |
| Missouri | 10,513 | 68,206 | 8.6 | 28,749 | 25,796 | 13,661 | 8,694 | 2,297 | 59,184 | 35,892 | 377,551 |
| Montana | 2,409 | 12,077 | 10.6 | 5,305 | 2,920 | 3,852 | 1,609 | 403 | 13,368 | 7,867 | 74,900 |
| Nebraska | 4,107 | 21,757 | -0.3 | 8,606 | 7,848 | 5,303 | 3,608 | 701 | 17,013 | 12,677 | 146,014 |
| Nevada | 4,032 | 28,825 | 13.5 | 10,478 | 13,898 | 4,449 | 3,281 | 914 | 19,777 | 18,932 | 141,472 |
| New Hampshire | 3,011 | 17,864 | 7.6 | 5,908 | 7,533 | 4,423 | 2,096 | 619 | 7,953 | 4,560 | 81,952 |
| New Jersey | 19,162 | 113,846 | 5.2 | 33,372 | 62,881 | 17,593 | 14,026 | 3,679 | 48,566 | 25,097 | 539,901 |
| New Mexico | 2,963 | 17,547 | 0.5 | 7,616 | 5,413 | 4,518 | 2,085 | 558 | 29,082 | 17,874 | 160,992 |
| New York | 48,436 | 299,209 | 10.1 | 60,170 | 124,431 | 114,608 | 51,750 | 11,846 | 115,237 | 55,853 | 1,333,389 |
| North Carolina | 15,118 | 93,642 | 16.0 | 38,832 | 36,439 | 18,371 | 13,022 | 3,153 | 74,376 | 127,121 | 663,871 |
| North Dakota | 1,763 | 9,706 | 5.1 | 3,931 | 3,215 | 2,560 | 1,246 | 318 | 9,320 | 11,872 | 67,818 |
| Ohio | 18,425 | 126,378 | -0.8 | 44,156 | 54,558 | 27,664 | 15,268 | 4,044 | 49,767 | 34,740 | 698,228 |
| Oklahoma | 5,565 | 31,947 | -1.4 | 13,379 | 12,296 | 6,272 | 4,934 | 1,084 | 28,548 | 11,388 | 289,272 |
| Oregon | 7,414 | 43,543 | 14.8 | 17,324 | 15,210 | 11,009 | 6,289 | 1,555 | 98,276 | 34,958 | 256,378 |
| Pennsylvania | 26,075 | 161,337 | 4.5 | 51,547 | 69,740 | 40,050 | 22,982 | 5,137 | 98,276 | 34,958 | 639,029 |
| Rhode Island | 2,318 | 13,882 | 6.4 | 4,509 | 6,091 | 3,282 | 1,773 | 471 | 11,193 | 6,734 | 56,329 |
| South Carolina | 7,068 | 46,847 | 5.6 | 19,764 | 17,125 | 9,958 | 5,653 | 1,529 | 35,084 | 53,366 | 328,637 |
| South Dakota | 1,860 | 8,713 | 4.1 | 3,855 | 2,767 | 2,091 | 1,224 | 278 | 11,316 | 8,216 | 67,551 |
| Tennessee | 8,570 | 61,744 | 10.3 | 23,225 | 26,719 | 11,800 | 8,650 | 2,149 | 50,594 | 20,508 | 382,946 |
| Texas | 37,506 | 285,777 | 10.3 | 134,543 | 107,153 | 44,081 | 39,008 | 10,468 | 204,813 | 178,270 | 1,712,188 |
| Utah | 4,763 | 29,903 | 14.9 | 13,638 | 11,151 | 5,114 | 3,527 | 957 | 37,303 | 16,661 | 211,410 |
| Vermont | 1,603 | 7,339 | 1.8 | 2,312 | 2,073 | 2,954 | 922 | 244 | 7,158 | 3,953 | 46,951 |
| Virginia | 15,700 | 118,537 | 5.4 | 35,053 | 44,371 | 39,113 | 20,344 | 5,188 | 201,455 | 138,183 | 553,258 |
| Washington | 13,444 | 78,480 | 12.2 | 26,178 | 34,251 | 18,051 | 14,514 | 3,005 | 75,860 | 75,113 | 514,147 |
| West Virginia | 2,524 | 14,894 | -10.2 | 6,234 | 4,989 | 3,671 | 1,897 | 465 | 24,116 | 8,508 | 120,883 |
| Wisconsin | 10,351 | 63,120 | 1.7 | 22,858 | 27,547 | 12,715 | 9,241 | 2,069 | 29,389 | 15,890 | 390,441 |
| Wyoming | 1,368 | 6,245 | -6.2 | 2,927 | 1,758 | 1,560 | 931 | 221 | 7,579 | 6,137 | 61,592 |

# Table A. States — State Government Employment and Payroll

| | State government employment and payroll, 2019 | | | | | | | | | | | |
| STATE | State government employment, 2019 | | | State government payroll, 2019 | | Full-time equivalent payroll (1,000 dollars) | | | | | | |
| | | | | | | | Percent of total for: | | | | | |
| | Full-time equivalent employees | Full-time employees | Part-time employees | Full-time March payroll (1,000 dollars) | Part-time March payroll (1,000 dollars) | Total March payroll (1,000 dollars)¹ | Administration | Judicial and legal | Police | Corrections | Highways and transportation | Welfare |
| | 312 | 313 | 314 | 315 | 316 | 317 | 318 | 319 | 320 | 321 | 322 | 323 |
| United States | 4,459,084 | 3,867,028 | 1,611,064 | 22,248,831,840 | 2,460,536,377 | 24,709,368 | 5.3 | 4.8 | 3.2 | 3.2 | 5.0 | 4.8 |
| Alabama | 94,644 | 82,063 | 35,345 | 413,997,906 | 41,402,363 | 455,400 | 3.2 | 3.4 | 1.9 | 1.9 | 4.2 | 3.8 |
| Alaska | 24,457 | 22,395 | 5,662 | 138,985,134 | 8,891,648 | 147,877 | 7.6 | 7.5 | 3.6 | 3.6 | 13.4 | 6.5 |
| Arizona | 74,127 | 61,539 | 33,146 | 325,688,617 | 36,244,033 | 361,933 | 3.8 | 3.6 | 3.2 | 3.2 | 3.4 | 7.1 |
| Arkansas | 62,409 | 56,943 | 16,143 | 253,402,002 | 16,670,269 | 270,072 | 5.2 | 2.7 | 2.3 | 2.3 | 5.2 | 6.1 |
| California | 439,267 | 364,283 | 191,063 | 2,873,703,847 | 464,914,727 | 3,338,619 | 6.2 | 1.7 | 4.1 | 4.1 | 6.1 | 1.7 |
| Colorado | 94,421 | 70,367 | 45,025 | 446,155,273 | 68,985,773 | 515,141 | 4.6 | 6.7 | 2.2 | 2.2 | 3.4 | 2.5 |
| Connecticut | 60,627 | 50,354 | 26,232 | 340,011,440 | 43,700,286 | 383,712 | 5.6 | 10.7 | 4.5 | 4.5 | 5.4 | 9.3 |
| Delaware | 25,447 | 22,444 | 7,622 | 114,842,477 | 17,562,228 | 132,405 | 4.1 | 7.8 | 7.1 | 7.1 | 5.0 | 4.3 |
| District of Columbia | X | X | X | X | X | X | X | X | X | X | X | X |
| Florida | 182,112 | 162,738 | 50,774 | 772,154,002 | 68,424,583 | 840,579 | 5.2 | 11.0 | 2.3 | 2.3 | 3.8 | 3.8 |
| Georgia | 131,128 | 115,230 | 53,087 | 543,364,621 | 54,076,399 | 597,441 | 3.6 | 3.3 | 2.2 | 2.2 | 2.7 | 3.9 |
| Hawaii | 56,571 | 51,031 | 18,274 | 269,022,845 | 23,763,184 | 292,786 | 3.0 | 5.3 | 0.0 | 0.0 | 1.6 | 0.7 |
| Idaho | 24,893 | 21,400 | 10,775 | 123,486,425 | 11,280,406 | 134,767 | 7.4 | 4.8 | 2.4 | 2.4 | 5.0 | 6.2 |
| Illinois | 123,673 | 103,391 | 49,982 | 652,359,234 | 92,078,187 | 744,437 | 5.6 | 4.6 | 3.4 | 3.4 | 6.0 | 9.3 |
| Indiana | 97,137 | 83,032 | 44,196 | 384,164,299 | 38,636,558 | 422,801 | 4.3 | 3.1 | 2.7 | 2.7 | 3.8 | 6.5 |
| Iowa | 51,032 | 40,995 | 29,343 | 278,887,106 | 24,891,356 | 303,778 | 3.1 | 4.6 | 2.2 | 2.2 | 4.5 | 4.4 |
| Kansas | 54,456 | 48,518 | 18,755 | 241,148,934 | 20,736,005 | 261,885 | 3.8 | 4.1 | 2.3 | 2.3 | 3.9 | 3.3 |
| Kentucky | 81,509 | 70,887 | 24,200 | 344,614,097 | 26,837,651 | 371,452 | 4.0 | 5.5 | 2.5 | 2.5 | 4.5 | 6.7 |
| Louisiana | 76,689 | 68,490 | 22,849 | 334,479,879 | 24,661,050 | 359,141 | 6.9 | 2.7 | 4.0 | 4.0 | 5.7 | 6.5 |
| Maine | 20,776 | 17,826 | 8,667 | 85,879,646 | 8,908,705 | 94,788 | 8.3 | 5.1 | 4.0 | 4.0 | 10.7 | 9.9 |
| Maryland | 84,213 | 75,074 | 14,052 | 423,564,352 | 46,192,351 | 469,757 | 5.3 | 6.3 | 3.9 | 3.9 | 5.9 | 6.4 |
| Massachusetts | 105,911 | 96,017 | 33,930 | 627,474,280 | 50,169,718 | 677,644 | 5.3 | 10.1 | 5.0 | 5.0 | 2.9 | 7.6 |
| Michigan | 150,483 | 120,317 | 73,998 | 753,397,772 | 143,624,566 | 897,022 | 4.4 | 1.5 | 2.0 | 2.0 | 1.9 | 7.4 |
| Minnesota | 84,449 | 71,299 | 33,150 | 453,148,346 | 55,881,835 | 509,030 | 9.0 | 5.4 | 1.3 | 1.3 | 6.3 | 2.6 |
| Mississippi | 53,994 | 49,123 | 13,150 | 219,232,176 | 15,538,110 | 234,770 | 3.7 | 1.2 | 2.2 | 2.2 | 4.6 | 5.6 |
| Missouri | 84,702 | 75,321 | 27,983 | 316,623,171 | 28,241,244 | 344,864 | 4.4 | 5.5 | 3.6 | 3.6 | 6.0 | 5.8 |
| Montana | 24,648 | 20,515 | 12,421 | 88,382,366 | 13,498,238 | 101,881 | 6.6 | 5.0 | 2.6 | 2.6 | 11.5 | 6.1 |
| Nebraska | 35,201 | 28,892 | 13,070 | 144,914,397 | 13,659,092 | 158,573 | 2.9 | 3.2 | 2.6 | 2.6 | 5.8 | 6.0 |
| Nevada | 30,167 | 26,980 | 9,988 | 154,171,506 | 13,078,469 | 167,250 | 10.5 | 3.3 | 3.6 | 3.6 | 5.7 | 6.8 |
| New Hampshire | 18,873 | 15,143 | 10,392 | 88,082,212 | 13,418,392 | 101,501 | 6.7 | 4.7 | 3.6 | 3.6 | 8.8 | 10.7 |
| New Jersey | 145,736 | 132,425 | 31,500 | 842,223,433 | 59,591,331 | 901,815 | 4.3 | 9.7 | 3.8 | 3.8 | 3.9 | 6.5 |
| New Mexico | 46,585 | 41,691 | 13,742 | 206,723,186 | 20,130,632 | 226,854 | 3.8 | 8.3 | 1.7 | 1.7 | 4.2 | 3.1 |
| New York | 252,429 | 236,307 | 43,676 | 1,543,878,430 | 83,163,643 | 1,627,042 | 7.9 | 10.1 | 3.8 | 3.8 | 4.2 | 1.6 |
| North Carolina | 149,437 | 132,927 | 44,791 | 686,734,362 | 55,402,055 | 742,136 | 4.0 | 4.9 | 2.9 | 2.9 | 5.9 | 0.7 |
| North Dakota | 18,259 | 15,517 | 8,940 | 80,896,674 | 10,731,056 | 91,628 | 5.4 | 4.8 | 1.4 | 1.4 | 6.9 | 3.5 |
| Ohio | 135,613 | 108,563 | 72,490 | 632,944,922 | 86,672,212 | 719,617 | 6.7 | 3.0 | 2.5 | 2.5 | 4.9 | 2.6 |
| Oklahoma | 64,235 | 56,938 | 23,944 | 263,739,633 | 22,814,069 | 286,554 | 4.6 | 4.7 | 4.0 | 4.0 | 4.7 | 8.8 |
| Oregon | 74,268 | 65,073 | 27,349 | 378,614,977 | 54,905,862 | 433,521 | 8.3 | 5.7 | 2.6 | 2.6 | 6.1 | 6.3 |
| Pennsylvania | 154,995 | 140,075 | 48,031 | 837,535,190 | 92,100,726 | 929,636 | 4.9 | 3.8 | 5.3 | 5.3 | 7.8 | 5.9 |
| Rhode Island | 19,068 | 17,569 | 7,093 | 110,314,976 | 8,067,422 | 118,382 | 7.8 | 7.0 | 2.7 | 2.7 | 4.5 | 8.3 |
| South Carolina | 84,751 | 76,168 | 22,676 | 341,361,630 | 26,108,419 | 367,470 | 4.7 | 1.5 | 2.5 | 2.5 | 4.7 | 5.1 |
| South Dakota | 14,281 | 12,474 | 6,285 | 59,123,241 | 5,836,172 | 64,959 | 6.4 | 6.3 | 2.8 | 2.8 | 8.5 | 10.5 |
| Tennessee | 79,800 | 71,174 | 27,010 | 347,720,064 | 26,667,009 | 374,387 | 7.4 | 4.8 | 2.7 | 2.7 | 5.3 | 8.6 |
| Texas | 314,952 | 280,095 | 90,619 | 1,540,446,536 | 129,677,689 | 1,670,124 | 3.5 | 2.2 | 3.4 | 3.4 | 4.0 | 6.2 |
| Utah | 63,247 | 53,939 | 29,901 | 301,401,427 | 35,991,946 | 337,393 | 5.0 | 3.0 | 1.4 | 1.4 | 2.7 | 3.6 |
| Vermont | 13,932 | 12,667 | 4,317 | 74,089,341 | 8,536,849 | 82,626 | 9.7 | 5.3 | 5.6 | 5.6 | 7.5 | 11.1 |
| Virginia | 126,239 | 106,343 | 61,567 | 585,974,706 | 77,165,482 | 663,140 | 4.1 | 3.5 | 2.9 | 2.9 | 6.7 | 2.5 |
| Washington | 129,855 | 109,790 | 52,894 | 669,958,622 | 106,391,102 | 776,350 | 3.4 | 2.2 | 1.9 | 1.9 | 5.8 | 8.9 |
| West Virginia | 38,877 | 35,257 | 11,785 | 153,938,852 | 11,999,806 | 165,939 | 5.0 | 5.1 | 2.9 | 2.9 | 12.9 | 6.4 |
| Wisconsin | 71,640 | 57,765 | 45,714 | 330,993,341 | 49,257,444 | 380,251 | 4.5 | 4.6 | 1.3 | 1.3 | 2.4 | 3.1 |
| Wyoming | 12,869 | 11,664 | 3,466 | 54,879,935 | 3,358,025 | 58,238 | 7.7 | 6.1 | 2.5 | 2.5 | 12.7 | 4.0 |

1. Includes program categories not shown separately.

# Table A. States — State Government Finances and Employment and Payroll

| | State government employment and payroll, 2019 (cont.) | | | | | | | State government finances, 2019 | | | | | | | |
| | Full-time equivalent payroll (1,000 dollars) (cont.) | | | | | | | General revenue (mil dol) | | | | | | | |
| | Percent of total for: | | | | | | | | From federal government | | From own sources | | | Taxes per capita[1] (dollars) | |
| STATE | Health | Hospitals | Social Insurance administration | Natural resources and parks | Utilities, sewerge, and waste management | Elementary and secondary education and libraries | Higher education | Total | Total | Per capita[1] (dollars) | Total | Taxes Total | Sales and gross receipts | Total | Sales and gross receipts |
| | 324 | 325 | 326 | 327 | 328 | 329 | 330 | 331 | 332 | 333 | 334 | 335 | 336 | 337 | 338 |
| United States ............... | 4.3 | 8.7 | 1.4 | 3.3 | 0.1 | 1.1 | 40.4 | X | X | X | X | X | X | X | X |
| Alabama............... | 7.4 | 14.0 | 0.9 | 2.2 | - | 0.0 | 47.8 | 35,488 | 11,279 | 2,300 | 18,918,472 | 11,577 | 5,619 | 2,361 | 1,146 |
| Alaska............... | 3.1 | 1.0 | 0.9 | 9.2 | - | 9.0 | 17.8 | 10,478 | 3,725 | 5,092 | 4,975,188 | 1,781 | 279 | 2,434 | 381 |
| Arizona............... | 3.4 | 0.8 | 2.3 | 2.3 | - | 0.0 | 52.9 | 49,668 | 16,861 | 2,317 | 24,410,837 | 18,164 | 10,444 | 2,495 | 1,435 |
| Arkansas............... | 5.2 | 10.5 | 1.8 | 4.0 | - | 0.1 | 44.7 | 25,034 | 7,897 | 2,617 | 14,265,468 | 10,218 | 4,920 | 3,386 | 1,630 |
| California............... | 3.8 | 10.0 | 2.6 | 4.2 | 0.2 | 0.0 | 36.8 | 436,479 | 102,638 | 2,598 | 233,905,211 | 188,235 | 60,234 | 4,764 | 1,524 |
| Colorado............... | 1.9 | 4.2 | 1.4 | 2.6 | - | 0.0 | 60.4 | 34,897 | 9,087 | 1,578 | 22,919,087 | 15,870 | 5,989 | 2,756 | 1,040 |
| Connecticut............... | 5.4 | 9.1 | 0.9 | 1.5 | 0.1 | 0.0 | 27.3 | 38,248 | 8,786 | 2,464 | 22,220,669 | 17,994 | 7,788 | 5,047 | 2,184 |
| Delaware............... | 6.2 | 3.4 | 0.8 | 2.7 | 0.4 | 0.0 | 35.3 | 10,746 | 2,584 | 2,654 | 7,167,298 | 4,596 | 603 | 4,719 | 619 |
| District of Columbia........... | X | X | X | X | X | X | X | X | X | X | X | X | X | X | X |
| Florida............... | 7.2 | 1.8 | 1.1 | 4.1 | - | 0.0 | 45.3 | 106,810 | 29,278 | 1,363 | 63,386,613 | 44,800 | 36,492 | 2,086 | 1,699 |
| Georgia............... | 3.1 | 6.0 | 0.6 | 4.6 | - | 0.0 | 55.0 | 58,717 | 15,789 | 1,487 | 31,559,127 | 24,713 | 9,509 | 2,328 | 896 |
| Hawaii............... | 4.2 | 6.4 | 0.3 | 1.9 | - | 41.8 | 20.2 | 16,686 | 2,887 | 2,039 | 11,430,729 | 8,208 | 5,062 | 5,797 | 3,575 |
| Idaho............... | 7.9 | 1.7 | 2.6 | 9.2 | - | 0.0 | 31.5 | 12,009 | 3,106 | 1,738 | 6,299,521 | 4,884 | 2,534 | 2,733 | 1,418 |
| Illinois............... | 2.2 | 9.2 | 1.2 | 1.9 | - | 0.0 | 34.7 | 91,966 | 20,972 | 1,655 | 53,393,758 | 42,501 | 19,534 | 3,354 | 1,542 |
| Indiana............... | 1.9 | 1.4 | 1.0 | 2.2 | - | 0.0 | 63.9 | 47,718 | 15,618 | 2,320 | 27,099,713 | 20,171 | 12,579 | 2,996 | 1,869 |
| Iowa............... | 0.9 | 17.0 | 1.0 | 3.0 | - | 0.0 | 47.7 | 31,009 | 6,382 | 2,023 | 19,513,384 | 10,584 | 4,858 | 3,355 | 1,540 |
| Kansas............... | 2.0 | 22.7 | 0.1 | 2.2 | - | 0.0 | 46.0 | 23,339 | 4,358 | 1,496 | 15,775,131 | 10,030 | 4,524 | 3,443 | 1,553 |
| Kentucky............... | 4.8 | 11.3 | 0.7 | 2.9 | - | 0.0 | 48.1 | 36,757 | 12,259 | 2,744 | 18,417,344 | 12,896 | 6,311 | 2,886 | 1,413 |
| Louisiana............... | 5.8 | 13.7 | 1.1 | 5.8 | - | 0.1 | 32.4 | 36,750 | 14,174 | 3,049 | 16,361,369 | 11,749 | 6,400 | 2,527 | 1,377 |
| Maine............... | 5.1 | 2.7 | 1.6 | 6.4 | - | 0.2 | 30.3 | 11,284 | 3,116 | 2,318 | 6,116,004 | 4,674 | 2,328 | 3,477 | 1,732 |
| Maryland............... | 6.6 | 4.0 | 0.8 | 2.9 | - | 0.0 | 32.4 | 51,389 | 13,456 | 2,226 | 30,430,036 | 23,606 | 9,896 | 3,905 | 1,637 |
| Massachusetts............... | 6.4 | 4.5 | 1.0 | 1.9 | 1.0 | 1.9 | 26.3 | 71,719 | 17,754 | 2,576 | 43,580,933 | 31,805 | 9,558 | 4,614 | 1,387 |
| Michigan............... | 3.9 | 11.1 | 0.5 | 2.4 | - | 0.0 | 52.0 | 57,264 | 11,803 | 2,093 | 33,913,296 | 28,176 | 11,170 | 4,996 | 1,981 |
| Minnesota............... | 4.3 | 5.2 | 1.1 | 4.3 | 1.0 | 0.0 | 44.7 | 23,697 | 8,330 | 2,799 | 11,401,455 | 8,289 | 5,233 | 2,785 | 1,758 |
| Mississippi............... | 11.2 | 16.5 | 0.9 | 4.5 | - | 0.0 | 39.5 | 23,339 | 12,012 | 1,957 | 19,336,866 | 13,181 | 5,549 | 2,148 | 904 |
| Missouri............... | 3.2 | 12.6 | 0.3 | 2.8 | - | 0.0 | 39.9 | 39,519 | 3,226 | 3,019 | 4,152,074 | 3,169 | 663 | 2,965 | 620 |
| Montana............... | 4.2 | 2.5 | 3.7 | 7.5 | - | 0.0 | 35.2 | 9,071 | 3,308 | 1,710 | 7,550,143 | 5,755 | 2,580 | 2,975 | 1,334 |
| Nebraska............... | 1.7 | 5.7 | 0.6 | 2.9 | - | 0.0 | 53.1 | 12,291 | 5,416 | 1,758 | 11,373,566 | 9,745 | 7,836 | 3,164 | 2,544 |
| Nevada............... | 4.9 | 4.8 | 1.2 | 3.6 | - | 0.0 | 35.4 | 23,008 | 2,822 | 2,075 | 4,750,850 | 2,969 | 994 | 2,184 | 731 |
| New Hampshire............... | 5.0 | 3.2 | 1.2 | 3.0 | - | 0.0 | 34.8 | 9,525 | 2,822 | 2,075 | 4,750,850 | 2,969 | 994 | 2,184 | 731 |
| New Jersey............... | 2.3 | 7.8 | 0.9 | 2.1 | 0.4 | 11.7 | 26.6 | 86,720 | 18,663 | 2,101 | 51,800,791 | 38,844 | 16,043 | 4,373 | 1,806 |
| New Mexico............... | 6.6 | 17.8 | 0.6 | 3.1 | - | 0.0 | 37.1 | 24,262 | 7,530 | 3,591 | 13,039,448 | 7,428 | 3,684 | 3,542 | 1,757 |
| New York............... | 3.3 | 15.2 | 2.8 | 2.0 | - | 0.0 | 19.8 | 224,820 | 67,090 | 3,449 | 117,008,537 | 91,621 | 27,231 | 4,710 | 1,400 |
| North Carolina............... | 1.1 | 16.3 | 0.7 | 3.0 | 0.1 | 0.0 | 44.8 | 71,552 | 19,608 | 1,870 | 39,917,035 | 29,316 | 12,804 | 2,795 | 1,221 |
| North Dakota............... | 10.3 | 3.9 | 0.9 | 5.9 | - | 0.0 | 40.5 | 9,521 | 1,795 | 2,356 | 6,713,278 | 4,970 | 1,564 | 6,521 | 2,052 |
| Ohio............... | 2.7 | 10.4 | 1.2 | 2.2 | - | 0.0 | 47.3 | 87,537 | 25,201 | 2,156 | 46,485,555 | 30,147 | 18,529 | 2,579 | 1,585 |
| Oklahoma............... | 7.7 | 1.5 | 1.5 | 2.9 | 0.2 | 0.0 | 44.5 | 28,409 | 11,057 | 1,869 | 23,788,074 | 10,732 | 4,644 | 2,712 | 1,174 |
| Oregon............... | 2.7 | 10.3 | 4.4 | 4.7 | - | 0.0 | 36.7 | 44,854 | 11,057 | 2,622 | 23,788,074 | 13,960 | 4,644 | 3,310 | 442 |
| Pennsylvania............... | 2.9 | 5.6 | 1.0 | 3.5 | - | 0.0 | 39.4 | 109,493 | 31,474 | 2,459 | 61,859,269 | 43,132 | 22,140 | 3,369 | 1,729 |
| Rhode Island............... | 5.4 | 3.7 | 1.9 | 2.3 | 2.0 | 2.3 | 23.0 | 10,460 | 2,932 | 2,768 | 5,824,995 | 3,724 | 1,831 | 3,515 | 1,728 |
| South Carolina............... | 5.9 | 7.0 | 1.0 | 2.8 | 0.1 | 0.0 | 45.5 | 36,196 | 9,617 | 1,868 | 19,528,955 | 11,221 | 5,241 | 2,179 | 1,018 |
| South Dakota............... | 5.2 | 1.7 | 1.3 | 6.5 | - | 0.0 | 36.8 | 5,370 | 1,525 | 1,724 | 2,914,919 | 1,940 | 1,615 | 2,193 | 1,825 |
| Tennessee............... | 5.9 | 3.5 | 0.6 | 5.4 | - | 0.0 | 43.0 | 40,248 | 11,934 | 1,747 | 19,133,321 | 14,827 | 10,729 | 2,171 | 1,571 |
| Texas............... | 6.3 | 6.8 | 1.2 | 3.5 | - | 0.0 | 50.4 | 168,759 | 50,031 | 1,725 | 96,294,729 | 63,330 | 53,905 | 2,184 | 1,859 |
| Utah............... | 3.2 | 16.5 | 1.3 | 2.2 | - | 0.0 | 51.7 | 23,038 | 4,779 | 1,491 | 16,501,720 | 9,968 | 4,057 | 3,109 | 1,266 |
| Vermont............... | 3.7 | 1.7 | 1.5 | 5.6 | - | 0.0 | 29.6 | 7,221 | 2,008 | 3,218 | 4,434,978 | 3,429 | 1,123 | 5,495 | 1,800 |
| Virginia............... | 5.6 | 9.3 | 0.8 | 2.9 | - | 0.0 | 47.1 | 65,939 | 12,218 | 1,431 | 43,305,047 | 26,286 | 8,596 | 3,080 | 1,007 |
| Washington............... | 5.8 | 9.9 | 2.8 | 4.1 | 0.4 | 0.0 | 43.0 | 68,489 | 13,963 | 1,834 | 38,141,983 | 27,992 | 21,374 | 3,676 | 2,807 |
| West Virginia............... | 1.6 | 2.9 | 1.1 | 4.9 | - | 0.0 | 39.9 | 15,955 | 5,168 | 2,884 | 8,632,690 | 5,938 | 2,919 | 3,313 | 1,629 |
| Wisconsin............... | 2.4 | 4.7 | 1.1 | 3.4 | - | 0.0 | 51.4 | 37,993 | 10,515 | 1,806 | 28,329,861 | 20,039 | 8,446 | 3,442 | 1,451 |
| Wyoming............... | 7.2 | 4.1 | 0.9 | 9.6 | - | 0.0 | 26.2 | 6,131 | 2,291 | 3,959 | 3,234,085 | 2,111 | 954 | 3,647 | 1,648 |

1. Based on resident population estimated as of July 1 of the year shown.   2. Copyright Election Data Services.

# Table A. States — **State Government Finances and Voting**

| STATE | State government finances, 2019 (cont.) General expenditures (mil dol) (cont.) Total | To local govern- ments | Direct expenditures Total | Per capita[1] (dollars) | By selected function Educa- tion | Health and hospitals | Highways | Public safety | Public welfare | Natural resources, parks, and recreation | Debt outstanding Total (mil dol) | Per capita[1] | Voting and registra- tion, November 2020 Percent registered | Percent voted | Presidential election, 2020 (percent of vote cast) Demo- cratic | Repub- lican | All other |
|---|---|---|---|---|---|---|---|---|---|---|---|---|---|---|---|---|---|
| | 339 | 340 | 341 | 342 | 343 | 344 | 345 | 346 | 347 | 348 | 349 | 350 | 351 | 352 | 353 | 354 | 355 |
| United States .......... | X | X | X | X | X | X | X | X | X | X | X | X | 72.7 | 66.8 | 51.0 | 46.6 | 2.3 |
| Alabama.............. | 30,621 | 7,604 | 26,817 | 5,469 | 12,597 | 3,790 | 1,971 | 873 | 8,050 | 300 | 10,098 | 2,059 | 68.0 | 60.5 | 36.6 | 62.0 | 1.4 |
| Alaska................ | 9,977 | 1,806 | 9,719 | 13,286 | 2,506 | 246 | 1,093 | 489 | 2,933 | 449 | 6,061 | 8,285 | 74.2 | 63.8 | 42.8 | 52.8 | 4.4 |
| Arizona............... | 37,843 | 10,584 | 31,607 | 4,342 | 13,134 | 591 | 2,271 | 1,410 | 16,055 | 435 | 13,616 | 1,871 | 76.4 | 71.9 | 49.4 | 49.1 | 1.6 |
| Arkansas............. | 21,190 | 5,507 | 17,833 | 5,909 | 8,076 | 1,354 | 1,480 | 467 | 7,753 | 350 | 7,184 | 2,381 | 62.0 | 54.0 | 34.8 | 62.4 | 2.8 |
| California............. | 337,836 | 115,185 | 280,538 | 7,100 | 111,656 | 23,010 | 13,558 | 10,722 | 143,428 | 7,328 | 145,293 | 3,677 | 69.4 | 65.1 | 63.5 | 34.3 | 2.2 |
| Colorado ............. | 32,011 | 8,175 | 29,569 | 5,135 | 12,723 | 1,807 | 1,946 | 1,413 | 9,704 | 525 | 19,241 | 3,341 | 71.3 | 67.6 | 55.4 | 41.9 | 2.7 |
| Connecticut.......... | 23,223 | 5,982 | 22,787 | 6,391 | 7,474 | 2,547 | 1,564 | 913 | 4,877 | 272 | 41,822 | 11,730 | 73.3 | 66.6 | 59.3 | 39.2 | 1.6 |
| Delaware............. | 9,639 | 1,710 | 8,860 | 9,098 | 3,606 | 548 | 651 | 513 | 2,653 | 134 | 4,964 | 5,098 | 75.1 | 67.7 | 58.7 | 39.8 | 1.5 |
| District of Columbia........... | X | X | X | X | X | X | X | X | X | X | X | X | 86.9 | 84.0 | 92.6 | 5.4 | 2.0 |
| Florida................ | 92,058 | 19,186 | 83,905 | 3,907 | 30,051 | 6,478 | 9,122 | 3,436 | 27,686 | 1,964 | 26,032 | 1,212 | 67.1 | 62.1 | 47.9 | 51.2 | 0.9 |
| Georgia ............... | 46,186 | 13,399 | 39,959 | 3,764 | 20,918 | 2,785 | 2,850 | 1,806 | 13,596 | 828 | 13,624 | 1,283 | 70.7 | 66.1 | 49.5 | 49.2 | 1.3 |
| Hawaii................ | 12,617 | 329 | 13,918 | 9,830 | 3,790 | 1,148 | 555 | 290 | 2,872 | 211 | 10,002 | 7,064 | 68.7 | 64.3 | 64.2 | 34.5 | 1.3 |
| Idaho................. | 9,770 | 2,708 | 8,445 | 4,725 | 3,511 | 237 | 927 | 391 | 2,882 | 354 | 3,381 | 1,892 | 69.3 | 64.9 | 33.1 | 63.8 | 3.1 |
| Illinois................ | 74,752 | 21,932 | 68,706 | 5,422 | 18,466 | 2,083 | 4,913 | 2,207 | 26,294 | 576 | 65,272 | 5,151 | 74.4 | 68.4 | 57.5 | 40.6 | 1.9 |
| Indiana............... | 40,597 | 10,484 | 32,810 | 4,874 | 16,204 | 748 | 3,013 | 1,023 | 15,043 | 592 | 22,436 | 3,333 | 69.3 | 61.0 | 41.0 | 57.0 | 2.0 |
| Iowa.................. | 21,469 | 5,393 | 19,076 | 6,046 | 7,422 | 2,580 | 2,347 | 404 | 6,654 | 329 | 6,185 | 1,960 | 76.0 | 70.5 | 44.6 | 52.8 | 2.6 |
| Kansas............... | 18,716 | 5,252 | 15,371 | 5,276 | 7,687 | 3,311 | 1,009 | 487 | 4,735 | 268 | 7,418 | 2,546 | 70.8 | 65.7 | 41.6 | 56.2 | 2.2 |
| Kentucky............. | 30,516 | 4,922 | 30,402 | 6,805 | 9,988 | 2,458 | 1,850 | 861 | 11,907 | 579 | 15,347 | 3,435 | 75.9 | 68.5 | 36.2 | 62.1 | 1.8 |
| Louisiana............. | 31,365 | 6,492 | 29,596 | 6,366 | 9,771 | 825 | 1,469 | 1,079 | 12,570 | 1,150 | 17,447 | 3,753 | 69.3 | 61.9 | 39.9 | 58.5 | 1.7 |
| Maine................. | 8,471 | 1,448 | 8,272 | 6,154 | 2,256 | 347 | 734 | 262 | 3,496 | 157 | 4,695 | 3,493 | 77.4 | 71.3 | 52.5 | 43.6 | 3.9 |
| Maryland ............. | 42,271 | 9,549 | 39,050 | 6,459 | 13,439 | 2,977 | 2,864 | 2,219 | 13,811 | 832 | 28,920 | 4,784 | 78.6 | 73.6 | 65.4 | 32.2 | 2.5 |
| Massachusetts....... | 58,584 | 9,908 | 59,366 | 8,613 | 13,405 | 2,442 | 2,433 | 2,208 | 24,819 | 513 | 78,663 | 11,413 | 72.4 | 66.3 | 65.1 | 31.9 | 3.0 |
| Michigan.............. | 71,242 | 23,597 | 56,878 | 5,695 | 27,643 | 7,127 | 3,951 | 2,494 | 21,020 | 648 | 31,964 | 3,201 | 73.8 | 66.9 | 50.6 | 47.8 | 1.5 |
| Minnesota............ | 44,757 | 14,965 | 35,980 | 6,380 | 16,239 | 917 | 3,128 | 1,267 | 16,126 | 1,015 | 17,114 | 3,035 | 82.9 | 77.9 | 52.4 | 45.3 | 2.3 |
| Mississippi........... | 19,209 | 5,176 | 17,186 | 5,775 | 5,906 | 1,968 | 1,129 | 483 | 6,665 | 326 | 7,225 | 2,428 | 80.4 | 70.3 | 41.1 | 57.6 | 1.3 |
| Missouri.............. | 30,465 | 6,381 | 28,915 | 4,711 | 10,157 | 4,602 | 1,468 | 1,061 | 9,652 | 230 | 17,068 | 2,781 | 75.7 | 66.8 | 41.4 | 56.8 | 1.8 |
| Montana.............. | 6,818 | 1,152 | 6,882 | 6,439 | 1,917 | 244 | 747 | 274 | 2,258 | 301 | 2,659 | 2,488 | 77.5 | 73.5 | 40.6 | 56.9 | 2.5 |
| Nebraska............. | 10,396 | 2,665 | 8,485 | 4,387 | 3,898 | 489 | 812 | 474 | 2,814 | 277 | 2,268 | 1,173 | 70.9 | 65.2 | 39.2 | 58.2 | 2.6 |
| Nevada............... | 14,809 | 4,925 | 12,845 | 4,170 | 5,629 | 566 | 903 | 429 | 4,309 | 179 | 3,307 | 1,074 | 66.2 | 61.5 | 50.1 | 47.7 | 2.3 |
| New Hampshire ................ | 7,644 | 1,765 | 7,285 | 5,358 | 2,305 | 212 | 522 | 204 | 2,647 | 84 | 7,600 | 5,589 | 78.3 | 74.0 | 52.7 | 45.4 | 1.9 |
| New Jersey ......... | 63,184 | 15,644 | 65,024 | 7,321 | 22,685 | 4,755 | 3,912 | 2,423 | 19,618 | 700 | 63,927 | 7,197 | 84.6 | 78.3 | 57.3 | 41.4 | 1.3 |
| New Mexico .......... | 19,688 | 5,351 | 16,806 | 8,015 | 6,299 | 1,640 | 766 | 665 | 6,215 | 296 | 7,310 | 3,486 | 68.6 | 62.6 | 54.3 | 43.5 | 2.2 |
| New York............. | 178,181 | 63,294 | 155,023 | 7,969 | 48,595 | 16,077 | 7,877 | 4,286 | 75,185 | 1,472 | 150,745 | 7,749 | 70.5 | 64.7 | 55.9 | 34.0 | 10.0 |
| North Carolina....... | 58,035 | 15,911 | 48,593 | 4,633 | 22,858 | 4,280 | 6,185 | 2,101 | 15,481 | 1,000 | 14,944 | 1,425 | 69.8 | 64.7 | 48.6 | 49.9 | 1.5 |
| North Dakota......... | 7,037 | 2,265 | 5,456 | 7,160 | 2,450 | 196 | 767 | 143 | 1,566 | 183 | 3,375 | 4,429 | 77.3 | 67.1 | 31.8 | 65.1 | 3.1 |
| Ohio.................. | 71,378 | 18,811 | 70,516 | 6,033 | 22,856 | 6,124 | 4,118 | 2,447 | 28,743 | 547 | 27,966 | 2,392 | 77.0 | 70.1 | 45.2 | 53.3 | 1.5 |
| Oklahoma............ | 21,989 | 4,611 | 20,741 | 5,242 | 8,227 | 1,156 | 2,400 | 806 | 7,051 | 262 | 9,551 | 2,414 | 67.3 | 58.3 | 32.3 | 65.4 | 2.3 |
| Oregon............... | 35,071 | 6,608 | 34,912 | 8,277 | 10,885 | 3,607 | 1,276 | 1,212 | 12,133 | 656 | 15,333 | 3,635 | 79.9 | 74.1 | 56.5 | 40.4 | 3.2 |
| Pennsylvania......... | 94,507 | 22,760 | 85,884 | 6,709 | 28,344 | 9,652 | 9,730 | 3,436 | 33,364 | 1,114 | 48,959 | 3,824 | 76.3 | 70.2 | 50.0 | 48.8 | 1.1 |
| Rhode Island........ | 8,409 | 1,374 | 8,466 | 7,992 | 2,219 | 293 | 522 | 325 | 3,362 | 117 | 9,275 | 8,756 | 74.1 | 66.3 | 59.4 | 38.6 | 2.0 |
| South Carolina ...... | 29,309 | 6,294 | 28,279 | 5,492 | 11,490 | 3,522 | 1,889 | 708 | 8,258 | 362 | 12,851 | 2,496 | 70.0 | 63.4 | 43.4 | 55.1 | 1.5 |
| South Dakota ........ | 4,744 | 906 | 4,437 | 5,016 | 1,627 | 180 | 660 | 178 | 1,193 | 202 | 3,509 | 3,967 | 67.4 | 58.5 | 35.6 | 61.8 | 2.6 |
| Tennessee ........... | 32,366 | 7,784 | 27,414 | 4,014 | 10,992 | 1,141 | 1,868 | 1,203 | 13,178 | 494 | 6,731 | 986 | 74.3 | 66.4 | 37.5 | 60.7 | 1.9 |
| Texas ................ | 145,442 | 32,072 | 131,353 | 4,530 | 58,083 | 11,908 | 17,751 | 5,228 | 41,491 | 1,094 | 53,794 | 1,855 | 71.8 | 63.9 | 46.5 | 52.1 | 1.5 |
| Utah.................. | 20,307 | 4,235 | 18,216 | 5,682 | 9,108 | 2,608 | 1,462 | 542 | 4,152 | 307 | 6,895 | 2,151 | 67.4 | 63.6 | 37.6 | 58.1 | 4.2 |
| Vermont.............. | 6,452 | 1,868 | 5,087 | 8,152 | 2,823 | 413 | 444 | 241 | 1,861 | 137 | 3,490 | 5,594 | 73.0 | 68.4 | 65.5 | 30.4 | 4.2 |
| Virginia............... | 53,170 | 12,751 | 46,566 | 5,456 | 18,237 | 7,165 | 5,318 | 2,358 | 13,263 | 488 | 29,111 | 3,411 | 76.0 | 71.5 | 54.1 | 44.0 | 1.9 |
| Washington.......... | 53,312 | 16,833 | 44,209 | 5,806 | 22,661 | 6,977 | 3,151 | 1,516 | 12,489 | 1,224 | 35,585 | 4,673 | 74.8 | 71.5 | 58.0 | 38.8 | 3.3 |
| West Virginia......... | 13,467 | 2,421 | 12,755 | 7,117 | 4,216 | 449 | 1,450 | 428 | 4,854 | 332 | 10,571 | 5,898 | 67.3 | 56.1 | 30.2 | 69.8 | 0.0 |
| Wisconsin............ | 37,774 | 11,610 | 32,090 | 5,511 | 13,063 | 3,041 | 2,868 | 1,072 | 11,864 | 599 | 22,664 | 3,893 | 76.7 | 73.6 | 49.5 | 48.8 | 1.7 |
| Wyoming............. | 5,225 | 1,320 | 4,839 | 8,362 | 1,955 | 300 | 494 | 182 | 858 | 245 | 883 | 1,525 | 69.3 | 65.5 | 26.4 | 69.5 | 4.1 |

1. Based on resident population estimated as of July 1 of the year shown.

# States and Counties

(For explanation of symbols, see page viii)

Page

| | |
|---|---|
| 53 | County Highlights and Rankings |
| 69 | State and County Column Headings |
| 73 | Table B |
| 73 | **AL**(Autauga)—**AL**(Walker) |
| 87 | **AL**(Washington)—**AR**(Cleburne) |
| 101 | **AR**(Cleveland)—**AR**(Yell) |
| 115 | **CA**(Alameda)—**CO**(Bent) |
| 129 | **CO**(Boulder)—**CT**(New London) |
| 143 | **CT**(Tolland)—**FL**(Polk) |
| 157 | **FL**(Putnam)—**GA**(Echols) |
| 171 | **GA**(Effingham)—**GA**(Pulaski) |
| 185 | **GA**(Putnam)—**ID**(Canyon) |
| 199 | **ID**(Caribou)—**IL**(Hancock) |
| 213 | **IL**(Hardin)—**IL**(Williamson) |
| 227 | **IL**(Winnebago)—**IN**(Perry) |
| 241 | **IN**(Pike)—**IA**(Floyd) |
| 255 | **IA**(Franklin)—**IA**(Wright) |
| 269 | **KS**(Allen)—**KS**(Nemaha) |
| 283 | **KS**(Neosho)—**KY**(Clark) |
| 297 | **KY**(Clay)—**KY**(Nicholas) |
| 311 | **KY**(Ohio)—**LA**(Natchitoches) |
| 325 | **LA**(Orleans)—**MD**(Queen Anne's) |
| 339 | **MD**(St. Mary's)—**MI**(Kent) |
| 353 | **MI**(Keweenaw)—**MN**(Faribault) |
| 367 | **MN**(Fillmore)—**MN**(Yellow Medicine) |
| 381 | **MS**(Adams)—**MS**(Stone) |
| 395 | **MS**(Sunflower)—**MO**(Jackson) |
| 409 | **MO**(Jasper)—**MO**(Wright) |
| 423 | **MO**(St. Louis city)—**NE**(Banner) |
| 437 | **NE**(Blaine)—**NE**(Pierce) |
| 451 | **NE**(Platte)—**NJ**(Hunterdon) |
| 465 | **NJ**(Mercer)—**NY**(Fulton) |
| 479 | **NY**(Genesee)—**NC**(Cherokee) |
| 493 | **NC**(Chowan)—**NC**(Surry) |
| 507 | **NC**(Swain)—**ND**(Walsh) |
| 521 | **ND**(Ward)—**OH**(Noble) |
| 535 | **OH**(Ottawa)—**OK**(Kingfisher) |
| 549 | **OK**(Kiowa)—**OR**(Marion) |
| 563 | **OR**(Morrow)—**PA**(Pike) |
| 577 | **PA**(Potter)—**SC**(Spartanburg) |
| 591 | **SC**(Sumter)—**SD**(Todd) |
| 605 | **SD**(Tripp)—**TN**(Marion) |
| 619 | **TN**(Marshall)—**TX**(Burnet) |
| 633 | **TX**(Caldwell)—**TX**(Grimes) |
| 647 | **TX**(Guadalupe)—**TX**(Martin) |
| 661 | **TX**(Mason)—**TX**(Titus) |
| 675 | **TX**(Tom Green)—**VT**(Caledonia) |
| 689 | **VT**(Chittenden)—**VA**(Loudoun) |
| 703 | **VA**(Louisa)—**VA**(Manassas Park city) |
| 717 | **VA**(Martinsville city)—**WV**(Cabell) |
| 731 | **WV**(Calhoun)—**WI**(Door) |
| 745 | **WI**(Douglas)—**WY**(Fremont) |
| 759 | **WY**(Goshen)—**WY**(Weston) |

# County Highlights and Rankings

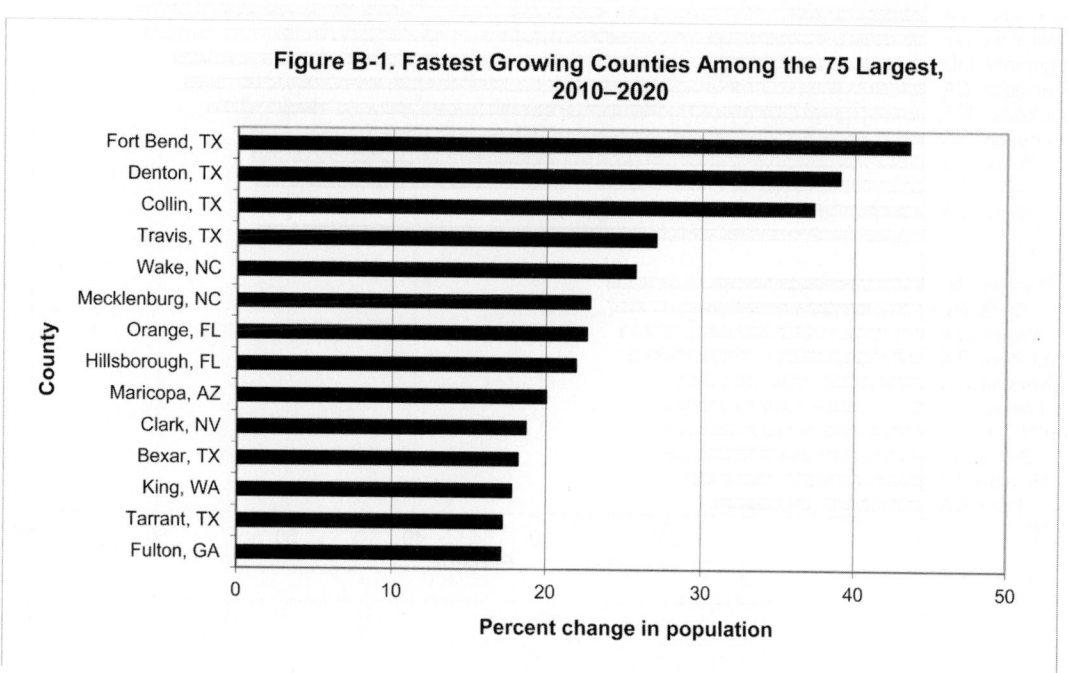

**Figure B-1. Fastest Growing Counties Among the 75 Largest, 2010–2020**

In the final year before the 2020 census, the 2020 population estimates show that Los Angeles County, CA remains, by far, the most populous county, with nearly 10 million residents. Next is Cook County, IL, which includes Chicago, with over 5.1 million people. Its population declined 3.4 percent from 2000 to 2010 and 1.7 percent from 2010 to 2020. New York City consists of five counties (the five boroughs), with Kings (Brooklyn) and Queens each with over 2 million residents, New York (Manhattan) with slightly over 1.6 million residents, and the Bronx with about 1.4 million. Queens county moved out of the top 10 in 2014 but Brooklyn's 1.4 percent growth keeps Kings the 9th most populous county in the nation.

Among the 75 most populous counties, the highest growth rates from 2010 to 2020 were found in the South, with the top four growth rates in Texas. From 2010 to 2020, the fastest-growing of these large counties was Fort Bend, TX, in the Houston metropolitan area. During the past nine years, Fort Bend County's population increased by 43.6 percent followed by Denton County, TX, in the Dallas-Fort Worth metropolitan area, at 39.0 percent. In 2020, Fort Bend County ranked 71ˢᵗ among the most populous counties. Seven other counties in Texas also had growth rates over 10 percent: Colin (also Dallas) Travis (Austin), Bexar (San Antonio), Harris (Houston), Tarrant (Fort Worth.), Hidalgo (McAllen) and Dallas. Other large counties that experienced more than 10 percent growth in the nine-year period were found mainly in the South and the West. Nearly 1,700 counties lost population during this period. The largest proportional losses were in counties with very small populations. Among the largest counties, Wayne, MI; Cuyahoga, OH; Suffolk, NY; Cook, IL; New Haven, CT;

Allegheny, PA; St. Louis, MO; Hartford, CT; Milwaukee, WI; Queens, NY; and Erie, NY declined in population between 2010 and 2020. Four hundred and thirty-six had population growth rates at or above 10 percent from 2010 to 2020. Fifty-five of these fast-growing counties had more than 50,000 residents and twenty-one counties had more than 100,000 residents.

Within states, the number and physical size of counties varied considerably: Delaware had three counties while Texas had 254 counties. For the 3,142 counties (and county equivalents—see Appendix A) in the United States, population in 2019 ranged from nearly 10 million in Los Angeles, CA, to 87 in Kalawao County, HI. Other particularly large counties in terms of population are Cook County, IL (over 5.1 million people), encompassing Chicago and its suburbs, Harris County, TX (containing Houston) with 4.7 million people, and Maricopa County, AZ (containing Phoenix), with nearly 4.6 million people. There were 46 counties with a population of 1,000,000 or more; these counties combined contain more than one-fourth of the U.S. population. Over half of the U.S. population lived in the 157 largest counties, those with a population of 450,000 or more. At the other extreme, there were 36 counties with fewer than 1,000 people in 2020. The median county population size was 25,583.

In terms of land area, counties range from the nearly 145,575 square miles of Yukon-Koyukuk Census Area, AK; to Kalawao County, HI, with 12 square miles; New York County, NY (Manhattan), with 22.7 square miles; Bristol County, RI, with 24.1 square miles; and Arlington County, VA, with 26 square miles.[1] Counties tend to be larger in the western United States

---

[1] Several independent cities in Virginia, which are treated as counties for tabulation purposes, were excluded here.

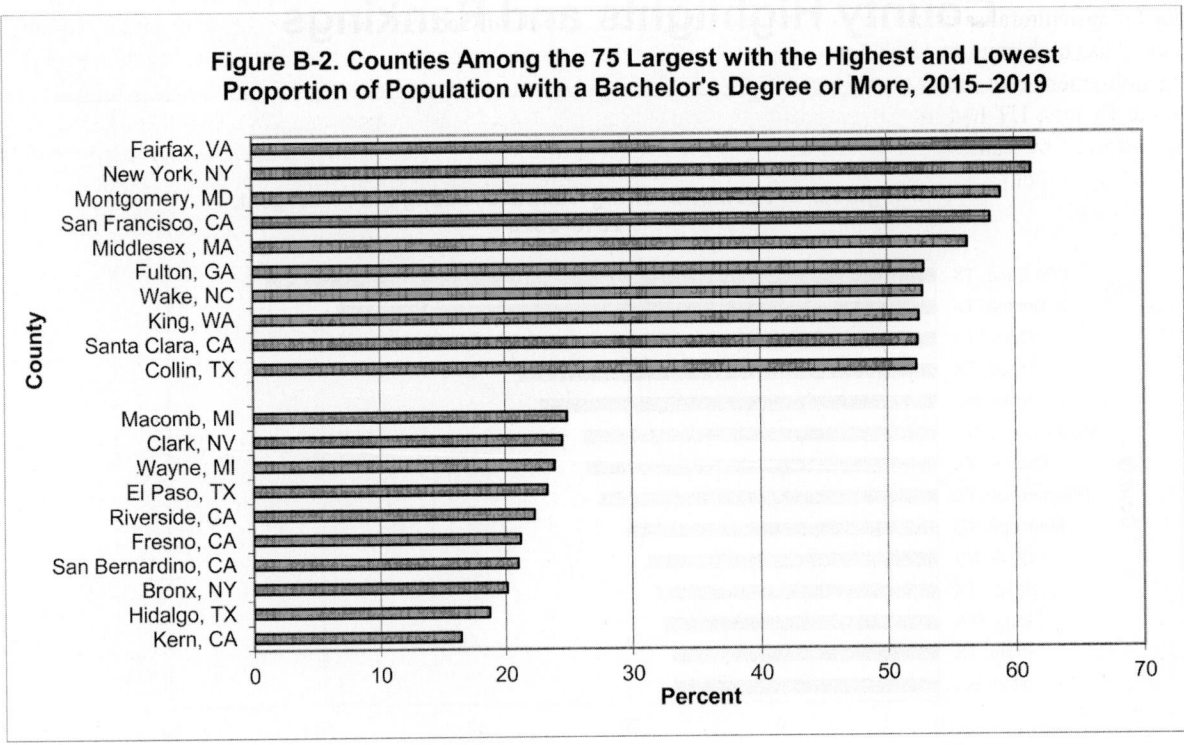

Figure B-2. Counties Among the 75 Largest with the Highest and Lowest Proportion of Population with a Bachelor's Degree or More, 2015–2019

(most of the largest 50 in size are in that region). The median land area for all U.S. counties was about 618 square miles in 2020.

While New York County, NY, had one of the smallest land areas, it had by far the highest population density among U.S. counties in 2020, with nearly 72,000 persons per square mile. No other county approached that density (although three other New York City boroughs were among the top five counties in population density). San Francisco had the highest population density outside of New York City, with Suffolk County, MA (Boston); Philadelphia County, PA; and Washington, DC, also among the top 10 counties. In 2020, the median county had nearly 45 persons per square mile, with 269 counties having more than 500 persons per square mile. The nation's largest county in terms of population (Los Angeles) had a population density of 2,449.4 persons per square mile. This density ranked 21st among the 75 most populous U.S. counties.

Proportionally large year-to-year labor force changes are not unusual for counties with small populations. The 2020 annual averages reflect a national labor force that decreased by 1.7 percent. Among the 75 most populous counties, McAllen-Edinburg-Mission, TX was the only county whose labor force grew by more than 2.0 percent while Phoenix-Mesa-Chandler, AZ and Salt Lake City, UT were the only two counties among the 75 most populous which experienced a labor force growth above 1.0 percent. Over 2.300 counties experienced declines in their labor forces from 2019 to 2020, with 182 counties losing 5 percent or more. This was unusually high and caused in large part by the coronavirus (COVID-19).

Among the most populous 75 counties, 59 counties had decreases in their labor forces, led by New York, NY at 6.4 percent, Miami-Dade, FL at 5.8 percent, and Oakland, MI at 4.9 percent.

The national annual average unemployment rate was 8.1 percent in 2020, reversing a steady drop since 2010. Unemployment increased drastically as businesses were forced to close due to stay-at-home orders instituted by the states because of COVID-19. The unemployment rate was up from 3.7 percent in 2019, 3.9 percent in 2018, 4.4 percent in 2017, and 4.9 percent in 2016. In 2020, 731 counties had unemployment rates above the national average of 8.1 percent and 224 counties had unemployment rates greater than 10 percent, down from 1,100 counties in 2010, but up significantly from 35 counties in 2017 and 20 counties in 2018. Of the 10 counties with the highest unemployment rates, Imperial County, CA, Maui County, HI, Atlantic County, NJ and Yuma County, AZ had populations over 100,000. Among the 75 most populous counties, 40 exceeded the national unemployment rate of 8.1 percent and one county equaled it, with the highest in Bronx county, NY at 16.0 percent and 14.7 percent in Clark county, NV. Salt Lake County, UT had the lowest unemployment rate at 5.1 percent, followed by Fairfax, VA at 5.8 percent and Gwinnett county, GA at 6.3 percent. Among all smaller counties in 2020, three counties in Nebraska, two counties each in Kansas and Texas, and one county in Hawaii had unemployment rates of less than 2 percent.

Among the 75 largest counties, two counties with high unemployment rates, Fresno and Kern counties in CA, ranked among the top counties for agricultural sales. Meanwhile, Wayne County, MI, which typically has high manufacturing employment, had the third highest unemployment rate among the 75 most populous counties. Salt Lake County, UT had the lowest unemployment rate of 5.1 percent followed by Fairfax, VA at 5.8 percent and Gwinnett county, GA at 6.3 percent.

Typically, counties with higher levels of educational attainment have lower unemployment rates and higher median household incomes. Of the top ten counties with the highest educational attainment, nine had an unemployment rate lower than the national average and all had median incomes higher than average. Nationally, 33.1 percent of the population held bachelor's degrees or higher in 2019.

# 75 Largest Counties by 2020 Population
## Selected Rankings

| Population, 2020 | | | Land area, 2020 | | | | Population density, 2020 | | | |
|---|---|---|---|---|---|---|---|---|---|---|
| Population rank | | Population [col 2] | Population rank | Land area rank | County | Land area (Square miles) [col 1] | Population rank | Density rank | County | Density (per square kilometer) [col 4] |
| 1 | Los Angeles, CA | 9,943,046 | 14 | 1 | San Bernardino, CA | 20,068 | 21 | 1 | New York, NY | 71,012.7 |
| 2 | Cook, IL | 5,108,284 | 4 | 2 | Maricopa, AZ | 9,202 | 9 | 2 | Kings, NY | 36,584.1 |
| 3 | Harris, TX | 4,738,253 | 44 | 3 | Pima, AZ | 9,189 | 29 | 3 | Bronx, NY | 33,202.4 |
| 4 | Maricopa, AZ | 4,579,081 | 63 | 4 | Kern, CA | 8,135 | 13 | 4 | Queens, NY | 20,476.7 |
| 5 | San Diego, CA | 3,332,427 | 11 | 5 | Clark, NV | 7,892 | 67 | 5 | San Francisco, CA | 18,477.7 |
| 6 | Orange, CA | 3,166,857 | 10 | 6 | Riverside, CA | 7,209 | 23 | 6 | Philadelphia, PA | 11,744.7 |
| 7 | Miami-Dade, FL | 2,707,303 | 46 | 7 | Fresno, CA | 5,958 | 2 | 7 | Cook, IL | 5,406.2 |
| 8 | Dallas, TX | 2,635,888 | 5 | 8 | San Diego, CA | 4,210 | 30 | 8 | Nassau, NY | 4,749.9 |
| 9 | Kings, NY | 2,538,934 | 1 | 9 | Los Angeles, CA | 4,059 | 57 | 9 | Bergen, NJ | 3,996.5 |
| 10 | Riverside, CA | 2,489,188 | 12 | 10 | King, WA | 2,115 | 6 | 10 | Orange, CA | 3,994.5 |
| 11 | Clark, NV | 2,315,963 | 73 | 11 | Snohomish, WA | 2,087 | 53 | 11 | Milwaukee, WI | 3,913.1 |
| 12 | King, WA | 2,274,315 | 25 | 12 | Palm Beach, FL | 1,964 | 48 | 12 | Pinellas, FL | 3,568.9 |
| 13 | Queens, NY | 2,225,821 | 7 | 13 | Miami-Dade, FL | 1,900 | 8 | 13 | Dallas, TX | 3,019.0 |
| 14 | San Bernardino, CA | 2,189,183 | 69 | 14 | Ventura, CA | 1,841 | 39 | 14 | Fairfax, VA | 2,943.3 |
| 15 | Tarrant, TX | 2,123,347 | 3 | 15 | Harris, TX | 1,707 | 19 | 15 | Wayne, MI | 2,845.1 |
| 16 | Bexar, TX | 2,026,823 | 61 | 16 | Pierce, WA | 1,668 | 59 | 16 | Du Page, IL | 2,798.9 |
| 17 | Broward, FL | 1,958,105 | 65 | 17 | Hidalgo, TX | 1,571 | 3 | 17 | Harris, TX | 2,775.8 |
| 18 | Santa Clara, CA | 1,907,105 | 74 | 18 | Worcester, MA | 1,511 | 35 | 18 | Cuyahoga, OH | 2,685.7 |
| 19 | Wayne, MI | 1,740,623 | 18 | 19 | Santa Clara, CA | 1,291 | 31 | 19 | Franklin, OH | 2,488.0 |
| 20 | Alameda, CA | 1,662,323 | 16 | 20 | Bexar, TX | 1,240 | 15 | 20 | Tarrant, TX | 2,453.9 |
| 21 | New York, NY | 1,611,989 | 17 | 21 | Broward, FL | 1,203 | 1 | 21 | Los Angeles, CA | 2,449.4 |
| 22 | Middlesex, MA | 1,609,379 | 60 | 22 | Erie, NY | 1,043 | 50 | 22 | Marion, IN | 2,436.2 |
| 23 | Philadelphia, PA | 1,578,487 | 26 | 23 | Hillsborough, FL | 1,022 | 33 | 23 | Hennepin, MN | 2,289.5 |
| 24 | Sacramento, CA | 1,559,146 | 70 | 24 | El Paso, TX | 1,013 | 20 | 24 | Alameda, CA | 2,254.0 |
| 25 | Palm Beach, FL | 1,507,600 | 32 | 25 | Travis, TX | 994 | 51 | 25 | Westchester, NY | 2,242.4 |
| 26 | Hillsborough, FL | 1,497,957 | 24 | 26 | Sacramento, CA | 965 | 54 | 26 | Gwinnett, GA | 2,188.1 |
| 27 | Suffolk, NY | 1,474,273 | 2 | 27 | Cook, IL | 945 | 41 | 27 | Mecklenburg, NC | 2,156.1 |
| 28 | Orange, FL | 1,404,396 | 27 | 28 | Suffolk, NY | 911 | 45 | 28 | Montgomery, MD | 2,133.1 |
| 29 | Bronx, NY | 1,401,142 | 28 | 29 | Orange, FL | 902 | 42 | 29 | Fulton, GA | 2,045.6 |
| 30 | Nassau, NY | 1,351,334 | 58 | 30 | Denton, TX | 879 | 22 | 30 | Middlesex, MA | 1,967.7 |
| 31 | Franklin, OH | 1,324,624 | 8 | 31 | Dallas, TX | 873 | 47 | 31 | St. Louis, MO | 1,957.1 |
| 32 | Travis, TX | 1,300,503 | 34 | 32 | Oakland, MI | 867 | 62 | 32 | Prince George's, MD | 1,884.4 |
| 33 | Hennepin, MN | 1,268,408 | 15 | 33 | Tarrant, TX | 865 | 66 | 33 | Macomb, MI | 1,816.8 |
| 34 | Oakland, MI | 1,253,459 | 71 | 34 | Fort Bend, TX | 862 | 72 | 34 | Montgomery, PA | 1,726.4 |
| 35 | Cuyahoga, OH | 1,227,883 | 43 | 35 | Collin, TX | 841 | 36 | 35 | Allegheny, PA | 1,659.4 |
| 36 | Allegheny, PA | 1,211,358 | 40 | 36 | Wake, NC | 835 | 16 | 36 | Bexar, TX | 1,634.1 |
| 37 | Salt Lake, UT | 1,165,517 | 22 | 37 | Middlesex, MA | 818 | 17 | 37 | Broward, FL | 1,628.1 |
| 38 | Contra Costa, CA | 1,152,333 | 6 | 38 | Orange, CA | 793 | 27 | 38 | Suffolk, NY | 1,617.9 |
| 39 | Fairfax, VA | 1,150,847 | 49 | 39 | Duval, FL | 763 | 24 | 39 | Sacramento, CA | 1,615.2 |
| 40 | Wake, NC | 1,132,271 | 56 | 40 | Shelby, TN | 761 | 38 | 40 | Contra Costa, CA | 1,607.4 |
| 41 | Mecklenburg, NC | 1,128,945 | 37 | 41 | Salt Lake, UT | 742 | 52 | 41 | Honolulu, HI | 1,604.8 |
| 42 | Fulton, GA | 1,077,402 | 20 | 42 | Alameda, CA | 738 | 37 | 42 | Salt Lake, UT | 1,570.6 |
| 43 | Collin, TX | 1,072,069 | 64 | 43 | Hartford, CT | 735 | 28 | 43 | Orange, FL | 1,557.0 |
| 44 | Pima, AZ | 1,061,175 | 36 | 44 | Allegheny, PA | 730 | 55 | 44 | Fairfield, CT | 1,507.9 |
| 45 | Montgomery, MD | 1,051,816 | 38 | 45 | Contra Costa, CA | 717 | 18 | 45 | Santa Clara, CA | 1,477.1 |
| 46 | Fresno, CA | 1,000,918 | 55 | 46 | Fairfield, CT | 625 | 26 | 46 | Hillsborough, FL | 1,465.9 |
| 47 | St. Louis, MO | 994,020 | 19 | 47 | Wayne, MI | 612 | 34 | 47 | Oakland, MI | 1,445.2 |
| 48 | Pinellas, FL | 976,802 | 68 | 48 | New Haven, CT | 604 | 7 | 48 | Miami-Dade, FL | 1,425.0 |
| 49 | Duval, FL | 966,728 | 52 | 49 | Honolulu, HI | 601 | 68 | 49 | New Haven, CT | 1,409.8 |
| 50 | Marion, IN | 966,183 | 75 | 50 | Baltimore, MD | 598 | 75 | 50 | Baltimore, MD | 1,380.4 |
| 51 | Westchester, NY | 965,802 | 33 | 51 | Hennepin, MN | 554 | 40 | 51 | Wake, NC | 1,356.7 |
| 52 | Honolulu, HI | 963,826 | 31 | 52 | Franklin, OH | 532 | 32 | 52 | Travis, TX | 1,308.2 |
| 53 | Milwaukee, WI | 945,016 | 42 | 53 | Fulton, GA | 527 | 43 | 53 | Collin, TX | 1,274.3 |
| 54 | Gwinnett, GA | 942,627 | 41 | 54 | Mecklenburg, NC | 524 | 49 | 54 | Duval, FL | 1,267.7 |
| 55 | Fairfield, CT | 942,426 | 47 | 55 | St. Louis, MO | 508 | 56 | 55 | Shelby, TN | 1,230.6 |
| 56 | Shelby, TN | 936,017 | 45 | 56 | Montgomery, MD | 493 | 64 | 56 | Hartford, CT | 1,210.0 |
| 57 | Bergen, NJ | 930,394 | 72 | 57 | Montgomery, PA | 483 | 12 | 57 | King, WA | 1,075.2 |
| 58 | Denton, TX | 919,324 | 62 | 58 | Prince George's, MD | 483 | 58 | 58 | Denton, TX | 1,046.5 |
| 59 | Du Page, IL | 917,481 | 66 | 59 | Macomb, MI | 479 | 71 | 59 | Fort Bend, TX | 974.5 |
| 60 | Erie, NY | 917,241 | 35 | 60 | Cuyahoga, OH | 457 | 60 | 60 | Erie, NY | 879.7 |
| 61 | Pierce, WA | 913,890 | 54 | 61 | Gwinnett, GA | 431 | 70 | 61 | El Paso, TX | 830.3 |
| 62 | Prince George's, MD | 909,612 | 51 | 62 | Westchester, NY | 431 | 5 | 62 | San Diego, CA | 791.5 |
| 63 | Kern, CA | 901,362 | 50 | 63 | Marion, IN | 397 | 25 | 63 | Palm Beach, FL | 767.5 |
| 64 | Hartford, CT | 889,226 | 39 | 64 | Fairfax, VA | 391 | 65 | 64 | Hidalgo, TX | 557.1 |
| 65 | Hidalgo, TX | 875,200 | 59 | 65 | Du Page, IL | 328 | 74 | 65 | Worcester, MA | 548.9 |
| 66 | Macomb, MI | 870,791 | 30 | 66 | Nassau, NY | 285 | 61 | 66 | Pierce, WA | 547.9 |
| 67 | San Francisco, CA | 866,606 | 48 | 67 | Pinellas, FL | 274 | 4 | 67 | Maricopa, AZ | 497.6 |
| 68 | New Haven, CT | 851,948 | 53 | 68 | Milwaukee, WI | 242 | 69 | 68 | Ventura, CA | 457.1 |
| 69 | Ventura, CA | 841,387 | 57 | 69 | Bergen, NJ | 233 | 73 | 69 | Snohomish, WA | 398.0 |
| 70 | El Paso, TX | 841,286 | 23 | 70 | Philadelphia, PA | 134 | 10 | 70 | Riverside, CA | 345.3 |
| 71 | Fort Bend, TX | 839,706 | 13 | 71 | Queens, NY | 109 | 11 | 71 | Clark, NV | 293.5 |
| 72 | Montgomery, PA | 833,869 | 9 | 72 | Kings, NY | 69 | 46 | 72 | Fresno, CA | 168.0 |
| 73 | Snohomish, WA | 830,393 | 67 | 73 | San Francisco, CA | 47 | 44 | 73 | Pima, AZ | 115.5 |
| 74 | Worcester, MA | 829,212 | 29 | 74 | Bronx, NY | 42 | 63 | 74 | Kern, CA | 110.8 |
| 75 | Baltimore, MD | 826,017 | 21 | 75 | New York, NY | 23 | 14 | 75 | San Bernardino, CA | 109.1 |

# 75 Largest Counties by 2020 Population
## Selected Rankings

| Percent population change, 2010–2020 | | | | Employment/residence ratio, 2015–2019 | | | | Percent White, not Hispanic or Latino, alone or in combination, 2020 | | | |
|---|---|---|---|---|---|---|---|---|---|---|---|
| Popu-lation rank | Percent change rank | County | Percent change [col 23] | Popu-lation rank | Number of employees per resident rank | County | Number of employees per resident [col 34] | Popu-lation rank | White rank | County | Percent white [col 5] |
| 71 | 1 | Fort Bend, TX | 43.6 | 21 | 1 | New York, NY | 2.94 | 36 | 1 | Allegheny, PA | 79.9 |
| 58 | 2 | Denton, TX | 39.0 | 42 | 2 | Fulton, GA | 1.78 | 66 | 2 | Macomb, MI | 79.5 |
| 43 | 3 | Collin, TX | 37.3 | 67 | 3 | San Francisco, CA | 1.48 | 74 | 3 | Worcester, MA | 76.9 |
| 32 | 4 | Travis, TX | 27.0 | 33 | 4 | Hennepin, MN | 1.33 | 72 | 4 | Montgomery, PA | 76.4 |
| 40 | 5 | Wake, NC | 25.7 | 50 | 5 | Marion, IN | 1.28 | 60 | 5 | Erie, NY | 76.2 |
| 41 | 6 | Mecklenburg, NC | 22.8 | 41 | 6 | Mecklenburg, NC | 1.27 | 48 | 6 | Pinellas, FL | 75.3 |
| 28 | 7 | Orange, FL | 22.6 | 8 | 7 | Dallas, TX | 1.26 | 34 | 7 | Oakland, MI | 73.1 |
| 26 | 8 | Hillsborough, FL | 21.9 | 28 | 7 | Orange, FL | 1.26 | 22 | 8 | Middlesex, MA | 72.6 |
| 4 | 9 | Maricopa, AZ | 20.0 | 35 | 9 | Cuyahoga, OH | 1.22 | 37 | 9 | Salt Lake, UT | 72.3 |
| 11 | 10 | Clark, NV | 18.7 | 47 | 9 | St. Louis, MO | 1.22 | 73 | 10 | Snohomish, WA | 71.0 |
| 16 | 11 | Bexar, TX | 18.2 | 59 | 9 | Du Page, IL | 1.22 | 33 | 11 | Hennepin, MN | 70.8 |
| 12 | 12 | King, WA | 17.8 | 32 | 12 | Travis, TX | 1.20 | 61 | 12 | Pierce, WA | 70.6 |
| 15 | 13 | Tarrant, TX | 17.2 | 72 | 13 | Montgomery, PA | 1.19 | 27 | 13 | Suffolk, NY | 67.2 |
| 42 | 14 | Fulton, GA | 17.1 | 49 | 14 | Duval, FL | 1.18 | 59 | 14 | Du Page, IL | 67.1 |
| 54 | 14 | Gwinnett, GA | 17.1 | 12 | 15 | King, WA | 1.17 | 47 | 15 | St. Louis, MO | 66.8 |
| 73 | 16 | Snohomish, WA | 16.4 | 56 | 15 | Shelby, TN | 1.17 | 31 | 16 | Franklin, OH | 64.2 |
| 3 | 17 | Harris, TX | 15.8 | 64 | 15 | Hartford, CT | 1.17 | 68 | 17 | New Haven, CT | 62.5 |
| 61 | 18 | Pierce, WA | 14.9 | 31 | 18 | Franklin, OH | 1.16 | 55 | 18 | Fairfield, CT | 61.6 |
| 25 | 19 | Palm Beach, FL | 14.2 | 18 | 19 | Santa Clara, CA | 1.15 | 40 | 19 | Wake, NC | 61.3 |
| 31 | 20 | Franklin, OH | 13.9 | 23 | 19 | Philadelphia, PA | 1.15 | 12 | 20 | King, WA | 61.2 |
| 10 | 21 | Riverside, CA | 13.7 | 37 | 19 | Salt Lake, UT | 1.15 | 64 | 21 | Hartford, CT | 60.9 |
| 37 | 22 | Salt Lake, UT | 13.2 | 3 | 22 | Harris, TX | 1.14 | 35 | 22 | Cuyahoga, OH | 60.1 |
| 65 | 23 | Hidalgo, TX | 13.0 | 34 | 22 | Oakland, MI | 1.14 | 30 | 23 | Nassau, NY | 58.9 |
| 17 | 24 | Broward, FL | 12.0 | 36 | 22 | Allegheny, PA | 1.14 | 58 | 24 | Denton, TX | 58.8 |
| 49 | 25 | Duval, FL | 11.9 | 19 | 25 | Wayne, MI | 1.09 | 75 | 25 | Baltimore, MD | 56.8 |
| 8 | 26 | Dallas, TX | 11.3 | 22 | 26 | Middlesex, MA | 1.08 | 4 | 26 | Maricopa, AZ | 56.2 |
| 20 | 27 | Alameda, CA | 10.1 | 26 | 26 | Hillsborough, FL | 1.08 | 43 | 27 | Collin, TX | 56.0 |
| 33 | 27 | Hennepin, MN | 10.1 | 40 | 26 | Wake, NC | 1.08 | 50 | 28 | Marion, IN | 55.9 |
| 24 | 29 | Sacramento, CA | 9.9 | 2 | 29 | Cook, IL | 1.07 | 57 | 29 | Bergen, NJ | 55.6 |
| 38 | 30 | Contra Costa, CA | 9.8 | 60 | 29 | Erie, NY | 1.07 | 25 | 30 | Palm Beach, FL | 54.4 |
| 7 | 31 | Miami-Dade, FL | 8.4 | 6 | 31 | Orange, CA | 1.06 | 49 | 31 | Duval, FL | 53.6 |
| 44 | 32 | Pima, AZ | 8.3 | 7 | 31 | Miami-Dade, FL | 1.06 | 51 | 31 | Westchester, NY | 53.6 |
| 45 | 32 | Montgomery, MD | 8.3 | 39 | 31 | Fairfax, VA | 1.06 | 39 | 33 | Fairfax, VA | 52.7 |
| 5 | 34 | San Diego, CA | 7.7 | 53 | 31 | Milwaukee, WI | 1.06 | 44 | 33 | Pima, AZ | 52.7 |
| 14 | 35 | San Bernardino, CA | 7.6 | 16 | 35 | Bexar, TX | 1.05 | 53 | 35 | Milwaukee, WI | 52.1 |
| 46 | 35 | Fresno, CA | 7.6 | 1 | 36 | Los Angeles, CA | 1.03 | 19 | 36 | Wayne, MI | 51.3 |
| 67 | 35 | San Francisco, CA | 7.6 | 25 | 36 | Palm Beach, FL | 1.03 | 32 | 37 | Travis, TX | 50.6 |
| 63 | 38 | Kern, CA | 7.4 | 48 | 36 | Pinellas, FL | 1.03 | 26 | 38 | Hillsborough, FL | 48.9 |
| 22 | 39 | Middlesex, MA | 7.1 | 4 | 39 | Maricopa, AZ | 1.02 | 21 | 39 | New York, NY | 48.8 |
| 18 | 40 | Santa Clara, CA | 7.0 | 5 | 39 | San Diego, CA | 1.02 | 5 | 40 | San Diego, CA | 47.5 |
| 50 | 40 | Marion, IN | 7.0 | 11 | 41 | Clark, NV | 1.01 | 41 | 40 | Mecklenburg, NC | 47.5 |
| 48 | 42 | Pinellas, FL | 6.5 | 63 | 41 | Kern, CA | 1.01 | 24 | 42 | Sacramento, CA | 47.2 |
| 39 | 43 | Fairfax, VA | 6.4 | 15 | 43 | Tarrant, TX | 1.00 | 69 | 43 | Ventura, CA | 46.7 |
| 62 | 44 | Prince George's, MD | 5.3 | 44 | 43 | Pima, AZ | 1.00 | 15 | 44 | Tarrant, TX | 46.4 |
| 6 | 45 | Orange, CA | 5.2 | 46 | 43 | Fresno, CA | 1.00 | 38 | 45 | Contra Costa, CA | 45.5 |
| 70 | 46 | El Paso, TX | 5.1 | 52 | 43 | Honolulu, HI | 1.00 | 45 | 46 | Montgomery, MD | 44.9 |
| 72 | 47 | Montgomery, PA | 4.3 | 55 | 43 | Fairfield, CT | 1.00 | 11 | 47 | Clark, NV | 44.2 |
| 34 | 48 | Oakland, MI | 4.2 | 70 | 43 | El Paso, TX | 1.00 | 2 | 48 | Cook, IL | 43.1 |
| 74 | 49 | Worcester, MA | 3.9 | 24 | 49 | Sacramento, CA | 0.99 | 67 | 49 | San Francisco, CA | 42.8 |
| 66 | 50 | Macomb, MI | 3.5 | 20 | 50 | Alameda, CA | 0.97 | 6 | 50 | Orange, CA | 41.8 |
| 23 | 51 | Philadelphia, PA | 3.4 | 43 | 51 | Collin, TX | 0.95 | 42 | 51 | Fulton, GA | 40.8 |
| 55 | 52 | Fairfield, CT | 2.8 | 45 | 51 | Montgomery, MD | 0.95 | 28 | 52 | Orange, FL | 40.6 |
| 57 | 52 | Bergen, NJ | 2.8 | 51 | 51 | Westchester, NY | 0.95 | 9 | 53 | Kings, NY | 38.4 |
| 75 | 54 | Baltimore, MD | 2.6 | 65 | 51 | Hidalgo, TX | 0.95 | 54 | 54 | Gwinnett, GA | 36.3 |
| 69 | 55 | Ventura, CA | 2.2 | 14 | 55 | San Bernardino, CA | 0.92 | 56 | 55 | Shelby, TN | 36.2 |
| 51 | 56 | Westchester, NY | 1.7 | 57 | 55 | Bergen, NJ | 0.92 | 23 | 56 | Philadelphia, PA | 35.8 |
| 21 | 57 | New York, NY | 1.6 | 68 | 55 | New Haven, CT | 0.92 | 17 | 57 | Broward, FL | 35.5 |
| 9 | 58 | Kings, NY | 1.4 | 17 | 58 | Broward, FL | 0.91 | 10 | 58 | Riverside, CA | 35.3 |
| 1 | 59 | Los Angeles, CA | 1.3 | 69 | 59 | Ventura, CA | 0.89 | 63 | 59 | Kern, CA | 33.9 |
| 29 | 60 | Bronx, NY | 1.2 | 27 | 60 | Suffolk, NY | 0.88 | 20 | 60 | Alameda, CA | 33.6 |
| 52 | 61 | Honolulu, HI | 1.1 | 74 | 60 | Worcester, MA | 0.88 | 18 | 61 | Santa Clara, CA | 32.6 |
| 30 | 62 | Nassau, NY | 0.9 | 75 | 60 | Baltimore, MD | 0.88 | 71 | 62 | Fort Bend, TX | 32.5 |
| 56 | 62 | Shelby, TN | 0.9 | 54 | 63 | Gwinnett, GA | 0.87 | 52 | 63 | Honolulu, HI | 31.7 |
| 59 | 64 | Du Page, IL | 0.1 | 30 | 64 | Nassau, NY | 0.86 | 46 | 64 | Fresno, CA | 29.6 |
| 13 | 65 | Queens, NY | -0.2 | 61 | 65 | Pierce, WA | 0.85 | 3 | 65 | Harris, TX | 29.5 |
| 60 | 65 | Erie, NY | -0.2 | 10 | 66 | Riverside, CA | 0.83 | 8 | 66 | Dallas, TX | 29.1 |
| 53 | 67 | Milwaukee, WI | -0.3 | 66 | 66 | Macomb, MI | 0.83 | 14 | 67 | San Bernardino, CA | 28.3 |
| 47 | 68 | St. Louis, MO | -0.5 | 73 | 68 | Snohomish, WA | 0.78 | 16 | 68 | Bexar, TX | 28.0 |
| 64 | 68 | Hartford, CT | -0.5 | 38 | 69 | Contra Costa, CA | 0.74 | 1 | 69 | Los Angeles, CA | 27.8 |
| 36 | 70 | Allegheny, PA | -1.0 | 62 | 69 | Prince George's, MD | 0.74 | 13 | 70 | Queens, NY | 25.8 |
| 68 | 71 | New Haven, CT | -1.2 | 9 | 71 | Kings, NY | 0.73 | 7 | 71 | Miami-Dade, FL | 14.0 |
| 27 | 72 | Suffolk, NY | -1.3 | 13 | 72 | Queens, NY | 0.67 | 62 | 72 | Prince George's, MD | 13.4 |
| 2 | 73 | Cook, IL | -1.7 | 29 | 73 | Bronx, NY | 0.66 | 70 | 73 | El Paso, TX | 12.4 |
| 35 | 74 | Cuyahoga, OH | -4.1 | 58 | 73 | Denton, TX | 0.66 | 29 | 74 | Bronx, NY | 9.6 |
| 19 | 75 | Wayne, MI | -4.4 | 71 | 75 | Fort Bend, TX | 0.57 | 65 | 75 | Hidalgo, TX | 6.0 |

# 75 Largest Counties by 2020 Population
## Selected Rankings

| Percent Black, not Hispanic or Latino, alone or in combination, 2020 | | | | Percent American Indian, Alaska Native, alone or in combination, 2020 | | | | Percent Asian or Pacific Islander, alone or in combination, 2020 | | | |
|---|---|---|---|---|---|---|---|---|---|---|---|
| Population rank | Black rank | County | Percent black [col 6] | Population rank | American Indian Alaska native rank | County | Percent American Indian, Alaska native [col 7] | Population rank | Asian or Pacific Islander rank | County | Percent Asian or Pacific Islander [col 8] |
| 62 | 1 | Prince George's, MD | 63.3 | 44 | 1 | Pima, AZ | 3.0 | 52 | 1 | Honolulu, HI | 79.0 |
| 56 | 2 | Shelby, TN | 54.8 | 61 | 2 | Pierce, WA | 2.6 | 18 | 2 | Santa Clara, CA | 42.5 |
| 42 | 3 | Fulton, GA | 44.8 | 4 | 3 | Maricopa, AZ | 2.2 | 67 | 3 | San Francisco, CA | 39.2 |
| 23 | 4 | Philadelphia, PA | 41.7 | 73 | 3 | Snohomish, WA | 2.2 | 20 | 4 | Alameda, CA | 36.7 |
| 19 | 5 | Wayne, MI | 39.6 | 5 | 5 | King, WA | 1.6 | 13 | 5 | Queens, NY | 28.6 |
| 41 | 6 | Mecklenburg, NC | 32.9 | 24 | 6 | Sacramento, CA | 1.5 | 6 | 6 | Orange, CA | 24.6 |
| 75 | 7 | Baltimore, MD | 31.7 | 52 | 6 | Honolulu, HI | 1.5 | 12 | 7 | King, WA | 24.1 |
| 35 | 8 | Cuyahoga, OH | 31.1 | 33 | 8 | Hennepin, MN | 1.4 | 39 | 8 | Fairfax, VA | 22.8 |
| 49 | 8 | Duval, FL | 31.1 | 53 | 9 | Milwaukee, WI | 1.2 | 71 | 9 | Fort Bend, TX | 22.4 |
| 50 | 10 | Marion, IN | 30.7 | 63 | 9 | Kern, CA | 1.2 | 38 | 10 | Contra Costa, CA | 21.9 |
| 9 | 11 | Kings, NY | 30.5 | 11 | 11 | Clark, NV | 1.1 | 24 | 11 | Sacramento, CA | 21.2 |
| 54 | 12 | Gwinnett, GA | 30.0 | 19 | 11 | Wayne, MI | 1.1 | 43 | 12 | Collin, TX | 18.4 |
| 29 | 13 | Bronx, NY | 29.9 | 37 | 11 | Salt Lake, UT | 1.1 | 57 | 13 | Bergen, NJ | 18.0 |
| 17 | 14 | Broward, FL | 29.8 | 46 | 11 | Fresno, CA | 1.1 | 45 | 14 | Montgomery, MD | 17.3 |
| 53 | 15 | Milwaukee, WI | 27.9 | 10 | 15 | Riverside, CA | 1.0 | 1 | 15 | Los Angeles, CA | 16.8 |
| 47 | 16 | St. Louis, MO | 26.1 | 38 | 15 | Contra Costa, CA | 1.0 | 73 | 16 | Snohomish, WA | 15.8 |
| 31 | 17 | Franklin, OH | 25.6 | 58 | 15 | Denton, TX | 1.0 | 5 | 17 | San Diego, CA | 15.3 |
| 2 | 18 | Cook, IL | 23.8 | 5 | 18 | San Diego, CA | 0.9 | 22 | 18 | Middlesex, MA | 14.8 |
| 8 | 19 | Dallas, TX | 23.6 | 14 | 18 | San Bernardino, CA | 0.9 | 59 | 19 | Du Page, IL | 14.0 |
| 71 | 20 | Fort Bend, TX | 21.5 | 15 | 18 | Tarrant, TX | 0.9 | 9 | 20 | Kings, NY | 13.9 |
| 28 | 21 | Orange, FL | 21.4 | 31 | 18 | Franklin, OH | 0.9 | 21 | 20 | New York, NY | 13.9 |
| 40 | 22 | Wake, NC | 21.0 | 41 | 18 | Mecklenburg, NC | 0.9 | 54 | 20 | Gwinnett, GA | 13.9 |
| 45 | 23 | Montgomery, MD | 20.2 | 43 | 18 | Collin, TX | 0.9 | 11 | 23 | Clark, NV | 13.7 |
| 3 | 24 | Harris, TX | 19.7 | 60 | 18 | Erie, NY | 0.9 | 30 | 24 | Nassau, NY | 12.4 |
| 25 | 25 | Palm Beach, FL | 19.6 | 62 | 18 | Prince George's, MD | 0.9 | 61 | 24 | Pierce, WA | 12.4 |
| 13 | 26 | Queens, NY | 19.0 | 66 | 18 | Macomb, MI | 0.9 | 46 | 26 | Fresno, CA | 12.0 |
| 15 | 27 | Tarrant, TX | 18.4 | 13 | 27 | Queens, NY | 0.8 | 58 | 27 | Denton, TX | 11.1 |
| 26 | 28 | Hillsborough, FL | 17.4 | 20 | 27 | Alameda, CA | 0.8 | 69 | 28 | Ventura, CA | 9.4 |
| 7 | 29 | Miami-Dade, FL | 15.7 | 23 | 27 | Philadelphia, PA | 0.8 | 14 | 29 | San Bernardino, CA | 9.3 |
| 33 | 30 | Hennepin, MN | 15.5 | 34 | 27 | Oakland, MI | 0.8 | 34 | 30 | Oakland, MI | 9.2 |
| 36 | 31 | Allegheny, PA | 14.7 | 40 | 27 | Wake, NC | 0.8 | 40 | 31 | Wake, NC | 9.0 |
| 51 | 31 | Westchester, NY | 14.7 | 49 | 27 | Duval, FL | 0.8 | 72 | 31 | Montgomery, PA | 9.0 |
| 64 | 31 | Hartford, CT | 14.7 | 67 | 27 | San Francisco, CA | 0.8 | 2 | 33 | Cook, IL | 8.8 |
| 34 | 34 | Oakland, MI | 14.6 | 69 | 27 | Ventura, CA | 0.8 | 42 | 34 | Fulton, GA | 8.7 |
| 68 | 35 | New Haven, CT | 14.4 | 75 | 27 | Baltimore, MD | 0.8 | 32 | 35 | Travis, TX | 8.6 |
| 60 | 36 | Erie, NY | 14.2 | 8 | 36 | Dallas, TX | 0.7 | 33 | 35 | Hennepin, MN | 8.6 |
| 66 | 37 | Macomb, MI | 14.0 | 18 | 36 | Santa Clara, CA | 0.7 | 10 | 37 | Riverside, CA | 8.5 |
| 11 | 38 | Clark, NV | 13.6 | 32 | 36 | Travis, TX | 0.7 | 23 | 37 | Philadelphia, PA | 8.5 |
| 21 | 39 | New York, NY | 13.4 | 35 | 36 | Cuyahoga, OH | 0.7 | 3 | 39 | Harris, TX | 8.0 |
| 30 | 40 | Nassau, NY | 12.4 | 47 | 36 | St. Louis, MO | 0.7 | 37 | 40 | Salt Lake, UT | 7.7 |
| 55 | 41 | Fairfield, CT | 12.1 | 48 | 36 | Pinellas, FL | 0.7 | 8 | 41 | Dallas, TX | 7.4 |
| 58 | 42 | Denton, TX | 11.9 | 50 | 36 | Marion, IN | 0.7 | 51 | 42 | Westchester, NY | 7.3 |
| 24 | 43 | Sacramento, CA | 11.8 | 54 | 36 | Gwinnett, GA | 0.7 | 75 | 42 | Baltimore, MD | 7.3 |
| 43 | 44 | Collin, TX | 11.7 | 1 | 44 | Los Angeles, CA | 0.6 | 41 | 44 | Mecklenburg, NC | 7.2 |
| 20 | 45 | Alameda, CA | 11.4 | 6 | 44 | Orange, CA | 0.6 | 15 | 45 | Tarrant, TX | 6.9 |
| 48 | 45 | Pinellas, FL | 11.4 | 9 | 44 | Kings, NY | 0.6 | 31 | 46 | Franklin, OH | 6.8 |
| 39 | 47 | Fairfax, VA | 11.1 | 16 | 44 | Bexar, TX | 0.6 | 64 | 47 | Hartford, CT | 6.7 |
| 72 | 48 | Montgomery, PA | 10.7 | 26 | 44 | Hillsborough, FL | 0.6 | 28 | 48 | Orange, FL | 6.6 |
| 38 | 49 | Contra Costa, CA | 10.0 | 28 | 44 | Orange, FL | 0.6 | 55 | 48 | Fairfield, CT | 6.6 |
| 61 | 49 | Pierce, WA | 10.0 | 29 | 44 | Bronx, NY | 0.6 | 49 | 50 | Duval, FL | 6.3 |
| 14 | 51 | San Bernardino, CA | 9.1 | 36 | 44 | Allegheny, PA | 0.6 | 74 | 51 | Worcester, MA | 6.1 |
| 32 | 52 | Travis, TX | 8.9 | 39 | 44 | Fairfax, VA | 0.6 | 63 | 52 | Kern, CA | 6.0 |
| 1 | 53 | Los Angeles, CA | 8.7 | 42 | 44 | Fulton, GA | 0.6 | 4 | 53 | Maricopa, AZ | 5.8 |
| 27 | 54 | Suffolk, NY | 8.4 | 45 | 44 | Montgomery, MD | 0.6 | 47 | 54 | St. Louis, MO | 5.7 |
| 12 | 55 | King, WA | 8.3 | 56 | 44 | Shelby, TN | 0.6 | 53 | 55 | Milwaukee, WI | 5.5 |
| 16 | 56 | Bexar, TX | 8.2 | 64 | 44 | Hartford, CT | 0.6 | 26 | 56 | Hillsborough, FL | 5.4 |
| 10 | 57 | Riverside, CA | 7.2 | 68 | 44 | New Haven, CT | 0.6 | 66 | 56 | Macomb, MI | 5.4 |
| 4 | 58 | Maricopa, AZ | 6.8 | 71 | 44 | Fort Bend, TX | 0.6 | 36 | 58 | Allegheny, PA | 5.0 |
| 22 | 59 | Middlesex, MA | 6.1 | 74 | 44 | Worcester, MA | 0.6 | 27 | 59 | Suffolk, NY | 4.9 |
| 57 | 59 | Bergen, NJ | 6.1 | 2 | 60 | Cook, IL | 0.5 | 62 | 59 | Prince George's, MD | 4.9 |
| 67 | 59 | San Francisco, CA | 6.1 | 3 | 60 | Harris, TX | 0.5 | 68 | 59 | New Haven, CT | 4.9 |
| 63 | 62 | Kern, CA | 5.9 | 21 | 60 | New York, NY | 0.5 | 17 | 62 | Broward, FL | 4.7 |
| 74 | 62 | Worcester, MA | 5.9 | 27 | 60 | Suffolk, NY | 0.5 | 50 | 62 | Marion, IN | 4.7 |
| 5 | 64 | San Diego, CA | 5.8 | 70 | 60 | El Paso, TX | 0.5 | 60 | 62 | Erie, NY | 4.7 |
| 59 | 65 | Du Page, IL | 5.7 | 72 | 60 | Montgomery, PA | 0.5 | 29 | 65 | Bronx, NY | 4.5 |
| 46 | 66 | Fresno, CA | 5.3 | 17 | 66 | Broward, FL | 0.4 | 48 | 65 | Pinellas, FL | 4.5 |
| 73 | 67 | Snohomish, WA | 5.0 | 22 | 66 | Middlesex, MA | 0.4 | 19 | 67 | Wayne, MI | 4.4 |
| 44 | 68 | Pima, AZ | 4.3 | 25 | 66 | Palm Beach, FL | 0.4 | 44 | 68 | Pima, AZ | 4.2 |
| 52 | 69 | Honolulu, HI | 3.7 | 30 | 66 | Nassau, NY | 0.4 | 16 | 69 | Bexar, TX | 4.0 |
| 70 | 70 | El Paso, TX | 3.6 | 51 | 66 | Westchester, NY | 0.4 | 35 | 69 | Cuyahoga, OH | 4.0 |
| 18 | 71 | Santa Clara, CA | 3.1 | 55 | 66 | Fairfield, CT | 0.4 | 25 | 71 | Palm Beach, FL | 3.6 |
| 37 | 72 | Salt Lake, UT | 2.5 | 59 | 66 | Du Page, IL | 0.4 | 56 | 72 | Shelby, TN | 3.4 |
| 69 | 73 | Ventura, CA | 2.4 | 57 | 73 | Bergen, NJ | 0.3 | 7 | 73 | Miami-Dade, FL | 1.9 |
| 6 | 74 | Orange, CA | 2.2 | 7 | 74 | Miami-Dade, FL | 0.2 | 70 | 74 | El Paso, TX | 1.8 |
| 65 | 75 | Hidalgo, TX | 0.5 | 65 | 75 | Hidalgo, TX | 0.1 | 65 | 75 | Hidalgo, TX | 1.0 |

# 75 Largest Counties by 2020 Population
## Selected Rankings

| Percent Hispanic or Latino,[1] 2020 | | | | Percent under 15 years old, 2020 | | | | Percent 65 years old and over, 2020 | | | |
|---|---|---|---|---|---|---|---|---|---|---|---|
| Population rank | Hispanic or Latino rank | County | Percent Hispanic or Latino [col 9] | Population rank | Under 18 years old rank | County | Percent under 18 years old [cols 10 and 11] | Population rank | 65 years old and over rank | County | Percent 65 years old and over [cols 17 and 18] |
| 65 | 1 | Hidalgo, TX | 92.5 | 65 | 1 | Hidalgo, TX | 26.4 | 48 | 1 | Pinellas, FL | 25.9 |
| 70 | 2 | El Paso, TX | 82.8 | 63 | 2 | Kern, CA | 23.9 | 25 | 2 | Palm Beach, FL | 24.8 |
| 7 | 3 | Miami-Dade, FL | 68.9 | 46 | 3 | Fresno, CA | 23.5 | 44 | 3 | Pima, AZ | 20.9 |
| 16 | 4 | Bexar, TX | 60.9 | 71 | 4 | Fort Bend, TX | 22.2 | 36 | 4 | Allegheny, PA | 19.8 |
| 29 | 5 | Bronx, NY | 56.4 | 70 | 5 | El Paso, TX | 22.0 | 35 | 5 | Cuyahoga, OH | 19.1 |
| 63 | 6 | Kern, CA | 55.2 | 3 | 6 | Harris, TX | 21.9 | 47 | 6 | St. Louis, MO | 19.0 |
| 14 | 7 | San Bernardino, CA | 54.9 | 37 | 7 | Salt Lake, UT | 21.7 | 52 | 7 | Honolulu, HI | 18.8 |
| 46 | 8 | Fresno, CA | 54.1 | 14 | 8 | San Bernardino, CA | 21.5 | 60 | 7 | Erie, NY | 18.8 |
| 10 | 9 | Riverside, CA | 50.6 | 8 | 9 | Dallas, TX | 21.4 | 30 | 9 | Nassau, NY | 18.6 |
| 1 | 10 | Los Angeles, CA | 48.5 | 54 | 9 | Gwinnett, GA | 21.4 | 72 | 9 | Montgomery, PA | 18.6 |
| 3 | 11 | Harris, TX | 43.8 | 15 | 11 | Tarrant, TX | 21.3 | 68 | 11 | New Haven, CT | 18.3 |
| 69 | 12 | Ventura, CA | 43.4 | 16 | 12 | Bexar, TX | 20.9 | 57 | 12 | Bergen, NJ | 18.0 |
| 8 | 13 | Dallas, TX | 40.8 | 56 | 12 | Shelby, TN | 20.9 | 75 | 12 | Baltimore, MD | 18.0 |
| 44 | 14 | Pima, AZ | 38.0 | 50 | 14 | Marion, IN | 20.8 | 34 | 14 | Oakland, MI | 17.9 |
| 5 | 15 | San Diego, CA | 34.3 | 29 | 15 | Bronx, NY | 20.6 | 66 | 14 | Macomb, MI | 17.9 |
| 6 | 16 | Orange, CA | 33.8 | 43 | 15 | Collin, TX | 20.6 | 27 | 16 | Suffolk, NY | 17.8 |
| 32 | 17 | Travis, TX | 33.3 | 10 | 17 | Riverside, CA | 20.2 | 51 | 16 | Westchester, NY | 17.8 |
| 28 | 18 | Orange, FL | 32.9 | 53 | 18 | Milwaukee, WI | 19.9 | 64 | 16 | Hartford, CT | 17.8 |
| 11 | 19 | Clark, NV | 31.8 | 31 | 19 | Franklin, OH | 19.6 | 17 | 19 | Broward, FL | 17.5 |
| 4 | 20 | Maricopa, AZ | 31.5 | 19 | 20 | Wayne, MI | 19.5 | 21 | 19 | New York, NY | 17.5 |
| 17 | 21 | Broward, FL | 31.4 | 58 | 20 | Denton, TX | 19.5 | 7 | 21 | Miami-Dade, FL | 17.0 |
| 26 | 22 | Hillsborough, FL | 29.9 | 61 | 20 | Pierce, WA | 19.5 | 13 | 22 | Queens, NY | 16.9 |
| 15 | 23 | Tarrant, TX | 29.7 | 24 | 23 | Sacramento, CA | 19.3 | 55 | 23 | Fairfield, CT | 16.8 |
| 13 | 24 | Queens, NY | 27.9 | 9 | 24 | Kings, NY | 19.2 | 38 | 24 | Contra Costa, CA | 16.7 |
| 38 | 25 | Contra Costa, CA | 26.3 | 39 | 24 | Fairfax, VA | 19.2 | 59 | 24 | Du Page, IL | 16.7 |
| 51 | 26 | Westchester, NY | 25.8 | 40 | 24 | Wake, NC | 19.2 | 69 | 24 | Ventura, CA | 16.7 |
| 2 | 27 | Cook, IL | 25.6 | 41 | 24 | Mecklenburg, NC | 19.2 | 67 | 27 | San Francisco, CA | 16.6 |
| 21 | 27 | New York, NY | 25.6 | 4 | 28 | Maricopa, AZ | 19.1 | 74 | 27 | Worcester, MA | 16.6 |
| 71 | 29 | Fort Bend, TX | 25.0 | 49 | 28 | Duval, FL | 19.1 | 45 | 29 | Montgomery, MD | 16.5 |
| 18 | 30 | Santa Clara, CA | 24.7 | 11 | 30 | Clark, NV | 18.9 | 19 | 30 | Wayne, MI | 16.3 |
| 24 | 31 | Sacramento, CA | 23.9 | 45 | 30 | Montgomery, MD | 18.9 | 4 | 31 | Maricopa, AZ | 16.0 |
| 25 | 32 | Palm Beach, FL | 23.4 | 62 | 32 | Prince George's, MD | 18.7 | 22 | 32 | Middlesex, MA | 15.9 |
| 20 | 33 | Alameda, CA | 22.2 | 73 | 32 | Snohomish, WA | 18.7 | 6 | 33 | Orange, CA | 15.7 |
| 54 | 34 | Gwinnett, GA | 21.8 | 26 | 34 | Hillsborough, FL | 18.3 | 2 | 34 | Cook, IL | 15.5 |
| 57 | 35 | Bergen, NJ | 21.5 | 59 | 34 | Du Page, IL | 18.3 | 11 | 34 | Clark, NV | 15.5 |
| 55 | 36 | Fairfield, CT | 20.9 | 69 | 34 | Ventura, CA | 18.3 | 10 | 36 | Riverside, CA | 15.1 |
| 27 | 37 | Suffolk, NY | 20.5 | 23 | 37 | Philadelphia, PA | 18.2 | 5 | 37 | San Diego, CA | 14.9 |
| 45 | 38 | Montgomery, MD | 20.0 | 33 | 37 | Hennepin, MN | 18.2 | 26 | 37 | Hillsborough, FL | 14.9 |
| 62 | 39 | Prince George's, MD | 19.8 | 38 | 37 | Contra Costa, CA | 18.2 | 33 | 37 | Hennepin, MN | 14.9 |
| 58 | 40 | Denton, TX | 19.6 | 47 | 40 | St. Louis, MO | 18.1 | 49 | 37 | Duval, FL | 14.9 |
| 68 | 40 | New Haven, CT | 19.6 | 28 | 41 | Orange, FL | 18.0 | 9 | 41 | Kings, NY | 14.8 |
| 64 | 42 | Hartford, CT | 19.1 | 75 | 41 | Baltimore, MD | 18.0 | 20 | 41 | Alameda, CA | 14.8 |
| 37 | 43 | Salt Lake, UT | 19.0 | 55 | 43 | Fairfield, CT | 17.9 | 24 | 41 | Sacramento, CA | 14.8 |
| 9 | 44 | Kings, NY | 18.8 | 2 | 44 | Cook, IL | 17.8 | 1 | 44 | Los Angeles, CA | 14.5 |
| 30 | 45 | Nassau, NY | 17.5 | 5 | 44 | San Diego, CA | 17.8 | 56 | 44 | Shelby, TN | 14.5 |
| 39 | 46 | Fairfax, VA | 16.5 | 6 | 46 | Orange, CA | 17.7 | 23 | 46 | Philadelphia, PA | 14.4 |
| 53 | 47 | Milwaukee, WI | 16.0 | 52 | 46 | Honolulu, HI | 17.7 | 39 | 46 | Fairfax, VA | 14.4 |
| 23 | 48 | Philadelphia, PA | 15.6 | 18 | 48 | Santa Clara, CA | 17.6 | 53 | 46 | Milwaukee, WI | 14.4 |
| 43 | 48 | Collin, TX | 15.6 | 1 | 49 | Los Angeles, CA | 17.5 | 61 | 46 | Pierce, WA | 14.4 |
| 67 | 50 | San Francisco, CA | 15.2 | 30 | 49 | Nassau, NY | 17.5 | 73 | 46 | Snohomish, WA | 14.4 |
| 59 | 51 | Du Page, IL | 14.7 | 32 | 49 | Travis, TX | 17.5 | 18 | 51 | Santa Clara, CA | 14.3 |
| 41 | 52 | Mecklenburg, NC | 13.8 | 51 | 49 | Westchester, NY | 17.5 | 62 | 52 | Prince George's, MD | 14.2 |
| 74 | 53 | Worcester, MA | 12.4 | 72 | 49 | Montgomery, PA | 17.5 | 29 | 53 | Bronx, NY | 13.8 |
| 61 | 54 | Pierce, WA | 11.8 | 17 | 54 | Broward, FL | 17.4 | 12 | 54 | King, WA | 13.7 |
| 49 | 55 | Duval, FL | 11.0 | 42 | 55 | Fulton, GA | 17.3 | 50 | 55 | Marion, IN | 13.1 |
| 50 | 55 | Marion, IN | 11.0 | 57 | 56 | Bergen, NJ | 17.1 | 46 | 56 | Fresno, CA | 12.9 |
| 73 | 55 | Snohomish, WA | 11.0 | 20 | 57 | Alameda, CA | 16.9 | 70 | 57 | El Paso, TX | 12.8 |
| 40 | 58 | Wake, NC | 10.3 | 35 | 57 | Cuyahoga, OH | 16.9 | 16 | 58 | Bexar, TX | 12.6 |
| 48 | 58 | Pinellas, FL | 10.3 | 64 | 57 | Hartford, CT | 16.9 | 28 | 58 | Orange, FL | 12.6 |
| 52 | 60 | Honolulu, HI | 10.2 | 66 | 57 | Macomb, MI | 16.9 | 31 | 58 | Franklin, OH | 12.6 |
| 12 | 61 | King, WA | 10.1 | 74 | 57 | Worcester, MA | 16.9 | 40 | 61 | Wake, NC | 12.5 |
| 22 | 62 | Middlesex, MA | 8.3 | 7 | 62 | Miami-Dade, FL | 16.8 | 42 | 62 | Fulton, GA | 12.3 |
| 42 | 63 | Fulton, GA | 7.2 | 13 | 62 | Queens, NY | 16.8 | 14 | 63 | San Bernardino, CA | 12.2 |
| 33 | 64 | Hennepin, MN | 7.0 | 27 | 62 | Suffolk, NY | 16.8 | 71 | 64 | Fort Bend, TX | 12.1 |
| 56 | 65 | Shelby, TN | 6.7 | 44 | 62 | Pima, AZ | 16.8 | 15 | 65 | Tarrant, TX | 12.0 |
| 35 | 66 | Cuyahoga, OH | 6.5 | 12 | 66 | King, WA | 16.6 | 41 | 66 | Mecklenburg, NC | 11.8 |
| 19 | 67 | Wayne, MI | 6.3 | 34 | 66 | Oakland, MI | 16.6 | 37 | 67 | Salt Lake, UT | 11.5 |
| 31 | 68 | Franklin, OH | 6.0 | 60 | 66 | Erie, NY | 16.6 | 43 | 67 | Collin, TX | 11.5 |
| 60 | 68 | Erie, NY | 6.0 | 68 | 69 | New Haven, CT | 16.2 | 63 | 67 | Kern, CA | 11.5 |
| 75 | 68 | Baltimore, MD | 6.0 | 22 | 70 | Middlesex, MA | 16.1 | 65 | 67 | Hidalgo, TX | 11.5 |
| 72 | 71 | Montgomery, PA | 5.6 | 25 | 71 | Palm Beach, FL | 15.5 | 8 | 71 | Dallas, TX | 11.4 |
| 34 | 72 | Oakland, MI | 4.6 | 36 | 71 | Allegheny, PA | 15.5 | 3 | 72 | Harris, TX | 11.3 |
| 47 | 73 | St. Louis, MO | 3.1 | 48 | 73 | Pinellas, FL | 12.9 | 58 | 73 | Denton, TX | 11.0 |
| 66 | 74 | Macomb, MI | 2.8 | 21 | 74 | New York, NY | 12.2 | 54 | 74 | Gwinnett, GA | 10.9 |
| 36 | 75 | Allegheny, PA | 2.3 | 67 | 75 | San Francisco, CA | 11.5 | 32 | 75 | Travis, TX | 10.5 |

1. May be of any race.

# 75 Largest Counties by 2020 Population
## Selected Rankings

| Percent female-headed family households, 2015–2019 | | | | Birth rate, 2020 | | | | Percent under 65 who have no health insurance, 2019 | | | |
|---|---|---|---|---|---|---|---|---|---|---|---|
| Population rank | Female households rank | County | Percent female households [col 30] | Population rank | Live birth rate rank | County | Birth rate [col 36] | Population rank | No health insurance rank | County | Percent with no health insurance [col 40] |
| 29 | 1 | Bronx, NY | 29.0 | 65 | 1 | Hidalgo, TX | 16.1 | 65 | 1 | Hidalgo, TX | 33.2 |
| 65 | 2 | Hidalgo, TX | 21.9 | 9 | 2 | Kings, NY | 14.7 | 8 | 2 | Dallas, TX | 25.1 |
| 23 | 3 | Philadelphia, PA | 20.1 | 50 | 3 | Marion, IN | 14.4 | 70 | 3 | El Paso, TX | 24.3 |
| 56 | 4 | Shelby, TN | 19.5 | 8 | 4 | Dallas, TX | 14.3 | 3 | 4 | Harris, TX | 24.2 |
| 62 | 5 | Prince George's, MD | 18.9 | 46 | 5 | Fresno, CA | 14.2 | 7 | 5 | Miami-Dade, FL | 19.4 |
| 19 | 6 | Wayne, MI | 18.6 | 63 | 5 | Kern, CA | 14.2 | 16 | 6 | Bexar, TX | 19.1 |
| 70 | 6 | El Paso, TX | 18.6 | 3 | 7 | Harris, TX | 14.0 | 15 | 7 | Tarrant, TX | 18.9 |
| 7 | 8 | Miami-Dade, FL | 18.3 | 37 | 8 | Salt Lake, UT | 13.8 | 25 | 8 | Palm Beach, FL | 17.9 |
| 9 | 9 | Kings, NY | 17.9 | 70 | 8 | El Paso, TX | 13.8 | 17 | 9 | Broward, FL | 17.7 |
| 46 | 10 | Fresno, CA | 17.6 | 29 | 10 | Bronx, NY | 13.7 | 54 | 9 | Gwinnett, GA | 17.7 |
| 14 | 11 | San Bernardino, CA | 16.9 | 31 | 10 | Franklin, OH | 13.7 | 32 | 11 | Travis, TX | 16.5 |
| 53 | 12 | Milwaukee, WI | 16.2 | 53 | 10 | Milwaukee, WI | 13.7 | 28 | 12 | Orange, FL | 15.4 |
| 63 | 12 | Kern, CA | 16.2 | 56 | 10 | Shelby, TN | 13.7 | 48 | 12 | Pinellas, FL | 15.4 |
| 13 | 14 | Queens, NY | 15.8 | 49 | 14 | Duval, FL | 13.4 | 71 | 14 | Fort Bend, TX | 15.0 |
| 16 | 14 | Bexar, TX | 15.6 | 14 | 15 | San Bernardino, CA | 13.2 | 26 | 15 | Hillsborough, FL | 14.9 |
| 49 | 16 | Duval, FL | 15.6 | 16 | 16 | Bexar, TX | 13.1 | 49 | 16 | Duval, FL | 14.3 |
| 3 | 17 | Harris, TX | 15.5 | 23 | 17 | Philadelphia, PA | 13.0 | 11 | 17 | Clark, NV | 13.7 |
| 35 | 18 | Cuyahoga, OH | 15.3 | 41 | 17 | Mecklenburg, NC | 13.0 | 58 | 18 | Denton, TX | 13.4 |
| 8 | 19 | Dallas, TX | 15.2 | 62 | 17 | Prince George's, MD | 13.0 | 4 | 19 | Maricopa, AZ | 13.3 |
| 17 | 20 | Broward, FL | 15.1 | 19 | 20 | Wayne, MI | 12.9 | 44 | 19 | Pima, AZ | 13.3 |
| 28 | 21 | Orange, FL | 15.0 | 15 | 21 | Tarrant, TX | 12.8 | 41 | 21 | Mecklenburg, NC | 13.2 |
| 1 | 22 | Los Angeles, CA | 14.7 | 61 | 22 | Pierce, WA | 12.4 | 56 | 21 | Shelby, TN | 13.2 |
| 50 | 22 | Marion, IN | 14.7 | 33 | 23 | Hennepin, MN | 12.3 | 42 | 23 | Fulton, GA | 12.6 |
| 2 | 24 | Cook, IL | 14.2 | 24 | 24 | Sacramento, CA | 12.2 | 43 | 24 | Collin, TX | 12.5 |
| 15 | 24 | Tarrant, TX | 14.2 | 52 | 24 | Honolulu, HI | 12.2 | 50 | 25 | Marion, IN | 12.3 |
| 68 | 24 | New Haven, CT | 14.2 | 13 | 26 | Queens, NY | 12.0 | 37 | 26 | Salt Lake, UT | 11.4 |
| 24 | 27 | Sacramento, CA | 14.1 | 28 | 27 | Orange, FL | 11.9 | 1 | 27 | Los Angeles, CA | 11.1 |
| 64 | 27 | Hartford, CT | 14.1 | 32 | 27 | Travis, TX | 11.9 | 2 | 28 | Cook, IL | 10.9 |
| 75 | 27 | Baltimore, MD | 14.1 | 2 | 29 | Cook, IL | 11.8 | 69 | 29 | Ventura, CA | 10.4 |
| 11 | 30 | Clark, NV | 13.9 | 39 | 29 | Fairfax, VA | 11.8 | 13 | 30 | Queens, NY | 10.3 |
| 31 | 30 | Franklin, OH | 13.9 | 54 | 29 | Gwinnett, GA | 11.8 | 62 | 31 | Prince George's, MD | 10.2 |
| 54 | 32 | Gwinnett, GA | 13.8 | 73 | 29 | Snohomish, WA | 11.8 | 40 | 32 | Wake, NC | 10.1 |
| 26 | 33 | Hillsborough, FL | 13.7 | 5 | 33 | San Diego, CA | 11.7 | 14 | 33 | San Bernardino, CA | 9.9 |
| 42 | 33 | Fulton, GA | 13.7 | 11 | 34 | Clark, NV | 11.6 | 10 | 34 | Riverside, CA | 9.8 |
| 47 | 35 | St. Louis, MO | 13.6 | 75 | 34 | Baltimore, MD | 11.6 | 46 | 35 | Fresno, CA | 9.7 |
| 41 | 36 | Mecklenburg, NC | 13.1 | 4 | 36 | Maricopa, AZ | 11.5 | 55 | 36 | Fairfield, CT | 9.3 |
| 10 | 37 | Riverside, CA | 13.0 | 26 | 36 | Hillsborough, FL | 11.5 | 63 | 36 | Kern, CA | 9.3 |
| 60 | 37 | Erie, NY | 13.0 | 10 | 38 | Riverside, CA | 11.4 | 39 | 38 | Fairfax, VA | 9.1 |
| 66 | 39 | Macomb, MI | 12.8 | 42 | 38 | Fulton, GA | 11.4 | 23 | 39 | Philadelphia, PA | 9.0 |
| 44 | 40 | Pima, AZ | 12.5 | 45 | 38 | Montgomery, MD | 11.4 | 5 | 40 | San Diego, CA | 8.9 |
| 51 | 41 | Westchester, NY | 12.4 | 47 | 41 | St. Louis, MO | 11.3 | 31 | 40 | Franklin, OH | 8.9 |
| 4 | 42 | Maricopa, AZ | 12.2 | 71 | 41 | Fort Bend, TX | 11.3 | 6 | 42 | Orange, CA | 8.8 |
| 38 | 42 | Contra Costa, CA | 12.2 | 35 | 43 | Cuyahoga, OH | 11.2 | 29 | 43 | Bronx, NY | 8.5 |
| 52 | 44 | Honolulu, HI | 12.1 | 40 | 43 | Wake, NC | 11.2 | 53 | 43 | Milwaukee, WI | 8.5 |
| 69 | 45 | Ventura, CA | 11.9 | 6 | 45 | Orange, CA | 11.1 | 57 | 45 | Bergen, NJ | 8.4 |
| 71 | 45 | Fort Bend, TX | 11.9 | 7 | 45 | Miami-Dade, FL | 11.1 | 47 | 46 | St. Louis, MO | 8.3 |
| 74 | 45 | Worcester, MA | 11.9 | 17 | 47 | Broward, FL | 11.0 | 45 | 47 | Montgomery, MD | 8.1 |
| 55 | 48 | Fairfield, CT | 11.8 | 18 | 47 | Santa Clara, CA | 11.0 | 19 | 48 | Wayne, MI | 7.6 |
| 5 | 49 | San Diego, CA | 11.7 | 58 | 47 | Denton, TX | 11.0 | 61 | 48 | Pierce, WA | 7.6 |
| 6 | 50 | Orange, CA | 11.5 | 20 | 50 | Alameda, CA | 10.9 | 35 | 50 | Cuyahoga, OH | 7.3 |
| 25 | 51 | Palm Beach, FL | 11.4 | 59 | 50 | Du Page, IL | 10.9 | 66 | 50 | Macomb, MI | 7.3 |
| 61 | 51 | Pierce, WA | 11.4 | 1 | 52 | Los Angeles, CA | 10.8 | 9 | 52 | Kings, NY | 7.2 |
| 30 | 53 | Nassau, NY | 11.2 | 12 | 52 | King, WA | 10.8 | 73 | 52 | Snohomish, WA | 7.2 |
| 40 | 53 | Wake, NC | 11.2 | 43 | 54 | Collin, TX | 10.6 | 68 | 54 | New Haven, CT | 7.0 |
| 20 | 55 | Alameda, CA | 11.1 | 60 | 54 | Erie, NY | 10.6 | 75 | 55 | Baltimore, MD | 6.9 |
| 27 | 55 | Suffolk, NY | 11.1 | 36 | 56 | Allegheny, PA | 10.5 | 24 | 56 | Sacramento, CA | 6.8 |
| 45 | 57 | Montgomery, MD | 11.0 | 66 | 56 | Macomb, MI | 10.5 | 38 | 57 | Contra Costa, CA | 6.5 |
| 21 | 58 | New York, NY | 10.9 | 69 | 56 | Ventura, CA | 10.5 | 59 | 58 | Du Page, IL | 6.4 |
| 48 | 59 | Pinellas, FL | 10.7 | 34 | 59 | Oakland, MI | 10.4 | 12 | 59 | King, WA | 6.3 |
| 57 | 60 | Bergen, NJ | 10.6 | 21 | 60 | New York, NY | 10.3 | 64 | 60 | Hartford, CT | 6.0 |
| 36 | 61 | Allegheny, PA | 10.5 | 27 | 60 | Suffolk, NY | 10.3 | 33 | 61 | Hennepin, MN | 5.8 |
| 37 | 62 | Salt Lake, UT | 10.3 | 30 | 60 | Nassau, NY | 10.3 | 34 | 62 | Oakland, MI | 5.7 |
| 58 | 63 | Denton, TX | 9.9 | 51 | 60 | Westchester, NY | 10.3 | 51 | 62 | Westchester, NY | 5.7 |
| 18 | 64 | Santa Clara, CA | 9.8 | 22 | 64 | Middlesex, MA | 10.2 | 21 | 64 | New York, NY | 5.6 |
| 34 | 64 | Oakland, MI | 9.8 | 38 | 64 | Contra Costa, CA | 10.2 | 18 | 65 | Santa Clara, CA | 5.5 |
| 73 | 64 | Snohomish, WA | 9.8 | 64 | 64 | Hartford, CT | 10.2 | 27 | 66 | Suffolk, NY | 5.2 |
| 32 | 67 | Travis, TX | 9.6 | 72 | 64 | Montgomery, PA | 10.2 | 67 | 66 | San Francisco, CA | 5.2 |
| 43 | 67 | Collin, TX | 9.6 | 44 | 68 | Pima, AZ | 10.1 | 20 | 68 | Alameda, CA | 5.1 |
| 72 | 67 | Montgomery, PA | 9.6 | 55 | 69 | Fairfield, CT | 10.0 | 36 | 68 | Allegheny, PA | 5.1 |
| 22 | 70 | Middlesex, MA | 9.4 | 74 | 69 | Worcester, MA | 10.0 | 72 | 70 | Montgomery, PA | 4.9 |
| 39 | 71 | Fairfax, VA | 9.1 | 25 | 71 | Palm Beach, FL | 9.9 | 30 | 71 | Nassau, NY | 4.8 |
| 59 | 71 | Du Page, IL | 9.1 | 67 | 71 | San Francisco, CA | 9.9 | 52 | 72 | Honolulu, HI | 4.4 |
| 33 | 73 | Hennepin, MN | 9.0 | 68 | 71 | New Haven, CT | 9.9 | 60 | 73 | Erie, NY | 4.1 |
| 12 | 74 | King, WA | 8.3 | 57 | 74 | Bergen, NJ | 9.6 | 74 | 74 | Worcester, MA | 3.5 |
| 67 | 75 | San Francisco, CA | 7.9 | 48 | 75 | Pinellas, FL | 8.1 | 22 | 75 | Middlesex, MA | 3.1 |

# 75 Largest Counties by 2020 Population
## Selected Rankings

| Percent college graduates (bachelor's degree or more), 2015–2019 | | | | Expenditures per student, 2017–2018 | | | | Per capita personal income, 2019 | | | |
|---|---|---|---|---|---|---|---|---|---|---|---|
| Population rank | College graduates rank | County | Percent college graduates [col 51] | Population rank | Expenditures rank | County | Expenditures per student (dollars) [col 53] | Population rank | Per capita income rank | County | Per capita income (dollars) [col 64] |
| 39 | 1 | Fairfax, VA | 61.6 | 30 | 1 | Nassau, NY | 26,492 | 21 | 1 | New York, NY | 197,847 |
| 21 | 2 | New York, NY | 61.3 | 51 | 2 | Westchester, NY | 25,952 | 67 | 2 | San Francisco, CA | 139,405 |
| 45 | 3 | Montgomery, MD | 58.9 | 21 | 3 | New York, NY | 25,791 | 55 | 3 | Fairfield, CT | 121,397 |
| 67 | 4 | San Francisco, CA | 58.1 | 27 | 3 | Kings, NY | 25,791 | 18 | 4 | Santa Clara, CA | 115,997 |
| 22 | 5 | Middlesex, MA | 56.3 | 64 | 3 | Queens, NY | 25,791 | 51 | 5 | Westchester, NY | 113,477 |
| 42 | 6 | Fulton, GA | 52.9 | 57 | 3 | Bronx, NY | 25,791 | 12 | 6 | King, WA | 94,974 |
| 40 | 7 | Wake, NC | 52.8 | 55 | 7 | Suffolk, NY | 25,319 | 30 | 7 | Nassau, NY | 93,241 |
| 12 | 8 | King, WA | 52.5 | 68 | 8 | Hartford, CT | 21,093 | 45 | 8 | Montgomery, MD | 90,139 |
| 18 | 9 | Santa Clara, CA | 52.4 | 72 | 9 | Bergen, NJ | 20,892 | 57 | 9 | Bergen, NJ | 89,456 |
| 43 | 10 | Collin, TX | 52.3 | 59 | 10 | Fairfield, CT | 20,186 | 42 | 10 | Fulton, GA | 88,832 |
| 33 | 11 | Hennepin, MN | 50.1 | 36 | 11 | New Haven, CT | 19,903 | 22 | 11 | Middlesex, MA | 87,192 |
| 32 | 12 | Travis, TX | 50.0 | 22 | 12 | Montgomery, PA | 18,902 | 39 | 12 | Fairfax, VA | 86,141 |
| 59 | 13 | Du Page, IL | 49.4 | 60 | 13 | Du Page, IL | 18,500 | 38 | 13 | Contra Costa, CA | 85,324 |
| 57 | 14 | Bergen, NJ | 49.3 | 2 | 14 | Allegheny, PA | 17,999 | 25 | 14 | Palm Beach, FL | 83,268 |
| 72 | 14 | Montgomery, PA | 49.3 | 45 | 15 | Middlesex, MA | 17,655 | 72 | 15 | Montgomery, PA | 82,037 |
| 51 | 16 | Westchester, NY | 48.9 | 35 | 16 | Erie, NY | 17,371 | 20 | 16 | Alameda, CA | 81,171 |
| 55 | 17 | Fairfield, CT | 47.9 | 62 | 17 | Cook, IL | 16,846 | 33 | 17 | Hennepin, MN | 76,552 |
| 20 | 18 | Alameda, CA | 47.4 | 67 | 18 | Montgomery, MD | 16,005 | 59 | 18 | Du Page, IL | 75,137 |
| 34 | 19 | Oakland, MI | 47.2 | 52 | 19 | Cuyahoga, OH | 15,670 | 34 | 19 | Oakland, MI | 73,271 |
| 71 | 20 | Fort Bend, TX | 46.2 | 23 | 20 | Prince George's, MD | 15,378 | 47 | 20 | St. Louis, MO | 73,016 |
| 30 | 21 | Nassau, NY | 46.0 | 74 | 21 | San Francisco, CA | 15,330 | 27 | 21 | Suffolk, NY | 71,911 |
| 41 | 22 | Mecklenburg, NC | 45.4 | 39 | 22 | Honolulu, HI | 15,242 | 6 | 22 | Orange, CA | 71,711 |
| 58 | 23 | Denton, TX | 45.1 | 33 | 23 | Philadelphia, PA | 15,238 | 32 | 23 | Travis, TX | 71,666 |
| 47 | 24 | St. Louis, MO | 43.6 | 75 | 24 | Worcester, MA | 15,164 | 43 | 24 | Collin, TX | 68,474 |
| 38 | 25 | Contra Costa, CA | 42.4 | 47 | 25 | Fairfax, VA | 14,937 | 36 | 25 | Allegheny, PA | 65,784 |
| 36 | 26 | Allegheny, PA | 41.6 | 12 | 26 | Hennepin, MN | 14,231 | 64 | 26 | Hartford, CT | 65,720 |
| 6 | 27 | Orange, CA | 40.6 | 42 | 27 | Baltimore, MD | 14,122 | 2 | 27 | Cook, IL | 65,306 |
| 31 | 28 | Franklin, OH | 40.0 | 46 | 28 | St. Louis, MO | 13,975 | 1 | 28 | Los Angeles, CA | 65,094 |
| 75 | 29 | Baltimore, MD | 39.4 | 18 | 29 | King, WA | 13,320 | 69 | 29 | Ventura, CA | 64,715 |
| 2 | 30 | Cook, IL | 38.8 | 61 | 30 | Fulton, GA | 13,189 | 5 | 30 | San Diego, CA | 63,729 |
| 5 | 30 | San Diego, CA | 38.8 | 73 | 31 | Fresno, CA | 13,178 | 75 | 31 | Baltimore, MD | 62,976 |
| 64 | 32 | Hartford, CT | 38.1 | 1 | 32 | Santa Clara, CA | 13,067 | 41 | 32 | Mecklenburg, NC | 62,890 |
| 9 | 33 | Kings, NY | 37.5 | 10 | 33 | Pierce, WA | 12,936 | 8 | 33 | Dallas, TX | 62,782 |
| 54 | 34 | Gwinnett, GA | 36.9 | 53 | 34 | Snohomish, WA | 12,822 | 40 | 34 | Wake, NC | 62,264 |
| 25 | 35 | Palm Beach, FL | 36.7 | 63 | 35 | Los Angeles, CA | 12,758 | 52 | 35 | Honolulu, HI | 61,174 |
| 74 | 36 | Worcester, MA | 36.4 | 34 | 36 | Riverside, CA | 12,726 | 3 | 36 | Harris, TX | 60,002 |
| 27 | 37 | Suffolk, NY | 36.3 | 31 | 37 | Milwaukee, WI | 12,541 | 71 | 37 | Fort Bend, TX | 59,653 |
| 37 | 38 | Salt Lake, UT | 35.6 | 69 | 38 | Kern, CA | 12,490 | 58 | 38 | Denton, TX | 59,414 |
| 52 | 39 | Honolulu, HI | 35.0 | 14 | 39 | Oakland, MI | 12,395 | 73 | 39 | Snohomish, WA | 58,729 |
| 68 | 39 | New Haven, CT | 35.0 | 6 | 40 | Franklin, OH | 12,361 | 74 | 40 | Worcester, MA | 58,563 |
| 28 | 41 | Orange, FL | 34.6 | 24 | 41 | Ventura, CA | 12,341 | 68 | 41 | New Haven, CT | 57,748 |
| 60 | 42 | Erie, NY | 34.4 | 38 | 42 | San Bernardino, CA | 12,199 | 23 | 42 | Philadelphia, PA | 57,265 |
| 69 | 43 | Ventura, CA | 33.8 | 66 | 43 | Orange, CA | 12,014 | 35 | 43 | Cuyahoga, OH | 56,502 |
| 26 | 44 | Hillsborough, FL | 33.5 | 20 | 44 | Sacramento, CA | 12,012 | 9 | 44 | Kings, NY | 56,080 |
| 62 | 45 | Prince George's, MD | 33.1 | 5 | 45 | Contra Costa, CA | 11,956 | 48 | 45 | Pinellas, FL | 55,607 |
| 73 | 46 | Snohomish, WA | 32.8 | 19 | 46 | Macomb, MI | 11,742 | 37 | 46 | Salt Lake, UT | 55,446 |
| 4 | 47 | Maricopa, AZ | 32.7 | 50 | 47 | Alameda, CA | 11,711 | 24 | 47 | Sacramento, CA | 55,266 |
| 1 | 48 | Los Angeles, CA | 32.5 | 56 | 48 | San Diego, CA | 11,551 | 7 | 48 | Miami-Dade, FL | 54,902 |
| 35 | 48 | Cuyahoga, OH | 32.5 | 65 | 49 | Wayne, MI | 11,520 | 50 | 49 | Marion, IN | 54,405 |
| 17 | 50 | Broward, FL | 32.4 | 54 | 50 | Marion, IN | 11,033 | 61 | 50 | Pierce, WA | 53,572 |
| 44 | 50 | Pima, AZ | 32.4 | 32 | 51 | Shelby, TN | 10,603 | 60 | 51 | Erie, NY | 53,498 |
| 15 | 52 | Tarrant, TX | 32.3 | 25 | 52 | Hidalgo, TX | 10,424 | 15 | 52 | Tarrant, TX | 53,292 |
| 13 | 53 | Queens, NY | 32.2 | 70 | 53 | Gwinnett, GA | 10,243 | 13 | 53 | Queens, NY | 53,008 |
| 48 | 54 | Pinellas, FL | 31.7 | 16 | 54 | Travis, TX | 10,006 | 17 | 54 | Broward, FL | 52,308 |
| 56 | 55 | Shelby, TN | 31.6 | 48 | 55 | Palm Beach, FL | 9,942 | 31 | 55 | Franklin, OH | 51,644 |
| 3 | 56 | Harris, TX | 31.5 | 3 | 56 | El Paso, TX | 9,933 | 56 | 56 | Shelby, TN | 50,744 |
| 8 | 56 | Dallas, TX | 31.5 | 7 | 57 | Bexar, TX | 9,771 | 62 | 57 | Prince George's, MD | 50,625 |
| 53 | 58 | Milwaukee, WI | 31.0 | 17 | 58 | Pinellas, FL | 9,621 | 4 | 58 | Maricopa, AZ | 49,704 |
| 24 | 59 | Sacramento, CA | 30.9 | 8 | 59 | Harris, TX | 9,544 | 53 | 59 | Milwaukee, WI | 49,098 |
| 50 | 59 | Marion, IN | 30.9 | 71 | 60 | Miami-Dade, FL | 9,542 | 11 | 60 | Clark, NV | 48,806 |
| 49 | 61 | Duval, FL | 30.0 | 28 | 61 | Broward, FL | 9,464 | 26 | 61 | Hillsborough, FL | 48,452 |
| 7 | 62 | Miami-Dade, FL | 29.8 | 15 | 62 | Dallas, TX | 9,460 | 16 | 62 | Bexar, TX | 47,830 |
| 23 | 63 | Philadelphia, PA | 29.7 | 41 | 63 | Fort Bend, TX | 9,343 | 66 | 63 | Macomb, MI | 47,487 |
| 16 | 64 | Bexar, TX | 28.0 | 58 | 64 | Orange, FL | 9,295 | 49 | 64 | Duval, FL | 47,475 |
| 61 | 65 | Pierce, WA | 27.2 | 49 | 65 | Tarrant, TX | 9,266 | 28 | 65 | Orange, FL | 46,250 |
| 66 | 66 | Macomb, MI | 24.9 | 43 | 66 | Mecklenburg, NC | 9,246 | 46 | 66 | Fresno, CA | 45,487 |
| 11 | 67 | Clark, NV | 24.5 | 75 | 67 | Denton, TX | 9,213 | 44 | 67 | Pima, AZ | 45,456 |
| 19 | 68 | Wayne, MI | 23.9 | 26 | 68 | Duval, FL | 9,163 | 19 | 68 | Wayne, MI | 44,512 |
| 70 | 69 | El Paso, TX | 23.3 | 40 | 69 | Collin, TX | 8,947 | 54 | 69 | Gwinnett, GA | 42,902 |
| 10 | 70 | Riverside, CA | 22.3 | 44 | 70 | Clark, NV | 8,943 | 10 | 70 | Riverside, CA | 42,418 |
| 46 | 71 | Fresno, CA | 21.2 | 4 | 71 | Hillsborough, FL | 8,889 | 14 | 71 | San Bernardino, CA | 42,043 |
| 14 | 72 | San Bernardino, CA | 21.0 | 37 | 72 | Wake, NC | 8,838 | 63 | 72 | Kern, CA | 41,843 |
| 29 | 73 | Bronx, NY | 20.1 | 9 | 73 | Pima, AZ | 8,536 | 29 | 73 | Bronx, NY | 39,711 |
| 65 | 74 | Hidalgo, TX | 18.7 | 13 | 74 | Maricopa, AZ | 8,083 | 70 | 74 | El Paso, TX | 37,715 |
| 63 | 75 | Kern, CA | 16.4 | 29 | 75 | Salt Lake, UT | 7,757 | 65 | 75 | Hidalgo, TX | 27,415 |

# 75 Largest Counties by 2020 Population
## Selected Rankings

| | Median household income, 2019 | | | | Median value of owner-occupied housing units, 2015–2019 | | | | Median gross rent of renter-occupied housing units, 2015–2019 | | |
|---|---|---|---|---|---|---|---|---|---|---|---|
| Popu-lation rank | Median income rank | County | Median income (dollars) [col 58] | Popu-lation rank | Median value rank | County | Median value (dollars) [col 91] | Popu-lation rank | Median rent rank | County | Median rent (dollars) [col 94] |
| 18 | 1 | Santa Clara, CA | 132,444 | 67 | 1 | San Francisco, CA | 1,097,800 | 18 | 1 | Santa Clara, CA | 2,268 |
| 39 | 2 | Fairfax, VA | 127,898 | 21 | 2 | New York, NY | 987,700 | 67 | 2 | San Francisco, CA | 1,895 |
| 67 | 3 | San Francisco, CA | 121,795 | 18 | 3 | Santa Clara, CA | 984,000 | 39 | 3 | Fairfax, VA | 1,881 |
| 30 | 4 | Nassau, NY | 117,767 | 20 | 4 | Alameda, CA | 769,300 | 6 | 4 | Orange, CA | 1,854 |
| 45 | 5 | Montgomery, MD | 110,012 | 9 | 5 | Kings, NY | 706,000 | 38 | 5 | Contra Costa, CA | 1,819 |
| 57 | 6 | Bergen, NJ | 107,971 | 6 | 6 | Orange, CA | 679,300 | 20 | 6 | Alameda, CA | 1,797 |
| 20 | 7 | Alameda, CA | 107,589 | 52 | 7 | Honolulu, HI | 678,200 | 69 | 7 | Ventura, CA | 1,776 |
| 38 | 8 | Contra Costa, CA | 106,555 | 38 | 8 | Contra Costa, CA | 625,800 | 30 | 8 | Nassau, NY | 1,772 |
| 22 | 9 | Middlesex, MA | 106,543 | 69 | 9 | Ventura, CA | 588,400 | 45 | 9 | Montgomery, MD | 1,768 |
| 27 | 10 | Suffolk, NY | 105,241 | 1 | 10 | Los Angeles, CA | 583,200 | 52 | 10 | Honolulu, HI | 1,745 |
| 12 | 11 | King, WA | 102,338 | 5 | 11 | San Diego, CA | 563,700 | 27 | 11 | Suffolk, NY | 1,742 |
| 51 | 12 | Westchester, NY | 101,741 | 39 | 12 | Fairfax, VA | 563,100 | 21 | 12 | New York, NY | 1,740 |
| 71 | 13 | Fort Bend, TX | 101,361 | 12 | 13 | King, WA | 549,200 | 5 | 13 | San Diego, CA | 1,658 |
| 55 | 14 | Fairfield, CT | 96,966 | 13 | 14 | Queens, NY | 543,800 | 22 | 14 | Middlesex, MA | 1,636 |
| 43 | 15 | Collin, TX | 96,847 | 51 | 15 | Westchester, NY | 540,600 | 12 | 15 | King, WA | 1,606 |
| 59 | 16 | Du Page, IL | 96,354 | 22 | 16 | Middlesex, MA | 500,700 | 13 | 16 | Queens, NY | 1,583 |
| 6 | 17 | Orange, CA | 95,761 | 30 | 17 | Nassau, NY | 493,500 | 51 | 17 | Westchester, NY | 1,537 |
| 21 | 18 | New York, NY | 93,007 | 45 | 18 | Montgomery, MD | 484,900 | 57 | 18 | Bergen, NJ | 1,506 |
| 72 | 19 | Montgomery, PA | 92,340 | 57 | 19 | Bergen, NJ | 469,500 | 55 | 19 | Fairfield, CT | 1,499 |
| 69 | 20 | Ventura, CA | 91,446 | 55 | 20 | Fairfield, CT | 428,500 | 62 | 20 | Prince George's, MD | 1,475 |
| 58 | 21 | Denton, TX | 90,910 | 29 | 21 | Bronx, NY | 404,700 | 1 | 21 | Los Angeles, CA | 1,460 |
| 73 | 22 | Snohomish, WA | 89,119 | 73 | 22 | Snohomish, WA | 403,000 | 73 | 22 | Snohomish, WA | 1,438 |
| 52 | 23 | Honolulu, HI | 86,946 | 27 | 23 | Suffolk, NY | 397,400 | 71 | 23 | Fort Bend, TX | 1,431 |
| 62 | 24 | Prince George's, MD | 85,357 | 24 | 24 | Sacramento, CA | 351,900 | 9 | 24 | Kings, NY | 1,426 |
| 40 | 25 | Wake, NC | 84,377 | 10 | 25 | Riverside, CA | 350,100 | 25 | 25 | Palm Beach, FL | 1,398 |
| 5 | 26 | San Diego, CA | 83,576 | 14 | 26 | San Bernardino, CA | 328,200 | 17 | 26 | Broward, FL | 1,392 |
| 33 | 27 | Hennepin, MN | 82,323 | 32 | 27 | Travis, TX | 324,800 | 43 | 27 | Collin, TX | 1,389 |
| 34 | 28 | Oakland, MI | 81,257 | 72 | 28 | Montgomery, PA | 316,100 | 10 | 28 | Riverside, CA | 1,375 |
| 32 | 29 | Travis, TX | 80,690 | 43 | 29 | Collin, TX | 315,300 | 7 | 29 | Miami-Dade, FL | 1,328 |
| 37 | 30 | Salt Lake, UT | 79,941 | 42 | 30 | Fulton, GA | 313,300 | 59 | 30 | Du Page, IL | 1,316 |
| 42 | 31 | Fulton, GA | 79,235 | 59 | 31 | Du Page, IL | 308,500 | 75 | 31 | Baltimore, MD | 1,302 |
| 61 | 32 | Pierce, WA | 78,779 | 37 | 32 | Salt Lake, UT | 305,700 | 72 | 32 | Montgomery, PA | 1,295 |
| 74 | 33 | Worcester, MA | 77,795 | 61 | 33 | Pierce, WA | 303,200 | 32 | 33 | Travis, TX | 1,289 |
| 75 | 34 | Baltimore, MD | 76,972 | 62 | 34 | Prince George's, MD | 302,800 | 14 | 34 | San Bernardino, CA | 1,283 |
| 64 | 35 | Hartford, CT | 75,336 | 7 | 35 | Miami-Dade, FL | 289,600 | 54 | 35 | Gwinnett, GA | 1,272 |
| 13 | 36 | Queens, NY | 72,975 | 25 | 36 | Palm Beach, FL | 283,600 | 24 | 36 | Sacramento, CA | 1,252 |
| 10 | 37 | Riverside, CA | 72,905 | 40 | 37 | Wake, NC | 281,700 | 61 | 37 | Pierce, WA | 1,250 |
| 1 | 38 | Los Angeles, CA | 72,721 | 74 | 38 | Worcester, MA | 280,600 | 58 | 38 | Denton, TX | 1,218 |
| 54 | 39 | Gwinnett, GA | 72,225 | 58 | 39 | Denton, TX | 277,800 | 28 | 39 | Orange, FL | 1,215 |
| 24 | 40 | Sacramento, CA | 71,891 | 33 | 40 | Hennepin, MN | 276,900 | 29 | 40 | Bronx, NY | 1,212 |
| 47 | 41 | St. Louis, MO | 70,161 | 71 | 41 | Fort Bend, TX | 265,900 | 42 | 41 | Fulton, GA | 1,205 |
| 15 | 42 | Tarrant, TX | 70,130 | 17 | 42 | Broward, FL | 265,000 | 68 | 42 | New Haven, CT | 1,153 |
| 68 | 43 | New Haven, CT | 69,687 | 11 | 43 | Clark, NV | 262,700 | 40 | 43 | Wake, NC | 1,150 |
| 41 | 44 | Mecklenburg, NC | 69,455 | 75 | 44 | Baltimore, MD | 261,500 | 41 | 44 | Mecklenburg, NC | 1,146 |
| 2 | 45 | Cook, IL | 69,375 | 4 | 45 | Maricopa, AZ | 260,200 | 26 | 45 | Hillsborough, FL | 1,142 |
| 4 | 46 | Maricopa, AZ | 68,634 | 46 | 46 | Fresno, CA | 255,000 | 33 | 46 | Hennepin, MN | 1,135 |
| 14 | 47 | San Bernardino, CA | 67,398 | 68 | 47 | New Haven, CT | 248,600 | 11 | 47 | Clark, NV | 1,132 |
| 25 | 48 | Palm Beach, FL | 66,621 | 2 | 48 | Cook, IL | 246,600 | 4 | 48 | Maricopa, AZ | 1,127 |
| 9 | 49 | Kings, NY | 66,501 | 34 | 49 | Oakland, MI | 242,700 | 2 | 49 | Cook, IL | 1,123 |
| 66 | 50 | Macomb, MI | 65,051 | 64 | 50 | Hartford, CT | 240,600 | 37 | 50 | Salt Lake, UT | 1,118 |
| 36 | 51 | Allegheny, PA | 64,799 | 41 | 51 | Mecklenburg, NC | 238,000 | 48 | 51 | Pinellas, FL | 1,112 |
| 31 | 52 | Franklin, OH | 64,648 | 28 | 52 | Orange, FL | 235,800 | 64 | 52 | Hartford, CT | 1,106 |
| 28 | 53 | Orange, FL | 63,133 | 54 | 53 | Gwinnett, GA | 217,900 | 8 | 53 | Dallas, TX | 1,105 |
| 11 | 54 | Clark, NV | 62,131 | 26 | 54 | Hillsborough, FL | 216,300 | 15 | 54 | Tarrant, TX | 1,095 |
| 8 | 55 | Dallas, TX | 61,807 | 63 | 55 | Kern, CA | 213,900 | 34 | 55 | Oakland, MI | 1,080 |
| 3 | 56 | Harris, TX | 61,638 | 48 | 56 | Pinellas, FL | 201,200 | 3 | 56 | Harris, TX | 1,078 |
| 17 | 57 | Broward, FL | 61,429 | 47 | 57 | St. Louis, MO | 198,800 | 49 | 57 | Duval, FL | 1,074 |
| 26 | 58 | Hillsborough, FL | 61,133 | 15 | 58 | Tarrant, TX | 188,500 | 74 | 58 | Worcester, MA | 1,060 |
| 60 | 59 | Erie, NY | 60,620 | 44 | 59 | Pima, AZ | 184,100 | 23 | 59 | Philadelphia, PA | 1,042 |
| 16 | 60 | Bexar, TX | 58,956 | 49 | 60 | Duval, FL | 180,700 | 16 | 60 | Bexar, TX | 1,015 |
| 49 | 61 | Duval, FL | 58,339 | 3 | 61 | Harris, TX | 175,700 | 46 | 61 | Fresno, CA | 998 |
| 46 | 62 | Fresno, CA | 56,926 | 31 | 62 | Franklin, OH | 175,100 | 47 | 62 | St. Louis, MO | 983 |
| 48 | 63 | Pinellas, FL | 56,708 | 8 | 63 | Dallas, TX | 174,900 | 63 | 63 | Kern, CA | 978 |
| 44 | 64 | Pima, AZ | 56,150 | 66 | 64 | Macomb, MI | 166,800 | 31 | 64 | Franklin, OH | 974 |
| 7 | 65 | Miami-Dade, FL | 54,991 | 23 | 65 | Philadelphia, PA | 163,000 | 66 | 65 | Macomb, MI | 962 |
| 53 | 66 | Milwaukee, WI | 53,505 | 16 | 66 | Bexar, TX | 161,800 | 56 | 66 | Shelby, TN | 942 |
| 63 | 67 | Kern, CA | 53,245 | 53 | 67 | Milwaukee, WI | 158,300 | 44 | 67 | Pima, AZ | 907 |
| 56 | 68 | Shelby, TN | 52,659 | 36 | 68 | Allegheny, PA | 154,700 | 36 | 68 | Allegheny, PA | 890 |
| 35 | 69 | Cuyahoga, OH | 52,503 | 60 | 69 | Erie, NY | 153,400 | 50 | 69 | Marion, IN | 889 |
| 19 | 70 | Wayne, MI | 50,753 | 56 | 70 | Shelby, TN | 150,400 | 53 | 70 | Milwaukee, WI | 880 |
| 50 | 71 | Marion, IN | 50,707 | 50 | 71 | Marion, IN | 136,700 | 19 | 71 | Wayne, MI | 875 |
| 70 | 72 | El Paso, TX | 48,629 | 35 | 72 | Cuyahoga, OH | 132,800 | 70 | 72 | El Paso, TX | 838 |
| 23 | 73 | Philadelphia, PA | 47,598 | 70 | 73 | El Paso, TX | 121,500 | 60 | 73 | Erie, NY | 829 |
| 65 | 74 | Hidalgo, TX | 41,656 | 19 | 74 | Wayne, MI | 113,000 | 35 | 74 | Cuyahoga, OH | 809 |
| 29 | 75 | Bronx, NY | 41,470 | 65 | 75 | Hidalgo, TX | 87,100 | 65 | 75 | Hidalgo, TX | 734 |

# 75 Largest Counties by 2020 Population
## Selected Rankings

| Percent of population below the poverty level, 2019 | | | | Percent under 18 years old below the poverty level, 2019 | | | | Unemployment rate, 2020 | | | |
|---|---|---|---|---|---|---|---|---|---|---|---|
| Population rank | Poverty rate rank | County | Poverty rate [col 59] | Population rank | Poverty rate for children rank | County | Poverty rate for children under 18 years [col 60] | Population rank | Unemployment rate rank | County | Unemployment rate [col 100] |
| 65 | 1 | Hidalgo, TX | 26.9 | 65 | 1 | Hidalgo, TX | 37.4 | 29 | 1 | Bronx, NY | 16.0 |
| 29 | 2 | Bronx, NY | 26.2 | 29 | 2 | Bronx, NY | 36.5 | 11 | 2 | Clark, NV | 14.7 |
| 23 | 3 | Philadelphia, PA | 23.0 | 23 | 3 | Philadelphia, PA | 31.6 | 19 | 3 | Wayne, MI | 13.8 |
| 46 | 4 | Fresno, CA | 20.5 | 19 | 4 | Wayne, MI | 29.1 | 1 | 4 | Los Angeles, CA | 12.8 |
| 19 | 5 | Wayne, MI | 19.8 | 46 | 5 | Fresno, CA | 29.0 | 9 | 5 | Kings, NY | 12.5 |
| 63 | 6 | Kern, CA | 19.0 | 70 | 6 | El Paso, TX | 26.2 | 13 | 5 | Queens, NY | 12.5 |
| 70 | 7 | El Paso, TX | 18.8 | 63 | 7 | Kern, CA | 26.0 | 63 | 5 | Kern, CA | 12.5 |
| 9 | 8 | Kings, NY | 17.7 | 56 | 8 | Shelby, TN | 25.7 | 23 | 8 | Philadelphia, PA | 12.4 |
| 56 | 9 | Shelby, TN | 17.2 | 9 | 9 | Kings, NY | 24.7 | 66 | 9 | Macomb, MI | 12.0 |
| 53 | 10 | Milwaukee, WI | 16.9 | 53 | 10 | Milwaukee, WI | 24.0 | 65 | 10 | Hidalgo, TX | 11.6 |
| 35 | 11 | Cuyahoga, OH | 16.2 | 35 | 11 | Cuyahoga, OH | 23.6 | 46 | 11 | Fresno, CA | 11.3 |
| 7 | 12 | Miami-Dade, FL | 15.7 | 16 | 12 | Bexar, TX | 22.4 | 2 | 12 | Cook, IL | 11.1 |
| 16 | 13 | Bexar, TX | 15.2 | 42 | 13 | Fulton, GA | 21.8 | 28 | 13 | Orange, FL | 10.6 |
| 50 | 13 | Marion, IN | 15.2 | 3 | 14 | Harris, TX | 21.3 | 35 | 14 | Cuyahoga, OH | 10.4 |
| 3 | 15 | Harris, TX | 15.0 | 8 | 15 | Dallas, TX | 20.9 | 52 | 15 | Honolulu, HI | 10.2 |
| 21 | 16 | New York, NY | 14.1 | 7 | 16 | Miami-Dade, FL | 20.8 | 10 | 16 | Riverside, CA | 9.9 |
| 8 | 17 | Dallas, TX | 14.0 | 49 | 17 | Duval, FL | 19.5 | 56 | 17 | Shelby, TN | 9.7 |
| 44 | 17 | Pima, AZ | 14.0 | 50 | 18 | Marion, IN | 19.4 | 57 | 18 | Bergen, NJ | 9.6 |
| 42 | 19 | Fulton, GA | 13.8 | 60 | 19 | Erie, NY | 19.3 | 61 | 18 | Pierce, WA | 9.6 |
| 26 | 20 | Hillsborough, FL | 13.5 | 44 | 20 | Pima, AZ | 19.0 | 21 | 20 | New York, NY | 9.5 |
| 31 | 20 | Franklin, OH | 13.5 | 11 | 21 | Clark, NV | 18.7 | 60 | 20 | Erie, NY | 9.5 |
| 49 | 20 | Duval, FL | 13.5 | 31 | 22 | Franklin, OH | 18.6 | 14 | 22 | San Bernardino, CA | 9.4 |
| 1 | 23 | Los Angeles, CA | 13.4 | 14 | 23 | San Bernardino, CA | 18.5 | 24 | 23 | Sacramento, CA | 9.3 |
| 11 | 24 | Clark, NV | 13.3 | 68 | 23 | New Haven, CT | 18.5 | 34 | 23 | Oakland, MI | 9.3 |
| 14 | 24 | San Bernardino, CA | 13.3 | 26 | 25 | Hillsborough, FL | 18.4 | 5 | 25 | San Diego, CA | 9.2 |
| 60 | 24 | Erie, NY | 13.3 | 1 | 26 | Los Angeles, CA | 18.3 | 36 | 26 | Allegheny, PA | 9.0 |
| 2 | 27 | Cook, IL | 13.0 | 2 | 27 | Cook, IL | 18.2 | 3 | 27 | Harris, TX | 8.9 |
| 24 | 28 | Sacramento, CA | 12.6 | 28 | 28 | Orange, FL | 17.9 | 38 | 27 | Contra Costa, CA | 8.9 |
| 28 | 28 | Orange, FL | 12.6 | 4 | 29 | Maricopa, AZ | 17.5 | 6 | 29 | Orange, CA | 8.8 |
| 17 | 30 | Broward, FL | 12.3 | 21 | 30 | New York, NY | 17.3 | 17 | 29 | Broward, FL | 8.8 |
| 4 | 31 | Maricopa, AZ | 12.2 | 17 | 31 | Broward, FL | 16.4 | 20 | 29 | Alameda, CA | 8.8 |
| 68 | 32 | New Haven, CT | 12.0 | 24 | 32 | Sacramento, CA | 16.0 | 74 | 29 | Worcester, MA | 8.8 |
| 25 | 33 | Palm Beach, FL | 11.4 | 25 | 32 | Palm Beach, FL | 16.0 | 69 | 33 | Ventura, CA | 8.6 |
| 48 | 33 | Pinellas, FL | 11.4 | 48 | 34 | Pinellas, FL | 14.9 | 27 | 34 | Suffolk, NY | 8.5 |
| 10 | 35 | Riverside, CA | 11.3 | 13 | 35 | Queens, NY | 14.4 | 73 | 34 | Snohomish, WA | 8.5 |
| 13 | 36 | Queens, NY | 11.0 | 10 | 36 | Riverside, CA | 14.1 | 30 | 36 | Nassau, NY | 8.4 |
| 32 | 37 | Travis, TX | 10.8 | 36 | 36 | Allegheny, PA | 14.1 | 51 | 36 | Westchester, NY | 8.4 |
| 36 | 37 | Allegheny, PA | 10.8 | 64 | 38 | Hartford, CT | 14.0 | 70 | 38 | El Paso, TX | 8.3 |
| 64 | 37 | Hartford, CT | 10.8 | 15 | 39 | Tarrant, TX | 13.9 | 53 | 39 | Milwaukee, WI | 8.2 |
| 5 | 40 | San Diego, CA | 10.3 | 41 | 40 | Mecklenburg, NC | 13.8 | 62 | 39 | Prince George's, MD | 8.2 |
| 41 | 40 | Mecklenburg, NC | 10.3 | 32 | 41 | Travis, TX | 13.6 | 64 | 41 | Hartford, CT | 8.1 |
| 15 | 42 | Tarrant, TX | 10.2 | 54 | 42 | Gwinnett, GA | 13.1 | 7 | 42 | Miami-Dade, FL | 8.0 |
| 33 | 43 | Hennepin, MN | 9.7 | 62 | 43 | Prince George's, MD | 13.0 | 42 | 42 | Fulton, GA | 8.0 |
| 6 | 44 | Orange, CA | 9.5 | 5 | 44 | San Diego, CA | 12.8 | 68 | 42 | New Haven, CT | 8.0 |
| 67 | 44 | San Francisco, CA | 9.5 | 6 | 45 | Orange, CA | 12.6 | 50 | 45 | Marion, IN | 7.9 |
| 61 | 46 | Pierce, WA | 9.4 | 47 | 46 | St. Louis, MO | 12.3 | 55 | 45 | Fairfield, CT | 7.9 |
| 74 | 46 | Worcester, MA | 9.4 | 33 | 47 | Hennepin, MN | 11.9 | 59 | 45 | Du Page, IL | 7.9 |
| 47 | 48 | St. Louis, MO | 9.3 | 74 | 47 | Worcester, MA | 11.9 | 41 | 48 | Mecklenburg, NC | 7.8 |
| 54 | 49 | Gwinnett, GA | 9.2 | 55 | 49 | Fairfield, CT | 11.8 | 67 | 48 | San Francisco, CA | 7.8 |
| 37 | 50 | Salt Lake, UT | 9.0 | 61 | 49 | Pierce, WA | 11.8 | 8 | 50 | Dallas, TX | 7.7 |
| 55 | 50 | Fairfield, CT | 9.0 | 66 | 51 | Macomb, MI | 11.7 | 44 | 50 | Pima, AZ | 7.7 |
| 20 | 52 | Alameda, CA | 8.9 | 69 | 52 | Ventura, CA | 11.4 | 71 | 50 | Fort Bend, TX | 7.7 |
| 75 | 52 | Baltimore, MD | 8.9 | 75 | 53 | Baltimore, MD | 11.0 | 25 | 53 | Palm Beach, FL | 7.6 |
| 62 | 54 | Prince George's, MD | 8.7 | 37 | 54 | Salt Lake, UT | 10.6 | 72 | 53 | Montgomery, PA | 7.6 |
| 66 | 55 | Macomb, MI | 8.6 | 51 | 55 | Westchester, NY | 10.1 | 12 | 55 | King, WA | 7.5 |
| 51 | 56 | Westchester, NY | 8.4 | 38 | 56 | Contra Costa, CA | 9.9 | 16 | 55 | Bexar, TX | 7.5 |
| 69 | 57 | Ventura, CA | 8.2 | 20 | 57 | Alameda, CA | 9.8 | 4 | 57 | Maricopa, AZ | 7.4 |
| 40 | 58 | Wake, NC | 8.0 | 40 | 58 | Wake, NC | 9.4 | 31 | 57 | Franklin, OH | 7.4 |
| 38 | 59 | Contra Costa, CA | 7.9 | 27 | 59 | Suffolk, NY | 9.3 | 15 | 59 | Tarrant, TX | 7.3 |
| 52 | 59 | Honolulu, HI | 7.9 | 45 | 59 | Montgomery, MD | 9.3 | 22 | 59 | Middlesex, MA | 7.3 |
| 34 | 61 | Oakland, MI | 7.8 | 67 | 59 | San Francisco, CA | 9.3 | 26 | 61 | Hillsborough, FL | 7.1 |
| 12 | 62 | King, WA | 7.7 | 34 | 62 | Oakland, MI | 9.2 | 18 | 62 | Santa Clara, CA | 7.0 |
| 45 | 63 | Montgomery, MD | 7.3 | 52 | 62 | Honolulu, HI | 9.2 | 48 | 62 | Pinellas, FL | 7.0 |
| 73 | 64 | Snohomish, WA | 7.0 | 73 | 64 | Snohomish, WA | 8.1 | 75 | 64 | Baltimore, MD | 6.8 |
| 22 | 65 | Middlesex, MA | 6.9 | 12 | 65 | King, WA | 7.9 | 49 | 65 | Duval, FL | 6.7 |
| 27 | 66 | Suffolk, NY | 6.8 | 39 | 65 | Fairfax, VA | 7.9 | 33 | 66 | Hennepin, MN | 6.6 |
| 71 | 67 | Fort Bend, TX | 6.6 | 59 | 67 | Du Page, IL | 7.4 | 58 | 67 | Denton, TX | 6.5 |
| 58 | 68 | Denton, TX | 6.5 | 71 | 67 | Fort Bend, TX | 7.4 | 40 | 68 | Wake, NC | 6.4 |
| 18 | 69 | Santa Clara, CA | 6.1 | 22 | 69 | Middlesex, MA | 7.1 | 32 | 69 | Travis, TX | 6.3 |
| 43 | 69 | Collin, TX | 6.1 | 58 | 69 | Denton, TX | 7.1 | 43 | 69 | Collin, TX | 6.3 |
| 39 | 71 | Fairfax, VA | 6.0 | 43 | 71 | Collin, TX | 6.8 | 45 | 69 | Montgomery, MD | 6.3 |
| 59 | 71 | Du Page, IL | 6.0 | 72 | 72 | Montgomery, PA | 6.7 | 47 | 69 | St. Louis, MO | 6.3 |
| 72 | 71 | Montgomery, PA | 6.0 | 30 | 73 | Nassau, NY | 6.5 | 54 | 69 | Gwinnett, GA | 6.3 |
| 57 | 74 | Bergen, NJ | 5.7 | 57 | 74 | Bergen, NJ | 5.8 | 39 | 74 | Fairfax, VA | 5.8 |
| 30 | 75 | Nassau, NY | 5.6 | 18 | 75 | Santa Clara, CA | 5.5 | 37 | 75 | Salt Lake, UT | 5.1 |

# 75 Largest Counties by 2020 Population
## Selected Rankings

| Population rank | Manufacturing rank | County | Percent employed in manufacturing [col 107/col 105] | Population rank | Professional services rank | County | Percent employed in professional services [col 110/col 105] | Population rank | Local taxes rank | County | Per capita local taxes (dollars) [col 183] |
|---|---|---|---|---|---|---|---|---|---|---|---|
| Manufacturing employment as a percent of total nonfarm employment, 2019 | | | | Professional, scientific and technical employment as a percent of total nonfarm employment, 2019 | | | | Per capita local government taxes, 2017 | | | |
| 66 | 1 | Macomb, MI | 23.9 | 39 | 1 | Fairfax, VA | 33.3 | 21 | 1 | New York, NY | 6,557 |
| 73 | 2 | Snohomish, WA | 22.8 | 45 | 2 | Montgomery, MD | 18.4 | 9 | 1 | Kings, NY | 6,557 |
| 19 | 3 | Wayne, MI | 13.7 | 67 | 3 | San Francisco, CA | 17.2 | 13 | 1 | Queens, NY | 6,557 |
| 20 | 4 | Alameda, CA | 12.1 | 22 | 4 | Middlesex, MA | 15.1 | 29 | 1 | Bronx, NY | 6,557 |
| 64 | 5 | Hartford, CT | 11.6 | 34 | 5 | Oakland, MI | 14.8 | 30 | 5 | Nassau, NY | 5,661 |
| 74 | 6 | Worcester, MA | 11.0 | 21 | 6 | New York, NY | 14.3 | 51 | 6 | Westchester, NY | 5,592 |
| 53 | 7 | Milwaukee, WI | 10.4 | 32 | 7 | Travis, TX | 13.5 | 67 | 7 | San Francisco, CA | 5,199 |
| 15 | 8 | Tarrant, TX | 10.3 | 42 | 8 | Fulton, GA | 13.3 | 27 | 8 | Suffolk, NY | 4,897 |
| 60 | 8 | Erie, NY | 10.3 | 18 | 9 | Santa Clara, CA | 13.2 | 57 | 9 | Bergen, NJ | 4,364 |
| 35 | 10 | Cuyahoga, OH | 10.0 | 5 | 10 | San Diego, CA | 12.1 | 45 | 10 | Montgomery, MD | 4,149 |
| 6 | 11 | Orange, CA | 9.6 | 40 | 10 | Wake, NC | 12.1 | 55 | 11 | Fairfield, CT | 3,926 |
| 46 | 11 | Fresno, CA | 9.6 | 62 | 12 | Prince George's, MD | 11.1 | 18 | 12 | Santa Clara, CA | 3,576 |
| 69 | 13 | Ventura, CA | 9.5 | 66 | 13 | Macomb, MI | 10.8 | 2 | 13 | Cook, IL | 3,471 |
| 59 | 14 | Du Page, IL | 9.1 | 43 | 14 | Collin, TX | 10.7 | 59 | 14 | Du Page, IL | 3,400 |
| 27 | 15 | Suffolk, NY | 9.0 | 20 | 15 | Alameda, CA | 10.6 | 39 | 15 | Fairfax, VA | 3,331 |
| 37 | 16 | Salt Lake, UT | 8.7 | 12 | 16 | King, WA | 9.8 | 32 | 16 | Travis, TX | 3,283 |
| 14 | 17 | San Bernardino, CA | 8.6 | 26 | 16 | Hillsborough, FL | 9.8 | 12 | 17 | King, WA | 3,282 |
| 72 | 18 | Montgomery, PA | 8.4 | 2 | 18 | Cook, IL | 9.7 | 35 | 18 | Cuyahoga, OH | 3,224 |
| 33 | 19 | Hennepin, MN | 8.3 | 8 | 19 | Dallas, TX | 9.2 | 20 | 19 | Alameda, CA | 3,118 |
| 68 | 19 | New Haven, CT | 8.3 | 55 | 19 | Fairfield, CT | 9.2 | 23 | 20 | Philadelphia, PA | 3,079 |
| 1 | 21 | Los Angeles, CA | 8.2 | 33 | 21 | Hennepin, MN | 9.0 | 31 | 21 | Franklin, OH | 3,069 |
| 55 | 21 | Fairfield, CT | 8.2 | 6 | 22 | Orange, CA | 8.9 | 72 | 22 | Montgomery, PA | 2,963 |
| 5 | 23 | San Diego, CA | 8.1 | 54 | 22 | Gwinnett, GA | 8.9 | 43 | 23 | Collin, TX | 2,953 |
| 48 | 24 | Pinellas, FL | 8.0 | 3 | 24 | Harris, TX | 8.6 | 22 | 24 | Middlesex, MA | 2,907 |
| 50 | 25 | Marion, IN | 7.9 | 24 | 24 | Sacramento, CA | 8.6 | 64 | 25 | Hartford, CT | 2,897 |
| 10 | 26 | Riverside, CA | 7.8 | 59 | 24 | Du Page, IL | 8.6 | 8 | 26 | Dallas, TX | 2,810 |
| 18 | 27 | Santa Clara, CA | 7.7 | 72 | 24 | Montgomery, PA | 8.6 | 60 | 27 | Erie, NY | 2,775 |
| 34 | 28 | Oakland, MI | 7.6 | 57 | 28 | Bergen, NJ | 8.5 | 3 | 28 | Harris, TX | 2,756 |
| 58 | 29 | Denton, TX | 7.4 | 75 | 28 | Baltimore, MD | 8.5 | 42 | 29 | Fulton, GA | 2,723 |
| 2 | 30 | Cook, IL | 7.3 | 47 | 30 | St. Louis, MO | 8.4 | 68 | 30 | New Haven, CT | 2,680 |
| 3 | 30 | Harris, TX | 7.3 | 69 | 31 | Ventura, CA | 8.1 | 25 | 31 | Palm Beach, FL | 2,672 |
| 47 | 32 | St. Louis, MO | 7.2 | 17 | 32 | Broward, FL | 8.0 | 36 | 32 | Allegheny, PA | 2,572 |
| 71 | 32 | Fort Bend, TX | 7.2 | 41 | 32 | Mecklenburg, NC | 8.0 | 47 | 33 | St. Louis, MO | 2,529 |
| 44 | 34 | Pima, AZ | 7.1 | 23 | 34 | Philadelphia, PA | 7.9 | 62 | 34 | Prince George's, MD | 2,422 |
| 57 | 35 | Bergen, NJ | 7.0 | 25 | 34 | Palm Beach, FL | 7.9 | 1 | 35 | Los Angeles, CA | 2,401 |
| 12 | 36 | King, WA | 6.9 | 28 | 34 | Orange, FL | 7.9 | 15 | 36 | Tarrant, TX | 2,302 |
| 54 | 37 | Gwinnett, GA | 6.8 | 1 | 37 | Los Angeles, CA | 7.7 | 38 | 37 | Contra Costa, CA | 2,278 |
| 22 | 38 | Middlesex, MA | 6.7 | 36 | 37 | Allegheny, PA | 7.7 | 7 | 38 | Miami-Dade, FL | 2,255 |
| 61 | 39 | Pierce, WA | 6.6 | 37 | 37 | Salt Lake, UT | 7.7 | 6 | 39 | Orange, CA | 2,214 |
| 63 | 39 | Kern, CA | 6.6 | 38 | 40 | Contra Costa, CA | 7.5 | 41 | 40 | Mecklenburg, NC | 2,128 |
| 70 | 39 | El Paso, TX | 6.6 | 50 | 40 | Marion, IN | 7.5 | 75 | 41 | Baltimore, MD | 2,116 |
| 8 | 42 | Dallas, TX | 6.5 | 7 | 42 | Miami-Dade, FL | 7.4 | 16 | 42 | Bexar, TX | 2,095 |
| 4 | 43 | Maricopa, AZ | 6.2 | 48 | 43 | Pinellas, FL | 7.2 | 5 | 43 | San Diego, CA | 2,082 |
| 56 | 44 | Shelby, TN | 6.0 | 31 | 44 | Franklin, OH | 7.1 | 58 | 44 | Denton, TX | 2,080 |
| 36 | 45 | Allegheny, PA | 5.1 | 4 | 45 | Maricopa, AZ | 6.9 | 33 | 45 | Hennepin, MN | 2,054 |
| 49 | 45 | Duval, FL | 5.1 | 19 | 45 | Wayne, MI | 6.9 | 28 | 46 | Orange, FL | 2,022 |
| 31 | 47 | Franklin, OH | 4.8 | 27 | 45 | Suffolk, NY | 6.9 | 17 | 47 | Broward, FL | 1,974 |
| 38 | 48 | Contra Costa, CA | 4.7 | 30 | 45 | Nassau, NY | 6.9 | 74 | 48 | Worcester, MA | 1,950 |
| 16 | 49 | Bexar, TX | 4.5 | 64 | 49 | Hartford, CT | 6.7 | 69 | 49 | Ventura, CA | 1,866 |
| 32 | 49 | Travis, TX | 4.5 | 35 | 50 | Cuyahoga, OH | 6.5 | 56 | 50 | Shelby, TN | 1,861 |
| 43 | 51 | Collin, TX | 4.3 | 60 | 50 | Erie, NY | 6.5 | 37 | 51 | Salt Lake, UT | 1,809 |
| 75 | 51 | Baltimore, MD | 4.3 | 71 | 50 | Fort Bend, TX | 6.5 | 61 | 52 | Pierce, WA | 1,795 |
| 24 | 53 | Sacramento, CA | 4.2 | 74 | 53 | Worcester, MA | 6.4 | 53 | 53 | Milwaukee, WI | 1,787 |
| 41 | 54 | Mecklenburg, NC | 4.0 | 49 | 54 | Duval, FL | 6.3 | 48 | 54 | Pinellas, FL | 1,781 |
| 26 | 55 | Hillsborough, FL | 3.8 | 16 | 55 | Bexar, TX | 6.2 | 71 | 55 | Fort Bend, TX | 1,761 |
| 7 | 56 | Miami-Dade, FL | 3.7 | 51 | 56 | Westchester, NY | 6.1 | 40 | 56 | Wake, NC | 1,755 |
| 28 | 56 | Orange, FL | 3.7 | 44 | 57 | Pima, AZ | 5.3 | 73 | 57 | Snohomish, WA | 1,726 |
| 17 | 58 | Broward, FL | 3.3 | 52 | 58 | Honolulu, HI | 5.1 | 52 | 58 | Honolulu, HI | 1,662 |
| 23 | 59 | Philadelphia, PA | 3.2 | 53 | 58 | Milwaukee, WI | 5.1 | 11 | 59 | Clark, NV | 1,657 |
| 13 | 60 | Queens, NY | 3.1 | 58 | 60 | Denton, TX | 5.0 | 24 | 60 | Sacramento, CA | 1,623 |
| 65 | 60 | Hidalgo, TX | 3.1 | 15 | 61 | Tarrant, TX | 4.9 | 54 | 60 | Gwinnett, GA | 1,623 |
| 40 | 62 | Wake, NC | 2.9 | 73 | 61 | Snohomish, WA | 4.9 | 4 | 62 | Maricopa, AZ | 1,612 |
| 62 | 63 | Prince George's, MD | 2.8 | 63 | 63 | Kern, CA | 4.8 | 70 | 63 | El Paso, TX | 1,585 |
| 9 | 64 | Kings, NY | 2.7 | 11 | 64 | Clark, NV | 4.7 | 34 | 64 | Oakland, MI | 1,572 |
| 25 | 65 | Palm Beach, FL | 2.6 | 56 | 65 | Shelby, TN | 4.4 | 49 | 65 | Duval, FL | 1,519 |
| 30 | 65 | Nassau, NY | 2.6 | 68 | 66 | New Haven, CT | 4.2 | 10 | 66 | Riverside, CA | 1,516 |
| 51 | 65 | Westchester, NY | 2.6 | 46 | 67 | Fresno, CA | 3.9 | 19 | 67 | Wayne, MI | 1,513 |
| 52 | 65 | Honolulu, HI | 2.6 | 70 | 68 | El Paso, TX | 3.7 | 14 | 68 | San Bernardino, CA | 1,502 |
| 11 | 69 | Clark, NV | 2.5 | 9 | 69 | Kings, NY | 3.6 | 44 | 69 | Pima, AZ | 1,500 |
| 42 | 69 | Fulton, GA | 2.5 | 10 | 69 | Riverside, CA | 3.6 | 26 | 70 | Hillsborough, FL | 1,459 |
| 29 | 71 | Bronx, NY | 2.1 | 61 | 69 | Pierce, WA | 3.6 | 63 | 71 | Kern, CA | 1,378 |
| 45 | 72 | Montgomery, MD | 1.4 | 65 | 72 | Hidalgo, TX | 3.1 | 46 | 72 | Fresno, CA | 1,304 |
| 67 | 73 | San Francisco, CA | 1.2 | 14 | 73 | San Bernardino, CA | 2.9 | 65 | 73 | Hidalgo, TX | 1,162 |
| 39 | 74 | Fairfax, VA | 0.9 | 13 | 74 | Queens, NY | 2.8 | 66 | 74 | Macomb, MI | 1,112 |
| 21 | 75 | New York, NY | 0.6 | 29 | 75 | Bronx, NY | 1.8 | 50 | 75 | Marion, IN | 969 |

# 75 Largest Counties by 2020 Population
## Selected Rankings

### Social Security beneficiaries rate per 1,000 population, 2019

| Population rank | Social Security rate rank | County | Social Security rate per 1000 population [col 85] |
|---|---|---|---|
| 48 | 1 | Pinellas, FL | 261 |
| 60 | 2 | Erie, NY | 227 |
| 44 | 3 | Pima, AZ | 225 |
| 36 | 4 | Allegheny, PA | 223 |
| 25 | 5 | Palm Beach, FL | 220 |
| 66 | 6 | Macomb, MI | 216 |
| 47 | 7 | St. Louis, MO | 210 |
| 35 | 8 | Cuyahoga, OH | 207 |
| 19 | 9 | Wayne, MI | 204 |
| 27 | 9 | Suffolk, NY | 204 |
| 34 | 11 | Oakland, MI | 198 |
| 72 | 11 | Montgomery, PA | 198 |
| 64 | 13 | Hartford, CT | 197 |
| 68 | 13 | New Haven, CT | 197 |
| 30 | 15 | Nassau, NY | 195 |
| 75 | 16 | Baltimore, MD | 194 |
| 74 | 17 | Worcester, MA | 188 |
| 52 | 18 | Honolulu, HI | 187 |
| 49 | 19 | Duval, FL | 182 |
| 51 | 20 | Westchester, NY | 178 |
| 57 | 20 | Bergen, NJ | 178 |
| 53 | 22 | Milwaukee, WI | 176 |
| 56 | 23 | Shelby, TN | 174 |
| 61 | 24 | Pierce, WA | 173 |
| 69 | 25 | Ventura, CA | 172 |
| 17 | 26 | Broward, FL | 171 |
| 26 | 27 | Hillsborough, FL | 169 |
| 11 | 28 | Clark, NV | 166 |
| 23 | 29 | Philadelphia, PA | 165 |
| 38 | 29 | Contra Costa, CA | 165 |
| 50 | 29 | Marion, IN | 165 |
| 59 | 29 | Du Page, IL | 165 |
| 55 | 33 | Fairfield, CT | 164 |
| 70 | 33 | El Paso, TX | 164 |
| 4 | 35 | Maricopa, AZ | 163 |
| 24 | 35 | Sacramento, CA | 163 |
| 7 | 37 | Miami-Dade, FL | 161 |
| 21 | 37 | New York, NY | 161 |
| 10 | 39 | Riverside, CA | 160 |
| 2 | 40 | Cook, IL | 157 |
| 22 | 41 | Middlesex, MA | 156 |
| 5 | 42 | San Diego, CA | 155 |
| 16 | 42 | Bexar, TX | 155 |
| 73 | 42 | Snohomish, WA | 155 |
| 13 | 45 | Queens, NY | 154 |
| 33 | 45 | Hennepin, MN | 154 |
| 6 | 47 | Orange, CA | 148 |
| 29 | 48 | Bronx, NY | 146 |
| 28 | 49 | Orange, FL | 144 |
| 46 | 49 | Fresno, CA | 144 |
| 67 | 49 | San Francisco, CA | 144 |
| 14 | 52 | San Bernardino, CA | 143 |
| 63 | 53 | Kern, CA | 141 |
| 1 | 54 | Los Angeles, CA | 139 |
| 45 | 55 | Montgomery, MD | 138 |
| 62 | 56 | Prince George's, MD | 137 |
| 9 | 57 | Kings, NY | 136 |
| 31 | 57 | Franklin, OH | 136 |
| 40 | 59 | Wake, NC | 134 |
| 12 | 60 | King, WA | 133 |
| 15 | 60 | Tarrant, TX | 133 |
| 20 | 62 | Alameda, CA | 132 |
| 65 | 62 | Hidalgo, TX | 132 |
| 41 | 64 | Mecklenburg, NC | 131 |
| 42 | 65 | Fulton, GA | 129 |
| 37 | 66 | Salt Lake, UT | 126 |
| 8 | 67 | Dallas, TX | 124 |
| 18 | 67 | Santa Clara, CA | 124 |
| 3 | 69 | Harris, TX | 119 |
| 54 | 69 | Gwinnett, GA | 119 |
| 39 | 71 | Fairfax, VA | 118 |
| 71 | 72 | Fort Bend, TX | 116 |
| 43 | 73 | Collin, TX | 113 |
| 58 | 74 | Denton, TX | 112 |
| 32 | 75 | Travis, TX | 106 |

### Military as a percent of all federal employment, 2019

| Population rank | Military employment rank | County | Percent military federal employment [col 195/ col 194+195] |
|---|---|---|---|
| 73 | 1 | Snohomish, WA | 74.6 |
| 61 | 2 | Pierce, WA | 72.4 |
| 70 | 3 | El Paso, TX | 68.8 |
| 5 | 4 | San Diego, CA | 67.2 |
| 71 | 5 | Fort Bend, TX | 64.6 |
| 52 | 6 | Honolulu, HI | 62.5 |
| 14 | 7 | San Bernardino, CA | 56.9 |
| 11 | 8 | Clark, NV | 54.3 |
| 16 | 9 | Bexar, TX | 51.4 |
| 43 | 10 | Collin, TX | 51.2 |
| 54 | 11 | Gwinnett, GA | 50.4 |
| 72 | 12 | Montgomery, PA | 47.2 |
| 58 | 13 | Denton, TX | 46.7 |
| 57 | 14 | Bergen, NJ | 43.5 |
| 49 | 15 | Duval, FL | 41.6 |
| 4 | 16 | Maricopa, AZ | 40.6 |
| 55 | 17 | Fairfield, CT | 39.8 |
| 69 | 17 | Ventura, CA | 39.8 |
| 74 | 19 | Worcester, MA | 39.1 |
| 44 | 20 | Pima, AZ | 38.7 |
| 29 | 21 | Bronx, NY | 36.3 |
| 47 | 22 | St. Louis, MO | 36.1 |
| 17 | 23 | Broward, FL | 35.4 |
| 10 | 24 | Riverside, CA | 35.1 |
| 9 | 25 | Kings, NY | 34.9 |
| 26 | 26 | Hillsborough, FL | 34.7 |
| 40 | 27 | Wake, NC | 34.5 |
| 30 | 28 | Nassau, NY | 33.6 |
| 6 | 29 | Orange, CA | 31.2 |
| 20 | 30 | Alameda, CA | 30.9 |
| 41 | 31 | Mecklenburg, NC | 30.3 |
| 34 | 32 | Oakland, MI | 30.2 |
| 22 | 33 | Middlesex, MA | 29.5 |
| 37 | 34 | Salt Lake, UT | 29.0 |
| 59 | 35 | Du Page, IL | 28.7 |
| 3 | 36 | Harris, TX | 28.1 |
| 25 | 37 | Palm Beach, FL | 27.7 |
| 65 | 38 | Hidalgo, TX | 27.5 |
| 12 | 39 | King, WA | 27.1 |
| 15 | 40 | Tarrant, TX | 26.8 |
| 38 | 40 | Contra Costa, CA | 26.8 |
| 1 | 42 | Los Angeles, CA | 26.4 |
| 48 | 43 | Pinellas, FL | 26.3 |
| 7 | 44 | Miami-Dade, FL | 25.9 |
| 68 | 45 | New Haven, CT | 25.8 |
| 33 | 46 | Hennepin, MN | 25.7 |
| 63 | 47 | Kern, CA | 25.5 |
| 51 | 48 | Westchester, NY | 25.4 |
| 18 | 49 | Santa Clara, CA | 24.9 |
| 64 | 50 | Hartford, CT | 24.6 |
| 24 | 51 | Sacramento, CA | 24.3 |
| 53 | 52 | Milwaukee, WI | 22.5 |
| 62 | 53 | Prince George's, MD | 22.4 |
| 31 | 54 | Franklin, OH | 22.1 |
| 2 | 55 | Cook, IL | 21.9 |
| 36 | 56 | Allegheny, PA | 21.1 |
| 13 | 57 | Queens, NY | 20.4 |
| 56 | 58 | Shelby, TN | 20.3 |
| 27 | 59 | Suffolk, NY | 19.1 |
| 39 | 60 | Fairfax, VA | 18.9 |
| 19 | 61 | Wayne, MI | 18.7 |
| 32 | 62 | Travis, TX | 18.4 |
| 8 | 63 | Dallas, TX | 18.2 |
| 66 | 64 | Macomb, MI | 17.5 |
| 28 | 65 | Orange, FL | 17.3 |
| 35 | 65 | Cuyahoga, OH | 17.3 |
| 50 | 67 | Marion, IN | 16.8 |
| 60 | 68 | Erie, NY | 16.1 |
| 75 | 69 | Baltimore, MD | 15.9 |
| 23 | 70 | Philadelphia, PA | 14.4 |
| 45 | 70 | Montgomery, MD | 14.4 |
| 46 | 72 | Fresno, CA | 14.0 |
| 67 | 73 | San Francisco, CA | 11.0 |
| 21 | 74 | New York, NY | 10.9 |
| 42 | 75 | Fulton, GA | 10.7 |

### Mean income tax, 2018

| Population rank | Mean income tax rate rank | County | Mean income tax rate [col 199] |
|---|---|---|---|
| 21 | 1 | New York, NY | 51,360 |
| 67 | 2 | San Francisco, CA | 34,303 |
| 55 | 3 | Fairfield, CT | 33,039 |
| 51 | 4 | Westchester, NY | 32,426 |
| 18 | 5 | Santa Clara, CA | 31,912 |
| 12 | 6 | King, WA | 21,255 |
| 30 | 7 | Nassau, NY | 21,129 |
| 22 | 8 | Middlesex, MA | 21,116 |
| 39 | 9 | Fairfax, VA | 20,715 |
| 42 | 10 | Fulton, GA | 20,513 |
| 57 | 11 | Bergen, NJ | 20,351 |
| 25 | 12 | Palm Beach, FL | 19,587 |
| 72 | 13 | Montgomery, PA | 18,779 |
| 45 | 14 | Montgomery, MD | 18,584 |
| 38 | 15 | Contra Costa, CA | 18,429 |
| 32 | 16 | Travis, TX | 17,331 |
| 20 | 17 | Alameda, CA | 16,571 |
| 59 | 18 | Du Page, IL | 15,866 |
| 43 | 19 | Collin, TX | 15,175 |
| 34 | 20 | Oakland, MI | 15,164 |
| 33 | 21 | Hennepin, MN | 14,993 |
| 47 | 22 | St. Louis, MO | 14,896 |
| 6 | 23 | Orange, CA | 14,034 |
| 27 | 24 | Suffolk, NY | 13,646 |
| 71 | 25 | Fort Bend, TX | 13,068 |
| 41 | 26 | Mecklenburg, NC | 13,040 |
| 40 | 27 | Wake, NC | 12,608 |
| 2 | 28 | Cook, IL | 12,585 |
| 58 | 29 | Denton, TX | 12,368 |
| 1 | 30 | Los Angeles, CA | 12,179 |
| 8 | 31 | Dallas, TX | 12,146 |
| 69 | 32 | Ventura, CA | 11,252 |
| 3 | 33 | Harris, TX | 11,110 |
| 5 | 34 | San Diego, CA | 10,997 |
| 64 | 35 | Hartford, CT | 10,791 |
| 7 | 36 | Miami-Dade, FL | 10,780 |
| 36 | 37 | Allegheny, PA | 10,356 |
| 75 | 38 | Baltimore, MD | 10,266 |
| 17 | 39 | Broward, FL | 10,195 |
| 73 | 40 | Snohomish, WA | 10,122 |
| 48 | 41 | Pinellas, FL | 10,065 |
| 15 | 42 | Tarrant, TX | 9,953 |
| 74 | 43 | Worcester, MA | 9,488 |
| 4 | 44 | Maricopa, AZ | 9,419 |
| 11 | 45 | Clark, NV | 9,346 |
| 68 | 46 | New Haven, CT | 9,283 |
| 26 | 47 | Hillsborough, FL | 9,146 |
| 9 | 48 | Kings, NY | 8,989 |
| 35 | 49 | Cuyahoga, OH | 8,866 |
| 37 | 50 | Salt Lake, UT | 8,699 |
| 28 | 51 | Orange, FL | 8,354 |
| 56 | 52 | Shelby, TN | 8,346 |
| 61 | 53 | Pierce, WA | 8,079 |
| 31 | 54 | Franklin, OH | 7,980 |
| 52 | 55 | Honolulu, HI | 7,898 |
| 60 | 56 | Erie, NY | 7,596 |
| 49 | 57 | Duval, FL | 7,530 |
| 24 | 58 | Sacramento, CA | 7,408 |
| 16 | 59 | Bexar, TX | 7,323 |
| 19 | 60 | Wayne, MI | 6,786 |
| 44 | 61 | Pima, AZ | 6,738 |
| 53 | 62 | Milwaukee, WI | 6,707 |
| 54 | 63 | Gwinnett, GA | 6,624 |
| 50 | 64 | Marion, IN | 6,405 |
| 23 | 65 | Philadelphia, PA | 6,326 |
| 66 | 66 | Macomb, MI | 6,041 |
| 46 | 67 | Fresno, CA | 5,844 |
| 13 | 68 | Queens, NY | 5,833 |
| 62 | 69 | Prince George's, MD | 5,817 |
| 10 | 70 | Riverside, CA | 5,807 |
| 63 | 71 | Kern, CA | 5,120 |
| 14 | 72 | San Bernardino, CA | 5,104 |
| 70 | 73 | El Paso, TX | 3,998 |
| 29 | 74 | Bronx, NY | 3,253 |
| 65 | 75 | Hidalgo, TX | 3,050 |

# 75 Largest Counties by 2020 Population
## Selected Rankings

### Nonemployer businesses, 2018

| Popu-lation rank | Nonemployer Businesses rank | County | Nonemployer businesses [col 167] |
|---|---|---|---|
| 1 | 1 | Los Angeles, CA | 1,107,080 |
| 7 | 2 | Miami-Dade, FL | 557,833 |
| 2 | 3 | Cook, IL | 492,996 |
| 3 | 4 | Harris, TX | 456,877 |
| 4 | 5 | Maricopa, AZ | 336,089 |
| 6 | 6 | Orange, CA | 323,593 |
| 17 | 7 | Broward, FL | 297,518 |
| 5 | 8 | San Diego, CA | 292,941 |
| 9 | 9 | Kings, NY | 288,852 |
| 13 | 10 | Queens, NY | 266,193 |
| 8 | 11 | Dallas, TX | 252,821 |
| 21 | 12 | New York, NY | 232,503 |
| 25 | 13 | Palm Beach, FL | 197,199 |
| 11 | 14 | Clark, NV | 189,386 |
| 15 | 15 | Tarrant, TX | 186,406 |
| 12 | 16 | King, WA | 183,205 |
| 10 | 17 | Riverside, CA | 176,665 |
| 28 | 18 | Orange, FL | 155,760 |
| 14 | 19 | San Bernardino, CA | 154,048 |
| 16 | 20 | Bexar, TX | 151,882 |
| 20 | 21 | Alameda, CA | 150,445 |
| 30 | 22 | Nassau, NY | 150,025 |
| 18 | 23 | Santa Clara, CA | 149,189 |
| 22 | 24 | Middlesex, MA | 148,625 |
| 26 | 25 | Hillsborough, FL | 140,515 |
| 27 | 26 | Suffolk, NY | 137,044 |
| 32 | 27 | Travis, TX | 132,806 |
| 19 | 28 | Wayne, MI | 132,583 |
| 29 | 29 | Bronx, NY | 120,174 |
| 45 | 30 | Montgomery, MD | 118,612 |
| 34 | 31 | Oakland, MI | 118,561 |
| 42 | 32 | Fulton, GA | 116,982 |
| 54 | 33 | Gwinnett, GA | 116,146 |
| 39 | 34 | Fairfax, VA | 114,600 |
| 24 | 35 | Sacramento, CA | 113,355 |
| 33 | 36 | Hennepin, MN | 110,717 |
| 23 | 37 | Philadelphia, PA | 106,028 |
| 41 | 38 | Mecklenburg, NC | 103,292 |
| 31 | 39 | Franklin, OH | 103,069 |
| 51 | 40 | Westchester, NY | 102,917 |
| 57 | 41 | Bergen, NJ | 101,881 |
| 43 | 42 | Collin, TX | 101,380 |
| 38 | 43 | Contra Costa, CA | 100,779 |
| 67 | 44 | San Francisco, CA | 100,598 |
| 55 | 45 | Fairfield, CT | 97,854 |
| 35 | 46 | Cuyahoga, OH | 97,290 |
| 40 | 47 | Wake, NC | 95,345 |
| 37 | 48 | Salt Lake, UT | 93,048 |
| 48 | 49 | Pinellas, FL | 91,910 |
| 59 | 50 | Du Page, IL | 86,929 |
| 36 | 51 | Allegheny, PA | 85,700 |
| 56 | 52 | Shelby, TN | 84,799 |
| 62 | 53 | Prince George's, MD | 82,444 |
| 71 | 54 | Fort Bend, TX | 80,262 |
| 47 | 55 | St. Louis, MO | 79,706 |
| 58 | 56 | Denton, TX | 79,667 |
| 49 | 57 | Duval, FL | 78,330 |
| 65 | 58 | Hidalgo, TX | 76,358 |
| 72 | 59 | Montgomery, PA | 74,145 |
| 75 | 60 | Baltimore, MD | 70,590 |
| 69 | 61 | Ventura, CA | 70,479 |
| 52 | 62 | Honolulu, HI | 69,723 |
| 44 | 63 | Pima, AZ | 69,496 |
| 66 | 64 | Macomb, MI | 68,207 |
| 50 | 65 | Marion, IN | 65,885 |
| 70 | 66 | El Paso, TX | 64,533 |
| 68 | 67 | New Haven, CT | 63,139 |
| 64 | 68 | Hartford, CT | 60,999 |
| 74 | 69 | Worcester, MA | 57,923 |
| 46 | 70 | Fresno, CA | 55,668 |
| 60 | 71 | Erie, NY | 52,711 |
| 73 | 72 | Snohomish, WA | 51,383 |
| 53 | 73 | Milwaukee, WI | 50,963 |
| 63 | 74 | Kern, CA | 49,553 |
| 61 | 75 | Pierce, WA | 48,267 |

### Value of residential construction authorized by building permits, 2020

| Popu-lation rank | Value ($1,000) rank | County | Value ($1,000) rank [col 169] |
|---|---|---|---|
| 4 | 1 | Maricopa, AZ | 9,546,985 |
| 3 | 2 | Harris, TX | 5,716,826 |
| 1 | 3 | Los Angeles, CA | 4,754,173 |
| 32 | 4 | Travis, TX | 3,994,543 |
| 43 | 5 | Collin, TX | 3,611,185 |
| 26 | 6 | Hillsborough, FL | 3,553,555 |
| 71 | 7 | Fort Bend, TX | 3,001,054 |
| 11 | 8 | Clark, NV | 2,824,227 |
| 15 | 9 | Tarrant, TX | 2,736,916 |
| 28 | 10 | Orange, FL | 2,718,839 |
| 40 | 11 | Wake, NC | 2,665,544 |
| 12 | 12 | King, WA | 2,507,262 |
| 10 | 13 | Riverside, CA | 2,491,108 |
| 25 | 14 | Palm Beach, FL | 2,212,910 |
| 58 | 15 | Denton, TX | 2,182,350 |
| 8 | 16 | Dallas, TX | 2,109,862 |
| 41 | 17 | Mecklenburg, NC | 2,024,696 |
| 37 | 18 | Salt Lake, UT | 2,014,033 |
| 5 | 19 | San Diego, CA | 1,997,781 |
| 16 | 20 | Bexar, TX | 1,782,553 |
| 7 | 21 | Miami-Dade, FL | 1,772,918 |
| 33 | 22 | Hennepin, MN | 1,761,533 |
| 6 | 23 | Orange, CA | 1,471,688 |
| 44 | 24 | Pima, AZ | 1,394,415 |
| 24 | 25 | Sacramento, CA | 1,298,819 |
| 49 | 26 | Duval, FL | 1,248,861 |
| 73 | 27 | Snohomish, WA | 1,244,700 |
| 14 | 28 | San Bernardino, CA | 1,182,727 |
| 20 | 29 | Alameda, CA | 1,157,616 |
| 31 | 30 | Franklin, OH | 1,154,609 |
| 61 | 31 | Pierce, WA | 1,111,475 |
| 54 | 32 | Gwinnett, GA | 1,054,068 |
| 18 | 33 | Santa Clara, CA | 1,011,537 |
| 42 | 34 | Fulton, GA | 985,112 |
| 2 | 35 | Cook, IL | 961,806 |
| 46 | 36 | Fresno, CA | 862,554 |
| 23 | 37 | Philadelphia, PA | 823,775 |
| 22 | 38 | Middlesex, MA | 820,959 |
| 67 | 39 | San Francisco, CA | 786,074 |
| 17 | 40 | Broward, FL | 776,309 |
| 65 | 41 | Hidalgo, TX | 774,933 |
| 9 | 42 | Kings, NY | 753,530 |
| 13 | 43 | Queens, NY | 715,879 |
| 38 | 44 | Contra Costa, CA | 694,892 |
| 34 | 45 | Oakland, MI | 638,506 |
| 62 | 46 | Prince George's, MD | 586,356 |
| 63 | 47 | Kern, CA | 571,906 |
| 70 | 48 | El Paso, TX | 571,276 |
| 51 | 49 | Westchester, NY | 553,914 |
| 57 | 50 | Bergen, NJ | 529,852 |
| 29 | 51 | Bronx, NY | 517,626 |
| 27 | 52 | Suffolk, NY | 501,651 |
| 55 | 53 | Fairfield, CT | 492,509 |
| 52 | 54 | Honolulu, HI | 486,429 |
| 50 | 55 | Marion, IN | 472,768 |
| 36 | 56 | Allegheny, PA | 468,936 |
| 66 | 57 | Macomb, MI | 460,853 |
| 19 | 58 | Wayne, MI | 435,269 |
| 30 | 59 | Nassau, NY | 430,676 |
| 47 | 60 | St. Louis, MO | 430,183 |
| 59 | 61 | Du Page, IL | 428,242 |
| 72 | 62 | Montgomery, PA | 426,776 |
| 48 | 63 | Pinellas, FL | 392,761 |
| 74 | 64 | Worcester, MA | 353,361 |
| 69 | 65 | Ventura, CA | 341,012 |
| 60 | 66 | Erie, NY | 340,877 |
| 56 | 67 | Shelby, TN | 339,871 |
| 75 | 68 | Baltimore, MD | 326,528 |
| 45 | 69 | Montgomery, MD | 326,527 |
| 39 | 70 | Fairfax, VA | 270,826 |
| 35 | 71 | Cuyahoga, OH | 239,544 |
| 21 | 72 | New York, NY | 224,779 |
| 64 | 73 | Hartford, CT | 149,573 |
| 53 | 74 | Milwaukee, WI | 145,159 |
| 68 | 75 | New Haven, CT | 139,929 |

### Full-time equivalent government employees, 2017

| Popu-lation rank | Government employees rank | County | Government employees rank [col 171] |
|---|---|---|---|
| 21 | 1 | New York, NY | 442,294 |
| 9 | 1 | Kings, NY | included in |
| 13 | 1 | Queens, NY | New York |
| 29 | 1 | Bronx, NY | |
| 1 | 5 | Los Angeles, CA | 400,857 |
| 2 | 6 | Cook, IL | 205,528 |
| 3 | 7 | Harris, TX | 187,579 |
| 4 | 8 | Maricopa, AZ | 124,148 |
| 8 | 9 | Dallas, TX | 115,153 |
| 5 | 10 | San Diego, CA | 106,292 |
| 7 | 11 | Miami-Dade, FL | 93,991 |
| 6 | 12 | Orange, CA | 92,158 |
| 16 | 13 | Bexar, TX | 83,813 |
| 15 | 14 | Tarrant, TX | 79,300 |
| 17 | 15 | Broward, FL | 78,792 |
| 10 | 16 | Riverside, CA | 78,634 |
| 12 | 17 | King, WA | 78,547 |
| 14 | 18 | San Bernardino, CA | 74,301 |
| 41 | 19 | Mecklenburg, NC | 70,012 |
| 27 | 20 | Suffolk, NY | 65,773 |
| 18 | 21 | Santa Clara, CA | 64,447 |
| 20 | 22 | Alameda, CA | 64,149 |
| 35 | 23 | Cuyahoga, OH | 63,287 |
| 30 | 24 | Nassau, NY | 61,582 |
| 11 | 25 | Clark, NV | 61,549 |
| 23 | 26 | Philadelphia, PA | 61,107 |
| 22 | 27 | Middlesex, MA | 56,125 |
| 24 | 28 | Sacramento, CA | 54,054 |
| 32 | 29 | Travis, TX | 51,463 |
| 25 | 30 | Palm Beach, FL | 49,998 |
| 19 | 31 | Wayne, MI | 48,307 |
| 28 | 32 | Orange, FL | 47,356 |
| 67 | 33 | San Francisco, CA | 46,523 |
| 26 | 34 | Hillsborough, FL | 46,480 |
| 31 | 35 | Franklin, OH | 46,320 |
| 51 | 36 | Westchester, NY | 45,049 |
| 39 | 37 | Fairfax, VA | 44,022 |
| 33 | 38 | Hennepin, MN | 43,507 |
| 42 | 39 | Fulton, GA | 42,801 |
| 36 | 40 | Allegheny, PA | 42,657 |
| 65 | 41 | Hidalgo, TX | 42,017 |
| 70 | 42 | El Paso, TX | 41,955 |
| 45 | 43 | Montgomery, MD | 38,557 |
| 60 | 44 | Erie, NY | 38,244 |
| 46 | 45 | Fresno, CA | 36,934 |
| 43 | 46 | Collin, TX | 36,296 |
| 37 | 47 | Salt Lake, UT | 35,948 |
| 53 | 48 | Milwaukee, WI | 35,769 |
| 59 | 49 | Du Page, IL | 35,243 |
| 47 | 50 | St. Louis, MO | 34,807 |
| 63 | 51 | Kern, CA | 34,360 |
| 40 | 52 | Wake, NC | 33,845 |
| 64 | 53 | Hartford, CT | 33,842 |
| 57 | 54 | Bergen, NJ | 33,707 |
| 56 | 55 | Shelby, TN | 33,448 |
| 50 | 56 | Marion, IN | 33,241 |
| 62 | 57 | Prince George's, MD | 32,780 |
| 55 | 58 | Fairfield, CT | 32,456 |
| 38 | 59 | Contra Costa, CA | 31,854 |
| 44 | 60 | Pima, AZ | 31,752 |
| 48 | 61 | Pinellas, FL | 31,247 |
| 34 | 62 | Oakland, MI | 31,208 |
| 69 | 63 | Ventura, CA | 30,149 |
| 68 | 64 | New Haven, CT | 29,926 |
| 54 | 65 | Gwinnett, GA | 29,474 |
| 75 | 66 | Baltimore, MD | 28,278 |
| 61 | 67 | Pierce, WA | 25,902 |
| 49 | 68 | Duval, FL | 24,856 |
| 74 | 69 | Worcester, MA | 24,532 |
| 72 | 70 | Montgomery, PA | 23,842 |
| 58 | 71 | Denton, TX | 23,475 |
| 66 | 72 | Macomb, MI | 20,652 |
| 73 | 73 | Snohomish, WA | 20,647 |
| 71 | 74 | Fort Bend, TX | 20,284 |
| 52 | 75 | Honolulu, HI | 10,620 |

# 75 Largest Counties by Value of Agricultural Sales
## Selected Rankings

### Value of agricultural sales, 2017

| Value of sales rank | County | Value of sales (millions of dollars) [col 125] |
|---|---|---|
| 1 | Fresno, CA | 5,743 |
| 2 | Tulare, CA | 4,475 |
| 3 | Monterey, CA | 4,116 |
| 4 | Kern, CA | 4,077 |
| 5 | Merced, CA | 2,938 |
| 6 | Stanislaus, CA | 2,526 |
| 7 | San Joaquin, CA | 2,176 |
| 8 | Weld, CO | 2,047 |
| 9 | Yakima, WA | 1,988 |
| 10 | Grant, WA | 1,939 |
| 11 | Imperial, CA | 1,860 |
| 12 | Sioux, IA | 1,696 |
| 13 | Kings, CA | 1,649 |
| 14 | Deaf Smith, TX | 1,639 |
| 15 | Ventura, CA | 1,633 |
| 16 | Santa Barbara, CA | 1,520 |
| 17 | Lancaster, PA | 1,507 |
| 18 | Madera, CA | 1,493 |
| 19 | Duplin, NC | 1,262 |
| 20 | Sampson, NC | 1,249 |
| 21 | Hartley, TX | 1,222 |
| 22 | Maricopa, AZ | 1,209 |
| 23 | Haskell, KS | 1,159 |
| 24 | Texas, OK | 1,136 |
| 25 | Scott, KS | 1,135 |
| 26 | Cuming, NE | 1,132 |
| 27 | Castro, TX | 1,122 |
| 28 | Sussex, DE | 1,013 |
| 29 | Benton, WA | 1,005 |
| 30 | Gray, KS | 991 |
| 31 | Riverside, CA | 932 |
| 32 | Cassia, ID | 927 |
| 33 | Lyon, IA | 924 |
| 34 | Sonoma, CA | 919 |
| 35 | Yuma, CO | 919 |
| 36 | Palm Beach, FL | 902 |
| 37 | Parmer, TX | 893 |
| 38 | Pinal, AZ | 862 |
| 39 | Sherman, TX | 838 |
| 40 | Miami-Dade, FL | 838 |
| 41 | San Diego, CA | 831 |
| 42 | Finney, KS | 823 |
| 43 | Grant, KS | 814 |
| 44 | Rockingham, VA | 796 |
| 45 | Gooding, ID | 783 |
| 46 | Custer, NE | 781 |
| 47 | Lincoln, NE | 755 |
| 48 | Dawson, NE | 748 |
| 49 | Stearns, MN | 748 |
| 50 | Plymouth, IA | 738 |
| 51 | Hansford, TX | 737 |
| 52 | Chester, PA | 713 |
| 53 | San Luis Obispo, CA | 702 |
| 53 | Marion, OR | 702 |
| 55 | Platte, NE | 689 |
| 56 | Twin Falls, ID | 680 |
| 57 | Washington, IA | 672 |
| 58 | Jerome, ID | 640 |
| 59 | Martin, MN | 636 |
| 60 | Dallam, TX | 635 |
| 61 | Mercer, OH | 632 |
| 61 | Franklin, WA | 632 |
| 63 | Glenn, CA | 629 |
| 64 | Swisher, TX | 624 |
| 65 | Logan, CO | 618 |
| 66 | Huron, MI | 611 |
| 67 | Renville, MN | 609 |
| 68 | Santa Cruz, CA | 607 |
| 69 | Morrow, OR | 597 |
| 70 | Benton, AR | 593 |
| 71 | Wayne, NC | 592 |
| 72 | Kossuth, IA | 588 |
| 73 | Hamilton, IA | 587 |
| 74 | Allegan, MI | 584 |
| 75 | Phelps, NE | 578 |

### Average agricultural sales per farm, 2017

| Value of sales rank | Average sales rank | County | Average sales per farm (dollars) [col 126] |
|---|---|---|---|
| 21 | 1 | Hartley, TX | 5,930,437 |
| 23 | 2 | Haskell, KS | 5,599,502 |
| 25 | 3 | Scott, KS | 4,809,487 |
| 11 | 4 | Imperial, CA | 4,696,157 |
| 51 | 5 | Hansford, TX | 4,074,127 |
| 3 | 6 | Monterey, CA | 3,728,396 |
| 39 | 7 | Sherman, TX | 3,420,641 |
| 14 | 8 | Deaf Smith, TX | 2,916,004 |
| 27 | 9 | Castro, TX | 2,728,944 |
| 43 | 10 | Grant, KS | 2,584,575 |
| 4 | 11 | Kern, CA | 2,355,161 |
| 30 | 12 | Gray, KS | 2,347,519 |
| 37 | 13 | Parmer, TX | 1,925,302 |
| 60 | 14 | Dallam, TX | 1,867,426 |
| 42 | 15 | Finney, KS | 1,829,091 |
| 13 | 16 | Kings, CA | 1,712,640 |
| 69 | 17 | Morrow, OR | 1,590,632 |
| 32 | 18 | Cassia, ID | 1,584,137 |
| 75 | 19 | Phelps, NE | 1,558,601 |
| 19 | 20 | Duplin, NC | 1,538,648 |
| 45 | 21 | Gooding, ID | 1,456,113 |
| 64 | 22 | Swisher, TX | 1,444,259 |
| 26 | 23 | Cuming, NE | 1,407,956 |
| 10 | 24 | Grant, WA | 1,400,936 |
| 24 | 25 | Texas, OK | 1,371,593 |
| 58 | 26 | Jerome, ID | 1,316,016 |
| 20 | 27 | Sampson, NC | 1,301,188 |
| 5 | 28 | Merced, CA | 1,257,337 |
| 1 | 29 | Fresno, CA | 1,202,926 |
| 35 | 30 | Yuma, CO | 1,186,972 |
| 38 | 31 | Pinal, AZ | 1,131,154 |
| 48 | 32 | Dawson, NE | 1,091,000 |
| 18 | 33 | Madera, CA | 1,076,903 |
| 71 | 34 | Wayne, NC | 1,074,539 |
| 2 | 35 | Tulare, CA | 1,068,739 |
| 16 | 36 | Santa Barbara, CA | 1,036,090 |
| 12 | 37 | Sioux, IA | 983,814 |
| 68 | 38 | Santa Cruz, CA | 970,464 |
| 28 | 39 | Sussex, DE | 904,900 |
| 55 | 40 | Platte, NE | 823,639 |
| 33 | 41 | Lyon, IA | 823,148 |
| 61 | 42 | Franklin, WA | 818,132 |
| 73 | 43 | Hamilton, IA | 801,623 |
| 15 | 44 | Ventura, CA | 765,008 |
| 47 | 45 | Lincoln, NE | 726,188 |
| 65 | 46 | Logan, CO | 717,686 |
| 46 | 47 | Custer, NE | 705,014 |
| 6 | 48 | Stanislaus, CA | 697,690 |
| 59 | 49 | Martin, MN | 697,610 |
| 36 | 50 | Palm Beach, FL | 694,699 |
| 9 | 51 | Yakima, WA | 673,451 |
| 29 | 52 | Benton, WA | 661,374 |
| 22 | 53 | Maricopa, AZ | 645,215 |
| 7 | 54 | San Joaquin, CA | 634,404 |
| 50 | 55 | Plymouth, IA | 605,578 |
| 57 | 56 | Washington, IA | 595,126 |
| 67 | 57 | Renville, MN | 593,752 |
| 56 | 58 | Twin Falls, ID | 561,716 |
| 63 | 59 | Glenn, CA | 536,012 |
| 66 | 60 | Huron, MI | 529,729 |
| 61 | 61 | Mercer, OH | 513,089 |
| 8 | 62 | Weld, CO | 503,982 |
| 74 | 63 | Allegan, MI | 498,612 |
| 72 | 64 | Kossuth, IA | 436,566 |
| 52 | 65 | Chester, PA | 432,848 |
| 44 | 66 | Rockingham, VA | 392,852 |
| 31 | 67 | Riverside, CA | 349,450 |
| 70 | 68 | Benton, AR | 306,494 |
| 40 | 69 | Miami-Dade, FL | 304,409 |
| 53 | 70 | San Luis Obispo, CA | 298,676 |
| 17 | 71 | Lancaster, PA | 295,068 |
| 34 | 72 | Sonoma, CA | 255,719 |
| 53 | 73 | Marion, OR | 254,104 |
| 49 | 74 | Stearns, MN | 253,466 |
| 41 | 75 | San Diego, CA | 163,601 |

### Number of farms, 2017

| Value of sales rank | Number of farms rank | County | Number of farms [col 113] |
|---|---|---|---|
| 17 | 1 | Lancaster, PA | 5,108 |
| 41 | 2 | San Diego, CA | 5,082 |
| 1 | 3 | Fresno, CA | 4,774 |
| 2 | 4 | Tulare, CA | 4,187 |
| 8 | 5 | Weld, CO | 4,062 |
| 6 | 6 | Stanislaus, CA | 3,621 |
| 34 | 7 | Sonoma, CA | 3,594 |
| 7 | 8 | San Joaquin, CA | 3,430 |
| 9 | 9 | Yakima, WA | 2,952 |
| 49 | 10 | Stearns, MN | 2,951 |
| 53 | 11 | Marion, OR | 2,761 |
| 40 | 12 | Miami-Dade, FL | 2,752 |
| 31 | 13 | Riverside, CA | 2,667 |
| 53 | 14 | San Luis Obispo, CA | 2,349 |
| 5 | 15 | Merced, CA | 2,337 |
| 15 | 16 | Ventura, CA | 2,135 |
| 44 | 17 | Rockingham, VA | 2,026 |
| 70 | 18 | Benton, AR | 1,936 |
| 22 | 19 | Maricopa, AZ | 1,874 |
| 4 | 20 | Kern, CA | 1,731 |
| 12 | 21 | Sioux, IA | 1,724 |
| 52 | 22 | Chester, PA | 1,646 |
| 29 | 23 | Benton, WA | 1,520 |
| 16 | 24 | Santa Barbara, CA | 1,467 |
| 18 | 25 | Madera, CA | 1,386 |
| 10 | 26 | Grant, WA | 1,384 |
| 72 | 27 | Kossuth, IA | 1,347 |
| 36 | 28 | Palm Beach, FL | 1,298 |
| 61 | 29 | Mercer, OH | 1,231 |
| 50 | 30 | Plymouth, IA | 1,219 |
| 56 | 31 | Twin Falls, ID | 1,211 |
| 63 | 32 | Glenn, CA | 1,173 |
| 74 | 33 | Allegan, MI | 1,172 |
| 66 | 34 | Huron, MI | 1,153 |
| 57 | 35 | Washington, IA | 1,129 |
| 33 | 36 | Lyon, IA | 1,122 |
| 28 | 37 | Sussex, DE | 1,119 |
| 46 | 38 | Custer, NE | 1,108 |
| 3 | 39 | Monterey, CA | 1,104 |
| 47 | 40 | Lincoln, NE | 1,040 |
| 67 | 41 | Renville, MN | 1,026 |
| 13 | 42 | Kings, CA | 963 |
| 20 | 43 | Sampson, NC | 960 |
| 59 | 44 | Martin, MN | 911 |
| 45 | 45 | Logan, CO | 861 |
| 55 | 46 | Platte, NE | 836 |
| 24 | 47 | Texas, OK | 828 |
| 19 | 48 | Duplin, NC | 820 |
| 26 | 49 | Cuming, NE | 804 |
| 35 | 50 | Yuma, CO | 774 |
| 61 | 51 | Franklin, WA | 772 |
| 38 | 52 | Pinal, AZ | 762 |
| 73 | 53 | Hamilton, IA | 732 |
| 48 | 54 | Dawson, NE | 686 |
| 68 | 55 | Santa Cruz, CA | 625 |
| 32 | 56 | Cassia, ID | 585 |
| 14 | 57 | Deaf Smith, TX | 562 |
| 71 | 58 | Wayne, NC | 551 |
| 45 | 59 | Gooding, ID | 538 |
| 58 | 60 | Jerome, ID | 486 |
| 37 | 61 | Parmer, TX | 464 |
| 42 | 62 | Finney, KS | 450 |
| 64 | 63 | Swisher, TX | 432 |
| 30 | 64 | Gray, KS | 422 |
| 27 | 65 | Castro, TX | 411 |
| 11 | 66 | Imperial, CA | 396 |
| 69 | 67 | Morrow, OR | 375 |
| 75 | 68 | Phelps, NE | 371 |
| 60 | 69 | Dallam, TX | 340 |
| 43 | 70 | Grant, KS | 315 |
| 39 | 71 | Sherman, TX | 245 |
| 25 | 72 | Scott, KS | 236 |
| 23 | 73 | Haskell, KS | 207 |
| 21 | 74 | Hartley, TX | 206 |
| 51 | 75 | Hansford, TX | 181 |

# 75 Largest Counties by Value of Agricultural Sales
## Selected Rankings

| Average size of farm, 2017 | | | | Average value of land and buildings per farm, 2017 | | | | Average value of land and buildings per acre, 2017 | | | |
|---|---|---|---|---|---|---|---|---|---|---|---|
| Value of sales rank | Size of farm rank | County | Average size of farm (acres) [col 119] | Value of sales rank | Value of land and buildings per farm rank | County | Average value per farm (dollars) [col 122] | Value of sales rank | Value of land and buildings per acre rank | County | Average value per acre (dollars) [col 123] |
| 21 | 1 | Hartley, TX | 4,052 | 11 | 1 | Imperial, CA | 14,670,312 | 40 | 1 | Miami-Dade, FL | 37,450 |
| 51 | 2 | Hansford, TX | 3,243 | 4 | 2 | Kern, CA | 9,787,244 | 15 | 2 | Ventura, CA | 25,498 |
| 69 | 3 | Morrow, OR | 3,003 | 3 | 3 | Monterey, CA | 8,944,364 | 41 | 3 | San Diego, CA | 23,209 |
| 60 | 4 | Dallam, TX | 2,633 | 13 | 4 | Kings, CA | 6,917,469 | 34 | 4 | Sonoma, CA | 22,186 |
| 39 | 5 | Sherman, TX | 2,409 | 5 | 5 | Merced, CA | 5,299,308 | 68 | 5 | Santa Cruz, CA | 21,352 |
| 25 | 6 | Scott, KS | 1,951 | 18 | 6 | Madera, CA | 5,102,429 | 17 | 6 | Lancaster, PA | 18,285 |
| 35 | 7 | Yuma, CO | 1,809 | 16 | 7 | Santa Barbara, CA | 5,060,150 | 31 | 7 | Riverside, CA | 18,144 |
| 23 | 8 | Haskell, KS | 1,757 | 21 | 8 | Hartley, TX | 5,028,050 | 6 | 8 | Stanislaus, CA | 15,619 |
| 42 | 8 | Finney, KS | 1,757 | 75 | 9 | Phelps, NE | 4,707,223 | 7 | 9 | San Joaquin, CA | 15,020 |
| 14 | 10 | Deaf Smith, TX | 1,722 | 51 | 10 | Hansford, TX | 4,353,971 | 5 | 10 | Merced, CA | 13,086 |
| 24 | 11 | Texas, OK | 1,544 | 38 | 11 | Pinal, AZ | 4,065,306 | 53 | 11 | Marion, OR | 12,367 |
| 38 | 12 | Pinal, AZ | 1,471 | 67 | 12 | Renville, MN | 3,990,664 | 52 | 12 | Chester, PA | 12,140 |
| 46 | 13 | Custer, NE | 1,358 | 1 | 13 | Fresno, CA | 3,917,829 | 2 | 13 | Tulare, CA | 11,726 |
| 27 | 14 | Castro, TX | 1,350 | 61 | 14 | Franklin, WA | 3,662,394 | 1 | 14 | Fresno, CA | 11,359 |
| 4 | 15 | Kern, CA | 1,326 | 63 | 15 | Glenn, CA | 3,547,184 | 11 | 15 | Imperial, CA | 11,135 |
| 65 | 16 | Logan, CO | 1,322 | 50 | 16 | Plymouth, IA | 3,535,476 | 18 | 16 | Madera, CA | 10,958 |
| 30 | 17 | Gray, KS | 1,318 | 32 | 17 | Cassia, ID | 3,508,509 | 13 | 17 | Kings, CA | 10,815 |
| 11 | 18 | Imperial, CA | 1,317 | 34 | 18 | Sonoma, CA | 3,501,852 | 12 | 18 | Sioux, IA | 10,405 |
| 47 | 19 | Lincoln, NE | 1,305 | 2 | 19 | Tulare, CA | 3,501,053 | 16 | 19 | Santa Barbara, CA | 10,381 |
| 64 | 20 | Swisher, TX | 1,304 | 73 | 20 | Hamilton, IA | 3,495,669 | 33 | 20 | Lyon, IA | 9,988 |
| 3 | 21 | Monterey, CA | 1,214 | 72 | 21 | Kossuth, IA | 3,480,288 | 61 | 21 | Mercer, OH | 9,673 |
| 37 | 22 | Parmer, TX | 1,183 | 39 | 22 | Sherman, TX | 3,446,896 | 22 | 22 | Maricopa, AZ | 9,236 |
| 43 | 23 | Grant, KS | 1,139 | 69 | 23 | Morrow, OR | 3,385,469 | 63 | 23 | Glenn, CA | 8,915 |
| 32 | 24 | Cassia, ID | 1,100 | 7 | 24 | San Joaquin, CA | 3,384,002 | 44 | 24 | Rockingham, VA | 8,844 |
| 75 | 25 | Phelps, NE | 921 | 46 | 25 | Custer, NE | 3,361,912 | 50 | 25 | Plymouth, IA | 8,561 |
| 48 | 26 | Dawson, NE | 889 | 59 | 26 | Martin, MN | 3,308,478 | 28 | 26 | Sussex, DE | 8,473 |
| 61 | 27 | Franklin, WA | 797 | 36 | 27 | Palm Beach, FL | 3,149,296 | 36 | 27 | Palm Beach, FL | 8,379 |
| 10 | 28 | Grant, WA | 753 | 6 | 28 | Stanislaus, CA | 3,116,617 | 73 | 28 | Hamilton, IA | 8,115 |
| 13 | 29 | Kings, CA | 640 | 15 | 29 | Ventura, CA | 3,106,351 | 72 | 29 | Kossuth, IA | 7,892 |
| 67 | 30 | Renville, MN | 608 | 33 | 30 | Lyon, IA | 3,082,560 | 53 | 30 | San Luis Obispo, CA | 7,547 |
| 9 | 31 | Yakima, WA | 603 | 48 | 31 | Dawson, NE | 3,034,516 | 4 | 31 | Kern, CA | 7,380 |
| 8 | 32 | Weld, CO | 517 | 53 | 32 | San Luis Obispo, CA | 2,991,939 | 3 | 32 | Monterey, CA | 7,368 |
| 59 | 33 | Martin, MN | 493 | 60 | 33 | Dallam, TX | 2,986,690 | 57 | 33 | Washington, IA | 6,889 |
| 16 | 34 | Santa Barbara, CA | 487 | 25 | 34 | Scott, KS | 2,955,605 | 59 | 34 | Martin, MN | 6,712 |
| 18 | 35 | Madera, CA | 466 | 35 | 35 | Yuma, CO | 2,953,997 | 67 | 35 | Renville, MN | 6,560 |
| 55 | 36 | Platte, NE | 459 | 12 | 36 | Sioux, IA | 2,918,016 | 70 | 36 | Benton, AR | 6,478 |
| 26 | 37 | Cuming, NE | 452 | 55 | 37 | Platte, NE | 2,903,960 | 55 | 37 | Platte, NE | 6,328 |
| 72 | 38 | Kossuth, IA | 441 | 26 | 38 | Cuming, NE | 2,752,249 | 26 | 38 | Cuming, NE | 6,087 |
| 73 | 39 | Hamilton, IA | 431 | 42 | 39 | Finney, KS | 2,749,764 | 66 | 39 | Huron, MI | 6,048 |
| 66 | 40 | Huron, MI | 430 | 23 | 40 | Haskell, KS | 2,682,671 | 58 | 40 | Jerome, ID | 6,039 |
| 50 | 41 | Plymouth, IA | 413 | 66 | 41 | Huron, MI | 2,597,983 | 45 | 41 | Gooding, ID | 6,016 |
| 5 | 42 | Merced, CA | 405 | 10 | 42 | Grant, WA | 2,574,272 | 74 | 42 | Allegan, MI | 5,981 |
| 29 | 43 | Benton, WA | 404 | 22 | 43 | Maricopa, AZ | 2,338,321 | 49 | 43 | Stearns, MN | 5,149 |
| 63 | 44 | Glenn, CA | 398 | 47 | 44 | Lincoln, NE | 2,184,451 | 75 | 44 | Phelps, NE | 5,114 |
| 53 | 45 | San Luis Obispo, CA | 396 | 68 | 45 | Santa Cruz, CA | 2,183,024 | 71 | 45 | Wayne, NC | 4,990 |
| 56 | 46 | Twin Falls, ID | 387 | 58 | 46 | Jerome, ID | 2,132,761 | 20 | 46 | Sampson, NC | 4,891 |
| 36 | 47 | Palm Beach, FL | 376 | 61 | 47 | Mercer, OH | 2,113,330 | 19 | 47 | Duplin, NC | 4,798 |
| 58 | 48 | Jerome, ID | 353 | 45 | 48 | Gooding, ID | 2,106,203 | 61 | 48 | Franklin, WA | 4,595 |
| 45 | 49 | Gooding, ID | 350 | 30 | 49 | Gray, KS | 2,103,461 | 56 | 49 | Twin Falls, ID | 4,439 |
| 1 | 50 | Fresno, CA | 345 | 28 | 50 | Sussex, DE | 2,085,980 | 29 | 50 | Benton, WA | 3,898 |
| 20 | 51 | Sampson, NC | 314 | 14 | 51 | Deaf Smith, TX | 1,942,201 | 10 | 51 | Grant, WA | 3,421 |
| 33 | 52 | Lyon, IA | 309 | 57 | 52 | Washington, IA | 1,894,242 | 48 | 52 | Dawson, NE | 3,412 |
| 71 | 53 | Wayne, NC | 300 | 27 | 53 | Castro, TX | 1,863,235 | 32 | 53 | Cassia, ID | 3,190 |
| 2 | 54 | Tulare, CA | 299 | 43 | 54 | Grant, KS | 1,794,738 | 38 | 54 | Pinal, AZ | 2,764 |
| 19 | 55 | Duplin, NC | 296 | 31 | 55 | Riverside, CA | 1,794,629 | 9 | 55 | Yakima, WA | 2,755 |
| 12 | 56 | Sioux, IA | 280 | 56 | 56 | Twin Falls, ID | 1,718,569 | 8 | 56 | Weld, CO | 2,586 |
| 57 | 57 | Washington, IA | 275 | 9 | 57 | Yakima, WA | 1,662,430 | 46 | 57 | Custer, NE | 2,475 |
| 22 | 58 | Maricopa, AZ | 253 | 29 | 58 | Benton, WA | 1,573,284 | 47 | 58 | Lincoln, NE | 1,674 |
| 28 | 59 | Sussex, DE | 246 | 65 | 59 | Logan, CO | 1,570,536 | 35 | 59 | Yuma, CO | 1,633 |
| 7 | 60 | San Joaquin, CA | 225 | 24 | 60 | Texas, OK | 1,563,361 | 30 | 60 | Gray, KS | 1,596 |
| 49 | 61 | Stearns, MN | 221 | 20 | 61 | Sampson, NC | 1,534,742 | 43 | 61 | Grant, KS | 1,576 |
| 61 | 62 | Mercer, OH | 218 | 71 | 62 | Wayne, NC | 1,497,288 | 42 | 62 | Finney, KS | 1,565 |
| 6 | 63 | Stanislaus, CA | 200 | 19 | 63 | Duplin, NC | 1,422,426 | 23 | 63 | Haskell, KS | 1,527 |
| 74 | 64 | Allegan, MI | 196 | 17 | 64 | Lancaster, PA | 1,410,238 | 25 | 64 | Scott, KS | 1,515 |
| 34 | 65 | Sonoma, CA | 158 | 64 | 65 | Swisher, TX | 1,364,672 | 39 | 65 | Sherman, TX | 1,431 |
| 70 | 66 | Benton, AR | 126 | 8 | 66 | Weld, CO | 1,335,953 | 27 | 66 | Castro, TX | 1,380 |
| 15 | 67 | Ventura, CA | 122 | 37 | 67 | Parmer, TX | 1,330,531 | 51 | 67 | Hansford, TX | 1,342 |
| 44 | 68 | Rockingham, VA | 113 | 53 | 68 | Marion, OR | 1,292,998 | 21 | 68 | Hartley, TX | 1,241 |
| 53 | 69 | Marion, OR | 105 | 74 | 69 | Allegan, MI | 1,172,374 | 65 | 69 | Logan, CO | 1,188 |
| 68 | 70 | Santa Cruz, CA | 102 | 49 | 70 | Stearns, MN | 1,135,611 | 60 | 70 | Dallam, TX | 1,134 |
| 31 | 71 | Riverside, CA | 99 | 52 | 71 | Chester, PA | 1,110,075 | 14 | 71 | Deaf Smith, TX | 1,128 |
| 52 | 72 | Chester, PA | 91 | 40 | 72 | Miami-Dade, FL | 1,068,826 | 69 | 72 | Morrow, OR | 1,127 |
| 17 | 73 | Lancaster, PA | 77 | 41 | 73 | San Diego, CA | 1,014,281 | 37 | 73 | Parmer, TX | 1,125 |
| 41 | 74 | San Diego, CA | 44 | 44 | 74 | Rockingham, VA | 997,634 | 64 | 74 | Swisher, TX | 1,046 |
| 40 | 75 | Miami-Dade, FL | 29 | 70 | 75 | Benton, AR | 815,643 | 24 | 75 | Texas, OK | 1,013 |

## Table B. States and Counties — **Land Area and Population**

| State / county code | CBSA code[1] | County code[2] | STATE County | Population, 2020 | | | | Population and population characteristics, 2020 | | | | | | | | | | |
|---|---|---|---|---|---|---|---|---|---|---|---|---|---|---|---|---|---|---|
| | | | | | | | | Race alone or in combination, not Hispanic or Latino (percent) | | | | | Age (percent) | | | | | |
| | | | | Land area[3] (sq. mi) | Total persons 2019 | Rank | Per square mile | White | Black | American Indian, Alaska Native | Asian and Pacific Islancer | Percent Hispanic or Latino[4] | Under 5 years | 5 to 14 years | 15 to 24 years | 25 to 34 years | 35 to 44 years | 45 to 54 years |
| | | | | 1 | 2 | 3 | 4 | 5 | 6 | 7 | 8 | 9 | 10 | 11 | 12 | 13 | 14 | 15 |

1. CBSA = Core Based Statistical Area. See Appendix A for explanation. See Appendix B for list of metropolitan areas with component counties.   2. County type code from the Economic Research Service of USDA Rural-Urban Continuum Codes. See Appendix A for definition.   3. Dry land or land partially or temporarily covered by water.   4. May be of any race.

## Table B. States and Counties — **Population and Households**

| STATE County | Population, 2020 (cont.) | | | | Population change, 2000-2020 | | | | | | | Households, 2015-2019 | | | | |
|---|---|---|---|---|---|---|---|---|---|---|---|---|---|---|---|---|
| | Age (percent) (cont.) | | | | Total persons | | Percent change | | Components of change, 2010-2020 | | | | | | Percent | |
| | 55 to 64 years | 65 to 74 years | 75 years and over | Percent female | 2000 | 2010 | 2000-2010 | 2010-2020 | Births | Deaths | Net Migration | Number | Persons per household | Family house-holds | Female family house-holder[1] | One person |
| | 16 | 17 | 18 | 19 | 20 | 21 | 22 | 23 | 24 | 25 | 26 | 27 | 28 | 29 | 30 | 31 |

1. No spouse present.

## Table B. States and Counties — **Population, Vital Statistics, Health, and Crime**

| STATE County | Persons in group quarters, 2020 | Daytime Population, 2015-2019 | | Births, 2020 | | Deaths, 2020 | | Persons under 65 with no health insurance, 2019 | | Medicare, 2020 | | | Crimes reported by county police or sheriff, 2019 | |
|---|---|---|---|---|---|---|---|---|---|---|---|---|---|---|
| | | Number | Employment/ residence ratio | Total | Rate[1] | Number | Rate[1] | Number | Percent | Total beneficiaries | Enrolled in Original Medicare | Enrolled in Medicare Advantage | Violent | Property |
| | 32 | 33 | 34 | 35 | 36 | 37 | 38 | 39 | 40 | 41 | 42 | 43 | 44 | 45 |

1. Per 1,000 estimated resident population.

## Table B. States and Counties — **Crime, Education, Money Income, and Poverty**

| STATE County | County law enforcement employment, 2019 | | Education | | | | | | Money income, 2015-2019 | | | | Income and poverty, 2019 | | |
|---|---|---|---|---|---|---|---|---|---|---|---|---|---|---|---|
| | | | School enrollment and attainment, 2015-2019 | | | | Local government expenditures,[3] 2017–2018 | | | Households | | | | Percent below poverty level | |
| | | | Enrollment[1] | | Attainment[2] (percent) | | | | | | Percent | | | | |
| | Officers | Civilians | Total | Percent private | High school graduate or less | Bachelor's degree or more | Total current spending (mil dol) | Current spending per student (dollars) | Per capita income[4] | Median income (dollars) | with income of less than $50,000 | with income of $200,000 or more | Median household income (dollars) | All persons | Children under 18 years | Children 5 to 17 years in families |
| | 46 | 47 | 48 | 49 | 50 | 51 | 52 | 53 | 54 | 55 | 56 | 57 | 58 | 59 | 60 | 61 |

1. All persons 3 years old and over enrolled in nursery school through college.   2. Persons 25 years old and over.   3. Elementary and secondary education expenditures.   4. Based on population estimated by the American Community Survey, 2014–2018.

## Table B. States and Counties — **Personal Income**

| STATE County | Personal income, 2019 | | | | | | | | | | Earnings, 2019 | | |
|---|---|---|---|---|---|---|---|---|---|---|---|---|---|
| | | | Per capita[1] | | Wages and salaries (mil dol) | Supplements to wages and salaries, employer contributions (mil dol) | | Proprietors' income (mil dol) | Dividends, interest, and rent (mil dol) | Personal transfer receipts (mil dol) | | Contributions for government social insurance (mil dol) | |
| | Total (mil dol) | Percent change 2018–2019 | Dollars | Rank | | Pension and insurance | Government social insurance | | | | Total (mil dol) | From employee and self-employed | From employer |
| | 62 | 63 | 64 | 65 | 66 | 67 | 68 | 69 | 70 | 71 | 72 | 73 | 74 |

1. Based on the resident population estimated as of July 1 of the year shown.

## Table B. States and Counties — **Earnings, Social Security, and Housing**

| STATE County | Earnings, 2019 (cont.) | | | | | | | | | Social Security beneficiaries, December 2019 | | Supple-mental Security Income recipients, 2019 | Housing units, 2020 | |
|---|---|---|---|---|---|---|---|---|---|---|---|---|---|---|
| | Percent by selected industries | | | | | | | | | | | | | |
| | Farm | Mining, quarrying, and extractions | Construction | Manu-facturing | Information; professional, scientific, technical services | Retail trade | Finance, insurance, real estate, and leasing | Health care and social assistance | Govern-ment | Number | Rate[1] | | Total | Percent change, 2010-2020 |
| | 75 | 76 | 77 | 78 | 79 | 80 | 81 | 82 | 83 | 84 | 85 | 86 | 87 | 88 |

1. Per 1,000 resident population estimated as of July 1 of the year shown.

## Table B. States and Counties — **Housing, Labor Force, and Employment**

| STATE County | Housing units, 2015-2019 | | | | | | | | Civilian labor force, 2020 | | | | Civilian employment[6], 2015-2019 | | |
|---|---|---|---|---|---|---|---|---|---|---|---|---|---|---|---|
| | | Occupied units | | | | | | | | | Unemployment | | | Percent | |
| | | | Owner-occupied | | | Renter-occupied | | | | | | | | | |
| | | | | Median owner cost as a percent of income | | | Median rent as a percent of income[2] | Sub-standard units[4] (percent) | | Percent change, 2019-2020 | | | | Management, business, science, and arts | Construction, production, and maintenance occupations |
| | Total | Percent | Median value[1] | With a mort-gage | Without a mort-gage[2] | Median rent[3] | | | Total | | Total | Rate[5] | Total | | |
| | 89 | 90 | 91 | 92 | 93 | 94 | 95 | 96 | 97 | 98 | 99 | 100 | 101 | 102 | 103 |

1. Specified owner-occupied units.   2. A value of 10.0 represents 10 percent or less; a value of 50.0 represents 50 percent or more.   3. Specified renter-occupied units.   4. Overcrowded or lacking complete plumbing facilities.   5. Percent of civilian labor force.   6. Civilian employed persons 16 years old and over.

## Table B. States and Counties — **Nonfarm Employment and Agriculture**

| STATE County | Private nonfarm establishments, employment and payroll, 2019 | | | | | | | | | Agriculture, 2017 | | | Farm producers whose primary occupation is farming (percent) |
|---|---|---|---|---|---|---|---|---|---|---|---|---|---|
| | | Employment | | | | | | Annual payroll | | Farms | | | |
| | | | | | | | | | | | Percent with: | | |
| | Number of establish-ments | Total | Health care and social assistance | Manufac-turing | Retail trade | Finance and insurance | Professional, scientific, and technical services | Total (mil dol) | Average per employee (dollars) | Number | Fewer than 50 acres | 1000 acres or more | |
| | 104 | 105 | 106 | 107 | 108 | 109 | 110 | 111 | 112 | 113 | 114 | 115 | 116 |

## Table B. States and Counties — **Agriculture**

| STATE County | Land in farms | | | | | Value of land and buildings (dollars) | | Value of machinery and equiopmnet, average per farm (dollars) | Value of products sold: | | | | Organic farms (number) | Farms with internet access (percent) | Government payments | |
|---|---|---|---|---|---|---|---|---|---|---|---|---|---|---|---|---|
| | | | Acres | | | | | | | | Percent from: | | | | | |
| | Acreage (1,000) | Percent change, 2012-2017 | Average size of farm | Total irrigated (1,000) | Total cropland (1,000) | Average per farm | Average per acre | | Total (mil dol) | Average per farm (acres) | Crops | Livestock and poultry products | | | Total ($1,000) | Percent of farms |
| | 117 | 118 | 119 | 120 | 121 | 122 | 123 | 124 | 125 | 126 | 127 | 128 | 129 | 130 | 131 | 132 |

## Table B. States and Counties — **Water Use, Wholesale Trade, Retail Trade, and Real Estate**

| STATE County | Water use, 2015 | | Wholesale Trade[1], 2017 | | | | Retail Trade[2], 2017 | | | | Real estate and rental and leasing[2], 2017 | | | |
|---|---|---|---|---|---|---|---|---|---|---|---|---|---|---|
| | Public supply water withdrawn (mil gal/ day) | Public supply gallons withdrawn per person per day | Number of establish-ments | Number of employees | Sales (mil dol) | Average payroll (mil dol) | Number of establish-ments | Number of employees | Sales (mil dol) | Average payroll (mil dol) | Number of establish-ments | Number of employees | Sales (mil dol) | Average payroll (mil dol) |
| | 133 | 134 | 135 | 136 | 137 | 138 | 139 | 140 | 141 | 142 | 143 | 144 | 145 | 146 |

1   Merchant wholesalers, except manufacturers' sales branches and offices.     2. Employer establishments.    2. Employer establishments.

## Table B. States and Counties — **Professional Services, Manufacturing, and Accommodation and Food Services**

| STATE County | Professional, scientific, and technical services, 2017 | | | | Manufacturing, 2017 | | | | Accommodation and food services, 2017 | | | |
|---|---|---|---|---|---|---|---|---|---|---|---|---|
| | Number of establish-ments | Number of employees | Sales (mil dol) | Average payroll (mil dol) | Number of establish-ments | Number of employees | Sales (mil dol) | Average payroll (mil dol) | Number of establish-hments | Number of employees | Sales (mil dol) | Annual payroll (mil dol) |
| | 147 | 148 | 149 | 150 | 151 | 152 | 153 | 154 | 155 | 156 | 157 | 158 |

## Table B. States and Counties — **Health Care and Social Assistance, Other Services, Nonemployer Businesses, and Residential Construction**

| STATE County | Health care and social assistance, 2017 | | | | Other services, 2017 | | | | Nonemployer businesses, 2018 | | Value of residential construction authorized by building permits, 2020 | |
|---|---|---|---|---|---|---|---|---|---|---|---|---|
| | Number of establish-ments | Number of employees | Receipts (mil dol) | Annual payroll (mil dol) | Number of establish-ments | Number of employees | Receipts (mil dol) | Annual payroll (mil dol) | Number | Receipts (mil dol) | New construction ($1,000) | Number of housing units |
| | 159 | 160 | 161 | 162 | 163 | 164 | 165 | 166 | 167 | 168 | 169 | 170 |

## Table B. States and Counties — Government Employment and Payroll, and Local Government Finances

| STATE County | Government employment and payroll, 2017 | | | | | | | | | Local government finances 2017 | | | | |
| | | | March payroll (percent of total) | | | | | | | General revenue | | | | |
| | | | | | | | | | | | | | Taxes | |
| | | | | | | | | | | | Inter-govern-mental (mil dol) | | Per capita[1] (dollars) | |
| | Full-time equivalent employees | March payroll (dollars) | Adminis-tration, judicial, and legal | Police and corrections | Fire protection | Highways and transpor-tation | Health and welfare | Natural resources and utilities | Education and libraries | Total (mil dol) | | Total (mil dol) | Total | Property |
| | 171 | 172 | 173 | 174 | 175 | 176 | 177 | 178 | 179 | 180 | 181 | 182 | 183 | 184 |

1. Based on the resident population estimated as of July 1 of the year shown.

## Table B. States and Counties — Local Government Finances, Government Employment, and Income Taxes

| STATE County | Local government finances, 2017 (cont.) | | | | | | | Debt outstanding | | Government employment, 2019 | | | Individual income tax returns, 2018 | | |
| | Direct general expenditure | | | | | | | | | | | | | | |
| | | | Percent of total for: | | | | | | | | | | | | |
| | Total (mil dol) | Per capita[1] (dollars) | Education | Health and hospitals | Police protection | Public welfare | Highways | Total (mil dol) | Per capita[1] (dollars) | Federal civilian | Federal military | State and local | Number of returns | Mean adjusted gross income | Mean income tax |
| | 185 | 186 | 187 | 188 | 189 | 190 | 191 | 192 | 193 | 194 | 195 | 196 | 197 | 198 | 199 |

1. Based on the resident population estimated as of July 1 of the year shown.

# Table B. States and Counties — **Land Area and Population**

| State / county code | CBSA code[1] | County Type code[2] | STATE County | Land area[3] (sq. mi) | Population, 2020 Total persons 2019 | Rank | Per square mile | White | Black | American Indian, Alaska Native | Asian and Pacific Islancer | Percent Hispanic or Latino[4] | Under 5 years | 5 to 14 years | 15 to 24 years | 25 to 34 years | 35 to 44 years | 45 to 54 years |
|---|---|---|---|---|---|---|---|---|---|---|---|---|---|---|---|---|---|---|
| | | | | 1 | 2 | 3 | 4 | 5 | 6 | 7 | 8 | 9 | 10 | 11 | 12 | 13 | 14 | 15 |
| 00000 | | 0 | UNITED STATES.......... | 3,533,044 | 329,484,123 | X | 93.3 | 61.8 | 13.6 | 1.3 | 7.2 | 18.6 | 5.9 | 12.4 | 12.9 | 14.0 | 12.8 | 12.3 |
| 01000 | | 0 | ALABAMA...................... | 50,647 | 4,921,532 | X | 97.2 | 66.6 | 27.3 | 1.2 | 2.0 | 4.6 | 5.9 | 12.4 | 12.9 | 13.2 | 12.1 | 12.4 |
| 01001 | 33860 | 2 | Autauga.................. | 594.5 | 56,145 | 918 | 94.4 | 74.8 | 21.3 | 0.9 | 1.8 | 3.1 | 6.0 | 13.0 | 12.0 | 13.3 | 13.1 | 13.2 |
| 01003 | 19300 | 3 | Baldwin .................. | 1,589.8 | 229,287 | 300 | 144.2 | 85.0 | 9.3 | 1.4 | 1.6 | 4.6 | 5.3 | 12.1 | 10.8 | 11.1 | 12.0 | 12.6 |
| 01005 | 21640 | 6 | Barbour .................. | 885.0 | 24,589 | 1,625 | 27.8 | 46.4 | 48.7 | 0.8 | 0.8 | 4.7 | 5.4 | 11.6 | 11.6 | 13.8 | 12.4 | 12.2 |
| 01007 | 13820 | 1 | Bibb...................... | 622.5 | 22,136 | 1,712 | 35.6 | 75.3 | 21.7 | 0.9 | 0.5 | 2.9 | 5.5 | 10.9 | 11.4 | 14.9 | 13.1 | 13.9 |
| 01009 | 13820 | 1 | Blount.................... | 644.9 | 57,879 | 897 | 89.7 | 88.0 | 2.0 | 1.1 | 0.6 | 9.5 | 5.9 | 12.8 | 11.7 | 12.1 | 12.0 | 13.2 |
| 01011 | | 6 | Bullock................... | 622.8 | 9,976 | 2,419 | 16.0 | 21.9 | 69.2 | 0.6 | 0.4 | 8.9 | 5.7 | 12.0 | 10.7 | 14.7 | 13.8 | 11.8 |
| 01013 | | 6 | Butler.................... | 776.8 | 19,504 | 1,854 | 25.1 | 51.9 | 45.2 | 0.7 | 1.6 | 1.7 | 5.6 | 12.7 | 11.5 | 11.3 | 12.2 | 11.6 |
| 01015 | 11500 | 3 | Calhoun .................. | 605.9 | 113,469 | 543 | 187.3 | 73.2 | 22.3 | 1.0 | 1.5 | 4.1 | 5.7 | 12.0 | 13.2 | 12.9 | 12.0 | 12.9 |
| 01017 | 29300 | 6 | Chambers ................. | 596.6 | 32,865 | 1,360 | 55.1 | 55.8 | 40.7 | 0.6 | 1.3 | 2.8 | 5.4 | 11.7 | 11.0 | 12.9 | 11.0 | 12.9 |
| 01019 | | 6 | Cherokee ................. | 553.5 | 26,294 | 1,557 | 47.5 | 92.9 | 4.8 | 1.4 | 0.6 | 1.8 | 4.6 | 10.5 | 10.5 | 11.0 | 9.8 | 13.3 |
| 01021 | 13820 | 1 | Chilton .................. | 692.8 | 44,397 | 1,088 | 64.1 | 81.2 | 10.6 | 0.8 | 0.7 | 7.9 | 6.1 | 13.4 | 12.1 | 12.6 | 12.0 | 12.8 |
| 01023 | | 9 | Choctaw .................. | 913.5 | 12,418 | 2,254 | 13.6 | 57.1 | 41.7 | 0.5 | 0.4 | 1.2 | 5.2 | 10.8 | 10.8 | 10.6 | 10.4 | 12.4 |
| 01025 | | 7 | Clarke ................... | 1,238.4 | 23,291 | 1,668 | 18.8 | 52.7 | 45.5 | 0.8 | 0.7 | 1.4 | 5.5 | 11.8 | 12.1 | 11.7 | 11.1 | 12.5 |
| 01027 | | 9 | Clay...................... | 604.0 | 13,112 | 2,213 | 21.7 | 82.5 | 14.8 | 1.1 | 0.5 | 3.1 | 5.5 | 11.5 | 11.1 | 11.2 | 11.1 | 13.3 |
| 01029 | | 8 | Cleburne ................. | 560.1 | 14,967 | 2,094 | 26.7 | 93.7 | 3.6 | 0.9 | 0.4 | 2.8 | 6.2 | 12.7 | 10.9 | 11.6 | 12.7 | 13.6 |
| 01031 | 21460 | 4 | Coffee ................... | 679.0 | 53,230 | 951 | 78.4 | 71.6 | 18.3 | 2.1 | 2.7 | 8.3 | 6.0 | 13.4 | 12.0 | 12.8 | 12.7 | 13.1 |
| 01033 | 22520 | 6 | Colbert .................. | 593.0 | 55,411 | 927 | 93.4 | 79.4 | 17.1 | 1.3 | 1.0 | 3.2 | 5.7 | 11.8 | 10.9 | 13.0 | 11.4 | 12.6 |
| 01035 | | 7 | Conecuh .................. | 850.3 | 11,851 | 2,301 | 13.9 | 51.2 | 46.1 | 0.9 | 0.6 | 2.5 | 5.1 | 11.9 | 10.8 | 10.7 | 10.3 | 12.1 |
| 01037 | 10760 | 8 | Coosa .................... | 650.9 | 10,650 | 2,369 | 16.4 | 66.6 | 30.6 | 1.0 | 0.5 | 2.5 | 4.3 | 9.0 | 9.2 | 10.6 | 10.2 | 14.3 |
| 01039 | | 6 | Covington ................ | 1,030.6 | 36,930 | 1,251 | 35.8 | 84.4 | 13.4 | 1.2 | 0.7 | 1.9 | 5.7 | 12.2 | 10.8 | 11.6 | 11.1 | 12.1 |
| 01041 | | 8 | Crenshaw ................. | 608.9 | 13,681 | 2,178 | 22.5 | 71.9 | 25.0 | 1.3 | 1.4 | 2.4 | 5.5 | 13.0 | 11.4 | 12.4 | 11.0 | 12.3 |
| 01043 | 18980 | 4 | Cullman .................. | 734.7 | 84,515 | 680 | 115.0 | 92.9 | 1.7 | 1.3 | 0.9 | 4.5 | 5.9 | 12.5 | 11.4 | 13.0 | 11.9 | 12.6 |
| 01045 | 37120 | 4 | Dale ..................... | 561.1 | 48,959 | 1,011 | 87.3 | 70.2 | 21.6 | 1.5 | 2.3 | 6.9 | 6.6 | 12.6 | 12.3 | 11.4 | 11.5 | 11.5 |
| 01047 | 42820 | 4 | Dallas ................... | 978.7 | 36,098 | 1,272 | 36.9 | 27.5 | 71.0 | 0.5 | 0.7 | 1.2 | 5.8 | 13.3 | 12.7 | 11.4 | 11.0 | 11.4 |
| 01049 | 22840 | 6 | DeKalb ................... | 777.1 | 71,658 | 762 | 92.2 | 81.5 | 2.0 | 2.7 | 0.6 | 15.3 | 6.0 | 13.6 | 12.5 | 12.0 | 12.0 | 13.0 |
| 01051 | 33860 | 2 | Elmore ................... | 618.5 | 82,158 | 694 | 132.8 | 74.6 | 21.8 | 1.0 | 1.3 | 3.0 | 5.6 | 12.4 | 12.0 | 14.1 | 12.9 | 13.1 |
| 01053 | 12120 | 6 | Escambia ................. | 945.4 | 36,281 | 1,268 | 38.4 | 62.0 | 32.3 | 4.6 | 0.7 | 2.6 | 6.1 | 12.4 | 11.4 | 13.7 | 12.0 | 12.6 |
| 01055 | 23460 | 6 | Etowah ................... | 535.1 | 102,371 | 594 | 191.3 | 78.7 | 16.5 | 1.1 | 1.1 | 4.3 | 5.8 | 11.8 | 11.8 | 12.3 | 11.8 | 13.2 |
| 01057 | | 6 | Fayette .................. | 627.7 | 16,241 | 2,020 | 25.9 | 85.4 | 12.6 | 0.9 | 0.6 | 2.1 | 5.5 | 11.8 | 11.2 | 11.4 | 11.3 | 12.5 |
| 01059 | | 6 | Franklin ................. | 633.9 | 31,507 | 1,389 | 49.7 | 76.9 | 4.5 | 1.1 | 0.5 | 18.4 | 7.0 | 13.6 | 12.3 | 12.5 | 12.0 | 12.9 |
| 01061 | 20020 | 3 | Geneva ................... | 574.5 | 26,411 | 1,551 | 46.0 | 85.1 | 10.2 | 1.9 | 0.7 | 4.3 | 5.5 | 12.2 | 10.9 | 11.4 | 11.5 | 12.3 |
| 01063 | 46220 | 8 | Greene ................... | 647.0 | 7,990 | 2,588 | 12.3 | 18.0 | 80.1 | 0.7 | 0.3 | 1.8 | 5.5 | 12.5 | 11.0 | 11.4 | 10.2 | 10.8 |
| 01065 | 46220 | 6 | Hale ..................... | 644.0 | 14,670 | 2,110 | 22.8 | 40.2 | 57.9 | 0.6 | 0.6 | 1.5 | 6.5 | 13.0 | 11.7 | 12.1 | 11.1 | 10.7 |
| 01067 | 20020 | 3 | Henry .................... | 561.8 | 17,223 | 1,961 | 30.7 | 71.2 | 25.9 | 1.0 | 0.7 | 2.7 | 5.0 | 11.5 | 10.9 | 12.0 | 12.0 | 12.6 |
| 01069 | 20020 | 3 | Houston................... | 579.8 | 106,580 | 575 | 183.8 | 67.3 | 28.6 | 1.1 | 1.6 | 3.5 | 6.3 | 12.6 | 11.5 | 12.9 | 12.3 | 12.4 |
| 01071 | 42420 | 6 | Jackson................... | 1,078.0 | 51,582 | 971 | 47.8 | 91.6 | 4.0 | 3.2 | 0.8 | 3.1 | 5.4 | 11.3 | 11.2 | 11.5 | 11.6 | 13.2 |
| 01073 | 13820 | 1 | Jefferson ................ | 1,111.5 | 655,342 | 104 | 589.6 | 50.3 | 44.1 | 0.6 | 2.2 | 4.1 | 6.3 | 12.7 | 12.4 | 14.6 | 12.8 | 11.8 |
| 01075 | | 8 | Lamar .................... | 604.8 | 13,764 | 2,172 | 22.8 | 87.4 | 11.0 | 0.9 | 0.4 | 2.0 | 5.9 | 11.8 | 11.0 | 10.9 | 11.0 | 12.8 |
| 01077 | 22520 | 3 | Lauderdale ............... | 668.0 | 93,368 | 637 | 139.8 | 85.8 | 10.8 | 0.9 | 1.2 | 3.0 | 5.1 | 10.9 | 14.8 | 12.1 | 10.7 | 11.9 |
| 01079 | 19460 | 3 | Lawrence ................. | 690.7 | 32,857 | 1,361 | 47.6 | 81.0 | 11.4 | 9.4 | 0.6 | 2.5 | 5.4 | 12.4 | 11.0 | 12.3 | 11.1 | 13.6 |
| 01081 | 12220 | 3 | Lee ...................... | 607.5 | 166,831 | 400 | 274.6 | 69.0 | 23.4 | 0.7 | 4.8 | 3.8 | 5.6 | 11.8 | 20.9 | 13.9 | 12.2 | 11.6 |
| 01083 | 26620 | 2 | Limestone ................ | 560.0 | 102,228 | 595 | 182.6 | 77.7 | 14.6 | 1.4 | 2.3 | 6.2 | 5.5 | 12.6 | 11.6 | 13.1 | 13.7 | 14.1 |
| 01085 | 33860 | 2 | Lowndes.................. | 716.1 | 9,641 | 2,443 | 13.5 | 25.1 | 72.4 | 0.6 | 0.4 | 2.0 | 6.1 | 12.5 | 11.1 | 12.7 | 10.5 | 11.6 |
| 01087 | | 6 | Macon .................... | 608.7 | 17,895 | 1,928 | 29.4 | 17.6 | 80.0 | 0.8 | 0.9 | 2.2 | 5.0 | 9.5 | 20.1 | 11.5 | 8.7 | 10.7 |
| 01089 | 26620 | 2 | Madison .................. | 801.6 | 379,453 | 190 | 473.4 | 66.7 | 25.6 | 1.5 | 3.6 | 5.2 | 5.8 | 12.1 | 12.9 | 14.4 | 12.5 | 12.5 |
| 01091 | | 7 | Marengo .................. | 976.9 | 18,733 | 1,886 | 19.2 | 45.1 | 51.9 | 0.6 | 0.5 | 2.8 | 5.9 | 12.5 | 11.6 | 12.4 | 10.9 | 12.5 |
| 01093 | | 7 | Marion ................... | 742.3 | 29,703 | 1,443 | 40.0 | 92.8 | 4.4 | 1.0 | 0.5 | 2.6 | 5.3 | 11.6 | 11.0 | 11.7 | 11.1 | 13.6 |
| 01095 | 10700 | 4 | Marshall ................. | 565.8 | 96,990 | 621 | 171.4 | 81.0 | 3.3 | 1.3 | 1.0 | 15.0 | 7.1 | 13.8 | 12.1 | 12.4 | 11.5 | 12.5 |
| 01097 | 33660 | 2 | Mobile ................... | 1,229.4 | 412,716 | 174 | 335.7 | 57.7 | 37.0 | 1.5 | 2.6 | 3.0 | 6.5 | 12.9 | 12.5 | 14.1 | 12.0 | 11.7 |
| 01099 | | 7 | Monroe ................... | 1,025.7 | 20,459 | 1,796 | 19.9 | 55.5 | 41.9 | 1.9 | 0.7 | 1.7 | 4.9 | 11.7 | 12.4 | 10.9 | 10.8 | 12.6 |
| 01101 | 33860 | 2 | Montgomery ............... | 785.4 | 224,639 | 307 | 286.0 | 32.9 | 60.3 | 0.6 | 3.9 | 3.7 | 6.8 | 12.9 | 13.3 | 14.4 | 12.5 | 11.7 |
| 01103 | 19460 | 2 | Morgan ................... | 579.7 | 119,883 | 527 | 206.8 | 76.9 | 13.8 | 1.7 | 1.0 | 8.7 | 6.1 | 12.8 | 11.8 | 12.2 | 12.2 | 12.8 |
| 01105 | | 8 | Perry .................... | 719.7 | 8,687 | 2,521 | 12.1 | 30.3 | 67.5 | 0.5 | 0.9 | 1.7 | 5.3 | 11.3 | 17.7 | 10.9 | 10.1 | 10.8 |
| 01107 | 46220 | 3 | Pickens .................. | 881.4 | 19,793 | 1,836 | 22.5 | 54.3 | 40.4 | 0.5 | 0.5 | 5.4 | 5.3 | 10.5 | 12.0 | 13.2 | 12.1 | 12.8 |
| 01109 | 45980 | 6 | Pike ..................... | 672.1 | 32,966 | 1,357 | 49.0 | 57.1 | 38.7 | 1.2 | 2.3 | 2.4 | 5.4 | 10.6 | 24.8 | 12.3 | 10.0 | 10.1 |
| 01111 | | 6 | Randolph.................. | 580.6 | 22,920 | 1,684 | 39.5 | 77.1 | 19.7 | 1.0 | 0.7 | 3.0 | 5.4 | 11.9 | 11.7 | 11.1 | 10.5 | 13.1 |
| 01113 | 17980 | 2 | Russell .................. | 641.2 | 58,237 | 894 | 90.8 | 47.4 | 46.2 | 1.1 | 1.8 | 5.9 | 6.8 | 13.4 | 11.7 | 15.3 | 12.5 | 12.1 |
| 01115 | 13820 | 1 | St. Clair ................ | 631.6 | 90,739 | 654 | 143.7 | 86.3 | 10.5 | 0.9 | 1.3 | 2.5 | 5.7 | 12.8 | 10.9 | 13.1 | 13.0 | 13.4 |
| 01117 | 13820 | 1 | Shelby ................... | 785.4 | 221,428 | 311 | 281.9 | 77.9 | 14.2 | 0.7 | 2.9 | 5.8 | 5.4 | 13.2 | 12.6 | 12.1 | 13.6 | 13.8 |
| 01119 | | 8 | Sumter ................... | 903.8 | 12,225 | 2,272 | 13.5 | 26.2 | 71.0 | 0.5 | 1.9 | 1.3 | 5.7 | 10.4 | 20.4 | 11.8 | 8.4 | 10.2 |
| 01121 | 45180 | 4 | Talladega ................ | 736.8 | 79,985 | 708 | 108.6 | 63.6 | 33.9 | 0.9 | 0.9 | 2.5 | 5.4 | 11.6 | 12.3 | 13.0 | 11.8 | 13.2 |
| 01123 | 10760 | 6 | Tallapoosa ............... | 716.5 | 40,133 | 1,180 | 56.0 | 69.7 | 27.4 | 0.8 | 0.8 | 2.5 | 5.1 | 11.8 | 10.5 | 11.6 | 10.1 | 12.7 |
| 01125 | 46220 | 2 | Tuscaloosa ............... | 1,320.8 | 210,758 | 322 | 159.6 | 61.4 | 33.2 | 0.6 | 2.1 | 4.0 | 5.8 | 11.7 | 19.4 | 14.7 | 11.9 | 11.2 |
| 01127 | 27530 | 1 | Walker ................... | 791.0 | 63,143 | 847 | 79.8 | 90.1 | 6.9 | 1.1 | 0.7 | 2.8 | 6.1 | 12.3 | 11.4 | 12.3 | 11.1 | 12.7 |

1. CBSA = Core Based Statistical Area. See Appendix A for explanation. See Appendix B for list of metropolitan areas with component counties.  2. County type code from the Economic Research
Service of USDA Rural-Urban Continuum Codes. See Appendix A for definition.   3. Dry land or land partially or temporarily covered by water.   4. May be of any race.

Items 1—15

AL(Autauga)—AL(Walker)  73

| STATE County | Age (percent) (cont.) 55 to 64 years | 65 to 74 years | 75 years and over | Percent female | Total persons 2000 | 2010 | Percent change 2000-2010 | 2010-2020 | Births | Deaths | Net Migration | Number | Persons per household | Family households | Female family householder[1] | One person |
|---|---|---|---|---|---|---|---|---|---|---|---|---|---|---|---|---|
| | 16 | 17 | 18 | 19 | 20 | 21 | 22 | 23 | 24 | 25 | 26 | 27 | 28 | 29 | 30 | 31 |
| UNITED STATES.............. | 12.9 | 9.9 | 7.0 | 50.8 | 281,421,906 | 308,758,105 | 9.7 | 6.7 | 40,009,421 | 27,751,753 | 8,468,350 | 120,756,048 | 2.62 | 65.5 | 12.4 | 27.9 |
| ALABAMA........................ | 13.4 | 10.5 | 7.2 | 51.7 | 4,447,100 | 4,780,118 | 7.5 | 3.0 | 600,479 | 532,587 | 74,102 | 1,867,893 | 2.55 | 65.7 | 14.2 | 29.8 |
| Autauga.......................... | 13.1 | 9.3 | 6.9 | 51.5 | 43,671 | 54,582 | 25.0 | 2.9 | 6,455 | 5,790 | 892 | 21,397 | 2.56 | 70.5 | 13.0 | 25.6 |
| Baldwin......................... | 14.6 | 13.0 | 8.6 | 51.6 | 140,415 | 182,263 | 29.8 | 25.8 | 22,965 | 21,613 | 45,301 | 80,930 | 2.59 | 66.1 | 9.7 | 29.6 |
| Barbour......................... | 12.8 | 11.8 | 8.5 | 47.2 | 29,038 | 27,454 | -5.5 | -10.4 | 2,860 | 3,241 | -2,497 | 9,345 | 2.41 | 66.2 | 18.4 | 31.9 |
| Bibb............................... | 13.2 | 10.0 | 7.1 | 46.8 | 20,826 | 22,904 | 10.0 | -3.4 | 2,613 | 2,643 | -736 | 6,891 | 2.99 | 69.5 | 11.4 | 28.5 |
| Blount........................... | 13.3 | 11.1 | 7.9 | 50.6 | 51,024 | 57,322 | 12.3 | 1.0 | 6,987 | 6,540 | 153 | 20,847 | 2.74 | 71.3 | 9.3 | 26.2 |
| Bullock.......................... | 13.7 | 10.3 | 7.4 | 45.3 | 11,714 | 10,913 | -6.8 | -8.6 | 1,320 | 1,276 | -992 | 3,521 | 2.79 | 71.1 | 22.4 | 25.8 |
| Butler............................ | 13.7 | 11.7 | 9.6 | 53.4 | 21,399 | 20,940 | -2.1 | -4.3 | 2,411 | 2,761 | -1,082 | 6,506 | 3.00 | 66.6 | 16.6 | 30.7 |
| Calhoun......................... | 13.7 | 11.2 | 7.3 | 51.9 | 112,249 | 118,533 | 5.6 | -4.3 | 13,556 | 14,808 | -3,766 | 44,605 | 2.50 | 67.2 | 15.9 | 28.1 |
| Chambers....................... | 14.5 | 12.0 | 8.6 | 52.3 | 36,583 | 34,154 | -6.6 | -3.8 | 3,934 | 4,664 | -542 | 13,448 | 2.46 | 63.6 | 16.0 | 30.5 |
| Cherokee ....................... | 16.1 | 14.4 | 9.9 | 50.4 | 23,988 | 25,984 | 8.3 | 1.2 | 2,386 | 3,602 | 1,547 | 10,737 | 2.39 | 68.1 | 8.5 | 28.4 |
| Chilton.......................... | 13.4 | 10.5 | 7.0 | 51.0 | 39,593 | 43,625 | 10.2 | 1.8 | 5,156 | 5,156 | 339 | 16,927 | 2.58 | 70.3 | 14.9 | 25.6 |
| Choctaw........................ | 15.8 | 13.1 | 11.0 | 52.7 | 15,922 | 13,858 | -13.0 | -10.4 | 1,382 | 1,791 | -1,025 | 5,300 | 2.40 | 64.9 | 18.1 | 32.6 |
| Clarke........................... | 14.4 | 11.3 | 9.6 | 52.7 | 27,867 | 25,834 | -7.3 | -9.8 | 2,804 | 3,069 | -2,295 | 9,090 | 2.61 | 62.3 | 11.8 | 36.1 |
| Clay............................... | 14.9 | 11.7 | 9.7 | 51.3 | 14,254 | 13,930 | -2.3 | -5.9 | 1,482 | 1,913 | -384 | 5,198 | 2.52 | 71.3 | 15.2 | 25.9 |
| Cleburne........................ | 13.6 | 11.7 | 8.6 | 50.9 | 14,123 | 14,972 | 6.0 | 0.0 | 1,814 | 1,929 | 116 | 5,680 | 2.60 | 67.0 | 8.2 | 29.2 |
| Coffee........................... | 12.5 | 10.3 | 7.3 | 50.7 | 43,615 | 49,951 | 14.5 | 6.6 | 6,389 | 5,330 | 2,203 | 19,924 | 2.56 | 69.0 | 13.7 | 27.4 |
| Colbert.......................... | 13.9 | 11.7 | 8.9 | 52.1 | 54,984 | 54,430 | -1.0 | 1.8 | 6,342 | 7,289 | 1,977 | 21,880 | 2.48 | 67.6 | 12.8 | 28.4 |
| Conecuh........................ | 15.1 | 13.5 | 10.5 | 52.2 | 14,089 | 13,232 | -6.1 | -10.4 | 1,417 | 1,799 | -1,000 | 4,553 | 2.71 | 65.8 | 19.7 | 33.0 |
| Coosa............................ | 17.7 | 14.5 | 10.2 | 49.9 | 12,202 | 11,756 | -3.7 | -9.4 | 939 | 1,434 | -612 | 4,032 | 2.57 | 65.9 | 9.7 | 32.1 |
| Covington....................... | 14.6 | 12.2 | 9.7 | 51.7 | 37,631 | 37,762 | 0.3 | -2.2 | 4,457 | 5,339 | 85 | 14,852 | 2.46 | 66.8 | 13.7 | 28.7 |
| Crenshaw....................... | 14.4 | 12.0 | 8.0 | 51.4 | 13,665 | 13,891 | 1.7 | -1.5 | 1,515 | 1,822 | 103 | 4,943 | 2.77 | 68.5 | 14.5 | 28.4 |
| Cullman......................... | 13.6 | 11.3 | 7.8 | 50.6 | 77,483 | 80,404 | 3.8 | 5.1 | 10,046 | 10,264 | 4,389 | 31,034 | 2.63 | 69.5 | 11.2 | 26.8 |
| Dale............................... | 13.0 | 10.3 | 7.5 | 51.0 | 49,129 | 50,253 | 2.3 | -2.6 | 6,790 | 5,262 | -2,828 | 18,806 | 2.55 | 66.5 | 14.8 | 28.4 |
| Dallas............................ | 14.8 | 11.8 | 7.8 | 53.8 | 46,365 | 43,813 | -5.5 | -17.6 | 5,148 | 5,519 | -7,381 | 15,910 | 2.42 | 64.9 | 23.7 | 32.1 |
| DeKalb .......................... | 13.0 | 10.7 | 7.2 | 50.3 | 64,452 | 71,090 | 10.3 | 0.8 | 8,600 | 8,124 | 132 | 26,040 | 2.70 | 70.5 | 11.2 | 26.6 |
| Elmore........................... | 13.6 | 9.9 | 6.5 | 51.7 | 65,874 | 79,279 | 20.3 | 3.6 | 9,620 | 7,918 | 1,209 | 29,708 | 2.58 | 72.5 | 12.7 | 24.9 |
| Escambia....................... | 12.9 | 10.6 | 8.2 | 49.8 | 38,440 | 38,325 | -0.3 | -5.3 | 4,551 | 4,842 | -1,740 | 13,089 | 2.61 | 61.3 | 14.3 | 34.8 |
| Etowah.......................... | 13.6 | 11.8 | 7.8 | 51.6 | 103,459 | 104,412 | 0.9 | -2.0 | 12,174 | 14,514 | 389 | 38,942 | 2.60 | 66.4 | 14.9 | 29.9 |
| Fayette.......................... | 14.2 | 12.5 | 9.5 | 51.1 | 18,495 | 17,237 | -6.8 | -5.8 | 1,791 | 2,420 | -355 | 6,802 | 2.38 | 70.2 | 10.3 | 26.6 |
| Franklin ........................ | 12.5 | 10.1 | 7.2 | 50.0 | 31,223 | 31,708 | 1.6 | -0.6 | 4,360 | 3,963 | -590 | 11,048 | 2.83 | 69.4 | 13.2 | 27.4 |
| Geneva .......................... | 14.2 | 12.2 | 9.0 | 51.3 | 25,764 | 26,783 | 4.0 | -1.4 | 2,998 | 3,710 | 355 | 10,383 | 2.52 | 66.5 | 11.5 | 28.3 |
| Greene .......................... | 14.3 | 14.4 | 10.0 | 53.5 | 9,974 | 9,039 | -9.4 | -11.6 | 1,016 | 1,119 | -955 | 2,951 | 2.80 | 52.3 | 18.5 | 45.2 |
| Hale............................... | 14.7 | 11.6 | 8.5 | 52.8 | 17,185 | 15,762 | -8.3 | -6.9 | 1,997 | 1,885 | -1,217 | 5,650 | 2.59 | 63.9 | 17.7 | 33.1 |
| Henry ............................ | 13.9 | 14.0 | 9.6 | 52.0 | 16,310 | 17,296 | 6.0 | -0.4 | 1,760 | 2,258 | 439 | 6,630 | 2.53 | 67.9 | 10.9 | 28.5 |
| Houston......................... | 13.3 | 10.8 | 7.9 | 52.2 | 88,787 | 101,562 | 14.4 | 4.9 | 13,519 | 11,212 | 2,785 | 39,311 | 2.64 | 65.5 | 15.4 | 30.5 |
| Jackson.......................... | 14.7 | 12.4 | 8.8 | 50.9 | 53,926 | 53,248 | -1.3 | -3.1 | 5,805 | 7,135 | -297 | 20,695 | 2.47 | 69.9 | 11.6 | 28.0 |
| Jefferson ....................... | 12.7 | 10.0 | 6.7 | 52.8 | 662,047 | 658,564 | -0.5 | -0.5 | 89,005 | 73,118 | -18,650 | 261,231 | 2.47 | 62.7 | 17.0 | 32.1 |
| Lamar............................ | 14.2 | 12.5 | 10.0 | 50.9 | 15,904 | 14,564 | -8.4 | -5.5 | 1,570 | 2,037 | -321 | 5,856 | 2.34 | 66.5 | 12.3 | 31.3 |
| Lauderdale...................... | 13.6 | 11.9 | 8.9 | 52.1 | 87,966 | 92,709 | 5.4 | 0.7 | 9,473 | 11,416 | 2,672 | 38,471 | 2.35 | 64.3 | 10.8 | 29.2 |
| Lawrence ....................... | 14.8 | 11.3 | 7.9 | 51.2 | 34,803 | 34,308 | -1.4 | -4.2 | 3,678 | 4,209 | -910 | 12,677 | 2.59 | 71.8 | 11.5 | 26.7 |
| Lee................................ | 11.1 | 8.0 | 4.9 | 50.9 | 115,092 | 140,300 | 21.9 | 18.9 | 18,526 | 10,814 | 18,627 | 60,237 | 2.58 | 61.6 | 12.2 | 29.7 |
| Limestone ...................... | 13.8 | 9.4 | 6.3 | 50.1 | 65,676 | 82,775 | 26.0 | 23.5 | 10,346 | 7,963 | 16,976 | 32,143 | 2.86 | 70.6 | 10.3 | 25.9 |
| Lowndes......................... | 15.1 | 11.4 | 8.9 | 53.4 | 13,473 | 11,302 | -16.1 | -14.7 | 1,311 | 1,402 | -1,589 | 4,251 | 2.35 | 64.5 | 27.9 | 33.0 |
| Macon............................ | 13.3 | 12.6 | 8.7 | 54.7 | 24,105 | 21,459 | -11.0 | -16.6 | 1,955 | 2,501 | -3,080 | 7,474 | 2.20 | 57.3 | 19.4 | 36.9 |
| Madison......................... | 14.3 | 9.1 | 6.5 | 51.1 | 276,700 | 334,812 | 21.0 | 13.3 | 42,947 | 31,108 | 32,892 | 148,189 | 2.38 | 64.5 | 12.4 | 30.3 |
| Marengo......................... | 13.7 | 11.8 | 8.8 | 53.1 | 22,539 | 21,047 | -6.6 | -11.0 | 2,367 | 2,712 | -1,984 | 7,361 | 2.59 | 52.1 | 13.2 | 45.8 |
| Marion........................... | 14.1 | 11.8 | 9.9 | 50.5 | 31,214 | 30,780 | -1.4 | -3.5 | 3,138 | 4,216 | 26 | 11,997 | 2.45 | 66.9 | 10.9 | 30.6 |
| Marshall......................... | 13.1 | 10.1 | 7.3 | 50.7 | 82,231 | 93,017 | 13.1 | 4.3 | 13,688 | 11,537 | 1,908 | 34,883 | 2.71 | 68.8 | 11.8 | 26.8 |
| Mobile........................... | 13.2 | 10.2 | 6.9 | 52.5 | 399,843 | 413,139 | 3.3 | -0.1 | 56,987 | 44,731 | -12,469 | 156,251 | 2.60 | 64.3 | 16.9 | 31.3 |
| Monroe.......................... | 15.0 | 12.4 | 9.3 | 52.6 | 24,324 | 23,066 | -5.2 | -11.3 | 2,249 | 2,755 | -2,120 | 8,166 | 2.57 | 62.7 | 17.7 | 35.4 |
| Montgomery.................... | 12.4 | 9.6 | 6.4 | 53.0 | 223,510 | 229,365 | 2.6 | -2.1 | 32,118 | 21,750 | -15,162 | 89,527 | 2.46 | 61.6 | 19.5 | 33.5 |
| Morgan........................... | 13.9 | 10.6 | 7.7 | 50.9 | 111,064 | 119,519 | 7.6 | 0.3 | 14,608 | 13,259 | -922 | 45,918 | 2.55 | 67.9 | 12.4 | 27.8 |
| Perry ............................. | 13.0 | 11.1 | 9.8 | 53.6 | 11,861 | 10,577 | -10.8 | -17.9 | 1,143 | 1,361 | -1,684 | 3,070 | 2.76 | 48.1 | 22.4 | 51.1 |
| Pickens .......................... | 14.3 | 11.3 | 8.5 | 50.0 | 20,949 | 19,746 | -5.7 | 0.2 | 2,246 | 2,537 | 276 | 7,637 | 2.41 | 66.4 | 14.2 | 30.6 |
| Pike............................... | 11.1 | 8.9 | 6.9 | 52.1 | 29,605 | 32,902 | 11.1 | 0.2 | 3,833 | 3,352 | -452 | 11,601 | 2.71 | 65.4 | 13.6 | 37.4 |
| Randolph........................ | 15.0 | 12.3 | 9.0 | 51.5 | 22,380 | 22,916 | 2.4 | 0.0 | 2,554 | 3,013 | 466 | 8,702 | 2.55 | 65.7 | 12.0 | 31.2 |
| Russell........................... | 13.1 | 9.2 | 5.9 | 52.3 | 49,756 | 52,966 | 6.5 | 10.0 | 8,594 | 6,103 | 2,586 | 23,262 | 2.47 | 64.3 | 16.2 | 32.1 |
| St. Clair ......................... | 13.8 | 10.5 | 7.0 | 50.6 | 64,742 | 83,366 | 28.8 | 8.8 | 10,495 | 9,401 | 6,310 | 32,036 | 2.68 | 73.0 | 10.5 | 24.2 |
| Shelby ........................... | 13.0 | 10.0 | 6.4 | 51.7 | 143,293 | 195,257 | 36.3 | 13.4 | 24,124 | 14,809 | 16,904 | 79,630 | 2.64 | 71.2 | 8.8 | 25.0 |
| Sumter........................... | 13.2 | 11.2 | 8.6 | 54.3 | 14,798 | 13,763 | -7.0 | -11.2 | 1,452 | 1,620 | -1,375 | 5,202 | 2.29 | 53.1 | 21.3 | 42.6 |
| Talladega........................ | 13.9 | 11.5 | 7.4 | 51.8 | 80,321 | 82,348 | 2.5 | -2.9 | 8,850 | 10,320 | -858 | 31,395 | 2.45 | 66.5 | 17.3 | 29.4 |
| Tallapoosa...................... | 15.4 | 13.7 | 9.1 | 51.5 | 41,475 | 41,621 | 0.4 | -3.6 | 4,519 | 5,535 | -460 | 16,205 | 2.46 | 69.0 | 13.8 | 28.1 |
| Tuscaloosa...................... | 11.2 | 8.7 | 5.4 | 52.0 | 164,875 | 194,678 | 18.1 | 8.3 | 25,164 | 17,915 | 8,832 | 72,461 | 2.72 | 64.8 | 14.9 | 28.5 |
| Walker........................... | 14.0 | 11.9 | 8.2 | 51.3 | 70,713 | 67,019 | -5.2 | -5.8 | 8,097 | 10,271 | -1,658 | 25,019 | 2.53 | 68.6 | 12.4 | 26.6 |

1. No spouse present.

# Table B. States and Counties — Population, Vital Statistics, Health, and Crime

| STATE County | Persons in group quarters, 2020 | Daytime Population, 2015-2019 Number | Employment/residence ratio | Births, 2020 Total | Rate[1] | Deaths, 2020 Number | Rate[1] | Persons under 65 with no health insurance, 2019 Number | Percent | Medicare, 2020 Total beneficiaries | Enrolled in Original Medicare | Enrolled in Medicare Advantage | Crimes reported by county police or sheriff, 2019 Violent | Property |
|---|---|---|---|---|---|---|---|---|---|---|---|---|---|---|
| | 32 | 33 | 34 | 35 | 36 | 37 | 38 | 39 | 40 | 41 | 42 | 43 | 44 | 45 |
| UNITED STATES............. | 8,024,233 | 324,697,795 | 1.00 | 3,748,000 | 11.4 | 3,070,859 | 9.3 | 28,980,723 | 10.8 | 61,521,688 | 37,063,987 | 24,457,709 | X | X |
| ALABAMA....................... | 116,767 | 4,831,069 | 0.98 | 56,739 | 11.5 | 58,354 | 11.9 | 457,718 | 11.6 | 1,062,205 | 576,373 | 485,833 | X | X |
| Autauga........................ | 455 | 44,451 | 0.56 | 606 | 10.8 | 582 | 10.4 | 4,366 | 9.4 | 11,247 | 5,593 | 5,654 | NA | NA |
| Baldwin......................... | 2,268 | 198,399 | 0.85 | 2,317 | 10.1 | 2,543 | 11.1 | 19,085 | 10.9 | 55,742 | 29,149 | 26,592 | NA | NA |
| Barbour........................ | 2,777 | 25,373 | 1.00 | 250 | 10.2 | 334 | 13.6 | 2,194 | 13.0 | 6,189 | 3,048 | 3,142 | NA | NA |
| Bibb............................. | 2,121 | 18,935 | 0.56 | 249 | 11.2 | 266 | 12.0 | 1,824 | 11.0 | 4,736 | 2,058 | 2,678 | NA | NA |
| Blount.......................... | 489 | 44,822 | 0.41 | 665 | 11.5 | 697 | 12.0 | 6,663 | 14.3 | 13,182 | 5,739 | 7,443 | NA | NA |
| Bullock......................... | 1,704 | 9,583 | 0.85 | 114 | 11.4 | 116 | 11.6 | 752 | 11.1 | 2,295 | 1,079 | 1,216 | NA | NA |
| Butler........................... | 333 | 19,320 | 0.94 | 200 | 10.3 | 266 | 13.6 | 1,755 | 11.6 | 4,964 | 3,106 | 1,858 | NA | NA |
| Calhoun........................ | 3,285 | 113,789 | 0.98 | 1,230 | 10.8 | 1,553 | 13.7 | 10,767 | 11.9 | 26,991 | 17,248 | 9,743 | NA | NA |
| Chambers..................... | 458 | 28,250 | 0.62 | 334 | 10.2 | 519 | 15.8 | 3,178 | 12.2 | 8,826 | 5,023 | 3,803 | NA | NA |
| Cherokee...................... | 290 | 22,523 | 0.66 | 236 | 9.0 | 406 | 15.4 | 2,580 | 13.0 | 7,359 | 4,272 | 3,087 | NA | NA |
| Chilton......................... | 393 | 36,995 | 0.59 | 533 | 12.0 | 554 | 12.5 | 4,995 | 13.7 | 9,666 | 3,861 | 5,805 | NA | NA |
| Choctaw....................... | 129 | 12,345 | 0.86 | 136 | 11.0 | 183 | 14.7 | 1,158 | 12.2 | 3,820 | 2,651 | 1,169 | NA | NA |
| Clarke.......................... | 280 | 23,928 | 0.97 | 265 | 11.4 | 348 | 14.9 | 2,209 | 12.0 | 6,418 | 3,084 | 3,334 | NA | NA |
| Clay............................. | 255 | 12,544 | 0.85 | 158 | 12.1 | 224 | 17.1 | 1,480 | 14.4 | 3,678 | 2,144 | 1,534 | NA | NA |
| Cleburne...................... | 178 | 12,175 | 0.50 | 177 | 11.8 | 216 | 14.4 | 1,506 | 12.8 | 3,541 | 2,439 | 1,103 | NA | NA |
| Coffee......................... | 600 | 47,690 | 0.82 | 568 | 10.7 | 590 | 11.1 | 5,604 | 13.0 | 10,922 | 7,241 | 3,681 | NA | NA |
| Colbert........................ | 474 | 57,592 | 1.13 | 631 | 11.4 | 724 | 13.1 | 4,721 | 10.8 | 14,149 | 9,175 | 4,974 | NA | NA |
| Conecuh....................... | 45 | 11,688 | 0.83 | 119 | 10.0 | 199 | 16.8 | 1,112 | 12.2 | 3,549 | 2,320 | 1,229 | NA | NA |
| Coosa.......................... | 259 | 8,391 | 0.42 | 79 | 7.4 | 143 | 13.4 | 891 | 11.3 | 3,112 | 1,697 | 1,415 | NA | NA |
| Covington..................... | 575 | 35,650 | 0.89 | 395 | 10.7 | 555 | 15.0 | 3,608 | 12.5 | 9,953 | 6,776 | 3,177 | NA | NA |
| Crenshaw..................... | 132 | 12,462 | 0.76 | 135 | 9.9 | 201 | 14.7 | 1,471 | 13.4 | 3,413 | 1,875 | 1,538 | NA | NA |
| Cullman........................ | 1,040 | 79,325 | 0.90 | 948 | 11.2 | 1,115 | 13.2 | 9,008 | 13.4 | 20,104 | 11,332 | 8,772 | NA | NA |
| Dale............................. | 927 | 50,641 | 1.07 | 642 | 13.1 | 622 | 12.7 | 4,777 | 12.0 | 10,821 | 6,917 | 3,904 | NA | NA |
| Dallas.......................... | 853 | 38,936 | 0.98 | 398 | 11.0 | 573 | 15.9 | 3,263 | 11.1 | 9,737 | 4,791 | 4,946 | NA | NA |
| DeKalb......................... | 781 | 66,877 | 0.85 | 816 | 11.4 | 893 | 12.5 | 10,814 | 18.7 | 15,511 | 9,619 | 5,892 | NA | NA |
| Elmore......................... | 4,172 | 66,679 | 0.58 | 902 | 11.0 | 876 | 10.7 | 5,781 | 8.9 | 17,506 | 9,194 | 8,312 | NA | NA |
| Escambia...................... | 2,392 | 37,342 | 1.02 | 434 | 12.0 | 503 | 13.9 | 3,875 | 14.2 | 8,832 | 4,913 | 3,919 | NA | NA |
| Etowah........................ | 2,085 | 97,550 | 0.87 | 1,173 | 11.5 | 1,609 | 15.7 | 10,281 | 12.7 | 26,225 | 13,840 | 12,385 | NA | NA |
| Fayette........................ | 291 | 14,565 | 0.69 | 160 | 9.9 | 247 | 15.2 | 1,508 | 12.0 | 4,610 | 3,042 | 1,568 | NA | NA |
| Franklin........................ | 272 | 30,153 | 0.90 | 428 | 13.6 | 416 | 13.2 | 3,952 | 15.4 | 6,771 | 4,836 | 1,936 | NA | NA |
| Geneva......................... | 229 | 21,973 | 0.56 | 264 | 10.0 | 379 | 14.4 | 3,009 | 14.6 | 6,946 | 4,322 | 2,624 | NA | NA |
| Greene......................... | 45 | 7,663 | 0.72 | 83 | 10.4 | 91 | 11.4 | 708 | 11.6 | 2,311 | 1,421 | 890 | NA | NA |
| Hale............................. | 221 | 12,403 | 0.54 | 178 | 12.1 | 203 | 13.8 | 1,335 | 11.6 | 4,009 | 2,611 | 1,398 | NA | NA |
| Henry........................... | 199 | 14,555 | 0.62 | 149 | 8.7 | 229 | 13.3 | 1,632 | 12.5 | 4,861 | 2,515 | 2,346 | NA | NA |
| Houston........................ | 1,418 | 112,625 | 1.18 | 1,376 | 12.9 | 1,253 | 11.8 | 11,329 | 13.3 | 24,387 | 14,042 | 10,345 | NA | NA |
| Jackson........................ | 597 | 48,542 | 0.83 | 559 | 10.8 | 736 | 14.3 | 5,584 | 13.8 | 13,310 | 8,549 | 4,761 | NA | NA |
| Jefferson...................... | 16,627 | 721,785 | 1.21 | 8,257 | 12.6 | 7,768 | 11.9 | 57,398 | 10.7 | 132,456 | 58,127 | 74,328 | NA | NA |
| Lamar.......................... | 217 | 12,232 | 0.68 | 175 | 12.7 | 223 | 16.2 | 1,257 | 11.9 | 3,855 | 2,892 | 963 | NA | NA |
| Lauderdale.................... | 2,411 | 87,269 | 0.87 | 916 | 9.8 | 1,217 | 13.0 | 8,571 | 11.9 | 22,437 | 14,448 | 7,989 | NA | NA |
| Lawrence...................... | 229 | 25,990 | 0.42 | 332 | 10.1 | 476 | 14.5 | 3,906 | 14.8 | 8,029 | 4,750 | 3,279 | NA | NA |
| Lee.............................. | 4,960 | 150,530 | 0.86 | 1,754 | 10.5 | 1,272 | 7.6 | 14,928 | 10.7 | 25,287 | 14,775 | 10,513 | NA | NA |
| Limestone..................... | 2,734 | 82,831 | 0.70 | 1,037 | 10.1 | 931 | 9.1 | 9,428 | 11.6 | 18,632 | 11,659 | 6,973 | NA | NA |
| Lowndes....................... | 96 | 8,667 | 0.58 | 110 | 11.4 | 161 | 16.7 | 826 | 10.8 | 2,924 | 1,102 | 1,822 | NA | NA |
| Macon.......................... | 1,582 | 17,739 | 0.86 | 170 | 9.5 | 283 | 15.8 | 1,349 | 10.7 | 4,621 | 2,301 | 2,320 | NA | NA |
| Madison........................ | 8,186 | 396,179 | 1.19 | 4,315 | 11.4 | 3,585 | 9.4 | 28,344 | 9.2 | 67,811 | 45,412 | 22,400 | NA | NA |
| Marengo........................ | 254 | 19,105 | 0.97 | 190 | 10.1 | 259 | 13.8 | 1,651 | 11.1 | 5,427 | 3,466 | 1,961 | NA | NA |
| Marion.......................... | 813 | 29,195 | 0.94 | 312 | 10.5 | 448 | 15.1 | 2,816 | 12.3 | 7,452 | 5,390 | 2,063 | NA | NA |
| Marshall....................... | 1,044 | 96,599 | 1.02 | 1,371 | 14.1 | 1,263 | 13.0 | 12,136 | 15.4 | 20,865 | 13,509 | 7,356 | NA | NA |
| Mobile.......................... | 7,386 | 424,641 | 1.06 | 5,393 | 13.1 | 5,011 | 12.1 | 42,767 | 12.7 | 85,179 | 34,793 | 50,386 | NA | NA |
| Monroe......................... | 230 | 21,114 | 0.98 | 182 | 8.9 | 286 | 14.0 | 2,083 | 12.9 | 5,322 | 2,828 | 2,494 | NA | NA |
| Montgomery................... | 8,755 | 260,549 | 1.34 | 3,064 | 13.6 | 2,419 | 10.8 | 21,123 | 11.7 | 43,719 | 21,322 | 22,398 | NA | NA |
| Morgan......................... | 2,007 | 119,408 | 1.00 | 1,416 | 11.8 | 1,383 | 11.5 | 12,379 | 12.8 | 26,371 | 16,890 | 9,481 | NA | NA |
| Perry............................ | 723 | 8,679 | 0.76 | 88 | 10.1 | 138 | 15.9 | 776 | 12.1 | 2,616 | 1,360 | 1,256 | NA | NA |
| Pickens........................ | 1,875 | 17,793 | 0.64 | 207 | 10.5 | 283 | 14.3 | 1,755 | 12.4 | 4,983 | 3,382 | 1,601 | NA | NA |
| Pike............................. | 1,912 | 35,864 | 1.18 | 353 | 10.7 | 376 | 11.4 | 3,498 | 13.5 | 6,497 | 3,526 | 2,970 | NA | NA |
| Randolph....................... | 390 | 19,959 | 0.69 | 243 | 10.6 | 306 | 13.4 | 2,165 | 12.3 | 5,910 | 3,712 | 2,199 | NA | NA |
| Russell......................... | 574 | 48,883 | 0.63 | 747 | 12.8 | 701 | 12.0 | 5,473 | 11.3 | 11,455 | 6,511 | 4,944 | NA | NA |
| St. Clair....................... | 1,515 | 73,531 | 0.62 | 946 | 10.4 | 1,046 | 11.5 | 7,465 | 10.2 | 19,289 | 8,607 | 10,682 | NA | NA |
| Shelby.......................... | 2,541 | 196,841 | 0.84 | 2,223 | 10.0 | 1,868 | 8.4 | 15,878 | 8.7 | 38,893 | 19,441 | 19,452 | NA | NA |
| Sumter......................... | 784 | 12,588 | 0.95 | 137 | 11.2 | 160 | 13.1 | 1,218 | 13.2 | 3,163 | 2,007 | 1,156 | NA | NA |
| Talladega...................... | 3,251 | 83,536 | 1.10 | 845 | 10.6 | 1,155 | 14.4 | 6,969 | 11.2 | 19,827 | 9,823 | 10,004 | NA | NA |
| Tallapoosa.................... | 576 | 37,913 | 0.82 | 375 | 9.3 | 585 | 14.6 | 3,761 | 12.1 | 11,481 | 6,215 | 5,267 | NA | NA |
| Tuscaloosa.................... | 10,665 | 215,687 | 1.09 | 2,362 | 11.2 | 2,063 | 9.8 | 17,245 | 10.1 | 36,950 | 20,254 | 16,697 | NA | NA |
| Walker.......................... | 852 | 59,961 | 0.83 | 767 | 12.1 | 988 | 15.6 | 6,674 | 13.2 | 17,532 | 8,085 | 9,447 | NA | NA |

1. Per 1,000 estimated resident population.

| STATE County | County law enforcement employment, 2019 | | Education — School enrollment and attainment, 2015-2019 | | | | Local government expenditures,[3] 2017-2018 | | Money income, 2015-2019 | | | | Income and poverty, 2019 | | | |
|---|---|---|---|---|---|---|---|---|---|---|---|---|---|---|---|---|
| | | | Enrollment[1] | | Attainment[2] (percent) | | | | | Households | | | | Percent below poverty level | | |
| | | | | | High school graduate or less | Bachelor's degree or more | Total current spending (mil dol) | Current spending per student (dollars) | Per capita income[4] | Median income (dollars) | Percent | | Median household income (dollars) | All persons | Children under 18 years | Children 5 to 17 years in families |
| | Officers | Civilians | Total | Percent private | | | | | | | with income of less than $50,000 | with income of $200,000 or more | | | | |
| | 46 | 47 | 48 | 49 | 50 | 51 | 52 | 53 | 54 | 55 | 56 | 57 | 58 | 59 | 60 | 61 |
| UNITED STATES | X | X | 81,084,866 | 16.3 | 39.0 | 32.1 | 629,620 | 12,485 | 34,103 | 62,843 | 40.5 | 7.7 | 65,712 | 12.3 | 16.8 | 15.8 |
| ALABAMA | X | X | 1,175,774 | 14.3 | 44.6 | 25.5 | 7,199 | 9,697 | 27,928 | 50,536 | 49.5 | 4.1 | 51,771 | 15.6 | 21.9 | 20.7 |
| Autauga | 32 | 11 | 13,131 | 15.5 | 45.1 | 26.6 | 76 | 8,202 | 29,819 | 58,731 | 43.1 | 3.2 | 58,233 | 12.1 | 15.9 | 14.4 |
| Baldwin | 135 | 184 | 45,187 | 19.1 | 36.9 | 31.9 | 295 | 9,284 | 32,626 | 58,320 | 42.6 | 5.7 | 59,871 | 10.1 | 13.5 | 13.3 |
| Barbour | 26 | 32 | 5,150 | 11.8 | 62.4 | 11.6 | 37 | 6,825 | 18,473 | 32,525 | 66.9 | 1.6 | 35,972 | 27.1 | 41.0 | 39.5 |
| Bibb | 11 | 1 | 4,106 | 11.7 | 65.8 | 10.4 | 31 | 9,491 | 20,778 | 47,542 | 53.0 | 2.2 | 47,918 | 20.3 | 25.9 | 25.2 |
| Blount | 44 | 4 | 12,102 | 10.2 | 52.9 | 13.1 | 84 | 9,033 | 24,747 | 49,358 | 50.7 | 2.5 | 52,902 | 16.3 | 21.0 | 21.4 |
| Bullock | 6 | 7 | 2,047 | 12.1 | 65.6 | 12.1 | 15 | 9,984 | 20,877 | 37,785 | 60.3 | 1.8 | 31,906 | 30.0 | 39.9 | 39.0 |
| Butler | 15 | 3 | 4,039 | 13.4 | 60.1 | 16.1 | 28 | 9,225 | 21,038 | 40,688 | 59.5 | 2.2 | 39,944 | 21.6 | 34.6 | 32.8 |
| Calhoun | NA | NA | 26,978 | 11.3 | 48.4 | 18.5 | 175 | 9,981 | 25,345 | 47,255 | 52.0 | 2.2 | 47,747 | 17.2 | 24.6 | 22.8 |
| Chambers | NA | NA | 6,807 | 12.1 | 55.1 | 13.3 | 44 | 9,567 | 22,729 | 42,289 | 56.2 | 1.1 | 42,015 | 19.6 | 30.4 | 31.1 |
| Cherokee | NA | NA | 5,061 | 5.6 | 58.7 | 12.8 | 42 | 10,617 | 24,301 | 41,919 | 55.2 | 2.4 | 45,982 | 15.9 | 23.3 | 21.6 |
| Chilton | NA | NA | 9,970 | 6.8 | 62.6 | 12.7 | 66 | 8,618 | 24,658 | 47,468 | 51.7 | 2.5 | 49,692 | 16.0 | 23.1 | 21.2 |
| Choctaw | NA | NA | 2,477 | 26.6 | 56.3 | 11.9 | 15 | 10,558 | 22,308 | 35,892 | 62.7 | 0.9 | 39,808 | 22.6 | 30.4 | 28.3 |
| Clarke | 14 | 33 | 5,299 | 13.4 | 61.2 | 12.2 | 39 | 10,281 | 22,225 | 37,404 | 61.5 | 2.1 | 43,822 | 18.4 | 28.6 | 27.9 |
| Clay | 11 | 16 | 2,798 | 13.8 | 61.1 | 10.6 | 17 | 9,094 | 23,519 | 40,845 | 58.9 | 2.5 | 40,562 | 16.6 | 26.0 | 22.3 |
| Cleburne | 10 | 18 | 3,209 | 12.1 | 58.9 | 15.4 | 24 | 9,306 | 23,142 | 44,741 | 56.1 | 1.4 | 51,276 | 13.3 | 17.9 | 17.7 |
| Coffee | 29 | 31 | 12,502 | 10.0 | 41.3 | 22.8 | 90 | 9,083 | 27,806 | 55,637 | 46.0 | 3.1 | 57,299 | 15.0 | 21.8 | 21.3 |
| Colbert | 31 | 26 | 10,710 | 10.2 | 52.7 | 18.1 | 87 | 10,705 | 25,536 | 48,065 | 51.7 | 1.8 | 46,683 | 14.6 | 19.4 | 18.6 |
| Conecuh | 11 | 27 | 2,467 | 10.9 | 66.8 | 12.8 | 21 | 10,444 | 19,769 | 37,837 | 63.1 | 1.2 | 41,539 | 22.2 | 34.7 | 33.0 |
| Coosa | 10 | 13 | 1,674 | 12.4 | 60.3 | 12.8 | 10 | 11,055 | 23,694 | 38,990 | 61.4 | 1.6 | 42,024 | 17.1 | 27.8 | 26.9 |
| Covington | NA | NA | 7,682 | 4.8 | 53.9 | 15.3 | 58 | 9,452 | 23,687 | 42,189 | 56.6 | 1.8 | 44,836 | 17.2 | 26.6 | 25.6 |
| Crenshaw | NA | NA | 2,938 | 7.6 | 60.3 | 17.4 | 21 | 9,100 | 24,563 | 43,163 | 56.6 | 2.0 | 43,309 | 16.8 | 25.0 | 23.5 |
| Cullman | NA | NA | 17,167 | 9.6 | 52.5 | 13.9 | 121 | 9,497 | 23,468 | 44,918 | 54.6 | 2.2 | 50,897 | 12.2 | 17.0 | 16.4 |
| Dale | NA | NA | 11,173 | 7.6 | 46.2 | 18.1 | 59 | 9,528 | 24,645 | 47,214 | 52.8 | 2.4 | 45,120 | 18.1 | 27.8 | 26.7 |
| Dallas | NA | NA | 9,584 | 13.2 | 54.1 | 14.7 | 68 | 10,433 | 19,612 | 33,845 | 64.7 | 1.1 | 34,034 | 26.0 | 41.0 | 37.1 |
| DeKalb | NA | NA | 16,645 | 9.6 | 59.4 | 13.0 | 112 | 9,183 | 21,907 | 40,440 | 58.9 | 2.1 | 44,277 | 18.2 | 29.4 | 30.5 |
| Elmore | NA | NA | 18,063 | 21.7 | 45.9 | 25.1 | 112 | 8,495 | 29,052 | 60,891 | 40.9 | 4.5 | 62,310 | 11.4 | 17.7 | 17.8 |
| Escambia | 33 | 56 | 7,668 | 9.7 | 62.6 | 12.5 | 56 | 10,371 | 19,028 | 36,275 | 64.0 | 1.6 | 42,712 | 20.5 | 27.1 | 24.4 |
| Etowah | 67 | 111 | 22,201 | 8.8 | 48.3 | 17.8 | 141 | 9,214 | 24,883 | 44,637 | 55.1 | 2.4 | 43,047 | 18.8 | 27.0 | 25.6 |
| Fayette | NA | NA | 3,588 | 8.9 | 58.7 | 12.4 | 22 | 9,499 | 22,483 | 39,856 | 59.0 | 1.5 | 44,212 | 17.0 | 23.0 | 22.2 |
| Franklin | NA | NA | 7,183 | 3.4 | 59.4 | 14.0 | 60 | 9,739 | 21,641 | 43,514 | 56.6 | 1.9 | 44,874 | 16.1 | 24.5 | 23.5 |
| Geneva | 12 | 13 | 5,708 | 10.0 | 55.7 | 12.5 | 38 | 9,382 | 21,711 | 41,732 | 59.2 | 1.1 | 43,089 | 16.1 | 21.3 | 23.0 |
| Greene | 13 | 20 | 1,699 | 19.6 | 59.4 | 10.1 | 13 | 12,455 | 14,884 | 24,145 | 74.0 | 0.9 | 28,699 | 31.7 | 45.8 | 44.1 |
| Hale | 11 | 20 | 3,593 | 12.2 | 59.4 | 14.2 | 24 | 9,502 | 21,191 | 34,046 | 61.7 | 1.5 | 40,745 | 20.5 | 30.8 | 30.9 |
| Henry | 15 | 4 | 3,587 | 17.4 | 53.2 | 17.1 | 22 | 8,969 | 25,276 | 50,017 | 50.0 | 1.8 | 48,597 | 17.2 | 29.2 | 28.0 |
| Houston | 72 | 99 | 24,742 | 15.7 | 45.2 | 21.7 | 142 | 9,089 | 27,097 | 47,580 | 52.0 | 4.0 | 50,138 | 19.4 | 27.8 | 25.7 |
| Jackson | 35 | 46 | 9,896 | 8.5 | 57.8 | 14.0 | 79 | 9,902 | 22,596 | 41,769 | 58.2 | 1.8 | 44,322 | 14.7 | 19.6 | 19.4 |
| Jefferson | 530 | 160 | 163,818 | 16.4 | 36.3 | 33.4 | 1,093 | 10,518 | 32,098 | 53,901 | 46.7 | 6.1 | 54,127 | 16.2 | 22.5 | 20.9 |
| Lamar | NA | NA | 2,979 | 5.2 | 57.4 | 11.9 | 21 | 8,879 | 22,636 | 38,364 | 60.0 | 1.2 | 42,686 | 15.3 | 21.6 | 19.8 |
| Lauderdale | 51 | 23 | 21,984 | 9.7 | 48.6 | 23.5 | 127 | 9,951 | 28,033 | 48,094 | 51.7 | 3.0 | 48,188 | 15.7 | 20.4 | 18.8 |
| Lawrence | 26 | 33 | 7,207 | 4.8 | 61.9 | 12.6 | 45 | 9,535 | 23,450 | 44,886 | 55.6 | 1.3 | 47,797 | 17.7 | 22.8 | 21.2 |
| Lee | 81 | 89 | 55,955 | 11.5 | 32.8 | 35.3 | 224 | 9,839 | 27,860 | 51,463 | 48.9 | 4.5 | 54,160 | 15.5 | 14.2 | 13.5 |
| Limestone | 48 | 72 | 22,669 | 15.8 | 45.7 | 25.7 | 125 | 8,823 | 28,411 | 59,686 | 42.3 | 4.7 | 64,729 | 12.0 | 16.3 | 14.3 |
| Lowndes | 17 | 25 | 2,194 | 14.8 | 61.3 | 14.7 | 20 | 13,819 | 20,209 | 30,036 | 68.1 | 1.1 | 33,930 | 26.6 | 36.1 | 35.7 |
| Macon | 16 | 12 | 5,256 | 49.3 | 48.4 | 20.4 | 23 | 12,026 | 20,419 | 33,370 | 66.3 | 1.8 | 34,281 | 29.3 | 39.2 | 39.0 |
| Madison | 131 | 208 | 95,021 | 17.5 | 28.9 | 42.8 | 533 | 9,862 | 36,620 | 65,449 | 39.4 | 7.5 | 68,609 | 11.5 | 16.8 | 14.7 |
| Marengo | 10 | 15 | 4,505 | 12.4 | 54.1 | 16.1 | 37 | 9,597 | 25,635 | 33,241 | 62.2 | 1.5 | 38,838 | 24.8 | 36.3 | 35.4 |
| Marion | 15 | 16 | 5,899 | 7.7 | 55.3 | 12.8 | 44 | 9,598 | 21,311 | 35,930 | 63.8 | 1.7 | 44,675 | 16.9 | 24.3 | 21.4 |
| Marshall | 38 | 36 | 23,321 | 4.2 | 49.5 | 19.4 | 173 | 9,504 | 24,453 | 45,983 | 53.4 | 3.4 | 48,913 | 15.7 | 21.5 | 21.9 |
| Mobile | 180 | 372 | 100,026 | 18.3 | 47.5 | 23.2 | 579 | 9,460 | 25,861 | 47,583 | 51.8 | 3.3 | 49,492 | 17.7 | 26.4 | 24.1 |
| Monroe | 18 | 33 | 4,076 | 9.8 | 61.6 | 13.5 | 33 | 9,401 | 20,527 | 30,441 | 65.1 | 2.0 | 40,005 | 23.3 | 31.4 | 30.7 |
| Montgomery | 127 | 48 | 58,437 | 23.5 | 38.3 | 33.4 | 290 | 9,354 | 29,071 | 50,124 | 49.9 | 4.5 | 52,711 | 15.8 | 23.3 | 24.3 |
| Morgan | 55 | 127 | 26,724 | 13.1 | 46.3 | 22.4 | 199 | 10,151 | 27,367 | 52,156 | 48.0 | 3.1 | 54,355 | 13.7 | 18.8 | 17.1 |
| Perry | 7 | 12 | 2,837 | 17.3 | 54.1 | 22.8 | 14 | 10,684 | 15,055 | 23,447 | 79.3 | 0.2 | 29,572 | 33.9 | 46.2 | 45.6 |
| Pickens | 10 | 25 | 4,030 | 18.5 | 55.7 | 14.1 | 26 | 10,099 | 23,024 | 39,848 | 57.8 | 2.1 | 40,249 | 24.3 | 34.0 | 32.7 |
| Pike | 21 | 14 | 11,272 | 9.7 | 50.8 | 23.7 | 42 | 10,197 | 22,453 | 37,446 | 61.0 | 2.2 | 41,271 | 21.8 | 29.8 | 26.2 |
| Randolph | 16 | 19 | 4,710 | 13.2 | 55.2 | 16.5 | 36 | 9,507 | 24,721 | 43,395 | 56.1 | 2.4 | 42,922 | 17.5 | 28.4 | 28.1 |
| Russell | 39 | 78 | 14,635 | 13.7 | 47.6 | 17.8 | 96 | 8,693 | 23,470 | 42,443 | 55.7 | 2.2 | 43,670 | 18.9 | 25.4 | 26.2 |
| St. Clair | 51 | 57 | 18,647 | 10.2 | 51.0 | 16.1 | 115 | 9,991 | 26,409 | 58,308 | 41.9 | 2.6 | 65,403 | 11.2 | 14.9 | 13.9 |
| Shelby | 131 | 90 | 55,659 | 18.2 | 27.2 | 42.9 | 298 | 9,991 | 38,549 | 77,799 | 30.7 | 8.8 | 77,801 | 6.2 | 7.7 | 7.0 |
| Sumter | 8 | 17 | 3,943 | 7.8 | 48.7 | 21.6 | 18 | 11,491 | 16,799 | 24,320 | 72.4 | 0.9 | 29,209 | 36.4 | 45.3 | 44.4 |
| Talladega | 42 | 53 | 17,839 | 9.8 | 53.4 | 15.0 | 115 | 9,721 | 23,305 | 41,325 | 56.8 | 1.7 | 47,719 | 17.4 | 25.1 | 24.0 |
| Tallapoosa | 23 | 32 | 8,271 | 8.2 | 52.1 | 18.4 | 57 | 9,533 | 25,217 | 45,828 | 53.3 | 2.8 | 47,100 | 16.0 | 25.9 | 24.5 |
| Tuscaloosa | 105 | 99 | 60,995 | 9.4 | 41.5 | 30.7 | 292 | 10,000 | 26,789 | 53,326 | 46.8 | 4.1 | 52,307 | 16.2 | 19.7 | 20.0 |
| Walker | 32 | 44 | 13,405 | 7.4 | 55.7 | 11.3 | 101 | 9,642 | 23,882 | 43,629 | 54.5 | 1.8 | 45,991 | 17.3 | 23.7 | 23.4 |

1. All persons 3 years old and over enrolled in nursery school through college.     2. Persons 25 years old and over.     3. Elementary and secondary education expenditures.     4. Based on population estimated by the American Community Survey, 2014–2018.

# Table B. States and Counties — **Personal Income**

| STATE County | Personal income, 2019 | | | | | | | | | | Earnings, 2019 | | |
|---|---|---|---|---|---|---|---|---|---|---|---|---|---|
| | Total (mil dol) | Percent change 2018-2019 | Per capita[1] Dollars | Per capita[1] Rank | Wages and salaries (mil dol) | Supplements to wages and salaries, employer contributions (mil dol) Pension and insurance | Supplements, Government social insurance | Proprietors' income (mil dol) | Dividends, interest, and rent (mil dol) | Personal transfer receipts (mil dol) | Total (mil dol) | Contributions for government social insurance (mil dol) From employee and self-employed | From employer |
| | 62 | 63 | 64 | 65 | 66 | 67 | 68 | 69 | 70 | 71 | 72 | 73 | 74 |
| UNITED STATES............. | 18,540,000 | 3.9 | 56,474 | X | 9,300,811 | 1,466,990 | 648,023 | 1,664,741 | 3,750,076 | 3,125,174 | 13,080,565 | 768,643 | 648,023 |
| ALABAMA ...................... | 216,449 | 3.7 | 44,102 | X | 102,465 | 16,974 | 7,408 | 14,345 | 39,056 | 50,500 | 141,193 | 9,440 | 7,408 |
| Autauga..................... | 2,454 | 4.0 | 43,917 | 1,501 | 509 | 88 | 36 | 106 | 356 | 542 | 739 | 57 | 36 |
| Baldwin ..................... | 10,600 | 5.3 | 47,485 | 1,079 | 3,263 | 488 | 237 | 673 | 2,113 | 2,308 | 4,662 | 345 | 237 |
| Barbour ..................... | 883 | 1.2 | 35,763 | 2,589 | 331 | 64 | 25 | 58 | 139 | 308 | 478 | 35 | 25 |
| Bibb........................... | 710 | 3.9 | 31,725 | 2,958 | 222 | 40 | 16 | 21 | 78 | 222 | 298 | 24 | 16 |
| Blount........................ | 2,106 | 2.2 | 36,412 | 2,523 | 357 | 66 | 26 | 116 | 257 | 550 | 565 | 50 | 26 |
| Bullock ...................... | 294 | 3.5 | 29,080 | 3,061 | 109 | 22 | 8 | 22 | 45 | 106 | 161 | 11 | 8 |
| Butler........................ | 730 | 1.1 | 37,523 | 2,384 | 265 | 51 | 20 | 36 | 113 | 240 | 372 | 28 | 20 |
| Calhoun ..................... | 4,362 | 3.1 | 38,394 | 2,280 | 1,983 | 417 | 147 | 188 | 775 | 1,284 | 2,734 | 189 | 147 |
| Chambers .................. | 1,194 | 3.3 | 35,900 | 2,580 | 359 | 66 | 26 | 31 | 177 | 403 | 482 | 40 | 26 |
| Cherokee ................... | 954 | 1.8 | 36,432 | 2,516 | 201 | 40 | 15 | 64 | 146 | 310 | 320 | 28 | 15 |
| Chilton....................... | 1,608 | 3.9 | 36,188 | 2,546 | 413 | 73 | 30 | 85 | 199 | 454 | 601 | 48 | 30 |
| Choctaw..................... | 505 | 1.2 | 40,145 | 2,034 | 204 | 37 | 15 | 30 | 62 | 177 | 285 | 22 | 15 |
| Clarke........................ | 897 | 3.2 | 37,965 | 2,329 | 357 | 64 | 26 | 36 | 140 | 303 | 483 | 37 | 26 |
| Clay........................... | 454 | 1.7 | 34,322 | 2,768 | 163 | 30 | 13 | 32 | 66 | 152 | 238 | 19 | 13 |
| Cleburne.................... | 534 | -1.9 | 35,791 | 2,586 | 98 | 19 | 7 | 41 | 63 | 153 | 165 | 14 | 7 |
| Coffee ....................... | 2,302 | 3.4 | 43,979 | 1,489 | 637 | 116 | 48 | 158 | 424 | 577 | 959 | 69 | 48 |
| Colbert....................... | 2,192 | 4.0 | 39,679 | 2,090 | 1,193 | 213 | 90 | 110 | 364 | 642 | 1,606 | 112 | 90 |
| Conecuh..................... | 421 | 2.1 | 34,892 | 2,700 | 139 | 27 | 11 | 32 | 64 | 168 | 209 | 16 | 11 |
| Coosa........................ | 347 | 4.0 | 32,523 | 2,918 | 59 | 13 | 5 | 9 | 48 | 117 | 86 | 9 | 5 |
| Covington................... | 1,362 | 1.9 | 36,756 | 2,476 | 490 | 97 | 37 | 78 | 206 | 466 | 702 | 53 | 37 |
| Crenshaw................... | 511 | 0.3 | 37,118 | 2,431 | 159 | 29 | 12 | 40 | 70 | 167 | 240 | 18 | 12 |
| Cullman...................... | 3,419 | 2.4 | 40,818 | 1,948 | 1,254 | 210 | 92 | 241 | 479 | 908 | 1,797 | 131 | 92 |
| Dale........................... | 1,861 | 2.2 | 37,853 | 2,339 | 1,321 | 324 | 110 | 70 | 324 | 547 | 1,825 | 104 | 110 |
| Dallas........................ | 1,427 | 3.0 | 38,362 | 2,283 | 534 | 96 | 40 | 69 | 219 | 542 | 739 | 55 | 40 |
| DeKalb....................... | 2,377 | 2.0 | 33,242 | 2,872 | 923 | 165 | 69 | 128 | 353 | 726 | 1,285 | 95 | 69 |
| Elmore....................... | 3,599 | 4.0 | 44,316 | 1,442 | 805 | 145 | 59 | 165 | 595 | 828 | 1,175 | 91 | 59 |
| Escambia ................... | 1,261 | 4.7 | 34,432 | 2,751 | 565 | 105 | 41 | 58 | 210 | 415 | 769 | 56 | 41 |
| Etowah....................... | 3,919 | 2.6 | 38,326 | 2,288 | 1,458 | 248 | 107 | 248 | 575 | 1,269 | 2,061 | 156 | 107 |
| Fayette ...................... | 594 | 3.2 | 36,465 | 2,514 | 151 | 33 | 11 | 17 | 80 | 222 | 212 | 19 | 11 |
| Franklin ..................... | 1,107 | 1.9 | 35,292 | 2,643 | 439 | 86 | 32 | 57 | 162 | 317 | 615 | 44 | 32 |
| Geneva ...................... | 968 | 3.6 | 36,848 | 2,462 | 193 | 39 | 14 | 86 | 138 | 319 | 332 | 27 | 14 |
| Greene....................... | 271 | 3.9 | 33,386 | 2,856 | 69 | 15 | 5 | 11 | 40 | 112 | 100 | 9 | 5 |
| Hale........................... | 560 | 3.7 | 38,207 | 2,305 | 111 | 22 | 8 | 28 | 79 | 196 | 170 | 15 | 8 |
| Henry ........................ | 748 | 2.7 | 43,502 | 1,545 | 164 | 27 | 11 | 68 | 120 | 208 | 271 | 21 | 11 |
| Houston...................... | 4,727 | 4.4 | 44,647 | 1,399 | 2,324 | 387 | 167 | 275 | 871 | 1,174 | 3,153 | 211 | 167 |
| Jackson...................... | 1,978 | 3.1 | 38,315 | 2,290 | 632 | 117 | 48 | 116 | 345 | 571 | 914 | 70 | 48 |
| Jefferson.................... | 37,756 | 3.0 | 57,329 | 393 | 22,380 | 3,284 | 1,567 | 4,247 | 8,273 | 6,611 | 31,478 | 1,979 | 1,567 |
| Lamar........................ | 485 | 3.4 | 35,148 | 2,669 | 155 | 28 | 11 | 21 | 70 | 172 | 216 | 18 | 11 |
| Lauderdale ................. | 3,594 | 3.5 | 38,758 | 2,228 | 1,179 | 207 | 86 | 179 | 724 | 956 | 1,652 | 127 | 86 |
| Lawrence ................... | 1,154 | 2.7 | 35,039 | 2,680 | 205 | 40 | 15 | 25 | 141 | 353 | 285 | 30 | 15 |
| Lee............................ | 6,546 | 4.2 | 39,781 | 2,079 | 2,711 | 493 | 193 | 301 | 1,269 | 1,216 | 3,699 | 238 | 193 |
| Limestone .................. | 4,446 | 6.8 | 44,946 | 1,362 | 1,313 | 244 | 98 | 220 | 600 | 871 | 1,875 | 131 | 98 |
| Lowndes..................... | 397 | -0.1 | 40,785 | 1,950 | 121 | 26 | 9 | 28 | 56 | 135 | 184 | 14 | 9 |
| Macon ....................... | 631 | 3.2 | 34,901 | 2,699 | 222 | 60 | 17 | 7 | 104 | 225 | 306 | 23 | 17 |
| Madison ..................... | 20,141 | 5.5 | 54,010 | 544 | 13,906 | 2,187 | 1,021 | 943 | 3,959 | 3,226 | 18,058 | 1,107 | 1,021 |
| Marengo ..................... | 784 | 2.3 | 41,567 | 1,835 | 325 | 58 | 24 | 46 | 120 | 254 | 453 | 33 | 24 |
| Marion ....................... | 1,010 | 2.3 | 34,000 | 2,802 | 393 | 73 | 30 | 58 | 162 | 322 | 554 | 41 | 30 |
| Marshall ..................... | 3,614 | 3.7 | 37,345 | 2,405 | 1,539 | 276 | 114 | 203 | 618 | 957 | 2,132 | 150 | 114 |
| Mobile ....................... | 16,575 | 3.2 | 40,112 | 2,038 | 9,201 | 1,456 | 664 | 1,207 | 2,762 | 4,301 | 12,528 | 829 | 664 |
| Monroe....................... | 729 | 3.1 | 35,174 | 2,664 | 296 | 52 | 22 | 35 | 113 | 252 | 405 | 30 | 22 |
| Montgomery ............... | 10,300 | 3.4 | 45,479 | 1,299 | 7,042 | 1,287 | 520 | 735 | 2,193 | 2,361 | 9,583 | 585 | 520 |
| Morgan....................... | 4,996 | 4.2 | 41,741 | 1,814 | 2,545 | 432 | 185 | 276 | 834 | 1,207 | 3,438 | 232 | 185 |
| Perry ......................... | 299 | 3.7 | 33,529 | 2,846 | 79 | 16 | 6 | 30 | 39 | 124 | 132 | 10 | 6 |
| Pickens ...................... | 675 | 2.2 | 33,890 | 2,814 | 161 | 35 | 12 | 31 | 92 | 239 | 238 | 20 | 12 |
| Pike........................... | 1,274 | 2.1 | 38,473 | 2,266 | 638 | 116 | 47 | 96 | 224 | 347 | 897 | 59 | 47 |
| Randolph.................... | 768 | 1.1 | 33,804 | 2,822 | 182 | 36 | 14 | 49 | 117 | 270 | 280 | 24 | 14 |
| Russell ...................... | 1,970 | 4.4 | 33,986 | 2,806 | 601 | 106 | 44 | 46 | 299 | 652 | 797 | 62 | 44 |
| St. Clair ..................... | 3,649 | 4.7 | 40,766 | 1,953 | 887 | 144 | 64 | 164 | 432 | 870 | 1,260 | 99 | 64 |
| Shelby ....................... | 12,372 | 4.4 | 56,830 | 411 | 5,336 | 718 | 364 | 888 | 2,209 | 1,673 | 7,305 | 473 | 364 |
| Sumter ...................... | 410 | 2.5 | 33,009 | 2,888 | 130 | 27 | 9 | 29 | 59 | 158 | 195 | 15 | 9 |
| Talladega ................... | 2,889 | 3.9 | 36,125 | 2,554 | 1,505 | 257 | 109 | 101 | 386 | 905 | 1,972 | 140 | 109 |
| Tallapoosa.................. | 1,716 | 3.4 | 42,509 | 1,690 | 521 | 93 | 38 | 103 | 321 | 515 | 755 | 60 | 38 |
| Tuscaloosa................. | 8,510 | 3.9 | 40,648 | 1,969 | 4,920 | 868 | 348 | 416 | 1,613 | 1,878 | 6,552 | 416 | 348 |
| Walker....................... | 2,698 | 3.0 | 42,477 | 1,693 | 794 | 138 | 59 | 132 | 511 | 832 | 1,124 | 90 | 59 |

1. Based on the resident population estimated as of July 1 of the year shown.

| STATE County | Farm | Mining, quarrying, and extractions | Construction | Manu-facturing | Information; professional, scientific, technical services | Retail trade | Finance, insurance, real estate, and leasing | Health care and social assistance | Govern-ment | Social Security beneficiaries, December 2019 Number | Rate[1] | Supplemental Security Income recipients, 2019 | Housing units, 2020 Total | Percent change, 2010-2020 |
|---|---|---|---|---|---|---|---|---|---|---|---|---|---|---|
| | 75 | 76 | 77 | 78 | 79 | 80 | 81 | 82 | 83 | 84 | 85 | 86 | 87 | 88 |
| UNITED STATES.................. | 0.7 | 1.3 | 6.1 | 9.1 | 14.2 | 5.6 | 9.4 | 11.0 | 15.7 | 62,502,630 | 190 | 8,075,821 | 140,775,530 | 6.9 |
| ALABAMA............................. | 0.4 | 0.7 | 6.2 | 14.1 | 10.1 | 6.4 | 6.6 | 11.2 | 19.9 | 1,159,320 | 236 | 160,187 | 2,302,643 | 6.0 |
| Autauga............................ | 1.8 | 0.9 | 7.8 | 16.5 | D | 7.7 | 5.1 | 9.8 | 17.9 | 12,205 | 219 | 1,468 | 24,165 | 9.2 |
| Baldwin............................. | 0.4 | 0.2 | 10.1 | 6.7 | 6.2 | 11.6 | 7.6 | 12.7 | 13.1 | 58,505 | 262 | 3,517 | 122,518 | 17.7 |
| Barbour............................ | 3.8 | 2.8 | 3.4 | 27.3 | D | 7.8 | 2.8 | D | 19.5 | 7,185 | 291 | 1,409 | 12,120 | 2.5 |
| Bibb.................................. | 0.0 | D | 24.8 | 11.9 | D | 5.6 | 2.0 | 8.6 | 23.9 | 5,475 | 245 | 869 | 9,340 | 4.0 |
| Blount............................... | 3.2 | D | 10.8 | 13.1 | 5.1 | 7.4 | 3.9 | D | 20.0 | 13,895 | 240 | 1,207 | 24,625 | 3.1 |
| Bullock............................. | 9.9 | 0.0 | D | D | D | 3.5 | 1.6 | 16.9 | 23.3 | 2,155 | 212 | 537 | 4,625 | 3.0 |
| Butler............................... | 2.4 | 0.0 | 5.5 | 22.6 | 3.8 | 8.0 | 2.8 | D | 13.3 | 5,535 | 284 | 1,013 | 10,177 | 2.2 |
| Calhoun............................ | 0.2 | D | 3.6 | 16.5 | 4.5 | 7.6 | 3.6 | 8.9 | 33.5 | 30,360 | 266 | 4,434 | 53,933 | 1.3 |
| Chambers......................... | 1.8 | 0.1 | 4.1 | 28.7 | D | 7.0 | 2.0 | D | 22.7 | 9,970 | 300 | 1,412 | 17,055 | 0.4 |
| Cherokee.......................... | 6.7 | 0.2 | 4.0 | 15.9 | 2.8 | 10.2 | 3.0 | D | 22.1 | 8,055 | 307 | 818 | 16,761 | 3.0 |
| Chilton............................. | -0.1 | D | 11.1 | 21.7 | D | 9.6 | 3.2 | 9.1 | 17.2 | 10,735 | 242 | 1,355 | 20,157 | 4.6 |
| Choctaw............................ | 2.3 | 0.2 | D | D | 2.0 | 4.8 | 2.2 | D | 8.4 | 4,270 | 340 | 792 | 7,459 | 2.6 |
| Clarke.............................. | 0.5 | 0.5 | 2.7 | 28.6 | D | 9.1 | 5.1 | D | 19.8 | 7,350 | 311 | 1,346 | 12,881 | 1.9 |
| Clay................................. | 0.7 | 0.0 | 4.0 | 44.9 | 1.1 | 4.2 | 3.4 | D | 19.5 | 3,990 | 305 | 489 | 6,858 | 1.2 |
| Cleburne........................... | 6.8 | D | 20.7 | 13.3 | D | 6.5 | 2.0 | 2.4 | 26.1 | 3,990 | 267 | 507 | 6,903 | 2.8 |
| Coffee.............................. | 2.1 | D | 3.3 | 17.8 | D | 9.4 | 4.2 | 11.8 | 17.3 | 11,910 | 226 | 1,410 | 23,719 | 6.2 |
| Colbert............................. | -0.2 | 0.6 | 8.7 | 28.0 | D | 7.8 | 4.0 | 7.8 | 20.5 | 15,525 | 281 | 1,871 | 27,007 | 4.8 |
| Conecuh........................... | 5.0 | D | 2.6 | 13.2 | 1.8 | 4.2 | 1.2 | D | 19.1 | 3,890 | 323 | 682 | 7,247 | 2.2 |
| Coosa.............................. | 0.2 | 0.4 | 4.1 | 37.5 | D | 3.1 | D | 5.1 | 19.9 | 2,955 | 277 | 451 | 6,661 | 2.7 |
| Covington......................... | 2.7 | 0.2 | 5.7 | 14.6 | 3.4 | 9.6 | 4.5 | D | 16.3 | 11,000 | 297 | 1,407 | 19,174 | 1.8 |
| Crenshaw.......................... | 7.2 | 0.3 | 4.7 | 24.7 | 2.2 | 3.0 | 3.0 | D | 14.0 | 4,025 | 291 | 579 | 6,887 | 2.3 |
| Cullman............................ | 0.7 | D | 7.9 | 21.6 | 3.6 | 8.3 | 4.6 | 12.7 | 13.6 | 22,315 | 267 | 2,565 | 38,565 | 4.1 |
| Dale................................. | 1.1 | D | 2.5 | 20.8 | 2.8 | 2.4 | 1.4 | 2.4 | 48.0 | 11,905 | 242 | 1,905 | 23,276 | 2.6 |
| Dallas.............................. | 3.1 | D | 6.0 | 25.8 | 2.9 | 7.2 | 3.8 | 13.9 | 18.6 | 11,110 | 299 | 3,793 | 20,459 | 1.3 |
| DeKalb............................. | 0.9 | 0.3 | 6.4 | 30.8 | 3.4 | 7.2 | 2.2 | 10.1 | 14.9 | 19,610 | 247 | 2,218 | 31,694 | 1.9 |
| Elmore.............................. | 0.2 | D | 9.2 | 15.3 | 4.7 | 10.7 | 4.3 | 9.4 | 22.1 | 19,610 | 240 | 2,218 | 34,618 | 6.0 |
| Escambia.......................... | 0.3 | 3.0 | 5.6 | 18.4 | D | 7.7 | 4.1 | 8.9 | 30.7 | 9,950 | 272 | 1,364 | 16,683 | 1.2 |
| Etowah............................. | 0.0 | D | 4.9 | 15.4 | 3.8 | 8.5 | 4.9 | D | 15.1 | 29,045 | 283 | 4,410 | 47,921 | 1.0 |
| Fayette............................. | 1.2 | 0.8 | 3.8 | 24.4 | 1.9 | 8.7 | 3.2 | D | 29.1 | 5,425 | 333 | 698 | 8,542 | 1.2 |
| Franklin............................ | 0.3 | 0.8 | 3.7 | 43.2 | 2.0 | 5.1 | 3.7 | D | 18.1 | 7,555 | 239 | 1,103 | 14,135 | 0.8 |
| Geneva............................. | 11.6 | D | 5.4 | 9.6 | D | 7.8 | 4.7 | 4.4 | 21.6 | 7,610 | 289 | 1,158 | 12,994 | 2.4 |
| Greene............................. | 5.2 | 0.0 | 4.7 | 26.7 | D | 6.3 | D | D | 29.2 | 2,600 | 321 | 717 | 5,191 | 3.7 |
| Hale................................. | 7.5 | D | 7.2 | 20.2 | D | 5.3 | 4.0 | 5.5 | 26.1 | 4,605 | 314 | 1,001 | 7,943 | 3.7 |
| Henry............................... | 10.5 | 0.2 | 5.8 | 10.2 | D | 4.0 | 2.5 | D | 14.2 | 5,340 | 312 | 608 | 9,298 | 4.6 |
| Houston............................ | 0.6 | D | 4.8 | 7.4 | 5.0 | 10.4 | 4.6 | 17.5 | 18.5 | 26,540 | 250 | 4,007 | 48,463 | 6.9 |
| Jackson............................ | 1.7 | 0.2 | 6.3 | 31.6 | 2.3 | 8.3 | 3.7 | D | 19.3 | 14,755 | 286 | 1,622 | 25,293 | 2.0 |
| Jefferson.......................... | 0.0 | 1.3 | 6.4 | 6.4 | 11.5 | 4.6 | 10.8 | 16.7 | 15.2 | 143,970 | 219 | 22,538 | 310,489 | 3.3 |
| Lamar............................... | 0.2 | D | 3.6 | 38.8 | 1.9 | 4.4 | 3.3 | 6.4 | 14.6 | 4,325 | 313 | 608 | 7,437 | 1.1 |
| Lauderdale......................... | -0.2 | D | 8.4 | 9.3 | 5.2 | 10.6 | 5.2 | 17.5 | 19.5 | 24,525 | 264 | 2,547 | 45,722 | 4.4 |
| Lawrence ......................... | -6.0 | D | 11.6 | 3.6 | 2.4 | 9.5 | 6.8 | D | 26.7 | 9,030 | 275 | 1,246 | 15,414 | 1.3 |
| Lee.................................. | 0.3 | D | 6.6 | 11.0 | 5.4 | 6.8 | 3.7 | 6.2 | 33.1 | 26,955 | 163 | 3,115 | 73,371 | 17.6 |
| Limestone ........................ | 0.7 | D | 13.3 | 15.0 | 4.8 | 9.0 | 2.8 | 4.2 | 31.4 | 20,100 | 203 | 2,067 | 36,869 | 5.4 |
| Lowndes ........................... | 8.0 | D | 4.5 | 41.9 | D | 5.1 | 3.1 | D | 15.9 | 3,115 | 320 | 702 | 5,305 | 3.2 |
| Macon.............................. | -0.4 | D | 2.3 | D | D | 3.7 | 0.6 | D | 66.3 | 4,930 | 273 | 1,027 | 10,356 | 0.9 |
| Madison............................ | 0.0 | 0.1 | 3.3 | 11.7 | 28.2 | 5.0 | 3.5 | 7.3 | 26.5 | 69,900 | 188 | 6,749 | 169,528 | 15.8 |
| Marengo............................ | 1.5 | 1.2 | 7.3 | 22.4 | 3.0 | 6.1 | 4.6 | D | 18.3 | 6,040 | 319 | 1,406 | 10,441 | 1.9 |
| Marion.............................. | 1.3 | D | 3.0 | 34.3 | 1.3 | 6.2 | 4.5 | 6.7 | 18.1 | 22,955 | 237 | 2,851 | 14,855 | 0.8 |
| Marshall........................... | -0.2 | D | 6.2 | 29.9 | 4.2 | 9.0 | 3.4 | 17.5 | 17.5 | 22,955 | 237 | 2,851 | 40,887 | 1.3 |
| Mobile.............................. | 0.3 | 0.3 | 7.2 | 15.0 | 9.1 | 6.3 | 7.2 | 12.6 | 14.4 | 93,555 | 226 | 15,111 | 185,524 | 4.1 |
| Monroe............................. | 2.4 | D | 4.0 | 26.5 | D | 5.7 | 3.3 | D | 17.7 | 6,035 | 292 | 876 | 11,494 | 1.4 |
| Montgomery...................... | 0.2 | D | 5.5 | 10.1 | 9.0 | 5.4 | 5.8 | 11.7 | 29.7 | 47,745 | 211 | 9,578 | 105,575 | 3.9 |
| Morgan............................. | -0.1 | 0.3 | 8.7 | 33.9 | 5.2 | 5.8 | 4.3 | 6.8 | 12.6 | 29,170 | 244 | 3,328 | 52,572 | 2.7 |
| Perry................................ | 11.6 | 0.0 | 2.6 | 16.6 | D | 3.5 | 2.4 | 7.5 | 19.9 | 2,725 | 305 | 965 | 4,776 | 1.0 |
| Pickens............................ | 5.2 | D | 3.3 | 14.3 | 2.4 | 5.0 | D | D | 33.6 | 5,515 | 277 | 1,070 | 9,693 | 2.2 |
| Pike................................. | 0.6 | 0.6 | 3.2 | 18.3 | 10.5 | 5.9 | 4.7 | D | 21.9 | 7,035 | 213 | 1,454 | 16,422 | 7.6 |
| Randolph.......................... | 5.9 | 1.1 | 6.5 | 15.9 | 2.9 | 8.8 | 4.0 | D | 21.0 | 6,475 | 284 | 846 | 12,215 | 1.9 |
| Russell............................. | 0.4 | 0.2 | 5.7 | 26.2 | 3.1 | 8.0 | 5.4 | 12.2 | 20.5 | 13,400 | 231 | 1,918 | 28,264 | 14.9 |
| St. Clair............................ | 0.1 | D | 12.1 | 18.9 | 5.1 | 8.7 | 4.1 | 8.4 | 14.4 | 20,570 | 229 | 1,923 | 37,639 | 6.2 |
| Shelby.............................. | 0.1 | 1.6 | 8.3 | 5.8 | 10.4 | 7.5 | 17.3 | 8.4 | 7.7 | 40,450 | 185 | 2,468 | 90,258 | 11.4 |
| Sumter............................. | 9.5 | 0.0 | 2.7 | 10.6 | D | 4.0 | 2.5 | D | 35.6 | 3,430 | 278 | 1,008 | 6,973 | 2.8 |
| Talladega.......................... | 0.2 | D | 5.4 | 40.8 | 2.3 | 4.4 | 2.2 | D | 14.7 | 21,840 | 272 | 3,679 | 38,360 | 3.3 |
| Tallapoosa......................... | 0.5 | D | 5.9 | 16.7 | 5.2 | 7.6 | 3.5 | 17.4 | 15.7 | 12,690 | 315 | 1,527 | 22,823 | 3.2 |
| Tuscaloosa........................ | 0.0 | 2.9 | 5.8 | 20.7 | 6.4 | 5.8 | 5.1 | 7.1 | 26.5 | 40,625 | 193 | 5,845 | 94,220 | 11.0 |
| Walker.............................. | 0.3 | 1.8 | 6.5 | 12.5 | 3.8 | 12.1 | 4.8 | 19.1 | 15.8 | 19,935 | 313 | 3,038 | 31,482 | 2.2 |

1. Per 1,000 resident population estimated as of July 1 of the year shown.

# Table B. States and Counties — Housing, Labor Force, and Employment

| STATE County | Housing units, 2015-2019 | | | | | | | | Civilian labor force, 2020 | | Civilian employment[6], 2015-2019 | | | |
|---|---|---|---|---|---|---|---|---|---|---|---|---|---|---|
| | Occupied units | | | | | | | | | | | | | |
| | Owner-occupied | | | | Renter-occupied | | | | | | Unemployment | | Percent | |
| | | | | Median owner cost as a percent of income | | Median rent as a percent of income[2] | | | | Percent change, 2019-2020 | | | | |
| | Total | Percent | Median value[1] | With a mortgage | Without a mortgage[2] | Median rent[3] | Median rent as % income[2] | Sub-standard units[4] (percent) | Total | Percent change 2019-2020 | Total | Rate[5] | Total | Management, business, science, and arts | Construction, production, and maintenance occupations |
| | 89 | 90 | 91 | 92 | 93 | 94 | 95 | 96 | 97 | 98 | 99 | 100 | 101 | 102 | 103 |
| UNITED STATES | 120,756,048 | 64.0 | 217,500 | 21.3 | 11.3 | 1,062 | 29.8 | 3.7 | 160,611,064 | -1.6 | 12,933,704 | 8.1 | 154,842,185 | 38.5 | 22.1 |
| ALABAMA | 1,867,893 | 68.8 | 142,700 | 19.2 | 10.0 | 792 | 29.0 | 2.1 | 2,230,118 | -0.3 | 131,056 | 5.9 | 2,097,384 | 34.9 | 27.1 |
| Autauga | 21,397 | 73.3 | 154,500 | 19.2 | 10.0 | 986 | 29.4 | 1.5 | 25,838 | -1.7 | 1,262 | 4.9 | 24,522 | 37.8 | 23.3 |
| Baldwin | 80,930 | 75.2 | 197,900 | 19.9 | 10.0 | 1,020 | 28.7 | 1.6 | 96,763 | -0.7 | 5,425 | 5.6 | 95,091 | 37.1 | 21.3 |
| Barbour | 9,345 | 60.9 | 90,700 | 19.4 | 12.3 | 576 | 29.6 | 4.0 | 8,587 | 1.0 | 605 | 7.0 | 8,413 | 26.9 | 38.7 |
| Bibb | 6,891 | 74.4 | 92,800 | 18.4 | 10.0 | 734 | 29.4 | 1.5 | 8,640 | 0.0 | 573 | 6.6 | 8,387 | 20.4 | 42.7 |
| Blount | 20,847 | 78.8 | 127,800 | 19.0 | 10.0 | 667 | 27.3 | 2.3 | 24,661 | -2.1 | 1,008 | 4.1 | 21,917 | 30.2 | 36.5 |
| Bullock | 3,521 | 71.1 | 70,300 | 22.2 | 11.5 | 654 | 32.2 | 0.9 | 4,818 | 0.7 | 265 | 5.5 | 4,319 | 20.5 | 45.4 |
| Butler | 6,506 | 70.0 | 95,900 | 21.8 | 11.7 | 620 | 27.1 | 1.8 | 9,056 | -1.3 | 801 | 8.8 | 7,930 | 27.3 | 36.3 |
| Calhoun | 44,605 | 70.1 | 118,000 | 18.7 | 10.0 | 716 | 27.6 | 2.0 | 46,240 | 0.3 | 3,260 | 7.1 | 47,896 | 28.2 | 31.7 |
| Chambers | 13,448 | 67.5 | 97,100 | 21.2 | 11.4 | 749 | 22.7 | 6.2 | 15,865 | 0.7 | 1,078 | 6.8 | 14,877 | 22.9 | 42.2 |
| Cherokee | 10,737 | 77.3 | 123,400 | 19.3 | 12.1 | 620 | 24.0 | 1.8 | 11,513 | -1.8 | 530 | 4.6 | 9,980 | 28.1 | 41.1 |
| Chilton | 16,927 | 74.5 | 107,300 | 19.7 | 10.0 | 726 | 28.4 | 3.4 | 19,592 | -1.3 | 974 | 5.0 | 17,526 | 22.2 | 40.9 |
| Choctaw | 5,300 | 80.9 | 74,100 | 19.2 | 11.3 | 525 | 35.5 | 2.7 | 4,621 | -0.5 | 308 | 6.7 | 4,173 | 26.4 | 37.6 |
| Clarke | 9,090 | 70.1 | 112,500 | 19.3 | 12.0 | 555 | 26.5 | 2.6 | 7,769 | 2.4 | 700 | 9.0 | 7,770 | 23.3 | 43.1 |
| Clay | 5,198 | 75.5 | 107,800 | 20.0 | 10.0 | 535 | 24.0 | 1.8 | 6,083 | 0.2 | 245 | 4.0 | 5,527 | 24.2 | 45.3 |
| Cleburne | 5,680 | 76.1 | 116,500 | 20.5 | 11.0 | 585 | 23.4 | 2.1 | 5,737 | -0.6 | 262 | 4.6 | 5,707 | 28.3 | 39.1 |
| Coffee | 19,924 | 65.7 | 147,400 | 17.9 | 10.0 | 803 | 25.0 | 1.8 | 21,502 | -1.1 | 944 | 4.4 | 21,907 | 34.2 | 31.3 |
| Colbert | 21,880 | 72.2 | 120,800 | 19.0 | 10.4 | 697 | 29.9 | 0.8 | 23,558 | -0.6 | 1,550 | 6.6 | 22,832 | 29.1 | 33.8 |
| Conecuh | 4,553 | 76.1 | 81,000 | 17.8 | 10.0 | 536 | 26.5 | 0.4 | 4,529 | 0.7 | 331 | 7.3 | 4,232 | 16.1 | 44.5 |
| Coosa | 4,032 | 79.2 | 84,000 | 20.4 | 10.4 | 589 | 22.9 | 3.5 | 4,392 | -2.5 | 242 | 5.5 | 4,202 | 22.0 | 36.6 |
| Covington | 14,852 | 74.5 | 97,600 | 18.3 | 10.0 | 612 | 28.9 | 3.0 | 15,172 | -2.4 | 738 | 4.9 | 14,822 | 29.7 | 36.2 |
| Crenshaw | 4,943 | 75.5 | 81,000 | 17.2 | 11.5 | 537 | 21.1 | 1.2 | 6,253 | -2.8 | 357 | 5.7 | 5,668 | 23.7 | 37.1 |
| Cullman | 31,034 | 74.8 | 128,600 | 19.4 | 11.2 | 706 | 27.1 | 2.3 | 38,310 | -0.6 | 1,650 | 4.3 | 34,228 | 26.7 | 36.8 |
| Dale | 18,806 | 60.1 | 112,000 | 18.9 | 10.0 | 730 | 25.0 | 1.7 | 20,710 | 0.5 | 1,009 | 4.9 | 17,994 | 26.9 | 33.4 |
| Dallas | 15,910 | 59.3 | 83,300 | 19.3 | 12.1 | 643 | 33.5 | 2.7 | 14,619 | -0.2 | 1,570 | 10.7 | 14,347 | 25.4 | 39.1 |
| DeKalb | 26,040 | 71.5 | 106,100 | 18.7 | 10.6 | 628 | 26.6 | 3.0 | 30,736 | -1.3 | 1,388 | 4.5 | 29,639 | 23.8 | 41.4 |
| Elmore | 29,708 | 73.3 | 166,600 | 18.8 | 10.0 | 892 | 31.2 | 1.9 | 36,620 | -1.7 | 1,777 | 4.9 | 34,180 | 38.1 | 26.4 |
| Escambia | 13,089 | 68.3 | 99,400 | 19.5 | 11.1 | 616 | 28.5 | 2.2 | 14,453 | -1.1 | 864 | 6.0 | 12,786 | 25.1 | 30.6 |
| Etowah | 38,942 | 72.6 | 113,700 | 19.1 | 10.8 | 669 | 27.7 | 1.6 | 40,724 | -4.1 | 3,126 | 7.7 | 42,650 | 29.0 | 31.0 |
| Fayette | 6,802 | 77.3 | 88,500 | 19.9 | 10.0 | 527 | 31.5 | 1.6 | 6,562 | -0.3 | 326 | 5.0 | 6,337 | 25.2 | 38.8 |
| Franklin | 11,048 | 70.8 | 95,000 | 17.6 | 10.5 | 581 | 20.6 | 3.6 | 14,424 | -1.2 | 638 | 4.4 | 12,840 | 24.1 | 43.6 |
| Geneva | 10,383 | 72.0 | 100,900 | 19.4 | 10.4 | 625 | 26.3 | 1.7 | 10,835 | -0.9 | 444 | 4.1 | 10,294 | 26.5 | 35.8 |
| Greene | 2,951 | 67.1 | 80,400 | 31.9 | 13.5 | 612 | 38.5 | 3.8 | 3,015 | 4.4 | 328 | 10.9 | 2,396 | 21.8 | 41.1 |
| Hale | 5,650 | 75.8 | 89,300 | 24.3 | 12.8 | 577 | 40.5 | 4.3 | 6,075 | 0.0 | 572 | 9.4 | 5,355 | 27.3 | 31.0 |
| Henry | 6,630 | 83.3 | 118,800 | 19.2 | 10.0 | 627 | 27.5 | 0.8 | 6,814 | | 327 | 4.8 | 6,920 | 30.1 | 31.7 |
| Houston | 39,311 | 65.7 | 137,600 | 19.2 | 10.0 | 754 | 28.0 | 1.4 | 46,358 | 0.1 | 2,497 | 5.4 | 44,374 | 32.1 | 25.1 |
| Jackson | 20,695 | 75.5 | 104,900 | 19.7 | 10.0 | 590 | 27.8 | 2.7 | 22,840 | -2.1 | 1,203 | 5.3 | 20,352 | 29.0 | 40.0 |
| Jefferson | 261,231 | 62.7 | 159,100 | 19.7 | 10.0 | 900 | 29.9 | 1.8 | 315,957 | -0.3 | 19,675 | 6.2 | 304,994 | 39.7 | 20.5 |
| Lamar | 5,856 | 75.6 | 80,900 | 15.8 | 11.3 | 425 | 33.2 | 2.0 | 5,730 | -1.2 | 278 | 4.9 | 5,244 | 29.4 | 39.7 |
| Lauderdale | 38,471 | 68.1 | 145,000 | 19.2 | 10.0 | 663 | 27.1 | 1.6 | 41,931 | -1.2 | 2,321 | 5.5 | 40,287 | 30.3 | 28.3 |
| Lawrence | 12,677 | 77.7 | 110,500 | 19.0 | 10.0 | 551 | 26.0 | 1.9 | 14,363 | 0.9 | 703 | 4.9 | 12,445 | 26.7 | 42.5 |
| Lee | 60,237 | 61.6 | 169,300 | 19.2 | 10.0 | 856 | 31.4 | 3.4 | 75,564 | -1.1 | 3,935 | 5.2 | 74,770 | 41.1 | 21.6 |
| Limestone | 32,143 | 77.6 | 158,800 | 18.0 | 10.0 | 691 | 25.9 | 2.4 | 44,624 | 0.8 | 1,900 | 4.3 | 40,510 | 35.2 | 30.0 |
| Lowndes | 4,251 | 74.3 | 70,000 | 26.8 | 16.6 | 586 | 42.4 | 2.4 | 3,661 | 4.6 | 488 | 13.3 | 3,430 | 25.1 | 40.6 |
| Macon | 7,474 | 65.5 | 80,800 | 21.7 | 13.7 | 641 | 31.1 | 2.1 | 8,063 | 2.6 | 773 | 9.6 | 7,064 | 31.3 | 26.5 |
| Madison | 148,189 | 66.8 | 186,100 | 17.6 | 10.0 | 857 | 27.9 | 1.6 | 187,247 | 1.2 | 8,668 | 4.6 | 176,910 | 47.9 | 17.3 |
| Marengo | 7,361 | 71.5 | 84,400 | 21.0 | 12.4 | 563 | 34.0 | 1.2 | 7,766 | 1.1 | 485 | 6.2 | 6,778 | 29.4 | 32.4 |
| Marion | 11,997 | 75.1 | 92,800 | 20.3 | 11.7 | 512 | 27.1 | 2.2 | 13,034 | 2.2 | 661 | 5.1 | 11,851 | 25.7 | 38.3 |
| Marshall | 34,883 | 70.2 | 128,800 | 20.1 | 10.0 | 649 | 27.7 | 2.2 | 44,275 | 0.5 | 1,845 | 4.2 | 39,193 | 30.5 | 34.6 |
| Mobile | 156,251 | 64.5 | 130,200 | 20.3 | 10.6 | 853 | 31.6 | 2.2 | 190,882 | 1.1 | 15,121 | 7.9 | 174,519 | 32.9 | 27.0 |
| Monroe | 8,166 | 63.6 | 102,200 | 19.2 | 11.6 | 582 | 31.6 | 0.8 | 7,474 | 2.1 | 610 | 8.2 | 6,423 | 27.1 | 34.8 |
| Montgomery | 89,527 | 58.4 | 129,800 | 19.3 | 10.0 | 884 | 30.6 | 2.4 | 107,103 | 1.0 | 8,368 | 7.8 | 101,013 | 37.6 | 21.4 |
| Morgan | 45,918 | 72.7 | 137,000 | 18.7 | 10.0 | 658 | 27.4 | 2.5 | 58,899 | 1.0 | 2,634 | 4.5 | 51,903 | 32.1 | 30.5 |
| Perry | 3,070 | 71.9 | 65,000 | 43.4 | 17.1 | 507 | 33.9 | 2.9 | 3,437 | 1.7 | 373 | 10.9 | 2,571 | 27.7 | 24.0 |
| Pickens | 7,637 | 75.3 | 96,000 | 19.4 | 10.7 | 457 | 31.1 | 1.3 | 7,660 | -3.1 | 499 | 6.5 | 7,017 | 30.0 | 36.4 |
| Pike | 11,601 | 62.3 | 120,000 | 19.6 | 10.0 | 669 | 36.0 | 1.3 | 15,636 | 2.0 | 772 | 4.9 | 13,992 | 30.5 | 26.1 |
| Randolph | 8,702 | 76.6 | 96,100 | 20.7 | 10.0 | 620 | 26.8 | 3.6 | 9,414 | -0.9 | 447 | 4.7 | 8,764 | 26.4 | 39.9 |
| Russell | 23,262 | 60.2 | 122,000 | 21.2 | 11.2 | 803 | 30.9 | 3.2 | 23,500 | -0.9 | 1,210 | 5.1 | 23,616 | 30.2 | 26.8 |
| St. Clair | 32,036 | 79.9 | 154,900 | 19.1 | 10.0 | 819 | 25.6 | 1.7 | 40,132 | -1.4 | 1,986 | 4.9 | 38,507 | 30.3 | 31.2 |
| Shelby | 79,630 | 80.5 | 211,900 | 18.9 | 10.0 | 1,047 | 26.3 | 1.3 | 114,048 | -2.0 | 4,402 | 3.9 | 106,839 | 46.3 | 16.5 |
| Sumter | 5,202 | 64.1 | 74,400 | 23.2 | 14.1 | 518 | 36.9 | 1.3 | 4,745 | -0.2 | 331 | 7.0 | 4,291 | 28.6 | 26.0 |
| Talladega | 31,395 | 71.3 | 105,300 | 19.8 | 10.5 | 628 | 28.9 | 2.2 | 36,123 | 1.3 | 2,532 | 7.0 | 31,166 | 28.6 | 37.3 |
| Tallapoosa | 16,205 | 74.9 | 109,300 | 19.2 | 10.0 | 665 | 30.5 | 3.4 | 17,864 | -3.2 | 1,239 | 6.9 | 15,119 | 29.9 | 34.6 |
| Tuscaloosa | 72,461 | 63.7 | 170,400 | 19.9 | 10.0 | 846 | 31.0 | 2.0 | 102,222 | -2.6 | 6,749 | 6.6 | 93,667 | 35.8 | 27.2 |
| Walker | 25,019 | 77.2 | 104,300 | 19.6 | 10.0 | 626 | 28.7 | 2.0 | 24,877 | -1.1 | 1,446 | 5.8 | 24,414 | 27.1 | 34.3 |

1. Specified owner-occupied units. 2. A value of 10.0 represents 10 percent or less; a value of 50.0 represents 50 percent or more. 3. Specified renter-occupied units. 4. Overcrowded or lacking complete plumbing facilities. 5. Percent of civilian labor force. 6. Civilian employed persons 16 years old and over.

# Table B. States and Counties — Nonfarm Employment and Agriculture

| | Private nonfarm establishments, employment and payroll, 2019 | | | | | | | | | Agriculture, 2017 | | | |
|---|---|---|---|---|---|---|---|---|---|---|---|---|---|
| | | Employment | | | | | | Annual payroll | | Farms | | | Farm producers whose primary occupation is farming (percent) |
| | | | | | | | | | | | Percent with: | | |
| STATE County | Number of establishments | Total | Health care and social assistance | Manufacturing | Retail trade | Finance and insurance | Professional, scientific, and technical services | Total (mil dol) | Average per employee (dollars) | Number | Fewer than 50 acres | 1000 acres or more | |
| | 104 | 105 | 106 | 107 | 108 | 109 | 110 | 111 | 112 | 113 | 114 | 115 | 116 |
| UNITED STATES | 7,959,103 | 132,989,428 | 20,864,810 | 12,109,803 | 15,693,290 | 6,553,166 | 9,284,491 | 7,428,554 | 55,858 | 2,042,220 | 41.9 | 8.5 | 41.1 |
| ALABAMA | 100,731 | 1,758,609 | 264,559 | 264,061 | 228,122 | 70,276 | 108,157 | 79,548 | 45,234 | 40,592 | 40.1 | 4.2 | 37.4 |
| Autauga | 867 | 11,510 | 1,585 | 1,185 | 2,639 | 361 | 268 | 386 | 33,515 | 371 | 31.5 | 6.7 | 37.1 |
| Baldwin | 5,644 | 67,084 | 8,814 | 4,564 | 13,553 | 1,700 | 2,185 | 2,373 | 35,369 | 842 | 53.8 | 4.5 | 45.0 |
| Barbour | 449 | 6,893 | 576 | 2,724 | 785 | 155 | 95 | 244 | 35,406 | 498 | 20.3 | 5.2 | 41.4 |
| Bibb | 285 | 3,951 | 722 | 713 | 513 | 71 | 81 | 164 | 41,559 | 205 | 35.6 | 5.9 | 32.3 |
| Blount | 695 | 6,825 | 985 | 1,103 | 1,209 | 215 | 270 | 236 | 34,603 | 1,146 | 40.8 | 1.4 | 43.1 |
| Bullock | 113 | 2,061 | 355 | 726 | 228 | 44 | 28 | 66 | 32,213 | 255 | 18.4 | 9.4 | 33.6 |
| Butler | 421 | 6,203 | 1,110 | 1,423 | 901 | 154 | 53 | 207 | 33,343 | 420 | 30.0 | 2.4 | 37.7 |
| Calhoun | 2,245 | 36,752 | 6,297 | 6,889 | 6,222 | 874 | 985 | 1,327 | 36,109 | 643 | 43.4 | 1.4 | 34.5 |
| Chambers | 550 | 7,118 | 768 | 2,231 | 1,108 | 131 | 186 | 252 | 35,353 | 331 | 25.4 | 11.5 | 28.5 |
| Cherokee | 388 | 4,570 | 687 | 977 | 784 | 106 | 52 | 137 | 30,043 | 530 | 40.4 | 4.7 | 38.0 |
| Chilton | 782 | 7,905 | 885 | 1,920 | 1,446 | 204 | 103 | 293 | 37,128 | 463 | 41.9 | 1.3 | 42.3 |
| Choctaw | 256 | 3,006 | 378 | 1,058 | 338 | 74 | 51 | 159 | 52,933 | 188 | 30.9 | 7.4 | 35.1 |
| Clarke | 592 | 6,429 | 989 | 1,297 | 1,571 | 332 | 82 | 236 | 36,757 | 320 | 41.3 | 5.0 | 24.4 |
| Clay | 189 | 3,609 | 550 | 2,139 | 274 | 84 | 47 | 126 | 34,885 | 381 | 24.7 | 2.4 | 46.5 |
| Cleburne | 173 | 1,669 | 65 | 369 | 342 | 59 | 13 | 67 | 40,222 | 319 | 31.0 | 0.9 | 44.7 |
| Coffee | 966 | 13,792 | 1,996 | 3,628 | 2,618 | 447 | 732 | 471 | 34,126 | 788 | 31.5 | 4.3 | 41.8 |
| Colbert | 1,231 | 24,190 | 3,107 | 6,992 | 2,829 | 659 | 326 | 988 | 40,835 | 591 | 39.9 | 8.5 | 43.1 |
| Conecuh | 194 | 2,679 | 510 | 428 | 326 | 49 | 20 | 91 | 33,929 | 344 | 27.9 | 6.7 | 40.7 |
| Coosa | 95 | 1,085 | 135 | 400 | 70 | 11 | 33 | 43 | 39,289 | 215 | 17.2 | 2.8 | 32.7 |
| Covington | 807 | 10,218 | 1,642 | 2,249 | 1,734 | 337 | 377 | 363 | 35,569 | 907 | 34.5 | 3.0 | 34.4 |
| Crenshaw | 207 | 3,117 | 428 | 1,016 | 246 | 99 | 19 | 121 | 38,785 | 543 | 28.2 | 4.1 | 39.7 |
| Cullman | 1,761 | 25,889 | 4,084 | 5,285 | 4,113 | 780 | 431 | 984 | 38,024 | 1,781 | 46.7 | 0.2 | 40.4 |
| Dale | 698 | 10,951 | 1,232 | 486 | 1,340 | 380 | 723 | 559 | 51,090 | 469 | 27.7 | 6.6 | 38.1 |
| Dallas | 673 | 10,030 | 1,727 | 3,117 | 1,376 | 281 | 395 | 385 | 38,347 | 528 | 28.0 | 16.7 | 33.8 |
| DeKalb | 1,102 | 18,973 | 2,674 | 7,100 | 2,321 | 418 | 438 | 669 | 35,286 | 1,939 | 44.8 | 1.5 | 43.2 |
| Elmore | 1,200 | 14,884 | 2,131 | 2,904 | 2,929 | 385 | 404 | 516 | 34,659 | 538 | 47.2 | 4.5 | 33.8 |
| Escambia | 741 | 10,006 | 1,255 | 2,019 | 1,766 | 327 | 194 | 366 | 36,617 | 437 | 46.7 | 8.0 | 37.2 |
| Etowah | 1,917 | 30,302 | 7,153 | 5,544 | 4,822 | 1,013 | 634 | 1,069 | 35,286 | 817 | 55.2 | 0.5 | 38.2 |
| Fayette | 281 | 3,177 | 689 | 929 | 573 | 74 | 46 | 106 | 33,361 | 324 | 28.4 | 3.4 | 37.2 |
| Franklin | 520 | 10,718 | 929 | 6,221 | 956 | 391 | 165 | 387 | 36,133 | 729 | 26.3 | 2.6 | 41.1 |
| Geneva | 406 | 3,909 | 570 | 751 | 727 | 142 | 103 | 127 | 32,547 | 820 | 33.7 | 4.9 | 42.5 |
| Greene | 97 | 1,369 | 262 | 589 | 155 | 36 | 7 | 50 | 36,458 | 325 | 24.9 | 16.9 | 47.6 |
| Hale | 185 | 1,906 | 445 | 522 | 281 | 69 | 23 | 76 | 39,708 | 393 | 27.7 | 9.4 | 37.6 |
| Henry | 285 | 3,509 | 164 | 355 | 267 | 94 | 164 | 243 | 69,208 | 455 | 25.5 | 9.7 | 36.9 |
| Houston | 2,701 | 45,246 | 9,669 | 4,712 | 7,474 | 1,186 | 1,082 | 1,777 | 39,269 | 698 | 42.8 | 5.0 | 34.4 |
| Jackson | 833 | 13,101 | 1,569 | 5,298 | 1,792 | 347 | 286 | 459 | 35,049 | 1,355 | 44.8 | 2.4 | 31.8 |
| Jefferson | 16,498 | 334,197 | 62,932 | 23,995 | 38,094 | 21,434 | 16,561 | 18,222 | 54,525 | 387 | 63.3 | 1.0 | 29.1 |
| Lamar | 247 | 3,185 | 358 | 1,604 | 315 | 114 | 49 | 126 | 39,529 | 269 | 29.0 | 2.6 | 24.7 |
| Lauderdale | 1,966 | 26,623 | 5,138 | 3,038 | 5,127 | 1,010 | 696 | 843 | 31,659 | 1,309 | 48.1 | 3.4 | 33.7 |
| Lawrence | 416 | 3,568 | 791 | 225 | 784 | 109 | 131 | 111 | 31,165 | 1,252 | 46.5 | 2.4 | 39.1 |
| Lee | 2,952 | 48,188 | 6,842 | 7,399 | 7,233 | 955 | 1,781 | 1,700 | 35,281 | 314 | 43.9 | 5.7 | 35.5 |
| Limestone | 1,423 | 18,202 | 2,328 | 4,182 | 2,670 | 406 | 897 | 762 | 41,856 | 1,156 | 52.2 | 3.5 | 35.5 |
| Lowndes | 108 | 1,765 | 130 | 815 | 196 | 25 | 12 | 86 | 48,998 | 512 | 35.0 | 9.2 | 38.1 |
| Macon | 197 | 4,922 | 1,605 | 386 | 429 | 42 | 28 | 214 | 43,538 | 373 | 26.5 | 7.5 | 34.0 |
| Madison | 8,539 | 173,625 | 26,646 | 17,082 | 20,250 | 3,819 | 43,484 | 9,927 | 57,175 | 1,021 | 56.4 | 4.4 | 36.3 |
| Marengo | 454 | 6,281 | 1,030 | 1,560 | 863 | 215 | 80 | 254 | 40,427 | 471 | 29.9 | 5.5 | 31.5 |
| Marion | 548 | 8,491 | 1,241 | 2,966 | 1,061 | 411 | 78 | 289 | 34,031 | 582 | 30.6 | 0.9 | 33.4 |
| Marshall | 1,859 | 34,461 | 4,166 | 12,912 | 4,723 | 852 | 697 | 1,228 | 35,631 | 1,444 | 51.7 | 0.6 | 38.7 |
| Mobile | 8,755 | 154,973 | 22,228 | 15,604 | 20,580 | 6,557 | 9,571 | 7,035 | 45,395 | 653 | 60.9 | 3.1 | 39.3 |
| Monroe | 392 | 5,394 | 748 | 1,385 | 798 | 155 | 42 | 235 | 43,544 | 477 | 41.5 | 7.1 | 34.0 |
| Montgomery | 5,528 | 105,054 | 18,328 | 12,289 | 12,107 | 4,197 | 5,859 | 4,627 | 44,042 | 575 | 27.0 | 9.7 | 41.3 |
| Morgan | 2,604 | 45,554 | 5,846 | 12,371 | 5,268 | 1,343 | 2,450 | 2,122 | 46,573 | 1,164 | 51.6 | 1.1 | 32.7 |
| Perry | 111 | 1,536 | 148 | 454 | 161 | 68 | 16 | 49 | 32,216 | 349 | 32.1 | 14.3 | 41.3 |
| Pickens | 254 | 2,756 | 564 | 635 | 417 | 144 | 96 | 95 | 34,575 | 377 | 31.0 | 6.9 | 37.5 |
| Pike | 652 | 12,108 | 1,141 | 2,159 | 1,623 | 357 | 1,037 | 458 | 37,830 | 594 | 26.4 | 5.2 | 35.5 |
| Randolph | 341 | 3,825 | 628 | 970 | 736 | 138 | 52 | 120 | 31,263 | 597 | 33.3 | 3.2 | 38.0 |
| Russell | 808 | 11,116 | 1,528 | 3,037 | 1,866 | 373 | 200 | 412 | 37,108 | 296 | 31.1 | 10.1 | 29.6 |
| St. Clair | 1,330 | 18,076 | 2,143 | 3,919 | 2,881 | 449 | 588 | 660 | 36,498 | 490 | 42.0 | 1.0 | 34.4 |
| Shelby | 5,338 | 85,296 | 7,801 | 5,791 | 10,252 | 9,495 | 5,226 | 4,672 | 54,777 | 447 | 50.6 | 1.8 | 26.3 |
| Sumter | 194 | 2,899 | 334 | 218 | 270 | 59 | 14 | 117 | 40,491 | 367 | 20.2 | 12.5 | 39.3 |
| Talladega | 1,250 | 25,477 | 3,196 | 9,726 | 2,810 | 535 | 416 | 1,131 | 44,401 | 566 | 42.2 | 1.1 | 40.2 |
| Tallapoosa | 777 | 10,487 | 1,999 | 2,553 | 1,608 | 280 | 210 | 352 | 33,524 | 347 | 25.9 | 3.2 | 33.9 |
| Tuscaloosa | 4,087 | 80,583 | 12,677 | 17,107 | 9,777 | 1,790 | 2,222 | 3,681 | 45,675 | 557 | 43.8 | 3.1 | 31.9 |
| Walker | 1,214 | 14,857 | 3,061 | 2,084 | 3,090 | 505 | 487 | 554 | 37,301 | 501 | 49.9 | 2.0 | 43.2 |

# Table B. States and Counties — Agriculture

|  | Land in farms | | | | | Value of land and buildings (dollars) | | Value of machinery and equipment, average per farm (dollars) | Value of products sold: | | | | Organic farms (number) | Farms with internet access (per-cent) | Government payments | |
| STATE County | Acreage (1,000) | Percent change, 2012-2017 | Average size of farm | Total irrigated (1,000) | Total cropland (1,000) | Average per farm | Average per acre |  | Total (mil dol) | Average per farm (acres) | Percent from: Crops | Percent from: Livestock and poultry products |  |  | Total ($1,000) | Percent of farms |
|---|---|---|---|---|---|---|---|---|---|---|---|---|---|---|---|---|
|  | 117 | 118 | 119 | 120 | 121 | 122 | 123 | 124 | 125 | 126 | 127 | 128 | 129 | 130 | 131 | 132 |
| UNITED STATES............ | 900,218 | -1.6 | 441 | 58,013.9 | 396,433.8 | 1,311,808 | 2,976 | 133,363 | 388,522.7 | 190,245 | 49.8 | 50.2 | 20,806 | 75.4 | 8,943,574 | 31.5 |
| ALABAMA...................... | 8,581 | -3.6 | 211 | 142.0 | 2,818.8 | 630,736 | 2,984 | 88,528 | 5,980.6 | 147,334 | 20.3 | 79.7 | 57 | 72.6 | 134,654 | 37.3 |
| Autauga...................... | 113 | 1.6 | 305 | 1.4 | 36.9 | 672,902 | 2,205 | 96,682 | 21.5 | 57,844 | 58.4 | 41.6 | NA | 76.8 | 1,069 | 37.5 |
| Baldwin ...................... | 175 | -9.1 | 208 | 7.4 | 110.4 | 1,208,662 | 5,822 | 136,496 | 120.4 | 142,973 | 84.4 | 15.6 | NA | 78.1 | 6,316 | 27.3 |
| Barbour ...................... | 153 | -25.2 | 307 | 3.6 | 37.3 | 684,780 | 2,233 | 87,026 | 105.6 | 211,978 | 11.5 | 88.5 | 2 | 60.6 | 3,003 | 62.9 |
| Bibb........................... | 56 | -0.6 | 273 | 0.1 | 15.8 | 766,325 | 2,807 | 68,656 | 4.2 | 20,483 | 53.5 | 46.5 | NA | 73.2 | 395 | 26.3 |
| Blount........................ | 148 | 1.4 | 129 | 0.9 | 43.8 | 465,684 | 3,602 | 87,452 | 242.9 | 211,915 | 5.5 | 94.5 | 3 | 74.6 | 3,166 | 32.3 |
| Bullock ...................... | 115 | -30.0 | 452 | 0.1 | 20.0 | 968,327 | 2,142 | 83,263 | D | D | D | D | NA | 69.0 | 1,277 | 47.1 |
| Butler ........................ | 84 | -4.5 | 201 | 0.6 | 17.7 | 492,971 | 2,454 | 83,737 | 132.0 | 314,355 | 3.5 | 96.5 | NA | 73.8 | 1,160 | 40.2 |
| Calhoun...................... | 89 | 9.5 | 138 | 1.1 | 28.5 | 587,727 | 4,246 | 74,079 | 86.8 | 135,003 | 15.0 | 85.0 | NA | 77.1 | 913 | 28.8 |
| Chambers ................... | 129 | 33.5 | 389 | 0.2 | 12.5 | 976,835 | 2,513 | 61,548 | 9.1 | 27,369 | 18.6 | 81.4 | NA | 71.9 | 822 | 38.4 |
| Cherokee ................... | 121 | -2.3 | 229 | 0.8 | 72.8 | 725,287 | 3,167 | 125,506 | 152.1 | 286,940 | 25.9 | 74.1 | 1 | 73.2 | 2,145 | 38.1 |
| Chilton....................... | 76 | -16.5 | 165 | 0.9 | 22.0 | 581,653 | 3,529 | 76,741 | 14.5 | 31,298 | 46.2 | 53.8 | NA | 73.9 | 761 | 19.7 |
| Choctaw..................... | 82 | 20.4 | 439 | 0.0 | 6.6 | 899,167 | 2,050 | 68,483 | 8.3 | 43,941 | 9.6 | 90.4 | NA | 66.5 | 275 | 41.0 |
| Clarke........................ | 65 | 36.0 | 202 | 0.0 | 11.2 | 460,092 | 2,279 | 61,402 | 2.9 | 8,994 | 42.7 | 57.3 | NA | 56.6 | 264 | 27.5 |
| Clay........................... | 71 | -8.5 | 187 | 0.0 | 12.2 | 525,095 | 2,807 | 85,885 | 81.3 | 213,462 | 1.2 | 98.8 | NA | 71.7 | 1,328 | 42.5 |
| Cleburne .................... | 48 | -3.6 | 152 | 0.4 | 8.9 | 524,930 | 3,460 | 71,623 | 117.5 | 368,197 | 2.6 | 97.4 | 7 | 76.8 | 366 | 21.0 |
| Coffee ....................... | 177 | -12.4 | 225 | 4.0 | 64.3 | 592,097 | 2,633 | 89,163 | 199.5 | 253,216 | 8.6 | 91.4 | 4 | 70.9 | 5,525 | 56.9 |
| Colbert ...................... | 150 | -1.6 | 254 | D | 85.8 | 677,872 | 2,665 | 105,238 | 60.2 | 101,846 | 62.9 | 37.1 | NA | 73.8 | 3,500 | 39.6 |
| Conecuh ..................... | 104 | 11.0 | 302 | 0.1 | 26.7 | 647,835 | 2,142 | 75,294 | 21.0 | 61,125 | 41.5 | 58.5 | NA | 56.7 | 693 | 44.2 |
| Coosa........................ | 46 | 27.7 | 214 | 0.0 | 5.2 | 489,452 | 2,292 | 62,564 | 1.7 | 7,940 | 17.6 | 82.4 | NA | 72.6 | 270 | 27.9 |
| Covington.................... | 161 | -22.6 | 178 | 1.3 | 54.3 | 477,739 | 2,684 | 79,394 | 126.5 | 139,485 | 16.6 | 83.4 | 6 | 71.6 | 4,147 | 47.0 |
| Crenshaw ................... | 114 | -12.0 | 211 | 0.0 | 23.6 | 525,225 | 2,494 | 97,983 | 164.9 | 303,591 | 2.3 | 97.7 | NA | 67.6 | 1,603 | 49.7 |
| Cullman ...................... | 209 | 7.7 | 117 | 1.9 | 60.9 | 466,918 | 3,977 | 77,508 | 470.7 | 264,303 | 4.0 | 96.0 | NA | 75.5 | 3,139 | 31.3 |
| Dale........................... | 137 | 5.6 | 292 | 2.4 | 52.5 | 879,051 | 3,009 | 103,692 | 172.0 | 366,825 | 9.9 | 90.1 | NA | 77.2 | 2,946 | 47.3 |
| Dallas........................ | 263 | 3.1 | 498 | 6.4 | 76.7 | 934,297 | 1,875 | 124,270 | 64.0 | 121,212 | 43.6 | 56.4 | NA | 69.3 | 3,427 | 51.1 |
| DeKalb....................... | 247 | 7.9 | 128 | 1.1 | 91.5 | 502,086 | 3,935 | 88,244 | 573.3 | 295,658 | 4.4 | 95.6 | 2 | 75.8 | 4,176 | 34.6 |
| Elmore........................ | 94 | 3.7 | 174 | 2.6 | 34.9 | 521,814 | 2,996 | 75,563 | 27.6 | 51,290 | 63.0 | 37.0 | 3 | 74.3 | 855 | 24.7 |
| Escambia .................... | 106 | -1.5 | 242 | 3.1 | 55.3 | 594,861 | 2,458 | 92,069 | 31.0 | 70,886 | 88.5 | 11.5 | NA | 72.5 | 3,842 | 40.7 |
| Etowah ...................... | 89 | 3.7 | 109 | 0.6 | 24.8 | 419,478 | 3,836 | 78,981 | 93.4 | 114,356 | 6.7 | 93.3 | NA | 73.8 | 804 | 21.5 |
| Fayette ...................... | 68 | -15.9 | 211 | 0.2 | 21.0 | 464,587 | 2,206 | 79,352 | 24.6 | 75,957 | 31.9 | 68.1 | NA | 75.9 | 660 | 48.1 |
| Franklin...................... | 130 | -13.9 | 178 | D | 32.4 | 435,546 | 2,451 | 84,184 | 139.4 | 191,202 | 4.5 | 95.5 | NA | 65.8 | 1,531 | 40.5 |
| Geneva ...................... | 183 | -16.2 | 224 | 5.5 | 81.7 | 553,557 | 2,476 | 99,070 | 137.6 | 167,778 | 25.0 | 75.0 | 1 | 71.8 | 4,527 | 51.7 |
| Greene....................... | 153 | 27.4 | 472 | 0.1 | 20.4 | 920,549 | 1,951 | 95,162 | 25.9 | 79,828 | 4.5 | 95.5 | NA | 67.7 | 1,127 | 49.2 |
| Hale........................... | 158 | -1.7 | 402 | 1.1 | 20.7 | 907,427 | 2,260 | 101,638 | 63.5 | 161,539 | 6.6 | 93.4 | 2 | 66.7 | 1,946 | 51.1 |
| Henry ........................ | 174 | 2.5 | 382 | 12.8 | 74.2 | 986,630 | 2,580 | 147,514 | 122.9 | 270,119 | 22.9 | 77.1 | NA | 76.3 | 3,937 | 60.2 |
| Houston ..................... | 149 | -25.0 | 213 | 10.0 | 85.9 | 751,866 | 3,533 | 102,777 | 69.7 | 99,887 | 68.1 | 31.9 | NA | 71.8 | 7,557 | 50.4 |
| Jackson ...................... | 219 | -5.7 | 161 | 1.4 | 95.5 | 490,606 | 3,042 | 86,387 | 153.5 | 113,276 | 19.9 | 80.1 | 3 | 72.5 | 3,198 | 36.0 |
| Jefferson .................... | 38 | -2.8 | 98 | 0.2 | 7.0 | 432,444 | 4,413 | 53,634 | 4.9 | 12,770 | 27.6 | 72.4 | NA | 80.1 | 220 | 11.1 |
| Lamar ........................ | 74 | -10.5 | 274 | 0.3 | 12.0 | 508,777 | 1,859 | 63,797 | 10.4 | 38,509 | 30.5 | 69.5 | NA | 68.4 | 394 | 32.7 |
| Lauderdale .................. | 211 | -0.5 | 161 | 0.9 | 109.0 | 583,196 | 3,626 | 85,411 | 72.3 | 55,264 | 55.7 | 44.3 | NA | 69.4 | 3,934 | 41.5 |
| Lawrence .................... | 214 | -12.3 | 171 | 6.4 | 120.8 | 648,125 | 3,796 | 112,499 | 212.9 | 170,043 | 27.7 | 72.3 | NA | 70.0 | 4,147 | 39.1 |
| Lee............................ | 68 | 14.8 | 216 | 0.2 | 9.0 | 895,686 | 4,150 | 95,632 | D | D | D | D | 4 | 79.9 | 432 | 26.8 |
| Limestone ................... | 225 | -8.9 | 194 | 10.4 | 151.1 | 803,618 | 4,132 | 109,978 | 131.4 | 113,669 | 59.8 | 40.2 | NA | 75.3 | 3,378 | 32.4 |
| Lowndes ..................... | 203 | -6.8 | 396 | 6.2 | 46.5 | 752,311 | 1,898 | 105,508 | 80.9 | 157,920 | 15.7 | 84.3 | NA | 60.4 | 2,573 | 53.3 |
| Macon ........................ | 120 | 16.3 | 322 | 2.6 | 40.8 | 756,523 | 2,347 | 98,817 | 19.6 | 52,458 | 84.2 | 15.8 | NA | 66.8 | 1,238 | 37.0 |
| Madison ...................... | 185 | -11.7 | 181 | 10.7 | 131.5 | 961,058 | 5,306 | 110,872 | 70.4 | 68,913 | 91.8 | 8.2 | 2 | 77.7 | 2,670 | 29.3 |
| Marengo ..................... | 147 | -10.9 | 313 | 0.8 | 28.0 | 627,909 | 2,007 | 79,112 | 15.9 | 33,735 | 32.9 | 67.1 | NA | 68.4 | 1,680 | 58.2 |
| Marion........................ | 83 | -27.4 | 142 | D | 20.3 | 331,932 | 2,339 | 62,866 | 105.3 | 181,003 | 3.1 | 96.9 | NA | 69.9 | 881 | 45.2 |
| Marshall ..................... | 145 | -11.0 | 101 | 0.3 | 44.9 | 445,314 | 4,431 | 86,970 | 283.6 | 196,403 | 2.9 | 97.1 | 3 | 78.3 | 1,729 | 24.9 |
| Mobile ....................... | 95 | 6.9 | 145 | 3.1 | 37.7 | 688,889 | 4,738 | 83,715 | 90.6 | 138,712 | 88.8 | 11.2 | NA | 77.0 | 1,537 | 15.8 |
| Monroe ...................... | 141 | 0.6 | 297 | 1.3 | 42.1 | 710,642 | 2,396 | 90,050 | 35.4 | 74,298 | 59.3 | 40.7 | NA | 58.5 | 2,156 | 51.4 |
| Montgomery................ | 233 | 5.9 | 405 | 0.1 | 57.0 | 1,078,207 | 2,660 | 81,751 | 46.9 | 81,496 | 30.8 | 69.2 | NA | 73.0 | 2,808 | 38.1 |
| Morgan....................... | 135 | -11.6 | 116 | 0.4 | 45.9 | 510,162 | 4,403 | 65,751 | 99.7 | 85,684 | 10.8 | 89.2 | 2 | 78.4 | 1,622 | 26.1 |
| Perry ......................... | 163 | 3.8 | 468 | 3.0 | 41.2 | 1,032,601 | 2,208 | 96,333 | 37.2 | 106,536 | 29.9 | 70.1 | 1 | 60.2 | 1,415 | 57.3 |
| Pickens ...................... | 105 | 2.1 | 278 | 2.4 | 27.1 | 702,898 | 2,528 | 96,970 | 88.8 | 235,520 | 8.9 | 91.1 | NA | 68.2 | 1,346 | 41.6 |
| Pike........................... | 166 | | 279 | 2.9 | 40.6 | 730,910 | 2,620 | 87,152 | 138.8 | 233,722 | 7.2 | 92.8 | NA | 70.9 | 3,305 | 54.4 |
| Randolph..................... | 120 | 5.6 | 202 | 0.2 | 19.7 | 636,356 | 3,157 | 97,755 | 152.7 | 255,797 | 1.8 | 98.2 | 3 | 77.2 | 1,680 | 36.2 |
| Russell....................... | 99 | -15.9 | 333 | 3.7 | 31.2 | 847,048 | 2,541 | 90,179 | 26.1 | 88,118 | 47.3 | 52.7 | NA | 66.9 | 2,666 | 44.9 |
| St. Clair ...................... | 59 | -12.6 | 120 | 0.8 | 14.4 | 514,787 | 4,277 | 77,055 | 58.3 | 118,880 | 13.3 | 86.7 | NA | 83.1 | 565 | 21.6 |
| Shelby ....................... | 67 | 15.0 | 150 | 2.0 | 26.2 | 597,813 | 3,976 | 69,175 | 16.5 | 37,007 | 84.0 | 16.0 | 6 | 77.0 | 399 | 17.7 |
| Sumter....................... | 180 | -24.8 | 491 | 1.1 | 31.1 | 919,441 | 1,872 | 81,695 | 21.9 | 59,594 | 22.1 | 77.9 | NA | 75.7 | 1,987 | 55.0 |
| Talladega .................... | 80 | -20.4 | 140 | 4.4 | 30.0 | 464,811 | 3,309 | 89,072 | 40.4 | 71,350 | 29.0 | 71.0 | NA | 75.4 | 1,080 | 38.0 |
| Tallapoosa .................. | 66 | 8.0 | 190 | D | 8.5 | 541,614 | 2,855 | 60,191 | 16.5 | 47,640 | 12.1 | 87.9 | NA | 75.5 | 494 | 23.6 |
| Tuscaloosa.................. | 95 | 10.4 | 170 | 0.4 | 23.5 | 573,374 | 3,364 | 62,210 | 38.6 | 69,284 | 17.9 | 82.1 | 2 | 77.2 | 919 | 30.9 |
| Walker........................ | 65 | 19.2 | 130 | 0.1 | 16.0 | 391,595 | 3,019 | 87,298 | 55.1 | 109,928 | 7.9 | 92.1 | NA | 67.5 | 431 | 15.8 |

| STATE County | Water use, 2015 | | Wholesale Trade[1], 2017 | | | | Retail Trade[2], 2017 | | | | Real estate and rental and leasing,[2] 2017 | | | |
|---|---|---|---|---|---|---|---|---|---|---|---|---|---|---|
| | Public supply water withdrawn (mil gal/day) | Public supply gallons withdrawn per person per day | Number of establishments | Number of employees | Sales (mil dol) | Average payroll (mil dol) | Number of establishments | Number of employees | Sales (mil dol) | Average payroll (mil dol) | Number of establishments | Number of employees | Sales (mil dol) | Average payroll (mil dol) |
| | 133 | 134 | 135 | 136 | 137 | 138 | 139 | 140 | 141 | 142 | 143 | 144 | 145 | 146 |
| UNITED STATES.............. | 36,595.83 | 120.1 | 352,065 | 5,156,359 | 5,700,966.9 | 337,908.8 | 1,064,087 | 15,938,821 | 4,949,601.5 | 441,795.6 | 410,820 | 2,194,885 | 674,147.0 | 113,409.6 |
| ALABAMA........................ | 761.53 | 156.7 | 4,410 | 63,468 | 66,906.3 | 3,461.1 | 17,958 | 230,069 | 63,740.4 | 5,819.1 | 4,362 | 22,778 | 5,785.1 | 968.2 |
| Autauga.......................... | 3.64 | 65.8 | 24 | 229 | 206.2 | 13.7 | 170 | 2,701 | 666.3 | 60.1 | 44 | 107 | 22.4 | 4.0 |
| Baldwin.......................... | 23.67 | 116.2 | 194 | 2,404 | 1,447.4 | 124.6 | 989 | 13,735 | 3,703.4 | 340.2 | 365 | 1,842 | 369.7 | 65.2 |
| Barbour.......................... | 3.23 | 121.9 | 12 | 123 | 65.1 | 3.0 | 91 | 898 | 205.8 | 19.4 | D | D | D | D |
| Bibb.............................. | 5.18 | 229.4 | 13 | 91 | 148.3 | 4.4 | 48 | 514 | 119.8 | 12.2 | 9 | 15 | 2.0 | 0.3 |
| Blount............................ | 56.86 | 985.9 | 36 | 379 | 173.4 | 15.8 | 131 | 1,227 | 340.9 | 30.5 | 9 | 146 | 11.7 | 2.0 |
| Bullock.......................... | 2.11 | 197.3 | 6 | 21 | 38.8 | 0.8 | 26 | 236 | 58.7 | 4.5 | 3 | 9 | 0.5 | 0.1 |
| Butler............................ | 2.28 | 113.1 | 13 | 70 | 60.3 | 3.2 | 100 | 945 | 220.6 | 21.1 | 21 | 34 | 5.4 | 0.8 |
| Calhoun.......................... | 25.25 | 218.4 | 90 | 1,614 | 1,347.9 | 69.2 | 466 | 6,380 | 1,652.3 | 151.6 | 76 | 307 | 58.6 | 10.8 |
| Chambers ....................... | 4.20 | 123.1 | D | D | D | D | 114 | 1,124 | 297.1 | 25.3 | 17 | 44 | 5.8 | 1.3 |
| Cherokee........................ | 3.20 | 123.7 | 15 | 119 | 68.9 | 4.3 | 85 | 894 | 207.8 | 21.7 | 9 | 27 | 6.3 | 1.4 |
| Chilton........................... | 4.40 | 100.1 | 24 | 391 | 303.3 | 17.7 | 148 | 1,502 | 510.6 | 38.7 | 25 | 62 | 8.0 | 1.4 |
| Choctaw.......................... | 1.21 | 91.9 | 5 | D | 12.4 | D | 52 | 326 | 90.3 | 7.0 | 5 | D | 0.7 | D |
| Clarke............................ | 2.58 | 104.6 | 16 | 142 | 108.9 | 6.7 | 134 | 1,336 | 348.2 | 33.4 | 20 | 96 | 10.6 | 2.2 |
| Clay.............................. | 1.78 | 131.3 | NA | NA | NA | NA | 36 | 286 | 63.0 | 5.1 | 7 | D | 0.8 | D |
| Cleburne........................ | 0.48 | 32.0 | 6 | D | 8.5 | D | 41 | 333 | 136.1 | 9.7 | 6 | 8 | 3.0 | 0.5 |
| Coffee............................ | 6.34 | 123.8 | 22 | 299 | 196.3 | 13.7 | 202 | 2,456 | 789.4 | 65.1 | 38 | 167 | 23.2 | 6.1 |
| Colbert........................... | 8.26 | 152.0 | 75 | 748 | 547.0 | 39.4 | 229 | 3,037 | 841.2 | 83.2 | 38 | 115 | 28.3 | 4.3 |
| Conecuh.......................... | 1.32 | 104.2 | 6 | 333 | 104.8 | 11.3 | 37 | 304 | 106.9 | 5.6 | 5 | 5 | 1.0 | 0.1 |
| Coosa............................ | 0.27 | 25.2 | D | D | D | 1.4 | 18 | 84 | 21.9 | 1.6 | NA | NA | NA | NA |
| Covington........................ | 4.09 | 108.1 | 31 | 426 | 463.6 | 17.1 | 186 | 1,827 | 461.8 | 44.3 | 24 | 92 | 10.4 | 2.2 |
| Crenshaw........................ | 1.93 | 138.2 | 10 | 472 | 192.2 | 23.6 | 34 | 274 | 67.0 | 6.0 | D | D | D | D |
| Cullman.......................... | 23.24 | 283.4 | 79 | 1,008 | 1,006.9 | 53.3 | 335 | 3,996 | 1,191.3 | 102.3 | 57 | 176 | 30.1 | 6.7 |
| Dale.............................. | 5.90 | 119.0 | 16 | 196 | 40.1 | 5.8 | 145 | 1,334 | 373.9 | 31.8 | 33 | 178 | 38.5 | 6.9 |
| Dallas............................ | 5.91 | 143.7 | 21 | 189 | 97.6 | 8.0 | 159 | 1,493 | 360.0 | 35.3 | 26 | 86 | 10.0 | 2.0 |
| DeKalb........................... | 5.91 | 83.1 | 43 | 595 | 252.0 | 26.0 | 239 | 2,265 | 695.9 | 57.6 | 24 | 150 | 17.8 | 5.5 |
| Elmore........................... | 12.19 | 149.6 | D | D | D | D | 206 | 2,939 | 774.4 | 72.1 | 59 | 133 | 25.3 | 4.1 |
| Escambia........................ | 5.00 | 132.3 | 27 | 131 | 102.3 | 7.6 | 166 | 1,693 | 426.5 | 40.1 | 26 | 62 | 10.4 | 1.7 |
| Etowah.......................... | 16.85 | 163.5 | 70 | 758 | 497.3 | 32.4 | 393 | 4,777 | 1,319.8 | 114.3 | 61 | 232 | 64.4 | 7.8 |
| Fayette........................... | 1.70 | 101.4 | 4 | 17 | 8.6 | 0.7 | 57 | 559 | 134.0 | 12.8 | 7 | 17 | 2.7 | 0.4 |
| Franklin.......................... | 5.94 | 187.4 | 16 | 71 | 100.5 | 3.6 | 99 | 933 | 236.4 | 20.2 | D | D | D | 0.9 |
| Geneva ......................... | 1.78 | 66.5 | 19 | 410 | 417.1 | 26.7 | 87 | 714 | 162.0 | 16.0 | 8 | 17 | 3.0 | 0.4 |
| Greene.......................... | 1.39 | 163.9 | NA | NA | NA | NA | 25 | 166 | 36.0 | 3.0 | NA | NA | NA | NA |
| Hale............................. | 3.10 | 205.7 | D | D | D | 1.3 | 40 | 279 | 73.3 | 7.0 | 4 | 6 | 0.5 | 0.2 |
| Henry............................ | 1.66 | 96.4 | 11 | 223 | 433.3 | 11.4 | 50 | 316 | 85.8 | 6.7 | 5 | 5 | 1.3 | 0.1 |
| Houston.......................... | 18.94 | 181.8 | 158 | 2,898 | 10,907.0 | 156.7 | 536 | 7,838 | 2,218.9 | 203.0 | 95 | 488 | 99.5 | 20.1 |
| Jackson.......................... | 11.79 | 224.9 | D | D | D | D | 174 | 1,832 | 498.6 | 45.0 | 28 | 76 | 16.9 | 2.6 |
| Jefferson........................ | 48.98 | 74.2 | 911 | 15,607 | 17,676.4 | 921.3 | 2,666 | 38,321 | 10,348.5 | 1,008.5 | 779 | 5,567 | 1,931.3 | 313.2 |
| Lamar........................... | 1.42 | 102.3 | D | D | D | 1.9 | 49 | 336 | 82.7 | 7.0 | 5 | D | 2.2 | D |
| Lauderdale...................... | 12.39 | 133.8 | 72 | 1,129 | 528.8 | 50.2 | 400 | 5,252 | 1,279.8 | 124.4 | 105 | 413 | 67.9 | 13.1 |
| Lawrence........................ | 7.73 | 233.4 | 16 | 107 | 62.3 | 3.3 | 82 | 768 | 202.9 | 18.4 | 7 | 25 | 2.0 | 0.5 |
| Lee.............................. | 15.83 | 100.8 | 83 | 999 | 465.5 | 48.1 | 478 | 7,212 | 1,901.9 | 169.0 | 157 | 678 | 139.6 | 25.1 |
| Limestone....................... | 11.52 | 125.7 | 48 | 751 | 455.6 | 33.9 | 266 | 2,790 | 794.7 | 72.9 | 60 | 183 | 29.5 | 6.0 |
| Lowndes......................... | 0.95 | 90.8 | NA | NA | NA | NA | 27 | 163 | 76.3 | 3.4 | 6 | 6 | 1.6 | 0.1 |
| Macon ........................... | 3.31 | 173.3 | D | D | D | 0.1 | 43 | 469 | 134.9 | 7.8 | 12 | 31 | 7.1 | 0.9 |
| Madison.......................... | 67.15 | 190.2 | 298 | 4,652 | 4,906.4 | 293.3 | 1,298 | 19,547 | 5,427.4 | 527.9 | 417 | 1,804 | 566.0 | 72.5 |
| Marengo......................... | 2.77 | 138.3 | 14 | 84 | 51.6 | 4.4 | 104 | 878 | 220.7 | 20.5 | 9 | 91 | 29.4 | 3.1 |
| Marion........................... | 6.06 | 200.9 | 24 | 219 | 199.2 | 9.6 | 109 | 1,096 | 257.2 | 25.1 | 9 | 16 | 3.7 | 0.5 |
| Marshall......................... | 23.76 | 250.8 | 82 | 1,238 | 1,284.8 | 51.8 | 396 | 4,752 | 1,535.8 | 122.3 | 65 | 574 | 47.8 | 11.7 |
| Mobile........................... | 66.10 | 159.1 | 508 | 6,103 | 3,430.8 | 317.7 | 1,493 | 20,361 | 5,607.8 | 527.7 | 438 | 2,424 | 565.7 | 98.4 |
| Monroe.......................... | 2.31 | 106.6 | 18 | 159 | 122.1 | 8.4 | 91 | 900 | 225.4 | 19.4 | D | D | D | D |
| Montgomery..................... | 28.00 | 123.6 | 274 | 5,628 | 4,338.6 | 308.4 | 916 | 12,601 | 3,448.9 | 324.8 | 257 | 1,579 | 341.3 | 67.9 |
| Morgan.......................... | 25.60 | 214.1 | D | D | D | 95.6 | 464 | 5,545 | 1,774.9 | 144.3 | 93 | 397 | 107.0 | 17.8 |
| Perry............................ | 2.02 | 209.3 | 4 | 38 | 18.6 | 1.1 | 29 | 156 | 33.2 | 3.2 | D | D | D | D |
| Pickens.......................... | 3.00 | 143.8 | D | D | D | 1.8 | 52 | 409 | 94.5 | 8.7 | 3 | 7 | 0.5 | 0.3 |
| Pike.............................. | 4.58 | 138.6 | 31 | 467 | 270.2 | 17.2 | 136 | 1,629 | 416.3 | 37.3 | 32 | 115 | 21.5 | 2.8 |
| Randolph........................ | 1.22 | 53.8 | 6 | 37 | 16.2 | 1.2 | 82 | 726 | 183.4 | 16.6 | D | D | D | 0.7 |
| Russell.......................... | 8.45 | 141.6 | D | D | D | 3.5 | 156 | 2,091 | 514.4 | 45.5 | D | D | D | D |
| St. Clair......................... | 8.93 | 102.6 | 60 | 1,154 | 510.8 | 50.3 | 230 | 2,862 | 811.5 | 69.6 | 42 | 124 | 45.6 | 5.5 |
| Shelby........................... | 14.21 | 68.1 | 333 | 4,810 | 8,118.3 | 345.4 | 700 | 10,386 | 3,607.4 | 292.3 | 236 | 1,294 | 559.8 | 77.0 |
| Sumter........................... | 2.06 | 157.2 | 10 | 137 | 49.5 | 4.9 | 50 | 226 | 54.0 | 4.8 | D | D | D | D |
| Talladega........................ | 16.00 | 197.9 | D | D | D | 27.1 | 273 | 2,880 | 781.9 | 66.6 | D | D | D | D |
| Tallapoosa....................... | 11.87 | 290.6 | 15 | 109 | 44.9 | 4.6 | 156 | 1,533 | 415.1 | 37.9 | 29 | 96 | 30.3 | 4.2 |
| Tuscaloosa...................... | 31.43 | 154.1 | 152 | 1,532 | 926.0 | 80.7 | 719 | 10,120 | 2,760.1 | 248.5 | 208 | 1,857 | 250.1 | 61.9 |
| Walker........................... | 43.88 | 672.0 | 37 | 332 | 234.2 | 16.3 | 281 | 3,293 | 949.1 | 83.9 | 44 | 188 | 32.8 | 6.6 |

1 Merchant wholesalers, except manufacturers' sales branches and offices. 2. Employer establishments.

# Professional Services, Manufacturing, and Accommodation and Food Services

| STATE County | Professional, scientific, and technical services, 2017 | | | | Manufacturing, 2017 | | | | Accommodation and food services, 2017 | | | |
|---|---|---|---|---|---|---|---|---|---|---|---|---|
| | Number of establish-ments | Number of employees | Sales (mil dol) | Average payroll (mil dol) | Number of establish-ments | Number of employees | Sales (mil dol) | Average payroll (mil dol) | Number of establis-hments | Number of employees | Sales (mil dol) | Annual payroll (mil dol) |
| | 147 | 148 | 149 | 150 | 151 | 152 | 153 | 154 | 155 | 156 | 157 | 158 |
| UNITED STATES.............. | 908,547 | 8,759,257 | 1,795,589 | 707,772.8 | 291,586 | 11,522,039 | 5,548,796.8 | 670,678.4 | 726,081 | 14,002,624 | 938,237.1 | 264,603.6 |
| ALABAMA ........................ | 9,452 | 99,578 | 20,371 | 7,087.4 | 4,133 | 245,725 | 137,441.2 | 12,916.0 | 9,085 | 183,590 | 10,527.7 | 2,761.5 |
| Autauga.......... | 51 | 266 | 29 | 10.2 | 23 | 1,152 | 760.1 | 79.7 | D | D | D | D |
| Baldwin .......... | D | D | D | D | 165 | 4,505 | 2,616.0 | 215.9 | 572 | 12,977 | 794.1 | 228.8 |
| Barbour .......... | 32 | 97 | 15 | 4.0 | 28 | 2,514 | 809.0 | 94.9 | 56 | 631 | 29.3 | 7.5 |
| Bibb.............. | 14 | 64 | 7 | 2.6 | 18 | 309 | 134.5 | 14.5 | 18 | 267 | 18.7 | 3.9 |
| Blount.......... | 43 | 219 | 34 | 9.6 | 37 | 1,041 | 321.3 | 37.9 | 49 | 677 | 35.9 | 9.4 |
| Bullock.......... | D | D | D | 1.1 | D | D | D | D | 9 | 79 | 4.6 | 1.1 |
| Butler.......... | 19 | 61 | 7 | 2.0 | 17 | 1,270 | 865.6 | 55.4 | D | D | D | D |
| Calhoun.......... | D | D | D | D | 106 | 5,986 | 2,699.2 | 308.1 | 225 | 4,642 | 247.0 | 66.2 |
| Chambers .......... | 32 | 174 | 18 | 5.5 | 24 | 1,810 | 880.2 | 85.5 | 59 | 950 | 43.8 | 10.6 |
| Cherokee ......... | 21 | 71 | 7 | 2.1 | 16 | 955 | 328.8 | 35.6 | 35 | 481 | 21.9 | 5.6 |
| Chilton.......... | 32 | 88 | 10 | 3.1 | 41 | 1,969 | 483.0 | 87.6 | 55 | 972 | 47.1 | 12.6 |
| Choctaw.......... | 20 | 48 | 6 | 1.4 | D | D | D | D | D | D | D | D |
| Clarke.......... | 29 | 94 | 13 | 2.5 | 23 | 1,521 | 648.1 | 85.4 | D | D | D | D |
| Clay.............. | 11 | 50 | 3 | 1.1 | 9 | 2,153 | 296.3 | 55.2 | 11 | 125 | 5.4 | 8.9 |
| Cleburne .......... | 8 | 12 | 1 | 0.3 | 9 | 288 | 125.1 | 13.7 | D | D | D | 1.3 |
| Coffee.......... | 67 | 758 | 138 | 55.0 | 38 | 3,020 | 938.9 | 111.6 | 106 | 1,672 | 82.9 | 22.1 |
| Colbert.......... | 69 | 278 | 32 | 10.9 | 92 | 6,620 | 3,691.0 | 358.0 | 107 | 1,966 | 88.2 | 23.3 |
| Conecuh.......... | 10 | 21 | 2 | 0.7 | 13 | 463 | 132.0 | 19.5 | D | D | D | 3.0 |
| Coosa.......... | 7 | 24 | 3 | 1.3 | 4 | 471 | 82.7 | 18.4 | D | D | D | D |
| Covington.......... | 62 | 283 | 32 | 11.4 | 32 | 2,115 | 689.7 | 89.0 | D | D | D | D |
| Crenshaw.......... | 14 | 31 | 4 | 1.0 | 6 | 1,050 | 308.9 | 36.7 | D | D | D | D |
| Cullman.......... | 101 | 435 | 47 | 16.2 | 98 | 5,617 | 2,109.7 | 257.2 | 148 | 2,829 | 151.5 | 40.0 |
| Dale.......... | D | D | D | D | 29 | 470 | 152.7 | 23.1 | 77 | 1,107 | 58.6 | 15.0 |
| Dallas.......... | D | D | D | D | 41 | 2,714 | 1,079.8 | 131.3 | 57 | 853 | 44.2 | 10.8 |
| DeKalb.......... | 72 | 324 | 37 | 11.7 | 98 | 6,493 | 1,442.4 | 286.2 | 93 | 1,576 | 97.9 | 25.8 |
| Elmore.......... | 99 | 440 | 92 | 22.0 | 57 | 2,718 | 890.2 | 147.6 | 104 | 2,773 | 454.8 | 59.0 |
| Escambia.......... | 42 | 170 | 24 | 6.5 | 36 | 1,936 | 758.0 | 110.3 | 62 | 2,113 | 303.2 | 46.4 |
| Etowah.......... | 127 | 656 | 81 | 28.2 | 89 | 5,358 | 1,607.9 | 231.1 | 187 | 3,587 | 187.6 | 50.4 |
| Fayette.......... | 14 | 35 | 4 | 1.3 | 22 | 828 | 170.5 | 35.0 | D | D | D | D |
| Franklin.......... | 25 | 143 | 13 | 5.1 | 40 | 5,365 | 1,553.9 | 206.7 | D | D | D | 7.1 |
| Geneva.......... | 29 | 98 | 9 | 2.8 | 22 | 698 | 120.2 | 23.5 | D | D | D | D |
| Greene.......... | 3 | 8 | 1 | 0.2 | 8 | 408 | 142.8 | 16.8 | 6 | 41 | 2.7 | 0.6 |
| Hale.......... | 8 | 25 | 2 | 0.4 | 14 | 474 | 148.3 | 26.0 | 15 | 153 | 7.1 | 1.6 |
| Henry.......... | 25 | 140 | 19 | 7.0 | 13 | 307 | 212.2 | 13.1 | D | D | D | D |
| Houston.......... | D | D | D | D | 104 | 4,667 | 1,377.1 | 193.1 | 256 | 5,206 | 280.2 | 73.1 |
| Jackson.......... | 59 | 242 | 26 | 9.0 | 66 | 5,895 | 1,515.2 | 218.0 | 83 | 1,295 | 64.1 | 16.2 |
| Jefferson.......... | D | D | D | D | 543 | 21,831 | 8,416.5 | 1,261.7 | 1,481 | 32,014 | 1,810.1 | 520.2 |
| Lamar.......... | D | D | D | 2.1 | 22 | 1,395 | 440.7 | 71.9 | D | D | D | D |
| Lauderdale.......... | 148 | 714 | 74 | 27.5 | 84 | 2,392 | 943.1 | 100.7 | 184 | 4,347 | 197.4 | 59.0 |
| Lawrence.......... | 31 | 139 | 21 | 7.1 | 23 | 149 | 33.3 | 6.6 | D | D | D | D |
| Lee.......... | 254 | 1,277 | 164 | 56.6 | 118 | 7,008 | 2,707.0 | 317.6 | 363 | 7,591 | 371.4 | 100.0 |
| Limestone.......... | D | D | D | D | 66 | 3,505 | 1,444.4 | 160.7 | 112 | 2,241 | 106.7 | 27.9 |
| Lowndes.......... | 5 | 11 | 1 | 0.5 | 11 | 938 | 1,169.7 | 52.3 | D | D | D | D |
| Macon.......... | 11 | 34 | 6 | 2.2 | 6 | 645 | 456.8 | 38.2 | 21 | 385 | 20.2 | 5.1 |
| Madison.......... | 1,469 | 37,711 | 10,167 | 3,380.3 | 265 | 15,834 | 6,741.9 | 1,004.9 | 786 | 16,885 | 886.0 | 253.3 |
| Marengo.......... | 16 | 86 | 14 | 3.4 | 22 | 1,276 | 578.5 | 73.6 | D | D | D | D |
| Marion.......... | 25 | 76 | 6 | 2.2 | 35 | 2,482 | 605.1 | 103.1 | D | D | D | D |
| Marshall.......... | 126 | 665 | 84 | 26.8 | 112 | 10,255 | 3,624.4 | 380.9 | 185 | 3,178 | 166.3 | 43.4 |
| Mobile .......... | D | D | D | D | 323 | 15,793 | 10,650.8 | 1,016.3 | 705 | 15,002 | 791.9 | 218.0 |
| Monroe.......... | 16 | 41 | 6 | 1.6 | 18 | 1,115 | 468.4 | 77.8 | D | D | D | D |
| Montgomery.......... | D | D | D | D | 169 | 12,464 | 11,183.8 | 660.0 | 516 | 11,673 | 861.7 | 186.5 |
| Morgan.......... | 193 | 2,421 | 529 | 144.8 | 158 | 11,931 | 10,358.9 | 741.9 | D | D | D | D |
| Perry.......... | 6 | 9 | 2 | 0.2 | 5 | 438 | 160.0 | 13.7 | 11 | 102 | 5.8 | 1.6 |
| Pickens.......... | 14 | 91 | 8 | 2.5 | 18 | 583 | 114.4 | 21.1 | D | D | D | D |
| Pike.......... | 34 | 440 | 39 | 18.5 | 26 | 2,306 | 876.7 | 122.5 | 81 | 1,568 | 70.4 | 18.3 |
| Randolph.......... | 16 | 48 | 5 | 1.3 | 24 | 1,060 | 339.4 | 40.3 | D | D | D | 4.3 |
| Russell.......... | 46 | 215 | 20 | 6.9 | 34 | 2,661 | 1,239.3 | 142.9 | 91 | 1,664 | 90.4 | 21.8 |
| St. Clair.......... | 101 | 652 | 79 | 21.0 | 71 | 4,123 | 1,344.5 | 194.0 | 123 | 2,299 | 118.1 | 30.9 |
| Shelby.......... | D | D | D | 321.6 | 158 | 5,452 | 1,816.8 | 305.8 | 405 | 8,120 | 451.8 | 130.6 |
| Sumter.......... | 6 | 13 | 2 | 0.4 | 7 | 213 | 67.3 | 9.2 | D | D | D | 4.0 |
| Talladega.......... | 66 | 382 | 42 | 17.7 | 86 | 9,360 | 11,590.0 | 617.8 | D | D | D | D |
| Tallapoosa.......... | 60 | 206 | 28 | 9.2 | 30 | 2,346 | 530.3 | 82.8 | 69 | 1,035 | 52.9 | 14.3 |
| Tuscaloosa.......... | D | D | D | D | 135 | 14,565 | 20,691.6 | 913.3 | D | D | D | D |
| Walker.......... | 79 | 432 | 39 | 21.7 | 48 | 1,890 | 618.8 | 77.6 | 92 | 1,688 | 91.2 | 23.7 |

# Health Care and Social Assistance, Other Services, Nonemployer Businesses, and Residential Construction

| STATE County | Health care and social assistance, 2017 | | | | Other services, 2017 | | | | Nonemployer businesses, 2018 | | Value of residential construction authorized by building permits, 2020 | |
|---|---|---|---|---|---|---|---|---|---|---|---|---|
| | Number of establish-ments | Number of employees | Receipts (mil dol) | Annual payroll (mil dol) | Number of establish-ments | Number of employees | Receipts (mil dol) | Annual payroll (mil dol) | Number | Receipts (mil dol) | New construction ($1,000) | Number of housing units |
| | 159 | 160 | 161 | 162 | 163 | 164 | 165 | 166 | 167 | 168 | 169 | 170 |
| UNITED STATES.............. | 892,245 | 20,506,502 | 2,527,903.3 | 990,056.2 | 560,845 | 3,696,831 | 544,127.7 | 133,751.1 | 26,485,532 | 1,292,866.7 | 307,209,904 | 1,471,141 |
| ALABAMA........................ | 10,636 | 258,399 | 31,238.9 | 11,948.8 | 6,149 | 38,476 | 5,844.4 | 1,302.6 | 336,445 | 14,797.0 | 4,358,692 | 19,982 |
| Autauga.......................... | D | D | D | D | 59 | 256 | 28.7 | 6.8 | 3,350 | 130.2 | 105,453 | 332 |
| Baldwin .......................... | 503 | 8,049 | 848.4 | 327.8 | 323 | 1,544 | 187.4 | 46.3 | 20,047 | 1,018.7 | 854,638 | 3,614 |
| Barbour.......................... | D | D | D | D | D | D | D | D | 1,329 | 47.5 | 2,881 | 12 |
| Bibb................................ | 24 | 601 | 55.5 | 25.4 | D | D | D | D | 3,710 | 158.8 | 9,632 | 37 |
| Blount............................. | 61 | 976 | 83.4 | 35.3 | D | D | D | D | 498 | 17.3 | 3,326 | 23 |
| Bullock ........................... | 12 | 422 | 41.1 | 14.9 | D | D | D | D | 1,081 | 42.7 | 0 | 0 |
| Butler............................. | 42 | 873 | 84.9 | 32.3 | D | D | D | D | 6,811 | 263.3 | 455 | 2 |
| Calhoun.......................... | 275 | 6,358 | 622.7 | 249.8 | 166 | 718 | 79.2 | 19.7 | 2,108 | 65.6 | 14,985 | 102 |
| Chambers ...................... | 55 | 778 | 60.6 | 26.3 | 37 | 120 | 11.9 | 3.5 | 1,686 | 85.8 | 1,995 | 19 |
| Cherokee ....................... | 35 | 456 | 36.8 | 16.0 | 20 | 157 | 9.1 | 2.5 | | | 4,736 | 29 |
| Chilton........................... | 72 | 813 | 83.8 | 33.1 | D | D | D | D | 2,873 | 124.5 | 6,011 | 41 |
| Choctaw.......................... | 24 | 356 | 35.8 | 13.7 | 16 | 58 | 6.9 | 2.2 | 745 | 26.2 | 209 | 1 |
| Clarke............................. | 58 | 897 | 67.1 | 29.3 | D | D | D | D | 1,446 | 59.3 | 430 | 10 |
| Clay................................. | 20 | 515 | 37.3 | 17.7 | 8 | 37 | 2.9 | 0.8 | 829 | 38.8 | 375 | 3 |
| Cleburne......................... | 12 | 60 | 5.0 | 2.1 | D | D | D | D | 997 | 40.7 | 170 | 1 |
| Coffee ............................ | 104 | 1,856 | 173.2 | 68.4 | 72 | 298 | 30.9 | 8.8 | 2,764 | 101.9 | 37,269 | 269 |
| Colbert ........................... | 118 | 3,062 | 333.4 | 133.5 | D | D | D | 18.1 | 3,599 | 143.6 | 23,948 | 197 |
| Conecuh.......................... | 18 | 465 | 34.1 | 14.1 | D | D | D | D | 734 | 24.4 | 154 | 11 |
| Coosa............................. | 3 | D | 7.7 | D | 7 | 18 | 2.7 | 0.5 | 426 | 16.0 | 0 | 0 |
| Covington........................ | 88 | 1,607 | 148.4 | 59.2 | 50 | 304 | 28.5 | 7.9 | 2,312 | 83.8 | 1,088 | 7 |
| Crenshaw........................ | 15 | 459 | 26.9 | 14.1 | 13 | 41 | 5.4 | 1.2 | 874 | 28.9 | 905 | 5 |
| Cullman.......................... | 194 | 4,235 | 398.7 | 153.0 | 111 | 605 | 66.3 | 19.9 | 5,989 | 287.4 | 26,468 | 146 |
| Dale................................ | 52 | 1,150 | 90.9 | 40.2 | 42 | 185 | 12.2 | 4.1 | 2,650 | 89.5 | 5,729 | 39 |
| Dallas............................. | 98 | 1,765 | 173.3 | 66.0 | 47 | 322 | 34.3 | 8.7 | 2,126 | 70.9 | 2,868 | 20 |
| DeKalb............................ | 134 | 2,527 | 216.5 | 90.6 | 52 | 172 | 17.6 | 5.0 | 4,910 | 246.5 | 7,360 | 78 |
| Elmore............................ | 122 | 2,170 | 176.6 | 67.5 | D | D | D | D | 5,400 | 245.5 | 25,416 | 153 |
| Escambia ........................ | 66 | 1,065 | 116.0 | 36.9 | 40 | 171 | 16.3 | 3.7 | 2,096 | 69.6 | 1,426 | 9 |
| Etowah............................ | 300 | 7,233 | 761.6 | 303.7 | 114 | 779 | 125.7 | 24.3 | 7,520 | 345.0 | 13,933 | 113 |
| Fayette ........................... | 43 | 690 | 56.4 | 27.3 | D | D | D | D | 1,028 | 39.9 | 0 | 0 |
| Franklin .......................... | 48 | 1,033 | 80.8 | 35.1 | D | D | D | D | 1,914 | 82.4 | 2,208 | 20 |
| Geneva ........................... | 32 | 578 | 44.6 | 18.1 | 21 | 48 | 6.0 | 1.2 | 1,633 | 68.2 | 3,643 | 22 |
| Greene............................ | 11 | 266 | 16.0 | 8.1 | 4 | 11 | 1.4 | 0.3 | 441 | 13.8 | 835 | 8 |
| Hale................................ | 17 | 385 | 27.7 | 11.9 | D | D | D | 0.2 | 896 | 28.6 | 5,137 | 31 |
| Henry.............................. | 21 | 168 | 11.8 | 5.4 | D | D | D | D | 1,090 | 43.8 | 11,418 | 65 |
| Houston.......................... | 322 | 9,826 | 1,180.0 | 494.0 | 185 | 926 | 110.0 | 26.3 | 7,419 | 347.0 | 102,338 | 401 |
| Jackson........................... | 103 | 1,550 | 142.8 | 57.4 | D | D | 22.0 | D | 3,215 | 128.9 | 11,539 | 46 |
| Jefferson......................... | 1,761 | 60,705 | 9,862.6 | 3,510.6 | 1,044 | 9,360 | 2,377.4 | 388.6 | 50,156 | 2,337.5 | 541,673 | 1,868 |
| Lamar............................. | 24 | 368 | 18.7 | 8.7 | D | D | D | D | 896 | 38.2 | 0 | 0 |
| Lauderdale ...................... | 241 | 5,150 | 578.7 | 203.1 | D | D | D | 15.1 | 6,645 | 337.1 | 12,661 | 99 |
| Lawrence......................... | 33 | 787 | 46.3 | 21.9 | D | D | D | D | 1,980 | 72.0 | 1,399 | 6 |
| Lee.................................. | 258 | 6,926 | 697.5 | 276.4 | 177 | 1,068 | 91.5 | 26.2 | 10,729 | 485.6 | 488,948 | 2,221 |
| Limestone ....................... | 140 | 2,302 | 215.8 | 88.9 | 78 | 491 | 58.5 | 16.7 | 6,237 | 269.2 | 91,746 | 423 |
| Lowndes.......................... | D | D | D | D | D | D | D | D | 623 | 23.2 | 0 | 0 |
| Macon ............................ | 22 | 1,719 | 275.4 | 117.5 | 17 | 43 | 5.0 | 1.2 | 995 | 25.3 | 1,085 | 5 |
| Madison.......................... | 1,021 | 25,276 | 3,441.1 | 1,233.5 | 500 | 3,917 | 498.5 | 137.8 | 25,693 | 1,086.5 | 700,850 | 3,918 |
| Marengo.......................... | 56 | 1,075 | 82.6 | 34.7 | 25 | 99 | 8.8 | 2.3 | 1,087 | 40.2 | 50 | 1 |
| Marion ............................ | 91 | 1,285 | 123.1 | 47.8 | D | D | D | D | 1,806 | 90.8 | 1,045 | 5 |
| Marshall ......................... | 196 | 3,982 | 382.4 | 155.8 | 87 | 364 | 29.5 | 7.7 | 7,199 | 351.9 | 30,971 | 173 |
| Mobile ............................ | 796 | 21,337 | 2,651.9 | 988.4 | 547 | 4,061 | 453.1 | 121.2 | 30,478 | 1,181.7 | 195,559 | 934 |
| Monroe ........................... | 30 | 708 | 55.9 | 24.1 | D | D | D | D | 1,343 | 46.2 | 325 | 2 |
| Montgomery..................... | 666 | 18,304 | 2,123.7 | 888.7 | 439 | 2,735 | 412.0 | 110.1 | 15,765 | 676.6 | 92,436 | 494 |
| Morgan........................... | 349 | 5,577 | 517.9 | 217.6 | D | D | D | D | 7,704 | 328.3 | 45,731 | 218 |
| Perry .............................. | 7 | 128 | 8.5 | 3.9 | 6 | 21 | 1.8 | 0.8 | 481 | 16.0 | 0 | 0 |
| Pickens........................... | 26 | 543 | 40.3 | 16.5 | D | D | D | 0.7 | 1,102 | 33.0 | 898 | 7 |
| Pike................................ | 55 | 1,138 | 110.5 | 44.9 | 35 | 148 | 11.9 | 3.4 | 1,734 | 72.8 | 10,037 | 64 |
| Randolph......................... | 27 | 563 | 39.8 | 18.1 | D | D | D | D | 1,427 | 52.2 | 0 | 0 |
| Russell............................ | D | D | D | D | 55 | 268 | 28.9 | 8.2 | 3,307 | 108.3 | 35,609 | 222 |
| St. Clair .......................... | 100 | 2,370 | 206.9 | 81.5 | 89 | 377 | 42.4 | 11.1 | 5,822 | 240.8 | 64,150 | 393 |
| Shelby............................. | 494 | 7,656 | 795.7 | 330.8 | 319 | 1,942 | 244.4 | 70.1 | 18,361 | 973.8 | 411,821 | 1,483 |
| Sumter............................ | 19 | 353 | 19.8 | 9.6 | 5 | 33 | 4.0 | 1.1 | 709 | 26.4 | 330 | 3 |
| Talladega ........................ | 143 | D | 232.1 | D | 59 | 228 | 22.8 | 8.8 | 4,128 | 174.9 | 18,960 | 125 |
| Tallapoosa ...................... | 79 | 2,116 | 194.9 | 83.4 | 52 | 236 | 26.9 | 6.8 | 2,668 | 125.6 | 68,576 | 167 |
| Tuscaloosa...................... | 449 | 13,463 | 1,478.5 | 619.0 | 237 | 1,452 | 166.9 | 46.7 | 12,880 | 633.8 | 244,944 | 1,223 |
| Walker............................ | 172 | 3,441 | 316.5 | 129.9 | 88 | 401 | 52.2 | 13.7 | 3,673 | 147.1 | 8,740 | 53 |

# Table B. States and Counties — Government Employment and Payroll, and Local Government Finances

| | Government employment and payroll, 2017 | | | | | | | | | Local government finances, 2017 | | | | |
| | | | March payroll (percent of total) | | | | | | | General revenue | | | | |
| | | | | | | | | | | | | Taxes | | |
| STATE County | Full-time equivalent employees | March payroll (dollars) | Adminis-tration, judicial, and legal | Police and corrections | Fire protection | Highways and transpor-tation | Health and welfare | Natural resources and utilities | Education and libraries | Total (mil dol) | Inter-govern-mental (mil dol) | Total (mil dol) | Per capita[1] (dollars) | |
| | | | | | | | | | | | | | Total | Property |
| | 171 | 172 | 173 | 174 | 175 | 176 | 177 | 178 | 179 | 180 | 181 | 182 | 183 | 184 |
| UNITED STATES | X | X | X | X | X | X | X | X | X | X | X | X | X | X |
| ALABAMA | X | X | X | X | X | X | X | X | X | X | X | X | X | X |
| Autauga | 1,734 | 5,695,272 | 5.1 | 10.7 | 6.8 | 2.4 | 0.8 | 7.3 | 65.2 | 140.0 | 69.5 | 50.6 | 914 | 271 |
| Baldwin | 7,998 | 27,159,727 | 6.9 | 8.8 | 3.3 | 4.6 | 13.2 | 8.1 | 51.4 | 778.0 | 221.7 | 271.7 | 1,278 | 507 |
| Barbour | 1,177 | 3,772,026 | 3.8 | 7.2 | 4.6 | 2.1 | 28.5 | 6.5 | 45.5 | 94.1 | 54.3 | 22.6 | 899 | 300 |
| Bibb | 873 | 2,675,441 | 2.8 | 5.2 | 0.0 | 2.9 | 29.7 | 2.2 | 55.0 | 44.7 | 28.2 | 8.7 | 385 | 216 |
| Blount | 1,544 | 5,111,998 | 5.5 | 6.9 | 1.7 | 3.3 | 1.6 | 5.1 | 74.3 | 114.0 | 70.5 | 26.7 | 462 | 209 |
| Bullock | 405 | 1,232,720 | 4.9 | 8.0 | 0.0 | 3.6 | 22.8 | 2.2 | 57.1 | 26.6 | 17.6 | 5.7 | 561 | 430 |
| Butler | 856 | 3,182,150 | 4.1 | 8.4 | 2.2 | 4.8 | 12.8 | 5.4 | 60.4 | 66.9 | 30.4 | 18.5 | 933 | 370 |
| Calhoun | 5,653 | 18,814,816 | 3.4 | 5.0 | 2.7 | 2.5 | 37.3 | 5.5 | 41.1 | 531.6 | 161.2 | 130.2 | 1,135 | 389 |
| Chambers | 1,105 | 3,409,335 | 5.5 | 11.8 | 4.8 | 6.0 | 4.5 | 8.6 | 57.1 | 77.4 | 41.1 | 25.1 | 744 | 257 |
| Cherokee | 1,032 | 3,120,377 | 5.3 | 7.0 | 0.1 | 3.1 | 21.5 | 2.9 | 58.8 | 71.9 | 30.6 | 19.6 | 760 | 379 |
| Chilton | 1,252 | 3,945,983 | 7.9 | 9.4 | 1.1 | 4.5 | 0.8 | 7.0 | 68.9 | 109.4 | 60.9 | 31.9 | 724 | 360 |
| Choctaw | 331 | 963,611 | 15.7 | 6.6 | 0.0 | 5.9 | 0.0 | 1.2 | 70.6 | 25.2 | 14.0 | 8.9 | 689 | 415 |
| Clarke | 1,138 | 3,692,713 | 5.0 | 9.2 | 0.0 | 4.0 | 18.5 | 11.4 | 50.9 | 73.4 | 36.3 | 30.5 | 1,269 | 384 |
| Clay | 702 | 2,190,795 | 3.6 | 7.5 | 0.1 | 4.5 | 38.7 | 5.7 | 39.2 | 48.7 | 18.7 | 6.5 | 490 | 236 |
| Cleburne | 580 | 2,297,353 | 2.5 | 8.1 | 0.2 | 3.5 | 10.8 | 5.4 | 68.0 | 51.2 | 26.0 | 7.5 | 505 | 263 |
| Coffee | 1,961 | 6,438,467 | 4.6 | 7.3 | 2.2 | 2.8 | 12.1 | 6.5 | 62.2 | 160.9 | 85.9 | 47.8 | 922 | 318 |
| Colbert | 2,129 | 7,472,138 | 3.6 | 7.5 | 3.1 | 3.6 | 8.2 | 18.7 | 52.9 | 169.6 | 79.1 | 46.2 | 845 | 406 |
| Conecuh | 565 | 1,621,775 | 4.7 | 13.5 | 0.0 | 6.6 | 2.0 | 21.2 | 52.0 | 35.7 | 23.1 | 8.7 | 704 | 354 |
| Coosa | 240 | 686,957 | 8.8 | 10.9 | 1.1 | 6.2 | 0.8 | 1.4 | 69.3 | 18.3 | 10.3 | 5.0 | 463 | 301 |
| Covington | 1,698 | 5,041,571 | 7.2 | 6.8 | 1.5 | 5.8 | 2.0 | 21.4 | 53.5 | 110.5 | 57.0 | 33.2 | 897 | 315 |
| Crenshaw | 475 | 1,722,202 | 4.5 | 5.4 | 0.0 | 6.6 | 2.1 | 7.3 | 73.9 | 31.3 | 17.9 | 8.0 | 579 | 232 |
| Cullman | 3,867 | 12,941,467 | 3.5 | 6.2 | 1.7 | 2.8 | 30.4 | 9.8 | 44.2 | 239.7 | 113.1 | 69.8 | 842 | 337 |
| Dale | 1,658 | 5,099,569 | 3.7 | 7.4 | 3.8 | 2.8 | 25.6 | 5.1 | 50.6 | 133.8 | 57.8 | 32.4 | 656 | 286 |
| Dallas | 1,571 | 5,179,439 | 3.8 | 7.9 | 2.1 | 2.6 | 2.3 | 5.2 | 73.0 | 117.2 | 70.1 | 34.1 | 870 | 328 |
| DeKalb | 2,270 | 8,027,316 | 3.3 | 7.8 | 2.1 | 3.3 | 4.0 | 11.1 | 66.6 | 188.6 | 111.4 | 45.0 | 630 | 261 |
| Elmore | 2,007 | 6,738,762 | 3.3 | 12.1 | 2.2 | 3.7 | 0.3 | 6.1 | 69.3 | 168.8 | 93.7 | 48.1 | 590 | 256 |
| Escambia | 1,615 | 4,788,979 | 4.1 | 10.1 | 2.9 | 4.9 | 17.9 | 2.1 | 55.3 | 107.4 | 56.1 | 34.2 | 926 | 410 |
| Etowah | 3,617 | 13,009,490 | 5.6 | 11.4 | 6.0 | 3.8 | 7.4 | 7.0 | 56.2 | 298.2 | 139.3 | 104.1 | 1,011 | 317 |
| Fayette | 482 | 1,481,148 | 6.5 | 7.7 | 1.5 | 4.5 | 2.9 | 6.2 | 69.1 | 45.3 | 26.3 | 12.5 | 762 | 310 |
| Franklin | 1,303 | 4,088,311 | 4.9 | 6.2 | 1.7 | 3.2 | 2.4 | 14.6 | 65.9 | 95.2 | 59.5 | 17.6 | 557 | 267 |
| Geneva | 1,068 | 3,268,649 | 4.0 | 5.8 | 0.3 | 4.1 | 29.0 | 4.7 | 51.7 | 83.1 | 37.0 | 15.4 | 585 | 310 |
| Greene | 316 | 869,694 | 5.7 | 10.2 | 0.0 | 2.7 | 3.8 | 5.6 | 71.0 | 29.8 | 11.6 | 6.4 | 770 | 470 |
| Hale | 599 | 1,933,870 | 3.9 | 6.7 | 0.0 | 3.4 | 20.1 | 4.4 | 61.4 | 43.3 | 22.3 | 8.2 | 557 | 289 |
| Henry | 506 | 1,532,820 | 6.1 | 10.0 | 0.0 | 5.4 | 3.6 | 6.3 | 68.2 | 37.6 | 19.3 | 10.5 | 612 | 285 |
| Houston | 5,705 | 32,358,632 | 2.5 | 4.7 | 2.4 | 1.5 | 31.7 | 4.8 | 51.0 | 687.0 | 134.1 | 125.4 | 1,201 | 386 |
| Jackson | 2,502 | 9,081,304 | 3.2 | 5.8 | 2.1 | 3.2 | 26.3 | 8.2 | 50.3 | 221.9 | 77.8 | 53.4 | 1,030 | 326 |
| Jefferson | 26,397 | 106,388,611 | 7.8 | 14.5 | 7.7 | 4.3 | 4.5 | 10.8 | 47.5 | 3110.8 | 944.3 | 1568.3 | 2,378 | 968 |
| Lamar | 464 | 1,463,915 | 7.4 | 8.4 | 1.2 | 4.7 | 4.3 | 9.0 | 63.8 | 47.7 | 26.0 | 13.0 | 938 | 588 |
| Lauderdale | 2,917 | 10,134,952 | 3.5 | 8.9 | 5.4 | 4.2 | 1.3 | 18.4 | 55.8 | 237.9 | 103.7 | 96.6 | 1,044 | 456 |
| Lawrence | 1,030 | 3,487,299 | 4.1 | 7.1 | 0.5 | 2.3 | 12.8 | 4.9 | 66.7 | 107.4 | 44.4 | 42.4 | 1,282 | 372 |
| Lee | 7,104 | 27,907,786 | 2.7 | 6.0 | 2.4 | 1.6 | 44.2 | 5.5 | 36.9 | 792.6 | 189.0 | 215.4 | 1,334 | 592 |
| Limestone | 2,212 | 9,373,150 | 4.8 | 8.1 | 1.7 | 3.2 | 0.2 | 10.7 | 70.8 | 247.5 | 102.6 | 47.4 | 504 | 236 |
| Lowndes | 405 | 1,193,381 | 10.1 | 10.0 | 0.0 | 5.6 | 0.4 | 2.4 | 70.9 | 32.1 | 17.7 | 10.4 | 1,029 | 521 |
| Macon | 705 | 2,144,009 | 7.4 | 10.8 | 3.6 | 4.4 | 4.8 | 7.0 | 58.0 | 58.7 | 28.1 | 22.7 | 1,211 | 373 |
| Madison | 18,821 | 86,685,197 | 2.9 | 5.5 | 2.9 | 2.6 | 51.1 | 5.2 | 26.3 | 2610.2 | 1033.3 | 518.3 | 1,433 | 614 |
| Marengo | 1,543 | 6,195,458 | 2.8 | 3.9 | 2.0 | 2.7 | 50.3 | 2.6 | 34.7 | 86.1 | 37.8 | 19.9 | 1,025 | 430 |
| Marion | 956 | 3,067,580 | 4.4 | 8.5 | 1.8 | 3.0 | 4.8 | 12.8 | 64.4 | 71.9 | 39.7 | 21.6 | 726 | 197 |
| Marshall | 4,648 | 16,573,051 | 2.4 | 5.3 | 2.4 | 1.9 | 32.4 | 8.8 | 45.7 | 452.0 | 154.7 | 88.7 | 928 | 392 |
| Mobile | 15,404 | 52,418,321 | 6.1 | 13.0 | 4.7 | 3.7 | 11.6 | 6.9 | 51.2 | 1481.8 | 632.7 | 614.0 | 1,483 | 580 |
| Monroe | 1,076 | 3,495,389 | 4.0 | 7.4 | 0.7 | 2.9 | 35.3 | 2.3 | 46.8 | 78.2 | 34.3 | 15.6 | 733 | 330 |
| Montgomery | 7,810 | 28,640,171 | 5.1 | 14.6 | 6.9 | 3.1 | 3.4 | 11.5 | 47.9 | 1337.9 | 303.7 | 303.3 | 1,334 | 364 |
| Morgan | 3,972 | 14,335,679 | 4.1 | 9.7 | 4.0 | 2.8 | 1.7 | 12.3 | 62.7 | 385.0 | 163.8 | 137.5 | 1,156 | 555 |
| Perry | 448 | 1,621,598 | 1.6 | 4.3 | 0.1 | 5.0 | 24.2 | 2.4 | 61.7 | 35.5 | 25.7 | 6.7 | 720 | 385 |
| Pickens | 578 | 1,769,470 | 7.0 | 9.8 | 0.0 | 5.1 | 3.7 | 5.5 | 65.7 | 55.3 | 26.7 | 8.7 | 428 | 270 |
| Pike | 1,030 | 3,548,434 | 7.9 | 12.1 | 3.1 | 5.2 | 5.5 | 10.4 | 55.4 | 108.7 | 43.0 | 24.0 | 721 | 275 |
| Randolph | 755 | 2,605,786 | 4.4 | 7.7 | 0.5 | 4.9 | 10.3 | 2.6 | 69.2 | 49.3 | 30.3 | 13.4 | 590 | 298 |
| Russell | 2,161 | 6,843,660 | 4.4 | 11.1 | 4.0 | 3.9 | 2.1 | 7.4 | 63.7 | 167.8 | 86.5 | 60.5 | 1,062 | 432 |
| St. Clair | 2,104 | 7,818,999 | 6.7 | 10.8 | 4.3 | 4.3 | 0.5 | 4.0 | 68.8 | 193.1 | 100.3 | 69.3 | 788 | 296 |
| Shelby | 4,679 | 17,149,718 | 6.6 | 15.8 | 7.0 | 3.3 | 3.1 | 7.2 | 55.7 | 539.4 | 208.6 | 236.0 | 1,105 | 580 |
| Sumter | 966 | 3,406,938 | 2.7 | 3.5 | 0.9 | 2.4 | 62.4 | 3.2 | 24.3 | 34.8 | 17.9 | 11.6 | 911 | 387 |
| Talladega | 3,181 | 10,111,425 | 4.4 | 7.2 | 2.3 | 2.6 | 21.6 | 7.2 | 52.4 | 208.0 | 100.4 | 70.1 | 874 | 414 |
| Tallapoosa | 1,719 | 5,539,492 | 4.7 | 8.6 | 4.1 | 2.4 | 24.5 | 7.1 | 47.0 | 126.9 | 73.3 | 32.8 | 808 | 402 |
| Tuscaloosa | 10,171 | 39,065,639 | 3.6 | 7.7 | 3.8 | 3.6 | 44.9 | 4.8 | 30.1 | 1109.3 | 251.9 | 212.1 | 1,021 | 440 |
| Walker | 2,038 | 7,363,701 | 4.1 | 8.3 | 1.5 | 3.6 | 1.8 | 7.2 | 72.8 | 170.1 | 83.9 | 58.9 | 922 | 286 |

1. Based on the resident population estimated as of July 1 of the year shown.

# Table B. States and Counties — Local Government Finances, Government Employment, and Income Taxes

| STATE County | Direct general expenditure — Total (mil dol) [185] | Per capita[1] (dollars) [186] | Percent of total for: Education [187] | Health and hospitals [188] | Police protection [189] | Public welfare [190] | Highways [191] | Debt outstanding — Total (mil dol) [192] | Per capita[1] (dollars) [193] | Government employment, 2019 — Federal civilian [194] | Federal military [195] | State and local [196] | Individual income tax returns, 2018 — Number of returns [197] | Mean adjusted gross income [198] | Mean income tax [199] |
|---|---|---|---|---|---|---|---|---|---|---|---|---|---|---|---|
| UNITED STATES | X | X | X | X | X | X | X | X | X | 2,879,000 | 1,946,000 | 19,911,000 | 152,624,340 | 75,676 | 9,888 |
| ALABAMA | X | X | X | X | X | X | X | X | X | 54,134 | 29,000 | 325,987 | 2,062,010 | 60,311 | 6,349 |
| Autauga | 137 | 2,470 | 59.0 | 0.1 | 6.7 | 0.5 | 5.1 | 156.4 | 2,824 | 83 | 271 | 2,163 | 24,360 | 58,334 | 5,017 |
| Baldwin | 834 | 3,922 | 37.7 | 18.0 | 5.7 | 0.1 | 8.0 | 1,143.7 | 5,382 | 366 | 969 | 9,606 | 100,970 | 68,660 | 7,884 |
| Barbour | 96 | 3,816 | 43.9 | 21.4 | 6.6 | 0.0 | 1.7 | 41.8 | 1,662 | 55 | 104 | 1,609 | 9,050 | 43,874 | 3,649 |
| Bibb | 50 | 2,231 | 63.9 | 5.4 | 7.4 | 0.0 | 10.0 | 42.0 | 1,862 | 65 | 85 | 1,220 | 7,830 | 51,474 | 4,502 |
| Blount | 113 | 1,948 | 77.7 | 0.5 | 4.9 | 0.0 | 5.7 | 65.1 | 1,127 | 85 | 241 | 2,020 | 22,620 | 51,897 | 4,056 |
| Bullock | 26 | 2,556 | 59.1 | 10.7 | 2.3 | 1.0 | 11.8 | 9.8 | 963 | 40 | 35 | 604 | 3,800 | 33,908 | 2,171 |
| Butler | 73 | 3,659 | 45.1 | 13.0 | 6.1 | 0.2 | 10.7 | 76.0 | 3,819 | 44 | 80 | 877 | 7,930 | 42,552 | 2,976 |
| Calhoun | 618 | 5,387 | 32.3 | 36.7 | 5.3 | 0.0 | 3.4 | 461.6 | 4,024 | 4,174 | 515 | 8,593 | 47,160 | 48,504 | 3,924 |
| Chambers | 74 | 2,202 | 60.5 | 1.0 | 6.3 | 0.3 | 3.6 | 58.9 | 1,746 | 54 | 138 | 1,828 | 13,900 | 41,147 | 2,744 |
| Cherokee | 76 | 2,940 | 55.1 | 0.0 | 5.1 | 18.8 | 5.9 | 57.6 | 2,234 | 48 | 109 | 1,282 | 9,700 | 47,822 | 3,756 |
| Chilton | 107 | 2,428 | 64.9 | 0.2 | 7.0 | 0.4 | 5.8 | 38.8 | 880 | 66 | 185 | 1,809 | 17,230 | 50,136 | 4,053 |
| Choctaw | 27 | 2,070 | 58.4 | 0.6 | 8.1 | 0.0 | 15.0 | 18.9 | 1,464 | 27 | 52 | 446 | 4,890 | 48,457 | 3,733 |
| Clarke | 78 | 3,229 | 61.9 | 0.3 | 7.4 | 0.2 | 5.5 | 116.6 | 4,847 | 73 | 98 | 1,721 | 9,720 | 49,307 | 4,052 |
| Clay | 49 | 3,682 | 36.1 | 38.7 | 3.4 | 0.1 | 3.0 | 18.9 | 1,418 | 51 | 55 | 916 | 5,630 | 46,696 | 3,710 |
| Cleburne | 53 | 3,575 | 44.2 | 0.1 | 3.7 | 12.7 | 9.5 | 25.5 | 1,709 | 50 | 62 | 729 | 5,720 | 50,423 | 3,772 |
| Coffee | 162 | 3,114 | 60.4 | 0.1 | 5.2 | 0.4 | 4.4 | 138.4 | 2,669 | 307 | 218 | 2,415 | 21,670 | 56,420 | 5,031 |
| Colbert | 180 | 3,285 | 51.5 | 7.7 | 5.0 | 0.2 | 5.6 | 186.0 | 3,400 | 598 | 230 | 4,109 | 24,000 | 53,404 | 5,081 |
| Conecuh | 43 | 3,442 | 55.2 | 0.3 | 6.0 | 0.0 | 8.4 | 62.0 | 4,985 | 26 | 50 | 723 | 4,630 | 39,851 | 2,825 |
| Coosa | 19 | 1,734 | 54.8 | 0.2 | 7.2 | 0.2 | 14.4 | 7.8 | 728 | 18 | 44 | 322 | 3,940 | 45,480 | 3,487 |
| Covington | 125 | 3,363 | 49.7 | 0.6 | 6.1 | 0.0 | 12.8 | 193.3 | 5,217 | 124 | 153 | 1,983 | 14,640 | 50,133 | 4,404 |
| Crenshaw | 36 | 2,622 | 67.1 | 6.3 | 8.8 | 0.0 | 3.0 | 24.5 | 1,769 | 42 | 57 | 589 | 5,690 | 42,995 | 3,143 |
| Cullman | 262 | 3,163 | 51.1 | 2.8 | 8.5 | 0.9 | 5.9 | 332.1 | 4,008 | 213 | 347 | 4,310 | 34,200 | 53,649 | 4,870 |
| Dale | 139 | 2,814 | 47.7 | 23.3 | 5.2 | 0.2 | 3.5 | 62.8 | 1,272 | 2,917 | 3,911 | 2,124 | 19,880 | 47,712 | 3,839 |
| Dallas | 121 | 3,074 | 58.0 | 0.2 | 6.3 | 0.1 | 5.9 | 70.0 | 1,785 | 152 | 153 | 2,360 | 14,970 | 39,248 | 2,890 |
| DeKalb | 190 | 2,664 | 61.8 | 4.3 | 7.1 | 0.3 | 7.2 | 110.2 | 1,542 | 169 | 297 | 3,354 | 27,130 | 44,879 | 3,517 |
| Elmore | 179 | 2,192 | 67.8 | 0.4 | 8.0 | 0.0 | 6.1 | 152.5 | 1,873 | 171 | 324 | 4,362 | 35,750 | 58,999 | 5,294 |
| Escambia | 109 | 2,955 | 57.2 | 4.1 | 8.9 | 0.6 | 6.5 | 95.6 | 2,584 | 59 | 143 | 3,997 | 13,960 | 45,823 | 3,639 |
| Etowah | 303 | 2,939 | 49.8 | 2.2 | 10.0 | 0.2 | 4.2 | 235.4 | 2,285 | 308 | 422 | 5,134 | 42,110 | 49,188 | 4,227 |
| Fayette | 44 | 2,685 | 55.3 | 1.7 | 5.0 | 0.0 | 12.1 | 29.4 | 1,786 | 36 | 69 | 1,224 | 6,260 | 45,084 | 3,065 |
| Franklin | 104 | 3,287 | 60.6 | 0.1 | 6.8 | 0.0 | 7.7 | 82.0 | 2,599 | 87 | 131 | 1,875 | 11,850 | 44,311 | 3,358 |
| Geneva | 85 | 3,234 | 47.4 | 26.9 | 4.0 | 0.2 | 5.9 | 23.3 | 885 | 57 | 109 | 1,383 | 10,180 | 42,800 | 3,161 |
| Greene | 32 | 3,850 | 43.5 | 29.6 | 2.7 | 0.6 | 4.4 | 24.8 | 2,980 | 27 | 34 | 543 | 3,240 | 32,716 | 1,927 |
| Hale | 46 | 3,078 | 55.3 | 19.3 | 5.9 | 0.0 | 5.6 | 21.3 | 1,437 | 50 | 61 | 838 | 6,250 | 41,111 | 2,882 |
| Henry | 38 | 2,225 | 59.9 | 0.0 | 8.5 | 0.4 | 8.4 | 44.1 | 2,578 | 43 | 71 | 676 | 7,150 | 49,775 | 4,491 |
| Houston | 740 | 7,094 | 20.2 | 53.1 | 3.3 | 0.1 | 3.1 | 511.0 | 4,896 | 302 | 486 | 8,457 | 45,090 | 57,063 | 6,338 |
| Jackson | 221 | 4,269 | 36.9 | 30.6 | 5.3 | 0.3 | 3.9 | 138.2 | 2,666 | 139 | 214 | 2,981 | 20,830 | 49,292 | 4,075 |
| Jefferson | 2,939 | 4,556 | 38.8 | 3.1 | 8.0 | 0.0 | 5.0 | 7,200.4 | 10,916 | 8,194 | 3,075 | 54,109 | 290,920 | 74,837 | 9,816 |
| Lamar | 44 | 3,197 | 53.6 | 0.1 | 5.5 | 0.3 | 5.6 | 35.5 | 2,556 | 37 | 57 | 554 | 5,290 | 47,119 | 3,445 |
| Lauderdale | 231 | 2,498 | 57.3 | 0.6 | 6.4 | 0.1 | 7.4 | 225.3 | 2,434 | 283 | 386 | 5,013 | 38,430 | 56,408 | 5,440 |
| Lawrence | 128 | 3,864 | 40.1 | 7.1 | 8.4 | 0.0 | 1.1 | 189.5 | 5,732 | 85 | 137 | 1,270 | 13,410 | 47,975 | 3,562 |
| Lee | 790 | 4,895 | 32.9 | 41.4 | 3.9 | 0.0 | 2.8 | 800.7 | 4,960 | 313 | 732 | 17,985 | 64,470 | 63,008 | 6,527 |
| Limestone | 311 | 3,306 | 55.3 | 25.6 | 5.2 | 0.3 | 2.1 | 507.5 | 5,391 | 1,483 | 404 | 5,593 | 41,930 | 66,903 | 6,698 |
| Lowndes | 34 | 3,350 | 59.7 | 0.0 | 6.8 | 0.0 | 12.7 | 30.0 | 2,968 | 32 | 40 | 497 | 4,630 | 39,227 | 2,602 |
| Macon | 56 | 2,967 | 50.6 | 1.6 | 5.9 | 0.2 | 12.3 | 72.0 | 3,834 | 907 | 76 | 1,871 | 7,320 | 36,332 | 2,415 |
| Madison | 2,623 | 7,251 | 22.4 | 52.5 | 2.8 | 0.0 | 2.6 | 2,849.0 | 7,875 | 18,244 | 2,248 | 26,544 | 173,430 | 74,997 | 8,891 |
| Marengo | 112 | 5,781 | 34.1 | 27.3 | 4.3 | 0.1 | 6.3 | 44.7 | 2,302 | 67 | 96 | 1,517 | 10,840 | 44,493 | 3,243 |
| Marion | 63 | 2,098 | 74.6 | 0.0 | 3.0 | 0.0 | 1.6 | 71.7 | 2,408 | 80 | 122 | 1,719 | 10,840 | 50,854 | 4,618 |
| Marshall | 446 | 4,665 | 40.7 | 35.3 | 3.5 | 0.1 | 3.6 | 275.9 | 2,887 | 269 | 402 | 6,066 | 39,240 | 50,854 | 4,618 |
| Mobile | 1,516 | 3,662 | 40.6 | 9.9 | 7.4 | 0.2 | 5.0 | 1,381.7 | 3,337 | 2,686 | 2,635 | 22,538 | 172,290 | 54,402 | 5,541 |
| Monroe | 75 | 3,533 | 46.0 | 28.6 | 2.0 | 0.2 | 6.7 | 44.7 | 2,100 | 53 | 86 | 1,377 | 7,830 | 45,580 | 3,526 |
| Montgomery | 1,303 | 5,734 | 23.5 | 50.1 | 4.4 | 0.1 | 2.0 | 952.0 | 4,189 | 6,261 | 3,735 | 26,218 | 96,560 | 56,801 | 6,035 |
| Morgan | 430 | 3,618 | 57.7 | 0.3 | 6.2 | 0.5 | 3.3 | 573.0 | 4,819 | 288 | 497 | 6,735 | 52,400 | 57,836 | 5,518 |
| Perry | 35 | 3,772 | 45.6 | 26.8 | 5.2 | 0.4 | 6.7 | 29.2 | 3,142 | 21 | 65 | 447 | 3,490 | 33,186 | 1,863 |
| Pickens | 62 | 3,068 | 49.5 | 22.0 | 6.8 | 0.1 | 5.2 | 31.3 | 1,551 | 312 | 76 | 856 | 7,210 | 43,666 | 3,204 |
| Pike | 149 | 4,469 | 30.6 | 19.4 | 7.0 | 0.0 | 10.4 | 146.5 | 4,393 | 94 | 137 | 3,288 | 12,530 | 48,823 | 4,814 |
| Randolph | 53 | 2,338 | 68.2 | 1.7 | 6.7 | 0.0 | 2.8 | 37.2 | 1,639 | 57 | 94 | 1,040 | 8,970 | 45,635 | 3,490 |
| Russell | 170 | 2,991 | 59.0 | 0.1 | 6.9 | 0.0 | 1.6 | 259.7 | 4,558 | 105 | 241 | 2,838 | 23,080 | 41,012 | 2,859 |
| St. Clair | 201 | 2,287 | 60.1 | 0.9 | 5.1 | 0.1 | 4.0 | 279.0 | 3,173 | 117 | 369 | 2,923 | 37,330 | 58,284 | 5,210 |
| Shelby | 614 | 2,873 | 62.2 | 1.5 | 6.4 | 0.1 | 3.4 | 917.6 | 4,295 | 404 | 904 | 8,604 | 96,620 | 86,714 | 10,879 |
| Sumter | 38 | 2,942 | 49.2 | 0.6 | 8.4 | 0.0 | 13.5 | 42.7 | 3,350 | 35 | 48 | 1,219 | 4,730 | 35,588 | 2,380 |
| Talladega | 215 | 2,681 | 59.5 | 0.5 | 7.9 | 0.3 | 5.4 | 265.2 | 3,309 | 439 | 323 | 4,435 | 32,640 | 46,230 | 3,550 |
| Tallapoosa | 129 | 3,172 | 47.1 | 19.9 | 5.2 | 0.0 | 3.6 | 76.8 | 1,891 | 97 | 167 | 2,011 | 16,990 | 52,808 | 5,155 |
| Tuscaloosa | 1,130 | 5,443 | 29.9 | 43.2 | 4.7 | 0.0 | 5.0 | 1,068.7 | 5,148 | 1,677 | 852 | 23,888 | 83,000 | 60,642 | 6,478 |
| Walker | 173 | 2,715 | 65.4 | 0.2 | 5.3 | 0.0 | 7.2 | 107.3 | 1,679 | 200 | 264 | 2,996 | 25,070 | 61,588 | 5,086 |

1. Based on the resident population estimated as of July 1 of the year shown.

# Table B. States and Counties — Land Area and Population

| State / county code | CBSA code[1] | County Type code[2] | STATE County | Population, 2020 | | | | Population and population characteristics, 2020 | | | | | | | | | | |
|---|---|---|---|---|---|---|---|---|---|---|---|---|---|---|---|---|---|---|
| | | | | | | | | Race alone or in combination, not Hispanic or Latino (percent) | | | | | Age (percent) | | | | | |
| | | | | Land area[3] (sq. mi) | Total persons 2019 | Rank | Per square mile | White | Black | American Indian, Alaska Native | Asian and Pacific Islancer | Percent Hispanic or Latino[4] | Under 5 years | 5 to 14 years | 15 to 24 years | 25 to 34 years | 35 to 44 years | 45 to 54 years |
| | | | | 1 | 2 | 3 | 4 | 5 | 6 | 7 | 8 | 9 | 10 | 11 | 12 | 13 | 14 | 15 |
| | | | ALABAMA—Cont'd | | | | | | | | | | | | | | | |
| 01129 | 33660 | 8 | Washington | 1,080.2 | 15,976 | 2,035 | 14.8 | 66.7 | 23.8 | 8.2 | 1.2 | 1.6 | 5.7 | 11.5 | 11.9 | 11.4 | 10.9 | 13.1 |
| 01131 | | 9 | Wilcox | 887.9 | 10,206 | 2,404 | 11.5 | 28.0 | 70.5 | 0.8 | 0.5 | 1.6 | 6.3 | 12.2 | 13.3 | 11.7 | 10.7 | 11.4 |
| 01133 | | 6 | Winston | 613.0 | 23,508 | 1,657 | 38.3 | 94.8 | 1.4 | 1.4 | 0.6 | 3.3 | 5.3 | 11.0 | 11.1 | 10.7 | 10.8 | 13.2 |
| 02000 | | 0 | ALASKA | 571,016.9 | 731,158 | X | 1.3 | 65.9 | 4.7 | 19.2 | 10.6 | 7.3 | 6.8 | 13.8 | 13.0 | 15.9 | 13.4 | 11.4 |
| 02013 | | 9 | Aleutians East | 6,985.2 | 3,401 | 2,940 | 0.5 | 12.4 | 9.6 | 20.7 | 46.1 | 14.9 | 1.7 | 4.1 | 9.9 | 15.8 | 19.2 | 23.0 |
| 02016 | | 9 | Aleutians West | 4,393.0 | 5,680 | 2,773 | 1.3 | 25.8 | 7.3 | 13.0 | 44.3 | 13.7 | 3.3 | 6.6 | 9.7 | 18.1 | 18.6 | 20.2 |
| 02020 | 11260 | 2 | Anchorage | 1,706.8 | 287,095 | 245 | 168.2 | 63.1 | 7.5 | 12.2 | 16.3 | 9.5 | 6.8 | 13.2 | 13.3 | 17.2 | 13.8 | 11.5 |
| 02050 | | 7 | Bethel | 40,627.1 | 18,437 | 1,902 | 0.5 | 12.1 | 1.4 | 85.9 | 2.1 | 2.1 | 10.9 | 19.3 | 15.7 | 15.4 | 11.0 | 9.3 |
| 02060 | | 9 | Bristol Bay | 482.0 | 788 | 3,118 | 1.6 | 56.6 | 3.9 | 41.8 | 7.7 | 8.4 | 5.7 | 11.9 | 8.3 | 13.6 | 11.8 | 11.2 |
| 02063 | | 9 | Chugach | 9,529.8 | 6,427 | 2,712 | 0.7 | 74.3 | 1.4 | 16.4 | 8.9 | 5.8 | 6.1 | 13.0 | 10.5 | 13.8 | 13.2 | 12.2 |
| 02066 | | 9 | Copper River | 24,692.1 | 2,919 | 2,969 | 0.1 | 73.9 | 1.2 | 22.7 | 3.9 | 5.2 | 6.1 | 15.1 | 11.8 | 11.8 | 12.1 | 10.3 |
| 02068 | | 8 | Denali | 12,641.0 | 2,081 | 3,043 | 0.2 | 81.8 | 3.7 | 7.8 | 6.6 | 4.0 | 4.0 | 10.4 | 9.0 | 17.6 | 15.5 | 15.4 |
| 02070 | | 9 | Dillingham | 18,334.0 | 4,833 | 2,838 | 0.3 | 22.3 | 1.4 | 79.6 | 2.8 | 3.1 | 9.1 | 17.9 | 13.7 | 15.9 | 11.0 | 9.1 |
| 02090 | 21820 | 3 | Fairbanks North Star | 7,334.8 | 95,651 | 627 | 13.0 | 75.1 | 6.3 | 11.3 | 6.2 | 8.1 | 7.1 | 13.1 | 15.9 | 18.3 | 13.2 | 9.8 |
| 02100 | | 9 | Haines | 2,343.4 | 2,614 | 2,992 | 1.1 | 83.8 | 1.4 | 15.9 | 2.1 | 3.6 | 4.3 | 9.4 | 10.0 | 10.1 | 12.1 | 12.0 |
| 02105 | | 9 | Hoonah-Angoon | 6,555.3 | 2,141 | 3,033 | 0.3 | 55.1 | 3.5 | 42.8 | 2.5 | 6.0 | 4.7 | 10.6 | 8.4 | 10.2 | 10.1 | 12.4 |
| 02110 | 27940 | 5 | Juneau | 2,704.0 | 31,849 | 1,385 | 11.8 | 71.2 | 2.2 | 17.7 | 12.1 | 7.0 | 5.3 | 11.9 | 11.7 | 15.2 | 14.2 | 12.9 |
| 02122 | | 7 | Kenai Peninsula | 16,017.4 | 59,414 | 876 | 3.7 | 84.6 | 1.3 | 11.9 | 3.6 | 4.4 | 5.9 | 12.6 | 10.9 | 12.6 | 12.4 | 11.6 |
| 02130 | 28540 | 7 | Ketchikan Gateway | 4,856.9 | 13,747 | 2,175 | 2.8 | 71.1 | 1.6 | 20.0 | 10.9 | 5.6 | 5.2 | 11.9 | 11.8 | 13.7 | 13.6 | 11.8 |
| 02150 | | 7 | Kodiak Island | 6,688.9 | 12,992 | 2,220 | 1.9 | 54.0 | 1.6 | 16.5 | 26.5 | 9.0 | 6.5 | 14.3 | 12.2 | 16.3 | 15.2 | 10.3 |
| 02158 | | 0 | Kusilvak | 17,077.1 | 8,328 | 2,554 | 0.5 | 5.2 | 0.7 | 94.2 | 1.2 | 1.6 | 13.1 | 22.8 | 16.2 | 15.4 | 9.0 | 8.0 |
| 02164 | | 9 | Lake and Peninsula | 23,831.8 | 1,493 | 3,079 | 0.1 | 32.1 | 2.0 | 67.9 | 6.2 | 2.9 | 9.0 | 14.9 | 13.8 | 13.3 | 12.7 | 11.9 |
| 02170 | 11260 | 2 | Matanuska-Susitna | 24,707.0 | 110,213 | 553 | 4.5 | 84.5 | 2.3 | 11.2 | 4.2 | 5.5 | 6.6 | 15.1 | 12.3 | 14.1 | 13.9 | 11.9 |
| 02180 | | 7 | Nome | 22,969.5 | 9,909 | 2,427 | 0.4 | 20.0 | 1.1 | 81.0 | 2.8 | 2.0 | 9.2 | 20.0 | 15.2 | 15.4 | 12.0 | 9.4 |
| 02185 | | 7 | North Slope | 88,823.6 | 9,294 | 2,474 | 0.1 | 32.8 | 2.6 | 57.0 | 9.6 | 4.2 | 7.8 | 16.0 | 12.1 | 16.0 | 13.5 | 12.9 |
| 02188 | | 7 | Northwest Arctic | 35,663.3 | 7,644 | 2,615 | 0.2 | 15.5 | 1.8 | 83.8 | 2.4 | 2.6 | 10.4 | 20.9 | 14.2 | 15.9 | 10.6 | 9.9 |
| 02195 | | 9 | Petersburg | 2,900.7 | 3,296 | 2,948 | 1.1 | 75.0 | 3.1 | 15.7 | 8.4 | 5.7 | 5.3 | 13.0 | 10.4 | 12.6 | 12.0 | 11.8 |
| 02198 | | 9 | Prince of Wales-Hyder | 5,267.7 | 6,147 | 2,734 | 1.2 | 51.9 | 1.6 | 47.3 | 4.4 | 4.0 | 5.2 | 13.3 | 10.5 | 11.4 | 11.6 | 12.3 |
| 02220 | | 7 | Sitka | 2,870.1 | 8,405 | 2,542 | 2.9 | 69.1 | 1.5 | 20.6 | 10.8 | 7.3 | 5.1 | 11.8 | 10.8 | 14.5 | 14.2 | 11.8 |
| 02230 | | 9 | Skagway | 433.9 | 1,179 | 3,102 | 2.7 | 83.8 | 2.2 | 7.4 | 5.2 | 6.4 | 4.4 | 9.4 | 5.4 | 17.4 | 22.2 | 13.7 |
| 02240 | | 9 | Southeast Fairbanks | 24,831.1 | 6,957 | 2,663 | 0.3 | 76.1 | 2.5 | 15.0 | 4.0 | 7.4 | 7.2 | 14.7 | 10.4 | 13.7 | 12.8 | 11.9 |
| 02261 | | 9 | Valdez-Cordova | NA | NA | 0 | NA | 74.3 | 1.4 | 18.1 | 6.7 | 5.6 | 6.7 | 13.0 | 10.7 | 13.7 | 12.8 | 12.0 |
| 02275 | | 9 | Wrangell | 2,556.0 | 2,510 | 3,002 | 1.0 | 75.3 | 2.4 | 24.6 | 5.3 | 3.4 | 4.4 | 11.4 | 9.4 | 11.1 | 9.7 | 10.2 |
| 02282 | | 9 | Yakutat | 7,623.3 | 637 | 3,134 | 0.1 | 45.8 | 5.7 | 48.4 | 11.1 | 4.7 | 5.3 | 12.1 | 9.4 | 14.1 | 10.7 | 13.7 |
| 02290 | | 9 | Yukon-Koyukuk | 145,575.5 | 5,077 | 2,821 | 0.0 | 27.1 | 1.6 | 73.0 | 1.7 | 2.8 | 7.7 | 16.6 | 11.9 | 12.2 | 11.9 | 9.6 |
| 04000 | | 0 | ARIZONA | 113,653.1 | 7,421,401 | X | 65.3 | 55.8 | 5.5 | 4.4 | 4.7 | 31.9 | 5.7 | 12.6 | 13.3 | 13.9 | 12.3 | 11.6 |
| 04001 | | 6 | Apache | 11,198.3 | 71,875 | 757 | 6.4 | 19.2 | 1.1 | 73.6 | 0.8 | 6.8 | 6.4 | 15.3 | 13.6 | 13.3 | 11.1 | 10.8 |
| 04003 | 43420 | 3 | Cochise | 6,209.8 | 127,450 | 505 | 20.5 | 57.1 | 4.9 | 1.6 | 3.7 | 35.5 | 5.5 | 12.0 | 11.9 | 12.5 | 11.3 | 10.1 |
| 04005 | 22380 | 3 | Coconino | 18,616.4 | 142,481 | 462 | 7.7 | 56.2 | 2.0 | 26.6 | 3.2 | 14.6 | 5.1 | 11.4 | 23.3 | 14.2 | 11.2 | 9.8 |
| 04007 | 37740 | 4 | Gila | 4,757.3 | 54,303 | 942 | 11.4 | 62.9 | 0.9 | 17.2 | 1.2 | 19.2 | 4.9 | 11.4 | 9.3 | 10.0 | 9.0 | 9.7 |
| 04009 | 40940 | 7 | Graham | 4,621.9 | 39,211 | 1,198 | 8.5 | 52.2 | 2.0 | 12.4 | 1.3 | 33.5 | 7.0 | 15.2 | 14.6 | 14.9 | 13.3 | 10.7 |
| 04011 | | 7 | Greenlee | 1,842.1 | 9,341 | 2,469 | 5.1 | 46.3 | 2.2 | 3.6 | 1.2 | 48.3 | 7.5 | 15.3 | 12.7 | 14.9 | 14.1 | 11.0 |
| 04012 | | 6 | La Paz | 4,496.6 | 21,480 | 1,747 | 4.8 | 58.8 | 1.3 | 12.7 | 1.2 | 28.0 | 4.2 | 8.8 | 8.0 | 8.6 | 8.3 | 8.2 |
| 04013 | 38060 | 1 | Maricopa | 9,201.8 | 4,579,081 | 4 | 497.6 | 56.2 | 6.8 | 2.2 | 5.8 | 31.5 | 6.0 | 13.1 | 13.2 | 14.7 | 13.0 | 12.3 |
| 04015 | 29420 | 3 | Mohave | 13,332.1 | 217,206 | 315 | 16.3 | 78.1 | 1.6 | 2.9 | 2.2 | 17.2 | 4.2 | 9.3 | 8.8 | 10.3 | 9.1 | 10.2 |
| 04017 | 43320 | 4 | Navajo | 9,949.5 | 112,112 | 548 | 11.3 | 43.4 | 1.4 | 44.0 | 1.1 | 11.9 | 6.4 | 14.8 | 12.5 | 12.3 | 10.8 | 10.4 |
| 04019 | 46060 | 2 | Pima | 9,188.7 | 1,061,175 | 44 | 115.5 | 52.7 | 4.3 | 3.0 | 4.2 | 38.0 | 5.3 | 11.5 | 15.0 | 13.1 | 11.5 | 10.6 |
| 04021 | 38060 | 1 | Pinal | 5,366.0 | 480,828 | 147 | 89.6 | 57.7 | 5.8 | 5.0 | 2.9 | 31.0 | 5.4 | 12.7 | 11.5 | 13.1 | 12.8 | 11.1 |
| 04023 | 35700 | 4 | Santa Cruz | 1,236.2 | 46,808 | 1,039 | 37.9 | 15.7 | 0.6 | 0.5 | 0.7 | 82.8 | 6.9 | 14.9 | 13.6 | 11.6 | 10.8 | 11.2 |
| 04025 | 39150 | 3 | Yavapai | 8,122.9 | 240,226 | 286 | 29.6 | 81.5 | 1.2 | 2.3 | 1.9 | 15.0 | 3.9 | 8.9 | 9.1 | 9.1 | 8.7 | 10.0 |
| 04027 | 49740 | 3 | Yuma | 5,513.8 | 217,824 | 314 | 39.5 | 31.0 | 2.3 | 1.4 | 1.9 | 64.7 | 7.0 | 13.7 | 14.9 | 14.0 | 11.0 | 9.7 |
| 05000 | | 0 | ARKANSAS | 52,037.5 | 3,030,522 | X | 58.2 | 73.6 | 16.3 | 1.7 | 2.5 | 8.0 | 6.1 | 13.0 | 13.1 | 13.2 | 12.3 | 11.8 |
| 05001 | | 6 | Arkansas | 992.1 | 17,383 | 1,952 | 17.5 | 70.2 | 26.2 | 0.9 | 0.9 | 3.8 | 6.4 | 13.0 | 10.4 | 11.9 | 11.7 | 12.0 |
| 05003 | | 7 | Ashley | 925.5 | 19,339 | 1,860 | 20.9 | 69.2 | 24.7 | 0.8 | 0.4 | 6.0 | 5.7 | 12.9 | 11.3 | 11.0 | 10.8 | 12.6 |
| 05005 | 34260 | 7 | Baxter | 553.9 | 42,242 | 1,134 | 76.3 | 95.6 | 0.7 | 1.7 | 0.8 | 2.6 | 4.4 | 10.0 | 9.1 | 9.4 | 9.6 | 10.7 |
| 05007 | 22220 | 2 | Benton | 847.7 | 288,774 | 244 | 340.7 | 74.2 | 2.4 | 2.7 | 6.0 | 17.3 | 6.9 | 14.6 | 12.4 | 15.0 | 14.5 | 12.1 |
| 05009 | 25460 | 7 | Boone | 589.7 | 37,625 | 1,234 | 63.8 | 95.3 | 0.8 | 2.1 | 1.0 | 2.7 | 5.6 | 12.8 | 11.2 | 11.5 | 11.5 | 12.1 |
| 05011 | | 7 | Bradley | 649.2 | 10,639 | 2,374 | 16.4 | 56.1 | 28.3 | 0.9 | 0.5 | 15.8 | 6.4 | 14.1 | 11.2 | 11.9 | 11.6 | 12.2 |
| 05013 | 15780 | 9 | Calhoun | 628.6 | 5,113 | 2,818 | 8.1 | 73.9 | 21.9 | 1.1 | 0.7 | 4.4 | 3.9 | 11.2 | 10.5 | 12.1 | 10.9 | 13.1 |
| 05015 | | 6 | Carroll | 629.9 | 28,276 | 1,487 | 44.9 | 79.6 | 1.0 | 2.4 | 3.5 | 15.5 | 5.7 | 12.2 | 11.0 | 10.5 | 10.7 | 11.3 |
| 05017 | | 7 | Chicot | 637.0 | 9,924 | 2,426 | 15.6 | 39.7 | 53.7 | 0.6 | 0.9 | 6.1 | 5.6 | 12.5 | 11.9 | 10.4 | 10.2 | 12.0 |
| 05019 | 11660 | 7 | Clark | 866.0 | 22,103 | 1,713 | 25.5 | 69.5 | 24.9 | 0.9 | 1.3 | 4.9 | 5.1 | 10.7 | 25.1 | 11.0 | 9.5 | 9.9 |
| 05021 | | 7 | Clay | 639.2 | 14,375 | 2,132 | 22.5 | 95.9 | 1.5 | 1.2 | 0.5 | 2.4 | 5.6 | 11.7 | 11.2 | 12.1 | 10.9 | 12.5 |
| 05023 | | 6 | Cleburne | 554.0 | 24,935 | 1,605 | 45.0 | 95.9 | 0.9 | 1.7 | 0.6 | 2.5 | 4.4 | 10.7 | 9.6 | 9.7 | 10.4 | 11.5 |

1. CBSA = Core Based Statistical Area. See Appendix A for explanation. See Appendix B for list of metropolitan areas with component counties.    2. County type code from the Economic Research Service of USDA Rural-Urban Continuum Codes. See Appendix A for definition.    3. Dry land or land partially or temporarily covered by water.    4. May be of any race.

| STATE County | 55 to 64 years | 65 to 74 years | 75 years and over | Percent female | 2000 | 2010 | 2000-2010 | 2010-2020 | Births | Deaths | Net Migration | Number | Persons per household | Family households | Female family householder[1] | One person |
|---|---|---|---|---|---|---|---|---|---|---|---|---|---|---|---|---|
| | 16 | 17 | 18 | 19 | 20 | 21 | 22 | 23 | 24 | 25 | 26 | 27 | 28 | 29 | 30 | 31 |
| **ALABAMA—Cont'd** | | | | | | | | | | | | | | | | |
| Washington | 14.9 | 12.0 | 8.5 | 51.2 | 18,097 | 17,580 | -2.9 | -9.1 | 1,755 | 2,096 | -1,266 | 5,990 | 2.74 | 70.3 | 12.8 | 26.8 |
| Wilcox | 13.4 | 12.3 | 8.8 | 53.0 | 13,183 | 11,664 | -11.5 | -12.5 | 1,378 | 1,537 | -1,317 | 3,854 | 2.71 | 59.3 | 22.1 | 35.2 |
| Winston | 15.3 | 12.7 | 9.9 | 50.9 | 24,843 | 24,488 | -1.4 | -4.0 | 2,488 | 3,235 | -216 | 9,592 | 2.44 | 65.3 | 10.8 | 30.7 |
| **ALASKA** | 12.5 | 8.8 | 4.3 | 47.8 | 626,932 | 710,246 | 13.3 | 2.9 | 111,891 | 44,413 | -47,260 | 253,346 | 2.80 | 65.7 | 10.6 | 26.1 |
| Aleutians East | 15.3 | 6.5 | 4.5 | 32.6 | 2,697 | 3,141 | 16.5 | 8.3 | 129 | 74 | 197 | 890 | 2.61 | 61.8 | 12.2 | 27.5 |
| Aleutians West | 15.8 | 6.5 | 1.3 | 33.7 | 5,465 | 5,561 | 1.8 | 2.1 | 340 | 126 | -105 | 1,187 | 3.74 | 61.7 | 9.0 | 28.9 |
| Anchorage | 11.9 | 8.0 | 4.3 | 49.1 | 260,283 | 292,252 | 12.3 | -1.8 | 45,399 | 17,327 | -33,456 | 106,567 | 2.69 | 64.4 | 11.3 | 26.3 |
| Bethel | 10.1 | 5.7 | 2.5 | 48.0 | 16,006 | 17,013 | 6.3 | 8.4 | 4,404 | 1,245 | -1,744 | 4,489 | 3.83 | 76.1 | 19.7 | 17.5 |
| Bristol Bay | 17.6 | 12.6 | 6.7 | 48.9 | 1,258 | 997 | -20.7 | -21.0 | 106 | 78 | -241 | 314 | 2.51 | 61.8 | 9.2 | 30.9 |
| Chugach | 16.9 | 10.5 | 3.8 | 47.1 | NA | 6,684 | NA | -3.8 | 855 | 316 | -811 | NA | NA | NA | NA | NA |
| Copper River | 14.4 | 12.6 | 6.4 | 46.8 | NA | 2,952 | NA | -1.1 | 370 | 90 | -320 | NA | NA | NA | NA | NA |
| Denali | 15.4 | 9.1 | 3.6 | 45.4 | 1,893 | 1,827 | -3.5 | 13.9 | 191 | 76 | 136 | 613 | 2.32 | 52.4 | 4.7 | 38.8 |
| Dillingham | 12.5 | 7.3 | 3.6 | 48.6 | 4,922 | 4,847 | -1.5 | -0.3 | 993 | 392 | -618 | 1,427 | 3.20 | 73.0 | 20.0 | 22.1 |
| Fairbanks North Star | 10.7 | 8.1 | 3.7 | 46.1 | 82,840 | 97,585 | 17.8 | -2.0 | 16,769 | 4,930 | -13,939 | 36,188 | 2.62 | 65.8 | 8.5 | 25.2 |
| Haines | 18.1 | 16.5 | 7.4 | 49.8 | 2,392 | 2,508 | 4.8 | 4.2 | 207 | 168 | 68 | 1,007 | 2.46 | 61.6 | 5.0 | 28.6 |
| Hoonah-Angoon | 18.4 | 16.5 | 8.5 | 45.9 | NA | 2,095 | NA | 2.2 | 214 | 149 | -21 | 791 | 2.51 | 60.4 | 11.0 | 33.4 |
| Juneau | 13.9 | 10.1 | 4.7 | 49.1 | 30,711 | 31,276 | 1.8 | 1.8 | 3,772 | 1,917 | -1,287 | 12,676 | 2.50 | 61.3 | 9.0 | 29.1 |
| Kenai Peninsula | 14.8 | 12.8 | 6.4 | 47.9 | 49,691 | 55,397 | 11.5 | 7.3 | 7,239 | 4,498 | 1,282 | 21,630 | 2.63 | 62.9 | 7.1 | 30.5 |
| Ketchikan Gateway | 14.8 | 11.0 | 6.1 | 49.0 | 14,070 | 13,534 | -3.8 | 1.6 | 1,668 | 977 | -480 | 5,397 | 2.51 | 64.0 | 10.8 | 27.6 |
| Kodiak Island | 12.9 | 8.4 | 4.0 | 46.7 | 13,913 | 13,606 | -2.2 | -4.5 | 2,048 | 616 | -2,075 | 4,261 | 3.08 | 69.6 | 12.4 | 23.4 |
| Kusilvak | 9.0 | 4.7 | 1.8 | 46.8 | 7,028 | 7,459 | 6.1 | 11.7 | 2,391 | 683 | -844 | 1,724 | 4.60 | 83.5 | 28.2 | 12.7 |
| Lake and Peninsula | 13.1 | 7.5 | 3.8 | 48.6 | 1,823 | 1,635 | -10.3 | -8.7 | 321 | 129 | -337 | 421 | 3.03 | 66.5 | 20.9 | 28.3 |
| Matanuska-Susitna | 13.1 | 9.0 | 4.1 | 48.1 | 59,322 | 88,981 | 50.0 | 23.9 | 14,306 | 6,012 | 12,795 | 31,217 | 3.31 | 70.8 | 8.6 | 22.8 |
| Nome | 10.3 | 6.0 | 2.5 | 47.4 | 9,196 | 9,492 | 3.2 | 4.4 | 2,180 | 805 | -966 | 2,844 | 3.30 | 74.8 | 21.2 | 19.8 |
| North Slope | 13.8 | 5.6 | 2.2 | 38.8 | 7,385 | 9,018 | 22.1 | 3.1 | 1,688 | 503 | -917 | 1,979 | 3.36 | 77.0 | 25.7 | 16.6 |
| Northwest Arctic | 10.2 | 4.7 | 3.1 | 45.9 | 7,208 | 7,523 | 4.4 | 1.6 | 1,843 | 557 | -1,175 | 1,795 | 4.06 | 73.4 | 22.2 | 20.0 |
| Petersburg | 14.8 | 13.3 | 6.9 | 48.5 | NA | 3,207 | NA | 2.8 | 367 | 213 | -68 | 1,187 | 2.68 | 64.5 | 11.6 | 28.4 |
| Prince of Wales-Hyder | 16.3 | 13.8 | 5.6 | 45.0 | NA | 6,168 | NA | -0.3 | 756 | 510 | -265 | 2,371 | 2.66 | 64.3 | 11.6 | 30.3 |
| Sitka | 14.7 | 10.6 | 6.4 | 48.8 | 8,835 | 8,881 | 0.5 | -5.4 | 931 | 647 | -767 | 3,547 | 2.36 | 59.9 | 10.3 | 31.9 |
| Skagway | 12.0 | 9.6 | 5.9 | 47.8 | NA | 968 | NA | 21.8 | 105 | 43 | 144 | 375 | 2.51 | 53.3 | 7.7 | 31.2 |
| Southeast Fairbanks | 13.3 | 10.7 | 5.3 | 45.5 | 6,174 | 7,029 | 13.8 | -9.1 | 1,124 | 476 | -734 | 2,170 | 3.05 | 67.8 | 7.8 | 27.1 |
| Valdez-Cordova | 16.8 | 10.6 | 3.9 | 47.0 | 10,195 | NA | NA | NA | NA | NA | NA | 3,177 | 2.83 | 62.4 | 5.4 | 32.1 |
| Wrangell | 17.9 | 16.4 | 9.5 | 47.5 | NA | 2,365 | NA | 6.1 | 239 | 220 | 128 | 1,027 | 2.38 | 63.7 | 7.7 | 31.6 |
| Yakutat | 14.8 | 12.6 | 7.4 | 45.7 | 808 | 662 | -18.1 | -3.8 | 78 | 39 | -65 | 219 | 2.65 | 58.0 | 10.5 | 36.5 |
| Yukon-Koyukuk | 13.5 | 11.2 | 5.5 | 47.2 | 6,551 | 5,583 | -14.8 | -9.1 | 858 | 597 | -775 | 1,856 | 2.69 | 58.4 | 17.0 | 35.0 |
| **ARIZONA** | 12.1 | 10.6 | 7.9 | 50.3 | 5,130,632 | 6,392,292 | 24.6 | 16.1 | 863,381 | 561,112 | 725,176 | 2,571,268 | 2.68 | 65.3 | 12.2 | 27.3 |
| Apache | 13.1 | 9.8 | 6.7 | 50.7 | 69,423 | 71,514 | 3.0 | 0.5 | 9,987 | 6,481 | -3,172 | 20,867 | 3.37 | 66.2 | 20.6 | 31.4 |
| Cochise | 13.0 | 13.5 | 10.3 | 49.2 | 117,755 | 131,359 | 11.6 | -3.0 | 15,790 | 13,365 | -6,564 | 50,163 | 2.34 | 64.7 | 11.5 | 30.0 |
| Coconino | 11.2 | 8.3 | 4.9 | 50.7 | 116,320 | 134,435 | 15.6 | 6.0 | 16,520 | 8,280 | -213 | 47,447 | 2.67 | 60.8 | 11.9 | 26.2 |
| Gila | 15.7 | 17.4 | 12.7 | 50.3 | 51,335 | 53,592 | 4.4 | 1.3 | 5,917 | 7,887 | 2,724 | 21,945 | 2.40 | 64.3 | 12.2 | 30.5 |
| Graham | 10.1 | 8.3 | 5.9 | 46.5 | 33,489 | 37,212 | 11.1 | 5.4 | 5,677 | 2,970 | -736 | 11,017 | 3.10 | 70.7 | 15.5 | 24.6 |
| Greenlee | 10.7 | 8.6 | 4.8 | 48.1 | 8,547 | 8,444 | -1.2 | 10.6 | 1,351 | 615 | 135 | 3,132 | 3.01 | 74.7 | 8.6 | 21.6 |
| La Paz | 12.2 | 18.3 | 23.4 | 48.9 | 19,715 | 20,489 | 3.9 | 4.8 | 2,034 | 2,741 | 1,724 | 9,346 | 2.20 | 63.4 | 10.3 | 31.2 |
| Maricopa | 11.7 | 9.2 | 6.8 | 50.6 | 3,072,149 | 3,817,371 | 24.3 | 20.0 | 550,333 | 303,538 | 513,596 | 1,552,096 | 2.75 | 65.6 | 12.2 | 26.7 |
| Mohave | 16.3 | 18.1 | 13.9 | 49.6 | 155,032 | 200,181 | 29.1 | 8.5 | 18,687 | 30,248 | 28,568 | 86,889 | 2.35 | 64.8 | 10.7 | 29.0 |
| Navajo | 13.3 | 11.9 | 7.6 | 49.8 | 97,470 | 107,488 | 10.3 | 4.3 | 15,776 | 10,400 | -723 | 34,990 | 3.04 | 69.6 | 18.1 | 26.0 |
| Pima | 12.2 | 11.8 | 9.1 | 50.8 | 843,746 | 980,266 | 16.2 | 8.3 | 117,433 | 96,509 | 60,898 | 404,739 | 2.46 | 61.5 | 12.5 | 30.5 |
| Pinal | 11.7 | 12.7 | 8.9 | 48.3 | 179,727 | 375,766 | 109.1 | 28.0 | 46,946 | 29,809 | 87,463 | 141,300 | 2.87 | 70.1 | 10.9 | 24.7 |
| Santa Cruz | 11.9 | 11.0 | 8.1 | 51.8 | 38,381 | 47,417 | 23.5 | -1.3 | 6,591 | 3,036 | -4,242 | 15,853 | 2.91 | 75.8 | 16.3 | 21.4 |
| Yavapai | 16.7 | 20.0 | 13.7 | 51.1 | 167,517 | 211,008 | 26.0 | 13.8 | 18,829 | 29,931 | 40,097 | 98,386 | 2.26 | 64.2 | 8.0 | 30.8 |
| Yuma | 9.7 | 9.7 | 10.4 | 48.4 | 160,026 | 195,750 | 22.3 | 11.3 | 31,510 | 15,302 | 5,621 | 73,098 | 2.79 | 73.9 | 14.3 | 21.0 |
| **ARKANSAS** | 12.7 | 10.2 | 7.5 | 50.9 | 2,673,400 | 2,916,029 | 9.1 | 3.9 | 388,332 | 320,002 | 45,657 | 1,158,071 | 2.52 | 66.1 | 13.2 | 28.6 |
| Arkansas | 14.0 | 12.0 | 8.5 | 51.8 | 20,749 | 19,045 | -8.2 | -8.7 | 2,434 | 2,441 | -1,662 | 7,491 | 2.36 | 63.0 | 12.8 | 30.1 |
| Ashley | 14.1 | 12.0 | 9.6 | 51.5 | 24,209 | 21,845 | -9.8 | -11.5 | 2,566 | 2,626 | -2,453 | 7,757 | 2.59 | 70.3 | 12.6 | 27.2 |
| Baxter | 15.3 | 16.4 | 15.0 | 51.6 | 38,386 | 41,511 | 8.1 | 1.8 | 3,590 | 6,984 | 4,148 | 18,435 | 2.22 | 64.9 | 6.5 | 30.5 |
| Benton | 10.8 | 8.1 | 5.7 | 50.1 | 153,406 | 221,348 | 44.3 | 30.5 | 36,510 | 18,040 | 48,661 | 97,249 | 2.70 | 72.4 | 10.0 | 22.1 |
| Boone | 13.8 | 11.8 | 9.6 | 50.7 | 33,948 | 36,918 | 8.7 | 1.9 | 4,468 | 4,571 | 838 | 15,034 | 2.45 | 64.0 | 9.0 | 26.5 |
| Bradley | 13.1 | 10.8 | 8.6 | 51.5 | 12,600 | 11,505 | -8.7 | -7.5 | 1,403 | 1,539 | -734 | 4,419 | 2.42 | 64.0 | 13.5 | 30.9 |
| Calhoun | 15.8 | 12.8 | 9.6 | 49.2 | 5,744 | 5,368 | -6.5 | -4.8 | 432 | 638 | -48 | 1,850 | 2.73 | 70.4 | 8.8 | 28.6 |
| Carroll | 14.5 | 14.5 | 9.6 | 50.2 | 25,357 | 27,451 | 8.3 | 3.0 | 3,237 | 3,206 | 808 | 11,139 | 2.49 | 68.3 | 7.4 | 28.0 |
| Chicot | 15.0 | 12.2 | 10.3 | 49.8 | 14,117 | 11,800 | -16.4 | -15.9 | 1,377 | 1,566 | -1,692 | 4,068 | 2.43 | 64.8 | 17.4 | 32.7 |
| Clark | 11.6 | 9.4 | 7.7 | 52.3 | 23,546 | 22,992 | -2.4 | -3.9 | 2,412 | 2,571 | -735 | 8,446 | 2.34 | 62.2 | 13.8 | 28.8 |
| Clay | 14.1 | 11.8 | 10.1 | 51.0 | 17,609 | 16,083 | -8.7 | -10.6 | 1,696 | 2,452 | -947 | 6,444 | 2.29 | 64.2 | 11.8 | 31.8 |
| Cleburne | 15.9 | 14.7 | 13.1 | 50.8 | 24,046 | 25,967 | 8.0 | -4.0 | 2,430 | 3,849 | 414 | 10,783 | 2.30 | 67.0 | 8.2 | 29.1 |

1. No spouse present.

# Table B. States and Counties — Population, Vital Statistics, Health, and Crime

| STATE County | Persons in group quarters, 2020 | Daytime Population, 2015-2019 | | Births, 2020 | | Deaths, 2020 | | Persons under 65 with no health insurance, 2019 | | Medicare, 2020 | | | Crimes reported by county police or sheriff, 2019 | |
|---|---|---|---|---|---|---|---|---|---|---|---|---|---|---|
| | | Number | Employment/residence ratio | Total | Rate[1] | Number | Rate[1] | Number | Percent | Total beneficiaries | Enrolled in Original Medicare | Enrolled in Medicare Advantage | Violent | Property |
| | 32 | 33 | 34 | 35 | 36 | 37 | 38 | 39 | 40 | 41 | 42 | 43 | 44 | 45 |
| ALABAMA—Cont'd | | | | | | | | | | | | | | |
| Washington | 147 | 15,540 | 0.82 | 181 | 11.3 | 231 | 14.5 | 1,781 | 13.8 | 4,232 | 2,599 | 1,634 | NA | NA |
| Wilcox | 108 | 10,950 | 1.09 | 123 | 12.1 | 143 | 14.0 | 934 | 11.6 | 3,165 | 1,941 | 1,224 | NA | NA |
| Winston | 301 | 22,777 | 0.89 | 241 | 10.3 | 362 | 15.4 | 2,389 | 13.0 | 6,131 | 3,727 | 2,403 | NA | NA |
| ALASKA | 27,015 | 745,138 | 1.02 | 9,733 | 13.3 | 5,249 | 7.2 | 79,260 | 12.7 | 104,362 | 102,526 | 1,836 | X | X |
| Aleutians East | 1,745 | 3,448 | 1.03 | 4 | 1.2 | 7 | 2.1 | 691 | 23.3 | 163 | 163 | 0 | NA | NA |
| Aleutians West | 2,542 | 6,453 | 1.20 | 33 | 5.8 | 7 | 1.2 | 884 | 17.7 | 235 | D | D | NA | NA |
| Anchorage | 8,022 | 303,422 | 1.06 | 3,861 | 13.4 | 2,047 | 7.1 | 27,810 | 11.1 | 39,937 | 39,199 | 738 | NA | NA |
| Bethel | 360 | 18,536 | 1.07 | 426 | 23.1 | 165 | 8.9 | 2,755 | 16.8 | 1,536 | D | D | NA | NA |
| Bristol Bay | 15 | 1,015 | 1.30 | 8 | 10.2 | 6 | 7.6 | 103 | 15.1 | 157 | D | D | NA | NA |
| Chugach | 183 | NA | NA | 73 | 11.4 | 34 | 5.3 | NA | NA | 0 | NA | NA | NA | NA |
| Copper River | 18 | NA | NA | 32 | 11.0 | 2 | 0.7 | NA | NA | 0 | NA | NA | NA | NA |
| Denali | 101 | 2,778 | 1.34 | 14 | 6.7 | 0 | 0.0 | 206 | 11.5 | 274 | D | D | NA | NA |
| Dillingham | 52 | 5,061 | 1.05 | 81 | 16.8 | 54 | 11.2 | 678 | 15.5 | 530 | D | D | NA | NA |
| Fairbanks North Star | 4,821 | 98,418 | 0.99 | 1,408 | 14.7 | 540 | 5.6 | 9,282 | 11.3 | 11,962 | 11,838 | 124 | NA | NA |
| Haines | 0 | 2,548 | 1.03 | 19 | 7.3 | 5 | 1.9 | 252 | 12.6 | 627 | 606 | 21 | NA | NA |
| Hoonah-Angoon | 0 | 2,158 | 1.02 | 19 | 8.9 | 23 | 10.7 | 323 | 20.0 | 517 | D | D | NA | NA |
| Juneau | 995 | 32,675 | 1.03 | 315 | 9.9 | 200 | 6.3 | 3,070 | 11.4 | 5,283 | 5,233 | 50 | NA | NA |
| Kenai Peninsula | 1,704 | 57,480 | 0.96 | 680 | 11.4 | 550 | 9.3 | 6,432 | 13.8 | 12,419 | 12,048 | 371 | NA | NA |
| Ketchikan Gateway | 264 | 13,942 | 1.02 | 135 | 9.8 | 101 | 7.3 | 1,547 | 13.5 | 2,570 | 2,542 | 28 | NA | NA |
| Kodiak Island | 388 | 13,704 | 1.04 | 145 | 11.2 | 64 | 4.9 | 2,047 | 18.2 | 1,728 | 1,710 | 18 | NA | NA |
| Kusilvak | 9 | 8,333 | 1.04 | 234 | 28.1 | 109 | 13.1 | 1,197 | 15.6 | 565 | D | D | NA | NA |
| Lake and Peninsula | 37 | 1,434 | 1.07 | 26 | 17.4 | 16 | 10.7 | 266 | 19.1 | 161 | 161 | 0 | NA | NA |
| Matanuska-Susitna | 2,101 | 92,801 | 0.71 | 1,372 | 12.4 | 721 | 6.5 | 11,867 | 12.8 | 15,595 | 15,238 | 357 | NA | NA |
| Nome | 188 | 10,087 | 1.03 | 167 | 16.9 | 113 | 11.4 | 1,533 | 17.2 | 902 | D | D | NA | NA |
| North Slope | 2,240 | 17,359 | 2.42 | 144 | 15.5 | 71 | 7.6 | 1,192 | 13.2 | 547 | D | D | NA | NA |
| Northwest Arctic | 400 | 7,975 | 1.1 | 158 | 20.7 | 89 | 11.6 | 1,201 | 17.5 | 586 | D | D | NA | NA |
| Petersburg | 43 | 3,369 | 1.07 | 26 | 7.9 | 38 | 11.5 | 430 | 16.7 | 845 | 834 | 11 | NA | NA |
| Prince of Wales-Hyder | 50 | 6,445 | 1.01 | 64 | 10.4 | 57 | 9.3 | 945 | 18.9 | 1,056 | D | D | NA | NA |
| Sitka | 257 | 8,618 | 1.00 | 80 | 9.5 | 51 | 6.1 | 951 | 13.4 | 1,496 | 1,483 | 13 | NA | NA |
| Skagway | 32 | 1,128 | 1.03 | 6 | 5.1 | 16 | 13.6 | 112 | 11.1 | 194 | D | D | NA | NA |
| Southeast Fairbanks | 364 | 7,255 | 1.14 | 96 | 13.8 | 48 | 6.9 | 1,044 | 18.3 | 1,345 | 1,316 | 29 | NA | NA |
| Valdez-Cordova | NA | 9,962 | 1.15 | NA | NA | NA | NA | 1,151 | 14.7 | 1,539 | 1,516 | 22 | NA | NA |
| Wrangell | 17 | 2,502 | 1.00 | 23 | 9.2 | 40 | 15.9 | 344 | 18.4 | 545 | D | D | NA | NA |
| Yakutat | 18 | 728 | 1.25 | 6 | 9.4 | 6 | 9.4 | 83 | 18.2 | 114 | 114 | 0 | NA | NA |
| Yukon-Koyukuk | 49 | 5,504 | 1.06 | 78 | 15.4 | 69 | 13.6 | 864 | 20.0 | 931 | D | D | NA | NA |
| ARIZONA | 155,824 | 7,018,338 | 0.99 | 81,451 | 11.0 | 66,385 | 8.9 | 789,960 | 13.6 | 1,369,035 | 787,095 | 581,940 | X | X |
| Apache | 953 | 70,196 | 0.93 | 891 | 12.4 | 774 | 10.8 | 12,371 | 21.0 | 13,310 | 11,902 | 1,407 | NA | NA |
| Cochise | 5,393 | 124,826 | 0.98 | 1,335 | 10.5 | 1,442 | 11.3 | 10,599 | 11.5 | 33,438 | 22,351 | 11,088 | 47 | 531 |
| Coconino | 11,916 | 141,338 | 1.00 | 1,440 | 10.1 | 963 | 6.8 | 17,084 | 15.2 | 21,301 | 17,234 | 4,067 | 127 | 264 |
| Gila | 846 | 54,038 | 1.03 | 481 | 8.9 | 910 | 16.8 | 5,535 | 14.8 | 17,194 | 13,162 | 4,031 | 176 | 384 |
| Graham | 2,831 | 37,210 | 0.94 | 514 | 13.1 | 321 | 8.2 | 3,591 | 11.8 | 5,929 | 3,443 | 2,486 | 61 | 104 |
| Greenlee | 35 | 11,162 | 1.43 | 123 | 13.2 | 57 | 6.1 | 640 | 7.9 | 1,301 | 1,026 | 276 | NA | NA |
| Maricopa | 65,297 | 4,373,052 | 1.02 | 52,573 | 11.5 | 36,247 | 7.9 | 493,829 | 13.3 | 722,323 | 397,785 | 324,538 | 40 | 310 |
| Mohave | 4,152 | 196,389 | 0.84 | 1,780 | 8.2 | 3,656 | 16.8 | 19,205 | 13.6 | 68,799 | 44,048 | 24,751 | NA | NA |
| Navajo | 2,214 | 109,289 | 1.00 | 1,363 | 12.2 | 1,227 | 10.9 | 15,219 | 17.4 | 24,710 | 19,199 | 5,511 | 43 | 341 |
| Pima | 24,584 | 1,026,976 | 1.00 | 10,667 | 10.1 | 11,132 | 10.5 | 107,748 | 13.3 | 236,346 | 118,826 | 117,520 | NA | NA |
| Pinal | 23,582 | 376,488 | 0.64 | 4,678 | 9.7 | 3,777 | 7.9 | 43,484 | 12.8 | 85,159 | 45,241 | 39,919 | 197 | 1,469 |
| Santa Cruz | 293 | 45,516 | 0.94 | 612 | 13.1 | 371 | 7.9 | 6,241 | 16.9 | 10,198 | 4,460 | 5,738 | 1 | 117 |
| Yavapai | 3,976 | 223,821 | 0.95 | 1,798 | 7.5 | 3,298 | 13.7 | 23,125 | 14.8 | 85,203 | 57,812 | 27,391 | 193 | 860 |
| Yuma | 9,391 | 206,702 | 0.96 | 3,008 | 13.8 | 1,880 | 8.6 | 28,673 | 17.7 | 37,910 | 26,323 | 11,587 | 136 | 807 |
| ARKANSAS | 82,365 | 2,998,060 | 1.00 | 36,332 | 12.0 | 33,832 | 11.2 | 261,617 | 10.8 | 646,559 | 453,618 | 192,942 | X | X |
| Arkansas | 247 | 19,997 | 1.27 | 208 | 12.0 | 236 | 13.6 | 1,338 | 9.8 | 4,219 | 3,376 | 843 | 15 | 58 |
| Ashley | 191 | 19,700 | 0.93 | 222 | 11.5 | 271 | 14.0 | 1,710 | 11.2 | 5,190 | 4,250 | 940 | 22 | 153 |
| Baxter | 595 | 42,974 | 1.10 | 350 | 8.3 | 753 | 17.8 | 2,924 | 10.2 | 15,264 | 10,164 | 5,100 | 70 | 616 |
| Benton | 2,034 | 272,089 | 1.05 | 3,875 | 13.4 | 1,998 | 6.9 | 26,888 | 11.1 | 44,485 | 27,144 | 17,341 | 159 | NA |
| Boone | 465 | 38,035 | 1.04 | 372 | 9.9 | 493 | 13.1 | 2,919 | 10.0 | 10,335 | 7,174 | 3,162 | 199 | 269 |
| Bradley | 224 | 10,400 | 0.88 | 136 | 12.8 | 131 | 12.3 | 1,114 | 13.1 | 2,538 | 2,032 | 507 | 8 | 51 |
| Calhoun | 175 | 5,449 | 1.12 | 29 | 5.7 | 53 | 10.4 | 372 | 9.6 | 1,185 | 889 | 296 | 11 | 68 |
| Carroll | 230 | 27,848 | 0.99 | 322 | 11.4 | 331 | 11.7 | 3,225 | 15.1 | 7,584 | 4,723 | 2,860 | 45 | 209 |
| Chicot | 734 | 10,872 | 1.08 | 114 | 11.5 | 161 | 16.2 | 678 | 9.4 | 2,647 | 1,923 | 724 | 13 | 76 |
| Clark | 2,976 | 21,582 | 0.92 | 218 | 9.9 | 259 | 11.7 | 1,471 | 9.4 | 4,637 | 3,303 | 1,334 | 11 | 70 |
| Clay | 125 | 13,510 | 0.78 | 153 | 10.6 | 208 | 14.5 | 1,131 | 10.1 | 4,005 | 3,122 | 883 | NA | NA |
| Cleburne | 339 | 23,717 | 0.86 | 215 | 8.6 | 413 | 16.6 | 1,897 | 10.6 | 8,190 | 6,290 | 1,900 | 142 | 297 |

1. Per 1,000 estimated resident population.

# Table B. States and Counties — Crime, Education, Money Income, and Poverty

| STATE County | County law enforcement employment, 2019 | | Education | | | | | | Money income, 2015-2019 | | | | Income and poverty, 2019 | | | |
| | | | School enrollment and attainment, 2015-2019 | | | | Local government expenditures[3] 2017-2018 | | | Households | | | | Percent below poverty level | | |
| | | | Enrollment[1] | | Attainment[2] (percent) | | | | | | | Percent | | | | | |
| | Officers | Civilians | Total | Percent private | High school graduate or less | Bachelor's degree or more | Total current spending (mil dol) | Current spending per student (dollars) | Per capita income[4] | Median income (dollars) | with income of less than $50,000 | with income of $200,000 or more | Median household income (dollars) | All persons | Children under 18 years | Children 5 to 17 years in families |
| | 46 | 47 | 48 | 49 | 50 | 51 | 52 | 53 | 54 | 55 | 56 | 57 | 58 | 59 | 60 | 61 |
| **ALABAMA—Cont'd** | | | | | | | | | | | | | | | | |
| Washington | 10 | 12 | 3,712 | 13.9 | 60.4 | 12.7 | 26 | 9,567 | 26,752 | 41,370 | 57.4 | 1.6 | 48,864 | 18.6 | 28.2 | 26.5 |
| Wilcox | 10 | 12 | 2,566 | 14.3 | 63.0 | 12.5 | 19 | 10,969 | 16,841 | 31,014 | 72.2 | 1.6 | 30,998 | 32.5 | 48.9 | 46.0 |
| Winston | 10 | 19 | 4,351 | 4.1 | 59.4 | 12.8 | 41 | 10,104 | 22,952 | 35,788 | 62.5 | 2.0 | 40,827 | 16.7 | 22.5 | 21.5 |
| **ALASKA** | X | X | 183,894 | 12.4 | 35.2 | 29.6 | 2,355 | 17,726 | 36,787 | 77,640 | 30.7 | 8.6 | 77,203 | 10.2 | 13.2 | 11.9 |
| Aleutians East | NA | NA | 497 | 7.2 | 58.0 | 11.5 | 9 | 37,335 | 33,939 | 69,250 | 37.8 | 5.7 | 66,923 | 14.8 | 13.4 | 11.6 |
| Aleutians West | NA | NA | 1,011 | 6.5 | 49.6 | 17.0 | 11 | 23,453 | 39,647 | 87,466 | 20.3 | 8.9 | 84,726 | 7.6 | 5.4 | 5.2 |
| Anchorage | NA | NA | 75,221 | 12.0 | 29.5 | 36.1 | 709 | 14,754 | 41,415 | 84,928 | 26.6 | 10.8 | 82,512 | 9.5 | 11.0 | 9.9 |
| Bethel | NA | NA | 5,419 | 0.9 | 65.5 | 11.9 | 166 | 32,052 | 19,885 | 53,553 | 46.6 | 4.3 | 48,262 | 23.5 | 31.6 | 29.9 |
| Bristol Bay | NA | NA | 171 | 7.0 | 43.0 | 22.5 | 21 | 44,698 | 44,168 | 79,808 | 27.1 | 7.0 | 87,950 | 9.8 | 11.0 | 12.6 |
| Chugach | NA | NA | NA | NA | NA | NA | NA | NA | NA | NA | NA | NA | NA | NA | NA | NA |
| Copper River | NA | NA | NA | NA | NA | NA | NA | NA | NA | NA | NA | NA | NA | NA | NA | NA |
| Denali | NA | NA | 339 | 20.4 | 27.5 | 41.1 | 10 | 10,430 | 34,365 | 81,719 | 32.0 | 3.8 | 78,434 | 6.6 | 8.6 | 7.6 |
| Dillingham | NA | NA | 1,299 | 3.5 | 50.9 | 22.1 | 34 | 30,671 | 25,993 | 56,898 | 45.6 | 3.4 | 57,436 | 18.1 | 27.1 | 26.4 |
| Fairbanks North Star | NA | NA | 26,104 | 15.6 | 28.1 | 31.4 | 255 | 16,344 | 37,025 | 76,992 | 29.7 | 8.1 | 74,076 | 7.2 | 7.9 | 7.9 |
| Haines | NA | NA | 493 | 0.4 | 32.6 | 33.8 | 5 | 19,900 | 31,731 | 58,059 | 42.7 | 2.9 | 62,548 | 9.8 | 14.4 | 13.6 |
| Hoonah-Angoon | NA | NA | 328 | 13.7 | 44.8 | 21.3 | 9 | 29,548 | 36,091 | 59,803 | 40.7 | 3.9 | 53,141 | 15.7 | 23.3 | 22.3 |
| Juneau | NA | NA | 7,710 | 11.2 | 25.2 | 38.4 | 81 | 17,035 | 43,680 | 88,390 | 23.0 | 9.5 | 85,419 | 6.3 | 8.0 | 7.1 |
| Kenai Peninsula | NA | NA | 12,564 | 18.0 | 38.6 | 24.6 | 156 | 17,417 | 34,226 | 66,064 | 37.7 | 6.5 | 68,625 | 10.4 | 12.5 | 11.7 |
| Ketchikan Gateway | NA | NA | 2,946 | 10.4 | 36.9 | 25.3 | 41 | 17,203 | 37,590 | 72,728 | 34.0 | 7.5 | 75,706 | 9.4 | 12.0 | 10.3 |
| Kodiak Island | NA | NA | 3,337 | 12.8 | 39.9 | 28.0 | 50 | 20,499 | 32,876 | 85,839 | 25.3 | 6.8 | 79,669 | 7.1 | 8.7 | 8.0 |
| Kusilvak | NA | NA | 2,626 | 1.1 | 77.1 | 3.9 | 75 | 29,855 | 13,762 | 36,754 | 64.4 | 1.2 | 35,369 | 26.8 | 35.5 | 36.2 |
| Lake and Peninsula | NA | NA | 288 | 1.4 | 56.2 | 16.1 | NA | NA | 24,836 | 44,135 | 55.3 | 1.7 | 51,693 | 17.4 | 25.3 | 26.5 |
| Matanuska-Susitna | NA | NA | 26,480 | 15.7 | 40.1 | 21.6 | 275 | 14,282 | 31,232 | 75,493 | 33.0 | 7.2 | 75,856 | 9.6 | 12.8 | 9.8 |
| Nome | NA | NA | 2,937 | 1.9 | 56.6 | 16.1 | 81 | 30,241 | 23,581 | 61,048 | 40.8 | 5.5 | 58,464 | 20.7 | 26.1 | 24.9 |
| North Slope | NA | NA | 2,199 | 2.6 | 53.2 | 16.5 | 78 | 36,058 | 48,730 | 79,306 | 32.9 | 9.4 | 84,784 | 11.4 | 15.3 | 14.3 |
| Northwest Arctic | NA | NA | 2,253 | 3.5 | 64.1 | 13.2 | 74 | 34,590 | 23,780 | 60,906 | 43.3 | 6.2 | 61,660 | 24.0 | 29.8 | 28.5 |
| Petersburg | NA | NA | 532 | 11.1 | 42.1 | 28.4 | 9 | 19,779 | 36,112 | 69,948 | 33.8 | 9.8 | 71,296 | 6.7 | 8.6 | 8.2 |
| Prince of Wales-Hyder | NA | NA | 1,417 | 8.3 | 47.1 | 18.3 | 35 | 26,325 | 27,080 | 52,379 | 48.8 | 2.6 | 50,049 | 17.6 | 23.3 | 22.0 |
| Sitka | NA | NA | 1,999 | 13.2 | 32.7 | 33.4 | 38 | 22,347 | 39,615 | 73,682 | 31.4 | 5.7 | 80,660 | 8.1 | 9.3 | 8.1 |
| Skagway | NA | NA | 144 | 6.3 | 34.5 | 29.6 | 3 | 24,138 | 39,453 | 73,906 | 25.3 | 1.9 | 74,898 | 4.0 | 7.3 | 7.2 |
| Southeast Fairbanks | NA | NA | 1,712 | 19.0 | 44.4 | 19.7 | 28 | 22,676 | 32,193 | 70,056 | 36.3 | 4.7 | 69,646 | 12.0 | 15.8 | 15.4 |
| Valdez-Cordova | NA | NA | 1,934 | 15.3 | 35.4 | 30.8 | 31 | 21,703 | 38,907 | 79,867 | 30.8 | 10.8 | 74,574 | 8.6 | 10.1 | 9.2 |
| Wrangell | NA | NA | 427 | 5.9 | 48.7 | 19.4 | 6 | 21,434 | 32,286 | 53,894 | 45.8 | 3.1 | 62,496 | 9.7 | 11.0 | 9.6 |
| Yakutat | NA | NA | 128 | 12.5 | 42.3 | 22.6 | 2 | 26,382 | 32,198 | 71,607 | 32.0 | 2.3 | 61,587 | 14.7 | 26.7 | 28.6 |
| Yukon-Koyukuk | NA | NA | 1,379 | 2.5 | 54.2 | 13.6 | 64 | 10,121 | 22,718 | 41,413 | 57.8 | 1.2 | 42,063 | 24.4 | 30.8 | 28.2 |
| **ARIZONA** | X | X | 1,765,901 | 11.6 | 36.7 | 29.5 | 9,097 | 8,296 | 30,694 | 58,945 | 42.6 | 5.9 | 62,027 | 13.5 | 19.2 | 18.1 |
| Apache | 25 | 47 | 19,612 | 3.4 | 51.8 | 12.3 | 139 | 13,183 | 15,128 | 32,508 | 66.3 | 0.7 | 35,386 | 33.4 | 44.7 | 47.6 |
| Cochise | 86 | 94 | 28,431 | 10.2 | 36.7 | 23.1 | 164 | 9,076 | 26,819 | 49,260 | 50.6 | 2.9 | 48,650 | 16.6 | 26.2 | 24.6 |
| Coconino | 61 | 178 | 49,204 | 6.0 | 31.5 | 36.2 | 181 | 9,827 | 27,339 | 59,460 | 42.9 | 5.3 | 58,301 | 15.9 | 15.4 | 14.6 |
| Gila | 51 | 79 | 9,790 | 11.2 | 41.5 | 18.7 | 76 | 10,533 | 24,251 | 43,524 | 56.2 | 1.9 | 45,980 | 21.1 | 32.2 | 31.0 |
| Graham | 23 | 59 | 10,275 | 6.3 | 43.5 | 15.2 | 54 | 8,429 | 18,683 | 51,353 | 48.6 | 1.5 | 50,685 | 20.1 | 25.2 | 23.9 |
| Greenlee | 16 | 26 | 2,290 | 8.7 | 46.7 | 13.5 | 17 | 10,048 | 25,780 | 63,473 | 38.2 | 1.4 | 68,066 | 9.3 | 11.1 | 10.5 |
| La Paz | 38 | 52 | 3,249 | 7.2 | 54.2 | 12.0 | 25 | 10,465 | 22,879 | 34,643 | 62.6 | 1.5 | 39,407 | 22.1 | 32.6 | 32.4 |
| Maricopa | 670 | 2,773 | 1,109,038 | 12.2 | 34.7 | 32.7 | 5,899 | 8,083 | 33,229 | 64,468 | 38.4 | 7.5 | 68,634 | 12.2 | 17.5 | 16.3 |
| Mohave | 83 | 163 | 35,195 | 15.6 | 49.3 | 12.9 | 181 | 7,779 | 26,166 | 45,587 | 54.0 | 2.3 | 49,712 | 16.0 | 28.2 | 27.3 |
| Navajo | 47 | 88 | 27,566 | 6.6 | 46.9 | 15.3 | 184 | 10,582 | 18,935 | 40,067 | 58.6 | 1.6 | 40,676 | 25.2 | 35.0 | 33.1 |
| Pima | 478 | 939 | 262,276 | 10.9 | 33.9 | 32.4 | 1,196 | 8,536 | 29,707 | 53,379 | 47.1 | 4.7 | 56,150 | 14.0 | 19.0 | 17.5 |
| Pinal | 207 | 272 | 99,143 | 13.0 | 43.5 | 19.5 | 427 | 8,352 | 24,959 | 58,174 | 42.5 | 2.9 | 62,067 | 12.1 | 17.0 | 16.2 |
| Santa Cruz | 35 | 48 | 12,308 | 8.0 | 52.1 | 20.4 | 81 | 8,054 | 21,480 | 41,259 | 57.6 | 3.0 | 47,779 | 18.8 | 27.8 | 26.1 |
| Yavapai | 120 | 73 | 43,786 | 19.3 | 34.8 | 25.9 | 198 | 8,527 | 30,341 | 52,451 | 47.7 | 3.2 | 54,242 | 12.0 | 16.7 | 16.4 |
| Yuma | 82 | 249 | 53,738 | 5.3 | 52.5 | 15.0 | 275 | 7,471 | 21,758 | 45,243 | 54.4 | 1.8 | 47,024 | 20.4 | 29.7 | 28.6 |
| **ARKANSAS** | X | X | 733,114 | 11.7 | 47.5 | 23.0 | 4,997 | 10,070 | 26,577 | 47,597 | 52.2 | 3.6 | 49,020 | 16.0 | 21.7 | 19.9 |
| Arkansas | 11 | 0 | 3,990 | 13.8 | 55.0 | 15.8 | 29 | 9,969 | 26,386 | 46,696 | 53.3 | 2.8 | 52,539 | 15.8 | 24.1 | 22.8 |
| Ashley | 19 | 3 | 4,619 | 9.4 | 60.4 | 12.3 | 35 | 9,256 | 23,391 | 44,744 | 55.2 | 2.2 | 47,744 | 18.5 | 24.8 | 23.8 |
| Baxter | 33 | 27 | 7,458 | 5.8 | 47.6 | 17.9 | 47 | 9,261 | 26,863 | 42,260 | 58.8 | 1.7 | 41,950 | 12.7 | 18.1 | 16.8 |
| Benton | 163 | 78 | 66,143 | 14.3 | 40.6 | 33.4 | 440 | 9,646 | 34,442 | 66,362 | 36.2 | 8.2 | 70,775 | 8.9 | 12.2 | 10.9 |
| Boone | 26 | 30 | 7,893 | 9.0 | 46.0 | 16.1 | 64 | 10,537 | 24,727 | 45,374 | 55.4 | 2.7 | 50,083 | 15.1 | 18.8 | 16.1 |
| Bradley | 5 | 2 | 2,544 | 5.3 | 60.3 | 14.3 | 22 | 10,514 | 22,945 | 43,184 | 56.7 | 1.8 | 39,982 | 22.9 | 31.3 | 29.6 |
| Calhoun | 6 | 6 | 1,095 | 11.5 | 60.5 | 11.4 | 7 | 10,391 | 23,688 | 46,417 | 52.9 | 0.3 | 48,415 | 15.4 | 20.2 | 18.7 |
| Carroll | 20 | 44 | 5,791 | 10.0 | 49.7 | 21.2 | 40 | 10,151 | 25,295 | 46,110 | 53.9 | 3.1 | 43,841 | 14.7 | 19.2 | 18.4 |
| Chicot | 7 | 1 | 2,500 | 10.2 | 57.3 | 15.9 | 17 | 11,693 | 20,232 | 34,147 | 67.1 | 2.5 | 37,014 | 31.0 | 45.1 | 44.2 |
| Clark | 17 | 16 | 7,354 | 11.4 | 40.2 | 26.6 | 41 | 15,690 | 21,337 | 41,620 | 56.6 | 1.2 | 44,105 | 20.6 | 23.1 | 21.4 |
| Clay | 10 | 10 | 3,066 | 7.7 | 59.4 | 12.6 | 22 | 9,147 | 22,505 | 37,233 | 61.8 | 1.7 | 40,686 | 18.7 | 25.5 | 23.3 |
| Cleburne | 26 | 23 | 4,084 | 14.1 | 56.9 | 15.5 | 33 | 10,109 | 27,227 | 44,483 | 54.3 | 2.6 | 46,214 | 14.1 | 21.9 | 19.9 |

1. All persons 3 years old and over enrolled in nursery school through college.  2. Persons 25 years old and over.  3. Elementary and secondary education expenditures.  4. Based on population estimated by the American Community Survey, 2014–2018.

# Table B. States and Counties — **Personal Income**

| STATE County | Personal income, 2019 | | | | | | | | | | Earnings, 2019 | | |
| | Total (mil dol) | Percent change 2018-2019 | Per capita[1] | | Wages and salaries (mil dol) | Supplements to wages and salaries, employer contributions (mil dol) | | Proprietors' income (mil dol) | Dividends, interest, and rent (mil dol) | Personal transfer receipts (mil dol) | Total (mil dol) | Contributions for government social insurance (mil dol) | |
| | | | Dollars | Rank | | Pension and insurance | Government social insurance | | | | | From employee and self-employed | From employer |
|---|---|---|---|---|---|---|---|---|---|---|---|---|---|
| | 62 | 63 | 64 | 65 | 66 | 67 | 68 | 69 | 70 | 71 | 72 | 73 | 74 |
| **ALABAMA—Cont'd** | | | | | | | | | | | | | |
| Washington | 618 | 2.8 | 37,837 | 2,340 | 251 | 52 | 17 | 24 | 69 | 192 | 344 | 25 | 17 |
| Wilcox | 362 | 2.3 | 34,903 | 2,698 | 141 | 27 | 10 | 31 | 55 | 157 | 209 | 16 | 10 |
| Winston | 845 | 3.4 | 35,776 | 2,587 | 332 | 61 | 27 | 68 | 140 | 269 | 487 | 37 | 27 |
| **ALASKA** | 45,945 | 3.6 | 62,629 | X | 21,815 | 5,865 | 1,557 | 3,457 | 8,745 | 7,939 | 32,693 | 1,676 | 1,557 |
| Aleutians East | 209 | 7.7 | 62,537 | 237 | 148 | 31 | 10 | 16 | 16 | 12 | 206 | 11 | 10 |
| Aleutians West | 327 | 2.9 | 58,008 | 368 | 232 | 53 | 16 | 21 | 32 | 24 | 321 | 16 | 16 |
| Anchorage | 20,202 | 3.4 | 70,145 | 127 | 10,518 | 2,469 | 773 | 1,695 | 4,206 | 3,163 | 15,455 | 813 | 773 |
| Bethel | 852 | 3.4 | 46,348 | 1,207 | 349 | 155 | 22 | 26 | 85 | 262 | 552 | 23 | 22 |
| Bristol Bay | 127 | 1.2 | 151,900 | 4 | 81 | 19 | 6 | 14 | 16 | 10 | 120 | 6 | 6 |
| Chugach | NA | NA | NA | 3,114 | NA | NA | NA | NA | NA | NA | NA | NA | NA |
| Copper River | NA | NA | NA | 3,114 | NA | NA | NA | NA | NA | NA | NA | NA | NA |
| Denali | 176 | 2.7 | 84,166 | 49 | 112 | 24 | 9 | 10 | 25 | 42 | 155 | 8 | 9 |
| Dillingham | 304 | 1.8 | 61,922 | 250 | 134 | 44 | 9 | 29 | 51 | 56 | 216 | 10 | 9 |
| Fairbanks North Star | 5,807 | 2.1 | 59,958 | 306 | 2,759 | 825 | 215 | 218 | 1,124 | 1,003 | 4,017 | 194 | 215 |
| Haines | 181 | 2.1 | 71,711 | 113 | 47 | 13 | 3 | 29 | 52 | 34 | 93 | 5 | 3 |
| Hoonah-Angoon | 139 | 4.0 | 64,905 | 188 | 34 | 15 | 2 | 17 | 27 | 36 | 68 | 4 | 2 |
| Juneau | 2,345 | 2.9 | 73,335 | 99 | 1,077 | 395 | 68 | 197 | 507 | 292 | 1,736 | 77 | 68 |
| Kenai Peninsula | 3,221 | 4.6 | 54,870 | 501 | 1,130 | 350 | 75 | 218 | 719 | 704 | 1,773 | 98 | 75 |
| Ketchikan Gateway | 987 | 3.3 | 70,985 | 120 | 413 | 131 | 29 | 144 | 180 | 173 | 717 | 35 | 29 |
| Kodiak Island | 863 | 3.7 | 66,382 | 164 | 377 | 120 | 28 | 131 | 152 | 124 | 656 | 31 | 28 |
| Kusilvak | 266 | 4.1 | 31,944 | 2,946 | 77 | 47 | 4 | 5 | 25 | 121 | 134 | 5 | 4 |
| Lake and Peninsula | 105 | -1.9 | 65,898 | 171 | 44 | 18 | 3 | 10 | 20 | 20 | 75 | 3 | 3 |
| Matanuska-Susitna | 5,301 | 5.2 | 48,943 | 928 | 1,263 | 368 | 91 | 357 | 782 | 939 | 2,080 | 118 | 91 |
| Nome | 543 | 4.1 | 54,272 | 529 | 232 | 91 | 15 | 17 | 66 | 158 | 355 | 15 | 15 |
| North Slope | 884 | 7.1 | 89,863 | 35 | 1,424 | 246 | 86 | 11 | 82 | 71 | 1,767 | 95 | 86 |
| Northwest Arctic | 393 | 3.2 | 51,526 | 723 | 237 | 68 | 16 | 13 | 40 | 126 | 333 | 16 | 16 |
| Petersburg | 242 | 4.2 | 74,021 | 96 | 63 | 26 | 4 | 55 | 59 | 49 | 147 | 7 | 4 |
| Prince of Wales-Hyder | 305 | 4.8 | 49,177 | 910 | 111 | 49 | 7 | 24 | 57 | 75 | 191 | 9 | 7 |
| Sitka | 632 | 3.6 | 74,440 | 93 | 243 | 73 | 17 | 83 | 149 | 94 | 415 | 21 | 17 |
| Skagway | 102 | 5.9 | 86,519 | 40 | 43 | 12 | 3 | 19 | 19 | 12 | 78 | 4 | 3 |
| Southeast Fairbanks | 362 | 3.3 | 52,458 | 649 | 191 | 51 | 13 | 24 | 62 | 87 | 278 | 16 | 13 |
| Valdez-Cordova | 591 | 4.1 | 64,202 | 203 | 320 | 98 | 22 | 51 | 107 | 92 | 491 | 24 | 22 |
| Wrangell | 124 | 0.1 | 49,549 | 871 | 39 | 14 | 3 | 9 | 28 | 35 | 65 | 4 | 3 |
| Yakutat | 39 | 7.9 | 66,967 | 154 | 15 | 5 | 1 | 4 | 8 | 8 | 25 | 1 | 1 |
| Yukon-Koyukuk | 317 | 1.5 | 60,579 | 285 | 101 | 58 | 5 | 10 | 49 | 114 | 174 | 7 | 5 |
| **ARIZONA** | 335,243 | 5.0 | 45,975 | X | 167,113 | 25,033 | 11,744 | 23,250 | 66,140 | 66,453 | 227,139 | 14,580 | 11,744 |
| Apache | 2,530 | 2.9 | 35,189 | 2,661 | 872 | 228 | 66 | 39 | 377 | 1,090 | 1,205 | 83 | 66 |
| Cochise | 5,259 | 4.2 | 41,766 | 1,809 | 2,005 | 477 | 159 | 242 | 1,051 | 1,749 | 2,883 | 198 | 159 |
| Coconino | 7,057 | 1.3 | 49,189 | 908 | 2,998 | 617 | 219 | 612 | 1,649 | 1,276 | 4,447 | 267 | 219 |
| Gila | 2,226 | 3.6 | 41,217 | 1,889 | 693 | 146 | 49 | 84 | 474 | 912 | 972 | 81 | 49 |
| Graham | 1,239 | 4.5 | 31,895 | 2,949 | 449 | 98 | 32 | 26 | 167 | 449 | 606 | 41 | 32 |
| Greenlee | 402 | 5.4 | 42,296 | 1,720 | 379 | 60 | 24 | 14 | 39 | 99 | 477 | 28 | 24 |
| La Paz | 718 | 5.7 | 34,006 | 2,801 | 256 | 52 | 20 | 47 | 139 | 276 | 375 | 28 | 20 |
| Maricopa | 222,943 | 5.3 | 49,704 | 858 | 125,773 | 17,137 | 8,687 | 15,421 | 43,379 | 35,575 | 167,018 | 10,378 | 8,687 |
| Mohave | 7,298 | 4.5 | 34,393 | 2,758 | 2,193 | 381 | 161 | 422 | 1,312 | 2,593 | 3,156 | 287 | 161 |
| Navajo | 3,690 | 4.2 | 33,268 | 2,866 | 1,225 | 266 | 99 | 191 | 620 | 1,548 | 1,781 | 136 | 99 |
| Pima | 47,605 | 4.3 | 45,456 | 1,309 | 20,428 | 3,658 | 1,467 | 3,231 | 10,814 | 11,294 | 28,784 | 1,924 | 1,467 |
| Pinal | 14,893 | 7.4 | 32,182 | 2,934 | 3,037 | 602 | 219 | 784 | 1,951 | 4,211 | 4,642 | 380 | 219 |
| Santa Cruz | 1,833 | 4.1 | 39,427 | 2,124 | 709 | 149 | 56 | 282 | 332 | 453 | 1,195 | 79 | 56 |
| Yavapai | 9,731 | 3.8 | 41,393 | 1,863 | 2,900 | 498 | 210 | 675 | 2,664 | 3,081 | 4,282 | 373 | 210 |
| Yuma | 7,818 | 5.5 | 36,570 | 2,500 | 3,196 | 664 | 275 | 1,180 | 1,173 | 1,847 | 5,316 | 297 | 275 |
| **ARKANSAS** | 134,683 | 3.3 | 44,582 | X | 60,093 | 8,854 | 4,429 | 7,551 | 32,547 | 31,559 | 80,928 | 5,551 | 4,429 |
| Arkansas | 838 | 3.9 | 47,899 | 1,029 | 480 | 71 | 37 | 97 | 117 | 202 | 685 | 43 | 37 |
| Ashley | 746 | 3.4 | 37,944 | 2,331 | 334 | 50 | 26 | 54 | 88 | 261 | 464 | 33 | 26 |
| Baxter | 1,583 | 4.0 | 37,749 | 2,351 | 608 | 101 | 46 | 83 | 307 | 605 | 839 | 72 | 46 |
| Benton | 25,876 | 2.9 | 92,697 | 30 | 8,486 | 918 | 559 | 415 | 14,711 | 1,980 | 10,378 | 686 | 559 |
| Boone | 1,363 | 2.3 | 36,414 | 2,521 | 632 | 102 | 47 | 58 | 230 | 435 | 839 | 64 | 47 |
| Bradley | 430 | 2.2 | 39,980 | 2,053 | 157 | 26 | 12 | 34 | 52 | 149 | 229 | 17 | 12 |
| Calhoun | 185 | 4.1 | 35,686 | 2,597 | 185 | 29 | 14 | 2 | 20 | 55 | 231 | 15 | 14 |
| Carroll | 937 | 0.3 | 33,021 | 2,887 | 383 | 64 | 30 | 39 | 185 | 293 | 516 | 41 | 30 |
| Chicot | 409 | 4.1 | 40,389 | 2,002 | 119 | 19 | 9 | 43 | 58 | 159 | 190 | 13 | 9 |
| Clark | 822 | 4.1 | 36,832 | 2,468 | 379 | 67 | 30 | 33 | 123 | 256 | 509 | 35 | 30 |
| Clay | 531 | 1.5 | 36,459 | 2,515 | 120 | 21 | 10 | 50 | 73 | 191 | 200 | 16 | 10 |
| Cleburne | 970 | 2.9 | 38,942 | 2,201 | 259 | 45 | 21 | 48 | 191 | 322 | 372 | 34 | 21 |

1. Based on the resident population estimated as of July 1 of the year shown.

# Table B. States and Counties — Earnings, Social Security, and Housing

| STATE County | Earnings, 2019 (cont.) Percent by selected industries | | | | | | | | | Social Security beneficiaries, December 2019 | | Supplemental Security Income recipients, 2019 | Housing units, 2020 | |
|---|---|---|---|---|---|---|---|---|---|---|---|---|---|---|
| | Farm | Mining, quarrying, and extractions | Construction | Manufacturing | Information; professional, scientific, technical services | Retail trade | Finance, insurance, real estate, and leasing | Health care and social assistance | Government | Number | Rate[1] | | Total | Percent change, 2010-2020 |
| | 75 | 76 | 77 | 78 | 79 | 80 | 81 | 82 | 83 | 84 | 85 | 86 | 87 | 88 |
| ALABAMA—Cont'd | | | | | | | | | | | | | | |
| Washington | 2.1 | D | 4.4 | 41.6 | D | 2.4 | D | D | 12.7 | 4,855 | 298 | 653 | 8,664 | 3.1 |
| Wilcox | 7.0 | 0 | 3.5 | 37.4 | 3.2 | 4.7 | 2.4 | D | 18.5 | 3,720 | 358 | 1,321 | 5,873 | 4.0 |
| Winston | -0.6 | D | 4.1 | 41.3 | D | 5.7 | 2.7 | 7.2 | 12.5 | 6,770 | 286 | 957 | 13,795 | 2.4 |
| ALASKA | 0.1 | 7.3 | 6.9 | 3.1 | 6.3 | 5.4 | 3.9 | 12.5 | 32.0 | 104,568 | 143 | 12,471 | 321,424 | 4.7 |
| Aleutians East | 0.0 | 0 | D | 74.8 | D | 0.8 | D | D | 11.6 | 165 | 49 | 15 | 755 | 1.1 |
| Aleutians West | 1.5 | D | 3.4 | 38.2 | D | D | D | 2.6 | 17.2 | 245 | 43 | 12 | 1,970 | 2.1 |
| Anchorage | 0.0 | 4.7 | 7.0 | 1.0 | 9.2 | 5.5 | 5.3 | 14.8 | 28.1 | 38,835 | 134 | 5,778 | 120,299 | 6.4 |
| Bethel | 0.0 | D | 1.2 | D | D | 5.3 | 4.1 | D | 46.3 | 1,830 | 99 | 393 | 6,048 | 2.2 |
| Bristol Bay | 0.0 | D | 5.1 | 42.5 | D | D | D | D | 19.9 | 135 | 161 | 11 | 985 | 1.8 |
| Chugach | NA | NA | NA | NA | NA | NA | NA | NA | NA | NA | NA | NA | 3,588 | 3.4 |
| Copper River | NA | NA | NA | NA | NA | NA | NA | NA | NA | NA | NA | NA | 2,632 | 0.0 |
| Denali | 0.0 | D | D | 0.6 | D | D | D | 0.7 | 25.8 | 270 | 127 | 10 | 1,768 | -0.3 |
| Dillingham | 0.0 | D | D | 19.9 | D | 3.5 | D | D | 28.7 | 575 | 117 | 119 | 2,457 | 1.2 |
| Fairbanks North Star | 0.1 | 2.8 | 7.8 | 1.0 | 4.4 | 5.6 | 2.8 | 10.6 | 47.7 | 12,055 | 124 | 1,162 | 44,311 | 6.0 |
| Haines | 0.0 | D | 14.5 | D | 3.5 | 7.4 | D | 8.4 | 19.1 | 610 | 238 | 38 | 1,695 | 3.9 |
| Hoonah-Angoon | 0.0 | 0.0 | D | 5.8 | D | 5.8 | D | 3.9 | 45.0 | 505 | 236 | 40 | 1,767 | 1.6 |
| Juneau | -0.2 | 6.8 | 5.0 | 1.5 | 4.8 | 5.4 | 3.4 | 7.5 | 47.1 | 4,870 | 152 | 490 | 14,027 | 7.4 |
| Kenai Peninsula | 0.1 | 7.3 | 6.6 | 6.1 | 4.3 | 6.4 | 2.9 | 13.1 | 31.4 | 12,585 | 214 | 1,005 | 31,542 | 3.2 |
| Ketchikan Gateway | 0.0 | D | 5.2 | 4.1 | 2.6 | 9.0 | 4.4 | 11.1 | 33.0 | 2,570 | 185 | 260 | 6,508 | 5.1 |
| Kodiak Island | 0.0 | D | 5.7 | 14.9 | 1.7 | 3.1 | 2.5 | D | 37.3 | 1,755 | 133 | 154 | 5,527 | 4.2 |
| Kusilvak | 0.0 | 0.0 | D | D | D | 6.6 | D | D | 67.9 | 800 | 95 | 204 | 2,237 | 2.5 |
| Lake and Peninsula | 0.0 | 0.0 | 1.8 | D | D | 2.7 | D | D | 37.3 | 175 | 111 | 19 | 1,510 | 0.3 |
| Matanuska-Susitna | 0.5 | 0.3 | 17.4 | 1.0 | 6.4 | 9.1 | 3.9 | 15.4 | 25.9 | 15,940 | 147 | 1,696 | 42,202 | 2.2 |
| Nome | 0.0 | 0.5 | 2.0 | D | D | 4.3 | 2.5 | D | 45.1 | 1,150 | 115 | 218 | 4,104 | 2.4 |
| North Slope | 0.0 | 60.9 | D | D | 2.4 | 0.7 | D | D | 13.4 | 695 | 74 | 36 | 2,630 | 5.2 |
| Northwest Arctic | 0.0 | D | D | 0.3 | D | D | D | D | 30.2 | 795 | 104 | 92 | 2,763 | 2.1 |
| Petersburg | 0.0 | D | 5.8 | 8.9 | D | 6.4 | 1.7 | D | 35.0 | 800 | 243 | 47 | 1,721 | 4.7 |
| Prince of Wales-Hyder | 0.0 | D | 4.5 | 2.8 | D | 6.9 | 3.0 | 5.2 | 48.1 | 1,085 | 175 | 103 | 3,457 | 3.0 |
| Sitka | 0.0 | 0.1 | 6.2 | 10.2 | 3.2 | 6.4 | 2.6 | 14.5 | 31.3 | 1,435 | 169 | 84 | 4,262 | 3.9 |
| Skagway | 0.0 | 0.0 | 3.2 | 6.5 | D | 14.2 | D | 0.7 | 23.9 | 140 | 117 | 0 | 734 | 15.4 |
| Southeast Fairbanks | 0.0 | D | 4.5 | 1.1 | 7 | 4.3 | 0.8 | 5.8 | 31.2 | 1,405 | 204 | 142 | 3,904 | -0.3 |
| Valdez-Cordova | 0.0 | 0.8 | 6.3 | 9.8 | 3.8 | 3.9 | D | 5.3 | 29.8 | 1,510 | ******** | 115 | NA | NA |
| Wrangell | 0.0 | 0.0 | 5.6 | 5.2 | D | 8.0 | 1.9 | D | 33.7 | 550 | 219 | 40 | 1,467 | 3.3 |
| Yakutat | 0.0 | 0.0 | D | 2.0 | D | 5.6 | D | 12.5 | 40.9 | 105 | 182 | 0 | 476 | 5.8 |
| Yukon-Koyukuk | 0.0 | 0.7 | 6.3 | D | D | 2.8 | 0.2 | D | 65.2 | 980 | 187 | 180 | 4,078 | 1.1 |
| ARIZONA | 0.7 | 0.9 | 6.7 | 7.7 | 10.6 | 6.7 | 11.0 | 12.5 | 15.5 | 1,399,403 | 192 | 119,532 | 3,120,343 | 9.7 |
| Apache | -1.1 | 1.0 | 2.1 | 1.0 | D | 3.4 | 2.3 | 13.3 | 60.5 | 13,820 | 192 | 3,966 | 33,097 | 1.8 |
| Cochise | 0.8 | 0.5 | 4.6 | 1.2 | 10.5 | 6.5 | 2.2 | 9.4 | 48.3 | 34,710 | 273 | 3,158 | 61,982 | 5.0 |
| Coconino | 0.2 | 0.1 | 4.6 | 8.1 | 5.4 | 7.0 | 4.1 | 15.5 | 29.7 | 21,590 | 150 | 2,390 | 68,354 | 7.9 |
| Gila | 0.1 | 9.8 | 5.0 | 12.2 | 3.3 | 7.1 | 2.8 | 9.1 | 36.0 | 18,070 | 334 | 1,377 | 34,024 | 4.1 |
| Graham | 2.0 | D | 4.0 | 1.9 | 1.9 | 9.0 | D | 11.2 | 33.8 | 6,585 | 169 | 760 | 13,952 | 7.5 |
| Greenlee | 0.8 | D | 1.7 | 0.1 | D | 2.3 | D | D | 7.2 | 1,465 | 155 | 109 | 4,486 | 2.6 |
| La Paz | 7.4 | 0.6 | 2.2 | 2.5 | 3.5 | 13.4 | D | D | 38.7 | 6,180 | 291 | 457 | 16,389 | 2.1 |
| Maricopa | 0.2 | 0.5 | 7.2 | 7.6 | 11.7 | 6.4 | 13.3 | 12.3 | 11.3 | 732,645 | 163 | 61,003 | 1,817,097 | 10.8 |
| Mohave | 0.8 | 0.4 | 7.5 | 6.6 | 4.8 | 12.9 | 5.2 | 18.8 | 17.6 | 72,295 | 338 | 4,646 | 117,650 | 6.1 |
| Navajo | 0.3 | 2.0 | 4.6 | 1.3 | 3.3 | 9.7 | 2.6 | 15.4 | 37.0 | 25,955 | 234 | 4,420 | 58,942 | 3.5 |
| Pima | 0.1 | 0.9 | 5.3 | 11.3 | 9.8 | 6.8 | 6.2 | 14.0 | 23.5 | 235,840 | 225 | 21,131 | 470,626 | 6.7 |
| Pinal | 5.6 | 2.6 | 4.8 | 6.3 | 5.7 | 8.1 | 3.2 | 7.0 | 32.5 | 93,450 | 202 | 6,325 | 186,629 | 17.2 |
| Santa Cruz | 0.9 | D | 2.2 | 2.3 | 2.4 | 9.4 | 1.7 | 4.4 | 33.0 | 10,690 | 229 | 1,396 | 18,601 | 3.3 |
| Yavapai | 0.2 | 2.7 | 9.4 | 5.9 | 6.7 | 10.3 | 5.0 | 15.5 | 19.0 | 85,265 | 361 | 3,583 | 122,590 | 11.0 |
| Yuma | 13.4 | 0.0 | 3.8 | 4.2 | 5.1 | 8.9 | 3.4 | 10.6 | 28.1 | 40,845 | 191 | 4,811 | 95,924 | 9.2 |
| ARKANSAS | 1.5 | 0.4 | 5.9 | 13.0 | 6.3 | 6.4 | 6.0 | 13.0 | 16.8 | 703,896 | 233 | 104,312 | 1,401,101 | 6.4 |
| Arkansas | 6.6 | D | 3.4 | 36.0 | 3.4 | 6.3 | 4.1 | D | 8.9 | 4,615 | 263 | 670 | 9,507 | 0.7 |
| Ashley | 6.5 | D | 9.5 | 32.7 | D | 5.1 | 2.7 | D | 11.1 | 5,785 | 293 | 878 | 10,189 | 0.5 |
| Baxter | -0.2 | 0.3 | 5.3 | 16.1 | D | 9.3 | 7.8 | 28.4 | 11.0 | 16,140 | 384 | 1,018 | 23,052 | 2.1 |
| Benton | 0.4 | 0.0 | 4.9 | 6.9 | 9.6 | 4.8 | 1.8 | 6.1 | 6.3 | 47,005 | 168 | 3,386 | 115,971 | 24.6 |
| Boone | -2.3 | 0 | D | 12.9 | 4.4 | 8.8 | 4.1 | 9.7 | 21.2 | 11,285 | 300 | 1,184 | 17,075 | 1.4 |
| Bradley | 1.5 | D | 5.7 | 23.9 | 1.6 | 4.6 | 3.7 | 16.2 | 17.3 | 2,795 | 259 | 490 | 5,801 | |
| Calhoun | 0.2 | D | 3.4 | 77.9 | D | D | D | 1.5 | 6.0 | 1,355 | 264 | 153 | 2,942 | 1.6 |
| Carroll | -1.5 | 0.2 | 7.5 | 31.1 | D | 6.7 | 5.5 | 10.0 | 12.5 | 8,170 | 288 | 576 | 13,897 | 2.5 |
| Chicot | 20.1 | 0.0 | 5.6 | 0.8 | D | 5.3 | 8.0 | 14.3 | 25.4 | 2,915 | 286 | 831 | 5,428 | 0.1 |
| Clark | 0.0 | D | 2.8 | 19.4 | 3.3 | 9.7 | 3.6 | D | 23.3 | 5,090 | 228 | 720 | 10,601 | 2.1 |
| Clay | 13.8 | 0.0 | 5.1 | 2.3 | D | 6.3 | 3.8 | 11.5 | 22.1 | 4,550 | 314 | 613 | 8,029 | 0.0 |
| Cleburne | -0.1 | 3.2 | 8.9 | 18.7 | 3.8 | 9.9 | 5.9 | D | 14.2 | 8,760 | 350 | 761 | 16,206 | 2.4 |

1. Per 1,000 resident population estimated as of July 1 of the year shown.

# Table B. States and Counties — Housing, Labor Force, and Employment

| STATE County | Housing units, 2015-2019 | | | | | | | | Civilian labor force, 2020 | | | | Civilian employment[6], 2015-2019 | | |
| | Occupied units | | | | | | | Sub-standard units[4] (percent) | | | Unemployment | | | Percent | |
| | Owner-occupied | | | | | Renter-occupied | | | | | | | | | |
| | | | | Median owner cost as a percent of income | | | Median rent as a percent of income[2] | | | | | | | Management, business, science, and arts | Construction, production, and maintenance occupations |
| | Total | Percent | Median value[1] | With a mortgage | Without a mortgage[2] | Median rent[3] | | | Total | Percent change, 2019-2020 | Total | Rate[5] | Total | | |
| | 89 | 90 | 91 | 92 | 93 | 94 | 95 | 96 | 97 | 98 | 99 | 100 | 101 | 102 | 103 |

**ALABAMA—Cont'd**

| STATE County | 89 | 90 | 91 | 92 | 93 | 94 | 95 | 96 | 97 | 98 | 99 | 100 | 101 | 102 | 103 |
|---|---|---|---|---|---|---|---|---|---|---|---|---|---|---|---|
| Washington | 5,990 | 82.7 | 89,100 | 18.0 | 10.7 | 730 | 23.2 | 2.8 | 6,729 | 0.4 | 539 | 8.0 | 5,582 | 25.3 | 38.9 |
| Wilcox | 3,854 | 67.9 | 79,400 | 21.6 | 17.3 | 509 | 28.8 | 3.0 | 2,871 | 6.1 | 423 | 14.7 | 2,995 | 22.5 | 37.1 |
| Winston | 9,592 | 79.4 | 86,100 | 21.7 | 12.2 | 529 | 25.7 | 1.9 | 10,017 | | 475 | 4.7 | 9,120 | 24.5 | 42.0 |
| | | | | | | | | | | | | | | | |
| ALASKA | 253,346 | 64.3 | 270,400 | 22.0 | 10.1 | 1,244 | 27.7 | 9.6 | 347,414 | -1.8 | 27,195 | 7.8 | 347,774 | 37.6 | 24.3 |
| | | | | | | | | | | | | | | | |
| Aleutians East | 890 | 63.8 | 119,900 | 16.9 | 10.0 | 939 | 19.6 | 5.6 | 2,523 | -1.4 | 100 | 4.0 | 2,306 | 16.8 | 58.5 |
| Aleutians West | 1,187 | 28.2 | 279,200 | 21.6 | 11.6 | 1,341 | 18.9 | 12.7 | 3,732 | 2.1 | 169 | 4.5 | 3,830 | 18.3 | 49.7 |
| Anchorage | 106,567 | 61.3 | 314,800 | 22.1 | 11.0 | 1,320 | 28.3 | 5.2 | 148,392 | -0.5 | 10,971 | 7.4 | 148,021 | 41.1 | 19.4 |
| Bethel | 4,489 | 58.2 | 83,600 | 18.3 | 10.9 | 1,352 | 22.2 | 53 | 6,963 | -1.9 | 805 | 11.6 | 6,243 | 35.2 | 21.8 |
| Bristol Bay | 314 | 53.8 | 188,800 | 18.8 | 10.0 | 1,120 | 19.1 | 6.7 | 417 | -7.5 | 31 | 7.4 | 468 | 31.8 | 32.5 |
| Chugach | NA | NA | NA | NA | NA | NA | NA | NA | NA | NA | NA | NA | NA | NA | NA |
| Copper River | NA | NA | NA | NA | NA | NA | NA | NA | NA | NA | NA | NA | NA | NA | NA |
| Denali | 613 | 81.6 | 223,000 | 18.7 | 10.0 | 845 | 16.7 | 19.2 | 680 | -38.3 | 107 | 15.7 | 1,508 | 19.1 | 25.1 |
| Dillingham | 1,427 | 62.0 | 183,500 | 19.0 | 12.5 | 1,020 | 23.4 | 32.0 | 1,742 | -12.3 | 143 | 8.2 | 2,030 | 48.0 | 17.9 |
| Fairbanks North Star | 36,188 | 58.9 | 240,000 | 21.9 | 10.0 | 1,305 | 28.0 | 10.9 | 44,679 | -1.8 | 2,841 | 6.4 | 45,538 | 35.8 | 26.0 |
| | | | | | | | | | | | | | | | |
| Haines | 1,007 | 66.4 | 254,000 | 25.5 | 11.0 | 951 | 22.9 | 20.9 | 1,018 | -10.2 | 159 | 15.6 | 1,181 | 38.8 | 17.1 |
| Hoonah-Angoon | 791 | 75.5 | 223,600 | 22.7 | 10.0 | 894 | 21.0 | 11.8 | 1,050 | -12.0 | 142 | 13.5 | 1,030 | 33.5 | 28.6 |
| Juneau | 12,676 | 64.8 | 345,900 | 21.9 | 10.0 | 1,310 | 25.9 | 4.2 | 16,374 | -5.4 | 1,081 | 6.6 | 17,132 | 44.5 | 22.1 |
| Kenai Peninsula | 21,630 | 72.8 | 242,200 | 22.1 | 10.0 | 987 | 27.5 | 7.9 | 26,341 | -0.8 | 2,393 | 9.1 | 25,314 | 35.6 | 26.7 |
| Ketchikan Gateway | 5,397 | 61.6 | 289,900 | 21.3 | 10.5 | 1,169 | 31.3 | 5.4 | 6,520 | -7.1 | 627 | 9.6 | 7,116 | 31.0 | 25.9 |
| Kodiak Island | 4,261 | 54.6 | 282,100 | 23.6 | 12.7 | 1,458 | 25.1 | 9.4 | 6,090 | 1.7 | 380 | 6.2 | 6,680 | 29.0 | 39.0 |
| Kusilvak | 1,724 | 71.7 | 71,500 | 16.9 | 13.8 | 632 | 20.1 | 62.6 | 2,299 | -13.3 | 445 | 19.4 | 2,192 | 28.5 | 26.1 |
| Lake and Peninsula | 421 | 71.5 | 114,900 | 26.7 | 11.6 | 743 | 14.9 | 24.7 | 689 | -3.0 | 65 | 9.4 | 653 | 37.4 | 27.6 |
| Matanuska-Susitna | 31,217 | 77.1 | 249,000 | 22.6 | 10.0 | 1,127 | 31.1 | 8.0 | 47,848 | -1.1 | 3,990 | 8.3 | 43,782 | 33.4 | 28.9 |
| Nome | 2,844 | 60.5 | 154,600 | 18.4 | 12.8 | 1,287 | 24.4 | 43.4 | 3,927 | -2.6 | 411 | 10.5 | 3,693 | 40.3 | 18.2 |
| | | | | | | | | | | | | | | | |
| North Slope | 1,979 | 52.7 | 152,500 | 16.5 | 10.0 | 1,000 | 16.5 | 36.8 | 3,200 | -10.1 | 189 | 5.9 | 5,404 | 36.1 | 36.5 |
| Northwest Arctic | 1,795 | 56.2 | 141,400 | 18.5 | 13.7 | 1,304 | 22.4 | 42.6 | 2,942 | 1.2 | 352 | 12.0 | 2,619 | 39.1 | 26.0 |
| Petersburg | 1,187 | 68.8 | 227,400 | 18.7 | 10.0 | 954 | 27.5 | 3.3 | 1,415 | 0.1 | 123 | 8.7 | 1,653 | 26.7 | 43.8 |
| Prince of Wales-Hyder | 2,371 | 72.8 | 174,900 | 19.3 | 10.0 | 783 | 23.4 | 12.1 | 2,828 | -2.1 | 260 | 9.2 | 2,753 | 34.2 | 31.6 |
| Sitka | 3,547 | 59.3 | 358,600 | 23.0 | 10.4 | 1,131 | 29.5 | 4.1 | 4,132 | -6.5 | 285 | 6.9 | 4,629 | 43.4 | 21.8 |
| Skagway | 375 | 53.3 | 338,500 | 22.4 | 10.0 | 1,004 | 19.7 | 7.5 | 606 | -22.7 | 130 | 21.5 | 738 | 28.5 | 21.5 |
| Southeast Fairbanks | 2,170 | 73.5 | 170,900 | 17.1 | 10.0 | 1,042 | 22.0 | 13.7 | 2,934 | 1.9 | 217 | 7.4 | 2,801 | 32.2 | 31.0 |
| Valdez-Cordova | 3,177 | 74.6 | 222,200 | 17.7 | 10.0 | 1,067 | 20.5 | 7.6 | 4,522 | -7.3 | 395 | 8.7 | 4,846 | 32.3 | 35.5 |
| Wrangell | 1,027 | 67.9 | 216,200 | 23.4 | 11.9 | 844 | 27.7 | 4.6 | 984 | -4.8 | 83 | 8.4 | 1,085 | 35.4 | 27.2 |
| Yakutat | 219 | 66.2 | 202,300 | 19.4 | 10.0 | 900 | 20.0 | 11.0 | 260 | 3.2 | 22 | 8.5 | 339 | 27.1 | 28.6 |
| | | | | | | | | | | | | | | | |
| Yukon-Koyukuk | 1,856 | 72.4 | 76,500 | 16.5 | 10.7 | 741 | 21.7 | 45.0 | 2,308 | -5.6 | 280 | 12.1 | 2,190 | 35.8 | 26.8 |
| | | | | | | | | | | | | | | | |
| ARIZONA | 2,571,268 | 64.4 | 225,500 | 21.1 | 10.0 | 1,052 | 29.1 | 5.0 | 3,570,220 | 0.9 | 282,070 | 7.9 | 3,130,658 | 36.5 | 20.1 |
| | | | | | | | | | | | | | | | |
| Apache | 20,867 | 78.9 | 59,900 | 20.5 | 10.0 | 506 | 16.6 | 23.0 | 20,195 | -1.2 | 2,581 | 12.8 | 18,400 | 31.2 | 25.3 |
| Cochise | 50,163 | 68.9 | 146,800 | 20.2 | 10.0 | 783 | 27.4 | 3.2 | 51,995 | 3.4 | 3,627 | 7.0 | 43,760 | 33.2 | 19.4 |
| Coconino | 47,447 | 61.0 | 280,900 | 21.7 | 10.0 | 1,181 | 31.3 | 8.9 | 75,280 | -2.4 | 7,295 | 9.7 | 66,820 | 37.9 | 17.6 |
| Gila | 21,945 | 75.6 | 165,800 | 22.8 | 11.6 | 816 | 25.8 | 5.7 | 21,008 | -1.9 | 1,621 | 7.7 | 18,695 | 33.7 | 23.2 |
| Graham | 11,017 | 70.3 | 138,700 | 20.2 | 10.0 | 769 | 25.6 | 8.3 | 15,919 | 3.2 | 1,008 | 6.3 | 13,104 | 25.9 | 29.7 |
| Greenlee | 3,132 | 48.6 | 83,800 | 16.4 | 10.0 | 466 | 10.0 | 6.3 | 4,454 | -0.6 | 244 | 5.5 | 3,870 | 22.6 | 49.4 |
| La Paz | 9,346 | 71.5 | 79,200 | 22.2 | 10.0 | 565 | 25.0 | 5.9 | 8,979 | -3.3 | 636 | 7.1 | 6,160 | 22.5 | 32.4 |
| Maricopa | 1,552,096 | 62.2 | 260,200 | 20.8 | 10.0 | 1,127 | 29.1 | 4.8 | 2,331,628 | 1.4 | 172,361 | 7.4 | 2,068,218 | 38.3 | 19.3 |
| Mohave | 86,889 | 69.3 | 160,500 | 22.5 | 10.0 | 826 | 28.2 | 4.7 | 88,743 | 1.8 | 8,854 | 10.0 | 72,932 | 24.8 | 24.9 |
| Navajo | 34,990 | 68.8 | 126,100 | 21.3 | 10.0 | 720 | 26.5 | 13.2 | 40,419 | -0.7 | 4,097 | 10.1 | 33,415 | 29.8 | 24.0 |
| | | | | | | | | | | | | | | | |
| Pima | 404,739 | 63.2 | 184,100 | 21.1 | 10.2 | 907 | 30.5 | 4.0 | 495,991 | -0.3 | 38,308 | 7.7 | 446,383 | 37.0 | 18.2 |
| Pinal | 141,300 | 76.7 | 183,100 | 21.6 | 10.0 | 1,064 | 29.2 | 4.5 | 191,074 | 0.7 | 14,400 | 7.5 | 157,798 | 31.0 | 23.6 |
| Santa Cruz | 15,853 | 67.7 | 151,200 | 23.5 | 11.1 | 675 | 30.6 | 6.2 | 19,429 | -0.9 | 2,220 | 11.4 | 17,163 | 26.5 | 24.0 |
| Yavapai | 98,386 | 72.0 | 252,000 | 23.8 | 10.8 | 947 | 29.7 | 2.5 | 106,985 | 0.4 | 8,007 | 7.5 | 88,037 | 32.1 | 21.5 |
| Yuma | 73,098 | 67.1 | 127,900 | 23.0 | 10.0 | 839 | 28.2 | 7.8 | 98,120 | -2.1 | 16,810 | 17.1 | 75,903 | 26.3 | 31.8 |
| | | | | | | | | | | | | | | | |
| ARKANSAS | 1,158,071 | 65.6 | 127,800 | 18.8 | 10.0 | 745 | 27.8 | 3.3 | 1,354,296 | -0.8 | 81,952 | 6.1 | 1,303,490 | 33.7 | 27.9 |
| | | | | | | | | | | | | | | | |
| Arkansas | 7,491 | 65.2 | 88,600 | 18.6 | 10.0 | 648 | 26.8 | 1.5 | 9,422 | -0.6 | 378 | 4.0 | 7,864 | 28.7 | 34.5 |
| Ashley | 7,757 | 77.5 | 76,400 | 17.5 | 10.0 | 533 | 30.2 | 2.1 | 7,186 | -6.8 | 668 | 9.3 | 7,796 | 26.0 | 39.6 |
| Baxter | 18,435 | 76.1 | 133,400 | 21.6 | 10.2 | 703 | 26.7 | 2.5 | 16,528 | 1.2 | 981 | 5.9 | 16,051 | 32.1 | 30.9 |
| Benton | 97,249 | 66.6 | 183,400 | 18.0 | 10.0 | 903 | 23.1 | 3.8 | 140,015 | 0.6 | 6,231 | 4.5 | 129,785 | 37.9 | 25.3 |
| Boone | 15,034 | 71.8 | 124,300 | 19.8 | 10.0 | 636 | 25.9 | 2.6 | 15,764 | -1.9 | 830 | 5.3 | 16,174 | 30.4 | 27.3 |
| Bradley | 4,419 | 65.2 | 89,800 | 17.1 | 11.8 | 585 | 23.2 | 2.9 | 4,386 | -2.1 | 256 | 5.8 | 4,091 | 25.4 | 43.6 |
| Calhoun | 1,850 | 84.8 | 70,000 | 16.4 | 10.0 | 640 | 30.4 | 1.9 | 2,426 | -0.4 | 125 | 5.2 | 2,111 | 25.5 | 45.2 |
| Carroll | 11,139 | 75.7 | 136,900 | 20.1 | 10.2 | 610 | 23.9 | 6.8 | 12,709 | -0.9 | 720 | 5.7 | 12,013 | 29.1 | 32.8 |
| Chicot | 4,068 | 61.8 | 69,700 | 21.3 | 12.9 | 601 | 28.1 | 2.5 | 3,250 | -0.2 | 336 | 10.3 | 3,282 | 30.2 | 27.7 |
| Clark | 8,446 | 57.5 | 111,600 | 16.8 | 10.0 | 606 | 28.6 | 3.3 | 9,314 | -4.0 | 549 | 5.9 | 10,575 | 30.9 | 26.5 |
| | | | | | | | | | | | | | | | |
| Clay | 6,444 | 67.4 | 77,400 | 18.6 | 11.5 | 618 | 28.5 | 2.5 | 5,519 | -2.7 | 333 | 6.0 | 6,391 | 27.4 | 33.8 |
| Cleburne | 10,783 | 75.2 | 129,900 | 19.6 | 10.3 | 704 | 24.3 | 1.5 | 9,157 | -0.8 | 697 | 7.6 | 10,314 | 27.6 | 32.3 |

1. Specified owner-occupied units.   2. A value of 10.0 represents 10 percent or less; a value of 50.0 represents 50 percent or more.   3. Specified renter-occupied units.   4. Overcrowded or lacking complete plumbing facilities.   5. Percent of civilian labor force.   6. Civilian employed persons 16 years old and over.

# Table B. States and Counties — Nonfarm Employment and Agriculture

| STATE County | Private nonfarm establishments, employment and payroll, 2019 | | | | | | | | | Agriculture, 2017 | | | |
|---|---|---|---|---|---|---|---|---|---|---|---|---|---|
| | Number of establish-ments | Employment | | | | | | Annual payroll | | Farms | | | Farm producers whose primary occupation is farming (percent) |
| | | Total | Health care and social assistance | Manufac-turing | Retail trade | Finance and insurance | Professional, scientific, and technical services | Total (mil dol) | Average per employee (dollars) | Number | Percent with: | | |
| | | | | | | | | | | | Fewer than 50 acres | 1000 acres or more | |
| | 104 | 105 | 106 | 107 | 108 | 109 | 110 | 111 | 112 | 113 | 114 | 115 | 116 |

| ALABAMA—Cont'd | | | | | | | | | | | | | |
|---|---|---|---|---|---|---|---|---|---|---|---|---|---|
| Washington | 208 | 5,614 | 272 | 3,510 | 286 | 40 | 27 | 434 | 77,355 | 435 | 36.6 | 5.3 | 34.1 |
| Wilcox | 164 | 2,133 | 236 | 994 | 252 | 82 | 18 | 102 | 48,035 | 318 | 25.8 | 19.8 | 39.1 |
| Winston | 442 | 6,856 | 731 | 3,228 | 793 | 171 | 74 | 251 | 36,676 | 484 | 40.3 | 1.0 | 37.1 |
| | | | | | | | | | | | | | |
| ALASKA | 21,399 | 264,971 | 51,584 | 12,334 | 33,461 | 7,271 | 19,332 | 16,572 | 62,542 | 990 | 67.0 | 4.3 | 41.9 |
| | | | | | | | | | | | | | |
| Aleutians East | 60 | 2,064 | 48 | 1,878 | 50 | NA | NA | 106 | 51,406 | 46 | 65.2 | 28.3 | 31.2 |
| Aleutians West | 116 | 3,454 | 101 | 2,261 | 109 | NA | 13 | 161 | 46,753 | (1) | (1) | (1) | (1) |
| Anchorage | 8,763 | 144,931 | 26,964 | 1,991 | 15,162 | 4,732 | 13,852 | 9,471 | 65,346 | 350 | 64.3 | 0.9 | 44.4 |
| Bethel | 217 | 2,422 | 563 | NA | 898 | 32 | 22 | 124 | 51,104 | (1) | (1) | (1) | (1) |
| Bristol Bay | 96 | 358 | 15 | 91 | 49 | NA | NA | 58 | 160,690 | (1) | (1) | (1) | (1) |
| Chugach | NA | NA | NA | NA | NA | NA | NA | NA | NA | | | | |
| Copper River | NA | NA | NA | NA | NA | NA | NA | NA | NA | | | | |
| Denali | 119 | 1,018 | 13 | NA | 56 | NA | 405 | 96 | 94,534 | (1) | (1) | (1) | (1) |
| Dillingham | 109 | 1,178 | 675 | 11 | 232 | 10 | 36 | 76 | 64,883 | (1) | (1) | (1) | (1) |
| Fairbanks North Star | 2,515 | 27,029 | 6,347 | 513 | 4,568 | 629 | 1,260 | 1,534 | 56,745 | 274 | 49.6 | 8.0 | 45.0 |
| | | | | | | | | | | | | | |
| Haines | 137 | 658 | 122 | 22 | 144 | NA | 14 | 35 | 53,862 | (1) | (1) | (1) | (1) |
| Hoonah-Angoon | 86 | 261 | 43 | 12 | 70 | NA | NA | 16 | 60,215 | (1) | (1) | (1) | (1) |
| Juneau | 1,130 | 11,097 | 2,348 | 272 | 1,644 | 330 | 483 | 640 | 57,655 | 60 | 96.7 | 0.0 | 41.1 |
| Kenai Peninsula | 2,216 | 15,060 | 4,035 | 753 | 2,576 | 292 | 558 | 788 | 52,336 | 260 | 82.7 | 1.9 | 37.9 |
| Ketchikan Gateway | 609 | 4,557 | 806 | 287 | 793 | 194 | 124 | 279 | 61,302 | (1) | (1) | (1) | (1) |
| Kodiak Island | 474 | 5,057 | 715 | 2,251 | 443 | 75 | 68 | 208 | 41,096 | (1) | (1) | (1) | (1) |
| Kusilvak | 74 | 650 | 22 | NA | 351 | NA | NA | 16 | 24,840 | (1) | (1) | (1) | (1) |
| Lake and Peninsula | 50 | 313 | NA | 52 | 11 | NA | NA | 19 | 61,741 | (1) | (1) | (1) | (1) |
| Matanuska-Susitna | 2,398 | 19,923 | 4,645 | 643 | 3,765 | 522 | 826 | 985 | 49,457 | (1) | (1) | (1) | (1) |
| Nome | 178 | 2,198 | 1,112 | NA | 365 | 35 | 15 | 134 | 60,973 | (1) | (1) | (1) | (1) |
| | | | | | | | | | | | | | |
| North Slope | 153 | 4,744 | 373 | NA | 285 | NA | 1,136 | 540 | 113,763 | (1) | (1) | (1) | (1) |
| Northwest Arctic | 68 | 1,822 | 452 | NA | 170 | NA | NA | 182 | 99,783 | (1) | (1) | (1) | (1) |
| Petersburg | 161 | 903 | 175 | 165 | 191 | 22 | 11 | 41 | 44,862 | (1) | (1) | (1) | (1) |
| Prince of Wales-Hyder | 172 | 1,013 | 159 | 132 | 234 | 26 | NA | 56 | 55,315 | (1) | (1) | (1) | (1) |
| Sitka | 367 | 3,010 | 904 | 361 | 389 | 65 | 73 | 163 | 54,111 | (1) | (1) | (1) | (1) |
| Skagway | 116 | 338 | 4 | NA | 90 | NA | NA | 32 | 93,598 | (1) | (1) | (1) | (1) |
| Southeast Fairbanks | 162 | 1,277 | 101 | 9 | 213 | 20 | 41 | 106 | 82,755 | (1) | (1) | (1) | (1) |
| Valdez-Cordova | 496 | 2,595 | 564 | 198 | 340 | 34 | 98 | 185 | 71,225 | (1) | (1) | (1) | (1) |
| Wrangell | 96 | 484 | 175 | 17 | 102 | 15 | 18 | 23 | 47,686 | (1) | (1) | (1) | (1) |
| Yakutat | 27 | 121 | 18 | NA | 38 | NA | NA | 8 | 68,017 | (1) | (1) | (1) | (1) |
| | | | | | | | | | | | | | |
| Yukon-Koyukuk | 95 | 340 | 26 | NA | 123 | NA | 14 | 13 | 36,859 | (1) | (1) | (1) | (1) |
| | | | | | | | | | | | | | |
| ARIZONA | 147,163 | 2,614,641 | 385,662 | 154,946 | 327,721 | 176,872 | 161,153 | 129,708 | 49,608 | 19,086 | 69.1 | 11.0 | 53.3 |
| | | | | | | | | | | | | | |
| Apache | 453 | 6,240 | 2,390 | 80 | 1,223 | 78 | 68 | 262 | 41,964 | 5,551 | 72.3 | 10.5 | 55.1 |
| Cochise | 2,136 | 26,186 | 5,066 | 308 | 5,045 | 487 | 3,025 | 989 | 37,757 | 1,083 | 39.6 | 17.7 | 47.4 |
| Coconino | 3,744 | 55,692 | 9,660 | 4,891 | 8,006 | 925 | 1,800 | 2,292 | 41,163 | 2,142 | 77.8 | 11.6 | 55.7 |
| Gila | 1,014 | 11,231 | 2,142 | 169 | 2,014 | 182 | 251 | 463 | 41,187 | 298 | 67.1 | 5.0 | 43.8 |
| Graham | 521 | 7,535 | 2,094 | 253 | 1,530 | 117 | 492 | 305 | 40,451 | 448 | 64.3 | 10.0 | 34.8 |
| Greenlee | 76 | 3,509 | 80 | NA | 203 | NA | 19 | 241 | 68,760 | 123 | 52.0 | 8.1 | 55.4 |
| La Paz | 337 | 3,737 | 415 | 174 | 1,134 | 58 | 15 | 108 | 28,926 | 97 | 28.9 | 26.8 | 54.5 |
| Maricopa | 97,970 | 1,796,017 | 252,407 | 110,471 | 215,276 | 148,446 | 124,185 | 95,246 | 53,032 | 1,874 | 81.7 | 4.7 | 44.0 |
| Mohave | 3,907 | 45,840 | 8,916 | 3,319 | 10,616 | 1,112 | 1,257 | 1,648 | 35,960 | 317 | 60.6 | 17.4 | 40.7 |
| Navajo | 1,791 | 19,469 | 4,834 | 387 | 3,994 | 339 | 342 | 742 | 38,120 | 4,205 | 67.0 | 11.8 | 61.4 |
| | | | | | | | | | | | | | |
| Pima | 20,400 | 331,057 | 63,841 | 23,565 | 46,672 | 13,284 | 17,635 | 14,743 | 44,534 | 661 | 79.7 | 7.9 | 42.9 |
| Pinal | 3,827 | 53,795 | 9,388 | 4,062 | 9,255 | 1,110 | 1,385 | 2,078 | 38,623 | 762 | 56.2 | 15.9 | 55.9 |
| Santa Cruz | 1,133 | 11,432 | 1,608 | 369 | 2,169 | 185 | 188 | 433 | 37,869 | 219 | 47.9 | 16.4 | 49.6 |
| Yavapai | 6,002 | 62,017 | 13,859 | 3,641 | 10,497 | 1,136 | 1,918 | 2,318 | 37,377 | 850 | 73.2 | 8.5 | 46.2 |
| Yuma | 3,113 | 45,290 | 8,297 | 3,257 | 8,225 | 1,261 | 1,237 | 1,678 | 37,055 | 456 | 60.7 | 13.2 | 50.4 |
| | | | | | | | | | | | | | |
| ARKANSAS | 67,243 | 1,053,453 | 186,435 | 161,716 | 138,148 | 39,953 | 39,670 | 46,260 | 43,912 | 42,625 | 30.3 | 7.2 | 41.7 |
| | | | | | | | | | | | | | |
| Arkansas | 484 | 9,152 | 985 | 4,081 | 984 | 257 | 69 | 376 | 41,126 | 488 | 16.2 | 32.2 | 43.6 |
| Ashley | 372 | 5,359 | 845 | 1,750 | 670 | 144 | 48 | 260 | 48,509 | 353 | 43.3 | 11.6 | 35.8 |
| Baxter | 1,045 | 13,624 | 4,094 | 2,617 | 2,175 | 530 | 434 | 570 | 41,846 | 479 | 35.5 | 3.8 | 36.9 |
| Benton | 6,834 | 122,863 | 10,347 | 12,201 | 12,909 | 3,216 | 9,145 | 7,596 | 61,824 | 1,936 | 48.2 | 1.4 | 38.7 |
| Boone | 849 | 11,799 | 2,409 | 1,781 | 2,046 | 335 | 224 | 435 | 36,861 | 1,313 | 33.0 | 4.3 | 41.9 |
| Bradley | 247 | 2,472 | 531 | 577 | 263 | 114 | 32 | 86 | 34,984 | 181 | 36.5 | 1.7 | 32.3 |
| Calhoun | 57 | 406 | 36 | NA | 49 | 14 | NA | 16 | 39,894 | 101 | 25.7 | 4.0 | 35.1 |
| Carroll | 688 | 8,382 | 670 | 2,940 | 1,067 | 230 | 160 | 258 | 30,743 | 1,169 | 25.7 | 4.0 | 41.1 |
| Chicot | 199 | 1,939 | 777 | 32 | 273 | 86 | 27 | 65 | 33,718 | 291 | 19.9 | 36.1 | 51.3 |
| Clark | 533 | 6,868 | 1,047 | 1,328 | 1,262 | 209 | 198 | 240 | 34,891 | 377 | 27.6 | 5.8 | 32.1 |
| | | | | | | | | | | | | | |
| Clay | 275 | 2,353 | 776 | 96 | 397 | 72 | 40 | 75 | 31,755 | 542 | 27.1 | 15.7 | 42.9 |
| Cleburne | 578 | 5,679 | 775 | 1,391 | 967 | 214 | 143 | 185 | 32,606 | 676 | 27.1 | 1.8 | 41.6 |

1. Agriculture Census data for Alaska is combined into regional groups. Aleutians West, Bethel, Bristol Bay, Dillingham, Kodiak Island, Kusilvak, Lake and Peninsula, Nome, North Slope, and Northwest Arctic are included with Aleutians East. Matanuska-Susitna and Valdez-Cordova are included with Anchorage. Denali, Southeast Fairbanks, and Yukon-Koyukuk are included with Fairbanks North Star. Haines, Hoonah-Angoon, Ketchikan Gateway, Petersburg, Prince of Wales-Hyder, Sitka, Skagway, Wrangell, and Yakutat are included with Juneau

| STATE County | Land in farms — Acreage (1,000) | Land in farms — Percent change, 2012-2017 | Acres — Average size of farm | Acres — Total irrigated (1,000) | Acres — Total cropland (1,000) | Value of land and buildings (dollars) — Average per farm | Value of land and buildings (dollars) — Average per acre | Value of machinery and equiopmnet, average per farm (dollars) | Value of products sold: Total (mil dol) | Value of products sold: Average per farm (acres) | Percent from: Crops | Percent from: Livestock and poultry products | Organic farms (number) | Farms with internet access (per-cent) | Government payments — Total ($1,000) | Government payments — Percent of farms |
|---|---|---|---|---|---|---|---|---|---|---|---|---|---|---|---|---|
| | 117 | 118 | 119 | 120 | 121 | 122 | 123 | 124 | 125 | 126 | 127 | 128 | 129 | 130 | 131 | 132 |
| **ALABAMA—Cont'd** | | | | | | | | | | | | | | | | |
| Washington .................. | 120 | 25.1 | 275 | 0.1 | 22.3 | 556,827 | 2,026 | 72,905 | 31.3 | 71,885 | 23.1 | 76.9 | NA | 65.1 | 1,739 | 33.8 |
| Wilcox ....................... | 165 | 38.2 | 520 | D | 26.9 | 945,109 | 1,818 | 66,889 | 9.1 | 28,736 | 26.7 | 73.3 | NA | 64.2 | 1,920 | 57.5 |
| Winston ...................... | 57 | -1.4 | 118 | 0.6 | 11.9 | 326,386 | 2,765 | 59,645 | 27.7 | 57,295 | 3.5 | 96.5 | NA | 70.5 | 631 | 43.2 |
| **ALASKA** ................... | 850 | 1.9 | 858 | 2.4 | 83.7 | 616,112 | 718 | 91,623 | 70.5 | 71,171 | 42.1 | 57.9 | 18 | 87.9 | 2,091 | 22.7 |
| Aleutians East .............. | 681 | 2.0 | 14,811 | 0.0 | 1.1 | 573,623 | 39 | 90,364 | 3.0 | 64,774 | 13.1 | 86.9 | 1 | 89.1 | 193 | 28.3 |
| Aleutians West ............. | (1) | (1) | (1) | (1) | (1) | (1) | (1) | (1) | (1) | (1) | (1) | (1) | (1) | (1) | (1) | (1) |
| Anchorage ................... | 34 | -5.5 | 98 | 1.1 | 15.6 | 809,796 | 8,242 | 106,917 | 37.5 | 107,245 | 45.6 | 54.4 | 4 | 87.7 | 262 | 20.6 |
| Bethel ........................ | (1) | (1) | (1) | (1) | (1) | (1) | (1) | (1) | (1) | (1) | (1) | (1) | (1) | (1) | (1) | (1) |
| Bristol Bay .................. | (1) | (1) | (1) | (1) | (1) | (1) | (1) | (1) | (1) | (1) | (1) | (1) | (1) | (1) | (1) | (1) |
| Chugach ..................... | | | | | | | | | | | | | | | | |
| Copper River ................ | | | | | | | | | | | | | | | | |
| Denali ........................ | (1) | (1) | (1) | (1) | (1) | (1) | (1) | (1) | (1) | (1) | (1) | (1) | (1) | (1) | (1) | (1) |
| Dillingham ................... | (1) | (1) | (1) | (1) | (1) | (1) | (1) | (1) | (1) | (1) | (1) | (1) | (1) | (1) | (1) | (1) |
| Fairbanks North Star ......... | 102 | 2.4 | 372 | 1.1 | 63.2 | 610,607 | 1,640 | 82,638 | 10.4 | 37,927 | 81.6 | 18.4 | 5 | 86.9 | 1,124 | 21.9 |
| Haines ....................... | (1) | (1) | (1) | (1) | (1) | (1) | (1) | (1) | (1) | (1) | (1) | (1) | (1) | (1) | (1) | (1) |
| Hoonah-Angoon .............. | (1) | (1) | (1) | (1) | (1) | (1) | (1) | (1) | (1) | (1) | (1) | (1) | (1) | (1) | (1) | (1) |
| Juneau ....................... | 1 | -0.3 | 9 | 0.0 | 0.0 | 881,816 | 100,397 | 225,293 | 14.1 | 235,489 | 6.5 | 93.5 | 1 | 85.0 | 11 | 11.7 |
| Kenai Peninsula ............. | 32 | 8.1 | 121 | 0.1 | 0.2 | 307,385 | 2,537 | 49,878 | 5.4 | 20,856 | 50.0 | 50.0 | 6 | 89.6 | 502 | 28.1 |
| Kodiak Island ............... | (1) | (1) | (1) | (1) | (1) | (1) | (1) | (1) | (1) | (1) | (1) | (1) | (1) | (1) | (1) | (1) |
| Kusilvak ..................... | (1) | (1) | (1) | (1) | (1) | (1) | (1) | (1) | (1) | (1) | (1) | (1) | (1) | (1) | (1) | (1) |
| Lake and Peninsula ......... | (1) | (1) | (1) | (1) | (1) | (1) | (1) | (1) | (1) | (1) | (1) | (1) | (1) | (1) | (1) | (1) |
| Matanuska-Susitna ........... | (1) | (1) | (1) | (1) | (1) | (1) | (1) | (1) | (1) | (1) | (1) | (1) | (1) | (1) | (1) | (1) |
| Nome ......................... | (1) | (1) | (1) | (1) | (1) | (1) | (1) | (1) | (1) | (1) | (1) | (1) | (1) | (1) | (1) | (1) |
| North Slope .................. | (1) | (1) | (1) | (1) | (1) | (1) | (1) | (1) | (1) | (1) | (1) | (1) | (1) | (1) | (1) | (1) |
| Northwest Arctic ............ | (1) | (1) | (1) | (1) | (1) | (1) | (1) | (1) | (1) | (1) | (1) | (1) | (1) | (1) | (1) | (1) |
| Petersburg ................... | (1) | (1) | (1) | (1) | (1) | (1) | (1) | (1) | (1) | (1) | (1) | (1) | (1) | (1) | (1) | (1) |
| Prince of Wales-Hyder ....... | (1) | (1) | (1) | (1) | (1) | (1) | (1) | (1) | (1) | (1) | (1) | (1) | (1) | (1) | (1) | (1) |
| Sitka ......................... | (1) | (1) | (1) | (1) | (1) | (1) | (1) | (1) | (1) | (1) | (1) | (1) | (1) | (1) | (1) | (1) |
| Skagway ..................... | (1) | (1) | (1) | (1) | (1) | (1) | (1) | (1) | (1) | (1) | (1) | (1) | (1) | (1) | (1) | (1) |
| Southeast Fairbanks ......... | (1) | (1) | (1) | (1) | (1) | (1) | (1) | (1) | (1) | (1) | (1) | (1) | (1) | (1) | (1) | (1) |
| Valdez-Cordova .............. | (1) | (1) | (1) | (1) | (1) | (1) | (1) | (1) | (1) | (1) | (1) | (1) | (1) | (1) | (1) | (1) |
| Wrangell ..................... | (1) | (1) | (1) | (1) | (1) | (1) | (1) | (1) | (1) | (1) | (1) | (1) | (1) | (1) | (1) | (1) |
| Yakutat ...................... | (1) | (1) | (1) | (1) | (1) | (1) | (1) | (1) | (1) | (1) | (1) | (1) | (1) | (1) | (1) | (1) |
| Yukon-Koyukuk ............... | (1) | (1) | (1) | (1) | (1) | (1) | (1) | (1) | (1) | (1) | (1) | (1) | (1) | (1) | (1) | (1) |
| **ARIZONA** .................. | 26,126 | -0.5 | 1,369 | 910.9 | 1,286.6 | 1,110,303 | 811 | 77,604 | 3,852.0 | 201,824 | 54.4 | 45.6 | 84 | 57.4 | 22,331 | 3.9 |
| Apache ....................... | 5,555 | -0.8 | 1,001 | 11.9 | 30.4 | 275,085 | 275 | 30,785 | 18.0 | 3,243 | 20.1 | 79.9 | 4 | 43.6 | 278 | 0.6 |
| Cochise ...................... | 973 | 6.2 | 899 | 86.0 | 152.9 | 1,802,553 | 2,005 | 99,038 | 144.7 | 133,648 | 56.9 | 43.1 | 4 | 78.9 | 3,119 | 10.6 |
| Coconino ..................... | 6,139 | 5.6 | 2,866 | 1.3 | 6.9 | 607,821 | 212 | 34,544 | 23.9 | 11,162 | 4.1 | 95.9 | 2 | 43.9 | 807 | 0.8 |
| Gila .......................... | 1,214 | 2.1 | 4,074 | 1.3 | 2.3 | 1,583,528 | 389 | 51,499 | 7.3 | 24,362 | 7.7 | 92.3 | NA | 75.5 | 143 | 1.7 |
| Graham ....................... | 1,290 | 3.1 | 2,880 | 46.7 | 52.4 | 1,866,459 | 648 | 107,676 | 62.1 | 138,558 | 88.2 | 11.8 | 9 | 73.0 | 1,083 | 12.7 |
| Greenlee ..................... | 66 | 25.9 | 536 | 5.1 | 5.0 | 835,412 | 1,559 | 76,520 | 8.7 | 70,650 | 24.9 | 75.1 | NA | 84.6 | 87 | 5.7 |
| La Paz ....................... | 250 | D | 2,574 | 97.1 | 102.6 | 5,190,405 | 2,016 | 484,196 | D | D | D | D | 1 | 71.1 | 1,280 | 34.0 |
| Maricopa ..................... | 474 | -0.3 | 253 | 180.2 | 257.2 | 2,338,321 | 9,236 | 171,562 | 1,209.1 | 645,215 | 39.3 | 60.7 | 18 | 85.0 | 5,310 | 9.4 |
| Mohave ....................... | 745 | -40.1 | 2,351 | 20.9 | 30.7 | 1,972,873 | 839 | 96,071 | 32.3 | 101,871 | 71.1 | 28.9 | NA | 87.4 | 390 | 2.8 |
| Navajo ....................... | 4,413 | 2.1 | 1,049 | 6.7 | 66.0 | 231,932 | 221 | 26,006 | 49.9 | 11,871 | 8.6 | 91.4 | NA | 40.4 | 340 | 0.3 |
| Pima ......................... | 2,618 | D | 3,960 | 30.0 | 40.7 | 2,088,810 | 527 | 79,142 | 75.5 | 114,174 | 84.3 | 15.7 | 3 | 72.0 | 1,234 | 5.4 |
| Pinal ......................... | 1,121 | -4.6 | 1,471 | 232.2 | 294.1 | 4,065,306 | 2,764 | 282,808 | 861.9 | 1,131,154 | 35.7 | 64.3 | 4 | 85.4 | 3,943 | 20.9 |
| Santa Cruz ................... | 198 | -8.0 | 903 | 2.6 | 3.2 | 1,800,767 | 1,994 | 62,799 | 19.6 | 89,639 | 48.8 | 51.2 | 1 | 88.1 | 413 | 5.0 |
| Yavapai ...................... | 822 | -0.3 | 967 | 7.5 | 8.0 | 1,596,867 | 1,651 | 63,905 | 35.7 | 42,036 | 40.1 | 59.9 | 4 | 86.4 | 115 | 1.5 |
| Yuma ......................... | 247 | 15.2 | 542 | 181.4 | 234.3 | 5,006,516 | 9,235 | 468,849 | D | D | D | D | 34 | 84.4 | 3,789 | 14.9 |
| **ARKANSAS** ................ | 13,889 | 0.6 | 326 | 4,855.1 | 7,825.9 | 1,030,741 | 3,163 | 126,667 | 9,651.2 | 226,420 | 37.6 | 62.4 | 81 | 73.6 | 321,742 | 19.5 |
| Arkansas ..................... | 414 | 3.0 | 849 | 314.3 | 359.0 | 2,675,627 | 3,151 | 312,157 | 214.3 | 439,098 | 99.6 | 0.4 | NA | 68.6 | 22,110 | 85.9 |
| Ashley ....................... | 132 | 18.3 | 374 | 88.9 | 103.9 | 1,081,824 | 2,894 | 201,843 | 70.2 | 198,856 | 90.4 | 9.6 | NA | 64.9 | 5,447 | 32.9 |
| Baxter ....................... | 101 | 10.4 | 212 | 0.0 | 13.3 | 569,724 | 2,690 | 58,020 | 31.8 | 66,309 | 2.7 | 97.3 | NA | 80.6 | 259 | 9.6 |
| Benton ....................... | 244 | -20.0 | 126 | 0.2 | 72.9 | 815,643 | 6,478 | 77,246 | 593.4 | 306,494 | 1.3 | 98.7 | 8 | 74.5 | 456 | 3.0 |
| Boone ........................ | 306 | 19.0 | 233 | 0.3 | 42.3 | 614,116 | 2,636 | 69,650 | 164.0 | 124,918 | 1.9 | 98.1 | NA | 76.6 | 1,492 | 7.6 |
| Bradley ...................... | 30 | 50.4 | 164 | 0.5 | 6.2 | 539,607 | 3,287 | 82,166 | 42.1 | 232,503 | 15.8 | 84.2 | NA | 74.0 | 21 | 7.2 |
| Calhoun ...................... | 19 | 35.1 | 186 | D | 3.1 | 480,703 | 2,591 | 61,897 | 5.5 | 54,683 | 6.3 | 93.7 | NA | 74.3 | 48 | 16.8 |
| Carroll ....................... | 290 | 13.3 | 248 | 0.0 | 54.6 | 683,129 | 2,750 | 83,790 | 363.6 | 311,050 | 0.9 | 99.1 | 9 | 76.5 | 878 | 8.2 |
| Chicot ....................... | 307 | 6.2 | 1,057 | 221.3 | 273.3 | 3,139,169 | 2,971 | 320,397 | 148.3 | 509,519 | 96.6 | 3.4 | NA | 72.5 | 11,746 | 76.6 |
| Clark ......................... | 103 | 14.1 | 273 | 2.3 | 29.6 | 661,236 | 2,421 | 94,828 | 17.5 | 46,414 | 32.4 | 67.6 | NA | 70.3 | 1,237 | 15.4 |
| Clay ......................... | 286 | -13.6 | 529 | 208.8 | 258.2 | 2,278,119 | 4,310 | 271,792 | 197.6 | 364,603 | 83.2 | 16.8 | NA | 70.3 | 12,965 | 59.8 |
| Cleburne ..................... | 129 | -17.8 | 192 | 0.2 | 29.2 | 562,588 | 2,938 | 64,986 | 57.6 | 85,138 | 2.6 | 97.4 | NA | 73.7 | 292 | 8.9 |

1.  Agriculture Census data for Alaska is combined into regional groups. Aleutians West, Bethel, Bristol Bay, Dillingham, Kodiak Island, Kusilvak, Lake and Peninsula, Nome, North Slope, and Northwest Arctic are included with Aleutians East.  Matanuska-Susitna and Valdez-Cordova are included with Anchorage. Denali, Southeast Fairbanks, and Yukon-Koyukuk are included with Fairbanks North Star. Haines, Hoonah-Angoon, Ketchican Gateway, Petersburg, Prince of Wales-Hyder, Sitka, Skagway, Wrangell, and Yakutat are included with Juneau.

| STATE County | Water use, 2015 Public supply water withdrawn (mil gal/day) | Public supply gallons withdrawn per person per day | Wholesale Trade[1], 2017 Number of establishments | Number of employees | Sales (mil dol) | Average payroll (mil dol) | Retail Trade[2], 2017 Number of establishments | Number of employees | Sales (mil dol) | Average payroll (mil dol) | Real estate and rental and leasing,[2] 2017 Number of establishments | Number of employees | Sales (mil dol) | Average payroll (mil dol) |
|---|---|---|---|---|---|---|---|---|---|---|---|---|---|---|
| | 133 | 134 | 135 | 136 | 137 | 138 | 139 | 140 | 141 | 142 | 143 | 144 | 145 | 146 |
| **ALABAMA—Cont'd** | | | | | | | | | | | | | | |
| Washington | 2.81 | 167.2 | 7 | 76 | 94.0 | 4.0 | 32 | 237 | 70.5 | 5.6 | NA | NA | NA | NA |
| Wilcox | 2.80 | 253.2 | 5 | 29 | 23.9 | 1.6 | 43 | 240 | 75.4 | 5.5 | 7 | D | 2.6 | D |
| Winston | 0.81 | 33.9 | 29 | 263 | 294.5 | 11.7 | 89 | 737 | 158.3 | 15.8 | 12 | 30 | 12.6 | 1.6 |
| **ALASKA** | 99.18 | 134.3 | 643 | 7,529 | 5,557.3 | 457.5 | 2,480 | 34,498 | 10,384.8 | 1,095.1 | 965 | 4,622 | 1,188.3 | 222.8 |
| Aleutians East | 1.22 | 365.2 | NA | NA | NA | NA | D | D | D | D | 3 | 41 | 0.4 | 0.4 |
| Aleutians West | 2.27 | 398.1 | 14 | 142 | 192.0 | 10.0 | 12 | 124 | 50.3 | 4.9 | D | D | D | 2.1 |
| Anchorage | 44.49 | 148.9 | 338 | 5,162 | 3,807.4 | 314.0 | 849 | 15,445 | 4,903.2 | 507.6 | 437 | 2,519 | 681.0 | 124.4 |
| Bethel | 0.30 | 16.7 | 4 | 18 | 17.8 | 1.0 | 51 | 832 | 156.2 | 16.2 | 9 | 49 | 7.2 | 2.0 |
| Bristol Bay | 0.07 | 78.5 | D | D | D | D | 10 | 51 | 15.1 | 1.7 | D | D | D | 0.2 |
| Chugach | | | | | | | | | | | | | | |
| Copper River | | | | | | | | | | | | | | |
| Denali | 0.04 | 20.8 | NA | NA | NA | NA | 18 | 34 | 18.9 | 2.0 | NA | NA | NA | NA |
| Dillingham | 0.25 | 50.0 | NA | NA | NA | NA | 17 | 202 | 48.1 | 5.6 | D | D | D | 0.8 |
| Fairbanks North Star | 14.12 | 141.7 | 70 | 693 | 470.6 | 38.4 | 294 | 4,749 | 1,484.8 | 159.0 | 158 | 764 | 188.3 | 42.1 |
| Haines | 0.29 | 114.4 | NA | NA | NA | NA | 19 | 114 | 24.6 | 3.6 | 6 | D | 0.5 | D |
| Hoonah-Angoon | 0.53 | 248.5 | NA | NA | NA | NA | 16 | 84 | 16.4 | 1.7 | NA | NA | NA | NA |
| Juneau | 4.80 | 146.5 | 33 | 278 | 203.1 | 16.0 | 148 | 1,790 | 464.8 | 53.3 | 64 | 195 | 60.0 | 8.0 |
| Kenai Peninsula | 3.63 | 62.5 | 53 | 334 | 207.3 | 19.6 | 253 | 2,607 | 817.1 | 84.7 | 75 | 267 | 82.1 | 14.5 |
| Ketchikan Gateway | 5.70 | 415.8 | 13 | 122 | 99.5 | 9.0 | 108 | 1,037 | 243.6 | 33.2 | 28 | 91 | 29.3 | 4.4 |
| Kodiak Island | 7.64 | 550.1 | 17 | 92 | 44.6 | 3.3 | 37 | 500 | 125.4 | 14.1 | D | D | D | 0.3 |
| Kusilvak | 0.48 | 59.1 | NA | NA | NA | NA | 27 | 338 | 65.4 | 5.9 | D | 4 | 0.2 | 0.0 |
| Lake and Peninsula | 0.12 | 76.8 | NA | NA | NA | NA | D | D | D | 0.4 | 3 | 4 | 0.2 | 0.0 |
| Matanuska-Susitna | 2.69 | 26.6 | 42 | 201 | 91.3 | 11.5 | 257 | 3,909 | 1,255.5 | 124.6 | 81 | 255 | 49.4 | 7.7 |
| Nome | 0.68 | 69.1 | D | D | D | 2.1 | 32 | 336 | 83.2 | 7.8 | 10 | 50 | 8.5 | 0.7 |
| North Slope | 0.38 | 39.2 | 7 | 171 | 133.3 | 13.3 | 21 | 301 | 85.2 | 9.1 | 8 | 85 | 20.7 | 6.1 |
| Northwest Arctic | 0.57 | 73.5 | NA | NA | NA | NA | 13 | 187 | 46.6 | 4.9 | D | D | D | D |
| Petersburg | 0.65 | 204.6 | 8 | 33 | 18.2 | 1.9 | 27 | 167 | 38.4 | 4.9 | 6 | 31 | 5.1 | 0.9 |
| Prince of Wales-Hyder | 0.34 | 53.6 | NA | NA | NA | NA | 24 | 222 | 55.0 | 6.3 | 16 | 46 | 8.4 | 2.4 |
| Sitka | 1.22 | 137.7 | 7 | 39 | 27.6 | 1.8 | 55 | 387 | 107.2 | 12.4 | NA | NA | NA | NA |
| Skagway | 0.24 | 227.1 | NA | NA | NA | NA | 46 | 137 | 43.0 | 5.0 | NA | NA | NA | NA |
| Southeast Fairbanks | 0.24 | 35.1 | 3 | 68 | 24.7 | 2.3 | 34 | 253 | 67.6 | 7.6 | 3 | 8 | 1.1 | 0.4 |
| Valdez-Cordova | 4.83 | 515.9 | 9 | 35 | 33.1 | 2.3 | 51 | 348 | 101.5 | 10.4 | 9 | 20 | 3.1 | 0.9 |
| Wrangell | 0.64 | 268.7 | NA | NA | NA | NA | 15 | 136 | 23.4 | 4.0 | 4 | 5 | 1.2 | 0.1 |
| Yakutat | 0.51 | 832.0 | NA | NA | NA | NA | 4 | D | 5.1 | D | NA | NA | NA | NA |
| Yukon-Koyukuk | 0.24 | 43.4 | 4 | 19 | 21.7 | 1.1 | 31 | 101 | 27.8 | 2.1 | NA | NA | NA | NA |
| **ARIZONA** | 1,195.15 | 175.0 | 5,508 | 79,541 | 71,206.1 | 4,826.6 | 17,918 | 324,912 | 104,365.6 | 9,505.5 | 9,937 | 48,305 | 12,992.2 | 2,267.2 |
| Apache | 5.77 | 80.7 | D | D | D | D | 94 | 1,129 | 280.5 | 24.6 | 13 | 31 | 4.3 | 0.7 |
| Cochise | 16.15 | 127.7 | D | D | D | 11.4 | 386 | 5,160 | 1,324.4 | 126.6 | 107 | 391 | 57.8 | 10.8 |
| Coconino | 19.87 | 142.8 | 96 | 832 | 365.3 | 37.9 | 580 | 8,004 | 2,197.1 | 207.2 | 224 | 884 | 201.3 | 30.5 |
| Gila | 4.83 | 90.9 | 20 | 140 | 53.2 | 7.4 | 150 | 2,082 | 553.0 | 52.5 | 69 | 159 | 35.8 | 6.1 |
| Graham | 3.80 | 100.9 | D | D | D | D | 101 | 1,546 | 408.4 | 40.7 | 21 | 65 | 11.9 | 2.0 |
| Greenlee | 1.70 | 178.4 | D | D | D | 0.5 | 18 | 188 | 64.1 | 5.2 | 4 | D | 3.8 | D |
| La Paz | 2.89 | 143.4 | D | D | D | 5.6 | 84 | 951 | 450.3 | 24.3 | 16 | D | 9.1 | D |
| Maricopa | 776.54 | 186.3 | 3,956 | 64,055 | 61,779.1 | 4,059.4 | 10,871 | 215,452 | 74,333.1 | 6,605.7 | 6,935 | 35,958 | 10,576.8 | 1,830.1 |
| Mohave | 47.86 | 233.8 | 101 | 1,217 | 455.1 | 53.7 | 626 | 9,435 | 3,063.4 | 262.1 | 241 | 736 | 150.2 | 20.1 |
| Navajo | 14.50 | 133.9 | D | D | D | D | 294 | 3,991 | 1,197.2 | 101.5 | 102 | 220 | 48.1 | 7.2 |
| Pima | 175.4 | 173.7 | 666 | 6,524 | 3,723.2 | 321.7 | 2,721 | 46,749 | 12,230.9 | 1,271.8 | 1,347 | 6,396 | 1,271.3 | 243.8 |
| Pinal | 63.24 | 155.5 | 109 | 1,179 | 605.9 | 62.5 | 507 | 8,908 | 2,578.1 | 227.6 | 223 | 1,425 | 207.4 | 44.7 |
| Santa Cruz | 5.17 | 111.3 | D | D | D | 63.8 | 199 | 2,205 | 513.2 | 51.7 | 44 | 127 | 21.1 | 3.9 |
| Yavapai | 21.14 | 95.1 | 141 | 1,447 | 952.8 | 72.9 | 815 | 10,820 | 2,952.7 | 294.1 | 414 | 1,075 | 245.0 | 42.3 |
| Yuma | 36.29 | 177.7 | 128 | 1,856 | 1,438.4 | 102.1 | 472 | 8,292 | 2,219.1 | 210.0 | 177 | 761 | 148.3 | 23.3 |
| **ARKANSAS** | 363.06 | 121.9 | 2,812 | 36,528 | 29,242.1 | 1,882.8 | 10,925 | 142,857 | 40,174.1 | 3,665.3 | 3,115 | 12,776 | 2,588.2 | 473.3 |
| Arkansas | 1.18 | 64.0 | 31 | 370 | 173.6 | 19.7 | 87 | 1,021 | 302.4 | 28.7 | 18 | 131 | 16.6 | 3.7 |
| Ashley | 1.20 | 57.6 | 16 | 124 | 141.7 | 6.3 | 67 | 724 | 183.0 | 17.1 | 8 | 17 | 2.7 | 0.6 |
| Baxter | 3.46 | 84.3 | D | D | D | D | 190 | 2,231 | 571.1 | 54.3 | 45 | 217 | 26.8 | 6.2 |
| Benton | 64.74 | 259.3 | 256 | 2,849 | 2,400.1 | 177.6 | 755 | 12,916 | 3,717.4 | 345.7 | 319 | 1,155 | 302.9 | 47.3 |
| Boone | 0.99 | 26.6 | D | D | D | D | 167 | 2,062 | 622.4 | 54.6 | 41 | 100 | 18.4 | 2.6 |
| Bradley | 0.21 | 18.9 | 10 | 128 | 75.2 | 5.0 | 33 | 268 | 71.4 | 5.9 | 5 | D | 0.6 | D |
| Calhoun | 0.24 | 45.9 | NA | NA | NA | NA | 8 | 47 | 12.6 | 1.0 | NA | NA | NA | NA |
| Carroll | 8.60 | 310.4 | 9 | 200 | 52.3 | 8.0 | 149 | 1,143 | 250.1 | 27.1 | 26 | 67 | 10.6 | 1.6 |
| Chicot | 0.62 | 56.2 | 12 | 130 | 177.8 | 6.9 | 43 | 307 | 58.3 | 6.4 | 9 | 20 | 2.2 | 0.3 |
| Clark | 1.84 | 81.3 | D | D | D | 2.3 | 87 | 1,269 | 376.9 | 35.6 | 28 | 75 | 10.7 | 1.5 |
| Clay | 0.91 | 60.2 | 15 | 255 | 211.7 | 8.1 | 50 | 496 | 127.7 | 11.9 | 6 | D | 1.3 | D |
| Cleburne | 11.15 | 437.8 | 18 | 106 | 35.7 | 3.5 | 92 | 1,068 | 412.5 | 30.1 | 30 | 53 | 8.6 | 1.5 |

1 Merchant wholesalers, except manufacturers' sales branches and offices.  2. Employer establishments.

Table B. States and Counties — **Professional Services, Manufacturing, and Accommodation and Food Services**

| STATE County | Professional, scientific, and technical services, 2017 | | | | Manufacturing, 2017 | | | | Accommodation and food services, 2017 | | | |
|---|---|---|---|---|---|---|---|---|---|---|---|---|
| | Number of establish-ments | Number of employees | Sales (mil dol) | Average payroll (mil dol) | Number of establish-ments | Number of employees | Sales (mil dol) | Average payroll (mil dol) | Number of establis-hments | Number of employees | Sales (mil dol) | Annual payroll (mil dol) |
| | 147 | 148 | 149 | 150 | 151 | 152 | 153 | 154 | 155 | 156 | 157 | 158 |
| ALABAMA—Cont'd | | | | | | | | | | | | |
| Washington | 10 | 29 | 7 | 2.2 | 14 | 3,610 | 5,633.1 | 326.9 | D | D | D | D |
| Wilcox | 10 | 18 | 2 | 0.5 | D | 856 | D | 62.8 | D | D | D | 1.4 |
| Winston | 25 | 76 | 7 | 2.4 | 46 | 2,556 | 738.2 | 96.6 | D | D | D | D |
| ALASKA | 1,932 | 17,595 | 3,193 | 1,252.8 | 535 | 12,515 | 8,616.9 | 585.3 | 2,214 | 28,853 | 2,536.5 | 741.7 |
| Aleutians East | NA | NA | NA | NA | D | D | D | D | 9 | 62 | 2.6 | 0.4 |
| Aleutians West | D | D | D | D | 7 | 2,528 | 472.7 | 73.0 | 7 | 109 | 14.8 | 3.8 |
| Anchorage | 1,178 | 14,141 | 2,567 | 1,039.8 | 171 | 1,675 | 412.5 | 89.9 | 798 | 15,618 | 1,224.8 | 386.2 |
| Bethel | D | D | D | D | D | 13 | D | 0.6 | 18 | 40 | 6.3 | 0.9 |
| Bristol Bay | NA | NA | NA | NA | 9 | 232 | 280.7 | 23.9 | 21 | 113 | 20.1 | 5.3 |
| Chugach | | | | | | | | | | | | |
| Copper River | | | | | | | | | | | | |
| Denali | D | D | D | D | NA | NA | NA | NA | 38 | 507 | 103.4 | 20.7 |
| Dillingham | NA | NA | NA | NA | 4 | 48 | 71.9 | 8.1 | 12 | 45 | 5.4 | 0.9 |
| Fairbanks North Star | 212 | 1,356 | 313 | 91.0 | 69 | 454 | 226.6 | 24.4 | 233 | 3,306 | 250.7 | 70.7 |
| Haines | 6 | 8 | 1 | 0.3 | 8 | 83 | 47.9 | 4.6 | 18 | 84 | 6.0 | 1.8 |
| Hoonah-Angoon | NA | NA | NA | NA | NA | NA | NA | NA | 15 | 61 | 11.0 | 1.6 |
| Juneau | D | D | D | D | 32 | 258 | 72.7 | 14.4 | 124 | 1,390 | 112.9 | 31.2 |
| Kenai Peninsula | D | D | D | D | D | 760 | D | 67.7 | 289 | 1,890 | 174.8 | 47.5 |
| Ketchikan Gateway | D | D | D | D | 15 | 380 | 217.8 | 25.8 | 61 | 554 | 56.3 | 15.1 |
| Kodiak Island | 22 | 53 | 6 | 2.3 | 25 | 2,319 | 591.3 | 73.3 | 48 | 429 | 34.5 | 9.6 |
| Kusilvak | NA | NA | NA | NA | NA | NA | NA | NA | NA | NA | NA | NA |
| Lake and Peninsula | NA | NA | NA | NA | D | D | D | D | 10 | 37 | 12.6 | 3.5 |
| Matanuska-Susitna | 164 | 699 | 105 | 41.0 | 57 | 346 | 64.0 | 16.3 | 236 | 2,490 | 215.4 | 51.8 |
| Nome | D | D | D | D | NA | NA | NA | NA | 19 | 156 | 16.3 | 4.8 |
| North Slope | D | D | D | D | NA | NA | NA | NA | 31 | 689 | 115.6 | 42.0 |
| Northwest Arctic | NA | NA | NA | NA | NA | NA | NA | NA | D | D | D | D |
| Petersburg | 6 | 14 | 1 | 0.5 | 5 | D | 91.7 | 9.9 | 13 | 69 | 6.8 | 1.6 |
| Prince of Wales-Hyder | NA | NA | NA | NA | D | 221 | D | 10.7 | 22 | 76 | 14.1 | 3.5 |
| Sitka | D | D | D | D | 16 | 448 | 154.5 | 16.5 | 41 | 378 | 44.0 | 11.9 |
| Skagway | NA | NA | NA | NA | D | 14 | D | 0.3 | 23 | 78 | 16.2 | 5.7 |
| Southeast Fairbanks | 11 | 50 | 11 | 3.4 | D | 10 | D | D | 28 | 141 | 14.3 | 3.9 |
| Valdez-Cordova | D | D | D | D | 13 | 306 | 534.7 | 25.2 | 70 | 340 | 39.0 | 11.4 |
| Wrangell | D | D | D | D | D | D | D | D | 5 | 37 | 3.5 | 1.0 |
| Yakutat | NA | NA | NA | NA | NA | NA | NA | NA | 6 | D | 3.3 | D |
| Yukon-Koyukuk | D | D | D | D | NA | NA | NA | NA | D | D | D | 0.3 |
| ARIZONA | 18,020 | 145,604 | 23,839 | 10,455.2 | 4,343 | 142,809 | 56,635.9 | 9,148.7 | 13,079 | 299,628 | 19,848.7 | 5,636.1 |
| Apache | D | D | D | D | D | D | D | D | 58 | 933 | 53.7 | 14.9 |
| Cochise | 223 | 3,433 | 502 | 227.8 | 41 | 268 | 118.0 | 13.6 | 261 | 3,522 | 183.7 | 55.3 |
| Coconino | 349 | 1,471 | 191 | 68.8 | 88 | 4,893 | 2,250.3 | 425.7 | 601 | 15,061 | 1,200.0 | 323.8 |
| Gila | 81 | 260 | 31 | 9.4 | D | 155 | D | 6.4 | 139 | 2,019 | 124.9 | 35.4 |
| Graham | 29 | 535 | 8 | 15.5 | D | 257 | D | 9.5 | 57 | 1,010 | 40.4 | 12.9 |
| Greenlee | 4 | 20 | 2 | 0.6 | NA | NA | NA | NA | 10 | 117 | 11.9 | 2.7 |
| La Paz | 11 | 22 | 2 | 0.7 | D | D | D | D | 81 | 1,410 | 93.8 | 25.3 |
| Maricopa | D | D | D | D | 2,945 | 100,076 | 38,022.0 | 6,055.9 | 8,004 | 195,797 | 13,229.4 | 3,765.8 |
| Mohave | D | D | D | D | 141 | 2,855 | 1,027.6 | 130.0 | 401 | 6,670 | 397.4 | 110.7 |
| Navajo | D | D | D | D | 40 | 317 | 83.3 | 12.8 | 229 | 3,924 | 294.9 | 73.3 |
| Pima | 2,554 | 17,645 | 2,637 | 1,069.3 | 617 | 23,392 | 9,958.9 | 1,965.6 | 1,852 | 44,608 | 2,877.5 | 808.1 |
| Pinal | D | D | D | D | 120 | 3,613 | 3,100.9 | 213.1 | 369 | 6,644 | 328.9 | 94.9 |
| Santa Cruz | 59 | 187 | 25 | 8.3 | 35 | 388 | 256.5 | 15.1 | 97 | 1,395 | 76.4 | 23.8 |
| Yavapai | 565 | 1,909 | 236 | 87.9 | 190 | 3,261 | 716.3 | 160.0 | 575 | 9,695 | 563.9 | 187.5 |
| Yuma | D | D | D | D | 75 | 3,095 | 967.4 | 132.4 | 345 | 6,823 | 371.6 | 101.7 |
| ARKANSAS | D | D | 5,516 | D | 2,571 | 152,696 | 58,363.7 | 7,024.8 | 5,977 | 106,280 | 5,484.8 | 1,572.8 |
| Arkansas | 23 | 75 | 10 | 2.7 | 30 | 3,894 | 1,656.0 | 142.6 | 42 | 385 | 20.5 | 5.0 |
| Ashley | 25 | 66 | 6 | 2.0 | D | 1,600 | D | 116.6 | 27 | 391 | 18.9 | 5.2 |
| Baxter | 71 | 449 | 39 | 15.2 | 48 | 2,083 | 613.3 | 95.6 | 106 | 1,618 | 73.6 | 21.2 |
| Benton | D | D | D | D | 173 | 11,608 | 3,579.3 | 478.9 | 512 | 10,860 | 610.3 | 177.9 |
| Boone | D | D | D | D | 50 | 1,697 | 438.5 | 70.6 | 72 | 1,135 | 58.1 | 16.8 |
| Bradley | 13 | 28 | 3 | 0.7 | 6 | 535 | 146.2 | 24.3 | 18 | 223 | 9.1 | 2.3 |
| Calhoun | NA | NA | NA | NA | NA | NA | NA | NA | 4 | 24 | 1.2 | 0.3 |
| Carroll | 38 | 692 | 198 | 67.0 | 32 | 3,118 | 796.6 | 98.9 | 144 | 1,345 | 69.9 | 21.2 |
| Chicot | 13 | 38 | 6 | 1.7 | 3 | 7 | 4.0 | 0.5 | 13 | 126 | 6.9 | 1.7 |
| Clark | D | D | D | D | 24 | 1,209 | 433.1 | 60.8 | 62 | 1,176 | 55.9 | 17.3 |
| Clay | 14 | 49 | 4 | 1.3 | 13 | 88 | 13.7 | 3.5 | 23 | 219 | 8.6 | 2.5 |
| Cleburne | 36 | 116 | 9 | 3.0 | 30 | 1,224 | 256.3 | 51.9 | 64 | 723 | 35.7 | 10.2 |

# Health Care and Social Assistance, Other Services, Nonemployer Businesses, and Residential Construction

| STATE County | Health care and social assistance, 2017 | | | | Other services, 2017 | | | | Nonemployer businesses, 2018 | | Value of residential construction authorized by building permits, 2020 | |
|---|---|---|---|---|---|---|---|---|---|---|---|---|
| | Number of establish-ments | Number of employees | Receipts (mil dol) | Annual payroll (mil dol) | Number of establish-ments | Number of employees | Receipts (mil dol) | Annual payroll (mil dol) | Number | Receipts (mil dol) | New construction ($1,000) | Number of housing units |
| | 159 | 160 | 161 | 162 | 163 | 164 | 165 | 166 | 167 | 168 | 169 | 170 |
| ALABAMA—Cont'd | | | | | | | | | | | | |
| Washington | 13 | 268 | 18.5 | 8.4 | D | D | D | D | 992 | 30.4 | 535 | 3 |
| Wilcox | 18 | 269 | 25.0 | 8.1 | 7 | 15 | 1.6 | 0.4 | 530 | 19.7 | 50 | 1 |
| Winston | 45 | 735 | 54.7 | 22.9 | D | D | D | D | 1,648 | 81.2 | 588 | 5 |
| ALASKA | 2,679 | 52,125 | 7,856.2 | 3,048.1 | 1,357 | 7,271 | 1,351.3 | 269.7 | 57,391 | 2,793.8 | 381,356 | 1,420 |
| Aleutians East | NA | NA | NA | NA | NA | NA | NA | NA | 216 | 17.4 | 0 | 0 |
| Aleutians West | 13 | 108 | 13.7 | 5.3 | 6 | 28 | 5.4 | 1.9 | 216 | 12.4 | 988 | 5 |
| Anchorage | 1,291 | 27,333 | 4,482.5 | 1,673.1 | 571 | 3,928 | 886.5 | 150.5 | 20,976 | 1,147.1 | 271,706 | 970 |
| Bethel | 52 | D | 127.5 | D | 24 | 66 | 7.8 | 1.7 | 667 | 19.9 | 3,957 | 15 |
| Bristol Bay | D | D | D | D | D | D | D | D | 212 | 10.5 | 155 | 1 |
| Chugach | | | | | | | | | | | NA | NA |
| Copper River | | | | | | | | | | | NA | NA |
| Denali | 3 | 11 | 0.8 | 0.5 | NA | NA | NA | NA | 195 | 8.7 | NA | NA |
| Dillingham | D | D | D | 39.7 | D | D | D | 0.4 | 863 | 41.3 | 0 | 0 |
| Fairbanks North Star | 313 | 6,507 | 890.8 | 371.6 | 178 | 896 | 121.0 | 32.9 | 5,992 | 236.2 | 4,982 | 19 |
| Haines | 10 | 133 | 8.7 | 3.9 | D | D | D | 0.2 | 390 | 16.8 | 487 | 4 |
| Hoonah-Angoon | 7 | D | 1.8 | D | NA | NA | NA | NA | 289 | 11.2 | 8,484 | 21 |
| Juneau | 133 | 2,288 | 310.2 | 134.4 | 76 | 393 | 62.4 | 16.6 | 2,860 | 141.5 | 10,110 | 46 |
| Kenai Peninsula | 252 | 4,062 | 493.6 | 198.8 | 142 | 533 | 64.9 | 18.4 | 6,770 | 295.4 | 24,992 | 95 |
| Ketchikan Gateway | 46 | 716 | 109.9 | 36.0 | 37 | 116 | 14.6 | 3.9 | 1,375 | 82.4 | 11,399 | 43 |
| Kodiak Island | 43 | 687 | 78.0 | 35.6 | 24 | 114 | 13.8 | 4.4 | 1,485 | 86.5 | 3,682 | 19 |
| Kusilvak | D | D | D | D | NA | NA | NA | NA | 532 | 5.3 | 0 | 0 |
| Lake and Peninsula | NA | NA | NA | NA | NA | NA | NA | NA | 240 | 8.0 | NA | NA |
| Matanuska-Susitna | 331 | 4,518 | 569.4 | 219.5 | 145 | 595 | 101.1 | 20.7 | 7,656 | 343.5 | 13,020 | 60 |
| Nome | D | D | D | D | 15 | 67 | 9.3 | 1.7 | 551 | 15.9 | 1,440 | 8 |
| North Slope | 9 | D | 78.9 | D | 7 | 39 | 7.9 | 1.6 | 276 | 7.5 | 5,182 | 38 |
| Northwest Arctic | D | D | D | D | D | D | D | D | 235 | 7.9 | 3,868 | 12 |
| Petersburg | 9 | 172 | 18.7 | 11.4 | D | D | D | D | 770 | 62.8 | 8,727 | 29 |
| Prince of Wales-Hyder | 12 | 175 | 13.5 | 6.2 | 12 | 37 | 2.1 | 0.5 | 579 | 26.2 | 0 | 0 |
| Sitka | 22 | 758 | 144.0 | 52.0 | 25 | 168 | 18.6 | 7.1 | 1,320 | 78.7 | 2,067 | 10 |
| Skagway | 5 | D | 0.4 | D | 4 | 19 | 1.4 | 0.4 | 141 | 7.0 | 300 | 1 |
| Southeast Fairbanks | 10 | 101 | 9.7 | 4.2 | D | D | D | 0.2 | 539 | 20.6 | NA | NA |
| Valdez-Cordova | 27 | 485 | 52.3 | 23.0 | 25 | 85 | 10.4 | 3.0 | 1,266 | 54.5 | 3,497 | 14 |
| Wrangell | 8 | 288 | 28.5 | 9.8 | D | D | D | 0.7 | 319 | 14.6 | 615 | 3 |
| Yakutat | D | D | D | 0.9 | NA | NA | NA | NA | 110 | 6.2 | 1,107 | 5 |
| Yukon-Koyukuk | 9 | D | 2.9 | D | D | D | D | 0.4 | 351 | 7.9 | 594 | 2 |
| ARIZONA | 18,816 | 379,015 | 48,130.1 | 18,409.1 | 9,271 | 68,350 | 8,727.5 | 2,305.9 | 502,593 | 23,719.1 | 13,798,358 | 60,342 |
| Apache | 71 | 2,832 | 370.2 | 142.0 | D | D | D | D | 2,599 | 66.2 | 22,980 | 82 |
| Cochise | 283 | 4,921 | 454.5 | 188.2 | 160 | 661 | 66.2 | 16.9 | 6,885 | 221.8 | 56,264 | 337 |
| Coconino | 422 | 8,994 | 1,229.8 | 448.2 | 249 | 1,422 | 143.0 | 43.3 | 9,691 | 406.4 | 215,702 | 941 |
| Gila | 143 | 2,046 | 251.6 | 103.3 | 56 | 185 | 23.5 | 5.0 | 3,315 | 133.4 | 45,597 | 203 |
| Graham | 72 | 1,974 | 219.3 | 88.0 | 32 | 230 | 28.4 | 9.8 | 1,563 | 66.1 | 35,298 | 187 |
| Greenlee | D | D | D | D | NA | NA | NA | NA | 267 | 4.6 | 2,508 | 11 |
| La Paz | D | D | D | D | 17 | 66 | 9.2 | 1.7 | 800 | 36.4 | 6,194 | 31 |
| Maricopa | 12,396 | 248,818 | 32,793.7 | 12,407.6 | 5,924 | 48,349 | 6,599.8 | 1,746.5 | 336,089 | 16,975.7 | 9,546,985 | 41,895 |
| Mohave | 500 | 8,679 | 1,115.4 | 403.1 | 293 | 1,375 | 133.7 | 35.5 | 11,375 | 557.6 | 281,431 | 1,403 |
| Navajo | 253 | 4,449 | 623.3 | 246.5 | 117 | 529 | 49.3 | 13.3 | 5,593 | 199.9 | 143,262 | 577 |
| Pima | 2,862 | 61,632 | 7,137.9 | 2,861.4 | 1,485 | 10,873 | 1,229.4 | 305.4 | 69,496 | 2,808.0 | 1,394,415 | 4,958 |
| Pinal | 504 | 9,830 | 1,010.9 | 444.0 | 254 | 1,534 | 131.5 | 40.3 | 20,866 | 768.2 | 1,268,850 | 6,324 |
| Santa Cruz | 73 | 1,783 | 165.4 | 82.9 | 54 | 221 | 19.0 | 5.1 | 4,263 | 215.0 | 47,570 | 212 |
| Yavapai | 790 | 13,844 | 1,568.4 | 579.8 | 390 | 1,672 | 178.9 | 49.5 | 19,895 | 868.6 | 522,371 | 1,891 |
| Yuma | 411 | 8,367 | 1,074.7 | 365.5 | 214 | 1,124 | 102.6 | 30.6 | 9,896 | 391.3 | 208,931 | 1,290 |
| ARKANSAS | 7,878 | 182,126 | 19,845.4 | 7,960.7 | 3,910 | 21,132 | 2,873.8 | 655.4 | 209,179 | 9,171.9 | 2,353,746 | 12,493 |
| Arkansas | 48 | 817 | 63.3 | 25.4 | 28 | 84 | 5.8 | 2.6 | 1,395 | 68.8 | 1,205 | 17 |
| Ashley | 40 | 893 | 67.7 | 31.3 | 22 | 91 | 11.3 | 2.9 | 1,063 | 41.9 | 0 | 0 |
| Baxter | 176 | 4,115 | 441.7 | 166.9 | 76 | 362 | 27.3 | 7.5 | 3,198 | 143.1 | 8,916 | 70 |
| Benton | 580 | 10,426 | 1,214.8 | 454.5 | 296 | 1,919 | 287.9 | 65.2 | 20,076 | 924.3 | 939,969 | 4,300 |
| Boone | 128 | 2,572 | 232.2 | 93.5 | D | D | D | D | 2,887 | 128.7 | 3,469 | 15 |
| Bradley | 31 | 527 | 44.0 | 16.9 | D | D | D | D | 512 | 27.6 | 21 | 1 |
| Calhoun | D | D | D | D | D | D | D | D | 162 | 7.7 | 0 | 0 |
| Carroll | 53 | 742 | 75.1 | 29.4 | 36 | 147 | 14.2 | 3.4 | 2,628 | 92.5 | 4,633 | 29 |
| Chicot | 33 | 604 | 43.4 | 20.0 | 14 | 64 | 7.4 | 1.5 | 556 | 20.2 | 170 | 2 |
| Clark | 55 | 1,126 | 76.9 | 33.9 | 30 | 84 | 10.7 | 2.5 | 1,326 | 51.1 | 1,768 | 15 |
| Clay | 28 | 707 | 42.8 | 20.9 | D | D | D | D | 843 | 38.3 | 672 | 3 |
| Cleburne | 50 | 892 | 60.8 | 25.3 | 31 | 105 | 11.1 | 2.7 | 2,160 | 90.5 | 3,973 | 90 |

# Table B. States and Counties — Government Employment and Payroll, and Local Government Finances

| | Government employment and payroll, 2017 | | | | | | | | | Local government finances, 2017 | | | | |
| STATE County | | | March payroll (percent of total) | | | | | | | General revenue | | | | |
| | | | | | | | | | | | | Taxes | | |
| | | | | | | | | | | | | | Per capita[1] (dollars) | |
| | Full-time equivalent employees | March payroll (dollars) | Administration, judicial, and legal | Police and corrections | Fire protection | Highways and transportation | Health and welfare | Natural resources and utilities | Education and libraries | Total (mil dol) | Intergovernmental (mil dol) | Total (mil dol) | Total | Property |
| | 171 | 172 | 173 | 174 | 175 | 176 | 177 | 178 | 179 | 180 | 181 | 182 | 183 | 184 |
| **ALABAMA—Cont'd** | | | | | | | | | | | | | | |
| Washington | 754 | 2,362,291 | 4.0 | 6.3 | 0.0 | 12.8 | 20.3 | 3.3 | 53.2 | 69.6 | 24.9 | 13.2 | 800 | 536 |
| Wilcox | 526 | 1,505,521 | 8.9 | 8.3 | 0.0 | 4.0 | 15.2 | 4.0 | 57.5 | 32.1 | 17.1 | 9.6 | 896 | 490 |
| Winston | 1,347 | 3,592,965 | 2.4 | 4.3 | 1.3 | 2.5 | 31.6 | 4.2 | 52.6 | 86.2 | 54.4 | 17.5 | 738 | 309 |
| **ALASKA** | X | X | X | X | X | X | X | X | X | X | X | X | X | X |
| Aleutians East | 141 | 727,422 | 20.6 | 9.0 | 0.0 | 14.7 | 0.0 | 10.3 | 38.5 | 39.1 | 19.9 | 12.8 | 3,714 | 445 |
| Aleutians West | 273 | 1,510,076 | 17.0 | 9.3 | 4.5 | 8.2 | 0.0 | 20.3 | 26.6 | 65.9 | 23.9 | 24.5 | 4,275 | 1,053 |
| Anchorage | 9,451 | 53,838,637 | 4.7 | 8.9 | 6.8 | 4.7 | 3.5 | 9.8 | 59.6 | 1,465.8 | 643.6 | 609.3 | 2,070 | 1,797 |
| Bethel | 302 | 1,271,943 | 24.4 | 15.0 | 3.4 | 5.0 | 23.8 | 11.3 | 0.0 | 48.2 | 24.2 | 10.4 | 578 | 5 |
| Bristol Bay | 68 | 336,805 | 14.4 | 14.0 | 4.2 | 12.3 | 3.1 | 12.7 | 38.9 | 16.8 | 6.2 | 7.2 | 8,307 | 5,524 |
| Chugach | | | | | | | | | | | | | | |
| Copper River | | | | | | | | | | | | | | |
| Denali | 98 | 453,460 | 8.3 | 0.0 | 0.0 | 0.0 | 0.0 | 3.9 | 86.6 | 15.1 | 8.3 | 6.1 | 2,929 | 1,138 |
| Dillingham | 229 | 1,071,597 | 16.0 | 10.5 | 0.8 | 6.1 | 17.1 | 4.7 | 42.0 | 35.8 | 23.2 | 5.7 | 1,148 | 467 |
| Fairbanks North Star | 3,002 | 16,676,902 | 7.1 | 2.7 | 3.0 | 4.7 | 1.3 | 2.3 | 77.7 | 420.1 | 226.7 | 155.1 | 1,556 | 1,370 |
| Haines | 101 | 485,115 | 11.9 | 8.4 | 1.6 | 5.0 | 0.0 | 5.3 | 60.7 | 18.0 | 9.4 | 6.2 | 2,474 | 1,197 |
| Hoonah-Angoon | 85 | 381,510 | 17.4 | 9.1 | 1.4 | 4.6 | 0.0 | 16.3 | 47.1 | 10.1 | 5.2 | 3.1 | 1,461 | 545 |
| Juneau | 1,882 | 11,995,964 | 5.2 | 5.5 | 2.7 | 6.8 | 32.3 | 6.9 | 40.5 | 355.5 | 97.8 | 102.9 | 3,206 | 1,520 |
| Kenai Peninsula | 2,016 | 10,480,980 | 10.5 | 4.7 | 7.0 | 3.3 | 0.5 | 4.6 | 62.5 | 558.4 | 145.6 | 137.3 | 2,345 | 1,334 |
| Ketchikan Gateway | 749 | 3,822,292 | 9.1 | 5.4 | 5.0 | 10.8 | 0.0 | 18.5 | 46.2 | 145.4 | 53.3 | 37.3 | 2,695 | 1,103 |
| Kodiak Island | 684 | 3,420,400 | 10.4 | 6.3 | 2.6 | 3.4 | 5.0 | 5.6 | 65.5 | 116.6 | 55.0 | 29.7 | 2,198 | 1,247 |
| Kusilvak | 213 | 728,558 | 22.4 | 13.1 | 0.6 | 8.1 | 0.0 | 22.8 | 20.3 | 21.4 | 14.2 | 1.8 | 221 | 18 |
| Lake and Peninsula | 179 | 756,201 | 17.1 | 0.0 | 0.1 | 3.5 | 0.9 | 3.6 | 72.3 | 29.8 | 17.8 | 8.9 | 5,519 | 3,213 |
| Matanuska-Susitna | 3,030 | 15,877,820 | 6.6 | 2.8 | 1.9 | 1.5 | 3.5 | 2.6 | 79.2 | 470.9 | 262.6 | 161.1 | 1,517 | 1,188 |
| Nome | 408 | 3,363,400 | 6.1 | 6.9 | 0.8 | 2.1 | 7.3 | 11.2 | 61.7 | 58.9 | 33.5 | 10.3 | 1,031 | 323 |
| North Slope | 1,780 | 10,868,287 | 23.3 | 6.4 | 2.8 | 3.4 | 12.3 | 13.3 | 29.3 | 602.0 | 65.8 | 400.2 | 40,860 | 40,759 |
| Northwest Arctic | 616 | 3,233,356 | 9.6 | 4.6 | 2.2 | 3.3 | 4.0 | 4.2 | 71.8 | 115.6 | 81.3 | 9.5 | 1,223 | 610 |
| Petersburg | 361 | 1,882,582 | 3.3 | 4.3 | 0.7 | 9.4 | 38.0 | 7.7 | 34.2 | 30.7 | 14.4 | 7.2 | 2,203 | 1,071 |
| Prince of Wales-Hyder | 236 | 1,050,369 | 8.2 | 5.0 | 0.0 | 3.7 | 15.7 | 9.2 | 54.7 | 31.7 | 20.4 | 3.9 | 606 | 158 |
| Sitka | 538 | 2,929,099 | 6.3 | 5.3 | 0.4 | 1.9 | 32.3 | 10.7 | 38.7 | 85.5 | 24.3 | 20.2 | 2,341 | 757 |
| Skagway | 81 | 480,789 | 12.4 | 12.6 | 4.5 | 3.4 | 22.5 | 8.6 | 30.1 | 21.3 | 7.0 | 9.9 | 8,445 | 1,711 |
| Southeast Fairbanks | 11 | 48,016 | 49.4 | 0.0 | 0.0 | 4.8 | 0.0 | 11.8 | 22.5 | 2.6 | 1.9 | 0.1 | 12 | 2 |
| Valdez-Cordova | 485 | 2,484,899 | 8.5 | 9.0 | 3.6 | 10.7 | 17.5 | 9.9 | 37.3 | 165.6 | 35.0 | 51.8 | 5,621 | 5,101 |
| Wrangell | NA | NA | NA | NA | NA | NA | NA | NA | NA | 28.5 | 8.8 | 4.5 | 1,790 | 683 |
| Yakutat | 41 | 136,787 | 21.6 | 29.3 | 0.0 | 11.9 | 0.0 | 13.8 | 22.0 | 4.8 | 2.3 | 1.7 | 2,752 | 837 |
| Yukon-Koyukuk | 399 | 1,696,302 | 7.5 | 1.6 | 1.4 | 2.3 | 0.0 | 8.9 | 74.8 | 57.4 | 51.6 | 1.5 | 281 | 154 |
| **ARIZONA** | X | X | X | X | X | X | X | X | X | X | X | X | X | X |
| Apache | 2,561 | 9,092,298 | 6.2 | 5.5 | 3.1 | 3.6 | 2.5 | 0.7 | 77.8 | 239.3 | 176.1 | 30.3 | 424 | 375 |
| Cochise | 4,703 | 17,332,719 | 11.3 | 11.0 | 5.9 | 3.1 | 2.2 | 4.0 | 60.1 | 436.0 | 208.8 | 154.2 | 1,235 | 895 |
| Coconino | 4,429 | 19,039,894 | 12.5 | 13.9 | 10.9 | 3.0 | 3.7 | 6.5 | 46.5 | 547.2 | 192.8 | 256.6 | 1,820 | 982 |
| Gila | 1,844 | 6,875,652 | 8.5 | 14.5 | 8.8 | 4.1 | 3.4 | 5.1 | 55.3 | 226.4 | 109.7 | 87.1 | 1,625 | 1,157 |
| Graham | 1,780 | 6,231,954 | 9.2 | 10.8 | 0.3 | 3.4 | 1.3 | 8.0 | 61.4 | 146.5 | 86.8 | 34.3 | 915 | 532 |
| Greenlee | 438 | 1,474,480 | 19.3 | 16.0 | 0.0 | 7.8 | 4.3 | 1.9 | 48.5 | 39.3 | 16.0 | 15.6 | 1,656 | 1,465 |
| La Paz | 819 | 2,764,287 | 12.2 | 19.2 | 8.2 | 4.0 | 6.2 | 4.1 | 43.2 | 79.6 | 41.4 | 22.7 | 1,099 | 679 |
| Maricopa | 124,148 | 610,024,911 | 8.9 | 14.9 | 6.5 | 3.1 | 6.4 | 14.3 | 44.7 | 16,182.0 | 5,622.0 | 6,973.8 | 1,612 | 981 |
| Mohave | 5,008 | 18,841,384 | 14.4 | 15.1 | 13.1 | 5.0 | 2.0 | 7.1 | 41.4 | 586.3 | 225.7 | 230.7 | 1,114 | 793 |
| Navajo | 3,430 | 12,806,832 | 2.5 | 10.3 | 6.3 | 1.5 | 0.5 | 4.3 | 74.1 | 527.2 | 226.3 | 112.3 | 1,030 | 712 |
| Pima | 31,752 | 127,346,777 | 11.2 | 14.7 | 7.2 | 3.8 | 2.7 | 7.1 | 50.9 | 3,446.6 | 1,195.6 | 1,540.0 | 1,500 | 1,074 |
| Pinal | 9,478 | 34,939,021 | 12.0 | 15.5 | 6.3 | 3.5 | 2.5 | 6.3 | 52.6 | 994.8 | 427.9 | 401.7 | 931 | 680 |
| Santa Cruz | 1,862 | 6,396,808 | 10.0 | 14.5 | 12.3 | 1.9 | 2.7 | 3.1 | 53.0 | 165.6 | 95.5 | 54.0 | 1,160 | 779 |
| Yavapai | 6,183 | 24,573,588 | 10.4 | 15.0 | 8.4 | 6.0 | 4.2 | 5.0 | 41.7 | 671.5 | 233.0 | 333.0 | 1,460 | 889 |
| Yuma | 7,776 | 27,599,788 | 11.8 | 12.2 | 3.2 | 2.3 | 2.4 | 7.4 | 59.2 | 761.8 | 338.4 | 336.9 | 1,608 | 1,179 |
| **ARKANSAS** | X | X | X | X | X | X | X | X | X | X | X | X | X | X |
| Arkansas | 744 | 2,133,165 | 6.2 | 10.7 | 2.2 | 4.5 | 0.9 | 6.3 | 66.6 | 71.5 | 40.2 | 20.0 | 1,121 | 431 |
| Ashley | 816 | 2,164,062 | 2.6 | 8.5 | 3.5 | 4.8 | 1.0 | 6.3 | 72.8 | 66.7 | 37.1 | 18.1 | 892 | 481 |
| Baxter | 1,084 | 3,322,357 | 5.6 | 9.8 | 4.0 | 6.0 | 0.2 | 8.6 | 64.0 | 67.4 | 43.0 | 14.6 | 353 | 180 |
| Benton | 7,857 | 28,010,843 | 6.2 | 10.0 | 5.6 | 2.4 | 0.3 | 7.5 | 66.3 | 800.1 | 441.1 | 239.3 | 898 | 389 |
| Boone | 1,418 | 3,852,255 | 5.0 | 7.5 | 2.9 | 3.1 | 2.0 | 5.1 | 73.1 | 95.2 | 64.6 | 17.6 | 470 | 215 |
| Bradley | 488 | 1,310,975 | 4.4 | 5.2 | 0.8 | 4.0 | 1.8 | 4.1 | 79.7 | 36.4 | 25.4 | 5.5 | 511 | 207 |
| Calhoun | 149 | 456,734 | 15.6 | 6.5 | 4.1 | 8.8 | 0.1 | 4.7 | 59.1 | 16.9 | 7.8 | 3.2 | 609 | 357 |
| Carroll | 1,025 | 2,774,615 | 5.7 | 8.8 | 2.2 | 5.7 | 0.2 | 11.6 | 63.2 | 78.5 | 42.7 | 20.1 | 722 | 340 |
| Chicot | 580 | 1,712,945 | 7.4 | 11.3 | 0.1 | 3.7 | 39.3 | 5.0 | 32.5 | 51.7 | 19.8 | 10.5 | 987 | 412 |
| Clark | 937 | 2,704,581 | 5.2 | 7.9 | 1.5 | 3.7 | 0.9 | 5.3 | 74.6 | 66.1 | 39.5 | 15.3 | 690 | 267 |
| Clay | 755 | 2,196,317 | 5.3 | 6.9 | 0.0 | 4.3 | 32.0 | 5.8 | 45.5 | 52.6 | 24.9 | 8.6 | 576 | 281 |
| Cleburne | 805 | 2,290,053 | 7.4 | 10.0 | 1.1 | 4.9 | 0.6 | 8.1 | 67.4 | 63.1 | 34.0 | 17.4 | 693 | 398 |

1. Based on the resident population estimated as of July 1 of the year shown.

# Table B. States and Counties — Local Government Finances, Government Employment, and Income Taxes

| STATE County | Direct general expenditure | | | | | | | Debt outstanding | | Government employment, 2019 | | | Individual income tax returns, 2018 | | |
|---|---|---|---|---|---|---|---|---|---|---|---|---|---|---|---|
| | Total (mil dol) | Per capita[1] (dollars) | Percent of total for: Education | Health and hospitals | Police protection | Public welfare | Highways | Total (mil dol) | Per capita[1] (dollars) | Federal civilian | Federal military | State and local | Number of returns | Mean adjusted gross income | Mean income tax |
| | 185 | 186 | 187 | 188 | 189 | 190 | 191 | 192 | 193 | 194 | 195 | 196 | 197 | 198 | 199 |
| ALABAMA—Cont'd | | | | | | | | | | | | | | | |
| Washington | 72 | 4,338 | 40.6 | 35.0 | 3.5 | 0.0 | 7.1 | 114.9 | 6,953 | 31 | 68 | 874 | 6,370 | 52,523 | 4,219 |
| Wilcox | 37 | 3,441 | 55.9 | 2.3 | 5.4 | 0.0 | 6.9 | 65.0 | 6,081 | 65 | 43 | 691 | 4,210 | 37,210 | 2,529 |
| Winston | 81 | 3,399 | 51.1 | 26.3 | 2.4 | 0.0 | 4.6 | 55.3 | 2,326 | 72 | 98 | 1,089 | 9,000 | 50,271 | 4,330 |
| ALASKA | X | X | X | X | X | X | X | X | X | 14,866 | 25,806 | 61,699 | 348,960 | 72,043 | 8,467 |
| Aleutians East | 102 | 29,450 | 8.1 | 0.0 | 1.0 | 0.1 | 1.3 | 16.0 | 4,635 | 14 | 11 | 257 | 810 | 45,274 | 3,484 |
| Aleutians West | 58 | 10,185 | 15.2 | 0.3 | 6.4 | 0.4 | 12.1 | 119.5 | 20,857 | 13 | 29 | 491 | 2,020 | 63,158 | 6,558 |
| Anchorage | 1,453 | 4,937 | 47.8 | 1.4 | 9.1 | 0.1 | 8.2 | 1,638.0 | 5,565 | 8,324 | 12,475 | 18,911 | 145,600 | 80,328 | 10,430 |
| Bethel | 52 | 2,897 | 5.6 | 0.1 | 6.2 | 0.0 | 4.5 | 16.7 | 925 | 86 | 122 | 2,898 | 7,130 | 39,972 | 3,369 |
| Bristol Bay | 13 | 14,455 | 32.2 | 0.0 | 8.5 | 0.0 | 7.7 | 0.4 | 488 | 42 | 6 | 162 | 490 | 65,235 | 7,012 |
| Chugach | | | | | | | | | | NA | NA | NA | NA | NA | NA |
| Copper River | | | | | | | | | | NA | NA | NA | NA | NA | NA |
| Denali | 13 | 6,094 | 76.6 | 0.1 | 0.0 | 0.0 | 0.0 | 0.0 | 0 | 198 | 46 | 158 | 1,230 | 69,965 | 7,760 |
| Dillingham | 35 | 7,070 | 34.6 | 0.5 | 4.4 | 0.0 | 2.3 | 19.4 | 3,919 | 45 | 33 | 671 | 2,280 | 46,760 | 4,196 |
| Fairbanks North Star | 388 | 3,888 | 55.4 | 1.1 | 1.9 | 0.0 | 6.2 | 171.3 | 1,719 | 3,134 | 9,131 | 7,220 | 46,180 | 69,879 | 7,702 |
| Haines | 14 | 5,376 | 39.2 | 0.7 | 4.2 | 0.0 | 4.2 | 0.0 | 0 | 11 | 17 | 169 | 1,300 | 67,294 | 6,518 |
| Hoonah-Angoon | 10 | 4,489 | 46.3 | 1.6 | 6.4 | 0.0 | 12.6 | 2.2 | 1,026 | 101 | 14 | 291 | 1,030 | 48,495 | 4,502 |
| Juneau | 344 | 10,707 | 21.5 | 30.3 | 4.1 | 0.0 | 3.8 | 179.4 | 5,588 | 670 | 454 | 5,925 | 16,850 | 79,118 | 9,168 |
| Kenai Peninsula | 541 | 9,244 | 26.5 | 44.6 | 1.7 | 0.0 | 2.6 | 214.8 | 3,669 | 367 | 482 | 4,492 | 28,510 | 69,312 | 7,748 |
| Ketchikan Gateway | 159 | 11,477 | 22.5 | 8.2 | 3.1 | 0.0 | 3.4 | 189.3 | 13,681 | 219 | 292 | 1,682 | 7,130 | 67,204 | 7,598 |
| Kodiak Island | 114 | 8,441 | 44.4 | 0.5 | 4.6 | 0.0 | 2.5 | 155.6 | 11,516 | 265 | 1,079 | 1,236 | 6,670 | 60,882 | 5,849 |
| Kusilvak | 19 | 2,256 | 44.7 | 0.6 | 6.5 | 0.0 | 2.9 | 0.1 | 10 | 21 | 56 | 1,606 | 3,010 | 23,657 | 1,131 |
| Lake and Peninsula | 24 | 14,839 | 72.0 | 0.0 | 0.5 | 0.0 | 2.8 | 32.8 | 20,372 | 42 | 10 | 366 | 680 | 39,797 | 2,946 |
| Matanuska-Susitna | 462 | 4,347 | 60.6 | 3.2 | 2.4 | 0.0 | 7.2 | 425.6 | 4,007 | 232 | 717 | 4,842 | 46,810 | 70,186 | 7,439 |
| Nome | 59 | 5,841 | 30.8 | 0.6 | 9.0 | 0.0 | 4.4 | 27.2 | 2,717 | 56 | 66 | 1,621 | 3,730 | 52,410 | 5,031 |
| North Slope | 403 | 41,116 | 22.0 | 6.2 | 6.3 | 1.5 | 4.1 | 381.7 | 38,965 | 18 | 48 | 2,036 | 3,620 | 56,197 | 6,048 |
| Northwest Arctic | 111 | 14,194 | 66.1 | 0.1 | 1.9 | 0.0 | 4.0 | 51.9 | 6,656 | 50 | 49 | 956 | 2,460 | 57,599 | 5,374 |
| Petersburg | 31 | 9,564 | 39.2 | 0.0 | 4.5 | 5.0 | 16.9 | 15.2 | 4,667 | 81 | 52 | 404 | 1,930 | 61,506 | 5,556 |
| Prince of Wales-Hyder | 32 | 5,024 | 42.2 | 2.8 | 4.6 | 0.0 | 7.0 | 8.6 | 1,340 | 95 | 41 | 966 | 2,250 | 55,584 | 5,608 |
| Sitka | 88 | 10,227 | 27.0 | 40.4 | 5.0 | 0.0 | 2.4 | 165.1 | 19,102 | 109 | 251 | 895 | 4,570 | 68,673 | 7,777 |
| Skagway | 29 | 25,028 | 11.1 | 8.4 | 2.5 | 0.1 | 1.7 | 24.2 | 20,739 | 52 | 8 | 126 | 760 | 62,153 | 6,507 |
| Southeast Fairbanks | 2 | 353 | 21.5 | 0.0 | 0.0 | 0.0 | 14.7 | 0.6 | 87 | 340 | 44 | 456 | 3,240 | 58,754 | 5,403 |
| Valdez-Cordova | 132 | 14,322 | 16.8 | 21.1 | 5.6 | 0.0 | 4.6 | 41.2 | 4,468 | 139 | 217 | 1,128 | 4,950 | 65,744 | 6,870 |
| Wrangell | 26 | 10,456 | 21.0 | 44.0 | 6.0 | 0.0 | 4.1 | 1.9 | 743 | 41 | 17 | 195 | 1,090 | 55,311 | 5,000 |
| Yakutat | 6 | 9,877 | 34.2 | 4.0 | 19.3 | 4.6 | 2.5 | 0.0 | 0 | 24 | 4 | 83 | 240 | 54,425 | 4,863 |
| Yukon-Koyukuk | 55 | 10,195 | 87.6 | 0.1 | 0.6 | 0.0 | 1.6 | 0.1 | 18 | 77 | 35 | 1,456 | 2,450 | 38,113 | 2,775 |
| ARIZONA | X | X | X | X | X | X | X | X | X | 56,554 | 34,735 | 367,505 | 3,093,860 | 67,797 | 7,960 |
| Apache | 199 | 2,777 | 76.5 | 1.6 | 2.6 | 0.1 | 4.2 | 186.3 | 2,603 | 2,598 | 153 | 7,894 | 24,860 | 39,986 | 2,629 |
| Cochise | 399 | 3,196 | 48.4 | 1.3 | 10.9 | 2.7 | 7.5 | 162.4 | 1,301 | 4,899 | 4,315 | 6,006 | 53,160 | 50,605 | 4,304 |
| Coconino | 510 | 3,619 | 36.3 | 2.7 | 7.6 | 0.9 | 8.5 | 201.0 | 1,426 | 2,687 | 291 | 15,858 | 61,230 | 61,614 | 6,615 |
| Gila | 180 | 3,356 | 47.7 | 1.1 | 10.2 | 1.9 | 5.4 | 58.4 | 1,091 | 281 | 115 | 5,317 | 22,020 | 50,746 | 4,491 |
| Graham | 166 | 4,425 | 54.1 | 1.3 | 7.8 | 1.5 | 3.7 | 60.5 | 1,615 | 391 | 77 | 2,403 | 12,120 | 50,868 | 3,965 |
| Greenlee | 37 | 3,879 | 42.4 | 5.8 | 14.7 | 0.7 | 6.1 | 1.4 | 147 | 34 | 20 | 521 | 3,490 | 60,853 | 4,909 |
| La Paz | 71 | 3,434 | 35.4 | 3.2 | 10.5 | 1.1 | 6.0 | 45.6 | 2,200 | 357 | 45 | 1,928 | 6,640 | 39,793 | 2,864 |
| Maricopa | 15,035 | 3,474 | 40.0 | 5.1 | 9.3 | 1.7 | 3.6 | 27,893.7 | 6,446 | 21,366 | 14,632 | 201,990 | 1,945,070 | 74,540 | 9,419 |
| Mohave | 532 | 2,570 | 33.3 | 4.1 | 9.9 | 0.1 | 6.2 | 543.6 | 2,626 | 477 | 447 | 7,839 | 86,120 | 47,845 | 4,288 |
| Navajo | 525 | 4,812 | 41.3 | 29.9 | 5.3 | 2.1 | 3.7 | 185.5 | 1,701 | 1,886 | 235 | 7,787 | 40,760 | 44,898 | 3,352 |
| Pima | 3,167 | 3,086 | 39.3 | 2.2 | 12.8 | 1.8 | 6.9 | 5,952.8 | 5,800 | 13,077 | 8,268 | 69,102 | 458,940 | 62,820 | 6,738 |
| Pinal | 970 | 2,247 | 41.5 | 3.3 | 11.4 | 0.1 | 7.7 | 1,061.2 | 2,459 | 1,528 | 946 | 18,571 | 160,860 | 53,789 | 4,416 |
| Santa Cruz | 164 | 3,523 | 47.3 | 1.0 | 7.1 | 1.8 | 4.0 | 93.5 | 2,009 | 1,730 | 99 | 2,094 | 20,930 | 46,355 | 3,821 |
| Yavapai | 710 | 3,114 | 35.4 | 3.3 | 9.4 | 0.1 | 10.8 | 559.6 | 2,454 | 1,554 | 507 | 9,440 | 110,410 | 61,135 | 6,549 |
| Yuma | 662 | 3,161 | 48.5 | 1.4 | 7.5 | 2.1 | 4.6 | 438.5 | 2,093 | 3,689 | 4,585 | 10,755 | 87,250 | 45,251 | 3,565 |
| ARKANSAS | X | X | X | X | X | X | X | X | X | 20,352 | 15,022 | 190,156 | 1,235,830 | 61,285 | 6,359 |
| Arkansas | 73 | 4,079 | 44.7 | 2.0 | 6.4 | 0.4 | 6.6 | 99.4 | 5,564 | 173 | 67 | 922 | 7,440 | 52,253 | 5,191 |
| Ashley | 62 | 3,074 | 62.3 | 0.1 | 6.8 | 0.1 | 6.3 | 112.9 | 5,559 | 68 | 76 | 940 | 7,800 | 46,918 | 3,614 |
| Baxter | 71 | 1,719 | 64.6 | 0.0 | 5.3 | 0.0 | 5.6 | 175.6 | 4,250 | 147 | 161 | 1,555 | 18,020 | 49,849 | 4,609 |
| Benton | 780 | 2,924 | 57.6 | 0.4 | 5.7 | 0.0 | 9.0 | 1,059.6 | 3,975 | 491 | 1,077 | 9,894 | 120,540 | 125,636 | 16,079 |
| Boone | 109 | 2,914 | 70.6 | 0.1 | 6.5 | 0.1 | 5.1 | 92.1 | 2,458 | 139 | 144 | 3,066 | 15,530 | 49,334 | 4,162 |
| Bradley | 35 | 3,199 | 71.2 | 0.6 | 5.4 | 0.1 | 6.6 | 20.7 | 1,915 | 28 | 41 | 802 | 4,080 | 42,756 | 2,923 |
| Calhoun | 15 | 2,963 | 37.2 | 2.1 | 5.7 | 0.9 | 11.2 | 98.1 | 18,854 | 10 | 19 | 263 | 1,870 | 49,963 | 3,604 |
| Carroll | 73 | 2,617 | 52.9 | 0.1 | 7.0 | 0.1 | 6.3 | 55.5 | 1,993 | 86 | 109 | 1,148 | 12,230 | 43,005 | 3,357 |
| Chicot | 49 | 4,589 | 32.9 | 35.9 | 4.3 | 0.1 | 6.5 | 11.5 | 1,079 | 46 | 36 | 902 | 3,750 | 42,040 | 3,915 |
| Clark | 65 | 2,937 | 63.5 | 0.1 | 8.4 | 0.2 | 7.2 | 40.9 | 1,842 | 92 | 77 | 2,443 | 8,210 | 48,291 | 3,940 |
| Clay | 52 | 3,469 | 44.6 | 28.0 | 6.4 | 0.3 | 6.4 | 20.9 | 1,404 | 52 | 56 | 863 | 5,700 | 40,541 | 3,102 |
| Cleburne | 56 | 2,252 | 59.6 | 0.0 | 8.6 | 0.3 | 10.2 | 63.9 | 2,549 | 89 | 96 | 894 | 10,350 | 49,353 | 4,189 |

1. Based on the resident population estimated as of July 1 of the year shown.

# Table B. States and Counties — Land Area and Population

| State / county code | CBSA code[1] | County Type code[2] | STATE County | Land area[3] (sq. mi) | Total persons 2019 | Rank | Per square mile | White | Black | American Indian, Alaska Native | Asian and Pacific Islander | Percent Hispanic or Latino[4] | Under 5 years | 5 to 14 years | 15 to 24 years | 25 to 34 years | 35 to 44 years | 45 to 54 years |
|---|---|---|---|---|---|---|---|---|---|---|---|---|---|---|---|---|---|---|
| | | | | 1 | 2 | 3 | 4 | 5 | 6 | 7 | 8 | 9 | 10 | 11 | 12 | 13 | 14 | 15 |
| | | | **ARKANSAS—Cont'd** | | | | | | | | | | | | | | | |
| 05025 | 38220 | 3 | Cleveland | 597.9 | 7,957 | 2,594 | 13.3 | 85.2 | 12.0 | 0.9 | 0.4 | 2.9 | 4.9 | 12.4 | 11.2 | 10.6 | 11.6 | 12.8 |
| 05027 | 31620 | 7 | Columbia | 766.1 | 23,331 | 1,666 | 30.5 | 60.6 | 35.9 | 0.8 | 1.2 | 2.9 | 5.9 | 11.5 | 20.9 | 11.2 | 10.1 | 10.6 |
| 05029 | | 6 | Conway | 551.9 | 21,037 | 1,768 | 38.1 | 83.0 | 12.5 | 1.6 | 0.8 | 4.3 | 5.8 | 12.9 | 10.8 | 11.7 | 11.4 | 12.3 |
| 05031 | 27860 | 3 | Craighead | 707.2 | 112,245 | 547 | 158.7 | 75.9 | 18.0 | 0.8 | 1.7 | 5.3 | 6.8 | 13.9 | 14.9 | 15.3 | 12.9 | 11.1 |
| 05033 | 22900 | 2 | Crawford | 595.3 | 63,409 | 842 | 106.5 | 86.2 | 2.4 | 4.0 | 2.1 | 8.4 | 6.0 | 13.7 | 11.9 | 12.4 | 12.0 | 12.6 |
| 05035 | 32820 | 1 | Crittenden | 612.9 | 47,616 | 1,029 | 77.7 | 41.3 | 55.5 | 0.6 | 1.0 | 3.0 | 7.5 | 14.9 | 13.4 | 13.3 | 11.9 | 11.7 |
| 05037 | | 6 | Cross | 616.3 | 16,142 | 2,024 | 26.2 | 73.2 | 24.2 | 0.9 | 1.1 | 2.2 | 6.2 | 13.3 | 11.1 | 12.0 | 12.1 | 12.5 |
| 05039 | | 6 | Dallas | 667.3 | 6,802 | 2,679 | 10.2 | 54.2 | 42.1 | 1.2 | 0.6 | 3.9 | 4.9 | 12.2 | 10.7 | 10.3 | 10.6 | 11.2 |
| 05041 | | 6 | Desha | 741.2 | 11,110 | 2,338 | 15.0 | 44.3 | 48.1 | 0.9 | 1.1 | 6.9 | 6.6 | 14.3 | 12.3 | 11.1 | 11.1 | 10.6 |
| 05043 | | 6 | Drew | 828.7 | 17,977 | 1,924 | 21.7 | 67.3 | 28.5 | 0.8 | 0.9 | 3.9 | 6.4 | 11.7 | 16.6 | 12.3 | 11.2 | 10.7 |
| 05045 | 30780 | 2 | Faulkner | 647.2 | 126,919 | 507 | 196.1 | 81.4 | 13.2 | 1.3 | 2.0 | 4.3 | 5.9 | 12.8 | 18.6 | 14.2 | 12.6 | 11.4 |
| 05047 | 22900 | 6 | Franklin | 609.0 | 17,897 | 1,927 | 29.4 | 93.0 | 1.4 | 2.4 | 1.4 | 3.4 | 5.4 | 12.7 | 12.3 | 11.3 | 11.8 | 12.1 |
| 05049 | | 9 | Fulton | 618.0 | 12,381 | 2,259 | 20.0 | 96.1 | 1.4 | 1.9 | 0.6 | 2.0 | 5.1 | 11.9 | 10.1 | 9.5 | 10.0 | 11.6 |
| 05051 | 26300 | 3 | Garland | 677.6 | 99,789 | 604 | 147.3 | 83.6 | 9.8 | 1.6 | 1.3 | 6.1 | 5.3 | 10.9 | 10.6 | 11.4 | 11.2 | 11.7 |
| 05053 | 30780 | 2 | Grant | 631.9 | 18,449 | 1,901 | 29.2 | 93.3 | 3.0 | 1.1 | 0.6 | 3.2 | 5.3 | 12.5 | 11.3 | 12.4 | 12.8 | 12.8 |
| 05055 | 37500 | 4 | Greene | 577.3 | 45,597 | 1,064 | 79.0 | 93.7 | 2.5 | 1.1 | 1.0 | 3.0 | 6.3 | 13.4 | 12.5 | 13.5 | 12.5 | 12.9 |
| 05057 | 26260 | 6 | Hempstead | 727.2 | 21,253 | 1,757 | 29.2 | 55.3 | 31.3 | 1.1 | 0.9 | 13.4 | 6.7 | 14.6 | 11.8 | 11.5 | 10.9 | 11.3 |
| 05059 | 31680 | 6 | Hot Spring | 614.9 | 33,787 | 1,332 | 54.9 | 84.2 | 12.0 | 1.4 | 0.8 | 3.6 | 5.1 | 11.3 | 11.8 | 12.7 | 12.4 | 12.7 |
| 05061 | | 6 | Howard | 587.3 | 13,109 | 2,214 | 22.3 | 64.4 | 21.6 | 1.6 | 0.9 | 13.2 | 7.0 | 14.6 | 12.1 | 11.5 | 11.5 | 11.3 |
| 05063 | 12900 | 7 | Independence | 763.7 | 37,757 | 1,229 | 49.4 | 89.4 | 3.1 | 1.2 | 1.2 | 6.9 | 6.1 | 13.7 | 12.9 | 12.0 | 11.9 | 12.1 |
| 05065 | | 9 | Izard | 580.2 | 13,613 | 2,184 | 23.5 | 93.9 | 2.6 | 1.5 | 0.7 | 2.6 | 4.4 | 10.3 | 10.1 | 10.7 | 10.9 | 12.3 |
| 05067 | | 6 | Jackson | 633.6 | 16,636 | 2,000 | 26.3 | 78.4 | 17.9 | 1.4 | 0.9 | 3.2 | 5.8 | 10.8 | 11.4 | 14.7 | 13.0 | 12.0 |
| 05069 | 38220 | 3 | Jefferson | 872.4 | 65,377 | 816 | 74.9 | 38.8 | 58.1 | 0.8 | 1.3 | 2.3 | 5.9 | 11.9 | 13.7 | 12.8 | 11.7 | 11.8 |
| 05071 | | 7 | Johnson | 661.0 | 26,513 | 1,548 | 40.1 | 80.0 | 2.3 | 2.1 | 3.3 | 14.4 | 6.5 | 13.7 | 13.4 | 12.0 | 12.2 | 11.6 |
| 05073 | | 8 | Lafayette | 529.5 | 6,596 | 2,693 | 12.5 | 60.2 | 36.0 | 1.0 | 0.9 | 3.2 | 5.0 | 9.9 | 10.2 | 11.2 | 10.2 | 11.9 |
| 05075 | | 6 | Lawrence | 587.5 | 16,410 | 2,012 | 27.9 | 96.1 | 1.7 | 1.6 | 0.6 | 1.7 | 6.1 | 12.2 | 13.5 | 11.8 | 11.3 | 11.7 |
| 05077 | | 7 | Lee | 599.8 | 8,513 | 2,535 | 14.2 | 42.2 | 54.3 | 1.2 | 0.8 | 3.0 | 5.1 | 9.5 | 11.0 | 15.4 | 12.0 | 13.0 |
| 05079 | 38220 | 3 | Lincoln | 559.5 | 12,944 | 2,226 | 23.1 | 65.3 | 30.0 | 1.0 | 0.4 | 4.6 | 4.6 | 8.7 | 13.5 | 17.3 | 14.4 | 13.5 |
| 05081 | 45500 | 3 | Little River | 531.3 | 12,180 | 2,279 | 22.9 | 74.8 | 20.6 | 2.4 | 0.8 | 3.7 | 6.1 | 12.2 | 11.1 | 11.5 | 11.1 | 12.4 |
| 05083 | | 6 | Logan | 709.1 | 21,410 | 1,751 | 30.2 | 92.2 | 1.9 | 2.1 | 2.4 | 3.1 | 5.8 | 11.4 | 11.4 | 11.4 | 10.9 | 13.4 |
| 05085 | 30780 | 2 | Lonoke | 771.4 | 73,921 | 744 | 95.8 | 87.5 | 7.1 | 1.2 | 1.6 | 4.7 | 6.3 | 14.2 | 12.4 | 14.0 | 13.5 | 12.7 |
| 05087 | 22220 | 2 | Madison | 834.2 | 16,644 | 1,999 | 20.0 | 90.5 | 0.9 | 2.6 | 1.8 | 6.1 | 6.2 | 12.5 | 11.3 | 11.9 | 11.5 | 12.0 |
| 05089 | | 9 | Marion | 596.7 | 16,790 | 1,985 | 28.1 | 95.2 | 1.0 | 2.2 | 1.0 | 2.6 | 4.8 | 10.9 | 8.4 | 9.1 | 9.5 | 11.1 |
| 05091 | 45500 | 3 | Miller | 624.0 | 43,177 | 1,116 | 69.2 | 69.5 | 26.5 | 1.4 | 1.0 | 3.7 | 6.3 | 13.3 | 11.7 | 13.0 | 12.7 | 12.4 |
| 05093 | 14180 | 4 | Mississippi | 902.1 | 40,066 | 1,183 | 44.4 | 58.7 | 36.5 | 0.7 | 0.7 | 4.7 | 7.3 | 14.4 | 13.3 | 13.1 | 12.0 | 11.5 |
| 05095 | | 7 | Monroe | 607.5 | 6,584 | 2,695 | 10.8 | 55.6 | 41.1 | 1.3 | 1.2 | 2.9 | 6.9 | 11.6 | 10.3 | 10.6 | 9.1 | 11.4 |
| 05097 | | 8 | Montgomery | 780.5 | 9,006 | 2,499 | 11.5 | 92.6 | 0.8 | 2.7 | 1.6 | 4.4 | 4.9 | 10.0 | 9.8 | 9.2 | 8.8 | 12.5 |
| 05099 | 26260 | 7 | Nevada | 617.8 | 8,099 | 2,577 | 13.1 | 64.2 | 31.3 | 0.9 | 0.7 | 4.7 | 5.8 | 12.5 | 11.4 | 10.6 | 10.9 | 12.9 |
| 05101 | 25460 | 9 | Newton | 821.1 | 7,602 | 2,619 | 9.3 | 95.4 | 1.0 | 3.1 | 0.7 | 2.2 | 4.8 | 10.2 | 10.3 | 9.5 | 10.4 | 11.3 |
| 05103 | 15780 | 7 | Ouachita | 732.7 | 23,167 | 1,676 | 31.6 | 56.7 | 41.1 | 1.0 | 0.8 | 2.6 | 5.6 | 12.7 | 11.3 | 11.3 | 10.7 | 11.6 |
| 05105 | 30780 | 2 | Perry | 551.4 | 10,327 | 2,398 | 18.7 | 93.6 | 2.7 | 2.0 | 0.7 | 3.1 | 5.4 | 12.3 | 11.1 | 10.7 | 11.3 | 13.2 |
| 05107 | 25760 | 6 | Phillips | 690.4 | 17,299 | 1,956 | 25.1 | 34.9 | 62.8 | 0.7 | 0.6 | 2.0 | 7.6 | 14.1 | 12.5 | 10.7 | 10.3 | 10.6 |
| 05109 | | 9 | Pike | 600.3 | 10,643 | 2,373 | 17.7 | 88.3 | 4.2 | 1.7 | 0.9 | 6.9 | 5.4 | 12.4 | 11.6 | 11.0 | 11.2 | 13.4 |
| 05111 | 27860 | 3 | Poinsett | 758.3 | 23,283 | 1,669 | 30.7 | 87.6 | 9.1 | 0.9 | 0.6 | 3.6 | 6.5 | 13.2 | 11.7 | 12.7 | 11.3 | 12.5 |
| 05113 | | 7 | Polk | 856.8 | 19,707 | 1,839 | 23.0 | 90.2 | 0.9 | 3.6 | 1.2 | 6.4 | 5.3 | 12.7 | 11.7 | 10.2 | 10.2 | 11.9 |
| 05115 | 40780 | 4 | Pope | 811.1 | 64,334 | 833 | 79.3 | 85.2 | 3.8 | 1.7 | 1.6 | 9.7 | 6.2 | 12.7 | 16.0 | 13.5 | 11.4 | 11.4 |
| 05117 | | 8 | Prairie | 647.5 | 7,966 | 2,591 | 12.3 | 86.4 | 11.5 | 0.9 | 0.4 | 2.0 | 5.5 | 11.5 | 10.2 | 10.9 | 10.3 | 12.4 |
| 05119 | 30780 | 2 | Pulaski | 758.3 | 392,980 | 180 | 518.2 | 52.9 | 39.2 | 1.0 | 2.9 | 6.3 | 6.4 | 13.0 | 12.3 | 14.6 | 12.8 | 11.7 |
| 05121 | | 7 | Randolph | 651.9 | 18,247 | 1,913 | 28.0 | 93.0 | 1.9 | 1.8 | 2.9 | 2.5 | 6.9 | 13.3 | 11.2 | 12.7 | 11.8 | 11.4 |
| 05123 | 22620 | 6 | St. Francis | 634.8 | 24,682 | 1,621 | 38.9 | 40.5 | 53.3 | 1.0 | 1.0 | 5.7 | 6.2 | 11.5 | 11.0 | 15.1 | 14.3 | 12.1 |
| 05125 | 30780 | 2 | Saline | 723.5 | 123,968 | 513 | 171.3 | 84.2 | 9.1 | 1.1 | 1.8 | 5.5 | 5.6 | 13.3 | 11.5 | 12.7 | 13.5 | 12.1 |
| 05127 | | 6 | Scott | 892.9 | 10,164 | 2,405 | 11.4 | 85.9 | 1.3 | 3.6 | 3.5 | 8.2 | 6.2 | 13.1 | 11.1 | 11.7 | 13.5 | 11.8 |
| 05129 | | 9 | Searcy | 665.8 | 7,842 | 2,600 | 11.8 | 94.6 | 0.9 | 3.1 | 0.7 | 3.1 | 4.8 | 11.3 | 9.8 | 10.1 | 10.0 | 12.0 |
| 05131 | 22900 | 2 | Sebastian | 531.2 | 127,590 | 504 | 240.2 | 71.6 | 8.1 | 3.5 | 5.4 | 14.9 | 6.5 | 13.2 | 12.8 | 13.5 | 12.1 | 12.1 |
| 05133 | | 6 | Sevier | 563.6 | 16,702 | 1,989 | 29.6 | 58.4 | 4.6 | 3.3 | 1.7 | 34.2 | 7.3 | 16.2 | 13.8 | 12.4 | 11.0 | 12.1 |
| 05135 | 12900 | 7 | Sharp | 604.6 | 17,424 | 1,949 | 28.8 | 94.7 | 1.5 | 2.3 | 0.8 | 2.7 | 5.5 | 11.2 | 10.4 | 10.0 | 9.7 | 11.8 |
| 05137 | | 9 | Stone | 606.4 | 12,674 | 2,238 | 20.9 | 96.3 | 0.9 | 2.0 | 0.7 | 1.9 | 5.1 | 10.7 | 9.8 | 9.7 | 9.8 | 10.5 |
| 05139 | 20980 | 7 | Union | 1,039.1 | 38,219 | 1,220 | 36.8 | 61.8 | 33.5 | 1.0 | 1.0 | 4.1 | 6.3 | 13.5 | 11.8 | 11.9 | 11.9 | 11.8 |
| 05141 | | 8 | Van Buren | 709.6 | 16,541 | 2,003 | 23.3 | 94.6 | 1.2 | 2.1 | 0.6 | 3.4 | 4.5 | 10.8 | 10.2 | 8.8 | 10.3 | 12.0 |
| 05143 | 22220 | 2 | Washington | 941.5 | 243,216 | 282 | 258.3 | 72.6 | 4.4 | 2.3 | 6.2 | 17.2 | 6.5 | 13.4 | 18.3 | 15.1 | 13.3 | 10.8 |
| 05145 | 42620 | 4 | White | 1,034.1 | 78,729 | 717 | 76.1 | 89.4 | 5.5 | 1.3 | 1.2 | 4.6 | 5.7 | 13.3 | 15.0 | 12.4 | 11.7 | 12.2 |
| 05147 | | 9 | Woodruff | 586.0 | 6,264 | 2,723 | 10.7 | 71.6 | 27.0 | 1.0 | 0.5 | 1.9 | 5.4 | 11.6 | 10.5 | 10.1 | 10.7 | 12.0 |
| 05149 | 40780 | 6 | Yell | 930.2 | 21,181 | 1,761 | 22.8 | 75.1 | 2.3 | 1.3 | 1.5 | 21.0 | 6.1 | 13.6 | 12.0 | 12.2 | 11.3 | 12.5 |

1. CBSA = Core Based Statistical Area. See Appendix A for explanation. See Appendix B for list of metropolitan areas with component counties. 2. County type code from the Economic Research Service of USDA Rural-Urban Continuum Codes. See Appendix A for definition. 3. Dry land or land partially or temporarily covered by water. 4. May be of any race.

# Table B. States and Counties — Population and Households

| STATE County | 55 to 64 years | 65 to 74 years | 75 years and over | Percent female | 2000 | 2010 | 2000-2010 | 2010-2020 | Births | Deaths | Net Migration | Number | Persons per household | Family households | Female family householder[1] | One person |
|---|---|---|---|---|---|---|---|---|---|---|---|---|---|---|---|---|
| | 16 | 17 | 18 | 19 | 20 | 21 | 22 | 23 | 24 | 25 | 26 | 27 | 28 | 29 | 30 | 31 |
| **ARKANSAS—Cont'd** | | | | | | | | | | | | | | | | |
| Cleveland | 14.9 | 12.6 | 9.0 | 50.9 | 8,571 | 8,692 | 1.4 | -8.5 | 834 | 1,030 | -541 | 3,217 | 2.50 | 70.1 | 13.3 | 29.0 |
| Columbia | 12.0 | 9.7 | 8.1 | 51.8 | 25,603 | 24,552 | -4.1 | -5.0 | 2,981 | 3,117 | -1,114 | 8,562 | 2.57 | 64.9 | 17.1 | 31.8 |
| Conway | 15.0 | 11.2 | 8.9 | 50.8 | 20,336 | 21,267 | 4.6 | -1.1 | 2,556 | 2,668 | -99 | 8,309 | 2.49 | 67.4 | 13.6 | 29.1 |
| Craighead | 10.9 | 8.2 | 5.9 | 51.6 | 82,148 | 96,443 | 17.4 | 16.4 | 15,141 | 9,517 | 10,155 | 41,375 | 2.50 | 64.7 | 15.5 | 28.7 |
| Crawford | 13.6 | 10.3 | 7.4 | 50.9 | 53,247 | 61,935 | 16.3 | 2.4 | 7,808 | 6,475 | 174 | 23,958 | 2.59 | 71.3 | 12.1 | 25.2 |
| Crittenden | 12.7 | 8.7 | 5.9 | 52.6 | 50,866 | 50,912 | 0.1 | -6.5 | 7,863 | 5,276 | -5,937 | 19,074 | 2.52 | 62.7 | 21.8 | 30.7 |
| Cross | 13.6 | 10.8 | 8.4 | 51.3 | 19,526 | 17,885 | -8.4 | -9.7 | 2,220 | 2,326 | -1,649 | 6,653 | 2.49 | 65.8 | 15.8 | 24.0 |
| Dallas | 15.0 | 13.9 | 11.2 | 50.8 | 9,210 | 8,124 | -11.8 | -16.3 | 768 | 1,112 | -982 | 2,969 | 2.21 | 60.5 | 14.7 | 37.6 |
| Desha | 14.4 | 11.6 | 8.0 | 53.5 | 15,341 | 13,000 | -15.3 | -14.5 | 1,643 | 1,662 | -1,889 | 5,204 | 2.23 | 68.1 | 19.9 | 28.2 |
| Drew | 12.8 | 9.9 | 8.3 | 51.2 | 18,723 | 18,517 | -1.1 | -2.9 | 2,438 | 2,075 | -921 | 7,157 | 2.43 | 65.0 | 11.3 | 27.0 |
| Faulkner | 11.2 | 8.1 | 5.3 | 51.3 | 86,014 | 113,242 | 31.7 | 12.1 | 15,679 | 9,461 | 7,393 | 44,575 | 2.66 | 64.7 | 11.3 | 27.2 |
| Franklin | 14.0 | 11.2 | 9.2 | 50.0 | 17,771 | 18,131 | 2.0 | -1.3 | 1,963 | 2,293 | 103 | 6,723 | 2.58 | 67.5 | 9.0 | 28.1 |
| Fulton | 15.2 | 14.6 | 12.1 | 51.0 | 11,642 | 12,231 | 5.1 | 1.2 | 1,182 | 1,842 | 816 | 4,973 | 2.42 | 65.9 | 10.7 | 28.5 |
| Garland | 14.2 | 14.0 | 10.7 | 51.9 | 88,068 | 96,003 | 9.0 | 3.9 | 11,205 | 14,174 | 6,792 | 40,209 | 2.40 | 62.8 | 12.7 | 31.9 |
| Grant | 14.1 | 10.6 | 8.1 | 50.6 | 16,464 | 17,842 | 8.4 | 3.4 | 1,940 | 1,952 | 626 | 6,863 | 2.62 | 73.0 | 11.5 | 24.2 |
| Greene | 12.3 | 9.6 | 7.1 | 50.8 | 37,331 | 42,090 | 12.7 | 8.3 | 5,773 | 4,984 | 2,716 | 17,254 | 2.56 | 71.2 | 13.5 | 23.8 |
| Hempstead | 14.0 | 10.6 | 8.5 | 51.7 | 23,587 | 22,593 | -4.2 | -5.9 | 3,216 | 2,508 | -2,053 | 8,094 | 2.65 | 70.6 | 15.6 | 24.5 |
| Hot Spring | 14.0 | 11.9 | 8.0 | 48.0 | 30,353 | 33,016 | 8.8 | 2.3 | 3,658 | 4,095 | 1,188 | 12,599 | 2.50 | 70.3 | 11.2 | 25.9 |
| Howard | 13.3 | 10.3 | 8.2 | 51.7 | 14,300 | 13,780 | -3.6 | -4.9 | 1,855 | 1,717 | -813 | 5,142 | 2.55 | 68.1 | 16.5 | 29.4 |
| Independence | 12.8 | 10.4 | 8.0 | 51.0 | 34,233 | 36,643 | 7.0 | 3.0 | 4,746 | 4,320 | 704 | 14,322 | 2.54 | 66.3 | 11.9 | 26.5 |
| Izard | 15.1 | 14.2 | 12.0 | 48.1 | 13,249 | 13,708 | 3.5 | -0.7 | 1,212 | 1,973 | 668 | 4,851 | 2.60 | 64.4 | 6.8 | 31.7 |
| Jackson | 13.4 | 10.7 | 7.9 | 50.4 | 18,418 | 18,005 | -2.2 | -7.6 | 1,993 | 2,348 | -1,023 | 6,137 | 2.34 | 61.2 | 15.9 | 34.1 |
| Jefferson | 13.7 | 11.1 | 7.5 | 50.8 | 84,278 | 77,429 | -8.1 | -15.6 | 9,191 | 8,653 | -12,706 | 26,801 | 2.36 | 60.8 | 18.4 | 34.2 |
| Johnson | 13.3 | 10.0 | 7.3 | 50.5 | 22,781 | 25,544 | 12.1 | 3.8 | 3,585 | 2,726 | 127 | 9,682 | 2.64 | 68.7 | 12.1 | 27.7 |
| Lafayette | 16.6 | 13.6 | 11.5 | 51.2 | 8,559 | 7,634 | -10.8 | -13.6 | 732 | 1,027 | -747 | 2,784 | 2.40 | 64.7 | 15.7 | 32.4 |
| Lawrence | 13.7 | 10.6 | 9.1 | 50.8 | 17,774 | 17,410 | -2 | -5.7 | 1,944 | 2,488 | -464 | 6,463 | 2.48 | 69.1 | 10.7 | 26.1 |
| Lee | 13.0 | 11.4 | 9.5 | 44.1 | 12,580 | 10,430 | -17.1 | -18.4 | 988 | 1,231 | -1,683 | 3,206 | 2.36 | 62.2 | 17.0 | 34.1 |
| Lincoln | 11.9 | 9.2 | 7.0 | 38.2 | 14,492 | 14,141 | -2.4 | -8.5 | 1,219 | 1,397 | -1,033 | 3,773 | 2.22 | 71.9 | 12.4 | 23.8 |
| Little River | 14.0 | 12.1 | 9.5 | 51.9 | 13,628 | 13,168 | -3.4 | -7.5 | 1,477 | 1,713 | -756 | 5,363 | 2.27 | 65.7 | 16.1 | 30.5 |
| Logan | 14.9 | 11.8 | 9.0 | 50.4 | 22,486 | 22,361 | -0.6 | -4.3 | 2,672 | 2,929 | -690 | 8,417 | 2.50 | 69.4 | 10.6 | 26.6 |
| Lonoke | 12.3 | 8.6 | 5.9 | 50.8 | 52,828 | 68,375 | 29.4 | 8.1 | 9,620 | 6,700 | 2,660 | 26,895 | 2.67 | 68.8 | 11.4 | 22.6 |
| Madison | 14.6 | 11.8 | 8.2 | 50.1 | 14,243 | 15,725 | 10.4 | 5.8 | 2,039 | 1,839 | 723 | 6,279 | 2.57 | 68.8 | 10.1 | 27.7 |
| Marion | 17.3 | 17.2 | 11.8 | 50.8 | 16,140 | 16,644 | 3.1 | 0.9 | 1,559 | 2,637 | 1,237 | 6,782 | 2.40 | 65.4 | 10.5 | 29.8 |
| Miller | 13.0 | 10.2 | 7.4 | 51.0 | 40,443 | 43,462 | 7.5 | -0.7 | 6,020 | 4,892 | -1,388 | 16,426 | 2.57 | 66.6 | 18.1 | 27.9 |
| Mississippi | 13.1 | 9.3 | 6.1 | 51.2 | 51,979 | 46,480 | -10.6 | -13.8 | 6,474 | 5,312 | -7,644 | 16,389 | 2.52 | 64.9 | 20.0 | 29.3 |
| Monroe | 16.0 | 13.0 | 11.0 | 52.9 | 10,254 | 8,155 | -20.5 | -19.3 | 992 | 1,189 | -1,385 | 3,333 | 2.09 | 59.2 | 21.3 | 38.5 |
| Montgomery | 17.2 | 15.2 | 12.4 | 50.6 | 9,245 | 9,501 | 2.8 | -5.2 | 879 | 1,357 | -10 | 3,754 | 2.35 | 68.3 | 7.8 | 29.8 |
| Nevada | 14.3 | 11.8 | 9.7 | 50.7 | 9,955 | 9,015 | -9.4 | -10.2 | 1,037 | 1,201 | -753 | 3,397 | 2.41 | 66.6 | 20.0 | 28.6 |
| Newton | 15.7 | 16.1 | 11.7 | 49.6 | 8,608 | 8,315 | -3.4 | -8.6 | 760 | 1,012 | -461 | 2,936 | 2.64 | 61.3 | 7.7 | 35.6 |
| Ouachita | 15.2 | 12.7 | 8.9 | 52.4 | 28,790 | 26,128 | -9.2 | -11.3 | 2,882 | 3,632 | -2,216 | 9,658 | 2.43 | 62.2 | 18.5 | 34.6 |
| Perry | 15.0 | 12.1 | 9.0 | 50.2 | 10,209 | 10,444 | 2.3 | -1.1 | 1,103 | 1,301 | 89 | 3,668 | 2.78 | 77.0 | 10.8 | 20.2 |
| Phillips | 13.9 | 12.0 | 8.2 | 53.4 | 26,445 | 21,755 | -17.7 | -20.5 | 2,989 | 2,744 | -4,755 | 7,485 | 2.46 | 64.6 | 23.9 | 32.3 |
| Pike | 14.1 | 11.9 | 9.1 | 50.2 | 11,303 | 11,286 | -0.2 | -5.7 | 1,146 | 1,453 | -331 | 4,280 | 2.47 | 69.4 | 7.6 | 28.5 |
| Poinsett | 13.4 | 10.6 | 8.0 | 51.4 | 25,614 | 24,570 | -4.1 | -5.2 | 3,128 | 3,516 | -886 | 9,526 | 2.47 | 70.4 | 16.1 | 26.2 |
| Polk | 14.3 | 13.3 | 10.4 | 51.2 | 20,229 | 20,654 | 2.1 | -4.6 | 2,321 | 2,809 | -447 | 8,243 | 2.42 | 68.8 | 11.5 | 26.7 |
| Pope | 12.2 | 9.5 | 7.1 | 50.5 | 54,469 | 61,754 | 13.4 | 4.2 | 8,093 | 6,086 | 572 | 22,579 | 2.67 | 65.9 | 11.2 | 28.4 |
| Prairie | 15.1 | 12.3 | 12.0 | 50.0 | 9,539 | 8,720 | -8.6 | -8.6 | 913 | 1,127 | -540 | 3,776 | 2.13 | 66.1 | 11.0 | 32.2 |
| Pulaski | 12.7 | 10.0 | 6.5 | 52.4 | 361,474 | 382,752 | 5.9 | 2.7 | 55,589 | 37,508 | -7,609 | 158,051 | 2.44 | 60.6 | 16.5 | 33.6 |
| Randolph | 12.9 | 10.7 | 9.0 | 50.5 | 18,195 | 17,969 | -1.2 | 1.5 | 2,320 | 2,523 | 487 | 7,355 | 2.36 | 67.9 | 12.5 | 29.7 |
| St. Francis | 12.8 | 10.3 | 6.7 | 44.5 | 29,329 | 28,246 | -3.7 | -12.6 | 3,434 | 2,964 | -4,069 | 9,388 | 2.41 | 62.5 | 21.6 | 33.6 |
| Saline | 12.6 | 10.9 | 7.7 | 51.2 | 83,529 | 107,126 | 28.3 | 15.7 | 13,684 | 11,063 | 14,129 | 44,075 | 2.68 | 71.0 | 11.2 | 25.2 |
| Scott | 14.7 | 11.5 | 9.6 | 49.3 | 10,996 | 11,217 | 2.0 | -9.4 | 1,276 | 1,305 | -1,040 | 3,944 | 2.61 | 65.1 | 9.1 | 32.3 |
| Searcy | 16.0 | 14.6 | 11.3 | 49.4 | 8,261 | 8,195 | -0.8 | -4.3 | 780 | 1,194 | 69 | 3,327 | 2.35 | 70.1 | 9.8 | 26.4 |
| Sebastian | 12.9 | 9.8 | 7.0 | 51.0 | 115,071 | 125,768 | 9.3 | 1.4 | 17,337 | 13,104 | -2,363 | 51,228 | 2.45 | 64.2 | 13.6 | 29.8 |
| Sevier | 11.7 | 9.1 | 6.4 | 49.7 | 15,757 | 17,061 | 8.3 | -2.1 | 2,684 | 1,631 | -1,408 | 5,885 | 2.88 | 72.7 | 11.8 | 23.6 |
| Sharp | 15.1 | 14.5 | 11.7 | 50.8 | 17,119 | 17,258 | 0.8 | 1.0 | 1,814 | 2,722 | 1,080 | 7,447 | 2.27 | 59.4 | 11.1 | 36.2 |
| Stone | 16.0 | 15.6 | 12.8 | 50.6 | 11,499 | 12,396 | 7.8 | 2.2 | 1,287 | 1,783 | 780 | 4,787 | 2.58 | 64.9 | 6.1 | 31.2 |
| Union | 14.0 | 11.0 | 7.8 | 51.5 | 45,629 | 41,634 | -8.8 | -8.2 | 5,146 | 5,474 | -3,110 | 15,726 | 2.48 | 67.2 | 15.9 | 30.0 |
| Van Buren | 16.3 | 15.0 | 12.0 | 50.4 | 16,192 | 17,294 | 6.8 | -4.4 | 1,636 | 2,398 | 28 | 6,813 | 2.41 | 68.0 | 10.5 | 29.1 |
| Washington | 10.1 | 7.4 | 5.0 | 50.0 | 157,715 | 203,045 | 28.7 | 19.8 | 32,933 | 15,690 | 22,752 | 87,325 | 2.57 | 63.0 | 9.9 | 27.0 |
| White | 12.9 | 9.6 | 7.3 | 51.4 | 67,165 | 77,075 | 14.8 | 2.1 | 9,778 | 8,564 | 449 | 29,055 | 2.58 | 68.1 | 11.8 | 28.3 |
| Woodruff | 14.1 | 13.9 | 10.8 | 51.5 | 8,741 | 7,264 | -16.9 | -13.8 | 822 | 1,037 | -790 | 2,932 | 2.19 | 63.3 | 16.9 | 30.2 |
| Yell | 14.1 | 10.2 | 8.1 | 49.7 | 21,139 | 22,185 | 4.9 | -4.5 | 2,820 | 2,663 | -1,153 | 7,503 | 2.82 | 73.9 | 14.4 | 22.6 |

1. No spouse present.

# Table B. States and Counties — Population, Vital Statistics, Health, and Crime

| STATE County | Persons in group quarters, 2020 | Daytime Population, 2015-2019 | | Births, 2020 | | Deaths, 2020 | | Persons under 65 with no health insurance, 2019 | | Medicare, 2020 | | | Crimes reported by county police or sheriff, 2019 | |
|---|---|---|---|---|---|---|---|---|---|---|---|---|---|---|
| | | Number | Employment/ residence ratio | Total | Rate[1] | Number | Rate[1] | Number | Percent | Total beneficiaries | Enrolled in Original Medicare | Enrolled in Medicare Advantage | Violent | Property |
| | 32 | 33 | 34 | 35 | 36 | 37 | 38 | 39 | 40 | 41 | 42 | 43 | 44 | 45 |
| ARKANSAS—Cont'd | | | | | | | | | | | | | | |
| Cleveland | 52 | 6,238 | 0.37 | 65 | 8.2 | 94 | 11.8 | 550 | 8.8 | 1,983 | 1,544 | 439 | 20 | 162 |
| Columbia | 2,204 | 23,750 | 1.00 | 281 | 12.0 | 300 | 12.9 | 1,678 | 9.7 | 5,168 | 4,135 | 1,033 | 23 | 67 |
| Conway | 286 | 19,384 | 0.83 | 239 | 11.4 | 238 | 11.3 | 1,603 | 9.7 | 5,163 | 3,748 | 1,415 | 46 | 234 |
| Craighead | 3,943 | 112,449 | 1.10 | 1,544 | 13.8 | 1,123 | 10.0 | 9,347 | 10.2 | 19,074 | 14,005 | 5,069 | 48 | 315 |
| Crawford | 540 | 55,608 | 0.73 | 714 | 11.3 | 719 | 11.3 | 5,805 | 11.2 | 13,975 | 7,958 | 6,017 | 55 | 254 |
| Crittenden | 1,011 | 45,562 | 0.85 | 689 | 14.5 | 567 | 11.9 | 3,647 | 9.2 | 9,243 | 6,531 | 2,712 | 123 | 286 |
| Cross | 225 | 15,225 | 0.79 | 199 | 12.3 | 225 | 13.9 | 1,269 | 9.7 | 3,974 | 2,828 | 1,146 | NA | NA |
| Dallas | 401 | 7,251 | 0.99 | 53 | 7.8 | 87 | 12.8 | 433 | 8.8 | 1,896 | 1,288 | 608 | NA | NA |
| Desha | 56 | 12,174 | 1.10 | 139 | 12.5 | 178 | 16.0 | 973 | 10.9 | 2,788 | 2,017 | 771 | NA | NA |
| Drew | 781 | 17,621 | 0.90 | 247 | 13.7 | 213 | 11.8 | 1,222 | 8.7 | 3,830 | 2,973 | 857 | 16 | 66 |
| Faulkner | 4,046 | 112,091 | 0.80 | 1,441 | 11.4 | 1,012 | 8.0 | 9,860 | 9.3 | 20,522 | 16,019 | 4,503 | 116 | 749 |
| Franklin | 447 | 16,427 | 0.80 | 173 | 9.7 | 203 | 11.3 | 1,518 | 10.8 | 4,338 | 2,754 | 1,584 | NA | NA |
| Fulton | 165 | 11,270 | 0.76 | 119 | 9.6 | 223 | 18.0 | 1,061 | 11.7 | 3,690 | 2,576 | 1,113 | 29 | 51 |
| Garland | 2,164 | 99,621 | 1.03 | 1,043 | 10.5 | 1,445 | 14.5 | 8,848 | 12.0 | 28,940 | 19,585 | 9,356 | 279 | NA |
| Grant | 143 | 15,363 | 0.64 | 182 | 9.9 | 201 | 10.9 | 1,128 | 7.6 | 3,975 | 3,032 | 943 | 28 | 167 |
| Greene | 601 | 43,355 | 0.91 | 534 | 11.7 | 539 | 11.8 | 3,253 | 8.7 | 9,636 | 6,866 | 2,770 | NA | NA |
| Hempstead | 301 | 22,067 | 1.02 | 274 | 12.9 | 251 | 11.8 | 2,170 | 12.6 | 4,666 | 3,284 | 1,381 | 29 | 143 |
| Hot Spring | 2,162 | 30,554 | 0.77 | 361 | 10.7 | 410 | 12.1 | 2,291 | 9.2 | 7,941 | 5,559 | 2,382 | NA | NA |
| Howard | 181 | 15,214 | 1.35 | 165 | 12.6 | 177 | 13.5 | 1,392 | 13.1 | 3,044 | 2,331 | 713 | 20 | 57 |
| Independence | 957 | 39,158 | 1.11 | 431 | 11.4 | 461 | 12.2 | 3,028 | 10.0 | 8,745 | 7,012 | 1,733 | 153 | 367 |
| Izard | 944 | 12,777 | 0.83 | 125 | 9.2 | 202 | 14.8 | 991 | 10.8 | 4,135 | 3,053 | 1,083 | 33 | 197 |
| Jackson | 2,117 | 17,061 | 1.01 | 187 | 11.2 | 227 | 13.6 | 1,080 | 9.4 | 3,806 | 2,860 | 946 | 45 | 119 |
| Jefferson | 5,351 | 71,288 | 1.08 | 789 | 12.1 | 880 | 13.5 | 3,927 | 8.0 | 15,417 | 10,498 | 4,920 | 78 | 414 |
| Johnson | 626 | 25,678 | 0.93 | 329 | 12.4 | 271 | 10.2 | 2,990 | 14.1 | 5,851 | 3,820 | 2,031 | 40 | 114 |
| Lafayette | 82 | 6,261 | 0.78 | 66 | 10.0 | 103 | 15.6 | 583 | 12.0 | 1,817 | 1,429 | 389 | NA | NA |
| Lawrence | 691 | 15,243 | 0.80 | 185 | 11.3 | 276 | 16.8 | 1,110 | 8.8 | 4,302 | 2,973 | 1,329 | NA | NA |
| Lee | 1,624 | 8,709 | 0.82 | 81 | 9.5 | 121 | 14.2 | 494 | 9.4 | 2,014 | 1,265 | 749 | 13 | 101 |
| Lincoln | 3,519 | 12,812 | 0.79 | 126 | 9.7 | 167 | 12.9 | 745 | 9.9 | 2,422 | 1,826 | 596 | 14 | 40 |
| Little River | 123 | 11,291 | 0.78 | 155 | 12.7 | 182 | 14.9 | 925 | 9.6 | 3,156 | 2,346 | 810 | 10 | 93 |
| Logan | 575 | 20,170 | 0.83 | 243 | 11.3 | 312 | 14.6 | 1,683 | 10.1 | 5,739 | 3,786 | 1,953 | 36 | 162 |
| Lonoke | 572 | 55,581 | 0.48 | 888 | 12.0 | 755 | 10.2 | 5,225 | 8.3 | 13,593 | 9,914 | 3,679 | 75 | 374 |
| Madison | 79 | 13,455 | 0.58 | 195 | 11.7 | 157 | 9.4 | 1,918 | 14.5 | 4,101 | 2,380 | 1,720 | 44 | 169 |
| Marion | 141 | 15,770 | 0.87 | 141 | 8.4 | 300 | 17.9 | 1,275 | 10.9 | 5,420 | 3,401 | 2,019 | 32 | 243 |
| Miller | 1,339 | 39,696 | 0.78 | 524 | 12.1 | 495 | 11.5 | 3,224 | 9.4 | 8,880 | 6,591 | 2,290 | 57 | 199 |
| Mississippi | 767 | 45,602 | 1.21 | 563 | 14.1 | 532 | 13.3 | 3,156 | 9.4 | 8,365 | 5,486 | 2,879 | 23 | 247 |
| Monroe | 84 | 6,591 | 0.84 | 104 | 15.8 | 124 | 18.8 | 474 | 9.6 | 1,979 | 1,406 | 573 | 1 | 32 |
| Montgomery | 121 | 7,911 | 0.67 | 86 | 9.5 | 160 | 17.8 | 973 | 15.0 | 2,659 | 1,960 | 700 | NA | NA |
| Nevada | 157 | 7,687 | 0.79 | 79 | 9.8 | 114 | 14.1 | 542 | 8.5 | 2,178 | 1,576 | 602 | NA | NA |
| Newton | 51 | 6,685 | 0.60 | 66 | 8.7 | 115 | 15.1 | 567 | 10.2 | 2,305 | 1,579 | 726 | NA | NA |
| Ouachita | 352 | 22,804 | 0.89 | 257 | 11.1 | 360 | 15.5 | 1,547 | 8.6 | 6,217 | 4,122 | 2,095 | NA | NA |
| Perry | 152 | 8,169 | 0.41 | 96 | 9.3 | 143 | 13.8 | 809 | 9.8 | 2,639 | 1,936 | 704 | 29 | 95 |
| Phillips | 246 | 18,467 | 0.98 | 247 | 14.3 | 275 | 15.9 | 1,245 | 9.0 | 4,244 | 2,764 | 1,480 | NA | NA |
| Pike | 196 | 9,728 | 0.78 | 104 | 9.8 | 148 | 13.9 | 1,174 | 14.1 | 2,804 | 2,087 | 717 | 10 | 73 |
| Poinsett | 341 | 20,908 | 0.69 | 278 | 11.9 | 362 | 15.5 | 2,010 | 10.7 | 5,761 | 3,708 | 2,053 | 42 | 138 |
| Polk | 155 | 20,011 | 0.99 | 192 | 9.7 | 316 | 16.0 | 2,220 | 14.8 | 5,453 | 4,179 | 1,274 | NA | NA |
| Pope | 3,345 | 66,061 | 1.09 | 767 | 11.9 | 636 | 9.9 | 6,019 | 11.9 | 13,007 | 9,105 | 3,903 | 91 | 309 |
| Prairie | 131 | 7,040 | 0.66 | 86 | 10.8 | 130 | 16.3 | 582 | 9.6 | 2,203 | 1,666 | 538 | 3 | 67 |
| Pulaski | 8,613 | 465,726 | 1.39 | 5,010 | 12.7 | 4,104 | 10.4 | 31,959 | 9.9 | 78,662 | 56,803 | 21,858 | 428 | 1,638 |
| Randolph | 349 | 17,413 | 0.96 | 269 | 14.7 | 272 | 14.9 | 1,641 | 11.6 | 4,552 | 3,060 | 1,492 | 40 | 112 |
| St. Francis | 4,095 | 25,918 | 1.00 | 287 | 11.6 | 334 | 13.5 | 1,482 | 9.0 | 4,921 | 3,239 | 1,682 | 40 | 286 |
| Saline | 1,407 | 92,409 | 0.53 | 1,339 | 10.8 | 1,201 | 9.7 | 8,511 | 8.5 | 26,769 | 19,872 | 6,897 | 187 | NA |
| Scott | 79 | 10,100 | 0.92 | 114 | 11.2 | 129 | 12.7 | 968 | 12.2 | 2,603 | 1,677 | 926 | 22 | 65 |
| Searcy | 71 | 7,012 | 0.70 | 68 | 8.7 | 121 | 15.4 | 652 | 11.4 | 2,540 | 1,729 | 811 | NA | NA |
| Sebastian | 2,185 | 142,226 | 1.26 | 1,615 | 12.7 | 1,383 | 10.8 | 14,914 | 14.2 | 26,667 | 16,671 | 9,996 | 45 | 223 |
| Sevier | 173 | 15,901 | 0.84 | 233 | 14.0 | 166 | 9.9 | 2,617 | 18.4 | 2,963 | 2,340 | 623 | 26 | 88 |
| Sharp | 189 | 15,718 | 0.76 | 176 | 10.1 | 250 | 14.3 | 1,504 | 11.9 | 5,807 | 4,178 | 1,628 | 44 | 141 |
| Stone | 151 | 11,837 | 0.86 | 135 | 10.7 | 184 | 14.5 | 1,167 | 13.2 | 4,220 | 3,237 | 984 | NA | NA |
| Union | 524 | 42,077 | 1.17 | 434 | 11.4 | 573 | 15.0 | 2,847 | 9.2 | 9,534 | 7,270 | 2,263 | 60 | 273 |
| Van Buren | 189 | 15,763 | 0.84 | 145 | 8.8 | 262 | 15.8 | 1,442 | 12.0 | 5,213 | 3,703 | 1,510 | NA | NA |
| Washington | 7,996 | 235,607 | 1.03 | 3,220 | 13.2 | 1,761 | 7.2 | 28,563 | 14.1 | 35,382 | 24,431 | 10,951 | 177 | 572 |
| White | 3,033 | 74,773 | 0.88 | 880 | 11.2 | 899 | 11.4 | 6,290 | 10.1 | 16,846 | 12,592 | 4,254 | 200 | 598 |
| Woodruff | 106 | 6,162 | 0.86 | 91 | 14.5 | 105 | 16.8 | 466 | 9.8 | 1,810 | 1,287 | 523 | 6 | 62 |
| Yell | 323 | 19,407 | 0.77 | 250 | 11.8 | 252 | 11.9 | 2,363 | 13.7 | 4,733 | 3,383 | 1,349 | 30 | 177 |

1. Per 1,000 estimated resident population.

# Table B. States and Counties — Crime, Education, Money Income, and Poverty

| STATE County | County law enforcement employment, 2019 | | Education — School enrollment and attainment, 2015-2019 | | | | Local government expenditures,[3] 2017-2018 | | Money income, 2015-2019 — Households | | | | Income and poverty, 2019 | | | |
|---|---|---|---|---|---|---|---|---|---|---|---|---|---|---|---|---|
| | | | Enrollment[1] | | Attainment[2] (percent) | | | | | | Percent | | Median household income (dollars) | Percent below poverty level | | |
| | Officers | Civilians | Total | Percent private | High school graduate or less | Bachelor's degree or more | Total current spending (mil dol) | Current spending per student (dollars) | Per capita income[4] | Median income (dollars) | with income of less than $50,000 | with income of $200,000 or more | | All persons | Children under 18 years | Children 5 to 17 years in families |
| | 46 | 47 | 48 | 49 | 50 | 51 | 52 | 53 | 54 | 55 | 56 | 57 | 58 | 59 | 60 | 61 |
| **ARKANSAS—Cont'd** | | | | | | | | | | | | | | | | |
| Cleveland | 8 | 4 | 1,828 | 6.9 | 57.0 | 16.2 | 14 | 8,985 | 24,225 | 46,684 | 53.9 | 2.5 | 53,773 | 13.7 | 19.8 | 18.2 |
| Columbia | 22 | 20 | 6,549 | 9.3 | 55.8 | 19.6 | 39 | 10,062 | 21,189 | 36,193 | 61.3 | 2.3 | 42,964 | 21.2 | 26.5 | 26.0 |
| Conway | 21 | 23 | 4,611 | 12.4 | 56.8 | 15.6 | 53 | 16,089 | 25,158 | 42,802 | 55.7 | 2.3 | 46,924 | 15.6 | 22.7 | 21.8 |
| Craighead | 40 | 73 | 30,472 | 6.0 | 45.3 | 26.2 | 180 | 9,258 | 26,948 | 47,286 | 52.3 | 3.5 | 47,766 | 16.9 | 25.0 | 21.9 |
| Crawford | 35 | 43 | 14,837 | 4.7 | 52.3 | 16.0 | 103 | 9,359 | 24,547 | 46,828 | 53.8 | 1.6 | 44,600 | 18.3 | 28.1 | 23.4 |
| Crittenden | 39 | 105 | 12,976 | 10.0 | 52.2 | 17.4 | 98 | 9,773 | 23,172 | 40,161 | 57.9 | 2.5 | 41,976 | 22.4 | 32.1 | 29.4 |
| Cross | 18 | 22 | 3,807 | 6.7 | 57.9 | 14.4 | 33 | 9,927 | 25,380 | 46,787 | 52.2 | 2.7 | 45,273 | 17.8 | 26.3 | 25.7 |
| Dallas | 9 | 19 | 1,715 | 28.6 | 60.1 | 11.9 | 9 | 11,041 | 20,517 | 38,072 | 63.7 | 0.7 | 41,314 | 18.8 | 31.7 | 30.3 |
| Desha | 7 | 1 | 2,781 | 4.9 | 63.5 | 14.1 | 27 | 10,568 | 19,235 | 31,893 | 66.9 | 2.1 | 37,005 | 25.4 | 38.6 | 34.6 |
| Drew | 14 | 1 | 5,461 | 12.6 | 48.4 | 22.8 | 40 | 12,842 | 25,543 | 46,997 | 51.3 | 2.8 | 44,162 | 18.5 | 25.0 | 23.4 |
| Faulkner | 50 | 113 | 36,743 | 13.5 | 38.6 | 30.6 | 177 | 9,180 | 26,913 | 52,827 | 47.8 | 3.4 | 58,482 | 14.4 | 15.9 | 14.2 |
| Franklin | 12 | 13 | 3,912 | 10.3 | 57.3 | 12.1 | 31 | 10,842 | 20,480 | 38,923 | 60.3 | 1.2 | 45,670 | 16.2 | 23.2 | 21.3 |
| Fulton | 7 | 11 | 2,350 | 9.6 | 50.2 | 12.2 | 17 | 9,927 | 19,697 | 35,405 | 62.8 | 0.4 | 32,926 | 17.7 | 26.5 | 24.0 |
| Garland | 54 | 127 | 19,526 | 7.3 | 42.3 | 21.6 | 151 | 9,841 | 27,171 | 44,777 | 54.4 | 2.8 | 45,786 | 16.9 | 23.9 | 20.9 |
| Grant | 16 | 2 | 3,739 | 8.5 | 53.0 | 18.7 | 40 | 8,497 | 28,153 | 55,388 | 45.3 | 1.3 | 58,566 | 12.1 | 15.6 | 14.6 |
| Greene | 23 | 3 | 10,930 | 8.2 | 57.6 | 16.2 | 71 | 8,941 | 23,494 | 47,056 | 52.2 | 1.8 | 47,098 | 15.5 | 20.0 | 18.1 |
| Hempstead | 25 | 33 | 5,381 | 8.4 | 54.2 | 15.1 | 41 | 11,850 | 21,521 | 42,860 | 54.7 | 1.3 | 40,867 | 18.5 | 26.3 | 25.5 |
| Hot Spring | 19 | 15 | 7,001 | 6.2 | 53.2 | 13.3 | 52 | 9,559 | 21,866 | 43,889 | 55.4 | 1.0 | 45,860 | 20.2 | 26.7 | 26.3 |
| Howard | 11 | 13 | 3,216 | 9.4 | 53.2 | 13.6 | 31 | 10,626 | 24,481 | 36,059 | 62.5 | 2.8 | 41,427 | 16.8 | 24.9 | 23.3 |
| Independence | 37 | 4 | 9,095 | 10.1 | 51.6 | 17.5 | 64 | 9,570 | 23,733 | 44,319 | 55.1 | 2.2 | 47,903 | 15.9 | 20.1 | 19.9 |
| Izard | 22 | 15 | 2,400 | 5.3 | 54.2 | 13.1 | 22 | 12,175 | 21,164 | 42,876 | 57.2 | 0.8 | 38,636 | 22.9 | 32.4 | 31.4 |
| Jackson | 14 | 18 | 3,374 | 3.9 | 62.3 | 10.1 | 22 | 10,088 | 20,295 | 34,109 | 67.3 | 2.7 | 40,555 | 24.4 | 33.1 | 29.4 |
| Jefferson | 48 | 54 | 17,162 | 13.6 | 52.7 | 18.0 | 124 | 11,483 | 21,010 | 40,726 | 59.4 | 1.9 | 40,396 | 17.5 | 23.5 | 21.4 |
| Johnson | 12 | 23 | 5,882 | 15.3 | 61.8 | 15.4 | 43 | 9,228 | 25,876 | 32,397 | 63.9 | 2.4 | 37,164 | 25.5 | 37.1 | 34.1 |
| Lafayette | 9 | 18 | 1,159 | 14.8 | 60.5 | 15.8 | 7 | 11,907 | 21,473 | 39,993 | 60.0 | 1.2 | 39,604 | 18.4 | 27.0 | 26.9 |
| Lawrence | 16 | 21 | 4,048 | 12.8 | 54.5 | 15.4 | 35 | 10,926 | 21,964 | 40,726 | 60.0 | 1.0 | 40,700 | 18.8 | 27.0 | 25.0 |
| Lee | 5 | 5 | 1,649 | 18.9 | 64.6 | 10.7 | 9 | 11,931 | 16,968 | 29,681 | 64.8 | 0.7 | 30,421 | 35.4 | 42.1 | 39.7 |
| Lincoln | 11 | 15 | 1,934 | 27.0 | 66.2 | 9.7 | 15 | 9,092 | 14,764 | 46,596 | 55.0 | 3.0 | 44,272 | 27.1 | 31.8 | 28.4 |
| Little River | 15 | 19 | 2,474 | 3.0 | 55.7 | 12.5 | 19 | 10,092 | 25,759 | 45,113 | 55.6 | 2.1 | 43,257 | 17.8 | 26.1 | 24.6 |
| Logan | 14 | 32 | 4,714 | 8.6 | 58.9 | 13.3 | 39 | 9,960 | 21,491 | 41,466 | 60.3 | 1.6 | 42,249 | 17.4 | 22.2 | 21.3 |
| Lonoke | 35 | 34 | 17,798 | 8.8 | 47.6 | 20.6 | 126 | 8,935 | 27,293 | 59,094 | 42.2 | 3.3 | 62,782 | 10.7 | 14.9 | 14.6 |
| Madison | 19 | 2 | 3,309 | 15.2 | 63.2 | 14.2 | 23 | 9,763 | 24,356 | 41,682 | 56.4 | 2.6 | 48,016 | 16.4 | 24.6 | 23.7 |
| Marion | 20 | 2 | 2,541 | 6.3 | 51.0 | 17.1 | 16 | 9,378 | 21,395 | 36,719 | 63.9 | 1.8 | 38,760 | 16.3 | 27.1 | 25.3 |
| Miller | 25 | 11 | 10,008 | 8.7 | 48.9 | 16.8 | 67 | 9,951 | 23,729 | 43,371 | 55.6 | 1.9 | 50,905 | 14.1 | 24.0 | 22.4 |
| Mississippi | 42 | 52 | 10,235 | 4.2 | 57.7 | 13.8 | 79 | 10,732 | 22,750 | 39,962 | 58.7 | 2.1 | 39,990 | 23.0 | 30.0 | 27.8 |
| Monroe | 6 | 8 | 1,514 | 13.7 | 62.4 | 12.7 | 13 | 12,377 | 23,538 | 38,468 | 62.0 | 1.7 | 34,289 | 25.5 | 40.8 | 39.3 |
| Montgomery | 11 | 13 | 1,552 | 3.1 | 56.6 | 13.5 | 11 | 10,600 | 22,631 | 35,741 | 64.6 | 2.1 | 39,591 | 19.9 | 31.0 | 29.7 |
| Nevada | 6 | 24 | 1,880 | 9.6 | 52.6 | 14.6 | 14 | 10,301 | 19,912 | 38,042 | 61.8 | 0.5 | 36,528 | 24.1 | 38.3 | 35.9 |
| Newton | 12 | 9 | 1,630 | 17.4 | 48.9 | 20.5 | 14 | 11,087 | 19,735 | 38,082 | 63.3 | 0.6 | 37,648 | 18.1 | 31.1 | 29.8 |
| Ouachita | 17 | 31 | 5,184 | 5.6 | 55.6 | 14.5 | 44 | 10,677 | 21,562 | 35,425 | 62.4 | 1.4 | 46,393 | 17.3 | 27.6 | 25.5 |
| Perry | 12 | 12 | 2,335 | 9.7 | 56.8 | 12.7 | 15 | 9,495 | 23,164 | 48,667 | 51.7 | 1.7 | 50,560 | 16.5 | 25.1 | 23.4 |
| Phillips | 14 | 6 | 4,904 | 6.9 | 51.7 | 13.8 | 53 | 13,633 | 18,651 | 29,320 | 66.5 | 1.4 | 31,521 | 33.3 | 49.9 | 48.2 |
| Pike | 19 | 5 | 2,290 | 10.9 | 50.7 | 17.3 | 21 | 10,006 | 23,362 | 40,401 | 59.8 | 1.7 | 40,536 | 23.1 | 32.9 | 30.0 |
| Poinsett | 16 | 28 | 5,229 | 6.3 | 62.7 | 11.1 | 47 | 11,405 | 20,963 | 40,921 | 58.7 | 0.8 | 39,722 | 20.2 | 28.9 | 24.2 |
| Polk | 16 | 13 | 4,142 | 10.4 | 51.6 | 14.1 | 35 | 9,945 | 23,103 | 37,974 | 64.2 | 2.9 | 47,519 | 16.9 | 19.9 | 17.4 |
| Pope | 35 | 53 | 17,363 | 6.3 | 50.7 | 22.8 | 101 | 9,784 | 23,391 | 43,462 | 56.0 | 3.0 | 46,205 | 15.0 | 22.1 | 20.8 |
| Prairie | 9 | 22 | 1,460 | 6.8 | 56.7 | 15.4 | 11 | 9,545 | 25,789 | 42,660 | 57.6 | 3.1 | 52,684 | 14.6 | 19.9 | 18.6 |
| Pulaski | 410 | 124 | 96,449 | 18.3 | 35.6 | 34.3 | 661 | 10,821 | 32,692 | 51,749 | 48.5 | 5.8 | 52,875 | 16.8 | 24.8 | 22.6 |
| Randolph | 12 | 13 | 3,869 | 12.2 | 53.2 | 13.8 | 22 | 8,374 | 23,248 | 37,218 | 64.3 | 1.8 | 42,875 | 17.3 | 25.8 | 24.1 |
| St. Francis | 19 | 23 | 5,885 | 9.0 | 55.7 | 10.4 | 37 | 10,655 | 18,447 | 35,348 | 67.2 | 2.2 | 34,389 | 32.0 | 42.3 | 39.7 |
| Saline | 57 | 50 | 28,536 | 11.5 | 40.7 | 26.7 | 147 | 8,339 | 30,553 | 64,412 | 38.0 | 3.9 | 66,944 | 9.6 | 14.9 | 13.4 |
| Scott | 9 | 19 | 2,178 | 1.1 | 61.6 | 9.3 | 24 | 10,114 | 18,281 | 36,092 | 64.9 | 1.2 | 38,065 | 19.6 | 29.8 | 27.1 |
| Searcy | 7 | 11 | 1,490 | 7.9 | 58.4 | 11.9 | 17 | 11,658 | 20,410 | 36,438 | 68.9 | 0.9 | 34,712 | 22.4 | 34.1 | 33.3 |
| Sebastian | 45 | 102 | 31,498 | 11.6 | 47.0 | 21.3 | 211 | 10,265 | 25,961 | 46,228 | 53.7 | 2.8 | 48,945 | 15.9 | 21.7 | 19.6 |
| Sevier | 30 | 5 | 4,613 | 2.9 | 60.9 | 12.8 | 42 | 12,668 | 22,320 | 47,704 | 52.8 | 1.8 | 43,493 | 17.7 | 25.0 | 23.7 |
| Sharp | 14 | 12 | 3,469 | 4.1 | 57.4 | 10.2 | 27 | 9,338 | 21,423 | 34,671 | 67.3 | 1.5 | 37,591 | 23.3 | 35.9 | 32.5 |
| Stone | 12 | 13 | 2,414 | 5.8 | 53.7 | 13.9 | 17 | 9,804 | 20,695 | 38,188 | 64.1 | 1.1 | 36,930 | 19.3 | 30.8 | 28.9 |
| Union | 30 | 36 | 9,319 | 10.6 | 50.6 | 21.1 | 71 | 9,315 | 25,193 | 44,663 | 54.8 | 3.2 | 46,602 | 18.9 | 25.7 | 22.2 |
| Van Buren | 21 | 16 | 3,030 | 13.4 | 57.5 | 14.5 | 27 | 11,671 | 22,168 | 38,499 | 61.5 | 1.3 | 38,096 | 19.9 | 29.6 | 25.0 |
| Washington | 156 | 147 | 71,546 | 10.3 | 40.7 | 33.3 | 428 | 9,921 | 27,790 | 50,451 | 49.5 | 5.1 | 52,158 | 15.1 | 16.0 | 15.3 |
| White | 53 | 51 | 20,911 | 30.4 | 54.9 | 20.9 | 126 | 9,782 | 23,702 | 44,029 | 54.0 | 3.1 | 46,388 | 15.4 | 20.4 | 18.9 |
| Woodruff | 7 | 4 | 1,223 | 6.7 | 62.1 | 14.0 | 12 | 11,071 | 23,423 | 32,845 | 61.7 | 1.9 | 36,568 | 27.1 | 38.1 | 34.4 |
| Yell | 16 | 13 | 5,047 | 3.7 | 61.8 | 12.8 | 42 | 9,861 | 22,015 | 43,923 | 55.5 | 1.6 | 44,469 | 15.6 | 23.7 | 22.1 |

1. All persons 3 years old and over enrolled in nursery school through college.   2. Persons 25 years old and over.   3. Elementary and secondary education expenditures.   4. Based on population estimated by the American Community Survey, 2014–2018.

Table B. States and Counties — **Personal Income**

| | Personal income, 2019 | | | | | Supplements to wages and salaries, employer contributions (mil dol) | | | | | Earnings, 2019 | Contributions for government social insurance (mil dol) | |
|---|---|---|---|---|---|---|---|---|---|---|---|---|---|
| STATE County | Total (mil dol) | Percent change 2018-2019 | Per capita[1] Dollars | Per capita[1] Rank | Wages and salaries (mil dol) | Pension and insurance | Government social insurance | Proprietors' income (mil dol) | Dividends, interest, and rent (mil dol) | Personal transfer receipts (mil dol) | Total (mil dol) | From employee and self-employed | From employer |
| | 62 | 63 | 64 | 65 | 66 | 67 | 68 | 69 | 70 | 71 | 72 | 73 | 74 |
| ARKANSAS—Cont'd | | | | | | | | | | | | | |
| Cleveland | 317 | -2.3 | 39,847 | 2,069 | 40 | 8 | 3 | 21 | 35 | 100 | 71 | 6 | 3 |
| Columbia | 879 | 2.0 | 37,456 | 2,394 | 373 | 65 | 28 | 25 | 148 | 267 | 492 | 35 | 28 |
| Conway | 811 | 1.9 | 38,892 | 2,210 | 290 | 48 | 22 | 60 | 120 | 248 | 420 | 30 | 22 |
| Craighead | 4,266 | 5.3 | 38,669 | 2,242 | 2,380 | 356 | 178 | 334 | 584 | 1,052 | 3,248 | 212 | 178 |
| Crawford | 2,145 | 3.7 | 33,904 | 2,813 | 833 | 124 | 65 | 79 | 260 | 651 | 1,102 | 84 | 65 |
| Crittenden | 1,872 | 3.8 | 39,041 | 2,184 | 698 | 111 | 54 | 196 | 225 | 545 | 1,059 | 71 | 54 |
| Cross | 635 | 7.7 | 38,693 | 2,238 | 191 | 30 | 14 | 65 | 82 | 199 | 300 | 21 | 14 |
| Dallas | 260 | 4.1 | 37,101 | 2,433 | 101 | 16 | 8 | 10 | 35 | 104 | 135 | 11 | 8 |
| Desha | 465 | 6.0 | 40,943 | 1,922 | 199 | 31 | 15 | 71 | 65 | 159 | 316 | 19 | 15 |
| Drew | 688 | 3.3 | 37,742 | 2,352 | 259 | 47 | 20 | 69 | 107 | 208 | 394 | 26 | 20 |
| Faulkner | 4,908 | 4.0 | 38,949 | 2,199 | 1,956 | 283 | 143 | 186 | 647 | 1,122 | 2,568 | 178 | 143 |
| Franklin | 592 | 0.3 | 33,404 | 2,852 | 195 | 40 | 15 | 6 | 93 | 191 | 257 | 21 | 15 |
| Fulton | 362 | 5.0 | 29,020 | 3,063 | 80 | 15 | 6 | 7 | 57 | 163 | 108 | 13 | 6 |
| Garland | 4,082 | 4.2 | 41,069 | 1,913 | 1,581 | 227 | 120 | 294 | 816 | 1,315 | 2,222 | 173 | 120 |
| Grant | 714 | 3.9 | 39,087 | 2,179 | 198 | 29 | 14 | 22 | 84 | 185 | 263 | 21 | 14 |
| Greene | 1,621 | 3.6 | 35,770 | 2,588 | 671 | 108 | 54 | 117 | 205 | 479 | 951 | 67 | 54 |
| Hempstead | 694 | -0.1 | 32,238 | 2,930 | 316 | 54 | 24 | 33 | 85 | 250 | 427 | 31 | 24 |
| Hot Spring | 1,062 | 3.5 | 31,443 | 2,981 | 351 | 61 | 27 | 47 | 143 | 374 | 486 | 40 | 27 |
| Howard | 459 | -5.0 | 34,762 | 2,721 | 252 | 43 | 20 | 43 | 59 | 148 | 358 | 23 | 20 |
| Independence | 1,380 | 4.2 | 36,476 | 2,512 | 732 | 115 | 57 | 113 | 205 | 423 | 1,018 | 70 | 57 |
| Izard | 421 | 2.2 | 30,923 | 3,011 | 116 | 22 | 9 | 22 | 65 | 184 | 169 | 16 | 9 |
| Jackson | 681 | 6.9 | 40,703 | 1,963 | 222 | 37 | 17 | 173 | 74 | 197 | 448 | 27 | 17 |
| Jefferson | 2,445 | 2.5 | 36,586 | 2,499 | 1,312 | 236 | 104 | 119 | 355 | 836 | 1,771 | 120 | 104 |
| Johnson | 769 | 0.3 | 28,920 | 3,069 | 310 | 50 | 24 | 34 | 95 | 268 | 419 | 33 | 24 |
| Lafayette | 242 | 1.5 | 36,524 | 2,506 | 46 | 9 | 4 | 18 | 35 | 84 | 76 | 6 | 4 |
| Lawrence | 564 | 2.7 | 34,400 | 2,755 | 154 | 27 | 12 | 49 | 78 | 214 | 242 | 19 | 12 |
| Lee | 284 | 9.9 | 32,099 | 2,938 | 79 | 14 | 6 | 37 | 41 | 107 | 136 | 9 | 6 |
| Lincoln | 365 | 2.0 | 28,024 | 3,080 | 119 | 23 | 9 | 29 | 44 | 122 | 181 | 13 | 9 |
| Little River | 434 | 2.2 | 35,383 | 2,638 | 194 | 28 | 14 | 24 | 56 | 144 | 259 | 19 | 14 |
| Logan | 717 | -2.2 | 33,386 | 2,856 | 195 | 35 | 15 | 40 | 96 | 288 | 285 | 24 | 15 |
| Lonoke | 2,987 | 3.5 | 40,749 | 1,956 | 571 | 84 | 43 | 131 | 425 | 691 | 829 | 67 | 43 |
| Madison | 568 | -1.1 | 34,245 | 2,775 | 142 | 24 | 11 | 38 | 78 | 165 | 215 | 18 | 11 |
| Marion | 524 | 3.4 | 31,405 | 2,984 | 143 | 30 | 12 | 19 | 94 | 221 | 205 | 20 | 12 |
| Miller | 1,516 | 3.7 | 35,035 | 2,682 | 603 | 90 | 47 | 93 | 223 | 443 | 832 | 61 | 47 |
| Mississippi | 1,499 | 7.5 | 36,882 | 2,456 | 1,018 | 139 | 74 | 64 | 192 | 472 | 1,294 | 87 | 74 |
| Monroe | 246 | 6.7 | 36,710 | 2,482 | 80 | 12 | 6 | 26 | 36 | 96 | 125 | 9 | 6 |
| Montgomery | 268 | 2.5 | 29,858 | 3,040 | 45 | 9 | 3 | 15 | 45 | 108 | 73 | 8 | 3 |
| Nevada | 283 | 1.3 | 34,274 | 2,773 | 108 | 17 | 9 | 16 | 37 | 113 | 150 | 12 | 9 |
| Newton | 233 | 3.5 | 30,096 | 3,033 | 35 | 8 | 3 | 10 | 41 | 95 | 56 | 7 | 3 |
| Ouachita | 907 | 3.3 | 38,801 | 2,223 | 284 | 52 | 21 | 20 | 130 | 293 | 377 | 31 | 21 |
| Perry | 379 | 3.3 | 36,248 | 2,543 | 47 | 8 | 4 | 25 | 52 | 118 | 84 | 8 | 4 |
| Phillips | 630 | 6.2 | 35,438 | 2,629 | 212 | 38 | 17 | 60 | 86 | 253 | 327 | 23 | 17 |
| Pike | 348 | 0.0 | 32,499 | 2,919 | 101 | 17 | 8 | 18 | 54 | 127 | 143 | 13 | 8 |
| Poinsett | 837 | 5.1 | 35,558 | 2,610 | 233 | 35 | 17 | 100 | 88 | 300 | 386 | 28 | 17 |
| Polk | 619 | 1.0 | 30,984 | 3,003 | 217 | 41 | 17 | 58 | 97 | 234 | 333 | 26 | 17 |
| Pope | 2,272 | 2.1 | 35,464 | 2,622 | 1,234 | 199 | 94 | 87 | 321 | 631 | 1,615 | 110 | 94 |
| Prairie | 301 | 0.6 | 37,330 | 2,409 | 60 | 10 | 5 | 30 | 44 | 99 | 105 | 8 | 5 |
| Pulaski | 20,351 | 3.3 | 51,927 | 685 | 14,782 | 2,205 | 1,091 | 1,302 | 4,404 | 4,129 | 19,380 | 1,216 | 1,091 |
| Randolph | 617 | 2.4 | 34,352 | 2,764 | 208 | 36 | 17 | 42 | 79 | 229 | 303 | 24 | 17 |
| St. Francis | 729 | 2.6 | 29,154 | 3,059 | 319 | 60 | 25 | 56 | 95 | 285 | 460 | 32 | 25 |
| Saline | 5,227 | 4.4 | 42,688 | 1,658 | 1,058 | 148 | 79 | 229 | 660 | 1,252 | 1,513 | 128 | 79 |
| Scott | 320 | -1.5 | 31,159 | 2,996 | 106 | 20 | 8 | 30 | 47 | 117 | 165 | 13 | 8 |
| Searcy | 222 | 3.1 | 28,143 | 3,077 | 51 | 9 | 4 | 7 | 38 | 106 | 71 | 9 | 4 |
| Sebastian | 5,474 | 3.6 | 42,821 | 1,646 | 3,177 | 459 | 237 | 513 | 1,004 | 1,310 | 4,386 | 291 | 237 |
| Sevier | 544 | -2.9 | 32,001 | 2,940 | 193 | 33 | 16 | 62 | 61 | 151 | 303 | 19 | 16 |
| Sharp | 600 | 1.9 | 34,403 | 2,754 | 120 | 21 | 9 | 58 | 87 | 261 | 207 | 20 | 9 |
| Stone | 395 | 3.3 | 31,593 | 2,970 | 94 | 17 | 7 | 28 | 83 | 173 | 146 | 15 | 7 |
| Union | 1,751 | 2.1 | 45,261 | 1,327 | 953 | 157 | 69 | 78 | 368 | 459 | 1,257 | 86 | 69 |
| Van Buren | 524 | 3.7 | 31,697 | 2,960 | 143 | 24 | 10 | 14 | 100 | 220 | 191 | 19 | 10 |
| Washington | 9,808 | 4.0 | 41,005 | 1,919 | 5,818 | 843 | 419 | 775 | 1,745 | 1,736 | 7,855 | 498 | 419 |
| White | 2,920 | 4.0 | 37,082 | 2,438 | 1,036 | 154 | 79 | 157 | 446 | 825 | 1,425 | 107 | 79 |
| Woodruff | 252 | 3.9 | 39,902 | 2,061 | 74 | 13 | 5 | 37 | 35 | 89 | 129 | 8 | 5 |
| Yell | 710 | -2.8 | 33,260 | 2,867 | 241 | 42 | 19 | 53 | 99 | 224 | 354 | 25 | 19 |

1. Based on the resident population estimated as of July 1 of the year shown.

| STATE County | Earnings, 2019 (cont.) — Percent by selected industries | | | | | | | | | Social Security beneficiaries, December 2019 | | Supplemental Security Income recipients, 2019 | Housing units, 2020 | |
|---|---|---|---|---|---|---|---|---|---|---|---|---|---|---|
| | Farm | Mining, quarrying, and extractions | Construction | Manufacturing | Information; professional, scientific, technical services | Retail trade | Finance, insurance, real estate, and leasing | Health care and social assistance | Government | Number | Rate[1] | | Total | Percent change, 2010-2020 |
| | 75 | 76 | 77 | 78 | 79 | 80 | 81 | 82 | 83 | 84 | 85 | 86 | 87 | 88 |
| **ARKANSAS—Cont'd** | | | | | | | | | | | | | | |
| Cleveland | 21.4 | 0.0 | 4.3 | 9.8 | D | 2.2 | D | 8.7 | 24.0 | 2,135 | 269 | 247 | 4,141 | 1.8 |
| Columbia | 1.4 | 5.9 | 2.7 | 30.5 | 2.2 | 6.2 | 4.7 | 7.1 | 21.1 | 5,745 | 245 | 1,158 | 11,722 | 1.1 |
| Conway | 6.7 | 0.4 | 10.5 | 17.5 | D | 7.4 | 3.1 | D | 19.9 | 5,725 | 274 | 925 | 9,903 | 1.9 |
| Craighead | 1.0 | D | 6.1 | 14.0 | 4.3 | 7.6 | 5.1 | 23.8 | 14.7 | 21,180 | 191 | 4,092 | 47,752 | 17.9 |
| Crawford | 0.4 | 0.6 | 7.5 | 23.3 | D | 6.6 | 3.1 | 6.9 | 11.9 | 15,770 | 249 | 1,982 | 27,191 | 4.1 |
| Crittenden | 5.2 | D | 5.2 | 14.0 | 2.4 | 6.6 | 3.9 | 10.3 | 14.1 | 10,655 | 222 | 3,096 | 21,827 | 1.6 |
| Cross | 11.6 | 0.0 | 2.2 | 4.6 | 2.6 | 11.7 | 5.2 | D | 20.7 | 4,395 | 269 | 796 | 7,980 | 1.6 |
| Dallas | 1.4 | 0.0 | 3.1 | 23.2 | 1.3 | 8.0 | 3.5 | D | 11.8 | 2,130 | 306 | 398 | 4,353 | 1.0 |
| Desha | 20.8 | 0.0 | 2.1 | 24.0 | 1.9 | 5.0 | 3.3 | D | 15.3 | 3,160 | 277 | 707 | 6,304 | 0.7 |
| Drew | 7.6 | 0.0 | 7.0 | 12.9 | 3.4 | 7.3 | 4.1 | 10.0 | 26.4 | 4,225 | 234 | 673 | 8,683 | 3.2 |
| Faulkner | -0.5 | 1.0 | 9.4 | 8.8 | 13.3 | 8.4 | 4.3 | 15.2 | 17.6 | 22,185 | 176 | 2,779 | 51,020 | 9.5 |
| Franklin | -4.2 | 4.5 | 5.1 | 20.2 | D | 7.5 | 1.7 | 8.3 | 24.5 | 4,795 | 270 | 527 | 8,044 | 0.2 |
| Fulton | -9.6 | D | 17.1 | D | 3.3 | 5.2 | 5.0 | D | 27.2 | 4,070 | 325 | 432 | 6,872 | 1.4 |
| Garland | 0.1 | 0.4 | 9.8 | 7.6 | 5.1 | 10.9 | 5.4 | 22.6 | 12.9 | 30,900 | 311 | 3,816 | 51,145 | 1.2 |
| Grant | -0.9 | D | 9.0 | 33.4 | D | 7.7 | 4.4 | D | 17.5 | 4,220 | 231 | 405 | 8,127 | 4.8 |
| Greene | 3.0 | D | 5.2 | 34.2 | D | 7.5 | 4.8 | 11.8 | 12.8 | 10,845 | 239 | 1,700 | 19,566 | 9.4 |
| Hempstead | 1.9 | D | 3.3 | 29.1 | 2.8 | 6.8 | 2.9 | D | 21.4 | 5,195 | 241 | 981 | 10,568 | 1.4 |
| Hot Spring | 0.0 | 1.0 | 4.6 | 19.1 | D | 5.9 | 3.2 | D | 20.9 | 8,745 | 259 | 1,136 | 14,536 | 1.1 |
| Howard | 6.8 | D | 3.1 | 40.9 | D | 5.4 | 2.2 | 10.9 | 12.1 | 3,375 | 255 | 502 | 6,260 | 0.5 |
| Independence | 2.8 | D | 5.3 | 24.1 | D | 6.4 | 5.1 | D | 11.2 | 9,730 | 257 | 1,164 | 16,704 | 3.2 |
| Izard | 0.9 | D | 6.3 | 8.6 | 1.6 | 6.5 | 6.8 | 18.0 | 28.9 | 4,605 | 338 | 526 | 7,319 | 1.1 |
| Jackson | 18.3 | 0.5 | 2.2 | 25.3 | 2.9 | 5.5 | 2.5 | D | 16.0 | 4,310 | 257 | 698 | 7,632 | 0.4 |
| Jefferson | 2.8 | 0.0 | 2.6 | 19.7 | D | 5.8 | 4.0 | 14.6 | 27.8 | 16,545 | 246 | 4,145 | 33,383 | 1.2 |
| Johnson | -0.6 | D | 5.8 | 28.3 | 1.8 | 6.7 | 3.2 | D | 15.4 | 6,500 | 244 | 928 | 11,613 | 2.7 |
| Lafayette | 16.1 | 0.7 | 10.9 | 3.2 | D | D | D | D | 24.7 | 1,975 | 297 | 409 | 4,408 | 1.3 |
| Lawrence | 10.3 | 1.1 | 5.8 | 7.3 | 1.5 | 8.4 | 3.1 | D | 24.0 | 4,785 | 291 | 833 | 8,099 | 1.3 |
| Lee | 21.7 | 0.0 | D | D | D | 4.7 | D | D | 26.0 | 2,200 | 247 | 646 | 4,356 | 0.0 |
| Lincoln | 9.3 | D | 8.0 | 8.6 | D | 2.9 | 2.6 | 7.8 | 36.6 | 2,610 | 201 | 475 | 4,984 | 2.5 |
| Little River | 5.7 | D | 4.1 | 49.2 | D | 5.2 | 1.2 | 3.4 | 16.0 | 3,510 | 284 | 423 | 6,553 | 1.5 |
| Logan | 5.8 | 1.4 | 6.7 | 18.9 | 2.6 | 8.9 | 4.7 | 11.1 | 23.3 | 6,360 | 296 | 883 | 10,179 | 0.7 |
| Lonoke | 3.4 | D | 12.0 | 11.3 | D | 8.6 | 6.8 | 9.7 | 19.1 | 14,785 | 201 | 1,869 | 30,471 | 11.8 |
| Madison | 5.0 | 0.0 | 7.3 | 28.0 | D | 7.9 | 2.8 | 5.8 | 16.0 | 4,415 | 267 | 442 | 7,636 | 2.0 |
| Marion | -2.7 | D | D | 44.5 | D | 7.3 | 4.9 | 7.0 | 13.9 | 5,995 | 360 | 500 | 9,675 | 3.5 |
| Miller | 1.1 | 1.6 | 7.3 | 23.0 | D | 5.0 | 3.4 | 7.6 | 14.8 | 9,690 | 223 | 2,031 | 19,603 | 1.7 |
| Mississippi | 4.7 | D | 5.5 | 47.4 | 1.1 | 3.8 | 2.4 | D | 10.5 | 9,795 | 241 | 2,780 | 20,591 | 0.6 |
| Monroe | 15.9 | 0.0 | 1.9 | 5.9 | D | 7.2 | 5.2 | D | 16.5 | 2,120 | 315 | 489 | 4,435 | -0.5 |
| Montgomery | -1.3 | D | 11.2 | 3.4 | 1.4 | 6.2 | 3.0 | 4.4 | 34.1 | 2,885 | 319 | 307 | 5,905 | 2.4 |
| Nevada | 3.2 | 0 | D | D | D | 6.5 | 2.4 | 9.4 | 14.3 | 2,365 | 288 | 440 | 4,600 | 0.7 |
| Newton | -8.8 | 0.0 | 13.8 | 4.2 | D | 6.7 | D | D | 36.6 | 2,740 | 355 | 343 | 4,710 | 1.2 |
| Ouachita | 0.3 | 1.5 | 4.5 | 10.4 | D | 7.4 | 3.8 | D | 24.6 | 6,775 | 290 | 1,260 | 13,099 | -0.2 |
| Perry | 12.4 | D | 13.3 | 3.2 | D | 3.6 | D | 22.8 | 19.1 | 2,885 | 277 | 389 | 5,035 | 1.2 |
| Phillips | 12.1 | -0.1 | 1.8 | 8.0 | D | 6.8 | 4.6 | 14.6 | 21.0 | 4,850 | 273 | 1,585 | 10,311 | 1.8 |
| Pike | -0.6 | D | 7.6 | 15.4 | D | 8.2 | D | D | 15.8 | 3,045 | 284 | 336 | 5,649 | 1.2 |
| Poinsett | 16.8 | D | 4.1 | 12.2 | D | 5.1 | 3.2 | 9.4 | 16.7 | 6,410 | 273 | 1,401 | 11,050 | 1.2 |
| Polk | 4.6 | D | 6.3 | 20.1 | 2.3 | 7.0 | 4.2 | 15.3 | 15.6 | 6,005 | 301 | 661 | 10,153 | 1.6 |
| Pope | 0.5 | 0.4 | 6.9 | 17.3 | 3.4 | 6.8 | 4.1 | 10.1 | 18.6 | 14,325 | 223 | 1,957 | 26,599 | 4.1 |
| Prairie | 22.1 | 0.1 | 4.0 | 2.7 | D | 6.3 | 3.6 | 9.2 | 23.3 | 2,365 | 292 | 290 | 4,573 | 1.5 |
| Pulaski | 0.1 | 0.2 | 4.8 | 5.2 | 9.7 | 5.4 | 10.7 | 13.7 | 18.3 | 83,120 | 212 | 15,961 | 187,931 | 7.1 |
| Randolph | 1.1 | D | 5.1 | 26.8 | D | 6.4 | 4.2 | D | 18.3 | 5,130 | 282 | 668 | 8,660 | 1.7 |
| St. Francis | 4.6 | 0.0 | D | D | 2.0 | 7.1 | 3.6 | D | 31.3 | 5,580 | 223 | 1,718 | 10,947 | 0.5 |
| Saline | -0.2 | D | 13.2 | 6.1 | 5.2 | 12.8 | 4.3 | 16.6 | 15.7 | 28,045 | 229 | 2,778 | 49,950 | 11.5 |
| Scott | 8.6 | D | 3.6 | 30.0 | D | 3.8 | 1.9 | D | 19.4 | 2,975 | 291 | 412 | 5,290 | 2.0 |
| Searcy | -12.3 | 0.0 | D | 9.4 | D | 8.5 | 4.0 | 18.1 | 26.5 | 2,865 | 364 | 399 | 4,960 | 1.2 |
| Sebastian | 0.2 | 0.7 | 3.9 | 16.3 | 5.0 | 6.7 | 9.8 | 18.3 | 11.4 | 29,410 | 230 | 4,283 | 57,688 | 5.5 |
| Sevier | 14.1 | D | D | D | D | 7.1 | 3.7 | 6.7 | 19.2 | 3,245 | 191 | 431 | 7,003 | 1.7 |
| Sharp | 10.8 | D | 7.4 | 3.4 | D | 11.4 | 5.5 | D | 18.5 | 6,445 | 373 | 866 | 9,906 | 0.9 |
| Stone | 2.2 | 0.0 | 6.9 | 6.5 | D | 11.8 | 9.0 | D | 20.4 | 4,635 | 367 | 566 | 6,977 | 3.9 |
| Union | 0.0 | 2.7 | 10.1 | 21.9 | 1.9 | 5.7 | 3.8 | 9.6 | 10.7 | 10,515 | 272 | 1,691 | 20,132 | 2.4 |
| Van Buren | -4.5 | 5.2 | 6.1 | 4.0 | 4.0 | 10.7 | 6.1 | D | 17.8 | 5,720 | 346 | 613 | 10,550 | 2.0 |
| Washington | 0.1 | D | 7.5 | 11.0 | 6.5 | 6.2 | 5.9 | 13.5 | 13.2 | 37,700 | 157 | 4,305 | 99,170 | 13.0 |
| White | 0.7 | 1.3 | 7.6 | 9.1 | 3.2 | 8.7 | 5.1 | 18.2 | 13.2 | 18,750 | 238 | 2,642 | 34,175 | 5.2 |
| Woodruff | 29.6 | D | 2.1 | 8.2 | D | 3.9 | 1.8 | D | 18.1 | 2,000 | 311 | 412 | 3,877 | -0.4 |
| Yell | 8.6 | D | 5.9 | 29.4 | 2.1 | 4.5 | 1.9 | D | 19.8 | 5,265 | 246 | 725 | 9,908 | 1.6 |

1. Per 1,000 resident population estimated as of July 1 of the year shown.

# Table B. States and Counties — Housing, Labor Force, and Employment

| STATE County | Total | Percent | Median value[1] | With a mortgage | Without a mortgage[2] | Median rent[3] | Median rent as a percent of income[2] | Sub-standard units[4] (percent) | Total | Percent change, 2019-2020 | Total | Rate[5] | Total | Management, business, science, and arts | Construction, production, and maintenance occupations |
|---|---|---|---|---|---|---|---|---|---|---|---|---|---|---|---|
| | 89 | 90 | 91 | 92 | 93 | 94 | 95 | 96 | 97 | 98 | 99 | 100 | 101 | 102 | 103 |
| **ARKANSAS—Cont'd** | | | | | | | | | | | | | | | |
| Cleveland | 3,217 | 76.0 | 97,100 | 18.0 | 10.0 | 630 | 21.6 | 1.8 | 3,203 | -1.3 | 195 | 6.1 | 3,061 | 32.4 | 35.6 |
| Columbia | 8,562 | 70.4 | 81,900 | 21.3 | 11.9 | 664 | 27.5 | 5.0 | 9,053 | -1.6 | 620 | 6.8 | 8,944 | 31.0 | 30.7 |
| Conway | 8,309 | 67.3 | 107,600 | 19.4 | 10.0 | 647 | 26.8 | 3.1 | 8,392 | 1.1 | 504 | 6.0 | 8,927 | 25.8 | 37.0 |
| Craighead | 41,375 | 57.1 | 140,700 | 17.3 | 10.0 | 757 | 27.9 | 3.1 | 55,775 | -0.7 | 2,972 | 5.3 | 49,761 | 34.0 | 27.1 |
| Crawford | 23,958 | 76.0 | 115,200 | 19.0 | 10.9 | 701 | 28.3 | 3.5 | 26,436 | -1.7 | 1,438 | 5.4 | 26,041 | 30.0 | 30.8 |
| Crittenden | 19,074 | 57.4 | 114,600 | 19.2 | 11.0 | 732 | 31.0 | 3.7 | 21,209 | -0.2 | 1,750 | 8.3 | 20,601 | 29.2 | 30.3 |
| Cross | 6,653 | 66.8 | 86,100 | 17.3 | 10.9 | 711 | 27.6 | 3.3 | 7,291 | -1.5 | 444 | 6.1 | 7,622 | 25.0 | 34.1 |
| Dallas | 2,969 | 67.7 | 70,400 | 18.8 | 10.8 | 673 | 26.6 | 0.7 | 2,832 | 0.2 | 151 | 5.3 | 2,952 | 19.7 | 35.8 |
| Desha | 5,204 | 61.5 | 70,300 | 19.0 | 12.4 | 495 | 27.2 | 2.2 | 5,321 | -1.6 | 373 | 7.0 | 4,563 | 28.2 | 33.0 |
| Drew | 7,157 | 67.7 | 99,800 | 16.6 | 10.0 | 658 | 27.2 | 2.3 | 7,730 | -1.3 | 534 | 6.9 | 7,919 | 35.5 | 25.3 |
| Faulkner | 44,575 | 62.3 | 161,500 | 18.9 | 10.0 | 826 | 30.5 | 3.3 | 61,682 | -2.0 | 3,361 | 5.4 | 58,112 | 37.7 | 22.6 |
| Franklin | 6,723 | 70.5 | 100,900 | 16.0 | 10.0 | 625 | 32.0 | 3.4 | 7,478 | -0.6 | 419 | 5.6 | 6,681 | 26.8 | 36.9 |
| Fulton | 4,973 | 79.7 | 86,400 | 24.3 | 10.6 | 545 | 28.2 | 5.5 | 4,924 | -1.1 | 253 | 5.1 | 4,077 | 26.9 | 28.8 |
| Garland | 40,209 | 67.8 | 139,900 | 20.5 | 10.0 | 771 | 29.9 | 2.5 | 41,347 | -0.3 | 3,295 | 8.0 | 41,221 | 31.7 | 21.2 |
| Grant | 6,863 | 80.1 | 124,600 | 16.5 | 10.0 | 736 | 27.8 | 2.6 | 8,222 | -2.4 | 426 | 5.2 | 7,760 | 32.4 | 33.7 |
| Greene | 17,254 | 67.0 | 118,800 | 17.6 | 11.2 | 698 | 26.4 | 3.0 | 19,450 | -3.0 | 1,158 | 6.0 | 18,761 | 28.4 | 36.4 |
| Hempstead | 8,094 | 69.3 | 81,100 | 18.4 | 10.0 | 694 | 30.4 | 3.1 | 9,593 | -3.3 | 499 | 5.2 | 9,218 | 25.8 | 37.2 |
| Hot Spring | 12,599 | 77.9 | 96,700 | 19.0 | 10.1 | 619 | 28.8 | 3.7 | 14,100 | 0.3 | 815 | 5.8 | 13,127 | 28.2 | 32.3 |
| Howard | 5,142 | 68.9 | 95,900 | 21.8 | 10.0 | 601 | 29.1 | 1.8 | 5,533 | -2.7 | 276 | 5.0 | 5,601 | 24.3 | 46.0 |
| Independence | 14,322 | 71.6 | 98,900 | 18.6 | 10.5 | 631 | 27.6 | 3.9 | 16,670 | 0.5 | 959 | 5.8 | 15,902 | 27.3 | 34.5 |
| Izard | 4,851 | 76.1 | 82,700 | 18.8 | 10.8 | 570 | 25.8 | 1.3 | 4,640 | -5.5 | 391 | 8.4 | 4,725 | 30.7 | 32.8 |
| Jackson | 6,137 | 69.4 | 66,600 | 18.6 | 10.0 | 588 | 29.3 | 2.7 | 5,863 | 0.8 | 419 | 7.1 | 5,542 | 24.5 | 32.9 |
| Jefferson | 26,801 | 62.1 | 83,500 | 19.5 | 10.2 | 740 | 29.0 | 2.5 | 27,398 | | 2,204 | 8.0 | 26,139 | 30.1 | 29.5 |
| Johnson | 9,682 | 70.4 | 103,500 | 21.0 | 10.2 | 620 | 30.3 | 4.3 | 10,120 | -1.5 | 638 | 6.3 | 10,448 | 26.2 | 39.5 |
| Lafayette | 2,784 | 79.1 | 64,600 | 20.4 | 10.0 | 473 | 36.8 | 3.0 | 2,421 | -1.5 | 185 | 7.6 | 2,511 | 29.6 | 32.8 |
| Lawrence | 6,463 | 67.3 | 76,800 | 18.3 | 10.0 | 549 | 25.7 | 2.2 | 6,786 | -0.8 | 417 | 6.1 | 6,600 | 27.6 | 34.9 |
| Lee | 3,206 | 59.8 | 74,000 | 18.8 | 12.6 | 525 | 46.8 | 2.7 | 2,805 | -0.6 | 210 | 7.5 | 2,790 | 23.4 | 34.3 |
| Lincoln | 3,773 | 77.1 | 76,300 | 17.2 | 10.8 | 534 | 22.3 | 3.6 | 3,834 | -1.7 | 229 | 6.0 | 3,209 | 28.8 | 33.8 |
| Little River | 5,363 | 71.5 | 77,900 | 15.9 | 10.0 | 601 | 24.3 | 2.0 | 5,385 | -2.6 | 347 | 6.4 | 4,915 | 29.2 | 40.3 |
| Logan | 8,417 | 73.7 | 99,000 | 19.4 | 10.0 | 581 | 23.0 | 3.6 | 8,656 | 1.5 | 575 | 6.6 | 9,059 | 23.6 | 39.8 |
| Lonoke | 26,895 | 69.2 | 142,500 | 18.6 | 10.0 | 796 | 27.1 | 2.4 | 33,550 | -2.1 | 1,858 | 5.5 | 32,568 | 36.3 | 25.6 |
| Madison | 6,279 | 78.0 | 120,400 | 19.6 | 10.0 | 615 | 24.0 | 4.0 | 7,413 | 0.1 | 287 | 3.9 | 6,674 | 30.6 | 34.8 |
| Marion | 6,782 | 78.9 | 120,700 | 21.3 | 10.1 | 632 | 28.0 | 2.3 | 6,251 | -6.2 | 403 | 6.4 | 5,744 | 29.7 | 32.6 |
| Miller | 16,426 | 64.6 | 109,100 | 17.8 | 10.0 | 737 | 29.7 | 5.3 | 19,379 | -1.8 | 1,332 | 6.9 | 17,792 | 31.5 | 29.4 |
| Mississippi | 16,389 | 56.4 | 84,500 | 18.1 | 10.0 | 632 | 25.4 | 4.8 | 17,264 | -0.7 | 1,559 | 9.0 | 17,365 | 25.9 | 38.3 |
| Monroe | 3,333 | 59.9 | 61,400 | 17.0 | 10.0 | 517 | 27.5 | 5.0 | 2,721 | -1.2 | 182 | 6.7 | 2,978 | 24.2 | 34.5 |
| Montgomery | 3,754 | 81.3 | 102,800 | 23.5 | 11.7 | 509 | 24.4 | 2.9 | 2,981 | 0.0 | 203 | 6.8 | 3,262 | 27.7 | 36.1 |
| Nevada | 3,397 | 66.9 | 66,300 | 19.0 | 10.0 | 696 | 28.0 | 2.1 | 3,550 | -2.8 | 208 | 5.9 | 3,204 | 30.6 | 36.2 |
| Newton | 2,936 | 84.2 | 114,700 | 22.9 | 10.0 | 547 | 31.7 | 5.6 | 3,246 | -2.3 | 162 | 5.0 | 2,878 | 30.4 | 30.6 |
| Ouachita | 9,658 | 69.1 | 75,300 | 19.1 | 10.9 | 525 | 27.4 | 4.2 | 9,821 | -0.2 | 592 | 6.0 | 9,487 | 32.2 | 33.7 |
| Perry | 3,668 | 76.4 | 106,300 | 17.7 | 10.0 | 631 | 23.5 | 5.3 | 4,188 | -2.9 | 232 | 5.5 | 3,793 | 29.2 | 32.1 |
| Phillips | 7,485 | 48.2 | 70,700 | 16.3 | 10.3 | 586 | 29.9 | 1.2 | 6,402 | 0.6 | 580 | 9.1 | 6,708 | 27.7 | 30.1 |
| Pike | 4,280 | 78.8 | 84,900 | 19.3 | 10.0 | 601 | 27.3 | 2.8 | 4,272 | 0.6 | 252 | 5.9 | 4,609 | 27.9 | 36.0 |
| Poinsett | 9,526 | 62.9 | 78,800 | 18.3 | 10.2 | 603 | 28.4 | 3.2 | 9,919 | -1.1 | 545 | 5.5 | 9,802 | 23.3 | 37.3 |
| Polk | 8,243 | 76.7 | 99,300 | 24.2 | 10.0 | 639 | 33.0 | 5.3 | 8,013 | -0.8 | 531 | 6.6 | 7,211 | 30.6 | 35.7 |
| Pope | 22,579 | 68.9 | 126,400 | 18.6 | 10.0 | 681 | 30.0 | 2.4 | 27,775 | -4.4 | 1,637 | 5.9 | 26,754 | 30.4 | 30.9 |
| Prairie | 3,776 | 73.9 | 67,400 | 19.0 | 10.7 | 619 | 23.1 | 3.1 | 3,608 | -0.1 | 193 | 5.3 | 3,450 | 32.6 | 31.5 |
| Pulaski | 158,051 | 58.8 | 157,200 | 19.2 | 10.0 | 861 | 29.3 | 2.6 | 188,928 | -0.4 | 14,044 | 7.4 | 184,202 | 41.3 | 18.9 |
| Randolph | 7,355 | 71.7 | 87,200 | 18.8 | 11.1 | 619 | 28.9 | 3.8 | 7,444 | 1.8 | 439 | 5.9 | 6,976 | 27.8 | 39.9 |
| St. Francis | 9,388 | 54.7 | 64,600 | 20.8 | 10.2 | 633 | 31.3 | 3.4 | 8,028 | -2.0 | 634 | 7.9 | 8,542 | 25.3 | 37.1 |
| Saline | 44,075 | 78.1 | 156,800 | 18.2 | 10.0 | 847 | 25.6 | 1.8 | 58,181 | -2.2 | 2,940 | 5.1 | 57,691 | 36.6 | 24.3 |
| Scott | 3,944 | 73.8 | 76,300 | 18.6 | 10.0 | 564 | 34.8 | 4.0 | 4,354 | 0.6 | 195 | 4.5 | 3,758 | 25.3 | 39.6 |
| Searcy | 3,327 | 73.9 | 103,500 | 19.2 | 11.4 | 535 | 28.8 | 1.8 | 2,885 | -0.8 | 184 | 6.4 | 3,097 | 25.6 | 35.5 |
| Sebastian | 51,228 | 58.8 | 123,100 | 18.9 | 10.0 | 696 | 26.3 | 3.5 | 56,158 | -1.1 | 3,355 | 6.0 | 57,296 | 31.2 | 29.6 |
| Sevier | 5,885 | 71.4 | 84,300 | 19.5 | 10.0 | 581 | 23.0 | 8.3 | 5,524 | 0.0 | 322 | 5.8 | 7,338 | 20.7 | 49.0 |
| Sharp | 7,447 | 77.2 | 80,700 | 21.0 | 10.4 | 527 | 29.5 | 3.8 | 5,717 | -0.8 | 408 | 7.1 | 5,957 | 27.6 | 32.8 |
| Stone | 4,787 | 74.6 | 121,100 | 21.4 | 10.0 | 604 | 33.0 | 4.7 | 4,618 | -2.5 | 328 | 7.1 | 4,526 | 30.9 | 34.6 |
| Union | 15,726 | 75.4 | 86,500 | 18.4 | 11.0 | 674 | 30.0 | 4.3 | 15,622 | -1.1 | 1,297 | 8.3 | 15,590 | 33.5 | 27.9 |
| Van Buren | 6,813 | 78.6 | 101,700 | 22.8 | 11.2 | 644 | 32.5 | 4.9 | 5,758 | -0.4 | 409 | 7.1 | 5,670 | 29.6 | 31.0 |
| Washington | 87,325 | 52.6 | 173,400 | 18.2 | 10.0 | 794 | 27.4 | 4.5 | 126,141 | 1.0 | 5,795 | 4.6 | 111,363 | 38.7 | 24.1 |
| White | 29,055 | 66.1 | 120,100 | 18.3 | 10.0 | 702 | 28.2 | 2.3 | 33,972 | 0.1 | 2,117 | 6.2 | 32,213 | 32.9 | 29.9 |
| Woodruff | 2,932 | 65.4 | 66,400 | 16.2 | 11.1 | 504 | 30.0 | 2.4 | 2,794 | 0.9 | 163 | 5.8 | 2,653 | 34.0 | 28.8 |
| Yell | 7,503 | 71.5 | 109,400 | 20.4 | 10.0 | 573 | 24.6 | 6.2 | 7,997 | -4.7 | 405 | 5.1 | 9,103 | 24.7 | 38.8 |

1. Specified owner-occupied units.   2. A value of 10.0 represents 10 percent or less; a value of 50.0 represents 50 percent or more.   3. Specified renter-occupied units.   4. Overcrowded or lacking complete plumbing facilities.   5. Percent of civilian labor force.   6. Civilian employed persons 16 years old and over.

| | Private nonfarm establishments, employment and payroll, 2019 | | | | | | | | | Agriculture, 2017 | | | |
| | | | | | | | | Annual payroll | | Farms | | | Farm producers whose primary occupation is farming (percent) |
| | | Employment | | | | | Professional, scientific, and technical services | | | | Percent with: | | |
| STATE County | Number of establishments | Total | Health care and social assistance | Manufacturing | Retail trade | Finance and insurance | | Total (mil dol) | Average per employee (dollars) | Number | Fewer than 50 acres | 1000 acres or more | |
| | 104 | 105 | 106 | 107 | 108 | 109 | 110 | 111 | 112 | 113 | 114 | 115 | 116 |
|---|---|---|---|---|---|---|---|---|---|---|---|---|---|
| ARKANSAS—Cont'd | | | | | | | | | | | | | |
| Cleveland | 71 | 608 | 199 | NA | 63 | NA | NA | 17 | 28,135 | 205 | 28.8 | 1.0 | 53.8 |
| Columbia | 543 | 7,142 | 1,132 | 1,862 | 896 | 280 | 403 | 282 | 39,483 | 297 | 32.3 | 2.0 | 33.3 |
| Conway | 410 | 4,732 | 789 | 906 | 756 | 120 | 72 | 192 | 40,627 | 768 | 28.9 | 3.4 | 43.2 |
| Craighead | 2,547 | 45,244 | 11,235 | 7,433 | 6,742 | 1,174 | 1,575 | 1,842 | 40,704 | 523 | 32.3 | 20.8 | 43.1 |
| Crawford | 1,072 | 15,019 | 1,455 | 4,905 | 2,017 | 343 | 259 | 519 | 34,533 | 799 | 44.2 | 2.1 | 35.4 |
| Crittenden | 858 | 13,738 | 1,793 | 1,941 | 2,041 | 197 | 203 | 487 | 35,454 | 262 | 26.7 | 34.7 | 53.6 |
| Cross | 361 | 3,820 | 906 | 570 | 760 | 247 | 107 | 119 | 31,275 | 300 | 18.3 | 28.0 | 57.8 |
| Dallas | 184 | 2,400 | 748 | 600 | 309 | 64 | 15 | 81 | 33,944 | 126 | 19.8 | 3.2 | 42.2 |
| Desha | 291 | 3,339 | 616 | 785 | 509 | 123 | 41 | 122 | 36,419 | 275 | 19.6 | 34.9 | 45.0 |
| Drew | 402 | 5,031 | 1,236 | 951 | 874 | 148 | 79 | 170 | 33,697 | 318 | 23.6 | 10.1 | 32.7 |
| Faulkner | 2,673 | 36,425 | 6,456 | 3,686 | 5,951 | 1,198 | 1,028 | 1,437 | 39,448 | 1,191 | 43.7 | 2.9 | 35.9 |
| Franklin | 289 | 3,605 | 502 | 1,090 | 583 | 359 | 100 | 127 | 35,223 | 752 | 25.5 | 5.5 | 40.9 |
| Fulton | 168 | 1,500 | 646 | 112 | 174 | 59 | 48 | 43 | 28,874 | 795 | 19.2 | 4.2 | 40.2 |
| Garland | 2,765 | 35,253 | 8,677 | 2,854 | 5,925 | 1,013 | 1,084 | 1,229 | 34,854 | 357 | 47.6 | 0.3 | 38.5 |
| Grant | 273 | 3,522 | 290 | 1,365 | 452 | 85 | 137 | 125 | 35,436 | 281 | 44.1 | 0.7 | 31.1 |
| Greene | 778 | 15,456 | 2,074 | 5,311 | 2,136 | 385 | 372 | 585 | 37,832 | 631 | 37.6 | 10.3 | 36.5 |
| Hempstead | 382 | 6,038 | 961 | 2,292 | 876 | 118 | 234 | 197 | 32,603 | 613 | 19.7 | 6.0 | 47.7 |
| Hot Spring | 495 | 5,916 | 1,089 | 1,521 | 786 | 202 | 89 | 223 | 37,624 | 563 | 44.2 | 1.4 | 36.7 |
| Howard | 268 | 6,003 | 1,013 | 3,024 | 520 | 126 | 35 | 203 | 33,882 | 586 | 27.8 | 1.9 | 44.2 |
| Independence | 787 | 15,779 | 3,489 | 5,016 | 1,817 | 438 | 171 | 625 | 39,597 | 890 | 23.6 | 4.7 | 37.4 |
| Izard | 207 | 2,091 | 606 | 386 | 349 | 85 | 29 | 62 | 29,733 | 632 | 19.1 | 6.6 | 44.3 |
| Jackson | 310 | 3,738 | 879 | 991 | 693 | 115 | 78 | 140 | 37,579 | 424 | 22.9 | 18.9 | 54.5 |
| Jefferson | 1,264 | 19,351 | 3,828 | 4,546 | 3,114 | 697 | 285 | 758 | 39,157 | 436 | 36.0 | 17.4 | 43.8 |
| Johnson | 390 | 7,345 | 1,046 | 2,523 | 884 | 121 | 85 | 223 | 30,425 | 577 | 26.7 | 1.9 | 46.8 |
| Lafayette | 106 | 682 | 108 | 54 | 120 | 58 | 30 | 23 | 33,416 | 287 | 17.4 | 10.8 | 43.0 |
| Lawrence | 283 | 2,993 | 704 | 361 | 592 | 83 | 41 | 93 | 30,909 | 535 | 24.3 | 12.5 | 43.2 |
| Lee | 119 | 925 | 283 | 11 | 164 | 39 | 28 | 33 | 35,242 | 221 | 14.5 | 27.1 | 60.2 |
| Lincoln | 154 | 1,478 | 347 | 338 | 194 | 42 | 14 | 55 | 37,513 | 369 | 33.6 | 16.3 | 46.6 |
| Little River | 174 | 2,677 | 310 | 1,208 | 394 | 52 | 38 | 173 | 64,636 | 414 | 17.9 | 6.5 | 39.2 |
| Logan | 382 | 3,821 | 668 | 972 | 736 | 147 | 138 | 126 | 32,930 | 873 | 24.6 | 3.2 | 42.7 |
| Lonoke | 1,056 | 11,619 | 1,585 | 1,858 | 2,102 | 524 | 472 | 395 | 34,024 | 702 | 35.6 | 13.8 | 42.7 |
| Madison | 213 | 2,518 | 220 | 1,045 | 386 | 85 | 53 | 90 | 35,564 | 1,229 | 21.2 | 3.3 | 45.9 |
| Marion | 213 | 2,042 | 258 | 688 | 420 | 104 | 61 | 62 | 30,366 | 587 | 25.4 | 6.5 | 43.6 |
| Miller | 704 | 12,155 | 991 | 2,443 | 1,264 | 229 | 217 | 478 | 39,292 | 513 | 37.2 | 6.0 | 36.9 |
| Mississippi | 759 | 14,618 | 1,702 | 6,633 | 1,627 | 193 | 90 | 801 | 54,784 | 284 | 16.5 | 48.6 | 73.2 |
| Monroe | 178 | 1,725 | 445 | 109 | 235 | 56 | 16 | 54 | 31,159 | 186 | 20.4 | 35.5 | 38.3 |
| Montgomery | 131 | 860 | 63 | 45 | 144 | 27 | 8 | 27 | 31,934 | 428 | 30.4 | 1.2 | 46.8 |
| Nevada | 114 | 1,969 | 394 | 604 | 287 | 65 | 23 | 72 | 36,816 | 355 | 26.5 | 3.1 | 36.6 |
| Newton | 84 | 683 | 221 | 73 | 104 | 18 | 6 | 17 | 24,701 | 537 | 19.4 | 1.5 | 38.5 |
| Ouachita | 505 | 7,499 | 1,408 | 2,742 | 958 | 175 | 146 | 313 | 41,776 | 205 | 42.9 | 0.5 | 26.9 |
| Perry | 111 | 697 | 186 | 19 | 124 | 21 | 13 | 36 | 51,482 | 397 | 29.2 | 1.3 | 42.7 |
| Phillips | 365 | 3,784 | 948 | 383 | 672 | 120 | 70 | 125 | 33,116 | 231 | 10.8 | 45.5 | 68.2 |
| Pike | 209 | 2,008 | 264 | 422 | 290 | 90 | 47 | 63 | 31,341 | 394 | 19.5 | 2.5 | 44.0 |
| Poinsett | 344 | 3,745 | 760 | 742 | 689 | 112 | 48 | 127 | 34,031 | 363 | 23.1 | 30.9 | 59.7 |
| Polk | 431 | 4,932 | 1,128 | 1,198 | 776 | 153 | 73 | 155 | 31,526 | 793 | 32.4 | 1.1 | 44.6 |
| Pope | 1,567 | 23,086 | 3,565 | 4,211 | 3,024 | 721 | 651 | 961 | 41,613 | 919 | 39.0 | 1.7 | 39.6 |
| Prairie | 148 | 986 | 198 | NA | 203 | 39 | 37 | 31 | 31,716 | 350 | 13.7 | 26.3 | 49.9 |
| Pulaski | 12,218 | 219,046 | 48,856 | 13,290 | 25,323 | 15,911 | 11,046 | 11,034 | 50,373 | 411 | 56.4 | 5.4 | 37.7 |
| Randolph | 307 | 3,713 | 890 | 512 | 649 | 129 | 63 | 111 | 29,806 | 657 | 21.0 | 7.8 | 44.3 |
| St. Francis | 441 | 5,474 | 1,461 | 757 | 1,014 | 171 | 101 | 175 | 31,892 | 278 | 28.1 | 29.1 | 55.4 |
| Saline | 2,070 | 22,623 | 4,773 | 1,357 | 4,452 | 516 | 770 | 761 | 33,635 | 371 | 52.3 | 0.5 | 32.0 |
| Scott | 141 | 1,985 | 359 | 852 | 195 | 38 | 15 | 62 | 31,113 | 528 | 23.9 | 1.3 | 45.0 |
| Searcy | 110 | 1,066 | 372 | 141 | 210 | 32 | 9 | 39 | 36,365 | 631 | 15.1 | 7.3 | 44.2 |
| Sebastian | 3,393 | 61,734 | 11,029 | 12,266 | 7,884 | 1,521 | 1,675 | 2,528 | 40,955 | 706 | 43.9 | 2.3 | 35.8 |
| Sevier | 249 | 3,704 | 374 | 1,391 | 592 | 146 | 50 | 129 | 34,724 | 540 | 27.0 | 3.0 | 44.2 |
| Sharp | 309 | 2,536 | 758 | 59 | 652 | 171 | 35 | 70 | 27,597 | 622 | 18.6 | 2.9 | 43.8 |
| Stone | 226 | 2,066 | 600 | 155 | 477 | 129 | 36 | 56 | 27,140 | 526 | 19.0 | 3.4 | 40.6 |
| Union | 1,064 | 16,180 | 1,967 | 3,115 | 2,073 | 410 | 363 | 888 | 54,863 | 268 | 39.6 | 0.4 | 32.3 |
| Van Buren | 306 | 3,239 | 737 | 137 | 558 | 101 | 73 | 122 | 37,801 | 611 | 22.3 | 2.1 | 36.1 |
| Washington | 5,711 | 93,767 | 16,220 | 13,907 | 12,121 | 2,459 | 4,410 | 4,200 | 44,796 | 2,279 | 42.7 | 1.1 | 37.6 |
| White | 1,552 | 22,593 | 4,567 | 2,177 | 3,464 | 595 | 532 | 796 | 35,242 | 1,613 | 34.2 | 3.7 | 37.7 |
| Woodruff | 127 | 1,126 | 200 | 175 | 177 | 38 | 6 | 39 | 34,606 | 187 | 16.0 | 36.9 | 52.5 |
| Yell | 319 | 4,863 | 1,070 | 1,678 | 492 | 165 | 88 | 164 | 33,803 | 718 | 21.2 | 4.3 | 46.7 |

| STATE County | Acreage (1,000) | Percent change, 2012-2017 | Average size of farm | Total irrigated (1,000) | Total cropland (1,000) | Value of land and buildings (dollars) Average per farm | Average per acre | Value of machinery and equiopmnet, average per farm (dollars) | Total (mil dol) | Average per farm (acres) | Crops | Livestock and poultry products | Organic farms (number) | Farms with internet access (percent) | Government payments Total ($1,000) | Percent of farms |
|---|---|---|---|---|---|---|---|---|---|---|---|---|---|---|---|---|
| | 117 | 118 | 119 | 120 | 121 | 122 | 123 | 124 | 125 | 126 | 127 | 128 | 129 | 130 | 131 | 132 |
| ARKANSAS—Cont'd | | | | | | | | | | | | | | | | |
| Cleveland | 32 | 16.2 | 154 | 0.1 | 6.7 | 652,384 | 4,232 | 161,950 | 128.9 | 628,927 | 0.4 | 99.6 | NA | 69.3 | 228 | 10.7 |
| Columbia | 54 | 19.2 | 183 | 0.3 | 12.1 | 425,028 | 2,323 | 83,478 | 49.0 | 164,848 | 8.0 | 92.0 | NA | 66.0 | 45 | 4.7 |
| Conway | 172 | -4.1 | 224 | 18.7 | 79.9 | 568,722 | 2,541 | 89,668 | 172.3 | 224,285 | 8.7 | 91.3 | NA | 68.1 | 838 | 11.8 |
| Craighead | 321 | -4.8 | 615 | 253.9 | 293.5 | 2,835,272 | 4,613 | 347,738 | 195.2 | 373,256 | 98.3 | 1.7 | NA | 78.4 | 14,200 | 54.9 |
| Crawford | 122 | -2.4 | 153 | 2.9 | 52.2 | 469,607 | 3,068 | 69,950 | 53.9 | 67,469 | 32.7 | 67.3 | 3 | 69.0 | 416 | 5.3 |
| Crittenden | 315 | -6.9 | 1,201 | 187.2 | 292.5 | 4,625,254 | 3,851 | 408,836 | 165.3 | 630,737 | 99.5 | 0.5 | 1 | 69.8 | 12,248 | 81.3 |
| Cross | 272 | -2.4 | 908 | 208.6 | 251.8 | 3,058,715 | 3,370 | 390,962 | 149.8 | 499,310 | 99.6 | 0.4 | 1 | 73.3 | 17,873 | 76.7 |
| Dallas | 29 | 38.4 | 233 | 0.0 | 4.2 | 454,901 | 1,949 | 49,777 | 1.7 | 13,825 | 67.5 | 32.5 | NA | 66.7 | 44 | 10.3 |
| Desha | 311 | 16.6 | 1,133 | 249.4 | 279.1 | 3,685,791 | 3,254 | 552,926 | 168.1 | 611,418 | 95.8 | 4.2 | NA | 75.6 | 9,994 | 79.3 |
| Drew | 134 | 3.1 | 421 | 57.2 | 77.0 | 1,217,911 | 2,893 | 194,702 | 60.1 | 188,858 | 67.1 | 32.9 | NA | 71.4 | 3,310 | 34.0 |
| Faulkner | 201 | 8.4 | 168 | 7.0 | 70 | 646,522 | 3,840 | 72,429 | 27.0 | 22,665 | 44.0 | 56.0 | 1 | 76.7 | 2,565 | 9.2 |
| Franklin | 184 | 15.0 | 244 | 1.2 | 56.5 | 626,738 | 2,564 | 86,566 | 181.2 | 240,979 | 2.8 | 97.2 | 2 | 76.3 | 520 | 5.9 |
| Fulton | 224 | 9.6 | 282 | 1.4 | 27.5 | 526,315 | 1,867 | 60,535 | 29.0 | 36,434 | 8.1 | 91.9 | NA | 73.2 | 672 | 10.9 |
| Garland | 34 | -4.9 | 95 | 0.1 | 8.2 | 496,951 | 5,226 | 46,231 | 10.0 | 27,972 | 30.5 | 69.5 | NA | 74.5 | 93 | 5.9 |
| Grant | 49 | -23.9 | 175 | 0.3 | 7.6 | 632,688 | 3,612 | 70,223 | 6.3 | 22,438 | 15.1 | 84.9 | NA | 82.2 | 36 | 4.3 |
| Greene | 261 | 0.3 | 414 | 164.8 | 214.7 | 1,754,440 | 4,239 | 169,942 | 139.3 | 220,792 | 91.9 | 8.1 | NA | 73.1 | 10,936 | 42.6 |
| Hempstead | 186 | -5.2 | 304 | D | 51.5 | 760,967 | 2,504 | 111,077 | 216.0 | 352,426 | 2.4 | 97.6 | NA | 70.3 | 512 | 10.9 |
| Hot Spring | 77 | 12.5 | 137 | 0.4 | 20.4 | 382,205 | 2,783 | 59,745 | 29.1 | 51,675 | 4.1 | 95.9 | NA | 75.0 | 180 | 4.3 |
| Howard | 150 | 2.2 | 256 | 0.4 | 34.5 | 731,438 | 2,855 | 98,674 | 262.1 | 447,229 | 0.7 | 99.3 | NA | 69.1 | 83 | 3.9 |
| Independence | 268 | 7.9 | 301 | 27.5 | 83.4 | 691,058 | 2,299 | 104,678 | 172.9 | 194,303 | 12.2 | 87.8 | NA | 73.3 | 2,947 | 16.3 |
| Izard | 190 | 9.8 | 300 | 0.3 | 31.7 | 613,810 | 2,047 | 74,547 | 65.9 | 104,286 | 2.0 | 98.0 | 1 | 75.6 | 438 | 11.1 |
| Jackson | 271 | -11.8 | 639 | 187.9 | 243.4 | 2,151,881 | 3,369 | 220,281 | 125.7 | 296,552 | 92.9 | 7.1 | 1 | 71.5 | 14,540 | 66.7 |
| Jefferson | 292 | 0.2 | 671 | 212.8 | 253.4 | 2,190,461 | 3,267 | 283,777 | 169.6 | 389,021 | 84.9 | 15.1 | NA | 70.0 | 14,205 | 60.1 |
| Johnson | 104 | -12.5 | 179 | 1.4 | 38.8 | 501,184 | 2,792 | 70,205 | 139.0 | 240,931 | 3.0 | 97.0 | 2 | 74.4 | 116 | 2.1 |
| Lafayette | 133 | 15.6 | 464 | 24.3 | 67.7 | 1,182,237 | 2,547 | 136,321 | 119.2 | 415,411 | 15.6 | 84.4 | NA | 80.5 | 3,646 | 33.4 |
| Lawrence | 270 | 6.7 | 504 | 147.2 | 204.1 | 1,807,457 | 3,586 | 203,207 | 150.6 | 281,490 | 66.3 | 33.7 | 1 | 76.4 | 13,714 | 40.6 |
| Lee | 258 | -1.1 | 1,167 | 169.5 | 246.7 | 3,702,289 | 3,173 | 414,909 | 130.1 | 588,670 | 99.8 | 0.2 | NA | 69.7 | 6,534 | 72.4 |
| Lincoln | 209 | 4.5 | 566 | 135.5 | 161.5 | 1,743,946 | 3,082 | 282,169 | 179.0 | 485,000 | 52.5 | 47.5 | NA | 62.6 | 7,490 | 51.8 |
| Little River | 159 | -7.1 | 385 | 10.3 | 58.2 | 929,079 | 2,414 | 114,187 | 85.5 | 206,606 | 19.4 | 80.6 | NA | 69.3 | 1,229 | 15.5 |
| Logan | 187 | -5.4 | 214 | D | 63.0 | 535,423 | 2,499 | 88,687 | 225.5 | 258,293 | 2.8 | 97.2 | 1 | 73.4 | 420 | 6.1 |
| Lonoke | 367 | 8.3 | 523 | 224.3 | 284.9 | 1,925,684 | 3,683 | 222,351 | 184.2 | 262,422 | 85.5 | 14.5 | 1 | 73.9 | 14,638 | 40.6 |
| Madison | 278 | 3.1 | 226 | 0.7 | 60.7 | 707,947 | 3,133 | 81,624 | 279.3 | 227,291 | 2.2 | 97.8 | 1 | 74.4 | 830 | 19.4 |
| Marion | 179 | 31.1 | 305 | 0.4 | 23.2 | 744,565 | 2,438 | 80,728 | 52.0 | 88,608 | 1.9 | 98.1 | 2 | 78.5 | 210 | 9.0 |
| Miller | 127 | -22.8 | 247 | 6.7 | 58.2 | 636,224 | 2,578 | 93,153 | 34.9 | 67,973 | 45.2 | 54.8 | NA | 73.5 | 1,786 | 9.7 |
| Mississippi | 477 | 0.2 | 1,678 | 350.2 | 469.3 | 5,765,623 | 3,436 | 743,149 | 300.9 | 1,059,602 | 100.0 | 0.0 | NA | 78.2 | 12,422 | 78.2 |
| Monroe | 203 | -24.4 | 1,091 | 159.9 | 189.0 | 3,078,477 | 2,823 | 366,858 | D | D | D | D | NA | 63.4 | 8,684 | 85.5 |
| Montgomery | 77 | 3.8 | 181 | 0.3 | 22.4 | 469,348 | 2,599 | 77,836 | 27.8 | 64,895 | 3.3 | 96.7 | NA | 64.3 | 9 | 2.1 |
| Nevada | 78 | 16.3 | 218 | 0.0 | 21.5 | 516,831 | 2,366 | 76,092 | 55.7 | 156,955 | 2.8 | 97.2 | NA | 67.3 | 158 | 9.3 |
| Newton | 101 | -11.6 | 188 | 0 | 14.6 | 458,102 | 2,436 | 55,239 | 24.0 | 44,665 | 3.6 | 96.4 | 6 | 76.0 | 92 | 6.9 |
| Ouachita | 28 | 8.8 | 138 | NA | 5.4 | 308,205 | 2,231 | 49,231 | 11.3 | 55,312 | 3.1 | 96.9 | NA | 63.9 | 22 | 3.4 |
| Perry | 78 | 10.3 | 195 | 6.6 | 31.7 | 539,122 | 2,758 | 96,815 | 56.1 | 141,340 | 21.8 | 78.2 | 2 | 75.6 | 1,013 | 18.6 |
| Phillips | 363 | 3.1 | 1,573 | 273.8 | 355.3 | 5,012,409 | 3,186 | 499,656 | 189.7 | 821,251 | 99.8 | 0.2 | NA | 68.8 | 11,787 | 90.0 |
| Pike | 82 | 17.8 | 209 | 0.5 | 23.1 | 545,769 | 2,608 | 79,626 | 106.5 | 270,330 | 0.9 | 99.1 | NA | 80.2 | 177 | 11.4 |
| Poinsett | 317 | -17.8 | 872 | 256.1 | 298.8 | 3,331,838 | 3,821 | 442,107 | 185.8 | 511,826 | 99.4 | 0.6 | NA | 71.1 | 16,877 | 75.2 |
| Polk | 140 | 18.0 | 177 | 0.1 | 37.6 | 523,224 | 2,959 | 79,598 | 159.0 | 200,503 | 1.1 | 98.9 | NA | 71.2 | 331 | 2.4 |
| Pope | 159 | 3.1 | 173 | 6.7 | 56.8 | 553,957 | 3,210 | 74,378 | 169.8 | 184,764 | 5.8 | 94.2 | 2 | 77.9 | 909 | 7.1 |
| Prairie | 273 | -0.9 | 779 | 171.3 | 213.1 | 2,274,086 | 2,920 | 291,350 | 127.1 | 363,197 | 93.0 | 7.0 | 2 | 76.3 | 13,847 | 76.9 |
| Pulaski | 79 | -5.8 | 193 | 20.7 | 44.6 | 757,344 | 3,929 | 71,002 | 20.3 | 49,333 | 81.1 | 18.9 | 7 | 78.1 | 1,580 | 12.7 |
| Randolph | 223 | 5.8 | 339 | 59.5 | 104.3 | 995,660 | 2,936 | 144,033 | 157.0 | 238,977 | 25.7 | 74.3 | 5 | 62.6 | 7,048 | 20.2 |
| St. Francis | 259 | -13.5 | 933 | 184.9 | 231.5 | 2,899,979 | 3,108 | 308,713 | 125.9 | 452,763 | 99.7 | 0.3 | NA | 69.1 | 9,903 | 61.5 |
| Saline | 42 | -6.5 | 113 | 0.1 | 12.8 | 386,927 | 3,424 | 54,372 | 8.2 | 21,989 | 13.1 | 86.9 | 7 | 78.4 | 46 | 3.0 |
| Scott | 93 | -6.8 | 177 | D | 25.9 | 434,292 | 2,454 | 71,379 | 136.7 | 258,813 | 1.2 | 98.8 | NA | 72.0 | 15 | 0.8 |
| Searcy | 195 | 15.8 | 310 | 0.1 | 30.1 | 583,263 | 1,883 | 64,568 | 18.0 | 28,564 | 7.4 | 92.6 | 2 | 68.9 | 579 | 11.9 |
| Sebastian | 101 | -15.2 | 143 | 0.2 | 29.1 | 477,234 | 3,343 | 58,679 | 118.4 | 167,670 | 1.4 | 98.6 | NA | 80.6 | 74 | 2.7 |
| Sevier | 142 | 10.1 | 263 | 0.2 | 30.0 | 702,812 | 2,668 | 96,124 | 230.8 | 427,398 | 0.5 | 99.5 | NA | 84.8 | 184 | 4.4 |
| Sharp | 162 | 2.2 | 260 | 0.3 | 30.1 | 554,504 | 2,132 | 65,348 | 146.9 | 236,125 | 1.1 | 98.9 | NA | 69.9 | 663 | 15.6 |
| Stone | 160 | 18.0 | 303 | 0.2 | 31.2 | 655,942 | 2,163 | 74,294 | 60.2 | 114,435 | 1.6 | 98.4 | NA | 71.9 | 370 | 8.4 |
| Union | 36 | -14.5 | 136 | 0.0 | 8.2 | 375,081 | 2,765 | 67,177 | 16.2 | 60,321 | 3.7 | 96.3 | NA | 74.6 | 13 | 2.6 |
| Van Buren | 128 | 3.9 | 209 | 0.1 | 21.6 | 575,172 | 2,754 | 63,590 | 16.3 | 26,755 | 6.6 | 93.4 | NA | 77.1 | 172 | 7.2 |
| Washington | 317 | 1.6 | 139 | 0.3 | 85.1 | 713,220 | 5,132 | 70,892 | 509.3 | 223,456 | 1.5 | 98.5 | 9 | 78.6 | 696 | 3.8 |
| White | 344 | -3.2 | 213 | 41.7 | 153.2 | 593,386 | 2,779 | 74,101 | 124.7 | 77,319 | 27.4 | 72.6 | 2 | 71.5 | 7,172 | 19.5 |
| Woodruff | 255 | -6.9 | 1,365 | 177.8 | 228.0 | 4,440,411 | 3,254 | 503,337 | D | D | D | D | 2 | 64.7 | 11,420 | 85.6 |
| Yell | 194 | 21.3 | 271 | 1.5 | 57.4 | 729,033 | 2,695 | 91,819 | 268.1 | 373,436 | 2.2 | 97.8 | NA | 73.7 | 989 | 13.2 |

# Table B. States and Counties — Water Use, Wholesale Trade, Retail Trade, and Real Estate

| STATE County | Water use, 2015 | | Wholesale Trade[1], 2017 | | | | Retail Trade[2], 2017 | | | | Real estate and rental and leasing,[2] 2017 | | | |
|---|---|---|---|---|---|---|---|---|---|---|---|---|---|---|
| | Public supply water withdrawn (mil gal/day) | Public supply gallons withdrawn per person per day | Number of establishments | Number of employees | Sales (mil dol) | Average payroll (mil dol) | Number of establishments | Number of employees | Sales (mil dol) | Average payroll (mil dol) | Number of establishments | Number of employees | Sales (mil dol) | Average payroll (mil dol) |
| | 133 | 134 | 135 | 136 | 137 | 138 | 139 | 140 | 141 | 142 | 143 | 144 | 145 | 146 |
| ARKANSAS—Cont'd | | | | | | | | | | | | | | |
| Cleveland | 0.50 | 60.2 | NA | NA | NA | NA | 16 | 59 | 16.2 | 1.0 | NA | NA | NA | NA |
| Columbia | 2.15 | 89.2 | D | D | D | D | 92 | 944 | 199.1 | 22.3 | 25 | 65 | 12.2 | 2.2 |
| Conway | 4.16 | 197.9 | 13 | 111 | 62.5 | 3.9 | 75 | 804 | 241.7 | 20.9 | 9 | 16 | 2.2 | 0.6 |
| Craighead | 14.12 | 135.3 | 127 | 1,524 | 1,403.2 | 69.3 | 462 | 6,646 | 1,828.9 | 163.8 | 124 | 498 | 119.0 | 19.0 |
| Crawford | 28.86 | 467.7 | 64 | 956 | 556.6 | 46.9 | 160 | 2,018 | 500.7 | 50.9 | 43 | 163 | 25.8 | 6.1 |
| Crittenden | 7.13 | 145.6 | 50 | 1,042 | 1,469.8 | 46.2 | 142 | 2,125 | 665.2 | 49.3 | 43 | 221 | 35.9 | 6.8 |
| Cross | 1.81 | 104.7 | 18 | 276 | 155.4 | 13.4 | 64 | 735 | 213.5 | 19.4 | 19 | 77 | 6.5 | 1.8 |
| Dallas | 0.18 | 23.7 | 5 | 34 | 5.8 | 0.6 | 42 | 325 | 77.3 | 8.3 | D | D | D | D |
| Desha | 0.43 | 35.9 | 21 | 241 | 365.5 | 12.1 | 63 | 566 | 123.7 | 11.7 | 9 | 63 | 3.8 | 1.2 |
| Drew | 2.22 | 118.2 | 9 | 127 | 159.6 | 7.9 | 79 | 953 | 254.2 | 22.8 | 20 | 61 | 8.1 | 1.9 |
| Faulkner | 0.15 | 1.2 | 79 | 776 | 554.8 | 37.8 | 408 | 6,226 | 1,880.1 | 165.7 | 141 | 514 | 97.7 | 16.7 |
| Franklin | 3.00 | 169.5 | 4 | 15 | 8.5 | 0.6 | 52 | 614 | 203.4 | 15.0 | 4 | 11 | 2.0 | 0.4 |
| Fulton | 1.40 | 114.7 | D | D | D | 0.9 | 31 | 180 | 55.4 | 4.2 | 4 | D | 0.6 | D |
| Garland | 16.07 | 165.4 | 75 | 659 | 460.1 | 39.1 | 496 | 6,194 | 1,684.3 | 157.7 | 130 | 490 | 77.5 | 17.1 |
| Grant | 1.01 | 55.8 | D | D | D | 4.7 | 49 | 463 | 131.6 | 12.0 | D | D | D | D |
| Greene | 4.42 | 100.0 | D | D | D | D | 157 | 2,022 | 579.5 | 52.9 | 35 | 75 | 13.4 | 1.9 |
| Hempstead | 2.80 | 126.8 | D | D | D | 3.2 | 78 | 915 | 217.2 | 21.3 | 18 | 58 | 7.3 | 1.6 |
| Hot Spring | 2.27 | 67.9 | D | D | D | D | 89 | 805 | 248.2 | 20.5 | 12 | 46 | 3.9 | 1.1 |
| Howard | 6.09 | 457.9 | 11 | 87 | 93.0 | 3.3 | 53 | 577 | 157.0 | 14.9 | 6 | 19 | 2.1 | 0.4 |
| Independence | 5.82 | 157.1 | D | D | D | D | 158 | 1,871 | 491.5 | 44.8 | 24 | 96 | 13.5 | 3.5 |
| Izard | 0.79 | 58.8 | 8 | 35 | 6.9 | 0.8 | 45 | 371 | 110.3 | 9.4 | 4 | 37 | 2.8 | 0.6 |
| Jackson | 0.39 | 22.5 | 16 | 189 | 135.4 | 9.3 | 67 | 731 | 229.7 | 19.3 | 12 | 43 | 3.5 | 0.9 |
| Jefferson | 11.88 | 166.0 | 57 | 465 | 314.3 | 21.8 | 282 | 3,133 | 776.8 | 77.0 | 56 | 202 | 34.5 | 6.5 |
| Johnson | 4.79 | 183.2 | 6 | 24 | 8.1 | 0.7 | 78 | 948 | 265.9 | 24.1 | D | D | D | 1.2 |
| Lafayette | 0.37 | 52.9 | NA | NA | NA | NA | 19 | 144 | 26.9 | 2.6 | 3 | D | 1.3 | D |
| Lawrence | 1.09 | 65.0 | 16 | 153 | 129.8 | 6.5 | 63 | 688 | 206.4 | 18.4 | 6 | 27 | 3.3 | 0.8 |
| Lee | 1.28 | 132.6 | 11 | 136 | 122.8 | 7.1 | 15 | 158 | 40.1 | 3.8 | 8 | D | 3.3 | D |
| Lincoln | 0.51 | 36.9 | D | D | D | D | 28 | 207 | 66.8 | 4.5 | D | D | D | D |
| Little River | 2.80 | 224.5 | NA | NA | NA | NA | 39 | 448 | 129.5 | 11.3 | NA | NA | NA | NA |
| Logan | 3.39 | 156.1 | D | D | D | 0.5 | 68 | 747 | 164.8 | 17.0 | 14 | 26 | 3.6 | 0.4 |
| Lonoke | 6.01 | 83.9 | 32 | 309 | 218.5 | 14.0 | 169 | 2,061 | 557.2 | 52.3 | 52 | 138 | 22.4 | 5.5 |
| Madison | 0.00 | 0.0 | D | D | D | D | 32 | 444 | 95.9 | 10.3 | D | D | D | D |
| Marion | 0.81 | 50.0 | 7 | D | 4.6 | D | 39 | 586 | 112.8 | 12.9 | 10 | D | 1.4 | D |
| Miller | 0.09 | 2.0 | 41 | 407 | 251.5 | 17.9 | 132 | 1,315 | 378.8 | 32.0 | D | D | D | D |
| Mississippi | 1.44 | 32.9 | 48 | 487 | 477.6 | 25.3 | 145 | 1,754 | 419.5 | 41.3 | 36 | 179 | 26.4 | 5.5 |
| Monroe | 1.52 | 205.4 | 12 | 177 | 81.9 | 8.5 | 30 | 266 | 63.0 | 6.4 | D | D | D | 0.2 |
| Montgomery | 0.52 | 58.0 | 5 | 45 | 27.4 | 1.5 | 21 | 120 | 28.9 | 3.1 | D | D | D | 0.2 |
| Nevada | 0.22 | 25.7 | D | D | D | 0.2 | 18 | 273 | 101.8 | 5.5 | 4 | 11 | 0.5 | 0.2 |
| Newton | 0.02 | 2.5 | NA | NA | NA | NA | 12 | 132 | 20.6 | 1.6 | NA | NA | NA | NA |
| Ouachita | 3.26 | 133.8 | D | D | D | D | 103 | 966 | 242.2 | 22.6 | 21 | 139 | 20.6 | 4.1 |
| Perry | 0.06 | 5.9 | D | D | D | 0.2 | 17 | 166 | 33.7 | 3.0 | D | D | D | D |
| Phillips | 6.10 | 312.6 | 29 | 344 | 643.5 | 16.0 | 73 | 772 | 148.1 | 17.7 | 17 | 63 | 10.0 | 2.0 |
| Pike | 1.13 | 104.4 | 7 | 141 | 31.7 | 3.3 | 34 | 338 | 79.8 | 7.6 | 6 | D | 6.8 | D |
| Poinsett | 2.18 | 90.7 | 24 | 609 | 379.0 | 21.5 | 66 | 694 | 166.2 | 14.6 | 8 | 33 | 3.9 | 0.8 |
| Polk | 1.69 | 83.6 | 15 | 83 | 58.2 | 3.8 | 80 | 829 | 193.8 | 19.3 | 14 | 46 | 5.7 | 1.2 |
| Pope | 11.00 | 173.5 | 64 | 626 | 392.0 | 27.4 | 272 | 3,261 | 895.2 | 79.6 | 82 | 242 | 46.5 | 7.2 |
| Prairie | 0.39 | 47.0 | D | D | D | D | 35 | 199 | 51.3 | 4.4 | D | D | D | 0.5 |
| Pulaski | 39.60 | 100.8 | 643 | 12,040 | 9,205.7 | 682.6 | 1,776 | 26,282 | 7,998.1 | 720.7 | 703 | 3,336 | 776.1 | 148.8 |
| Randolph | 1.30 | 74.4 | 10 | 267 | 155.7 | 7.5 | 62 | 691 | 189.5 | 18.8 | 15 | 41 | 4.1 | 0.9 |
| St. Francis | 3.84 | 144.4 | 24 | 462 | 512.3 | 21.8 | 95 | 978 | 254.2 | 21.7 | 26 | 139 | 21.9 | 3.3 |
| Saline | 29.64 | 252.3 | 67 | 523 | 558.2 | 22.1 | 300 | 4,187 | 1,276.9 | 109.1 | 86 | 285 | 70.9 | 9.8 |
| Scott | 1.55 | 147.4 | D | D | D | 3.2 | 21 | 191 | 44.8 | 4.5 | 5 | 16 | 1.3 | 0.4 |
| Searcy | 0.31 | 39.4 | D | D | D | 0.1 | 28 | 196 | 51.3 | 4.5 | 7 | 21 | 1.8 | 0.5 |
| Sebastian | 0.13 | 1.0 | 184 | 2,729 | 1,657.8 | 140.1 | 573 | 8,405 | 2,250.5 | 206.8 | 164 | 1,198 | 193.5 | 46.6 |
| Sevier | 0.74 | 42.8 | 7 | 56 | 18.9 | 3.0 | 62 | 654 | 202.3 | 16.3 | 7 | 21 | 2.4 | 0.7 |
| Sharp | 0.98 | 57.9 | 5 | D | 2.2 | D | 59 | 709 | 169.4 | 16.0 | 8 | 26 | 2.3 | 0.5 |
| Stone | 1.77 | 142.1 | 5 | 58 | 30.3 | 1.4 | 45 | 467 | 126.4 | 12.8 | 6 | 12 | 0.9 | 0.2 |
| Union | 4.29 | 106.9 | D | D | D | D | 178 | 2,036 | 560.1 | 52.6 | 36 | 212 | 70.7 | 9.2 |
| Van Buren | 2.19 | 130.6 | 13 | 74 | 36.3 | 2.9 | 59 | 596 | 159.6 | 13.7 | 6 | D | 1.5 | D |
| Washington | 0.19 | 0.8 | 226 | 2,680 | 2,389.2 | 148.2 | 787 | 12,569 | 3,587.8 | 335.9 | 303 | 1,219 | 319.1 | 46.3 |
| White | 7.72 | 97.5 | 60 | 480 | 350.8 | 23.8 | 325 | 3,777 | 1,051.9 | 96.8 | 69 | 187 | 33.1 | 6.2 |
| Woodruff | 0.53 | 78.6 | 14 | 151 | 105.8 | 7.4 | 26 | 200 | 44.8 | 3.6 | D | D | D | 0.1 |
| Yell | 4.82 | 222.0 | D | D | D | D | 53 | 574 | 125.9 | 11.9 | 8 | 18 | 2.4 | 0.3 |

1 Merchant wholesalers, except manufacturers' sales branches and offices.    2. Employer establishments.

# Table B. States and Counties — Professional Services, Manufacturing, and Accommodation and Food Services

| STATE County | Professional, scientific, and technical services, 2017 | | | | Manufacturing, 2017 | | | | Accommodation and food services, 2017 | | | |
|---|---|---|---|---|---|---|---|---|---|---|---|---|
| | Number of establishments | Number of employees | Sales (mil dol) | Average payroll (mil dol) | Number of establishments | Number of employees | Sales (mil dol) | Average payroll (mil dol) | Number of establishments | Number of employees | Sales (mil dol) | Annual payroll (mil dol) |
| | 147 | 148 | 149 | 150 | 151 | 152 | 153 | 154 | 155 | 156 | 157 | 158 |
| ARKANSAS—Cont'd | | | | | | | | | | | | |
| Cleveland | NA | NA | NA | NA | 4 | 61 | 11.6 | 2.3 | NA | NA | NA | NA |
| Columbia | 32 | 313 | 36 | 13.4 | 28 | 1,613 | 864.8 | 102.1 | 41 | 720 | 43.8 | 10.3 |
| Conway | 26 | 159 | 9 | 2.5 | 19 | 687 | 475.6 | 46.9 | 29 | 439 | 29.9 | 7.6 |
| Craighead | D | D | D | D | 91 | 6,341 | 2,536.8 | 301.7 | 242 | 5,080 | 251.5 | 73.0 |
| Crawford | 86 | 280 | 33 | 11.8 | 53 | 4,525 | 1,266.8 | 163.9 | 89 | 1,527 | 73.1 | 21.5 |
| Crittenden | D | D | D | D | 35 | 1,674 | 657.3 | 76.4 | 93 | 1,613 | 90.2 | 24.6 |
| Cross | 25 | 90 | 8 | 2.9 | 11 | 508 | 124.1 | 16.8 | D | D | D | D |
| Dallas | 7 | 34 | 3 | 0.9 | 10 | 532 | 219.8 | 24.7 | D | D | D | D |
| Desha | 17 | 47 | 5 | 1.7 | 7 | 696 | 391.2 | 45.5 | 28 | 312 | 14.6 | 3.9 |
| Drew | 19 | 104 | 9 | 3.5 | 23 | 855 | 228.8 | 37.9 | 36 | 598 | 29.7 | 8.5 |
| Faulkner | D | D | D | D | 91 | 2,949 | 838.0 | 144.3 | 227 | 4,784 | 234.3 | 70.1 |
| Franklin | 19 | 78 | 9 | 2.3 | 14 | 1,158 | 438.2 | 48.1 | 27 | 316 | 13.5 | 3.8 |
| Fulton | 12 | 30 | 3 | 0.8 | 8 | 73 | 7.1 | 2.3 | 15 | 213 | 9.5 | 2.6 |
| Garland | D | D | D | D | 88 | 2,427 | 694.6 | 115.8 | 306 | 5,510 | 291.2 | 85.8 |
| Grant | 18 | 109 | 13 | 5.7 | 13 | D | 798.4 | D | D | D | D | D |
| Greene | 52 | 346 | 32 | 12.8 | 52 | 4,933 | 1,753.3 | 223.1 | 70 | 1,236 | 59.8 | 16.5 |
| Hempstead | 14 | 120 | 10 | 4.6 | D | D | D | D | 36 | 554 | 27.7 | 7.2 |
| Hot Spring | 30 | 75 | 6 | 2.0 | 30 | 1,391 | 821.5 | 69.3 | 42 | 592 | 27.7 | 7.2 |
| Howard | 8 | 38 | 3 | 1.1 | 18 | 3,126 | 1,469.3 | 105.8 | 17 | 308 | 20.8 | 4.1 |
| Independence | 59 | 194 | 36 | 7.8 | 34 | 3,693 | 1,467.9 | 151.6 | 64 | 1,267 | 52.3 | 13.7 |
| Izard | D | D | 2 | D | 6 | D | 45.8 | 9.9 | 14 | D | 4.9 | D |
| Jackson | D | D | D | D | 17 | 917 | 340.2 | 44.4 | 26 | 332 | 16.1 | 5.0 |
| Jefferson | D | D | D | D | 53 | 4,781 | 1,582.0 | 233.8 | D | D | D | 23.9 |
| Johnson | 24 | 80 | 8 | 2.9 | 33 | 2,641 | 659.0 | 83.1 | 40 | 595 | 27.0 | 7.6 |
| Lafayette | D | D | 1 | D | 5 | 81 | 16.6 | 3.3 | 6 | 49 | 4.0 | 0.8 |
| Lawrence | 13 | 33 | 5 | 1.0 | 19 | 390 | 75.5 | 14.4 | 25 | 304 | 14.6 | 3.8 |
| Lee | 8 | 27 | 3 | 1.0 | NA | NA | NA | NA | 4 | 46 | 2.7 | 0.6 |
| Lincoln | D | D | D | D | 8 | 365 | 85.8 | 16.1 | D | D | D | D |
| Little River | 8 | 39 | 6 | 1.4 | 16 | 1,098 | 639.0 | 84.7 | D | D | D | D |
| Logan | 29 | 122 | 13 | 4.9 | 15 | 1,081 | 248.5 | 46.7 | 30 | 350 | 13.6 | 3.7 |
| Lonoke | 89 | 501 | 69 | 25.9 | 28 | 1,917 | 407.4 | 104.0 | 93 | 1,562 | 87.4 | 23.1 |
| Madison | D | D | D | D | 23 | 957 | 315.2 | 36.7 | 17 | 191 | 10.0 | 2.2 |
| Marion | 15 | 63 | 4 | 1.6 | 15 | 1,138 | 353.9 | 43.8 | 27 | 198 | 9.4 | 2.6 |
| Miller | 40 | 183 | 18 | 6.5 | 24 | 2,265 | 1,075.4 | 128.6 | D | D | D | D |
| Mississippi | 30 | 106 | 11 | 3.5 | 47 | 6,325 | 4,801.8 | 410.6 | 80 | 1,156 | 60.9 | 15.4 |
| Monroe | 13 | 19 | 2 | 0.4 | 5 | 94 | 39.0 | 3.7 | 17 | 241 | 11.1 | 3.2 |
| Montgomery | D | D | D | 0.3 | 8 | 42 | 21.0 | D | 14 | 304 | 28.9 | 6.8 |
| Nevada | 6 | 30 | 1 | 0.4 | D | D | D | D | 9 | 90 | 4.1 | 1.2 |
| Newton | 4 | 5 | 1 | 0.1 | 12 | 63 | 13.0 | 2.3 | 13 | 128 | 7.6 | 2.2 |
| Ouachita | D | D | D | D | D | D | D | D | 35 | 507 | 23.2 | 6.0 |
| Perry | 8 | 13 | 3 | 1.0 | NA | NA | NA | NA | D | D | D | D |
| Phillips | D | D | D | D | 11 | 284 | 85.9 | 11.8 | 26 | 273 | 14.6 | 3.5 |
| Pike | 9 | 52 | 5 | 1.4 | 17 | 274 | 84.1 | 10.9 | 20 | 193 | 10.6 | 2.7 |
| Poinsett | 14 | 56 | 4 | 1.3 | 20 | 648 | 264.1 | 26.7 | 38 | 434 | 20.0 | 5.7 |
| Polk | D | D | D | D | 27 | 1,103 | 195.1 | 43.5 | 36 | 411 | 18.7 | 5.5 |
| Pope | D | D | D | D | 59 | 4,327 | 1,729.0 | 195.7 | 141 | 2,846 | 136.8 | 40.2 |
| Prairie | 7 | 35 | 3 | 0.8 | 3 | 48 | 6.5 | D | D | D | D | 0.8 |
| Pulaski | D | D | D | D | 311 | 12,027 | 5,425.2 | 663.9 | 1,060 | 20,953 | 1,127.3 | 333.2 |
| Randolph | 21 | 62 | 6 | 2.5 | 22 | 231 | 67.4 | 10.3 | 24 | 387 | 16.5 | 4.6 |
| St. Francis | 32 | 112 | 11 | 3.6 | 8 | 722 | 212.5 | 24.3 | 40 | 603 | 30.1 | 7.9 |
| Saline | 165 | 766 | 120 | 36.7 | D | 1,401 | D | 66.1 | 160 | 3,083 | 164.6 | 47.5 |
| Scott | 9 | 10 | 1 | 0.3 | 10 | 1,040 | 358.3 | 35.1 | 13 | 221 | 6.4 | 3.0 |
| Searcy | 5 | 12 | 1 | 0.4 | 12 | 92 | 23.6 | 3.7 | D | D | D | 1.5 |
| Sebastian | D | D | D | D | 163 | 13,257 | 4,207.1 | 564.7 | 298 | 6,051 | 304.3 | 90.6 |
| Sevier | 12 | 63 | 4 | 1.4 | D | D | D | D | 20 | 267 | 16.1 | 3.1 |
| Sharp | 12 | 33 | 2 | 0.6 | 14 | 100 | 19.7 | 2.3 | 28 | 284 | 13.2 | 3.7 |
| Stone | 10 | 41 | 2 | 0.7 | 18 | 99 | 18.3 | 3.6 | 27 | 444 | 19.7 | 5.3 |
| Union | D | D | D | D | 46 | 3,082 | 1,304.0 | 178.1 | 77 | 1,079 | 60.0 | 15.5 |
| Van Buren | 17 | 59 | 6 | 2.1 | 8 | 102 | 11.9 | 2.8 | 20 | 281 | 14.5 | 4.4 |
| Washington | D | D | D | D | 203 | 13,101 | 4,188.1 | 555.1 | 571 | 11,310 | 581.7 | 169.3 |
| White | 104 | 500 | 54 | 19.9 | 62 | 1,858 | 761.1 | 83.2 | 143 | 2,535 | 126.2 | 36.0 |
| Woodruff | D | D | D | 0.2 | 6 | 190 | 69.6 | 7.3 | 8 | 75 | 3.2 | 0.9 |
| Yell | 22 | 89 | 8 | 3.2 | 18 | 1,786 | 609.9 | 65.5 | 23 | 350 | 15.5 | 4.4 |

| STATE County | Health care and social assistance, 2017 | | | | Other services, 2017 | | | | Nonemployer businesses, 2018 | | Value of residential construction authorized by building permits, 2020 | |
|---|---|---|---|---|---|---|---|---|---|---|---|---|
| | Number of establishments | Number of employees | Receipts (mil dol) | Annual payroll (mil dol) | Number of establishments | Number of employees | Receipts (mil dol) | Annual payroll (mil dol) | Number | Receipts (mil dol) | New construction ($1,000) | Number of housing units |
| | 159 | 160 | 161 | 162 | 163 | 164 | 165 | 166 | 167 | 168 | 169 | 170 |
| ARKANSAS—Cont'd | | | | | | | | | | | | |
| Cleveland | 5 | 150 | 3.9 | 2.5 | 6 | 20 | 2.8 | 0.8 | 364 | 13.5 | 0 | 0 |
| Columbia | 68 | 1,318 | 78.6 | 34.0 | 33 | 112 | 10.3 | 2.8 | 1,304 | 55.1 | 605 | 2 |
| Conway | 48 | 716 | 55.1 | 23.2 | D | D | D | D | 1,382 | 54.2 | 1,428 | 11 |
| Craighead | 354 | 10,046 | 1,184.9 | 475.1 | 125 | 776 | 81.1 | 20.5 | 8,431 | 403.1 | 116,811 | 988 |
| Crawford | 111 | 1,584 | 128.4 | 57.7 | 63 | 281 | 30.4 | 8.1 | 3,947 | 181.1 | 16,959 | 116 |
| Crittenden | 126 | 1,744 | 129.4 | 53.2 | 48 | 333 | 31.0 | 8.8 | 3,734 | 136.5 | 15,964 | 77 |
| Cross | 44 | 982 | 55.5 | 27.2 | D | D | D | D | 1,349 | 55.4 | 1,518 | 11 |
| Dallas | 19 | 813 | 42.1 | 19.5 | 11 | 27 | 3.3 | 0.9 | 342 | 13.5 | 0 | 0 |
| Desha | 24 | 489 | 29.2 | 11.5 | 13 | 38 | 3.2 | 0.7 | 715 | 29.2 | 750 | 3 |
| Drew | 43 | 1,272 | 104.4 | 37.2 | 17 | 52 | 7.3 | 2.0 | 1,129 | 55.4 | 675 | 3 |
| Faulkner | 317 | 5,737 | 535.5 | 217.6 | 159 | 734 | 74.4 | 19.5 | 8,835 | 369.7 | 128,407 | 891 |
| Franklin | 34 | 565 | 47.4 | 18.6 | 14 | 44 | 3.9 | 1.1 | 986 | 40.0 | 1,381 | 17 |
| Fulton | 31 | 581 | 44.5 | 17.3 | D | D | D | 0.3 | 868 | 32.0 | 0 | 0 |
| Garland | 343 | 8,736 | 1,069.9 | 380.5 | 186 | 762 | 67.3 | 19.5 | 8,116 | 361.2 | 18,872 | 88 |
| Grant | 26 | 259 | 15.4 | 6.9 | D | D | 4.9 | D | 1,178 | 49.6 | 4,405 | 29 |
| Greene | 103 | 1,895 | 187.9 | 68.3 | 39 | 175 | 16.5 | 4.3 | 2,929 | 135.0 | 21,021 | 157 |
| Hempstead | 46 | 1,121 | 67.0 | 27.1 | 28 | 133 | 14.1 | 3.4 | 1,037 | 40.4 | 0 | 0 |
| Hot Spring | 54 | 1,188 | 92.9 | 40.5 | 24 | 84 | 6.6 | 2.0 | 1,946 | 71.8 | 4,600 | 71 |
| Howard | 34 | 918 | 49.0 | 24.3 | 21 | 100 | 11.4 | 2.8 | 784 | 34.7 | 0 | 0 |
| Independence | 99 | 3,494 | 343.3 | 151.3 | 54 | 250 | 19.3 | 6.1 | 2,481 | 104.6 | 1,810 | 31 |
| Izard | 34 | 539 | 30.3 | 13.4 | 11 | 46 | 3.6 | 0.9 | 900 | 37.5 | 1,013 | 7 |
| Jackson | 43 | 808 | 63.6 | 25.9 | 16 | 88 | 10.8 | 1.7 | 811 | 30.2 | 1,738 | 15 |
| Jefferson | 203 | 3,893 | 399.0 | 166.1 | 70 | 382 | 39.5 | 10.4 | 3,601 | 134.7 | 6,492 | 68 |
| Johnson | 38 | 907 | 79.1 | 35.3 | 29 | 123 | 10.6 | 3.7 | 1,334 | 51.6 | 6,151 | 39 |
| Lafayette | 7 | 141 | 6.3 | 2.9 | 3 | 35 | 2.0 | 0.8 | 326 | 11.4 | 418 | 2 |
| Lawrence | 28 | 656 | 47.6 | 23.0 | D | D | D | D | 1,186 | 54.8 | 5,626 | 64 |
| Lee | 18 | 314 | 23.2 | 9.2 | 6 | 12 | 0.9 | 0.3 | 656 | 24.5 | 0 | 0 |
| Lincoln | 18 | 320 | 25.3 | 10.9 | 11 | 33 | 4.8 | 1.0 | 540 | 27.4 | 877 | 5 |
| Little River | 21 | 331 | 23.5 | 10.8 | D | D | D | D | 562 | 21.9 | 780 | 3 |
| Logan | 44 | 674 | 58.6 | 24.0 | 27 | 74 | 9.2 | 2.0 | 1,235 | 42.3 | 184 | 1 |
| Lonoke | 116 | 1,459 | 106.6 | 43.4 | 58 | 261 | 22.5 | 6.3 | 4,706 | 205.1 | 47,959 | 405 |
| Madison | 15 | 226 | 14.6 | 7.2 | D | D | D | 1.9 | 1,283 | 50.5 | 8,364 | 54 |
| Marion | 14 | 297 | 14.8 | 5.9 | 12 | 28 | 2.8 | 0.5 | 1,150 | 38.0 | 10,700 | 78 |
| Miller | 78 | 1,117 | 77.0 | 28.3 | D | D | D | D | 2,607 | 119.2 | 4,987 | 21 |
| Mississippi | 112 | 1,750 | 149.4 | 62.7 | 43 | 181 | 16.3 | 4.9 | 2,309 | 75.3 | 5,408 | 37 |
| Monroe | 23 | 526 | 23.4 | 11.2 | D | D | 5.7 | D | 569 | 24.8 | 0 | 0 |
| Montgomery | 11 | 72 | 6.2 | 2.8 | D | D | 3.1 | D | 793 | 31.5 | NA | NA |
| Nevada | 15 | 381 | 19.6 | 9.9 | 11 | 23 | 2.4 | 0.5 | 416 | 17.2 | 0 | 0 |
| Newton | 10 | 201 | 8.1 | 3.7 | NA | NA | NA | NA | 649 | 25.9 | 125 | 1 |
| Ouachita | D | D | D | D | D | D | 10.0 | D | 1,157 | 41.6 | 1,400 | 7 |
| Perry | 13 | 184 | 10.4 | 4.1 | D | D | 1.9 | D | 721 | 33.3 | 124 | 8 |
| Phillips | 57 | 934 | 75.4 | 29.9 | 25 | 81 | 6.8 | 2.0 | 1,340 | 43.8 | 1,185 | 51 |
| Pike | 18 | 304 | 19.5 | 8.2 | 3 | 26 | 1.5 | 0.5 | 892 | 39.4 | NA | NA |
| Poinsett | 43 | 783 | 49.1 | 22.0 | D | D | D | D | 1,449 | 54.8 | 7,053 | 70 |
| Polk | 45 | 1,009 | 64.7 | 32.9 | 27 | 121 | 10.4 | 2.6 | 1,524 | 55.1 | 627 | 3 |
| Pope | D | D | D | 121.1 | 94 | 483 | 38.8 | 13.5 | 3,792 | 152.6 | 11,920 | 120 |
| Prairie | 9 | 188 | 8.1 | 4.0 | D | D | 3.1 | D | 537 | 21.3 | 100 | 2 |
| Pulaski | 1,515 | 46,881 | 6,685.8 | 2,627.9 | 809 | 5,543 | 951.3 | 210.5 | 29,614 | 1,369.0 | 265,502 | 1,315 |
| Randolph | 43 | 855 | 56.2 | 26.3 | 15 | 40 | 3.8 | 0.9 | 1,290 | 61.4 | 2,280 | 29 |
| St. Francis | 62 | 1,494 | 112.3 | 45.4 | 18 | 53 | 4.7 | 1.2 | 1,707 | 60.5 | 644 | 3 |
| Saline | 241 | 4,292 | 379.8 | 150.6 | 135 | 1,204 | 116.4 | 34.7 | 8,448 | 376.2 | 94,927 | 433 |
| Scott | 23 | 359 | 21.9 | 11.8 | D | D | D | D | 619 | 21.7 | 0 | 0 |
| Searcy | 13 | 276 | 13.8 | 5.7 | NA | NA | NA | NA | 781 | 33.7 | 0 | 0 |
| Sebastian | 428 | 11,389 | 1,422.6 | 551.1 | 199 | 906 | 90.1 | 25.0 | 8,902 | 461.4 | 70,572 | 410 |
| Sevier | 39 | 556 | 47.4 | 16.7 | D | D | D | D | 912 | 40.6 | 167 | 5 |
| Sharp | 50 | 643 | 34.6 | 16.2 | 26 | 157 | 13.3 | 3.5 | 1,187 | 43.3 | 442 | 4 |
| Stone | 35 | 610 | 39.5 | 17.7 | 14 | 49 | 4.4 | 1.3 | 1,162 | 43.3 | 470 | 9 |
| Union | 122 | 2,265 | 203.3 | 82.1 | 66 | 340 | 41.5 | 10.8 | 2,832 | 121.3 | 1,858 | 14 |
| Van Buren | 30 | 798 | 55.5 | 25.8 | D | D | D | D | 1,229 | 40.6 | 2,387 | 10 |
| Washington | 607 | 15,329 | 1,795.3 | 803.6 | 332 | 2,017 | 488.5 | 64.3 | 17,827 | 871.2 | 472,970 | 2,011 |
| White | 179 | 4,860 | 425.0 | 165.6 | 94 | 534 | 53.5 | 16.3 | 5,399 | 227.1 | 16,858 | 137 |
| Woodruff | 13 | 138 | 10.2 | 4.9 | 4 | 14 | 1.1 | 0.3 | 418 | 18.0 | 240 | 2 |
| Yell | D | D | D | 28.3 | 13 | 50 | 5.3 | 1.4 | 1,144 | 45.9 | 1,198 | 13 |

# Table B. States and Counties — Government Employment and Payroll, and Local Government Finances

| | Government employment and payroll, 2017 | | | | | | | | | Local government finances, 2017 | | | | |
| | | | March payroll (percent of total) | | | | | | | General revenue | | | | |
| | | | | | | | | | | | | Taxes | | |
| STATE County | Full-time equivalent employees | March payroll (dollars) | Administration, judicial, and legal | Police and corrections | Fire protection | Highways and transportation | Health and welfare | Natural resources and utilities | Education and libraries | Total (mil dol) | Intergovernmental (mil dol) | Total (mil dol) | Per capita[1] (dollars) Total | Property |
| | 171 | 172 | 173 | 174 | 175 | 176 | 177 | 178 | 179 | 180 | 181 | 182 | 183 | 184 |
|---|---|---|---|---|---|---|---|---|---|---|---|---|---|---|
| **ARKANSAS—Cont'd** | | | | | | | | | | | | | | |
| Cleveland | 305 | 798,054 | 7.6 | 5.8 | 0.6 | 4.1 | 0.7 | 2.3 | 78.6 | 21.7 | 15.0 | 5.5 | 674 | 565 |
| Columbia | 844 | 2,253,957 | 2.7 | 8.3 | 2.5 | 5.0 | 2.0 | 6.6 | 71.9 | 89.8 | 41.0 | 20.1 | 847 | 327 |
| Conway | 853 | 2,670,239 | 4.5 | 9.4 | 0.9 | 4.5 | 1.8 | 5.3 | 72.2 | 77.7 | 43.6 | 19.2 | 924 | 533 |
| Craighead | 3,526 | 12,146,858 | 5.4 | 9.1 | 2.8 | 3.9 | 1.3 | 10.5 | 66.8 | 330.8 | 193.4 | 87.6 | 817 | 400 |
| Crawford | 2,030 | 6,952,646 | 3.7 | 6.8 | 2.6 | 2.5 | 0.9 | 4.1 | 78.8 | 170.0 | 106.8 | 44.9 | 714 | 287 |
| Crittenden | 2,206 | 7,455,061 | 5.4 | 11.1 | 4.5 | 2.4 | 1.1 | 6.8 | 65.9 | 213.8 | 120.6 | 63.5 | 1,303 | 537 |
| Cross | 689 | 2,089,801 | 5.7 | 9.2 | 1.3 | 2.8 | 1.0 | 1.8 | 77.0 | 51.9 | 35.4 | 8.5 | 505 | 272 |
| Dallas | 258 | 586,128 | 10.9 | 16.4 | 0.3 | 4.8 | 0.2 | 7.3 | 59.7 | 21.1 | 13.0 | 4.4 | 597 | 161 |
| Desha | 761 | 2,240,254 | 5.2 | 7.5 | 2.4 | 1.0 | 24.3 | 4.1 | 55.6 | 64.1 | 32.8 | 15.1 | 1,282 | 580 |
| Drew | 1,073 | 3,599,821 | 3.8 | 2.7 | 0.9 | 1.7 | 33.3 | 2.0 | 55.4 | 97.2 | 43.3 | 14.6 | 794 | 258 |
| Faulkner | 3,353 | 11,828,937 | 4.8 | 9.3 | 4.1 | 2.6 | 0.4 | 4.8 | 73.3 | 352.7 | 181.7 | 85.9 | 695 | 316 |
| Franklin | 664 | 2,103,759 | 6.3 | 5.1 | 1.2 | 3.4 | 0.3 | 3.7 | 79.9 | 54.1 | 38.0 | 10.8 | 609 | 313 |
| Fulton | 479 | 1,338,215 | 3.1 | 4.6 | 0.6 | 4.5 | 30.2 | 2.0 | 54.7 | 38.2 | 17.6 | 4.5 | 370 | 184 |
| Garland | 2,866 | 10,381,906 | 5.0 | 10.8 | 3.8 | 3.3 | 1.5 | 7.3 | 67.5 | 304.8 | 156.1 | 87.2 | 886 | 270 |
| Grant | 747 | 2,335,417 | 6.2 | 6.6 | 0.0 | 2.7 | 0.0 | 3.4 | 80.1 | 59.1 | 43.4 | 9.3 | 513 | 249 |
| Greene | 1,572 | 5,019,874 | 3.6 | 6.2 | 3.1 | 3.3 | 0.9 | 9.6 | 67.4 | 131.4 | 72.9 | 26.5 | 589 | 229 |
| Hempstead | 883 | 2,974,446 | 3.4 | 6.9 | 1.6 | 2.9 | 1.8 | 8.7 | 70.1 | 68.8 | 45.3 | 13.9 | 636 | 226 |
| Hot Spring | 1,015 | 3,410,812 | 9.2 | 2.3 | 2.1 | 0.8 | 1.4 | 7.7 | 76.4 | 85.2 | 55.2 | 20.3 | 605 | 266 |
| Howard | 644 | 1,981,608 | 6.0 | 5.8 | 0.8 | 2.9 | 1.4 | 10.5 | 72.6 | 44.4 | 30.9 | 5.4 | 407 | 239 |
| Independence | 1,475 | 4,100,818 | 4.9 | 7.1 | 1.2 | 3.7 | 3.8 | 8.7 | 68.8 | 137.4 | 83.0 | 30.6 | 817 | 304 |
| Izard | 513 | 1,366,925 | 8.2 | 5.1 | 0.9 | 4.7 | 1.2 | 3.2 | 76.5 | 31.3 | 23.1 | 4.4 | 321 | 230 |
| Jackson | 508 | 1,537,477 | 7.9 | 9.3 | 3.4 | 4.6 | 3.0 | 6.1 | 65.1 | 49.3 | 26.4 | 15.5 | 913 | 506 |
| Jefferson | 2,423 | 7,522,146 | 6.1 | 11.6 | 3.8 | 3.7 | 1.8 | 1.5 | 71.4 | 218.7 | 128.4 | 62.7 | 906 | 391 |
| Johnson | 886 | 2,790,408 | 4.5 | 6.7 | 0.9 | 2.5 | 1.1 | 8.4 | 71.0 | 71.1 | 47.4 | 15.5 | 587 | 218 |
| Lafayette | 259 | 833,135 | 9.0 | 8.5 | 0.7 | 7.1 | 0.1 | 5.7 | 68.8 | 13.9 | 8.3 | 3.0 | 435 | 213 |
| Lawrence | 929 | 2,618,570 | 4.5 | 3.7 | 1.4 | 1.9 | 24.3 | 3.2 | 60.5 | 72.5 | 40.7 | 10.4 | 626 | 255 |
| Lee | 295 | 770,474 | 10.0 | 9.8 | 3.4 | 7.8 | 5.7 | 9.9 | 53.1 | 19.8 | 12.9 | 4.4 | 481 | 227 |
| Lincoln | 318 | 967,097 | 6.0 | 10.1 | 0.1 | 4.2 | 1.7 | 3.7 | 73.2 | 30.0 | 22.0 | 4.7 | 348 | 172 |
| Little River | 656 | 1,951,349 | 6.5 | 4.0 | 0.5 | 4.0 | 37.3 | 4.2 | 43.3 | 76.3 | 24.2 | 28.8 | 2,327 | 1,888 |
| Logan | 753 | 2,180,723 | 8.0 | 8.0 | 0.4 | 3.9 | 4.0 | 3.3 | 71.9 | 51.2 | 35.9 | 9.7 | 444 | 237 |
| Lonoke | 2,432 | 7,248,544 | 4.3 | 6.9 | 2.1 | 2.2 | 1.1 | 5.8 | 76.7 | 190.4 | 128.3 | 41.3 | 568 | 251 |
| Madison | 471 | 1,252,806 | 6.4 | 6.5 | 0.0 | 8.9 | 4.8 | 5.8 | 67.3 | 34.2 | 23.0 | 6.8 | 415 | 202 |
| Marion | 395 | 1,135,636 | 9.2 | 10.7 | 0.1 | 8.3 | 0.5 | 7.0 | 63.4 | 29.6 | 18.9 | 7.5 | 457 | 245 |
| Miller | 1,433 | 4,198,444 | 5.8 | 18.1 | 6.5 | 4.2 | 1.3 | 0.6 | 61.1 | 126.3 | 79.2 | 28.0 | 641 | 283 |
| Mississippi | 1,946 | 6,023,869 | 6.2 | 10.2 | 3.6 | 2.9 | 2.3 | 7.6 | 66.3 | 182.4 | 89.0 | 37.4 | 888 | 345 |
| Monroe | 365 | 969,748 | 9.0 | 9.3 | 1.3 | 5.0 | 4.2 | 1.3 | 69.6 | 23.7 | 14.9 | 5.1 | 730 | 364 |
| Montgomery | 308 | 758,187 | 8.9 | 7.6 | 0.0 | 7.0 | 0.2 | 7.2 | 68.3 | 18.7 | 13.5 | 3.2 | 361 | 232 |
| Nevada | 364 | 973,929 | 7.1 | 8.2 | 1.0 | 4.7 | 1.8 | 10.8 | 66.2 | 23.1 | 15.5 | 5.6 | 672 | 538 |
| Newton | 464 | 1,314,007 | 2.7 | 2.9 | 0.0 | 3.2 | 0.0 | 2.7 | 88.0 | 24.7 | 21.9 | 1.3 | 170 | 158 |
| Ouachita | 1,035 | 3,046,677 | 8.8 | 7.0 | 3.4 | 4.0 | 3.0 | 4.7 | 68.1 | 80.4 | 49.8 | 20.5 | 862 | 227 |
| Perry | 290 | 964,106 | 7.6 | 7.2 | 0.3 | 3.0 | 0.1 | 3.7 | 78.1 | 17.4 | 14.0 | 1.9 | 180 | 139 |
| Phillips | 876 | 2,488,292 | 6.7 | 8.8 | 3.4 | 3.5 | 2.2 | 6.4 | 67.4 | 71.4 | 48.6 | 15.7 | 846 | 290 |
| Pike | 434 | 1,240,094 | 5.3 | 5.4 | 0.0 | 3.7 | 0.6 | 4.6 | 79.7 | 32.9 | 23.1 | 6.6 | 617 | 241 |
| Poinsett | 1,057 | 3,162,684 | 5.0 | 8.8 | 0.8 | 2.6 | 2.7 | 3.4 | 76.2 | 71.6 | 52.3 | 9.5 | 394 | 185 |
| Polk | 827 | 2,358,009 | 4.5 | 7.3 | 0.8 | 4.7 | 1.2 | 3.3 | 76.9 | 85.1 | 37.9 | 12.0 | 597 | 188 |
| Pope | 2,123 | 6,616,052 | 3.1 | 9.1 | 3.7 | 2.3 | 2.5 | 6.1 | 70.2 | 179.0 | 102.7 | 50.3 | 790 | 324 |
| Prairie | 327 | 898,490 | 11.0 | 10.0 | 0.4 | 3.6 | 3.8 | 7.6 | 61.0 | 24.2 | 12.9 | 6.1 | 741 | 388 |
| Pulaski | 14,961 | 57,806,343 | 6.2 | 13.0 | 6.7 | 5.4 | 3.9 | 11.0 | 51.8 | 1554.1 | 690.2 | 522.7 | 1,329 | 629 |
| Randolph | 727 | 2,362,081 | 6.6 | 4.3 | 1.1 | 4.7 | 35.3 | 4.0 | 41.9 | 57.3 | 27.0 | 7.9 | 447 | 123 |
| St. Francis | 878 | 2,624,352 | 6.2 | 9.3 | 2.8 | 2.4 | 2.4 | 7.7 | 68.1 | 69.9 | 43.3 | 17.5 | 673 | 177 |
| Saline | 2,670 | 9,187,746 | 5.6 | 9.4 | 5.4 | 2.3 | 0.5 | 3.9 | 71.3 | 262.1 | 159.4 | 72.6 | 608 | 307 |
| Scott | 367 | 1,012,757 | 4.6 | 5.2 | 0.0 | 4.4 | 1.1 | 9.1 | 72.6 | 18.1 | 15.0 | 1.5 | 145 | 72 |
| Searcy | 270 | 681,711 | 7.8 | 9.3 | 0.0 | 6.3 | 0.1 | 4.7 | 70.7 | 17.5 | 12.9 | 3.1 | 390 | 218 |
| Sebastian | 4,499 | 16,611,357 | 5.6 | 9.0 | 4.9 | 3.6 | 1.5 | 9.1 | 66.0 | 471.2 | 238.0 | 139.0 | 1,088 | 387 |
| Sevier | 821 | 2,336,559 | 4.8 | 6.0 | 0.5 | 2.3 | 0.9 | 4.4 | 81.0 | 55.5 | 42.1 | 6.3 | 371 | 150 |
| Sharp | 793 | 1,880,777 | 7.2 | 7.5 | 3.4 | 6.1 | 0.1 | 6.2 | 67.1 | 39.8 | 28.0 | 6.5 | 377 | 178 |
| Stone | 396 | 1,118,159 | 5.7 | 7.4 | 0.0 | 4.7 | 0.1 | 9.5 | 71.6 | 25.2 | 17.4 | 5.5 | 441 | 194 |
| Union | 1,547 | 4,466,114 | 2.1 | 10.0 | 4.5 | 4.3 | 0.4 | 7.3 | 68.1 | 128.0 | 70.7 | 38.7 | 981 | 364 |
| Van Buren | 633 | 1,658,575 | 8.9 | 10.9 | 0.2 | 4.8 | 0.3 | 2.2 | 71.0 | 49.4 | 28.4 | 14.6 | 879 | 535 |
| Washington | 7,736 | 25,196,414 | 5.5 | 11.1 | 4.4 | 4.8 | 0.7 | 5.0 | 66.9 | 804.5 | 444.4 | 225.6 | 969 | 333 |
| White | 2,591 | 8,236,317 | 4.7 | 8.2 | 2.4 | 2.9 | 0.8 | 5.0 | 74.6 | 210.9 | 132.2 | 49.9 | 633 | 236 |
| Woodruff | 312 | 811,976 | 12.0 | 6.6 | 0.3 | 6.3 | 3.0 | 7.7 | 62.8 | 25.7 | 16.4 | 5.1 | 779 | 470 |
| Yell | 887 | 2,628,007 | 3.2 | 7.0 | 0.0 | 3.7 | 1.2 | 3.5 | 80.6 | 54.1 | 42.2 | 8.4 | 392 | 228 |

1. Based on the resident population estimated as of July 1 of the year shown.

| STATE County | Total (mil dol) | Per capita[1] (dollars) | Education | Health and hospitals | Police protection | Public welfare | Highways | Total (mil dol) | Per capita[1] (dollars) | Federal civilian | Federal military | State and local | Number of returns | Mean adjusted gross income | Mean income tax |
|---|---|---|---|---|---|---|---|---|---|---|---|---|---|---|---|
| | 185 | 186 | 187 | 188 | 189 | 190 | 191 | 192 | 193 | 194 | 195 | 196 | 197 | 198 | 199 |
| **ARKANSAS—Cont'd** | | | | | | | | | | | | | | | |
| Cleveland | 26 | 3,145 | 55.7 | 0.1 | 2.8 | 0.0 | 5.1 | 7.3 | 890 | 12 | 31 | 326 | 3,080 | 52,913 | 4,081 |
| Columbia | 88 | 3,716 | 43.8 | 25.0 | 3.8 | 0.0 | 6.8 | 73.9 | 3,120 | 37 | 83 | 2,035 | 8,730 | 52,286 | 4,503 |
| Conway | 95 | 4,555 | 72.3 | 0.8 | 4.9 | 0.1 | 5.3 | 52.1 | 2,504 | 57 | 80 | 1,583 | 8,220 | 50,488 | 4,134 |
| Craighead | 317 | 2,959 | 58.7 | 0.4 | 5.2 | 0.1 | 5.3 | 496.1 | 4,629 | 335 | 418 | 8,266 | 42,900 | 57,553 | 5,812 |
| Crawford | 168 | 2,664 | 60.0 | 0.0 | 5.8 | 0.1 | 6.8 | 237.5 | 3,776 | 92 | 244 | 2,180 | 24,130 | 49,026 | 3,743 |
| Crittenden | 216 | 4,431 | 60.5 | 0.7 | 6.5 | 0.0 | 4.5 | 235.2 | 4,828 | 93 | 182 | 2,587 | 20,140 | 45,560 | 3,827 |
| Cross | 52 | 3,072 | 63.9 | 4.6 | 7.3 | 0.0 | 5.0 | 18.4 | 1,093 | 56 | 87 | 1,039 | 6,850 | 44,007 | 3,389 |
| Dallas | 19 | 2,662 | 54.2 | 0.1 | 11.9 | 0.4 | 8.2 | 8.1 | 1,106 | 20 | 26 | 365 | 2,650 | 43,471 | 3,214 |
| Desha | 61 | 5,184 | 44.3 | 24.7 | 6.5 | 0.1 | 3.5 | 33.0 | 2,807 | 74 | 44 | 915 | 4,720 | 39,006 | 3,266 |
| Drew | 102 | 5,574 | 42.3 | 40.6 | 3.0 | 0.1 | 3.7 | 57.9 | 3,152 | 51 | 67 | 1,969 | 6,960 | 50,300 | 4,411 |
| Faulkner | 364 | 2,948 | 49.5 | 0.1 | 5.0 | 0.1 | 9.7 | 829.8 | 6,717 | 226 | 476 | 7,510 | 50,670 | 59,870 | 5,748 |
| Franklin | 63 | 3,509 | 74.9 | 2.9 | 3.4 | 0.3 | 4.7 | 71.5 | 4,016 | 162 | 67 | 930 | 6,680 | 47,064 | 3,567 |
| Fulton | 41 | 3,375 | 41.5 | 24.6 | 2.3 | 0.2 | 4.0 | 19.3 | 1,594 | 27 | 48 | 609 | 4,430 | 39,328 | 2,779 |
| Garland | 292 | 2,972 | 52.9 | 0.3 | 5.9 | 0.1 | 3.2 | 550.8 | 5,602 | 479 | 379 | 4,025 | 43,450 | 55,037 | 5,732 |
| Grant | 57 | 3,156 | 75.4 | 0.3 | 3.8 | 0.1 | 4.5 | 88.4 | 4,883 | 29 | 70 | 784 | 7,430 | 55,764 | 4,551 |
| Greene | 127 | 2,823 | 57.7 | 0.2 | 5.3 | 0.1 | 5.3 | 109.7 | 2,438 | 95 | 174 | 2,008 | 17,510 | 49,314 | 3,964 |
| Hempstead | 68 | 3,090 | 68.2 | 0.2 | 4.8 | 0.1 | 6.3 | 47.4 | 2,165 | 61 | 82 | 1,677 | 8,370 | 39,390 | 2,421 |
| Hot Spring | 79 | 2,340 | 66.5 | 0.0 | 3.9 | 0.1 | 5.6 | 90.0 | 2,678 | 64 | 122 | 1,915 | 12,420 | 46,446 | 3,383 |
| Howard | 44 | 3,259 | 75.2 | 0.0 | 3.4 | 0.0 | 1.3 | 111.4 | 8,332 | 51 | 51 | 782 | 5,420 | 42,329 | 3,111 |
| Independence | 125 | 3,334 | 56.9 | 0.1 | 3.8 | 0.1 | 4.0 | 241.3 | 6,453 | 101 | 143 | 2,100 | 14,940 | 50,406 | 4,477 |
| Izard | 32 | 2,358 | 71.3 | 1.2 | 5.6 | 0.2 | 5.8 | 23.2 | 1,695 | 27 | 49 | 996 | 4,790 | 42,639 | 3,181 |
| Jackson | 54 | 3,143 | 41.1 | 0.4 | 6.7 | 0.5 | 4.9 | 46.9 | 2,755 | 42 | 56 | 1,455 | 5,660 | 41,791 | 3,209 |
| Jefferson | 208 | 3,005 | 55.9 | 0.1 | 9.7 | 0.2 | 4.2 | 198.7 | 2,870 | 1,375 | 266 | 6,576 | 27,620 | 44,488 | 3,574 |
| Johnson | 65 | 2,478 | 64.2 | 0.2 | 4.8 | 0.0 | 8.4 | 75.7 | 2,865 | 82 | 101 | 1,072 | 9,810 | 43,097 | 3,083 |
| Lafayette | 14 | 2,049 | 48.9 | 0.3 | 10.9 | 0.2 | 10.9 | 9.5 | 1,401 | 27 | 25 | 378 | 2,440 | 39,657 | 2,885 |
| Lawrence | 70 | 4,233 | 47.3 | 24.8 | 3.4 | 5.1 | 3.7 | 51.3 | 3,097 | 49 | 61 | 1,124 | 6,140 | 40,770 | 2,744 |
| Lee | 19 | 2,094 | 52.8 | 0.4 | 14.4 | 0.3 | 13.8 | 5.9 | 642 | 36 | 28 | 663 | 2,690 | 39,443 | 3,212 |
| Lincoln | 21 | 1,590 | 65.2 | 0.1 | 5.7 | 0.0 | 10.2 | 10.7 | 793 | 26 | 37 | 1,180 | 3,810 | 44,667 | 3,214 |
| Little River | 73 | 5,881 | 27.5 | 18.4 | 3.0 | 7.0 | 3.0 | 58.5 | 4,729 | 44 | 47 | 795 | 4,870 | 44,787 | 3,127 |
| Logan | 49 | 2,272 | 66.5 | 3.1 | 5.8 | 0.0 | 7.3 | 64.3 | 2,958 | 94 | 81 | 1,304 | 8,370 | 42,413 | 2,826 |
| Lonoke | 182 | 2,506 | 67.8 | 0.2 | 5.7 | 0.1 | 4.3 | 182.9 | 2,513 | 134 | 283 | 2,806 | 30,400 | 57,184 | 4,822 |
| Madison | 35 | 2,154 | 61.7 | 3.8 | 6.0 | 0.0 | 13.0 | 33.7 | 2,069 | 43 | 64 | 586 | 6,520 | 45,164 | 3,655 |
| Marion | 33 | 1,984 | 63.1 | 0.2 | 7.4 | 0.2 | 10.4 | 23.1 | 1,408 | 34 | 64 | 527 | 6,480 | 41,949 | 2,984 |
| Miller | 125 | 2,857 | 53.5 | 0.4 | 9.3 | 0.1 | 3.8 | 96.8 | 2,212 | 74 | 162 | 1,918 | 17,410 | 48,428 | 4,143 |
| Mississippi | 179 | 4,254 | 47.8 | 2.4 | 5.9 | 0.1 | 3.2 | 811.3 | 19,265 | 103 | 155 | 2,528 | 15,770 | 45,630 | 3,653 |
| Monroe | 24 | 3,387 | 53.5 | 1.4 | 7.6 | 0.3 | 8.8 | 5.6 | 804 | 31 | 26 | 393 | 2,840 | 37,592 | 3,148 |
| Montgomery | 18 | 1,992 | 59.9 | 0.1 | 4.6 | 0.0 | 12.5 | 12.1 | 1,362 | 49 | 34 | 450 | 3,230 | 40,733 | 2,946 |
| Nevada | 24 | 2,927 | 60.1 | 0.3 | 5.2 | 0.0 | 2.6 | 12.6 | 1,515 | 26 | 31 | 446 | 3,240 | 39,610 | 2,517 |
| Newton | 23 | 2,938 | 96.5 | 0.0 | 0.5 | 0.0 | 0.3 | 35.1 | 4,486 | 41 | 30 | 371 | 2,880 | 38,083 | 2,280 |
| Ouachita | 85 | 3,569 | 55.6 | 9.2 | 4.1 | 0.1 | 4.8 | 67.9 | 2,849 | 98 | 89 | 1,837 | 9,420 | 47,083 | 3,764 |
| Perry | 17 | 1,668 | 82.5 | 0.0 | 1.7 | 0.0 | 1.4 | 10.8 | 1,045 | 23 | 40 | 357 | 3,870 | 47,212 | 3,432 |
| Phillips | 67 | 3,622 | 63.4 | 0.2 | 7.6 | 0.1 | 4.7 | 68.1 | 3,664 | 57 | 68 | 1,169 | 6,750 | 36,336 | 2,709 |
| Pike | 32 | 2,935 | 75.3 | 0.1 | 3.9 | 0.2 | 5.4 | 21.0 | 1,953 | 48 | 41 | 572 | 4,100 | 41,993 | 2,995 |
| Poinsett | 76 | 3,149 | 72.4 | 0.1 | 6.4 | 0.1 | 5.5 | 39.8 | 1,655 | 65 | 90 | 1,151 | 8,750 | 41,664 | 2,971 |
| Polk | 83 | 4,120 | 42.5 | 37.6 | 2.7 | 0.0 | 5.1 | 57.4 | 2,849 | 108 | 77 | 1,065 | 7,440 | 41,899 | 3,034 |
| Pope | 180 | 2,833 | 60.5 | 1.7 | 6.2 | 0.1 | 5.6 | 260.0 | 4,085 | 280 | 237 | 4,358 | 25,580 | 54,123 | 4,841 |
| Prairie | 21 | 2,595 | 52.7 | 1.8 | 10.0 | 0.0 | 8.2 | 13.7 | 1,663 | 42 | 31 | 336 | 3,240 | 46,691 | 3,361 |
| Pulaski | 1,618 | 4,114 | 42.2 | 2.6 | 7.4 | 0.0 | 4.9 | 2,355.1 | 5,988 | 9,173 | 4,990 | 43,326 | 177,030 | 68,102 | 8,586 |
| Randolph | 56 | 3,164 | 40.2 | 27.6 | 5.5 | 0.1 | 5.2 | 21.8 | 1,237 | 49 | 68 | 1,213 | 7,090 | 39,898 | 2,837 |
| St. Francis | 64 | 2,449 | 56.7 | 0.5 | 7.4 | 0.1 | 5.2 | 34.0 | 1,307 | 623 | 81 | 1,373 | 8,380 | 37,471 | 2,690 |
| Saline | 255 | 2,135 | 57.8 | 0.2 | 6.4 | 0.1 | 8.9 | 400.0 | 3,347 | 99 | 470 | 3,784 | 52,360 | 62,821 | 5,816 |
| Scott | 19 | 1,851 | 81.5 | 0.5 | 2.5 | 0.0 | 1.5 | 22.8 | 2,199 | 72 | 40 | 547 | 3,840 | 38,687 | 2,457 |
| Searcy | 20 | 2,474 | 68.3 | 0.1 | 3.8 | 0.2 | 8.9 | 17.5 | 2,209 | 36 | 30 | 348 | 2,960 | 33,165 | 2,034 |
| Sebastian | 452 | 3,537 | 49.5 | 0.4 | 5.0 | 0.1 | 9.6 | 1,147.3 | 8,978 | 918 | 528 | 6,791 | 53,130 | 55,219 | 5,659 |
| Sevier | 54 | 3,161 | 74.3 | 0.2 | 2.2 | 0.1 | 4.9 | 18.7 | 1,093 | 70 | 65 | 1,147 | 6,080 | 41,102 | 2,567 |
| Sharp | 40 | 2,326 | 68.6 | 0.1 | 5.5 | 0.0 | 8.6 | 21.5 | 1,254 | 49 | 67 | 784 | 6,440 | 39,179 | 2,747 |
| Stone | 24 | 1,903 | 66.2 | 0.1 | 6.7 | 0.3 | 9.0 | 14.9 | 1,189 | 69 | 48 | 499 | 4,650 | 39,428 | 2,735 |
| Union | 127 | 3,229 | 58.5 | 0.1 | 5.6 | 0.0 | 6.8 | 128.1 | 3,245 | 148 | 149 | 2,435 | 16,840 | 60,832 | 6,567 |
| Van Buren | 46 | 2,806 | 57.3 | 1.1 | 10.0 | 0.5 | 10.0 | 44.1 | 2,662 | 33 | 64 | 653 | 6,190 | 44,716 | 3,392 |
| Washington | 754 | 3,240 | 56.2 | 0.5 | 5.5 | 0.0 | 5.6 | 1,274.3 | 5,475 | 2,111 | 913 | 18,671 | 101,060 | 63,181 | 7,094 |
| White | 200 | 2,534 | 64.7 | 0.1 | 6.7 | 0.2 | 7.4 | 183.1 | 2,322 | 169 | 297 | 3,281 | 29,920 | 53,930 | 4,980 |
| Woodruff | 29 | 4,409 | 44.9 | 0.3 | 4.4 | 21.8 | 6.7 | 21.3 | 3,245 | 37 | 24 | 425 | 2,530 | 39,310 | 3,109 |
| Yell | 54 | 2,524 | 81.4 | 0.0 | 2.5 | 0.0 | 1.9 | 55.6 | 2,585 | 95 | 82 | 1,169 | 8,280 | 42,652 | 3,040 |

1. Based on the resident population estimated as of July 1 of the year shown.

# Table B. States and Counties — **Land Area and Population**

| State / county code | CBSA code[1] | County Type code[2] | STATE County | Land area[3] (sq. mi) | Total persons 2019 | Rank | Per square mile | White | Black | American Indian, Alaska Native | Asian and Pacific Islancer | Percent Hispanic or Latino[4] | Under 5 years | 5 to 14 years | 15 to 24 years | 25 to 34 years | 35 to 44 years | 45 to 54 years |
|---|---|---|---|---|---|---|---|---|---|---|---|---|---|---|---|---|---|---|
| | | | | 1 | 2 | 3 | 4 | 5 | 6 | 7 | 8 | 9 | 10 | 11 | 12 | 13 | 14 | 15 |
| 06000 | | 0 | CALIFORNIA ............... | 155,854.0 | 39,368,078 | X | 252.6 | 38.5 | 6.5 | 1.0 | 17.6 | 39.5 | 5.9 | 12.6 | 13.0 | 15.3 | 13.5 | 12.5 |
| 06001 | 41860 | 1 | Alameda .................. | 737.5 | 1,662,323 | 20 | 2,254.0 | 33.6 | 11.4 | 0.8 | 36.7 | 22.2 | 5.6 | 11.3 | 11.2 | 16.5 | 15.5 | 13.1 |
| 06003 | | 8 | Alpine ..................... | 738.3 | 1,119 | 3,105 | 1.5 | 63.8 | 1.6 | 22.0 | 2.7 | 13.0 | 4.5 | 10.1 | 12.8 | 11.0 | 8.0 | 11.4 |
| 06005 | | 6 | Amador .................... | 594.6 | 40,083 | 1,182 | 67.4 | 79.1 | 3.2 | 2.7 | 2.7 | 14.9 | 4.0 | 8.4 | 8.1 | 11.3 | 12.0 | 12.4 |
| 06007 | 17020 | 3 | Butte ...................... | 1,636.5 | 212,744 | 321 | 130.0 | 73.5 | 2.8 | 3.0 | 7.0 | 17.9 | 5.4 | 11.4 | 18.0 | 13.1 | 11.5 | 10.2 |
| 06009 | | 6 | Calaveras ................. | 1,020.0 | 46,308 | 1,051 | 45.4 | 82.5 | 1.6 | 2.9 | 3.1 | 13.3 | 4.4 | 9.3 | 8.9 | 9.8 | 10.0 | 11.1 |
| 06011 | | 6 | Colusa ..................... | 1,150.7 | 21,558 | 1,745 | 18.7 | 35.2 | 1.8 | 1.9 | 2.3 | 60.4 | 6.8 | 15.4 | 13.6 | 13.4 | 12.6 | 11.1 |
| 06013 | 41860 | 1 | Contra Costa .............. | 716.9 | 1,152,333 | 38 | 1,607.4 | 45.5 | 10.0 | 1.0 | 21.9 | 26.3 | 5.5 | 12.7 | 12.0 | 12.6 | 13.7 | 13.6 |
| 06015 | 18860 | 7 | Del Norte ................. | 1,006.2 | 27,968 | 1,499 | 27.8 | 65.4 | 4.2 | 9.6 | 4.5 | 20.4 | 5.1 | 12.2 | 10.8 | 13.8 | 13.9 | 11.2 |
| 06017 | 40900 | 1 | El Dorado ................. | 1,707.9 | 192,925 | 350 | 113.0 | 79.8 | 1.5 | 1.8 | 7.0 | 13.3 | 4.5 | 11.3 | 10.6 | 10.0 | 11.8 | 12.5 |
| 06019 | 23420 | 2 | Fresno .................... | 5,958.4 | 1,000,918 | 46 | 168.0 | 29.6 | 5.3 | 1.1 | 12.0 | 54.1 | 7.3 | 16.2 | 14.2 | 15.3 | 13.0 | 10.9 |
| 06021 | | 6 | Glenn ..................... | 1,314.0 | 28,283 | 1,486 | 21.5 | 51.6 | 1.5 | 2.4 | 3.2 | 43.1 | 6.8 | 15.4 | 13.4 | 13.1 | 11.7 | 10.6 |
| 06023 | 21700 | 5 | Humboldt .................. | 3,568.2 | 134,977 | 481 | 37.8 | 77.7 | 2.5 | 7.5 | 5.1 | 12.5 | 5.1 | 10.6 | 15.6 | 13.2 | 13.1 | 10.7 |
| 06025 | 20940 | 3 | Imperial .................. | 4,175.5 | 180,267 | 368 | 43.2 | 10.1 | 2.6 | 0.9 | 1.6 | 85.4 | 7.6 | 16.2 | 14.5 | 14.9 | 12.4 | 10.7 |
| 06027 | | 7 | Inyo ...................... | 10,197.3 | 18,046 | 1,918 | 1.8 | 63.0 | 1.6 | 12.0 | 2.5 | 23.7 | 5.4 | 11.9 | 9.4 | 10.8 | 12.4 | 10.5 |
| 06029 | 12540 | 2 | Kern ...................... | 8,134.6 | 901,362 | 63 | 110.8 | 33.9 | 5.9 | 1.2 | 6.0 | 55.2 | 7.4 | 16.5 | 14.5 | 15.6 | 13.0 | 10.9 |
| 06031 | 25260 | 3 | Kings ..................... | 1,391.0 | 152,692 | 439 | 109.8 | 32.9 | 7.1 | 1.4 | 5.3 | 55.6 | 7.4 | 15.3 | 14.9 | 16.9 | 14.1 | 11.0 |
| 06033 | 17340 | 4 | Lake ...................... | 1,256.6 | 64,479 | 831 | 51.3 | 71.0 | 2.8 | 4.3 | 2.8 | 22.8 | 6.0 | 12.0 | 9.8 | 11.4 | 11.0 | 11.1 |
| 06035 | 45000 | 7 | Lassen .................... | 4,541.3 | 30,016 | 1,435 | 6.6 | 67.5 | 8.6 | 4.3 | 3.3 | 19.6 | 5.0 | 9.9 | 12.1 | 18.3 | 14.6 | 12.2 |
| 06037 | 31080 | 1 | Los Angeles ............... | 4,059.3 | 9,943,046 | 1 | 2,449.4 | 28.8 | 8.7 | 0.6 | 16.8 | 48.5 | 5.6 | 11.9 | 12.7 | 16.3 | 13.7 | 13.1 |
| 06039 | 31460 | 3 | Madera .................... | 2,136.9 | 157,761 | 427 | 73.8 | 34.0 | 3.6 | 1.7 | 2.9 | 59.5 | 7.0 | 15.8 | 13.7 | 13.9 | 12.8 | 11.2 |
| 06041 | 41860 | 1 | Marin ..................... | 520.4 | 257,332 | 271 | 494.5 | 73.9 | 3.4 | 0.8 | 9.1 | 16.4 | 4.4 | 11.2 | 10.6 | 8.7 | 11.6 | 15.1 |
| 06043 | | 8 | Mariposa .................. | 1,448.8 | 17,160 | 1,966 | 11.8 | 81.4 | 1.9 | 4.6 | 2.6 | 12.7 | 4.3 | 9.3 | 8.7 | 11.3 | 10.1 | 10.3 |
| 06045 | 46380 | 4 | Mendocino ................. | 3,506.8 | 86,061 | 671 | 24.5 | 66.4 | 1.5 | 5.4 | 3.2 | 26.4 | 5.5 | 12.0 | 10.5 | 11.1 | 12.2 | 11.4 |
| 06047 | 32900 | 2 | Merced .................... | 1,938.0 | 279,252 | 250 | 144.1 | 27.4 | 3.6 | 0.9 | 8.6 | 61.5 | 7.4 | 16.6 | 16.1 | 14.7 | 12.5 | 10.9 |
| 06049 | | 6 | Modoc ..................... | 3,948.2 | 8,763 | 2,514 | 2.2 | 79.3 | 2.0 | 4.6 | 2.0 | 15.0 | 4.7 | 11.3 | 9.0 | 10.1 | 10.4 | 10.4 |
| 06051 | | 7 | Mono ...................... | 3,048.9 | 14,534 | 2,119 | 4.8 | 67.2 | 1.4 | 2.5 | 3.6 | 27.3 | 4.5 | 9.8 | 11.7 | 15.5 | 13.2 | 12.8 |
| 06053 | 41500 | 2 | Monterey .................. | 3,281.7 | 430,906 | 165 | 131.3 | 31.0 | 3.1 | 0.8 | 7.9 | 59.7 | 6.8 | 14.8 | 14.0 | 14.1 | 13.1 | 11.6 |
| 06055 | 34900 | 3 | Napa ...................... | 748.3 | 135,965 | 477 | 181.7 | 53.6 | 2.7 | 1.1 | 10.4 | 34.8 | 4.6 | 11.4 | 12.2 | 12.3 | 12.7 | 12.9 |
| 06057 | 46020 | 4 | Nevada .................... | 957.8 | 99,606 | 605 | 104.0 | 87.1 | 1.1 | 2.1 | 2.8 | 9.8 | 4.2 | 9.6 | 8.8 | 9.8 | 11.6 | 11.0 |
| 06059 | 31080 | 1 | Orange .................... | 792.8 | 3,166,857 | 6 | 3,994.5 | 41.8 | 2.2 | 0.6 | 24.6 | 33.8 | 5.7 | 12.0 | 12.6 | 14.5 | 12.9 | 13.6 |
| 06061 | 40900 | 1 | Placer .................... | 1,407.1 | 402,950 | 179 | 286.4 | 74.2 | 2.6 | 1.3 | 11.3 | 14.7 | 5.2 | 12.8 | 11.2 | 10.8 | 13.4 | 13.0 |
| 06063 | | 7 | Plumas .................... | 2,553.1 | 18,967 | 1,876 | 7.4 | 85.8 | 1.9 | 3.9 | 2.2 | 9.7 | 4.7 | 9.6 | 8.7 | 9.6 | 10.2 | 9.8 |
| 06065 | 40140 | 1 | Riverside ................. | 7,209.3 | 2,489,188 | 10 | 345.3 | 35.3 | 7.2 | 1.0 | 8.5 | 50.6 | 6.1 | 14.1 | 13.8 | 14.0 | 13.1 | 12.1 |
| 06067 | 40900 | 1 | Sacramento ................ | 965.3 | 1,559,146 | 24 | 1,615.2 | 47.2 | 11.8 | 1.5 | 21.2 | 23.9 | 6.2 | 13.1 | 12.1 | 15.8 | 13.8 | 12.1 |
| 06069 | 41940 | 1 | San Benito ................ | 1,388.7 | 64,055 | 838 | 46.1 | 34.3 | 1.5 | 1.0 | 4.7 | 61.2 | 6.5 | 14.6 | 13.3 | 13.7 | 14.0 | 12.4 |
| 06071 | 40140 | 1 | San Bernardino ............ | 20,068.0 | 2,189,183 | 14 | 109.1 | 28.3 | 9.1 | 0.9 | 9.3 | 54.9 | 6.8 | 14.7 | 14.3 | 15.5 | 13.2 | 11.9 |
| 06073 | 41740 | 1 | San Diego ................. | 4,210.2 | 3,332,427 | 5 | 791.5 | 47.5 | 5.8 | 0.9 | 15.3 | 34.3 | 5.9 | 11.9 | 13.4 | 16.4 | 13.7 | 12.0 |
| 06075 | 41860 | 1 | San Francisco ............. | 46.9 | 866,606 | 67 | 18,477.7 | 42.8 | 6.1 | 0.8 | 39.2 | 15.2 | 4.4 | 7.1 | 8.8 | 22.9 | 16.0 | 12.7 |
| 06077 | 44700 | 2 | San Joaquin ............... | 1,392.4 | 767,967 | 84 | 551.5 | 32.4 | 8.3 | 1.2 | 19.8 | 42.3 | 6.8 | 15.3 | 14.0 | 14.1 | 13.4 | 11.9 |
| 06079 | 42020 | 2 | San Luis Obispo ........... | 3,300.9 | 282,249 | 247 | 85.5 | 70.5 | 2.4 | 1.2 | 5.4 | 23.2 | 4.4 | 9.9 | 17.9 | 11.3 | 11.5 | 10.4 |
| 06081 | 41860 | 1 | San Mateo ................. | 448.6 | 758,308 | 88 | 1,690.4 | 41.4 | 3.0 | 0.6 | 35.4 | 23.8 | 5.4 | 11.1 | 10.6 | 15.0 | 14.5 | 13.4 |
| 06083 | 42200 | 2 | Santa Barbara ............. | 2,733.9 | 444,766 | 160 | 162.7 | 45.6 | 2.4 | 1.0 | 7.2 | 46.4 | 6.1 | 12.4 | 19.0 | 13.5 | 11.5 | 10.4 |
| 06085 | 41940 | 2 | Santa Clara ............... | 1,291.1 | 1,907,105 | 18 | 1,477.1 | 32.6 | 3.1 | 0.7 | 42.5 | 24.7 | 5.7 | 11.9 | 11.9 | 16.4 | 14.4 | 13.4 |
| 06087 | 42100 | 2 | Santa Cruz ................ | 445.1 | 269,925 | 256 | 606.4 | 59.6 | 1.7 | 1.2 | 6.9 | 34.0 | 4.7 | 10.7 | 18.0 | 12.1 | 11.7 | 11.8 |
| 06089 | 39820 | 3 | Shasta .................... | 3,775.5 | 179,027 | 372 | 47.4 | 82.4 | 2.1 | 4.1 | 4.8 | 10.9 | 5.7 | 12.2 | 11.1 | 12.9 | 11.9 | 10.9 |
| 06091 | | 8 | Sierra .................... | 953.2 | 2,920 | 2,968 | 3.1 | 83.8 | 1.1 | 2.9 | 1.5 | 12.8 | 4.7 | 9.4 | 7.1 | 8.0 | 11.1 | 10.9 |
| 06093 | | 6 | Siskiyou .................. | 6,278.8 | 43,245 | 1,113 | 6.9 | 79.0 | 2.5 | 6.7 | 3.3 | 13.4 | 5.0 | 11.7 | 10.0 | 10.0 | 10.6 | 10.4 |
| 06095 | 46700 | 2 | Solano .................... | 821.8 | 446,935 | 158 | 543.8 | 40.8 | 15.9 | 1.4 | 20.5 | 27.8 | 5.9 | 12.3 | 12.0 | 14.6 | 13.2 | 12.0 |
| 06097 | 42220 | 2 | Sonoma .................... | 1,575.6 | 489,819 | 145 | 310.9 | 65.1 | 2.5 | 1.6 | 6.5 | 27.6 | 4.7 | 10.9 | 11.4 | 12.6 | 12.8 | 12.3 |
| 06099 | 33700 | 2 | Stanislaus ................ | 1,496.0 | 550,081 | 126 | 367.7 | 41.8 | 3.6 | 1.2 | 8.1 | 48.3 | 7.0 | 15.3 | 13.9 | 14.5 | 12.9 | 11.5 |
| 06101 | 49700 | 3 | Sutter .................... | 602.7 | 96,385 | 625 | 159.9 | 47.3 | 3.1 | 2.0 | 19.4 | 32.1 | 6.7 | 14.3 | 13.0 | 14.0 | 12.6 | 11.3 |
| 06103 | 39780 | 4 | Tehama .................... | 2,949.1 | 64,494 | 830 | 21.9 | 69.1 | 1.5 | 3.4 | 2.4 | 26.6 | 6.0 | 13.7 | 11.4 | 12.2 | 11.4 | 10.9 |
| 06105 | | 8 | Trinity ................... | 3,179.3 | 12,216 | 2,275 | 3.8 | 85.4 | 1.5 | 7.1 | 3.2 | 7.6 | 4.5 | 9.7 | 8.3 | 8.9 | 10.8 | 11.2 |
| 06107 | 47300 | 2 | Tulare .................... | 4,823.9 | 468,680 | 153 | 97.2 | 28.4 | 1.7 | 1.2 | 4.3 | 66.0 | 7.6 | 17.5 | 15.1 | 14.3 | 12.9 | 10.8 |
| 06109 | 43760 | 4 | Tuolumne .................. | 2,220.9 | 54,515 | 938 | 24.5 | 81.6 | 2.6 | 2.9 | 2.6 | 13.2 | 4.4 | 9.5 | 9.3 | 12.2 | 11.5 | 10.6 |
| 06111 | 37100 | 2 | Ventura ................... | 1,840.8 | 841,387 | 69 | 457.1 | 46.7 | 2.4 | 0.8 | 9.4 | 43.4 | 5.5 | 12.8 | 13.0 | 13.5 | 12.5 | 12.7 |
| 06113 | 40900 | 1 | Yolo ...................... | 1,014.7 | 219,728 | 313 | 216.5 | 49.4 | 3.5 | 1.3 | 18.4 | 32.1 | 5.1 | 11.7 | 23.7 | 13.5 | 12.0 | 10.4 |
| 06115 | 49700 | 1 | Yuba ...................... | 632.0 | 80,160 | 706 | 126.8 | 57.2 | 5.4 | 3.3 | 9.7 | 29.7 | 7.7 | 15.5 | 13.3 | 15.8 | 13.1 | 10.2 |
| 08000 | | 0 | COLORADO ................ | 103,637.5 | 5,807,719 | X | 56.0 | 69.7 | 5.0 | 1.3 | 4.7 | 21.9 | 5.6 | 12.2 | 12.9 | 15.9 | 14.1 | 12.1 |
| 08001 | 19740 | 1 | Adams ..................... | 1,166.7 | 519,883 | 138 | 445.6 | 50.4 | 4.3 | 1.1 | 5.3 | 41.2 | 6.6 | 14.5 | 13.4 | 16.4 | 15.0 | 12.2 |
| 08003 | | 7 | Alamosa ................... | 722.6 | 16,180 | 2,022 | 22.4 | 48.4 | 1.9 | 2.0 | 1.6 | 47.8 | 6.2 | 14.2 | 19.5 | 13.9 | 11.6 | 9.3 |
| 08005 | 19740 | 1 | Arapahoe .................. | 797.9 | 657,452 | 103 | 824.0 | 61.8 | 12.5 | 1.1 | 8.2 | 20.0 | 5.9 | 12.9 | 12.1 | 15.9 | 14.6 | 12.7 |
| 08007 | | 7 | Archuleta ................. | 1,350.1 | 14,196 | 2,144 | 10.5 | 78.6 | 1.1 | 2.5 | 1.5 | 18.2 | 4.1 | 10.3 | 8.4 | 9.2 | 11.4 | 11.0 |
| 08009 | | 9 | Baca ...................... | 2,555.0 | 3,555 | 2,930 | 1.4 | 85.3 | 1.7 | 2.5 | 0.8 | 11.9 | 5.4 | 13.1 | 10.2 | 10.3 | 10.9 | 9.9 |
| 08011 | | 7 | Bent ...................... | 1,512.8 | 5,356 | 2,797 | 3.5 | 57.8 | 7.0 | 1.8 | 1.6 | 32.6 | 3.8 | 8.9 | 11.2 | 15.7 | 14.3 | 13.5 |

1. CBSA = Core Based Statistical Area. See Appendix A for explanation. See Appendix B for list of metropolitan areas with component counties.   2. County type code from the Economic Research Service of USDA Rural-Urban Continuum Codes. See Appendix A for definition.   3. Dry land or land partially or temporarily covered by water.   4. May be of any race.

# Table B. States and Counties — **Population and Households**

| STATE County | 55 to 64 years | 65 to 74 years | 75 years and over | Percent female | Total persons 2000 | Total persons 2010 | Percent change 2000-2010 | Percent change 2010-2020 | Births | Deaths | Net Migration | Number | Persons per household | Family households | Female family householder[1] | One person |
|---|---|---|---|---|---|---|---|---|---|---|---|---|---|---|---|---|
| | 16 | 17 | 18 | 19 | 20 | 21 | 22 | 23 | 24 | 25 | 26 | 27 | 28 | 29 | 30 | 31 |
| CALIFORNIA | 12.1 | 8.8 | 6.4 | 50.3 | 33,871,648 | 37,254,522 | 10.0 | 5.7 | 4,963,885 | 2,658,968 | -177,201 | 13,044,266 | 2.95 | 68.7 | 13.0 | 23.8 |
| Alameda | 12.1 | 8.7 | 6.1 | 50.8 | 1,443,741 | 1,510,283 | 4.6 | 10.1 | 195,466 | 100,835 | 57,544 | 577,177 | 2.82 | 66.6 | 11.1 | 24.4 |
| Alpine | 16.6 | 18.1 | 7.6 | 47.7 | 1,208 | 1,175 | -2.7 | -4.8 | 68 | 109 | -19 | 350 | 2.87 | 66.0 | 7.7 | 28.9 |
| Amador | 15.9 | 16.9 | 11.1 | 45.7 | 35,100 | 38,091 | 8.5 | 5.2 | 3,000 | 4,503 | 3,452 | 14,594 | 2.38 | 67.6 | 7.7 | 27.5 |
| Butte | 12.1 | 10.9 | 7.5 | 50.5 | 203,171 | 220,005 | 8.3 | -3.3 | 24,618 | 23,905 | -7,835 | 85,320 | 2.57 | 60.3 | 11.5 | 27.6 |
| Calaveras | 17.6 | 17.4 | 11.5 | 50.1 | 40,554 | 45,577 | 12.4 | 1.6 | 3,762 | 5,117 | 2,115 | 16,942 | 2.66 | 69.0 | 10.1 | 25.5 |
| Colusa | 11.7 | 9.0 | 6.4 | 49.1 | 18,804 | 21,407 | 13.8 | 0.7 | 3,030 | 1,523 | -1,364 | 7,227 | 2.94 | 73.7 | 11.4 | 21.5 |
| Contra Costa | 13.2 | 9.7 | 7.0 | 51.1 | 948,816 | 1,049,188 | 10.6 | 9.8 | 124,894 | 79,024 | 57,637 | 394,769 | 2.87 | 72.1 | 12.2 | 21.8 |
| Del Norte | 14.0 | 11.9 | 7.1 | 45.4 | 27,507 | 28,610 | 4.0 | -2.2 | 3,111 | 2,968 | -827 | 9,945 | 2.47 | 61.7 | 13.1 | 31.4 |
| El Dorado | 16.6 | 13.9 | 8.8 | 50.3 | 156,299 | 181,079 | 15.9 | 6.5 | 16,332 | 15,422 | 11,054 | 70,974 | 2.63 | 70.1 | 7.9 | 23.7 |
| Fresno | 10.3 | 7.6 | 5.3 | 50.1 | 799,407 | 930,503 | 16.4 | 7.6 | 156,504 | 69,429 | -16,383 | 307,906 | 3.14 | 72.2 | 17.6 | 22.2 |
| Glenn | 12.4 | 9.6 | 7.0 | 49.1 | 26,453 | 28,121 | 6.3 | 0.6 | 3,987 | 2,561 | -1,273 | 10,085 | 2.74 | 73.3 | 12.3 | 21.3 |
| Humboldt | 12.6 | 12.1 | 6.9 | 50.5 | 126,518 | 134,613 | 6.4 | 0.3 | 14,896 | 13,697 | -800 | 54,679 | 2.41 | 55.2 | 10.7 | 32.1 |
| Imperial | 10.2 | 7.7 | 6.0 | 48.8 | 142,361 | 174,522 | 22.6 | 3.3 | 30,275 | 10,616 | -14,046 | 44,829 | 3.81 | 76.2 | 19.1 | 20.6 |
| Inyo | 15.1 | 13.7 | 10.6 | 49.9 | 17,945 | 18,533 | 3.3 | -2.6 | 2,055 | 2,041 | -495 | 7,950 | 2.18 | 54.0 | 9.8 | 39.9 |
| Kern | 10.4 | 7.0 | 4.5 | 48.9 | 661,645 | 839,619 | 26.9 | 7.4 | 140,814 | 60,952 | -17,936 | 270,282 | 3.17 | 74.2 | 16.2 | 20.7 |
| Kings | 9.5 | 6.3 | 4.6 | 45.0 | 129,461 | 152,990 | 18.2 | -0.2 | 23,918 | 8,594 | -15,992 | 43,452 | 3.13 | 76.9 | 17.2 | 17.6 |
| Lake | 15.2 | 14.2 | 9.3 | 50.1 | 58,309 | 64,662 | 10.9 | -0.3 | 7,553 | 8,751 | 1,058 | 25,660 | 2.46 | 59.2 | 11.5 | 32.4 |
| Lassen | 11.6 | 10.1 | 6.2 | 38.6 | 33,828 | 34,895 | 3.2 | -14.0 | 3,095 | 2,302 | -5,792 | 9,280 | 2.26 | 66.5 | 10.3 | 28.2 |
| Los Angeles | 12.1 | 8.3 | 6.2 | 50.7 | 9,519,338 | 9,818,646 | 3.1 | 1.3 | 1,256,001 | 639,278 | -489,743 | 3,316,795 | 2.99 | 66.7 | 14.7 | 25.7 |
| Madera | 11.0 | 8.8 | 5.8 | 51.6 | 123,109 | 150,834 | 22.5 | 4.6 | 22,963 | 11,118 | -4,992 | 44,881 | 3.28 | 78.5 | 13.8 | 17.5 |
| Marin | 15.0 | 13.3 | 10.3 | 51.1 | 247,289 | 252,425 | 2.1 | 1.9 | 23,233 | 19,889 | 1,714 | 105,432 | 2.40 | 62.6 | 7.7 | 29.9 |
| Mariposa | 17.2 | 16.6 | 12.1 | 49.5 | 17,130 | 18,271 | 6.7 | -6.1 | 1,481 | 1,904 | -683 | 7,643 | 2.13 | 62.8 | 9.1 | 31.4 |
| Mendocino | 13.5 | 14.9 | 8.9 | 50.4 | 86,265 | 87,850 | 1.8 | -2.0 | 10,379 | 8,825 | -3,275 | 34,408 | 2.47 | 60.8 | 12.3 | 30.9 |
| Merced | 10.1 | 6.9 | 4.8 | 49.5 | 210,554 | 255,797 | 21.5 | 9.2 | 42,237 | 17,487 | -1,251 | 80,008 | 3.32 | 76.3 | 16.6 | 18.6 |
| Modoc | 15.4 | 16.6 | 12.1 | 49.8 | 9,449 | 9,686 | 2.5 | -9.5 | 849 | 1,062 | -710 | 3,616 | 2.33 | 58.9 | 7.5 | 35.0 |
| Mono | 15.4 | 11.2 | 6.0 | 46.8 | 12,853 | 14,204 | 10.5 | 2.3 | 1,432 | 484 | -639 | 4,765 | 2.93 | 52.5 | 6.2 | 34.9 |
| Monterey | 11.0 | 8.5 | 6.0 | 49.2 | 401,762 | 415,058 | 3.3 | 3.8 | 64,707 | 25,951 | -22,977 | 127,155 | 3.27 | 72.2 | 13.3 | 21.6 |
| Napa | 13.6 | 11.4 | 8.9 | 50.2 | 124,279 | 136,535 | 9.9 | -0.4 | 14,281 | 12,639 | -2,103 | 48,705 | 2.78 | 68.0 | 9.9 | 25.9 |
| Nevada | 16.3 | 17.7 | 11.2 | 50.9 | 92,033 | 98,749 | 7.3 | 0.9 | 8,242 | 10,248 | 2,955 | 40,855 | 2.40 | 64.0 | 8.8 | 29.1 |
| Orange | 13.0 | 8.9 | 6.8 | 50.7 | 2,846,289 | 3,010,258 | 5.8 | 5.2 | 382,227 | 200,199 | -23,093 | 1,037,492 | 3.01 | 71.7 | 11.5 | 21.1 |
| Placer | 13.3 | 11.4 | 8.9 | 51.2 | 248,399 | 348,481 | 40.3 | 15.6 | 38,055 | 32,025 | 48,512 | 142,855 | 2.67 | 70.3 | 8.6 | 24.1 |
| Plumas | 17.0 | 18.9 | 11.6 | 50.2 | 20,824 | 20,007 | -3.9 | -5.2 | 1,640 | 2,247 | -439 | 8,450 | 2.16 | 59.6 | 8.9 | 32.0 |
| Riverside | 11.6 | 8.6 | 6.5 | 50.1 | 1,545,387 | 2,189,785 | 41.7 | 13.7 | 306,180 | 165,842 | 159,537 | 724,893 | 3.28 | 72.7 | 13.0 | 22.0 |
| Sacramento | 12.1 | 8.9 | 5.9 | 51.1 | 1,223,499 | 1,418,736 | 16.0 | 9.9 | 199,908 | 116,032 | 57,463 | 543,025 | 2.76 | 66.0 | 14.1 | 26.2 |
| San Benito | 12.0 | 8.2 | 5.3 | 49.7 | 53,234 | 55,265 | 3.8 | 15.9 | 7,706 | 3,291 | 4,401 | 18,135 | 3.31 | 80.4 | 13.5 | 14.4 |
| San Bernardino | 11.3 | 7.5 | 4.7 | 50.2 | 1,709,434 | 2,035,172 | 19.1 | 7.6 | 309,050 | 140,804 | -14,352 | 636,041 | 3.29 | 76.5 | 16.9 | 18.7 |
| San Diego | 11.8 | 8.7 | 6.2 | 49.6 | 2,813,833 | 3,095,352 | 10.0 | 7.7 | 436,256 | 217,725 | 21,049 | 1,125,286 | 2.87 | 67.3 | 11.7 | 23.9 |
| San Francisco | 11.5 | 9.3 | 7.3 | 49.0 | 776,733 | 805,184 | 3.7 | 7.6 | 90,880 | 59,821 | 30,059 | 362,354 | 2.36 | 47.9 | 7.9 | 35.6 |
| San Joaquin | 11.2 | 7.9 | 5.4 | 50.1 | 563,598 | 685,298 | 21.6 | 12.1 | 103,227 | 55,216 | 34,946 | 228,567 | 3.17 | 74.6 | 15.1 | 20.3 |
| San Luis Obispo | 13.1 | 12.7 | 8.7 | 49.5 | 246,681 | 269,599 | 9.3 | 4.7 | 26,424 | 24,329 | 10,603 | 105,981 | 2.51 | 63.4 | 8.5 | 26.0 |
| San Mateo | 13.0 | 9.6 | 7.5 | 50.4 | 707,161 | 718,528 | 1.6 | 5.5 | 90,332 | 49,160 | -1,021 | 263,543 | 2.87 | 70.1 | 10.1 | 22.3 |
| Santa Barbara | 11.1 | 8.9 | 7.1 | 50.0 | 399,347 | 423,947 | 6.2 | 4.9 | 57,494 | 31,893 | -4,565 | 145,856 | 2.91 | 65.8 | 11.2 | 23.7 |
| Santa Clara | 12.0 | 7.9 | 6.4 | 49.3 | 1,682,585 | 1,781,680 | 5.9 | 7.0 | 234,493 | 102,786 | -5,084 | 640,215 | 2.95 | 71.8 | 9.8 | 20.3 |
| Santa Cruz | 13.0 | 11.4 | 6.6 | 50.6 | 255,602 | 262,345 | 2.6 | 2.9 | 29,038 | 18,157 | -3,207 | 95,818 | 2.72 | 63.3 | 10.3 | 26.2 |
| Shasta | 14.1 | 12.5 | 8.8 | 50.9 | 163,256 | 177,222 | 8.6 | 1.0 | 20,960 | 22,660 | 3,697 | 71,181 | 2.47 | 65.1 | 11.3 | 27.8 |
| Sierra | 17.7 | 19.2 | 11.9 | 49.2 | 3,555 | 3,239 | -8.9 | -9.8 | 239 | 366 | -192 | 1,319 | 2.28 | 63.5 | 6.1 | 33.4 |
| Siskiyou | 15.3 | 16.4 | 10.4 | 50.3 | 44,301 | 44,899 | 1.3 | -3.7 | 4,553 | 5,744 | -452 | 19,240 | 2.23 | 59.2 | 9.3 | 33.4 |
| Solano | 13.3 | 10.3 | 6.5 | 50.2 | 394,542 | 413,343 | 4.8 | 8.1 | 52,637 | 33,374 | 14,593 | 149,865 | 2.88 | 71.7 | 14.2 | 22.2 |
| Sonoma | 14.1 | 12.8 | 8.4 | 51.2 | 458,614 | 483,866 | 5.5 | 1.2 | 50,154 | 42,154 | -1,784 | 189,374 | 2.59 | 63.6 | 10.1 | 27.5 |
| Stanislaus | 11.2 | 8.0 | 5.6 | 50.4 | 446,997 | 514,451 | 15.1 | 6.9 | 77,693 | 42,529 | 729 | 173,898 | 3.09 | 74.1 | 14.8 | 20.5 |
| Sutter | 12.0 | 9.1 | 7.1 | 50.2 | 78,930 | 94,751 | 20.0 | 1.7 | 13,256 | 8,230 | -3,431 | 32,636 | 2.91 | 73.0 | 13.3 | 21.7 |
| Tehama | 14.1 | 11.6 | 8.7 | 50.4 | 56,039 | 63,441 | 13.2 | 1.7 | 7,821 | 6,703 | -18 | 24,189 | 2.61 | 66.6 | 12.4 | 26.6 |
| Trinity | 17.2 | 18.5 | 10.8 | 48.8 | 13,022 | 13,782 | 5.8 | -11.4 | 1,129 | 1,661 | -1,028 | 5,939 | 2.09 | 57.7 | 10.0 | 34.1 |
| Tulare | 9.9 | 7.0 | 4.8 | 50.1 | 368,021 | 442,176 | 20.1 | 6.0 | 75,989 | 30,822 | -18,658 | 138,238 | 3.30 | 77.0 | 16.4 | 18.4 |
| Tuolumne | 15.2 | 16.2 | 11.2 | 47.9 | 54,501 | 55,344 | 1.5 | -1.5 | 4,727 | 6,688 | 1,151 | 22,502 | 2.24 | 61.1 | 8.4 | 31.9 |
| Ventura | 13.3 | 9.6 | 7.1 | 50.5 | 753,197 | 823,446 | 9.3 | 2.2 | 101,757 | 57,839 | -25,563 | 271,040 | 3.08 | 72.5 | 11.9 | 21.7 |
| Yolo | 10.2 | 7.8 | 5.5 | 51.6 | 168,660 | 200,855 | 19.1 | 9.4 | 23,815 | 13,254 | 8,331 | 74,296 | 2.81 | 63.4 | 10.7 | 22.6 |
| Yuba | 11.1 | 8.4 | 4.9 | 49.4 | 60,219 | 72,142 | 19.8 | 11.1 | 12,192 | 6,183 | 2,041 | 26,354 | 2.84 | 69.8 | 12.1 | 22.9 |
| COLORADO | 12.2 | 9.4 | 5.7 | 49.6 | 4,301,261 | 5,029,319 | 16.9 | 15.5 | 666,380 | 370,913 | 477,921 | 2,148,994 | 2.56 | 63.7 | 9.4 | 27.4 |
| Adams | 10.8 | 7.0 | 4.0 | 49.5 | 348,618 | 441,701 | 26.7 | 17.7 | 72,249 | 30,148 | 35,594 | 166,450 | 3.00 | 70.3 | 12.9 | 22.3 |
| Alamosa | 10.7 | 8.9 | 5.8 | 50.6 | 14,966 | 15,440 | 3.2 | 4.8 | 2,166 | 1,381 | -43 | 6,162 | 2.44 | 61.1 | 9.7 | 30.5 |
| Arapahoe | 11.9 | 8.7 | 5.3 | 50.3 | 487,967 | 572,477 | 17.3 | 14.8 | 80,049 | 39,016 | 43,713 | 240,304 | 2.66 | 65.6 | 10.7 | 27.1 |
| Archuleta | 18.0 | 18.8 | 8.7 | 50.0 | 9,898 | 12,084 | 22.1 | 17.5 | 1,194 | 962 | 1,868 | 5,858 | 2.23 | 65.5 | 5.2 | 26.9 |
| Baca | 14.2 | 13.1 | 12.9 | 49.9 | 4,517 | 3,787 | -16.2 | -6.1 | 418 | 564 | -87 | 1,662 | 2.09 | 64.7 | 8.4 | 31.6 |
| Bent | 12.5 | 11.1 | 8.9 | 36.5 | 5,998 | 6,499 | 8.4 | -17.6 | 410 | 627 | -963 | 1,767 | 2.23 | 63.3 | 11.4 | 30.7 |

1. No spouse present.

# Table B. States and Counties — **Population, Vital Statistics, Health, and Crime**

| STATE County | Persons in group quarters, 2020 | Daytime Population, 2015-2019 Number | Employment/ residence ratio | Births, 2020 Total | Rate[1] | Deaths, 2020 Number | Rate[1] | Persons under 65 with no health insurance, 2019 Number | Percent | Medicare, 2020 Total beneficiaries | Enrolled in Original Medicare | Enrolled in Medicare Advantage | Crimes reported by county police or sheriff, 2019 Violent | Property |
|---|---|---|---|---|---|---|---|---|---|---|---|---|---|---|
| | 32 | 33 | 34 | 35 | 36 | 37 | 38 | 39 | 40 | 41 | 42 | 43 | 44 | 45 |
| CALIFORNIA ................... | 829,974 | 39,286,280 | 1.00 | 448,160 | 11.4 | 304,961 | 7.7 | 2,947,957 | 8.9 | 6,406,280 | 3,499,131 | 2,907,149 | X | X |
| Alameda......................... | 36,710 | 1,629,986 | 0.97 | 18,186 | 10.9 | 11,948 | 7.2 | 72,422 | 5.1 | 247,693 | 135,017 | 112,676 | 633 | 2,169 |
| Alpine.......................... | 24 | 1,458 | 2.27 | 10 | 8.9 | 13 | 11.6 | 72 | 8.7 | 265 | 248 | 17 | 13 | 29 |
| Amador......................... | 4,390 | 37,925 | 0.96 | 314 | 7.8 | 494 | 12.3 | 1,851 | 7.5 | 11,753 | 8,920 | 2,834 | 66 | 416 |
| Butte........................... | 5,443 | 225,363 | 1.00 | 2,236 | 10.5 | 2,371 | 11.1 | 13,710 | 7.8 | 45,888 | 42,635 | 3,253 | 283 | 1,290 |
| Calaveras...................... | 493 | 39,562 | 0.65 | 390 | 8.4 | 569 | 12.3 | 2,579 | 7.9 | 13,379 | 12,147 | 1,233 | 139 | 668 |
| Colusa......................... | 262 | 21,399 | 0.99 | 260 | 12.1 | 134 | 6.2 | 2,195 | 12.2 | 3,792 | 3,529 | 263 | 26 | 169 |
| Contra Costa................... | 10,284 | 1,000,874 | 0.74 | 11,771 | 10.2 | 9,228 | 8.0 | 62,447 | 6.5 | 202,459 | 103,340 | 99,118 | 332 | 1,499 |
| Del Norte...................... | 3,082 | 28,141 | 1.07 | 267 | 9.5 | 306 | 10.9 | 1,652 | 8.4 | 6,358 | 6,181 | 178 | 84 | 424 |
| El Dorado...................... | 1,643 | 165,160 | 0.72 | 1,608 | 8.3 | 1,846 | 9.6 | 9,349 | 6.2 | 45,839 | 29,073 | 16,766 | 173 | 1,695 |
| Fresno......................... | 17,672 | 985,853 | 1.00 | 14,251 | 14.2 | 7,941 | 7.9 | 82,348 | 9.7 | 145,804 | 93,653 | 52,151 | 1,082 | 3,281 |
| Glenn .......................... | 319 | 27,156 | 0.93 | 385 | 13.6 | 249 | 8.8 | 2,905 | 12.4 | 5,754 | 5,492 | 262 | 55 | 362 |
| Humboldt....................... | 5,258 | 136,162 | 1.00 | 1,370 | 10.1 | 1,523 | 11.3 | 10,146 | 9.5 | 29,511 | 28,269 | 1,242 | 310 | 1,092 |
| Imperial........................ | 8,535 | 178,769 | 0.97 | 2,541 | 14.1 | 1,120 | 6.2 | 14,070 | 9.6 | 32,249 | 24,264 | 7,985 | 211 | 588 |
| Inyo............................ | 410 | 18,306 | 1.04 | 181 | 10.0 | 192 | 10.6 | 1,234 | 9.1 | 4,548 | 4,238 | 310 | 60 | 192 |
| Kern............................ | 30,547 | 891,435 | 1.01 | 12,826 | 14.2 | 6,980 | 7.7 | 70,588 | 9.3 | 120,883 | 71,492 | 49,391 | 2,864 | 9,449 |
| Kings........................... | 15,835 | 151,285 | 1.01 | 2,184 | 14.3 | 986 | 6.5 | 11,212 | 9.4 | 17,921 | 13,576 | 4,345 | 125 | 432 |
| Lake............................ | 1,101 | 59,826 | 0.81 | 732 | 11.4 | 917 | 14.2 | 4,557 | 9.4 | 17,316 | 16,272 | 1,044 | 193 | 435 |
| Lassen ......................... | 7,045 | 32,068 | 1.15 | 291 | 9.7 | 208 | 6.9 | 914 | 5.0 | 5,239 | 5,085 | 154 | 78 | 172 |
| Los Angeles ................... | 180,231 | 10,210,030 | 1.03 | 107,789 | 10.8 | 73,513 | 7.4 | 940,376 | 11.1 | 1,523,583 | 735,371 | 788,212 | 5,564 | 15,040 |
| Madera......................... | 7,131 | 149,672 | 0.90 | 2,042 | 12.9 | 1,178 | 7.5 | 13,807 | 11.0 | 24,911 | 15,639 | 9,272 | 457 | 1,017 |
| Marin........................... | 8,416 | 256,349 | 0.97 | 2,065 | 8.0 | 2,183 | 8.5 | 10,406 | 5.3 | 59,640 | 34,800 | 24,841 | 63 | 738 |
| Mariposa....................... | 726 | 17,952 | 1.08 | 136 | 7.9 | 185 | 10.8 | 1,190 | 9.9 | 4,947 | 4,536 | 411 | 79 | 171 |
| Mendocino ..................... | 1,970 | 87,721 | 1.01 | 890 | 10.3 | 979 | 11.4 | 6,908 | 10.6 | 23,347 | 21,226 | 2,121 | 178 | 414 |
| Merced ......................... | 6,943 | 254,066 | 0.83 | 3,855 | 13.8 | 1,922 | 6.9 | 26,939 | 11.4 | 37,596 | 31,828 | 5,768 | 500 | 1,515 |
| Modoc.......................... | 333 | 8,677 | 0.92 | 68 | 7.8 | 104 | 11.9 | 709 | 11.5 | 2,568 | 2,496 | 72 | 31 | 29 |
| Mono........................... | 337 | 14,521 | 1.03 | 119 | 8.2 | 42 | 2.9 | 1,258 | 10.6 | 2,017 | 1,902 | 115 | 25 | 53 |
| Monterey....................... | 17,092 | 429,449 | 0.98 | 5,799 | 13.5 | 2,892 | 6.7 | 42,273 | 11.9 | 66,756 | 59,648 | 7,108 | 258 | 1,125 |
| Napa........................... | 5,079 | 149,335 | 1.14 | 1,226 | 9.0 | 1,415 | 10.4 | 9,391 | 8.7 | 29,865 | 17,885 | 11,980 | 69 | 269 |
| Nevada......................... | 1,172 | 95,772 | 0.92 | 790 | 7.9 | 1,065 | 10.7 | 4,976 | 7.0 | 29,958 | 24,696 | 5,263 | 118 | 424 |
| Orange......................... | 43,650 | 3,269,476 | 1.06 | 35,163 | 11.1 | 23,233 | 7.3 | 235,378 | 8.8 | 519,799 | 249,496 | 270,303 | 247 | 881 |
| Placer.......................... | 3,932 | 389,983 | 1.03 | 3,678 | 9.1 | 3,692 | 9.2 | 15,593 | 4.9 | 87,193 | 43,225 | 43,968 | 183 | 1,077 |
| Plumas......................... | 280 | 19,015 | 1.05 | 154 | 8.1 | 212 | 11.2 | 886 | 6.7 | 6,361 | 6,010 | 352 | 128 | 294 |
| Riverside....................... | 34,027 | 2,237,528 | 0.83 | 28,383 | 11.4 | 19,696 | 7.9 | 203,605 | 9.8 | 388,660 | 160,703 | 227,957 | 932 | 7,220 |
| Sacramento.................... | 23,150 | 1,520,068 | 0.99 | 19,084 | 12.2 | 13,340 | 8.6 | 89,530 | 6.8 | 264,354 | 126,787 | 137,566 | 2,408 | 8,067 |
| San Benito..................... | 397 | 49,189 | 0.61 | 808 | 12.6 | 383 | 6.0 | 4,869 | 9.0 | 9,249 | 7,859 | 1,390 | 33 | 152 |
| San Bernardino................ | 38,664 | 2,079,740 | 0.92 | 28,808 | 13.2 | 16,533 | 7.6 | 184,915 | 9.9 | 306,339 | 118,403 | 187,936 | 1,503 | 3,945 |
| San Diego...................... | 110,390 | 3,348,048 | 1.02 | 39,080 | 11.7 | 25,040 | 7.5 | 247,271 | 8.9 | 544,476 | 274,208 | 270,268 | 1,499 | 4,300 |
| San Francisco.................. | 25,704 | 1,118,860 | 1.48 | 8,547 | 9.9 | 6,898 | 8.0 | 37,808 | 5.2 | 149,174 | 85,542 | 63,632 | NA | NA |
| San Joaquin.................... | 15,882 | 706,150 | 0.88 | 9,957 | 13.0 | 6,339 | 8.3 | 53,036 | 8.2 | 114,638 | 64,537 | 50,101 | 903 | 2,439 |
| San Luis Obispo................ | 16,154 | 281,867 | 1.00 | 2,423 | 8.6 | 2,641 | 9.4 | 14,763 | 7.0 | 65,649 | 53,715 | 11,934 | 181 | 963 |
| San Mateo...................... | 9,323 | 777,997 | 1.03 | 8,226 | 10.8 | 5,556 | 7.3 | 37,100 | 5.8 | 128,316 | 68,619 | 59,697 | 324 | 3,245 |
| Santa Barbara................. | 19,942 | 456,816 | 1.06 | 5,390 | 12.1 | 3,550 | 8.0 | 43,058 | 12.0 | 78,006 | 64,839 | 13,167 | 232 | 1,257 |
| Santa Clara.................... | 30,408 | 2,070,092 | 1.15 | 21,070 | 11.0 | 12,110 | 6.3 | 91,263 | 5.5 | 273,330 | 148,270 | 125,060 | 356 | 1,518 |
| Santa Cruz..................... | 13,687 | 261,635 | 0.91 | 2,402 | 8.9 | 2,096 | 7.8 | 17,026 | 7.9 | 52,198 | 43,837 | 8,362 | 331 | 1,630 |
| Shasta.......................... | 2,924 | 179,780 | 1.01 | 1,918 | 10.7 | 2,381 | 13.3 | 11,387 | 8.1 | 47,823 | 44,986 | 2,838 | 386 | 824 |
| Sierra.......................... | 33 | 2,646 | 0.67 | 18 | 6.2 | 40 | 13.7 | 144 | 7.1 | 906 | 869 | 37 | 1 | 31 |
| Siskiyou........................ | 596 | 44,213 | 1.05 | 383 | 8.9 | 607 | 14.0 | 2,910 | 9.3 | 13,715 | 13,326 | 389 | 70 | 146 |
| Solano.......................... | 11,825 | 388,545 | 0.74 | 5,023 | 11.2 | 3,953 | 8.8 | 21,935 | 6.0 | 81,509 | 43,610 | 37,899 | 109 | 377 |
| Sonoma......................... | 10,527 | 485,255 | 0.94 | 4,353 | 8.9 | 4,751 | 9.7 | 30,207 | 7.8 | 108,116 | 56,246 | 51,870 | 497 | 1,089 |
| Stanislaus...................... | 6,536 | 517,859 | 0.89 | 7,335 | 13.3 | 4,676 | 8.5 | 39,386 | 8.4 | 88,425 | 45,388 | 43,037 | 370 | 1,602 |
| Sutter.......................... | 736 | 90,259 | 0.85 | 1,244 | 12.9 | 899 | 9.3 | 7,213 | 8.9 | 17,814 | 16,607 | 1,207 | 77 | 576 |
| Tehama......................... | 795 | 61,133 | 0.88 | 765 | 11.9 | 735 | 11.4 | 4,448 | 8.7 | 15,149 | 14,191 | 958 | 95 | 472 |
| Trinity.......................... | 375 | 13,019 | 1.07 | 96 | 7.9 | 180 | 14.7 | 867 | 10.1 | 3,550 | 3,438 | 111 | 61 | 190 |
| Tulare.......................... | 4,977 | 452,618 | 0.95 | 6,787 | 14.5 | 3,404 | 7.3 | 38,062 | 9.5 | 64,059 | 48,976 | 15,083 | 334 | 1,893 |
| Tuolumne....................... | 3,941 | 53,994 | 1.00 | 457 | 8.4 | 679 | 12.5 | 2,498 | 6.8 | 16,030 | 14,871 | 1,159 | 169 | 669 |
| Ventura ........................ | 11,350 | 803,045 | 0.89 | 8,820 | 10.5 | 6,741 | 8.0 | 72,745 | 10.4 | 153,894 | 97,896 | 55,998 | 159 | 543 |
| Yolo............................ | 10,093 | 233,142 | 1.16 | 2,068 | 9.4 | 1,489 | 6.8 | 12,146 | 6.7 | 32,665 | 17,797 | 14,868 | 69 | 294 |
| Yuba........................... | 1,193 | 70,036 | 0.79 | 1,138 | 14.2 | 604 | 7.5 | 5,424 | 8.1 | 13,032 | 11,412 | 1,620 | 204 | 847 |
| COLORADO ................... | 117,920 | 5,601,991 | 1.00 | 63,285 | 10.9 | 42,735 | 7.4 | 453,100 | 9.4 | 938,457 | 527,427 | 411,031 | X | X |
| Adams.......................... | 4,035 | 446,083 | 0.78 | 6,565 | 12.6 | 3,458 | 6.7 | 56,098 | 12.3 | 63,404 | 27,326 | 36,078 | 513 | 2,845 |
| Alamosa ........................ | 852 | 17,259 | 1.17 | 191 | 11.8 | 127 | 7.8 | 1,686 | 13.1 | 3,027 | 2,157 | 870 | NA | NA |
| Arapahoe ...................... | 5,061 | 639,867 | 0.99 | 7,501 | 11.4 | 4,531 | 6.9 | 54,044 | 9.6 | 95,381 | 48,022 | 47,359 | NA | NA |
| Archuleta....................... | 129 | 12,861 | 0.93 | 109 | 7.7 | 103 | 7.3 | 1,386 | 13.8 | 4,103 | 3,278 | 825 | 46 | 125 |
| Baca........................... | 82 | 3,506 | 0.97 | 39 | 11.0 | 49 | 13.8 | 413 | 15.8 | 961 | 905 | 56 | 3 | 2 |
| Bent........................... | 1,518 | 5,724 | 0.96 | 41 | 7.7 | 54 | 10.1 | 315 | 10.9 | 1,109 | 1,034 | 75 | 17 | 166 |

1. Per 1,000 estimated resident population.

# Table B. States and Counties — Crime, Education, Money Income, and Poverty

| | County law enforcement employment, 2019 | | School enrollment and attainment, 2015-2019 | | | | Local government expenditures,[3] 2017-2018 | | Money income, 2015-2019 | | | | Income and poverty, 2019 | | | |
| | | | Enrollment[1] | | Attainment[2] (percent) | | | | | | Households | | | Percent below poverty level | | |
| | | | | | | | | | | | | Percent | | | | |
| STATE County | Officers | Civilians | Total | Percent private | High school graduate or less | Bachelor's degree or more | Total current spending (mil dol) | Current spending per student (dollars) | Per capita income[4] | Median income (dollars) | with income of less than $50,000 | with income of $200,000 or more | Median household income (dollars) | All persons | Children under 18 years | Children 5 to 17 years in families |
| | 46 | 47 | 48 | 49 | 50 | 51 | 52 | 53 | 54 | 55 | 56 | 57 | 58 | 59 | 60 | 61 |
| CALIFORNIA .................... | X | X | 10,423,307 | 13.9 | 37.2 | 33.9 | 77,639 | 12,510 | 36,955 | 75,235 | 34.4 | 12.2 | 80,423 | 11.8 | 15.6 | 15.2 |
| Alameda......... | 950 | 546 | 414,966 | 14.7 | 29.1 | 47.4 | 2,674 | 11,711 | 47,314 | 99,406 | 26.2 | 19.4 | 107,589 | 8.9 | 9.8 | 9.2 |
| Alpine.......... | 13 | 3 | 239 | 15.9 | 36.9 | 34.5 | 4 | 48,263 | 36,739 | 63,750 | 38.3 | 8.9 | 58,112 | 17.2 | 29.6 | 25.5 |
| Amador.......... | 49 | 45 | 6,856 | 12.6 | 37.5 | 19.3 | 50 | 11,937 | 31,987 | 62,772 | 39.5 | 5.2 | 62,640 | 9.8 | 12.5 | 12.0 |
| Butte............ | 101 | 151 | 64,726 | 8.5 | 33.3 | 27.2 | 394 | 12,417 | 29,506 | 52,537 | 48.0 | 5.6 | 58,394 | 16.1 | 16.1 | 15.1 |
| Calaveras........ | 58 | 55 | 8,130 | 9.7 | 37.8 | 18.3 | 78 | 14,237 | 33,456 | 63,158 | 38.4 | 5.6 | 68,248 | 12.1 | 19.6 | 19.1 |
| Colusa........... | 40 | 19 | 5,717 | 3.7 | 55.0 | 15.0 | 65 | 13,701 | 26,932 | 59,401 | 42.8 | 3.9 | 59,048 | 12.0 | 15.3 | 13.7 |
| Contra Costa..... | 612 | 316 | 295,019 | 15.7 | 27.8 | 42.4 | 2,129 | 11,956 | 48,178 | 99,716 | 24.6 | 19.4 | 106,555 | 7.9 | 9.9 | 9.3 |
| Del Norte........ | 21 | 31 | 5,889 | 9.3 | 51.0 | 14.8 | 54 | 12,752 | 23,674 | 45,283 | 52.7 | 3.4 | 48,979 | 17.9 | 25.8 | 23.8 |
| El Dorado........ | 154 | 190 | 41,882 | 12.1 | 27.9 | 34.3 | 328 | 11,770 | 42,749 | 83,377 | 30.3 | 13.0 | 86,202 | 8.7 | 9.1 | 7.8 |
| Fresno........... | 414 | 768 | 294,428 | 7.8 | 46.8 | 21.2 | 2,655 | 13,178 | 24,422 | 53,969 | 46.8 | 4.9 | 56,926 | 20.5 | 29.0 | 28.8 |
| Glenn............ | 33 | 40 | 6,877 | 9.5 | 52.4 | 14.2 | 72 | 13,338 | 22,668 | 49,633 | 50.3 | 2.7 | 55,682 | 12.1 | 18.1 | 17.4 |
| Humboldt......... | 187 | 58 | 34,601 | 8.2 | 31.7 | 30.4 | 253 | 14,019 | 28,769 | 48,041 | 51.6 | 4.1 | 51,134 | 19.1 | 20.5 | 20.8 |
| Imperial......... | 187 | 85 | 55,145 | 5.2 | 54.6 | 15.2 | 532 | 14,102 | 18,018 | 47,622 | 52.5 | 2.9 | 48,102 | 22.0 | 30.7 | 28.5 |
| Inyo............. | 38 | 40 | 3,491 | 9.8 | 40.0 | 27.2 | 69 | 15,237 | 32,590 | 57,316 | 44.2 | 3.2 | 59,605 | 11.7 | 15.4 | 14.5 |
| Kern............. | 779 | 376 | 258,576 | 9.2 | 53.7 | 16.4 | 2,373 | 12,490 | 23,326 | 53,350 | 47.1 | 4.4 | 53,245 | 19.0 | 26.0 | 26.3 |
| Kings............ | 85 | 204 | 44,298 | 13.5 | 51.9 | 14.7 | 367 | 13,899 | 22,373 | 57,848 | 43.7 | 3.7 | 57,297 | 16.0 | 22.3 | 20.5 |
| Lake............. | 91 | 31 | 13,658 | 9.8 | 44.7 | 15.5 | 127 | 13,368 | 27,362 | 47,040 | 53.7 | 3.2 | 46,897 | 18.3 | 25.2 | 24.7 |
| Lassen........... | 58 | 22 | 5,067 | 20.1 | 51.3 | 12.9 | 47 | 12,304 | 22,134 | 56,352 | 43.1 | 3.0 | 53,613 | 16.5 | 15.0 | 13.8 |
| Los Angeles...... | 9,565 | 6,365 | 2,638,441 | 15.4 | 41.5 | 32.5 | 19,045 | 12,758 | 34,156 | 68,044 | 38.0 | 10.2 | 72,721 | 13.4 | 18.3 | 18.1 |
| Madera........... | 86 | 35 | 43,940 | 8.6 | 53.0 | 14.6 | 399 | 12,562 | 22,853 | 57,585 | 44.6 | 4.0 | 61,105 | 17.6 | 26.1 | 24.4 |
| Marin............ | 194 | 104 | 59,457 | 23.8 | 17.2 | 59.5 | 515 | 15,249 | 72,466 | 115,246 | 23.1 | 27.9 | 112,069 | 6.9 | 7.2 | 6.2 |
| Mariposa......... | 62 | 16 | 2,900 | 10.9 | 33.8 | 24.6 | 25 | 13,191 | 28,757 | 48,820 | 51.0 | 1.6 | 52,760 | 15.1 | 21.9 | 22.0 |
| Mendocino........ | 131 | 71 | 19,971 | 9.8 | 38.5 | 24.4 | 190 | 14,443 | 29,035 | 51,416 | 48.5 | 4.0 | 52,309 | 14.0 | 20.4 | 17.9 |
| Merced........... | 122 | 133 | 86,919 | 5.0 | 56.1 | 13.8 | 755 | 12,839 | 23,011 | 53,672 | 47.1 | 4.5 | 59,733 | 17.0 | 22.5 | 22.4 |
| Modoc............ | 20 | 10 | 1,636 | 15.9 | 48.2 | 15.2 | 17 | 16,888 | 22,843 | 45,507 | 53.2 | 0.2 | 45,311 | 20.5 | 31.0 | 29.2 |
| Mono............. | 20 | 21 | 2,596 | 1.7 | 35.6 | 28.8 | 33 | 17,597 | 34,552 | 62,260 | 34.1 | 5.9 | 66,226 | 10.1 | 12.0 | 11.2 |
| Monterey......... | 295 | 135 | 122,014 | 8.6 | 48.8 | 24.7 | 1,038 | 13,332 | 30,073 | 71,015 | 33.7 | 8.9 | 76,509 | 12.6 | 18.4 | 17.1 |
| Napa............. | 100 | 34 | 33,586 | 17.7 | 32.3 | 35.7 | 285 | 13,975 | 45,195 | 88,596 | 27.4 | 15.0 | 90,230 | 7.4 | 8.4 | 7.7 |
| Nevada........... | 59 | 93 | 18,741 | 12.2 | 24.0 | 37.2 | 192 | 12,358 | 39,233 | 66,096 | 39.3 | 8.8 | 69,550 | 9.4 | 13.5 | 12.8 |
| Orange........... | 1,888 | 1,835 | 852,709 | 15.1 | 31.8 | 40.6 | 5,837 | 12,014 | 41,514 | 90,234 | 27.4 | 15.5 | 95,761 | 9.5 | 12.6 | 12.5 |
| Placer........... | 248 | 286 | 94,405 | 12.2 | 23.4 | 39.7 | 746 | 10,758 | 43,759 | 89,691 | 27.4 | 13.3 | 97,688 | 6.6 | 7.5 | 7.5 |
| Plumas........... | 36 | 30 | 3,595 | 16.0 | 31.0 | 23.7 | 32 | 14,737 | 35,198 | 55,359 | 45.1 | 4.2 | 59,376 | 13.2 | 19.3 | 19.2 |
| Riverside........ | 1,788 | 1,962 | 673,258 | 11.6 | 44.8 | 22.3 | 5,459 | 12,726 | 28,596 | 67,005 | 37.9 | 7.1 | 72,905 | 11.3 | 14.1 | 13.2 |
| Sacramento....... | 1,348 | 644 | 401,026 | 11.0 | 34.7 | 30.9 | 2,926 | 12,012 | 32,751 | 67,151 | 37.4 | 7.3 | 71,891 | 12.6 | 16.0 | 14.9 |
| San Benito....... | 23 | 36 | 17,021 | 10.3 | 44.3 | 19.7 | 130 | 11,548 | 33,174 | 86,958 | 26.5 | 10.8 | 84,209 | 10.4 | 13.9 | 13.8 |
| San Bernardino... | 1,927 | 1,415 | 617,368 | 10.9 | 46.3 | 21.0 | 4,917 | 12,199 | 25,215 | 63,362 | 39.5 | 5.4 | 67,398 | 13.3 | 18.5 | 17.9 |
| San Diego........ | 2,601 | 1,743 | 868,720 | 14.4 | 30.8 | 38.8 | 5,870 | 11,551 | 38,073 | 78,980 | 31.7 | 11.7 | 83,576 | 10.3 | 12.8 | 12.5 |
| San Francisco.... | 857 | 187 | 163,183 | 31.7 | 23.6 | 58.1 | 934 | 15,330 | 68,883 | 112,449 | 26.6 | 27.0 | 121,795 | 9.5 | 9.3 | 9.8 |
| San Joaquin...... | 293 | 444 | 210,473 | 11.7 | 49.1 | 18.8 | 1,719 | 11,932 | 27,521 | 64,432 | 38.9 | 6.4 | 68,458 | 13.6 | 18.2 | 16.4 |
| San Luis Obispo.. | 303 | 115 | 77,507 | 8.3 | 28.7 | 35.4 | 432 | 12,432 | 37,233 | 73,518 | 33.8 | 9.1 | 76,599 | 11.6 | 11.2 | 10.7 |
| San Mateo........ | 334 | 412 | 182,198 | 22.8 | 25.3 | 51.0 | 1,349 | 14,178 | 61,545 | 122,641 | 20.0 | 28.5 | 135,234 | 6.1 | 6.3 | 5.9 |
| Santa Barbara.... | 477 | 197 | 135,829 | 9.0 | 36.9 | 34.2 | 879 | 12,735 | 36,039 | 74,624 | 33.7 | 11.9 | 74,530 | 12.0 | 13.1 | 12.8 |
| Santa Clara...... | 1,316 | 400 | 507,312 | 22.3 | 25.7 | 52.4 | 3,550 | 13,067 | 56,248 | 124,055 | 20.6 | 27.9 | 132,444 | 6.1 | 5.5 | 5.4 |
| Santa Cruz....... | 148 | 186 | 80,037 | 12.5 | 29.2 | 40.8 | 522 | 12,762 | 41,312 | 82,234 | 31.7 | 15.4 | 85,770 | 10.6 | 10.4 | 10.3 |
| Shasta........... | 138 | 60 | 40,696 | 14.9 | 34.3 | 22.2 | 348 | 12,909 | 29,720 | 54,667 | 46.3 | 4.5 | 61,464 | 13.3 | 16.5 | 15.3 |
| Sierra........... | 6 | 6 | 486 | 14.8 | 39.3 | 17.9 | 8 | 19,290 | 33,747 | 52,148 | 47.9 | 2.3 | 55,862 | 12.7 | 18.4 | 18.5 |
| Siskiyou......... | 81 | 22 | 8,968 | 8.7 | 35.3 | 23.2 | 95 | 14,935 | 25,241 | 45,241 | 54.0 | 2.4 | 45,954 | 17.4 | 25.5 | 25.4 |
| Solano........... | 122 | 390 | 108,022 | 13.3 | 34.1 | 26.9 | 745 | 11,321 | 35,400 | 81,472 | 29.5 | 9.4 | 85,704 | 9.0 | 11.2 | 11.2 |
| Sonoma........... | 223 | 358 | 117,903 | 12.3 | 30.0 | 35.5 | 930 | 13,201 | 42,178 | 81,018 | 29.6 | 11.0 | 87,084 | 7.2 | 7.7 | 6.8 |
| Stanislaus....... | 527 | 179 | 152,625 | 7.7 | 50.4 | 17.1 | 1,335 | 12,294 | 26,258 | 60,704 | 41.2 | 4.9 | 62,761 | 13.0 | 17.1 | 15.8 |
| Sutter........... | 105 | 23 | 26,084 | 7.2 | 45.2 | 18.2 | 239 | 10,083 | 27,371 | 59,050 | 42.5 | 4.8 | 60,910 | 12.8 | 17.4 | 16.9 |
| Tehama........... | 73 | 34 | 15,653 | 8.6 | 45.6 | 15.7 | 148 | 13,465 | 24,301 | 44,551 | 55.0 | 3.5 | 51,672 | 16.3 | 23.7 | 22.9 |
| Trinity.......... | 36 | 14 | 1,999 | 6.6 | 42.4 | 19.5 | 28 | 17,774 | 26,523 | 40,846 | 59.9 | 1.3 | 43,881 | 16.5 | 25.3 | 25.6 |
| Tulare........... | 563 | 228 | 140,603 | 7.1 | 54.7 | 14.6 | 1,360 | 13,071 | 21,380 | 49,687 | 50.3 | 3.9 | 64,729 | 11.3 | 15.2 | 14.5 |
| Tuolumne......... | 60 | 71 | 9,911 | 13.4 | 37.6 | 20.5 | 80 | 13,179 | 34,702 | 60,108 | 42.4 | 6.1 | 64,729 | 11.3 | 15.2 | 14.5 |
| Ventura.......... | 761 | 470 | 222,739 | 14.4 | 34.1 | 33.8 | 1,700 | 12,341 | 38,595 | 88,131 | 27.3 | 13.3 | 91,446 | 8.2 | 11.4 | 11.1 |
| Yolo............. | 84 | 178 | 81,173 | 7.5 | 31.6 | 41.4 | 379 | 12,616 | 34,515 | 70,228 | 36.7 | 10.7 | 70,951 | 16.9 | 13.0 | 12.9 |
| Yuba............. | 126 | 37 | 22,038 | 10.8 | 41.3 | 17.1 | 185 | 12,631 | 25,583 | 58,054 | 44.0 | 3.6 | 56,607 | 15.2 | 18.8 | 18.0 |
| COLORADO ...... | X | X | 1,395,735 | 12.6 | 29.6 | 40.9 | 9,216 | 10,157 | 38,226 | 72,331 | 34.2 | 8.8 | 77,104 | 9.4 | 11.2 | 10.7 |
| Adams............ | 400 | 183 | 131,969 | 10.2 | 44.8 | 24.3 | 801 | 9,456 | 30,313 | 71,202 | 33.3 | 5.5 | 75,341 | 9.3 | 12.1 | 11.1 |
| Alamosa.......... | 23 | 23 | 4,994 | 5.1 | 41.6 | 25.0 | 29 | 10,791 | 22,131 | 37,515 | 62.9 | 2.7 | 38,213 | 19.6 | 24.9 | 23.8 |
| Arapahoe......... | 243 | 325 | 161,819 | 13.5 | 28.2 | 42.8 | 1,268 | 10,608 | 40,443 | 77,469 | 30.1 | 9.9 | 82,364 | 7.2 | 9.0 | 8.9 |
| Archuleta........ | 17 | 22 | 2,120 | 18.8 | 28.7 | 37.2 | 14 | 8,548 | 31,629 | 52,221 | 45.8 | 4.7 | 57,604 | 10.8 | 20.1 | 19.6 |
| Baca............. | 5 | 8 | 727 | 2.6 | 41.8 | 21.4 | 9 | 13,804 | 24,061 | 35,878 | 63.3 | 1.2 | 38,945 | 19.3 | 28.7 | 26.6 |
| Bent............. | 13 | 17 | 851 | 6.8 | 64.0 | 11.4 | 16 | 7,924 | 13,930 | 30,900 | 69.1 | 0.7 | 35,764 | 34.4 | 31.0 | 29.9 |

1. All persons 3 years old and over enrolled in nursery school through college.  2. Persons 25 years old and over.  3. Elementary and secondary education expenditures.  4. Based on population estimated by the American Community Survey, 2014–2018.

# Table B. States and Counties — **Personal Income**

| STATE County | Personal income, 2019 | | | | | | | | | | Earnings, 2019 | | |
| | Total (mil dol) | Percent change 2018-2019 | Per capita[1] | | Wages and salaries (mil dol) | Supplements to wages and salaries, employer contributions (mil dol) | | Proprietors' income (mil dol) | Dividends, interest, and rent (mil dol) | Personal transfer receipts (mil dol) | Total (mil dol) | Contributions for government social insurance (mil dol) | |
| | | | Dollars | Rank | | Pension and insurance | Government social insurance | | | | | From employee and self-employed | From employer |
| | 62 | 63 | 64 | 65 | 66 | 67 | 68 | 69 | 70 | 71 | 72 | 73 | 74 |
|---|---|---|---|---|---|---|---|---|---|---|---|---|---|
| CALIFORNIA ................ | 2,632,280 | 4.7 | 66,745 | X | 1,333,110 | 205,335 | 85,315 | 253,079 | 576,065 | 373,130 | 1,876,840 | 106,351 | 85,315 |
| Alameda.......................... | 135,664 | 5.4 | 81,171 | 62 | 68,234 | 10,216 | 4,462 | 12,211 | 24,780 | 15,172 | 95,122 | 5,269 | 4,462 |
| Alpine............................. | 81 | 3.8 | 72,155 | 109 | 52 | 9 | 3 | 2 | 25 | 18 | 67 | 4 | 3 |
| Amador.......................... | 1,825 | 4.6 | 45,920 | 1,258 | 619 | 175 | 38 | 151 | 428 | 468 | 982 | 64 | 38 |
| Butte............................. | 10,490 | 2.2 | 47,860 | 1,034 | 4,052 | 871 | 289 | 1,020 | 2,166 | 2,628 | 6,232 | 383 | 289 |
| Calaveras....................... | 2,347 | 4.3 | 51,131 | 743 | 453 | 119 | 30 | 201 | 579 | 593 | 803 | 61 | 30 |
| Colusa........................... | 1,106 | 3.4 | 51,342 | 733 | 444 | 97 | 37 | 193 | 266 | 193 | 772 | 35 | 37 |
| Contra Costa.................. | 98,423 | 5.0 | 85,324 | 45 | 28,666 | 4,582 | 1,950 | 7,593 | 23,827 | 10,329 | 42,792 | 2,517 | 1,950 |
| Del Norte....................... | 1,069 | 4.6 | 38,445 | 2,274 | 361 | 116 | 23 | 89 | 204 | 375 | 588 | 36 | 23 |
| El Dorado....................... | 13,479 | 5.1 | 69,895 | 130 | 3,291 | 608 | 229 | 1,032 | 2,748 | 1,952 | 5,160 | 337 | 229 |
| Fresno........................... | 45,446 | 5.5 | 45,487 | 1,297 | 19,924 | 4,262 | 1,463 | 4,970 | 7,841 | 10,492 | 30,619 | 1,681 | 1,463 |
| Glenn ............................ | 1,397 | 6.0 | 49,194 | 906 | 442 | 103 | 35 | 246 | 282 | 313 | 826 | 40 | 35 |
| Humboldt........................ | 6,731 | 4.2 | 49,654 | 863 | 2,339 | 578 | 161 | 941 | 1,495 | 1,622 | 4,020 | 239 | 161 |
| Imperial......................... | 7,330 | 7.4 | 40,447 | 1,994 | 2,854 | 802 | 212 | 1,228 | 994 | 1,884 | 5,096 | 247 | 212 |
| Inyo............................... | 1,092 | 4.4 | 60,513 | 287 | 379 | 114 | 23 | 211 | 220 | 239 | 727 | 40 | 23 |
| Kern............................... | 37,667 | 5.9 | 41,843 | 1,791 | 17,845 | 3,894 | 1,299 | 4,226 | 5,992 | 7,605 | 27,263 | 1,460 | 1,299 |
| Kings............................. | 6,031 | 7.8 | 39,433 | 2,120 | 2,723 | 746 | 206 | 661 | 1,048 | 1,256 | 4,336 | 196 | 206 |
| Lake............................... | 2,850 | 3.9 | 44,259 | 1,451 | 737 | 185 | 53 | 242 | 546 | 921 | 1,216 | 88 | 53 |
| Lassen ........................... | 1,166 | 2.9 | 38,130 | 2,316 | 515 | 190 | 34 | 56 | 239 | 310 | 795 | 41 | 34 |
| Los Angeles ................... | 653,483 | 4.1 | 65,094 | 186 | 329,573 | 52,054 | 21,879 | 69,552 | 145,177 | 107,044 | 473,058 | 26,855 | 21,879 |
| Madera........................... | 6,492 | 5.9 | 41,267 | 1,885 | 2,465 | 574 | 180 | 1,056 | 1,107 | 1,389 | 4,275 | 216 | 180 |
| Marin............................. | 36,685 | 3.8 | 141,735 | 6 | 9,332 | 1,388 | 599 | 4,804 | 12,688 | 2,445 | 16,123 | 901 | 599 |
| Mariposa........................ | 969 | 4.3 | 56,354 | 431 | 259 | 74 | 17 | 107 | 198 | 232 | 457 | 29 | 17 |
| Mendocino ..................... | 4,596 | 4.2 | 52,976 | 617 | 1,471 | 334 | 105 | 645 | 1,094 | 1,220 | 2,555 | 163 | 105 |
| Merced........................... | 11,406 | 6.6 | 41,077 | 1,911 | 3,879 | 920 | 295 | 1,440 | 1,810 | 2,785 | 6,533 | 328 | 295 |
| Modoc............................ | 452 | 8.6 | 51,088 | 751 | 111 | 36 | 8 | 86 | 92 | 137 | 241 | 12 | 8 |
| Mono ............................. | 773 | 6.4 | 53,500 | 572 | 355 | 76 | 25 | 110 | 241 | 83 | 567 | 30 | 25 |
| Monterey........................ | 25,973 | 5.7 | 59,838 | 310 | 10,889 | 2,244 | 847 | 4,205 | 5,875 | 3,671 | 18,184 | 891 | 847 |
| Napa.............................. | 10,430 | 4.3 | 75,717 | 86 | 4,969 | 893 | 363 | 1,324 | 2,774 | 1,340 | 7,549 | 420 | 363 |
| Nevada........................... | 6,339 | 4.2 | 63,542 | 218 | 1,655 | 343 | 116 | 749 | 1,687 | 1,205 | 2,862 | 192 | 116 |
| Orange........................... | 227,733 | 4.0 | 71,711 | 113 | 117,087 | 16,944 | 8,147 | 22,868 | 55,784 | 24,906 | 165,046 | 9,536 | 8,147 |
| Placer............................ | 27,459 | 5.2 | 68,936 | 140 | 11,014 | 1,680 | 786 | 2,093 | 5,403 | 3,678 | 15,573 | 965 | 786 |
| Plumas........................... | 1,051 | 4.9 | 55,910 | 449 | 318 | 94 | 22 | 97 | 294 | 292 | 531 | 36 | 22 |
| Riverside........................ | 104,795 | 5.6 | 42,418 | 1,700 | 39,176 | 7,834 | 2,774 | 7,881 | 17,330 | 20,159 | 57,665 | 3,530 | 2,774 |
| Sacramento.................... | 85,776 | 5.1 | 55,266 | 482 | 46,650 | 10,095 | 2,987 | 6,960 | 15,114 | 16,623 | 66,694 | 3,569 | 2,987 |
| San Benito..................... | 3,471 | 7.3 | 55,261 | 483 | 950 | 199 | 71 | 274 | 552 | 467 | 1,494 | 84 | 71 |
| San Bernardino ............... | 91,658 | 5.4 | 42,043 | 1,755 | 42,525 | 8,365 | 3,120 | 6,598 | 13,799 | 18,893 | 60,607 | 3,594 | 3,120 |
| San Diego....................... | 212,749 | 4.4 | 63,729 | 215 | 110,232 | 19,530 | 7,805 | 16,487 | 48,053 | 29,821 | 154,055 | 8,692 | 7,805 |
| San Francisco ................. | 122,892 | 4.7 | 139,405 | 7 | 107,656 | 11,075 | 5,420 | 14,582 | 33,067 | 8,419 | 138,733 | 7,601 | 5,420 |
| San Joaquin ................... | 35,927 | 6.8 | 47,139 | 1,114 | 13,780 | 2,759 | 1,018 | 3,239 | 5,482 | 7,890 | 20,796 | 1,189 | 1,018 |
| San Luis Obispo.............. | 17,271 | 4.9 | 61,004 | 276 | 6,648 | 1,393 | 451 | 2,094 | 4,478 | 2,602 | 10,586 | 622 | 451 |
| San Mateo...................... | 102,803 | 4.3 | 134,107 | 9 | 57,441 | 5,385 | 2,885 | 9,069 | 29,701 | 5,517 | 74,781 | 4,250 | 2,885 |
| Santa Barbara................. | 29,503 | 4.5 | 66,076 | 168 | 12,614 | 2,344 | 899 | 3,927 | 8,940 | 3,671 | 19,784 | 1,073 | 899 |
| Santa Clara.................... | 223,625 | 4.9 | 115,997 | 14 | 163,502 | 14,433 | 7,953 | 16,836 | 47,160 | 14,443 | 202,723 | 11,533 | 7,953 |
| Santa Cruz..................... | 19,560 | 4.6 | 71,592 | 116 | 6,400 | 1,183 | 435 | 2,322 | 4,633 | 2,433 | 10,340 | 577 | 435 |
| Shasta........................... | 8,723 | 4.2 | 48,438 | 979 | 3,365 | 709 | 243 | 701 | 1,710 | 2,565 | 5,017 | 335 | 243 |
| Sierra ............................ | 142 | 4.3 | 47,245 | 1,103 | 25 | 11 | 1 | 11 | 42 | 32 | 48 | 3 | 1 |
| Siskiyou......................... | 2,091 | 4.1 | 48,021 | 1,012 | 642 | 168 | 48 | 211 | 498 | 678 | 1,069 | 71 | 48 |
| Solano........................... | 23,951 | 5.9 | 53,505 | 570 | 10,041 | 1,950 | 699 | 1,246 | 4,066 | 4,358 | 13,936 | 827 | 699 |
| Sonoma.......................... | 32,972 | 4.3 | 66,700 | 160 | 13,247 | 2,208 | 945 | 3,794 | 8,504 | 4,866 | 20,195 | 1,204 | 945 |
| Stanislaus ...................... | 25,188 | 5.3 | 45,742 | 1,281 | 10,340 | 2,008 | 768 | 2,378 | 4,181 | 5,317 | 15,495 | 888 | 768 |
| Sutter............................ | 4,616 | 5.6 | 47,605 | 1,068 | 1,494 | 301 | 113 | 532 | 817 | 989 | 2,440 | 144 | 113 |
| Tehama.......................... | 2,816 | 6.1 | 43,268 | 1,580 | 912 | 199 | 69 | 238 | 610 | 768 | 1,418 | 91 | 69 |
| Trinity............................ | 534 | 3.2 | 43,471 | 1,549 | 122 | 38 | 8 | 54 | 146 | 189 | 222 | 16 | 8 |
| Tulare............................ | 19,974 | 6.9 | 42,845 | 1,643 | 7,454 | 1,740 | 558 | 2,798 | 3,097 | 4,718 | 12,549 | 630 | 558 |
| Tuolumne ....................... | 2,661 | 3.9 | 48,841 | 938 | 881 | 215 | 59 | 195 | 663 | 722 | 1,350 | 93 | 59 |
| Ventura .......................... | 54,749 | 4.3 | 64,715 | 192 | 21,113 | 3,613 | 1,496 | 4,926 | 12,244 | 7,387 | 31,148 | 1,797 | 1,496 |
| Yolo............................... | 12,375 | 4.7 | 56,123 | 442 | 7,265 | 1,876 | 455 | 1,092 | 2,784 | 1,796 | 10,689 | 516 | 455 |
| Yuba.............................. | 3,425 | 6.3 | 43,536 | 1,542 | 1,327 | 383 | 98 | 227 | 491 | 1,011 | 2,036 | 109 | 98 |
| COLORADO .................... | 352,185 | 5.1 | 61,159 | X | 182,087 | 23,794 | 12,573 | 35,866 | 77,890 | 45,459 | 254,320 | 14,294 | 12,573 |
| Adams............................ | 23,533 | 5.7 | 45,481 | 1,298 | 13,813 | 1,947 | 942 | 2,229 | 2,987 | 3,557 | 18,931 | 997 | 942 |
| Alamosa......................... | 651 | 4.9 | 40,076 | 2,040 | 356 | 64 | 26 | 63 | 116 | 166 | 509 | 27 | 26 |
| Arapahoe ....................... | 42,335 | 4.9 | 64,477 | 196 | 25,031 | 2,819 | 1,722 | 4,099 | 9,594 | 4,950 | 33,672 | 1,932 | 1,722 |
| Archuleta........................ | 604 | 3.6 | 43,088 | 1,604 | 174 | 28 | 12 | 83 | 189 | 149 | 297 | 20 | 12 |
| Baca.............................. | 155 | -5.3 | 43,297 | 1,570 | 45 | 11 | 3 | 16 | 40 | 46 | 74 | 4 | 3 |
| Bent.............................. | 175 | 3.8 | 31,374 | 2,986 | 46 | 9 | 4 | 18 | 47 | 56 | 77 | 4 | 4 |

1. Based on the resident population estimated as of July 1 of the year shown.

# Earnings, Social Security, and Housing

| STATE County | Earnings, 2019 (cont.) | | | | | | | | | Social Security beneficiaries, December 2019 | | Supplemental Security Income recipients, 2019 | Housing units, 2020 | |
|---|---|---|---|---|---|---|---|---|---|---|---|---|---|---|
| | Percent by selected industries | | | | | | | | | | | | | |
| | Farm | Mining, quarrying, and extractions | Construction | Manu-facturing | Information; professional, scientific, technical services | Retail trade | Finance, insurance, real estate, and leasing | Health care and social assistance | Govern-ment | Number | Rate[1] | | Total | Percent change, 2010-2020 |
| | 75 | 76 | 77 | 78 | 79 | 80 | 81 | 82 | 83 | 84 | 85 | 86 | 87 | 88 |
| CALIFORNIA ............ | 1.2 | 0.2 | 5.5 | 8.7 | 19.7 | 5.0 | 8.7 | 9.4 | 15.8 | 6,070,395 | 154 | 1,222,078 | 14,383,358 | 5.1 |
| Alameda............... | 0.1 | 0.0 | 7.1 | 11.0 | 21.4 | 4.9 | 5.6 | 10.0 | 15.2 | 220,590 | 132 | 48,016 | 616,277 | 5.8 |
| Alpine.................. | 0.0 | 0.0 | D | D | D | D | 2.5 | 1.2 | 28.5 | 260 | 234 | 31 | 1,788 | 1.6 |
| Amador................ | 0.8 | 1.5 | 7.1 | 4.3 | 6.1 | 7.3 | 3.5 | D | 41.6 | 11,860 | 299 | 687 | 18,362 | 1.8 |
| Butte.................. | 3.9 | 0.1 | 7.5 | 4.8 | 5.9 | 9.1 | 5.5 | 19.1 | 21.7 | 47,060 | 215 | 9,947 | 86,511 | -9.7 |
| Calaveras............. | 0.2 | D | 14.5 | 2.4 | D | 8.3 | 4.6 | 7.0 | 30.3 | 13,630 | 296 | 991 | 28,279 | 1.3 |
| Colusa................. | 25.4 | 0.1 | 3.4 | 11.7 | 1.8 | 5.5 | 4.4 | D | 21.2 | 3,890 | 181 | 561 | 8,244 | 4.7 |
| Contra Costa.......... | 0.1 | 0.5 | 8.5 | 5.7 | 14.2 | 5.2 | 12.2 | 14.2 | 12.4 | 189,825 | 165 | 25,047 | 419,341 | 4.8 |
| Del Norte............. | 3.0 | 0.0 | 3.5 | 1.5 | 3.0 | 7.4 | 2.2 | D | 47.1 | 6,665 | 239 | 1,878 | 11,437 | 2.2 |
| El Dorado............. | 0.0 | 0.1 | 14.9 | 3.9 | 10.8 | 6.7 | 9.9 | 11.7 | 18.9 | 45,465 | 236 | 2,910 | 92,659 | 5.1 |
| Fresno................. | 7.3 | 0.2 | 5.7 | 5.8 | D | 6.2 | 5.6 | 14.2 | 21.9 | 143,880 | 144 | 43,771 | 338,360 | 7.2 |
| Glenn.................. | 26.5 | D | 4.7 | 7.1 | 2.9 | 5.1 | 2.3 | D | 21.3 | 6,085 | 214 | 1,092 | 11,346 | 5.3 |
| Humboldt.............. | 2.5 | D | 8.5 | 3.6 | 7.2 | 9.6 | 5.0 | 13.4 | 26.9 | 29,170 | 215 | 5,697 | 63,469 | 3.1 |
| Imperial............... | 18.6 | 0.8 | 2.6 | 2.2 | 2.2 | 7.1 | 2.3 | 6.5 | 35.4 | 34,575 | 192 | 10,166 | 58,467 | 4.3 |
| Inyo................... | 1.0 | D | 4.4 | 16.0 | 2.9 | 6.9 | 1.9 | D | 38.7 | 4,645 | 258 | 390 | 9,543 | 0.7 |
| Kern................... | 8.3 | 6.2 | 5.8 | 3.8 | 4.8 | 6.1 | 3.1 | 9.1 | 25.0 | 127,085 | 141 | 33,754 | 303,109 | 6.6 |
| Kings.................. | 15.7 | D | 2.1 | 7.7 | 1.6 | 4.2 | 1.8 | 8.5 | 42.1 | 18,960 | 124 | 4,695 | 46,949 | 7.0 |
| Lake................... | 2.5 | 1.0 | 9.3 | 2.0 | 4.5 | 9.2 | 3.2 | 19.3 | 25.8 | 18,305 | 284 | 3,618 | 34,470 | -2.9 |
| Lassen................. | 5.4 | 0.0 | 2.2 | 0.1 | D | 4.6 | 1.6 | D | 68.0 | 5,510 | 180 | 982 | 12,776 | 0.5 |
| Los Angeles........... | 0.0 | 0.1 | 3.9 | 7.1 | 19.4 | 5 | 10.3 | 9.9 | 14.6 | 1,392,425 | 139 | 389,075 | 3,599,420 | 4.5 |
| Madera................ | 16.8 | D | 4.5 | 6.5 | 2.5 | 4.9 | 2.1 | 14.9 | 22.5 | 26,225 | 167 | 4,663 | 51,063 | 3.9 |
| Marin.................. | 0.3 | D | 7.6 | 6.3 | 19.7 | 6.3 | 11.1 | 10.4 | 11.4 | 53,265 | 206 | 3,122 | 112,665 | 1.3 |
| Mariposa.............. | 3.0 | D | 6.2 | 1.5 | D | 8.3 | D | 4.0 | 41.0 | 5,025 | 294 | 459 | 10,587 | 3.8 |
| Mendocino ............ | 2.7 | 0.1 | 8.5 | 7.5 | D | 10.5 | 3.9 | 13.0 | 21.1 | 22,405 | 258 | 3,345 | 41,035 | 1.8 |
| Merced................ | 17.2 | D | 4.2 | 10.1 | 2.3 | 5.9 | 3.1 | 9.2 | 24.7 | 39,405 | 142 | 10,887 | 89,306 | 6.7 |
| Modoc................. | 30.0 | 0.0 | 4.6 | 0.5 | 2.2 | 4.1 | 2.6 | D | 34.0 | 2,605 | 294 | 432 | 5,266 | 1.5 |
| Mono.................. | 2.1 | D | 8.2 | 1.5 | D | 5.8 | 5.8 | 2.6 | 30.8 | 1,610 | 111 | 94 | 14,190 | 2.0 |
| Monterey.............. | 13.6 | 0.3 | 4.2 | 2.2 | 4.9 | 5.2 | 4.7 | 7.6 | 23.2 | 65,270 | 151 | 8,074 | 143,139 | 2.9 |
| Napa................... | 3.0 | D | 7.8 | 19.9 | 6.5 | 4.8 | 5.5 | 9.4 | 15.1 | 27,850 | 202 | 2,084 | 55,292 | 1.0 |
| Nevada................ | 0.1 | 0.1 | 14.4 | 4.3 | 10.7 | 8.6 | 5.9 | 12.4 | 20.7 | 29,035 | 291 | 1,885 | 54,503 | 3.6 |
| Orange................ | 0.1 | 0.0 | 7.6 | 10.5 | 14.4 | 5.5 | 13.7 | 8.9 | 10.3 | 470,315 | 148 | 72,767 | 1,117,849 | 6.6 |
| Placer................. | 0.2 | D | 12.2 | 3.3 | 10.3 | 7.8 | 11.6 | 17.1 | 12.5 | 83,350 | 209 | 5,928 | 170,594 | 11.8 |
| Plumas................ | 0.2 | D | 8.3 | 8.0 | 4.4 | 6.8 | 3.4 | 5.8 | 37.8 | 6,195 | 325 | 615 | 15,944 | 2.4 |
| Riverside.............. | 1.0 | 0.4 | 10.8 | 6.0 | 5.3 | 8.2 | 4.3 | 11.2 | 23.0 | 394,965 | 160 | 62,354 | 859,850 | 7.4 |
| Sacramento............ | 0.3 | 0.0 | 6.7 | 3.0 | 11.4 | 4.9 | 7.2 | 12.3 | 33.5 | 253,330 | 163 | 64,774 | 580,865 | 4.5 |
| San Benito............ | 6.5 | D | 11.8 | 19.4 | D | 4.9 | 4.7 | 4.2 | 21.1 | 9,170 | 146 | 939 | 20,084 | 12.4 |
| San Bernardino........ | 0.3 | 0.1 | 7.0 | 7.2 | 5.0 | 7.2 | 4.2 | 12.5 | 22.7 | 310,510 | 143 | 71,516 | 728,299 | 4.1 |
| San Diego ............ | 0.4 | 0.0 | 5.5 | 8.7 | 18.9 | 4.8 | 7.5 | 8.6 | 23.2 | 517,095 | 155 | 81,167 | 1,232,928 | 5.8 |
| San Francisco ........ | 0.0 | D | 2.8 | 1.5 | 37.0 | 3.1 | 15.3 | 4.3 | 11.2 | 126,675 | 144 | 40,086 | 405,241 | 7.5 |
| San Joaquin .......... | 5.1 | 0.1 | 6.6 | 7.1 | 3.4 | 6.5 | 4.7 | 11.4 | 20.7 | 115,645 | 152 | 27,610 | 250,918 | 7.3 |
| San Luis Obispo....... | 2.5 | 0.1 | 10.4 | 5.8 | 10.9 | 8.2 | 5.6 | 10.4 | 19.9 | 62,275 | 220 | 4,277 | 124,014 | 5.7 |
| San Mateo............. | 0.1 | 0.0 | 4.2 | 8.0 | 41.5 | 3.3 | 11.9 | 6.0 | 5.4 | 115,800 | 151 | 9,536 | 281,131 | 3.7 |
| Santa Barbara......... | 5.1 | 1.1 | 4.9 | 6.9 | 14.5 | 5.3 | 7.4 | 10.5 | 19.3 | 75,035 | 168 | 8,350 | 161,754 | 5.8 |
| Santa Clara........... | 0.1 | 0.0 | 3.5 | 21.6 | 37.8 | 2.9 | 5.2 | 6.1 | 5.8 | 238,495 | 124 | 43,143 | 677,305 | 7.2 |
| Santa Cruz............ | 4.5 | 0.0 | 7.0 | 5.9 | 11.8 | 7.2 | 6.6 | 12.7 | 18.4 | 49,295 | 180 | 5,440 | 106,249 | 1.7 |
| Shasta................. | 1.1 | D | 7.3 | 3.7 | 6.4 | 9.2 | 5.6 | 19.1 | 22.2 | 49,295 | 274 | 9,270 | 78,280 | 1.3 |
| Sierra................. | 3.1 | D | D | D | D | D | D | D | 58.3 | 905 | 306 | 75 | 2,353 | 1.1 |
| Siskiyou............... | 9.2 | D | 5.3 | 5.9 | 4.6 | 7.0 | 2.6 | 13.4 | 31.2 | 13,885 | 318 | 2,534 | 24,267 | 1.5 |
| Solano................. | 1.0 | 0.4 | 12.2 | 12.8 | 4.3 | 6.0 | 3.5 | 16.2 | 23.7 | 81,465 | 182 | 11,490 | 161,002 | 5.4 |
| Sonoma................ | 1.4 | 0.1 | 10.8 | 12.4 | 10.3 | 6.6 | 6.7 | 13.4 | 13.3 | 101,715 | 207 | 8,533 | 206,444 | 0.9 |
| Stanislaus............. | 7.0 | 0.0 | 6.5 | 11.5 | 4.3 | 7.1 | 4.3 | 17.1 | 18.2 | 91,100 | 166 | 20,722 | 183,550 | 2.3 |
| Sutter................. | 8.6 | 0.3 | 6.5 | 4.7 | 3.7 | 9.6 | 5.7 | 12.3 | 17.8 | 18,045 | 186 | 3,929 | 34,567 | 2.1 |
| Tehama................ | 8.8 | D | 8.5 | 8.4 | 2.5 | 8.1 | 2.5 | 12.0 | 22.6 | 15,895 | 244 | 3,020 | 27,708 | 2.7 |
| Trinity................. | 4.5 | D | 7.2 | 7.4 | 5.4 | 8.0 | D | D | 38.9 | 3,745 | 304 | 615 | 8,974 | 3.4 |
| Tulare................. | 16.0 | D | 4.7 | 7.7 | 3.5 | 6.0 | 3.3 | 7.1 | 23.4 | 67,340 | 145 | 18,834 | 152,855 | 7.9 |
| Tuolumne.............. | -1.1 | 0.5 | 8.6 | 4.8 | D | 8.1 | 4.2 | 16.1 | 32.9 | 16,420 | 302 | 1,613 | 31,712 | 1.5 |
| Ventura............... | 4.3 | 0.7 | 6.1 | 8.4 | 10.8 | 7.1 | 8.1 | 9.8 | 16.7 | 145,535 | 172 | 15,803 | 291,563 | 3.5 |
| Yolo................... | 2.5 | 0.1 | 5.0 | 5.3 | 6.1 | 4.1 | 3.5 | 6.6 | 43.7 | 30,750 | 140 | 5,049 | 79,875 | 6.4 |
| Yuba.................. | 3.6 | 0.5 | 7.5 | 3.2 | D | 3.2 | 1.4 | D | 51.3 | 13,860 | 176 | 3,736 | 29,330 | 6.2 |
| COLORADO ............ | 0.5 | 5.1 | 7.8 | 5.7 | 16.7 | 5.0 | 8.6 | 8.7 | 15.2 | 897,886 | 156 | 72,761 | 2,432,829 | 9.9 |
| Adams................. | 0.2 | 0.9 | 14.1 | 6.1 | 5.8 | 5.0 | 3.7 | 7.7 | 23.5 | 63,340 | 123 | 6,967 | 178,451 | 9.4 |
| Alamosa............... | 7.1 | 0.5 | 5.9 | 1.4 | 3.1 | 8.8 | 7.4 | 21.1 | 23.8 | 3,120 | 193 | 631 | 7,010 | 7.0 |
| Arapahoe ............. | 0.0 | 2.1 | 7.9 | 2.0 | 22.4 | 5.2 | 12.9 | 10.3 | 9.2 | 90,260 | 138 | 7,647 | 259,444 | 8.8 |
| Archuleta.............. | -0.6 | 0.5 | 17.2 | 1.6 | 8.1 | 10.7 | 6.5 | 7.3 | 21.2 | 3,930 | 282 | 111 | 9,701 | 10.7 |
| Baca................... | 9.8 | D | 4.1 | 0.8 | D | 5.7 | 4.4 | 1.2 | 42.9 | 955 | 264 | 65 | 2,262 | 0.7 |
| Bent................... | 18.6 | D | D | D | 0.5 | 4.1 | D | 3.5 | 30.7 | 1,095 | 196 | 257 | 2,270 | 1.2 |

1. Per 1,000 resident population estimated as of July 1 of the year shown.

# Table B. States and Counties — Housing, Labor Force, and Employment

| | Housing units, 2015-2019 | | | | | | | | Civilian labor force, 2020 | | | | Civilian employment[6], 2015-2019 | | |
| | Occupied units | | | | | | | | | | Unemployment | | | Percent | |
| | Owner-occupied | | | | | Renter-occupied | | | | | | | | | |
| STATE County | | | | Median owner cost as a percent of income | | | Median rent as a percent of income[2] | Sub-standard units[4] (percent) | | Percent change, 2019-2020 | | | | Management, business, science, and arts | Construction, production, and maintenance occupations |
| | Total | Percent | Median value[1] | With a mort-gage | Without a mort-gage[2] | Median rent[3] | | | Total | | Total | Rate[5] | Total | | |
| | 89 | 90 | 91 | 92 | 93 | 94 | 95 | 96 | 97 | 98 | 99 | 100 | 101 | 102 | 103 |
|---|---|---|---|---|---|---|---|---|---|---|---|---|---|---|---|
| CALIFORNIA | 13,044,266 | 54.8 | 505,000 | 25.2 | 10.9 | 1,503 | 32.5 | 8.6 | 18,821,167 | -2.8 | 1,908,089 | 10.1 | 18,591,241 | 39.3 | 21.0 |
| Alameda | 577,177 | 53.5 | 769,300 | 23.8 | 10.0 | 1,797 | 29.8 | 8.2 | 813,807 | -3.2 | 71,396 | 8.8 | 862,449 | 50.2 | 16.1 |
| Alpine | 350 | 84.0 | 365,300 | 30.7 | 10.2 | 538 | 17.2 | 2.9 | 517 | -4.4 | 58 | 11.2 | 343 | 37.3 | 26.8 |
| Amador | 14,594 | 76.5 | 313,700 | 25.2 | 12.8 | 1,103 | 32.4 | 2.9 | 14,424 | -2.5 | 1,317 | 9.1 | 13,665 | 33.9 | 21.7 |
| Butte | 85,320 | 59.0 | 271,700 | 23.2 | 12.5 | 1,060 | 34.8 | 3.8 | 92,604 | -5.4 | 8,509 | 9.2 | 96,095 | 36.8 | 20.7 |
| Calaveras | 16,942 | 77.8 | 319,500 | 28.8 | 13.4 | 1,339 | 34.8 | 2.3 | 21,207 | -0.5 | 1,619 | 7.6 | 17,223 | 32.3 | 23.0 |
| Colusa | 7,227 | 61.0 | 265,200 | 23.5 | 10.7 | 889 | 27.6 | 8.1 | 10,499 | -4.3 | 1,675 | 16.0 | 9,692 | 24.0 | 43.6 |
| Contra Costa | 394,769 | 65.9 | 625,800 | 24.3 | 10.9 | 1,819 | 31.3 | 5.4 | 541,256 | -3.3 | 48,020 | 8.9 | 559,366 | 44.1 | 16.8 |
| Del Norte | 9,945 | 63.1 | 218,800 | 25.4 | 11.7 | 878 | 32.0 | 5.2 | 9,350 | -2.8 | 884 | 9.5 | 9,015 | 35.0 | 19.8 |
| El Dorado | 70,974 | 75.7 | 460,900 | 24.6 | 12.5 | 1,260 | 30.9 | 2.6 | 90,752 | -1.5 | 7,490 | 8.3 | 85,101 | 41.2 | 17.2 |
| Fresno | 307,906 | 53.3 | 255,000 | 23.3 | 10.4 | 998 | 33.6 | 9.8 | 445,518 | -1.1 | 50,260 | 11.3 | 407,511 | 30.3 | 29.0 |
| Glenn | 10,085 | 59.4 | 233,200 | 24.4 | 12.1 | 835 | 30.0 | 6.1 | 12,568 | -3.2 | 1,113 | 8.9 | 11,279 | 22.7 | 41.2 |
| Humboldt | 54,679 | 56.8 | 313,200 | 26.8 | 11.5 | 981 | 36.1 | 4.9 | 59,411 | -4.3 | 4,965 | 8.4 | 62,030 | 34.7 | 20.5 |
| Imperial | 44,829 | 58.3 | 195,800 | 23.7 | 11.8 | 830 | 32.4 | 10.7 | 69,602 | -5.5 | 15,653 | 22.5 | 60,074 | 25.2 | 27.0 |
| Inyo | 7,950 | 65.5 | 269,100 | 25.6 | 11.7 | 919 | 24.6 | 3.1 | 8,305 | -5.1 | 647 | 7.8 | 8,238 | 36.4 | 21.7 |
| Kern | 270,282 | 58.3 | 213,900 | 23.3 | 11.2 | 978 | 32.8 | 9.6 | 383,769 | -1.1 | 48,062 | 12.5 | 343,445 | 27.1 | 35.2 |
| Kings | 43,452 | 52.3 | 215,900 | 21.8 | 10.0 | 990 | 29.2 | 8.3 | 56,416 | -2.4 | 6,533 | 11.6 | 54,408 | 25.2 | 34.4 |
| Lake | 25,660 | 66.4 | 219,400 | 27.0 | 14.2 | 978 | 32.0 | 3.6 | 28,267 | -2.5 | 2,700 | 9.6 | 24,304 | 27.5 | 26.1 |
| Lassen | 9,280 | 67.6 | 203,000 | 20.7 | 10.0 | 956 | 29.3 | 2.1 | 9,461 | -2.6 | 676 | 7.1 | 8,611 | 31.2 | 22.5 |
| Los Angeles | 3,316,795 | 45.8 | 583,200 | 27.3 | 11.0 | 1,460 | 34.0 | 11.7 | 4,921,499 | -3.9 | 629,811 | 12.8 | 4,929,863 | 37.6 | 21.3 |
| Madera | 44,881 | 64.1 | 251,200 | 22.8 | 10.6 | 1,014 | 32.1 | 10.0 | 61,738 | -0.6 | 6,649 | 10.8 | 58,268 | 23.9 | 37.1 |
| Marin | 105,432 | 63.7 | 995,800 | 25.5 | 12.1 | 2,069 | 30.8 | 5.0 | 130,138 | -6.1 | 8,662 | 6.7 | 130,747 | 55.3 | 10.3 |
| Mariposa | 7,643 | 68.4 | 275,700 | 27.2 | 12.9 | 929 | 35.7 | 3.8 | 7,162 | -6.0 | 781 | 10.9 | 7,038 | 29.5 | 19.6 |
| Mendocino | 34,408 | 59.9 | 377,500 | 29.5 | 12.2 | 1,146 | 33.5 | 6.0 | 37,015 | -4.2 | 3,292 | 8.9 | 37,689 | 32.3 | 24.1 |
| Merced | 80,008 | 52.2 | 252,700 | 23.5 | 10.2 | 1,021 | 30.3 | 9.1 | 114,736 | -0.9 | 13,981 | 12.2 | 104,983 | 23.4 | 38.3 |
| Modoc | 3,616 | 74.9 | 140,600 | 22.5 | 12.2 | 717 | 25.6 | 3.0 | 3,170 | -0.1 | 270 | 8.5 | 2,957 | 33.3 | 25.8 |
| Mono | 4,765 | 64.4 | 312,500 | 24.0 | 16.9 | 1,206 | 24.9 | 1.5 | 7,846 | -7.7 | 929 | 11.8 | 8,328 | 35.5 | 15.5 |
| Monterey | 127,155 | 51.0 | 516,600 | 26.2 | 10.5 | 1,495 | 32.5 | 14.1 | 213,471 | -3.5 | 23,082 | 10.8 | 187,708 | 29.1 | 32.4 |
| Napa | 48,705 | 64.2 | 635,900 | 24.5 | 10.9 | 1,700 | 31.4 | 6.4 | 68,867 | -5.9 | 6,014 | 8.7 | 71,141 | 36.6 | 22.2 |
| Nevada | 40,855 | 74.2 | 431,000 | 27.8 | 14.0 | 1,273 | 35.9 | 2.4 | 46,489 | -3.9 | 3,702 | 8.0 | 44,059 | 41.5 | 18.0 |
| Orange | 1,037,492 | 57.4 | 679,300 | 25.6 | 10.0 | 1,854 | 33.2 | 9.1 | 1,553,301 | -3.7 | 136,563 | 8.8 | 1,592,151 | 42.5 | 17.3 |
| Placer | 142,855 | 71.9 | 471,500 | 24.1 | 12.3 | 1,542 | 31.1 | 2.2 | 184,892 | -1.9 | 13,573 | 7.3 | 178,325 | 46.2 | 14.4 |
| Plumas | 8,450 | 72.9 | 242,300 | 22.6 | 13.3 | 916 | 28.3 | 3.2 | 7,559 | -4.0 | 806 | 10.7 | 7,733 | 37.4 | 23.1 |
| Riverside | 724,893 | 66.3 | 350,100 | 26.1 | 12.7 | 1,375 | 34.5 | 7.2 | 1,107,723 | 0.2 | 109,998 | 9.9 | 1,032,734 | 30.2 | 26.1 |
| Sacramento | 543,025 | 56.4 | 351,900 | 23.6 | 10.1 | 1,252 | 32.5 | 5.2 | 707,182 | -0.4 | 65,558 | 9.3 | 704,151 | 39.1 | 18.7 |
| San Benito | 18,135 | 63.7 | 551,500 | 26.6 | 11.2 | 1,535 | 29.4 | 7.7 | 31,528 | -2.3 | 3,119 | 9.9 | 29,347 | 29.3 | 31.8 |
| San Bernardino | 636,041 | 59.8 | 328,200 | 25.0 | 11.1 | 1,283 | 34.0 | 9.1 | 966,206 | 0.1 | 91,295 | 9.4 | 914,514 | 29.5 | 29.0 |
| San Diego | 1,125,286 | 53.3 | 563,700 | 26.0 | 10.8 | 1,658 | 33.3 | 7.1 | 1,538,361 | -2.6 | 141,814 | 9.2 | 1,588,732 | 42.2 | 17.0 |
| San Francisco | 362,354 | 37.6 | 1,097,800 | 24.6 | 10.0 | 1,895 | 24.1 | 8.5 | 556,056 | -4.1 | 43,525 | 7.8 | 523,501 | 57.6 | 9.0 |
| San Joaquin | 228,567 | 56.6 | 342,100 | 24.1 | 10.9 | 1,208 | 32.3 | 7.7 | 331,828 | 1.6 | 37,374 | 11.3 | 313,129 | 27.9 | 32.1 |
| San Luis Obispo | 105,981 | 61.6 | 574,000 | 26.3 | 11.6 | 1,476 | 32.3 | 3.4 | 132,690 | -4.9 | 10,195 | 7.7 | 131,679 | 40.1 | 18.5 |
| San Mateo | 263,543 | 60.2 | 1,089,400 | 24.3 | 10.0 | 2,316 | 29.3 | 8.2 | 433,876 | -4.9 | 29,748 | 6.9 | 414,747 | 49.6 | 13.9 |
| Santa Barbara | 145,856 | 52.1 | 577,400 | 24.8 | 10.9 | 1,643 | 33.4 | 11.0 | 217,510 | -2.1 | 17,335 | 8.0 | 213,438 | 35.9 | 23.6 |
| Santa Clara | 640,215 | 56.4 | 984,000 | 23.9 | 10.0 | 2,268 | 28.4 | 8.4 | 1,020,693 | -2.6 | 71,340 | 7.0 | 999,094 | 54.2 | 14.2 |
| Santa Cruz | 95,818 | 60.1 | 756,600 | 25.9 | 11.8 | 1,717 | 35.2 | 7.5 | 133,646 | -5.3 | 12,714 | 9.5 | 136,406 | 41.6 | 19.8 |
| Shasta | 71,181 | 64.0 | 252,300 | 24.2 | 13.0 | 1,039 | 31.6 | 4.4 | 72,827 | 0.0 | 6,343 | 8.7 | 74,134 | 33.6 | 21.3 |
| Sierra | 1,319 | 75.4 | 231,400 | 25.0 | 13.7 | 972 | 28.1 | 1.8 | 1,282 | -3.4 | 103 | 8.0 | 1,268 | 29.1 | 32.7 |
| Siskiyou | 19,240 | 65.0 | 198,900 | 24.5 | 12.4 | 856 | 31.1 | 3.6 | 16,923 | -1.4 | 1,637 | 9.7 | 16,538 | 35.2 | 25.5 |
| Solano | 149,865 | 61.5 | 406,900 | 24.1 | 10.3 | 1,592 | 31.8 | 5.7 | 202,812 | -2.5 | 19,205 | 9.5 | 206,978 | 33.0 | 24.7 |
| Sonoma | 189,374 | 61.5 | 609,600 | 25.6 | 11.4 | 1,621 | 32.4 | 5.4 | 245,091 | -4.7 | 19,300 | 7.9 | 256,074 | 38.2 | 20.7 |
| Stanislaus | 173,898 | 57.8 | 291,600 | 24.0 | 11.3 | 1,155 | 32.0 | 7.1 | 240,641 | -0.7 | 25,756 | 10.7 | 228,175 | 28.0 | 32.9 |
| Sutter | 32,636 | 57.7 | 279,400 | 23.3 | 10.4 | 1,033 | 31.0 | 7.1 | 45,467 | 0.0 | 4,995 | 11.0 | 39,045 | 28.5 | 31.9 |
| Tehama | 24,189 | 65.4 | 215,900 | 25.3 | 12.7 | 872 | 33.8 | 5.2 | 25,579 | 0.1 | 2,262 | 8.8 | 24,633 | 25.7 | 30.3 |
| Trinity | 5,939 | 68.9 | 287,700 | 28.1 | 10.0 | 799 | 34.6 | 3.2 | 4,515 | -2.8 | 360 | 8.0 | 4,465 | 30.2 | 25.6 |
| Tulare | 138,238 | 57.1 | 205,000 | 24.3 | 10.0 | 942 | 33.1 | 10.0 | 199,151 | -1.5 | 26,299 | 13.2 | 178,687 | 26.9 | 36.0 |
| Tuolumne | 22,502 | 71.3 | 289,200 | 24.7 | 13.6 | 1,009 | 31.9 | 3.9 | 20,038 | -4.3 | 2,084 | 10.4 | 20,993 | 33.3 | 21.7 |
| Ventura | 271,040 | 63.2 | 588,400 | 25.7 | 10.1 | 1,776 | 33.8 | 6.6 | 408,867 | -3.0 | 34,964 | 8.6 | 415,752 | 38.2 | 22.6 |
| Yolo | 74,296 | 51.6 | 424,900 | 22.6 | 10.0 | 1,324 | 32.4 | 5.9 | 105,026 | -2.7 | 7,915 | 7.5 | 99,367 | 44.7 | 19.9 |
| Yuba | 26,354 | 60.5 | 257,100 | 23.8 | 10.8 | 1,002 | 31.4 | 7.2 | 30,042 | 1.0 | 3,163 | 10.5 | 29,821 | 27.9 | 29.0 |
| COLORADO | 2,148,994 | 65.2 | 343,300 | 21.4 | 10.0 | 1,271 | 30.5 | 2.9 | 3,122,237 | -0.1 | 226,764 | 7.3 | 2,904,589 | 42.8 | 19.1 |
| Adams | 166,450 | 66.2 | 307,600 | 22.3 | 10.6 | 1,346 | 32.3 | 5.6 | 274,498 | 0.4 | 22,029 | 8.0 | 261,893 | 30.7 | 28.5 |
| Alamosa | 6,162 | 57.8 | 157,500 | 21.4 | 10.4 | 700 | 29.2 | 4.7 | 8,403 | | 535 | 6.4 | 6,934 | 33.4 | 23.6 |
| Arapahoe | 240,304 | 63.5 | 358,200 | 21.4 | 10.0 | 1,390 | 31.3 | 3.1 | 366,768 | 0.5 | 28,825 | 7.9 | 347,236 | 42.7 | 18.1 |
| Archuleta | 5,858 | 72.2 | 314,400 | 25.7 | 10.9 | 961 | 30.0 | 3.1 | 7,123 | 3.2 | 538 | 7.6 | 6,088 | 33.3 | 25.3 |
| Baca | 1,662 | 71.0 | 83,500 | 23.4 | 12.0 | 443 | 25.1 | 2.3 | 2,132 | -1.1 | 50 | 2.3 | 1,622 | 40.6 | 30.1 |
| Bent | 1,767 | 59.6 | 81,000 | 18.6 | 15.1 | 643 | 33.0 | 2.7 | 1,828 | -2.0 | 80 | 4.4 | 1,454 | 30.4 | 27.4 |

1. Specified owner-occupied units.    2. A value of 10.0 represents 10 percent or less; a value of 50.0 represents 50 percent or more.    3. Specified renter-occupied units.    4. Overcrowded or lacking complete plumbing facilities.    5. Percent of civilian labor force.    6. Civilian employed persons 16 years old and over.

Items 89—103

# Table B. States and Counties — Nonfarm Employment and Agriculture

| STATE County | Private nonfarm establishments, employment and payroll, 2019 | | | | | | | | | Agriculture, 2017 | | | |
|---|---|---|---|---|---|---|---|---|---|---|---|---|---|
| | Employment | | | | | | | Annual payroll | | Farms | | | Farm producers whose primary occupation is farming (percent) |
| | Number of establishments | Total | Health care and social assistance | Manufacturing | Retail trade | Finance and insurance | Professional, scientific, and technical services | Total (mil dol) | Average per employee (dollars) | Number | Percent with: | | |
| | | | | | | | | | | | Fewer than 50 acres | 1000 acres or more | |
| | 104 | 105 | 106 | 107 | 108 | 109 | 110 | 111 | 112 | 113 | 114 | 115 | 116 |
| CALIFORNIA .............. | 966,224 | 15,516,824 | 2,100,768 | 1,174,938 | 1,676,677 | 626,332 | 1,270,184 | 1,077,176 | 69,420 | 70,521 | 64.0 | 6.3 | 46.6 |
| Alameda.................. | 40,767 | 712,218 | 101,316 | 86,057 | 70,556 | 20,670 | 75,838 | 54,003 | 75,824 | 446 | 62.6 | 8.1 | 35.8 |
| Alpine..................... | 42 | 551 | 26 | NA | NA | NA | NA | 10 | 17,858 | 6 | 33.3 | 16.7 | 40.0 |
| Amador................... | 828 | 8,385 | 1,411 | 786 | 1,556 | 192 | 224 | 338 | 40,279 | 482 | 54.6 | 6.2 | 44.5 |
| Butte...................... | 4,637 | 63,691 | 15,291 | 4,305 | 11,017 | 2,572 | 2,612 | 2,705 | 42,467 | 1,912 | 65.2 | 2.9 | 51.3 |
| Calaveras............... | 910 | 6,407 | 1,040 | 400 | 1,152 | 108 | 331 | 240 | 37,536 | 699 | 59.7 | 6.3 | 35.9 |
| Colusa.................... | 376 | 3,873 | 429 | 586 | 522 | 93 | 48 | 186 | 48,078 | 459 | 75.6 | 6.1 | 54.7 |
| Contra Costa........... | 24,558 | 344,558 | 60,364 | 16,254 | 44,576 | 22,589 | 25,977 | 24,904 | 72,279 | 459 | 75.6 | 6.1 | 46.7 |
| Del Norte................ | 393 | 4,155 | 1,266 | 154 | 902 | 80 | 102 | 148 | 35,625 | 90 | 58.9 | 4.4 | 41.6 |
| El Dorado............... | 4,626 | 49,977 | 7,060 | 3,205 | 6,493 | 3,082 | 2,560 | 2,560 | 51,226 | 1,390 | 83.4 | 1.3 | 39.6 |
| Fresno.................... | 17,351 | 277,632 | 49,427 | 26,580 | 38,020 | 9,477 | 10,933 | 12,531 | 45,135 | 4,774 | 54.1 | 6.6 | 57.1 |
| Glenn ..................... | 472 | 5,003 | 524 | 841 | 870 | 118 | 111 | 272 | 54,389 | 1,173 | 48.3 | 8.5 | 53.9 |
| Humboldt................. | 3,214 | 35,465 | 7,309 | 2,257 | 6,824 | 1,106 | 2,037 | 1,413 | 39,831 | 849 | 46.6 | 14.6 | 52.4 |
| Imperial ................. | 2,545 | 33,174 | 6,085 | 2,890 | 8,116 | 758 | 908 | 1,240 | 37,379 | 396 | 24.7 | 35.1 | 73.7 |
| Inyo ....................... | 488 | 5,397 | 1,187 | 189 | 1,021 | 74 | 156 | 222 | 41,102 | 85 | 41.2 | 25.9 | 37.9 |
| Kern ...................... | 13,197 | 199,982 | 30,947 | 13,138 | 30,814 | 5,219 | 9,548 | 9,521 | 47,611 | 1,731 | 40.3 | 17.6 | 51.1 |
| Kings ..................... | 1,682 | 25,541 | 5,008 | 4,337 | 4,333 | 492 | 896 | 1,066 | 41,730 | 963 | 48.9 | 9.2 | 54.1 |
| Lake ...................... | 1,111 | 10,001 | 2,513 | 283 | 2,175 | 229 | 252 | 416 | 41,562 | 636 | 69.3 | 2.8 | 41.4 |
| Lassen ................... | 405 | 3,540 | 885 | 49 | 758 | 89 | 327 | 141 | 39,777 | 377 | 37.9 | 17.0 | 47.7 |
| Los Angeles ........... | 285,383 | 3,900,772 | 587,230 | 317,988 | 414,073 | 154,508 | 301,529 | 246,653 | 63,232 | 1,035 | 89.9 | 1.0 | 46.1 |
| Madera................... | 2,039 | 27,690 | 6,453 | 3,832 | 3,733 | 458 | 484 | 1,308 | 47,242 | 1,386 | 41.6 | 8.7 | 52.7 |
| Marin ..................... | 10,028 | 102,686 | 16,868 | 4,117 | 15,116 | 3,887 | 8,608 | 7,526 | 73,293 | 343 | 44.9 | 14.0 | 42.6 |
| Mariposa................ | 321 | 2,491 | 426 | 82 | 312 | 23 | 77 | 97 | 38,749 | 299 | 34.8 | 23.1 | 46.1 |
| Mendocino.............. | 2,445 | 23,543 | 4,389 | 2,468 | 4,463 | 484 | 704 | 967 | 41,077 | 1,128 | 44.6 | 11.1 | 43.2 |
| Merced................... | 3,177 | 46,799 | 7,339 | 10,037 | 8,249 | 1,150 | 900 | 1,883 | 40,238 | 2,337 | 53.5 | 7.4 | 56.3 |
| Modoc.................... | 149 | 1,108 | 388 | 12 | 171 | 30 | 36 | 48 | 43,323 | 423 | 25.1 | 27.9 | 55.4 |
| Mono ..................... | 621 | 6,670 | 527 | 100 | 657 | 21 | 100 | 242 | 36,291 | 65 | 55.4 | 23.1 | 41.7 |
| Monterey ................ | 8,830 | 119,472 | 17,563 | 8,917 | 17,687 | 2,760 | 8,232 | 5,928 | 49,622 | 1,104 | 45.6 | 20.2 | 44.3 |
| Napa ..................... | 4,328 | 64,170 | 9,905 | 11,708 | 6,733 | 1,428 | 1,995 | 3,679 | 57,328 | 1,866 | 74.9 | 2.8 | 32.2 |
| Nevada................... | 3,119 | 29,332 | 5,681 | 1,418 | 4,217 | 964 | 1,399 | 1,256 | 42,830 | 673 | 81.4 | 0.9 | 44.8 |
| Orange................... | 99,577 | 1,536,500 | 177,947 | 148,163 | 149,310 | 90,182 | 136,382 | 95,656 | 62,256 | 193 | 83.4 | 1.6 | 42.6 |
| Placer.................... | 11,279 | 160,796 | 24,514 | 4,845 | 23,155 | 10,082 | 11,858 | 8,833 | 54,932 | 1,237 | 85.0 | 1.1 | 36.8 |
| Plumas................... | 605 | 3,830 | 903 | 516 | 596 | 124 | 180 | 187 | 48,760 | 162 | 45.1 | 16 | 44.1 |
| Riverside................ | 40,608 | 611,156 | 86,261 | 47,536 | 96,358 | 11,411 | 22,052 | 26,004 | 42,549 | 2,667 | 87.1 | 2.1 | 40.9 |
| Sacramento............ | 30,796 | 505,255 | 90,693 | 21,192 | 62,352 | 30,342 | 43,370 | 30,064 | 59,502 | 1,161 | 66.8 | 5.9 | 48.5 |
| San Benito ............. | 1,008 | 13,517 | 1,493 | 3,423 | 1,454 | 229 | 281 | 667 | 49,316 | 610 | 60.0 | 10.8 | 45.5 |
| San Bernardino........ | 37,018 | 625,872 | 101,894 | 53,533 | 92,024 | 15,750 | 18,052 | 28,991 | 46,321 | 1,062 | 84.4 | 1.5 | 41.8 |
| San Diego .............. | 87,334 | 1,313,497 | 187,372 | 105,850 | 148,910 | 58,696 | 159,564 | 81,122 | 61,761 | 5,082 | 90.8 | 0.8 | 36.2 |
| San Francisco ......... | 34,863 | 706,852 | 71,700 | 8,291 | 52,723 | 63,981 | 121,844 | 85,768 | 121,338 | 10 | 100.0 | NA | 27.3 |
| San Joaquin ........... | 11,989 | 195,234 | 29,698 | 17,872 | 26,907 | 6,270 | 5,066 | 9,144 | 46,836 | 3,430 | 62.1 | 5.4 | 55.0 |
| San Luis Obispo....... | 8,538 | 96,358 | 15,676 | 7,219 | 14,544 | 2,061 | 5,368 | 4,785 | 49,659 | 2,349 | 56.4 | 7.8 | 40.8 |
| San Mateo.............. | 21,528 | 416,263 | 38,216 | 24,489 | 36,370 | 19,136 | 44,085 | 55,894 | 134,276 | 241 | 66.4 | 6.2 | 48.2 |
| Santa Barbara......... | 11,859 | 156,903 | 23,462 | 13,364 | 18,491 | 4,245 | 11,703 | 8,972 | 57,182 | 1,467 | 60.7 | 9.5 | 44.0 |
| Santa Clara............ | 49,035 | 1,102,219 | 118,072 | 85,171 | 81,928 | 26,371 | 145,806 | 148,408 | 134,644 | 890 | 78.1 | 5.2 | 44.9 |
| Santa Cruz............. | 7,067 | 82,282 | 14,124 | 5,446 | 12,691 | 2,467 | 4,545 | 4,412 | 53,615 | 625 | 80.5 | 2.1 | 43.2 |
| Shasta.................... | 4,337 | 50,946 | 10,801 | 2,188 | 9,423 | 2,061 | 2,248 | 2,252 | 44,209 | 1,337 | 71.4 | 4.6 | 31.5 |
| Sierra .................... | 68 | 212 | 47 | NA | 22 | NA | NA | 8 | 39,472 | 38 | 23.7 | 26.3 | 56.9 |
| Siskiyou.................. | 1,043 | 8,441 | 1,751 | 812 | 1,505 | 206 | 288 | 326 | 38,596 | 745 | 36.4 | 15.3 | 49.3 |
| Solano.................... | 7,250 | 116,890 | 22,790 | 9,703 | 18,956 | 3,548 | 4,048 | 6,267 | 53,611 | 849 | 65.3 | 8.1 | 44.6 |
| Sonoma.................. | 14,297 | 178,775 | 27,440 | 22,213 | 24,905 | 6,139 | 8,611 | 10,109 | 56,548 | 3,594 | 73.7 | 3.5 | 40.6 |
| Stanislaus............... | 9,279 | 145,805 | 26,936 | 19,321 | 23,873 | 3,466 | 6,214 | 7,026 | 48,189 | 3,621 | 65.3 | 3.5 | 51.3 |
| Sutter .................... | 1,849 | 22,192 | 3,787 | 1,737 | 4,462 | 881 | 624 | 993 | 44,747 | 1,157 | 49.3 | 6.1 | 54.3 |
| Tehama.................. | 994 | 13,098 | 1,923 | 1,861 | 2,095 | 199 | 228 | 627 | 47,864 | 1,479 | 61.9 | 7.6 | 44.7 |
| Trinity ................... | 244 | 1,504 | 320 | 194 | 327 | 35 | 78 | 60 | 39,933 | 185 | 64.3 | 2.2 | 48.0 |
| Tulare.................... | 6,531 | 99,961 | 17,063 | 13,594 | 16,231 | 3,046 | 2,275 | 4,149 | 41,505 | 4,187 | 57.6 | 5.4 | 52.3 |
| Tuolumne................ | 1,266 | 12,806 | 2,747 | 810 | 2,381 | 219 | 456 | 565 | 44,141 | 417 | 57.3 | 8.9 | 37.7 |
| Ventura .................. | 21,519 | 265,180 | 40,627 | 25,321 | 39,683 | 11,695 | 21,439 | 15,093 | 56,915 | 2,135 | 78.4 | 2.2 | 43.0 |
| Yolo....................... | 4,234 | 70,273 | 7,557 | 6,839 | 7,969 | 1,113 | 3,811 | 3,569 | 50,792 | 949 | 49.5 | 9.2 | 53.5 |
| Yuba...................... | 828 | 10,834 | 2,788 | 445 | 1,437 | 153 | 523 | 532 | 49,061 | 764 | 59.8 | 3.8 | 51.0 |
| COLORADO ............. | 174,258 | 2,473,192 | 321,090 | 128,547 | 280,500 | 120,339 | 206,059 | 142,461 | 57,602 | 38,893 | 46.2 | 14.9 | 38.2 |
| Adams.................... | 10,497 | 184,813 | 25,297 | 14,691 | 21,637 | 3,615 | 6,533 | 9,330 | 50,485 | 905 | 62.0 | 10.9 | 32.2 |
| Alamosa.................. | 496 | 5,834 | 1,747 | 112 | 1,121 | 360 | 136 | 221 | 37,851 | 280 | 23.2 | 20.7 | 46.9 |
| Arapahoe................ | 19,233 | 296,968 | 39,427 | 6,495 | 33,388 | 25,859 | 25,102 | 19,475 | 65,579 | 121 | 41.4 | 14.3 | 33.7 |
| Archuleta................ | 524 | 3,733 | 576 | 111 | 757 | 87 | 110 | 121 | 32,384 | 399 | 41.4 | 14.3 | 49.6 |
| Baca...................... | 81 | 520 | 226 | NA | 98 | 21 | 6 | 18 | 34,883 | 667 | 3.3 | 48.4 | 47.4 |
| Bent....................... | 54 | 574 | 84 | NA | 107 | 50 | NA | 22 | 38,373 | 274 | 12.8 | 33.9 | 47.4 |

| STATE County | Acreage (1,000) | Percent change, 2012-2017 | Average size of farm | Total irrigated (1,000) | Total cropland (1,000) | Value of land and buildings (dollars) Average per farm | Average per acre | Value of machinery and equipment, average per farm (dollars) | Total (mil dol) | Average per farm (acres) | Crops | Livestock and poultry products | Organic farms (number) | Farms with internet access (percent) | Government payments Total ($1,000) | Percent of farms |
|---|---|---|---|---|---|---|---|---|---|---|---|---|---|---|---|---|
| | 117 | 118 | 119 | 120 | 121 | 122 | 123 | 124 | 125 | 126 | 127 | 128 | 129 | 130 | 131 | 132 |
| CALIFORNIA | 24,523 | -4.1 | 348 | 7,833.6 | 9,597.4 | 3,252,414 | 9,353 | 165,070 | 45,154.4 | 640,297 | 73.9 | 26.1 | 3,794 | 82.0 | 127,938 | 7.5 |
| Alameda | 183 | 3.1 | 411 | 7.5 | 17.3 | 2,920,388 | 7,106 | 58,016 | 46.2 | 103,509 | 77.4 | 22.6 | 13 | 84.5 | 201 | 6.3 |
| Alpine | 3 | D | 529 | D | D | 2,418,390 | 4,573 | 104,011 | D | D | D | D | NA | 83.3 | D | 16.7 |
| Amador | 181 | 17.0 | 377 | 10.4 | 14.6 | 1,677,097 | 4,454 | 56,975 | 31.0 | 64,376 | 74.8 | 25.2 | 13 | 86.3 | 146 | 6.0 |
| Butte | 348 | -8.6 | 182 | 192.5 | 214.2 | 2,193,553 | 12,042 | 163,066 | 524.2 | 274,140 | 96.8 | 3.2 | 100 | 80.4 | 6,677 | 6.1 |
| Calaveras | 240 | 12.9 | 343 | 5.8 | 14.8 | 1,218,142 | 3,554 | 52,629 | 27.4 | 39,249 | 35.0 | 65.0 | 11 | 79.3 | 1,050 | 9.3 |
| Colusa | 457 | 0.8 | 608 | 233.7 | 277.5 | 4,846,803 | 7,967 | 320,815 | 553.9 | 737,571 | 98.4 | 1.6 | 28 | 83.1 | 9,646 | 31.0 |
| Contra Costa | 156 | 21.9 | 339 | 22.6 | 41.5 | 2,346,158 | 6,922 | 93,600 | 83.2 | 181,327 | 77.4 | 22.6 | 9 | 84.5 | 503 | 6.1 |
| Del Norte | 20 | D | 220 | 9.1 | 7.9 | 1,826,521 | 8,315 | 157,104 | 43.4 | 482,111 | 25.6 | 74.4 | 11 | 97.8 | D | 1.1 |
| El Dorado | 91 | -29.1 | 65 | 5.6 | 8.4 | 649,878 | 9,926 | 41,388 | 24.5 | 17,650 | 87.0 | 13.0 | 27 | 87.6 | 82 | 1.3 |
| Fresno | 1,647 | -4.3 | 345 | 972.6 | 1,142.7 | 3,917,829 | 11,359 | 234,782 | 5,742.8 | 1,202,926 | 71.1 | 28.9 | 183 | 75.6 | 8,894 | 7.7 |
| Glenn | 467 | -30.2 | 398 | 243.2 | 265.1 | 3,547,184 | 8,915 | 248,370 | 628.7 | 536,012 | 82.9 | 17.1 | 30 | 83.3 | 3,676 | 15.8 |
| Humboldt | 621 | 4.6 | 731 | 20.7 | 20.6 | 2,260,162 | 3,090 | 89,370 | D | D | D | D | 123 | 79.5 | 626 | 3.9 |
| Imperial | 522 | 1.2 | 1,317 | 456.1 | 504.0 | 14,670,312 | 11,135 | 884,987 | 1,859.7 | 4,696,157 | 65.7 | 34.3 | 45 | 90.7 | 3,640 | 27.5 |
| Inyo | 287 | -13.3 | 3,375 | 17.1 | 13.6 | 2,093,360 | 620 | 70,430 | 10.6 | 124,800 | 14.3 | 85.7 | 1 | 83.5 | 1,101 | 16.5 |
| Kern | 2,295 | -1.5 | 1,326 | 730.7 | 954.1 | 9,787,244 | 7,380 | 406,230 | 4,076.8 | 2,355,161 | 84.3 | 15.7 | 30 | 82.7 | 5,101 | 13.8 |
| Kings | 616 | -8.6 | 640 | 371.7 | 488.5 | 6,917,469 | 10,815 | 349,654 | 1,649.3 | 1,712,640 | 50.1 | 49.9 | 25 | 80.4 | 5,849 | 29.8 |
| Lake | 138 | -8.2 | 218 | 13.9 | 20.1 | 1,426,769 | 6,555 | 67,768 | 71.9 | 113,002 | 96.5 | 3.5 | 101 | 78.9 | 106 | 5.5 |
| Lassen | 473 | -1.9 | 1,256 | 53.6 | 56.3 | 2,685,457 | 2,139 | 126,819 | 46.0 | 121,923 | 45.0 | 55.0 | 16 | 77.7 | 1,473 | 12.5 |
| Los Angeles | 58 | -37.0 | 56 | 13.8 | 29.6 | 1,037,785 | 18,580 | 56,113 | 154.6 | 149,380 | 86.5 | 13.5 | 41 | 78.7 | 523 | 2.1 |
| Madera | 645 | -1.3 | 466 | 300.2 | 346.1 | 5,102,429 | 10,958 | 229,882 | 1,492.6 | 1,076,903 | 77.4 | 22.6 | 74 | 77.1 | 2,192 | 8.4 |
| Marin | 140 | -18.0 | 408 | 5.0 | 13.7 | 2,393,811 | 5,862 | 102,186 | 95.3 | 277,962 | 11.6 | 88.4 | 67 | 86.9 | 712 | 13.1 |
| Mariposa | 301 | 6.3 | 1,008 | 2.8 | 5.8 | 2,080,275 | 2,064 | 55,341 | 24.7 | 82,706 | 7.7 | 92.3 | 4 | 87.3 | 1,670 | 23.4 |
| Mendocino | 782 | 1.5 | 693 | 27.9 | 64.3 | 2,653,485 | 3,829 | 87,962 | 173.0 | 153,362 | 86.6 | 13.4 | 131 | 81.6 | 226 | 3.3 |
| Merced | 946 | -3.3 | 405 | 493.7 | 546.5 | 5,299,308 | 13,086 | 334,860 | 2,938.4 | 1,257,337 | 43.9 | 56.1 | 68 | 78.5 | 8,726 | 12.5 |
| Modoc | 571 | 9.1 | 1,350 | 142.1 | 159.9 | 2,640,981 | 1,956 | 195,540 | 114.8 | 271,357 | 70.2 | 29.8 | 24 | 82.5 | 2,537 | 31.2 |
| Mono | 73 | 29.5 | 1,124 | 41.7 | 7.9 | 2,158,060 | 1,921 | 140,666 | 9.6 | 147,277 | 65.6 | 34.4 | NA | 78.5 | D | 4.6 |
| Monterey | 1,340 | 5.7 | 1,214 | 294.6 | 366.7 | 8,944,364 | 7,368 | 805,557 | 4,116.1 | 3,728,396 | 99.1 | 0.9 | 191 | 75.5 | 1,269 | 7.4 |
| Napa | 256 | 1.0 | 137 | 60.9 | 67.7 | 6,052,361 | 44,154 | 94,303 | 573.2 | 307,198 | 97.7 | 2.3 | 117 | 84.7 | 924 | 2.1 |
| Nevada | 52 | 23.6 | 77 | 5.0 | 4.8 | 574,346 | 7,425 | 34,251 | 12.5 | 18,517 | 54.5 | 45.5 | 58 | 88.4 | 173 | 4.2 |
| Orange | 32 | -46.4 | 168 | 4.2 | 9.6 | 3,205,502 | 19,094 | 162,436 | 82.5 | 427,497 | 99.5 | 0.5 | 7 | 79.8 | 6 | 3.1 |
| Placer | 119 | 29.7 | 96 | 20.2 | 28.5 | 643,458 | 6,715 | 44,969 | 54.9 | 44,378 | 68.4 | 31.6 | 26 | 88.0 | 327 | 2.7 |
| Plumas | 191 | 9.6 | 1,179 | 18.3 | 29.2 | 2,707,681 | 2,298 | 86,429 | 9.6 | 59,235 | 35.1 | 64.9 | 4 | 82.1 | 120 | 6.8 |
| Riverside | 264 | -23.3 | 99 | 126.2 | 179.7 | 1,794,629 | 18,144 | 102,551 | 932.0 | 349,450 | 75.9 | 24.1 | 228 | 80.4 | 1,621 | 2.4 |
| Sacramento | 260 | 5.4 | 224 | 100.4 | 118.8 | 2,252,041 | 10,048 | 129,041 | 430.5 | 370,759 | 70.1 | 29.9 | 27 | 83.2 | 1,966 | 9.7 |
| San Benito | 520 | -13.9 | 853 | 18.1 | 35.6 | 3,043,028 | 3,569 | 98,840 | 162.9 | 267,057 | 79.9 | 20.1 | 75 | 81.3 | 575 | 9.8 |
| San Bernardino | 68 | -11.6 | 64 | 22.2 | 29.3 | 1,278,283 | 19,897 | 78,028 | 373.9 | 352,096 | 25.2 | 74.8 | 25 | 75.9 | 647 | 2.6 |
| San Diego | 222 | 0.3 | 44 | 42.7 | 64.1 | 1,014,281 | 23,209 | 43,529 | 831.4 | 163,601 | 93.5 | 6.5 | 419 | 84.2 | 558 | 1.1 |
| San Francisco | 0 | 650 | 9 | D | D | 699,840 | 77,760 | 28,864 | D | D | D | D | NA | 80.0 | NA | NA |
| San Joaquin | 773 | -1.8 | 225 | 487.1 | 524.4 | 3,384,002 | 15,020 | 189,084 | 2,176.0 | 634,404 | 74.8 | 25.2 | 35 | 81.7 | 5,788 | 6.9 |
| San Luis Obispo | 931 | -30.4 | 396 | 75.8 | 246.4 | 2,991,939 | 7,547 | 87,783 | 701.6 | 298,676 | 95.2 | 4.8 | 132 | 84.4 | 4,285 | 9.4 |
| San Mateo | 46 | -4.5 | 191 | 3.0 | 6.9 | 1,863,577 | 9,769 | 75,026 | 79.4 | 329,560 | 98.2 | 1.8 | 20 | 88.4 | D | 0.8 |
| Santa Barbara | 715 | 2.0 | 487 | 119.9 | 146.3 | 5,060,150 | 10,381 | 183,019 | 1,519.9 | 1,036,090 | 98.0 | 2.0 | 163 | 87.7 | 854 | 5.2 |
| Santa Clara | 288 | 25.3 | 324 | 19.2 | 34.3 | 2,633,548 | 8,136 | 89,041 | 310.2 | 348,525 | 94.7 | 5.3 | 34 | 86.2 | 518 | 3.6 |
| Santa Cruz | 64 | -36.1 | 102 | 20.1 | 23.5 | 2,183,024 | 21,352 | 116,842 | 606.5 | 970,464 | 98.7 | 1.3 | 130 | 85.1 | 37 | 1.9 |
| Shasta | 410 | 8.9 | 307 | 48.8 | 38.9 | 895,229 | 2,919 | 51,324 | 62.2 | 46,546 | 35.0 | 65.0 | 16 | 81.5 | 547 | 2.0 |
| Sierra | 57 | 45.7 | 1,501 | 10.2 | 3.7 | 2,555,225 | 1,702 | 115,102 | 4.0 | 106,000 | 17.2 | 82.8 | 4 | 84.2 | 84 | 21.1 |
| Siskiyou | 687 | -4.9 | 923 | 115.6 | 142.8 | 2,828,476 | 3,066 | 142,709 | 192.4 | 258,301 | 81.3 | 18.7 | 55 | 82.8 | 2,638 | 14.6 |
| Solano | 343 | -15.8 | 404 | 110.4 | 152.1 | 3,691,320 | 9,148 | 178,302 | 296.6 | 349,337 | 83.8 | 16.2 | 63 | 85.2 | 2,554 | 13.9 |
| Sonoma | 567 | -3.8 | 158 | 86.4 | 129.9 | 3,501,852 | 22,186 | 77,715 | 919.1 | 255,719 | 66.9 | 33.1 | 332 | 87.7 | 2,410 | 3.8 |
| Stanislaus | 723 | -5.9 | 200 | 380.6 | 404.7 | 3,116,617 | 15,619 | 182,695 | 2,526.3 | 697,690 | 53.0 | 47.0 | 35 | 82.2 | 5,006 | 7.5 |
| Sutter | 381 | 1.5 | 329 | 198.8 | 257.3 | 3,136,319 | 9,525 | 240,974 | 412.2 | 356,252 | 99.0 | 1.0 | 42 | 79.0 | 7,687 | 14.0 |
| Tehama | 614 | -0.5 | 415 | 72.8 | 77.2 | 1,789,414 | 4,313 | 80,756 | 218.4 | 147,640 | 74.5 | 25.5 | 18 | 81.6 | 1,168 | 4.7 |
| Trinity | 66 | -62.6 | 356 | D | 1.7 | 683,288 | 1,921 | 35,076 | D | D | D | 100.0 | 9 | 82.2 | D | 2.2 |
| Tulare | 1,250 | 0.9 | 299 | 568.2 | 721.4 | 3,501,053 | 11,726 | 218,462 | 4,474.8 | 1,068,739 | 49.7 | 50.3 | 104 | 78.2 | 13,824 | 10.4 |
| Tuolumne | 123 | 39.5 | 294 | 2.6 | 1.5 | 1,091,612 | 3,715 | 52,254 | 32.3 | 77,384 | 3.5 | 96.5 | 3 | 81.5 | 397 | 8.4 |
| Ventura | 260 | -7.5 | 122 | 98.1 | 123.4 | 3,106,351 | 25,498 | 127,982 | 1,633.3 | 765,008 | 99.2 | 0.8 | 128 | 80.6 | 830 | 2.5 |
| Yolo | 460 | -0.3 | 484 | 234.7 | 312.8 | 4,699,589 | 9,703 | 240,642 | 571.6 | 602,266 | 97.1 | 2.9 | 98 | 86.4 | 2,850 | 19.2 |
| Yuba | 180 | -4.3 | 235 | 72.5 | 79.3 | 1,931,854 | 8,221 | 157,292 | 179.1 | 234,438 | 88.9 | 11.1 | 21 | 82.5 | 2,828 | 4.6 |
| COLORADO | 31,821 | -0.2 | 818 | 2,761.2 | 11,056.3 | 1,315,440 | 1,608 | 117,337 | 7,491.7 | 192,623 | 29.9 | 70.1 | 323 | 81.4 | 198,697 | 23.0 |
| Adams | 705 | 2.1 | 779 | 21.6 | 586.8 | 1,256,762 | 1,613 | 129,421 | 126.5 | 139,779 | 81.9 | 18.1 | 17 | 80.3 | 5,037 | 26.1 |
| Alamosa | 192 | 5.3 | 686 | 79.5 | 81.1 | 1,453,990 | 2,120 | 272,790 | 89.3 | 319,050 | 90.9 | 9.1 | 13 | 83.2 | 836 | 22.9 |
| Arapahoe | 283 | -0.1 | 332 | 1.2 | 136.8 | 751,389 | 2,260 | 63,650 | 26.7 | 31,370 | 78.0 | 22.0 | 3 | 87.7 | 1,705 | 15.0 |
| Archuleta | 210 | 0.0 | 527 | 18.3 | 12.5 | 1,634,748 | 3,104 | 67,028 | 11.2 | 27,965 | 12.6 | 87.4 | 2 | 79.4 | 143 | 2.5 |
| Baca | 1,472 | -2.1 | 2,207 | 45.5 | 742.4 | 1,729,456 | 784 | 149,867 | 114.1 | 171,039 | 45.9 | 54.1 | 1 | 77.1 | 19,467 | 84.0 |
| Bent | 735 | 1.2 | 2,681 | 38.8 | 93.3 | 1,850,149 | 690 | 166,448 | 60.8 | 221,964 | 27.0 | 73.0 | NA | 69.0 | 1,854 | 51.5 |

# Water Use, Wholesale Trade, Retail Trade, and Real Estate

| STATE County | Water use, 2015 | | Wholesale Trade[1], 2017 | | | | Retail Trade[2], 2017 | | | | Real estate and rental and leasing,[2] 2017 | | | |
|---|---|---|---|---|---|---|---|---|---|---|---|---|---|---|
| | Public supply water withdrawn (mil gal/ day) | Public supply gallons withdrawn per person per day | Number of establish-ments | Number of employees | Sales (mil dol) | Average payroll (mil dol) | Number of establish-ments | Number of employees | Sales (mil dol) | Average payroll (mil dol) | Number of establish-ments | Number of employees | Sales (mil dol) | Average payroll (mil dol) |
| | 133 | 134 | 135 | 136 | 137 | 138 | 139 | 140 | 141 | 142 | 143 | 144 | 145 | 146 |
| CALIFORNIA | 5,147.74 | 131.5 | 52,861 | 735,916 | 806,282.5 | 55,597.1 | 108,233 | 1,723,278 | 594,861.4 | 54,229.4 | 57,434 | 314,273 | 110,821.5 | 18,275.0 |
| Alameda | 168.74 | 103.0 | 2,293 | 42,146 | 43,709.9 | 3,185.0 | 4,318 | 72,619 | 26,888.7 | 2,494.6 | 2,189 | 11,182 | 3,621.8 | 622.9 |
| Alpine | 0.10 | 90.1 | NA | NA | NA | NA | NA | NA | NA | NA | NA | NA | NA | NA |
| Amador | 8.64 | 233.5 | 20 | 87 | 31.5 | 4.0 | 124 | 1,474 | 367.6 | 40.4 | 45 | 109 | 29.4 | 3.8 |
| Butte | 28.81 | 127.8 | 144 | 2,073 | 975.0 | 102.1 | 693 | 10,652 | 3,137.5 | 307.3 | 262 | 1,136 | 207.2 | 36.2 |
| Calaveras | 6.59 | 147.0 | D | D | D | 1.8 | 135 | 1,213 | 327.7 | 34.1 | 43 | 96 | 25.7 | 3.8 |
| Colusa | 2.46 | 114.5 | 23 | 394 | 579.4 | 25.5 | 61 | 444 | 159.8 | 13.0 | 17 | 40 | 10.6 | 2.4 |
| Contra Costa | 139.97 | 124.2 | 780 | 8,912 | 10,017.1 | 629.2 | 2,489 | 45,970 | 14,986.4 | 1,521.1 | 1,478 | 7,402 | 2,767.7 | 456.2 |
| Del Norte | 2.40 | 88.1 | D | D | D | D | 59 | 913 | 229.9 | 25.4 | 31 | 100 | 18.7 | 2.5 |
| El Dorado | 27.80 | 150.7 | 117 | 548 | 272.0 | 36.3 | 514 | 6,612 | 1,921.0 | 209.2 | 269 | 1,190 | 277.9 | 49.4 |
| Fresno | 157.08 | 161.1 | 827 | 14,164 | 10,890.9 | 780.4 | 2,438 | 38,243 | 11,347.0 | 1,066.6 | 852 | 4,904 | 1,097.9 | 207.2 |
| Glenn | 2.55 | 91.0 | 19 | 212 | 148.0 | 11.3 | 71 | 846 | 262.6 | 23.5 | 15 | 29 | 3.9 | 1.2 |
| Humboldt | 11.49 | 84.7 | 100 | 992 | 499.8 | 48.1 | 596 | 7,173 | 2,055.5 | 214.6 | 161 | 608 | 135.0 | 22.3 |
| Imperial | 24.47 | 135.8 | 202 | 1,837 | 1,656.4 | 87.4 | 473 | 8,444 | 1,815.1 | 199.7 | 161 | 736 | 136.0 | 22.6 |
| Inyo | 2.42 | 132.5 | 18 | 156 | 56.2 | 5.9 | 90 | 855 | 250.4 | 26.3 | 28 | 100 | 13.8 | 2.4 |
| Kern | 168.87 | 191.4 | 561 | 8,010 | 6,451.9 | 443.1 | 1,960 | 31,869 | 9,382.8 | 860.7 | 670 | 3,474 | 759.1 | 147.4 |
| Kings | 29.07 | 192.6 | 61 | 704 | 969.0 | 42.0 | 292 | 4,155 | 1,272.1 | 117.0 | 64 | 169 | 30.1 | 5.6 |
| Lake | 2.93 | 45.4 | D | D | D | D | 166 | 2,167 | 614.1 | 58.5 | 27 | 56 | 9.4 | 1.4 |
| Lassen | 3.80 | 121.2 | 9 | 68 | 91.7 | 2.7 | 76 | 887 | 234.1 | 24.1 | 27 | 56 | 9.4 | 1.4 |
| Los Angeles | 1,256.44 | 123.5 | 21,107 | 241,372 | 208,449.7 | 14,003.0 | 29,356 | 434,013 | 150,362.8 | 13,607.1 | 16,872 | 96,996 | 37,782.5 | 5,682.8 |
| Madera | 15.63 | 100.8 | 87 | 1,032 | 709.1 | 59.7 | 315 | 3,648 | 1,157.5 | 101.3 | 103 | 390 | 68.6 | 12.7 |
| Marin | 28.19 | 107.9 | 317 | 2,332 | 1,830.8 | 163.2 | 1,011 | 15,549 | 5,625.2 | 645.1 | 642 | 2,969 | 1,874.6 | 229.9 |
| Mariposa | 0.68 | 38.8 | D | D | D | 0.4 | 42 | 339 | 80.0 | 8.4 | 20 | D | 7.0 | D |
| Mendocino | 8.83 | 100.7 | 75 | 860 | 606.8 | 39.6 | 436 | 4,628 | 1,270.7 | 135.9 | 136 | 546 | 97.4 | 18.0 |
| Merced | 50.17 | 186.9 | 105 | 1,759 | 1,948.0 | 87.1 | 544 | 8,453 | 2,366.8 | 214.4 | 156 | 569 | 116.3 | 18.4 |
| Modoc | 0.60 | 66.9 | 5 | 40 | 85.8 | 1.8 | 23 | 192 | 48.5 | 5.0 | 6 | 12 | 2.0 | 0.4 |
| Mono | 1.98 | 142.4 | 6 | 17 | 15.1 | 0.7 | 72 | 702 | 164.8 | 17.8 | 67 | 389 | 54.2 | 11.5 |
| Monterey | 38.23 | 88.1 | 349 | 5,528 | 8,040.7 | 353.6 | 1,292 | 17,597 | 5,629.5 | 529.7 | 474 | 1,987 | 587.0 | 90.9 |
| Napa | 15.28 | 107.3 | D | D | D | D | 497 | 6,742 | 2,005.4 | 219.8 | 219 | 1,041 | 256.2 | 50.0 |
| Nevada | 11.44 | 115.7 | 71 | 502 | 436.5 | 35.0 | 383 | 4,167 | 1,134.8 | 127.4 | 185 | 868 | 287.2 | 46.0 |
| Orange | 457.54 | 144.3 | 6,681 | 86,779 | 107,972.6 | 6,183.7 | 9,707 | 156,535 | 52,345.3 | 4,927.0 | 6,589 | 45,719 | 15,227.2 | 3,118.6 |
| Placer | 54.07 | 144.0 | 319 | 4,675 | 3,572.4 | 305.9 | 1,310 | 23,374 | 7,799.0 | 752.3 | 804 | 3,658 | 988.0 | 198.2 |
| Plumas | 2.15 | 116.8 | D | D | D | D | 85 | 583 | 152.2 | 16.7 | 28 | 69 | 11.2 | 1.9 |
| Riverside | 440.73 | 186.7 | 1,704 | 25,471 | 30,247.5 | 1,510.0 | 5,305 | 96,868 | 31,941.7 | 2,855.7 | 2,407 | 10,120 | 2,538.3 | 451.9 |
| Sacramento | 231.29 | 154.1 | 1,080 | 16,782 | 27,125.2 | 1,030.1 | 3,536 | 62,292 | 19,354.7 | 1,845.5 | 1,743 | 9,009 | 2,445.2 | 442.6 |
| San Benito | 5.24 | 89.1 | 35 | 543 | 271.8 | 43.8 | 102 | 1,407 | 460.5 | 46.6 | 55 | 126 | 29.7 | 5.0 |
| San Bernardino | 337.70 | 158.7 | 2,804 | 40,924 | 37,448.3 | 2,109.0 | 4,898 | 94,285 | 32,452.6 | 2,747.8 | 1,820 | 9,536 | 2,794.2 | 425.2 |
| San Diego | 408.82 | 123.9 | 3,905 | 46,390 | 38,374.0 | 3,219.4 | 9,455 | 152,542 | 46,665.9 | 4,557.3 | 6,381 | 29,423 | 11,062.2 | 1,670.4 |
| San Francisco | 67.54 | 78.1 | 963 | 14,014 | 24,706.6 | 1,245.9 | 3,396 | 49,883 | 19,362.3 | 2,119.1 | 2,132 | 19,351 | 9,399.2 | 1,564.4 |
| San Joaquin | 93.39 | 128.6 | 528 | 11,280 | 14,265.8 | 728.0 | 1,635 | 27,256 | 8,856.3 | 789.3 | 566 | 2,832 | 634.9 | 119.8 |
| San Luis Obispo | 29.10 | 103.4 | 276 | 2,519 | 1,282.1 | 131.0 | 1,130 | 14,708 | 4,372.0 | 442.0 | 494 | 2,024 | 420.2 | 80.3 |
| San Mateo | 68.39 | 89.4 | 969 | 15,805 | 16,383.0 | 1,587.4 | 2,018 | 37,611 | 20,064.3 | 1,354.4 | 1,265 | 7,710 | 3,367.9 | 484.2 |
| Santa Barbara | 49.94 | 112.3 | 405 | 5,968 | 5,075.3 | 490.4 | 1,479 | 18,889 | 5,267.8 | 547.6 | 747 | 3,072 | 712.4 | 130.7 |
| Santa Clara | 194.59 | 101.5 | 2,252 | 74,332 | 136,798.9 | 13,052.5 | 4,682 | 86,436 | 47,472.2 | 3,353.6 | 2,776 | 15,765 | 6,369.5 | 1,023.2 |
| Santa Cruz | 18.45 | 67.3 | D | D | D | D | 893 | 12,885 | 4,650.4 | 364.9 | 370 | 1,539 | 351.3 | 69.8 |
| Shasta | 33.47 | 186.4 | 157 | 1,664 | 1,056.6 | 85.1 | 634 | 9,391 | 2,847.7 | 274.5 | 220 | 887 | 188.2 | 32.5 |
| Sierra | 0.37 | 124.7 | NA | NA | NA | NA | D | D | D | D | NA | NA | NA | NA |
| Siskiyou | 4.61 | 105.8 | 27 | 237 | 128.7 | 7.4 | 171 | 1,546 | 476.4 | 41.3 | 57 | 106 | 13.8 | 2.9 |
| Solano | 52.18 | 119.7 | 253 | 4,043 | 3,119.4 | 261.9 | 1,061 | 20,104 | 6,244.8 | 601.4 | 393 | 1,603 | 503.6 | 68.8 |
| Sonoma | 40.49 | 80.6 | 560 | 8,903 | 6,419.4 | 684.6 | 1,815 | 25,679 | 7,760.5 | 843.1 | 712 | 3,334 | 817.5 | 152.7 |
| Stanislaus | 80.48 | 149.5 | 406 | 6,921 | 6,198.1 | 401.0 | 1,462 | 24,398 | 7,486.4 | 680.1 | 481 | 2,047 | 505.1 | 85.9 |
| Sutter | 13.61 | 141.1 | 66 | 1,325 | 1,101.1 | 94.9 | 287 | 4,473 | 1,306.0 | 124.0 | 84 | 369 | 76.5 | 12.4 |
| Tehama | 5.26 | 83.1 | 24 | 209 | 100.3 | 8.3 | 142 | 1,963 | 754.9 | 56.0 | 44 | 115 | 27.6 | 3.3 |
| Trinity | 0.99 | 75.8 | D | D | D | D | 48 | 402 | 98.2 | 10.1 | D | D | D | 0.3 |
| Tulare | 70.33 | 152.9 | 338 | 4,898 | 8,105.4 | 270.5 | 1,098 | 16,622 | 4,599.0 | 429.3 | 330 | 1,500 | 256.8 | 49.1 |
| Tuolumne | 5.47 | 101.8 | D | D | D | D | 175 | 2,484 | 660.0 | 69.5 | 74 | 202 | 51.8 | 6.7 |
| Ventura | 130.02 | 152.9 | 1,028 | 15,216 | 24,894.7 | 1,221.6 | 2,569 | 40,494 | 13,115.6 | 1,238.4 | 1,204 | 4,751 | 1,300.4 | 235.6 |
| Yolo | 26.91 | 126.3 | 253 | 6,068 | 8,608.1 | 349.2 | 485 | 8,138 | 2,701.8 | 249.8 | 305 | 1,689 | 338.2 | 71.0 |
| Yuba | 8.95 | 120.1 | 27 | 748 | 267.4 | 23.4 | 121 | 1,431 | 489.0 | 43.5 | 40 | 91 | 15.5 | 2.7 |
| COLORADO | 843.95 | 154.7 | 5,953 | 80,365 | 83,316.2 | 5,376.3 | 19,056 | 279,982 | 84,930.6 | 8,419.7 | 12,087 | 47,642 | 12,757.1 | 2,387.8 |
| Adams | 78.89 | 160.6 | 640 | 15,232 | 17,240.8 | 896.1 | 1,183 | 20,600 | 6,974.1 | 653.1 | 583 | 2,855 | 750.8 | 132.3 |
| Alamosa | 2.09 | 126.7 | 12 | 155 | 52.0 | 6.3 | 84 | 1,147 | 317.6 | 33.8 | 23 | 86 | 13.3 | 2.9 |
| Arapahoe | 141.39 | 224.0 | 630 | 9,086 | 19,344.0 | 751.8 | 1,925 | 34,509 | 12,075.7 | 1,129.1 | 1,378 | 5,732 | 1,674.9 | 323.8 |
| Archuleta | 1.51 | 122.2 | 6 | 20 | 5.1 | 0.7 | 87 | 805 | 192.1 | 22.6 | 40 | 184 | 46.2 | 6.8 |
| Baca | 0.45 | 124.5 | 13 | D | 49.9 | D | 14 | 105 | 29.8 | 2.3 | D | D | D | 0.0 |
| Bent | 0.81 | 138.9 | NA | NA | NA | NA | 15 | 72 | 20.8 | 1.4 | NA | NA | NA | NA |

1   Merchant wholesalers, except manufacturers' sales branches and offices.    2. Employer establishments.

| STATE County | Professional, scientific, and technical services, 2017 | | | | Manufacturing, 2017 | | | | Accommodation and food services, 2017 | | | |
|---|---|---|---|---|---|---|---|---|---|---|---|---|
| | Number of establishments | Number of employees | Sales (mil dol) | Average payroll (mil dol) | Number of establishments | Number of employees | Sales (mil dol) | Average payroll (mil dol) | Number of establishments | Number of employees | Sales (mil dol) | Annual payroll (mil dol) |
| | 147 | 148 | 149 | 150 | 151 | 152 | 153 | 154 | 155 | 156 | 157 | 158 |
| CALIFORNIA | 126,250 | 1,200,190 | 290,528 | 114,637.8 | 37,887 | 1,160,890 | 510,858.5 | 76,483.1 | 89,596 | 1,739,010 | 133,716.9 | 37,677.2 |
| Alameda | 6,087 | 66,185 | 16,814 | 6,898.5 | 1,821 | 80,007 | 28,157.5 | 6,799.9 | 4,191 | 66,352 | 5,042.0 | 1,433.6 |
| Alpine | NA | NA | NA | NA | NA | NA | NA | NA | 8 | 54 | 7.4 | 1.8 |
| Amador | 71 | 366 | 33 | 13.0 | 48 | 586 | 113.6 | 31.1 | 109 | 2,387 | 290.8 | 66.0 |
| Butte | D | D | D | D | 180 | 4,287 | 1,340.8 | 207.0 | 452 | 8,241 | 498.3 | 138.1 |
| Calaveras | D | D | D | D | 38 | 325 | 62.7 | 12.7 | 122 | 1,226 | 76.1 | 22.3 |
| Colusa | D | D | D | 5 | 23 | 695 | 450.4 | 36.1 | 36 | 910 | 96.3 | 23.4 |
| Contra Costa | 3,499 | 26,743 | 5,558 | 2,300.1 | 530 | 16,239 | 28,515.2 | 1,222.9 | 2,074 | 34,165 | 2,448.6 | 681.5 |
| Del Norte | 22 | 102 | 9 | 2.7 | 12 | 112 | 57.4 | 5.4 | 64 | 626 | 47.2 | 12.2 |
| El Dorado | D | D | D | D | 173 | 2,514 | 541.9 | 145.2 | 494 | 7,660 | 548.1 | 153.7 |
| Fresno | 1,581 | 10,913 | 1,577 | 610.6 | 579 | 22,774 | 9,073.4 | 1,091.3 | 1,647 | 30,013 | 1,783.4 | 501.2 |
| Glenn | D | D | 11 | D | 29 | 731 | 358.0 | 41.9 | 52 | 670 | 41.3 | 10.5 |
| Humboldt | D | D | D | D | 128 | 2,351 | 596.2 | 83.4 | 403 | 6,356 | 400.6 | 108.2 |
| Imperial | 178 | 826 | 88 | 34.7 | 65 | 2,679 | 1,291.9 | 121.7 | 281 | 4,212 | 246.3 | 69.9 |
| Inyo | D | D | D | D | 11 | 170 | 100.4 | 9.5 | 88 | 1,412 | 134.1 | 30.7 |
| Kern | 1,189 | 9,602 | 1,543 | 569.4 | 376 | 13,484 | 6,938.0 | 678.5 | 1,467 | 23,929 | 1,507.6 | 410.3 |
| Kings | 105 | 965 | 110 | 44.9 | 60 | 4,514 | 2,819.7 | 215.2 | 196 | 4,210 | 460.0 | 97.9 |
| Lake | 68 | 286 | 35 | 11.8 | 42 | 296 | 70.0 | 14.2 | 126 | 1,535 | 116.9 | 30.7 |
| Lassen | 23 | 277 | 23 | 8.8 | 4 | 10 | 2.9 | 0.7 | 49 | 679 | 43.2 | 12.8 |
| Los Angeles | 35,978 | 305,750 | 69,868 | 25,566.5 | 11,794 | 316,464 | 150,047.8 | 19,199.0 | 23,592 | 456,789 | 34,725.9 | 9,833.4 |
| Madera | 107 | 401 | 45 | 15.7 | 104 | 3,517 | 1,490.1 | 192.4 | 210 | 2,882 | 200.0 | 52.1 |
| Marin | 1,796 | 8,000 | 1,780 | 671.9 | 232 | 2,962 | 780.9 | 167.4 | 763 | 13,685 | 1,025.8 | 321.3 |
| Mariposa | D | D | D | D | 12 | 80 | 13.4 | 4.4 | 43 | 986 | 102.1 | 25.9 |
| Mendocino | D | D | D | D | 127 | 2,469 | 780.2 | 120.6 | 341 | 4,127 | 266.4 | 77.2 |
| Merced | 157 | 800 | 88 | 30.4 | 116 | 9,822 | 5,969.6 | 459.9 | 340 | 5,365 | 320.4 | 81.5 |
| Modoc | D | D | 5 | D | D | D | D | D | 16 | 108 | 8.5 | 2.6 |
| Mono | 34 | 106 | 15 | 4.5 | 9 | 68 | 12.2 | 2.3 | 144 | 4,192 | 451.9 | 123.4 |
| Monterey | D | D | D | D | 280 | 9,070 | 4,346.2 | 429.1 | 1,095 | 22,103 | 1,894.5 | 524.6 |
| Napa | 430 | 2,044 | 363 | 130.9 | 492 | 13,127 | 5,873.9 | 890.0 | 408 | 11,705 | 1,068.5 | 339.4 |
| Nevada | 343 | 1,358 | 253 | 94.5 | 138 | 1,532 | 340.8 | 92.2 | 245 | 6,165 | 339.9 | 104.3 |
| Orange | 15,867 | 136,767 | 27,615 | 10,958.5 | 4,572 | 146,707 | 46,800.8 | 9,553.6 | 8,230 | 176,590 | 13,293.4 | 3,749.0 |
| Placer | 1,352 | 10,437 | 1,687 | 670.6 | 261 | 4,288 | 1,136.9 | 244.7 | 951 | 22,782 | 2,120.4 | 491.1 |
| Plumas | D | D | D | D | 18 | 470 | 164.8 | 25.9 | 102 | 447 | 41.9 | 10.3 |
| Riverside | 3,610 | 20,461 | 2,938 | 1,075.2 | 1,547 | 43,177 | 16,073.4 | 2,229.1 | 3,934 | 92,264 | 6,997.9 | 1,971.1 |
| Sacramento | 3,873 | 41,600 | 7,781 | 4,041.1 | 761 | 20,006 | 7,899.3 | 1,226.6 | 2,906 | 55,359 | 3,462.8 | 993.9 |
| San Benito | 75 | 273 | 40 | 15.2 | 61 | 3,251 | 1,037.2 | 167.5 | 96 | 1,418 | 90.1 | 26.3 |
| San Bernardino | 2,749 | 19,069 | 2,751 | 978.9 | 1,795 | 50,781 | 20,023.9 | 2,719.3 | 3,652 | 66,336 | 4,102.2 | 1,139.9 |
| San Diego | 14,150 | 143,999 | 37,220 | 14,266.1 | 2,972 | 100,773 | 31,443.8 | 7,335.4 | 7,810 | 184,771 | 14,854.7 | 4,139.0 |
| San Francisco | 7,124 | 115,262 | 40,464 | 14,240.6 | 704 | 8,583 | 1,983.3 | 485.9 | 4,559 | 89,123 | 9,426.2 | 2,717.4 |
| San Joaquin | 808 | 4,858 | 616 | 259.4 | 533 | 19,833 | 9,574.9 | 1,001.7 | 1,185 | 19,118 | 1,206.4 | 332.1 |
| San Luis Obispo | 929 | 4,853 | 797 | 298.6 | 453 | 6,433 | 2,341.3 | 343.5 | 985 | 17,024 | 1,148.8 | 331.1 |
| San Mateo | 3,232 | 43,162 | 14,613 | 6,440.1 | 611 | 32,094 | 29,757.6 | 3,380.4 | 2,167 | 41,461 | 3,911.8 | 1,106.8 |
| Santa Barbara | 1,389 | 10,005 | 2,286 | 777.4 | 495 | 12,939 | 4,208.2 | 916.0 | 1,215 | 25,145 | 2,149.9 | 607.0 |
| Santa Clara | 8,759 | 144,020 | 42,452 | 18,186.4 | 2,197 | 92,225 | 36,391.1 | 8,114.1 | 5,107 | 90,824 | 7,261.7 | 2,143.6 |
| Santa Cruz | 878 | 4,262 | 648 | 315.9 | 314 | 5,049 | 1,707.6 | 301.1 | 711 | 12,032 | 831.9 | 251.5 |
| Shasta | 357 | 2,043 | 273 | 116.7 | 130 | 2,158 | 656.3 | 106.5 | 373 | 6,351 | 428.5 | 118.5 |
| Sierra | D | D | D | D | NA | NA | NA | NA | 18 | 83 | 6.9 | 2.6 |
| Siskiyou | 90 | 309 | 36 | 12.3 | 31 | 750 | 287.5 | 39.9 | 146 | 1,554 | 101.6 | 28.4 |
| Solano | 590 | 3,726 | 627 | 233.5 | 245 | 9,085 | 7,191.2 | 610.6 | 775 | 13,222 | 848.6 | 237.8 |
| Sonoma | 1,585 | 8,289 | 1,357 | 544.3 | 891 | 21,445 | 7,370.0 | 1,344.2 | 1,306 | 24,035 | 2,085.6 | 534.8 |
| Stanislaus | 682 | 5,805 | 762 | 314.5 | 406 | 19,982 | 12,685.7 | 1,202.1 | 926 | 16,562 | 971.6 | 273.6 |
| Sutter | 124 | 596 | 73 | 26.8 | 56 | 1,493 | 576.3 | 81.7 | 146 | 2,343 | 153.5 | 40.4 |
| Tehama | 55 | 217 | 22 | 7.4 | 43 | 1,783 | 605.5 | 85.5 | 126 | 1,986 | 163.8 | 43.9 |
| Trinity | D | D | 8 | D | 10 | 208 | 70.5 | 12.2 | 47 | 260 | 17.8 | 4.6 |
| Tulare | D | D | D | D | 231 | 13,534 | 8,119.7 | 719.1 | 640 | 9,898 | 641.4 | 173.1 |
| Tuolumne | 86 | 454 | 62 | 23.1 | 45 | 814 | 210.3 | 43.5 | 169 | 2,130 | 172.2 | 42.7 |
| Ventura | 2,816 | 23,034 | 3,557 | 2,558.7 | 872 | 25,626 | 9,751.9 | 1,585.0 | 1,718 | 32,483 | 2,147.4 | 629.8 |
| Yolo | D | D | D | D | 167 | 6,006 | 2,528.8 | 311.7 | 453 | 9,599 | 813.4 | 227.2 |
| Yuba | 86 | 517 | 93 | 35.6 | 36 | 478 | 109.8 | 20.6 | 87 | 1,066 | 72.3 | 19.2 |
| COLORADO | 26,285 | 187,406 | 38,430 | 15,299.1 | 5,111 | 121,372 | 50,809.1 | 7,356.9 | 14,121 | 290,915 | 19,455.8 | 5,804.1 |
| Adams | D | D | D | D | 425 | 13,055 | 6,223.2 | 698.1 | 814 | 16,036 | 1,019.0 | 296.4 |
| Alamosa | D | D | D | D | 16 | 103 | 21.1 | 4.0 | 52 | 877 | 47.7 | 12.9 |
| Arapahoe | D | D | D | D | 417 | 6,866 | 2,194.1 | 401.3 | 1,417 | 30,207 | 1,916.3 | 589.4 |
| Archuleta | 44 | 111 | 13 | 4.7 | 17 | 94 | 10.5 | 2.8 | 53 | 687 | 45.4 | 14.5 |
| Baca | D | D | D | D | NA | NA | NA | NA | D | D | D | 0.3 |
| Bent | D | D | D | 0.1 | NA | NA | NA | NA | D | D | D | 0.6 |

# Health Care and Social Assistance, Other Services, Nonemployer Businesses, and Residential Construction

| STATE County | Health care and social assistance, 2017 | | | | Other services, 2017 | | | | Nonemployer businesses, 2018 | | Value of residential construction authorized by building permits, 2020 | |
|---|---|---|---|---|---|---|---|---|---|---|---|---|
| | Number of establishments | Number of employees | Receipts (mil dol) | Annual payroll (mil dol) | Number of establishments | Number of employees | Receipts (mil dol) | Annual payroll (mil dol) | Number | Receipts (mil dol) | New construction ($1,000) | Number of housing units |
| | 159 | 160 | 161 | 162 | 163 | 164 | 165 | 166 | 167 | 168 | 169 | 170 |
| CALIFORNIA ............ | 113,390 | 2,043,117 | 311,312.2 | 119,500.6 | 62,302 | 437,372 | 64,253.9 | 15,637.9 | 3,453,769 | 189,304.5 | 25,423,120 | 106,075 |
| Alameda............. | 4,812 | 99,670 | 16,159.0 | 6,593.0 | 3,104 | 22,883 | 3,697.8 | 963.0 | 150,445 | 7,890.3 | 1,157,616 | 4,120 |
| Alpine................. | D | D | D | D | D | D | D | 0.3 | 105 | 3.8 | 2,299 | 8 |
| Amador.............. | 107 | 1,331 | 198.1 | 71.5 | 63 | 243 | 30.3 | 8.9 | 2,835 | 131.5 | 25,242 | 103 |
| Butte.................. | 720 | 15,210 | 1,977.3 | 749.0 | 334 | 3,579 | 221.1 | 71.1 | 13,121 | 646.2 | 345,503 | 1,837 |
| Calaveras.......... | 90 | 1,061 | 140.7 | 51.1 | 70 | 258 | 24.7 | 7.0 | 3,591 | 183.2 | 40,763 | 132 |
| Colusa............... | 26 | 245 | 19.5 | 9.0 | D | D | 6.7 | D | 1,051 | 62.9 | 2,672 | 10 |
| Contra Costa...... | 3,233 | 58,665 | 10,075.4 | 3,904.3 | 1,711 | 11,985 | 1,428.0 | 468.2 | 100,779 | 5,858.8 | 694,892 | 2,803 |
| Del Norte........... | 63 | 1,264 | 151.8 | 59.9 | 21 | 78 | 8.9 | 2.4 | 1,326 | 56.4 | 12,578 | 45 |
| El Dorado.......... | 457 | 6,597 | 878.6 | 351.2 | 317 | 1,595 | 161.4 | 49.0 | 17,441 | 918.6 | 214,419 | 619 |
| Fresno............... | 2,398 | 47,713 | 6,660.5 | 2,612.0 | 1,035 | 8,607 | 994.0 | 274.0 | 55,668 | 3,044.1 | 862,554 | 3,130 |
| Glenn ................ | 43 | 492 | 52.8 | 23.3 | 24 | 105 | 20.9 | 4.3 | 1,437 | 74.7 | 13,178 | 60 |
| Humboldt........... | 425 | 7,355 | 904.8 | 349.4 | 230 | 1,276 | 219.2 | 42.1 | 11,275 | 480.5 | 51,976 | 291 |
| Imperial............. | 288 | 5,327 | 652.5 | 237.9 | 143 | 695 | 73.4 | 20.1 | 10,098 | 375.2 | 77,094 | 480 |
| Inyo................... | 48 | 1,151 | 143.1 | 56.2 | 36 | 148 | 18.5 | 5.0 | 1,299 | 47.6 | 908 | 7 |
| Kern................... | 1,684 | 30,531 | 4,574.8 | 1,580.8 | 865 | 6,057 | 671.3 | 199.2 | 49,553 | 2,607.6 | 571,906 | 2,502 |
| Kings................. | 213 | 4,803 | 631.8 | 221.0 | 116 | 501 | 59.1 | 14.2 | 5,119 | 230.5 | 76,662 | 306 |
| Lake.................. | 151 | 2,462 | 350.6 | 123.3 | 69 | 278 | 29.4 | 7.3 | 3,898 | 155.4 | 22,678 | 104 |
| Lassen .............. | 55 | 818 | 98.7 | 41.8 | 33 | 85 | 8.0 | 2.1 | 1,116 | 36.3 | 3,111 | 21 |
| Los Angeles....... | 32,968 | 568,696 | 84,372.2 | 31,197.5 | 17,225 | 119,814 | 17,931.3 | 4,151.3 | 1,107,080 | 61,133.5 | 4,754,173 | 20,903 |
| Madera.............. | 210 | 6,171 | 999.5 | 387.3 | 117 | 576 | 59.2 | 16.5 | 7,504 | 404.3 | 214,136 | 852 |
| Marin................. | 1,158 | 16,799 | 2,350.7 | 1,064.8 | 703 | 4,890 | 847.4 | 203.5 | 38,207 | 2,985.1 | 73,588 | 155 |
| Mariposa........... | 38 | 411 | 39.3 | 18.6 | D | D | D | 1.0 | 1,376 | 52.4 | 9,826 | 39 |
| Mendocino......... | 252 | 4,475 | 528.3 | 209.7 | 170 | 709 | 107.0 | 22.8 | 8,390 | 348.3 | 15,912 | 139 |
| Merced.............. | 444 | 7,595 | 944.4 | 372.0 | 188 | 944 | 95.5 | 25.1 | 11,717 | 627.9 | 314,044 | 1,019 |
| Modoc............... | 20 | 406 | 34.1 | 13.7 | D | D | D | D | 590 | 25.1 | 793 | 7 |
| Mono................. | 36 | 510 | 91.7 | 30.5 | 51 | 206 | 25.5 | 6.7 | 1,443 | 81.4 | 26,637 | 85 |
| Monterey............ | 999 | 17,169 | 2,803.9 | 1,072.6 | 607 | 4,058 | 573.0 | 128.9 | 25,963 | 1,502.6 | 137,710 | 753 |
| Napa.................. | 433 | 10,558 | 1,653.4 | 671.2 | 264 | 1,645 | 203.6 | 54.3 | 12,334 | 792.1 | 159,548 | 770 |
| Nevada.............. | 331 | 5,790 | 662.3 | 266.7 | 198 | 1,239 | 184.1 | 45.1 | 12,499 | 624.0 | 99,975 | 440 |
| Orange.............. | 12,231 | 176,838 | 25,663.6 | 9,314.5 | 5,568 | 39,646 | 4,742.0 | 1,311.0 | 323,593 | 19,312.8 | 1,471,688 | 6,027 |
| Placer............... | 1,272 | 23,006 | 4,079.8 | 1,634.6 | 656 | 4,634 | 565.5 | 164.9 | 34,802 | 2,104.9 | 925,095 | 3,169 |
| Plumas.............. | 52 | 858 | 111.1 | 37.8 | 37 | 134 | 19.8 | 4.7 | 1,648 | 79.9 | 8,233 | 42 |
| Riverside........... | 4,754 | 82,483 | 10,754.6 | 4,057.5 | 2,707 | 16,881 | 1,744.7 | 484.1 | 176,665 | 8,195.5 | 2,491,108 | 10,103 |
| Sacramento....... | 3,612 | 89,537 | 15,276.4 | 6,122.5 | 2,348 | 17,404 | 2,349.3 | 698.7 | 113,355 | 5,503.0 | 1,298,819 | 6,170 |
| San Benito......... | 95 | 1,411 | 182.4 | 80.6 | 74 | 319 | 38.9 | 8.7 | 3,737 | 220.3 | 136,561 | 494 |
| San Bernardino... | 4,287 | 96,414 | 14,309.5 | 5,291.9 | 2,480 | 16,673 | 1,747.0 | 519.3 | 154,048 | 6,947.0 | 1,182,727 | 5,129 |
| San Diego ......... | 9,592 | 175,430 | 26,431.3 | 10,143.4 | 5,823 | 42,595 | 4,888.7 | 1,379.5 | 292,941 | 15,032.5 | 1,997,781 | 9,472 |
| San Francisco .... | 3,330 | 69,425 | 13,052.4 | 5,235.7 | 2,470 | 25,451 | 5,265.1 | 1,112.6 | 100,598 | 6,396.5 | 786,074 | 2,004 |
| San Joaquin....... | 1,513 | 32,650 | 4,236.8 | 1,600.6 | 890 | 5,434 | 565.0 | 163.8 | 26,143 | 1,417.6 | 1,001,931 | 3,914 |
| San Luis Obispo.. | 1,055 | 15,066 | 1,912.6 | 788.2 | 450 | 2,986 | 265.6 | 78.7 | 26,143 | 1,417.6 | 227,992 | 911 |
| San Mateo......... | 2,482 | 39,334 | 6,048.9 | 2,538.7 | 1,498 | 11,771 | 3,065.5 | 595.7 | 73,688 | 5,048.1 | 450,934 | 1,074 |
| Santa Barbara.... | 1,428 | 23,163 | 3,662.0 | 1,207.5 | 785 | 5,846 | 1,759.0 | 211.2 | 35,272 | 2,084.9 | 205,525 | 1,027 |
| Santa Clara....... | 5,898 | 115,234 | 22,181.0 | 9,057.1 | 3,306 | 22,499 | 5,467.5 | 974.4 | 149,189 | 9,101.9 | 1,011,537 | 5,357 |
| Santa Cruz........ | 940 | 15,312 | 2,175.2 | 831.2 | 473 | 3,232 | 413.6 | 127.6 | 25,378 | 1,337.0 | 39,102 | 244 |
| Shasta............... | 670 | 11,319 | 1,553.8 | 575.1 | 300 | 1,659 | 211.8 | 51.1 | 11,945 | 587.4 | 109,818 | 393 |
| Sierra................ | D | D | D | D | NA | NA | NA | NA | 259 | 9.5 | 2,784 | 8 |
| Siskiyou............. | 105 | 1,834 | 217.4 | 97.6 | 73 | 310 | 25.3 | 7.1 | 3,347 | 135.8 | 4,057 | 17 |
| Solano............... | 931 | 22,704 | 3,667.2 | 1,608.3 | 576 | 3,920 | 630.6 | 159.5 | 26,699 | 1,152.7 | 381,135 | 1,733 |
| Sonoma............. | 1,537 | 26,382 | 3,737.9 | 1,632.2 | 959 | 5,810 | 684.6 | 206.7 | 46,153 | 2,559.9 | 403,558 | 1,395 |
| Stanislaus......... | 1,150 | 26,140 | 4,257.6 | 1,674.9 | 691 | 4,587 | 592.0 | 171.6 | 30,181 | 1,594.8 | 130,340 | 552 |
| Sutter................ | 240 | 3,790 | 479.9 | 181.6 | 126 | 625 | 68.4 | 20.1 | 6,251 | 467.3 | 44,425 | 144 |
| Tehama.............. | 133 | 2,017 | 224.3 | 95.6 | 65 | 306 | 35.1 | 10.5 | 3,307 | 170.6 | 27,334 | 147 |
| Trinity................ | 22 | 303 | 29.7 | 12.3 | 10 | 36 | 3.9 | 1.0 | 915 | 35.2 | 2,545 | 29 |
| Tulare................ | 884 | 16,822 | 1,970.5 | 745.0 | 405 | 2,313 | 267.4 | 75.4 | 21,098 | 1,068.9 | 348,742 | 1,575 |
| Tuolumne........... | 171 | 2,734 | 399.0 | 154.7 | 78 | 379 | 54.6 | 13.1 | 4,246 | 189.9 | 14,886 | 56 |
| Ventura.............. | 2,785 | 39,199 | 5,017.7 | 1,883.6 | 1,319 | 6,955 | 794.9 | 205.1 | 70,479 | 3,935.7 | 341,012 | 1,501 |
| Yolo................... | 422 | 7,576 | 1,068.5 | 384.4 | 337 | 2,145 | 259.1 | 80.1 | 13,931 | 695.2 | 290,576 | 1,206 |
| Yuba.................. | 83 | 2,776 | 458.2 | 172.5 | 54 | 183 | 19.8 | 5.1 | 3,742 | 144.3 | 104,410 | 442 |
| COLORADO ........ | 16,659 | 320,813 | 40,055.8 | 15,719.7 | 11,667 | 74,317 | 10,973.6 | 2,785.1 | 535,299 | 26,587.4 | 10,266,541 | 40,469 |
| Adams............... | 780 | 25,410 | 3,232.4 | 1,364.4 | 762 | 5,516 | 789.8 | 208.2 | 36,594 | 1,743.4 | 873,086 | 4,027 |
| Alamosa............ | 78 | 1,579 | 157.9 | 61.7 | 36 | 182 | 15.2 | 4.1 | 1,163 | 50.9 | 11,348 | 62 |
| Arapahoe........... | 2,220 | 42,044 | 5,093.9 | 1,978.9 | 1,297 | 8,655 | 1,179.8 | 349.4 | 62,543 | 3,342.4 | 983,387 | 4,116 |
| Archuleta.......... | 36 | 534 | 56.5 | 24.8 | 27 | 125 | 12.2 | 3.8 | 2,054 | 93.0 | 37,737 | 153 |
| Baca................. | D | D | D | D | D | D | D | 1.8 | 346 | 19.0 | 428 | 2 |
| Bent.................. | 8 | D | 7.6 | D | D | D | D | 0.0 | 231 | 7.5 | 558 | 5 |

# Government Employment and Payroll, and Local Government Finances

| STATE County | Government employment and payroll, 2017 | | March payroll (percent of total) | | | | | | | Local government finances, 2017 | | | | |
|---|---|---|---|---|---|---|---|---|---|---|---|---|---|---|
| | | | | | | | | | | General revenue | | | | |
| | | | | | | | | | | | | Taxes | | |
| | Full-time equivalent employees | March payroll (dollars) | Administration, judicial, and legal | Police and corrections | Fire protection | Highways and transportation | Health and welfare | Natural resources and utilities | Education and libraries | Total (mil dol) | Inter-governmental (mil dol) | Total (mil dol) | Per capita[1] (dollars) | |
| | | | | | | | | | | | | | Total | Property |
| | 171 | 172 | 173 | 174 | 175 | 176 | 177 | 178 | 179 | 180 | 181 | 182 | 183 | 184 |
| CALIFORNIA ................. | X | X | X | X | X | X | X | X | X | X | X | X | X | X |
| Alameda....................... | 64,149 | 475,857,164 | 5.3 | 11.8 | 4.5 | 12.0 | 18.3 | 8.9 | 37.1 | 14373.8 | 5114.9 | 5176.4 | 3,118 | 1,931 |
| Alpine.......................... | 231 | 1,070,089 | 11.6 | 13.3 | 0.6 | 7.5 | 10.7 | 11.3 | 43.0 | 27.1 | 11.7 | 9.0 | 8,109 | 6,925 |
| Amador........................ | 1,294 | 7,214,202 | 10.5 | 17.2 | 2.0 | 9.5 | 7.2 | 5.6 | 41.4 | 169.1 | 77.7 | 60.2 | 1,561 | 1,297 |
| Butte........................... | 8,504 | 42,774,851 | 6.8 | 10.0 | 1.8 | 1.9 | 13.8 | 4.8 | 59.1 | 1192.9 | 763.9 | 283.0 | 1,237 | 929 |
| Calaveras..................... | 1,697 | 7,848,362 | 7.6 | 10.1 | 4.4 | 3.7 | 11.2 | 10.7 | 50.1 | 223.4 | 93.0 | 88.0 | 1,926 | 1,484 |
| Colusa......................... | 1,140 | 5,836,524 | 10.7 | 11.0 | 1.8 | 3.4 | 8.9 | 7.9 | 54.6 | 160.6 | 92.8 | 42.5 | 1,969 | 1,639 |
| Contra Costa................. | 31,854 | 210,481,302 | 5.0 | 11.4 | 4.6 | 3.2 | 13.8 | 6.4 | 53.5 | 7830.5 | 2442.7 | 2609.8 | 2,278 | 1,722 |
| Del Norte..................... | 1,235 | 5,038,894 | 9.9 | 10.2 | 0.4 | 3.4 | 13.6 | 3.9 | 56.0 | 142.7 | 99.3 | 24.8 | 907 | 687 |
| El Dorado..................... | 6,359 | 37,808,926 | 8.0 | 11.1 | 5.9 | 4.3 | 8.4 | 4.8 | 51.7 | 1119.1 | 502.7 | 376.1 | 1,993 | 1,580 |
| Fresno......................... | 36,934 | 197,609,273 | 3.7 | 9.8 | 2.2 | 3.1 | 10.2 | 4.8 | 65.2 | 7476.9 | 5082.5 | 1284.5 | 1,304 | 863 |
| Glenn.......................... | 1,414 | 6,777,443 | 5.3 | 8.2 | 0.6 | 2.8 | 13.1 | 9.2 | 57.4 | 177.4 | 112.0 | 37.3 | 1,335 | 1,074 |
| Humboldt...................... | 7,047 | 33,486,085 | 5.0 | 10.8 | 3.5 | 3.3 | 16.5 | 4.5 | 53.3 | 842.2 | 456.4 | 199.8 | 1,464 | 975 |
| Imperial....................... | 11,605 | 61,911,631 | 3.6 | 5.2 | 1.6 | 1.1 | 22.1 | 13.1 | 47.5 | 1462.1 | 823.5 | 203.7 | 1,121 | 778 |
| Inyo............................ | 1,450 | 7,655,248 | 8.3 | 9.1 | 0.4 | 5.3 | 40.1 | 3.4 | 32.6 | 240.0 | 76.8 | 60.1 | 3,361 | 2,278 |
| Kern............................ | 34,360 | 176,954,893 | 4.6 | 10.6 | 4.2 | 2.3 | 11.2 | 4.9 | 61.3 | 5452.6 | 3333.2 | 1223.1 | 1,378 | 1,104 |
| Kings........................... | 5,524 | 28,583,504 | 5.7 | 10.6 | 2.5 | 2.1 | 9.0 | 5.0 | 62.4 | 740.6 | 510.5 | 136.2 | 910 | 697 |
| Lake............................ | 2,574 | 13,586,601 | 12.1 | 11.3 | 3.0 | 2.5 | 19.1 | 8.3 | 42.5 | 326.6 | 186.0 | 82.3 | 1,283 | 1,054 |
| Lassen......................... | 1,151 | 5,516,096 | 9.6 | 10.3 | 1.0 | 3.4 | 11.8 | 8.6 | 53.7 | 139.9 | 87.7 | 32.1 | 1,039 | 769 |
| Los Angeles .................. | 400,857 | 2,602,396,081 | 9.3 | 12.0 | 4.1 | 6.0 | 15.8 | 8.9 | 41.0 | 76029.6 | 33818.3 | 24255.9 | 2,401 | 1,555 |
| Madera......................... | 5,262 | 26,791,038 | 8.4 | 9.4 | 1.2 | 1.9 | 12.2 | 3.1 | 62.3 | 765.1 | 488.6 | 169.1 | 1,088 | 800 |
| Marin........................... | 8,963 | 64,533,285 | 8.9 | 10.4 | 7.2 | 3.3 | 9.5 | 10.9 | 46.6 | 1844.1 | 429.4 | 992.4 | 3,820 | 2,667 |
| Mariposa...................... | 1,190 | 5,377,779 | 8.7 | 10.2 | 0.3 | 1.9 | 25.2 | 3.1 | 47.2 | 117.1 | 46.9 | 38.7 | 2,225 | 1,198 |
| Mendocino.................... | 4,443 | 21,862,401 | 7.5 | 9.3 | 1.6 | 3.4 | 17.1 | 4.7 | 54.7 | 598.3 | 288.5 | 169.2 | 1,932 | 1,421 |
| Merced......................... | 11,500 | 58,776,936 | 5.1 | 7.4 | 1.2 | 0.8 | 9.4 | 5.9 | 67.6 | 1626.6 | 1064.3 | 295.0 | 1,088 | 823 |
| Modoc.......................... | 652 | 2,830,084 | 5.3 | 7.5 | 1.0 | 4.7 | 39.8 | 3.2 | 35.5 | 85.4 | 45.7 | 12.4 | 1,400 | 1,190 |
| Mono........................... | 1,043 | 6,681,287 | 5.5 | 9.9 | 5.4 | 6.4 | 41.1 | 8.7 | 22.3 | 223.5 | 42.1 | 96.6 | 6,729 | 4,157 |
| Monterey...................... | 19,396 | 124,781,670 | 5.0 | 7.7 | 2.4 | 3.5 | 27.9 | 4.3 | 45.9 | 3511.6 | 1416.5 | 905.1 | 2,083 | 1,363 |
| Napa........................... | 5,188 | 34,539,364 | 11.5 | 13.0 | 3.1 | 2.6 | 10.1 | 4.7 | 51.6 | 1003.8 | 286.8 | 481.0 | 3,439 | 2,612 |
| Nevada........................ | 3,579 | 20,669,561 | 8.1 | 9.0 | 5.0 | 3.6 | 27.5 | 9.5 | 30.1 | 623.3 | 178.3 | 205.8 | 2,072 | 1,681 |
| Orange......................... | 92,158 | 640,816,778 | 6.4 | 12.9 | 4.5 | 3.4 | 7.1 | 5.6 | 58.9 | 17778.6 | 6803.8 | 7027.1 | 2,214 | 1,642 |
| Placer.......................... | 12,560 | 81,230,645 | 8.4 | 10.8 | 4.6 | 2.2 | 5.7 | 9.7 | 55.2 | 2568.8 | 781.5 | 956.1 | 2,483 | 2,009 |
| Plumas......................... | 1,602 | 7,016,012 | 5.4 | 6.8 | 2.2 | 4.2 | 38.8 | 2.8 | 37.1 | 178.4 | 57.4 | 41.0 | 2,196 | 1,842 |
| Riverside...................... | 78,634 | 510,029,742 | 9.1 | 10.8 | 1.6 | 2.1 | 13.2 | 5.3 | 56.2 | 13772.3 | 6716.3 | 3661.4 | 1,516 | 1,075 |
| Sacramento................... | 54,054 | 337,517,731 | 5.1 | 12.8 | 4.7 | 4.1 | 8.2 | 13.9 | 48.3 | 9780.9 | 4722.9 | 2478.2 | 1,623 | 1,100 |
| San Benito .................... | 2,271 | 13,908,888 | 4.2 | 7.1 | 3.0 | 2.5 | 29.0 | 4.1 | 48.5 | 389.8 | 134.3 | 98.3 | 1,634 | 1,266 |
| San Bernardino.............. | 74,301 | 438,536,700 | 5.0 | 10.4 | 4.1 | 2.2 | 13.6 | 4.7 | 57.9 | 17953.7 | 11376.5 | 3231.3 | 1,502 | 1,057 |
| San Diego..................... | 106,292 | 629,893,092 | 7.5 | 10.0 | 3.7 | 3.2 | 10.6 | 6.2 | 54.8 | 19787.8 | 7386.6 | 6914.4 | 2,082 | 1,537 |
| San Francisco................ | 46,523 | 341,836,622 | 9.1 | 13.1 | 6.3 | 19.8 | 19.6 | 9.1 | 21.6 | 11180.4 | 2928.8 | 4565.3 | 5,199 | 3,155 |
| San Joaquin.................. | 26,849 | 155,624,972 | 5.8 | 10.0 | 3.2 | 3.1 | 16.7 | 4.3 | 55.8 | 4670.9 | 2331.8 | 1193.2 | 1,605 | 1,058 |
| San Luis Obispo............. | 8,813 | 50,653,491 | 7.9 | 12.7 | 2.9 | 2.9 | 9.1 | 5.7 | 55.0 | 1483.1 | 506.8 | 703.3 | 2,490 | 1,941 |
| San Mateo..................... | 22,759 | 158,326,401 | 7.5 | 11.8 | 5.7 | 4.8 | 15.0 | 6.3 | 45.2 | 5510.9 | 1257.2 | 2768.5 | 3,601 | 2,566 |
| Santa Barbara................ | 21,940 | 144,599,650 | 4.5 | 8.3 | 4.1 | 2.5 | 11.8 | 5.0 | 62.2 | 3756.5 | 1907.2 | 1050.1 | 2,358 | 1,793 |
| Santa Clara................... | 64,447 | 499,870,139 | 7.4 | 9.0 | 2.9 | 4.9 | 23.8 | 5.0 | 45.0 | 16775.4 | 4010.7 | 6910.6 | 3,576 | 2,611 |
| Santa Cruz.................... | 9,852 | 61,097,077 | 8.1 | 9.9 | 4.0 | 7.1 | 13.0 | 8.3 | 47.4 | 1679.8 | 750.9 | 583.8 | 2,124 | 1,525 |
| Shasta......................... | 7,243 | 36,294,479 | 6.8 | 9.0 | 2.6 | 3.2 | 14.4 | 8.3 | 53.1 | 1172.7 | 746.1 | 233.3 | 1,300 | 1,022 |
| Sierra.......................... | 176 | 898,827 | 18.3 | 11.1 | 0.1 | 12.9 | 19.6 | 2.9 | 32.9 | 32.6 | 18.6 | 7.5 | 2,503 | 2,286 |
| Siskiyou....................... | 2,199 | 9,992,933 | 7.8 | 10.7 | 0.7 | 5.6 | 10.8 | 5.9 | 56.7 | 263.0 | 162.7 | 61.2 | 1,404 | 1,118 |
| Solano......................... | 14,181 | 85,279,873 | 8.8 | 16.3 | 3.9 | 2.8 | 11.4 | 6.1 | 48.9 | 2297.4 | 1058.2 | 776.7 | 1,751 | 1,193 |
| Sonoma........................ | 17,488 | 111,265,857 | 7.9 | 11.7 | 4.0 | 2.5 | 14.7 | 6.4 | 51.1 | 3029.9 | 1128.7 | 1172.0 | 2,332 | 1,749 |
| Stanislaus .................... | 21,814 | 117,420,075 | 5.4 | 8.3 | 2.5 | 1.5 | 10.8 | 8.8 | 60.3 | 3252.8 | 1916.6 | 608.7 | 1,117 | 790 |
| Sutter.......................... | 3,622 | 18,902,432 | 8.5 | 7.5 | 3.2 | 2.5 | 14.2 | 4.0 | 57.2 | 576.2 | 385.6 | 121.5 | 1,263 | 934 |
| Tehama........................ | 2,385 | 11,043,707 | 7.6 | 10.4 | 1.3 | 2.9 | 13.7 | 2.7 | 59.4 | 296.7 | 202.0 | 64.5 | 1,011 | 762 |
| Trinity.......................... | 758 | 3,725,402 | 13.0 | 8.5 | 0.2 | 4.1 | 26.3 | 11.0 | 31.4 | 169.3 | 52.8 | 18.0 | 1,416 | 1,146 |
| Tulare.......................... | 23,861 | 111,821,006 | 5.1 | 7.4 | 1.9 | 1.1 | 31.0 | 4.0 | 48.3 | 3798.6 | 1982.9 | 537.8 | 1,163 | 700 |
| Tuolumne...................... | 1,693 | 9,306,935 | 10.7 | 14.8 | 1.2 | 3.0 | 16.4 | 4.4 | 43.3 | 249.9 | 129.2 | 79.4 | 1,471 | 1,139 |
| Ventura........................ | 30,149 | 195,583,899 | 9.7 | 10.1 | 4.8 | 2.6 | 14.9 | 6.1 | 50.1 | 5323.1 | 2073.3 | 1584.4 | 1,866 | 1,524 |
| Yolo............................ | 6,637 | 38,527,487 | 9.8 | 12.8 | 4.7 | 1.8 | 10.7 | 5.7 | 50.1 | 1166.9 | 519.0 | 370.2 | 1,694 | 1,024 |
| Yuba........................... | 3,322 | 18,247,567 | 4.6 | 9.1 | 0.8 | 1.3 | 8.4 | 2.1 | 71.0 | 456.6 | 300.1 | 82.2 | 1,074 | 899 |
| COLORADO ................... | X | X | X | X | X | X | X | X | X | X | X | X | X | X |
| Adams.......................... | 14,992 | 69,690,300 | 8.7 | 12.9 | 5.2 | 3.2 | 6.3 | 8.0 | 54.1 | 2225.9 | 744.1 | 1078.7 | 2,142 | 1,310 |
| Alamosa........................ | 762 | 2,550,582 | 10.8 | 11.4 | 0.6 | 3.8 | 15.4 | 6.5 | 49.7 | 90.7 | 45.1 | 29.2 | 1,815 | 941 |
| Arapahoe...................... | 22,355 | 105,178,450 | 5.8 | 11.6 | 7.4 | 2.4 | 6.1 | 7.5 | 57.7 | 3305.1 | 951.2 | 1574.1 | 2,443 | 1,631 |
| Archuleta...................... | 641 | 2,961,826 | 6.0 | 5.7 | 1.9 | 3.4 | 33.2 | 8.1 | 24.0 | 90.9 | 23.2 | 31.0 | 2,328 | 1,600 |
| Baca............................ | 501 | 1,750,268 | 6.6 | 3.7 | 0.0 | 9.0 | 53.1 | 2.6 | 23.8 | 42.3 | 11.6 | 9.2 | 2,583 | 2,369 |
| Bent............................ | 316 | 859,960 | 6.9 | 8.2 | 0.0 | 5.1 | 27.0 | 9.8 | 39.8 | 66.7 | 18.9 | 6.5 | 1,114 | 938 |

1. Based on the resident population estimated as of July 1 of the year shown.

# Table B. States and Counties — Local Government Finances, Government Employment, and Income Taxes

| STATE County | Direct general expenditure Total (mil dol) [185] | Per capita[1] (dollars) [186] | Education [187] | Health and hospitals [188] | Police protection [189] | Public welfare [190] | Highways [191] | Debt outstanding Total (mil dol) [192] | Per capita[1] (dollars) [193] | Federal civilian [194] | Federal military [195] | State and local [196] | Number of returns [197] | Mean adjusted gross income [198] | Mean income tax [199] |
|---|---|---|---|---|---|---|---|---|---|---|---|---|---|---|---|
| CALIFORNIA ................. | X | X | X | X | X | X | X | X | X | 248,689 | 204,524 | 2,384,397 | 18,174,590 | 89,228 | 13,035 |
| Alameda........................ | 14,423 | 8,688 | 25.2 | 15.7 | 5.5 | 4.4 | 8.2 | 28,138.5 | 16,949 | 8,825 | 3,939 | 108,375 | 800,510 | 108,434 | 16,571 |
| Alpine.......................... | 26 | 23,218 | 16.5 | 11.9 | 11.9 | 6.7 | 6.6 | 5.6 | 5,076 | 1 | 2 | 269 | 460 | 60,063 | 5,583 |
| Amador........................ | 175 | 4,546 | 31.7 | 6.5 | 7.3 | 7.4 | 6.3 | 92.4 | 2,398 | 87 | 52 | 4,749 | 17,260 | 67,001 | 7,091 |
| Butte........................... | 1,304 | 5,701 | 50.9 | 7.0 | 4.3 | 9.8 | 3.6 | 710.4 | 3,106 | 572 | 319 | 16,405 | 87,730 | 61,448 | 6,301 |
| Calaveras...................... | 210 | 4,586 | 35.1 | 10.8 | 5.3 | 10.1 | 5.3 | 219.1 | 4,797 | 128 | 67 | 2,646 | 20,430 | 67,219 | 7,101 |
| Colusa......................... | 171 | 7,941 | 41.4 | 5.9 | 5.2 | 5.1 | 4.7 | 70.1 | 3,249 | 71 | 31 | 2,082 | 9,960 | 59,046 | 4,921 |
| Contra Costa.................. | 7,425 | 6,482 | 36.0 | 19.5 | 6.6 | 5.9 | 3.6 | 7,724.2 | 6,743 | 4,794 | 1,751 | 43,604 | 554,800 | 118,817 | 18,429 |
| Del Norte...................... | 150 | 5,498 | 34.9 | 7.8 | 3.7 | 15.2 | 2.9 | 77.2 | 2,821 | 148 | 47 | 3,529 | 9,990 | 50,457 | 4,292 |
| El Dorado...................... | 1,095 | 5,805 | 42.4 | 3.6 | 4.8 | 5.6 | 4.3 | 1,027.1 | 5,444 | 665 | 284 | 10,261 | 92,180 | 97,002 | 13,280 |
| Fresno......................... | 7,003 | 7,108 | 41.9 | 23.6 | 5.5 | 6.2 | 2.3 | 5,206.5 | 5,285 | 10,192 | 1,659 | 65,280 | 404,380 | 57,417 | 5,844 |
| Glenn.......................... | 183 | 6,538 | 46.3 | 8.9 | 5.1 | 11.3 | 5.1 | 81.7 | 2,924 | 197 | 67 | 1,927 | 12,520 | 51,356 | 4,295 |
| Humboldt...................... | 899 | 6,588 | 47.0 | 7.6 | 4.6 | 11.5 | 3.4 | 424.0 | 3,106 | 774 | 381 | 13,492 | 57,280 | 54,854 | 5,248 |
| Imperial....................... | 1,513 | 8,327 | 43.0 | 24.1 | 3.4 | 7.9 | 2.3 | 1,540.3 | 8,480 | 2,157 | 365 | 16,944 | 80,240 | 43,367 | 3,248 |
| Inyo............................ | 256 | 14,342 | 28.8 | 39.4 | 4.2 | 3.4 | 4.3 | 71.0 | 3,974 | 305 | 26 | 2,908 | 8,450 | 61,869 | 6,251 |
| Kern............................ | 7,088 | 7,988 | 40.8 | 3.7 | 3.6 | 5.7 | 1.6 | 4,158.1 | 4,686 | 11,070 | 3,798 | 56,967 | 343,720 | 55,249 | 5,120 |
| Kings........................... | 801 | 5,349 | 48.0 | 4.4 | 4.9 | 9.9 | 2.5 | 315.7 | 2,110 | 1,198 | 5,021 | 14,265 | 55,610 | 49,394 | 3,886 |
| Lake............................ | 323 | 5,034 | 41.9 | 6.6 | 5.7 | 13.2 | 3.8 | 137.8 | 2,150 | 163 | 93 | 4,063 | 25,730 | 49,407 | 4,352 |
| Lassen......................... | 161 | 5,218 | 44.3 | 7.4 | 5.1 | 8.9 | 9.3 | 102.1 | 3,301 | 1,833 | 40 | 4,365 | 9,860 | 58,772 | 5,063 |
| Los Angeles .................. | 70,277 | 6,956 | 35.1 | 12.0 | 8.8 | 8.4 | 2.5 | 101,862.8 | 10,082 | 48,007 | 17,224 | 560,255 | 4,702,060 | 81,685 | 12,179 |
| Madera......................... | 839 | 5,399 | 50.1 | 4.5 | 3.8 | 8.2 | 6.4 | 657.0 | 4,227 | 302 | 221 | 11,059 | 59,580 | 51,966 | 4,480 |
| Marin .......................... | 1,821 | 7,010 | 33.9 | 7.4 | 6.6 | 3.6 | 3.4 | 2,913.7 | 11,217 | 679 | 513 | 14,865 | 132,030 | 202,985 | 40,355 |
| Mariposa ...................... | 124 | 7,138 | 20.5 | 24.9 | 6.6 | 10.4 | 6.7 | 38.1 | 2,189 | 647 | 24 | 1,542 | 7,520 | 56,365 | 5,351 |
| Mendocino .................... | 1,357 | 15,496 | 17.1 | 6.6 | 2.4 | 4.4 | 1.9 | 535.6 | 6,116 | 275 | 163 | 6,669 | 39,550 | 55,747 | 5,511 |
| Merced ........................ | 1,617 | 5,964 | 57.0 | 4.2 | 4.0 | 8.9 | 3.7 | 974.0 | 3,593 | 742 | 400 | 17,133 | 107,320 | 48,237 | 3,887 |
| Modoc ......................... | 90 | 10,126 | 30.1 | 31.7 | 3.8 | 6.7 | 10.7 | 5.1 | 575 | 208 | 12 | 912 | 3,390 | 43,588 | 3,812 |
| Mono .......................... | 215 | 14,972 | 17.1 | 32.4 | 4.8 | 2.4 | 10.8 | 79.7 | 5,549 | 216 | 240 | 1,428 | 6,420 | 64,043 | 6,862 |
| Monterey ...................... | 3,623 | 8,338 | 38.0 | 22.1 | 4.4 | 5.0 | 2.3 | 2,222.3 | 5,114 | 5,239 | 5,309 | 29,711 | 197,090 | 71,269 | 8,424 |
| Napa ........................... | 965 | 6,897 | 37.4 | 6.8 | 8.1 | 4.4 | 5.0 | 1,190.6 | 8,512 | 227 | 195 | 10,728 | 68,540 | 99,868 | 14,637 |
| Nevada ........................ | 569 | 5,727 | 23.4 | 33.6 | 5.3 | 5.8 | 4.9 | 495.9 | 4,990 | 338 | 145 | 6,038 | 50,100 | 75,481 | 9,037 |
| Orange ........................ | 16,921 | 5,331 | 45.3 | 3.6 | 8.0 | 5.5 | 3.9 | 24,733.3 | 7,792 | 11,219 | 5,079 | 151,444 | 1,527,670 | 94,533 | 14,034 |
| Placer......................... | 2,370 | 6,153 | 45.4 | 3.4 | 5.7 | 4.8 | 2.6 | 3,219.6 | 8,360 | 706 | 611 | 19,128 | 188,660 | 97,092 | 12,898 |
| Plumas........................ | 203 | 10,853 | 25.1 | 37.2 | 3.4 | 4.8 | 5.8 | 67.3 | 3,605 | 337 | 27 | 2,053 | 8,900 | 59,255 | 5,654 |
| Riverside...................... | 14,495 | 6,002 | 43.7 | 8.4 | 7.1 | 6.8 | 3.4 | 18,182.0 | 7,529 | 7,364 | 3,984 | 129,609 | 1,026,690 | 59,495 | 5,807 |
| Sacramento................... | 9,386 | 6,146 | 39.1 | 6.7 | 6.0 | 6.9 | 3.7 | 20,813.8 | 13,628 | 9,522 | 3,062 | 170,603 | 708,500 | 66,774 | 7,408 |
| San Benito .................... | 408 | 6,784 | 42.6 | 29.5 | 3.2 | 5.4 | 1.3 | 350.4 | 5,824 | 119 | 92 | 2,954 | 29,290 | 77,140 | 8,968 |
| San Bernardino.............. | 17,614 | 8,189 | 33.3 | 29.0 | 5.1 | 6.1 | 3.7 | 10,972.1 | 5,101 | 14,026 | 18,491 | 114,073 | 930,510 | 55,326 | 5,104 |
| San Diego..................... | 20,668 | 6,223 | 39.3 | 9.4 | 5.6 | 4.4 | 3.1 | 41,671.1 | 12,547 | 48,266 | 98,986 | 206,249 | 1,598,530 | 83,078 | 10,997 |
| San Francisco................ | 12,046 | 13,719 | 12.8 | 28.1 | 4.4 | 11.8 | 1.0 | 22,442.2 | 25,559 | 13,250 | 1,645 | 97,303 | 483,190 | 172,985 | 34,303 |
| San Joaquin .................. | 4,399 | 5,918 | 46.3 | 10.7 | 6.0 | 7.0 | 4.7 | 5,064.4 | 6,813 | 3,212 | 1,168 | 41,374 | 324,270 | 62,118 | 6,146 |
| San Luis Obispo ............. | 1,450 | 5,132 | 41.1 | 6.4 | 6.0 | 7.8 | 4.0 | 1,526.6 | 5,404 | 566 | 527 | 22,523 | 132,030 | 78,930 | 9,761 |
| San Mateo .................... | 4,972 | 6,466 | 36.9 | 9.7 | 8.0 | 4.1 | 3.1 | 7,250.3 | 9,429 | 3,541 | 1,353 | 28,888 | 390,390 | 192,039 | 37,713 |
| Santa Barbara................ | 3,449 | 7,745 | 33.0 | 26.8 | 4.7 | 4.2 | 1.9 | 1,732.6 | 3,891 | 3,521 | 3,107 | 32,924 | 202,420 | 83,343 | 11,736 |
| Santa Clara................... | 15,564 | 8,054 | 30.8 | 24.8 | 4.9 | 5.6 | 1.9 | 20,395.8 | 10,554 | 9,985 | 3,307 | 87,220 | 934,040 | 171,602 | 31,912 |
| Santa Cruz.................... | 1,651 | 6,006 | 38.1 | 7.8 | 5.8 | 7.8 | 3.0 | 1,276.1 | 4,643 | 530 | 381 | 19,884 | 130,370 | 94,197 | 13,737 |
| Shasta......................... | 1,007 | 5,615 | 46.0 | 8.4 | 6.0 | 11.3 | 2.6 | 573.9 | 3,199 | 1,295 | 308 | 11,850 | 77,700 | 59,273 | 5,752 |
| Sierra.......................... | 29 | 9,599 | 27.0 | 11.6 | 10.3 | 8.4 | 14.5 | 9.6 | 3,181 | 44 | 4 | 342 | 1,260 | 55,978 | 5,112 |
| Siskiyou....................... | 279 | 6,392 | 47.4 | 6.1 | 5.7 | 7.5 | 4.7 | 222.6 | 5,102 | 769 | 63 | 3,570 | 19,160 | 49,535 | 4,467 |
| Solano......................... | 2,262 | 5,100 | 38.2 | 8.0 | 9.9 | 6.9 | 4.7 | 2,036.9 | 4,593 | 3,630 | 7,342 | 22,032 | 212,820 | 71,309 | 7,587 |
| Sonoma........................ | 3,042 | 6,054 | 40.6 | 9.0 | 6.5 | 6.9 | 3.8 | 3,352.6 | 6,672 | 1,363 | 1,487 | 26,682 | 245,380 | 85,708 | 11,379 |
| Stanislaus..................... | 3,266 | 5,996 | 49.6 | 7.6 | 5.3 | 9.3 | 2.5 | 6,781.2 | 12,449 | 831 | 807 | 30,271 | 231,670 | 58,213 | 5,722 |
| Sutter .......................... | 640 | 6,650 | 45.1 | 9.0 | 6.3 | 6.7 | 2.0 | 348.7 | 3,624 | 76 | 144 | 4,734 | 41,400 | 56,804 | 5,288 |
| Tehama........................ | 314 | 4,924 | 48.0 | 7.3 | 5.6 | 12.1 | 4.3 | 75.1 | 1,175 | 212 | 95 | 3,894 | 26,090 | 51,299 | 4,459 |
| Trinity.......................... | 176 | 13,793 | 18.3 | 13.3 | 2.3 | 6.2 | 8.8 | 54.7 | 4,301 | 213 | 18 | 858 | 4,440 | 48,898 | 4,291 |
| Tulare.......................... | 3,764 | 8,141 | 41.8 | 26.6 | 3.2 | 6.7 | 3.0 | 1,853.2 | 4,009 | 1,079 | 681 | 32,442 | 182,200 | 48,280 | 4,224 |
| Tuolumne...................... | 269 | 4,975 | 34.3 | 8.2 | 6.8 | 7.8 | 10.0 | 84.4 | 1,563 | 437 | 75 | 5,140 | 24,360 | 62,942 | 6,344 |
| Ventura........................ | 5,426 | 6,390 | 40.7 | 12.5 | 7.4 | 4.4 | 2.7 | 4,701.3 | 5,537 | 7,500 | 4,953 | 38,698 | 405,950 | 84,283 | 11,252 |
| Yolo............................ | 1,145 | 5,242 | 35.1 | 5.0 | 6.5 | 7.9 | 5.2 | 1,501.7 | 6,873 | 3,485 | 322 | 39,379 | 91,460 | 79,716 | 9,700 |
| Yuba............................ | 454 | 5,931 | 49.9 | 2.4 | 3.8 | 11.7 | 1.9 | 650.2 | 8,490 | 1,531 | 4,017 | 5,805 | 30,730 | 50,744 | 3,817 |
| COLORADO ................... | X | X | X | X | X | X | X | X | X | 53,124 | 53,846 | 403,856 | 2,764,910 | 82,387 | 10,703 |
| Adams.......................... | 2,273 | 4,513 | 40.5 | 0.4 | 5.8 | 4.2 | 4.7 | 3,879.8 | 7,704 | 3,109 | 1,626 | 45,773 | 242,060 | 60,439 | 5,800 |
| Alamosa........................ | 74 | 4,606 | 39.6 | 6.1 | 8.7 | 19.1 | 6.7 | 92.8 | 5,759 | 133 | 37 | 2,177 | 6,760 | 49,349 | 4,212 |
| Arapahoe...................... | 3,091 | 4,796 | 44.2 | 0.9 | 7.2 | 1.7 | 6.4 | 6,725.3 | 10,435 | 3,302 | 3,442 | 35,080 | 321,130 | 86,464 | 11,919 |
| Archuleta...................... | 88 | 6,637 | 16.1 | 41.2 | 2.8 | 5.1 | 9.9 | 74.0 | 5,568 | 53 | 34 | 870 | 6,620 | 62,118 | 6,395 |
| Baca............................ | 41 | 11,513 | 22.0 | 37.3 | 3.4 | 12.5 | 7.7 | 3.0 | 845 | 34 | 9 | 636 | 1,510 | 35,770 | 3,846 |
| Bent............................ | 63 | 10,742 | 21.0 | 2.0 | 3.6 | 8.1 | 3.2 | 6.4 | 1,097 | 34 | 9 | 437 | 1,610 | 38,622 | 2,901 |

1. Based on the resident population estimated as of July 1 of the year shown.

# Table B. States and Counties — Land Area and Population

| State / county code | CBSA code[1] | County Type code[2] | STATE County | Land area[3] (sq. mi) | Total persons 2019 | Rank | Per square mile | White | Black | American Indian, Alaska Native | Asian and Pacific Islander | Percent Hispanic or Latino[4] | Under 5 years | 5 to 14 years | 15 to 24 years | 25 to 34 years | 35 to 44 years | 45 to 54 years |
|---|---|---|---|---|---|---|---|---|---|---|---|---|---|---|---|---|---|---|
| | | | | 1 | 2 | 3 | 4 | 5 | 6 | 7 | 8 | 9 | 10 | 11 | 12 | 13 | 14 | 15 |
| | | | COLORADO—Cont'd | | | | | | | | | | | | | | | |
| 08013 | 14500 | 2 | Boulder | 726.4 | 327,171 | 218 | 450.4 | 79.8 | 1.6 | 0.9 | 6.4 | 13.8 | 4.2 | 10.6 | 18.0 | 14.1 | 12.5 | 12.6 |
| 08014 | 19740 | 1 | Broomfield | 33.0 | 72,236 | 755 | 2,189.0 | 78.1 | 2.0 | 1.0 | 8.5 | 12.9 | 4.9 | 12.5 | 11.8 | 16.2 | 14.4 | 13.6 |
| 08015 | | 7 | Chaffee | 1,013.4 | 20,661 | 1,785 | 20.4 | 86.5 | 2.1 | 1.6 | 1.4 | 9.8 | 3.7 | 8.5 | 8.6 | 12.8 | 12.9 | 11.5 |
| 08017 | | 9 | Cheyenne | 1,778.3 | 1,795 | 3,065 | 1.0 | 83.8 | 1.9 | 1.4 | 2.3 | 11.8 | 5.1 | 15.1 | 10.3 | 9.9 | 12.9 | 9.4 |
| 08019 | 19740 | 1 | Clear Creek | 395.1 | 9,586 | 2,448 | 24.3 | 89.6 | 1.6 | 1.5 | 1.8 | 7.4 | 4.0 | 7.8 | 7.8 | 12.3 | 13.3 | 14.7 |
| 08021 | | 9 | Conejos | 1,287.4 | 8,143 | 2,571 | 6.3 | 46.7 | 1.1 | 1.6 | 0.7 | 51.2 | 5.9 | 14.7 | 12.9 | 11.0 | 10.9 | 10.5 |
| 08023 | | 9 | Costilla | 1,227.6 | 3,921 | 2,902 | 3.2 | 35.4 | 1.4 | 2.2 | 1.8 | 60.6 | 4.8 | 11.0 | 10.0 | 9.7 | 10.0 | 10.4 |
| 08025 | | 8 | Crowley | 787.4 | 5,696 | 2,771 | 7.2 | 54.8 | 10.6 | 2.4 | 1.5 | 32.0 | 2.5 | 7.0 | 12.0 | 21.0 | 17.1 | 15.0 |
| 08027 | | 8 | Custer | 738.6 | 5,183 | 2,813 | 7.0 | 91.2 | 1.9 | 1.9 | 0.8 | 6.0 | 3.6 | 7.7 | 7.5 | 7.2 | 8.9 | 10.1 |
| 08029 | | 6 | Delta | 1,142.1 | 31,067 | 1,402 | 27.2 | 82.5 | 1.0 | 1.5 | 1.1 | 15.4 | 4.5 | 11.3 | 9.9 | 9.9 | 10.8 | 10.4 |
| 08031 | 19740 | 1 | Denver | 153.1 | 735,538 | 19 | 4,804.3 | 57.4 | 10.1 | 1.2 | 5.1 | 28.9 | 5.7 | 10.2 | 10.8 | 23.7 | 16.5 | 11.5 |
| 08033 | | 9 | Dolores | 1,067.2 | 2,096 | 3,041 | 2.0 | 88.8 | 1.0 | 4.1 | 1.0 | 7.5 | 3.6 | 11.6 | 10.2 | 9.1 | 11.2 | 11.3 |
| 08035 | 19740 | 1 | Douglas | 840.2 | 360,750 | 201 | 429.4 | 83.1 | 2.2 | 0.8 | 7.2 | 9.4 | 5.6 | 14.1 | 12.7 | 11.7 | 14.6 | 15.5 |
| 08037 | 20780 | 5 | Eagle | 1,684.5 | 54,929 | 934 | 32.6 | 68.1 | 1.2 | 0.7 | 1.8 | 29.3 | 5.3 | 12.1 | 11.0 | 15.3 | 15.6 | 14.8 |
| 08039 | 19740 | 1 | Elbert | 1,850.8 | 27,313 | 1,518 | 14.8 | 89.0 | 1.8 | 1.2 | 1.7 | 8.2 | 4.5 | 12.0 | 11.5 | 9.3 | 11.8 | 14.5 |
| 08041 | 17820 | 2 | El Paso | 2,126.4 | 728,310 | 92 | 342.5 | 71.8 | 7.8 | 1.5 | 5.3 | 18.0 | 6.4 | 13.2 | 14.6 | 16.3 | 13.2 | 11.2 |
| 08043 | 15860 | 4 | Fremont | 1,533.9 | 47,867 | 1,025 | 31.2 | 80.5 | 4.0 | 2.3 | 1.3 | 13.4 | 4.0 | 9.1 | 9.6 | 14.2 | 13.4 | 12.4 |
| 08045 | 24060 | 5 | Garfield | 2,947.4 | 60,366 | 869 | 20.5 | 68.5 | 1.0 | 1.2 | 1.3 | 29.4 | 6.7 | 14.1 | 12.1 | 13.6 | 14.1 | 12.3 |
| 08047 | 19740 | 1 | Gilpin | 150.0 | 6,235 | 2,727 | 41.6 | 88.5 | 1.9 | 1.9 | 2.4 | 7.6 | 3.0 | 8.6 | 7.8 | 10.7 | 13.6 | 16.3 |
| 08049 | | 7 | Grand | 1,846.4 | 15,794 | 2,043 | 8.6 | 87.7 | 1.2 | 1.2 | 1.4 | 9.9 | 3.6 | 9.4 | 9.8 | 14.0 | 13.3 | 12.9 |
| 08051 | | 7 | Gunnison | 3,239.1 | 17,593 | 1,939 | 5.4 | 88.1 | 0.9 | 1.5 | 1.4 | 9.7 | 3.9 | 9.2 | 20.2 | 14.5 | 13.6 | 12.3 |
| 08053 | | 9 | Hinsdale | 1,117.2 | 808 | 3,116 | 0.7 | 91.0 | 2.2 | 2.1 | 1.1 | 5.8 | 4.0 | 9.0 | 10.9 | 8.2 | 10.3 | 9.8 |
| 08055 | | 6 | Huerfano | 1,591.0 | 6,883 | 2,669 | 4.3 | 63.6 | 1.3 | 2.1 | 1.3 | 33.4 | 3.6 | 9.2 | 8.4 | 8.9 | 9.2 | 10.2 |
| 08057 | | 9 | Jackson | 1,613.7 | 1,389 | 3,086 | 0.9 | 84.9 | 0.4 | 1.6 | 0.8 | 13.3 | 5.2 | 9.7 | 7.6 | 11.2 | 11.7 | 13.3 |
| 08059 | 19740 | 1 | Jefferson | 764.4 | 583,283 | 116 | 763.1 | 79.4 | 1.8 | 1.1 | 4.1 | 15.7 | 4.9 | 10.7 | 10.9 | 15.4 | 14.4 | 12.4 |
| 08061 | | 9 | Kiowa | 1,767.8 | 1,458 | 3,083 | 0.8 | 89.1 | 0.8 | 2.0 | 0.9 | 9.1 | 8.0 | 12.2 | 10.0 | 10.8 | 9.9 | 8.8 |
| 08063 | | 7 | Kit Carson | 2,160.8 | 7,121 | 2,648 | 3.3 | 78.9 | 1.1 | 1.4 | 1.2 | 18.9 | 6.1 | 15.1 | 11.4 | 11.1 | 12.1 | 9.5 |
| 08065 | | 6 | Lake | 376.9 | 7,987 | 2,589 | 21.2 | 61.7 | 0.9 | 1.7 | 1.3 | 35.7 | 5.7 | 11.6 | 12.5 | 17.1 | 13.8 | 12.3 |
| 08067 | 20420 | 4 | La Plata | 1,689.7 | 56,564 | 913 | 33.5 | 80.2 | 0.9 | 6.6 | 1.3 | 12.9 | 4.4 | 10.5 | 12.1 | 13.1 | 13.5 | 12.2 |
| 08069 | 22660 | 2 | Larimer | 2,595.8 | 360,428 | 202 | 138.9 | 84.0 | 1.7 | 1.1 | 3.5 | 12.1 | 4.6 | 10.9 | 17.3 | 15.0 | 12.7 | 10.8 |
| 08071 | | 7 | Las Animas | 4,772.9 | 14,420 | 2,128 | 3.0 | 55.1 | 2.2 | 2.1 | 1.8 | 40.1 | 4.3 | 10.2 | 12.0 | 11.4 | 10.8 | 10.8 |
| 08073 | | 8 | Lincoln | 2,577.7 | 5,680 | 2,773 | 2.2 | 77.5 | 5.6 | 1.7 | 1.6 | 15.2 | 5.2 | 11.4 | 11.3 | 15.9 | 13.9 | 11.7 |
| 08075 | 44540 | 7 | Logan | 1,838.6 | 21,974 | 1,721 | 12.0 | 76.8 | 4.7 | 1.5 | 1.1 | 17.2 | 5.0 | 9.8 | 13.4 | 15.8 | 13.1 | 11.4 |
| 08077 | 24300 | 3 | Mesa | 3,328.7 | 155,603 | 432 | 46.7 | 82.5 | 1.3 | 1.4 | 1.7 | 15.0 | 5.3 | 11.9 | 12.3 | 13.1 | 12.5 | 10.7 |
| 08079 | | 9 | Mineral | 875.8 | 773 | 3,121 | 0.9 | 91.5 | 1.3 | 1.2 | 1.3 | 6.2 | 4.0 | 7.0 | 7.5 | 10.1 | 11.3 | 11.6 |
| 08081 | 18780 | 7 | Moffat | 4,743.2 | 13,144 | 2,210 | 2.8 | 81.0 | 1.3 | 1.6 | 1.5 | 16.6 | 6.3 | 14.6 | 11.7 | 12.6 | 12.9 | 11.2 |
| 08083 | | 6 | Montezuma | 2,029.3 | 26,408 | 1,552 | 13.0 | 74.2 | 0.8 | 13.3 | 1.2 | 12.7 | 5.0 | 12.5 | 10.2 | 11.1 | 11.1 | 10.8 |
| 08085 | 33940 | 4 | Montrose | 2,240.9 | 43,322 | 1,108 | 19.3 | 77.2 | 0.8 | 1.5 | 1.3 | 20.8 | 5.1 | 12.2 | 10.5 | 10.2 | 11.5 | 10.9 |
| 08087 | 22820 | 6 | Morgan | 1,280.5 | 28,941 | 1,461 | 22.6 | 58.4 | 4.4 | 0.9 | 1.1 | 36.3 | 7.3 | 14.5 | 12.4 | 14.4 | 12.4 | 10.9 |
| 08089 | | 6 | Otero | 1,261.9 | 18,201 | 1,914 | 14.4 | 54.5 | 1.4 | 1.6 | 1.3 | 42.7 | 5.9 | 13.1 | 13.3 | 11.4 | 11.9 | 10.4 |
| 08091 | 33940 | 9 | Ouray | 540.7 | 5,001 | 2,826 | 9.2 | 91.5 | 0.7 | 1.1 | 1.4 | 6.9 | 3.0 | 8.1 | 7.7 | 8.1 | 10.9 | 12.4 |
| 08093 | 19740 | 1 | Park | 2,193.5 | 18,955 | 1,877 | 8.6 | 90.4 | 1.3 | 1.7 | 1.7 | 6.9 | 3.5 | 8.5 | 7.3 | 9.7 | 12.1 | 14.0 |
| 08095 | | 9 | Phillips | 687.9 | 4,367 | 2,866 | 6.3 | 75.4 | 1.0 | 0.8 | 1.0 | 22.7 | 6.2 | 13.6 | 11.8 | 10.3 | 11.8 | 10.4 |
| 08097 | 24060 | 7 | Pitkin | 970.7 | 17,894 | 1,929 | 18.4 | 85.8 | 1.2 | 0.6 | 2.7 | 11.0 | 4.0 | 8.0 | 9.3 | 14.6 | 13.2 | 14.8 |
| 08099 | | 7 | Prowers | 1,638.4 | 12,106 | 2,288 | 7.4 | 58.8 | 1.3 | 1.5 | 0.9 | 39.0 | 6.8 | 14.7 | 14.0 | 11.2 | 11.6 | 10.1 |
| 08101 | 39380 | 3 | Pueblo | 2,386.6 | 169,823 | 393 | 71.2 | 53.0 | 2.4 | 1.5 | 1.6 | 43.4 | 5.6 | 12.7 | 12.4 | 13.3 | 12.3 | 11.4 |
| 08103 | | 9 | Rio Blanco | 3,221.0 | 6,342 | 2,718 | 2.0 | 86.0 | 1.8 | 1.9 | 1.7 | 10.8 | 5.7 | 14.7 | 12.9 | 12.0 | 13.1 | 10.0 |
| 08105 | | 9 | Rio Grande | 912.0 | 11,296 | 2,327 | 12.4 | 53.1 | 1.1 | 2.1 | 0.9 | 44.2 | 5.5 | 12.4 | 12.2 | 11.5 | 12.1 | 10.8 |
| 08107 | 44460 | 7 | Routt | 2,362.0 | 25,560 | 1,575 | 10.8 | 90.6 | 1.4 | 0.8 | 1.5 | 6.9 | 4.2 | 9.9 | 10.3 | 14.2 | 14.9 | 14.0 |
| 08109 | | 9 | Saguache | 3,168.6 | 6,938 | 2,666 | 2.2 | 61.0 | 1.3 | 2.8 | 1.4 | 35.5 | 4.6 | 11.2 | 9.8 | 8.5 | 10.6 | 11.8 |
| 08111 | | 9 | San Juan | 387.5 | 748 | 3,123 | 1.9 | 84.9 | 1.5 | 1.9 | 0.8 | 12.8 | 2.8 | 7.1 | 6.8 | 12.0 | 14.6 | 12.3 |
| 08113 | | 9 | San Miguel | 1,286.7 | 8,105 | 2,573 | 6.3 | 86.0 | 1.0 | 1.2 | 1.5 | 11.7 | 3.8 | 9.9 | 9.0 | 15.5 | 14.5 | 16.0 |
| 08115 | | 9 | Sedgwick | 548.0 | 2,260 | 3,021 | 4.1 | 80.7 | 1.6 | 1.2 | 1.2 | 17.1 | 4.7 | 10.8 | 9.7 | 10.0 | 10.8 | 9.1 |
| 08117 | 14720 | 5 | Summit | 608.3 | 30,631 | 1,415 | 50.4 | 82.1 | 1.6 | 0.7 | 2.0 | 15.0 | 4.4 | 8.9 | 8.7 | 19.9 | 15.2 | 14.3 |
| 08119 | 17820 | 2 | Teller | 557.1 | 25,529 | 1,577 | 45.8 | 89.0 | 1.4 | 1.9 | 1.9 | 8.0 | 3.7 | 9.9 | 8.6 | 9.6 | 10.9 | 12.8 |
| 08121 | | 9 | Washington | 2,518.1 | 4,875 | 2,834 | 1.9 | 87.0 | 1.8 | 0.9 | 0.6 | 11.1 | 5.5 | 13.0 | 10.9 | 11.8 | 12.1 | 10.8 |
| 08123 | 24540 | 2 | Weld | 3,984.9 | 333,983 | 213 | 83.8 | 66.1 | 1.7 | 1.1 | 2.5 | 30.2 | 7.0 | 14.5 | 13.2 | 15.4 | 14.2 | 11.9 |
| 08125 | | 7 | Yuma | 2,364.4 | 10,047 | 2,416 | 4.2 | 73.4 | 1.0 | 0.8 | 0.5 | 25.3 | 7.3 | 15.8 | 11.6 | 11.7 | 11.8 | 10.9 |
| 09000 | | 0 | CONNECTICUT | 4,842.7 | 3,557,006 | X | 734.5 | 66.9 | 11.6 | 0.7 | 5.6 | 17.2 | 5.0 | 11.4 | 13.3 | 12.6 | 12.1 | 13.0 |
| 09001 | 14860 | 2 | Fairfield | 625.0 | 942,426 | 55 | 1,507.9 | 61.6 | 12.1 | 0.4 | 6.6 | 20.9 | 5.4 | 12.5 | 13.2 | 11.5 | 12.5 | 13.9 |
| 09003 | 25540 | 1 | Hartford | 734.9 | 889,226 | 64 | 1,210.0 | 60.9 | 14.7 | 0.6 | 6.7 | 19.1 | 5.3 | 11.6 | 12.6 | 13.4 | 12.7 | 12.6 |
| 09005 | 45860 | 4 | Litchfield | 920.5 | 179,610 | 369 | 195.1 | 88.2 | 2.5 | 0.6 | 2.7 | 7.5 | 4.1 | 9.9 | 10.8 | 10.9 | 11.2 | 13.4 |
| 09007 | 25540 | 1 | Middlesex | 369.3 | 161,657 | 414 | 437.7 | 84.6 | 6.0 | 0.6 | 4.1 | 6.7 | 4.0 | 9.4 | 12.1 | 11.7 | 11.2 | 13.4 |
| 09009 | 35300 | 2 | New Haven | 604.3 | 851,948 | 68 | 1,409.8 | 62.5 | 14.4 | 0.6 | 4.9 | 19.6 | 5.1 | 11.1 | 13.4 | 13.5 | 12.1 | 12.6 |
| 09011 | 35980 | 2 | New London | 665.1 | 264,999 | 262 | 398.4 | 77.2 | 7.4 | 1.8 | 5.3 | 11.7 | 4.8 | 10.7 | 13.3 | 13.3 | 11.5 | 12.2 |

1. CBSA = Core Based Statistical Area. See Appendix A for explanation. See Appendix B for list of metropolitan areas with component counties. 2. County type code from the Economic Research
Service of USDA Rural-Urban Continuum Codes. See Appendix A for definition. 3. Dry land or land partially or temporarily covered by water. 4. May be of any race.

# Table B. States and Counties — Population and Households

| STATE County | 55 to 64 years | 65 to 74 years | 75 years and over | Percent female | 2000 | 2010 | 2000-2010 | 2010-2020 | Births | Deaths | Net Migration | Number | Persons per household | Family house-holds | Female family house-holder[1] | One person |
|---|---|---|---|---|---|---|---|---|---|---|---|---|---|---|---|---|
| | 16 | 17 | 18 | 19 | 20 | 21 | 22 | 23 | 24 | 25 | 26 | 27 | 28 | 29 | 30 | 31 |
| COLORADO—Cont'd | | | | | | | | | | | | | | | | |
| Boulder | 12.4 | 9.7 | 6.0 | 49.7 | 269,814 | 294,560 | 9.2 | 11.1 | 28,618 | 18,203 | 22,129 | 127,415 | 2.44 | 57.9 | 7.8 | 28.7 |
| Broomfield | 11.6 | 9.0 | 5.9 | 50.0 | 38,272 | 55,848 | 45.9 | 29.3 | 6,961 | 3,734 | 13,025 | 27,470 | 2.46 | 65.3 | 6.9 | 26.0 |
| Chaffee | 15.8 | 16.5 | 9.9 | 47.1 | 16,242 | 17,809 | 9.6 | 16.0 | 1,388 | 1,798 | 3,242 | 8,231 | 2.21 | 65.1 | 6.5 | 27.9 |
| Cheyenne | 14.8 | 13.0 | 9.5 | 49.1 | 2,231 | 1,833 | -17.8 | -2.1 | 234 | 195 | -77 | 774 | 2.56 | 62.7 | 5.3 | 37.0 |
| Clear Creek | 18.5 | 15.0 | 6.7 | 47.2 | 9,322 | 9,073 | -2.7 | 5.7 | 718 | 634 | 429 | 4,395 | 2.13 | 58.1 | 7.5 | 35.0 |
| Conejos | 13.6 | 12.1 | 8.3 | 49.6 | 8,400 | 8,256 | -1.7 | -1.4 | 1,087 | 855 | -345 | 3,183 | 2.53 | 62.2 | 9.8 | 35.7 |
| Costilla | 17.1 | 16.0 | 10.9 | 48.7 | 3,663 | 3,524 | -3.8 | 11.3 | 324 | 401 | 474 | 1,746 | 2.14 | 59.7 | 14.2 | 35.3 |
| Crowley | 11.3 | 8.3 | 5.8 | 27.0 | 5,518 | 5,823 | 5.5 | -2.2 | 342 | 437 | -85 | 1,301 | 3.69 | 69.6 | 12.9 | 26.4 |
| Custer | 20.9 | 23.7 | 10.4 | 48.6 | 3,503 | 4,251 | 21.4 | 21.9 | 279 | 395 | 1,040 | 2,214 | 2.15 | 67.9 | 2.8 | 28.5 |
| Delta | 15.6 | 16.4 | 11.1 | 50.0 | 27,834 | 30,954 | 11.2 | 0.4 | 3,030 | 3,866 | 961 | 11,883 | 2.46 | 65.1 | 7.1 | 30.3 |
| Denver | 9.5 | 7.4 | 4.7 | 49.9 | 554,636 | 599,454 | 8.1 | 22.7 | 94,551 | 45,101 | 84,583 | 301,501 | 2.29 | 48.4 | 9.4 | 38.2 |
| Dolores | 16.8 | 14.9 | 11.4 | 47.5 | 1,844 | 2,064 | 11.9 | 1.6 | 167 | 182 | 43 | 759 | 2.45 | 64.7 | 9.9 | 32.0 |
| Douglas | 12.8 | 8.3 | 4.8 | 50.0 | 175,766 | 285,480 | 62.4 | 26.4 | 35,738 | 12,895 | 52,175 | 120,709 | 2.78 | 77.6 | 6.6 | 17.7 |
| Eagle | 12.8 | 9.2 | 4.0 | 47.2 | 41,659 | 52,200 | 25.3 | 5.2 | 6,415 | 1,227 | -2,478 | 18,171 | 3.00 | 66.2 | 5.1 | 21.3 |
| Elbert | 18.6 | 12.2 | 5.6 | 49.2 | 19,872 | 23,084 | 16.2 | 18.3 | 1,956 | 1,373 | 3,648 | 9,280 | 2.76 | 79.8 | 6.7 | 15.9 |
| El Paso | 11.7 | 8.4 | 5.2 | 49.5 | 516,929 | 622,256 | 20.4 | 17.0 | 94,876 | 44,122 | 55,160 | 257,507 | 2.65 | 68.2 | 10.1 | 25.1 |
| Fremont | 14.5 | 13.5 | 9.4 | 42.8 | 46,145 | 46,825 | 1.5 | 2.2 | 3,854 | 5,672 | 2,838 | 17,136 | 2.11 | 66.8 | 10.2 | 27.3 |
| Garfield | 12.8 | 9.4 | 4.8 | 48.9 | 43,791 | 56,373 | 28.7 | 7.1 | 8,215 | 3,345 | -901 | 21,498 | 2.70 | 70.9 | 8.9 | 20.9 |
| Gilpin | 21.3 | 14.0 | 4.8 | 47.7 | 4,757 | 5,449 | 14.5 | 14.4 | 403 | 289 | 657 | 2,802 | 2.11 | 58.5 | 4.4 | 34.4 |
| Grand | 17.8 | 14.0 | 5.3 | 46.8 | 12,442 | 14,843 | 19.3 | 6.4 | 1,179 | 668 | 406 | 5,884 | 2.55 | 61.2 | 7.2 | 29.7 |
| Gunnison | 11.9 | 9.8 | 4.5 | 46.2 | 13,956 | 15,324 | 9.8 | 14.8 | 1,461 | 730 | 1,511 | 6,616 | 2.41 | 51.1 | 5.8 | 33.6 |
| Hinsdale | 17.1 | 19.3 | 11.5 | 48.9 | 790 | 843 | 6.7 | -4.2 | 52 | 58 | -32 | 377 | 2.27 | 67.6 | 4.0 | 28.1 |
| Huerfano | 17.7 | 19.3 | 13.3 | 49.0 | 7,862 | 6,712 | -14.6 | 2.5 | 475 | 1,058 | 753 | 3,225 | 2.02 | 57.8 | 9.1 | 37.4 |
| Jackson | 16.4 | 15.0 | 9.8 | 47.9 | 1,577 | 1,393 | -11.7 | -0.3 | 125 | 119 | -13 | 558 | 2.24 | 68.5 | 2.9 | 23.1 |
| Jefferson | 13.9 | 10.6 | 6.7 | 50.0 | 525,507 | 534,832 | 1.8 | 9.1 | 58,907 | 44,010 | 33,366 | 232,284 | 2.44 | 64.3 | 8.9 | 27.0 |
| Kiowa | 16.6 | 12.6 | 11.2 | 50.8 | 1,622 | 1,398 | -13.8 | 4.3 | 173 | 200 | 86 | 638 | 2.28 | 60.5 | 10.3 | 34.5 |
| Kit Carson | 14.4 | 10.7 | 9.6 | 49.7 | 8,011 | 8,278 | 3.3 | -14.0 | 962 | 759 | -1,403 | 3,004 | 2.30 | 66.7 | 6.6 | 31.5 |
| Lake | 12.6 | 9.7 | 4.7 | 46.4 | 7,812 | 7,310 | -6.4 | 9.3 | 865 | 394 | 198 | 3,392 | 2.24 | 57.1 | 7.5 | 26.9 |
| La Plata | 14.7 | 12.9 | 6.6 | 49.6 | 43,941 | 51,335 | 16.8 | 10.2 | 5,287 | 3,448 | 3,345 | 21,794 | 2.47 | 60.0 | 6.2 | 31.3 |
| Larimer | 12.0 | 10.3 | 6.4 | 50.2 | 251,494 | 299,628 | 19.1 | 20.3 | 34,394 | 22,266 | 48,125 | 137,021 | 2.45 | 61.6 | 7.1 | 24.3 |
| Las Animas | 15.0 | 15.4 | 10.1 | 47.7 | 15,207 | 15,506 | 2.0 | -7.0 | 1,332 | 1,798 | -643 | 6,670 | 2.01 | 57.1 | 9.1 | 35.9 |
| Lincoln | 12.6 | 9.1 | 8.8 | 41.0 | 6,087 | 5,469 | -10.2 | 3.9 | 601 | 547 | 155 | 1,518 | 2.31 | 60.5 | 8.9 | 34.5 |
| Logan | 12.7 | 10.5 | 8.3 | 43.3 | 20,504 | 22,709 | 10.8 | -3.2 | 2,336 | 2,131 | -946 | 8,393 | 2.54 | 60.1 | 7.3 | 33.5 |
| Mesa | 13.6 | 12.1 | 8.4 | 50.7 | 116,255 | 146,736 | 26.2 | 6.0 | 17,901 | 15,037 | 6,040 | 61,742 | 2.39 | 63.4 | 10.4 | 28.8 |
| Mineral | 16.9 | 20.6 | 11.0 | 49.9 | 831 | 712 | -14.3 | 8.6 | 59 | 72 | 73 | 367 | 2.20 | 71.7 | 5.2 | 23.2 |
| Moffat | 14.1 | 10.2 | 6.4 | 48.7 | 13,184 | 13,791 | 4.6 | -4.7 | 1,803 | 1,016 | -1,455 | 5,366 | 2.42 | 64.9 | 12.9 | 28.3 |
| Montezuma | 15.1 | 15.1 | 9.0 | 50.6 | 23,830 | 25,541 | 7.2 | 3.4 | 2,953 | 2,794 | 727 | 10,655 | 2.42 | 66.2 | 9.7 | 27.1 |
| Montrose | 14.8 | 14.4 | 10.6 | 50.8 | 33,432 | 41,272 | 23.5 | 5.0 | 4,540 | 4,420 | 1,941 | 17,140 | 2.40 | 68.3 | 8.9 | 28.1 |
| Morgan | 12.1 | 8.7 | 7.2 | 49.0 | 27,171 | 28,159 | 3.6 | 2.8 | 4,443 | 2,683 | -990 | 10,928 | 2.57 | 70.1 | 12.5 | 24.4 |
| Otero | 13.1 | 11.7 | 9.2 | 50.8 | 20,311 | 18,833 | -7.3 | -3.4 | 2,242 | 2,461 | -409 | 7,722 | 2.31 | 60.2 | 11.9 | 33.1 |
| Ouray | 19.9 | 20.3 | 9.5 | 49.7 | 3,742 | 4,442 | 18.7 | 12.6 | 295 | 295 | 551 | 2,163 | 2.22 | 63.3 | 4.8 | 25.6 |
| Park | 22.1 | 16.8 | 5.9 | 47.1 | 14,523 | 16,199 | 11.5 | 17.0 | 1,186 | 935 | 2,471 | 6,931 | 2.56 | 68.8 | 3.2 | 24.1 |
| Phillips | 13.0 | 11.0 | 11.8 | 50.5 | 4,480 | 4,442 | -0.8 | -1.7 | 547 | 520 | -108 | 1,689 | 2.49 | 66.0 | 6.7 | 31.4 |
| Pitkin | 15.0 | 13.3 | 7.9 | 47.8 | 14,872 | 17,145 | 15.3 | 4.4 | 1,422 | 547 | -128 | 7,467 | 2.38 | 51.6 | 8.5 | 32.9 |
| Prowers | 12.8 | 11.1 | 7.6 | 50.0 | 14,483 | 12,551 | -13.3 | -3.5 | 1,647 | 1,275 | -832 | 4,888 | 2.38 | 64.7 | 11.3 | 30.7 |
| Pueblo | 13.0 | 11.4 | 8.0 | 50.7 | 141,472 | 159,066 | 12.4 | 6.8 | 19,008 | 17,833 | 9,683 | 64,314 | 2.52 | 63.7 | 13.2 | 30.0 |
| Rio Blanco | 14.6 | 10.2 | 6.8 | 48.8 | 5,986 | 6,673 | 11.5 | -5.0 | 768 | 522 | -588 | 2,301 | 2.62 | 63.7 | 7.1 | 31.6 |
| Rio Grande | 13.6 | 13.3 | 8.7 | 50.2 | 12,413 | 11,982 | -3.5 | -5.7 | 1,373 | 1,286 | -775 | 4,823 | 2.30 | 61.1 | 9.9 | 34.3 |
| Routt | 15.0 | 12.2 | 5.2 | 48.1 | 19,690 | 23,506 | 19.4 | 8.7 | 2,205 | 1,121 | 942 | 9,603 | 2.57 | 62.3 | 5.6 | 26.4 |
| Saguache | 17.8 | 17.9 | 7.8 | 50.4 | 5,917 | 6,108 | 3.2 | 13.6 | 687 | 456 | 598 | 2,860 | 2.30 | 62.5 | 9.8 | 34.9 |
| San Juan | 18.0 | 17.1 | 9.2 | 44.8 | 558 | 699 | 25.3 | 7.0 | 46 | 43 | 44 | 289 | 2.04 | 46.0 | 2.4 | 40.1 |
| San Miguel | 14.7 | 12.1 | 4.4 | 46.8 | 6,594 | 7,359 | 11.6 | 10.1 | 669 | 239 | 317 | 3,552 | 2.26 | 53.3 | 5.9 | 36.0 |
| Sedgwick | 15.8 | 15.7 | 13.5 | 50.3 | 2,747 | 2,382 | -13.3 | -5.1 | 232 | 364 | 10 | 967 | 2.36 | 58.3 | 10.1 | 37.7 |
| Summit | 13.7 | 10.6 | 4.2 | 45.4 | 23,548 | 27,994 | 18.9 | 9.4 | 2,824 | 643 | 417 | 10,641 | 2.84 | 55.2 | 5.1 | 29.0 |
| Teller | 20.3 | 17.4 | 6.8 | 49.3 | 20,555 | 23,360 | 13.6 | 9.3 | 1,858 | 1,676 | 1,985 | 10,248 | 2.38 | 68.6 | 5.9 | 25.1 |
| Washington | 14.6 | 11.7 | 9.6 | 47.9 | 4,926 | 4,798 | -2.6 | 1.6 | 472 | 505 | 109 | 2,087 | 2.19 | 68.8 | 9.6 | 26.4 |
| Weld | 11.1 | 8.0 | 4.8 | 49.4 | 180,926 | 252,836 | 39.7 | 32.1 | 41,936 | 17,551 | 56,262 | 104,671 | 2.85 | 73.1 | 10.0 | 20.6 |
| Yuma | 12.2 | 10.2 | 8.5 | 49.7 | 9,841 | 10,049 | 2.1 | 0.0 | 1,473 | 1,011 | -472 | 4,028 | 2.45 | 70.3 | 11.2 | 24.7 |
| CONNECTICUT | 14.4 | 10.2 | 7.9 | 51.2 | 3,405,565 | 3,574,151 | 5.0 | -0.5 | 367,838 | 312,319 | -72,975 | 1,370,746 | 2.53 | 65.2 | 12.5 | 28.5 |
| Fairfield | 14.3 | 9.3 | 7.5 | 51.3 | 882,567 | 916,910 | 3.9 | 2.8 | 102,311 | 68,987 | -7,760 | 340,189 | 2.72 | 69.5 | 11.8 | 24.9 |
| Hartford | 13.8 | 9.9 | 7.9 | 51.4 | 857,183 | 894,016 | 4.3 | -0.5 | 96,172 | 82,497 | -18,646 | 350,408 | 2.48 | 65.4 | 14.1 | 29.9 |
| Litchfield | 17.3 | 13.3 | 9.5 | 50.5 | 182,193 | 189,876 | 4.2 | -5.4 | 14,714 | 18,539 | -6,434 | 74,143 | 2.41 | 65.4 | 8.5 | 28.4 |
| Middlesex | 16.4 | 12.4 | 9.4 | 51.2 | 155,071 | 165,673 | 6.8 | -2.4 | 13,875 | 15,674 | -2,137 | 66,971 | 2.34 | 63.6 | 9.4 | 29.3 |
| New Haven | 14.0 | 10.3 | 8.0 | 51.8 | 824,008 | 862,476 | 4.7 | -1.2 | 90,367 | 79,982 | -20,996 | 330,572 | 2.51 | 62.5 | 14.2 | 31.2 |
| New London | 14.8 | 11.1 | 8.3 | 49.9 | 259,088 | 274,076 | 5.8 | -3.3 | 27,091 | 25,087 | -11,140 | 107,827 | 2.37 | 64.6 | 11.9 | 28.8 |

1. No spouse present.

# Table B. States and Counties — Population, Vital Statistics, Health, and Crime

| STATE County | Persons in group quarters, 2020 | Daytime Population, 2015-2019 | | Births, 2020 | | Deaths, 2020 | | Persons under 65 with no health insurance, 2019 | | Medicare, 2020 | | | Crimes reported by county police or sheriff, 2019 | |
|---|---|---|---|---|---|---|---|---|---|---|---|---|---|---|
| | | Number | Employment/ residence ratio | Total | Rate[1] | Number | Rate[1] | Number | Percent | Total beneficiaries | Enrolled in Original Medicare | Enrolled in Medicare Advantage | Violent | Property |
| | 32 | 33 | 34 | 35 | 36 | 37 | 38 | 39 | 40 | 41 | 42 | 43 | 44 | 45 |
| COLORADO—Cont'd | | | | | | | | | | | | | | |
| Boulder | 10,463 | 358,000 | 1.21 | 2,517 | 7.7 | 2,164 | 6.6 | 20,712 | 7.7 | 53,602 | 29,926 | 23,676 | 85 | 617 |
| Broomfield | 282 | 72,777 | 1.13 | 672 | 9.3 | 445 | 6.2 | 3,245 | 5.3 | 10,980 | 5,089 | 5,891 | NA | NA |
| Chaffee | 1,395 | 19,773 | 1.03 | 138 | 6.7 | 191 | 9.2 | 1,539 | 11.3 | 5,527 | 4,163 | 1,364 | 3 | 57 |
| Cheyenne | 42 | 2,131 | 1.13 | 14 | 7.8 | 14 | 7.8 | 204 | 14.6 | 393 | 354 | 40 | 3 | 14 |
| Clear Creek | 84 | 7,943 | 0.69 | 63 | 6.6 | 81 | 8.4 | 547 | 7.2 | 1,925 | 1,086 | 840 | NA | NA |
| Conejos | 36 | 7,198 | 0.68 | 97 | 11.9 | 80 | 9.8 | 708 | 11.1 | 1,951 | 1,326 | 625 | NA | NA |
| Costilla | 0 | 3,466 | 0.79 | 36 | 9.2 | 29 | 7.4 | 419 | 15.3 | 1,194 | 825 | 369 | NA | NA |
| Crowley | 2,675 | 5,342 | 0.79 | 27 | 4.7 | 48 | 8.4 | 286 | 12.5 | 869 | 656 | 213 | 5 | 61 |
| Custer | 150 | 4,356 | 0.75 | 37 | 7.1 | 42 | 8.1 | 413 | 12.9 | 1,827 | 1,312 | 514 | 8 | 67 |
| Delta | 582 | 28,799 | 0.84 | 268 | 8.6 | 422 | 13.6 | 2,524 | 11.5 | 9,488 | 6,584 | 2,904 | NA | NA |
| Denver | 16,064 | 858,070 | 1.39 | 9,233 | 12.6 | 5,406 | 7.3 | 63,469 | 10.1 | 92,682 | 42,509 | 50,173 | NA | NA |
| Dolores | 0 | 1,696 | 0.79 | 16 | 7.6 | 11 | 5.2 | 157 | 10.5 | 568 | 496 | 72 | 1 | 22 |
| Douglas | 651 | 301,198 | 0.81 | 3,525 | 9.8 | 1,704 | 4.7 | 13,403 | 4.3 | 48,922 | 26,470 | 22,452 | 318 | 2,098 |
| Eagle | 55 | 53,487 | 0.96 | 559 | 10.2 | 163 | 3.0 | 6,582 | 13.7 | 6,419 | 5,852 | 567 | 28 | 190 |
| Elbert | 73 | 17,870 | 0.42 | 216 | 7.9 | 170 | 6.2 | 1,584 | 7.1 | 5,063 | 3,143 | 1,920 | 5 | 5 |
| El Paso | 19,818 | 693,547 | 0.98 | 9,089 | 12.5 | 5,259 | 7.2 | 54,443 | 9.0 | 111,603 | 67,502 | 44,101 | 484 | 2,351 |
| Fremont | 7,851 | 47,087 | 0.98 | 378 | 7.9 | 612 | 12.8 | 2,680 | 9.2 | 12,673 | 7,050 | 5,623 | 35 | 198 |
| Garfield | 865 | 54,681 | 0.86 | 763 | 12.6 | 405 | 6.7 | 7,965 | 15.7 | 8,910 | 7,776 | 1,134 | 70 | 205 |
| Gilpin | 49 | 8,360 | 1.62 | 31 | 5.0 | 38 | 6.1 | 218 | 4.3 | 1,071 | 646 | 425 | 25 | 83 |
| Grand | 222 | 14,898 | 0.95 | 110 | 7.0 | 70 | 4.4 | 1,427 | 11.2 | 2,693 | 2,104 | 588 | NA | NA |
| Gunnison | 990 | 16,888 | 1.01 | 131 | 7.4 | 77 | 4.4 | 1,398 | 9.9 | 2,522 | 2,207 | 315 | 6 | 21 |
| Hinsdale | 56 | 877 | 1.05 | 6 | 7.4 | 7 | 8.7 | 68 | 12.2 | 251 | 217 | 34 | NA | NA |
| Huerfano | 165 | 6,668 | 1.00 | 39 | 5.7 | 106 | 15.4 | 399 | 8.8 | 2,440 | 1,768 | 672 | 24 | 151 |
| Jackson | 2 | 1,222 | 0.94 | 12 | 8.6 | 12 | 8.6 | 159 | 15.2 | 342 | 303 | 39 | NA | NA |
| Jefferson | 8,123 | 523,299 | 0.83 | 5,701 | 9.8 | 5,002 | 8.6 | 35,897 | 7.5 | 108,278 | 45,558 | 62,720 | NA | NA |
| Kiowa | 14 | 1,465 | 0.96 | 19 | 13.0 | 14 | 9.6 | 111 | 10.5 | 348 | 326 | 22 | 2 | 14 |
| Kit Carson | 58 | 7,556 | 1.03 | 91 | 12.8 | 71 | 10.0 | 811 | 14.5 | 1,522 | 1,376 | 146 | 4 | 42 |
| Lake | 153 | 5,805 | 0.54 | 95 | 11.9 | 35 | 4.4 | 969 | 14.4 | 1,029 | 886 | 143 | NA | NA |
| La Plata | 1,809 | 57,260 | 1.06 | 439 | 7.8 | 399 | 7.1 | 5,001 | 11.3 | 11,336 | 9,135 | 2,201 | 50 | 197 |
| Larimer | 9,905 | 340,550 | 0.98 | 3,241 | 9.0 | 2,488 | 6.9 | 22,447 | 7.7 | 65,205 | 38,640 | 26,565 | 211 | 753 |
| Las Animas | 808 | 14,267 | 1.00 | 111 | 7.7 | 177 | 12.3 | 1,297 | 13.2 | 4,283 | 2,999 | 1,285 | 16 | 53 |
| Lincoln | 1,011 | 6,083 | 1.34 | 58 | 10.2 | 48 | 8.5 | 382 | 10.4 | 1,047 | 929 | 119 | NA | NA |
| Logan | 3,448 | 21,675 | 0.94 | 219 | 10.0 | 218 | 9.9 | 1,627 | 11.2 | 4,278 | 3,331 | 948 | 33 | 109 |
| Mesa | 4,434 | 149,552 | 0.98 | 1,596 | 10.3 | 1,688 | 10.8 | 13,094 | 10.9 | 35,695 | 22,687 | 13,008 | 164 | 1,059 |
| Mineral | 0 | 858 | 1.08 | 6 | 7.8 | 10 | 12.9 | 56 | 11.0 | 298 | 242 | 56 | NA | NA |
| Moffat | 102 | 11,546 | 0.74 | 154 | 11.7 | 99 | 7.5 | 1,103 | 10.0 | 2,368 | 2,124 | 244 | 15 | 10 |
| Montezuma | 243 | 25,732 | 0.97 | 253 | 9.6 | 323 | 12.2 | 2,751 | 13.9 | 6,921 | 5,535 | 1,386 | 15 | 136 |
| Montrose | 542 | 40,945 | 0.96 | 410 | 9.5 | 495 | 11.4 | 4,115 | 12.9 | 11,510 | 8,219 | 3,291 | 20 | 176 |
| Morgan | 563 | 28,712 | 1.01 | 428 | 14.8 | 268 | 9.3 | 3,547 | 14.9 | 5,099 | 4,108 | 992 | 21 | 127 |
| Otero | 489 | 18,398 | 1.02 | 218 | 12.0 | 241 | 13.2 | 1,572 | 11.4 | 4,707 | 3,438 | 1,269 | 21 | 92 |
| Ouray | 0 | 4,781 | 0.99 | 22 | 4.4 | 23 | 4.6 | 309 | 8.6 | 1,466 | 1,220 | 245 | NA | NA |
| Park | 92 | 13,635 | 0.52 | 138 | 7.3 | 105 | 5.5 | 1,289 | 8.8 | 3,824 | 2,715 | 1,108 | 41 | 65 |
| Phillips | 60 | 4,358 | 1.04 | 50 | 11.4 | 57 | 13.1 | 435 | 13.5 | 1,038 | 959 | 79 | 2 | 2 |
| Pitkin | 72 | 26,486 | 1.73 | 129 | 7.2 | 51 | 2.9 | 1,500 | 10.6 | 3,045 | 2,856 | 189 | 2 | 23 |
| Prowers | 315 | 12,086 | 1.01 | 145 | 12.0 | 123 | 10.2 | 1,383 | 14.6 | 2,515 | 2,299 | 216 | 2 | 16 |
| Pueblo | 4,077 | 164,205 | 0.97 | 1,803 | 10.6 | 2,000 | 11.8 | 12,543 | 9.6 | 39,116 | 19,892 | 19,224 | 45 | 1,058 |
| Rio Blanco | 262 | 6,779 | 1.14 | 58 | 9.1 | 49 | 7.7 | 479 | 9.5 | 1,221 | 1,031 | 190 | NA | NA |
| Rio Grande | 198 | 11,427 | 1.03 | 122 | 10.8 | 133 | 11.8 | 1,227 | 14.2 | 3,289 | 2,620 | 669 | 3 | 55 |
| Routt | 324 | 26,507 | 1.10 | 189 | 7.4 | 139 | 5.4 | 2,043 | 9.6 | 4,227 | 3,713 | 514 | 5 | 48 |
| Saguache | 18 | 6,368 | 0.92 | 54 | 7.8 | 60 | 8.6 | 899 | 18.1 | 1,400 | 1,165 | 235 | NA | NA |
| San Juan | 0 | 743 | 1.45 | 2 | 2.7 | 6 | 8.0 | 65 | 12.2 | 173 | 157 | 16 | 1 | 9 |
| San Miguel | 9 | 9,208 | 1.23 | 53 | 6.5 | 35 | 4.3 | 844 | 12.3 | 1,141 | 1,016 | 125 | 2 | 5 |
| Sedgwick | 35 | 2,212 | 0.88 | 18 | 8.0 | 25 | 11.1 | 199 | 12.4 | 730 | 669 | 61 | NA | NA |
| Summit | 273 | 33,131 | 1.13 | 252 | 8.2 | 75 | 2.4 | 3,217 | 12.1 | 3,669 | 3,018 | 651 | 36 | 237 |
| Teller | 132 | 22,093 | 0.79 | 173 | 6.8 | 218 | 8.5 | 1,544 | 8.0 | 6,698 | 3,948 | 2,751 | 13 | 53 |
| Washington | 184 | 4,362 | 0.77 | 46 | 9.4 | 49 | 10.1 | 386 | 10.6 | 1,079 | 991 | 88 | 3 | 74 |
| Weld | 5,724 | 277,913 | 0.82 | 4,428 | 13.3 | 2,042 | 6.1 | 30,273 | 10.9 | 46,091 | 25,768 | 20,323 | 136 | 664 |
| Yuma | 196 | 10,360 | 1.07 | 141 | 14.0 | 91 | 9.1 | 1,194 | 15.1 | 1,918 | 1,778 | 140 | 4 | 44 |
| CONNECTICUT | 110,606 | 3,536,304 | 0.98 | 34,258 | 9.6 | 33,477 | 9.4 | 201,146 | 7.0 | 691,541 | 381,673 | 309,868 | X | X |
| Fairfield | 19,496 | 943,974 | 1.00 | 9,461 | 10.0 | 7,423 | 7.9 | 72,442 | 9.3 | 159,488 | 103,589 | 55,899 | NA | NA |
| Hartford | 24,967 | 967,368 | 1.17 | 9,108 | 10.2 | 8,831 | 9.9 | 43,194 | 6.0 | 174,825 | 84,713 | 90,112 | NA | NA |
| Litchfield | 2,463 | 155,802 | 0.72 | 1,375 | 7.7 | 1,973 | 11.0 | 8,618 | 6.1 | 43,407 | 25,563 | 17,844 | NA | NA |
| Middlesex | 4,956 | 150,065 | 0.85 | 1,274 | 7.9 | 1,662 | 10.3 | 5,840 | 4.7 | 37,212 | 18,780 | 18,431 | NA | NA |
| New Haven | 28,515 | 825,683 | 0.92 | 8,397 | 9.9 | 8,512 | 10.0 | 47,465 | 7.0 | 167,724 | 89,020 | 78,705 | NA | NA |
| New London | 10,858 | 266,270 | 0.99 | 2,455 | 9.3 | 2,768 | 10.4 | 12,556 | 6.1 | 57,332 | 33,939 | 23,393 | NA | NA |

1. Per 1,000 estimated resident population.

# Table B. States and Counties — Crime, Education, Money Income, and Poverty

| STATE County | County law enforcement employment, 2019 | | School enrollment and attainment, 2015-2019 | | | | Local government expenditures,[3] 2017-2018 | | Money income, 2015-2019 | | | | Income and poverty, 2019 | | | |
| | | | Enrollment[1] | | Attainment[2] (percent) | | | | | Households | | | | Percent below poverty level | | |
| | | | | | | | | | | | Percent | | | | | |
| | Officers | Civilians | Total | Percent private | High school graduate or less | Bachelor's degree or more | Total current spending (mil dol) | Current spending per student (dollars) | Per capita income[4] | Median income (dollars) | with income of less than $50,000 | with income of $200,000 or more | Median household income (dollars) | All persons | Children under 18 years | Children 5 to 17 years in families |
| | 46 | 47 | 48 | 49 | 50 | 51 | 52 | 53 | 54 | 55 | 56 | 57 | 58 | 59 | 60 | 61 |

| COLORADO—Cont'd | | | | | | | | | | | | | | | | |
| Boulder | 108 | 288 | 97,364 | 11.5 | 16.7 | 62.1 | 693.4[5] | 10,884[5] | 46,826 | 83,019 | 31.4 | 14.2 | 88,341 | 10.7 | 8.1 | 7.0 |
| Broomfield | NA | NA | 17,330 | 12.9 | 17.5 | 55.7 | [5] | [5] | 50,863 | 96,416 | 23.0 | 15.0 | 106,892 | 4.1 | 4.6 | 3.8 |
| Chaffee | 21 | 37 | 2,791 | 10.0 | 33.5 | 33.8 | 25 | 10,937 | 29,827 | 55,771 | 44.6 | 2.8 | 60,331 | 11.0 | 13.1 | 11.4 |
| Cheyenne | 6 | 4 | 467 | 6.9 | 45.1 | 19.2 | 5 | 17,896 | 29,081 | 53,977 | 46.0 | 4.5 | 52,337 | 13.7 | 20.6 | 19.8 |
| Clear Creek | 25 | 42 | 1,570 | 9.0 | 23.3 | 48.1 | 10 | 12,632 | 39,203 | 67,060 | 38.0 | 7.5 | 79,900 | 7.4 | 9.7 | 8.8 |
| Conejos | 12 | 20 | 1,985 | 9.9 | 46.0 | 21.0 | 16 | 9,864 | 20,640 | 36,084 | 62.0 | 1.2 | 40,944 | 19.9 | 26.2 | 25.1 |
| Costilla | 8 | 6 | 650 | 11.1 | 43.5 | 19.3 | 7 | 13,841 | 21,732 | 30,965 | 71.3 | 2.3 | 34,731 | 24.6 | 38.9 | 36.6 |
| Crowley | 8 | 7 | 1,227 | 11.6 | 56.5 | 9.0 | 5 | 10,327 | 15,517 | 42,135 | 58.1 | 0.4 | 38,224 | 40.0 | 30.3 | 27.2 |
| Custer | 11 | 6 | 774 | 33.2 | 27.2 | 32.5 | 4 | 10,489 | 29,258 | 53,119 | 46.9 | 1.4 | 57,381 | 11.8 | 25.2 | 25.0 |
| Delta | 36 | 29 | 5,825 | 8.5 | 47.6 | 21.7 | 44 | 8,703 | 26,499 | 45,269 | 54.7 | 4.4 | 51,525 | 15.1 | 21.1 | 18.3 |
| Denver | NA | NA | 155,885 | 17.8 | 28.8 | 49.4 | 1,233 | 11,270 | 43,770 | 68,592 | 36.4 | 10.1 | 75,456 | 12.1 | 15.6 | 15.6 |
| Dolores | 5 | 4 | 327 | 16.2 | 42.3 | 24.1 | 4 | 14,048 | 26,323 | 45,972 | 53.9 | 2.8 | 50,232 | 12.8 | 15.4 | 13.7 |
| Douglas | 351 | 171 | 94,579 | 15.2 | 13.9 | 58.0 | 632 | 9,350 | 53,313 | 119,730 | 14.8 | 20.1 | 123,697 | 2.7 | 2.4 | 2.2 |
| Eagle | 46 | 36 | 11,841 | 11.0 | 26.9 | 47.2 | 81 | 11,671 | 44,391 | 84,790 | 27.9 | 13.3 | 89,268 | 7.9 | 8.2 | 6.5 |
| Elbert | 36 | 9 | 5,911 | 16.9 | 27.4 | 35.8 | 31 | 9,445 | 44,690 | 99,199 | 22.0 | 13.6 | 101,733 | 5.7 | 7.1 | 6.2 |
| El Paso | 525 | 317 | 189,593 | 12.8 | 26.2 | 38.5 | 1,152 | 9,526 | 33,728 | 72,675 | 35.3 | 6.6 | 72,675 | 8.8 | 10.4 | 10.2 |
| Fremont | NA | NA | 7,843 | 12.6 | 49.3 | 18.5 | 50 | 9,816 | 22,692 | 49,409 | 50.7 | 2.2 | 52,239 | 16.1 | 20.9 | 18.5 |
| Garfield | 96 | 32 | 14,277 | 12.5 | 39.0 | 31.0 | 125 | 10,606 | 33,393 | 75,937 | 31.3 | 5.4 | 73,788 | 9.6 | 11.9 | 10.7 |
| Gilpin | 35 | 23 | 931 | 11.5 | 23.0 | 39.6 | 6 | 11,678 | 49,641 | 76,429 | 26.4 | 6.8 | 74,806 | 6.8 | 8.5 | 5.7 |
| Grand | 24 | 27 | 2,499 | 6.3 | 32.2 | 38.0 | 20 | 11,800 | 37,876 | 71,198 | 35.4 | 6.4 | 71,029 | 8.1 | 9.5 | 8.5 |
| Gunnison | 14 | 15 | 4,916 | 12.0 | 18.5 | 56.2 | 21 | 9,972 | 32,191 | 56,577 | 43.0 | 5.5 | 60,408 | 11.6 | 10.1 | 8.8 |
| Hinsdale | 3 | 1 | 122 | 17.2 | 29.0 | 45.7 | 2 | 18,292 | 33,570 | 56,339 | 44.6 | 7.2 | 61,391 | 8.9 | 19.1 | 17.4 |
| Huerfano | 19 | 21 | 1,214 | 14.3 | 37.2 | 21.1 | 9 | 11,728 | 25,140 | 38,137 | 61.7 | 1.3 | 39,955 | 20.0 | 35.4 | 31.8 |
| Jackson | 5 | 7 | 217 | 14.3 | 40.8 | 20.0 | 3 | 16,392 | 28,199 | 53,300 | 48.6 | 1.1 | 53,577 | 12.6 | 18.5 | 18.3 |
| Jefferson | 532 | 229 | 128,059 | 13.9 | 25.8 | 45.2 | 841 | 9,756 | 44,119 | 82,986 | 28.2 | 10.8 | 89,696 | 7.2 | 9.0 | 8.7 |
| Kiowa | 5 | 3 | 304 | 5.9 | 40.0 | 21.4 | 3 | 14,742 | 25,937 | 41,731 | 57.8 | 3.6 | 46,901 | 13.5 | 17.6 | 17.7 |
| Kit Carson | 6 | 23 | 1,583 | 5.1 | 48.1 | 18.7 | 15 | 10,688 | 27,797 | 49,349 | 51.2 | 2.8 | 42,951 | 13.2 | 20.0 | 19.3 |
| Lake | 7 | 14 | 1,746 | 13.3 | 42.0 | 32.3 | 13 | 12,185 | 29,122 | 50,565 | 48.9 | 2.3 | 56,850 | 11.4 | 18.1 | 17.4 |
| La Plata | 91 | 25 | 12,819 | 14.5 | 24.5 | 44.2 | 88 | 11,923 | 39,493 | 68,685 | 36.6 | 7.9 | 77,829 | 11.3 | 14.2 | 13.3 |
| Larimer | 276 | 132 | 95,900 | 9.8 | 23.1 | 47.3 | 467 | 9,849 | 37,363 | 71,881 | 34.8 | 7.7 | 75,332 | 10.3 | 8.5 | 8.1 |
| Las Animas | 12 | 20 | 2,689 | 10.9 | 41.6 | 19.5 | 24 | 10,898 | 25,813 | 41,817 | 56.3 | 2.2 | 43,452 | 17.6 | 24.9 | 24.0 |
| Lincoln | 10 | 16 | 920 | 3.3 | 63.6 | 12.6 | 14 | 19,270 | 16,753 | 47,258 | 53.4 | 4.1 | 43,077 | 17.0 | 21.0 | 21.2 |
| Logan | 20 | 25 | 4,686 | 6.6 | 44.2 | 20.2 | 28 | 9,480 | 26,727 | 53,318 | 47.5 | 3.1 | 58,429 | 12.9 | 16.3 | 17.1 |
| Mesa | 131 | 101 | 35,877 | 10.5 | 37.1 | 28.2 | 226 | 9,967 | 29,596 | 55,379 | 45.5 | 3.6 | 60,249 | 11.2 | 14.3 | 12.5 |
| Mineral | 4 | 3 | 105 | 11.4 | 18.4 | 46.5 | 2 | 23,163 | 33,955 | 62,188 | 38.1 | 3.5 | 63,015 | 9.4 | 16.0 | 16.2 |
| Moffat | 18 | 20 | 2,788 | 10.3 | 50.1 | 17.6 | 20 | 8,944 | 29,100 | 57,229 | 44.4 | 4.0 | 63,232 | 12.2 | 14.5 | 12.0 |
| Montezuma | 27 | 40 | 5,257 | 8.5 | 39.2 | 29.6 | 36 | 8,972 | 26,072 | 49,470 | 50.5 | 1.9 | 52,641 | 14.9 | 24.6 | 21.5 |
| Montrose | 48 | 38 | 8,580 | 11.8 | 42.4 | 25.2 | 58 | 8,825 | 28,078 | 50,489 | 49.5 | 2.4 | 50,707 | 13.2 | 18.5 | 17.8 |
| Morgan | 43 | 5 | 6,684 | 8.7 | 52.4 | 17.5 | 52 | 9,284 | 25,064 | 53,682 | 46.6 | 1.7 | 57,535 | 11.9 | 15.0 | 14.3 |
| Otero | 22 | 1 | 4,465 | 10.3 | 44.7 | 18.6 | 36 | 10,747 | 21,110 | 38,169 | 61.6 | 1.3 | 43,723 | 24.2 | 33.6 | 33.3 |
| Ouray | 8 | 0 | 718 | 15.3 | 18.2 | 47.8 | 9 | 16,812 | 41,160 | 66,417 | 39.8 | 8.0 | 72,836 | 6.9 | 9.7 | 8.5 |
| Park | 30 | 25 | 2,851 | 16.3 | 33.2 | 29.8 | 17 | 10,253 | 36,500 | 73,622 | 31.2 | 3.5 | 77,221 | 8.6 | 13.0 | 11.9 |
| Phillips | 4 | 0 | 980 | 5.4 | 44.6 | 23.8 | 13 | 14,320 | 26,494 | 51,155 | 47.8 | 2.9 | 53,432 | 10.4 | 14.8 | 14.2 |
| Pitkin | 22 | 33 | 2,991 | 18.2 | 20.2 | 60.8 | 31 | 18,216 | 54,875 | 78,935 | 27.1 | 9.3 | 92,820 | 6.2 | 5.8 | 5.4 |
| Prowers | 10 | 19 | 2,897 | 8.4 | 45.6 | 18.3 | 24 | 10,193 | 23,698 | 41,929 | 58.6 | 2.2 | 47,051 | 18.2 | 25.5 | 24.5 |
| Pueblo | NA | NA | 40,312 | 7.0 | 39.4 | 21.8 | 244 | 9,096 | 25,051 | 46,783 | 52.7 | 2.4 | 51,075 | 17.8 | 24.1 | 23.9 |
| Rio Blanco | 13 | 14 | 1,741 | 7.1 | 39.6 | 22.7 | 15 | 11,715 | 26,487 | 54,357 | 44.7 | 2.0 | 65,960 | 10.3 | 12.8 | 11.3 |
| Rio Grande | 17 | 26 | 2,799 | 8.7 | 45.1 | 24.5 | 20 | 10,276 | 24,582 | 39,123 | 59.8 | 2.1 | 48,496 | 14.4 | 21.8 | 21.2 |
| Routt | 23 | 29 | 5,241 | 7.3 | 23.4 | 49.8 | 44 | 12,651 | 43,769 | 77,443 | 28.9 | 9.9 | 87,691 | 7.2 | 6.5 | 6.6 |
| Saguache | 10 | 7 | 1,348 | 9.1 | 47.5 | 22.1 | 14 | 14,473 | 22,511 | 38,571 | 61.1 | 1.2 | 37,004 | 25.4 | 38.8 | 34.3 |
| San Juan | 2 | 0 | 89 | 0.0 | 32.0 | 35.7 | 2 | 25,015 | 34,502 | 54,625 | 45.3 | 2.8 | 50,524 | 13.7 | 19.3 | 20.3 |
| San Miguel | 12 | 20 | 1,482 | 16.4 | 18.5 | 59.4 | 16 | 14,170 | 47,526 | 67,038 | 37.8 | 8.9 | 77,209 | 9.9 | 9.3 | 8.7 |
| Sedgwick | 2 | 1 | 493 | 2.4 | 42.1 | 19.5 | 6 | 9,636 | 27,527 | 43,150 | 53.9 | 4.7 | 42,808 | 14.6 | 24.5 | 23.0 |
| Summit | 53 | 21 | 5,117 | 12.9 | 25.1 | 52.4 | 42 | 11,614 | 41,281 | 79,277 | 25.5 | 7.2 | 86,570 | 6.4 | 6.9 | 6.1 |
| Teller | 47 | 28 | 4,800 | 11.7 | 26.0 | 37.7 | 29 | 9,969 | 35,703 | 66,592 | 34.3 | 6.1 | 70,599 | 7.9 | 12.1 | 11.2 |
| Washington | 10 | 21 | 996 | 7.2 | 42.7 | 15.4 | 12 | 13,168 | 27,860 | 50,094 | 49.9 | 2.6 | 53,263 | 11.8 | 17.2 | 16.3 |
| Weld | 129 | 276 | 83,564 | 9.7 | 39.2 | 27.5 | 423 | 9,645 | 31,793 | 74,150 | 32.8 | 6.1 | 78,160 | 8.4 | 9.8 | 9.8 |
| Yuma | 7 | 14 | 2,236 | 10.9 | 45.8 | 21.8 | 21 | 11,387 | 27,955 | 52,022 | 47.3 | 2.3 | 52,841 | 12.8 | 15.9 | 14.7 |
| CONNECTICUT | X | X | 898,354 | 20.0 | 36.2 | 39.3 | 10,304 | 19,939 | 44,496 | 78,444 | 32.9 | 12.6 | 78,920 | 9.9 | 13.5 | 12.9 |
| Fairfield | NA | NA | 250,634 | 23.6 | 31.5 | 47.9 | 2,923 | 20,186 | 57,263 | 95,645 | 28.3 | 21.5 | 96,966 | 9.0 | 11.8 | 11.9 |
| Hartford | NA | NA | 220,397 | 16.3 | 36.9 | 38.1 | 2,718 | 21,093 | 44,480 | 75,148 | 35.0 | 10.4 | 75,336 | 10.8 | 14.0 | 13.1 |
| Litchfield | NA | NA | 37,971 | 16.3 | 36.9 | 35.4 | 461 | 20,874 | 44,480 | 79,906 | 31.2 | 10.1 | 81,015 | 7.1 | 9.4 | 8.5 |
| Middlesex | NA | NA | 37,431 | 24.8 | 31.6 | 42.2 | 426 | 13,925 | 46,023 | 85,898 | 29.0 | 11.9 | 81,721 | 7.0 | 7.2 | 6.6 |
| New Haven | NA | NA | 215,964 | 23.6 | 40.6 | 35.0 | 2,384 | 19,903 | 38,009 | 69,905 | 36.9 | 9.1 | 69,687 | 12 | 18.5 | 17.8 |
| New London | NA | NA | 60,287 | 19.5 | 37.2 | 33.3 | 710 | 19,842 | 39,426 | 73,490 | 33.2 | 8.4 | 75,226 | 7.8 | 11.6 | 10.5 |

1. All persons 3 years old and over enrolled in nursery school through college.   2. Persons 25 years old and over.   3. Elementary and secondary education expenditures.   4. Based on population estimated by the American Community Survey, 2014–2018.   5. Broomfield county is included with Boulder county.

| STATE County | Personal income, 2019 | | | | | | | | | | Earnings, 2019 | | |
|---|---|---|---|---|---|---|---|---|---|---|---|---|---|
| | Total (mil dol) | Percent change 2018-2019 | Per capita[1] | | Wages and salaries (mil dol) | Supplements to wages and salaries, employer contributions (mil dol) | | Proprietors' income (mil dol) | Dividends, interest, and rent (mil dol) | Personal transfer receipts (mil dol) | Total (mil dol) | Contributions for government social insurance (mil dol) | |
| | | | Dollars | Rank | | Pension and insurance | Government social insurance | | | | | From employee and self-employed | From employer |
| | 62 | 63 | 64 | 65 | 66 | 67 | 68 | 69 | 70 | 71 | 72 | 73 | 74 |
| COLORADO—Cont'd | | | | | | | | | | | | | |
| Boulder | 24,963 | 4.6 | 76,527 | 79 | 14,832 | 1,741 | 950 | 1,718 | 7,445 | 2,247 | 19,241 | 1,078 | 950 |
| Broomfield | 5,003 | 6.1 | 70,996 | 119 | 3,902 | 377 | 238 | 342 | 898 | 450 | 4,860 | 288 | 238 |
| Chaffee | 998 | 4.3 | 49,036 | 922 | 355 | 56 | 24 | 106 | 312 | 202 | 541 | 34 | 24 |
| Cheyenne | 101 | 26.0 | 55,188 | 486 | 35 | 7 | 3 | 20 | 27 | 19 | 65 | 3 | 3 |
| Clear Creek | 600 | 4.6 | 61,816 | 257 | 170 | 23 | 12 | 43 | 154 | 74 | 248 | 15 | 12 |
| Conejos | 315 | 4.3 | 38,385 | 2,282 | 56 | 12 | 4 | 64 | 49 | 91 | 135 | 8 | 4 |
| Costilla | 138 | 5.4 | 35,621 | 2,602 | 33 | 7 | 3 | 13 | 29 | 53 | 56 | 4 | 3 |
| Crowley | 126 | 5.4 | 20,822 | 3,110 | 52 | 9 | 3 | 17 | 26 | 42 | 82 | 4 | 3 |
| Custer | 215 | 4.0 | 42,381 | 1,706 | 31 | 6 | 2 | 29 | 74 | 63 | 68 | 6 | 2 |
| Delta | 1,226 | 3.8 | 39,339 | 2,136 | 343 | 62 | 24 | 114 | 290 | 365 | 543 | 40 | 24 |
| Denver | 59,198 | 4.6 | 81,405 | 59 | 41,852 | 5,059 | 2,822 | 11,593 | 13,343 | 5,297 | 61,326 | 3,282 | 2,822 |
| Dolores | 78 | 3.8 | 38,005 | 2,324 | 22 | 5 | 1 | 4 | 18 | 23 | 32 | 2 | 1 |
| Douglas | 27,550 | 6.7 | 78,455 | 71 | 9,804 | 1,064 | 648 | 1,826 | 4,811 | 2,057 | 13,342 | 793 | 648 |
| Eagle | 4,673 | 3.3 | 84,765 | 47 | 1,817 | 192 | 134 | 606 | 1,799 | 270 | 2,749 | 149 | 134 |
| Elbert | 1,619 | 6.3 | 60,586 | 283 | 198 | 28 | 14 | 90 | 265 | 194 | 330 | 25 | 14 |
| El Paso | 36,825 | 5.5 | 51,117 | 746 | 18,393 | 3,092 | 1,380 | 2,173 | 7,467 | 6,531 | 25,038 | 1,381 | 1,380 |
| Fremont | 1,768 | 3.8 | 36,966 | 2,453 | 643 | 131 | 43 | 162 | 346 | 532 | 978 | 64 | 43 |
| Garfield | 3,621 | 3.7 | 60,285 | 293 | 1,439 | 202 | 102 | 293 | 1,346 | 429 | 2,035 | 113 | 102 |
| Gilpin | 314 | 6.3 | 50,346 | 807 | 248 | 19 | 17 | 24 | 62 | 45 | 309 | 18 | 17 |
| Grand | 816 | 4.1 | 51,891 | 688 | 328 | 45 | 23 | 80 | 254 | 108 | 477 | 27 | 23 |
| Gunnison | 853 | 3.6 | 48,853 | 937 | 388 | 63 | 26 | 110 | 266 | 106 | 588 | 31 | 26 |
| Hinsdale | 41 | 2.3 | 50,294 | 809 | 10 | 2 | 1 | 3 | 18 | 8 | 16 | 1 | 1 |
| Huerfano | 287 | 2.1 | 41,561 | 1,840 | 66 | 14 | 5 | 19 | 74 | 104 | 104 | 9 | 5 |
| Jackson | 78 | 1.3 | 55,715 | 463 | 24 | 4 | 2 | 13 | 21 | 13 | 43 | 2 | 2 |
| Jefferson | 38,480 | 4.9 | 66,017 | 169 | 15,992 | 2,132 | 1,141 | 2,911 | 8,041 | 4,686 | 22,176 | 1,313 | 1,141 |
| Kiowa | 77 | 16.5 | 54,765 | 508 | 25 | 5 | 2 | 14 | 17 | 16 | 46 | 2 | 2 |
| Kit Carson | 324 | 17.0 | 45,658 | 1,290 | 116 | 21 | 9 | 60 | 80 | 66 | 206 | 10 | 9 |
| Lake | 304 | 5.1 | 37,391 | 2,401 | 117 | 20 | 8 | 31 | 64 | 49 | 176 | 10 | 8 |
| La Plata | 3,273 | 2.0 | 58,216 | 364 | 1,354 | 193 | 92 | 330 | 1,097 | 444 | 1,969 | 113 | 92 |
| Larimer | 19,945 | 4.9 | 55,884 | 451 | 9,622 | 1,375 | 645 | 1,350 | 4,736 | 2,731 | 12,992 | 736 | 645 |
| Las Animas | 577 | -0.5 | 39,762 | 2,081 | 220 | 40 | 16 | 23 | 126 | 200 | 299 | 21 | 16 |
| Lincoln | 176 | 5.0 | 30,855 | 3,013 | 100 | 20 | 6 | 17 | 42 | 48 | 142 | 7 | 6 |
| Logan | 1,046 | 3.4 | 46,697 | 1,160 | 366 | 65 | 27 | 229 | 208 | 197 | 687 | 35 | 27 |
| Mesa | 7,205 | 3.7 | 46,719 | 1,158 | 3,164 | 471 | 234 | 569 | 1,415 | 1,608 | 4,438 | 279 | 234 |
| Mineral | 58 | 3.2 | 75,044 | 90 | 21 | 3 | 2 | 6 | 20 | 11 | 32 | 2 | 2 |
| Moffat | 582 | 4.5 | 43,842 | 1,508 | 260 | 47 | 19 | 62 | 94 | 126 | 388 | 22 | 19 |
| Montezuma | 1,140 | 3.0 | 43,542 | 1,541 | 377 | 70 | 27 | 75 | 275 | 296 | 549 | 37 | 27 |
| Montrose | 1,789 | 3.1 | 41,852 | 1,789 | 689 | 116 | 50 | 185 | 400 | 455 | 1,040 | 68 | 50 |
| Morgan | 1,389 | 5.5 | 47,774 | 1,046 | 627 | 95 | 48 | 214 | 203 | 246 | 983 | 51 | 48 |
| Otero | 726 | 3.9 | 39,708 | 2,085 | 275 | 51 | 22 | 49 | 140 | 240 | 397 | 26 | 22 |
| Ouray | 317 | 3.8 | 64,076 | 205 | 81 | 12 | 6 | 40 | 116 | 46 | 139 | 9 | 6 |
| Park | 872 | 5.8 | 46,261 | 1,215 | 116 | 19 | 8 | 66 | 194 | 153 | 209 | 16 | 8 |
| Phillips | 210 | 10.7 | 49,259 | 900 | 71 | 14 | 5 | 46 | 47 | 47 | 136 | 7 | 5 |
| Pitkin | 2,813 | 1.6 | 158,313 | 3 | 1,004 | 100 | 70 | 264 | 1,750 | 123 | 1,438 | 77 | 70 |
| Prowers | 543 | 6.2 | 44,623 | 1,404 | 173 | 34 | 13 | 112 | 102 | 131 | 331 | 16 | 13 |
| Pueblo | 6,852 | 4.2 | 40,680 | 1,965 | 3,066 | 477 | 228 | 337 | 1,230 | 2,023 | 4,108 | 267 | 228 |
| Rio Blanco | 315 | 4.6 | 49,846 | 840 | 161 | 30 | 11 | 13 | 69 | 60 | 215 | 11 | 11 |
| Rio Grande | 568 | 5.9 | 50,408 | 800 | 182 | 31 | 15 | 83 | 115 | 151 | 310 | 18 | 15 |
| Routt | 2,095 | 2.8 | 81,699 | 58 | 780 | 102 | 56 | 168 | 974 | 163 | 1,105 | 62 | 56 |
| Saguache | 253 | 9.3 | 37,022 | 2,447 | 66 | 13 | 6 | 49 | 60 | 56 | 134 | 6 | 6 |
| San Juan | 35 | 6.6 | 47,933 | 1,023 | 11 | 2 | 1 | 4 | 10 | 6 | 18 | 1 | 1 |
| San Miguel | 740 | 3.3 | 90,444 | 33 | 271 | 31 | 20 | 90 | 357 | 41 | 412 | 22 | 20 |
| Sedgwick | 110 | 10.6 | 48,824 | 940 | 31 | 7 | 2 | 13 | 25 | 31 | 53 | 3 | 2 |
| Summit | 2,217 | 3.6 | 71,479 | 117 | 1,031 | 115 | 75 | 320 | 735 | 147 | 1,541 | 83 | 75 |
| Teller | 1,300 | 5.4 | 51,206 | 739 | 344 | 49 | 24 | 73 | 283 | 277 | 490 | 35 | 24 |
| Washington | 215 | 9.3 | 43,765 | 1,520 | 61 | 13 | 5 | 26 | 45 | 45 | 104 | 6 | 5 |
| Weld | 16,289 | 7.7 | 50,198 | 816 | 6,614 | 893 | 473 | 1,962 | 2,344 | 2,202 | 9,942 | 543 | 473 |
| Yuma | 493 | 11.3 | 49,201 | 904 | 190 | 31 | 16 | 107 | 108 | 88 | 344 | 16 | 16 |
| CONNECTICUT | 275,557 | 3.1 | 77,273 | X | 120,355 | 18,357 | 8,226 | 30,472 | 61,031 | 36,204 | 177,410 | 9,844 | 8,226 |
| Fairfield | 114,518 | 3.3 | 121,397 | 11 | 39,659 | 4,960 | 2,413 | 15,053 | 34,275 | 8,582 | 62,085 | 3,342 | 2,413 |
| Hartford | 58,604 | 2.4 | 65,720 | 175 | 38,487 | 5,770 | 2,671 | 6,143 | 9,044 | 9,502 | 53,071 | 2,940 | 2,671 |
| Litchfield | 11,804 | 3.7 | 65,458 | 177 | 3,148 | 576 | 242 | 1,459 | 2,444 | 1,803 | 5,425 | 325 | 242 |
| Middlesex | 11,235 | 2.9 | 69,167 | 138 | 4,219 | 724 | 311 | 1,089 | 1,946 | 1,584 | 6,344 | 366 | 311 |
| New Haven | 49,360 | 3.3 | 57,748 | 381 | 23,061 | 3,798 | 1,736 | 4,382 | 8,257 | 9,464 | 32,977 | 1,893 | 1,736 |
| New London | 15,837 | 2.7 | 59,717 | 314 | 7,694 | 1,575 | 560 | 1,309 | 2,913 | 2,810 | 11,138 | 610 | 560 |

1. Based on the resident population estimated as of July 1 of the year shown.

# Table B. States and Counties — Earnings, Social Security, and Housing

| STATE County | Earnings, 2019 (cont.) | | | | | | | | | Social Security beneficiaries, December 2019 | | Supplemental Security Income recipients, 2019 | Housing units, 2020 | |
| | Percent by selected industries | | | | | | | | | | | | | |
| | Farm | Mining, quarrying, and extractions | Construction | Manufacturing | Information; professional, scientific, technical services | Retail trade | Finance, insurance, real estate, and leasing | Health care and social assistance | Government | Number | Rate[1] | | Total | Percent change, 2010-2020 |
| | 75 | 76 | 77 | 78 | 79 | 80 | 81 | 82 | 83 | 84 | 85 | 86 | 87 | 88 |
|---|---|---|---|---|---|---|---|---|---|---|---|---|---|---|
| COLORADO—Cont'd | | | | | | | | | | | | | | |
| Boulder | 0.1 | 0.3 | 3.6 | 11.0 | 34.2 | 4.2 | 5.8 | 9.0 | 14.3 | 48,010 | 147 | 2,203 | 138,578 | 9.1 |
| Broomfield | 0.0 | D | 6.4 | 14.2 | 36.4 | 3.6 | 5 | 3.4 | 2.5 | 9,885 | 140 | 424 | 29,261 | 29.3 |
| Chaffee | 0.9 | D | 14.2 | 1.7 | 7.2 | 10.4 | 6.5 | 5.1 | 25.7 | 5,385 | 265 | 206 | 11,351 | 13.2 |
| Cheyenne | 14.5 | 12.6 | D | D | D | 4.0 | D | D | 21.4 | 360 | 197 | 15 | 1,000 | 2.7 |
| Clear Creek | 0.0 | D | 6.4 | 1.7 | D | 5.1 | D | 2.3 | 15.5 | 1,815 | 189 | 75 | 5,826 | 2.6 |
| Conejos | 22.3 | 1.8 | 10.1 | 2.5 | D | 4.7 | D | 9.4 | 18.5 | 1,975 | 242 | 329 | 4,398 | 2.6 |
| Costilla | 8.3 | 0.3 | D | D | D | 6.4 | D | 4.9 | 27.3 | 1,265 | 327 | 235 | 2,742 | 4.9 |
| Crowley | 20.6 | D | 2.6 | 0.2 | D | 4.1 | D | D | 40.9 | 920 | 154 | 148 | 1,603 | 2.8 |
| Custer | 2.7 | D | 20.6 | 4.2 | 8.2 | 10.0 | 4.8 | D | 19.4 | 1,845 | 360 | 63 | 4,158 | 5.2 |
| Delta | 2.2 | 7.2 | 9.2 | 4.9 | 4.7 | 9.2 | 4.7 | 8.9 | 28.6 | 9,575 | 309 | 552 | 14,995 | 2.9 |
| Denver | 0.0 | 15.2 | 4.1 | 2.9 | 18.0 | 2.7 | 10.6 | 6.4 | 10.9 | 87,205 | 120 | 14,084 | 319,568 | 12.0 |
| Dolores | 5.4 | D | 5.2 | 8.9 | D | 6.2 | D | D | 35.6 | 585 | 286 | 24 | 1,537 | 4.8 |
| Douglas | 0.0 | 2.9 | 7.6 | 1.1 | 20.5 | 5.9 | 19.8 | 7.9 | 6.9 | 44,310 | 126 | 1,149 | 132,822 | 24.3 |
| Eagle | 0.1 | 0.2 | 14.0 | 0.8 | 7.3 | 7.3 | 7.5 | 13.6 | 8.9 | 5,655 | 102 | 110 | 33,022 | 5.5 |
| Elbert | -6.7 | D | 26.6 | 2.3 | 12.9 | 6.3 | 4.9 | 2.0 | 16.1 | 4,640 | 174 | 97 | 9,527 | 6.6 |
| El Paso | 0.0 | 0.0 | 7.7 | 3.9 | 15.2 | 5.8 | 7.3 | 10.2 | 30.6 | 111,760 | 155 | 9,127 | 286,047 | 13.1 |
| Fremont | 0.3 | 0.6 | 7.4 | 5.4 | D | 6.5 | 3.0 | 11.7 | 41.0 | 13,090 | 273 | 1,182 | 20,330 | 5.6 |
| Garfield | 0.1 | 6.1 | 17.7 | 1.3 | 7.2 | 7.2 | 7.2 | 11.3 | 17.7 | 8,650 | 144 | 439 | 24,413 | 4.8 |
| Gilpin | 0.0 | 0.1 | 2.6 | D | D | D | D | D | 11.4 | 1,055 | 169 | 29 | 3,679 | 3.1 |
| Grand | 0.6 | 0.7 | 16.5 | 1.3 | 5.1 | 7.7 | 8.1 | 2.7 | 20.1 | 2,490 | 158 | 55 | 17,272 | 7.5 |
| Gunnison | 0.2 | 8.0 | 13 | 1.2 | 7.6 | 6.9 | 4.8 | 3.6 | 24.1 | 2,345 | 134 | 54 | 12,116 | 6.2 |
| Hinsdale | -2.5 | D | 14.5 | D | D | D | D | D | 28.0 | 240 | 304 | 0 | 1,446 | 4.3 |
| Huerfano | -0.5 | D | 11.9 | 5.6 | 4.0 | 9.4 | 4.0 | D | 23.8 | 2,505 | 365 | 289 | 5,377 | 5.9 |
| Jackson | 16.3 | 6.5 | D | D | 11.0 | 6.7 | D | D | 19.5 | 320 | 229 | 12 | 1,321 | 2.8 |
| Jefferson | 0.0 | 0.5 | 8.8 | 12.4 | 17.2 | 6.4 | 5.8 | 10.3 | 14.0 | 100,395 | 172 | 5,319 | 240,855 | 4.7 |
| Kiowa | 32.6 | 0.5 | 1.8 | 1.8 | D | 6.8 | D | D | 26.0 | 335 | 236 | 18 | 813 | 1.0 |
| Kit Carson | 25.4 | D | 4.2 | 3.5 | 2.3 | 5.9 | 6.1 | 4.3 | 18.6 | 1,445 | 202 | 100 | 3,574 | 1.3 |
| Lake | 0.0 | D | 13.3 | D | D | 5.2 | 1.7 | D | 22.8 | 995 | 123 | 61 | 4,534 | 6.2 |
| La Plata | -0.2 | 4.0 | 10.8 | 2.3 | 10.3 | 6.7 | 8.3 | 12.8 | 19.9 | 10,810 | 192 | 433 | 27,411 | 6.0 |
| Larimer | 0.3 | 0.5 | 9.1 | 13.9 | 12.5 | 6.4 | 5.8 | 8.6 | 21.1 | 60,740 | 170 | 2,571 | 148,795 | 12.1 |
| Las Animas | 0.6 | 8.3 | 4.8 | 2.3 | 2.7 | 10.1 | 4.3 | 16.4 | 27.4 | 4,255 | 295 | 548 | 8,641 | 5.4 |
| Lincoln | 4.2 | 0.2 | 4.1 | 0.9 | 2.9 | 8.8 | 6.6 | 5.7 | 44.0 | 1,040 | 183 | 67 | 2,523 | 4.2 |
| Logan | 9.6 | 4.3 | 4.5 | 7.3 | 0.9 | 7.4 | 4.8 | 10.1 | 21.3 | 4,160 | 186 | 389 | 9,068 | 1.0 |
| Mesa | 0.2 | 6.7 | 10.8 | 4.2 | 5.6 | 7.6 | 7 | 16.9 | 16.4 | 34,995 | 226 | 2,652 | 65,771 | 5.0 |
| Mineral | 0.6 | 0.5 | D | D | D | D | D | D | 16.2 | 275 | 364 | 0 | 1,330 | 10.6 |
| Moffat | 0.6 | 13.9 | 5.1 | 0.9 | D | 7.3 | 3.5 | 15.4 | 17.1 | 2,465 | 187 | 186 | 6,244 | 0.8 |
| Montezuma | 1.5 | 2.7 | 7.9 | 3.4 | 4.0 | 10.2 | 3.9 | 14.7 | 26.5 | 6,985 | 265 | 463 | 12,381 | 2.4 |
| Montrose | 0.8 | 0.4 | 13.4 | 8.0 | 4.5 | 8.8 | 5.5 | 11.2 | 21.4 | 11,400 | 266 | 637 | 20,009 | 9.7 |
| Morgan | 12.1 | 4.6 | 5.7 | 21.1 | 4.7 | 4.5 | 2.8 | 7.3 | 12.5 | 5,100 | 176 | 435 | 12,039 | 4.8 |
| Otero | 6.0 | D | 4.8 | 8.4 | 4.7 | 6.0 | 3.6 | 17.1 | 23.6 | 4,580 | 251 | 859 | 8,946 | -0.3 |
| Ouray | 1.4 | 1.9 | 16.5 | 3.6 | 10.1 | 9.4 | 7.2 | 3.4 | 16.8 | 1,360 | 272 | 17 | 3,433 | 11.3 |
| Park | -0.5 | D | 19.3 | 2.4 | 10.5 | 7.6 | 4.1 | 2.2 | 22.1 | 3,830 | 203 | 128 | 14,317 | 2.7 |
| Phillips | 9.7 | D | 6.9 | 1.1 | D | 4.6 | D | D | 25.3 | 940 | 216 | 54 | 2,140 | 2.5 |
| Pitkin | -0.1 | D | 5.9 | 0.6 | 9.1 | 5.7 | 12.2 | 2.4 | 14.0 | 2,750 | 154 | 24 | 14,013 | 8.2 |
| Prowers | 17.1 | 0.9 | 4.9 | 5.5 | D | 6.8 | 5.0 | 6.6 | 23.1 | 2,500 | 207 | 365 | 5,937 | -0.1 |
| Pueblo | 0.2 | D | 8.2 | 9.0 | 7.2 | 8.2 | 3.6 | 20.6 | 20.5 | 38,335 | 228 | 6,045 | 71,999 | 3.6 |
| Rio Blanco | 0.9 | 29.8 | 7.2 | 1.1 | 1.6 | 2.7 | 2.8 | 1.1 | 32.5 | 1,165 | 183 | 53 | 3,364 | 1.6 |
| Rio Grande | 16.5 | D | 6.6 | 3.1 | 1.8 | 4.5 | 4.6 | D | 18.2 | 3,420 | 303 | 492 | 6,891 | 4.0 |
| Routt | 0.0 | 5.9 | 11.7 | 0.7 | 7.7 | 7.8 | 9.4 | 7.0 | 15.7 | 3,750 | 147 | 76 | 17,354 | 6.5 |
| Saguache | 34.9 | D | 5.1 | 2.1 | 2.4 | 2.9 | D | 2.6 | 20.0 | 1,400 | 205 | 100 | 4,142 | 7.8 |
| San Juan | 0.0 | 0.5 | D | D | D | 15.6 | D | D | 24.2 | 175 | 242 | 0 | 790 | 4.5 |
| San Miguel | -0.3 | 0.5 | 15.0 | 2.4 | 8.7 | 7.0 | 5.2 | 3.7 | 12.9 | 1,065 | 130 | 29 | 6,845 | 3.1 |
| Sedgwick | 17.3 | 0.3 | 2.0 | 2.8 | D | 4.1 | D | 3.0 | 31.2 | 720 | 317 | 65 | 1,428 | 0.8 |
| Summit | 0.0 | D | 16.0 | 1.0 | 7.4 | 8.0 | 10.0 | 7.0 | 11.3 | 3,180 | 103 | 36 | 31,773 | 6.5 |
| Teller | -0.3 | D | 8.1 | D | 7.5 | 7.7 | 4.7 | 4.2 | 17.1 | 6,720 | 264 | 227 | 13,472 | 6.5 |
| Washington | 12.7 | D | D | D | D | 6.1 | 4.1 | 0.8 | 22.0 | 1,045 | 215 | 64 | 2,496 | 2.7 |
| Weld | 5.9 | 9.8 | 14 | 11.0 | 4.5 | 5.5 | 4.6 | 6.5 | 10.9 | 45,050 | 139 | 3,564 | 113,437 | 17.8 |
| Yuma | 18.8 | 4.6 | 9.4 | 5.6 | 1.1 | 5.3 | 6.1 | D | 16.4 | 1,920 | 192 | 93 | 4,537 | 1.6 |
| CONNECTICUT | 0.1 | 0 | 5.5 | 10.9 | 12.3 | 5.4 | 16.8 | 11.5 | 12.6 | 689,996 | 193 | 66,783 | 1,530,096 | 2.8 |
| Fairfield | 0.0 | 0 | 4.6 | 8.3 | 16.2 | 5.1 | 24.8 | 8.8 | 7.4 | 155,230 | 164 | 12,790 | 377,497 | 4.5 |
| Hartford | 0.0 | 0 | 4.2 | 11.9 | 12.2 | 4.5 | 20.8 | 11.7 | 12.8 | 175,990 | 197 | 21,626 | 381,991 | 2.1 |
| Litchfield | 0.1 | 0.2 | D | D | 7.4 | 9.0 | 5.5 | 12.0 | 12.4 | 43,260 | 240 | 1,773 | 88,512 | 1.1 |
| Middlesex | 0.3 | D | 9.4 | 18.2 | 7.8 | 6.6 | 5.0 | 14.3 | 14.2 | 36,495 | 225 | 1,659 | 76,742 | 2.5 |
| New Haven | 0.1 | 0.1 | 6.5 | 8.7 | 9.6 | 6.1 | 6.4 | 15.7 | 14.1 | 168,330 | 197 | 20,933 | 369,956 | 2.2 |
| New London | 0.7 | 0.1 | 5.1 | 19.4 | 7.8 | 5.6 | 3.1 | 11.1 | 25.5 | 58,025 | 218 | 4,579 | 124,734 | 3.1 |

1. Per 1,000 resident population estimated as of July 1 of the year shown.

Table B. States and Counties — **Housing, Labor Force, and Employment**

| STATE County | Housing units, 2015-2019 | | | | | | | | Civilian labor force, 2020 | | | | Civilian employment[6], 2015-2019 | | |
| | Occupied units | | | | | | | | | | Unemployment | | | Percent | |
| | | Owner-occupied | | | | Renter-occupied | | | | | | | | | |
| | | | | Median owner cost as a percent of income | | | Median rent as a percent of income[2] | Sub-standard units[4] (percent) | | Percent change, 2019-2020 | | | | Management, business, science, and arts | Construction, production, and maintenance occupations |
| | Total | Percent | Median value[1] | With a mort-gage[1] | Without a mort-gage[2] | Median rent[3] | | | Total | | Total | Rate[5] | Total | | |
| | 89 | 90 | 91 | 92 | 93 | 94 | 95 | 96 | 97 | 98 | 99 | 100 | 101 | 102 | 103 |

COLORADO—Cont'd

| | | | | | | | | | | | | | | | |
|---|---|---|---|---|---|---|---|---|---|---|---|---|---|---|---|
| Boulder | 127,415 | 62.2 | 497,300 | 20.6 | 10.0 | 1,495 | 33.7 | 1.9 | 192,879 | -1.3 | 11,887 | 6.2 | 175,849 | 54.4 | 12.1 |
| Broomfield | 27,470 | 65.2 | 413,500 | 19.9 | 10.0 | 1,679 | 28.3 | 1.6 | 40,226 | -0.6 | 2,670 | 6.6 | 38,422 | 55.2 | 12.2 |
| Chaffee | 8,231 | 77.8 | 357,800 | 25.6 | 10.7 | 931 | 29.1 | 2.4 | 9,741 | -0.3 | 615 | 6.3 | 8,364 | 36.4 | 25.4 |
| Cheyenne | 774 | 71.4 | 108,000 | 15.9 | 10.0 | 525 | 16.2 | 0.3 | 1,103 | -3.1 | 27 | 2.4 | 849 | 41.3 | 29.7 |
| Clear Creek | 4,395 | 77.2 | 378,300 | 23.6 | 10.0 | 938 | 28.8 | 1.3 | 6,151 | 1.5 | 523 | 8.5 | 5,265 | 37.8 | 20.2 |
| Conejos | 3,183 | 79.4 | 120,500 | 23.6 | 11.3 | 543 | 29.8 | 5.6 | 4,071 | -2.2 | 227 | 5.6 | 3,029 | 32.6 | 31.9 |
| Costilla | 1,746 | 69.0 | 113,000 | 27.5 | 11.5 | 532 | 31.3 | 7.2 | 1,823 | -2.4 | 132 | 7.2 | 1,394 | 29.0 | 36.0 |
| Crowley | 1,301 | 72.5 | 79,400 | 20.8 | 12.6 | 914 | 25.5 | 4.2 | 1,528 | -1.7 | 83 | 5.4 | 1,924 | 27.8 | 17.9 |
| Custer | 2,214 | 87.2 | 281,900 | 27.1 | 10.0 | 620 | 32.9 | 4.2 | 2,087 | 1.2 | 105 | 5.0 | 1,725 | 37.0 | 34.0 |
| Delta | 11,883 | 73.6 | 241,500 | 24.6 | 10.1 | 834 | 36.4 | 2.0 | 14,340 | 0.0 | 940 | 6.6 | 11,012 | 31.9 | 28.9 |
| Denver | 301,501 | 49.9 | 390,600 | 20.9 | 10.0 | 1,311 | 28.8 | 3.4 | 423,824 | 1.0 | 34,900 | 8.2 | 402,046 | 48.6 | 16.0 |
| Dolores | 759 | 81.6 | 116,600 | 20.7 | 10.0 | 817 | 27.9 | 2.9 | 1,137 | -0.8 | 78 | 6.9 | 792 | 30.8 | 28.5 |
| Douglas | 120,709 | 79.0 | 468,700 | 19.9 | 10.0 | 1,725 | 27.5 | 1.3 | 194,649 | -1.4 | 11,237 | 5.8 | 182,234 | 54.6 | 10.0 |
| Eagle | 18,171 | 69.8 | 562,300 | 24.8 | 10.4 | 1,594 | 32.4 | 4.3 | 35,708 | -2.5 | 3,389 | 9.5 | 34,128 | 34.2 | 21.7 |
| Elbert | 9,280 | 90.0 | 463,600 | 23.3 | 10.2 | 1,182 | 24.4 | 1.9 | 14,961 | -2.4 | 697 | 4.7 | 13,760 | 44.5 | 22.1 |
| El Paso | 257,507 | 64.0 | 275,000 | 21.5 | 10.0 | 1,174 | 30.9 | 3.1 | 344,120 | 1.0 | 25,036 | 7.3 | 323,140 | 42.0 | 17.8 |
| Fremont | 17,136 | 75.0 | 174,000 | 22.5 | 10.0 | 838 | 29.5 | 2.2 | 15,393 | 2.3 | 1,190 | 7.7 | 14,327 | 33.3 | 21.5 |
| Garfield | 21,498 | 66.9 | 360,600 | 22.2 | 10.0 | 1,201 | 27.5 | 4.8 | 31,762 | -2.6 | 2,156 | 6.8 | 31,575 | 35.0 | 25.5 |
| Gilpin | 2,802 | 79.6 | 353,400 | 25.2 | 10.0 | 1,159 | 28.0 | 3.7 | 4,016 | 5.0 | 454 | 11.3 | 3,848 | 44.4 | 13.1 |
| Grand | 5,884 | 68.5 | 308,200 | 23.0 | 10.1 | 1,047 | 27.8 | 2.6 | 9,364 | -5.0 | 788 | 8.4 | 8,646 | 31.8 | 27.2 |
| Gunnison | 6,616 | 61.4 | 367,300 | 22.0 | 10.0 | 992 | 30.0 | 3.7 | 11,347 | -0.9 | 751 | 6.6 | 9,747 | 34.7 | 22.9 |
| Hinsdale | 377 | 71.6 | 350,000 | 23.9 | 10.0 | 756 | 23.2 | 4.2 | 426 | -3.0 | 18 | 4.2 | 384 | 40.6 | 19.5 |
| Huerfano | 3,225 | 71.8 | 161,600 | 26.2 | 14.2 | 626 | 31.3 | 1.2 | 2,644 | 4.6 | 249 | 9.4 | 2,560 | 31.4 | 26.5 |
| Jackson | 558 | 79.9 | 196,700 | 22.5 | 10.0 | 625 | 14.4 | 1.3 | 940 | -4.1 | 40 | 4.3 | 641 | 31.4 | 30.4 |
| Jefferson | 232,284 | 70.7 | 397,700 | 20.8 | 10.0 | 1,376 | 30.7 | 1.6 | 333,682 | -0.1 | 23,625 | 7.1 | 316,280 | 47.3 | 16.0 |
| Kiowa | 638 | 74.3 | 86,500 | 21.5 | 11.8 | 750 | 22.5 | 0.0 | 956 | 0.1 | 22 | 2.3 | 697 | 35.3 | 23.8 |
| Kit Carson | 3,004 | 66.4 | 122,400 | 23.3 | 10.0 | 736 | 23.8 | 1.3 | 4,249 | -1.6 | 119 | 2.8 | 3,514 | 33.6 | 30.4 |
| Lake | 3,392 | 68.4 | 232,100 | 24.0 | 13.0 | 982 | 33.7 | 1.6 | 4,853 | -2.1 | 399 | 8.2 | 4,249 | 35.5 | 25.1 |
| La Plata | 21,794 | 70.7 | 395,600 | 22.1 | 10.0 | 1,147 | 31.7 | 2.6 | 31,366 | -1.6 | 2,147 | 6.8 | 29,339 | 42.6 | 17.5 |
| Larimer | 137,021 | 65.0 | 363,800 | 21.0 | 10.0 | 1,297 | 33.4 | 1.7 | 203,683 | -0.5 | 12,841 | 6.3 | 182,498 | 45.4 | 17.4 |
| Las Animas | 6,670 | 67.4 | 151,100 | 23.6 | 11.3 | 712 | 30.0 | 2.1 | 6,497 | -0.6 | 483 | 7.4 | 5,844 | 31.5 | 22.7 |
| Lincoln | 1,518 | 70.6 | 147,100 | 21.3 | 10.0 | 780 | 30.7 | 1.9 | 2,375 | -1.8 | 106 | 4.5 | 1,499 | 40.2 | 14.7 |
| Logan | 8,393 | 65.2 | 157,900 | 18.4 | 10.9 | 786 | 29.9 | 3.8 | 11,017 | -1.1 | 522 | 4.7 | 11,271 | 30.0 | 30.9 |
| Mesa | 61,742 | 68.2 | 227,000 | 22.2 | 10.0 | 963 | 31.1 | 2.6 | 75,551 | -0.6 | 5,634 | 7.5 | 70,796 | 36.1 | 24.1 |
| Mineral | 367 | 86.6 | 308,600 | 25.9 | 10.2 | 720 | 24.4 | 0.5 | 469 | -9.8 | 27 | 5.8 | 418 | 41.1 | 15.3 |
| Moffat | 5,366 | 64.8 | 182,300 | 20.0 | 10.1 | 779 | 26.8 | 4.8 | 7,285 | -1.5 | 436 | 6.0 | 6,163 | 29.7 | 33.0 |
| Montezuma | 10,655 | 72.1 | 222,800 | 22.4 | 10.2 | 815 | 24.8 | 3.7 | 12,703 | -1.2 | 908 | 7.1 | 11,212 | 32.1 | 24.2 |
| Montrose | 17,140 | 72.7 | 224,400 | 23.2 | 10.0 | 901 | 28.6 | 2.5 | 21,639 | -0.6 | 1,431 | 6.6 | 17,659 | 31.9 | 27.0 |
| Morgan | 10,928 | 64.0 | 201,200 | 23.0 | 10.1 | 835 | 28.4 | 4.8 | 16,125 | -2.0 | 871 | 5.4 | 13,632 | 26.9 | 39.8 |
| Otero | 7,722 | 63.9 | 94,900 | 19.7 | 11.5 | 675 | 28.9 | 2.4 | 8,228 | -0.9 | 508 | 6.2 | 7,030 | 35.1 | 28.5 |
| Ouray | 2,163 | 72.3 | 449,000 | 28.2 | 10.0 | 1,223 | 31.0 | 0.9 | 2,493 | 2.3 | 201 | 8.1 | 2,329 | 51.1 | 14.2 |
| Park | 6,931 | 86.4 | 327,200 | 22.7 | 10.0 | 1,125 | 37.2 | 2.2 | 11,114 | -1.3 | 634 | 5.7 | 8,880 | 33.0 | 25.1 |
| Phillips | 1,689 | 71.9 | 158,500 | 22.0 | 10.0 | 751 | 18.5 | 5.0 | 2,467 | -5.1 | 66 | 2.7 | 1,948 | 35.6 | 35.1 |
| Pitkin | 7,467 | 65.0 | 615,900 | 26.4 | 10.0 | 1,473 | 30.2 | 3.3 | 11,253 | 1.2 | 1,176 | 10.5 | 11,937 | 43.8 | 13.6 |
| Prowers | 4,888 | 65.4 | 106,100 | 20.9 | 10.9 | 697 | 29.1 | 3.2 | 6,213 | -1.9 | 276 | 4.4 | 5,494 | 32.5 | 28.8 |
| Pueblo | 64,314 | 64.7 | 164,600 | 21.8 | 11.6 | 829 | 32.5 | 3.1 | 77,355 | 2.1 | 6,318 | 8.2 | 68,559 | 31.6 | 22.6 |
| Rio Blanco | 2,301 | 70.8 | 197,100 | 19.4 | 10.8 | 770 | 24.5 | 2.0 | 2,949 | -1.3 | 160 | 5.4 | 2,806 | 31.7 | 29.1 |
| Rio Grande | 4,823 | 63.2 | 172,000 | 18.5 | 10.3 | 634 | 26.6 | 4.0 | 5,453 | 0.1 | 385 | 7.1 | 4,614 | 27.2 | 31.0 |
| Routt | 9,603 | 71.2 | 535,300 | 24.1 | 10.0 | 1,282 | 27.7 | 2.6 | 16,206 | -0.6 | 1,277 | 7.9 | 14,464 | 37.6 | 18.6 |
| Saguache | 2,860 | 71.5 | 152,700 | 22.3 | 11.5 | 631 | 30.8 | 6.1 | 3,438 | -3.0 | 242 | 7.0 | 2,748 | 29.2 | 38.1 |
| San Juan | 289 | 62.6 | 322,400 | 25.0 | 13.9 | 1,019 | 23.1 | 4.2 | 509 | -8.5 | 34 | 6.7 | 344 | 28.2 | 18.3 |
| San Miguel | 3,552 | 61.0 | 479,300 | 25.3 | 11.6 | 1,140 | 30.0 | 3.8 | 5,410 | -6.0 | 592 | 10.9 | 5,100 | 42.5 | 15.4 |
| Sedgwick | 967 | 71.4 | 97,100 | 18.0 | 10.0 | 556 | 29.5 | 2.7 | 1,166 | -2.5 | 40 | 3.4 | 942 | 35.7 | 29.0 |
| Summit | 10,641 | 65.5 | 579,600 | 23.7 | 10.0 | 1,393 | 29.0 | 7.2 | 22,402 | -2.4 | 2,198 | 9.8 | 19,805 | 32.0 | 20.0 |
| Teller | 10,248 | 82.5 | 292,700 | 23.0 | 10.0 | 1,102 | 28.7 | 2.0 | 12,984 | 1.5 | 968 | 7.5 | 11,875 | 41.1 | 20.3 |
| Washington | 2,087 | 66.9 | 151,300 | 20.1 | 10.0 | 734 | 23.2 | 3.3 | 2,895 | -2.7 | 91 | 3.1 | 2,284 | 38.0 | 29.5 |
| Weld | 104,671 | 72.9 | 299,000 | 21.6 | 10.0 | 1,085 | 29.5 | 3.2 | 166,666 | -1.5 | 11,624 | 7.0 | 152,572 | 34.2 | 28.7 |
| Yuma | 4,028 | 66.9 | 191,800 | 21.8 | 10.0 | 823 | 24.4 | 4.7 | 5,604 | -8.3 | 159 | 2.8 | 4,899 | 33.5 | 32.1 |
| CONNECTICUT | 1,370,746 | 66.1 | 275,400 | 22.7 | 14.7 | 1,180 | 30.9 | 2.2 | 1,872,631 | -2.3 | 148,010 | 7.9 | 1,815,636 | 43.7 | 17.5 |
| Fairfield | 340,189 | 67.1 | 428,500 | 24.0 | 15.7 | 1,499 | 32.5 | 2.9 | 466,208 | -3.0 | 37,044 | 7.9 | 477,144 | 46.0 | 15.0 |
| Hartford | 350,408 | 64.1 | 240,600 | 21.8 | 14.7 | 1,106 | 29.9 | 2.1 | 476,115 | -1.8 | 38,682 | 8.1 | 451,120 | 44.2 | 17.5 |
| Litchfield | 74,143 | 76.5 | 251,000 | 22.6 | 14.2 | 1,065 | 29.9 | 1.4 | 102,109 | -2.9 | 7,007 | 6.9 | 97,180 | 41.6 | 20.9 |
| Middlesex | 66,971 | 73.5 | 286,800 | 22.2 | 14.5 | 1,162 | 30.7 | 1.2 | 92,221 | -2.0 | 6,010 | 6.5 | 88,260 | 45.8 | 16.2 |
| New Haven | 330,572 | 61.8 | 248,600 | 23.2 | 15.3 | 1,153 | 31.6 | 2.3 | 457,171 | -1.4 | 36,502 | 8.0 | 430,091 | 42.3 | 18.5 |
| New London | 107,827 | 66.4 | 241,700 | 22.2 | 13.9 | 1,130 | 29.5 | 1.5 | 131,992 | -4.0 | 12,679 | 9.6 | 132,898 | 41.0 | 17.8 |

1. Specified owner-occupied units.    2. A value of 10.0 represents 10 percent or less; a value of 50.0 represents 50 percent or more.    3. Specified renter-occupied units.    4. Overcrowded or lacking complete plumbing facilities.    5. Percent of civilian labor force.    6. Civilian employed persons 16 years old and over.

Items 89—103

# Table B. States and Counties — Nonfarm Employment and Agriculture

| STATE County | Private nonfarm establishments, employment and payroll, 2019 | | | | | | | | | Agriculture, 2017 | | | |
|---|---|---|---|---|---|---|---|---|---|---|---|---|---|
| | Number of establishments | Employment | | | | | | Annual payroll | | Farms | | | Farm producers whose primary occupation is farming (percent) |
| | | Total | Health care and social assistance | Manufacturing | Retail trade | Finance and insurance | Professional, scientific, and technical services | Total (mil dol) | Average per employee (dollars) | Number | Percent with: | | |
| | | | | | | | | | | | Fewer than 50 acres | 1000 acres or more | |
| | 104 | 105 | 106 | 107 | 108 | 109 | 110 | 111 | 112 | 113 | 114 | 115 | 116 |

| STATE County | 104 | 105 | 106 | 107 | 108 | 109 | 110 | 111 | 112 | 113 | 114 | 115 | 116 |
|---|---|---|---|---|---|---|---|---|---|---|---|---|---|
| COLORADO—Cont'd | | | | | | | | | | | | | |
| Boulder | 12,856 | 156,833 | 22,151 | 15,212 | 19,238 | 3,836 | 26,065 | 10,686 | 68,137 | 1,012 | 79.0 | 2.2 | 33.1 |
| Broomfield | 2,285 | 49,475 | 3,058 | 3,002 | 5,170 | 2,546 | 3,982 | 3,775 | 76,292 | 38 | 76.3 | 7.9 | 38.3 |
| Chaffee | 1,005 | 6,375 | 1,007 | 159 | 1,250 | 209 | 228 | 224 | 35,065 | 289 | 52.2 | 4.2 | 42.4 |
| Cheyenne | 65 | 644 | 84 | NA | 49 | 41 | 3 | 36 | 55,612 | 377 | 4.8 | 53.1 | 47.2 |
| Clear Creek | 346 | 3,111 | 67 | 36 | 338 | 17 | 46 | 109 | 35,091 | 33 | 42.4 | 9.1 | 27.8 |
| Conejos | 120 | 809 | 231 | 79 | 180 | NA | 17 | 30 | 37,578 | 524 | 33.4 | 14.3 | 49.2 |
| Costilla | 48 | 370 | 35 | 91 | 70 | NA | NA | 12 | 31,100 | 229 | 51.5 | 9.6 | 38.2 |
| Crowley | 39 | 495 | 77 | NA | 82 | 19 | NA | 21 | 43,226 | 246 | 21.1 | 30.9 | 51.7 |
| Custer | 148 | 559 | NA | 39 | 157 | 13 | 16 | 18 | 31,803 | 315 | 35.6 | 8.6 | 39.0 |
| Delta | 860 | 6,509 | 1,568 | 464 | 1,179 | 240 | 193 | 228 | 35,032 | 1,615 | 69.7 | 2.8 | 41.8 |
| Denver | 26,604 | 468,854 | 56,741 | 17,058 | 32,918 | 27,428 | 49,355 | 32,548 | 69,421 | 12 | 100.0 | NA | 60.0 |
| Dolores | 58 | 287 | 52 | 28 | 48 | NA | 5 | 13 | 44,864 | 313 | 21.7 | 10.5 | 38.1 |
| Douglas | 9,652 | 117,725 | 15,010 | 7,389 | 18,575 | 8,077 | 9,235 | 7,140 | 60,652 | 1,223 | 67.4 | 3.0 | 25.0 |
| Eagle | 3,518 | 34,864 | 2,405 | 353 | 4,662 | 864 | 1,588 | 1,471 | 42,189 | 257 | 44.7 | 7.8 | 34.2 |
| Elbert | 626 | 2,866 | 138 | 156 | 455 | 73 | 374 | 119 | 41,459 | 1,632 | 43.2 | 10.5 | 29.8 |
| El Paso | 17,950 | 250,547 | 40,352 | 10,506 | 31,437 | 12,254 | 24,252 | 12,129 | 48,410 | 1,345 | 45.5 | 10.6 | 34.9 |
| Fremont | 877 | 8,352 | 2,159 | 490 | 1,687 | 207 | 183 | 291 | 34,802 | 1,034 | 72.4 | 6.8 | 30.9 |
| Garfield | 2,525 | 21,262 | 3,083 | 205 | 3,160 | 770 | 1,106 | 1,061 | 49,884 | 661 | 49.5 | 13.2 | 45.1 |
| Gilpin | 131 | 4,842 | 20 | 13 | 15 | NA | 49 | 183 | 37,793 | 37 | 51.4 | NA | 31.3 |
| Grand | 871 | 8,954 | 475 | 181 | 730 | 105 | 940 | 293 | 32,720 | 290 | 47.9 | 13.1 | 31.6 |
| Gunnison | 1,177 | 6,984 | 682 | 173 | 1,045 | 147 | 348 | 247 | 35,425 | 309 | 40.5 | 18.1 | 40.0 |
| Hinsdale | 65 | 130 | NA | NA | 13 | NA | NA | 5 | 35,438 | 26 | 7.7 | 15.4 | 25.7 |
| Huerfano | 145 | 1,063 | 395 | NA | 224 | 19 | 43 | 40 | 38,022 | 437 | 21.3 | 21.7 | 30.5 |
| Jackson | 64 | 283 | NA | NA | 53 | NA | 13 | 11 | 38,406 | 131 | 15.3 | 37.4 | 46.2 |
| Jefferson | 17,576 | 199,991 | 27,922 | 8,599 | 30,721 | 13,938 | 23,410 | 10,271 | 51,358 | 597 | 77.7 | 2.7 | 28.6 |
| Kiowa | 36 | 275 | 114 | NA | 87 | NA | NA | 9 | 32,884 | 388 | 3.1 | 48.2 | 47.5 |
| Kit Carson | 273 | 1,936 | 319 | 99 | 372 | 115 | 45 | 72 | 37,312 | 574 | 11.0 | 40.4 | 47.0 |
| Lake | 224 | 1,443 | 164 | 90 | 209 | 18 | 34 | 44 | 30,507 | 33 | 42.4 | 6.1 | 25.0 |
| La Plata | 2,408 | 22,810 | 3,665 | 636 | 3,509 | 1,506 | 1,171 | 1,130 | 49,540 | 1,093 | 53.4 | 3.9 | 35.1 |
| Larimer | 11,236 | 133,269 | 20,253 | 11,755 | 19,749 | 3,647 | 9,802 | 6,446 | 48,369 | 2,043 | 71.0 | 4.4 | 31.6 |
| Las Animas | 361 | 3,234 | 813 | 73 | 725 | 119 | 49 | 115 | 35,563 | 549 | 19.1 | 37.7 | 44.8 |
| Lincoln | 136 | 1,380 | 255 | 14 | 321 | 63 | 34 | 52 | 37,374 | 489 | 8.4 | 51.5 | 48.9 |
| Logan | 569 | 5,073 | 1,183 | 366 | 1,035 | 206 | 91 | 187 | 36,884 | 861 | 16.6 | 33.1 | 45.7 |
| Mesa | 4,560 | 54,402 | 11,576 | 2,995 | 8,133 | 1,813 | 2,452 | 2,429 | 44,654 | 2,465 | 80.1 | 2.6 | 32.1 |
| Mineral | 68 | 197 | NA | NA | 54 | NA | 4 | 8 | 42,695 | 19 | NA | 21.1 | 21.9 |
| Moffat | 389 | 3,366 | 566 | 58 | 698 | 100 | 84 | 162 | 48,139 | 462 | 24.5 | 28.1 | 29.9 |
| Montezuma | 764 | 6,527 | 1,411 | 284 | 1,293 | 184 | 280 | 237 | 36,364 | 1,123 | 49.2 | 5.9 | 37.0 |
| Montrose | 1,291 | 12,795 | 2,781 | 1,209 | 2,215 | 326 | 578 | 515 | 40,289 | 1,135 | 55.9 | 6.1 | 40.4 |
| Morgan | 709 | 10,936 | 1,385 | 3,591 | 1,163 | 215 | 160 | 474 | 43,357 | 740 | 28.4 | 19.5 | 51.8 |
| Otero | 403 | 4,299 | 1,208 | 518 | 679 | 141 | 123 | 152 | 35,408 | 444 | 30.4 | 16.9 | 42.7 |
| Ouray | 301 | 1,326 | 62 | 81 | 178 | 42 | 69 | 52 | 39,418 | 122 | 55.7 | 9.8 | 44.0 |
| Park | 491 | 1,598 | 62 | 70 | 278 | 34 | 106 | 57 | 35,424 | 278 | 30.2 | 14.4 | 30.1 |
| Phillips | 136 | 1,073 | 292 | NA | 154 | 59 | 27 | 44 | 41,314 | 326 | 11.3 | 39.9 | 56.3 |
| Pitkin | 1,633 | 18,129 | 779 | 224 | 1,467 | 272 | 789 | 780 | 43,000 | 112 | 51.8 | 8.9 | 38.8 |
| Prowers | 337 | 2,988 | 730 | 222 | 648 | 183 | 71 | 95 | 31,643 | 472 | 10.2 | 36.9 | 47.6 |
| Pueblo | 3,212 | 51,786 | 13,462 | 5,200 | 8,327 | 1,157 | 2,513 | 2,174 | 41,973 | 839 | 41.7 | 14.2 | 37.6 |
| Rio Blanco | 196 | 1,842 | 376 | 31 | 235 | 37 | 40 | 111 | 60,161 | 320 | 35.3 | 22.2 | 41.4 |
| Rio Grande | 351 | 2,504 | 475 | 90 | 416 | 104 | 53 | 96 | 38,172 | 321 | 27.7 | 16.2 | 59.8 |
| Routt | 1,780 | 14,822 | 1,364 | 167 | 1,805 | 243 | 687 | 554 | 37,382 | 887 | 50.3 | 11.8 | 30.4 |
| Saguache | 132 | 1,003 | 89 | 117 | 133 | 14 | 9 | 33 | 32,802 | 288 | 23.6 | 25.3 | 52.4 |
| San Juan | 70 | 222 | NA | NA | 36 | NA | 2 | 7 | 33,410 | NA | NA | NA | NA |
| San Miguel | 664 | 5,522 | 163 | 126 | 579 | 66 | 162 | 193 | 34,935 | 133 | 30.1 | 17.3 | 35.0 |
| Sedgwick | 68 | 434 | 149 | 40 | 60 | 26 | 7 | 13 | 30,993 | 212 | 8.0 | 41.0 | 57.9 |
| Summit | 2,305 | 22,640 | 1,293 | 357 | 3,380 | 237 | 619 | 752 | 33,216 | 55 | 45.5 | 14.5 | 30.9 |
| Teller | 777 | 6,194 | 651 | 85 | 907 | 159 | 251 | 239 | 38,602 | 159 | 51.6 | 10.1 | 39.8 |
| Washington | 110 | 514 | 36 | 90 | 76 | 38 | 17 | 20 | 38,323 | 757 | 10.2 | 39.1 | 52.1 |
| Weld | 6,602 | 91,965 | 9,195 | 14,153 | 10,491 | 3,027 | 2,950 | 5,013 | 54,511 | 4,062 | 46.3 | 10.2 | 38.1 |
| Yuma | 359 | 2,451 | 581 | 92 | 397 | 153 | 79 | 97 | 39,719 | 774 | 16.0 | 42.5 | 57.9 |
| CONNECTICUT | 88,916 | 1,538,341 | 295,248 | 159,618 | 179,766 | 107,531 | 106,252 | 100,305 | 65,203 | 5,521 | 70.9 | 0.7 | 39.0 |
| Fairfield | 26,947 | 422,988 | 73,060 | 34,516 | 49,143 | 31,402 | 38,822 | 35,015 | 82,780 | 402 | 81.8 | 2.2 | 49.3 |
| Hartford | 22,701 | 461,158 | 86,677 | 53,319 | 48,270 | 53,707 | 31,013 | 30,659 | 66,483 | 786 | 70.9 | 0.5 | 39.3 |
| Litchfield | 4,693 | 52,519 | 9,851 | 9,403 | 8,521 | 1,332 | 1,745 | 2,358 | 44,893 | 1,217 | 67.7 | 0.7 | 33.8 |
| Middlesex | 4,215 | 63,621 | 15,168 | 9,488 | 8,439 | 1,437 | 2,802 | 3,360 | 52,807 | 441 | 79.4 | 0.2 | 42.1 |
| New Haven | 19,355 | 343,018 | 77,102 | 28,477 | 40,928 | 10,269 | 14,558 | 18,568 | 54,131 | 686 | 81.9 | 0.3 | 38.3 |
| New London | 5,924 | 107,959 | 17,922 | 15,038 | 14,801 | 1,955 | 7,994 | 5,614 | 51,998 | 823 | 63.1 | 0.6 | 43.3 |

| STATE County | Acreage (1,000) [117] | Percent change, 2012-2017 [118] | Average size of farm [119] | Total irrigated (1,000) [120] | Total cropland (1,000) [121] | Average per farm [122] | Average per acre [123] | Value of machinery and equipment, average per farm (dollars) [124] | Total (mil dol) [125] | Average per farm (acres) [126] | Crops [127] | Livestock and poultry products [128] | Organic farms (number) [129] | Farms with internet access (percent) [130] | Total ($1,000) [131] | Percent of farms [132] |
|---|---|---|---|---|---|---|---|---|---|---|---|---|---|---|---|---|
| COLORADO—Cont'd | | | | | | | | | | | | | | | | |
| Boulder | 107 | -19.5 | 106 | 27.2 | 38.1 | 1,329,691 | 12,571 | 67,305 | 43.9 | 43,378 | 87.4 | 12.6 | 29 | 90.7 | 501 | 5.6 |
| Broomfield | 9 | -23.8 | 224 | 1.2 | 7.7 | 1,223,811 | 5,467 | 48,312 | 0.6 | 16,132 | 77.5 | 22.5 | NA | 63.2 | D | 2.6 |
| Chaffee | 66 | -14.6 | 229 | 16.5 | 17.2 | 1,164,108 | 5,075 | 71,744 | 12.2 | 42,343 | 32.7 | 67.3 | NA | 83.7 | 56 | 3.5 |
| Cheyenne | 1,076 | 10.1 | 2,853 | 23.6 | 599.1 | 2,389,477 | 838 | 260,682 | 89.2 | 236,676 | 76.0 | 24.0 | NA | 73.7 | 10,125 | 82.0 |
| Clear Creek | 10 | 24.8 | 314 | 0.3 | 0.9 | 990,682 | 3,157 | 54,704 | 0.2 | 5,273 | 30.5 | 69.5 | NA | 81.8 | NA | NA |
| Conejos | 266 | 3.3 | 508 | 119.5 | 131.5 | 857,351 | 1,687 | 133,635 | 53.9 | 102,939 | 63.2 | 36.8 | 17 | 64.9 | 718 | 18.1 |
| Costilla | 358 | -4.9 | 1,562 | 39.3 | 50.6 | 1,898,119 | 1,215 | 154,192 | 22.1 | 96,328 | 87.7 | 12.3 | 2 | 59.4 | 308 | 9.2 |
| Crowley | 484 | -3.1 | 1,969 | 4.7 | 31.0 | 1,002,121 | 509 | 88,555 | 96.3 | 391,370 | 1.4 | 98.6 | 5 | 71.5 | 1,701 | 48.8 |
| Custer | 161 | -14.5 | 512 | 22.6 | 29.0 | 994,967 | 1,943 | 81,523 | 9.7 | 30,733 | 47.3 | 52.7 | NA | 79.7 | 173 | 3.2 |
| Delta | 237 | -5.5 | 147 | 66.8 | 69.4 | 705,824 | 4,813 | 67,981 | 67.1 | 41,559 | 45.2 | 54.8 | 33 | 83.7 | 897 | 6.3 |
| Denver | 0 | -9.8 | 11 | D | 0.0 | 685,079 | 63,728 | 26,067 | D | D | D | D | 1 | 75.0 | NA | NA |
| Dolores | 158 | -1.4 | 504 | 7.2 | 69.9 | 898,017 | 1,783 | 69,774 | 8.5 | 27,208 | 70.9 | 29.1 | 5 | 67.7 | 1,123 | 44.4 |
| Douglas | 202 | 0.8 | 165 | 2.3 | 26.5 | 1,112,035 | 6,747 | 50,523 | 18.9 | 15,428 | 62.0 | 38.0 | 3 | 87.3 | 428 | 3.4 |
| Eagle | 155 | 19.9 | 604 | 14.8 | 15.4 | 1,998,474 | 3,309 | 83,582 | 8.2 | 32,074 | 23.9 | 76.1 | 7 | 86.0 | 60 | 5.1 |
| Elbert | 1,018 | -2.4 | 624 | 6.7 | 189.9 | 952,293 | 1,526 | 61,221 | 35.4 | 21,675 | 21.2 | 78.8 | 1 | 87.1 | 3,286 | 11.9 |
| El Paso | 630 | -2.9 | 468 | 8.9 | 51.5 | 659,073 | 1,407 | 46,294 | 31.9 | 23,716 | 43.3 | 56.7 | 5 | 80.8 | 1,300 | 7.5 |
| Fremont | 278 | -4.3 | 269 | 11.3 | 15.1 | 628,195 | 2,336 | 48,150 | 21.8 | 21,089 | 23.6 | 76.4 | 3 | 81.1 | 244 | 2.0 |
| Garfield | 475 | 52.9 | 719 | 52.0 | 77.4 | 1,724,785 | 2,399 | 89,572 | 35.9 | 54,256 | 24.5 | 75.5 | 4 | 80.0 | 360 | 7.3 |
| Gilpin | 4 | -32.2 | 106 | D | 0.5 | 446,403 | 4,226 | 33,954 | 0.2 | 5,838 | 17.1 | 82.9 | NA | 81.1 | 28 | 16.2 |
| Grand | 241 | 6.2 | 831 | 36.3 | 36.5 | 1,827,034 | 2,199 | 92,591 | 14.4 | 49,793 | 23.5 | 76.5 | NA | 76.9 | 30 | 1.7 |
| Gunnison | 267 | 40.3 | 864 | 57.7 | 49.3 | 2,217,820 | 2,567 | 107,695 | 24.1 | 78,045 | 22.7 | 77.3 | NA | 83.2 | 51 | 3.9 |
| Hinsdale | 10 | 2.4 | 403 | 2.5 | 1.1 | 741,831 | 1,841 | 70,145 | D | D | D | D | NA | 76.9 | D | 3.8 |
| Huerfano | 582 | 0.1 | 1,331 | 8.6 | 19.4 | 1,280,062 | 962 | 58,781 | 13.2 | 30,174 | 10.8 | 89.2 | 1 | 70.7 | 311 | 7.8 |
| Jackson | 301 | -12.0 | 2,301 | 70.4 | 67.4 | 3,045,594 | 1,323 | 196,274 | 24.5 | 186,924 | 18.5 | 81.5 | 3 | 71.8 | 35 | 6.9 |
| Jefferson | 69 | 0.4 | 115 | 1.2 | 5.4 | 885,635 | 7,715 | 36,831 | 9.0 | 15,144 | 77.8 | 22.2 | 2 | 86.1 | 151 | 3.0 |
| Kiowa | 1,092 | -1.9 | 2,814 | 2.7 | 676.3 | 2,122,946 | 754 | 162,846 | 65.5 | 168,729 | 61.8 | 38.2 | NA | 72.2 | 10,450 | 82.0 |
| Kit Carson | 1,358 | -1.4 | 2,366 | 117.8 | 890.7 | 2,883,863 | 1,219 | 375,287 | 474.3 | 826,947 | 32.4 | 67.6 | NA | 81.2 | 13,382 | 71.4 |
| Lake | 12 | -1.9 | 362 | 1.9 | 0.5 | 743,849 | 2,055 | 46,972 | D | D | D | D | NA | 84.8 | NA | NA |
| La Plata | 549 | -7.0 | 503 | 58.5 | 89.7 | 1,135,305 | 2,259 | 74,568 | 24.4 | 22,281 | 38.3 | 61.7 | 3 | 87.5 | 1,066 | 8.4 |
| Larimer | 482 | 7.1 | 236 | 60.2 | 98.6 | 1,095,047 | 4,637 | 76,905 | 150.7 | 73,772 | 45.7 | 54.3 | 20 | 85.1 | 633 | 5.5 |
| Las Animas | 1,796 | -16.1 | 3,272 | 17.3 | 75.6 | 1,968,541 | 602 | 80,465 | 25.8 | 47,080 | 14.7 | 85.3 | 7 | 71.8 | 4,168 | 27.5 |
| Lincoln | 1,500 | 1.8 | 3,067 | 5.8 | 588.3 | 2,158,581 | 704 | 200,920 | 67.9 | 138,855 | 58.9 | 41.1 | 2 | 74.4 | 10,318 | 61.8 |
| Logan | 1,138 | 3.5 | 1,322 | 107.1 | 562.7 | 1,570,536 | 1,188 | 208,745 | 617.9 | 717,686 | 16.0 | 84.0 | 2 | 82.9 | 11,511 | 68.8 |
| Mesa | 343 | -11.5 | 139 | 76.2 | 79.8 | 767,492 | 5,523 | 57,249 | 94.2 | 38,209 | 48.7 | 51.3 | 12 | 86.2 | 730 | 3.9 |
| Mineral | 8 | 27.2 | 444 | 2.9 | 0.7 | 1,526,408 | 3,441 | 31,306 | D | D | D | 100.0 | NA | 84.2 | D | 10.5 |
| Moffat | 953 | 2.5 | 2,063 | 30.1 | 123.4 | 1,648,562 | 799 | 107,741 | 33.1 | 71,727 | 11.7 | 88.3 | 2 | 71.9 | 2,109 | 26.4 |
| Montezuma | 691 | 0.0 | 615 | 79 | 117.2 | 744,860 | 1,211 | 85,128 | 46.4 | 41,339 | 64.3 | 35.7 | 6 | 79.0 | 1,237 | 13.3 |
| Montrose | 331 | 0.3 | 291 | 79.8 | 65.6 | 900,826 | 3,093 | 79,919 | 81.2 | 71,565 | 22.7 | 77.3 | 10 | 83.6 | 792 | 10.2 |
| Morgan | 659 | 1.9 | 891 | 132.3 | 340.4 | 1,546,382 | 1,735 | 241,369 | 559.5 | 756,130 | 16.1 | 83.9 | 5 | 82.0 | 6,040 | 53.9 |
| Otero | 688 | -2.7 | 1,548 | 49.3 | 78.1 | 1,162,135 | 750 | 148,382 | 121.5 | 273,759 | 25.1 | 74.9 | 3 | 77.9 | 2,724 | 48.9 |
| Ouray | 85 | 4.7 | 698 | 8.6 | 8.6 | 2,130,534 | 3,054 | 88,951 | 4.2 | 34,459 | 17.2 | 82.8 | 1 | 89.3 | 64 | 4.1 |
| Park | 189 | 5.1 | 680 | 12.0 | 11.2 | 1,142,769 | 1,680 | 55,212 | 5.1 | 18,374 | 7.5 | 92.5 | NA | 78.1 | 26 | 2.5 |
| Phillips | 439 | 0.6 | 1,347 | 79.3 | 384.4 | 2,279,496 | 1,692 | 324,052 | 174.2 | 534,479 | 50.5 | 49.5 | NA | 86.2 | 8,706 | 79.4 |
| Pitkin | 33 | 1.9 | 292 | 10.1 | 5.9 | 2,225,232 | 7,617 | 88,246 | 2.9 | 26,000 | 34.9 | 65.1 | NA | 70.5 | NA | NA |
| Prowers | 1,011 | | 2,143 | 88.3 | 525.2 | 1,726,199 | 806 | 220,792 | 310.0 | 656,875 | 20.3 | 79.7 | NA | 65.7 | 9,185 | 75.6 |
| Pueblo | 896 | 0.0 | 1,067 | 18.1 | 72.0 | 1,097,156 | 1,028 | 76,147 | 52.0 | 62,033 | 41.0 | 59.0 | 7 | 77.5 | 2,117 | 13.9 |
| Rio Blanco | 411 | -19.0 | 1,284 | 27.1 | 43.7 | 1,763,734 | 1,373 | 101,073 | 18.8 | 58,597 | 14.8 | 85.2 | 2 | 83.1 | 390 | 13.4 |
| Rio Grande | 177 | -4.3 | 553 | 93.6 | 109.1 | 1,726,626 | 3,123 | 311,844 | 99.0 | 308,274 | 88.7 | 11.3 | 9 | 82.9 | 998 | 26.5 |
| Routt | 465 | -24.1 | 524 | 42.7 | 81.0 | 1,647,039 | 3,141 | 69,340 | 31.6 | 35,679 | 16.2 | 83.8 | 4 | 81.1 | 962 | 10.8 |
| Saguache | 314 | 0.8 | 1,090 | 118.4 | 132.7 | 2,068,512 | 1,898 | 218,769 | 105.4 | 365,983 | 85.8 | 14.2 | 15 | 76.4 | 843 | 18.8 |
| San Juan | NA | NA | NA | NA | NA | NA | NA | NA | NA | NA | NA | NA | NA | NA | NA | NA |
| San Miguel | 136 | 7.5 | 1,023 | 15.9 | 14.3 | 1,524,124 | 1,490 | 79,047 | 6.4 | 47,925 | 14.0 | 86.0 | NA | 77.4 | 322 | 15.0 |
| Sedgwick | 349 | 3.8 | 1,645 | 40.4 | 228.1 | 2,226,323 | 1,353 | 321,028 | 93.9 | 442,693 | 49.3 | 50.7 | 2 | 82.1 | 4,776 | 81.6 |
| Summit | 27 | 4.8 | 483 | 5.5 | 6.8 | 2,041,182 | 4,225 | 90,529 | 1.5 | 27,127 | 35.1 | 64.9 | NA | 94.5 | D | 1.8 |
| Teller | 71 | 0.6 | 449 | 0.9 | 5.6 | 908,034 | 2,023 | 40,497 | 1.2 | 7,811 | 12.2 | 87.8 | 3 | 83.0 | NA | NA |
| Washington | 1,358 | 11.6 | 1,794 | 40.4 | 836.2 | 1,969,481 | 1,098 | 221,219 | 184.6 | 243,802 | 50.9 | 49.1 | 5 | 74.4 | 15,207 | 72.7 |
| Weld | 2,099 | 7.3 | 517 | 323.4 | 923.0 | 1,335,953 | 2,586 | 159,664 | 2,047.2 | 503,982 | 17.0 | 83.0 | 42 | 82.9 | 19,375 | 26.0 |
| Yuma | 1,400 | 3.5 | 1,809 | 209.0 | 628.2 | 2,953,997 | 1,633 | 313,588 | 918.7 | 1,186,972 | 21.8 | 78.2 | 4 | 82.4 | 19,612 | 69.8 |
| CONNECTICUT | 382 | -12.6 | 69 | 7.4 | 148.6 | 862,636 | 12,483 | 62,250 | 580.1 | 105,074 | 72.4 | 27.6 | 113 | 82.4 | 1,850 | 4.4 |
| Fairfield | 52 | -3.2 | 130 | 0.2 | 4.6 | 1,402,826 | 10,794 | 57,696 | 42.1 | 104,649 | 52.7 | 47.3 | 9 | 87.1 | 33 | 3.5 |
| Hartford | 48 | -11.5 | 61 | 3.4 | 26.2 | 981,893 | 16,126 | 64,077 | 93.9 | 119,481 | 95.1 | 4.9 | 9 | 80.9 | 212 | 3.8 |
| Litchfield | 90 | -0.7 | 74 | 0.5 | 37.4 | 840,256 | 11,322 | 56,949 | 41.1 | 33,800 | 51.0 | 49.0 | 28 | 83.3 | 363 | 3.8 |
| Middlesex | 16 | -31.8 | 37 | 0.7 | 6.5 | 555,077 | 14,911 | 66,715 | 57.1 | 129,426 | 96.8 | 3.2 | 10 | 83.9 | 304 | 4.8 |
| New Haven | 27 | -36.3 | 39 | 1.3 | 11.4 | 922,400 | 23,490 | 67,524 | 111.6 | 162,720 | 95.9 | 4.1 | 26 | 78.4 | 323 | 5.2 |
| New London | 60 | -7.7 | 73 | 0.6 | 23.9 | 837,723 | 11,467 | 67,605 | 135.8 | 164,989 | 55.5 | 44.5 | 22 | 88.0 | 267 | 6.1 |

# Table B. States and Counties — Water Use, Wholesale Trade, Retail Trade, and Real Estate

| STATE County | Water use, 2015 Public supply water withdrawn (mil gal/day) | Water use, 2015 Public supply gallons withdrawn per person per day | Wholesale Trade[1], 2017 Number of establish-ments | Wholesale Trade[1], 2017 Number of employees | Wholesale Trade[1], 2017 Sales (mil dol) | Wholesale Trade[1], 2017 Average payroll (mil dol) | Retail Trade[2], 2017 Number of establish-ments | Retail Trade[2], 2017 Number of employees | Retail Trade[2], 2017 Sales (mil dol) | Retail Trade[2], 2017 Average payroll (mil dol) | Real estate and rental and leasing,[2] 2017 Number of establish-ments | Real estate and rental and leasing,[2] 2017 Number of employees | Real estate and rental and leasing,[2] 2017 Sales (mil dol) | Real estate and rental and leasing,[2] 2017 Average payroll (mil dol) |
|---|---|---|---|---|---|---|---|---|---|---|---|---|---|---|
| | 133 | 134 | 135 | 136 | 137 | 138 | 139 | 140 | 141 | 142 | 143 | 144 | 145 | 146 |
| COLORADO—Cont'd | | | | | | | | | | | | | | |
| Boulder | 48.30 | 151.2 | 428 | 6,066 | 4,884.6 | 516.0 | 1,187 | 19,230 | 5,635.2 | 602.1 | 775 | 2,894 | 754.0 | 161.8 |
| Broomfield | 5.67 | 87.1 | 70 | 845 | 430.1 | 55.9 | 286 | 5,338 | 1,280.6 | 139.7 | 150 | 425 | 170.4 | 22.9 |
| Chaffee | 3.96 | 212.2 | 15 | 117 | 53.6 | 4.2 | 141 | 1,234 | 371.1 | 34.3 | 79 | 159 | 26.9 | 4.7 |
| Cheyenne | 0.36 | 196.8 | 6 | D | 96.7 | D | 16 | 52 | 14.4 | 1.3 | NA | NA | NA | NA |
| Clear Creek | 1.17 | 125.8 | 15 | D | 14.7 | D | 51 | 311 | 84.4 | 7.7 | D | D | D | D |
| Conejos | 3.38 | 415.7 | D | D | D | 1.2 | 22 | 188 | 34.0 | 4.5 | NA | NA | NA | NA |
| Costilla | 0.45 | 125.6 | NA | NA | NA | NA | 14 | 73 | 14.7 | 1.1 | NA | NA | NA | NA |
| Crowley | 1.00 | 179.8 | NA | NA | NA | NA | 9 | 79 | 18.8 | 2.5 | NA | NA | NA | NA |
| Custer | 0.38 | 85.5 | NA | NA | NA | NA | 21 | 133 | 49.4 | 4.7 | 8 | D | 3.8 | D |
| Delta | 5.00 | 166.8 | 26 | 165 | 39.4 | 6.9 | 115 | 1,270 | 326.9 | 33.4 | 42 | 921 | 19.9 | 25.3 |
| Denver | 163.27 | 239.2 | 1,191 | 20,941 | 20,105.5 | 1,388.8 | 2,471 | 32,753 | 9,218.5 | 1,047.9 | 2,033 | 11,307 | 3,866.6 | 710.4 |
| Dolores | 0.25 | 126.4 | D | D | D | 1.4 | 5 | 43 | 10.4 | 0.8 | 4 | D | 0.6 | D |
| Douglas | 38.48 | 119.4 | 284 | 2,321 | 2,462.7 | 181.5 | 939 | 18,587 | 5,755.5 | 520.1 | 761 | 1,725 | 577.3 | 90.7 |
| Eagle | 10.31 | 192.3 | 78 | 492 | 216.9 | 23.8 | 410 | 4,299 | 1,072.6 | 143.2 | 420 | 2,082 | 445.9 | 79.1 |
| Elbert | 0.85 | 34.4 | 17 | 61 | 26.9 | 2.4 | 37 | 450 | 140.4 | 12.4 | D | D | D | D |
| El Paso | 88.35 | 131.0 | 459 | 4,834 | 3,005.6 | 320.3 | 2,096 | 31,935 | 9,701.0 | 926.1 | 1,326 | 4,447 | 1,015.8 | 188.2 |
| Fremont | 6.42 | 137.5 | 20 | 82 | 26.1 | 3.0 | 134 | 1,751 | 487.4 | 45.8 | 42 | 183 | 18.2 | 4.5 |
| Garfield | 11.64 | 200.4 | 55 | 342 | 173.3 | 16.9 | 277 | 3,209 | 1,169.5 | 109.9 | 152 | 582 | 116.8 | 25.5 |
| Gilpin | 0.68 | 116.7 | NA | NA | NA | NA | 10 | 29 | 7.6 | 0.9 | 8 | 11 | 3.6 | 0.7 |
| Grand | 2.31 | 158.1 | D | D | D | 0.8 | 96 | 706 | 210.4 | 21.4 | 91 | 457 | 67.7 | 14.4 |
| Gunnison | 1.88 | 117.0 | 7 | 16 | 4.4 | 0.5 | 134 | 1,081 | 243.5 | 25.7 | 104 | 235 | 45.8 | 7.6 |
| Hinsdale | 0.24 | 310.1 | NA | NA | NA | NA | 13 | 20 | 5.0 | 0.6 | 7 | D | 4.9 | D |
| Huerfano | 2.97 | 457.5 | 3 | 7 | 0.6 | 0.1 | 25 | 202 | 50.1 | 5.0 | D | D | D | D |
| Jackson | 0.32 | 236.0 | NA | NA | NA | NA | 10 | 54 | 16.2 | 1.3 | NA | NA | NA | NA |
| Jefferson | 22.99 | 40.7 | 548 | 4,685 | 2,929.7 | 307.6 | 1,864 | 29,948 | 8,899.4 | 882.2 | 1,096 | 3,036 | 807.1 | 145.0 |
| Kiowa | 0.23 | 161.6 | 5 | D | 16.4 | D | 8 | 92 | 29.8 | 1.9 | NA | NA | NA | NA |
| Kit Carson | 1.27 | 163.7 | 21 | 224 | 231.7 | 10.6 | 38 | 356 | 116.3 | 9.2 | 11 | 39 | 3.2 | 1.0 |
| Lake | 4.47 | 597.2 | 4 | 20 | 7.8 | 0.8 | 31 | 219 | 48.0 | 5.5 | D | D | D | D |
| La Plata | 5.74 | 105.0 | 51 | 548 | 232.6 | 27.7 | 306 | 3,396 | 989.1 | 104.5 | 173 | 438 | 105.1 | 17.2 |
| Larimer | 30.58 | 91.7 | 336 | 4,724 | 3,723.5 | 349.3 | 1,266 | 19,902 | 5,893.8 | 574.0 | 766 | 2,404 | 663.9 | 109.9 |
| Las Animas | 3.46 | 246.1 | 10 | 32 | 13.0 | 0.8 | 61 | 733 | 193.2 | 19.7 | 18 | 64 | 8.5 | 1.8 |
| Lincoln | 0.78 | 140.4 | NA | NA | NA | NA | 25 | 322 | 126.9 | 8.1 | 4 | 7 | 0.5 | 0.1 |
| Logan | 3.26 | 147.9 | 23 | 102 | 122.8 | 5.2 | 99 | 1,094 | 346.8 | 29.5 | 17 | 48 | 7.1 | 1.5 |
| Mesa | 17.51 | 117.9 | 200 | 1,731 | 942.8 | 90.2 | 572 | 8,305 | 2,463.2 | 237.0 | 303 | 936 | 208.8 | 39.0 |
| Mineral | 0.22 | 303.0 | NA | NA | NA | NA | 12 | 47 | 10.8 | 1.2 | 3 | 4 | 1.0 | 0.1 |
| Moffat | 1.74 | 134.5 | 24 | 161 | 188.4 | 9.7 | 63 | 687 | 198.0 | 21.2 | 11 | 17 | 3.8 | 0.6 |
| Montezuma | 4.37 | 167.0 | 25 | 226 | 131.3 | 10.9 | 112 | 1,309 | 363.9 | 37.6 | 36 | 127 | 27.4 | 4.4 |
| Montrose | 9.13 | 223.0 | 51 | 316 | 120.7 | 14.4 | 165 | 2,148 | 650.1 | 62.5 | 60 | 154 | 30.4 | 5.0 |
| Morgan | 5.96 | 210.2 | 36 | 299 | 236.7 | 14.2 | 95 | 1,064 | 303.9 | 28.4 | 30 | 74 | 11.1 | 2.0 |
| Otero | 3.99 | 217.5 | 20 | 126 | 63.0 | 4.8 | 67 | 700 | 180.6 | 17.1 | 13 | 78 | 7.4 | 2.4 |
| Ouray | 1.80 | 383.7 | NA | NA | NA | NA | 40 | 223 | 37.9 | 4.9 | 10 | D | 3.7 | D |
| Park | 0.40 | 24.2 | D | D | D | D | 49 | 254 | 63.7 | 7.0 | 27 | D | 8.8 | D |
| Phillips | 0.80 | 184.0 | D | D | D | 4.8 | 25 | 179 | 50.7 | 4.5 | NA | NA | NA | NA |
| Pitkin | 8.13 | 457.1 | 16 | 30 | 32.8 | 1.4 | 234 | 1,680 | 404.9 | 56.0 | 242 | 950 | 238.4 | 47.6 |
| Prowers | 1.91 | 159.8 | 15 | 110 | 67.4 | 4.3 | 57 | 679 | 167.3 | 16.0 | D | D | D | D |
| Pueblo | 41.29 | 252.4 | 94 | 1,037 | 684.5 | 57.8 | 506 | 8,124 | 2,144.4 | 214.1 | 148 | 658 | 116.3 | 19.4 |
| Rio Blanco | 1.89 | 287.6 | D | D | D | D | 30 | 207 | 54.6 | 5.2 | 8 | 19 | 1.8 | 0.6 |
| Rio Grande | 1.20 | 104.0 | 23 | 322 | 202.7 | 14.1 | 51 | 406 | 111.9 | 11.2 | D | D | D | D |
| Routt | 2.88 | 119.4 | 50 | 376 | 200.3 | 20.8 | 193 | 1,737 | 417.0 | 50.8 | 155 | 587 | 97.8 | 19.5 |
| Saguache | 0.69 | 110.4 | 11 | 253 | 72.4 | 8.9 | 21 | 138 | 23.9 | 2.6 | D | D | D | D |
| San Juan | 0.24 | 342.4 | NA | NA | NA | NA | 17 | 38 | 9.5 | 1.1 | D | D | D | 0.1 |
| San Miguel | 0.68 | 86.3 | 9 | D | 6.0 | D | 82 | 548 | 115.1 | 15.8 | 84 | 264 | 68.0 | 11.5 |
| Sedgwick | 0.69 | 287.6 | 5 | D | 12.4 | D | 11 | 68 | 26.1 | 1.8 | NA | NA | NA | NA |
| Summit | 6.63 | 219.1 | 31 | 79 | 29.7 | 3.9 | 321 | 3,423 | 794.4 | 90.6 | 339 | 1,388 | 297.6 | 64.3 |
| Teller | 6.17 | 263.8 | 11 | 37 | 13.7 | 1.2 | 70 | 898 | 261.3 | 23.5 | 54 | 101 | 17.4 | 3.3 |
| Washington | 0.35 | 72.0 | 5 | D | 34.2 | D | 12 | 79 | 28.6 | 2.1 | 5 | D | 0.6 | D |
| Weld | 28.30 | 99.2 | 274 | 3,365 | 3,733.5 | 201.6 | 668 | 10,289 | 3,768.8 | 327.5 | 314 | 1,616 | 358.6 | 79.6 |
| Yuma | 1.42 | 140.0 | 31 | 342 | 656.5 | 20.0 | 58 | 425 | 109.1 | 10.9 | 13 | 35 | 7.4 | 1.2 |
| CONNECTICUT | 239.93 | 66.8 | 3,504 | 62,298 | 102,896.5 | 5,173.7 | 12,391 | 186,297 | 55,404.5 | 5,560.8 | 3,459 | 20,224 | 6,691.3 | 1,114.8 |
| Fairfield | 87.98 | 92.8 | 1,131 | 19,293 | 66,731.7 | 1,916.5 | 3,368 | 51,210 | 16,458.8 | 1,717.6 | 1,165 | 7,605 | 2,604.5 | 536.9 |
| Hartford | 59.22 | 66.1 | 930 | 19,164 | 19,181.9 | 1,276.8 | 3,127 | 50,098 | 14,132.7 | 1,389.1 | 892 | 5,100 | 1,480.8 | 259.4 |
| Litchfield | 10.18 | 55.4 | 139 | 798 | 436.3 | 43.1 | 673 | 8,552 | 2,683.2 | 268.7 | 141 | 414 | 81.1 | 17.0 |
| Middlesex | 7.31 | 44.6 | 157 | 2,363 | 1,550.5 | 159.5 | 645 | 8,983 | 2,588.1 | 259.2 | 753 | 4,825 | 1,990.9 | 210.9 |
| New Haven | 54.49 | 63.4 | 880 | 16,151 | 12,689.6 | 1,516.0 | 2,870 | 42,459 | 12,606.7 | 1,225.2 | 228 | 818 | 236.0 | 32.3 |
| New London | 8.11 | 29.8 | 149 | 2,638 | 1,679.6 | 176.0 | 1,015 | 15,140 | 4,030.8 | 422.3 | 228 | 818 | 236.0 | 32.3 |

1   Merchant wholesalers, except manufacturers' sales branches and offices.   2. Employer establishments.

# Table B. States and Counties — Professional Services, Manufacturing, and Accommodation and Food Services

| STATE County | Professional, scientific, and technical services, 2017 | | | | Manufacturing, 2017 | | | | Accommodation and food services, 2017 | | | |
|---|---|---|---|---|---|---|---|---|---|---|---|---|
| | Number of establishments | Number of employees | Sales (mil dol) | Average payroll (mil dol) | Number of establishments | Number of employees | Sales (mil dol) | Average payroll (mil dol) | Number of establishments | Number of employees | Sales (mil dol) | Annual payroll (mil dol) |
| | 147 | 148 | 149 | 150 | 151 | 152 | 153 | 154 | 155 | 156 | 157 | 158 |
| COLORADO—Cont'd | | | | | | | | | | | | |
| Boulder | D | D | D | D | 561 | 14,472 | 5,400.1 | 939.4 | 939 | 19,279 | 1,157.9 | 360.1 |
| Broomfield | D | D | D | D | 82 | 3,161 | 3,441.8 | 210.6 | 178 | 3,689 | 261.1 | 77.9 |
| Chaffee | 104 | 203 | 24 | 8.5 | 30 | 131 | 33.7 | 6.2 | 114 | 1,502 | 87.3 | 29.1 |
| Cheyenne | 4 | 6 | 0 | 0.1 | NA | NA | NA | NA | 3 | 22 | 0.9 | 0.2 |
| Clear Creek | 46 | 64 | 13 | 4.2 | 10 | 32 | 4.5 | D | 54 | 628 | 42.8 | 14.0 |
| Conejos | D | D | 2 | D | 7 | 70 | 21.0 | 3.1 | 13 | 47 | 5.1 | 1.4 |
| Costilla | NA | NA | NA | NA | D | D | D | D | D | D | D | 0.8 |
| Crowley | NA | NA | NA | NA | NA | NA | NA | NA | NA | NA | NA | NA |
| Custer | 10 | 18 | 5 | 0.8 | 8 | 33 | 10.1 | 2.1 | 12 | 57 | 3.6 | 0.9 |
| Delta | 63 | 168 | 15 | 4.8 | 52 | 337 | 108.4 | 17.7 | 82 | 736 | 35.7 | 11.0 |
| Denver | 5,110 | 49,765 | 12,602 | 4,732.3 | 773 | 16,660 | 5,404.4 | 836.0 | 2,354 | 55,545 | 4,477.0 | 1,297.4 |
| Dolores | D | D | D | 0.2 | 4 | 12 | 6.3 | 1.5 | 7 | D | 5.2 | D |
| Douglas | D | D | D | D | 137 | 8,329 | 3,526.8 | 907.6 | 571 | 12,316 | 742.9 | 226.7 |
| Eagle | D | D | D | D | 58 | 335 | 131.9 | 18.2 | 295 | 8,137 | 680.6 | 204.9 |
| Elbert | 97 | 230 | 43 | 12.8 | 19 | 200 | 35.7 | 8.6 | 27 | 196 | 11.0 | 3.3 |
| El Paso | 2,632 | 22,760 | 4,501 | 1,766.2 | 454 | 11,012 | 3,335.0 | 673.2 | 1,377 | 30,711 | 1,909.3 | 547.3 |
| Fremont | 64 | 217 | 19 | 6.7 | 37 | 347 | 135.4 | 19.4 | 79 | 1,149 | 60.4 | 18.1 |
| Garfield | D | D | D | D | 39 | 172 | 45.4 | 10.5 | 215 | 3,223 | 218.0 | 69.5 |
| Gilpin | 20 | 36 | 7 | 2.0 | D | 36 | D | D | 13 | 2,672 | 433.2 | 100.9 |
| Grand | 81 | 911 | 70 | 46.8 | 17 | 157 | 25.6 | 6.0 | 136 | 2,141 | 130.2 | 46.7 |
| Gunnison | 143 | 276 | 40 | 13.2 | 28 | 141 | 24.6 | 5.1 | 135 | 2,531 | 130.6 | 41.9 |
| Hinsdale | NA | NA | NA | NA | NA | NA | NA | NA | 17 | 15 | 3.4 | 0.8 |
| Huerfano | 13 | 34 | 3 | 1.8 | NA | NA | NA | NA | 22 | 174 | 10.2 | 2.3 |
| Jackson | 6 | 13 | 2 | 0.4 | NA | NA | NA | NA | 11 | 55 | 3.7 | 1.1 |
| Jefferson | D | D | D | D | 463 | 8,742 | 4,128.7 | 555.3 | 1,245 | 24,777 | 1,535.2 | 462.9 |
| Kiowa | NA | NA | NA | NA | NA | NA | NA | NA | D | D | D | D |
| Kit Carson | 17 | 34 | 4 | 1.3 | 6 | 93 | 30.8 | 3.6 | 27 | 314 | 16.1 | 4.1 |
| Lake | 15 | 31 | 5 | 1.2 | 5 | 57 | 8.0 | D | 34 | 242 | 13.6 | 4.4 |
| La Plata | D | D | D | D | 65 | 567 | 109.2 | 23.7 | 227 | 4,528 | 256.3 | 84.7 |
| Larimer | 1,642 | 9,285 | 1,638 | 668.0 | 444 | 9,931 | 3,909.8 | 646.2 | 949 | 18,194 | 1,087.4 | 338.0 |
| Las Animas | 26 | 65 | 6 | 2.3 | 5 | 28 | 6.7 | 1.7 | 46 | 593 | 36.7 | 9.6 |
| Lincoln | 10 | 40 | 3 | 1.0 | NA | NA | NA | NA | 25 | 261 | 19.5 | 4.4 |
| Logan | 26 | 91 | 13 | 4.1 | 21 | 358 | 208.3 | 12.4 | 46 | 710 | 33.9 | 9.8 |
| Mesa | 497 | 3,416 | 527 | 209.4 | 150 | 2,740 | 552.2 | 124.2 | 335 | 6,462 | 356.5 | 117.5 |
| Mineral | D | D | D | 0.2 | NA | NA | NA | NA | 19 | 79 | 10.3 | 3.4 |
| Moffat | D | D | D | D | 12 | 49 | 18.2 | 2.0 | 33 | 383 | 19.2 | 5.7 |
| Montezuma | 73 | 262 | 30 | 11.3 | 34 | 277 | 53.0 | 10.6 | 89 | 1,368 | 98.0 | 25.7 |
| Montrose | D | D | D | D | 62 | 1,292 | 267.7 | 49.3 | 98 | 1,270 | 70.5 | 21.4 |
| Morgan | 38 | 126 | 15 | 4.8 | 31 | 3,061 | 2,644.4 | 143.9 | 65 | 988 | 66.1 | 13.3 |
| Otero | D | D | D | D | 16 | 465 | 92.0 | 23.0 | 49 | 572 | 28.9 | 8.1 |
| Ouray | 53 | 83 | 13 | 4.0 | 9 | 29 | 8.7 | 1.9 | 55 | 499 | 35.5 | 10.6 |
| Park | 62 | 102 | 12 | 4.3 | D | 49 | D | 1.7 | 43 | 250 | 24.5 | 5.0 |
| Phillips | 10 | 22 | 5 | 0.9 | D | 83 | D | 4.3 | 15 | D | 3.3 | D |
| Pitkin | D | D | D | D | 20 | 157 | 43.8 | 9.4 | 161 | 5,778 | 502.0 | 158.7 |
| Prowers | 26 | 98 | 7 | 2.5 | 15 | 184 | 39.0 | 6.7 | 34 | 360 | 19.3 | 4.7 |
| Pueblo | D | D | D | D | 94 | 4,380 | 2,103.4 | 256.3 | 336 | 5,761 | 289.7 | 84.7 |
| Rio Blanco | 10 | 38 | 3 | 1.4 | 3 | 21 | 3.5 | 1.0 | 23 | 192 | 9.0 | 3.8 |
| Rio Grande | 22 | 55 | 6 | 1.9 | 12 | 72 | 11.1 | 2.3 | 43 | 312 | 18.8 | 5.7 |
| Routt | 209 | 636 | 102 | 37.9 | 35 | 158 | 22.8 | 6.8 | 163 | 5,244 | 272.4 | 88.1 |
| Saguache | D | D | D | 0.4 | 5 | 132 | 17.9 | 2.9 | 7 | 39 | 2.7 | 0.8 |
| San Juan | D | D | 0 | D | NA | NA | NA | NA | 22 | 83 | 11.0 | 2.5 |
| San Miguel | 80 | 145 | 28 | 7.6 | 13 | 87 | 15.5 | 5.2 | 74 | 1,628 | 132.4 | 42.8 |
| Sedgwick | D | D | D | 0.3 | D | D | D | 0.8 | 6 | D | 1.4 | D |
| Summit | D | D | D | D | 25 | 292 | 76.7 | 14.2 | 271 | 6,898 | 491.2 | 143.6 |
| Teller | 122 | 269 | 32 | 13.0 | 10 | 70 | 10.4 | 3.1 | 65 | 1,992 | 129.7 | 43.8 |
| Washington | D | D | 2 | D | 6 | 67 | 37.2 | 3.1 | 5 | 29 | 1.3 | 0.4 |
| Weld | D | D | D | D | 305 | 11,968 | 5,959.4 | 657.2 | 472 | 7,843 | 406.0 | 117.9 |
| Yuma | D | D | D | D | 13 | 77 | 122.5 | 3.1 | 29 | 267 | 12.5 | 3.6 |
| CONNECTICUT | 9,126 | 107,375 | 21,343 | 9,625.6 | 3,986 | 156,822 | 57,882.0 | 11,021.6 | 8,762 | 146,456 | 10,791.5 | 3,069.7 |
| Fairfield | 3,519 | 47,964 | 10,736 | 4,661.4 | 779 | 34,375 | 12,255.6 | 2,634.2 | 2,438 | 35,172 | 2,597.3 | 767.7 |
| Hartford | D | D | D | D | 1,163 | 51,705 | 19,350.3 | 3,875.2 | 2,154 | 37,527 | 2,338.8 | 694.8 |
| Litchfield | D | D | D | D | 326 | 8,974 | 2,711.2 | 513.5 | 434 | 4,988 | 321.1 | 96.4 |
| Middlesex | D | D | D | D | 219 | 9,702 | 3,640.9 | 600.2 | 441 | 6,004 | 400.2 | 123.7 |
| New Haven | 1,814 | 14,470 | 2,620 | 1,218.0 | 1,048 | 29,004 | 10,764.2 | 1,792.6 | 2,047 | 29,719 | 1,854.2 | 534.1 |
| New London | 527 | 8,313 | 973 | 972.2 | 175 | 14,024 | 6,228.6 | 1,105.9 | 750 | 25,852 | 2,816.6 | 722.2 |

Table B. States and Counties — **Health Care and Social Assistance, Other Services, Nonemployer Businesses, and Residential Construction**

| STATE County | Health care and social assistance, 2017 | | | | Other services, 2017 | | | | Nonemployer businesses, 2018 | | Value of residential construction authorized by building permits, 2020 | |
|---|---|---|---|---|---|---|---|---|---|---|---|---|
| | Number of establish-ments | Number of employees | Receipts (mil dol) | Annual payroll (mil dol) | Number of establish-ments | Number of employees | Receipts (mil dol) | Annual payroll (mil dol) | Number | Receipts (mil dol) | New construction ($1,000) | Number of housing units |
| | 159 | 160 | 161 | 162 | 163 | 164 | 165 | 166 | 167 | 168 | 169 | 170 |
| COLORADO—Cont'd | | | | | | | | | | | | |
| Boulder | 1,432 | 21,332 | 2,566.9 | 1,012.1 | 823 | 5,173 | 671.5 | 202.2 | 40,829 | 2,155.5 | 405,860 | 1,634 |
| Broomfield | 197 | 2,472 | 287.5 | 117.3 | 145 | 732 | 158.2 | 36.2 | 6,398 | 298.3 | 128,723 | 427 |
| Chaffee | 77 | 966 | 114.5 | 46.4 | 51 | 188 | 17.7 | 5.7 | 2,622 | 111.3 | 45,289 | 186 |
| Cheyenne | D | D | D | D | NA | NA | NA | NA | 213 | 12.5 | 912 | 6 |
| Clear Creek | 18 | 92 | 6.6 | 2.6 | 11 | 83 | 4.6 | 2.4 | 1,030 | 54.1 | 14,936 | 94 |
| Conejos | 15 | 235 | 20.9 | 8.8 | D | D | D | 0.2 | 638 | 32.5 | 10,242 | 36 |
| Costilla | D | D | D | 1.1 | NA | NA | NA | NA | 275 | 11.2 | 0 | 0 |
| Crowley | 3 | 78 | 3.6 | 2.1 | NA | NA | NA | NA | 188 | 7.0 | 1,406 | 12 |
| Custer | NA | NA | NA | NA | 13 | 39 | 2.5 | 0.7 | 719 | 37.1 | 32,903 | 118 |
| Delta | 77 | 1,701 | 137.4 | 61.6 | 52 | 174 | 18.3 | 4.7 | 3,066 | 147.5 | 13,467 | 73 |
| Denver | 2,302 | 56,165 | 8,008.4 | 3,078.4 | 1,874 | 15,112 | 2,404.0 | 600.2 | 75,501 | 4,194.2 | 831,523 | 5,059 |
| Dolores | 4 | D | 2.3 | D | NA | NA | NA | NA | 161 | 6.2 | 1,466 | 6 |
| Douglas | 934 | 13,150 | 2,257.4 | 710.1 | 646 | 3,907 | 432.2 | 133.1 | 34,908 | 1,810.7 | 971,500 | 3,370 |
| Eagle | 193 | 2,058 | 452.0 | 137.4 | 208 | 1,267 | 132.6 | 43.1 | 7,578 | 468.3 | 151,637 | 403 |
| Elbert | 24 | 140 | 9.9 | 4.7 | 40 | 104 | 12.8 | 3.7 | 2,968 | 143.6 | 86,670 | 296 |
| El Paso | 2,126 | 39,317 | 4,488.9 | 1,848.1 | 1,264 | 10,636 | 2,375.2 | 452.2 | 54,516 | 2,307.0 | 2,098,524 | 6,819 |
| Fremont | 101 | 2,204 | 165.1 | 78.5 | 47 | 196 | 16.4 | 5.0 | 2,901 | 114.0 | 34,694 | 155 |
| Garfield | 162 | 3,170 | 509.6 | 202.8 | 150 | 677 | 77.6 | 23.9 | 6,458 | 349.0 | 110,826 | 334 |
| Gilpin | D | D | D | 0.6 | 7 | 51 | 2.7 | 0.8 | 622 | 24.7 | 5,554 | 20 |
| Grand | 37 | 420 | 53.2 | 19.2 | 45 | 116 | 17.1 | 4.1 | 2,056 | 108.3 | 77,530 | 234 |
| Gunnison | 61 | 561 | 77.2 | 27.5 | 78 | 251 | 31.4 | 6.6 | 2,593 | 104.7 | 77,640 | 228 |
| Hinsdale | NA | NA | NA | NA | D | D | D | 0.3 | 148 | 4.8 | 4,636 | 11 |
| Huerfano | 14 | 426 | 32.6 | 16.0 | 6 | 17 | 1.5 | 0.2 | 586 | 24.5 | 7,854 | 43 |
| Jackson | NA | NA | NA | NA | NA | NA | NA | NA | 183 | 8.2 | 1,024 | 5 |
| Jefferson | 1,821 | 32,117 | 3,636.5 | 1,458.9 | 1,216 | 6,781 | 804.7 | 231.7 | 58,970 | 2,715.1 | 494,525 | 2,161 |
| Kiowa | D | D | D | D | NA | NA | NA | NA | 169 | 6.1 | 853 | 5 |
| Kit Carson | 28 | 323 | 27.1 | 11.9 | 17 | 53 | 9.3 | 2.2 | 743 | 35.9 | 2,455 | 11 |
| Lake | 11 | 134 | 14.4 | 6.2 | 17 | 39 | 4.4 | 1.0 | 643 | 28.3 | 10,517 | 53 |
| La Plata | 232 | 3,351 | 524.9 | 176.4 | 157 | 643 | 60.4 | 19.1 | 6,793 | 322.1 | 115,453 | 253 |
| Larimer | 1,147 | 19,461 | 2,304.6 | 932.0 | 734 | 4,247 | 640.0 | 141.2 | 33,935 | 1,591.3 | 664,597 | 2,565 |
| Las Animas | 42 | 814 | 69.3 | 30.7 | 35 | 154 | 14.6 | 3.9 | 1,026 | 37.1 | 5,582 | 34 |
| Lincoln | 11 | 309 | 27.7 | 11.4 | D | D | 4.2 | D | 386 | 14.9 | 1,326 | 9 |
| Logan | 70 | 1,152 | 124.7 | 47.1 | 51 | 187 | 22.7 | 5.4 | 1,417 | 65.1 | 2,903 | 12 |
| Mesa | 438 | 11,405 | 1,429.2 | 530.6 | 315 | 1,662 | 188.0 | 55.8 | 12,271 | 576.3 | 144,225 | 1,004 |
| Mineral | NA | NA | NA | NA | D | D | D | 0.1 | 162 | 8.8 | 8,092 | 18 |
| Moffat | 44 | 547 | 72.8 | 29.0 | 32 | 157 | 11.0 | 3.6 | 1,015 | 43.1 | 3,267 | 17 |
| Montezuma | 83 | 1,456 | 131.0 | 56.0 | 54 | 183 | 19.9 | 4.9 | 2,403 | 91.0 | 5,453 | 27 |
| Montrose | 159 | 3,064 | 301.9 | 117.2 | 83 | 329 | 39.6 | 10.6 | 3,944 | 190.4 | 40,285 | 263 |
| Morgan | 58 | 1,447 | 156.6 | 63.7 | 44 | 155 | 16.9 | 4.5 | 2,020 | 114.8 | 25,145 | 147 |
| Otero | 55 | 1,296 | 92.1 | 46.0 | 22 | 83 | 8.2 | 2.3 | 1,017 | 32.3 | 1,332 | 6 |
| Ouray | D | D | D | 1.9 | 11 | 77 | 5.2 | 1.5 | 945 | 53.7 | 24,519 | 71 |
| Park | D | D | D | 2.6 | 19 | 55 | 5.9 | 1.6 | 2,088 | 87.5 | 41,851 | 162 |
| Phillips | 9 | 266 | 29.4 | 12.2 | D | D | D | 0.6 | 456 | 22.5 | 542 | 3 |
| Pitkin | 72 | 767 | 149.0 | 56.6 | 109 | 824 | 103.0 | 32.3 | 3,783 | 261.4 | 154,163 | 72 |
| Prowers | 31 | 621 | 55.6 | 24.4 | 24 | 87 | 11.2 | 2.5 | 881 | 44.6 | 3,699 | 15 |
| Pueblo | 407 | 13,427 | 1,403.4 | 602.5 | 235 | 1,348 | 125.1 | 36.1 | 8,929 | 376.3 | 93,369 | 579 |
| Rio Blanco | D | D | D | D | D | D | D | 0.9 | 582 | 21.6 | 2,224 | 8 |
| Rio Grande | 27 | 435 | 34.3 | 15.5 | 27 | 94 | 14.3 | 3.3 | 1,119 | 46.0 | 5,985 | 31 |
| Routt | 139 | 1,488 | 188.7 | 76.9 | 114 | 561 | 57.4 | 17.5 | 3,938 | 203.2 | 133,450 | 285 |
| Saguache | D | D | D | 2.1 | D | D | D | 1.7 | 598 | 22.8 | 7,628 | 54 |
| San Juan | D | D | D | D | D | D | D | 0.8 | 126 | 5.4 | 1,989 | 7 |
| San Miguel | 29 | 151 | 13.7 | 6.8 | 40 | 230 | 44.0 | 9.8 | 1,763 | 111.6 | 87,794 | 34 |
| Sedgwick | D | D | D | D | D | D | D | 0.2 | 195 | 8.5 | 0 | 0 |
| Summit | 115 | 1,211 | 181.3 | 65.2 | 144 | 467 | 58.1 | 15.5 | 4,502 | 268.8 | 158,092 | 318 |
| Teller | 62 | 546 | 54.5 | 19.5 | 51 | 166 | 19.2 | 5.4 | 2,735 | 119.7 | 35,791 | 113 |
| Washington | 7 | D | 3.2 | D | D | D | 1.5 | D | 394 | 19.4 | 1,466 | 7 |
| Weld | 551 | 8,971 | 1,127.3 | 446.2 | 418 | 2,212 | 278.9 | 74.4 | 24,243 | 1,206.3 | 965,414 | 4,185 |
| Yuma | 22 | 576 | 58.6 | 23.3 | 28 | 66 | 9.2 | 1.8 | 1,020 | 44.2 | 1,198 | 6 |
| CONNECTICUT | 11,065 | 295,083 | 35,302.4 | 14,069.0 | 7,507 | 46,490 | 6,055.5 | 1,645.5 | 286,874 | 17,158.9 | 1,061,756 | 5,471 |
| Fairfield | 3,071 | 72,290 | 10,942.1 | 3,753.5 | 2,180 | 14,442 | 2,103.2 | 530.8 | 97,854 | 7,149.9 | 492,509 | 1,862 |
| Hartford | 2,994 | 89,797 | 9,665.2 | 4,186.6 | 1,971 | 13,694 | 1,736.4 | 523.0 | 60,999 | 3,329.7 | 149,573 | 850 |
| Litchfield | 510 | 9,674 | 903.0 | 391.3 | 383 | 1,643 | 206.0 | 53.9 | 17,272 | 952.7 | 74,394 | 232 |
| Middlesex | 506 | 16,147 | 1,681.2 | 822.4 | 361 | 1,791 | 214.5 | 58.2 | 13,696 | 767.9 | 43,591 | 213 |
| New Haven | 2,585 | 76,203 | 9,164.6 | 3,610.4 | 1,772 | 10,522 | 1,229.8 | 347.7 | 63,139 | 3,324.7 | 139,929 | 1,373 |
| New London | 798 | 17,766 | 1,817.6 | 795.5 | 483 | 2,577 | 312.5 | 70.6 | 17,321 | 839.7 | 89,950 | 456 |

# Table B. States and Counties — Government Employment and Payroll, and Local Government Finances

| | Government employment and payroll, 2017 | | | | | | | | | Local government finances, 2017 | | | | |
| | | | March payroll (percent of total) | | | | | | | General revenue | | | | |
| | | | | | | | | | | | | Taxes | | |
| | | | | | | | | | | | | | Per capita[1] (dollars) | |
| STATE County | Full-time equivalent employees | March payroll (dollars) | Administration, judicial, and legal | Police and corrections | Fire protection | Highways and transportation | Health and welfare | Natural resources and utilities | Education and libraries | Total (mil dol) | Inter-governmental (mil dol) | Total (mil dol) | Total | Property |
| | 171 | 172 | 173 | 174 | 175 | 176 | 177 | 178 | 179 | 180 | 181 | 182 | 183 | 184 |
| COLORADO—Cont'd | | | | | | | | | | | | | | |
| Boulder | 12,053 | 61,752,806 | 8.2 | 10.5 | 5.2 | 2.6 | 6.2 | 8.9 | 54.2 | 1954.8 | 451.4 | 1117.5 | 3,466 | 2,375 |
| Broomfield | 765 | 4,448,301 | 17.0 | 35.1 | 0.0 | 2.2 | 18.5 | 18.7 | 3.6 | 273.2 | 16.5 | 152.4 | 2,232 | 881 |
| Chaffee | 1,135 | 4,646,831 | 6.2 | 7.5 | 2.1 | 3.2 | 53.5 | 3.6 | 23.4 | 132.3 | 22.6 | 39.0 | 1,981 | 1,067 |
| Cheyenne | 169 | 548,400 | 9.2 | 6.3 | 0.0 | 7.8 | 27.9 | 2.5 | 42.3 | 23.6 | 6.1 | 7.2 | 3,864 | 3,071 |
| Clear Creek | 427 | 1,724,550 | 16.5 | 24.0 | 1.6 | 8.8 | 8.7 | 6.7 | 32.1 | 66.0 | 11.4 | 42.6 | 4,445 | 4,026 |
| Conejos | 395 | 1,321,761 | 8.4 | 7.8 | 0.1 | 6.0 | 12.7 | 2.8 | 61.8 | 36.8 | 22.5 | 10.5 | 1,288 | 676 |
| Costilla | 251 | 708,815 | 12.3 | 6.7 | 0.4 | 13.7 | 16.4 | 4.5 | 40.8 | 30.5 | 15.2 | 11.1 | 2,964 | 2,907 |
| Crowley | 150 | 410,874 | 13.1 | 10.8 | 0.3 | 6.5 | 12.7 | 6.6 | 46.1 | 10.7 | 5.9 | 3.3 | 572 | 460 |
| Custer | 171 | 559,390 | 16.4 | 12.3 | 0.7 | 9.8 | 18.9 | 5.3 | 32.5 | 15.7 | 3.5 | 9.0 | 1,862 | 1,472 |
| Delta | 1,739 | 6,945,583 | 4.5 | 5.8 | 0.3 | 2.2 | 46.5 | 5.1 | 34.5 | 170.2 | 46.8 | 36.1 | 1,181 | 657 |
| Denver | 36,005 | 196,656,831 | 6.4 | 13.4 | 4.4 | 12.1 | 25.3 | 9.5 | 28.0 | 7076.0 | 1282.7 | 2796.9 | 3,968 | 1,774 |
| Dolores | 126 | 375,221 | 16.8 | 9.8 | 0.0 | 17.6 | 6.4 | 2.8 | 44.2 | 111.7 | 4.4 | 106.0 | 51,812 | 51,744 |
| Douglas | 9,267 | 41,903,758 | 7.5 | 11.2 | 2.2 | 3.1 | 1.3 | 6.0 | 64.9 | 1489.9 | 399.3 | 759.3 | 2,261 | 1,645 |
| Eagle | 2,056 | 10,948,776 | 10.1 | 15.1 | 8.5 | 10.0 | 7.7 | 12.6 | 29.1 | 436.2 | 48.2 | 284.6 | 5,180 | 3,292 |
| Elbert | 740 | 2,611,721 | 6.9 | 15.3 | 6.0 | 7.8 | 5.0 | 3.3 | 54.8 | 83.4 | 38.8 | 32.5 | 1,261 | 1,083 |
| El Paso | 21,142 | 92,207,013 | 5.4 | 12.5 | 5.0 | 2.8 | 3.7 | 16.5 | 52.2 | 2505.7 | 1031.1 | 1027.5 | 1,468 | 741 |
| Fremont | 1,458 | 4,946,191 | 3.4 | 13.1 | 3.6 | 4.9 | 8.8 | 8.1 | 49.4 | 144.6 | 65.4 | 53.0 | 1,115 | 620 |
| Garfield | 3,290 | 14,845,350 | 6.6 | 9.6 | 5.1 | 2.8 | 25.3 | 8.2 | 39.5 | 459.3 | 124.2 | 219.2 | 3,713 | 2,801 |
| Gilpin | 348 | 1,824,112 | 16.8 | 31.6 | 11.2 | 10.6 | 3.2 | 6.8 | 15.2 | 68.8 | 19.7 | 35.2 | 5,845 | 1,587 |
| Grand | 973 | 4,190,299 | 8.9 | 8.1 | 2.1 | 9.0 | 35.2 | 11.9 | 22.9 | 145.4 | 20.2 | 73.2 | 4,762 | 3,288 |
| Gunnison | 926 | 4,431,480 | 12.3 | 7.7 | 2.4 | 8.3 | 36.3 | 8.3 | 20.1 | 130.8 | 29.9 | 51.4 | 3,047 | 2,036 |
| Hinsdale | 71 | 261,394 | 18.6 | 8.1 | 0.0 | 16.5 | 21.2 | 2.7 | 27.3 | 7.4 | 2.1 | 3.9 | 4,889 | 3,826 |
| Huerfano | 700 | 2,721,294 | 8.5 | 5.1 | 0.1 | 4.9 | 51.8 | 4.8 | 22.2 | 60.2 | 14.5 | 14.4 | 2,177 | 1,713 |
| Jackson | 98 | 302,997 | 14.7 | 12.1 | 0.0 | 9.0 | 6.3 | 7.6 | 48.7 | 8.5 | 3.5 | 3.4 | 2,456 | 1,698 |
| Jefferson | 17,598 | 81,941,659 | 6.8 | 14.0 | 6.3 | 2.2 | 5.0 | 6.9 | 56.3 | 2003.7 | 586.2 | 1064.2 | 1,849 | 1,404 |
| Kiowa | 180 | 564,349 | 8.9 | 5.9 | 0.1 | 8.4 | 45.8 | 1.3 | 27.6 | 18.8 | 5.4 | 4.7 | 3,440 | 3,197 |
| Kit Carson | 608 | 1,994,649 | 6.2 | 6.5 | 2.1 | 6.0 | 35.6 | 5.2 | 36.1 | 54.1 | 17.4 | 15.5 | 2,177 | 1,921 |
| Lake | 1,118 | 4,657,023 | 2.0 | 2.1 | 1.2 | 1.5 | 9.5 | 1.8 | 80.4 | 128.0 | 30.5 | 64.7 | 8,345 | 8,023 |
| La Plata | 2,202 | 9,074,777 | 10.6 | 12.4 | 7.6 | 5.9 | 6.4 | 6.5 | 43.8 | 278.0 | 81.9 | 141.6 | 2,549 | 1,493 |
| Larimer | 11,808 | 54,787,902 | 10.2 | 11.8 | 1.5 | 3.1 | 9.4 | 17.5 | 42.7 | 1540.8 | 372.6 | 770.7 | 2,240 | 1,405 |
| Las Animas | 709 | 2,423,976 | 10.8 | 9.5 | 3.1 | 7.7 | 8.7 | 7.6 | 49.2 | 80.1 | 37.3 | 25.5 | 1,797 | 1,084 |
| Lincoln | 467 | 1,624,923 | 5.3 | 7.5 | 0.0 | 6.9 | 41.7 | 3.1 | 34.4 | 51.4 | 17.8 | 14.2 | 2,575 | 1,877 |
| Logan | 871 | 2,830,803 | 9.0 | 10.3 | 4.7 | 6.5 | 6.9 | 7.3 | 52.2 | 66.3 | 28.1 | 30.2 | 1,353 | 1,095 |
| Mesa | 5,190 | 22,403,036 | 4.9 | 11.8 | 4.6 | 3.6 | 5.4 | 9.0 | 57.7 | 577.7 | 216.3 | 242.8 | 1,606 | 965 |
| Mineral | 61 | 209,780 | 21.3 | 16.1 | 0.0 | 14.2 | 0.7 | 9.2 | 36.7 | 5.7 | 1.3 | 3.0 | 4,024 | 3,351 |
| Moffat | 604 | 2,569,544 | 8.6 | 13.4 | 0.2 | 9.9 | 5.3 | 7.6 | 50.6 | 71.8 | 26.9 | 35.0 | 2,675 | 1,998 |
| Montezuma | 1,062 | 3,219,203 | 7.9 | 15.7 | 2.3 | 5.9 | 5.3 | 9.5 | 45.2 | 146.6 | 48.6 | 63.7 | 2,439 | 1,516 |
| Montrose | 1,571 | 5,748,173 | 9.9 | 13.0 | 4.5 | 6.3 | 7.2 | 11.9 | 43.3 | 260.5 | 64.1 | 69.1 | 1,654 | 894 |
| Morgan | 1,324 | 4,512,710 | 7.0 | 9.1 | 0.3 | 4.3 | 9.2 | 11.2 | 56.4 | 132.7 | 42.6 | 64.3 | 2,272 | 1,843 |
| Otero | 918 | 2,842,292 | 7.6 | 6.5 | 2.0 | 3.0 | 11.2 | 10.6 | 57.7 | 111.8 | 43.9 | 49.9 | 2,721 | 768 |
| Ouray | 273 | 1,080,668 | 26.0 | 6.3 | 3.5 | 2.8 | 4.7 | 8.1 | 46.6 | 28.1 | 9.5 | 13.4 | 2,780 | 1,961 |
| Park | 500 | 1,877,930 | 8.4 | 11.9 | 11.3 | 9.1 | 8.0 | 3.3 | 41.3 | 57.8 | 20.8 | 27.9 | 1,562 | 1,305 |
| Phillips | 470 | 1,792,239 | 5.2 | 2.6 | 2.1 | 4.0 | 45.7 | 3.8 | 36.0 | 76.8 | 11.6 | 35.8 | 8,339 | 2,524 |
| Pitkin | 1,762 | 10,803,740 | 7.6 | 6.9 | 2.0 | 19.6 | 34.4 | 8.1 | 13.5 | 430.1 | 33.6 | 195.5 | 10,876 | 4,990 |
| Prowers | 905 | 3,180,058 | 6.5 | 5.7 | 1.3 | 3.3 | 33.2 | 9.6 | 37.5 | 94.4 | 31.5 | 21.9 | 1,823 | 990 |
| Pueblo | 5,844 | 27,151,176 | 5.3 | 12.1 | 10.8 | 2.9 | 7.3 | 6.0 | 53.1 | 632.0 | 278.7 | 262.3 | 1,578 | 1,002 |
| Rio Blanco | 833 | 3,496,255 | 6.4 | 6.4 | 0.8 | 4.4 | 51.0 | 8.7 | 19.3 | 126.6 | 22.5 | 61.2 | 9,630 | 8,689 |
| Rio Grande | 590 | 1,781,261 | 9.1 | 10.8 | 0.5 | 7.6 | 7.3 | 4.4 | 59.9 | 63.0 | 27.8 | 27.1 | 2,397 | 1,001 |
| Routt | 1,137 | 5,312,629 | 13.1 | 10.3 | 4.5 | 13.5 | 3.6 | 8.5 | 39.4 | 181.0 | 38.0 | 98.9 | 3,931 | 2,246 |
| Saguache | 371 | 1,125,818 | 10.1 | 11.3 | 0.8 | 6.9 | 8.3 | 6.3 | 54.7 | 29.9 | 20.0 | 6.5 | 984 | 803 |
| San Juan | 45 | 188,197 | 26.2 | 10.9 | 0.0 | 14.2 | 6.3 | 7.6 | 34.6 | 6.0 | 1.9 | 3.4 | 4,756 | 2,510 |
| San Miguel | 506 | 2,335,059 | 13.2 | 12.8 | 4.6 | 14.8 | 8.6 | 5.5 | 40.4 | 112.3 | 19.6 | 58.8 | 7,320 | 5,091 |
| Sedgwick | 248 | 987,639 | 12.6 | 1.7 | 1.9 | 4.2 | 48.6 | 4.7 | 24.4 | 18.5 | 5.4 | 9.3 | 4,045 | 3,830 |
| Summit | 1,494 | 7,358,035 | 10.1 | 11.2 | 11.6 | 11.3 | 10.6 | 9.1 | 28.1 | 270.6 | 23.2 | 178.7 | 5,799 | 3,071 |
| Teller | 911 | 3,287,414 | 10.6 | 16.0 | 5.2 | 5.8 | 11.0 | 4.7 | 41.0 | 108.2 | 39.5 | 52.4 | 2,124 | 1,450 |
| Washington | 317 | 920,735 | 9.5 | 11.0 | 0.0 | 10.1 | 6.2 | 4.2 | 57.4 | 35.7 | 13.9 | 11.5 | 2,336 | 2,094 |
| Weld | 9,283 | 38,633,927 | 9.4 | 12.0 | 4.9 | 4.5 | 6.1 | 8.0 | 51.9 | 1519.5 | 370.2 | 817.7 | 2,673 | 1,807 |
| Yuma | 811 | 3,063,690 | 3.5 | 4.9 | 0.0 | 5.3 | 41.4 | 5.1 | 39.2 | 99.2 | 29.6 | 23.9 | 2,395 | 2,047 |
| CONNECTICUT | X | X | X | X | X | X | X | X | X | X | X | X | X | X |
| Fairfield | 32,456 | 194,415,331 | 3.6 | 10.0 | 5.5 | 3.2 | 3.2 | 2.7 | 70.4 | 5258.8 | 1080.1 | 3702.6 | 3,926 | 3,839 |
| Hartford | 33,842 | 190,339,606 | 3.1 | 8.1 | 3.5 | 2.5 | 2.8 | 4.5 | 73.2 | 4687.9 | 1738.5 | 2587.4 | 2,897 | 2,869 |
| Litchfield | 5,925 | 29,964,946 | 5.0 | 6.9 | 1.8 | 5.1 | 1.7 | 2.4 | 75.6 | 786.3 | 156.1 | 562.8 | 3,098 | 3,078 |
| Middlesex | 5,401 | 28,429,699 | 4.8 | 7.3 | 4.5 | 3.8 | 3.4 | 3.0 | 71.2 | 732.7 | 138.0 | 509.0 | 3,124 | 3,103 |
| New Haven | 29,926 | 153,376,297 | 3.4 | 10.0 | 6.1 | 3.3 | 3.0 | 4.9 | 68.4 | 4281.3 | 1605.5 | 2298.8 | 2,680 | 2,641 |
| New London | 9,464 | 49,779,946 | 3.9 | 7.1 | 3.6 | 3.2 | 2.1 | 5.2 | 70.6 | 1247.5 | 411.9 | 703.0 | 2,629 | 2,606 |

1. Based on the resident population estimated as of July 1 of the year shown.

# Table B. States and Counties — Local Government Finances, Government Employment, and Income Taxes

| | Local government finances, 2017 (cont.) | | | | | | | | | Government employment, 2019 | | | Individual income tax returns, 2018 | | |
| STATE County | Direct general expenditure | | | | | | | Debt outstanding | | | | | | | |
| | Total (mil dol) | Per capita[1] (dollars) | Percent of total for: | | | | | Total (mil dol) | Per capita[1] (dollars) | Federal civilian | Federal military | State and local | Number of returns | Mean adjusted gross income | Mean income tax |
| | | | Education | Health and hospitals | Police protection | Public welfare | Highways | | | | | | | | |
| | 185 | 186 | 187 | 188 | 189 | 190 | 191 | 192 | 193 | 194 | 195 | 196 | 197 | 198 | 199 |
|---|---|---|---|---|---|---|---|---|---|---|---|---|---|---|---|
| COLORADO—Cont'd | | | | | | | | | | | | | | | |
| Boulder | 2,019 | 6,261 | 43.6 | 1.9 | 5.4 | 2.4 | 6.8 | 2,807.4 | 8,707 | 2,027 | 836 | 34,078 | 160,980 | 110,472 | 16,864 |
| Broomfield | 178 | 2,613 | 0.0 | 1.4 | 10.0 | 8.1 | 5.8 | 794.6 | 11,640 | 171 | 170 | 1,603 | 36,360 | 99,725 | 13,820 |
| Chaffee | 127 | 6,454 | 20.5 | 44.9 | 3.5 | 1.3 | 8.9 | 94.2 | 4,791 | 70 | 46 | 2,020 | 9,760 | 69,945 | 7,654 |
| Cheyenne | 22 | 11,717 | 27.0 | 26.8 | 3.6 | 12.6 | 8.1 | 0.6 | 317 | 12 | 4 | 273 | 800 | 41,858 | 5,330 |
| Clear Creek | 57 | 5,896 | 22.6 | 5.3 | 9.5 | 5.4 | 11.1 | 16.0 | 1,672 | 34 | 23 | 573 | 4,800 | 84,621 | 11,034 |
| Conejos | 37 | 4,597 | 43.4 | 4.7 | 3.3 | 3.1 | 5.0 | 9.3 | 1,139 | 43 | 20 | 511 | 3,060 | 39,458 | 2,423 |
| Costilla | 30 | 7,893 | 34.5 | 7.0 | 2.0 | 14.8 | 13.4 | 22.5 | 5,980 | 10 | 9 | 356 | 1,350 | 37,813 | 2,593 |
| Crowley | 12 | 2,012 | 37.1 | 0.0 | 4.1 | 13.1 | 9.5 | 0.6 | 97 | 10 | 8 | 512 | 1,180 | 39,409 | 2,638 |
| Custer | 15 | 3,120 | 28.6 | 14.7 | 6.3 | 3.1 | 11.1 | 3.6 | 736 | 11 | 12 | 250 | 2,250 | 62,399 | 6,563 |
| Delta | 171 | 5,618 | 30.3 | 42.9 | 3.5 | 2.2 | 3.9 | 61.4 | 2,012 | 166 | 76 | 2,401 | 13,580 | 47,744 | 3,995 |
| Denver | 6,027 | 8,550 | 21.7 | 18.7 | 3.9 | 2.3 | 2.5 | 15,683.0 | 22,247 | 13,156 | 2,523 | 59,920 | 368,310 | 93,695 | 14,043 |
| Dolores | 11 | 5,229 | 35.3 | 5.0 | 5.9 | 5.8 | 23.0 | 1.1 | 556 | 9 | 5 | 230 | 830 | 45,605 | 3,489 |
| Douglas | 1,412 | 4,203 | 45.4 | 0.6 | 5.3 | 1.9 | 10.2 | 1,862.0 | 5,545 | 419 | 851 | 13,518 | 168,690 | 122,753 | 18,232 |
| Eagle | 414 | 7,528 | 23.3 | 3.3 | 5.8 | 1.4 | 9.0 | 982.1 | 17,879 | 126 | 135 | 3,312 | 29,750 | 107,558 | 16,002 |
| Elbert | 87 | 3,391 | 39.0 | 4.7 | 4.7 | 5.8 | 9.3 | 141.0 | 5,473 | 31 | 65 | 990 | 12,680 | 95,541 | 12,537 |
| El Paso | 2,339 | 3,341 | 52.0 | 1.2 | 7.5 | 3.0 | 8.3 | 4,561.4 | 6,515 | 12,211 | 38,622 | 41,671 | 334,070 | 68,987 | 7,490 |
| Fremont | 124 | 2,597 | 42.7 | 0.6 | 11.8 | 6.5 | 5.9 | 93.7 | 1,970 | 254 | 144 | 4,995 | 18,490 | 53,213 | 4,710 |
| Garfield | 522 | 8,849 | 40.3 | 15.0 | 4.4 | 3.6 | 4.7 | 58.5 | 9,726 | 9 | 15 | 475 | 28,280 | 79,544 | 10,420 |
| Gilpin | 59 | 9,777 | 10.7 | 3.1 | 15.8 | 3.3 | 18.5 | 261.8 | 17,040 | 116 | 38 | 1,324 | 2,950 | 69,166 | 7,371 |
| Grand | 124 | 8,064 | 15.9 | 23.8 | 5.6 | 2.0 | 6.2 | 127.1 | 7,535 | 168 | 40 | 2,159 | 7,900 | 73,554 | 8,835 |
| Gunnison | 125 | 7,417 | 16.8 | 22.1 | 6.0 | 3.2 | 16.2 | 0.4 | 512 | 3 | 2 | 89 | 8,420 | 68,455 | 7,591 |
| Hinsdale | 7 | 9,005 | 27.8 | 19.2 | 7.2 | 1.0 | 18.7 | 45.6 | 6,873 | 15 | 16 | 457 | 380 | 65,074 | 5,771 |
| Huerfano | 62 | 9,332 | 14.0 | 46.5 | 2.7 | 8.6 | 2.6 | 3.6 | 2,615 | 33 | 3 | 129 | 2,720 | 43,923 | 3,556 |
| Jackson | 8 | 6,122 | 36.3 | 5.1 | 6.0 | 5.0 | 19.1 | | | | | | 660 | 47,589 | 4,083 |
| Jefferson | 2,037 | 3,541 | 45.2 | 1.2 | 6.5 | 2.4 | 4.7 | 1,389.2 | 2,414 | 8,142 | 1,466 | 29,047 | 301,970 | 88,039 | 11,417 |
| Kiowa | 15 | 11,116 | 25.5 | 41.1 | 3.4 | 2.7 | 10.8 | 9.2 | 6,739 | 17 | 3 | 248 | 600 | 47,107 | 4,142 |
| Kit Carson | 56 | 7,836 | 30.0 | 28.2 | 5.3 | 3.4 | 9.8 | 23.8 | 3,331 | 40 | 17 | 745 | 3,230 | 43,593 | 4,304 |
| Lake | 154 | 19,851 | 81.8 | 5.8 | 1.2 | 2.0 | 1.0 | 99.2 | 12,798 | 55 | 19 | 633 | 3,680 | 52,991 | 4,688 |
| La Plata | 266 | 4,789 | 34.5 | 0.5 | 6.7 | 3.8 | 6.9 | 1,497.6 | 4,352 | 2,525 | 864 | 37,343 | 173,750 | 80,261 | 10,074 |
| Larimer | 1,367 | 3,973 | 34.9 | 5.0 | 7.1 | 3.4 | 10.2 | 52.5 | 3,697 | 66 | 33 | 1,452 | 6,030 | 48,017 | 4,497 |
| Las Animas | 65 | 4,557 | 38.8 | 6.2 | 4.9 | 0.0 | 12.9 | 17.9 | 3,246 | 20 | 11 | 989 | 2,050 | 46,398 | 3,574 |
| Lincoln | 54 | 9,842 | 31.9 | 29.8 | 2.2 | 5.8 | 10.4 | 62.7 | 2,811 | 59 | 45 | 2,333 | 8,610 | 56,102 | 5,440 |
| Logan | 64 | 2,860 | 49.0 | 2.2 | 3.9 | 6.7 | 9.5 | 401.0 | 2,653 | 1,636 | 381 | 8,781 | 71,300 | 61,564 | 6,357 |
| Mesa | 591 | 3,907 | 38.0 | 2.0 | 11.7 | 5.4 | 9.0 | 12.3 | 16,271 | 5 | 2 | 118 | 430 | 57,286 | 5,391 |
| Mineral | 5 | 6,173 | 50.2 | 0.6 | 3.6 | 0.0 | 11.6 | 35.2 | 2,690 | 131 | 32 | 937 | 5,890 | 57,135 | 4,955 |
| Moffat | 70 | 5,354 | 28.9 | 0.9 | 8.7 | 7.8 | 17.7 | 61.1 | 2,339 | 337 | 63 | 2,283 | 11,800 | 54,749 | 5,646 |
| Montezuma | 128 | 4,908 | 29.0 | 4.7 | 8.1 | 9.2 | 9.3 | 104.8 | 2,510 | 308 | 102 | 2,924 | 19,580 | 54,322 | 4,918 |
| Montrose | 244 | 5,845 | 24.7 | 41.6 | 5.0 | 2.3 | 5.5 | 133.3 | 4,711 | 105 | 69 | 2,119 | 13,550 | 53,065 | 4,561 |
| Morgan | 119 | 4,190 | 50.6 | 1.4 | 5.9 | 4.3 | 8.6 | 41.9 | 2,284 | 101 | 43 | 1,815 | 7,500 | 38,750 | 2,972 |
| Otero | 106 | 5,764 | 35.5 | 3.5 | 4.9 | 5.1 | 6.5 | 9.3 | 1,941 | 13 | 12 | 394 | 2,650 | 88,549 | 12,004 |
| Ouray | 27 | 5,584 | 39.0 | 3.8 | 5.3 | 4.2 | 10.7 | 31.4 | 1,754 | 47 | 46 | 745 | 8,060 | 67,941 | 6,989 |
| Park | 53 | 2,945 | 34.5 | 3.3 | 6.8 | 9.1 | 10.5 | 29.9 | 6,969 | 23 | 10 | 592 | 2,020 | 55,462 | 5,007 |
| Phillips | 73 | 17,041 | 21.9 | 33.3 | 3.6 | 0.9 | 5.3 | 352.2 | 19,597 | 74 | 43 | 2,372 | 10,330 | 164,819 | 30,550 |
| Pitkin | 354 | 19,692 | 9.6 | 35.7 | 4.0 | 1.8 | 3.4 | 173.9 | 14,495 | 35 | 29 | 1,363 | 4,880 | 44,684 | 3,546 |
| Prowers | 90 | 7,469 | 28.7 | 33.7 | 5.9 | 4.8 | 5.3 | 578.0 | 3,476 | 1,083 | 415 | 11,641 | 72,210 | 52,552 | 4,740 |
| Pueblo | 625 | 3,759 | 39.5 | 1.4 | 7.3 | 4.1 | 4.1 | 83.3 | 13,109 | 63 | 15 | 1,163 | 2,710 | 60,301 | 5,666 |
| Rio Blanco | 118 | 18,501 | 14.5 | 29.0 | 3.1 | 2.2 | 7.8 | 27.7 | 2,450 | 114 | 27 | 846 | 5,840 | 49,770 | 4,325 |
| Rio Grande | 60 | 5,263 | 34.6 | 6.7 | 6.3 | 9.4 | 7.8 | 122.5 | 4,865 | 101 | 61 | 2,369 | 13,510 | 104,399 | 16,513 |
| Routt | 156 | 6,199 | 29.7 | 1.6 | 5.2 | 2.9 | 10.0 | 16.3 | 2,452 | 43 | 17 | 509 | 2,140 | 37,985 | 3,577 |
| Saguache | 30 | 4,587 | 48.1 | 1.4 | 5.4 | 15.0 | 10.4 | 1.4 | 1,996 | 4 | 2 | 83 | 360 | 48,008 | 4,156 |
| San Juan | 6 | 7,721 | 42.9 | 2.2 | 10.0 | 2.1 | 0.0 | 94.6 | 11,774 | 32 | 20 | 802 | 4,560 | 102,269 | 16,540 |
| San Miguel | 109 | 13,587 | 21.9 | 7.6 | 4.9 | 1.3 | 9.3 | 9.1 | 3,928 | 16 | 5 | 320 | 1,070 | 39,333 | 3,179 |
| Sedgwick | 17 | 7,132 | 44.0 | 0.0 | 3.4 | 6.6 | 9.6 | 621.2 | 20,155 | 52 | 75 | 2,471 | 17,240 | 94,414 | 13,594 |
| Summit | 245 | 7,947 | 19.8 | 4.6 | 5.0 | 0.9 | 6.2 | 95.2 | 3,860 | 59 | 61 | 1,389 | 12,410 | 66,808 | 6,866 |
| Teller | 106 | 4,279 | 29.5 | 6.2 | 7.5 | 2.7 | 7.9 | 24.2 | 4,910 | 41 | 11 | 456 | 1,980 | 49,930 | 3,789 |
| Washington | 31 | 6,382 | 40.2 | 3.2 | 3.4 | 10.0 | 11.8 | 1,730.4 | 5,657 | 598 | 775 | 16,960 | 143,900 | 71,647 | 7,795 |
| Weld | 1,347 | 4,403 | 41.0 | 1.2 | 6.1 | 2.3 | 8.2 | 37.3 | 3,744 | 44 | 24 | 1,005 | 4,280 | 47,445 | 4,291 |
| Yuma | 95 | 9,551 | 23.8 | 40.5 | 2.5 | 3.4 | 5.8 | | | | | | | | |
| CONNECTICUT | X | X | X | X | X | X | X | X | X | 18,251 | 13,638 | 220,785 | 1,768,610 | 100,776 | 15,929 |
| Fairfield | 4,968 | 5,268 | 53.4 | 1.0 | 6.9 | 0.9 | 3.8 | 3,859.8 | 4,093 | 2,854 | 1,885 | 44,011 | 462,580 | 165,244 | 33,039 |
| Hartford | 4,406 | 4,933 | 57.9 | 0.9 | 5.1 | 0.7 | 3.4 | 3,703.3 | 4,147 | 5,620 | 1,831 | 65,396 | 449,820 | 81,443 | 10,791 |
| Litchfield | 772 | 4,250 | 63.4 | 1.5 | 4.1 | 0.2 | 7.2 | 219.8 | 1,210 | 457 | 359 | 7,401 | 94,490 | 80,182 | 10,226 |
| Middlesex | 783 | 4,808 | 59.2 | 0.7 | 4.2 | 0.3 | 3.8 | 394.4 | 2,420 | 398 | 319 | 9,415 | 84,920 | 88,827 | 11,875 |
| New Haven | 4,329 | 5,047 | 54.3 | 0.8 | 5.3 | 0.3 | 3.3 | 3,393.1 | 3,956 | 5,522 | 1,917 | 43,653 | 416,540 | 74,374 | 9,283 |
| New London | 1,142 | 4,271 | 59.0 | 0.8 | 6.3 | 0.5 | 6.0 | 611.7 | 2,288 | 2,842 | 6,810 | 27,127 | 135,780 | 73,015 | 8,582 |

1. Based on the resident population estimated as of July 1 of the year shown.

# Table B. States and Counties — **Land Area and Population**

| State / county code | CBSA code[1] | County Type code[2] | STATE County | Land area[3] (sq. mi) | Total persons 2019 | Rank | Per square mile | White | Black | American Indian, Alaska Native | Asian and Pacific Islander | Percent Hispanic or Latino[4] | Under 5 years | 5 to 14 years | 15 to 24 years | 25 to 34 years | 35 to 44 years | 45 to 54 years |
|---|---|---|---|---|---|---|---|---|---|---|---|---|---|---|---|---|---|---|
| | | | | 1 | 2 | 3 | 4 | 5 | 6 | 7 | 8 | 9 | 10 | 11 | 12 | 13 | 14 | 15 |
| | | | CONNECTICUT—Cont'd | | | | | | | | | | | | | | | |
| 09013 | 25540 | 1 | Tolland | 410.4 | 150,600 | 447 | 367.0 | 84.9 | 4.3 | 0.6 | 6.0 | 6.2 | 4.0 | 9.5 | 21.9 | 11.4 | 10.7 | 11.7 |
| 09015 | 49340 | 2 | Windham | 512.9 | 116,540 | 535 | 227.2 | 83.4 | 2.9 | 1.1 | 1.9 | 12.6 | 4.7 | 11.1 | 13.2 | 12.8 | 12.2 | 13.0 |
| 10000 | | 0 | DELAWARE | 1,948.5 | 986,809 | X | 506.4 | 63.2 | 23.8 | 0.9 | 4.9 | 9.8 | 5.5 | 11.6 | 12.0 | 13.1 | 11.7 | 11.8 |
| 10001 | 20100 | 3 | Kent | 586.1 | 183,643 | 363 | 313.3 | 62.5 | 28.8 | 1.3 | 3.5 | 7.7 | 6.1 | 12.8 | 13.6 | 13.5 | 11.9 | 11.3 |
| 10003 | 37980 | 1 | New Castle | 426.3 | 561,531 | 122 | 1,317.2 | 57.4 | 27.0 | 0.7 | 6.7 | 10.7 | 5.6 | 11.9 | 12.7 | 14.4 | 12.7 | 12.6 |
| 10005 | 41540 | 2 | Sussex | 936.2 | 241,635 | 284 | 258.1 | 77.1 | 12.8 | 1.0 | 1.8 | 9.4 | 4.9 | 10.2 | 9.0 | 9.9 | 9.4 | 10.5 |
| 11000 | | 0 | DISTRICT OF COLUMBIA | 61.1 | 712,816 | X | 11,666.4 | 39.6 | 45.3 | 0.8 | 5.6 | 11.4 | 6.2 | 9.8 | 12.3 | 23.1 | 15.9 | 10.4 |
| 11001 | 47900 | 1 | District of Columbia | 61.1 | 712,816 | 93 | 11,666.4 | 39.6 | 45.3 | 0.8 | 5.6 | 11.4 | 6.2 | 9.8 | 12.3 | 23.1 | 15.9 | 10.4 |
| 12000 | | 0 | FLORIDA | 53,647.9 | 21,733,312 | X | 405.1 | 54.5 | 16.5 | 0.6 | 3.7 | 26.5 | 5.2 | 11.0 | 11.4 | 13.0 | 12.2 | 12.4 |
| 12001 | 23540 | 2 | Alachua | 875.6 | 271,218 | 255 | 309.8 | 62.7 | 21.3 | 0.6 | 7.3 | 10.7 | 5.0 | 10.1 | 23.0 | 15.3 | 11.4 | 9.6 |
| 12003 | 27260 | 1 | Baker | 585.2 | 29,566 | 1,447 | 50.5 | 81.7 | 15.0 | 1.0 | 1.1 | 2.9 | 6.1 | 13.5 | 12.2 | 14.7 | 13.3 | 13.2 |
| 12005 | 37460 | 3 | Bay | 758.6 | 171,322 | 386 | 225.8 | 79.0 | 12.1 | 1.4 | 3.8 | 6.9 | 5.5 | 11.5 | 10.7 | 13.3 | 12.4 | 12.6 |
| 12007 | | 6 | Bradford | 294.0 | 28,593 | 1,475 | 97.3 | 74.2 | 20.9 | 0.9 | 1.1 | 4.7 | 5.2 | 10.8 | 11.1 | 15.7 | 13.4 | 12.6 |
| 12009 | 37340 | 2 | Brevard | 1,015.0 | 608,459 | 112 | 599.5 | 75.6 | 11.3 | 0.8 | 3.7 | 11.2 | 4.5 | 10.3 | 10.2 | 11.5 | 10.8 | 11.8 |
| 12011 | 33100 | 1 | Broward | 1,202.7 | 1,958,105 | 17 | 1,628.1 | 35.5 | 29.8 | 0.4 | 4.7 | 31.4 | 5.7 | 11.7 | 11.1 | 13.3 | 13.5 | 13.5 |
| 12013 | | 6 | Calhoun | 567.3 | 14,078 | 2,153 | 24.8 | 79.5 | 13.3 | 2.1 | 1.2 | 6.1 | 4.8 | 11.6 | 10.8 | 14.1 | 12.7 | 13.6 |
| 12015 | 39460 | 3 | Charlotte | 681.1 | 194,711 | 348 | 285.9 | 84.7 | 6.2 | 0.7 | 1.9 | 8.0 | 2.8 | 6.7 | 7.0 | 7.8 | 7.5 | 9.8 |
| 12017 | 26140 | 3 | Citrus | 581.9 | 153,010 | 437 | 262.9 | 88.7 | 3.6 | 0.9 | 2.2 | 6.3 | 3.6 | 8.4 | 7.7 | 8.7 | 8.0 | 10.3 |
| 12019 | 27260 | 1 | Clay | 604.6 | 221,770 | 309 | 366.8 | 73.5 | 13.3 | 1.0 | 4.6 | 10.7 | 5.4 | 13.3 | 11.9 | 12.4 | 13.1 | 13.1 |
| 12021 | 34940 | 2 | Collier | 1,997.0 | 392,973 | 181 | 196.8 | 62.8 | 7.1 | 0.4 | 2.0 | 28.6 | 4.2 | 9.4 | 9.3 | 9.4 | 9.6 | 10.9 |
| 12023 | 29380 | 4 | Columbia | 797.6 | 72,654 | 752 | 91.1 | 73.5 | 19.2 | 1.0 | 1.5 | 6.7 | 5.6 | 12.3 | 12.2 | 13.0 | 11.9 | 11.6 |
| 12027 | 11580 | 6 | DeSoto | 636.7 | 38,520 | 1,211 | 60.5 | 55.4 | 12.5 | 0.5 | 0.7 | 31.9 | 5.0 | 10.2 | 12.1 | 13.8 | 12.7 | 11.3 |
| 12029 | | 6 | Dixie | 705.0 | 17,057 | 1,971 | 24.2 | 85.4 | 10.0 | 1.3 | 0.7 | 4.2 | 4.6 | 10.4 | 9.8 | 11.4 | 12.3 | 12.1 |
| 12031 | 27260 | 1 | Duval | 762.6 | 966,728 | 49 | 1,267.7 | 53.6 | 31.1 | 0.8 | 6.3 | 11.0 | 6.7 | 12.4 | 12.0 | 16.5 | 13.0 | 11.9 |
| 12033 | 37860 | 2 | Escambia | 657.0 | 322,364 | 222 | 490.7 | 66.7 | 24.2 | 1.6 | 5.0 | 6.1 | 5.9 | 11.6 | 14.5 | 14.8 | 11.3 | 10.9 |
| 12035 | 19660 | 2 | Flagler | 486.2 | 118,451 | 531 | 243.6 | 76.1 | 10.9 | 0.7 | 3.2 | 11.1 | 3.8 | 9.4 | 9.3 | 8.8 | 9.7 | 11.6 |
| 12037 | | 6 | Franklin | 545.0 | 12,201 | 2,277 | 22.4 | 80.2 | 14.2 | 1.0 | 0.7 | 5.9 | 4.1 | 8.7 | 8.9 | 14.9 | 12.0 | 11.8 |
| 12039 | 45220 | 2 | Gadsden | 516.4 | 45,277 | 1,071 | 87.7 | 33.2 | 55.4 | 0.5 | 0.9 | 10.9 | 5.8 | 12.0 | 11.6 | 11.9 | 12.1 | 13.1 |
| 12041 | 23540 | 2 | Gilchrist | 349.7 | 18,885 | 1,881 | 54.0 | 87.1 | 6.0 | 1.1 | 0.8 | 6.4 | 5.3 | 11.2 | 14.9 | 10.7 | 10.4 | 11.8 |
| 12043 | | 6 | Glades | 806.8 | 14,198 | 2,143 | 17.6 | 61.0 | 13.2 | 4.3 | 0.8 | 21.8 | 2.9 | 8.5 | 9.1 | 12.4 | 12.7 | 12.1 |
| 12045 | | 3 | Gulf | 553.5 | 13,534 | 2,189 | 24.5 | 84.5 | 12.4 | 1.5 | 1.0 | 2.9 | 4.7 | 10.8 | 9.5 | 10.6 | 9.6 | 12.1 |
| 12047 | | 6 | Hamilton | 514.3 | 14,521 | 2,120 | 28.2 | 55.7 | 33.3 | 1.3 | 1.0 | 10.4 | 5.3 | 10.6 | 14.5 | 14.0 | 11.4 | 12.3 |
| 12049 | 48100 | 6 | Hardee | 637.6 | 26,822 | 1,533 | 42.1 | 47.6 | 7.4 | 0.7 | 1.3 | 44.3 | 6.7 | 14.8 | 14.1 | 13.5 | 11.3 | 11.8 |
| 12051 | 17500 | 4 | Hendry | 1,156.0 | 42,813 | 1,123 | 37.0 | 31.3 | 11.1 | 1.5 | 1.1 | 55.7 | 7.2 | 15.1 | 13.4 | 14.0 | 12.6 | 12.3 |
| 12053 | 45300 | 1 | Hernando | 473.0 | 198,792 | 339 | 420.3 | 77.3 | 6.3 | 0.8 | 2.0 | 15.4 | 4.4 | 10.5 | 9.8 | 10.8 | 10.4 | 11.9 |
| 12055 | 42700 | 3 | Highlands | 1,017.7 | 106,639 | 573 | 104.8 | 67.0 | 10.4 | 0.8 | 1.9 | 21.3 | 4.2 | 9.6 | 8.7 | 9.6 | 8.8 | 9.4 |
| 12057 | 45300 | 1 | Hillsborough | 1,021.9 | 1,497,957 | 26 | 1,465.9 | 48.9 | 17.4 | 0.6 | 5.4 | 29.9 | 5.9 | 12.4 | 12.4 | 15.3 | 13.9 | 13.0 |
| 12059 | | 6 | Holmes | 478.9 | 19,594 | 1,847 | 40.9 | 88.2 | 7.4 | 2.2 | 1.1 | 3.2 | 5.2 | 11.3 | 11.7 | 12.9 | 11.6 | 13.1 |
| 12061 | 42680 | 3 | Indian River | 502.8 | 162,518 | 411 | 323.2 | 76.1 | 9.7 | 0.6 | 2.1 | 12.9 | 4.0 | 8.4 | 8.8 | 9.4 | 8.7 | 10.5 |
| 12063 | | 6 | Jackson | 918.2 | 46,085 | 1,055 | 50.2 | 67.3 | 27.1 | 1.4 | 1.2 | 5.1 | 5.2 | 10.3 | 11.5 | 13.1 | 13.5 | 12.5 |
| 12065 | 45220 | 2 | Jefferson | 598.1 | 14,543 | 2,118 | 24.3 | 62.6 | 32.5 | 1.0 | 1.0 | 4.4 | 4.1 | 9.9 | 9.0 | 11.6 | 11.5 | 14.0 |
| 12067 | | 9 | Lafayette | 543.3 | 8,482 | 2,538 | 15.6 | 72.7 | 13.0 | 0.8 | 0.7 | 14.3 | 4.2 | 11.1 | 14.6 | 13.4 | 13.9 | 13.0 |
| 12069 | 36740 | 1 | Lake | 951.6 | 375,492 | 191 | 394.6 | 69.3 | 11.7 | 0.8 | 2.9 | 17.1 | 4.8 | 10.8 | 9.7 | 11.1 | 11.0 | 11.7 |
| 12071 | 15980 | 2 | Lee | 781.0 | 790,767 | 83 | 1,012.5 | 66.9 | 8.8 | 0.5 | 2.3 | 23.0 | 4.5 | 9.6 | 9.8 | 10.9 | 10.3 | 11.2 |
| 12073 | 45220 | 2 | Leon | 668.4 | 295,460 | 234 | 442.0 | 57.4 | 32.9 | 0.8 | 4.5 | 6.8 | 5.1 | 10.3 | 24.2 | 14.2 | 11.3 | 10.0 |
| 12075 | 23540 | 6 | Levy | 1,118.2 | 42,214 | 1,135 | 37.8 | 80.5 | 9.7 | 1.3 | 1.2 | 9.1 | 5.0 | 11.2 | 9.4 | 11.3 | 10.0 | 11.5 |
| 12077 | | 8 | Liberty | 835.6 | 8,364 | 2,547 | 10.0 | 72.7 | 19.7 | 1.4 | 0.7 | 7.0 | 4.3 | 9.3 | 11.7 | 17.1 | 15.5 | 14.1 |
| 12079 | | 6 | Madison | 696.5 | 18,707 | 1,888 | 26.9 | 55.7 | 37.7 | 1.1 | 0.7 | 6.2 | 4.9 | 10.5 | 10.5 | 13.6 | 12.1 | 12.7 |
| 12081 | 35840 | 2 | Manatee | 742.8 | 411,219 | 175 | 553.6 | 72.1 | 9.4 | 0.6 | 2.8 | 16.8 | 4.4 | 10.1 | 9.5 | 10.3 | 10.3 | 11.6 |
| 12083 | 36100 | 2 | Marion | 1,588.4 | 373,513 | 192 | 235.2 | 70.5 | 13.5 | 0.9 | 2.3 | 14.7 | 4.9 | 10.4 | 9.7 | 11.0 | 10.0 | 10.8 |
| 12085 | 38940 | 2 | Martin | 543.8 | 162,088 | 412 | 298.1 | 79.0 | 5.7 | 0.6 | 2.0 | 14.1 | 4.0 | 9.0 | 8.8 | 9.1 | 9.4 | 11.4 |
| 12086 | 33100 | 1 | Miami-Dade | 1,899.9 | 2,707,303 | 7 | 1,425.0 | 14.0 | 15.7 | 0.2 | 1.9 | 68.9 | 5.7 | 11.1 | 11.5 | 13.9 | 13.6 | 14.2 |
| 12087 | 28580 | 4 | Monroe | 983.0 | 73,900 | 745 | 75.2 | 65.9 | 7.2 | 0.9 | 2.1 | 25.5 | 4.5 | 8.5 | 8.6 | 12.1 | 12.2 | 13.9 |
| 12089 | 27260 | 1 | Nassau | 648.7 | 91,113 | 652 | 140.5 | 87.7 | 6.5 | 0.8 | 1.7 | 4.9 | 5.0 | 11.2 | 9.8 | 11.3 | 11.5 | 12.4 |
| 12091 | 18880 | 3 | Okaloosa | 930.2 | 212,820 | 320 | 228.8 | 76.1 | 11.4 | 1.3 | 5.5 | 10.0 | 6.4 | 12.5 | 12.3 | 15.8 | 12.7 | 10.5 |
| 12093 | 36380 | 4 | Okeechobee | 769.2 | 42,297 | 1,133 | 55.0 | 63.1 | 9.0 | 1.2 | 1.3 | 26.5 | 6.1 | 11.5 | 11.0 | 13.2 | 12.3 | 12.5 |
| 12095 | 36740 | 1 | Orange | 902.0 | 1,404,396 | 28 | 1,557.0 | 40.6 | 21.4 | 0.6 | 6.6 | 32.9 | 5.9 | 12.1 | 13.7 | 16.7 | 14.6 | 12.9 |
| 12097 | 36740 | 1 | Osceola | 1,327.5 | 385,315 | 184 | 290.3 | 30.8 | 10.7 | 0.5 | 3.4 | 56.0 | 6.1 | 13.6 | 13.4 | 14.4 | 14.6 | 13.1 |
| 12099 | 33100 | 1 | Palm Beach | 1,964.3 | 1,507,600 | 25 | 767.5 | 54.4 | 19.6 | 0.4 | 3.6 | 23.4 | 5.0 | 10.5 | 10.6 | 11.8 | 11.5 | 12.3 |
| 12101 | 45300 | 1 | Pasco | 746.6 | 570,412 | 119 | 764.0 | 73.5 | 7.0 | 0.8 | 3.8 | 17.1 | 5.0 | 11.6 | 10.5 | 11.4 | 12.3 | 12.9 |
| 12103 | 45300 | 1 | Pinellas | 273.7 | 976,802 | 48 | 3,568.9 | 75.3 | 11.4 | 0.7 | 4.5 | 10.3 | 4.1 | 8.8 | 9.1 | 12.2 | 11.1 | 12.5 |
| 12105 | 29460 | 2 | Polk | 1,797.8 | 744,552 | 89 | 414.1 | 57.2 | 15.8 | 0.7 | 2.4 | 25.7 | 5.7 | 12.4 | 12.0 | 13.4 | 12.2 | 11.7 |

1.  CBSA = Core Based Statistical Area. See Appendix A for explanation. See Appendix B for list of metropolitan areas with component counties.
Service of USDA Rural-Urban Continuum Codes. See Appendix A for definition.  3.  Dry land or land partially or temporarily covered by water.
2.  County type code from the Economic Research
4.  May be of any race.

# Table B. States and Counties — Population and Households

| STATE County | Population, 2020 (cont.) — Age (percent) (cont.) | | | | Population change, 2000-2020 | | | | | | | Households, 2015-2019 | | | | |
| | 55 to 64 years | 65 to 74 years | 75 years and over | Percent female | Total persons 2000 | Total persons 2010 | Percent change 2000-2010 | Percent change 2010-2020 | Births | Deaths | Net Migration | Number | Persons per household | Family households | Female family householder[1] | One person |
|---|---|---|---|---|---|---|---|---|---|---|---|---|---|---|---|---|
| | 16 | 17 | 18 | 19 | 20 | 21 | 22 | 23 | 24 | 25 | 26 | 27 | 28 | 29 | 30 | 31 |
| **CONNECTICUT—Cont'd** | | | | | | | | | | | | | | | | |
| Tolland | 13.9 | 9.7 | 7.2 | 49.9 | 136,364 | 152,745 | 12.0 | -1.4 | 11,859 | 10,770 | -3,324 | 55,683 | 2.44 | 64.1 | 8.3 | 25.1 |
| Windham | 15.2 | 10.6 | 7.2 | 50.4 | 109,091 | 118,379 | 8.5 | -1.6 | 11,449 | 10,783 | -2,538 | 44,953 | 2.47 | 66.0 | 12.3 | 26.6 |
| **DELAWARE** | 14.2 | 12.0 | 8.0 | 51.7 | 783,600 | 897,947 | 14.6 | 9.9 | 112,071 | 89,006 | 65,890 | 363,322 | 2.57 | 66.1 | 12.8 | 27.5 |
| Kent | 12.9 | 10.5 | 7.4 | 51.9 | 126,697 | 162,352 | 28.1 | 13.1 | 22,616 | 16,251 | 14,926 | 65,796 | 2.61 | 68.6 | 15.4 | 25.1 |
| New Castle | 13.6 | 9.7 | 6.8 | 51.6 | 500,265 | 538,496 | 7.6 | 4.3 | 66,276 | 47,310 | 4,474 | 205,829 | 2.62 | 64.2 | 13.0 | 28.8 |
| Sussex | 16.4 | 18.5 | 11.2 | 51.8 | 156,638 | 197,099 | 25.8 | 22.6 | 23,179 | 25,445 | 46,490 | 91,697 | 2.41 | 68.4 | 10.4 | 26.3 |
| **DISTRICT OF COLUMBIA** | 9.8 | 7.2 | 5.4 | 52.6 | 572,059 | 601,767 | 5.2 | 18.5 | 96,247 | 51,168 | 64,816 | 284,386 | 2.30 | 43.5 | 14.0 | 44.1 |
| District of Columbia | 9.8 | 7.2 | 5.4 | 52.6 | 572,059 | 601,767 | 5.2 | 18.5 | 96,247 | 51,168 | 64,816 | 284,386 | 2.30 | 43.5 | 14.0 | 44.1 |
| **FLORIDA** | 13.5 | 11.7 | 9.7 | 51.1 | 15,982,378 | 18,804,589 | 17.7 | 15.6 | 2,244,645 | 1,985,189 | 2,661,420 | 7,736,311 | 2.65 | 64.6 | 12.9 | 28.6 |
| Alachua | 10.5 | 9.1 | 6.1 | 51.8 | 217,955 | 247,337 | 13.5 | 9.7 | 29,028 | 19,449 | 14,319 | 97,995 | 2.55 | 51.3 | 10.5 | 37.2 |
| Baker | 12.3 | 8.7 | 5.9 | 47.4 | 22,259 | 27,115 | 21.8 | 9.0 | 3,521 | 2,532 | 1,466 | 8,693 | 2.92 | 71.7 | 12.2 | 22.4 |
| Bay | 15.1 | 11.1 | 7.8 | 50.4 | 148,217 | 168,843 | 13.9 | 1.5 | 22,772 | 19,089 | -1,269 | 71,422 | 2.52 | 64.1 | 12.9 | 29.2 |
| Bradford | 12.8 | 10.1 | 8.3 | 44.5 | 26,088 | 28,519 | 9.3 | 0.3 | 3,083 | 3,136 | 40 | 9,115 | 2.57 | 64.6 | 14.1 | 28.5 |
| Brevard | 16.4 | 13.4 | 11.1 | 51.1 | 476,230 | 543,376 | 14.1 | 12.0 | 53,029 | 71,908 | 83,982 | 230,417 | 2.51 | 63.1 | 10.4 | 30.9 |
| Broward | 13.7 | 9.8 | 7.7 | 51.2 | 1,623,018 | 1,748,158 | 7.7 | 12.0 | 223,220 | 154,173 | 142,202 | 690,050 | 2.77 | 63.9 | 15.1 | 29.2 |
| Calhoun | 13.2 | 10.2 | 8.9 | 46.1 | 13,017 | 14,632 | 12.4 | -3.8 | 1,415 | 1,696 | -268 | 4,501 | 2.74 | 60.7 | 8.8 | 35.3 |
| Charlotte | 17.2 | 21.8 | 19.4 | 51.2 | 141,627 | 159,974 | 13.0 | 21.7 | 10,452 | 26,195 | 50,133 | 76,891 | 2.31 | 63.8 | 7.0 | 29.7 |
| Citrus | 16.2 | 19.8 | 17.3 | 51.6 | 118,085 | 141,229 | 19.6 | 8.3 | 10,688 | 25,947 | 27,055 | 63,681 | 2.24 | 62.2 | 9.4 | 32.2 |
| Clay | 14.0 | 10.4 | 6.3 | 50.8 | 140,814 | 190,872 | 35.5 | 16.2 | 22,080 | 17,476 | 26,357 | 74,333 | 2.83 | 76.1 | 12.5 | 20.0 |
| Collier | 13.7 | 16.0 | 17.5 | 50.8 | 251,377 | 321,514 | 27.9 | 22.2 | 32,959 | 33,701 | 72,076 | 142,979 | 2.57 | 68.2 | 8.4 | 26.5 |
| Columbia | 13.6 | 11.5 | 8.3 | 48.4 | 56,513 | 67,526 | 19.5 | 7.6 | 8,215 | 8,219 | 5,139 | 25,133 | 2.59 | 62.6 | 12.6 | 31.6 |
| DeSoto | 12.1 | 11.7 | 11.2 | 44.0 | 32,209 | 34,862 | 8.2 | 10.5 | 3,958 | 3,358 | 3,072 | 12,072 | 2.77 | 68.4 | 13.4 | 26.3 |
| Dixie | 15.1 | 13.7 | 10.6 | 45.0 | 13,827 | 16,422 | 18.8 | 3.9 | 1,567 | 2,249 | 1,316 | 6,504 | 2.31 | 67.3 | 12.3 | 26.2 |
| Duval | 12.7 | 9.1 | 5.8 | 51.6 | 778,879 | 864,253 | 11.0 | 11.9 | 131,243 | 82,659 | 54,206 | 359,544 | 2.55 | 62.1 | 15.6 | 30.6 |
| Escambia | 13.4 | 10.4 | 7.1 | 50.7 | 294,410 | 297,632 | 1.1 | 8.3 | 39,699 | 34,266 | 19,367 | 120,104 | 2.46 | 59.6 | 13.9 | 32.2 |
| Flagler | 15.8 | 18.2 | 13.5 | 51.9 | 49,832 | 95,691 | 92.0 | 23.8 | 8,357 | 12,998 | 27,247 | 42,121 | 2.59 | 69.6 | 10.3 | 24.1 |
| Franklin | 15.1 | 15.2 | 9.3 | 43.6 | 11,057 | 11,549 | 4.4 | 5.6 | 1,028 | 1,358 | 976 | 4,444 | 2.29 | 61.4 | 11.6 | 33.0 |
| Gadsden | 14.1 | 11.5 | 7.9 | 52.4 | 45,087 | 47,737 | 5.9 | -5.2 | 5,615 | 4,835 | -3,271 | 17,149 | 2.46 | 65.4 | 22.2 | 30.6 |
| Gilchrist | 14.5 | 12.0 | 9.2 | 48.3 | 14,437 | 16,944 | 17.4 | 11.5 | 1,950 | 1,953 | 1,944 | 6,464 | 2.57 | 67.3 | 12.8 | 28.1 |
| Glades | 13.3 | 13.9 | 15.2 | 44.4 | 10,576 | 12,890 | 21.9 | 10.1 | 679 | 1,187 | 1,803 | 4,700 | 2.57 | 66.2 | 10.5 | 28.4 |
| Gulf | 16.8 | 15.6 | 10.4 | 49.9 | 13,332 | 15,863 | 19.0 | -14.7 | 1,239 | 1,833 | -1,746 | 5,757 | 2.28 | 67.8 | 12.8 | 28.2 |
| Hamilton | 13.2 | 11.3 | 7.4 | 42.2 | 13,327 | 14,799 | 11.0 | -1.9 | 1,632 | 1,447 | -508 | 4,381 | 2.51 | 70.3 | 17.0 | 24.4 |
| Hardee | 10.7 | 9.3 | 7.9 | 47.3 | 26,938 | 27,737 | 3.0 | -3.3 | 3,861 | 2,188 | -2,616 | 7,863 | 3.17 | 71.4 | 12.8 | 24.2 |
| Hendry | 11.4 | 7.8 | 6.2 | 47.3 | 36,210 | 39,141 | 8.1 | 9.4 | 6,050 | 3,076 | 587 | 12,527 | 3.09 | 74.8 | 18.0 | 23.3 |
| Hernando | 14.6 | 14.8 | 12.7 | 51.7 | 130,802 | 172,778 | 32.1 | 15.1 | 15,764 | 27,343 | 37,515 | 75,348 | 2.44 | 67.6 | 11.6 | 26.7 |
| Highlands | 13.2 | 17.1 | 19.4 | 51.3 | 87,366 | 98,786 | 13.1 | 7.9 | 9,120 | 15,875 | 14,684 | 41,740 | 2.43 | 64.8 | 9.4 | 29.9 |
| Hillsborough | 12.2 | 8.9 | 6.0 | 51.0 | 998,948 | 1,229,207 | 23.1 | 21.9 | 173,670 | 106,438 | 200,501 | 526,175 | 2.66 | 63.1 | 13.7 | 28.9 |
| Holmes | 13.7 | 11.2 | 9.2 | 46.9 | 18,564 | 19,924 | 7.3 | -1.7 | 2,016 | 2,710 | 377 | 7,092 | 2.48 | 67.4 | 13.7 | 28.6 |
| Indian River | 15.6 | 18.0 | 16.5 | 52.0 | 112,947 | 138,028 | 22.2 | 17.7 | 12,969 | 20,372 | 31,790 | 58,612 | 2.60 | 62.5 | 7.2 | 32.2 |
| Jackson | 13.3 | 11.4 | 9.3 | 46.0 | 46,755 | 49,763 | 6.4 | -7.4 | 5,080 | 6,163 | -2,585 | 17,149 | 2.35 | 65.0 | 14.1 | 31.0 |
| Jefferson | 15.5 | 14.4 | 10.1 | 47.7 | 12,902 | 14,761 | 14.4 | -1.5 | 1,288 | 1,648 | 131 | 5,770 | 2.09 | 65.2 | 12.9 | 31.5 |
| Lafayette | 12.2 | 9.9 | 7.7 | 44.0 | 7,022 | 8,868 | 26.3 | -4.4 | 717 | 787 | -331 | 2,175 | 3.27 | 70.6 | 11.1 | 26.5 |
| Lake | 13.7 | 14.5 | 12.6 | 51.6 | 210,528 | 297,047 | 41.1 | 26.4 | 32,814 | 40,880 | 86,017 | 134,317 | 2.55 | 67.8 | 10.6 | 26.7 |
| Lee | 14.1 | 15.7 | 13.8 | 51.1 | 440,888 | 618,761 | 40.3 | 27.8 | 67,541 | 70,855 | 174,050 | 275,965 | 2.64 | 65.9 | 9.4 | 28.0 |
| Leon | 10.5 | 8.9 | 5.5 | 52.8 | 239,452 | 275,483 | 15.0 | 7.3 | 30,932 | 18,935 | 7,804 | 113,658 | 2.42 | 54.3 | 13.2 | 31.3 |
| Levy | 15.9 | 15.0 | 10.6 | 51.1 | 34,450 | 40,803 | 18.4 | 3.5 | 4,137 | 5,577 | 2,865 | 16,374 | 2.44 | 62.9 | 13.3 | 30.7 |
| Liberty | 12.3 | 9.2 | 6.5 | 38.1 | 7,021 | 8,367 | 19.2 | 0.0 | 783 | 685 | -110 | 2,459 | 2.81 | 65.6 | 10.3 | 30.2 |
| Madison | 14.1 | 12.9 | 8.7 | 47.1 | 18,733 | 19,228 | 2.6 | -2.7 | 2,055 | 2,335 | -240 | 6,778 | 2.41 | 62.4 | 16.2 | 32.3 |
| Manatee | 14.9 | 15.4 | 13.3 | 51.8 | 264,002 | 322,879 | 22.3 | 27.4 | 35,262 | 39,943 | 92,241 | 145,356 | 2.61 | 65.9 | 10.0 | 28.4 |
| Marion | 13.7 | 15.6 | 13.7 | 52.0 | 258,916 | 331,296 | 28.0 | 12.7 | 35,199 | 50,252 | 57,293 | 139,172 | 2.47 | 61.2 | 11.9 | 29.7 |
| Martin | 15.8 | 15.8 | 16.7 | 50.8 | 126,731 | 146,851 | 15.9 | 10.4 | 12,552 | 19,451 | 22,052 | 64,528 | 2.40 | 68.7 | 8.8 | 32.8 |
| Miami-Dade | 12.9 | 9.1 | 7.9 | 51.3 | 2,253,362 | 2,498,003 | 10.9 | 8.4 | 321,239 | 201,879 | 89,523 | 883,372 | 3.00 | 68.7 | 18.3 | 25.7 |
| Monroe | 16.5 | 14.3 | 9.4 | 48.1 | 79,589 | 73,090 | -8.2 | 1.1 | 7,368 | 7,228 | 651 | 32,068 | 2.30 | 58.6 | 7.5 | 30.2 |
| Nassau | 15.5 | 14.3 | 9.0 | 50.8 | 57,663 | 73,314 | 27.1 | 24.3 | 8,159 | 8,402 | 17,998 | 32,603 | 2.52 | 73.5 | 10.2 | 21.1 |
| Okaloosa | 13.4 | 9.8 | 6.7 | 49.2 | 170,498 | 180,824 | 6.1 | 17.7 | 27,837 | 17,878 | 21,809 | 77,962 | 2.54 | 64.5 | 10.0 | 29.0 |
| Okeechobee | 12.9 | 10.9 | 9.6 | 46.3 | 35,910 | 39,987 | 11.4 | 5.8 | 5,430 | 4,519 | 1,409 | 13,904 | 2.85 | 66.0 | 15.0 | 24.9 |
| Orange | 11.5 | 7.6 | 5.0 | 51.0 | 896,344 | 1,145,949 | 27.8 | 22.6 | 166,552 | 79,895 | 171,139 | 461,705 | 2.85 | 73.3 | 15.9 | 20.9 |
| Osceola | 11.1 | 8.1 | 5.5 | 50.7 | 172,493 | 268,685 | 55.8 | 43.4 | 42,546 | 21,853 | 95,598 | 103,141 | 3.39 | 62.3 | 11.4 | 31.0 |
| Palm Beach | 13.4 | 12.0 | 12.8 | 51.6 | 1,131,184 | 1,320,134 | 16.7 | 14.2 | 148,812 | 147,119 | 185,670 | 554,095 | 2.61 | 62.3 | 11.4 | 31.0 |
| Pasco | 13.5 | 12.4 | 10.4 | 51.4 | 344,765 | 464,707 | 34.8 | 22.7 | 50,812 | 61,649 | 116,153 | 204,198 | 2.53 | 66.2 | 12.3 | 27.6 |
| Pinellas | 16.2 | 13.9 | 12.0 | 52.0 | 921,482 | 916,799 | -0.5 | 6.5 | 85,520 | 122,048 | 96,368 | 407,546 | 2.32 | 56.2 | 10.7 | 35.9 |
| Polk | 12.2 | 11.3 | 9.2 | 51.0 | 483,924 | 602,068 | 24.4 | 23.7 | 77,998 | 68,161 | 132,537 | 235,283 | 2.86 | 68.6 | 13.5 | 25.3 |

1. No spouse present.

# Table B. States and Counties — Population, Vital Statistics, Health, and Crime

| STATE County | Persons in group quarters, 2020 | Daytime Population, 2015-2019 Number | Employment/ residence ratio | Births, 2020 Total | Rate[1] | Deaths, 2020 Number | Rate[1] | Persons under 65 with no health insurance, 2019 Number | Percent | Medicare, 2020 Total beneficiaries | Enrolled in Original Medicare | Enrolled in Medicare Advantage | Crimes reported by county police or sheriff, 2019 Violent | Property |
|---|---|---|---|---|---|---|---|---|---|---|---|---|---|---|
| | 32 | 33 | 34 | 35 | 36 | 37 | 38 | 39 | 40 | 41 | 42 | 43 | 44 | 45 |
| CONNECTICUT—Cont'd | | | | | | | | | | | | | | |
| Tolland | 14,683 | 124,695 | 0.66 | 1,156 | 7.7 | 1,151 | 7.6 | 5,013 | 4.4 | 27,810 | 13,094 | 14,716 | NA | NA |
| Windham | 4,668 | 102,447 | 0.75 | 1,032 | 8.9 | 1,157 | 9.9 | 6,018 | 6.5 | 23,743 | 12,974 | 10,769 | NA | NA |
| DELAWARE | 24,938 | 956,591 | 1.00 | 10,516 | 10.7 | 10,368 | 10.5 | 60,910 | 8.0 | 215,611 | 174,723 | 40,889 | X | X |
| Kent | 4,615 | 168,194 | 0.89 | 2,151 | 11.7 | 1,947 | 10.6 | 13,058 | 9.0 | 38,492 | 31,219 | 7,273 | NA | NA |
| New Castle | 17,477 | 572,848 | 1.06 | 6,097 | 10.9 | 5,242 | 9.3 | 30,759 | 6.8 | 102,830 | 81,083 | 21,747 | 912 | 4,125 |
| Sussex | 2,846 | 215,549 | 0.91 | 2,268 | 9.4 | 3,179 | 13.2 | 17,093 | 10.4 | 74,288 | 62,420 | 11,869 | NA | NA |
| DISTRICT OF COLUMBIA | 38,802 | 1,158,889 | 2.25 | 9,155 | 12.8 | 6,093 | 8.5 | 23,635 | 4.0 | 94,033 | 73,216 | 20,818 | X | X |
| District of Columbia | 38,802 | 1,158,889 | 2.25 | 9,155 | 12.8 | 6,093 | 8.5 | 23,635 | 4.0 | 94,032 | 73,215 | 20,818 | NA | NA |
| FLORIDA | 430,318 | 20,846,478 | 0.99 | 219,996 | 10.1 | 230,396 | 10.6 | 2,716,165 | 16.4 | 4,677,489 | 2,409,421 | 2,268,069 | X | X |
| Alachua | 14,597 | 281,016 | 1.13 | 2,680 | 9.9 | 2,236 | 8.2 | 26,882 | 12.4 | 45,840 | 32,658 | 13,182 | 738 | 1,817 |
| Baker | 2,831 | 24,905 | 0.69 | 359 | 12.1 | 284 | 9.6 | 2,484 | 11.1 | 4,993 | 3,267 | 1,726 | 121 | 275 |
| Bay | 3,645 | 184,282 | 1.03 | 1,894 | 11.1 | 2,065 | 12.1 | 20,640 | 14.8 | 37,013 | 26,004 | 11,009 | 360 | 2,192 |
| Bradford | 4,225 | 25,600 | 0.82 | 292 | 10.2 | 355 | 12.4 | 2,568 | 13.5 | 5,845 | 3,695 | 2,151 | 65 | 226 |
| Brevard | 6,768 | 576,593 | 0.96 | 5,236 | 8.6 | 8,105 | 13.3 | 62,753 | 13.9 | 159,117 | 90,617 | 68,500 | 540 | 3,548 |
| Broward | 16,287 | 1,843,538 | 0.91 | 21,491 | 11.0 | 17,369 | 8.9 | 282,874 | 17.7 | 331,717 | 137,131 | 194,586 | 349 | 694 |
| Calhoun | 1,688 | 12,859 | 0.68 | 121 | 8.6 | 159 | 11.3 | 1,559 | 15.8 | 3,006 | 1,721 | 1,286 | 30 | 78 |
| Charlotte | 2,885 | 173,592 | 0.88 | 1,002 | 5.1 | 3,117 | 16.0 | 19,097 | 17.4 | 71,649 | 42,202 | 29,447 | 301 | 1,833 |
| Citrus | 2,253 | 138,571 | 0.85 | 1,029 | 6.7 | 2,865 | 18.7 | 15,868 | 17.1 | 60,329 | 33,745 | 26,584 | 366 | 2,046 |
| Clay | 1,249 | 170,082 | 0.57 | 2,211 | 10.0 | 2,104 | 9.5 | 23,029 | 12.6 | 43,817 | 29,529 | 14,288 | 510 | 2,876 |
| Collier | 4,547 | 382,374 | 1.07 | 3,163 | 8.0 | 4,286 | 10.9 | 54,372 | 21.3 | 104,801 | 74,053 | 30,748 | 864 | 3,884 |
| Columbia | 5,441 | 70,216 | 1.01 | 789 | 10.9 | 926 | 12.7 | 7,488 | 14.2 | 16,604 | 10,682 | 5,923 | 158 | 894 |
| DeSoto | 3,583 | 37,862 | 1.07 | 403 | 10.5 | 371 | 9.6 | 6,914 | 26.5 | 7,295 | 4,412 | 2,883 | 104 | 512 |
| Dixie | 1,758 | 15,341 | 0.75 | 153 | 9.0 | 235 | 13.8 | 1,918 | 17.5 | 4,291 | 2,522 | 1,769 | 85 | 147 |
| Duval | 20,763 | 1,020,654 | 1.18 | 12,932 | 13.4 | 9,624 | 10.0 | 114,072 | 14.3 | 168,337 | 96,414 | 71,924 | NA | NA |
| Escambia | 18,012 | 334,145 | 1.14 | 3,812 | 11.8 | 3,925 | 12.2 | 34,595 | 14.0 | 71,316 | 44,657 | 26,660 | 1,451 | 7,196 |
| Flagler | 684 | 97,041 | 0.70 | 840 | 7.1 | 1,560 | 13.2 | 12,438 | 15.8 | 39,414 | 20,980 | 18,434 | 172 | 1,022 |
| Franklin | 1,765 | 11,551 | 0.94 | 88 | 7.2 | 141 | 11.6 | 1,324 | 17.9 | 3,140 | 1,933 | 1,207 | 59 | 211 |
| Gadsden | 3,267 | 41,994 | 0.76 | 492 | 10.9 | 540 | 11.9 | 5,549 | 16.3 | 10,425 | 4,244 | 6,181 | 77 | 264 |
| Gilchrist | 1,182 | 15,312 | 0.60 | 191 | 10.1 | 210 | 11.1 | 2,347 | 17.2 | 4,091 | 2,686 | 1,405 | 33 | 145 |
| Glades | 1,468 | 12,425 | 0.70 | 61 | 4.3 | 147 | 10.4 | 2,241 | 26.5 | 2,792 | 1,624 | 1,168 | 24 | 124 |
| Gulf | 398 | 14,929 | 0.89 | 115 | 8.5 | 196 | 14.5 | 1,314 | 13.5 | 3,834 | 2,657 | 1,177 | 29 | 156 |
| Hamilton | 2,705 | 14,124 | 0.95 | 167 | 11.5 | 146 | 10.1 | 1,369 | 15.2 | 2,956 | 1,936 | 1,020 | 41 | 186 |
| Hardee | 1,590 | 26,489 | 0.93 | 330 | 12.3 | 225 | 8.4 | 4,077 | 19.9 | 4,505 | 2,626 | 1,880 | 47 | 270 |
| Hendry | 696 | 39,095 | 0.91 | 593 | 13.9 | 363 | 8.5 | 9,450 | 26.9 | 6,016 | 3,579 | 2,438 | 126 | 630 |
| Hernando | 1,832 | 170,917 | 0.77 | 1,607 | 8.1 | 3,098 | 15.6 | 21,668 | 15.6 | 60,299 | 25,000 | 35,299 | 465 | 2,258 |
| Highlands | 1,734 | 102,258 | 0.96 | 835 | 7.8 | 1,807 | 16.9 | 13,536 | 20.4 | 34,125 | 19,311 | 14,814 | 209 | 1,755 |
| Hillsborough | 23,501 | 1,478,192 | 1.08 | 17,169 | 11.5 | 12,718 | 8.5 | 184,263 | 14.9 | 242,267 | 112,429 | 129,838 | 1,752 | 11,353 |
| Holmes | 1,760 | 17,036 | 0.65 | 192 | 9.8 | 277 | 14.1 | 2,563 | 18.4 | 4,429 | 2,899 | 1,530 | 72 | 158 |
| Indian River | 1,326 | 155,064 | 1.02 | 1,277 | 7.9 | 2,416 | 14.9 | 19,417 | 18.6 | 54,766 | 36,923 | 17,843 | 274 | 1,398 |
| Jackson | 6,313 | 47,070 | 0.95 | 468 | 10.2 | 668 | 14.5 | 4,557 | 14.9 | 11,273 | 7,520 | 3,753 | NA | NA |
| Jefferson | 1,273 | 11,798 | 0.52 | 110 | 7.6 | 170 | 11.7 | 1,417 | 14.9 | 3,695 | 1,755 | 1,941 | 69 | 157 |
| Lafayette | 1,259 | 7,725 | 0.62 | 57 | 6.7 | 82 | 9.7 | 1,115 | 19.6 | 1,480 | 998 | 482 | 13 | 37 |
| Lake | 3,948 | 308,380 | 0.73 | 3,456 | 9.2 | 4,968 | 13.2 | 41,405 | 15.7 | 107,014 | 58,966 | 48,048 | 445 | 2,231 |
| Lee | 8,476 | 724,406 | 0.96 | 6,919 | 8.7 | 8,819 | 11.2 | 96,245 | 17.9 | 201,807 | 121,956 | 79,851 | 1,236 | 4,632 |
| Leon | 15,011 | 308,808 | 1.13 | 2,919 | 9.9 | 2,175 | 7.4 | 28,530 | 11.9 | 46,687 | 22,293 | 24,394 | 322 | 1,447 |
| Levy | 335 | 35,045 | 0.64 | 431 | 10.2 | 604 | 14.3 | 5,879 | 19.3 | 12,433 | 7,312 | 5,121 | 462 | 527 |
| Liberty | 2,050 | 7,914 | 0.81 | 66 | 7.9 | 71 | 8.5 | 738 | 14.6 | 1,392 | 675 | 717 | 19 | 66 |
| Madison | 2,022 | 17,573 | 0.86 | 198 | 10.6 | 248 | 13.3 | 2,095 | 16.7 | 4,554 | 2,866 | 1,688 | 109 | 174 |
| Manatee | 4,814 | 356,360 | 0.83 | 3,496 | 8.5 | 4,839 | 11.8 | 48,162 | 16.8 | 104,964 | 62,059 | 42,906 | 1,368 | 4,650 |
| Marion | 9,218 | 344,900 | 0.93 | 3,532 | 9.5 | 5,608 | 15.0 | 42,094 | 16.9 | 120,124 | 60,208 | 59,916 | 1,014 | 3,987 |
| Martin | 4,216 | 163,640 | 1.07 | 1,230 | 7.6 | 2,210 | 13.6 | 17,411 | 16.2 | 47,211 | 31,601 | 15,610 | 293 | 1,528 |
| Miami-Dade | 42,231 | 2,775,778 | 1.06 | 30,086 | 11.1 | 23,189 | 8.6 | 428,190 | 19.4 | 475,765 | 138,854 | 336,912 | 5,353 | 31,822 |
| Monroe | 1,998 | 79,755 | 1.10 | 675 | 9.1 | 771 | 10.4 | 12,047 | 21.6 | 17,430 | 14,577 | 2,852 | 186 | 681 |
| Nassau | 543 | 69,996 | 0.65 | 859 | 9.4 | 1,074 | 11.8 | 8,114 | 11.8 | 23,249 | 15,344 | 7,904 | 174 | 919 |
| Okaloosa | 5,221 | 213,377 | 1.10 | 2,733 | 12.8 | 2,085 | 9.8 | 24,981 | 14.5 | 41,619 | 31,868 | 9,751 | 535 | 2,375 |
| Okeechobee | 3,024 | 40,282 | 0.94 | 529 | 12.5 | 498 | 11.8 | 6,699 | 22.0 | 8,577 | 4,741 | 3,836 | 162 | 780 |
| Orange | 35,766 | 1,523,521 | 1.26 | 16,684 | 11.9 | 9,545 | 6.8 | 183,226 | 15.4 | 192,564 | 90,443 | 102,122 | 4,874 | 20,832 |
| Osceola | 3,288 | 304,550 | 0.71 | 4,493 | 11.7 | 2,723 | 7.1 | 51,986 | 16.3 | 62,680 | 22,554 | 40,126 | 678 | 4,509 |
| Palm Beach | 21,850 | 1,488,446 | 1.03 | 14,865 | 9.9 | 16,781 | 11.1 | 199,201 | 17.9 | 332,132 | 194,156 | 137,977 | 1,589 | 9,137 |
| Pasco | 5,798 | 456,680 | 0.68 | 5,194 | 9.1 | 7,082 | 12.4 | 66,501 | 15.7 | 136,335 | 55,566 | 80,770 | 1,452 | 6,456 |
| Pinellas | 19,928 | 979,484 | 1.03 | 7,928 | 8.1 | 13,821 | 14.1 | 110,535 | 15.4 | 249,082 | 118,278 | 130,804 | 466 | 3,831 |
| Polk | 13,783 | 653,593 | 0.88 | 8,169 | 11.0 | 8,009 | 10.8 | 95,885 | 17.1 | 160,554 | 68,277 | 92,278 | 1,166 | 5,089 |

1. Per 1,000 estimated resident population.

# Table B. States and Counties — Crime, Education, Money Income, and Poverty

| STATE County | County law enforcement employment, 2019 | | Education | | | | | | Money income, 2015-2019 | | | | Income and poverty, 2019 | | |
| | | | School enrollment and attainment, 2015-2019 | | | | Local government expenditures,[3] 2017-2018 | | | Households | | | Percent below poverty level | | |
| | | | Enrollment[1] | | Attainment[2] (percent) | | | | | | Percent | | | | | |
| | Officers | Civilians | Total | Percent private | High school graduate or less | Bachelor's degree or more | Total current spending (mil dol) | Current spending per student (dollars) | Per capita income[4] | Median income (dollars) | with income of less than $50,000 | with income of $200,000 or more | Median household income (dollars) | All persons | Children under 18 years | Children 5 to 17 years in families |
| | 46 | 47 | 48 | 49 | 50 | 51 | 52 | 53 | 54 | 55 | 56 | 57 | 58 | 59 | 60 | 61 |

| CONNECTICUT—Cont'd | | | | | | | | | | | | | | | | |
| Tolland | NA | NA | 47,043 | 7.7 | 32.0 | 41.8 | 381 | 19,783 | 41,371 | 87,069 | 28.6 | 11.8 | 90,181 | 7.8 | 6.3 | 5.8 |
| Windham | NA | NA | 28,627 | 8.7 | 44.8 | 24.3 | 301 | 19,297 | 32,732 | 66,550 | 36.5 | 5.5 | 63,309 | 11.5 | 14.0 | 13.3 |
| DELAWARE | X | X | 228,567 | 16.1 | 41.3 | 32.0 | 2,084 | 15,289 | 35,450 | 68,287 | 36.8 | 7.0 | 70,348 | 11.2 | 16.3 | 15.2 |
| Kent | NA | NA | 45,885 | 13.4 | 45.5 | 23.7 | 373 | 14,053 | 28,845 | 60,910 | 41.9 | 3.6 | 58,804 | 12.7 | 17.8 | 17.3 |
| New Castle | 387 | 76 | 140,343 | 18.3 | 39.2 | 36.2 | 1,267 | 16,228 | 37,532 | 73,892 | 34.0 | 8.6 | 76,076 | 10.4 | 13.4 | 12.5 |
| Sussex | NA | NA | 42,339 | 11.5 | 43.1 | 28.3 | 444 | 14,014 | 35,491 | 63,162 | 39.2 | 5.9 | 64,839 | 12.1 | 22.8 | 20.8 |
| DISTRICT OF COLUMBIA | X | X | 168,553 | 40.5 | 25.9 | 58.5 | 1,930 | 22,311 | 56,147 | 86,420 | 32.6 | 18.3 | 90,395 | 14.1 | 20.8 | 20.1 |
| District of Columbia | NA | NA | 168,553 | 40.5 | 25.9 | 58.5 | 1,930 | 22,311 | 56,147 | 86,420 | 32.6 | 18.3 | 90,395 | 14.1 | 20.8 | 20.1 |
| FLORIDA | X | X | 4,758,186 | 18.0 | 40.4 | 29.9 | 26,305 | 9,280 | 31,619 | 55,660 | 45.0 | 6.0 | 59,198 | 12.7 | 18.2 | 16.9 |
| Alachua | 268 | 110 | 92,264 | 10.2 | 28.6 | 43.3 | 274 | 8,849 | 28,646 | 49,689 | 50.2 | 5.5 | 49,880 | 18.4 | 17.7 | 16.7 |
| Baker | 44 | 19 | 6,471 | 11.7 | 56.9 | 13.2 | 43 | 8,565 | 24,686 | 63,275 | 40.8 | 3.3 | 60,454 | 14.9 | 21.0 | 19.9 |
| Bay | 208 | 65 | 40,380 | 14.6 | 40.3 | 23.7 | 244 | 8,690 | 29,290 | 54,316 | 45.7 | 3.4 | 58,376 | 12.1 | 18.2 | 18.0 |
| Bradford | 41 | 47 | 5,315 | 12.0 | 60.8 | 9.9 | 35 | 10,724 | 20,529 | 45,921 | 52.8 | 1.1 | 47,509 | 21.0 | 27.2 | 28.8 |
| Brevard | 558 | 724 | 123,632 | 17.8 | 36.1 | 30.2 | 650 | 8,841 | 32,176 | 56,775 | 43.5 | 5.0 | 57,582 | 9.4 | 12.7 | 11.8 |
| Broward | 390 | 939 | 475,572 | 21.1 | 38.4 | 32.4 | 2,589 | 9,464 | 32,909 | 59,547 | 42.5 | 7.4 | 61,429 | 12.3 | 16.4 | 15.7 |
| Calhoun | 25 | 8 | 3,036 | 7.1 | 64.7 | 9.5 | 22 | 9,810 | 19,304 | 38,568 | 62.0 | 1.0 | 41,399 | 20.3 | 28.3 | 25.8 |
| Charlotte | 309 | 139 | 24,519 | 14.0 | 44.0 | 23.4 | 153 | 9,592 | 32,144 | 51,499 | 48.5 | 4.0 | 54,756 | 11.4 | 19.9 | 17.9 |
| Citrus | 205 | 123 | 22,187 | 15.3 | 48.8 | 18.3 | 142 | 9,165 | 26,989 | 44,237 | 56.3 | 2.1 | 49,762 | 15.2 | 24.5 | 18.4 |
| Clay | 277 | 302 | 54,685 | 14.0 | 39.2 | 24.9 | 306 | 8,155 | 30,483 | 65,740 | 36.4 | 4.5 | 75,776 | 8.3 | 12.1 | 10.6 |
| Collier | 570 | 334 | 65,701 | 17.1 | 38.7 | 36.4 | 515 | 10,999 | 45,567 | 69,653 | 35.5 | 11.9 | 76,415 | 9.4 | 14.0 | 13.1 |
| Columbia | 153 | 48 | 14,579 | 16.8 | 50.6 | 14.9 | 90 | 8,878 | 24,275 | 46,494 | 52.8 | 2.3 | 47,087 | 15.6 | 21.7 | 21.4 |
| DeSoto | 56 | 59 | 6,698 | 6.3 | 70.3 | 11.5 | 48 | 9,927 | 18,503 | 35,438 | 64.3 | 1.0 | 41,673 | 21.8 | 30.5 | 28.5 |
| Dixie | NA | NA | 2,892 | 19.3 | 64.8 | 10.4 | 20 | 9,291 | 21,144 | 39,828 | 64.5 | 3.4 | 39,793 | 22.2 | 30.5 | 28.3 |
| Duval | NA | NA | 226,037 | 20.0 | 38.0 | 30.0 | 1,187 | 9,163 | 31,327 | 55,807 | 44.4 | 5.0 | 58,339 | 13.5 | 19.5 | 19.4 |
| Escambia | 417 | 238 | 76,677 | 22.3 | 36.7 | 26.5 | 365 | 9,059 | 28,022 | 50,915 | 49.0 | 3.8 | 52,822 | 15.5 | 24.6 | 24.6 |
| Flagler | 212 | 77 | 20,895 | 12.6 | 42.1 | 25.1 | 113 | 8,636 | 29,383 | 54,514 | 45.1 | 3.4 | 62,831 | 9.7 | 16.5 | 14.7 |
| Franklin | NA | NA | 1,859 | 12.4 | 53.9 | 18.8 | 15 | 11,789 | 26,574 | 46,643 | 52.7 | 4.0 | 47,504 | 19.2 | 30.1 | 29.3 |
| Gadsden | 44 | 23 | 10,748 | 15.4 | 56.4 | 16.2 | 57 | 10,740 | 20,734 | 41,401 | 59.2 | 1.2 | 42,975 | 19.7 | 31.2 | 31.2 |
| Gilchrist | 49 | 17 | 3,894 | 7.8 | 56.7 | 13.4 | 30 | 11,111 | 21,578 | 43,640 | 57.0 | 1.7 | 46,171 | 15.0 | 20.8 | 20.5 |
| Glades | 32 | 98 | 2,493 | 16.3 | 63.7 | 11.2 | 18 | 10,478 | 21,953 | 40,977 | 58.9 | 2.4 | 45,412 | 19.3 | 25.7 | 21.6 |
| Gulf | 32 | 14 | 2,737 | 15.0 | 50.4 | 19.2 | 22 | 10,945 | 23,252 | 47,712 | 53.3 | 1.7 | 53,228 | 14.0 | 23.1 | 22.9 |
| Hamilton | 19 | 41 | 2,857 | 8.9 | 72.3 | 7.9 | 17 | 10,374 | 16,136 | 38,569 | 57.5 | 2.1 | 39,241 | 32.5 | 51.5 | 43.1 |
| Hardee | 76 | 29 | 6,499 | 7.4 | 71.9 | 10.2 | 50 | 9,598 | 17,754 | 38,682 | 60.7 | 1.5 | 41,430 | 22.1 | 28.3 | 25.9 |
| Hendry | 86 | 76 | 9,959 | 8.0 | 67.4 | 8.3 | 68 | 9,261 | 19,167 | 40,820 | 59.7 | 2.2 | 45,107 | 19.5 | 28.3 | 26.0 |
| Hernando | 252 | 305 | 35,749 | 16.4 | 47.3 | 18.4 | 193 | 8,618 | 25,860 | 48,812 | 51.2 | 2.1 | 50,705 | 12.4 | 19.1 | 17.8 |
| Highlands | 147 | 195 | 16,841 | 13.5 | 53.0 | 17.2 | 116 | 9,313 | 25,752 | 40,942 | 59.3 | 2.3 | 45,723 | 15.8 | 26.2 | 24.2 |
| Hillsborough | 1,245 | 2,068 | 359,686 | 18.1 | 38.5 | 33.5 | 1,930 | 8,889 | 32,343 | 58,884 | 42.7 | 7.0 | 61,133 | 13.5 | 18.4 | 17.6 |
| Holmes | NA | NA | 3,752 | 6.6 | 62.0 | 10.7 | 31 | 9,479 | 19,201 | 39,102 | 58.5 | 1.7 | 38,874 | 20.1 | 28.4 | 27.6 |
| Indian River | 190 | 289 | 26,962 | 15.5 | 41.1 | 29.1 | 171 | 9,624 | 36,516 | 54,740 | 45.6 | 6.7 | 61,173 | 11.5 | 20.1 | 16.7 |
| Jackson | 49 | 25 | 9,479 | 17.4 | 58.0 | 12.6 | 63 | 9,471 | 20,019 | 39,872 | 58.7 | 1.9 | 40,543 | 19.4 | 26.9 | 25.6 |
| Jefferson | 26 | 33 | 2,439 | 30.3 | 52.7 | 22.3 | 8 | 10,661 | 24,510 | 47,240 | 51.1 | 1.9 | 47,192 | 17.6 | 25.6 | 24.6 |
| Lafayette | 13 | 21 | 2,206 | 14.8 | 63.6 | 11.8 | 12 | 9,502 | 20,776 | 50,165 | 49.8 | 2.0 | 45,480 | 18.0 | 27.2 | 22.7 |
| Lake | 463 | 240 | 69,066 | 19.3 | 42.6 | 24.0 | 370 | 8,570 | 28,348 | 54,513 | 46.1 | 3.3 | 57,660 | 10.9 | 18.4 | 17.5 |
| Lee | 1,030 | 474 | 143,251 | 14.1 | 42.6 | 28.2 | 902 | 9,680 | 33,543 | 57,832 | 42.7 | 6.1 | 62,364 | 11.2 | 18.6 | 16.9 |
| Leon | 271 | 444 | 103,229 | 12.1 | 25.5 | 46.2 | 303 | 8,135 | 30,586 | 53,106 | 47.1 | 5.7 | 55,081 | 20.8 | 22.0 | 20.1 |
| Levy | 115 | 36 | 8,075 | 13.6 | 56.2 | 12.2 | 52 | 9,322 | 21,579 | 37,326 | 62.9 | 1.2 | 41,405 | 18.2 | 28.9 | 25.7 |
| Liberty | 25 | 7 | 1,734 | 4.3 | 64.4 | 14.4 | 15 | 10,460 | 17,872 | 38,015 | 63.7 | 0.6 | 44,236 | 23.0 | 26.8 | 25.0 |
| Madison | 64 | 17 | 3,569 | 12.0 | 58.7 | 14.0 | 27 | 9,506 | 19,510 | 37,037 | 63.7 | 1.2 | 37,919 | 22.7 | 32.6 | 32.7 |
| Manatee | 753 | 394 | 73,294 | 16.9 | 40.4 | 29.8 | 450 | 9,187 | 33,853 | 59,009 | 42.7 | 5.9 | 64,151 | 11.3 | 19.4 | 18.8 |
| Marion | 545 | 233 | 66,953 | 16.9 | 49.3 | 20.1 | 402 | 9,325 | 25,839 | 45,371 | 54.6 | 2.9 | 49,079 | 14.9 | 25.4 | 23.7 |
| Martin | 260 | 127 | 28,780 | 16.1 | 33.9 | 34.1 | 174 | 9,163 | 42,199 | 61,133 | 41.7 | 9.3 | 70,806 | 8.9 | 13.6 | 12.7 |
| Miami-Dade | 2,953 | 1,115 | 661,117 | 20.5 | 45.8 | 29.8 | 3,386 | 9,542 | 28,224 | 51,347 | 48.9 | 6.4 | 54,991 | 15.7 | 20.8 | 18.8 |
| Monroe | 196 | 312 | 11,999 | 18.7 | 35.3 | 34.4 | 114 | 13,384 | 46,028 | 70,033 | 35.8 | 10.0 | 67,528 | 9.9 | 14.8 | 15.4 |
| Nassau | 140 | 117 | 16,977 | 19.4 | 40.4 | 29.6 | 107 | 9,042 | 36,553 | 69,943 | 34.5 | 7.2 | 72,818 | 9.0 | 12.6 | 11.9 |
| Okaloosa | 323 | 121 | 46,808 | 13.0 | 32.3 | 31.1 | 287 | 9,065 | 33,019 | 63,412 | 38.7 | 5.3 | 63,520 | 10.6 | 16.4 | 15.6 |
| Okeechobee | NA | NA | 8,094 | 13.9 | 62.3 | 11.3 | 61 | 9,494 | 21,367 | 41,760 | 58.9 | 2.1 | 45,225 | 18.4 | 27.7 | 25.7 |
| Orange | 1,581 | 683 | 365,516 | 17.7 | 35.8 | 34.6 | 1,967 | 9,295 | 30,456 | 58,254 | 42.9 | 6.6 | 63,133 | 12.6 | 17.9 | 17.2 |
| Osceola | 411 | 226 | 90,723 | 17.0 | 44.5 | 21.8 | 555 | 8,405 | 22,196 | 52,279 | 47.9 | 2.7 | 52,270 | 13.4 | 19.1 | 17.9 |
| Palm Beach | 1,644 | 1,979 | 324,367 | 19.0 | 35.5 | 36.7 | 1,949 | 9,942 | 39,933 | 63,299 | 40.2 | 9.7 | 66,621 | 11.4 | 16.0 | 15.0 |
| Pasco | NA | NA | 110,153 | 16.5 | 43.7 | 24.0 | 640 | 8,692 | 29,001 | 52,828 | 47.8 | 3.9 | 55,745 | 11.3 | 16.3 | 14.2 |
| Pinellas | 827 | 1,558 | 181,115 | 19.0 | 36.7 | 31.7 | 980 | 9,621 | 35,196 | 54,090 | 46.2 | 5.5 | 56,708 | 11.4 | 14.9 | 14.4 |
| Polk | 652 | 518 | 155,378 | 17.3 | 49.7 | 20.2 | 962 | 9,239 | 24,864 | 50,584 | 49.4 | 3.2 | 51,817 | 14.0 | 21.1 | 18.1 |

1. All persons 3 years old and over enrolled in nursery school through college.　2. Persons 25 years old and over.　3. Elementary and secondary education expenditures.　4. Based on population estimated by the American Community Survey, 2014–2018.

# Table B. States and Counties — **Personal Income**

| STATE County | Personal income, 2019 | | | | | | | | | | Earnings, 2019 | | |
|---|---|---|---|---|---|---|---|---|---|---|---|---|---|
| | Total (mil dol) | Percent change 2018-2019 | Per capita¹ Dollars | Per capita¹ Rank | Wages and salaries (mil dol) | Supplements to wages and salaries, employer contributions (mil dol) Pension and insurance | Supplements to wages and salaries, employer contributions (mil dol) Government social insurance | Proprietors' income (mil dol) | Dividends, interest, and rent (mil dol) | Personal transfer receipts (mil dol) | Total (mil dol) | Contributions for government social insurance (mil dol) From employee and self-employed | Contributions for government social insurance (mil dol) From employer |
| | 62 | 63 | 64 | 65 | 66 | 67 | 68 | 69 | 70 | 71 | 72 | 73 | 74 |
| CONNECTICUT—Cont'd | | | | | | | | | | | | | |
| Tolland | 8,637 | 3.2 | 57,307 | 395 | 2,208 | 565 | 149 | 627 | 1,348 | 1,238 | 3,549 | 197 | 149 |
| Windham | 5,561 | 3.8 | 47,618 | 1,067 | 1,879 | 390 | 143 | 410 | 804 | 1,221 | 2,822 | 172 | 143 |
| DELAWARE | 53,055 | 3.8 | 54,323 | X | 27,709 | 4,884 | 1,980 | 4,540 | 9,759 | 11,071 | 39,113 | 2,405 | 1,980 |
| Kent | 7,791 | 4.8 | 43,097 | 1,603 | 3,501 | 879 | 268 | 452 | 1,223 | 2,014 | 5,099 | 309 | 268 |
| New Castle | 33,064 | 3.6 | 59,175 | 334 | 20,447 | 3,306 | 1,426 | 2,625 | 6,120 | 5,718 | 27,804 | 1,677 | 1,426 |
| Sussex | 12,200 | 3.7 | 52,085 | 676 | 3,762 | 699 | 287 | 1,463 | 2,416 | 3,339 | 6,210 | 419 | 287 |
| DISTRICT OF COLUMBIA | 58,864 | 3.3 | 83,111 | X | 78,460 | 14,816 | 5,621 | 6,946 | 10,606 | 7,062 | 105,843 | 5,458 | 5,621 |
| District of Columbia | 58,864 | 3.3 | 83,111 | 53 | 78,460 | 14,816 | 5,621 | 6,946 | 10,606 | 7,062 | 105,843 | 5,458 | 5,621 |
| FLORIDA | 1,125,984 | 4.0 | 52,391 | X | 497,685 | 71,975 | 33,817 | 68,977 | 311,365 | 216,742 | 672,455 | 44,515 | 33,817 |
| Alachua | 12,562 | 3.7 | 46,690 | 1,161 | 7,334 | 1,386 | 507 | 399 | 2,775 | 2,344 | 9,626 | 576 | 507 |
| Baker | 971 | 4.9 | 33,254 | 2,868 | 288 | 63 | 21 | 36 | 124 | 259 | 407 | 29 | 21 |
| Bay | 7,982 | -0.8 | 45,690 | 1,283 | 3,823 | 658 | 277 | 432 | 1,679 | 1,880 | 5,190 | 340 | 277 |
| Bradford | 964 | 3.0 | 34,179 | 2,784 | 273 | 56 | 19 | 23 | 149 | 285 | 371 | 29 | 19 |
| Brevard | 28,839 | 5.4 | 47,991 | 1,026 | 12,993 | 1,990 | 895 | 1,244 | 6,191 | 7,013 | 17,122 | 1,200 | 895 |
| Broward | 102,146 | 3.9 | 52,308 | 657 | 51,376 | 6,910 | 3,419 | 4,863 | 25,419 | 16,242 | 66,568 | 4,225 | 3,419 |
| Calhoun | 416 | 3.7 | 29,526 | 3,052 | 108 | 24 | 8 | 23 | 64 | 145 | 163 | 13 | 8 |
| Charlotte | 8,084 | 4.6 | 42,793 | 1,649 | 2,227 | 342 | 156 | 412 | 2,301 | 2,794 | 3,136 | 294 | 156 |
| Citrus | 5,845 | 4.1 | 39,055 | 2,182 | 1,410 | 243 | 98 | 304 | 1,412 | 2,331 | 2,055 | 218 | 98 |
| Clay | 9,656 | 4.0 | 44,042 | 1,478 | 2,465 | 392 | 171 | 363 | 1,694 | 2,140 | 3,392 | 258 | 171 |
| Collier | 38,252 | 2.9 | 99,382 | 23 | 8,613 | 1,079 | 576 | 2,001 | 21,973 | 4,051 | 12,269 | 856 | 576 |
| Columbia | 2,551 | 4.1 | 35,583 | 2,605 | 1,090 | 218 | 78 | 112 | 385 | 807 | 1,498 | 106 | 78 |
| DeSoto | 993 | 6.9 | 26,128 | 3,098 | 431 | 76 | 32 | 98 | 181 | 347 | 637 | 42 | 32 |
| Dixie | 476 | 2.5 | 28,314 | 3,074 | 106 | 24 | 8 | 20 | 95 | 193 | 157 | 15 | 8 |
| Duval | 45,470 | 3.3 | 47,475 | 1,081 | 33,787 | 4,898 | 2,352 | 2,915 | 8,653 | 8,901 | 43,953 | 2,673 | 2,352 |
| Escambia | 14,117 | 4.4 | 44,349 | 1,438 | 7,910 | 1,436 | 576 | 652 | 2,969 | 3,456 | 10,574 | 667 | 576 |
| Flagler | 5,473 | 5.3 | 47,558 | 1,072 | 1,008 | 169 | 71 | 285 | 1,407 | 1,509 | 1,533 | 150 | 71 |
| Franklin | 453 | 3.0 | 37,320 | 2,410 | 129 | 25 | 9 | 34 | 154 | 129 | 197 | 16 | 9 |
| Gadsden | 1,639 | 2.8 | 35,885 | 2,582 | 587 | 116 | 42 | 71 | 306 | 493 | 816 | 56 | 42 |
| Gilchrist | 646 | 5.5 | 34,756 | 2,722 | 156 | 31 | 11 | 51 | 88 | 188 | 249 | 18 | 11 |
| Glades | 360 | 4.3 | 26,102 | 3,099 | 118 | 23 | 9 | 30 | 64 | 107 | 179 | 12 | 9 |
| Gulf | 632 | 4.0 | 46,346 | 1,208 | 161 | 31 | 11 | 32 | 136 | 170 | 236 | 19 | 11 |
| Hamilton | 399 | 2.9 | 27,643 | 3,086 | 155 | 37 | 11 | 18 | 56 | 144 | 222 | 16 | 11 |
| Hardee | 796 | 4.5 | 29,545 | 3,050 | 301 | 62 | 22 | 81 | 125 | 216 | 467 | 28 | 22 |
| Hendry | 1,395 | 4.4 | 33,204 | 2,877 | 594 | 104 | 47 | 157 | 192 | 338 | 902 | 45 | 47 |
| Hernando | 7,412 | 5.1 | 38,223 | 2,303 | 1,949 | 319 | 138 | 260 | 1,275 | 2,631 | 2,667 | 252 | 138 |
| Highlands | 3,663 | 4.0 | 34,480 | 2,745 | 1,138 | 204 | 82 | 177 | 765 | 1,418 | 1,601 | 140 | 82 |
| Hillsborough | 71,320 | 4.9 | 48,452 | 975 | 46,184 | 6,407 | 3,104 | 4,066 | 12,939 | 12,668 | 59,762 | 3,672 | 3,104 |
| Holmes | 624 | 3.8 | 31,822 | 2,957 | 130 | 32 | 9 | 38 | 99 | 229 | 208 | 19 | 9 |
| Indian River | 12,925 | 3.6 | 80,818 | 65 | 2,679 | 384 | 184 | 661 | 6,121 | 2,151 | 3,908 | 305 | 184 |
| Jackson | 1,672 | 4.0 | 36,014 | 2,569 | 603 | 131 | 42 | 139 | 294 | 567 | 916 | 65 | 42 |
| Jefferson | 598 | 3.4 | 41,989 | 1,770 | 97 | 20 | 7 | 31 | 134 | 159 | 154 | 14 | 7 |
| Lafayette | 233 | 7.0 | 27,645 | 3,085 | 53 | 13 | 4 | 22 | 39 | 69 | 92 | 6 | 4 |
| Lake | 15,942 | 5.8 | 43,425 | 1,557 | 4,473 | 715 | 316 | 625 | 2,873 | 4,501 | 6,130 | 512 | 316 |
| Lee | 40,119 | 4.1 | 52,064 | 677 | 14,063 | 2,052 | 957 | 3,508 | 14,398 | 8,341 | 20,580 | 1,469 | 957 |
| Leon | 13,621 | 3.6 | 46,394 | 1,199 | 7,969 | 1,490 | 542 | 750 | 2,812 | 2,302 | 10,751 | 640 | 542 |
| Levy | 1,564 | 6.3 | 37,680 | 2,362 | 349 | 70 | 25 | 119 | 242 | 518 | 563 | 47 | 25 |
| Liberty | 217 | 4.2 | 25,973 | 3,100 | 81 | 19 | 6 | 11 | 35 | 68 | 118 | 8 | 6 |
| Madison | 626 | 2.7 | 33,859 | 2,817 | 169 | 40 | 12 | 52 | 119 | 211 | 273 | 21 | 12 |
| Manatee | 19,605 | 4.4 | 48,618 | 952 | 6,547 | 944 | 451 | 1,320 | 5,515 | 4,394 | 9,263 | 676 | 451 |
| Marion | 13,999 | 4.8 | 38,293 | 2,293 | 4,743 | 790 | 329 | 598 | 3,166 | 4,851 | 6,460 | 554 | 329 |
| Martin | 13,748 | 2.7 | 85,394 | 44 | 3,447 | 501 | 236 | 607 | 6,581 | 1,911 | 4,791 | 341 | 236 |
| Miami-Dade | 149,166 | 3.4 | 54,902 | 496 | 74,558 | 10,439 | 4,992 | 16,137 | 38,234 | 25,320 | 106,127 | 6,635 | 4,992 |
| Monroe | 7,516 | 2.8 | 101,262 | 20 | 2,158 | 318 | 157 | 410 | 4,181 | 727 | 3,042 | 193 | 157 |
| Nassau | 5,435 | 4.8 | 61,329 | 268 | 1,142 | 178 | 79 | 248 | 1,435 | 967 | 1,646 | 129 | 79 |
| Okaloosa | 10,680 | 4.8 | 50,681 | 781 | 5,876 | 1,211 | 460 | 676 | 2,877 | 2,148 | 8,223 | 477 | 460 |
| Okeechobee | 1,348 | 4.1 | 31,971 | 2,943 | 493 | 87 | 36 | 111 | 196 | 432 | 727 | 50 | 36 |
| Orange | 64,447 | 4.3 | 46,250 | 1,220 | 50,976 | 6,559 | 3,417 | 4,335 | 10,661 | 10,504 | 65,287 | 3,907 | 3,417 |
| Osceola | 13,248 | 5.5 | 35,258 | 2,649 | 4,410 | 652 | 312 | 708 | 1,527 | 3,138 | 6,082 | 435 | 312 |
| Palm Beach | 124,633 | 3.2 | 83,268 | 52 | 39,377 | 5,072 | 2,582 | 7,440 | 57,484 | 15,259 | 54,471 | 3,564 | 2,582 |
| Pasco | 23,305 | 6.0 | 42,070 | 1,752 | 5,667 | 907 | 395 | 991 | 3,344 | 6,132 | 7,960 | 667 | 395 |
| Pinellas | 54,217 | 3.6 | 55,607 | 467 | 25,480 | 3,692 | 1,710 | 2,201 | 13,576 | 11,572 | 33,084 | 2,237 | 1,710 |
| Polk | 26,563 | 5.0 | 36,649 | 2,491 | 11,447 | 1,802 | 789 | 1,357 | 5,185 | 7,332 | 15,395 | 1,102 | 789 |

1. Based on the resident population estimated as of July 1 of the year shown.

| STATE County | Earnings, 2019 (cont.) | | | | | | | | | Social Security beneficiaries, December 2019 | | Supplemental Security Income recipients, 2019 | Housing units, 2020 | |
|---|---|---|---|---|---|---|---|---|---|---|---|---|---|---|
| | Percent by selected industries | | | | | | | | | | | | | |
| | Farm | Mining, quarrying, and extractions | Construction | Manufacturing | Information; professional, scientific, technical services | Retail trade | Finance, insurance, real estate, and leasing | Health care and social assistance | Government | Number | Rate[1] | | Total | Percent change, 2010-2020 |
| | 75 | 76 | 77 | 78 | 79 | 80 | 81 | 82 | 83 | 84 | 85 | 86 | 87 | 88 |
| CONNECTICUT—Cont'd | | | | | | | | | | | | | | |
| Tolland | 0.6 | D | 8.4 | D | 6.4 | 6.6 | 4.4 | 10.2 | 37.4 | 27,665 | 183 | 1,202 | 60,598 | 4.5 |
| Windham | 0.2 | D | D | 17.9 | 4.4 | 8.0 | 3.3 | 15.3 | 18.6 | 25,000 | 214 | 2,221 | 50,066 | 2.1 |
| DELAWARE | 1.1 | D | 6.3 | 5.7 | 11.3 | 5.5 | 20.3 | 13.5 | 16.1 | 219,490 | 225 | 17,059 | 449,531 | 10.8 |
| Kent | 2.5 | D | 6.0 | D | 4.3 | 7.2 | 4.2 | 12.4 | 38.1 | 39,720 | 219 | 3,602 | 75,742 | 15.9 |
| New Castle | 0.1 | D | 5.3 | D | 13.9 | 4.6 | 25.7 | 13.4 | 13.0 | 105,295 | 188 | 10,162 | 226,377 | 4.1 |
| Sussex | 4.6 | 0.1 | 10.9 | 10.2 | 5.3 | 8.5 | 9.5 | 15.1 | 11.9 | 74,475 | 317 | 3,295 | 147,412 | 19.8 |
| DISTRICT OF COLUMBIA | 0.0 | 0.0 | 1.4 | 0.2 | 27.6 | 1.1 | 4.6 | 5.7 | 37.8 | 83,899 | 118 | 25,521 | 327,516 | 10.4 |
| District of Columbia | 0.0 | 0.0 | 1.4 | 0.2 | 27.6 | 1.1 | 4.6 | 5.7 | 37.8 | 83,899 | 118 | 25,521 | 327,516 | 10.4 |
| FLORIDA | 0.5 | 0.1 | 7.1 | 4.8 | 12.9 | 7.2 | 10.0 | 12.4 | 13.9 | 4,747,364 | 221 | 576,861 | 9,814,383 | 9.2 |
| Alachua | 0.6 | D | 3.7 | 3.7 | 8.8 | 5.4 | 5.9 | 18.1 | 36.0 | 47,055 | 175 | 6,581 | 122,344 | 8.5 |
| Baker | 1.1 | 0.0 | 7.1 | 1.6 | 2.3 | 9.8 | 2.7 | 9.5 | 39.0 | 5,525 | 189 | 641 | 10,367 | 7.0 |
| Bay | 0.0 | D | 8.7 | 5.0 | 9.7 | 9.3 | 7.0 | D | 23.3 | 40,005 | 230 | 4,077 | 106,426 | 6.8 |
| Bradford | 0.6 | D | 4.8 | 3.6 | 3.8 | 10.3 | 3.2 | 9.6 | 32.7 | 6,300 | 222 | 840 | 11,109 | 0.9 |
| Brevard | 0.1 | 0 | 6.5 | 18.6 | 11.9 | 6.7 | 4.8 | 13.1 | 14.8 | 166,545 | 277 | 11,920 | 286,637 | 6.2 |
| Broward | 0.0 | 0 | 7.0 | 3.7 | 14.7 | 7.9 | 9.9 | 10.2 | 13.4 | 333,670 | 171 | 45,801 | 831,992 | 2.7 |
| Calhoun | 6.1 | D | 9.4 | 0.2 | 2.5 | 5.9 | 2.6 | D | 32.1 | 3,235 | 229 | 543 | 5,976 | -0.5 |
| Charlotte | 1.1 | 0.3 | 9.2 | 1.7 | 8.4 | 12.3 | 6.8 | 21.8 | 13.8 | 72,580 | 383 | 2,878 | 109,385 | 8.7 |
| Citrus | 0.1 | 0.3 | 11.2 | 1.7 | 5.9 | 11.3 | 5.6 | 22.2 | 13.1 | 63,015 | 420 | 3,593 | 81,706 | 4.7 |
| Clay | 0.0 | D | 11.0 | 3.3 | 12.0 | 9.9 | 5.2 | 18.5 | 14.5 | 46,050 | 210 | 3,190 | 83,902 | 11.2 |
| Collier | 0.9 | 0.0 | 11.1 | 3.5 | 12.7 | 8.4 | 10.9 | 12.9 | 8.9 | 100,610 | 261 | 4,145 | 226,087 | 14.6 |
| Columbia | 0.4 | 0.2 | 3.4 | 10.7 | 4.3 | 9.6 | 4.3 | 12.6 | 26.5 | 17,885 | 249 | 2,703 | 29,487 | 3.0 |
| DeSoto | 12.9 | D | 6.8 | 4.5 | D | 6.2 | 2.4 | 10.9 | 17.4 | 7,880 | 208 | 1,057 | 15,490 | 6.2 |
| Dixie | 3.8 | 0.0 | 6.0 | 18.8 | D | 7.0 | D | D | 36.0 | 4,675 | 277 | 782 | 9,586 | 2.9 |
| Duval | 0.0 | 0.0 | 6.2 | 5.1 | 11.9 | 5.7 | 15.1 | 13.1 | 13.1 | 174,780 | 182 | 26,642 | 424,806 | 9.3 |
| Escambia | 0.1 | 0.1 | 5.7 | 5.1 | 8.0 | 6.5 | 10.3 | 15.7 | 25.2 | 75,215 | 236 | 10,115 | 144,621 | 5.8 |
| Flagler | 0.7 | D | 8.0 | 3.1 | 10.1 | 9.5 | 6.7 | 12.4 | 16.3 | 39,660 | 343 | 1,779 | 55,556 | 14.3 |
| Franklin | 0.0 | 0.2 | 7.7 | D | 6.0 | 10.5 | 7.5 | D | 28.7 | 3,250 | 268 | 317 | 8,902 | 2.8 |
| Gadsden | 11.1 | 1.7 | 10.4 | 7.3 | 2.5 | 4.7 | 1.3 | 3.5 | 30.6 | 11,460 | 251 | 2,542 | 21,787 | 11.7 |
| Gilchrist | 17.2 | 0.1 | 7.7 | 4.4 | 2.2 | 3.4 | 2.0 | 18.1 | 25.8 | 4,485 | 241 | 516 | 7,750 | 6.0 |
| Glades | 17.3 | D | 8.0 | 8.3 | D | 1.9 | D | D | 18.7 | 2,770 | 199 | 194 | 7,181 | 2.8 |
| Gulf | 0.0 | D | 10.6 | 2.2 | 10.4 | 8.2 | 6.4 | D | 25.7 | 4,100 | 304 | 365 | 9,888 | 8.6 |
| Hamilton | 5.3 | 0.0 | D | D | D | 5.5 | 1.3 | D | 29.4 | 3,275 | 226 | 653 | 5,860 | 1.4 |
| Hardee | 17.0 | D | 4.5 | 3.7 | 1.5 | 5.4 | 4.5 | D | 21.9 | 4,850 | 181 | 754 | 9,848 | 1.3 |
| Hendry | 27.3 | D | 5.4 | 2.7 | 4.6 | 6.1 | 2.1 | 4.2 | 16.8 | 6,645 | 158 | 1,430 | 15,227 | 4.6 |
| Hernando | 0.2 | 0.0 | 6.4 | 5.7 | D | 10.8 | 4.3 | 21.7 | 15.3 | 64,135 | 330 | 4,638 | 89,805 | 6.3 |
| Highlands | 6.6 | D | 5.3 | 2.3 | 4.0 | 10.6 | 4.2 | 22.9 | 16.2 | 35,675 | 337 | 2,952 | 56,320 | 1.7 |
| Hillsborough | 0.4 | 0.0 | 6.1 | 3.9 | 17.2 | 6.2 | 14.3 | 11.0 | 12.2 | 248,995 | 169 | 41,177 | 612,875 | 14.3 |
| Holmes | 4.9 | D | 11.0 | 2.3 | D | 6.2 | 3.0 | D | 39.2 | 5,565 | 283 | 852 | 8,743 | 1.2 |
| Indian River | 2.1 | 0.0 | 7.3 | 4.5 | 12.9 | 8.7 | 9.4 | 16.7 | 9.6 | 55,280 | 346 | 2,640 | 83,611 | 9.5 |
| Jackson | 5.8 | 1.1 | 7.1 | 7.1 | D | 7.8 | 3.1 | 8.5 | 33.8 | 12,620 | 273 | 1,878 | 21,168 | 0.8 |
| Jefferson | 8.8 | 0.1 | 9.6 | 0.2 | D | 6.2 | 4.4 | D | 21.1 | 3,900 | 273 | 553 | 6,918 | 4.2 |
| Lafayette | 19.7 | D | D | 3.7 | D | 5.7 | D | D | 38.6 | 1,585 | 189 | 153 | 3,417 | 2.7 |
| Lake | 1.9 | 0.2 | 11.8 | 4.0 | 5.8 | 11.4 | 5.4 | 20.1 | 13.0 | 110,655 | 302 | 7,497 | 167,010 | 15.2 |
| Lee | 0.3 | 0.1 | 11.3 | 2.2 | 11.3 | 8.9 | 7.3 | 11.1 | 16.2 | 202,045 | 262 | 12,811 | 417,359 | 12.5 |
| Leon | 0.0 | D | 4.4 | 1.3 | 15.9 | 5.7 | 6.2 | 13.9 | 34.8 | 48,125 | 164 | 6,464 | 134,575 | 8.4 |
| Levy | 13.3 | 0.6 | 12.6 | 7.8 | 3.6 | 9.1 | 4.0 | 5.5 | 18.8 | 13,200 | 317 | 1,457 | 21,157 | 5.1 |
| Liberty | 3.2 | 0.0 | 3.1 | 20.8 | D | D | D | D | 37.2 | 1,555 | 186 | 293 | 3,484 | 3.8 |
| Madison | 7.6 | D | 2.2 | 12.4 | 2.3 | 7.2 | 4.4 | 13.1 | 26.5 | 4,865 | 262 | 873 | 8,745 | 3.1 |
| Manatee | 2.0 | D | 10.1 | 7.3 | 8.7 | 9.2 | 7.8 | 13.3 | 10.7 | 105,930 | 263 | 6,029 | 203,442 | 17.8 |
| Marion | 0.4 | 0.2 | 8.0 | 9.1 | 6.3 | 10.5 | 5.5 | 18.1 | 15.2 | 122,830 | 336 | 10,591 | 175,527 | 7.0 |
| Martin | 1.0 | D | 9.2 | 5.6 | 10.2 | 8.0 | 8.5 | 18.5 | 9.2 | 47,460 | 294 | 1,585 | 81,396 | 3.9 |
| Miami-Dade | 0.4 | 0.1 | 7.0 | 2.9 | 14.8 | 6.3 | 12.2 | 10.8 | 12.4 | 436,570 | 161 | 160,488 | 1,048,012 | 5.9 |
| Monroe | 0.0 | D | 7.8 | 0.8 | 8.1 | 8.8 | 7.0 | 6.2 | 21.5 | 17,525 | 237 | 1,263 | 54,450 | 3.2 |
| Nassau | 0.2 | 0.0 | 5.3 | 7.9 | 7.8 | 7.5 | 4.3 | 9.0 | 18.4 | 23,210 | 262 | 1,197 | 42,673 | 21.9 |
| Okaloosa | 0.1 | D | 4.8 | 2.7 | 14.9 | 6.7 | 5.5 | 7.2 | 40.8 | 44,480 | 210 | 3,481 | 99,357 | 7.5 |
| Okeechobee | 10.1 | 0.1 | 6.2 | 3.7 | 6.2 | 8.5 | 3.1 | D | 20.0 | 9,305 | 221 | 1,163 | 18,855 | 1.9 |
| Orange | 0.2 | 0.0 | 6.1 | 5.2 | 14.4 | 5.8 | 8.2 | 9.9 | 9.9 | 200,780 | 144 | 34,423 | 566,693 | 16.2 |
| Osceola | 0.6 | 0.0 | 8.3 | 1.9 | 5.5 | 10.1 | 4.5 | 15.3 | 15.6 | 67,015 | 179 | 11,056 | 170,481 | 33 |
| Palm Beach | 0.8 | 0.0 | 6.3 | 3.7 | 14.2 | 6.6 | 11.4 | 12.5 | 10.5 | 329,185 | 220 | 23,656 | 697,426 | 4.9 |
| Pasco | 0.2 | 0.1 | 8.8 | 3.5 | 7.7 | 13.7 | 4.6 | 19.9 | 14.6 | 143,420 | 259 | 12,911 | 255,073 | 11.4 |
| Pinellas | 0.0 | 0.0 | 5.4 | 9.3 | 13.0 | 7.1 | 10.0 | 15.0 | 10.8 | 254,605 | 261 | 21,287 | 514,117 | 2.1 |
| Polk | 0.9 | 0.6 | 6.4 | 8.9 | 5.4 | 7.7 | 7.8 | 13.4 | 12.8 | 170,180 | 235 | 22,715 | 313,005 | 11.3 |

1. Per 1,000 resident population estimated as of July 1 of the year shown.

| STATE County | Total [89] | Percent [90] | Median value[1] [91] | With a mortgage [92] | Without a mortgage[2] [93] | Median rent[3] [94] | Median rent as a percent of income[2] [95] | Sub-standard units[4] (percent) [96] | Total [97] | Percent change, 2019-2020 [98] | Total [99] | Rate[5] [100] | Total [101] | Management, business, science, and arts [102] | Construction, production, and maintenance occupations [103] |
|---|---|---|---|---|---|---|---|---|---|---|---|---|---|---|---|
| **CONNECTICUT—Cont'd** | | | | | | | | | | | | | | | |
| Tolland | 55,683 | 71.8 | 253,100 | 20.7 | 12.2 | 1,155 | 29.8 | 1.3 | 84,598 | -3.1 | 5,236 | 6.2 | 79,361 | 45.6 | 17.8 |
| Windham | 44,953 | 69.0 | 204,400 | 22.4 | 13.0 | 946 | 29.5 | 1.7 | 62,218 | -3.1 | 4,849 | 7.8 | 59,582 | 34.0 | 24.3 |
| **DELAWARE** | 363,322 | 71.2 | 251,100 | 21.2 | 10.0 | 1,130 | 29.6 | 2.0 | 484,358 | -0.8 | 37,900 | 7.8 | 455,620 | 40.3 | 20.3 |
| Kent | 65,796 | 69.1 | 220,600 | 22.3 | 10.0 | 1,084 | 32.7 | 2.3 | 79,715 | 0.8 | 6,872 | 8.6 | 79,883 | 33.0 | 24.8 |
| New Castle | 205,829 | 67.9 | 260,800 | 20.5 | 10.0 | 1,163 | 28.6 | 1.6 | 298,640 | -0.7 | 23,039 | 7.7 | 276,420 | 44.3 | 17.7 |
| Sussex | 91,697 | 80.3 | 258,600 | 21.8 | 10.2 | 1,030 | 30.0 | 2.6 | 106,004 | -2.1 | 7,990 | 7.5 | 99,317 | 34.9 | 23.9 |
| **DISTRICT OF COLUMBIA** | 284,386 | 41.6 | 601,500 | 21.0 | 10.0 | 1,541 | 28.5 | 3.8 | 409,734 | -1.3 | 32,895 | 8.0 | 376,871 | 63.6 | 6.7 |
| District of Columbia | 284,386 | 41.6 | 601,500 | 21.0 | 10.0 | 1,541 | 28.5 | 3.8 | 409,734 | -1.3 | 32,895 | 8.0 | 376,871 | 63.6 | 6.7 |
| **FLORIDA** | 7,736,311 | 65.4 | 215,300 | 23.0 | 12.0 | 1,175 | 33.3 | 3.2 | 10,114,329 | -2.1 | 781,491 | 7.7 | 9,495,353 | 35.6 | 19.9 |
| Alachua | 97,995 | 55.0 | 186,700 | 20.1 | 10.8 | 983 | 36.4 | 2.0 | 134,931 | -2.9 | 7,235 | 5.4 | 126,034 | 47.0 | 12.0 |
| Baker | 8,693 | 75.3 | 138,900 | 18.5 | 10.0 | 782 | 24.2 | 2.5 | 11,759 | -2.1 | 580 | 4.9 | 11,039 | 31.6 | 27.1 |
| Bay | 71,422 | 65.1 | 178,400 | 22.6 | 10.4 | 1,049 | 30.5 | 2.3 | 82,763 | -2.0 | 5,017 | 6.1 | 82,767 | 32.9 | 22.6 |
| Bradford | 9,115 | 69.3 | 110,800 | 20.4 | 10.0 | 833 | 28.9 | 3.8 | 10,791 | -1.4 | 641 | 5.9 | 9,393 | 30.2 | 26.2 |
| Brevard | 230,417 | 74.3 | 196,400 | 22.1 | 10.8 | 1,068 | 30.9 | 1.8 | 281,591 | -0.8 | 18,724 | 6.6 | 252,483 | 39.3 | 19.2 |
| Broward | 690,050 | 62.1 | 265,000 | 25.6 | 14.8 | 1,392 | 36.0 | 4.3 | 1,020,586 | -2.0 | 90,289 | 8.8 | 966,496 | 36.1 | 18.7 |
| Calhoun | 4,501 | 81.4 | 85,000 | 20.3 | 10.0 | 596 | 31.2 | 2.2 | 4,754 | | 246 | 5.2 | 4,815 | 27.7 | 28.2 |
| Charlotte | 76,891 | 79.7 | 195,400 | 23.4 | 12.6 | 1,032 | 34.7 | 1.9 | 71,182 | -1.1 | 5,214 | 7.3 | 62,286 | 30.0 | 21.9 |
| Citrus | 63,681 | 81.9 | 131,500 | 22.7 | 10.0 | 843 | 33.0 | 1.6 | 46,500 | -1.9 | 3,918 | 8.4 | 46,529 | 30.4 | 24.2 |
| Clay | 74,333 | 75.0 | 185,400 | 20.4 | 10.0 | 1,148 | 29.5 | 1.8 | 105,658 | -1.7 | 5,654 | 5.4 | 96,866 | 35.2 | 22.9 |
| Collier | 142,979 | 73.3 | 360,800 | 23.8 | 12.4 | 1,317 | 32.9 | 4.3 | 177,497 | -1.7 | 12,210 | 6.9 | 158,452 | 31.0 | 22.7 |
| Columbia | 25,133 | 71.4 | 124,200 | 20.3 | 10.2 | 829 | 27.6 | 3.9 | 29,566 | -0.9 | 1,726 | 5.8 | 26,530 | 30.7 | 25.7 |
| DeSoto | 12,072 | 69.7 | 87,700 | 22.7 | 10.4 | 687 | 33.6 | 5.3 | 14,356 | 0.9 | 741 | 5.2 | 13,571 | 16.9 | 45.0 |
| Dixie | 6,504 | 78.6 | 78,500 | 17.7 | 10.0 | 631 | 26.1 | 6.0 | 5,886 | 0.5 | 321 | 5.5 | 5,143 | 25.5 | 36.9 |
| Duval | 359,544 | 56.7 | 180,700 | 21.6 | 10.5 | 1,074 | 29.8 | 2.6 | 485,114 | -0.5 | 32,633 | 6.7 | 453,177 | 37.3 | 20.1 |
| Escambia | 120,104 | 62.0 | 146,200 | 21.3 | 10.4 | 983 | 29.5 | 1.7 | 144,779 | -0.1 | 9,671 | 6.7 | 136,650 | 33.6 | 20.0 |
| Flagler | 42,121 | 76.0 | 217,500 | 23.9 | 12.0 | 1,255 | 33.1 | 1.4 | 46,636 | -2.3 | 3,553 | 7.6 | 43,163 | 32.6 | 21.2 |
| Franklin | 4,444 | 74.6 | 166,000 | 25.9 | 10.7 | 802 | 27.1 | 3.1 | 4,681 | 0.3 | 252 | 5.4 | 4,268 | 26.6 | 27.6 |
| Gadsden | 17,149 | 72.9 | 96,800 | 21.8 | 10.0 | 678 | 29.3 | 3.8 | 18,421 | -0.8 | 1,245 | 6.8 | 16,813 | 29.8 | 25.1 |
| Gilchrist | 6,464 | 83.0 | 104,800 | 20.8 | 10.0 | 645 | 26.6 | 2.8 | 7,028 | -3.8 | 345 | 4.9 | 6,698 | 22.2 | 33.5 |
| Glades | 4,700 | 79.6 | 76,700 | 23.1 | 10.0 | 781 | 32.1 | 4.0 | 5,192 | -3.0 | 289 | 5.6 | 3,842 | 23.3 | 40.7 |
| Gulf | 5,757 | 74.2 | 160,700 | 20.4 | 11.2 | 1,037 | 27.6 | 1.8 | 5,022 | -3.4 | 298 | 5.9 | 5,750 | 31.2 | 26.3 |
| Hamilton | 4,381 | 67.9 | 80,100 | 20.4 | 10.0 | 604 | 28.9 | 4.2 | 4,199 | -2.5 | 319 | 7.6 | 3,770 | 23.3 | 27.5 |
| Hardee | 7,863 | 64.9 | 83,400 | 19.4 | 11.1 | 740 | 28.2 | 7.6 | 8,466 | 1.1 | 551 | 6.5 | 9,752 | 21.3 | 39.5 |
| Hendry | 12,527 | 65.3 | 94,500 | 23.5 | 10.0 | 786 | 28.7 | 9.2 | 15,589 | 1.4 | 1,257 | 8.1 | 17,749 | 19.2 | 47.9 |
| Hernando | 75,348 | 78.1 | 142,500 | 21.4 | 10.6 | 984 | 32.2 | 1.7 | 71,534 | -0.6 | 5,769 | 8.1 | 68,907 | 32.1 | 21.5 |
| Highlands | 41,740 | 75.3 | 108,100 | 21.6 | 10.7 | 816 | 32.0 | 3.0 | 34,641 | -3.3 | 2,674 | 7.7 | 33,555 | 28.4 | 26.2 |
| Hillsborough | 526,175 | 58.6 | 216,300 | 21.5 | 10.7 | 1,142 | 31.1 | 3.4 | 754,552 | -0.3 | 53,788 | 7.1 | 696,246 | 39.1 | 18.7 |
| Holmes | 7,092 | 76.4 | 92,700 | 20.5 | 10.7 | 703 | 31.8 | 3.3 | 6,850 | -1.4 | 378 | 5.5 | 6,849 | 28.5 | 28.4 |
| Indian River | 58,612 | 79.2 | 202,300 | 22.6 | 11.7 | 959 | 32.9 | 1.6 | 64,233 | -2.1 | 4,751 | 7.4 | 58,626 | 33.1 | 21.0 |
| Jackson | 17,149 | 69.8 | 99,000 | 20.2 | 10.1 | 656 | 27.5 | 3.5 | 17,231 | 0.7 | 933 | 5.4 | 16,690 | 29.4 | 25.9 |
| Jefferson | 5,770 | 77.0 | 133,500 | 24.1 | 10.0 | 770 | 31.0 | 1.5 | 5,347 | -3.5 | 298 | 5.6 | 5,107 | 34.4 | 19.3 |
| Lafayette | 2,175 | 84.4 | 104,700 | 17.5 | 10.1 | 532 | 20.4 | 6.3 | 2,781 | -2.3 | 117 | 4.2 | 2,481 | 27.2 | 30.8 |
| Lake | 134,317 | 74.5 | 181,400 | 22.3 | 11.5 | 1,084 | 32.6 | 2.1 | 154,028 | -4.8 | 13,645 | 8.9 | 141,998 | 34.4 | 19.4 |
| Lee | 275,965 | 72.3 | 224,800 | 23.2 | 12.4 | 1,154 | 31.7 | 2.7 | 345,417 | | 25,256 | 7.3 | 308,959 | 31.2 | 22.3 |
| Leon | 113,658 | 53.0 | 203,100 | 20.6 | 10.0 | 1,024 | 34.6 | 2.1 | 150,626 | -2.8 | 8,743 | 5.8 | 146,627 | 46.1 | 11.6 |
| Levy | 16,374 | 77.7 | 105,700 | 23.2 | 11.5 | 720 | 30.3 | 2.9 | 16,569 | -0.9 | 1,001 | 6.0 | 15,416 | 24.1 | 29.3 |
| Liberty | 2,459 | 75.4 | 75,100 | 25.4 | 10.6 | 595 | 20.7 | 2.2 | 2,598 | -2.1 | 125 | 4.8 | 2,323 | 25.4 | 31.6 |
| Madison | 6,778 | 73.6 | 89,100 | 20.9 | 12.1 | 689 | 28.3 | 2.7 | 7,599 | 1.8 | 459 | 6.0 | 6,314 | 28.9 | 26.8 |
| Manatee | 145,356 | 72.9 | 238,500 | 22.7 | 11.9 | 1,144 | 32.0 | 2.3 | 176,496 | -2.3 | 11,925 | 6.8 | 162,556 | 34.8 | 20.7 |
| Marion | 139,172 | 75.0 | 139,900 | 22.1 | 11.0 | 896 | 29.9 | 2.5 | 138,469 | -0.6 | 9,760 | 7.0 | 125,900 | 30.1 | 21.9 |
| Martin | 64,528 | 78.0 | 273,700 | 23.5 | 12.7 | 1,111 | 32.3 | 1.9 | 72,180 | -1.6 | 4,384 | 6.1 | 65,826 | 38.1 | 18.0 |
| Miami-Dade | 883,372 | 51.2 | 289,600 | 27.3 | 14.4 | 1,328 | 38.1 | 6.6 | 1,291,854 | -5.8 | 103,681 | 8.0 | 1,310,482 | 33.0 | 21.0 |
| Monroe | 32,068 | 59.5 | 494,100 | 29.7 | 12.2 | 1,627 | 34.0 | 4.6 | 44,755 | -3.9 | 3,532 | 7.9 | 40,056 | 31.4 | 21.4 |
| Nassau | 32,603 | 80.0 | 230,900 | 20.6 | 10.0 | 1,046 | 30.6 | 1.4 | 41,338 | -1.2 | 2,347 | 5.7 | 37,539 | 35.1 | 23.6 |
| Okaloosa | 77,962 | 63.4 | 219,100 | 21.1 | 10.7 | 1,127 | 30.2 | 2.2 | 96,000 | -0.7 | 5,281 | 5.5 | 89,114 | 38.1 | 18.4 |
| Okeechobee | 13,904 | 72.2 | 106,000 | 21.3 | 10.1 | 772 | 31.1 | 5.1 | 17,138 | -3.5 | 921 | 5.4 | 15,304 | 21.8 | 38.8 |
| Orange | 461,705 | 55.4 | 235,800 | 22.3 | 11.2 | 1,215 | 33.1 | 3.5 | 737,546 | -2.9 | 78,496 | 10.6 | 688,631 | 37.2 | 18.3 |
| Osceola | 103,141 | 61.6 | 201,000 | 24.6 | 11.8 | 1,228 | 36.3 | 3.2 | 185,793 | 0.1 | 25,155 | 13.5 | 165,008 | 27.0 | 23.6 |
| Palm Beach | 554,095 | 68.9 | 283,600 | 24.3 | 14.3 | 1,398 | 34.9 | 3.4 | 717,379 | -2.4 | 54,434 | 7.6 | 684,112 | 37.3 | 17.3 |
| Pasco | 204,198 | 72.1 | 162,100 | 21.9 | 11.2 | 1,062 | 31.4 | 1.8 | 238,327 | -0.8 | 17,142 | 7.2 | 218,061 | 36.6 | 19.5 |
| Pinellas | 407,546 | 67.0 | 201,200 | 22.7 | 13.6 | 1,112 | 31.8 | 2.2 | 484,330 | -0.4 | 34,064 | 7.0 | 453,031 | 39.4 | 16.5 |
| Polk | 235,283 | 68.9 | 150,800 | 22.0 | 11.4 | 978 | 30.1 | 3.6 | 316,626 | 3.3 | 28,647 | 9.0 | 285,305 | 29.5 | 25.8 |

1. Specified owner-occupied units, lacking complete plumbing facilities.
2. A value of 10.0 represents 10 percent or less; a value of 50.0 represents 50 percent or more.
3. Specified renter-occupied units.
4. Overcrowded or
5. Percent of civilian labor force.
6. Civilian employed persons 16 years old and over.

| | Private nonfarm establishments, employment and payroll, 2019 | | | | | | | Agriculture, 2017 | | | | | |
| | Employment | | | | | | | Annual payroll | | Farms | | | Farm producers whose primary occupation is farming (percent) |
| | | | | | | | Professional, scientific, and technical services | | | | Percent with: | | |
| STATE County | Number of establish-ments | Total | Health care and social assistance | Manufac-turing | Retail trade | Finance and insurance | | | Total (mil dol) | Average per employee (dollars) | Number | Fewer than 50 acres | 1000 acres or more | |
| | 104 | 105 | 106 | 107 | 108 | 109 | 110 | 111 | 112 | 113 | 114 | 115 | 116 |
|---|---|---|---|---|---|---|---|---|---|---|---|---|---|
| CONNECTICUT—Cont'd | | | | | | | | | | | | | |
| Tolland | 2,405 | 27,887 | 5,923 | 3,502 | 4,641 | 488 | 1,217 | 1,183 | 42,426 | 520 | 70.6 | 0.8 | 34.4 |
| Windham | 2,111 | 30,243 | 6,981 | 5,875 | 4,991 | 673 | 601 | 1,290 | 42,661 | 646 | 63.3 | 0.6 | 38.4 |
| DELAWARE | 26,142 | 413,410 | 72,115 | 29,709 | 57,004 | 41,700 | 31,580 | 23,948 | 57,928 | 2,302 | 55.7 | 6.9 | 52.8 |
| Kent | 3,793 | 54,277 | 12,020 | 4,682 | 9,522 | 1,538 | 2,384 | 2,368 | 43,635 | 822 | 53.4 | 6.3 | 48.5 |
| New Castle | 16,235 | 275,091 | 47,646 | 12,838 | 34,070 | 36,637 | 24,512 | 18,074 | 65,703 | 361 | 66.5 | 5.8 | 43.5 |
| Sussex | 5,777 | 74,351 | 12,135 | 12,188 | 13,410 | 1,964 | 2,398 | 2,944 | 39,594 | 1,119 | 54.0 | 7.6 | 58.9 |
| DISTRICT OF COLUMBIA. | 23,993 | 528,826 | 73,308 | 1,102 | 21,578 | 20,242 | 102,250 | 44,758 | 84,636 | NA | NA | NA | NA |
| District of Columbia | 23,993 | 528,826 | 73,308 | 1,102 | 21,578 | 20,242 | 102,250 | 44,758 | 84,636 | NA | NA | NA | NA |
| FLORIDA | 574,512 | 8,860,042 | 1,172,376 | 324,315 | 1,090,399 | 375,949 | 535,691 | 426,908 | 48,184 | 47,590 | 71.0 | 3.2 | 41.3 |
| Alachua | 6,295 | 100,237 | 28,744 | 3,739 | 14,133 | 4,001 | 6,320 | 4,441 | 44,301 | 1,611 | 72.1 | 2.5 | 34.5 |
| Baker | 398 | 5,759 | 1,829 | 156 | 897 | 128 | 121 | 191 | 33,234 | 328 | 69.2 | 1.2 | 35.0 |
| Bay | 4,673 | 59,187 | 9,699 | 3,696 | 11,440 | 1,571 | 4,610 | 2,441 | 41,234 | 190 | 69.5 | 4.2 | 23.2 |
| Bradford | 430 | 4,468 | 797 | 359 | 1,070 | 102 | 92 | 149 | 33,303 | 490 | 73.9 | 3.1 | 37.2 |
| Brevard | 14,516 | 189,379 | 32,969 | 15,622 | 29,431 | 4,903 | 19,615 | 8,889 | 46,937 | 522 | 81.0 | 3.1 | 43.5 |
| Broward | 62,337 | 733,082 | 103,064 | 24,268 | 107,360 | 31,573 | 58,447 | 37,665 | 51,378 | 640 | 97.2 | 0.2 | 41.5 |
| Calhoun | 183 | 1,761 | 623 | 12 | 259 | 59 | 10 | 55 | 31,425 | 289 | 56.4 | 3.8 | 29.7 |
| Charlotte | 4,079 | 39,400 | 8,836 | 554 | 9,543 | 1,005 | 1,481 | 1,467 | 37,221 | 306 | 66.3 | 6.2 | 37.2 |
| Citrus | 2,772 | 30,318 | 10,256 | 343 | 5,617 | 676 | 821 | 1,128 | 37,211 | 609 | 75.2 | 2.1 | 43.4 |
| Clay | 3,993 | 44,049 | 9,009 | 1,137 | 8,612 | 999 | 2,290 | 1,661 | 37,706 | 361 | 83.4 | 0.6 | 41.1 |
| Collier | 12,563 | 136,239 | 20,517 | 3,739 | 23,084 | 4,006 | 5,480 | 6,260 | 45,949 | 322 | 77.3 | 7.1 | 36.2 |
| Columbia | 1,403 | 19,182 | 4,533 | 898 | 3,014 | 439 | 652 | 783 | 40,796 | 979 | 66.4 | 1.7 | 40.6 |
| DeSoto | 490 | 5,729 | 1,014 | 451 | 1,008 | 132 | 133 | 210 | 36,606 | 761 | 57.4 | 7.4 | 44.5 |
| Dixie | 187 | 1,334 | 114 | 362 | 235 | 33 | 40 | 45 | 33,603 | 235 | 60.9 | 5.5 | 51.6 |
| Duval | 26,277 | 461,408 | 65,881 | 23,371 | 53,260 | 42,477 | 29,110 | 23,757 | 51,489 | 366 | 81.7 | 2.7 | 42.4 |
| Escambia | 6,943 | 113,336 | 20,803 | 4,650 | 15,935 | 10,088 | 7,564 | 4,913 | 43,347 | 649 | 72.7 | 2.2 | 36.1 |
| Flagler | 2,269 | 18,774 | 3,298 | 470 | 3,925 | 564 | 705 | 631 | 33,636 | 116 | 56.9 | 15.5 | 43.0 |
| Franklin | 317 | 2,188 | 244 | 73 | 460 | 47 | 79 | 70 | 31,941 | 15 | 66.7 | 6.7 | 20.0 |
| Gadsden | 658 | 8,384 | 1,568 | 885 | 1,138 | 114 | 174 | 370 | 44,132 | 522 | 54.0 | 1.3 | 33.6 |
| Gilchrist | 248 | 1,879 | 601 | 167 | 213 | 41 | 41 | 66 | 35,230 | 565 | 67.3 | 2.1 | 45.7 |
| Glades | 119 | 909 | 55 | 147 | 115 | 15 | 6 | 33 | 36,252 | 354 | 53.7 | 7.1 | 48.1 |
| Gulf | 305 | 1,691 | 212 | 52 | 302 | 100 | 85 | 58 | 34,268 | 46 | 30.4 | 4.3 | 26.4 |
| Hamilton | 169 | 1,696 | 218 | 503 | 415 | 20 | 49 | 81 | 47,982 | 338 | 43.5 | 3.0 | 43.2 |
| Hardee | 400 | 4,614 | 1,026 | 364 | 692 | 262 | 80 | 169 | 36,733 | 1,038 | 56.9 | 5.5 | 41.4 |
| Hendry | 625 | 6,758 | 1,195 | 827 | 1,446 | 236 | 349 | 247 | 36,542 | 436 | 50.9 | 10.8 | 47.3 |
| Hernando | 3,434 | 37,090 | 9,323 | 2,113 | 7,772 | 798 | 1,394 | 1,288 | 34,730 | 747 | 80.6 | 0.9 | 42.9 |
| Highlands | 1,959 | 20,941 | 5,925 | 795 | 4,657 | 453 | 727 | 706 | 33,728 | 989 | 66.4 | 6.1 | 45.7 |
| Hillsborough | 38,343 | 628,921 | 92,138 | 23,739 | 73,505 | 55,140 | 61,868 | 33,260 | 52,884 | 2,265 | 83.6 | 1.0 | 40.7 |
| Holmes | 266 | 1,788 | 542 | 85 | 363 | 43 | 48 | 53 | 29,907 | 721 | 41.5 | 1.7 | 34.5 |
| Indian River | 4,409 | 47,529 | 9,545 | 2,105 | 8,957 | 1,512 | 2,357 | 1,928 | 40,569 | 450 | 76.2 | 5.1 | 42.6 |
| Jackson | 797 | 9,643 | 1,653 | 848 | 1,991 | 292 | 267 | 337 | 34,947 | 1,154 | 45.6 | 4.3 | 41.5 |
| Jefferson | 252 | 1,558 | 178 | 54 | 307 | 137 | 66 | 50 | 31,870 | 592 | 62.8 | 4.1 | 36.3 |
| Lafayette | 98 | 615 | 136 | 53 | 94 | 19 | 26 | 18 | 28,945 | 257 | 43.2 | 4.7 | 50.0 |
| Lake | 7,659 | 87,243 | 19,396 | 3,327 | 17,373 | 2,063 | 2,258 | 3,191 | 36,581 | 1,703 | 75.1 | 2.7 | 37.6 |
| Lee | 19,540 | 232,778 | 38,748 | 5,412 | 40,336 | 6,008 | 13,582 | 10,024 | 43,062 | 800 | 81.6 | 2.4 | 38.5 |
| Leon | 7,721 | 103,869 | 18,905 | 1,651 | 16,506 | 5,548 | 9,768 | 4,542 | 43,731 | 325 | 67.4 | 3.7 | 32.5 |
| Levy | 762 | 6,162 | 538 | 735 | 1,474 | 241 | 219 | 199 | 32,225 | 1,058 | 66.4 | 3.9 | 49.4 |
| Liberty | 83 | 1,032 | 263 | 321 | 151 | NA | 20 | 39 | 37,786 | 111 | 54.1 | 0.9 | 28.1 |
| Madison | 295 | 2,888 | 862 | 489 | 475 | 80 | 68 | 96 | 33,297 | 669 | 38.6 | 4.6 | 43.5 |
| Manatee | 9,512 | 107,967 | 16,427 | 8,194 | 20,225 | 3,213 | 5,681 | 4,397 | 40,726 | 753 | 66.3 | 7.0 | 45.7 |
| Marion | 7,360 | 89,254 | 18,251 | 8,221 | 17,513 | 1,989 | 2,998 | 3,326 | 37,266 | 3,985 | 81.2 | 1.2 | 43.8 |
| Martin | 5,784 | 62,750 | 14,093 | 2,531 | 10,231 | 1,764 | 3,102 | 2,631 | 41,922 | 594 | 79.0 | 5.9 | 40.5 |
| Miami-Dade | 86,855 | 995,962 | 145,933 | 36,882 | 137,727 | 48,733 | 73,675 | 52,226 | 52,438 | 2,752 | 93.2 | 0.3 | 47.1 |
| Monroe | 3,829 | 31,744 | 2,104 | 328 | 6,112 | 654 | 927 | 1,187 | 37,381 | 40 | 100.0 | NA | 26.8 |
| Nassau | 1,948 | 19,531 | 2,826 | 1,343 | 3,408 | 436 | 665 | 750 | 38,380 | 373 | 70.2 | 2.7 | 37.8 |
| Okaloosa | 5,528 | 63,186 | 8,804 | 2,380 | 12,091 | 1,845 | 7,271 | 2,621 | 41,487 | 481 | 64.4 | 1.0 | 33.4 |
| Okeechobee | 836 | 7,440 | 1,462 | 251 | 1,732 | 193 | 231 | 259 | 34,789 | 599 | 48.1 | 13.9 | 43.8 |
| Orange | 39,740 | 772,591 | 80,527 | 28,235 | 86,995 | 24,408 | 61,170 | 37,681 | 48,773 | 622 | 83.8 | 1.9 | 51.8 |
| Osceola | 6,835 | 87,036 | 13,319 | 1,668 | 16,286 | 1,398 | 1,949 | 2,930 | 33,667 | 392 | 65.1 | 10.2 | 37.7 |
| Palm Beach | 49,958 | 550,168 | 91,692 | 14,174 | 79,188 | 20,705 | 43,434 | 27,990 | 50,875 | 1,298 | 89.1 | 3.0 | 48.3 |
| Pasco | 10,158 | 104,038 | 22,046 | 3,514 | 23,899 | 2,434 | 4,755 | 3,992 | 38,366 | 1,165 | 71.7 | 4.0 | 36.5 |
| Pinellas | 29,449 | 399,445 | 73,881 | 31,928 | 52,847 | 28,480 | 28,941 | 19,981 | 50,022 | 148 | 97.3 | 0.7 | 31.2 |
| Polk | 12,161 | 195,939 | 29,740 | 16,967 | 28,222 | 11,374 | 6,505 | 8,217 | 41,936 | 2,080 | 63.2 | 4.8 | 37.6 |

# Table B. States and Counties — Agriculture

| STATE County | Land in farms | | | | | Value of land and buildings (dollars) | | Value of machinery and equiopmnet, average per farm (dollars) | Value of products sold: | | | | Organic farms (number) | Farms with internet access (per-cent) | Government payments | |
|---|---|---|---|---|---|---|---|---|---|---|---|---|---|---|---|---|
| | Acreage (1,000) | Percent change, 2012-2017 | Acres | | | Average per farm | Average per acre | | Total (mil dol) | Average per farm (acres) | Percent from: | | | | Total ($1,000) | Percent of farms |
| | | | Average size of farm | Total irrigated (1,000) | Total cropland (1,000) | | | | | | Crops | Livestock and poultry products | | | | |
| | 117 | 118 | 119 | 120 | 121 | 122 | 123 | 124 | 125 | 126 | 127 | 128 | 129 | 130 | 131 | 132 |
| CONNECTICUT—Cont'd | | | | | | | | | | | | | | | | |
| Tolland | 36 | -25.4 | 69 | 0.2 | 16.2 | 665,187 | 9,703 | 62,379 | 53.4 | 102,725 | 62.3 | 37.7 | 3 | 79.2 | 215 | 3.5 |
| Windham | 52 | -10.8 | 80 | 0.4 | 22.4 | 760,709 | 9,452 | 57,273 | 45.1 | 69,800 | 36.9 | 63.1 | 6 | 78.2 | 132 | 4.6 |
| DELAWARE | 525 | 3.3 | 228 | 163.3 | 452.2 | 1,920,109 | 8,414 | 198,096 | 1,466.0 | 636,826 | 22.2 | 77.8 | 13 | 78.7 | 15,162 | 35.4 |
| Kent | 182 | 5.9 | 222 | 57.8 | 155.7 | 1,758,029 | 7,923 | 179,332 | 391.3 | 476,038 | 27.9 | 72.1 | 4 | 76.6 | 4,531 | 34.3 |
| New Castle | 67 | 5.1 | 187 | 5.1 | 53.7 | 1,775,015 | 9,499 | 140,504 | 62.1 | 171,986 | 55.1 | 44.9 | 2 | 80.9 | 1,342 | 26.6 |
| Sussex | 275 | 1.2 | 246 | 100.3 | 242.8 | 2,085,980 | 8,473 | 230,459 | 1,012.6 | 904,900 | 18 | 82 | 7 | 79.4 | 9,289 | 39.1 |
| DISTRICT OF COLUMBIA. | NA | NA | NA | NA | NA | NA | NA | NA | NA | NA | NA | NA | NA | NA | NA | NA |
| District of Columbia | NA | NA | NA | NA | NA | NA | NA | NA | NA | NA | NA | NA | NA | NA | NA | NA |
| FLORIDA | 9,732 | 1.9 | 204 | 1,519.4 | 2,825.8 | 1,206,788 | 5,901 | 72,754 | 7,357.3 | 154,599 | 77.5 | 22.5 | 251 | 76.3 | 59,120 | 8.4 |
| Alachua | 178 | -5.2 | 111 | 10.4 | 55.9 | 866,214 | 7,832 | 55,524 | 99.9 | 62,019 | 75.3 | 24.7 | 23 | 79.1 | 661 | 5.6 |
| Baker | 33 | 1.1 | 102 | 0.2 | 3.6 | 465,756 | 4,588 | 38,815 | 13.2 | 40,256 | 16.1 | 83.9 | NA | 78.4 | 72 | 2.4 |
| Bay | 74 | 601.5 | 387 | 1.7 | 4.6 | 794,316 | 2,051 | 41,297 | 2.9 | 15,274 | 69.6 | 30.4 | NA | 73.2 | 23 | 3.2 |
| Bradford | 59 | 67.8 | 120 | 0.6 | 12.9 | 573,623 | 4,777 | 61,641 | 13.1 | 26,708 | 27.3 | 72.7 | NA | 75.3 | 85 | 1.6 |
| Brevard | 157 | 6.9 | 300 | 10.5 | 21.0 | 1,507,863 | 5,027 | 62,406 | 59.0 | 112,977 | 76.9 | 23.1 | 1 | 78.7 | 186 | 3.6 |
| Broward | 7 | -53.5 | 11 | 0.7 | 1.6 | 348,907 | 33,140 | 31,890 | 24.9 | 38,883 | 94.2 | 5.8 | NA | 85.5 | 18 | 0.6 |
| Calhoun | 118 | 175.5 | 409 | 3.3 | 25.4 | 880,123 | 2,154 | 63,913 | 22.0 | 76,163 | 85.5 | 14.5 | NA | 75.4 | 876 | 10.7 |
| Charlotte | 113 | -48.1 | 368 | 11.6 | 18.5 | 2,644,176 | 7,176 | 66,707 | 43.9 | 143,402 | 91.8 | 8.2 | 2 | 73.9 | 329 | 5.2 |
| Citrus | 56 | 37.5 | 92 | 0.9 | 9.6 | 726,032 | 7,929 | 49,552 | 13.5 | 22,118 | 67.3 | 32.7 | NA | 81.0 | 197 | 2.6 |
| Clay | D | D | D | 0.3 | 4.7 | 366,072 | 5,726 | 42,781 | 5.5 | 15,111 | 55.7 | 44.3 | NA | 72.9 | 23 | 2.2 |
| Collier | 148 | 20.1 | 461 | 37.3 | 82.7 | 2,189,706 | 4,749 | 121,437 | 189.7 | 588,994 | 96.8 | 3.2 | 2 | 83.5 | 141 | 5.3 |
| Columbia | 107 | 5.5 | 109 | 4.2 | 33.6 | 476,995 | 4,361 | 52,758 | 40.2 | 41,040 | 41.2 | 58.8 | 1 | 75.5 | 1,189 | 10.7 |
| DeSoto | 335 | 10.5 | 440 | 57.7 | 73.5 | 2,203,290 | 5,008 | 100,533 | 168.3 | 221,145 | 72.1 | 27.9 | 3 | 73.2 | 1,066 | 13.4 |
| Dixie | 56 | 24.8 | 240 | 3.3 | 16.3 | 892,927 | 3,722 | 65,150 | 10.8 | 46,034 | 84.6 | 15.4 | NA | 67.2 | 273 | 3.8 |
| Duval | 30 | 6.1 | 82 | 2.3 | 5.9 | 616,395 | 7,523 | 47,364 | 9.0 | 24,656 | 71.5 | 28.5 | 4 | 70.5 | 135 | 4.4 |
| Escambia | 59 | -21.1 | 91 | 2.0 | 33.1 | 562,575 | 6,206 | 75,174 | 26.9 | 41,507 | 91.2 | 8.8 | 4 | 76.9 | 1,957 | 11.7 |
| Flagler | 79 | 81.8 | 683 | 3.4 | 5.8 | 3,801,926 | 5,565 | 81,473 | 14.3 | 123,388 | 90.2 | 9.8 | 3 | 73.3 | 55 | 8.6 |
| Franklin | D | D | D | NA | 0.1 | D | D | 126,657 | D | D | D | D | 3 | 80.0 | 56 | 26.7 |
| Gadsden | 66 | 30.4 | 127 | 4.1 | 19.4 | 540,049 | 4,256 | 62,186 | 90.5 | 173,354 | 81.4 | 18.6 | 1 | 73.6 | 238 | 8.6 |
| Gilchrist | 82 | -2.0 | 146 | 10.9 | 42.0 | 774,215 | 5,321 | 88,720 | 89.7 | 158,837 | 33.9 | 66.1 | NA | 73.5 | 462 | 8.7 |
| Glades | 429 | -3.2 | 1,211 | 65.8 | 43.9 | 5,217,205 | 4,308 | 142,023 | 78.2 | 220,924 | 62.7 | 37.3 | NA | 66.1 | 869 | 12.7 |
| Gulf | D | D | D | 0.0 | 1.6 | 6,070,106 | 1,435 | 47,216 | D | D | D | D | NA | 56.5 | 17 | 8.7 |
| Hamilton | 88 | 22.9 | 261 | 8.9 | 33.6 | 1,009,713 | 3,865 | 157,590 | 42.4 | 125,364 | 69.9 | 30.1 | 3 | 76.6 | 1,825 | 18.6 |
| Hardee | 377 | 37.6 | 363 | 46.8 | 68.1 | 1,937,932 | 5,337 | 95,859 | 204.7 | 197,171 | 60.0 | 40.0 | 1 | 70.8 | 1,611 | 16.6 |
| Hendry | 433 | -12.6 | 993 | 183.3 | 203.5 | 4,867,433 | 4,900 | 262,368 | 329.5 | 755,716 | 91.6 | 8.4 | 4 | 78.4 | 1,163 | 11.7 |
| Hernando | 50 | -18.8 | 67 | 2.0 | 8.2 | 706,024 | 10,488 | 41,410 | 20.4 | 27,309 | 73.8 | 26.2 | 2 | 77.9 | 313 | 5.1 |
| Highlands | 376 | -23.3 | 380 | 73.3 | 87.6 | 1,484,269 | 3,906 | 92,897 | 196.7 | 198,865 | 72.5 | 27.5 | 2 | 71.1 | 2,471 | 11.0 |
| Hillsborough | 180 | -16.1 | 80 | 25.8 | 76.3 | 929,585 | 11,678 | 64,339 | 447.6 | 197,627 | 91.6 | 8.4 | 25 | 75.8 | 584 | 3.0 |
| Holmes | 101 | -4.1 | 140 | 1.1 | 36.3 | 468,245 | 3,334 | 53,453 | 28.2 | 39,122 | 34.7 | 65.3 | 2 | 73.0 | 1,500 | 32.6 |
| Indian River | 183 | 12.4 | 406 | 52.9 | 55.5 | 2,330,717 | 5,745 | 90,645 | 106.5 | 236,611 | 89.3 | 10.7 | 9 | 68.2 | 426 | 7.1 |
| Jackson | 275 | 4.8 | 238 | 18.4 | 152.3 | 880,430 | 3,694 | 101,848 | 93.3 | 80,821 | 87.0 | 13.0 | 8 | 76.4 | 9,120 | 28.8 |
| Jefferson | 168 | 29.7 | 284 | 2.8 | 35.6 | 983,884 | 3,468 | 56,238 | 36.1 | 60,990 | 24.2 | 75.8 | 4 | 77.7 | 1,147 | 10.8 |
| Lafayette | 94 | 2.6 | 364 | 17.0 | 32.4 | 1,102,316 | 3,024 | 186,306 | 85.9 | 334,218 | 24.6 | 75.4 | NA | 68.9 | 1,200 | 14.0 |
| Lake | 184 | 20.8 | 108 | 19.6 | 43.5 | 856,367 | 7,932 | 60,783 | 215.7 | 126,665 | 91.1 | 8.9 | 21 | 81.0 | 890 | 5.9 |
| Lee | 87 | 0.1 | 109 | 10.5 | 22.2 | 1,330,329 | 12,206 | 53,238 | 104.4 | 130,449 | 93.8 | 6.2 | 7 | 76.0 | 163 | 1.8 |
| Leon | 92 | 12.8 | 282 | 2.3 | 7.6 | 1,204,029 | 4,266 | 50,879 | 5.5 | 17,034 | 73.4 | 26.6 | 7 | 84.6 | 54 | 4.6 |
| Levy | 187 | 12.0 | 177 | 20.8 | 79.1 | 816,534 | 4,608 | 90,455 | 131.0 | 123,818 | 38.5 | 61.5 | 2 | 83.8 | 2,571 | 7.0 |
| Liberty | 35 | 145.3 | 313 | 0.0 | 1.4 | 1,227,282 | 3,916 | 64,185 | 2.5 | 22,126 | 6.8 | 93.2 | NA | 79.3 | 106 | 19.8 |
| Madison | 168 | 17.3 | 251 | 16.4 | 47.5 | 884,049 | 3,527 | 87,915 | 88.3 | 131,945 | 40.1 | 59.9 | 6 | 65.6 | 1,002 | 16.9 |
| Manatee | 193 | 3.4 | 256 | 56.5 | 71.2 | 1,991,183 | 7,784 | 116,597 | 360.1 | 478,246 | 89.3 | 10.7 | 3 | 76.2 | 151 | 3.6 |
| Marion | 331 | 2.9 | 83 | 13.2 | 79.3 | 922,892 | 11,114 | 47,624 | 145.5 | 36,501 | 41.1 | 58.9 | 5 | 82.3 | 813 | 1.4 |
| Martin | 154 | 10.4 | 259 | 28.8 | 49.6 | 1,137,728 | 4,396 | 88,036 | 112.6 | 189,505 | 91.4 | 8.6 | 3 | 81.5 | 487 | 4.2 |
| Miami-Dade | 79 | -3.4 | 29 | 36.8 | 55.2 | 1,068,826 | 37,450 | 51,638 | 837.7 | 304,409 | 98.8 | 1.2 | 35 | 70.6 | 1,733 | 1.8 |
| Monroe | 0 | -64.9 | 4 | 0.0 | 0.1 | D | D | 67,514 | 3.9 | 98,375 | 22.1 | 77.9 | NA | 90.0 | NA | NA |
| Nassau | 55 | 38.8 | 146 | 0.2 | 4.6 | 547,982 | 3,744 | 52,065 | 12.9 | 34,525 | 5.5 | 94.5 | NA | 73.2 | 236 | 12.6 |
| Okaloosa | 47 | -24.2 | 97 | 0.4 | 13.7 | 461,280 | 4,763 | 38,830 | 8.3 | 17,318 | 70.3 | 29.7 | NA | 71.9 | 992 | 14.3 |
| Okeechobee | 297 | -32.7 | 497 | 23.2 | 42.2 | 2,096,854 | 4,223 | 135,425 | 235.9 | 393,791 | 13.6 | 86.4 | NA | 70.1 | 2,671 | 24.4 |
| Orange | 109 | -17.5 | 176 | 8.1 | 15.7 | 1,482,493 | 8,432 | 70,712 | 232.0 | 372,932 | 96.6 | 3.4 | 2 | 77.0 | 205 | 4.5 |
| Osceola | 525 | -4.0 | 1,339 | 36.5 | 40.2 | 5,196,486 | 3,880 | 121,645 | 85.4 | 217,982 | 52.4 | 47.6 | 3 | 78.6 | 1,048 | 9.4 |
| Palm Beach | 488 | -5.1 | 376 | 370.7 | 438.9 | 3,149,296 | 8,379 | 179,904 | 901.7 | 694,699 | 98.3 | 1.7 | 10 | 84.2 | 114 | 1.1 |
| Pasco | 192 | 11.9 | 164 | 3.6 | 25.9 | 1,177,181 | 7,161 | 56,608 | 65.0 | 55,766 | 23.1 | 76.9 | NA | 77.8 | 876 | 11.8 |
| Pinellas | 2 | 64.3 | 16 | 0.1 | 0.7 | 943,499 | 57,464 | 29,146 | 1.9 | 13,068 | 73.5 | 26.5 | 3 | 81.8 | D | 1.4 |
| Polk | 487 | -6.5 | 234 | 88.6 | 131 | 1,420,064 | 6,064 | 69,094 | 297.7 | 143,136 | 81.0 | 19.0 | 5 | 71.5 | 2,242 | 11.5 |

Items 117—132

# Table B. States and Counties — Water Use, Wholesale Trade, Retail Trade, and Real Estate

| STATE County | Water use, 2015 | | Wholesale Trade[1], 2017 | | | | Retail Trade[2], 2017 | | | | Real estate and rental and leasing,[2] 2017 | | | |
|---|---|---|---|---|---|---|---|---|---|---|---|---|---|---|
| | Public supply water withdrawn (mil gal/day) | Public supply gallons withdrawn per person per day | Number of establishments | Number of employees | Sales (mil dol) | Average payroll (mil dol) | Number of establishments | Number of employees | Sales (mil dol) | Average payroll (mil dol) | Number of establishments | Number of employees | Sales (mil dol) | Average payroll (mil dol) |
| | 133 | 134 | 135 | 136 | 137 | 138 | 139 | 140 | 141 | 142 | 143 | 144 | 145 | 146 |
| CONNECTICUT—Cont'd | | | | | | | | | | | | | | |
| Tolland | 4.93 | 32.6 | 57 | 561 | 217.4 | 28.2 | 358 | 4,747 | 1,468.1 | 137.8 | 94 | 487 | 74.4 | 18.3 |
| Windham | 7.71 | 66.1 | 61 | 1,330 | 409.6 | 57.5 | 335 | 5,108 | 1,436.0 | 141.0 | 48 | 128 | 28.3 | 4.4 |
| DELAWARE | 86.35 | 91.3 | 962 | 9,315 | 7,224.6 | 522.6 | 3,648 | 58,201 | 17,667.9 | 1,558.0 | 1,286 | 6,096 | 5,163.7 | 287.7 |
| Kent | 11.81 | 68.1 | D | D | D | D | 600 | 9,606 | 3,059.7 | 263.9 | 161 | 697 | 225.2 | 26.9 |
| New Castle | 60.61 | 108.9 | D | D | D | D | 1,967 | 35,232 | 10,672.4 | 947.7 | 819 | 4,088 | 4,623.7 | 210.5 |
| Sussex | 13.93 | 64.6 | D | D | D | D | 1,081 | 13,363 | 3,935.9 | 346.3 | 306 | 1,311 | 314.8 | 50.3 |
| DISTRICT OF COLUMBIA. | 0 | 0 | 321 | 3,851 | 3,387.1 | 292.5 | 1,743 | 23,133 | 5,533.9 | 672.6 | 1,350 | 11,000 | 4,524.7 | 883.3 |
| District of Columbia | 0 | 0 | 321 | 3,851 | 3,387.1 | 292.5 | 1,743 | 23,133 | 5,533.9 | 672.6 | 1,350 | 11,000 | 4,524.7 | 883.3 |
| FLORIDA | 2,384.85 | 117.6 | 27,230 | 283,295 | 288,642.2 | 16,241.2 | 74,496 | 1,086,052 | 333,134.6 | 29,686.2 | 37,660 | 176,886 | 49,175.2 | 8,183.6 |
| Alachua | 23.37 | 89.9 | 187 | 1,947 | 1,652.3 | 112.9 | 914 | 14,317 | 3,726.7 | 339.1 | 365 | 2,033 | 348.9 | 74.3 |
| Baker | 0.90 | 32.8 | D | D | D | 2.4 | 78 | 862 | 283.5 | 21.8 | 6 | 2 | 0.7 | 0.1 |
| Bay | 47.62 | 262.2 | 155 | 1,728 | 746.4 | 81.2 | 826 | 11,802 | 3,219.3 | 305.2 | 313 | 1,138 | 272.3 | 37.7 |
| Bradford | 0.95 | 35.3 | D | D | D | 2.1 | 79 | 1,022 | 343.3 | 27.1 | 18 | 40 | 11.0 | 0.9 |
| Brevard | 31.45 | 55.4 | 454 | 3,798 | 2,494.8 | 207.9 | 2,026 | 29,154 | 7,884.4 | 747.1 | 860 | 2,623 | 512.5 | 92.3 |
| Broward | 233.65 | 123.2 | 3,886 | 38,460 | 43,614.1 | 2,218.3 | 7,220 | 107,986 | 35,031.3 | 3,159.0 | 4,042 | 19,924 | 6,193.7 | 936.6 |
| Calhoun | 0.58 | 40.1 | D | D | D | 0.7 | 35 | 281 | 72.1 | 6.6 | 4 | 7 | 1.5 | 0.2 |
| Charlotte | 7.51 | 43.4 | 102 | 1,037 | 578.8 | 53.3 | 599 | 9,574 | 2,757.2 | 254.8 | 304 | 963 | 208.8 | 29.7 |
| Citrus | 14.15 | 100.3 | 74 | 527 | 272.7 | 24.8 | 431 | 5,633 | 1,864.0 | 157.5 | 163 | 1,014 | 94.7 | 21.7 |
| Clay | 13.51 | 66.2 | D | D | D | 32.3 | 569 | 8,396 | 2,097.1 | 209.5 | 230 | 894 | 177.6 | 29.4 |
| Collier | 51.82 | 145.0 | 309 | 3,273 | 3,862.1 | 269.5 | 1,499 | 22,521 | 7,302.9 | 693.1 | 1,189 | 3,758 | 863.4 | 184.6 |
| Columbia | 3.27 | 47.8 | D | D | D | D | 260 | 2,694 | 823.8 | 71.7 | 56 | 158 | 29.1 | 5.0 |
| DeSoto | 33.20 | 936.3 | 14 | 72 | 21.9 | 2.5 | 83 | 1,112 | 282.6 | 27.8 | 20 | 68 | 8.5 | 1.3 |
| Dixie | 0.62 | 38.3 | D | D | D | D | 37 | 247 | 73.7 | 5.3 | NA | NA | NA | NA |
| Duval | 109.03 | 119.4 | 1,154 | 22,455 | 20,621.7 | 1,393.3 | 3,381 | 53,267 | 17,312.7 | 1,472.3 | 1,494 | 8,805 | 3,144.4 | 489.2 |
| Escambia | 37.52 | 120.6 | 274 | 2,557 | 1,547.2 | 133.8 | 1,081 | 15,794 | 4,534.5 | 417.2 | 376 | 1,679 | 403.5 | 60.3 |
| Flagler | 9.17 | 87.0 | 56 | 281 | 138.6 | 14.6 | 261 | 3,821 | 1,087.4 | 97.7 | 168 | 426 | 92.4 | 16.0 |
| Franklin | 1.97 | 167.5 | 9 | 162 | 99.6 | 5.5 | 66 | 433 | 100.9 | 9.9 | 17 | 90 | 11.8 | 3.1 |
| Gadsden | 4.07 | 88.4 | 33 | 644 | 475.8 | 32.1 | 127 | 1,366 | 386.2 | 29.5 | 17 | 42 | 6.8 | 1.2 |
| Gilchrist | 0.23 | 13.4 | 14 | 80 | 72.0 | 3.4 | 34 | 183 | 51.8 | 3.5 | 6 | 6 | 1.2 | 0.1 |
| Glades | 0.51 | 37.3 | NA | NA | NA | NA | 14 | 88 | 28.7 | 1.8 | 3 | D | 1.7 | D |
| Gulf | 1.88 | 118.5 | 7 | 17 | 10.2 | 0.5 | 46 | 326 | 79.3 | 7.1 | 15 | 50 | 12.5 | 1.9 |
| Hamilton | 0.88 | 61.6 | NA | NA | NA | NA | 44 | 365 | 190.8 | 8.0 | 7 | 22 | 1.2 | 0.4 |
| Hardee | 1.64 | 59.6 | 19 | 132 | 95.2 | 6.9 | 71 | 717 | 187.0 | 15.8 | 18 | 39 | 6.1 | 1.0 |
| Hendry | 3.31 | 84.6 | 28 | 259 | 177.5 | 11.2 | 110 | 1,481 | 344.1 | 31.6 | 25 | 66 | 11.7 | 1.9 |
| Hernando | 18.25 | 102.3 | 83 | 451 | 209.3 | 20.8 | 473 | 7,896 | 2,036.5 | 191.0 | 187 | 465 | 87.1 | 12.6 |
| Highlands | 7.49 | 75.3 | 53 | 434 | 200.2 | 20.7 | 329 | 4,589 | 1,261.2 | 109.9 | 98 | 302 | 60.5 | 8.3 |
| Hillsborough | 196.38 | 145.6 | 1,739 | 27,830 | 24,464.4 | 1,608.7 | 4,661 | 74,470 | 24,583.3 | 2,111.0 | 2,377 | 11,801 | 4,040.4 | 577.0 |
| Holmes | 1.01 | 52.3 | D | D | D | D | 54 | 390 | 108.3 | 9.7 | 6 | 16 | 1.3 | 0.2 |
| Indian River | 16.94 | 114.5 | 133 | 856 | 5,125.0 | 96.6 | 684 | 9,009 | 2,267.7 | 222.7 | 276 | 1,569 | 257.8 | 52.5 |
| Jackson | 2.46 | 50.6 | 16 | 145 | 60.0 | 4.6 | 178 | 1,991 | 602.0 | 48.0 | 30 | 90 | 12.6 | 3.2 |
| Jefferson | 0.65 | 46.2 | D | D | D | D | 46 | 381 | 69.8 | 6.3 | 7 | 21 | 1.1 | 0.3 |
| Lafayette | 0.17 | 19.6 | 7 | 83 | 51.0 | 3.8 | 16 | 83 | 25.8 | 1.9 | NA | NA | NA | NA |
| Lake | 49.28 | 151.2 | 234 | 1,534 | 713.1 | 65.7 | 1,061 | 16,287 | 4,938.9 | 441.6 | 486 | 2,065 | 356.5 | 69.4 |
| Lee | 64.55 | 92.0 | 638 | 5,832 | 3,164.5 | 278.6 | 2,651 | 39,847 | 12,227.8 | 1,071.9 | 1,562 | 6,205 | 1,535.0 | 384.0 |
| Leon | 28.51 | 99.6 | 214 | 2,547 | 1,942.0 | 167.0 | 1,019 | 16,228 | 4,081.0 | 396.5 | 441 | 2,832 | 450.3 | 97.0 |
| Levy | 1.48 | 37.2 | 29 | 164 | 103.7 | 6.2 | 148 | 1,474 | 421.9 | 34.1 | 28 | 57 | 7.5 | 1.2 |
| Liberty | 0.46 | 55.2 | D | D | D | D | 18 | 157 | 28.2 | 2.3 | NA | NA | NA | NA |
| Madison | 1.23 | 66.8 | 9 | 48 | 45.3 | 3.6 | 59 | 543 | 157.4 | 11.3 | 8 | 19 | 2.4 | 0.6 |
| Manatee | 42.83 | 117.9 | 332 | 3,190 | 2,148.9 | 187.2 | 1,301 | 19,601 | 5,421.8 | 508.6 | 635 | 2,122 | 523.3 | 82.5 |
| Marion | 27.71 | 80.7 | 280 | 3,742 | 2,243.1 | 188.1 | 1,152 | 16,765 | 5,080.9 | 454.7 | 416 | 1,614 | 301.9 | 47.7 |
| Martin | 15.61 | 99.9 | 170 | 1,165 | 641.5 | 62.6 | 768 | 10,396 | 3,089.8 | 274.5 | 343 | 1,781 | 429.1 | 71.3 |
| Miami-Dade | 351.92 | 130.7 | 8,150 | 67,693 | 80,939.0 | 3,590.8 | 10,680 | 141,982 | 45,110.7 | 3,964.1 | 6,073 | 25,412 | 8,315.8 | 1,227.9 |
| Monroe | 0.00 | 0.0 | 76 | 303 | 514.7 | 13.9 | 626 | 6,105 | 1,688.5 | 159.9 | 383 | 1,135 | 290.6 | 39.4 |
| Nassau | 7.00 | 89.2 | D | D | D | 11.9 | 257 | 3,197 | 791.3 | 77.6 | 117 | 270 | 69.7 | 11.5 |
| Okaloosa | 21.81 | 109.8 | 125 | 610 | 252.0 | 29.0 | 871 | 12,309 | 3,609.7 | 322.4 | 413 | 1,900 | 398.5 | 73.7 |
| Okeechobee | 2.72 | 68.9 | 25 | 158 | 101.8 | 8.9 | 139 | 1,765 | 625.1 | 45.4 | 35 | 98 | 13.1 | 2.7 |
| Orange | 221.92 | 172.3 | 1,693 | 26,493 | 32,960.3 | 1,847.7 | 5,020 | 86,080 | 26,388.5 | 2,238.7 | 2,865 | 24,857 | 9,060.4 | 1,425.3 |
| Osceola | 42.19 | 130.2 | 147 | 1,852 | 4,306.5 | 82.8 | 969 | 15,629 | 4,128.2 | 368.4 | 590 | 4,330 | 842.0 | 147.0 |
| Palm Beach | 238.73 | 167.8 | 1,992 | 18,587 | 17,481.6 | 1,181.9 | 5,680 | 79,430 | 24,445.7 | 2,335.5 | 3,149 | 14,516 | 3,696.6 | 733.4 |
| Pasco | 61.15 | 122.8 | 296 | 1,822 | 1,004.9 | 92.7 | 1,511 | 23,252 | 7,098.7 | 615.0 | 565 | 2,135 | 367.5 | 58.0 |
| Pinellas | 23.86 | 25.1 | 1,141 | 12,087 | 7,620.0 | 677.8 | 3,732 | 53,060 | 16,891.1 | 1,530.8 | 1,928 | 8,946 | 1,635.7 | 348.8 |
| Polk | 67.54 | 103.9 | 528 | 7,950 | 13,505.1 | 402.1 | 1,793 | 27,302 | 8,333.4 | 726.9 | 727 | 3,332 | 734.1 | 127.7 |

1  Merchant wholesalers, except manufacturers' sales branches and offices.  2. Employer establishments.

Table B. States and Counties — **Professional Services, Manufacturing, and Accommodation and Food Services**

| STATE County | Professional, scientific, and technical services, 2017 | | | | Manufacturing, 2017 | | | | Accommodation and food services, 2017 | | | |
|---|---|---|---|---|---|---|---|---|---|---|---|---|
| | Number of establish-ments | Number of employees | Sales (mil dol) | Average payroll (mil dol) | Number of establish-ments | Number of employees | Sales (mil dol) | Average payroll (mil dol) | Number of establis-hments | Number of employees | Sales (mil dol) | Annual payroll (mil dol) |
| | 147 | 148 | 149 | 150 | 151 | 152 | 153 | 154 | 155 | 156 | 157 | 158 |
| CONNECTICUT—Cont'd | | | | | | | | | | | | |
| Tolland | D | D | D | D | 116 | 3,378 | 835.7 | 192.2 | 245 | 3,859 | 259.9 | 70.0 |
| Windham | D | D | D | D | 160 | 5,660 | 2,095.6 | 307.9 | 253 | 3,335 | 203.3 | 60.8 |
| DELAWARE | 2,979 | 29,589 | 6,967 | 2,934.2 | 558 | 27,536 | 16,689.2 | 1,543.9 | 2,141 | 39,950 | 2,554.4 | 714.1 |
| Kent | D | D | D | D | 70 | 4,635 | 2,139.3 | 210.0 | 285 | 6,916 | 476.3 | 115.0 |
| New Castle | D | D | D | D | 354 | 12,407 | 10,948.5 | 902.6 | 1,214 | 23,051 | 1,360.6 | 387.0 |
| Sussex | D | D | D | D | 134 | 10,494 | 3,601.4 | 431.3 | 642 | 9,983 | 717.5 | 212.1 |
| DISTRICT OF COLUMBIA | 5,593 | 98,614 | 34,765 | 12,275.9 | 113 | 1,092 | 277.4 | 49.8 | 2,733 | 72,890 | 6,739.8 | 2,060.2 |
| District of Columbia | 5,593 | 98,614 | 34,765 | 12,275.9 | 113 | 1,092 | 277.4 | 49.8 | 2,733 | 72,890 | 6,739.8 | 2,060.2 |
| FLORIDA | 79,028 | 508,370 | 88,908 | 34,319.6 | 13,471 | 296,389 | 106,341.9 | 16,575.6 | 42,071 | 955,006 | 67,950.4 | 18,462.3 |
| Alachua | 863 | 5,461 | 780 | 311.2 | 170 | 3,869 | 1,122.6 | 250.3 | 591 | 13,043 | 682.2 | 200.8 |
| Baker | 19 | 98 | 10 | 3.7 | 6 | 153 | 111.3 | 8.4 | 34 | 773 | 31.6 | 8.8 |
| Bay | D | D | D | D | 106 | 3,728 | 1,488.1 | 208.0 | 515 | 11,608 | 757.4 | 215.8 |
| Bradford | 32 | 101 | 10 | 3.6 | 13 | 397 | 127.5 | 20.0 | 41 | 696 | 34.5 | 9.2 |
| Brevard | 1,920 | 19,218 | 3,795 | 1,440.5 | 447 | 19,579 | 5,939.2 | 1,622.5 | 1,202 | 23,254 | 1,343.2 | 384.3 |
| Broward | 10,546 | 54,062 | 9,520 | 3,632.7 | 1,432 | 22,969 | 6,516.5 | 1,183.8 | 4,113 | 83,312 | 6,473.1 | 1,691.1 |
| Calhoun | 6 | 6 | 1 | 0.3 | 4 | 28 | 5.9 | 0.7 | D | D | D | D |
| Charlotte | D | D | D | D | 78 | 512 | 101.9 | 21.2 | 294 | 6,067 | 336.5 | 96.2 |
| Citrus | D | D | D | D | 55 | 338 | 75.1 | 13.6 | 235 | 3,513 | 175.3 | 50.8 |
| Clay | D | D | D | D | 79 | 892 | 329.0 | 47.6 | 293 | 6,148 | 310.6 | 93.8 |
| Collier | 1,602 | 5,768 | 1,104 | 381.5 | 209 | 3,397 | 797.0 | 173.7 | 881 | 24,719 | 1,957.8 | 527.6 |
| Columbia | D | D | D | D | 38 | 709 | 336.5 | 39.3 | 126 | 2,936 | 150.4 | 43.1 |
| DeSoto | 32 | 113 | 17 | 3.9 | 14 | 451 | 246.1 | 24.0 | 40 | 569 | 33.0 | 8.1 |
| Dixie | 15 | 32 | 2 | 0.7 | 10 | 587 | 126.5 | 19.5 | D | D | D | D |
| Duval | 3,408 | 37,455 | 6,724 | 2,668.0 | 593 | 22,768 | 11,136.8 | 1,349.6 | 2,195 | 46,414 | 2,773.1 | 772.1 |
| Escambia | D | D | D | D | 157 | 3,655 | 2,866.8 | 260.8 | 589 | 14,086 | 805.8 | 223.2 |
| Flagler | 223 | 591 | 87 | 31.9 | 44 | 582 | 135.7 | 24.1 | 180 | 3,451 | 250.8 | 61.1 |
| Franklin | 23 | 74 | 9 | 3.1 | NA | NA | NA | NA | 46 | 602 | 43.6 | 14.1 |
| Gadsden | D | D | D | D | 25 | 880 | 182.0 | 41.2 | 51 | 791 | 46.1 | 13.9 |
| Gilchrist | 17 | 44 | 4 | 1.3 | 8 | 128 | 53.4 | 5.6 | 21 | 272 | 20.1 | 4.8 |
| Glades | 6 | 17 | 1 | 0.3 | 7 | 168 | 126.2 | 9.8 | 9 | 120 | 5.1 | 1.3 |
| Gulf | 30 | 162 | 21 | 6.4 | 7 | 49 | 6.8 | 1.5 | 35 | 421 | 25.9 | 7.9 |
| Hamilton | 11 | 40 | 3 | 1.5 | D | D | D | D | D | D | D | D |
| Hardee | 20 | 67 | 6 | 2.0 | 10 | 325 | 134.1 | 13.4 | 31 | 433 | 23.2 | 5.0 |
| Hendry | D | D | D | D | D | 830 | D | 44.5 | 60 | 886 | 47.8 | 13.0 |
| Hernando | 307 | 1,340 | 155 | 52.1 | 82 | 1,876 | 451.1 | 78.2 | 270 | 5,154 | 254.6 | 75.8 |
| Highlands | D | D | D | D | 46 | 707 | 284.3 | 33.2 | 143 | 2,450 | 125.6 | 35.8 |
| Hillsborough | 5,832 | 61,813 | 11,447 | 4,706.0 | 801 | 20,022 | 11,450.1 | 1,076.0 | 2,514 | 57,750 | 4,268.3 | 1,098.7 |
| Holmes | 14 | 40 | 5 | 1.6 | 10 | 75 | 11.0 | 3.4 | D | D | D | D |
| Indian River | 518 | 2,076 | 287 | 116.0 | 87 | 1,538 | 362.3 | 82.6 | 282 | 5,660 | 334.0 | 97.0 |
| Jackson | 47 | 258 | 29 | 11.3 | 22 | 693 | 309.4 | 36.7 | 80 | 1,090 | 56.8 | 14.0 |
| Jefferson | 21 | 135 | 13 | 5.0 | D | D | D | D | 20 | 164 | 8.9 | 2.2 |
| Lafayette | D | D | 2 | D | 5 | 21 | 5.5 | 1.0 | D | D | D | D |
| Lake | 661 | 2,303 | 287 | 100.8 | 161 | 2,836 | 999.7 | 120.1 | 552 | 11,402 | 646.2 | 189.2 |
| Lee | D | D | 2,197 | D | 366 | 5,117 | 1,176.3 | 249.1 | 1,432 | 31,343 | 1,925.7 | 566.6 |
| Leon | D | D | D | D | 111 | 1,442 | 468.4 | 77.3 | 722 | 15,606 | 830.6 | 227.1 |
| Levy | 53 | 209 | 25 | 7.9 | 20 | 482 | 123.9 | 21.5 | 65 | 818 | 37.4 | 11.1 |
| Liberty | NA | NA | NA | NA | 4 | 290 | 197.6 | 17.4 | 4 | D | 1.7 | D |
| Madison | 15 | 111 | 15 | 6.5 | 13 | 401 | 238.1 | 17.4 | 22 | 304 | 15.9 | 4.6 |
| Manatee | D | D | D | D | 275 | 7,264 | 2,110.5 | 388.8 | 679 | 13,593 | 766.2 | 235.3 |
| Marion | D | D | D | D | 188 | 7,109 | 3,330.8 | 337.8 | 516 | 9,478 | 531.8 | 146.7 |
| Martin | 763 | 3,789 | 515 | 182.1 | 193 | 2,558 | 749.1 | 141.6 | 384 | 7,302 | 439.7 | 126.5 |
| Miami-Dade | 14,033 | 71,986 | 15,104 | 5,282.6 | 2,157 | 31,558 | 7,916.7 | 1,515.7 | 5,693 | 131,647 | 11,060.0 | 2,971.1 |
| Monroe | 393 | 981 | 155 | 49.4 | 68 | 264 | 68.4 | 14.0 | 542 | 12,466 | 1,229.2 | 341.4 |
| Nassau | 208 | 681 | 92 | 34.1 | 35 | 1,316 | 905.4 | 88.9 | 175 | 4,663 | 324.7 | 102.4 |
| Okaloosa | D | D | D | D | 93 | 1,565 | 481.9 | 81.1 | 520 | 11,250 | 694.9 | 199.7 |
| Okeechobee | 57 | 256 | 28 | 9.1 | 22 | 373 | 192.3 | 17.1 | 55 | 875 | 56.7 | 13.0 |
| Orange | 5,659 | 50,796 | 8,331 | 3,635.1 | 835 | 26,099 | 8,135.6 | 1,652.1 | 3,221 | 124,474 | 11,132.8 | 2,814.9 |
| Osceola | 523 | 2,029 | 265 | 90.6 | 98 | 1,197 | 523.2 | 54.9 | 684 | 18,989 | 1,500.4 | 390.8 |
| Palm Beach | 8,348 | 43,426 | 8,585 | 3,051.8 | 966 | 13,242 | 4,082.6 | 782.3 | 3,064 | 68,960 | 4,656.3 | 1,349.2 |
| Pasco | D | D | D | D | 235 | 2,934 | 828.5 | 147.6 | 648 | 12,747 | 701.1 | 198.8 |
| Pinellas | D | D | D | D | 1,027 | 28,004 | 8,845.1 | 1,582.5 | 2,294 | 45,742 | 3,056.6 | 840.0 |
| Polk | 1,077 | 6,200 | 780 | 299.4 | 408 | 14,431 | 8,312.3 | 772.9 | 893 | 18,045 | 1,019.0 | 272.7 |

# Health Care and Social Assistance, Other Services, Nonemployer Businesses, and Residential Construction

| STATE County | Health care and social assistance, 2017 | | | | Other services, 2017 | | | | Nonemployer businesses, 2018 | | Value of residential construction authorized by building permits, 2020 | |
|---|---|---|---|---|---|---|---|---|---|---|---|---|
| | Number of establishments | Number of employees | Receipts (mil dol) | Annual payroll (mil dol) | Number of establishments | Number of employees | Receipts (mil dol) | Annual payroll (mil dol) | Number | Receipts (mil dol) | New construction ($1,000) | Number of housing units |
| | 159 | 160 | 161 | 162 | 163 | 164 | 165 | 166 | 167 | 168 | 169 | 170 |
| CONNECTICUT—Cont'd | | | | | | | | | | | | |
| Tolland | 320 | 5,852 | 511.8 | 213.3 | 190 | 1,144 | 176.4 | 43.8 | 9,675 | 469.8 | 45,631 | 336 |
| Windham | 281 | 7,354 | 616.9 | 296.0 | 167 | 677 | 76.7 | 17.6 | 6,918 | 324.3 | 26,181 | 149 |
| DELAWARE | 2,648 | 68,348 | 8,855.3 | 3,720.5 | 1,608 | 10,252 | 1,220.8 | 339.1 | 68,623 | 4,504.8 | 1,099,377 | 8,455 |
| Kent | 433 | 10,827 | 1,248.9 | 497.3 | 248 | 1,249 | 124.2 | 37.3 | 11,282 | 774.2 | 289,710 | 2,136 |
| New Castle | 1,651 | 45,968 | 6,099.3 | 2,652.7 | 1,014 | 7,184 | 906.7 | 247.1 | 39,630 | 2,711.0 | 185,574 | 1,796 |
| Sussex | 564 | 11,553 | 1,507.1 | 570.5 | 346 | 1,819 | 189.9 | 54.7 | 17,711 | 1,019.6 | 624,094 | 4,523 |
| DISTRICT OF COLUMBIA. | 2,203 | 74,134 | 11,400.2 | 4,376.0 | 3,502 | 73,756 | 24,225.6 | 5,459.1 | 62,583 | 3,285.4 | 886,594 | 7,370 |
| District of Columbia | 2,203 | 74,134 | 11,400.2 | 4,376.0 | 3,502 | 73,756 | 24,225.6 | 5,459.1 | 62,583 | 3,285.4 | 886,594 | 7,370 |
| FLORIDA | 61,554 | 1,127,155 | 155,283.6 | 55,472.2 | 38,932 | 222,512 | 29,616.9 | 6,946.2 | 2,388,050 | 106,471.4 | 36,884,176 | 164,074 |
| Alachua | 807 | 26,051 | 3,922.9 | 1,462.7 | 413 | 3,892 | 568.1 | 126.6 | 19,727 | 790.7 | 258,850 | 1,767 |
| Baker | 37 | 1,882 | 128.0 | 80.8 | D | D | D | 3.0 | 1,417 | 53.9 | 27,844 | 152 |
| Bay | 549 | 10,448 | 1,288.9 | 495.7 | 319 | 1,922 | 176.4 | 47.3 | 14,652 | 735.6 | 473,592 | 2,846 |
| Bradford | 40 | 822 | 83.6 | 26.9 | 26 | 89 | 9.3 | 2.4 | 1,272 | 47.0 | 16,308 | 74 |
| Brevard | 1,600 | 32,031 | 3,775.1 | 1,369.8 | 1,023 | 4,793 | 466.5 | 135.2 | 46,350 | 1,883.5 | 1,284,438 | 4,739 |
| Broward | 6,614 | 95,619 | 13,231.2 | 4,736.4 | 4,538 | 22,941 | 2,920.0 | 750.2 | 297,518 | 12,828.7 | 776,309 | 4,428 |
| Calhoun | 22 | 632 | 44.5 | 21.1 | D | D | D | D | 820 | 29.6 | 1,527 | 14 |
| Charlotte | 514 | 8,529 | 1,259.5 | 400.0 | 335 | 1,418 | 146.7 | 39.7 | 14,383 | 681.5 | 763,035 | 2,933 |
| Citrus | 393 | 9,117 | 940.2 | 381.0 | 225 | 964 | 70.3 | 19.9 | 10,082 | 402.8 | 184,925 | 927 |
| Clay | 512 | 9,445 | 1,174.4 | 437.7 | D | D | D | 36.5 | 14,765 | 583.1 | 278,574 | 1,432 |
| Collier | 1,143 | 20,535 | 2,704.8 | 980.5 | 1,011 | 6,309 | 662.8 | 191.7 | 43,851 | 2,768.9 | 1,471,129 | 4,473 |
| Columbia | 185 | 4,270 | 584.6 | 244.9 | 83 | 310 | 33.5 | 7.6 | 4,267 | 205.7 | 26,369 | 194 |
| DeSoto | 46 | 1,140 | 102.5 | 43.3 | 26 | 95 | 8.7 | 2.5 | 1,714 | 78.2 | 17,191 | 110 |
| Dixie | 11 | 148 | 9.2 | 3.7 | D | D | 4.4 | D | 858 | 34.8 | 4,055 | 43 |
| Duval | 2,861 | 71,476 | 10,405.2 | 3,809.1 | 1,791 | 11,787 | 1,591.2 | 408.2 | 78,330 | 3,132.1 | 1,248,861 | 9,134 |
| Escambia | 834 | 21,902 | 2,968.5 | 1,142.0 | 453 | 3,230 | 312.0 | 87.1 | 22,950 | 975.0 | 479,734 | 2,210 |
| Flagler | 230 | 3,762 | 499.9 | 166.6 | 162 | 670 | 61.5 | 16.4 | 10,029 | 445.3 | 487,565 | 1,862 |
| Franklin | 19 | 163 | 16.0 | 6.9 | 24 | 60 | 5.8 | 1.9 | 1,332 | 58.3 | 20,590 | 94 |
| Gadsden | 56 | 2,329 | 163.9 | 108.4 | 36 | 115 | 13.3 | 3.5 | 3,101 | 97.7 | 2,851 | 69 |
| Gilchrist | 21 | 621 | 61.7 | 24.9 | 13 | 30 | 4.3 | 1.0 | 1,052 | 44.4 | 12,278 | 83 |
| Glades | D | D | D | D | 6 | 18 | 1.4 | 0.3 | 600 | 20.8 | 7,486 | 43 |
| Gulf | 25 | 225 | 14.4 | 6.5 | 12 | 41 | 4.0 | 0.6 | 1,211 | 61.1 | 66,299 | 242 |
| Hamilton | D | D | D | 6.1 | D | D | D | D | 626 | 22.3 | 3,432 | 22 |
| Hardee | 64 | 1,096 | 84.1 | 38.3 | 26 | 63 | 5.6 | 1.7 | 1,411 | 56.6 | 9,069 | 50 |
| Hendry | 61 | 1,261 | 99.7 | 45.4 | 44 | 170 | 20.3 | 4.7 | 3,150 | 125.9 | 60,848 | 391 |
| Hernando | 550 | 8,989 | 1,247.9 | 394.8 | 237 | 1,223 | 96.9 | 30.4 | 12,819 | 512.7 | 249,272 | 1,100 |
| Highlands | 309 | 6,442 | 729.3 | 284.6 | 125 | 403 | 39.6 | 9.5 | 6,295 | 266.0 | 85,682 | 384 |
| Hillsborough | 4,107 | 83,163 | 13,642.4 | 4,492.1 | 2,364 | 17,128 | 2,235.3 | 505.5 | 140,515 | 6,246.8 | 3,553,555 | 12,392 |
| Holmes | 28 | 509 | 39.2 | 14.9 | D | D | D | D | 1,214 | 41.9 | 1,464 | 11 |
| Indian River | 495 | 9,718 | 1,155.2 | 454.1 | 313 | 1,726 | 167.0 | 51.2 | 14,992 | 815.1 | 462,031 | 1,162 |
| Jackson | 83 | 1,723 | 164.4 | 70.4 | 52 | 236 | 28.4 | 7.1 | 2,735 | 94.3 | 24,539 | 140 |
| Jefferson | D | D | D | D | 13 | 48 | 2.8 | 1.0 | 1,101 | 38.6 | 14,635 | 70 |
| Lafayette | 10 | 132 | 8.6 | 3.5 | D | D | 1.0 | D | 357 | 14.5 | 2,454 | 34 |
| Lake | 950 | 19,135 | 2,211.1 | 823.4 | 521 | 2,529 | 250.2 | 67.7 | 28,499 | 1,156.6 | 992,065 | 3,923 |
| Lee | 1,700 | 38,678 | 5,215.6 | 2,158.8 | 1,425 | 7,305 | 834.2 | 222.8 | 73,304 | 3,642.3 | 1,944,756 | 10,673 |
| Leon | 783 | 18,888 | 2,413.6 | 941.8 | 615 | 4,473 | 728.7 | 194.5 | 22,644 | 908.4 | 296,997 | 1,897 |
| Levy | 63 | 687 | 46.6 | 21.5 | 52 | 151 | 16.8 | 3.6 | 2,959 | 122.1 | 23,144 | 230 |
| Liberty | D | D | D | D | D | D | D | 0.4 | 420 | 14.7 | 100 | 1 |
| Madison | 41 | 684 | 58.3 | 24.0 | D | D | D | D | 1,090 | 43.1 | 8,780 | 43 |
| Manatee | 963 | 16,571 | 1,943.6 | 725.7 | 656 | 2,970 | 301.2 | 85.3 | 34,301 | 1,674.6 | 1,005,174 | 5,052 |
| Marion | 909 | 15,902 | 2,033.6 | 716.2 | 516 | 2,545 | 263.3 | 71.9 | 27,086 | 1,189.9 | 573,152 | 3,827 |
| Martin | 585 | 11,069 | 1,257.1 | 517.8 | 430 | 2,212 | 238.5 | 70.0 | 17,112 | 988.5 | 261,228 | 544 |
| Miami-Dade | 9,159 | 144,441 | 21,236.1 | 7,238.0 | 5,335 | 31,051 | 3,952.6 | 909.2 | 557,833 | 23,858.8 | 1,772,918 | 9,831 |
| Monroe | 228 | 2,093 | 341.7 | 110.2 | 284 | 1,121 | 160.4 | 39.1 | 13,401 | 811.1 | 123,189 | 377 |
| Nassau | 186 | 2,449 | 267.9 | 97.1 | 137 | 543 | 66.9 | 19.5 | 6,809 | 309.2 | 330,165 | 1,438 |
| Okaloosa | 590 | 8,400 | 1,141.5 | 397.9 | 371 | 1,496 | 158.2 | 43.4 | 17,467 | 882.8 | 308,202 | 993 |
| Okeechobee | 117 | 1,510 | 214.1 | 71.3 | 69 | 284 | 33.4 | 9.2 | 2,421 | 95.5 | 2,932 | 15 |
| Orange | 3,581 | 74,659 | 11,145.8 | 3,816.0 | 2,327 | 19,594 | 3,802.3 | 664.1 | 155,760 | 6,416.7 | 2,718,839 | 12,196 |
| Osceola | 671 | 12,936 | 2,032.1 | 663.3 | 424 | 2,078 | 219.4 | 59.9 | 40,299 | 1,448.6 | 1,172,213 | 5,848 |
| Palm Beach | 5,903 | 87,245 | 12,287.3 | 4,300.0 | 3,725 | 20,953 | 2,620.0 | 676.6 | 197,199 | 10,093.6 | 2,212,910 | 7,499 |
| Pasco | 1,320 | 21,001 | 2,968.7 | 1,027.3 | 702 | 3,131 | 327.1 | 91.7 | 41,689 | 1,672.5 | 1,103,940 | 5,373 |
| Pinellas | 3,506 | 72,361 | 9,734.2 | 3,602.0 | 2,132 | 11,092 | 1,237.4 | 342.0 | 91,910 | 4,447.9 | 392,761 | 1,483 |
| Polk | 1,148 | 27,693 | 3,505.9 | 1,282.2 | 717 | 3,805 | 460.2 | 122.5 | 49,619 | 1,941.6 | 1,749,382 | 9,492 |

# Table B. States and Counties — Government Employment and Payroll, and Local Government Finances

| | Government employment and payroll, 2017 | | | | | | | | | Local government finances, 2017 | | | | |
| | | | March payroll (percent of total) | | | | | | | General revenue | | | | |
| STATE County | Full-time equivalent employees | March payroll (dollars) | Administra-tion, judicial, and legal | Police and corrections | Fire protection | Highways and transpor-tation | Health and welfare | Natural resources and utilities | Education and libraries | Total (mil dol) | Inter-govern-mental (mil dol) | Taxes Total (mil dol) | Taxes Per capita[1] (dollars) Total | Per capita[1] (dollars) Property |
| | 171 | 172 | 173 | 174 | 175 | 176 | 177 | 178 | 179 | 180 | 181 | 182 | 183 | 184 |
| CONNECTICUT—Cont'd | | | | | | | | | | | | | | |
| Tolland | 4,812 | 22,573,034 | 5.3 | 3.0 | 1.3 | 4.4 | 2.1 | 2.9 | 79.5 | 565.4 | 166.0 | 355.5 | 2,354 | 2,328 |
| Windham | 4,280 | 19,365,768 | 4.5 | 2.2 | 1.3 | 3.7 | 1.6 | 2.2 | 82.2 | 470.0 | 202.8 | 232.7 | 1,999 | 1,987 |
| DELAWARE | X | X | X | X | X | X | X | X | X | X | X | X | X | X |
| Kent | 4,452 | 19,533,638 | 5.1 | 6.3 | 0.1 | 0.8 | 1.8 | 5.1 | 79.4 | 513.1 | 312.8 | 109.2 | 619 | 544 |
| New Castle | 13,758 | 68,584,293 | 5.3 | 10.4 | 1.5 | 3.5 | 2.2 | 4.2 | 71.3 | 2025.0 | 893.9 | 754.9 | 1,358 | 1,110 |
| Sussex | 5,620 | 25,192,097 | 5.1 | 4.7 | 0.0 | 1.2 | 3.0 | 4.1 | 80.6 | 778.5 | 394.2 | 223.2 | 995 | 758 |
| DISTRICT OF COLUMBIA. | X | X | X | X | X | X | X | X | X | X | X | X | X | X |
| District of Columbia | 52,851 | 365,723,937 | 11.0 | 13.7 | 4.0 | 26.1 | 14.2 | 7.0 | 16.9 | 13246.4 | 4230.7 | 7455.9 | 10,729 | 3,500 |
| FLORIDA | X | X | X | X | X | X | X | X | X | X | X | X | X | X |
| Alachua | 9,388 | 36,390,714 | 11.6 | 16.6 | 5.0 | 5.0 | 3.8 | 11.7 | 43.8 | 964.9 | 331.4 | 356.8 | 1,340 | 1,075 |
| Baker | 895 | 2,640,778 | 6.4 | 3.9 | 1.5 | 2.9 | 2.3 | 6.2 | 76.3 | 95.3 | 47.5 | 19.8 | 699 | 497 |
| Bay | 8,252 | 28,582,062 | 5.1 | 9.8 | 2.7 | 3.8 | 24.7 | 5.2 | 47.3 | 727.1 | 246.4 | 300.1 | 1,625 | 1,070 |
| Bradford | 829 | 2,600,500 | 11.9 | 12.6 | 2.6 | 2.7 | 5.3 | 3.1 | 58.3 | 70.2 | 37.7 | 20.1 | 742 | 535 |
| Brevard | 19,876 | 71,754,734 | 7.2 | 12.8 | 6.8 | 4.6 | 7.5 | 8.1 | 51.1 | 2102.4 | 685.7 | 749.1 | 1,275 | 944 |
| Broward | 78,792 | 389,220,424 | 5.0 | 13.2 | 5.4 | 3.7 | 32.3 | 6.1 | 32.6 | 12006.7 | 2460.8 | 3819.2 | 1,974 | 1,586 |
| Calhoun | 513 | 1,282,306 | 5.4 | 4.0 | 0.2 | 6.0 | 0.0 | 4.7 | 79.5 | 51.3 | 38.9 | 9.1 | 628 | 475 |
| Charlotte | 4,751 | 17,974,279 | 14.2 | 16.9 | 10.5 | 6.1 | 1.4 | 9.3 | 39.5 | 618.7 | 101.2 | 291.7 | 1,607 | 1,248 |
| Citrus | 3,578 | 11,361,255 | 9.3 | 10.8 | 0.3 | 3.5 | 2.1 | 4.9 | 64.4 | 336.2 | 116.8 | 150.9 | 1,038 | 922 |
| Clay | 6,503 | 21,317,220 | 4.6 | 11.7 | 4.3 | 1.2 | 0.2 | 4.6 | 71.4 | 605.5 | 265.1 | 186.8 | 880 | 650 |
| Collier | 10,447 | 46,883,097 | 6.7 | 17.1 | 7.6 | 3.2 | 5.8 | 8.0 | 48.7 | 1526.6 | 242.6 | 884.1 | 2,372 | 2,131 |
| Columbia | 2,321 | 6,980,424 | 2.7 | 2.9 | 4.1 | 4.8 | 0.6 | 4.2 | 74.5 | 216.5 | 115.2 | 60.4 | 863 | 586 |
| DeSoto | 1,357 | 4,556,679 | 3.8 | 10.1 | 5.2 | 2.6 | 26.0 | 5.2 | 46.1 | 131.5 | 47.3 | 34.1 | 916 | 685 |
| Dixie | 497 | 1,302,803 | 5.3 | 1.2 | 1.4 | 6.5 | 9.6 | 7.5 | 67.8 | 48.8 | 29.9 | 12.2 | 734 | 608 |
| Duval | 24,856 | 100,025,694 | 5.9 | 14.4 | 6.6 | 2.8 | 1.4 | 8.4 | 55.1 | 3573.5 | 1256.5 | 1424.5 | 1,519 | 1,040 |
| Escambia | 10,448 | 35,675,524 | 5.8 | 14.8 | 3.2 | 2.8 | 2.7 | 10.5 | 57.0 | 1196.2 | 415.4 | 375.5 | 1,199 | 750 |
| Flagler | 3,121 | 11,214,768 | 13.1 | 10.3 | 7.2 | 4.8 | 1.7 | 8.4 | 51.9 | 340.4 | 100.6 | 160.2 | 1,457 | 1,232 |
| Franklin | 406 | 1,170,228 | 7.7 | 10.5 | 1.8 | 8.5 | 3.3 | 13.4 | 50.7 | 60.5 | 18.7 | 27.8 | 2,370 | 1,981 |
| Gadsden | 1,599 | 4,603,867 | 10.2 | 11.9 | 2.1 | 7.2 | 5.1 | 4.2 | 56.9 | 110.0 | 62.3 | 31.7 | 690 | 492 |
| Gilchrist | 601 | 1,591,452 | 4.9 | 0.6 | 0.6 | 3.2 | 1.1 | 3.5 | 77.7 | 55.7 | 34.8 | 15.1 | 842 | 615 |
| Glades | 599 | 2,023,704 | 8.3 | 23.3 | 0.3 | 2.4 | 3.4 | 5.9 | 55.1 | 79.2 | 22.4 | 12.5 | 922 | 817 |
| Gulf | 574 | 1,787,849 | 13.1 | 12.7 | 0.2 | 4.9 | 5.8 | 8.7 | 51.0 | 59.9 | 21.9 | 28.8 | 1,790 | 1,416 |
| Hamilton | 590 | 1,857,007 | 5.1 | 15.9 | 0.6 | 5.0 | 3.6 | 21.4 | 45.0 | 68.3 | 36.7 | 24.4 | 1,695 | 1,312 |
| Hardee | 748 | 2,319,628 | 5.9 | 4.0 | 0.0 | 0.8 | 0.2 | 5.0 | 83.5 | 105.5 | 53.9 | 30.9 | 1,138 | 955 |
| Hendry | 1,627 | 5,855,789 | 10.9 | 8.7 | 1.1 | 3.8 | 23.3 | 9.8 | 40.2 | 185.4 | 78.0 | 54.4 | 1,327 | 982 |
| Hernando | 5,021 | 18,120,327 | 6.4 | 11.5 | 8.6 | 1.9 | 0.5 | 19.0 | 50.2 | 540.1 | 171.4 | 256.1 | 1,371 | 1,219 |
| Highlands | 3,221 | 10,134,596 | 7.5 | 15.0 | 2.0 | 3.4 | 1.9 | 4.9 | 63.1 | 302.8 | 127.4 | 101.8 | 980 | 722 |
| Hillsborough | 46,480 | 191,019,225 | 6.7 | 13.8 | 5.9 | 5.6 | 3.1 | 6.8 | 55.7 | 5912.8 | 2418.9 | 2081.3 | 1,459 | 1,059 |
| Holmes | 866 | 2,311,727 | 6.1 | 5.9 | 0.0 | 4.4 | 24.9 | 2.4 | 53.6 | 77.3 | 49.3 | 9.9 | 512 | 379 |
| Indian River | 4,114 | 16,229,719 | 5.2 | 17.3 | 4.8 | 4.9 | 3.8 | 12.9 | 48.2 | 515.4 | 108.8 | 297.9 | 1,931 | 1,523 |
| Jackson | 2,613 | 8,483,063 | 4.3 | 5.6 | 2.7 | 2.7 | 34.0 | 2.4 | 46.1 | 239.2 | 100.8 | 41.7 | 863 | 469 |
| Jefferson | 325 | 834,386 | 9.1 | 5.8 | 2.5 | 10.1 | 6.7 | 9.7 | 53.0 | 32.7 | 12.5 | 13.6 | 957 | 634 |
| Lafayette | 289 | 851,769 | 6.4 | 11.1 | 0.0 | 3.3 | 2.5 | 3.6 | 68.0 | 21.4 | 14.4 | 4.6 | 536 | 472 |
| Lake | 10,219 | 34,465,237 | 6.4 | 13.3 | 7.7 | 2.4 | 0.9 | 8.0 | 57.1 | 1031.6 | 329.0 | 396.8 | 1,149 | 840 |
| Lee | 33,112 | 154,572,301 | 3.8 | 7.4 | 4.9 | 3.0 | 47.3 | 4.2 | 28.2 | 4646.6 | 703.6 | 1284.1 | 1,735 | 1,464 |
| Leon | 10,881 | 43,682,389 | 11.3 | 12.5 | 4.1 | 4.4 | 3.2 | 16.6 | 43.8 | 1141.1 | 385.1 | 396.5 | 1,363 | 930 |
| Levy | 1,384 | 4,036,870 | 4.9 | 14.1 | 1.5 | 5.6 | 6.2 | 3.4 | 60.8 | 115.5 | 56.7 | 36.2 | 899 | 697 |
| Liberty | 361 | 949,943 | 2.1 | 0.1 | 0.2 | 4.3 | 5.4 | 3.4 | 83.8 | 28.2 | 20.4 | 4.9 | 596 | 460 |
| Madison | 966 | 2,812,310 | 4.8 | 8.6 | 1.3 | 3.2 | 16.6 | 4.4 | 57.9 | 81.8 | 42.0 | 17.7 | 958 | 662 |
| Manatee | 11,742 | 43,798,961 | 8.5 | 15.5 | 5.9 | 4.3 | 2.7 | 5.6 | 54.2 | 1621.6 | 387.7 | 743.5 | 1,929 | 1,635 |
| Marion | 10,693 | 35,122,480 | 4.8 | 11.0 | 9.8 | 2.1 | 0.8 | 6.7 | 62.3 | 962.8 | 383.9 | 311.7 | 882 | 738 |
| Martin | 4,598 | 17,470,072 | 9.1 | 12.8 | 13.8 | 2.5 | 0.8 | 6.9 | 49.2 | 590.0 | 114.2 | 354.1 | 2,217 | 1,997 |
| Miami-Dade | 93,991 | 469,157,165 | 7.3 | 16.4 | 6.7 | 8.4 | 14.5 | 8.1 | 35.3 | 14969.4 | 3265.1 | 6118.7 | 2,255 | 1,664 |
| Monroe | 3,903 | 18,227,562 | 9.3 | 19.0 | 8.7 | 3.9 | 4.4 | 19.6 | 31.8 | 657.8 | 128.3 | 317.3 | 4,149 | 2,611 |
| Nassau | 2,061 | 7,376,121 | 4.4 | 2.6 | 12.3 | 4.2 | 1.0 | 5.2 | 65.5 | 256.9 | 67.6 | 142.4 | 1,717 | 1,372 |
| Okaloosa | 7,072 | 25,149,599 | 5.4 | 11.0 | 5.4 | 3.1 | 2.3 | 7.6 | 62.0 | 694.5 | 252.3 | 258.8 | 1,272 | 1,009 |
| Okeechobee | 1,602 | 4,781,680 | 7.2 | 16.2 | 6.5 | 2.6 | 1.2 | 4.1 | 58.1 | 127.6 | 62.5 | 47.6 | 1,153 | 668 |
| Orange | 47,356 | 187,768,001 | 6.4 | 13.0 | 7.6 | 5.4 | 2.5 | 9.0 | 50.9 | 6687.2 | 1776.9 | 2741.4 | 2,022 | 1,417 |
| Osceola | 12,048 | 43,029,385 | 7.4 | 13.3 | 6.6 | 2.3 | 0.7 | 9.0 | 57.0 | 1423.1 | 490.2 | 507.4 | 1,435 | 910 |
| Palm Beach | 49,998 | 226,983,697 | 8.2 | 17.4 | 9.4 | 2.1 | 4.2 | 12.1 | 43.4 | 7208.5 | 1451.4 | 3928.2 | 2,672 | 2,261 |
| Pasco | 15,712 | 51,647,343 | 6.3 | 11.7 | 7.1 | 2.3 | 2.7 | 5.6 | 62.3 | 1509.9 | 621.2 | 521.2 | 993 | 718 |
| Pinellas | 31,247 | 129,067,444 | 9.1 | 18.1 | 6.3 | 3.6 | 2.9 | 12.1 | 43.4 | 3760.3 | 955.5 | 1724.8 | 1,781 | 1,359 |
| Polk | 24,085 | 82,688,018 | 8.2 | 12.8 | 5.3 | 2.2 | 2.2 | 10.4 | 55.7 | 2121.7 | 833.2 | 708.7 | 1,034 | 683 |

1. Based on the resident population estimated as of July 1 of the year shown.

# Local Government Finances, Government Employment, and Income Taxes

| STATE County | Local government finances, 2017 (cont.) | | | | | | | | | Government employment, 2019 | | | Individual income tax returns, 2018 | | |
|---|---|---|---|---|---|---|---|---|---|---|---|---|---|---|---|
| | Direct general expenditure | | | | | | | Debt outstanding | | | | | | | |
| | Total (mil dol) | Per capita[1] (dollars) | Percent of total for: | | | | | Total (mil dol) | Per capita[1] (dollars) | Federal civilian | Federal military | State and local | Number of returns | Mean adjusted gross income | Mean income tax |
| | | | Education | Health and hospitals | Police protection | Public welfare | Highways | | | | | | | | |
| | 185 | 186 | 187 | 188 | 189 | 190 | 191 | 192 | 193 | 194 | 195 | 196 | 197 | 198 | 199 |
| **CONNECTICUT—Cont'd** | | | | | | | | | | | | | | | |
| Tolland | 564 | 3,733 | 64.8 | 0.6 | 2.3 | 0.5 | 5.2 | 175.1 | 1,160 | 271 | 291 | 16,831 | 68,970 | 84,333 | 10,268 |
| Windham | 478 | 4,105 | 72.9 | 0.8 | 2.8 | 0.2 | 3.8 | 122.9 | 1,056 | 287 | 226 | 6,951 | 55,500 | 60,023 | 5,763 |
| **DELAWARE** | X | X | X | X | X | X | X | X | X | 5,751 | 8,681 | 60,601 | 469,840 | 70,372 | 8,195 |
| Kent | 521 | 2,950 | 74.0 | 1.3 | 5.4 | 0.0 | 1.4 | 371.1 | 2,103 | 1,679 | 4,429 | 17,451 | 83,340 | 54,201 | 4,807 |
| New Castle | 2,034 | 3,657 | 59.7 | 0.9 | 8.5 | 0.0 | 4.7 | 1,908.6 | 3,433 | 3,425 | 2,961 | 34,704 | 271,300 | 76,549 | 9,489 |
| Sussex | 757 | 3,375 | 71.3 | 1.9 | 4.1 | 0.1 | 1.5 | 432.8 | 1,929 | 647 | 1,291 | 8,446 | 115,210 | 67,516 | 7,597 |
| **DISTRICT OF COLUMBIA.** | X | X | X | X | X | X | X | X | X | 191,495 | 14,158 | 42,658 | 349,990 | 105,463 | 17,764 |
| District of Columbia | 13,255 | 19,075 | 21.9 | 5.0 | 4.8 | 29.6 | 3.4 | 17,640.3 | 25,385 | 191,495 | 14,158 | 42,658 | 349,990 | 105,463 | 17,764 |
| **FLORIDA** | X | X | X | X | X | X | X | X | X | 143,148 | 92,838 | 968,609 | 10,228,040 | 73,560 | 10,140 |
| Alachua | 994 | 3,733 | 37.3 | 2.9 | 7.9 | 1.7 | 4.3 | 2,434.8 | 9,143 | 4,627 | 527 | 37,122 | 115,560 | 66,619 | 8,237 |
| Baker | 101 | 3,561 | 44.1 | 4.7 | 9.3 | 0.8 | 4.2 | 33.7 | 1,194 | 66 | 46 | 2,673 | 11,110 | 52,848 | 4,253 |
| Bay | 719 | 3,891 | 43.3 | 2.6 | 8.1 | 0.0 | 4.5 | 596.2 | 3,227 | 3,787 | 2,626 | 8,801 | 82,060 | 59,521 | 6,572 |
| Bradford | 69 | 2,530 | 49.0 | 5.7 | 10.0 | 0.0 | 7.2 | 8.0 | 293 | 30 | 46 | 1,961 | 10,310 | 47,373 | 3,968 |
| Brevard | 2,232 | 3,797 | 35.9 | 9.2 | 7.0 | 0.2 | 4.6 | 1,844.7 | 3,138 | 6,576 | 2,901 | 22,506 | 292,710 | 65,100 | 7,575 |
| Broward | 11,894 | 6,148 | 25.1 | 28.0 | 9.9 | 1.2 | 1.3 | 11,449.4 | 5,919 | 7,150 | 3,920 | 100,925 | 970,530 | 72,413 | 10,195 |
| Calhoun | 62 | 4,282 | 56.7 | 1.5 | 4.7 | 0.5 | 19.2 | 0.6 | 43 | 24 | 22 | 907 | 4,880 | 45,432 | 3,497 |
| Charlotte | 621 | 3,423 | 26.1 | 3.8 | 10.9 | 1.3 | 15.4 | 578.0 | 3,184 | 350 | 325 | 5,770 | 88,530 | 62,218 | 6,737 |
| Citrus | 373 | 2,562 | 41.6 | 6.4 | 9.6 | 0.6 | 8.5 | 478.1 | 3,288 | 253 | 256 | 4,162 | 68,660 | 53,580 | 5,401 |
| Clay | 587 | 2,768 | 53.4 | 2.7 | 8.2 | 0.7 | 5.0 | 436.2 | 2,055 | 388 | 381 | 6,857 | 102,940 | 62,927 | 6,223 |
| Collier | 1,423 | 3,817 | 38.1 | 3.4 | 12.7 | 0.5 | 5.2 | 1,944.8 | 5,219 | 756 | 661 | 13,159 | 190,930 | 158,523 | 29,256 |
| Columbia | 209 | 2,992 | 55.4 | 2.3 | 6.7 | 0.2 | 8.3 | 130.0 | 1,857 | 1,229 | 116 | 4,318 | 28,620 | 47,008 | 4,014 |
| DeSoto | 134 | 3,610 | 36.2 | 25.9 | 5.7 | 0.4 | 3.7 | 54.6 | 1,465 | 56 | 60 | 1,727 | 12,660 | 44,093 | 3,563 |
| Dixie | 53 | 3,195 | 50.3 | 9.1 | 6.4 | 0.0 | 10.6 | 12.1 | 726 | 10 | 27 | 1,003 | 5,220 | 46,139 | 4,451 |
| Duval | 3,287 | 3,504 | 42.7 | 2.5 | 11.3 | 0.4 | 1.6 | 9,891.8 | 10,546 | 17,407 | 12,396 | 37,822 | 467,290 | 63,203 | 7,530 |
| Escambia | 1,281 | 4,088 | 37.7 | 1.8 | 6.4 | 0.0 | 3.6 | 4,176.3 | 13,332 | 5,923 | 10,828 | 15,418 | 148,140 | 57,976 | 6,463 |
| Flagler | 359 | 3,261 | 34.5 | 1.7 | 6.0 | 0.2 | 9.5 | 505.1 | 4,591 | 164 | 199 | 3,666 | 55,430 | 66,563 | 8,071 |
| Franklin | 58 | 4,946 | 27.5 | 17.3 | 8.2 | 0.2 | 9.3 | 25.0 | 2,129 | 16 | 18 | 984 | 4,690 | 64,978 | 8,214 |
| Gadsden | 124 | 2,689 | 49.6 | 4.8 | 7.9 | 1.3 | 8.7 | 49.8 | 1,083 | 92 | 74 | 4,192 | 20,000 | 42,223 | 3,204 |
| Gilchrist | 54 | 3,024 | 56.1 | 5.3 | 5.1 | 0.6 | 9.7 | 4.8 | 268 | 31 | 30 | 1,068 | 6,900 | 49,608 | 4,266 |
| Glades | 78 | 5,754 | 22.3 | 0.9 | 5.1 | 0.0 | 5.2 | 171.7 | 12,643 | 10 | 21 | 533 | 4,190 | 52,777 | 5,533 |
| Gulf | 79 | 4,912 | 27.1 | 29.7 | 4.9 | 0.1 | 10.6 | 36.5 | 2,265 | 12 | 23 | 975 | 5,810 | 60,439 | 6,103 |
| Hamilton | 88 | 6,146 | 45.8 | 2.8 | 3.9 | 0.2 | 3.2 | 3.3 | 227 | 30 | 20 | 1,082 | 4,740 | 43,147 | 3,131 |
| Hardee | 111 | 4,075 | 44.6 | 4.3 | 9.9 | 0.2 | 4.8 | 20.9 | 768 | 54 | 44 | 1,639 | 9,780 | 42,520 | 3,402 |
| Hendry | 192 | 4,691 | 37.5 | 18.0 | 6.8 | 0.4 | 7.5 | 76.9 | 1,875 | 78 | 148 | 2,152 | 16,590 | 45,617 | 4,323 |
| Hernando | 567 | 3,037 | 35.2 | 3.4 | 8.3 | 0.1 | 4.7 | 301.1 | 1,613 | 370 | 336 | 5,511 | 85,980 | 49,124 | 4,397 |
| Highlands | 318 | 3,058 | 44.6 | 3.2 | 10.1 | 0.6 | 5.4 | 109.0 | 1,050 | 285 | 183 | 3,969 | 43,010 | 45,263 | 3,849 |
| Hillsborough | 5,978 | 4,190 | 37.7 | 2.8 | 6.8 | 0.7 | 2.7 | 6,717.3 | 4,708 | 15,671 | 8,335 | 65,073 | 681,230 | 70,742 | 9,146 |
| Holmes | 101 | 5,220 | 56.8 | 13.7 | 3.9 | 0.1 | 6.1 | 24.5 | 1,260 | 66 | 35 | 1,339 | 6,910 | 41,885 | 3,061 |
| Indian River | 534 | 3,462 | 35.9 | 4.1 | 7.3 | 0.9 | 7.3 | 253.7 | 1,645 | 390 | 276 | 4,711 | 78,500 | 106,913 | 17,196 |
| Jackson | 232 | 4,797 | 36.3 | 34.9 | 3.4 | 0.0 | 8.5 | 47.9 | 992 | 427 | 70 | 4,583 | 18,350 | 45,867 | 3,869 |
| Jefferson | 33 | 2,315 | 29.8 | 7.0 | 11.4 | 0.8 | 7.6 | 12.4 | 874 | 33 | 23 | 522 | 6,010 | 56,015 | 5,341 |
| Lafayette | 21 | 2,485 | 56.6 | 5.8 | 7.8 | 0.1 | 11.7 | 0.2 | 24 | 13 | 12 | 583 | 2,540 | 42,744 | 2,978 |
| Lake | 1,055 | 3,055 | 41.1 | 4.6 | 9.0 | 0.5 | 3.7 | 940.7 | 2,723 | 633 | 632 | 11,899 | 174,150 | 59,214 | 6,100 |
| Lee | 4,462 | 6,030 | 24.7 | 37.4 | 5.2 | 0.2 | 2.4 | 4,561.8 | 6,165 | 2,432 | 1,400 | 40,901 | 359,450 | 84,592 | 11,892 |
| Leon | 1,119 | 3,844 | 39.5 | 2.4 | 8.5 | 0.3 | 7.3 | 4,855.4 | 16,687 | 1,935 | 541 | 50,854 | 129,860 | 67,496 | 8,289 |
| Levy | 116 | 2,876 | 48.0 | 6.2 | 8.3 | 0.9 | 4.9 | 21.9 | 544 | 72 | 103 | 1,807 | 17,670 | 42,840 | 3,586 |
| Liberty | 29 | 3,500 | 55.6 | 7.9 | 5.2 | 0.5 | 10.6 | 4.1 | 494 | 35 | 11 | 750 | 2,610 | 48,948 | 3,777 |
| Madison | 83 | 4,507 | 47.3 | 13.3 | 6.1 | 0.4 | 9.6 | 31.5 | 1,705 | 40 | 29 | 1,306 | 7,140 | 43,587 | 3,749 |
| Manatee | 1,401 | 3,634 | 41.4 | 2.2 | 8.3 | 0.5 | 4.9 | 1,651.1 | 4,283 | 1,029 | 729 | 12,577 | 187,490 | 78,260 | 10,652 |
| Marion | 966 | 2,733 | 49.5 | 3.7 | 7.4 | 0.9 | 5.5 | 629.5 | 1,782 | 752 | 623 | 14,753 | 166,940 | 51,068 | 5,269 |
| Martin | 620 | 3,884 | 34.0 | 6.0 | 8.9 | 1.9 | 4.7 | 239.7 | 1,501 | 310 | 275 | 5,798 | 78,950 | 127,132 | 21,100 |
| Miami-Dade | 14,469 | 5,333 | 28.2 | 13.3 | 9.2 | 3.4 | 1.6 | 28,281.1 | 10,423 | 20,979 | 7,336 | 122,728 | 1,355,670 | 69,516 | 10,780 |
| Monroe | 709 | 9,268 | 19.7 | 5.8 | 9.3 | 0.5 | 3.3 | 1,611.0 | 21,063 | 1,194 | 1,399 | 4,873 | 44,640 | 126,979 | 22,184 |
| Nassau | 270 | 3,251 | 48.8 | 4.8 | 8.8 | 0.2 | 3.8 | 449.9 | 2,211 | 534 | 153 | 3,084 | 42,930 | 92,093 | 13,144 |
| Okaloosa | 716 | 3,520 | 47.8 | 2.0 | 7.4 | 0.3 | 3.9 | 32.2 | 781 | 8,963 | 17,687 | 8,333 | 104,670 | 72,395 | 9,076 |
| Okeechobee | 126 | 3,046 | 50.5 | 3.0 | 8.7 | 0.6 | 5.3 | 164.6 | 1,985 | 76 | 68 | 2,204 | 15,720 | 47,302 | 4,208 |
| Orange | 6,836 | 5,042 | 38.6 | 1.6 | 6.7 | 0.2 | 3.0 | 10,178.6 | 7,507 | 12,527 | 2,620 | 69,400 | 672,180 | 64,747 | 8,354 |
| Osceola | 1,330 | 3,762 | 44.5 | 0.5 | 8.7 | 1.5 | 5.3 | 2,606.1 | 7,370 | 519 | 650 | 13,237 | 180,260 | 44,133 | 3,686 |
| Palm Beach | 6,985 | 4,750 | 30.3 | 4.6 | 10.4 | 1.7 | 2.1 | 5,584.1 | 3,798 | 7,087 | 2,709 | 58,447 | 737,630 | 112,802 | 19,587 |
| Pasco | 1,578 | 3,005 | 51.7 | 2.4 | 8.5 | 0.2 | 3.5 | 1,241.2 | 2,364 | 900 | 954 | 16,426 | 248,430 | 57,613 | 5,872 |
| Pinellas | 3,846 | 3,972 | 32.7 | 5.4 | 11.5 | 1.6 | 3.3 | 3,338.4 | 3,448 | 7,714 | 2,755 | 37,274 | 486,890 | 73,938 | 10,065 |
| Polk | 2,193 | 3,200 | 47.9 | 2.6 | 8.2 | 1.4 | 4.0 | 2,340.3 | 3,415 | 1,246 | 1,298 | 27,592 | 315,270 | 52,183 | 4,842 |

1. Based on the resident population estimated as of July 1 of the year shown.

# Land Area and Population

| State / county code | CBSA code[1] | County Type code[2] | STATE County | Land area[3] (sq. mi) | Total persons 2019 | Rank | Per square mile | White | Black | American Indian, Alaska Native | Asian and Pacific Islander | Percent Hispanic or Latino[4] | Under 5 years | 5 to 14 years | 15 to 24 years | 25 to 34 years | 35 to 44 years | 45 to 54 years |
|---|---|---|---|---|---|---|---|---|---|---|---|---|---|---|---|---|---|---|
| | | | | | 1 | 2 | 3 | 4 | 5 | 6 | 7 | 8 | 9 | 10 | 11 | 12 | 13 |

Header note: Columns 5–8 = Race alone or in combination, not Hispanic or Latino (percent). Columns 10–15 = Age (percent).

| State / county code | CBSA code[1] | County Type code[2] | STATE County | Land area (sq. mi) | Total persons 2019 | Rank | Per sq. mile | White | Black | Am. Ind., Alaska Native | Asian and Pacific Islander | Pct Hispanic or Latino | Under 5 yrs | 5 to 14 yrs | 15 to 24 yrs | 25 to 34 yrs | 35 to 44 yrs | 45 to 54 yrs |
|---|---|---|---|---|---|---|---|---|---|---|---|---|---|---|---|---|---|---|
| | | | **FLORIDA—Cont'd** | | | | | | | | | | | | | | | |
| 12107 | 37260 | 4 | Putnam | 728.4 | 74,815 | 741 | 102.7 | 72.5 | 16.7 | 1.0 | 1.0 | 10.4 | 5.7 | 12.0 | 10.6 | 11.4 | 10.0 | 11.0 |
| 12109 | 27260 | 1 | St. Johns | 600.6 | 278,715 | 251 | 464.1 | 83.1 | 6.0 | 0.6 | 4.5 | 7.9 | 4.8 | 12.7 | 10.8 | 9.4 | 13.4 | 13.7 |
| 12111 | 38940 | 2 | St. Lucie | 571.7 | 337,186 | 211 | 589.8 | 56.7 | 21.7 | 0.6 | 2.7 | 20.3 | 4.9 | 11.1 | 10.5 | 11.5 | 11.0 | 11.8 |
| 12113 | 37860 | 2 | Santa Rosa | 1,012.4 | 189,139 | 355 | 186.8 | 84.6 | 7.3 | 1.6 | 3.9 | 6.1 | 5.5 | 12.5 | 11.3 | 13.8 | 13.4 | 12.9 |
| 12115 | 35840 | 2 | Sarasota | 556.0 | 443,465 | 161 | 797.6 | 83.8 | 5.0 | 0.6 | 2.4 | 9.7 | 3.4 | 7.9 | 8.1 | 8.4 | 8.4 | 10.4 |
| 12117 | 36740 | 1 | Seminole | 309.4 | 474,171 | 152 | 1,532.6 | 60.0 | 12.8 | 0.6 | 6.0 | 22.9 | 5.1 | 11.8 | 11.8 | 14.6 | 14.2 | 13.2 |
| 12119 | 45540 | 3 | Sumter | 557.1 | 139,018 | 470 | 249.5 | 86.0 | 7.0 | 0.7 | 1.3 | 6.0 | 1.8 | 4.0 | 4.0 | 5.7 | 5.8 | 6.3 |
| 12121 | | 6 | Suwannee | 688.6 | 44,851 | 1,078 | 65.1 | 76.7 | 12.7 | 1.0 | 0.9 | 10.2 | 5.3 | 12.0 | 11.0 | 12.3 | 11.3 | 11.6 |
| 12123 | | 6 | Taylor | 1,043.3 | 21,600 | 1,742 | 20.7 | 74.3 | 20.4 | 1.6 | 1.3 | 4.3 | 5.4 | 11.3 | 10.2 | 14.4 | 11.9 | 12.5 |
| 12125 | | 6 | Union | 243.6 | 15,182 | 2,083 | 62.3 | 70.9 | 23.2 | 0.7 | 0.9 | 5.8 | 5.4 | 11.4 | 12.0 | 15.1 | 13.2 | 13.2 |
| 12127 | 19660 | 2 | Volusia | 1,101.3 | 561,497 | 123 | 509.8 | 71.7 | 11.5 | 0.8 | 2.5 | 15.5 | 4.6 | 9.8 | 10.8 | 11.8 | 10.6 | 11.7 |
| 12129 | 45220 | 2 | Wakulla | 606.4 | 34,319 | 1,322 | 56.6 | 80.9 | 14.5 | 1.3 | 1.2 | 4.1 | 5.1 | 11.6 | 10.7 | 14.0 | 13.9 | 14.3 |
| 12131 | 18880 | 3 | Walton | 1,038.3 | 76,648 | 726 | 73.8 | 86.4 | 5.8 | 1.7 | 2.1 | 6.4 | 5.6 | 11.5 | 9.3 | 11.6 | 12.6 | 12.7 |
| 12133 | | 6 | Washington | 584.7 | 25,932 | 1,566 | 44.4 | 79.3 | 15.6 | 2.0 | 1.5 | 4.1 | 5.2 | 11.2 | 11.9 | 14.1 | 12.7 | 13.4 |
| 13000 | | 0 | **GEORGIA** | 57,716.3 | 10,710,017 | X | 185.6 | 53.2 | 33.0 | 0.7 | 5.2 | 10.0 | 6.0 | 13.2 | 13.6 | 14.1 | 13.0 | 13.0 |
| 13001 | | 7 | Appling | 508.3 | 18,325 | 1,908 | 36.1 | 70.4 | 20.0 | 0.5 | 0.9 | 9.8 | 6.3 | 13.9 | 12.4 | 11.6 | 11.9 | 11.9 |
| 13003 | 20060 | 9 | Atkinson | 342.7 | 8,393 | 2,543 | 24.5 | 56.5 | 16.7 | 0.8 | 0.7 | 26.8 | 7.1 | 14.3 | 13.5 | 13.4 | 12.6 | 12.9 |
| 13005 | | 7 | Bacon | 284.1 | 11,036 | 2,344 | 38.8 | 74.8 | 17.4 | 0.4 | 0.8 | 8.5 | 6.7 | 14.2 | 12.8 | 12.4 | 12.2 | 12.8 |
| 13007 | | 3 | Baker | 342.0 | 2,971 | 2,964 | 8.7 | 49.9 | 43.1 | 0.8 | 1.4 | 6.3 | 5.4 | 10.2 | 9.9 | 11.4 | 9.7 | 12.0 |
| 13009 | 33300 | 4 | Baldwin | 258.7 | 45,099 | 1,073 | 174.3 | 53.4 | 42.9 | 0.5 | 2.0 | 2.5 | 4.5 | 10.6 | 21.8 | 11.7 | 10.5 | 11.1 |
| 13011 | | 8 | Banks | 232.6 | 19,352 | 1,858 | 83.2 | 88.5 | 3.4 | 0.9 | 1.4 | 7.5 | 5.3 | 12.0 | 12.3 | 11.9 | 12.2 | 13.7 |
| 13013 | 12060 | 1 | Barrow | 161.0 | 85,588 | 674 | 531.6 | 69.9 | 14.3 | 0.7 | 4.5 | 12.7 | 6.5 | 14.6 | 12.5 | 14.7 | 14.2 | 12.9 |
| 13015 | 12060 | 1 | Bartow | 458.9 | 109,426 | 559 | 238.5 | 78.1 | 12.0 | 0.8 | 1.5 | 9.4 | 6.1 | 13.2 | 12.5 | 14.0 | 12.5 | 13.8 |
| 13017 | 22340 | 7 | Ben Hill | 250.1 | 16,614 | 2,001 | 66.4 | 55.5 | 37.6 | 0.7 | 1.0 | 6.6 | 6.3 | 14.0 | 11.9 | 11.9 | 12.0 | 12.1 |
| 13019 | | 6 | Berrien | 453.4 | 19,408 | 1,857 | 42.8 | 82.1 | 11.4 | 0.7 | 1.4 | 5.8 | 6.1 | 13.1 | 12.0 | 12.6 | 11.7 | 12.8 |
| 13021 | 31420 | 3 | Bibb | 249.4 | 152,737 | 438 | 612.4 | 37.9 | 56.7 | 0.5 | 2.7 | 3.7 | 6.4 | 13.6 | 14.3 | 14.1 | 11.5 | 11.3 |
| 13023 | | 6 | Bleckley | 215.9 | 12,955 | 2,224 | 60.0 | 68.5 | 27.8 | 0.4 | 1.3 | 3.2 | 5.0 | 11.1 | 18.7 | 12.0 | 10.5 | 11.6 |
| 13025 | 15260 | 3 | Brantley | 443.2 | 19,202 | 1,867 | 43.3 | 93.2 | 4.4 | 1.2 | 0.6 | 2.4 | 6.1 | 12.9 | 11.8 | 12.5 | 11.4 | 13.7 |
| 13027 | 46660 | 6 | Brooks | 493.2 | 15,357 | 2,069 | 31.1 | 57.1 | 35.6 | 0.9 | 1.8 | 6.2 | 5.6 | 12.3 | 10.6 | 11.8 | 10.9 | 13.0 |
| 13029 | 42340 | 2 | Bryan | 437.6 | 40,755 | 1,161 | 93.1 | 74.1 | 16.6 | 1.0 | 3.5 | 8.1 | 7.2 | 17.2 | 12.4 | 13.8 | 16.4 | 12.0 |
| 13031 | 44340 | 4 | Bulloch | 676.0 | 80,839 | 704 | 119.6 | 64.6 | 30.2 | 0.6 | 1.9 | 4.5 | 5.6 | 10.9 | 27.1 | 13.5 | 10.8 | 10.0 |
| 13033 | 12260 | 2 | Burke | 827.0 | 22,648 | 1,699 | 27.4 | 49.1 | 47.2 | 0.9 | 0.9 | 3.5 | 6.6 | 14.3 | 12.5 | 12.3 | 11.4 | 12.0 |
| 13035 | 12060 | 1 | Butts | 183.7 | 25,426 | 1,583 | 138.4 | 67.3 | 28.9 | 0.6 | 1.0 | 3.6 | 5.6 | 11.3 | 12.9 | 15.1 | 13.0 | 13.2 |
| 13037 | | 8 | Calhoun | 280.4 | 6,231 | 2,729 | 22.2 | 33.0 | 61.4 | 0.4 | 1.1 | 5.2 | 3.8 | 9.5 | 13.0 | 15.1 | 15.4 | 14.0 |
| 13039 | 41220 | 4 | Camden | 630.4 | 55,388 | 928 | 87.9 | 72.0 | 20.1 | 1.1 | 2.9 | 7.1 | 6.9 | 13.5 | 14.7 | 16.4 | 11.7 | 10.3 |
| 13043 | | 7 | Candler | 243.1 | 10,985 | 2,349 | 45.2 | 61.8 | 26.1 | 0.4 | 0.9 | 11.9 | 6.3 | 14.3 | 12.4 | 12.0 | 10.9 | 12.3 |
| 13045 | 12060 | 1 | Carroll | 499.1 | 121,633 | 521 | 243.7 | 71.9 | 20.9 | 0.7 | 1.4 | 7.3 | 6.4 | 13.0 | 16.8 | 13.5 | 12.2 | 12.1 |
| 13047 | 16860 | 2 | Catoosa | 162.2 | 67,996 | 789 | 419.2 | 91.9 | 3.8 | 0.9 | 2.0 | 3.2 | 5.2 | 12.8 | 12.0 | 12.3 | 12.4 | 13.6 |
| 13049 | | 6 | Charlton | 780.1 | 13,430 | 2,194 | 17.2 | 62.3 | 30.4 | 2.2 | 1.1 | 5.9 | 4.4 | 10.7 | 11.2 | 17.6 | 13.7 | 13.4 |
| 13051 | 42340 | 2 | Chatham | 433.1 | 289,463 | 242 | 668.4 | 49.2 | 41.7 | 0.7 | 3.9 | 6.7 | 6.0 | 11.4 | 14.0 | 16.0 | 12.6 | 11.3 |
| 13053 | 17980 | 2 | Chattahoochee | 248.7 | 10,551 | 2,384 | 42.4 | 59.1 | 20.1 | 1.5 | 5.2 | 18.2 | 8.3 | 12.2 | 28.8 | 26.1 | 11.0 | 4.6 |
| 13055 | 44900 | 6 | Chattooga | 313.3 | 24,843 | 1,609 | 79.3 | 83.6 | 10.8 | 0.7 | 0.8 | 5.7 | 6.0 | 12.5 | 12.0 | 12.7 | 11.9 | 13.0 |
| 13057 | 12060 | 1 | Cherokee | 421.5 | 265,274 | 260 | 630.0 | 79.1 | 8.1 | 0.7 | 3.0 | 11.1 | 5.6 | 13.4 | 12.6 | 12.0 | 13.5 | 14.6 |
| 13059 | 12020 | 3 | Clarke | 119.2 | 127,795 | 503 | 1,072.1 | 57.1 | 28.6 | 0.5 | 4.9 | 10.9 | 4.9 | 9.5 | 28.1 | 16.2 | 11.3 | 9.2 |
| 13061 | | 9 | Clay | 195.4 | 2,866 | 2,974 | 14.7 | 36.9 | 60.2 | 0.5 | 1.5 | 1.8 | 5.6 | 10.7 | 10.1 | 10.5 | 11.1 | 9.0 |
| 13063 | 12060 | 1 | Clayton | 141.7 | 292,646 | 240 | 2,065.3 | 10.1 | 72.6 | 0.7 | 5.5 | 13.3 | 7.2 | 15.6 | 14.3 | 15.9 | 12.8 | 12.7 |
| 13065 | | 6 | Clinch | 815.0 | 6,582 | 2,696 | 8.1 | 66.7 | 27.8 | 1.0 | 0.6 | 5.7 | 6.5 | 14.4 | 12.5 | 12.8 | 11.5 | 12.6 |
| 13067 | 12060 | 1 | Cobb | 339.8 | 762,944 | 86 | 2,245.3 | 52.7 | 29.2 | 0.7 | 6.5 | 13.4 | 5.9 | 12.9 | 13.0 | 14.7 | 14.1 | 13.9 |
| 13069 | 20060 | 7 | Coffee | 592.3 | 43,218 | 1,115 | 73.0 | 58.2 | 29.5 | 0.6 | 1.1 | 12.1 | 6.7 | 13.6 | 13.7 | 14.6 | 12.8 | 12.6 |
| 13071 | 34220 | 6 | Colquitt | 547.0 | 45,542 | 1,067 | 83.3 | 55.5 | 24.0 | 0.6 | 1.1 | 19.9 | 6.5 | 14.7 | 12.9 | 12.8 | 12.8 | 12.3 |
| 13073 | 12260 | 2 | Columbia | 290.2 | 160,377 | 418 | 552.6 | 69.1 | 20.0 | 0.8 | 6.0 | 7.3 | 6.2 | 14.4 | 12.2 | 13.8 | 14.3 | 12.6 |
| 13075 | | 6 | Cook | 228.5 | 17,291 | 1,957 | 75.7 | 64.9 | 28.6 | 0.7 | 1.0 | 6.3 | 6.3 | 14.6 | 12.7 | 12.1 | 12.0 | 12.6 |
| 13077 | 12060 | 1 | Coweta | 441.1 | 150,849 | 445 | 342.0 | 71.8 | 19.1 | 0.7 | 2.9 | 7.3 | 5.7 | 13.3 | 12.6 | 12.5 | 13.1 | 14.5 |
| 13079 | 31420 | 3 | Crawford | 324.9 | 12,231 | 2,271 | 37.6 | 74.2 | 21.6 | 1.0 | 1.0 | 3.7 | 4.7 | 11.7 | 10.5 | 11.9 | 10.8 | 13.2 |
| 13081 | 18380 | 6 | Crisp | 272.7 | 22,034 | 1,717 | 80.8 | 50.4 | 45.2 | 0.5 | 1.2 | 3.9 | 5.9 | 13.4 | 12.0 | 11.9 | 11.4 | 11.9 |
| 13083 | 16860 | 2 | Dade | 174.0 | 16,057 | 2,029 | 92.3 | 94.6 | 1.9 | 1.4 | 1.4 | 2.4 | 5.0 | 10.8 | 14.7 | 11.5 | 11.5 | 11.9 |
| 13085 | 12060 | 1 | Dawson | 210.8 | 27,113 | 1,523 | 128.6 | 91.9 | 1.7 | 0.9 | 1.4 | 5.5 | 5.5 | 10.9 | 11.3 | 12.6 | 11.3 | 13.0 |
| 13087 | 12460 | 6 | Decatur | 597.2 | 26,457 | 1,550 | 44.3 | 50.3 | 43.1 | 0.7 | 0.9 | 6.5 | 7.1 | 13.1 | 12.9 | 12.7 | 11.8 | 12.3 |
| 13089 | 12060 | 1 | DeKalb | 267.7 | 762,009 | 87 | 2,846.5 | 30.9 | 54.7 | 0.7 | 7.4 | 8.4 | 6.8 | 12.5 | 11.8 | 16.9 | 14.2 | 12.7 |
| 13091 | | 6 | Dodge | 496.0 | 20,452 | 1,797 | 41.2 | 65.6 | 30.6 | 0.6 | 0.8 | 3.9 | 4.9 | 10.7 | 11.4 | 13.0 | 12.7 | 14.3 |
| 13093 | | 6 | Dooly | 392.6 | 13,174 | 2,207 | 33.6 | 43.3 | 49.4 | 0.5 | 0.9 | 7.1 | 4.0 | 9.0 | 11.2 | 12.0 | 13.2 | 13.4 |
| 13095 | 10500 | 3 | Dougherty | 328.6 | 86,477 | 670 | 263.2 | 24.7 | 71.7 | 0.5 | 1.3 | 3.1 | 6.4 | 13.1 | 15.6 | 13.4 | 11.2 | 11.0 |
| 13097 | 12060 | 1 | Douglas | 200.1 | 147,988 | 455 | 739.6 | 37.7 | 51.0 | 0.8 | 2.3 | 10.6 | 6.0 | 14.5 | 14.0 | 13.0 | 13.2 | 14.5 |
| 13099 | | 6 | Early | 512.6 | 10,037 | 2,417 | 19.6 | 44.6 | 52.2 | 0.9 | 1.1 | 2.4 | 6.6 | 13.7 | 12.7 | 11.2 | 10.8 | 12.3 |
| 13101 | 46660 | 3 | Echols | 420.4 | 4,002 | 2,897 | 9.5 | 62.9 | 6.3 | 2.3 | 1.1 | 29.4 | 6.1 | 16.6 | 12.2 | 12.5 | 14.1 | 12.5 |

1. CBSA = Core Based Statistical Area. See Appendix A for explanation. See Appendix B for list of metropolitan areas with component counties. Service of USDA Rural-Urban Continuum Codes. See Appendix A for definition. 3. Dry land or land partially or temporarily covered by water. 2. County type code from the Economic Research Service of USDA Rural-Urban Continuum Codes. 4. May be of any race.

# Table B. States and Counties — **Population and Households**

| STATE County | Age (percent) (cont.) 55 to 64 years | 65 to 74 years | 75 years and over | Percent female | Population change, 2000-2020 Total persons 2000 | Total persons 2010 | Percent change 2000-2010 | Percent change 2010-2020 | Components of change, 2010-2020 Births | Deaths | Net Migration | Households, 2015-2019 Number | Persons per household | Family households | Percent Female family householder[1] | One person |
|---|---|---|---|---|---|---|---|---|---|---|---|---|---|---|---|---|
| | 16 | 17 | 18 | 19 | 20 | 21 | 22 | 23 | 24 | 25 | 26 | 27 | 28 | 29 | 30 | 31 |
| **FLORIDA—Cont'd** | | | | | | | | | | | | | | | | |
| Putnam | 15.4 | 13.9 | 10.0 | 50.7 | 70,423 | 74,380 | 5.6 | 0.6 | 8,544 | 10,141 | 2,078 | 28,943 | 2.48 | 62.4 | 14.6 | 31.5 |
| St. Johns | 14.2 | 12.7 | 8.3 | 51.3 | 123,135 | 190,040 | 54.3 | 46.7 | 21,247 | 18,957 | 85,729 | 88,773 | 2.73 | 72.5 | 9.3 | 21.9 |
| St. Lucie | 14.1 | 13.7 | 11.4 | 51.2 | 192,695 | 277,257 | 43.9 | 21.6 | 31,170 | 32,154 | 60,786 | 114,761 | 2.70 | 67.5 | 12.0 | 26.1 |
| Santa Rosa | 14.2 | 10.2 | 6.3 | 49.0 | 117,743 | 151,371 | 28.6 | 25.0 | 19,341 | 14,692 | 32,905 | 63,514 | 2.65 | 73.8 | 11.2 | 20.9 |
| Sarasota | 15.5 | 18.9 | 19.0 | 52.4 | 325,957 | 379,427 | 16.4 | 16.9 | 29,564 | 56,762 | 90,895 | 182,842 | 2.26 | 61.2 | 8.4 | 31.9 |
| Seminole | 12.9 | 9.5 | 6.8 | 51.6 | 365,196 | 422,720 | 15.8 | 12.2 | 46,852 | 34,082 | 38,885 | 173,668 | 2.63 | 67.3 | 12.0 | 25.7 |
| Sumter | 13.3 | 32.1 | 27.0 | 50.4 | 53,345 | 93,420 | 75.1 | 48.8 | 4,768 | 17,198 | 57,449 | 56,230 | 2.06 | 66.2 | 4.1 | 30.0 |
| Suwannee | 14.4 | 12.2 | 9.8 | 48.7 | 34,844 | 41,552 | 19.3 | 7.9 | 4,732 | 5,713 | 4,201 | 14,888 | 2.85 | 71.8 | 11.8 | 23.7 |
| Taylor | 13.1 | 12.1 | 9.0 | 45.5 | 19,256 | 22,572 | 17.2 | -4.3 | 2,422 | 2,530 | -879 | 7,405 | 2.44 | 70.2 | 16.6 | 26.8 |
| Union | 14.2 | 9.1 | 6.2 | 35.5 | 13,442 | 15,537 | 15.6 | -2.3 | 1,661 | 2,291 | 273 | 3,960 | 2.42 | 71.0 | 19.4 | 23.8 |
| Volusia | 15.3 | 14.2 | 11.1 | 51.3 | 443,343 | 494,592 | 11.6 | 13.5 | 49,329 | 70,615 | 87,922 | 216,495 | 2.43 | 61.1 | 10.8 | 31.4 |
| Wakulla | 14.2 | 10.3 | 5.9 | 46.1 | 22,863 | 30,781 | 34.6 | 11.5 | 3,337 | 2,757 | 2,952 | 11,200 | 2.58 | 72.0 | 10.7 | 22.2 |
| Walton | 16.1 | 12.8 | 7.7 | 49.7 | 40,601 | 55,044 | 35.6 | 39.2 | 7,651 | 6,353 | 20,080 | 27,420 | 2.42 | 66.6 | 9.8 | 27.2 |
| Washington | 13.7 | 10.5 | 7.3 | 45.9 | 20,973 | 24,901 | 18.7 | 4.1 | 2,515 | 3,136 | 1,626 | 8,827 | 2.48 | 67.1 | 14.2 | 28.1 |
| **GEORGIA** | 12.4 | 9.0 | 5.7 | 51.4 | 8,186,453 | 9,688,737 | 18.4 | 10.5 | 1,327,232 | 814,702 | 505,498 | 3,758,798 | 2.70 | 67.2 | 14.9 | 27.2 |
| Appling | 13.7 | 10.8 | 7.5 | 50.2 | 17,419 | 18,237 | 4.7 | 0.5 | 2,526 | 2,074 | -368 | 6,656 | 2.69 | 73.2 | 14.8 | 24.2 |
| Atkinson | 12.2 | 8.5 | 5.5 | 49.2 | 7,609 | 8,380 | 10.1 | 0.2 | 1,292 | 770 | -517 | 2,880 | 2.85 | 68.6 | 16.1 | 28.2 |
| Bacon | 11.4 | 10.9 | 6.6 | 51.1 | 10,103 | 11,097 | 9.8 | -0.5 | 1,536 | 1,329 | -261 | 3,874 | 2.70 | 69.9 | 14.6 | 25.3 |
| Baker | 15.7 | 15.4 | 10.4 | 50.4 | 4,074 | 3,447 | -15.4 | -13.8 | 318 | 284 | -512 | 1,425 | 2.20 | 55.3 | 10.0 | 42.2 |
| Baldwin | 12.7 | 10.3 | 6.8 | 49.4 | 44,700 | 45,840 | 2.6 | -1.6 | 4,560 | 4,499 | -879 | 16,191 | 2.51 | 59.1 | 15.4 | 31.3 |
| Banks | 14.1 | 11.2 | 7.2 | 49.0 | 14,422 | 18,373 | 27.4 | 5.3 | 2,014 | 1,660 | 633 | 6,740 | 2.78 | 72.3 | 8.7 | 26.7 |
| Barrow | 11.6 | 8.1 | 4.9 | 51.0 | 46,144 | 69,349 | 50.3 | 23.4 | 10,754 | 6,010 | 11,417 | 26,305 | 2.99 | 75.6 | 12.7 | 20.8 |
| Bartow | 13.3 | 9.1 | 5.6 | 50.7 | 76,019 | 100,083 | 31.7 | 9.3 | 13,234 | 9,186 | 5,335 | 37,627 | 2.76 | 72.9 | 13.2 | 22.2 |
| Ben Hill | 13.5 | 11.1 | 7.2 | 53.0 | 17,484 | 17,662 | 1.0 | -5.9 | 2,336 | 2,276 | -1,106 | 6,443 | 2.59 | 62.4 | 19.4 | 33.2 |
| Berrien | 13.9 | 10.2 | 7.5 | 50.9 | 16,235 | 19,291 | 18.8 | 0.6 | 2,381 | 2,170 | -108 | 7,367 | 2.57 | 68.6 | 13.6 | 27.1 |
| Bibb | 12.4 | 9.8 | 6.7 | 53.2 | 153,887 | 155,844 | 1.3 | -2.0 | 21,880 | 17,004 | -7,988 | 58,116 | 2.53 | 60.1 | 19.5 | 35.0 |
| Bleckley | 13.3 | 9.3 | 8.5 | 52.3 | 11,666 | 13,060 | 11.9 | -0.8 | 1,368 | 1,440 | -44 | 4,176 | 2.71 | 65.3 | 12.5 | 30.9 |
| Brantley | 14.0 | 10.6 | 7.0 | 50.6 | 14,629 | 18,409 | 25.8 | 4.3 | 2,202 | 1,982 | 571 | 6,823 | 2.74 | 67.1 | 13.3 | 29.8 |
| Brooks | 15.4 | 12.4 | 8.0 | 51.8 | 16,450 | 16,314 | -0.8 | -5.9 | 1,964 | 2,055 | -886 | 6,335 | 2.44 | 63.4 | 11.9 | 33.3 |
| Bryan | 10.0 | 6.8 | 4.1 | 50.8 | 23,417 | 30,238 | 29.1 | 34.8 | 5,240 | 2,560 | 7,818 | 13,048 | 2.83 | 77.7 | 13.8 | 18.4 |
| Bulloch | 10.2 | 7.4 | 4.6 | 51.0 | 55,983 | 70,244 | 25.5 | 15.1 | 8,954 | 5,414 | 6,949 | 27,375 | 2.48 | 60.8 | 15.7 | 26.6 |
| Burke | 14.0 | 10.5 | 6.4 | 52.4 | 22,243 | 23,334 | 4.9 | -2.9 | 3,223 | 2,503 | -1,430 | 8,193 | 2.72 | 72.5 | 17.8 | 24.8 |
| Butts | 12.9 | 9.6 | 6.5 | 47.0 | 19,522 | 23,685 | 21.3 | 7.4 | 2,771 | 2,644 | 1,606 | 8,279 | 2.56 | 70.3 | 17.5 | 23.9 |
| Calhoun | 12.4 | 9.2 | 7.4 | 39.4 | 6,320 | 6,697 | 6.0 | -7.0 | 552 | 624 | -398 | 1,736 | 2.56 | 66.4 | 26.3 | 25.3 |
| Camden | 11.9 | 9.1 | 5.5 | 48.8 | 43,664 | 50,525 | 15.7 | 9.6 | 8,042 | 3,543 | 342 | 19,338 | 2.66 | 74.4 | 15.2 | 19.9 |
| Candler | 12.8 | 11.6 | 7.5 | 51.2 | 9,577 | 10,987 | 14.7 | 0.0 | 1,430 | 1,334 | -99 | 4,013 | 2.63 | 69.2 | 17.7 | 28.3 |
| Carroll | 11.8 | 8.6 | 5.6 | 51.4 | 87,268 | 110,565 | 26.7 | 10.0 | 15,307 | 10,587 | 6,366 | 41,903 | 2.71 | 71.0 | 14.6 | 22.3 |
| Catoosa | 13.2 | 10.7 | 7.8 | 51.6 | 53,282 | 63,928 | 20.0 | 6.4 | 7,275 | 6,305 | 3,145 | 24,778 | 2.67 | 70.3 | 10.9 | 25.0 |
| Charlton | 12.9 | 8.9 | 7.2 | 41.1 | 10,282 | 12,167 | 18.3 | 10.4 | 1,211 | 1,102 | 1,040 | 3,675 | 3.22 | 73.1 | 9.8 | 22.3 |
| Chatham | 12.2 | 9.8 | 6.6 | 52.0 | 232,048 | 265,112 | 14.2 | 9.2 | 39,133 | 24,516 | 9,651 | 108,568 | 2.55 | 61.8 | 16.6 | 30.1 |
| Chattahoochee | 4.0 | 2.6 | 2.3 | 35.4 | 14,882 | 11,263 | -24.3 | -6.3 | 2,319 | 350 | -2,746 | 2,570 | 2.79 | 73.4 | 16.8 | 21.6 |
| Chattooga | 13.5 | 10.8 | 7.6 | 49.1 | 25,470 | 26,022 | 2.2 | -4.5 | 2,991 | 3,273 | -905 | 9,260 | 2.49 | 63.2 | 10.8 | 31.9 |
| Cherokee | 13.1 | 9.7 | 5.6 | 50.8 | 141,903 | 214,400 | 51.1 | 23.7 | 28,646 | 14,992 | 37,051 | 88,137 | 2.79 | 75.1 | 8.5 | 20.0 |
| Clarke | 8.9 | 7.4 | 4.7 | 52.6 | 101,489 | 116,670 | 15.0 | 9.5 | 13,992 | 7,535 | 4,520 | 48,844 | 2.36 | 49.4 | 13.0 | 33.3 |
| Clay | 14.6 | 16.8 | 11.5 | 52.6 | 3,357 | 3,186 | -5.1 | -10.0 | 324 | 408 | -241 | 1,242 | 2.33 | 57.0 | 17.6 | 39.2 |
| Clayton | 11.4 | 6.9 | 3.3 | 53.4 | 236,517 | 259,639 | 9.8 | 12.7 | 43,599 | 17,114 | 6,386 | 94,279 | 2.95 | 66.6 | 25.7 | 28.2 |
| Clinch | 12.7 | 10.5 | 6.5 | 52.2 | 6,878 | 6,792 | -1.3 | -3.1 | 977 | 795 | -392 | 2,477 | 2.51 | 66.2 | 20.3 | 27.8 |
| Cobb | 12.4 | 8.3 | 4.9 | 51.5 | 607,751 | 688,076 | 13.2 | 10.9 | 94,924 | 43,006 | 23,236 | 280,374 | 2.64 | 67.6 | 12.9 | 25.3 |
| Coffee | 11.7 | 8.8 | 5.6 | 48.7 | 37,413 | 42,346 | 13.2 | 2.1 | 6,011 | 4,146 | -1,037 | 14,438 | 2.71 | 68.7 | 15.6 | 27.6 |
| Colquitt | 11.6 | 9.6 | 6.8 | 50.5 | 42,053 | 45,496 | 8.2 | 0.1 | 6,717 | 4,900 | -1,757 | 15,505 | 2.88 | 68.8 | 17.3 | 25.7 |
| Columbia | 12.0 | 9.0 | 5.4 | 51.2 | 89,288 | 124,013 | 38.9 | 29.3 | 17,823 | 9,280 | 27,609 | 47,215 | 3.18 | 74.2 | 10.0 | 22.1 |
| Cook | 12.5 | 9.8 | 7.1 | 51.9 | 15,771 | 17,205 | 9.1 | 0.5 | 2,239 | 1,968 | -195 | 6,217 | 2.74 | 68.2 | 15.1 | 27.3 |
| Coweta | 13.4 | 9.2 | 5.7 | 51.5 | 89,215 | 127,368 | 42.8 | 18.4 | 16,830 | 10,234 | 16,819 | 52,035 | 2.74 | 74.0 | 13.2 | 20.8 |
| Crawford | 17.3 | 12.5 | 7.5 | 49.7 | 12,495 | 12,593 | 0.8 | -2.9 | 1,300 | 1,320 | -344 | 4,510 | 2.70 | 69.2 | 12.9 | 27.2 |
| Crisp | 13.4 | 12.0 | 8.1 | 52.8 | 21,996 | 23,435 | 6.5 | -6.0 | 2,941 | 2,608 | -1,755 | 8,585 | 2.60 | 66.5 | 23.1 | 31.2 |
| Dade | 14.2 | 12.4 | 8.2 | 51.0 | 15,154 | 16,644 | 9.8 | -3.5 | 1,633 | 1,844 | -373 | 6,195 | 2.40 | 73.3 | 10.5 | 22.7 |
| Dawson | 14.7 | 12.8 | 7.8 | 50.3 | 15,999 | 22,377 | 39.9 | 21.2 | 2,540 | 2,112 | 4,304 | 9,041 | 2.70 | 71.8 | 8.3 | 24.8 |
| Decatur | 12.8 | 9.9 | 7.3 | 51.4 | 28,240 | 27,841 | -1.4 | -5.0 | 3,753 | 3,254 | -1,892 | 10,084 | 2.53 | 70.5 | 18.2 | 26.6 |
| DeKalb | 11.8 | 8.3 | 5.0 | 52.8 | 665,865 | 692,031 | 3.9 | 10.1 | 111,976 | 46,741 | 4,968 | 282,436 | 2.61 | 57.9 | 16.8 | 34.1 |
| Dodge | 13.7 | 10.9 | 8.3 | 47.8 | 19,171 | 21,792 | 13.7 | -6.1 | 2,362 | 2,370 | -1,330 | 7,628 | 2.46 | 67.7 | 15.5 | 28.8 |
| Dooly | 15.1 | 12.9 | 9.2 | 45.7 | 11,525 | 14,920 | 29.5 | -11.7 | 1,131 | 1,282 | -1,607 | 5,020 | 2.47 | 66.7 | 21.1 | 29.2 |
| Dougherty | 12.3 | 10.2 | 6.8 | 54.0 | 96,065 | 94,564 | -1.6 | -8.6 | 13,278 | 9,445 | -11,983 | 34,087 | 2.51 | 60.8 | 24.6 | 34.4 |
| Douglas | 12.4 | 7.8 | 4.5 | 52.6 | 92,174 | 132,281 | 43.5 | 11.9 | 17,669 | 9,654 | 7,694 | 49,187 | 2.89 | 70.8 | 17.9 | 23.5 |
| Early | 13.4 | 10.4 | 9.1 | 53.5 | 12,354 | 11,008 | -10.9 | -8.8 | 1,397 | 1,412 | -963 | 4,074 | 2.50 | 65.3 | 22.6 | 31.2 |
| Echols | 12.2 | 8.3 | 5.3 | 49.4 | 3,754 | 4,023 | 7.2 | -0.5 | 602 | 281 | -345 | 1,561 | 2.55 | 70.3 | 12.2 | 27.9 |

1. No spouse present.

# Table B. States and Counties — Population, Vital Statistics, Health, and Crime

| STATE County | Persons in group quarters, 2020 | Daytime Population, 2015-2019 | | Births, 2020 | | Deaths, 2020 | | Persons under 65 with no health insurance, 2019 | | Medicare, 2020 | | | Crimes reported by county police or sheriff, 2019 | |
|---|---|---|---|---|---|---|---|---|---|---|---|---|---|---|
| | | Number | Employment/residence ratio | Total | Rate[1] | Number | Rate[1] | Number | Percent | Total beneficiaries | Enrolled in Original Medicare | Enrolled in Medicare Advantage | Violent | Property |
| | 32 | 33 | 34 | 35 | 36 | 37 | 38 | 39 | 40 | 41 | 42 | 43 | 44 | 45 |
| FLORIDA—Cont'd | | | | | | | | | | | | | | |
| Putnam | 1,422 | 67,484 | 0.77 | 814 | 10.9 | 1,064 | 14.2 | 9,959 | 18.2 | 20,044 | 11,513 | 8,531 | 124 | 797 |
| St. Johns | 2,967 | 223,667 | 0.81 | 2,282 | 8.2 | 2,330 | 8.4 | 22,830 | 10.9 | 60,527 | 41,223 | 19,304 | 281 | 2,104 |
| St. Lucie | 3,041 | 279,627 | 0.74 | 3,152 | 9.3 | 3,913 | 11.6 | 44,138 | 18.1 | 82,954 | 44,890 | 38,064 | 259 | 1,160 |
| Santa Rosa | 6,169 | 141,189 | 0.56 | 1,967 | 10.4 | 1,855 | 9.8 | 17,956 | 12.0 | 36,535 | 23,331 | 13,204 | 209 | 1,360 |
| Sarasota | 6,003 | 447,058 | 1.16 | 2,895 | 6.5 | 6,732 | 15.2 | 43,250 | 16.0 | 149,539 | 98,404 | 51,135 | 475 | 4,183 |
| Seminole | 3,513 | 435,977 | 0.89 | 4,654 | 9.8 | 3,972 | 8.4 | 45,093 | 11.4 | 82,284 | 43,862 | 38,422 | 445 | 2,551 |
| Sumter | 8,264 | 134,939 | 1.40 | 469 | 3.4 | 2,269 | 16.3 | 5,957 | 12.5 | 74,807 | 40,166 | 34,641 | 176 | 884 |
| Suwannee | 2,513 | 40,459 | 0.78 | 458 | 10.2 | 617 | 13.8 | 5,457 | 17.0 | 11,010 | 7,110 | 3,900 | 148 | 432 |
| Taylor | 2,428 | 22,657 | 1.13 | 234 | 10.8 | 262 | 12.1 | 2,149 | 14.8 | 5,002 | 3,348 | 1,654 | 71 | 229 |
| Union | 4,814 | 16,115 | 1.22 | 163 | 10.7 | 257 | 16.9 | 1,046 | 12.5 | 2,196 | 1,430 | 766 | 47 | 83 |
| Volusia | 14,480 | 512,130 | 0.89 | 4,836 | 8.6 | 7,812 | 13.9 | 66,872 | 16.6 | 152,020 | 70,988 | 81,032 | 546 | 2,205 |
| Wakulla | 3,248 | 25,603 | 0.50 | 325 | 9.5 | 306 | 8.9 | 3,131 | 12.4 | 6,424 | 2,890 | 3,533 | 75 | 503 |
| Walton | 2,031 | 69,247 | 1.03 | 850 | 11.1 | 740 | 9.7 | 9,397 | 16.4 | 16,802 | 11,444 | 5,357 | 114 | 755 |
| Washington | 2,620 | 23,188 | 0.81 | 276 | 10.6 | 357 | 13.8 | 3,248 | 17.7 | 5,694 | 3,823 | 1,872 | 32 | 132 |
| GEORGIA | 268,022 | 10,406,270 | 1.00 | 125,818 | 11.7 | 94,638 | 8.8 | 1,387,604 | 15.7 | 1,772,282 | 1,011,229 | 761,053 | X | X |
| Appling | 416 | 18,891 | 1.06 | 213 | 11.6 | 224 | 12.2 | 3,001 | 20.6 | 3,895 | 2,247 | 1,648 | NA | NA |
| Atkinson | 21 | 7,560 | 0.79 | 134 | 16.0 | 81 | 9.7 | 1,615 | 23.8 | 1,378 | 739 | 640 | NA | NA |
| Bacon | 297 | 11,199 | 1.00 | 136 | 12.3 | 113 | 10.2 | 1,743 | 19.6 | 2,253 | 1,342 | 911 | NA | NA |
| Baker | 0 | 2,615 | 0.51 | 24 | 8.1 | 25 | 8.4 | 411 | 18.1 | 706 | 332 | 374 | NA | NA |
| Baldwin | 5,819 | 43,896 | 0.93 | 390 | 8.6 | 510 | 11.3 | 5,396 | 16.9 | 9,135 | 4,541 | 4,594 | NA | NA |
| Banks | 0 | 16,490 | 0.71 | 200 | 10.3 | 173 | 8.9 | 3,350 | 21.3 | 3,596 | 1,981 | 1,615 | NA | NA |
| Barrow | 360 | 64,830 | 0.61 | 1,094 | 12.8 | 725 | 8.5 | 12,872 | 17.8 | 13,495 | 7,413 | 6,082 | NA | NA |
| Bartow | 1,034 | 99,033 | 0.87 | 1,236 | 11.3 | 1,091 | 10.0 | 16,360 | 18.0 | 19,013 | 10,912 | 8,101 | NA | NA |
| Ben Hill | 298 | 17,504 | 1.08 | 214 | 12.9 | 250 | 15.0 | 2,279 | 17.1 | 3,846 | 2,056 | 1,791 | NA | NA |
| Berrien | 199 | 16,415 | 0.63 | 252 | 13.0 | 260 | 13.4 | 3,313 | 21.1 | 3,816 | 2,362 | 1,454 | 51 | 132 |
| Bibb | 6,719 | 174,476 | 1.34 | 1,993 | 13.0 | 1,891 | 12.4 | 17,632 | 14.6 | 31,208 | 16,991 | 14,216 | NA | NA |
| Bleckley | 1,413 | 10,879 | 0.57 | 135 | 10.4 | 121 | 9.3 | 1,343 | 14.5 | 2,570 | 1,746 | 823 | 21 | 100 |
| Brantley | 89 | 14,609 | 0.38 | 215 | 11.2 | 241 | 12.6 | 2,899 | 18.6 | 3,581 | 2,213 | 1,368 | NA | NA |
| Brooks | 51 | 12,474 | 0.48 | 155 | 10.1 | 215 | 14.0 | 2,317 | 19.2 | 3,996 | 2,373 | 1,623 | 50 | 196 |
| Bryan | 120 | 28,467 | 0.49 | 565 | 13.9 | 283 | 6.9 | 4,479 | 12.6 | 5,986 | 3,516 | 2,470 | 47 | 277 |
| Bulloch | 6,525 | 72,838 | 0.90 | 893 | 11.0 | 636 | 7.9 | 10,565 | 16.7 | 10,973 | 6,719 | 4,254 | NA | NA |
| Burke | 296 | 25,404 | 1.31 | 288 | 12.7 | 248 | 11.0 | 2,775 | 15.1 | 4,789 | 2,374 | 2,415 | 27 | 225 |
| Butts | 3,014 | 21,831 | 0.76 | 303 | 11.9 | 286 | 11.2 | 2,762 | 15.3 | 4,833 | 2,386 | 2,447 | 57 | 329 |
| Calhoun | 1,700 | 6,240 | 0.92 | 49 | 7.9 | 49 | 7.9 | 636 | 18.4 | 1,194 | 666 | 528 | NA | NA |
| Camden | 2,017 | 50,455 | 0.89 | 733 | 13.2 | 425 | 7.7 | 6,350 | 14.2 | 9,186 | 6,262 | 2,924 | NA | NA |
| Candler | 267 | 9,762 | 0.76 | 128 | 11.7 | 128 | 11.7 | 1,883 | 22.2 | 2,421 | 1,275 | 1,146 | 9 | 41 |
| Carroll | 4,114 | 111,609 | 0.89 | 1,545 | 12.7 | 1,186 | 9.8 | 16,536 | 16.7 | 21,544 | 12,138 | 9,407 | NA | NA |
| Catoosa | 468 | 52,445 | 0.54 | 616 | 9.1 | 722 | 10.6 | 6,996 | 12.7 | 14,120 | 8,926 | 5,193 | NA | NA |
| Charlton | 2,571 | 11,623 | 0.65 | 107 | 8.0 | 106 | 7.9 | 1,777 | 20.3 | 2,169 | 1,355 | 814 | NA | NA |
| Chatham | 14,736 | 322,744 | 1.25 | 3,528 | 12.2 | 2,836 | 9.8 | 37,624 | 16.4 | 51,262 | 28,463 | 22,800 | NA | NA |
| Chattahoochee | 2,697 | 19,359 | 2.39 | 198 | 18.8 | 32 | 3.0 | 826 | 10.7 | 611 | 336 | 275 | NA | NA |
| Chattooga | 1,274 | 23,155 | 0.81 | 307 | 12.4 | 312 | 12.6 | 3,309 | 17.3 | 5,799 | 3,162 | 2,637 | NA | NA |
| Cherokee | 1,624 | 198,194 | 0.61 | 2,695 | 10.2 | 1,855 | 7.0 | 31,738 | 14.3 | 43,730 | 26,217 | 17,513 | 74 | 1,145 |
| Clarke | 10,411 | 143,396 | 1.29 | 1,298 | 10.2 | 834 | 6.5 | 17,173 | 16.8 | 17,308 | 10,320 | 6,987 | NA | NA |
| Clay | 57 | 2,671 | 0.69 | 31 | 10.8 | 44 | 15.4 | 343 | 17.3 | 774 | 415 | 359 | NA | NA |
| Clayton | 4,775 | 272,485 | 0.91 | 4,046 | 13.8 | 1,982 | 6.8 | 46,395 | 18.2 | 37,074 | 16,716 | 20,358 | 1,440 | 7,364 |
| Clinch | 126 | 7,066 | 1.17 | 91 | 13.8 | 78 | 11.9 | 962 | 17.9 | 1,469 | 824 | 645 | NA | NA |
| Cobb | 9,693 | 737,286 | 0.96 | 8,769 | 11.5 | 5,210 | 6.8 | 94,503 | 14.4 | 106,424 | 64,235 | 42,188 | 6 | 74 |
| Coffee | 3,277 | 43,995 | 1.06 | 555 | 12.8 | 453 | 10.5 | 6,927 | 20.7 | 7,327 | 4,251 | 3,076 | NA | NA |
| Colquitt | 1,231 | 43,643 | 0.90 | 586 | 12.9 | 504 | 11.1 | 8,674 | 23.5 | 8,848 | 5,071 | 3,777 | NA | NA |
| Columbia | 715 | 119,955 | 0.57 | 1,789 | 11.2 | 1,138 | 7.1 | 16,534 | 12.2 | 24,439 | 16,221 | 8,218 | NA | NA |
| Cook | 143 | 15,754 | 0.80 | 211 | 12.2 | 211 | 12.2 | 2,638 | 18.7 | 3,495 | 2,052 | 1,442 | 36 | 131 |
| Coweta | 637 | 120,271 | 0.67 | 1,630 | 10.8 | 1,211 | 8.0 | 15,969 | 12.5 | 24,596 | 14,109 | 10,488 | 596 | 1,007 |
| Crawford | 142 | 9,616 | 0.41 | 104 | 8.5 | 159 | 13.0 | 1,717 | 17.4 | 2,751 | 1,473 | 1,278 | NA | NA |
| Crisp | 467 | 23,203 | 1.06 | 237 | 10.8 | 317 | 14.4 | 2,888 | 16.7 | 4,480 | 2,313 | 2,167 | NA | NA |
| Dade | 984 | 14,419 | 0.74 | 145 | 9.0 | 198 | 12.3 | 1,972 | 16.5 | 3,737 | 2,317 | 1,421 | NA | NA |
| Dawson | 110 | 21,899 | 0.78 | 291 | 10.7 | 252 | 9.3 | 3,496 | 16.8 | 5,412 | 3,304 | 2,108 | NA | NA |
| Decatur | 1,053 | 25,698 | 0.90 | 349 | 13.2 | 342 | 12.9 | 3,810 | 18.3 | 5,801 | 3,503 | 2,298 | NA | NA |
| DeKalb | 11,844 | 713,623 | 0.90 | 10,514 | 13.8 | 5,495 | 7.2 | 109,204 | 16.9 | 106,989 | 53,718 | 53,271 | NA | NA |
| Dodge | 1,818 | 19,293 | 0.80 | 206 | 10.1 | 295 | 14.4 | 2,518 | 16.8 | 4,188 | 2,570 | 1,618 | NA | NA |
| Dooly | 1,876 | 13,516 | 0.96 | 110 | 8.3 | 133 | 10.1 | 1,612 | 18.7 | 2,214 | 1,175 | 1,039 | NA | NA |
| Dougherty | 4,555 | 102,682 | 1.38 | 1,169 | 13.5 | 1,046 | 12.1 | 11,325 | 16.5 | 17,469 | 9,237 | 8,232 | 38 | 340 |
| Douglas | 1,252 | 124,122 | 0.72 | 1,669 | 11.3 | 1,167 | 7.9 | 18,875 | 14.8 | 21,224 | 11,123 | 10,101 | NA | NA |
| Early | 137 | 9,944 | 0.90 | 124 | 12.4 | 159 | 15.8 | 1,238 | 15.5 | 2,425 | 1,401 | 1,023 | NA | NA |
| Echols | 0 | 2,907 | 0.39 | 36 | 9.0 | 24 | 6.0 | 793 | 23.4 | 562 | 357 | 205 | 3 | 19 |

1. Per 1,000 estimated resident population.

# Table B. States and Counties — Crime, Education, Money Income, and Poverty

| STATE County | County law enforcement employment, 2019 | | Education | | | | | | Money income, 2015-2019 | | | | Income and poverty, 2019 | | | |
|---|---|---|---|---|---|---|---|---|---|---|---|---|---|---|---|---|
| | | | School enrollment and attainment, 2015-2019 | | | | Local government expenditures,[3] 2017-2018 | | Households | | | | | Percent below poverty level | | |
| | | | Enrollment[1] | | Attainment[2] (percent) | | | | | | | Percent | | | | |
| | Officers | Civilians | Total | Percent private | High school graduate or less | Bachelor's degree or more | Total current spending (mil dol) | Current spending per student (dollars) | Per capita income[4] | Median income (dollars) | with income of less than $50,000 | with income of $200,000 or more | Median household income (dollars) | All persons | Children under 18 years | Children 5 to 17 years in families |
| | 46 | 47 | 48 | 49 | 50 | 51 | 52 | 53 | 54 | 55 | 56 | 57 | 58 | 59 | 60 | 61 |
| FLORIDA—Cont'd | | | | | | | | | | | | | | | | |
| Putnam | 120 | 92 | 15,367 | 12.7 | 59.4 | 12.3 | 106 | 9,541 | 20,915 | 37,670 | 63.3 | 1.4 | 41,986 | 22.4 | 35.0 | 34.6 |
| St. Johns | 333 | 362 | 59,478 | 18.7 | 26.6 | 44.7 | 337 | 8,282 | 43,194 | 82,252 | 29.8 | 13.0 | 90,839 | 6.4 | 6.8 | 5.7 |
| St. Lucie | 316 | 425 | 67,262 | 17.3 | 46.9 | 21.6 | 373 | 9,120 | 27,121 | 52,322 | 47.8 | 3.5 | 57,292 | 10.5 | 15.7 | 15.1 |
| Santa Rosa | 220 | 219 | 41,353 | 11.8 | 35.5 | 27.7 | 240 | 8,582 | 31,691 | 67,949 | 35.8 | 5.3 | 66,139 | 9.8 | 12.2 | 11.0 |
| Sarasota | 414 | 560 | 65,198 | 16.2 | 35.8 | 35.4 | 484 | 11,282 | 41,081 | 62,236 | 39.6 | 7.8 | 66,258 | 7.8 | 12.0 | 11.3 |
| Seminole | 437 | 799 | 117,636 | 17.1 | 26.8 | 39.6 | 570 | 8,393 | 35,175 | 66,768 | 37.0 | 7.9 | 69,954 | 9.3 | 11.8 | 10.8 |
| Sumter | NA | NA | 9,424 | 20.2 | 38.2 | 31.8 | 84 | 9,741 | 35,560 | 57,226 | 43.3 | 3.5 | 63,464 | 8.9 | 22.1 | 22.1 |
| Suwannee | NA | NA | 9,591 | 17.7 | 56.6 | 15.5 | 56 | 9,160 | 22,808 | 47,839 | 52.7 | 2.8 | 49,453 | 17.1 | 26.3 | 23.5 |
| Taylor | 55 | 25 | 4,273 | 21.7 | 65.0 | 8.3 | 28 | 9,818 | 17,391 | 40,306 | 59.8 | 0.8 | 42,987 | 19.9 | 27.1 | 26.1 |
| Union | NA | NA | 2,458 | 8.1 | 60.9 | 9.2 | 21 | 9,024 | 17,010 | 44,270 | 54.5 | 1.7 | 50,893 | 19.9 | 20.2 | 18.7 |
| Volusia | 424 | 310 | 110,802 | 19.7 | 41.3 | 23.7 | 538 | 8,541 | 28,517 | 49,494 | 50.5 | 3.3 | 53,554 | 13.1 | 20.0 | 18.0 |
| Wakulla | 56 | 85 | 7,377 | 13.7 | 48.5 | 18.3 | 45 | 8,675 | 25,733 | 61,410 | 42.8 | 2.9 | 58,640 | 12.0 | 16.6 | 14.3 |
| Walton | 263 | 271 | 13,423 | 11.7 | 39.3 | 28.5 | 92 | 9,653 | 34,475 | 58,093 | 43.7 | 6.8 | 62,600 | 10.8 | 17.6 | 16.5 |
| Washington | 42 | 40 | 4,646 | 13.5 | 58.9 | 12.5 | 34 | 9,972 | 18,401 | 37,022 | 62.6 | 1.1 | 42,518 | 20.1 | 29.4 | 27.2 |
| GEORGIA | X | X | 2,749,494 | 14.5 | 40.6 | 31.3 | 18,995 | 10,741 | 31,067 | 58,700 | 43.2 | 6.5 | 61,950 | 13.5 | 19.5 | 18.6 |
| Appling | 29 | 39 | 4,188 | 7.0 | 63.8 | 9.8 | 38 | 10,507 | 21,300 | 40,304 | 57.4 | 1.8 | 43,821 | 20.9 | 31.0 | 29.9 |
| Atkinson | NA | NA | 2,104 | 1.1 | 67.5 | 12.2 | 18 | 10,713 | 20,338 | 37,197 | 59.7 | 2.5 | 38,581 | 23.2 | 36.8 | 36.7 |
| Bacon | 11 | 18 | 2,476 | 7.1 | 60.3 | 10.5 | 23 | 10,564 | 20,326 | 37,519 | 60.7 | 1.2 | 41,807 | 19.2 | 31.1 | 28.5 |
| Baker | 8 | 1 | 711 | 23.3 | 57.3 | 12.9 | 5 | 16,524 | 22,525 | 32,917 | 66.5 | 0.8 | 39,515 | 24.8 | 39.8 | 40.0 |
| Baldwin | 59 | 60 | 13,709 | 13.7 | 49.3 | 23.3 | 55 | 10,450 | 21,716 | 43,672 | 53.9 | 2.1 | 44,429 | 22.9 | 29.4 | 27.2 |
| Banks | 46 | 18 | 4,401 | 11.4 | 62.1 | 13.4 | 30 | 10,585 | 21,403 | 47,811 | 52.8 | 1.6 | 53,481 | 12.8 | 19.1 | 18.6 |
| Barrow | NA | NA | 20,429 | 15.6 | 49.6 | 19.0 | 138 | 10,020 | 25,266 | 62,345 | 39.8 | 2.3 | 68,924 | 9.7 | 13.5 | 13.5 |
| Bartow | 186 | 41 | 24,175 | 11.7 | 49.8 | 18.6 | 180 | 9,970 | 26,493 | 57,423 | 44.3 | 2.9 | 60,254 | 12.5 | 16.1 | 14.9 |
| Ben Hill | NA | NA | 3,925 | 10.3 | 61.7 | 11.2 | 33 | 10,354 | 17,774 | 32,229 | 68.2 | 0.6 | 35,931 | 22.8 | 32.0 | 29.1 |
| Berrien | 26 | 23 | 4,224 | 4.6 | 63.4 | 11.4 | 31 | 9,759 | 19,547 | 40,415 | 59.4 | 0.6 | 42,058 | 20.1 | 28.6 | 27.4 |
| Bibb | 408 | 87 | 41,393 | 22.8 | 44.5 | 25.4 | 274 | 11,112 | 25,640 | 41,334 | 57.0 | 3.9 | 43,066 | 23.2 | 36.4 | 35.5 |
| Bleckley | 17 | 26 | 3,712 | 3.6 | 52.4 | 18.7 | 25 | 9,891 | 20,458 | 48,174 | 52.5 | 1.2 | 48,115 | 18.5 | 25.5 | 24.4 |
| Brantley | NA | NA | 4,053 | 8.6 | 65.7 | 8.9 | 35 | 10,170 | 20,191 | 38,857 | 60.4 | 0.4 | 44,680 | 16.1 | 26.9 | 26.5 |
| Brooks | 24 | 21 | 3,392 | 11.7 | 49.9 | 17.8 | 25 | 11,585 | 25,700 | 38,285 | 61.7 | 3.6 | 39,554 | 21.9 | 35.3 | 34.8 |
| Bryan | 42 | 21 | 11,470 | 14.8 | 32.7 | 33.1 | 82 | 8,909 | 31,623 | 72,624 | 36.1 | 5.9 | 79,296 | 7.8 | 10.4 | 10.0 |
| Bulloch | NA | NA | 28,988 | 6.2 | 39.9 | 27.0 | 116 | 10,795 | 22,328 | 45,550 | 54.0 | 3.4 | 48,788 | 21.9 | 21.4 | 21.5 |
| Burke | 76 | 44 | 5,513 | 13.4 | 58.2 | 13.2 | 61 | 14,532 | 22,173 | 44,151 | 57.7 | 2.2 | 42,632 | 23.6 | 36.3 | 37.1 |
| Butts | NA | NA | 4,870 | 14.1 | 64.1 | 12.2 | 36 | 10,327 | 27,195 | 43,471 | 57.5 | 2.7 | 47,410 | 14.7 | 21.0 | 20.0 |
| Calhoun | NA | NA | 1,376 | 12.0 | 63.8 | 9.1 | 14 | 12,111 | 14,005 | 34,167 | 65.3 | 1.6 | 34,960 | 35.9 | 39.0 | 36.1 |
| Camden | NA | NA | 12,378 | 9.5 | 40.9 | 23.9 | 87 | 9,352 | 29,741 | 56,951 | 44.1 | 3.7 | 58,552 | 13.8 | 20.3 | 19.3 |
| Candler | 18 | 11 | 2,820 | 7.5 | 60.0 | 13.0 | 25 | 8,752 | 20,687 | 33,736 | 62.4 | 1.6 | 41,665 | 23.1 | 34.3 | 33.6 |
| Carroll | 113 | 81 | 33,546 | 9.0 | 48.8 | 21.1 | 196 | 9,810 | 24,993 | 53,737 | 47.0 | 2.9 | 59,193 | 14.9 | 21.8 | 18.2 |
| Catoosa | 72 | 48 | 15,995 | 14.5 | 45.4 | 21.4 | 120 | 11,073 | 27,308 | 56,235 | 43.9 | 3.1 | 60,753 | 10.0 | 15.2 | 13.4 |
| Charlton | 18 | 12 | 2,892 | 12.4 | 66.0 | 8.8 | 18 | 10,372 | 18,323 | 41,961 | 58.4 | 1.6 | 42,891 | 25.9 | 38.2 | 41.2 |
| Chatham | 125 | 19 | 77,973 | 25.3 | 34.3 | 33.6 | 442 | 11,769 | 32,229 | 56,842 | 44.4 | 5.8 | 57,122 | 14.8 | 21.6 | 21.6 |
| Chattahoochee | NA | NA | 2,434 | 10.8 | 34.1 | 29.5 | 10 | 11,931 | 23,769 | 47,096 | 53.5 | 1.9 | 46,106 | 18.5 | 23.5 | 23.9 |
| Chattooga | 28 | 18 | 5,164 | 8.5 | 66.3 | 8.9 | 44 | 10,447 | 18,715 | 36,807 | 65.7 | 1.1 | 41,561 | 17.4 | 22.8 | 22.1 |
| Cherokee | 282 | 76 | 66,150 | 15.5 | 30.3 | 38.2 | 424 | 9,815 | 36,503 | 82,740 | 27.8 | 8.6 | 88,635 | 6.5 | 8.3 | 7.5 |
| Clarke | NA | NA | 50,093 | 6.8 | 31.9 | 44.0 | 185 | 13,593 | 23,827 | 38,623 | 59.9 | 3.7 | 41,036 | 25.7 | 27.3 | 26.5 |
| Clay | 6 | 1 | 667 | 3.6 | 56.3 | 12.5 | 4 | 15,825 | 17,036 | 22,325 | 75.0 | 0.5 | 24,732 | 28.8 | 41.9 | 41.0 |
| Clayton | 347 | 133 | 82,519 | 10.9 | 48.8 | 19.5 | 671 | 9,614 | 20,970 | 47,864 | 52.4 | 1.4 | 50,737 | 16.0 | 24.4 | 25.0 |
| Clinch | 12 | 3 | 1,702 | 0.7 | 64.7 | 10.0 | 15 | 10,665 | 15,152 | 27,658 | 69.3 | 0.2 | 42,928 | 22.0 | 30.6 | 27.9 |
| Cobb | 1,066 | 368 | 202,778 | 16.0 | 26.4 | 47.4 | 1,289 | 10,604 | 40,031 | 77,932 | 30.8 | 10.7 | 79,622 | 8.3 | 11.7 | 10.4 |
| Coffee | 55 | 70 | 10,294 | 9.1 | 61.5 | 13.0 | 79 | 10,231 | 19,905 | 40,859 | 60.2 | 2.3 | 41,675 | 20.2 | 28.0 | 27.2 |
| Colquitt | NA | NA | 11,125 | 5.2 | 64.1 | 12.8 | 99 | 10,304 | 20,529 | 36,435 | 62.5 | 2.4 | 41,899 | 21.9 | 31.9 | 29.9 |
| Columbia | 276 | 62 | 40,051 | 12.8 | 31.1 | 36.4 | 259 | 9,404 | 34,579 | 82,339 | 26.6 | 8.6 | 88,285 | 5.6 | 7.4 | 7.2 |
| Cook | 28 | 24 | 4,221 | 6.6 | 57.2 | 13.5 | 33 | 10,369 | 20,128 | 41,854 | 59.4 | 0.9 | 40,670 | 21.4 | 32.2 | 32.7 |
| Coweta | 166 | 101 | 36,659 | 16.6 | 39.4 | 30.3 | 240 | 10,058 | 33,875 | 75,913 | 32.2 | 6.3 | 79,232 | 9.0 | 12.0 | 10.9 |
| Crawford | NA | NA | 2,712 | 9.1 | 56.9 | 10.5 | 18 | 10,440 | 23,204 | 46,283 | 54.6 | 1.7 | 50,411 | 15.3 | 24.2 | 21.1 |
| Crisp | 59 | 40 | 5,768 | 7.7 | 54.8 | 14.1 | 45 | 11,022 | 22,933 | 36,042 | 65.5 | 2.9 | 40,482 | 26.7 | 43.9 | 46.8 |
| Dade | NA | NA | 4,035 | 35.5 | 57.0 | 15.2 | 22 | 10,257 | 24,000 | 42,581 | 55.8 | 2.2 | 45,986 | 13.5 | 17.3 | 16.1 |
| Dawson | 71 | 42 | 5,017 | 9.1 | 41.9 | 30.6 | 46 | 13,065 | 34,239 | 66,281 | 36.1 | 7.5 | 69,155 | 8.8 | 12.0 | 11.4 |
| Decatur | 49 | 14 | 5,684 | 4.6 | 54.2 | 13.8 | 52 | 10,081 | 21,568 | 41,481 | 59.1 | 2.3 | 40,473 | 23.4 | 35.6 | 35.2 |
| DeKalb | 1,155 | 461 | 195,216 | 22.9 | 31.5 | 44.2 | 1,254 | 11,739 | 36,077 | 62,399 | 40.5 | 9.1 | 63,606 | 12.9 | 18.5 | 16.7 |
| Dodge | 23 | 33 | 4,244 | 6.1 | 58.6 | 14.3 | 38 | 11,997 | 20,216 | 39,302 | 61.1 | 2.0 | 41,118 | 20.6 | 28.8 | 25.6 |
| Dooly | 37 | 38 | 2,966 | 14.2 | 63.4 | 12.1 | 15 | 12,115 | 22,716 | 37,340 | 57.2 | 1.2 | 37,109 | 28.2 | 41.5 | 42.7 |
| Dougherty | 239 | 33 | 27,079 | 8.6 | 46.8 | 21.4 | 165 | 11,371 | 22,059 | 39,584 | 61.7 | 2.5 | 36,894 | 27.6 | 39.5 | 37.2 |
| Douglas | 298 | 39 | 41,238 | 13.2 | 42.6 | 28.2 | 272 | 10,271 | 27,743 | 63,835 | 38.8 | 3.6 | 62,915 | 10.9 | 16.6 | 15.7 |
| Early | 18 | 21 | 2,455 | 8.2 | 56.7 | 15.1 | 24 | 11,716 | 19,026 | 30,640 | 65.8 | 0.7 | 40,011 | 27.3 | 40.8 | 35.6 |
| Echols | 8 | 1 | 835 | 5.5 | 67.8 | 9.7 | 9 | 11,139 | 23,885 | 39,494 | 58.6 | 1.5 | 40,408 | 21.6 | 31.0 | 27.2 |

1. All persons 3 years old and over enrolled in nursery school through college.   2. Persons 25 years old and over.   3. Elementary and secondary education expenditures.   4. Based on population estimated by the American Community Survey, 2014–2018.

# Table B. States and Counties — **Personal Income**

| STATE County | Personal income, 2019 Total (mil dol) [62] | Percent change 2018-2019 [63] | Per capita[1] Dollars [64] | Per capita[1] Rank [65] | Wages and salaries (mil dol) [66] | Pension and insurance [67] | Government social insurance [68] | Proprietors' income (mil dol) [69] | Dividends, interest, and rent (mil dol) [70] | Personal transfer receipts (mil dol) [71] | Earnings, 2019 Total (mil dol) [72] | From employee and self-employed [73] | From employer [74] |
|---|---|---|---|---|---|---|---|---|---|---|---|---|---|
| FLORIDA—Cont'd | | | | | | | | | | | | | |
| Putnam | 2,455 | 3.8 | 32,949 | 2,894 | 726 | 143 | 51 | 107 | 397 | 950 | 1,026 | 88 | 51 |
| St. Johns | 18,659 | 5.5 | 70,498 | 124 | 4,122 | 585 | 281 | 728 | 4,896 | 2,418 | 5,716 | 410 | 281 |
| St. Lucie | 13,501 | 5.7 | 41,125 | 1,904 | 3,861 | 635 | 267 | 576 | 2,766 | 3,717 | 5,339 | 426 | 267 |
| Santa Rosa | 8,475 | 6.3 | 45,981 | 1,250 | 1,917 | 341 | 139 | 433 | 1,493 | 1,747 | 2,830 | 210 | 139 |
| Sarasota | 29,008 | 3.3 | 66,878 | 155 | 9,308 | 1,256 | 631 | 1,831 | 12,686 | 5,757 | 13,027 | 977 | 631 |
| Seminole | 24,137 | 4.3 | 51,156 | 742 | 11,614 | 1,555 | 781 | 1,000 | 4,467 | 3,841 | 14,951 | 960 | 781 |
| Sumter | 6,407 | 5.1 | 48,387 | 987 | 1,519 | 268 | 109 | 226 | 2,092 | 2,636 | 2,122 | 250 | 109 |
| Suwannee | 1,505 | 0.8 | 33,882 | 2,815 | 465 | 91 | 33 | 145 | 215 | 511 | 734 | 55 | 33 |
| Taylor | 670 | 1.4 | 31,054 | 3,000 | 296 | 55 | 20 | 32 | 117 | 233 | 404 | 30 | 20 |
| Union | 342 | 2.9 | 22,440 | 3,108 | 146 | 39 | 10 | 12 | 53 | 112 | 207 | 14 | 10 |
| Volusia | 24,444 | 4.4 | 44,180 | 1,459 | 8,230 | 1,275 | 572 | 1,143 | 5,521 | 6,595 | 11,220 | 867 | 572 |
| Wakulla | 1,271 | 5.5 | 37,660 | 2,364 | 253 | 56 | 17 | 53 | 183 | 276 | 379 | 30 | 17 |
| Walton | 4,776 | 5.6 | 64,481 | 195 | 1,277 | 178 | 89 | 399 | 1,665 | 684 | 1,943 | 133 | 89 |
| Washington | 771 | 3.8 | 30,268 | 3,027 | 251 | 51 | 17 | 37 | 113 | 261 | 356 | 28 | 17 |
| GEORGIA | 512,138 | 3.8 | 48,188 | X | 270,472 | 41,139 | 18,111 | 43,511 | 94,341 | 86,129 | 373,233 | 21,864 | 18,111 |
| Appling | 653 | 3.5 | 35,514 | 2,619 | 374 | 88 | 25 | 60 | 83 | 185 | 547 | 31 | 25 |
| Atkinson | 258 | 3.0 | 31,542 | 2,974 | 123 | 25 | 8 | 22 | 31 | 72 | 179 | 10 | 8 |
| Bacon | 373 | 4.1 | 33,393 | 2,854 | 157 | 33 | 11 | 31 | 48 | 114 | 232 | 14 | 11 |
| Baker | 126 | 10.4 | 41,323 | 1,876 | 22 | 4 | 1 | 19 | 24 | 33 | 47 | 3 | 1 |
| Baldwin | 1,532 | 3.8 | 34,128 | 2,786 | 618 | 150 | 41 | 56 | 277 | 507 | 865 | 55 | 41 |
| Banks | 676 | 0.0 | 35,151 | 2,667 | 161 | 30 | 11 | 55 | 101 | 164 | 257 | 18 | 11 |
| Barrow | 3,116 | 5.9 | 37,436 | 2,396 | 950 | 155 | 64 | 189 | 374 | 607 | 1,358 | 90 | 64 |
| Bartow | 4,256 | 4.4 | 39,505 | 2,114 | 2,020 | 341 | 141 | 220 | 595 | 904 | 2,721 | 169 | 141 |
| Ben Hill | 555 | 3.4 | 33,228 | 2,873 | 220 | 45 | 16 | 27 | 97 | 192 | 309 | 21 | 16 |
| Berrien | 639 | 5.9 | 32,935 | 2,895 | 150 | 36 | 11 | 38 | 109 | 193 | 234 | 16 | 11 |
| Bibb | 6,539 | 2.9 | 42,696 | 1,657 | 4,214 | 673 | 285 | 414 | 1,361 | 1,702 | 5,586 | 340 | 285 |
| Bleckley | 454 | 3.2 | 35,272 | 2,645 | 97 | 28 | 6 | 19 | 92 | 121 | 150 | 11 | 6 |
| Brantley | 540 | 3.6 | 28,248 | 3,075 | 101 | 25 | 7 | 40 | 57 | 174 | 173 | 14 | 7 |
| Brooks | 582 | 5.7 | 37,655 | 2,366 | 142 | 26 | 10 | 42 | 111 | 171 | 220 | 15 | 10 |
| Bryan | 2,134 | 5.4 | 53,853 | 554 | 381 | 77 | 26 | 91 | 302 | 327 | 575 | 37 | 26 |
| Bulloch | 2,605 | 4.2 | 32,724 | 2,907 | 1,056 | 236 | 70 | 132 | 459 | 591 | 1,494 | 87 | 70 |
| Burke | 861 | 7.0 | 38,481 | 2,263 | 1,343 | 183 | 81 | 61 | 114 | 239 | 1,668 | 95 | 81 |
| Butts | 873 | 4.5 | 35,017 | 2,684 | 302 | 59 | 21 | 31 | 130 | 234 | 413 | 28 | 21 |
| Calhoun | 170 | 8.7 | 27,390 | 3,090 | 41 | 11 | 3 | 26 | 29 | 59 | 81 | 5 | 3 |
| Camden | 2,033 | 5.1 | 37,184 | 2,426 | 998 | 263 | 81 | 44 | 395 | 468 | 1,385 | 76 | 81 |
| Candler | 369 | 5.0 | 34,192 | 2,783 | 109 | 25 | 7 | 34 | 58 | 120 | 175 | 12 | 7 |
| Carroll | 4,697 | 3.8 | 39,147 | 2,170 | 2,047 | 362 | 139 | 252 | 764 | 1,048 | 2,800 | 174 | 139 |
| Catoosa | 2,460 | 3.7 | 36,399 | 2,524 | 622 | 111 | 43 | 155 | 292 | 598 | 930 | 69 | 43 |
| Charlton | 335 | 5.8 | 25,020 | 3,103 | 85 | 16 | 6 | 12 | 41 | 108 | 120 | 9 | 6 |
| Chatham | 13,978 | 2.0 | 48,294 | 992 | 8,767 | 1,432 | 625 | 982 | 3,107 | 2,594 | 11,807 | 680 | 625 |
| Chattahoochee | 328 | 3.8 | 30,062 | 3,034 | 889 | 276 | 89 | 2 | 126 | 41 | 1,255 | 44 | 89 |
| Chattooga | 783 | 2.0 | 31,577 | 2,972 | 234 | 53 | 17 | 43 | 107 | 269 | 347 | 26 | 17 |
| Cherokee | 13,703 | 5.2 | 52,953 | 619 | 3,242 | 491 | 219 | 916 | 1,879 | 1,835 | 4,868 | 314 | 219 |
| Clarke | 4,475 | 3.3 | 34,869 | 2,704 | 3,680 | 783 | 235 | 197 | 1,112 | 886 | 4,895 | 260 | 235 |
| Clay | 100 | 5.8 | 35,142 | 2,670 | 20 | 5 | 2 | 5 | 21 | 39 | 32 | 3 | 2 |
| Clayton | 8,377 | 4.1 | 28,665 | 3,071 | 7,999 | 1,479 | 502 | 338 | 1,098 | 2,140 | 10,317 | 583 | 502 |
| Clinch | 259 | 4.8 | 39,136 | 2,171 | 94 | 19 | 7 | 38 | 30 | 78 | 158 | 10 | 7 |
| Cobb | 44,995 | 3.4 | 59,194 | 330 | 25,494 | 3,115 | 1,677 | 3,179 | 8,341 | 4,965 | 33,465 | 1,938 | 1,677 |
| Coffee | 1,474 | 4.3 | 34,073 | 2,794 | 720 | 136 | 51 | 78 | 213 | 399 | 985 | 60 | 51 |
| Colquitt | 1,586 | 4.7 | 34,780 | 2,719 | 593 | 127 | 41 | 158 | 254 | 458 | 920 | 54 | 41 |
| Columbia | 8,021 | 5.5 | 51,184 | 741 | 1,631 | 280 | 111 | 400 | 1,323 | 1,270 | 2,422 | 159 | 111 |
| Cook | 584 | 4.4 | 33,845 | 2,819 | 158 | 37 | 11 | 19 | 95 | 174 | 226 | 16 | 11 |
| Coweta | 7,191 | 4.6 | 48,424 | 982 | 1,939 | 323 | 134 | 228 | 1,061 | 1,082 | 2,625 | 171 | 134 |
| Crawford | 423 | 1.3 | 34,100 | 2,790 | 51 | 11 | 3 | 17 | 57 | 122 | 84 | 8 | 3 |
| Crisp | 792 | 6.3 | 35,421 | 2,632 | 354 | 66 | 25 | 56 | 134 | 235 | 501 | 31 | 25 |
| Dade | 558 | -4.1 | 34,639 | 2,733 | 157 | 32 | 11 | 40 | 88 | 154 | 240 | 18 | 11 |
| Dawson | 1,208 | 5.3 | 46,278 | 1,214 | 347 | 58 | 24 | 92 | 207 | 218 | 521 | 34 | 24 |
| Decatur | 1,014 | 3.9 | 38,393 | 2,281 | 339 | 73 | 23 | 88 | 195 | 283 | 523 | 32 | 23 |
| DeKalb | 40,308 | 3.4 | 53,086 | 608 | 20,051 | 2,869 | 1,361 | 2,913 | 8,147 | 5,365 | 27,194 | 1,579 | 1,361 |
| Dodge | 659 | 3.0 | 31,970 | 2,944 | 192 | 48 | 12 | 25 | 110 | 206 | 277 | 19 | 12 |
| Dooly | 460 | 10.3 | 34,331 | 2,766 | 151 | 31 | 11 | 69 | 77 | 118 | 262 | 14 | 11 |
| Dougherty | 3,259 | 2.9 | 37,050 | 2,444 | 2,304 | 425 | 161 | 150 | 642 | 1,003 | 3,040 | 182 | 161 |
| Douglas | 5,341 | 4.4 | 36,497 | 2,508 | 2,087 | 336 | 144 | 165 | 648 | 1,098 | 2,733 | 173 | 144 |
| Early | 412 | 8.7 | 40,465 | 1,991 | 220 | 45 | 15 | 35 | 67 | 121 | 315 | 18 | 15 |
| Echols | 113 | 1.7 | 28,211 | 3,076 | 18 | 5 | 1 | 3 | 17 | 26 | 27 | 2 | 1 |

1. Based on the resident population estimated as of July 1 of the year shown.

# Table B. States and Counties — Earnings, Social Security, and Housing

| | Earnings, 2019 (cont.) | | | | | | | | | Social Security beneficiaries, December 2019 | | Supplemental Security Income recipients, 2019 | Housing units, 2020 | |
| | Percent by selected industries | | | | | | | | | | | | | |
| STATE County | Farm | Mining, quarrying, and extractions | Construction | Manufacturing | Information; professional, scientific, technical services | Retail trade | Finance, insurance, real estate, and leasing | Health care and social assistance | Government | Number | Rate[1] | | Total | Percent change, 2010-2020 |
|---|---|---|---|---|---|---|---|---|---|---|---|---|---|---|
| | 75 | 76 | 77 | 78 | 79 | 80 | 81 | 82 | 83 | 84 | 85 | 86 | 87 | 88 |
| FLORIDA—Cont'd | | | | | | | | | | | | | | |
| Putnam | 2.3 | D | 5.6 | 13.6 | D | 9.9 | 3.6 | 13.4 | 24.5 | 21,795 | 293 | 2,990 | 37,716 | 1.0 |
| St. Johns | 0.6 | D | 7.5 | 7.2 | 10.5 | 9.2 | 9.0 | 10.9 | 12.6 | 58,625 | 221 | 2,478 | 121,167 | 34.9 |
| St. Lucie | 1.3 | 0.0 | 7.4 | 5.0 | 6.8 | 8.5 | 5.0 | 16.1 | 18.7 | 86,035 | 262 | 7,178 | 149,485 | 9.3 |
| Santa Rosa | 0.7 | 0.3 | 8.7 | 2.5 | 9.3 | 8.5 | 6.7 | 13.3 | 22.8 | 38,795 | 210 | 2,617 | 75,925 | 17.2 |
| Sarasota | 0.1 | 0.1 | 9.8 | 4.7 | 13.3 | 8.4 | 9.5 | 17.7 | 8.8 | 147,050 | 338 | 4,687 | 254,303 | 11.3 |
| Seminole | 0.1 | 0 | 11.6 | 3.8 | 17.7 | 8.1 | 12.4 | 9.9 | 7.9 | 84,730 | 180 | 7,485 | 196,076 | 8.1 |
| Sumter | 0.8 | 0.5 | 10.5 | 5.8 | 6.3 | 8.6 | 7.8 | 16.2 | 18.9 | 74,860 | 560 | 1,680 | 77,152 | 45.5 |
| Suwannee | 7.7 | 0.7 | 6.5 | 14.4 | D | 11.2 | 2.3 | 8.3 | 21.7 | 12,145 | 273 | 1,508 | 19,656 | 2.6 |
| Taylor | 0.7 | D | 7.3 | 28.5 | D | 8.4 | 1.8 | 8.1 | 20.2 | 5,600 | 260 | 763 | 11,209 | 1.9 |
| Union | 1.2 | 0.0 | 5.9 | 0.6 | D | D | 1.3 | D | 57.1 | 2,500 | 164 | 389 | 4,721 | 4.7 |
| Volusia | 1.0 | 0.1 | 7.8 | 7.2 | 7.9 | 9.7 | 6.3 | 18 | 13.1 | 159,085 | 287 | 12,486 | 266,709 | 4.9 |
| Wakulla | 0.2 | 0.1 | 10.7 | 13.7 | 7.0 | 7.4 | 6.5 | 5.2 | 30.5 | 6,965 | 207 | 641 | 14,175 | 10.7 |
| Walton | 0.5 | D | 12.1 | 1.1 | 10.1 | 10.9 | 10.3 | 8.6 | 12.3 | 16,660 | 225 | 966 | 57,416 | 27.2 |
| Washington | 2.9 | D | 7.8 | 2.6 | D | 7.9 | 2.6 | D | 33.4 | 6,305 | 246 | 852 | 11,059 | 2.4 |
| GEORGIA | 0.5 | 0.3 | 5.9 | 8.4 | 15.3 | 5.8 | 9.4 | 9.7 | 15.5 | 1,871,862 | 176 | 259,199 | 4,426,846 | 8.3 |
| Appling | 8.0 | 0.0 | 4.3 | 8.1 | 1.5 | 8.5 | 3.6 | D | 13.7 | 4,250 | 231 | 639 | 8,630 | 1.4 |
| Atkinson | 9.0 | D | 1.7 | 32.9 | D | 2.3 | D | 3.8 | 12.7 | 1,545 | 187 | 357 | 3,513 | -0.3 |
| Bacon | 8.7 | D | D | 16.9 | D | 4.0 | D | D | 15.0 | 2,405 | 217 | 373 | 4,831 | 0.6 |
| Baker | 44.6 | 0.0 | 0.1 | 0.3 | D | 1.8 | D | 3.2 | 13.6 | 780 | 256 | 122 | 1,665 | 0.9 |
| Baldwin | -0.1 | D | 3.6 | 11.3 | D | 9.5 | 3.9 | 14.5 | 36.7 | 9,900 | 220 | 1,525 | 20,796 | 3.0 |
| Banks | 8.4 | 0.1 | 9.0 | 10.7 | D | 8.1 | D | 3.9 | 19.0 | 4,110 | 214 | 316 | 7,911 | 4.3 |
| Barrow | -0.8 | D | 12.5 | 11.1 | 3.9 | 13.4 | 5.3 | 8.1 | 15.1 | 14,515 | 174 | 1,884 | 30,534 | 15.7 |
| Bartow | 0.2 | 0.4 | 7.5 | 26.6 | 4.7 | 7.6 | 3.2 | 7.9 | 12.6 | 20,790 | 193 | 2,283 | 43,093 | 8.3 |
| Ben Hill | 3.6 | 0.0 | 1.8 | 21.1 | 1.9 | 7.5 | 3.3 | D | 18.0 | 4,030 | 241 | 874 | 8,060 | 1.3 |
| Berrien | 9.6 | 0.0 | 3.6 | 30.3 | D | 5.8 | 4.7 | D | 22.4 | 4,240 | 219 | 788 | 8,873 | 1.9 |
| Bibb | 0.0 | 0.6 | 3.1 | 7.9 | 6.7 | 7.0 | 14.9 | 22.4 | 11.9 | 33,495 | 219 | 8,078 | 70,009 | 0.2 |
| Bleckley | 5.8 | 0.0 | 6.8 | 2.1 | 2.4 | 9.8 | D | 5.2 | 43.2 | 2,750 | 213 | 403 | 5,336 | 0.6 |
| Brantley | | D | 10.4 | 11.5 | D | 4.4 | D | 3.8 | 25.3 | 3,800 | 199 | 521 | 8,240 | 1.9 |
| Brooks | 16.7 | 0.0 | 2.7 | 5.8 | D | 5.1 | 3 | 7.6 | 16.4 | 3,910 | 252 | 771 | 7,871 | 2.2 |
| Bryan | 0.4 | 0.1 | 10.1 | 10.3 | D | 9.1 | 8.1 | 7.8 | 25.9 | 6,495 | 164 | 656 | 15,845 | 33.9 |
| Bulloch | 1.4 | 0.0 | 6.4 | 7.9 | D | 7.8 | 6.5 | 12.5 | 32.6 | 11,690 | 147 | 1,713 | 32,327 | 12.2 |
| Burke | 2.2 | 0.0 | D | 2.1 | 3.2 | 1.9 | 1.2 | D | 5.9 | 5,385 | 239 | 1,030 | 10,221 | 3.6 |
| Butts | 0.5 | 0.0 | D | 13.5 | D | 7.6 | 3.2 | D | 23.5 | 5,495 | 221 | 761 | 9,596 | 2.5 |
| Calhoun | 30.5 | D | D | D | D | 4.9 | D | D | 31.4 | 1,290 | 209 | 304 | 2,404 | -0.2 |
| Camden | 0.0 | 0.1 | D | 6.6 | 3.6 | 5.3 | 3.8 | 4.8 | 59.4 | 9,955 | 182 | 843 | 22,874 | 8.3 |
| Candler | 6.5 | 0.2 | 8.3 | 3.2 | D | 9.5 | D | D | 23.3 | 2,605 | 240 | 464 | 4,809 | 1.1 |
| Carroll | 0.8 | D | 7.9 | 20.2 | D | 6.4 | 4.3 | 16.5 | 16.9 | 23,630 | 197 | 3,446 | 46,599 | 4.4 |
| Catoosa | 0.5 | D | 8.6 | 10.2 | D | 15.0 | 7.2 | 10.1 | 17.9 | 14,885 | 220 | 851 | 28,025 | 5.4 |
| Charlton | 0.8 | D | 6.2 | 16.2 | D | 5.7 | D | 0.7 | 22.8 | 2,430 | 183 | 374 | 4,585 | 2.5 |
| Chatham | 0.0 | 0.0 | 4.6 | 16.2 | 5.7 | 6.6 | 6.1 | 13 | 16.8 | 53,780 | 185 | 6,959 | 128,473 | 7.7 |
| Chattahoochee | -0.1 | 0.0 | 0.6 | 0.0 | D | 0.1 | D | 0.2 | 89.8 | 740 | 67 | 150 | 3,329 | -1.4 |
| Chattooga | 0.5 | 0.2 | 5.2 | 38.1 | 2.2 | 6.9 | 5 | D | 21.9 | 6,495 | 261 | 1,003 | 11,053 | 0.6 |
| Cherokee | 0.1 | 0.1 | 14.9 | 9.4 | 8.3 | 8.7 | 7.2 | 10.3 | 12.9 | 44,105 | 170 | 2,450 | 99,866 | 21.2 |
| Clarke | 0.0 | D | 3.3 | 9.4 | 3.8 | 6.6 | 7.4 | 17 | 34.9 | 18,355 | 143 | 3,113 | 54,805 | 7.4 |
| Clay | 12.2 | 0.0 | D | 0.8 | 0.2 | 4.5 | D | 10.9 | 38.5 | 815 | 281 | 153 | 2,134 | 1.5 |
| Clayton | 0.0 | D | 4.2 | 3.6 | 1.7 | 4.4 | 2.3 | 5.5 | 10.2 | 40,905 | 140 | 9,123 | 107,099 | 2.1 |
| Clinch | 5.8 | 0.0 | 1.2 | 34.0 | D | 4.7 | D | 2.4 | 17.8 | 1,675 | 252 | 370 | 3,020 | 0.5 |
| Cobb | 0.0 | 0.1 | 10.1 | 5.6 | 18.2 | 7.3 | 9.5 | 9.2 | 8.2 | 106,860 | 140 | 9,408 | 307,899 | 7.5 |
| Coffee | 3.6 | 0.0 | 5.9 | 19.7 | 2.4 | 9.1 | 3.9 | D | 15.8 | 8,160 | 189 | 1,589 | 17,483 | 2.5 |
| Colquitt | 10.4 | 0.1 | 3.8 | 15.8 | D | 8.1 | 6.3 | D | 25.9 | 9,765 | 214 | 1,934 | 18,837 | 2.9 |
| Columbia | 0.0 | D | 9.3 | 10.4 | 8.7 | 11.9 | 7.8 | 12.2 | 17.9 | 26,080 | 166 | 1,780 | 61,324 | 26.2 |
| Cook | 2.9 | D | 7.6 | 13.7 | D | 8.4 | D | D | 31.5 | 3,910 | 227 | 719 | 7,539 | 3.5 |
| Coweta | 0.1 | D | 6.9 | 14.6 | 5.0 | 10.1 | 5.1 | 18.5 | 14.7 | 25,775 | 174 | 2,201 | 57,771 | 15.1 |
| Crawford | 15.4 | D | 10.6 | 1.1 | D | 3.0 | D | 8.6 | 28.5 | 2,905 | 235 | 248 | 5,453 | 3.4 |
| Crisp | 4.7 | D | 2.2 | 14.1 | D | 8.9 | 4.4 | D | 17.5 | 4,915 | 220 | 1,024 | 10,829 | 0.9 |
| Dade | 1.4 | 0.1 | D | 26.6 | D | 6.4 | 5.2 | 7.2 | 13.7 | 4,095 | 254 | 352 | 7,349 | 0.6 |
| Dawson | 1.7 | 0.3 | 6.5 | 12.7 | 5.4 | 21.2 | 6.9 | 9.1 | 14.5 | 5,580 | 214 | 434 | 12,578 | 20.4 |
| Decatur | 9.5 | D | 3.8 | 9.5 | 3.1 | 9.0 | 7.5 | D | 24.6 | 6,335 | 239 | 1,358 | 12,268 | 1.2 |
| DeKalb | 0.0 | D | 4.9 | 3.9 | 13.9 | 6.0 | 7.9 | 11.8 | 14.1 | 107,020 | 141 | 18,214 | 319,238 | 4.7 |
| Dodge | 4.2 | 0.1 | 3.0 | 12.5 | D | 8.3 | 3.2 | D | 35.5 | 4,610 | 223 | 824 | 9,927 | 0.7 |
| Dooly | 20.9 | 0.0 | D | D | 0.3 | 3.3 | 3.2 | D | 17.0 | 2,675 | 200 | 698 | 6,220 | -1.7 |
| Dougherty | 0.7 | 0.0 | 4.0 | 9.4 | 6.1 | 7.0 | 5.1 | 19.2 | 23.3 | 19,185 | 218 | 5,018 | 40,639 | -0.4 |
| Douglas | 0.0 | D | 6.6 | 10.1 | 3.8 | 10.3 | 4.8 | 10.2 | 15.4 | 23,265 | 159 | 2,976 | 53,860 | 4.3 |
| Early | 8.9 | 0.0 | D | D | D | 6.3 | 3.6 | D | 23.5 | 2,615 | 257 | 583 | 4,985 | 0.2 |
| Echols | -2.4 | 0.0 | D | D | 0.1 | D | D | D | 37.4 | 590 | 149 | 34 | 1,621 | 4.2 |

1. Per 1,000 resident population estimated as of July 1 of the year shown.

| STATE County | Total (89) | Percent (90) | Median value[1] (91) | With a mortgage (92) | Without a mortgage[2] (93) | Median rent[3] (94) | Median rent as a percent of income[2] (95) | Sub-standard units[4] (percent) (96) | Total (97) | Percent change, 2019-2020 (98) | Total (99) | Rate[5] (100) | Total (101) | Management, business, science, and arts (102) | Construction, production, and maintenance occupations (103) |
|---|---|---|---|---|---|---|---|---|---|---|---|---|---|---|---|
| **FLORIDA—Cont'd** | | | | | | | | | | | | | | | |
| Putnam | 28,943 | 70.7 | 89,100 | 21.8 | 10.0 | 728 | 31.6 | 2.7 | 26,470 | -0.4 | 2,092 | 7.9 | 25,868 | 24.3 | 31.9 |
| St. Johns | 88,773 | 80.4 | 304,700 | 21.4 | 10.0 | 1,312 | 30.6 | 1.1 | 133,843 | -1.4 | 7,071 | 5.3 | 112,316 | 46.4 | 13.5 |
| St. Lucie | 114,761 | 73.2 | 186,900 | 24.6 | 13.1 | 1,185 | 33.7 | 2.5 | 145,363 | -0.8 | 11,303 | 7.8 | 132,350 | 28.7 | 24.2 |
| Santa Rosa | 63,514 | 76.0 | 199,200 | 20.7 | 10.0 | 1,129 | 28.2 | 2.0 | 82,647 | -1.4 | 4,299 | 5.2 | 75,427 | 38.3 | 21.0 |
| Sarasota | 182,842 | 74.9 | 251,600 | 23.6 | 12.0 | 1,277 | 31.6 | 1.6 | 184,464 | -2.4 | 12,589 | 6.8 | 170,059 | 35.2 | 18.3 |
| Seminole | 173,668 | 64.6 | 242,600 | 21.3 | 10.5 | 1,242 | 30.0 | 2.5 | 237,769 | -6.4 | 17,167 | 7.2 | 232,498 | 45.5 | 14.7 |
| Sumter | 56,230 | 89.7 | 258,600 | 23.6 | 10.0 | 914 | 37.0 | 1.0 | 32,225 | -1.3 | 2,534 | 7.9 | 25,261 | 30.6 | 20.7 |
| Suwannee | 14,888 | 74.3 | 101,300 | 20.3 | 10.0 | 722 | 23.0 | 3.9 | 17,758 | -2.9 | 1,000 | 5.6 | 16,778 | 27.1 | 29.8 |
| Taylor | 7,405 | 77.4 | 85,100 | 20.1 | 11.2 | 666 | 30.2 | 4.5 | 8,169 | -0.8 | 510 | 6.2 | 6,426 | 24.9 | 30.6 |
| Union | 3,960 | 66.1 | 118,400 | 21.2 | 10.0 | 657 | 23.2 | 3.4 | 4,552 | -1.2 | 216 | 4.7 | 3,671 | 29.8 | 24.8 |
| Volusia | 216,495 | 70.3 | 175,000 | 23.1 | 12.4 | 1,046 | 34.0 | 1.4 | 249,560 | -1.8 | 18,975 | 7.6 | 226,928 | 31.8 | 21.8 |
| Wakulla | 11,200 | 82.0 | 144,000 | 19.2 | 10.0 | 896 | 30.3 | 3.3 | 14,885 | -3.4 | 669 | 4.5 | 13,623 | 38.3 | 19.4 |
| Walton | 27,420 | 74.6 | 229,500 | 22.5 | 10.0 | 979 | 32.6 | 3.3 | 31,957 | -0.5 | 1,887 | 5.9 | 30,611 | 35.6 | 18.2 |
| Washington | 8,827 | 77.7 | 113,200 | 23.3 | 11.6 | 717 | 29.8 | 3.1 | 9,478 | -2.4 | 549 | 5.8 | 8,481 | 28.8 | 29.0 |
| **GEORGIA** | 3,758,798 | 63.3 | 176,000 | 20.1 | 10.0 | 1,006 | 29.6 | 2.5 | 5,072,155 | -1.5 | 330,964 | 6.5 | 4,834,622 | 37.6 | 23.8 |
| Appling | 6,656 | 76.4 | 80,200 | 18.7 | 11.8 | 560 | 20.1 | 3.1 | 9,105 | -1.8 | 431 | 4.7 | 7,174 | 27.7 | 40.4 |
| Atkinson | 2,880 | 73.3 | 62,700 | 19.1 | 10.0 | 454 | 25.8 | 4.1 | 4,434 | -2.8 | 172 | 3.9 | 3,256 | 24.4 | 47.2 |
| Bacon | 3,874 | 72.4 | 81,100 | 18.8 | 10.0 | 526 | 26.0 | 2.9 | 4,652 | -6.2 | 227 | 4.9 | 4,108 | 27.8 | 36.6 |
| Baker | 1,425 | 65.3 | 81,800 | 24.1 | 13.5 | 513 | 19.4 | 3.4 | 1,145 | -1.2 | 69 | 6.0 | 1,092 | 28.8 | 35.6 |
| Baldwin | 16,191 | 58.5 | 120,300 | 19.6 | 11.6 | 703 | 34.2 | 1.8 | 17,607 | -0.6 | 1,219 | 6.9 | 18,464 | 31.1 | 24.9 |
| Banks | 6,740 | 75.4 | 151,100 | 22.5 | 11.6 | 731 | 26.0 | 3.5 | 9,778 | -0.2 | 419 | 4.3 | 7,713 | 26.7 | 36.2 |
| Barrow | 26,305 | 75.6 | 159,700 | 19.9 | 10.0 | 994 | 27.6 | 2.5 | 40,112 | -2.9 | 2,055 | 5.1 | 37,291 | 30.3 | 30.6 |
| Bartow | 37,627 | 66.3 | 163,900 | 20.1 | 10.7 | 937 | 27.6 | 3.2 | 49,719 | -2.2 | 3,107 | 6.2 | 48,127 | 30.3 | 31.6 |
| Ben Hill | 6,443 | 58.8 | 83,400 | 24.1 | 12.8 | 656 | 27.3 | 0.2 | 5,460 | 0.0 | 377 | 6.9 | 6,130 | 22.9 | 37.8 |
| Berrien | 7,367 | 64.2 | 88,000 | 19.5 | 10.0 | 653 | 25.3 | 2.9 | 7,405 | -2.5 | 428 | 5.8 | 7,370 | 23.4 | 40.8 |
| Bibb | 58,116 | 52.3 | 120,200 | 19.9 | 11.3 | 818 | 33.0 | 2.4 | 67,421 | -0.8 | 5,050 | 7.5 | 63,504 | 33.8 | 20.4 |
| Bleckley | 4,176 | 74.3 | 102,300 | 18.1 | 11.6 | 620 | 25.9 | 1.3 | 4,501 | -2.0 | 292 | 6.5 | 4,535 | 29.1 | 33.4 |
| Brantley | 6,823 | 80.3 | 72,300 | 19.9 | 14.4 | 622 | 28.9 | 3.2 | 6,995 | -3.5 | 377 | 5.4 | 6,769 | 25.1 | 39.3 |
| Brooks | 6,335 | 74.0 | 107,800 | 19.1 | 13.4 | 685 | 29.9 | 2.6 | 6,842 | -0.5 | 354 | 5.2 | 6,074 | 25.6 | 29.9 |
| Bryan | 13,048 | 70.1 | 223,900 | 20.3 | 10.0 | 1,242 | 32.1 | 1.2 | 18,718 | -2.5 | 986 | 5.3 | 16,208 | 41.3 | 22.3 |
| Bulloch | 27,375 | 53.9 | 140,700 | 19.4 | 10.0 | 805 | 31.8 | 2.3 | 37,020 | -1.2 | 2,419 | 6.5 | 34,290 | 32.0 | 27.5 |
| Burke | 8,193 | 70.8 | 88,700 | 18.2 | 10.2 | 616 | 30.3 | 4.5 | 9,026 | -1.2 | 647 | 7.2 | 9,366 | 22.9 | 41.7 |
| Butts | 8,279 | 69.8 | 132,600 | 20.8 | 11.7 | 850 | 31.0 | 2.8 | 10,739 | -2.3 | 664 | 6.2 | 9,503 | 28.3 | 30.9 |
| Calhoun | 1,736 | 65.6 | 51,500 | 22.7 | 14.7 | 543 | 22.0 | 1.7 | 2,237 | -1.2 | 113 | 5.1 | 1,677 | 29.6 | 30.8 |
| Camden | 19,338 | 62.5 | 165,700 | 20.2 | 10.0 | 945 | 28.5 | 2.5 | 21,089 | -0.9 | 997 | 4.7 | 21,519 | 33.1 | 25.3 |
| Candler | 4,013 | 57.6 | 96,100 | 21.2 | 10.0 | 597 | 35.8 | 6.8 | 5,633 | 0.2 | 250 | 4.4 | 4,389 | 20.2 | 36.4 |
| Carroll | 41,903 | 66.9 | 137,800 | 20.3 | 10.0 | 877 | 26.1 | 2.5 | 54,475 | -2.2 | 3,466 | 6.4 | 53,011 | 30.2 | 30.8 |
| Catoosa | 24,778 | 74.0 | 144,300 | 19.1 | 10.0 | 775 | 26.5 | 1.9 | 32,611 | -2.7 | 1,417 | 4.3 | 31,272 | 30.5 | 28.6 |
| Charlton | 3,675 | 71.3 | 83,300 | 19.2 | 10.0 | 557 | 23.5 | 1.3 | 4,842 | -1.5 | 182 | 3.8 | 4,109 | 24.3 | 38.0 |
| Chatham | 108,568 | 54.7 | 194,500 | 21.1 | 11.8 | 1,085 | 29.8 | 2.1 | 140,631 | -0.4 | 10,713 | 7.6 | 138,578 | 36.6 | 21.1 |
| Chattahoochee | 2,570 | 24.6 | 71,300 | 18.1 | 12.2 | 1,192 | 28.5 | 3.7 | 1,988 | -2.1 | 124 | 6.2 | 2,219 | 36.1 | 27.1 |
| Chattooga | 9,260 | 68.2 | 72,200 | 19.9 | 11.4 | 608 | 26.6 | 7.6 | 9,767 | -6.0 | 820 | 8.4 | 9,098 | 20.4 | 46.5 |
| Cherokee | 88,137 | 76.8 | 253,500 | 19.6 | 10.0 | 1,241 | 28.6 | 2.0 | 133,045 | -3.1 | 6,464 | 4.9 | 127,265 | 43.0 | 18.0 |
| Clarke | 48,844 | 39.0 | 170,700 | 19.8 | 10.0 | 856 | 35.4 | 2.5 | 58,206 | -2.6 | 3,753 | 6.4 | 61,423 | 40.7 | 19.5 |
| Clay | 1,242 | 72.9 | 71,400 | 29.9 | 17.9 | 340 | 29.9 | 1.5 | 896 | 1.6 | 111 | 12.4 | 916 | 25.7 | 27.4 |
| Clayton | 94,279 | 49.5 | 108,500 | 22.2 | 10.0 | 991 | 30.9 | 4.0 | 139,790 | 1.2 | 14,433 | 10.3 | 129,683 | 24.5 | 31.0 |
| Clinch | 2,477 | 74.9 | 61,000 | 22.8 | 16.5 | 439 | 27.8 | 8.3 | 2,943 | 4.3 | 117 | 4.0 | 2,232 | 27.4 | 38.2 |
| Cobb | 280,374 | 64.5 | 253,900 | 19.2 | 10.0 | 1,202 | 28.1 | 2.1 | 414,357 | -2.3 | 24,453 | 5.9 | 396,869 | 46.9 | 16.7 |
| Coffee | 14,438 | 63.1 | 100,600 | 21.8 | 11.1 | 623 | 24.4 | 2.6 | 18,938 | -1.5 | 1,103 | 5.8 | 16,896 | 29.9 | 34.4 |
| Colquitt | 15,505 | 63.0 | 85,600 | 19.6 | 11.5 | 647 | 31.9 | 3.5 | 21,917 | 0.7 | 979 | 4.5 | 18,685 | 23.2 | 39.4 |
| Columbia | 47,215 | 79.5 | 203,400 | 19.3 | 10.0 | 1,149 | 25.6 | 1.3 | 74,411 | -2.3 | 3,145 | 4.2 | 68,738 | 44.3 | 18.3 |
| Cook | 6,217 | 66.4 | 87,500 | 22.0 | 14.0 | 737 | 26.8 | 2.7 | 8,102 | -0.8 | 384 | 4.7 | 7,128 | 25.7 | 36.5 |
| Coweta | 52,035 | 72.8 | 211,600 | 19.4 | 10.0 | 1,096 | 29.4 | 2.1 | 73,981 | -1.9 | 4,584 | 6.2 | 70,819 | 35.4 | 26.3 |
| Crawford | 4,510 | 77.9 | 93,400 | 20.7 | 10.6 | 709 | 26.3 | 1.2 | 5,488 | -2.8 | 277 | 5.0 | 4,707 | 23.7 | 35.0 |
| Crisp | 8,585 | 54.5 | 83,100 | 19.9 | 11.0 | 649 | 33.4 | 3.1 | 9,515 | -0.9 | 748 | 7.9 | 9,049 | 28.0 | 28.4 |
| Dade | 6,195 | 71.5 | 123,500 | 20.7 | 10.6 | 724 | 22.6 | 2.5 | 7,860 | -3.3 | 322 | 4.1 | 7,057 | 24.5 | 34.4 |
| Dawson | 9,041 | 81.5 | 230,300 | 19.7 | 10.0 | 918 | 25.8 | 2.0 | 12,319 | -3.2 | 584 | 4.7 | 12,035 | 38.7 | 24.2 |
| Decatur | 10,084 | 57.5 | 110,900 | 20.5 | 11.2 | 677 | 25.7 | 6.7 | 11,304 | -3.2 | 541 | 4.8 | 10,317 | 25.8 | 31.5 |
| DeKalb | 282,436 | 54.6 | 215,600 | 20.6 | 10.3 | 1,169 | 31.3 | 2.6 | 395,683 | -0.6 | 31,474 | 8.0 | 382,970 | 45.3 | 17.8 |
| Dodge | 7,628 | 67.8 | 76,300 | 18.9 | 11.1 | 620 | 24.6 | 3.1 | 6,957 | -2.9 | 415 | 6.0 | 7,878 | 30.2 | 32.3 |
| Dooly | 5,020 | 69.6 | 83,300 | 18.6 | 12.3 | 543 | 25.5 | 2.3 | 4,925 | -3.0 | 378 | 7.7 | 5,089 | 24.3 | 37.0 |
| Dougherty | 34,087 | 46.0 | 103,500 | 21.2 | 12.4 | 746 | 30.7 | 2.7 | 37,712 | 0.0 | 3,243 | 8.6 | 34,980 | 29.3 | 26.3 |
| Douglas | 49,187 | 64.3 | 161,600 | 19.7 | 10.0 | 1,087 | 28.6 | 2.1 | 72,297 | -1.3 | 5,367 | 7.4 | 69,595 | 35.3 | 26.4 |
| Early | 4,074 | 64.0 | 86,900 | 27.2 | 14.2 | 709 | 28.6 | 6.5 | 4,424 | 0.9 | 256 | 5.8 | 3,703 | 28.7 | 40.3 |
| Echols | 1,561 | 67.8 | 59,100 | 19.1 | 10.0 | 669 | 25.3 | 2.8 | 1,899 | -0.7 | 74 | 3.9 | 1,758 | 17.7 | 38.3 |

1. Specified owner-occupied units.    2. A value of 10.0 represents 10 percent or less; a value of 50.0 represents 50 percent or more.    3. Specified renter-occupied units.    4. Overcrowded or lacking complete plumbing facilities.    5. Percent of civilian labor force.    6. Civilian employed persons 16 years old and over.

| | Private nonfarm establishments, employment and payroll, 2019 | | | | | | | | | Agriculture, 2017 | | | |
|---|---|---|---|---|---|---|---|---|---|---|---|---|---|
| | | Employment | | | | | | Annual payroll | | Farms | | | Farm producers whose primary occupation is farming (percent) |
| | | | | | | | Professional, scientific, and technical services | | | | | Percent with: | | |
| STATE County | Number of establishments | Total | Health care and social assistance | Manufacturing | Retail trade | Finance and insurance | | Total (mil dol) | Average per employee (dollars) | Number | Fewer than 50 acres | 1000 acres or more | |
| | 104 | 105 | 106 | 107 | 108 | 109 | 110 | 111 | 112 | 113 | 114 | 115 | 116 |

| FLORIDA—Cont'd | | | | | | | | | | | | | |
|---|---|---|---|---|---|---|---|---|---|---|---|---|---|
| Putnam | 1,234 | 12,733 | 2,623 | 1,917 | 2,756 | 299 | 240 | 462 | 36,278 | 564 | 68.1 | 2.8 | 38.2 |
| St. Johns | 6,804 | 64,619 | 8,429 | 2,118 | 11,684 | 1,811 | 3,589 | 2,613 | 40,431 | 253 | 68.8 | 3.2 | 47.9 |
| St. Lucie | 6,124 | 65,928 | 13,187 | 4,330 | 11,886 | 1,208 | 2,701 | 2,508 | 38,036 | 415 | 55.7 | 13.3 | 49.2 |
| Santa Rosa | 2,933 | 28,587 | 4,507 | 845 | 5,759 | 1,065 | 1,590 | 1,049 | 36,704 | 699 | 63.2 | 2.9 | 38.6 |
| Sarasota | 14,331 | 149,363 | 29,165 | 7,000 | 24,650 | 4,931 | 9,972 | 6,848 | 45,849 | 292 | 71.2 | 5.8 | 41.7 |
| Seminole | 14,014 | 185,304 | 21,443 | 7,358 | 27,689 | 16,658 | 13,853 | 8,913 | 48,099 | 403 | 86.4 | 0.7 | 40.1 |
| Sumter | 1,665 | 26,554 | 5,118 | 1,040 | 4,320 | 672 | 666 | 1,014 | 38,168 | 1,307 | 75.7 | 1.9 | 37.0 |
| Suwannee | 660 | 8,791 | 1,450 | 2,048 | 1,933 | 146 | 172 | 288 | 32,749 | 1,079 | 56.3 | 2.8 | 48.2 |
| Taylor | 400 | 4,652 | 547 | 1,564 | 892 | 65 | 98 | 187 | 40,272 | 240 | 41.3 | 9.2 | 39.5 |
| Union | 125 | 1,609 | 319 | 148 | 239 | 17 | 2 | 63 | 38,973 | 308 | 62.7 | 3.6 | 40.5 |
| Volusia | 13,224 | 156,337 | 28,221 | 9,710 | 27,457 | 5,328 | 7,976 | 5,943 | 38,015 | 1,575 | 83.1 | 1.5 | 39.7 |
| Wakulla | 453 | 3,656 | 179 | 501 | 853 | 71 | 159 | 121 | 33,145 | 209 | 71.8 | 1.0 | 32.9 |
| Walton | 2,569 | 22,400 | 2,018 | 264 | 4,685 | 475 | 753 | 837 | 37,345 | 598 | 50.0 | 3.3 | 42.8 |
| Washington | 410 | 4,186 | 701 | 284 | 717 | 81 | 493 | 136 | 32,473 | 437 | 59.7 | 0.9 | 34.4 |
| GEORGIA | 239,034 | 4,040,559 | 529,248 | 393,260 | 477,602 | 182,902 | 294,597 | 212,578 | 52,611 | 42,439 | 42.3 | 5.3 | 39.4 |
| Appling | 381 | 5,608 | 937 | 598 | 981 | 142 | 37 | 298 | 53,157 | 548 | 43.8 | 6.6 | 39.2 |
| Atkinson | 92 | 1,566 | 33 | 709 | 156 | 59 | NA | 54 | 34,353 | 215 | 34.4 | 11.6 | 45.3 |
| Bacon | 238 | 3,280 | 558 | 807 | 315 | 138 | 59 | 109 | 33,093 | 273 | 38.1 | 4.8 | 39.5 |
| Baker | 30 | 262 | 24 | NA | 31 | 5 | NA | 10 | 36,561 | 147 | 29.3 | 21.8 | 56.1 |
| Baldwin | 811 | 12,959 | 3,781 | 1,108 | 2,382 | 322 | 233 | 463 | 35,765 | 139 | 36.0 | 6.5 | 23.4 |
| Banks | 287 | 3,388 | 164 | 294 | 790 | 26 | 54 | 97 | 28,733 | 463 | 40.0 | 0.9 | 48.9 |
| Barrow | 1,314 | 18,540 | 1,600 | 3,404 | 2,915 | 314 | 675 | 690 | 37,226 | 288 | 58.0 | 0.3 | 37.4 |
| Bartow | 2,163 | 36,040 | 3,191 | 9,478 | 4,247 | 655 | 737 | 1,528 | 42,387 | 469 | 56.5 | 1.9 | 39.9 |
| Ben Hill | 295 | 4,708 | 477 | 1,977 | 724 | 125 | 41 | 158 | 33,578 | 217 | 41.0 | 5.5 | 40.9 |
| Berrien | 247 | 3,100 | 310 | 1,562 | 371 | 141 | 32 | 134 | 43,220 | 349 | 33.5 | 8.6 | 48.4 |
| Bibb | 4,108 | 74,800 | 15,795 | 5,722 | 9,764 | 9,723 | 2,407 | 3,279 | 43,840 | 98 | 55.1 | 1.0 | 44.6 |
| Bleckley | 169 | 1,590 | 271 | 74 | 398 | 76 | 24 | 47 | 29,539 | 231 | 43.7 | 5.2 | 21.6 |
| Brantley | 203 | 1,397 | 159 | 143 | 314 | 33 | 20 | 50 | 35,580 | 235 | 49.4 | NA | 33.9 |
| Brooks | 210 | 2,088 | 563 | 158 | 341 | 69 | 40 | 74 | 35,225 | 360 | 31.1 | 11.9 | 43.8 |
| Bryan | 708 | 6,376 | 751 | 346 | 1,255 | 168 | 297 | 209 | 32,837 | 95 | 58.9 | 8.4 | 35.8 |
| Bulloch | 1,495 | 18,826 | 3,586 | 1,570 | 3,480 | 528 | 720 | 603 | 32,027 | 478 | 30.5 | 11.9 | 38.3 |
| Burke | 346 | 9,936 | 637 | 305 | 855 | 124 | 115 | 840 | 84,558 | 467 | 25.3 | 14.6 | 43.7 |
| Butts | 374 | 5,337 | 586 | 894 | 911 | 91 | 73 | 198 | 37,075 | 173 | 37.0 | 1.7 | 44.6 |
| Calhoun | 61 | 605 | 74 | NA | 62 | 31 | 4 | 24 | 39,398 | 169 | 20.1 | 16.0 | 40.1 |
| Camden | 857 | 9,529 | 1,029 | 228 | 1,857 | 288 | 1,368 | 322 | 33,770 | 47 | 70.2 | 2.1 | 28.4 |
| Candler | 226 | 2,341 | 438 | 288 | 387 | 100 | 104 | 73 | 31,178 | 197 | 32.0 | 6.1 | 30.5 |
| Carroll | 2,131 | 38,833 | 5,757 | 10,604 | 5,237 | 686 | 550 | 1,711 | 44,055 | 867 | 53.1 | 0.9 | 38.8 |
| Catoosa | 953 | 12,738 | 1,543 | 1,551 | 3,442 | 471 | 308 | 427 | 33,535 | 250 | 48.8 | 0.4 | 39.5 |
| Charlton | 150 | 1,708 | 74 | 277 | 262 | 39 | 79 | 63 | 36,861 | 120 | 50.8 | 5.0 | 28.1 |
| Chatham | 8,078 | 147,681 | 19,069 | 16,979 | 19,149 | 2,393 | 5,187 | 6,607 | 44,740 | 67 | 73.1 | NA | 44.0 |
| Chattahoochee | 100 | 1,429 | 44 | NA | 79 | 16 | 640 | 72 | 50,694 | 12 | 58.3 | NA | 22.7 |
| Chattooga | 318 | 4,762 | 382 | 2,189 | 667 | 118 | 74 | 158 | 33,121 | 323 | 36.8 | 1.2 | 43.1 |
| Cherokee | 5,706 | 58,791 | 7,808 | 6,741 | 10,984 | 1,490 | 3,790 | 2,367 | 40,254 | 430 | 73.3 | NA | 39.3 |
| Clarke | 3,104 | 49,838 | 10,595 | 5,918 | 7,821 | 1,075 | 1,613 | 1,969 | 39,504 | 91 | 62.6 | 1.1 | 28.4 |
| Clay | 39 | 277 | 11 | NA | 69 | NA | NA | 9 | 31,069 | 67 | 14.9 | 20.9 | 42.5 |
| Clayton | 3,781 | 71,159 | 7,993 | 4,451 | 10,297 | 1,343 | 840 | 3,012 | 42,321 | 19 | 89.5 | NA | 47.2 |
| Clinch | 127 | 1,641 | 227 | 696 | 133 | 53 | 20 | 61 | 37,155 | 113 | 39.8 | 8.8 | 47.1 |
| Cobb | 20,869 | 360,023 | 38,321 | 19,686 | 38,224 | 18,113 | 42,890 | 21,889 | 60,798 | 116 | 92.2 | NA | 37.5 |
| Coffee | 823 | 14,227 | 2,230 | 3,919 | 1,835 | 289 | 243 | 507 | 35,665 | 608 | 34.7 | 9.2 | 40.2 |
| Colquitt | 895 | 11,646 | 2,438 | 3,127 | 2,043 | 290 | 199 | 402 | 34,488 | 498 | 31.9 | 8.8 | 47.7 |
| Columbia | 2,529 | 33,186 | 4,129 | 2,936 | 6,483 | 737 | 1,562 | 1,273 | 38,347 | 183 | 56.3 | 2.7 | 29.7 |
| Cook | 312 | 3,302 | 431 | 773 | 661 | 122 | 83 | 106 | 32,245 | 239 | 45.6 | 14.6 | 41.7 |
| Coweta | 2,603 | 38,164 | 5,418 | 5,542 | 6,305 | 800 | 1,598 | 1,462 | 38,304 | 368 | 56.3 | 1.6 | 35.0 |
| Crawford | 107 | 508 | 37 | 43 | 110 | 25 | NA | 19 | 36,811 | 192 | 35.4 | 3.6 | 37.1 |
| Crisp | 501 | 7,333 | 1,293 | 1,348 | 1,211 | 174 | 144 | 262 | 35,733 | 236 | 32.6 | 14.0 | 33.1 |
| Dade | 213 | 2,956 | 235 | 670 | 517 | 94 | 40 | 105 | 35,659 | 198 | 41.9 | 1.5 | 40.6 |
| Dawson | 736 | 8,353 | 520 | 994 | 3,360 | 236 | 203 | 240 | 28,745 | 192 | 60.4 | NA | 38.4 |
| Decatur | 601 | 6,969 | 911 | 1,276 | 1,400 | 238 | 109 | 244 | 34,961 | 337 | 24.3 | 21.7 | 45.1 |
| DeKalb | 17,293 | 289,015 | 47,967 | 12,059 | 32,115 | 17,836 | 20,855 | 16,038 | 55,493 | 34 | 97.1 | NA | 42.3 |
| Dodge | 307 | 3,489 | 990 | 685 | 644 | 137 | 58 | 114 | 32,579 | 391 | 24.8 | 6.1 | 33.3 |
| Dooly | 151 | 2,792 | 44 | 1,578 | 223 | 54 | 17 | 85 | 30,601 | 297 | 30.3 | 18.9 | 45.8 |
| Dougherty | 2,222 | 36,694 | 8,526 | 2,810 | 5,790 | 843 | 2,256 | 1,435 | 39,099 | 110 | 40.0 | 10.0 | 38.9 |
| Douglas | 2,670 | 38,685 | 4,893 | 4,093 | 8,013 | 660 | 1,060 | 1,508 | 38,976 | 93 | 67.7 | NA | 34.7 |
| Early | 214 | 2,969 | 372 | 954 | 360 | 105 | 84 | 138 | 46,591 | 321 | 14.6 | 17.4 | 44.7 |
| Echols | 28 | 105 | 5 | NA | 10 | NA | NA | 5 | 46,219 | 66 | 37.9 | 13.6 | 43.4 |

# Table B. States and Counties — **Agriculture**

| STATE County | Acreage (1,000) | Percent change, 2012-2017 | Acres — Average size of farm | Acres — Total irrigated (1,000) | Acres — Total cropland (1,000) | Value of land and buildings (dollars) — Average per farm | Value of land and buildings (dollars) — Average per acre | Value of machinery and equipment, average per farm (dollars) | Value of products sold — Total (mil dol) | Value of products sold — Average per farm (acres) | Percent from: Crops | Percent from: Livestock and poultry products | Organic farms (number) | Farms with internet access (percent) | Government payments — Total ($1,000) | Government payments — Percent of farms |
|---|---|---|---|---|---|---|---|---|---|---|---|---|---|---|---|---|
| | 117 | 118 | 119 | 120 | 121 | 122 | 123 | 124 | 125 | 126 | 127 | 128 | 129 | 130 | 131 | 132 |
| **FLORIDA—Cont'd** | | | | | | | | | | | | | | | | |
| Putnam | 85 | 20.3 | 150 | 6.7 | 13.5 | 596,965 | 3,977 | 65,825 | 46.1 | 81,681 | 76.0 | 24.0 | NA | 74.8 | 466 | 6.4 |
| St. Johns | 34 | 2.3 | 136 | 14.4 | 17.5 | 1,156,661 | 8,507 | 130,543 | 61.4 | 242,605 | 95.3 | 4.7 | 2 | 85.4 | D | 2.8 |
| St. Lucie | 226 | 15.8 | 545 | 48.2 | 68.7 | 3,254,221 | 5,976 | 109,557 | 139.6 | 336,446 | 89.3 | 10.7 | 2 | 68.9 | 947 | 14.5 |
| Santa Rosa | 85 | -12.9 | 122 | 1.2 | 52.8 | 652,601 | 5,362 | 61,267 | 38.5 | 55,117 | 89.1 | 10.9 | NA | 79.3 | 4,549 | 23.2 |
| Sarasota | 71 | -11.2 | 244 | 5.2 | 11.0 | 2,240,920 | 9,195 | 63,666 | 23.1 | 79,092 | 77.4 | 22.6 | 1 | 76.7 | 348 | 5.1 |
| Seminole | 35 | 61.0 | 87 | 1.2 | 7.1 | 716,815 | 8,271 | 38,395 | 21.3 | 52,965 | 91.2 | 8.8 | 5 | 78.7 | 136 | 3.2 |
| Sumter | 177 | -3.4 | 135 | 2.0 | 21.3 | 816,314 | 6,025 | 50,217 | 54.5 | 41,666 | 36.7 | 63.3 | 8 | 74.7 | 1,099 | 16.7 |
| Suwannee | 170 | -12.2 | 157 | 27.4 | 78.5 | 695,473 | 4,418 | 83,821 | 258.9 | 239,981 | 21.5 | 78.5 | 9 | 76.6 | 1,988 | 6.8 |
| Taylor | 59 | 56.8 | 245 | 0.2 | 4.5 | 892,861 | 3,650 | 55,626 | 11.8 | 49,113 | 21.1 | 78.9 | NA | 75.0 | 10 | 3.8 |
| Union | 54 | 16.3 | 175 | 1.5 | 10.3 | 649,755 | 3,722 | 54,492 | 7.7 | 25,010 | 47.7 | 52.3 | NA | 74.0 | 93 | 3.6 |
| Volusia | 114 | 7.8 | 73 | 9.2 | 25.1 | 748,025 | 10,309 | 51,317 | 196.4 | 124,693 | 93.4 | 6.6 | 6 | 76.3 | 1,139 | 5.8 |
| Wakulla | 24 | -23.4 | 113 | 0.1 | 2.0 | 387,249 | 3,420 | 38,983 | 2.4 | 11,431 | 22.9 | 77.1 | NA | 65.6 | 161 | 7.7 |
| Walton | 89 | -39.7 | 149 | 1.2 | 23.1 | 562,514 | 3,771 | 45,630 | 30.6 | 51,186 | 19.6 | 80.4 | NA | 80.1 | 1,007 | 16.7 |
| Washington | 45 | -22.4 | 104 | 0.9 | 16.0 | 395,545 | 3,820 | 53,771 | 8.9 | 20,471 | 66.5 | 33.5 | 2 | 71.9 | 458 | 18.1 |
| **GEORGIA** | 9,954 | 3.5 | 235 | 1,287.5 | 4,372.1 | 822,958 | 3,509 | 115,773 | 9,573.3 | 225,577 | 34.2 | 65.8 | 139 | 76.0 | 247,428 | 31.8 |
| Appling | 128 | 4.4 | 234 | 13.7 | 85.1 | 760,709 | 3,249 | 136,171 | 166.6 | 303,936 | 36.6 | 63.4 | 7 | 71.5 | 2,462 | 31.9 |
| Atkinson | 72 | -17.2 | 334 | 9.1 | 40.8 | 971,298 | 2,904 | 192,878 | 71.1 | 330,563 | 38.6 | 61.4 | NA | 72.1 | 2,132 | 34.0 |
| Bacon | 62 | 7.6 | 228 | 6.1 | 38.6 | 937,359 | 4,116 | 144,034 | 63.2 | 231,440 | 50.5 | 49.5 | 1 | 80.2 | 1,141 | 20.9 |
| Baker | 131 | -10.6 | 891 | 26.2 | 53.2 | 1,980,216 | 2,222 | 343,799 | 57.0 | 387,469 | 63.5 | 36.5 | 2 | 62.6 | 5,182 | 59.9 |
| Baldwin | 34 | 81.5 | 244 | 0.1 | 8.0 | 527,772 | 2,165 | 66,743 | 1.4 | 10,216 | 33.7 | 66.3 | NA | 79.9 | 93 | 12.9 |
| Banks | 56 | -5.5 | 122 | 0.5 | 12.1 | 655,689 | 5,381 | 95,826 | 169.5 | 366,194 | 2.3 | 97.7 | NA | 76.9 | 943 | 27.2 |
| Barrow | 22 | -25.0 | 78 | 0.4 | 6.1 | 508,793 | 6,557 | 50,503 | 36.0 | 124,844 | 2.4 | 97.6 | 5 | 78.8 | 161 | 12.5 |
| Bartow | 77 | 21.2 | 165 | 1.0 | 16.9 | 994,446 | 6,025 | 65,838 | 71.4 | 152,292 | 12.5 | 87.5 | NA | 77.8 | 919 | 22.2 |
| Ben Hill | 53 | -8.3 | 242 | 5.3 | 22.4 | 597,934 | 2,469 | 111,377 | 20.9 | 96,244 | 69.0 | 31.0 | NA | 71.0 | 891 | 44.2 |
| Berrien | 117 | -18.6 | 335 | 21.4 | 61.2 | 1,046,386 | 3,128 | 170,691 | 85.5 | 244,931 | 70.7 | 29.3 | NA | 77.7 | 5,470 | 44.1 |
| Bibb | 9 | -38.2 | 93 | 0.1 | 4.1 | 371,840 | 4,006 | 77,778 | 4.8 | 49,429 | 6.4 | 93.6 | NA | 83.7 | 27 | 9.2 |
| Bleckley | 48 | -26.8 | 209 | 8.1 | 23.2 | 539,162 | 2,580 | 59,016 | 12.4 | 53,654 | 93.6 | 6.4 | NA | 58.9 | 638 | 53.7 |
| Brantley | 24 | 4.5 | 104 | 0.9 | 5.7 | 304,091 | 2,919 | 47,411 | D | D | D | D | NA | 78.7 | 127 | 17.4 |
| Brooks | 178 | 20.4 | 496 | 22.1 | 91.6 | 1,733,285 | 3,496 | 215,239 | 118.9 | 330,253 | 63.2 | 36.8 | NA | 72.5 | 3,288 | 55.0 |
| Bryan | 26 | 69.8 | 272 | 0.0 | 3.5 | 837,457 | 3,077 | 67,866 | 3.0 | 31,400 | 33.1 | 66.9 | NA | 74.7 | 33 | 7.4 |
| Bulloch | 197 | 9.4 | 413 | 17.7 | 126.1 | 1,181,592 | 2,864 | 202,153 | 89.9 | 187,992 | 82.2 | 17.8 | 1 | 76.8 | 9,661 | 46.9 |
| Burke | 223 | 38.3 | 478 | 43.2 | 123.3 | 1,357,912 | 2,843 | 266,151 | 118.1 | 252,972 | 57.2 | 42.8 | 3 | 76.4 | 7,072 | 45.6 |
| Butts | 31 | 48.0 | 181 | 0.0 | 6.4 | 704,674 | 3,897 | 61,476 | 4.3 | 24,694 | 37.7 | 62.3 | NA | 68.8 | 112 | 16.8 |
| Calhoun | 116 | 7.8 | 686 | 38 | 66.3 | 2,067,724 | 3,014 | 298,570 | 63.5 | 375,917 | 79.9 | 20.1 | NA | 66.3 | 6,347 | 72.2 |
| Camden | 6 | -64.8 | 118 | 0.1 | 2.1 | 337,997 | 2,865 | 59,431 | 0.7 | 15,809 | 32.6 | 67.4 | 1 | 80.9 | 22 | 10.6 |
| Candler | 55 | 2.4 | 278 | 3.9 | 24.5 | 879,763 | 3,160 | 95,491 | 21.8 | 110,853 | 72.9 | 27.1 | 1 | 69.5 | 796 | 37.1 |
| Carroll | 85 | -0.8 | 98 | D | 18.2 | 470,633 | 4,786 | 69,536 | 186.0 | 214,525 | 4.0 | 96.0 | 1 | 81.2 | 519 | 11.6 |
| Catoosa | 24 | 17.0 | 97 | 0.1 | 7.5 | 531,318 | 5,503 | 69,075 | 26.7 | 106,880 | 16.5 | 83.5 | NA | 78.0 | 360 | 25.2 |
| Charlton | 21 | 54.9 | 173 | D | 2.3 | 418,803 | 2,427 | 55,039 | 3.8 | 31,842 | 13.9 | 86.1 | NA | 66.7 | 41 | 18.3 |
| Chatham | 5 | 22.0 | 70 | 0.7 | 1.7 | 354,509 | 5,078 | 129,996 | 12.2 | 182,448 | 95.7 | 4.3 | 1 | 85.1 | NA | NA |
| Chattahoochee | 2 | -57.4 | 145 | D | 0.1 | 428,469 | 2,958 | 70,000 | D | D | 100.0 | D | NA | 75.0 | D | 16.7 |
| Chattooga | 55 | 10.3 | 171 | 0.0 | 15.9 | 605,993 | 3,542 | 72,635 | 74.2 | 229,836 | 2.8 | 97.2 | NA | 78.0 | 403 | 25.4 |
| Cherokee | 24 | -4.5 | 56 | 0.1 | 6.6 | 484,880 | 8,675 | 56,805 | 21.7 | 50,484 | 22.8 | 77.2 | NA | 89.8 | 182 | 10.0 |
| Clarke | 8 | -9.5 | 88 | 0.1 | 1.6 | 683,725 | 7,735 | 52,373 | 44.7 | 491,385 | 9.5 | 90.5 | NA | 90.1 | 59 | 9.9 |
| Clay | 45 | 12.9 | 674 | 6.6 | 23.0 | 1,596,179 | 2,367 | 277,136 | 15.8 | 236,239 | 88.1 | 11.9 | 1 | 83.6 | 1,883 | 58.2 |
| Clayton | 1 | -29.3 | 31 | 0.0 | 0.2 | 259,453 | 8,355 | 26,985 | 0.2 | 12,842 | 48.4 | 51.6 | NA | 89.5 | D | 5.3 |
| Clinch | 27 | 3.0 | 243 | 3.7 | 8.9 | 967,887 | 3,983 | 179,959 | 33.9 | 299,876 | 89.8 | 10.2 | NA | 72.6 | 290 | 31.0 |
| Cobb | 3 | -50.9 | 22 | 0.0 | 0.4 | 248,584 | 11,425 | 38,141 | D | D | D | D | NA | 92.2 | D | 1.7 |
| Coffee | 189 | 12.6 | 311 | 35.3 | 103.5 | 938,419 | 3,016 | 187,887 | 185.5 | 305,051 | 39.0 | 61.0 | 4 | 75.8 | 8,386 | 37.2 |
| Colquitt | 186 | -1.4 | 373 | 45.4 | 120.8 | 1,235,184 | 3,308 | 224,282 | 295.9 | 594,273 | 66.5 | 33.5 | NA | 72.7 | 8,659 | 55.6 |
| Columbia | 23 | 74.5 | 125 | 0.1 | 3.4 | 555,982 | 4,452 | 39,103 | 2.8 | 15,208 | 51.0 | 49.0 | 2 | 84.7 | 50 | 8.7 |
| Cook | 79 | 15.4 | 330 | 15.9 | 48.9 | 1,004,035 | 3,038 | 167,271 | 88.1 | 368,544 | 63.1 | 36.9 | NA | 70.7 | 4,636 | 50.2 |
| Coweta | 53 | -3.8 | 145 | 0.3 | 12.5 | 699,043 | 4,825 | 67,682 | 11.7 | 31,793 | 36.4 | 63.6 | NA | 81.0 | 136 | 9.5 |
| Crawford | 35 | 3.9 | 184 | 2.7 | 10.7 | 739,529 | 4,029 | 103,828 | 61.0 | 317,745 | 20.6 | 79.4 | 3 | 82.8 | 243 | 18.8 |
| Crisp | 108 | -7.8 | 458 | 20.2 | 66.9 | 1,403,992 | 3,065 | 198,462 | 60.0 | 254,085 | 82.0 | 18.0 | NA | 77.5 | 3,815 | 55.9 |
| Dade | 29 | -10.6 | 147 | 0.0 | 4.8 | 584,184 | 3,981 | 58,107 | 25.1 | 126,631 | 3.2 | 96.8 | 3 | 68.7 | 198 | 24.7 |
| Dawson | 19 | 49.1 | 99 | 0.0 | 3.9 | 689,160 | 6,983 | 66,115 | 46.8 | 243,880 | 2.0 | 98.0 | 2 | 77.6 | 134 | 12.0 |
| Decatur | 192 | -3.6 | 569 | 82.9 | 133.4 | 1,855,602 | 3,260 | 315,713 | 179.5 | 532,591 | 83.7 | 16.3 | 3 | 68.2 | 15,914 | 70.0 |
| DeKalb | 0 | -84.1 | 14 | 0.0 | 0.0 | 646,015 | 46,933 | 17,575 | 0.5 | 16,059 | 71.1 | 28.9 | NA | 97.1 | NA | NA |
| Dodge | 103 | 14.4 | 264 | 11.6 | 36.3 | 685,800 | 2,600 | 81,427 | 30.5 | 78,115 | 73.9 | 26.1 | NA | 73.7 | 1,767 | 52.7 |
| Dooly | 186 | 46.6 | 626 | 29.7 | 121.2 | 1,830,024 | 2,922 | 240,010 | 99.2 | 334,020 | 86.0 | 14.0 | 1 | 76.1 | 4,446 | 67.3 |
| Dougherty | 64 | -1.4 | 586 | 18.4 | 24.0 | 2,267,228 | 3,869 | 167,695 | 40.3 | 366,355 | 89.6 | 10.4 | NA | 85.5 | 1,747 | 38.2 |
| Douglas | 7 | -15.1 | 76 | 0.2 | 1.9 | 509,840 | 6,695 | 49,702 | 0.7 | 7,398 | 46.4 | 53.6 | NA | 67.7 | 9 | 5.4 |
| Early | 168 | | 522 | 32.2 | 80.0 | 1,341,069 | 2,569 | 186,003 | 59.3 | 184,629 | 91.7 | 8.3 | NA | 73.5 | 4,866 | 74.5 |
| Echols | 23 | 71.6 | 346 | 3.4 | 9.2 | 965,776 | 2,791 | 147,334 | 17.9 | 271,697 | 82.0 | 18.0 | NA | 74.2 | 215 | 37.9 |

# Table B. States and Counties — Water Use, Wholesale Trade, Retail Trade, and Real Estate

| STATE County | Water use, 2015 — Public supply water withdrawn (mil gal/day) [133] | Public supply gallons withdrawn per person per day [134] | Wholesale Trade[1], 2017 — Number of establishments [135] | Number of employees [136] | Sales (mil dol) [137] | Average payroll (mil dol) [138] | Retail Trade[2], 2017 — Number of establishments [139] | Number of employees [140] | Sales (mil dol) [141] | Average payroll (mil dol) [142] | Real estate and rental and leasing,[2] 2017 — Number of establishments [143] | Number of employees [144] | Sales (mil dol) [145] | Average payroll (mil dol) [146] |
|---|---|---|---|---|---|---|---|---|---|---|---|---|---|---|
| **FLORIDA—Cont'd** | | | | | | | | | | | | | | |
| Putnam | 2.10 | 29.2 | D | D | D | D | 227 | 2,867 | 737.4 | 65.9 | 45 | 118 | 17.1 | 3.3 |
| St. Johns | 16.83 | 74.3 | 184 | 1,792 | 1,550.3 | 116.2 | 882 | 11,538 | 3,214.8 | 285.3 | 470 | 1,405 | 327.4 | 56.1 |
| St. Lucie | 29.37 | 98.4 | 211 | 1,683 | 790.9 | 78.2 | 746 | 11,839 | 3,794.9 | 332.7 | 319 | 1,341 | 241.1 | 40.7 |
| Santa Rosa | 14.96 | 89.6 | 63 | 521 | 427.3 | 23.1 | 373 | 5,294 | 1,508.1 | 132.7 | 200 | 441 | 101.3 | 13.8 |
| Sarasota | 20.17 | 49.7 | 428 | 4,165 | 2,809.4 | 211.7 | 1,753 | 23,507 | 7,397.3 | 680.2 | 1,028 | 3,337 | 734.5 | 151.3 |
| Seminole | 57.59 | 128.2 | 614 | 5,162 | 2,685.3 | 282.7 | 1,706 | 27,111 | 9,607.0 | 765.7 | 866 | 3,732 | 824.7 | 154.4 |
| Sumter | 24.13 | 203.0 | 49 | 450 | 267.0 | 22.4 | 241 | 3,726 | 1,089.7 | 88.2 | 137 | 652 | 90.5 | 18.3 |
| Suwannee | 1.19 | 27.2 | 34 | 359 | 238.0 | 12.1 | 136 | 2,007 | 572.4 | 51.4 | 26 | 61 | 12.6 | 1.7 |
| Taylor | 1.75 | 77.8 | 17 | 51 | 110.7 | 2.6 | 87 | 872 | 268.1 | 20.2 | 12 | 24 | 6.0 | 0.6 |
| Union | 0.22 | 14.4 | D | D | D | 0.8 | 29 | 191 | 63.9 | 5.6 | D | D | D | 0.3 |
| Volusia | 55.45 | 107.1 | 399 | 2,842 | 1,426.2 | 136.7 | 1,981 | 26,960 | 7,739.6 | 706.8 | 754 | 2,752 | 597.5 | 97.4 |
| Wakulla | 2.31 | 73.3 | D | D | D | D | 65 | 865 | 236.5 | 18.3 | 21 | 27 | 6.8 | 0.9 |
| Walton | 11.17 | 175.9 | 55 | 665 | 418.3 | 29.2 | 416 | 4,802 | 1,122.1 | 114.6 | 305 | 1,211 | 303.4 | 51.9 |
| Washington | 0.95 | 38.5 | 12 | 145 | 147.3 | 7.0 | 67 | 813 | 203.9 | 18.2 | 9 | 13 | 2.4 | 0.5 |
| **GEORGIA** | 1,069.91 | 104.7 | 10,832 | 165,088 | 188,899.2 | 10,594.4 | 34,100 | 485,505 | 148,624.6 | 12,560.5 | 12,426 | 63,935 | 23,009.0 | 3,795.4 |
| Appling | 0.76 | 41.2 | 19 | 170 | 129.7 | 8.3 | 79 | 1,173 | 578.7 | 36.8 | 10 | 25 | 3.2 | 0.5 |
| Atkinson | 0.43 | 51.2 | 5 | 94 | 47.4 | 3.6 | 23 | 141 | 41.7 | 3.2 | NA | NA | NA | NA |
| Bacon | 0.98 | 86.7 | 14 | 201 | 180.0 | 8.6 | 44 | 324 | 81.2 | 6.6 | NA | NA | NA | NA |
| Baker | 0.15 | 47.2 | 3 | D | 7.7 | D | 4 | 28 | 4.1 | 0.4 | NA | NA | NA | NA |
| Baldwin | 3.82 | 84.0 | D | D | D | D | 182 | 2,401 | 674.2 | 58.5 | D | D | D | D |
| Banks | 2.61 | 141.1 | 12 | 98 | 88.1 | 3.9 | 49 | 838 | 223.6 | 19.1 | 7 | 31 | 3.3 | 0.6 |
| Barrow | 4.46 | 59.2 | 61 | 1,232 | 765.4 | 63.1 | 168 | 2,870 | 993.6 | 82.7 | 56 | 116 | 32.5 | 4.2 |
| Bartow | 52.41 | 510.1 | 128 | 2,375 | 1,066.5 | 111.0 | 330 | 4,163 | 1,330.7 | 104.8 | 106 | 435 | 87.7 | 16.3 |
| Ben Hill | 2.46 | 141.4 | 7 | 38 | 18.9 | 1.8 | 74 | 769 | 207.0 | 17.4 | 13 | 26 | 4.0 | 0.7 |
| Berrien | 0.78 | 41.1 | 19 | 194 | 96.7 | 6.8 | 52 | 439 | 100.3 | 8.9 | 7 | 12 | 2.0 | 0.2 |
| Bibb | 24.13 | 157.0 | 178 | 2,913 | 1,943.5 | 149.2 | 761 | 10,330 | 2,685.4 | 259.1 | 203 | 1,050 | 208.6 | 40.3 |
| Bleckley | 0.20 | 16.3 | 6 | 38 | 22.4 | 1.6 | 41 | 458 | 93.8 | 9.5 | 4 | D | 1.5 | D |
| Brantley | 0.24 | 13.0 | 5 | 13 | 1.6 | 0.5 | 40 | 261 | 70.9 | 5.4 | NA | NA | NA | NA |
| Brooks | 0.90 | 57.5 | 11 | 103 | 107.3 | 4.9 | 40 | 350 | 93.4 | 7.6 | D | D | D | 0.3 |
| Bryan | 2.59 | 73.7 | 19 | 299 | 512.8 | 20.1 | 104 | 1,221 | 373.4 | 25.8 | 39 | 101 | 41.3 | 4.7 |
| Bulloch | 4.77 | 65.7 | 46 | 257 | 207.6 | 12.6 | 264 | 3,407 | 957.5 | 83.5 | 81 | 319 | 50.5 | 10.2 |
| Burke | 0.85 | 37.4 | 26 | 546 | 395.9 | 23.9 | 58 | 818 | 221.2 | 20.0 | 12 | 21 | 6.0 | 0.7 |
| Butts | 2.74 | 116.1 | 9 | 253 | 402.3 | 11.6 | 75 | 868 | 316.0 | 19.9 | 17 | 24 | 6.1 | 0.7 |
| Calhoun | 0.44 | 67.9 | 3 | 36 | 28.0 | 2.0 | 13 | 90 | 13.9 | 1.6 | NA | NA | NA | NA |
| Camden | 3.94 | 75.6 | D | D | D | D | 151 | 1,941 | 600.9 | 45.0 | 27 | 127 | 27.3 | 3.2 |
| Candler | 0.41 | 37.7 | 7 | 63 | 27.0 | 1.9 | 49 | 401 | 145.7 | 10.1 | 3 | 5 | 0.4 | 0.1 |
| Carroll | 11.32 | 98.8 | 82 | 824 | 538.0 | 38.7 | 376 | 5,369 | 1,556.9 | 129.8 | 97 | 347 | 66.9 | 9.7 |
| Catoosa | 4.68 | 70.9 | 43 | 487 | 412.7 | 23.0 | 185 | 3,050 | 948.3 | 82.9 | 37 | 148 | 27.7 | 4.5 |
| Charlton | 0.76 | 58.6 | 6 | 82 | 37.3 | 3.2 | 33 | 247 | 61.0 | 4.7 | 4 | 32 | 1.9 | 0.5 |
| Chatham | 28.08 | 97.9 | 332 | 4,916 | 5,884.0 | 288.9 | 1,321 | 19,102 | 5,006.2 | 473.7 | 464 | 2,376 | 567.2 | 91.6 |
| Chattahoochee | 0.31 | 27.3 | NA | NA | NA | NA | 16 | 94 | 14.8 | 1.6 | NA | NA | NA | NA |
| Chattooga | 2.80 | 112.4 | D | D | D | D | 74 | 703 | 159.9 | 15.3 | 10 | 34 | 3.9 | 0.7 |
| Cherokee | 19.29 | 81.8 | 254 | 2,350 | 1,347.2 | 145.0 | 660 | 10,814 | 2,977.0 | 270.7 | 292 | 707 | 210.9 | 31.1 |
| Clarke | 11.48 | 92.6 | 94 | 2,042 | 2,980.2 | 95.9 | 513 | 8,212 | 1,982.0 | 195.5 | 202 | 1,392 | 242.9 | 54.2 |
| Clay | 0.25 | 79.6 | NA | NA | NA | NA | 11 | 52 | 7.8 | 0.9 | NA | NA | NA | NA |
| Clayton | 10.76 | 39.3 | 214 | 4,698 | 4,813.8 | 262.0 | 700 | 10,714 | 3,454.9 | 285.6 | 180 | 1,019 | 331.3 | 50.6 |
| Clinch | 0.45 | 65.3 | 7 | 120 | 19.8 | 3.2 | 27 | 153 | 36.9 | 3.0 | NA | NA | NA | NA |
| Cobb | 42.22 | 57.0 | 1,049 | 17,683 | 25,968.6 | 1,263.9 | 2,261 | 39,473 | 18,543.7 | 1,066.1 | 1,257 | 6,174 | 1,911.4 | 343.9 |
| Coffee | 4.05 | 94.0 | 45 | 503 | 467.9 | 25.3 | 188 | 1,910 | 542.1 | 46.2 | 28 | 74 | 10.6 | 2.0 |
| Colquitt | 3.11 | 67.8 | 49 | 453 | 456.1 | 21.3 | 198 | 2,047 | 528.2 | 50.9 | 30 | 92 | 14.1 | 2.7 |
| Columbia | 15.92 | 110.5 | 84 | 1,038 | 477.8 | 51.1 | 323 | 6,455 | 2,111.5 | 191.8 | 120 | 499 | 147.6 | 21.4 |
| Cook | 1.40 | 81.8 | 10 | 72 | 23.0 | 3.4 | 65 | 692 | 200.1 | 16.0 | 7 | 46 | 4.2 | 1.3 |
| Coweta | 7.45 | 53.8 | 99 | 1,171 | 1,899.0 | 59.7 | 390 | 6,192 | 1,943.1 | 155.3 | 170 | 370 | 109.3 | 18.8 |
| Crawford | 0.40 | 32.3 | NA | NA | NA | NA | 21 | 95 | 21.8 | 2.3 | NA | NA | NA | NA |
| Crisp | 2.01 | 87.8 | 31 | 347 | 376.2 | 18.4 | 115 | 1,237 | 315.9 | 28.6 | 22 | 194 | 15.1 | 5.2 |
| Dade | 2.02 | 124.2 | NA | NA | NA | NA | 61 | 510 | 153.9 | 11.1 | 5 | 5 | 1.4 | 0.2 |
| Dawson | 1.51 | 64.8 | 18 | 158 | 38.6 | 5.5 | 200 | 3,364 | 747.6 | 66.7 | 25 | 55 | 18.4 | 2.1 |
| Decatur | 2.43 | 89.4 | 28 | 251 | 186.0 | 10.3 | 119 | 1,341 | 378.4 | 32.1 | 26 | 68 | 10.0 | 1.9 |
| DeKalb | 0.00 | 0.0 | 717 | 10,312 | 13,007.6 | 592.9 | 2,231 | 34,345 | 10,567.2 | 946.9 | 1,064 | 4,685 | 1,298.4 | 245.6 |
| Dodge | 1.39 | 66.6 | 11 | 44 | 9.3 | 1.1 | 62 | 708 | 160.7 | 14.1 | D | D | D | D |
| Dooly | 2.28 | 162.5 | 12 | 84 | 47.1 | 3.7 | 35 | 307 | 127.6 | 8.6 | 3 | 4 | 0.2 | 0.1 |
| Dougherty | 11.57 | 126.7 | 114 | 1,997 | 1,189.7 | 93.7 | 443 | 5,959 | 1,578.7 | 135.0 | 117 | 484 | 106.1 | 17.0 |
| Douglas | 17.52 | 124.5 | 116 | 2,209 | 2,721.6 | 139.7 | 416 | 8,318 | 2,760.6 | 221.2 | 140 | 605 | 141.6 | 22.4 |
| Early | 1.21 | 114.4 | 12 | 69 | 66.3 | 3.2 | 52 | 413 | 92.4 | 7.6 | 4 | 9 | 0.9 | 0.2 |
| Echols | 0.05 | 12.4 | NA | NA | NA | NA | 5 | 12 | 4.2 | 0.2 | NA | NA | NA | NA |

1  Merchant wholesalers, except manufacturers' sales branches and offices.  2. Employer establishments.

# Table B. States and Counties — Professional Services, Manufacturing, and Accommodation and Food Services

| STATE County | Professional, scientific, and technical services, 2017 | | | | Manufacturing, 2017 | | | | Accommodation and food services, 2017 | | | |
|---|---|---|---|---|---|---|---|---|---|---|---|---|
| | Number of establishments | Number of employees | Sales (mil dol) | Average payroll (mil dol) | Number of establishments | Number of employees | Sales (mil dol) | Average payroll (mil dol) | Number of establishments | Number of employees | Sales (mil dol) | Annual payroll (mil dol) |
| | 147 | 148 | 149 | 150 | 151 | 152 | 153 | 154 | 155 | 156 | 157 | 158 |
| FLORIDA—Cont'd | | | | | | | | | | | | |
| Putnam | D | D | D | D | 38 | 1,514 | 932.0 | 94.3 | 95 | 1,313 | 69.4 | 18.0 |
| St. Johns | D | D | D | D | 116 | 1,933 | 570.9 | 86.8 | 600 | 12,316 | 824.6 | 232.7 |
| St. Lucie | 549 | 2,367 | 314 | 105.4 | 152 | 3,100 | 817.1 | 142.4 | 444 | 8,381 | 456.7 | 127.8 |
| Santa Rosa | 324 | 1,687 | 259 | 101.6 | 54 | 665 | 547.9 | 36.2 | 218 | 4,863 | 238.4 | 65.6 |
| Sarasota | D | D | D | D | 353 | 6,921 | 1,667.2 | 367.9 | 957 | 18,847 | 1,151.7 | 343.2 |
| Seminole | 2,038 | 12,586 | 1,979 | 786.7 | 375 | 6,585 | 1,902.1 | 317.5 | 877 | 17,286 | 1,038.1 | 290.9 |
| Sumter | 160 | 753 | 104 | 40.1 | 39 | 943 | 474.7 | 49.1 | 143 | 3,614 | 177.1 | 56.6 |
| Suwannee | 43 | 165 | 18 | 5.4 | 16 | 1,959 | 597.7 | 63.0 | 56 | 963 | 51.0 | 12.3 |
| Taylor | 28 | 77 | 8 | 2.9 | 20 | 1,617 | 546.7 | 94.1 | 36 | 429 | 21.7 | 5.9 |
| Union | D | D | D | 0.3 | D | D | D | D | 4 | 55 | 3.0 | 0.9 |
| Volusia | D | D | D | D | 372 | 9,074 | 2,391.4 | 440.2 | 1,187 | 22,605 | 1,293.3 | 366.1 |
| Wakulla | D | D | D | D | D | D | D | D | 44 | 576 | 32.6 | 8.0 |
| Walton | 265 | 655 | 115 | 32.6 | 40 | 202 | 63.9 | 8.7 | 237 | 6,326 | 526.7 | 146.7 |
| Washington | 31 | 290 | 40 | 14.7 | D | 257 | D | 6.7 | 37 | 436 | 23.1 | 5.9 |
| GEORGIA | 29,635 | 265,189 | 54,904 | 20,100.5 | 7,510 | 368,836 | 169,058.5 | 19,071.6 | 21,201 | 426,884 | 26,010.1 | 6,999.7 |
| Appling | 16 | 41 | 5 | 1.3 | 26 | 597 | 222.3 | 30.0 | 34 | 508 | 25.6 | 6.5 |
| Atkinson | NA | NA | NA | NA | 12 | 647 | 249.8 | 29.0 | 8 | 66 | 2.5 | 0.6 |
| Bacon | 16 | 59 | 6 | 2.0 | 13 | 947 | 220.3 | 33.6 | D | D | D | D |
| Baker | NA | NA | NA | NA | NA | NA | NA | NA | NA | NA | NA | NA |
| Baldwin | D | D | D | D | D | D | D | D | D | D | D | D |
| Banks | 14 | 43 | 6 | 2.6 | 11 | 262 | 47.1 | 11.6 | 47 | 1,005 | 59.8 | 16.1 |
| Barrow | 115 | 785 | 78 | 29.5 | 72 | 2,350 | 1,008.2 | 113.9 | 100 | 2,385 | 177.5 | 38.7 |
| Bartow | D | D | D | D | 117 | 7,853 | 4,466.2 | 416.8 | 197 | 4,081 | 234.5 | 64.2 |
| Ben Hill | D | D | D | D | 28 | 1,348 | 508.7 | 63.1 | 29 | 298 | 17.8 | 4.9 |
| Berrien | 14 | 38 | 6 | 1.4 | D | 1,237 | D | 52.8 | 24 | 225 | 11.5 | 2.8 |
| Bibb | D | D | D | D | 119 | 5,179 | 2,443.3 | 297.1 | 444 | 8,531 | 451.0 | 122.9 |
| Bleckley | 8 | 14 | 2 | 1.0 | 7 | 66 | 12.9 | 3.4 | 16 | D | 12.4 | D |
| Brantley | 9 | 17 | 2 | 0.6 | 9 | 91 | 53.7 | 5.2 | 13 | 132 | 5.7 | 1.7 |
| Brooks | 13 | 45 | 4 | 1.2 | 6 | 186 | 95.7 | 9.2 | 14 | 166 | 7.7 | 2.1 |
| Bryan | 74 | 262 | 40 | 12.4 | 12 | 270 | 160.1 | 16.6 | 81 | 1,243 | 65.1 | 16.7 |
| Bulloch | D | D | D | D | 43 | 1,332 | 416.1 | 56.4 | 153 | 4,021 | 173.6 | 47.4 |
| Burke | D | D | D | D | 9 | 291 | 65.6 | 13.0 | 26 | 402 | 22.0 | 5.8 |
| Butts | 22 | 71 | 7 | 2.5 | 12 | 842 | 420.7 | 29.8 | 37 | 426 | 26.4 | 6.7 |
| Calhoun | 3 | 2 | 0 | 0.1 | NA | NA | NA | NA | 4 | D | 0.4 | D |
| Camden | D | D | D | D | 14 | D | 111.9 | D | 118 | 2,108 | 107.1 | 29.7 |
| Candler | 26 | 90 | 12 | 3.4 | 8 | 163 | 47.3 | 4.4 | D | D | D | D |
| Carroll | 145 | 595 | 102 | 30.3 | 108 | 8,686 | 5,937.5 | 417.7 | 215 | 4,122 | 215.8 | 55.7 |
| Catoosa | D | D | D | D | 47 | 1,403 | 484.2 | 72.5 | 100 | 2,225 | 122.0 | 32.7 |
| Charlton | D | D | D | D | 6 | 237 | 66.7 | 11.3 | 13 | 178 | 8.8 | 2.3 |
| Chatham | D | D | D | D | 182 | 16,361 | 10,193.1 | 1,407.2 | 1,018 | 22,190 | 1,492.3 | 383.6 |
| Chattahoochee | D | D | D | D | NA | NA | NA | NA | 6 | 73 | 3.1 | 0.8 |
| Chattooga | D | D | 9 | D | 14 | 2,412 | 614.2 | 84.8 | 27 | 314 | 16.9 | 3.5 |
| Cherokee | 791 | 3,015 | 546 | 164.1 | 170 | 4,510 | 1,221.0 | 209.1 | 380 | 7,874 | 420.4 | 122.8 |
| Clarke | 318 | 1,510 | 192 | 71.2 | 92 | 5,698 | 2,376.5 | 261.8 | 360 | 7,878 | 396.0 | 110.0 |
| Clay | NA | NA | NA | NA | NA | NA | NA | NA | NA | NA | NA | NA |
| Clayton | 221 | 952 | 145 | 48.3 | 90 | 4,457 | 1,893.6 | 221.2 | 410 | 8,801 | 585.6 | 153.4 |
| Clinch | D | D | D | D | D | 672 | D | D | 13 | D | 4.0 | D |
| Cobb | 3,463 | 32,471 | 6,501 | 2,369.0 | 497 | 19,472 | 8,818.9 | 1,362.6 | 1,649 | 34,784 | 2,056.6 | 565.8 |
| Coffee | 51 | 241 | 27 | 11.2 | 45 | 3,178 | 910.0 | 127.5 | 60 | 1,625 | 75.0 | 19.4 |
| Colquitt | 64 | 214 | 26 | 7.3 | 40 | 2,754 | 944.9 | 94.7 | D | D | D | D |
| Columbia | 216 | 1,224 | 204 | 67.0 | 58 | 3,168 | 2,485.7 | 146.8 | 225 | 4,871 | 242.2 | 65.2 |
| Cook | 17 | 60 | 7 | 2.2 | 30 | 637 | 516.5 | 29.7 | 36 | 550 | 30.9 | 6.6 |
| Coweta | D | D | D | D | 83 | 5,044 | 2,426.7 | 258.7 | 188 | 4,500 | 235.2 | 68.2 |
| Crawford | D | D | D | 0.5 | D | 30 | D | 1.6 | D | D | D | D |
| Crisp | 26 | 153 | 19 | 7.5 | 19 | 1,012 | 538.3 | 52.4 | 50 | 1,030 | 59.2 | 12.4 |
| Dade | 11 | 39 | 3 | 1.0 | 11 | 452 | 185.8 | 24.2 | D | D | D | D |
| Dawson | 55 | 198 | 25 | 8.5 | 24 | 1,115 | 211.1 | 33.0 | 58 | 1,191 | 66.4 | 17.3 |
| Decatur | 30 | 107 | 13 | 3.7 | 28 | 1,191 | 724.3 | 59.1 | 47 | 689 | 34.9 | 8.9 |
| DeKalb | 2,951 | 22,022 | 3,899 | 1,594.3 | 408 | 11,374 | 4,013.1 | 573.4 | 1,561 | 27,441 | 1,768.0 | 490.7 |
| Dodge | 18 | 72 | 10 | 3.9 | 14 | 294 | 133.9 | 12.9 | 25 | 416 | 16.1 | 4.3 |
| Dooly | D | D | D | D | 8 | D | 398.9 | D | 13 | 158 | 9.2 | 1.7 |
| Dougherty | 212 | 2,000 | 269 | 104.0 | 57 | 2,910 | 2,729.2 | 176.2 | 221 | 4,431 | 235.8 | 63.6 |
| Douglas | 218 | 970 | 123 | 43.8 | 91 | 3,804 | 1,043.9 | 162.1 | 241 | 5,171 | 287.3 | 74.6 |
| Early | 15 | 74 | 9 | 3.2 | 5 | 785 | 647.9 | 60.4 | D | D | D | D |
| Echols | NA | NA | NA | NA | NA | NA | NA | NA | NA | NA | NA | NA |

# Health Care and Social Assistance, Other Services, Nonemployer Businesses, and Residential Construction

| STATE County | Health care and social assistance, 2017 | | | | Other services, 2017 | | | | Nonemployer businesses, 2018 | | Value of residential construction authorized by building permits, 2020 | |
|---|---|---|---|---|---|---|---|---|---|---|---|---|
| | Number of establishments | Number of employees | Receipts (mil dol) | Annual payroll (mil dol) | Number of establishments | Number of employees | Receipts (mil dol) | Annual payroll (mil dol) | Number | Receipts (mil dol) | New construction ($1,000) | Number of housing units |
| | 159 | 160 | 161 | 162 | 163 | 164 | 165 | 166 | 167 | 168 | 169 | 170 |
| FLORIDA—Cont'd | | | | | | | | | | | | |
| Putnam | 159 | 2,213 | 236.4 | 83.8 | 104 | 372 | 38.5 | 9.3 | 4,248 | 153.6 | 21,255 | 128 |
| St. Johns | 660 | 8,990 | 1,013.8 | 367.9 | 413 | 5,020 | 1,828.8 | 192.6 | 24,012 | 1,220.8 | 1,377,131 | 5,090 |
| St. Lucie | 788 | 13,396 | 1,853.0 | 616.2 | 421 | 1,685 | 166.5 | 46.6 | 28,819 | 1,084.3 | 928,065 | 4,694 |
| Santa Rosa | 285 | 3,931 | 512.9 | 173.0 | 165 | 598 | 63.8 | 16.0 | 13,056 | 565.4 | 404,820 | 2,181 |
| Sarasota | 1,589 | 28,231 | 3,766.3 | 1,300.3 | 1,015 | 5,093 | 717.0 | 157.2 | 45,398 | 2,470.0 | 1,336,384 | 4,437 |
| Seminole | 1,411 | 20,780 | 2,625.8 | 933.6 | 912 | 4,645 | 478.0 | 139.9 | 46,371 | 1,967.3 | 595,070 | 2,532 |
| Sumter | 202 | 4,682 | 635.1 | 217.7 | 82 | 436 | 34.6 | 11.4 | 7,013 | 298.5 | 875,645 | 3,698 |
| Suwannee | 70 | 1,357 | 114.5 | 45.3 | 39 | 188 | 23.7 | 5.4 | 2,704 | 110.4 | 13,886 | 75 |
| Taylor | 37 | 572 | 50.0 | 20.9 | D | D | D | 3.6 | 1,007 | 38.1 | 5,379 | 29 |
| Union | 10 | 350 | 29.6 | 12.8 | D | D | 1.5 | D | 594 | 29.9 | 413 | 38 |
| Volusia | 1,452 | 28,599 | 3,465.8 | 1,273.6 | 1,049 | 4,675 | 697.6 | 150.2 | 44,735 | 1,880.9 | 1,168,352 | 4,581 |
| Wakulla | D | D | D | D | 37 | 108 | 9.9 | 2.7 | 2,163 | 86.1 | 69,320 | 409 |
| Walton | 135 | 2,054 | 250.5 | 96.9 | 112 | 525 | 58.1 | 16.3 | 8,981 | 595.9 | 708,516 | 1,762 |
| Washington | 44 | 753 | 72.2 | 26.5 | D | D | D | 2.8 | 1,631 | 58.6 | 10,300 | 60 |
| GEORGIA | 25,260 | 521,146 | 68,759.7 | 25,771.5 | 14,976 | 97,721 | 14,342.5 | 3,347.6 | 955,621 | 39,863.1 | 11,592,571 | 55,827 |
| Appling | D | D | D | D | 23 | 74 | 7.9 | 2.1 | 1,197 | 49.8 | 444 | 4 |
| Atkinson | 5 | 23 | 1.0 | 0.6 | 3 | 14 | 0.9 | 0.2 | 568 | 25.6 | 3,779 | 36 |
| Bacon | 19 | 513 | 44.7 | 17.5 | D | D | D | D | 634 | 27.0 | 0 | 0 |
| Baker | D | D | D | 1.4 | D | D | D | D | 193 | 6.5 | 0 | 0 |
| Baldwin | 101 | 3,547 | 345.2 | 137.4 | 53 | 268 | 21.5 | 6.3 | 3,229 | 115.7 | 26,574 | 107 |
| Banks | 15 | 150 | 17.6 | 7.8 | D | D | D | D | 1,329 | 61.1 | 11,983 | 70 |
| Barrow | 104 | 1,471 | 135.7 | 55.9 | 93 | 475 | 60.0 | 15.2 | 6,921 | 299.3 | 124,575 | 935 |
| Bartow | 214 | 3,349 | 449.3 | 159.6 | 131 | 718 | 110.1 | 26.3 | 8,562 | 397.8 | 133,994 | 861 |
| Ben Hill | 27 | 468 | 36.1 | 15.8 | 23 | 87 | 7.8 | 1.6 | 1,031 | 36.7 | 1,328 | 10 |
| Berrien | 25 | 286 | 22.0 | 8.5 | 11 | 37 | 4.0 | 0.8 | 1,142 | 55.8 | 9,500 | 45 |
| Bibb | 619 | 16,894 | 2,123.3 | 756.1 | 245 | 1,633 | 217.1 | 58.6 | 13,187 | 443.7 | 27,542 | 157 |
| Bleckley | 17 | 248 | 17.5 | 6.8 | D | D | D | D | 843 | 24.5 | 3,512 | 14 |
| Brantley | D | D | D | D | 12 | 28 | 2.9 | 0.7 | 1,053 | 37.9 | 6,690 | 32 |
| Brooks | 20 | 558 | 39.9 | 19.2 | D | D | D | D | 964 | 36.9 | 7,980 | 38 |
| Bryan | 62 | 688 | 57.2 | 21.3 | 48 | 488 | 37.1 | 13.0 | 3,089 | 134.7 | 149,227 | 516 |
| Bulloch | 204 | 3,328 | 354.0 | 125.2 | 95 | 469 | 50.8 | 12.4 | 5,056 | 204.5 | 77,126 | 515 |
| Burke | 31 | 648 | 38.9 | 19.1 | D | D | D | D | 1,535 | 51.4 | 10,409 | 49 |
| Butts | 31 | 487 | 42.9 | 19.0 | D | D | D | D | 1,778 | 71.6 | 51,153 | 170 |
| Calhoun | 9 | 69 | 6.3 | 2.3 | NA | NA | NA | NA | 283 | 13.5 | 1,518 | 5 |
| Camden | 100 | 1,023 | 105.4 | 38.2 | 60 | 277 | 22.7 | 6.6 | 2,956 | 93.6 | 87,661 | 441 |
| Candler | 20 | 512 | 30.7 | 15.2 | 7 | 33 | 2.3 | 0.4 | 819 | 38.9 | 245 | 2 |
| Carroll | 214 | 5,294 | 785.2 | 314.3 | 119 | 610 | 69.1 | 17.2 | 8,815 | 339.9 | 194,302 | 817 |
| Catoosa | 108 | 2,325 | 231.8 | 75.2 | 49 | 303 | 31.0 | 8.2 | 4,334 | 202.2 | 51,551 | 280 |
| Charlton | 13 | 73 | 4.4 | 2.6 | D | D | D | D | 507 | 14.7 | 6,927 | 25 |
| Chatham | 761 | 21,255 | 2,541.9 | 977.6 | 467 | 3,587 | 436.1 | 128.4 | 22,971 | 1,039.6 | 241,431 | 1,754 |
| Chattahoochee | D | D | D | D | D | D | D | D | 275 | 4.7 | 950 | 3 |
| Chattooga | 27 | 385 | 31.3 | 12.7 | D | D | D | D | 1,420 | 83.4 | 450 | 4 |
| Cherokee | 507 | 7,011 | 980.3 | 315.9 | 356 | 1,838 | 203.2 | 61.2 | 26,324 | 1,162.7 | 669,764 | 2,394 |
| Clarke | 456 | 10,263 | 1,666.8 | 621.3 | 196 | 1,403 | 275.9 | 48.3 | 9,267 | 342.5 | 99,497 | 635 |
| Clay | 4 | 20 | 0.9 | 0.4 | NA | NA | NA | NA | 161 | 5.2 | 540 | 3 |
| Clayton | 413 | 8,536 | 924.3 | 397.7 | 281 | 1,678 | 201.4 | 57.1 | 31,318 | 794.3 | 231,127 | 954 |
| Clinch | 12 | 222 | 16.2 | 6.4 | D | D | 1.5 | D | 380 | 11.9 | 1,719 | 9 |
| Cobb | 2,134 | 39,978 | 5,569.5 | 2,444.9 | 1,384 | 10,745 | 1,045.6 | 353.6 | 83,097 | 3,749.5 | 581,617 | 2,494 |
| Coffee | 107 | 2,342 | 281.5 | 98.4 | 46 | 233 | 21.3 | 7.4 | 3,086 | 135.8 | 7,280 | 52 |
| Colquitt | 93 | 1,961 | 239.7 | 88.8 | 53 | 221 | 21.3 | 5.5 | 2,975 | 112.3 | 15,975 | 150 |
| Columbia | 268 | 4,211 | 387.6 | 162.8 | 183 | 1,097 | 110.9 | 34.5 | 11,171 | 483.8 | 256,622 | 1,402 |
| Cook | 40 | 427 | 36.2 | 15.6 | D | D | D | D | 1,167 | 48.9 | 6,174 | 47 |
| Coweta | 264 | 5,616 | 1,103.4 | 286.8 | 150 | 783 | 77.1 | 22.6 | 12,141 | 446.0 | 364,986 | 1,014 |
| Crawford | D | D | D | D | 10 | 45 | 5.3 | 1.6 | 787 | 25.1 | 4,485 | 39 |
| Crisp | 63 | 1,147 | 170.4 | 47.4 | 32 | 151 | 20.2 | 5.1 | 1,489 | 58.9 | 3,556 | 27 |
| Dade | D | D | D | D | D | D | D | D | 1,064 | 48.9 | 1,245 | 8 |
| Dawson | 48 | 521 | 62.9 | 19.5 | 36 | 177 | 18.6 | 5.4 | 2,438 | 125.6 | 93,206 | 430 |
| Decatur | 53 | 840 | 72.2 | 32.5 | 43 | 151 | 14.2 | 3.1 | 1,836 | 67.6 | 5,447 | 30 |
| DeKalb | 2,078 | 47,637 | 5,974.3 | 2,416.6 | 1,142 | 7,483 | 1,120.3 | 296.6 | 83,364 | 2,969.8 | 563,691 | 2,225 |
| Dodge | 52 | 974 | 71.3 | 27.2 | D | D | 9.5 | D | 1,501 | 47.2 | 4,182 | 20 |
| Dooly | 11 | 46 | 3.3 | 1.3 | 9 | 30 | 3.1 | 0.6 | 675 | 27.1 | 0 | 0 |
| Dougherty | 313 | 8,489 | 1,158.0 | 457.9 | 134 | 906 | 83.3 | 25.5 | 6,660 | 205.7 | 25,013 | 165 |
| Douglas | 283 | 4,710 | 557.8 | 205.3 | 189 | 978 | 116.4 | 32.0 | 13,714 | 469.8 | 141,843 | 747 |
| Early | D | D | D | D | 13 | 30 | 3.9 | 0.9 | 636 | 27.7 | 2,458 | 15 |
| Echols | NA | NA | NA | NA | NA | NA | NA | NA | 219 | 9.2 | 1,116 | 7 |

# Table B. States and Counties — Government Employment and Payroll, and Local Government Finances

| STATE County | Full-time equivalent employees | March payroll (dollars) | Administration, judicial, and legal | Police and corrections | Fire protection | Highways and transportation | Health and welfare | Natural resources and utilities | Education and libraries | Total (mil dol) | Intergovernmental (mil dol) | Taxes Total (mil dol) | Per capita Total | Per capita Property |
|---|---|---|---|---|---|---|---|---|---|---|---|---|---|---|
| | 171 | 172 | 173 | 174 | 175 | 176 | 177 | 178 | 179 | 180 | 181 | 182 | 183 | 184 |
| **FLORIDA—Cont'd** | | | | | | | | | | | | | | |
| Putnam | 3,479 | 13,112,988 | 7.6 | 7.0 | 1.1 | 1.6 | 2.9 | 25.2 | 53.4 | 376.7 | 160.4 | 158.0 | 2,153 | 1,944 |
| St. Johns | 7,186 | 26,903,374 | 6.5 | 13.1 | 7.2 | 1.8 | 1.5 | 5.5 | 58.9 | 892.9 | 224.0 | 373.4 | 1,532 | 1,301 |
| St. Lucie | 9,995 | 39,089,749 | 5.6 | 13.6 | 8.1 | 1.8 | 2.1 | 7.9 | 56.2 | 1233.7 | 354.0 | 481.5 | 1,537 | 1,305 |
| Santa Rosa | 4,207 | 14,416,778 | 9.4 | 14.5 | 1.9 | 1.2 | 0.5 | 3.0 | 69.1 | 437.9 | 208.1 | 149.9 | 861 | 674 |
| Sarasota | 17,288 | 78,763,540 | 6.7 | 9.0 | 5.7 | 3.0 | 37.1 | 5.9 | 29.2 | 2415.6 | 267.3 | 869.1 | 2,071 | 1,600 |
| Seminole | 13,454 | 52,063,875 | 6.8 | 15.3 | 6.8 | 2.3 | 0.9 | 6.6 | 59.4 | 1543.0 | 567.5 | 635.9 | 1,374 | 997 |
| Sumter | 1,805 | 6,376,591 | 5.2 | 22.2 | 6.8 | 3.6 | 0.3 | 3.5 | 57.8 | 324.6 | 40.6 | 147.5 | 1,180 | 970 |
| Suwannee | 1,197 | 3,723,646 | 3.6 | 1.8 | 4.9 | 6.2 | 0.6 | 2.5 | 75.4 | 133.6 | 77.9 | 36.7 | 831 | 608 |
| Taylor | 789 | 2,322,312 | 10.4 | 15.2 | 3.4 | 3.6 | 0.6 | 5.4 | 59.3 | 65.8 | 29.8 | 26.6 | 1,221 | 958 |
| Union | 458 | 1,208,109 | 4.9 | 0.0 | 0.0 | 3.5 | 6.5 | 4.7 | 79.3 | 37.5 | 26.0 | 6.6 | 430 | 264 |
| Volusia | 15,886 | 61,497,255 | 8.2 | 16.3 | 5.4 | 3.3 | 3.7 | 9.0 | 50.9 | 2518.3 | 603.8 | 818.9 | 1,522 | 1,139 |
| Wakulla | 828 | 2,436,692 | 6.0 | 0.3 | 5.9 | 0.0 | 1.4 | 1.9 | 82.7 | 81.8 | 44.9 | 23.9 | 747 | 540 |
| Walton | 2,607 | 9,053,017 | 9.8 | 15.2 | 13.3 | 6.1 | 0.9 | 2.3 | 45.9 | 299.8 | 56.1 | 203.9 | 2,998 | 2,197 |
| Washington | 914 | 2,754,142 | 8.1 | 9.4 | 0.3 | 3.8 | 0.6 | 3.3 | 73.2 | 80.8 | 48.3 | 20.8 | 846 | 605 |
| **GEORGIA** | X | X | X | X | X | X | X | X | X | X | X | X | X | X |
| Appling | 1,086 | 3,342,587 | 6.1 | 8.6 | 0.2 | 3.5 | 32.0 | 1.7 | 46.6 | 102.4 | 29.0 | 37.2 | 2,017 | 1,350 |
| Atkinson | 387 | 1,735,046 | 12.3 | 1.5 | 0.2 | 1.4 | 0.2 | 2.5 | 79.5 | 29.6 | 15.8 | 6.7 | 823 | 591 |
| Bacon | 409 | 1,529,007 | 7.4 | 6.4 | 0.4 | 3.3 | 2.1 | 0.9 | 78.7 | 36.3 | 19.0 | 12.0 | 1,074 | 685 |
| Baker | 108 | 297,903 | 11.0 | 14.3 | 0.0 | 3.6 | 0.0 | 0.3 | 70.0 | 8.8 | 3.9 | 4.0 | 1,269 | 1,032 |
| Baldwin | 1,778 | 5,432,600 | 6.3 | 9.3 | 4.4 | 2.0 | 33.9 | 4.7 | 37.6 | 127.0 | 50.2 | 53.5 | 1,189 | 682 |
| Banks | 698 | 2,400,762 | 6.9 | 7.3 | 3.3 | 1.4 | 0.4 | 2.7 | 75.9 | 49.6 | 20.3 | 22.5 | 1,209 | 685 |
| Barrow | 2,202 | 8,733,548 | 5.0 | 11.5 | 5.9 | 1.2 | 1.0 | 2.7 | 70.8 | 231.3 | 102.4 | 91.5 | 1,159 | 715 |
| Bartow | 3,757 | 13,895,285 | 7.5 | 8.7 | 5.1 | 2.0 | 2.4 | 7.7 | 65.5 | 366.7 | 121.3 | 185.2 | 1,762 | 1,084 |
| Ben Hill | 984 | 3,377,220 | 3.2 | 13.7 | 2.1 | 2.6 | 23.0 | 7.1 | 44.3 | 80.3 | 27.8 | 21.4 | 1,255 | 746 |
| Berrien | 769 | 2,162,144 | 8.7 | 10.1 | 0.2 | 4.4 | 5.2 | 2.5 | 66.4 | 51.5 | 27.3 | 17.6 | 919 | 626 |
| Bibb | 5,503 | 21,076,493 | 6.9 | 10.9 | 6.6 | 2.0 | 4.6 | 6.0 | 60.1 | 496.0 | 236.8 | 147.4 | 965 | 668 |
| Bleckley | 506 | 1,546,372 | 5.4 | 6.5 | 1.9 | 2.9 | 1.1 | 2.0 | 79.2 | 50.5 | 22.9 | 12.3 | 960 | 650 |
| Brantley | 727 | 2,183,628 | 4.6 | 7.2 | 1.1 | 3.6 | 3.8 | 0.6 | 79.0 | 51.0 | 30.1 | 15.1 | 800 | 553 |
| Brooks | 526 | 1,838,263 | 7.0 | 9.5 | 2.6 | 3.3 | 1.3 | 2.0 | 71.7 | 43.0 | 20.0 | 16.8 | 1,074 | 805 |
| Bryan | 1,278 | 4,501,314 | 6.6 | 8.0 | 2.8 | 1.4 | 2.9 | 1.3 | 74.8 | 126.1 | 52.7 | 55.4 | 1,496 | 916 |
| Bulloch | 2,971 | 9,601,219 | 4.9 | 9.6 | 1.3 | 2.5 | 11.3 | 7.2 | 61.5 | 217.6 | 83.3 | 89.1 | 1,173 | 620 |
| Burke | 1,327 | 4,425,013 | 6.6 | 9.5 | 11.8 | 4.3 | 1.1 | 3.2 | 61.4 | 121.9 | 26.0 | 85.6 | 3,798 | 3,136 |
| Butts | 814 | 2,725,650 | 9.8 | 12.2 | 6.4 | 3.5 | 0.8 | 6.4 | 60.4 | 64.4 | 23.7 | 30.9 | 1,285 | 754 |
| Calhoun | 218 | 687,826 | 14.0 | 11.0 | 0.3 | 3.3 | 3.3 | 5.6 | 58.7 | 20.6 | 8.3 | 5.1 | 799 | 543 |
| Camden | 1,951 | 7,933,054 | 5.7 | 7.5 | 6.1 | 2.1 | 0.1 | 3.8 | 71.6 | 167.0 | 62.6 | 70.9 | 1,337 | 865 |
| Candler | 593 | 1,891,822 | 5.5 | 6.0 | 1.3 | 1.5 | 30.0 | 4.0 | 50.9 | 52.5 | 21.2 | 11.3 | 1,051 | 581 |
| Carroll | 3,751 | 13,069,947 | 4.9 | 10.1 | 4.6 | 1.9 | 0.8 | 7.6 | 68.4 | 347.9 | 148.3 | 137.3 | 1,170 | 667 |
| Catoosa | 2,056 | 7,580,949 | 4.8 | 6.3 | 2.5 | 2.2 | 0.7 | 3.3 | 78.6 | 375.1 | 78.8 | 72.4 | 1,089 | 625 |
| Charlton | 565 | 1,709,347 | 8.1 | 7.2 | 0.2 | 2.9 | 28.1 | 3.2 | 48.9 | 37.1 | 14.5 | 16.2 | 1,264 | 946 |
| Chatham | 10,284 | 40,221,481 | 9.9 | 15.3 | 4.9 | 6.0 | 3.4 | 8.4 | 47.8 | 1791.9 | 329.5 | 666.3 | 2,303 | 1,474 |
| Chattahoochee | 222 | 618,747 | 7.8 | 5.5 | 0.0 | 4.0 | 0.5 | 3.7 | 77.5 | 14.0 | 8.5 | 3.7 | 366 | 173 |
| Chattooga | 833 | 2,858,771 | 7.1 | 8.6 | 0.5 | 3.2 | 1.5 | 10.5 | 67.1 | 73.5 | 35.6 | 21.2 | 856 | 506 |
| Cherokee | 7,585 | 27,818,955 | 5.2 | 8.4 | 6.7 | 1.4 | 0.7 | 4.0 | 72.1 | 720.1 | 235.5 | 354.2 | 1,430 | 1,009 |
| Clarke | 4,590 | 17,021,391 | 7.3 | 12.9 | 4.9 | 2.6 | 7.2 | 9.0 | 51.4 | 858.6 | 139.6 | 176.8 | 1,394 | 995 |
| Clay | 123 | 336,328 | 16.3 | 9.1 | 0.0 | 6.1 | 6.9 | 5.7 | 53.2 | 15.6 | 4.1 | 6.1 | 2,064 | 1,723 |
| Clayton | 10,560 | 39,755,068 | 6.7 | 13.9 | 5.5 | 1.0 | 2.3 | 3.3 | 64.1 | 970.9 | 408.0 | 420.7 | 1,480 | 853 |
| Clinch | 411 | 1,464,990 | 8.8 | 6.4 | 1.6 | 2.4 | 27.5 | 2.8 | 50.6 | 38.1 | 11.4 | 10.6 | 1,590 | 1,274 |
| Cobb | 23,528 | 98,376,437 | 7.2 | 8.7 | 4.8 | 1.1 | 1.6 | 5.3 | 70.1 | 2465.8 | 764.5 | 1279.6 | 1,700 | 1,123 |
| Coffee | 1,553 | 4,931,558 | 4.3 | 8.1 | 2.8 | 2.2 | 1.1 | 3.4 | 75.2 | 131.2 | 66.7 | 41.0 | 955 | 516 |
| Colquitt | 3,145 | 10,521,532 | 3.5 | 3.9 | 1.3 | 1.5 | 41.9 | 1.8 | 44.2 | 266.6 | 83.0 | 51.7 | 1,137 | 661 |
| Columbia | 4,609 | 16,921,276 | 5.1 | 8.7 | 4.4 | 2.0 | 0.6 | 4.9 | 72.4 | 427.3 | 162.2 | 200.6 | 1,324 | 802 |
| Cook | 785 | 2,201,477 | 8.0 | 11.1 | 2.5 | 3.0 | 0.0 | 5.7 | 66.9 | 55.1 | 24.4 | 19.4 | 1,129 | 640 |
| Coweta | 4,327 | 18,142,647 | 5.4 | 10.0 | 5.4 | 1.9 | 0.6 | 4.7 | 67.7 | 380.3 | 133.8 | 197.7 | 1,381 | 870 |
| Crawford | 405 | 906,421 | 23.0 | 8.0 | 1.8 | 4.0 | 1.4 | 1.1 | 59.3 | 25.9 | 13.0 | 10.3 | 838 | 642 |
| Crisp | 1,102 | 3,736,914 | 6.6 | 10.8 | 3.9 | 2.5 | 2.3 | 12.4 | 58.3 | 92.9 | 39.0 | 35.2 | 1,552 | 868 |
| Dade | 469 | 1,393,274 | 2.2 | 10.4 | 0.3 | 2.0 | 1.6 | 7.0 | 75.5 | 35.2 | 15.9 | 16.2 | 999 | 585 |
| Dawson | 830 | 3,226,303 | 9.9 | 10.6 | 5.8 | 1.7 | 1.2 | 3.9 | 64.0 | 82.4 | 20.5 | 52.8 | 2,165 | 1,173 |
| Decatur | 1,626 | 5,616,306 | 4.4 | 8.5 | 1.3 | 2.1 | 31.4 | 2.8 | 47.4 | 142.0 | 40.5 | 41.4 | 1,550 | 1,012 |
| DeKalb | 26,965 | 181,019,159 | 4.5 | 4.7 | 1.6 | 0.4 | 23.7 | 2.6 | 61.7 | 3353.9 | 815.6 | 1202.8 | 1,601 | 1,144 |
| Dodge | 1,053 | 3,799,900 | 2.7 | 4.3 | 0.8 | 1.4 | 27.3 | 1.4 | 61.2 | 124.3 | 32.9 | 18.1 | 872 | 513 |
| Dooly | 452 | 1,485,388 | 12.1 | 18.9 | 0.5 | 5.1 | 6.1 | 5.1 | 52.1 | 39.7 | 12.3 | 16.4 | 1,200 | 806 |
| Dougherty | 4,459 | 14,873,418 | 6.7 | 11.0 | 4.7 | 2.8 | 7.3 | 9.4 | 55.6 | 360.7 | 141.6 | 134.9 | 1,508 | 932 |
| Douglas | 4,718 | 18,148,063 | 7.4 | 10.3 | 4.0 | 1.5 | 0.8 | 4.6 | 68.7 | 459.2 | 185.2 | 194.5 | 1,357 | 891 |
| Early | 535 | 1,742,905 | 6.9 | 3.7 | 2.8 | 3.9 | 5.1 | 4.7 | 70.3 | 36.7 | 17.5 | 15.4 | 1,492 | 1,063 |
| Echols | 160 | 484,584 | 5.8 | 5.7 | 0.0 | 3.8 | 1.3 | 2.1 | 81.2 | 12.3 | 7.1 | 4.5 | 1,144 | 846 |

1. Based on the resident population estimated as of July 1 of the year shown.

# Table B. States and Counties — Local Government Finances, Government Employment, and Income Taxes

| STATE County | Direct general expenditure Total (mil dol) | Per capita[1] (dollars) | Percent of total for: Education | Health and hospitals | Police protection | Public welfare | Highways | Debt outstanding Total (mil dol) | Per capita[1] (dollars) | Government employment, 2019 Federal civilian | Federal military | State and local | Individual income tax returns, 2018 Number of returns | Mean adjusted gross income | Mean income tax |
|---|---|---|---|---|---|---|---|---|---|---|---|---|---|---|---|
| | 185 | 186 | 187 | 188 | 189 | 190 | 191 | 192 | 193 | 194 | 195 | 196 | 197 | 198 | 199 |
| **FLORIDA—Cont'd** | | | | | | | | | | | | | | | |
| Putnam | 363 | 4,952 | 40.6 | 2.6 | 5.0 | 0.0 | 3.7 | 76.6 | 1,044 | 126 | 127 | 3,682 | 29,720 | 42,279 | 3,291 |
| St. Johns | 912 | 3,742 | 44.8 | 2.1 | 8.8 | 0.9 | 4.7 | 1,335.0 | 5,478 | 627 | 459 | 9,757 | 124,620 | 112,459 | 17,241 |
| St. Lucie | 1,195 | 3,816 | 40.7 | 0.9 | 8.5 | 1.1 | 9.1 | 2,143.9 | 6,846 | 755 | 645 | 13,079 | 152,600 | 54,061 | 5,428 |
| Santa Rosa | 426 | 2,446 | 57.6 | 1.6 | 10.2 | 0.0 | 3.8 | 320.3 | 1,840 | 826 | 1,377 | 5,965 | 81,550 | 66,985 | 7,112 |
| Sarasota | 2,329 | 5,550 | 24.9 | 33.8 | 5.7 | 0.0 | 3.9 | 1,745.3 | 4,159 | 1,048 | 798 | 13,759 | 218,340 | 98,766 | 15,395 |
| Seminole | 1,549 | 3,346 | 46.8 | 1.8 | 8.7 | 0.6 | 7.8 | 836.5 | 1,807 | 965 | 814 | 16,063 | 230,430 | 72,124 | 9,108 |
| Sumter | 308 | 2,461 | 27.6 | 1.1 | 7.3 | 0.5 | 7.7 | 579.9 | 4,640 | 1,646 | 216 | 3,191 | 61,940 | 89,940 | 11,552 |
| Suwannee | 133 | 3,012 | 48.3 | 4.6 | 5.7 | 0.7 | 11.8 | 25.2 | 571 | 102 | 73 | 2,403 | 17,080 | 43,313 | 3,644 |
| Taylor | 72 | 3,310 | 46.7 | 2.7 | 8.8 | 0.5 | 6.5 | 7.4 | 340 | 34 | 33 | 1,366 | 7,960 | 48,639 | 3,987 |
| Union | 36 | 2,332 | 59.1 | 5.4 | 6.1 | 0.7 | 5.8 | 2.6 | 169 | 20 | 18 | 1,931 | 4,710 | 47,180 | 3,511 |
| Volusia | 2,354 | 4,377 | 29.1 | 27.0 | 7.0 | 0.3 | 3.9 | 2,352.3 | 4,373 | 1,373 | 1,066 | 18,428 | 262,540 | 57,243 | 6,251 |
| Wakulla | 81 | 2,540 | 54.9 | 4.5 | 14.2 | 0.1 | 3.1 | 5.9 | 183 | 85 | 53 | 1,835 | 13,880 | 55,710 | 4,834 |
| Walton | 291 | 4,277 | 36.2 | 5.4 | 11.8 | 0.1 | 8.1 | 108.8 | 1,599 | 156 | 162 | 3,338 | 33,880 | 107,269 | 17,731 |
| Washington | 92 | 3,755 | 55.1 | 3.4 | 4.3 | 0.0 | 10.7 | 11.1 | 453 | 34 | 40 | 1,856 | 9,710 | 44,513 | 3,340 |
| **GEORGIA** | X | X | X | X | X | X | X | X | X | 103,398 | 93,172 | 587,914 | 4,580,240 | 67,877 | 8,108 |
| Appling | 102 | 5,538 | 40.4 | 32.1 | 3.5 | 0.0 | 7.3 | 31.6 | 1,714 | 48 | 48 | 1,277 | 7,180 | 45,834 | 3,921 |
| Atkinson | 31 | 3,822 | 57.4 | 0.9 | 3.8 | 0.1 | 4.9 | 6.5 | 790 | 16 | 24 | 409 | 3,040 | 32,027 | 1,886 |
| Bacon | 37 | 3,288 | 59.8 | 2.0 | 5.7 | 0.1 | 6.7 | 35.9 | 3,203 | 26 | 29 | 580 | 4,120 | 43,406 | 3,351 |
| Baker | 10 | 3,032 | 62.8 | 1.2 | 7.0 | 0.0 | 11.2 | 0.0 | 0 | 1 | 8 | 119 | 1,150 | 38,521 | 3,000 |
| Baldwin | 117 | 2,604 | 51.0 | 8.3 | 8.3 | 0.0 | 3.5 | 44.2 | 983 | 78 | 133 | 5,373 | 16,860 | 46,943 | 4,052 |
| Banks | 45 | 2,416 | 70.1 | 0.3 | 8.7 | 0.0 | 3.3 | 28.6 | 1,536 | 19 | 51 | 879 | 7,480 | 48,634 | 3,545 |
| Barrow | 229 | 2,900 | 63.2 | 2.3 | 6.4 | 0.2 | 5.1 | 237.0 | 3,003 | 170 | 221 | 2,827 | 35,980 | 52,430 | 4,474 |
| Bartow | 343 | 3,265 | 56.3 | 1.7 | 8.8 | 0.2 | 5.5 | 214.2 | 2,038 | 215 | 285 | 4,853 | 46,440 | 56,314 | 5,211 |
| Ben Hill | 85 | 4,966 | 41.8 | 30.5 | 4.5 | 0.3 | 3.3 | 9.0 | 530 | 27 | 44 | 979 | 6,450 | 39,311 | 2,682 |
| Berrien | 50 | 2,624 | 61.1 | 2.3 | 5.4 | 0.3 | 4.8 | 20.1 | 1,052 | 38 | 51 | 911 | 6,870 | 41,618 | 3,480 |
| Bibb | 530 | 3,466 | 54.8 | 7.6 | 5.1 | 0.0 | 0.8 | 365.4 | 2,391 | 1,016 | 396 | 8,894 | 64,560 | 53,239 | 5,586 |
| Bleckley | 49 | 3,802 | 51.6 | 21.8 | 3.6 | 0.0 | 5.5 | 14.3 | 1,117 | 31 | 31 | 1,258 | 4,730 | 48,578 | 3,868 |
| Brantley | 54 | 2,869 | 73.8 | 3.4 | 3.1 | 0.1 | 5.1 | 7.1 | 379 | 27 | 51 | 809 | 6,680 | 40,798 | 2,582 |
| Brooks | 44 | 2,811 | 53.8 | 1.1 | 7.4 | 0.1 | 9.7 | 5.4 | 344 | 24 | 41 | 659 | 6,120 | 43,662 | 3,808 |
| Bryan | 124 | 3,343 | 64.8 | 2.3 | 6.8 | 0.0 | 3.9 | 64.1 | 1,732 | 203 | 106 | 1,986 | 18,190 | 70,401 | 6,930 |
| Bulloch | 221 | 2,914 | 51.0 | 11.6 | 6.6 | 0.0 | 3.8 | 271.8 | 3,577 | 142 | 208 | 7,960 | 27,840 | 50,514 | 4,811 |
| Burke | 118 | 5,240 | 54.9 | 8.9 | 3.7 | 0.0 | 6.5 | 19.9 | 883 | 40 | 59 | 1,478 | 9,360 | 47,731 | 4,261 |
| Butts | 64 | 2,672 | 55.0 | 1.1 | 9.5 | 0.1 | 6.4 | 25.9 | 1,076 | 57 | 59 | 1,569 | 9,740 | 51,187 | 4,591 |
| Calhoun | 21 | 3,312 | 49.5 | 15.0 | 5.2 | 0.8 | 5.3 | 11.8 | 1,851 | 21 | 12 | 471 | 1,850 | 36,688 | 2,469 |
| Camden | 162 | 3,061 | 53.2 | 3.7 | 5.6 | 0.1 | 5.8 | 134.7 | 2,542 | 2,636 | 4,281 | 2,278 | 22,730 | 53,108 | 4,301 |
| Candler | 53 | 4,905 | 38.5 | 30.5 | 5.4 | 0.0 | 4.9 | 23.6 | 2,200 | 22 | 28 | 698 | 4,200 | 42,401 | 3,075 |
| Carroll | 368 | 3,131 | 62.2 | 0.5 | 6.2 | 0.0 | 2.9 | 446.1 | 3,800 | 228 | 310 | 7,206 | 48,500 | 56,260 | 5,276 |
| Catoosa | 245 | 3,677 | 50.4 | 28.3 | 3.3 | 0.2 | 2.5 | 151.2 | 2,274 | 85 | 179 | 2,665 | 27,640 | 53,827 | 4,519 |
| Charlton | 40 | 3,130 | 43.5 | 5.7 | 7.3 | 0.1 | 7.0 | 6.8 | 527 | 38 | 29 | 413 | 3,620 | 43,331 | 2,842 |
| Chatham | 1,847 | 6,382 | 27.6 | 30.7 | 6.4 | 0.8 | 5.3 | 535.1 | 1,850 | 2,677 | 5,229 | 16,504 | 128,480 | 63,585 | 7,181 |
| Chattahoochee | 15 | 1,473 | 66.7 | 0.0 | 0.0 | 0.0 | 3.0 | 4.4 | 428 | 106 | 14,338 | 248 | 3,280 | 38,115 | 1,777 |
| Chattooga | 77 | 3,128 | 60.8 | 0.4 | 4.9 | 0.2 | 4.9 | 61.9 | 2,500 | 40 | 63 | 1,253 | 9,310 | 42,259 | 2,894 |
| Cherokee | 666 | 2,689 | 64.8 | 0.5 | 5.2 | 0.1 | 4.5 | 788.3 | 3,182 | 378 | 687 | 8,490 | 119,480 | 79,061 | 9,124 |
| Clarke | 893 | 7,041 | 22.4 | 50.5 | 3.5 | 0.1 | 1.4 | 638.1 | 5,031 | 943 | 385 | 24,090 | 47,880 | 56,757 | 6,147 |
| Clay | 16 | 5,472 | 25.4 | 32.0 | 5.2 | 0.5 | 4.8 | 7.1 | 2,405 | 41 | 7 | 149 | 1,140 | 33,664 | 2,430 |
| Clayton | 1,004 | 3,532 | 58.4 | 1.4 | 7.7 | 0.2 | 4.0 | 348.4 | 1,226 | 1,483 | 816 | 13,061 | 129,190 | 35,630 | 2,332 |
| Clinch | 33 | 4,969 | 46.2 | 28.1 | 3.6 | 0.4 | 3.8 | 14.1 | 2,117 | 14 | 17 | 508 | 2,380 | 41,683 | 3,072 |
| Cobb | 2,829 | 3,758 | 50.1 | 1.4 | 5.8 | 0.0 | 5.8 | 3,892.4 | 5,172 | 2,616 | 2,351 | 33,935 | 355,230 | 86,207 | 11,564 |
| Coffee | 134 | 3,132 | 62.4 | 1.1 | 5.8 | 0.0 | 6.5 | 19.3 | 449 | 112 | 107 | 2,699 | 15,530 | 42,226 | 3,865 |
| Colquitt | 176 | 3,860 | 63.1 | 5.1 | 4.3 | 0.1 | 3.8 | 108.7 | 2,389 | 119 | 119 | 3,636 | 16,990 | 43,889 | 3,635 |
| Columbia | 423 | 2,791 | 66.2 | 0.5 | 4.4 | 0.0 | 5.5 | 366.8 | 2,421 | 434 | 417 | 5,637 | 66,080 | 76,455 | 8,249 |
| Cook | 53 | 3,083 | 53.2 | 1.3 | 8.7 | 0.5 | 7.0 | 26.8 | 1,558 | 30 | 46 | 1,258 | 6,610 | 40,083 | 3,197 |
| Coweta | 402 | 2,805 | 60.7 | 1.5 | 7.3 | 0.0 | 6.3 | 203.0 | 1,418 | 276 | 395 | 5,145 | 65,390 | 72,132 | 7,875 |
| Crawford | 26 | 2,125 | 65.2 | 0.4 | 5.3 | 0.0 | 5.5 | 9.4 | 766 | 8 | 33 | 442 | 5,000 | 44,300 | 3,224 |
| Crisp | 96 | 4,230 | 50.5 | 1.0 | 7.0 | 0.0 | 9.0 | 34.4 | 1,516 | 51 | 59 | 1,411 | 8,120 | 42,878 | 3,392 |
| Dade | 39 | 2,378 | 64.4 | 1.2 | 6.3 | 0.2 | 5.1 | 11.5 | 707 | 24 | 40 | 589 | 6,070 | 49,795 | 4,052 |
| Dawson | 81 | 3,302 | 58.5 | 4.0 | 5.7 | 0.0 | 5.9 | 83.1 | 3,405 | 44 | 69 | 1,210 | 11,370 | 66,679 | 7,053 |
| Decatur | 144 | 5,406 | 36.2 | 32.4 | 5.5 | 0.0 | 3.1 | 71.6 | 2,680 | 48 | 68 | 2,287 | 11,010 | 43,264 | 3,747 |
| DeKalb | 3,396 | 4,521 | 39.0 | 25.5 | 5.0 | 0.3 | 1.8 | 2,576.9 | 3,430 | 11,231 | 2,069 | 33,928 | 349,050 | 72,200 | 9,424 |
| Dodge | 91 | 4,394 | 43.9 | 25.4 | 3.0 | 0.0 | 3.3 | 15.2 | 734 | 38 | 50 | 1,718 | 7,180 | 40,156 | 2,803 |
| Dooly | 40 | 2,894 | 42.0 | 3.1 | 11.5 | 1.0 | 8.5 | 26.0 | 1,898 | 58 | 31 | 706 | 3,700 | 40,811 | 3,127 |
| Dougherty | 384 | 4,290 | 47.5 | 3.3 | 6.7 | 0.0 | 2.7 | 142.2 | 1,589 | 2,633 | 529 | 6,483 | 35,360 | 44,298 | 4,091 |
| Douglas | 433 | 3,019 | 62.7 | 1.7 | 5.6 | 0.0 | 2.9 | 355.3 | 2,478 | 217 | 388 | 5,680 | 64,830 | 51,819 | 4,655 |
| Early | 42 | 4,090 | 56.3 | 6.1 | 5.0 | 0.0 | 5.4 | 8.5 | 820 | 36 | 27 | 1,258 | 4,090 | 39,138 | 3,004 |
| Echols | 12 | 3,072 | 74.2 | 2.0 | 1.9 | 0.0 | 8.6 | 5.6 | 1,431 | 1 | 11 | 204 | 1,410 | 40,223 | 3,021 |

1. Based on the resident population estimated as of July 1 of the year shown.

# Table B. States and Counties — Land Area and Population

| State / county code | CBSA code[1] | County Type code[2] | STATE County | Land area[3] (sq. mi) | Total persons 2019 | Rank | Per square mile | White | Black | American Indian, Alaska Native | Asian and Pacific Islancer | Percent Hispanic or Latino[4] | Under 5 years | 5 to 14 years | 15 to 24 years | 25 to 34 years | 35 to 44 years | 45 to 54 years |
|---|---|---|---|---|---|---|---|---|---|---|---|---|---|---|---|---|---|---|
| | | | | 1 | 2 | 3 | 4 | 5 | 6 | 7 | 8 | 9 | 10 | 11 | 12 | 13 | 14 | 15 |
| | | | GEORGIA—Cont'd | | | | | | | | | | | | | | | |
| 13103 | 42340 | 2 | Effingham | 478.8 | 65,765 | 811 | 137.4 | 78.7 | 15.5 | 0.9 | 1.8 | 5.2 | 6.5 | 15.3 | 12.2 | 14.0 | 14.5 | 12.9 |
| 13105 | | 6 | Elbert | 351.1 | 19,335 | 1,861 | 55.1 | 64.3 | 29.0 | 0.7 | 1.3 | 6.1 | 5.8 | 12.6 | 10.9 | 12.5 | 11.0 | 11.9 |
| 13107 | | 7 | Emanuel | 680.6 | 22,507 | 1,704 | 33.1 | 59.8 | 34.9 | 0.6 | 1.0 | 4.7 | 6.7 | 14.1 | 13.0 | 13.2 | 12.5 | 11.1 |
| 13109 | | 6 | Evans | 182.9 | 10,638 | 2,376 | 58.2 | 57.0 | 30.6 | 0.8 | 1.1 | 11.9 | 6.2 | 15.3 | 13.2 | 12.4 | 12.0 | 11.9 |
| 13111 | | 8 | Fannin | 387.1 | 26,521 | 1,547 | 68.5 | 95.5 | 1.1 | 1.1 | 0.9 | 2.7 | 3.8 | 9.0 | 9.0 | 8.9 | 9.4 | 11.4 |
| 13113 | 12060 | 1 | Fayette | 194.6 | 115,821 | 539 | 595.2 | 61.4 | 26.5 | 0.8 | 6.0 | 7.7 | 4.5 | 13.3 | 13.4 | 9.0 | 11.5 | 13.9 |
| 13115 | 40660 | 3 | Floyd | 509.8 | 98,604 | 612 | 193.4 | 71.9 | 15.6 | 0.7 | 1.9 | 11.8 | 5.9 | 12.9 | 14.1 | 13.0 | 12.0 | 12.3 |
| 13117 | 12060 | 1 | Forsyth | 224.6 | 250,847 | 276 | 1,116.9 | 68.9 | 4.9 | 0.6 | 17.7 | 9.6 | 5.7 | 15.7 | 12.8 | 9.4 | 16.1 | 16.3 |
| 13119 | | 8 | Franklin | 261.4 | 23,504 | 1,658 | 89.9 | 83.7 | 10.1 | 0.7 | 1.8 | 5.3 | 5.9 | 12.0 | 13.6 | 12.6 | 10.5 | 12.5 |
| 13121 | 12060 | 1 | Fulton | 526.7 | 1,077,402 | 42 | 2,045.6 | 40.8 | 44.8 | 0.6 | 8.7 | 7.2 | 5.5 | 11.8 | 13.7 | 17.5 | 14.2 | 13.5 |
| 13123 | | 6 | Gilmer | 426.2 | 31,978 | 1,383 | 75.0 | 86.3 | 1.0 | 1.0 | 0.8 | 12.0 | 4.9 | 10.5 | 9.7 | 9.5 | 10.3 | 12.2 |
| 13125 | | 9 | Glascock | 143.7 | 2,984 | 2,962 | 20.8 | 88.3 | 10.0 | 1.1 | 0.4 | 1.8 | 4.9 | 12.2 | 12.9 | 10.2 | 12.1 | 14.5 |
| 13127 | 15260 | 3 | Glynn | 419.6 | 85,568 | 675 | 203.9 | 65.0 | 27.0 | 0.7 | 2.2 | 6.9 | 5.4 | 12.0 | 11.8 | 11.8 | 11.6 | 12.0 |
| 13129 | 15660 | 4 | Gordon | 356.4 | 58,780 | 880 | 164.9 | 77.8 | 4.9 | 0.7 | 1.6 | 16.6 | 5.9 | 13.3 | 13.4 | 12.8 | 12.6 | 13.9 |
| 13131 | | 6 | Grady | 454.5 | 24,491 | 1,627 | 53.9 | 58.9 | 28.7 | 1.1 | 0.9 | 11.7 | 6.5 | 13.9 | 11.9 | 11.3 | 11.8 | 12.0 |
| 13133 | | 6 | Greene | 387.5 | 18,837 | 1,884 | 48.6 | 60.2 | 32.5 | 0.8 | 1.4 | 6.5 | 4.7 | 10.6 | 9.4 | 9.2 | 9.9 | 10.2 |
| 13135 | 12060 | 1 | Gwinnett | 430.8 | 942,627 | 54 | 2,188.1 | 36.3 | 30.0 | 0.7 | 13.9 | 21.8 | 6.3 | 15.1 | 14.1 | 13.2 | 14.1 | 14.4 |
| 13137 | 18460 | 6 | Habersham | 276.9 | 46,047 | 1,057 | 166.3 | 77.4 | 4.3 | 0.8 | 2.7 | 16.4 | 5.8 | 12.4 | 13.5 | 12.6 | 11.8 | 12.1 |
| 13139 | 23580 | 3 | Hall | 393.0 | 206,591 | 330 | 525.7 | 61.1 | 8.0 | 0.6 | 2.5 | 29.1 | 6.3 | 13.6 | 13.7 | 12.9 | 12.2 | 13.1 |
| 13141 | 33300 | 7 | Hancock | 471.1 | 8,494 | 2,537 | 18.0 | 25.8 | 70.2 | 0.7 | 1.3 | 2.9 | 3.4 | 8.9 | 10.3 | 14.5 | 11.5 | 11.8 |
| 13143 | 12060 | 1 | Haralson | 282.2 | 30,383 | 1,420 | 107.7 | 92.1 | 5.3 | 1.0 | 1.2 | 2.3 | 6.4 | 13.4 | 12.3 | 13.1 | 12.2 | 13.7 |
| 13145 | 17980 | 2 | Harris | 463.8 | 36,080 | 1,273 | 77.8 | 77.9 | 16.9 | 0.9 | 2.1 | 4.2 | 4.9 | 12.5 | 11.5 | 10.9 | 12.2 | 13.9 |
| 13147 | | 6 | Hart | 232.4 | 26,406 | 1,553 | 113.6 | 76.0 | 19.9 | 0.5 | 1.5 | 3.8 | 5.1 | 11.6 | 10.9 | 11.4 | 11.0 | 12.8 |
| 13149 | 12060 | 1 | Heard | 296.0 | 11,973 | 2,296 | 40.4 | 86.7 | 10.6 | 1.0 | 0.9 | 2.8 | 5.3 | 13.3 | 12.0 | 12.6 | 11.5 | 13.7 |
| 13151 | 12060 | 1 | Henry | 318.7 | 239,139 | 288 | 750.4 | 39.6 | 50.5 | 0.8 | 4.2 | 7.6 | 5.8 | 14.3 | 14.5 | 12.6 | 13.2 | 14.8 |
| 13153 | 47580 | 3 | Houston | 376.0 | 160,110 | 420 | 425.8 | 56.8 | 34.3 | 0.9 | 4.4 | 6.8 | 6.5 | 14.7 | 13.0 | 14.5 | 13.7 | 11.8 |
| 13155 | | 7 | Irwin | 354.4 | 9,387 | 2,466 | 26.5 | 67.5 | 27.8 | 0.5 | 1.0 | 4.3 | 5.5 | 11.5 | 12.8 | 14.6 | 12.6 | 12.7 |
| 13157 | 27600 | 4 | Jackson | 339.7 | 76,199 | 730 | 224.3 | 81.3 | 8.1 | 0.7 | 2.6 | 9.0 | 6.3 | 14.4 | 11.9 | 13.3 | 13.9 | 13.4 |
| 13159 | 12060 | 1 | Jasper | 368.4 | 14,483 | 2,123 | 39.3 | 77.0 | 19.2 | 0.8 | 0.6 | 3.9 | 5.3 | 13.5 | 11.6 | 11.5 | 11.6 | 13.4 |
| 13161 | | 7 | Jeff Davis | 330.9 | 15,213 | 2,082 | 46.0 | 71.2 | 15.6 | 0.6 | 0.8 | 13.0 | 6.6 | 14.7 | 13.0 | 11.8 | 12.1 | 12.6 |
| 13163 | | 6 | Jefferson | 526.6 | 15,267 | 2,074 | 29.0 | 43.4 | 52.5 | 0.5 | 0.8 | 3.9 | 6.5 | 12.8 | 12.3 | 12.5 | 11.7 | 12.1 |
| 13165 | | 6 | Jenkins | 347.3 | 8,746 | 2,515 | 25.2 | 50.7 | 41.6 | 0.7 | 1.3 | 6.8 | 5.5 | 11.1 | 13.6 | 15.2 | 12.5 | 11.9 |
| 13167 | 20140 | 7 | Johnson | 303.0 | 9,667 | 2,498 | 31.9 | 62.3 | 34.7 | 0.5 | 0.7 | 2.8 | 5.4 | 9.8 | 11.2 | 13.1 | 13.2 | 13.9 |
| 13169 | 31420 | 3 | Jones | 393.9 | 28,787 | 1,467 | 73.1 | 71.6 | 26.3 | 0.7 | 0.9 | 1.9 | 5.1 | 13.2 | 12.3 | 11.6 | 12.3 | 13.3 |
| 13171 | 12060 | 1 | Lamar | 183.5 | 19,261 | 1,864 | 105.0 | 67.0 | 30.3 | 0.8 | 1.1 | 2.7 | 5.9 | 11.3 | 16.4 | 13.0 | 10.6 | 12.4 |
| 13173 | 46660 | 3 | Lanier | 196.5 | 10,737 | 2,364 | 54.6 | 69.1 | 22.8 | 1.4 | 2.0 | 7.4 | 6.4 | 13.6 | 12.7 | 14.2 | 13.3 | 11.9 |
| 13175 | 20140 | 5 | Laurens | 807.3 | 47,512 | 1,030 | 58.9 | 58.5 | 38.3 | 0.6 | 1.3 | 2.8 | 6.6 | 13.8 | 12.1 | 12.5 | 11.9 | 12.1 |
| 13177 | 10500 | 3 | Lee | 355.9 | 30,234 | 1,426 | 85.0 | 70.4 | 23.9 | 0.7 | 3.1 | 3.3 | 6.1 | 15.2 | 12.7 | 13.0 | 14.5 | 13.5 |
| 13179 | 25980 | 3 | Liberty | 516.5 | 63,004 | 849 | 122.0 | 41.1 | 45.7 | 1.2 | 3.8 | 12.7 | 10.1 | 14.4 | 17.3 | 19.8 | 11.1 | 8.3 |
| 13181 | 12260 | 2 | Lincoln | 210.4 | 8,031 | 2,586 | 38.2 | 68.4 | 29.2 | 1.1 | 0.9 | 2.2 | 5.1 | 10.5 | 9.4 | 11.7 | 9.8 | 11.9 |
| 13183 | 25980 | 3 | Long | 400.4 | 20,171 | 1,815 | 50.4 | 59.5 | 28.9 | 1.3 | 2.9 | 11.2 | 7.3 | 15.4 | 12.5 | 15.8 | 15.0 | 12.2 |
| 13185 | 46660 | 3 | Lowndes | 497.2 | 118,268 | 533 | 237.9 | 54.2 | 38.3 | 0.7 | 2.8 | 6.1 | 6.7 | 13.7 | 19.9 | 14.9 | 11.6 | 10.0 |
| 13187 | | 6 | Lumpkin | 282.9 | 34,186 | 1,325 | 120.8 | 91.7 | 2.1 | 1.4 | 1.3 | 5.1 | 4.4 | 9.6 | 21.9 | 11.4 | 10.0 | 11.1 |
| 13189 | 12260 | 2 | McDuffie | 257.4 | 21,162 | 1,763 | 82.2 | 54.6 | 42.2 | 0.7 | 0.9 | 3.5 | 6.4 | 14.3 | 12.1 | 11.9 | 11.1 | 11.7 |
| 13191 | 15260 | 3 | McIntosh | 431.4 | 14,387 | 2,131 | 33.3 | 63.5 | 33.9 | 0.8 | 0.9 | 2.4 | 3.9 | 8.5 | 9.0 | 10.6 | 9.1 | 13.0 |
| 13193 | | 3 | Macon | 400.7 | 12,712 | 2,232 | 31.7 | 33.6 | 60.6 | 0.5 | 1.8 | 4.5 | 5.1 | 10.1 | 12.1 | 15.1 | 12.5 | 12.2 |
| 13195 | 12020 | 3 | Madison | 282.3 | 30,457 | 1,417 | 107.9 | 81.7 | 10.4 | 0.8 | 2.4 | 6.5 | 6.0 | 13.1 | 11.4 | 12.8 | 12.0 | 12.9 |
| 13197 | 17980 | 2 | Marion | 365.7 | 8,516 | 2,534 | 23.3 | 60.4 | 30.5 | 1.4 | 1.6 | 7.9 | 4.9 | 12.0 | 11.3 | 11.2 | 10.2 | 13.0 |
| 13199 | 12060 | 1 | Meriwether | 500.9 | 21,164 | 1,762 | 42.3 | 58.2 | 39.0 | 0.8 | 0.9 | 2.7 | 5.7 | 11.7 | 11.5 | 12.1 | 10.8 | 11.9 |
| 13201 | | 8 | Miller | 282.4 | 5,622 | 2,779 | 19.9 | 68.1 | 28.8 | 0.8 | 1.0 | 2.9 | 6.2 | 12.9 | 11.0 | 10.8 | 11.1 | 11.8 |
| 13205 | | 6 | Mitchell | 512.2 | 21,602 | 1,741 | 42.2 | 46.4 | 47.9 | 0.7 | 1.2 | 5.0 | 5.4 | 13.0 | 12.1 | 13.5 | 12.4 | 12.8 |
| 13207 | 31420 | 3 | Monroe | 396.1 | 28,042 | 1,498 | 70.8 | 73.2 | 22.9 | 0.7 | 1.3 | 2.5 | 5.0 | 11.8 | 11.8 | 11.9 | 12.0 | 13.2 |
| 13209 | 47080 | 9 | Montgomery | 241.1 | 9,012 | 2,498 | 37.4 | 66.9 | 25.8 | 0.6 | 0.9 | 7.2 | 5.2 | 11.2 | 15.3 | 12.6 | 11.5 | 12.7 |
| 13211 | 12060 | 1 | Morgan | 347.4 | 19,636 | 1,843 | 56.5 | 74.9 | 21.8 | 0.7 | 0.9 | 3.2 | 5.8 | 12.5 | 11.5 | 10.7 | 11.6 | 12.8 |
| 13213 | 19140 | 3 | Murray | 344.5 | 40,032 | 1,184 | 116.2 | 86.2 | 1.5 | 0.8 | 0.7 | 15.7 | 6.2 | 13.4 | 13.0 | 13.1 | 11.9 | 13.9 |
| 13215 | 17980 | 2 | Muscogee | 216.5 | 196,442 | 343 | 907.4 | 41.8 | 48.4 | 0.9 | 4.0 | 8.0 | 7.0 | 13.9 | 13.4 | 16.3 | 12.9 | 10.9 |
| 13217 | 12060 | 1 | Newton | 273.8 | 113,295 | 545 | 413.8 | 44.5 | 49.0 | 0.7 | 1.8 | 6.2 | 6.3 | 14.4 | 14.3 | 12.8 | 12.5 | 13.9 |
| 13219 | 12020 | 3 | Oconee | 184.3 | 41,124 | 1,154 | 223.1 | 84.8 | 5.5 | 0.5 | 5.1 | 5.8 | 5.3 | 15.7 | 13.1 | 8.8 | 14.1 | 14.1 |
| 13221 | 12020 | 3 | Oglethorpe | 439.1 | 15,383 | 2,067 | 35.0 | 76.3 | 18.2 | 0.8 | 1.3 | 5.5 | 5.2 | 11.6 | 10.9 | 12.4 | 12.1 | 13.4 |
| 13223 | 12060 | 1 | Paulding | 312.4 | 173,359 | 380 | 554.9 | 68.8 | 23.4 | 0.8 | 1.9 | 7.4 | 6.2 | 14.7 | 13.1 | 13.7 | 14.0 | 15.0 |
| 13225 | 47580 | 3 | Peach | 150.3 | 27,950 | 1,500 | 186.0 | 46.3 | 45.0 | 0.9 | 1.5 | 8.2 | 5.6 | 11.7 | 17.1 | 12.3 | 11.0 | 11.5 |
| 13227 | 12060 | 3 | Pickens | 232.1 | 33,127 | 1,353 | 142.7 | 94.4 | 1.8 | 1.0 | 1.1 | 3.3 | 4.9 | 10.7 | 10.7 | 11.5 | 10.6 | 13.0 |
| 13229 | 48180 | 6 | Pierce | 340.5 | 19,522 | 1,851 | 57.3 | 85.5 | 9.3 | 0.9 | 1.1 | 5.0 | 6.5 | 13.6 | 12.2 | 11.9 | 12.3 | 13.2 |
| 13231 | 12060 | 1 | Pike | 216.1 | 19,121 | 1,871 | 88.5 | 88.5 | 9.6 | 0.8 | 0.9 | 1.8 | 5.3 | 13.1 | 12.8 | 11.7 | 12.7 | 14.6 |
| 13233 | 16340 | 4 | Polk | 310.3 | 42,840 | 1,121 | 138.1 | 72.9 | 13.4 | 0.7 | 0.9 | 13.8 | 6.7 | 14.2 | 12.9 | 13.2 | 12.1 | 12.3 |
| 13235 | | 3 | Pulaski | 249.3 | 11,191 | 2,334 | 44.9 | 63.4 | 32.7 | 0.7 | 1.6 | 3.6 | 4.3 | 10.3 | 11.0 | 12.1 | 12.1 | 12.9 |

1. CBSA = Core Based Statistical Area. See Appendix A for explanation. See Appendix B for list of metropolitan areas with component counties.
Service of USDA Rural-Urban Continuum Codes. See Appendix A for definition.  3. Dry land or land partially or temporarily covered by water.
2. County type code from the Economic Research
4. May be of any race.

| STATE County | 55 to 64 years | 65 to 74 years | 75 years and over | Percent female | 2000 | 2010 | 2000-2010 | 2010-2020 | Births | Deaths | Net Migration | Number | Persons per household | Family house-holds | Female family house-holder[1] | One person |
|---|---|---|---|---|---|---|---|---|---|---|---|---|---|---|---|---|
| | 16 | 17 | 18 | 19 | 20 | 21 | 22 | 23 | 24 | 25 | 26 | 27 | 28 | 29 | 30 | 31 |
| **GEORGIA—Cont'd** | | | | | | | | | | | | | | | | |
| Effingham | 12.2 | 7.7 | 4.5 | 50.3 | 37,535 | 52,258 | 39.2 | 25.8 | 7,731 | 4,367 | 10,116 | 21,172 | 2.84 | 72.9 | 10.3 | 23.2 |
| Elbert | 14.0 | 12.3 | 9.2 | 52.4 | 20,511 | 20,165 | -1.7 | -4.1 | 2,290 | 2,651 | -463 | 7,559 | 2.50 | 67.0 | 17.4 | 28.9 |
| Emanuel | 12.7 | 10.0 | 6.8 | 50.5 | 21,837 | 22,592 | 3.5 | -0.4 | 3,179 | 2,970 | -300 | 8,387 | 2.58 | 67.8 | 20.0 | 27.7 |
| Evans | 12.6 | 9.5 | 7.1 | 51.3 | 10,495 | 11,008 | 4.9 | -3.4 | 1,523 | 1,159 | -753 | 4,020 | 2.53 | 68.3 | 22.2 | 26.0 |
| Fannin | 17.7 | 18.7 | 12.2 | 51.1 | 19,798 | 23,706 | 19.7 | 11.9 | 2,060 | 3,355 | 4,103 | 10,408 | 2.42 | 68.1 | 7.4 | 27.1 |
| Fayette | 14.9 | 11.6 | 7.8 | 51.7 | 91,263 | 106,556 | 16.8 | 8.7 | 8,671 | 8,629 | 9,258 | 40,285 | 2.77 | 77.2 | 8.9 | 19.9 |
| Floyd | 12.5 | 10.0 | 7.3 | 51.5 | 90,565 | 96,318 | 6.4 | 2.4 | 12,044 | 11,091 | 1,390 | 35,679 | 2.61 | 69.9 | 14.2 | 26.0 |
| Forsyth | 11.5 | 7.4 | 5.1 | 50.4 | 98,407 | 175,498 | 78.3 | 42.9 | 23,878 | 11,285 | 62,196 | 76,753 | 2.97 | 81.7 | 7.5 | 14.5 |
| Franklin | 13.3 | 11.3 | 8.3 | 50.8 | 20,285 | 22,098 | 8.9 | 6.4 | 2,779 | 2,976 | 1,608 | 8,460 | 2.62 | 72.4 | 13.0 | 25.0 |
| Fulton | 11.4 | 7.5 | 4.8 | 51.7 | 816,006 | 920,393 | 12.8 | 17.1 | 127,279 | 64,807 | 93,668 | 410,576 | 2.44 | 53.7 | 13.7 | 38.2 |
| Gilmer | 16.4 | 16.6 | 10.0 | 50.0 | 23,456 | 28,299 | 20.6 | 13.0 | 3,282 | 3,132 | 3,517 | 12,021 | 2.51 | 66.8 | 7.8 | 26.6 |
| Glascock | 14.5 | 10.2 | 8.5 | 50.8 | 2,556 | 3,086 | 20.7 | -3.3 | 282 | 399 | 15 | 1,108 | 2.63 | 65.5 | 14.1 | 30.5 |
| Glynn | 14.0 | 12.6 | 8.8 | 53.2 | 67,568 | 79,630 | 17.9 | 7.5 | 9,862 | 8,819 | 4,914 | 34,119 | 2.43 | 65.2 | 15.3 | 29.8 |
| Gordon | 12.6 | 9.3 | 6.3 | 50.8 | 44,104 | 55,221 | 25.2 | 6.4 | 7,090 | 5,383 | 1,880 | 20,561 | 2.75 | 73.0 | 13.2 | 23.1 |
| Grady | 13.5 | 11.1 | 8.1 | 51.6 | 23,659 | 25,019 | 5.7 | -2.1 | 3,434 | 2,649 | -1,311 | 9,136 | 2.70 | 69.2 | 14.7 | 26.7 |
| Greene | 16.1 | 18.0 | 11.9 | 51.1 | 14,406 | 15,996 | 11.0 | 17.8 | 1,734 | 2,074 | 3,177 | 7,132 | 2.40 | 69.8 | 13.5 | 26.9 |
| Gwinnett | 11.9 | 7.0 | 3.9 | 51.2 | 588,448 | 805,241 | 36.8 | 17.1 | 118,457 | 42,232 | 61,468 | 293,330 | 3.10 | 75.7 | 13.8 | 19.9 |
| Habersham | 12.9 | 10.9 | 7.9 | 52.1 | 35,902 | 43,028 | 19.8 | 7.0 | 5,139 | 4,540 | 2,412 | 15,139 | 2.80 | 69.0 | 8.9 | 26.8 |
| Hall | 12.2 | 9.3 | 6.7 | 50.3 | 139,277 | 179,765 | 29.1 | 14.9 | 26,275 | 14,787 | 15,362 | 64,352 | 3.06 | 74.8 | 12.6 | 20.5 |
| Hancock | 15.6 | 14.1 | 10.0 | 44.1 | 10,076 | 9,401 | -6.7 | -9.6 | 701 | 1,030 | -612 | 2,974 | 2.03 | 59.0 | 23.1 | 37.6 |
| Haralson | 12.6 | 9.5 | 7.0 | 51.2 | 25,690 | 28,780 | 12.0 | 5.6 | 3,574 | 3,630 | 1,667 | 11,259 | 2.56 | 70.7 | 13.6 | 24.7 |
| Harris | 15.0 | 12.2 | 7.1 | 50.0 | 23,695 | 31,992 | 35.0 | 12.8 | 3,054 | 2,791 | 3,851 | 12,156 | 2.76 | 78.8 | 8.4 | 19.2 |
| Hart | 14.5 | 13.1 | 9.8 | 50.7 | 22,997 | 25,205 | 9.6 | 4.8 | 2,779 | 3,028 | 1,461 | 9,853 | 2.52 | 68.4 | 12.5 | 27.3 |
| Heard | 14.1 | 10.6 | 6.9 | 50.6 | 11,012 | 11,828 | 7.4 | 1.2 | 1,318 | 1,310 | 139 | 4,502 | 2.57 | 72.3 | 15.1 | 23.4 |
| Henry | 12.5 | 7.9 | 4.4 | 52.6 | 119,341 | 203,777 | 70.8 | 17.4 | 25,018 | 14,710 | 25,114 | 75,984 | 2.95 | 77.4 | 17.0 | 19.2 |
| Houston | 12.5 | 8.1 | 5.2 | 51.7 | 110,765 | 139,573 | 26.0 | 14.7 | 20,778 | 11,809 | 11,620 | 57,056 | 2.66 | 71.1 | 15.4 | 23.8 |
| Irwin | 11.9 | 10.2 | 8.3 | 47.2 | 9,931 | 9,513 | -4.2 | -1.3 | 1,052 | 1,095 | -109 | 3,329 | 2.75 | 62.8 | 14.0 | 35.9 |
| Jackson | 12.1 | 8.9 | 5.8 | 50.5 | 41,589 | 60,470 | 45.4 | 26.0 | 8,387 | 5,889 | 13,218 | 23,166 | 2.89 | 77.0 | 11.1 | 18.3 |
| Jasper | 14.6 | 11.3 | 6.8 | 50.9 | 11,426 | 13,894 | 21.6 | 4.2 | 1,686 | 1,410 | 308 | 5,171 | 2.67 | 73.6 | 12.0 | 22.4 |
| Jeff Davis | 12.7 | 9.8 | 6.6 | 50.7 | 12,684 | 15,074 | 18.8 | 0.9 | 2,104 | 1,638 | -323 | 5,279 | 2.82 | 71.8 | 19.1 | 24.2 |
| Jefferson | 13.2 | 11.5 | 7.4 | 51.5 | 17,266 | 16,916 | -2.0 | -9.7 | 2,092 | 2,301 | -1,456 | 5,664 | 2.66 | 63.5 | 20.8 | 30.7 |
| Jenkins | 12.8 | 10.4 | 6.9 | 46.6 | 8,575 | 8,333 | -2.8 | 5.0 | 1,028 | 1,060 | 374 | 3,443 | 2.53 | 60.8 | 14.0 | 33.0 |
| Johnson | 13.6 | 11.1 | 8.8 | 43.7 | 8,560 | 9,978 | 16.6 | -3.1 | 986 | 1,043 | -251 | 3,393 | 2.68 | 65.1 | 17.9 | 33.6 |
| Jones | 13.9 | 11.1 | 7.2 | 51.9 | 23,639 | 28,648 | 21.2 | 0.5 | 2,999 | 2,730 | -127 | 10,701 | 2.65 | 71.7 | 15.8 | 24.2 |
| Lamar | 13.0 | 10.5 | 7.0 | 52.1 | 15,912 | 18,307 | 15.1 | 5.2 | 2,082 | 2,139 | 989 | 6,494 | 2.69 | 57.7 | 12.1 | 37.8 |
| Lanier | 13.2 | 9.1 | 5.7 | 49.9 | 7,241 | 10,076 | 39.2 | 6.6 | 1,405 | 945 | 194 | 3,714 | 2.71 | 68.3 | 11.0 | 29.4 |
| Laurens | 12.6 | 10.5 | 7.9 | 52.6 | 44,874 | 48,436 | 7.9 | -1.9 | 6,576 | 5,790 | -1,700 | 17,142 | 2.71 | 67.4 | 18.2 | 28.9 |
| Lee | 11.6 | 8.9 | 4.5 | 49.9 | 24,757 | 28,295 | 14.3 | 6.9 | 3,597 | 2,201 | 526 | 10,226 | 2.78 | 77.0 | 13.9 | 20.4 |
| Liberty | 9.1 | 6.4 | 3.6 | 49.4 | 61,610 | 63,584 | 3.2 | -0.9 | 14,611 | 3,596 | -12,011 | 23,485 | 2.53 | 70.9 | 19.4 | 24.0 |
| Lincoln | 16.3 | 14.9 | 10.3 | 51.6 | 8,348 | 7,996 | -4.2 | 0.4 | 764 | 918 | 188 | 3,475 | 2.23 | 61.6 | 12.4 | 33.3 |
| Long | 11.3 | 7.0 | 3.5 | 49.6 | 10,304 | 14,345 | 39.2 | 40.6 | 2,674 | 958 | 4,040 | 5,695 | 3.26 | 72.8 | 16.5 | 22.8 |
| Lowndes | 10.2 | 7.8 | 5.2 | 51.9 | 92,115 | 109,257 | 18.6 | 8.2 | 16,698 | 9,024 | 1,164 | 41,282 | 2.69 | 62.8 | 16.2 | 29.5 |
| Lumpkin | 13.2 | 11.4 | 7.0 | 50.7 | 21,016 | 29,947 | 42.5 | 14.2 | 3,077 | 2,763 | 3,871 | 11,570 | 2.67 | 67.4 | 10.1 | 24.4 |
| McDuffie | 13.8 | 11.2 | 7.5 | 53.2 | 21,231 | 21,865 | 3.0 | -3.2 | 2,964 | 2,521 | -1,140 | 8,153 | 2.59 | 70.8 | 20.4 | 23.2 |
| McIntosh | 16.9 | 16.9 | 12.1 | 51.6 | 10,847 | 14,332 | 32.1 | 0.4 | 1,193 | 1,301 | 150 | 6,042 | 2.33 | 67.3 | 13.0 | 27.8 |
| Macon | 14.2 | 11.9 | 6.8 | 44.7 | 14,074 | 14,743 | 4.8 | -13.8 | 1,402 | 1,607 | -1,849 | 4,696 | 2.40 | 64.6 | 19.2 | 33.8 |
| Madison | 13.9 | 10.7 | 7.1 | 50.8 | 25,730 | 28,174 | 9.5 | 8.1 | 3,470 | 2,997 | 1,826 | 10,744 | 2.70 | 75.9 | 13.4 | 20.9 |
| Marion | 16.5 | 12.4 | 8.6 | 50.9 | 7,144 | 8,738 | 22.3 | -2.5 | 918 | 850 | -294 | 3,408 | 2.45 | 70.3 | 13.3 | 28.4 |
| Meriwether | 15.2 | 12.3 | 8.8 | 52.3 | 22,534 | 21,976 | -2.5 | -3.7 | 2,557 | 2,769 | -600 | 8,051 | 2.58 | 68.4 | 22.5 | 28.6 |
| Miller | 13.9 | 11.0 | 11.3 | 52.1 | 6,383 | 6,127 | -4.0 | -8.2 | 700 | 849 | -357 | 2,333 | 2.40 | 66.7 | 16.5 | 29.0 |
| Mitchell | 13.3 | 10.0 | 7.4 | 48.1 | 23,932 | 23,500 | -1.8 | -8.1 | 2,706 | 2,570 | -2,050 | 7,982 | 2.50 | 68.3 | 24.2 | 27.1 |
| Monroe | 14.6 | 11.7 | 8.1 | 50.3 | 21,757 | 26,121 | 20.1 | 7.4 | 2,766 | 2,864 | 2,041 | 9,760 | 2.67 | 63.3 | 9.6 | 32.7 |
| Montgomery | 13.4 | 10.8 | 7.3 | 49.1 | 8,270 | 9,177 | 11.0 | -1.8 | 1,012 | 903 | -282 | 3,097 | 2.66 | 67.9 | 14.6 | 26.7 |
| Morgan | 14.0 | 12.0 | 8.9 | 52.1 | 15,457 | 17,867 | 15.6 | 9.9 | 2,014 | 1,851 | 1,610 | 6,942 | 2.74 | 73.4 | 12.5 | 22.0 |
| Murray | 13.1 | 9.4 | 6.2 | 50.7 | 36,506 | 39,628 | 8.6 | 1.0 | 5,066 | 3,857 | -803 | 14,385 | 2.74 | 73.4 | 11.7 | 21.1 |
| Muscogee | 11.4 | 8.4 | 5.8 | 51.5 | 186,291 | 190,570 | 2.3 | 3.1 | 30,688 | 19,397 | -5,776 | 72,759 | 2.59 | 61.6 | 18.4 | 33.8 |
| Newton | 12.0 | 8.5 | 5.1 | 52.8 | 62,001 | 99,964 | 61.2 | 13.3 | 13,808 | 8,779 | 8,352 | 37,018 | 2.88 | 71.9 | 18.9 | 22.6 |
| Oconee | 12.6 | 9.7 | 6.5 | 51.1 | 26,225 | 32,840 | 25.2 | 25.2 | 3,491 | 2,472 | 7,272 | 13,423 | 2.82 | 79.9 | 9.7 | 18.3 |
| Oglethorpe | 14.7 | 11.5 | 8.2 | 50.7 | 12,635 | 14,874 | 17.7 | 3.4 | 1,608 | 1,489 | 390 | 5,651 | 2.62 | 65.7 | 11.7 | 31.6 |
| Paulding | 11.9 | 7.3 | 4.1 | 51.3 | 81,678 | 142,385 | 74.3 | 21.8 | 19,464 | 9,277 | 20,799 | 53,299 | 2.99 | 77.8 | 13.3 | 18.3 |
| Peach | 14.1 | 10.2 | 6.5 | 52.0 | 23,668 | 28,054 | 18.5 | -0.4 | 3,122 | 2,651 | -599 | 10,136 | 2.49 | 65.1 | 19.8 | 30.3 |
| Pickens | 15.9 | 14.3 | 8.5 | 50.9 | 22,983 | 29,402 | 27.9 | 12.7 | 3,109 | 3,346 | 3,946 | 11,868 | 2.62 | 71.9 | 8.9 | 23.4 |
| Pierce | 12.7 | 10.3 | 7.3 | 50.4 | 15,636 | 18,760 | 20 | 4.1 | 2,484 | 2,122 | 407 | 7,048 | 2.71 | 75.5 | 19.2 | 21.8 |
| Pike | 13.4 | 9.8 | 6.6 | 51.0 | 13,688 | 17,879 | 30.6 | 6.9 | 1,781 | 1,790 | 1,266 | 6,143 | 2.94 | 73.7 | 8.0 | 24.0 |
| Polk | 12.7 | 9.3 | 6.6 | 51.1 | 38,127 | 41,487 | 8.8 | 3.3 | 5,778 | 4,970 | 549 | 15,038 | 2.77 | 67.5 | 14.8 | 26.0 |
| Pulaski | 14.0 | 12.6 | 10.7 | 56.8 | 9,588 | 12,014 | 25.3 | -6.9 | 940 | 1,138 | -633 | 3,687 | 2.62 | 67.2 | 17.6 | 28.3 |

1. No spouse present.

# Table B. States and Counties — Population, Vital Statistics, Health, and Crime

| STATE County | Persons in group quarters, 2020 | Daytime Population, 2015-2019 | | Births, 2020 | | Deaths, 2020 | | Persons under 65 with no health insurance, 2019 | | Medicare, 2020 | | | Crimes reported by county police or sheriff, 2019 | |
|---|---|---|---|---|---|---|---|---|---|---|---|---|---|---|
| | | Number | Employment/ residence ratio | Total | Rate[1] | Number | Rate[1] | Number | Percent | Total beneficiaries | Enrolled in Original Medicare | Enrolled in Medicare Advantage | Violent | Property |
| | 32 | 33 | 34 | 35 | 36 | 37 | 38 | 39 | 40 | 41 | 42 | 43 | 44 | 45 |
| GEORGIA—Cont'd | | | | | | | | | | | | | | |
| Effingham | 607 | 43,797 | 0.42 | 806 | 12.3 | 525 | 8.0 | 7,337 | 13.0 | 9,711 | 5,645 | 4,066 | NA | NA |
| Elbert | 275 | 17,893 | 0.83 | 214 | 11.1 | 282 | 14.6 | 2,691 | 18.2 | 5,185 | 2,857 | 2,328 | NA | NA |
| Emanuel | 1,368 | 21,951 | 0.93 | 288 | 12.8 | 303 | 13.5 | 3,044 | 17.4 | 4,959 | 2,632 | 2,328 | NA | NA |
| Evans | 404 | 10,837 | 1.04 | 130 | 12.2 | 125 | 11.8 | 1,575 | 18.9 | 2,147 | 1,171 | 976 | NA | NA |
| Fannin | 160 | 25,178 | 0.98 | 199 | 7.5 | 390 | 14.7 | 3,493 | 19.4 | 8,416 | 5,548 | 2,868 | 31 | 274 |
| Fayette | 491 | 113,736 | 1.03 | 868 | 7.5 | 990 | 8.5 | 10,537 | 11.3 | 24,228 | 15,127 | 9,101 | 50 | 346 |
| Floyd | 3,725 | 100,798 | 1.08 | 1,088 | 11.0 | 1,252 | 12.7 | 14,101 | 18.1 | 20,813 | 12,918 | 7,895 | NA | NA |
| Forsyth | 652 | 205,955 | 0.80 | 2,366 | 9.4 | 1,522 | 6.1 | 23,650 | 10.9 | 32,265 | 19,470 | 12,795 | 123 | 1,377 |
| Franklin | 816 | 22,205 | 0.94 | 280 | 11.9 | 308 | 13.1 | 3,484 | 19.3 | 5,475 | 3,070 | 2,406 | 16 | 221 |
| Fulton | 34,052 | 1,445,383 | 1.78 | 12,264 | 11.4 | 8,114 | 7.5 | 114,040 | 12.6 | 139,258 | 76,162 | 63,096 | NA | NA |
| Gilmer | 171 | 26,753 | 0.71 | 326 | 10.2 | 411 | 12.9 | 5,046 | 21.8 | 8,912 | 5,900 | 3,012 | 72 | 327 |
| Glascock | 85 | 2,219 | 0.40 | 26 | 8.7 | 31 | 10.4 | 338 | 14.0 | 599 | 383 | 217 | NA | NA |
| Glynn | 1,531 | 90,762 | 1.16 | 898 | 10.5 | 1,007 | 11.8 | 12,217 | 18.5 | 19,413 | 12,635 | 6,778 | NA | NA |
| Gordon | 669 | 57,020 | 0.99 | 684 | 11.6 | 641 | 10.9 | 9,421 | 19.6 | 10,906 | 7,043 | 3,864 | NA | NA |
| Grady | 200 | 21,478 | 0.67 | 289 | 11.8 | 285 | 11.6 | 3,997 | 20.4 | 5,457 | 3,179 | 2,278 | NA | NA |
| Greene | 177 | 18,440 | 1.17 | 178 | 9.4 | 252 | 13.4 | 2,295 | 18.0 | 6,167 | 3,928 | 2,238 | NA | NA |
| Gwinnett | 4,945 | 857,059 | 0.87 | 11,123 | 11.8 | 5,160 | 5.5 | 147,535 | 17.7 | 110,107 | 58,245 | 51,862 | NA | NA |
| Habersham | 2,380 | 41,755 | 0.84 | 473 | 10.3 | 506 | 11.0 | 7,023 | 20.3 | 9,947 | 5,638 | 4,309 | NA | NA |
| Hall | 2,962 | 202,189 | 1.04 | 2,546 | 12.3 | 1,773 | 8.6 | 36,780 | 21.6 | 36,023 | 21,578 | 14,444 | NA | NA |
| Hancock | 1,301 | 7,950 | 0.73 | 53 | 6.2 | 136 | 16.0 | 669 | 13.1 | 2,307 | 1,061 | 1,246 | 252 | 1,315 |
| Haralson | 350 | 26,854 | 0.81 | 360 | 11.8 | 361 | 11.9 | 4,091 | 16.6 | 6,275 | 3,480 | 2,795 | 15 | 50 |
| Harris | 446 | 25,121 | 0.42 | 314 | 8.7 | 385 | 10.7 | 3,082 | 10.9 | 7,271 | 4,673 | 2,598 | 146 | 384 |
| Hart | 662 | 23,955 | 0.81 | 246 | 9.3 | 349 | 13.2 | 3,845 | 19.5 | 6,760 | 4,115 | 2,645 | NA | NA |
| Heard | 130 | 9,319 | 0.51 | 117 | 9.8 | 135 | 11.3 | 1,536 | 16.0 | 2,277 | 1,230 | 1,047 | NA | NA |
| Henry | 982 | 193,168 | 0.70 | 2,657 | 11.1 | 1,824 | 7.6 | 25,109 | 12.2 | 34,973 | 17,890 | 17,083 | 18 | 96 |
| Houston | 1,546 | 154,139 | 1.01 | 1,995 | 12.5 | 1,415 | 8.8 | 17,854 | 13.2 | 26,029 | 18,311 | 7,717 | NA | NA |
| Irwin | 1,147 | 8,169 | 0.68 | 105 | 11.2 | 116 | 12.4 | 1,037 | 15.7 | 1,955 | 1,121 | 834 | NA | NA |
| Jackson | 716 | 65,587 | 0.93 | 889 | 11.7 | 683 | 9.0 | 9,513 | 15.3 | 13,733 | 7,836 | 5,896 | NA | NA |
| Jasper | 95 | 11,052 | 0.49 | 140 | 9.7 | 134 | 9.3 | 2,005 | 17.3 | 3,029 | 1,611 | 1,418 | 69 | 193 |
| Jeff Davis | 112 | 14,291 | 0.86 | 194 | 12.8 | 165 | 10.8 | 2,410 | 19.5 | 2,809 | 1,727 | 1,082 | NA | NA |
| Jefferson | 599 | 14,630 | 0.83 | 188 | 12.3 | 237 | 15.5 | 2,187 | 18.5 | 3,891 | 2,057 | 1,834 | 27 | 70 |
| Jenkins | 1,184 | 7,983 | 0.75 | 98 | 11.2 | 111 | 12.7 | 1,053 | 17.7 | 1,791 | 912 | 880 | NA | NA |
| Johnson | 1,692 | 8,176 | 0.61 | 105 | 10.9 | 144 | 14.9 | 1,031 | 16.6 | 1,899 | 1,043 | 856 | NA | NA |
| Jones | 306 | 21,279 | 0.39 | 231 | 8.0 | 322 | 11.2 | 3,307 | 14.2 | 5,892 | 3,192 | 2,701 | 22 | 303 |
| Lamar | 1,175 | 16,941 | 0.75 | 216 | 11.2 | 235 | 12.2 | 1,920 | 13.1 | 4,016 | 2,056 | 1,960 | NA | NA |
| Lanier | 260 | 8,366 | 0.46 | 127 | 11.8 | 123 | 11.5 | 1,298 | 15.1 | 1,642 | 1,078 | 564 | 22 | 117 |
| Laurens | 1,184 | 48,944 | 1.09 | 620 | 13.0 | 634 | 13.3 | 5,843 | 15.5 | 10,946 | 6,399 | 4,547 | 92 | 523 |
| Lee | 907 | 22,191 | 0.44 | 334 | 11.0 | 277 | 9.2 | 2,969 | 11.7 | 5,029 | 3,157 | 1,872 | 41 | 494 |
| Liberty | 2,349 | 67,245 | 1.22 | 1,429 | 22.7 | 390 | 6.2 | 6,540 | 12.2 | 7,811 | 5,022 | 2,789 | NA | NA |
| Lincoln | 67 | 6,166 | 0.47 | 78 | 9.7 | 110 | 13.7 | 1,029 | 17.5 | 2,173 | 1,242 | 931 | NA | NA |
| Long | 330 | 13,609 | 0.29 | 261 | 12.9 | 116 | 5.8 | 2,987 | 17.4 | 1,831 | 1,196 | 635 | NA | NA |
| Lowndes | 6,067 | 122,061 | 1.13 | 1,639 | 13.9 | 1,046 | 8.8 | 16,602 | 17.3 | 18,731 | 12,654 | 6,077 | NA | NA |
| Lumpkin | 3,054 | 28,156 | 0.73 | 286 | 8.4 | 328 | 9.6 | 4,668 | 19.0 | 6,783 | 4,288 | 2,495 | NA | NA |
| McDuffie | 265 | 20,713 | 0.91 | 270 | 12.8 | 288 | 13.6 | 2,770 | 16.3 | 4,974 | 2,304 | 2,670 | 28 | 178 |
| McIntosh | 71 | 11,429 | 0.55 | 103 | 7.2 | 143 | 9.9 | 1,668 | 16.4 | 3,251 | 1,746 | 1,505 | NA | NA |
| Macon | 1,893 | 12,491 | 0.81 | 131 | 10.3 | 180 | 14.2 | 1,522 | 17.6 | 2,565 | 1,269 | 1,296 | NA | NA |
| Madison | 248 | 21,292 | 0.37 | 355 | 11.7 | 320 | 10.5 | 4,470 | 18.3 | 6,560 | 3,692 | 2,868 | 99 | 363 |
| Marion | 75 | 6,816 | 0.49 | 83 | 9.7 | 108 | 12.7 | 1,231 | 19.0 | 1,682 | 928 | 754 | 2 | 29 |
| Meriwether | 245 | 17,989 | 0.61 | 234 | 11.1 | 307 | 14.5 | 2,851 | 17.4 | 5,408 | 2,574 | 2,834 | 42 | 257 |
| Miller | 127 | 5,451 | 0.86 | 66 | 11.7 | 91 | 16.2 | 749 | 17.2 | 1,390 | 863 | 527 | NA | NA |
| Mitchell | 2,108 | 21,945 | 0.96 | 215 | 10.0 | 259 | 12.0 | 2,932 | 18.4 | 4,592 | 2,468 | 2,124 | 28 | 265 |
| Monroe | 1,153 | 24,671 | 0.77 | 245 | 8.7 | 346 | 12.3 | 2,895 | 13.5 | 6,220 | 3,707 | 2,513 | NA | NA |
| Montgomery | 806 | 7,411 | 0.52 | 98 | 10.9 | 98 | 10.9 | 1,187 | 17.8 | 1,778 | 1,018 | 760 | NA | NA |
| Morgan | 163 | 20,463 | 1.24 | 210 | 10.7 | 168 | 8.6 | 2,266 | 14.8 | 4,729 | 2,571 | 2,157 | NA | NA |
| Murray | 273 | 33,460 | 0.64 | 465 | 11.6 | 455 | 11.4 | 6,817 | 20.4 | 7,818 | 5,492 | 2,326 | 47 | 85 |
| Muscogee | 6,861 | 218,890 | 1.28 | 2,794 | 14.2 | 2,102 | 10.7 | 21,415 | 13.3 | 35,598 | 21,237 | 14,361 | NA | NA |
| Newton | 1,649 | 90,069 | 0.62 | 1,401 | 12.4 | 1,049 | 9.3 | 15,821 | 16.7 | 18,967 | 9,344 | 9,623 | 81 | 821 |
| Oconee | 147 | 37,120 | 0.95 | 345 | 8.4 | 293 | 7.1 | 3,401 | 9.9 | 7,092 | 4,597 | 2,495 | NA | NA |
| Oglethorpe | 218 | 10,561 | 0.32 | 146 | 9.5 | 189 | 12.3 | 2,374 | 19.6 | 3,368 | 1,958 | 1,411 | NA | NA |
| Paulding | 529 | 116,311 | 0.44 | 1,887 | 10.9 | 1,235 | 7.1 | 20,564 | 13.7 | 22,485 | 12,711 | 9,774 | 320 | 1,862 |
| Peach | 1,440 | 25,394 | 0.83 | 333 | 11.9 | 302 | 10.8 | 3,800 | 17.7 | 5,671 | 3,482 | 2,189 | 25 | 156 |
| Pickens | 398 | 27,758 | 0.74 | 301 | 9.1 | 398 | 12.0 | 4,086 | 16.2 | 9,356 | 5,816 | 3,540 | NA | NA |
| Pierce | 149 | 16,555 | 0.64 | 250 | 12.8 | 253 | 13.0 | 2,854 | 18.0 | 4,187 | 2,610 | 1,577 | 31 | 321 |
| Pike | 257 | 13,726 | 0.42 | 209 | 10.9 | 217 | 11.3 | 2,361 | 15.0 | 3,665 | 2,035 | 1,630 | 20 | 193 |
| Polk | 386 | 39,717 | 0.88 | 574 | 13.4 | 557 | 13.0 | 6,337 | 17.9 | 8,840 | 5,084 | 3,756 | NA | NA |
| Pulaski | 1,237 | 10,763 | 0.88 | 102 | 9.1 | 145 | 13.0 | 1,161 | 15.8 | 2,106 | 1,323 | 783 | 50 | 209 |

1. Per 1,000 estimated resident population.

| STATE County | County law enforcement employment, 2019 | | School enrollment and attainment, 2015-2019 | | | | Local government expenditures,[3] 2017-2018 | | Money income, 2015-2019 | | | | Income and poverty, 2019 | | | |
|---|---|---|---|---|---|---|---|---|---|---|---|---|---|---|---|---|
| | | | Enrollment[1] | | Attainment[2] (percent) | | | | | | Households | | | Percent below poverty level | | |
| | | | | | | | | | | | | Percent | | | | |
| | Officers | Civilians | Total | Percent private | High school graduate or less | Bachelor's degree or more | Total current spending (mil dol) | Current spending per student (dollars) | Per capita income[4] | Median income (dollars) | with income of less than $50,000 | with income of $200,000 or more | Median household income (dollars) | All persons | Children under 18 years | Children 5 to 17 years in families |
| | 46 | 47 | 48 | 49 | 50 | 51 | 52 | 53 | 54 | 55 | 56 | 57 | 58 | 59 | 60 | 61 |
| GEORGIA—Cont'd | | | | | | | | | | | | | | | | |
| Effingham | 80 | 76 | 15,747 | 9.6 | 49.2 | 20.0 | 121 | 9,790 | 31,704 | 66,822 | 36.0 | 3.9 | 71,628 | 9.1 | 11.9 | 11.5 |
| Elbert | 33 | 29 | 4,290 | 7.3 | 62.5 | 11.6 | 33 | 10,759 | 22,355 | 38,678 | 60.7 | 1.2 | 40,962 | 20.6 | 29.5 | 27.9 |
| Emanuel | NA | NA | 5,509 | 13.7 | 62.2 | 13.0 | 44 | 10,130 | 19,389 | 37,453 | 63.4 | 2.0 | 38,529 | 20.9 | 30.6 | 30.0 |
| Evans | NA | NA | 2,695 | 18.6 | 62.7 | 15.8 | 19 | 10,025 | 20,725 | 42,607 | 57.9 | 1.0 | 41,425 | 24.1 | 36.0 | 33.9 |
| Fannin | 40 | 16 | 4,154 | 15.6 | 46.9 | 21.8 | 38 | 12,695 | 27,940 | 47,997 | 51.5 | 3.3 | 47,980 | 12.1 | 22.7 | 21.4 |
| Fayette | 133 | 68 | 30,036 | 15.3 | 25.6 | 46.2 | 220 | 10,720 | 44,061 | 90,145 | 24.5 | 15.3 | 94,563 | 5.4 | 7.1 | 6.4 |
| Floyd | 160 | 94 | 23,872 | 20.2 | 51.9 | 20.9 | 182 | 11,170 | 25,776 | 48,336 | 51.4 | 3.4 | 50,520 | 16.4 | 24.1 | 22.2 |
| Forsyth | 360 | 81 | 67,251 | 16.1 | 22.6 | 53.1 | 445 | 9,315 | 43,832 | 107,218 | 20.7 | 18.1 | 113,332 | 4.5 | 5.1 | 4.4 |
| Franklin | 36 | 21 | 5,224 | 17.0 | 57.1 | 15.9 | 39 | 10,584 | 23,187 | 42,488 | 55.6 | 2.9 | 47,691 | 17.3 | 25.7 | 27.4 |
| Fulton | 750 | 240 | 276,351 | 21.4 | 25.1 | 52.9 | 1,999 | 13,189 | 47,163 | 69,673 | 37.5 | 14.1 | 79,235 | 13.8 | 21.8 | 21.1 |
| Gilmer | 61 | 46 | 5,460 | 7.1 | 53.7 | 20.4 | 49 | 11,914 | 27,282 | 52,625 | 48.4 | 4.2 | 57,007 | 14.6 | 25.7 | 25.2 |
| Glascock | NA | NA | 747 | 4.0 | 62.7 | 11.6 | 7 | 13,484 | 23,541 | 45,000 | 53.6 | 2.8 | 49,350 | 16.9 | 20.8 | 18.5 |
| Glynn | 114 | 6 | 19,034 | 12.5 | 38.6 | 29.9 | 143 | 10,740 | 33,848 | 52,977 | 47.0 | 6.1 | 58,017 | 15.4 | 25.9 | 25.4 |
| Gordon | 82 | 40 | 13,595 | 7.8 | 57.0 | 14.7 | 106 | 9,824 | 24,092 | 45,865 | 55.3 | 2.5 | 53,373 | 12.3 | 16.2 | 15.3 |
| Grady | 19 | 4 | 6,424 | 7.2 | 59.0 | 13.9 | 45 | 9,604 | 22,543 | 43,531 | 55.8 | 1.6 | 42,603 | 21.7 | 29.1 | 28.2 |
| Greene | NA | NA | 3,211 | 8.4 | 48.1 | 26.9 | 35 | 13,888 | 37,818 | 52,129 | 48.8 | 8.1 | 60,973 | 15.0 | 27.6 | 25.8 |
| Gwinnett | 764 | 243 | 264,189 | 13.4 | 34.7 | 36.9 | 1,931 | 10,243 | 30,636 | 71,026 | 33.5 | 7.2 | 72,225 | 9.2 | 13.1 | 13.0 |
| Habersham | 52 | 35 | 10,462 | 20.0 | 55.1 | 19.3 | 78 | 11,297 | 23,046 | 50,563 | 49.4 | 2.9 | 51,296 | 15.2 | 20.0 | 21.0 |
| Hall | 293 | 138 | 51,124 | 10.0 | 48.6 | 24.5 | 353 | 9,978 | 29,680 | 62,984 | 39.2 | 6.3 | 66,698 | 13.5 | 21.8 | 21.6 |
| Hancock | 26 | 20 | 1,142 | 7.4 | 69.0 | 8.6 | 15 | 17,033 | 16,704 | 31,860 | 70.0 | 1.1 | 31,715 | 31.2 | 43.5 | 40.2 |
| Haralson | 75 | 2 | 6,870 | 9.5 | 54.8 | 14.8 | 60 | 10,410 | 25,074 | 48,550 | 51.4 | 1.0 | 49,020 | 14.4 | 20.2 | 18.6 |
| Harris | 79 | 2 | 7,832 | 12.0 | 31.5 | 31.4 | 56 | 10,487 | 36,273 | 76,319 | 32.6 | 7.5 | 74,052 | 6.5 | 10.5 | 9.8 |
| Hart | 27 | 5 | 5,611 | 14.2 | 55.5 | 15.7 | 39 | 10,808 | 23,181 | 43,204 | 55.8 | 2.4 | 50,647 | 14.6 | 21.2 | 21.1 |
| Heard | 22 | 14 | 2,575 | 9.3 | 65.5 | 8.7 | 23 | 10,747 | 22,714 | 51,358 | 48.5 | 1.6 | 48,984 | 16.7 | 24.7 | 22.2 |
| Henry | 473 | 99 | 62,860 | 14.6 | 40.4 | 28.5 | 454 | 10,797 | 29,889 | 71,288 | 33.2 | 5.2 | 72,927 | 8.1 | 12.8 | 11.8 |
| Houston | 133 | 202 | 44,130 | 10.8 | 35.5 | 28.3 | 304 | 10,295 | 29,462 | 61,723 | 41.2 | 4.4 | 65,641 | 11.0 | 16.0 | 16.2 |
| Irwin | 15 | 8 | 2,347 | 6.0 | 58.7 | 15.1 | 20 | 11,606 | 19,878 | 37,736 | 60.4 | 4.4 | 42,317 | 20.5 | 29.2 | 28.6 |
| Jackson | 98 | 14 | 17,020 | 12.3 | 48.1 | 22.7 | 131 | 10,018 | 28,915 | 65,385 | 36.2 | 4.8 | 66,645 | 10.3 | 12.3 | 11.4 |
| Jasper | 29 | 15 | 3,045 | 14.8 | 55.4 | 14.2 | 25 | 10,510 | 22,814 | 51,250 | 49.4 | 1.8 | 59,207 | 12.4 | 19.8 | 18.5 |
| Jeff Davis | NA | NA | 4,008 | 5.0 | 59.4 | 10.3 | 30 | 9,714 | 17,833 | 36,669 | 64.8 | 0.3 | 40,603 | 21.2 | 32.4 | 33.0 |
| Jefferson | 47 | 3 | 3,502 | 10.5 | 65.4 | 9.9 | 27 | 10,381 | 19,259 | 33,944 | 66.4 | 0.7 | 35,379 | 25.1 | 35.6 | 34.8 |
| Jenkins | NA | NA | 1,760 | 7.1 | 65.5 | 7.9 | 15 | 12,558 | 15,531 | 27,375 | 71.4 | 0.5 | 32,808 | 29.0 | 37.0 | 34.8 |
| Johnson | NA | NA | 1,870 | 4.2 | 70.8 | 9.1 | 12 | 10,600 | 21,087 | 42,446 | 57.2 | 0.1 | 37,771 | 24.2 | 31.6 | 31.3 |
| Jones | 39 | 39 | 7,456 | 13.1 | 50.1 | 19.6 | 52 | 9,684 | 25,854 | 56,020 | 44.9 | 2.4 | 56,819 | 12.8 | 17.0 | 16.1 |
| Lamar | 35 | 27 | 5,343 | 7.2 | 52.4 | 15.8 | 26 | 9,637 | 22,413 | 44,846 | 53.8 | 2.5 | 57,402 | 14.8 | 21.4 | 21.0 |
| Lanier | 18 | 3 | 2,465 | 5.2 | 57.9 | 16.5 | 19 | 11,136 | 19,207 | 40,986 | 60.3 | 1.5 | 42,667 | 18.5 | 28.6 | 25.3 |
| Laurens | 67 | 41 | 11,098 | 10.4 | 57.2 | 16.2 | 89 | 9,988 | 21,665 | 39,120 | 60.9 | 3.0 | 42,018 | 23.9 | 31.4 | 29.2 |
| Lee | NA | NA | 8,782 | 11.7 | 38.9 | 25.4 | 59 | 9,098 | 29,348 | 69,280 | 35.0 | 3.9 | 72,425 | 9.3 | 13.5 | 12.4 |
| Liberty | NA | NA | 17,174 | 11.3 | 40.9 | 18.3 | 109 | 10,948 | 22,811 | 48,007 | 51.3 | 1.5 | 51,410 | 14.5 | 21.7 | 24.2 |
| Lincoln | NA | NA | 1,467 | 4.6 | 57.5 | 15.9 | 14 | 12,348 | 26,918 | 39,742 | 57.0 | 4.1 | 45,299 | 17.3 | 26.4 | 26.6 |
| Long | 28 | 2 | 4,872 | 6.2 | 47.1 | 17.4 | 32 | 8,700 | 22,639 | 54,605 | 43.6 | 1.7 | 52,051 | 14.2 | 21.7 | 21.2 |
| Lowndes | NA | NA | 34,720 | 9.1 | 44.8 | 25.3 | 185 | 9,493 | 23,348 | 42,441 | 55.7 | 2.7 | 46,450 | 20.4 | 28.6 | 32.1 |
| Lumpkin | NA | NA | 8,857 | 8.0 | 45.6 | 26.4 | 43 | 11,135 | 27,155 | 51,790 | 48.1 | 4.7 | 57,078 | 14.5 | 17.2 | 16.0 |
| McDuffie | NA | NA | 4,896 | 6.6 | 58.3 | 14.0 | 42 | 10,680 | 21,625 | 43,468 | 58.3 | 1.6 | 44,473 | 18.0 | 29.4 | 28.2 |
| McIntosh | 52 | 28 | 2,606 | 17.7 | 52.2 | 17.6 | 16 | 11,650 | 26,892 | 49,504 | 50.5 | 1.7 | 46,682 | 18.7 | 31.5 | 29.3 |
| Macon | NA | NA | 3,028 | 12.4 | 61.2 | 8.8 | 16 | 11,731 | 17,883 | 32,161 | 69.2 | 1.3 | 34,634 | 29.4 | 41.0 | 38.9 |
| Madison | 52 | 43 | 6,471 | 14.3 | 56.6 | 18.5 | 71 | 11,435 | 24,511 | 52,500 | 47.5 | 2.3 | 52,326 | 15.6 | 25.1 | 26.2 |
| Marion | 6 | 8 | 1,774 | 7.4 | 57.7 | 13.4 | 15 | 10,609 | 22,541 | 44,643 | 54.5 | 2.3 | 41,190 | 21.1 | 32.4 | 29.6 |
| Meriwether | 27 | 22 | 4,588 | 10.2 | 60.1 | 10.4 | 30 | 10,970 | 21,038 | 37,121 | 60.8 | 1.9 | 43,571 | 22.2 | 31.1 | 31.0 |
| Miller | NA | NA | 1,147 | 15.2 | 53.9 | 14.0 | 11 | 10,840 | 23,870 | 44,542 | 53.2 | 2.2 | 41,769 | 21.3 | 31.3 | 28.6 |
| Mitchell | NA | NA | 5,271 | 9.3 | 62.3 | 9.7 | 46 | 12,469 | 18,619 | 38,116 | 63.2 | 2.5 | 37,383 | 30.7 | 42.3 | 39.6 |
| Monroe | NA | NA | 6,173 | 28.6 | 47.6 | 26.8 | 49 | 11,994 | 33,290 | 54,754 | 46.2 | 8.1 | 70,488 | 10.6 | 15.6 | 14.5 |
| Montgomery | 11 | 10 | 2,104 | 27.5 | 56.3 | 16.0 | 10 | 11,150 | 22,477 | 41,996 | 57.1 | 2.9 | 43,657 | 19.4 | 27.2 | 26.4 |
| Morgan | NA | NA | 3,832 | 9.7 | 48.4 | 20.7 | 37 | 12,069 | 33,571 | 66,178 | 36.6 | 6.6 | 67,089 | 11.1 | 17.2 | 16.4 |
| Murray | 45 | 27 | 9,346 | 5.6 | 66.7 | 10.1 | 66 | 8,905 | 21,989 | 48,033 | 51.6 | 1.5 | 50,061 | 14.5 | 20.8 | 19.3 |
| Muscogee | NA | NA | 55,688 | 10.9 | 39.3 | 25.8 | 328 | 10,330 | 26,097 | 46,408 | 53.3 | 3.9 | 47,035 | 18.4 | 25.4 | 24.2 |
| Newton | 168 | 122 | 29,438 | 14.3 | 47.1 | 19.8 | 202 | 10,280 | 25,247 | 56,316 | 44.5 | 3.5 | 58,222 | 11.2 | 17.5 | 17.1 |
| Oconee | 62 | 34 | 11,146 | 16.6 | 24.8 | 52.3 | 75 | 9,653 | 44,336 | 90,751 | 27.9 | 13.1 | 100,878 | 5.6 | 6.6 | 5.8 |
| Oglethorpe | NA | NA | 3,453 | 10.3 | 59.1 | 17.1 | 23 | 10,892 | 23,523 | 47,120 | 53.5 | 1.3 | 54,346 | 12.8 | 18.9 | 18.6 |
| Paulding | 226 | 39 | 43,090 | 11.9 | 43.5 | 22.6 | 284 | 9,635 | 27,979 | 68,370 | 34.1 | 3.4 | 76,580 | 6.8 | 9.3 | 9.0 |
| Peach | 30 | 30 | 8,138 | 6.7 | 46.8 | 20.5 | 37 | 10,034 | 24,486 | 46,636 | 52.6 | 3.3 | 49,815 | 19.8 | 27.0 | 24.4 |
| Pickens | NA | NA | 6,256 | 9.7 | 46.2 | 23.1 | 52 | 11,761 | 31,965 | 67,631 | 36.8 | 4.5 | 66,180 | 10.3 | 17.8 | 15.9 |
| Pierce | 21 | 32 | 3,958 | 7.2 | 59.4 | 12.6 | 37 | 10,118 | 22,268 | 42,804 | 54.4 | 1.6 | 46,701 | 16.4 | 23.7 | 22.8 |
| Pike | 28 | 23 | 4,458 | 12.3 | 53.6 | 17.1 | 31 | 9,144 | 27,639 | 64,878 | 38.8 | 4.5 | 74,905 | 8.7 | 12.0 | 10.8 |
| Polk | NA | NA | 9,877 | 11.9 | 59.1 | 13.7 | 74 | 9,429 | 23,857 | 44,891 | 53.2 | 1.7 | 47,896 | 16.2 | 22.7 | 21.6 |
| Pulaski | NA | NA | 2,005 | 6.7 | 55.7 | 13.2 | 14 | 10,505 | 20,746 | 45,148 | 55.0 | 0.2 | 42,274 | 20.7 | 27.6 | 26.1 |

1. All persons 3 years old and over enrolled in nursery school through college.  2. Persons 25 years old and over.  3. Elementary and secondary education expenditures.  4. Based on population estimated by the American Community Survey, 2014–2018.

# Table B. States and Counties — **Personal Income**

| STATE County | Personal income, 2019 Total (mil dol) | Percent change 2018-2019 | Per capita[1] Dollars | Per capita[1] Rank | Wages and salaries (mil dol) | Supplements to wages and salaries, employer contributions (mil dol) Pension and insurance | Government social insurance | Proprietors' income (mil dol) | Dividends, interest, and rent (mil dol) | Personal transfer receipts (mil dol) | Earnings, 2019 Total (mil dol) | Contributions for government social insurance (mil dol) From employee and self-employed | From employer |
|---|---|---|---|---|---|---|---|---|---|---|---|---|---|
| | 62 | 63 | 64 | 65 | 66 | 67 | 68 | 69 | 70 | 71 | 72 | 73 | 74 |
| GEORGIA—Cont'd | | | | | | | | | | | | | |
| Effingham | 2,718 | 5.0 | 42,271 | 1,726 | 530 | 109 | 35 | 86 | 275 | 464 | 759 | 51 | 35 |
| Elbert | 682 | 1.0 | 35,550 | 2,612 | 233 | 51 | 16 | 45 | 124 | 229 | 345 | 26 | 16 |
| Emanuel | 768 | 4.3 | 33,921 | 2,811 | 262 | 59 | 18 | 46 | 117 | 261 | 386 | 26 | 18 |
| Evans | 374 | 1.5 | 35,109 | 2,673 | 167 | 32 | 12 | 25 | 60 | 102 | 237 | 15 | 12 |
| Fannin | 979 | 3.4 | 37,385 | 2,402 | 261 | 45 | 18 | 116 | 210 | 325 | 440 | 36 | 18 |
| Fayette | 7,720 | 4.1 | 67,467 | 152 | 2,423 | 359 | 162 | 374 | 1,504 | 995 | 3,317 | 209 | 162 |
| Floyd | 3,872 | 3.0 | 39,314 | 2,146 | 1,891 | 319 | 132 | 313 | 661 | 1,042 | 2,655 | 168 | 132 |
| Forsyth | 15,599 | 5.8 | 63,864 | 211 | 4,492 | 616 | 300 | 851 | 2,230 | 1,261 | 6,259 | 373 | 300 |
| Franklin | 794 | -3.9 | 34,017 | 2,800 | 330 | 57 | 23 | 88 | 125 | 243 | 497 | 33 | 23 |
| Fulton | 94,512 | 3.1 | 88,832 | 37 | 78,802 | 9,072 | 5,018 | 17,248 | 22,425 | 7,146 | 110,140 | 6,050 | 5,018 |
| Gilmer | 1,116 | 1.9 | 35,576 | 2,606 | 263 | 54 | 18 | 122 | 217 | 358 | 457 | 35 | 18 |
| Glascock | 103 | 4.7 | 34,786 | 2,718 | 13 | 4 | 1 | 3 | 15 | 29 | 21 | 2 | 1 |
| Glynn | 4,109 | 2.4 | 48,170 | 1,004 | 1,930 | 334 | 135 | 246 | 1,166 | 863 | 2,645 | 167 | 135 |
| Gordon | 2,061 | 1.6 | 35,557 | 2,611 | 1,139 | 180 | 80 | 136 | 259 | 492 | 1,534 | 95 | 80 |
| Grady | 900 | 4.1 | 36,527 | 2,505 | 249 | 51 | 17 | 126 | 135 | 235 | 443 | 28 | 17 |
| Greene | 1,076 | 1.8 | 58,708 | 347 | 286 | 45 | 19 | 89 | 450 | 241 | 439 | 32 | 19 |
| Gwinnett | 40,167 | 4.2 | 42,902 | 1,632 | 22,273 | 2,965 | 1,465 | 3,564 | 5,543 | 5,241 | 30,267 | 1,750 | 1,465 |
| Habersham | 1,579 | 2.0 | 34,831 | 2,710 | 614 | 133 | 43 | 95 | 286 | 420 | 884 | 58 | 43 |
| Hall | 9,318 | 4.0 | 45,576 | 1,295 | 4,877 | 740 | 321 | 724 | 1,722 | 1,644 | 6,662 | 404 | 321 |
| Hancock | 265 | 4.1 | 31,297 | 2,990 | 62 | 17 | 4 | 7 | 40 | 106 | 90 | 8 | 4 |
| Haralson | 1,120 | 2.8 | 37,582 | 2,377 | 329 | 64 | 22 | 60 | 147 | 305 | 475 | 34 | 22 |
| Harris | 1,743 | 5.0 | 49,471 | 881 | 206 | 43 | 14 | 60 | 324 | 318 | 324 | 27 | 14 |
| Hart | 988 | -0.5 | 37,701 | 2,358 | 279 | 56 | 20 | 109 | 179 | 275 | 464 | 33 | 20 |
| Heard | 381 | 1.6 | 31,965 | 2,945 | 110 | 28 | 7 | 24 | 45 | 105 | 169 | 11 | 7 |
| Henry | 9,425 | 5.4 | 40,182 | 2,031 | 2,981 | 506 | 209 | 392 | 1,116 | 1,737 | 4,088 | 262 | 209 |
| Houston | 6,955 | 4.8 | 44,059 | 1,476 | 3,485 | 893 | 265 | 243 | 1,275 | 1,411 | 4,885 | 276 | 265 |
| Irwin | 339 | 6.4 | 36,036 | 2,562 | 84 | 19 | 6 | 40 | 58 | 99 | 150 | 9 | 6 |
| Jackson | 3,175 | 5.6 | 43,508 | 1,544 | 1,253 | 206 | 86 | 229 | 373 | 582 | 1,774 | 110 | 86 |
| Jasper | 548 | 4.0 | 38,568 | 2,249 | 86 | 20 | 6 | 36 | 75 | 134 | 148 | 12 | 6 |
| Jeff Davis | 485 | 6.3 | 32,069 | 2,939 | 177 | 37 | 12 | 33 | 71 | 146 | 260 | 17 | 12 |
| Jefferson | 554 | 6.5 | 36,033 | 2,565 | 212 | 43 | 15 | 35 | 93 | 195 | 305 | 21 | 15 |
| Jenkins | 274 | 6.0 | 31,635 | 2,966 | 56 | 13 | 4 | 22 | 45 | 88 | 95 | 7 | 4 |
| Johnson | 254 | 2.0 | 26,341 | 3,096 | 56 | 15 | 4 | 8 | 38 | 94 | 83 | 7 | 4 |
| Jones | 1,107 | 4.0 | 38,526 | 2,257 | 181 | 37 | 12 | 31 | 155 | 268 | 261 | 22 | 12 |
| Lamar | 656 | 3.8 | 34,375 | 2,760 | 165 | 36 | 11 | 29 | 87 | 187 | 241 | 18 | 11 |
| Lanier | 302 | 5.0 | 28,991 | 3,064 | 56 | 16 | 4 | 11 | 42 | 93 | 87 | 6 | 4 |
| Laurens | 1,854 | 2.6 | 38,995 | 2,194 | 834 | 176 | 59 | 120 | 300 | 525 | 1,189 | 76 | 59 |
| Lee | 1,478 | 5.1 | 49,274 | 899 | 282 | 52 | 19 | 51 | 198 | 242 | 404 | 26 | 19 |
| Liberty | 2,299 | 3.5 | 37,424 | 2,398 | 1,851 | 546 | 164 | 27 | 449 | 552 | 2,588 | 115 | 164 |
| Lincoln | 305 | 4.1 | 38,527 | 2,256 | 47 | 11 | 3 | 18 | 51 | 89 | 80 | 7 | 3 |
| Long | 533 | 2.9 | 27,232 | 3,092 | 47 | 15 | 3 | 13 | 75 | 135 | 78 | 6 | 3 |
| Lowndes | 4,544 | 3.9 | 38,706 | 2,236 | 2,434 | 523 | 175 | 244 | 872 | 1,018 | 3,376 | 188 | 175 |
| Lumpkin | 1,292 | 4.3 | 38,450 | 2,271 | 344 | 85 | 23 | 73 | 226 | 311 | 526 | 35 | 23 |
| McDuffie | 820 | 2.9 | 38,456 | 2,270 | 284 | 56 | 21 | 38 | 129 | 235 | 399 | 26 | 21 |
| McIntosh | 445 | 3.4 | 30,981 | 3,004 | 75 | 17 | 5 | 15 | 90 | 139 | 112 | 11 | 5 |
| Macon | 419 | 2.9 | 32,343 | 2,925 | 128 | 27 | 8 | 64 | 71 | 124 | 227 | 13 | 8 |
| Madison | 1,102 | 0.9 | 36,879 | 2,457 | 146 | 35 | 10 | 82 | 158 | 286 | 273 | 22 | 10 |
| Marion | 250 | 2.3 | 29,849 | 3,041 | 39 | 10 | 3 | 6 | 56 | 76 | 58 | 5 | 3 |
| Meriwether | 763 | 3.3 | 36,050 | 2,561 | 205 | 45 | 15 | 33 | 114 | 245 | 299 | 24 | 15 |
| Miller | 275 | 7.8 | 48,011 | 1,013 | 72 | 20 | 4 | 41 | 48 | 66 | 138 | 7 | 4 |
| Mitchell | 822 | 4.6 | 37,608 | 2,372 | 266 | 56 | 19 | 109 | 157 | 222 | 450 | 25 | 19 |
| Monroe | 1,284 | 3.8 | 46,563 | 1,180 | 369 | 89 | 24 | 56 | 206 | 269 | 537 | 35 | 24 |
| Montgomery | 285 | 5.1 | 31,037 | 3,002 | 62 | 13 | 4 | 14 | 39 | 84 | 93 | 7 | 4 |
| Morgan | 996 | 3.8 | 51,677 | 708 | 323 | 54 | 22 | 46 | 218 | 193 | 445 | 30 | 22 |
| Murray | 1,242 | 2.3 | 30,964 | 3,008 | 338 | 64 | 24 | 85 | 133 | 349 | 510 | 37 | 24 |
| Muscogee | 8,878 | 3.5 | 45,352 | 1,316 | 5,334 | 1,039 | 379 | 318 | 2,374 | 2,090 | 7,070 | 406 | 379 |
| Newton | 3,877 | 5.1 | 34,696 | 2,727 | 1,248 | 225 | 84 | 110 | 475 | 925 | 1,667 | 114 | 84 |
| Oconee | 2,733 | 4.2 | 67,845 | 149 | 636 | 100 | 42 | 207 | 646 | 283 | 985 | 61 | 42 |
| Oglethorpe | 560 | 0.2 | 36,698 | 2,483 | 74 | 17 | 5 | 50 | 88 | 138 | 146 | 11 | 5 |
| Paulding | 6,574 | 6.2 | 38,978 | 2,195 | 1,104 | 203 | 74 | 207 | 660 | 1,051 | 1,587 | 113 | 74 |
| Peach | 1,084 | 5.4 | 39,336 | 2,137 | 441 | 88 | 32 | 61 | 203 | 290 | 621 | 38 | 32 |
| Pickens | 1,580 | 3.8 | 48,493 | 972 | 406 | 65 | 27 | 92 | 307 | 375 | 590 | 45 | 27 |
| Pierce | 730 | 4.4 | 37,486 | 2,392 | 171 | 33 | 12 | 34 | 117 | 212 | 250 | 19 | 12 |
| Pike | 785 | 4.3 | 41,401 | 1,861 | 128 | 27 | 9 | 37 | 110 | 156 | 200 | 16 | 9 |
| Polk | 1,451 | 4.4 | 34,055 | 2,795 | 493 | 92 | 35 | 51 | 191 | 446 | 670 | 48 | 35 |
| Pulaski | 352 | 3.6 | 31,603 | 2,968 | 120 | 24 | 8 | 15 | 76 | 101 | 167 | 11 | 8 |

1. Based on the resident population estimated as of July 1 of the year shown.

| STATE County | Earnings, 2019 (cont.) — Percent by selected industries | | | | | | | | | Social Security beneficiaries, December 2019 | | Supplemental Security Income recipients, 2019 | Housing units, 2020 | |
|---|---|---|---|---|---|---|---|---|---|---|---|---|---|---|
| | Farm | Mining, quarrying, and extractions | Construction | Manufacturing | Information; professional, scientific, technical services | Retail trade | Finance, insurance, real estate, and leasing | Health care and social assistance | Government | Number | Rate[1] | | Total | Percent change, 2010-2020 |
| | 75 | 76 | 77 | 78 | 79 | 80 | 81 | 82 | 83 | 84 | 85 | 86 | 87 | 88 |
| **GEORGIA—Cont'd** | | | | | | | | | | | | | | |
| Effingham | 0.8 | D | 6.8 | 18.7 | 6.6 | 5.3 | 3.6 | D | 27.1 | 10,495 | 163 | 878 | 24,737 | 24.4 |
| Elbert | -0.1 | 3.0 | 4.0 | 28.1 | 2.3 | 6.7 | D | 5.6 | 23.1 | 5,755 | 300 | 846 | 9,690 | 1.1 |
| Emanuel | 2.8 | 0 | 3.5 | 23.2 | D | 6.8 | 3.6 | D | 29.3 | 5,510 | 244 | 1,011 | 9,941 | -0.2 |
| Evans | 4.1 | 0 | 5.5 | 37.8 | 2.2 | 5.7 | D | D | 16.1 | 2,375 | 223 | 422 | 4,803 | 2.9 |
| Fannin | 0.8 | 0 | 13.4 | D | 4.8 | 12.5 | 9.4 | D | 13.5 | 8,975 | 344 | 673 | 18,016 | 11.1 |
| Fayette | 0.0 | D | 9.6 | 14.0 | 9.1 | 6.6 | 7.4 | 15.2 | 12.9 | 24,115 | 210 | 1,136 | 43,924 | 7.7 |
| Floyd | 0.3 | D | 2.8 | 17.1 | | 6.0 | 4.7 | 26.2 | 13.6 | 22,795 | 232 | 3,244 | 40,964 | 1.0 |
| Forsyth | 0.2 | 0.1 | 11.3 | 10.4 | 14.3 | 6.1 | 6.7 | D | 9.9 | 30,575 | 125 | 1,095 | 89,364 | 39.5 |
| Franklin | 7.1 | 0.3 | 4.4 | 18.0 | 3.9 | 7.2 | D | D | 13.3 | 6,200 | 266 | 870 | 10,855 | 2.8 |
| Fulton | 0.0 | 0.2 | 2.7 | 2.6 | 30.8 | 3.4 | 15.2 | 7.6 | 9.2 | 137,775 | 129 | 26,989 | 493,854 | 13.0 |
| Gilmer | 7.4 | D | 9.9 | 14.8 | D | 9.7 | 5.8 | 4.7 | 16.9 | 9,385 | 297 | 605 | 17,656 | 6.6 |
| Glascock | -1.9 | 0.0 | D | 0.2 | D | 4.1 | D | D | 41.6 | 685 | 231 | 105 | 1,540 | 1.2 |
| Glynn | 0.0 | D | 4.4 | 5.8 | 5.6 | 7.1 | 4.9 | 15.4 | 23.9 | 20,385 | 239 | 1,906 | 44,326 | 8.9 |
| Gordon | 1.9 | 0.0 | 4.8 | 41.2 | D | 4.8 | 4.0 | 9.3 | 11.6 | 11,930 | 206 | 1,523 | 23,088 | 3.6 |
| Grady | 9.1 | 0.7 | 5.1 | 15.9 | D | 5.8 | 3.7 | D | 17.2 | 5,605 | 228 | 1,040 | 10,975 | 2.0 |
| Greene | 2.1 | D | 13.1 | 8.1 | 6.3 | 6.4 | 12.3 | 9.4 | 12.1 | 6,180 | 337 | 593 | 10,952 | 26.1 |
| Gwinnett | 0.0 | D | 10.1 | 9.1 | 14.5 | 7.6 | 8.5 | 7.2 | 9.4 | 111,635 | 119 | 12,486 | 320,312 | 9.9 |
| Habersham | 2.1 | D | 4.5 | 27.4 | D | 7.6 | 4.9 | 5.3 | 20.2 | 10,755 | 236 | 967 | 18,917 | 4.3 |
| Hall | -0.1 | D | 6.6 | 20.0 | 4.5 | 6.1 | 6.8 | 16.7 | 10.8 | 38,550 | 189 | 3,130 | 77,669 | 12.8 |
| Hancock | 2.6 | D | D | D | D | 4.1 | D | D | 43.9 | 2,585 | 305 | 446 | 5,456 | 2.2 |
| Haralson | 1.5 | 0.3 | 10.1 | 27.3 | 3.6 | 5.9 | 3.9 | D | 20.3 | 7,000 | 234 | 1,061 | 12,663 | 3.1 |
| Harris | 0.0 | 0.1 | 11.2 | 19.8 | D | 2.3 | 7.6 | D | 22.6 | 7,445 | 212 | 474 | 14,745 | 10.1 |
| Hart | 5.4 | 0.0 | 5.0 | 32.0 | D | 6.9 | D | 5.3 | 15.3 | 7,070 | 269 | 586 | 13,417 | 3.2 |
| Heard | 4.6 | 0.0 | 9.5 | 20.4 | D | 1.6 | 2.0 | 2.3 | 22.8 | 2,550 | 216 | 355 | 5,283 | 2.6 |
| Henry | 0.0 | 1.1 | 6.0 | 6.7 | 7.1 | 9.2 | 3.8 | 13.4 | 18.6 | 38,595 | 164 | 4,492 | 86,287 | 12.8 |
| Houston | 0.1 | D | 3.2 | 7.2 | 9.3 | 5.2 | 2.8 | 5.9 | 55.4 | 27,410 | 174 | 3,723 | 66,039 | 13.6 |
| Irwin | 19.6 | 0.0 | 9.3 | 7.3 | D | 4.7 | D | 3.1 | 23.8 | 2,260 | 239 | 369 | 4,126 | 2.6 |
| Jackson | 1.0 | D | 7.8 | 26.1 | D | 8.5 | 6.7 | 3.2 | 11.3 | 14,440 | 198 | 1,787 | 27,653 | 16.4 |
| Jasper | 0.4 | 0.0 | 11.7 | 15.5 | D | 4.3 | 3.9 | D | 24.3 | 3,350 | 236 | 354 | 6,638 | 7.9 |
| Jeff Davis | 4.5 | 0.0 | 2.0 | 25.5 | D | 6.9 | 2.6 | D | 19.4 | 3,220 | 213 | 495 | 6,549 | 0.9 |
| Jefferson | 5.9 | 10.3 | 7.1 | 18.4 | D | 6.1 | 3.0 | D | 18.2 | 4,260 | 278 | 838 | 7,280 | -0.2 |
| Jenkins | 16.4 | 0.0 | 3.6 | 3.3 | D | 4.8 | D | 10.5 | 23.2 | 1,915 | 219 | 456 | 4,284 | 1.5 |
| Johnson | -3.9 | 0.0 | 5.9 | 6.5 | D | 6.6 | D | 17.0 | 37.8 | 2,160 | 222 | 550 | 4,133 | 0.3 |
| Jones | -0.2 | 0.1 | 17.5 | 0.7 | D | 7.3 | 3.8 | D | 26.2 | 6,430 | 224 | 367 | 12,049 | 3.8 |
| Lamar | 3.9 | D | 4.1 | 18.5 | D | 7.3 | 3.8 | 5.1 | 28.2 | 4,380 | 231 | 579 | 7,755 | 5.7 |
| Lanier | 3.4 | 0.2 | 5.4 | 11.0 | D | D | D | 2.6 | 42.7 | 1,830 | 173 | 328 | 4,492 | 1.5 |
| Laurens | 1.6 | D | 5.3 | 14.5 | D | 7.4 | 4.2 | 13.6 | 29.6 | 11,900 | 250 | 2,073 | 21,678 | 1.5 |
| Lee | 8.3 | D | 12.6 | 5.6 | D | 8.8 | 3.4 | D | 21.2 | 5,550 | 185 | 521 | 11,660 | 13.5 |
| Liberty | 0.0 | D | 1.0 | 8.6 | D | 2.9 | 1.7 | 1.9 | 72.9 | 8,750 | 139 | 1,360 | 29,490 | 10.1 |
| Lincoln | 1.1 | 0.0 | 21.3 | 2.2 | 3.0 | 6.9 | 7.3 | 2.3 | 26.6 | 2,380 | 299 | 225 | 4,960 | 3.6 |
| Long | 3.1 | D | 8.4 | D | 6.8 | 7.0 | 4.8 | 8.5 | 34.8 | 2,115 | 108 | 269 | 7,379 | 23.2 |
| Lowndes | 0.3 | D | 7.8 | 8.0 | 8.0 | 6.5 | 5.6 | 7.9 | 41.2 | 20,645 | 176 | 4,258 | 50,604 | 15.2 |
| Lumpkin | -0.2 | D | 8.5 | 8.2 | 3.5 | 6.5 | 4.8 | D | 21.8 | 7,215 | 214 | 504 | 14,062 | 8.9 |
| McDuffie | 5.2 | D | 6.8 | 22.9 | 4.1 | 8.1 | 4.8 | D | 32.3 | 5,295 | 248 | 932 | 9,767 | 1.1 |
| McIntosh | 0.6 | D | 11.2 | 0.5 | D | 7.6 | 1.8 | 2.3 | 20.5 | 3,460 | 242 | 376 | 9,414 | 5.9 |
| Macon | 20.6 | D | 2.6 | 25.8 | D | 4.1 | 2.4 | D | 25.7 | 2,535 | 196 | 403 | 6,109 | -0.5 |
| Madison | 12.7 | D | 14.7 | 6.5 | D | 4.8 | D | D | 31.5 | 7,210 | 241 | 973 | 12,049 | 2.1 |
| Marion | -1.3 | D | D | D | D | 4.8 | D | D | 23.5 | 1,815 | 216 | 265 | 4,258 | 2.5 |
| Meriwether | 0.8 | 0.0 | 15.5 | 18.8 | D | 5.1 | 2.7 | D | 40.8 | 5,730 | 271 | 816 | 10,137 | 1.9 |
| Miller | 25.1 | 0.0 | 2.4 | 0.9 | D | 5.3 | 6.0 | 3.6 | 40.8 | 1,450 | 256 | 316 | 2,753 | -1.4 |
| Mitchell | 20.4 | 0.0 | 4.4 | 23.2 | D | 5.4 | 5.1 | D | 17.7 | 4,965 | 227 | 992 | 9,091 | 1.0 |
| Monroe | 0.1 | D | 6.9 | 2.0 | 5.6 | 4.9 | 3.4 | 5.7 | 34.3 | 6,465 | 234 | 537 | 11,461 | 9.0 |
| Montgomery | 6.2 | 0.0 | 4.1 | 6.8 | D | 4.3 | 4.1 | D | 22.1 | 1,975 | 216 | 324 | 4,019 | 2.0 |
| Morgan | 0.0 | 0.1 | 5.9 | 23.1 | D | 7.3 | 9.0 | D | 14.9 | 4,995 | 260 | 408 | 8,123 | 8.7 |
| Murray | 3.4 | D | 3.7 | 33.4 | D | 7.8 | 4.2 | D | 17.1 | 8,505 | 213 | 1,156 | 16,339 | 2.2 |
| Muscogee | 0.0 | 0.1 | 3.1 | 7.3 | 7.1 | 5.0 | 17.5 | 12.3 | 28.8 | 39,705 | 203 | 7,737 | 85,681 | 3.6 |
| Newton | -0.5 | 0.0 | 9.1 | 20.0 | 4.3 | 6.7 | 3.8 | 8.9 | 19.0 | 20,810 | 186 | 3,165 | 40,835 | 6.5 |
| Oconee | 0.7 | D | 8.9 | 4.2 | 10.4 | 6.8 | 11.6 | 11.6 | 10.8 | 7,325 | 182 | 324 | 15,188 | 22.5 |
| Oglethorpe | 24.6 | 3.1 | 13.9 | 7.3 | D | 2.8 | 5.4 | D | 20.6 | 3,560 | 234 | 284 | 6,807 | 5.1 |
| Paulding | 0.0 | 0.2 | 13.4 | 5.2 | 5.1 | 10.8 | 5.6 | 12.7 | 24.1 | 24,435 | 144 | 1,680 | 61,228 | 17.4 |
| Peach | 4.3 | 0.0 | D | D | 2.2 | 6.7 | D | 5.2 | 20.6 | 5,865 | 212 | 1,126 | 11,955 | 6.2 |
| Pickens | 0.2 | D | 8.5 | 10.5 | 5.2 | 7.8 | 7.9 | D | 14.4 | 9,700 | 297 | 590 | 14,369 | 5.1 |
| Pierce | 5.8 | 0.0 | 8.7 | 7.9 | 2.8 | 7.4 | 3.7 | D | 19.1 | 4,330 | 223 | 708 | 8,353 | 4.6 |
| Pike | 0.4 | D | 22.6 | 13.6 | D | 5.0 | 5.1 | D | 21.7 | 4,040 | 213 | 377 | 7,290 | 6.8 |
| Polk | 1.7 | D | 5.5 | 33.4 | 2.1 | 9.3 | 2.1 | 8.6 | 17.1 | 9,925 | 233 | 1,669 | 17,204 | 1.7 |
| Pulaski | 8.0 | 0.0 | D | D | 3.6 | 6.7 | D | 30.2 | 20.3 | 2,270 | 203 | 384 | 5,183 | 0.6 |

1. Per 1,000 resident population estimated as of July 1 of the year shown.

| STATE County | Housing units, 2015-2019 | | | | | | | | Civilian labor force, 2020 | | | | Civilian employment[6], 2015-2019 | | |
|---|---|---|---|---|---|---|---|---|---|---|---|---|---|---|---|
| | Occupied units | | | | | | | | | | Unemployment | | | Percent | |
| | Owner-occupied | | | | | Renter-occupied | | | | | | | | | |
| | | | | Median owner cost as a percent of income | | | Median rent as a percent of income[2] | Sub-standard units[4] (percent) | | Percent change, 2019-2020 | | | | Management, business, science, and arts | Construction, production, and maintenance occupations |
| | | | Median value[1] | With a mortgage | Without a mortgage[2] | Median rent[3] | | | | | | | | | |
| | Total | Percent | | | | | | | Total | | Total | Rate[5] | Total | | |
| | 89 | 90 | 91 | 92 | 93 | 94 | 95 | 96 | 97 | 98 | 99 | 100 | 101 | 102 | 103 |
| GEORGIA—Cont'd | | | | | | | | | | | | | | | |
| Effingham | 21,172 | 75.9 | 163,600 | 19.4 | 10.0 | 1,023 | 28.3 | 1.8 | 31,400 | -2.6 | 1,573 | 5.0 | 28,887 | 32.6 | 33.2 |
| Elbert | 7,559 | 72.2 | 84,700 | 22.2 | 12.2 | 685 | 27.3 | 2.1 | 7,969 | 1.5 | 551 | 6.9 | 7,769 | 24.5 | 36.6 |
| Emanuel | 8,387 | 63.8 | 73,100 | 21.6 | 10.1 | 619 | 25.0 | 2.7 | 8,518 | 1.2 | 603 | 7.1 | 8,551 | 28.3 | 37.1 |
| Evans | 4,020 | 68.0 | 87,900 | 20.1 | 11.8 | 662 | 25.1 | 4.8 | 5,002 | 1.1 | 257 | 5.1 | 4,400 | 24.5 | 36.8 |
| Fannin | 10,408 | 79.0 | 197,800 | 23.5 | 11.2 | 794 | 27.4 | 1.4 | 11,066 | -3.7 | 524 | 4.7 | 9,998 | 29.8 | 26.6 |
| Fayette | 40,285 | 82.0 | 281,400 | 20.0 | 10.0 | 1,326 | 27.5 | 0.9 | 56,444 | -2.8 | 3,055 | 5.4 | 53,105 | 45.6 | 20.2 |
| Floyd | 35,679 | 61.0 | 132,400 | 19.7 | 10.0 | 762 | 29.4 | 3.2 | 43,234 | -2.1 | 2,712 | 6.3 | 41,723 | 31.4 | 30.3 |
| Forsyth | 76,753 | 84.0 | 339,700 | 19.0 | 10.0 | 1,380 | 27.0 | 1.3 | 120,904 | -3.4 | 5,533 | 4.6 | 115,231 | 52.8 | 12.7 |
| Franklin | 8,460 | 66.4 | 108,700 | 19.6 | 11.1 | 627 | 27.8 | 1.1 | 9,892 | -3.1 | 584 | 5.9 | 9,263 | 30.7 | 35.1 |
| Fulton | 410,576 | 51.6 | 313,300 | 19.8 | 10.0 | 1,205 | 29.4 | 2.1 | 557,232 | -0.6 | 44,639 | 8.0 | 535,127 | 51.7 | 12.6 |
| Gilmer | 12,021 | 74.5 | 177,500 | 21.5 | 10.0 | 765 | 26.7 | 3.0 | 11,963 | -0.6 | 562 | 4.7 | 13,076 | 27.1 | 32.8 |
| Glascock | 1,108 | 74.7 | 81,700 | 17.7 | 10.0 | 625 | 25.3 | 5.1 | 1,244 | -2.5 | 51 | 4.1 | 1,322 | 28.6 | 31.7 |
| Glynn | 34,119 | 62.5 | 179,000 | 20.0 | 10.0 | 890 | 29.2 | 2.5 | 38,215 | -1.6 | 2,605 | 6.8 | 39,037 | 34.5 | 21.6 |
| Gordon | 20,561 | 65.1 | 136,000 | 19.3 | 11.3 | 731 | 28.6 | 3.9 | 28,062 | -1.5 | 1,531 | 5.5 | 26,262 | 24.0 | 38.6 |
| Grady | 9,136 | 64.3 | 119,800 | 21.2 | 13.6 | 778 | 34.4 | 2.9 | 10,601 | -0.6 | 462 | 4.4 | 10,520 | 30.2 | 35.6 |
| Greene | 7,132 | 74.6 | 210,300 | 26.1 | 12.7 | 727 | 28.7 | 0.8 | 7,080 | -3.1 | 417 | 5.9 | 6,482 | 28.1 | 23.9 |
| Gwinnett | 293,330 | 66.3 | 217,900 | 20.9 | 10.0 | 1,272 | 30.8 | 3.2 | 481,453 | -1.9 | 30,544 | 6.3 | 459,927 | 38.0 | 22.3 |
| Habersham | 15,139 | 78.3 | 150,500 | 20.5 | 12.0 | 772 | 25.0 | 2.5 | 18,691 | -3.9 | 1,016 | 5.4 | 18,747 | 31.2 | 33.0 |
| Hall | 64,352 | 69.5 | 201,300 | 20.3 | 10.0 | 963 | 28.1 | 6.1 | 101,949 | -0.8 | 4,816 | 4.7 | 93,448 | 30.2 | 34.3 |
| Hancock | 2,974 | 70.6 | 85,000 | 29.7 | 15.7 | 678 | 36.2 | 0.8 | 2,532 | 0.7 | 223 | 8.8 | 2,152 | 25.2 | 15.0 |
| Haralson | 11,259 | 68.4 | 124,200 | 18.1 | 12.1 | 677 | 29.8 | 1.8 | 12,245 | -2.7 | 720 | 5.9 | 12,814 | 28.7 | 37.0 |
| Harris | 12,156 | 87.6 | 217,000 | 20.2 | 10.9 | 863 | 27.1 | 1.3 | 16,527 | -2.4 | 764 | 4.6 | 15,470 | 40.3 | 23.8 |
| Hart | 9,853 | 74.8 | 137,700 | 22.4 | 10.5 | 720 | 31.3 | 3.0 | 11,373 | -2.0 | 616 | 5.4 | 9,819 | 28.6 | 32.7 |
| Heard | 4,502 | 71.1 | 109,000 | 20.2 | 11.0 | 630 | 23.9 | 2.0 | 5,132 | -1.8 | 334 | 6.5 | 5,137 | 21.0 | 44.4 |
| Henry | 75,984 | 70.4 | 175,300 | 20.1 | 10.0 | 1,181 | 28.1 | 1.1 | 115,352 | -1.3 | 8,521 | 7.4 | 109,797 | 35.2 | 25.5 |
| Houston | 57,056 | 63.6 | 142,600 | 18.8 | 10.0 | 920 | 28.2 | 3.0 | 70,418 | -1.1 | 3,649 | 5.2 | 71,363 | 36.0 | 23.4 |
| Irwin | 3,329 | 73.6 | 79,800 | 19.3 | 11.2 | 559 | 28.5 | 4.3 | 3,467 | 1.9 | 191 | 5.5 | 3,630 | 31.5 | 34.4 |
| Jackson | 23,166 | 77.9 | 189,700 | 19.1 | 10.0 | 823 | 28.0 | 2.3 | 39,648 | 5.0 | 1,738 | 4.4 | 31,077 | 35.6 | 26.8 |
| Jasper | 5,171 | 76.8 | 139,400 | 21.8 | 12.2 | 848 | 26.3 | 2.3 | 6,756 | -3.2 | 334 | 4.9 | 5,680 | 26.7 | 38.0 |
| Jeff Davis | 5,279 | 65.1 | 79,800 | 19.4 | 12.1 | 584 | 22.9 | 5.3 | 6,099 | -1.4 | 314 | 5.1 | 5,309 | 24.2 | 36.2 |
| Jefferson | 5,664 | 63.2 | 72,500 | 18.3 | 12.2 | 561 | 29.4 | 4.0 | 6,658 | 0.3 | 449 | 6.7 | 5,977 | 23.5 | 45.5 |
| Jenkins | 3,443 | 71.4 | 74,500 | 21.6 | 15.7 | 548 | 29.9 | 2.6 | 3,178 | | 200 | 6.3 | 3,319 | 26.4 | 32.1 |
| Johnson | 3,393 | 67.3 | 64,100 | 17.9 | 10.0 | 533 | 33.0 | 4.1 | 4,006 | -1.7 | 198 | 4.9 | 3,958 | 22.4 | 43.0 |
| Jones | 10,701 | 80.2 | 124,100 | 21.5 | 11.9 | 775 | 33.6 | 1.9 | 13,346 | -3.0 | 609 | 4.6 | 12,306 | 33.6 | 29.1 |
| Lamar | 6,494 | 69.2 | 145,600 | 22.6 | 10.0 | 706 | 34.8 | 1.1 | 7,980 | -2.8 | 504 | 6.3 | 6,932 | 27.8 | 34.5 |
| Lanier | 3,714 | 64.4 | 114,900 | 21.2 | 10.0 | 712 | 31.5 | 1.0 | 3,723 | -1.5 | 194 | 5.2 | 3,574 | 36.1 | 26.1 |
| Laurens | 17,142 | 64.2 | 94,600 | 20.9 | 10.0 | 615 | 29.8 | 2.9 | 19,367 | | 1,292 | 6.7 | 17,458 | 31.8 | 29.8 |
| Lee | 10,226 | 72.8 | 152,800 | 18.8 | 10.0 | 881 | 23.6 | 1.7 | 14,763 | -2.5 | 682 | 4.6 | 13,314 | 40.0 | 25.1 |
| Liberty | 23,485 | 45.7 | 124,500 | 21.4 | 10.9 | 1,067 | 29.8 | 3.2 | 25,741 | -0.9 | 1,601 | 6.2 | 22,248 | 28.1 | 30.2 |
| Lincoln | 3,475 | 75.0 | 127,700 | 23.3 | 13.0 | 708 | 31.5 | 1.5 | 3,497 | -1.5 | 190 | 5.4 | 3,289 | 23.4 | 35.9 |
| Long | 5,695 | 69.1 | 122,100 | 20.2 | 10.0 | 799 | 32.9 | 3.3 | 8,139 | -2.5 | 357 | 4.4 | 6,546 | 29.1 | 33.6 |
| Lowndes | 41,282 | 52.3 | 140,600 | 21.1 | 10.0 | 802 | 30.5 | 2.7 | 51,599 | -0.2 | 3,115 | 6.0 | 47,834 | 31.4 | 23.4 |
| Lumpkin | 11,570 | 70.7 | 185,800 | 20.8 | 10.2 | 891 | 33.7 | 1.4 | 16,463 | -2.9 | 792 | 4.8 | 15,740 | 26.9 | 26.3 |
| McDuffie | 8,153 | 61.3 | 109,500 | 21.3 | 10.3 | 695 | 31.7 | 1.7 | 8,695 | -0.7 | 679 | 7.8 | 8,481 | 26.3 | 34.6 |
| McIntosh | 6,042 | 79.8 | 128,000 | 23.1 | 11.8 | 813 | 32.1 | 0.9 | 5,949 | -2.5 | 340 | 5.7 | 6,069 | 25.8 | 30.0 |
| Macon | 4,696 | 64.0 | 63,800 | 21.7 | 13.4 | 592 | 27.6 | 1.8 | 4,614 | -2.3 | 311 | 6.7 | 4,416 | 20.0 | 43.2 |
| Madison | 10,744 | 74.7 | 131,300 | 19.5 | 10.0 | 715 | 25.5 | 3.5 | 12,972 | -3.3 | 670 | 5.2 | 12,821 | 28.2 | 32.8 |
| Marion | 3,408 | 78.4 | 90,000 | 25.6 | 10.9 | 536 | 20.8 | 1.8 | 3,192 | -3.1 | 163 | 5.1 | 3,218 | 24.2 | 39.2 |
| Meriwether | 8,051 | 68.5 | 103,300 | 22.5 | 14.7 | 714 | 31.7 | 2.0 | 8,816 | -0.9 | 755 | 8.6 | 8,071 | 26.1 | 43.9 |
| Miller | 2,333 | 65.5 | 89,400 | 19.3 | 12.4 | 652 | 22.7 | 8.1 | 2,772 | -1.1 | 112 | 4.0 | 2,426 | 33.8 | 27.6 |
| Mitchell | 7,982 | 62.2 | 85,400 | 26.0 | 15.1 | 637 | 25.4 | 3.3 | 8,289 | -2.4 | 538 | 6.5 | 8,288 | 26.0 | 33.5 |
| Monroe | 9,760 | 79.1 | 166,900 | 19.0 | 11.0 | 704 | 28.5 | 1.4 | 12,700 | -2.5 | 650 | 5.1 | 11,331 | 39.3 | 23.6 |
| Montgomery | 3,097 | 69.2 | 84,000 | 19.3 | 10.1 | 627 | 30.0 | 1.8 | 3,891 | -0.9 | 210 | 5.4 | 3,532 | 32.4 | 31.2 |
| Morgan | 6,942 | 73.2 | 229,000 | 21.5 | 10.0 | 832 | 21.1 | 1.7 | 9,142 | -2.6 | 497 | 5.4 | 8,432 | 33.6 | 33.0 |
| Murray | 14,385 | 69.6 | 108,300 | 19.2 | 10.0 | 688 | 24.2 | 3.9 | 15,350 | -1.8 | 1,159 | 7.6 | 17,531 | 18.1 | 45.3 |
| Muscogee | 72,759 | 48.0 | 141,300 | 21.7 | 10.3 | 906 | 30.4 | 2.3 | 77,258 | -0.7 | 6,049 | 7.8 | 77,701 | 32.9 | 23.1 |
| Newton | 37,018 | 67.6 | 151,800 | 21.0 | 10.0 | 1,010 | 32.2 | 2.2 | 51,901 | -1.9 | 3,699 | 7.1 | 48,967 | 28.4 | 31.7 |
| Oconee | 13,423 | 82.9 | 286,600 | 19.4 | 10.0 | 1,001 | 26.7 | 1.6 | 19,208 | -4.2 | 721 | 3.8 | 18,706 | 55.4 | 14.1 |
| Oglethorpe | 5,651 | 77.2 | 118,500 | 22.2 | 11.1 | 725 | 28.9 | 0.7 | 6,791 | -3.5 | 327 | 4.8 | 6,518 | 32.3 | 36.6 |
| Paulding | 53,299 | 75.8 | 173,600 | 20.3 | 10.0 | 1,163 | 28.7 | 1.9 | 85,313 | -2.7 | 4,595 | 5.4 | 78,024 | 34.0 | 25.5 |
| Peach | 10,136 | 65.4 | 135,100 | 20.7 | 10.0 | 698 | 31.8 | 2.0 | 11,814 | -0.7 | 730 | 6.2 | 10,625 | 28.5 | 33.1 |
| Pickens | 11,868 | 75.9 | 200,200 | 20.1 | 10.1 | 856 | 25.1 | 2.7 | 14,927 | -3.4 | 717 | 4.8 | 14,170 | 29.2 | 30.2 |
| Pierce | 7,048 | 77.5 | 104,400 | 19.5 | 12.2 | 669 | 30.3 | 1.6 | 8,500 | -2.2 | 385 | 4.5 | 7,535 | 26.8 | 40.6 |
| Pike | 6,143 | 84.0 | 173,000 | 21.2 | 10.9 | 828 | 23.7 | 1.1 | 8,705 | -3.2 | 432 | 5.0 | 8,032 | 35.3 | 32.3 |
| Polk | 15,038 | 64.6 | 110,700 | 20.2 | 11.2 | 718 | 30.6 | 3.1 | 18,292 | -1.5 | 1,092 | 6.0 | 18,213 | 21.9 | 44.4 |
| Pulaski | 3,687 | 64.3 | 120,100 | 17.4 | 10.0 | 722 | 31.6 | 2.0 | 3,976 | -2.2 | 179 | 4.5 | 3,692 | 28.1 | 28.3 |

1. Specified owner-occupied units.    2. A value of 10.0 represents 10 percent or less; a value of 50.0 represents 50 percent or more.    3. Specified renter-occupied units.    4. Overcrowded or lacking complete plumbing facilities.    5. Percent of civilian labor force.    6. Civilian employed persons 16 years old and over.

| STATE County | Number of establishments | Total | Health care and social assistance | Manufacturing | Retail trade | Finance and insurance | Professional, scientific, and technical services | Total (mil dol) | Average per employee (dollars) | Number | Fewer than 50 acres | 1000 acres or more | Farm producers whose primary occupation is farming (percent) |
|---|---|---|---|---|---|---|---|---|---|---|---|---|---|
| | 104 | 105 | 106 | 107 | 108 | 109 | 110 | 111 | 112 | 113 | 114 | 115 | 116 |
| **GEORGIA—Cont'd** | | | | | | | | | | | | | |
| Effingham | 781 | 8,229 | 977 | 1,665 | 1,495 | 147 | 241 | 354 | 43,021 | 254 | 55.5 | 6.7 | 35.4 |
| Elbert | 437 | 4,515 | 463 | 1,898 | 581 | 191 | 54 | 157 | 34,877 | 453 | 34.0 | 4.2 | 39.4 |
| Emanuel | 376 | 4,951 | 888 | 1,688 | 778 | 173 | 87 | 168 | 33,836 | 465 | 28.4 | 8.4 | 34.9 |
| Evans | 226 | 4,052 | 423 | 1,949 | 439 | 77 | 57 | 116 | 28,667 | 143 | 31.5 | 6.3 | 39.8 |
| Fannin | 634 | 5,453 | 874 | 319 | 1,453 | 147 | 232 | 172 | 31,478 | 211 | 64.0 | 0.9 | 44.2 |
| Fayette | 3,608 | 43,278 | 7,064 | 2,907 | 7,300 | 1,088 | 2,439 | 1,880 | 43,435 | 148 | 58.1 | NA | 53.4 |
| Floyd | 2,012 | 37,509 | 8,352 | 6,710 | 4,249 | 720 | 758 | 1,498 | 39,945 | 547 | 51.7 | 2.0 | 35.2 |
| Forsyth | 6,783 | 77,736 | 9,486 | 8,517 | 10,796 | 1,598 | 6,314 | 3,812 | 49,042 | 291 | 63.2 | NA | 49.9 |
| Franklin | 461 | 6,781 | 625 | 1,571 | 860 | 145 | 131 | 240 | 35,340 | 753 | 47.7 | 0.4 | 52.4 |
| Fulton | 37,892 | 840,711 | 91,211 | 21,329 | 54,345 | 59,037 | 111,773 | 65,363 | 77,748 | 195 | 75.9 | NA | 38.8 |
| Gilmer | 589 | 6,271 | 464 | 1,654 | 1,226 | 123 | 202 | 192 | 30,607 | 330 | 56.4 | 0.3 | 49.7 |
| Glascock | 27 | 213 | 92 | NA | 31 | NA | NA | 6 | 28,685 | 76 | 35.5 | 3.9 | 38.2 |
| Glynn | 2,595 | 32,263 | 5,277 | 2,101 | 5,205 | 718 | 994 | 1,229 | 38,106 | 53 | 71.7 | NA | 25.0 |
| Gordon | 1,027 | 19,993 | 2,215 | 5,967 | 2,297 | 294 | 231 | 902 | 45,132 | 740 | 52.6 | 1.2 | 39.5 |
| Grady | 409 | 4,764 | 472 | 1,155 | 714 | 134 | 110 | 166 | 34,820 | 415 | 29.9 | 9.2 | 33.7 |
| Greene | 457 | 5,475 | 768 | 488 | 816 | 186 | 316 | 232 | 42,439 | 248 | 27.8 | 7.3 | 40.2 |
| Gwinnett | 24,903 | 337,848 | 31,588 | 23,032 | 45,866 | 14,942 | 29,986 | 18,027 | 53,359 | 177 | 67.2 | 0.6 | 39.7 |
| Habersham | 861 | 11,972 | 1,203 | 3,815 | 1,918 | 373 | 277 | 439 | 36,674 | 379 | 63.6 | NA | 42.2 |
| Hall | 4,569 | 83,924 | 14,580 | 22,554 | 9,269 | 2,022 | 2,080 | 3,961 | 47,197 | 551 | 63.5 | 0.4 | 38.1 |
| Hancock | 63 | 618 | 141 | 152 | 99 | 25 | 17 | 21 | 34,379 | 145 | 20.0 | 3.4 | 36.2 |
| Haralson | 467 | 5,742 | 708 | 1,792 | 1,052 | 139 | 70 | 224 | 39,047 | 321 | 49.2 | NA | 38.2 |
| Harris | 447 | 4,199 | 241 | 1,027 | 390 | 79 | 77 | 123 | 29,285 | 289 | 47.4 | 2.4 | 40.0 |
| Hart | 406 | 6,485 | 350 | 3,119 | 964 | 123 | 151 | 228 | 35,140 | 516 | 40.3 | 0.6 | 50.7 |
| Heard | 109 | 1,262 | NA | 502 | 104 | 17 | 16 | 66 | 52,023 | 227 | 30.0 | 1.8 | 37.2 |
| Henry | 3,931 | 55,750 | 9,091 | 2,920 | 10,402 | 1,180 | 1,961 | 2,110 | 37,845 | 240 | 72.9 | NA | 39.7 |
| Houston | 2,587 | 41,427 | 6,951 | 4,698 | 7,571 | 1,125 | 4,219 | 1,482 | 35,782 | 277 | 55.2 | 2.9 | 21.5 |
| Irwin | 118 | 1,521 | 424 | 207 | 112 | 29 | 39 | 57 | 37,554 | 348 | 29.9 | 10.1 | 46.2 |
| Jackson | 1,366 | 26,556 | 1,266 | 6,541 | 3,230 | 237 | 464 | 1,131 | 42,595 | 734 | 52.7 | 0.8 | 38.2 |
| Jasper | 183 | 1,567 | 225 | 376 | 247 | 62 | 24 | 49 | 31,027 | 251 | 51.4 | 3.6 | 37.0 |
| Jeff Davis | 242 | 3,810 | 342 | 1,550 | 617 | 74 | 38 | 136 | 35,569 | 197 | 41.1 | 11.7 | 37.5 |
| Jefferson | 307 | 3,753 | 401 | 895 | 608 | 156 | 62 | 147 | 39,281 | 318 | 25.2 | 12.6 | 43.5 |
| Jenkins | 101 | 1,058 | 232 | 16 | 154 | 29 | 11 | 35 | 33,026 | 210 | 35.2 | 11.4 | 33.9 |
| Johnson | 123 | 879 | 195 | 119 | 206 | 32 | 22 | 26 | 29,878 | 284 | 28.9 | 7.4 | 28.5 |
| Jones | 328 | 2,737 | 371 | 41 | 509 | 78 | 78 | 94 | 34,164 | 165 | 33.3 | 5.5 | 29.3 |
| Lamar | 259 | 3,179 | 340 | 747 | 517 | 104 | 70 | 122 | 38,484 | 220 | 52.3 | 0.9 | 38.7 |
| Lanier | 91 | 899 | 188 | 188 | 186 | 64 | 19 | 27 | 30,102 | 103 | 45.6 | 19.4 | 38.1 |
| Laurens | 1,056 | 16,005 | 3,114 | 2,963 | 2,511 | 505 | 334 | 607 | 37,946 | 626 | 30.4 | 4.2 | 29.0 |
| Lee | 414 | 4,382 | 444 | 470 | 740 | 148 | 105 | 216 | 49,190 | 206 | 40.8 | 13.6 | 33.7 |
| Liberty | 852 | 13,781 | 2,104 | 2,607 | 2,234 | 339 | 409 | 600 | 43,535 | 69 | 63.8 | 1.4 | 23.3 |
| Lincoln | 145 | 818 | 36 | 65 | 150 | 40 | 37 | 27 | 32,707 | 104 | 40.4 | 1.9 | 25.6 |
| Long | 82 | 410 | 32 | NA | 92 | 21 | NA | 11 | 26,007 | 85 | 44.7 | NA | 28.9 |
| Lowndes | 2,770 | 42,174 | 7,897 | 3,455 | 6,606 | 1,190 | 1,244 | 1,522 | 36,082 | 380 | 52.6 | 3.2 | 30.9 |
| Lumpkin | 529 | 4,916 | 661 | 592 | 827 | 97 | 207 | 172 | 34,980 | 240 | 60.8 | 1.3 | 45.3 |
| McDuffie | 414 | 6,056 | 847 | 1,707 | 965 | 155 | 104 | 214 | 35,395 | 269 | 46.8 | 3.0 | 31.1 |
| McIntosh | 171 | 1,147 | 40 | 14 | 285 | 73 | 37 | 33 | 28,671 | 32 | 68.8 | 6.3 | 33.3 |
| Macon | 173 | 1,807 | 423 | 582 | 262 | 43 | 21 | 77 | 42,621 | 339 | 34.2 | 5.9 | 43.1 |
| Madison | 379 | 2,364 | 221 | 210 | 450 | 76 | 178 | 77 | 32,747 | 673 | 51.0 | 1.0 | 44.4 |
| Marion | 80 | 1,218 | 144 | 28 | 111 | 13 | 13 | 28 | 22,789 | 222 | 29.3 | 6.8 | 46.1 |
| Meriwether | 284 | 2,925 | 853 | 278 | 547 | 52 | 38 | 105 | 36,027 | 344 | 34.9 | 4.1 | 39.0 |
| Miller | 135 | 1,303 | 593 | 16 | 160 | 47 | 45 | 50 | 38,065 | 144 | 27.1 | 15.3 | 45.7 |
| Mitchell | 358 | 4,983 | 510 | 2,186 | 663 | 128 | 148 | 159 | 31,814 | 425 | 31.5 | 11.8 | 48.9 |
| Monroe | 529 | 6,265 | 1,038 | 670 | 969 | 156 | 242 | 248 | 39,628 | 219 | 38.8 | 4.1 | 35.9 |
| Montgomery | 105 | 1,152 | 118 | 212 | 151 | 47 | 20 | 32 | 28,073 | 179 | 27.4 | 10.1 | 32.3 |
| Morgan | 510 | 6,862 | 529 | 1,870 | 979 | 180 | 239 | 246 | 35,866 | 513 | 37.8 | 1.8 | 46.6 |
| Murray | 408 | 6,381 | 552 | 2,596 | 922 | 94 | 70 | 229 | 35,829 | 278 | 45.0 | 1.4 | 37.9 |
| Muscogee | 4,359 | 81,640 | 15,632 | 7,376 | 10,933 | 12,713 | 2,635 | 3,837 | 46,999 | 37 | 35.1 | 2.7 | 40.8 |
| Newton | 1,457 | 19,983 | 2,527 | 3,816 | 3,223 | 528 | 351 | 915 | 45,812 | 292 | 50.3 | 1.4 | 36.5 |
| Oconee | 1,234 | 14,201 | 1,973 | 2,010 | 2,275 | 477 | 1,018 | 587 | 41,370 | 329 | 49.2 | 1.2 | 29.7 |
| Oglethorpe | 176 | 1,157 | 107 | 91 | 121 | 26 | 36 | 40 | 34,324 | 427 | 38.6 | 2.8 | 39.9 |
| Paulding | 2,072 | 20,304 | 3,303 | 919 | 4,987 | 427 | 761 | 696 | 34,260 | 212 | 60.4 | NA | 26.3 |
| Peach | 468 | 7,086 | 593 | 2,720 | 1,014 | 117 | 160 | 293 | 41,354 | 228 | 51.3 | 3.5 | 43.9 |
| Pickens | 717 | 6,851 | 1,443 | 816 | 1,267 | 256 | 200 | 286 | 41,748 | 258 | 62.4 | 0.4 | 45.1 |
| Pierce | 329 | 3,097 | 123 | 492 | 439 | 79 | 52 | 111 | 35,758 | 352 | 46.6 | 6.5 | 36.6 |
| Pike | 260 | 1,930 | 237 | 179 | 189 | 164 | 62 | 66 | 34,336 | 286 | 45.1 | 1.0 | 30.2 |
| Polk | 609 | 10,588 | 1,266 | 3,943 | 1,443 | 150 | 109 | 394 | 37,194 | 401 | 51.4 | 3.0 | 38.8 |
| Pulaski | 171 | 2,082 | 894 | 119 | 302 | 72 | 43 | 78 | 37,582 | 189 | 41.8 | 7.9 | 40.5 |

# Table B. States and Counties — **Agriculture**

| STATE County | Acreage (1,000) | Percent change, 2012-2017 | Average size of farm | Total irrigated (1,000) | Total cropland (1,000) | Value of land and buildings (dollars) Average per farm | Average per acre | Value of machinery and equiopmnet, average per farm (dollars) | Total (mil dol) | Average per farm (acres) | Crops | Livestock and poultry products | Organic farms (number) | Farms with internet access (percent) | Government payments Total ($1,000) | Percent of farms |
|---|---|---|---|---|---|---|---|---|---|---|---|---|---|---|---|---|
| | 117 | 118 | 119 | 120 | 121 | 122 | 123 | 124 | 125 | 126 | 127 | 128 | 129 | 130 | 131 | 132 |
| GEORGIA—Cont'd | | | | | | | | | | | | | | | | |
| Effingham | 50 | 24.9 | 199 | 2.8 | 28.4 | 646,043 | 3,250 | 104,350 | 16.3 | 64,063 | 92.0 | 8.0 | 4 | 82.7 | 369 | 22.0 |
| Elbert | 79 | 39.3 | 175 | 1.0 | 19.2 | 687,827 | 3,925 | 83,075 | 107.1 | 236,494 | 4.9 | 95.1 | 4 | 64.7 | 919 | 28.3 |
| Emanuel | 139 | -8.7 | 298 | 4.8 | 45.7 | 622,886 | 2,091 | 95,023 | 33.0 | 71,013 | 85.0 | 15.0 | 1 | 74.4 | 2,565 | 40.9 |
| Evans | 36 | -1.2 | 249 | 5.6 | 14.4 | 671,882 | 2,699 | 104,565 | 32.2 | 224,979 | 60.7 | 39.3 | NA | 79.0 | 567 | 29.4 |
| Fannin | 16 | 17.9 | 78 | 0.2 | 4.0 | 457,536 | 5,902 | 48,477 | 23.0 | 109,175 | 21.7 | 78.3 | NA | 84.4 | 239 | 16.6 |
| Fayette | 11 | -1.8 | 76 | 0.1 | 3.5 | 415,120 | 5,441 | 42,880 | 4.1 | 27,432 | 79.0 | 21.0 | 1 | 86.5 | 13 | 5.4 |
| Floyd | 75 | 6.7 | 137 | 0.7 | 21.7 | 665,897 | 4,866 | 88,749 | 53.4 | 97,700 | 10.2 | 89.8 | NA | 77.3 | 442 | 13.5 |
| Forsyth | 18 | 12.1 | 62 | 0.3 | 5.3 | 488,028 | 7,879 | 64,873 | 45.9 | 157,732 | 11.1 | 88.9 | 1 | 74.2 | 93 | 9.6 |
| Franklin | 79 | 1.9 | 105 | 0.2 | 21.0 | 601,312 | 5,748 | 100,018 | 371.8 | 493,734 | 0.6 | 99.4 | 1 | 78.6 | 928 | 27.5 |
| Fulton | 12 | -13.3 | 63 | 0.2 | 2.2 | 1,195,368 | 19,063 | 42,330 | 2.3 | 11,641 | 39.9 | 60.1 | 7 | 91.8 | 110 | 10.8 |
| Gilmer | 28 | 11.0 | 86 | 0.1 | 6.9 | 658,366 | 7,657 | 86,301 | 205.4 | 622,530 | 1.5 | 98.5 | NA | 87.3 | 402 | 30.6 |
| Glascock | 21 | -10.6 | 283 | D | 6.1 | 449,699 | 1,592 | 52,828 | 2.0 | 26,947 | 82.0 | 18.0 | NA | 86.8 | 187 | 23.7 |
| Glynn | 2 | -46.5 | 36 | 0.1 | 0.2 | 559,751 | 15,411 | 68,477 | 0.3 | 5,887 | 26.0 | 74.0 | 1 | 54.7 | 8 | 9.4 |
| Gordon | 75 | -12.1 | 101 | 1.0 | 29.8 | 660,422 | 6,545 | 73,663 | 294.2 | 397,519 | 2.9 | 97.1 | 1 | 73.1 | 1,580 | 29.5 |
| Grady | 124 | -5.0 | 298 | 13.2 | 69.6 | 1,120,918 | 3,760 | 138,658 | 100.7 | 242,636 | 66.8 | 33.2 | NA | 69.2 | 5,132 | 52.0 |
| Greene | 76 | 55.8 | 305 | 0.2 | 9.0 | 1,094,410 | 3,586 | 98,364 | 79.1 | 319,073 | 3.0 | 97.0 | NA | 74.2 | 1,429 | 12.9 |
| Gwinnett | 11 | 1.9 | 60 | 0.1 | 3.0 | 563,058 | 9,339 | 53,633 | 16.8 | 95,068 | 96.5 | 3.5 | NA | 78.5 | D | 4.5 |
| Habersham | 26 | -32.2 | 68 | 0.1 | 9.7 | 485,967 | 7,160 | 99,560 | 123.0 | 324,485 | 1.5 | 98.5 | 1 | 81.3 | 269 | 16.1 |
| Hall | 41 | -21.7 | 74 | 0.1 | 13.8 | 689,173 | 9,332 | 79,541 | 128.5 | 233,156 | 2.3 | 97.7 | 1 | 80.9 | 789 | 20.3 |
| Hancock | 39 | 21.3 | 267 | 0.5 | 5.7 | 559,388 | 2,092 | 60,754 | 4.4 | 30,297 | 29.7 | 70.3 | 1 | 74.5 | 134 | 19.3 |
| Haralson | 27 | 0.5 | 84 | 0.2 | 7.7 | 398,850 | 4,750 | 59,412 | 75.4 | 234,754 | 1.8 | 98.2 | NA | 69.8 | 194 | 15.3 |
| Harris | 42 | 29.2 | 145 | 0.3 | 12.2 | 626,701 | 4,322 | 54,841 | 5.1 | 17,550 | 72.5 | 27.5 | 1 | 72.0 | 68 | 9.3 |
| Hart | 66 | -2.9 | 129 | 2.2 | 26.1 | 722,492 | 5,622 | 116,647 | 215.1 | 416,953 | 6.2 | 93.8 | NA | 75.8 | 1,103 | 38.2 |
| Heard | 38 | 42.0 | 169 | 0.3 | 7.6 | 611,638 | 3,611 | 77,480 | 43.3 | 190,767 | 8.4 | 91.6 | NA | 84.1 | 346 | 18.9 |
| Henry | 12 | -42.4 | 52 | 0.1 | 3.4 | 389,536 | 7,482 | 33,481 | 2.8 | 11,654 | 59.0 | 41.0 | 8 | 80.0 | 126 | 4.2 |
| Houston | 39 | -17.5 | 141 | 7.6 | 18.3 | 567,998 | 4,021 | 86,680 | 18.2 | 65,588 | 55.0 | 45.0 | NA | 82.7 | 445 | 19.9 |
| Irwin | 123 | -17.3 | 353 | 31.4 | 73.3 | 1,146,611 | 3,249 | 180,086 | 63.1 | 181,417 | 81.2 | 18.8 | 2 | 75.0 | 5,680 | 59.2 |
| Jackson | 75 | -3.5 | 102 | 1.4 | 19.8 | 647,798 | 6,370 | 76,784 | 197.6 | 269,181 | 5.2 | 94.8 | 1 | 76.6 | 711 | 19.2 |
| Jasper | 43 | -2.2 | 171 | 0.3 | 8.0 | 594,493 | 3,481 | 58,915 | 27.0 | 107,641 | 7.2 | 92.8 | 1 | 77.3 | 40 | 8.8 |
| Jeff Davis | 72 | -9.2 | 363 | 10.3 | 46.3 | 910,689 | 2,505 | 215,420 | 40.6 | 205,914 | 70.9 | 29.1 | NA | 80.2 | 1,831 | 36.5 |
| Jefferson | 125 | -14.2 | 393 | 33.6 | 69.7 | 1,017,736 | 2,590 | 140,667 | 58.5 | 183,912 | 73 | 27.0 | NA | 65.7 | 2,413 | 45.9 |
| Jenkins | 79 | -13.2 | 378 | 7.3 | 33.4 | 989,382 | 2,620 | 96,519 | 21.6 | 103,071 | 87.2 | 12.8 | NA | 79.5 | 2,721 | 46.2 |
| Johnson | 75 | 31.2 | 263 | 9.1 | 25.0 | 551,537 | 2,093 | 59,306 | 12.3 | 43,447 | 82.2 | 17.8 | NA | 71.5 | 1,340 | 35.2 |
| Jones | 36 | 58.3 | 221 | D | 7.1 | 619,706 | 2,806 | 67,284 | 5.5 | 33,448 | 14.3 | 85.7 | 3 | 81.8 | 115 | 15.8 |
| Lamar | 32 | -8.9 | 147 | 0.2 | 10.2 | 514,385 | 3,501 | 87,640 | 46.5 | 211,136 | 8.5 | 91.5 | 2 | 79.5 | 84 | 10.0 |
| Lanier | 47 | 12.4 | 454 | 7.3 | 25.4 | 1,376,047 | 3,030 | 220,587 | 22.9 | 222,184 | 98.9 | 1.1 | NA | 74.8 | 1,027 | 33.0 |
| Laurens | 155 | -16.0 | 247 | 9.3 | 42.7 | 518,267 | 2,096 | 73,721 | 25.7 | 41,003 | 71.6 | 28.4 | NA | 74.3 | 2,465 | 57.2 |
| Lee | 120 | 14.1 | 584 | 19.1 | 60.1 | 1,976,595 | 3,384 | 150,525 | 60.4 | 293,097 | 77.3 | 22.7 | NA | 69.9 | 3,101 | 46.1 |
| Liberty | 6 | 2.5 | 92 | 0.1 | 1.0 | 488,481 | 5,289 | 36,656 | 0.5 | 6,986 | 61.0 | 39.0 | NA | 75.4 | 19 | 11.6 |
| Lincoln | 18 | -22.5 | 176 | D | 3.3 | 549,595 | 3,125 | 64,987 | 4.2 | 40,346 | 17.0 | 83.0 | NA | 67.3 | 177 | 20.2 |
| Long | 10 | -0.8 | 120 | 0.1 | 3.0 | 402,578 | 3,361 | 71,266 | 7.3 | 85,518 | 8.8 | 91.2 | NA | 85.9 | 63 | 9.4 |
| Lowndes | 62 | -4.8 | 163 | 7.0 | 24.4 | 886,787 | 5,451 | 79,945 | 35.5 | 93,363 | 85.7 | 14.3 | 3 | 73.4 | 1,121 | 31.8 |
| Lumpkin | 27 | 55.1 | 112 | 0.2 | 6.9 | 665,565 | 5,925 | 82,836 | 51.3 | 213,658 | 7.6 | 92.4 | NA | 80.4 | 227 | 19.2 |
| McDuffie | 44 | 15.0 | 162 | 0.6 | 11.0 | 486,078 | 2,994 | 72,590 | D | D | D | D | NA | 74.0 | 255 | 20.4 |
| McIntosh | 10 | -42.9 | 305 | D | 1.0 | 722,542 | 2,372 | 84,945 | D | D | D | D | 1 | 87.5 | NA | NA |
| Macon | 111 | 10.1 | 328 | 25.2 | 59.5 | 1,056,552 | 3,217 | 182,358 | 271.6 | 801,212 | 20.0 | 80.0 | 1 | 71.1 | 2,615 | 48.7 |
| Madison | 69 | -3.7 | 102 | 0.3 | 23.0 | 577,100 | 5,664 | 83,713 | 239.6 | 355,947 | 1.7 | 98.3 | 1 | 77.4 | 569 | 21.2 |
| Marion | 64 | 35.0 | 288 | 1.8 | 14.4 | 684,034 | 2,375 | 99,628 | 20.6 | 92,599 | 24.6 | 75.4 | NA | 72.1 | 777 | 42.3 |
| Meriwether | 71 | 14.5 | 206 | 1.0 | 13.9 | 673,529 | 3,262 | 85,309 | 12.5 | 36,445 | 75.5 | 24.5 | NA | 75.9 | 261 | 15.7 |
| Miller | 80 | -16.3 | 557 | 26.6 | 54.6 | 1,805,529 | 3,242 | 327,966 | 47.9 | 332,486 | 87.9 | 12.1 | NA | 75.0 | 6,028 | 70.1 |
| Mitchell | 190 | -0.8 | 446 | 66.6 | 121.8 | 1,651,849 | 3,702 | 281,683 | 262.7 | 618,111 | 44.7 | 55.3 | NA | 72.7 | 6,948 | 51.1 |
| Monroe | 49 | 40.9 | 222 | 0.0 | 5.6 | 727,734 | 3,279 | 79,453 | 51.2 | 233,685 | 1.9 | 98.1 | NA | 84.0 | 94 | 10.5 |
| Montgomery | 60 | 3.2 | 333 | 7.2 | 22.5 | 859,810 | 2,584 | 96,589 | 15.5 | 86,732 | 87.2 | 12.8 | NA | 58.1 | 1,018 | 39.1 |
| Morgan | 88 | -6.6 | 172 | 1.1 | 26 | 803,842 | 4,673 | 74,639 | 121.0 | 235,889 | 4.8 | 95.2 | NA | 76.8 | 922 | 19.1 |
| Murray | 47 | 0.5 | 170 | 0.9 | 11.9 | 860,300 | 5,068 | 105,708 | 122.7 | 441,428 | 3.5 | 96.5 | NA | 76.3 | 470 | 31.7 |
| Muscogee | 9 | 117.3 | 251 | D | 2.9 | 1,255,529 | 5,003 | 50,172 | 0.2 | 5,324 | 79.2 | 20.8 | NA | 94.6 | D | 2.7 |
| Newton | 43 | 5.0 | 146 | 0.1 | 8.5 | 655,436 | 4,475 | 74,446 | 12.4 | 42,308 | 15.1 | 84.9 | 2 | 79.8 | 245 | 12.0 |
| Oconee | 36 | -21.2 | 108 | 0.6 | 7.9 | 784,386 | 7,235 | 69,487 | 42.2 | 128,210 | 33.7 | 66.3 | 1 | 79.6 | 383 | 26.1 |
| Oglethorpe | 73 | -9.7 | 171 | 1.6 | 17.9 | 743,571 | 4,353 | 76,951 | 198.4 | 464,553 | 7.1 | 92.9 | 4 | 84.8 | 934 | 31.4 |
| Paulding | 15 | 77.2 | 70 | 0.1 | 4.1 | 404,436 | 5,781 | 52,413 | 9.5 | 44,967 | 5.7 | 94.3 | NA | 77.8 | 34 | 10.4 |
| Peach | 58 | 64.5 | 255 | 16.4 | 33.8 | 1,469,850 | 5,766 | 191,004 | 65.4 | 286,654 | 88.1 | 11.9 | NA | 81.6 | 254 | 14.5 |
| Pickens | 17 | -1.5 | 64 | 0.1 | 4.0 | 548,090 | 8,529 | 101,132 | 77.1 | 298,841 | 1.2 | 98.8 | NA | 71.7 | 215 | 20.5 |
| Pierce | 81 | 3.4 | 230 | 8.3 | 47.6 | 661,340 | 2,877 | 151,848 | 42.1 | 119,608 | 82.5 | 17.5 | NA | 70.7 | 1,653 | 34.1 |
| Pike | 41 | 7.4 | 143 | 0.7 | 8.6 | 699,823 | 4,897 | 65,136 | 18.8 | 65,881 | 14.2 | 85.8 | NA | 78.7 | 225 | 18.2 |
| Polk | 62 | 38.4 | 155 | 0.1 | 28.7 | 561,506 | 3,616 | 116,950 | 46.0 | 114,691 | 29.2 | 70.8 | 1 | 73.3 | 232 | 15.7 |
| Pulaski | 53 | -15.8 | 278 | 14.0 | 28.8 | 743,749 | 2,671 | 115,184 | 49.7 | 262,937 | 38.9 | 61.1 | NA | 87.3 | 703 | 57.1 |

# Table B. States and Counties — Water Use, Wholesale Trade, Retail Trade, and Real Estate

| STATE County | Water use, 2015 | | Wholesale Trade[1], 2017 | | | | Retail Trade[2], 2017 | | | | Real estate and rental and leasing,[2] 2017 | | | |
|---|---|---|---|---|---|---|---|---|---|---|---|---|---|---|
| | Public supply water withdrawn (mil gal/day) | Public supply gallons withdrawn per person per day | Number of establishments | Number of employees | Sales (mil dol) | Average payroll (mil dol) | Number of establishments | Number of employees | Sales (mil dol) | Average payroll (mil dol) | Number of establishments | Number of employees | Sales (mil dol) | Average payroll (mil dol) |
| | 133 | 134 | 135 | 136 | 137 | 138 | 139 | 140 | 141 | 142 | 143 | 144 | 145 | 146 |
| GEORGIA—Cont'd | | | | | | | | | | | | | | |
| Effingham | 35.80 | 626.9 | 18 | 108 | 129.2 | 7.1 | 103 | 1,626 | 444.5 | 34.9 | 32 | 90 | 23.5 | 2.5 |
| Elbert | 1.59 | 82.1 | 40 | 229 | 97.6 | 8.2 | 72 | 581 | 126.0 | 12.6 | NA | NA | NA | NA |
| Emanuel | 1.59 | 70.0 | 17 | 83 | 96.7 | 3.6 | 81 | 792 | 245.6 | 19.5 | 13 | 32 | 13.8 | 0.9 |
| Evans | 0.54 | 50.1 | 5 | 19 | 6.8 | 0.7 | 61 | 454 | 166.0 | 11.8 | 4 | 14 | 1.5 | 0.3 |
| Fannin | 1.75 | 72.0 | 18 | 122 | 45.4 | 3.7 | 127 | 1,400 | 377.7 | 31.6 | D | D | D | D |
| Fayette | 9.46 | 85.4 | 171 | 2,181 | 1,692.4 | 144.5 | 435 | 7,502 | 1,725.0 | 163.0 | 225 | 596 | 151.8 | 25.1 |
| Floyd | 11.80 | 122.3 | 67 | 1,010 | 655.3 | 50.1 | 389 | 4,269 | 1,214.8 | 106.6 | 83 | 260 | 56.2 | 9.3 |
| Forsyth | 20.61 | 97.0 | 417 | 5,823 | 4,457.7 | 361.9 | 677 | 9,901 | 3,037.0 | 273.0 | 344 | 908 | 216.0 | 37.6 |
| Franklin | 1.95 | 87.4 | 22 | 188 | 90.0 | 8.6 | 91 | 893 | 293.1 | 20.3 | 11 | 23 | 14.7 | 0.9 |
| Fulton | 196.32 | 194.3 | 1,514 | 29,384 | 40,998.6 | 2,272.3 | 3,599 | 55,155 | 16,815.9 | 1,592.5 | 2,598 | 23,538 | 12,563.1 | 1,968.0 |
| Gilmer | 2.50 | 85.0 | 19 | 99 | 44.3 | 3.5 | 99 | 1,192 | 328.0 | 30.1 | 38 | 143 | 21.0 | 4.0 |
| Glascock | 0.07 | 22.8 | NA | NA | NA | NA | 4 | 19 | 6.1 | 0.3 | NA | NA | NA | NA |
| Glynn | 9.31 | 111.4 | 81 | 747 | 769.0 | 31.3 | 468 | 5,607 | 1,498.9 | 132.5 | 165 | 510 | 98.0 | 18.0 |
| Gordon | 10.43 | 184.4 | 75 | 1,096 | 519.9 | 60.7 | 230 | 2,520 | 660.1 | 54.5 | 41 | 126 | 42.3 | 4.4 |
| Grady | 1.73 | 68.6 | 29 | 365 | 443.5 | 18.4 | 69 | 778 | 214.6 | 18.5 | 20 | 39 | 5.4 | 1.2 |
| Greene | 1.23 | 73.6 | 18 | 101 | 57.3 | 5.5 | 74 | 780 | 271.9 | 20.5 | 25 | 55 | 17.2 | 2.2 |
| Gwinnett | 0.64 | 0.7 | 1,772 | 30,291 | 39,243.9 | 2,351.1 | 2,821 | 46,440 | 15,083.3 | 1,342.0 | 1,206 | 5,857 | 1,636.6 | 306.9 |
| Habersham | 5.66 | 128.6 | D | D | D | D | 174 | 1,906 | 512.0 | 45.3 | D | D | D | D |
| Hall | 88.18 | 455.6 | 261 | 4,205 | 7,901.5 | 236.0 | 619 | 9,240 | 2,948.9 | 268.3 | 220 | 549 | 180.7 | 25.3 |
| Hancock | 1.20 | 140.3 | NA | NA | NA | NA | 22 | 78 | 22.7 | 2.0 | NA | NA | NA | NA |
| Haralson | 2.53 | 87.7 | 18 | 126 | 36.6 | 7.4 | 88 | 992 | 398.6 | 29.2 | 12 | 11 | 1.4 | 0.3 |
| Harris | 6.26 | 187.5 | D | D | D | D | 47 | 308 | 73.4 | 5.6 | D | D | D | D |
| Hart | 1.46 | 57.2 | 12 | 71 | 43.6 | 2.8 | 76 | 1,032 | 213.6 | 20.3 | 9 | 62 | 11.7 | 1.3 |
| Heard | 1.22 | 105.7 | NA | NA | NA | NA | 20 | 99 | 23.0 | 2.0 | 3 | D | 0.2 | D |
| Henry | 34.13 | 156.7 | 123 | 1,918 | 1,313.0 | 110.8 | 589 | 10,479 | 2,947.9 | 245.3 | 177 | 567 | 164.4 | 24.7 |
| Houston | 23.75 | 158.3 | 43 | 320 | 175.8 | 13.7 | 486 | 7,535 | 2,075.9 | 182.5 | 124 | 455 | 114.3 | 15.7 |
| Irwin | 0.44 | 47.6 | 8 | 68 | 29.3 | 2.5 | 27 | 161 | 32.0 | 2.8 | D | D | D | 0.3 |
| Jackson | 9.52 | 150.3 | 74 | 1,686 | 1,935.7 | 101.1 | 242 | 3,590 | 1,578.6 | 109.4 | 68 | 159 | 42.5 | 4.8 |
| Jasper | 0.75 | 55.0 | NA | NA | NA | NA | 22 | 207 | 54.2 | 4.0 | 5 | 5 | 2.1 | 0.4 |
| Jeff Davis | 0.90 | 60.3 | 13 | 104 | 78.8 | 4.9 | 66 | 678 | 164.3 | 14.8 | D | D | D | D |
| Jefferson | 1.32 | 82.0 | 10 | 204 | 191.7 | 5.1 | 64 | 584 | 136.8 | 12.6 | 9 | 47 | 2.2 | 1.2 |
| Jenkins | 0.45 | 50.2 | D | D | D | D | 18 | 146 | 32.9 | 2.7 | NA | NA | NA | NA |
| Johnson | 0.53 | 54.9 | 3 | 14 | 11.8 | 0.6 | 25 | 158 | 40.9 | 2.8 | NA | NA | NA | NA |
| Jones | 2.73 | 95.8 | D | D | D | D | 44 | 518 | 107.0 | 10.3 | D | D | D | D |
| Lamar | 1.87 | 102.7 | 7 | 260 | 143.1 | 12.3 | 45 | 525 | 116.9 | 11.0 | 4 | 8 | 1.0 | 0.1 |
| Lanier | 0.36 | 34.9 | NA | NA | NA | NA | 18 | 154 | 33.7 | 2.9 | NA | NA | NA | NA |
| Laurens | 3.54 | 74.2 | 40 | 368 | 278.1 | 15.7 | 233 | 2,579 | 754.2 | 56.3 | D | D | D | D |
| Lee | 1.87 | 64.0 | 17 | 291 | 768.6 | 35.6 | 65 | 726 | 224.9 | 17.7 | 20 | 53 | 12.5 | 2.0 |
| Liberty | 6.42 | 102.8 | D | D | D | D | 185 | 2,122 | 598.6 | 52.2 | D | D | D | 5.3 |
| Lincoln | 0.01 | 1.3 | 5 | D | 7.4 | D | 20 | 169 | 41.1 | 3.5 | 6 | 12 | 2.1 | 0.4 |
| Long | 0.51 | 28.8 | NA | NA | NA | NA | 14 | 74 | 19.1 | 1.5 | D | D | D | 0.3 |
| Lowndes | 13.16 | 116.6 | 109 | 1,113 | 904.7 | 51.5 | 523 | 6,980 | 2,098.9 | 167.1 | 143 | 701 | 124.2 | 22.7 |
| Lumpkin | 1.27 | 40.4 | 13 | 234 | 73.2 | 15.6 | 77 | 886 | 236.4 | 22.1 | D | D | D | D |
| McDuffie | 2.59 | 120.2 | D | D | D | D | 86 | 1,106 | 335.5 | 27.1 | 13 | 46 | 5.5 | 1.3 |
| McIntosh | 1.04 | 74.5 | 7 | 97 | 13.9 | 2.0 | 43 | 315 | 90.9 | 6.6 | D | D | D | D |
| Macon | 1.03 | 75.6 | 10 | 99 | 264.0 | 5.0 | 40 | 267 | 43.7 | 5.2 | 4 | D | 0.3 | D |
| Madison | 0.53 | 18.6 | D | D | D | 2.6 | 58 | 450 | 116.1 | 8.8 | D | D | D | D |
| Marion | 1.41 | 160.9 | NA | NA | NA | NA | 24 | 118 | 28.8 | 2.6 | NA | NA | NA | NA |
| Meriwether | 0.31 | 14.6 | D | D | D | 0.8 | 70 | 546 | 104.7 | 11.2 | NA | NA | NA | NA |
| Miller | 0.26 | 44.4 | 9 | 97 | 61.1 | 3.9 | 22 | 171 | 57.6 | 4.7 | 3 | 3 | 0.2 | 0.0 |
| Mitchell | 3.77 | 167.0 | 21 | 194 | 222.4 | 7.9 | 82 | 677 | 139.5 | 13.6 | D | D | D | D |
| Monroe | 1.58 | 58.3 | D | D | D | 5.0 | 78 | 903 | 226.7 | 20.3 | 21 | 39 | 7.6 | 1.2 |
| Montgomery | 0.18 | 20.1 | NA | NA | NA | NA | 24 | 165 | 32.2 | 3.2 | NA | NA | NA | NA |
| Morgan | 1.34 | 74.3 | 16 | 134 | 139.1 | 5.8 | 69 | 984 | 258.8 | 22.3 | D | D | D | D |
| Murray | 1.89 | 47.8 | 27 | 556 | 332.8 | 38.6 | 92 | 867 | 190.1 | 17.1 | D | D | D | 7.2 |
| Muscogee | 36.77 | 183.3 | 152 | 1,978 | 1,683.1 | 109.0 | 794 | 11,547 | 2,979.2 | 268.9 | 242 | 1,314 | 342.8 | 54.5 |
| Newton | 13.73 | 130.2 | 59 | 924 | 2,501.7 | 76.5 | 218 | 3,266 | 950.7 | 80.4 | 64 | 233 | 54.1 | 8.6 |
| Oconee | 0.35 | 9.7 | 29 | 215 | 110.3 | 12.4 | 116 | 2,246 | 566.9 | 51.9 | 88 | 479 | 96.6 | 29.5 |
| Oglethorpe | 0.26 | 17.5 | D | D | D | 4.0 | 19 | 118 | 31.3 | 2.7 | D | D | D | D |
| Paulding | 0.10 | 0.7 | 62 | 282 | 146.7 | 14.0 | 271 | 4,863 | 1,472.7 | 121.0 | 83 | 427 | 92.3 | 25.4 |
| Peach | 2.31 | 86.5 | D | D | D | D | 101 | 1,020 | 356.7 | 26.6 | 24 | 59 | 13.0 | 2.5 |
| Pickens | 2.77 | 91.4 | 34 | 144 | 72.4 | 4.7 | 100 | 1,223 | 417.4 | 33.1 | 35 | 54 | 10.3 | 1.8 |
| Pierce | 0.57 | 29.8 | 13 | 360 | 330.9 | 14.0 | 66 | 409 | 123.5 | 9.1 | 9 | 59 | 32.7 | 1.9 |
| Pike | 2.97 | 165.5 | 8 | 172 | 41.6 | 8.0 | 36 | 173 | 39.3 | 3.9 | 9 | 12 | 3.6 | 0.4 |
| Polk | 5.60 | 134.9 | 19 | 225 | 332.4 | 9.6 | 132 | 1,485 | 377.8 | 36.6 | D | D | D | D |
| Pulaski | 1.09 | 95.6 | NA | NA | NA | NA | 38 | 306 | 82.4 | 7.8 | 5 | 9 | 0.7 | 0.2 |

1  Merchant wholesalers, except manufacturers' sales branches and offices.   2. Employer establishments.

# Table B. States and Counties — Professional Services, Manufacturing, and Accommodation and Food Services

| STATE County | Professional, scientific, and technical services, 2017 | | | | Manufacturing, 2017 | | | | Accommodation and food services, 2017 | | | |
|---|---|---|---|---|---|---|---|---|---|---|---|---|
| | Number of establishments | Number of employees | Sales (mil dol) | Average payroll (mil dol) | Number of establishments | Number of employees | Sales (mil dol) | Average payroll (mil dol) | Number of establishments | Number of employees | Sales (mil dol) | Annual payroll (mil dol) |
| | 147 | 148 | 149 | 150 | 151 | 152 | 153 | 154 | 155 | 156 | 157 | 158 |
| GEORGIA—Cont'd | | | | | | | | | | | | |
| Effingham | D | D | D | D | 24 | 1,669 | 832.2 | 106.9 | 59 | 809 | 40.2 | 9.9 |
| Elbert | 20 | 67 | 5 | 1.6 | 83 | 1,803 | 311.1 | 77.6 | D | D | D | 4.1 |
| Emanuel | 21 | 92 | 13 | 4.1 | 33 | 1,827 | 790.6 | 69.5 | 37 | 422 | 23.1 | 5.6 |
| Evans | 15 | 59 | 12 | 2.5 | 13 | 1,838 | 471.7 | 64.4 | 12 | 204 | 8.3 | 2.4 |
| Fannin | 43 | 219 | 19 | 7.5 | 28 | 290 | 39.8 | 9.6 | 64 | 914 | 52.2 | 13.9 |
| Fayette | D | D | D | D | 100 | 2,836 | 1,269.6 | 144.8 | 258 | 6,095 | 325.9 | 94.9 |
| Floyd | 178 | 808 | 123 | 31.6 | 98 | 6,153 | 3,872.4 | 358.9 | 202 | 4,241 | 227.3 | 64.8 |
| Forsyth | 1,325 | 6,253 | 1,142 | 388.1 | 231 | 8,010 | 2,375.0 | 375.9 | 349 | 6,999 | 364.8 | 101.5 |
| Franklin | 28 | 150 | 13 | 4.5 | 32 | 1,570 | 419.5 | 69.5 | 44 | 755 | 43.1 | 10.6 |
| Fulton | 7,614 | 110,359 | 28,925 | 10,416.8 | 602 | 19,877 | 10,427.0 | 1,149.7 | 3,407 | 85,340 | 6,380.6 | 1,745.6 |
| Gilmer | D | D | D | D | 31 | 1,252 | 346.1 | 40.3 | 63 | 922 | 48.1 | 13.4 |
| Glascock | NA | NA | NA | NA | NA | NA | NA | NA | 3 | 33 | 0.7 | 0.2 |
| Glynn | D | D | D | D | 50 | 2,026 | 969.5 | 140.7 | 287 | 7,637 | 594.2 | 165.7 |
| Gordon | 50 | 357 | 34 | 10.2 | 94 | 6,483 | 2,280.3 | 303.9 | 91 | 1,572 | 89.2 | 22.7 |
| Grady | 17 | 79 | 7 | 2.2 | 19 | 981 | 242.5 | 37.3 | 30 | 345 | 19.4 | 4.2 |
| Greene | 42 | 268 | 55 | 22.9 | 12 | 449 | 516.4 | 23.5 | 37 | 1,639 | 197.7 | 46.7 |
| Gwinnett | 3,312 | 28,939 | 5,504 | 2,013.5 | 705 | 22,058 | 7,837.2 | 1,211.4 | 1,942 | 31,382 | 1,910.3 | 483.3 |
| Habersham | 66 | 274 | 42 | 10.7 | 54 | 4,158 | 1,191.9 | 171.5 | 72 | 1,413 | 71.0 | 19.7 |
| Hall | 410 | 1,919 | 331 | 107.0 | 223 | 20,766 | 8,353.0 | 932.3 | 312 | 5,862 | 338.5 | 93.2 |
| Hancock | NA | NA | NA | NA | D | D | D | D | D | D | D | D |
| Haralson | 30 | 78 | 8 | 2.7 | 31 | 1,553 | 957.8 | 99.9 | 41 | 513 | 27.8 | 6.3 |
| Harris | 37 | 76 | 12 | 3.1 | 14 | 1,204 | 436.9 | 40.7 | D | D | D | 8.8 |
| Hart | 40 | 167 | 21 | 7.9 | 28 | 2,582 | 621.8 | 114.5 | D | D | D | D |
| Heard | D | D | D | D | 11 | 531 | 193.2 | 23.3 | D | D | D | D |
| Henry | 314 | 1,460 | 257 | 69.3 | 66 | 2,729 | 1,962.3 | 149.3 | 379 | 7,164 | 396.5 | 105.2 |
| Houston | D | D | D | D | 56 | 4,714 | 2,558.9 | 200.4 | 329 | 7,252 | 347.1 | 92.4 |
| Irwin | D | D | D | D | NA | NA | NA | NA | 5 | 68 | 2.1 | 0.5 |
| Jackson | 101 | 400 | 59 | 23.0 | 76 | 6,367 | 2,925.5 | 311.5 | 78 | 1,318 | 68.9 | 18.3 |
| Jasper | 11 | 25 | 3 | 1.0 | 15 | 433 | 141.6 | 20.2 | 11 | 146 | 6.2 | 2.0 |
| Jeff Davis | 12 | 32 | 3 | 1.4 | 28 | 1,324 | 323.2 | 54.2 | 24 | 335 | 16.2 | 4.3 |
| Jefferson | 11 | 62 | 7 | 2.0 | 25 | 979 | 277.1 | 50.0 | 31 | 303 | 15.3 | 3.4 |
| Jenkins | 6 | 13 | 1 | 0.5 | 5 | 20 | 20.6 | 1.6 | 14 | 145 | 7.5 | 1.7 |
| Johnson | 6 | 12 | 1 | 0.3 | 5 | 83 | 17.1 | 3.6 | 9 | 46 | 1.6 | 0.4 |
| Jones | 13 | 68 | 5 | 1.6 | D | 38 | D | 2.2 | D | D | D | D |
| Lamar | 15 | 67 | 6 | 1.9 | 12 | 550 | 299.6 | 29.5 | 43 | 524 | 21.3 | 4.9 |
| Lanier | 5 | 21 | 1 | 0.3 | D | D | D | D | D | D | D | D |
| Laurens | 59 | 386 | 55 | 20.3 | 41 | 2,253 | 925.2 | 116.4 | 104 | 2,025 | 101.1 | 26.7 |
| Lee | 30 | 254 | 17 | 6.4 | 14 | 423 | 100.4 | 14.8 | 25 | 322 | 19.4 | 4.7 |
| Liberty | D | D | D | D | 19 | 1,904 | 1,276.2 | 123.5 | 114 | 2,078 | 104.3 | 26.4 |
| Lincoln | D | D | D | D | 4 | 77 | 19.1 | 3.3 | 10 | 90 | 3.7 | 1.1 |
| Long | NA | NA | NA | NA | NA | NA | NA | NA | D | D | D | D |
| Lowndes | D | D | D | D | 91 | 3,150 | 2,777.5 | 162.7 | 276 | 5,921 | 307.6 | 81.4 |
| Lumpkin | 66 | 192 | 23 | 8.2 | 30 | 537 | 108.8 | 23.6 | 63 | 1,165 | 63.8 | 18.4 |
| McDuffie | D | D | 11 | D | 28 | 1,595 | 585.5 | 69.3 | 41 | 679 | 34.6 | 8.1 |
| McIntosh | 14 | 39 | 4 | 1.3 | 4 | 8 | 2.3 | 0.6 | 31 | 390 | 22.9 | 5.7 |
| Macon | 8 | 19 | 2 | 0.5 | 13 | 514 | 338.4 | 39.1 | 15 | 177 | 8.8 | 2.1 |
| Madison | 26 | 87 | 9 | 2.7 | 23 | 166 | 23.0 | 6.8 | 23 | 146 | 9.4 | 2.5 |
| Marion | 4 | 10 | 1 | 0.4 | D | D | D | D | D | D | D | 0.4 |
| Meriwether | 11 | 43 | 6 | 1.7 | 17 | 627 | 208.1 | 29.7 | 27 | 239 | 11.3 | 2.8 |
| Miller | D | D | 4 | D | 3 | 11 | 2.4 | 0.6 | 8 | D | 3.5 | D |
| Mitchell | 22 | 208 | 21 | 9.1 | 13 | 2,176 | 881.4 | 78.2 | 29 | 360 | 20.1 | 5.6 |
| Monroe | 47 | 208 | 27 | 10.3 | 15 | 584 | 164.0 | 19.3 | 42 | 643 | 32.8 | 8.9 |
| Montgomery | 6 | 20 | 2 | 0.6 | 8 | 156 | 50.9 | 5.9 | 9 | 79 | 3.5 | 0.9 |
| Morgan | 46 | 282 | 52 | 15.7 | 27 | 1,315 | 428.5 | 67.9 | 57 | 884 | 47.0 | 14.6 |
| Murray | 16 | 76 | 18 | 6.6 | 57 | 3,485 | 1,012.0 | 132.7 | 43 | 622 | 29.5 | 7.8 |
| Muscogee | D | D | D | D | 116 | 6,683 | 2,208.9 | 353.0 | 474 | 10,818 | 574.5 | 162.6 |
| Newton | 102 | 405 | 54 | 15.6 | 78 | 3,694 | 2,412.4 | 199.5 | 133 | 2,251 | 131.8 | 36.4 |
| Oconee | 169 | 915 | 122 | 44.3 | 26 | 2,258 | 584.6 | 117.7 | 79 | 1,749 | 78.1 | 21.7 |
| Oglethorpe | 12 | 32 | 3 | 0.8 | 10 | 81 | 9.2 | 2.8 | 6 | 142 | 4.9 | 2.2 |
| Paulding | 203 | 756 | 146 | 36.8 | 50 | 801 | 283.0 | 42.6 | 158 | 3,743 | 205.5 | 55.7 |
| Peach | 28 | 134 | 16 | 5.8 | D | D | D | D | D | D | D | 11.4 |
| Pickens | 80 | 180 | 25 | 6.9 | 42 | 776 | 139.0 | 36.4 | 49 | 909 | 52.8 | 15.1 |
| Pierce | 17 | 48 | 4 | 1.1 | 13 | 436 | 181.6 | 20.0 | 26 | 340 | 16.1 | 4.4 |
| Pike | 25 | 66 | 8 | 2.3 | 10 | 195 | 85.8 | 6.9 | D | D | D | D |
| Polk | 31 | 108 | 12 | 3.0 | 29 | 3,841 | 1,100.6 | 171.4 | 66 | 1,223 | 61.1 | 16.3 |
| Pulaski | 13 | 38 | 5 | 2.1 | NA | NA | NA | NA | D | D | D | 3.4 |

Items 147—158

## Table B. States and Counties — Health Care and Social Assistance, Other Services, Nonemployer Businesses, and Residential Construction

| STATE County | Health care and social assistance, 2017 | | | | Other services, 2017 | | | | Nonemployer businesses, 2018 | | Value of residential construction authorized by building permits, 2020 | |
|---|---|---|---|---|---|---|---|---|---|---|---|---|
| | Number of establishments | Number of employees | Receipts (mil dol) | Annual payroll (mil dol) | Number of establishments | Number of employees | Receipts (mil dol) | Annual payroll (mil dol) | Number | Receipts (mil dol) | New construction ($1,000) | Number of housing units |
| | 159 | 160 | 161 | 162 | 163 | 164 | 165 | 166 | 167 | 168 | 169 | 170 |
| GEORGIA—Cont'd | | | | | | | | | | | | |
| Effingham | 70 | 890 | 95.5 | 38.6 | 56 | 503 | 63.1 | 21.1 | 4,143 | 165.9 | 172,262 | 756 |
| Elbert | 31 | 512 | 38.0 | 16.0 | D | D | D | D | 1,417 | 55.9 | 10,116 | 83 |
| Emanuel | 44 | 870 | 66.8 | 29.9 | 24 | 96 | 12.2 | 2.6 | 1,797 | 75.5 | 423 | 4 |
| Evans | 16 | 426 | 31.8 | 14.3 | 14 | 88 | 8.6 | 3.0 | 731 | 21.3 | 2,886 | 18 |
| Fannin | 61 | 902 | 112.4 | 40.0 | 33 | 218 | 19.0 | 5.5 | 2,929 | 162.0 | 79,792 | 382 |
| Fayette | 430 | 6,561 | 880.0 | 327.6 | 249 | 1,732 | 155.9 | 51.4 | 11,534 | 540.8 | 175,752 | 564 |
| Floyd | 322 | 8,395 | 1,147.7 | 397.0 | 115 | 808 | 79.6 | 23.5 | 6,962 | 270.4 | 46,376 | 282 |
| Forsyth | 562 | 9,270 | 1,334.7 | 495.4 | 388 | 2,365 | 298.0 | 88.4 | 22,850 | 1,184.0 | 413,289 | 2,485 |
| Franklin | 38 | 626 | 54.5 | 23.4 | 29 | 189 | 35.7 | 7.1 | 1,706 | 73.4 | 16,707 | 119 |
| Fulton | 3,950 | 86,258 | 14,367.1 | 4,845.8 | 2,372 | 22,826 | 5,613.3 | 900.1 | 116,982 | 6,137.5 | 985,112 | 4,289 |
| Gilmer | 48 | 507 | 41.0 | 16.8 | 36 | 196 | 19.8 | 5.3 | 2,917 | 130.6 | 49,006 | 327 |
| Glascock | D | D | D | D | 3 | 5 | 0.4 | 0.1 | 192 | 6.4 | NA | NA |
| Glynn | 257 | 5,461 | 670.2 | 260.3 | 147 | 913 | 99.2 | 23.3 | 7,136 | 335.6 | 147,110 | 439 |
| Gordon | 78 | 2,313 | 247.0 | 90.7 | 45 | 234 | 24.6 | 6.9 | 3,543 | 177.8 | 41,981 | 256 |
| Grady | 35 | 504 | 57.1 | 18.0 | 30 | 115 | 12.2 | 2.7 | 1,557 | 60.4 | 5,919 | 32 |
| Greene | 45 | 745 | 70.3 | 25.9 | 29 | 177 | 13.7 | 3.9 | 1,693 | 98.6 | 151,990 | 243 |
| Gwinnett | 2,114 | 31,297 | 3,716.0 | 1,507.7 | 1,565 | 8,763 | 1,068.5 | 300.4 | 116,146 | 4,968.9 | 1,054,068 | 4,551 |
| Habersham | 91 | 1,311 | 120.0 | 52.2 | 62 | 252 | 24.6 | 7.2 | 3,281 | 136.0 | 45,204 | 262 |
| Hall | 488 | 13,913 | 1,965.5 | 721.0 | 282 | 1,514 | 177.5 | 51.0 | 16,827 | 799.5 | 347,411 | 1,706 |
| Hancock | 7 | 136 | 9.6 | 4.6 | 5 | 10 | 0.9 | 0.3 | 622 | 14.9 | 2,968 | 17 |
| Haralson | 43 | 583 | 184.1 | 74.1 | D | D | D | D | 2,201 | 93.0 | 19,310 | 83 |
| Harris | 30 | 246 | 16.9 | 7.0 | D | D | D | 1.0 | 2,631 | 121.0 | 57,374 | 237 |
| Hart | 33 | 340 | 33.7 | 13.9 | D | D | D | D | 1,891 | 72.6 | 35,576 | 181 |
| Heard | D | D | D | D | D | D | D | 0.6 | 791 | 31.3 | 7,132 | 32 |
| Henry | 477 | 8,458 | 1,079.1 | 432.0 | 256 | 1,186 | 128.0 | 35.1 | 22,449 | 786.1 | 451,888 | 1,867 |
| Houston | 341 | 7,395 | 750.1 | 269.9 | 166 | 976 | 86.8 | 25.7 | 10,494 | 342.9 | 236,202 | 1,219 |
| Irwin | 11 | 343 | 24.1 | 9.4 | D | D | D | D | 636 | 23.8 | 2,352 | 27 |
| Jackson | 112 | 1,473 | 102.8 | 45.2 | 84 | 513 | 37.8 | 21.0 | 5,949 | 246.6 | 261,117 | 1,422 |
| Jasper | D | D | D | D | D | D | D | D | 1,177 | 48.6 | 18,876 | 107 |
| Jeff Davis | 23 | 327 | 21.6 | 8.9 | D | D | D | D | 938 | 33.0 | 250 | 3 |
| Jefferson | 23 | 480 | 30.7 | 13.8 | D | D | D | D | 1,404 | 44.1 | 0 | 0 |
| Jenkins | 12 | 258 | 18.7 | 7.2 | 6 | 15 | 5.2 | 0.6 | 541 | 19.8 | 2,320 | 23 |
| Johnson | 14 | 206 | 15.5 | 7.1 | D | D | D | D | 556 | 21.6 | 0 | 0 |
| Jones | 38 | 446 | 33.2 | 13.4 | 22 | 58 | 9.8 | 1.7 | 2,026 | 76.5 | 17,577 | 114 |
| Lamar | D | D | D | 8.5 | 16 | 173 | 16.4 | 5.3 | 1,216 | 42.6 | 27,000 | 127 |
| Lanier | D | D | D | D | NA | NA | NA | NA | 606 | 24.8 | 4,797 | 45 |
| Laurens | 152 | 3,081 | 498.3 | 184.0 | D | D | D | D | 3,929 | 180.6 | 4,419 | 25 |
| Lee | 41 | 379 | 33.3 | 13.7 | 28 | 234 | 27.1 | 8.2 | 2,308 | 88.4 | 14,393 | 172 |
| Liberty | 91 | 2,157 | 259.7 | 104.0 | D | D | D | D | 3,214 | 92.7 | 84,423 | 421 |
| Lincoln | 6 | 43 | 3.3 | 1.7 | D | D | D | D | 627 | 24.5 | 9,888 | 44 |
| Long | 5 | 25 | 2.7 | 1.1 | D | D | D | D | 790 | 24.8 | 69,939 | 300 |
| Lowndes | 459 | 7,871 | 901.8 | 327.8 | 147 | 670 | 70.0 | 17.4 | 7,548 | 376.7 | 224,579 | 1,276 |
| Lumpkin | 56 | 725 | 96.7 | 26.1 | D | D | D | D | 2,613 | 109.2 | 47,635 | 228 |
| McDuffie | 51 | 955 | 73.9 | 29.9 | D | D | 22.4 | D | 1,754 | 53.1 | 11,116 | 53 |
| McIntosh | D | D | D | D | 13 | 71 | 5.8 | 1.4 | 968 | 37.6 | 12,193 | 71 |
| Macon | 21 | 463 | 31.8 | 12.5 | 11 | 50 | 5.3 | 1.4 | 765 | 23.0 | 2,300 | 11 |
| Madison | 23 | 201 | 13.0 | 5.4 | D | D | D | 0.8 | 2,348 | 90.1 | 53,514 | 235 |
| Marion | D | D | D | D | D | D | D | D | 447 | 17.0 | 3,174 | 16 |
| Meriwether | 23 | 965 | 74.7 | 30.4 | D | D | D | D | 1,618 | 53.7 | 12,586 | 71 |
| Miller | 15 | 572 | 57.3 | 22.4 | D | D | 1.9 | D | 334 | 12.1 | 1,578 | 6 |
| Mitchell | 34 | 556 | 46.4 | 17.9 | 24 | 80 | 8.3 | 2.1 | 1,443 | 53.3 | 1,266 | 6 |
| Monroe | D | D | D | D | 36 | 141 | 19.5 | 5.6 | 2,243 | 99.5 | 46,063 | 203 |
| Montgomery | D | D | D | D | D | D | D | D | 569 | 19.5 | 2,703 | 20 |
| Morgan | D | D | D | D | 30 | 126 | 14.0 | 3.3 | 2,103 | 94.7 | 95,348 | 449 |
| Murray | D | D | D | D | 27 | 109 | 13.5 | 3.3 | 1,919 | 81.4 | 14,687 | 63 |
| Muscogee | 666 | 15,924 | 1,829.0 | 701.9 | D | D | D | 62.2 | 13,277 | 454.5 | 155,057 | 1,342 |
| Newton | 137 | 2,426 | 252.1 | 109.5 | 79 | 318 | 35.4 | 9.5 | 10,350 | 321.1 | 151,206 | 857 |
| Oconee | 161 | 1,830 | 188.6 | 81.5 | 55 | 445 | 65.4 | 18.3 | 4,301 | 229.5 | 112,724 | 281 |
| Oglethorpe | 10 | 117 | 7.5 | 3.4 | D | D | D | 0.7 | 1,096 | 40.6 | 1,005 | 67 |
| Paulding | 177 | 3,538 | 427.9 | 167.3 | 153 | 652 | 59.5 | 16.6 | 13,997 | 495.5 | 188,939 | 1,893 |
| Peach | 41 | 565 | 46.9 | 16.8 | 25 | 127 | 17.3 | 3.7 | 1,898 | 70.2 | 15,258 | 88 |
| Pickens | 69 | 1,327 | 173.6 | 69.4 | 51 | 433 | 43.4 | 12.9 | 3,259 | 149.4 | 64,174 | 230 |
| Pierce | 22 | 194 | 10.0 | 4.5 | 18 | 50 | 4.6 | 1.1 | 1,248 | 47.2 | 6,290 | 64 |
| Pike | D | D | D | 7.9 | D | D | D | 1.0 | 1,593 | 67.0 | 32,517 | 139 |
| Polk | 60 | 1,086 | 106.2 | 40.0 | 39 | 324 | 33.0 | 9.1 | 2,651 | 95.0 | 1,136 | 8 |
| Pulaski | 24 | 866 | 69.9 | 37.2 | 8 | 23 | 1.7 | 0.5 | 701 | 21.1 | 3,159 | 23 |

# Table B. States and Counties — Government Employment and Payroll, and Local Government Finances

| STATE County | Full-time equivalent employees | March payroll (dollars) | March payroll (percent of total) Administration, judicial, and legal | Police and corrections | Fire protection | Highways and transportation | Health and welfare | Natural resources and utilities | Education and libraries | Local government finances, 2017 General revenue Total (mil dol) | Inter-govern-mental (mil dol) | Taxes Total (mil dol) | Taxes Per capita[1] (dollars) Total | Per capita[1] Property |
|---|---|---|---|---|---|---|---|---|---|---|---|---|---|---|
| | 171 | 172 | 173 | 174 | 175 | 176 | 177 | 178 | 179 | 180 | 181 | 182 | 183 | 184 |
| GEORGIA—Cont'd | | | | | | | | | | | | | | |
| Effingham | 2,530 | 9,080,471 | 3.6 | 7.9 | 3.3 | 0.3 | 16.3 | 1.7 | 65.9 | 309.6 | 84.7 | 77.1 | 1,284 | 814 |
| Elbert | 896 | 2,970,610 | 6.2 | 7.8 | 2.0 | 1.7 | 20.2 | 7.3 | 52.5 | 75.9 | 26.0 | 23.1 | 1,209 | 835 |
| Emanuel | 1,261 | 4,150,130 | 4.9 | 7.6 | 1.3 | 3.5 | 30.0 | 2.6 | 48.4 | 102.6 | 38.6 | 24.9 | 1,104 | 674 |
| Evans | 422 | 1,286,946 | 7.8 | 7.6 | 0.9 | 3.0 | 5.2 | 6.0 | 68.8 | 45.3 | 17.6 | 11.0 | 1,024 | 575 |
| Fannin | 769 | 2,585,142 | 6.7 | 7.4 | 1.9 | 5.6 | 5.3 | 4.4 | 68.0 | 70.6 | 21.5 | 40.9 | 1,613 | 900 |
| Fayette | 4,117 | 16,713,945 | 6.0 | 9.2 | 7.0 | 1.7 | 0.2 | 2.8 | 69.9 | 390.4 | 113.5 | 204.2 | 1,814 | 1,311 |
| Floyd | 3,978 | 13,144,682 | 5.5 | 10.4 | 4.8 | 3.7 | 2.2 | 6.5 | 65.2 | 377.8 | 154.9 | 149.8 | 1,537 | 984 |
| Forsyth | 7,135 | 29,839,562 | 5.0 | 7.0 | 3.1 | 1.1 | 0.7 | 3.7 | 77.7 | 787.1 | 261.5 | 401.3 | 1,753 | 1,192 |
| Franklin | 813 | 2,679,239 | 7.3 | 11.5 | 0.2 | 2.5 | 5.3 | 5.3 | 66.3 | 69.1 | 29.8 | 28.8 | 1,259 | 769 |
| Fulton | 42,801 | 177,707,903 | 10.3 | 12.6 | 4.9 | 15.0 | 2.3 | 8.6 | 45.6 | 5704.3 | 1290.5 | 2829.1 | 2,723 | 1,993 |
| Gilmer | 1,014 | 3,339,978 | 6.6 | 11.8 | 6.5 | 2.2 | 0.4 | 4.9 | 65.8 | 82.8 | 30.0 | 41.8 | 1,374 | 907 |
| Glascock | 155 | 449,339 | 13.2 | 0.0 | 5.4 | 4.1 | 0.6 | 1.9 | 72.9 | 9.2 | 5.0 | 3.4 | 1,121 | 779 |
| Glynn | 5,449 | 23,453,893 | 3.7 | 5.6 | 3.0 | 1.4 | 51.6 | 2.8 | 30.6 | 646.3 | 95.8 | 166.8 | 1,970 | 1,270 |
| Gordon | 1,956 | 7,118,689 | 6.8 | 9.3 | 4.7 | 2.8 | 0.9 | 5.4 | 67.0 | 188.7 | 76.4 | 80.5 | 1,408 | 848 |
| Grady | 1,026 | 3,285,875 | 5.2 | 8.1 | 1.9 | 2.9 | 0.9 | 6.2 | 68.7 | 77.2 | 39.9 | 24.9 | 1,004 | 660 |
| Greene | 705 | 2,544,629 | 9.2 | 9.3 | 0.0 | 1.9 | 5.3 | 3.1 | 68.1 | 70.6 | 17.0 | 44.4 | 2,584 | 1,740 |
| Gwinnett | 29,474 | 120,526,583 | 6.0 | 7.6 | 4.0 | 0.7 | 1.6 | 4.1 | 75.1 | 3209.4 | 1189.1 | 1490.4 | 1,623 | 1,143 |
| Habersham | 2,098 | 7,199,487 | 4.7 | 5.7 | 2.3 | 1.5 | 28.7 | 4.0 | 50.7 | 178.4 | 56.4 | 53.7 | 1,206 | 708 |
| Hall | 6,731 | 25,935,061 | 6.3 | 9.9 | 7.4 | 1.6 | 4.5 | 5.5 | 63.2 | 690.5 | 254.6 | 301.9 | 1,518 | 931 |
| Hancock | 373 | 1,041,916 | 6.1 | 11.4 | 0.0 | 2.9 | 4.6 | 5.2 | 67.2 | 36.2 | 10.8 | 16.3 | 1,906 | 1,560 |
| Haralson | 1,199 | 4,193,572 | 7.0 | 8.5 | 4.9 | 2.4 | 3.1 | 4.4 | 69.1 | 96.5 | 47.8 | 37.0 | 1,264 | 843 |
| Harris | 1,107 | 3,723,319 | 4.3 | 10.2 | 0.0 | 1.9 | 3.4 | 5.1 | 72.6 | 87.1 | 31.4 | 43.6 | 1,286 | 931 |
| Hart | 810 | 2,529,596 | 5.8 | 8.0 | 1.1 | 3.8 | 6.2 | 4.4 | 69.2 | 70.8 | 30.6 | 30.2 | 1,175 | 769 |
| Heard | 481 | 1,604,121 | 7.1 | 7.8 | 9.0 | 2.5 | 2.4 | 4.7 | 63.9 | 42.1 | 13.8 | 24.7 | 2,105 | 1,020 |
| Henry | 8,159 | 29,035,197 | 7.8 | 7.6 | 5.5 | 1.3 | 1.3 | 3.9 | 69.5 | 734.4 | 268.8 | 359.7 | 1,597 | 1,063 |
| Houston | 5,858 | 21,180,489 | 4.8 | 8.7 | 3.3 | 1.8 | 1.4 | 3.6 | 73.8 | 499.9 | 208.4 | 210.1 | 1,371 | 786 |
| Irwin | 532 | 1,604,996 | 6.2 | 3.8 | 1.0 | 1.9 | 21.1 | 1.0 | 58.1 | 68.2 | 16.3 | 9.7 | 1,039 | 768 |
| Jackson | 2,489 | 8,810,842 | 5.3 | 10.7 | 0.4 | 1.8 | 2.7 | 4.9 | 70.6 | 239.9 | 80.4 | 118.3 | 1,749 | 1,150 |
| Jasper | 495 | 1,474,019 | 6.3 | 7.0 | 0.8 | 3.9 | 4.2 | 4.1 | 73.1 | 50.0 | 18.7 | 17.1 | 1,232 | 974 |
| Jeff Davis | 678 | 2,258,153 | 5.7 | 5.5 | 2.2 | 2.0 | 20.6 | 3.4 | 59.6 | 47.0 | 23.8 | 17.4 | 1,157 | 586 |
| Jefferson | 787 | 2,446,812 | 1.8 | 11.2 | 1.8 | 3.1 | 19.6 | 4.3 | 56.1 | 62.2 | 24.2 | 20.8 | 1,328 | 875 |
| Jenkins | 336 | 1,040,123 | 7.2 | 9.2 | 2.4 | 4.7 | 3.6 | 5.3 | 67.0 | 32.6 | 19.0 | 9.6 | 1,089 | 691 |
| Johnson | 458 | 1,259,519 | 5.0 | 6.4 | 1.0 | 1.5 | 5.3 | 1.3 | 79.2 | 23.5 | 12.1 | 7.6 | 777 | 502 |
| Jones | 1,120 | 3,183,020 | 5.1 | 7.9 | 0.3 | 3.1 | 0.3 | 4.3 | 78.9 | 78.9 | 38.3 | 32.9 | 1,154 | 879 |
| Lamar | 619 | 1,781,815 | 9.3 | 13.6 | 1.7 | 3.3 | 1.3 | 6.4 | 61.9 | 50.0 | 20.3 | 20.0 | 1,077 | 784 |
| Lanier | 268 | 837,866 | 6.2 | 5.8 | 0.1 | 3.0 | 0.4 | 1.4 | 82.5 | 24.9 | 15.2 | 8.3 | 796 | 590 |
| Laurens | 2,010 | 7,072,670 | 5.6 | 8.0 | 2.4 | 2.6 | 6.2 | 4.7 | 68.5 | 181.6 | 91.0 | 62.7 | 1,324 | 703 |
| Lee | 1,242 | 3,793,199 | 5.8 | 8.2 | 2.9 | 2.2 | 3.3 | 1.8 | 73.6 | 94.7 | 39.7 | 41.9 | 1,423 | 1,002 |
| Liberty | 2,654 | 8,935,784 | 5.5 | 9.6 | 2.1 | 0.9 | 19.3 | 1.8 | 56.9 | 235.8 | 92.9 | 64.3 | 1,044 | 724 |
| Lincoln | 331 | 996,570 | 9.2 | 9.9 | 0.0 | 2.7 | 7.4 | 2.8 | 65.3 | 28.4 | 11.5 | 11.5 | 1,461 | 1,059 |
| Long | 530 | 1,540,782 | 5.8 | 5.9 | 1.3 | 2.4 | 0.0 | 2.0 | 82.6 | 50.4 | 34.5 | 12.8 | 679 | 534 |
| Lowndes | 6,117 | 22,966,286 | 3.4 | 6.7 | 2.0 | 1.2 | 45.2 | 3.0 | 38.1 | 682.2 | 148.6 | 151.5 | 1,313 | 833 |
| Lumpkin | 901 | 2,978,044 | 9.0 | 9.7 | 5.0 | 2.6 | 0.3 | 4.7 | 67.0 | 71.5 | 24.4 | 36.5 | 1,313 | 833 |
| McDuffie | 951 | 2,939,097 | 6.2 | 7.6 | 4.1 | 1.8 | 2.2 | 5.4 | 70.7 | 71.6 | 33.9 | 28.4 | 1,113 | 737 |
| McIntosh | 456 | 1,345,123 | 11.3 | 17.8 | 0.0 | 3.1 | 4.1 | 4.9 | 56.7 | 34.8 | 10.9 | 16.8 | 1,322 | 763 |
| Macon | 396 | 1,181,723 | 8.9 | 11.4 | 0.9 | 3.8 | 4.8 | 6.8 | 62.4 | 34.1 | 12.7 | 15.8 | 1,191 | 842 |
| Madison | 1,010 | 3,188,661 | 5.5 | 7.3 | 0.0 | 1.9 | 4.0 | 1.2 | 77.9 | 76.6 | 42.4 | 25.0 | 1,188 | 881 |
| Marion | 270 | 959,139 | 6.9 | 7.5 | 0.0 | 4.4 | 3.8 | 3.8 | 72.7 | 21.1 | 10.8 | 7.7 | 854 | 611 |
| Meriwether | 805 | 2,343,467 | 7.3 | 11.0 | 1.1 | 3.7 | 6.2 | 2.5 | 65.8 | 61.4 | 26.6 | 24.5 | 914 | 622 |
| Miller | 730 | 2,625,098 | 4.8 | 3.8 | 0.2 | 0.8 | 63.4 | 2.2 | 24.3 | 70.1 | 44.9 | 9.3 | 1,165 | 855 |
| Mitchell | 956 | 2,922,809 | 8.0 | 11.9 | 2.7 | 2.3 | 3.1 | 4.4 | 64.5 | 80.6 | 33.4 | 30.4 | 1,606 | 1,250 |
| Monroe | 1,099 | 3,677,674 | 4.3 | 11.0 | 1.8 | 3.3 | 15.3 | 3.5 | 58.0 | 65.8 | 21.0 | 26.6 | 1,363 | 983 |
| Montgomery | 232 | 670,711 | 6.8 | 6.5 | 0.5 | 3.4 | 2.2 | 2.5 | 76.4 | 21.1 | 11.2 | 7.4 | 977 | 779 |
| Morgan | 924 | 3,191,062 | 7.4 | 6.4 | 0.9 | 2.5 | 17.6 | 6.5 | 56.9 | 84.7 | 24.5 | 36.0 | 825 | 549 |
| Murray | 1,158 | 4,056,208 | 4.4 | 7.6 | 3.3 | 2.4 | 0.9 | 1.8 | 74.6 | 99.6 | 55.2 | 33.1 | 1,960 | 1,286 |
| Muscogee | 8,905 | 30,532,915 | 4.4 | 12.3 | 5.0 | 2.3 | 10.2 | 7.3 | 55.0 | 735.2 | 271.0 | 279.8 | 833 | 437 |
| Newton | 3,939 | 13,566,867 | 6.2 | 9.2 | 3.9 | 1.5 | 0.8 | 5.3 | 70.7 | 407.0 | 153.8 | 127.3 | 1,443 | 978 |
| Oconee | 1,258 | 4,271,841 | 7.0 | 7.4 | 0.4 | 2.2 | 0.4 | 5.1 | 73.5 | 124.4 | 43.1 | 65.8 | 1,181 | 787 |
| Oglethorpe | 444 | 1,472,805 | 4.5 | 9.3 | 0.0 | 4.3 | 3.4 | 3.2 | 75.3 | 33.8 | 15.7 | 14.4 | 1,725 | 1,077 |
| Paulding | 4,614 | 16,994,666 | 3.6 | 7.4 | 3.1 | 1.4 | 0.5 | 2.7 | 80.5 | 418.6 | 199.9 | 173.4 | 967 | 746 |
| | | | | | | | | | | | | | 1,087 | 714 |
| Peach | 1,076 | 3,658,633 | 7.8 | 11.0 | 3.6 | 0.7 | 2.3 | 7.2 | 59.7 | 107.4 | 25.9 | 37.8 | 1,401 | 910 |
| Pickens | 1,035 | 3,467,071 | 11.4 | 10.9 | 8.0 | 2.5 | 0.7 | 4.3 | 61.7 | 96.1 | 31.4 | 50.0 | 1,586 | 1,085 |
| Pierce | 645 | 2,462,438 | 6.8 | 4.5 | 0.2 | 1.7 | 3.4 | 1.9 | 77.6 | 52.7 | 28.4 | 18.6 | 967 | 633 |
| Pike | 616 | 1,844,937 | 6.8 | 10.0 | 0.0 | 3.6 | 0.2 | 1.7 | 76.4 | 43.0 | 21.4 | 18.5 | 1,017 | 780 |
| Polk | 1,416 | 4,844,573 | 4.5 | 11.6 | 2.6 | 2.3 | 0.7 | 3.6 | 71.5 | 129.6 | 63.4 | 45.6 | 1,090 | 687 |
| Pulaski | 330 | 1,035,204 | 12.7 | 7.8 | 3.8 | 4.0 | 1.3 | 5.4 | 65.0 | 24.6 | 11.1 | 10.7 | 957 | 655 |

1. Based on the resident population estimated as of July 1 of the year shown.

| STATE County | Local government finances, 2017 (cont.) | | | | | | | Debt outstanding | | Government employment, 2019 | | | Individual income tax returns, 2018 | | |
|---|---|---|---|---|---|---|---|---|---|---|---|---|---|---|---|
| | Direct general expenditure | | | | | | | | | | | | | | |
| | | | Percent of total for: | | | | | | | | | | | Mean adjusted gross income | Mean income tax |
| | Total (mil dol) | Per capita[1] (dollars) | Education | Health and hospitals | Police protection | Public welfare | Highways | Total (mil dol) | Per capita[1] (dollars) | Federal civilian | Federal military | State and local | Number of returns | | |
| | 185 | 186 | 187 | 188 | 189 | 190 | 191 | 192 | 193 | 194 | 195 | 196 | 197 | 198 | 199 |
| GEORGIA—Cont'd | | | | | | | | | | | | | | | |
| Effingham | 247 | 4,109 | 51.8 | 22.6 | 4.0 | 0.0 | 2.9 | 244.8 | 4,074 | 74 | 170 | 3,067 | 27,280 | 61,353 | 5,161 |
| Elbert | 75 | 3,911 | 43.2 | 19.6 | 4.2 | 0.7 | 2.3 | 11.1 | 580 | 123 | 51 | 1,152 | 8,060 | 42,326 | 3,093 |
| Emanuel | 103 | 4,578 | 42.1 | 32.1 | 2.8 | 0.0 | 7.5 | 21.3 | 947 | 60 | 57 | 2,009 | 8,690 | 42,400 | 2,987 |
| Evans | 46 | 4,311 | 41.0 | 26.8 | 4.7 | 0.0 | 7.2 | 11.8 | 1,097 | 44 | 27 | 618 | 4,210 | 46,167 | 4,221 |
| Fannin | 66 | 2,589 | 58.3 | 3.9 | 4.8 | 0.1 | 8.7 | 23.4 | 924 | 58 | 70 | 930 | 10,890 | 54,414 | 5,405 |
| Fayette | 390 | 3,460 | 62.8 | 1.0 | 6.8 | 0.0 | 5.4 | 197.3 | 1,753 | 543 | 352 | 4,728 | 55,040 | 98,060 | 13,059 |
| Floyd | 407 | 4,176 | 48.2 | 8.6 | 5.5 | 0.1 | 2.8 | 185.7 | 1,904 | 215 | 255 | 5,467 | 40,330 | 56,323 | 5,283 |
| Forsyth | 809 | 3,532 | 63.1 | 0.4 | 4.5 | 0.0 | 2.0 | 715.3 | 3,125 | 237 | 651 | 8,148 | 105,503 | 111,292 | 15,250 |
| Franklin | 65 | 2,852 | 59.6 | 3.3 | 10.4 | 0.0 | 5.7 | 33.0 | 1,444 | 49 | 60 | 1,074 | 9,350 | 46,094 | 3,540 |
| Fulton | 5,538 | 5,330 | 38.8 | 2.5 | 6.5 | 0.5 | 3.8 | 20,584.7 | 19,813 | 24,975 | 2,989 | 80,842 | 480,910 | 118,916 | 20,513 |
| Gilmer | 82 | 2,694 | 64.1 | 0.3 | 5.9 | 0.0 | 2.9 | 91.4 | 3,002 | 87 | 83 | 1,218 | 13,000 | 52,236 | 4,980 |
| Glascock | 10 | 3,420 | 63.8 | 1.4 | 6.9 | 0.0 | 13.2 | 0.2 | 61 | 2 | 8 | 188 | 1,090 | 45,310 | 3,044 |
| Glynn | 656 | 7,741 | 24.1 | 51.9 | 3.0 | 0.0 | 3.7 | 373.4 | 4,410 | 1,981 | 298 | 5,406 | 37,760 | 67,803 | 7,762 |
| Gordon | 172 | 3,008 | 64.1 | 1.3 | 6.3 | 0.3 | 4.1 | 115.4 | 2,019 | 101 | 153 | 2,592 | 23,010 | 50,184 | 4,091 |
| Grady | 101 | 4,058 | 53.3 | 2.0 | 3.9 | 0.0 | 6.2 | 43.1 | 1,739 | 78 | 65 | 1,150 | 9,630 | 43,820 | 3,401 |
| Greene | 67 | 3,875 | 51.4 | 1.0 | 7.4 | 0.2 | 10.9 | 33.8 | 1,965 | 49 | 48 | 827 | 8,250 | 104,935 | 16,163 |
| Gwinnett | 3,194 | 3,478 | 59.1 | 1.5 | 5.3 | 0.1 | 4.2 | 3,329.3 | 3,626 | 2,554 | 2,595 | 35,875 | 424,940 | 62,022 | 6,624 |
| Habersham | 134 | 3,011 | 55.4 | 3.9 | 5.9 | 0.0 | 3.6 | 143.8 | 3,229 | 101 | 115 | 2,779 | 18,550 | 52,640 | 4,562 |
| Hall | 705 | 3,544 | 54.9 | 6.3 | 5.7 | 0.3 | 3.3 | 1,794.7 | 9,021 | 456 | 539 | 10,548 | 90,510 | 64,891 | 7,124 |
| Hancock | 38 | 4,457 | 42.8 | 12.2 | 3.7 | 1.5 | 8.9 | 21.4 | 2,506 | 15 | 19 | 696 | 3,130 | 36,869 | 2,600 |
| Haralson | 97 | 3,324 | 64.5 | 2.9 | 6.2 | 0.1 | 2.7 | 40.2 | 1,373 | 49 | 79 | 1,513 | 11,800 | 51,776 | 4,158 |
| Harris | 81 | 2,399 | 67.2 | 3.3 | 5.5 | 0.0 | 2.2 | 39.6 | 1,166 | 60 | 93 | 1,249 | 15,040 | 76,092 | 8,390 |
| Hart | 75 | 2,907 | 62.7 | 3.6 | 5.1 | 0.1 | 4.6 | 446.1 | 17,353 | 82 | 68 | 1,099 | 10,530 | 50,489 | 4,777 |
| Heard | 40 | 3,441 | 56.1 | 8.0 | 4.9 | 0.1 | 6.4 | 9.3 | 788 | 15 | 32 | 635 | 4,220 | 45,775 | 3,198 |
| Henry | 765 | 3,395 | 62.2 | 0.2 | 7.6 | 0.0 | 6.2 | 723.5 | 3,213 | 1,017 | 626 | 8,661 | 107,030 | 55,587 | 5,066 |
| Houston | 482 | 3,144 | 61.8 | 1.7 | 6.5 | 0.0 | 3.4 | 144.8 | 945 | 15,953 | 3,780 | 9,751 | 70,760 | 57,262 | 5,163 |
| Irwin | 44 | 4,714 | 46.1 | 32.8 | 5.2 | 0.1 | 4.1 | 7.8 | 834 | 26 | 22 | 620 | 3,530 | 41,714 | 2,984 |
| Jackson | 249 | 3,677 | 51.6 | 3.1 | 5.1 | 0.0 | 3.4 | 406.8 | 6,014 | 166 | 193 | 3,045 | 31,560 | 61,082 | 5,796 |
| Jasper | 50 | 3,594 | 51.7 | 23.7 | 4.2 | 0.3 | 5.4 | 6.9 | 495 | 22 | 38 | 664 | 5,860 | 52,885 | 4,279 |
| Jeff Davis | 49 | 3,270 | 62.4 | 2.7 | 4.7 | 0.1 | 11.4 | 19.6 | 1,306 | 31 | 40 | 864 | 5,380 | 41,005 | 3,031 |
| Jefferson | 65 | 4,172 | 42.7 | 21.7 | 4.8 | 0.0 | 8.1 | 48.6 | 3,112 | 52 | 39 | 959 | 6,630 | 39,187 | 2,774 |
| Jenkins | 33 | 3,682 | 66.2 | 3.4 | 4.3 | 0.5 | 7.6 | 14.2 | 1,612 | 22 | 20 | 408 | 3,040 | 38,512 | 2,595 |
| Johnson | 24 | 2,489 | 56.4 | 2.6 | 5.2 | 0.4 | 3.7 | 1.7 | 176 | 17 | 21 | 605 | 3,010 | 38,703 | 2,379 |
| Jones | 77 | 2,704 | 66.9 | 0.5 | 5.4 | 0.2 | 4.6 | 18.2 | 640 | 36 | 76 | 1,099 | 12,030 | 53,888 | 4,432 |
| Lamar | 51 | 2,734 | 51.9 | 0.2 | 7.1 | 0.0 | 4.9 | 66.4 | 3,570 | 44 | 48 | 1,023 | 7,360 | 48,524 | 3,768 |
| Lanier | 25 | 2,424 | 72.1 | 0.8 | 7.3 | 0.1 | 6.0 | 6.3 | 605 | 15 | 27 | 663 | 3,470 | 50,023 | 4,645 |
| Laurens | 182 | 3,841 | 50.9 | 12.5 | 5.5 | 0.0 | 7.7 | 45.5 | 960 | 1,632 | 125 | 2,818 | 20,150 | 50,023 | 4,645 |
| Lee | 100 | 3,404 | 69.2 | 0.6 | 9.7 | 0.0 | 3.6 | 68.0 | 2,311 | 49 | 78 | 1,482 | 14,010 | 62,373 | 5,842 |
| Liberty | 250 | 4,049 | 46.7 | 20.7 | 5.4 | 0.1 | 1.9 | 125.9 | 2,043 | 3,750 | 15,342 | 3,266 | 27,270 | 41,100 | 2,596 |
| Lincoln | 29 | 3,637 | 49.5 | 4.2 | 3.8 | 0.2 | 5.6 | 33.9 | 4,318 | 17 | 21 | 404 | 3,350 | 49,914 | 4,020 |
| Long | 50 | 2,660 | 78.1 | 0.7 | 5.3 | 0.0 | 1.9 | 11.2 | 596 | 12 | 51 | 738 | 5,780 | 41,531 | 2,347 |
| Lowndes | 769 | 6,664 | 29.8 | 48.5 | 3.8 | 0.0 | 2.2 | 530.1 | 4,594 | 1,178 | 4,866 | 10,307 | 46,090 | 52,528 | 5,302 |
| Lumpkin | 72 | 2,206 | 56.0 | 1.0 | 7.1 | 0.0 | 5.5 | 51.1 | 1,558 | 78 | 304 | 2,979 | 12,840 | 58,034 | 5,560 |
| McDuffie | 69 | 3,190 | 61.3 | 8.0 | 6.4 | 0.2 | 2.5 | 23.2 | 1,076 | 27 | 229 | 1,185 | 9,240 | 45,367 | 3,445 |
| McIntosh | 36 | 2,539 | 53.0 | 3.6 | 9.4 | 0.2 | 3.2 | 17.7 | 1,257 | 53 | 40 | 550 | 5,060 | 47,277 | 4,140 |
| Macon | 38 | 2,882 | 62.5 | 3.1 | 4.8 | 0.3 | 5.3 | 18.5 | 1,398 | 27 | 30 | 823 | 4,270 | 37,506 | 2,525 |
| Madison | 78 | 2,648 | 69.5 | 2.9 | 3.8 | 0.1 | 3.9 | 39.5 | 1,350 | 43 | 79 | 1,178 | 12,500 | 48,636 | 3,804 |
| Marion | 22 | 2,553 | 67.8 | 2.5 | 3.3 | 1.2 | 2.9 | 20.9 | 2,486 | 15 | 22 | 337 | 2,720 | 42,509 | 3,123 |
| Meriwether | 63 | 2,999 | 49.4 | 3.1 | 8.1 | 0.0 | 3.5 | 62.3 | 2,964 | 47 | 56 | 1,205 | 8,790 | 45,953 | 3,938 |
| Miller | 66 | 11,343 | 16.2 | 64.6 | 1.9 | 5.0 | 2.1 | 24.1 | 4,149 | 19 | 15 | 897 | 2,280 | 42,194 | 3,870 |
| Mitchell | 87 | 3,895 | 57.2 | 1.0 | 6.4 | 0.0 | 7.0 | 41.4 | 1,854 | 77 | 53 | 1,337 | 8,150 | 40,588 | 2,913 |
| Monroe | 96 | 3,537 | 51.0 | 13.5 | 8.3 | 0.1 | 4.1 | 33.9 | 1,248 | 34 | 72 | 2,778 | 11,810 | 69,736 | 7,672 |
| Montgomery | 23 | 2,489 | 55.2 | 1.1 | 5.6 | 0.2 | 8.2 | 3.7 | 412 | 14 | 22 | 379 | 3,100 | 46,583 | 3,707 |
| Morgan | 83 | 4,496 | 53.2 | 19.8 | 2.1 | 0.9 | 3.5 | 183.3 | 9,974 | 41 | 51 | 1,145 | 8,850 | 67,642 | 7,512 |
| Murray | 96 | 2,412 | 69.0 | 3.3 | 4.1 | 0.0 | 3.8 | 47.3 | 1,188 | 96 | 106 | 1,273 | 15,360 | 42,683 | 2,764 |
| Muscogee | 809 | 4,171 | 45.6 | 8.7 | 6.3 | 4.6 | 4.5 | 618.0 | 3,188 | 6,392 | 6,930 | 12,960 | 83,960 | 54,508 | 5,749 |
| Newton | 393 | 3,647 | 50.9 | 18.8 | 5.2 | 0.0 | 3.7 | 151.9 | 1,409 | 264 | 294 | 4,353 | 48,500 | 47,290 | 3,839 |
| Oconee | 121 | 3,165 | 64.3 | 0.5 | 4.1 | 0.2 | 7.8 | 152.0 | 3,986 | 73 | 107 | 1,538 | 17,900 | 112,710 | 15,571 |
| Oglethorpe | 34 | 2,252 | 66.1 | 3.3 | 3.8 | 0.1 | 6.9 | 15.3 | 1,030 | 15 | 40 | 558 | 6,320 | 48,033 | 3,752 |
| Paulding | 393 | 2,465 | 73.0 | 0.2 | 5.0 | 0.0 | 6.4 | 747.8 | 4,688 | 152 | 497 | 5,192 | 72,720 | 58,485 | 5,180 |
| Peach | 87 | 3,236 | 42.9 | 13.8 | 7.5 | 0.0 | 3.2 | 12.3 | 457 | 121 | 75 | 2,006 | 11,500 | 50,380 | 4,102 |
| Pickens | 88 | 2,801 | 58.0 | 0.5 | 5.9 | 0.0 | 6.6 | 32.2 | 1,022 | 61 | 86 | 1,226 | 15,100 | 66,468 | 6,928 |
| Pierce | 53 | 2,772 | 70.4 | 2.1 | 4.8 | 0.1 | 4.8 | 4.5 | 235 | 53 | 52 | 709 | 7,710 | 49,731 | 4,377 |
| Pike | 49 | 2,699 | 73.6 | 0.9 | 3.6 | 0.1 | 8.5 | 20.9 | 1,146 | 34 | 50 | 732 | 7,870 | 63,411 | 5,736 |
| Polk | 128 | 3,051 | 65.6 | 0.5 | 6.5 | 0.1 | 4.6 | 116.9 | 2,795 | 75 | 113 | 1,662 | 16,950 | 44,213 | 3,209 |
| Pulaski | 26 | 2,287 | 54.0 | 2.6 | 10.5 | 0.0 | 5.4 | 2.3 | 209 | 16 | 26 | 592 | 3,600 | 59,534 | 5,744 |

1. Based on the resident population estimated as of July 1 of the year shown.

# Table B. States and Counties — Land Area and Population

| State / county code | CBSA code[1] | County Type code[2] | STATE County | Land area[3] (sq. mi) | Total persons 2019 | Rank | Per square mile | White | Black | American Indian, Alaska Native | Asian and Pacific Islancer | Percent Hispanic or Latino[4] | Under 5 years | 5 to 14 years | 15 to 24 years | 25 to 34 years | 35 to 44 years | 45 to 54 years |
|---|---|---|---|---|---|---|---|---|---|---|---|---|---|---|---|---|---|---|
| | | | | 1 | 2 | 3 | 4 | 5 | 6 | 7 | 8 | 9 | 10 | 11 | 12 | 13 | 14 | 15 |
| | | | GEORGIA—Cont'd | | | | | | | | | | | | | | | |
| 13237 | | 6 | Putnam | 344.7 | 22,520 | 1,703 | 65.3 | 67.1 | 26.6 | 0.5 | 0.9 | 6.3 | 5.0 | 11.1 | 10.2 | 10.7 | 9.9 | 12.0 |
| 13239 | 21640 | 9 | Quitman | 151.2 | 2,271 | 3,020 | 15.0 | 49.9 | 47.5 | 1.1 | 1.0 | 1.8 | 5.4 | 9.8 | 9.0 | 8.6 | 8.4 | 10.7 |
| 13241 | | 7 | Rabun | 370.1 | 17,273 | 1,959 | 46.7 | 89.4 | 2.1 | 1.2 | 1.3 | 7.6 | 4.2 | 8.6 | 9.8 | 10.5 | 9.6 | 12.7 |
| 13243 | | 6 | Randolph | 428.2 | 6,682 | 2,689 | 15.6 | 35.9 | 61.3 | 0.3 | 0.6 | 2.7 | 4.8 | 12.1 | 12.0 | 10.0 | 9.7 | 10.2 |
| 13245 | 12260 | 2 | Richmond | 324.3 | 202,079 | 337 | 623.1 | 35.1 | 58.7 | 0.9 | 2.8 | 5.3 | 6.7 | 12.4 | 14.6 | 16.5 | 11.9 | 10.5 |
| 13247 | 12060 | 1 | Rockdale | 129.8 | 90,939 | 653 | 700.6 | 28.8 | 60.0 | 0.7 | 2.5 | 10.3 | 5.7 | 13.7 | 13.7 | 12.3 | 11.6 | 13.8 |
| 13249 | 11140 | 8 | Schley | 166.9 | 5,196 | 2,811 | 31.1 | 72.9 | 20.9 | 0.5 | 1.3 | 5.9 | 5.2 | 13.2 | 12.9 | 10.3 | 11.9 | 16.0 |
| 13251 | | 6 | Screven | 645.8 | 14,012 | 2,157 | 21.7 | 55.9 | 41.1 | 0.8 | 1.0 | 2.7 | 6.0 | 11.1 | 10.8 | 13.2 | 11.3 | 12.3 |
| 13253 | | 6 | Seminole | 237.5 | 8,060 | 2,583 | 33.9 | 62.4 | 33.4 | 0.7 | 1.1 | 3.8 | 5.0 | 11.4 | 11.4 | 10.8 | 10.3 | 12.2 |
| 13255 | 12060 | 1 | Spalding | 196.0 | 67,414 | 796 | 343.9 | 58.2 | 36.3 | 0.8 | 1.4 | 5.2 | 6.4 | 13.1 | 11.8 | 13.8 | 11.5 | 11.9 |
| 13257 | 45740 | 7 | Stephens | 178.9 | 26,107 | 1,564 | 145.9 | 84.0 | 12.3 | 0.9 | 1.5 | 4.0 | 6.2 | 12.6 | 13.1 | 11.8 | 10.8 | 11.8 |
| 13259 | 17980 | 8 | Stewart | 458.7 | 6,689 | 2,688 | 14.6 | 22.5 | 40.3 | 0.8 | 3.0 | 34.5 | 2.6 | 6.1 | 14.4 | 27.1 | 15.4 | 10.0 |
| 13261 | 11140 | 6 | Sumter | 482.9 | 29,282 | 1,455 | 60.6 | 39.6 | 53.3 | 0.6 | 1.6 | 6.0 | 6.1 | 12.7 | 17.4 | 12.1 | 10.6 | 11.4 |
| 13263 | 17980 | 8 | Talbot | 391.4 | 6,143 | 2,735 | 15.7 | 41.7 | 55.0 | 1.0 | 0.6 | 3.1 | 4.7 | 8.5 | 9.7 | 10.5 | 9.5 | 12.3 |
| 13265 | | 8 | Taliaferro | 194.6 | 1,562 | 3,076 | 8.0 | 40.3 | 54.5 | 0.8 | 2.2 | 4.7 | 5.1 | 9.2 | 9.2 | 10.4 | 9.6 | 11.5 |
| 13267 | | 6 | Tattnall | 480.8 | 25,365 | 1,589 | 52.8 | 58.3 | 28.9 | 0.5 | 0.9 | 12.6 | 4.9 | 11.8 | 13.0 | 16.0 | 13.6 | 13.3 |
| 13269 | | 8 | Taylor | 376.7 | 8,074 | 2,580 | 21.4 | 59.2 | 37.7 | 0.6 | 1.2 | 2.8 | 5.3 | 11.1 | 10.8 | 11.8 | 10.2 | 13.8 |
| 13271 | | 7 | Telfair | 437.3 | 15,781 | 2,044 | 36.1 | 49.8 | 34.5 | 0.4 | 1.0 | 15.4 | 4.1 | 8.6 | 10.5 | 15.3 | 15.6 | 13.6 |
| 13273 | 10500 | 3 | Terrell | 335.8 | 8,523 | 2,532 | 25.4 | 37.0 | 59.7 | 0.7 | 0.9 | 3.0 | 5.9 | 12.7 | 12.4 | 12.7 | 10.9 | 10.7 |
| 13275 | 45620 | 4 | Thomas | 544.6 | 44,372 | 1,089 | 81.5 | 58.5 | 36.9 | 0.8 | 1.3 | 3.8 | 6.2 | 13.4 | 11.8 | 12.0 | 11.8 | 12.0 |
| 13277 | 45700 | 5 | Tift | 260.9 | 40,719 | 1,164 | 156.1 | 55.6 | 31.1 | 0.5 | 1.8 | 12.5 | 6.8 | 13.9 | 14.8 | 13.4 | 12.2 | 11.5 |
| 13279 | 47080 | 7 | Toombs | 364.0 | 26,973 | 1,527 | 74.1 | 61.0 | 27.2 | 0.5 | 1.2 | 11.5 | 6.9 | 14.9 | 12.9 | 12.5 | 11.9 | 11.4 |
| 13281 | | 9 | Towns | 166.5 | 12,247 | 2,269 | 73.6 | 94.4 | 1.6 | 0.7 | 1.3 | 3.1 | 3.7 | 6.7 | 15.5 | 7.7 | 6.7 | 9.3 |
| 13283 | 20140 | 7 | Treutlen | 199.4 | 6,822 | 2,674 | 34.2 | 64.9 | 31.7 | 0.5 | 0.5 | 3.6 | 5.8 | 13.2 | 13.4 | 12.7 | 12.3 | 10.9 |
| 13285 | 29300 | 4 | Troup | 414.0 | 70,214 | 774 | 169.6 | 57.1 | 37.5 | 0.6 | 2.8 | 3.8 | 6.5 | 13.6 | 13.5 | 13.9 | 11.9 | 12.4 |
| 13287 | | 6 | Turner | 285.4 | 7,882 | 2,599 | 27.6 | 54.2 | 40.0 | 0.7 | 1.3 | 5.0 | 7.3 | 13.5 | 13.2 | 12.6 | 11.4 | 11.0 |
| 13289 | 31420 | 3 | Twiggs | 359.3 | 8,103 | 2,575 | 22.6 | 56.6 | 40.3 | 1.0 | 0.7 | 3.0 | 4.7 | 11.0 | 9.4 | 11.1 | 10.6 | 12.4 |
| 13291 | | 9 | Union | 322.1 | 25,358 | 1,592 | 78.7 | 94.5 | 1.2 | 1.0 | 0.8 | 3.7 | 3.6 | 8.6 | 8.5 | 8.3 | 8.4 | 10.7 |
| 13293 | 45580 | 4 | Upson | 323.5 | 26,527 | 1,546 | 82.0 | 68.4 | 29.0 | 0.8 | 1.0 | 2.5 | 6.0 | 12.8 | 11.5 | 12.8 | 11.1 | 12.6 |
| 13295 | 16860 | 2 | Walker | 446.4 | 70,116 | 775 | 157.1 | 91.9 | 5.3 | 0.8 | 1.0 | 2.7 | 5.5 | 12.1 | 11.3 | 12.5 | 12.6 | 12.8 |
| 13297 | 12060 | 1 | Walton | 326.8 | 96,875 | 623 | 296.4 | 73.6 | 19.9 | 0.7 | 2.2 | 5.3 | 5.9 | 13.7 | 12.8 | 12.5 | 12.4 | 13.7 |
| 13299 | 48180 | 5 | Ware | 899.2 | 35,826 | 1,282 | 39.8 | 64.0 | 31.2 | 0.8 | 1.6 | 4.4 | 6.8 | 13.9 | 12.0 | 13.6 | 11.6 | 11.9 |
| 13301 | | 8 | Warren | 284.4 | 5,232 | 2,807 | 18.4 | 39.5 | 58.3 | 0.6 | 0.8 | 2.1 | 5.4 | 11.0 | 10.5 | 11.9 | 10.1 | 12.2 |
| 13303 | | 7 | Washington | 678.5 | 20,150 | 1,816 | 29.7 | 43.2 | 54.2 | 0.4 | 0.8 | 2.8 | 5.5 | 12.3 | 11.6 | 13.8 | 11.9 | 12.2 |
| 13305 | 27700 | 6 | Wayne | 641.8 | 30,023 | 1,434 | 46.8 | 72.6 | 20.7 | 0.9 | 1.1 | 6.6 | 6.1 | 13.6 | 11.8 | 13.7 | 13.0 | 12.8 |
| 13307 | | 8 | Webster | 209.7 | 2,595 | 2,995 | 12.4 | 52.2 | 42.5 | 0.7 | 0.7 | 5.2 | 4.7 | 9.7 | 12.6 | 9.7 | 8.7 | 14.0 |
| 13309 | | 9 | Wheeler | 295.5 | 7,751 | 2,612 | 26.2 | 56.7 | 38.2 | 0.6 | 0.5 | 5.5 | 4.4 | 8.9 | 13.1 | 17.8 | 14.7 | 14.2 |
| 13311 | | 6 | White | 240.7 | 31,094 | 1,401 | 129.2 | 93.5 | 2.7 | 1.1 | 0.8 | 3.5 | 4.9 | 10.9 | 12.9 | 10.7 | 10.2 | 12.4 |
| 13313 | 19140 | 3 | Whitfield | 290.4 | 103,837 | 586 | 357.6 | 57.9 | 4.3 | 0.5 | 1.8 | 36.7 | 6.4 | 14.4 | 14.2 | 13.2 | 12.6 | 12.9 |
| 13315 | | 8 | Wilcox | 377.8 | 8,502 | 2,536 | 22.5 | 60.0 | 34.7 | 0.8 | 1.2 | 4.8 | 4.8 | 10.6 | 12.0 | 14.8 | 13.9 | 13.1 |
| 13317 | | 6 | Wilkes | 469.5 | 9,694 | 2,436 | 20.6 | 53.4 | 41.4 | 0.6 | 0.9 | 5.6 | 5.1 | 11.8 | 10.9 | 10.7 | 10.2 | 12.0 |
| 13319 | | 8 | Wilkinson | 449.2 | 8,812 | 2,508 | 19.6 | 58.3 | 38.7 | 0.9 | 0.7 | 3.1 | 5.6 | 12.5 | 11.7 | 10.8 | 11.4 | 11.4 |
| 13321 | 10500 | 3 | Worth | 570.7 | 19,972 | 1,825 | 35.0 | 68.4 | 28.7 | 0.7 | 1.1 | 2.4 | 5.5 | 12.4 | 11.4 | 12.4 | 11.0 | 12.6 |
| 15000 | | 0 | HAWAII | 6,422.4 | 1,407,006 | X | 219.1 | 36.2 | 3.0 | 1.7 | 75.6 | 10.9 | 6.0 | 11.8 | 11.5 | 14.0 | 12.9 | 11.7 |
| 15001 | 25900 | 5 | Hawaii | 4,028.4 | 203,340 | 335 | 50.5 | 48.0 | 1.8 | 2.3 | 68.1 | 13.5 | 5.5 | 12.2 | 10.3 | 11.7 | 12.2 | 11.3 |
| 15003 | 46520 | 2 | Honolulu | 600.6 | 963,826 | 52 | 1,604.8 | 31.7 | 3.7 | 1.5 | 79.0 | 10.2 | 6.1 | 11.6 | 12.1 | 14.9 | 12.9 | 11.6 |
| 15005 | | 3 | Kalawao | 12.0 | 87 | 3,143 | 7.3 | 40.2 | 2.3 | 5.7 | 66.7 | 1.1 | 0.0 | 0.0 | 0.0 | 5.7 | 9.2 | 14.9 |
| 15007 | 28180 | 5 | Kauai | 619.9 | 71,851 | 759 | 115.9 | 44.7 | 1.4 | 1.8 | 68.4 | 11.6 | 5.6 | 12.4 | 10.0 | 11.9 | 13.0 | 12.0 |
| 15009 | 27980 | 3 | Maui | 1,161.5 | 167,902 | 398 | 144.6 | 44.0 | 1.5 | 1.8 | 68.0 | 11.9 | 5.7 | 12.2 | 10.2 | 12.0 | 13.6 | 12.6 |
| 16000 | | 0 | IDAHO | 82,645.1 | 1,826,913 | X | 22.1 | 83.5 | 1.2 | 1.9 | 2.7 | 13.0 | 6.2 | 14.1 | 13.5 | 13.2 | 12.8 | 11.2 |
| 16001 | 14260 | 2 | Ada | 1,052.0 | 494,399 | 142 | 470.0 | 86.5 | 1.9 | 1.1 | 4.5 | 8.8 | 5.4 | 13.1 | 12.8 | 14.2 | 14.3 | 12.6 |
| 16003 | | 9 | Adams | 1,362.8 | 4,447 | 2,862 | 3.3 | 89.3 | 0.7 | 2.4 | 1.8 | 3.7 | 3.9 | 9.7 | 8.2 | 8.8 | 9.1 | 10.1 |
| 16005 | 38540 | 3 | Bannock | 1,112.5 | 88,795 | 661 | 79.8 | 85.1 | 1.4 | 3.6 | 2.9 | 9.2 | 6.6 | 14.5 | 14.8 | 14.3 | 13.3 | 10.3 |
| 16007 | | 9 | Bear Lake | 975.7 | 6,143 | 2,735 | 6.3 | 94.2 | 0.6 | 1.2 | 0.9 | 4.3 | 6.6 | 15.2 | 11.9 | 10.6 | 11.2 | 10.0 |
| 16009 | | 6 | Benewah | 776.9 | 9,430 | 2,462 | 12.1 | 88.0 | 1.1 | 9.6 | 1.1 | 4.2 | 5.3 | 12.8 | 10.1 | 9.9 | 10.3 | 11.6 |
| 16011 | 13940 | 4 | Bingham | 2,093.7 | 47,202 | 1,032 | 22.5 | 75.4 | 0.6 | 5.9 | 1.5 | 18.1 | 7.1 | 17.6 | 14.0 | 12.0 | 12.7 | 11.3 |
| 16013 | 25200 | 7 | Blaine | 2,637.7 | 23,426 | 1,663 | 8.9 | 74.6 | 0.9 | 0.7 | 1.5 | 23.5 | 4.5 | 11.8 | 11.1 | 10.7 | 12.4 | 13.0 |
| 16015 | 14260 | 2 | Boise | 1,899.6 | 8,065 | 2,581 | 4.2 | 91.6 | 0.8 | 2.2 | 2.0 | 5.4 | 3.4 | 8.5 | 8.6 | 7.9 | 10.7 | 12.6 |
| 16017 | 41760 | 6 | Bonner | 1,733.2 | 46,817 | 1,038 | 27.0 | 94.6 | 0.5 | 2.1 | 1.4 | 3.6 | 4.7 | 11.0 | 9.3 | 9.5 | 11.1 | 11.5 |
| 16019 | 26820 | 3 | Bonneville | 1,866.0 | 122,134 | 518 | 65.5 | 83.9 | 1.0 | 1.3 | 2.0 | 13.6 | 8.0 | 17.4 | 13.3 | 13.9 | 13.3 | 10.1 |
| 16021 | | 7 | Boundary | 1,268.7 | 12,656 | 2,240 | 10.0 | 91.3 | 1.0 | 2.6 | 1.8 | 5.3 | 6.2 | 12.7 | 11.0 | 9.5 | 10.8 | 11.3 |
| 16023 | 26820 | 3 | Butte | 2,236.5 | 2,646 | 2,990 | 1.2 | 93.1 | 0.6 | 1.8 | 1.1 | 5.3 | 5.1 | 13.8 | 11.5 | 9.2 | 10.9 | 9.4 |
| 16025 | 25200 | 9 | Camas | 1,074.2 | 1,130 | 3,104 | 1.1 | 91.7 | 0.7 | 3.0 | 1.2 | 6.4 | 3.7 | 14.2 | 9.0 | 7.3 | 13.9 | 11.9 |
| 16027 | 14260 | 2 | Canyon | 587.1 | 237,053 | 291 | 403.8 | 72.1 | 1.1 | 1.5 | 2.1 | 25.5 | 6.9 | 15.6 | 14.2 | 13.8 | 12.9 | 11.3 |

1. CBSA = Core Based Statistical Area. See Appendix A for explanation. See Appendix B for list of metropolitan areas with component counties.
Service of USDA Rural-Urban Continuum Codes. See Appendix A for definition.
3. Dry land or land partially or temporarily covered by water.
2. County type code from the Economic Research
4. May be of any race.

# Table B. States and Counties — Population and Households

| STATE County | Age (percent) (cont.) 55 to 64 years | 65 to 74 years | 75 years and over | Percent female | Total persons 2000 | 2010 | Percent change 2000-2010 | 2010-2020 | Components of change, 2010-2020 Births | Deaths | Net Migration | Households, 2015-2019 Number | Persons per household | Family house-holds | Female family house-holder[1] | One person |
|---|---|---|---|---|---|---|---|---|---|---|---|---|---|---|---|---|
| | 16 | 17 | 18 | 19 | 20 | 21 | 22 | 23 | 24 | 25 | 26 | 27 | 28 | 29 | 30 | 31 |
| **GEORGIA—Cont'd** | | | | | | | | | | | | | | | | |
| Putnam............... | 16.0 | 15.4 | 9.7 | 52.0 | 18,812 | 21,218 | 12.8 | 6.1 | 2,399 | 2,498 | 1,404 | 8,937 | 2.41 | 70.3 | 13.5 | 24.7 |
| Quitman ............. | 16.1 | 18.2 | 13.9 | 52.4 | 2,598 | 2,510 | -3.4 | -9.5 | 239 | 330 | -148 | 842 | 2.72 | 68.5 | 24.2 | 29.1 |
| Rabun ................. | 15.8 | 16.9 | 11.8 | 51.3 | 15,050 | 16,273 | 8.1 | 6.1 | 1,537 | 2,109 | 1,586 | 6,662 | 2.43 | 63.3 | 9.0 | 30.8 |
| Randolph............. | 15.1 | 14.1 | 12.1 | 54.0 | 7,791 | 7,721 | -0.9 | -13.5 | 796 | 931 | -921 | 2,553 | 2.66 | 63.1 | 23.9 | 30.1 |
| Richmond............. | 12.5 | 9.1 | 5.8 | 51.5 | 199,775 | 200,591 | 0.4 | 0.7 | 29,913 | 20,254 | -8,096 | 71,400 | 2.69 | 60.7 | 22.0 | 33.8 |
| Rockdale............. | 14.0 | 9.7 | 5.6 | 53.1 | 70,111 | 85,173 | 21.5 | 6.8 | 10,091 | 6,985 | 2,671 | 31,076 | 2.86 | 73.3 | 18.7 | 22.7 |
| Schley................. | 12.6 | 9.5 | 8.4 | 52.0 | 3,766 | 5,013 | 33.1 | 3.7 | 518 | 376 | 39 | 1,864 | 2.80 | 77.9 | 15.3 | 21.9 |
| Screven.............. | 15.0 | 12.4 | 7.8 | 50.7 | 15,374 | 14,592 | -5.1 | -4.0 | 1,886 | 1,790 | -687 | 5,098 | 2.66 | 66.6 | 16.7 | 29.3 |
| Seminole............. | 14.8 | 13.5 | 10.6 | 52.0 | 9,369 | 8,729 | -6.8 | -7.7 | 984 | 1,243 | -412 | 3,363 | 2.44 | 64.3 | 18.6 | 32.0 |
| Spalding............. | 12.7 | 11.2 | 7.7 | 52.1 | 58,417 | 64,107 | 9.7 | 5.2 | 8,614 | 7,694 | 2,398 | 24,336 | 2.63 | 67.6 | 20.4 | 26.9 |
| Stephens............. | 13.5 | 12.2 | 8.0 | 51.9 | 25,435 | 26,162 | 2.9 | -0.2 | 3,157 | 3,619 | 419 | 9,543 | 2.62 | 71.1 | 9.8 | 26.4 |
| Stewart............... | 9.5 | 8.2 | 6.7 | 31.0 | 5,252 | 6,060 | 15.4 | 10.4 | 448 | 685 | 817 | 1,816 | 2.26 | 62.7 | 24.3 | 36.1 |
| Sumter................ | 12.1 | 10.5 | 7.1 | 53.0 | 33,200 | 32,742 | -1.4 | -10.6 | 3,960 | 3,660 | -3,802 | 11,510 | 2.43 | 63.0 | 18.4 | 33.1 |
| Talbot................. | 18.3 | 16.6 | 9.9 | 52.3 | 6,498 | 6,892 | 6.1 | -10.9 | 572 | 832 | -493 | 2,809 | 2.24 | 65.8 | 18.3 | 29.9 |
| Taliaferro............ | 16.1 | 15.7 | 13.3 | 50.1 | 2,077 | 1,717 | -17.3 | -9.0 | 148 | 244 | -59 | 593 | 2.71 | 67.3 | 22.8 | 28.0 |
| Tattnall.............. | 12.2 | 9.2 | 6.1 | 42.5 | 22,305 | 25,504 | 14.3 | -0.5 | 2,848 | 2,482 | -521 | 8,241 | 2.31 | 71.3 | 17.9 | 25.1 |
| Taylor................. | 15.2 | 12.1 | 9.8 | 52.6 | 8,815 | 8,982 | 1.9 | -10.1 | 886 | 1,019 | -795 | 3,473 | 2.30 | 63.6 | 15.1 | 32.7 |
| Telfair................ | 12.8 | 10.8 | 8.6 | 41.2 | 11,794 | 16,492 | 39.8 | -4.3 | 1,352 | 1,529 | -556 | 4,668 | 2.74 | 69.8 | 19.3 | 28.0 |
| Terrell................ | 14.3 | 11.9 | 8.6 | 52.4 | 10,970 | 9,507 | -13.3 | -10.4 | 1,185 | 1,097 | -1,081 | 3,399 | 2.47 | 69.1 | 26.6 | 26.6 |
| Thomas............... | 13.7 | 11.0 | 8.2 | 52.5 | 42,737 | 44,719 | 4.6 | -0.8 | 5,754 | 5,307 | -768 | 17,595 | 2.49 | 69.1 | 18.6 | 24.7 |
| Tift.................... | 11.7 | 9.2 | 6.5 | 51.8 | 38,407 | 40,132 | 4.5 | 1.5 | 5,743 | 4,227 | -978 | 15,144 | 2.54 | 70.7 | 18.9 | 24.7 |
| Toombs................ | 12.2 | 9.7 | 7.6 | 52.7 | 26,067 | 27,174 | 4.2 | -0.7 | 3,971 | 3,186 | -973 | 10,030 | 2.65 | 65.2 | 16.6 | 30.3 |
| Towns................. | 15.0 | 19.4 | 15.9 | 52.5 | 9,319 | 10,474 | 12.4 | 16.9 | 890 | 1,776 | 2,628 | 4,898 | 2.2 | 66.1 | 8.7 | 27.3 |
| Treutlen............. | 13.3 | 10.7 | 7.8 | 49.9 | 6,854 | 6,883 | 0.4 | -0.9 | 773 | 745 | -89 | 2,490 | 2.53 | 66.4 | 13.6 | 32.6 |
| Troup.................. | 12.7 | 9.4 | 6.2 | 52.2 | 58,779 | 67,042 | 14.1 | 4.7 | 9,278 | 7,145 | 1,077 | 24,928 | 2.74 | 68.0 | 18.7 | 28.9 |
| Turner................. | 11.8 | 10.5 | 8.7 | 51.8 | 9,504 | 8,931 | -6.0 | -11.7 | 1,166 | 1,195 | -1,059 | 3,169 | 2.39 | 72.5 | 17.0 | 24.0 |
| Twiggs................ | 17.1 | 14.0 | 9.7 | 51.0 | 10,590 | 9,035 | -14.7 | -10.3 | 901 | 1,154 | -689 | 3,044 | 2.68 | 60.4 | 14.4 | 38.2 |
| Union................. | 17.3 | 20.2 | 14.4 | 51.5 | 17,289 | 21,370 | 23.6 | 18.7 | 1,696 | 3,160 | 5,439 | 9,743 | 2.33 | 71.4 | 10.0 | 25.3 |
| Upson................. | 14.2 | 11.2 | 7.8 | 52.6 | 27,597 | 27,147 | -1.6 | -2.3 | 3,297 | 3,852 | -63 | 10,154 | 2.53 | 60.1 | 13.1 | 34.5 |
| Walker................. | 14.0 | 11.1 | 8.2 | 50.9 | 61,053 | 68,739 | 12.6 | 2.0 | 7,576 | 7,922 | 1,767 | 25,975 | 2.60 | 70.9 | 12.9 | 25.1 |
| Walton................ | 12.8 | 9.6 | 6.6 | 51.3 | 60,687 | 83,808 | 38.1 | 15.6 | 10,955 | 8,323 | 10,454 | 31,670 | 2.86 | 77.1 | 14.3 | 19.3 |
| Ware................... | 12.3 | 10.0 | 7.8 | 50.0 | 35,483 | 36,298 | 2.3 | -1.3 | 5,029 | 4,717 | -793 | 13,823 | 2.39 | 64.5 | 15.0 | 32.1 |
| Warren................ | 15.9 | 12.6 | 10.5 | 53.4 | 6,336 | 5,834 | -7.9 | -10.3 | 611 | 794 | -423 | 2,244 | 2.32 | 64.9 | 23.4 | 33.0 |
| Washington.......... | 14.5 | 10.5 | 7.7 | 48.8 | 21,176 | 21,181 | 0.0 | -4.9 | 2,376 | 2,332 | -1,100 | 7,503 | 2.46 | 70.8 | 22.8 | 25.4 |
| Wayne................. | 12.5 | 9.9 | 6.6 | 49.0 | 26,565 | 30,102 | 13.3 | -0.3 | 3,979 | 3,463 | -618 | 10,400 | 2.66 | 68.9 | 13.5 | 25.9 |
| Webster............... | 16.2 | 13.6 | 10.6 | 50.8 | 2,390 | 2,799 | 17.1 | -7.3 | 227 | 226 | -209 | 1,140 | 2.29 | 63.5 | 11.1 | 31.9 |
| Wheeler.............. | 11.4 | 8.9 | 6.7 | 35.2 | 6,179 | 7,431 | 20.3 | 4.3 | 647 | 622 | 241 | 1,862 | 4.14 | 62.2 | 10.6 | 36.2 |
| White................. | 14.7 | 13.5 | 9.8 | 51.1 | 19,944 | 27,139 | 36.1 | 14.6 | 2,750 | 2,902 | 4,106 | 11,695 | 2.49 | 73.4 | 10.4 | 23.5 |
| Whitfield............. | 11.7 | 8.4 | 6.3 | 50.3 | 83,525 | 102,602 | 22.8 | 1.2 | 14,025 | 8,455 | -4,301 | 35,629 | 2.88 | 72.2 | 13.7 | 23.4 |
| Wilcox................ | 11.8 | 10.4 | 8.5 | 40.0 | 8,577 | 9,251 | 7.9 | -8.1 | 898 | 1,072 | -594 | 2,575 | 2.61 | 70.2 | 17.8 | 26.6 |
| Wilkes................ | 15.1 | 12.8 | 11.4 | 52.1 | 10,687 | 10,593 | -0.9 | -8.5 | 1,082 | 1,454 | -557 | 3,979 | 2.45 | 60.8 | 15.4 | 37.6 |
| Wilkinson............ | 15.8 | 11.9 | 9.0 | 51.8 | 10,220 | 9,569 | -6.4 | -7.9 | 1,078 | 1,225 | -611 | 3,185 | 2.80 | 65.3 | 16.9 | 29.9 |
| Worth................. | 14.7 | 11.6 | 8.5 | 52.3 | 21,967 | 21,667 | -1.4 | -7.8 | 2,450 | 2,393 | -1,768 | 8,002 | 2.54 | 73.7 | 19.3 | 21.5 |
| **HAWAII** ............. | 12.6 | 11.0 | 8.6 | 49.9 | 1,211,537 | 1,360,304 | 12.3 | 3.4 | 185,462 | 114,918 | -22,986 | 459,424 | 3.00 | 69.5 | 12.1 | 24.2 |
| Hawaii................ | 14.2 | 14.2 | 8.3 | 50.5 | 148,677 | 185,076 | 24.5 | 9.9 | 23,915 | 17,545 | 11,962 | 69,453 | 2.82 | 66.6 | 12.9 | 26.7 |
| Honolulu.............. | 11.9 | 10.0 | 9.7 | 49.7 | 876,156 | 953,203 | 8.8 | 1.1 | 133,124 | 79,117 | -42,716 | 312,795 | 3.03 | 70.0 | 12.1 | 24.0 |
| Kalawao............... | 19.5 | 23.0 | 27.6 | 52.9 | 147 | 90 | -38.8 | -3.3 | 0 | 3 | 0 | 39 | 1.41 | 23.1 | 0.0 | 66.7 |
| Kauai.................. | 13.8 | 12.8 | 8.5 | 50.5 | 58,463 | 67,095 | 14.8 | 7.1 | 8,626 | 6,054 | 2,237 | 22,658 | 3.13 | 70.3 | 10.3 | 22.9 |
| Maui................... | 14.0 | 11.9 | 7.6 | 50.4 | 128,094 | 154,840 | 20.9 | 8.4 | 19,797 | 12,199 | 5,531 | 54,479 | 3.00 | 70.2 | 12.2 | 22.8 |
| **IDAHO**................ | 12.1 | 10.2 | 6.6 | 49.9 | 1,293,953 | 1,567,658 | 21.2 | 16.5 | 229,433 | 133,664 | 162,950 | 630,008 | 2.68 | 67.6 | 9.1 | 26.3 |
| Ada.................... | 12.0 | 9.6 | 5.9 | 49.9 | 300,904 | 392,364 | 30.4 | 26.0 | 51,321 | 29,449 | 79,865 | 173,353 | 2.58 | 63.5 | 8.4 | 29.6 |
| Adams................. | 19.4 | 20.0 | 10.9 | 48.1 | 3,476 | 3,978 | 14.4 | 11.8 | 300 | 368 | 533 | 1,757 | 2.29 | 68.5 | 7.8 | 29.2 |
| Bannock.............. | 11.1 | 9.5 | 5.7 | 50.2 | 75,565 | 82,842 | 9.6 | 7.2 | 12,731 | 7,199 | 422 | 31,155 | 2.68 | 64.4 | 9.5 | 27.6 |
| Bear Lake............ | 13.2 | 12.2 | 9.1 | 49.9 | 6,411 | 5,989 | -6.6 | 2.6 | 816 | 646 | -17 | 2,424 | 2.46 | 74.6 | 8.2 | 22.6 |
| Benewah.............. | 16.5 | 14.5 | 8.9 | 49.1 | 9,171 | 9,283 | 1.2 | 1.6 | 1,075 | 1,243 | 317 | 3,418 | 2.65 | 67.5 | 8.2 | 25.2 |
| Bingham.............. | 11.6 | 8.9 | 5.8 | 50.0 | 41,735 | 45,605 | 9.3 | 3.5 | 7,045 | 3,709 | -1,777 | 14,989 | 3.04 | 74.7 | 10.2 | 22.2 |
| Blaine................ | 15.8 | 12.8 | 7.9 | 50.2 | 18,991 | 21,377 | 12.6 | 9.6 | 2,172 | 1,112 | 988 | 7,995 | 2.75 | 62.4 | 9.0 | 34.5 |
| Boise................. | 20.6 | 19.0 | 8.8 | 48.4 | 6,670 | 7,027 | 5.4 | 14.8 | 447 | 559 | 1,137 | 3,205 | 2.28 | 70.0 | 6.2 | 22.8 |
| Bonner................ | 16.7 | 16.6 | 9.4 | 49.9 | 36,835 | 40,877 | 11.0 | 14.5 | 4,169 | 4,430 | 6,186 | 17,537 | 2.46 | 68.3 | 7.3 | 26.7 |
| Bonneville........... | 10.4 | 8.2 | 5.5 | 50.1 | 82,522 | 104,294 | 26.4 | 17.1 | 19,476 | 8,615 | 7,035 | 39,768 | 2.84 | 71.8 | 10.0 | 24.0 |
| Boundary............. | 14.5 | 15.0 | 9.0 | 49.5 | 9,871 | 10,972 | 11.2 | 15.3 | 1,384 | 1,158 | 1,458 | 4,653 | 2.52 | 61.8 | 4.5 | 31.6 |
| Butte.................. | 14.8 | 15.0 | 10.4 | 49.0 | 2,899 | 2,893 | -0.2 | -8.5 | 285 | 280 | -257 | 967 | 2.62 | 66.9 | 7.8 | 31.4 |
| Camas................. | 16.5 | 15.8 | 7.5 | 49.1 | 991 | 1,117 | 12.7 | 1.2 | 96 | 79 | -9 | 394 | 2.66 | 60.9 | 6.6 | 35.0 |
| Canyon................ | 10.9 | 8.8 | 5.7 | 50.3 | 131,441 | 188,930 | 43.7 | 25.5 | 31,604 | 14,882 | 31,325 | 72,807 | 2.94 | 70.2 | 11.8 | 24.0 |

1. No spouse present.

## Table B. States and Counties — Population, Vital Statistics, Health, and Crime

| STATE County | Persons in group quarters, 2020 | Daytime Population, 2015-2019 Number | Daytime Population, 2015-2019 Employment/residence ratio | Births, 2020 Total | Births, 2020 Rate[1] | Deaths, 2020 Number | Deaths, 2020 Rate[1] | Persons under 65 with no health insurance, 2019 Number | Persons under 65 with no health insurance, 2019 Percent | Medicare, 2020 Total beneficiaries | Medicare, 2020 Enrolled in Original Medicare | Medicare, 2020 Enrolled in Medicare Advantage | Crimes reported by county police or sheriff, 2019 Violent | Crimes reported by county police or sheriff, 2019 Property |
|---|---|---|---|---|---|---|---|---|---|---|---|---|---|---|
| | 32 | 33 | 34 | 35 | 36 | 37 | 38 | 39 | 40 | 41 | 42 | 43 | 44 | 45 |
| **GEORGIA—Cont'd** | | | | | | | | | | | | | | |
| Putnam | 157 | 19,059 | 0.72 | 230 | 10.2 | 307 | 13.6 | 2,988 | 18.2 | 6,026 | 3,627 | 2,398 | NA | NA |
| Quitman | 0 | 1,982 | 0.61 | 18 | 7.9 | 39 | 17.2 | 308 | 19.7 | 721 | 384 | 336 | NA | NA |
| Rabun | 448 | 17,109 | 1.08 | 154 | 8.9 | 254 | 14.7 | 2,592 | 21.5 | 5,298 | 3,609 | 1,689 | NA | NA |
| Randolph | 350 | 6,890 | 0.97 | 58 | 8.7 | 83 | 12.4 | 777 | 16.4 | 1,624 | 813 | 811 | NA | NA |
| Richmond | 12,830 | 242,973 | 1.48 | 2,817 | 13.9 | 2,254 | 11.2 | 22,751 | 14.3 | 37,523 | 20,606 | 16,917 | NA | NA |
| Rockdale | 766 | 91,537 | 1.05 | 954 | 10.5 | 805 | 8.9 | 12,249 | 16.1 | 16,014 | 7,824 | 8,190 | NA | NA |
| Schley | 0 | 4,588 | 0.70 | 50 | 9.6 | 45 | 8.7 | 634 | 14.8 | 813 | 425 | 388 | NA | NA |
| Screven | 479 | 12,387 | 0.70 | 175 | 12.5 | 190 | 13.6 | 1,658 | 15.5 | 3,410 | 1,759 | 1,652 | NA | NA |
| Seminole | 52 | 7,910 | 0.86 | 86 | 10.7 | 115 | 14.3 | 1,026 | 16.9 | 2,209 | 1,376 | 833 | 37 | 112 |
| Spalding | 1,379 | 61,836 | 0.87 | 848 | 12.6 | 872 | 12.9 | 8,962 | 17.0 | 15,321 | 7,919 | 7,402 | NA | NA |
| Stephens | 567 | 26,356 | 1.06 | 311 | 11.9 | 385 | 14.7 | 3,227 | 16.1 | 6,900 | 4,085 | 2,815 | NA | NA |
| Stewart | 1,699 | 6,029 | 0.82 | 40 | 6.0 | 70 | 10.5 | 726 | 18.5 | 1,103 | 502 | 601 | NA | NA |
| Sumter | 1,604 | 31,115 | 1.09 | 338 | 11.5 | 363 | 12.4 | 3,632 | 16.2 | 6,202 | 3,101 | 3,101 | NA | NA |
| Talbot | 0 | 4,519 | 0.29 | 56 | 9.1 | 101 | 16.4 | 733 | 16.1 | 1,662 | 797 | 865 | NA | NA |
| Taliaferro | 7 | 1,215 | 0.34 | 13 | 8.3 | 30 | 19.2 | 189 | 17.5 | 491 | 233 | 258 | NA | NA |
| Tattnall | 4,446 | 24,488 | 0.88 | 247 | 9.7 | 280 | 11.0 | 3,326 | 19.7 | 4,096 | 2,250 | 1,846 | NA | NA |
| Taylor | 214 | 7,312 | 0.73 | 84 | 10.4 | 88 | 10.9 | 978 | 15.8 | 1,825 | 945 | 880 | NA | NA |
| Telfair | 3,133 | 15,855 | 0.96 | 124 | 7.9 | 178 | 11.3 | 1,637 | 16.8 | 2,414 | 1,388 | 1,026 | NA | NA |
| Terrell | 267 | 8,229 | 0.85 | 107 | 12.6 | 109 | 12.8 | 1,106 | 17.1 | 2,215 | 1,088 | 1,127 | NA | NA |
| Thomas | 651 | 47,091 | 1.13 | 521 | 11.7 | 577 | 13.0 | 5,679 | 16.1 | 10,317 | 6,006 | 4,311 | NA | NA |
| Tift | 1,588 | 46,321 | 1.32 | 537 | 13.2 | 466 | 11.4 | 6,242 | 19.2 | 7,783 | 4,417 | 3,366 | 60 | 477 |
| Toombs | 390 | 28,771 | 1.18 | 364 | 13.5 | 350 | 13.0 | 4,238 | 19.4 | 5,734 | 3,436 | 2,298 | NA | NA |
| Towns | 1,167 | 11,316 | 0.93 | 100 | 8.2 | 193 | 15.8 | 1,272 | 18.9 | 5,012 | 3,158 | 1,854 | 23 | 88 |
| Treutlen | 494 | 5,862 | 0.61 | 69 | 10.1 | 82 | 12.0 | 986 | 19.0 | 1,399 | 714 | 684 | 5 | 34 |
| Troup | 2,162 | 78,514 | 1.30 | 910 | 13.0 | 810 | 11.5 | 9,452 | 16.6 | 13,185 | 7,623 | 5,562 | NA | NA |
| Turner | 393 | 7,404 | 0.82 | 124 | 15.7 | 127 | 16.1 | 902 | 15.1 | 1,962 | 910 | 1,052 | NA | NA |
| Twiggs | 106 | 7,749 | 0.83 | 69 | 8.5 | 148 | 18.3 | 1,026 | 16.8 | 2,349 | 1,158 | 1,191 | NA | NA |
| Union | 296 | 23,549 | 1.03 | 169 | 6.7 | 316 | 12.5 | 3,005 | 18.8 | 9,124 | 5,825 | 3,300 | NA | NA |
| Upson | 414 | 25,290 | 0.91 | 298 | 11.2 | 425 | 16.0 | 2,876 | 13.7 | 6,674 | 3,304 | 3,370 | NA | NA |
| Walker | 1,353 | 55,700 | 0.55 | 731 | 10.4 | 887 | 12.7 | 8,359 | 15.1 | 14,959 | 9,254 | 5,715 | NA | NA |
| Walton | 792 | 76,050 | 0.63 | 1,157 | 11.9 | 985 | 10.2 | 12,001 | 15.2 | 17,779 | 9,770 | 8,010 | NA | NA |
| Ware | 2,471 | 39,988 | 1.34 | 503 | 14.0 | 488 | 13.6 | 4,500 | 16.7 | 8,147 | 4,860 | 3,287 | 39 | 444 |
| Warren | 89 | 4,379 | 0.56 | 59 | 11.3 | 72 | 13.8 | 596 | 15.1 | 1,452 | 687 | 765 | 83 | 432 |
| Washington | 1,879 | 20,837 | 1.06 | 222 | 11.0 | 257 | 12.8 | 2,312 | 15.6 | 4,562 | 2,313 | 2,248 | NA | NA |
| Wayne | 2,062 | 29,081 | 0.93 | 348 | 11.6 | 362 | 12.1 | 4,052 | 17.6 | 6,083 | 3,594 | 2,489 | NA | NA |
| Webster | 0 | 2,130 | 0.55 | 20 | 7.7 | 24 | 9.2 | 330 | 16.8 | 578 | 285 | 293 | NA | NA |
| Wheeler | 2,510 | 7,341 | 0.76 | 63 | 8.1 | 67 | 8.6 | 645 | 15.5 | 1,164 | 640 | 524 | NA | NA |
| White | 823 | 25,904 | 0.72 | 280 | 9.0 | 375 | 12.1 | 4,283 | 18.8 | 7,434 | 4,159 | 3,275 | 22 | 136 |
| Whitfield | 1,126 | 114,985 | 1.23 | 1,250 | 12.0 | 941 | 9.1 | 20,150 | 22.9 | 17,308 | 13,003 | 4,305 | NA | NA |
| Wilcox | 1,849 | 7,990 | 0.68 | 76 | 8.9 | 110 | 12.9 | 816 | 15.6 | 1,718 | 985 | 733 | NA | NA |
| Wilkes | 182 | 9,082 | 0.81 | 95 | 9.8 | 163 | 16.8 | 1,429 | 19.7 | 2,720 | 1,565 | 1,155 | 22 | 30 |
| Wilkinson | 179 | 8,981 | 0.99 | 96 | 10.9 | 138 | 15.7 | 1,134 | 16.3 | 2,337 | 1,313 | 1,024 | 10 | 34 |
| Worth | 177 | 16,370 | 0.53 | 196 | 9.8 | 275 | 13.8 | 2,746 | 17.2 | 4,380 | 2,464 | 1,916 | NA | NA |
| **HAWAII** | 42,318 | 1,423,351 | 1.00 | 16,420 | 11.7 | 13,768 | 9.8 | 53,289 | 4.8 | 280,940 | 147,885 | 133,055 | X | X |
| Hawaii | 3,929 | 199,419 | 1.00 | 2,078 | 10.2 | 2,227 | 11.0 | 8,771 | 5.6 | 46,305 | 29,323 | 16,982 | NA | NA |
| Honolulu | 34,383 | 986,282 | 1.00 | 11,793 | 12.2 | 9,226 | 9.6 | 33,662 | 4.4 | 187,165 | 94,183 | 92,982 | NA | NA |
| Kalawao | 3 | 60 | 0.88 | 0 | 0.0 | 0 | 0.0 | 0 | 0.0 | 0 | D | D | NA | NA |
| Kauai | 1,158 | 71,753 | 1.00 | 755 | 10.5 | 760 | 10.6 | 2,952 | 5.2 | 15,825 | 9,052 | 6,773 | NA | NA |
| Maui | 2,845 | 165,837 | 1.00 | 1,794 | 10.7 | 1,555 | 9.3 | 7,904 | 5.9 | 31,643 | 15,326 | 16,317 | 449 | 4,984 |
| **IDAHO** | 30,633 | 1,695,098 | 0.97 | 21,656 | 11.9 | 14,637 | 8.0 | 187,336 | 12.7 | 349,014 | 219,432 | 129,582 | X | X |
| Ada | 11,296 | 480,085 | 1.10 | 4,912 | 9.9 | 3,426 | 6.9 | 38,528 | 9.5 | 85,153 | 41,856 | 43,297 | 257 | 595 |
| Adams | 21 | 3,600 | 0.68 | 32 | 7.2 | 28 | 6.3 | 487 | 16.4 | 1,405 | 1,052 | 353 | 10 | 43 |
| Bannock | 2,010 | 83,886 | 0.95 | 1,092 | 12.3 | 734 | 8.3 | 8,817 | 12.1 | 15,611 | 10,004 | 5,607 | 5 | 37 |
| Bear Lake | 31 | 5,331 | 0.73 | 73 | 11.9 | 62 | 10.1 | 532 | 11.0 | 1,471 | 1,422 | 49 | NA | NA |
| Benewah | 71 | 9,563 | 1.12 | 94 | 10.0 | 126 | 13.4 | 1,055 | 14.9 | 2,774 | 2,657 | 117 | 25 | 36 |
| Bingham | 338 | 42,447 | 0.82 | 609 | 12.9 | 392 | 8.3 | 5,725 | 14.6 | 8,195 | 5,869 | 2,325 | 27 | 154 |
| Blaine | 258 | 23,150 | 1.06 | 186 | 7.9 | 136 | 5.8 | 3,511 | 19.0 | 5,164 | 4,198 | 966 | 5 | 18 |
| Boise | 34 | 6,489 | 0.70 | 47 | 5.8 | 66 | 8.2 | 696 | 12.1 | 2,294 | 1,248 | 1,046 | 17 | 62 |
| Bonner | 406 | 41,795 | 0.90 | 418 | 8.9 | 502 | 10.7 | 4,535 | 13.4 | 13,283 | 8,868 | 4,415 | 26 | 342 |
| Bonneville | 1,285 | 120,885 | 1.13 | 1,841 | 15.1 | 936 | 7.7 | 11,095 | 10.9 | 19,475 | 13,824 | 5,650 | 109 | 607 |
| Boundary | 67 | 11,595 | 0.94 | 153 | 12.1 | 137 | 10.8 | 1,526 | 16.5 | 3,532 | 2,481 | 1,051 | NA | NA |
| Butte | 18 | 4,909 | 3.45 | 21 | 7.9 | 25 | 9.4 | 206 | 10.5 | 679 | 654 | 24 | 7 | 1 |
| Camas | 0 | 799 | 0.49 | 6 | 5.3 | 11 | 9.7 | 95 | 11.4 | 259 | 203 | 56 | 3 | 8 |
| Canyon | 3,509 | 195,893 | 0.77 | 3,007 | 12.7 | 1,683 | 7.1 | 29,380 | 15.3 | 39,962 | 18,080 | 21,882 | 117 | 430 |

1. Per 1,000 estimated resident population.

Items 32—45

# Table B. States and Counties — Crime, Education, Money Income, and Poverty

| STATE County | County law enforcement employment, 2019 | | Education — School enrollment and attainment, 2015-2019 | | | | Local government expenditures[3] 2017-2018 | | Money income, 2015-2019 | | | | Income and poverty, 2019 | | | |
|---|---|---|---|---|---|---|---|---|---|---|---|---|---|---|---|---|
| | | | Enrollment[1] | | Attainment[2] (percent) | | | | | Households | | | | Percent below poverty level | | |
| | Officers | Civilians | Total | Percent private | High school graduate or less | Bachelor's degree or more | Total current spending (mil dol) | Current spending per student (dollars) | Per capita income[4] | Median income (dollars) | Percent with income of less than $50,000 | Percent with income of $200,000 or more | Median household income (dollars) | All persons | Children under 18 years | Children 5 to 17 years in families |
| | 46 | 47 | 48 | 49 | 50 | 51 | 52 | 53 | 54 | 55 | 56 | 57 | 58 | 59 | 60 | 61 |
| **GEORGIA—Cont'd** | | | | | | | | | | | | | | | | |
| Putnam | NA | NA | 4,460 | 19.6 | 50.0 | 23.4 | 37 | 12,602 | 31,504 | 51,606 | 47.4 | 4.5 | 62,282 | 15.1 | 27.5 | 24.3 |
| Quitman | NA | NA | 421 | 8.8 | 70.8 | 7.1 | 7 | 20,429 | 18,639 | 26,667 | 67.5 | 0.5 | 35,522 | 22.8 | 36.0 | 37.2 |
| Rabun | NA | NA | 2,750 | 12.3 | 48.3 | 24.2 | 31 | 13,956 | 28,266 | 42,500 | 57.1 | 3.6 | 46,639 | 13.0 | 20.9 | 20.4 |
| Randolph | NA | NA | 1,877 | 7.5 | 61.2 | 10.3 | 15 | 13,275 | 16,691 | 31,699 | 77.3 | 1.1 | 33,505 | 25.3 | 37.5 | 36.2 |
| Richmond | NA | NA | 50,767 | 13.7 | 47.2 | 21.4 | 323 | 10,234 | 22,787 | 42,728 | 56.2 | 2.4 | 44,810 | 21.7 | 34.4 | 34.7 |
| Rockdale | 246 | 24 | 24,206 | 12.2 | 42.7 | 26.3 | 189 | 11,332 | 28,320 | 60,189 | 40.5 | 4.7 | 60,281 | 12.1 | 18.8 | 16.9 |
| Schley | NA | NA | 1,498 | 4.2 | 47.3 | 13.3 | 16 | 12,214 | 20,965 | 44,448 | 55.5 | 0.9 | 51,129 | 15.9 | 23.3 | 21.0 |
| Screven | 8 | 14 | 3,036 | 9.7 | 58.0 | 15.4 | 24 | 10,344 | 22,213 | 39,842 | 56.6 | 2.0 | 42,331 | 24.1 | 33.0 | 33.0 |
| Seminole | NA | NA | 1,632 | 11.3 | 52.3 | 16.2 | 17 | 11,083 | 29,614 | 33,357 | 66.2 | 1.5 | 38,618 | 22.6 | 36.2 | 33.8 |
| Spalding | 95 | 81 | 14,480 | 10.8 | 58.0 | 17.8 | 117 | 11,227 | 24,582 | 47,111 | 52.2 | 3.4 | 49,237 | 17.7 | 29.0 | 28.6 |
| Stephens | NA | NA | 6,119 | 14.0 | 56.5 | 21.0 | 41 | 10,146 | 23,546 | 47,164 | 51.7 | 2.4 | 45,416 | 15.0 | 23.2 | 20.7 |
| Stewart | NA | NA | 786 | 9.0 | 68.3 | 12.2 | 8 | 16,361 | 16,705 | 29,732 | 68.6 | 2.8 | 34,566 | 34.7 | 36.4 | 34.2 |
| Sumter | 74 | 24 | 8,755 | 6.1 | 53.2 | 18.3 | 50 | 10,885 | 20,654 | 36,681 | 60.6 | 1.6 | 40,250 | 26.7 | 38.4 | 38.8 |
| Talbot | NA | NA | 1,105 | 15.1 | 63.2 | 12.1 | 0 | 0 | 21,357 | 38,549 | 61.9 | 0.9 | 41,045 | 19.6 | 33.1 | 35.1 |
| Taliaferro | NA | NA | 238 | 8.4 | 75.1 | 8.1 | 5 | NA | 21,416 | 37,446 | 67.1 | 0.8 | 37,046 | 22.5 | 35.3 | 35.7 |
| Tattnall | NA | NA | 5,785 | 12.4 | 62.3 | 13.2 | 36 | 9,374 | 18,319 | 40,730 | 57.1 | 2.8 | 41,457 | 26.5 | 33.6 | 35.9 |
| Taylor | 11 | 11 | 1,835 | 19.0 | 59.0 | 12.6 | 16 | 11,245 | 21,298 | 33,832 | 67.4 | 3.1 | 28,234 | 22.9 | 32.1 | 30.8 |
| Telfair | NA | NA | 2,425 | 3.3 | 76.9 | 8.6 | 19 | 10,563 | 17,202 | 32,405 | 70.1 | 1.6 | 36,264 | 27.7 | 34.0 | 30.2 |
| Terrell | NA | NA | 2,110 | 13.9 | 54.9 | 11.2 | 15 | 11,204 | 20,045 | 34,768 | 60.8 | 1.8 | 37,122 | 28.2 | 39.8 | 37.6 |
| Thomas | NA | NA | 11,105 | 12.2 | 46.7 | 24.7 | 97 | 11,115 | 27,454 | 43,740 | 54.8 | 3.9 | 47,978 | 18.0 | 25.0 | 24.2 |
| Tift | 51 | 52 | 10,796 | 6.5 | 52.7 | 18.1 | 82 | 10,442 | 22,974 | 45,639 | 53.9 | 2.5 | 47,764 | 21.5 | 31.0 | 30.6 |
| Toombs | NA | NA | 7,029 | 10.4 | 57.4 | 16.2 | 56 | 10,171 | 24,399 | 40,175 | 58.7 | 2.4 | 41,442 | 19.2 | 29.7 | 29.3 |
| Towns | 19 | 19 | 2,481 | 27.8 | 40.0 | 30.1 | 15 | 15,122 | 25,972 | 48,219 | 53.3 | 2.2 | 49,023 | 12.7 | 19.2 | 19.4 |
| Treutlen | 8 | 12 | 1,417 | 8.4 | 67.1 | 12.4 | 12 | 10,227 | 21,830 | 32,241 | 63.0 | 1.7 | 37,478 | 31.6 | 47.8 | 50.0 |
| Troup | NA | NA | 17,805 | 11.0 | 51.7 | 19.0 | 127 | 10,380 | 23,124 | 45,649 | 53.0 | 2.7 | 47,296 | 19.2 | 27.3 | 26.2 |
| Turner | NA | NA | 1,679 | 5.7 | 56.8 | 11.7 | 15 | 11,868 | 19,527 | 37,039 | 60.6 | 1.1 | 35,400 | 28.0 | 43.0 | 43.6 |
| Twiggs | NA | NA | 1,531 | 16.3 | 64.3 | 11.8 | 12 | 15,327 | 20,979 | 34,879 | 64.9 | 1.1 | 44,079 | 19.0 | 30.4 | 29.0 |
| Union | 41 | 5 | 3,491 | 19.0 | 45.0 | 22.5 | 34 | 11,787 | 28,953 | 50,021 | 50.0 | 3.9 | 57,049 | 11.6 | 19.4 | 18.7 |
| Upson | NA | NA | 5,949 | 10.6 | 56.1 | 13.4 | 44 | 10,506 | 21,747 | 37,640 | 62.2 | 1.2 | 40,926 | 19.1 | 28.6 | 27.3 |
| Walker | NA | NA | 14,835 | 12.8 | 53.3 | 16.6 | 113 | 11,193 | 24,247 | 46,157 | 53.8 | 2.8 | 47,548 | 14.7 | 20.1 | 17.4 |
| Walton | 188 | 18 | 23,641 | 14.7 | 49.1 | 21.7 | 158 | 9,956 | 27,889 | 61,599 | 39.6 | 4.6 | 67,007 | 10.8 | 16.7 | 16.2 |
| Ware | NA | NA | 7,619 | 8.0 | 57.8 | 15.2 | 71 | 11,546 | 20,411 | 36,869 | 61.0 | 2.1 | 37,642 | 26.3 | 41.9 | 40.3 |
| Warren | NA | NA | 977 | 13.7 | 66.6 | 12.3 | 9 | 14,502 | 23,448 | 37,203 | 63.0 | 1.8 | 37,313 | 26.5 | 40.3 | 40.5 |
| Washington | NA | NA | 4,735 | 10.2 | 63.4 | 12.3 | 37 | 11,788 | 20,303 | 38,068 | 60.6 | 1.3 | 41,674 | 21.4 | 29.2 | 28.6 |
| Wayne | NA | NA | 7,044 | 5.8 | 58.6 | 13.0 | 54 | 10,044 | 21,090 | 45,483 | 54.7 | 1.5 | 47,382 | 18.4 | 26.6 | 26.1 |
| Webster | 7 | 0 | 738 | 9.1 | 61.4 | 14.7 | 5 | 13,308 | 24,102 | 37,297 | 62.1 | 1.3 | 41,547 | 18.4 | 28.4 | 26.8 |
| Wheeler | NA | NA | 1,712 | 7.1 | 69.7 | 13.6 | 12 | 11,905 | 14,654 | 30,192 | 63.9 | 0.2 | 35,927 | 34.2 | 33.6 | 30.7 |
| White | NA | NA | 6,069 | 14.3 | 45.7 | 22.7 | 71 | 11,700 | 27,406 | 52,493 | 48.4 | 2.7 | 54,786 | 10.9 | 17.6 | 16.2 |
| Whitfield | NA | NA | 27,141 | 3.7 | 60.0 | 15.7 | 216 | 10,313 | 23,974 | 48,623 | 51.4 | 4.1 | 53,115 | 12.7 | 17.6 | 17.6 |
| Wilcox | NA | NA | 1,714 | 5.4 | 63.4 | 12.2 | 14 | 11,775 | 16,183 | 36,964 | 59.3 | 1.0 | 38,261 | 29.4 | 38.9 | 34.4 |
| Wilkes | NA | NA | 2,153 | 8.8 | 60.3 | 13.8 | 18 | 11,671 | 24,674 | 37,838 | 57.6 | 2.1 | 39,716 | 19.6 | 31.2 | 29.7 |
| Wilkinson | 15 | 13 | 2,216 | 15.6 | 65.9 | 10.9 | 19 | 13,297 | 22,141 | 38,996 | 59.2 | 1.6 | 43,902 | 17.0 | 29.4 | 27.7 |
| Worth | 27 | 15 | 4,589 | 7.4 | 59.8 | 11.0 | 32 | 9,877 | 24,415 | 45,398 | 53.3 | 1.8 | 43,142 | 20.3 | 31.5 | 30.4 |
| **HAWAII** | X | X | 325,063 | 22.9 | 35.4 | 33.0 | 2,756 | 15,242 | 35,567 | 81,275 | 30.1 | 10.3 | 83,734 | 9.0 | 11.2 | 10.5 |
| Hawaii | 434 | 130 | 42,786 | 17.4 | 37.8 | 29.4 | (7) | (7) | 30,542 | 62,409 | 41.4 | 5.6 | 64,929 | 13.1 | 18.8 | 18.1 |
| Honolulu | NA | NA | 234,111 | 24.6 | 34.0 | 35.0 | (7)2,756.3 | (7)15,242 | 36,816 | 85,857 | 27.7 | 11.8 | 86,946 | 7.9 | 9.2 | 8.7 |
| Kalawao | NA | NA | 4 | 75.0 | 37.1 | 24.2 | (7) | (7) | 49,118 | 69,375 | 23.1 | 0.0 | 0 | 0.0 | 0.0 | 0.0 |
| Kauai | 132 | 54 | 14,008 | 17.0 | 37.5 | 29.2 | (7) | (7) | 33,143 | 83,554 | 31.4 | 7.6 | 73,029 | 9.3 | 11.7 | 10.8 |
| Maui | 334 | 99 | 34,154 | 20.6 | 39.3 | 27.2 | (7) | (7) | 35,241 | 80,948 | 29.4 | 9.5 | 77,375 | 10.7 | 13.4 | 11.7 |
| **IDAHO** | X | X | 451,807 | 14.6 | 36.6 | 27.6 | 2,319 | 7,704 | 27,970 | 55,785 | 44.4 | 4.0 | 60,830 | 11.0 | 12.7 | 11.2 |
| Ada | 167 | 553 | 119,298 | 13.6 | 27.2 | 38.5 | 591 | 7,555 | 34,919 | 66,293 | 37.3 | 6.8 | 72,295 | 9.4 | 9.6 | 8.2 |
| Adams | 8 | 16 | 644 | 9.2 | 44.6 | 21.2 | 4 | 9,891 | 28,026 | 48,856 | 51.8 | 3.6 | 50,042 | 12.7 | 20.9 | 20.0 |
| Bannock | 42 | 83 | 25,570 | 9.6 | 34.0 | 27.8 | 103 | 6,887 | 25,076 | 51,734 | 48.6 | 2.3 | 53,306 | 13.5 | 15.0 | 13.2 |
| Bear Lake | 8 | 6 | 1,304 | 9.2 | 44.9 | 18.9 | 9 | 7,395 | 26,378 | 54,167 | 45.3 | 2.4 | 54,550 | 11.8 | 15.3 | 14.4 |
| Benewah | 10 | 18 | 1,774 | 12.9 | 50.9 | 15.2 | 15 | 10,836 | 23,255 | 47,556 | 53.8 | 1.6 | 49,303 | 15.4 | 23.1 | 21.7 |
| Bingham | 37 | 48 | 12,830 | 9.6 | 42.2 | 21.2 | 74 | 7,109 | 23,059 | 55,472 | 44.2 | 2.1 | 59,959 | 11.1 | 13.5 | 11.8 |
| Blaine | 50 | 14 | 4,571 | 10.5 | 36.9 | 35.4 | 54 | 15,157 | 33,645 | 56,694 | 42.9 | 5.4 | 75,778 | 7.2 | 7.9 | 6.8 |
| Boise | 13 | 10 | 1,352 | 11.5 | 36.2 | 27.1 | 9 | 10,632 | 31,975 | 59,646 | 43.7 | 3.7 | 64,443 | 11.6 | 17.0 | 15.6 |
| Bonner | 41 | 58 | 8,168 | 19.2 | 39.0 | 23.8 | 49 | 9,638 | 27,918 | 50,256 | 49.8 | 3.1 | 56,729 | 12.6 | 18.0 | 16.6 |
| Bonneville | 71 | 120 | 33,152 | 12.5 | 33.5 | 31.3 | 166 | 6,757 | 28,671 | 60,615 | 41.7 | 5.0 | 65,165 | 9.3 | 10.8 | 9.9 |
| Boundary | 8 | 19 | 2,311 | 27.5 | 50.9 | 19.8 | 13 | 8,386 | 25,782 | 43,423 | 55.0 | 4.4 | 50,709 | 14.8 | 21.7 | 21.8 |
| Butte | 4 | 10 | 559 | 11.8 | 43.3 | 16.2 | 4 | 8,573 | 30,294 | 42,132 | 55.1 | 3.8 | 51,742 | 14.6 | 18.0 | 15.5 |
| Camas | 4 | 2 | 244 | 12.3 | 49.9 | 16.3 | 2 | 14,370 | 30,209 | 39,688 | 55.8 | 7.9 | 56,014 | 9.0 | 6.9 | 6.9 |
| Canyon | 70 | 212 | 59,627 | 13.8 | 47.0 | 18.7 | 297 | 7,272 | 22,232 | 52,134 | 47.3 | 2.1 | 58,670 | 10.1 | 11.6 | 10.0 |

1. All persons 3 years old and over enrolled in nursery school through college.   2. Persons 25 years old and over.   3. Elementary and secondary education expenditures.   4. Based on population estimated by the American Community Survey, 2014–2018.   7. Hawaii, Kalawao, Kauai, and Maui counties are included with Honolulu county.

# Table B. States and Counties — **Personal Income**

| STATE County | Personal income, 2019 Total (mil dol) 62 | Percent change 2018-2019 63 | Per capita[1] Dollars 64 | Rank 65 | Wages and salaries (mil dol) 66 | Pension and insurance 67 | Government social insurance 68 | Proprietors' income (mil dol) 69 | Dividends, interest, and rent (mil dol) 70 | Personal transfer receipts (mil dol) 71 | Earnings, 2019 Total (mil dol) 72 | From employee and self-employed 73 | From employer 74 |
|---|---|---|---|---|---|---|---|---|---|---|---|---|---|
| **GEORGIA—Cont'd** | | | | | | | | | | | | | |
| Putnam | 987 | 3.6 | 44,643 | 1,402 | 218 | 47 | 15 | 37 | 244 | 250 | 317 | 25 | 15 |
| Quitman | 72 | 5.1 | 31,466 | 2,980 | 14 | 4 | 1 | 2 | 11 | 31 | 21 | 2 | 1 |
| Rabun | 704 | 3.3 | 41,058 | 1,915 | 190 | 36 | 14 | 53 | 234 | 201 | 293 | 23 | 14 |
| Randolph | 238 | 7.3 | 35,186 | 2,662 | 77 | 18 | 6 | 21 | 41 | 80 | 121 | 8 | 6 |
| Richmond | 7,973 | 3.3 | 39,370 | 2,131 | 6,278 | 1,410 | 459 | 562 | 1,571 | 2,138 | 8,710 | 466 | 459 |
| Rockdale | 3,285 | 3.8 | 36,138 | 2,553 | 1,731 | 266 | 116 | 124 | 474 | 786 | 2,237 | 141 | 116 |
| Schley | 174 | 4.3 | 33,044 | 2,885 | 42 | 12 | 3 | 12 | 25 | 38 | 69 | 4 | 3 |
| Screven | 494 | 4.3 | 35,346 | 2,641 | 124 | 28 | 9 | 17 | 77 | 159 | 177 | 14 | 9 |
| Seminole | 384 | 10.8 | 47,527 | 1,074 | 93 | 19 | 7 | 96 | 55 | 108 | 215 | 12 | 7 |
| Spalding | 2,384 | 3.7 | 35,740 | 2,592 | 978 | 196 | 66 | 97 | 386 | 710 | 1,337 | 90 | 66 |
| Stephens | 1,020 | 1.2 | 39,335 | 2,138 | 399 | 73 | 27 | 49 | 173 | 299 | 548 | 38 | 27 |
| Stewart | 153 | 7.7 | 23,081 | 3,106 | 75 | 13 | 5 | 7 | 26 | 56 | 101 | 7 | 5 |
| Sumter | 1,120 | 4.9 | 37,936 | 2,332 | 482 | 109 | 34 | 69 | 215 | 330 | 694 | 42 | 34 |
| Talbot | 221 | 3.7 | 35,705 | 2,595 | 33 | 8 | 2 | 8 | 38 | 76 | 51 | 5 | 2 |
| Taliaferro | 57 | 3.7 | 37,088 | 2,436 | 10 | 3 | 1 | 3 | 11 | 19 | 16 | 1 | 1 |
| Tattnall | 736 | -2.1 | 29,106 | 3,060 | 247 | 57 | 17 | 89 | 117 | 207 | 410 | 22 | 17 |
| Taylor | 263 | 6.1 | 32,801 | 2,906 | 82 | 17 | 6 | 10 | 45 | 88 | 115 | 9 | 6 |
| Telfair | 342 | 2.1 | 21,535 | 3,109 | 104 | 23 | 7 | 33 | 53 | 133 | 168 | 12 | 7 |
| Terrell | 372 | 7.0 | 43,654 | 1,530 | 82 | 18 | 5 | 32 | 77 | 110 | 138 | 9 | 5 |
| Thomas | 2,055 | 2.9 | 46,223 | 1,225 | 964 | 162 | 67 | 145 | 449 | 523 | 1,339 | 85 | 67 |
| Tift | 1,598 | 4.0 | 39,317 | 2,143 | 902 | 191 | 58 | 161 | 274 | 398 | 1,312 | 74 | 58 |
| Toombs | 997 | 3.5 | 37,143 | 2,429 | 471 | 89 | 33 | 55 | 150 | 291 | 649 | 41 | 33 |
| Towns | 472 | 4.4 | 39,181 | 2,164 | 132 | 29 | 9 | 29 | 131 | 183 | 198 | 18 | 9 |
| Treutlen | 205 | 2.7 | 29,682 | 3,046 | 40 | 11 | 3 | 12 | 29 | 69 | 66 | 5 | 3 |
| Troup | 2,635 | 3.8 | 37,690 | 2,360 | 1,995 | 308 | 138 | 97 | 446 | 673 | 2,539 | 153 | 138 |
| Turner | 294 | 7.9 | 36,836 | 2,465 | 78 | 17 | 5 | 28 | 42 | 100 | 128 | 8 | 5 |
| Twiggs | 350 | 5.2 | 43,163 | 1,591 | 109 | 16 | 9 | 52 | 38 | 111 | 186 | 13 | 9 |
| Union | 921 | 4.9 | 37,588 | 2,376 | 293 | 62 | 19 | 59 | 224 | 333 | 433 | 36 | 19 |
| Upson | 965 | 3.4 | 36,653 | 2,489 | 286 | 54 | 20 | 42 | 160 | 299 | 402 | 31 | 20 |
| Walker | 2,377 | 2.9 | 34,078 | 2,793 | 544 | 124 | 40 | 146 | 351 | 695 | 854 | 65 | 40 |
| Walton | 4,011 | 4.6 | 42,406 | 1,704 | 1,155 | 192 | 78 | 251 | 585 | 806 | 1,676 | 112 | 78 |
| Ware | 1,248 | 3.2 | 34,915 | 2,697 | 686 | 132 | 56 | 65 | 212 | 438 | 939 | 62 | 56 |
| Warren | 191 | 5.0 | 36,427 | 2,517 | 81 | 14 | 6 | 6 | 29 | 68 | 106 | 8 | 6 |
| Washington | 740 | 2.5 | 36,321 | 2,533 | 285 | 64 | 19 | 24 | 170 | 224 | 392 | 26 | 19 |
| Wayne | 977 | 2.9 | 32,633 | 2,913 | 385 | 80 | 26 | 49 | 132 | 301 | 540 | 36 | 26 |
| Webster | 90 | 7.8 | 34,700 | 2,726 | 26 | 5 | 2 | 8 | 17 | 22 | 41 | 3 | 2 |
| Wheeler | 153 | 2.4 | 19,472 | 3,113 | 45 | 10 | 3 | 12 | 20 | 57 | 70 | 5 | 3 |
| White | 1,085 | 3.7 | 35,238 | 2,654 | 327 | 61 | 23 | 69 | 191 | 303 | 480 | 36 | 23 |
| Whitfield | 4,177 | 2.4 | 39,927 | 2,057 | 2,758 | 423 | 194 | 439 | 914 | 834 | 3,815 | 224 | 194 |
| Wilcox | 253 | 4.5 | 29,334 | 3,056 | 44 | 12 | 3 | 43 | 41 | 83 | 102 | 6 | 3 |
| Wilkes | 417 | -0.9 | 42,644 | 1,666 | 116 | 26 | 8 | 42 | 77 | 124 | 191 | 12 | 8 |
| Wilkinson | 331 | 3.1 | 37,011 | 2,448 | 170 | 31 | 13 | 19 | 46 | 108 | 232 | 16 | 13 |
| Worth | 749 | 6.6 | 36,974 | 2,452 | 136 | 28 | 9 | 50 | 119 | 195 | 223 | 16 | 9 |
| **HAWAII** | 80,727 | 2.9 | 57,026 | X | 39,570 | 8,077 | 3,111 | 6,709 | 17,168 | 12,673 | 57,466 | 3,470 | 3,111 |
| Hawaii | 8,782 | 3.2 | 43,578 | 1,536 | 3,463 | 711 | 266 | 774 | 2,049 | 2,104 | 5,213 | 350 | 266 |
| Honolulu | 59,618 | 2.6 | 61,174 | 273 | 30,330 | 6,413 | 2,412 | 4,615 | 12,400 | 8,483 | 43,770 | 2,575 | 2,412 |
| Kalawao | [2]8,601 | [2]4.7 | [2]51,348 | [2]732 | [2]4,081 | [2]655 | [2]305 | [2]972 | [2]1,866 | [2] 1,381 | [2]6,013 | [2]384 | [2]305 |
| Kauai | 3,726 | 3.9 | 51,545 | 718 | 1,696 | 298 | 128 | 347 | 853 | 705 | 2,470 | 161 | 128 |
| Maui | [2] | [2] | [2] | [2] | [2] | [2] | [2] | [2] | [2] | [2] | [2] | [2] | [2] |
| **IDAHO** | 82,148 | 5.3 | 45,917 | X | 35,649 | 5,877 | 2,965 | 9,208 | 18,343 | 14,970 | 53,699 | 3,401 | 2,965 |
| Ada | 26,437 | 5.1 | 54,896 | 497 | 14,429 | 2,106 | 1,140 | 2,653 | 6,022 | 3,609 | 20,329 | 1,280 | 1,140 |
| Adams | 168 | 5.1 | 39,038 | 2,186 | 51 | 10 | 4 | 12 | 55 | 48 | 78 | 6 | 4 |
| Bannock | 3,446 | 4.0 | 39,246 | 2,155 | 1,487 | 292 | 129 | 148 | 600 | 794 | 2,055 | 139 | 129 |
| Bear Lake | 264 | 6.2 | 43,103 | 1,600 | 62 | 15 | 6 | 25 | 51 | 64 | 108 | 8 | 6 |
| Benewah | 369 | 4.8 | 39,695 | 2,088 | 159 | 31 | 14 | 31 | 74 | 64 | 236 | 17 | 14 |
| Bingham | 1,775 | 6.6 | 37,909 | 2,334 | 608 | 118 | 52 | 193 | 323 | 382 | 971 | 61 | 52 |
| Blaine | 2,696 | 1.2 | 117,097 | 12 | 622 | 82 | 52 | 323 | 1,564 | 186 | 1,078 | 66 | 52 |
| Boise | 339 | 5.8 | 43,320 | 1,568 | 56 | 13 | 5 | 21 | 78 | 83 | 94 | 9 | 5 |
| Bonner | 1,934 | 4.1 | 42,277 | 1,723 | 601 | 111 | 52 | 134 | 616 | 474 | 898 | 71 | 52 |
| Bonneville | 5,967 | 4.6 | 50,114 | 824 | 2,347 | 380 | 199 | 922 | 1,407 | 950 | 3,848 | 237 | 199 |
| Boundary | 467 | 5.4 | 38,142 | 2,313 | 160 | 31 | 15 | 47 | 103 | 129 | 253 | 19 | 15 |
| Butte | 116 | 6.4 | 44,828 | 1,381 | 830 | 76 | 62 | 13 | 22 | 28 | 981 | 60 | 62 |
| Camas | 52 | 5.9 | 47,417 | 1,090 | 18 | 3 | 1 | 13 | 13 | 10 | 35 | 2 | 1 |
| Canyon | 7,917 | 6.5 | 34,446 | 2,750 | 2,945 | 504 | 257 | 738 | 1,146 | 1,806 | 4,443 | 300 | 257 |

1. Based on the resident population estimated as of July 1 of the year shown.  2. Kalawao county is included with Maui county.

# Table B. States and Counties — Earnings, Social Security, and Housing

| STATE County | Earnings, 2019 (cont.) Percent by selected industries | | | | | | | | | Social Security beneficiaries, December 2019 | | Supplemental Security Income recipients, 2019 | Housing units, 2020 | |
|---|---|---|---|---|---|---|---|---|---|---|---|---|---|---|
| | Farm | Mining, quarrying, and extractions | Construction | Manufacturing | Information; professional, scientific, technical services | Retail trade | Finance, insurance, real estate, and leasing | Health care and social assistance | Government | Number | Rate[1] | | Total | Percent change, 2010-2020 |
| | 75 | 76 | 77 | 78 | 79 | 80 | 81 | 82 | 83 | 84 | 85 | 86 | 87 | 88 |

| | | | | | | | | | | | | | | |
|---|---|---|---|---|---|---|---|---|---|---|---|---|---|---|
| **GEORGIA—Cont'd** | | | | | | | | | | | | | | |
| Putnam | 1.0 | 0.0 | 9.2 | 11.8 | D | 8.8 | 8.3 | D | 27.4 | 6,255 | 283 | 541 | 13,480 | 5.3 |
| Quitman | 2.6 | 0.0 | 0.6 | D | D | 4.6 | D | D | 36.3 | 800 | 348 | 130 | 2,085 | 1.9 |
| Rabun | -0.1 | D | 12.5 | 6.4 | 4.2 | 13.3 | 6.3 | D | 17.9 | 5,565 | 325 | 409 | 12,792 | 3.9 |
| Randolph | 8.2 | 0.0 | D | D | D | 5.6 | D | D | 24.9 | 1,765 | 259 | 395 | 4,096 | -1.4 |
| Richmond | 0.0 | D | 2.9 | 8.0 | 6.7 | 4.4 | 5.8 | 13.4 | 41.9 | 41,455 | 205 | 8,309 | 89,713 | 3.9 |
| Rockdale | 0.0 | 0.0 | D | 17.1 | 7.8 | 8.1 | 7.6 | 12.0 | 13.7 | 17,410 | 192 | 2,065 | 34,324 | 3.2 |
| Schley | 5.2 | 0.0 | D | 31.5 | D | 3.9 | D | D | 29.4 | 930 | 177 | 141 | 2,240 | 1.4 |
| Screven | 3.2 | 0.0 | 4.6 | 26.7 | D | 6.7 | 4.2 | D | 30.4 | 3,690 | 265 | 658 | 6,852 | 1.7 |
| Seminole | 16.9 | 0.0 | 1.3 | 19.0 | D | 3.7 | 5.3 | 12.3 | 11.9 | 2,560 | 316 | 422 | 4,845 | 1.0 |
| Spalding | 0.0 | 0.1 | 4.1 | 16.4 | D | 7.0 | 4.6 | 15.4 | 24.5 | 16,530 | 247 | 2,750 | 28,150 | 5.0 |
| Stephens | 1.7 | D | 5.0 | 24.4 | 2.5 | 8.1 | 3.0 | D | 18.4 | 7,485 | 288 | 1,211 | 12,622 | -0.3 |
| Stewart | 4.0 | 0.0 | 0.4 | 1.0 | D | 2.1 | D | D | 26.6 | 1,135 | 171 | 253 | 2,341 | -1.8 |
| Sumter | 5.7 | D | 2.1 | 19.6 | D | 6.2 | 6.0 | D | 21.9 | 6,740 | 229 | 1,281 | 13,880 | -0.2 |
| Talbot | 3.5 | 17.3 | 25.7 | 0.1 | D | 1.7 | 6.6 | D | 25.9 | 1,755 | 284 | 291 | 3,449 | 1.2 |
| Taliaferro | 15.5 | 0.0 | D | D | 0.0 | D | D | D | 45.3 | 495 | 320 | 100 | 1,027 | 1.1 |
| Tattnall | 19.5 | 0.0 | 5.4 | 2.4 | D | 4.1 | 3.0 | D | 25.9 | 4,550 | 180 | 924 | 10,110 | 1.5 |
| Taylor | 1.4 | D | 17.6 | 2.9 | D | 6.1 | D | 12.9 | 20.6 | 2,005 | 249 | 417 | 4,637 | 1.6 |
| Telfair | 6.4 | 0.0 | D | D | 1.6 | 6.0 | 4.1 | 5.0 | 25.3 | 2,755 | 174 | 564 | 7,307 | 0.2 |
| Terrell | 16.8 | 0.0 | 3.1 | 3.9 | D | 11.6 | 6.9 | D | 23.5 | 2,460 | 287 | 503 | 4,163 | 0.0 |
| Thomas | 0.8 | D | 3.2 | 17.5 | 3.5 | 6.2 | 8.9 | 21.4 | 15.4 | 11,470 | 258 | 2,170 | 20,950 | 3.8 |
| Tift | 1.8 | 0.1 | 4.2 | 6.4 | 3.1 | 8.2 | 7.6 | 6.6 | 33.4 | 8,505 | 209 | 1,623 | 17,133 | 4.2 |
| Toombs | 3.1 | 0.0 | 4.7 | 14.4 | D | 9.0 | 5.0 | D | 14.2 | 6,295 | 235 | 1,262 | 12,268 | 1.2 |
| Towns | -0.6 | D | 8.2 | D | 9.8 | 5.4 | 7.1 | D | 14.6 | 5,195 | 433 | 219 | 8,418 | 8.9 |
| Treutlen | 5.4 | 0.0 | 3.3 | 8.4 | 0.5 | 9.2 | D | D | 29.7 | 1,570 | 229 | 314 | 3,037 | 1.5 |
| Troup | 0.0 | D | 5.5 | 34.0 | 2.7 | 7.4 | 4.8 | 8.7 | 10.1 | 14,855 | 212 | 2,373 | 29,102 | 3.8 |
| Turner | 18.0 | 0.0 | 0.7 | 12.6 | D | 7.7 | D | 4.1 | 23.1 | 2,120 | 267 | 447 | 3,943 | 2.6 |
| Twiggs | 1.2 | D | D | D | D | D | D | D | 9.1 | 2,625 | 322 | 397 | 4,280 | 1.0 |
| Union | 0.5 | D | 7.2 | 4.8 | D | 8.4 | 4.8 | D | 27.8 | 9,540 | 388 | 478 | 15,222 | 8.3 |
| Upson | 3.2 | 0.1 | 4.7 | 19.6 | 3.8 | 7.3 | 4.0 | 20.5 | 20.7 | 7,365 | 279 | 1,150 | 12,172 | 0.1 |
| Walker | 1.8 | 0.1 | 6.6 | 30.2 | D | 5.5 | 6.0 | 5.5 | 21.5 | 16,800 | 241 | 2,101 | 30,711 | 2.1 |
| Walton | 0.1 | 0.1 | 15.6 | 15.3 | 3.2 | 6.1 | 5.3 | 8.6 | 15.9 | 19,120 | 202 | 2,560 | 35,517 | 9.5 |
| Ware | 0.8 | 0.0 | 3.4 | 12.6 | D | 10.0 | 3.7 | D | 20.5 | 8,250 | 230 | 1,743 | 16,909 | 3.6 |
| Warren | 0.8 | D | 1.9 | 30.0 | D | 3.4 | D | D | 13.3 | 1,570 | 301 | 252 | 2,993 | 0.3 |
| Washington | 0.2 | 3.7 | 5.3 | 8.5 | 3.3 | 5.4 | 4.6 | D | 27.9 | 5,095 | 250 | 853 | 9,588 | 6.0 |
| Wayne | 1.3 | 0.0 | 12.4 | 20.1 | D | 8.2 | 3.1 | D | 29.9 | 6,760 | 225 | 1,086 | 12,399 | 1.6 |
| Webster | 12.5 | D | 0.5 | D | 0.1 | 4.7 | D | 0.2 | 16.1 | 590 | 229 | 86 | 1,551 | 1.8 |
| Wheeler | 7.9 | 0.0 | D | D | D | 2.5 | D | D | 22.8 | 1,275 | 162 | 203 | 2,637 | 0.3 |
| White | 1.9 | 0.1 | 13.7 | 11.1 | 3.8 | 9.4 | 4.1 | D | 15.7 | 7,930 | 259 | 603 | 16,546 | 3.1 |
| Whitfield | 0.0 | D | 1.9 | 34.2 | 8.8 | 5.3 | 6.5 | 10.8 | 9.8 | 19,425 | 186 | 2,505 | 40,612 | 1.8 |
| Wilcox | 35.6 | 0.0 | D | D | D | D | D | D | 3.7 | 1,800 | 209 | 329 | 3,542 | 0.9 |
| Wilkes | 16.7 | D | 10.3 | 13.7 | D | 6.2 | 4.1 | 4.7 | 23.6 | 3,035 | 311 | 437 | 5,177 | 0.4 |
| Wilkinson | 0.3 | 36.1 | 9.7 | 6.1 | D | 2.1 | D | 3.1 | 10.2 | 2,650 | 297 | 367 | 4,501 | 0.3 |
| Worth | 20.1 | 0.0 | 4.1 | 6.5 | D | 7.7 | D | D | 23.3 | 4,590 | 227 | 655 | 9,401 | 1.7 |
| **HAWAII** | 0.4 | 0.1 | 7.8 | 1.7 | 7.1 | 5.9 | 6.4 | 10.3 | 28.8 | NA | NA | NA | 554,088 | 6.7 |
| Hawaii | 2.2 | 0.0 | D | D | 5.7 | 8.7 | D | D | 24.5 | 47,130 | 233 | 4,979 | 90,258 | 9.6 |
| Honolulu | 0.2 | 0.1 | 7.5 | 1.8 | 7.8 | 5.3 | 6.7 | 10.5 | 31.9 | 181,575 | 187 | 14,894 | 356,788 | 5.9 |
| Kalawao | (2) | (2) | (2) | (2) | (2) | (2) | (2) | (2) | (2) | 10 | 115 | 0 | 112 | -0.9 |
| Kauai | 0.3 | D | D | D | 4.4 | 7.9 | D | D | 19.0 | 16,235 | 225 | 0 | 31,747 | 6.5 |
| Maui | (2)0.6 | (2)D | (2)9.0 | (2)1.2 | (2)4.8 | (2)7.8 | (2)5.8 | (2)10.2 | (2)13.4 | 32,060 | 191 | 1,976 | 75,183 | 6.8 |
| **IDAHO** | 4.3 | 0.5 | 7.8 | 11.3 | 8.8 | 8.2 | 6.3 | 11.9 | 15.7 | 358,633 | 200 | 31,007 | 768,815 | 15.1 |
| Ada | 0.3 | 0.2 | 7.5 | 12.4 | 10.7 | 7.6 | 8.0 | 13.4 | 13.3 | 85,370 | 177 | 6,755 | 199,614 | 25.2 |
| Adams | 3.9 | 0.1 | 8.6 | 10.3 | 5.6 | 6.3 | 2.2 | D | 26.7 | 1,420 | 333 | 64 | 2,709 | 2.7 |
| Bannock | 0.3 | D | 5.7 | 7.9 | 4.7 | 7.8 | 6.0 | 15.8 | 24.4 | 15,775 | 180 | 2,001 | 35,166 | 5.9 |
| Bear Lake | 10.8 | D | 2.9 | 3.7 | D | 7.9 | 4.1 | D | 36.0 | 1,560 | 254 | 94 | 4,257 | 8.8 |
| Benewah | 0.3 | D | 4.3 | 17.2 | D | 4.1 | 2.5 | D | 33.0 | 2,920 | 314 | 236 | 4,831 | 4.4 |
| Bingham | 9.2 | 0.1 | 7.2 | 13.4 | 2.6 | 5.3 | 6.2 | 10.6 | 22.7 | 8,920 | 191 | 961 | 17,254 | 6.9 |
| Blaine | 0.5 | D | 15.0 | 2.3 | 14.4 | 6.4 | 13.7 | 7.2 | 9.1 | 4,790 | 208 | 92 | 15,703 | 4.3 |
| Boise | 0.6 | 0.4 | 11.9 | 2.6 | D | 4.0 | D | D | 31.7 | 2,410 | 305 | 107 | 5,775 | 9.1 |
| Bonner | -0.1 | 1.4 | 9.2 | 15.6 | 7.8 | 9.3 | 6.0 | 18.2 | 13,610 | 297 | 831 | 25,489 | 3.3 | |
| Bonneville | 1.2 | D | 6.6 | 9.4 | 7.1 | 18.6 | 5.2 | 16.4 | 11.4 | 20,580 | 173 | 2,363 | 45,393 | 14.3 |
| Boundary | 3.3 | D | 11.8 | 12.3 | 4.6 | 6.4 | 6.0 | 6.1 | 26.4 | 3,605 | 292 | 259 | 5,661 | 9.4 |
| Butte | 1.2 | D | D | D | D | 0.2 | D | 0.7 | 1.7 | 720 | 275 | 86 | 1,383 | 2.1 |
| Camas | 28.0 | 0.2 | 9.3 | D | D | D | D | D | 18.5 | 265 | 243 | 10 | 882 | 6.1 |
| Canyon | 3.6 | 0.1 | 14.3 | 15.6 | 4.6 | 8.7 | 4.2 | 11.2 | 12.8 | 41,485 | 180 | 4,495 | 83,359 | 20.1 |

1. Per 1,000 resident population estimated as of July 1 of the year shown.   2. Kalawao county is included with Maui county.

# Table B. States and Counties — Housing, Labor Force, and Employment

| STATE County | Housing units, 2015-2019 Occupied units Owner-occupied Total | Percent | Median value[1] | Median owner cost as a percent of income With a mortgage | Without a mortgage[2] | Renter-occupied Median rent[3] | Median rent as a percent of income[2] | Sub-standard units[4] (percent) | Civilian labor force, 2020 Total | Percent change, 2019-2020 | Unemployment Total | Rate[5] | Civilian employment[6], 2015-2019 Total | Percent Management, business, science, and arts | Construction, production, and maintenance occupations |
|---|---|---|---|---|---|---|---|---|---|---|---|---|---|---|---|
| | 89 | 90 | 91 | 92 | 93 | 94 | 95 | 96 | 97 | 98 | 99 | 100 | 101 | 102 | 103 |
| **GEORGIA—Cont'd** | | | | | | | | | | | | | | | |
| Putnam | 8,937 | 77 | 160,400 | 21.4 | 10.3 | 788 | 30.2 | 1.3 | 8,210 | -1.4 | 539 | 6.6 | 9,501 | 25.9 | 34.6 |
| Quitman | 842 | 70.2 | 59,900 | 25.4 | 17.4 | 709 | 40.3 | 5.0 | 800 | -1.5 | 52 | 6.5 | 817 | 14.2 | 39.7 |
| Rabun | 6,662 | 74.3 | 174,500 | 24.6 | 10.2 | 738 | 33.1 | 2.8 | 7,215 | -1.4 | 384 | 5.3 | 6,150 | 29.2 | 27.0 |
| Randolph | 2,553 | 57.2 | 71,900 | 27.1 | 13.9 | 583 | 24.4 | 4.9 | 2,490 | -2.6 | 147 | 5.9 | 2,397 | 29.7 | 36.1 |
| Richmond | 71,400 | 52.6 | 108,000 | 21.5 | 11.0 | 888 | 33.8 | 1.9 | 85,112 | -0.3 | 6,463 | 7.6 | 82,032 | 29.9 | 23.6 |
| Rockdale | 31,076 | 66.4 | 163,900 | 21.4 | 10.0 | 1,056 | 28.9 | 2.8 | 43,989 | -1.4 | 3,332 | 7.6 | 40,693 | 34.1 | 26.7 |
| Schley | 1,864 | 71.5 | 125,300 | 19.7 | 13.0 | 626 | 33.6 | 1.7 | 2,145 | -5.0 | 93 | 4.3 | 2,245 | 32.1 | 31.1 |
| Screven | 5,098 | 73.5 | 81,200 | 22.4 | 10.3 | 663 | 29.1 | 4.4 | 5,047 | -2.2 | 412 | 8.2 | 5,473 | 28.1 | 37.8 |
| Seminole | 3,363 | 65.3 | 81,100 | 20.6 | 12.9 | 721 | 29.9 | 5.6 | 3,093 | -0.4 | 155 | 5.0 | 2,967 | 33.4 | 26.4 |
| Spalding | 24,336 | 62.5 | 122,800 | 21.2 | 11.8 | 872 | 29.6 | 1.9 | 28,305 | -1.5 | 2,149 | 7.6 | 26,767 | 27.8 | 34.2 |
| Stephens | 9,543 | 69.8 | 105,300 | 21.2 | 10.9 | 750 | 30.2 | 2.3 | 10,414 | -3.9 | 669 | 6.4 | 10,507 | 27.5 | 33.1 |
| Stewart | 1,816 | 73.5 | 49,200 | 24.9 | 13.8 | 529 | 31.8 | 4.0 | 2,509 | 0.5 | 142 | 5.7 | 1,485 | 29.6 | 30.0 |
| Sumter | 11,510 | 55.5 | 95,700 | 21.3 | 13.2 | 736 | 31.5 | 3.9 | 12,525 | -2.7 | 924 | 7.4 | 11,813 | 30.6 | 31.3 |
| Talbot | 2,809 | 79 | 86,400 | 26.0 | 13.0 | 652 | 24.4 | 2.4 | 2,757 | -2.5 | 189 | 6.9 | 2,590 | 21.4 | 41.0 |
| Taliaferro | 593 | 69 | 80,200 | 32.6 | 13.4 | 547 | 26.4 | 2.7 | 530 | -2.9 | 38 | 7.2 | 598 | 19.2 | 41.3 |
| Tattnall | 8,241 | 67.2 | 89,200 | 18.8 | 10.5 | 543 | 21.5 | 5.6 | 9,651 | -1.1 | 414 | 4.3 | 7,974 | 31.5 | 32.7 |
| Taylor | 3,473 | 70.5 | 83,600 | 21.1 | 13.4 | 619 | 36.0 | 2.2 | 2,746 | -9.6 | 183 | 6.7 | 3,090 | 28.3 | 34.6 |
| Telfair | 4,668 | 67 | 64,600 | 23.5 | 10.0 | 599 | 28.8 | 4.0 | 3,782 | -12.0 | 374 | 9.9 | 5,055 | 24.4 | 46.7 |
| Terrell | 3,399 | 55 | 92,100 | 21.4 | 10.8 | 638 | 34.1 | 4.3 | 3,497 | -1.5 | 262 | 7.5 | 3,356 | 22.1 | 32.5 |
| Thomas | 17,595 | 61.2 | 146,800 | 20.6 | 13.1 | 861 | 34.0 | 3.4 | 16,930 | -0.5 | 1,056 | 6.2 | 19,498 | 35.6 | 24.8 |
| Tift | 15,144 | 60 | 118,100 | 23.0 | 10.0 | 689 | 27.4 | 1.6 | 20,500 | 0.9 | 1,037 | 5.1 | 18,424 | 30.6 | 29.7 |
| Toombs | 10,030 | 60.5 | 95,300 | 19.7 | 10.9 | 598 | 26.6 | 4.8 | 12,081 | 0.5 | 794 | 6.6 | 10,475 | 26.0 | 34.3 |
| Towns | 4,898 | 78.3 | 230,500 | 24.3 | 10.8 | 717 | 27.2 | 1.1 | 3,841 | -3.4 | 256 | 6.7 | 4,438 | 31.5 | 19.7 |
| Treutlen | 2,490 | 66.9 | 69,400 | 27.8 | 14.1 | 585 | 18.4 | 3.7 | 2,710 | -0.7 | 163 | 6.0 | 2,443 | 26.6 | 36.1 |
| Troup | 24,928 | 57.4 | 136,700 | 21.3 | 11.3 | 823 | 32.3 | 2.6 | 37,448 | -1.5 | 2,721 | 7.3 | 29,625 | 28.7 | 35.2 |
| Turner | 3,169 | 69.1 | 79,700 | 22.0 | 13.6 | 538 | 26.8 | 1.7 | 3,264 | 1.3 | 236 | 7.2 | 3,100 | 29.9 | 26.8 |
| Twiggs | 3,044 | 80.9 | 65,100 | 22.9 | 13.7 | 613 | 29.2 | 2.1 | 2,790 | -2.4 | 207 | 7.4 | 2,895 | 19.3 | 38.7 |
| Union | 9,743 | 77.2 | 196,500 | 21.9 | 10.0 | 780 | 30.6 | 1.0 | 10,464 | -0.4 | 458 | 4.4 | 8,606 | 37.2 | 25.2 |
| Upson | 10,154 | 67.6 | 93,300 | 19.9 | 11.4 | 639 | 30.3 | 1.8 | 11,328 | -0.8 | 750 | 6.6 | 10,270 | 30.2 | 31.3 |
| Walker | 25,975 | 71.8 | 120,600 | 20.5 | 10.8 | 753 | 29.9 | 3.0 | 30,601 | -2.6 | 1,495 | 4.9 | 29,880 | 27.3 | 32.6 |
| Walton | 31,670 | 74.3 | 190,300 | 20.4 | 10.0 | 1,000 | 33.4 | 1.7 | 44,917 | -2.8 | 2,441 | 5.4 | 41,903 | 31.4 | 29.4 |
| Ware | 13,823 | 63.6 | 83,100 | 19.3 | 11.4 | 658 | 31.8 | 1.4 | 15,284 | -1.4 | 852 | 5.6 | 13,292 | 26.2 | 31.6 |
| Warren | 2,244 | 66.6 | 66,200 | 22.4 | 14.4 | 631 | 29.2 | 2.7 | 2,635 | -4.6 | 179 | 6.8 | 2,115 | 20.7 | 40.4 |
| Washington | 7,503 | 66.6 | 84,800 | 22.7 | 13.3 | 646 | 25.5 | 4.2 | 6,931 | -1.9 | 445 | 6.4 | 7,341 | 27.1 | 30.9 |
| Wayne | 10,400 | 63.5 | 105,400 | 19.7 | 10.9 | 655 | 22.6 | 3.5 | 11,540 | 0.1 | 646 | 5.6 | 10,919 | 26.2 | 32.7 |
| Webster | 1,140 | 83.9 | 65,700 | 20.9 | 10.3 | 381 | 33.7 | 0.7 | 1,008 | -1.9 | 57 | 5.7 | 1,080 | 32.7 | 31.2 |
| Wheeler | 1,862 | 63.3 | 58,800 | 19.9 | 12.3 | 450 | 34.7 | 3.1 | 1,608 | -4.1 | 116 | 7.2 | 2,491 | 20.6 | 48.1 |
| White | 11,695 | 75.9 | 176,300 | 21.3 | 10.0 | 827 | 27.8 | 1.8 | 16,197 | -2.3 | 706 | 4.4 | 12,767 | 34.3 | 23.4 |
| Whitfield | 35,629 | 64.6 | 133,400 | 19.2 | 10.1 | 720 | 24.5 | 6.6 | 43,275 | -1.5 | 3,138 | 7.3 | 47,674 | 23.4 | 45.2 |
| Wilcox | 2,575 | 76.1 | 61,800 | 19.2 | 10.0 | 513 | 25.6 | 2.6 | 2,737 | 0.8 | 156 | 5.7 | 2,624 | 28.0 | 36.8 |
| Wilkes | 3,979 | 67.5 | 87,200 | 21.9 | 15.8 | 762 | 32.8 | 1.0 | 3,712 | -2.6 | 248 | 6.7 | 4,092 | 25.5 | 32.5 |
| Wilkinson | 3,185 | 75.2 | 70,700 | 17.6 | 11.5 | 635 | 34.4 | 3.0 | 3,790 | -6.5 | 203 | 5.4 | 3,276 | 25.0 | 36.5 |
| Worth | 8,002 | 65.4 | 98,700 | 21.0 | 11.4 | 722 | 30.6 | 1.0 | 8,844 | -2.3 | 475 | 5.4 | 8,829 | 25.7 | 37.4 |
| **HAWAII** | 459,424 | 58.9 | 615,300 | 25.6 | 10.0 | 1,617 | 32.5 | 9.6 | 648,191 | -3.1 | 75,395 | 11.6 | 680,258 | 34.7 | 18.9 |
| Hawaii | 69,453 | 67.7 | 350,000 | 24.4 | 10.0 | 1,180 | 29.6 | 8.9 | 89,119 | -2.6 | 10,133 | 11.4 | 88,098 | 32.4 | 21.1 |
| Honolulu | 312,795 | 56.2 | 678,200 | 25.5 | 10.0 | 1,745 | 33.6 | 9.7 | 440,915 | -2.8 | 44,779 | 10.2 | 470,749 | 36.5 | 18.3 |
| Kalawao | 39 | NA | NA | NA | NA | 1,063 | 13.2 | NA | NA | NA | NA | NA | 52 | 23.1 | 13.5 |
| Kauai | 22,658 | 63.2 | 570,700 | 26.5 | 10.0 | 1,375 | 28.2 | 8.5 | 35,337 | -4.1 | 5,719 | 16.2 | 36,460 | 27.8 | 20.3 |
| Maui | 54,479 | 61 | 633,500 | 27.1 | 10.0 | 1,510 | 29.5 | 10.8 | 82,820 | -5.3 | 14,764 | 17.8 | 84,899 | 30.2 | 19.4 |
| **IDAHO** | 630,008 | 70 | 212,300 | 20.9 | 10.0 | 853 | 28.3 | 3.1 | 892,151 | 1.4 | 47,786 | 5.4 | 792,237 | 34.9 | 25.3 |
| Ada | 173,353 | 69.3 | 270,800 | 20.2 | 10.0 | 995 | 28.6 | 1.9 | 257,677 | 1.6 | 13,893 | 5.4 | 229,029 | 44.0 | 16.1 |
| Adams | 1,757 | 80.3 | 215,500 | 22.1 | 10.0 | 685 | 26.8 | 3.1 | 1,758 | 2.7 | 164 | 9.3 | 1,573 | 29.9 | 28.4 |
| Bannock | 31,155 | 67.5 | 160,000 | 19.3 | 10.0 | 691 | 27.4 | 2.6 | 42,560 | 0.2 | 2,065 | 4.9 | 37,593 | 36.5 | 22.4 |
| Bear Lake | 2,424 | 76.6 | 149,800 | 18.2 | 10.0 | 589 | 24.8 | 3.3 | 3,075 | 3.7 | 129 | 4.2 | 2,585 | 27.2 | 32.3 |
| Benewah | 3,418 | 73.3 | 167,000 | 24.4 | 10.2 | 644 | 23.9 | 5.6 | 4,140 | 1.7 | 310 | 7.5 | 3,482 | 23.4 | 35.4 |
| Bingham | 14,989 | 76.8 | 160,100 | 20.0 | 10.0 | 662 | 25.1 | 3.8 | 23,776 | 0.2 | 916 | 3.9 | 19,732 | 31.6 | 31.2 |
| Blaine | 7,995 | 70.1 | 428,900 | 24.6 | 11.6 | 933 | 28.7 | 2.7 | 12,666 | 2.9 | 987 | 7.8 | 12,325 | 27.9 | 24.9 |
| Boise | 3,205 | 82.8 | 232,300 | 22.4 | 10.0 | 754 | 26.6 | 2.4 | 3,609 | 2.9 | 288 | 8.0 | 2,947 | 41.4 | 23.5 |
| Bonner | 17,537 | 75.7 | 254,200 | 24.7 | 10.0 | 853 | 28.7 | 3.6 | 21,119 | 5.6 | 1,592 | 7.5 | 18,299 | 28.6 | 33.1 |
| Bonneville | 39,768 | 69.6 | 181,200 | 19.2 | 10.0 | 819 | 28.2 | 3.5 | 58,136 | 2.1 | 2,347 | 4.0 | 51,226 | 37.7 | 22.8 |
| Boundary | 4,653 | 76.7 | 205,000 | 21.5 | 10.7 | 613 | 26.1 | 6.0 | 5,502 | 3.0 | 339 | 6.2 | 4,670 | 26.1 | 31.7 |
| Butte | 967 | 82.3 | 128,900 | 22.2 | 10.3 | 656 | 24.5 | 1.2 | 1,393 | 1.6 | 50 | 3.6 | 951 | 34.9 | 30.4 |
| Camas | 394 | 75.6 | 183,300 | 24.2 | 10.0 | 853 | 32.2 | 1.3 | 684 | 0.6 | 33 | 4.8 | 493 | 36.5 | 24.9 |
| Canyon | 72,807 | 69.5 | 178,300 | 21.2 | 10.0 | 896 | 29.0 | 4.4 | 105,981 | 1.4 | 6,218 | 5.9 | 96,436 | 26.2 | 31.7 |

1. Specified owner-occupied units. lacking complete plumbing facilities.   2. A value of 10.0 represents 10 percent or less; a value of 50.0 represents 50 percent or more.   3. Specified renter-occupied units.   4. Overcrowded or   5. Percent of civilian labor force.   6. Civilian employed persons 16 years old and over.

# Table B. States and Counties — Nonfarm Employment and Agriculture

| STATE County | Private nonfarm establishments, employment and payroll, 2019 | | | | | | | Annual payroll | | Agriculture, 2017 | | | Farm producers whose primary occupation is farming (percent) |
|---|---|---|---|---|---|---|---|---|---|---|---|---|---|
| | | Employment | | | | | | | | Farms | | | |
| | | | | | | | Professional, scientific, and technical services | | | | Percent with: | | |
| | Number of establish-ments | Total | Health care and social assistance | Manufac-turing | Retail trade | Finance and insurance | | Total (mil dol) | Average per employee (dollars) | Number | Fewer than 50 acres | 1000 acres or more | |
| | 104 | 105 | 106 | 107 | 108 | 109 | 110 | 111 | 112 | 113 | 114 | 115 | 116 |
| GEORGIA—Cont'd | | | | | | | | | | | | | |
| Putnam | 460 | 4,789 | 501 | 1,208 | 752 | 129 | 88 | 168 | 35,085 | 186 | 24.7 | 2.2 | 42.6 |
| Quitman | 34 | 197 | 12 | NA | 32 | 9 | NA | 6 | 29,376 | 37 | 8.1 | 10.8 | 22.2 |
| Rabun | 497 | 4,873 | 541 | 308 | 1,084 | 119 | 133 | 147 | 30,193 | 135 | 68.9 | 0.7 | 34.2 |
| Randolph | 129 | 1,443 | 269 | 175 | 154 | 37 | 25 | 50 | 34,769 | 153 | 5.9 | 21.6 | 42.5 |
| Richmond | 4,261 | 85,150 | 24,230 | 7,843 | 10,883 | 1,456 | 3,935 | 3,845 | 45,154 | 118 | 63.6 | 0.8 | 34.2 |
| Rockdale | 2,073 | 33,671 | 4,092 | 5,323 | 5,049 | 640 | 833 | 1,607 | 47,734 | 74 | 75.7 | 1.4 | 42.6 |
| Schley | 60 | 741 | 27 | 415 | 85 | NA | 5 | 28 | 38,197 | 89 | 12.4 | 7.9 | 30.9 |
| Screven | 221 | 2,324 | 284 | 848 | 375 | 102 | 46 | 83 | 35,601 | 352 | 22.2 | 12.8 | 41.4 |
| Seminole | 178 | 1,587 | 440 | 23 | 282 | 72 | 56 | 58 | 36,376 | 157 | 25.5 | 12.1 | 68.4 |
| Spalding | 1,137 | 16,786 | 3,816 | 3,021 | 2,894 | 416 | 401 | 665 | 39,611 | 225 | 66.2 | 0.4 | 36.9 |
| Stephens | 541 | 8,468 | 1,221 | 2,790 | 1,234 | 182 | 128 | 314 | 37,080 | 227 | 45.8 | NA | 44.5 |
| Stewart | 55 | 731 | 103 | NA | 70 | 13 | 8 | 31 | 41,860 | 104 | 13.5 | 8.7 | 22.9 |
| Sumter | 604 | 8,744 | 2,691 | 1,148 | 1,406 | 203 | 287 | 294 | 33,596 | 371 | 18.9 | 12.4 | 35.2 |
| Talbot | 63 | 585 | 15 | NA | 68 | 67 | NA | 23 | 39,444 | 102 | 22.5 | 4.9 | 38.3 |
| Taliaferro | 14 | 40 | NA | NA | 17 | NA | NA | 1 | 28,875 | 48 | 4.2 | 4.2 | 40.5 |
| Tattnall | 303 | 3,648 | 862 | 111 | 428 | 125 | 52 | 130 | 35,590 | 547 | 37.7 | 3.3 | 44.8 |
| Taylor | 126 | 1,690 | 222 | 64 | 119 | 42 | NA | 41 | 24,398 | 224 | 13.4 | 5.4 | 32.4 |
| Telfair | 176 | 2,968 | 146 | 1,337 | 329 | 70 | 23 | 79 | 26,714 | 255 | 25.5 | 3.5 | 41.6 |
| Terrell | 173 | 1,808 | 152 | 360 | 389 | 66 | 41 | 61 | 33,920 | 256 | 22.3 | 16.4 | 42.5 |
| Thomas | 1,129 | 17,052 | 3,666 | 2,260 | 2,349 | 713 | 416 | 748 | 43,851 | 408 | 31.1 | 12.5 | 41.3 |
| Tift | 1,078 | 18,014 | 3,681 | 1,553 | 2,841 | 427 | 651 | 702 | 38,962 | 306 | 38.2 | 16.3 | 44.0 |
| Toombs | 655 | 9,338 | 2,010 | 1,438 | 1,773 | 275 | 186 | 321 | 34,397 | 320 | 36.6 | 6.9 | 35.3 |
| Towns | 281 | 2,748 | 562 | 39 | 410 | 62 | 69 | 87 | 31,639 | 105 | 61.0 | NA | 46.8 |
| Treutlen | 85 | 669 | 138 | 98 | 213 | 39 | NA | 19 | 28,857 | 148 | 26.4 | 4.7 | 26.8 |
| Troup | 1,458 | 34,576 | 3,684 | 11,300 | 3,363 | 887 | 485 | 1,557 | 45,035 | 261 | 46.7 | 1.5 | 36.0 |
| Turner | 153 | 1,614 | 184 | 420 | 282 | 59 | 33 | 53 | 32,672 | 246 | 30.1 | 9.3 | 35.2 |
| Twiggs | 74 | 1,600 | 154 | NA | 72 | 9 | 7 | 61 | 38,162 | 116 | 43.1 | 12.1 | 30.8 |
| Union | 564 | 6,214 | 1,450 | 451 | 1,176 | 149 | 184 | 246 | 39,549 | 251 | 61.8 | NA | 36.0 |
| Upson | 457 | 5,503 | 1,409 | 1,216 | 1,004 | 186 | 103 | 204 | 37,016 | 235 | 46.0 | 1.7 | 34.2 |
| Walker | 680 | 11,267 | 959 | 4,577 | 1,371 | 260 | 154 | 356 | 31,607 | 624 | 44.9 | 1.6 | 35.7 |
| Walton | 1,754 | 18,603 | 2,690 | 2,759 | 2,734 | 412 | 470 | 727 | 39,104 | 437 | 55.8 | 0.7 | 41.5 |
| Ware | 866 | 12,007 | 2,384 | 1,789 | 2,413 | 250 | 224 | 420 | 34,988 | 248 | 50.4 | 7.7 | 40.3 |
| Warren | 79 | 905 | 130 | 389 | 68 | 6 | 11 | 41 | 45,699 | 135 | 19.3 | 4.4 | 42.4 |
| Washington | 346 | 5,227 | 862 | 536 | 741 | 147 | 240 | 225 | 43,132 | 383 | 26.4 | 6.8 | 35.7 |
| Wayne | 578 | 6,172 | 1,188 | 896 | 1,142 | 150 | 147 | 245 | 39,726 | 316 | 46.8 | 3.8 | 36.9 |
| Webster | 32 | 367 | NA | NA | 52 | 6 | NA | 16 | 43,003 | 109 | 18.3 | 20.2 | 45.2 |
| Wheeler | 59 | 725 | NA | NA | 76 | NA | NA | 26 | 36,051 | 143 | 21.7 | 10.5 | 33.5 |
| White | 649 | 6,523 | 820 | 640 | 1,229 | 172 | 136 | 217 | 33,215 | 301 | 67.4 | 0.3 | 40.4 |
| Whitfield | 2,178 | 48,524 | 5,491 | 17,624 | 4,909 | 679 | 1,508 | 2,181 | 44,953 | 386 | 52.3 | 0.5 | 36.9 |
| Wilcox | 80 | 671 | 111 | 14 | 82 | 61 | 5 | 18 | 27,522 | 287 | 30.0 | 9.1 | 41.0 |
| Wilkes | 194 | 1,880 | 300 | 321 | 370 | 77 | 32 | 67 | 35,612 | 277 | 25.3 | 5.1 | 39.3 |
| Wilkinson | 136 | 2,030 | 135 | 243 | 182 | 41 | 29 | 104 | 51,281 | 140 | 22.1 | 3.6 | 24.0 |
| Worth | 265 | 2,917 | 423 | 630 | 503 | 119 | 71 | 98 | 33,598 | 469 | 32.0 | 15.4 | 44.9 |
| HAWAII | 32,889 | 553,206 | 73,182 | 12,416 | 71,593 | 21,448 | 23,056 | 25,839 | 46,707 | 7,328 | 89.5 | 1.7 | 46.7 |
| Hawaii | 4,196 | 56,111 | 7,988 | 1,669 | 10,068 | 1,139 | 1,413 | 2,294 | 40,876 | 4,220 | 89.4 | 1.4 | 46.4 |
| Honolulu | 21,420 | 368,904 | 54,323 | 9,411 | 47,291 | 18,468 | 18,773 | 17,932 | 48,610 | 927 | 91.3 | 1.7 | 55.4 |
| Kalawao | NA | NA | NA | NA | NA | NA | NA | NA | NA | NA | NA | NA | NA |
| Kauai | 2,196 | 28,272 | 3,254 | 385 | 4,212 | 407 | 719 | 1,193 | 42,204 | 773 | 87.8 | 2.2 | 44.0 |
| Maui | 4,838 | 67,706 | 7,605 | 951 | 9,968 | 892 | 1,794 | 2,993 | 44,200 | 1,408 | 89.6 | 2.3 | 43.8 |
| IDAHO | 50,547 | 616,778 | 102,857 | 63,485 | 84,201 | 23,403 | 36,701 | 26,955 | 43,703 | 24,996 | 56.0 | 9.7 | 40.8 |
| Ada | 14,788 | 214,169 | 37,708 | 15,713 | 25,971 | 10,116 | 14,827 | 11,027 | 51,486 | 1,304 | 86.8 | 2.1 | 28.8 |
| Adams | 110 | 535 | 43 | 133 | 105 | 7 | 21 | 18 | 33,273 | 232 | 42.2 | 13.8 | 49.2 |
| Bannock | 2,154 | 25,109 | 5,341 | 2,105 | 4,568 | 2,475 | 902 | 914 | 36,386 | 757 | 51.9 | 9.0 | 35.1 |
| Bear Lake | 130 | 998 | 287 | 44 | 235 | 37 | 16 | 31 | 30,595 | 395 | 25.8 | 14.4 | 39.3 |
| Benewah | 252 | 2,187 | 345 | 547 | 296 | 50 | 61 | 91 | 41,412 | 288 | 42.7 | 8.7 | 41.2 |
| Bingham | 908 | 10,970 | 1,890 | 2,438 | 1,362 | 232 | 329 | 433 | 39,516 | 1,177 | 59.8 | 13.7 | 42.9 |
| Blaine | 1,532 | 11,872 | 777 | 318 | 1,401 | 247 | 903 | 507 | 42,734 | 190 | 47.4 | 23.2 | 42.8 |
| Boise | 175 | 770 | 18 | 28 | 100 | 8 | 8 | 19 | 24,870 | 90 | 68.9 | 7.8 | 45.0 |
| Bonner | 1,656 | 13,063 | 1,763 | 1,806 | 2,203 | 294 | 509 | 488 | 37,329 | 1,213 | 72.2 | 0.6 | 29.7 |
| Bonneville | 3,681 | 54,154 | 9,930 | 3,365 | 7,539 | 1,400 | 9,110 | 2,656 | 49,049 | 1,109 | 67.1 | 8.4 | 31.0 |
| Boundary | 404 | 2,641 | 523 | 494 | 457 | 52 | 129 | 98 | 37,120 | 348 | 48.9 | 4.6 | 39.1 |
| Butte | 54 | 334 | 116 | NA | 112 | 10 | 7 | 10 | 30,347 | 189 | 24.3 | 27.0 | 58.1 |
| Camas | 30 | 190 | 6 | NA | 22 | NA | NA | 5 | 26,337 | 151 | 18.5 | 27.2 | 47.0 |
| Canyon | 4,785 | 59,146 | 8,457 | 10,616 | 8,377 | 1,146 | 1,492 | 2,210 | 37,362 | 2,289 | 77.7 | 2.6 | 36.8 |

# Table B. States and Counties — Agriculture

| | | | Land in farms | | | Value of land and buildings (dollars) | | Value of machinery and equiopmnet, average per farm (dollars) | Value of products sold: | | | | Organic farms (number) | Farms with internet access (percent) | Government payments | |
|---|---|---|---|---|---|---|---|---|---|---|---|---|---|---|---|---|
| | | | Acres | | | | | | | | Percent from: | | | | | |
| STATE County | Acreage (1,000) | Percent change, 2012-2017 | Average size of farm | Total irrigated (1,000) | Total cropland (1,000) | Average per farm | Average per acre | | Total (mil dol) | Average per farm (acres) | Crops | Livestock and poultry products | | | Total ($1,000) | Percent of farms |
| | 117 | 118 | 119 | 120 | 121 | 122 | 123 | 124 | 125 | 126 | 127 | 128 | 129 | 130 | 131 | 132 |
| GEORGIA—Cont'd | | | | | | | | | | | | | | | | |
| Putnam | 38 | 34.3 | 206 | 1.8 | 14.7 | 865,498 | 4,206 | 100,518 | 34.7 | 186,780 | 20.0 | 80.0 | NA | 71.5 | 333 | 16.1 |
| Quitman | 19 | 111.8 | 521 | D | 4.2 | 1,201,990 | 2,307 | 64,583 | D | D | D | D | NA | 56.8 | 265 | 56.8 |
| Rabun | 8 | -5.7 | 56 | D | 3.1 | 464,254 | 8,240 | 94,314 | D | D | D | D | 3 | 84.4 | 149 | 11.9 |
| Randolph | 122 | 2.2 | 797 | 33.6 | 66.3 | 2,256,188 | 2,832 | 275,608 | 43.4 | 283,784 | 85.0 | 15.0 | NA | 57.5 | 4,997 | 77.8 |
| Richmond | 13 | -4.4 | 113 | 0.1 | 4.1 | 414,735 | 3,680 | 64,945 | D | D | 100.0 | D | 3 | 76.3 | D | 1.7 |
| Rockdale | 4 | -22.8 | 57 | 0.0 | 1.0 | 402,413 | 7,060 | 38,569 | 0.5 | 6,108 | 49.6 | 50.4 | NA | 78.4 | D | 1.4 |
| Schley | 35 | -1.5 | 392 | D | 8.5 | 1,116,611 | 2,848 | 60,512 | 14.8 | 166,843 | 18.9 | 81.1 | NA | 51.7 | 474 | 70.8 |
| Screven | 187 | 3.7 | 532 | 32.4 | 80.9 | 1,169,455 | 2,200 | 206,003 | 50.6 | 143,838 | 95.7 | 4.3 | 2 | 80.4 | 4,807 | 58.8 |
| Seminole | 105 | 19.0 | 669 | 41.0 | 79.5 | 2,100,309 | 3,141 | 286,662 | 61.9 | 394,401 | 92.3 | 7.7 | 1 | 69.4 | 6,536 | 65.6 |
| Spalding | 17 | -9.5 | 76 | 0.1 | 4.8 | 387,738 | 5,121 | 32,050 | 9.3 | 41,338 | 34.9 | 65.1 | NA | 78.7 | 107 | 5.3 |
| Stephens | 20 | 6.0 | 86 | 0.0 | 4.8 | 496,202 | 5,774 | 112,819 | 114.3 | 503,595 | 0.3 | 99.7 | NA | 76.7 | 359 | 37.4 |
| Stewart | 51 | -13.8 | 491 | 0.3 | 7.6 | 1,016,280 | 2,069 | 71,304 | 5.1 | 48,692 | 23.4 | 76.6 | NA | 59.6 | 534 | 56.7 |
| Sumter | 175 | 9.0 | 471 | 41.3 | 96.7 | 1,346,756 | 2,859 | 258,509 | 133.2 | 359,005 | 54.5 | 45.5 | 1 | 76.3 | 4,274 | 58.8 |
| Talbot | 30 | -10.9 | 296 | 0.0 | 4.8 | 809,499 | 2,735 | 70,159 | 1.1 | 10,520 | 35.3 | 64.7 | NA | 79.4 | 242 | 18.6 |
| Taliaferro | 18 | 30.2 | 374 | 0.1 | 2.2 | 1,231,989 | 3,292 | 81,632 | 24.3 | 505,479 | 0.9 | 99.1 | NA | 70.8 | 323 | 39.6 |
| Tattnall | 114 | 5.7 | 208 | 14.6 | 52.5 | 804,395 | 3,874 | 140,385 | 387.7 | 708,722 | 28.7 | 71.3 | 5 | 77.3 | 1,663 | 32.7 |
| Taylor | 64 | 4.3 | 286 | 4.1 | 20.2 | 595,317 | 2,079 | 82,999 | 27.7 | 123,790 | 37.0 | 63.0 | NA | 72.8 | 686 | 52.2 |
| Telfair | 52 | -21.7 | 205 | 6.4 | 17.5 | 452,315 | 2,210 | 59,880 | 10.3 | 40,475 | 91.7 | 8.3 | NA | 71.8 | 725 | 48.2 |
| Terrell | 134 | 11.1 | 524 | 32.0 | 88.2 | 1,468,061 | 2,803 | 285,046 | 53.1 | 207,543 | 92.9 | 7.1 | NA | 78.9 | 7,191 | 74.2 |
| Thomas | 187 | 8.1 | 459 | 11.4 | 85.3 | 1,670,886 | 3,640 | 166,382 | 78.7 | 192,958 | 68.1 | 31.9 | 1 | 78.9 | 5,807 | 52.2 |
| Tift | 121 | 42.8 | 394 | 33.3 | 77.0 | 1,251,034 | 3,174 | 203,433 | 84.0 | 274,425 | 95.4 | 4.6 | 3 | 79.4 | 6,522 | 56.2 |
| Toombs | 81 | 10.1 | 252 | 16.5 | 39.4 | 700,382 | 2,780 | 153,211 | 83.2 | 260,081 | 88.9 | 11.1 | 4 | 70.6 | 572 | 37.2 |
| Towns | 7 | -20.0 | 64 | 0.0 | 2.0 | 352,577 | 5,497 | 70,921 | 2.2 | 21,257 | 70.8 | 29.2 | NA | 80.0 | 85 | 27.6 |
| Treutlen | 37 | 5.2 | 250 | 1.4 | 11.5 | 470,130 | 1,883 | 73,055 | 6.1 | 40,939 | 95.1 | 4.9 | NA | 62.2 | 233 | 34.5 |
| Troup | 45 | 38.2 | 172 | 0.2 | 6.4 | 557,147 | 3,246 | 64,542 | 5.9 | 22,778 | 46.6 | 53.4 | 6 | 83.1 | 288 | 15.7 |
| Turner | 92 | 6.4 | 376 | 27.9 | 58.2 | 1,159,576 | 3,084 | 220,779 | 65.2 | 265,240 | 57.1 | 42.9 | 1 | 76.8 | 4,558 | 58.1 |
| Twiggs | 39 | 1.4 | 338 | 3.5 | 14.0 | 817,257 | 2,417 | 82,020 | 7.1 | 61,362 | 88.8 | 11.2 | 1 | 69.8 | 495 | 35.3 |
| Union | 19 | -6.1 | 77 | 0.0 | 7.3 | 446,988 | 5,771 | 77,107 | 37.4 | 148,940 | 15.0 | 85.0 | 1 | 74.9 | 158 | 27.9 |
| Upson | 32 | -29.3 | 135 | 0.6 | 9.5 | 532,497 | 3,954 | 76,595 | 42.9 | 182,468 | 6.2 | 93.8 | 5 | 77.4 | 113 | 10.2 |
| Walker | 91 | 13.9 | 145 | 0.2 | 26.1 | 616,558 | 4,239 | 72,803 | 152.4 | 244,223 | 3.1 | 96.9 | 1 | 80.9 | 1,100 | 27.7 |
| Walton | 47 | -9.6 | 109 | 0.8 | 13.4 | 614,553 | 5,664 | 57,774 | 26.6 | 60,899 | 23.0 | 77.0 | 1 | 75.7 | 276 | 19.7 |
| Ware | 63 | 12.0 | 256 | 4.3 | 22.6 | 656,805 | 2,565 | 103,573 | 31.7 | 127,879 | 74.3 | 25.7 | NA | 71.0 | 1,210 | 20.6 |
| Warren | 38 | 10.7 | 282 | D | 11.3 | 841,802 | 2,982 | 57,368 | 3.1 | 22,807 | 48.2 | 51.8 | NA | 63.7 | 444 | 25.2 |
| Washington | 96 | -3.8 | 251 | 8.8 | 35.1 | 528,574 | 2,106 | 87,028 | 19.9 | 51,893 | 65.0 | 35.0 | 1 | 70.2 | 1,684 | 47.5 |
| Wayne | 63 | 0.4 | 198 | 5.8 | 20.3 | 411,476 | 2,080 | 89,301 | 27.5 | 86,889 | 48.4 | 51.6 | NA | 72.2 | 539 | 19.0 |
| Webster | 60 | 24.3 | 548 | 10.4 | 33.3 | 1,338,185 | 2,442 | 377,331 | 22.8 | 209,394 | 79.6 | 20.4 | NA | 67.0 | 1,529 | 79.8 |
| Wheeler | 57 | 11.8 | 396 | 0.7 | 7.3 | 724,035 | 1,829 | 51,312 | 3.4 | 23,853 | 90.7 | 9.3 | NA | 74.1 | 269 | 54.5 |
| White | 19 | -20.3 | 62 | 0.1 | 6.2 | 505,000 | 8,126 | 77,920 | 92.8 | 308,399 | 1.8 | 98.2 | NA | 84.1 | 388 | 13.6 |
| Whitfield | 37 | -6.5 | 95 | D | 9.1 | 585,687 | 6,185 | 64,119 | 136.8 | 354,433 | 1.3 | 98.7 | NA | 79.0 | 565 | 33.7 |
| Wilcox | 91 | -21.4 | 316 | 19.0 | 54.1 | 824,117 | 2,608 | 132,330 | 98.6 | 343,700 | 37.7 | 62.3 | NA | 72.1 | 3,486 | 71.1 |
| Wilkes | 91 | -2.8 | 329 | 0.6 | 18.9 | 1,133,797 | 3,447 | 67,477 | 154.8 | 558,939 | 3.5 | 96.5 | 1 | 72.6 | 536 | 26.4 |
| Wilkinson | 30 | 90.8 | 217 | 0.4 | 11.0 | 444,442 | 2,050 | 50,675 | 6.2 | 44,186 | 40.0 | 60.0 | 1 | 52.9 | 183 | 20.7 |
| Worth | 218 | -5.1 | 464 | 49.2 | 129.9 | 1,394,353 | 3,003 | 249,754 | 104.3 | 222,354 | 83.9 | 16.1 | NA | 77.0 | 9,709 | 54.2 |
| HAWAII | 1,135 | 0.5 | 155 | 45.5 | 191.2 | 1,445,188 | 9,328 | 50,701 | 563.8 | 76,938 | 74.0 | 26.0 | 167 | 76.1 | 8,362 | 9.0 |
| Hawaii | 664 | -3.3 | 157 | 6.7 | 82.3 | 1,091,739 | 6,934 | 41,883 | 269.2 | 63,789 | 59.5 | 40.5 | 89 | 75.9 | 5,339 | 7.9 |
| Honolulu | 72 | 3.8 | 77 | 11.7 | 23.1 | 1,920,259 | 24,794 | 92,036 | 151.4 | 163,305 | 90.4 | 9.6 | 25 | 71.1 | 350 | 10.8 |
| Kalawao | NA | NA | NA | NA | NA | NA | NA | NA | NA | NA | NA | NA | NA | NA | NA | NA |
| Kauai | 150 | 4.2 | 194 | 22.3 | 29.3 | 1,744,739 | 8,982 | 47,852 | 61.0 | 78,946 | 75.2 | 24.8 | 13 | 70.4 | 499 | 12.5 |
| Maui | 249 | 8.6 | 177 | 4.8 | 56.6 | 2,027,299 | 11,466 | 51,466 | 82.2 | 58,385 | 90.3 | 9.7 | 40 | 83.5 | 2,174 | 9.4 |
| IDAHO | 11,692 | -0.6 | 468 | 3,398.3 | 5,894.7 | 1,340,738 | 2,866 | 175,951 | 7,567.4 | 302,746 | 42.4 | 57.6 | 295 | 83.9 | 129,605 | 24.3 |
| Ada | 112 | -22.0 | 86 | 57.3 | 62.9 | 819,575 | 9,511 | 80,276 | 131.6 | 100,936 | 32.5 | 67.5 | 9 | 88.3 | 471 | 4.4 |
| Adams | 163 | 19.7 | 703 | 22.2 | 16.9 | 1,089,369 | 1,550 | 83,372 | 12.6 | 54,306 | 25.4 | 74.6 | NA | 78.9 | 93 | 7.8 |
| Bannock | 315 | 6.8 | 416 | 40.0 | 181.5 | 812,963 | 1,953 | 97,293 | 37.8 | 49,943 | 56.8 | 43.2 | 1 | 81.6 | 5,606 | 33.9 |
| Bear Lake | 297 | 15.2 | 752 | 54.7 | 108.3 | 1,081,929 | 1,439 | 141,605 | 36.5 | 92,443 | 47.9 | 52.1 | 15 | 76.5 | 1,549 | 37.5 |
| Benewah | 140 | -4.6 | 486 | 0.2 | 82.1 | 940,817 | 1,936 | 128,882 | 19.1 | 66,358 | 96.6 | 3.4 | 2 | 76.7 | 1,931 | 34.4 |
| Bingham | 933 | 7.3 | 793 | 333.9 | 397.7 | 2,016,632 | 2,544 | 258,056 | 453.1 | 384,999 | 77.8 | 22.2 | 6 | 84.9 | 13,317 | 24.8 |
| Blaine | 211 | 17.9 | 1,112 | 37.3 | 52.1 | 2,813,206 | 2,530 | 179,893 | 27.2 | 142,905 | 61.4 | 38.6 | 15 | 90.0 | 800 | 20.5 |
| Boise | 53 | D | 591 | 1.3 | 1.7 | 847,323 | 1,433 | 52,205 | 2.6 | 28,733 | 59.8 | 40.2 | NA | 75.6 | 6 | 3.3 |
| Bonner | 89 | 10.8 | 74 | 1.2 | 32.8 | 370,858 | 5,036 | 42,446 | 10.2 | 8,406 | 60.2 | 39.8 | 10 | 79.7 | 224 | 2.9 |
| Bonneville | 419 | 2.3 | 378 | 131.6 | 260.6 | 1,100,960 | 2,915 | 130,606 | 167.9 | 151,363 | 66.0 | 34.0 | 2 | 82.8 | 7,179 | 22.4 |
| Boundary | 69 | -8.4 | 198 | 2.3 | 44.5 | 946,934 | 4,784 | 100,250 | 30.8 | 88,509 | 87.4 | 12.6 | 4 | 79.3 | 1,129 | 16.7 |
| Butte | 130 | 4.1 | 690 | 69.4 | 78.6 | 1,415,857 | 2,053 | 212,368 | 42.2 | 223,169 | 84.6 | 15.4 | 6 | 92.6 | 1,877 | 52.9 |
| Camas | 193 | 14.9 | 1,276 | 27.4 | 98.6 | 1,710,551 | 1,341 | 171,377 | 24.7 | 163,490 | 82.9 | 17.1 | 39 | 74.2 | 826 | 42.4 |
| Canyon | 275 | -9.5 | 120 | 213.4 | 219.4 | 989,782 | 8,240 | 161,943 | 574.8 | 251,095 | 54.7 | 45.3 | 11 | 86.4 | 2,463 | 8.0 |

# Table B. States and Counties — Water Use, Wholesale Trade, Retail Trade, and Real Estate

| STATE County | Water use, 2015 | | Wholesale Trade[1], 2017 | | | | Retail Trade[2], 2017 | | | | Real estate and rental and leasing,[2] 2017 | | | |
|---|---|---|---|---|---|---|---|---|---|---|---|---|---|---|
| | Public supply water withdrawn (mil gal/day) | Public supply gallons withdrawn per person per day | Number of establishments | Number of employees | Sales (mil dol) | Average payroll (mil dol) | Number of establishments | Number of employees | Sales (mil dol) | Average payroll (mil dol) | Number of establishments | Number of employees | Sales (mil dol) | Average payroll (mil dol) |
| | 133 | 134 | 135 | 136 | 137 | 138 | 139 | 140 | 141 | 142 | 143 | 144 | 145 | 146 |
| GEORGIA—Cont'd | | | | | | | | | | | | | | |
| Putnam | 3.76 | 176.1 | 22 | 141 | 92.5 | 7.0 | 79 | 1,004 | 219.6 | 20.6 | 20 | 46 | 11.8 | 2.7 |
| Quitman | 0.17 | 73.8 | NA | NA | NA | NA | 13 | 62 | 13.2 | 1.1 | NA | NA | NA | NA |
| Rabun | 1.60 | 98.3 | NA | NA | NA | NA | 75 | 1,039 | 272.8 | 27.8 | D | D | D | D |
| Randolph | 0.88 | 122.3 | 9 | 77 | 86.0 | 3.6 | 25 | 130 | 30.3 | 2.9 | 3 | 4 | 1.3 | 0.1 |
| Richmond | 39.17 | 194.1 | 172 | 1,995 | 1,162.3 | 104.9 | 769 | 11,085 | 2,842.6 | 265.1 | 197 | 837 | 230.7 | 37.6 |
| Rockdale | 11.76 | 132.3 | 76 | 574 | 456.6 | 36.0 | 303 | 4,896 | 1,285.0 | 121.2 | 83 | 505 | 156.8 | 28.8 |
| Schley | 0.45 | 87.1 | 4 | 80 | 22.5 | 3.0 | 10 | 73 | 13.5 | 1.5 | NA | NA | NA | NA |
| Screven | 0.74 | 52.3 | 4 | 12 | 5.2 | 0.4 | 50 | 384 | 95.4 | 8.3 | 4 | 8 | 1.1 | 0.4 |
| Seminole | 0.50 | 57.8 | 11 | 68 | 78.3 | 4.1 | 41 | 286 | 69.8 | 5.5 | 42 | 141 | 23.0 | 3.9 |
| Spalding | 6.83 | 106.6 | 45 | 619 | 372.5 | 38.9 | 225 | 3,025 | 850.0 | 72.9 | 12 | 47 | 13.2 | 1.8 |
| Stephens | 3.38 | 132.1 | D | D | D | D | 103 | 1,157 | 273.3 | 28.3 | NA | NA | NA | NA |
| Stewart | 0.78 | 133.3 | NA | NA | NA | NA | 13 | 66 | 19.8 | 1.5 | D | D | D | D |
| Sumter | 2.48 | 80.6 | 29 | 423 | 125.6 | 13.5 | 134 | 1,402 | 295.3 | 30.1 | NA | NA | NA | NA |
| Talbot | 1.33 | 209.9 | NA | NA | NA | NA | 12 | 57 | 16.3 | 1.2 | NA | NA | NA | NA |
| Taliaferro | 0.05 | 30.5 | NA | NA | NA | NA | 3 | 15 | 2.8 | 0.3 | NA | NA | NA | NA |
| Tattnall | 1.22 | 48.4 | 17 | 495 | 244.4 | 23.2 | 63 | 460 | 104.8 | 9.2 | 6 | 18 | 1.5 | 0.3 |
| Taylor | 0.62 | 74.4 | D | D | D | D | 21 | 113 | 45.1 | 3.2 | 3 | 8 | 0.7 | 0.1 |
| Telfair | 1.22 | 74.4 | 6 | 33 | 18.9 | 1.2 | 43 | 324 | 78.7 | 5.9 | NA | NA | NA | NA |
| Terrell | 1.44 | 158.0 | 8 | D | 347.0 | D | 41 | 368 | 92.6 | 10.0 | D | D | D | D |
| Thomas | 5.64 | 125.2 | D | D | D | 27.8 | 212 | 2,332 | 642.5 | 56.8 | 50 | 144 | 36.3 | 4.8 |
| Tift | 4.78 | 117.3 | 66 | 1,436 | 1,230.2 | 69.8 | 231 | 2,894 | 882.4 | 71.2 | 45 | 179 | 34.7 | 7.1 |
| Toombs | 2.84 | 104.3 | D | D | D | D | 148 | 1,764 | 510.2 | 44.1 | D | D | D | D |
| Towns | 1.54 | 137.7 | 8 | 15 | 2.2 | 0.3 | 62 | 406 | 83.3 | 7.6 | 17 | 47 | 9.5 | 1.7 |
| Treutlen | 0.30 | 44.2 | NA | NA | NA | NA | 23 | 222 | 42.6 | 4.1 | NA | NA | NA | NA |
| Troup | 8.54 | 122.4 | 53 | 583 | 495.4 | 26.5 | 264 | 3,278 | 923.8 | 83.2 | 73 | 202 | 56.6 | 7.8 |
| Turner | 0.79 | 96.2 | 14 | 214 | 92.5 | 7.7 | 40 | 270 | 91.9 | 6.8 | D | D | D | 1.3 |
| Twiggs | 0.45 | 53.6 | NA | NA | NA | NA | 13 | 65 | 21.3 | 1.1 | NA | NA | NA | NA |
| Union | 1.82 | 81.7 | 9 | 76 | 42.3 | 3.1 | 95 | 1,250 | 321.4 | 27.8 | 34 | 56 | 11.8 | 2.0 |
| Upson | 3.02 | 114.5 | 7 | 15 | 9.6 | 0.5 | 89 | 1,048 | 223.9 | 23.2 | D | D | D | D |
| Walker | 6.62 | 97.3 | 36 | 229 | 123.5 | 10.4 | 148 | 1,391 | 335.6 | 28.9 | 15 | 30 | 3.8 | 0.7 |
| Walton | 3.16 | 35.7 | 60 | 636 | 329.6 | 36.7 | 199 | 2,740 | 894.1 | 71.5 | 84 | 218 | 44.1 | 10.0 |
| Ware | 2.74 | 77.5 | 36 | 269 | 452.5 | 11.0 | 199 | 2,463 | 699.7 | 60.8 | 32 | 101 | 14.5 | 2.4 |
| Warren | 0.32 | 58.6 | 4 | 52 | 25.0 | 2.2 | 12 | 69 | 13.4 | 1.7 | 5 | D | 2.7 | D |
| Washington | 2.00 | 96.1 | 13 | 101 | 168.4 | 4.5 | 70 | 788 | 171.6 | 17.0 | 13 | 46 | 20.5 | 2.6 |
| Wayne | 1.50 | 50.8 | D | D | D | D | 116 | 1,135 | 290.5 | 25.6 | 12 | 55 | 6.6 | 1.3 |
| Webster | 0.10 | 37.8 | 3 | D | 1.8 | D | 6 | 46 | 9.2 | 1.1 | NA | NA | NA | NA |
| Wheeler | 0.18 | 22.8 | NA | NA | NA | NA | 14 | 61 | 17.5 | 1.2 | NA | NA | NA | NA |
| White | 1.47 | 51.9 | 13 | 78 | 36.4 | 2.3 | 114 | 1,222 | 346.7 | 31.2 | 28 | 83 | 14.8 | 2.8 |
| Whitfield | 23.96 | 229.9 | 218 | 3,518 | 1,201.4 | 150.5 | 396 | 4,678 | 1,413.3 | 119.1 | D | D | D | 8.6 |
| Wilcox | 0.44 | 49.7 | 5 | D | 18.2 | D | 23 | 80 | 19.3 | 1.4 | NA | NA | NA | NA |
| Wilkes | 1.01 | 102.4 | 7 | 70 | 31.3 | 2.1 | 41 | 386 | 73.7 | 7.7 | 3 | 7 | 0.4 | 0.1 |
| Wilkinson | 0.70 | 76.5 | 5 | D | 6.2 | D | 25 | 165 | 32.6 | 2.9 | 4 | D | 0.3 | D |
| Worth | 1.01 | 48.8 | 27 | 257 | 213.7 | 10.0 | 47 | 567 | 135.0 | 13.7 | 5 | 11 | 5.1 | 0.2 |
| HAWAII | 266.92 | 186.4 | 1,423 | 16,829 | 11,342.6 | 858.8 | 4,644 | 72,908 | 21,658.9 | 2,184.7 | 2,069 | 13,101 | 4,409.0 | 675.2 |
| Hawaii | 39.70 | 202.1 | 169 | 1,696 | 760.7 | 72.8 | 629 | 10,232 | 3,086.0 | 307.8 | 281 | 1,280 | 339.4 | 56.5 |
| Honolulu | 168.78 | 169.0 | 1,036 | 13,217 | 9,376.4 | 689.4 | 2,856 | 48,609 | 14,266.1 | 1,437.5 | 1,288 | 8,864 | 3,150.9 | 483.1 |
| Kalawao | 0.01 | 112.4 | NA | NA | NA | NA | NA | NA | NA | NA | NA | NA | NA | NA |
| Kauai | 16.34 | 227.8 | 72 | 562 | 376.8 | 33.0 | 356 | 4,170 | 1,253.1 | 127.4 | 157 | 973 | 267.3 | 48.1 |
| Maui | 42.09 | 255.7 | 146 | 1,354 | 828.7 | 63.7 | D | D | D | D | 343 | 1,984 | 651.4 | 87.5 |
| IDAHO | 275.79 | 166.6 | 1,870 | 23,878 | 23,736.5 | 1,259.7 | 6,133 | 82,312 | 24,936.1 | 2,312.2 | 2,644 | 7,521 | 1,635.4 | 276.7 |
| Ada | 72.71 | 167.5 | 587 | 9,095 | 9,627.7 | 558.3 | 1,505 | 24,520 | 7,401.4 | 742.7 | 968 | 3,285 | 764.3 | 134.2 |
| Adams | 0.53 | 137.9 | NA | NA | NA | NA | D | D | D | D | 4 | D | 0.9 | D |
| Bannock | 16.91 | 201.9 | 82 | 859 | 616.8 | 37.9 | 321 | 4,601 | 1,290.2 | 114.2 | 99 | 270 | 55.0 | 7.6 |
| Bear Lake | 0.37 | 62.5 | D | D | D | D | 27 | 241 | 58.3 | 4.9 | 5 | 12 | 1.3 | 0.2 |
| Benewah | 0.52 | 57.4 | D | D | D | D | 32 | 301 | 73.6 | 7.1 | D | D | D | 0.1 |
| Bingham | 1.99 | 44.2 | 48 | 811 | 473.1 | 35.7 | 118 | 1,203 | 312.1 | 29.2 | 30 | 71 | 9.8 | 2.0 |
| Blaine | 5.44 | 251.9 | 35 | 299 | 252.5 | 17.3 | 176 | 1,427 | 375.1 | 49.4 | D | D | D | 0.4 |
| Boise | 0.48 | 68.0 | NA | NA | NA | NA | 16 | 100 | 22.4 | 1.8 | D | D | D | 9.8 |
| Bonner | 1.71 | 40.9 | 33 | 203 | 68.7 | 8.5 | 216 | 2,194 | 585.9 | 58.7 | 93 | 251 | 43.9 | 9.8 |
| Bonneville | 36.62 | 332.6 | 184 | 2,305 | 2,841.8 | 112.5 | 515 | 7,783 | 2,247.5 | 201.7 | 151 | 477 | 124.9 | 19.4 |
| Boundary | 0.39 | 34.5 | 8 | 56 | 31.0 | 2.7 | 48 | 469 | 116.9 | 10.6 | 10 | 15 | 2.0 | 0.3 |
| Butte | 1.43 | 571.8 | 5 | D | 37.8 | D | 11 | 116 | 19.2 | 2.3 | NA | NA | NA | NA |
| Camas | 0.14 | 131.3 | NA | NA | NA | NA | 3 | 23 | 7.0 | 0.5 | NA | NA | NA | NA |
| Canyon | 15.92 | 76.7 | 166 | 1,881 | 1,445.1 | 89.5 | 509 | 7,976 | 2,764.2 | 234.7 | 183 | 507 | 91.7 | 17.2 |

1  Merchant wholesalers, except manufacturers' sales branches and offices.    2. Employer establishments.

| STATE County | Professional, scientific, and technical services, 2017 | | | | Manufacturing, 2017 | | | | Accommodation and food services, 2017 | | | |
|---|---|---|---|---|---|---|---|---|---|---|---|---|
| | Number of establishments | Number of employees | Sales (mil dol) | Average payroll (mil dol) | Number of establishments | Number of employees | Sales (mil dol) | Average payroll (mil dol) | Number of establishments | Number of employees | Sales (mil dol) | Annual payroll (mil dol) |
| | 147 | 148 | 149 | 150 | 151 | 152 | 153 | 154 | 155 | 156 | 157 | 158 |
| GEORGIA—Cont'd | | | | | | | | | | | | |
| Putnam | 31 | 62 | 14 | 2.8 | 30 | 701 | 193.1 | 35.3 | D | D | D | D |
| Quitman | NA | NA | NA | NA | NA | NA | NA | NA | NA | NA | NA | NA |
| Rabun | 48 | 137 | 15 | 5.1 | 21 | 397 | 184.9 | 15.6 | 69 | 1,173 | 74.0 | 20.9 |
| Randolph | 6 | 28 | 3 | 0.8 | D | D | D | D | 14 | 74 | 4.4 | 1.2 |
| Richmond | 464 | 4,935 | 877 | 282.2 | 104 | 7,900 | 4,281.4 | 501.2 | 466 | 10,315 | 562.8 | 149.2 |
| Rockdale | D | D | D | D | 81 | 4,621 | 2,375.3 | 250.4 | 208 | 4,733 | 253.2 | 72.7 |
| Schley | NA | NA | NA | NA | 5 | 255 | 100.1 | 13.6 | 3 | 22 | 1.0 | 0.3 |
| Screven | 12 | 46 | 5 | 2.1 | 9 | 707 | 148.7 | 33.0 | 16 | 178 | 8.9 | 2.1 |
| Seminole | D | D | D | D | 4 | 9 | 5.9 | 0.7 | 17 | 119 | 8.0 | 1.3 |
| Spalding | 82 | 396 | 60 | 19.7 | 53 | 2,833 | 1,467.7 | 152.4 | 108 | 1,799 | 102.0 | 26.0 |
| Stephens | 37 | 129 | 16 | 4.7 | 49 | 2,075 | 711.6 | 85.8 | 38 | 736 | 30.7 | 8.5 |
| Stewart | D | D | D | D | D | 11 | D | 0.5 | NA | NA | NA | NA |
| Sumter | D | D | D | D | 31 | 1,025 | 330.3 | 36.9 | 51 | 855 | 40.6 | 11.6 |
| Talbot | NA | NA | NA | NA | NA | NA | NA | NA | NA | NA | NA | NA |
| Taliaferro | NA | NA | NA | NA | NA | NA | NA | NA | NA | NA | NA | NA |
| Tattnall | 17 | 57 | 5 | 1.3 | 8 | 91 | 10.8 | 3.7 | 23 | 275 | 12.7 | 3.3 |
| Taylor | NA | NA | NA | NA | 6 | 62 | 40.0 | 2.1 | 6 | 36 | 2.1 | 0.4 |
| Telfair | 7 | 28 | 2 | 0.7 | D | D | D | D | 20 | 149 | 7.9 | 1.8 |
| Terrell | 13 | 39 | 4 | 1.3 | 11 | 373 | 162.0 | 14.2 | D | D | D | D |
| Thomas | 76 | 417 | 58 | 16.7 | 46 | 2,174 | 718.1 | 114.4 | 104 | 1,784 | 91.3 | 24.5 |
| Tift | 79 | 623 | 70 | 29.9 | 33 | 1,318 | 487.3 | 60.2 | 100 | 2,463 | 119.1 | 33.1 |
| Toombs | 53 | 197 | 18 | 6.8 | 28 | 1,120 | 682.5 | 43.0 | 70 | 1,132 | 59.7 | 15.5 |
| Towns | D | D | 6 | D | 11 | 33 | 5.9 | 1.5 | 38 | 481 | 39.3 | 6.6 |
| Treutlen | NA | NA | NA | NA | 4 | D | 8.7 | D | 5 | 37 | 1.7 | 0.5 |
| Troup | 98 | 501 | 72 | 25.9 | 91 | 12,465 | 11,966.6 | 631.1 | 135 | 2,373 | 121.7 | 32.3 |
| Turner | 9 | 66 | 5 | 1.5 | 9 | 328 | 159.1 | 14.8 | 17 | 222 | 9.5 | 2.4 |
| Twiggs | D | D | D | D | NA | NA | NA | NA | 6 | D | 1.6 | D |
| Union | 42 | 144 | 20 | 6.2 | 29 | 344 | 83.3 | 16.6 | 50 | 687 | 36.2 | 9.8 |
| Upson | 26 | 84 | 18 | 3.7 | 20 | 1,199 | 441.9 | 51.2 | 47 | 603 | 31.3 | 8.4 |
| Walker | 47 | 181 | 22 | 7.9 | 41 | 4,047 | 1,513.4 | 156.2 | 57 | 703 | 32.7 | 8.5 |
| Walton | 150 | 451 | 53 | 20.1 | 47 | 2,514 | 1,107.1 | 135.9 | 118 | 2,100 | 109.0 | 31.4 |
| Ware | 56 | 275 | 22 | 7.9 | 30 | 1,347 | 470.2 | 59.0 | 75 | 1,537 | 78.3 | 20.0 |
| Warren | D | D | 1 | D | 5 | 439 | 106.6 | 19.4 | NA | NA | NA | NA |
| Washington | 20 | 225 | 19 | 11.6 | 17 | 596 | 269.9 | 28.0 | 29 | 432 | 18.1 | 5.3 |
| Wayne | 32 | 146 | 13 | 6.8 | 24 | 943 | 751.8 | 72.9 | 51 | 720 | 41.8 | 9.7 |
| Webster | NA | NA | NA | NA | NA | NA | NA | NA | NA | NA | NA | NA |
| Wheeler | NA | NA | NA | NA | NA | NA | NA | NA | NA | NA | NA | NA |
| White | 52 | 136 | 16 | 5.0 | 35 | 736 | 131.4 | 33.6 | 3 | 20 | 0.2 | 0.1 |
| Whitfield | D | D | D | D | 246 | 17,470 | 6,817.6 | 735.6 | 90 | 1,138 | 82.8 | 22.7 |
| Wilcox | 3 | 4 | 0 | 0.1 | D | 11 | D | D | 178 | 3,533 | 212.2 | 55.4 |
| Wilkes | 11 | 32 | 3 | 1.0 | 13 | 411 | 211.6 | 18.4 | NA | NA | NA | NA |
| Wilkinson | D | D | 3 | D | 10 | 424 | 129.4 | 26.8 | 16 | 170 | 7.8 | 2.1 |
| Worth | 15 | 75 | 6 | 2.3 | 9 | 534 | 315.9 | 23.7 | 4 | 15 | 0.6 | 0.1 |
| HAWAII | 3,347 | 22,058 | 3,701 | 1,423.2 | 783 | 11,850 | 6,055.9 | 558.5 | 3,865 | 112,743 | 12,101.8 | 3,296.8 |
| Hawaii | D | D | D | D | 110 | 1,433 | 447.3 | 66.7 | 477 | 13,731 | 1,363.8 | 396.8 |
| Honolulu | 2,473 | 18,527 | 3,232 | 1,238.8 | 501 | 9,240 | 5,325.4 | 439.0 | 2,569 | 67,611 | 6,805.4 | 1,767.8 |
| Kalawao | NA | NA | NA | NA | NA | NA | NA | NA | NA | NA | NA | NA |
| Kauai | D | D | D | D | 52 | 332 | 62.7 | 12.0 | 273 | 9,430 | 1,002.9 | 319.8 |
| Maui | D | D | D | D | 120 | 845 | 220.4 | 40.8 | 546 | 21,971 | 2,929.7 | 812.5 |
| IDAHO | D | D | D | D | 1,877 | 58,746 | 20,263.4 | 3,486.1 | 3,856 | 65,463 | 3,598.1 | 1,006.2 |
| Ada | 1,882 | 13,732 | 2,180 | 815.1 | 423 | 14,550 | 3,307.2 | 1,400.9 | 1,049 | 21,248 | 1,132.3 | 336.7 |
| Adams | 7 | 11 | 1 | 0.3 | 7 | 113 | 13.5 | 4.7 | 15 | 42 | 2.5 | 0.7 |
| Bannock | D | D | D | D | 51 | 2,082 | 921.9 | 120.5 | 201 | 3,462 | 176.7 | 47.4 |
| Bear Lake | 7 | 8 | 1 | 0.3 | 4 | 25 | 3.7 | 0.8 | 21 | 159 | 8.2 | 1.7 |
| Benewah | D | D | D | 1.0 | D | 532 | D | D | 24 | 152 | 7.6 | 2.0 |
| Bingham | 62 | 225 | 27 | 8.0 | 38 | 2,156 | 655.0 | 93.8 | 48 | 639 | 28.5 | 7.3 |
| Blaine | 169 | 831 | 116 | 60.7 | 39 | 306 | 63.8 | 14.7 | 113 | 3,128 | 178.0 | 60.5 |
| Boise | D | D | D | D | D | D | D | 1.1 | D | D | D | D |
| Bonner | 155 | 437 | 57 | 20.8 | 75 | 1,501 | 485.6 | 65.4 | 127 | 1,993 | 89.6 | 27.2 |
| Bonneville | D | D | D | D | 140 | 3,026 | 810.7 | 129.2 | 237 | 5,065 | 257.2 | 77.1 |
| Boundary | D | D | 10 | D | 27 | 451 | 172.9 | 20.4 | D | D | D | D |
| Butte | 4 | 6 | 0 | 0.1 | NA | NA | NA | NA | D | D | D | D |
| Camas | NA | NA | NA | NA | NA | NA | NA | NA | D | D | D | 0.4 |
| Canyon | D | D | D | D | 225 | 8,812 | 3,186.0 | 399.0 | 282 | 5,348 | 265.2 | 72.5 |

Table B. States and Counties — **Health Care and Social Assistance, Other Services, Nonemployer Businesses, and Residential Construction**

| STATE County | Health care and social assistance, 2017 | | | | Other services, 2017 | | | | Nonemployer businesses, 2018 | | Value of residential construction authorized by building permits, 2020 | |
|---|---|---|---|---|---|---|---|---|---|---|---|---|
| | Number of establish-ments | Number of employees | Receipts (mil dol) | Annual payroll (mil dol) | Number of establish-ments | Number of employees | Receipts (mil dol) | Annual payroll (mil dol) | Number | Receipts (mil dol) | New construction ($1,000) | Number of housing units |
| | 159 | 160 | 161 | 162 | 163 | 164 | 165 | 166 | 167 | 168 | 169 | 170 |
| GEORGIA—Cont'd | | | | | | | | | | | | |
| Putnam | 35 | 418 | 35.0 | 13.9 | 23 | 85 | 10.6 | 2.0 | 2,000 | 91.1 | 55,173 | 200 |
| Quitman | NA | NA | NA | NA | NA | NA | NA | NA | 133 | 3.7 | 0 | 0 |
| Rabun | 38 | 586 | 43.2 | 21.2 | 38 | 111 | 10.5 | 3.2 | 1,942 | 83.5 | 32,924 | 98 |
| Randolph | 10 | 288 | 25.4 | 11.7 | 6 | 12 | 1.1 | 0.3 | 444 | 13.0 | 1,134 | 12 |
| Richmond | 608 | 24,655 | 3,883.8 | 1,371.0 | 293 | 1,799 | 220.3 | 59.1 | 13,144 | 416.9 | 108,022 | 644 |
| Rockdale | 262 | 4,060 | 484.6 | 169.9 | 144 | 733 | 105.1 | 24.5 | 9,297 | 300.3 | 74,882 | 269 |
| Schley | 5 | 27 | 4.4 | 0.7 | D | D | D | D | 335 | 12.7 | 3,192 | 15 |
| Screven | 15 | 302 | 23.3 | 9.9 | 24 | 75 | 11.0 | 2.2 | 976 | 35.8 | 465 | 21 |
| Seminole | 25 | 439 | 39.0 | 19.3 | D | D | D | D | 555 | 24.0 | 0 | 0 |
| Spalding | 141 | 3,050 | 318.1 | 130.2 | 72 | 385 | 39.5 | 11.5 | 5,166 | 178.8 | 91,468 | 430 |
| Stephens | 55 | 1,232 | 110.3 | 46.6 | 30 | 151 | 14.1 | 4.9 | 1,701 | 56.4 | 1,807 | 17 |
| Stewart | 8 | D | 8.4 | D | D | D | D | 0.1 | 242 | 4.7 | 466 | 2 |
| Sumter | 81 | 2,645 | 428.5 | 107.0 | D | D | D | D | 1,858 | 67.3 | 4,336 | 21 |
| Talbot | 4 | D | 0.6 | D | 3 | 3 | 0.2 | 0.1 | 417 | 12.4 | 1,399 | 6 |
| Taliaferro | NA | NA | NA | NA | NA | NA | NA | NA | 118 | 4.7 | NA | NA |
| Tattnall | 28 | 787 | 117.0 | 29.5 | D | D | D | D | 1,319 | 55.3 | 4,512 | 33 |
| Taylor | 19 | 214 | 11.8 | 5.4 | D | D | D | 0.8 | 495 | 19.8 | 1,579 | 9 |
| Telfair | 15 | 205 | 11.9 | 5.3 | 16 | 63 | 8.3 | 1.8 | 702 | 30.2 | 390 | 1 |
| Terrell | D | D | D | D | D | D | D | D | 690 | 25.4 | 760 | 6 |
| Thomas | 141 | 3,748 | 502.3 | 181.6 | 64 | 462 | 53.7 | 10.7 | 3,313 | 144.9 | 33,219 | 175 |
| Tift | 101 | 3,260 | 517.7 | 197.4 | 51 | 319 | 31.9 | 10.4 | 3,088 | 134.5 | 13,745 | 129 |
| Toombs | D | D | D | D | D | D | D | D | 1,908 | 84.6 | 2,092 | 19 |
| Towns | 30 | 573 | 43.0 | 19.4 | 8 | 24 | 2.0 | 0.5 | 1,266 | 53.2 | 20,046 | 83 |
| Treutlen | 8 | 128 | 9.3 | 3.9 | D | D | D | D | 481 | 20.5 | 2,654 | 16 |
| Troup | 133 | 2,671 | 396.0 | 171.6 | 83 | 349 | 44.4 | 10.8 | 5,166 | 172.3 | 102,395 | 634 |
| Turner | 10 | 142 | 8.4 | 3.6 | 5 | 14 | 1.3 | 0.3 | 698 | 21.9 | 1,280 | 7 |
| Twiggs | D | D | D | D | 4 | 9 | 1.7 | 0.3 | 619 | 21.1 | 339 | 9 |
| Union | 69 | 1,380 | 121.0 | 52.6 | 37 | 173 | 24.3 | 4.7 | 2,443 | 103.2 | 65,630 | 229 |
| Upson | 63 | 1,350 | 139.1 | 57.1 | 45 | 168 | 21.8 | 5.5 | 1,704 | 59.0 | 15,791 | 106 |
| Walker | 52 | 922 | 66.2 | 27.3 | 40 | 212 | 20.8 | 6.4 | 4,184 | 223.4 | 43,414 | 231 |
| Walton | 154 | 2,131 | 240.9 | 82.7 | 110 | 396 | 40.7 | 10.5 | 8,713 | 358.1 | 143,048 | 862 |
| Ware | 125 | 2,727 | 186.3 | 85.7 | 52 | 307 | 31.0 | 9.0 | 1,926 | 79.2 | 6,202 | 45 |
| Warren | 7 | 135 | 7.3 | 3.8 | 3 | 4 | 0.2 | 0.1 | 442 | 12.7 | 0 | 0 |
| Washington | 38 | 914 | 50.7 | 24.7 | 27 | 84 | 12.5 | 2.7 | 1,271 | 43.4 | 10,327 | 94 |
| Wayne | 74 | 1,202 | 131.4 | 46.0 | 36 | 133 | 13.3 | 3.2 | 1,710 | 67.5 | 14,372 | 84 |
| Webster | NA | NA | NA | NA | NA | NA | NA | NA | 158 | 6.9 | 835 | 4 |
| Wheeler | NA | NA | NA | NA | 3 | 3 | 0.5 | 0.1 | 375 | 15.8 | 0 | 0 |
| White | 39 | 921 | 66.5 | 27.1 | 46 | 182 | 18.2 | 4.7 | 2,724 | 118.4 | 29,244 | 138 |
| Whitfield | D | D | D | D | 125 | 857 | 114.2 | 31.1 | 6,147 | 318.0 | 48,722 | 382 |
| Wilcox | 9 | 64 | 3.1 | 1.4 | D | D | 2.1 | D | 532 | 19.2 | NA | NA |
| Wilkes | 18 | 331 | 12.7 | 5.6 | D | D | D | D | 652 | 24.0 | 3,771 | 42 |
| Wilkinson | 14 | 154 | 10.4 | 4.4 | D | D | 2.5 | D | 667 | 19.3 | 1,882 | 9 |
| Worth | D | D | D | D | 22 | 69 | 7.9 | 2.1 | 1,357 | 52.6 | 5,253 | 24 |
| HAWAII | 3,677 | 73,551 | 9,785.7 | 3,998.4 | 2,912 | 20,219 | 2,561.2 | 652.4 | 112,589 | 5,556.1 | 1,169,770 | 3,164 |
| Hawaii | 476 | 8,222 | 917.2 | 420.1 | 290 | 1,379 | 204.4 | 46.5 | 18,139 | 847.5 | 337,836 | 870 |
| Honolulu | 2,578 | 54,227 | 7,522.5 | 2,994.2 | 2,064 | 15,687 | 1,965.5 | 500.9 | 69,723 | 3,464.2 | 486,429 | 1,621 |
| Kalawao | NA | NA | NA | NA | NA | NA | NA | NA | NA | NA | NA | NA |
| Kauai | 192 | 3,324 | 413.3 | 180.2 | 149 | 761 | 91.9 | 25.7 | 7,150 | 343.3 | 140,525 | 147 |
| Maui | 431 | 7,778 | 932.7 | 403.9 | 409 | 2,392 | 299.4 | 79.2 | 17,577 | 901.0 | 204,980 | 526 |
| IDAHO | 5,310 | 98,100 | 10,469.0 | 4,292.3 | 2,749 | 14,016 | 1,574.0 | 434.6 | 135,986 | 6,336.3 | 3,948,590 | 19,130 |
| Ada | 1,558 | 35,758 | 4,249.2 | 1,814.5 | 850 | 4,988 | 599.6 | 170.9 | 41,464 | 2,011.0 | 1,576,192 | 6,060 |
| Adams | 6 | D | 3.4 | D | D | D | D | 0.3 | 361 | 14.7 | 6,723 | 11 |
| Bannock | 386 | 5,589 | 621.3 | 214.2 | 114 | 506 | 70.3 | 15.9 | 5,392 | 229.5 | 76,274 | 763 |
| Bear Lake | 11 | 305 | 42.3 | 12.6 | 3 | 11 | 1.3 | 0.2 | 486 | 20.4 | 18,410 | 63 |
| Benewah | 14 | 360 | 31.3 | 14.3 | 17 | 78 | 10.1 | 2.6 | 653 | 28.5 | 8,500 | 48 |
| Bingham | 115 | 1,855 | 210.0 | 90.0 | 54 | 289 | 40.2 | 10.0 | 3,000 | 141.5 | 41,820 | 304 |
| Blaine | 79 | 858 | 105.9 | 45.8 | 95 | 382 | 54.1 | 14.2 | 3,735 | 255.6 | 164,451 | 221 |
| Boise | 6 | 22 | 2.0 | 0.7 | D | D | D | D | 725 | 29.3 | 34,532 | 122 |
| Bonner | 144 | 1,830 | 156.0 | 68.4 | 88 | 364 | 37.0 | 9.5 | 4,351 | 188.8 | 37,400 | 203 |
| Bonneville | 578 | 9,005 | 1,077.5 | 365.9 | D | D | D | D | 9,129 | 452.8 | 143,835 | 880 |
| Boundary | 27 | 686 | 49.5 | 25.6 | 25 | 56 | 5.0 | 1.3 | 1,033 | 45.4 | 18,397 | 76 |
| Butte | D | D | D | 4.2 | D | D | D | D | 195 | 5.4 | 339 | 2 |
| Camas | D | D | D | D | NA | NA | NA | NA | 105 | 3.4 | 1,091 | 7 |
| Canyon | 404 | 7,256 | 631.9 | 270.9 | 226 | 1,253 | 118.9 | 36.3 | 13,941 | 587.3 | 501,684 | 3,395 |

# Table B. States and Counties — Government Employment and Payroll, and Local Government Finances

| STATE County | Government employment and payroll, 2017 | | | | | | | | | Local government finances, 2017 | | | | |
| | | | March payroll (percent of total) | | | | | | | General revenue | | | | |
| | | | | | | | | | | | | Taxes | | |
| | | | Administration, judicial, and legal | Police and corrections | Fire protection | Highways and transportation | Health and welfare | Natural resources and utilities | Education and libraries | Total (mil dol) | Inter-govern-mental (mil dol) | Total (mil dol) | Per capita[1] (dollars) | |
| | Full-time equivalent employees | March payroll (dollars) | | | | | | | | | | | Total | Property |
| | 171 | 172 | 173 | 174 | 175 | 176 | 177 | 178 | 179 | 180 | 181 | 182 | 183 | 184 |
| **GEORGIA—Cont'd** | | | | | | | | | | | | | | |
| Putnam | 881 | 3,538,894 | 6.2 | 7.9 | 2.1 | 1.8 | 12.8 | 3.2 | 61.7 | 65.5 | 18.6 | 38.9 | 1,793 | 1,367 |
| Quitman | 122 | 349,836 | 12.7 | 5.9 | 0.0 | 5.7 | 5.9 | 5.3 | 64.1 | 7.3 | 5.0 | 1.1 | 490 | 418 |
| Rabun | 673 | 2,344,388 | 8.2 | 11.1 | 2.4 | 3.9 | 6.0 | 6.9 | 60.0 | 77.5 | 24.3 | 44.9 | 2,713 | 2,004 |
| Randolph | 512 | 1,450,960 | 7.1 | 7.3 | 2.0 | 3.0 | 35.0 | 2.3 | 39.4 | 36.2 | 11.4 | 8.9 | 1,281 | 947 |
| Richmond | 8,236 | 26,032,382 | 7.5 | 11.6 | 4.8 | 2.7 | 4.9 | 6.8 | 60.7 | 725.4 | 288.5 | 251.4 | 1,247 | 793 |
| Rockdale | 3,741 | 12,706,963 | 7.1 | 9.9 | 4.1 | 1.3 | 0.9 | 2.4 | 72.2 | 308.4 | 121.5 | 123.7 | 1,377 | 966 |
| Schley | 218 | 765,828 | 2.2 | 3.2 | 0.9 | 3.3 | 0.0 | 6.5 | 83.8 | 20.0 | 12.0 | 5.2 | 985 | 710 |
| Screven | 593 | 1,723,859 | 7.4 | 9.7 | 3.7 | 3.4 | 3.3 | 6.4 | 63.2 | 42.9 | 19.2 | 16.9 | 1,208 | 865 |
| Seminole | 314 | 1,210,782 | 6.2 | 8.6 | 2.7 | 4.3 | 0.0 | 4.9 | 73.1 | 27.7 | 12.0 | 12.8 | 1,553 | 1,128 |
| Spalding | 2,703 | 9,580,057 | 6.0 | 13.0 | 5.1 | 1.7 | 7.6 | 6.7 | 57.9 | 233.8 | 102.5 | 81.5 | 1,247 | 851 |
| Stephens | 918 | 3,145,718 | 6.2 | 9.5 | 2.5 | 2.4 | 2.6 | 5.3 | 69.4 | 120.8 | 37.3 | 32.5 | 1,264 | 810 |
| Stewart | 214 | 662,289 | 14.8 | 9.7 | 7.8 | 3.7 | 2.1 | 2.3 | 59.6 | 19.2 | 9.9 | 5.9 | 933 | 670 |
| Sumter | 1,277 | 4,584,282 | 6.1 | 12.1 | 4.9 | 2.1 | 9.6 | 2.2 | 60.6 | 119.4 | 52.9 | 39.7 | 1,328 | 835 |
| Talbot | 170 | 542,906 | 15.2 | 11.0 | 0.9 | 7.1 | 5.1 | 4.8 | 55.9 | 15.4 | 4.7 | 8.8 | 1,403 | 1,018 |
| Taliaferro | 102 | 290,207 | 10.8 | 14.2 | 0.0 | 4.0 | 0.2 | 1.8 | 65.6 | 11.9 | 5.6 | 3.8 | 2,326 | 1,895 |
| Tattnall | 773 | 2,319,640 | 7.4 | 5.8 | 5.9 | 2.7 | 1.4 | 2.5 | 73.1 | 63.0 | 31.8 | 23.1 | 910 | 521 |
| Taylor | 334 | 1,035,828 | 7.3 | 9.5 | 0.5 | 4.9 | 1.7 | 3.0 | 72.5 | 26.7 | 13.2 | 10.0 | 1,228 | 793 |
| Telfair | 451 | 1,273,226 | 9.4 | 9.9 | 1.3 | 3.4 | 5.9 | 4.4 | 64.8 | 35.2 | 15.3 | 14.4 | 901 | 595 |
| Terrell | 402 | 1,272,609 | 2.9 | 15.0 | 3.6 | 4.1 | 4.0 | 2.8 | 66.6 | 29.3 | 12.4 | 12.0 | 1,382 | 972 |
| Thomas | 2,432 | 8,226,021 | 5.8 | 8.0 | 3.5 | 3.5 | 11.5 | 6.8 | 54.1 | 239.8 | 81.1 | 70.4 | 1,578 | 1,018 |
| Tift | 3,567 | 17,141,046 | 2.2 | 3.3 | 0.8 | 0.9 | 69.8 | 0.7 | 21.2 | 437.6 | 62.7 | 53.0 | 1,311 | 580 |
| Toombs | 1,296 | 4,145,016 | 4.7 | 11.1 | 1.5 | 1.8 | 3.4 | 2.0 | 75.0 | 255.8 | 52.0 | 33.8 | 1,259 | 672 |
| Towns | 545 | 1,581,012 | 6.2 | 7.5 | 0.8 | 2.1 | 0.0 | 25.2 | 49.2 | 30.0 | 6.1 | 18.0 | 1,556 | 885 |
| Treutlen | 236 | 727,124 | 8.1 | 7.8 | 2.8 | 3.1 | 3.0 | 2.6 | 72.1 | 18.5 | 10.3 | 4.9 | 725 | 473 |
| Troup | 3,172 | 11,325,815 | 5.0 | 12.0 | 4.6 | 1.8 | 10.2 | 7.2 | 54.8 | 269.0 | 96.5 | 100.4 | 1,434 | 889 |
| Turner | 381 | 1,245,789 | 15.4 | 9.4 | 1.8 | 6.0 | 8.2 | 3.6 | 54.5 | 33.1 | 12.6 | 11.3 | 1,425 | 984 |
| Twiggs | 316 | 948,450 | 8.4 | 21.9 | 0.0 | 6.9 | 0.0 | 1.8 | 56.2 | 21.0 | 8.0 | 10.4 | 1,263 | 1,014 |
| Union | 1,436 | 4,910,401 | 2.3 | 5.6 | 1.7 | 1.3 | 49.4 | 3.8 | 34.8 | 64.9 | 23.0 | 37.4 | 1,599 | 965 |
| Upson | 994 | 3,137,963 | 10.3 | 9.5 | 1.5 | 2.7 | 1.8 | 3.4 | 70.0 | 80.2 | 39.9 | 29.0 | 1,109 | 739 |
| Walker | 2,516 | 8,497,959 | 5.8 | 5.3 | 3.2 | 1.4 | 9.0 | 4.8 | 68.8 | 168.2 | 109.7 | 29.4 | 427 | 318 |
| Walton | 3,207 | 11,641,649 | 7.0 | 8.5 | 4.5 | 1.9 | 2.7 | 5.6 | 66.7 | 285.0 | 104.0 | 126.2 | 1,383 | 959 |
| Ware | 1,754 | 5,525,840 | 5.7 | 9.9 | 3.8 | 2.0 | 14.3 | 2.5 | 57.0 | 142.9 | 64.6 | 47.5 | 1,332 | 772 |
| Warren | 178 | 576,699 | 9.1 | 7.1 | 0.0 | 3.2 | 5.8 | 6.2 | 63.4 | 29.2 | 16.1 | 8.4 | 1,588 | 1,130 |
| Washington | 1,050 | 3,510,831 | 5.7 | 8.4 | 1.0 | 3.5 | 29.8 | 3.4 | 46.3 | 114.5 | 25.8 | 32.3 | 1,589 | 1,046 |
| Wayne | 1,443 | 5,278,781 | 3.6 | 5.9 | 0.8 | 3.1 | 32.7 | 2.1 | 50.2 | 167.8 | 43.9 | 41.7 | 1,401 | 782 |
| Webster | 89 | 326,252 | 19.3 | 0.0 | 0.0 | 3.3 | 3.2 | 3.6 | 70.6 | 7.5 | 4.2 | 2.7 | 1,061 | 847 |
| Wheeler | 239 | 548,888 | 9.8 | 9.1 | 0.0 | 4.6 | 4.2 | 3.0 | 67.8 | 21.2 | 12.0 | 6.9 | 865 | 649 |
| White | 959 | 3,298,378 | 5.4 | 10.6 | 1.9 | 1.7 | 0.4 | 3.8 | 72.9 | 90.8 | 30.9 | 41.0 | 1,391 | 850 |
| Whitfield | 4,150 | 15,747,724 | 3.6 | 7.3 | 4.9 | 3.2 | 3.2 | 6.1 | 63.8 | 438.7 | 191.0 | 143.1 | 1,373 | 862 |
| Wilcox | 265 | 790,584 | 7.2 | 6.4 | 0.0 | 1.9 | 0.7 | 2.1 | 81.4 | 28.3 | 17.9 | 7.9 | 902 | 667 |
| Wilkes | 651 | 2,382,391 | 4.7 | 3.5 | 1.5 | 3.1 | 27.4 | 5.1 | 54.2 | 40.2 | 15.5 | 15.8 | 1,598 | 1,124 |
| Wilkinson | 402 | 1,138,739 | 9.9 | 16.2 | 0.7 | 7.7 | 0.0 | 1.3 | 63.2 | 33.5 | 13.8 | 17.3 | 1,936 | 1,288 |
| Worth | 737 | 2,288,460 | 5.9 | 9.4 | 4.8 | 4.0 | 1.6 | 2.7 | 70.4 | 60.4 | 28.8 | 20.7 | 1,009 | 744 |
| **HAWAII** | X | X | X | X | X | X | X | X | X | X | X | X | X | X |
| Hawaii | 2,603 | 13,993,658 | 20.0 | 24.2 | 18.6 | 5.1 | 5.4 | 20.3 | 0.0 | 431.6 | 77.9 | 313.8 | 1,569 | 1,333 |
| Honolulu | 10,620 | 59,188,205 | 13.8 | 33.1 | 16.0 | 4.1 | 4.2 | 22.8 | 0.0 | 2,959.6 | 435.8 | 1,639.6 | 1,662 | 1,115 |
| Kalawao | NA | NA | NA | NA | NA | NA | NA | NA | NA | NA | NA | NA | NA | NA |
| Kauai | 1,280 | 7,098,138 | 18.8 | 20.4 | 16.1 | 10.7 | 2.7 | 26.7 | 0.0 | 220.1 | 47.9 | 143.6 | 2,000 | 1,680 |
| Maui | 2,489 | 15,157,755 | 20.1 | 26.1 | 17.0 | 6.5 | 3.3 | 24.9 | 0.0 | 479.7 | 60.6 | 331.2 | 1,994 | 1,641 |
| **IDAHO** | X | X | X | X | X | X | X | X | X | X | X | X | X | X |
| Ada | 13,487 | 49,826,035 | 10.4 | 13.0 | 6.1 | 4.5 | 2.9 | 5.2 | 55.2 | 1,472.9 | 569.1 | 542.9 | 1,189 | 1,098 |
| Adams | 143 | 495,412 | 15.9 | 18.3 | 0.0 | 12.1 | 2.2 | 5.2 | 42.2 | 22.5 | 7.2 | 13.5 | 3,280 | 3,070 |
| Bannock | 2,787 | 10,399,001 | 7.8 | 15.0 | 5.6 | 5.2 | 3.7 | 7.9 | 51.7 | 252.4 | 111.2 | 86.8 | 1,015 | 978 |
| Bear Lake | 236 | 719,522 | 10.0 | 11.5 | 0.2 | 8.7 | 0.1 | 8.6 | 54.9 | 23.6 | 9.9 | 4.4 | 729 | 697 |
| Benewah | 499 | 1,677,766 | 4.3 | 6.0 | 2.1 | 4.0 | 42.0 | 2.7 | 38.0 | 27.5 | 15.4 | 8.7 | 945 | 894 |
| Bingham | 1,540 | 4,733,371 | 6.4 | 10.9 | 3.1 | 4.3 | 0.3 | 5.2 | 68.9 | 127.9 | 76.7 | 32.1 | 699 | 678 |
| Blaine | 949 | 4,578,178 | 11.2 | 9.9 | 6.3 | 3.5 | 0.9 | 6.1 | 59.1 | 132.4 | 29.3 | 71.1 | 3,176 | 2,943 |
| Boise | 267 | 908,437 | 14.7 | 9.2 | 1.9 | 5.4 | 0.8 | 2.8 | 64.1 | 25.1 | 11.9 | 8.2 | 1,122 | 1,112 |
| Bonner | 1,321 | 4,417,721 | 11.7 | 16.3 | 3.8 | 6.0 | 0.2 | 7.5 | 51.5 | 122.9 | 45.7 | 54.1 | 1,240 | 1,164 |
| Bonneville | 3,793 | 12,587,349 | 7.5 | 11.1 | 6.4 | 4.9 | 2.2 | 11.1 | 55.5 | 352.2 | 180.0 | 100.2 | 876 | 830 |
| Boundary | 552 | 1,761,941 | 6.8 | 7.1 | 0.6 | 3.5 | 38.3 | 7.0 | 36.0 | 44.8 | 16.2 | 10.2 | 849 | 840 |
| Butte | 234 | 1,127,536 | 3.1 | 3.5 | 0.0 | 3.3 | 62.6 | 3.9 | 23.5 | 19.0 | 5.6 | 3.9 | 1,508 | 1,438 |
| Camas | 60 | 195,129 | 17.6 | 11.0 | 0.0 | 20.5 | 0.0 | 7.2 | 43.6 | 6.2 | 3.7 | 1.8 | 1,623 | 1,599 |
| Canyon | 6,362 | 22,467,284 | 8.4 | 13.1 | 3.7 | 2.1 | 3.4 | 4.6 | 61.4 | 589.3 | 301.3 | 187.6 | 865 | 795 |

1. Based on the resident population estimated as of July 1 of the year shown.

# Local Government Finances, Government Employment, and Income Taxes

| STATE County | Local government finances, 2017 (cont.) | | | | | | | Debt outstanding | | Government employment, 2019 | | | Individual income tax returns, 2018 | | |
|---|---|---|---|---|---|---|---|---|---|---|---|---|---|---|---|
| | Direct general expenditure | | | | | | | | | | | | | | |
| | | | Percent of total for: | | | | | | | | | | | | |
| | Total (mil dol) | Per capita[1] (dollars) | Education | Health and hospitals | Police protection | Public welfare | Highways | Total (mil dol) | Per capita[1] (dollars) | Federal civilian | Federal military | State and local | Number of returns | Mean adjusted gross income | Mean income tax |
| | 185 | 186 | 187 | 188 | 189 | 190 | 191 | 192 | 193 | 194 | 195 | 196 | 197 | 198 | 199 |
| GEORGIA—Cont'd | | | | | | | | | | | | | | | |
| Putnam | 63 | 2,909 | 58.1 | 5.0 | 7.1 | 0.3 | 4.8 | 22.4 | 1,032 | 65 | 59 | 1,411 | 9,660 | 63,973 | 7,203 |
| Quitman | 9 | 3,946 | 62.4 | 3.8 | 4.3 | 0.0 | 5.7 | 13.3 | 5,684 | 7 | 6 | 140 | 830 | 36,828 | 2,325 |
| Rabun | 66 | 3,983 | 56.3 | 5.2 | 5.2 | 0.2 | 4.5 | 51.1 | 3,087 | 59 | 45 | 799 | 7,790 | 56,009 | 6,067 |
| Randolph | 37 | 5,275 | 31.9 | 37.1 | 4.3 | 0.0 | 4.9 | 12.3 | 1,760 | 22 | 17 | 581 | 2,460 | 36,811 | 2,559 |
| Richmond | 766 | 3,796 | 46.4 | 4.0 | 5.7 | 0.2 | 5.3 | 1,279.5 | 6,344 | 7,825 | 11,718 | 24,377 | 85,790 | 45,593 | 4,052 |
| Rockdale | 316 | 3,521 | 58.9 | 0.6 | 6.4 | 0.4 | 6.6 | 180.5 | 2,009 | 121 | 243 | 4,449 | 40,390 | 48,464 | 4,061 |
| Schley | 24 | 4,529 | 70.4 | 2.6 | 3.6 | 0.1 | 6.4 | 16.3 | 3,115 | 8 | 14 | 335 | 1,780 | 44,173 | 2,989 |
| Screven | 42 | 3,032 | 56.2 | 3.8 | 4.8 | 0.0 | 7.4 | 9.1 | 655 | 35 | 36 | 952 | 5,470 | 45,709 | 3,570 |
| Seminole | 30 | 3,671 | 56.3 | 0.5 | 7.2 | 0.0 | 12.0 | 5.5 | 669 | 24 | 21 | 453 | 3,300 | 43,734 | 3,868 |
| Spalding | 233 | 3,569 | 54.7 | 10.1 | 6.3 | 0.1 | 2.8 | 109.7 | 1,679 | 140 | 175 | 5,276 | 27,750 | 47,741 | 4,023 |
| Stephens | 119 | 4,638 | 33.3 | 41.4 | 3.6 | 0.4 | 2.9 | 46.3 | 1,798 | 70 | 68 | 1,637 | 10,750 | 47,400 | 4,038 |
| Stewart | 19 | 2,939 | 62.3 | 0.3 | 9.4 | 0.0 | 8.4 | 8.2 | 1,299 | 105 | 13 | 253 | 1,500 | 38,083 | 2,615 |
| Sumter | 115 | 3,854 | 47.1 | 13.3 | 6.2 | 0.0 | 3.8 | 25.7 | 858 | 119 | 74 | 2,500 | 11,420 | 42,453 | 3,629 |
| Talbot | 14 | 2,245 | 50.1 | 3.5 | 9.0 | 0.1 | 9.1 | 3.5 | 556 | 9 | 17 | 270 | 2,490 | 38,953 | 2,761 |
| Taliaferro | 12 | 7,514 | 35.3 | 0.4 | 9.7 | 0.0 | 4.4 | 0.7 | 436 | 4 | 4 | 143 | 600 | 37,767 | 2,315 |
| Tattnall | 64 | 2,519 | 58.9 | 2.4 | 4.8 | 0.0 | 12.0 | 17.5 | 688 | 51 | 56 | 1,789 | 7,880 | 44,239 | 3,471 |
| Taylor | 26 | 3,257 | 60.9 | 2.2 | 7.4 | 0.1 | 7.7 | 4.3 | 533 | 20 | 21 | 414 | 3,080 | 44,881 | 3,616 |
| Telfair | 36 | 2,267 | 50.1 | 5.2 | 6.2 | 0.0 | 7.7 | 9.4 | 590 | 29 | 34 | 737 | 3,720 | 36,886 | 2,526 |
| Terrell | 31 | 3,590 | 52.3 | 3.6 | 7.6 | 0.1 | 6.1 | 6.7 | 772 | 54 | 22 | 562 | 3,760 | 40,728 | 3,389 |
| Thomas | 224 | 5,008 | 45.9 | 12.6 | 5.5 | 0.0 | 5.1 | 284.7 | 6,382 | 170 | 118 | 3,136 | 18,710 | 56,056 | 5,810 |
| Tift | 447 | 11,066 | 19.8 | 67.1 | 2.8 | 0.0 | 0.9 | 152.7 | 3,777 | 190 | 104 | 5,902 | 16,680 | 49,392 | 4,501 |
| Toombs | 274 | 10,199 | 28.0 | 57.6 | 1.9 | 0.0 | 2.4 | 180.9 | 6,735 | 68 | 71 | 1,643 | 10,590 | 48,144 | 4,222 |
| Towns | 28 | 2,429 | 50.7 | 4.8 | 5.7 | 0.0 | 5.2 | 27.1 | 2,348 | 28 | 29 | 509 | 5,540 | 54,367 | 5,196 |
| Treutlen | 19 | 2,796 | 59.4 | 2.6 | 6.0 | 2.2 | 5.1 | 8.9 | 1,319 | 14 | 17 | 350 | 2,440 | 40,314 | 2,640 |
| Troup | 275 | 3,929 | 49.6 | 8.8 | 7.3 | 0.1 | 3.4 | 83.9 | 1,197 | 143 | 181 | 3,852 | 28,560 | 50,982 | 4,720 |
| Turner | 32 | 4,038 | 43.9 | 4.1 | 10.4 | 0.4 | 10.0 | 7.4 | 943 | 29 | 20 | 465 | 3,430 | 36,227 | 2,463 |
| Twiggs | 22 | 2,653 | 58.3 | 0.3 | 12.4 | 0.3 | 7.1 | 9.4 | 1,141 | 11 | 21 | 311 | 3,470 | 41,767 | 3,267 |
| Union | 54 | 2,310 | 69.0 | 2.7 | 6.6 | 0.0 | 0.5 | 22.0 | 943 | 62 | 65 | 1,939 | 10,950 | 54,675 | 5,026 |
| Upson | 82 | 3,122 | 54.0 | 3.4 | 5.3 | 0.2 | 2.9 | 25.1 | 957 | 42 | 69 | 1,314 | 10,860 | 45,085 | 3,452 |
| Walker | 212 | 3,067 | 55.4 | 11.0 | 4.7 | 0.0 | 3.4 | 112.8 | 1,635 | 113 | 183 | 2,903 | 26,440 | 47,823 | 3,823 |
| Walton | 269 | 2,944 | 57.7 | 0.5 | 6.1 | 0.2 | 5.0 | 266.7 | 2,922 | 163 | 251 | 3,626 | 41,160 | 60,126 | 5,651 |
| Ware | 148 | 4,157 | 48.8 | 17.2 | 4.9 | 0.4 | 4.5 | 42.6 | 1,195 | 84 | 90 | 3,276 | 13,580 | 43,367 | 3,430 |
| Warren | 23 | 4,349 | 36.1 | 2.7 | 4.8 | 0.5 | 8.9 | 4.9 | 934 | 22 | 14 | 261 | 2,270 | 37,919 | 2,756 |
| Washington | 95 | 4,653 | 38.1 | 28.9 | 4.5 | 0.0 | 6.0 | 59.7 | 2,937 | 48 | 49 | 2,017 | 7,750 | 47,159 | 3,829 |
| Wayne | 171 | 5,750 | 36.4 | 39.7 | 3.6 | 0.0 | 7.6 | 53.7 | 1,803 | 366 | 74 | 1,871 | 10,920 | 47,160 | 3,680 |
| Webster | 7 | 2,657 | 69.0 | 6.0 | 4.7 | 0.3 | 5.4 | 1.7 | 664 | 5 | 7 | 119 | 910 | 35,268 | 2,612 |
| Wheeler | 22 | 2,718 | 54.7 | 2.3 | 5.0 | 0.0 | 4.7 | 8.5 | 1,073 | 11 | 14 | 306 | 1,810 | 35,693 | 2,256 |
| White | 90 | 3,057 | 65.9 | 1.4 | 5.8 | 0.1 | 6.0 | 33.3 | 1,132 | 53 | 80 | 1,135 | 12,210 | 51,020 | 4,348 |
| Whitfield | 437 | 4,196 | 54.7 | 12.4 | 4.0 | 0.0 | 4.5 | 83.0 | 796 | 159 | 277 | 5,472 | 42,730 | 61,324 | 6,666 |
| Wilcox | 28 | 3,228 | 70.1 | 3.2 | 4.7 | 0.6 | 6.5 | 9.0 | 1,020 | 26 | 18 | 540 | 2,620 | 37,877 | 2,529 |
| Wilkes | 39 | 3,946 | 48.9 | 3.4 | 7.9 | 0.0 | 4.2 | 35.7 | 3,625 | 29 | 26 | 764 | 4,000 | 43,343 | 3,528 |
| Wilkinson | 34 | 3,841 | 54.6 | 1.9 | 9.4 | 0.5 | 8.9 | 22.6 | 2,524 | 20 | 23 | 443 | 3,920 | 42,027 | 3,063 |
| Worth | 65 | 3,150 | 60.2 | 0.3 | 6.9 | 0.1 | 7.3 | 53.5 | 2,604 | 35 | 54 | 821 | 8,250 | 44,011 | 3,356 |
| HAWAII | X | X | X | X | X | X | X | X | X | 34,305 | 55,796 | 92,748 | 694,850 | 67,444 | 7,424 |
| Hawaii | 497 | 2,484 | 0.0 | 0.0 | 12.8 | 1.5 | 7.8 | 510.8 | 2,554 | 1,295 | 1,399 | 13,619 | 90,210 | 55,999 | 5,660 |
| Honolulu | 1,982 | 2,009 | 0.0 | 2.2 | 14.5 | 1.1 | 9.1 | 6,208.2 | 6,294 | [2]31,577 | [2]52,560 | [2]66,184 | 486,000 | 70,504 | 7,898 |
| Kalawao | NA | NA | NA | NA | NA | NA | NA | NA | NA | [2]872 | [2]1,208 | [2]8,359 | NA | NA | NA |
| Kauai | 221 | 3,070 | 0.0 | 0.0 | 15.0 | 5.2 | 11.1 | 204.1 | 2,842 | 561 | 629 | 4,586 | 36,070 | 64,632 | 6,920 |
| Maui | 468 | 2,816 | 0.0 | 0.0 | 11.5 | 13.0 | 10.1 | 336.6 | 2,026 | [2] | [2] | [2] | 82,600 | 63,140 | 6,780 |
| IDAHO | X | X | X | X | X | X | X | X | X | 13,234 | 9,033 | 114,765 | 783,110 | 63,054 | 6,578 |
| Ada | 1,348 | 2,953 | 40.4 | 2.4 | 9.0 | 0.6 | 6.9 | 674.3 | 1,477 | 5,943 | 1,562 | 31,006 | 223,710 | 79,838 | 9,877 |
| Adams | 12 | 2,950 | 37.5 | 0.0 | 12.4 | 0.0 | 13.3 | 3.5 | 836 | 113 | 13 | 218 | 1,900 | 49,759 | 4,306 |
| Bannock | 232 | 2,710 | 41.8 | 2.2 | 8.8 | 0.6 | 6.9 | 66.9 | 783 | 578 | 270 | 8,043 | 35,740 | 53,629 | 4,432 |
| Bear Lake | 26 | 4,316 | 35.0 | 25.0 | 2.5 | 0.5 | 0.9 | 8.0 | 1,322 | 48 | 19 | 653 | 2,630 | 49,500 | 3,662 |
| Benewah | 51 | 5,558 | 28.1 | 43.0 | 3.5 | 0.5 | 5.1 | 38.9 | 4,252 | 58 | 29 | 1,296 | 4,050 | 47,177 | 3,522 |
| Bingham | 123 | 2,676 | 55.4 | 0.3 | 8.7 | 0.4 | 6.5 | 74.3 | 1,619 | 207 | 145 | 3,979 | 19,210 | 51,526 | 4,403 |
| Blaine | 119 | 5,311 | 46.3 | 2.3 | 8.2 | 2.7 | 5.2 | 51.8 | 2,316 | 98 | 71 | 1,266 | 12,530 | 129,344 | 21,480 |
| Boise | 20 | 2,672 | 44.5 | 1.3 | 0.9 | 1.0 | 10.3 | 11.1 | 1,512 | 154 | 24 | 352 | 3,430 | 61,732 | 6,381 |
| Bonner | 124 | 2,831 | 36.9 | 5.9 | 16.1 | 0.1 | 10.9 | 30.9 | 707 | 198 | 142 | 2,290 | 20,920 | 61,050 | 6,257 |
| Bonneville | 323 | 2,824 | 45.1 | 4.3 | 8.3 | 0.3 | 4.5 | 206.1 | 1,800 | 726 | 373 | 6,139 | 50,580 | 64,880 | 6,579 |
| Boundary | 46 | 3,865 | 26.9 | 28.9 | 10.0 | 1.0 | 8.7 | 14.7 | 1,231 | 150 | 38 | 954 | 5,440 | 49,309 | 3,881 |
| Butte | 17 | 6,693 | 20.8 | 37.5 | 4.0 | 0.0 | 4.9 | 5.0 | 1,946 | 59 | 26 | 166 | 1,050 | 48,038 | 3,556 |
| Camas | 5 | 4,955 | 37.8 | 1.4 | 8.0 | 1.6 | 17.6 | 2.6 | 2,354 | 23 | 3 | 95 | 460 | 58,761 | 4,709 |
| Canyon | 542 | 2,500 | 50.2 | 1.7 | 7.5 | 0.6 | 4.4 | 345.4 | 1,593 | 376 | 713 | 9,311 | 93,890 | 51,771 | 4,073 |

1. Based on the resident population estimated as of July 1 of the year shown.    2. Kalawao county is included with Maui county.

# Table B. States and Counties — Land Area and Population

| State / county code | CBSA code[1] | County Type code[2] | STATE County | Land area[3] (sq. mi) | Total persons 2019 | Rank | Per square mile | White | Black | American Indian, Alaska Native | Asian and Pacific Islander | Percent Hispanic or Latino[4] | Under 5 years | 5 to 14 years | 15 to 24 years | 25 to 34 years | 35 to 44 years | 45 to 54 years |
|---|---|---|---|---|---|---|---|---|---|---|---|---|---|---|---|---|---|---|
| | | | | 1 | 2 | 3 | 4 | 5 | 6 | 7 | 8 | 9 | 10 | 11 | 12 | 13 | 14 | 15 |
| | | | IDAHO—Cont'd | | | | | | | | | | | | | | | |
| 16029 | | 6 | Caribou | 1,764.2 | 7,123 | 2,647 | 4.0 | 91.8 | 0.3 | 1.2 | 1.0 | 6.9 | 7.0 | 16.2 | 12.5 | 10.6 | 13.7 | 10.1 |
| 16031 | 15420 | 7 | Cassia | 2,565.6 | 24,277 | 1,637 | 9.5 | 70.5 | 0.6 | 1.2 | 1.3 | 27.6 | 8.0 | 17.5 | 14.8 | 12.0 | 12.4 | 9.9 |
| 16033 | | 9 | Clark | 1,763.1 | 852 | 3,113 | 0.5 | 55.4 | 2.1 | 0.8 | 1.3 | 41.7 | 6.5 | 11.7 | 12.9 | 12.7 | 12.8 | 11.7 |
| 16035 | | 6 | Clearwater | 2,457.3 | 8,846 | 2,505 | 3.6 | 92.2 | 0.9 | 3.7 | 1.6 | 4.0 | 3.6 | 8.2 | 9.2 | 11.0 | 11.0 | 11.7 |
| 16037 | | 8 | Custer | 4,922.2 | 4,249 | 2,876 | 0.9 | 93.5 | 0.8 | 2.1 | 0.9 | 4.9 | 4.3 | 9.5 | 7.9 | 8.4 | 10.0 | 10.9 |
| 16039 | 34300 | 6 | Elmore | 3,075.1 | 27,448 | 1,513 | 8.9 | 75.4 | 3.6 | 2.0 | 4.7 | 17.8 | 7.7 | 13.5 | 15.3 | 16.8 | 11.6 | 9.5 |
| 16041 | 30860 | 3 | Franklin | 663.0 | 14,215 | 2,142 | 21.4 | 91.8 | 0.5 | 1.3 | 0.7 | 6.8 | 7.4 | 17.6 | 14.8 | 11.1 | 12.9 | 10.6 |
| 16043 | 39940 | 6 | Fremont | 1,864.0 | 13,218 | 2,206 | 7.1 | 86.2 | 0.5 | 1.3 | 1.0 | 12.0 | 6.1 | 13.1 | 14.0 | 11.8 | 12.7 | 11.4 |
| 16045 | 14260 | 2 | Gem | 559.8 | 18,703 | 1,889 | 33.4 | 89.0 | 0.5 | 1.9 | 1.8 | 8.8 | 5.6 | 13.1 | 10.8 | 10.5 | 11.1 | 11.0 |
| 16047 | | 7 | Gooding | 729.3 | 15,618 | 2,050 | 21.4 | 68.7 | 0.5 | 1.5 | 1.2 | 29.5 | 6.6 | 15.5 | 13.0 | 12.0 | 11.4 | 11.3 |
| 16049 | | 6 | Idaho | 8,477.5 | 16,823 | 1,983 | 2.0 | 92.5 | 0.9 | 4.2 | 1.1 | 3.7 | 4.8 | 11.2 | 10.1 | 8.8 | 9.9 | 10.0 |
| 16051 | 26820 | 3 | Jefferson | 1,093.7 | 30,581 | 1,416 | 28.0 | 87.8 | 0.5 | 1.2 | 1.3 | 10.6 | 7.8 | 19.5 | 14.3 | 11.6 | 13.9 | 10.3 |
| 16053 | 46300 | 7 | Jerome | 597.5 | 24,578 | 1,626 | 41.1 | 60.8 | 0.5 | 1.1 | 0.8 | 37.8 | 7.6 | 17.2 | 13.5 | 13.4 | 12.7 | 10.8 |
| 16055 | 17660 | 3 | Kootenai | 1,237.8 | 170,628 | 389 | 137.8 | 92.3 | 0.8 | 2.2 | 2.0 | 5.1 | 5.7 | 12.8 | 11.2 | 12.9 | 12.4 | 11.7 |
| 16057 | 34140 | 4 | Latah | 1,075.9 | 40,830 | 1,159 | 37.9 | 91.0 | 1.5 | 1.8 | 3.9 | 4.6 | 5.1 | 10.3 | 25.4 | 14.7 | 10.8 | 9.0 |
| 16059 | | 7 | Lemhi | 4,563.7 | 8,054 | 2,585 | 1.8 | 95.1 | 0.7 | 2.1 | 0.8 | 3.3 | 4.9 | 10.0 | 8.2 | 8.8 | 10.6 | 9.4 |
| 16061 | | 8 | Lewis | 478.8 | 3,838 | 2,906 | 8.0 | 86.5 | 1.2 | 7.6 | 2.6 | 5.5 | 4.8 | 12.8 | 10.0 | 8.6 | 10.2 | 9.5 |
| 16063 | | 9 | Lincoln | 1,201.4 | 5,358 | 2,796 | 4.5 | 66.1 | 0.8 | 1.6 | 1.1 | 32.1 | 6.4 | 15.3 | 14.0 | 12.4 | 13.1 | 10.9 |
| 16065 | 39940 | 3 | Madison | 469.3 | 40,318 | 1,173 | 85.9 | 89.0 | 0.9 | 0.8 | 2.8 | 8.1 | 9.6 | 12.7 | 35.4 | 15.1 | 7.6 | 6.0 |
| 16067 | 15420 | 7 | Minidoka | 757.0 | 21,216 | 1,759 | 28.0 | 62.7 | 0.6 | 1.4 | 0.9 | 35.7 | 7.4 | 16.7 | 13.1 | 13.0 | 11.6 | 10.3 |
| 16069 | 30300 | 3 | Nez Perce | 848.3 | 40,755 | 1,161 | 48.0 | 88.9 | 0.9 | 6.5 | 1.8 | 4.4 | 5.7 | 11.9 | 12.0 | 13.1 | 11.5 | 11.5 |
| 16071 | | 8 | Oneida | 1,199.0 | 4,520 | 2,857 | 3.8 | 94.2 | 0.7 | 1.0 | 1.0 | 4.2 | 6.7 | 15.5 | 12.2 | 9.2 | 12.0 | 10.2 |
| 16073 | 14260 | 2 | Owyhee | 7,668.2 | 12,133 | 2,283 | 1.6 | 70.8 | 0.7 | 3.5 | 1.1 | 25.5 | 6.3 | 14.1 | 13.2 | 11.4 | 11.7 | 11.5 |
| 16075 | 36620 | 6 | Payette | 406.9 | 24,771 | 1,614 | 60.9 | 80.0 | 0.7 | 2.1 | 1.9 | 17.6 | 6.7 | 14.6 | 11.9 | 11.9 | 11.4 | 11.8 |
| 16077 | 38540 | 6 | Power | 1,403.8 | 7,643 | 2,616 | 5.4 | 62.4 | 0.8 | 3.2 | 1.1 | 34.2 | 7.7 | 18.1 | 13.0 | 12.1 | 11.4 | 9.2 |
| 16079 | | 6 | Shoshone | 2,637.4 | 12,911 | 2,228 | 4.9 | 93.5 | 1.0 | 3.1 | 1.1 | 3.7 | 5.9 | 10.8 | 9.9 | 11.4 | 10.2 | 11.8 |
| 16081 | 27220 | 9 | Teton | 449.1 | 12,501 | 2,246 | 27.8 | 82.2 | 0.5 | 0.8 | 1.0 | 16.5 | 5.5 | 14.0 | 10.5 | 12.8 | 16.6 | 14.8 |
| 16083 | 46300 | 5 | Twin Falls | 1,921.7 | 88,411 | 662 | 46.0 | 79.3 | 1.1 | 1.3 | 2.6 | 17.3 | 6.9 | 15.6 | 12.8 | 13.9 | 13.1 | 10.3 |
| 16085 | | 8 | Valley | 3,665.1 | 11,792 | 2,306 | 3.2 | 93.4 | 0.7 | 1.5 | 1.1 | 4.8 | 4.3 | 10.3 | 8.3 | 9.8 | 12.6 | 11.8 |
| 16087 | | 6 | Washington | 1,452.9 | 10,360 | 2,396 | 7.1 | 81.2 | 0.8 | 2.2 | 1.3 | 16.4 | 5.2 | 12.6 | 11.7 | 9.0 | 10.1 | 10.8 |
| 17000 | | 0 | ILLINOIS | 55,513.7 | 12,587,530 | X | 226.7 | 62.0 | 15.0 | 0.5 | 6.7 | 17.6 | 5.8 | 12.4 | 13.0 | 13.8 | 13.0 | 12.4 |
| 17001 | 39500 | 5 | Adams | 855.1 | 64,783 | 822 | 75.8 | 93.1 | 5.5 | 0.4 | 1.3 | 1.8 | 6.1 | 12.6 | 11.4 | 12.0 | 11.7 | 11.3 |
| 17003 | 16020 | 3 | Alexander | 235.4 | 5,497 | 2,788 | 23.4 | 66.3 | 32.4 | 0.9 | 0.8 | 2.0 | 4.7 | 12.3 | 9.5 | 9.7 | 10.2 | 12.0 |
| 17005 | 41180 | 1 | Bond | 380.3 | 16,262 | 2,018 | 42.8 | 88.6 | 7.2 | 0.8 | 1.2 | 3.8 | 4.5 | 10.7 | 10.9 | 14.0 | 13.4 | 11.9 |
| 17007 | 40420 | 2 | Boone | 280.7 | 52,777 | 961 | 188.0 | 72.9 | 3.0 | 0.6 | 1.8 | 23.2 | 5.4 | 13.8 | 13.8 | 11.1 | 12.0 | 13.8 |
| 17009 | | 7 | Brown | 305.7 | 6,546 | 2,699 | 21.4 | 74.0 | 18.9 | 0.4 | 0.4 | 6.9 | 4.9 | 8.4 | 13.6 | 19.0 | 15.6 | 12.5 |
| 17011 | 36837 | 7 | Bureau | 869.1 | 32,303 | 1,371 | 37.2 | 88.2 | 1.6 | 0.5 | 1.3 | 9.6 | 5.1 | 11.8 | 11.4 | 10.6 | 11.3 | 12.1 |
| 17013 | 41180 | 1 | Calhoun | 253.8 | 4,616 | 2,852 | 18.2 | 94.7 | 0.7 | 0.5 | 0.5 | 1.4 | 4.5 | 12.0 | 9.5 | 9.0 | 10.4 | 13.2 |
| 17015 | | 7 | Carroll | 445.4 | 14,241 | 2,140 | 32.0 | 93.9 | 1.9 | 0.6 | 1.1 | 4.0 | 5.0 | 10.9 | 10.2 | 10.1 | 10.6 | 11.4 |
| 17017 | | 6 | Cass | 375.8 | 11,925 | 2,298 | 31.7 | 74.6 | 5.0 | 0.3 | 0.9 | 20.0 | 6.9 | 13.3 | 11.9 | 11.2 | 12.8 | 12.2 |
| 17019 | 16580 | 3 | Champaign | 996.1 | 209,192 | 323 | 210.0 | 68.8 | 15.2 | 0.5 | 12.1 | 6.4 | 5.3 | 10.4 | 25.6 | 13.9 | 11.3 | 9.5 |
| 17021 | 45380 | 6 | Christian | 709.5 | 32,075 | 1,379 | 45.2 | 95.8 | 2.2 | 0.5 | 1.0 | 1.6 | 5.2 | 11.1 | 11.2 | 12.2 | 12.1 | 12.7 |
| 17023 | | 6 | Clark | 501.4 | 15,268 | 2,073 | 30.5 | 97.3 | 0.8 | 0.5 | 0.6 | 1.6 | 5.9 | 12.6 | 11.1 | 11.0 | 11.6 | 12.3 |
| 17025 | | 7 | Clay | 468.4 | 13,079 | 2,215 | 27.9 | 96.6 | 1.1 | 0.6 | 1.1 | 1.8 | 6.0 | 12.9 | 11.0 | 11.7 | 11.2 | 12.1 |
| 17027 | 41180 | 1 | Clinton | 473.9 | 37,398 | 1,241 | 78.9 | 92.3 | 4.2 | 0.4 | 1.0 | 3.3 | 5.6 | 11.9 | 11.0 | 13.0 | 13.2 | 12.4 |
| 17029 | 16660 | 5 | Coles | 508.3 | 50,383 | 984 | 99.1 | 91.9 | 4.8 | 0.5 | 1.4 | 3.0 | 4.8 | 9.8 | 19.9 | 12.7 | 11.1 | 10.6 |
| 17031 | 16980 | 1 | Cook | 944.9 | 5,108,284 | 2 | 5,406.2 | 43.1 | 23.8 | 0.5 | 8.8 | 25.6 | 5.9 | 11.9 | 12.2 | 16.3 | 13.7 | 12.3 |
| 17033 | | 7 | Crawford | 443.6 | 18,512 | 1,899 | 41.7 | 91.5 | 5.7 | 0.5 | 0.9 | 2.4 | 5.4 | 11.2 | 10.6 | 13.3 | 13.1 | 12.1 |
| 17035 | 16660 | 9 | Cumberland | 345.9 | 10,649 | 2,370 | 30.8 | 97.4 | 1.0 | 0.7 | 0.9 | 1.2 | 5.8 | 12.7 | 10.6 | 11.0 | 12.2 | 11.9 |
| 17037 | 16980 | 1 | DeKalb | 631.3 | 104,491 | 582 | 165.5 | 76.8 | 9.3 | 0.5 | 3.4 | 12.0 | 5.5 | 12.1 | 22.7 | 11.8 | 11.1 | 10.4 |
| 17039 | | 3 | De Witt | 397.6 | 15,368 | 2,068 | 38.7 | 95.4 | 1.5 | 0.6 | 0.8 | 2.9 | 5.6 | 11.4 | 11.1 | 11.4 | 12.2 | 12.6 |
| 17041 | | 6 | Douglas | 416.6 | 19,510 | 1,853 | 46.8 | 91.0 | 1.3 | 0.7 | 1.0 | 7.4 | 6.5 | 13.7 | 12.4 | 12.4 | 11.9 | 11.3 |
| 17043 | 16980 | 1 | DuPage | 327.8 | 917,481 | 59 | 2,798.9 | 67.1 | 5.7 | 0.4 | 14.0 | 14.7 | 5.8 | 12.5 | 12.4 | 12.5 | 13.3 | 12.9 |
| 17045 | | 6 | Edgar | 623.3 | 16,858 | 1,980 | 27.0 | 97.5 | 1.1 | 0.6 | 0.5 | 1.3 | 5.2 | 10.8 | 10.6 | 10.7 | 11.7 | 12.2 |
| 17047 | | 9 | Edwards | 222.4 | 6,356 | 2,717 | 28.6 | 97.4 | 1.2 | 0.6 | 0.8 | 1.3 | 6.3 | 12.2 | 10.9 | 10.0 | 11.5 | 12.3 |
| 17049 | 20820 | 7 | Effingham | 478.8 | 34,065 | 1,326 | 71.1 | 96.3 | 1.0 | 0.3 | 1.1 | 2.1 | 6.6 | 13.3 | 11.3 | 12.8 | 11.9 | 11.1 |
| 17051 | | 6 | Fayette | 716.4 | 21,264 | 1,756 | 29.7 | 93.0 | 4.9 | 0.6 | 0.7 | 2.0 | 5.6 | 11.2 | 12.0 | 12.8 | 12.7 | 12.1 |
| 17053 | | 3 | Ford | 485.6 | 12,949 | 2,225 | 26.7 | 94.0 | 1.9 | 0.7 | 1.1 | 4.0 | 6.0 | 12.4 | 11.8 | 11.5 | 12.0 | 11.8 |
| 17055 | | 4 | Franklin | 408.9 | 38,060 | 1,224 | 93.1 | 96.7 | 1.2 | 0.8 | 0.7 | 1.9 | 5.9 | 12.1 | 10.9 | 11.5 | 11.4 | 12.8 |
| 17057 | 37900 | 7 | Fulton | 865.7 | 33,690 | 1,334 | 38.9 | 92.2 | 4.5 | 0.6 | 0.6 | 3.2 | 4.8 | 10.9 | 11.2 | 12.1 | 12.8 | 12.7 |
| 17059 | | 8 | Gallatin | 322.9 | 4,793 | 2,842 | 14.8 | 96.9 | 1.4 | 1.2 | 0.7 | 1.6 | 4.5 | 11.7 | 9.8 | 10.7 | 11.5 | 12.8 |
| 17061 | | 6 | Greene | 543.0 | 12,702 | 2,235 | 23.4 | 96.8 | 1.6 | 0.5 | 0.4 | 1.3 | 5.4 | 11.9 | 11.1 | 11.4 | 11.8 | 12.6 |
| 17063 | 16980 | 1 | Grundy | 418.1 | 50,993 | 977 | 122.0 | 86.4 | 2.2 | 0.4 | 1.4 | 10.7 | 6.0 | 14.2 | 12.5 | 12.4 | 13.9 | 13.2 |
| 17065 | | 7 | Hamilton | 434.6 | 8,084 | 2,579 | 18.6 | 96.3 | 1.2 | 0.7 | 1.0 | 1.9 | 5.4 | 12.3 | 11.1 | 10.6 | 11.9 | 11.6 |
| 17067 | 22800 | 7 | Hancock | 793.7 | 17,422 | 1,950 | 22.0 | 96.9 | 1.3 | 0.6 | 0.7 | 1.7 | 5.2 | 12.1 | 10.2 | 10.3 | 11.0 | 11.5 |

1. CBSA = Core Based Statistical Area. See Appendix A for explanation. See Appendix B for list of metropolitan areas with component counties.
Service of USDA Rural-Urban Continuum Codes. See Appendix A for definition. 3. Dry land or land partially or temporarily covered by water.
2. County type code from the Economic Research
4. May be of any race.

Items 1—15

# Table B. States and Counties — Population and Households

| STATE County | 55 to 64 years | 65 to 74 years | 75 years and over | Percent female | Total persons 2000 | Total persons 2010 | Percent change 2000-2010 | Percent change 2010-2020 | Births | Deaths | Net Migration | Number | Persons per household | Family households | Female family householder[1] | One person |
|---|---|---|---|---|---|---|---|---|---|---|---|---|---|---|---|---|
| | 16 | 17 | 18 | 19 | 20 | 21 | 22 | 23 | 24 | 25 | 26 | 27 | 28 | 29 | 30 | 31 |
| **IDAHO—Cont'd** | | | | | | | | | | | | | | | | |
| Caribou | 12.0 | 10.3 | 7.6 | 49.3 | 7,304 | 6,960 | -4.7 | 2.3 | 938 | 667 | -110 | 2,538 | 2.71 | 72.2 | 5.3 | 24.5 |
| Cassia | 10.7 | 8.4 | 6.3 | 49.0 | 21,416 | 22,964 | 7.2 | 5.7 | 3,931 | 2,038 | -584 | 7,711 | 3.02 | 78.4 | 9.6 | 18.2 |
| Clark | 12.1 | 10.3 | 9.3 | 47.9 | 1,022 | 982 | -3.9 | -13.2 | 100 | 67 | -167 | 290 | 3.07 | 65.5 | 1.0 | 33.8 |
| Clearwater | 16.6 | 16.4 | 12.1 | 44.6 | 8,930 | 8,761 | -1.9 | 1.0 | 663 | 1,107 | 534 | 3,555 | 2.17 | 59.7 | 4.2 | 33.0 |
| Custer | 17.4 | 20.2 | 11.5 | 47.7 | 4,342 | 4,366 | 0.6 | -2.7 | 370 | 438 | -50 | 1,774 | 2.29 | 62.7 | 4.9 | 36.0 |
| Elmore | 11.4 | 8.6 | 5.7 | 47.7 | 29,130 | 27,040 | -7.2 | 1.5 | 4,687 | 1,926 | -2,409 | 10,606 | 2.45 | 66.5 | 7.8 | 25.3 |
| Franklin | 11.0 | 8.4 | 6.3 | 48.9 | 11,329 | 12,786 | 12.9 | 11.2 | 2,047 | 1,046 | 428 | 4,397 | 3.05 | 79.4 | 9.2 | 17.6 |
| Fremont | 13.1 | 10.6 | 7.4 | 47.6 | 11,819 | 13,243 | 12.0 | -0.2 | 1,924 | 1,083 | -879 | 4,347 | 2.87 | 82.3 | 10.6 | 15.3 |
| Gem | 15.1 | 13.4 | 9.4 | 49.8 | 15,181 | 16,719 | 10.1 | 11.9 | 2,016 | 2,087 | 2,057 | 6,683 | 2.57 | 68.6 | 12.0 | 27.5 |
| Gooding | 11.8 | 10.6 | 7.9 | 49.0 | 14,155 | 15,471 | 9.3 | 1.0 | 2,080 | 1,457 | -483 | 5,469 | 2.75 | 70.2 | 10.7 | 23.1 |
| Idaho | 16.1 | 17.2 | 11.9 | 47.5 | 15,511 | 16,269 | 4.9 | 3.4 | 1,587 | 1,884 | 869 | 6,407 | 2.45 | 68.0 | 5.4 | 27.4 |
| Jefferson | 10.4 | 7.7 | 4.5 | 48.8 | 19,155 | 26,144 | 36.5 | 17.0 | 4,839 | 1,615 | 1,212 | 8,791 | 3.24 | 79.9 | 5.2 | 15.7 |
| Jerome | 11.2 | 8.1 | 5.4 | 48.5 | 18,342 | 22,355 | 21.9 | 9.9 | 3,982 | 1,662 | -103 | 8,022 | 2.94 | 71.1 | 12.5 | 19.9 |
| Kootenai | 13.5 | 12.1 | 7.7 | 50.5 | 108,685 | 138,466 | 27.4 | 23.2 | 18,081 | 14,096 | 28,134 | 62,304 | 2.5 | 68.1 | 9.4 | 25.5 |
| Latah | 10.1 | 8.9 | 5.8 | 49.1 | 34,935 | 37,243 | 6.6 | 9.6 | 4,455 | 2,293 | 1,436 | 15,422 | 2.35 | 54.9 | 5.4 | 29.8 |
| Lemhi | 16.3 | 18.6 | 13.3 | 49.7 | 7,806 | 7,936 | 1.7 | 1.5 | 754 | 1,027 | 389 | 3,582 | 2.13 | 58.0 | 5.5 | 39.3 |
| Lewis | 15.7 | 16.2 | 12.1 | 49.6 | 3,747 | 3,821 | 2.0 | 0.4 | 393 | 422 | 49 | 1,646 | 2.31 | 55.8 | 5.8 | 36.9 |
| Lincoln | 12.7 | 9.5 | 5.7 | 48.4 | 4,044 | 5,207 | 28.8 | 2.9 | 733 | 393 | -190 | 1,724 | 3.08 | 73.3 | 8.5 | 23.3 |
| Madison | 6.0 | 4.4 | 3.1 | 49.1 | 27,467 | 37,542 | 36.7 | 7.4 | 11,234 | 1,405 | -7,116 | 11,016 | 3.51 | 82.4 | 3.7 | 11.8 |
| Minidoka | 11.6 | 9.0 | 7.3 | 50.0 | 20,174 | 20,055 | -0.6 | 5.8 | 3,255 | 1,859 | -239 | 7,250 | 2.85 | 75.6 | 12.5 | 20.8 |
| Nez Perce | 13.5 | 11.5 | 9.3 | 50.7 | 37,410 | 39,270 | 5.0 | 3.8 | 4,841 | 5,148 | 1,819 | 16,384 | 2.4 | 65.5 | 10.0 | 28.4 |
| Oneida | 13.4 | 11.4 | 9.2 | 49.9 | 4,125 | 4,286 | 3.9 | 5.5 | 532 | 403 | 103 | 1,615 | 2.7 | 71.6 | 7.3 | 26.3 |
| Owyhee | 13.6 | 10.8 | 7.3 | 49.1 | 10,644 | 11,526 | 8.3 | 5.3 | 1,437 | 985 | 155 | 4,321 | 2.64 | 72.0 | 9.8 | 22.8 |
| Payette | 12.6 | 10.8 | 8.3 | 49.9 | 20,578 | 22,622 | 9.9 | 9.5 | 3,127 | 2,206 | 1,244 | 8,876 | 2.6 | 67.8 | 9.3 | 27.3 |
| Power | 11.9 | 9.8 | 6.9 | 49.6 | 7,538 | 7,819 | 3.7 | -2.3 | 1,283 | 626 | -846 | 2,633 | 2.89 | 67.7 | 9.5 | 29.9 |
| Shoshone | 15.7 | 14.8 | 9.5 | 49.6 | 13,771 | 12,800 | -7.1 | 0.9 | 1,425 | 1,899 | 578 | 5,466 | 2.25 | 63.5 | 9.6 | 30.5 |
| Teton | 12.7 | 9.4 | 3.7 | 47.8 | 5,999 | 10,165 | 69.4 | 23.0 | 1,498 | 446 | 1,267 | 3,707 | 3.08 | 66.6 | 6.6 | 23.9 |
| Twin Falls | 11.3 | 9.4 | 6.9 | 50.7 | 64,284 | 77,240 | 20.2 | 14.5 | 12,295 | 7,679 | 6,590 | 31,146 | 2.69 | 69.3 | 9.8 | 25.2 |
| Valley | 16.0 | 18.1 | 8.8 | 48.6 | 7,651 | 9,854 | 28.8 | 19.7 | 932 | 741 | 1,731 | 3,869 | 2.72 | 71.5 | 7.1 | 23.3 |
| Washington | 14.2 | 15.2 | 11.2 | 50.2 | 9,977 | 10,198 | 2.2 | 1.6 | 1,073 | 1,230 | 325 | 4,035 | 2.46 | 67.1 | 12.5 | 28.0 |
| **ILLINOIS** | 13.0 | 9.6 | 7.0 | 50.9 | 12,419,293 | 12,831,572 | 3.3 | -1.9 | 1,578,161 | 1,094,778 | -728,322 | 4,846,134 | 2.57 | 64.2 | 12.2 | 29.6 |
| Adams | 13.9 | 11.1 | 10.0 | 50.8 | 68,277 | 67,095 | -1.7 | -3.4 | 8,265 | 8,288 | -2,255 | 27,112 | 2.38 | 64.9 | 10.3 | 29.5 |
| Alexander | 16.4 | 14.6 | 10.6 | 51.4 | 9,590 | 8,238 | -14.1 | -33.3 | 852 | 984 | -2,645 | 2,154 | 2.85 | 63.0 | 14.5 | 33.7 |
| Bond | 14.6 | 11.6 | 8.5 | 47.8 | 17,633 | 17,770 | 0.8 | -8.5 | 1,598 | 1,756 | -1,369 | 6,299 | 2.35 | 65.2 | 6.9 | 27.5 |
| Boone | 13.4 | 9.6 | 7.0 | 50.0 | 41,786 | 54,161 | 29.6 | -2.6 | 5,828 | 4,152 | -3,080 | 18,571 | 2.86 | 74.8 | 11.0 | 20.8 |
| Brown | 11.5 | 8.1 | 6.5 | 35.5 | 6,950 | 6,937 | -0.2 | -5.6 | 589 | 568 | -415 | 2,055 | 2.31 | 63.4 | 7.8 | 28.7 |
| Bureau | 14.6 | 12.6 | 10.5 | 50.9 | 35,503 | 34,980 | -1.5 | -7.7 | 3,424 | 3,982 | -2,129 | 13,698 | 2.39 | 67.8 | 9.8 | 27.8 |
| Calhoun | 16.4 | 13.4 | 11.4 | 49.9 | 5,084 | 5,087 | 0.1 | -9.3 | 472 | 546 | -400 | 1,664 | 2.83 | 69.4 | 4.9 | 28.2 |
| Carroll | 15.3 | 14.9 | 11.6 | 50.1 | 16,674 | 15,394 | -7.7 | -7.5 | 1,430 | 1,929 | -646 | 6,508 | 2.17 | 61.8 | 8.4 | 32.4 |
| Cass | 13.2 | 10.5 | 8.0 | 50.0 | 13,695 | 13,641 | -0.4 | -12.6 | 1,764 | 1,496 | -2,007 | 5,043 | 2.44 | 66.2 | 12.5 | 28.4 |
| Champaign | 10.2 | 8.2 | 5.5 | 50.4 | 179,669 | 201,087 | 11.9 | 4.0 | 24,032 | 13,406 | -2,649 | 82,369 | 2.36 | 51.4 | 9.2 | 36.5 |
| Christian | 14.7 | 11.3 | 9.5 | 49.0 | 35,372 | 34,797 | -1.6 | -7.8 | 3,552 | 4,228 | -2,043 | 13,901 | 2.24 | 61.7 | 10.4 | 31.4 |
| Clark | 14.9 | 11.3 | 9.4 | 50.6 | 17,008 | 16,335 | -4.0 | -6.5 | 1,922 | 2,085 | -896 | 6,720 | 2.31 | 71.7 | 11.1 | 22.7 |
| Clay | 14.1 | 12.0 | 9.1 | 50.7 | 14,560 | 13,812 | -5.1 | -5.3 | 1,605 | 1,776 | -558 | 5,696 | 2.27 | 67.0 | 10.7 | 28.5 |
| Clinton | 14.7 | 10.3 | 7.8 | 48.1 | 35,535 | 37,752 | 6.2 | -0.9 | 4,238 | 3,622 | -960 | 14,353 | 2.47 | 66.8 | 7.9 | 27.1 |
| Coles | 12.8 | 10.3 | 7.9 | 51.4 | 53,196 | 53,876 | 1.3 | -6.5 | 5,092 | 5,217 | -3,400 | 20,926 | 2.27 | 57.3 | 11.9 | 32.9 |
| Cook | 12.3 | 9.0 | 6.5 | 51.4 | 5,376,741 | 5,195,024 | -3.4 | -1.7 | 686,879 | 418,298 | -354,916 | 1,972,108 | 2.59 | 60.0 | 14.2 | 32.8 |
| Crawford | 14.4 | 11.4 | 8.5 | 47.6 | 20,452 | 19,818 | -3.1 | -6.6 | 2,127 | 2,402 | -1,031 | 7,666 | 2.16 | 64.9 | 10.1 | 30.3 |
| Cumberland | 15.1 | 11.8 | 8.9 | 49.9 | 11,253 | 11,045 | -1.8 | -3.6 | 1,245 | 1,178 | -460 | 4,299 | 2.49 | 68.6 | 7.0 | 26.4 |
| DeKalb | 11.2 | 7.9 | 5.5 | 50.4 | 88,969 | 105,162 | 18.2 | -0.6 | 11,950 | 7,363 | -5,340 | 38,150 | 2.62 | 60.7 | 11.4 | 27.3 |
| De Witt | 15.5 | 11.4 | 8.7 | 50.4 | 16,798 | 16,558 | -1.4 | -7.2 | 1,792 | 1,945 | -1,032 | 6,694 | 2.34 | 64.9 | 8.1 | 32.5 |
| Douglas | 13.2 | 10.3 | 8.5 | 50.4 | 19,922 | 19,971 | 0.2 | -2.3 | 2,604 | 2,002 | -1,059 | 7,613 | 2.55 | 66.5 | 8.6 | 27.4 |
| DuPage | 13.9 | 9.9 | 6.8 | 50.8 | 904,161 | 916,741 | 1.4 | 0.1 | 108,206 | 63,877 | -43,882 | 342,791 | 2.67 | 69.8 | 9.1 | 25.0 |
| Edgar | 15.4 | 13.0 | 10.3 | 50.9 | 19,704 | 18,576 | -5.7 | -9.2 | 1,835 | 2,443 | -1,107 | 7,542 | 2.26 | 68.1 | 12.0 | 28.6 |
| Edwards | 14.0 | 12.8 | 9.8 | 51.2 | 6,971 | 6,726 | -3.5 | -5.5 | 816 | 749 | -435 | 2,773 | 2.31 | 65.0 | 8.8 | 30.0 |
| Effingham | 14.4 | 10.5 | 8.0 | 50.1 | 34,264 | 34,244 | -0.1 | -0.5 | 4,623 | 3,672 | -1,123 | 13,877 | 2.43 | 66.1 | 8.0 | 29.0 |
| Fayette | 13.8 | 10.7 | 9.0 | 47.1 | 21,802 | 22,152 | 1.6 | -4.0 | 2,459 | 2,276 | -1,077 | 7,737 | 2.57 | 68.5 | 8.6 | 26.9 |
| Ford | 14.5 | 10.5 | 9.4 | 50.9 | 14,241 | 14,075 | -1.2 | -8.0 | 1,481 | 2,020 | -588 | 5,771 | 2.23 | 61.5 | 9.3 | 34.4 |
| Franklin | 14.0 | 11.8 | 9.5 | 50.6 | 39,018 | 39,996 | 2.5 | -4.8 | 4,825 | 5,458 | -1,270 | 16,235 | 2.37 | 60.7 | 12.1 | 31.4 |
| Fulton | 14.0 | 12.0 | 9.5 | 47.9 | 38,250 | 37,070 | -3.1 | -9.1 | 3,465 | 4,706 | -2,122 | 13,940 | 2.33 | 63.3 | 9.0 | 31.8 |
| Gallatin | 14.4 | 13.4 | 11.1 | 51.1 | 6,445 | 5,589 | -13.3 | -14.2 | 523 | 823 | -496 | 2,293 | 2.2 | 65.7 | 16.2 | 31.8 |
| Greene | 15.5 | 11.4 | 8.9 | 49.4 | 14,761 | 13,886 | -5.9 | -8.5 | 1,414 | 1,621 | -984 | 4,949 | 2.6 | 66.1 | 10.2 | 29.9 |
| Grundy | 12.7 | 8.9 | 6.1 | 50.3 | 37,535 | 50,079 | 33.4 | 1.8 | 6,193 | 4,103 | -1,170 | 19,676 | 2.56 | 71.2 | 10.0 | 23.6 |
| Hamilton | 14.6 | 12.2 | 10.3 | 51.1 | 8,621 | 8,457 | -1.9 | -4.4 | 877 | 1,111 | -128 | 3,400 | 2.37 | 69.5 | 9.4 | 24.3 |
| Hancock | 15.3 | 13.6 | 10.7 | 50.5 | 20,121 | 19,104 | -5.1 | -8.8 | 1,933 | 2,119 | -1,497 | 7,409 | 2.39 | 66.6 | 8.5 | 29.6 |

1. No spouse present.

# Table B. States and Counties — Population, Vital Statistics, Health, and Crime

| STATE County | Persons in group quarters, 2020 | Daytime Population, 2015-2019 | | Births, 2020 | | Deaths, 2020 | | Persons under 65 with no health insurance, 2019 | | Medicare, 2020 | | | Crimes reported by county police or sheriff, 2019 | |
|---|---|---|---|---|---|---|---|---|---|---|---|---|---|---|
| | | Number | Employment/ residence ratio | Total | Rate[1] | Number | Rate[1] | Number | Percent | Total beneficiaries | Enrolled in Original Medicare | Enrolled in Medicare Advantage | Violent | Property |
| | 32 | 33 | 34 | 35 | 36 | 37 | 38 | 39 | 40 | 41 | 42 | 43 | 44 | 45 |
| **IDAHO—Cont'd** | | | | | | | | | | | | | | |
| Caribou | 32 | 7,653 | 1.24 | 92 | 12.9 | 67 | 9.4 | 620 | 10.5 | 1,404 | 1,355 | 49 | NA | NA |
| Cassia | 286 | 24,272 | 1.06 | 384 | 15.8 | 201 | 8.3 | 3,414 | 17.1 | 4,065 | 3,063 | 1,002 | 34 | 257 |
| Clark | 2 | 807 | 0.74 | 12 | 14.1 | 2 | 2.3 | 196 | 29.7 | 143 | 101 | 42 | 1 | 5 |
| Clearwater | 740 | 8,572 | 0.96 | 64 | 7.2 | 99 | 11.2 | 805 | 14.4 | 2,834 | 2,755 | 79 | NA | NA |
| Custer | 21 | 4,164 | 1.00 | 27 | 6.4 | 38 | 8.9 | 436 | 14.6 | 1,368 | 1,333 | 34 | 4 | 16 |
| Elmore | 699 | 25,789 | 0.92 | 419 | 15.3 | 211 | 7.7 | 3,386 | 15.0 | 4,735 | 3,383 | 1,352 | 29 | 55 |
| Franklin | 102 | 11,133 | 0.60 | 203 | 14.3 | 98 | 6.9 | 1,546 | 13.2 | 2,385 | 1,957 | 428 | 8 | 34 |
| Fremont | 458 | 11,362 | 0.69 | 180 | 13.6 | 97 | 7.3 | 1,742 | 17.0 | 2,608 | 2,018 | 590 | 2 | 42 |
| Gem | 147 | 14,822 | 0.64 | 190 | 10.2 | 211 | 11.3 | 2,034 | 14.7 | 4,987 | 2,448 | 2,539 | 22 | 64 |
| Gooding | 31 | 14,035 | 0.84 | 188 | 12.0 | 119 | 7.6 | 2,424 | 19.9 | 3,133 | 2,195 | 937 | 19 | 41 |
| Idaho | 524 | 15,917 | 0.92 | 152 | 9.0 | 192 | 11.4 | 1,834 | 16.2 | 5,113 | 4,957 | 156 | 26 | 63 |
| Jefferson | 175 | 24,036 | 0.65 | 439 | 14.4 | 163 | 5.3 | 3,555 | 13.6 | 4,276 | 3,195 | 1,081 | 19 | 68 |
| Jerome | 152 | 23,983 | 1.02 | 356 | 14.5 | 170 | 6.9 | 4,349 | 21.1 | 3,702 | 2,378 | 1,324 | 19 | 95 |
| Kootenai | 1,467 | 150,649 | 0.91 | 1,785 | 10.5 | 1,675 | 9.8 | 16,442 | 12.4 | 40,392 | 26,300 | 14,092 | 149 | 707 |
| Latah | 2,774 | 36,576 | 0.85 | 394 | 9.6 | 229 | 5.6 | 3,462 | 10.8 | 6,623 | 5,623 | 1,000 | 14 | 89 |
| Lemhi | 80 | 7,590 | 0.92 | 75 | 9.3 | 103 | 12.8 | 738 | 13.5 | 2,720 | 2,650 | 70 | 6 | 9 |
| Lewis | 74 | 4,051 | 1.15 | 34 | 8.9 | 41 | 10.7 | 417 | 15.2 | 1,446 | 1,404 | 42 | 5 | 49 |
| Lincoln | 37 | 4,722 | 0.75 | 63 | 11.8 | 37 | 6.9 | 918 | 20.4 | 899 | 698 | 202 | 17 | 26 |
| Madison | 624 | 39,732 | 1.04 | 1,131 | 28.1 | 150 | 3.7 | 3,395 | 9.4 | 3,349 | 2,489 | 860 | 22 | 54 |
| Minidoka | 82 | 19,841 | 0.91 | 306 | 14.4 | 209 | 9.9 | 3,090 | 17.9 | 3,929 | 2,822 | 1,107 | 12 | 82 |
| Nez Perce | 1,062 | 43,105 | 1.15 | 444 | 10.9 | 489 | 12.0 | 3,851 | 12.2 | 10,412 | 7,777 | 2,636 | 8 | 83 |
| Oneida | 50 | 4,024 | 0.79 | 56 | 12.4 | 38 | 8.4 | 464 | 13.1 | 1,035 | 1,014 | 22 | 5 | 19 |
| Owyhee | 160 | 10,345 | 0.76 | 145 | 12.0 | 109 | 9.0 | 2,167 | 23.1 | 2,398 | 1,281 | 1,117 | 30 | 60 |
| Payette | 149 | 21,108 | 0.78 | 304 | 12.3 | 248 | 10.0 | 3,063 | 16.0 | 5,805 | 3,253 | 2,552 | 14 | 85 |
| Power | 47 | 8,083 | 1.13 | 110 | 14.4 | 67 | 8.8 | 1,164 | 18.4 | 1,420 | 1,055 | 365 | 10 | 29 |
| Shoshone | 160 | 12,415 | 0.96 | 141 | 10.9 | 189 | 14.6 | 1,374 | 14.2 | 3,899 | 3,523 | 376 | 14 | 106 |
| Teton | 7 | 9,604 | 0.69 | 128 | 10.2 | 67 | 5.4 | 1,663 | 15.8 | 1,572 | 1,503 | 68 | 15 | 33 |
| Twin Falls | 962 | 85,402 | 1.01 | 1,155 | 13.1 | 850 | 9.6 | 10,179 | 14.1 | 16,836 | 10,345 | 6,491 | 42 | 136 |
| Valley | 66 | 11,504 | 1.17 | 96 | 8.1 | 94 | 8.0 | 1,163 | 13.8 | 3,125 | 2,025 | 1,100 | 22 | 58 |
| Washington | 119 | 9,375 | 0.83 | 92 | 8.9 | 110 | 10.6 | 1,257 | 17.1 | 3,147 | 2,116 | 1,031 | 5 | 16 |
| **ILLINOIS** | 296,880 | 12,755,165 | 1.00 | 140,833 | 11.2 | 118,232 | 9.4 | 907,738 | 8.7 | 2,265,099 | 1,581,792 | 683,306 | X | X |
| Adams | 2,018 | 68,590 | 1.08 | 737 | 11.4 | 900 | 13.9 | 3,118 | 6.1 | 15,041 | 12,272 | 2,769 | NA | NA |
| Alexander | 132 | 5,355 | 0.56 | 43 | 7.8 | 84 | 15.3 | 336 | 7.7 | 1,620 | 1,393 | 228 | 6 | 53 |
| Bond | 1,602 | 15,014 | 0.79 | 147 | 9.0 | 177 | 10.9 | 835 | 7.1 | 3,614 | 2,750 | 865 | NA | NA |
| Boone | 310 | 46,191 | 0.71 | 501 | 9.5 | 414 | 7.8 | 3,785 | 8.5 | 9,935 | 6,479 | 3,456 | 27 | 146 |
| Brown | 1,985 | 8,537 | 1.80 | 64 | 9.8 | 61 | 9.3 | 158 | 4.3 | 1,080 | 837 | 244 | 1 | 10 |
| Bureau | 457 | 30,499 | 0.83 | 298 | 9.2 | 406 | 12.6 | 1,837 | 7.3 | 8,150 | 6,637 | 1,514 | 29 | 113 |
| Calhoun | 90 | 3,756 | 0.46 | 36 | 7.8 | 68 | 14.7 | 254 | 7.1 | 1,189 | 925 | 264 | NA | NA |
| Carroll | 223 | 13,485 | 0.85 | 129 | 9.1 | 201 | 14.1 | 778 | 7.4 | 4,232 | 3,006 | 1,226 | 9 | 21 |
| Cass | 184 | 11,640 | 0.85 | 159 | 13.3 | 145 | 12.2 | 997 | 10.2 | 2,521 | 1,923 | 598 | 2 | 30 |
| Champaign | 16,384 | 218,543 | 1.08 | 2,200 | 10.5 | 1,528 | 7.3 | 11,815 | 7.1 | 30,217 | 15,258 | 14,959 | 48 | 405 |
| Christian | 1,596 | 29,625 | 0.77 | 340 | 10.6 | 395 | 12.3 | 1,663 | 6.8 | 7,674 | 5,574 | 2,100 | NA | NA |
| Clark | 221 | 13,791 | 0.74 | 185 | 12.1 | 214 | 14.0 | 792 | 6.5 | 3,700 | 2,826 | 874 | NA | NA |
| Clay | 305 | 13,497 | 1.04 | 153 | 11.7 | 192 | 14.7 | 734 | 7.1 | 3,272 | 2,952 | 319 | 5 | 43 |
| Clinton | 2,071 | 31,378 | 0.66 | 388 | 10.4 | 422 | 11.3 | 1,737 | 6.0 | 7,575 | 5,908 | 1,667 | 9 | 73 |
| Coles | 3,296 | 52,560 | 1.05 | 461 | 9.1 | 518 | 10.3 | 2,904 | 7.4 | 9,851 | 7,506 | 2,345 | 16 | 98 |
| Cook | 90,933 | 5,366,343 | 1.07 | 60,224 | 11.8 | 46,387 | 9.1 | 466,676 | 10.9 | 830,121 | 560,562 | 269,558 | 231 | 1,035 |
| Crawford | 1,493 | 19,768 | 1.11 | 190 | 10.3 | 247 | 13.3 | 939 | 6.8 | 4,457 | 3,896 | 562 | 6 | 76 |
| Cumberland | 117 | 8,580 | 0.57 | 105 | 9.9 | 121 | 11.4 | 607 | 7.1 | 2,459 | 1,822 | 637 | 1 | 13 |
| De Witt | 249 | 15,031 | 0.89 | 169 | 11.0 | 228 | 14.8 | 806 | 6.5 | 3,596 | 2,567 | 1,029 | NA | NA |
| Douglas | 168 | 19,037 | 0.94 | 230 | 11.8 | 186 | 9.5 | 1,718 | 10.9 | 3,952 | 2,542 | 1,410 | 8 | 30 |
| DuPage | 12,825 | 1,036,635 | 1.22 | 9,987 | 10.9 | 7,223 | 7.9 | 49,519 | 6.4 | 159,928 | 116,854 | 43,074 | 81 | 434 |
| Edgar | 283 | 17,695 | 1.04 | 165 | 9.8 | 242 | 14.4 | 921 | 7.0 | 4,322 | 3,435 | 887 | 4 | 32 |
| Edwards | 52 | 6,415 | 0.99 | 92 | 14.5 | 70 | 11.0 | 306 | 6.2 | 1,499 | 1,336 | 164 | NA | NA |
| Effingham | 440 | 38,972 | 1.28 | 435 | 12.8 | 355 | 10.4 | 1,666 | 6.0 | 7,468 | 6,420 | 1,049 | 16 | 107 |
| Fayette | 1,651 | 19,489 | 0.76 | 220 | 10.3 | 223 | 10.5 | 1,463 | 9.3 | 4,513 | 3,777 | 736 | 14 | 50 |
| Ford | 411 | 12,343 | 0.85 | 141 | 10.9 | 180 | 13.9 | 737 | 7.1 | 3,089 | 2,264 | 824 | 11 | 52 |
| Franklin | 517 | 33,978 | 0.68 | 455 | 12.0 | 557 | 14.6 | 2,400 | 8.0 | 9,668 | 7,408 | 2,260 | 14 | 217 |
| Fulton | 2,685 | 30,511 | 0.69 | 297 | 8.8 | 479 | 14.2 | 1,946 | 7.9 | 8,231 | 6,040 | 2,190 | 29 | 155 |
| Gallatin | 25 | 4,392 | 0.68 | 33 | 6.9 | 78 | 16.3 | 256 | 7.1 | 1,410 | 1,193 | 217 | 5 | 38 |
| Greene | 284 | 10,472 | 0.56 | 133 | 10.5 | 177 | 13.9 | 711 | 7.0 | 2,982 | 2,423 | 558 | 9 | 20 |
| Grundy | 254 | 45,797 | 0.80 | 560 | 11.0 | 451 | 8.8 | 2,385 | 5.5 | 8,971 | 7,667 | 1,304 | 20 | 79 |
| Hamilton | 117 | 7,293 | 0.74 | 81 | 10.0 | 117 | 14.5 | 419 | 6.7 | 1,926 | 1,653 | 273 | 7 | 42 |
| Hancock | 223 | 14,665 | 0.59 | 179 | 10.3 | 218 | 12.5 | 921 | 6.9 | 4,706 | 3,956 | 750 | NA | NA |

1. Per 1,000 estimated resident population.

# Table B. States and Counties — Crime, Education, Money Income, and Poverty

| STATE County | County law enforcement employment, 2019 Officers | Civilians | Enrollment[1] Total | Percent private | Attainment[2] (percent) High school graduate or less | Bachelor's degree or more | Local government expenditures,[3] 2017-2018 Total current spending (mil dol) | Current spending per student (dollars) | Per capita income[4] | Median income (dollars) | Households Percent with income of less than $50,000 | with income of $200,000 or more | Median household income (dollars) | Percent below poverty level All persons | Children under 18 years | Children 5 to 17 years in families |
|---|---|---|---|---|---|---|---|---|---|---|---|---|---|---|---|---|
| | 46 | 47 | 48 | 49 | 50 | 51 | 52 | 53 | 54 | 55 | 56 | 57 | 58 | 59 | 60 | 61 |
| **IDAHO—Cont'd** | | | | | | | | | | | | | | | | |
| Caribou | 9 | 17 | 1,949 | 7.1 | 41.1 | 19.0 | 13 | 8,689 | 26,221 | 59,053 | 40.4 | 2.6 | 61,518 | 10.1 | 13.1 | 11.7 |
| Cassia | 33 | 45 | 6,862 | 6.3 | 46.2 | 18.6 | 39 | 7,111 | 21,687 | 52,935 | 45.5 | 2.8 | 53,284 | 12.0 | 13.9 | 12.9 |
| Clark | 3 | 4 | 230 | 1.3 | 68.2 | 12.2 | 2 | 16,526 | 18,860 | 46,154 | 57.2 | 0.0 | 48,726 | 14.5 | 17.9 | 16.2 |
| Clearwater | 15 | 17 | 1,323 | 10.8 | 45.0 | 17.9 | 14 | 12,614 | 23,412 | 43,915 | 54.7 | 1.8 | 49,816 | 13.5 | 23.1 | 22.7 |
| Custer | 9 | 8 | 576 | 2.8 | 39.2 | 24.4 | 6 | 10,000 | 23,805 | 40,875 | 62.8 | 0.2 | 49,055 | 13.0 | 17.7 | 16.9 |
| Elmore | 22 | 46 | 6,227 | 14.5 | 36.8 | 18.7 | 34 | 7,240 | 23,028 | 46,855 | 54.2 | 0.6 | 47,796 | 14.1 | 16.9 | 16.7 |
| Franklin | 10 | 5 | 3,965 | 8.5 | 47.5 | 17.7 | 21 | 6,638 | 24,185 | 57,215 | 41.0 | 4.2 | 55,421 | 8.2 | 10.0 | 8.9 |
| Fremont | 21 | 17 | 3,330 | 16.6 | 43.6 | 21.7 | 17 | 8,013 | 23,577 | 58,065 | 43.2 | 1.7 | 56,162 | 12.7 | 17.4 | 16.5 |
| Gem | 13 | 21 | 3,776 | 14.8 | 49.1 | 17.5 | 21 | 8,213 | 22,661 | 45,492 | 53.9 | 1.1 | 53,733 | 12.7 | 18.0 | 15.7 |
| Gooding | 16 | 18 | 3,665 | 6.1 | 56.0 | 17.2 | 25 | 7,496 | 23,607 | 47,204 | 51.5 | 1.9 | 50,205 | 15.1 | 19.6 | 18.3 |
| Idaho | 18 | 20 | 3,120 | 14.0 | 45.6 | 19.8 | 20 | 10,855 | 22,672 | 41,516 | 58.3 | 1.1 | 48,866 | 13.8 | 18.8 | 16.5 |
| Jefferson | 22 | 38 | 9,086 | 18.0 | 35.3 | 24.7 | 45 | 6,272 | 24,677 | 63,048 | 33.5 | 3.9 | 63,078 | 9.2 | 9.9 | 8.2 |
| Jerome | 22 | 33 | 6,271 | 11.1 | 53.4 | 16.1 | 33 | 6,897 | 21,461 | 52,921 | 46.2 | 1.4 | 56,320 | 14.2 | 19.5 | 15.8 |
| Kootenai | 97 | 200 | 35,556 | 17.0 | 34.1 | 25.2 | 171 | 7,575 | 29,948 | 57,242 | 42.8 | 3.7 | 62,764 | 10.0 | 13.1 | 11.4 |
| Latah | 29 | 24 | 15,566 | 9.6 | 23.9 | 45.4 | 50 | 10,568 | 27,011 | 49,158 | 50.6 | 4.6 | 54,999 | 11.7 | 9.8 | 9.0 |
| Lemhi | 7 | 16 | 1,359 | 9.8 | 38.2 | 22.9 | 9 | 9,320 | 25,075 | 39,324 | 58.3 | 2.1 | 46,062 | 13.5 | 21.3 | 19.8 |
| Lewis | 8 | 8 | 792 | 15.0 | 41.2 | 17.9 | 9 | 12,223 | 24,094 | 41,198 | 59.1 | 1.6 | 45,342 | 13.2 | 20.5 | 18.2 |
| Lincoln | 9 | 3 | 1,367 | 5.3 | 56.7 | 10.3 | 9 | 9,668 | 21,656 | 50,053 | 49.9 | 2.8 | 50,608 | 11.2 | 14.5 | 12.7 |
| Madison | 23 | 39 | 19,272 | 47.9 | 19.3 | 36.4 | 46 | 6,617 | 16,585 | 39,160 | 59.5 | 1.6 | 46,177 | 27.4 | 17.7 | 15.6 |
| Minidoka | 21 | 11 | 5,207 | 8.7 | 52.9 | 14.1 | 32 | 7,525 | 24,262 | 53,370 | 47.4 | 2.4 | 56,173 | 11.3 | 14.4 | 13.1 |
| Nez Perce | 22 | 55 | 8,649 | 12.1 | 40.1 | 22.9 | 54 | 10,156 | 29,544 | 58,107 | 43.6 | 3.3 | 67,645 | 11.1 | 13.6 | 11.9 |
| Oneida | 9 | 5 | 1,182 | 4.0 | 44.7 | 17.1 | 9 | 5,832 | 22,120 | 53,841 | 47.7 | 0.3 | 55,833 | 11.3 | 14.0 | 12.4 |
| Owyhee | 12 | 16 | 2,579 | 11.8 | 57.2 | 12.0 | 19 | 8,281 | 22,806 | 43,798 | 55.1 | 2.9 | 48,395 | 14.9 | 19.7 | 18.7 |
| Payette | 18 | 31 | 5,423 | 13.8 | 49.1 | 15.7 | 31 | 7,068 | 25,768 | 50,579 | 49.5 | 3.8 | 53,643 | 11.0 | 15.2 | 14.1 |
| Power | 10 | 16 | 2,142 | 7.3 | 53.4 | 15.1 | 16 | 9,806 | 23,343 | 48,823 | 52.1 | 1.9 | 50,981 | 13.1 | 17.0 | 16.2 |
| Shoshone | 15 | 22 | 2,048 | 7.2 | 50.8 | 11.5 | 19 | 11,390 | 23,987 | 39,386 | 59.1 | 1.0 | 39,239 | 19.1 | 26.3 | 26.2 |
| Teton | 11 | 9 | 2,763 | 22.7 | 22.8 | 41.0 | 16 | 8,627 | 31,191 | 74,216 | 32.5 | 3.5 | 73,637 | 7.7 | 10.8 | 9.7 |
| Twin Falls | 45 | 73 | 22,375 | 14.2 | 40.1 | 20.6 | 120 | 7,405 | 25,352 | 52,919 | 47.2 | 2.8 | 56,141 | 11.0 | 13.3 | 11.6 |
| Valley | 17 | 24 | 1,617 | 11.9 | 32.1 | 30.4 | 16 | 11,351 | 33,709 | 64,475 | 39.0 | 7.5 | 64,041 | 8.8 | 12.0 | 11.5 |
| Washington | 14 | 21 | 2,126 | 12.3 | 48.8 | 16.3 | 15 | 8,401 | 22,650 | 40,992 | 61.0 | 1.7 | 45,813 | 13.4 | 19.8 | 17.4 |
| **ILLINOIS** | X | X | 3,228,360 | 18.6 | 36.7 | 34.7 | 31,466 | 15,781 | 36,038 | 65,886 | 38.7 | 8.3 | 69,212 | 11.4 | 15.6 | 14.6 |
| Adams | NA | NA | 14,764 | 21.5 | 42.9 | 25.0 | 111 | 12,078 | 30,343 | 52,993 | 46.5 | 3.7 | 55,942 | 12.3 | 18.3 | 16.1 |
| Alexander | 7 | 6 | 1,273 | 2.5 | 54.2 | 12.9 | 14 | 16,880 | 19,519 | 36,806 | 64.6 | 0.4 | 37,408 | 24.0 | 38.9 | 37.3 |
| Bond | 12 | 11 | 4,116 | 20.7 | 43.9 | 22.9 | 27 | 11,740 | 28,119 | 57,289 | 42.6 | 3.3 | 53,447 | 13.0 | 16.9 | 15.6 |
| Boone | NA | NA | 14,186 | 16.3 | 48.6 | 23.5 | 139 | 14,445 | 31,943 | 69,272 | 35.7 | 7.3 | 72,426 | 8.9 | 12.2 | 10.6 |
| Brown | 7 | 1 | 1,318 | 17.1 | 54.5 | 14.1 | 10 | 13,010 | 23,070 | 61,655 | 39.3 | 2.2 | 60,254 | 14.8 | 13.4 | 13.1 |
| Bureau | NA | NA | 7,125 | 13.8 | 45.8 | 20.6 | 74 | 14,807 | 30,255 | 57,436 | 44.4 | 3.9 | 57,935 | 11.2 | 14.0 | 13.5 |
| Calhoun | NA | NA | 1,125 | 22.8 | 51.9 | 14.2 | 8 | 14,043 | 28,822 | 63,009 | 40.3 | 3.1 | 63,373 | 9.7 | 12.7 | 11.3 |
| Carroll | NA | NA | 2,606 | 9.2 | 47.0 | 19.5 | 24 | 10,797 | 29,166 | 52,410 | 48.4 | 2.0 | 56,468 | 11.1 | 17.0 | 16.1 |
| Cass | 7 | 1 | 2,823 | 11.3 | 59.0 | 15.1 | 26 | 11,271 | 26,992 | 52,373 | 47.2 | 2.4 | 51,011 | 13.0 | 15.7 | 14.8 |
| Champaign | 66 | 88 | 79,180 | 8.6 | 27.0 | 45.0 | 378 | 15,027 | 30,578 | 52,797 | 47.7 | 5.6 | 53,485 | 19.9 | 18.8 | 18.8 |
| Christian | NA | NA | 6,998 | 14.4 | 52.1 | 17.1 | 80 | 14,506 | 27,361 | 52,834 | 47.0 | 2.2 | 51,573 | 12.3 | 17.2 | 14.8 |
| Clark | NA | NA | 3,365 | 7.7 | 47.1 | 19.9 | 27 | 10,278 | 30,796 | 56,531 | 43.7 | 2.9 | 58,323 | 10.8 | 15.5 | 14.1 |
| Clay | 12 | 6 | 2,578 | 3.6 | 50.6 | 15.0 | 22 | 9,282 | 26,815 | 48,500 | 52.9 | 2.3 | 49,377 | 13.8 | 18.4 | 17.9 |
| Clinton | NA | NA | 8,080 | 15.8 | 41.8 | 22.3 | 58 | 11,132 | 31,783 | 66,639 | 36.7 | 3.8 | 67,201 | 7.5 | 8.8 | 8.2 |
| Coles | 21 | 8 | 15,284 | 6.0 | 39.1 | 26.0 | 97 | 15,324 | 28,084 | 46,202 | 53.3 | 2.7 | 50,467 | 16.8 | 18.4 | 17.5 |
| Cook | 1,849 | 3,569 | 1,285,634 | 23.3 | 36.0 | 38.8 | 12,609 | 16,846 | 37,552 | 64,660 | 39.9 | 9.4 | 69,375 | 13.0 | 18.2 | 17.7 |
| Crawford | 17 | 7 | 3,941 | 19.4 | 43.6 | 17.1 | 34 | 11,885 | 26,546 | 49,779 | 50.2 | 3.1 | 50,352 | 12.6 | 15.8 | 14.8 |
| Cumberland | 5 | 10 | 2,313 | 6.9 | 49.5 | 15.4 | 18 | 10,821 | 27,812 | 56,206 | 44.5 | 3.1 | 53,991 | 10.1 | 14.4 | 12.9 |
| DeKalb | NA | NA | 35,172 | 9.5 | 32.6 | 32.3 | 254 | 15,129 | 29,430 | 63,317 | 40.3 | 4.4 | 64,676 | 13.0 | 11.9 | 10.7 |
| De Witt | NA | NA | 3,422 | 12.0 | 46.2 | 21.3 | 35 | 13,398 | 30,729 | 55,587 | 45.2 | 4.0 | 59,521 | 11.0 | 16.6 | 14.9 |
| Douglas | 12 | 13 | 4,162 | 15.4 | 50.2 | 18.9 | 45 | 12,752 | 28,139 | 56,714 | 43.7 | 2.8 | 64,283 | 8.0 | 11.4 | 10.6 |
| DuPage | 391 | 95 | 242,749 | 22.5 | 25.2 | 49.4 | 2,848 | 18,500 | 46,272 | 92,809 | 25.9 | 14.8 | 96,354 | 6.0 | 7.4 | 6.6 |
| Edgar | 6 | 0 | 3,291 | 7.6 | 47.2 | 17.2 | 33 | 11,145 | 27,797 | 53,647 | 47.0 | 1.8 | 51,333 | 13.2 | 20.6 | 19.4 |
| Edwards | NA | NA | 1,362 | 6.2 | 42.6 | 13.9 | 10 | 11,258 | 28,255 | 51,080 | 49.0 | 1.9 | 54,930 | 11.8 | 15.2 | 14.1 |
| Effingham | 20 | 26 | 7,583 | 18.6 | 42.7 | 22.7 | 54 | 10,822 | 31,665 | 56,685 | 43.6 | 4.1 | 58,943 | 9.2 | 11.6 | 10.0 |
| Fayette | 11 | 6 | 4,550 | 16.2 | 56.4 | 11.3 | 35 | 12,917 | 23,194 | 46,650 | 53.3 | 1.9 | 48,913 | 15.1 | 21.7 | 20.6 |
| Ford | 20 | 6 | 2,957 | 9.5 | 45.1 | 21.4 | 40 | 14,493 | 28,957 | 52,092 | 47.8 | 3.0 | 55,369 | 10.0 | 14.3 | 12.6 |
| Franklin | 16 | 24 | 8,020 | 7.3 | 44.1 | 16.7 | 68 | 11,098 | 24,098 | 42,769 | 56.8 | 1.5 | 43,695 | 16.4 | 23.4 | 21.8 |
| Fulton | 20 | 22 | 7,567 | 5.7 | 45.9 | 17.6 | 54 | 11,744 | 27,741 | 51,643 | 48.3 | 2.7 | 50,461 | 13.3 | 18.1 | 17.3 |
| Gallatin | 4 | 0 | 1,035 | 4.3 | 52.9 | 10.9 | 9 | 12,394 | 33,377 | 44,076 | 54.4 | 4.1 | 46,515 | 19.0 | 26.2 | 24.5 |
| Greene | 6 | 0 | 2,643 | 12.8 | 57.0 | 12.8 | 23 | 12,437 | 25,414 | 49,885 | 50.1 | 1.8 | 52,116 | 13.7 | 19.1 | 18.4 |
| Grundy | 30 | 15 | 13,087 | 8.9 | 40.6 | 22.9 | 168 | 12,447 | 35,208 | 77,350 | 29.7 | 5.9 | 83,135 | 5.6 | 7.0 | 6.4 |
| Hamilton | NA | NA | 1,744 | 15.0 | 47.8 | 17.0 | 14 | 11,359 | 27,702 | 54,046 | 45.6 | 3.0 | 51,031 | 12.7 | 17.4 | 17.2 |
| Hancock | 11 | 12 | 3,684 | 9.0 | 43.1 | 21.0 | 37 | 12,686 | 28,576 | 52,561 | 46.2 | 2.9 | 53,724 | 12.3 | 18.1 | 17.2 |

1. All persons 3 years old and over enrolled in nursery school through college.   2. Persons 25 years old and over.   3. Elementary and secondary education expenditures.   4. Based on population estimated by the American Community Survey, 2014–2018.

# Table B. States and Counties — Personal Income

| STATE County | Personal income, 2019 | | | | | | | | | | Earnings, 2019 | | |
|---|---|---|---|---|---|---|---|---|---|---|---|---|---|
| | Total (mil dol) | Percent change 2018-2019 | Per capita¹ | | Wages and salaries (mil dol) | Supplements to wages and salaries, employer contributions (mil dol) | | Proprietors' income (mil dol) | Dividends, interest, and rent (mil dol) | Personal transfer receipts (mil dol) | Total (mil dol) | Contributions for government social insurance (mil dol) | |
| | | | Dollars | Rank | | Pension and insurance | Government social insurance | | | | | From employee and self-employed | From employer |
| | 62 | 63 | 64 | 65 | 66 | 67 | 68 | 69 | 70 | 71 | 72 | 73 | 74 |
| IDAHO—Cont'd | | | | | | | | | | | | | |
| Caribou | 304 | 7.5 | 42,527 | 1,684 | 213 | 43 | 18 | 25 | 57 | 63 | 299 | 18 | 18 |
| Cassia | 1,195 | 8.5 | 49,712 | 856 | 464 | 80 | 41 | 366 | 193 | 182 | 951 | 42 | 41 |
| Clark | 35 | -9.7 | 41,622 | 1,829 | 15 | 3 | 1 | 7 | 7 | 5 | 25 | 1 | 1 |
| Clearwater | 322 | 2.3 | 36,805 | 2,471 | 110 | 25 | 9 | 27 | 75 | 109 | 172 | 14 | 9 |
| Custer | 190 | 2.9 | 43,920 | 1,498 | 56 | 13 | 5 | 18 | 57 | 49 | 91 | 7 | 5 |
| Elmore | 1,117 | 7.1 | 40,585 | 1,975 | 483 | 132 | 46 | 145 | 255 | 239 | 806 | 38 | 46 |
| Franklin | 517 | 6.1 | 37,293 | 2,415 | 126 | 29 | 11 | 60 | 85 | 99 | 226 | 14 | 11 |
| Fremont | 497 | 6.4 | 37,953 | 2,330 | 133 | 27 | 12 | 70 | 113 | 107 | 242 | 15 | 12 |
| Gem | 713 | 6.1 | 39,382 | 2,129 | 152 | 30 | 14 | 37 | 159 | 195 | 233 | 21 | 14 |
| Gooding | 937 | 13.0 | 61,715 | 260 | 243 | 42 | 22 | 357 | 122 | 140 | 663 | 21 | 22 |
| Idaho | 599 | 3.5 | 35,946 | 2,576 | 182 | 40 | 16 | 48 | 162 | 170 | 286 | 22 | 16 |
| Jefferson | 1,106 | 7.3 | 37,025 | 2,446 | 260 | 50 | 22 | 154 | 180 | 191 | 486 | 22 | 16 |
| Jerome | 981 | 10.2 | 40,190 | 2,029 | 380 | 64 | 33 | 267 | 144 | 165 | 745 | 34 | 22 |
| Kootenai | 7,729 | 5.6 | 46,645 | 1,167 | 2,920 | 485 | 250 | 578 | 1,729 | 1,588 | 4,233 | 299 | 33 |
| Latah | 1,678 | 3.4 | 41,833 | 1,792 | 602 | 148 | 50 | 127 | 422 | 287 | 927 | 58 | 250 |
| Lemhi | 355 | 5.1 | 44,199 | 1,456 | 104 | 23 | 9 | 28 | 107 | 101 | 165 | 13 | 50 |
| Lewis | 191 | 3.1 | 49,643 | 864 | 63 | 14 | 5 | 18 | 35 | 79 | 100 | 8 | 9 |
| Lincoln | 223 | 12.0 | 41,610 | 1,832 | 65 | 14 | 6 | 79 | 29 | 41 | 164 | 6 | 5 |
| Madison | 1,149 | 5.9 | 28,780 | 3,070 | 589 | 113 | 51 | 165 | 179 | 263 | 918 | 53 | 6 |
| Minidoka | 871 | 6.6 | 41,421 | 1,859 | 335 | 60 | 29 | 121 | 170 | 162 | 545 | 31 | 51 |
| Nez Perce | 1,907 | 3.3 | 47,200 | 1,109 | 980 | 174 | 85 | 160 | 414 | 433 | 1,400 | 94 | 29 |
| Oneida | 179 | 8.6 | 39,523 | 2,112 | 41 | 10 | 3 | 20 | 31 | 43 | 74 | 5 | 85 |
| Owyhee | 466 | 8.7 | 39,409 | 2,125 | 104 | 20 | 9 | 105 | 79 | 98 | 239 | 11 | 3 |
| Payette | 1,003 | 6.8 | 41,890 | 1,786 | 282 | 51 | 25 | 143 | 179 | 232 | 500 | 33 | 9 |
| Power | 333 | 10.3 | 43,313 | 1,569 | 162 | 31 | 14 | 65 | 58 | 63 | 272 | 14 | 25 |
| Shoshone | 474 | 2.2 | 36,793 | 2,473 | 184 | 34 | 16 | 15 | 103 | 157 | 248 | 20 | 14 |
| Teton | 497 | 7.0 | 40,911 | 1,930 | 148 | 23 | 13 | 53 | 140 | 63 | 237 | 15 | 16 |
| Twin Falls | 3,682 | 5.0 | 42,379 | 1,707 | 1,576 | 269 | 136 | 598 | 689 | 745 | 2,579 | 155 | 13 |
| Valley | 582 | 4.6 | 51,088 | 751 | 200 | 35 | 17 | 56 | 218 | 112 | 308 | 22 | 136 |
| Washington | 370 | 4.4 | 36,292 | 2,540 | 103 | 22 | 9 | 23 | 82 | 119 | 156 | 13 | 17 |
| ILLINOIS | 744,641 | 2.4 | 58,786 | X | 394,896 | 64,441 | 26,093 | 56,788 | 150,488 | 112,157 | 542,219 | 30,610 | 9 |
| Adams | 3,134 | 1.9 | 47,898 | 1,030 | 1,585 | 305 | 110 | 248 | 618 | 674 | 2,246 | 133 | 26,093 |
| Alexander | 217 | 1.5 | 37,723 | 2,355 | 49 | 14 | 3 | 16 | 30 | 97 | 83 | 6 | 110 |
| Bond | 622 | 1.3 | 37,871 | 2,336 | 220 | 55 | 15 | 31 | 108 | 163 | 321 | 21 | 3 |
| Boone | 2,553 | 1.6 | 47,682 | 1,058 | 973 | 186 | 73 | 95 | 355 | 422 | 1,327 | 81 | 15 |
| Brown | 247 | 6.1 | 37,500 | 2,388 | 232 | 43 | 15 | 14 | 43 | 47 | 305 | 17 | 73 |
| Bureau | 1,404 | 1.7 | 43,040 | 1,612 | 544 | 111 | 40 | 24 | 270 | 322 | 718 | 49 | 15 |
| Calhoun | 206 | 1.2 | 43,368 | 1,560 | 27 | 8 | 2 | 14 | 34 | 52 | 50 | 4 | 40 |
| Carroll | 649 | 1.8 | 45,342 | 1,318 | 200 | 50 | 15 | 55 | 134 | 166 | 320 | 22 | 2 |
| Cass | 510 | 1.2 | 41,954 | 1,773 | 249 | 53 | 18 | 30 | 75 | 122 | 350 | 21 | 15 |
| Champaign | 9,639 | 3.0 | 45,967 | 1,253 | 5,361 | 1,330 | 322 | 628 | 2,022 | 1,351 | 7,641 | 365 | 18 |
| Christian | 1,375 | 0.0 | 42,576 | 1,676 | 439 | 97 | 31 | 109 | 224 | 361 | 675 | 43 | 322 |
| Clark | 658 | 1.1 | 42,593 | 1,673 | 207 | 51 | 14 | 35 | 115 | 164 | 307 | 20 | 31 |
| Clay | 531 | -1.8 | 40,308 | 2,013 | 219 | 55 | 15 | 19 | 88 | 162 | 308 | 20 | 14 |
| Clinton | 1,790 | 2.8 | 47,653 | 1,063 | 475 | 107 | 33 | 117 | 294 | 342 | 732 | 45 | 15 |
| Coles | 2,113 | 2.2 | 41,751 | 1,811 | 1,114 | 257 | 73 | 140 | 398 | 475 | 1,584 | 87 | 33 |
| Cook | 336,342 | 2.5 | 65,306 | 180 | 200,483 | 29,404 | 13,033 | 30,003 | 76,615 | 47,683 | 272,922 | 15,225 | 73 |
| Crawford | 895 | 3.4 | 47,935 | 1,021 | 397 | 121 | 27 | 86 | 149 | 188 | 631 | 35 | 13,033 |
| Cumberland | 904 | 6.6 | 83,954 | 50 | 115 | 26 | 8 | 460 | 66 | 100 | 610 | 12 | 27 |
| DeKalb | 4,418 | 3.7 | 42,120 | 1,748 | 1,915 | 452 | 125 | 241 | 771 | 755 | 2,733 | 144 | 8 |
| De Witt | 735 | -1.7 | 47,023 | 1,132 | 296 | 69 | 19 | 56 | 124 | 160 | 440 | 26 | 125 |
| Douglas | 1,009 | 1.3 | 51,859 | 692 | 399 | 84 | 29 | 167 | 153 | 171 | 679 | 38 | 19 |
| DuPage | 69,346 | 2.7 | 75,137 | 89 | 45,083 | 6,301 | 3,038 | 6,010 | 13,637 | 6,767 | 60,431 | 3,351 | 29 |
| Edgar | 732 | 0.9 | 42,665 | 1,662 | 321 | 69 | 22 | 36 | 120 | 199 | 449 | 28 | 3,038 |
| Edwards | 262 | 0.8 | 40,991 | 1,920 | 99 | 23 | 7 | 26 | 45 | 61 | 155 | 10 | 22 |
| Effingham | 1,744 | 2.6 | 51,296 | 736 | 990 | 176 | 71 | 178 | 350 | 314 | 1,416 | 82 | 7 |
| Fayette | 775 | 1.6 | 36,310 | 2,535 | 214 | 52 | 14 | 52 | 133 | 212 | 332 | 22 | 71 |
| Ford | 656 | 0.4 | 50,606 | 790 | 230 | 51 | 15 | 77 | 101 | 138 | 373 | 22 | 14 |
| Franklin | 1,473 | 1.6 | 38,289 | 2,295 | 411 | 93 | 29 | 73 | 220 | 468 | 607 | 44 | 15 |
| Fulton | 1,321 | -0.3 | 38,465 | 2,268 | 346 | 92 | 22 | 63 | 222 | 371 | 523 | 36 | 29 |
| Gallatin | 205 | -0.2 | 42,451 | 1,697 | 62 | 13 | 5 | 15 | 36 | 67 | 95 | 6 | 22 |
| Greene | 475 | 0.0 | 36,652 | 2,490 | 97 | 25 | 6 | 39 | 72 | 134 | 168 | 11 | 5 |
| Grundy | 2,846 | 3.5 | 55,741 | 460 | 1,325 | 265 | 91 | 406 | 364 | 386 | 2,087 | 115 | 6 |
| Hamilton | 354 | 1.1 | 43,609 | 1,534 | 98 | 23 | 7 | 34 | 66 | 93 | 161 | 10 | 91 |
| Hancock | 798 | 0.1 | 45,052 | 1,355 | 159 | 41 | 11 | 87 | 136 | 195 | 298 | 20 | 7 |
| | | | | | | | | | | | | | 11 |

1. Based on the resident population estimated as of July 1 of the year shown.

Items 62—74

## Table B. States and Counties — Earnings, Social Security, and Housing

| STATE County | Earnings, 2019 (cont.) — Percent by selected industries | | | | | | | | | Social Security beneficiaries, December 2019 | | Supplemental Security Income recipients, 2019 | Housing units, 2020 | |
|---|---|---|---|---|---|---|---|---|---|---|---|---|---|---|
| | Farm | Mining, quarrying, and extractions | Construction | Manufacturing | Information; professional, scientific, technical services | Retail trade | Finance, insurance, real estate, and leasing | Health care and social assistance | Government | Number | Rate[1] | | Total | Percent change 2010-2020 |
| | 75 | 76 | 77 | 78 | 79 | 80 | 81 | 82 | 83 | 84 | 85 | 86 | 87 | 88 |
| **IDAHO—Cont'd** | | | | | | | | | | | | | | |
| Caribou | 8.1 | 20.9 | 8.3 | 28.9 | 3.4 | 3.0 | 1.1 | D | 14.5 | 1,485 | 208 | 78 | 3,346 | 3.8 |
| Cassia | 32.0 | 0.8 | 5.6 | 9.1 | 2.3 | 6.1 | 4.7 | 7.9 | 8.7 | 4,405 | 183 | 414 | 9,006 | 7.5 |
| Clark | 19.9 | 0.8 | D | D | D | D | D | D | 27.1 | 140 | 167 | 0 | 566 | 6.6 |
| Clearwater | 1.5 | D | 6.3 | 4.7 | 2.3 | 5.5 | 2.7 | D | 33.4 | 2,880 | 329 | 262 | 4,698 | 5.5 |
| Custer | 11.2 | 7.7 | 5.7 | D | 6.6 | 5.0 | 2.6 | D | 28.2 | 1,430 | 335 | 66 | 3,189 | 2.8 |
| Elmore | 15.3 | 0.1 | 2.5 | 4.6 | 1.8 | 4.4 | 2.3 | 5.0 | 52.8 | 5,110 | 187 | 614 | 12,653 | 4.0 |
| Franklin | 12.7 | D | 8.5 | 7.5 | D | 8.6 | 5.0 | D | 23.3 | 2,550 | 183 | 168 | 5,033 | 11.2 |
| Fremont | 15.4 | D | 14.3 | 2.6 | D | 5.4 | D | 12.5 | 24.7 | 2,715 | 207 | 173 | 9,317 | 9.3 |
| Gem | -0.5 | D | 12.5 | 6.1 | 4.9 | 7.6 | 5.2 | 12.5 | 25.1 | 5,200 | 286 | 422 | 7,649 | 7.7 |
| Gooding | 56.3 | D | 2.4 | 8.1 | D | 2.1 | D | D | 7.3 | 3,225 | 211 | 298 | 6,334 | 3.9 |
| Idaho | 1.3 | 1.9 | 9.6 | 11.1 | 3.5 | 5.6 | 4.4 | 9.4 | 29.2 | 4,585 | 276 | 332 | 8,790 | 0.5 |
| Jefferson | 9.9 | 0.1 | 12.1 | 14.0 | 3.8 | 8.7 | 5.0 | 4.7 | 13.0 | 4,570 | 153 | 335 | 10,031 | 15.0 |
| Jerome | 30.6 | 0.0 | 4.3 | 13.9 | D | 6.0 | D | 3.8 | 7.9 | 3,910 | 160 | 390 | 8,751 | 8.1 |
| Kootenai | 0.0 | 0.7 | 10.0 | 8.0 | 7.5 | 9.9 | 7.1 | 13.3 | 19.2 | 40,935 | 247 | 2,602 | 77,143 | 22.1 |
| Latah | 0.3 | D | 5.2 | 2.5 | 7.5 | 6.8 | 4.3 | 11.0 | 43.6 | 6,805 | 169 | 539 | 17,482 | 9.3 |
| Lemhi | 6.9 | 2.0 | 8.5 | 3.0 | 7.8 | 6.0 | 3.3 | D | 35.9 | 2,785 | 345 | 182 | 4,999 | 5.7 |
| Lewis | 6.4 | D | 6.5 | 12.1 | 4.5 | 6.6 | 6.9 | 6.4 | 24.7 | 2,255 | 586 | 252 | 1,996 | 6.1 |
| Lincoln | 48.3 | 0.0 | 4.8 | 7.6 | D | D | D | 4.0 | 16.9 | 970 | 182 | 72 | 2,034 | 2.9 |
| Madison | 3.8 | D | 5.3 | 6.2 | 6.3 | 7.9 | 4.0 | D | 14.0 | 3,380 | 84 | 337 | 15,724 | 39.3 |
| Minidoka | 16.5 | D | 5.4 | 16.1 | 3.8 | 6.2 | 4.0 | D | 15.6 | 3,975 | 190 | 375 | 8,560 | 11.7 |
| Nez Perce | 0.2 | D | 5.8 | 19.7 | 4.4 | 7.4 | 8.7 | 16.8 | 18.3 | 10,820 | 266 | 983 | 18,095 | 3.8 |
| Oneida | 19.2 | D | D | 6.4 | D | 6.1 | D | 3.9 | 29.1 | 1,080 | 240 | 61 | 2,036 | 6.8 |
| Owyhee | 42.5 | 0.7 | 5.1 | 4.3 | 4.1 | 4.0 | 2.3 | D | 14.2 | 2,585 | 219 | 249 | 5,047 | 5.6 |
| Payette | 12.1 | 0.3 | 5.7 | 13.5 | D | 4.4 | D | 13.8 | 11.0 | 6,235 | 259 | 570 | 9,692 | 8.4 |
| Power | 20.5 | 0.0 | 1.6 | 30.0 | 2.6 | D | 5.1 | D | 12.3 | 1,510 | 198 | 134 | 3,065 | 4.1 |
| Shoshone | -0.2 | 16.9 | 6.8 | 3.3 | 5.4 | 19.2 | 3.0 | 7.8 | 20.6 | 4,165 | 324 | 442 | 7,196 | 1.7 |
| Teton | 3.3 | D | 20.4 | 3.0 | 10.7 | 7.0 | 3.2 | 8.3 | 13.4 | 1,490 | 123 | 49 | 6,322 | 15.4 |
| Twin Falls | 9.3 | D | 4.9 | 15.0 | 4.8 | 8.0 | 5.1 | 16.6 | 11.5 | 17,675 | 203 | 1,842 | 34,941 | 12.4 |
| Valley | 1.5 | D | 13.3 | 1.8 | D | 9.3 | 6.5 | 11.2 | 22.8 | 3,075 | 269 | 122 | 12,911 | 9.6 |
| Washington | 6.6 | 0.1 | 5.3 | 19.1 | 7.1 | 6.5 | 3.3 | D | 26.5 | 3,260 | 322 | 284 | 4,733 | 4.5 |
| **ILLINOIS** | 0.4 | 0.2 | 4.7 | 10.6 | 14.2 | 4.9 | 11.4 | 10.3 | 13.5 | 2,267,082 | 179 | 264,282 | 5,400,159 | 1.9 |
| Adams | 1.8 | 0.9 | 5.8 | 16.5 | 4.4 | 7.1 | 7.3 | 19.8 | 11.7 | 15,585 | 238 | 1,348 | 30,261 | 1.4 |
| Alexander | 12.2 | D | 1.1 | 11.7 | D | 2.7 | D | D | 30.9 | 1,645 | 284 | 341 | 3,957 | -1.2 |
| Bond | 2.6 | 0.0 | 5.9 | 23.8 | 2.9 | 3.2 | 2.6 | 4.4 | 12.7 | 3,800 | 231 | 297 | 7,288 | 2.8 |
| Boone | 0.0 | D | 9.1 | 49.1 | 2.3 | 3.5 | 2.4 | D | 11.9 | 10,460 | 196 | 580 | 20,123 | 0.8 |
| Brown | 0.8 | D | 5.9 | 0.8 | D | D | D | 3.9 | 20.7 | 1,095 | 165 | 68 | 2,453 | -0.4 |
| Bureau | -4.7 | D | 6.3 | 16.8 | 3.1 | 9.3 | 3.7 | D | 31.3 | 8,245 | 253 | 432 | 15,655 | -0.4 |
| Calhoun | 11.9 | D | 10.2 | D | D | 6.5 | 6.5 | D | 26.8 | 1,290 | 272 | 77 | 2,906 | 2.5 |
| Carroll | 6.2 | D | 7.1 | 16.7 | D | 4.0 | 6.1 | D | 26.8 | 4,200 | 293 | 226 | 8,476 | 0.5 |
| Cass | 5.0 | 0.1 | D | D | D | 4.1 | 4.7 | D | 14.6 | 2,660 | 219 | 237 | 5,822 | -0.2 |
| Champaign | 0.7 | D | 4.8 | 5.9 | 7.5 | 5.4 | 5.3 | 15.0 | 38.8 | 28,570 | 136 | 3,215 | 95,758 | 9.4 |
| Christian | 6.4 | D | 5.6 | 14.5 | D | 6.8 | 4.2 | D | 17.3 | 8,250 | 254 | 626 | 15,595 | 0.2 |
| Clark | 2.7 | 1.9 | 7.1 | 31.5 | 6.7 | 5.7 | 4.4 | 4.4 | 16.4 | 3,975 | 257 | 300 | 7,822 | 0.6 |
| Clay | -0.7 | 2.4 | 2.4 | 39.1 | D | 5.5 | 3.5 | D | 20.7 | 3,440 | 261 | 301 | 6,479 | 1.2 |
| Clinton | 3.3 | 0.2 | 12.5 | 10.9 | 2.8 | 9.4 | 5.2 | 11.0 | 23.0 | 7,650 | 204 | 305 | 16,100 | 5.2 |
| Coles | 1.2 | 0.2 | 5.0 | 8.6 | D | 6.3 | 7.7 | 17.9 | 25.0 | 10,015 | 197 | 1,220 | 23,436 | 0.0 |
| Cook | 0.0 | 0.1 | 3.3 | 6.4 | 19.5 | 4.0 | 15.6 | 9.7 | 11.4 | 807,915 | 157 | 145,411 | 2,209,260 | 1.3 |
| Crawford | 1.5 | 1.5 | 7.7 | 42.8 | 7.5 | 4.3 | 3.5 | 3.2 | 16.9 | 4,675 | 250 | 352 | 8,691 | 0.3 |
| Cumberland | 73.7 | 0.0 | D | D | 0.2 | 2.4 | D | 3.4 | 5.3 | 2,530 | 236 | 150 | 4,867 | -0.1 |
| DeKalb | 1.5 | D | 9.9 | 12.4 | D | 7.9 | 3.7 | 11.6 | 29.2 | 15,805 | 150 | 1,115 | 41,358 | 0.7 |
| De Witt | 5.2 | D | 8.9 | 9.2 | D | 5.5 | 2.6 | 2.0 | 16.7 | 3,690 | 236 | 244 | 7,584 | 0.9 |
| Douglas | 4.4 | D | 10.7 | 43.2 | D | 4.8 | 3.2 | 3.2 | 10.0 | 4,080 | 210 | 241 | 8,487 | 1.2 |
| DuPage | 0.0 | 0.1 | 6.2 | 9.3 | 15.4 | 4.6 | 9.6 | 10.1 | 7.8 | 152,360 | 165 | 7,953 | 362,769 | 1.9 |
| Edgar | 2.6 | 0.0 | 3.5 | 35.7 | 3.0 | 4.4 | 9.9 | D | 14.7 | 4,505 | 262 | 392 | 8,811 | 0.1 |
| Edwards | 7.2 | D | D | D | D | 4.2 | 4.2 | 1.5 | 10.8 | 1,595 | 249 | 68 | 3,196 | 0.3 |
| Effingham | 2.0 | D | 7.2 | 14.9 | D | 7.8 | 5.5 | 18.9 | 8.5 | 7,785 | 228 | 448 | 15,038 | 3.2 |
| Fayette | 6.2 | 1.8 | 4.9 | 6.0 | 3.5 | 7.4 | 6.6 | D | 24.2 | 4,870 | 228 | 409 | 9,305 | 0.0 |
| Ford | 6.9 | D | 7.4 | 15.0 | 2.9 | 4.3 | 2.1 | D | 13.5 | 3,150 | 243 | 205 | 6,355 | 1.2 |
| Franklin | 2.1 | 10.1 | 8.2 | 6.7 | D | 7.6 | 3.5 | 12.4 | 24.9 | 10,455 | 271 | 1,301 | 18,625 | -0.4 |
| Fulton | 3.5 | 0.0 | 5.3 | 3.5 | 4.2 | 8.1 | 5.6 | D | 28.5 | 8,685 | 254 | 679 | 16,376 | 1.1 |
| Gallatin | 10.7 | D | 1.7 | 1.7 | D | 3.0 | 1.2 | D | 14.8 | 1,515 | 313 | 205 | 2,755 | 0.3 |
| Greene | 13.8 | 0.1 | 5.4 | 2.8 | 3.1 | 6.7 | D | D | 25.9 | 3,080 | 238 | 349 | 6,422 | 0.5 |
| Grundy | 0.7 | 2.9 | 11.1 | 19.0 | D | 8.0 | 2.5 | 7.6 | 11.0 | 9,690 | 190 | 414 | 21,525 | 7.6 |
| Hamilton | 10.9 | D | 7.1 | 3.9 | D | 3.7 | 4.0 | 4.3 | 19.6 | 1,990 | 245 | 180 | 4,114 | 0.2 |
| Hancock | 13.8 | 0.7 | 7.1 | 6.2 | 6.1 | 4.6 | 5.5 | D | 20.5 | 4,905 | 277 | 288 | 9,223 | -0.5 |

1. Per 1,000 resident population estimated as of July 1 of the year shown.

| STATE County | Housing units, 2015-2019 Occupied units Owner-occupied Total | Percent | Median value[1] | Median owner cost as a percent of income With a mortgage | Without a mortgage[2] | Renter-occupied Median rent[3] | Median rent as a percent of income[2] | Sub-standard units[4] (percent) | Civilian labor force, 2020 Total | Percent change, 2019-2020 | Unemployment Total | Rate[5] | Civilian employment[6], 2015-2019 Total | Percent Management, business, science, and arts | Construction, production, and maintenance occupations |
|---|---|---|---|---|---|---|---|---|---|---|---|---|---|---|---|
| | 89 | 90 | 91 | 92 | 93 | 94 | 95 | 96 | 97 | 98 | 99 | 100 | 101 | 102 | 103 |
| **IDAHO—Cont'd** | | | | | | | | | | | | | | | |
| Caribou | 2,538 | 79.3 | 145,900 | 18.9 | 10.0 | 570 | 21.7 | 3.4 | 4,110 | 0.2 | 148 | 3.6 | 2,983 | 28.1 | 37.7 |
| Cassia | 7,711 | 70.3 | 162,100 | 19.3 | 10.0 | 647 | 21.3 | 5.0 | 12,384 | 2.2 | 454 | 3.7 | 10,141 | 27.2 | 40.9 |
| Clark | 290 | 62.4 | 98,100 | 26.9 | 10.0 | 592 | 19.0 | 5.9 | 378 | 1.1 | 15 | 4.0 | 409 | 15.6 | 63.3 |
| Clearwater | 3,555 | 78.1 | 155,900 | 23.0 | 10.6 | 719 | 27.7 | 2.8 | 3,021 | 4.7 | 249 | 8.2 | 2,842 | 28.5 | 31.8 |
| Custer | 1,774 | 78.6 | 191,600 | 30.5 | 10.0 | 608 | 28.5 | 1.7 | 2,218 | 0.2 | 112 | 5.0 | 1,482 | 37.6 | 28.3 |
| Elmore | 10,606 | 59.8 | 153,800 | 21.9 | 10.0 | 833 | 29.0 | 3.0 | 11,897 | 3.2 | 576 | 4.8 | 9,539 | 25.2 | 35.4 |
| Franklin | 4,397 | 81.3 | 211,900 | 21.3 | 10.0 | 680 | 23.8 | 2.6 | 7,060 | 0.0 | 253 | 3.6 | 5,915 | 28.9 | 41.0 |
| Fremont | 4,347 | 81.2 | 169,900 | 20.6 | 10.2 | 798 | 24.2 | 4.5 | 7,894 | 2.0 | 288 | 3.6 | 5,356 | 27.1 | 37.0 |
| Gem | 6,683 | 74.8 | 168,700 | 23.1 | 10.8 | 745 | 28.0 | 3.1 | 8,422 | 1.4 | 498 | 5.9 | 7,233 | 28.0 | 33.5 |
| Gooding | 5,469 | 70.5 | 157,800 | 21.0 | 10.0 | 704 | 25.5 | 5.2 | 8,256 | 0.7 | 312 | 3.8 | 6,984 | 28.5 | 39.3 |
| Idaho | 6,407 | 76.7 | 173,200 | 27.1 | 10.0 | 682 | 22.6 | 4.8 | 6,709 | 2.3 | 449 | 6.7 | 6,214 | 28.8 | 29.7 |
| Jefferson | 8,791 | 80.7 | 203,700 | 19.9 | 10.0 | 796 | 21.2 | 4.6 | 14,142 | 1.3 | 483 | 3.4 | 12,917 | 34.4 | 30.7 |
| Jerome | 8,022 | 64.9 | 164,600 | 21.9 | 10.0 | 772 | 26.7 | 4.3 | 12,198 | -0.1 | 487 | 4.0 | 11,055 | 26.1 | 41.6 |
| Kootenai | 62,304 | 70.4 | 260,300 | 22.6 | 10.0 | 993 | 29.3 | 2.3 | 81,286 | 1.8 | 5,632 | 6.9 | 73,191 | 32.9 | 23.4 |
| Latah | 15,422 | 54.3 | 228,200 | 19.4 | 10.0 | 737 | 29.8 | 2.7 | 19,756 | -3.2 | 938 | 4.7 | 20,190 | 43.9 | 19.3 |
| Lemhi | 3,582 | 77.6 | 177,000 | 22.5 | 11.2 | 657 | 25.7 | 4.0 | 3,651 | 3.1 | 243 | 6.7 | 3,194 | 32.5 | 28.3 |
| Lewis | 1,646 | 69.4 | 132,300 | 21.6 | 10.6 | 593 | 27.1 | 3.8 | 1,662 | 2.2 | 128 | 7.7 | 1,454 | 29.6 | 32.3 |
| Lincoln | 1,724 | 68.2 | 145,600 | 24.4 | 10.0 | 639 | 23.4 | 5.2 | 2,667 | 0.0 | 156 | 5.8 | 2,559 | 24.2 | 41.7 |
| Madison | 11,016 | 44.6 | 219,100 | 21.1 | 10.0 | 723 | 42.1 | 10.2 | 22,306 | 1.6 | 606 | 2.7 | 18,393 | 39.5 | 20.6 |
| Minidoka | 7,250 | 73.4 | 149,600 | 18.1 | 10.0 | 701 | 23.0 | 3.4 | 11,642 | 2.3 | 448 | 3.8 | 9,547 | 21.7 | 45.5 |
| Nez Perce | 16,384 | 72.8 | 191,500 | 19.4 | 10.0 | 727 | 29.7 | 2.5 | 20,793 | -0.6 | 1,013 | 4.9 | 19,215 | 32.6 | 28.0 |
| Oneida | 1,615 | 79.9 | 167,100 | 23.5 | 10.6 | 748 | 30.3 | 2.3 | 2,342 | 1.2 | 98 | 4.2 | 1,725 | 30.0 | 37.2 |
| Owyhee | 4,321 | 70.9 | 147,800 | 23.6 | 10.0 | 668 | 24.7 | 7.1 | 5,493 | 1.6 | 327 | 6.0 | 5,168 | 24.4 | 42.0 |
| Payette | 8,876 | 72.7 | 165,600 | 21.4 | 10.0 | 735 | 25.3 | 6.5 | 11,941 | 2.3 | 632 | 5.3 | 9,876 | 27.5 | 35.7 |
| Power | 2,633 | 74.4 | 151,300 | 21.2 | 11.1 | 636 | 18.3 | 2.3 | 4,132 | 1.5 | 168 | 4.1 | 3,304 | 27.3 | 43.9 |
| Shoshone | 5,466 | 70.0 | 127,900 | 19.9 | 12.6 | 670 | 25.6 | 2.5 | 5,444 | 6.1 | 498 | 9.1 | 5,025 | 22.1 | 31.1 |
| Teton | 3,707 | 77.7 | 346,600 | 21.9 | 12.7 | 881 | 26.4 | 2.4 | 6,632 | 0.4 | 376 | 5.7 | 5,930 | 34.7 | 23.2 |
| Twin Falls | 31,146 | 68.9 | 166,800 | 21.2 | 11.2 | 800 | 27.9 | 2.5 | 41,250 | 0.2 | 2,167 | 5.3 | 40,237 | 30.8 | 30.6 |
| Valley | 3,869 | 80.8 | 314,800 | 22.0 | 10.0 | 773 | 38.5 | 2.1 | 5,643 | 1.9 | 448 | 7.9 | 4,714 | 32.5 | 24.6 |
| Washington | 4,035 | 72.5 | 148,900 | 21.3 | 11.1 | 683 | 33.4 | 2.7 | 4,752 | 3.5 | 256 | 5.4 | 4,104 | 25.1 | 33.4 |
| **ILLINOIS** | 4,846,134 | 66.1 | 194,500 | 21.3 | 12.8 | 1,010 | 29.1 | 2.7 | 6,249,147 | -3.1 | 591,615 | 9.5 | 6,250,862 | 38.7 | 22.2 |
| Adams | 27,112 | 71.8 | 127,800 | 18.8 | 10.4 | 713 | 27.7 | 2.2 | 31,259 | -3.2 | 1,972 | 6.3 | 32,677 | 32.4 | 26.2 |
| Alexander | 2,154 | 73.7 | 56,000 | 24.1 | 13.4 | 526 | 23.2 | 1.5 | 1,918 | -1.5 | 188 | 9.8 | 2,102 | 25.5 | 28.0 |
| Bond | 6,299 | 75.9 | 119,700 | 18.3 | 12.7 | 609 | 26.2 | 4.9 | 7,484 | -3.9 | 512 | 6.8 | 7,417 | 30.0 | 29.5 |
| Boone | 18,571 | 81.0 | 153,400 | 21.4 | 12.2 | 854 | 27.6 | 2.8 | 25,580 | -3.2 | 2,860 | 11.2 | 25,777 | 29.3 | 33.7 |
| Brown | 2,055 | 74.8 | 96,200 | 15.6 | 10.0 | 615 | 21.8 | 2.6 | 2,996 | -0.6 | 115 | 3.8 | 2,423 | 29.5 | 36.1 |
| Bureau | 13,698 | 76.6 | 111,700 | 18.7 | 11.2 | 680 | 23.7 | 1.5 | 16,363 | -5.1 | 1,239 | 7.6 | 15,890 | 28.4 | 32.2 |
| Calhoun | 1,664 | 87.3 | 121,000 | 19.1 | 10.8 | 617 | 30.4 | 1.7 | 2,228 | -3.8 | 169 | 7.6 | 1,996 | 36.5 | 26.8 |
| Carroll | 6,508 | 76.6 | 106,100 | 18.6 | 11.6 | 631 | 24.3 | 0.8 | 7,376 | -0.2 | 532 | 7.2 | 6,570 | 31.8 | 33.7 |
| Cass | 5,043 | 74.6 | 80,800 | 17.5 | 11.3 | 598 | 22.7 | 2.1 | 5,813 | -4.5 | 387 | 6.7 | 5,978 | 21.0 | 41.8 |
| Champaign | 82,369 | 53.0 | 162,100 | 19.0 | 11.0 | 882 | 31.8 | 5.3 | 108,725 | | 6,993 | 6.4 | 103,206 | 46.2 | 17.4 |
| Christian | 13,901 | 75.1 | 90,900 | 17.7 | 11.7 | 674 | 26.7 | 1.3 | 13,598 | -6.4 | 1,101 | 8.1 | 14,704 | 32.1 | 30.5 |
| Clark | 6,720 | 76.6 | 92,500 | 17.8 | 11.1 | 687 | 24.0 | 1.4 | 7,331 | -4.3 | 579 | 7.9 | 7,572 | 27.7 | 35.0 |
| Clay | 5,696 | 74.5 | 84,500 | 19.1 | 10.8 | 604 | 19.4 | 3.7 | 6,231 | -3.9 | 585 | 9.4 | 6,030 | 25.5 | 40.6 |
| Clinton | 14,353 | 79.6 | 146,200 | 18.7 | 12.1 | 808 | 23.8 | 1.8 | 19,796 | -4.1 | 1,154 | 5.8 | 18,656 | 37.2 | 29.3 |
| Coles | 20,926 | 61.5 | 100,500 | 18.6 | 11.6 | 697 | 28.9 | 1.6 | 23,148 | -3.1 | 1,716 | 7.4 | 24,994 | 33.2 | 26.4 |
| Cook | 1,972,108 | 56.9 | 246,600 | 23.4 | 14.4 | 1,123 | 29.6 | 3.6 | 2,539,907 | -2.2 | 281,454 | 11.1 | 2,567,330 | 41.0 | 19.8 |
| Crawford | 7,666 | 77.7 | 87,800 | 16.7 | 11.8 | 638 | 27.5 | 2.3 | 8,472 | -2.5 | 590 | 7.0 | 7,645 | 30.5 | 31.2 |
| Cumberland | 4,299 | 78.7 | 105,800 | 17.6 | 12.1 | 582 | 19.4 | 2.1 | 5,944 | -2.9 | 338 | 5.7 | 5,396 | 27.2 | 36.8 |
| DeKalb | 38,150 | 57.0 | 173,100 | 20.4 | 13.1 | 930 | 30.4 | 1.6 | 53,031 | -4.8 | 4,353 | 8.2 | 53,520 | 34.1 | 24.3 |
| De Witt | 6,694 | 78.7 | 99,500 | 18.1 | 10.7 | 629 | 25.8 | 1.5 | 7,120 | -4.6 | 464 | 6.5 | 7,926 | 32.2 | 29.2 |
| Douglas | 7,613 | 74.0 | 111,400 | 18.3 | 10.0 | 724 | 26.1 | 3.1 | 10,036 | -2.4 | 562 | 5.6 | 9,522 | 25.6 | 35.1 |
| DuPage | 342,791 | 73.4 | 308,500 | 21.6 | 13.2 | 1,316 | 28.5 | 2.4 | 484,215 | -4.7 | 38,391 | 7.9 | 491,547 | 46.7 | 16.9 |
| Edgar | 7,542 | 76.5 | 82,800 | 17.3 | 10.7 | 611 | 27.9 | 1.3 | 8,899 | -0.7 | 588 | 6.6 | 8,218 | 28.6 | 36.1 |
| Edwards | 2,773 | 80.2 | 79,300 | 16.9 | 10.0 | 495 | 22.7 | 1.6 | 2,681 | -1.4 | 219 | 8.2 | 2,929 | 26.5 | 44.6 |
| Effingham | 13,877 | 75.4 | 143,600 | 18.8 | 10.6 | 650 | 23.0 | 2.0 | 19,265 | -0.4 | 1,277 | 6.6 | 17,279 | 32.0 | 30.0 |
| Fayette | 7,737 | 80.5 | 88,300 | 18.1 | 11.9 | 612 | 26.8 | 2.1 | 9,700 | -2.4 | 723 | 7.5 | 8,760 | 26.4 | 32.6 |
| Ford | 5,771 | 72.2 | 98,300 | 18.9 | 11.5 | 676 | 27.8 | 0.8 | 6,236 | -2.4 | 402 | 6.4 | 6,306 | 33.7 | 31.0 |
| Franklin | 16,235 | 74.2 | 76,100 | 19.5 | 11.4 | 637 | 28.8 | 1.9 | 15,985 | -2.7 | 1,686 | 10.5 | 15,804 | 27.3 | 31.2 |
| Fulton | 13,940 | 78.3 | 88,600 | 18.3 | 11.5 | 646 | 26.2 | 1.4 | 14,615 | -3.9 | 1,251 | 8.6 | 15,016 | 32.1 | 29.5 |
| Gallatin | 2,293 | 77.1 | 73,900 | 15.4 | 11.2 | 439 | 26.3 | 2.0 | 2,032 | -13.2 | 186 | 9.2 | 2,124 | 26.6 | 36.8 |
| Greene | 4,949 | 77.6 | 81,000 | 18.7 | 11.5 | 638 | 27.1 | 1.4 | 5,904 | -2.6 | 382 | 6.5 | 6,153 | 25.4 | 31.4 |
| Grundy | 19,676 | 71.3 | 199,300 | 19.9 | 12.0 | 999 | 23.9 | 2.1 | 24,269 | -5.4 | 2,131 | 8.8 | 25,357 | 30.5 | 31.6 |
| Hamilton | 3,400 | 80.0 | 93,500 | 18.3 | 11.1 | 609 | 25.2 | 1.0 | 4,225 | -4.6 | 281 | 6.7 | 3,559 | 32.9 | 36.1 |
| Hancock | 7,409 | 81.4 | 85,600 | 17.9 | 10.1 | 617 | 21.8 | 1.5 | 8,081 | -4.1 | 560 | 6.9 | 8,133 | 29.9 | 33.9 |

1. Specified owner-occupied units. lacking complete plumbing facilities.  2. A value of 10.0 represents 10 percent or less; a value of 50.0 represents 50 percent or more.  3. Specified renter-occupied units.  4. Overcrowded or
5. Percent of civilian labor force.  6. Civilian employed persons 16 years old and over.

# Table B. States and Counties — Nonfarm Employment and Agriculture

| STATE County | Private nonfarm establishments, employment and payroll, 2019 | | | | | | | | | Agriculture, 2017 | | | |
|---|---|---|---|---|---|---|---|---|---|---|---|---|---|
| | Number of establishments | Employment | | | | | | Annual payroll | | Farms | | | Farm producers whose primary occupation is farming (percent) |
| | | Total | Health care and social assistance | Manufacturing | Retail trade | Finance and insurance | Professional, scientific, and technical services | Total (mil dol) | Average per employee (dollars) | Number | Percent with: | | |
| | | | | | | | | | | | Fewer than 50 acres | 1000 acres or more | |
| | 104 | 105 | 106 | 107 | 108 | 109 | 110 | 111 | 112 | 113 | 114 | 115 | 116 |
| IDAHO—Cont'd | | | | | | | | | | | | | |
| Caribou | 183 | 2,086 | 334 | 736 | 295 | 51 | 23 | 114 | 54,618 | 411 | 28.2 | 23.6 | 48.2 |
| Cassia | 734 | 8,440 | 1,305 | 1,269 | 1,221 | 169 | 218 | 314 | 37,172 | 585 | 44.3 | 25.3 | 51.7 |
| Clark | 15 | 81 | NA | NA | NA | NA | NA | 3 | 34,444 | 68 | 19.1 | 38.2 | 52.1 |
| Clearwater | 226 | 1,664 | 490 | 224 | 231 | 34 | 42 | 73 | 43,629 | 312 | 50.6 | 3.8 | 36.9 |
| Custer | 175 | 766 | 86 | 18 | 200 | 20 | 25 | 29 | 37,617 | 267 | 34.8 | 12.0 | 50.6 |
| Elmore | 457 | 4,346 | 746 | 730 | 1,011 | 179 | 141 | 142 | 32,776 | 340 | 59.4 | 14.1 | 41.5 |
| Franklin | 334 | 2,416 | 439 | 366 | 541 | 71 | 57 | 81 | 33,339 | 787 | 44.7 | 7.6 | 33.9 |
| Fremont | 329 | 1,784 | 245 | 110 | 253 | 36 | 48 | 72 | 40,213 | 513 | 37.4 | 9.7 | 36.9 |
| Gem | 407 | 2,886 | 837 | 222 | 461 | 64 | 113 | 95 | 32,789 | 860 | 77.4 | 3.4 | 39.7 |
| Gooding | 363 | 2,845 | 668 | 703 | 339 | 69 | 63 | 115 | 40,461 | 538 | 58.6 | 7.6 | 50.9 |
| Idaho | 495 | 3,555 | 685 | 647 | 508 | 129 | 60 | 132 | 37,181 | 708 | 32.1 | 17.1 | 45.5 |
| Jefferson | 561 | 4,287 | 291 | 927 | 645 | 97 | 114 | 155 | 36,054 | 750 | 60.0 | 10.0 | 41.7 |
| Jerome | 557 | 6,474 | 476 | 1,623 | 958 | 104 | 103 | 262 | 40,547 | 486 | 49.8 | 7.8 | 55.3 |
| Kootenai | 5,177 | 55,398 | 11,066 | 5,021 | 8,714 | 2,331 | 2,814 | 2,383 | 43,019 | 1,073 | 68.7 | 2.1 | 32.1 |
| Latah | 933 | 9,401 | 1,672 | 328 | 1,805 | 218 | 593 | 292 | 31,074 | 1,041 | 48.7 | 9.1 | 34.3 |
| Lemhi | 298 | 1,914 | 548 | 88 | 382 | 37 | 71 | 63 | 32,707 | 351 | 48.1 | 8.8 | 42.8 |
| Lewis | 110 | 872 | 117 | 162 | 200 | 43 | 4 | 29 | 33,229 | 197 | 26.9 | 27.9 | 50.0 |
| Lincoln | 93 | 574 | 104 | 154 | 87 | 9 | 23 | 21 | 36,361 | 276 | 32.6 | 13.0 | 59.4 |
| Madison | 971 | 19,380 | 2,033 | 1,074 | 1,724 | 275 | 770 | 484 | 24,954 | 454 | 50.7 | 10.8 | 47.3 |
| Minidoka | 447 | 5,596 | 602 | 1,116 | 705 | 154 | 134 | 254 | 45,473 | 620 | 51.8 | 9.0 | 49.5 |
| Nez Perce | 1,139 | 16,753 | 3,137 | 3,405 | 2,533 | 1,065 | 503 | 712 | 42,487 | 446 | 47.3 | 24.2 | 42.4 |
| Oneida | 102 | 754 | 193 | 51 | 176 | 51 | 7 | 23 | 30,565 | 422 | 28.7 | 23.0 | 36.1 |
| Owyhee | 198 | 1,945 | 246 | 157 | 245 | 27 | 25 | 61 | 31,301 | 565 | 46.2 | 15.2 | 55.6 |
| Payette | 558 | 5,731 | 867 | 1,435 | 501 | 170 | 243 | 228 | 39,763 | 640 | 70.3 | 3.1 | 40.6 |
| Power | 178 | 2,298 | 165 | 728 | 174 | 45 | 34 | 87 | 37,731 | 295 | 29.2 | 39.3 | 52.6 |
| Shoshone | 355 | 3,733 | 614 | 144 | 1,015 | 64 | 155 | 141 | 37,716 | 48 | 75.0 | NA | 30.0 |
| Teton | 545 | 2,700 | 338 | 146 | 381 | 43 | 118 | 118 | 43,707 | 277 | 38.6 | 12.3 | 44.0 |
| Twin Falls | 2,716 | 31,286 | 6,344 | 3,558 | 5,216 | 948 | 1,171 | 1,132 | 36,186 | 1,211 | 51.6 | 5.9 | 49.6 |
| Valley | 672 | 4,241 | 540 | 106 | 582 | 58 | 99 | 141 | 33,333 | 188 | 57.4 | 6.4 | 33.4 |
| Washington | 208 | 1,985 | 371 | 597 | 290 | 36 | 62 | 70 | 35,226 | 535 | 49.7 | 15.0 | 50.2 |
| ILLINOIS | 320,417 | 5,530,388 | 818,916 | 547,133 | 581,767 | 341,539 | 401,094 | 325,435 | 58,845 | 72,651 | 35.6 | 10.8 | 43.4 |
| Adams | 1,774 | 32,059 | 6,642 | 5,675 | 4,865 | 1,498 | 691 | 1,397 | 43,573 | 1,308 | 26.8 | 9.9 | 42.8 |
| Alexander | 91 | 892 | 189 | 58 | 88 | 15 | 28 | 35 | 39,760 | 126 | 30.2 | 11.9 | 36.5 |
| Bond | 314 | 3,898 | 509 | 759 | 333 | 115 | 58 | 133 | 34,096 | 637 | 46.3 | 7.4 | 36.7 |
| Boone | 844 | 16,653 | 925 | 9,181 | 1,343 | 202 | 532 | 891 | 53,496 | 457 | 54.5 | 5.5 | 47.4 |
| Brown | 110 | 3,404 | 165 | 150 | 133 | 44 | 23 | 178 | 52,251 | 419 | 25.3 | 7.2 | 30.2 |
| Bureau | 722 | 9,803 | 2,194 | 1,680 | 1,044 | 313 | 168 | 426 | 43,450 | 1,038 | 28.9 | 13.8 | 46.5 |
| Calhoun | 74 | 490 | 80 | NA | 102 | 53 | 10 | 12 | 24,288 | 474 | 25.7 | 5.3 | 34.2 |
| Carroll | 380 | 3,334 | 468 | 952 | 426 | 186 | 57 | 126 | 37,869 | 627 | 31.6 | 9.4 | 47.3 |
| Cass | 238 | 4,316 | 308 | 1,874 | 445 | 166 | 59 | 181 | 41,854 | 429 | 29.1 | 14.5 | 36.9 |
| Champaign | 4,232 | 69,618 | 13,993 | 5,926 | 9,592 | 3,261 | 2,775 | 3,161 | 45,399 | 1,214 | 33.1 | 16.0 | 50.0 |
| Christian | 703 | 7,917 | 1,547 | 934 | 1,289 | 329 | 156 | 327 | 41,352 | 794 | 33.8 | 16.5 | 48.2 |
| Clark | 343 | 3,936 | 320 | 1,170 | 439 | 155 | 179 | 146 | 37,102 | 733 | 37.8 | 10.8 | 38.2 |
| Clay | 346 | 5,068 | 698 | 2,363 | 532 | 134 | 113 | 210 | 41,414 | 732 | 37.8 | 8.6 | 37.4 |
| Clinton | 849 | 8,912 | 1,801 | 930 | 1,724 | 292 | 193 | 310 | 34,776 | 831 | 32.4 | 6.0 | 39.1 |
| Coles | 1,143 | 18,983 | 4,600 | 2,371 | 2,382 | 749 | 439 | 726 | 38,224 | 701 | 39.8 | 10.4 | 39.1 |
| Cook | 134,204 | 2,437,240 | 378,314 | 177,985 | 223,175 | 196,653 | 236,693 | 164,451 | 67,474 | 182 | 78.6 | 0.5 | 34.7 |
| Crawford | 416 | 6,060 | 786 | 1,947 | 749 | 221 | 271 | 271 | 44,796 | 566 | 38.0 | 11.3 | 41.4 |
| Cumberland | 193 | 2,010 | 287 | 620 | 263 | 90 | 18 | 61 | 30,574 | 724 | 40.2 | 4.8 | 34.3 |
| DeKalb | 1,968 | 28,336 | 5,013 | 4,003 | 4,327 | 780 | 1,116 | 1,203 | 42,469 | 779 | 35.3 | 14.2 | 53.7 |
| De Witt | 364 | 4,476 | 504 | 505 | 662 | 101 | 188 | 245 | 54,763 | 504 | 44.6 | 12.5 | 43.4 |
| Douglas | 582 | 6,768 | 294 | 2,869 | 906 | 204 | 97 | 299 | 44,144 | 600 | 45.5 | 12.5 | 40.5 |
| DuPage | 34,049 | 585,492 | 67,127 | 53,408 | 55,353 | 32,908 | 50,406 | 37,300 | 63,707 | 77 | 87.0 | NA | 41.8 |
| Edgar | 337 | 6,212 | 1,026 | 2,413 | 648 | 188 | 87 | 229 | 36,918 | 637 | 30.3 | 16.6 | 48.2 |
| Edwards | 150 | 2,109 | 83 | 1,311 | 161 | 68 | 25 | 90 | 42,826 | 291 | 31.3 | 11.3 | 40.8 |
| Effingham | 1,255 | 26,864 | 2,813 | 3,477 | 2,771 | 434 | 516 | 1,073 | 39,934 | 1,193 | 37.6 | 6.1 | 35.5 |
| Fayette | 473 | 4,600 | 901 | 313 | 744 | 237 | 123 | 145 | 31,595 | 1,239 | 38.7 | 8.2 | 37.1 |
| Ford | 350 | 4,338 | 1,261 | 786 | 535 | 113 | 100 | 176 | 40,650 | 564 | 29.8 | 13.8 | 49.5 |
| Franklin | 687 | 6,947 | 896 | 592 | 1,423 | 233 | 178 | 253 | 36,367 | 596 | 41.4 | 7.6 | 38.1 |
| Fulton | 608 | 5,791 | 1,692 | 140 | 1,262 | 329 | 206 | 199 | 34,335 | 973 | 27.9 | 12.7 | 45.4 |
| Gallatin | 83 | 556 | 94 | 21 | 49 | 23 | 19 | 24 | 42,567 | 165 | 20.0 | 21.8 | 60.6 |
| Greene | 210 | 1,580 | 320 | 90 | 369 | 124 | 72 | 55 | 35,084 | 733 | 25.9 | 14.2 | 43.0 |
| Grundy | 1,184 | 17,241 | 2,627 | 1,575 | 2,248 | 294 | 715 | 958 | 55,546 | 412 | 29.6 | 19.2 | 49.7 |
| Hamilton | 190 | 1,597 | 434 | 75 | 143 | 42 | 33 | 67 | 41,910 | 552 | 30.4 | 10.5 | 32.8 |
| Hancock | 373 | 2,887 | 533 | 410 | 466 | 159 | 198 | 116 | 40,147 | 1,109 | 26.1 | 12.4 | 47.5 |

# Table B. States and Counties — Agriculture

| | Agriculture, 2017 (cont.) | | | | | | | | | | | | | | |
|---|---|---|---|---|---|---|---|---|---|---|---|---|---|---|---|
| | Land in farms | | | | Value of land and buildings (dollars) | | Value of machinery and equioopmnet, average per farm (dollars) | Value of products sold: | | | | | Farms with internet access (percent) | Government payments | |
| | | Acres | | | | | | | | Percent from: | | | | | |
| STATE County | Acreage (1,000) | Percent change, 2012-2017 | Average size of farm | Total irrigated (1,000) | Total cropland (1,000) | Average per farm | Average per acre | | Total (mil dol) | Average per farm (acres) | Crops | Livestock and poultry products | Organic farms (number) | | Total ($1,000) | Percent of farms |
| | 117 | 118 | 119 | 120 | 121 | 122 | 123 | 124 | 125 | 126 | 127 | 128 | 129 | 130 | 131 | 132 |

| STATE County | 117 | 118 | 119 | 120 | 121 | 122 | 123 | 124 | 125 | 126 | 127 | 128 | 129 | 130 | 131 | 132 |
|---|---|---|---|---|---|---|---|---|---|---|---|---|---|---|---|---|
| IDAHO—Cont'd | | | | | | | | | | | | | | | | |
| Caribou | 366 | -7.1 | 892 | 61.1 | 217.1 | 1,654,989 | 1,856 | 219,066 | 90.3 | 219,757 | 62.2 | 37.8 | 1 | 81.8 | 3,905 | 52.8 |
| Cassia | 643 | 5.3 | 1,100 | 259.3 | 385.0 | 3,508,509 | 3,190 | 512,076 | 926.7 | 1,584,137 | 27.6 | 72.4 | 4 | 87.9 | 11,088 | 38.6 |
| Clark | 149 | -1.2 | 2,197 | 31.6 | 40.7 | 3,856,994 | 1,755 | 320,235 | 25.9 | 380,309 | 74.4 | 25.6 | 2 | 75.0 | 1,490 | 55.9 |
| Clearwater | 57 | -22.1 | 181 | D | 24.8 | 447,579 | 2,469 | 54,776 | 7.3 | 23,487 | 46.8 | 53.2 | NA | 79.5 | 639 | 18.3 |
| Custer | 148 | 3.5 | 554 | 60.1 | 49.7 | 1,509,320 | 2,726 | 169,068 | 36.4 | 136,468 | 23.7 | 76.3 | 3 | 90.3 | 343 | 13.1 |
| Elmore | 358 | 4.0 | 1,054 | 113.2 | 148.9 | 2,626,980 | 2,492 | 400,675 | 429.9 | 1,264,462 | 27.8 | 72.2 | 7 | 82.6 | 684 | 11.5 |
| Franklin | 228 | -13 | 290 | 65.3 | 132.1 | 674,740 | 2,325 | 128,763 | 82.8 | 105,187 | 31.1 | 68.9 | 26 | 83.5 | 2,730 | 39.6 |
| Fremont | 280 | -11.6 | 545 | 101.9 | 171.0 | 1,507,845 | 2,767 | 225,633 | 138.2 | 269,419 | 84.7 | 15.3 | NA | 80.7 | 4,130 | 38.8 |
| Gem | 183 | 2.3 | 213 | 29.9 | 28.3 | 599,509 | 2,815 | 69,848 | 39.2 | 45,560 | 39.6 | 60.4 | 8 | 85.2 | 834 | 10.9 |
| Gooding | 188 | -21.4 | 350 | 121.8 | 125.3 | 2,106,203 | 6,016 | 286,549 | 783.4 | 1,456,113 | 9.3 | 90.7 | 9 | 83.3 | 2,416 | 16.5 |
| Idaho | 537 | -15.9 | 759 | 2.7 | 180.6 | 1,248,054 | 1,644 | 112,824 | 43.7 | 61,689 | 54.8 | 45.2 | NA | 79.2 | 3,060 | 36.4 |
| Jefferson | 334 | 3.3 | 445 | 198.3 | 228.3 | 1,566,390 | 3,522 | 264,821 | 294.6 | 392,743 | 58.7 | 41.3 | 6 | 86.9 | 4,603 | 24.4 |
| Jerome | 172 | -8.7 | 353 | 134.9 | 134.6 | 2,132,761 | 6,039 | 418,233 | 639.6 | 1,316,016 | 18.9 | 81.1 | 16 | 84.2 | 2,050 | 33.5 |
| Kootenai | 140 | 12.4 | 130 | 13.7 | 62.2 | 719,383 | 5,525 | 49,847 | 21.5 | 20,057 | 81.0 | 19.0 | 3 | 81.7 | 1,211 | 9.9 |
| Latah | 350 | -16.1 | 336 | 0.2 | 254.7 | 853,388 | 2,542 | 126,243 | 78.0 | 74,900 | 94.3 | 5.7 | 7 | 82.9 | 7,066 | 35.4 |
| Lemhi | 174 | -7.2 | 496 | 72.2 | 46.4 | 1,251,971 | 2,526 | 116,297 | 33.3 | 94,818 | 16.6 | 83.4 | 3 | 89.7 | 295 | 10.0 |
| Lewis | 200 | -9.4 | 1,017 | D | 151.1 | 1,997,866 | 1,964 | 238,202 | 37.8 | 191,787 | 90.2 | 9.8 | NA | 86.3 | 2,929 | 70.1 |
| Lincoln | 135 | 4.0 | 489 | 77.3 | 84.9 | 1,784,513 | 3,651 | 226,469 | 203.1 | 735,830 | 21.0 | 79.0 | 8 | 83.7 | 1,163 | 35.9 |
| Madison | 196 | -2.6 | 432 | 125.9 | 160.0 | 1,797,444 | 4,162 | 289,620 | 157.0 | 345,855 | 93.2 | 6.8 | 1 | 90.3 | 3,919 | 39.0 |
| Minidoka | 268 | 9.6 | 432 | 232.7 | 242.6 | 1,958,696 | 4,539 | 300,481 | 354.4 | 571,692 | 73.4 | 26.6 | 6 | 81.9 | 4,319 | 40.2 |
| Nez Perce | 382 | 18.4 | 856 | 1.1 | 233.8 | 1,779,079 | 2,079 | 230,280 | 74.3 | 166,632 | 81.4 | 18.6 | 1 | 84.1 | 7,319 | 43.0 |
| Oneida | 320 | -2.7 | 758 | 25.7 | 151.3 | 1,256,611 | 1,658 | 138,236 | 36.2 | 85,863 | 44.9 | 55.1 | 2 | 80.6 | 5,063 | 55.2 |
| Owyhee | 727 | -2.9 | 1,287 | 119.0 | 137.8 | 2,096,160 | 1,628 | 232,705 | 273.4 | 483,851 | 22.6 | 77.4 | 7 | 82.8 | 2,365 | 21.4 |
| Payette | 163 | 3.5 | 254 | 59.2 | 57.4 | 890,516 | 3,505 | 141,684 | 167.4 | 261,563 | 30.0 | 70.0 | 3 | 88.1 | 900 | 12.0 |
| Power | 486 | 4.1 | 1,649 | 147.7 | 381.4 | 4,068,136 | 2,467 | 438,509 | 235.4 | 798,108 | 89.6 | 10.4 | 3 | 86.1 | 10,617 | 59.3 |
| Shoshone | 2 | D | 51 | 0.1 | 0.6 | 340,665 | 6,715 | 42,286 | 0.2 | 4,479 | 30.2 | 69.8 | NA | 77.1 | NA | NA |
| Teton | 117 | -11.9 | 424 | 48.9 | 84.4 | 1,675,941 | 3,954 | 180,866 | 45.3 | 163,679 | 90.4 | 9.6 | 16 | 89.9 | 865 | 28.9 |
| Twin Falls | 469 | -3.1 | 387 | 241.5 | 257.3 | 1,718,569 | 4,439 | 225,875 | 680.2 | 561,716 | 24.8 | 75.2 | 26 | 88.2 | 4,456 | 29.7 |
| Valley | 51 | -16.8 | 271 | 22.1 | 4.3 | 703,011 | 2,594 | 62,528 | 10.5 | 56,069 | 8.7 | 91.3 | 1 | 67.6 | 84 | 4.3 |
| Washington | 468 | 9.8 | 876 | 38.8 | 80.4 | 1,185,859 | 1,354 | 132,560 | 50.2 | 93,918 | 41.8 | 58.2 | 6 | 80.7 | 1,591 | 28.4 |
| ILLINOIS | 27,006 | 0.3 | 372 | 612.5 | 24,003.1 | 2,705,291 | 7,278 | 220,485 | 17,010 | 234,133 | 81.4 | 18.6 | 328 | 76.9 | 521,229 | 66.9 |
| Adams | 478 | 22.9 | 365 | 11.0 | 382.8 | 2,376,117 | 6,506 | 227,632 | 269.4 | 205,979 | 73.2 | 26.8 | 9 | 75.7 | 4,870 | 60.0 |
| Alexander | 50 | -19.1 | 401 | 2.6 | 43.1 | 1,371,640 | 3,423 | 144,673 | 16.6 | 131,889 | 98.5 | 1.5 | NA | 81.0 | 1,130 | 61.9 |
| Bond | 173 | -12.9 | 271 | 0.0 | 149.8 | 1,865,147 | 6,874 | 180,014 | 84.8 | 133,047 | 89.7 | 10.3 | 2 | 76.9 | 3,877 | 60.3 |
| Boone | 114 | -15.8 | 248 | 0.9 | 106.0 | 1,939,825 | 7,811 | 171,009 | 78.4 | 171,589 | 75.5 | 24.5 | 6 | 84.2 | 3,291 | 38.5 |
| Brown | 142 | 3.0 | 338 | 0.2 | 88.2 | 1,892,515 | 5,598 | 125,319 | 49.8 | 118,947 | 78.6 | 21.4 | 4 | 70.4 | 3,224 | 83.1 |
| Bureau | 437 | -2.9 | 421 | 13.5 | 408.2 | 3,460,532 | 8,219 | 290,184 | 360.0 | 346,795 | 89.4 | 10.6 | 12 | 79.3 | 17,896 | 79.8 |
| Calhoun | 115 | 30.6 | 242 | D | 68.7 | 1,013,115 | 4,189 | 98,750 | 38.7 | 81,584 | 72.8 | 27.2 | NA | 63.5 | 2,393 | 68.6 |
| Carroll | 246 | -4.1 | 392 | 10.6 | 210.0 | 3,150,456 | 8,039 | 269,522 | 216.8 | 345,845 | 69.0 | 31.0 | NA | 80.1 | 4,490 | 75.6 |
| Cass | 198 | 8.1 | 461 | 30.5 | 169.8 | 2,946,206 | 6,398 | 268,331 | 121.9 | 284,096 | 79.2 | 20.8 | 2 | 73.7 | 5,151 | 84.4 |
| Champaign | 583 | -5.5 | 480 | 14.2 | 566.4 | 4,471,916 | 9,317 | 322,994 | 375.6 | 309,349 | 96.2 | 3.8 | 5 | 80.3 | 5,861 | 82.3 |
| Christian | 403 | 7.8 | 507 | 0.0 | 382.6 | 4,381,139 | 8,638 | 311,771 | 278.7 | 351,029 | 90.7 | 9.3 | 2 | 76.8 | 10,174 | 79.5 |
| Clark | 261 | -2.1 | 356 | 11.8 | 227.7 | 2,032,961 | 5,708 | 205,302 | 163.3 | 222,795 | 73.9 | 26.1 | 1 | 71.8 | 4,408 | 77.8 |
| Clay | 294 | 8.9 | 402 | D | 258.7 | 2,058,866 | 5,121 | 214,709 | 116.1 | 158,626 | 86.9 | 13.1 | NA | 73.1 | 5,966 | 76.2 |
| Clinton | 236 | -17.4 | 284 | 1.2 | 214.9 | 1,965,321 | 6,928 | 263,352 | 247.0 | 297,197 | 45.1 | 54.9 | NA | 78.1 | 6,067 | 74.8 |
| Coles | 237 | -11.2 | 338 | 0.1 | 219.9 | 2,660,833 | 7,875 | 196,323 | 133.9 | 191,073 | 95.7 | 4.3 | 1 | 77.5 | 4,071 | 75.5 |
| Cook | 12 | 40.1 | 65 | 0.1 | 10.8 | 1,349,568 | 20,635 | 77,634 | 19.7 | 108,187 | 89.5 | 10.5 | 3 | 89.0 | 81 | 17.0 |
| Crawford | 220 | 2.2 | 388 | 9.1 | 193.5 | 2,039,041 | 5,254 | 213,282 | 108.4 | 191,594 | 90.4 | 9.6 | NA | 77.0 | 5,598 | 77.0 |
| Cumberland | 172 | 0.9 | 237 | 1.0 | 145.7 | 1,504,990 | 6,344 | 141,060 | 120.6 | 166,609 | 65.3 | 34.7 | 1 | 71.7 | 6,191 | 82.6 |
| DeKalb | 372 | -6.5 | 477 | 0.5 | 362.6 | 4,495,730 | 9,420 | 299,674 | 384.2 | 493,178 | 60.9 | 39.1 | 11 | 87.2 | 8,773 | 71.6 |
| De Witt | 186 | -4.9 | 369 | 0.2 | 177.1 | 3,094,024 | 8,387 | 202,905 | 120.4 | 238,875 | 91.6 | 8.4 | 5 | 79.6 | 4,331 | 69.0 |
| Douglas | 245 | -6.9 | 408 | 0.0 | 236.6 | 3,701,396 | 9,071 | 203,298 | 159.5 | 265,888 | 93.9 | 6.1 | 7 | 64.3 | 3,836 | 57.2 |
| DuPage | 2 | -70.2 | 28 | 0.0 | 1.6 | 471,456 | 16,807 | 38,061 | 3.9 | 50,377 | 97.6 | 2.4 | NA | 92.2 | 89 | 10.4 |
| Edgar | 318 | -9.5 | 499 | D | 299.5 | 3,708,748 | 7,425 | 289,955 | D | D | D | D | NA | 80.4 | 7,464 | 82.4 |
| Edwards | 112 | 4.7 | 384 | D | 98.6 | 2,062,863 | 5,372 | 228,647 | 61.1 | 209,842 | 76.2 | 23.8 | NA | 78.0 | 3,203 | 80.4 |
| Effingham | 299 | 4.3 | 251 | 0.4 | 260.7 | 1,657,000 | 6,603 | 179,393 | 195.1 | 163,505 | 64.8 | 35.2 | NA | 79.2 | 9,626 | 73.8 |
| Fayette | 349 | 15.1 | 282 | 0.2 | 297.8 | 1,520,838 | 5,398 | 170,568 | 164.9 | 133,117 | 88.9 | 11.1 | NA | 70.2 | 6,528 | 63.9 |
| Ford | 270 | -12.3 | 479 | 1.0 | 262.7 | 3,722,553 | 7,769 | 260,649 | 190.7 | 338,188 | 83.1 | 16.9 | 4 | 80.1 | 3,095 | 70.0 |
| Franklin | 174 | -4.2 | 292 | 0.1 | 148.8 | 1,300,260 | 4,460 | 160,135 | 83.7 | 140,460 | 77.6 | 22.4 | NA | 67.6 | 3,569 | 54.4 |
| Fulton | 402 | 13.4 | 414 | 1.0 | 305.5 | 2,561,328 | 6,193 | 194,999 | 220.4 | 226,489 | 77.0 | 23.0 | 2 | 76.1 | 4,859 | 59.8 |
| Gallatin | 178 | -4.5 | 1,078 | 28.0 | 161.3 | 5,719,128 | 5,307 | 451,878 | 92.4 | 559,933 | 97.9 | 2.1 | NA | 79.4 | 4,534 | 80.6 |
| Greene | 328 | 13.1 | 448 | 0.8 | 257.8 | 2,768,122 | 6,184 | 222,206 | 183.4 | 250,139 | 78.7 | 21.3 | 5 | 77.5 | 4,745 | 72.7 |
| Grundy | 233 | 7.4 | 566 | D | 224.9 | 5,018,301 | 8,868 | 344,786 | 135.5 | 328,791 | 97.6 | 2.4 | NA | 84.7 | 1,375 | 49.3 |
| Hamilton | 201 | -10.2 | 363 | NA | 174.4 | 1,747,933 | 4,810 | 179,660 | 86.2 | 156,243 | 95.1 | 4.9 | 1 | 61.4 | 6,053 | 83.5 |
| Hancock | 455 | 17.9 | 411 | 6.8 | 378.2 | 2,911,061 | 7,091 | 222,490 | 320.2 | 288,772 | 69.7 | 30.3 | NA | 71.6 | 4,399 | 59.0 |

# Table B. States and Counties — Water Use, Wholesale Trade, Retail Trade, and Real Estate

| STATE County | Water use, 2015 | | Wholesale Trade[1], 2017 | | | | Retail Trade[2], 2017 | | | | Real estate and rental and leasing,[2] 2017 | | | |
|---|---|---|---|---|---|---|---|---|---|---|---|---|---|---|
| | Public supply water withdrawn (mil gal/day) | Public supply gallons withdrawn per person per day | Number of establishments | Number of employees | Sales (mil dol) | Average payroll (mil dol) | Number of establishments | Number of employees | Sales (mil dol) | Average payroll (mil dol) | Number of establishments | Number of employees | Sales (mil dol) | Average payroll (mil dol) |
| | 133 | 134 | 135 | 136 | 137 | 138 | 139 | 140 | 141 | 142 | 143 | 144 | 145 | 146 |
| **IDAHO—Cont'd** | | | | | | | | | | | | | | |
| Caribou | 1.56 | 230.4 | 13 | 61 | 169.4 | 2.5 | 33 | 290 | 92.4 | 7.7 | 5 | D | 1.5 | D |
| Cassia | 3.66 | 155.7 | 27 | 451 | 909.5 | 24.7 | 121 | 1,180 | 341.0 | 29.7 | 28 | 51 | 9.2 | 0.9 |
| Clark | 0.17 | 193.2 | NA | NA | NA | NA | NA | NA | NA | NA | NA | NA | NA | NA |
| Clearwater | 1.04 | 122.4 | 6 | 126 | 60.2 | 7.6 | 37 | 279 | 71.0 | 6.6 | D | D | D | 0.4 |
| Custer | 0.95 | 232.4 | NA | NA | NA | NA | 27 | 221 | 40.5 | 3.8 | 9 | 9 | 3.4 | 0.2 |
| Elmore | 5.62 | 217.2 | 17 | 104 | 61.3 | 4.2 | 77 | 991 | 285.4 | 25.7 | 17 | 34 | 5.8 | 0.9 |
| Franklin | 11.62 | 888.8 | 12 | 142 | 293.7 | 7.5 | 45 | 575 | 136.8 | 11.3 | 13 | 17 | 1.9 | 0.2 |
| Fremont | 1.34 | 104.5 | 13 | 115 | 108.5 | 4.8 | 43 | 290 | 91.3 | 7.8 | 16 | 18 | 4.0 | 0.7 |
| Gem | 0.86 | 51.0 | D | D | D | 1.5 | 46 | 467 | 123.3 | 11.7 | 16 | 26 | 4.8 | 0.7 |
| Gooding | 3.59 | 234.9 | 20 | 163 | 277.2 | 9.1 | 45 | 351 | 95.3 | 8.3 | 10 | 18 | 4.4 | 0.4 |
| Idaho | 0.93 | 57.2 | 14 | 58 | 57.1 | 2.5 | 65 | 529 | 98.0 | 12.0 | 13 | D | 1.3 | D |
| Jefferson | 1.07 | 39.4 | 27 | D | 1,597.0 | D | 60 | 644 | 209.6 | 17.7 | D | D | D | D |
| Jerome | 3.67 | 160.9 | 31 | 319 | 496.7 | 18.3 | 72 | 881 | 281.7 | 26.3 | 15 | 18 | 4.7 | 0.5 |
| Kootenai | 34.69 | 230.7 | 135 | 1,540 | 1,066.9 | 77.1 | 600 | 8,570 | 2,942.5 | 255.1 | 301 | 731 | 181.1 | 28.7 |
| Latah | 4.20 | 108.3 | 18 | 377 | 247.0 | 22.0 | 131 | 1,838 | 352.6 | 41.5 | 50 | 165 | 19.6 | 3.9 |
| Lemhi | 1.63 | 210.7 | 6 | 54 | 45.5 | 2.4 | 39 | 396 | 109.3 | 11.0 | 13 | 22 | 3.0 | 0.7 |
| Lewis | 0.53 | 139.9 | 7 | D | 81.2 | D | 20 | 202 | 39.5 | 3.6 | 4 | 11 | 1.2 | 0.3 |
| Lincoln | 0.40 | 75.5 | NA | NA | NA | NA | 13 | 47 | 16.2 | 1.4 | NA | NA | NA | NA |
| Madison | 3.39 | 88.6 | 42 | 729 | 343.7 | 27.3 | 191 | 1,749 | 496.0 | 45.6 | 59 | 261 | 32.8 | 5.4 |
| Minidoka | 2.81 | 137.3 | 49 | 774 | 533.2 | 38.3 | 61 | 751 | 215.4 | 18.5 | 12 | 32 | 7.4 | 1.5 |
| Nez Perce | 10.07 | 251.4 | 48 | 560 | 487.5 | 25.4 | 205 | 2,557 | 717.6 | 71.2 | 45 | 139 | 27.2 | 4.6 |
| Oneida | 0.72 | 168.2 | NA | NA | NA | NA | 18 | 162 | 37.9 | 2.7 | 5 | D | 1.6 | D |
| Owyhee | 0.73 | 64.5 | D | D | D | D | 24 | 160 | 41.2 | 4.6 | D | D | D | 0.1 |
| Payette | 1.94 | 84.7 | 22 | 167 | 86.4 | 6.6 | 56 | 480 | 158.5 | 13.1 | 29 | 41 | 7.3 | 1.1 |
| Power | 2.15 | 281.1 | 14 | 220 | 200.9 | 9.0 | 17 | 175 | 31.9 | 3.5 | 9 | D | 1.9 | D |
| Shoshone | 2.48 | 199.5 | 10 | D | 36.5 | D | 56 | 1,096 | 730.0 | 44.3 | D | D | D | 0.7 |
| Teton | 0.78 | 73.8 | 7 | 42 | 33.8 | 1.6 | 44 | 385 | 99.4 | 9.5 | 38 | 45 | 14.1 | 2.1 |
| Twin Falls | 15.74 | 191.1 | 132 | 1,309 | 836.6 | 63.1 | 371 | 5,189 | 1,549.2 | 138.9 | 142 | 352 | 70.5 | 10.5 |
| Valley | 1.55 | 153.4 | 13 | 59 | 24.2 | 2.9 | 76 | 457 | 123.8 | 11.8 | 56 | 122 | 28.4 | 3.7 |
| Washington | 0.74 | 74.1 | 8 | 147 | 183.4 | 4.5 | 31 | 289 | 102.2 | 6.8 | 7 | 8 | 2.0 | 0.4 |
| **ILLINOIS** | 1,475.66 | 114.7 | 15,409 | 272,391 | 311,140.6 | 19,071.8 | 38,189 | 629,878 | 173,473.9 | 16,246.9 | 13,589 | 82,763 | 32,355.2 | 4,824.5 |
| Adams | 10.35 | 154.4 | D | D | D | D | 286 | 4,946 | 1,095.9 | 115.6 | 64 | 263 | 42.9 | 7.3 |
| Alexander | 1.61 | 237.5 | D | D | D | 0.7 | 18 | 87 | 18.6 | 1.6 | NA | NA | NA | NA |
| Bond | 1.30 | 76.7 | 19 | 267 | 211.9 | 11.1 | 37 | 377 | 101.3 | 7.1 | 7 | 16 | 1.2 | 0.2 |
| Boone | 3.56 | 66.4 | 29 | 280 | 132.8 | 13.7 | 99 | 1,326 | 395.8 | 34.4 | 25 | 60 | 12.3 | 1.8 |
| Brown | 0.04 | 5.9 | D | D | D | D | 15 | 133 | 29.2 | 2.9 | NA | NA | NA | NA |
| Bureau | 2.96 | 88.1 | 48 | 950 | 1,461.4 | 45.3 | 89 | 1,067 | 262.9 | 25.0 | D | D | D | 0.7 |
| Calhoun | 0.41 | 83.7 | D | D | D | 0.2 | 14 | 95 | 24.5 | 2.5 | NA | NA | NA | NA |
| Carroll | 1.03 | 70.5 | 29 | 211 | 362.7 | 11.3 | 57 | 495 | 98.7 | 10.5 | 10 | D | 2.6 | D |
| Cass | 1.10 | 85.6 | 14 | 147 | 279.6 | 8.0 | 44 | 440 | 114.1 | 10.2 | 5 | 19 | 3.2 | 0.6 |
| Champaign | 24.42 | 116.9 | 165 | 2,935 | 2,433.6 | 134.6 | 615 | 9,880 | 2,470.0 | 232.5 | 231 | 2,083 | 388.5 | 85.4 |
| Christian | 2.55 | 75.8 | 42 | 621 | 792.1 | 32.2 | 116 | 1,437 | 425.3 | 35.7 | 19 | 106 | 12.3 | 3.1 |
| Clark | 1.60 | 100.1 | 20 | 236 | 225.3 | 8.9 | 45 | 443 | 137.0 | 11.3 | 6 | 34 | 5.9 | 1.1 |
| Clay | 0.01 | 0.7 | 24 | 217 | 163.2 | 8.5 | 53 | 547 | 125.2 | 12.5 | 7 | 50 | 8.6 | 0.9 |
| Clinton | 6.98 | 184.7 | 44 | 385 | 298.3 | 15.9 | 131 | 1,745 | 485.5 | 49.4 | 19 | 198 | 46.1 | 7.4 |
| Coles | 4.10 | 78.1 | 46 | 474 | 692.8 | 24.8 | 188 | 2,860 | 861.6 | 69.8 | D | D | D | D |
| Cook | 834.08 | 159.2 | 5,775 | 96,186 | 104,269.8 | 7,100.3 | 14,620 | 245,527 | 66,513.5 | 6,535.4 | 6,605 | 47,645 | 19,438.6 | 3,040.5 |
| Crawford | 2.18 | 112.3 | 18 | 184 | 186.7 | 7.7 | 71 | 748 | 183.4 | 18.2 | 18 | 75 | 9.8 | 2.6 |
| Cumberland | 0.27 | 24.8 | D | D | D | D | 29 | 284 | 63.5 | 4.3 | NA | NA | NA | NA |
| DeKalb | 7.64 | 73.2 | 75 | 884 | 1,039.8 | 45.0 | 251 | 4,426 | 1,043.9 | 97.4 | 76 | 391 | 80.3 | 13.4 |
| De Witt | 1.31 | 80.6 | 26 | 325 | 849.6 | 21.1 | 55 | 704 | 232.0 | 19.4 | 9 | 24 | 1.3 | 0.3 |
| Douglas | 0.67 | 33.8 | 31 | 327 | 388.5 | 16.4 | 126 | 1,194 | 204.7 | 20.8 | 9 | 34 | 4.5 | 1.2 |
| DuPage | 6.80 | 7.3 | 2,325 | 49,609 | 66,305.4 | 3,795.2 | 3,144 | 63,797 | 19,876.7 | 1,790.7 | 1,508 | 10,695 | 6,075.0 | 727.4 |
| Edgar | 1.58 | 89.4 | 10 | 651 | 277.1 | 22.8 | 54 | 672 | 183.5 | 16.6 | 10 | 28 | 3.4 | 0.8 |
| Edwards | 0.04 | 6.1 | 13 | 174 | 135.5 | 6.3 | 24 | 179 | 42.5 | 3.5 | 3 | D | 0.8 | D |
| Effingham | 2.06 | 59.9 | 63 | 1,416 | 853.5 | 70.1 | 201 | 2,976 | 875.6 | 73.8 | 35 | 374 | 50.1 | 13.2 |
| Fayette | 1.27 | 57.6 | 26 | 361 | 635.8 | 14.9 | 78 | 798 | 224.2 | 20.1 | 8 | 25 | 2.3 | 0.7 |
| Ford | 1.44 | 104.8 | 33 | 372 | 381.4 | 16.7 | 49 | 561 | 125.9 | 11.8 | D | D | D | D |
| Franklin | 15.39 | 389.8 | 23 | 229 | 105.4 | 9.5 | 131 | 1,537 | 392.7 | 36.0 | 13 | 52 | 6.1 | 1.5 |
| Fulton | 2.55 | 71.4 | 21 | 204 | 221.2 | 9.5 | 105 | 1,396 | 305.4 | 29.3 | 14 | 52 | 5.8 | 1.0 |
| Gallatin | 3.25 | 617.3 | 5 | D | 24.2 | D | 12 | 51 | 9.4 | 1.1 | NA | NA | NA | NA |
| Greene | 0.70 | 52.9 | 20 | 183 | 174.0 | 7.9 | 42 | 391 | 117.2 | 8.4 | 3 | 1 | 0.2 | 0.0 |
| Grundy | 3.41 | 67.5 | 42 | 503 | 562.7 | 29.5 | 146 | 2,498 | 711.4 | 56.9 | 33 | 99 | 18.5 | 3.1 |
| Hamilton | 0.00 | 0.0 | 8 | 83 | 79.1 | 4.0 | 32 | 160 | 43.8 | 3.5 | 5 | 20 | 2.0 | 0.7 |
| Hancock | 1.32 | 71.2 | 31 | 315 | 352.5 | 15.2 | 59 | 545 | 100.8 | 11.3 | 5 | 20 | 1.4 | 0.2 |

1   Merchant wholesalers, except manufacturers' sales branches and offices.   2. Employer establishments.

# Table B. States and Counties — Professional Services, Manufacturing, and Accommodation and Food Services

| STATE County | Professional, scientific, and technical services, 2017 | | | | Manufacturing, 2017 | | | | Accommodation and food services, 2017 | | | |
|---|---|---|---|---|---|---|---|---|---|---|---|---|
| | Number of establish-ments | Number of employees | Sales (mil dol) | Average payroll (mil dol) | Number of establish-ments | Number of employees | Sales (mil dol) | Average payroll (mil dol) | Number of establis-hments | Number of employees | Sales (mil dol) | Annual payroll (mil dol) |
| | 147 | 148 | 149 | 150 | 151 | 152 | 153 | 154 | 155 | 156 | 157 | 158 |
| IDAHO—Cont'd | | | | | | | | | | | | |
| Caribou | 8 | 29 | 2 | 0.7 | 8 | D | 524.2 | D | 18 | 112 | 3.9 | 1.2 |
| Cassia | 49 | 168 | 15 | 5.4 | 32 | 1,208 | 909.5 | 61.5 | D | D | D | D |
| Clark | NA | NA | NA | NA | NA | NA | NA | NA | NA | NA | NA | NA |
| Clearwater | 10 | 37 | 2 | 1.0 | 17 | 219 | 42.5 | 7.1 | 28 | 191 | 8.4 | 2.4 |
| Custer | 5 | 24 | 2 | 0.4 | 3 | D | 2.4 | 0.5 | 33 | 162 | 14.2 | 4.1 |
| Elmore | 23 | 110 | 14 | 4.0 | 15 | 581 | 301.3 | 22.2 | 52 | 665 | 34.6 | 9.4 |
| Franklin | 24 | 50 | 5 | 1.7 | 21 | 347 | 56.3 | 16.6 | 19 | 264 | 7.2 | 2.3 |
| Fremont | 10 | 39 | 3 | 1.1 | 10 | 24 | 7.2 | 1.4 | 35 | 186 | 17.5 | 4.1 |
| Gem | 31 | 95 | 12 | 4.2 | 16 | 102 | 9.7 | 3.1 | 24 | 267 | 13.6 | 3.6 |
| Gooding | 33 | 117 | 14 | 6.2 | 23 | 602 | 833.5 | 40.3 | 32 | 239 | 10.1 | 2.7 |
| Idaho | 24 | 64 | 6 | 1.7 | 26 | 524 | 222.3 | 24.4 | 44 | 243 | 13.5 | 3.6 |
| Jefferson | 35 | 119 | 15 | 3.8 | D | D | D | D | 34 | 392 | 23.8 | D |
| Jerome | 28 | 98 | 15 | 4.4 | 23 | 1,467 | 1,017.6 | 62.8 | 34 | 392 | 23.8 | 5.7 |
| Kootenai | 499 | 2,576 | 389 | 138.5 | 262 | 4,995 | 1,241.2 | 238.2 | 416 | 7,789 | 502.1 | 141.2 |
| Latah | 68 | 478 | 51 | 18.4 | 30 | 306 | 97.3 | 15.1 | 117 | 1,870 | 82.3 | 24.2 |
| Lemhi | 27 | 59 | 4 | 1.4 | 7 | 75 | 25.9 | 3.4 | 35 | 283 | 14.1 | 4.0 |
| Lewis | D | D | 0 | D | 5 | 147 | 56.6 | 5.4 | 17 | D | 6.6 | D |
| Lincoln | D | D | D | D | D | D | D | D | D | D | D | 0.3 |
| Madison | D | D | D | D | 30 | 1,076 | 217.5 | 33.6 | 65 | 1,301 | 53.5 | 14.1 |
| Minidoka | D | D | D | D | 30 | 989 | 488.4 | 52.3 | D | D | D | D |
| Nez Perce | D | D | D | D | 39 | 3,838 | 1,483.2 | 219.4 | 108 | 1,935 | 100.2 | 28.9 |
| Oneida | 5 | 9 | 1 | 0.3 | 5 | D | 9.7 | 2.4 | D | D | D | 0.9 |
| Owyhee | 9 | 22 | 4 | 0.7 | D | D | D | 6.2 | D | D | D | D |
| Payette | 41 | 175 | 16 | 6.4 | 33 | 1,459 | 300.0 | 42.1 | 30 | 306 | 11.3 | 3.1 |
| Power | 10 | 28 | 3 | 1.1 | D | D | D | D | 14 | 61 | 3.0 | 0.6 |
| Shoshone | 23 | 138 | 14 | 6.0 | 14 | 144 | 33.6 | 5.5 | D | D | D | 8.2 |
| Teton | 47 | 111 | 18 | 5.4 | 14 | 102 | 20.9 | 4.3 | 45 | 300 | 19.6 | 5.8 |
| Twin Falls | 231 | 1,073 | 139 | 47.0 | 93 | 3,691 | 1,605.6 | 172.2 | 202 | 3,687 | 180.1 | 50.1 |
| Valley | 44 | 96 | 13 | 4.3 | 16 | 78 | 5.7 | 1.9 | 79 | 995 | 73.7 | 23.4 |
| Washington | 16 | 60 | 5 | 1.5 | 14 | 537 | 128.2 | 21.5 | 23 | 221 | 7.9 | 2.1 |
| ILLINOIS | 38,649 | 394,230 | 85,307 | 33,131.2 | 13,162 | 527,862 | 241,484.3 | 30,116.3 | 29,025 | 536,245 | 35,314.4 | 10,188.5 |
| Adams | D | D | 74 | D | 80 | D | 3,070.5 | D | D | D | D | D |
| Alexander | D | D | D | D | D | 142 | D | 6.8 | D | D | D | D |
| Bond | 26 | 77 | 7 | 2.4 | 16 | 861 | 286.3 | 45.4 | 35 | 386 | 17.7 | 4.6 |
| Boone | D | D | D | D | 79 | 7,262 | 4,699.6 | 571.6 | 64 | 932 | 46.2 | 11.6 |
| Brown | 5 | 16 | 1 | 0.5 | 6 | 129 | 21.7 | 6.1 | D | D | D | D |
| Bureau | 43 | 156 | 19 | 6.1 | 39 | 1,589 | 580.6 | 86.4 | 73 | 727 | 33.5 | 9.1 |
| Calhoun | NA | NA | NA | NA | 3 | 8 | 2.0 | 0.5 | D | D | D | D |
| Carroll | 20 | 56 | 9 | 2.3 | 21 | 925 | 178.0 | 32.8 | 41 | 302 | 20.9 | 3.9 |
| Cass | 8 | 51 | 9 | 1.7 | D | D | D | D | 32 | 256 | 10.8 | 3.1 |
| Champaign | D | D | D | D | 118 | 6,336 | 2,578.8 | 313.1 | 570 | 11,228 | 574.4 | 169.5 |
| Christian | 37 | 498 | 22 | 8.1 | 20 | 710 | 471.1 | 44.0 | 74 | 948 | 39.3 | 11.2 |
| Clark | 20 | 176 | 23 | 12.3 | 17 | 1,202 | 768.9 | 54.9 | 38 | 427 | 18.6 | 5.0 |
| Clay | 15 | 85 | 5 | 2.0 | 19 | 2,029 | 973.0 | 102.9 | 26 | 259 | 11.8 | 3.1 |
| Clinton | D | D | D | D | 36 | 756 | 198.1 | 31.2 | 93 | 1,082 | 49.8 | 13.0 |
| Coles | 61 | 411 | 39 | 15.6 | D | D | D | D | 128 | 2,124 | 89.9 | 25.1 |
| Cook | 19,322 | 245,667 | 60,907 | 23,107.3 | 4,745 | 181,875 | 74,966.9 | 10,425.4 | 12,521 | 250,110 | 19,616.0 | 5,708.6 |
| Crawford | 29 | 265 | 23 | 9.3 | D | 1,871 | D | 139.3 | 36 | 555 | 23.0 | 5.6 |
| Cumberland | 6 | 17 | 1 | 0.4 | D | D | D | D | 13 | 122 | 3.3 | 1.1 |
| DeKalb | 153 | 1,071 | 180 | 53.8 | 113 | 3,564 | 1,308.4 | 181.3 | 210 | 3,331 | 166.4 | 46.2 |
| De Witt | 27 | 259 | 32 | 13.1 | 12 | 567 | 275.8 | 31.3 | 47 | 435 | 23.0 | 6.3 |
| Douglas | 28 | 96 | 8 | 2.6 | 76 | 2,647 | 1,010.2 | 139.4 | 66 | 760 | 36.7 | 9.7 |
| DuPage | D | D | D | D | 1,640 | 55,728 | 18,792.6 | 3,300.0 | 2,231 | 44,640 | 2,932.1 | 850.4 |
| Edgar | 28 | 96 | 11 | 3.5 | 20 | 2,884 | 936.8 | 110.8 | 38 | 364 | 16.8 | 4.5 |
| Edwards | D | D | 2 | D | D | D | D | D | D | D | D | 0.4 |
| Effingham | 68 | 425 | 52 | 18.7 | 63 | 3,204 | 791.0 | 145.1 | 116 | 2,099 | 109.1 | 30.2 |
| Fayette | 24 | 119 | 10 | 4.1 | 16 | 276 | 89.3 | 10.4 | 44 | 606 | 24.3 | 7.6 |
| Ford | 17 | 106 | 12 | 4.2 | 19 | 727 | 529.7 | 34.5 | 34 | 321 | 14.1 | 4.3 |
| Franklin | 45 | 213 | 28 | 7.9 | 39 | 510 | 122.4 | 23.1 | 83 | 966 | 42.6 | 10.7 |
| Fulton | 36 | 233 | 23 | 9.9 | 16 | D | 19.5 | 4.6 | 76 | 829 | 33.4 | 9.4 |
| Gallatin | 8 | 21 | 2 | 0.8 | 5 | 10 | 1.9 | 0.5 | 5 | D | 1.9 | D |
| Greene | 11 | 98 | 9 | 4.7 | 11 | 79 | 38.4 | 4.9 | 23 | 186 | 6.8 | 2.0 |
| Grundy | 91 | 1,008 | 119 | 58.7 | 48 | 1,387 | 1,696.9 | 117.7 | 133 | 1,891 | 101.6 | 26.4 |
| Hamilton | 14 | 37 | 3 | 1.0 | 7 | 58 | 11.7 | 2.4 | 9 | 130 | 4.8 | 1.5 |
| Hancock | 20 | 166 | 52 | 12.4 | 21 | 622 | 400.2 | 26.9 | 35 | 184 | 10.3 | 2.9 |

# Health Care and Social Assistance, Other Services, Nonemployer Businesses, and Residential Construction

| STATE County | Health care and social assistance, 2017 | | | | Other services, 2017 | | | | Nonemployer businesses, 2018 | | Value of residential construction authorized by building permits, 2020 | |
|---|---|---|---|---|---|---|---|---|---|---|---|---|
| | Number of establishments | Number of employees | Receipts (mil dol) | Annual payroll (mil dol) | Number of establishments | Number of employees | Receipts (mil dol) | Annual payroll (mil dol) | Number | Receipts (mil dol) | New construction ($1,000) | Number of housing units |
| | 159 | 160 | 161 | 162 | 163 | 164 | 165 | 166 | 167 | 168 | 169 | 170 |
| IDAHO—Cont'd | | | | | | | | | | | | |
| Caribou | 21 | 310 | 27.7 | 12.9 | D | D | D | D | 466 | 16.9 | 4,011 | 19 |
| Cassia | 81 | 1,317 | 123.7 | 43.5 | D | D | D | D | 1,581 | 92.2 | 20,357 | 112 |
| Clark | NA | NA | NA | NA | NA | NA | NA | NA | 76 | 2.6 | 0 | 0 |
| Clearwater | 23 | 527 | 49.3 | 25.9 | 10 | 26 | 3.5 | 1.0 | 511 | 21.1 | 3,489 | 35 |
| Custer | 11 | D | 4.0 | D | D | D | 1.3 | D | 464 | 15.2 | 2,855 | 11 |
| Elmore | 45 | 759 | 73.7 | 28.1 | 31 | 106 | 8.7 | 2.3 | 1,354 | 50.0 | 16,665 | 86 |
| Franklin | 27 | 437 | 33.1 | 15.0 | D | D | D | D | 1,094 | 51.6 | 16,996 | 91 |
| Fremont | 22 | 254 | 19.2 | 8.5 | 15 | 53 | 8.4 | 1.9 | 1,230 | 61.5 | 26,304 | 136 |
| Gem | 38 | 757 | 56.8 | 26.7 | 25 | 78 | 8.3 | 2.2 | 1,353 | 54.6 | 23,104 | 148 |
| Gooding | 35 | 537 | 45.1 | 20.3 | 22 | 95 | 11.1 | 2.8 | 903 | 42.3 | 6,433 | 32 |
| Idaho | 28 | 822 | 67.7 | 31.0 | D | D | D | D | 1,306 | 54.4 | 1,285 | 8 |
| Jefferson | D | D | D | 7.7 | D | D | D | D | 2,488 | 121.5 | 53,910 | 291 |
| Jerome | 39 | 506 | 40.8 | 18.3 | 40 | 185 | 21.4 | 5.1 | 1,204 | 58.4 | 14,024 | 94 |
| Kootenai | 528 | 10,600 | 1,069.8 | 492.9 | 284 | 1,450 | 139.3 | 41.4 | 13,392 | 632.3 | 561,941 | 2,929 |
| Latah | 93 | 1,569 | 137.4 | 56.8 | 64 | 384 | 34.1 | 10.1 | 2,695 | 103.9 | 29,275 | 122 |
| Lemhi | 29 | 494 | 46.6 | 20.4 | 20 | 73 | 7.8 | 2.1 | 795 | 32.0 | 4,439 | 36 |
| Lewis | 13 | D | 4.0 | D | NA | NA | NA | NA | 268 | 10.8 | 1,659 | 9 |
| Lincoln | D | D | D | D | D | D | D | D | 260 | 12.6 | 2,268 | 15 |
| Madison | 111 | 1,846 | 170.0 | 55.9 | 35 | 163 | 13.8 | 3.5 | 3,199 | 135.6 | 109,363 | 648 |
| Minidoka | 32 | 462 | 39.1 | 17.3 | D | D | D | D | 1,250 | 66.8 | 18,088 | 173 |
| Nez Perce | 139 | 3,288 | 309.4 | 124.4 | 71 | 473 | 39.9 | 13.3 | 2,115 | 89.8 | 21,518 | 99 |
| Oneida | 11 | 176 | 13.9 | 5.4 | D | D | D | D | 379 | 14.3 | 3,520 | 22 |
| Owyhee | 20 | 229 | 12.5 | 5.7 | D | D | D | D | 753 | 33.5 | 8,205 | 37 |
| Payette | 53 | 628 | 52.5 | 23.6 | D | D | D | D | 1,465 | 68.2 | 34,723 | 250 |
| Power | 15 | 162 | 13.6 | 5.7 | 11 | 41 | 5.1 | 1.2 | 375 | 18.4 | 2,250 | 9 |
| Shoshone | 31 | 531 | 41.0 | 17.3 | 17 | 47 | 3.9 | 1.3 | 803 | 25.0 | 3,993 | 29 |
| Teton | 35 | 291 | 29.0 | 12.7 | 24 | 82 | 10.9 | 3.4 | 1,647 | 62.7 | 66,469 | 247 |
| Twin Falls | 393 | 6,474 | 682.8 | 245.1 | 154 | 921 | 88.8 | 27.2 | 6,019 | 289.8 | 156,247 | 959 |
| Valley | 31 | 499 | 54.2 | 23.8 | 31 | 138 | 15.0 | 3.2 | 1,363 | 62.0 | 100,849 | 286 |
| Washington | 24 | 421 | 30.7 | 11.6 | D | D | D | D | 608 | 22.9 | 4,705 | 42 |
| ILLINOIS | 34,235 | 817,733 | 97,626.1 | 38,088.6 | 24,296 | 171,261 | 29,599.4 | 7,164.3 | 996,670 | 44,899.3 | 3,646,298 | 18,058 |
| Adams | D | D | D | 313.6 | D | D | 95.2 | D | 3,922 | 158.7 | 14,388 | 66 |
| Alexander | D | D | D | D | D | D | D | D | 255 | 8.0 | 0 | 0 |
| Bond | D | D | D | D | D | D | D | D | 932 | 31.3 | 1,457 | 9 |
| Boone | 64 | 947 | 87.7 | 32.0 | 71 | 343 | 38.7 | 10.2 | 3,036 | 126.3 | 7,588 | 43 |
| Brown | 10 | 185 | 12.5 | 6.4 | 11 | 21 | 2.4 | 0.5 | 324 | 12.7 | 0 | 0 |
| Bureau | 70 | 2,049 | 200.8 | 83.5 | D | D | D | D | 1,817 | 71.9 | 14,128 | 99 |
| Calhoun | D | D | D | D | NA | NA | NA | NA | 309 | 11.8 | 965 | 6 |
| Carroll | 23 | 489 | 24.4 | 9.9 | 29 | 161 | 14.1 | 4.0 | 957 | 35.7 | 9,221 | 34 |
| Cass | 13 | 333 | 14.2 | 7.2 | D | D | 7.5 | D | 691 | 20.0 | 1,102 | 8 |
| Champaign | D | D | D | D | 301 | 2,176 | 749.9 | 84.3 | 12,462 | 490.4 | 251,520 | 1,509 |
| Christian | 55 | 1,685 | 136.8 | 52.9 | 57 | 277 | 18.5 | 5.3 | 1,728 | 64.4 | 1,911 | 13 |
| Clark | 24 | 301 | 18.8 | 7.6 | 19 | 71 | 5.2 | 1.1 | 1,003 | 38.9 | 350 | 2 |
| Clay | 44 | 787 | 54.7 | 26.8 | 31 | 88 | 8.8 | 1.8 | 936 | 36.2 | 1,213 | 5 |
| Clinton | 92 | 1,854 | 157.6 | 60.6 | 78 | 276 | 27.6 | 7.2 | 2,356 | 86.8 | 29,020 | 102 |
| Coles | 161 | 4,918 | 485.4 | 193.3 | 86 | 374 | 32.0 | 10.2 | 2,581 | 101.0 | 612 | 4 |
| Cook | 15,121 | 382,756 | 45,827.4 | 18,285.6 | 10,518 | 83,813 | 18,747.7 | 4,054.0 | 492,996 | 22,554.9 | 961,806 | 5,579 |
| Crawford | 39 | 853 | 67.0 | 31.4 | 31 | 151 | 13.0 | 3.5 | 1,305 | 52.4 | 0 | 0 |
| Cumberland | 17 | 519 | 18.4 | 8.4 | 18 | 60 | 4.7 | 1.1 | 723 | 22.9 | 150 | 1 |
| DeKalb | 220 | 4,937 | 570.2 | 203.4 | 152 | 766 | 69.9 | 19.3 | 6,184 | 236.9 | 25,647 | 152 |
| De Witt | 25 | 508 | 41.0 | 16.5 | 20 | 70 | 7.6 | 1.5 | 904 | 28.8 | 4,594 | 19 |
| Douglas | 31 | 285 | 21.2 | 9.6 | 34 | 160 | 16.5 | 5.3 | 1,580 | 65.4 | 5,575 | 24 |
| DuPage | 3,509 | 66,417 | 8,717.6 | 3,477.7 | 2,142 | 18,378 | 2,422.5 | 809.5 | 86,929 | 4,703.1 | 428,242 | 1,671 |
| Edgar | 31 | 926 | 76.6 | 37.2 | 24 | 93 | 10.1 | 2.5 | 883 | 29.9 | 0 | 0 |
| Edwards | 12 | 96 | 4.3 | 2.3 | 17 | 54 | 5.5 | 1.3 | 423 | 14.2 | NA | NA |
| Effingham | 134 | 2,859 | 393.6 | 120.6 | 106 | 1,034 | 75.8 | 39.1 | 2,729 | 124.4 | 11,172 | 52 |
| Fayette | 47 | 833 | 54.4 | 23.9 | 34 | 120 | 14.6 | 3.2 | 1,224 | 48.6 | 710 | 3 |
| Ford | 35 | 1,162 | 126.8 | 50.9 | 30 | 110 | 11.1 | 2.6 | 910 | 33.9 | 2,878 | 12 |
| Franklin | 67 | 975 | 87.1 | 33.1 | 54 | 187 | 23.7 | 4.7 | 2,099 | 74.5 | 1,028 | 12 |
| Fulton | 69 | 1,904 | 160.5 | 71.9 | 45 | 179 | 17.0 | 4.4 | 1,614 | 49.6 | 5,015 | 38 |
| Gallatin | 8 | 88 | 4.3 | 2.0 | D | D | 3.2 | D | 303 | 12.8 | NA | NA |
| Greene | 17 | 334 | 22.4 | 11.2 | 15 | 44 | 4.8 | 1.0 | 767 | 27.5 | 475 | 2 |
| Grundy | 145 | 2,271 | 298.0 | 107.2 | 102 | 1,140 | 149.6 | 59.9 | 3,007 | 130.2 | 26,755 | 127 |
| Hamilton | 22 | 444 | 30.7 | 13.8 | 20 | 74 | 7.3 | 2.1 | 567 | 18.8 | NA | NA |
| Hancock | 29 | 507 | 46.6 | 19.7 | D | D | D | D | 1,235 | 46.3 | 644 | 3 |

# Table B. States and Counties — Government Employment and Payroll, and Local Government Finances

| | Government employment and payroll, 2017 | | | | | | | | | Local government finances, 2017 | | | | |
| | | | March payroll (percent of total) | | | | | | | General revenue | | | | |
| | | | | | | | | | | | | Taxes | | |
| STATE County | Full-time equivalent employees | March payroll (dollars) | Administration, judicial, and legal | Police and corrections | Fire protection | Highways and transportation | Health and welfare | Natural resources and utilities | Education and libraries | Total (mil dol) | Intergovernmental (mil dol) | Total (mil dol) | Per capita[1] (dollars) Total | Per capita[1] (dollars) Property |
| | 171 | 172 | 173 | 174 | 175 | 176 | 177 | 178 | 179 | 180 | 181 | 182 | 183 | 184 |
|---|---|---|---|---|---|---|---|---|---|---|---|---|---|---|
| **IDAHO—Cont'd** | | | | | | | | | | | | | | |
| Caribou | 572 | 1,914,185 | 6.3 | 6.5 | 0.8 | 4.7 | 45.7 | 2.6 | 32.8 | 34.1 | 14.6 | 11.8 | 1,696 | 1,650 |
| Cassia | 991 | 2,996,775 | 7.0 | 10.5 | 1.8 | 4.2 | 0.1 | 7.9 | 64.5 | 99.4 | 43.8 | 23.6 | 999 | 781 |
| Clark | 75 | 232,218 | 8.2 | 10.3 | 0.0 | 11.7 | 0.0 | 4.2 | 47.4 | 5.5 | 2.2 | 3.0 | 3,416 | 3,206 |
| Clearwater | 394 | 1,125,348 | 14.2 | 11.9 | 0.2 | 6.6 | 2.3 | 7.2 | 50.6 | 37.5 | 15.0 | 16.8 | 1,943 | 1,848 |
| Custer | 196 | 574,965 | 18.3 | 8.6 | 0.2 | 10.2 | 1.3 | 7.2 | 49.7 | 16.0 | 9.3 | 4.0 | 957 | 942 |
| Elmore | 863 | 2,481,291 | 9.8 | 13.7 | 0.8 | 6.9 | 0.7 | 7.6 | 57.9 | 76.7 | 36.9 | 23.5 | 873 | 845 |
| Franklin | 650 | 2,076,070 | 4.6 | 5.0 | 0.0 | 3.0 | 34.8 | 3.4 | 48.5 | 35.5 | 24.0 | 7.3 | 542 | 514 |
| Fremont | 535 | 1,694,793 | 11.3 | 18.2 | 2.5 | 5.7 | 2.5 | 7.2 | 51.5 | 41.5 | 17.2 | 19.2 | 1,461 | 1,373 |
| Gem | 552 | 1,556,250 | 10.0 | 12.9 | 1.5 | 3.8 | 0.2 | 9.2 | 61.5 | 43.9 | 21.6 | 17.2 | 993 | 961 |
| Gooding | 630 | 1,677,452 | 10.4 | 10.2 | 2.1 | 4.8 | 3.5 | 7.5 | 59.6 | 45.5 | 26.9 | 12.0 | 793 | 760 |
| Idaho | 583 | 2,063,196 | 5.1 | 6.8 | 0.0 | 8.7 | 29.3 | 3.0 | 42.7 | 53.6 | 22.5 | 12.6 | 772 | 765 |
| Jefferson | 1,004 | 2,656,177 | 5.5 | 8.7 | 2.3 | 3.0 | 0.2 | 2.3 | 77.1 | 62.2 | 44.8 | 13.8 | 484 | 481 |
| Jerome | 735 | 2,359,346 | 10.4 | 10.8 | 3.0 | 5.1 | 0.5 | 5.7 | 63.0 | 71.5 | 36.6 | 19.9 | 837 | 801 |
| Kootenai | 7,757 | 33,973,530 | 5.6 | 7.5 | 3.1 | 2.0 | 43.3 | 3.4 | 34.7 | 903.2 | 235.7 | 172.7 | 1,098 | 1,032 |
| Latah | 1,054 | 4,213,007 | 10.8 | 11.5 | 0.8 | 5.6 | 1.2 | 7.5 | 60.1 | 101.0 | 40.2 | 38.9 | 979 | 922 |
| Lemhi | 405 | 1,523,806 | 6.3 | 7.1 | 1.2 | 4.3 | 47.3 | 5.0 | 24.7 | 41.9 | 15.7 | 4.9 | 628 | 557 |
| Lewis | 247 | 708,412 | 12.5 | 8.3 | 0.0 | 8.9 | 0.3 | 5.5 | 63.9 | 18.5 | 11.0 | 4.5 | 1,160 | 1,151 |
| Lincoln | 212 | 611,303 | 11.7 | 7.9 | 0.7 | 8.8 | 0.3 | 5.8 | 61.8 | 16.1 | 9.9 | 3.8 | 710 | 679 |
| Madison | 1,187 | 3,602,076 | 7.3 | 15.1 | 0.5 | 4.2 | 3.8 | 4.2 | 62.9 | 188.3 | 82.1 | 37.3 | 948 | 912 |
| Minidoka | 985 | 3,305,590 | 5.5 | 7.1 | 2.1 | 0.9 | 28.7 | 11.8 | 42.3 | 69.4 | 33.4 | 12.6 | 610 | 586 |
| Nez Perce | 1,420 | 5,752,990 | 8.6 | 12.6 | 6.6 | 3.8 | 3.9 | 7.3 | 51.8 | 146.3 | 62.4 | 52.8 | 1,310 | 1,245 |
| Oneida | 171 | 533,003 | 12.7 | 8.6 | 0.2 | 5.9 | 0.4 | 10.8 | 59.3 | 19.5 | 8.3 | 7.4 | 1,676 | 1,648 |
| Owyhee | 397 | 1,244,002 | 10.1 | 10.8 | 1.6 | 5.5 | 0.2 | 4.6 | 66.7 | 41.2 | 24.2 | 10.6 | 912 | 896 |
| Payette | 839 | 2,498,050 | 9.3 | 11.4 | 0.9 | 4.6 | 0.4 | 6.3 | 66.0 | 63.2 | 32.4 | 21.6 | 932 | 816 |
| Power | 503 | 1,520,346 | 4.3 | 5.9 | 0.2 | 7.8 | 24.2 | 4.4 | 52.2 | 46.0 | 19.4 | 16.7 | 2,197 | 2,168 |
| Shoshone | 634 | 2,055,612 | 7.9 | 7.7 | 5.0 | 7.5 | 21.1 | 6.6 | 43.5 | 57.1 | 21.8 | 16.0 | 1,277 | 1,232 |
| Teton | 337 | 1,217,589 | 12.3 | 5.8 | 7.6 | 3.9 | 0.0 | 5.1 | 63.1 | 51.0 | 16.6 | 15.6 | 1,360 | 1,244 |
| Twin Falls | 3,370 | 11,792,355 | 9.0 | 10.1 | 2.7 | 3.0 | 3.4 | 4.8 | 65.7 | 334.9 | 158.4 | 104.5 | 1,225 | 1,142 |
| Valley | 584 | 2,091,242 | 14.1 | 12.9 | 9.0 | 8.7 | 8.2 | 9.3 | 36.7 | 51.5 | 15.1 | 26.6 | 2,491 | 2,362 |
| Washington | 495 | 1,799,937 | 7.5 | 9.3 | 0.2 | 5.2 | 27.4 | 6.8 | 41.8 | 53.7 | 19.9 | 12.0 | 1,197 | 1,141 |
| **ILLINOIS** | X | X | X | X | X | X | X | X | X | X | X | X | X | X |
| Adams | 2,643 | 9,098,939 | 5.8 | 9.6 | 4.6 | 4.8 | 5.3 | 5.0 | 63.7 | 224.4 | 101.1 | 88.0 | 1,333 | 1,136 |
| Alexander | 305 | 979,817 | 8.5 | 11.0 | 0.0 | 13.3 | 6.9 | 2.0 | 54.8 | 21.8 | 13.9 | 5.1 | 806 | 733 |
| Bond | 530 | 1,889,362 | 10.2 | 12.9 | 1.5 | 5.9 | 5.8 | 4.7 | 58.5 | 47.0 | 18.6 | 18.2 | 1,093 | 1,021 |
| Boone | 1,618 | 7,647,159 | 6.0 | 10.3 | 2.5 | 1.9 | 1.1 | 3.9 | 72.4 | 181.0 | 69.6 | 95.1 | 1,778 | 1,649 |
| Brown | 225 | 848,596 | 12.4 | 7.8 | 0.6 | 8.0 | 5.8 | 1.8 | 63.4 | 16.3 | 5.9 | 7.5 | 1,140 | 937 |
| Bureau | 1,597 | 6,419,293 | 6.4 | 7.1 | 2.1 | 3.5 | 24.1 | 5.7 | 50.1 | 161.3 | 47.6 | 55.3 | 1,668 | 1,625 |
| Calhoun | 153 | 520,148 | 10.7 | 6.9 | 5.2 | 9.0 | 6.9 | 0.7 | 60.6 | 14.2 | 7.3 | 5.4 | 1,103 | 1,073 |
| Carroll | 510 | 1,913,321 | 10.1 | 11.2 | 3.0 | 5.1 | 1.3 | 4.8 | 63.6 | 54.7 | 16.8 | 30.8 | 2,125 | 2,082 |
| Cass | 521 | 1,645,754 | 9.1 | 6.5 | 0.9 | 6.3 | 1.2 | 4.4 | 68.7 | 49.2 | 25.0 | 16.0 | 1,283 | 1,219 |
| Champaign | 7,428 | 31,761,851 | 6.9 | 10.0 | 4.4 | 6.9 | 6.0 | 5.4 | 58.4 | 800.2 | 271.0 | 407.2 | 1,935 | 1,617 |
| Christian | 1,367 | 4,737,664 | 7.8 | 8.5 | 2.2 | 4.6 | 1.5 | 4.9 | 69.5 | 100.5 | 45.5 | 41.4 | 1,254 | 1,213 |
| Clark | 603 | 2,100,872 | 6.9 | 8.2 | 4.0 | 6.3 | 2.0 | 11.0 | 58.8 | 56.7 | 30.5 | 17.4 | 1,103 | 998 |
| Clay | 738 | 3,020,248 | 4.9 | 4.4 | 0.0 | 3.5 | 40.7 | 5.3 | 40.6 | 47.0 | 27.8 | 13.0 | 982 | 931 |
| Clinton | 1,017 | 4,028,845 | 9.3 | 11.1 | 0.2 | 4.9 | 3.6 | 7.4 | 62.7 | 88.7 | 31.2 | 43.5 | 1,155 | 1,129 |
| Coles | 2,736 | 10,830,347 | 5.2 | 7.0 | 3.9 | 2.1 | 1.7 | 2.8 | 76.5 | 177.8 | 65.9 | 63.8 | 1,243 | 1,137 |
| Cook | 205,528 | 1,216,640,423 | 5.4 | 14.9 | 5.4 | 10.0 | 5.1 | 6.3 | 49.8 | 36045.1 | 11878.6 | 18051.5 | 3,471 | 2,414 |
| Crawford | 965 | 3,932,672 | 3.5 | 5.7 | 1.2 | 4.0 | 45.2 | 1.6 | 38.0 | 104.9 | 20.9 | 30.2 | 1,593 | 1,577 |
| Cumberland | 341 | 1,506,830 | 9.1 | 5.3 | 3.0 | 5.7 | 2.3 | 3.2 | 71.0 | 28.6 | 14.4 | 10.4 | 954 | 932 |
| DeKalb | 3,861 | 16,075,945 | 5.7 | 12.8 | 5.5 | 3.7 | 5.5 | 4.7 | 61.0 | 445.5 | 134.4 | 243.6 | 2,336 | 1,993 |
| De Witt | 693 | 2,635,487 | 7.4 | 7.0 | 2.2 | 3.0 | 21.6 | 3.6 | 53.1 | 78.9 | 16.5 | 36.8 | 2,315 | 2,262 |
| Douglas | 708 | 2,533,662 | 12.4 | 7.5 | 0.9 | 3.4 | 0.8 | 2.7 | 72.2 | 70.7 | 25.4 | 38.5 | 1,962 | 1,878 |
| DuPage | 35,243 | 186,878,707 | 4.3 | 9.7 | 4.6 | 2.4 | 2.8 | 8.2 | 66.2 | 4840.5 | 983.3 | 3163.1 | 3,400 | 3,022 |
| Edgar | 693 | 2,517,272 | 6.7 | 9.4 | 3.4 | 4.5 | 3.8 | 3.6 | 68.2 | 55.7 | 25.2 | 24.6 | 1,415 | 1,273 |
| Edwards | 194 | 602,546 | 7.0 | 7.5 | 6.7 | 2.6 | 2.8 | 4.1 | 69.0 | 14.1 | 7.0 | 4.8 | 750 | 734 |
| Effingham | 1,157 | 4,609,810 | 7.7 | 10.5 | 2.7 | 4.5 | 1.6 | 3.9 | 68.0 | 106.9 | 44.0 | 46.0 | 1,347 | 1,283 |
| Fayette | 744 | 2,354,663 | 9.7 | 10.9 | 0.0 | 4.6 | 4.4 | 4.7 | 65.0 | 56.9 | 26.8 | 18.6 | 865 | 848 |
| Ford | 637 | 2,097,747 | 4.4 | 7.0 | 0.0 | 3.8 | 3.4 | 2.7 | 75.8 | 57.3 | 20.0 | 31.5 | 2,367 | 2,302 |
| Franklin | 1,590 | 6,048,883 | 2.8 | 5.4 | 11.6 | 2.4 | 15.2 | 7.4 | 54.4 | 152.4 | 70.8 | 35.4 | 908 | 797 |
| Fulton | 1,476 | 5,334,986 | 6.7 | 6.2 | 2.9 | 3.5 | 6.7 | 4.3 | 69.2 | 134.6 | 56.2 | 52.4 | 1,492 | 1,444 |
| Gallatin | 212 | 610,109 | 22.1 | 6.7 | 1.9 | 4.3 | 3.1 | 2.6 | 58.1 | 14.7 | 7.9 | 5.2 | 1,024 | 974 |
| Greene | 497 | 1,565,591 | 6.9 | 9.6 | 1.8 | 10.1 | 3.8 | 8.9 | 57.9 | 35.1 | 15.6 | 14.1 | 1,075 | 1,024 |
| Grundy | 2,319 | 8,628,263 | 6.9 | 8.5 | 3.4 | 2.8 | 2.3 | 2.4 | 71.3 | 245.3 | 63.3 | 156.2 | 3,085 | 2,993 |
| Hamilton | 399 | 1,349,402 | 7.5 | 3.5 | 0.0 | 3.5 | 38.5 | 4.4 | 41.2 | 38.1 | 11.6 | 6.7 | 825 | 824 |
| Hancock | 727 | 2,287,973 | 8.0 | 5.8 | 2.3 | 5.6 | 6.9 | 4.2 | 65.4 | 56.2 | 20.2 | 27.9 | 1,552 | 1,498 |

1. Based on the resident population estimated as of July 1 of the year shown.

# Table B. States and Counties — Local Government Finances, Government Employment, and Income Taxes

| STATE County | Direct general expenditure Total (mil dol) | Per capita[1] (dollars) | Percent of total for: Education | Health and hospitals | Police protection | Public welfare | Highways | Debt outstanding Total (mil dol) | Per capita[1] (dollars) | Government employment, 2019 Federal civilian | Federal military | State and local | Individual income tax returns, 2018 Number of returns | Mean adjusted gross income | Mean income tax |
|---|---|---|---|---|---|---|---|---|---|---|---|---|---|---|---|
| | 185 | 186 | 187 | 188 | 189 | 190 | 191 | 192 | 193 | 194 | 195 | 196 | 197 | 198 | 199 |
| IDAHO—Cont'd | | | | | | | | | | | | | | | |
| Caribou | 32 | 4,598 | 41.8 | 0.5 | 10.5 | 0.5 | 14.0 | 11.3 | 1,618 | 40 | 22 | 738 | 2,810 | 53,025 | 4,330 |
| Cassia | 98 | 4,157 | 58.8 | 0.2 | 3.5 | 0.6 | 4.5 | 106.2 | 4,489 | 135 | 74 | 1,462 | 9,860 | 49,485 | 4,152 |
| Clark | 3 | 3,341 | 78.8 | 0.0 | 1.1 | 0.0 | 2.7 | 1.0 | 1,091 | 25 | 3 | 105 | 310 | 26,203 | 1,713 |
| Clearwater | 65 | 7,495 | 23.2 | 0.8 | 3.3 | 0.4 | 45.1 | 8.7 | 1,011 | 143 | 25 | 787 | 3,430 | 51,255 | 4,239 |
| Custer | 17 | 4,152 | 39.7 | 1.9 | 5.8 | 0.0 | 18.3 | 0.9 | 223 | 143 | 13 | 296 | 1,990 | 49,649 | 4,133 |
| Elmore | 69 | 2,563 | 47.0 | 1.5 | 8.9 | 1.0 | 9.1 | 15.8 | 589 | 868 | 3,474 | 1,098 | 11,960 | 43,719 | 3,123 |
| Franklin | 34 | 2,550 | 63.9 | 1.7 | 7.1 | 0.5 | 6.8 | 6.6 | 491 | 33 | 43 | 1,066 | 5,610 | 51,223 | 3,444 |
| Fremont | 45 | 3,389 | 36.4 | 1.4 | 6.5 | 0.2 | 12.9 | 28.8 | 2,191 | 73 | 40 | 962 | 5,560 | 44,409 | 3,382 |
| Gem | 37 | 2,139 | 56.4 | 2.0 | 6.0 | 1.6 | 5.0 | 19.8 | 1,141 | 82 | 56 | 885 | 8,050 | 49,443 | 3,775 |
| Gooding | 49 | 3,238 | 46.2 | 1.9 | 4.6 | 0.5 | 9.2 | 45.1 | 2,981 | 50 | 47 | 910 | 6,370 | 35,404 | 3,491 |
| Idaho | 51 | 3,128 | 39.5 | 5.3 | 4.5 | 1.4 | 15.7 | 7.3 | 444 | 327 | 51 | 904 | 6,490 | 49,612 | 4,359 |
| Jefferson | 51 | 1,797 | 83.9 | 0.0 | 1.3 | 0.0 | 0.6 | 64.0 | 2,252 | 52 | 93 | 1,348 | 11,600 | 56,179 | 4,076 |
| Jerome | 78 | 3,293 | 50.9 | 0.7 | 5.6 | 0.7 | 1.7 | 87.2 | 3,670 | 47 | 76 | 1,037 | 9,590 | 43,785 | 3,349 |
| Kootenai | 910 | 5,783 | 24.0 | 51.5 | 3.8 | 0.2 | 2.5 | 282.6 | 1,796 | 671 | 514 | 11,075 | 79,490 | 63,810 | 6,669 |
| Latah | 100 | 2,511 | 45.7 | 0.3 | 8.4 | 0.4 | 9.1 | 40.2 | 1,012 | 167 | 140 | 6,716 | 16,470 | 60,361 | 5,666 |
| Lemhi | 35 | 4,523 | 21.8 | 46.0 | 4.2 | 0.8 | 6.2 | 16.3 | 2,079 | 214 | 25 | 630 | 3,690 | 48,183 | 4,379 |
| Lewis | 19 | 4,818 | 50.4 | 0.9 | 6.2 | 0.9 | 14.0 | 4.9 | 1,260 | 73 | 12 | 390 | 1,710 | 38,219 | 3,388 |
| Lincoln | 17 | 3,123 | 54.2 | 2.1 | 3.2 | 0.5 | 10.7 | 7.8 | 1,456 | 82 | 17 | 406 | 2,040 | 42,012 | 2,721 |
| Madison | 164 | 4,163 | 27.0 | 42.0 | 4.0 | 0.0 | 5.2 | 95.1 | 2,416 | 58 | 125 | 2,188 | 14,370 | 44,336 | 2,991 |
| Minidoka | 66 | 3,178 | 47.5 | 0.0 | 5.8 | 0.3 | 1.8 | 47.1 | 2,271 | 91 | 66 | 1,415 | 9,050 | 46,892 | 3,367 |
| Nez Perce | 143 | 3,557 | 37.2 | 6.2 | 8.4 | 0.2 | 7.5 | 115.4 | 2,863 | 201 | 125 | 4,059 | 18,690 | 56,680 | 4,917 |
| Oneida | 17 | 3,945 | 39.7 | 2.8 | 6.6 | 0.2 | 6.0 | 2.7 | 612 | 18 | 14 | 428 | 1,900 | 29,264 | 2,848 |
| Owyhee | 39 | 3,318 | 51.6 | 0.2 | 5.5 | 0.8 | 5.6 | 14.2 | 1,222 | 43 | 36 | 649 | 4,860 | 42,234 | 3,316 |
| Payette | 59 | 2,558 | 53.5 | 1.9 | 6.9 | 0.7 | 6.7 | 17.1 | 737 | 35 | 75 | 1,090 | 10,310 | 56,822 | 5,250 |
| Power | 41 | 5,366 | 39.9 | 19.6 | 3.6 | 0.2 | 8.5 | 4.7 | 615 | 19 | 24 | 651 | 3,220 | 45,211 | 3,438 |
| Shoshone | 55 | 4,415 | 35.2 | 21.4 | 4.9 | 0.2 | 11.2 | 30.8 | 2,459 | 69 | 40 | 839 | 5,570 | 44,826 | 3,596 |
| Teton | 44 | 3,826 | 35.2 | 2.4 | 2.7 | 0.1 | 5.6 | 19.6 | 1,710 | 42 | 38 | 457 | 5,260 | 65,351 | 6,824 |
| Twin Falls | 340 | 3,979 | 56.0 | 2.5 | 4.2 | 1.1 | 6.8 | 321.2 | 3,763 | 390 | 270 | 4,958 | 37,610 | 50,456 | 4,533 |
| Valley | 54 | 5,039 | 30.4 | 0.2 | 7.5 | 0.1 | 14.5 | 39.8 | 3,735 | 257 | 35 | 727 | 5,540 | 74,036 | 8,407 |
| Washington | 51 | 5,105 | 28.9 | 33.1 | 6.1 | 0.3 | 5.7 | 18.5 | 1,843 | 55 | 32 | 721 | 4,290 | 45,202 | 3,286 |
| ILLINOIS | X | X | X | X | X | X | X | X | X | 79,611 | 44,379 | 751,365 | 6,116,290 | 79,956 | 10,859 |
| Adams | 271 | 4,103 | 48.8 | 2.9 | 5.2 | 0.3 | 3.9 | 180.9 | 2,738 | 232 | 129 | 3,975 | 31,520 | 61,448 | 6,319 |
| Alexander | 25 | 3,936 | 49.0 | 1.0 | 6.6 | 0.0 | 7.1 | 4.5 | 711 | 22 | 11 | 319 | 2,280 | 39,582 | 2,634 |
| Bond | 43 | 2,554 | 48.0 | 7.0 | 7.3 | 0.2 | 7.9 | 44.9 | 2,698 | 287 | 30 | 736 | 6,880 | 53,498 | 4,540 |
| Boone | 176 | 3,286 | 58.6 | 1.5 | 6.6 | 0.7 | 6.7 | 124.3 | 2,323 | 66 | 107 | 2,213 | 25,820 | 66,624 | 7,253 |
| Brown | 18 | 2,713 | 52.8 | 1.1 | 6.3 | 0.0 | 11.5 | 11.0 | 1,658 | 42 | 9 | 420 | 2,290 | 53,841 | 4,583 |
| Bureau | 167 | 5,035 | 36.8 | 25.8 | 4.4 | 0.9 | 5.6 | 96.1 | 2,899 | 104 | 65 | 2,179 | 16,540 | 55,986 | 5,175 |
| Calhoun | 14 | 2,939 | 43.1 | 0.2 | 3.3 | 0.0 | 29.2 | 7.5 | 1,553 | 19 | 9 | 248 | 2,080 | 57,629 | 5,055 |
| Carroll | 54 | 3,702 | 50.3 | 0.6 | 5.3 | 1.2 | 10.3 | 42.1 | 2,911 | 441 | 28 | 708 | 7,300 | 55,145 | 4,964 |
| Cass | 47 | 3,734 | 46.3 | 11.0 | 4.5 | 0.2 | 5.9 | 28.6 | 2,288 | 55 | 24 | 857 | 6,190 | 48,398 | 3,607 |
| Champaign | 798 | 3,794 | 50.3 | 2.7 | 6.2 | 2.4 | 5.6 | 724.3 | 3,442 | 1,250 | 419 | 36,257 | 84,710 | 67,776 | 7,675 |
| Christian | 116 | 3,518 | 58.5 | 1.2 | 5.5 | 0.1 | 7.1 | 64.4 | 1,952 | 72 | 62 | 1,704 | 15,080 | 54,738 | 4,734 |
| Clark | 58 | 3,679 | 39.6 | 1.9 | 6.0 | 0.2 | 18.6 | 44.3 | 2,801 | 47 | 31 | 836 | 7,300 | 54,942 | 4,617 |
| Clay | 39 | 2,908 | 48.9 | 0.0 | 7.1 | 0.4 | 13.5 | 36.2 | 2,729 | 40 | 26 | 938 | 6,180 | 47,863 | 3,593 |
| Clinton | 92 | 2,444 | 54.3 | 2.8 | 8.3 | 0.0 | 7.8 | 42.9 | 1,139 | 101 | 72 | 2,176 | 17,440 | 66,075 | 6,521 |
| Coles | 236 | 4,596 | 59.4 | 0.7 | 8.1 | 0.6 | 5.5 | 151.0 | 2,940 | 143 | 100 | 5,334 | 20,800 | 54,559 | 5,176 |
| Cook | 32,810 | 6,309 | 38.3 | 5.9 | 7.7 | 1.1 | 3.7 | 68,769.6 | 13,223 | 38,121 | 10,720 | 281,222 | 2,529,070 | 84,862 | 12,585 |
| Crawford | 105 | 5,541 | 24.9 | 52.0 | 3.6 | 0.2 | 5.8 | 50.4 | 2,656 | 57 | 35 | 1,830 | 8,440 | 56,315 | 5,087 |
| Cumberland | 29 | 2,656 | 46.3 | 1.3 | 5.1 | 0.1 | 20.5 | 8.0 | 740 | 30 | 32 | 445 | 4,870 | 53,805 | 4,248 |
| DeKalb | 464 | 4,451 | 48.5 | 4.5 | 8.1 | 0.4 | 4.8 | 351.9 | 3,373 | 194 | 203 | 11,402 | 46,670 | 60,864 | 5,917 |
| De Witt | 73 | 4,559 | 38.9 | 23.3 | 5.5 | 0.1 | 6.6 | 35.7 | 2,246 | 41 | 31 | 1,071 | 7,570 | 59,718 | 5,618 |
| Douglas | 67 | 3,399 | 52.3 | 2.7 | 5.8 | 0.3 | 9.0 | 17.5 | 893 | 57 | 39 | 1,042 | 9,940 | 55,260 | 4,683 |
| DuPage | 4,927 | 5,296 | 52.3 | 0.7 | 7.3 | 1.0 | 6.9 | 4,689.0 | 5,040 | 4,698 | 1,891 | 48,357 | 469,520 | 104,308 | 15,866 |
| Edgar | 55 | 3,146 | 55.3 | 3.2 | 6.6 | 0.1 | 9.7 | 32.2 | 1,850 | 49 | 34 | 950 | 7,850 | 52,293 | 4,547 |
| Edwards | 14 | 2,228 | 53.1 | 1.6 | 5.7 | 0.1 | 8.2 | 4.8 | 740 | 22 | 13 | 292 | 2,880 | 51,545 | 3,701 |
| Effingham | 108 | 3,169 | 46.3 | 2.2 | 8.2 | 0.9 | 11.3 | 67.0 | 1,965 | 151 | 69 | 1,560 | 17,520 | 63,568 | 6,624 |
| Fayette | 58 | 2,685 | 52.3 | 5.9 | 7.7 | 0.0 | 6.5 | 45.6 | 2,119 | 51 | 40 | 1,077 | 8,940 | 50,078 | 4,052 |
| Ford | 56 | 4,199 | 59.5 | 1.3 | 6.6 | 0.1 | 11.4 | 80.8 | 6,076 | 48 | 25 | 782 | 6,440 | 56,748 | 5,099 |
| Franklin | 160 | 4,095 | 42.2 | 12.3 | 6.5 | 0.2 | 5.1 | 41.2 | 1,057 | 186 | 77 | 1,937 | 16,240 | 48,024 | 3,755 |
| Fulton | 128 | 3,655 | 57.1 | 5.7 | 5.3 | 0.3 | 5.1 | 65.1 | 1,854 | 95 | 64 | 2,184 | 15,270 | 51,267 | 4,369 |
| Gallatin | 14 | 2,741 | 49.4 | 0.0 | 2.6 | 0.7 | 11.7 | 7.8 | 1,533 | 18 | 10 | 230 | 2,170 | 53,271 | 4,370 |
| Greene | 34 | 2,618 | 54.4 | 4.4 | 7.0 | 0.4 | 11.9 | 20.3 | 1,544 | 47 | 26 | 704 | 5,560 | 47,890 | 3,559 |
| Grundy | 261 | 5,154 | 61.7 | 0.8 | 5.3 | 0.5 | 5.6 | 283.3 | 5,597 | 110 | 103 | 3,071 | 25,520 | 71,692 | 7,893 |
| Hamilton | 39 | 4,749 | 28.6 | 44.1 | 2.7 | 0.4 | 7.4 | 27.9 | 3,408 | 38 | 16 | 512 | 3,610 | 52,093 | 4,245 |
| Hancock | 55 | 3,053 | 57.3 | 0.0 | 4.8 | 6.7 | 8.9 | 27.9 | 1,556 | 72 | 35 | 988 | 8,450 | 54,009 | 4,480 |

1. Based on the resident population estimated as of July 1 of the year shown.

# Table B. States and Counties — **Land Area and Population**

| State / county code | CBSA code[1] | County Type code[2] | STATE County | Land area[3] (sq. mi) | Total persons 2019 | Rank | Per square mile | White | Black | American Indian, Alaska Native | Asian and Pacific Islander | Percent Hispanic or Latino[4] | Under 5 years | 5 to 14 years | 15 to 24 years | 25 to 34 years | 35 to 44 years | 45 to 54 years |
|---|---|---|---|---|---|---|---|---|---|---|---|---|---|---|---|---|---|---|
| | | | | 1 | 2 | 3 | 4 | 5 | 6 | 7 | 8 | 9 | 10 | 11 | 12 | 13 | 14 | 15 |
| | | | ILLINOIS—Cont'd | | | | | | | | | | | | | | | |
| 17069 | | 9 | Hardin | 177.4 | 3,808 | 2,910 | 21.5 | 96.1 | 1.2 | 1.0 | 1.1 | 2.0 | 4.4 | 9.2 | 10.1 | 8.8 | 10.6 | 13.3 |
| 17071 | 15460 | 9 | Henderson | 378.8 | 6,535 | 2,700 | 17.3 | 96.8 | 1.1 | 0.6 | 0.9 | 1.8 | 4.7 | 10.3 | 9.1 | 9.6 | 10.0 | 11.9 |
| 17073 | 19340 | 2 | Henry | 823.1 | 48,411 | 1,015 | 58.8 | 91.2 | 2.8 | 0.5 | 0.8 | 6.1 | 5.2 | 12.4 | 11.2 | 10.7 | 12.2 | 12.4 |
| 17075 | | 6 | Iroquois | 1,117.4 | 26,711 | 1,539 | 23.9 | 90.6 | 1.7 | 0.6 | 0.8 | 7.5 | 5.7 | 11.7 | 11.4 | 10.4 | 11.2 | 11.6 |
| 17077 | 16060 | 3 | Jackson | 583.6 | 56,675 | 912 | 97.1 | 76.2 | 16.7 | 0.9 | 4.3 | 4.6 | 5.3 | 10.1 | 24.1 | 13.0 | 10.1 | 9.4 |
| 17079 | | 7 | Jasper | 494.6 | 9,465 | 2,460 | 19.1 | 97.6 | 0.7 | 0.5 | 0.4 | 1.4 | 5.9 | 12.5 | 10.9 | 10.4 | 11.9 | 11.5 |
| 17081 | 34500 | 7 | Jefferson | 571.2 | 37,235 | 1,243 | 65.2 | 87.0 | 9.7 | 0.6 | 1.6 | 3.0 | 6.1 | 12.4 | 11.0 | 12.7 | 12.4 | 12.1 |
| 17083 | 41180 | 1 | Jersey | 369.7 | 21,616 | 1,739 | 58.5 | 96.7 | 1.3 | 0.8 | 1.1 | 1.5 | 5.0 | 11.1 | 13.6 | 10.5 | 11.5 | 12.2 |
| 17085 | | 6 | Jo Daviess | 600.9 | 21,239 | 1,758 | 35.3 | 95.2 | 1.2 | 0.6 | 0.8 | 3.2 | 4.2 | 10.7 | 9.8 | 8.7 | 9.9 | 11.2 |
| 17087 | 16060 | 8 | Johnson | 343.7 | 12,358 | 2,263 | 36.0 | 89.5 | 7.3 | 0.8 | 0.6 | 3.1 | 4.6 | 10.4 | 11.4 | 12.8 | 12.3 | 12.3 |
| 17089 | 16980 | 1 | Kane | 519.4 | 531,010 | 133 | 1,022.4 | 57.7 | 6.0 | 0.4 | 4.9 | 32.5 | 6.1 | 13.8 | 13.7 | 12.0 | 13.2 | 13.6 |
| 17091 | 28100 | 3 | Kankakee | 676.5 | 108,594 | 561 | 160.5 | 72.8 | 16.0 | 0.5 | 1.4 | 11.2 | 5.8 | 12.4 | 15.0 | 12.3 | 11.8 | 12.0 |
| 17093 | 16980 | 1 | Kendall | 320.2 | 130,638 | 493 | 408.0 | 67.7 | 8.6 | 0.4 | 4.3 | 20.9 | 6.3 | 16.3 | 13.6 | 12.2 | 16.0 | 14.4 |
| 17095 | 23660 | 4 | Knox | 716.4 | 49,053 | 1,007 | 68.5 | 84.4 | 10.0 | 0.6 | 1.3 | 6.2 | 5.7 | 10.7 | 13.0 | 11.9 | 11.4 | 11.1 |
| 17097 | 16980 | 1 | Lake | 443.6 | 693,593 | 97 | 1,563.6 | 61.5 | 7.6 | 0.5 | 9.6 | 22.7 | 5.6 | 13.5 | 14.6 | 11.3 | 12.6 | 13.4 |
| 17099 | 36837 | 4 | LaSalle | 1,135.2 | 107,571 | 567 | 94.8 | 85.8 | 3.3 | 0.5 | 1.3 | 10.3 | 5.4 | 11.8 | 11.9 | 12.0 | 12.1 | 12.1 |
| 17101 | | 7 | Lawrence | 372.2 | 15,467 | 2,063 | 41.6 | 85.1 | 10.6 | 0.6 | 0.6 | 4.0 | 4.6 | 11.0 | 11.9 | 15.4 | 13.8 | 12.1 |
| 17103 | 19940 | 6 | Lee | 724.8 | 33,647 | 1,336 | 46.4 | 86.6 | 6.2 | 0.6 | 1.1 | 6.8 | 5.0 | 11.1 | 10.5 | 12.9 | 12.2 | 12.6 |
| 17105 | 38700 | 4 | Livingston | 1,043.6 | 35,414 | 1,291 | 33.9 | 89.9 | 4.9 | 0.5 | 1.0 | 5.0 | 5.8 | 11.8 | 11.8 | 12.5 | 12.1 | 11.7 |
| 17107 | 30660 | 6 | Logan | 618.1 | 28,383 | 1,485 | 45.9 | 87.1 | 9.2 | 0.6 | 1.1 | 3.6 | 5.0 | 10.6 | 13.5 | 13.9 | 13.4 | 11.6 |
| 17109 | 31380 | 5 | McDonough | 589.3 | 29,295 | 1,454 | 49.7 | 88.9 | 6.2 | 0.7 | 2.8 | 3.0 | 4.5 | 9.3 | 25.5 | 10.9 | 9.9 | 9.2 |
| 17111 | 16980 | 1 | McHenry | 603.4 | 305,888 | 229 | 506.9 | 81.1 | 2.1 | 0.5 | 3.7 | 14.2 | 5.3 | 13.1 | 12.6 | 11.6 | 12.7 | 14.1 |
| 17113 | 14010 | 3 | McLean | 1,183.2 | 171,256 | 387 | 144.7 | 81.3 | 9.7 | 0.5 | 5.7 | 5.2 | 5.6 | 11.9 | 21.1 | 12.7 | 12.1 | 10.9 |
| 17115 | 19500 | 3 | Macon | 580.6 | 103,015 | 590 | 177.4 | 78.1 | 20.3 | 0.6 | 1.7 | 2.5 | 6.1 | 12.3 | 12.3 | 12.0 | 11.4 | 11.2 |
| 17117 | 41180 | 1 | Macoupin | 863.0 | 44,567 | 1,084 | 51.6 | 96.9 | 1.6 | 0.8 | 0.7 | 1.3 | 5.3 | 11.7 | 11.5 | 10.7 | 12.0 | 12.3 |
| 17119 | 41180 | 1 | Madison | 715.5 | 262,635 | 268 | 367.1 | 86.2 | 10.1 | 0.7 | 1.7 | 3.5 | 5.6 | 12.2 | 11.8 | 13.2 | 12.5 | 12.3 |
| 17121 | 16460 | 4 | Marion | 572.5 | 37,045 | 1,247 | 64.7 | 92.6 | 5.5 | 0.6 | 0.9 | 2.2 | 6.3 | 13.3 | 10.7 | 12.0 | 11.5 | 11.6 |
| 17123 | 37900 | 2 | Marshall | 386.8 | 11,309 | 2,326 | 29.2 | 95.3 | 1.3 | 0.7 | 0.9 | 3.1 | 5.4 | 11.6 | 9.9 | 11.1 | 10.7 | 11.9 |
| 17125 | | 6 | Mason | 539.4 | 13,173 | 2,208 | 24.4 | 97.1 | 1.2 | 0.8 | 0.9 | 1.3 | 5.0 | 11.6 | 10.9 | 10.7 | 11.4 | 12.3 |
| 17127 | 37140 | 7 | Massac | 237.2 | 13,636 | 2,182 | 57.5 | 89.3 | 7.9 | 1.1 | 0.8 | 3.3 | 5.4 | 11.9 | 11.2 | 10.4 | 11.2 | 12.8 |
| 17129 | 44100 | 3 | Menard | 314.4 | 12,068 | 2,291 | 38.4 | 96.6 | 1.7 | 0.9 | 0.8 | 1.6 | 5.4 | 12.2 | 10.7 | 11.0 | 12.7 | 11.8 |
| 17131 | 19340 | 2 | Mercer | 561.2 | 15,225 | 2,080 | 27.1 | 95.9 | 1.2 | 0.6 | 0.7 | 2.7 | 5.0 | 12.1 | 10.8 | 10.2 | 12.0 | 12.5 |
| 17133 | 41180 | 1 | Monroe | 385.3 | 34,739 | 1,311 | 90.2 | 97.0 | 0.8 | 0.6 | 1.0 | 1.7 | 5.7 | 12.4 | 10.6 | 10.5 | 13.2 | 12.8 |
| 17135 | | 6 | Montgomery | 703.8 | 28,045 | 1,497 | 39.8 | 93.7 | 4.1 | 0.5 | 0.8 | 1.9 | 5.3 | 11.4 | 10.8 | 12.2 | 12.5 | 12.1 |
| 17137 | 27300 | 4 | Morgan | 568.9 | 33,400 | 1,345 | 58.7 | 89.3 | 8.0 | 0.6 | 1.1 | 2.8 | 4.9 | 10.3 | 13.6 | 12.3 | 12.4 | 11.6 |
| 17139 | | 6 | Moultrie | 336.0 | 14,347 | 2,134 | 42.7 | 97.0 | 1.2 | 0.6 | 0.6 | 1.7 | 6.0 | 14.3 | 12.1 | 11.3 | 12.1 | 11.1 |
| 17141 | 40300 | 4 | Ogle | 758.7 | 50,306 | 987 | 66.3 | 87.2 | 1.9 | 0.5 | 0.9 | 10.9 | 5.5 | 12.7 | 11.7 | 11.5 | 11.9 | 12.4 |
| 17143 | 37900 | 2 | Peoria | 618.7 | 177,652 | 376 | 287.1 | 71.6 | 20.7 | 0.6 | 4.9 | 5.2 | 6.8 | 13.1 | 12.5 | 13.5 | 12.2 | 11.5 |
| 17145 | | 6 | Perry | 441.9 | 20,664 | 1,784 | 46.8 | 87.0 | 10.1 | 0.6 | 0.9 | 3.3 | 4.7 | 10.7 | 12.7 | 13.6 | 13.0 | 12.2 |
| 17147 | 16580 | 3 | Piatt | 439.2 | 16,355 | 2,015 | 37.2 | 96.9 | 1.3 | 0.6 | 1.0 | 1.5 | 5.7 | 12.3 | 11.1 | 11.1 | 12.4 | 11.7 |
| 17149 | | 7 | Pike | 831.4 | 15,239 | 2,078 | 18.3 | 96.4 | 1.9 | 0.6 | 0.5 | 1.5 | 6.3 | 12.3 | 10.6 | 11.6 | 11.3 | 11.8 |
| 17151 | | 8 | Pope | 368.9 | 4,142 | 2,884 | 11.2 | 91.7 | 6.0 | 1.1 | 0.6 | 2.0 | 2.5 | 5.7 | 11.4 | 7.7 | 10.6 | 13.2 |
| 17153 | | 8 | Pulaski | 199.3 | 5,201 | 2,809 | 26.1 | 66.8 | 31.1 | 1.5 | 1.1 | 2.6 | 5.2 | 11.7 | 11.2 | 11.0 | 10.7 | 11.0 |
| 17155 | 36837 | 8 | Putnam | 160.1 | 5,716 | 2,765 | 35.7 | 92.3 | 1.1 | 0.6 | 0.9 | 6.2 | 4.3 | 11.5 | 10.7 | 10.3 | 11.3 | 11.8 |
| 17157 | | 4 | Randolph | 575.4 | 31,351 | 1,395 | 54.5 | 85.2 | 11.2 | 0.5 | 0.8 | 3.5 | 5.0 | 10.5 | 11.0 | 13.8 | 13.3 | 12.5 |
| 17159 | | 7 | Richland | 360.0 | 15,507 | 2,058 | 43.1 | 95.9 | 1.5 | 0.7 | 1.3 | 1.9 | 5.9 | 12.7 | 10.7 | 11.8 | 11.6 | 11.5 |
| 17161 | 19340 | 2 | Rock Island | 427.5 | 140,907 | 464 | 329.6 | 72.7 | 12.8 | 0.6 | 3.2 | 13.4 | 6.0 | 12.5 | 12.4 | 12.3 | 11.9 | 11.3 |
| 17163 | 41180 | 1 | St. Clair | 657.7 | 258,046 | 270 | 392.3 | 63.1 | 31.9 | 0.7 | 2.6 | 4.5 | 6.1 | 13.1 | 11.9 | 13.1 | 12.9 | 12.1 |
| 17165 | | 6 | Saline | 380.0 | 23,182 | 1,675 | 61.0 | 93.2 | 4.9 | 1.0 | 1.1 | 1.8 | 5.9 | 11.1 | 11.4 | 11.9 | 11.5 | 12.6 |
| 17167 | 44100 | 3 | Sangamon | 868.3 | 193,882 | 349 | 223.3 | 81.9 | 14.9 | 0.6 | 2.8 | 2.5 | 5.7 | 12.3 | 11.9 | 12.4 | 12.6 | 12.1 |
| 17169 | | 7 | Schuyler | 437.3 | 6,738 | 2,684 | 15.4 | 92.6 | 4.8 | 0.5 | 0.5 | 2.3 | 4.3 | 10.3 | 9.6 | 11.1 | 12.5 | 13.8 |
| 17171 | 27300 | 9 | Scott | 250.8 | 4,950 | 2,828 | 19.7 | 96.5 | 1.3 | 0.7 | 1.3 | 1.4 | 4.7 | 12.6 | 11.1 | 10.5 | 12.0 | 12.1 |
| 17173 | | 6 | Shelby | 758.5 | 21,299 | 1,754 | 28.1 | 97.6 | 0.9 | 0.5 | 0.7 | 1.3 | 5.5 | 11.9 | 10.3 | 10.8 | 11.2 | 11.7 |
| 17175 | 37900 | 2 | Stark | 288.1 | 5,262 | 2,804 | 18.3 | 95.9 | 1.6 | 0.6 | 1.2 | 2.3 | 5.5 | 12.1 | 11.2 | 10.2 | 11.4 | 12.0 |
| 17177 | 23300 | 4 | Stephenson | 564.2 | 43,831 | 1,100 | 77.7 | 84.2 | 12.7 | 0.6 | 1.3 | 4.5 | 5.5 | 11.9 | 11.0 | 10.5 | 10.4 | 11.8 |
| 17179 | 37900 | 2 | Tazewell | 646.5 | 130,777 | 492 | 202.3 | 94.9 | 2.0 | 0.6 | 1.4 | 2.6 | 5.4 | 12.8 | 11.2 | 11.8 | 13.1 | 12.4 |
| 17181 | | 6 | Union | 413.4 | 16,498 | 2,008 | 39.9 | 92.2 | 2.1 | 1.0 | 0.9 | 5.4 | 5.5 | 11.6 | 10.8 | 11.6 | 11.4 | 12.6 |
| 17183 | 19180 | 3 | Vermilion | 898.3 | 74,855 | 740 | 83.3 | 79.9 | 15.3 | 0.6 | 1.3 | 5.4 | 6.2 | 13.2 | 11.6 | 11.8 | 11.6 | 11.5 |
| 17185 | | 7 | Wabash | 223.3 | 11,190 | 2,335 | 50.1 | 95.3 | 1.6 | 0.7 | 1.6 | 2.3 | 6.0 | 12.2 | 10.8 | 11.3 | 11.1 | 11.2 |
| 17187 | | 6 | Warren | 542.4 | 16,696 | 1,993 | 30.8 | 84.2 | 3.7 | 0.7 | 2.8 | 10.3 | 6.1 | 11.9 | 15.7 | 10.7 | 11.1 | 10.8 |
| 17189 | | 6 | Washington | 562.6 | 13,764 | 2,172 | 24.5 | 96.9 | 1.4 | 0.5 | 0.6 | 1.4 | 5.5 | 11.5 | 10.7 | 11.4 | 11.7 | 12.2 |
| 17191 | | 7 | Wayne | 713.8 | 16,031 | 2,032 | 22.5 | 96.9 | 1.0 | 0.5 | 0.8 | 1.7 | 6.1 | 12.6 | 10.6 | 11.2 | 11.4 | 11.8 |
| 17193 | | 6 | White | 495.0 | 13,364 | 2,196 | 27.0 | 96.9 | 0.9 | 0.7 | 0.8 | 1.6 | 5.3 | 12.9 | 10.0 | 10.8 | 12.1 | 11.4 |
| 17195 | 44580 | 4 | Whiteside | 684.1 | 54,656 | 937 | 79.9 | 85.1 | 2.4 | 0.5 | 0.6 | 12.6 | 5.6 | 12.2 | 11.4 | 11.2 | 10.9 | 12.2 |
| 17197 | 16980 | 1 | Will | 835.9 | 688,726 | 98 | 823.9 | 63.3 | 12.7 | 0.4 | 6.8 | 18.5 | 5.7 | 13.7 | 13.8 | 12.0 | 13.4 | 14.4 |
| 17199 | 16060 | 3 | Williamson | 420.2 | 66,415 | 806 | 158.1 | 90.7 | 5.8 | 0.7 | 1.8 | 2.9 | 5.6 | 12.1 | 10.6 | 12.8 | 13.0 | 12.4 |

1. CBSA = Core Based Statistical Area. See Appendix A for explanation. See Appendix B for list of metropolitan areas with component counties. 2. County type code from the Economic Research Service of USDA Rural-Urban Continuum Codes. See Appendix A for definition. 3. Dry land or land partially or temporarily covered by water. 4. May be of any race.

# Table B. States and Counties — Population and Households

| STATE County | \*55 to 64 years | 65 to 74 years | 75 years and over | Percent female | Total persons 2000 | Total persons 2010 | Percent change 2000-2010 | Percent change 2010-2020 | Births | Deaths | Net Migration | Number | Persons per household | Family households | Female family householder[1] | One person |
|---|---|---|---|---|---|---|---|---|---|---|---|---|---|---|---|---|
| | 16 | 17 | 18 | 19 | 20 | 21 | 22 | 23 | 24 | 25 | 26 | 27 | 28 | 29 | 30 | 31 |
| **ILLINOIS—Cont'd** | | | | | | | | | | | | | | | | |
| Hardin | 15.9 | 16.3 | 11.4 | 49.4 | 4,800 | 4,319 | -10.0 | -11.8 | 340 | 650 | -196 | 1,363 | 2.80 | 68.2 | 6.6 | 29.1 |
| Henderson | 17.6 | 14.0 | 12.9 | 50.6 | 8,213 | 7,329 | -10.8 | -10.8 | 650 | 872 | -577 | 2,977 | 2.27 | 67.4 | 8.3 | 26.3 |
| Henry | 14.5 | 12.0 | 9.4 | 50.3 | 51,020 | 50,487 | | -4.1 | 5,388 | 5,531 | -1,922 | 19,856 | 2.45 | 67.9 | 8.0 | 28.6 |
| Iroquois | 15.4 | 12.1 | 10.4 | 51.1 | 31,334 | 29,719 | -5.2 | -10.1 | 3,122 | 3,843 | -2,286 | 11,741 | 2.33 | 65.2 | 8.4 | 29.2 |
| Jackson | 11.2 | 9.7 | 7.0 | 50.5 | 59,612 | 60,204 | 1.0 | -5.9 | 6,818 | 4,890 | -5,593 | 23,883 | 2.26 | 51.4 | 11.8 | 37.4 |
| Jasper | 15.7 | 11.9 | 9.1 | 49.9 | 10,117 | 9,701 | -4.1 | -2.4 | 1,181 | 1,041 | -374 | 3,711 | 2.57 | 65.3 | 7.1 | 29.8 |
| Jefferson | 13.4 | 11.3 | 8.5 | 48.6 | 40,045 | 38,825 | -3.0 | -4.1 | 4,822 | 4,641 | -1,752 | 14,985 | 2.37 | 65.5 | 12.1 | 30.3 |
| Jersey | 15.8 | 11.3 | 8.9 | 51.3 | 21,668 | 23,011 | 6.2 | -6.1 | 2,180 | 2,576 | -1,008 | 8,499 | 2.47 | 67.7 | 8.5 | 27.6 |
| Jo Daviess | 16.0 | 16.9 | 12.5 | 49.9 | 22,289 | 22,681 | 1.8 | -6.4 | 1,820 | 2,544 | -703 | 9,970 | 2.15 | 62.8 | 6.9 | 31.3 |
| Johnson | 14.2 | 12.1 | 9.8 | 45.4 | 12,878 | 12,571 | -2.4 | -1.7 | 1,108 | 1,321 | | 4,303 | 2.38 | 70.7 | 9.5 | 25.2 |
| Kane | 12.8 | 8.9 | 5.9 | 50.2 | 404,119 | 515,318 | 27.5 | 3.0 | 68,272 | 32,624 | -19,868 | 179,637 | 2.93 | 73.3 | 11.1 | 21.9 |
| Kankakee | 13.1 | 10.0 | 7.5 | 50.6 | 103,833 | 113,450 | 9.3 | -4.3 | 13,470 | 11,623 | -6,722 | 39,796 | 2.64 | 66.0 | 12.8 | 28.2 |
| Kendall | 10.2 | 6.9 | 4.2 | 50.4 | 54,544 | 114,800 | 110.5 | 13.8 | 16,579 | 5,765 | 5,119 | 40,721 | 3.09 | 78.7 | 11.0 | 17.3 |
| Knox | 13.7 | 12.4 | 10.1 | 49.4 | 55,836 | 52,923 | -5.2 | -7.3 | 5,670 | 6,902 | -2,607 | 20,680 | 2.23 | 57.0 | 11.1 | 36.3 |
| Lake | 13.7 | 9.2 | 6.2 | 50.0 | 644,356 | 703,396 | 9.2 | -1.4 | 78,712 | 46,294 | -42,702 | 246,122 | 2.78 | 73.2 | 10.3 | 22.4 |
| LaSalle | 15.0 | 11.1 | 8.7 | 49.5 | 111,509 | 113,911 | 2.2 | -5.6 | 12,054 | 13,376 | -4,997 | 45,095 | 2.36 | 65.0 | 11.4 | 29.9 |
| Lawrence | 12.9 | 10.2 | 8.1 | 44.1 | 15,452 | 16,903 | 9.4 | -8.5 | 1,652 | 2,095 | -995 | 6,306 | 2.16 | 65.1 | 10.6 | 30.3 |
| Lee | 14.9 | 11.9 | 8.8 | 46.9 | 36,062 | 36,034 | -0.1 | -6.6 | 3,544 | 3,806 | -2,125 | 13,788 | 2.25 | 60.8 | 9.3 | 32.5 |
| Livingston | 14.5 | 10.9 | 9.0 | 49.5 | 39,678 | 38,951 | -1.8 | -9.1 | 4,301 | 4,386 | -3,514 | 14,307 | 2.36 | 66.0 | 11.8 | 29.7 |
| Logan | 12.9 | 10.3 | 8.7 | 49.0 | 31,183 | 30,305 | -2.8 | -6.3 | 3,067 | 3,425 | -1,574 | 10,797 | 2.31 | 62.7 | 7.9 | 30.9 |
| McDonough | 11.8 | 10.5 | 8.4 | 51.0 | 32,913 | 32,616 | -0.9 | -10.2 | 2,875 | 3,154 | -3,076 | 11,408 | 2.38 | 51.8 | 7.0 | 39.0 |
| McHenry | 14.8 | 9.7 | 6.1 | 50.2 | 260,077 | 308,886 | 18.8 | -1.0 | 32,574 | 21,334 | -14,336 | 112,453 | 2.72 | 74.1 | 9.6 | 21.0 |
| McLean | 11.5 | 8.4 | 5.8 | 51.5 | 150,433 | 169,578 | 12.7 | 1.0 | 20,671 | 12,100 | -6,983 | 65,845 | 2.50 | 61.1 | 8.8 | 28.9 |
| Macon | 13.7 | 11.7 | 9.2 | 52.1 | 114,706 | 110,777 | -3.4 | -7.0 | 13,606 | 12,558 | -8,838 | 43,912 | 2.32 | 59.4 | 12.8 | 35.6 |
| Macoupin | 15.3 | 12.3 | 8.9 | 50.8 | 49,019 | 47,764 | -2.6 | -6.7 | 4,698 | 5,787 | -2,111 | 18,875 | 2.36 | 68.1 | 11.1 | 26.7 |
| Madison | 14.3 | 10.5 | 7.7 | 51.3 | 258,941 | 269,296 | 4.0 | -2.5 | 31,044 | 28,998 | -8,648 | 107,659 | 2.41 | 65.0 | 12.2 | 28.2 |
| Marion | 14.3 | 11.4 | 9.1 | 51.0 | 41,691 | 39,437 | -5.4 | -6.1 | 4,989 | 5,259 | -2,112 | 15,946 | 2.32 | 62.6 | 12.3 | 31.8 |
| Marshall | 15.1 | 13.2 | 11.1 | 50.2 | 13,180 | 12,647 | -4.0 | -10.6 | 1,239 | 1,636 | -941 | 4,884 | 2.33 | 63.9 | 7.8 | 31.7 |
| Mason | 14.9 | 12.9 | 10.3 | 50.4 | 16,038 | 14,666 | -8.6 | -10.2 | 1,384 | 1,921 | -955 | 5,977 | 2.24 | 63.5 | 10.0 | 28.5 |
| Massac | 15.3 | 11.1 | 10.7 | 52.0 | 15,161 | 15,429 | 1.8 | -11.6 | 1,602 | 2,147 | -1,251 | 5,822 | 2.39 | 67.1 | 9.3 | 29.6 |
| Menard | 16.0 | 11.7 | 8.5 | 51.2 | 12,486 | 12,705 | 1.8 | -5.0 | 1,231 | 1,334 | -535 | 5,188 | 2.35 | 72.1 | 9.6 | 24.3 |
| Mercer | 15.0 | 12.3 | 10.1 | 50.1 | 16,957 | 16,436 | -3.1 | -7.4 | 1,497 | 1,911 | -795 | 6,516 | 2.36 | 67.3 | 6.9 | 28.2 |
| Monroe | 15.7 | 10.9 | 8.2 | 50.3 | 27,619 | 33,006 | 19.5 | 5.3 | 3,548 | 3,001 | 1,217 | 13,586 | 2.50 | 74.4 | 8.0 | 22.3 |
| Montgomery | 14.7 | 11.4 | 9.6 | 47.5 | 30,652 | 30,103 | -1.8 | -6.8 | 3,050 | 3,676 | -1,443 | 11,522 | 2.24 | 64.8 | 9.9 | 30.1 |
| Morgan | 14.2 | 11.3 | 9.4 | 49.3 | 36,616 | 35,542 | -2.9 | -6.0 | 3,686 | 4,046 | -1,768 | 13,719 | 2.25 | 61.2 | 12.1 | 33.6 |
| Moultrie | 13.6 | 10.6 | 8.9 | 51.0 | 14,287 | 14,856 | 4.0 | -3.4 | 1,854 | 1,890 | -468 | 5,959 | 2.40 | 66.2 | 7.3 | 27.9 |
| Ogle | 14.8 | 10.9 | 8.6 | 50.4 | 51,032 | 53,494 | 4.8 | -6.0 | 5,454 | 5,165 | -3,508 | 21,021 | 2.40 | 67.1 | 9.3 | 28.2 |
| Peoria | 12.3 | 10.4 | 7.6 | 51.5 | 183,433 | 186,496 | 1.7 | -4.7 | 26,544 | 18,706 | -16,711 | 73,253 | 2.42 | 59.7 | 13.5 | 34.6 |
| Perry | 13.3 | 11.3 | 8.5 | 44.9 | 23,094 | 22,346 | -3.2 | -7.5 | 2,079 | 2,471 | -1,297 | 8,433 | 2.21 | 67.1 | 13.6 | 29.0 |
| Piatt | 15.3 | 11.8 | 8.6 | 50.3 | 16,365 | 16,727 | 2.2 | -2.2 | 1,800 | 1,765 | -405 | 6,692 | 2.44 | 70.6 | 7.6 | 27.0 |
| Pike | 13.8 | 11.9 | 10.3 | 49.9 | 17,384 | 16,432 | -5.5 | -7.3 | 1,907 | 2,065 | -1,026 | 6,309 | 2.41 | 64.6 | 9.4 | 31.5 |
| Pope | 19.9 | 16.5 | 12.5 | 48.6 | 4,413 | 4,478 | 1.5 | -7.5 | 293 | 502 | -128 | 1,694 | 2.34 | 57.1 | 5.3 | 42.1 |
| Pulaski | 15.8 | 14.0 | 9.4 | 51.2 | 7,348 | 6,165 | -16.1 | -15.6 | 617 | 811 | -772 | 2,095 | 2.62 | 61.5 | 12.8 | 34.0 |
| Putnam | 16.1 | 14.5 | 9.5 | 49.4 | 6,086 | 6,008 | -1.3 | -4.9 | 494 | 586 | -199 | 2,413 | 2.37 | 67.5 | 7.4 | 27.4 |
| Randolph | 14.3 | 10.8 | 7.9 | 44.8 | 33,893 | 33,427 | -1.4 | -6.2 | 3,410 | 3,862 | -1,620 | 11,883 | 2.39 | 65.5 | 10.7 | 30.4 |
| Richland | 14.4 | 10.9 | 10.4 | 50.9 | 16,149 | 16,233 | 0.5 | -4.5 | 1,933 | 2,099 | -552 | 6,452 | 2.40 | 63.1 | 10.0 | 31.5 |
| Rock Island | 13.3 | 11.5 | 8.7 | 50.8 | 149,374 | 147,542 | -1.2 | -4.5 | 18,503 | 16,001 | -9,120 | 60,546 | 2.30 | 62.0 | 13.2 | 32.9 |
| St. Clair | 13.9 | 10.0 | 6.9 | 51.8 | 256,082 | 270,075 | 5.5 | -4.5 | 33,796 | 26,398 | -19,551 | 104,105 | 2.48 | 62.6 | 14.7 | 32.4 |
| Saline | 14.6 | 11.7 | 9.3 | 50.8 | 26,733 | 24,913 | -6.8 | -6.9 | 3,151 | 3,695 | -1,174 | 9,972 | 2.36 | 66.7 | 12.5 | 30.0 |
| Sangamon | 14.1 | 11.2 | 7.7 | 52.1 | 188,951 | 197,463 | 4.5 | -1.8 | 23,441 | 20,207 | -6,740 | 83,711 | 2.30 | 60.0 | 13.0 | 33.5 |
| Schuyler | 14.8 | 13.1 | 10.6 | 47.7 | 7,189 | 7,543 | 4.9 | -10.7 | 636 | 909 | -533 | 2,775 | 2.19 | 60.6 | 11.6 | 30.6 |
| Scott | 16.4 | 10.9 | 9.6 | 50.7 | 5,537 | 5,358 | -3.2 | -7.6 | 473 | 603 | -279 | 1,939 | 2.55 | 71.3 | 10.0 | 26.0 |
| Shelby | 15.0 | 12.9 | 10.7 | 50.3 | 22,893 | 22,355 | -2.4 | -4.7 | 2,471 | 2,459 | -1,055 | 9,189 | 2.34 | 67.3 | 8.3 | 27.4 |
| Stark | 14.8 | 12.7 | 10.1 | 50.1 | 6,332 | 5,992 | -5.4 | -12.2 | 621 | 805 | -551 | 2,315 | 2.31 | 65.9 | 10.5 | 30.9 |
| Stephenson | 15.2 | 12.8 | 11.0 | 51.4 | 48,979 | 47,704 | -2.6 | -8.1 | 5,034 | 5,614 | -3,304 | 19,739 | 2.24 | 64.0 | 12.5 | 31.7 |
| Tazewell | 13.6 | 11.1 | 8.6 | 50.8 | 128,485 | 135,392 | 5.4 | -3.4 | 15,598 | 14,532 | -5,640 | 54,291 | 2.41 | 66.6 | 9.7 | 28.8 |
| Union | 14.3 | 12.4 | 9.8 | 50.0 | 18,293 | 17,804 | -2.7 | -7.3 | 1,878 | 2,329 | -846 | 6,654 | 2.48 | 68.5 | 10.6 | 27.0 |
| Vermilion | 13.8 | 11.4 | 8.8 | 50.1 | 83,919 | 81,625 | -2.7 | -8.3 | 10,333 | 9,944 | -7,166 | 31,151 | 2.40 | 62.3 | 14.0 | 32.6 |
| Wabash | 15.1 | 12.4 | 9.8 | 50.2 | 12,937 | 11,947 | -7.7 | -6.3 | 1,423 | 1,407 | -771 | 4,839 | 2.37 | 66.6 | 8.6 | 30.1 |
| Warren | 13.0 | 11.7 | 9.0 | 50.5 | 18,735 | 17,701 | -5.5 | -5.7 | 2,161 | 1,885 | -1,285 | 6,862 | 2.33 | 66.4 | 13.1 | 29.3 |
| Washington | 15.8 | 11.7 | 9.7 | 50.0 | 15,148 | 14,716 | -2.9 | -6.5 | 1,574 | 1,559 | -967 | 6,020 | 2.29 | 67.4 | 8.7 | 27.5 |
| Wayne | 14.1 | 11.8 | 10.4 | 50.6 | 17,151 | 16,760 | -2.3 | -4.3 | 2,083 | 2,156 | -639 | 7,051 | 2.32 | 64.2 | 9.2 | 31.8 |
| White | 15.0 | 12.1 | 10.4 | 50.7 | 15,371 | 14,665 | -4.6 | -8.9 | 1,559 | 2,269 | -584 | 6,041 | 2.23 | 66.3 | 11.2 | 31.0 |
| Whiteside | 14.7 | 12.2 | 9.7 | 50.6 | 60,653 | 58,505 | -3.5 | -6.6 | 6,432 | 6,721 | -3,560 | 23,084 | 2.38 | 64.8 | 10.4 | 29.8 |
| Will | 13.0 | 8.5 | 5.5 | 50.4 | 502,266 | 677,590 | 34.9 | 1.6 | 79,232 | 44,899 | -23,264 | 229,498 | 2.96 | 74.6 | 11.0 | 21.2 |
| Williamson | 13.6 | 11.2 | 8.6 | 50.2 | 61,296 | 66,373 | 8.3 | 0.1 | 7,869 | 7,851 | 63 | 27,029 | 2.41 | 62.7 | 11.5 | 31.9 |

1. No spouse present.

# Table B. States and Counties — Population, Vital Statistics, Health, and Crime

| STATE County | Persons in group quarters, 2020 | Daytime Population, 2015-2019 Number | Employment/ residence ratio | Births, 2020 Total | Rate[1] | Deaths, 2020 Number | Rate[1] | Persons under 65 with no health insurance, 2019 Number | Percent | Medicare, 2020 Total beneficiaries | Enrolled in Original Medicare | Enrolled in Medicare Advantage | Crimes reported by county police or sheriff, 2019 Violent | Property |
|---|---|---|---|---|---|---|---|---|---|---|---|---|---|---|
| | 32 | 33 | 34 | 35 | 36 | 37 | 38 | 39 | 40 | 41 | 42 | 43 | 44 | 45 |
| ILLINOIS—Cont'd | | | | | | | | | | | | | | |
| Hardin | 17 | 3,536 | 0.69 | 33 | 8.7 | 68 | 17.9 | 225 | 8.1 | 1,090 | 919 | 171 | NA | NA |
| Henderson | 51 | 5,258 | 0.52 | 62 | 9.5 | 92 | 14.1 | 385 | 7.8 | 1,678 | 1,333 | 344 | NA | NA |
| Henry | 722 | 41,876 | 0.68 | 479 | 9.9 | 614 | 12.7 | 2,607 | 6.8 | 11,595 | 7,818 | 3,777 | 6 | 51 |
| Iroquois | 468 | 24,268 | 0.72 | 292 | 10.9 | 341 | 12.8 | 1,823 | 8.7 | 6,894 | 5,665 | 1,229 | 19 | 142 |
| Jackson | 3,587 | 60,268 | 1.09 | 605 | 10.7 | 509 | 9.0 | 3,691 | 8.3 | 10,281 | 7,373 | 2,908 | 29 | 299 |
| Jasper | 54 | 7,904 | 0.61 | 116 | 12.3 | 97 | 10.2 | 530 | 6.9 | 2,243 | 1,880 | 363 | 48 | 56 |
| Jefferson | 2,143 | 43,207 | 1.33 | 423 | 11.4 | 467 | 12.5 | 2,066 | 7.3 | 8,460 | 7,104 | 1,355 | 59 | 286 |
| Jersey | 795 | 17,676 | 0.58 | 211 | 9.8 | 220 | 10.2 | 995 | 5.9 | 5,159 | 3,771 | 1,388 | 25 | 115 |
| Jo Daviess | 167 | 19,195 | 0.78 | 156 | 7.3 | 267 | 12.6 | 1,210 | 8.0 | 6,540 | 3,234 | 3,306 | 6 | 74 |
| Johnson | 1,272 | 11,838 | 0.83 | 100 | 8.1 | 154 | 12.5 | 617 | 7.3 | 2,993 | 2,197 | 796 | 26 | 93 |
| Kane | 6,819 | 495,522 | 0.86 | 6,211 | 11.7 | 3,815 | 7.2 | 46,080 | 10.2 | 81,566 | 57,986 | 23,580 | 87 | 280 |
| Kankakee | 6,270 | 106,768 | 0.92 | 1,205 | 11.1 | 1,280 | 11.8 | 6,322 | 7.4 | 21,977 | 16,455 | 5,523 | 56 | 250 |
| Kendall | 208 | 98,609 | 0.57 | 1,482 | 11.3 | 660 | 5.1 | 7,377 | 6.4 | 16,166 | 11,958 | 4,208 | 52 | 225 |
| Knox | 3,931 | 50,331 | 0.99 | 530 | 10.8 | 617 | 12.6 | 2,578 | 7.3 | 12,239 | 7,914 | 4,326 | 51 | 171 |
| Lake | 18,830 | 716,944 | 1.04 | 6,927 | 10.0 | 5,168 | 7.5 | 47,402 | 8.2 | 113,825 | 90,291 | 23,534 | 104 | 771 |
| LaSalle | 3,384 | 105,165 | 0.91 | 1,059 | 9.8 | 1,270 | 11.8 | 5,551 | 6.5 | 24,391 | 20,238 | 4,153 | 33 | 167 |
| Lawrence | 2,419 | 14,965 | 0.83 | 143 | 9.2 | 220 | 14.2 | 786 | 7.4 | 3,253 | 2,896 | 357 | NA | NA |
| Lee | 3,033 | 34,101 | 0.98 | 326 | 9.7 | 428 | 12.7 | 1,536 | 6.2 | 7,791 | 6,021 | 1,770 | 21 | 81 |
| Livingston | 2,310 | 36,900 | 1.05 | 427 | 12.1 | 389 | 11.0 | 1,752 | 6.5 | 7,945 | 5,895 | 2,051 | 28 | 96 |
| Logan | 4,142 | 27,469 | 0.87 | 287 | 10.1 | 317 | 11.2 | 1,082 | 5.5 | 6,056 | 4,205 | 1,852 | 11 | 63 |
| McDonough | 3,326 | 31,186 | 1.05 | 246 | 8.4 | 308 | 10.5 | 1,563 | 7.4 | 5,854 | 4,140 | 1,714 | 8 | 19 |
| McHenry | 1,648 | 258,682 | 0.70 | 2,957 | 9.7 | 2,389 | 7.8 | 18,546 | 7.1 | 53,469 | 42,414 | 11,055 | 65 | 257 |
| McLean | 10,506 | 176,845 | 1.05 | 1,752 | 10.2 | 1,318 | 7.7 | 7,970 | 5.8 | 27,167 | 18,010 | 9,157 | 30 | 117 |
| Macon | 4,009 | 111,972 | 1.14 | 1,234 | 12.0 | 1,245 | 12.1 | 4,788 | 6.0 | 24,263 | 19,173 | 5,090 | 38 | 282 |
| Macoupin | 863 | 37,214 | 0.59 | 451 | 10.1 | 603 | 13.5 | 2,355 | 6.7 | 11,110 | 8,702 | 2,437 | 10 | 250 |
| Madison | 3,950 | 246,225 | 0.85 | 2,754 | 10.5 | 3,027 | 11.5 | 13,796 | 6.4 | 55,288 | 33,224 | 22,064 | 126 | 1,059 |
| Marion | 759 | 36,976 | 0.95 | 455 | 12.3 | 418 | 11.3 | 2,057 | 7.0 | 9,342 | 8,175 | 1,167 | 43 | 202 |
| Marshall | 255 | 10,158 | 0.71 | 122 | 10.8 | 139 | 12.3 | 572 | 6.6 | 3,017 | 2,294 | 723 | 3 | 24 |
| Mason | 206 | 11,669 | 0.69 | 127 | 9.6 | 178 | 13.5 | 704 | 6.9 | 3,390 | 2,695 | 696 | 12 | 85 |
| Massac | 292 | 12,378 | 0.67 | 136 | 10.0 | 182 | 13.3 | 764 | 7.1 | 3,687 | 3,264 | 424 | NA | NA |
| Menard | 154 | 8,514 | 0.39 | 104 | 8.6 | 148 | 12.3 | 563 | 5.8 | 2,789 | 1,733 | 1,057 | 10 | 51 |
| Mercer | 185 | 11,779 | 0.49 | 139 | 9.1 | 179 | 11.8 | 764 | 6.4 | 3,822 | 2,399 | 1,424 | NA | NA |
| Monroe | 343 | 26,966 | 0.60 | 383 | 11.0 | 348 | 10.0 | 1,092 | 3.8 | 7,169 | 4,049 | 3,120 | 9 | 34 |
| Montgomery | 2,451 | 27,961 | 0.92 | 287 | 10.2 | 391 | 13.9 | 1,431 | 7.0 | 6,737 | 5,439 | 1,297 | 42 | 133 |
| Morgan | 3,328 | 35,073 | 1.05 | 318 | 9.5 | 386 | 11.6 | 1,470 | 6.1 | 7,774 | 5,809 | 1,965 | 11 | 108 |
| Moultrie | 395 | 13,467 | 0.83 | 160 | 11.2 | 152 | 10.6 | 993 | 8.6 | 3,125 | 2,591 | 533 | 8 | 38 |
| Ogle | 526 | 45,062 | 0.76 | 526 | 10.5 | 520 | 10.3 | 2,926 | 7.2 | 11,011 | 7,674 | 3,337 | 13 | 149 |
| Peoria | 4,811 | 202,220 | 1.24 | 2,411 | 13.6 | 1,920 | 10.8 | 10,443 | 7.3 | 36,351 | 22,814 | 13,537 | 55 | 866 |
| Perry | 2,472 | 19,506 | 0.78 | 173 | 8.4 | 244 | 11.8 | 1,008 | 6.9 | 4,632 | 3,401 | 1,230 | 7 | 67 |
| Piatt | 72 | 12,372 | 0.51 | 180 | 11.0 | 189 | 11.6 | 744 | 5.7 | 3,701 | 2,062 | 1,639 | 16 | 63 |
| Pike | 488 | 13,998 | 0.75 | 203 | 13.3 | 187 | 12.3 | 916 | 7.7 | 3,788 | 3,167 | 621 | 15 | 37 |
| Pope | 272 | 3,743 | 0.58 | 27 | 6.5 | 35 | 8.5 | 246 | 8.4 | 1,142 | 893 | 250 | NA | NA |
| Pulaski | 21 | 6,085 | 1.32 | 42 | 8.1 | 67 | 12.9 | 342 | 8.4 | 1,426 | 1,228 | 199 | NA | NA |
| Putnam | 2 | 4,694 | 0.64 | 37 | 6.5 | 65 | 11.4 | 276 | 6.3 | 1,413 | 1,071 | 342 | 1 | 19 |
| Randolph | 4,313 | 32,685 | 1.03 | 292 | 9.3 | 422 | 13.5 | 1,426 | 6.6 | 6,894 | 4,820 | 2,074 | 4 | 18 |
| Richland | 364 | 15,714 | 0.99 | 165 | 10.6 | 196 | 12.6 | 926 | 7.7 | 3,755 | 3,185 | 569 | NA | NA |
| Rock Island | 4,586 | 152,490 | 1.13 | 1,686 | 12.0 | 1,536 | 10.9 | 8,856 | 8.1 | 31,140 | 17,718 | 13,422 | 54 | 222 |
| St. Clair | 4,561 | 243,928 | 0.85 | 2,985 | 11.6 | 2,926 | 11.3 | 15,294 | 7.2 | 49,451 | 29,363 | 20,088 | NA | NA |
| Saline | 883 | 23,042 | 0.90 | 270 | 11.6 | 343 | 14.8 | 1,366 | 7.5 | 6,093 | 4,914 | 1,179 | NA | NA |
| Sangamon | 4,017 | 211,382 | 1.16 | 2,166 | 11.2 | 2,315 | 11.9 | 10,156 | 6.5 | 41,685 | 24,333 | 17,352 | 204 | 467 |
| Schuyler | 463 | 6,123 | 0.71 | 61 | 9.1 | 77 | 11.4 | 376 | 7.3 | 1,658 | 1,308 | 350 | 3 | 14 |
| Scott | 43 | 4,170 | 0.63 | 41 | 8.3 | 53 | 10.7 | 226 | 5.7 | 1,126 | 869 | 257 | 0 | 15 |
| Shelby | 205 | 17,926 | 0.63 | 235 | 11.0 | 276 | 13.0 | 1,004 | 6.1 | 5,360 | 4,616 | 744 | 1 | 49 |
| Stark | 92 | 4,643 | 0.67 | 51 | 9.7 | 68 | 12.9 | 293 | 7.2 | 1,304 | 950 | 355 | 9 | 33 |
| Stephenson | 836 | 43,436 | 0.92 | 471 | 10.7 | 597 | 13.6 | 2,211 | 6.6 | 11,688 | 6,430 | 5,258 | 36 | 87 |
| Tazewell | 2,565 | 126,344 | 0.89 | 1,339 | 10.2 | 1,556 | 11.9 | 5,767 | 5.5 | 28,902 | 19,554 | 9,348 | NA | NA |
| Union | 565 | 15,721 | 0.82 | 178 | 10.8 | 239 | 14.5 | 1,080 | 8.4 | 4,497 | 3,339 | 1,158 | NA | NA |
| Vermilion | 2,906 | 77,261 | 0.99 | 880 | 11.8 | 981 | 13.1 | 3,829 | 6.6 | 17,556 | 10,070 | 7,486 | 87 | 403 |
| Wabash | 88 | 10,005 | 0.71 | 135 | 12.1 | 134 | 12.0 | 606 | 6.7 | 2,813 | 2,445 | 368 | 0 | 2 |
| Warren | 978 | 16,517 | 0.92 | 196 | 11.7 | 189 | 11.3 | 1,059 | 8.5 | 3,940 | 3,004 | 937 | NA | NA |
| Washington | 246 | 14,256 | 1.03 | 146 | 10.6 | 147 | 10.7 | 611 | 5.6 | 3,222 | 2,633 | 589 | NA | NA |
| Wayne | 76 | 14,338 | 0.72 | 200 | 12.5 | 224 | 14.0 | 1,054 | 8.4 | 3,899 | 3,525 | 374 | 15 | 108 |
| White | 391 | 13,261 | 0.90 | 121 | 9.1 | 199 | 14.9 | 749 | 7.2 | 3,638 | 3,148 | 491 | NA | NA |
| Whiteside | 1,011 | 53,177 | 0.89 | 580 | 10.6 | 647 | 11.8 | 2,895 | 6.8 | 13,964 | 10,769 | 3,195 | 17 | 79 |
| Will | 8,743 | 617,365 | 0.79 | 7,169 | 10.4 | 5,185 | 7.5 | 40,663 | 6.9 | 107,466 | 76,580 | 30,886 | 125 | 658 |
| Williamson | 1,997 | 68,081 | 1.03 | 702 | 10.6 | 849 | 12.8 | 3,504 | 6.7 | 15,250 | 11,919 | 3,331 | 8 | 248 |

1. Per 1,000 estimated resident population.

| STATE County | County law enforcement employment, 2019 | | Education | | | | | | Money income, 2015-2019 | | | | Income and poverty, 2019 | | | |
|---|---|---|---|---|---|---|---|---|---|---|---|---|---|---|---|---|
| | | | School enrollment and attainment, 2015-2019 | | | | Local government expenditures,[3] 2017-2018 | | | | Households | | | Percent below poverty level | | |
| | | | Enrollment[1] | | Attainment[2] (percent) | | | | | | | Percent | | | | |
| | Officers | Civilians | Total | Percent private | High school graduate or less | Bachelor's degree or more | Total current spending (mil dol) | Current spending per student (dollars) | Per capita income[4] | Median income (dollars) | with income of less than $50,000 | with income of $200,000 or more | Median household income (dollars) | All persons | Children under 18 years | Children 5 to 17 years in families |
| | 46 | 47 | 48 | 49 | 50 | 51 | 52 | 53 | 54 | 55 | 56 | 57 | 58 | 59 | 60 | 61 |
| ILLINOIS—Cont'd | | | | | | | | | | | | | | | | |
| Hardin | NA | NA | 677 | 2.2 | 52.7 | 10.8 | 6 | 10,531 | 28,932 | 51,250 | 48.9 | 2.5 | 44,573 | 18.0 | 28.7 | 25.1 |
| Henderson | NA | NA | 1,194 | 18.8 | 50.5 | 18.3 | 10 | 13,032 | 28,950 | 53,676 | 45.3 | 1.3 | 55,535 | 11.0 | 16.9 | 16.8 |
| Henry | 22 | 43 | 11,037 | 9.4 | 42.5 | 23.3 | 109 | 12,957 | 30,345 | 59,933 | 42.1 | 2.9 | 61,680 | 8.7 | 12.5 | 9.9 |
| Iroquois | 13 | 2 | 6,036 | 11.0 | 50.3 | 15.9 | 67 | 15,423 | 27,574 | 52,700 | 48.0 | 2.2 | 51,542 | 11.6 | 15.2 | 14.8 |
| Jackson | 32 | 54 | 20,823 | 6.8 | 32.2 | 34.1 | 99 | 14,110 | 24,804 | 37,241 | 61.2 | 3.2 | 37,308 | 25.4 | 28.0 | 23.1 |
| Jasper | 7 | 8 | 2,011 | 13.9 | 49.4 | 16.6 | 22 | 16,577 | 25,849 | 54,256 | 45.7 | 2.0 | 59,980 | 9.9 | 14.3 | 14.0 |
| Jefferson | 20 | 41 | 8,116 | 13.2 | 43.2 | 17.9 | 73 | 12,830 | 26,466 | 49,896 | 50.1 | 3.0 | 50,531 | 14.9 | 23.8 | 23.1 |
| Jersey | NA | NA | 5,334 | 28.4 | 46.5 | 20.3 | 27 | 10,651 | 30,182 | 63,028 | 41.4 | 3.2 | 63,351 | 10.4 | 14.4 | 12.1 |
| Jo Daviess | 18 | 20 | 4,265 | 10.0 | 43.6 | 25.0 | 44 | 13,109 | 34,437 | 57,946 | 43.1 | 3.3 | 62,522 | 9.3 | 12.8 | 11.2 |
| Johnson | NA | NA | 2,305 | 4.3 | 49.3 | 16.0 | 21 | 11,529 | 23,140 | 52,774 | 47.9 | 2.1 | 55,022 | 12.9 | 16.8 | 14.5 |
| Kane | 87 | 35 | 147,104 | 15.0 | 38.1 | 33.1 | 1,410 | 14,273 | 36,270 | 79,394 | 30.4 | 10.3 | 84,311 | 8.2 | 13.0 | 10.9 |
| Kankakee | 49 | 142 | 28,329 | 20.7 | 46.2 | 20.8 | 230 | 12,901 | 27,984 | 58,902 | 42.6 | 3.4 | 60,763 | 12.7 | 18.3 | 16.8 |
| Kendall | 60 | 58 | 37,274 | 13.8 | 30.4 | 36.2 | 494 | 18,188 | 36,382 | 96,563 | 19.7 | 10.2 | 105,586 | 4.0 | 5.3 | 4.4 |
| Knox | NA | NA | 11,535 | 19.7 | 47.5 | 18.9 | 97 | 13,320 | 24,545 | 44,129 | 55.1 | 1.9 | 43,404 | 18.6 | 24.2 | 22.0 |
| Lake | 162 | 65 | 188,905 | 16.4 | 30.1 | 45.3 | 2,640 | 19,898 | 45,766 | 89,427 | 28.1 | 16.7 | 92,638 | 7.2 | 10.3 | 9.1 |
| LaSalle | NA | NA | 23,912 | 13.1 | 47.1 | 17.8 | 248 | 15,920 | 30,277 | 58,142 | 43.6 | 3.4 | 59,528 | 12.2 | 18.1 | 16.3 |
| Lawrence | NA | NA | 2,633 | 5.7 | 49.8 | 14.9 | 24 | 10,840 | 23,613 | 46,636 | 53.3 | 2.7 | 47,114 | 18.4 | 25.4 | 25.2 |
| Lee | NA | NA | 7,278 | 16.4 | 45.0 | 18.2 | 61 | 14,729 | 29,984 | 58,194 | 43.8 | 3.5 | 57,619 | 12.0 | 16.8 | 15.8 |
| Livingston | 30 | 36 | 7,279 | 11.0 | 54.8 | 14.8 | 92 | 15,659 | 27,925 | 55,160 | 45.9 | 2.3 | 54,422 | 13.0 | 18.0 | 17.5 |
| Logan | 23 | 3 | 6,816 | 22.8 | 46.6 | 19.9 | 48 | 14,392 | 27,546 | 57,308 | 42.5 | 3.7 | 53,806 | 12.8 | 16.3 | 15.6 |
| McDonough | 14 | 10 | 10,960 | 5.2 | 37.6 | 32.2 | 58 | 17,246 | 24,094 | 44,471 | 54.4 | 2.4 | 49,048 | 23.3 | 23.0 | 19.6 |
| McHenry | 98 | 248 | 78,407 | 12.6 | 33.4 | 34.4 | 1,091 | 15,550 | 39,006 | 86,799 | 26.0 | 9.8 | 89,379 | 5.4 | 6.6 | 5.9 |
| McLean | 51 | 79 | 57,327 | 12.8 | 28.7 | 44.8 | 338 | 13,267 | 34,532 | 67,675 | 38.0 | 7.1 | 69,230 | 14.3 | 9.6 | 8.1 |
| Macon | 48 | 103 | 23,713 | 18.0 | 45.0 | 22.1 | 202 | 12,533 | 29,644 | 50,480 | 49.6 | 4.2 | 52,259 | 16.7 | 21.5 | 21.8 |
| Macoupin | 19 | 30 | 10,505 | 12.4 | 48.5 | 18.4 | 96 | 11,570 | 28,809 | 55,159 | 46.4 | 2.9 | 54,174 | 13.5 | 18.3 | 16.2 |
| Madison | 85 | 82 | 65,753 | 14.2 | 37.4 | 26.8 | 472 | 11,720 | 32,427 | 60,738 | 41.3 | 4.6 | 66,591 | 10.9 | 14.9 | 14.5 |
| Marion | NA | NA | 8,061 | 11.0 | 46.2 | 15.2 | 85 | 12,588 | 25,563 | 47,519 | 52.5 | 1.6 | 52,291 | 15.0 | 22.8 | 22.2 |
| Marshall | 8 | 9 | 2,200 | 11.7 | 46.5 | 18.9 | 17 | 13,394 | 30,808 | 58,465 | 42.5 | 2.0 | 63,628 | 9.0 | 13.8 | 12.6 |
| Mason | 9 | 13 | 2,696 | 10.3 | 53.6 | 16.0 | 34 | 13,153 | 28,534 | 48,417 | 51.1 | 3.1 | 51,967 | 12.1 | 19.6 | 16.9 |
| Massac | NA | NA | 3,038 | 3.7 | 46.8 | 13.9 | 31 | 13,540 | 23,539 | 47,481 | 52.4 | 1.2 | 54,159 | 15.9 | 23.9 | 21.3 |
| Menard | NA | NA | 2,593 | 11.5 | 41.8 | 24.0 | 27 | 10,883 | 37,098 | 74,684 | 32.4 | 5.6 | 77,306 | 8.2 | 12.8 | 12.1 |
| Mercer | 12 | 18 | 3,323 | 8.4 | 47.7 | 17.4 | 35 | 12,463 | 29,268 | 59,787 | 40.5 | 1.6 | 62,241 | 8.1 | 12.4 | 10.8 |
| Monroe | NA | NA | 7,908 | 21.0 | 31.7 | 35.0 | 63 | 12,226 | 42,152 | 85,747 | 30.1 | 9.6 | 90,904 | 4.2 | 3.8 | 3.2 |
| Montgomery | 13 | 18 | 5,902 | 14.7 | 51.6 | 16.7 | 49 | 11,512 | 26,609 | 52,748 | 47.1 | 2.3 | 51,665 | 16.0 | 21.6 | 19.8 |
| Morgan | NA | NA | 8,146 | 29.6 | 49.3 | 19.3 | 70 | 15,223 | 26,755 | 51,437 | 48.8 | 1.8 | 55,937 | 13.8 | 17.3 | 16.6 |
| Moultrie | NA | NA | 3,077 | 15.6 | 48.6 | 18.6 | 19 | 11,337 | 29,538 | 61,456 | 41.0 | 3.0 | 65,472 | 8.6 | 12.6 | 11.9 |
| Ogle | NA | NA | 11,819 | 11.5 | 43.2 | 21.4 | 132 | 15,080 | 31,397 | 60,986 | 40.3 | 3.3 | 63,565 | 9.4 | 13.0 | 12.4 |
| Peoria | NA | NA | 45,975 | 24.5 | 36.7 | 30.6 | 390 | 13,846 | 32,041 | 55,842 | 45.1 | 5.2 | 57,717 | 14.8 | 21.9 | 19.5 |
| Perry | 6 | 15 | 4,104 | 7.4 | 53.3 | 12.5 | 28 | 10,312 | 26,035 | 52,428 | 47.8 | 1.8 | 56,880 | 14.4 | 18.3 | 16.2 |
| Piatt | 14 | 22 | 3,513 | 14.2 | 40.0 | 28.6 | 38 | 14,370 | 33,468 | 70,849 | 37.3 | 3.8 | 72,265 | 6.5 | 7.5 | 6.8 |
| Pike | 13 | 17 | 3,423 | 10.3 | 50.0 | 17.8 | 31 | 12,268 | 26,157 | 47,492 | 51.8 | 2.5 | 54,309 | 14.4 | 20.5 | 19.2 |
| Pope | NA | NA | 579 | 9.5 | 50.0 | 13.1 | 6 | 11,705 | 22,065 | 38,056 | 57.9 | 0.6 | 46,478 | 16.9 | 23.0 | 22.1 |
| Pulaski | NA | NA | 1,162 | 7.9 | 47.5 | 12.5 | 16 | 19,635 | 21,605 | 34,640 | 64.5 | 1.6 | 38,506 | 22.0 | 34.0 | 32.6 |
| Putnam | 8 | 5 | 1,114 | 9.8 | 45.4 | 16.6 | 13 | 14,974 | 34,438 | 63,638 | 38.7 | 3.2 | 74,367 | 7.6 | 13.6 | 12.4 |
| Randolph | NA | NA | 6,312 | 21.2 | 53.0 | 13.3 | 52 | 12,911 | 26,558 | 53,816 | 45.9 | 3.2 | 58,444 | 12.7 | 16.9 | 16.0 |
| Richland | NA | NA | 3,526 | 9.0 | 40.8 | 21.0 | 26 | 10,958 | 26,358 | 48,894 | 51.1 | 2.8 | 49,775 | 13.1 | 17.6 | 16.0 |
| Rock Island | 65 | 89 | 33,727 | 18.1 | 40.8 | 23.5 | 287 | 13,404 | 29,756 | 54,858 | 44.7 | 2.9 | 57,772 | 13.6 | 21.2 | 20.0 |
| St. Clair | NA | NA | 65,640 | 16.3 | 37.1 | 28.0 | 588 | 14,509 | 30,807 | 55,179 | 46.1 | 4.8 | 57,726 | 13.3 | 20.0 | 18.1 |
| Saline | NA | NA | 4,935 | 8.5 | 40.8 | 19.2 | 55 | 13,858 | 25,342 | 44,090 | 56.1 | 2.7 | 48,547 | 15.7 | 24.6 | 22.7 |
| Sangamon | 77 | 127 | 48,407 | 14.8 | 34.7 | 34.5 | 384 | 13,151 | 35,509 | 61,912 | 40.8 | 5.6 | 61,768 | 12.2 | 18.0 | 16.0 |
| Schuyler | NA | NA | 1,276 | 7.4 | 46.2 | 20.4 | 13 | 12,269 | 24,831 | 52,357 | 47.9 | 2.2 | 56,015 | 11.3 | 16.5 | 15.3 |
| Scott | 3 | 4 | 1,144 | 13.4 | 54.3 | 15.8 | 11 | 13,054 | 27,052 | 57,118 | 45.0 | 1.7 | 61,231 | 11.2 | 16.8 | 15.5 |
| Shelby | 14 | 14 | 4,412 | 16.0 | 52.1 | 16.8 | 27 | 11,742 | 27,029 | 52,953 | 47.1 | 1.9 | 61,376 | 9.4 | 12.9 | 11.8 |
| Stark | NA | NA | 1,085 | 15.9 | 50.3 | 17.0 | 12 | 13,159 | 30,018 | 54,907 | 45.1 | 3.0 | 54,804 | 11.2 | 17.4 | 16.1 |
| Stephenson | 28 | 10 | 9,538 | 12.1 | 42.5 | 19.6 | 90 | 14,081 | 27,911 | 48,805 | 51.6 | 2.1 | 51,786 | 11.5 | 18.4 | 15.2 |
| Tazewell | 41 | 19 | 31,132 | 15.0 | 39.0 | 26.3 | 267 | 13,442 | 33,507 | 63,454 | 38.2 | 5.1 | 66,206 | 8.2 | 9.9 | 9.1 |
| Union | NA | NA | 3,299 | 7.6 | 47.0 | 21.7 | 35 | 11,807 | 25,879 | 50,625 | 49.1 | 2.5 | 51,575 | 17.2 | 24.7 | 20.4 |
| Vermilion | 41 | 56 | 17,400 | 6.5 | 52.7 | 14.7 | 182 | 14,416 | 25,307 | 46,515 | 53.1 | 2.1 | 46,295 | 16.8 | 27.0 | 27.0 |
| Wabash | 5 | 7 | 2,418 | 17.1 | 40.1 | 18.2 | 15 | 9,444 | 27,781 | 50,770 | 49.4 | 2.7 | 54,249 | 13.5 | 17.7 | 16.7 |
| Warren | NA | NA | 4,657 | 27.7 | 45.3 | 22.8 | 31 | 11,574 | 25,730 | 50,310 | 49.5 | 2.5 | 53,621 | 11.4 | 14.9 | 14.2 |
| Washington | NA | NA | 2,871 | 29.1 | 41.2 | 20.9 | 22 | 12,332 | 32,838 | 61,763 | 40.3 | 2.3 | 64,297 | 8.1 | 11.1 | 10.4 |
| Wayne | 9 | 12 | 3,316 | 8.0 | 47.0 | 15.4 | 27 | 10,961 | 26,313 | 50,251 | 49.7 | 1.3 | 51,829 | 13.1 | 22.5 | 21.3 |
| White | NA | NA | 2,832 | 7.1 | 47.4 | 15.5 | 35 | 14,291 | 26,746 | 49,290 | 50.6 | 2.0 | 51,388 | 14.5 | 20.5 | 19.1 |
| Whiteside | 20 | 43 | 11,881 | 12.0 | 47.1 | 18.2 | 126 | 13,970 | 29,856 | 54,232 | 46.4 | 3.3 | 50,880 | 11.4 | 16.8 | 15.7 |
| Will | 238 | 358 | 188,034 | 15.0 | 35.9 | 34.3 | 1,623 | 14,389 | 36,524 | 86,961 | 26.4 | 10.2 | 90,249 | 6.5 | 9.0 | 7.9 |
| Williamson | NA | NA | 14,622 | 9.0 | 38.8 | 24.3 | 125 | 12,055 | 28,375 | 50,734 | 49.2 | 3.3 | 56,958 | 13.4 | 18.4 | 16.4 |

1. All persons 3 years old and over enrolled in nursery school through college.    2. Persons 25 years old and over.    3. Elementary and secondary education expenditures.    4. Based on population estimated by the American Community Survey, 2014–2018.

# Table B. States and Counties — **Personal Income**

| STATE County | Total (mil dol) | Percent change 2018-2019 | Per capita[1] Dollars | Per capita[1] Rank | Wages and salaries (mil dol) | Pension and insurance | Government social insurance | Proprietors' income (mil dol) | Dividends, interest, and rent (mil dol) | Personal transfer receipts (mil dol) | Earnings Total (mil dol) | From employee and self-employed | From employer |
|---|---|---|---|---|---|---|---|---|---|---|---|---|---|
| | 62 | 63 | 64 | 65 | 66 | 67 | 68 | 69 | 70 | 71 | 72 | 73 | 74 |
| ILLINOIS—Cont'd | | | | | | | | | | | | | |
| Hardin | 143 | 0.0 | 37,438 | 2,395 | 25 | 8 | 2 | 11 | 21 | 54 | 45 | 4 | 2 |
| Henderson | 282 | -1.7 | 42,414 | 1,702 | 45 | 13 | 3 | 22 | 52 | 70 | 82 | 6 | 3 |
| Henry | 2,284 | 2.1 | 46,700 | 1,159 | 617 | 145 | 41 | 124 | 425 | 454 | 928 | 61 | 41 |
| Iroquois | 1,220 | 0.6 | 44,994 | 1,360 | 327 | 71 | 24 | 131 | 219 | 308 | 552 | 34 | 24 |
| Jackson | 2,155 | 2.0 | 37,967 | 2,328 | 1,283 | 354 | 77 | 128 | 446 | 509 | 1,842 | 91 | 77 |
| Jasper | 402 | -2.0 | 41,811 | 1,795 | 93 | 27 | 6 | 41 | 82 | 93 | 167 | 10 | 6 |
| Jefferson | 1,527 | 2.6 | 40,516 | 1,988 | 974 | 184 | 68 | 88 | 240 | 414 | 1,314 | 78 | 68 |
| Jersey | 933 | 1.7 | 42,836 | 1,644 | 207 | 50 | 14 | 31 | 139 | 221 | 302 | 22 | 14 |
| Jo Daviess | 1,035 | 1.7 | 48,743 | 945 | 300 | 68 | 21 | 41 | 269 | 227 | 430 | 32 | 21 |
| Johnson | 456 | 2.3 | 36,717 | 2,480 | 92 | 28 | 5 | 18 | 66 | 125 | 143 | 11 | 5 |
| Kane | 27,772 | 3.1 | 52,163 | 666 | 12,019 | 2,196 | 815 | 1,602 | 4,361 | 3,681 | 16,632 | 929 | 815 |
| Kankakee | 4,690 | 3.1 | 42,687 | 1,659 | 2,197 | 457 | 154 | 194 | 660 | 1,117 | 3,002 | 178 | 154 |
| Kendall | 6,441 | 4.2 | 49,931 | 836 | 1,401 | 288 | 92 | 254 | 696 | 690 | 2,035 | 120 | 92 |
| Knox | 2,006 | 1.0 | 40,363 | 2,007 | 840 | 176 | 74 | 103 | 363 | 605 | 1,193 | 81 | 74 |
| Lake | 56,401 | 2.5 | 80,973 | 63 | 29,229 | 4,616 | 1,847 | 3,710 | 13,999 | 5,066 | 39,403 | 2,151 | 1,847 |
| LaSalle | 5,000 | 2.4 | 46,013 | 1,246 | 2,111 | 435 | 147 | 275 | 802 | 1,066 | 2,967 | 181 | 147 |
| Lawrence | 484 | -0.9 | 30,877 | 3,012 | 191 | 47 | 13 | 38 | 94 | 167 | 288 | 19 | 13 |
| Lee | 1,458 | 2.0 | 42,776 | 1,650 | 603 | 120 | 41 | 72 | 251 | 340 | 837 | 54 | 41 |
| Livingston | 1,607 | -0.7 | 45,090 | 1,353 | 664 | 138 | 45 | 201 | 254 | 338 | 1,049 | 59 | 45 |
| Logan | 1,078 | -0.3 | 37,660 | 2,364 | 402 | 90 | 27 | 77 | 179 | 273 | 597 | 37 | 27 |
| McDonough | 1,144 | | 38,550 | 2,253 | 512 | 150 | 30 | 78 | 243 | 262 | 771 | 38 | 30 |
| McHenry | 17,454 | 3.2 | 56,711 | 415 | 4,993 | 959 | 343 | 824 | 2,720 | 2,226 | 7,119 | 426 | 343 |
| McLean | 8,508 | 0.2 | 49,602 | 867 | 5,093 | 908 | 312 | 631 | 1,422 | 1,167 | 6,944 | 379 | 312 |
| Macon | 5,072 | 0.7 | 48,764 | 943 | 2,850 | 509 | 202 | 340 | 883 | 1,158 | 3,900 | 235 | 202 |
| Macoupin | 1,902 | 1.5 | 42,341 | 1,715 | 463 | 109 | 31 | 88 | 317 | 487 | 690 | 51 | 31 |
| Madison | 12,681 | 2.8 | 48,222 | 998 | 5,137 | 1,030 | 363 | 694 | 1,991 | 2,593 | 7,225 | 436 | 363 |
| Marion | 1,596 | 2.2 | 42,895 | 1,636 | 598 | 134 | 45 | 93 | 252 | 494 | 870 | 57 | 45 |
| Marshall | 525 | -1.1 | 45,901 | 1,261 | 128 | 30 | 9 | 31 | 93 | 128 | 198 | 14 | 9 |
| Mason | 550 | -2.0 | 41,137 | 1,901 | 137 | 39 | 8 | 40 | 95 | 155 | 225 | 15 | 8 |
| Massac | 536 | 2.0 | 38,926 | 2,206 | 161 | 38 | 10 | 14 | 89 | 182 | 223 | 17 | 10 |
| Menard | 573 | 0.9 | 47,012 | 1,133 | 70 | 21 | 4 | 32 | 100 | 114 | 127 | 10 | 4 |
| Mercer | 736 | 0.8 | 47,701 | 1,053 | 124 | 31 | 9 | 57 | 121 | 161 | 221 | 16 | 9 |
| Monroe | 2,048 | 4.0 | 59,125 | 335 | 366 | 75 | 25 | 103 | 334 | 297 | 569 | 38 | 25 |
| Montgomery | 1,032 | 0.3 | 36,324 | 2,532 | 373 | 86 | 26 | 75 | 205 | 302 | 559 | 37 | 26 |
| Morgan | 1,365 | 0.9 | 40,560 | 1,978 | 635 | 129 | 43 | 85 | 259 | 361 | 893 | 55 | 43 |
| Moultrie | 934 | 0.2 | 64,409 | 200 | 194 | 40 | 14 | 351 | 119 | 140 | 598 | 33 | 14 |
| Ogle | 2,399 | 1.2 | 47,369 | 1,093 | 816 | 177 | 54 | 94 | 395 | 457 | 1,141 | 73 | 54 |
| Peoria | 9,154 | 0.9 | 51,089 | 750 | 6,911 | 1,102 | 451 | 438 | 1,703 | 1,718 | 8,902 | 515 | 451 |
| Perry | 799 | 1.8 | 38,179 | 2,308 | 206 | 57 | 13 | 107 | 124 | 215 | 383 | 24 | 13 |
| Piatt | 898 | 0.2 | 54,940 | 494 | 153 | 39 | 10 | 63 | 145 | 141 | 265 | 17 | 10 |
| Pike | 653 | 0.8 | 41,957 | 1,772 | 168 | 41 | 12 | 77 | 111 | 171 | 298 | 18 | 12 |
| Pope | 123 | 0.9 | 29,510 | 3,053 | 20 | 7 | 1 | 5 | 22 | 47 | 32 | 3 | 1 |
| Pulaski | 202 | 1.8 | 37,936 | 2,332 | 98 | 26 | 6 | 9 | 30 | 81 | 140 | 8 | 6 |
| Putnam | 352 | 4.1 | 61,363 | 267 | 101 | 21 | 7 | 91 | 53 | 56 | 220 | 12 | 7 |
| Randolph | 1,207 | 3.2 | 37,988 | 2,325 | 525 | 125 | 35 | 67 | 205 | 313 | 751 | 45 | 35 |
| Richland | 656 | 0.8 | 42,275 | 1,725 | 260 | 55 | 18 | 41 | 122 | 172 | 375 | 24 | 18 |
| Rock Island | 6,485 | 2.3 | 45,711 | 1,282 | 4,770 | 874 | 315 | 399 | 1,265 | 1,362 | 6,358 | 370 | 315 |
| St. Clair | 11,976 | 2.7 | 46,116 | 1,234 | 5,120 | 1,144 | 382 | 492 | 2,107 | 2,718 | 7,139 | 417 | 382 |
| Saline | 961 | 2.0 | 40,891 | 1,931 | 342 | 80 | 22 | 67 | 157 | 318 | 511 | 34 | 22 |
| Sangamon | 9,625 | 2.4 | 49,445 | 886 | 5,434 | 1,098 | 357 | 533 | 1,873 | 1,847 | 7,423 | 412 | 357 |
| Schuyler | 302 | 2.6 | 44,586 | 1,410 | 77 | 20 | 5 | 34 | 46 | 66 | 135 | 8 | 5 |
| Scott | 208 | -1.7 | 41,942 | 1,775 | 45 | 13 | 3 | 22 | 32 | 43 | 83 | 5 | 3 |
| Shelby | 877 | -0.3 | 40,517 | 1,987 | 222 | 49 | 16 | 86 | 146 | 215 | 374 | 25 | 16 |
| Stark | 227 | -5.8 | 42,563 | 1,677 | 62 | 15 | 5 | 18 | 45 | 55 | 99 | 6 | 5 |
| Stephenson | 1,913 | 0.8 | 42,980 | 1,620 | 839 | 166 | 59 | 78 | 370 | 479 | 1,142 | 75 | 59 |
| Tazewell | 6,394 | 1.1 | 48,515 | 968 | 2,448 | 453 | 172 | 299 | 1,117 | 1,229 | 3,373 | 212 | 172 |
| Union | 724 | 1.6 | 43,491 | 1,546 | 195 | 52 | 13 | 35 | 115 | 222 | 295 | 20 | 13 |
| Vermilion | 3,039 | 1.1 | 40,109 | 2,039 | 1,321 | 302 | 93 | 180 | 476 | 850 | 1,896 | 117 | 93 |
| Wabash | 496 | 1.3 | 43,098 | 1,602 | 152 | 42 | 10 | 28 | 99 | 119 | 231 | 14 | 10 |
| Warren | 695 | -1.6 | 41,270 | 1,884 | 268 | 56 | 19 | 76 | 124 | 156 | 420 | 25 | 19 |
| Washington | 677 | 0.9 | 48,740 | 946 | 358 | 82 | 25 | 76 | 126 | 135 | 541 | 31 | 25 |
| Wayne | 642 | 2.3 | 39,592 | 2,105 | 164 | 43 | 12 | 84 | 109 | 170 | 302 | 19 | 12 |
| White | 684 | 1.5 | 50,560 | 794 | 192 | 41 | 14 | 110 | 121 | 168 | 356 | 21 | 14 |
| Whiteside | 2,478 | 2.4 | 44,916 | 1,370 | 984 | 237 | 67 | 124 | 452 | 612 | 1,412 | 85 | 67 |
| Will | 37,228 | 3.1 | 53,895 | 547 | 13,591 | 2,464 | 948 | 1,846 | 4,970 | 4,826 | 18,850 | 1,084 | 948 |
| Williamson | 3,055 | 2.7 | 45,878 | 1,265 | 1,317 | 309 | 92 | 166 | 473 | 695 | 1,885 | 114 | 92 |

1. Based on the resident population estimated as of July 1 of the year shown.

# Table B. States and Counties — Earnings, Social Security, and Housing

| STATE County | Earnings, 2019 (cont.) — Percent by selected industries | | | | | | | | | Social Security beneficiaries, December 2019 | | Supplemental Security Income recipients, 2019 | Housing units, 2020 | |
|---|---|---|---|---|---|---|---|---|---|---|---|---|---|---|
| | Farm | Mining, quarrying, and extractions | Construction | Manufacturing | Information; professional, scientific, technical services | Retail trade | Finance, insurance, real estate, and leasing | Health care and social assistance | Government | Number | Rate[1] | | Total | Percent change, 2010-2020 |
| | 75 | 76 | 77 | 78 | 79 | 80 | 81 | 82 | 83 | 84 | 85 | 86 | 87 | 88 |
| **ILLINOIS—Cont'd** | | | | | | | | | | | | | | |
| Hardin | 6.4 | D | D | D | D | 4.3 | D | 22.9 | 26.0 | 1,200 | 312 | 165 | 2,483 | -0.2 |
| Henderson | 18.3 | D | 7.4 | 1.7 | D | D | D | 6.5 | 27.3 | 1,780 | 265 | 117 | 3,869 | 1.1 |
| Henry | 2.6 | D | 11.7 | 9.0 | 3.9 | 7.2 | 6.1 | 11.3 | 24.0 | 11,855 | 242 | 604 | 22,166 | 0.0 |
| Iroquois | 14.2 | D | 8.6 | 7.2 | D | 8.1 | 6.5 | D | 16.0 | 7,150 | 264 | 565 | 13,507 | 0.4 |
| Jackson | 0.2 | D | 4.5 | 3.1 | 3.8 | 5.7 | 2.8 | 19.2 | 41.5 | 10,265 | 181 | 1,643 | 28,921 | 1.2 |
| Jasper | 14.3 | 3.0 | 7.2 | 6.1 | 1.4 | 5.6 | 5.4 | 1.2 | 23.7 | 2,410 | 251 | 133 | 4,345 | 0.0 |
| Jefferson | 0.2 | 1.1 | 3.1 | 25.2 | 3.8 | 6.8 | 5.1 | 18.6 | 13.3 | 8,940 | 238 | 920 | 17,065 | 0.7 |
| Jersey | -0.5 | D | 8.8 | 3.5 | D | 9.4 | 4.8 | D | 29.5 | 5,495 | 254 | 346 | 10,236 | 3.8 |
| Jo Daviess | -1.6 | D | 10.7 | 13.2 | D | 7.2 | 4.5 | D | 20.8 | 6,550 | 309 | 180 | 13,740 | 1.2 |
| Johnson | 0.8 | D | 5.6 | 2.0 | 2.8 | 4.1 | D | 8.2 | 50.4 | 3,230 | 261 | 274 | 5,633 | 0.7 |
| Kane | 0.0 | 0.1 | 7.8 | 17.0 | 8.9 | 5.1 | 5.5 | 10.2 | 17.4 | 79,865 | 150 | 5,197 | 191,891 | 5.4 |
| Kankakee | 1.3 | D | 4.8 | 23.2 | 3.9 | 6.2 | 4.0 | 16.4 | 15.8 | 22,945 | 209 | 2,646 | 45,772 | 1.2 |
| Kendall | 0.4 | 0.1 | 9.2 | 15.6 | 6.2 | 8.6 | 3.7 | 5.5 | 23.7 | 17,105 | 133 | 807 | 43,249 | 7.2 |
| Knox | 1.7 | D | 3.9 | 7.1 | D | 7.0 | 4.0 | D | 18.5 | 12,215 | 246 | 1,303 | 23,851 | -0.9 |
| Lake | 0.0 | 0.0 | 4.2 | 22.3 | 10.0 | 6.8 | 6.8 | 6.3 | 13.3 | 110,265 | 158 | 8,226 | 265,832 | 2.1 |
| LaSalle | 1.6 | 3.5 | 5.8 | 19.0 | 3.7 | 6.6 | 4.9 | 8.4 | 15.3 | 25,855 | 238 | 1,692 | 50,190 | 0.4 |
| Lawrence | 0.2 | 11.7 | D | D | 1.9 | 3.9 | 11.0 | 9.4 | 21.3 | 3,455 | 221 | 293 | 7,271 | 4.8 |
| Lee | -0.1 | 0.9 | 3.5 | 29.8 | 2.6 | 5.6 | 4.6 | D | 16.8 | 8,275 | 243 | 497 | 15,097 | 0.3 |
| Livingston | 7.8 | D | 5.9 | 19.9 | 6.4 | 8.7 | 3.8 | 9.3 | 16.1 | 8,295 | 233 | 495 | 15,913 | 0.1 |
| Logan | 5.3 | 0.0 | 3.6 | 19.8 | 3.2 | 6.0 | 4.3 | 12.4 | 18.1 | 6,395 | 223 | 390 | 12,019 | -0.7 |
| McDonough | 4.0 | 0.0 | 4.6 | 14.9 | 2.6 | 5.8 | 2.8 | 5.9 | 46.4 | 5,790 | 194 | 545 | 14,363 | -0.4 |
| McHenry | -0.1 | 0.1 | 11.6 | 17.4 | 6.6 | 7.9 | 3.8 | 8.8 | 16.5 | 53,540 | 174 | 2,109 | 119,583 | 3.0 |
| McLean | 0.9 | 0.0 | 3.6 | 3.2 | D | 5.9 | 35.9 | 10.0 | 15.8 | 27,920 | 163 | 1,901 | 72,874 | 4.6 |
| Macon | 0.6 | 0.0 | 6.8 | 30.1 | 4.3 | 5.1 | 4.8 | 13.1 | 11.2 | 25,160 | 242 | 3,249 | 50,245 | -0.5 |
| Macoupin | 1.0 | D | 9.3 | 6.8 | D | 8.0 | 6.0 | D | 22.9 | 11,910 | 264 | 988 | 21,844 | 1.2 |
| Madison | 0.1 | 0.1 | 8.6 | 16.9 | 6.5 | 6.3 | 4.4 | 12.1 | 16.5 | 57,605 | 219 | 5,562 | 120,298 | 2.7 |
| Marion | 1.6 | 2.6 | 6.0 | 19.0 | 2.7 | 4.7 | 4.1 | 16.5 | 18.1 | 9,535 | 256 | 1,206 | 18,292 | 0.0 |
| Marshall | 5.5 | D | 8.6 | 30.7 | 1.7 | 3.9 | 3.1 | 8.3 | 16.3 | 3,180 | 277 | 154 | 5,922 | 0.1 |
| Mason | 7.7 | D | 4.7 | 3.4 | 2.1 | 4.4 | 5.2 | 7.1 | 34.4 | 3,580 | 267 | 295 | 7,029 | -0.7 |
| Massac | -0.8 | D | 3.7 | D | 1.5 | 4.5 | 2.1 | D | 29.6 | 4,085 | 297 | 450 | 7,105 | -0.1 |
| Menard | 8.4 | 0.0 | 9.4 | 1.5 | 7.5 | 6.8 | 5.7 | D | 31.7 | 2,885 | 237 | 173 | 5,746 | 1.6 |
| Mercer | 13.3 | 0.0 | 7.3 | 14.7 | 2.5 | 6.2 | 5.4 | D | 20.8 | 4,225 | 273 | 154 | 7,407 | 0.7 |
| Monroe | 2.7 | 0.4 | 11.1 | 5.2 | D | 8.3 | 10.4 | 6.6 | 17.3 | 7,185 | 207 | 158 | 14,623 | 9.1 |
| Montgomery | 3.0 | 1.5 | 6.4 | 8.2 | 3.2 | 9.1 | 6.2 | 16.4 | 19.5 | 7,185 | 253 | 626 | 13,143 | 0.5 |
| Morgan | 1.7 | 0.0 | 4.6 | 22.3 | 3.6 | 7.2 | 6.8 | 12.4 | 16.7 | 8,190 | 243 | 852 | 15,455 | -0.4 |
| Moultrie | 1.8 | 7.3 | 7.8 | 49.2 | 7.1 | 2.1 | 1.9 | 4.0 | 5.9 | 3,240 | 224 | 166 | 6,613 | 5.6 |
| Ogle | 0.3 | D | 6.3 | 18.8 | D | 4.2 | 5.6 | D | 17.1 | 11,400 | 225 | 576 | 22,732 | 0.8 |
| Peoria | 0.2 | 0.0 | 4.0 | 25.9 | 9.1 | 4.1 | 5.2 | 19.1 | 9.4 | 38,175 | 213 | 5,018 | 83,548 | 0.6 |
| Perry | 0.5 | D | 4.0 | 21.9 | D | 4.8 | 7.7 | D | 29.0 | 4,970 | 237 | 440 | 9,692 | 2.8 |
| Piatt | 8.0 | 0.4 | 7.1 | 6.8 | 4.1 | 6.3 | 7.1 | 11.2 | 20.7 | 3,885 | 249 | 338 | 8,020 | 2.8 |
| Pike | 13.5 | D | 6.6 | 2.4 | D | 6.6 | 10.3 | D | 44.5 | 1,195 | 286 | 97 | 2,505 | 0.9 |
| Pope | -1.2 | D | 4.0 | D | D | 2.6 | D | D | 45.2 | 1,575 | 297 | 248 | 3,184 | 0.9 |
| Pulaski | 2.6 | D | D | D | D | 2.5 | D | 1.9 | 8.7 | 1,485 | 258 | 53 | 3,191 | 3.7 |
| Putnam | 6.0 | 3.6 | D | 28.2 | D | 4.4 | 5.5 | D | 24.5 | 7,315 | 231 | 498 | 13,960 | 2.0 |
| Randolph | 2.5 | D | 5.1 | 20.1 | D | 7.9 | 3.2 | D | 17.4 | 4,030 | 260 | 420 | 7,510 | 0.0 |
| Richland | 3.6 | 2.5 | 3.3 | 8.8 | 2.2 | 6.2 | 5.8 | 16.1 | 17.4 | 4,030 | 260 | 420 | 7,510 | 0.0 |
| Rock Island | 0.1 | 0.1 | 4.2 | 10.8 | 7.1 | 5.0 | 4.1 | 9.1 | 19.3 | 32,185 | 227 | 2,760 | 66,137 | 0.6 |
| St. Clair | 0.2 | 0.3 | 5.5 | 6.3 | D | 6.3 | 3.8 | 12.4 | 31.0 | 51,135 | 197 | 7,621 | 121,322 | 4.4 |
| Saline | 0.3 | 1.2 | 8.3 | 4.5 | 10.9 | 8.1 | 6.3 | 17.5 | 22.0 | 6,530 | 277 | 1,063 | 11,693 | 0.0 |
| Sangamon | 0.5 | D | 4.2 | 2.9 | 10.6 | 5.7 | 8.2 | 21.4 | 24.8 | 43,870 | 225 | 4,511 | 92,473 | 2.9 |
| Schuyler | 13.1 | D | 7.0 | 4.9 | 1.7 | 4.9 | D | 5.8 | 29.1 | 1,635 | 239 | 88 | 3,441 | -0.5 |
| Scott | 12.4 | 0.0 | 4.8 | D | D | 3.5 | D | 1.0 | 22.2 | 1,085 | 218 | 70 | 2,451 | -0.4 |
| Shelby | 10.5 | D | 5.5 | 22.6 | 8.7 | 5.5 | 4.3 | 7.3 | 13.9 | 5,470 | 253 | 313 | 10,666 | 2.6 |
| Stark | 9.3 | 0.0 | 13.4 | 20.4 | D | 6.5 | D | 4.7 | 17.6 | 1,390 | 262 | 71 | 2,652 | -0.8 |
| Stephenson | -0.4 | 0.1 | 13.9 | 20.1 | 4.3 | 5.5 | 8.4 | 13.9 | 16.3 | 12,100 | 273 | 1,079 | 21,858 | |
| Tazewell | 0.9 | 0.1 | 9.7 | 16.2 | 5.4 | 8.2 | 6.4 | 6.4 | 16.0 | 30,640 | 232 | 1,853 | 58,980 | 2.5 |
| Union | 2.4 | D | 4.3 | 5.3 | 3.4 | 7.7 | 3.8 | D | 33.2 | 4,800 | 288 | 601 | 8,018 | 1.2 |
| Vermilion | 2.1 | D | 3.4 | 21.5 | 1.9 | 5.6 | 4.4 | 9.3 | 23.9 | 18,595 | 246 | 2,650 | 36,054 | -0.7 |
| Wabash | 2.1 | 11.8 | 11.1 | 6.4 | D | 6.1 | 4.4 | D | 32.0 | 2,930 | 258 | 187 | 5,595 | 0.2 |
| Warren | 8.5 | D | 5.1 | 31.5 | 2.0 | 5.4 | 3.5 | D | 15.0 | 3,640 | 216 | 285 | 7,667 | -0.2 |
| Washington | 0.7 | D | 4.6 | D | D | 9.5 | 3.8 | D | 9.0 | 3,245 | 233 | 129 | 6,681 | 2.2 |
| Wayne | 11.4 | 7.3 | 5.8 | 1.6 | D | 7.1 | 3.8 | D | 20.5 | 4,055 | 250 | 285 | 8,003 | 0.4 |
| White | 10.0 | 19.6 | D | D | D | 7.5 | 4.3 | D | 14.5 | 3,985 | 294 | 409 | 7,161 | -0.3 |
| Whiteside | 2.0 | D | 4.3 | 21.7 | 5.3 | 6.0 | 3.9 | 6.9 | 27.5 | 14,655 | 265 | 1,021 | 25,780 | 0.0 |
| Will | 0.1 | 0.1 | 8.4 | 11.3 | 7.0 | 6.7 | 4.1 | 10.0 | 14.9 | 108,160 | 157 | 7,090 | 248,063 | 4.4 |
| Williamson | 0.0 | D | 4.6 | 11.8 | D | 7.6 | 7.7 | 17.8 | 26.3 | 15,930 | 239 | 1,557 | 31,733 | 4.5 |

1. Per 1,000 resident population estimated as of July 1 of the year shown.

# Table B. States and Counties — Housing, Labor Force, and Employment

| STATE County | Housing units, 2015-2019 Occupied units — Owner-occupied Total | Percent | Median value[1] | Median owner cost as a percent of income — With a mortgage | Without a mortgage[2] | Renter-occupied Median rent[3] | Median rent as a percent of income[2] | Sub-standard units[4] (percent) | Civilian labor force, 2020 Total | Percent change, 2019-2020 | Unemployment Total | Rate[5] | Civilian employment[6], 2015-2019 Total | Percent — Management, business, science, and arts | Construction, production, and maintenance occupations |
|---|---|---|---|---|---|---|---|---|---|---|---|---|---|---|---|
| | 89 | 90 | 91 | 92 | 93 | 94 | 95 | 96 | 97 | 98 | 99 | 100 | 101 | 102 | 103 |
| **ILLINOIS—Cont'd** | | | | | | | | | | | | | | | |
| Hardin | 1,363 | 82.0 | 80,500 | 18.4 | 10.5 | 401 | 25.2 | 3.3 | 1,289 | -1.4 | 118 | 9.2 | 1,310 | 27.6 | 39.5 |
| Henderson | 2,977 | 82.2 | 92,000 | 16.0 | 10.0 | 558 | 27.7 | 2.2 | 3,544 | -2.5 | 246 | 6.9 | 3,294 | 32.8 | 34.3 |
| Henry | 19,856 | 78.6 | 119,200 | 17.6 | 12.3 | 679 | 25.1 | 0.8 | 23,925 | -4.7 | 1,864 | 7.8 | 22,995 | 30.5 | 30.4 |
| Iroquois | 11,741 | 76.1 | 102,800 | 19.3 | 11.9 | 686 | 27.1 | 1.1 | 13,582 | -5.5 | 844 | 6.2 | 12,897 | 30.5 | 31.8 |
| Jackson | 23,883 | 51.5 | 108,600 | 19.0 | 11.1 | 691 | 35.5 | 1.7 | 27,251 | -2.7 | 2,105 | 7.7 | 24,955 | 39.4 | 20.9 |
| Jasper | 3,711 | 78.9 | 99,500 | 20.1 | 11.2 | 578 | 28.7 | 2.0 | 4,591 | -1.4 | 276 | 6.0 | 4,392 | 24.4 | 33.4 |
| Jefferson | 14,985 | 73.2 | 96,700 | 19.0 | 11.1 | 701 | 27.9 | 2.3 | 16,993 | -2.5 | 1,714 | 10.1 | 16,090 | 29.8 | 31.8 |
| Jersey | 8,499 | 81.9 | 147,900 | 19.5 | 11.5 | 645 | 28.5 | 0.9 | 10,678 | -3.6 | 795 | 7.4 | 10,401 | 33.8 | 30.4 |
| Jo Daviess | 9,970 | 78.4 | 149,100 | 20.6 | 12.8 | 669 | 25.0 | 2.0 | 10,426 | -3.8 | 866 | 8.3 | 10,689 | 29.9 | 28.4 |
| Johnson | 4,303 | 84.2 | 109,500 | 19.3 | 11.9 | 663 | 27.8 | 1.7 | 4,098 | -1.1 | 366 | 8.9 | 3,814 | 29.9 | 28.4 |
| Kane | 179,637 | 73.9 | 238,300 | 22.4 | 13.4 | 1,148 | 29.9 | 3.9 | 261,649 | -4.2 | 23,911 | 9.1 | 268,488 | 42.0 | 18.7 |
| Kankakee | 39,796 | 68.0 | 147,900 | 20.1 | 13.2 | 887 | 30.0 | 2.2 | 53,934 | -3.6 | 4,770 | 8.8 | 50,949 | 35.1 | 25.4 |
| Kendall | 40,721 | 83.6 | 235,600 | 22.1 | 13.2 | 1,414 | 29.1 | 2.4 | 66,386 | -4.7 | 5,547 | 8.4 | 65,645 | 30.8 | 30.6 |
| Knox | 20,680 | 66.7 | 83,000 | 18.3 | 11.9 | 634 | 30.8 | 1.9 | 21,110 | -3.1 | 1,825 | 8.6 | 21,091 | 40.3 | 21.3 |
| Lake | 246,122 | 73.0 | 265,100 | 21.8 | 13.7 | 1,197 | 29.1 | 2.5 | 364,544 | -3.5 | 29,411 | 8.1 | 351,926 | 43.1 | 18.9 |
| LaSalle | 45,095 | 71.8 | 127,200 | 19.6 | 11.3 | 777 | 28.3 | 1.6 | 52,663 | -3.9 | 4,964 | 9.4 | 52,003 | 27.9 | 32.0 |
| Lawrence | 6,306 | 74.2 | 78,400 | 17.3 | 10.0 | 628 | 25.8 | 2.8 | 5,781 | -2.4 | 505 | 8.7 | 6,166 | 33.1 | 31.0 |
| Lee | 13,788 | 72.3 | 119,900 | 18.6 | 11.6 | 694 | 26.1 | 1.2 | 17,154 | -2.8 | 1,172 | 6.8 | 15,438 | 34.3 | 29.7 |
| Livingston | 14,307 | 72.8 | 109,400 | 19.5 | 12.4 | 715 | 27.2 | 1.8 | 15,909 | -4.2 | 1,069 | 6.7 | 15,981 | 25.7 | 33.6 |
| Logan | 10,797 | 72.7 | 103,200 | 17.8 | 10.3 | 689 | 22.6 | 0.7 | 12,122 | -4.4 | 817 | 6.7 | 12,426 | 31.3 | 30.2 |
| McDonough | 11,408 | 66.3 | 91,800 | 18.6 | 11.0 | 662 | 32.1 | 2.1 | 12,819 | -4.7 | 911 | 7.1 | 13,558 | 34.6 | 23.2 |
| McHenry | 112,453 | 79.4 | 224,500 | 22.0 | 13.2 | 1,197 | 28.9 | 1.7 | 158,406 | -4.7 | 13,157 | 8.3 | 164,564 | 37.8 | 22.5 |
| McLean | 65,845 | 64.6 | 167,000 | 18.4 | 10.3 | 845 | 26.7 | 1.2 | 85,103 | -3.6 | 5,762 | 6.8 | 88,705 | 43.5 | 15.6 |
| Macon | 43,912 | 68.4 | 98,400 | 18.2 | 10.3 | 684 | 29.6 | 1.2 | 47,177 | -3.3 | 4,665 | 9.9 | 46,487 | 32.6 | 27.1 |
| Macoupin | 18,875 | 76.7 | 100,600 | 18.4 | 10.4 | 701 | 30.5 | 1.4 | 22,119 | -4.1 | 1,523 | 6.9 | 20,551 | 31.3 | 29.4 |
| Madison | 107,659 | 71.4 | 135,600 | 18.6 | 11.5 | 824 | 29.6 | 1.5 | 131,452 | -2.7 | 10,182 | 7.7 | 127,273 | 36.0 | 23.7 |
| Marion | 15,946 | 73.6 | 77,900 | 18.6 | 11.2 | 646 | 27.6 | 2.4 | 17,114 | -3.4 | 1,591 | 9.3 | 16,244 | 25.7 | 34.0 |
| Marshall | 4,884 | 81.8 | 104,900 | 18.4 | 11.2 | 623 | 24.9 | 0.9 | 5,085 | -4.9 | 375 | 7.4 | 5,338 | 30.8 | 31.4 |
| Mason | 5,977 | 77.8 | 83,000 | 18.2 | 12.7 | 664 | 24.0 | 1.1 | 5,916 | -6.1 | 476 | 8.0 | 6,339 | 28.4 | 31.7 |
| Massac | 5,822 | 75.5 | 83,600 | 19.9 | 11.5 | 692 | 30.9 | 3.2 | 5,424 | -3.4 | 541 | 10.0 | 5,636 | 26.4 | 27.4 |
| Menard | 5,188 | 78.1 | 146,900 | 17.7 | 10.0 | 783 | 25.0 | 1.3 | 6,101 | -4.9 | 391 | 6.4 | 6,223 | 38.2 | 20.9 |
| Mercer | 6,516 | 79.0 | 107,600 | 17.9 | 10.9 | 655 | 24.1 | 0.5 | 7,689 | -4.8 | 617 | 8.0 | 7,531 | 28.7 | 35.8 |
| Monroe | 13,586 | 83.4 | 211,300 | 18.9 | 11.4 | 872 | 31.2 | 0.4 | 18,386 | -4.2 | 1,023 | 5.6 | 18,127 | 44.1 | 20.4 |
| Montgomery | 11,522 | 76.4 | 84,100 | 17.5 | 10.6 | 644 | 24.7 | 1.3 | 11,453 | -2.8 | 998 | 8.7 | 11,571 | 33.1 | 27.1 |
| Morgan | 13,719 | 68.5 | 110,100 | 19.3 | 10.9 | 720 | 27.7 | 1.4 | 15,873 | -3.9 | 1,089 | 6.9 | 15,573 | 31.2 | 25.5 |
| Moultrie | 5,959 | 77.6 | 107,400 | 19.0 | 10.5 | 709 | 23.9 | 3.2 | 7,292 | -4.3 | 367 | 5.0 | 7,045 | 29.0 | 39.7 |
| Ogle | 21,021 | 73.2 | 146,900 | 19.3 | 11.3 | 745 | 24.6 | 2.2 | 24,161 | -3.4 | 1,962 | 8.1 | 25,032 | 31.7 | 31.5 |
| Peoria | 73,253 | 65.1 | 129,800 | 19.3 | 11.5 | 798 | 28.1 | 1.6 | 83,684 | -1.8 | 8,668 | 10.4 | 81,821 | 38.9 | 19.6 |
| Perry | 8,433 | 75.4 | 86,500 | 18.0 | 10.9 | 576 | 28.0 | 1.8 | 8,093 | -3.0 | 763 | 9.4 | 8,126 | 27.0 | 33.9 |
| Piatt | 6,692 | 82.5 | 138,600 | 18.6 | 11.1 | 851 | 26.7 | 0.6 | 8,462 | -2.5 | 453 | 5.4 | 8,268 | 37.8 | 25.4 |
| Pike | 6,309 | 78.6 | 85,400 | 18.8 | 11.8 | 568 | 26.6 | 1.9 | 6,753 | -4.3 | 405 | 6.0 | 6,815 | 29.8 | 31.4 |
| Pope | 1,694 | 82.5 | 93,000 | 20.6 | 11.2 | 313 | 31.3 | 2.1 | 1,643 | -2.6 | 130 | 7.9 | 1,138 | 25.7 | 37.4 |
| Pulaski | 2,095 | 76.4 | 61,300 | 20.6 | 12.2 | 519 | 31.3 | 2.1 | 1,768 | -9.6 | 206 | 11.7 | 1,830 | 29.6 | 26.7 |
| Putnam | 2,413 | 80.5 | 126,800 | 20.3 | 10.0 | 693 | 25.7 | 0.7 | 2,942 | -3.3 | 244 | 8.3 | 2,906 | 31.0 | 37.0 |
| Randolph | 11,883 | 74.0 | 110,200 | 18.2 | 10.0 | 680 | 23.8 | 1.4 | 13,764 | -4.1 | 978 | 7.1 | 13,190 | 25.2 | 33.7 |
| Richland | 6,452 | 71.2 | 89,600 | 16.5 | 10.0 | 566 | 23.1 | 1.8 | 7,362 | 1.4 | 498 | 6.8 | 7,165 | 28.5 | 34.8 |
| Rock Island | 60,546 | 68.0 | 119,400 | 18.9 | 12.0 | 748 | 26.1 | 2.0 | 68,915 | -3.0 | 6,472 | 9.4 | 67,890 | 30.8 | 29.4 |
| St. Clair | 104,105 | 65.2 | 128,000 | 19.7 | 11.8 | 861 | 30.2 | 1.6 | 125,155 | -1.8 | 11,319 | 9.0 | 120,965 | 36.1 | 22.4 |
| Saline | 9,972 | 74.5 | 73,800 | 17.4 | 10.8 | 622 | 28.5 | 2.9 | 9,534 | -1.4 | 929 | 9.7 | 9,902 | 30.9 | 27.3 |
| Sangamon | 83,711 | 68.9 | 141,200 | 18.7 | 10.6 | 820 | 28.5 | 1.8 | 96,869 | -3.1 | 7,953 | 8.2 | 94,128 | 43.0 | 16.2 |
| Schuyler | 2,775 | 75.7 | 86,600 | 17.8 | 11.3 | 672 | 20.9 | 1.4 | 3,126 | -4.3 | 193 | 6.2 | 2,893 | 30.0 | 28.0 |
| Scott | 1,939 | 79.9 | 89,200 | 16.3 | 10.5 | 542 | 21.9 | 1.5 | 2,349 | -6.6 | 155 | 6.6 | 2,311 | 32.3 | 32.5 |
| Shelby | 9,189 | 81.0 | 96,500 | 18.3 | 11.5 | 653 | 28.4 | 2.5 | 10,256 | -3.7 | 660 | 6.4 | 10,426 | 26.5 | 34.8 |
| Stark | 2,315 | 78.5 | 87,700 | 17.2 | 11.4 | 630 | 23.6 | 0.3 | 2,373 | -6.1 | 173 | 7.3 | 2,479 | 33.2 | 33.8 |
| Stephenson | 19,739 | 69.9 | 97,800 | 19.0 | 12.9 | 651 | 27.8 | 1.2 | 21,106 | -3.3 | 1,546 | 7.3 | 20,922 | 29.5 | 31.7 |
| Tazewell | 54,291 | 76.4 | 142,000 | 19.2 | 11.6 | 741 | 25.8 | 1.3 | 61,977 | -3.6 | 5,029 | 8.1 | 62,800 | 36.1 | 24.1 |
| Union | 6,654 | 77.3 | 109,900 | 19.4 | 11.3 | 612 | 21.8 | 2.7 | 7,192 | -3.7 | 582 | 8.1 | 6,962 | 27.7 | 30.5 |
| Vermilion | 31,151 | 69.6 | 79,900 | 17.6 | 10.9 | 685 | 26.4 | 1.9 | 32,542 | -2.0 | 2,871 | 8.8 | 31,285 | 27.1 | 31.9 |
| Wabash | 4,839 | 78.0 | 82,800 | 17.5 | 10.7 | 633 | 29.2 | 1.2 | 5,718 | -2.4 | 476 | 8.3 | 5,412 | 29.2 | 34.7 |
| Warren | 6,862 | 74.3 | 83,900 | 18.4 | 11.4 | 628 | 25.3 | 3.8 | 8,085 | -5.1 | 498 | 6.2 | 7,948 | 30.7 | 32.2 |
| Washington | 6,020 | 77.5 | 113,500 | 18.6 | 12.2 | 716 | 21.7 | 1.9 | 9,576 | -4.5 | 504 | 5.3 | 7,110 | 33.4 | 33.1 |
| Wayne | 7,051 | 76.6 | 86,400 | 17.7 | 11.3 | 626 | 26.3 | 4.3 | 6,795 | -3.0 | 565 | 8.3 | 7,370 | 27.4 | 38.9 |
| White | 6,041 | 80.6 | 71,500 | 17.9 | 11.0 | 552 | 25.8 | 1.4 | 6,137 | -7.4 | 466 | 7.6 | 5,939 | 29.4 | 35.6 |
| Whiteside | 23,084 | 74.9 | 106,700 | 18.1 | 11.6 | 695 | 26.1 | 1.6 | 27,673 | -3.8 | 2,212 | 8.0 | 26,264 | 25.8 | 33.5 |
| Will | 229,498 | 81.3 | 232,000 | 22.1 | 13.4 | 1,176 | 31.1 | 1.8 | 344,335 | -4.1 | 32,572 | 9.5 | 351,095 | 36.7 | 24.3 |
| Williamson | 27,029 | 70.1 | 114,900 | 18.6 | 11.2 | 735 | 26.2 | 1.0 | 31,533 | -2.1 | 2,739 | 8.7 | 28,866 | 32.6 | 23.7 |

1. Specified owner-occupied units.   2. A value of 10.0 represents 10 percent or less; a value of 50.0 represents 50 percent or more.   3. Specified renter-occupied units.   4. Overcrowded or
lacking complete plumbing facilities.   5. Percent of civilian labor force.   6. Civilian employed persons 16 years old and over.

# Table B. States and Counties — **Nonfarm Employment and Agriculture**

| STATE County | Private nonfarm establishments, employment and payroll, 2019 | | | | | | | | | Agriculture, 2017 | | | Farm producers whose primary occupation is farming (percent) |
|---|---|---|---|---|---|---|---|---|---|---|---|---|---|
| | Number of establishments | Employment | | | | | | Annual payroll | | Farms | | | |
| | | Total | Health care and social assistance | Manufacturing | Retail trade | Finance and insurance | Professional, scientific, and technical services | Total (mil dol) | Average per employee (dollars) | Number | Percent with: | | |
| | | | | | | | | | | | Fewer than 50 acres | 1000 acres or more | |
| | 104 | 105 | 106 | 107 | 108 | 109 | 110 | 111 | 112 | 113 | 114 | 115 | 116 |

ILLINOIS—Cont'd

| STATE County | 104 | 105 | 106 | 107 | 108 | 109 | 110 | 111 | 112 | 113 | 114 | 115 | 116 |
|---|---|---|---|---|---|---|---|---|---|---|---|---|---|
| Hardin | 60 | 590 | 303 | NA | 65 | 17 | 25 | 20 | 33,268 | 161 | 21.7 | 3.1 | 30.6 |
| Henderson | 102 | 630 | 126 | NA | 97 | 86 | 23 | 19 | 30,425 | 438 | 23.3 | 11.4 | 58.8 |
| Henry | 1,014 | 13,075 | 1,553 | 4,225 | 1,850 | 498 | 269 | 501 | 38,293 | 1,353 | 37.0 | 10.7 | 45.5 |
| Iroquois | 640 | 5,711 | 1,568 | 582 | 964 | 337 | 95 | 208 | 36,379 | 1,516 | 26.2 | 14.7 | 48.3 |
| Jackson | 1,257 | 16,190 | 3,815 | 1,008 | 3,374 | 601 | 621 | 621 | 38,374 | 772 | 34.7 | 6.9 | 40.4 |
| Jasper | 228 | 1,655 | 215 | 218 | 192 | 113 | 29 | 66 | 39,773 | 913 | 36.0 | 6.9 | 39.2 |
| Jefferson | 932 | 18,967 | 3,859 | 5,100 | 2,256 | 453 | 361 | 874 | 46,086 | 1,099 | 41.4 | 5.2 | 34.5 |
| Jersey | 412 | 4,421 | 874 | 151 | 842 | 197 | 108 | 140 | 31,637 | 519 | 35.5 | 9.2 | 43.2 |
| Jo Daviess | 699 | 6,214 | 740 | 756 | 980 | 201 | 166 | 230 | 37,012 | 947 | 31.0 | 6.0 | 42.3 |
| Johnson | 168 | 1,347 | 276 | 37 | 234 | 67 | 124 | 33 | 24,804 | 653 | 34.2 | 2.6 | 36.3 |
| Kane | 13,054 | 194,210 | 24,962 | 32,379 | 23,880 | 7,603 | 10,032 | 9,487 | 48,850 | 605 | 53.1 | 7.9 | 47.9 |
| Kankakee | 2,291 | 36,106 | 7,288 | 5,641 | 5,497 | 958 | 674 | 1,532 | 42,441 | 756 | 36.6 | 12.3 | 46.9 |
| Kendall | 2,244 | 23,758 | 1,794 | 2,343 | 5,385 | 659 | 938 | 894 | 37,617 | 313 | 39.9 | 13.7 | 52.5 |
| Knox | 991 | 15,184 | 3,288 | 1,205 | 3,347 | 420 | 302 | 494 | 32,524 | 853 | 22.5 | 14.3 | 50.7 |
| Lake | 19,832 | 317,680 | 37,495 | 38,288 | 35,938 | 14,487 | 30,518 | 24,159 | 76,048 | 302 | 77.8 | 2.6 | 40.6 |
| LaSalle | 2,598 | 37,530 | 4,874 | 5,582 | 6,022 | 1,249 | 1,231 | 1,633 | 43,521 | 1,496 | 33.4 | 11.4 | 45.9 |
| Lawrence | 234 | 3,370 | 391 | 1,089 | 351 | 336 | 40 | 123 | 36,622 | 426 | 34.7 | 15.3 | 44.0 |
| Lee | 704 | 10,724 | 2,096 | 3,402 | 1,322 | 255 | 285 | 450 | 41,971 | 832 | 33.5 | 14.5 | 50.8 |
| Livingston | 852 | 11,206 | 1,465 | 3,372 | 1,551 | 449 | 222 | 485 | 43,304 | 1,313 | 29.9 | 14.3 | 49.2 |
| Logan | 552 | 7,435 | 1,469 | 1,226 | 951 | 253 | 222 | 260 | 34,960 | 683 | 30.7 | 19.0 | 51.6 |
| McDonough | 639 | 8,573 | 1,929 | 1,578 | 1,386 | 301 | 215 | 287 | 33,476 | 760 | 29.3 | 13.2 | 47.2 |
| McHenry | 8,066 | 87,858 | 12,383 | 14,309 | 13,691 | 1,951 | 3,349 | 4,121 | 46,903 | 881 | 58.2 | 6.5 | 50.7 |
| McLean | 3,530 | 72,469 | 8,086 | 3,203 | 8,874 | 18,520 | 2,375 | 3,827 | 52,812 | 1,416 | 33.9 | 13.8 | 46.4 |
| Macon | 2,311 | 46,084 | 8,137 | 8,286 | 5,088 | 1,591 | 1,061 | 2,438 | 52,913 | 589 | 41.8 | 16.0 | 49.9 |
| Macoupin | 851 | 7,970 | 1,511 | 485 | 1,309 | 426 | 118 | 276 | 34,613 | 1,169 | 31.7 | 9.9 | 43.2 |
| Madison | 5,638 | 90,989 | 14,511 | 11,009 | 11,847 | 2,861 | 3,678 | 3,991 | 43,868 | 1,079 | 47.1 | 8.4 | 36.7 |
| Marion | 864 | 11,032 | 2,750 | 2,846 | 1,245 | 361 | 226 | 418 | 37,878 | 1,004 | 39.1 | 4.7 | 32.5 |
| Marshall | 244 | 2,529 | 464 | 862 | 252 | 91 | 36 | 97 | 38,170 | 472 | 21.2 | 11.0 | 45.6 |
| Mason | 262 | 2,064 | 519 | 101 | 399 | 130 | 23 | 71 | 34,315 | 548 | 25.2 | 18.4 | 47.8 |
| Massac | 230 | 3,036 | 743 | 342 | 325 | 100 | 118 | 118 | 38,786 | 417 | 34.3 | 7.9 | 50.3 |
| Menard | 194 | 1,174 | 79 | 52 | 258 | 98 | 91 | 42 | 35,922 | 386 | 32.4 | 12.7 | 46.5 |
| Mercer | 263 | 2,440 | 392 | 624 | 323 | 126 | 37 | 83 | 34,078 | 748 | 33.2 | 10.8 | 46.0 |
| Monroe | 825 | 8,186 | 798 | 369 | 1,336 | 360 | 669 | 303 | 37,025 | 568 | 46.0 | 10.9 | 41.2 |
| Montgomery | 689 | 7,067 | 1,449 | 685 | 1,345 | 401 | 178 | 244 | 34,569 | 1,067 | 33.3 | 13.8 | 47.9 |
| Morgan | 803 | 13,435 | 2,339 | 2,294 | 1,823 | 1,012 | 612 | 497 | 36,961 | 693 | 29.0 | 13.3 | 44.2 |
| Moultrie | 323 | 5,158 | 750 | 2,388 | 385 | 146 | 124 | 219 | 42,466 | 526 | 50.2 | 10.3 | 35.6 |
| Ogle | 1,037 | 13,315 | 1,572 | 2,893 | 1,409 | 484 | 296 | 623 | 46,810 | 1,011 | 43.0 | 10.8 | 46.7 |
| Peoria | 4,366 | 103,464 | 26,110 | 6,092 | 10,378 | 3,541 | 4,560 | 6,213 | 60,051 | 884 | 37.8 | 7.7 | 38.4 |
| Perry | 392 | 4,089 | 1,044 | 415 | 631 | 144 | 85 | 152 | 37,155 | 572 | 34.4 | 7.9 | 40.6 |
| Piatt | 316 | 2,410 | 452 | 225 | 435 | 155 | 108 | 86 | 35,698 | 422 | 36.0 | 21.1 | 53.6 |
| Pike | 349 | 2,995 | 696 | 142 | 551 | 204 | 61 | 104 | 34,654 | 956 | 24.7 | 15.2 | 40.4 |
| Pope | 48 | 334 | 95 | NA | 60 | 10 | 5 | 5 | 16,219 | 322 | 25.2 | 2.5 | 32.9 |
| Pulaski | 87 | 598 | 121 | NA | 129 | 40 | 20 | 17 | 27,997 | 222 | 26.1 | 15.3 | 38.6 |
| Putnam | 119 | 1,192 | 25 | 486 | 83 | 41 | NA | 67 | 56,024 | 147 | 19.7 | 10.2 | 43.1 |
| Randolph | 636 | 10,606 | 2,042 | 2,819 | 1,352 | 290 | 285 | 410 | 38,665 | 808 | 34.7 | 7.9 | 36.7 |
| Richland | 450 | 5,218 | 1,011 | 474 | 651 | 190 | 77 | 208 | 39,779 | 596 | 41.9 | 8.4 | 35.9 |
| Rock Island | 3,041 | 61,613 | 9,418 | 6,889 | 7,606 | 2,635 | 3,003 | 3,409 | 55,333 | 649 | 38.1 | 6.0 | 42.5 |
| St. Clair | 5,150 | 74,730 | 14,495 | 4,986 | 13,057 | 2,058 | 4,565 | 3,035 | 40,611 | 793 | 46.0 | 9.5 | 42.3 |
| Saline | 533 | 6,456 | 2,140 | 373 | 1,143 | 291 | 157 | 242 | 37,509 | 452 | 39.6 | 9.5 | 36.6 |
| Sangamon | 4,885 | 82,451 | 22,121 | 2,885 | 11,799 | 5,406 | 4,563 | 3,630 | 44,031 | 1,083 | 46.4 | 13.0 | 44.5 |
| Schuyler | 149 | 1,130 | 344 | 60 | 180 | 56 | 31 | 39 | 34,112 | 544 | 21.5 | 11.2 | 33.5 |
| Scott | 76 | 590 | 20 | NA | 70 | 53 | 7 | 26 | 43,861 | 300 | 25.0 | 15.0 | 45.3 |
| Shelby | 418 | 4,206 | 583 | 1,092 | 523 | 178 | 280 | 149 | 35,399 | 1,197 | 36.9 | 7.7 | 43.4 |
| Stark | 108 | 863 | 89 | 244 | 132 | 62 | 24 | 40 | 46,780 | 362 | 30.9 | 18.2 | 51.3 |
| Stephenson | 1,014 | 15,067 | 2,403 | 3,146 | 1,852 | 1,061 | 423 | 718 | 47,627 | 965 | 43.7 | 7.5 | 48.7 |
| Tazewell | 2,783 | 43,264 | 5,218 | 6,712 | 6,962 | 1,879 | 1,475 | 1,876 | 43,372 | 857 | 35.9 | 10.2 | 44.7 |
| Union | 359 | 3,476 | 1,287 | 243 | 669 | 132 | 98 | 103 | 29,734 | 590 | 30.7 | 5.4 | 37.2 |
| Vermilion | 1,342 | 23,295 | 4,766 | 4,608 | 3,424 | 1,091 | 430 | 1,010 | 43,352 | 1,049 | 38.9 | 14.1 | 47.8 |
| Wabash | 256 | 2,845 | 859 | 231 | 300 | 108 | 132 | 115 | 40,538 | 208 | 38.0 | 17.3 | 37.7 |
| Warren | 369 | 6,249 | 624 | 2,142 | 581 | 217 | 102 | 240 | 38,332 | 711 | 28.4 | 14.9 | 55.9 |
| Washington | 375 | 7,472 | 315 | 2,747 | 615 | 171 | 97 | 357 | 47,771 | 715 | 23.9 | 15.2 | 48.6 |
| Wayne | 382 | 3,032 | 999 | 73 | 588 | 135 | 57 | 95 | 31,176 | 1,025 | 31.6 | 11.0 | 42.8 |
| White | 356 | 3,222 | 689 | 300 | 505 | 147 | 77 | 110 | 33,997 | 496 | 34.9 | 17.3 | 46.3 |
| Whiteside | 1,181 | 17,838 | 3,351 | 4,049 | 2,433 | 549 | 401 | 718 | 40,232 | 959 | 33.5 | 10.4 | 45.6 |
| Will | 15,832 | 236,021 | 28,658 | 19,316 | 28,684 | 4,610 | 8,641 | 11,263 | 47,720 | 801 | 56.4 | 7.2 | 42.3 |
| Williamson | 1,572 | 24,460 | 6,590 | 3,245 | 3,702 | 1,530 | 633 | 970 | 39,669 | 610 | 45.7 | 4.6 | 28.7 |

| STATE County | Acreage (1,000) | Percent change, 2012-2017 | Average size of farm | Total irrigated (1,000) | Total cropland (1,000) | Average per farm | Average per acre | Value of machinery and equipment, average per farm (dollars) | Total (mil dol) | Average per farm (acres) | Crops | Livestock and poultry products | Organic farms (number) | Farms with internet access (percent) | Total ($1,000) | Percent of farms |
|---|---|---|---|---|---|---|---|---|---|---|---|---|---|---|---|---|
| | 117 | 118 | 119 | 120 | 121 | 122 | 123 | 124 | 125 | 126 | 127 | 128 | 129 | 130 | 131 | 132 |
| **ILLINOIS—Cont'd** | | | | | | | | | | | | | | | | |
| Hardin | 37 | 10.2 | 227 | NA | 21.2 | 745,489 | 3,281 | 84,639 | D | D | D | D | 1 | 90.1 | 679 | 38.5 |
| Henderson | 193 | 12.4 | 440 | 12.7 | 165.1 | 3,065,949 | 6,961 | 208,592 | 122.3 | 279,297 | 83.4 | 16.6 | 5 | 84.5 | 3,938 | 76.9 |
| Henry | 484 | 1.0 | 358 | 7.4 | 440.8 | 2,828,369 | 7,902 | 225,411 | 353.0 | 260,897 | 77.2 | 22.8 | 7 | 81.1 | 15,746 | 72.4 |
| Iroquois | 681 | 1.8 | 449 | 5.7 | 655.9 | 3,302,325 | 7,348 | 249,933 | 420.5 | 277,400 | 84.4 | 15.6 | 11 | 77.8 | 8,969 | 61.8 |
| Jackson | 222 | 3.5 | 287 | 1.7 | 176.8 | 1,546,191 | 5,386 | 149,783 | 87.2 | 112,920 | 85.9 | 14.1 | 3 | 69.3 | 4,446 | 47.8 |
| Jasper | 250 | -0.5 | 273 | D | 218.7 | 1,540,385 | 5,634 | 162,342 | 165.9 | 181,690 | 61.6 | 38.4 | 2 | 75.0 | 7,045 | 84.2 |
| Jefferson | 269 | 25.9 | 245 | 0.1 | 222.6 | 1,019,111 | 4,158 | 108,307 | 94.6 | 86,101 | 82.8 | 17.2 | 6 | 69.9 | 5,256 | 64.8 |
| Jersey | 190 | 22.0 | 366 | 0.0 | 152.1 | 2,314,178 | 6,330 | 220,839 | 82.1 | 158,143 | 95.3 | 4.7 | NA | 75.0 | 4,326 | 64.5 |
| Jo Daviess | 289 | 6.5 | 306 | 0.8 | 202.1 | 1,867,551 | 6,110 | 196,260 | 151.9 | 160,415 | 65.0 | 35.0 | 3 | 75.1 | 6,275 | 67.9 |
| Johnson | 105 | 17.6 | 162 | 0.5 | 55.8 | 597,733 | 3,701 | 66,612 | 18.2 | 27,824 | 80.0 | 20.0 | NA | 65.4 | 2,524 | 45.2 |
| Kane | 170 | 1.0 | 281 | 1.3 | 161.9 | 2,971,109 | 10,558 | 229,206 | 181.3 | 299,633 | 84.8 | 15.2 | 10 | 83.6 | 3,782 | 45.6 |
| Kankakee | 313 | -8.7 | 414 | 18.4 | 300.4 | 3,237,465 | 7,822 | 285,392 | 221.1 | 292,508 | 91.3 | 8.7 | 3 | 75.5 | 2,346 | 45.4 |
| Kendall | 138 | 6.3 | 441 | D | 133.6 | 3,991,102 | 9,059 | 260,029 | 101.6 | 324,655 | 92.4 | 7.6 | 3 | 85.6 | 2,128 | 60.7 |
| Knox | 414 | 19.1 | 485 | 0.0 | 355.5 | 3,553,390 | 7,319 | 247,401 | 284.4 | 333,419 | 78.5 | 21.5 | 2 | 82.6 | 7,150 | 73.0 |
| Lake | 31 | 1.8 | 101 | 0.4 | 23.9 | 1,230,772 | 12,149 | 102,194 | 39.1 | 129,364 | 85.2 | 14.8 | 8 | 77.8 | 423 | 10.9 |
| LaSalle | 573 | -4.9 | 383 | 6.1 | 545.4 | 3,495,212 | 9,125 | 265,453 | 370.9 | 247,958 | 94.1 | 5.9 | 3 | 81.6 | 12,749 | 72.1 |
| Lawrence | 225 | 22.2 | 528 | 29.1 | 205.5 | 2,983,608 | 5,650 | 327,369 | 156.4 | 367,045 | 71.4 | 28.6 | 1 | 71.1 | 7,260 | 71.8 |
| Lee | 392 | 6.2 | 471 | 23.9 | 374.4 | 4,085,091 | 8,668 | 320,174 | 278.9 | 335,184 | 89.8 | 10.2 | 6 | 81.3 | 9,291 | 70.9 |
| Livingston | 601 | -8.5 | 457 | 0.1 | 581.3 | 3,764,261 | 8,230 | 280,950 | 408.4 | 311,023 | 85.2 | 14.8 | 9 | 84.3 | 6,848 | 57.9 |
| Logan | 354 | -2.5 | 518 | 2.6 | 338.1 | 4,343,867 | 8,380 | 297,104 | 245.7 | 359,712 | 90.5 | 9.5 | 2 | 79.9 | 4,262 | 79.8 |
| McDonough | 315 | 7.8 | 414 | 0.1 | 272.5 | 3,037,172 | 7,334 | 233,227 | 214.1 | 281,663 | 82.9 | 17.1 | 1 | 79.7 | 3,359 | 66.8 |
| McHenry | 208 | -11.0 | 236 | 9.6 | 189.7 | 2,256,366 | 9,541 | 174,498 | 163.8 | 185,871 | 75.3 | 24.7 | 5 | 80.6 | 4,201 | 30.6 |
| McLean | 620 | -10.4 | 438 | 3.2 | 599.9 | 4,310,555 | 9,844 | 280,498 | 457.1 | 322,784 | 85.0 | 15.0 | 3 | 81.1 | 9,368 | 73.0 |
| Macon | 277 | -17.6 | 471 | 0.1 | 268 | 4,356,859 | 9,250 | 281,637 | 180.0 | 305,610 | 97.7 | 2.3 | 5 | 84.6 | 4,815 | 69.4 |
| Macoupin | 421 | -4.1 | 360 | 0.0 | 355.3 | 2,589,379 | 7,195 | 203,190 | 236.5 | 202,305 | 82.6 | 17.4 | 1 | 74.7 | 4,907 | 63.5 |
| Madison | 319 | 3.8 | 295 | 0.5 | 287.8 | 2,571,906 | 8,706 | 176,232 | 174.7 | 161,912 | 89.9 | 10.1 | 6 | 79.0 | 6,378 | 56.1 |
| Marion | 249 | -6.8 | 248 | 0.2 | 203.7 | 1,210,249 | 4,885 | 147,171 | 111.9 | 111,406 | 70.8 | 29.2 | 5 | 73.4 | 6,728 | 76.1 |
| Marshall | 199 | -5.0 | 421 | 2.9 | 180.7 | 3,396,974 | 8,075 | 245,161 | 119.5 | 253,275 | 96.7 | 3.3 | 3 | 82.4 | 4,124 | 79.4 |
| Mason | 312 | 7.6 | 569 | 136.9 | 289.3 | 3,974,753 | 6,983 | 352,375 | 192.9 | 352,035 | 88.9 | 11.1 | 1 | 73.9 | 6,447 | 86.7 |
| Massac | 119 | 16.0 | 284 | 10.4 | 95.5 | 1,100,621 | 3,871 | 176,306 | 45.8 | 109,928 | 86.5 | 13.5 | NA | 77.7 | 3,252 | 63.1 |
| Menard | 168 | 6.5 | 435 | 3.8 | 139 | 3,399,115 | 7,807 | 261,654 | 89.7 | 232,267 | 93.2 | 6.8 | NA | 80.6 | 3,801 | 71.0 |
| Mercer | 282 | 12.0 | 377 | 7.8 | 247.3 | 2,471,925 | 6,551 | 197,201 | 217.5 | 290,807 | 71.9 | 28.1 | 3 | 78.2 | 6,604 | 76.7 |
| Monroe | 176 | -8.8 | 310 | 3.2 | 151 | 2,065,789 | 6,659 | 215,520 | 88.2 | 155,195 | 74.5 | 25.5 | NA | 75.7 | 4,205 | 59.5 |
| Montgomery | 439 | 14.8 | 411 | D | 398.1 | 2,992,259 | 7,276 | 254,828 | 263.0 | 246,517 | 85.6 | 14.4 | NA | 79.9 | 6,978 | 76.9 |
| Morgan | 300 | -2.9 | 433 | 7.7 | 265.2 | 3,573,396 | 8,247 | 266,790 | 172.0 | 248,221 | 88.5 | 11.5 | 11 | 81.8 | 4,846 | 75.8 |
| Moultrie | 202 | -1.6 | 384 | 0.0 | 194.9 | 3,351,079 | 8,737 | 205,183 | 141.2 | 268,365 | 93.6 | 6.4 | 9 | 67.5 | 2,939 | 56.7 |
| Ogle | 355 | -5.8 | 351 | 0.8 | 326.8 | 3,015,903 | 8,599 | 246,650 | 276.4 | 273,371 | 75.4 | 24.6 | 1 | 85.3 | 3,708 | 61.4 |
| Peoria | 250 | -0.1 | 283 | 3.5 | 216.3 | 2,232,609 | 7,892 | 180,266 | 145.2 | 164,249 | 90.6 | 9.4 | 5 | 77.4 | 2,517 | 64.1 |
| Perry | 184 | 2.0 | 322 | 0.2 | 153.5 | 1,539,778 | 4,782 | 174,881 | 66.3 | 115,955 | 92.3 | 7.7 | NA | 62.9 | 3,806 | 73.8 |
| Piatt | 256 | -1.2 | 607 | 1.1 | 251.2 | 5,617,953 | 9,260 | 341,248 | 165.3 | 391,673 | 99.5 | 0.5 | 2 | 85.5 | 3,844 | 68.0 |
| Pike | 447 | 8.6 | 468 | 3.1 | 342.9 | 2,766,090 | 5,916 | 203,650 | 278.9 | 291,705 | 64.8 | 35.2 | NA | 72.4 | 5,868 | 74.4 |
| Pope | 66 | -15.3 | 205 | D | 30.9 | 715,861 | 3,491 | 75,796 | 8.9 | 27,503 | 76.5 | 23.5 | NA | 65.8 | 1,815 | 59.0 |
| Pulaski | 101 | 23.3 | 456 | 5.2 | 83.1 | 1,871,568 | 4,103 | 181,600 | 40.1 | 180,450 | 96.8 | 3.2 | NA | 68.9 | 2,466 | 64.9 |
| Putnam | 50 | -17.3 | 339 | 0.2 | 43.7 | 2,833,889 | 8,372 | 272,212 | D | D | D | D | NA | 76.2 | 837 | 80.3 |
| Randolph | 262 | -6.0 | 324 | D | 216.5 | 1,757,514 | 5,423 | 158,325 | 98.5 | 121,881 | 85.7 | 14.3 | NA | 73.0 | 5,839 | 66.2 |
| Richland | 178 | -5.5 | 299 | 1.0 | 158.2 | 1,581,240 | 5,280 | 158,267 | 119.8 | 201,000 | 57.6 | 42.4 | NA | 82.0 | 5,629 | 76.2 |
| Rock Island | 160 | 7.0 | 246 | 4.8 | 132.4 | 1,712,532 | 6,965 | 179,885 | 99.9 | 153,954 | 80.3 | 19.7 | 2 | 76.6 | 3,685 | 60.9 |
| St. Clair | 237 | -5.8 | 299 | 0.5 | 218.8 | 2,218,716 | 7,417 | 183,236 | 135.6 | 171,019 | 83.1 | 16.9 | 4 | 77.3 | 4,907 | 62.3 |
| Saline | 145 | 3.6 | 321 | D | 122.3 | 1,546,065 | 4,824 | 155,847 | 73.6 | 162,761 | 75.9 | 24.1 | NA | 72.1 | 3,646 | 51.5 |
| Sangamon | 531 | 3.4 | 491 | 0.8 | 496.8 | 4,381,113 | 8,931 | 259,349 | 352.6 | 325,599 | 92.4 | 7.6 | 3 | 78.8 | 6,498 | 63.3 |
| Schuyler | 212 | 16.3 | 389 | 2.4 | 153.8 | 2,135,913 | 5,484 | 176,634 | 116.3 | 213,836 | 64.6 | 35.4 | NA | 69.1 | 2,423 | 78.1 |
| Scott | 155 | 5.4 | 518 | 6.3 | 131.8 | 3,367,008 | 6,498 | 237,953 | 84.7 | 282,197 | 82.0 | 18.0 | NA | 70.7 | 1,621 | 65.3 |
| Shelby | 362 | -10.7 | 303 | D | 325.7 | 2,133,632 | 7,047 | 191,064 | 219.1 | 183,081 | 79.9 | 20.1 | 1 | 75.1 | 8,283 | 75.1 |
| Stark | 179 | 6.3 | 494 | D | 169.2 | 4,103,207 | 8,314 | 302,743 | 117.0 | 323,246 | 96.1 | 3.9 | 1 | 79.3 | 3,618 | 82.9 |
| Stephenson | 305 | -13.5 | 316 | 0.3 | 277.9 | 2,662,038 | 8,424 | 231,782 | 288.5 | 298,940 | 53.2 | 46.8 | 13 | 81.7 | 9,220 | 66.5 |
| Tazewell | 304 | -9.8 | 355 | 39.7 | 282.2 | 2,938,881 | 8,272 | 216,559 | 220.4 | 257,216 | 82.6 | 17.4 | 19 | 78.5 | 3,523 | 58.0 |
| Union | 151 | 24.3 | 255 | 4.5 | 110.7 | 1,084,536 | 4,248 | 113,939 | 47.9 | 81,264 | 90.3 | 9.7 | 1 | 72.5 | 3,589 | 55.4 |
| Vermilion | 471 | 8.5 | 449 | 0.8 | 443.7 | 3,686,595 | 8,203 | 261,347 | 283.0 | 269,783 | 95.2 | 4.8 | 6 | 77.8 | 5,125 | 56.7 |
| Wabash | 115 | 8.5 | 555 | 2.7 | 105.6 | 3,070,471 | 5,533 | 310,687 | 52.6 | 252,793 | 96.1 | 3.9 | NA | 76.0 | 2,768 | 75.0 |
| Warren | 341 | 0.8 | 480 | 0.2 | 301.4 | 3,880,818 | 8,092 | 284,489 | 253.6 | 356,689 | 79.1 | 20.9 | 3 | 81.6 | 2,469 | 74.5 |
| Washington | 349 | -1.7 | 488 | 1.8 | 325.7 | 3,134,730 | 6,422 | 325,140 | 203.8 | 284,987 | 69.0 | 31.0 | NA | 69.2 | 8,727 | 81.3 |
| Wayne | 368 | -0.1 | 359 | 1.2 | 319.6 | 1,710,457 | 4,764 | 188,596 | 188.3 | 183,664 | 72.6 | 27.4 | NA | 72.4 | 10,861 | 76.5 |
| White | 289 | -6.9 | 584 | 21.1 | 256.8 | 2,908,130 | 4,983 | 282,694 | 141.1 | 284,377 | 93.2 | 6.8 | NA | 73.6 | 7,825 | 71.8 |
| Whiteside | 371 | -8.1 | 387 | 60.4 | 340.4 | 3,051,253 | 7,892 | 283,093 | 301.0 | 313,911 | 69.9 | 30.1 | 8 | 78.6 | 12,909 | 74.1 |
| Will | 217 | -7.5 | 270 | 0.5 | 208.2 | 2,403,429 | 8,888 | 163,473 | 133.5 | 166,674 | 92.3 | 7.7 | 6 | 81.5 | 1,014 | 30.6 |
| Williamson | 104 | 0.4 | 170 | 0.0 | 73.3 | 673,017 | 3,953 | 94,199 | 34.7 | 56,907 | 74.4 | 25.6 | NA | 70.7 | 2,137 | 39.7 |

# Table B. States and Counties — Water Use, Wholesale Trade, Retail Trade, and Real Estate

| STATE County | Water use, 2015 Public supply water withdrawn (mil gal/day) | Public supply gallons withdrawn per person per day | Wholesale Trade[1], 2017 Number of establishments | Number of employees | Sales (mil dol) | Average payroll (mil dol) | Retail Trade[2], 2017 Number of establishments | Number of employees | Sales (mil dol) | Average payroll (mil dol) | Real estate and rental and leasing,[2] 2017 Number of establishments | Number of employees | Sales (mil dol) | Average payroll (mil dol) |
|---|---|---|---|---|---|---|---|---|---|---|---|---|---|---|
| | 133 | 134 | 135 | 136 | 137 | 138 | 139 | 140 | 141 | 142 | 143 | 144 | 145 | 146 |
| **ILLINOIS—Cont'd** | | | | | | | | | | | | | | |
| Hardin | 0.13 | 31.4 | NA | NA | NA | NA | 11 | 64 | 13.6 | 1.2 | 3 | 4 | 0.2 | 0.1 |
| Henderson | 8.82 | 1,260.9 | D | D | D | D | 13 | 97 | 19.0 | 1.5 | 3 | 2 | 0.2 | 0.0 |
| Henry | 3.97 | 80.2 | D | D | D | D | 153 | 1,836 | 481.2 | 45.8 | 24 | 65 | 12.3 | 1.4 |
| Iroquois | 1.99 | 69.4 | 55 | 512 | 518.7 | 23.6 | 78 | 994 | 291.3 | 22.2 | 20 | 45 | 9.1 | 1.2 |
| Jackson | 5.96 | 100.4 | 27 | 274 | 169.4 | 12.4 | 217 | 3,527 | 904.6 | 77.0 | 77 | 362 | 63.8 | 8.3 |
| Jasper | 1.77 | 184.2 | 21 | 219 | 180.4 | 7.4 | 24 | 166 | 65.0 | 4.7 | 4 | D | 0.5 | D |
| Jefferson | 0.00 | 0.0 | D | D | D | 36.6 | 158 | 2,165 | 596.2 | 53.1 | 28 | 77 | 18.7 | 2.6 |
| Jersey | 1.08 | 48.3 | 24 | 194 | 232.1 | 5.9 | 57 | 847 | 217.5 | 22.2 | 14 | 32 | 3.3 | 0.9 |
| Jo Daviess | 2.03 | 91.9 | 23 | 129 | 197.0 | 6.8 | 108 | 957 | 332.6 | 23.8 | 21 | 118 | 57.7 | 5.7 |
| Johnson | 0.86 | 67.4 | 5 | D | D | 14.2 | 32 | 238 | 91.3 | 5.4 | 4 | D | 1.1 | D |
| Kane | 61.75 | 116.3 | 742 | 11,620 | 13,711.5 | 761.0 | 1,495 | 25,639 | 6,686.0 | 632.4 | 479 | 2,585 | 1,265.4 | 137.2 |
| Kankakee | 13.26 | 119.6 | 116 | 2,199 | 1,384.6 | 110.9 | 366 | 5,932 | 1,516.0 | 134.1 | 89 | 331 | 68.4 | 12.2 |
| Kendall | 7.66 | 62.1 | 61 | 1,265 | 2,084.1 | 58.2 | 241 | 5,803 | 1,434.9 | 135.0 | 79 | 172 | 40.6 | 6.1 |
| Knox | 0.49 | 9.5 | 45 | 706 | 555.2 | 35.4 | 178 | 3,582 | 862.3 | 96.0 | 36 | 115 | 17.8 | 3.0 |
| Lake | 59.65 | 84.7 | 1,071 | 29,609 | 31,296.7 | 2,787.0 | 2,204 | 38,507 | 12,816.9 | 1,047.9 | 837 | 3,736 | 1,881.8 | 240.0 |
| LaSalle | 9.07 | 81.5 | 110 | 1,881 | 2,658.7 | 109.7 | 404 | 6,246 | 1,710.6 | 153.5 | 73 | 355 | 55.1 | 11.7 |
| Lawrence | 0.62 | 37.6 | 12 | 218 | 124.0 | 7.9 | 32 | 389 | 82.0 | 8.2 | D | D | D | D |
| Lee | 3.94 | 113.9 | D | D | D | D | 105 | 1,340 | 408.5 | 35.0 | 22 | 90 | 16.7 | 3.0 |
| Livingston | 5.66 | 154.3 | D | D | D | D | 124 | 1,487 | 418.9 | 39.7 | 19 | 70 | 9.4 | 2.6 |
| Logan | 2.88 | 97.6 | 37 | 508 | 694.9 | 30.1 | 84 | 1,028 | 302.8 | 25.0 | 20 | 61 | 6.5 | 1.6 |
| McDonough | 2.63 | 83.9 | 28 | 219 | 274.3 | 9.4 | 109 | 1,597 | 319.9 | 33.6 | 30 | 118 | 16.7 | 3.0 |
| McHenry | 18.70 | 60.8 | 381 | 5,182 | 3,114.1 | 316.5 | 849 | 14,300 | 3,898.2 | 368.9 | 240 | 704 | 142.3 | 26.8 |
| McLean | 10.50 | 60.6 | 165 | 2,108 | 7,165.7 | 150.4 | 523 | 9,562 | 2,301.8 | 211.7 | 154 | 754 | 151.1 | 26.2 |
| Macon | 20.02 | 186.6 | 95 | 1,303 | 1,363.5 | 74.8 | 395 | 5,594 | 1,515.3 | 143.5 | 88 | 484 | 78.7 | 14.9 |
| Macoupin | 2.96 | 64.3 | 43 | 584 | 831.3 | 28.1 | 132 | 1,370 | 385.8 | 34.4 | 15 | 45 | 4.5 | 1.0 |
| Madison | 52.40 | 196.8 | 204 | 3,258 | 2,977.0 | 181.2 | 772 | 12,202 | 3,599.7 | 317.1 | 235 | 963 | 220.4 | 36.5 |
| Marion | 0.99 | 25.8 | D | D | D | D | 131 | 1,334 | 340.1 | 34.8 | 23 | 108 | 11.4 | 2.3 |
| Marshall | 1.67 | 139.4 | D | D | D | 9.2 | 37 | 274 | 53.4 | 5.8 | 7 | 17 | 1.9 | 0.6 |
| Mason | 0.57 | 41.6 | 26 | 304 | 791.3 | 16.4 | 43 | 418 | 90.5 | 7.3 | D | D | D | 0.7 |
| Massac | 4.73 | 320.3 | D | D | D | 4.1 | 36 | 316 | 98.1 | 8.2 | 3 | 8 | 0.6 | 0.2 |
| Menard | 0.81 | 65.1 | 10 | 71 | 50.6 | 3.2 | 27 | 266 | 65.5 | 5.4 | 4 | 6 | 0.5 | 0.2 |
| Mercer | 0.88 | 55.5 | D | D | D | 17.0 | 37 | 347 | 73.7 | 7.2 | 36 | 163 | 57.3 | 8.7 |
| Monroe | 0.37 | 10.9 | D | D | D | 16.5 | 93 | 1,307 | 457.9 | 39.4 | 20 | 79 | 7.9 | 1.7 |
| Montgomery | 2.69 | 93.1 | 39 | 371 | 382.9 | 18.9 | 125 | 1,407 | 395.3 | 33.9 | 25 | 100 | 14.6 | 3.2 |
| Morgan | 0.31 | 8.9 | D | D | D | 4.8 | 141 | 1,884 | 473.5 | 42.8 | 3 | D | 1.3 | D |
| Moultrie | 0.17 | 11.4 | 11 | 80 | 240.7 | | 44 | 366 | 102.5 | 7.3 | | | | |
| Ogle | 5.09 | 98.5 | D | D | D | D | 123 | 1,514 | 430.5 | 35.3 | 36 | 103 | 23.8 | 4.5 |
| Peoria | 23.98 | 128.8 | 187 | 2,906 | 1,517.5 | 146.5 | 679 | 11,032 | 2,544.7 | 260.8 | 190 | 1,019 | 186.6 | 39.7 |
| Perry | 0.70 | 32.5 | 14 | 103 | 67.2 | 4.1 | 54 | 632 | 183.6 | 19.0 | 5 | 14 | 1.8 | 0.4 |
| Piatt | 1.12 | 68.3 | 24 | 286 | 286.1 | 15.9 | 40 | 405 | 131.7 | 11.4 | D | D | D | D |
| Pike | 1.90 | 118.8 | 26 | 244 | 254.7 | 12.5 | 53 | 625 | 141.8 | 12.9 | 8 | 95 | 39.6 | 3.4 |
| Pope | 0.00 | 0.0 | NA | NA | NA | NA | 10 | 46 | 9.0 | 0.9 | 4 | D | 0.8 | D |
| Pulaski | 0.17 | 29.9 | D | D | D | 4.5 | 15 | 88 | 38.1 | 2.3 | NA | NA | NA | NA |
| Putnam | 0.46 | 81.5 | 10 | 48 | 165.6 | 2.4 | 12 | 133 | 24.9 | 2.5 | D | D | D | 0.1 |
| Randolph | 2.43 | 74.0 | 26 | 513 | 473.4 | 25.6 | 101 | 1,305 | 360.2 | 33.9 | 11 | 27 | 8.5 | 1.1 |
| Richland | 1.42 | 88.6 | 24 | 294 | 198.8 | 13.2 | 60 | 754 | 171.5 | 16.0 | 7 | 31 | 3.6 | 1.0 |
| Rock Island | 17.39 | 119.0 | 140 | 2,769 | 2,567.4 | 163.3 | 426 | 7,912 | 1,993.9 | 199.8 | 125 | 520 | 103.1 | 17.1 |
| St. Clair | 17.32 | 65.6 | 180 | 2,236 | 2,868.2 | 109.3 | 863 | 13,465 | 3,272.6 | 333.3 | 229 | 936 | 185.2 | 35.4 |
| Saline | 0.00 | 0.0 | 12 | 159 | 97.1 | 9.0 | 101 | 1,266 | 344.3 | 31.4 | 13 | 67 | 21.0 | 2.7 |
| Sangamon | 23.21 | 116.8 | 175 | 2,446 | 3,128.9 | 129.9 | 723 | 12,124 | 3,024.0 | 289.8 | 243 | 887 | 227.4 | 29.6 |
| Schuyler | 0.92 | 130.8 | D | D | D | 2.1 | 22 | 210 | 33.6 | 4.8 | 3 | D | 3.1 | D |
| Scott | 4.82 | 946.6 | D | D | D | 3.5 | 7 | 64 | 22.0 | 1.7 | NA | NA | NA | NA |
| Shelby | 1.65 | 75.8 | 27 | 225 | 210.7 | 9.4 | 61 | 627 | 137.3 | 11.8 | D | D | D | D |
| Stark | 0.59 | 101.9 | D | D | D | 6.0 | 19 | 154 | 51.5 | 5.6 | NA | NA | NA | NA |
| Stephenson | 3.85 | 84.2 | D | D | D | D | 156 | 1,978 | 560.3 | 49.2 | 33 | 95 | 11.5 | 2.6 |
| Tazewell | 14.72 | 109.2 | 121 | 2,153 | 1,905.5 | 113.9 | 388 | 7,392 | 2,172.4 | 206.0 | 95 | 349 | 64.6 | 12.6 |
| Union | 1.28 | 73.5 | 10 | 94 | 46.7 | 4.0 | 59 | 687 | 177.2 | 17.3 | 5 | 19 | 2.8 | 0.6 |
| Vermilion | 8.95 | 112.9 | D | D | D | 91.7 | 247 | 3,709 | 934.8 | 84.3 | 46 | 153 | 38.3 | 5.9 |
| Wabash | 1.75 | 151.6 | 12 | 162 | 141.4 | 11.4 | 42 | 369 | 88.7 | 7.3 | 12 | 28 | 6.1 | 0.9 |
| Warren | 2.75 | 156.9 | 28 | 253 | 416.8 | 11.6 | 53 | 610 | 135.8 | 12.3 | 10 | 19 | 1.6 | 0.3 |
| Washington | 0.63 | 44.1 | 32 | 870 | 384.7 | 39.0 | 57 | 587 | 293.4 | 19.9 | 9 | 12 | 1.4 | 0.5 |
| Wayne | 1.12 | 68.2 | 14 | 125 | 172.4 | 5.3 | 61 | 595 | 155.1 | 14.9 | 7 | 24 | 7.9 | 0.5 |
| White | 1.08 | 75.4 | 30 | 265 | 139.2 | 10.0 | 59 | 544 | 140.2 | 12.3 | 10 | 24 | 5.7 | 0.8 |
| Whiteside | 3.62 | 63.4 | 67 | 629 | 732.5 | 28.6 | 180 | 2,643 | 623.0 | 62.4 | 36 | 95 | 14.7 | 2.4 |
| Will | 35.06 | 51.0 | 777 | 17,101 | 23,383.3 | 1,091.2 | 1,640 | 31,930 | 8,424.0 | 763.5 | 542 | 2,128 | 554.5 | 109.0 |
| Williamson | 1.22 | 18.1 | 57 | 650 | 250.6 | 27.9 | 249 | 3,669 | 1,175.0 | 99.2 | 60 | 185 | 34.1 | 5.7 |

1  Merchant wholesalers, except manufacturers' sales branches and offices.　　2. Employer establishments.

| STATE County | Professional, scientific, and technical services, 2017 | | | | Manufacturing, 2017 | | | | Accommodation and food services, 2017 | | | |
|---|---|---|---|---|---|---|---|---|---|---|---|---|
| | Number of establishments | Number of employees | Sales (mil dol) | Average payroll (mil dol) | Number of establishments | Number of employees | Sales (mil dol) | Average payroll (mil dol) | Number of establishments | Number of employees | Sales (mil dol) | Annual payroll (mil dol) |
| | 147 | 148 | 149 | 150 | 151 | 152 | 153 | 154 | 155 | 156 | 157 | 158 |
| ILLINOIS—Cont'd | | | | | | | | | | | | |
| Hardin | NA | NA | NA | NA | NA | NA | NA | NA | D | D | D | D |
| Henderson | 8 | 28 | 2 | 0.7 | NA | NA | NA | NA | 14 | 46 | 2.1 | 0.4 |
| Henry | 65 | 309 | 25 | 9.1 | 50 | 4,441 | 2,594.6 | 169.9 | 93 | 1,119 | 46.8 | 13.3 |
| Iroquois | 33 | 90 | 10 | 3.5 | 23 | 516 | 381.0 | 22.7 | 55 | 437 | 20.2 | 5.2 |
| Jackson | D | D | D | D | 45 | 890 | 208.6 | 33.4 | 149 | 2,640 | 104.6 | 30.6 |
| Jasper | 9 | 23 | 2 | 0.9 | 10 | 185 | 41.4 | 6.3 | 16 | 140 | 5.3 | 1.5 |
| Jefferson | 63 | 403 | 50 | 20.6 | D | 3,311 | D | 189.7 | 86 | 1,720 | 82.8 | 23.1 |
| Jersey | 17 | 100 | 11 | 5.5 | 15 | 131 | 28.4 | 5.3 | 58 | 704 | 33.1 | 10.2 |
| Jo Daviess | 49 | 182 | 29 | 8.5 | 40 | 871 | 354.8 | 45.0 | 99 | 1,333 | 82.1 | 24.6 |
| Johnson | 10 | 160 | 5 | 1.3 | 11 | 40 | 9.2 | 1.9 | D | D | D | D |
| Kane | D | D | D | D | 777 | 30,913 | 12,294.9 | 1,761.5 | 949 | 18,433 | 1,058.3 | 308.4 |
| Kankakee | 143 | 767 | 89 | 35.3 | 109 | 5,400 | 5,619.1 | 331.0 | 243 | 4,101 | 187.0 | 56.8 |
| Kendall | 222 | 858 | 117 | 39.9 | 78 | 2,311 | 683.0 | 123.2 | 197 | 3,504 | 189.3 | 50.9 |
| Knox | 59 | 290 | 34 | 12.3 | 33 | 1,015 | 344.9 | 43.0 | 128 | 1,768 | 83.7 | 22.8 |
| Lake | D | D | D | D | 817 | 28,356 | 10,863.9 | 1,751.5 | 1,563 | 28,143 | 1,677.7 | 489.3 |
| LaSalle | D | D | D | D | 144 | 5,277 | 2,708.6 | 312.1 | 310 | 4,079 | 216.9 | 58.3 |
| Lawrence | 8 | 39 | 5 | 1.8 | D | D | D | 42.9 | 20 | 252 | 10.6 | 3.0 |
| Lee | 29 | 284 | 40 | 14.0 | 35 | 3,240 | 1,432.6 | 163.7 | 74 | 895 | 42.2 | 11.4 |
| Livingston | 45 | 212 | 33 | 9.7 | 61 | 3,225 | 1,121.4 | 179.9 | 72 | 767 | 34.8 | 9.3 |
| Logan | 32 | 229 | 21 | 9.0 | 14 | 1,198 | 766.0 | 59.1 | 61 | 970 | 35.0 | 10.7 |
| McDonough | 44 | 231 | 22 | 8.0 | 16 | 1,399 | 349.9 | 69.4 | 88 | 1,309 | 52.9 | 14.6 |
| McHenry | D | D | D | D | 455 | 13,071 | 4,155.5 | 730.0 | 592 | 10,042 | 544.9 | 156.8 |
| McLean | D | D | D | D | 81 | 4,592 | 993.1 | 185.3 | 417 | 8,518 | 413.1 | 123.6 |
| Macon | 142 | 1,099 | 147 | 54.3 | 101 | 6,731 | 6,948.0 | 460.3 | 255 | 4,345 | 206.4 | 62.5 |
| Macoupin | 38 | 116 | 13 | 3.7 | 34 | 404 | 199.7 | 23.2 | D | D | D | D |
| Madison | D | D | D | D | 183 | 9,719 | 12,113.7 | 625.0 | 583 | 11,077 | 525.4 | 155.2 |
| Marion | 54 | 287 | 20 | 8.8 | 43 | 2,800 | 771.9 | 127.5 | 83 | 934 | 43.9 | 12.3 |
| Marshall | 12 | 73 | 8 | 2.5 | 9 | 840 | 251.1 | 45.6 | 29 | 226 | 7.5 | 2.2 |
| Mason | 11 | 27 | 3 | 0.8 | 9 | 91 | 28.3 | 4.7 | 34 | 249 | 9.8 | 2.8 |
| Massac | 10 | 185 | 26 | 10.6 | D | D | D | D | D | D | D | D |
| Menard | 15 | 115 | 10 | 3.8 | 8 | 43 | 7.3 | 1.4 | 20 | 161 | 6.6 | 1.9 |
| Mercer | D | D | D | D | 12 | 478 | 103.8 | 24.0 | 25 | 212 | 7.6 | 2.3 |
| Monroe | 73 | 655 | 109 | 46.9 | 21 | 226 | 57.0 | 12.5 | 76 | 1,338 | 60.8 | 17.9 |
| Montgomery | 36 | 204 | 19 | 8.5 | 23 | 660 | 351.1 | 33.1 | 71 | 1,013 | 45.6 | 13.5 |
| Morgan | 39 | 630 | 56 | 29.0 | D | D | D | D | 87 | 1,531 | 74.2 | 18.2 |
| Moultrie | 13 | 116 | 15 | 5.8 | 42 | 2,083 | 606.0 | 99.2 | 25 | 251 | 10.8 | 2.9 |
| Ogle | 74 | 386 | 55 | 22.5 | 65 | 3,388 | 1,165.8 | 163.4 | 115 | 1,251 | 60.0 | 15.4 |
| Peoria | D | D | D | D | 137 | 5,392 | 2,453.2 | 365.1 | 458 | 7,790 | 404.7 | 120.1 |
| Perry | 25 | 87 | 11 | 2.6 | 15 | 345 | 249.3 | 18.4 | 43 | 516 | 19.2 | 5.6 |
| Piatt | 24 | 114 | 10 | 4.8 | 17 | 190 | 37.0 | 9.2 | 37 | 307 | 13.3 | 4.0 |
| Pike | 18 | 98 | 13 | 4.8 | 12 | 131 | 43.4 | 5.2 | 28 | 343 | 14.8 | 4.0 |
| Pope | D | D | D | 0.2 | NA | NA | NA | NA | D | D | D | 0.2 |
| Pulaski | NA | NA | NA | NA | NA | NA | NA | NA | D | D | D | 0.4 |
| Putnam | D | D | D | D | 5 | 384 | 788.1 | 23.9 | 16 | 87 | 3.3 | 0.7 |
| Randolph | 28 | 276 | 33 | 10.6 | 28 | 2,607 | 639.1 | 81.2 | 69 | 980 | 41.5 | 11.8 |
| Richland | 29 | 81 | 10 | 3.5 | 23 | 421 | 160.1 | 20.4 | 37 | 536 | 22.2 | 6.5 |
| Rock Island | D | D | D | D | 134 | 8,809 | 2,958.2 | 381.9 | 351 | 6,301 | 342.6 | 96.8 |
| St. Clair | D | D | D | D | 141 | 5,127 | 2,607.9 | 273.0 | 542 | 10,972 | 612.0 | 167.8 |
| Saline | 42 | 163 | 17 | 5.7 | 22 | 308 | 55.0 | 12.2 | 49 | 877 | 42.3 | 9.8 |
| Sangamon | 497 | 5,245 | 755 | 320.4 | 89 | 2,699 | 847.3 | 138.9 | 552 | 10,006 | 490.8 | 148.3 |
| Schuyler | 8 | 36 | 4 | 1.3 | D | D | D | D | 12 | 111 | 5.8 | 1.8 |
| Scott | 4 | 7 | 1 | 0.2 | NA | NA | NA | NA | 8 | 42 | 2.1 | 0.4 |
| Shelby | 22 | 253 | 31 | 12.4 | 21 | 1,138 | 413.9 | 58.9 | 38 | 429 | 16.0 | 4.6 |
| Stark | 8 | 40 | 4 | 1.3 | 5 | 202 | 54.3 | 11.3 | 3 | 24 | 0.8 | 0.1 |
| Stephenson | 69 | 401 | 64 | 20.6 | 63 | 2,711 | 961.6 | 159.4 | 90 | 1,384 | 63.1 | 17.6 |
| Tazewell | 182 | 1,410 | 166 | 74.5 | 113 | 5,589 | 2,667.5 | 330.3 | 322 | 6,290 | 353.7 | 98.4 |
| Union | 22 | 88 | 9 | 3.3 | 13 | 234 | 78.4 | 10.7 | 37 | 421 | 13.3 | 4.0 |
| Vermilion | 76 | 385 | 58 | 17.6 | 80 | 4,496 | 1,881.2 | 244.5 | 152 | 2,338 | 101.6 | 28.0 |
| Wabash | 24 | 110 | 11 | 4.8 | 10 | 307 | 62.8 | 16.5 | 17 | 281 | 10.1 | 3.4 |
| Warren | 19 | 115 | 11 | 3.0 | 13 | 1,920 | 834.9 | 94.1 | 36 | 450 | 20.6 | 6.1 |
| Washington | 23 | 107 | 19 | 5.7 | 13 | 1,701 | 588.9 | 94.3 | 29 | 367 | 15.8 | 4.4 |
| Wayne | 20 | 83 | 8 | 2.9 | 17 | 122 | 48.7 | 5.8 | 26 | 345 | 15.4 | 4.2 |
| White | 16 | 68 | 8 | 2.7 | 13 | 273 | 77.6 | 13.2 | 25 | 321 | 13.4 | 4.1 |
| Whiteside | 75 | 397 | 62 | 19.3 | 87 | 3,540 | 1,334.1 | 207.5 | 135 | 1,769 | 74.6 | 21.4 |
| Will | D | D | D | D | 588 | 19,153 | 15,580.8 | 1,229.2 | 1,194 | 23,183 | 1,500.8 | 399.3 |
| Williamson | D | D | D | D | 42 | 3,226 | 1,212.7 | 147.3 | 159 | 2,979 | 147.1 | 41.1 |

# Table B. States and Counties — Health Care and Social Assistance, Other Services, Nonemployer Businesses, and Residential Construction

| STATE County | Health care and social assistance, 2017 | | | | Other services, 2017 | | | | Nonemployer businesses, 2018 | | Value of residential construction authorized by building permits, 2020 | |
|---|---|---|---|---|---|---|---|---|---|---|---|---|
| | Number of establishments | Number of employees | Receipts (mil dol) | Annual payroll (mil dol) | Number of establishments | Number of employees | Receipts (mil dol) | Annual payroll (mil dol) | Number | Receipts (mil dol) | New construction ($1,000) | Number of housing units |
| | 159 | 160 | 161 | 162 | 163 | 164 | 165 | 166 | 167 | 168 | 169 | 170 |
| ILLINOIS—Cont'd | | | | | | | | | | | | |
| Hardin | 11 | 277 | 18.1 | 9.5 | NA | NA | NA | NA | 166 | 4.5 | 0 | 0 |
| Henderson | 7 | 132 | 8.0 | 4.0 | 6 | 14 | 3.2 | 0.6 | 347 | 15.6 | 1,756 | 9 |
| Henry | 79 | 1,632 | 160.2 | 58.2 | 90 | 279 | 38.0 | 8.1 | 2,801 | 97.4 | 5,847 | 22 |
| Iroquois | 70 | 1,617 | 106.5 | 52.2 | 46 | 117 | 14.2 | 3.2 | 1,847 | 66.9 | 4,366 | 25 |
| Jackson | 154 | 4,373 | 665.8 | 192.2 | 93 | 534 | 57.6 | 11.2 | 3,125 | 122.3 | 690 | 5 |
| Jasper | 19 | 184 | 11.5 | 3.9 | 17 | 49 | 5.6 | 1.1 | 819 | 26.1 | 150 | 1 |
| Jefferson | 144 | 4,053 | 518.8 | 185.8 | 73 | 522 | 62.1 | 15.5 | 2,246 | 93.1 | 720 | 6 |
| Jersey | 35 | 1,144 | 85.1 | 36.4 | 37 | 121 | 11.2 | 3.2 | 1,328 | 43.9 | 24,524 | 102 |
| Jo Daviess | 44 | 793 | 53.9 | 22.8 | 53 | 240 | 28.1 | 7.1 | 1,920 | 79.9 | 10,253 | 37 |
| Johnson | 18 | 230 | 12.8 | 5.2 | D | D | 5.2 | D | 751 | 21.4 | 0 | 0 |
| Kane | 1,224 | 24,156 | 3,291.4 | 1,233.3 | 911 | 6,364 | 722.3 | 214.4 | 35,338 | 1,562.1 | 241,926 | 1,272 |
| Kankakee | 290 | 7,298 | 720.7 | 336.0 | 175 | 846 | 92.4 | 27.1 | 5,872 | 215.7 | 17,610 | 93 |
| Kendall | 181 | 1,555 | 144.4 | 54.7 | 181 | 774 | 76.9 | 22.2 | 8,888 | 350.0 | 90,801 | 559 |
| Knox | 134 | 3,593 | 353.0 | 116.0 | 70 | 469 | 132.7 | 15.3 | 2,169 | 69.7 | 3,457 | 19 |
| Lake | 2,077 | 36,794 | 5,359.2 | 1,818.9 | 1,403 | 8,027 | 918.2 | 271.2 | 56,075 | 3,167.6 | 233,042 | 780 |
| LaSalle | 272 | 5,359 | 521.3 | 191.1 | 225 | 1,040 | 122.5 | 33.3 | 5,667 | 215.0 | 21,950 | 93 |
| Lawrence | 23 | 526 | 34.8 | 14.6 | 19 | 73 | 7.6 | 1.7 | 747 | 29.3 | 9,660 | 56 |
| Lee | 88 | 2,381 | 262.7 | 96.0 | 60 | 341 | 32.8 | 9.4 | 1,915 | 71.6 | 4,220 | 25 |
| Livingston | 67 | 1,501 | 146.6 | 46.6 | 69 | 339 | 36.7 | 9.5 | 1,927 | 69.4 | 3,191 | 15 |
| Logan | 58 | 1,572 | 132.9 | 50.1 | 44 | 187 | 17.7 | 4.3 | 1,387 | 50.5 | 1,364 | 7 |
| McDonough | 66 | 1,822 | 170.9 | 71.5 | D | D | D | 5.1 | 1,549 | 51.7 | 500 | 1 |
| McHenry | 772 | 11,931 | 1,547.5 | 571.8 | 685 | 3,465 | 338.1 | 104.5 | 22,922 | 1,024.4 | 113,568 | 585 |
| McLean | 377 | 8,713 | 1,061.4 | 402.1 | 272 | 2,436 | 262.9 | 85.3 | 9,646 | 404.9 | 49,951 | 265 |
| Macon | 287 | 8,451 | 1,014.4 | 362.8 | 172 | 1,030 | 293.2 | 35.5 | 5,091 | 180.2 | 8,466 | 31 |
| Macoupin | 81 | 1,724 | 121.9 | 51.1 | 79 | 270 | 25.6 | 6.3 | 2,475 | 95.0 | 19,003 | 95 |
| Madison | 686 | 14,455 | 1,350.4 | 532.0 | 472 | 2,893 | 312.6 | 90.2 | 14,601 | 597.4 | 115,675 | 442 |
| Marion | 107 | 2,905 | 274.2 | 99.7 | 67 | 256 | 25.3 | 6.5 | 2,076 | 72.9 | 1,260 | 15 |
| Marshall | 17 | 454 | 20.9 | 11.6 | D | D | D | D | 619 | 22.5 | 951 | 4 |
| Mason | 19 | 505 | 38.6 | 17.9 | 19 | 84 | 8.3 | 1.9 | 641 | 24.8 | 1,426 | 6 |
| Massac | D | D | D | D | D | D | D | D | 819 | 25.5 | 1,336 | 31 |
| Menard | 8 | 135 | 6.4 | 4.0 | 16 | 67 | 10.0 | 2.3 | 783 | 25.3 | 3,027 | 13 |
| Mercer | 21 | 296 | 28.4 | 12.1 | 18 | 93 | 7.2 | 2.0 | 901 | 29.4 | 1,215 | 8 |
| Monroe | 91 | 1,049 | 71.5 | 30.7 | 76 | 396 | 35.7 | 13.6 | 2,284 | 94.4 | 33,313 | 126 |
| Montgomery | 68 | 1,451 | 158.1 | 50.4 | 47 | 194 | 24.9 | 5.5 | 1,520 | 53.2 | 5,944 | 25 |
| Morgan | 103 | 2,597 | 239.0 | 98.4 | D | D | D | D | 1,839 | 67.0 | 2,106 | 18 |
| Moultrie | 32 | 716 | 39.3 | 19.0 | 19 | 66 | 6.4 | 1.7 | 1,052 | 45.1 | 4,755 | 20 |
| Ogle | 84 | 1,526 | 115.4 | 49.4 | 81 | 329 | 37.5 | 9.7 | 3,129 | 126.2 | 9,239 | 42 |
| Peoria | 514 | 24,940 | 3,301.2 | 1,375.5 | 298 | 4,335 | 454.6 | 203.8 | 9,508 | 396.5 | 17,640 | 60 |
| Perry | 46 | 973 | 76.2 | 31.2 | 46 | 184 | 14.2 | 3.6 | 951 | 28.5 | 6,555 | 29 |
| Piatt | D | D | D | D | 20 | 60 | 5.7 | 1.7 | 1,152 | 44.0 | 12,122 | 38 |
| Pike | 23 | 702 | 62.1 | 24.3 | 27 | 68 | 7.6 | 1.7 | 1,069 | 42.9 | 3,812 | 23 |
| Pope | 9 | 74 | 3.6 | 1.5 | D | D | D | 0.2 | 265 | 7.7 | 0 | 0 |
| Pulaski | 13 | 148 | 5.0 | 1.7 | D | D | 3.8 | D | 274 | 7.7 | 1,000 | 10 |
| Putnam | 5 | 25 | 1.8 | 0.6 | D | D | D | D | 380 | 15.5 | 2,583 | 17 |
| Randolph | 76 | 1,886 | 182.1 | 77.8 | 65 | 322 | 49.0 | 12.2 | 1,371 | 46.5 | 6,139 | 29 |
| Richland | 54 | 1,170 | 88.2 | 40.1 | 43 | 164 | 20.6 | 4.9 | 1,088 | 44.1 | 238 | 1 |
| Rock Island | 406 | 9,287 | 980.8 | 387.9 | 220 | 1,514 | 149.2 | 46.7 | 6,836 | 267.3 | 21,446 | 113 |
| St. Clair | 603 | 14,766 | 1,501.4 | 626.2 | 384 | 2,244 | 209.2 | 66.1 | 14,016 | 473.1 | 127,666 | 561 |
| Saline | 83 | 2,720 | 225.0 | 82.3 | 40 | 168 | 20.7 | 4.5 | 1,425 | 51.8 | 0 | 0 |
| Sangamon | 470 | 22,478 | 3,032.8 | 1,050.0 | 467 | 3,250 | 402.8 | 131.2 | 12,140 | 475.0 | 70,791 | 322 |
| Schuyler | 12 | 338 | 29.6 | 12.5 | 13 | 30 | 4.9 | 0.9 | 458 | 16.0 | 0 | 0 |
| Scott | 7 | 14 | 1.3 | 0.5 | NA | NA | NA | NA | 335 | 11.5 | NA | NA |
| Shelby | 31 | 651 | 37.3 | 18.4 | 29 | 98 | 13.9 | 3.1 | 1,278 | 51.2 | 10,255 | 50 |
| Stark | 7 | 109 | 7.4 | 3.3 | NA | NA | NA | NA | 357 | 16.7 | 1,117 | 3 |
| Stephenson | 102 | 2,535 | 337.9 | 105.7 | 89 | 396 | 36.2 | 10.3 | 2,719 | 99.2 | 2,129 | 13 |
| Tazewell | 275 | 5,116 | 442.7 | 168.1 | 222 | 1,192 | 129.7 | 40.3 | 6,599 | 256.0 | 26,663 | 123 |
| Union | 61 | 1,300 | 80.3 | 32.9 | 26 | 104 | 9.9 | 3.0 | 1,009 | 37.6 | 1,934 | 18 |
| Vermilion | 146 | 5,046 | 578.9 | 265.4 | 112 | 509 | 53.0 | 14.3 | 3,880 | 134.9 | 1,277 | 7 |
| Wabash | 20 | 861 | 65.9 | 25.6 | 22 | 94 | 7.2 | 1.5 | 750 | 27.0 | 0 | 0 |
| Warren | 38 | 586 | 49.8 | 17.5 | 36 | 102 | 11.4 | 1.9 | 864 | 35.3 | 640 | 2 |
| Washington | 22 | 400 | 33.0 | 12.1 | 28 | 61 | 6.4 | 1.3 | 902 | 34.9 | 5,595 | 22 |
| Wayne | 42 | 947 | 62.3 | 29.2 | 24 | 113 | 14.1 | 3.5 | 1,306 | 47.4 | 0 | 0 |
| White | 38 | 707 | 57.4 | 19.6 | 29 | 132 | 18.4 | 4.3 | 1,081 | 45.2 | 238 | 1 |
| Whiteside | 98 | 3,423 | 253.6 | 112.9 | 99 | 542 | 49.0 | 12.9 | 2,757 | 103.2 | 3,631 | 22 |
| Will | 1,551 | 27,115 | 2,757.6 | 1,123.0 | 1,160 | 7,504 | 992.3 | 250.7 | 50,716 | 2,300.0 | 405,600 | 1,775 |
| Williamson | 212 | 7,024 | 913.4 | 304.4 | 96 | 501 | 49.1 | 13.3 | 4,188 | 159.7 | 24,047 | 178 |

# Table B. States and Counties — Government Employment and Payroll, and Local Government Finances

| STATE County | Government employment and payroll, 2017 | | | | | | | | | Local government finances, 2017 | | | | |
| | | | March payroll (percent of total) | | | | | | | General revenue | | | | |
| | | | | | | | | | | | | Taxes | | |
| | Full-time equivalent employees | March payroll (dollars) | Adminis-tration, judicial, and legal | Police and corrections | Fire protection | Highways and transpor-tation | Health and welfare | Natural resources and utilities | Education and libraries | Total (mil dol) | Inter-govern-mental (mil dol) | Total (mil dol) | Per capita[1] (dollars) | |
| | | | | | | | | | | | | | Total | Property |
| | 171 | 172 | 173 | 174 | 175 | 176 | 177 | 178 | 179 | 180 | 181 | 182 | 183 | 184 |

ILLINOIS—Cont'd

| STATE County | 171 | 172 | 173 | 174 | 175 | 176 | 177 | 178 | 179 | 180 | 181 | 182 | 183 | 184 |
|---|---|---|---|---|---|---|---|---|---|---|---|---|---|---|
| Hardin | 355 | 964,662 | 4.8 | 3.6 | 0.1 | 67.0 | 2.4 | 1.9 | 20.2 | 11.7 | 8.6 | 1.6 | 397 | 394 |
| Henderson | 259 | 794,157 | 14.4 | 7.2 | 0.6 | 7.5 | 12.6 | 1.1 | 56.2 | 20.1 | 7.1 | 9.7 | 1,426 | 1,331 |
| Henry | 2,305 | 8,261,504 | 5.7 | 8.4 | 1.8 | 3.2 | 22.2 | 5.0 | 53.1 | 213.8 | 62.0 | 79.6 | 1,618 | 1,542 |
| Iroquois | 913 | 3,709,326 | 9.8 | 7.1 | 1.2 | 5.1 | 0.1 | 2.7 | 72.9 | 88.2 | 33.1 | 43.8 | 1,579 | 1,529 |
| Jackson | 2,296 | 7,460,282 | 7.1 | 10.1 | 3.1 | 5.0 | 9.8 | 7.0 | 57.0 | 194.5 | 82.7 | 80.7 | 1,393 | 1,067 |
| Jasper | 432 | 1,484,947 | 9.8 | 6.5 | 0.5 | 5.8 | 9.2 | 4.1 | 63.9 | 35.3 | 16.7 | 14.8 | 1,553 | 1,523 |
| Jefferson | 1,644 | 6,130,923 | 6.1 | 7.1 | 3.2 | 2.8 | 2.3 | 2.1 | 75.4 | 160.2 | 85.3 | 51.9 | 1,365 | 1,060 |
| Jersey | 864 | 3,597,723 | 4.3 | 7.3 | 0.3 | 2.7 | 46.9 | 2.8 | 34.2 | 96.2 | 21.3 | 25.3 | 1,154 | 1,020 |
| Jo Daviess | 765 | 3,218,722 | 9.5 | 10.0 | 0.5 | 5.3 | 1.7 | 2.3 | 69.9 | 89.9 | 21.7 | 55.6 | 2,586 | 2,420 |
| Johnson | 315 | 1,048,861 | 8.9 | 6.6 | 0.3 | 2.5 | 1.7 | 5.0 | 75.0 | 31.6 | 15.9 | 8.6 | 694 | 693 |
| Kane | 23,307 | 119,685,773 | 5.1 | 8.6 | 4.9 | 1.9 | 1.5 | 6.0 | 70.9 | 3,136.7 | 933.2 | 1,815.4 | 3,411 | 3,119 |
| Kankakee | 4,432 | 18,124,649 | 7.1 | 11.6 | 4.2 | 3.3 | 1.1 | 3.4 | 67.9 | 414.1 | 162.1 | 176.4 | 1,596 | 1,513 |
| Kendall | 4,540 | 19,788,537 | 4.2 | 9.3 | 3.0 | 1.1 | 1.2 | 4.0 | 76.8 | 518.1 | 169.8 | 289.8 | 2,297 | 2,210 |
| Knox | 2,096 | 8,569,394 | 6.4 | 10.6 | 3.9 | 3.1 | 9.3 | 5.2 | 60.3 | 204.4 | 73.6 | 90.3 | 1,786 | 1,567 |
| Lake | 29,578 | 152,226,049 | 4.8 | 8.4 | 4.0 | 2.3 | 3.2 | 5.9 | 70.2 | 4,064.7 | 1,048.8 | 2,555.2 | 3,637 | 3,411 |
| LaSalle | 4,070 | 17,042,807 | 6.3 | 11.5 | 3.2 | 3.6 | 3.2 | 3.3 | 67.6 | 434.5 | 146.0 | 224.5 | 2,048 | 1,856 |
| Lawrence | 637 | 2,603,378 | 5.2 | 4.6 | 0.0 | 3.1 | 44.1 | 4.3 | 38.3 | 34.7 | 20.0 | 9.3 | 583 | 562 |
| Lee | 1,093 | 4,234,740 | 7.8 | 5.9 | 4.3 | 3.8 | 2.3 | 3.3 | 71.4 | 114.7 | 35.6 | 59.7 | 1,727 | 1,645 |
| Livingston | 1,606 | 6,248,617 | 6.4 | 8.5 | 1.8 | 4.4 | 3.1 | 2.8 | 72.2 | 148.2 | 51.3 | 69.8 | 1,931 | 1,905 |
| Logan | 857 | 3,071,770 | 8.8 | 11.1 | 4.2 | 5.4 | 2.9 | 2.8 | 62.2 | 82.8 | 30.5 | 35.3 | 1,216 | 1,141 |
| McDonough | 1,763 | 7,331,926 | 3.9 | 6.9 | 1.8 | 2.5 | 50.9 | 2.2 | 31.1 | 180.5 | 37.6 | 37.6 | 1,238 | 1,187 |
| McHenry | 11,450 | 51,968,532 | 6.0 | 10.6 | 5.1 | 2.8 | 2.4 | 5.1 | 66.5 | 1,368.0 | 327.1 | 853.5 | 2,773 | 2,571 |
| McLean | 6,093 | 27,474,898 | 6.9 | 11.6 | 5.5 | 5.0 | 3.2 | 7.0 | 58.0 | 710.8 | 161.0 | 421.6 | 2,440 | 1,942 |
| Macon | 3,903 | 16,584,624 | 6.5 | 13.5 | 5.1 | 3.4 | 3.3 | 7.1 | 58.7 | 459.7 | 187.3 | 179.9 | 1,707 | 1,439 |
| Macoupin | 1,653 | 6,375,505 | 7.8 | 8.5 | 1.0 | 2.5 | 2.9 | 4.3 | 72.3 | 134.0 | 67.7 | 45.9 | 1,008 | 981 |
| Madison | 8,747 | 37,036,497 | 7.7 | 11.8 | 4.2 | 3.9 | 2.2 | 6.4 | 61.9 | 1,040.8 | 419.6 | 447.2 | 1,686 | 1,525 |
| Marion | 1,918 | 7,344,248 | 5.0 | 6.5 | 1.7 | 7.3 | 2.0 | 3.7 | 72.8 | 172.0 | 96.7 | 48.9 | 1,296 | 1,243 |
| Marshall | 337 | 1,117,988 | 11.6 | 9.5 | 0.0 | 6.6 | 0.0 | 4.9 | 65.4 | 37.9 | 10.4 | 24.3 | 2,081 | 1,983 |
| Mason | 921 | 3,116,994 | 6.9 | 6.0 | 1.2 | 5.5 | 34.8 | 1.8 | 43.6 | 73.4 | 20.8 | 23.6 | 1,726 | 1,654 |
| Massac | 699 | 2,610,477 | 4.8 | 7.7 | 3.3 | 2.2 | 30.4 | 5.6 | 45.9 | 76.2 | 28.9 | 15.0 | 1,049 | 1,010 |
| Menard | 584 | 1,894,919 | 5.0 | 8.0 | 0.2 | 3.3 | 19.9 | 3.7 | 58.7 | 49.9 | 15.2 | 22.1 | 1,796 | 1,702 |
| Mercer | 579 | 2,003,713 | 12.5 | 9.1 | 0.0 | 4.3 | 3.1 | 2.9 | 66.8 | 58.0 | 19.1 | 31.4 | 2,015 | 1,977 |
| Monroe | 1,165 | 4,473,031 | 6.0 | 9.5 | 3.8 | 2.6 | 11.6 | 6.0 | 60.2 | 108.2 | 27.3 | 52.7 | 1,542 | 1,469 |
| Montgomery | 998 | 3,803,424 | 8.5 | 8.7 | 2.8 | 7.2 | 5.1 | 5.8 | 61.2 | 89.9 | 39.1 | 38.0 | 1,322 | 1,218 |
| Morgan | 1,605 | 5,143,648 | 4.6 | 8.3 | 1.8 | 3.1 | 3.9 | 3.8 | 73.9 | 104.5 | 45.3 | 48.0 | 1,401 | 1,282 |
| Moultrie | 430 | 1,479,487 | 14.0 | 9.8 | 4.2 | 4.1 | 1.2 | 10.0 | 51.6 | 37.3 | 13.6 | 19.1 | 1,299 | 1,259 |
| Ogle | 2,519 | 10,005,346 | 4.9 | 8.9 | 3.1 | 3.1 | 0.8 | 6.4 | 72.3 | 253.4 | 74.0 | 135.1 | 2,650 | 2,523 |
| Peoria | 7,199 | 29,894,387 | 5.5 | 13.1 | 5.3 | 5.3 | 4.3 | 7.0 | 57.7 | 789.4 | 287.3 | 349.9 | 1,917 | 1,574 |
| Perry | 795 | 3,003,562 | 3.7 | 8.6 | 1.5 | 3.4 | 31.0 | 5.3 | 44.9 | 76.1 | 26.8 | 15.7 | 740 | 705 |
| Piatt | 646 | 2,360,752 | 6.1 | 9.2 | 1.1 | 5.1 | 19.3 | 3.5 | 54.0 | 69.2 | 23.1 | 29.6 | 1,800 | 1,746 |
| Pike | 632 | 2,444,120 | 6.8 | 6.8 | 0.0 | 6.5 | 5.9 | 6.7 | 64.2 | 51.2 | 22.6 | 18.5 | 1,179 | 1,125 |
| Pope | 135 | 415,660 | 15.1 | 7.7 | 0.0 | 1.6 | 3.8 | 7.7 | 64.1 | 13.9 | 5.4 | 4.3 | 1,038 | 862 |
| Pulaski | 386 | 1,105,044 | 8.6 | 9.0 | 0.0 | 19.5 | 1.8 | 1.9 | 58.6 | 36.9 | 21.2 | 4.2 | 760 | 725 |
| Putnam | 233 | 671,429 | 16.2 | 13.2 | 0.0 | 6.9 | 3.9 | 4.1 | 54.4 | 23.1 | 11.0 | 9.7 | 1,704 | 1,671 |
| Randolph | 1,559 | 6,096,504 | 5.2 | 5.7 | 0.1 | 2.9 | 36.8 | 4.0 | 44.3 | 154.5 | 45.5 | 34.4 | 1,065 | 1,025 |
| Richland | 889 | 3,418,270 | 4.4 | 5.0 | 0.4 | 2.4 | 0.9 | 2.7 | 83.8 | 51.4 | 24.3 | 15.2 | 962 | 892 |
| Rock Island | 5,593 | 24,550,254 | 5.5 | 11.1 | 4.6 | 6.9 | 5.2 | 6.0 | 59.2 | 641.0 | 245.2 | 281.6 | 1,958 | 1,716 |
| St. Clair | 8,822 | 37,845,721 | 5.8 | 10.1 | 2.3 | 3.3 | 2.8 | 3.4 | 70.2 | 1,174.6 | 618.8 | 385.0 | 1,466 | 1,326 |
| Saline | 1,033 | 4,590,790 | 5.0 | 7.0 | 0.9 | 3.6 | 2.0 | 4.1 | 77.0 | 93.2 | 49.2 | 27.2 | 1,135 | 1,045 |
| Sangamon | 8,122 | 38,176,352 | 4.7 | 9.9 | 4.6 | 5.3 | 2.1 | 15.0 | 58.2 | 836.8 | 305.4 | 387.8 | 1,968 | 1,666 |
| Schuyler | 432 | 1,599,283 | 9.0 | 5.5 | 1.8 | 4.4 | 47.8 | 4.3 | 26.3 | 44.3 | 8.6 | 10.1 | 1,452 | 1,433 |
| Scott | 217 | 643,991 | 11.1 | 6.8 | 0.0 | 3.3 | 3.7 | 4.9 | 70.3 | 16.9 | 6.8 | 5.8 | 1,155 | 1,065 |
| Shelby | 566 | 2,072,677 | 12.4 | 9.0 | 2.0 | 8.3 | 3.6 | 4.7 | 57.8 | 49.8 | 22.8 | 20.4 | 936 | 905 |
| Stark | 211 | 748,179 | 18.3 | 1.8 | 0.0 | 6.6 | 0.0 | 1.4 | 71.8 | 20.9 | 5.7 | 12.7 | 2,333 | 2,301 |
| Stephenson | 2,026 | 7,213,358 | 5.0 | 9.1 | 4.2 | 3.0 | 5.7 | 3.7 | 68.9 | 183.6 | 67.5 | 81.4 | 1,810 | 1,664 |
| Tazewell | 4,866 | 20,502,398 | 5.1 | 10.5 | 3.8 | 2.6 | 2.2 | 5.8 | 69.1 | 562.7 | 186.8 | 261.2 | 1,957 | 1,719 |
| Union | 680 | 2,600,080 | 10.9 | 6.2 | 0.3 | 5.0 | 3.8 | 3.6 | 69.3 | 62.7 | 34.3 | 20.8 | 1,228 | 1,189 |
| Vermilion | 3,100 | 11,926,933 | 8.2 | 8.8 | 2.8 | 3.9 | 4.1 | 4.3 | 64.0 | 364.0 | 228.8 | 92.7 | 1,193 | 1,097 |
| Wabash | 712 | 2,806,713 | 2.5 | 4.1 | 1.4 | 1.6 | 51.1 | 2.6 | 36.5 | 70.4 | 14.6 | 10.3 | 894 | 854 |
| Warren | 588 | 2,061,899 | 7.5 | 10.1 | 4.3 | 4.6 | 2.8 | 2.3 | 68.1 | 52.8 | 22.0 | 24.5 | 1,429 | 1,322 |
| Washington | 516 | 2,031,972 | 7.3 | 8.5 | 1.3 | 4.6 | 28.1 | 4.1 | 44.5 | 56.8 | 14.1 | 18.9 | 1,354 | 1,343 |
| Wayne | 625 | 2,027,975 | 8.6 | 8.9 | 1.5 | 3.7 | 2.8 | 9.1 | 63.3 | 58.3 | 37.2 | 15.5 | 943 | 876 |
| White | 840 | 2,861,503 | 5.9 | 7.1 | 0.7 | 2.9 | 2.0 | 5.6 | 75.1 | 51.3 | 29.1 | 15.3 | 1,101 | 1,061 |
| Whiteside | 3,520 | 15,743,468 | 4.3 | 4.2 | 1.1 | 2.0 | 52.8 | 2.7 | 32.4 | 435.9 | 86.1 | 81.8 | 1,461 | 1,368 |
| Will | 23,275 | 114,026,103 | 4.8 | 11.3 | 6.0 | 2.1 | 2.5 | 4.7 | 67.5 | 2,958.3 | 810.0 | 1,726.1 | 2,500 | 2,290 |
| Williamson | 2,486 | 9,699,278 | 4.3 | 9.1 | 3.8 | 3.5 | 0.8 | 5.3 | 71.6 | 249.9 | 118.8 | 96.8 | 1,443 | 1,274 |

1. Based on the resident population estimated as of July 1 of the year shown.

| STATE County | Direct general expenditure Total (mil dol) 185 | Per capita[1] (dollars) 186 | Education 187 | Health and hospitals 188 | Police protection 189 | Public welfare 190 | Highways 191 | Debt outstanding Total (mil dol) 192 | Per capita[1] (dollars) 193 | Federal civilian 194 | Federal military 195 | State and local 196 | Number of returns 197 | Mean adjusted gross income 198 | Mean income tax 199 |
|---|---|---|---|---|---|---|---|---|---|---|---|---|---|---|---|
| ILLINOIS—Cont'd | | | | | | | | | | | | | | | |
| Hardin | 9 | 2,206 | 54.1 | 0.7 | 2.4 | 0.1 | 8.4 | 8.3 | 2,107 | 9 | 8 | 178 | 1,480 | 45,558 | 3,506 |
| Henderson | 19 | 2,793 | 45.2 | 5.0 | 7.3 | 0.1 | 9.4 | 6.5 | 957 | 33 | 13 | 371 | 3,130 | 50,687 | 4,259 |
| Henry | 239 | 4,853 | 43.6 | 22.3 | 5.6 | 0.2 | 4.9 | 164.0 | 3,331 | 119 | 97 | 3,570 | 23,810 | 62,181 | 6,026 |
| Iroquois | 93 | 3,340 | 65.1 | 1.7 | 4.0 | 0.1 | 6.9 | 50.4 | 1,815 | 95 | 54 | 1,389 | 13,130 | 55,028 | 4,769 |
| Jackson | 198 | 3,410 | 46.6 | 4.2 | 7.9 | 0.1 | 6.4 | 116.0 | 2,004 | 144 | 116 | 11,197 | 21,840 | 54,042 | 5,503 |
| Jasper | 34 | 3,551 | 50.2 | 4.5 | 8.6 | 0.4 | 12.8 | 9.3 | 973 | 36 | 19 | 595 | 4,630 | 54,248 | 4,272 |
| Jefferson | 150 | 3,934 | 62.3 | 0.6 | 5.8 | 0.1 | 7.4 | 162.8 | 4,278 | 144 | 72 | 2,214 | 16,330 | 52,180 | 4,671 |
| Jersey | 95 | 4,316 | 24.8 | 44.6 | 3.9 | 0.1 | 4.3 | 45.1 | 2,058 | 40 | 42 | 1,205 | 10,190 | 58,257 | 5,370 |
| Jo Daviess | 92 | 4,288 | 44.6 | 1.6 | 6.7 | 0.1 | 11.1 | 28.6 | 1,333 | 66 | 43 | 1,291 | 11,740 | 58,747 | 6,093 |
| Johnson | 29 | 2,327 | 57.8 | 1.4 | 5.0 | 0.0 | 6.9 | 22.0 | 1,776 | 71 | 22 | 815 | 4,920 | 52,473 | 4,016 |
| Kane | 2,962 | 5,564 | 57.5 | 0.2 | 7.1 | 0.7 | 4.7 | 5,354.6 | 10,060 | 1,615 | 1,063 | 31,299 | 250,080 | 77,317 | 9,371 |
| Kankakee | 447 | 4,046 | 52.8 | 0.5 | 6.7 | 0.1 | 5.4 | 491.0 | 4,442 | 258 | 214 | 6,058 | 50,160 | 58,111 | 5,506 |
| Kendall | 503 | 3,986 | 60.4 | 1.4 | 5.7 | 0.4 | 4.7 | 715.8 | 5,674 | 126 | 260 | 6,254 | 60,540 | 74,812 | 7,843 |
| Knox | 204 | 4,039 | 49.3 | 7.0 | 5.7 | 0.2 | 6.3 | 253.8 | 5,017 | 165 | 92 | 3,086 | 22,770 | 52,402 | 4,771 |
| Lake | 4,041 | 5,752 | 56.9 | 2.0 | 5.9 | 0.1 | 5.0 | 3,441.8 | 4,900 | 5,306 | 14,953 | 37,913 | 343,530 | 125,616 | 22,073 |
| LaSalle | 427 | 3,897 | 52.5 | 3.2 | 6.3 | 0.1 | 6.3 | 325.9 | 2,972 | 352 | 213 | 5,936 | 54,010 | 58,791 | 5,769 |
| Lawrence | 37 | 2,275 | 49.5 | 3.9 | 6.3 | 1.0 | 11.3 | 33.9 | 2,114 | 42 | 27 | 863 | 6,110 | 50,367 | 4,051 |
| Lee | 118 | 3,418 | 55.3 | 1.5 | 7.5 | 0.1 | 8.4 | 89.5 | 2,590 | 84 | 64 | 1,821 | 16,120 | 58,447 | 5,760 |
| Livingston | 152 | 4,198 | 53.8 | 3.2 | 5.8 | 0.1 | 9.7 | 49.4 | 1,366 | 88 | 67 | 2,176 | 16,890 | 59,672 | 5,593 |
| Logan | 81 | 2,803 | 46.7 | 3.3 | 5.4 | 0.1 | 6.9 | 37.8 | 1,304 | 91 | 49 | 1,402 | 12,310 | 55,811 | 4,880 |
| McDonough | 187 | 6,140 | 23.7 | 49.7 | 3.8 | 0.6 | 4.6 | 80.4 | 2,645 | 94 | 56 | 4,811 | 11,620 | 53,612 | 4,858 |
| McHenry | 1,398 | 4,543 | 51.3 | 2.3 | 7.2 | 0.4 | 6.6 | 1,098.7 | 3,569 | 482 | 618 | 14,893 | 156,390 | 77,818 | 9,221 |
| McLean | 716 | 4,145 | 43.6 | 1.2 | 7.8 | 1.4 | 6.3 | 932.9 | 5,400 | 435 | 330 | 15,164 | 76,040 | 75,144 | 8,730 |
| Macon | 452 | 4,288 | 47.3 | 1.2 | 10.4 | 1.1 | 5.4 | 425.2 | 4,034 | 317 | 212 | 5,570 | 47,630 | 61,080 | 6,419 |
| Macoupin | 132 | 2,903 | 58.4 | 3.6 | 6.0 | 0.1 | 6.2 | 89.6 | 1,969 | 120 | 89 | 2,357 | 21,000 | 54,492 | 4,698 |
| Madison | 971 | 3,659 | 46.4 | 1.1 | 8.2 | 1.9 | 5.8 | 909.3 | 3,428 | 581 | 528 | 16,017 | 125,310 | 65,094 | 6,957 |
| Marion | 161 | 4,268 | 65.7 | 0.8 | 3.2 | 0.1 | 3.3 | 97.1 | 2,573 | 104 | 74 | 2,163 | 17,490 | 49,043 | 4,024 |
| Marshall | 40 | 3,404 | 48.7 | 1.3 | 5.4 | 0.0 | 12.0 | 17.2 | 1,471 | 36 | 23 | 488 | 5,730 | 57,508 | 5,265 |
| Mason | 74 | 5,403 | 35.7 | 31.8 | 3.4 | 1.2 | 4.4 | 29.9 | 2,189 | 58 | 27 | 1,116 | 6,200 | 51,793 | 4,423 |
| Massac | 82 | 5,758 | 31.1 | 31.3 | 4.3 | 0.3 | 4.5 | 41.2 | 2,886 | 33 | 27 | 901 | 6,090 | 48,852 | 3,907 |
| Menard | 52 | 4,217 | 45.5 | 16.1 | 4.6 | 0.6 | 10.8 | 26.5 | 2,157 | 27 | 24 | 664 | 6,040 | 65,858 | 6,141 |
| Mercer | 55 | 3,550 | 49.3 | 1.7 | 5.5 | 0.1 | 10.0 | 33.3 | 2,134 | 55 | 31 | 747 | 7,690 | 58,828 | 5,272 |
| Monroe | 100 | 2,926 | 50.0 | 9.5 | 8.6 | 0.7 | 9.0 | 139.3 | 4,075 | 68 | 69 | 1,380 | 17,400 | 81,422 | 9,319 |
| Montgomery | 91 | 3,144 | 49.0 | 0.2 | 3.7 | 0.1 | 6.9 | 52.5 | 1,825 | 99 | 52 | 1,467 | 12,330 | 55,711 | 4,807 |
| Morgan | 125 | 3,634 | 61.3 | 1.2 | 5.9 | 1.9 | 6.5 | 106.7 | 3,115 | 73 | 61 | 1,980 | 15,260 | 55,968 | 5,115 |
| Moultrie | 38 | 2,556 | 46.8 | 1.8 | 4.9 | 0.0 | 13.9 | 14.9 | 1,014 | 31 | 28 | 510 | 6,620 | 58,899 | 5,424 |
| Ogle | 263 | 5,155 | 57.1 | 0.0 | 5.1 | 1.0 | 8.1 | 256.0 | 5,020 | 116 | 101 | 2,887 | 25,220 | 61,869 | 5,952 |
| Peoria | 762 | 4,174 | 41.8 | 0.9 | 7.5 | 1.9 | 6.9 | 757.5 | 4,150 | 1,489 | 406 | 9,278 | 85,880 | 70,779 | 8,772 |
| Perry | 74 | 3,455 | 33.6 | 35.0 | 7.1 | 0.1 | 5.0 | 48.2 | 2,264 | 44 | 37 | 1,326 | 8,630 | 50,621 | 3,853 |
| Piatt | 64 | 3,896 | 43.0 | 15.3 | 5.8 | 2.3 | 9.5 | 21.5 | 1,308 | 37 | 33 | 1,024 | 8,050 | 71,167 | 7,655 |
| Pike | 56 | 3,597 | 44.7 | 4.0 | 5.4 | 0.4 | 11.4 | 29.7 | 1,894 | 56 | 30 | 923 | 6,750 | 49,715 | 4,307 |
| Pope | 14 | 3,296 | 33.7 | 0.0 | 2.3 | 0.0 | 8.0 | 1.3 | 299 | 21 | 8 | 202 | 1,570 | 47,593 | 3,790 |
| Pulaski | 33 | 6,047 | 52.1 | 1.1 | 2.0 | 0.1 | 5.3 | 13.0 | 2,347 | 60 | 11 | 844 | 2,260 | 43,477 | 3,013 |
| Putnam | 24 | 4,151 | 45.0 | 2.6 | 5.2 | 0.0 | 15.5 | 4.2 | 746 | 13 | 12 | 332 | 2,940 | 63,436 | 6,173 |
| Randolph | 167 | 5,165 | 31.6 | 36.9 | 3.7 | 0.8 | 4.6 | 76.7 | 2,376 | 93 | 55 | 2,379 | 13,330 | 55,358 | 4,768 |
| Richland | 82 | 5,151 | 76.9 | 0.5 | 3.9 | 0.1 | 4.9 | 48.6 | 3,071 | 44 | 31 | 949 | 7,240 | 51,008 | 4,053 |
| Rock Island | 680 | 4,731 | 45.1 | 3.5 | 7.9 | 0.6 | 5.4 | 822.9 | 5,721 | 4,973 | 664 | 7,830 | 69,070 | 56,885 | 5,448 |
| St. Clair | 1,086 | 4,136 | 53.4 | 1.5 | 7.9 | 1.0 | 5.1 | 903.4 | 3,440 | 6,032 | 5,110 | 12,957 | 119,910 | 61,519 | 6,393 |
| Saline | 96 | 4,017 | 60.8 | 0.0 | 5.5 | 0.1 | 6.1 | 81.1 | 3,385 | 101 | 46 | 1,754 | 9,980 | 51,173 | 4,477 |
| Sangamon | 878 | 4,456 | 45.6 | 1.6 | 9.1 | 0.1 | 8.1 | 3,122.8 | 15,846 | 1,747 | 442 | 18,499 | 98,210 | 71,801 | 8,461 |
| Schuyler | 43 | 6,222 | 24.1 | 55.3 | 3.0 | 0.0 | 6.4 | 6.1 | 874 | 21 | 13 | 588 | 3,120 | 52,077 | 4,341 |
| Scott | 18 | 3,591 | 51.9 | 2.0 | 3.2 | 16.0 | 7.0 | 13.4 | 2,694 | 14 | 10 | 349 | 2,420 | 54,810 | 4,537 |
| Shelby | 50 | 2,312 | 40.5 | 3.4 | 6.6 | 0.0 | 16.1 | 19.1 | 879 | 88 | 43 | 791 | 10,150 | 54,970 | 4,481 |
| Stark | 20 | 3,641 | 53.8 | 0.5 | 4.8 | 0.8 | 11.9 | 6.4 | 1,170 | 26 | 11 | 299 | 2,610 | 57,540 | 5,284 |
| Stephenson | 180 | 3,998 | 54.4 | 5.2 | 6.4 | 0.2 | 5.4 | 109.7 | 2,438 | 119 | 88 | 2,746 | 21,920 | 52,079 | 4,574 |
| Tazewell | 569 | 4,263 | 58.3 | 1.5 | 6.6 | 0.2 | 6.6 | 369.4 | 2,767 | 492 | 289 | 7,224 | 63,460 | 67,453 | 7,824 |
| Union | 69 | 4,066 | 66.1 | 0.1 | 6.2 | 0.0 | 5.5 | 31.5 | 1,854 | 63 | 32 | 1,241 | 7,540 | 50,642 | 4,140 |
| Vermilion | 307 | 3,957 | 52.9 | 0.9 | 5.4 | 0.1 | 7.9 | 108.9 | 1,402 | 1,420 | 147 | 4,537 | 32,570 | 48,911 | 4,109 |
| Wabash | 71 | 6,114 | 22.8 | 51.8 | 2.3 | 0.0 | 2.4 | 10.6 | 917 | 23 | 23 | 1,096 | 5,040 | 58,818 | 5,693 |
| Warren | 54 | 3,167 | 49.0 | 2.2 | 5.3 | 1.7 | 10.3 | 49.6 | 2,893 | 66 | 32 | 965 | 7,700 | 53,205 | 4,497 |
| Washington | 57 | 4,075 | 31.7 | 27.3 | 3.6 | 2.7 | 11.1 | 36.8 | 2,635 | 45 | 28 | 756 | 6,790 | 64,370 | 6,366 |
| Wayne | 57 | 3,470 | 54.4 | 0.3 | 4.1 | 1.3 | 12.8 | 37.6 | 2,289 | 57 | 33 | 962 | 7,190 | 49,242 | 3,738 |
| White | 53 | 3,818 | 54.9 | 1.8 | 4.6 | 0.1 | 9.5 | 19.8 | 1,426 | 50 | 27 | 817 | 6,120 | 54,386 | 4,959 |
| Whiteside | 435 | 7,767 | 24.7 | 52.7 | 2.5 | 0.7 | 2.6 | 122.8 | 2,194 | 151 | 109 | 4,911 | 27,420 | 54,199 | 4,877 |
| Will | 2,808 | 4,066 | 52.3 | 2.1 | 7.9 | 0.0 | 5.4 | 4,430.4 | 6,416 | 1,051 | 1,390 | 34,873 | 337,760 | 77,630 | 9,159 |
| Williamson | 253 | 3,774 | 57.8 | 0.2 | 6.7 | 0.1 | 5.7 | 320.5 | 4,781 | 1,687 | 130 | 4,231 | 29,840 | 56,379 | 5,368 |

1. Based on the resident population estimated as of July 1 of the year shown.

# Table B. States and Counties — Land Area and Population

| State / county code | CBSA code[1] | County Type code[2] | STATE County | Land area[3] (sq. mi) | Population, 2020 Total persons 2019 | Rank | Per square mile | White | Black | American Indian, Alaska Native | Asian and Pacific Islancer | Percent Hispanic or Latino[4] | Under 5 years | 5 to 14 years | 15 to 24 years | 25 to 34 years | 35 to 44 years | 45 to 54 years |
|---|---|---|---|---|---|---|---|---|---|---|---|---|---|---|---|---|---|---|
| | | | | 1 | 2 | 3 | 4 | 5 | 6 | 7 | 8 | 9 | 10 | 11 | 12 | 13 | 14 | 15 |
| | | | ILLINOIS—Cont'd | | | | | | | | | | | | | | | |
| 17201 | 40420 | 2 | Winnebago | 513.1 | 281,295 | 248 | 548.2 | 69.4 | 15.4 | 0.7 | 3.5 | 13.8 | 6.3 | 12.9 | 12.3 | 12.7 | 11.7 | 12.2 |
| 17203 | 37900 | 2 | Woodford | 527.5 | 38,091 | 1,223 | 72.2 | 96.4 | 1.3 | 0.5 | 1.1 | 1.9 | 5.6 | 13.9 | 12.7 | 10.4 | 12.6 | 12.0 |
| 18000 | | 0 | INDIANA | 35,826.4 | 6,754,953 | X | 188.5 | 79.9 | 10.9 | 0.7 | 3.2 | 7.4 | 6.1 | 13.0 | 13.7 | 13.3 | 12.4 | 12.0 |
| 18001 | 19540 | 6 | Adams | 338.9 | 35,839 | 1,281 | 105.8 | 94.3 | 1.0 | 0.4 | 0.5 | 4.5 | 9.1 | 17.2 | 13.2 | 11.7 | 10.5 | 10.6 |
| 18003 | 23060 | 2 | Allen | 657.3 | 382,187 | 188 | 581.4 | 75.3 | 13.7 | 0.8 | 5.5 | 7.9 | 7.0 | 14.2 | 13.2 | 14.2 | 12.4 | 11.6 |
| 18005 | 18020 | 3 | Bartholomew | 406.9 | 84,447 | 681 | 207.5 | 81.4 | 3.0 | 0.6 | 8.7 | 7.8 | 6.4 | 13.5 | 11.8 | 14.4 | 12.6 | 12.3 |
| 18007 | 29200 | 6 | Benton | 406.4 | 8,741 | 2,517 | 21.5 | 92.9 | 1.4 | 0.6 | 0.4 | 5.9 | 6.2 | 13.8 | 12.3 | 11.4 | 12.0 | 12.0 |
| 18009 | | 6 | Blackford | 165.1 | 11,782 | 2,307 | 71.4 | 96.4 | 1.6 | 0.8 | 0.8 | 1.9 | 5.8 | 12.4 | 11.1 | 11.6 | 10.0 | 13.0 |
| 18011 | 26900 | 1 | Boone | 422.9 | 69,347 | 781 | 164.0 | 90.6 | 2.9 | 0.5 | 4.2 | 3.4 | 6.4 | 15.0 | 12.0 | 12.1 | 14.3 | 13.3 |
| 18013 | 26900 | 1 | Brown | 312.0 | 15,112 | 2,088 | 48.4 | 96.4 | 1.2 | 1.1 | 0.8 | 2.0 | 4.1 | 9.6 | 9.4 | 9.8 | 10.3 | 13.4 |
| 18015 | 29200 | 3 | Carroll | 372.2 | 20,228 | 1,808 | 54.3 | 94.1 | 1.3 | 0.5 | 0.4 | 4.7 | 5.6 | 12.5 | 11.3 | 11.7 | 11.5 | 12.7 |
| 18017 | 30900 | 4 | Cass | 412.1 | 37,388 | 1,242 | 90.7 | 79.2 | 2.5 | 0.7 | 2.1 | 16.7 | 5.7 | 12.5 | 12.4 | 11.5 | 12.3 | 12.7 |
| 18019 | 31140 | 1 | Clark | 372.8 | 119,266 | 528 | 319.9 | 84.5 | 9.9 | 0.7 | 1.7 | 5.8 | 6.1 | 12.4 | 11.3 | 14.2 | 13.2 | 13.0 |
| 18021 | 45460 | 3 | Clay | 357.6 | 26,246 | 1,559 | 73.4 | 96.9 | 1.4 | 0.7 | 0.6 | 1.7 | 6.2 | 12.7 | 11.2 | 12.1 | 12.2 | 12.4 |
| 18023 | 23140 | 6 | Clinton | 405.1 | 32,206 | 1,375 | 79.5 | 82.0 | 1.1 | 0.5 | 0.5 | 16.6 | 6.7 | 14.9 | 12.7 | 12.3 | 11.5 | 11.8 |
| 18025 | | 8 | Crawford | 305.6 | 10,629 | 2,378 | 34.8 | 96.9 | 1.2 | 1.0 | 0.6 | 1.7 | 5.7 | 12.0 | 10.8 | 10.6 | 11.2 | 12.8 |
| 18027 | 47780 | 7 | Daviess | 429.5 | 33,505 | 1,339 | 78.0 | 91.7 | 2.8 | 0.4 | 0.6 | 5.5 | 8.3 | 16.2 | 13.2 | 12.5 | 11.2 | 10.5 |
| 18029 | 17140 | 1 | Dearborn | 305.1 | 49,824 | 994 | 163.3 | 97.1 | 1.3 | 0.5 | 0.8 | 1.4 | 5.5 | 12.2 | 12.1 | 11.2 | 11.9 | 13.4 |
| 18031 | 24700 | 6 | Decatur | 372.6 | 26,584 | 1,542 | 71.3 | 95.7 | 1.0 | 0.5 | 1.6 | 2.1 | 6.3 | 13.1 | 12.4 | 12.4 | 11.8 | 12.5 |
| 18033 | 12140 | 4 | DeKalb | 362.8 | 43,670 | 1,104 | 120.4 | 95.6 | 1.1 | 0.5 | 0.8 | 3.0 | 6.5 | 13.3 | 12.3 | 12.9 | 12.0 | 12.4 |
| 18035 | 34620 | 3 | Delaware | 392.1 | 113,454 | 544 | 289.3 | 88.6 | 8.5 | 0.7 | 2.0 | 2.6 | 4.8 | 9.9 | 22.5 | 11.8 | 10.1 | 11.0 |
| 18037 | 27540 | 5 | Dubois | 427.3 | 42,542 | 1,127 | 99.6 | 89.7 | 1.1 | 0.3 | 0.8 | 8.7 | 6.4 | 13.6 | 12.0 | 11.1 | 11.7 | 12.0 |
| 18039 | 21140 | 3 | Elkhart | 463.2 | 206,161 | 332 | 445.1 | 75.8 | 7.1 | 0.7 | 1.7 | 17.1 | 7.4 | 15.2 | 13.6 | 13.0 | 12.0 | 11.7 |
| 18041 | 18220 | 6 | Fayette | 215.0 | 22,892 | 1,687 | 106.5 | 96.4 | 2.1 | 0.6 | 0.5 | 1.5 | 5.8 | 12.1 | 11.7 | 11.4 | 11.3 | 12.9 |
| 18043 | 31140 | 1 | Floyd | 148.5 | 78,936 | 715 | 531.6 | 89.2 | 6.9 | 0.7 | 1.8 | 3.7 | 5.9 | 12.6 | 12.0 | 12.9 | 12.9 | 12.6 |
| 18045 | | 6 | Fountain | 395.7 | 16,511 | 2,007 | 41.7 | 96.0 | 0.9 | 0.7 | 0.8 | 2.8 | 5.8 | 12.0 | 11.6 | 11.7 | 11.1 | 12.7 |
| 18047 | 17140 | 6 | Franklin | 384.4 | 22,761 | 1,692 | 59.2 | 97.5 | 0.6 | 0.5 | 0.9 | 1.3 | 6.1 | 12.3 | 11.9 | 10.7 | 11.4 | 13.1 |
| 18049 | | 7 | Fulton | 368.4 | 20,018 | 1,822 | 54.3 | 92.5 | 1.5 | 0.9 | 0.8 | 5.5 | 6.3 | 13.2 | 11.3 | 11.4 | 11.7 | 11.7 |
| 18051 | | 6 | Gibson | 487.4 | 33,825 | 1,331 | 69.4 | 94.6 | 4.0 | 0.6 | 0.9 | 1.9 | 6.3 | 13.3 | 12.0 | 12.2 | 12.0 | 12.2 |
| 18053 | 31980 | 4 | Grant | 414.1 | 65,225 | 817 | 157.5 | 86.8 | 8.9 | 0.8 | 1.3 | 4.7 | 5.8 | 11.6 | 16.1 | 11.1 | 10.3 | 11.5 |
| 18055 | | 6 | Greene | 542.5 | 32,203 | 1,376 | 59.4 | 97.0 | 0.6 | 0.8 | 0.6 | 1.9 | 5.6 | 12.0 | 11.1 | 11.7 | 11.3 | 12.9 |
| 18057 | 26900 | 1 | Hamilton | 394.4 | 344,238 | 209 | 872.8 | 84.1 | 5.4 | 0.4 | 7.8 | 4.4 | 6.2 | 15.2 | 12.5 | 11.9 | 14.9 | 14.1 |
| 18059 | 26900 | 1 | Hancock | 306.0 | 79,553 | 712 | 260.0 | 92.6 | 4.0 | 0.7 | 1.5 | 2.8 | 5.7 | 13.3 | 11.6 | 12.2 | 13.2 | 13.3 |
| 18061 | 31140 | 1 | Harrison | 484.1 | 40,682 | 1,165 | 84.0 | 96.3 | 1.2 | 0.7 | 0.9 | 2.2 | 5.4 | 12.7 | 11.0 | 11.4 | 12.9 | 12.7 |
| 18063 | 26900 | 1 | Hendricks | 406.9 | 173,251 | 381 | 425.8 | 83.9 | 9.2 | 0.5 | 3.9 | 4.5 | 5.6 | 14.3 | 12.6 | 12.7 | 14.5 | 13.3 |
| 18065 | 35220 | 4 | Henry | 391.9 | 48,033 | 1,021 | 122.6 | 94.4 | 3.4 | 0.6 | 0.8 | 2.1 | 5.1 | 11.3 | 11.8 | 12.8 | 11.9 | 13.3 |
| 18067 | 29020 | 3 | Howard | 293.1 | 82,732 | 689 | 282.3 | 86.6 | 9.7 | 0.9 | 1.9 | 3.7 | 6.1 | 12.7 | 11.9 | 12.4 | 11.1 | 11.9 |
| 18069 | 26540 | 6 | Huntington | 382.6 | 36,395 | 1,267 | 95.1 | 95.4 | 1.2 | 0.9 | 0.9 | 2.7 | 5.7 | 12.0 | 13.4 | 12.1 | 11.8 | 12.2 |
| 18071 | 42980 | 4 | Jackson | 510.0 | 44,222 | 1,093 | 86.7 | 87.7 | 1.8 | 0.7 | 2.9 | 8.1 | 6.8 | 13.6 | 11.7 | 12.1 | 12.8 | 12.7 |
| 18073 | 16980 | 1 | Jasper | 559.7 | 33,440 | 1,343 | 59.7 | 92.2 | 1.3 | 0.6 | 0.7 | 6.3 | 5.5 | 12.7 | 13.7 | 11.4 | 11.9 | 12.2 |
| 18075 | | 6 | Jay | 383.9 | 20,416 | 1,798 | 53.2 | 95.4 | 1.0 | 0.7 | 0.7 | 3.4 | 7.1 | 13.9 | 12.1 | 11.7 | 10.6 | 12.4 |
| 18077 | 31500 | 6 | Jefferson | 360.6 | 32,110 | 1,377 | 89.0 | 93.6 | 2.8 | 0.7 | 1.2 | 3.1 | 5.7 | 10.9 | 13.5 | 12.0 | 11.8 | 12.6 |
| 18079 | 35860 | 6 | Jennings | 376.6 | 27,515 | 1,512 | 73.1 | 95.7 | 1.4 | 0.9 | 0.6 | 2.8 | 5.9 | 13.1 | 12.1 | 12.8 | 10.8 | 14.2 |
| 18081 | 26900 | 1 | Johnson | 320.4 | 160,607 | 417 | 501.3 | 88.8 | 3.5 | 0.6 | 5.0 | 3.8 | 6.2 | 14.0 | 12.6 | 13.6 | 13.6 | 12.6 |
| 18083 | 47180 | 5 | Knox | 515.8 | 36,522 | 1,263 | 70.8 | 93.2 | 3.6 | 0.7 | 1.3 | 2.6 | 5.7 | 12.0 | 15.7 | 11.7 | 11.6 | 10.8 |
| 18085 | 47700 | 4 | Kosciusko | 531.4 | 78,988 | 714 | 148.6 | 88.6 | 1.6 | 0.6 | 2.1 | 8.3 | 6.3 | 13.1 | 12.9 | 13.0 | 11.7 | 11.6 |
| 18087 | | 6 | LaGrange | 379.6 | 40,119 | 1,181 | 105.7 | 94.5 | 0.7 | 0.5 | 0.7 | 4.4 | 9.1 | 17.5 | 14.9 | 11.8 | 10.9 | 10.6 |
| 18089 | 16980 | 1 | Lake | 498.7 | 487,536 | 146 | 977.6 | 54.7 | 24.3 | 0.6 | 2.0 | 20.0 | 5.9 | 13.1 | 12.7 | 12.4 | 12.7 | 12.2 |
| 18091 | 33140 | 3 | LaPorte | 598.3 | 109,663 | 558 | 183.3 | 80.5 | 13.0 | 0.8 | 1.1 | 7.1 | 5.8 | 12.0 | 11.6 | 13.4 | 12.1 | 12.4 |
| 18093 | 13260 | 6 | Lawrence | 449.1 | 45,496 | 1,068 | 101.3 | 96.6 | 0.9 | 1.0 | 0.9 | 1.8 | 5.6 | 11.7 | 11.5 | 11.3 | 11.8 | 12.9 |
| 18095 | 26900 | 1 | Madison | 451.9 | 129,681 | 497 | 287.0 | 86.2 | 9.6 | 0.7 | 1.0 | 4.5 | 5.4 | 12.0 | 12.3 | 12.9 | 12.2 | 12.8 |
| 18097 | 26900 | 1 | Marion | 396.6 | 966,183 | 50 | 2,436.2 | 55.9 | 30.7 | 0.7 | 4.7 | 11.0 | 7.1 | 13.7 | 13.1 | 17.0 | 13.1 | 11.2 |
| 18099 | 38500 | 6 | Marshall | 443.6 | 46,108 | 1,054 | 103.9 | 88.0 | 1.2 | 0.7 | 0.9 | 10.6 | 6.2 | 13.8 | 12.9 | 11.0 | 11.4 | 12.2 |
| 18101 | | 7 | Martin | 335.8 | 10,079 | 2,413 | 30.0 | 97.6 | 0.6 | 0.6 | 0.8 | 1.3 | 5.8 | 12.3 | 11.5 | 10.9 | 11.5 | 12.6 |
| 18103 | 37940 | 6 | Miami | 373.8 | 35,328 | 1,295 | 94.5 | 90.1 | 5.9 | 1.6 | 0.8 | 3.6 | 5.2 | 11.9 | 12.1 | 13.5 | 12.7 | 13.1 |
| 18105 | 14020 | 3 | Monroe | 394.5 | 148,219 | 452 | 375.7 | 85.4 | 4.8 | 0.7 | 8.1 | 3.7 | 4.3 | 8.6 | 28.4 | 14.3 | 10.9 | 9.3 |
| 18107 | 18820 | 6 | Montgomery | 504.6 | 38,365 | 1,215 | 76.0 | 92.6 | 1.7 | 0.7 | 1.2 | 5.1 | 5.8 | 13.0 | 12.9 | 12.0 | 11.2 | 12.0 |
| 18109 | 26900 | 1 | Morgan | 403.8 | 70,707 | 770 | 175.1 | 96.6 | 1.0 | 0.8 | 1.0 | 1.8 | 5.6 | 12.4 | 11.9 | 11.7 | 11.8 | 13.7 |
| 18111 | 16980 | 1 | Newton | 401.8 | 13,907 | 2,163 | 34.6 | 90.9 | 1.3 | 0.8 | 0.7 | 7.3 | 5.5 | 11.6 | 11.0 | 11.8 | 11.8 | 12.5 |
| 18113 | 28340 | 6 | Noble | 410.8 | 47,832 | 1,026 | 116.4 | 87.7 | 1.3 | 0.7 | 0.9 | 10.5 | 6.5 | 13.5 | 12.4 | 12.4 | 11.9 | 12.4 |
| 18115 | 17140 | 1 | Ohio | 86.2 | 5,892 | 2,753 | 68.4 | 96.7 | 1.3 | 0.8 | 0.6 | 1.9 | 5.0 | 11.3 | 9.7 | 10.4 | 10.5 | 12.5 |
| 18117 | | 6 | Orange | 398.4 | 19,651 | 1,842 | 49.3 | 95.7 | 2.2 | 1.0 | 0.7 | 1.8 | 6.4 | 12.6 | 11.5 | 11.2 | 11.2 | 11.8 |
| 18119 | 14020 | 3 | Owen | 385.3 | 20,833 | 1,778 | 54.1 | 97.1 | 1.0 | 1.0 | 1.0 | 1.3 | 5.4 | 11.3 | 11.1 | 11.1 | 11.1 | 13.1 |
| 18121 | 45460 | 6 | Parke | 444.7 | 16,871 | 1,979 | 37.9 | 95.1 | 3.1 | 0.7 | 0.3 | 1.6 | 5.9 | 11.6 | 11.1 | 13.0 | 11.4 | 12.2 |
| 18123 | | 6 | Perry | 381.7 | 19,154 | 1,870 | 50.2 | 94.4 | 3.4 | 0.6 | 1.0 | 1.5 | 5.6 | 12.0 | 11.3 | 13.2 | 12.7 | 11.8 |

1. CBSA = Core Based Statistical Area. See Appendix A for explanation. See Appendix B for list of metropolitan areas with component counties.   2. County type code from the Economic Research Service of USDA Rural-Urban Continuum Codes. See Appendix A for definition.   3. Dry land or land partially or temporarily covered by water.   4. May be of any race.

# Table B. States and Counties — Population and Households

| STATE County | Population, 2020 (cont.) Age (percent) (cont.) | | | | Population change, 2000-2020 Total persons | | Percent change | | Components of change, 2010-2020 | | | Households, 2015-2019 | | Percent | | |
|---|---|---|---|---|---|---|---|---|---|---|---|---|---|---|---|---|
| | 55 to 64 years | 65 to 74 years | 75 years and over | Percent female | 2000 | 2010 | 2000-2010 | 2010-2020 | Births | Deaths | Net Migration | Number | Persons per household | Family households | Female family householder[1] | One person |
| | 16 | 17 | 18 | 19 | 20 | 21 | 22 | 23 | 24 | 25 | 26 | 27 | 28 | 29 | 30 | 31 |
| ILLINOIS—Cont'd | | | | | | | | | | | | | | | | |
| Winnebago | 13.5 | 10.6 | 7.8 | 51.2 | 278,418 | 295,270 | 6.1 | -4.7 | 36,735 | 29,437 | -21,464 | 114,779 | 2.44 | 64.4 | 14.2 | 30.1 |
| Woodford | 13.9 | 10.7 | 8.2 | 50.1 | 35,469 | 38,656 | 9.0 | -1.5 | 4,419 | 3,908 | -1,065 | 14,499 | 2.60 | 72.4 | 8.5 | 23.9 |
| INDIANA | 12.9 | 9.8 | 6.8 | 50.7 | 6,080,485 | 6,484,050 | 6.6 | 4.2 | 849,083 | 637,936 | 61,821 | 2,570,419 | 2.52 | 64.7 | 11.7 | 29.0 |
| Adams | 11.4 | 8.9 | 7.3 | 50.1 | 33,625 | 34,384 | 2.3 | 4.2 | 6,819 | 3,165 | -2,208 | 12,481 | 2.79 | 70.4 | 9.9 | 25.8 |
| Allen | 12.0 | 9.2 | 6.1 | 51.0 | 331,849 | 355,339 | 7.1 | 7.6 | 53,562 | 32,475 | 6,018 | 145,065 | 2.53 | 64.5 | 12.5 | 29.2 |
| Bartholomew | 12.1 | 9.7 | 7.2 | 49.7 | 71,435 | 76,782 | 7.5 | 10.0 | 10,804 | 7,615 | 4,497 | 31,452 | 2.59 | 65.8 | 10.2 | 27.4 |
| Benton | 14.2 | 9.9 | 8.2 | 50.0 | 9,421 | 8,836 | -6.2 | -1.1 | 1,102 | 880 | -321 | 3,432 | 2.49 | 66.0 | 10.4 | 29.1 |
| Blackford | 14.3 | 12.2 | 9.6 | 50.8 | 14,048 | 12,766 | -9.1 | -7.7 | 1,399 | 1,619 | -768 | 5,235 | 2.26 | 66.4 | 11.2 | 29.1 |
| Boone | 12.7 | 8.5 | 5.7 | 50.2 | 46,107 | 56,647 | 22.9 | 22.4 | 7,756 | 5,207 | 10,095 | 25,485 | 2.55 | 72.6 | 8.0 | 22.1 |
| Brown | 17.7 | 16.3 | 9.4 | 50.2 | 14,957 | 15,247 | 1.9 | -0.9 | 1,195 | 1,619 | 299 | 6,189 | 2.41 | 72.8 | 8.0 | 22.7 |
| Carroll | 14.6 | 11.8 | 8.5 | 49.4 | 20,165 | 20,160 | 0.0 | 0.3 | 2,179 | 1,910 | -200 | 8,002 | 2.49 | 67.6 | 7.9 | 27.7 |
| Cass | 14.1 | 10.5 | 8.3 | 50.0 | 40,930 | 38,970 | -4.8 | -4.1 | 4,822 | 4,140 | -2,283 | 14,688 | 2.49 | 64.6 | 10.3 | 31.6 |
| Clark | 13.4 | 10.1 | 6.2 | 51.3 | 96,472 | 110,221 | 14.3 | 8.2 | 14,825 | 12,174 | 6,399 | 44,200 | 2.60 | 65.1 | 11.6 | 29.7 |
| Clay | 14.3 | 11.1 | 7.7 | 50.5 | 26,556 | 26,896 | 1.3 | -2.4 | 3,266 | 3,120 | -793 | 10,543 | 2.44 | 71.4 | 10.6 | 23.9 |
| Clinton | 13.0 | 9.5 | 7.6 | 50.3 | 33,866 | 33,219 | -1.9 | -3.0 | 4,494 | 3,717 | -1,804 | 12,033 | 2.63 | 68.0 | 10.5 | 27.0 |
| Crawford | 16.2 | 13.0 | 7.7 | 49.2 | 10,743 | 10,713 | -0.3 | -0.8 | 1,186 | 1,225 | -43 | 3,835 | 2.73 | 66.5 | 8.3 | 28.8 |
| Daviess | 12.1 | 9.0 | 6.9 | 49.6 | 29,820 | 31,654 | 6.2 | 5.8 | 5,572 | 3,461 | -242 | 11,227 | 2.90 | 71.8 | 7.5 | 25.5 |
| Dearborn | 15.0 | 11.4 | 7.2 | 50.1 | 46,109 | 50,025 | 8.5 | -0.4 | 5,328 | 4,842 | -672 | 18,870 | 2.59 | 72.7 | 9.2 | 21.8 |
| Decatur | 14.0 | 10.1 | 7.3 | 50.3 | 24,555 | 25,737 | 4.8 | 3.3 | 3,416 | 2,776 | 226 | 10,353 | 2.53 | 67.0 | 7.7 | 26.7 |
| DeKalb | 13.8 | 10.1 | 6.8 | 50.5 | 40,285 | 42,255 | 4.9 | 3.3 | 5,447 | 4,219 | 210 | 16,801 | 2.53 | 68.8 | 9.1 | 27.2 |
| Delaware | 12.2 | 9.8 | 7.9 | 51.9 | 118,769 | 117,674 | -0.9 | -3.6 | 12,104 | 12,799 | -3,518 | 46,026 | 2.34 | 58.9 | 12.6 | 30.8 |
| Dubois | 14.7 | 10.6 | 8.0 | 49.9 | 39,674 | 41,886 | 5.6 | 1.6 | 5,497 | 4,182 | -645 | 16,893 | 2.46 | 66.3 | 8.6 | 28.8 |
| Elkhart | 11.7 | 8.9 | 6.6 | 50.6 | 182,791 | 197,569 | 8.1 | 4.3 | 31,576 | 17,207 | -5,702 | 71,718 | 2.80 | 69.8 | 12.1 | 25.3 |
| Fayette | 13.8 | 12.4 | 8.6 | 50.8 | 25,588 | 24,302 | -5.0 | -5.8 | 2,580 | 3,280 | -699 | 9,584 | 2.38 | 67.2 | 14.5 | 26.8 |
| Floyd | 14.0 | 10.4 | 6.6 | 51.3 | 70,823 | 74,575 | 5.3 | 5.8 | 9,055 | 7,826 | 3,174 | 29,189 | 2.60 | 66.4 | 11.8 | 28.6 |
| Fountain | 14.5 | 11.2 | 9.4 | 50.3 | 17,954 | 17,242 | -4.0 | -4.2 | 1,961 | 2,055 | -640 | 6,974 | 2.33 | 67.3 | 9.7 | 27.3 |
| Franklin | 14.9 | 11.4 | 8.1 | 50.2 | 22,151 | 23,097 | 4.3 | -1.5 | 2,604 | 2,188 | -752 | 8,687 | 2.60 | 74.5 | 10.0 | 22.4 |
| Fulton | 13.8 | 11.8 | 8.7 | 50.1 | 20,511 | 20,856 | 1.7 | -4.0 | 2,538 | 2,461 | -910 | 7,799 | 2.55 | 64.1 | 7.6 | 31.0 |
| Gibson | 14.1 | 10.4 | 7.6 | 49.8 | 32,500 | 33,503 | 3.1 | 1.0 | 4,246 | 3,793 | -109 | 13,340 | 2.46 | 70.1 | 10.6 | 24.7 |
| Grant | 13.9 | 11.2 | 8.6 | 52.2 | 73,403 | 70,063 | -4.6 | -6.9 | 7,837 | 8,508 | -4,170 | 26,372 | 2.31 | 64.1 | 13.4 | 30.3 |
| Greene | 15.1 | 11.6 | 8.7 | 50.1 | 33,157 | 33,172 | 0.0 | -2.9 | 3,548 | 4,011 | -485 | 12,781 | 2.49 | 67.7 | 8.4 | 27.1 |
| Hamilton | 11.9 | 8.1 | 5.2 | 51.1 | 182,740 | 274,555 | 50.2 | 25.4 | 39,281 | 16,783 | 46,980 | 119,789 | 2.68 | 73.7 | 8.3 | 21.0 |
| Hancock | 13.7 | 10.2 | 6.9 | 50.8 | 55,391 | 70,060 | 26.5 | 13.5 | 8,069 | 6,537 | 8,006 | 28,740 | 2.59 | 70.4 | 7.8 | 23.1 |
| Harrison | 15.1 | 11.4 | 7.5 | 50.2 | 34,325 | 39,363 | 14.7 | 3.4 | 4,358 | 4,015 | 998 | 14,403 | 2.75 | 69.7 | 7.0 | 24.7 |
| Hendricks | 12.4 | 9.0 | 5.7 | 50.0 | 104,093 | 145,455 | 39.7 | 19.1 | 17,624 | 11,436 | 21,634 | 59,015 | 2.73 | 73.7 | 9.4 | 21.6 |
| Henry | 14.1 | 11.3 | 8.4 | 47.7 | 48,508 | 49,466 | 2.0 | -2.9 | 4,871 | 5,967 | -306 | 18,304 | 2.45 | 67.2 | 11.7 | 27.1 |
| Howard | 13.8 | 11.2 | 8.9 | 51.6 | 84,964 | 82,748 | -2.6 | 0.0 | 10,165 | 10,109 | -14 | 34,701 | 2.34 | 67.2 | 13.6 | 30.8 |
| Huntington | 14.7 | 10.4 | 7.6 | 50.6 | 38,075 | 37,123 | -2.5 | -2.0 | 4,331 | 4,205 | -848 | 14,742 | 2.37 | 67.2 | 9.0 | 27.5 |
| Jackson | 13.4 | 9.8 | 7.2 | 50.0 | 41,335 | 42,376 | 2.5 | 4.4 | 6,072 | 4,772 | 562 | 16,753 | 2.58 | 70.1 | 11.5 | 25.0 |
| Jasper | 14.0 | 10.8 | 7.8 | 50.1 | 30,043 | 33,481 | 11.4 | -0.1 | 3,882 | 3,373 | -537 | 12,533 | 2.60 | 69.7 | 8.7 | 25.5 |
| Jay | 13.4 | 10.9 | 7.7 | 50.1 | 21,806 | 21,253 | -2.5 | -3.9 | 3,079 | 2,406 | -1,514 | 8,174 | 2.52 | 65.5 | 9.8 | 28.9 |
| Jefferson | 14.8 | 11.3 | 7.5 | 51.8 | 31,705 | 32,400 | 2.2 | -0.9 | 3,748 | 3,864 | -154 | 12,632 | 2.35 | 65.2 | 11.5 | 29.7 |
| Jennings | 14.1 | 10.3 | 6.8 | 49.8 | 27,554 | 28,529 | 3.5 | -3.6 | 3,406 | 3,055 | -1,365 | 10,792 | 2.53 | 68.5 | 10.1 | 26.8 |
| Johnson | 12.3 | 9.0 | 6.3 | 50.6 | 115,209 | 139,856 | 21.4 | 14.8 | 18,935 | 13,201 | 15,061 | 56,628 | 2.67 | 71.4 | 10.2 | 23.2 |
| Knox | 13.7 | 10.6 | 8.2 | 49.5 | 39,256 | 38,440 | -2.1 | -5.0 | 4,560 | 4,531 | -1,951 | 15,038 | 2.32 | 63.7 | 11.4 | 30.2 |
| Kosciusko | 13.7 | 10.5 | 7.2 | 50.0 | 74,057 | 77,355 | 4.5 | 2.1 | 10,479 | 7,262 | -1,524 | 30,997 | 2.50 | 70.5 | 10.9 | 23.4 |
| LaGrange | 10.9 | 8.6 | 5.8 | 49.2 | 34,909 | 37,131 | 6.4 | 8.0 | 7,720 | 2,753 | -1,982 | 12,325 | 3.16 | 77.3 | 7.2 | 18.0 |
| Lake | 13.5 | 10.2 | 7.2 | 51.6 | 484,564 | 496,112 | 2.4 | -1.7 | 60,605 | 49,845 | -19,348 | 186,731 | 2.57 | 65.6 | 16.0 | 29.7 |
| LaPorte | 13.9 | 11.3 | 7.5 | 48.4 | 110,106 | 111,466 | 1.2 | -1.6 | 13,352 | 12,298 | -2,797 | 43,039 | 2.37 | 65.1 | 13.5 | 28.8 |
| Lawrence | 14.5 | 12.0 | 8.7 | 50.4 | 45,922 | 46,134 | 0.5 | -1.4 | 5,051 | 5,714 | 70 | 18,781 | 2.39 | 67.5 | 10.3 | 27.8 |
| Madison | 13.5 | 11.0 | 8.0 | 49.9 | 133,358 | 131,639 | -1.3 | -1.5 | 15,091 | 15,437 | -1,544 | 51,003 | 2.42 | 65.2 | 13.9 | 29.4 |
| Marion | 11.8 | 8.0 | 5.1 | 51.8 | 860,454 | 903,373 | 5.0 | 7.0 | 146,841 | 82,135 | -1,333 | 372,058 | 2.51 | 54.5 | 14.7 | 37.7 |
| Marshall | 13.5 | 10.8 | 8.2 | 50.1 | 45,128 | 47,050 | 4.3 | -2.0 | 5,868 | 4,760 | -2,051 | 17,304 | 2.64 | 69.1 | 9.2 | 26.1 |
| Martin | 14.7 | 12.4 | 8.4 | 49.5 | 10,369 | 10,380 | 0.1 | -2.9 | 1,243 | 1,116 | -425 | 4,187 | 2.37 | 60.0 | 7.8 | 32.9 |
| Miami | 13.2 | 10.8 | 7.6 | 46.1 | 36,082 | 36,905 | 2.3 | -4.3 | 3,776 | 3,797 | -1,565 | 13,611 | 2.47 | 67.3 | 10.4 | 28.2 |
| Monroe | 10.1 | 8.3 | 5.7 | 50.3 | 120,563 | 137,957 | 14.4 | 7.4 | 13,178 | 9,555 | 6,670 | 55,624 | 2.36 | 52.6 | 9.4 | 32.7 |
| Montgomery | 14.3 | 10.4 | 8.5 | 49.4 | 37,629 | 38,121 | 1.3 | 0.6 | 4,635 | 4,104 | -269 | 15,382 | 2.41 | 67.4 | 9.4 | 26.4 |
| Morgan | 15.1 | 10.9 | 6.9 | 50.4 | 66,689 | 68,926 | 3.4 | 2.6 | 7,893 | 6,949 | 895 | 26,034 | 2.65 | 72.0 | 9.6 | 22.7 |
| Newton | 15.4 | 11.6 | 8.8 | 49.3 | 14,566 | 14,239 | -2.2 | -2.3 | 1,490 | 1,641 | -173 | 5,573 | 2.47 | 68.8 | 11.5 | 27.0 |
| Noble | 13.8 | 10.6 | 6.5 | 49.7 | 46,275 | 47,538 | 2.7 | 0.6 | 6,270 | 4,630 | -1,333 | 18,276 | 2.56 | 68.8 | 9.4 | 25.9 |
| Ohio | 17.1 | 13.5 | 10.0 | 51.0 | 5,623 | 6,097 | 8.4 | -3.4 | 570 | 662 | -110 | 2,555 | 2.28 | 70.1 | 8.7 | 27.4 |
| Orange | 14.8 | 12.1 | 8.4 | 50.1 | 19,306 | 19,846 | 2.8 | -1.0 | 2,455 | 2,321 | -329 | 7,898 | 2.44 | 67.4 | 12.7 | 28.1 |
| Owen | 16.7 | 12.7 | 7.6 | 49.7 | 21,786 | 21,573 | -1.0 | -3.4 | 2,306 | 2,430 | -617 | 8,776 | 2.35 | 66.4 | 9.5 | 27.3 |
| Parke | 14.8 | 11.9 | 8.2 | 52.9 | 17,241 | 17,349 | 0.6 | -2.8 | 2,021 | 1,688 | -818 | 6,057 | 2.56 | 72.4 | 7.7 | 22.6 |
| Perry | 14.3 | 11.3 | 7.8 | 46.0 | 18,899 | 19,338 | 2.3 | -1.0 | 2,116 | 2,148 | -148 | 7,615 | 2.29 | 65.2 | 8.6 | 30.2 |

1. No spouse present.

# Table B. States and Counties — Population, Vital Statistics, Health, and Crime

| STATE County | Persons in group quarters, 2020 | Daytime Population, 2015-2019 | | Births, 2020 | | Deaths, 2020 | | Persons under 65 with no health insurance, 2019 | | Medicare, 2020 | | | Crimes reported by county police or sheriff, 2019 | |
|---|---|---|---|---|---|---|---|---|---|---|---|---|---|---|
| | | Number | Employment/ residence ratio | Total | Rate[1] | Number | Rate[1] | Number | Percent | Total beneficiaries | Enrolled in Original Medicare | Enrolled in Medicare Advantage | Violent | Property |
| | 32 | 33 | 34 | 35 | 36 | 37 | 38 | 39 | 40 | 41 | 42 | 43 | 44 | 45 |
| ILLINOIS—Cont'd | | | | | | | | | | | | | | |
| Winnebago | 4,742 | 289,202 | 1.03 | 3,440 | 12.2 | 2,972 | 10.6 | 18,485 | 8.1 | 59,232 | 35,022 | 24,210 | 123 | 517 |
| Woodford | 1,005 | 31,947 | 0.63 | 405 | 10.6 | 430 | 11.3 | 1,746 | 5.7 | 7,966 | 5,885 | 2,081 | 24 | 104 |
| INDIANA | 190,123 | 6,609,716 | 0.98 | 80,976 | 12.0 | 67,236 | 10.0 | 565,717 | 10.3 | 1,283,388 | 822,981 | 460,407 | X | X |
| Adams | 421 | 34,684 | 0.95 | 631 | 17.6 | 290 | 8.1 | 4,240 | 14.3 | 6,201 | 3,928 | 2,273 | NA | NA |
| Allen | 6,125 | 383,716 | 1.06 | 5,286 | 13.8 | 3,415 | 8.9 | 33,536 | 10.6 | 67,968 | 32,178 | 35,789 | 184 | 642 |
| Bartholomew | 1,148 | 92,508 | 1.25 | 1,051 | 12.4 | 790 | 9.4 | 6,827 | 9.8 | 15,717 | 11,487 | 4,230 | 51 | 402 |
| Benton | 92 | 7,421 | 0.69 | 112 | 12.8 | 92 | 10.5 | 816 | 11.5 | 1,884 | 1,424 | 460 | NA | NA |
| Blackford | 163 | 10,882 | 0.78 | 139 | 11.8 | 171 | 14.5 | 889 | 9.9 | 3,135 | 2,093 | 1,043 | 8 | 16 |
| Boone | 574 | 58,201 | 0.78 | 776 | 11.2 | 524 | 7.6 | 4,192 | 7.1 | 11,573 | 7,710 | 3,863 | NA | NA |
| Brown | 163 | 11,725 | 0.51 | 117 | 7.7 | 175 | 11.6 | 1,177 | 10.5 | 4,227 | 2,688 | 1,539 | NA | NA |
| Carroll | 106 | 16,056 | 0.57 | 198 | 9.8 | 214 | 10.6 | 1,720 | 10.7 | 4,462 | 3,166 | 1,296 | NA | NA |
| Cass | 1,052 | 36,009 | 0.89 | 402 | 10.8 | 439 | 11.7 | 4,250 | 14.1 | 8,088 | 5,469 | 2,618 | NA | NA |
| Clark | 1,328 | 110,796 | 0.90 | 1,396 | 11.7 | 1,299 | 10.9 | 8,738 | 8.9 | 23,864 | 16,364 | 7,500 | NA | NA |
| Clay | 341 | 22,300 | 0.67 | 324 | 12.3 | 303 | 11.5 | 1,862 | 8.8 | 5,999 | 4,320 | 1,679 | NA | NA |
| Clinton | 831 | 29,478 | 0.81 | 421 | 13.1 | 343 | 10.7 | 3,138 | 11.8 | 6,534 | 4,267 | 2,267 | 19 | 175 |
| Crawford | 62 | 8,778 | 0.57 | 118 | 11.1 | 128 | 12.0 | 1,039 | 12.5 | 2,674 | 1,920 | 754 | NA | NA |
| Daviess | 581 | 30,881 | 0.84 | 573 | 17.1 | 346 | 10.3 | 5,235 | 18.8 | 5,608 | 4,642 | 966 | NA | NA |
| Dearborn | 530 | 40,957 | 0.65 | 532 | 10.7 | 497 | 10.0 | 3,329 | 8.3 | 10,867 | 7,406 | 3,461 | NA | NA |
| Decatur | 377 | 29,113 | 1.19 | 315 | 11.8 | 297 | 11.2 | 1,938 | 8.9 | 5,536 | 3,848 | 1,688 | NA | NA |
| DeKalb | 615 | 44,659 | 1.08 | 541 | 12.4 | 431 | 9.9 | 3,316 | 9.2 | 8,897 | 4,244 | 4,653 | NA | NA |
| Delaware | 8,601 | 115,016 | 1.00 | 1,064 | 9.4 | 1,266 | 11.2 | 8,911 | 10.3 | 23,875 | 15,762 | 8,113 | NA | NA |
| Dubois | 894 | 48,135 | 1.25 | 503 | 11.8 | 446 | 10.5 | 3,026 | 8.7 | 8,959 | 7,527 | 1,433 | NA | NA |
| Elkhart | 3,691 | 234,209 | 1.31 | 3,098 | 15.0 | 1,817 | 8.8 | 25,976 | 15.1 | 35,101 | 21,881 | 13,220 | NA | NA |
| Fayette | 392 | 20,391 | 0.72 | 271 | 11.8 | 309 | 13.5 | 1,775 | 9.8 | 6,360 | 4,895 | 1,466 | NA | NA |
| Floyd | 1,329 | 69,414 | 0.79 | 853 | 10.8 | 817 | 10.4 | 5,223 | 8.0 | 16,291 | 11,613 | 4,678 | 16 | 281 |
| Fountain | 172 | 14,361 | 0.72 | 200 | 12.1 | 193 | 11.7 | 1,229 | 9.5 | 4,010 | 2,888 | 1,122 | NA | NA |
| Franklin | 187 | 17,779 | 0.55 | 264 | 11.6 | 231 | 10.1 | 1,719 | 9.3 | 5,075 | 3,452 | 1,624 | 4 | 137 |
| Fulton | 224 | 18,028 | 0.77 | 241 | 12.0 | 224 | 11.2 | 1,999 | 12.6 | 4,695 | 2,783 | 1,912 | NA | NA |
| Gibson | 738 | 39,859 | 1.38 | 412 | 12.2 | 412 | 12.2 | 2,099 | 7.7 | 7,062 | 4,827 | 2,235 | 32 | 116 |
| Grant | 4,798 | 67,835 | 1.05 | 765 | 11.7 | 873 | 13.4 | 4,732 | 9.8 | 15,843 | 10,401 | 5,442 | 36 | 254 |
| Greene | 281 | 25,973 | 0.56 | 360 | 11.2 | 443 | 13.8 | 2,644 | 10.5 | 7,396 | 5,532 | 1,864 | NA | NA |
| Hamilton | 1,871 | 301,724 | 0.87 | 3,760 | 10.9 | 2,124 | 6.2 | 17,843 | 6.0 | 48,779 | 32,363 | 16,416 | NA | NA |
| Hancock | 649 | 62,338 | 0.67 | 802 | 10.1 | 700 | 8.8 | 4,734 | 7.2 | 15,524 | 9,337 | 6,187 | 60 | 233 |
| Harrison | 461 | 33,476 | 0.63 | 414 | 10.2 | 395 | 9.7 | 2,981 | 9.1 | 8,785 | 6,429 | 2,356 | NA | NA |
| Hendricks | 4,067 | 148,560 | 0.82 | 1,647 | 9.5 | 1,262 | 7.3 | 11,029 | 7.7 | 28,404 | 17,537 | 10,866 | NA | NA |
| Henry | 3,557 | 42,538 | 0.71 | 480 | 10.0 | 577 | 12.0 | 3,180 | 9.0 | 11,476 | 6,975 | 4,501 | NA | NA |
| Howard | 1,281 | 85,552 | 1.09 | 995 | 12.0 | 1,016 | 12.3 | 6,419 | 9.9 | 19,535 | 12,687 | 6,848 | 56 | 126 |
| Huntington | 1,339 | 33,084 | 0.82 | 419 | 11.5 | 420 | 11.5 | 2,802 | 9.6 | 8,357 | 4,074 | 4,284 | NA | NA |
| Jackson | 595 | 46,472 | 1.12 | 606 | 13.7 | 484 | 10.9 | 4,071 | 11.2 | 9,095 | 5,725 | 3,370 | 12 | 277 |
| Jasper | 932 | 30,686 | 0.82 | 359 | 10.7 | 365 | 10.9 | 2,931 | 11.0 | 7,045 | 5,249 | 1,796 | NA | NA |
| Jay | 243 | 19,596 | 0.87 | 288 | 14.1 | 243 | 11.9 | 1,832 | 11.2 | 4,575 | 3,058 | 1,517 | 7 | 53 |
| Jefferson | 2,044 | 31,778 | 0.97 | 367 | 11.4 | 378 | 11.8 | 2,218 | 9.1 | 7,294 | 5,270 | 2,024 | NA | NA |
| Jennings | 298 | 23,549 | 0.70 | 313 | 11.4 | 342 | 12.4 | 2,219 | 9.8 | 6,036 | 3,951 | 2,085 | 10 | 89 |
| Johnson | 2,418 | 131,659 | 0.72 | 1,857 | 11.6 | 1,409 | 8.8 | 11,109 | 8.3 | 28,513 | 18,174 | 10,340 | NA | NA |
| Knox | 1,882 | 38,613 | 1.09 | 398 | 10.9 | 422 | 11.6 | 2,644 | 9.4 | 8,133 | 6,445 | 1,688 | NA | NA |
| Kosciusko | 1,574 | 79,951 | 1.02 | 991 | 12.5 | 786 | 10.0 | 8,085 | 12.5 | 15,876 | 8,336 | 7,540 | NA | NA |
| LaGrange | 325 | 38,851 | 0.98 | 781 | 19.5 | 283 | 7.1 | 9,514 | 27.9 | 5,820 | 3,450 | 2,370 | NA | NA |
| Lake | 6,490 | 465,129 | 0.90 | 5,722 | 11.7 | 5,384 | 11.0 | 39,323 | 9.9 | 95,353 | 62,504 | 32,849 | NA | NA |
| LaPorte | 6,387 | 104,251 | 0.87 | 1,249 | 11.4 | 1,310 | 11.9 | 8,550 | 10.3 | 23,751 | 17,238 | 6,513 | 57 | 416 |
| Lawrence | 653 | 40,504 | 0.76 | 514 | 11.3 | 603 | 13.3 | 3,570 | 10.0 | 10,832 | 7,577 | 3,255 | 27 | 142 |
| Madison | 6,213 | 117,121 | 0.78 | 1,377 | 10.6 | 1,554 | 12.0 | 10,641 | 10.7 | 29,612 | 16,188 | 13,424 | 37 | 369 |
| Marion | 18,370 | 1,081,355 | 1.28 | 13,959 | 14.4 | 8,808 | 9.1 | 100,212 | 12.3 | 147,109 | 85,426 | 61,683 | NA | NA |
| Marshall | 671 | 47,012 | 1.03 | 525 | 11.4 | 507 | 11.0 | 5,495 | 14.7 | 9,552 | 5,317 | 4,234 | NA | NA |
| Martin | 133 | 13,789 | 1.75 | 103 | 10.2 | 115 | 11.4 | 752 | 9.4 | 2,350 | 1,941 | 408 | NA | NA |
| Miami | 3,343 | 31,665 | 0.72 | 352 | 10.0 | 389 | 11.0 | 2,850 | 11.1 | 7,486 | 5,139 | 2,347 | NA | NA |
| Monroe | 15,006 | 154,837 | 1.12 | 1,234 | 8.3 | 1,029 | 6.9 | 10,733 | 9.4 | 21,863 | 15,825 | 6,038 | NA | NA |
| Montgomery | 1,175 | 36,985 | 0.93 | 431 | 11.2 | 420 | 10.9 | 3,368 | 11.1 | 8,129 | 5,240 | 2,889 | NA | NA |
| Morgan | 590 | 55,785 | 0.57 | 747 | 10.6 | 728 | 10.3 | 5,814 | 10.0 | 15,470 | 9,562 | 5,908 | NA | NA |
| Newton | 169 | 11,384 | 0.56 | 152 | 10.9 | 158 | 11.4 | 1,462 | 13.2 | 3,229 | 2,402 | 827 | NA | NA |
| Noble | 823 | 44,740 | 0.88 | 628 | 13.1 | 495 | 10.3 | 4,980 | 12.7 | 9,280 | 4,528 | 4,752 | NA | NA |
| Ohio | 52 | 4,284 | 0.45 | 56 | 9.5 | 82 | 13.9 | 387 | 8.5 | 1,526 | 986 | 540 | NA | NA |
| Orange | 258 | 19,036 | 0.94 | 264 | 13.4 | 251 | 12.8 | 1,746 | 11.3 | 4,806 | 3,406 | 1,400 | NA | NA |
| Owen | 197 | 17,355 | 0.63 | 237 | 11.4 | 281 | 13.5 | 2,091 | 12.7 | 5,016 | 3,330 | 1,686 | NA | NA |
| Parke | 1,603 | 14,866 | 0.68 | 187 | 11.1 | 193 | 11.4 | 1,494 | 12.5 | 3,681 | 2,584 | 1,098 | NA | NA |
| Perry | 1,583 | 17,784 | 0.84 | 206 | 10.8 | 204 | 10.7 | 1,135 | 8.1 | 4,309 | 3,265 | 1,044 | NA | NA |

1. Per 1,000 estimated resident population.

# Table B. States and Counties — Crime, Education, Money Income, and Poverty

| STATE County | County law enforcement employment, 2019 | | Education — School enrollment and attainment, 2015-2019 | | | | Local government expenditures[3], 2017-2018 | | Money income, 2015-2019 | | | | Income and poverty, 2019 | | | |
|---|---|---|---|---|---|---|---|---|---|---|---|---|---|---|---|---|
| | | | Enrollment[1] | | Attainment[2] (percent) | | | | | | Households Percent | | | Percent below poverty level | | |
| | Officers | Civilians | Total | Percent private | High school graduate or less | Bachelor's degree or more | Total current spending (mil dol) | Current spending per student (dollars) | Per capita income[4] | Median income (dollars) | with income of less than $50,000 | with income of $200,000 or more | Median household income (dollars) | All persons | Children under 18 years | Children 5 to 17 years in families |
| | 46 | 47 | 48 | 49 | 50 | 51 | 52 | 53 | 54 | 55 | 56 | 57 | 58 | 59 | 60 | 61 |
| **ILLINOIS—Cont'd** | | | | | | | | | | | | | | | | |
| Winnebago | 116 | 215 | 68,038 | 18.6 | 44.1 | 22.9 | 689 | 15,422 | 29,278 | 54,489 | 45.8 | 3.6 | 59,510 | 16.1 | 26.6 | 26.4 |
| Woodford | NA | NA | 9,960 | 16.4 | 36.5 | 33.2 | 103 | 13,589 | 37,170 | 72,808 | 31.1 | 7.7 | 77,117 | 6.1 | 6.9 | 6.0 |
| **INDIANA** | X | X | 1,668,419 | 16.4 | 44.6 | 26.5 | 10,663 | 10,135 | 29,777 | 56,303 | 44.4 | 4.2 | 57,617 | 11.9 | 15.1 | 13.9 |
| Adams | NA | NA | 8,634 | 27.5 | 55.9 | 15.5 | 46 | 10,580 | 23,316 | 52,504 | 48.0 | 2.6 | 53,532 | 12.6 | 21.3 | 21.2 |
| Allen | NA | NA | 96,178 | 23.7 | 38.9 | 28.5 | 557 | 10,195 | 29,160 | 54,857 | 45.3 | 3.9 | 56,838 | 10.5 | 13.2 | 12.3 |
| Bartholomew | 43 | 62 | 19,248 | 16.8 | 40.3 | 33.5 | 125 | 10,057 | 31,774 | 63,431 | 39.1 | 5.1 | 71,181 | 11.2 | 15.9 | 16.7 |
| Benton | NA | NA | 1,959 | 12.7 | 53.0 | 16.1 | 21 | 11,412 | 25,187 | 49,488 | 50.5 | 1.8 | 55,016 | 8.7 | 12.6 | 11.1 |
| Blackford | NA | NA | 2,384 | 9.6 | 60.2 | 13.5 | 17 | 10,202 | 24,302 | 43,505 | 57.2 | 1.0 | 46,811 | 12.8 | 18.4 | 17.1 |
| Boone | NA | NA | 16,934 | 13.6 | 25.8 | 49.3 | 118 | 9,576 | 48,295 | 83,077 | 27.5 | 14.5 | 84,137 | 6.9 | 6.3 | 5.6 |
| Brown | 16 | 28 | 2,783 | 20.6 | 40.8 | 28.9 | 24 | 11,935 | 34,372 | 61,030 | 40.9 | 6.1 | 59,617 | 10.5 | 14.5 | 13.3 |
| Carroll | NA | NA | 4,199 | 16.8 | 52.4 | 16.7 | 24 | 9,635 | 29,366 | 54,806 | 44.6 | 2.6 | 58,965 | 8.7 | 11.6 | 10.5 |
| Cass | NA | NA | 8,689 | 11.3 | 54.5 | 13.3 | 68 | 10,489 | 24,948 | 49,415 | 50.8 | 1.3 | 51,777 | 12.6 | 15.9 | 13.6 |
| Clark | NA | NA | 26,556 | 16.0 | 45.2 | 21.2 | 170 | 9,901 | 29,062 | 55,630 | 44.4 | 2.9 | 59,441 | 9.7 | 12.7 | 12.5 |
| Clay | NA | NA | 5,641 | 9.3 | 52.4 | 17.0 | 42 | 10,069 | 26,257 | 55,637 | 43.8 | 1.2 | 54,349 | 11.3 | 15.5 | 14.6 |
| Clinton | 20 | 38 | 7,601 | 13.6 | 57.6 | 16.3 | 58 | 9,335 | 24,489 | 54,286 | 45.9 | 1.5 | 55,336 | 11.3 | 16.0 | 14.6 |
| Crawford | NA | NA | 2,231 | 5.7 | 65.5 | 10.3 | 16 | 10,152 | 22,443 | 41,662 | 56.0 | 1.2 | 44,332 | 16.5 | 24.4 | 23.0 |
| Daviess | NA | NA | 7,195 | 31.6 | 60.2 | 13.2 | 48 | 10,218 | 23,848 | 53,629 | 45.5 | 2.4 | 56,611 | 11.3 | 16.5 | 16.2 |
| Dearborn | NA | NA | 11,191 | 11.1 | 49.3 | 22.9 | 79 | 9,452 | 30,800 | 68,658 | 35.1 | 4.9 | 71,351 | 9.3 | 13.4 | 12.0 |
| Decatur | NA | NA | 5,995 | 13.6 | 53.0 | 20.6 | 43 | 10,427 | 27,006 | 57,949 | 41.8 | 1.9 | 60,891 | 9.6 | 13.6 | 12.5 |
| DeKalb | NA | NA | 10,470 | 19.1 | 52.0 | 18.0 | 91 | 12,910 | 29,073 | 56,421 | 43.7 | 3.4 | 59,714 | 9.5 | 15.4 | 14.5 |
| Delaware | NA | NA | 34,571 | 6.5 | 45.4 | 23.7 | 157 | 10,091 | 25,107 | 43,512 | 55.7 | 2.9 | 45,912 | 21.5 | 23.0 | 23.3 |
| Dubois | NA | NA | 10,227 | 11.6 | 49.1 | 23.7 | 73 | 10,156 | 30,366 | 60,666 | 40.5 | 2.8 | 63,869 | 5.9 | 6.8 | 6.7 |
| Elkhart | NA | NA | 50,071 | 14.9 | 54.0 | 19.8 | 368 | 10,107 | 26,150 | 57,021 | 43.1 | 3.4 | 55,782 | 9.6 | 12.3 | 11.7 |
| Fayette | NA | NA | 4,780 | 12.4 | 60.0 | 13.3 | 40 | 11,290 | 24,015 | 46,175 | 53.5 | 1.0 | 48,499 | 13.5 | 17.5 | 16.3 |
| Floyd | NA | NA | 19,003 | 14.9 | 40.1 | 29.6 | 123 | 10,241 | 33,478 | 64,468 | 38.5 | 5.2 | 64,614 | 10.6 | 14.8 | 13.4 |
| Fountain | NA | NA | 3,408 | 10.8 | 54.7 | 14.9 | 26 | 9,887 | 28,058 | 52,874 | 46.9 | 2.3 | 55,337 | 12.5 | 17.4 | 13.6 |
| Franklin | 11 | 0 | 5,069 | 20.6 | 55.2 | 21.1 | 46 | 9,993 | 29,166 | 62,462 | 38.9 | 3.2 | 67,315 | 9.2 | 13.2 | 12.5 |
| Fulton | NA | NA | 4,649 | 8.3 | 52.9 | 15.8 | 26 | 10,373 | 26,427 | 52,034 | 47.5 | 1.4 | 55,011 | 9.6 | 14.4 | 13.7 |
| Gibson | 16 | 30 | 7,785 | 18.5 | 49.0 | 18.0 | 50 | 10,055 | 28,076 | 55,462 | 44.1 | 3.0 | 61,872 | 9.3 | 11.7 | 10.9 |
| Grant | NA | NA | 17,576 | 34.7 | 53.2 | 17.9 | 109 | 9,921 | 23,046 | 44,356 | 55.2 | 1.4 | 47,509 | 16.0 | 22.7 | 21.5 |
| Greene | 20 | 40 | 6,762 | 11.4 | 53.2 | 14.9 | 51 | 10,117 | 25,453 | 51,613 | 47.9 | 1.8 | 51,918 | 13.5 | 19.9 | 18.3 |
| Hamilton | NA | NA | 89,551 | 16.7 | 18.1 | 59.3 | 627 | 9,352 | 49,287 | 98,173 | 21.7 | 16.2 | 107,710 | 4.2 | 4.3 | 4.0 |
| Hancock | 41 | 41 | 17,815 | 13.8 | 39.9 | 30.8 | 127 | 9,336 | 34,119 | 74,072 | 32.2 | 5.5 | 77,905 | 5.2 | 5.7 | 5.2 |
| Harrison | NA | NA | 8,888 | 14.4 | 51.2 | 18.7 | 59 | 9,569 | 28,119 | 57,712 | 43.9 | 3.3 | 64,270 | 8.2 | 11.8 | 11.1 |
| Hendricks | NA | NA | 43,475 | 14.8 | 33.5 | 37.0 | 269 | 8,947 | 36,569 | 81,933 | 26.5 | 6.1 | 85,827 | 4.9 | 5.1 | 4.7 |
| Henry | NA | NA | 9,816 | 7.4 | 54.4 | 16.4 | 73 | 10,187 | 24,687 | 49,832 | 50.2 | 1.5 | 49,293 | 12.9 | 16.4 | 14.0 |
| Howard | 35 | 113 | 18,712 | 9.4 | 46.4 | 21.6 | 135 | 9,773 | 28,129 | 52,373 | 47.7 | 2.7 | 54,048 | 12.2 | 18.3 | 16.0 |
| Huntington | NA | NA | 8,718 | 17.4 | 49.5 | 20.1 | 51 | 9,827 | 26,502 | 53,632 | 46.4 | 2.0 | 54,565 | 9.7 | 12.6 | 11.5 |
| Jackson | 17 | 57 | 9,871 | 17.7 | 54.6 | 16.7 | 68 | 9,804 | 25,840 | 51,250 | 49.2 | 3.0 | 59,472 | 10.1 | 12.4 | 11.9 |
| Jasper | NA | NA | 7,940 | 14.9 | 55.0 | 14.3 | 49 | 9,879 | 28,240 | 63,892 | 38.0 | 2.4 | 65,352 | 8.8 | 10.9 | 10.5 |
| Jay | 13 | 34 | 4,735 | 9.6 | 60.9 | 11.4 | 35 | 10,701 | 23,443 | 47,658 | 52.5 | 1.2 | 49,447 | 12.5 | 19.3 | 18.1 |
| Jefferson | NA | NA | 7,172 | 27.5 | 52.6 | 17.9 | 46 | 10,803 | 27,152 | 52,718 | 47.7 | 2.3 | 53,378 | 11.6 | 17.1 | 15.9 |
| Jennings | 17 | 3 | 6,397 | 11.3 | 58.0 | 12.0 | 47 | 11,120 | 26,017 | 54,191 | 46.5 | 1.8 | 55,022 | 11.6 | 16.4 | 15.7 |
| Johnson | 123 | 16 | 38,304 | 17.5 | 39.0 | 32.7 | 248 | 9,231 | 34,408 | 72,440 | 32.1 | 6.3 | 74,427 | 6.0 | 7.6 | 6.8 |
| Knox | 28 | 36 | 9,180 | 7.9 | 47.1 | 16.9 | 53 | 10,080 | 26,041 | 47,380 | 51.7 | 2.7 | 48,760 | 12.7 | 16.0 | 16.3 |
| Kosciusko | 32 | 68 | 17,947 | 18.8 | 49.1 | 23.3 | 121 | 10,224 | 29,786 | 61,366 | 40.3 | 3.5 | 64,677 | 9.0 | 11.1 | 10.5 |
| LaGrange | NA | NA | 8,344 | 32.4 | 68.7 | 10.5 | 58 | 10,651 | 24,549 | 64,498 | 37.2 | 3.8 | 70,746 | 7.5 | 11.3 | 10.7 |
| Lake | 157 | 284 | 119,745 | 13.6 | 46.4 | 22.6 | 814 | 10,248 | 28,923 | 56,128 | 44.6 | 3.7 | 57,280 | 14.5 | 20.3 | 18.8 |
| LaPorte | 68 | 91 | 22,573 | 14.7 | 51.1 | 18.0 | 177 | 10,131 | 27,264 | 53,658 | 46.9 | 3.2 | 56,427 | 13.4 | 23.2 | 23.7 |
| Lawrence | NA | NA | 9,740 | 11.9 | 54.0 | 15.8 | 67 | 10,292 | 25,557 | 53,610 | 46.4 | 1.9 | 56,261 | 10.5 | 14.5 | 13.6 |
| Madison | 46 | 55 | 28,258 | 16.1 | 51.0 | 18.4 | 185 | 9,705 | 25,557 | 49,522 | 50.4 | 2.0 | 54,145 | 13.1 | 18.6 | 16.0 |
| Marion | NA | NA | 239,411 | 17.2 | 41.9 | 30.9 | 1,751 | 11,033 | 28,566 | 48,316 | 51.4 | 3.9 | 50,707 | 15.2 | 19.4 | 18.5 |
| Marshall | NA | NA | 10,668 | 17.9 | 53.8 | 19.7 | 71 | 9,591 | 25,761 | 52,658 | 47.4 | 2.4 | 53,695 | 10.9 | 12.9 | 11.7 |
| Martin | NA | NA | 2,255 | 6.8 | 54.6 | 13.7 | 14 | 10,157 | 26,758 | 52,726 | 46.4 | 1.3 | 56,735 | 10.9 | 14.6 | 13.0 |
| Miami | NA | NA | 8,140 | 5.9 | 55.9 | 13.0 | 49 | 9,269 | 24,028 | 50,657 | 49.3 | 2.3 | 53,658 | 13.0 | 17.6 | 16.4 |
| Monroe | 46 | 14 | 58,638 | 7.0 | 29.5 | 45.7 | 154 | 10,692 | 28,514 | 49,839 | 50.1 | 4.4 | 53,113 | 20.8 | 15.0 | 12.6 |
| Montgomery | 25 | 46 | 8,543 | 18.2 | 51.6 | 18.2 | 62 | 10,161 | 27,162 | 55,522 | 43.9 | 2.6 | 54,366 | 12.1 | 16.3 | 15.5 |
| Morgan | NA | NA | 16,050 | 11.8 | 51.2 | 17.3 | 99 | 9,124 | 30,721 | 64,335 | 36.3 | 4.0 | 63,414 | 8.9 | 12.7 | 11.3 |
| Newton | NA | NA | 2,843 | 7.1 | 57.0 | 11.9 | 23 | 10,296 | 26,007 | 55,356 | 45.0 | 1.9 | 61,457 | 9.5 | 12.5 | 11.5 |
| Noble | 48 | 26 | 10,637 | 11.2 | 58.4 | 13.7 | 70 | 9,588 | 26,832 | 56,789 | 43.6 | 2.4 | 60,980 | 7.8 | 11.0 | 10.0 |
| Ohio | NA | NA | 1,083 | 3.3 | 55.5 | 14.7 | 9 | 10,924 | 30,517 | 60,128 | 39.2 | 1.3 | 58,294 | 9.6 | 14.8 | 13.3 |
| Orange | NA | NA | 4,004 | 9.0 | 63.4 | 10.9 | 35 | 11,372 | 24,080 | 47,917 | 52.2 | 1.8 | 47,931 | 14.1 | 21.3 | 21.0 |
| Owen | NA | NA | 4,209 | 17.9 | 59.0 | 13.1 | 25 | 9,848 | 25,229 | 49,543 | 50.4 | 1.3 | 53,387 | 12.0 | 18.1 | 17.0 |
| Parke | NA | NA | 3,057 | 14.3 | 54.3 | 12.0 | 23 | 10,044 | 23,574 | 52,618 | 47.0 | 1.3 | 51,554 | 15.1 | 22.0 | 21.3 |
| Perry | NA | NA | 3,813 | 10.9 | 54.4 | 16.1 | 28 | 9,403 | 25,707 | 52,348 | 47.5 | 2.3 | 53,908 | 12.1 | 14.7 | 13.7 |

1. All persons 3 years old and over enrolled in nursery school through college.  2. Persons 25 years old and over.  3. Elementary and secondary education expenditures.  4. Based on population estimated by the American Community Survey, 2014–2018.

# Table B. States and Counties — **Personal Income**

| STATE County | Personal income, 2019 | | | | | Supplements to wages and salaries, employer contributions (mil dol) | | | | | Earnings, 2019 | | |
|---|---|---|---|---|---|---|---|---|---|---|---|---|---|
| | Total (mil dol) | Percent change 2018-2019 | Per capita[1] Dollars | Per capita[1] Rank | Wages and salaries (mil dol) | Pension and insurance | Government social insurance | Proprietors' income (mil dol) | Dividends, interest, and rent (mil dol) | Personal transfer receipts (mil dol) | Total (mil dol) | Contributions for government social insurance (mil dol) From employee and self-employed | From employer |
| | 62 | 63 | 64 | 65 | 66 | 67 | 68 | 69 | 70 | 71 | 72 | 73 | 74 |
| ILLINOIS—Cont'd | | | | | | | | | | | | | |
| Winnebago | 12,499 | 1.4 | 44,234 | 1,454 | 6,891 | 1,257 | 475 | 543 | 1,921 | 2,755 | 9,165 | 550 | 475 |
| Woodford | 2,045 | 0.7 | 53,163 | 599 | 458 | 96 | 32 | 120 | 387 | 315 | 707 | 45 | 32 |
| INDIANA | 327,712 | 3.4 | 48,687 | X | 160,989 | 26,363 | 11,684 | 30,859 | 53,008 | 63,282 | 229,896 | 14,143 | 11,684 |
| Adams | 1,452 | 2.6 | 40,584 | 1,976 | 608 | 113 | 45 | 296 | 218 | 277 | 1,063 | 60 | 45 |
| Allen | 18,055 | 3.3 | 47,602 | 1,069 | 10,128 | 1,576 | 745 | 1,459 | 3,132 | 3,415 | 13,908 | 855 | 745 |
| Bartholomew | 4,402 | 3.9 | 52,546 | 642 | 3,167 | 473 | 223 | 274 | 759 | 728 | 4,137 | 251 | 223 |
| Benton | 369 | 2.8 | 42,135 | 1,746 | 106 | 21 | 8 | 31 | 65 | 83 | 166 | 11 | 8 |
| Blackford | 456 | 1.5 | 38,813 | 2,221 | 124 | 26 | 9 | 24 | 68 | 155 | 183 | 14 | 9 |
| Boone | 5,172 | 3.9 | 76,241 | 80 | 1,569 | 225 | 118 | 286 | 1,102 | 495 | 2,198 | 135 | 118 |
| Brown | 737 | 2.7 | 48,862 | 936 | 96 | 21 | 7 | 52 | 135 | 167 | 176 | 15 | 7 |
| Carroll | 880 | 1.0 | 43,432 | 1,552 | 236 | 41 | 17 | 70 | 137 | 184 | 364 | 25 | 17 |
| Cass | 1,534 | 3.5 | 40,693 | 1,964 | 624 | 119 | 45 | 81 | 234 | 444 | 869 | 58 | 45 |
| Clark | 5,356 | 4.5 | 45,270 | 1,326 | 2,565 | 444 | 190 | 313 | 669 | 1,139 | 3,512 | 225 | 190 |
| Clay | 1,023 | 3.3 | 39,004 | 2,192 | 307 | 63 | 22 | 46 | 146 | 285 | 439 | 33 | 22 |
| Clinton | 1,293 | 2.1 | 39,905 | 2,060 | 508 | 90 | 37 | 59 | 194 | 310 | 694 | 47 | 37 |
| Crawford | 371 | 4.2 | 35,054 | 2,678 | 71 | 16 | 5 | 22 | 43 | 129 | 114 | 10 | 5 |
| Daviess | 1,467 | 2.6 | 43,998 | 1,485 | 525 | 93 | 39 | 242 | 234 | 300 | 898 | 52 | 39 |
| Dearborn | 2,470 | 4.4 | 49,949 | 833 | 638 | 113 | 48 | 112 | 385 | 485 | 910 | 66 | 48 |
| Decatur | 1,173 | 2.7 | 44,168 | 1,463 | 727 | 120 | 52 | 79 | 178 | 255 | 978 | 61 | 52 |
| DeKalb | 1,970 | 3.7 | 45,324 | 1,323 | 1,175 | 187 | 87 | 151 | 309 | 408 | 1,600 | 100 | 87 |
| Delaware | 4,352 | 3.1 | 38,129 | 2,317 | 2,133 | 400 | 157 | 197 | 722 | 1,306 | 2,887 | 193 | 157 |
| Dubois | 2,445 | 2.2 | 57,217 | 398 | 1,384 | 213 | 100 | 184 | 615 | 389 | 1,882 | 114 | 100 |
| Elkhart | 9,693 | 1.7 | 46,975 | 1,136 | 7,009 | 1,136 | 506 | 1,083 | 1,539 | 1,675 | 9,735 | 573 | 506 |
| Fayette | 970 | 3.4 | 41,993 | 1,768 | 262 | 50 | 20 | 48 | 134 | 334 | 380 | 31 | 20 |
| Floyd | 4,487 | 4.1 | 57,145 | 399 | 1,483 | 252 | 110 | 277 | 927 | 796 | 2,121 | 138 | 110 |
| Fountain | 677 | 2.2 | 41,412 | 1,860 | 193 | 41 | 15 | 36 | 104 | 182 | 285 | 21 | 15 |
| Franklin | 1,085 | 3.4 | 47,664 | 1,062 | 174 | 37 | 13 | 76 | 195 | 234 | 300 | 24 | 13 |
| Fulton | 874 | 2.9 | 43,775 | 1,519 | 278 | 52 | 20 | 109 | 151 | 206 | 459 | 30 | 20 |
| Gibson | 1,550 | 2.9 | 46,049 | 1,242 | 1,208 | 204 | 87 | 102 | 223 | 333 | 1,601 | 97 | 87 |
| Grant | 2,729 | 2.1 | 41,491 | 1,850 | 1,229 | 219 | 93 | 153 | 410 | 885 | 1,694 | 116 | 93 |
| Greene | 1,333 | 5.4 | 41,747 | 1,812 | 263 | 56 | 20 | 75 | 202 | 362 | 414 | 32 | 20 |
| Hamilton | 26,116 | 4.5 | 77,263 | 76 | 9,086 | 1,180 | 636 | 2,036 | 5,225 | 2,010 | 12,937 | 766 | 636 |
| Hancock | 4,075 | 4.9 | 52,134 | 671 | 1,323 | 224 | 93 | 210 | 625 | 680 | 1,849 | 123 | 93 |
| Harrison | 1,760 | 4.3 | 43,431 | 1,553 | 446 | 85 | 32 | 81 | 242 | 387 | 644 | 47 | 32 |
| Hendricks | 8,682 | 4.9 | 50,979 | 761 | 3,351 | 513 | 252 | 449 | 1,135 | 1,179 | 4,566 | 289 | 252 |
| Henry | 1,912 | 3.4 | 39,863 | 2,067 | 540 | 105 | 39 | 92 | 274 | 570 | 776 | 59 | 39 |
| Howard | 3,542 | 2.2 | 42,911 | 1,630 | 2,103 | 337 | 150 | 151 | 535 | 973 | 2,741 | 183 | 150 |
| Huntington | 1,564 | 3.0 | 42,824 | 1,645 | 597 | 110 | 44 | 52 | 270 | 375 | 804 | 56 | 44 |
| Jackson | 1,955 | 3.3 | 44,192 | 1,458 | 1,120 | 207 | 83 | 77 | 296 | 400 | 1,488 | 93 | 83 |
| Jasper | 1,536 | 4.5 | 45,777 | 1,277 | 506 | 95 | 38 | 149 | 218 | 322 | 788 | 50 | 38 |
| Jay | 803 | -2.1 | 39,317 | 2,143 | 301 | 57 | 22 | 126 | 104 | 205 | 505 | 30 | 22 |
| Jefferson | 1,425 | 4.4 | 44,102 | 1,473 | 563 | 115 | 41 | 101 | 208 | 383 | 820 | 55 | 41 |
| Jennings | 1,142 | 4.5 | 41,169 | 1,898 | 338 | 60 | 26 | 51 | 133 | 359 | 474 | 35 | 26 |
| Johnson | 7,911 | 4.4 | 50,018 | 832 | 2,618 | 410 | 195 | 475 | 1,195 | 1,275 | 3,698 | 241 | 195 |
| Knox | 1,679 | 2.6 | 45,882 | 1,263 | 780 | 162 | 56 | 118 | 269 | 473 | 1,116 | 70 | 56 |
| Kosciusko | 3,848 | 1.9 | 48,425 | 981 | 2,142 | 429 | 158 | 228 | 632 | 684 | 2,956 | 179 | 158 |
| LaGrange | 1,714 | 1.4 | 43,275 | 1,576 | 679 | 121 | 51 | 315 | 220 | 246 | 1,166 | 64 | 51 |
| Lake | 22,690 | 3.6 | 46,736 | 1,157 | 10,238 | 1,683 | 749 | 1,486 | 3,138 | 5,203 | 14,155 | 911 | 749 |
| LaPorte | 4,825 | 3.6 | 43,910 | 1,502 | 1,882 | 328 | 143 | 292 | 786 | 1,139 | 2,646 | 178 | 143 |
| Lawrence | 1,930 | 3.6 | 42,538 | 1,681 | 628 | 108 | 46 | 77 | 266 | 520 | 859 | 64 | 46 |
| Madison | 5,109 | 3.2 | 39,428 | 2,123 | 1,795 | 306 | 132 | 247 | 718 | 1,490 | 2,481 | 182 | 132 |
| Marion | 52,478 | 3.6 | 54,405 | 523 | 39,842 | 5,881 | 2,827 | 11,799 | 7,591 | 8,601 | 60,349 | 3,374 | 2,827 |
| Marshall | 1,952 | 1.1 | 42,196 | 1,738 | 787 | 139 | 58 | 114 | 314 | 419 | 1,098 | 73 | 58 |
| Martin | 402 | 3.0 | 39,207 | 2,162 | 594 | 171 | 49 | 26 | 69 | 103 | 839 | 47 | 49 |
| Miami | 1,278 | 2.1 | 35,970 | 2,574 | 408 | 90 | 34 | 90 | 205 | 346 | 622 | 42 | 34 |
| Monroe | 6,481 | 4.4 | 43,660 | 1,528 | 3,300 | 744 | 236 | 463 | 1,465 | 1,055 | 4,743 | 275 | 236 |
| Montgomery | 1,571 | 3.0 | 40,979 | 1,921 | 752 | 127 | 56 | 74 | 232 | 383 | 1,008 | 67 | 56 |
| Morgan | 3,197 | 3.9 | 45,352 | 1,316 | 689 | 124 | 51 | 165 | 403 | 695 | 1,029 | 78 | 51 |
| Newton | 580 | 3.2 | 41,454 | 1,853 | 142 | 26 | 11 | 36 | 77 | 133 | 215 | 15 | 11 |
| Noble | 1,948 | 2.9 | 40,799 | 1,949 | 829 | 151 | 62 | 156 | 252 | 411 | 1,197 | 77 | 62 |
| Ohio | 243 | 4.2 | 41,339 | 1,870 | 46 | 9 | 3 | 11 | 31 | 57 | 69 | 6 | 3 |
| Orange | 755 | 1.9 | 38,432 | 2,275 | 298 | 47 | 22 | 70 | 103 | 230 | 438 | 31 | 22 |
| Owen | 866 | 4.0 | 41,632 | 1,826 | 237 | 63 | 18 | 48 | 121 | 238 | 365 | 26 | 18 |
| Parke | 636 | 4.1 | 37,570 | 2,379 | 119 | 28 | 9 | 66 | 104 | 185 | 222 | 17 | 9 |
| Perry | 734 | 1.8 | 38,272 | 2,300 | 293 | 58 | 20 | 60 | 110 | 196 | 431 | 29 | 20 |

1. Based on the resident population estimated as of July 1 of the year shown.

| STATE County | Farm | Mining, quarrying, and extractions | Construction | Manu-facturing | Information; professional, scientific, technical services | Retail trade | Finance, insurance, real estate, and leasing | Health care and social assistance | Govern-ment | Social Security beneficiaries, December 2019 Number | Rate[1] | Supplemental Security Income recipients, 2019 | Housing units, 2020 Total | Percent change, 2010-2020 |
|---|---|---|---|---|---|---|---|---|---|---|---|---|---|---|
| | 75 | 76 | 77 | 78 | 79 | 80 | 81 | 82 | 83 | 84 | 85 | 86 | 87 | 88 |
| ILLINOIS—Cont'd | | | | | | | | | | | | | | |
| Winnebago | 0.0 | 0.0 | 4.9 | 22.4 | 4.8 | 6.0 | 6.6 | 18.4 | 12.7 | 62,335 | 221 | 7,533 | 125,754 | -0.2 |
| Woodford | 3.9 | 0.2 | 10.8 | 16.4 | D | 7.9 | 4.6 | 8.5 | 18.1 | 8,120 | 212 | 249 | 15,679 | 3.5 |
| INDIANA | 0.6 | 0.3 | 6.3 | 19.8 | 7.2 | 5.7 | 9.3 | 12.8 | 12.3 | 1,370,231 | 204 | 127,800 | 2,940,441 | 5.2 |
| Adams | 6.3 | D | 17.0 | 31.9 | D | 6.1 | 3.2 | 4.0 | 12.3 | 6,555 | 184 | 376 | 13,441 | 3.3 |
| Allen | 0.2 | D | 7.6 | 17.3 | 6.2 | 6.6 | 7.9 | 20.0 | 9.1 | 72,465 | 191 | 8,115 | 162,340 | 6.7 |
| Bartholomew | 0.5 | D | 3.6 | 47.5 | 4.9 | 4.5 | 3.6 | 7.0 | 9.1 | 16,685 | 198 | 1,238 | 34,810 | 5.2 |
| Benton | 10.6 | 0.0 | 7.2 | 14.0 | D | 3.7 | 10.7 | 2.8 | 18.8 | 2,085 | 238 | 177 | 3,935 | 0.2 |
| Blackford | 4.6 | D | 6.5 | 31.1 | D | 5.3 | 4.6 | 9.4 | 14.9 | 3,455 | 293 | 272 | 6,003 | -0.8 |
| Boone | 0.3 | D | 9.8 | 8.3 | 13.1 | 6.1 | 5.8 | 6.5 | 12.5 | 11,645 | 171 | 486 | 28,714 | 26.2 |
| Brown | -0.1 | 0.0 | 12.5 | 7.2 | 8.8 | 7.6 | 3.8 | 8.5 | 21.8 | 4,440 | 295 | 195 | 8,877 | 7.1 |
| Carroll | 8.2 | D | 7.9 | D | 2.3 | 4.5 | 4.5 | D | 12.0 | 4,840 | 239 | 210 | 9,728 | 2.7 |
| Cass | 2.6 | D | 6.3 | 29.5 | 2.1 | 5.6 | 3.7 | D | 21.9 | 8,735 | 232 | 794 | 16,357 | -0.7 |
| Clark | 0.2 | D | 7.0 | 17.7 | 4.0 | 7.8 | 6.9 | 12.0 | 11.6 | 25,470 | 215 | 2,178 | 52,562 | 10.0 |
| Clay | 1.3 | D | 4.5 | 40.4 | D | 7.2 | 3.2 | 7.6 | 15.0 | 6,690 | 254 | 637 | 11,778 | 0.7 |
| Clinton | 3.0 | 0.0 | 5.2 | 38.9 | D | 4.8 | 5.2 | 7.2 | 13.7 | 7,100 | 220 | 528 | 13,452 | 1.0 |
| Crawford | 0.2 | D | D | D | 1.8 | D | D | 5.4 | 21.0 | 3,040 | 287 | 347 | 5,592 | 1.3 |
| Daviess | 4.6 | D | 21.2 | 16.3 | 4.7 | 9.2 | 4.2 | D | 12.7 | 5,850 | 175 | 510 | 12,583 | 0.9 |
| Dearborn | 0.0 | D | 8.5 | 14.5 | D | 9.2 | 4.8 | 8.4 | 21.4 | 11,580 | 234 | 636 | 20,810 | 3.2 |
| Decatur | 2.1 | D | 3.5 | 44.7 | D | 4.5 | 3.6 | 3.8 | 11.0 | 5,995 | 226 | 415 | 11,520 | 2.8 |
| DeKalb | 1.0 | 0.3 | 3.7 | 52.5 | 2.5 | 3.5 | 2.4 | 5.8 | 7.7 | 9,635 | 221 | 649 | 18,419 | 4.8 |
| Delaware | 0.3 | D | 4.7 | 9.7 | 7.7 | 8.2 | 6.7 | 19.8 | 21.3 | 25,925 | 228 | 2,986 | 52,700 | 0.7 |
| Dubois | 1.3 | D | 4.2 | 37.2 | 4.2 | 6.8 | 3.9 | D | 6.7 | 9,350 | 220 | 385 | 18,132 | 4.3 |
| Elkhart | 0.7 | D | 4.1 | 50.2 | 2.3 | 4.2 | 3.1 | 7.5 | 5.5 | 36,720 | 178 | 3,022 | 80,082 | 3.0 |
| Fayette | 0.6 | D | 3.6 | 26.1 | 3.9 | 7.3 | 4.0 | 19.4 | 15.6 | 7,020 | 305 | 897 | 10,813 | -0.8 |
| Floyd | -0.1 | D | 7.3 | 21.2 | 7.9 | 5.1 | 4.4 | 21.6 | 12.3 | 17,185 | 219 | 1,422 | 33,325 | 4.2 |
| Fountain | 2.7 | 0.0 | 3.5 | 40.7 | 2.8 | 6.2 | 4.9 | D | 14.9 | 4,435 | 270 | 332 | 9,931 | 2.2 |
| Franklin | 3.9 | 0.3 | 10.7 | 15.6 | D | 9.5 | 4.9 | D | 17.6 | 6,265 | 275 | 429 | 9,759 | 4.1 |
| Fulton | 1.8 | D | 14.1 | 28.2 | D | 7.1 | 5.0 | D | 18.1 | 5,020 | 250 | 362 | 15,607 | 0.5 |
| Gibson | 1.3 | D | 1.8 | 51.6 | D | 3.3 | 1.8 | D | 4.8 | 7,515 | 223 | 469 | 30,549 | 6.6 |
| Grant | 0.7 | D | 3.1 | 21.8 | 3.2 | 6.7 | 4.2 | D | 15.1 | 17,225 | 262 | 2,120 | 15,319 | 0.3 |
| Greene | 6.5 | 1.1 | 7.9 | 7.4 | 4.5 | 7.6 | 3.1 | D | 27.5 | 8,110 | 253 | 728 | 135,395 | 0.7 |
| Hamilton | 0.3 | 0.2 | 7.9 | 4.4 | 14.3 | 6.5 | 18.9 | 10.2 | 7.5 | 48,130 | 142 | 1,663 | 31,706 | 26.8 |
| Hancock | 0.7 | 0.0 | 9.7 | 27.0 | 8.1 | 5.3 | 4.2 | 6.2 | 15.0 | 16,295 | 208 | 682 | 17,403 | 12.7 |
| Harrison | 0.9 | 1.1 | 7.1 | 16.7 | 2.7 | 7.5 | 4.4 | D | 19.3 | 9,530 | 235 | 613 | 64,967 | 5.3 |
| Hendricks | 0.0 | D | 6.8 | 6.6 | 3.4 | 11.0 | 3.5 | 7.1 | 14.5 | 28,770 | 169 | 979 | 21,285 | 17.2 |
| Henry | 2.2 | D | 6.8 | 23.4 | 2.5 | 8.1 | 4.7 | D | 24.1 | 12,600 | 262 | 1,067 | 40,033 | 0.0 |
| Howard | 0.3 | D | 3.6 | 45.3 | 2.6 | 6.0 | 3.8 | 12.6 | 10.0 | 21,645 | 262 | 2,249 | 16,214 | 3.5 |
| Huntington | 0.9 | D | 6.0 | 28.3 | D | 5.2 | 5.8 | 9.2 | 10.6 | 8,945 | 245 | 627 | 19,607 | 2.6 |
| Jackson | 1.0 | D | 3.9 | 40.4 | 1.9 | 5.5 | 3.2 | D | 14.9 | 9,825 | 222 | 765 | 13,815 | 7.7 |
| Jasper | 6.3 | D | 8.2 | 16.1 | D | 6.2 | 3.9 | 6.2 | 10.6 | 7,940 | 237 | 469 | 9,262 | 4.9 |
| Jay | 14.9 | D | 7.8 | 34.7 | 1.9 | 4.0 | 2.8 | D | 9.9 | 4,985 | 244 | 371 | 14,595 | 0.4 |
| Jefferson | -0.2 | 0.0 | 4.1 | 26.8 | D | 12.3 | 3.3 | 13.1 | 15.2 | 8,140 | 252 | 715 | 12,515 | 2.0 |
| Jennings | 2.9 | D | 14.2 | 26.1 | 2.0 | 5.3 | 2.2 | 6.7 | 15.6 | 6,655 | 241 | 669 | | 3.7 |
| Johnson | -0.1 | D | 8.3 | 11.4 | 5.8 | 9.3 | 5.9 | 12.5 | 13.6 | 29,980 | 189 | 1,701 | 63,311 | 11.8 |
| Knox | 3.0 | 7.3 | 5.0 | 10.8 | 2.8 | 5.9 | 3.3 | 12.9 | 25.1 | 8,710 | 238 | 946 | 17,186 | 0.9 |
| Kosciusko | 1.4 | D | 4.3 | 49.4 | 2.8 | 5.1 | 3.6 | 6.5 | 6.0 | 16,770 | 211 | 879 | 39,382 | 6.3 |
| LaGrange | 6.7 | D | 6.5 | 46.7 | 1.8 | 7.0 | 2.4 | D | 6.4 | 6,120 | 154 | 294 | 15,194 | 7.8 |
| Lake | 0.1 | 0.1 | 8.5 | 18.9 | 4.5 | 6.2 | 4.0 | 16.4 | 10.7 | 102,455 | 211 | 12,613 | 215,597 | 3.3 |
| LaPorte | 0.9 | 0.1 | 8.4 | 21.8 | 3.4 | 6.6 | 4.3 | 13.5 | 15.0 | 25,535 | 232 | 2,174 | 49,355 | 1.9 |
| Lawrence | 0.6 | 1.2 | 6.4 | 26.2 | 8.1 | 7.8 | 5.1 | 14.7 | 11.9 | 11,790 | 260 | 1,060 | 21,227 | 0.7 |
| Madison | 0.5 | D | 6.8 | 15.6 | 3.4 | 6.4 | 6.2 | 16.7 | 15.4 | 32,075 | 248 | 3,388 | 59,382 | 0.5 |
| Marion | 0.0 | 0.0 | 5.1 | 11.1 | D | 4.1 | 19.4 | 14.1 | 10.9 | 158,800 | 165 | 26,113 | 427,900 | 2.4 |
| Marshall | 2.4 | D | 4.7 | 38.7 | 3.5 | 6.2 | 4.7 | 8.4 | 10.6 | 10,235 | 222 | 591 | 20,460 | 3.1 |
| Martin | 0.4 | 0.0 | 1.1 | 3.8 | 9.9 | 1.2 | 0.6 | 0.8 | 76.7 | 2,490 | 245 | 205 | 4,826 | 0.8 |
| Miami | 5.9 | 0.7 | 5.9 | 17.3 | 2.3 | 5.4 | 4.6 | D | 26.9 | 7,525 | 212 | 757 | 15,419 | -0.4 |
| Monroe | 0.0 | 0.4 | 4.4 | 11.7 | D | 5.1 | 4.5 | 13.0 | 33.2 | 22,695 | 153 | 1,850 | 63,090 | 6.7 |
| Montgomery | 0.4 | D | 3.7 | 39.0 | D | 5.5 | 2.9 | D | 12.0 | 8,855 | 231 | 636 | 16,782 | 1.5 |
| Morgan | 0.8 | 0.2 | 13.1 | 16.6 | 5.0 | 7.3 | 6.4 | 9.7 | 14.4 | 16,800 | 239 | 951 | 28,898 | 4.1 |
| Newton | 11.2 | D | 8.2 | 16.8 | D | 3.1 | D | 5.6 | 16.5 | 3,185 | 228 | 196 | 6,120 | 1.5 |
| Noble | 1.8 | D | 4.7 | 52.5 | D | 5.6 | 2.7 | 5.4 | 9.0 | 10,165 | 212 | 652 | 20,875 | 3.8 |
| Ohio | -0.2 | 0.0 | D | D | D | 1.7 | D | 4.5 | 25.8 | 1,530 | 259 | 80 | 2,911 | 5.2 |
| Orange | 0.4 | D | 22.4 | 12.8 | D | 4.3 | 2.5 | D | 11.1 | 5,205 | 265 | 495 | 9,305 | 1.4 |
| Owen | 0.3 | 0.0 | 6.7 | 49.2 | D | 4.7 | 2.7 | D | 11.6 | 5,555 | 266 | 370 | 10,338 | 2.4 |
| Parke | 0.3 | D | 13.0 | 16.3 | D | 7.3 | 5.6 | D | 24.7 | 3,935 | 233 | 252 | 8,326 | 2.9 |
| Perry | 1.5 | 0.5 | 4.6 | 37.4 | D | 5.1 | 8.0 | 5.4 | 19.9 | 4,725 | 246 | 342 | 8,778 | 3.3 |

1. Per 1,000 resident population estimated as of July 1 of the year shown.

# Table B. States and Counties — Housing, Labor Force, and Employment

| STATE County | Housing units, 2015-2019 | | | | | | | | Civilian labor force, 2020 | | | | Civilian employment[6], 2015-2019 | | |
|---|---|---|---|---|---|---|---|---|---|---|---|---|---|---|---|
| | Occupied units | | | | | | | | | | Unemployment | | | Percent | |
| | Owner-occupied | | | | | Renter-occupied | | | | | | | | | |
| | | | | Median owner cost as a percent of income | | Median rent as a percent of income[2] | | | | | | | | | Construction, production, and maintenance occupations |
| | | | | With a mortgage | Without a mortgage[2] | Median rent[3] | | Substandard units[4] (percent) | Total | Percent change, 2019-2020 | Total | Rate[5] | Total | Management, business, science, and arts | |
| | Total | Percent | Median value[1] | | | | | | | | | | | | |
| | 89 | 90 | 91 | 92 | 93 | 94 | 95 | 96 | 97 | 98 | 99 | 100 | 101 | 102 | 103 |
| ILLINOIS—Cont'd | | | | | | | | | | | | | | | |
| Winnebago | 114,779 | 65.9 | 120,400 | 19.7 | 11.8 | 799 | 28.8 | 2.2 | 135,759 | -2.9 | 15,506 | 11.4 | 132,720 | 30.6 | 29.3 |
| Woodford | 14,499 | 81.8 | 168,700 | 19.4 | 10.8 | 758 | 27.6 | 1.1 | 18,012 | -5.1 | 1,095 | 6.1 | 18,653 | 39.3 | 21.8 |
| INDIANA | 2,570,419 | 69.1 | 141,700 | 18.3 | 10.0 | 826 | 28.6 | 1.9 | 3,319,010 | -1.9 | 236,028 | 7.1 | 3,202,509 | 34.2 | 28.5 |
| Adams | 12,481 | 77.1 | 129,300 | 19.4 | 10.0 | 624 | 25.2 | 9.9 | 16,771 | -3.3 | 935 | 5.6 | 15,495 | 24.1 | 41.3 |
| Allen | 145,065 | 68.4 | 129,300 | 17.7 | 10.0 | 773 | 27.5 | 1.9 | 184,986 | -1.1 | 14,203 | 7.7 | 182,422 | 34.5 | 27.1 |
| Bartholomew | 31,452 | 71.4 | 153,500 | 18.3 | 10.0 | 937 | 23.7 | 3.1 | 43,802 | -2.8 | 2,827 | 6.5 | 40,151 | 42.9 | 26.4 |
| Benton | 3,432 | 74.0 | 92,500 | 18.4 | 10.0 | 708 | 28.8 | 1.6 | 4,268 | -5.7 | 221 | 5.2 | 4,177 | 23.4 | 38.4 |
| Blackford | 5,235 | 74.7 | 70,600 | 19.4 | 11.8 | 644 | 28.6 | 1.9 | 4,826 | -3.4 | 384 | 8.0 | 5,194 | 26.0 | 41.4 |
| Boone | 25,485 | 77.1 | 233,700 | 18.4 | 10.0 | 1,011 | 25.3 | 1.0 | 35,586 | -2.9 | 1,522 | 4.3 | 34,202 | 50.3 | 18.7 |
| Brown | 6,189 | 84.4 | 191,700 | 19.7 | 11.7 | 841 | 30.1 | 2.9 | 7,568 | -0.7 | 506 | 6.7 | 7,042 | 34.8 | 24.2 |
| Carroll | 8,002 | 80.7 | 131,700 | 17.7 | 11.0 | 691 | 20.6 | 2.3 | 9,659 | -4.3 | 583 | 6.0 | 9,367 | 27.4 | 42.0 |
| Cass | 14,688 | 74.0 | 89,300 | 18.0 | 10.0 | 650 | 24.4 | 2.4 | 17,551 | -2.7 | 1,191 | 6.8 | 17,738 | 24.0 | 42.0 |
| Clark | 44,200 | 71.1 | 142,800 | 18.8 | 10.1 | 845 | 28.6 | 1.3 | 61,138 | -1.9 | 4,412 | 7.2 | 58,218 | 32.3 | 28.8 |
| Clay | 10,543 | 77.6 | 101,400 | 17.6 | 10.0 | 670 | 24.1 | 2.7 | 11,450 | -2.8 | 761 | 6.6 | 12,009 | 26.8 | 37.8 |
| Clinton | 12,033 | 70.6 | 111,500 | 18.1 | 10.0 | 726 | 26.8 | 2.4 | 16,825 | -2.3 | 939 | 5.6 | 15,183 | 24.1 | 43.4 |
| Crawford | 3,835 | 82.6 | 91,900 | 19.1 | 12.6 | 673 | 27.2 | 3.6 | 4,849 | -0.5 | 343 | 7.1 | 4,247 | 20.9 | 41.3 |
| Daviess | 11,227 | 73.5 | 130,000 | 17.4 | 10.0 | 692 | 26.9 | 3.7 | 16,523 | -1.4 | 632 | 3.8 | 14,545 | 26.1 | 39.8 |
| Dearborn | 18,870 | 79.6 | 170,400 | 19.2 | 10.0 | 744 | 29.3 | 1.6 | 25,224 | -2.5 | 1,742 | 6.9 | 24,757 | 32.2 | 30.3 |
| Decatur | 10,353 | 69.8 | 134,400 | 17.1 | 10.0 | 791 | 22.9 | 1.0 | 14,928 | -3.5 | 1,115 | 7.5 | 13,247 | 30.4 | 39.1 |
| DeKalb | 16,801 | 76.1 | 124,300 | 18.4 | 10.0 | 701 | 23.9 | 1.5 | 22,116 | -2.5 | 1,525 | 6.9 | 21,541 | 27.8 | 42.0 |
| Delaware | 46,026 | 63.5 | 93,600 | 17.4 | 11.3 | 727 | 32.4 | 1.3 | 52,466 | -2.0 | 3,833 | 7.3 | 52,763 | 33.3 | 23.4 |
| Dubois | 16,893 | 77.9 | 156,500 | 18.5 | 10.0 | 660 | 26.6 | 1.5 | 22,595 | -2.5 | 1,026 | 4.5 | 22,529 | 31.1 | 34.7 |
| Elkhart | 71,718 | 69.8 | 144,300 | 17.9 | 10.0 | 810 | 26.8 | 2.6 | 110,494 | -1.9 | 8,339 | 7.5 | 97,025 | 27.5 | 38.2 |
| Fayette | 9,584 | 69.5 | 81,900 | 17.9 | 11.2 | 705 | 29.1 | 2.3 | 8,543 | -2.6 | 843 | 9.9 | 10,070 | 24.4 | 38.4 |
| Floyd | 29,189 | 73.0 | 171,400 | 18.1 | 10.0 | 796 | 27.8 | 0.9 | 41,093 | -2.5 | 2,671 | 6.5 | 38,812 | 37.5 | 24.7 |
| Fountain | 6,974 | 76.2 | 103,000 | 18.0 | 10.0 | 674 | 25.2 | 2.0 | 7,752 | -3.4 | 505 | 6.5 | 7,601 | 27.6 | 42.4 |
| Franklin | 8,687 | 84.0 | 160,200 | 19.4 | 10.0 | 676 | 24.9 | 2.4 | 11,270 | -1.8 | 662 | 5.9 | 11,301 | 29.6 | 34.4 |
| Fulton | 7,799 | 74.7 | 103,600 | 18.2 | 10.0 | 690 | 28.1 | 1.1 | 9,680 | -2.1 | 640 | 6.6 | 9,330 | 25.0 | 39.2 |
| Gibson | 13,340 | 75.5 | 116,100 | 17.0 | 10.1 | 659 | 22.8 | 2.6 | 19,537 | -0.1 | 1,205 | 6.2 | 16,591 | 27.5 | 40.9 |
| Grant | 26,372 | 68.4 | 92,700 | 18.8 | 10.3 | 700 | 28.5 | 0.9 | 31,534 | -0.9 | 1,996 | 6.3 | 29,082 | 28.9 | 30.8 |
| Greene | 12,781 | 79.6 | 97,600 | 18.0 | 11.2 | 636 | 27.6 | 4.4 | 13,264 | -2.8 | 822 | 6.2 | 14,333 | 30.0 | 32.0 |
| Hamilton | 119,789 | 76.9 | 263,300 | 17.7 | 10.0 | 1,176 | 24.9 | 1.0 | 182,979 | -2.5 | 8,524 | 4.7 | 172,742 | 54.9 | 11.0 |
| Hancock | 28,740 | 79.2 | 169,300 | 18.8 | 10.0 | 895 | 26.1 | 1.1 | 40,363 | -1.8 | 2,279 | 5.6 | 39,099 | 39.4 | 23.4 |
| Harrison | 14,403 | 83.0 | 151,900 | 19.3 | 10.0 | 732 | 29.1 | 2.3 | 19,874 | -2.8 | 1,248 | 6.3 | 17,546 | 31.6 | 30.7 |
| Hendricks | 59,015 | 78.5 | 191,700 | 17.9 | 10.0 | 1,076 | 26.0 | 1.2 | 89,386 | -2.3 | 4,501 | 5.0 | 86,609 | 41.7 | 20.8 |
| Henry | 18,304 | 73.5 | 100,400 | 17.9 | 10.4 | 721 | 28.2 | 1.5 | 21,866 | -1.4 | 1,442 | 6.6 | 19,851 | 29.3 | 31.1 |
| Howard | 34,701 | 70.2 | 109,500 | 17.0 | 10.0 | 700 | 27.2 | 1.3 | 36,931 | 1.3 | 4,010 | 10.9 | 37,078 | 30.5 | 33.8 |
| Huntington | 14,742 | 75.5 | 107,200 | 18.2 | 10.1 | 703 | 27.7 | 1.4 | 18,032 | -1.4 | 1,186 | 6.6 | 18,586 | 28.7 | 34.6 |
| Jackson | 16,753 | 72.1 | 125,300 | 18.6 | 10.9 | 767 | 27.5 | 1.8 | 23,066 | 0.5 | 1,650 | 7.2 | 20,694 | 28.3 | 35.9 |
| Jasper | 12,533 | 76.6 | 159,700 | 17.4 | 10.0 | 796 | 27.4 | 1.4 | 15,583 | -4.0 | 1,093 | 7.0 | 15,558 | 25.0 | 39.0 |
| Jay | 8,174 | 73.0 | 91,400 | 18.0 | 10.1 | 677 | 23.5 | 3.3 | 8,801 | -6.6 | 577 | 6.6 | 9,616 | 24.5 | 43.9 |
| Jefferson | 12,632 | 71.9 | 132,900 | 18.3 | 11.8 | 742 | 27.3 | 3.1 | 14,965 | -0.9 | 1,107 | 7.4 | 14,357 | 30.2 | 36.4 |
| Jennings | 10,792 | 77.7 | 109,800 | 19.7 | 10.0 | 714 | 22.7 | 1.9 | 13,582 | -0.2 | 1,033 | 7.6 | 13,854 | 23.4 | 40.0 |
| Johnson | 56,628 | 72.7 | 164,600 | 17.1 | 10.0 | 955 | 26.4 | 1.3 | 82,335 | -1.7 | 4,623 | 5.6 | 78,779 | 40.7 | 21.5 |
| Knox | 15,038 | 64.6 | 96,800 | 17.0 | 10.0 | 659 | 26.4 | 1.4 | 18,228 | -1.6 | 1,080 | 5.9 | 18,220 | 27.6 | 32.5 |
| Kosciusko | 30,997 | 74.8 | 150,000 | 18.0 | 10.0 | 788 | 24.5 | 2.1 | 41,420 | -2.6 | 2,541 | 6.1 | 39,704 | 29.8 | 39.0 |
| LaGrange | 12,325 | 82.1 | 195,900 | 18.0 | 10.0 | 709 | 22.6 | 4.2 | 19,178 | -4.3 | 1,233 | 6.4 | 16,834 | 21.1 | 51.3 |
| Lake | 186,731 | 69.7 | 149,500 | 19.3 | 11.1 | 886 | 29.4 | 2.3 | 225,185 | -1.4 | 23,196 | 10.3 | 219,897 | 30.4 | 28.4 |
| LaPorte | 43,039 | 72.3 | 132,600 | 18.6 | 10.0 | 747 | 29.0 | 2.7 | 47,444 | 0.0 | 4,520 | 9.5 | 47,646 | 26.8 | 31.0 |
| Lawrence | 18,781 | 77.0 | 117,400 | 19.0 | 10.0 | 686 | 26.6 | 2.1 | 20,687 | -2.1 | 1,435 | 6.9 | 20,914 | 27.3 | 34.4 |
| Madison | 51,003 | 69.6 | 97,300 | 18.0 | 10.4 | 774 | 29.9 | 1.5 | 58,758 | -0.4 | 4,621 | 7.9 | 57,537 | 29.9 | 27.2 |
| Marion | 372,358 | 53.9 | 136,700 | 19.0 | 10.6 | 889 | 31.1 | 2.2 | 495,063 | 0.2 | 38,918 | 7.9 | 464,897 | 35.7 | 24.6 |
| Marshall | 17,304 | 74.7 | 138,700 | 18.8 | 10.2 | 754 | 29.9 | 3.4 | 22,646 | -2.8 | 1,456 | 6.4 | 21,088 | 24.9 | 41.5 |
| Martin | 4,187 | 78.2 | 117,100 | 17.6 | 10.0 | 540 | 22.6 | 2.2 | 5,410 | 0.5 | 229 | 4.2 | 4,705 | 29.8 | 41.5 |
| Miami | 13,611 | 71.5 | 88,700 | 17.8 | 10.0 | 714 | 27.3 | 1.6 | 15,009 | -2.0 | 1,237 | 8.2 | 15,293 | 21.6 | 42.2 |
| Monroe | 55,624 | 55.9 | 175,600 | 18.7 | 10.0 | 920 | 35.9 | 1.7 | 68,155 | -3.1 | 3,765 | 5.5 | 73,499 | 44.9 | 15.6 |
| Montgomery | 15,382 | 71.5 | 125,700 | 17.8 | 10.3 | 684 | 25.0 | 1.6 | 18,538 | -1.7 | 958 | 5.2 | 18,944 | 25.8 | 39.2 |
| Morgan | 26,034 | 77.1 | 160,700 | 18.5 | 10.0 | 837 | 24.5 | 1.8 | 35,605 | -2.3 | 1,912 | 5.4 | 33,408 | 29.3 | 33.5 |
| Newton | 5,573 | 77.9 | 118,400 | 19.9 | 10.0 | 707 | 21.0 | 1.8 | 6,591 | -4.4 | 452 | 6.9 | 6,139 | 20.7 | 40.7 |
| Noble | 18,276 | 76.6 | 125,800 | 18.4 | 10.0 | 712 | 24.8 | 2.3 | 22,709 | -4.3 | 1,905 | 8.4 | 22,957 | 23.8 | 46.5 |
| Ohio | 2,555 | 76.2 | 151,200 | 21.4 | 10.8 | 669 | 21.4 | 1.8 | 3,157 | -1.9 | 233 | 7.4 | 2,932 | 26.2 | 33.6 |
| Orange | 7,898 | 74.4 | 94,700 | 18.6 | 10.5 | 623 | 25.9 | 1.9 | 8,499 | -2.0 | 941 | 11.1 | 8,706 | 24.2 | 37.6 |
| Owen | 8,776 | 79.2 | 117,600 | 20.4 | 11.6 | 696 | 28.5 | 3.8 | 9,015 | -3.1 | 568 | 6.3 | 9,683 | 24.5 | 40.3 |
| Parke | 6,057 | 78.8 | 88,900 | 18.3 | 10.0 | 619 | 23.1 | 4.1 | 6,844 | -2.7 | 376 | 5.5 | 6,623 | 22.9 | 39.0 |
| Perry | 7,615 | 77.7 | 111,500 | 16.9 | 10.0 | 571 | 25.3 | 1.8 | 9,089 | -1.7 | 594 | 6.5 | 8,372 | 27.8 | 39.7 |

1. Specified owner-occupied units, lacking complete plumbing facilities.   2. A value of 10.0 represents 10 percent or less; a value of 50.0 represents 50 percent or more.   3. Specified renter-occupied units.   4. Overcrowded or
5. Percent of civilian labor force.   6. Civilian employed persons 16 years old and over.

# Table B. States and Counties — Nonfarm Employment and Agriculture

| STATE County | Private nonfarm establishments, employment and payroll, 2019 | | | | | | | | | Agriculture, 2017 | | | |
|---|---|---|---|---|---|---|---|---|---|---|---|---|---|
| | Number of establishments | Employment | | | | | | Annual payroll | | Farms | | | Farm producers whose primary occupation is farming (percent) |
| | | Total | Health care and social assistance | Manufacturing | Retail trade | Finance and insurance | Professional, scientific, and technical services | Total (mil dol) | Average per employee (dollars) | Number | Percent with: | | |
| | | | | | | | | | | | Fewer than 50 acres | 1000 acres or more | |
| | 104 | 105 | 106 | 107 | 108 | 109 | 110 | 111 | 112 | 113 | 114 | 115 | 116 |
| ILLINOIS—Cont'd | | | | | | | | | | | | | |
| Winnebago | 6,290 | 118,139 | 20,393 | 25,464 | 14,072 | 3,313 | 4,184 | 5,423 | 45,905 | 736 | 50.5 | 6.5 | 39.6 |
| Woodford | 754 | 8,371 | 1,227 | 1,946 | 1,104 | 247 | 198 | 339 | 40,489 | 920 | 35.8 | 8.2 | 41.4 |
| INDIANA | 148,917 | 2,834,056 | 443,718 | 522,442 | 328,113 | 110,133 | 132,893 | 132,502 | 46,753 | 56,649 | 46.4 | 7.1 | 38.1 |
| Adams | 726 | 12,132 | 1,592 | 5,129 | 1,495 | 284 | 223 | 472 | 38,890 | 1,450 | 62.8 | 2.8 | 34.2 |
| Allen | 9,318 | 185,463 | 37,359 | 28,348 | 23,040 | 9,036 | 6,213 | 8,496 | 45,809 | 1,548 | 58.1 | 4.5 | 33.0 |
| Bartholomew | 1,890 | 45,897 | 5,203 | 12,730 | 4,942 | 941 | 2,822 | 2,464 | 53,689 | 564 | 47.0 | 8.7 | 36.6 |
| Benton | 170 | 1,276 | 113 | 359 | 131 | 97 | 29 | 49 | 38,440 | 358 | 30.2 | 23.2 | 52.2 |
| Blackford | 240 | 2,826 | 526 | 946 | 270 | 61 | 87 | 110 | 38,936 | 234 | 30.8 | 9.0 | 44.2 |
| Boone | 1,601 | 26,598 | 2,701 | 2,795 | 5,722 | 461 | 806 | 1,119 | 42,064 | 626 | 57.5 | 11.3 | 40.3 |
| Brown | 340 | 1,999 | 239 | 117 | 339 | 30 | 88 | 56 | 27,870 | 182 | 46.7 | NA | 31.2 |
| Carroll | 400 | 5,042 | 314 | 2,484 | 367 | 93 | 148 | 188 | 37,217 | 573 | 43.8 | 14.0 | 43.1 |
| Cass | 664 | 12,789 | 2,528 | 4,226 | 1,440 | 278 | 187 | 448 | 35,056 | 642 | 40.5 | 8.1 | 38.0 |
| Clark | 2,531 | 49,604 | 6,572 | 7,154 | 9,847 | 2,550 | 1,130 | 2,120 | 42,735 | 483 | 44.9 | 4.6 | 32.1 |
| Clay | 470 | 5,990 | 618 | 2,238 | 969 | 126 | 60 | 202 | 33,650 | 585 | 49.6 | 8.9 | 37.6 |
| Clinton | 611 | 10,658 | 1,291 | 4,745 | 984 | 209 | 120 | 390 | 36,559 | 560 | 41.6 | 15.4 | 48.7 |
| Crawford | 131 | 1,349 | 126 | 365 | 144 | 26 | 20 | 51 | 38,164 | 391 | 33.8 | 0.8 | 26.8 |
| Daviess | 920 | 11,255 | 1,444 | 2,533 | 1,439 | 317 | 411 | 417 | 37,040 | 1,230 | 62.4 | 4.7 | 28.3 |
| Dearborn | 936 | 13,136 | 2,613 | 2,149 | 1,911 | 283 | 447 | 473 | 36,020 | 598 | 38.5 | 0.8 | 30.6 |
| Decatur | 647 | 11,731 | 1,465 | 4,569 | 1,186 | 383 | 126 | 501 | 42,741 | 581 | 31.3 | 8.6 | 48.0 |
| DeKalb | 966 | 20,984 | 1,249 | 10,320 | 1,435 | 291 | 499 | 1,029 | 49,049 | 771 | 48.6 | 6.9 | 33.3 |
| Delaware | 2,398 | 41,543 | 10,693 | 4,033 | 5,738 | 2,482 | 1,297 | 1,599 | 38,501 | 546 | 52.9 | 8.1 | 45.0 |
| Dubois | 1,266 | 28,901 | 3,883 | 10,993 | 3,221 | 560 | 378 | 1,306 | 45,204 | 757 | 37.1 | 4.8 | 37.5 |
| Elkhart | 4,951 | 135,841 | 11,540 | 75,097 | 10,186 | 2,003 | 2,078 | 6,267 | 46,132 | 1,667 | 64.1 | 1.9 | 34.8 |
| Fayette | 427 | 5,077 | 1,456 | 1,069 | 769 | 155 | 98 | 183 | 35,968 | 343 | 39.9 | 7.3 | 39.3 |
| Floyd | 1,872 | 29,578 | 7,425 | 5,992 | 3,038 | 755 | 1,331 | 1,187 | 40,115 | 229 | 61.6 | 1.7 | 32.6 |
| Fountain | 309 | 3,732 | 309 | 1,744 | 533 | 142 | 76 | 142 | 37,926 | 497 | 35.4 | 12.5 | 42.5 |
| Franklin | 427 | 5,849 | 736 | 886 | 749 | 175 | 57 | 253 | 43,234 | 704 | 34.5 | 2.6 | 36.2 |
| Fulton | 450 | 5,119 | 652 | 1,652 | 1,020 | 167 | 91 | 187 | 36,601 | 635 | 42.8 | 11.5 | 44.4 |
| Gibson | 687 | 18,028 | 1,372 | 8,892 | 1,507 | 139 | 348 | 943 | 52,280 | 513 | 38.2 | 12.7 | 40.3 |
| Grant | 1,269 | 28,260 | 5,355 | 4,975 | 2,783 | 677 | 317 | 1,022 | 36,156 | 494 | 40.5 | 13.0 | 47.5 |
| Greene | 538 | 4,680 | 970 | 555 | 1,029 | 147 | 295 | 153 | 32,663 | 828 | 43.4 | 4.5 | 37.4 |
| Hamilton | 9,354 | 152,333 | 20,851 | 5,226 | 16,473 | 21,025 | 18,236 | 8,379 | 55,005 | 585 | 62.6 | 4.8 | 33.8 |
| Hancock | 1,455 | 22,339 | 2,906 | 3,932 | 2,801 | 404 | 2,492 | 1,067 | 47,756 | 551 | 56.4 | 10.2 | 34.7 |
| Harrison | 685 | 9,086 | 1,672 | 1,721 | 1,266 | 249 | 155 | 314 | 34,529 | 1,054 | 51.0 | 2.7 | 30.5 |
| Hendricks | 3,387 | 64,409 | 8,415 | 3,364 | 12,863 | 1,265 | 1,428 | 2,685 | 41,681 | 658 | 62.9 | 6.4 | 41.6 |
| Henry | 846 | 11,369 | 2,474 | 2,596 | 1,631 | 283 | 149 | 400 | 35,184 | 636 | 50.2 | 6.3 | 40.3 |
| Howard | 1,812 | 31,965 | 5,236 | 9,388 | 4,840 | 807 | 796 | 1,423 | 44,522 | 422 | 36.3 | 10.0 | 49.1 |
| Huntington | 852 | 13,178 | 1,855 | 4,175 | 1,336 | 317 | 346 | 460 | 34,873 | 611 | 41.6 | 10.8 | 33.7 |
| Jackson | 1,000 | 21,718 | 2,689 | 8,998 | 1,880 | 529 | 323 | 1,054 | 48,531 | 665 | 38.3 | 8.4 | 40.1 |
| Jasper | 745 | 8,657 | 1,017 | 1,528 | 1,548 | 294 | 211 | 348 | 40,170 | 611 | 39.1 | 13.7 | 47.5 |
| Jay | 417 | 5,766 | 893 | 2,611 | 575 | 143 | 135 | 233 | 40,331 | 770 | 45.3 | 8.3 | 37.4 |
| Jefferson | 645 | 11,531 | 1,981 | 3,321 | 1,680 | 175 | 220 | 457 | 39,622 | 684 | 39.8 | 2.2 | 30.5 |
| Jennings | 417 | 6,240 | 825 | 1,848 | 613 | 99 | 95 | 271 | 43,503 | 510 | 46.3 | 5.3 | 34.4 |
| Johnson | 3,280 | 50,203 | 7,689 | 6,436 | 10,295 | 1,376 | 1,272 | 1,903 | 37,903 | 642 | 62.9 | 5.9 | 37.8 |
| Knox | 843 | 14,047 | 3,408 | 2,034 | 1,772 | 294 | 272 | 567 | 40,390 | 496 | 30.0 | 19.8 | 53.4 |
| Kosciusko | 1,970 | 37,988 | 4,153 | 13,269 | 3,796 | 805 | 613 | 1,963 | 51,667 | 1,042 | 55.6 | 6.9 | 32.2 |
| LaGrange | 886 | 14,342 | 917 | 8,024 | 1,409 | 235 | 204 | 624 | 43,487 | 2,144 | 62.0 | 1.7 | 34.0 |
| Lake | 10,094 | 168,185 | 35,277 | 21,771 | 23,779 | 4,007 | 6,121 | 8,177 | 48,619 | 384 | 52.3 | 6.3 | 42.0 |
| LaPorte | 2,247 | 35,649 | 5,478 | 7,904 | 5,624 | 779 | 773 | 1,430 | 40,118 | 740 | 42.4 | 9.5 | 45.4 |
| Lawrence | 918 | 12,104 | 2,707 | 2,737 | 1,869 | 326 | 660 | 503 | 41,583 | 840 | 42.4 | 3.0 | 31.1 |
| Madison | 2,245 | 35,252 | 6,613 | 4,775 | 4,569 | 882 | 680 | 1,370 | 38,864 | 667 | 51.3 | 9.7 | 41.7 |
| Marion | 23,540 | 533,265 | 90,670 | 41,988 | 47,358 | 24,813 | 39,802 | 29,674 | 55,647 | 192 | 79.7 | 3.1 | 26.2 |
| Marshall | 1,085 | 18,014 | 2,374 | 6,588 | 2,004 | 492 | 314 | 711 | 39,473 | 829 | 47.8 | 6.4 | 39.3 |
| Martin | 204 | 2,303 | 136 | 396 | 302 | 39 | 786 | 116 | 50,241 | 260 | 39.2 | 7.3 | 29.2 |
| Miami | 567 | 6,869 | 908 | 1,894 | 960 | 347 | 110 | 232 | 33,829 | 629 | 37.8 | 7.9 | 43.5 |
| Monroe | 3,151 | 52,369 | 10,834 | 7,962 | 6,609 | 1,490 | 1,952 | 2,189 | 41,804 | 490 | 50.0 | 0.6 | 30.0 |
| Montgomery | 851 | 13,625 | 1,371 | 5,223 | 1,646 | 239 | 174 | 612 | 44,914 | 634 | 38.0 | 14.5 | 44.3 |
| Morgan | 1,202 | 13,014 | 1,814 | 2,374 | 2,367 | 390 | 412 | 481 | 36,959 | 501 | 49.1 | 7.0 | 38.1 |
| Newton | 242 | 2,284 | 219 | 484 | 286 | 106 | 54 | 84 | 36,657 | 358 | 28.2 | 17.3 | 50.7 |
| Noble | 888 | 17,958 | 1,299 | 10,284 | 1,736 | 272 | 235 | 713 | 39,723 | 1,015 | 52.4 | 5.6 | 33.6 |
| Ohio | 69 | 883 | 87 | 17 | 23 | 12 | 15 | 25 | 28,211 | 158 | 31.6 | 1.3 | 27.7 |
| Orange | 380 | 6,940 | 744 | 1,430 | 564 | 122 | 67 | 288 | 41,570 | 448 | 37.1 | 5.1 | 37.2 |
| Owen | 315 | 3,375 | 595 | 1,156 | 500 | 98 | 96 | 114 | 33,717 | 649 | 45.6 | 3.7 | 31.9 |
| Parke | 272 | 2,256 | 142 | 323 | 425 | 52 | 146 | 73 | 32,451 | 597 | 37.0 | 8.7 | 42.5 |
| Perry | 353 | 5,622 | 932 | 2,190 | 652 | 118 | 236 | 230 | 40,984 | 445 | 29.4 | 2.5 | 28.4 |

# Table B. States and Counties — Agriculture

| STATE County | Land in farms — Acreage (1,000) | Percent change, 2012-2017 | Acres — Average size of farm | Acres — Total irrigated (1,000) | Total cropland (1,000) | Value of land and buildings (dollars) — Average per farm | Average per acre | Value of machinery and equipmnet, average per farm (dollars) | Value of products sold: Total (mil dol) | Average per farm (acres) | Percent from: Crops | Livestock and poultry products | Organic farms (number) | Farms with internet access (percent) | Government payments — Total ($1,000) | Percent of farms |
|---|---|---|---|---|---|---|---|---|---|---|---|---|---|---|---|---|
| | 117 | 118 | 119 | 120 | 121 | 122 | 123 | 124 | 125 | 126 | 127 | 128 | 129 | 130 | 131 | 132 |
| **ILLINOIS—Cont'd** | | | | | | | | | | | | | | | | |
| Winnebago | 179 | -2.3 | 243 | 1.6 | 161.0 | 1,732,196 | 7,137 | 145,947 | 107.2 | 145,644 | 78.1 | 21.9 | 9 | 81.7 | 6,961 | 54.9 |
| Woodford | 283 | -12.3 | 308 | D | 259.0 | 2,761,299 | 8,972 | 205,867 | 216.1 | 234,897 | 75.2 | 24.8 | 23 | 77.6 | 3,479 | 73.6 |
| **INDIANA** | 14,970 | 1.7 | 264 | 555.4 | 12,909.7 | 1,737,741 | 6,576 | 163,136 | 11,107.3 | 196,073 | 64.1 | 35.9 | 657 | 71.9 | 342,914 | 47.9 |
| Adams | 213 | 1.3 | 147 | 0.5 | 194.4 | 1,215,197 | 8,274 | 101,610 | 283.1 | 195,266 | 34.9 | 65.1 | 15 | 50.8 | 3,926 | 35.4 |
| Allen | 282 | 4.0 | 182 | 0.4 | 249.5 | 1,446,442 | 7,950 | 113,983 | 175.8 | 113,581 | 71.1 | 28.9 | 19 | 65.6 | 5,389 | 47.2 |
| Bartholomew | 160 | -6.5 | 284 | 15.5 | 142.2 | 1,902,512 | 6,688 | 159,738 | 93 | 164,929 | 88.3 | 11.7 | NA | 79.3 | 3,919 | 55.7 |
| Benton | 251 | -1.3 | 701 | 6.6 | 246.4 | 5,422,824 | 7,734 | 405,595 | 178.9 | 499,765 | 86.9 | 13.1 | NA | 80.7 | 4,100 | 81.6 |
| Blackford | 92 | 4.7 | 394 | NA | 86.3 | 2,601,618 | 6,605 | 219,419 | 63.5 | 271,265 | 74.8 | 25.2 | NA | 76.5 | 1,567 | 67.9 |
| Boone | 230 | 3.8 | 367 | 0.1 | 219.9 | 2,800,238 | 7,621 | 206,909 | 134.6 | 215,045 | 92.9 | 7.1 | 3 | 89.8 | 6,471 | 42.0 |
| Brown | 15 | 1.4 | 81 | 0.1 | 5.7 | 356,748 | 4,390 | 44,535 | 1.9 | 10,527 | 85.4 | 14.6 | 1 | 81.9 | 162 | 18.7 |
| Carroll | 224 | 9.9 | 391 | 1.6 | 213.4 | 2,949,885 | 7,539 | 249,168 | 250 | 436,283 | 55.0 | 45.0 | 8 | 71.6 | 4,972 | 59.9 |
| Cass | 199 | -0.4 | 311 | 0.8 | 182.7 | 1,939,195 | 6,243 | 212,168 | 149.5 | 232,807 | 66.4 | 33.6 | 2 | 77.4 | 3,933 | 66.5 |
| Clark | 94 | 19.6 | 195 | 0.7 | 68.6 | 1,072,932 | 5,515 | 113,762 | 41.7 | 86,244 | 84.3 | 15.7 | NA | 71.4 | 1,746 | 37.7 |
| Clay | 162 | -0.5 | 277 | 0.6 | 139.1 | 1,412,114 | 5,100 | 142,144 | 76.2 | 130,309 | 96.4 | 3.6 | 1 | 72.6 | 3,616 | 63.8 |
| Clinton | 247 | 10.6 | 441 | D | 238.8 | 3,162,178 | 7,169 | 282,883 | 184.7 | 329,836 | 78.4 | 21.6 | 4 | 86.3 | 7,274 | 65.5 |
| Crawford | 53 | 13.4 | 135 | D | 18.7 | 416,192 | 3,094 | 51,196 | 9.3 | 23,754 | 41.1 | 58.9 | NA | 73.4 | 475 | 23.0 |
| Daviess | 225 | 0.0 | 183 | 4.8 | 193.2 | 1,395,457 | 7,620 | 144,028 | 270.9 | 220,211 | 41.6 | 58.4 | 1 | 52.3 | 6,725 | 23.3 |
| Dearborn | 65 | 14.2 | 108 | 0.1 | 29.8 | 525,303 | 4,861 | 56,517 | 12.2 | 20,390 | 69.6 | 30.4 | NA | 80.4 | 278 | 20.2 |
| Decatur | 202 | 8.3 | 348 | 0.0 | 179.2 | 2,461,834 | 7,083 | 231,518 | 178.9 | 307,904 | 57.8 | 42.2 | NA | 76.4 | 5,753 | 67.6 |
| DeKalb | 159 | -1.2 | 206 | 3.6 | 140.1 | 1,143,430 | 5,547 | 121,735 | 93.8 | 121,624 | 65.3 | 34.7 | NA | 74.8 | 5,313 | 60.3 |
| Delaware | 168 | -4.3 | 307 | 0.6 | 157.6 | 2,038,398 | 6,633 | 188,657 | 92.2 | 168,773 | 92.9 | 7.1 | NA | 83.5 | 4,044 | 56.4 |
| Dubois | 179 | 2.4 | 236 | 0.7 | 133.4 | 1,289,755 | 5,455 | 169,974 | 248.8 | 328,690 | 28.1 | 71.9 | NA | 71.9 | 6,327 | 60.1 |
| Elkhart | 175 | 1.2 | 105 | 25.0 | 146.5 | 1,173,021 | 11,178 | 98,422 | 298.3 | 178,936 | 23.4 | 76.6 | 86 | 45.5 | 3,082 | 16.3 |
| Fayette | 86 | 9.8 | 251 | 0.2 | 72.2 | 1,416,832 | 5,656 | 164,773 | 45.8 | 133,458 | 85.0 | 15.0 | NA | 73.8 | 2,463 | 51.9 |
| Floyd | 25 | 16.5 | 109 | 0.1 | 13.8 | 595,936 | 5,456 | 71,967 | 7.1 | 31,162 | 90.8 | 9.2 | NA | 72.5 | 335 | 18.3 |
| Fountain | 212 | -1.1 | 427 | D | 187.4 | 2,871,619 | 6,732 | 224,601 | 112.7 | 226,809 | 94.0 | 6.0 | 3 | 77.7 | 4,181 | 66.6 |
| Franklin | 133 | 6.5 | 189 | D | 92.2 | 1,071,362 | 5,670 | 102,977 | 67.1 | 95,261 | 76.8 | 23.2 | NA | 69.2 | 3,190 | 50.4 |
| Fulton | 214 | 13.8 | 338 | 25.1 | 197.5 | 2,131,615 | 6,312 | 233,337 | 140.2 | 220,855 | 78.1 | 21.9 | 2 | 75.0 | 5,859 | 58.9 |
| Gibson | 220 | -17.8 | 430 | 7.8 | 203.1 | 2,785,651 | 6,484 | 297,121 | 149 | 290,458 | 85.6 | 14.4 | NA | 76.4 | 6,226 | 64.9 |
| Grant | 190 | 3.7 | 385 | D | 181.4 | 2,716,186 | 7,059 | 216,027 | 118.6 | 240,095 | 83.3 | 16.7 | 3 | 82.4 | 1,747 | 60.1 |
| Greene | 169 | -6.7 | 204 | 2.2 | 124.2 | 1,010,049 | 4,952 | 110,938 | 108.4 | 130,965 | 56.6 | 43.4 | 10 | 78.9 | 4,026 | 30.9 |
| Hamilton | 127 | -2.7 | 218 | D | 117.4 | 1,903,729 | 8,750 | 168,283 | 104.6 | 178,853 | 97.5 | 2.5 | 2 | 83.9 | 1,209 | 39.5 |
| Hancock | 170 | 2.3 | 308 | 0.0 | 161.2 | 2,322,612 | 7,543 | 185,610 | 115.7 | 210,031 | 72.1 | 27.9 | 3 | 89.3 | 4,498 | 45.2 |
| Harrison | 149 | 10.1 | 141 | 0.3 | 100.7 | 644,972 | 4,575 | 84,004 | 88.7 | 84,114 | 51.9 | 48.1 | NA | 71.5 | 2,627 | 26.7 |
| Hendricks | 153 | -30 | 232 | 0.1 | 141.5 | 1,774,141 | 7,638 | 128,330 | 84.4 | 128,293 | 96.2 | 3.8 | 6 | 82.1 | 3,541 | 36.9 |
| Henry | 168 | -4.5 | 265 | 0.1 | 151.9 | 1,586,242 | 5,987 | 174,623 | 122.9 | 193,259 | 65.0 | 35.0 | NA | 80.3 | 5,297 | 51.3 |
| Howard | 146 | 1.1 | 345 | D | 138.0 | 2,681,236 | 7,765 | 219,981 | 97.1 | 230,078 | 83.1 | 16.9 | 3 | 79.9 | 4,286 | 68.2 |
| Huntington | 197 | 4.4 | 323 | D | 181.4 | 2,212,369 | 6,854 | 221,979 | 160.2 | 262,116 | 56.5 | 43.5 | NA | 81.7 | 2,019 | 62.7 |
| Jackson | 201 | 9.5 | 303 | 11.1 | 155.9 | 1,602,696 | 5,295 | 202,939 | 186.5 | 280,451 | 40.6 | 59.4 | NA | 73.1 | 4,533 | 56.2 |
| Jasper | 270 | -4.5 | 442 | 25.9 | 251.6 | 2,854,205 | 6,456 | 251,325 | 298.7 | 488,939 | 47.7 | 52.3 | 1 | 78.4 | 4,951 | 67.6 |
| Jay | 208 | 18.1 | 270 | D | 190.3 | 2,044,970 | 7,585 | 180,334 | 372.6 | 483,891 | 23.7 | 76.3 | 2 | 75.2 | 3,566 | 60.8 |
| Jefferson | 107 | 11.7 | 156 | 0.0 | 66.9 | 645,429 | 4,141 | 87,775 | 31.4 | 45,975 | 84.6 | 15.4 | 1 | 74.4 | 1,417 | 33.0 |
| Jennings | 128 | 3.8 | 251 | 0.5 | 91.1 | 1,146,513 | 4,564 | 142,906 | 63.3 | 124,141 | 70.2 | 29.8 | NA | 73.1 | 2,427 | 39.8 |
| Johnson | 141 | -2.4 | 220 | 2.4 | 122.0 | 1,482,614 | 6,745 | 148,278 | 75.3 | 117,271 | 88.6 | 11.4 | 2 | 82.7 | 4,292 | 37.9 |
| Knox | 311 | -5.5 | 627 | 32.1 | 289.9 | 3,905,102 | 6,224 | 365,221 | 245.3 | 494,635 | 79.7 | 20.3 | 3 | 85.7 | 10,316 | 70.8 |
| Kosciusko | 262 | 2.7 | 251 | 30.1 | 229.7 | 1,788,066 | 7,120 | 177,429 | 298 | 286,019 | 42.0 | 58.0 | 7 | 68.2 | 8,572 | 43.7 |
| LaGrange | 195 | -4.3 | 91 | 38.9 | 144.1 | 842,807 | 9,249 | 75,820 | 275.6 | 128,536 | 29.9 | 70.1 | 271 | 19.8 | 2,970 | 9.9 |
| Lake | 112 | -15.5 | 293 | 4.7 | 106.0 | 2,040,013 | 6,966 | 155,636 | 65.4 | 170,253 | 93.0 | 7.0 | NA | 78.4 | 2,032 | 44.3 |
| LaPorte | 249 | 9.2 | 336 | 68.5 | 230.0 | 2,473,619 | 7,355 | 210,852 | 166.4 | 224,842 | 82.2 | 17.8 | 7 | 79.6 | 8,644 | 58.4 |
| Lawrence | 147 | 9.4 | 175 | 0.1 | 82.3 | 651,657 | 3,716 | 72,451 | 43.5 | 51,751 | 69.4 | 30.6 | 2 | 71.0 | 3,141 | 35.6 |
| Madison | 208 | 1.3 | 312 | 1.5 | 198.0 | 2,526,028 | 8,108 | 205,713 | 129.5 | 194,096 | 92.6 | 7.4 | 2 | 81.3 | 5,002 | 53.4 |
| Marion | 17 | -13.5 | 90 | 0.1 | 15.5 | 955,725 | 10,564 | 62,277 | 12.1 | 62,844 | 92.2 | 7.8 | 3 | 81.8 | 423 | 19.3 |
| Marshall | 199 | -3.5 | 240 | 16.5 | 177.3 | 1,503,166 | 6,259 | 151,905 | 145.2 | 175,111 | 65.2 | 34.8 | 21 | 63.3 | 5,212 | 45.7 |
| Martin | 62 | -0.5 | 239 | D | 35.3 | 930,286 | 3,886 | 144,218 | 60.8 | 233,777 | 29.6 | 70.4 | NA | 62.3 | 879 | 35.4 |
| Miami | 194 | 10.4 | 308 | 3.4 | 175.1 | 2,076,888 | 6,750 | 165,044 | 179.5 | 285,297 | 51.5 | 48.5 | 1 | 80.1 | 5,973 | 71.1 |
| Monroe | 48 | -9.5 | 97 | 0.1 | 22.2 | 640,303 | 6,569 | 62,244 | 9.4 | 19,196 | 73.7 | 26.3 | 16 | 83.3 | 493 | 19.8 |
| Montgomery | 282 | -1.7 | 445 | 2.1 | 260.0 | 3,108,274 | 6,984 | 270,937 | 162.7 | 256,550 | 94.4 | 5.6 | 3 | 81.7 | 6,451 | 66.6 |
| Morgan | 137 | -0.5 | 273 | D | 111.4 | 1,713,947 | 6,289 | 140,341 | 65.5 | 130,812 | 87.4 | 12.6 | NA | 79.0 | 2,760 | 41.9 |
| Newton | 181 | -5.8 | 506 | 4.7 | 161.3 | 3,403,798 | 6,733 | 335,472 | 196.2 | 547,925 | 48.6 | 51.4 | NA | 84.4 | 1,666 | 57.0 |
| Noble | 200 | 10.2 | 197 | 21.1 | 172.3 | 1,259,845 | 6,394 | 141,041 | 158.5 | 156,156 | 53.2 | 46.8 | 35 | 69.1 | 5,315 | 46.5 |
| Ohio | 24 | 11.9 | 152 | 0.0 | 11.1 | 663,843 | 4,368 | 71,293 | 4.6 | 29,051 | 76.2 | 23.8 | NA | 75.3 | 445 | 17.7 |
| Orange | 102 | 4.3 | 229 | 0.1 | 72.2 | 1,026,470 | 4,489 | 136,262 | 122.5 | 273,498 | 31.1 | 68.9 | NA | 77.2 | 4,352 | 41.3 |
| Owen | 112 | 17.2 | 172 | 0.2 | 73.6 | 783,546 | 4,543 | 73,005 | 34.1 | 52,582 | 84.7 | 15.3 | 3 | 71.0 | 1,857 | 33.4 |
| Parke | 181 | 2.4 | 303 | 3.9 | 139.9 | 1,768,164 | 5,840 | 173,611 | 92.7 | 155,295 | 84.4 | 15.6 | 23 | 62.1 | 3,317 | 47.6 |
| Perry | 77 | 16.0 | 173 | 0.5 | 42.2 | 693,181 | 4,013 | 91,796 | 40.1 | 90,036 | 39.3 | 60.7 | 1 | 71.2 | 1,039 | 34.2 |

| STATE County | Water use, 2015 | | Wholesale Trade[1], 2017 | | | | Retail Trade[2], 2017 | | | | Real estate and rental and leasing,[2] 2017 | | | |
|---|---|---|---|---|---|---|---|---|---|---|---|---|---|---|
| | Public supply water withdrawn (mil gal/day) | Public supply gallons withdrawn per person per day | Number of establishments | Number of employees | Sales (mil dol) | Average payroll (mil dol) | Number of establishments | Number of employees | Sales (mil dol) | Average payroll (mil dol) | Number of establishments | Number of employees | Sales (mil dol) | Average payroll (mil dol) |
| | 133 | 134 | 135 | 136 | 137 | 138 | 139 | 140 | 141 | 142 | 143 | 144 | 145 | 146 |
| ILLINOIS—Cont'd | | | | | | | | | | | | | | |
| Winnebago | 29.23 | 101.8 | 308 | 4,281 | 2,788.2 | 230.7 | 927 | 14,085 | 3,946.2 | 349.8 | 231 | 1,316 | 281.1 | 57.3 |
| Woodford | 7.26 | 185.1 | D | D | D | 36.4 | 88 | 1,014 | 414.5 | 31.3 | 15 | 32 | 2.5 | 0.9 |
| INDIANA | 627.84 | 94.8 | 6,271 | 96,880 | 87,596.7 | 5,505.8 | 21,327 | 336,615 | 102,106.0 | 8,660.1 | 6,686 | 34,736 | 8,392.2 | 1,519.7 |
| Adams | 2.31 | 66.0 | 32 | 285 | 176.6 | 10.8 | 141 | 1,511 | 413.2 | 35.4 | 21 | 81 | 8.1 | 2.9 |
| Allen | 35.70 | 96.9 | 473 | 7,968 | 8,489.9 | 424.7 | 1,270 | 22,962 | 6,972.1 | 632.0 | 445 | 2,215 | 541.5 | 89.4 |
| Bartholomew | 9.49 | 116.9 | D | D | D | D | 315 | 5,106 | 1,311.6 | 112.8 | 75 | 355 | 91.4 | 12.4 |
| Benton | 0.40 | 46.1 | 12 | 243 | 279.1 | 10.6 | 29 | 157 | 46.4 | 3.5 | 3 | 4 | 0.4 | 0.1 |
| Blackford | 1.04 | 84.6 | 9 | 90 | 156.3 | 3.9 | 33 | 310 | 80.3 | 6.1 | 12 | 22 | 3.7 | 0.5 |
| Boone | 2.01 | 31.7 | 69 | 1,149 | 1,920.7 | 71.8 | 167 | 5,585 | 2,803.6 | 179.0 | 82 | 305 | 57.6 | 12.4 |
| Brown | 0.00 | 0.0 | 5 | 18 | 10.0 | 0.8 | 75 | 331 | 54.5 | 6.6 | 14 | 66 | 5.0 | 1.7 |
| Carroll | 1.31 | 66.0 | 24 | 171 | 153.4 | 7.3 | 51 | 376 | 97.8 | 8.7 | 11 | 16 | 2.8 | 0.6 |
| Cass | 5.94 | 156.4 | 35 | 418 | 465.8 | 19.1 | 113 | 1,560 | 360.9 | 33.5 | 17 | 59 | 11.7 | 1.8 |
| Clark | 22.11 | 191.6 | 98 | 1,339 | 1,699.3 | 72.7 | 421 | 10,327 | 3,354.3 | 277.9 | 111 | 467 | 110.8 | 18.2 |
| Clay | 0.33 | 12.5 | 11 | 84 | 39.3 | 3.4 | 79 | 993 | 292.8 | 22.2 | 12 | 31 | 3.2 | 0.7 |
| Clinton | 3.68 | 112.9 | 28 | 179 | 202.4 | 8.3 | 97 | 1,007 | 270.4 | 23.9 | 4 | 5 | 1.4 | 0.2 |
| Crawford | 1.64 | 156.4 | NA | NA | NA | NA | 25 | 183 | 71.8 | 4.2 | 19 | 56 | 7.0 | 1.3 |
| Daviess | 3.55 | 107.9 | 31 | 341 | 162.5 | 14.9 | 124 | 1,498 | 586.4 | 44.5 | 35 | 139 | 20.6 | 3.4 |
| Dearborn | 4.41 | 89.2 | 34 | D | 281.1 | D | 140 | 2,079 | 682.0 | 52.5 | 16 | 46 | 8.5 | 1.5 |
| Decatur | 2.59 | 97.7 | D | D | D | D | 109 | 1,306 | 348.5 | 29.7 | 32 | 128 | 19.6 | 5.2 |
| DeKalb | 3.17 | 74.4 | D | D | D | D | 109 | 1,476 | 465.4 | 38.8 | 22 | 77 | — | — |
| Delaware | 9.58 | 82.0 | 89 | 829 | 520.8 | 35.7 | 430 | 5,983 | 1,662.1 | 144.3 | 104 | 430 | 91.5 | 16.7 |
| Dubois | 5.84 | 137.5 | 76 | 1,204 | 594.6 | 56.8 | 228 | 3,336 | 1,005.8 | 91.3 | 33 | 115 | 14.0 | 3.0 |
| Elkhart | 12.95 | 63.6 | 317 | 6,495 | 4,827.8 | 332.6 | 705 | 9,603 | 2,889.9 | 262.5 | 181 | 767 | 167.3 | 28.7 |
| Fayette | 2.42 | 103.3 | 10 | 143 | 104.3 | 6.6 | 72 | 1,041 | 230.4 | 21.6 | 20 | 53 | 8.9 | 1.2 |
| Floyd | 1.21 | 15.8 | 59 | 562 | 299.1 | 26.6 | 201 | 3,402 | 908.2 | 83.1 | 95 | 280 | 54.9 | 9.1 |
| Fountain | 0.88 | 53.0 | 13 | 81 | 244.3 | 4.0 | 62 | 572 | 141.7 | 11.0 | 6 | 22 | 1.4 | 0.4 |
| Franklin | 2.62 | 114.6 | D | D | D | D | 63 | 882 | 205.0 | 17.6 | 13 | 31 | 3.9 | 0.5 |
| Fulton | 1.07 | 52.7 | 18 | 162 | 146.7 | 7.7 | 83 | 1,037 | 289.5 | 24.3 | 12 | 20 | 3.3 | 0.6 |
| Gibson | 1.71 | 50.6 | 23 | 349 | 569.0 | 16.1 | 117 | 1,597 | 501.7 | 39.3 | 24 | 73 | 14.1 | 2.1 |
| Grant | 3.95 | 58.1 | 40 | 385 | 287.2 | 18.5 | 222 | 2,810 | 763.8 | 63.1 | 53 | 196 | 23.6 | 4.7 |
| Greene | 2.87 | 88.5 | 19 | 90 | 46.1 | 3.5 | 93 | 996 | 293.6 | 23.2 | 10 | 26 | 3.6 | 0.5 |
| Hamilton | 35.69 | 115.2 | 388 | 5,238 | 3,744.7 | 372.1 | 896 | 17,614 | 5,631.6 | 478.8 | 537 | 2,995 | 1,144.7 | 219.1 |
| Hancock | 3.18 | 43.8 | 43 | 1,030 | 1,863.7 | 51.2 | 161 | 2,267 | 800.0 | 63.0 | 55 | 156 | 42.4 | 7.1 |
| Harrison | 2.64 | 66.7 | D | D | D | 10.8 | 108 | 1,287 | 557.3 | 34.4 | 26 | 59 | 8.1 | 1.4 |
| Hendricks | 5.10 | 32.2 | 117 | 3,332 | 4,285.6 | 177.3 | 442 | 12,750 | 4,672.0 | 382.4 | 148 | 410 | 113.3 | 16.1 |
| Henry | 4.68 | 95.5 | 29 | D | 846.0 | D | 142 | 1,668 | 538.8 | 40.8 | 29 | 70 | 8.7 | 1.6 |
| Howard | 8.47 | 102.6 | D | D | D | D | 341 | 5,141 | 1,382.5 | 117.8 | 78 | 300 | 62.9 | 9.9 |
| Huntington | 3.73 | 101.8 | 41 | 374 | 292.8 | 12.2 | 135 | 1,335 | 324.0 | 30.9 | 36 | 96 | 14.1 | 2.3 |
| Jackson | 5.39 | 122.3 | 35 | 641 | 879.4 | 30.0 | 179 | 1,998 | 621.9 | 53.1 | 53 | 163 | 27.4 | 4.2 |
| Jasper | 0.91 | 27.2 | 34 | 225 | 391.8 | 12.5 | 126 | 1,584 | 507.6 | 33.9 | 22 | 103 | 26.7 | 2.5 |
| Jay | 1.52 | 72.0 | 16 | 168 | 233.7 | 7.7 | 57 | 756 | 145.8 | 14.6 | 13 | 21 | 2.9 | 0.4 |
| Jefferson | 5.63 | 173.7 | 17 | 253 | 334.2 | 11.3 | 119 | 1,633 | 437.6 | 45.1 | 31 | 98 | 10.9 | 2.7 |
| Jennings | 1.03 | 36.9 | 16 | 234 | 99.1 | 9.6 | 57 | 716 | 194.3 | 18.7 | 9 | 54 | 8.9 | 2.2 |
| Johnson | 10.04 | 67.1 | 92 | 2,269 | 1,579.8 | 135.2 | 487 | 9,952 | 3,248.9 | 255.9 | 151 | 461 | 127.5 | 16.9 |
| Knox | 4.20 | 110.7 | 51 | 545 | 331.0 | 24.6 | 154 | 1,936 | 507.6 | 44.9 | 40 | 189 | 30.7 | 5.2 |
| Kosciusko | 3.33 | 42.4 | D | D | D | D | 278 | 4,012 | 1,099.7 | 99.4 | 83 | 198 | 35.4 | 6.2 |
| LaGrange | 0.97 | 25.0 | 31 | 326 | 137.6 | 12.6 | 152 | 1,297 | 343.9 | 32.3 | 31 | 95 | 11.6 | 1.7 |
| Lake | 76.05 | 155.9 | 370 | 4,017 | 3,769.5 | 227.1 | 1,539 | 24,996 | 7,473.7 | 625.4 | 411 | 1,921 | 396.3 | 76.4 |
| LaPorte | 9.35 | 84.3 | 92 | 1,254 | 756.2 | 53.1 | 428 | 5,969 | 1,507.0 | 131.8 | 84 | 343 | 67.0 | 13.6 |
| Lawrence | 5.05 | 111.0 | 24 | 178 | 76.9 | 7.4 | 170 | 1,970 | 589.9 | 50.5 | 27 | 90 | 14.2 | 3.1 |
| Madison | 14.98 | 115.5 | 77 | 914 | 1,423.1 | 53.9 | 317 | 4,897 | 1,508.0 | 116.2 | 92 | 390 | 86.7 | 11.7 |
| Marion | 111.30 | 118.5 | 1,225 | 26,017 | 22,159.4 | 1,717.1 | 2,933 | 49,975 | 15,372.5 | 1,347.2 | 1,429 | 11,817 | 3,291.7 | 591.8 |
| Marshall | 2.85 | 60.8 | 55 | 697 | 405.4 | 34.1 | 164 | 2,094 | 666.1 | 50.7 | 29 | 98 | 15.6 | 2.9 |
| Martin | 0.63 | 61.6 | D | D | D | 1.2 | 33 | 325 | 154.8 | 8.0 | 4 | D | 4.0 | D |
| Miami | 2.66 | 74.2 | D | D | D | D | 93 | 925 | 263.9 | 21.3 | 20 | 54 | 7.2 | 1.8 |
| Monroe | 15.78 | 109.0 | D | D | D | D | 436 | 6,467 | 1,853.5 | 157.5 | 193 | 1,037 | 193.6 | 36.8 |
| Montgomery | 3.61 | 94.4 | D | D | D | D | 134 | 1,675 | 489.3 | 40.3 | 31 | 107 | 14.8 | 2.3 |
| Morgan | 10.52 | 151.0 | 38 | 502 | 243.4 | 23.5 | 169 | 2,205 | 671.2 | 58.0 | 62 | 140 | 24.0 | 4.9 |
| Newton | 0.56 | 40.0 | 18 | 132 | 218.8 | 7.4 | 36 | 256 | 72.6 | 5.0 | 6 | 14 | 1.4 | 0.3 |
| Noble | 2.30 | 48.2 | 38 | 469 | 756.7 | 22.9 | 139 | 1,650 | 475.4 | 41.7 | 35 | 102 | 14.7 | 3.2 |
| Ohio | 0.74 | 124.6 | D | D | D | D | 8 | 56 | 11.7 | 1.0 | 3 | 8 | 0.8 | 0.1 |
| Orange | 0.00 | 0.0 | 12 | 96 | 80.4 | 3.1 | 64 | 622 | 168.1 | 14.8 | 10 | 30 | 3.5 | 0.6 |
| Owen | 1.40 | 67.1 | D | D | D | D | 47 | 571 | 128.9 | 12.3 | 6 | 12 | 1.3 | 0.2 |
| Parke | 0.93 | 55.0 | 13 | 101 | 63.2 | 3.9 | 39 | 350 | 92.9 | 7.1 | 10 | 21 | 2.5 | 0.6 |
| Perry | 0.33 | 17.1 | 8 | 59 | 11.9 | 2.0 | 69 | 726 | 166.6 | 15.1 | 13 | 34 | 5.0 | 0.9 |

1   Merchant wholesalers, except manufacturers' sales branches and offices.    2. Employer establishments.

# Table B. States and Counties — Professional Services, Manufacturing, and Accommodation and Food Services

| STATE County | Professional, scientific, and technical services, 2017 | | | | Manufacturing, 2017 | | | | Accommodation and food services, 2017 | | | |
|---|---|---|---|---|---|---|---|---|---|---|---|---|
| | Number of establish-ments | Number of employees | Sales (mil dol) | Average payroll (mil dol) | Number of establish-ments | Number of employees | Sales (mil dol) | Average payroll (mil dol) | Number of establis-hments | Number of employees | Sales (mil dol) | Annual payroll (mil dol) |
| | 147 | 148 | 149 | 150 | 151 | 152 | 153 | 154 | 155 | 156 | 157 | 158 |
| ILLINOIS—Cont'd | | | | | | | | | | | | |
| Winnebago | 577 | 5,044 | 818 | 286.8 | 564 | 24,052 | 9,218.4 | 1,493.9 | 579 | 11,359 | 587.5 | 171.5 |
| Woodford | 55 | 218 | 22 | 9.1 | 42 | 1,777 | 622.6 | 106.7 | 63 | 875 | 32.9 | 9.8 |
| INDIANA | 13,003 | 121,271 | 22,318 | 8,504.6 | 8,064 | 496,083 | 246,671.7 | 28,131.2 | 13,647 | 280,340 | 15,250.0 | 4,288.5 |
| Adams | 45 | 245 | 23 | 10.8 | 65 | 5,498 | 2,290.9 | 255.0 | 54 | 952 | 35.4 | 10.1 |
| Allen | 817 | 5,643 | 872 | 307.3 | 468 | 26,533 | 18,954.5 | 1,548.3 | 746 | 16,830 | 822.2 | 244.1 |
| Bartholomew | D | D | D | D | 140 | 12,239 | 5,531.9 | 637.4 | 176 | 4,414 | 216.3 | 62.5 |
| Benton | 9 | 27 | 2 | 0.9 | D | D | D | D | 6 | 46 | 2.0 | 0.4 |
| Blackford | 11 | 104 | 15 | 4.2 | 20 | 976 | 325.8 | 47.7 | D | D | D | D |
| Boone | 185 | 803 | 175 | 47.2 | 74 | 2,037 | 674.6 | 96.7 | 130 | 2,103 | 103.5 | 30.3 |
| Brown | D | D | D | D | 17 | 88 | 14.4 | 3.0 | 44 | 533 | 23.6 | 7.9 |
| Carroll | 18 | 152 | 9 | 2.9 | D | D | D | D | 35 | 528 | 25.6 | 7.7 |
| Cass | 38 | 190 | 18 | 5.8 | 45 | 4,304 | 1,658.9 | 175.0 | 70 | 1,083 | 45.6 | 12.8 |
| Clark | D | D | D | D | 142 | 7,569 | 2,468.5 | 360.5 | 235 | 5,966 | 297.8 | 88.0 |
| Clay | 23 | 71 | 7 | 2.5 | 28 | 2,266 | 650.2 | 107.4 | 46 | 615 | 25.7 | 6.9 |
| Clinton | 35 | 127 | 12 | 4.2 | 40 | 3,638 | 3,253.4 | 186.0 | 54 | 779 | 33.5 | 9.8 |
| Crawford | D | D | 1 | D | D | 393 | D | D | D | D | D | D |
| Daviess | 50 | 475 | 50 | 22.2 | 96 | 2,126 | 793.9 | 81.1 | 62 | 934 | 40.2 | 10.9 |
| Dearborn | 59 | 334 | 27 | 10.5 | 46 | 1,455 | 503.9 | 78.2 | D | D | D | D |
| Decatur | 31 | 114 | 14 | 4.9 | 49 | 4,661 | 5,622.6 | 273.6 | 50 | 947 | 48.4 | 14.4 |
| DeKalb | 67 | 469 | 55 | 21.9 | 120 | 9,664 | 5,787.2 | 555.4 | 89 | 1,388 | 66.8 | 18.7 |
| Delaware | D | D | D | D | 118 | 4,234 | 1,723.0 | 219.5 | 220 | 5,143 | 225.4 | 66.9 |
| Dubois | 82 | 345 | 39 | 13.7 | 95 | 11,152 | 2,846.1 | 447.5 | 104 | 1,727 | 72.9 | 22.1 |
| Elkhart | D | D | D | D | 819 | 70,059 | 24,520.9 | 3,763.1 | 378 | 7,145 | 369.0 | 97.5 |
| Fayette | 28 | 172 | 17 | 5.0 | 28 | 1,484 | 438.1 | 74.9 | 45 | 605 | 28.1 | 7.8 |
| Floyd | D | D | D | D | 106 | 5,128 | 1,664.3 | 290.4 | 159 | 3,166 | 150.3 | 43.1 |
| Fountain | 17 | 76 | 7 | 2.2 | 23 | 1,325 | 290.2 | 67.3 | D | D | D | D |
| Franklin | 23 | 47 | 5 | 1.6 | 20 | 782 | 359.1 | 38.3 | 41 | 808 | 42.2 | 11.4 |
| Fulton | 21 | 95 | 25 | 3.5 | 39 | 1,503 | 355.7 | 71.8 | 38 | 447 | 20.2 | 5.4 |
| Gibson | 49 | 329 | 58 | 16.7 | 42 | 8,454 | 11,193.6 | 565.6 | 62 | 1,195 | 52.6 | 16.4 |
| Grant | 69 | 318 | 37 | 12.5 | 63 | 4,443 | 1,807.4 | 268.8 | 133 | 2,343 | 107.2 | 31.9 |
| Greene | D | D | D | D | 24 | 452 | 121.3 | 20.7 | 49 | 612 | 25.2 | 7.4 |
| Hamilton | D | D | D | D | 201 | 4,536 | 1,552.9 | 259.6 | 702 | 17,827 | 975.4 | 300.3 |
| Hancock | 141 | 1,949 | 159 | 216.2 | 67 | 3,175 | 1,217.2 | 161.6 | 119 | 2,543 | 116.8 | 34.0 |
| Harrison | 44 | 158 | 14 | 5.9 | 42 | 1,685 | 561.6 | 82.0 | 55 | 2,416 | 310.0 | 60.3 |
| Hendricks | 300 | 1,526 | 179 | 62.0 | 96 | 3,068 | 1,402.4 | 175.4 | 330 | 7,632 | 371.7 | 109.1 |
| Henry | 44 | 138 | 15 | 4.7 | 47 | 2,217 | 723.3 | 121.2 | 73 | 1,199 | 54.8 | 16.1 |
| Howard | 118 | 714 | 93 | 29.8 | 62 | 9,251 | 4,005.0 | 629.8 | 195 | 4,095 | 185.4 | 54.0 |
| Huntington | 53 | 214 | 27 | 8.4 | 61 | 3,835 | 2,051.1 | 180.6 | 89 | 1,206 | 54.3 | 15.2 |
| Jackson | 53 | 277 | 33 | 10.1 | 62 | 7,969 | 3,037.8 | 468.9 | 88 | 1,431 | 71.3 | 20.4 |
| Jasper | D | D | D | D | 37 | 1,428 | 737.9 | 73.8 | 66 | 931 | 42.3 | 10.8 |
| Jay | 26 | 98 | 7 | 2.9 | 32 | 2,856 | 923.3 | 127.5 | 31 | 483 | 20.5 | 5.4 |
| Jefferson | 37 | 230 | 24 | 8.7 | 47 | 3,209 | 1,575.2 | 175.4 | 73 | 1,143 | 53.3 | 14.6 |
| Jennings | 20 | 93 | 13 | 4.4 | 41 | 1,802 | 439.6 | 79.0 | 24 | 450 | 19.9 | 5.3 |
| Johnson | 289 | 1,575 | 212 | 71.7 | 134 | 6,191 | 2,223.5 | 324.9 | 295 | 6,596 | 337.8 | 97.9 |
| Knox | 42 | 236 | 25 | 9.3 | 36 | 2,122 | 708.8 | 90.6 | 80 | 1,562 | 67.9 | 20.4 |
| Kosciusko | 145 | 547 | 69 | 22.3 | 177 | 10,861 | 6,565.7 | 635.5 | 179 | 2,704 | 140.5 | 40.1 |
| LaGrange | 32 | 170 | 17 | 7.1 | 166 | 7,087 | 2,108.6 | 378.8 | 67 | 771 | 44.3 | 12.0 |
| Lake | D | D | D | D | 341 | 22,241 | 20,585.0 | 1,714.1 | 1,019 | 20,007 | 1,206.1 | 321.3 |
| LaPorte | 145 | 706 | 90 | 32.7 | 166 | 7,181 | 2,445.0 | 373.8 | 228 | 5,006 | 367.7 | 88.1 |
| Lawrence | 62 | 718 | 95 | 38.1 | 59 | 3,040 | 719.1 | 177.8 | 64 | 1,284 | 59.5 | 16.8 |
| Madison | 163 | 641 | 69 | 24.2 | 101 | 4,446 | 2,222.6 | 260.2 | 229 | 4,301 | 194.2 | 58.8 |
| Marion | 2,742 | 50,117 | 11,155 | 4,510.8 | 839 | 43,459 | 21,231.2 | 3,074.1 | 2,308 | 53,348 | 3,276.2 | 931.4 |
| Marshall | 65 | 298 | 28 | 10.0 | 139 | 5,622 | 1,536.6 | 263.3 | 87 | 1,596 | 77.1 | 22.9 |
| Martin | 36 | 874 | 147 | 57.3 | 10 | 353 | 150.9 | 20.6 | 23 | 199 | 8.0 | 2.4 |
| Miami | 32 | 112 | 16 | 6.5 | 36 | 1,919 | 598.4 | 74.4 | 53 | 901 | 32.7 | 9.8 |
| Monroe | 275 | 1,830 | 241 | 93.7 | 97 | 7,432 | 1,424.5 | 375.5 | 395 | 8,100 | 423.3 | 119.2 |
| Montgomery | 46 | 167 | 16 | 5.2 | 52 | 4,928 | 3,316.4 | 311.4 | 82 | 1,365 | 66.5 | 18.0 |
| Morgan | 99 | 420 | 65 | 24.1 | 66 | 2,628 | 808.1 | 111.9 | 79 | 1,673 | 78.2 | 24.0 |
| Newton | 14 | 50 | 5 | 1.6 | 18 | 549 | 134.3 | 23.3 | 27 | 365 | 23.7 | 6.4 |
| Noble | 53 | 196 | 32 | 8.9 | 117 | 8,594 | 2,823.7 | 400.8 | 74 | 1,143 | 49.7 | 13.5 |
| Ohio | D | D | 2 | D | D | 15 | D | 0.5 | D | D | D | D |
| Orange | 16 | 61 | 5 | 1.8 | 22 | 1,257 | 300.0 | 55.1 | 39 | 640 | 30.4 | 7.9 |
| Owen | 23 | 107 | 10 | 4.6 | 28 | 1,055 | 194.8 | 47.2 | 22 | 281 | 13.9 | 3.8 |
| Parke | 10 | 93 | 21 | 8.4 | 16 | 365 | 124.8 | 16.3 | D | D | D | 2.9 |
| Perry | 26 | 194 | 19 | 8.8 | 23 | 2,084 | 941.8 | 126.5 | 39 | 574 | 25.5 | 7.5 |

# Table B. States and Counties — Health Care and Social Assistance, Other Services, Nonemployer Businesses, and Residential Construction

| STATE County | Health care and social assistance, 2017 | | | | Other services, 2017 | | | | Nonemployer businesses, 2018 | | Value of residential construction authorized by building permits, 2020 | |
|---|---|---|---|---|---|---|---|---|---|---|---|---|
| | Number of establishments | Number of employees | Receipts (mil dol) | Annual payroll (mil dol) | Number of establishments | Number of employees | Receipts (mil dol) | Annual payroll (mil dol) | Number | Receipts (mil dol) | New construction ($1,000) | Number of housing units |
| | 159 | 160 | 161 | 162 | 163 | 164 | 165 | 166 | 167 | 168 | 169 | 170 |
| ILLINOIS—Cont'd | | | | | | | | | | | | |
| Winnebago | 655 | 20,279 | 2,635.3 | 1,015.1 | 499 | 3,092 | 343.3 | 90.1 | 17,456 | 643.9 | 26,205 | 149 |
| Woodford | 51 | 1,335 | 85.8 | 37.5 | 45 | 203 | 28.6 | 7.7 | 2,567 | 98.2 | 11,552 | 44 |
| INDIANA | 16,668 | 436,616 | 51,837.3 | 19,641.2 | 10,777 | 75,446 | 11,120.7 | 2,561.7 | 422,619 | 18,127.0 | 5,872,298 | 24,919 |
| Adams | 54 | 1,611 | 113.2 | 54.0 | 75 | 359 | 34.7 | 9.2 | 3,294 | 185.4 | 16,264 | 83 |
| Allen | 1,043 | 36,047 | 4,501.6 | 1,698.5 | 668 | 5,046 | 550.3 | 167.3 | 24,346 | 1,072.6 | 413,550 | 1,694 |
| Bartholomew | 217 | 5,146 | 608.5 | 229.4 | 112 | 729 | 87.2 | 21.5 | 4,312 | 183.9 | 43,019 | 169 |
| Benton | 12 | 69 | 4.8 | 2.0 | D | D | D | D | 604 | 25.9 | 393 | 3 |
| Blackford | 35 | 508 | 55.2 | 19.0 | 20 | 55 | 5.4 | 1.5 | 569 | 22.2 | 470 | 3 |
| Boone | 131 | 2,420 | 244.1 | 103.7 | 94 | 623 | 46.5 | 14.6 | 5,607 | 284.3 | 254,890 | 692 |
| Brown | 20 | 258 | 10.4 | 6.3 | 13 | 66 | 7.4 | 2.0 | 1,319 | 56.4 | 8,664 | 46 |
| Carroll | 34 | 266 | 20.6 | 8.1 | D | D | D | D | 1,785 | 68.7 | 5,710 | 29 |
| Cass | 69 | 2,619 | 200.9 | 91.9 | 56 | 259 | 28.6 | 6.4 | 7,223 | 314.5 | 238,117 | 1,414 |
| Clark | 261 | 6,386 | 693.7 | 277.7 | 167 | 1,318 | 154.8 | 41.8 | | | | |
| Clay | 49 | 592 | 56.8 | 20.1 | 34 | 169 | 23.9 | 6.3 | 1,406 | 49.2 | 1,956 | 9 |
| Clinton | 58 | 1,114 | 77.3 | 33.9 | 42 | 316 | 37.3 | 9.9 | 1,732 | 65.7 | 8,937 | 48 |
| Crawford | 18 | 126 | 8.0 | 3.3 | 8 | 31 | 3.6 | 0.7 | 635 | 27.8 | 0 | 0 |
| Daviess | 80 | 1,373 | 113.9 | 47.3 | 67 | 412 | 121.8 | 17.7 | 2,259 | 98.0 | 9,515 | 34 |
| Dearborn | 126 | 2,718 | 248.2 | 104.4 | 66 | 267 | 28.2 | 7.4 | 3,051 | 133.7 | 36,859 | 142 |
| Decatur | 58 | 1,425 | 159.2 | 52.3 | 45 | 271 | 36.4 | 6.5 | 1,574 | 64.6 | 15,576 | 96 |
| DeKalb | 83 | 1,798 | 149.2 | 61.4 | 70 | 292 | 25.3 | 6.3 | 2,625 | 109.2 | 33,633 | 133 |
| Delaware | 379 | 10,377 | 1,061.6 | 418.1 | 169 | 1,102 | 133.5 | 31.5 | 5,665 | 214.4 | 35,867 | 171 |
| Dubois | 132 | 3,672 | 417.2 | 163.7 | 91 | 468 | 57.7 | 15.4 | 2,908 | 117.1 | 22,199 | 85 |
| Elkhart | 373 | 11,322 | 1,394.3 | 488.3 | 345 | 2,646 | 334.3 | 98.6 | 13,005 | 605.0 | 113,306 | 488 |
| Fayette | 54 | 1,437 | 128.3 | 47.0 | 35 | 151 | 13.1 | 3.8 | 1,109 | 46.8 | 2,254 | 19 |
| Floyd | 269 | 6,982 | 728.4 | 281.2 | 122 | 673 | 77.1 | 19.6 | 5,557 | 245.0 | 85,892 | 276 |
| Fountain | 19 | 456 | 33.5 | 12.0 | 29 | 83 | 9.7 | 2.0 | 1,600 | 70.8 | 17,615 | 58 |
| Franklin | 52 | 552 | 48.5 | 23.3 | 46 | 142 | 16.1 | 3.7 | 1,332 | 55.2 | 3,542 | 15 |
| Fulton | 41 | 1,049 | 87.3 | 39.0 | 40 | 166 | 20.1 | 4.3 | 1,637 | 58.3 | 23,186 | 159 |
| Gibson | 69 | 1,357 | 99.9 | 40.9 | 50 | 259 | 25.6 | 7.1 | 3,224 | 115.1 | 13,498 | 76 |
| Grant | 173 | 4,669 | 416.2 | 163.5 | 103 | 471 | 47.2 | 14.0 | 1,760 | 53.8 | NA | NA |
| Greene | 61 | 996 | 87.3 | 34.5 | 42 | 174 | 17.7 | 4.1 | 30,975 | 1,639.6 | 999,981 | 3,807 |
| Hamilton | 1,101 | 20,604 | 2,519.0 | 968.3 | 555 | 4,097 | 373.3 | 118.9 | 5,385 | 214.7 | 198,390 | 761 |
| Hancock | 151 | 2,734 | 298.6 | 115.6 | 96 | 369 | 37.7 | 10.1 | | | | |
| Harrison | 80 | 1,432 | 144.8 | 52.2 | 34 | 163 | 19.9 | 5.0 | 2,607 | 107.2 | 21,638 | 83 |
| Hendricks | 333 | 8,433 | 973.6 | 386.1 | 247 | 1,924 | 180.8 | 62.7 | 12,097 | 505.3 | 366,068 | 1,371 |
| Henry | 97 | 2,316 | 226.9 | 81.5 | 65 | 330 | 45.7 | 8.0 | 2,371 | 96.7 | 10,858 | 57 |
| Howard | 233 | 5,323 | 589.4 | 222.2 | 126 | 705 | 88.2 | 17.4 | 4,225 | 155.5 | 33,740 | 162 |
| Huntington | 70 | 1,823 | 145.4 | 62.5 | 76 | 395 | 32.3 | 9.1 | 1,913 | 69.9 | 19,299 | 75 |
| Jackson | 98 | 2,690 | 266.3 | 111.7 | 76 | 440 | 40.1 | 11.8 | 2,182 | 91.2 | 35,466 | 181 |
| Jasper | 51 | 1,430 | 91.4 | 43.3 | D | D | D | D | 1,890 | 81.8 | 39,061 | 149 |
| Jay | 32 | 914 | 83.1 | 33.8 | 35 | 118 | 11.6 | 2.1 | 1,355 | 62.0 | 4,655 | 20 |
| Jefferson | 86 | 2,073 | 245.7 | 90.6 | 47 | 239 | 19.6 | 5.8 | 1,818 | 81.1 | 9,372 | 50 |
| Jennings | 82 | 827 | 65.1 | 28.2 | 24 | 105 | 9.4 | 2.2 | 1,474 | 55.5 | 11,925 | 77 |
| Johnson | 338 | 8,458 | 758.5 | 333.1 | 236 | 1,461 | 147.1 | 47.2 | 11,003 | 520.4 | 314,589 | 1,114 |
| Knox | 92 | 3,462 | 359.0 | 147.7 | 60 | 320 | 32.5 | 8.5 | 1,877 | 70.4 | 14,502 | 61 |
| Kosciusko | 159 | 4,448 | 357.3 | 146.6 | 137 | 1,001 | 138.5 | 35.4 | 5,090 | 217.4 | 68,209 | 341 |
| LaGrange | 40 | 1,072 | 86.2 | 32.8 | 51 | 262 | 33.2 | 8.7 | 3,847 | 169.8 | 27,149 | 124 |
| Lake | 1,305 | 33,653 | 4,638.8 | 1,608.7 | 857 | 6,614 | 690.6 | 226.2 | 29,393 | 1,195.5 | 400,217 | 1,512 |
| LaPorte | 229 | 5,904 | 777.2 | 299.5 | 198 | 1,031 | 133.0 | 29.0 | 5,909 | 224.1 | 34,612 | 120 |
| Lawrence | 130 | 2,896 | 232.9 | 100.3 | 61 | 338 | 31.8 | 9.0 | 2,577 | 93.3 | 6,689 | 31 |
| Madison | 280 | 6,683 | 729.3 | 273.2 | 170 | 967 | 89.8 | 23.6 | 6,743 | 251.6 | 65,211 | 274 |
| Marion | 2,837 | 90,422 | 12,707.3 | 4,909.0 | 1,658 | 17,871 | 4,220.9 | 783.5 | 65,885 | 2,811.7 | 472,768 | 2,344 |
| Marshall | 84 | 2,285 | 207.3 | 74.8 | 77 | 770 | 83.6 | 24.3 | 2,903 | 114.3 | 31,953 | 254 |
| Martin | 11 | 149 | 7.3 | 3.5 | 13 | 41 | 5.2 | 0.9 | 630 | 21.1 | 0 | 0 |
| Miami | 42 | 970 | 103.3 | 39.8 | 46 | 215 | 26.8 | 5.9 | 1,639 | 66.7 | 5,478 | 32 |
| Monroe | 416 | 10,565 | 977.0 | 419.6 | 212 | 1,533 | 429.8 | 55.5 | 9,711 | 384.0 | 111,219 | 619 |
| Montgomery | 95 | 1,380 | 125.2 | 54.3 | 68 | 268 | 27.8 | 6.5 | 2,230 | 90.3 | 15,031 | 76 |
| Morgan | 121 | 2,236 | 251.6 | 93.5 | 97 | 454 | 47.1 | 13.0 | 4,649 | 191.2 | 75,181 | 329 |
| Newton | 12 | 238 | 13.6 | 7.3 | D | D | D | D | 745 | 31.2 | 5,710 | 28 |
| Noble | 67 | 1,371 | 148.6 | 52.6 | 73 | 280 | 40.5 | 6.5 | 2,668 | 118.7 | 31,547 | 142 |
| Ohio | 10 | 97 | 7.5 | 2.6 | D | D | D | 0.4 | 297 | 9.8 | 5,400 | 20 |
| Orange | 51 | 802 | 71.1 | 25.9 | 29 | 148 | 14.3 | 3.6 | 1,215 | 47.7 | 1,080 | 9 |
| Owen | 37 | 547 | 36.8 | 16.0 | 27 | 66 | 6.2 | 1.2 | 1,415 | 61.1 | 8,228 | 31 |
| Parke | 18 | 180 | 10.6 | 4.6 | 27 | 83 | 10.0 | 2.1 | 1,153 | 58.9 | 6,484 | 41 |
| Perry | 40 | 872 | 81.9 | 28.6 | 22 | 79 | 6.2 | 2.0 | 837 | 36.0 | 7,956 | 38 |

# Table B. States and Counties — Government Employment and Payroll, and Local Government Finances

| STATE County | Full-time equivalent employees | March payroll (dollars) | Administration, judicial, and legal | Police and corrections | Fire protection | Highways and transportation | Health and welfare | Natural resources and utilities | Education and libraries | Total (mil dol) | Inter-governmental (mil dol) | Total (mil dol) | Per capita[1] Total | Per capita[1] Property |
|---|---|---|---|---|---|---|---|---|---|---|---|---|---|---|
| | 171 | 172 | 173 | 174 | 175 | 176 | 177 | 178 | 179 | 180 | 181 | 182 | 183 | 184 |
| **ILLINOIS—Cont'd** | | | | | | | | | | | | | | |
| Winnebago | 10,108 | 44,836,553 | 5.1 | 12.7 | 6.4 | 3.8 | 4.4 | 7.2 | 59.3 | 1,312.2 | 577.3 | 543.2 | 1,908 | 1,647 |
| Woodford | 1,499 | 5,765,466 | 4.7 | 6.6 | 0.6 | 4.4 | 1.8 | 3.1 | 78.6 | 134.5 | 43.5 | 75.8 | 1,961 | 1,944 |
| **INDIANA** | X | X | X | X | X | X | X | X | X | X | X | X | X | X |
| Adams | 1,650 | 6,028,628 | 4.5 | 5.6 | 0.8 | 2.0 | 48.1 | 3.5 | 34.8 | 155.3 | 42.9 | 31.8 | 899 | 864 |
| Allen | 11,137 | 44,759,235 | 6.4 | 13.3 | 4.4 | 4.4 | 2.2 | 5.4 | 62.8 | 1,165.8 | 549.6 | 465.5 | 1,252 | 957 |
| Bartholomew | 4,111 | 16,129,167 | 3.8 | 7.4 | 3.0 | 1.4 | 41.3 | 4.3 | 37.9 | 614.4 | 104.8 | 80.3 | 977 | 940 |
| Benton | 510 | 1,372,200 | 8.0 | 7.0 | 0.0 | 6.2 | 4.0 | 3.9 | 70.0 | 35.0 | 18.5 | 12.5 | 1,449 | 1,427 |
| Blackford | 455 | 1,477,437 | 7.4 | 11.6 | 2.1 | 4.5 | 0.9 | 4.9 | 66.6 | 35.8 | 18.6 | 10.7 | 890 | 831 |
| Boone | 2,824 | 12,455,140 | 4.4 | 5.8 | 4.6 | 1.5 | 32.3 | 3.8 | 47.0 | 336.7 | 101.6 | 82.4 | 1,254 | 1,158 |
| Brown | 531 | 1,549,624 | 11.9 | 12.7 | 0.1 | 3.8 | 2.4 | 3.2 | 62.6 | 42.4 | 20.6 | 15.9 | 1,056 | 985 |
| Carroll | 617 | 1,867,416 | 8.6 | 11.2 | 0.0 | 4.7 | 2.0 | 6.1 | 66.9 | 64.3 | 26.5 | 31.5 | 1,572 | 1,512 |
| Cass | 2,218 | 7,871,513 | 4.7 | 5.4 | 1.7 | 2.2 | 35.1 | 8.4 | 41.2 | 214.1 | 68.1 | 52.9 | 1,399 | 1,338 |
| Clark | 4,970 | 19,005,482 | 3.8 | 8.7 | 3.3 | 2.3 | 35.3 | 3.7 | 42.3 | 443.8 | 214.5 | 135.9 | 1,166 | 990 |
| Clay | 895 | 2,956,288 | 7.5 | 9.7 | 2.1 | 3.0 | 2.0 | 3.4 | 71.3 | 77.5 | 39.9 | 23.7 | 907 | 594 |
| Clinton | 1,276 | 4,458,361 | 4.8 | 9.9 | 7.6 | 3.3 | 0.9 | 11.1 | 60.4 | 144.6 | 65.7 | 59.9 | 1,860 | 1,773 |
| Crawford | 366 | 927,209 | 12.9 | 8.0 | 0.0 | 7.1 | 3.7 | 3.7 | 62.7 | 54.6 | 18.3 | 32.3 | 3,069 | 3,040 |
| Daviess | 1,409 | 5,500,456 | 5.1 | 7.3 | 1.2 | 2.8 | 41.3 | 6.9 | 34.7 | 114.7 | 43.9 | 48.4 | 1,459 | 801 |
| Dearborn | 1,792 | 6,094,969 | 8.8 | 11.8 | 2.4 | 2.5 | 0.8 | 6.4 | 64.5 | 387.3 | 101.7 | 92.6 | 1,869 | 1,838 |
| Decatur | 1,159 | 4,643,596 | 4.4 | 5.1 | 2.2 | 2.7 | 44.5 | 3.6 | 37.4 | 180.5 | 44.4 | 48.6 | 1,825 | 1,648 |
| DeKalb | 1,668 | 5,500,156 | 7.7 | 7.8 | 2.3 | 3.4 | 0.8 | 8.7 | 60.7 | 165.9 | 78.1 | 47.1 | 1,101 | 1,079 |
| Delaware | 3,844 | 13,695,351 | 5.6 | 8.5 | 3.6 | 3.7 | 3.0 | 4.5 | 69.0 | 411.1 | 185.8 | 88.7 | 769 | 723 |
| Dubois | 1,617 | 5,303,292 | 8.0 | 8.8 | 0.6 | 5.9 | 2.2 | 17.8 | 55.8 | 432.2 | 98.9 | 73.0 | 1,717 | 1,055 |
| Elkhart | 7,096 | 28,545,949 | 5.0 | 8.8 | 4.0 | 2.1 | 1.6 | 3.1 | 72.4 | 664.7 | 333.5 | 219.8 | 1,077 | 1,000 |
| Fayette | 775 | 2,716,489 | 5.6 | 12.0 | 4.3 | 2.8 | 1.3 | 6.9 | 61.2 | 53.5 | 36.5 | 12.0 | 519 | 495 |
| Floyd | 2,198 | 8,020,054 | 6.9 | 11.6 | 6.7 | 2.5 | 3.5 | 3.7 | 64.5 | 251.0 | 102.4 | 82.1 | 1,066 | 941 |
| Fountain | 608 | 1,865,542 | 8.9 | 8.6 | 1.8 | 5.6 | 5.6 | 6.3 | 63.0 | 46.3 | 26.3 | 14.0 | 854 | 831 |
| Franklin | 488 | 1,612,210 | 9.1 | 10.3 | 0.2 | 5.2 | 1.7 | 2.7 | 69.5 | 38.3 | 21.4 | 12.3 | 544 | 497 |
| Fulton | 1,010 | 4,020,897 | 4.0 | 4.9 | 1.5 | 2.2 | 50.7 | 2.2 | 33.0 | 52.2 | 27.7 | 13.8 | 691 | 638 |
| Gibson | 1,180 | 3,779,902 | 8.4 | 7.7 | 2.8 | 4.0 | 4.0 | 5.2 | 66.7 | 118.5 | 54.8 | 43.6 | 1,294 | 1,255 |
| Grant | 2,178 | 6,806,498 | 7.1 | 16.1 | 4.4 | 4.3 | 2.0 | 2.2 | 63.5 | 191.2 | 108.8 | 56.0 | 844 | 823 |
| Greene | 1,388 | 4,795,988 | 5.2 | 6.2 | 1.5 | 2.3 | 22.7 | 3.8 | 57.0 | 89.8 | 46.1 | 26.0 | 808 | 614 |
| Hamilton | 11,180 | 46,901,560 | 5.6 | 7.3 | 7.0 | 1.8 | 13.0 | 4.3 | 59.4 | 1,579.9 | 431.8 | 523.3 | 1,621 | 1,294 |
| Hancock | 2,973 | 11,331,967 | 5.7 | 7.4 | 4.9 | 1.6 | 31.4 | 3.5 | 44.0 | 346.2 | 106.5 | 80.0 | 1,066 | 861 |
| Harrison | 1,449 | 5,427,289 | 5.5 | 4.9 | 0.0 | 2.3 | 38.3 | 3.0 | 44.6 | 242.7 | 198.7 | 22.0 | 552 | 496 |
| Hendricks | 6,012 | 25,797,353 | 1.8 | 3.2 | 5.6 | 0.5 | 32.4 | 2.5 | 53.8 | 1,123.8 | 250.6 | 246.2 | 1,505 | 1,160 |
| Henry | 2,426 | 9,309,783 | 4.2 | 5.2 | 2.5 | 2.2 | 42.6 | 3.4 | 38.9 | 260.9 | 93.7 | 94.6 | 1,964 | 1,746 |
| Howard | 3,042 | 10,199,579 | 6.3 | 12.7 | 4.2 | 3.2 | 3.2 | 3.4 | 65.3 | 262.2 | 141.0 | 92.8 | 1,129 | 1,108 |
| Huntington | 1,056 | 3,701,086 | 6.5 | 8.3 | 4.1 | 4.8 | 1.2 | 4.1 | 67.3 | 86.1 | 41.5 | 30.0 | 827 | 803 |
| Jackson | 2,284 | 9,879,858 | 3.1 | 6.8 | 3.4 | 1.4 | 47.6 | 1.9 | 34.7 | 269.6 | 56.5 | 34.4 | 784 | 656 |
| Jasper | 1,147 | 3,521,843 | 9.2 | 13.1 | 0.1 | 4.0 | 1.3 | 8.4 | 63.1 | 112.3 | 45.4 | 55.2 | 1,651 | 1,295 |
| Jay | 1,038 | 3,710,595 | 6.1 | 6.3 | 1.7 | 2.7 | 34.7 | 2.2 | 45.1 | 67.0 | 33.0 | 21.9 | 1,048 | 1,000 |
| Jefferson | 937 | 3,151,949 | 10.0 | 11.3 | 0.2 | 4.3 | 1.1 | 5.5 | 67.5 | 87.0 | 43.5 | 27.6 | 862 | 841 |
| Jennings | 904 | 2,910,827 | 5.9 | 9.9 | 0.7 | 3.9 | 2.4 | 5.6 | 69.8 | 72.6 | 39.4 | 15.5 | 562 | 556 |
| Johnson | 5,177 | 19,308,878 | 5.2 | 8.2 | 3.1 | 2.1 | 23.1 | 3.5 | 52.9 | 828.1 | 232.2 | 162.0 | 1,053 | 946 |
| Knox | 2,859 | 12,155,381 | 2.7 | 4.6 | 1.3 | 1.1 | 72.7 | 1.4 | 16.1 | 376.5 | 57.1 | 39.6 | 1,068 | 1,009 |
| Kosciusko | 2,847 | 9,243,126 | 6.2 | 8.7 | 2.3 | 3.3 | 1.3 | 3.7 | 73.7 | 283.8 | 145.5 | 87.8 | 1,111 | 1,055 |
| LaGrange | 1,108 | 3,552,148 | 5.6 | 6.9 | 0.4 | 3.1 | 0.6 | 2.9 | 78.8 | 88.7 | 49.7 | 28.5 | 727 | 656 |
| Lake | 17,781 | 66,212,636 | 10.8 | 11.0 | 4.9 | 5.6 | 2.6 | 8.5 | 55.1 | 1,999.5 | 909.7 | 681.5 | 1,406 | 1,358 |
| LaPorte | 4,410 | 14,500,631 | 6.4 | 12.0 | 4.0 | 3.7 | 2.8 | 6.2 | 63.3 | 390.5 | 185.2 | 125.9 | 1,146 | 1,100 |
| Lawrence | 1,538 | 5,349,008 | 5.2 | 9.5 | 3.3 | 3.7 | 9.0 | 4.9 | 63.7 | 156.2 | 61.6 | 34.8 | 764 | 692 |
| Madison | 3,950 | 14,479,297 | 8.4 | 13.9 | 4.7 | 4.1 | 2.2 | 10.8 | 54.4 | 433.7 | 172.4 | 193.6 | 1,496 | 1,073 |
| Marion | 33,241 | 142,174,649 | 4.1 | 7.5 | 7.3 | 2.7 | 18.2 | 7.4 | 51.9 | 4,655.4 | 1,870.7 | 920.8 | 969 | 812 |
| Marshall | 1,537 | 5,374,932 | 7.2 | 11.1 | 1.8 | 4.2 | 1.1 | 5.4 | 67.8 | 196.2 | 62.4 | 105.1 | 2,264 | 2,219 |
| Martin | 333 | 1,020,519 | 9.9 | 14.7 | 1.2 | 4.5 | 0.1 | 5.2 | 60.6 | 31.8 | 17.3 | 9.3 | 914 | 878 |
| Miami | 1,211 | 4,312,290 | 4.8 | 8.7 | 3.0 | 4.2 | 1.2 | 5.8 | 69.5 | 118.3 | 65.7 | 27.4 | 765 | 729 |
| Monroe | 3,742 | 13,600,022 | 11.4 | 13.3 | 5.5 | 5.1 | 2.5 | 7.1 | 53.1 | 404.3 | 152.3 | 174.8 | 1,192 | 1,133 |
| Montgomery | 1,253 | 4,400,029 | 6.4 | 11.3 | 6.6 | 3.5 | 1.0 | 8.6 | 61.8 | 112.8 | 56.1 | 39.4 | 1,029 | 1,009 |
| Morgan | 2,165 | 6,661,029 | 6.7 | 10.1 | 5.0 | 2.9 | 1.2 | 3.0 | 70.4 | 177.2 | 96.3 | 60.2 | 863 | 804 |
| Newton | 555 | 1,682,981 | 12.2 | 10.2 | 0.0 | 3.3 | 4.6 | 1.7 | 66.6 | 83.5 | 23.4 | 45.5 | 3,242 | 3,206 |
| Noble | 1,436 | 4,855,396 | 9.2 | 11.3 | 1.4 | 3.6 | 1.0 | 5.0 | 68.2 | 126.3 | 62.2 | 45.8 | 968 | 951 |
| Ohio | 222 | 764,712 | 11.3 | 9.1 | 0.0 | 5.9 | 2.0 | 8.5 | 56.9 | 26.7 | 14.4 | 7.4 | 1,257 | 1,008 |
| Orange | 659 | 2,044,620 | 9.1 | 9.3 | 0.5 | 6.1 | 0.5 | 5.4 | 68.7 | 52.2 | 30.4 | 17.2 | 885 | 753 |
| Owen | 516 | 1,573,362 | 5.9 | 12.0 | 4.0 | 5.2 | 0.5 | 1.7 | 70.8 | 37.9 | 22.1 | 12.1 | 583 | 543 |
| Parke | 613 | 1,570,005 | 9.0 | 10.2 | 0.9 | 5.0 | 4.2 | 4.3 | 64.2 | 40.9 | 25.5 | 10.9 | 644 | 602 |
| Perry | 912 | 3,316,555 | 5.3 | 5.5 | 0.5 | 4.7 | 39.6 | 5.6 | 38.2 | 61.5 | 32.0 | 15.6 | 823 | 743 |

1. Based on the resident population estimated as of July 1 of the year shown.

Items 171—184

# Local Government Finances, Government Employment, and Income Taxes

| STATE County | Local government finances, 2017 (cont.) | | | | | | | | | Government employment, 2019 | | | Individual income tax returns, 2018 | | |
|---|---|---|---|---|---|---|---|---|---|---|---|---|---|---|---|
| | Direct general expenditure | | | | | | | Debt outstanding | | | | | | Mean | Mean |
| | | | Percent of total for: | | | | | | | | | | | adjusted | income |
| | Total (mil dol) | Per capita¹ (dollars) | Education | Health and hospitals | Police protection | Public welfare | Highways | Total (mil dol) | Per capita¹ (dollars) | Federal civilian | Federal military | State and local | Number of returns | gross income | tax |
| | 185 | 186 | 187 | 188 | 189 | 190 | 191 | 192 | 193 | 194 | 195 | 196 | 197 | 198 | 199 |
| ILLINOIS—Cont'd | | | | | | | | | | | | | | | |
| Winnebago | 1,377 | 4,836 | 51.3 | 0.8 | 9.5 | 2.6 | 6.4 | 965.8 | 3,393 | 838 | 590 | 13,413 | 135,990 | 57,013 | 5,729 |
| Woodford | 141 | 3,654 | 63.8 | 1.1 | 4.0 | 0.0 | 7.7 | 66.6 | 1,722 | 71 | 76 | 1,965 | 18,140 | 81,549 | 9,447 |
| INDIANA | X | X | X | X | X | X | X | X | X | 39,033 | 20,441 | 388,680 | 3,150,090 | 62,080 | 6,632 |
| Adams | 161 | 4,543 | 27.3 | 44.3 | 1.8 | 0.4 | 3.5 | 134.5 | 3,802 | 62 | 105 | 2,208 | 15,340 | 54,385 | 4,938 |
| Allen | 1,076 | 2,895 | 51.0 | 1.2 | 8.3 | 0.0 | 6.4 | 1,152.5 | 3,100 | 2,060 | 1,148 | 17,051 | 182,340 | 61,625 | 6,804 |
| Bartholomew | 630 | 7,658 | 19.8 | 63.2 | 2.4 | 0.1 | 2.0 | 281.4 | 3,422 | 174 | 246 | 6,279 | 39,640 | 71,231 | 7,970 |
| Benton | 34 | 3,942 | 58.6 | 1.0 | 3.2 | 0.6 | 8.1 | 44.5 | 5,158 | 26 | 26 | 581 | 4,180 | 49,308 | 3,904 |
| Blackford | 30 | 2,520 | 51.2 | 0.6 | 4.4 | 0.1 | 8.2 | 28.5 | 2,371 | 24 | 34 | 495 | 5,720 | 44,997 | 3,324 |
| Boone | 324 | 4,926 | 35.0 | 33.6 | 4.2 | 0.1 | 1.6 | 414.9 | 6,313 | 110 | 200 | 3,873 | 32,670 | 114,753 | 17,547 |
| Brown | 44 | 2,911 | 57.8 | 1.4 | 1.9 | 0.1 | 4.4 | 27.4 | 1,826 | 15 | 44 | 718 | 7,740 | 58,858 | 5,939 |
| Carroll | 64 | 3,169 | 37.8 | 2.0 | 3.0 | 0.8 | 7.8 | 20.3 | 1,012 | 66 | 60 | 776 | 9,810 | 54,330 | 4,557 |
| Cass | 195 | 5,149 | 36.0 | 38.5 | 2.7 | 0.0 | 4.5 | 90.8 | 2,399 | 87 | 109 | 2,988 | 17,810 | 46,989 | 3,622 |
| Clark | 378 | 3,244 | 43.2 | 0.5 | 5.3 | 0.1 | 3.3 | 505.8 | 4,340 | 1,874 | 350 | 4,487 | 57,970 | 55,384 | 4,987 |
| Clay | 87 | 3,320 | 51.5 | 0.5 | 2.1 | 0.0 | 8.4 | 89.3 | 3,413 | 63 | 77 | 1,215 | 12,040 | 49,413 | 3,928 |
| Clinton | 112 | 3,481 | 50.7 | 1.4 | 5.1 | 0.4 | 7.5 | 104.2 | 3,238 | 58 | 94 | 1,602 | 15,390 | 49,857 | 3,895 |
| Crawford | 30 | 2,833 | 49.9 | 2.7 | 1.7 | 0.0 | 13.5 | 12.9 | 1,221 | 24 | 31 | 469 | 4,670 | 42,181 | 2,782 |
| Daviess | 117 | 3,538 | 43.1 | 0.6 | 2.6 | 0.0 | 9.6 | 106.9 | 3,226 | 74 | 97 | 1,882 | 14,670 | 53,124 | 4,510 |
| Dearborn | 339 | 6,840 | 23.3 | 51.1 | 2.3 | 0.0 | 2.3 | 156.3 | 3,154 | 103 | 146 | 2,853 | 25,020 | 62,356 | 5,918 |
| Decatur | 151 | 5,672 | 27.4 | 52.7 | 1.6 | 0.0 | 3.0 | 138.4 | 5,197 | 63 | 78 | 1,609 | 13,000 | 52,994 | 4,520 |
| DeKalb | 157 | 3,674 | 56.5 | 0.2 | 5.0 | 0.2 | 3.0 | 96.5 | 2,254 | 83 | 128 | 2,005 | 21,010 | 55,785 | 5,036 |
| Delaware | 644 | 5,590 | 24.6 | 0.6 | 2.4 | 0.0 | 1.8 | 330.8 | 2,870 | 307 | 320 | 10,444 | 47,870 | 49,385 | 4,443 |
| Dubois | 332 | 7,812 | 21.5 | 62.2 | 1.5 | 0.0 | 2.9 | 242.2 | 5,693 | 103 | 125 | 2,189 | 22,490 | 66,753 | 7,386 |
| Elkhart | 694 | 3,396 | 57.5 | 1.1 | 4.4 | 0.0 | 4.3 | 677.6 | 3,318 | 284 | 603 | 8,391 | 96,080 | 61,165 | 6,485 |
| Fayette | 74 | 3,215 | 55.0 | 3.5 | 5.6 | 1.8 | 6.4 | 27.9 | 1,203 | 41 | 68 | 1,020 | 10,220 | 44,164 | 3,256 |
| Floyd | 196 | 2,546 | 59.9 | 0.3 | 4.0 | 0.0 | 2.5 | 338.1 | 4,391 | 229 | 230 | 4,149 | 38,940 | 67,937 | 7,490 |
| Fountain | 43 | 2,591 | 60.2 | 0.9 | 2.9 | 0.8 | 9.8 | 24.1 | 1,469 | 56 | 48 | 718 | 7,850 | 49,384 | 3,809 |
| Franklin | 46 | 2,014 | 57.1 | 0.7 | 4.9 | 0.0 | 6.7 | 17.0 | 751 | 43 | 67 | 927 | 10,820 | 62,841 | 6,097 |
| Fulton | 54 | 2,705 | 54.1 | 0.9 | 2.4 | 0.0 | 8.2 | 54.9 | 2,744 | 42 | 59 | 1,340 | 9,460 | 51,389 | 4,450 |
| Gibson | 99 | 2,936 | 51.2 | 2.2 | 4.1 | 0.0 | 6.7 | 140.6 | 4,170 | 90 | 98 | 1,304 | 15,940 | 55,271 | 4,597 |
| Grant | 174 | 2,622 | 55.2 | 0.7 | 7.8 | 0.1 | 5.9 | 108.9 | 1,642 | 1,102 | 181 | 2,785 | 29,030 | 46,068 | 3,673 |
| Greene | 86 | 2,670 | 61.3 | 2.4 | 3.8 | 0.0 | 5.8 | 80.4 | 2,498 | 78 | 94 | 1,875 | 14,360 | 50,109 | 4,012 |
| Hamilton | 1,475 | 4,568 | 37.8 | 29.2 | 4.5 | 0.0 | 3.6 | 2,786.9 | 8,630 | 425 | 1,000 | 14,100 | 160,610 | 117,401 | 17,764 |
| Hancock | 314 | 4,180 | 38.5 | 33.8 | 2.7 | 0.0 | 3.3 | 302.0 | 4,027 | 125 | 231 | 4,173 | 38,570 | 69,816 | 7,196 |
| Harrison | 100 | 2,520 | 57.3 | 1.5 | 1.4 | 0.0 | 4.1 | 97.7 | 2,457 | 98 | 119 | 2,044 | 18,590 | 55,604 | 4,793 |
| Hendricks | 946 | 5,782 | 28.0 | 50.3 | 2.0 | 0.1 | 1.4 | 911.9 | 5,573 | 273 | 496 | 9,272 | 82,160 | 71,980 | 7,428 |
| Henry | 203 | 4,223 | 36.5 | 36.9 | 2.9 | 0.0 | 2.5 | 114.6 | 2,378 | 98 | 132 | 2,871 | 21,670 | 48,751 | 3,880 |
| Howard | 263 | 3,202 | 53.5 | 0.4 | 6.4 | 0.1 | 2.3 | 246.6 | 2,999 | 189 | 243 | 4,898 | 40,740 | 53,268 | 4,688 |
| Huntington | 68 | 1,862 | 75.3 | 0.0 | 5.4 | 0.0 | 1.1 | 34.2 | 944 | 88 | 105 | 1,366 | 17,710 | 52,379 | 4,517 |
| Jackson | 245 | 5,568 | 28.7 | 57.2 | 2.0 | 0.0 | 1.3 | 132.7 | 3,018 | 96 | 130 | 3,145 | 21,360 | 53,926 | 4,704 |
| Jasper | 87 | 2,615 | 57.8 | 2.0 | 2.1 | 0.1 | 1.2 | 184.1 | 5,509 | 102 | 97 | 1,449 | 15,520 | 56,920 | 4,898 |
| Jay | 73 | 3,500 | 47.9 | 2.2 | 3.1 | 0.6 | 5.9 | 65.4 | 3,130 | 44 | 60 | 864 | 9,170 | 45,709 | 3,967 |
| Jefferson | 84 | 2,614 | 57.0 | 1.1 | 3.4 | 0.0 | 5.5 | 38.0 | 1,187 | 78 | 90 | 2,333 | 14,610 | 53,218 | 4,596 |
| Jennings | 74 | 2,672 | 60.1 | 1.2 | 4.5 | 0.6 | 6.1 | 58.5 | 2,114 | 84 | 82 | 1,182 | 12,790 | 46,680 | 3,597 |
| Johnson | 899 | 5,842 | 28.6 | 39.7 | 1.4 | 0.1 | 2.2 | 571.6 | 3,716 | 422 | 762 | 7,017 | 76,150 | 69,282 | 7,366 |
| Knox | 348 | 9,399 | 17.5 | 64.7 | 1.4 | 0.0 | 2.1 | 305.6 | 8,249 | 155 | 103 | 4,771 | 16,490 | 53,417 | 4,837 |
| Kosciusko | 270 | 3,422 | 53.4 | 0.4 | 2.9 | 0.1 | 5.0 | 320.9 | 4,061 | 156 | 232 | 2,881 | 38,840 | 63,769 | 6,643 |
| LaGrange | 92 | 2,345 | 62.9 | 0.5 | 3.0 | 0.6 | 4.5 | 53.9 | 1,374 | 55 | 117 | 1,246 | 17,700 | 57,196 | 4,748 |
| Lake | 1,737 | 3,584 | 44.1 | 0.5 | 6.6 | 0.1 | 3.3 | 2,491.0 | 5,140 | 1,441 | 1,429 | 23,239 | 231,160 | 58,629 | 5,991 |
| LaPorte | 386 | 3,511 | 46.8 | 0.7 | 5.0 | 0.1 | 3.5 | 349.2 | 3,179 | 180 | 328 | 6,403 | 51,730 | 55,109 | 5,366 |
| Lawrence | 127 | 2,781 | 49.9 | 17.8 | 4.6 | 0.0 | 1.2 | 90.5 | 1,984 | 127 | 133 | 1,814 | 21,160 | 49,910 | 4,131 |
| Madison | 454 | 3,508 | 36.0 | 0.4 | 4.6 | 0.0 | 5.2 | 559.3 | 4,322 | 249 | 368 | 6,044 | 59,420 | 49,990 | 4,783 |
| Marion | 4,334 | 4,559 | 37.1 | 30.7 | 5.4 | 0.0 | 0.1 | 7,535.8 | 7,928 | 15,492 | 3,135 | 67,209 | 457,540 | 57,537 | 6,405 |
| Marshall | 111 | 2,391 | 64.4 | 0.8 | 4.8 | 0.6 | 1.7 | 87.0 | 1,874 | 100 | 136 | 2,040 | 21,990 | 53,906 | 4,864 |
| Martin | 31 | 3,065 | 46.7 | 0.4 | 3.5 | 0.1 | 6.4 | 16.8 | 1,654 | 4,774 | 67 | 465 | 4,730 | 51,143 | 3,908 |
| Miami | 112 | 3,109 | 59.5 | 0.4 | 2.2 | 0.0 | 6.3 | 111.1 | 3,097 | 598 | 113 | 1,912 | 15,350 | 46,349 | 3,424 |
| Monroe | 506 | 3,453 | 29.3 | 0.4 | 3.2 | 0.1 | 3.1 | 595.1 | 4,058 | 328 | 418 | 23,649 | 58,030 | 67,209 | 7,923 |
| Montgomery | 101 | 2,621 | 62.9 | 0.3 | 4.8 | 0.0 | 2.1 | 156.9 | 4,092 | 74 | 111 | 1,961 | 18,030 | 50,978 | 4,188 |
| Morgan | 172 | 2,461 | 60.6 | 1.0 | 2.9 | 0.2 | 3.5 | 153.2 | 2,195 | 100 | 208 | 2,588 | 34,190 | 59,683 | 5,559 |
| Newton | 63 | 4,481 | 40.2 | 0.4 | 2.0 | 0.9 | 3.9 | 52.1 | 3,713 | 25 | 41 | 681 | 6,690 | 51,395 | 4,259 |
| Noble | 116 | 2,459 | 61.4 | 0.0 | 3.5 | 0.1 | 4.8 | 186.4 | 3,938 | 84 | 140 | 1,790 | 22,400 | 52,213 | 4,195 |
| Ohio | 22 | 3,749 | 37.9 | 0.7 | 5.0 | 0.0 | 4.7 | 8.9 | 1,525 | 13 | 17 | 317 | 2,720 | 53,909 | 4,708 |
| Orange | 48 | 2,453 | 69.9 | 0.3 | 2.3 | 0.0 | 6.2 | 71.6 | 3,688 | 44 | 58 | 843 | 8,770 | 43,708 | 3,226 |
| Owen | 39 | 1,887 | 65.8 | 3.4 | 0.1 | 0.0 | 8.1 | 32.4 | 1,557 | 33 | 61 | 764 | 9,670 | 46,242 | 3,342 |
| Parke | 39 | 2,304 | 58.2 | 1.3 | 3.5 | 0.0 | 10.1 | 31.1 | 1,841 | 51 | 46 | 1,028 | 6,810 | 50,227 | 4,115 |
| Perry | 48 | 2,520 | 54.8 | 0.5 | 3.1 | 0.0 | 9.8 | 44.5 | 2,347 | 73 | 52 | 1,410 | 8,470 | 49,647 | 3,970 |

1. Based on the resident population estimated as of July 1 of the year shown.

| State / county code | CBSA code[1] | County Type code[2] | STATE County | Land area[3] (sq. mi) | Total persons 2019 | Rank | Per square mile | White | Black | American Indian, Alaska Native | Asian and Pacific Islander | Percent Hispanic or Latino[4] | Under 5 years | 5 to 14 years | 15 to 24 years | 25 to 34 years | 35 to 44 years | 45 to 54 years |
|---|---|---|---|---|---|---|---|---|---|---|---|---|---|---|---|---|---|---|
| | | | | 1 | 2 | 3 | 4 | 5 | 6 | 7 | 8 | 9 | 10 | 11 | 12 | 13 | 14 | 15 |
| | | | INDIANA—Cont'd | | | | | | | | | | | | | | | |
| 18125 | 27540 | 8 | Pike | 334.3 | 12,378 | 2,260 | 37.0 | 96.8 | 1.0 | 0.7 | 0.8 | 1.6 | 6.0 | 12.5 | 10.5 | 11.7 | 11.0 | 12.5 |
| 18127 | 16980 | 1 | Porter | 418.1 | 170,980 | 388 | 408.9 | 83.4 | 4.9 | 0.6 | 2.0 | 10.7 | 5.3 | 12.3 | 13.0 | 12.1 | 13.4 | 12.7 |
| 18129 | 21780 | 2 | Posey | 409.4 | 25,275 | 1,596 | 61.7 | 96.8 | 1.8 | 0.6 | 0.8 | 1.3 | 5.4 | 12.7 | 10.7 | 11.6 | 11.7 | 12.2 |
| 18131 | | 6 | Pulaski | 433.7 | 12,388 | 2,258 | 28.6 | 94.3 | 1.3 | 1.0 | 1.0 | 3.5 | 5.3 | 12.4 | 11.7 | 11.5 | 11.5 | 12.1 |
| 18133 | 26900 | 6 | Putnam | 480.5 | 37,469 | 1,238 | 78.0 | 92.6 | 4.4 | 0.7 | 1.7 | 2.1 | 5.0 | 10.8 | 16.7 | 12.9 | 11.5 | 12.2 |
| 18135 | | 6 | Randolph | 452.4 | 24,191 | 1,641 | 53.5 | 94.6 | 1.3 | 0.7 | 0.7 | 3.9 | 5.6 | 12.5 | 11.8 | 10.9 | 11.2 | 12.7 |
| 18137 | | 6 | Ripley | 446.4 | 28,448 | 1,481 | 63.7 | 96.6 | 0.8 | 0.7 | 1.1 | 1.9 | 6.5 | 12.8 | 12.5 | 11.3 | 11.5 | 12.9 |
| 18139 | | 6 | Rush | 408.1 | 16,649 | 1,997 | 40.8 | 96.5 | 1.6 | 0.6 | 0.6 | 1.8 | 5.7 | 12.2 | 12.1 | 11.8 | 11.3 | 12.8 |
| 18141 | 43780 | 2 | St. Joseph | 457.8 | 271,484 | 254 | 593.0 | 74.2 | 15.2 | 1.0 | 3.4 | 9.4 | 6.4 | 12.9 | 14.9 | 13.6 | 12.0 | 11.4 |
| 18143 | 42500 | 1 | Scott | 190.4 | 23,788 | 1,651 | 124.9 | 95.7 | 1.0 | 0.6 | 1.2 | 2.5 | 5.7 | 12.8 | 11.8 | 12.7 | 11.9 | 13.8 |
| 18145 | 26900 | 1 | Shelby | 411.1 | 44,871 | 1,077 | 109.1 | 92.7 | 1.9 | 0.6 | 1.1 | 4.9 | 5.7 | 12.6 | 11.7 | 12.2 | 11.8 | 13.0 |
| 18147 | | 8 | Spencer | 396.9 | 20,225 | 1,809 | 51.0 | 95.4 | 1.3 | 0.6 | 0.6 | 3.2 | 5.0 | 12.4 | 11.3 | 10.7 | 11.6 | 12.7 |
| 18149 | | 6 | Starke | 309.1 | 23,049 | 1,679 | 74.6 | 94.5 | 0.8 | 0.9 | 0.6 | 4.4 | 5.8 | 12.8 | 11.4 | 11.8 | 11.5 | 12.4 |
| 18151 | 11420 | 7 | Steuben | 308.8 | 34,831 | 1,308 | 112.8 | 94.3 | 1.2 | 0.6 | 1.1 | 3.9 | 5.6 | 10.9 | 13.6 | 10.6 | 10.1 | 12.2 |
| 18153 | 45460 | 3 | Sullivan | 447.2 | 20,578 | 1,791 | 46.0 | 92.8 | 5.4 | 0.9 | 0.5 | 1.8 | 5.0 | 10.4 | 11.8 | 14.2 | 13.3 | 13.4 |
| 18155 | | 8 | Switzerland | 220.7 | 10,724 | 2,367 | 48.6 | 96.3 | 1.5 | 0.5 | 0.5 | 2.0 | 5.6 | 14.1 | 11.1 | 10.6 | 12.2 | 13.0 |
| 18157 | 29200 | 3 | Tippecanoe | 498.9 | 196,115 | 344 | 393.1 | 76.7 | 6.9 | 0.6 | 9.1 | 8.9 | 5.6 | 11.3 | 26.8 | 14.3 | 10.8 | 9.5 |
| 18159 | | 6 | Tipton | 260.5 | 15,227 | 2,079 | 58.5 | 95.4 | 1.2 | 0.6 | 0.9 | 3.1 | 5.4 | 11.3 | 11.3 | 11.8 | 10.4 | 13.3 |
| 18161 | 17140 | 1 | Union | 161.2 | 7,119 | 2,649 | 44.2 | 95.7 | 1.5 | 0.8 | 1.0 | 2.5 | 5.5 | 11.3 | 11.8 | 11.6 | 11.3 | 13.2 |
| 18163 | 21780 | 2 | Vanderburgh | 233.4 | 182,447 | 366 | 781.7 | 85.2 | 11.9 | 0.6 | 2.2 | 3.0 | 5.9 | 12.0 | 13.2 | 14.2 | 12.1 | 11.2 |
| 18165 | 45460 | 3 | Vermillion | 256.9 | 15,329 | 2,070 | 59.7 | 97.3 | 1.2 | 0.7 | 0.7 | 1.5 | 5.5 | 12.3 | 11.9 | 11.5 | 11.5 | 12.6 |
| 18167 | 45460 | 3 | Vigo | 403.6 | 106,608 | 574 | 264.1 | 87.3 | 8.7 | 0.8 | 2.7 | 2.9 | 5.4 | 11.3 | 18.2 | 13.2 | 11.6 | 11.3 |
| 18169 | 47340 | 6 | Wabash | 412.5 | 30,784 | 1,411 | 74.6 | 95.1 | 1.4 | 1.1 | 0.8 | 2.8 | 5.3 | 11.4 | 13.8 | 11.0 | 10.9 | 11.7 |
| 18171 | 29200 | 8 | Warren | 364.7 | 8,194 | 2,565 | 22.5 | 96.1 | 1.2 | 0.7 | 0.9 | 2.4 | 5.0 | 12.9 | 10.8 | 11.1 | 11.2 | 12.8 |
| 18173 | 21780 | 2 | Warrick | 384.8 | 63,269 | 845 | 164.4 | 93.1 | 2.6 | 0.6 | 3.3 | 2.1 | 5.3 | 13.6 | 11.7 | 11.5 | 13.1 | 12.8 |
| 18175 | 31140 | 1 | Washington | 513.7 | 28,213 | 1,489 | 54.9 | 97.5 | 0.9 | 0.6 | 0.5 | 1.3 | 5.6 | 12.7 | 11.7 | 12.3 | 11.9 | 12.9 |
| 18177 | 39980 | 5 | Wayne | 401.8 | 65,778 | 810 | 163.7 | 90.2 | 6.9 | 0.8 | 1.8 | 3.3 | 5.9 | 12.3 | 13.2 | 11.7 | 11.2 | 12.3 |
| 18179 | | 2 | Wells | 368.1 | 28,142 | 1,493 | 76.5 | 94.9 | 1.5 | 0.6 | 0.9 | 3.3 | 6.3 | 14.0 | 11.5 | 11.8 | 11.8 | 11.4 |
| 18181 | | 6 | White | 505.1 | 24,165 | 1,642 | 47.8 | 89.6 | 1.0 | 0.8 | 0.7 | 9.0 | 6.2 | 12.9 | 11.6 | 11.2 | 10.7 | 12.0 |
| 18183 | 23060 | 2 | Whitley | 335.6 | 34,378 | 1,319 | 102.4 | 96.3 | 1.1 | 0.8 | 0.9 | 2.3 | 5.8 | 12.8 | 11.7 | 12.2 | 11.5 | 12.6 |
| 19000 | | 0 | IOWA | 55,853.7 | 3,163,561 | X | 56.6 | 86.2 | 5.0 | 0.7 | 3.4 | 6.5 | 6.1 | 12.9 | 13.8 | 12.6 | 12.3 | 11.3 |
| 19001 | | 8 | Adair | 569.3 | 7,059 | 2,654 | 12.4 | 95.8 | 1.3 | 0.6 | 0.8 | 2.6 | 6.5 | 12.5 | 9.4 | 11.8 | 10.8 | 10.7 |
| 19003 | | 9 | Adams | 423.4 | 3,588 | 2,927 | 8.5 | 96.7 | 1.0 | 0.8 | 0.8 | 1.6 | 5.2 | 12.6 | 9.7 | 10.6 | 10.6 | 10.1 |
| 19005 | | 6 | Allamakee | 639.0 | 13,642 | 2,181 | 21.3 | 89.9 | 2.0 | 0.5 | 0.7 | 7.6 | 6.9 | 12.7 | 10.4 | 10.3 | 10.3 | 10.7 |
| 19007 | | 7 | Appanoose | 497.3 | 12,430 | 2,253 | 25.0 | 95.7 | 1.3 | 0.8 | 1.1 | 2.4 | 5.6 | 12.7 | 10.7 | 10.0 | 10.8 | 11.4 |
| 19009 | | 8 | Audubon | 443.0 | 5,481 | 2,789 | 12.4 | 96.3 | 1.0 | 0.5 | 0.9 | 2.2 | 5.9 | 11.8 | 9.7 | 10.1 | 10.4 | 10.7 |
| 19011 | 16300 | 2 | Benton | 716.1 | 25,414 | 1,585 | 35.5 | 96.8 | 1.4 | 0.5 | 0.7 | 1.7 | 6.1 | 12.8 | 11.1 | 10.8 | 11.8 | 12.7 |
| 19013 | 47940 | 3 | Black Hawk | 565.8 | 130,786 | 491 | 231.2 | 82.5 | 11.1 | 0.6 | 3.7 | 4.7 | 6.2 | 12.1 | 17.6 | 13.1 | 11.7 | 10.1 |
| 19015 | 11180 | 2 | Boone | 570.5 | 26,277 | 1,558 | 46.1 | 95.0 | 1.7 | 0.7 | 0.9 | 2.9 | 5.3 | 11.7 | 11.1 | 12.4 | 13.0 | 12.1 |
| 19017 | 47940 | 3 | Bremer | 435.5 | 25,311 | 1,595 | 58.1 | 95.5 | 1.9 | 0.4 | 1.5 | 1.8 | 5.7 | 12.8 | 15.8 | 10.3 | 12.3 | 10.9 |
| 19019 | | 6 | Buchanan | 571.1 | 21,287 | 1,755 | 37.3 | 96.9 | 1.2 | 0.4 | 0.9 | 1.7 | 6.6 | 14.5 | 11.9 | 10.9 | 12.5 | 11.2 |
| 19021 | 44740 | 7 | Buena Vista | 574.9 | 19,772 | 1,837 | 34.4 | 56.9 | 3.6 | 0.4 | 13.2 | 27.1 | 7.4 | 14.5 | 14.9 | 12.5 | 11.8 | 9.8 |
| 19023 | | 8 | Butler | 580.1 | 14,333 | 2,135 | 24.7 | 97.0 | 0.8 | 0.4 | 1.3 | 1.6 | 5.0 | 13.3 | 11.0 | 10.0 | 11.9 | 11.6 |
| 19025 | | 9 | Calhoun | 568.9 | 9,473 | 2,457 | 16.6 | 94.6 | 3.2 | 0.7 | 0.6 | 2.0 | 5.4 | 12.3 | 11.0 | 12.1 | 11.1 | 10.3 |
| 19027 | 16140 | 7 | Carroll | 569.5 | 19,914 | 1,828 | 35.0 | 95.0 | 1.8 | 0.4 | 0.8 | 3.0 | 6.1 | 14.1 | 11.1 | 10.8 | 11.2 | 10.8 |
| 19029 | | 6 | Cass | 564.3 | 12,817 | 2,229 | 22.7 | 95.2 | 0.9 | 0.4 | 1.2 | 3.1 | 5.4 | 12.8 | 10.5 | 10.4 | 11.4 | 10.6 |
| 19031 | | 6 | Cedar | 579.5 | 18,485 | 1,900 | 31.9 | 96.0 | 1.4 | 0.6 | 1.0 | 2.4 | 5.0 | 12.3 | 10.9 | 10.6 | 12.6 | 12.3 |
| 19033 | 32380 | 5 | Cerro Gordo | 568.3 | 42,103 | 1,138 | 74.1 | 91.1 | 3.1 | 0.6 | 1.8 | 5.3 | 5.5 | 12.0 | 11.1 | 11.2 | 11.2 | 10.9 |
| 19035 | | 6 | Cherokee | 576.9 | 11,190 | 2,335 | 19.4 | 92.9 | 1.6 | 0.6 | 1.0 | 4.8 | 5.5 | 12.7 | 9.7 | 10.7 | 11.0 | 10.2 |
| 19037 | | 6 | Chickasaw | 504.3 | 11,834 | 2,303 | 23.5 | 95.6 | 1.1 | 0.3 | 0.6 | 3.3 | 6.0 | 13.3 | 11.2 | 10.3 | 11.0 | 11.0 |
| 19039 | | 6 | Clarke | 431.1 | 9,353 | 2,468 | 21.7 | 81.3 | 1.2 | 0.8 | 1.5 | 16.5 | 6.3 | 14.8 | 12.0 | 10.7 | 11.6 | 11.9 |
| 19041 | 43980 | 7 | Clay | 567.2 | 15,976 | 2,035 | 28.2 | 94.1 | 1.3 | 0.6 | 1.1 | 4.1 | 5.9 | 12.8 | 10.9 | 11.3 | 12.0 | 10.8 |
| 19043 | | 8 | Clayton | 778.5 | 17,321 | 1,955 | 22.2 | 96.6 | 1.3 | 0.4 | 0.7 | 2.1 | 5.3 | 11.9 | 10.1 | 10.0 | 10.3 | 11.1 |
| 19045 | 17540 | 4 | Clinton | 695.0 | 46,392 | 1,048 | 66.8 | 92.8 | 4.2 | 0.7 | 1.1 | 3.3 | 6.0 | 12.9 | 11.2 | 11.3 | 11.6 | 11.7 |
| 19047 | | 7 | Crawford | 714.2 | 16,834 | 1,982 | 23.6 | 64.1 | 3.5 | 0.5 | 2.7 | 29.9 | 6.9 | 13.9 | 13.5 | 12.2 | 11.1 | 11.2 |
| 19049 | 19780 | 2 | Dallas | 588.3 | 96,963 | 622 | 164.8 | 85.5 | 3.4 | 0.5 | 5.8 | 6.3 | 7.1 | 15.6 | 11.6 | 14.5 | 16.2 | 12.3 |
| 19051 | | 9 | Davis | 502.2 | 9,051 | 2,492 | 18.0 | 97.4 | 0.6 | 0.6 | 0.7 | 1.7 | 9.0 | 16.0 | 12.6 | 10.7 | 10.5 | 10.4 |
| 19053 | | 9 | Decatur | 531.9 | 7,769 | 2,608 | 14.6 | 93.1 | 2.5 | 1.0 | 1.4 | 3.4 | 6.1 | 12.4 | 20.0 | 9.7 | 9.5 | 9.2 |
| 19055 | | 6 | Delaware | 577.7 | 16,937 | 1,977 | 29.3 | 97.0 | 1.2 | 0.4 | 0.5 | 1.6 | 5.9 | 13.2 | 11.7 | 9.6 | 11.1 | 11.7 |
| 19057 | 15460 | 5 | Des Moines | 416.1 | 38,708 | 1,207 | 93.0 | 88.8 | 7.9 | 0.8 | 1.6 | 3.5 | 5.7 | 12.9 | 11.0 | 11.3 | 11.9 | 11.6 |
| 19059 | 44020 | 7 | Dickinson | 380.5 | 17,549 | 1,943 | 46.1 | 95.8 | 1.1 | 0.5 | 1.3 | 2.4 | 4.9 | 10.9 | 10.1 | 10.2 | 10.5 | 10.9 |
| 19061 | 20220 | 3 | Dubuque | 608.3 | 97,590 | 619 | 160.4 | 91.3 | 4.6 | 0.5 | 2.5 | 2.8 | 6.2 | 12.7 | 13.7 | 12.9 | 11.4 | 10.9 |
| 19063 | | 7 | Emmet | 395.9 | 9,095 | 2,487 | 23.0 | 87.5 | 1.6 | 0.8 | 1.0 | 10.1 | 5.1 | 10.3 | 13.8 | 10.8 | 11.9 | 10.6 |
| 19065 | | 6 | Fayette | 730.8 | 19,258 | 1,865 | 26.4 | 94.2 | 2.2 | 0.6 | 1.3 | 3.0 | 5.5 | 12.0 | 12.6 | 10.9 | 10.5 | 11.0 |
| 19067 | | 7 | Floyd | 500.6 | 15,480 | 2,062 | 30.9 | 91.5 | 3.3 | 0.5 | 1.8 | 4.3 | 6.1 | 12.8 | 11.2 | 10.6 | 10.8 | 11.6 |

1. CBSA = Core Based Statistical Area. See Appendix A for explanation. See Appendix B for list of metropolitan areas with component counties.    2. County type code from the Economic Research Service of USDA Rural-Urban Continuum Codes. See Appendix A for definition.    3. Dry land or land partially or temporarily covered by water.    4. May be of any race.

# Table B. States and Counties — Population and Households

| STATE County | 55 to 64 years | 65 to 74 years | 75 years and over | Percent female | 2000 | 2010 | 2000-2010 | 2010-2020 | Births | Deaths | Net Migration | Number | Persons per household | Family households | Female family householder[1] | One person |
|---|---|---|---|---|---|---|---|---|---|---|---|---|---|---|---|---|
| | 16 | 17 | 18 | 19 | 20 | 21 | 22 | 23 | 24 | 25 | 26 | 27 | 28 | 29 | 30 | 31 |
| INDIANA—Cont'd | | | | | | | | | | | | | | | | |
| Pike | 15.3 | 11.8 | 8.8 | 49.8 | 12,837 | 12,695 | -1.1 | -2.5 | 1,459 | 1,485 | -287 | 5,129 | 2.37 | 63.9 | 8.0 | 31.0 |
| Porter | 13.8 | 10.8 | 6.7 | 50.7 | 146,798 | 164,288 | 11.9 | 4.1 | 17,738 | 15,187 | 4,279 | 64,517 | 2.55 | 68.2 | 10.6 | 25.3 |
| Posey | 15.5 | 12.0 | 8.1 | 50.2 | 27,061 | 25,912 | -4.2 | -2.5 | 2,813 | 2,526 | -927 | 10,155 | 2.49 | 71.5 | 9.9 | 24.6 |
| Pulaski | 14.7 | 11.7 | 9.2 | 49.0 | 13,755 | 13,379 | -2.7 | -7.4 | 1,360 | 1,550 | -809 | 5,151 | 2.41 | 65.0 | 11.3 | 30.3 |
| Putnam | 13.6 | 9.9 | 7.4 | 47.5 | 36,019 | 37,941 | 5.3 | -1.2 | 3,726 | 3,640 | -555 | 13,543 | 2.37 | 68.3 | 10.5 | 25.3 |
| Randolph | 14.3 | 11.4 | 9.6 | 50.5 | 27,401 | 26,170 | -4.5 | -7.6 | 3,003 | 3,060 | -1,930 | 10,229 | 2.40 | 63.5 | 11.3 | 30.6 |
| Ripley | 13.7 | 10.8 | 8.0 | 50.3 | 26,523 | 28,814 | 8.6 | -1.3 | 3,594 | 3,194 | -755 | 10,992 | 2.54 | 68.1 | 9.8 | 27.7 |
| Rush | 14.8 | 10.8 | 8.4 | 50.8 | 18,261 | 17,391 | -4.8 | -4.3 | 1,897 | 1,987 | -655 | 6,667 | 2.46 | 70.7 | 9.7 | 24.3 |
| St. Joseph | 12.3 | 9.8 | 6.8 | 51.2 | 265,559 | 266,914 | 0.5 | 1.7 | 36,085 | 26,871 | -4,568 | 101,872 | 2.53 | 62.8 | 12.5 | 31.5 |
| Scott | 14.1 | 10.6 | 6.7 | 51.0 | 22,960 | 24,195 | 5.4 | -1.7 | 2,855 | 3,198 | -59 | 8,971 | 2.60 | 69.3 | 10.7 | 25.7 |
| Shelby | 14.8 | 10.6 | 7.4 | 50.4 | 43,445 | 44,379 | 2.1 | 1.1 | 5,145 | 4,690 | 64 | 17,823 | 2.45 | 67.0 | 9.7 | 26.5 |
| Spencer | 15.8 | 12.1 | 8.5 | 49.6 | 20,391 | 20,955 | 2.8 | -3.5 | 2,152 | 2,063 | -818 | 8,127 | 2.48 | 69.9 | 7.2 | 26.2 |
| Starke | 14.7 | 11.9 | 7.7 | 50.0 | 23,556 | 23,362 | -0.8 | -1.3 | 2,719 | 2,853 | -163 | 8,549 | 2.68 | 68.3 | 10.5 | 24.4 |
| Steuben | 15.3 | 13.1 | 8.5 | 49.5 | 33,214 | 34,147 | 2.8 | 2.0 | 3,881 | 3,286 | 121 | 14,261 | 2.32 | 66.3 | 7.9 | 26.4 |
| Sullivan | 13.3 | 10.8 | 7.7 | 45.3 | 21,751 | 21,473 | -1.3 | -4.2 | 2,287 | 2,441 | -740 | 7,644 | 2.43 | 69.5 | 12.3 | 26.1 |
| Switzerland | 14.9 | 10.7 | 7.9 | 48.1 | 9,065 | 10,688 | 17.9 | 0.3 | 1,257 | 1,099 | -128 | 4,349 | 2.43 | 70.3 | 10.2 | 24.5 |
| Tippecanoe | 9.5 | 7.2 | 5.0 | 49.0 | 148,955 | 172,801 | 16.0 | 13.5 | 23,037 | 12,174 | 12,500 | 70,526 | 2.48 | 55.0 | 9.6 | 31.7 |
| Tipton | 14.7 | 12.1 | 9.7 | 50.0 | 16,577 | 15,936 | -3.9 | -4.4 | 1,526 | 1,787 | -457 | 6,376 | 2.34 | 68.2 | 7.3 | 25.0 |
| Union | 15.0 | 12.1 | 8.2 | 51.3 | 7,349 | 7,516 | 2.3 | -5.3 | 738 | 734 | -404 | 2,823 | 2.50 | 66.2 | 8.3 | 29.3 |
| Vanderburgh | 13.5 | 10.4 | 7.4 | 51.5 | 171,922 | 179,701 | 4.5 | 1.5 | 22,717 | 19,986 | 169 | 75,607 | 2.31 | 59.2 | 12.8 | 33.7 |
| Vermillion | 14.1 | 12.2 | 8.5 | 50.4 | 16,788 | 16,210 | -3.4 | -5.4 | 1,690 | 2,178 | -394 | 6,664 | 2.30 | 66.9 | 12.4 | 27.2 |
| Vigo | 11.9 | 9.8 | 7.3 | 49.5 | 105,848 | 107,850 | 1.9 | -1.2 | 12,774 | 11,955 | -2,016 | 41,994 | 2.36 | 58.8 | 12.8 | 32.4 |
| Wabash | 14.2 | 11.6 | 10.2 | 51.4 | 34,960 | 32,888 | -5.9 | -6.4 | 3,429 | 4,359 | -1,169 | 12,862 | 2.29 | 66.0 | 9.2 | 28.4 |
| Warren | 14.9 | 11.5 | 9.6 | 50.2 | 8,419 | 8,511 | 1.1 | -3.7 | 878 | 810 | -386 | 3,357 | 2.43 | 74.9 | 9.0 | 22.5 |
| Warrick | 13.8 | 10.8 | 7.4 | 50.8 | 52,383 | 59,689 | 13.9 | 6.0 | 6,561 | 5,815 | 2,871 | 24,514 | 2.51 | 72.7 | 8.5 | 22.4 |
| Washington | 14.8 | 11.0 | 7.0 | 50.3 | 27,223 | 28,260 | 3.8 | -0.2 | 3,215 | 3,150 | -102 | 10,940 | 2.52 | 70.5 | 10.7 | 26.9 |
| Wayne | 13.8 | 11.1 | 8.6 | 51.7 | 71,097 | 68,996 | -3.0 | -4.7 | 8,043 | 8,897 | -2,350 | 26,644 | 2.39 | 64.6 | 14.9 | 29.4 |
| Wells | 14.2 | 10.6 | 8.2 | 50.6 | 27,600 | 27,637 | 0.1 | 1.8 | 3,509 | 2,925 | -66 | 10,888 | 2.52 | 71.4 | 10.1 | 24.9 |
| White | 14.4 | 12.3 | 8.6 | 50.1 | 25,267 | 24,641 | -2.5 | -1.9 | 2,985 | 2,873 | -576 | 9,879 | 2.41 | 68.9 | 9.7 | 25.9 |
| Whitley | 14.6 | 11.3 | 7.6 | 50.1 | 30,707 | 33,287 | 8.4 | 3.3 | 3,951 | 3,190 | 356 | 13,742 | 2.42 | 69.9 | 9.1 | 25.5 |
| IOWA | 13.0 | 10.2 | 7.7 | 50.2 | 2,926,324 | 3,046,877 | 4.1 | 3.8 | 395,193 | 300,959 | 22,741 | 1,265,473 | 2.40 | 63.4 | 9.2 | 29.3 |
| Adair | 15.4 | 11.9 | 11.1 | 50.1 | 8,243 | 7,682 | -6.8 | -8.1 | 790 | 1,074 | -338 | 3,200 | 2.17 | 63.2 | 7.4 | 30.7 |
| Adams | 17.6 | 12.7 | 10.9 | 50.1 | 4,482 | 4,029 | -10.1 | -10.9 | 424 | 480 | -391 | 1,598 | 2.21 | 64.6 | 4.9 | 28.8 |
| Allamakee | 15.1 | 13.4 | 10.2 | 49.1 | 14,675 | 14,328 | -2.4 | -4.8 | 1,854 | 1,722 | -819 | 5,953 | 2.27 | 63.8 | 5.8 | 30.5 |
| Appanoose | 14.9 | 13.4 | 10.6 | 50.3 | 13,721 | 12,892 | -6.0 | -3.6 | 1,488 | 1,744 | -196 | 5,353 | 2.30 | 59.6 | 8.3 | 37.1 |
| Audubon | 16.2 | 12.8 | 12.5 | 51.4 | 6,830 | 6,115 | -10.5 | -10.4 | 621 | 833 | -423 | 2,649 | 2.06 | 60.2 | 7.1 | 34.3 |
| Benton | 15.6 | 10.5 | 8.7 | 49.9 | 25,308 | 26,071 | 3.0 | -2.5 | 2,961 | 2,442 | -1,182 | 10,306 | 2.45 | 71.0 | 7.7 | 24.2 |
| Black Hawk | 11.8 | 10.3 | 7.2 | 50.8 | 128,012 | 131,086 | 2.4 | -0.2 | 17,344 | 12,480 | -5,082 | 52,905 | 2.41 | 59.4 | 10.0 | 30.9 |
| Boone | 15.1 | 11.7 | 7.7 | 49.3 | 26,224 | 26,308 | 0.3 | -0.1 | 2,952 | 3,069 | 101 | 10,954 | 2.35 | 66.8 | 7.9 | 28.5 |
| Bremer | 12.2 | 10.6 | 9.5 | 50.5 | 23,325 | 24,280 | 4.1 | 4.2 | 2,666 | 2,245 | 629 | 9,676 | 2.42 | 69.6 | 5.5 | 25.6 |
| Buchanan | 13.9 | 10.7 | 7.9 | 50.2 | 21,093 | 20,953 | -0.7 | 1.6 | 2,899 | 2,132 | -420 | 8,029 | 2.59 | 69.4 | 8.7 | 26.0 |
| Buena Vista | 12.4 | 9.4 | 7.2 | 49.2 | 20,411 | 20,265 | -0.7 | -2.4 | 3,172 | 1,858 | -1,817 | 7,524 | 2.53 | 66.6 | 9.3 | 28.7 |
| Butler | 14.0 | 12.6 | 10.6 | 50.4 | 15,305 | 14,869 | -2.8 | -3.6 | 1,553 | 1,917 | -163 | 6,254 | 2.30 | 68.0 | 6.2 | 27.7 |
| Calhoun | 14.1 | 12.4 | 11.2 | 47.9 | 11,115 | 10,177 | -8.4 | -6.9 | 1,108 | 1,434 | -373 | 4,146 | 2.16 | 64.4 | 9.2 | 30.6 |
| Carroll | 14.7 | 11.0 | 10.3 | 50.3 | 21,421 | 20,816 | -2.8 | -4.3 | 2,585 | 2,468 | -1,019 | 8,726 | 2.28 | 65.1 | 6.8 | 32.0 |
| Cass | 15.1 | 12.7 | 11.0 | 50.7 | 14,684 | 13,952 | -5.0 | -8.1 | 1,470 | 2,003 | -594 | 5,921 | 2.16 | 56.9 | 7.0 | 37.3 |
| Cedar | 15.5 | 11.6 | 9.1 | 50.2 | 18,187 | 18,445 | 1.4 | 0.2 | 1,860 | 1,835 | 23 | 7,391 | 2.45 | 65.2 | 5.8 | 29.5 |
| Cerro Gordo | 15.2 | 12.8 | 10.1 | 51.2 | 46,447 | 44,151 | -4.9 | -4.6 | 4,797 | 5,237 | -1,592 | 19,241 | 2.16 | 59.5 | 10.2 | 34.9 |
| Cherokee | 15.3 | 12.9 | 11.9 | 50.2 | 13,035 | 12,067 | -7.4 | -7.3 | 1,266 | 1,644 | -504 | 5,319 | 2.04 | 59.6 | 8.5 | 34.1 |
| Chickasaw | 15.5 | 11.8 | 9.9 | 49.2 | 13,095 | 12,442 | -5.0 | -4.9 | 1,435 | 1,395 | -647 | 5,173 | 2.29 | 62.1 | 3.9 | 32.0 |
| Clarke | 13.5 | 11.1 | 8.0 | 48.9 | 9,133 | 9,286 | 1.7 | 0.7 | 1,270 | 1,126 | -70 | 3,895 | 2.35 | 61.6 | 5.0 | 30.3 |
| Clay | 14.3 | 11.9 | 10.1 | 50.6 | 17,372 | 16,667 | -4.1 | -4.1 | 2,007 | 1,962 | -729 | 7,212 | 2.20 | 62.4 | 9.4 | 32.0 |
| Clayton | 16.5 | 13.7 | 11.1 | 49.4 | 18,678 | 18,128 | -2.9 | -4.5 | 1,916 | 2,035 | -681 | 7,510 | 2.30 | 61.9 | 7.2 | 32.1 |
| Clinton | 15.0 | 11.3 | 9.1 | 50.9 | 50,149 | 49,113 | -2.1 | -5.5 | 5,717 | 5,692 | -2,764 | 19,635 | 2.35 | 62.6 | 10.8 | 30.9 |
| Crawford | 13.1 | 9.9 | 8.1 | 49.0 | 16,942 | 17,091 | 0.9 | -1.5 | 2,389 | 1,689 | -953 | 6,437 | 2.55 | 69.7 | 11.2 | 24.1 |
| Dallas | 10.0 | 7.4 | 5.1 | 50.6 | 40,750 | 66,139 | 62.3 | 46.6 | 12,508 | 4,519 | 22,612 | 34,399 | 2.52 | 68.4 | 6.9 | 25.3 |
| Davis | 12.6 | 10.2 | 7.9 | 50.0 | 8,541 | 8,753 | 2.5 | 3.4 | 1,492 | 884 | -307 | 3,176 | 2.77 | 71.9 | 6.5 | 25.1 |
| Decatur | 12.0 | 11.3 | 9.9 | 49.4 | 8,689 | 8,456 | -2.7 | -8.1 | 969 | 946 | -710 | 6,907 | 2.45 | 63.1 | 8.2 | 30.1 |
| Delaware | 16.2 | 11.5 | 9.1 | 49.5 | 18,404 | 17,770 | -3.4 | -4.7 | 2,059 | 1,780 | -1,115 | 6,907 | 2.45 | 70.7 | 5.2 | 25.6 |
| Des Moines | 14.1 | 12.1 | 9.5 | 51.3 | 42,351 | 40,325 | -4.8 | -4.0 | 4,699 | 4,776 | -1,522 | 16,891 | 2.29 | 61.4 | 11.6 | 32.5 |
| Dickinson | 15.1 | 16.0 | 11.3 | 50.3 | 16,424 | 16,667 | 1.5 | 5.3 | 1,631 | 2,053 | 1,320 | 8,174 | 2.07 | 61.6 | 9.2 | 33.6 |
| Dubuque | 13.4 | 10.4 | 8.2 | 50.7 | 89,143 | 93,643 | 5.0 | 4.2 | 12,260 | 9,288 | 1,043 | 38,210 | 2.43 | 65.1 | 8.9 | 28.6 |
| Emmet | 14.9 | 12.5 | 10.1 | 49.8 | 11,027 | 10,302 | -6.6 | -11.7 | 1,119 | 1,272 | -1,062 | 4,036 | 2.10 | 60.0 | 8.3 | 33.9 |
| Fayette | 15.3 | 12.0 | 10.2 | 49.6 | 22,008 | 20,882 | -5.1 | -7.8 | 2,171 | 2,563 | -1,234 | 8,200 | 2.33 | 65.0 | 8.3 | 29.6 |
| Floyd | 14.2 | 12.2 | 10.6 | 50.5 | 16,900 | 16,293 | -3.6 | -5.0 | 1,927 | 1,909 | -822 | 6,903 | 2.24 | 60.9 | 9.0 | 34.8 |

1. No spouse present.

# Table B. States and Counties — Population, Vital Statistics, Health, and Crime

| STATE County | Persons in group quarters, 2020 | Daytime Population, 2015-2019 | | Births, 2020 | | Deaths, 2020 | | Persons under 65 with no health insurance, 2019 | | Medicare, 2020 | | | Crimes reported by county police or sheriff, 2019 | |
|---|---|---|---|---|---|---|---|---|---|---|---|---|---|---|
| | | Number | Employment/ residence ratio | Total | Rate[1] | Number | Rate[1] | Number | Percent | Total beneficiaries | Enrolled in Original Medicare | Enrolled in Medicare Advantage | Violent | Property |
| | 32 | 33 | 34 | 35 | 36 | 37 | 38 | 39 | 40 | 41 | 42 | 43 | 44 | 45 |
| INDIANA—Cont'd | | | | | | | | | | | | | | |
| Pike | 213 | 10,417 | 0.65 | 152 | 12.3 | 160 | 12.9 | 996 | 10.2 | 2,927 | 1,968 | 960 | NA | NA |
| Porter | 3,421 | 156,876 | 0.85 | 1,656 | 9.7 | 1,798 | 10.5 | 10,367 | 7.4 | 34,221 | 23,553 | 10,668 | 38 | 350 |
| Posey | 242 | 23,600 | 0.84 | 276 | 10.9 | 290 | 11.5 | 1,483 | 7.3 | 5,649 | 3,925 | 1,724 | NA | NA |
| Pulaski | 193 | 12,017 | 0.90 | 119 | 9.6 | 145 | 11.7 | 1,049 | 10.8 | 3,074 | 2,255 | 819 | 41 | 118 |
| Putnam | 4,827 | 36,211 | 0.93 | 376 | 10.0 | 378 | 10.1 | 2,249 | 8.5 | 7,450 | 4,804 | 2,645 | NA | NA |
| Randolph | 329 | 22,368 | 0.77 | 265 | 11.0 | 294 | 12.2 | 2,242 | 11.7 | 5,926 | 4,487 | 1,439 | NA | NA |
| Ripley | 449 | 26,152 | 0.84 | 394 | 13.8 | 327 | 11.5 | 2,288 | 10.0 | 6,370 | 4,521 | 1,849 | NA | NA |
| Rush | 178 | 14,740 | 0.74 | 186 | 11.2 | 197 | 11.8 | 1,464 | 11.0 | 3,724 | 2,633 | 1,091 | NA | NA |
| St. Joseph | 11,614 | 270,837 | 1.00 | 3,508 | 12.9 | 2,640 | 9.7 | 22,075 | 10.2 | 50,362 | 30,049 | 20,313 | NA | NA |
| Scott | 312 | 21,445 | 0.77 | 264 | 11.1 | 331 | 13.9 | 1,794 | 9.2 | 5,570 | 3,792 | 1,778 | 6 | 219 |
| Shelby | 683 | 41,203 | 0.85 | 488 | 10.9 | 477 | 10.6 | 3,296 | 9.0 | 9,377 | 5,762 | 3,615 | 58 | 154 |
| Spencer | 368 | 18,455 | 0.80 | 182 | 9.0 | 250 | 12.4 | 1,438 | 8.9 | 4,585 | 3,436 | 1,149 | NA | NA |
| Starke | 16 | 18,362 | 0.52 | 270 | 11.7 | 315 | 13.7 | 1,949 | 10.6 | 5,804 | 3,938 | 1,866 | 36 | 196 |
| Steuben | 1,311 | 33,259 | 0.93 | 381 | 10.9 | 379 | 10.9 | 2,729 | 10.5 | 8,133 | 4,155 | 3,978 | 12 | 322 |
| Sullivan | 2,260 | 19,348 | 0.84 | 211 | 10.3 | 226 | 11.0 | 1,379 | 9.4 | 4,447 | 3,228 | 1,220 | NA | NA |
| Switzerland | 107 | 8,598 | 0.54 | 113 | 10.5 | 124 | 11.6 | 1,092 | 12.5 | 2,090 | 1,489 | 600 | NA | NA |
| Tippecanoe | 16,552 | 203,554 | 1.13 | 2,142 | 10.9 | 1,304 | 6.6 | 15,778 | 10.1 | 25,622 | 18,302 | 7,321 | 41 | 500 |
| Tipton | 205 | 13,831 | 0.82 | 145 | 9.5 | 168 | 11.0 | 1,080 | 9.1 | 3,629 | 2,370 | 1,259 | NA | NA |
| Union | 67 | 5,308 | 0.46 | 70 | 9.8 | 70 | 9.8 | 603 | 10.7 | 1,684 | 1,258 | 426 | NA | NA |
| Vanderburgh | 7,069 | 201,922 | 1.24 | 2,069 | 11.3 | 2,032 | 11.1 | 15,281 | 10.6 | 37,708 | 24,742 | 12,966 | 51 | 482 |
| Vermillion | 209 | 13,697 | 0.72 | 172 | 11.2 | 215 | 14.0 | 1,070 | 8.7 | 3,873 | 2,932 | 941 | NA | NA |
| Vigo | 9,595 | 114,220 | 1.14 | 1,115 | 10.5 | 1,254 | 11.8 | 8,240 | 10.3 | 22,187 | 16,182 | 6,005 | 218 | NA |
| Wabash | 1,843 | 30,380 | 0.93 | 303 | 9.8 | 427 | 13.9 | 2,421 | 10.5 | 8,055 | 4,459 | 3,596 | NA | NA |
| Warren | 84 | 6,787 | 0.62 | 75 | 9.2 | 95 | 11.6 | 539 | 8.3 | 1,920 | 1,496 | 424 | NA | NA |
| Warrick | 727 | 49,816 | 0.60 | 612 | 9.7 | 675 | 10.7 | 3,948 | 7.6 | 13,199 | 9,165 | 4,034 | 88 | 508 |
| Washington | 261 | 22,658 | 0.57 | 301 | 10.7 | 324 | 11.5 | 2,491 | 10.9 | 6,234 | 4,025 | 2,210 | NA | NA |
| Wayne | 2,705 | 68,352 | 1.07 | 770 | 11.7 | 935 | 14.2 | 5,788 | 11.4 | 15,964 | 13,142 | 2,821 | 6 | 147 |
| | 472 | 26,229 | 0.87 | 335 | 11.9 | 267 | 9.5 | 2,000 | 8.8 | 6,049 | 3,217 | 2,832 | NA | NA |
| White | 307 | 23,203 | 0.91 | 287 | 11.9 | 298 | 12.3 | 2,413 | 12.7 | 5,753 | 4,211 | 1,543 | 12 | 44 |
| Whitley | 436 | 30,398 | 0.80 | 356 | 10.4 | 339 | 9.9 | 2,386 | 8.7 | 7,307 | 3,255 | 4,052 | NA | NA |
| IOWA | 96,247 | 3,140,742 | 1.00 | 37,068 | 11.7 | 30,857 | 9.8 | 144,986 | 5.7 | 637,150 | 477,159 | 159,991 | X | X |
| Adair | 154 | 6,510 | 0.83 | 98 | 13.9 | 96 | 13.6 | 280 | 5.1 | 1,901 | 1,613 | 288 | 0 | 24 |
| Adams | 69 | 3,460 | 0.87 | 36 | 10.0 | 39 | 10.9 | 163 | 6.0 | 1,017 | 988 | 29 | 3 | 20 |
| Allamakee | 215 | 12,755 | 0.84 | 174 | 12.8 | 192 | 14.1 | 900 | 8.7 | 3,397 | 2,840 | 557 | 7 | 20 |
| Appanoose | 118 | 12,119 | 0.94 | 143 | 11.5 | 172 | 13.8 | 598 | 6.3 | 3,373 | 2,451 | 922 | 8 | 70 |
| Audubon | 116 | 5,024 | 0.81 | 58 | 10.6 | 74 | 13.5 | 283 | 6.9 | 1,587 | 1,508 | 79 | NA | NA |
| Benton | 311 | 19,498 | 0.53 | 284 | 11.2 | 259 | 10.2 | 930 | 4.5 | 5,376 | 3,860 | 1,517 | NA | NA |
| Black Hawk | 4,256 | 141,063 | 1.13 | 1,594 | 12.2 | 1,282 | 9.8 | 5,945 | 5.7 | 26,841 | 16,309 | 10,532 | 20 | 123 |
| Boone | 722 | 22,829 | 0.74 | 262 | 10.0 | 318 | 12.1 | 945 | 4.5 | 6,008 | 4,669 | 1,339 | 17 | 48 |
| Bremer | 1,581 | 23,447 | 0.89 | 264 | 10.4 | 239 | 9.4 | 701 | 3.7 | 5,445 | 3,997 | 1,448 | 4 | 24 |
| Buchanan | 310 | 18,582 | 0.76 | 274 | 12.9 | 203 | 9.5 | 828 | 4.8 | 4,314 | 2,910 | 1,404 | 33 | 115 |
| Buena Vista | 796 | 20,904 | 1.08 | 288 | 14.6 | 147 | 7.4 | 1,581 | 10.2 | 3,538 | 3,212 | 326 | 12 | 41 |
| Butler | 211 | 11,876 | 0.61 | 152 | 10.6 | 192 | 13.4 | 585 | 5.3 | 3,746 | 2,912 | 834 | 7 | 25 |
| Calhoun | 723 | 8,764 | 0.77 | 103 | 10.9 | 142 | 15.0 | 355 | 5.2 | 2,576 | 2,327 | 249 | 8 | 31 |
| Carroll | 398 | 21,212 | 1.09 | 225 | 11.3 | 242 | 12.2 | 715 | 4.5 | 4,905 | 4,295 | 610 | NA | NA |
| Cass | 291 | 13,177 | 1.01 | 136 | 10.6 | 184 | 14.4 | 705 | 7.3 | 3,635 | 3,140 | 496 | 10 | 38 |
| Cedar | 319 | 14,036 | 0.53 | 172 | 9.3 | 195 | 10.5 | 621 | 4.2 | 4,065 | 3,063 | 1,002 | 12 | 43 |
| Cerro Gordo | 1,132 | 45,626 | 1.13 | 462 | 11.0 | 541 | 12.8 | 1,537 | 4.8 | 11,145 | 9,907 | 1,238 | 10 | 98 |
| Cherokee | 361 | 10,927 | 0.93 | 114 | 10.2 | 170 | 15.2 | 449 | 5.4 | 3,053 | 2,672 | 381 | 2 | 12 |
| Chickasaw | 90 | 11,623 | 0.94 | 139 | 11.7 | 131 | 11.1 | 625 | 6.7 | 2,831 | 2,492 | 338 | 10 | 29 |
| Clarke | 156 | 9,539 | 1.04 | 121 | 12.9 | 111 | 11.9 | 535 | 7.2 | 2,110 | 1,772 | 339 | 7 | 36 |
| Clay | 215 | 16,711 | 1.06 | 186 | 11.6 | 191 | 12.0 | 737 | 5.9 | 4,001 | 3,788 | 214 | 13 | 11 |
| Clayton | 255 | 15,811 | 0.80 | 176 | 10.2 | 210 | 12.1 | 964 | 7.3 | 4,648 | 3,399 | 1,249 | 11 | 41 |
| Clinton | 633 | 46,045 | 0.96 | 516 | 11.1 | 529 | 11.4 | 1,928 | 5.2 | 10,903 | 7,918 | 2,985 | 17 | 92 |
| Crawford | 535 | 16,729 | 0.96 | 217 | 12.9 | 157 | 9.3 | 1,569 | 11.5 | 3,356 | 2,933 | 423 | 3 | 3 |
| Dallas | 692 | 78,807 | 0.82 | 1,309 | 13.5 | 538 | 5.5 | 3,257 | 3.9 | 13,145 | 9,702 | 3,443 | NA | NA |
| Davis | 92 | 8,173 | 0.82 | 158 | 17.5 | 84 | 9.3 | 746 | 10.2 | 1,731 | 1,353 | 377 | 4 | 1 |
| Decatur | 652 | 7,279 | 0.81 | 90 | 11.6 | 78 | 10.0 | 399 | 7.1 | 1,815 | 1,498 | 317 | NA | NA |
| Delaware | 218 | 15,765 | 0.85 | 189 | 11.2 | 190 | 11.2 | 753 | 5.5 | 3,812 | 2,798 | 1,014 | 4 | 46 |
| Des Moines | 630 | 42,071 | 1.14 | 445 | 11.5 | 475 | 12.3 | 1,531 | 5.1 | 9,619 | 7,955 | 1,664 | 14 | 118 |
| Dickinson | 180 | 17,327 | 1.02 | 152 | 8.7 | 186 | 10.6 | 581 | 4.6 | 5,250 | 4,845 | 406 | 5 | 29 |
| Dubuque | 4,139 | 105,435 | 1.17 | 1,196 | 12.3 | 970 | 9.9 | 3,726 | 4.9 | 20,642 | 9,402 | 11,240 | 13 | 114 |
| Emmet | 431 | 8,839 | 0.87 | 106 | 11.7 | 113 | 12.4 | 490 | 7.1 | 2,343 | 2,201 | 143 | 0 | 7 |
| Fayette | 832 | 18,738 | 0.89 | 202 | 10.5 | 233 | 12.1 | 839 | 5.7 | 4,947 | 3,873 | 1,074 | NA | NA |
| Floyd | 228 | 15,233 | 0.93 | 171 | 11.0 | 176 | 11.4 | 760 | 6.3 | 3,955 | 3,416 | 539 | NA | NA |

1. Per 1,000 estimated resident population.

# Table B. States and Counties — Crime, Education, Money Income, and Poverty

Group headers: **County law enforcement employment, 2019** (Officers, Civilians) · **Education — School enrollment and attainment, 2015-2019** (Enrollment: Total, Percent private; Attainment (percent): High school graduate or less, Bachelor's degree or more) · **Local government expenditures, 2017-2018** (Total current spending (mil dol), Current spending per student (dollars)) · **Money income, 2015-2019** (Per capita income; Households: Median income (dollars), Percent with income of less than $50,000, Percent with income of $200,000 or more) · **Income and poverty, 2019** (Median household income (dollars); Percent below poverty level: All persons, Children under 18 years, Children 5 to 17 years in families)

| STATE County | Officers (46) | Civilians (47) | Enroll. Total (48) | Enroll. % private (49) | HS grad or less (50) | Bachelor's+ (51) | Total cur. spend (mil dol) (52) | Cur. spend per student $ (53) | Per capita income (54) | Median income $ (55) | % income < $50,000 (56) | % income ≥ $200,000 (57) | Median HH income $ (58) | Poverty All persons (59) | Children under 18 (60) | Children 5–17 in families (61) |
|---|---|---|---|---|---|---|---|---|---|---|---|---|---|---|---|---|
| **INDIANA—Cont'd** | | | | | | | | | | | | | | | | |
| Pike | NA | NA | 2,792 | 5.7 | 55.0 | 14.1 | 19 | 10,187 | 25,624 | 50,194 | 49.9 | 0.6 | 52,378 | 9.4 | 12.2 | 10.9 |
| Porter | 66 | 9 | 41,090 | 19.2 | 41.5 | 28.4 | 265 | 9,865 | 34,595 | 71,152 | 35.4 | 5.9 | 74,064 | 8.1 | 8.5 | 7.1 |
| Posey | NA | NA | 6,007 | 23.9 | 44.8 | 21.9 | 38 | 10,865 | 32,091 | 64,196 | 38.2 | 2.9 | 68,435 | 8.6 | 10.1 | 8.7 |
| Pulaski | NA | NA | 2,859 | 9.2 | 57.6 | 12.8 | 22 | 10,583 | 24,992 | 49,580 | 50.4 | 1.1 | 52,432 | 11.2 | 15.6 | 14.7 |
| Putnam | NA | NA | 9,390 | 30.9 | 53.5 | 16.8 | 60 | 10,573 | 26,120 | 61,047 | 39.9 | 2.6 | 64,098 | 10.5 | 13.6 | 12.4 |
| Randolph | NA | NA | 5,317 | 9.1 | 55.9 | 14.3 | 43 | 8,618 | 24,964 | 48,036 | 52.6 | 1.4 | 50,496 | 13.3 | 18.9 | 16.6 |
| Ripley | NA | NA | 6,423 | 14.6 | 54.3 | 18.0 | 39 | 12,187 | 28,312 | 56,332 | 44.7 | 3.1 | 61,219 | 9.1 | 12.0 | 11.3 |
| Rush | NA | NA | 3,499 | 7.4 | 59.1 | 13.9 | 24 | 10,546 | 26,301 | 54,346 | 44.7 | 2.3 | 54,371 | 11.0 | 15.1 | 13.9 |
| St. Joseph | 116 | 168 | 73,259 | 33.7 | 42.2 | 29.6 | 399 | 10,147 | 28,215 | 52,769 | 47.9 | 4.1 | 53,881 | 15.3 | 21.5 | 19.6 |
| Scott | NA | NA | 4,780 | 20.3 | 60.3 | 12.0 | 39 | 9,917 | 23,646 | 48,700 | 51.3 | 1.0 | 50,571 | 13.6 | 18.7 | 17.5 |
| Shelby | 33 | 65 | 10,068 | 9.3 | 55.1 | 18.7 | 71 | 9,550 | 29,583 | 60,404 | 39.5 | 3.9 | 59,798 | 10.1 | 13.4 | 11.9 |
| Spencer | NA | NA | 4,298 | 6.9 | 55.6 | 15.2 | 30 | 9,284 | 31,287 | 57,305 | 42.5 | 3.1 | 59,183 | 8.7 | 10.2 | 9.9 |
| Starke | 13 | 23 | 5,003 | 13.7 | 57.2 | 11.9 | 35 | 10,152 | 24,673 | 51,190 | 48.6 | 1.6 | 54,001 | 13.3 | 18.7 | 16.1 |
| Steuben | 23 | 30 | 7,852 | 25.5 | 48.5 | 21.8 | 39 | 10,307 | 29,949 | 58,279 | 41.1 | 2.8 | 59,444 | 8.6 | 12.0 | 11.1 |
| Sullivan | NA | NA | 4,062 | 10.4 | 55.5 | 12.1 | 31 | 9,803 | 24,356 | 50,245 | 49.8 | 1.4 | 53,375 | 14.4 | 19.0 | 17.0 |
| Switzerland | NA | NA | 2,384 | 6.0 | 62.5 | 10.9 | 15 | 10,145 | 24,944 | 49,383 | 50.4 | 2.5 | 52,482 | 14.5 | 23.7 | 21.8 |
| Tippecanoe | NA | NA | 73,012 | 9.2 | 33.8 | 38.7 | 239 | 10,038 | 27,451 | 53,130 | 47.7 | 4.2 | 51,228 | 16.1 | 14.0 | 13.1 |
| Tipton | NA | NA | 3,030 | 9.8 | 51.5 | 24.7 | 22 | 9,626 | 31,025 | 58,118 | 40.3 | 2.9 | 69,960 | 7.8 | 9.5 | 8.4 |
| Union | NA | NA | 1,381 | 6.0 | 55.9 | 18.5 | 15 | 11,190 | 25,661 | 50,375 | 49.4 | 1.4 | 53,866 | 9.6 | 13.4 | 12.5 |
| Vanderburgh | NA | NA | 43,316 | 21.1 | 42.4 | 27.0 | 249 | 10,654 | 29,638 | 49,708 | 50.2 | 3.2 | 51,603 | 14.4 | 19.4 | 16.5 |
| Vermillion | NA | NA | 3,350 | 6.7 | 52.4 | 14.2 | 24 | 9,512 | 26,493 | 50,243 | 49.7 | 1.7 | 50,723 | 12.2 | 16.3 | 15.2 |
| Vigo | NA | NA | 29,254 | 14.3 | 43.6 | 25.1 | 160 | 11,010 | 25,113 | 45,230 | 54.1 | 2.6 | 48,082 | 20.8 | 24.8 | 23.0 |
| Wabash | NA | NA | 7,073 | 18.6 | 53.8 | 18.1 | 52 | 9,680 | 27,149 | 54,259 | 46.0 | 2.3 | 59,874 | 10.9 | 14.7 | 12.9 |
| Warren | NA | NA | 1,696 | 5.2 | 49.8 | 19.4 | 13 | 9,607 | 31,104 | 60,553 | 40.5 | 2.5 | 64,109 | 9.4 | 14.1 | 12.5 |
| Warrick | NA | NA | 15,014 | 14.9 | 38.0 | 30.3 | 96 | 9,397 | 37,466 | 73,482 | 33.6 | 7.4 | 79,644 | 6.3 | 8.0 | 7.0 |
| Washington | NA | NA | 5,915 | 12.7 | 58.0 | 13.4 | 42 | 10,225 | 25,015 | 47,983 | 51.8 | 2.7 | 51,948 | 12.1 | 17.7 | 16.9 |
| Wayne | NA | NA | 14,935 | 17.7 | 50.8 | 18.2 | 98 | 9,411 | 26,510 | 46,516 | 53.4 | 3.0 | 47,885 | 15.2 | 23.3 | 18.5 |
| Wells | 16 | 0 | 6,545 | 12.3 | 48.0 | 18.2 | 46 | 9,264 | 28,663 | 59,237 | 42.4 | 3.2 | 61,636 | 7.4 | 10.3 | 9.3 |
| White | 16 | 29 | 5,299 | 8.9 | 52.8 | 16.4 | 47 | 9,909 | 27,461 | 54,576 | 46.7 | 1.9 | 58,295 | 10.3 | 13.3 | 12.5 |
| Whitley | NA | NA | 7,195 | 18.8 | 47.3 | 20.7 | 56 | 9,113 | 31,843 | 61,741 | 39.7 | 3.5 | 67,347 | 7.7 | 9.4 | 9.0 |
| **IOWA** | X | X | 797,393 | 15.1 | 38.9 | 28.6 | 6,005 | 11,732 | 32,176 | 60,523 | 41.3 | 4.6 | 61,807 | 11.0 | 12.8 | 11.7 |
| Adair | 4 | 9 | 1,500 | 8.6 | 46.3 | 18.5 | 10 | 10,862 | 30,822 | 53,363 | 47.1 | 2.1 | 57,977 | 10.7 | 12.9 | 12.0 |
| Adams | 8 | 0 | 656 | 9.1 | 45.4 | 15.6 | 6 | 14,087 | 28,440 | 49,255 | 50.9 | 3.2 | 53,385 | 12.5 | 19.6 | 18.4 |
| Allamakee | 10 | 11 | 3,136 | 21.0 | 51.1 | 17.8 | 26 | 11,376 | 27,892 | 52,216 | 48.2 | 3.2 | 56,240 | 13.0 | 15.6 | 15.2 |
| Appanoose | 8 | 8 | 2,551 | 12.2 | 46.4 | 16.3 | 21 | 10,260 | 26,012 | 40,167 | 59.4 | 2.0 | 41,627 | 19.3 | 24.1 | 22.9 |
| Audubon | NA | NA | 994 | 7.2 | 47.6 | 17.1 | 6 | 11,597 | 30,344 | 52,055 | 48.2 | 2.6 | 70,023 | 7.8 | 8.2 | 7.7 |
| Benton | 14 | 21 | 5,625 | 10.0 | 42.8 | 22.1 | 40 | 10,571 | 33,301 | 67,729 | 36.4 | 4.2 | 60,010 | 13.7 | 15.2 | 13.7 |
| Black Hawk | 74 | 49 | 37,807 | 11.8 | 39.1 | 29.1 | 251 | 13,227 | 30,563 | 53,539 | 46.4 | 3.8 | 63,975 | 8.7 | 10.5 | 9.5 |
| Boone | 13 | 18 | 5,874 | 7.3 | 37.6 | 25.3 | 42 | 10,538 | 32,889 | 64,000 | 39.2 | 3.8 | 67,229 | 6.7 | 6.3 | 5.9 |
| Bremer | 13 | 25 | 7,150 | 31.9 | 36.5 | 30.7 | 58 | 10,632 | 32,814 | 70,395 | 33.2 | 3.1 | 67,906 | 9.5 | 14.4 | 13.3 |
| Buchanan | 13 | 17 | 5,147 | 14.3 | 44.4 | 21.5 | 33 | 10,533 | 33,373 | 64,837 | 39.2 | 4.0 |  |  |  |  |
| Buena Vista | 14 | 22 | 5,462 | 12.2 | 51.9 | 19.7 | 50 | 11,525 | 26,526 | 57,125 | 43.7 | 2.4 | 57,048 | 10.2 | 13.5 | 12.9 |
| Butler | 12 | 6 | 3,192 | 7.1 | 45.5 | 17.3 | 20 | 10,702 | 30,036 | 57,115 | 41.5 | 2.5 | 57,080 | 10.5 | 12.8 | 10.1 |
| Calhoun | 7 | 5 | 1,988 | 7.5 | 44.2 | 18.7 | 18 | 10,320 | 28,434 | 52,323 | 47.8 | 3.9 | 57,007 | 11.5 | 14.2 | 13.8 |
| Carroll | NA | NA | 4,643 | 25.6 | 43.1 | 22.4 | 35 | 10,374 | 31,648 | 59,212 | 44.6 | 3.1 | 61,379 | 8.2 | 9.0 | 8.5 |
| Cass | 9 | 3 | 2,716 | 4.2 | 49.9 | 20.1 | 33 | 12,920 | 28,302 | 50,187 | 49.8 | 2.2 | 53,692 | 14.3 | 19.0 | 17.1 |
| Cedar | 14 | 25 | 4,177 | 5.3 | 42.0 | 22.2 | 36 | 10,428 | 32,268 | 67,795 | 34.6 | 3.8 | 71,029 | 7.8 | 8.8 | 8.0 |
| Cerro Gordo | 20 | 53 | 9,033 | 14.2 | 37.9 | 23.6 | 69 | 11,613 | 32,757 | 53,963 | 46.1 | 3.8 | 54,589 | 8.5 | 12.7 | 12.0 |
| Cherokee | 8 | 13 | 2,156 | 11.4 | 42.7 | 22.2 | 18 | 10,584 | 32,999 | 56,550 | 42.3 | 4.6 | 59,901 | 10.2 | 12.2 | 11.1 |
| Chickasaw | 10 | 6 | 2,354 | 8.5 | 50.8 | 17.2 | 18 | 10,716 | 31,095 | 60,034 | 42.1 | 2.1 | 60,111 | 8.5 | 12.2 | 12.0 |
| Clarke | 7 | 15 | 2,040 | 10.5 | 51.1 | 16.0 | 19 | 10,515 | 27,416 | 54,427 | 45.4 | 2.1 | 57,708 | 11.7 | 17.5 | 15.7 |
| Clay | 11 | 14 | 3,540 | 13.1 | 40.9 | 22.7 | 27 | 11,174 | 28,747 | 50,521 | 49.7 | 1.6 | 57,174 | 10.3 | 13.9 | 13.3 |
| Clayton | 13 | 16 | 3,522 | 14.7 | 53.3 | 17.7 | 49 | 21,137 | 28,294 | 53,152 | 46.8 | 2.8 | 56,865 | 11.0 | 14.0 | 13.1 |
| Clinton | 25 | 22 | 10,321 | 10.2 | 47.9 | 17.9 | 88 | 11,579 | 28,483 | 51,688 | 48.7 | 2.0 | 58,077 | 13.1 | 18.3 | 16.7 |
| Crawford | 11 | 10 | 4,293 | 11.4 | 57.2 | 14.3 | 33 | 10,725 | 29,196 | 55,755 | 45.0 | 3.3 | 58,077 | 12.5 | 17.2 | 16.3 |
| Dallas | 29 | 43 | 23,384 | 16.0 | 22.1 | 50.4 | 193 | 10,103 | 45,090 | 88,479 | 24.5 | 11.5 | 97,535 | 4.3 | 4.4 | 4.1 |
| Davis | 6 | 11 | 1,623 | 25.0 | 49.4 | 19.5 | 14 | 11,926 | 29,387 | 63,404 | 37.4 | 5.8 | 57,150 | 11.5 | 16.5 | 16.9 |
| Decatur | 6 | 10 | 2,278 | 32.7 | 48.4 | 25.0 | 13 | 12,712 | 22,875 | 44,462 | 55.0 | 1.8 | 44,873 | 18.3 | 24.3 | 22.6 |
| Delaware | 14 | 7 | 3,926 | 20.2 | 51.1 | 16.6 | 24 | 10,611 | 32,112 | 63,750 | 37.9 | 2.7 | 70,091 | 8.6 | 10.9 | 10.5 |
| Des Moines | 22 | 4 | 8,851 | 12.3 | 41.4 | 20.1 | 70 | 11,116 | 29,941 | 51,267 | 48.4 | 2.8 | 53,547 | 14.9 | 21.2 | 19.5 |
| Dickinson | 9 | 11 | 3,336 | 4.6 | 30.3 | 33.6 | 29 | 11,354 | 39,052 | 59,969 | 41.0 | 6.0 | 66,211 | 7.0 | 8.3 | 7.3 |
| Dubuque | 79 | 12 | 25,158 | 30.8 | 40.0 | 30.8 | 170 | 11,601 | 32,905 | 63,031 | 39.2 | 4.9 | 62,481 | 10.7 | 11.4 | 10.5 |
| Emmet | 9 | 9 | 2,384 | 6.3 | 42.2 | 18.8 | 20 | 10,887 | 28,651 | 51,563 | 49.0 | 3.2 | 55,635 | 11.4 | 15.3 | 15.0 |
| Fayette | 12 | 26 | 4,515 | 18.3 | 47.9 | 18.3 | 42 | 11,848 | 28,643 | 51,128 | 48.8 | 3.4 | 53,142 | 13.2 | 16.8 | 16.4 |
| Floyd | 11 | 8 | 3,420 | 13.1 | 45.3 | 19.8 | 24 | 11,810 | 29,622 | 50,406 | 49.5 | 3.4 | 55,369 | 11.7 | 15.2 | 14.5 |

Column item numbers: 46 · 47 · 48 · 49 · 50 · 51 · 52 · 53 · 54 · 55 · 56 · 57 · 58 · 59 · 60 · 61

1. All persons 3 years old and over enrolled in nursery school through college. 2. Persons 25 years old and over. 3. Elementary and secondary education expenditures. 4. Based on population estimated by the American Community Survey, 2014–2018.

| STATE County | Personal income, 2019 | | | | | | | | | | Earnings, 2019 | | |
|---|---|---|---|---|---|---|---|---|---|---|---|---|---|
| | Total (mil dol) | Percent change 2018-2019 | Per capita[1] Dollars | Rank | Wages and salaries (mil dol) | Supplements to wages and salaries, employer contributions (mil dol) Pension and insurance | Government social insurance | Proprietors' income (mil dol) | Dividends, interest, and rent (mil dol) | Personal transfer receipts (mil dol) | Total (mil dol) | Contributions for government social insurance (mil dol) From employee and self-employed | From employer |
| | 62 | 63 | 64 | 65 | 66 | 67 | 68 | 69 | 70 | 71 | 72 | 73 | 74 |
| INDIANA—Cont'd | | | | | | | | | | | | | |
| Pike | 518 | 4.0 | 41,827 | 1,793 | 151 | 34 | 11 | 19 | 68 | 145 | 215 | 16 | 11 |
| Porter | 9,324 | 3.9 | 54,723 | 511 | 3,185 | 501 | 235 | 570 | 1,466 | 1,574 | 4,491 | 293 | 235 |
| Posey | 1,282 | 4.2 | 50,407 | 801 | 564 | 119 | 39 | 77 | 209 | 253 | 800 | 50 | 39 |
| Pulaski | 562 | 2.6 | 45,473 | 1,302 | 219 | 41 | 16 | 43 | 100 | 136 | 319 | 20 | 16 |
| Putnam | 1,439 | 3.1 | 38,303 | 2,292 | 570 | 106 | 43 | 65 | 208 | 334 | 784 | 53 | 43 |
| Randolph | 967 | 2.4 | 39,222 | 2,159 | 269 | 54 | 20 | 84 | 146 | 282 | 428 | 30 | 20 |
| Ripley | 1,226 | 3.0 | 43,282 | 1,575 | 620 | 111 | 44 | 68 | 211 | 263 | 843 | 54 | 44 |
| Rush | 787 | 3.5 | 47,445 | 1,086 | 225 | 45 | 16 | 78 | 117 | 179 | 364 | 23 | 16 |
| St. Joseph | 13,193 | 1.9 | 48,535 | 965 | 6,469 | 1,023 | 478 | 1,388 | 2,325 | 2,536 | 9,359 | 573 | 478 |
| Scott | 930 | 4.4 | 38,948 | 2,200 | 340 | 65 | 25 | 37 | 109 | 271 | 467 | 34 | 25 |
| Shelby | 2,004 | 3.4 | 44,799 | 1,383 | 904 | 149 | 66 | 114 | 291 | 459 | 1,233 | 81 | 66 |
| Spencer | 941 | 5.3 | 46,393 | 1,200 | 318 | 61 | 23 | 72 | 134 | 198 | 473 | 31 | 23 |
| Starke | 820 | 2.1 | 35,648 | 2,599 | 162 | 34 | 13 | 47 | 102 | 281 | 257 | 22 | 13 |
| Steuben | 1,552 | 3.6 | 44,854 | 1,379 | 671 | 123 | 50 | 82 | 303 | 344 | 926 | 62 | 50 |
| Sullivan | 762 | 3.0 | 36,863 | 2,460 | 289 | 63 | 20 | 34 | 117 | 228 | 405 | 28 | 20 |
| Switzerland | 346 | 3.5 | 32,174 | 2,935 | 85 | 14 | 6 | 20 | 49 | 87 | 124 | 9 | 6 |
| Tippecanoe | 7,736 | 2.4 | 39,525 | 2,110 | 4,737 | 914 | 339 | 481 | 1,478 | 1,247 | 6,471 | 379 | 339 |
| Tipton | 718 | 1.8 | 47,392 | 1,092 | 273 | 47 | 20 | 40 | 135 | 164 | 381 | 26 | 20 |
| Union | 287 | 3.2 | 40,659 | 1,968 | 51 | 11 | 4 | 23 | 44 | 70 | 89 | 7 | 4 |
| Vanderburgh | 8,571 | 3.9 | 47,233 | 1,105 | 5,694 | 884 | 417 | 512 | 1,584 | 1,936 | 7,507 | 470 | 417 |
| Vermillion | 599 | 3.1 | 38,633 | 2,246 | 241 | 44 | 17 | 16 | 90 | 183 | 318 | 24 | 17 |
| Vigo | 4,221 | 2.9 | 39,432 | 2,121 | 2,227 | 414 | 166 | 201 | 761 | 1,199 | 3,007 | 196 | 166 |
| Wabash | 1,362 | 2.7 | 43,955 | 1,495 | 501 | 90 | 37 | 86 | 269 | 404 | 715 | 51 | 37 |
| Warren | 374 | 1.4 | 45,202 | 1,339 | 88 | 16 | 7 | 33 | 55 | 81 | 143 | 9 | 7 |
| Warrick | 3,688 | 4.4 | 58,539 | 354 | 884 | 142 | 64 | 260 | 623 | 601 | 1,351 | 91 | 64 |
| Washington | 1,115 | 3.0 | 39,759 | 2,082 | 248 | 48 | 19 | 80 | 133 | 286 | 394 | 29 | 19 |
| Wayne | 2,824 | 3.1 | 42,857 | 1,641 | 1,339 | 245 | 99 | 172 | 427 | 837 | 1,855 | 124 | 99 |
| Wells | 1,213 | 2.3 | 42,857 | 1,641 | 471 | 83 | 35 | 82 | 213 | 256 | 670 | 44 | 35 |
| White | 1,061 | 1.0 | 44,008 | 1,483 | 403 | 73 | 31 | 58 | 193 | 255 | 566 | 38 | 31 |
| Whitley | 1,581 | 2.6 | 46,555 | 1,181 | 630 | 116 | 45 | 89 | 247 | 309 | 880 | 58 | 45 |
| IOWA | 163,639 | 3.6 | 51,791 | X | 78,469 | 14,136 | 6,003 | 16,122 | 31,476 | 29,100 | 114,731 | 7,086 | 6,003 |
| Adair | 395 | 5.8 | 55,251 | 485 | 118 | 24 | 10 | 46 | 68 | 76 | 198 | 12 | 10 |
| Adams | 240 | 8.0 | 66,683 | 161 | 54 | 12 | 4 | 83 | 34 | 42 | 153 | 8 | 4 |
| Allamakee | 684 | 5.4 | 49,945 | 835 | 191 | 45 | 15 | 117 | 130 | 138 | 368 | 22 | 15 |
| Appanoose | 497 | 3.8 | 40,000 | 2,051 | 175 | 39 | 14 | 38 | 82 | 155 | 267 | 20 | 14 |
| Audubon | 321 | 11.6 | 58,359 | 360 | 66 | 14 | 5 | 97 | 62 | 61 | 183 | 8 | 5 |
| Benton | 1,424 | 4.1 | 55,520 | 472 | 254 | 55 | 20 | 259 | 231 | 234 | 588 | 37 | 20 |
| Black Hawk | 6,079 | 1.8 | 46,326 | 1,210 | 3,708 | 644 | 288 | 311 | 1,207 | 1,291 | 4,951 | 318 | 288 |
| Boone | 1,306 | 3.5 | 49,794 | 844 | 415 | 89 | 35 | 100 | 237 | 267 | 638 | 43 | 35 |
| Bremer | 1,273 | 3.1 | 50,810 | 770 | 446 | 89 | 34 | 93 | 256 | 224 | 662 | 43 | 34 |
| Buchanan | 1,076 | 4.0 | 50,823 | 768 | 275 | 59 | 22 | 160 | 221 | 189 | 516 | 30 | 22 |
| Buena Vista | 949 | 0.8 | 48,394 | 986 | 470 | 91 | 35 | 157 | 169 | 166 | 753 | 42 | 35 |
| Butler | 715 | 4.3 | 49,509 | 877 | 145 | 33 | 12 | 90 | 139 | 154 | 281 | 19 | 12 |
| Calhoun | 525 | 7.3 | 54,273 | 528 | 115 | 26 | 9 | 104 | 100 | 106 | 253 | 14 | 9 |
| Carroll | 1,179 | 4.6 | 58,482 | 356 | 481 | 90 | 36 | 249 | 234 | 205 | 856 | 47 | 36 |
| Cass | 644 | 3.3 | 50,153 | 823 | 233 | 52 | 18 | 74 | 134 | 159 | 377 | 24 | 18 |
| Cedar | 1,051 | 3.9 | 56,419 | 425 | 223 | 46 | 18 | 119 | 207 | 157 | 406 | 26 | 18 |
| Cerro Gordo | 2,303 | 3.6 | 54,253 | 531 | 1,114 | 193 | 87 | 283 | 462 | 494 | 1,676 | 111 | 87 |
| Cherokee | 738 | 5.5 | 65,727 | 174 | 207 | 42 | 16 | 241 | 126 | 123 | 507 | 29 | 16 |
| Chickasaw | 746 | 5.5 | 62,483 | 239 | 220 | 41 | 18 | 207 | 136 | 119 | 486 | 29 | 18 |
| Clarke | 395 | -1.3 | 42,031 | 1,756 | 183 | 36 | 14 | 26 | 61 | 99 | 260 | 17 | 14 |
| Clay | 869 | 3.9 | 54,270 | 530 | 376 | 71 | 29 | 159 | 177 | 165 | 635 | 37 | 29 |
| Clayton | 912 | 5.3 | 51,982 | 682 | 293 | 60 | 25 | 146 | 195 | 191 | 523 | 31 | 25 |
| Clinton | 2,157 | 4.4 | 46,466 | 1,188 | 905 | 171 | 73 | 147 | 363 | 516 | 1,295 | 88 | 73 |
| Crawford | 771 | 6.9 | 45,817 | 1,268 | 313 | 61 | 23 | 129 | 141 | 154 | 527 | 30 | 23 |
| Dallas | 6,378 | 5.1 | 68,246 | 147 | 2,785 | 384 | 202 | 336 | 1,120 | 537 | 3,707 | 229 | 202 |
| Davis | 363 | 3.0 | 40,289 | 2,015 | 80 | 19 | 6 | 86 | 53 | 76 | 191 | 12 | 6 |
| Decatur | 287 | 3.6 | 36,494 | 2,509 | 88 | 21 | 8 | 24 | 49 | 81 | 140 | 10 | 8 |
| Delaware | 933 | 4.9 | 54,823 | 505 | 303 | 64 | 24 | 193 | 175 | 150 | 584 | 32 | 24 |
| Des Moines | 1,977 | 1.9 | 50,725 | 777 | 955 | 175 | 79 | 235 | 372 | 459 | 1,443 | 95 | 79 |
| Dickinson | 1,076 | 4.2 | 62,364 | 240 | 381 | 74 | 32 | 132 | 308 | 191 | 618 | 40 | 32 |
| Dubuque | 5,014 | 3.2 | 51,525 | 724 | 2,882 | 472 | 222 | 339 | 1,144 | 911 | 3,914 | 251 | 222 |
| Emmet | 451 | 5.6 | 48,926 | 930 | 153 | 32 | 12 | 68 | 82 | 101 | 265 | 16 | 12 |
| Fayette | 916 | 3.3 | 46,621 | 1,170 | 276 | 60 | 23 | 135 | 183 | 229 | 494 | 31 | 23 |
| Floyd | 785 | 4.8 | 50,203 | 814 | 276 | 62 | 21 | 93 | 147 | 175 | 453 | 28 | 21 |

1. Based on the resident population estimated as of July 1 of the year shown.

# Table B. States and Counties — Earnings, Social Security, and Housing

Column groups: Earnings, 2019 (cont.) — Percent by selected industries (items 75–83) · Social Security beneficiaries, December 2019 (items 84–85) · Supplemental Security Income recipients, 2019 (item 86) · Housing units, 2020 (items 87–88)

| STATE County | Farm | Mining, quarrying, and extractions | Construction | Manufacturing | Information; professional, scientific, technical services | Retail trade | Finance, insurance, real estate, and leasing | Health care and social assistance | Government | Number | Rate[1] | Supplemental Security Income recipients, 2019 | Total | Percent change, 2010-2020 |
|---|---|---|---|---|---|---|---|---|---|---|---|---|---|---|
| | 75 | 76 | 77 | 78 | 79 | 80 | 81 | 82 | 83 | 84 | 85 | 86 | 87 | 88 |
| **INDIANA—Cont'd** | | | | | | | | | | | | | | |
| Pike | 2.6 | 2.3 | 2.0 | 6.0 | D | 3.4 | 2.8 | D | 15.9 | 3,285 | 265 | 231 | 5,811 | 2.6 |
| Porter | 0.2 | 0.1 | 11.7 | 20.5 | 6.1 | 6.7 | 4.5 | 13.2 | 9.5 | 36,320 | 213 | 2,043 | 70,359 | 6.4 |
| Posey | 2.6 | 0.2 | 5.5 | 46.3 | D | 5.0 | 2.6 | D | 8.1 | 6,035 | 238 | 342 | 11,624 | 3.7 |
| Pulaski | 8.0 | D | 3.8 | 33.2 | D | 5.0 | 4.1 | D | 19.7 | 3,320 | 268 | 269 | 6,150 | 1.6 |
| Putnam | 0.1 | 0.7 | 7.5 | 21.3 | 3.0 | 5.2 | 3.8 | D | 17.8 | 7,930 | 211 | 480 | 15,480 | 5.3 |
| Randolph | 6.1 | D | 10.9 | 25.9 | 6.2 | 3.9 | 3.1 | D | 14.8 | 6,485 | 264 | 489 | 11,698 | -0.4 |
| Ripley | 2.2 | 0.2 | 5.5 | 15.6 | D | 4.0 | 3.2 | D | 8.6 | 6,240 | 219 | 322 | 12,590 | 5.4 |
| Rush | 7.5 | D | 8.2 | 22.3 | D | 6.2 | D | D | 17.4 | 4,010 | 241 | 276 | 7,521 | 0.2 |
| St. Joseph | 0.1 | D | 5.4 | 14.5 | 10.2 | 6.4 | 5.4 | 15.4 | 9.1 | 53,440 | 197 | 5,680 | 117,655 | 2.4 |
| Scott | 0.4 | D | 3.6 | 39.8 | D | 7.0 | D | D | 14.5 | 6,270 | 263 | 838 | 10,739 | 2.8 |
| Shelby | 0.6 | 0.3 | 8.2 | 32.2 | 3.7 | 4.7 | 3.2 | 6.3 | 14.4 | 9,980 | 223 | 690 | 19,523 | 2.4 |
| Spencer | 3.4 | D | 7.5 | 21.5 | 4.5 | 3.0 | 3.3 | D | 11.0 | 4,905 | 242 | 268 | 9,233 | 4.1 |
| Starke | 4.3 | 0.0 | 4.7 | 17.8 | 2.7 | 8.6 | 3.1 | D | 19.3 | 6,465 | 281 | 542 | 11,274 | 2.9 |
| Steuben | 0.6 | 0.0 | 5.1 | 36.3 | 3.8 | 7.1 | 2.8 | D | 8.7 | 8,630 | 248 | 419 | 20,424 | 5.5 |
| Sullivan | 1.6 | D | 2.7 | 9.5 | 3.1 | 4.8 | 2.4 | D | 26.6 | 4,925 | 239 | 414 | 9,009 | 0.8 |
| Switzerland | 0.6 | D | D | D | 1.8 | D | D | D | 20.6 | 2,005 | 185 | 202 | 5,531 | 10.7 |
| Tippecanoe | 0.1 | D | 4.7 | 23.8 | D | 5.1 | 4.2 | 13.4 | 26.6 | 26,885 | 138 | 2,406 | 79,534 | 11.9 |
| Tipton | 3.5 | 0.0 | 7.4 | 41.7 | 3.3 | 5.1 | 3.9 | D | 10.7 | 3,825 | 253 | 156 | 7,050 | 0.7 |
| Union | 1.2 | 0.0 | 8.2 | 16.5 | D | 8.5 | D | D | 22.7 | 1,665 | 233 | 129 | 3,286 | 1.5 |
| Vanderburgh | 0.1 | 0.0 | 7.3 | 13.6 | 6.6 | 6.3 | 5.2 | 18.9 | 9.3 | 40,310 | 222 | 4,543 | 84,834 | 2.2 |
| Vermillion | -0.4 | D | 16.3 | 23.5 | D | 7.1 | 2.3 | D | 11.7 | 4,245 | 274 | 336 | 7,503 | 0.2 |
| Vigo | 0.2 | 1.1 | 6.2 | 14.7 | 4.1 | 7.4 | 5.1 | 19.5 | 18.8 | 23,840 | 223 | 3,190 | 47,483 | 3.2 |
| Wabash | 2.8 | D | 6.9 | 27.4 | 3.6 | 7.6 | 4.6 | D | 11.4 | 8,865 | 286 | 576 | 14,200 | 0.2 |
| Warren | 15.4 | D | 5.3 | 23.3 | 2.0 | D | D | 11.2 | 13.7 | 2,045 | 248 | 87 | 3,788 | 2.9 |
| Warrick | 0.3 | 0.7 | 8.4 | 18.0 | 6.1 | 4.6 | 6.4 | 27.3 | 9.7 | 13,970 | 222 | 691 | 27,054 | 11.8 |
| Washington | 5.2 | D | 11.4 | 24.9 | D | 9.8 | 3.8 | D | 16.4 | 6,780 | 242 | 608 | 12,562 | 2.8 |
| Wayne | 1.1 | 0.1 | 4.2 | 21.5 | D | 6.8 | 4.7 | 22.1 | 13.6 | 17,520 | 266 | 2,205 | 31,513 | 0.9 |
| Wells | 3.8 | D | 5.7 | 26.3 | D | 5.6 | D | D | 10.0 | 6,415 | 228 | 306 | 12,006 | 3.0 |
| White | 4.2 | D | 10.1 | 29.4 | D | 7.0 | 4.5 | 6.7 | 12.3 | 6,130 | 253 | 342 | 13,216 | 1.9 |
| Whitley | 2.0 | 0.0 | 6.9 | 47.0 | D | 5.6 | D | D | 9.5 | 7,860 | 232 | 391 | 15,152 | 6.1 |
| **IOWA** | 5.4 | 0.2 | 6.9 | 16.1 | 6.4 | 5.6 | 9.8 | 10.2 | 16.1 | 657,019 | 208 | 51,633 | 1,429,008 | 6.9 |
| Adair | 15.7 | D | 8.1 | 16.1 | D | 6.0 | 3.8 | 6.3 | 13.9 | 1,840 | 259 | 92 | 3,721 | 0.6 |
| Adams | 21.6 | D | 3.8 | 28.6 | D | 3.8 | D | 10.5 | 8.5 | 1,010 | 280 | 62 | 2,011 | 0.0 |
| Allamakee | 17.5 | 0.3 | 7.1 | 16.1 | 2.2 | 6.6 | 3.6 | 6.9 | 18.2 | 3,575 | 262 | 152 | 7,897 | 3.7 |
| Appanoose | 2.3 | D | 3.3 | 21.6 | D | 7.2 | 3.6 | D | 16.0 | 3,600 | 287 | 429 | 6,650 | 0.2 |
| Audubon | 46.4 | 0.0 | 4.3 | 6.5 | 3.0 | 3.1 | 3.2 | D | 12.5 | 1,580 | 288 | 67 | 3,013 | 1.4 |
| Benton | 12.9 | D | 9.1 | 25.0 | 3.0 | 6.4 | 4.8 | D | 16.6 | 5,685 | 223 | 316 | 11,210 | 1.1 |
| Black Hawk | 0.9 | D | 4.5 | 25.7 | 5.1 | 6.5 | 5.5 | 13.5 | 15.2 | 27,870 | 213 | 3,290 | 58,795 | 5.2 |
| Boone | 5.9 | D | 8.6 | 5.5 | D | 6.3 | 4.1 | 7.5 | 24.8 | 6,135 | 233 | 315 | 12,099 | 2.9 |
| Bremer | 6.0 | D | 6.3 | 15.5 | 3.3 | 6.4 | 16.9 | D | 19.3 | 5,650 | 224 | 196 | 10,655 | 7.5 |
| Buchanan | 19.1 | 0.2 | 8.6 | 18.9 | 2.9 | 6.1 | D | 4.0 | 18.4 | 4,560 | 216 | 286 | 9,161 | 2.2 |
| Buena Vista | 12.7 | 0.0 | 4.8 | 30.3 | 2.8 | 5.1 | 3.8 | D | 15.5 | 3,670 | 186 | 259 | 8,289 | 0.6 |
| Butler | 18.6 | D | 5.3 | 17.8 | 2.4 | 4.6 | 5.0 | D | 16.1 | 3,900 | 271 | 173 | 6,830 | 2.2 |
| Calhoun | 32.9 | 0.0 | 3.4 | 2.3 | D | 5.1 | 3.9 | D | 15.9 | 2,705 | 281 | 160 | 5,153 | 0.9 |
| Carroll | 17.8 | D | 6.4 | 12.2 | D | 6.3 | 8.0 | D | 8.9 | 4,970 | 247 | 261 | 9,577 | 2.1 |
| Cass | 12.0 | D | 8.0 | 10.1 | 4.0 | 7.1 | 6.0 | 8.5 | 25.2 | 3,790 | 296 | 288 | 6,576 | -0.2 |
| Cedar | 15.2 | D | 8.1 | 10.8 | 5.4 | 4.5 | 3.6 | D | 15.9 | 4,195 | 226 | 159 | 8,316 | 3.4 |
| Cerro Gordo | 2.2 | D | 5.7 | 16.6 | 5.7 | 6.9 | 6.0 | 23.3 | 11.9 | 11,570 | 272 | 793 | 22,360 | 0.9 |
| Cherokee | 13.3 | 0.2 | 5.4 | 29.1 | 4.8 | 3.8 | 2.8 | D | 12.6 | 3,060 | 272 | 119 | 5,778 | 0.0 |
| Chickasaw | 12.1 | D | 5.3 | 35.7 | D | 5.6 | 3.2 | D | 8.3 | 2,940 | 246 | 123 | 5,700 | 0.4 |
| Clarke | 3.9 | D | D | 34.0 | 1.5 | 6.5 | 2.6 | D | 20.3 | 2,200 | 234 | 136 | 4,502 | 10.2 |
| Clay | 13.6 | D | 6.3 | 6.1 | D | 12.2 | 3.8 | 10.9 | 17.1 | 4,125 | 256 | 223 | 8,412 | 4.3 |
| Clayton | 19.1 | D | 14.0 | 11.9 | 2.7 | 4.9 | 3.7 | D | 15.7 | 4,855 | 277 | 210 | 9,177 | 2.0 |
| Clinton | 3.1 | D | 5.5 | 28.3 | 3.9 | 6.6 | 4.5 | 14.0 | 12.2 | 11,400 | 245 | 1,105 | 22,275 | 2.5 |
| Crawford | 15.1 | 0.0 | 5.5 | 26.9 | 2.5 | 4.5 | 4.0 | 5.4 | 17.2 | 3,550 | 210 | 218 | 7,066 | 1.8 |
| Dallas | 1.5 | D | 4.8 | 4.0 | 6.7 | 6.3 | 40.7 | 10.3 | 7.9 | 12,770 | 136 | 452 | 39,908 | 46.4 |
| Davis | 9.5 | D | 14.4 | 13.6 | D | 6.4 | 3.7 | 7.1 | 21.5 | 1,805 | 200 | 129 | 3,597 | -0.1 |
| Decatur | 7.9 | D | D | D | 2.9 | 5.9 | D | D | 23.2 | 1,850 | 236 | 186 | 3,846 | 0.3 |
| Delaware | 18.6 | 0.1 | 7.4 | 26.4 | 2.4 | 4.1 | 4.3 | D | 11.7 | 3,855 | 226 | 227 | 8,141 | 1.4 |
| Des Moines | 1.5 | D | 5.0 | 31.1 | 3.6 | 6.2 | 3.5 | 16.1 | 13.1 | 9,845 | 252 | 1,041 | 18,632 | 0.5 |
| Dickinson | 8.0 | D | 8.6 | 19.6 | 4.3 | 9.2 | 5.4 | 6.5 | | 5,320 | 308 | 176 | 14,465 | 12.6 |
| Dubuque | 2.2 | D | 5.6 | 20.5 | 7.1 | 6.0 | 11.7 | 13.5 | 8.1 | 21,725 | 223 | 1,636 | 41,963 | 7.7 |
| Emmet | 17.2 | D | 5.5 | 15.6 | D | 5.3 | D | D | 16.6 | 2,400 | 260 | 133 | 4,822 | 1.3 |
| Fayette | 17.4 | D | 7.6 | 10.7 | 2.4 | 5.2 | 3.6 | D | 14.2 | 5,135 | 262 | 520 | 9,602 | 0.4 |
| Floyd | 13.5 | 0.0 | 4.4 | 30.9 | 2.4 | 6.3 | 4.7 | D | 14.8 | 4,125 | 263 | 292 | 7,582 | 0.8 |

1. Per 1,000 resident population estimated as of July 1 of the year shown.

# Table B. States and Counties — Housing, Labor Force, and Employment

| STATE County | Housing units, 2015-2019 Owner-occupied Total | Percent | Median value[1] | Median owner cost as a percent of income With a mortgage | Without a mortgage[2] | Renter-occupied Median rent[3] | Median rent as a percent of income[2] | Sub-standard units[4] (percent) | Civilian labor force, 2020 Total | Percent change, 2019-2020 | Unemployment Total | Rate[5] | Civilian employment[6], 2015-2019 Total | Percent Management, business, science, and arts | Construction, production, and maintenance occupations |
|---|---|---|---|---|---|---|---|---|---|---|---|---|---|---|---|
| | 89 | 90 | 91 | 92 | 93 | 94 | 95 | 96 | 97 | 98 | 99 | 100 | 101 | 102 | 103 |
| INDIANA—Cont'd | | | | | | | | | | | | | | | |
| Pike | 5,129 | 82.4 | 93,400 | 16.7 | 11.1 | 576 | 25.1 | 2.0 | 6,144 | 0.3 | 362 | 5.9 | 5,650 | 24.3 | 43.1 |
| Porter | 64,517 | 75.1 | 185,400 | 18.8 | 10.1 | 933 | 28.6 | 1.5 | 83,987 | -2.7 | 6,767 | 8.1 | 79,993 | 36.7 | 28.0 |
| Posey | 10,155 | 81.7 | 144,200 | 17.0 | 10.0 | 688 | 27.7 | 1.1 | 12,834 | -4.7 | 632 | 4.9 | 12,401 | 31.4 | 30.2 |
| Pulaski | 5,151 | 75.9 | 102,700 | 19.1 | 10.0 | 652 | 24.6 | 1.4 | 6,395 | -2.3 | 388 | 6.1 | 5,450 | 27.7 | 41.6 |
| Putnam | 13,543 | 71.6 | 128,800 | 18.4 | 10.5 | 787 | 21.9 | 2.7 | 16,336 | -1.9 | 1,018 | 6.2 | 16,558 | 29.8 | 34.1 |
| Randolph | 10,229 | 76.3 | 81,200 | 17.4 | 11.8 | 685 | 24.3 | 1.8 | 11,191 | -1.1 | 735 | 6.6 | 11,372 | 28.1 | 37.6 |
| Ripley | 10,992 | 75.9 | 152,100 | 19.5 | 10.1 | 693 | 25.0 | 4.1 | 13,875 | -3.2 | 884 | 6.4 | 13,966 | 27.7 | 37.3 |
| Rush | 6,667 | 71.4 | 112,600 | 16.7 | 10.8 | 709 | 24.1 | 1.9 | 8,738 | -3.2 | 547 | 6.3 | 7,563 | 23.9 | 41.9 |
| St. Joseph | 101,872 | 68.2 | 129,000 | 17.8 | 10.0 | 793 | 28.8 | 2.1 | 133,385 | -3.4 | 11,188 | 8.4 | 128,744 | 36.3 | 25.4 |
| Scott | 8,971 | 72.0 | 103,000 | 19.8 | 11.4 | 789 | 29.2 | 1.9 | 10,484 | | 878 | 8.4 | 10,135 | 24.0 | 42.7 |
| Shelby | 17,823 | 73.3 | 133,200 | 17.9 | 10.0 | 801 | 26.5 | 2.0 | 23,086 | -0.4 | 1,643 | 7.1 | 22,227 | 31.8 | 33.1 |
| Spencer | 8,127 | 81.9 | 127,700 | 17.8 | 10.0 | 708 | 25.5 | 1.4 | 10,333 | -4.8 | 579 | 5.6 | 9,938 | 29.8 | 39.6 |
| Starke | 8,549 | 80.1 | 115,500 | 19.0 | 10.5 | 697 | 26.8 | 1.8 | 9,730 | -1.8 | 783 | 8.0 | 9,751 | 25.1 | 40.4 |
| Steuben | 14,261 | 77.8 | 148,800 | 19.1 | 10.0 | 783 | 22.1 | 1.1 | 19,915 | -3.7 | 1,242 | 6.2 | 17,584 | 27.6 | 36.2 |
| Sullivan | 7,644 | 74.4 | 86,000 | 16.5 | 11.4 | 667 | 29.1 | 3.2 | 7,903 | -3.4 | 551 | 7.0 | 8,591 | 26.6 | 35.7 |
| Switzerland | 4,349 | 74.8 | 116,400 | 22.1 | 13.7 | 679 | 23.9 | 4.5 | 4,782 | -4.9 | 388 | 8.1 | 4,519 | 18.3 | 45.5 |
| Tippecanoe | 70,526 | 55.1 | 153,000 | 17.7 | 10.0 | 851 | 32.2 | 1.6 | 94,147 | -3.9 | 5,936 | 6.3 | 96,929 | 40.2 | 24.1 |
| Tipton | 6,376 | 81.7 | 116,000 | 17.9 | 10.0 | 798 | 29.5 | 1.7 | 9,019 | -3.4 | 554 | 6.1 | 7,530 | 35.6 | 36.5 |
| Union | 2,823 | 72.0 | 116,300 | 20.4 | 10.1 | 797 | 26.7 | 2.1 | 3,443 | -3.9 | 189 | 5.5 | 3,395 | 22.1 | 35.7 |
| Vanderburgh | 75,607 | 64.8 | 129,000 | 19.0 | 11.8 | 789 | 30.1 | 1.5 | 91,573 | -2.7 | 6,559 | 7.2 | 88,298 | 34.4 | 25.6 |
| Vermillion | 6,664 | 73.1 | 79,700 | 16.7 | 10.0 | 667 | 27.7 | 2.9 | 6,590 | -3.7 | 496 | 7.5 | 6,771 | 26.7 | 38.7 |
| Vigo | 41,994 | 61.8 | 97,500 | 17.8 | 10.1 | 746 | 32.7 | 2.1 | 45,814 | -2.2 | 3,616 | 7.9 | 48,625 | 33.2 | 25.8 |
| Wabash | 12,862 | 74.0 | 100,500 | 16.6 | 10.0 | 706 | 26.5 | 1.2 | 14,499 | -3.5 | 884 | 6.1 | 14,712 | 28.4 | 34.4 |
| Warren | 3,357 | 83.9 | 122,100 | 17.9 | 10.0 | 740 | 32.9 | 1.9 | 4,001 | -4.7 | 216 | 5.4 | 3,974 | 31.9 | 38.8 |
| Warrick | 24,514 | 78.5 | 168,200 | 17.3 | 10.0 | 865 | 25.3 | 1.7 | 31,654 | -4.2 | 1,724 | 5.4 | 31,695 | 41.1 | 25.0 |
| Washington | 10,940 | 77.8 | 118,100 | 18.5 | 11.9 | 674 | 30.6 | 2.8 | 13,465 | -2.4 | 932 | 6.9 | 12,238 | 30.0 | 41.5 |
| Wayne | 26,644 | 66.9 | 100,500 | 18.7 | 10.2 | 688 | 26.7 | 1.9 | 29,989 | -3.1 | 2,038 | 6.8 | 30,108 | 29.6 | 31.4 |
| | 10,888 | 78.9 | 127,600 | 17.2 | 10.0 | 702 | 25.7 | 1.2 | 14,006 | -3.0 | 795 | 5.7 | 13,966 | 31.2 | 33.1 |
| Wells | | | | | | | | | | | | | | | |
| White | 9,879 | 77.5 | 113,500 | 18.3 | 10.3 | 706 | 24.7 | 1.3 | 13,254 | -2.3 | 668 | 5.0 | 11,283 | 27.5 | 40.6 |
| Whitley | 13,742 | 82.2 | 143,100 | 18.4 | 10.0 | 722 | 25.4 | 1.0 | 17,155 | -2.9 | 1,016 | 5.9 | 17,204 | 28.9 | 37.7 |
| IOWA | 1,265,473 | 71.1 | 147,800 | 18.8 | 11.1 | 789 | 26.9 | 1.9 | 1,666,420 | -3.9 | 87,655 | 5.3 | 1,613,902 | 36.2 | 26.8 |
| Adair | 3,200 | 74.3 | 104,600 | 17.9 | 10.0 | 585 | 22.9 | 0.7 | 4,179 | -2.9 | 166 | 4.0 | 3,506 | 33.0 | 34.6 |
| Adams | 1,598 | 81.7 | 87,500 | 17.5 | 12.1 | 549 | 20.6 | 2.3 | 2,029 | -7.7 | 77 | 3.8 | 1,674 | 32.6 | 33.0 |
| Allamakee | 5,953 | 79.3 | 130,400 | 20.6 | 11.9 | 620 | 22.8 | 2.6 | 7,034 | -5.6 | 359 | 5.1 | 6,752 | 28.8 | 37.2 |
| Appanoose | 5,353 | 71.0 | 81,400 | 19.8 | 13.9 | 624 | 25.9 | 2.7 | 5,914 | -4.4 | 313 | 5.3 | 5,263 | 26.3 | 34.9 |
| Audubon | 2,649 | 76.8 | 79,500 | 16.5 | 10.0 | 582 | 22.3 | 1.2 | 3,011 | -4.4 | 103 | 3.4 | 2,906 | 28.3 | 33.8 |
| Benton | 10,306 | 82.8 | 157,200 | 19.1 | 11.0 | 640 | 25.3 | 1.1 | 12,924 | -5.5 | 652 | 5.0 | 13,269 | 31.2 | 34.3 |
| Black Hawk | 52,905 | 65.5 | 147,500 | 19.0 | 10.9 | 801 | 28.8 | 1.7 | 67,318 | -4.0 | 4,145 | 6.2 | 69,185 | 31.8 | 27.5 |
| Boone | 10,954 | 78.1 | 132,700 | 19.0 | 11.7 | 742 | 24.3 | 1.8 | 14,468 | -4.1 | 588 | 4.1 | 14,064 | 32.4 | 29.6 |
| Bremer | 9,676 | 82.7 | 165,700 | 19.1 | 10.3 | 661 | 25.1 | 0.7 | 13,500 | -5.3 | 533 | 3.9 | 13,158 | 36.0 | 28.0 |
| Buchanan | 8,029 | 79.3 | 142,500 | 17.9 | 11.1 | 678 | 23.8 | 2.2 | 11,095 | -3.6 | 507 | 4.6 | 10,518 | 33.6 | 31.9 |
| Buena Vista | 7,524 | 64.6 | 115,600 | 17.5 | 10.0 | 706 | 23.9 | 5.7 | 11,643 | -1.6 | 426 | 3.7 | 10,343 | 28.2 | 41.1 |
| Butler | 6,254 | 77.9 | 116,700 | 18.9 | 11.5 | 598 | 22.1 | 0.9 | 7,675 | -3.9 | 358 | 4.7 | 7,202 | 29.5 | 32.8 |
| Calhoun | 4,146 | 78.8 | 82,000 | 16.8 | 10.0 | 588 | 23.6 | 2.1 | 4,096 | -4.7 | 168 | 4.1 | 4,314 | 33.8 | 28.9 |
| Carroll | 8,726 | 75.8 | 138,900 | 16.7 | 10.0 | 621 | 28.0 | 0.9 | 10,546 | -4.6 | 417 | 4.0 | 10,628 | 33.1 | 28.9 |
| Cass | 5,921 | 69.2 | 102,100 | 18.4 | 11.4 | 646 | 25.9 | 0.5 | 6,752 | -3.4 | 263 | 3.9 | 6,370 | 29.2 | 31.7 |
| Cedar | 7,391 | 81.2 | 156,000 | 18.6 | 11.7 | 719 | 19.3 | 0.7 | 10,368 | -5.3 | 453 | 4.4 | 9,634 | 32.8 | 31.3 |
| Cerro Gordo | 19,241 | 69.1 | 128,600 | 18.3 | 11.3 | 707 | 26.4 | 1.4 | 23,096 | -3.1 | 1,203 | 5.2 | 22,181 | 31.6 | 30.7 |
| Cherokee | 5,319 | 74.6 | 106,700 | 15.5 | 10.0 | 550 | 22.9 | 0.8 | 6,192 | -0.4 | 234 | 3.8 | 5,822 | 30.0 | 34.7 |
| Chickasaw | 5,173 | 80.7 | 119,100 | 17.7 | 10.8 | 630 | 22.1 | 0.8 | 6,271 | -4.3 | 295 | 4.7 | 6,519 | 29.0 | 37.6 |
| Clarke | 3,895 | 70.4 | 110,300 | 19.4 | 13.9 | 783 | 23.1 | 4.8 | 4,761 | -3.6 | 257 | 5.4 | 4,615 | 32.8 | 37.4 |
| Clay | 7,212 | 67.7 | 124,100 | 19.0 | 11.7 | 700 | 25.7 | 1.2 | 8,336 | -2.9 | 378 | 4.5 | 8,111 | 31.5 | 31.8 |
| Clayton | 7,510 | 74.5 | 121,700 | 20.3 | 10.8 | 648 | 23.1 | 2.8 | 9,703 | -5.4 | 553 | 5.7 | 9,087 | 28.1 | 38.1 |
| Clinton | 19,635 | 73.3 | 118,100 | 19.9 | 13.2 | 683 | 29.7 | 1.0 | 21,989 | -4.7 | 1,480 | 6.7 | 22,152 | 27.7 | 34.2 |
| Crawford | 6,437 | 69.1 | 93,400 | 17.4 | 10.0 | 631 | 22.6 | 6.1 | 7,931 | -3.5 | 416 | 5.2 | 7,824 | 25.2 | 47.7 |
| Dallas | 34,399 | 73.7 | 248,100 | 18.1 | 10.6 | 1,045 | 24.8 | 2.1 | 49,406 | -4.3 | 1,785 | 3.6 | 47,496 | 54.7 | 13.6 |
| Davis | 3,176 | 85.2 | 115,000 | 19.3 | 10.3 | 631 | 25.1 | 7.1 | 4,108 | -3.7 | 165 | 4.0 | 4,202 | 33.7 | 32.3 |
| Decatur | 3,200 | 67.0 | 83,200 | 19.3 | 11.9 | 567 | 29.2 | 5.2 | 4,266 | -3.4 | 162 | 3.8 | 3,666 | 32.8 | 26.4 |
| Delaware | 6,907 | 84.4 | 136,100 | 17.7 | 12.1 | 621 | 22.3 | 1.9 | 10,381 | -3.0 | 412 | 4.0 | 9,258 | 33.7 | 34.1 |
| Des Moines | 16,891 | 70.3 | 108,500 | 18.6 | 12.9 | 778 | 30.8 | 1.5 | 18,609 | -4.9 | 1,387 | 7.5 | 19,045 | 30.0 | 31.4 |
| Dickinson | 8,174 | 79.4 | 194,300 | 19.4 | 11.3 | 844 | 28.4 | 0.6 | 9,752 | -4.1 | 497 | 5.1 | 9,204 | 35.5 | 28.4 |
| Dubuque | 38,210 | 73.3 | 166,800 | 19.2 | 11.5 | 783 | 26.4 | 1.5 | 54,877 | -3.3 | 3,211 | 5.9 | 50,533 | 37.0 | 24.3 |
| Emmet | 4,036 | 78.1 | 89,300 | 19.2 | 11.8 | 694 | 27.5 | 3.4 | 5,047 | -5.0 | 250 | 5.0 | 4,721 | 32.5 | 35.4 |
| Fayette | 8,200 | 76.6 | 104,100 | 18.8 | 12.7 | 612 | 26.8 | 0.9 | 10,299 | -3.9 | 564 | 5.5 | 9,796 | 29.9 | 33.0 |
| Floyd | 6,903 | 71.2 | 106,200 | 17.8 | 10.7 | 539 | 28.4 | 1.6 | 8,250 | -7.6 | 499 | 6.0 | 7,857 | 35.1 | 30.5 |

1. Specified owner-occupied units, lacking complete plumbing facilities.  2. A value of 10.0 represents 10 percent or less; a value of 50.0 represents 50 percent or more.  3. Specified renter-occupied units.  4. Overcrowded or
5. Percent of civilian labor force.  6. Civilian employed persons 16 years old and over.

Items 89—103

# Table B. States and Counties — Nonfarm Employment and Agriculture

| STATE County | Private nonfarm establishments, employment and payroll, 2019 | | | | | | | Annual payroll | | Agriculture, 2017 | | | Farm producers whose primary occupation is farming (percent) |
| | Number of establishments | Employment | | | | | | | | Farms | Percent with: | | |
| | | Total | Health care and social assistance | Manufacturing | Retail trade | Finance and insurance | Professional, scientific, and technical services | Total (mil dol) | Average per employee (dollars) | Number | Fewer than 50 acres | 1000 acres or more | |
| | 104 | 105 | 106 | 107 | 108 | 109 | 110 | 111 | 112 | 113 | 114 | 115 | 116 |
| INDIANA—Cont'd | | | | | | | | | | | | | |
| Pike | 189 | 2,006 | 346 | 178 | 191 | 29 | 31 | 97 | 48,224 | 327 | 36.4 | 5.8 | 24.1 |
| Porter | 3,630 | 56,373 | 8,950 | 10,587 | 7,627 | 1,111 | 2,135 | 2,645 | 46,922 | 445 | 48.1 | 7.0 | 42.6 |
| Posey | 497 | 9,136 | 395 | 2,974 | 739 | 142 | 298 | 560 | 61,261 | 491 | 38.1 | 13.6 | 49.3 |
| Pulaski | 310 | 3,999 | 632 | 1,559 | 455 | 123 | 72 | 176 | 44,077 | 547 | 41.0 | 13.9 | 44.4 |
| Putnam | 710 | 11,799 | 1,907 | 2,680 | 1,231 | 235 | 188 | 423 | 35,827 | 828 | 44.8 | 4.5 | 30.8 |
| Randolph | 475 | 5,444 | 808 | 1,707 | 592 | 130 | 134 | 205 | 37,739 | 754 | 43.1 | 9.4 | 46.6 |
| Ripley | 640 | 9,407 | 1,577 | 1,578 | 879 | 275 | 139 | 396 | 42,122 | 879 | 36.9 | 4.7 | 34.3 |
| Rush | 371 | 4,131 | 588 | 646 | 434 | 90 | 167 | 173 | 41,850 | 557 | 30.9 | 9.2 | 51.3 |
| St. Joseph | 5,847 | 123,110 | 19,725 | 15,048 | 16,414 | 3,853 | 5,079 | 5,411 | 43,952 | 629 | 56.0 | 5.7 | 41.1 |
| Scott | 400 | 6,885 | 1,402 | 2,294 | 830 | 110 | 113 | 253 | 36,692 | 315 | 46.7 | 4.1 | 34.1 |
| Shelby | 937 | 16,198 | 2,270 | 5,331 | 1,536 | 280 | 239 | 759 | 46,831 | 567 | 44.4 | 11.8 | 44.8 |
| Spencer | 428 | 5,009 | 213 | 1,519 | 592 | 169 | 142 | 240 | 47,829 | 665 | 37.4 | 5.4 | 40.5 |
| Starke | 305 | 3,120 | 601 | 757 | 585 | 77 | 72 | 111 | 35,737 | 507 | 47.1 | 8.9 | 37.3 |
| Steuben | 955 | 15,036 | 1,644 | 5,247 | 2,123 | 202 | 203 | 550 | 36,589 | 472 | 40.3 | 6.1 | 35.4 |
| Sullivan | 327 | 3,929 | 539 | 415 | 572 | 94 | 113 | 173 | 44,029 | 450 | 30.7 | 9.3 | 43.5 |
| Switzerland | 130 | 1,601 | 118 | 108 | 186 | 30 | 15 | 50 | 31,476 | 410 | 38.3 | 1.0 | 30.5 |
| Tippecanoe | 3,684 | 73,250 | 13,466 | 17,469 | 9,425 | 1,954 | 3,032 | 3,351 | 45,744 | 693 | 53.7 | 9.1 | 35.4 |
| Tipton | 316 | 4,600 | 681 | 2,058 | 431 | 68 | 55 | 191 | 41,624 | 404 | 43.8 | 11.1 | 43.4 |
| Union | 120 | 1,013 | 94 | 312 | 222 | 22 | 69 | 31 | 30,361 | 238 | 26.5 | 9.2 | 51.2 |
| Vanderburgh | 5,052 | 105,146 | 20,707 | 12,514 | 12,423 | 4,517 | 3,925 | 4,773 | 45,391 | 251 | 47.8 | 6.4 | 43.4 |
| Vermillion | 259 | 3,606 | 597 | 693 | 605 | 49 | 38 | 205 | 56,854 | 283 | 36.7 | 14.5 | 49.8 |
| Vigo | 2,405 | 43,228 | 9,681 | 6,658 | 6,434 | 1,133 | 1,023 | 1,688 | 39,046 | 477 | 55.6 | 6.9 | 37.7 |
| Wabash | 731 | 10,829 | 2,080 | 2,491 | 1,392 | 300 | 196 | 403 | 37,235 | 724 | 41.0 | 7.6 | 40.6 |
| Warren | 129 | 1,612 | 241 | 782 | 87 | 31 | 19 | 59 | 36,659 | 417 | 37.9 | 11.5 | 36.4 |
| Warrick | 1,186 | 15,273 | 4,050 | 2,702 | 1,521 | 422 | 444 | 770 | 50,440 | 364 | 53.0 | 9.1 | 41.7 |
| Washington | 465 | 4,643 | 662 | 1,557 | 924 | 135 | 123 | 163 | 35,214 | 865 | 36.8 | 5.4 | 40.1 |
| Wayne | 1,430 | 25,712 | 5,902 | 5,587 | 3,617 | 779 | 373 | 1,016 | 39,495 | 768 | 38.5 | 4.2 | 40.4 |
| Wells | 635 | 10,790 | 1,576 | 2,990 | 1,009 | 158 | 291 | 423 | 39,233 | 581 | 39.1 | 14.3 | 44.9 |
| White | 587 | 7,794 | 771 | 3,316 | 1,084 | 175 | 126 | 295 | 37,805 | 539 | 37.3 | 14.8 | 48.8 |
| Whitley | 682 | 12,112 | 1,243 | 5,403 | 1,416 | 224 | 241 | 548 | 45,217 | 696 | 48.3 | 7.8 | 31.6 |
| IOWA | 82,770 | 1,380,747 | 217,669 | 218,714 | 187,226 | 93,220 | 59,816 | 63,671 | 46,113 | 86,104 | 31.7 | 9.8 | 45.0 |
| Adair | 200 | 1,923 | 337 | 382 | 267 | 72 | 43 | 63 | 32,699 | 738 | 22.2 | 14.5 | 44.4 |
| Adams | 108 | 936 | 244 | 205 | 121 | 20 | 24 | 35 | 36,990 | 509 | 21.2 | 10.4 | 51.2 |
| Allamakee | 391 | 3,906 | 937 | 927 | 578 | 216 | 94 | 142 | 36,322 | 997 | 26.7 | 6.0 | 38.2 |
| Appanoose | 278 | 3,674 | 622 | 1,125 | 602 | 105 | 67 | 134 | 36,562 | 675 | 35.9 | 6.4 | 36.4 |
| Audubon | 181 | 1,339 | 322 | 156 | 181 | 56 | 21 | 49 | 37,302 | 1,148 | 33.3 | 11.8 | 47.8 |
| Benton | 548 | 4,456 | 827 | 838 | 811 | 188 | 81 | 166 | 37,302 | 968 | 42.3 | 8.9 | 49.3 |
| Black Hawk | 3,174 | 63,661 | 11,203 | 12,006 | 9,224 | 2,006 | 3,016 | 2,853 | 44,820 | 967 | 44.0 | 10.2 | 42.5 |
| Boone | 595 | 6,826 | 1,542 | 503 | 1,037 | 175 | 157 | 277 | 40,643 | 963 | 35.7 | 6.3 | 36.5 |
| Bremer | 584 | 9,195 | 1,905 | 1,720 | 1,288 | 1,006 | 232 | 404 | 43,969 | 1,057 | 35.3 | 8.0 | 40.5 |
| Buchanan | 494 | 5,348 | 993 | 1,469 | 797 | 216 | 133 | 204 | 38,236 | 802 | 24.9 | 12.7 | 49.5 |
| Buena Vista | 540 | 8,973 | 1,208 | 3,480 | 1,064 | 246 | 226 | 342 | 38,104 | 1,074 | 40.3 | 8.8 | 52.4 |
| Butler | 335 | 2,476 | 472 | 718 | 446 | 142 | 45 | 87 | 35,153 | 813 | 33.1 | 16.1 | 46.9 |
| Calhoun | 276 | 2,395 | 651 | 110 | 356 | 108 | 38 | 93 | 38,772 | 1,074 | 35.2 | 7.3 | 52.0 |
| Carroll | 857 | 10,264 | 2,161 | 1,420 | 1,687 | 782 | 157 | 382 | 37,194 | 643 | 28.1 | 14.3 | 45.6 |
| Cass | 462 | 4,775 | 1,001 | 719 | 889 | 198 | 147 | 174 | 36,339 | 933 | 37.6 | 9.4 | 48.0 |
| Cedar | 454 | 4,099 | 607 | 604 | 565 | 136 | 113 | 150 | 36,658 | 760 | 33.6 | 12.6 | 45.9 |
| Cerro Gordo | 1,425 | 20,349 | 4,234 | 3,080 | 3,656 | 943 | 667 | 879 | 43,174 | 863 | 21.7 | 8.9 | 46.2 |
| Cherokee | 340 | 3,961 | 803 | 402 | 657 | 145 | 76 | 154 | 38,766 | 973 | 33.4 | 7.5 | 55.4 |
| Chickasaw | 389 | 4,302 | 583 | 1,679 | 473 | 159 | 63 | 180 | 41,883 | 624 | 23.2 | 6.3 | 44.4 |
| Clarke | 171 | 3,469 | 507 | 1,217 | 548 | 75 | 52 | 127 | 36,569 | 716 | 26.0 | 15.4 | 31.6 |
| Clay | 581 | 7,225 | 1,432 | 526 | 1,456 | 305 | 174 | 277 | 38,274 | 1,525 | 26.8 | 5.0 | 50.4 |
| Clayton | 503 | 5,460 | 1,087 | 946 | 767 | 182 | 88 | 220 | 40,228 | 1,169 | 30.8 | 9.8 | 42.2 |
| Clinton | 1,139 | 17,023 | 2,835 | 4,172 | 2,472 | 621 | 320 | 693 | 40,691 | 915 | 26.6 | 12.0 | 44.7 |
| Crawford | 415 | 5,715 | 950 | 2,174 | 651 | 203 | 90 | 236 | 41,359 | 924 | 41.0 | 8.2 | 49.3 |
| Dallas | 2,183 | 40,360 | 4,196 | 2,066 | 6,870 | 13,594 | 1,365 | 2,268 | 56,190 | 826 | 30.4 | 4.6 | 35.9 |
| Davis | 170 | 1,497 | 431 | 241 | 245 | 62 | 38 | 59 | 39,598 | 659 | 25.2 | 9.0 | 35.8 |
| Decatur | 137 | 1,786 | 374 | 57 | 261 | 38 | 38 | 46 | 25,553 | 1,331 | 30.2 | 5.2 | 39.1 |
| Delaware | 510 | 6,229 | 1,031 | 2,316 | 675 | 268 | 90 | 258 | 41,348 | 593 | 32.5 | 7.6 | 49.7 |
| Des Moines | 1,100 | 19,924 | 3,761 | 4,903 | 3,328 | 515 | 324 | 828 | 41,576 | 411 | 29.7 | 15.8 | 40.7 |
| Dickinson | 782 | 7,729 | 1,166 | 1,861 | 1,233 | 216 | 207 | 314 | 40,688 | | | | 51.7 |
| Dubuque | 2,767 | 56,000 | 8,466 | 10,098 | 7,325 | 4,008 | 1,767 | 2,506 | 44,749 | 1,402 | 30.5 | 3.4 | 42.8 |
| Emmet | 289 | 2,953 | 683 | 704 | 462 | 139 | 54 | 104 | 35,372 | 488 | 26.2 | 13.9 | 56.7 |
| Fayette | 557 | 5,743 | 1,029 | 526 | 822 | 201 | 118 | 192 | 33,366 | 1,265 | 31.2 | 6.7 | 46.9 |
| Floyd | 392 | 5,430 | 1,160 | 1,719 | 835 | 206 | 69 | 232 | 42,689 | 917 | 33.9 | 10.9 | 45.2 |

# Table B. States and Counties — **Agriculture**

| STATE County | Land in farms — Acreage (1,000) | Percent change, 2012-2017 | Acres — Average size of farm | Total irrigated (1,000) | Total cropland (1,000) | Value of land and buildings (dollars) — Average per farm | Average per acre | Value of machinery and equipment, average per farm (dollars) | Value of products sold: Total (mil dol) | Average per farm (acres) | Percent from: Crops | Livestock and poultry products | Organic farms (number) | Farms with internet access (per-cent) | Government payments Total ($1,000) | Percent of farms |
|---|---|---|---|---|---|---|---|---|---|---|---|---|---|---|---|---|
| | 117 | 118 | 119 | 120 | 121 | 122 | 123 | 124 | 125 | 126 | 127 | 128 | 129 | 130 | 131 | 132 |
| **INDIANA—Cont'd** | | | | | | | | | | | | | | | | |
| Pike | 81 | 0.7 | 246 | D | 63.3 | 1,214,424 | 4,931 | 138,768 | 42.6 | 130,343 | 68.1 | 31.9 | NA | 75.8 | 2,207 | 65.1 |
| Porter | 123 | 1.6 | 275 | 10.4 | 114.7 | 1,862,858 | 6,766 | 198,333 | 77.3 | 173,742 | 88.4 | 11.6 | NA | 79.6 | 3,698 | 53.9 |
| Posey | 194 | -15.3 | 395 | 13.0 | 175.2 | 2,396,308 | 6,073 | 274,785 | 118.2 | 240,717 | 92.7 | 7.3 | 2 | 74.7 | 7,423 | 68.2 |
| Pulaski | 232 | 7.1 | 424 | 29.9 | 217.7 | 2,607,043 | 6,150 | 185,625 | 188.2 | 344,108 | 68.0 | 32.0 | NA | 73.9 | 5,770 | 72.2 |
| Putnam | 185 | -6.4 | 223 | 0.5 | 140.4 | 1,437,116 | 6,435 | 103,171 | 85.1 | 102,793 | 84.7 | 15.3 | NA | 78.1 | 5,065 | 47.3 |
| Randolph | 242 | 0.3 | 321 | 0.2 | 224.5 | 2,057,555 | 6,417 | 207,262 | 213.1 | 282,635 | 55.4 | 44.6 | 10 | 75.9 | 5,720 | 59.3 |
| Ripley | 176 | 5.6 | 200 | 0.1 | 132.1 | 1,034,225 | 5,163 | 123,313 | 86.5 | 98,396 | 78.3 | 21.7 | 8 | 72.8 | 3,808 | 58.7 |
| Rush | 211 | 1.8 | 379 | 0.1 | 195.5 | 2,752,265 | 7,255 | 243,945 | 165.1 | 296,497 | 67.1 | 32.9 | NA | 79.4 | 4,579 | 63.4 |
| St. Joseph | 150 | -1.3 | 238 | 28.1 | 133.8 | 1,928,345 | 8,087 | 164,410 | 103.9 | 165,261 | 80.0 | 20.0 | 3 | 74.4 | 4,335 | 44.7 |
| Scott | 59 | 15.2 | 188 | 3.7 | 45.7 | 1,010,422 | 5,368 | 116,543 | 23.8 | 75,695 | 88.9 | 11.1 | NA | 74.0 | 1,843 | 38.7 |
| Shelby | 220 | -5.4 | 389 | 5.2 | 207.4 | 2,700,637 | 6,945 | 229,128 | 135.6 | 239,138 | 88.3 | 11.7 | 1 | 76.4 | 6,523 | 56.1 |
| Spencer | 169 | -0.6 | 255 | 0.7 | 133.5 | 1,174,320 | 4,609 | 174,957 | 92.9 | 139,705 | 75.0 | 25.0 | 1 | 73.1 | 4,531 | 62.0 |
| Starke | 146 | 9.2 | 287 | 38.7 | 130.6 | 1,438,032 | 5,004 | 146,138 | 69.2 | 136,578 | 94.9 | 5.1 | 1 | 68.6 | 6,179 | 73.8 |
| Steuben | 120 | 15.1 | 255 | 13.0 | 101.7 | 1,411,595 | 5,537 | 144,619 | 64.0 | 135,672 | 71.0 | 29.0 | NA | 74.4 | 2,336 | 63.1 |
| Sullivan | 160 | -6 | 356 | 7.1 | 139.6 | 1,770,624 | 4,977 | 190,307 | 88.9 | 197,533 | 90.3 | 9.7 | 2 | 82.2 | 4,194 | 69.3 |
| Switzerland | 55 | 8.9 | 134 | D | 27.8 | 549,933 | 4,098 | 67,288 | 14.8 | 36,127 | 74.6 | 25.4 | 9 | 68.0 | 711 | 24.4 |
| Tippecanoe | 222 | 0.9 | 321 | 4.6 | 197 | 2,893,652 | 9,027 | 188,551 | 143.0 | 206,338 | 83.1 | 16.9 | 1 | 85.6 | 3,070 | 51.9 |
| Tipton | 161 | 11.1 | 399 | D | 155.9 | 3,261,470 | 8,167 | 274,856 | 114.9 | 284,366 | 82.1 | 17.9 | NA | 85.6 | 2,684 | 67.6 |
| Union | 83 | 11.5 | 349 | NA | 71 | 2,211,666 | 6,340 | 200,911 | 41.9 | 175,992 | 91.5 | 8.5 | NA | 79.0 | 2,457 | 69.3 |
| Vanderburgh | 64 | -16.5 | 255 | 1.3 | 61 | 2,013,091 | 7,906 | 184,217 | 39.0 | 155,339 | 98.0 | 2.0 | NA | 74.9 | 1,620 | 61.4 |
| Vermillion | 123 | 4.0 | 435 | 0.3 | 104.4 | 2,467,818 | 5,674 | 233,387 | 76.8 | 271,247 | 76.8 | 23.2 | NA | 71.0 | 1,888 | 53.7 |
| Vigo | 120 | 2.1 | 252 | 1.6 | 100.9 | 1,364,518 | 5,425 | 135,673 | 55.4 | 116,216 | 98.9 | 1.1 | 2 | 75.7 | 1,454 | 55.1 |
| Wabash | 211 | 6.9 | 292 | 1.8 | 190.3 | 1,835,188 | 6,290 | 192,429 | 161.2 | 222,664 | 59.1 | 40.9 | 2 | 76.8 | 3,814 | 62.3 |
| Warren | 188 | 6.8 | 450 | 7.1 | 167.9 | 3,057,183 | 6,786 | 239,741 | 130.6 | 313,293 | 79.7 | 20.3 | 2 | 78.9 | 4,075 | 62.8 |
| Warrick | 105 | 5.6 | 289 | 0.1 | 86.4 | 1,569,892 | 5,429 | 175,673 | 49.4 | 135,821 | 92.6 | 7.4 | NA | 77.2 | 2,445 | 46.4 |
| Washington | 212 | 6.2 | 245 | 0.2 | 147 | 1,102,858 | 4,501 | 129,743 | 160.6 | 185,609 | 42.5 | 57.5 | 2 | 70.4 | 3,910 | 36.0 |
| Wayne | 163 | 4.9 | 213 | 0.9 | 131.5 | 1,169,226 | 5,492 | 110,988 | 100.1 | 130,385 | 68.6 | 31.4 | 31 | 71.1 | 3,878 | 48.6 |
| Wells | 225 | 12.3 | 387 | 0.0 | 215.4 | 2,922,012 | 7,544 | 255,414 | 192.1 | 330,707 | 62.5 | 37.5 | 4 | 79.9 | 3,224 | 70.7 |
| White | 283 | -1.8 | 525 | 5.0 | 266.5 | 4,101,696 | 7,815 | 319,127 | 256.8 | 476,384 | 61.8 | 38.2 | NA | 86.3 | 3,750 | 67.7 |
| Whitley | 176 | 25.8 | 253 | 1.6 | 154.1 | 1,606,248 | 6,343 | 187,714 | 120.9 | 173,774 | 65.3 | 34.7 | NA | 78.4 | 3,103 | 51.1 |
| **IOWA** | 30,564 | -0.2 | 355 | 222.0 | 26,546 | 2,506,812 | 7,062 | 230,716 | 28,956.5 | 336,296 | 47.8 | 52.2 | 785 | 79.6 | 682,995 | 71.2 |
| Adair | 335 | 3.5 | 454 | 0.1 | 261.5 | 2,372,891 | 5,230 | 223,107 | 188.0 | 254,749 | 54.9 | 45.1 | 9 | 77.8 | 7,583 | 75.7 |
| Adams | 223 | -2.5 | 439 | D | 174.1 | 2,189,535 | 4,987 | 197,021 | 108.3 | 212,764 | 59.2 | 40.8 | NA | 76.8 | 6,239 | 79.8 |
| Allamakee | 292 | 0.9 | 293 | 0.0 | 188.8 | 1,564,078 | 5,345 | 172,462 | 200.5 | 201,062 | 33.3 | 66.7 | 18 | 75.4 | 6,176 | 81.4 |
| Appanoose | 179 | -4.5 | 266 | NA | 108.3 | 1,079,752 | 4,065 | 87,526 | 44.5 | 65,901 | 59.4 | 40.6 | 1 | 76.3 | 3,894 | 55.9 |
| Audubon | 276 | -1.7 | 439 | D | 253.4 | 3,012,527 | 6,866 | 285,066 | 276.6 | 440,482 | 47.0 | 53.0 | 2 | 76.9 | 9,364 | 83.9 |
| Benton | 421 | -0.4 | 366 | 0.2 | 384.2 | 2,706,668 | 7,387 | 245,776 | 347.7 | 302,895 | 62.8 | 37.2 | 11 | 80.7 | 7,127 | 77.0 |
| Black Hawk | 292 | -1.6 | 302 | 0.4 | 275.6 | 2,632,645 | 8,723 | 250,907 | 261.2 | 269,871 | 61.9 | 38.1 | 5 | 84.2 | 6,532 | 73.1 |
| Boone | 315 | 0.5 | 326 | 0.2 | 288.7 | 2,595,336 | 7,966 | 205,330 | 218.4 | 225,806 | 73.9 | 26.1 | 4 | 83.1 | 5,716 | 70.4 |
| Bremer | 262 | -3.5 | 272 | 0.5 | 240.8 | 2,174,810 | 7,991 | 227,554 | 229.9 | 238,769 | 59.4 | 40.6 | 10 | 81.0 | 5,871 | 77.8 |
| Buchanan | 330 | -3.5 | 312 | 0.1 | 309.5 | 2,459,772 | 7,884 | 252,896 | 366.4 | 346,651 | 50.9 | 49.1 | 33 | 79.6 | 8,083 | 70.3 |
| Buena Vista | 357 | -1.2 | 445 | D | 334.9 | 3,577,469 | 8,045 | 302,722 | 537.2 | 669,864 | 32.9 | 67.1 | 4 | 81.9 | 6,426 | 84.4 |
| Butler | 359 | | 334 | 1.6 | 333.5 | 2,397,864 | 7,169 | 202,064 | 291.5 | 271,395 | 62.8 | 37.2 | 3 | 81.3 | 11,840 | 80.4 |
| Calhoun | 351 | -2 | 432 | 0.7 | 334 | 3,429,727 | 7,942 | 290,880 | 321.6 | 395,549 | 54.8 | 45.2 | 7 | 79.1 | 5,250 | 65.6 |
| Carroll | 349 | -2.8 | 325 | 0.5 | 319.5 | 2,608,918 | 8,030 | 239,347 | 573.0 | 533,488 | 31.1 | 68.9 | 10 | 80.4 | 5,341 | 55.5 |
| Cass | 285 | -1.8 | 443 | D | 241.9 | 2,610,781 | 5,895 | 236,918 | 203.4 | 316,264 | 56.9 | 43.1 | 7 | 77.1 | 5,293 | 73.4 |
| Cedar | 340 | 8.9 | 365 | 0.3 | 307.8 | 2,866,790 | 7,858 | 275,263 | 321.5 | 344,610 | 58.5 | 41.5 | 4 | 80.6 | 4,563 | 73.1 |
| Cerro Gordo | 320 | -2.2 | 421 | 0.5 | 305.2 | 3,088,779 | 7,341 | 250,319 | 226.6 | 298,208 | 74.6 | 25.4 | 1 | 89.1 | 7,612 | 80.0 |
| Cherokee | 339 | 0.4 | 392 | 0.0 | 299.4 | 3,105,879 | 7,914 | 300,575 | 387.1 | 448,583 | 41.9 | 58.1 | 1 | 83.2 | 5,329 | 47.6 |
| Chickasaw | 293 | -2 | 301 | 0.3 | 269.2 | 2,248,034 | 7,463 | 227,181 | 303.0 | 311,375 | 48.9 | 51.1 | 1 | 71.2 | 6,716 | 63.7 |
| Clarke | 193 | 14.1 | 309 | D | 112.2 | 1,319,833 | 4,271 | 98,535 | 123.2 | 197,500 | 17.4 | 82.6 | 1 | 71.8 | 7,147 | 68.6 |
| Clay | 329 | 3.2 | 460 | 0.7 | 306.8 | 3,670,430 | 7,985 | 310,588 | 348.7 | 487,042 | 46.9 | 53.1 | 1 | 85.9 | 9,744 | 84.8 |
| Clayton | 413 | 3.7 | 271 | 0.4 | 305.5 | 1,500,568 | 5,545 | 166,778 | 364.2 | 238,852 | 39.1 | 60.9 | 30 | 72.0 | 12,580 | 81.6 |
| Clinton | 403 | -3.5 | 345 | 0.4 | 358.9 | 2,599,062 | 7,544 | 252,926 | 339.8 | 290,686 | 61.9 | 38.1 | 2 | 80.8 | 9,749 | 76.6 |
| Crawford | 440 | -2.5 | 481 | 0.5 | 391 | 3,255,634 | 6,774 | 269,409 | 400.3 | 437,461 | 55.3 | 44.7 | 4 | 76.6 | 8,356 | 59.8 |
| Dallas | 293 | -4.2 | 318 | 0.8 | 257.6 | 2,594,337 | 8,169 | 186,706 | 237.6 | 257,187 | 60.5 | 39.5 | 20 | 82.1 | 4,422 | 48.6 |
| Davis | 199 | -7.2 | 240 | 0.1 | 121.8 | 941,546 | 3,916 | 125,282 | 91.9 | 111,274 | 31.0 | 69.0 | 39 | 59.9 | 2,701 | 50.6 |
| Decatur | 236 | 2.0 | 358 | D | 142.6 | 1,377,822 | 3,845 | 129,758 | 103.4 | 156,869 | 34.6 | 65.4 | 3 | 72.1 | 7,942 | 67.1 |
| Delaware | 365 | -0.3 | 274 | 0.3 | 321.6 | 2,129,527 | 7,773 | 261,745 | 534.6 | 401,616 | 35.2 | 64.8 | 8 | 81.1 | 13,135 | 83.1 |
| Des Moines | 175 | 1.1 | 295 | 4.5 | 143.9 | 2,035,386 | 6,910 | 183,368 | 108.3 | 182,639 | 80.8 | 19.2 | 2 | 79.4 | 3,495 | 65.6 |
| Dickinson | 187 | 0.0 | 456 | 1.0 | 176.5 | 3,389,715 | 7,439 | 285,670 | 179.3 | 436,195 | 49.2 | 50.8 | 2 | 83.7 | 4,070 | 76.6 |
| Dubuque | 313 | 7.6 | 224 | 0.4 | 258.5 | 1,930,243 | 8,633 | 188,850 | 440.1 | 313,898 | 29.2 | 70.8 | 15 | 82.0 | 5,667 | 73.5 |
| Emmet | 230 | 4.9 | 471 | 0.8 | 211.9 | 3,382,133 | 7,182 | 356,936 | 234.9 | 481,395 | 49.9 | 50.1 | 1 | 86.3 | 4,554 | 83.4 |
| Fayette | 385 | -0.9 | 304 | 1.3 | 337.6 | 2,183,246 | 7,176 | 222,654 | 372.9 | 294,761 | 46.1 | 53.9 | 9 | 79.5 | 14,587 | 84.3 |
| Floyd | 311 | -2.2 | 339 | 2.1 | 286.3 | 2,493,726 | 7,359 | 238,169 | 281.0 | 306,482 | 55.1 | 44.9 | 6 | 79.6 | 10,269 | 77.1 |

# Table B. States and Counties — Water Use, Wholesale Trade, Retail Trade, and Real Estate

| STATE County | Water use, 2015 | | Wholesale Trade[1], 2017 | | | | Retail Trade[2], 2017 | | | | Real estate and rental and leasing,[2] 2017 | | | |
|---|---|---|---|---|---|---|---|---|---|---|---|---|---|---|
| | Public supply water withdrawn (mil gal/day) | Public supply gallons withdrawn per person per day | Number of establishments | Number of employees | Sales (mil dol) | Average payroll (mil dol) | Number of establishments | Number of employees | Sales (mil dol) | Average payroll (mil dol) | Number of establishments | Number of employees | Sales (mil dol) | Average payroll (mil dol) |
| | 133 | 134 | 135 | 136 | 137 | 138 | 139 | 140 | 141 | 142 | 143 | 144 | 145 | 146 |
| INDIANA—Cont'd | | | | | | | | | | | | | | |
| Pike | 1.01 | 80.2 | 7 | D | 64.7 | D | 28 | 214 | 63.6 | 4.5 | 8 | 58 | 4.7 | 1.8 |
| Porter | 15.84 | 94.5 | 148 | 2,215 | 1,822.8 | 128.1 | 460 | 7,862 | 2,183.2 | 191.8 | 164 | 762 | 155.7 | 28.6 |
| Posey | 1.00 | 39.2 | 15 | 214 | 297.5 | 11.7 | 58 | 729 | 342.7 | 20.6 | 13 | 56 | 10.1 | 2.6 |
| Pulaski | 0.38 | 29.5 | 25 | 176 | 529.5 | 8.1 | 47 | 367 | 161.4 | 10.6 | D | D | D | D |
| Putnam | 4.03 | 107.2 | 18 | 115 | 50.8 | 6.7 | 97 | 1,281 | 400.0 | 33.8 | 30 | 115 | 12.0 | 2.5 |
| Randolph | 2.18 | 86.6 | 14 | 121 | 110.9 | 4.5 | 71 | 647 | 206.2 | 14.3 | 12 | 46 | 4.7 | 0.6 |
| Ripley | 1.19 | 41.5 | 24 | 189 | 101.9 | 6.5 | 103 | 764 | 233.5 | 22.3 | 20 | 31 | 4.8 | 0.8 |
| Rush | 0.99 | 59.4 | 30 | 689 | 515.7 | 39.1 | 45 | 430 | 118.4 | 9.9 | D | D | D | D |
| St. Joseph | 21.76 | 81.1 | 292 | 4,437 | 3,881.1 | 253.9 | 914 | 16,377 | 5,381.5 | 439.8 | 236 | 1,604 | 282.0 | 60.4 |
| Scott | 3.33 | 140.2 | 9 | 166 | 39.2 | 8.0 | 79 | 826 | 237.0 | 20.0 | 22 | 55 | 7.5 | 1.4 |
| Shelby | 5.66 | 127.3 | 35 | 504 | 293.9 | 22.9 | 124 | 1,589 | 512.5 | 41.2 | 44 | 148 | 23.2 | 4.8 |
| Spencer | 1.97 | 95.1 | 18 | 304 | 184.7 | 12.1 | 67 | 676 | 155.9 | 14.1 | 9 | 31 | 4.3 | 1.5 |
| Starke | 0.61 | 26.6 | 10 | 75 | 52.1 | 2.8 | 60 | 606 | 174.1 | 15.6 | 13 | 25 | 1.9 | 0.4 |
| Steuben | 1.36 | 39.6 | 42 | 336 | 191.6 | 13.8 | 174 | 2,157 | 623.0 | 45.3 | 43 | 107 | 15.3 | 3.3 |
| Sullivan | 1.36 | 65.0 | 18 | 146 | 144.5 | 6.5 | 48 | 527 | 173.5 | 12.7 | 6 | 14 | 0.9 | 0.2 |
| Switzerland | 1.04 | 98.8 | D | D | D | D | 19 | 143 | 43.8 | 3.0 | NA | NA | NA | NA |
| Tippecanoe | 12.84 | 69.1 | 112 | 1,434 | 1,211.3 | 74.9 | 553 | 9,689 | 2,707.4 | 234.5 | 209 | 1,218 | 209.6 | 45.1 |
| Tipton | 0.91 | 59.6 | 16 | 133 | 173.0 | 6.9 | 42 | 430 | 140.7 | 10.3 | 10 | 22 | 2.4 | 0.5 |
| Union | 0.35 | 48.7 | NA | NA | NA | NA | 22 | 199 | 33.9 | 3.7 | NA | NA | NA | NA |
| Vanderburgh | 17.40 | 95.7 | 255 | 3,880 | 2,689.5 | 208.8 | 760 | 13,106 | 3,570.8 | 330.0 | 220 | 1,274 | 297.1 | 48.4 |
| Vermillion | 1.25 | 79.7 | 10 | 68 | 136.5 | 2.7 | 50 | 612 | 191.1 | 14.9 | D | D | D | D |
| Vigo | 12.20 | 113.1 | 91 | 1,152 | 450.4 | 48.5 | 417 | 6,558 | 1,683.5 | 152.6 | 92 | 576 | 103.7 | 18.2 |
| Wabash | 3.48 | 108.3 | 26 | 328 | 317.3 | 17.6 | 116 | 1,359 | 345.4 | 32.7 | 27 | 120 | 8.6 | 2.6 |
| Warren | 0.43 | 52.0 | 8 | 109 | 100.3 | 5.0 | 15 | 102 | 19.5 | 1.2 | 3 | 6 | 0.3 | 0.0 |
| Warrick | 2.70 | 43.6 | 36 | 256 | 306.6 | 12.7 | 129 | 1,519 | 425.1 | 37.3 | 51 | 165 | 33.8 | 4.3 |
| Washington | 2.44 | 87.7 | 12 | 45 | 16.2 | 1.6 | 76 | 909 | 274.4 | 22.7 | 15 | 55 | 5.3 | 1.0 |
| Wayne | 5.01 | 74.8 | 50 | 802 | 984.3 | 36.3 | 251 | 3,768 | 957.1 | 85.8 | 56 | 212 | 42.4 | 8.2 |
| Wells | 2.44 | 87.3 | 28 | 529 | 941.9 | 38.2 | 80 | 985 | 250.6 | 24.5 | 31 | 102 | 11.8 | 2.6 |
| White | 0.97 | 39.9 | 23 | 210 | 308.7 | 10.0 | 93 | 1,116 | 316.4 | 28.3 | 23 | 35 | 6.6 | 1.0 |
| Whitley | 1.38 | 41.3 | 24 | 443 | 271.8 | 20.6 | 103 | 1,367 | 397.0 | 39.8 | 19 | 87 | 27.7 | 5.1 |
| IOWA | 390.38 | 125.0 | 4,426 | 62,658 | 64,162.6 | 3,389.3 | 11,479 | 181,416 | 50,063.1 | 4,518.5 | 3,130 | 13,999 | 2,890.1 | 575.6 |
| Adair | 0.67 | 92.7 | D | D | D | 5.3 | 29 | 259 | 49.6 | 4.7 | 7 | D | 4.8 | D |
| Adams | 1.00 | 263.4 | D | D | D | D | 13 | 118 | 22.2 | 2.4 | NA | NA | NA | NA |
| Allamakee | 0.91 | 65.5 | 29 | 281 | 217.3 | 13.2 | 62 | 627 | 141.7 | 13.6 | 14 | 36 | 2.7 | 0.7 |
| Appanoose | 6.65 | 530.8 | 15 | 105 | 67.5 | 3.5 | 51 | 651 | 140.6 | 15.9 | 6 | 10 | 1.8 | 0.2 |
| Audubon | 0.24 | 41.6 | D | D | D | 5.7 | 15 | 155 | 39.0 | 3.4 | NA | NA | NA | NA |
| Benton | 1.26 | 49.1 | 34 | 418 | 403.5 | 20.0 | 83 | 811 | 205.7 | 19.5 | D | D | D | D |
| Black Hawk | 16.56 | 124.1 | 144 | 2,591 | 2,150.4 | 137.7 | 506 | 8,863 | 2,213.9 | 222.1 | 152 | 786 | 143.2 | 26.8 |
| Boone | 2.39 | 89.7 | D | D | D | D | 64 | 1,037 | 261.0 | 26.2 | 18 | 53 | 5.2 | 1.2 |
| Bremer | 1.51 | 61.1 | 23 | 198 | 165.0 | 9.8 | 80 | 1,404 | 290.4 | 33.5 | D | D | D | D |
| Buchanan | 1.20 | 57.0 | 31 | 303 | 214.1 | 14.3 | 74 | 853 | 218.7 | 20.2 | 10 | D | 3.4 | D |
| Buena Vista | 3.57 | 174.2 | D | D | D | D | 83 | 1,073 | 241.2 | 25.6 | 18 | 74 | 11.0 | 2.7 |
| Butler | 0.83 | 55.6 | 27 | 186 | 222.6 | 8.9 | 51 | 411 | 102.4 | 8.6 | NA | NA | NA | NA |
| Calhoun | 1.02 | 103.9 | 20 | 228 | 247.8 | 10.9 | 45 | 350 | 160.9 | 10.3 | D | D | D | D |
| Carroll | 2.04 | 99.5 | 57 | 1,303 | 1,001.3 | 51.8 | 140 | 1,559 | 372.1 | 40.8 | 20 | 107 | 23.7 | 4.2 |
| Cass | 1.23 | 91.6 | 20 | 232 | 318.5 | 9.5 | 72 | 927 | 199.6 | 21.1 | 15 | 71 | 14.8 | 3.2 |
| Cedar | 1.04 | 56.7 | 27 | 443 | 367.0 | 22.0 | 58 | 597 | 189.0 | 13.1 | 13 | 21 | 3.6 | 0.7 |
| Cerro Gordo | 6.50 | 151.1 | 93 | 1,071 | 924.7 | 56.1 | 212 | 3,596 | 912.4 | 85.1 | 66 | 160 | 32.7 | 4.8 |
| Cherokee | 1.93 | 166.8 | 19 | 175 | 201.3 | 8.5 | 52 | 602 | 133.3 | 14.5 | D | D | D | 0.1 |
| Chickasaw | 1.06 | 87.6 | 35 | 382 | 363.7 | 18.8 | 52 | 454 | 114.2 | 10.0 | 8 | 10 | 1.1 | 0.3 |
| Clarke | 1.19 | 128.5 | 8 | 72 | 48.8 | 3.0 | 29 | 563 | 130.3 | 12.6 | 5 | 11 | 1.1 | 0.1 |
| Clay | 2.27 | 137.5 | 49 | 658 | 450.4 | 23.8 | 100 | 1,419 | 478.2 | 37.9 | 23 | 102 | 14.3 | 2.9 |
| Clayton | 0.73 | 41.4 | 22 | 336 | 529.0 | 17.3 | 86 | 747 | 225.8 | 18.5 | 10 | D | 4.2 | D |
| Clinton | 3.87 | 81.0 | 38 | 329 | 302.9 | 17.3 | 181 | 2,570 | 591.0 | 58.1 | 36 | 112 | 23.8 | 4.2 |
| Crawford | 2.96 | 173.2 | 21 | 251 | 326.9 | 13.0 | 47 | 788 | 174.2 | 17.0 | 10 | D | 2.5 | D |
| Dallas | 1.47 | 18.3 | 64 | 841 | 491.1 | 55.2 | 281 | 6,246 | 1,527.1 | 145.9 | 126 | 973 | 152.3 | 61.1 |
| Davis | 0.00 | 0.0 | 10 | 82 | 38.8 | 2.9 | 29 | 282 | 66.2 | 6.4 | 6 | 11 | 1.7 | 0.3 |
| Decatur | 0.55 | 66.9 | 9 | 83 | 36.1 | 2.5 | 22 | 224 | 40.2 | 4.7 | 4 | 8 | 1.4 | 0.3 |
| Delaware | 0.88 | 50.6 | 29 | 332 | 385.5 | 17.4 | 60 | 682 | 197.8 | 17.3 | 2 | 35 | 4.4 | 1.1 |
| Des Moines | 5.76 | 143.8 | 53 | 657 | 723.3 | 26.8 | 197 | 3,285 | 911.2 | 83.4 | 47 | 639 | 105.2 | 27.2 |
| Dickinson | 2.07 | 121.0 | D | D | D | D | 109 | 1,168 | 340.4 | 34.7 | 36 | 50 | 16.5 | 1.7 |
| Dubuque | 7.71 | 79.4 | 156 | 2,348 | 2,185.6 | 122.0 | 424 | 7,233 | 1,763.9 | 175.4 | 121 | 400 | 94.9 | 14.4 |
| Emmet | 1.17 | 119.8 | 19 | 152 | 132.7 | 8.8 | 41 | 440 | 87.0 | 9.1 | NA | NA | NA | NA |
| Fayette | 1.12 | 55.3 | 39 | 369 | 317.9 | 17.8 | 98 | 866 | 193.9 | 19.4 | 15 | 28 | 15.7 | 1.0 |
| Floyd | 2.22 | 139.1 | 23 | 382 | 276.9 | 20.4 | 59 | 810 | 215.0 | 18.1 | 9 | 22 | 2.6 | 0.5 |

1 Merchant wholesalers, except manufacturers' sales branches and offices.  2. Employer establishments.

— **Professional Services, Manufacturing, and Accommodation and Food Services**

| STATE County | Professional, scientific, and technical services, 2017 | | | | Manufacturing, 2017 | | | | Accommodation and food services, 2017 | | | |
|---|---|---|---|---|---|---|---|---|---|---|---|---|
| | Number of establish-ments | Number of employees | Sales (mil dol) | Average payroll (mil dol) | Number of establish-ments | Number of employees | Sales (mil dol) | Average payroll (mil dol) | Number of establis-hments | Number of employees | Sales (mil dol) | Annual payroll (mil dol) |
| | 147 | 148 | 149 | 150 | 151 | 152 | 153 | 154 | 155 | 156 | 157 | 158 |
| INDIANA—Cont'd | | | | | | | | | | | | |
| Pike | 8 | 31 | 3 | 1.0 | 6 | 178 | 85.5 | 9.9 | 14 | 232 | 8.0 | 2.3 |
| Porter | D | D | D | D | 153 | 10,337 | 9,156.2 | 860.1 | 321 | 6,263 | 315.5 | 91.2 |
| Posey | 39 | 307 | 45 | 16.8 | 31 | 3,322 | 3,816.0 | 291.5 | 31 | 558 | 21.7 | 6.6 |
| Pulaski | 23 | 69 | 7 | 1.5 | 17 | 1,387 | 499.6 | 79.1 | D | D | D | D |
| Putnam | 41 | 170 | 14 | 5.8 | 28 | 3,011 | 1,192.7 | 146.6 | 82 | 1,299 | 58.0 | 16.1 |
| Randolph | 34 | 128 | 21 | 4.4 | 47 | 1,671 | 560.0 | 81.8 | 39 | 514 | 21.7 | 6.2 |
| Ripley | 35 | 126 | 12 | 4.7 | 37 | 1,496 | 730.5 | 69.8 | 43 | 641 | 25.5 | 8.6 |
| Rush | 32 | 138 | 15 | 4.9 | 25 | 662 | 252.5 | 26.2 | 24 | 350 | 17.9 | 5.1 |
| St. Joseph | 501 | 4,779 | 1,569 | 296.4 | 336 | 15,133 | 6,184.5 | 845.7 | 568 | 11,392 | 591.2 | 166.6 |
| Scott | 22 | 117 | 11 | 3.8 | 19 | 2,076 | 962.5 | 110.2 | D | D | D | D |
| Shelby | 57 | 235 | 56 | 11.9 | 81 | 5,560 | 2,796.9 | 289.3 | 80 | 1,579 | 69.4 | 20.4 |
| Spencer | 24 | 223 | 33 | 13.1 | 23 | 1,488 | 1,296.3 | 77.1 | 28 | 334 | 16.8 | 4.5 |
| Starke | 17 | 106 | 9 | 3.2 | 22 | 734 | 225.5 | 31.1 | D | D | D | D |
| Steuben | 62 | 217 | 20 | 6.7 | 89 | 5,078 | 1,435.2 | 229.1 | 90 | 1,400 | 72.2 | 21.2 |
| Sullivan | 21 | 98 | 8 | 3.6 | 18 | 497 | 103.5 | 18.3 | 36 | 486 | 19.4 | 6.4 |
| Switzerland | 10 | 15 | 1 | 0.5 | D | 81 | D | D | D | D | D | D |
| Tippecanoe | 335 | 2,869 | 439 | 173.9 | 124 | 17,647 | 14,562.9 | 1,054.5 | 451 | 9,044 | 475.7 | 128.8 |
| Tipton | D | D | D | D | 16 | 1,034 | 326.7 | 50.9 | 25 | 443 | 16.7 | 5.0 |
| Union | D | D | 2 | D | D | 238 | D | 12.1 | D | D | D | 1.5 |
| Vanderburgh | D | D | D | D | 240 | 11,605 | 4,073.2 | 537.1 | 471 | 11,279 | 640.0 | 174.4 |
| Vermillion | 10 | 30 | 2 | 0.6 | 15 | 748 | 708.5 | 58.8 | 29 | 389 | 22.2 | 5.4 |
| Vigo | D | D | D | D | 111 | 6,828 | 3,641.3 | 363.8 | 276 | 5,623 | 276.8 | 78.8 |
| Wabash | 49 | 204 | 36 | 8.6 | 57 | 2,545 | 1,407.5 | 148.3 | 68 | 1,008 | 45.6 | 12.4 |
| Warren | D | D | 1 | D | 12 | 668 | 154.9 | 28.1 | D | D | D | D |
| Warrick | 103 | 416 | 61 | 22.0 | 48 | 2,057 | 1,099.8 | 143.0 | 75 | 1,210 | 56.4 | 15.9 |
| Washington | 37 | 121 | 10 | 3.6 | 34 | 1,478 | 388.1 | 66.7 | 30 | 452 | 18.4 | 5.5 |
| Wayne | D | D | D | D | 96 | 5,100 | 1,989.4 | 253.7 | 149 | 2,834 | 136.3 | 41.9 |
| Wells | 35 | 262 | 29 | 10.6 | 54 | 2,732 | 1,118.8 | 122.4 | 41 | 572 | 24.1 | 6.6 |
| White | 29 | 99 | 12 | 3.7 | 42 | 2,695 | 1,227.0 | 111.0 | 61 | 665 | 28.9 | 7.8 |
| Whitley | 41 | 212 | 21 | 7.4 | 63 | 4,744 | 2,143.9 | 267.3 | 60 | 985 | 37.8 | 10.8 |
| IOWA | 6,426 | 52,390 | 7,907 | 3,160.8 | 3,489 | 210,722 | 109,727.8 | 11,373.9 | 7,283 | 123,866 | 7,110.7 | 1,897.4 |
| Adair | 12 | 42 | 3 | 1.2 | NA | NA | NA | NA | 18 | 218 | 8.3 | 2.2 |
| Adams | 7 | 26 | 3 | 0.8 | D | 201 | D | 9.9 | D | D | D | 0.4 |
| Allamakee | 19 | 79 | 14 | 3.4 | 24 | 956 | 271.4 | 42.3 | 35 | 235 | 10.4 | 2.5 |
| Appanoose | 21 | 89 | 8 | 3.2 | 19 | 1,088 | 336.2 | 46.3 | 32 | 547 | 26.3 | 7.7 |
| Audubon | D | D | D | D | 9 | 153 | 57.9 | 7.2 | 9 | 62 | 2.0 | 0.5 |
| Benton | 28 | 72 | 8 | 2.7 | 24 | 631 | 301.4 | 33.9 | 33 | 277 | 11.5 | 3.3 |
| Black Hawk | D | D | D | D | 144 | 13,259 | 5,936.3 | 650.3 | 318 | 6,704 | 363.7 | 97.4 |
| Boone | 38 | 134 | 17 | 6.3 | 34 | 476 | 129.9 | 28.1 | 42 | 528 | 22.5 | 6.4 |
| Bremer | 33 | 225 | 23 | 7.1 | 34 | 1,352 | 575.3 | 78.8 | 52 | 571 | 25.3 | 6.8 |
| Buchanan | 22 | 114 | 24 | 4.0 | 39 | 1,331 | 582.8 | 61.8 | 38 | 348 | 14.8 | 3.9 |
| Buena Vista | 35 | 211 | 28 | 10.7 | 27 | 3,601 | 1,757.3 | 155.4 | 42 | 677 | 28.5 | 7.4 |
| Butler | 19 | 49 | 5 | 1.4 | 24 | 732 | 379.9 | 29.0 | 19 | 122 | 3.6 | 0.8 |
| Calhoun | 13 | 35 | 6 | 1.4 | 13 | 82 | 11.2 | 3.9 | 14 | 110 | 3.6 | 1.0 |
| Carroll | 40 | 174 | 25 | 8.0 | 40 | 1,411 | 769.0 | 70.7 | 53 | 719 | 30.3 | 9.0 |
| Cass | 24 | 140 | 15 | 5.9 | 22 | 506 | 138.1 | 23.3 | 30 | 346 | 14.5 | 4.0 |
| Cedar | 23 | 101 | 13 | 4.1 | 27 | 506 | 176.6 | 23.9 | 35 | 300 | 11.4 | 3.2 |
| Cerro Gordo | D | D | D | D | 59 | 3,321 | 1,561.0 | 161.5 | 126 | 2,062 | 100.0 | 29.8 |
| Cherokee | 16 | 79 | 12 | 2.9 | 16 | 328 | 359.6 | 14.5 | 27 | 293 | 13.2 | 3.4 |
| Chickasaw | 19 | 60 | 8 | 2.3 | 39 | 1,391 | 662.1 | 67.9 | 27 | 247 | 9.0 | 2.5 |
| Clarke | 11 | 37 | 9 | 1.9 | 9 | 1,309 | 854.8 | 62.2 | 19 | 538 | 86.3 | 11.3 |
| Clay | 37 | 180 | 25 | 9.3 | 29 | 636 | 285.5 | 31.2 | 53 | 738 | 32.5 | 9.2 |
| Clayton | 24 | 94 | 9 | 3.0 | 28 | 924 | 202.3 | 38.1 | 52 | 341 | 13.1 | 3.4 |
| Clinton | 60 | 273 | 29 | 12.2 | 54 | 4,230 | 3,596.9 | 251.0 | 107 | 1,812 | 100.9 | 23.7 |
| Crawford | 21 | 97 | 9 | 3.0 | 18 | 2,171 | 943.2 | 104.2 | 35 | 412 | 17.9 | 4.6 |
| Dallas | 248 | 1,425 | 244 | 91.6 | 39 | 2,042 | 608.1 | 87.9 | 170 | 3,245 | 192.7 | 56.5 |
| Davis | 12 | 54 | 7 | 1.9 | 10 | 119 | 44.0 | 5.9 | D | D | D | D |
| Decatur | 7 | 30 | 2 | 0.6 | 5 | D | 11.9 | D | 9 | 88 | 3.8 | 1.1 |
| Delaware | 25 | 99 | 12 | 4.9 | 43 | 2,204 | 807.0 | 116.7 | D | D | D | D |
| Des Moines | 53 | 362 | 38 | 13.9 | 48 | 4,568 | 1,629.2 | 249.9 | 106 | 2,077 | 93.3 | 29.7 |
| Dickinson | 50 | 179 | 22 | 7.3 | 35 | 1,537 | 698.1 | 72.5 | 105 | 1,019 | 73.9 | 23.0 |
| Dubuque | D | D | D | D | 148 | 9,356 | 4,987.4 | 463.1 | 251 | 4,953 | 270.8 | 78.0 |
| Emmet | 16 | 60 | 7 | 2.9 | 17 | 614 | 188.8 | 29.0 | 20 | 240 | 9.9 | 2.9 |
| Fayette | 31 | 116 | 11 | 4.4 | 22 | 459 | 123.8 | 20.5 | 42 | 411 | 17.6 | 4.9 |
| Floyd | 24 | 75 | 7 | 2.9 | 19 | 1,264 | 809.9 | 73.3 | 31 | 350 | 16.1 | 3.9 |

# Table B. States and Counties — Health Care and Social Assistance, Other Services, Nonemployer Businesses, and Residential Construction

| STATE County | Health care and social assistance, 2017 | | | | Other services, 2017 | | | | Nonemployer businesses, 2018 | | Value of residential construction authorized by building permits, 2020 | |
|---|---|---|---|---|---|---|---|---|---|---|---|---|
| | Number of establishments | Number of employees | Receipts (mil dol) | Annual payroll (mil dol) | Number of establishments | Number of employees | Receipts (mil dol) | Annual payroll (mil dol) | Number | Receipts (mil dol) | New construction ($1,000) | Number of housing units |
| | 159 | 160 | 161 | 162 | 163 | 164 | 165 | 166 | 167 | 168 | 169 | 170 |
| INDIANA—Cont'd | | | | | | | | | | | | |
| Pike | 14 | 299 | 17.1 | 5.8 | 20 | 64 | 8.2 | 2.6 | 566 | 17.9 | 0 | 0 |
| Porter | 389 | 8,715 | 959.2 | 369.0 | 292 | 1,837 | 191.5 | 54.7 | 10,392 | 455.9 | 163,111 | 580 |
| Posey | 44 | 395 | 27.2 | 11.1 | 49 | 177 | 20.8 | 5.3 | 1,439 | 52.9 | 12,082 | 52 |
| Pulaski | 17 | 535 | 43.8 | 19.3 | 28 | 86 | 8.6 | 2.1 | 805 | 30.3 | 608 | 15 |
| Putnam | 74 | 1,909 | 133.8 | 58.5 | 58 | 253 | 23.4 | 5.8 | 2,089 | 89.1 | 44,099 | 164 |
| Randolph | 47 | 801 | 58.4 | 24.5 | 37 | 124 | 12.7 | 2.9 | 1,385 | 55.1 | 2,790 | 17 |
| Ripley | 79 | 1,548 | 148.7 | 61.5 | 60 | 185 | 21.8 | 5.0 | 1,780 | 73.8 | 17,791 | 113 |
| Rush | 33 | 622 | 61.3 | 23.7 | 24 | 107 | 12.2 | 2.4 | 1,156 | 45.9 | 4,416 | 14 |
| St. Joseph | 663 | 19,436 | 2,404.8 | 829.9 | 431 | 3,025 | 354.6 | 99.3 | 16,013 | 658.6 | 89,385 | 286 |
| Scott | 59 | 1,306 | 91.2 | 34.4 | D | D | D | D | 1,161 | 40.8 | 11,404 | 40 |
| Shelby | 89 | 1,979 | 236.8 | 90.0 | 61 | 405 | 52.0 | 13.1 | 2,853 | 132.4 | 19,945 | 82 |
| Spencer | 29 | 220 | 22.3 | 6.6 | 30 | 188 | 20.9 | 6.6 | 1,181 | 48.1 | 13,777 | 71 |
| Starke | 26 | 424 | 55.5 | 20.0 | 26 | 85 | 14.7 | 2.4 | 1,181 | 48.2 | 13,501 | 59 |
| Steuben | 75 | 1,359 | 132.6 | 48.8 | 85 | 439 | 62.0 | 16.4 | 2,182 | 109.7 | 53,220 | 163 |
| Sullivan | 29 | 614 | 55.8 | 22.1 | 28 | 119 | 8.2 | 2.0 | 921 | 28.7 | 550 | 5 |
| Switzerland | 15 | 124 | 7.6 | 3.4 | 10 | 42 | 3.8 | 0.9 | 611 | 26.9 | 3,904 | 67 |
| Tippecanoe | 478 | 12,733 | 1,627.6 | 612.3 | 249 | 2,110 | 491.8 | 74.0 | 9,707 | 420.6 | 213,635 | 1,317 |
| Tipton | 34 | 593 | 73.0 | 25.2 | 26 | 92 | 15.3 | 2.6 | 1,036 | 40.7 | 5,831 | 28 |
| Union | 11 | 90 | 4.7 | 2.6 | D | D | D | 1.2 | 458 | 21.2 | 1,858 | 12 |
| Vanderburgh | 617 | 20,158 | 2,479.5 | 879.0 | 357 | 2,901 | 334.5 | 89.0 | 9,790 | 415.7 | 116,618 | 675 |
| Vermillion | 24 | 574 | 75.2 | 25.3 | 23 | 63 | 5.5 | 1.2 | 741 | 24.7 | 4,347 | 27 |
| Vigo | 371 | 9,441 | 1,268.8 | 440.5 | 172 | 1,053 | 124.8 | 29.2 | 4,892 | 189.7 | 9,480 | 58 |
| Wabash | 75 | 2,225 | 156.0 | 64.2 | 55 | 321 | 39.1 | 6.9 | 1,757 | 63.0 | 8,662 | 39 |
| Warren | 13 | 269 | 26.4 | 11.1 | D | D | D | D | 504 | 22.6 | 8,632 | 31 |
| Warrick | 151 | 3,806 | 485.9 | 172.3 | 98 | 524 | 51.4 | 14.6 | 3,992 | 181.9 | 102,351 | 331 |
| Washington | 57 | 680 | 52.7 | 21.4 | 31 | 142 | 15.1 | 3.2 | 1,749 | 69.8 | 9,331 | 59 |
| Wayne | 201 | 5,563 | 728.7 | 250.2 | 105 | 542 | 54.9 | 13.4 | 3,644 | 160.8 | 6,675 | 34 |
| Wells | 57 | 1,367 | 109.3 | 42.5 | 60 | 301 | 31.3 | 8.6 | 1,745 | 66.5 | 12,683 | 57 |
| White | 37 | 874 | 66.8 | 25.7 | 36 | 138 | 14.4 | 3.4 | 1,457 | 65.1 | 8,360 | 38 |
| Whitley | 62 | 1,323 | 120.9 | 45.6 | 66 | 408 | 43.4 | 11.6 | 2,187 | 82.3 | 26,823 | 92 |
| IOWA | 8,610 | 216,965 | 22,419.8 | 9,241.6 | 5,975 | 31,306 | 4,186.5 | 1,021.5 | 212,333 | 9,723.7 | 2,701,682 | 12,623 |
| Adair | 26 | 336 | 22.6 | 9.6 | D | D | 2.1 | D | 609 | 27.4 | 2,618 | 12 |
| Adams | 14 | 423 | 67.2 | 22.2 | 11 | 26 | 2.2 | 0.5 | 345 | 17.1 | 1,575 | 7 |
| Allamakee | 43 | 894 | 57.8 | 29.0 | D | D | D | D | 1,143 | 58.4 | 6,355 | 38 |
| Appanoose | 31 | 688 | 60.2 | 25.3 | 22 | 82 | 8.2 | 1.9 | 883 | 42.1 | 1,146 | 12 |
| Audubon | 17 | 305 | 24.0 | 10.2 | 18 | 48 | 7.0 | 1.4 | 474 | 22.5 | 2,803 | 6 |
| Benton | 51 | 815 | 73.0 | 26.3 | D | D | D | 3.7 | 1,791 | 73.0 | 5,258 | 20 |
| Black Hawk | 373 | 11,246 | 1,179.9 | 485.6 | 211 | 1,507 | 154.9 | 41.9 | 7,420 | 367.1 | 62,134 | 268 |
| Boone | 53 | 1,613 | 114.6 | 52.7 | 46 | 146 | 19.5 | 5.3 | 1,812 | 67.0 | 9,120 | 50 |
| Bremer | 63 | 1,756 | 133.0 | 63.2 | 55 | 206 | 21.5 | 6.2 | 1,773 | 79.9 | 14,970 | 63 |
| Buchanan | 40 | 937 | 72.4 | 35.5 | 24 | 83 | 6.8 | 2.0 | 1,523 | 70.5 | 5,900 | 53 |
| Buena Vista | 51 | 1,312 | 108.9 | 44.7 | 34 | 129 | 17.0 | 3.6 | 1,377 | 58.5 | 4,915 | 26 |
| Butler | 28 | 514 | 23.4 | 12.5 | D | D | D | 2.6 | 1,066 | 43.6 | 5,382 | 29 |
| Calhoun | 31 | 658 | 56.9 | 22.5 | D | D | D | 2.1 | 754 | 29.1 | 5,430 | 34 |
| Carroll | 110 | 2,246 | 199.8 | 83.2 | 65 | 285 | 27.0 | 6.4 | 1,863 | 99.7 | 7,535 | 34 |
| Cass | 47 | 1,174 | 98.7 | 42.3 | 40 | 252 | 18.6 | 4.6 | 1,168 | 44.7 | 2,283 | 10 |
| Cedar | 52 | 606 | 33.0 | 16.0 | 36 | 89 | 13.4 | 2.7 | 1,321 | 53.3 | 11,394 | 54 |
| Cerro Gordo | 138 | 4,072 | 391.7 | 181.4 | D | D | D | D | 2,842 | 117.7 | 26,692 | 176 |
| Cherokee | 43 | 861 | 78.0 | 38.0 | 23 | 57 | 7.8 | 1.7 | 835 | 32.4 | 875 | 8 |
| Chickasaw | 33 | 647 | 62.2 | 22.4 | 26 | 90 | 10.0 | 2.3 | 890 | 40.6 | 1,310 | 7 |
| Clarke | 27 | 521 | 47.3 | 18.2 | 15 | 53 | 6.4 | 1.1 | 547 | 23.1 | 13,955 | 109 |
| Clay | 63 | 1,485 | 171.9 | 68.5 | 42 | 227 | 20.9 | 5.7 | 1,330 | 69.0 | 5,884 | 23 |
| Clayton | 69 | 1,145 | 69.1 | 32.4 | D | D | 11.3 | D | 1,465 | 59.0 | 5,877 | 25 |
| Clinton | 128 | 3,359 | 255.0 | 107.2 | 97 | 317 | 33.8 | 8.3 | 2,613 | 106.4 | 9,787 | 56 |
| Crawford | 38 | 941 | 77.4 | 33.6 | 27 | 106 | 12.3 | 2.9 | 897 | 43.5 | 1,282 | 5 |
| Dallas | 202 | 3,989 | 402.1 | 170.7 | 119 | 476 | 39.5 | 12.2 | 6,881 | 363.0 | 285,302 | 1,155 |
| Davis | 20 | 435 | 40.0 | 16.7 | 11 | 29 | 2.8 | 0.6 | 754 | 52.4 | 848 | 4 |
| Decatur | 16 | 343 | 29.2 | 13.4 | D | D | D | D | 591 | 26.4 | 713 | 3 |
| Delaware | 37 | 1,109 | 89.9 | 40.9 | 41 | 147 | 16.5 | 4.1 | 1,338 | 69.7 | 5,975 | 30 |
| Des Moines | 148 | 3,534 | 402.5 | 166.8 | 84 | 396 | 44.5 | 10.2 | 2,337 | 85.9 | 5,292 | 43 |
| Dickinson | 63 | 951 | 84.7 | 31.9 | 41 | 196 | 18.9 | 5.1 | 1,727 | 85.2 | 52,655 | 198 |
| Dubuque | 283 | 8,469 | 943.9 | 405.4 | 205 | 1,118 | 135.7 | 31.4 | 6,400 | 306.5 | 60,978 | 277 |
| Emmet | 34 | 781 | 54.0 | 20.8 | 26 | 83 | 9.0 | 2.9 | 655 | 24.3 | 2,722 | 10 |
| Fayette | 53 | 1,302 | 91.3 | 43.1 | 49 | 219 | 28.1 | 7.0 | 1,369 | 55.0 | 4,070 | 19 |
| Floyd | 49 | 929 | 65.6 | 30.5 | 38 | 106 | 13.0 | 2.6 | 1,106 | 40.5 | 3,557 | 15 |

# Table B. States and Counties — Government Employment and Payroll, and Local Government Finances

| STATE County | Full-time equivalent employees | March payroll (dollars) | Adminis-tration, judicial, and legal | Police and corrections | Fire protection | Highways and transpor-tation | Health and welfare | Natural resources and utilities | Education and libraries | Total (mil dol) | Inter-govern-mental (mil dol) | Taxes Total (mil dol) | Per capita[1] (dollars) Total | Per capita[1] (dollars) Property |
|---|---|---|---|---|---|---|---|---|---|---|---|---|---|---|
| | 171 | 172 | 173 | 174 | 175 | 176 | 177 | 178 | 179 | 180 | 181 | 182 | 183 | 184 |
| **INDIANA—Cont'd** | | | | | | | | | | | | | | |
| Pike | 376 | 1,267,623 | 15.7 | 9.9 | 0.0 | 4.7 | 4.6 | 3.9 | 61.1 | 47.2 | 17.8 | 26.4 | 2,144 | 1,901 |
| Porter | 5,027 | 18,091,151 | 5.3 | 10.3 | 4.0 | 3.2 | 0.9 | 5.1 | 69.2 | 545.8 | 244.5 | 213.1 | 1,265 | 1,187 |
| Posey | 873 | 2,782,425 | 8.1 | 8.2 | 1.0 | 6.0 | 5.3 | 5.9 | 63.3 | 71.4 | 31.2 | 29.1 | 1,138 | 1,119 |
| Pulaski | 761 | 2,900,049 | 6.3 | 6.4 | 0.7 | 2.1 | 48.4 | 2.8 | 33.0 | 106.3 | 21.3 | 11.1 | 889 | 851 |
| Putnam | 1,541 | 5,175,859 | 5.5 | 6.3 | 1.4 | 2.7 | 25.6 | 1.8 | 54.9 | 116.6 | 69.2 | 30.9 | 828 | 720 |
| Randolph | 917 | 2,927,259 | 9.3 | 9.1 | 3.5 | 3.4 | 2.7 | 2.9 | 67.3 | 82.0 | 36.9 | 36.8 | 1,479 | 1,469 |
| Ripley | 1,144 | 4,018,061 | 6.8 | 6.0 | 0.7 | 3.0 | 2.2 | 4.7 | 75.7 | 122.0 | 61.2 | 46.6 | 1,638 | 1,477 |
| Rush | 959 | 2,631,447 | 6.7 | 4.4 | 4.5 | 4.2 | 39.4 | 2.5 | 38.0 | 82.4 | 33.5 | 35.4 | 2,127 | 1,980 |
| St. Joseph | 8,426 | 31,242,839 | 6.0 | 13.1 | 8.2 | 4.8 | 2.4 | 7.1 | 57.0 | 1,052.2 | 450.4 | 395.1 | 1,463 | 1,159 |
| Scott | 971 | 3,502,719 | 8.3 | 7.3 | 0.2 | 1.8 | 20.4 | 9.5 | 51.0 | 63.9 | 36.2 | 18.1 | 761 | 717 |
| Shelby | 2,332 | 8,734,101 | 3.2 | 7.2 | 4.0 | 1.7 | 40.8 | 2.0 | 40.6 | 305.0 | 133.0 | 46.2 | 1,042 | 907 |
| Spencer | 735 | 2,727,729 | 6.8 | 7.5 | 0.1 | 2.9 | 1.4 | 4.6 | 75.5 | 51.8 | 27.7 | 14.1 | 691 | 661 |
| Starke | 720 | 2,236,687 | 9.2 | 10.2 | 0.2 | 4.9 | 4.1 | 2.5 | 67.4 | 73.3 | 34.9 | 30.3 | 1,323 | 1,279 |
| Steuben | 1,033 | 3,461,764 | 11.8 | 11.4 | 2.4 | 4.2 | 5.4 | 6.5 | 57.0 | 103.3 | 50.0 | 32.9 | 957 | 917 |
| Sullivan | 901 | 2,822,687 | 6.5 | 5.8 | 2.0 | 3.7 | 34.8 | 2.7 | 43.9 | 99.4 | 25.8 | 35.7 | 1,722 | 1,714 |
| Switzerland | 376 | 1,180,396 | 9.7 | 9.0 | 0.0 | 4.8 | 1.6 | 4.0 | 61.5 | 44.8 | 31.4 | 8.4 | 781 | 541 |
| Tippecanoe | 4,813 | 18,962,006 | 7.7 | 11.8 | 5.4 | 4.6 | 1.8 | 5.5 | 61.8 | 572.0 | 262.5 | 175.4 | 917 | 873 |
| Tipton | 976 | 3,778,449 | 5.5 | 4.2 | 1.6 | 1.6 | 46.3 | 5.3 | 34.7 | 87.2 | 23.4 | 50.7 | 3,348 | 3,121 |
| Union | 369 | 1,190,233 | 12.2 | 4.3 | 12.1 | 3.3 | 1.2 | 1.9 | 64.3 | 28.8 | 13.0 | 10.7 | 1,501 | 1,096 |
| Vanderburgh | 5,246 | 21,088,289 | 6.4 | 14.4 | 6.5 | 5.0 | 2.1 | 6.8 | 56.5 | 623.8 | 279.8 | 199.4 | 1,102 | 1,043 |
| Vermillion | 522 | 1,635,207 | 9.3 | 8.6 | 1.4 | 4.1 | 1.6 | 3.0 | 71.3 | 45.1 | 21.8 | 17.1 | 1,103 | 1,073 |
| Vigo | 3,041 | 12,168,829 | 7.9 | 11.1 | 6.8 | 4.5 | 2.2 | 4.0 | 62.3 | 354.2 | 164.1 | 107.7 | 1,001 | 869 |
| Wabash | 1,039 | 2,928,878 | 7.5 | 12.2 | 6.2 | 3.8 | 0.8 | 4.8 | 62.8 | 100.0 | 49.4 | 34.6 | 1,101 | 905 |
| Warren | 249 | 839,058 | 11.8 | 10.1 | 0.0 | 8.0 | 0.0 | 7.1 | 62.6 | 33.2 | 13.6 | 14.1 | 1,715 | 1,025 |
| Warrick | 1,874 | 6,732,032 | 7.6 | 6.2 | 0.8 | 2.6 | 1.0 | 4.4 | 76.9 | 189.5 | 92.4 | 54.2 | 869 | 821 |
| Washington | 1,007 | 3,213,506 | 6.6 | 6.9 | 1.0 | 2.9 | 2.0 | 2.9 | 76.9 | 84.9 | 37.8 | 41.7 | 1,502 | 1,312 |
| Wayne | 2,582 | 8,382,351 | 8.8 | 11.8 | 3.5 | 3.3 | 3.9 | 12.8 | 54.5 | 214.4 | 101.3 | 72.0 | 1,089 | 1,056 |
| Wells | 1,007 | 3,432,160 | 6.4 | 11.5 | 1.0 | 3.1 | 0.8 | 5.3 | 71.5 | 83.6 | 38.4 | 31.6 | 1,129 | 743 |
| White | 1,085 | 3,734,602 | 6.7 | 8.0 | 3.4 | 3.8 | 0.5 | 5.7 | 70.0 | 97.2 | 44.5 | 28.3 | 1,171 | 1,137 |
| Whitley | 934 | 3,049,878 | 6.9 | 15.4 | 0.0 | 6.3 | 1.7 | 2.6 | 65.1 | 91.5 | 44.6 | 27.9 | 827 | 782 |
| **IOWA** | X | X | X | X | X | X | X | X | X | X | X | X | X | X |
| Adair | 354 | 1,285,189 | 7.2 | 8.9 | 0.0 | 9.0 | 29.6 | 2.9 | 42.0 | 47.0 | 12.3 | 16.6 | 2,349 | 2,103 |
| Adams | 157 | 536,007 | 6.4 | 8.7 | 0.0 | 14.6 | 5.9 | 3.8 | 58.4 | 16.9 | 7.2 | 8.0 | 2,191 | 1,957 |
| Allamakee | 696 | 2,613,749 | 5.1 | 4.8 | 0.1 | 6.7 | 28.3 | 2.0 | 51.8 | 70.8 | 34.8 | 22.0 | 1,593 | 1,327 |
| Appanoose | 487 | 1,695,790 | 8.2 | 7.5 | 1.0 | 7.1 | 1.5 | 2.5 | 71.5 | 40.1 | 22.4 | 14.2 | 1,150 | 952 |
| Audubon | 329 | 1,358,335 | 6.5 | 4.5 | 0.0 | 6.9 | 37.5 | 1.4 | 42.6 | 28.3 | 11.2 | 12.0 | 2,168 | 1,909 |
| Benton | 947 | 3,311,970 | 5.1 | 7.1 | 0.3 | 6.3 | 1.2 | 8.0 | 71.6 | 93.9 | 44.1 | 39.5 | 1,539 | 1,375 |
| Black Hawk | 5,888 | 25,785,248 | 3.3 | 8.3 | 3.8 | 3.9 | 4.2 | 8.5 | 66.6 | 632.3 | 290.7 | 225.2 | 1,703 | 1,428 |
| Boone | 1,215 | 5,162,400 | 4.1 | 5.3 | 1.0 | 4.6 | 35.3 | 4.8 | 44.5 | 129.6 | 40.4 | 39.8 | 1,505 | 1,282 |
| Bremer | 1,085 | 4,122,991 | 6.2 | 7.2 | 0.0 | 5.1 | 2.4 | 3.7 | 73.9 | 174.1 | 50.8 | 47.5 | 1,914 | 1,644 |
| Buchanan | 918 | 3,620,791 | 4.4 | 5.5 | 0.6 | 4.6 | 28.2 | 4.7 | 51.2 | 96.9 | 31.7 | 29.3 | 1,388 | 1,210 |
| Buena Vista | 1,352 | 5,090,475 | 4.6 | 4.7 | 0.3 | 3.9 | 32.3 | 3.9 | 49.5 | 147.2 | 44.9 | 38.8 | 1,926 | 1,633 |
| Butler | 294 | 1,055,541 | 15.9 | 11.9 | 0.1 | 16.9 | 5.6 | 6.5 | 39.7 | 77.3 | 47.4 | 20.5 | 1,408 | 1,249 |
| Calhoun | 436 | 1,312,852 | 8.3 | 8.1 | 0.0 | 11.1 | 8.7 | 4.7 | 58.1 | 41.9 | 17.4 | 18.7 | 1,924 | 1,721 |
| Carroll | 863 | 3,050,468 | 6.9 | 6.8 | 0.1 | 7.3 | 4.3 | 8.9 | 63.6 | 94.9 | 34.5 | 38.2 | 1,883 | 1,533 |
| Cass | 847 | 3,327,200 | 4.3 | 3.9 | 0.5 | 4.6 | 48.0 | 2.5 | 35.0 | 109.5 | 27.2 | 30.5 | 2,320 | 2,039 |
| Cedar | 836 | 2,809,400 | 6.7 | 7.4 | 0.1 | 6.4 | 2.8 | 6.1 | 70.4 | 74.0 | 32.0 | 34.1 | 1,842 | 1,622 |
| Cerro Gordo | 1,648 | 6,969,938 | 4.9 | 9.0 | 3.3 | 8.9 | 4.0 | 6.8 | 61.9 | 207.4 | 72.8 | 81.2 | 1,889 | 1,558 |
| Cherokee | 412 | 1,513,267 | 10.4 | 8.1 | 0.6 | 9.1 | 0.4 | 6.7 | 64.0 | 44.6 | 18.5 | 20.5 | 1,809 | 1,506 |
| Chickasaw | 414 | 1,499,116 | 7.6 | 5.1 | 2.5 | 8.2 | 4.9 | 6.4 | 63.5 | 43.6 | 19.2 | 19.4 | 1,615 | 1,288 |
| Clarke | 569 | 1,988,284 | 4.8 | 6.3 | 0.0 | 5.1 | 30.2 | 3.3 | 49.5 | 64.1 | 20.2 | 16.8 | 1,784 | 1,545 |
| Clay | 1,189 | 4,750,090 | 3.8 | 4.5 | 0.6 | 6.4 | 41.5 | 8.7 | 31.9 | 171.3 | 30.6 | 31.5 | 1,951 | 1,618 |
| Clayton | 933 | 3,473,585 | 4.8 | 4.9 | 0.0 | 5.3 | 15.5 | 3.1 | 65.7 | 113.5 | 40.9 | 28.7 | 1,625 | 1,427 |
| Clinton | 1,789 | 6,779,889 | 5.3 | 8.4 | 3.6 | 6.0 | 2.3 | 4.4 | 66.8 | 199.2 | 83.7 | 85.7 | 1,828 | 1,486 |
| Crawford | 911 | 3,895,378 | 3.8 | 3.8 | 0.1 | 5.3 | 35.1 | 5.8 | 44.0 | 110.7 | 39.5 | 30.7 | 1,800 | 1,623 |
| Dallas | 2,920 | 12,020,235 | 4.6 | 4.2 | 0.7 | 2.3 | 5.1 | 2.6 | 79.3 | 348.1 | 145.8 | 153.2 | 1,757 | 1,655 |
| Davis | 430 | 1,804,127 | 4.8 | 4.9 | 0.4 | 5.0 | 39.5 | 2.6 | 42.5 | 54.3 | 17.0 | 11.2 | 1,252 | 1,119 |
| Decatur | 459 | 1,645,151 | 4.6 | 6.2 | 0.0 | 7.5 | 24.3 | 2.3 | 54.3 | 45.9 | 17.3 | 11.5 | 1,451 | 1,299 |
| Delaware | 1,012 | 4,298,102 | 3.6 | 3.6 | 0.0 | 4.1 | 47.3 | 1.6 | 39.5 | 120.6 | 28.4 | 33.4 | 1,949 | 1,687 |
| Des Moines | 1,812 | 7,541,692 | 4.6 | 7.2 | 3.2 | 3.8 | 1.6 | 3.4 | 74.8 | 195.4 | 87.2 | 67.3 | 1,711 | 1,383 |
| Dickinson | 882 | 3,384,380 | 8.1 | 6.7 | 0.1 | 4.5 | 28.3 | 6.2 | 45.3 | 120.7 | 17.3 | 54.4 | 3,173 | 2,769 |
| Dubuque | 3,536 | 14,203,131 | 6.2 | 8.7 | 3.8 | 7.3 | 5.2 | 4.6 | 62.8 | 432.9 | 186.4 | 173.0 | 1,782 | 1,463 |
| Emmet | 710 | 2,757,375 | 4.6 | 5.0 | 0.2 | 5.0 | 1.5 | 3.4 | 79.4 | 79.3 | 33.9 | 25.5 | 2,707 | 2,459 |
| Fayette | 867 | 2,840,894 | 4.8 | 9.5 | 0.4 | 6.9 | 1.6 | 4.8 | 70.6 | 80.6 | 37.9 | 31.0 | 1,576 | 1,356 |
| Floyd | 760 | 3,203,579 | 6.3 | 5.2 | 0.6 | 6.4 | 30.1 | 2.7 | 48.1 | 102.8 | 36.5 | 22.7 | 1,438 | 1,248 |

1. Based on the resident population estimated as of July 1 of the year shown.

# Table B. States and Counties — Local Government Finances, Government Employment, and Income Taxes

| STATE County | Total (mil dol) [185] | Per capita[1] (dollars) [186] | Education [187] | Health and hospitals [188] | Police protection [189] | Public welfare [190] | Highways [191] | Total (mil dol) [192] | Per capita[1] (dollars) [193] | Federal civilian [194] | Federal military [195] | State and local [196] | Number of returns [197] | Mean adjusted gross income [198] | Mean income tax [199] |
|---|---|---|---|---|---|---|---|---|---|---|---|---|---|---|---|
| INDIANA—Cont'd | | | | | | | | | | | | | | | |
| Pike | 37 | 3,019 | 46.8 | 0.5 | 2.6 | 0.0 | 6.5 | 18.4 | 1,492 | 32 | 36 | 633 | 5,660 | 51,885 | 4,040 |
| Porter | 489 | 2,899 | 57.1 | 0.9 | 4.2 | 0.1 | 4.8 | 656.5 | 3,896 | 458 | 497 | 6,625 | 84,060 | 73,500 | 8,508 |
| Posey | 55 | 2,144 | 72.1 | 0.1 | 2.0 | 0.1 | 1.9 | 55.3 | 2,161 | 69 | 75 | 1,076 | 12,180 | 66,627 | 6,472 |
| Pulaski | 56 | 4,431 | 45.6 | 25.5 | 3.3 | 0.0 | 5.7 | 39.3 | 3,136 | 41 | 36 | 1,014 | 5,880 | 49,794 | 4,021 |
| Putnam | 95 | 2,537 | 64.8 | 1.3 | 2.6 | 0.1 | 7.1 | 89.2 | 2,387 | 74 | 97 | 2,343 | 16,160 | 53,115 | 4,488 |
| Randolph | 70 | 2,812 | 62.6 | 0.2 | 3.6 | 2.2 | 2.4 | 48.1 | 1,931 | 57 | 72 | 1,204 | 11,450 | 44,181 | 3,212 |
| Ripley | 88 | 3,101 | 71.0 | 0.1 | 1.9 | 0.0 | 4.7 | 91.0 | 3,201 | 76 | 83 | 1,220 | 13,740 | 56,885 | 5,297 |
| Rush | 76 | 4,553 | 28.1 | 0.8 | 2.8 | 0.1 | 12.1 | 48.4 | 2,911 | 43 | 49 | 1,056 | 7,940 | 49,885 | 3,928 |
| St. Joseph | 872 | 3,227 | 43.1 | 1.4 | 6.4 | 0.0 | 4.4 | 651.7 | 2,413 | 875 | 855 | 12,771 | 127,170 | 59,916 | 6,383 |
| Scott | 85 | 3,553 | 45.4 | 7.2 | 3.5 | 0.0 | 5.0 | 44.7 | 1,879 | 85 | 70 | 1,092 | 10,620 | 45,213 | 3,342 |
| Shelby | 261 | 5,886 | 26.4 | 60.4 | 2.8 | 0.0 | 0.4 | 219.7 | 4,951 | 125 | 131 | 2,610 | 21,580 | 56,197 | 5,028 |
| Spencer | 55 | 2,669 | 57.3 | 0.6 | 2.8 | 0.0 | 5.5 | 29.0 | 1,420 | 67 | 59 | 942 | 9,680 | 56,057 | 4,794 |
| Starke | 58 | 2,513 | 60.1 | 2.4 | 1.8 | 0.0 | 6.5 | 46.1 | 2,009 | 41 | 68 | 913 | 10,480 | 47,655 | 3,799 |
| Steuben | 102 | 2,966 | 46.3 | 3.9 | 4.9 | 0.0 | 3.7 | 87.7 | 2,550 | 60 | 99 | 1,422 | 16,510 | 58,949 | 5,961 |
| Sullivan | 139 | 6,712 | 19.4 | 24.0 | 2.7 | 0.0 | 3.6 | 80.8 | 3,903 | 52 | 55 | 1,852 | 8,660 | 48,895 | 3,811 |
| Switzerland | 35 | 3,288 | 47.6 | 1.9 | 2.7 | 0.0 | 7.0 | 9.0 | 838 | 17 | 32 | 459 | 4,190 | 49,880 | 4,006 |
| Tippecanoe | 495 | 2,591 | 50.5 | 0.5 | 5.0 | 0.6 | 7.0 | 455.2 | 2,381 | 428 | 604 | 24,335 | 76,220 | 64,778 | 7,122 |
| Tipton | 84 | 5,551 | 30.7 | 0.3 | 1.2 | 0.0 | 4.7 | 75.1 | 4,964 | 36 | 44 | 735 | 7,580 | 57,170 | 4,979 |
| Union | 30 | 4,187 | 54.0 | 0.8 | 3.0 | 0.0 | 7.3 | 14.6 | 2,036 | 13 | 21 | 398 | 3,390 | 47,109 | 3,336 |
| Vanderburgh | 561 | 3,103 | 43.3 | 1.8 | 8.0 | 0.3 | 3.0 | 993.1 | 5,488 | 1,039 | 554 | 10,009 | 85,940 | 57,378 | 5,919 |
| Vermillion | 58 | 3,737 | 51.7 | 0.3 | 3.0 | 0.1 | 3.9 | 44.2 | 2,849 | 30 | 45 | 676 | 7,270 | 47,296 | 3,598 |
| Vigo | 323 | 3,007 | 46.1 | 0.8 | 6.9 | 0.0 | 2.8 | 295.7 | 2,749 | 1,092 | 311 | 7,745 | 45,980 | 51,603 | 4,883 |
| Wabash | 101 | 3,209 | 50.3 | 0.8 | 5.3 | 0.0 | 9.4 | 64.3 | 2,050 | 71 | 87 | 1,465 | 14,990 | 57,483 | 5,920 |
| Warren | 26 | 3,184 | 52.4 | 1.7 | 4.0 | 0.1 | 7.3 | 17.1 | 2,087 | 11 | 24 | 339 | 3,990 | 57,437 | 5,097 |
| Warrick | 181 | 2,902 | 55.1 | 0.5 | 3.3 | 0.0 | 3.6 | 86.1 | 1,380 | 113 | 185 | 2,010 | 30,480 | 78,596 | 9,622 |
| Washington | 69 | 2,498 | 70.8 | 0.4 | 3.9 | 0.0 | 5.8 | 55.1 | 1,986 | 63 | 83 | 1,104 | 12,590 | 46,087 | 3,417 |
| Wayne | 211 | 3,192 | 45.6 | 3.7 | 5.2 | 0.0 | 3.8 | 140.7 | 2,127 | 165 | 189 | 4,625 | 30,050 | 47,757 | 4,099 |
| Wells | 84 | 3,007 | 52.4 | 0.6 | 5.4 | 0.0 | 5.2 | 62.5 | 2,234 | 51 | 83 | 1,268 | 13,300 | 59,499 | 5,307 |
| White | 99 | 4,095 | 50.3 | 0.3 | 2.4 | 0.0 | 5.3 | 74.4 | 3,081 | 54 | 71 | 1,334 | 12,110 | 51,282 | 4,496 |
| Whitley | 83 | 2,463 | 48.8 | 0.6 | 3.8 | 0.0 | 6.6 | 74.7 | 2,216 | 80 | 100 | 1,403 | 17,000 | 60,128 | 5,699 |
| IOWA | X | X | X | X | X | X | X | X | X | 17,645 | 11,391 | 242,922 | 1,458,830 | 66,543 | 7,041 |
| Adair | 41 | 5,850 | 43.7 | 17.2 | 3.6 | 0.3 | 20.8 | 45.4 | 6,434 | 27 | 25 | 452 | 3,470 | 50,701 | 4,051 |
| Adams | 21 | 5,618 | 34.7 | 6.1 | 7.5 | 0.9 | 20.0 | 12.1 | 3,299 | 27 | 13 | 196 | 1,730 | 49,005 | 4,066 |
| Allamakee | 72 | 5,235 | 39.4 | 27.9 | 3.8 | 0.3 | 12.9 | 30.9 | 2,241 | 72 | 49 | 1,088 | 6,530 | 47,671 | 3,830 |
| Appanoose | 44 | 3,557 | 57.5 | 2.3 | 8.2 | 1.5 | 11.9 | 14.2 | 1,149 | 49 | 45 | 639 | 5,390 | 45,109 | 3,572 |
| Audubon | 40 | 7,137 | 31.4 | 36.7 | 4.0 | 0.2 | 14.0 | 11.5 | 2,063 | 24 | 20 | 339 | 2,730 | 51,344 | 4,197 |
| Benton | 92 | 3,590 | 47.8 | 3.7 | 4.5 | 0.2 | 13.0 | 114.3 | 4,456 | 62 | 92 | 1,502 | 11,790 | 66,270 | 5,908 |
| Black Hawk | 672 | 5,082 | 54.9 | 3.3 | 4.9 | 1.3 | 5.4 | 488.7 | 3,696 | 537 | 486 | 10,837 | 58,940 | 61,946 | 6,229 |
| Boone | 134 | 5,075 | 36.6 | 33.4 | 4.0 | 1.0 | 10.1 | 189.6 | 7,171 | 118 | 93 | 2,203 | 12,520 | 64,459 | 5,958 |
| Bremer | 118 | 4,756 | 58.0 | 5.4 | 4.4 | 0.1 | 7.0 | 96.1 | 3,870 | 62 | 85 | 1,886 | 11,040 | 72,225 | 7,130 |
| Buchanan | 111 | 5,252 | 33.0 | 36.1 | 3.4 | 0.1 | 8.5 | 86.2 | 4,079 | 49 | 76 | 1,408 | 9,420 | 60,557 | 5,421 |
| Buena Vista | 149 | 7,408 | 38.7 | 30.8 | 3.1 | 0.1 | 5.9 | 129.6 | 6,435 | 86 | 70 | 1,710 | 9,560 | 59,962 | 5,537 |
| Butler | 77 | 5,255 | 28.8 | 37.1 | 3.8 | 0.3 | 10.8 | 30.6 | 2,094 | 42 | 52 | 750 | 6,630 | 57,370 | 4,477 |
| Calhoun | 45 | 4,582 | 48.7 | 7.7 | 4.5 | 0.8 | 14.2 | 16.8 | 1,722 | 35 | 32 | 642 | 4,440 | 55,241 | 4,587 |
| Carroll | 86 | 4,239 | 44.9 | 3.1 | 4.3 | 0.4 | 14.0 | 44.7 | 2,206 | 72 | 72 | 1,243 | 9,930 | 62,854 | 5,770 |
| Cass | 112 | 8,551 | 32.4 | 42.6 | 2.7 | 0.2 | 7.5 | 108.9 | 8,282 | 65 | 46 | 1,359 | 6,130 | 53,295 | 4,580 |
| Cedar | 77 | 4,167 | 52.1 | 5.8 | 4.8 | 0.1 | 15.0 | 47.0 | 2,540 | 80 | 66 | 1,019 | 8,860 | 63,591 | 5,812 |
| Cerro Gordo | 207 | 4,813 | 51.8 | 4.7 | 6.4 | 0.4 | 6.5 | 139.0 | 3,232 | 165 | 150 | 2,769 | 20,400 | 62,187 | 6,194 |
| Cherokee | 46 | 4,068 | 48.3 | 3.5 | 4.4 | 0.1 | 16.7 | 23.4 | 2,070 | 42 | 39 | 968 | 5,590 | 57,396 | 4,825 |
| Chickasaw | 42 | 3,536 | 49.6 | 5.1 | 4.9 | 0.1 | 17.4 | 34.4 | 2,867 | 53 | 43 | 606 | 5,690 | 58,701 | 5,342 |
| Clarke | 68 | 7,197 | 32.5 | 38.8 | 3.5 | 0.3 | 9.0 | 27.6 | 2,935 | 33 | 34 | 804 | 4,350 | 50,139 | 3,856 |
| Clay | 174 | 10,801 | 19.4 | 52.3 | 2.0 | 0.3 | 5.7 | 160.9 | 9,965 | 56 | 57 | 1,530 | 7,920 | 62,358 | 6,147 |
| Clayton | 126 | 7,161 | 43.9 | 25.9 | 3.1 | 0.3 | 9.6 | 40.7 | 2,309 | 75 | 63 | 1,208 | 8,290 | 49,726 | 4,163 |
| Clinton | 192 | 4,087 | 50.0 | 0.3 | 5.7 | 0.0 | 10.3 | 260.8 | 5,563 | 112 | 166 | 2,481 | 22,050 | 54,770 | 4,695 |
| Crawford | 104 | 6,113 | 35.2 | 35.6 | 2.9 | 0.1 | 9.5 | 56.6 | 3,313 | 55 | 59 | 1,251 | 7,660 | 54,008 | 4,395 |
| Dallas | 356 | 4,078 | 66.2 | 6.3 | 3.3 | 0.3 | 6.5 | 470.3 | 5,392 | 113 | 341 | 4,144 | 44,750 | 106,081 | 15,420 |
| Davis | 53 | 5,925 | 28.2 | 47.1 | 3.3 | 0.3 | 9.6 | 19.1 | 2,129 | 37 | 32 | 552 | 3,560 | 48,872 | 3,904 |
| Decatur | 47 | 5,867 | 41.5 | 29.8 | 3.9 | 0.2 | 10.4 | 30.7 | 3,867 | 29 | 26 | 523 | 3,010 | 49,022 | 4,386 |
| Delaware | 119 | 6,911 | 28.2 | 46.0 | 2.8 | 0.1 | 7.9 | 68.5 | 3,996 | 33 | 61 | 1,271 | 8,160 | 56,735 | 4,796 |
| Des Moines | 213 | 5,423 | 66.0 | 2.1 | 4.4 | 0.2 | 5.0 | 106.4 | 2,705 | 146 | 140 | 2,452 | 18,660 | 55,455 | 5,078 |
| Dickinson | 109 | 6,372 | 28.9 | 34.5 | 4.2 | 0.3 | 8.5 | 128.4 | 7,489 | 76 | 62 | 1,167 | 8,990 | 74,181 | 8,368 |
| Dubuque | 417 | 4,296 | 46.7 | 1.8 | 5.2 | 0.3 | 5.1 | 473.0 | 4,873 | 257 | 357 | 4,552 | 47,480 | 67,516 | 7,326 |
| Emmet | 88 | 9,365 | 72.7 | 3.1 | 3.0 | 0.2 | 6.1 | 67.0 | 7,109 | 35 | 32 | 684 | 4,290 | 51,292 | 4,040 |
| Fayette | 86 | 4,360 | 57.0 | 2.9 | 5.0 | 0.2 | 12.3 | 49.5 | 2,513 | 73 | 68 | 1,129 | 8,750 | 49,579 | 4,120 |
| Floyd | 86 | 5,465 | 36.1 | 33.2 | 3.6 | 0.5 | 9.4 | 42.8 | 2,713 | 41 | 56 | 994 | 7,260 | 55,386 | 4,729 |

1. Based on the resident population estimated as of July 1 of the year shown.

Table B. States and Counties — **Land Area and Population**

| State / county code | CBSA code[1] | County Type code[2] | STATE County | Land area[3] (sq. mi) | Total persons 2019 | Rank | Per square mile | White | Black | American Indian, Alaska Native | Asian and Pacific Islander | Percent Hispanic or Latino[4] | Under 5 years | 5 to 14 years | 15 to 24 years | 25 to 34 years | 35 to 44 years | 45 to 54 years |
|---|---|---|---|---|---|---|---|---|---|---|---|---|---|---|---|---|---|---|
| | | | | 1 | 2 | 3 | 4 | 5 | 6 | 7 | 8 | 9 | 10 | 11 | 12 | 13 | 14 | 15 |
| | | | IOWA—Cont'd | | | | | | | | | | | | | | | |
| 19069 | | 7 | Franklin | 582.0 | 9,971 | 2,420 | 17.1 | 85.1 | 1.1 | 0.7 | 0.8 | 13.3 | 6.0 | 12.8 | 11.2 | 9.9 | 11.9 | 10.8 |
| 19071 | | 8 | Fremont | 511.2 | 6,729 | 2,685 | 13.2 | 95.4 | 1.4 | 1.0 | 0.6 | 3.2 | 5.8 | 12.6 | 10.1 | 9.9 | 11.0 | 11.6 |
| 19073 | | 6 | Greene | 569.6 | 8,795 | 2,510 | 15.4 | 95.1 | 1.1 | 0.9 | 0.9 | 3.0 | 6.0 | 12.6 | 11.1 | 10.1 | 10.8 | 11.1 |
| 19075 | 47940 | 3 | Grundy | 501.9 | 12,217 | 2,274 | 24.3 | 97.1 | 1.2 | 0.5 | 0.6 | 1.6 | 5.5 | 13.4 | 11.1 | 11.0 | 12.0 | 11.5 |
| 19077 | 19780 | 2 | Guthrie | 590.6 | 10,737 | 2,364 | 18.2 | 95.8 | 1.2 | 0.9 | 0.6 | 2.8 | 5.1 | 11.8 | 11.0 | 9.4 | 11.4 | 11.6 |
| 19079 | | 6 | Hamilton | 576.7 | 14,716 | 2,105 | 25.5 | 89.5 | 1.7 | 0.7 | 2.9 | 6.7 | 6.3 | 12.7 | 11.1 | 10.5 | 11.5 | 11.6 |
| 19081 | | 7 | Hancock | 571.0 | 10,507 | 2,388 | 18.4 | 93.1 | 1.2 | 0.5 | 0.9 | 5.2 | 5.5 | 11.8 | 11.3 | 10.1 | 11.7 | 10.4 |
| 19083 | | 6 | Hardin | 569.3 | 16,575 | 2,002 | 29.1 | 92.6 | 2.0 | 0.7 | 1.2 | 4.6 | 4.9 | 10.4 | 11.9 | 10.8 | 11.7 | 11.4 |
| 19085 | 36540 | 2 | Harrison | 696.9 | 13,928 | 2,161 | 20.0 | 96.8 | 0.8 | 0.9 | 0.7 | 2.0 | 5.8 | 13.2 | 10.8 | 10.6 | 11.5 | 12.3 |
| 19087 | | 6 | Henry | 434.3 | 19,697 | 1,840 | 45.4 | 89.2 | 3.5 | 0.8 | 3.1 | 5.2 | 5.4 | 11.9 | 13.7 | 11.8 | 11.7 | 12.0 |
| 19089 | | 6 | Howard | 473.2 | 9,176 | 2,481 | 19.4 | 96.8 | 1.2 | 0.7 | 0.6 | 1.8 | 6.7 | 14.0 | 11.4 | 10.6 | 11.6 | 10.3 |
| 19091 | | 7 | Humboldt | 434.4 | 9,473 | 2,457 | 21.8 | 93.8 | 1.3 | 0.7 | 0.7 | 4.8 | 6.1 | 13.6 | 10.9 | 11.0 | 11.7 | 9.9 |
| 19093 | | 8 | Ida | 431.5 | 6,833 | 2,672 | 15.8 | 94.9 | 1.0 | 0.7 | 1.0 | 3.7 | 5.9 | 14.5 | 11.9 | 10.1 | 11.2 | 10.6 |
| 19095 | | 6 | Iowa | 586.5 | 16,138 | 2,025 | 27.5 | 95.4 | 1.3 | 0.5 | 0.8 | 3.2 | 5.7 | 13.5 | 11.0 | 10.9 | 12.1 | 11.7 |
| 19097 | | 6 | Jackson | 636.0 | 19,205 | 1,866 | 30.2 | 96.2 | 1.3 | 0.6 | 1.5 | 1.6 | 5.7 | 12.4 | 11.0 | 10.4 | 10.7 | 11.9 |
| 19099 | 19780 | 6 | Jasper | 730.4 | 37,148 | 1,246 | 50.9 | 93.7 | 2.7 | 0.7 | 1.1 | 3.0 | 5.6 | 12.3 | 11.2 | 12.4 | 12.5 | 12.3 |
| 19101 | 21840 | 7 | Jefferson | 435.5 | 18,347 | 1,907 | 42.1 | 82.5 | 2.9 | 0.9 | 11.9 | 3.5 | 4.2 | 8.5 | 13.6 | 14.5 | 11.0 | 10.2 |
| 19103 | 26980 | 3 | Johnson | 613.0 | 153,740 | 434 | 250.8 | 79.4 | 8.9 | 0.7 | 7.6 | 5.9 | 5.6 | 10.9 | 23.7 | 15.3 | 12.2 | 9.8 |
| 19105 | 16300 | 2 | Jones | 575.6 | 20,617 | 1,788 | 35.8 | 94.4 | 2.9 | 0.6 | 0.8 | 2.2 | 5.0 | 11.9 | 11.0 | 11.2 | 12.5 | 12.1 |
| 19107 | | 8 | Keokuk | 579.2 | 10,085 | 2,411 | 17.4 | 96.5 | 1.1 | 0.5 | 0.5 | 2.4 | 5.8 | 12.5 | 11.4 | 10.3 | 11.8 | 11.0 |
| 19109 | | 7 | Kossuth | 972.7 | 14,680 | 2,109 | 15.1 | 94.0 | 1.4 | 0.4 | 0.9 | 4.2 | 5.8 | 12.3 | 10.6 | 10.2 | 10.7 | 10.1 |
| 19111 | 22800 | 5 | Lee | 517.6 | 33,480 | 1,341 | 64.7 | 92.2 | 4.1 | 0.8 | 1.1 | 3.8 | 5.6 | 12.0 | 11.0 | 11.7 | 11.8 | 11.7 |
| 19113 | 16300 | 2 | Linn | 717.1 | 227,854 | 302 | 317.7 | 87.2 | 7.9 | 0.7 | 3.5 | 3.6 | 6.1 | 12.7 | 13.0 | 13.5 | 13.3 | 12.1 |
| 19115 | | 8 | Louisa | 401.8 | 11,011 | 2,347 | 27.4 | 78.6 | 1.6 | 0.8 | 4.0 | 16.1 | 6.0 | 12.3 | 11.5 | 11.2 | 12.7 | 11.6 |
| 19117 | | 6 | Lucas | 430.7 | 8,518 | 2,533 | 19.8 | 96.3 | 0.6 | 0.6 | 0.7 | 2.6 | 6.3 | 12.9 | 10.5 | 10.9 | 10.2 | 11.2 |
| 19119 | | 8 | Lyon | 587.6 | 11,756 | 2,310 | 20.0 | 95.8 | 0.8 | 0.7 | 0.9 | 3.1 | 7.6 | 15.5 | 12.2 | 10.4 | 12.5 | 10.4 |
| 19121 | 19780 | 2 | Madison | 561.0 | 16,521 | 2,006 | 29.4 | 95.8 | 1.0 | 0.7 | 1.5 | 2.4 | 5.5 | 14.6 | 12.1 | 10.1 | 13.2 | 13.5 |
| 19123 | 36820 | 7 | Mahaska | 570.9 | 22,370 | 1,707 | 39.2 | 93.2 | 2.7 | 0.5 | 2.3 | 2.5 | 6.5 | 12.9 | 14.5 | 11.4 | 11.9 | 11.0 |
| 19125 | 37800 | 6 | Marion | 554.5 | 33,168 | 1,352 | 59.8 | 95.2 | 1.5 | 0.6 | 1.9 | 2.2 | 5.6 | 13.4 | 14.3 | 10.7 | 12.1 | 11.4 |
| 19127 | 32260 | 4 | Marshall | 572.5 | 39,495 | 1,194 | 69.0 | 69.7 | 2.6 | 0.7 | 4.3 | 24.2 | 6.3 | 14.4 | 12.9 | 11.8 | 11.5 | 11.3 |
| 19129 | 36540 | 2 | Mills | 437.4 | 14,766 | 2,103 | 33.8 | 94.8 | 1.0 | 1.0 | 0.8 | 3.5 | 4.9 | 13.6 | 11.4 | 9.6 | 12.4 | 13.3 |
| 19131 | | 7 | Mitchell | 469.3 | 10,647 | 2,371 | 22.7 | 97.1 | 0.7 | 0.4 | 0.8 | 1.7 | 6.0 | 13.4 | 11.8 | 10.7 | 10.7 | 10.8 |
| 19133 | | 6 | Monona | 694.1 | 8,598 | 2,525 | 12.4 | 94.8 | 1.4 | 2.0 | 0.6 | 2.6 | 5.9 | 12.2 | 10.9 | 10.3 | 10.3 | 10.5 |
| 19135 | | 7 | Monroe | 433.7 | 7,770 | 2,607 | 17.9 | 96.2 | 1.4 | 0.5 | 0.7 | 2.4 | 5.9 | 12.9 | 12.0 | 10.6 | 11.6 | 12.1 |
| 19137 | | 6 | Montgomery | 424.1 | 9,935 | 2,423 | 23.4 | 94.2 | 1.1 | 1.0 | 0.8 | 4.2 | 6.1 | 12.4 | 11.4 | 10.4 | 10.6 | 11.7 |
| 19139 | 34700 | 4 | Muscatine | 437.4 | 42,394 | 1,132 | 96.9 | 77.3 | 3.3 | 0.6 | 1.5 | 18.5 | 6.3 | 14.0 | 12.3 | 12.6 | 12.1 | 11.8 |
| 19141 | | 7 | O'Brien | 573.0 | 13,679 | 2,179 | 23.9 | 92.2 | 1.6 | 0.4 | 1.1 | 5.6 | 6.2 | 13.2 | 11.3 | 11.3 | 11.7 | 10.2 |
| 19143 | | 7 | Osceola | 398.7 | 5,987 | 2,746 | 15.0 | 89.0 | 1.3 | 0.9 | 1.4 | 8.7 | 5.7 | 13.2 | 10.9 | 10.4 | 10.6 | 10.5 |
| 19145 | | 6 | Page | 534.9 | 15,073 | 2,091 | 28.2 | 91.5 | 3.4 | 1.1 | 1.6 | 3.9 | 4.7 | 10.1 | 10.9 | 12.3 | 12.9 | 11.7 |
| 19147 | | 7 | Palo Alto | 563.9 | 8,845 | 2,506 | 15.7 | 93.5 | 2.2 | 0.7 | 1.4 | 3.3 | 6.2 | 12.8 | 12.1 | 11.1 | 11.0 | 10.1 |
| 19149 | | 3 | Plymouth | 862.8 | 25,219 | 1,598 | 29.2 | 91.1 | 2.3 | 0.7 | 1.3 | 5.7 | 6.1 | 14.3 | 12.1 | 10.6 | 12.3 | 11.4 |
| 19151 | | 6 | Pocahontas | 577.2 | 6,607 | 2,692 | 11.4 | 93.2 | 1.6 | 0.7 | 1.2 | 4.6 | 5.9 | 12.8 | 10.4 | 10.0 | 10.5 | 9.8 |
| 19153 | 19780 | 2 | Polk | 572.2 | 494,281 | 143 | 863.8 | 78.6 | 8.5 | 0.6 | 6.0 | 8.8 | 6.8 | 13.7 | 12.6 | 15.3 | 14.0 | 12.0 |
| 19155 | 36540 | 2 | Pottawattamie | 951.3 | 93,328 | 638 | 98.1 | 88.4 | 2.7 | 1.0 | 1.5 | 8.4 | 6.1 | 13.1 | 12.2 | 12.1 | 12.3 | 11.8 |
| 19157 | | 7 | Poweshiek | 584.9 | 18,381 | 1,904 | 31.4 | 92.5 | 2.0 | 0.7 | 2.8 | 3.4 | 4.7 | 11.4 | 18.1 | 9.6 | 10.7 | 10.5 |
| 19159 | | 9 | Ringgold | 535.5 | 4,801 | 2,841 | 9.0 | 95.6 | 1.0 | 0.6 | 1.0 | 2.6 | 5.6 | 12.9 | 12.0 | 8.7 | 11.1 | 10.0 |
| 19161 | | 9 | Sac | 575.0 | 9,603 | 2,445 | 16.7 | 94.7 | 1.1 | 0.4 | 0.8 | 4.1 | 6.0 | 12.8 | 10.5 | 10.2 | 10.2 | 10.9 |
| 19163 | 19340 | 2 | Scott | 458.1 | 173,216 | 382 | 378.1 | 81.7 | 9.7 | 0.8 | 3.6 | 7.2 | 6.1 | 13.3 | 12.2 | 12.9 | 13.3 | 12.1 |
| 19165 | | 6 | Shelby | 590.8 | 11,430 | 2,323 | 19.3 | 94.6 | 1.3 | 0.8 | 0.9 | 3.5 | 5.2 | 13.1 | 11.3 | 9.4 | 9.8 | 11.7 |
| 19167 | | 7 | Sioux | 767.9 | 35,043 | 1,303 | 45.6 | 86.9 | 0.9 | 0.4 | 1.0 | 11.4 | 7.0 | 15.2 | 17.8 | 10.6 | 11.7 | 9.6 |
| 19169 | 11180 | 3 | Story | 572.5 | 98,237 | 616 | 171.6 | 85.0 | 3.5 | 0.5 | 8.9 | 3.9 | 4.3 | 9.2 | 32.5 | 13.3 | 10.1 | 8.3 |
| 19171 | | 6 | Tama | 721.0 | 16,801 | 1,984 | 23.3 | 80.7 | 1.5 | 7.5 | 1.0 | 11.5 | 6.5 | 13.4 | 12.0 | 10.6 | 10.8 | 11.8 |
| 19173 | | 9 | Taylor | 531.9 | 6,092 | 2,739 | 11.5 | 90.6 | 0.9 | 0.5 | 0.6 | 8.4 | 5.6 | 13.6 | 11.0 | 9.9 | 11.9 | 10.3 |
| 19175 | | 6 | Union | 423.6 | 12,157 | 2,280 | 28.7 | 94.4 | 1.7 | 0.6 | 1.1 | 3.1 | 5.5 | 12.7 | 13.4 | 10.8 | 11.8 | 11.0 |
| 19177 | | 9 | Van Buren | 484.8 | 7,069 | 2,653 | 14.6 | 97.3 | 0.7 | 0.6 | 0.9 | 1.7 | 6.6 | 12.8 | 11.0 | 9.8 | 10.9 | 11.5 |
| 19179 | 36900 | 5 | Wapello | 431.8 | 34,985 | 1,305 | 81.0 | 81.1 | 4.6 | 0.7 | 3.5 | 11.5 | 6.5 | 12.5 | 12.3 | 12.9 | 12.2 | 11.3 |
| 19181 | 19780 | 2 | Warren | 569.8 | 52,265 | 966 | 91.7 | 94.7 | 1.5 | 0.6 | 1.4 | 3.2 | 6.1 | 14.1 | 13.3 | 11.6 | 13.1 | 12.4 |
| 19183 | 26980 | 3 | Washington | 568.8 | 21,992 | 1,720 | 38.7 | 90.9 | 1.7 | 0.5 | 1.0 | 7.1 | 6.5 | 14.1 | 11.4 | 11.7 | 11.4 | 11.2 |
| 19185 | | 9 | Wayne | 525.5 | 6,415 | 2,713 | 12.2 | 97.1 | 1.2 | 0.6 | 0.8 | 3.4 | 7.1 | 14.6 | 11.2 | 10.0 | 10.2 | 10.0 |
| 19187 | 22700 | 5 | Webster | 715.7 | 35,934 | 1,277 | 50.2 | 88.0 | 6.1 | 0.7 | 1.6 | 5.9 | 5.9 | 11.9 | 15.3 | 12.4 | 11.1 | 10.4 |
| 19189 | | 7 | Winnebago | 400.5 | 10,277 | 2,401 | 25.7 | 92.2 | 2.1 | 0.6 | 1.3 | 5.2 | 5.1 | 12.8 | 13.5 | 10.5 | 11.0 | 10.6 |
| 19191 | | 7 | Winneshiek | 689.8 | 19,862 | 1,833 | 28.8 | 95.6 | 1.2 | 0.3 | 1.3 | 2.4 | 4.5 | 10.3 | 18.5 | 9.1 | 9.9 | 10.5 |
| 19193 | 43580 | 3 | Woodbury | 872.9 | 103,138 | 589 | 118.2 | 72.2 | 6.1 | 2.2 | 4.1 | 18.0 | 6.9 | 14.6 | 14.2 | 13.0 | 12.6 | 11.3 |
| 19195 | 32380 | 9 | Worth | 400.1 | 7,359 | 2,630 | 18.4 | 95.0 | 1.2 | 0.6 | 1.0 | 3.2 | 6.1 | 11.3 | 10.2 | 11.7 | 11.9 | 12.1 |
| 19197 | | 7 | Wright | 580.4 | 12,416 | 2,255 | 21.4 | 84.8 | 1.4 | 0.6 | 1.0 | 13.5 | 6.5 | 13.4 | 11.2 | 10.0 | 10.9 | 10.4 |

1. CBSA = Core Based Statistical Area. See Appendix A for explanation. See Appendix B for list of metropolitan areas with component counties. Service of USDA Rural-Urban Continuum Codes. See Appendix A for definition.   2. County type code from the Economic Research Service of USDA Rural-Urban Continuum Codes. See Appendix A for definition.   3. Dry land or land partially or temporarily covered by water.   4. May be of any race.

# Table B. States and Counties — **Population and Households**

| STATE County | Population, 2020 (cont.) Age (percent) (cont.) | | | | Population change, 2000-2020 Total persons | | Percent change | | Components of change, 2010-2020 | | | Households, 2015-2019 | | | Percent | |
|---|---|---|---|---|---|---|---|---|---|---|---|---|---|---|---|---|
| | 55 to 64 years | 65 to 74 years | 75 years and over | Percent female | 2000 | 2010 | 2000-2010 | 2010-2020 | Births | Deaths | Net Migration | Number | Persons per household | Family house-holds | Female family house-holder[1] | One person |
| | 16 | 17 | 18 | 19 | 20 | 21 | 22 | 23 | 24 | 25 | 26 | 27 | 28 | 29 | 30 | 31 |
| IOWA—Cont'd | | | | | | | | | | | | | | | | |
| Franklin | 14.7 | 12.4 | 10.2 | 49.3 | 10,704 | 10,680 | -0.2 | -6.6 | 1,230 | 1,228 | -717 | 4,192 | 2.38 | 70.5 | 9.2 | 23.5 |
| Fremont | 15.7 | 13.2 | 10.2 | 49.4 | 8,010 | 7,438 | -7.1 | -9.5 | 776 | 937 | -553 | 2,945 | 2.31 | 66.2 | 7.5 | 25.9 |
| Greene | 15.2 | 12.6 | 10.7 | 50.5 | 10,366 | 9,337 | -9.9 | -5.8 | 1,040 | 1,220 | -361 | 3,971 | 2.22 | 62.6 | 8.1 | 30.2 |
| Grundy | 14.1 | 11.5 | 9.9 | 50.4 | 12,369 | 12,453 | 0.7 | -1.9 | 1,332 | 1,396 | -168 | 5,155 | 2.35 | 69.0 | 5.6 | 26.4 |
| Guthrie | 15.9 | 13.0 | 10.7 | 50.0 | 11,353 | 10,959 | -3.5 | -2.0 | 1,085 | 1,274 | -30 | 4,452 | 2.36 | 69.4 | 6.7 | 26.0 |
| Hamilton | 15.2 | 11.4 | 9.8 | 49.9 | 16,438 | 15,675 | -4.6 | -6.1 | 1,837 | 1,792 | -1,008 | 6,259 | 2.36 | 66.0 | 8.0 | 30.9 |
| Hancock | 15.3 | 13.1 | 10.8 | 49.9 | 12,100 | 11,340 | -6.3 | -7.3 | 1,150 | 1,262 | -720 | 4,790 | 2.21 | 69.0 | 7.3 | 26.1 |
| Hardin | 15.7 | 12.4 | 11.0 | 50.1 | 18,812 | 17,534 | -6.8 | -5.5 | 1,834 | 2,246 | -529 | 7,148 | 2.27 | 61.9 | 7.2 | 33.5 |
| Harrison | 15.7 | 11.3 | 8.8 | 49.7 | 15,666 | 14,937 | -4.7 | -6.8 | 1,616 | 1,885 | -748 | 6,019 | 2.29 | 68.0 | 9.1 | 26.0 |
| Henry | 13.6 | 11.3 | 8.7 | 48.4 | 20,336 | 20,150 | -0.9 | -2.2 | 2,305 | 2,223 | -545 | 7,743 | 2.39 | 68.8 | 10.0 | 26.9 |
| Howard | 14.6 | 10.9 | 9.8 | 50.0 | 9,932 | 9,564 | -3.7 | -4.1 | 1,204 | 1,152 | -438 | 3,810 | 2.36 | 63.6 | 6.9 | 32.8 |
| Humboldt | 15.4 | 11.4 | 10.0 | 50.9 | 10,381 | 9,814 | -5.5 | -3.5 | 1,105 | 1,207 | -234 | 4,174 | 2.26 | 62.7 | 7.6 | 32.2 |
| Ida | 14.0 | 11.7 | 10.1 | 50.0 | 7,837 | 7,089 | -9.5 | -3.6 | 777 | 1,010 | -21 | 3,011 | 2.25 | 62.7 | 5.6 | 29.6 |
| Iowa | 15.5 | 10.5 | 9.2 | 49.8 | 15,671 | 16,356 | 4.4 | -1.3 | 1,873 | 1,880 | -206 | 6,780 | 2.34 | 68.3 | 7.6 | 26.1 |
| Jackson | 16.1 | 12.3 | 9.5 | 50.0 | 20,296 | 19,855 | -2.2 | -3.3 | 2,163 | 2,334 | -469 | 8,239 | 2.32 | 65.9 | 9.0 | 28.3 |
| Jasper | 13.9 | 10.8 | 8.9 | 48.7 | 37,213 | 36,842 | -1.0 | 0.8 | 4,219 | 3,917 | 23 | 14,574 | 2.42 | 68.9 | 9.6 | 26.2 |
| Jefferson | 13.5 | 16.5 | 8.1 | 45.4 | 16,181 | 16,836 | 4.0 | 9.0 | 1,534 | 1,703 | 1,679 | 6,754 | 2.54 | 60.3 | 8.1 | 35.6 |
| Johnson | 9.7 | 7.9 | 4.9 | 50.4 | 111,006 | 130,882 | 17.9 | 17.5 | 18,276 | 7,217 | 11,786 | 59,134 | 2.38 | 54.8 | 7.3 | 28.8 |
| Jones | 14.8 | 11.9 | 9.7 | 48.1 | 20,221 | 20,634 | 2.0 | -0.1 | 2,114 | 2,079 | -45 | 8,130 | 2.38 | 65.7 | 7.3 | 30.4 |
| Keokuk | 14.7 | 12.2 | 10.3 | 49.4 | 11,400 | 10,510 | -7.8 | -4.0 | 1,220 | 1,182 | -463 | 4,394 | 2.29 | 63.6 | 5.5 | 31.1 |
| Kossuth | 15.6 | 12.7 | 12.0 | 49.5 | 17,163 | 15,545 | -9.4 | -5.6 | 1,702 | 1,879 | -680 | 6,646 | 2.21 | 64.0 | 6.9 | 31.3 |
| Lee | 15.0 | 12.5 | 8.8 | 49.7 | 38,052 | 35,862 | -5.8 | -6.6 | 4,047 | 4,460 | -1,963 | 14,233 | 2.32 | 65.8 | 9.9 | 29.1 |
| Linn | 12.7 | 9.6 | 7.1 | 50.8 | 191,701 | 211,242 | 10.2 | 7.9 | 28,253 | 17,783 | 6,335 | 90,877 | 2.41 | 61.8 | 9.2 | 30.2 |
| Louisa | 15.1 | 11.1 | 8.5 | 49.1 | 12,183 | 11,387 | -6.5 | -3.3 | 1,379 | 1,102 | -662 | 4,335 | 2.55 | 68.6 | 7.8 | 26.1 |
| Lucas | 15.4 | 12.2 | 10.3 | 49.2 | 9,422 | 8,900 | -5.5 | -4.3 | 1,055 | 1,102 | -333 | 3,701 | 2.29 | 67.3 | 10.1 | 28.2 |
| Lyon | 12.2 | 10.3 | 8.9 | 49.2 | 11,763 | 11,581 | -1.5 | 1.5 | 1,755 | 1,140 | -441 | 4,548 | 2.55 | 73.7 | 5.3 | 22.8 |
| Madison | 13.7 | 10.2 | 7.2 | 50.2 | 14,019 | 15,681 | 11.9 | 5.4 | 1,720 | 1,539 | 661 | 6,462 | 2.45 | 72.7 | 5.1 | 25.5 |
| Mahaska | 13.3 | 10.3 | 8.2 | 49.1 | 22,335 | 22,384 | 0.2 | -0.1 | 2,838 | 2,384 | -457 | 8,927 | 2.41 | 66.9 | 10.1 | 27.3 |
| Marion | 13.7 | 10.4 | 8.3 | 49.9 | 32,052 | 33,307 | 3.9 | -0.4 | 3,717 | 3,490 | -360 | 13,365 | 2.35 | 66.5 | 7.6 | 28.2 |
| Marshall | 13.2 | 10.4 | 8.2 | 49.4 | 39,311 | 40,648 | 3.4 | -2.8 | 5,444 | 4,905 | -1,700 | 15,419 | 2.52 | 64.5 | 8.6 | 28.1 |
| Mills | 14.9 | 12.4 | 7.5 | 49.9 | 14,547 | 15,059 | 3.5 | -1.9 | 1,512 | 1,401 | -402 | 5,554 | 2.61 | 70.5 | 8.0 | 25.2 |
| Mitchell | 15.3 | 10.6 | 10.7 | 50.2 | 10,874 | 10,776 | -0.9 | -1.2 | 1,270 | 1,382 | -6 | 4,355 | 2.38 | 65.3 | 8.7 | 32.1 |
| Monona | 15.1 | 12.2 | 12.7 | 51.6 | 10,020 | 9,243 | -7.8 | -7.0 | 897 | 1,486 | -50 | 4,000 | 2.13 | 60.7 | 8.8 | 33.4 |
| Monroe | 14.2 | 11.8 | 8.9 | 50.6 | 8,016 | 7,967 | -0.6 | -2.5 | 888 | 1,040 | -38 | 3,294 | 2.33 | 65.9 | 6.6 | 30.2 |
| Montgomery | 15.2 | 12.2 | 10.0 | 51.2 | 11,771 | 10,740 | -8.8 | -7.5 | 1,169 | 1,542 | -435 | 4,554 | 2.16 | 63.7 | 10.8 | 30.0 |
| Muscatine | 13.3 | 10.5 | 7.2 | 50.1 | 41,722 | 42,803 | 2.6 | -1.0 | 5,515 | 4,102 | -1,818 | 16,660 | 2.54 | 67.1 | 11.1 | 26.8 |
| O'Brien | 13.9 | 11.1 | 10.4 | 49.5 | 15,102 | 14,398 | -4.7 | -5.0 | 1,690 | 1,946 | -458 | 6,169 | 2.18 | 59.2 | 5.5 | 35.1 |
| Osceola | 15.6 | 11.5 | 11.7 | 49.4 | 7,003 | 6,462 | -7.7 | -7.4 | 705 | 723 | -462 | 2,633 | 2.26 | 62.0 | 6.2 | 30.0 |
| Page | 14.2 | 12.8 | 10.4 | 46.9 | 16,976 | 15,946 | -6.1 | -5.5 | 1,641 | 2,102 | -407 | 6,280 | 2.25 | 61.0 | 6.3 | 35.4 |
| Palo Alto | 14.2 | 11.7 | 10.8 | 50.0 | 10,147 | 9,421 | -7.2 | -6.1 | 1,055 | 1,319 | -312 | 3,702 | 2.34 | 62.7 | 7.7 | 32.3 |
| Plymouth | 13.8 | 10.8 | 8.6 | 49.9 | 24,849 | 24,986 | 0.6 | 0.9 | 2,897 | 2,499 | -158 | 10,250 | 2.41 | 71.2 | 9.0 | 24.8 |
| Pocahontas | 16.6 | 12.6 | 11.4 | 50.1 | 8,662 | 7,310 | -15.6 | -9.6 | 778 | 969 | -518 | 3,215 | 2.07 | 61.2 | 5.3 | 33.7 |
| Polk | 11.7 | 8.4 | 5.4 | 50.7 | 374,601 | 430,631 | 15.0 | 14.8 | 69,207 | 35,253 | 29,574 | 187,798 | 2.51 | 62.5 | 11.5 | 29.2 |
| Pottawattamie | 14.0 | 10.8 | 7.6 | 50.7 | 87,704 | 93,149 | 6.2 | 0.2 | 11,965 | 9,439 | -2,315 | 36,799 | 2.48 | 65.5 | 12.5 | 28.4 |
| Poweshiek | 13.7 | 11.4 | 9.8 | 51.3 | 18,815 | 18,915 | 0.5 | -2.8 | 1,789 | 2,267 | -55 | 7,767 | 2.15 | 61.4 | 9.7 | 31.4 |
| Ringgold | 15.5 | 12.1 | 12.2 | 50.7 | 5,469 | 5,130 | -6.2 | -6.4 | 548 | 801 | -77 | 1,971 | 2.41 | 67.1 | 7.1 | 28.3 |
| Sac | 15.3 | 13.2 | 11.0 | 50.4 | 11,529 | 10,350 | -10.2 | -7.2 | 1,126 | 1,416 | -452 | 4,313 | 2.23 | 66.8 | 7.3 | 29.4 |
| Scott | 13.0 | 10.1 | 7.1 | 50.9 | 158,668 | 165,227 | 4.1 | 4.8 | 22,344 | 15,267 | 972 | 67,021 | 2.52 | 62.8 | 10.6 | 30.6 |
| Shelby | 15.9 | 12.1 | 11.4 | 50.5 | 13,173 | 12,174 | -7.6 | -6.1 | 1,226 | 1,496 | -469 | 5,030 | 2.27 | 63.1 | 6.9 | 31.1 |
| Sioux | 11.5 | 8.7 | 7.9 | 49.9 | 31,589 | 33,704 | 6.7 | 4.0 | 5,158 | 2,576 | -1,251 | 12,248 | 2.66 | 73.7 | 4.0 | 22.3 |
| Story | 9.3 | 7.6 | 5.5 | 47.8 | 79,981 | 89,542 | 12.0 | 9.7 | 9,301 | 5,291 | 4,630 | 37,811 | 2.28 | 51.0 | 5.2 | 28.8 |
| Tama | 14.4 | 11.3 | 9.2 | 50.2 | 18,103 | 17,767 | -1.9 | -5.4 | 2,214 | 2,133 | -1,045 | 6,767 | 2.46 | 65.8 | 8.5 | 28.4 |
| Taylor | 14.2 | 12.6 | 10.8 | 49.0 | 6,958 | 6,317 | -9.2 | -3.6 | 759 | 782 | -199 | 2,616 | 2.32 | 62.3 | 6.1 | 30.3 |
| Union | 13.6 | 11.4 | 9.7 | 51.5 | 12,309 | 12,534 | 1.8 | -3.0 | 1,382 | 1,510 | -250 | 5,176 | 2.33 | 61.6 | 9.8 | 30.6 |
| Van Buren | 15.0 | 13.3 | 9.0 | 49.3 | 7,809 | 7,571 | -3.0 | -6.6 | 932 | 942 | -488 | 2,894 | 2.44 | 66.5 | 7.4 | 30.2 |
| Wapello | 13.5 | 10.8 | 7.9 | 50.0 | 36,051 | 35,622 | -1.2 | -1.8 | 4,495 | 4,212 | -901 | 14,481 | 2.37 | 65.1 | 11.0 | 30.1 |
| Warren | 12.7 | 9.8 | 6.8 | 50.6 | 40,671 | 46,226 | 13.7 | 13.1 | 5,688 | 4,194 | 4,576 | 19,258 | 2.51 | 72.1 | 8.5 | 23.5 |
| Washington | 13.6 | 11.1 | 9.0 | 50.4 | 20,670 | 21,703 | 5.0 | 1.3 | 2,988 | 2,490 | -192 | 8,832 | 2.46 | 69.8 | 9.4 | 24.2 |
| Wayne | 14.5 | 11.5 | 10.9 | 50.4 | 6,730 | 6,403 | -4.9 | 0.2 | 909 | 875 | -16 | 2,689 | 2.35 | 63.4 | 7.2 | 32.3 |
| Webster | 13.9 | 10.8 | 8.3 | 48.3 | 40,235 | 38,011 | -5.5 | -5.5 | 4,515 | 4,574 | -2,025 | 15,306 | 2.2 | 57.7 | 10.1 | 35.9 |
| Winnebago | 14.0 | 12.1 | 10.3 | 50.5 | 11,723 | 10,874 | -7.2 | -5.5 | 1,138 | 1,319 | -419 | 4,577 | 2.16 | 59.4 | 10.3 | 35.0 |
| Winneshiek | 15.1 | 12.1 | 9.9 | 50.2 | 21,310 | 21,058 | -1.2 | -5.7 | 1,788 | 1,874 | -1,109 | 8,257 | 2.18 | 64.8 | 7.2 | 29.4 |
| Woodbury | 11.9 | 9.3 | 6.3 | 50.1 | 103,877 | 102,173 | -1.6 | 0.9 | 15,358 | 9,743 | -4,668 | 39,016 | 2.56 | 66.1 | 13.9 | 27.4 |
| Worth | 14.8 | 12.4 | 9.6 | 49.2 | 7,909 | 7,591 | -4.0 | -3.1 | 801 | 853 | -179 | 3,172 | 2.31 | 64.8 | 8.2 | 29.4 |
| Wright | 13.9 | 12.3 | 11.3 | 49.7 | 14,334 | 13,229 | -7.7 | -6.1 | 1,566 | 1,707 | -669 | 5,718 | 2.19 | 62.1 | 9.0 | 32.1 |

1. No spouse present.

# Table B. States and Counties — Population, Vital Statistics, Health, and Crime

| STATE County | Persons in group quarters, 2020 | Daytime Population, 2015-2019 Number | Employment/residence ratio | Births, 2020 Total | Rate[1] | Deaths, 2020 Number | Rate[1] | Persons under 65 with no health insurance, 2019 Number | Percent | Medicare, 2020 Total beneficiaries | Enrolled in Original Medicare | Enrolled in Medicare Advantage | Crimes reported by county police or sheriff, 2019 Violent | Property |
|---|---|---|---|---|---|---|---|---|---|---|---|---|---|---|
| | 32 | 33 | 34 | 35 | 36 | 37 | 38 | 39 | 40 | 41 | 42 | 43 | 44 | 45 |
| **IOWA—Cont'd** | | | | | | | | | | | | | | |
| Franklin | 185 | 9,827 | 0.93 | 108 | 10.8 | 114 | 11.4 | 557 | 7.1 | 2,388 | 2,216 | 172 | 0 | 1 |
| Fremont | 131 | 6,194 | 0.77 | 64 | 9.5 | 96 | 14.3 | 275 | 5.2 | 1,757 | 1,529 | 228 | 11 | 43 |
| Greene | 88 | 8,448 | 0.89 | 90 | 10.2 | 110 | 12.5 | 345 | 5.1 | 2,333 | 1,863 | 469 | NA | NA |
| Grundy | 145 | 10,272 | 0.67 | 127 | 10.4 | 140 | 11.5 | 397 | 4.1 | 2,895 | 2,211 | 684 | 7 | 63 |
| Guthrie | 168 | 9,034 | 0.69 | 97 | 9.0 | 129 | 12.0 | 521 | 6.4 | 2,756 | 2,124 | 633 | 2 | 45 |
| Hamilton | 198 | 14,206 | 0.89 | 176 | 12.0 | 178 | 12.1 | 627 | 5.4 | 3,401 | 2,887 | 514 | 6 | 54 |
| Hancock | 184 | 10,803 | 1.0 | 110 | 10.5 | 119 | 11.3 | 420 | 5.2 | 2,692 | 2,572 | 120 | 6 | 31 |
| Hardin | 769 | 16,824 | 0.97 | 160 | 9.7 | 234 | 14.1 | 732 | 5.9 | 4,180 | 3,535 | 645 | 11 | 83 |
| Harrison | 281 | 11,198 | 0.6 | 157 | 11.3 | 190 | 13.6 | 576 | 5.2 | 3,311 | 2,742 | 569 | 7 | 43 |
| Henry | 1,576 | 20,431 | 1.05 | 202 | 10.3 | 211 | 10.7 | 758 | 5.2 | 4,498 | 3,925 | 573 | 20 | 69 |
| Howard | 190 | 8,949 | 0.94 | 119 | 13.0 | 124 | 13.5 | 435 | 6.1 | 2,230 | 2,096 | 134 | 0 | 35 |
| Humboldt | 104 | 9,130 | 0.91 | 105 | 11.1 | 124 | 13.1 | 420 | 5.7 | 2,233 | 2,093 | 140 | 0 | 41 |
| Ida | 107 | 7,642 | 1.22 | 68 | 10.0 | 108 | 15.8 | 276 | 5.2 | 1,680 | 1,482 | 198 | 10 | 68 |
| Iowa | 283 | 16,704 | 1.06 | 165 | 10.2 | 214 | 13.3 | 535 | 4.1 | 3,593 | 2,795 | 798 | 8 | 55 |
| Jackson | 192 | 16,774 | 0.72 | 211 | 11.0 | 242 | 12.6 | 820 | 5.4 | 4,904 | 3,199 | 1,705 | 3 | 23 |
| Jasper | 1,809 | 31,185 | 0.67 | 400 | 10.8 | 401 | 10.8 | 1,260 | 4.4 | 8,259 | 6,238 | 2,021 | 6 | 44 |
| Jefferson | 1,704 | 18,913 | 1.09 | 141 | 7.7 | 191 | 10.4 | 1,134 | 9.1 | 4,614 | 3,496 | 1,118 | 2 | 26 |
| Johnson | 8,388 | 155,357 | 1.08 | 1,719 | 11.2 | 835 | 5.4 | 7,387 | 5.9 | 21,338 | 16,424 | 4,913 | 65 | 196 |
| Jones | 1,196 | 17,680 | 0.7 | 198 | 9.6 | 230 | 11.2 | 844 | 5.5 | 4,706 | 3,413 | 1,293 | NA | NA |
| Keokuk | 136 | 8,367 | 0.62 | 105 | 10.4 | 114 | 11.3 | 481 | 6.1 | 2,472 | 1,914 | 558 | NA | NA |
| Kossuth | 219 | 15,136 | 1.02 | 173 | 11.8 | 178 | 12.1 | 536 | 4.8 | 3,895 | 3,705 | 189 | 6 | 32 |
| Lee | 1,271 | 36,413 | 1.14 | 328 | 9.8 | 462 | 13.8 | 1,351 | 5.3 | 8,512 | 7,752 | 761 | 15 | 108 |
| Linn | 5,133 | 235,233 | 1.1 | 2,674 | 11.7 | 1,921 | 8.4 | 8,883 | 4.8 | 42,398 | 27,356 | 15,043 | NA | NA |
| Louisa | 117 | 10,139 | 0.81 | 123 | 11.2 | 125 | 11.4 | 786 | 8.9 | 2,375 | 1,852 | 524 | 23 | 58 |
| Lucas | 63 | 8,485 | 0.98 | 100 | 11.7 | 96 | 11.3 | 372 | 5.6 | 2,129 | 1,639 | 490 | 12 | 68 |
| Lyon | 166 | 10,717 | 0.82 | 174 | 14.8 | 108 | 9.2 | 618 | 6.5 | 2,344 | 2,110 | 234 | 11 | 57 |
| Madison | 148 | 12,596 | 0.59 | 173 | 10.5 | 156 | 9.4 | 661 | 4.9 | 3,294 | 2,387 | 907 | 2 | 26 |
| Mahaska | 995 | 19,916 | 0.79 | 263 | 11.8 | 250 | 11.2 | 963 | 5.5 | 4,669 | 3,608 | 1,061 | 23 | 64 |
| Marion | 1,326 | 35,335 | 1.13 | 336 | 10.1 | 365 | 11.0 | 1,090 | 4.2 | 7,206 | 5,812 | 1,394 | 19 | 44 |
| Marshall | 1,247 | 40,360 | 1.02 | 490 | 12.4 | 476 | 12.1 | 2,374 | 7.6 | 8,589 | 6,366 | 2,223 | 42 | 94 |
| Mills | 537 | 12,191 | 0.61 | 135 | 9.1 | 151 | 10.2 | 570 | 4.8 | 3,296 | 2,658 | 638 | 17 | 98 |
| Mitchell | 214 | 9,993 | 0.88 | 131 | 12.3 | 129 | 12.1 | 627 | 7.6 | 2,450 | 2,328 | 122 | 0 | 5 |
| Monona | 193 | 7,928 | 0.81 | 92 | 10.7 | 147 | 17.1 | 381 | 5.9 | 2,360 | 1,820 | 540 | 16 | 37 |
| Monroe | 107 | 7,786 | 0.99 | 85 | 10.9 | 77 | 9.9 | 361 | 5.9 | 1,787 | 1,348 | 439 | 5 | 20 |
| Montgomery | 188 | 10,196 | 1.03 | 107 | 10.8 | 150 | 15.1 | 468 | 6.1 | 2,727 | 2,299 | 429 | NA | NA |
| Muscatine | 553 | 45,546 | 1.13 | 512 | 12.1 | 392 | 9.2 | 2,119 | 6.1 | 8,642 | 6,063 | 2,579 | 21 | 56 |
| O'Brien | 410 | 13,376 | 0.94 | 160 | 11.7 | 181 | 13.2 | 666 | 6.2 | 3,336 | 3,089 | 247 | 21 | 57 |
| Osceola | 107 | 5,235 | 0.73 | 63 | 10.5 | 75 | 12.5 | 347 | 7.6 | 1,421 | 1,321 | 100 | 14 | 73 |
| Page | 1,523 | 15,586 | 1.05 | 131 | 8.7 | 206 | 13.7 | 604 | 5.9 | 3,961 | 3,482 | 479 | 3 | 26 |
| Palo Alto | 251 | 8,598 | 0.91 | 98 | 11.1 | 123 | 13.9 | 389 | 5.7 | 2,267 | 2,166 | 101 | 15 | 48 |
| Plymouth | 284 | 23,514 | 0.88 | 277 | 11.0 | 236 | 9.4 | 1,079 | 5.3 | 5,286 | 4,307 | 979 | 13 | 45 |
| Pocahontas | 95 | 6,631 | 0.95 | 71 | 10.7 | 85 | 12.9 | 352 | 7.0 | 1,810 | 1,680 | 131 | NA | NA |
| Polk | 9,364 | 517,856 | 1.15 | 6,524 | 13.2 | 3,966 | 8.0 | 24,317 | 5.8 | 77,265 | 52,988 | 24,276 | 112 | 822 |
| Pottawattamie | 1,981 | 85,719 | 0.84 | 1,058 | 11.3 | 1,042 | 11.2 | 4,487 | 6.0 | 19,937 | 13,284 | 6,653 | 57 | 251 |
| Poweshiek | 1,639 | 19,577 | 1.13 | 161 | 8.8 | 221 | 12.0 | 683 | 5.2 | 4,246 | 3,476 | 770 | NA | NA |
| Ringgold | 177 | 4,496 | 0.78 | 47 | 9.8 | 66 | 13.7 | 261 | 7.2 | 1,291 | 1,176 | 115 | 4 | 21 |
| Sac | 167 | 8,693 | 0.77 | 108 | 11.2 | 119 | 12.4 | 411 | 5.6 | 2,575 | 2,445 | 129 | 6 | 32 |
| Scott | 3,377 | 178,411 | 1.07 | 2,046 | 11.8 | 1,571 | 9.1 | 7,743 | 5.5 | 33,556 | 22,284 | 11,272 | 40 | 170 |
| Shelby | 200 | 11,747 | 1.02 | 126 | 11.0 | 138 | 12.1 | 490 | 5.6 | 3,058 | 2,700 | 358 | 3 | 14 |
| Sioux | 2,242 | 37,549 | 1.15 | 500 | 14.3 | 263 | 7.5 | 1,775 | 6.5 | 6,085 | 5,637 | 448 | 7 | 50 |
| Story | 10,514 | 101,714 | 1.09 | 826 | 8.4 | 564 | 5.7 | 4,788 | 6.4 | 14,002 | 11,341 | 2,660 | 13 | 105 |
| Tama | 296 | 14,682 | 0.72 | 200 | 11.9 | 188 | 11.2 | 1,121 | 8.4 | 3,896 | 2,877 | 1,019 | 13 | 57 |
| Taylor | 53 | 5,636 | 0.82 | 65 | 10.7 | 69 | 11.3 | 390 | 8.3 | 1,533 | 1,508 | 25 | 8 | 14 |
| Union | 456 | 12,926 | 1.1 | 121 | 10.0 | 139 | 11.4 | 590 | 6.3 | 2,941 | 2,511 | 430 | 4 | 46 |
| Van Buren | 57 | 6,550 | 0.8 | 96 | 13.6 | 97 | 13.7 | 410 | 7.6 | 1,828 | 1,448 | 380 | 4 | 58 |
| Wapello | 913 | 36,099 | 1.06 | 435 | 12.4 | 405 | 11.6 | 2,229 | 8.1 | 7,775 | 5,316 | 2,459 | 25 | 91 |
| Warren | 1,440 | 36,868 | 0.49 | 578 | 11.1 | 463 | 8.9 | 1,614 | 3.8 | 9,658 | 6,882 | 2,776 | 54 | 136 |
| Washington | 292 | 20,412 | 0.85 | 285 | 13.0 | 226 | 10.3 | 1,141 | 6.5 | 4,964 | 4,060 | 903 | 35 | 61 |
| Wayne | 93 | 5,960 | 0.83 | 90 | 14.0 | 92 | 14.3 | 382 | 7.8 | 1,536 | 1,178 | 358 | 8 | 28 |
| Webster | 2,742 | 38,354 | 1.11 | 429 | 11.9 | 442 | 12.3 | 1,732 | 6.5 | 8,119 | 6,486 | 1,633 | 27 | 113 |
| Winnebago | 486 | 11,014 | 1.09 | 97 | 9.4 | 128 | 12.5 | 386 | 5.0 | 2,547 | 2,348 | 199 | NA | NA |
| Winneshiek | 1,918 | 20,852 | 1.05 | 187 | 9.4 | 198 | 10.0 | 666 | 4.8 | 4,740 | 3,970 | 771 | 9 | 12 |
| Woodbury | 2,513 | 101,973 | 0.99 | 1,421 | 13.8 | 968 | 9.4 | 6,474 | 7.6 | 18,786 | 12,951 | 5,835 | 22 | 186 |
| Worth | 80 | 5,882 | 0.6 | 80 | 10.9 | 66 | 9.0 | 299 | 5.2 | 1,740 | 1,559 | 181 | 6 | 27 |
| Wright | 205 | 12,990 | 1.04 | 158 | 12.7 | 174 | 14.0 | 639 | 6.7 | 3,051 | 2,815 | 236 | 4 | 21 |

1. Per 1,000 estimated resident population.

# Table B. States and Counties — Crime, Education, Money Income, and Poverty

| STATE County | County law enforcement employment, 2019 | | Education — School enrollment and attainment, 2015-2019 | | | | Local government expenditures,[3] 2017-2018 | | Money income, 2015-2019 | | | | Income and poverty, 2019 | | | |
| | | | Enrollment[1] | | Attainment[2] (percent) | | | | Per capita income[4] | Households | | | Percent below poverty level | | | |
| | | | | | | | | | | | Percent | | | | | |
| | Officers | Civilians | Total | Percent private | High school graduate or less | Bachelor's degree or more | Total current spending (mil dol) | Current spending per student (dollars) | Per capita income[4] | Median income (dollars) | with income of less than $50,000 | with income of $200,000 or more | Median household income (dollars) | All persons | Children under 18 years | Children 5 to 17 years in families |
| | 46 | 47 | 48 | 49 | 50 | 51 | 52 | 53 | 54 | 55 | 56 | 57 | 58 | 59 | 60 | 61 |
| **IOWA—Cont'd** | | | | | | | | | | | | | | | | |
| Franklin | 7 | 3 | 2,118 | 3.2 | 44.8 | 18.0 | 18 | 11,434 | 29,546 | 56,419 | 43.3 | 2.6 | 55,847 | 11.2 | 17.1 | 16.4 |
| Fremont | 10 | 17 | 1,565 | 10.0 | 41.5 | 18.9 | 14 | 11,364 | 29,261 | 56,750 | 44.2 | 2.9 | 60,498 | 11.6 | 14.7 | 13.9 |
| Greene | 8 | 10 | 2,033 | 13.3 | 46.3 | 18.7 | 17 | 11,071 | 28,451 | 53,050 | 47.5 | 1.9 | 55,562 | 11.2 | 14.1 | 12.9 |
| Grundy | 12 | 4 | 2,767 | 9.0 | 37.0 | 27.9 | 30 | 10,589 | 36,231 | 68,038 | 34.7 | 5.2 | 73,683 | 5.5 | 7.1 | 6.4 |
| Guthrie | 9 | 5 | 2,272 | 5.4 | 45.8 | 19.8 | 27 | 10,923 | 31,128 | 61,161 | 41.7 | 3.6 | 64,965 | 8.8 | 11.6 | 11.5 |
| Hamilton | 11 | 20 | 3,340 | 6.5 | 39.1 | 24.3 | 31 | 11,846 | 30,127 | 60,910 | 39.8 | 2.9 | 63,234 | 9.1 | 12.4 | 12.5 |
| Hancock | 8 | 1 | 2,223 | 5.7 | 43.0 | 23.8 | 19 | 11,825 | 30,309 | 61,761 | 39.2 | 2.8 | 62,030 | 9.0 | 10.9 | 10.3 |
| Hardin | 11 | 19 | 3,726 | 9.4 | 42.3 | 20.8 | 38 | 11,468 | 28,770 | 54,196 | 45.0 | 2.5 | 57,103 | 10.6 | 15.2 | 14.5 |
| Harrison | 10 | 1 | 3,120 | 7.9 | 45.7 | 19.5 | 30 | 10,454 | 32,443 | 63,854 | 37.9 | 3.1 | 63,901 | 8.7 | 9.9 | 9.5 |
| Henry | NA | NA | 4,602 | 14.4 | 43.4 | 19.6 | 37 | 10,981 | 26,408 | 53,501 | 45.3 | 2.3 | 57,747 | 11.7 | 15.3 | 15.0 |
| Howard | 9 | 8 | 2,038 | 24.4 | 49.1 | 15.8 | 18 | 11,438 | 29,032 | 55,060 | 43.5 | 1.7 | 58,028 | 9.6 | 13.4 | 12.9 |
| Humboldt | 8 | 13 | 2,110 | 14.6 | 44.2 | 21.2 | 19 | 11,725 | 30,480 | 54,575 | 44.3 | 3.1 | 56,695 | 10.8 | 12.7 | 11.8 |
| Ida | 9 | 7 | 1,481 | 3.7 | 45.6 | 18.6 | 14 | 11,557 | 33,789 | 58,427 | 44.7 | 4.5 | 58,249 | 10.5 | 11.5 | 10.5 |
| Iowa | 14 | 15 | 3,541 | 9.5 | 43.3 | 18.8 | 28 | 10,479 | 33,755 | 61,443 | 39.9 | 3.7 | 66,861 | 6.7 | 8.2 | 7.2 |
| Jackson | 10 | 9 | 3,989 | 12.7 | 49.4 | 19.2 | 33 | 10,982 | 29,660 | 55,967 | 44.7 | 2.9 | 57,687 | 12.3 | 17.3 | 16.7 |
| Jasper | 17 | 30 | 7,929 | 11.6 | 47.5 | 18.8 | 61 | 10,287 | 28,604 | 58,952 | 41.3 | 2.8 | 67,317 | 8.4 | 10.3 | 9.5 |
| Jefferson | 12 | 2 | 4,113 | 33.8 | 34.8 | 31.5 | 26 | 11,488 | 28,611 | 47,708 | 51.3 | 3.4 | 51,741 | 14.2 | 17.6 | 17.2 |
| Johnson | 72 | 20 | 53,954 | 8.9 | 21.3 | 52.8 | 213 | 11,319 | 35,108 | 62,542 | 41.3 | 7.6 | 67,295 | 16.0 | 10.5 | 10.2 |
| Jones | 11 | 16 | 4,377 | 13.8 | 47.2 | 19.5 | 33 | 10,502 | 29,453 | 57,549 | 43.1 | 4.2 | 63,399 | 9.5 | 11.9 | 10.4 |
| Keokuk | 6 | 6 | 2,227 | 6.8 | 47.5 | 18.3 | 12 | 9,691 | 29,957 | 52,881 | 47.5 | 3.3 | 54,937 | 11.4 | 15.9 | 15.2 |
| Kossuth | 10 | 15 | 3,140 | 17.6 | 43.5 | 17.5 | 23 | 12,022 | 31,225 | 56,073 | 43.6 | 3.1 | 59,568 | 10.3 | 13.6 | 13.0 |
| Lee | 18 | 19 | 7,020 | 15.8 | 47.9 | 16.4 | 54 | 10,809 | 26,710 | 49,564 | 50.4 | 1.9 | 51,089 | 15.0 | 22.1 | 20.6 |
| Linn | 129 | 66 | 55,897 | 16.6 | 32.5 | 33.1 | 474 | 12,432 | 34,545 | 64,903 | 37.7 | 5.2 | 64,783 | 11.1 | 13.0 | 11.4 |
| Louisa | 11 | 20 | 2,406 | 3.4 | 53.6 | 14.9 | 27 | 10,631 | 29,021 | 56,673 | 44.9 | 3.3 | 60,544 | 9.2 | 12.2 | 11.7 |
| Lucas | 5 | 8 | 1,862 | 10.2 | 49.2 | 17.0 | 14 | 10,564 | 31,069 | 55,205 | 45.4 | 3.8 | 53,085 | 14.9 | 22.5 | 22.7 |
| Lyon | 11 | 11 | 2,927 | 15.1 | 46.5 | 20.7 | 23 | 10,414 | 29,124 | 64,982 | 34.3 | 2.5 | 66,139 | 7.4 | 9.2 | 8.8 |
| Madison | 8 | 10 | 4,006 | 9.4 | 42.3 | 24.5 | 34 | 10,203 | 35,470 | 66,316 | 37.0 | 6.0 | 72,578 | 6.9 | 7.6 | 6.9 |
| Mahaska | 10 | 16 | 5,618 | 23.7 | 47.4 | 22.0 | 32 | 10,782 | 28,174 | 54,825 | 44.7 | 2.4 | 59,242 | 10.8 | 12.9 | 12.5 |
| Marion | 17 | 21 | 8,453 | 30.7 | 40.8 | 28.8 | 59 | 10,708 | 30,644 | 61,038 | 38.0 | 3.1 | 66,381 | 8.2 | 9.6 | 9.0 |
| Marshall | 19 | 36 | 9,126 | 7.9 | 50.4 | 18.8 | 78 | 10,736 | 27,196 | 56,437 | 43.3 | 1.9 | 55,631 | 13.0 | 15.2 | 13.1 |
| Mills | 12 | 15 | 3,692 | 11.4 | 41.3 | 24.1 | 27 | 10,131 | 34,850 | 72,079 | 33.3 | 5.9 | 72,679 | 8.8 | 10.6 | 9.7 |
| Mitchell | 10 | 10 | 2,286 | 15.0 | 46.6 | 19.0 | 17 | 10,771 | 29,676 | 58,302 | 41.7 | 3.6 | 61,247 | 9.3 | 13.1 | 12.2 |
| Monona | 8 | 10 | 1,781 | 6.9 | 46.0 | 15.3 | 18 | 11,591 | 29,591 | 47,605 | 52.3 | 2.8 | 53,104 | 12.3 | 15.4 | 14.6 |
| Monroe | NA | NA | 1,560 | 5.9 | 51.4 | 17.4 | 12 | 10,004 | 28,999 | 58,269 | 43.9 | 1.9 | 56,397 | 10.7 | 16.0 | 14.8 |
| Montgomery | NA | NA | 2,021 | 4.6 | 44.8 | 18.5 | 19 | 11,381 | 29,781 | 51,696 | 47.9 | 3.1 | 52,715 | 13.2 | 18.3 | 17.0 |
| Muscatine | 21 | 60 | 10,403 | 6.2 | 46.3 | 21.5 | 79 | 10,472 | 29,511 | 58,453 | 41.9 | 2.8 | 61,767 | 10.1 | 13.6 | 11.6 |
| O'Brien | 9 | 19 | 2,987 | 19.5 | 47.8 | 18.8 | 27 | 10,912 | 30,564 | 53,703 | 46.1 | 2.2 | 60,298 | 10.1 | 12.4 | 11.7 |
| Osceola | 9 | 5 | 1,348 | 10.2 | 48.1 | 17.9 | 8 | 10,010 | 30,955 | 59,725 | 40.5 | 4.0 | 60,702 | 9.3 | 13.2 | 12.5 |
| Page | 9 | 8 | 3,052 | 8.6 | 44.9 | 20.2 | 29 | 11,183 | 27,390 | 51,867 | 48.0 | 1.9 | 52,179 | 14.9 | 20.4 | 18.5 |
| Palo Alto | 8 | 9 | 2,204 | 11.9 | 36.5 | 25.5 | 20 | 11,634 | 28,445 | 52,932 | 46.2 | 2.9 | 55,074 | 11.1 | 13.6 | 12.6 |
| Plymouth | 12 | 22 | 6,052 | 17.5 | 40.8 | 23.2 | 47 | 11,120 | 33,982 | 67,297 | 34.7 | 5.0 | 69,165 | 7.2 | 8.7 | 7.6 |
| Pocahontas | 7 | 10 | 1,268 | 19.1 | 47.2 | 15.0 | 33 | 31,900 | 30,013 | 52,448 | 47.8 | 2.9 | 56,146 | 10.1 | 15.9 | 15.8 |
| Polk | 157 | 368 | 121,602 | 17.8 | 32.9 | 36.7 | 994 | 12,586 | 35,433 | 67,637 | 36.4 | 6.5 | 67,331 | 10.1 | 11.9 | 10.0 |
| Pottawattamie | 53 | 148 | 22,483 | 10.3 | 43.2 | 21.5 | 215 | 13,636 | 30,318 | 60,065 | 41.9 | 3.5 | 60,322 | 11.0 | 13.4 | 12.0 |
| Poweshiek | 12 | 1 | 4,915 | 39.9 | 44.0 | 26.1 | 31 | 10,892 | 31,827 | 53,852 | 46.3 | 4.9 | 57,741 | 11.6 | 12.4 | 11.4 |
| Ringgold | 6 | 7 | 1,156 | 6.9 | 44.1 | 21.4 | 10 | 12,082 | 28,756 | 55,271 | 46.7 | 3.7 | 50,890 | 15.1 | 25.1 | 23.9 |
| Sac | 9 | 13 | 2,031 | 6.4 | 44.1 | 20.0 | 19 | 11,561 | 32,274 | 58,232 | 42.4 | 3.3 | 59,392 | 9.5 | 13.1 | 12.3 |
| Scott | 49 | 117 | 42,682 | 17.7 | 36.8 | 31.8 | 356 | 12,368 | 33,643 | 61,183 | 40.9 | 6.0 | 65,724 | 11.8 | 14.5 | 13.0 |
| Shelby | 9 | 1 | 2,654 | 11.9 | 42.9 | 22.2 | 21 | 10,903 | 32,703 | 58,000 | 43.6 | 5.0 | 60,498 | 8.7 | 10.3 | 9.0 |
| Sioux | 14 | 27 | 10,350 | 44.7 | 39.9 | 29.1 | 58 | 10,754 | 29,360 | 69,844 | 33.2 | 3.9 | 71,566 | 7.0 | 7.4 | 7.1 |
| Story | 32 | 57 | 43,029 | 4.0 | 20.3 | 51.6 | 125 | 10,863 | 29,539 | 56,756 | 44.9 | 4.9 | 64,151 | 15.8 | 8.0 | 6.9 |
| Tama | 13 | 10 | 3,910 | 7.4 | 48.8 | 15.8 | 27 | 11,262 | 28,585 | 56,037 | 43.3 | 2.3 | 56,281 | 13.1 | 17.6 | 16.6 |
| Taylor | 8 | 5 | 1,383 | 4.3 | 48.3 | 14.5 | 11 | 10,999 | 29,778 | 50,431 | 49.6 | 3.5 | 51,510 | 12.5 | 17.1 | 16.5 |
| Union | 6 | 7 | 2,861 | 17.1 | 45.3 | 18.0 | 22 | 10,624 | 27,750 | 49,900 | 50.1 | 2.8 | 52,914 | 12.1 | 16.1 | 14.6 |
| Van Buren | 6 | 6 | 1,453 | 26.0 | 49.7 | 14.4 | 10 | 9,845 | 27,252 | 48,591 | 52.5 | 2.4 | 50,425 | 12.8 | 18.8 | 18.2 |
| Wapello | 13 | 31 | 7,893 | 11.3 | 46.8 | 18.8 | 91 | 14,117 | 26,424 | 46,246 | 53.8 | 1.9 | 50,533 | 14.2 | 18.5 | 14.9 |
| Warren | 20 | 16 | 13,272 | 20.3 | 35.4 | 31.9 | 97 | 10,010 | 36,496 | 77,048 | 30.8 | 6.7 | 79,557 | 7.3 | 7.9 | 6.6 |
| Washington | 17 | 28 | 4,932 | 10.4 | 47.8 | 21.8 | 43 | 11,027 | 30,547 | 62,556 | 40.1 | 3.6 | 63,230 | 9.1 | 12.1 | 11.6 |
| Wayne | 8 | 8 | 1,247 | 16.0 | 56.5 | 12.7 | 13 | 10,507 | 25,628 | 44,768 | 54.1 | 2.3 | 49,962 | 19.1 | 28.9 | 25.9 |
| Webster | 17 | 19 | 8,438 | 12.2 | 43.4 | 19.9 | 57 | 10,970 | 26,498 | 47,466 | 52.4 | 2.3 | 53,344 | 12.1 | 15.5 | 14.6 |
| Winnebago | 7 | 1 | 2,524 | 14.9 | 41.7 | 21.8 | 27 | 11,758 | 28,342 | 49,870 | 50.2 | 1.9 | 50,258 | 10.2 | 13.6 | 12.3 |
| Winneshiek | 11 | 15 | 5,959 | 43.4 | 37.8 | 30.6 | 32 | 12,226 | 32,075 | 65,263 | 36.6 | 4.4 | 65,482 | 8.6 | 8.5 | 8.0 |
| Woodbury | 40 | 73 | 26,910 | 16.9 | 45.1 | 23.1 | 233 | 12,282 | 28,688 | 59,224 | 41.8 | 3.5 | 58,419 | 12.8 | 16.1 | 15.8 |
| Worth | 13 | 12 | 1,617 | 5.2 | 43.1 | 17.5 | 15 | 10,562 | 31,949 | 57,130 | 42.7 | 3.9 | 60,139 | 9.8 | 14.5 | 14.1 |
| Wright | 10 | 15 | 2,706 | 2.9 | 44.5 | 17.2 | 31 | 11,224 | 29,622 | 53,333 | 46.8 | 2.2 | 54,251 | 11.0 | 14.9 | 14.9 |

1. All persons 3 years old and over enrolled in nursery school through college.  2. Persons 25 years old and over.  3. Elementary and secondary education expenditures.  4. Based on population estimated by the American Community Survey, 2014–2018.

# Table B. States and Counties — Personal Income

| | Personal income, 2019 | | | | | | | | | | Earnings, 2019 | | |
|---|---|---|---|---|---|---|---|---|---|---|---|---|---|
| | | | Per capita[1] | | | Supplements to wages and salaries, employer contributions (mil dol) | | | | | | Contributions for government social insurance (mil dol) | |
| STATE County | Total (mil dol) | Percent change 2018-2019 | Dollars | Rank | Wages and salaries (mil dol) | Pension and insurance | Government social insurance | Proprietors' income (mil dol) | Dividends, interest, and rent (mil dol) | Personal transfer receipts (mil dol) | Total (mil dol) | From employee and self-employed | From employer |
| | 62 | 63 | 64 | 65 | 66 | 67 | 68 | 69 | 70 | 71 | 72 | 73 | 74 |

IOWA—Cont'd

| | | | | | | | | | | | | | |
|---|---|---|---|---|---|---|---|---|---|---|---|---|---|
| Franklin | 536 | 7.1 | 53,241 | 595 | 173 | 38 | 13 | 113 | 109 | 103 | 337 | 17 | 13 |
| Fremont | 351 | 12.7 | 50,441 | 799 | 108 | 21 | 9 | 57 | 58 | 82 | 195 | 11 | 9 |
| Greene | 471 | 6.6 | 53,039 | 612 | 152 | 34 | 12 | 76 | 98 | 101 | 274 | 16 | 12 |
| Grundy | 673 | 4.1 | 55,017 | 491 | 199 | 36 | 16 | 76 | 140 | 113 | 327 | 21 | 16 |
| Guthrie | 586 | 2.7 | 54,843 | 504 | 135 | 30 | 11 | 84 | 125 | 115 | 260 | 17 | 11 |
| Hamilton | 863 | 3.9 | 58,399 | 358 | 240 | 50 | 18 | 222 | 156 | 151 | 530 | 27 | 18 |
| Hancock | 613 | 4.4 | 57,702 | 382 | 290 | 57 | 23 | 108 | 115 | 107 | 477 | 25 | 23 |
| Hardin | 890 | 4.1 | 52,844 | 630 | 283 | 62 | 22 | 168 | 193 | 179 | 535 | 30 | 22 |
| Harrison | 690 | 7.4 | 49,085 | 918 | 172 | 36 | 13 | 65 | 108 | 154 | 286 | 19 | 13 |
| Henry | 897 | 2.5 | 44,932 | 1,365 | 408 | 80 | 31 | 72 | 173 | 195 | 592 | 38 | 31 |
| Howard | 487 | 5.9 | 53,184 | 598 | 159 | 36 | 13 | 93 | 101 | 91 | 301 | 16 | 13 |
| Humboldt | 504 | 5.4 | 52,742 | 634 | 166 | 37 | 13 | 78 | 94 | 94 | 293 | 17 | 13 |
| Ida | 385 | 6.7 | 56,128 | 441 | 183 | 34 | 14 | 48 | 100 | 72 | 279 | 16 | 14 |
| Iowa | 912 | 3.7 | 56,382 | 427 | 415 | 102 | 39 | 100 | 193 | 150 | 656 | 38 | 39 |
| Jackson | 915 | 5.1 | 47,052 | 1,128 | 235 | 49 | 19 | 93 | 165 | 213 | 396 | 28 | 19 |
| Jasper | 1,636 | 3.7 | 43,987 | 1,486 | 492 | 99 | 39 | 108 | 283 | 353 | 738 | 52 | 39 |
| Jefferson | 767 | 3.6 | 41,936 | 1,776 | 329 | 75 | 25 | 62 | 228 | 174 | 491 | 33 | 25 |
| Johnson | 8,300 | 1.9 | 54,917 | 495 | 4,565 | 1,229 | 332 | 515 | 1,975 | 965 | 6,642 | 371 | 332 |
| Jones | 950 | 4.3 | 45,927 | 1,257 | 260 | 55 | 21 | 91 | 183 | 201 | 426 | 29 | 21 |
| Keokuk | 493 | 5.6 | 48,084 | 1,009 | 94 | 21 | 7 | 93 | 88 | 104 | 214 | 13 | 7 |
| Kossuth | 843 | 5.6 | 56,902 | 410 | 297 | 57 | 22 | 190 | 174 | 158 | 566 | 31 | 22 |
| Lee | 1,462 | 3.2 | 43,430 | 1,554 | 749 | 141 | 65 | 85 | 288 | 389 | 1,040 | 72 | 65 |
| Linn | 12,136 | 1.6 | 53,533 | 569 | 7,399 | 1,181 | 566 | 549 | 2,387 | 2,004 | 9,695 | 617 | 566 |
| Louisa | 510 | 5.6 | 46,211 | 1,228 | 166 | 35 | 12 | 81 | 73 | 103 | 294 | 16 | 12 |
| Lucas | 370 | 3.1 | 43,061 | 1,608 | 159 | 29 | 12 | 35 | 70 | 92 | 235 | 16 | 12 |
| Lyon | 785 | 6.4 | 66,756 | 158 | 188 | 36 | 15 | 279 | 130 | 90 | 518 | 19 | 15 |
| Madison | 824 | 3.1 | 50,460 | 798 | 154 | 34 | 13 | 56 | 136 | 140 | 258 | 19 | 13 |
| Mahaska | 1,028 | 4.8 | 46,516 | 1,183 | 350 | 73 | 28 | 111 | 190 | 211 | 563 | 35 | 28 |
| Marion | 1,691 | 4.3 | 50,862 | 765 | 933 | 160 | 69 | 89 | 347 | 307 | 1,251 | 81 | 69 |
| Marshall | 1,755 | 2.8 | 44,589 | 1,409 | 790 | 149 | 62 | 122 | 330 | 430 | 1,122 | 74 | 62 |
| Mills | 875 | 5.1 | 57,897 | 374 | 171 | 41 | 13 | 163 | 126 | 166 | 387 | 25 | 13 |
| Mitchell | 738 | 8.6 | 69,700 | 132 | 236 | 43 | 19 | 227 | 118 | 98 | 524 | 30 | 19 |
| Monona | 432 | 8.4 | 50,103 | 825 | 100 | 21 | 8 | 61 | 81 | 103 | 191 | 12 | 8 |
| Monroe | 357 | 5.7 | 46,356 | 1,206 | 197 | 34 | 15 | 26 | 67 | 79 | 272 | 17 | 15 |
| Montgomery | 445 | 6.3 | 44,918 | 1,369 | 177 | 38 | 14 | 40 | 86 | 131 | 270 | 18 | 14 |
| Muscatine | 2,072 | 3.0 | 48,563 | 963 | 1,182 | 200 | 90 | 143 | 385 | 407 | 1,616 | 103 | 90 |
| O'Brien | 842 | 5.9 | 61,246 | 272 | 263 | 54 | 20 | 180 | 173 | 145 | 516 | 26 | 20 |
| Osceola | 365 | 6.9 | 61,277 | 271 | 96 | 20 | 8 | 116 | 60 | 57 | 240 | 10 | 8 |
| Page | 671 | 5.7 | 44,442 | 1,426 | 245 | 55 | 19 | 66 | 141 | 180 | 385 | 25 | 19 |
| Palo Alto | 524 | 6.0 | 58,989 | 338 | 150 | 35 | 12 | 152 | 89 | 96 | 349 | 18 | 12 |
| Plymouth | 1,512 | 5.3 | 60,040 | 302 | 604 | 103 | 47 | 238 | 310 | 214 | 992 | 55 | 47 |
| Pocahontas | 410 | 8.2 | 61,992 | 249 | 163 | 28 | 15 | 117 | 65 | 76 | 323 | 15 | 15 |
| Polk | 26,399 | 3.0 | 53,859 | 552 | 18,413 | 2,688 | 1,355 | 1,635 | 4,882 | 3,896 | 24,091 | 1,490 | 1,355 |
| Pottawattamie | 4,395 | 4.5 | 47,150 | 1,113 | 1,876 | 322 | 151 | 337 | 690 | 1,000 | 2,686 | 180 | 151 |
| Poweshiek | 873 | 2.5 | 47,199 | 1,110 | 474 | 80 | 37 | 70 | 182 | 170 | 661 | 42 | 37 |
| Ringgold | 247 | 7.1 | 50,405 | 802 | 54 | 14 | 4 | 55 | 50 | 59 | 127 | 7 | 4 |
| Sac | 574 | 4.8 | 59,085 | 337 | 127 | 27 | 11 | 162 | 112 | 104 | 327 | 16 | 11 |
| Scott | 9,643 | 3.0 | 55,758 | 459 | 4,441 | 716 | 347 | 627 | 1,880 | 1,700 | 6,131 | 398 | 347 |
| Shelby | 645 | 5.6 | 56,323 | 434 | 242 | 50 | 18 | 102 | 133 | 133 | 413 | 24 | 18 |
| Sioux | 2,042 | 3.9 | 58,585 | 350 | 926 | 174 | 72 | 476 | 406 | 254 | 1,647 | 81 | 72 |
| Story | 4,239 | 2.7 | 43,643 | 1,531 | 2,479 | 610 | 179 | 280 | 1,017 | 603 | 3,548 | 206 | 179 |
| Tama | 825 | 5.9 | 48,923 | 932 | 233 | 52 | 18 | 89 | 155 | 169 | 391 | 25 | 18 |
| Taylor | 276 | 6.9 | 45,093 | 1,351 | 80 | 18 | 6 | 53 | 44 | 68 | 158 | 9 | 6 |
| Union | 521 | 3.3 | 42,530 | 1,682 | 252 | 57 | 21 | 43 | 95 | 139 | 372 | 24 | 21 |
| Van Buren | 302 | 4.4 | 42,908 | 1,631 | 77 | 19 | 6 | 43 | 60 | 78 | 145 | 10 | 6 |
| Wapello | 1,411 | 4.0 | 40,344 | 2,008 | 701 | 134 | 56 | 83 | 222 | 399 | 975 | 67 | 56 |
| Warren | 2,665 | 3.4 | 51,776 | 698 | 502 | 101 | 39 | 161 | 407 | 405 | 803 | 59 | 39 |
| Washington | 1,365 | 3.8 | 62,145 | 246 | 331 | 68 | 26 | 392 | 232 | 216 | 817 | 43 | 26 |
| Wayne | 287 | 4.4 | 44,614 | 1,405 | 79 | 21 | 6 | 54 | 53 | 65 | 161 | 10 | 6 |
| Webster | 1,744 | 4.5 | 48,588 | 956 | 888 | 173 | 69 | 195 | 306 | 388 | 1,324 | 82 | 69 |
| Winnebago | 474 | 2.9 | 45,793 | 1,274 | 169 | 36 | 14 | 55 | 95 | 110 | 273 | 17 | 14 |
| Winneshiek | 1,038 | 4.9 | 51,904 | 687 | 445 | 91 | 36 | 140 | 229 | 183 | 712 | 44 | 36 |
| Woodbury | 4,653 | 4.3 | 45,132 | 1,346 | 2,332 | 412 | 181 | 382 | 726 | 960 | 3,308 | 212 | 181 |
| Worth | 330 | 7.3 | 44,675 | 1,397 | 96 | 19 | 7 | 37 | 61 | 72 | 160 | 11 | 7 |
| Wright | 694 | 0.5 | 55,258 | 484 | 267 | 57 | 21 | 130 | 126 | 142 | 475 | 28 | 21 |

1. Based on the resident population estimated as of July 1 of the year shown.

Items 62—74

# Table B. States and Counties — Earnings, Social Security, and Housing

| STATE County | Earnings, 2019 (cont.) Percent by selected industries | | | | | | | | | Social Security beneficiaries, December 2019 | | Supplemental Security Income recipients, 2019 | Housing units, 2020 | |
| | Farm | Mining, quarrying, and extractions | Construction | Manufacturing | Information; professional, scientific, technical services | Retail trade | Finance, insurance, real estate, and leasing | Health care and social assistance | Government | Number | Rate[1] | | Total | Percent change, 2010-2020 |
|---|---|---|---|---|---|---|---|---|---|---|---|---|---|---|
| | 75 | 76 | 77 | 78 | 79 | 80 | 81 | 82 | 83 | 84 | 85 | 86 | 87 | 88 |
| **IOWA—Cont'd** | | | | | | | | | | | | | | |
| Franklin | 26.6 | D | 4.6 | 19.5 | 2.1 | 3.5 | 4.2 | 4.2 | 13.4 | 2,450 | 243 | 113 | 4,842 | -1.1 |
| Fremont | 24.0 | 0.0 | 4.0 | 21.1 | D | 6.4 | 3.2 | D | 12.8 | 1,840 | 266 | 114 | 3,430 | 0.1 |
| Greene | 20.9 | D | 3.3 | 21.0 | D | 4.1 | 5.2 | D | 18.6 | 2,415 | 272 | 153 | 4,558 | 0.2 |
| Grundy | 14.4 | 0.0 | 18.6 | 5.8 | 3.1 | 3.5 | D | D | 12.8 | 3,000 | 245 | 99 | 5,602 | 1.3 |
| Guthrie | 18.2 | 0.0 | 8.9 | 9.0 | 5.2 | 2.7 | 11.0 | D | 21.2 | 2,990 | 278 | 143 | 5,843 | 1.5 |
| Hamilton | 28.2 | D | 4.2 | 14.9 | 3.0 | 4.3 | 5.7 | D | 15.0 | 3,580 | 242 | 182 | 7,234 | 0.2 |
| Hancock | 19.1 | D | 2.4 | 39.9 | 1.5 | 2.8 | 1.9 | D | 8.9 | 2,630 | 247 | 74 | 5,356 | 0.5 |
| Hardin | 20.8 | D | 7.7 | 6.8 | 3.3 | 5.6 | 4.6 | 6.4 | 18.5 | 4,330 | 258 | 251 | 8,163 | -0.7 |
| Harrison | 12.2 | 0.0 | 4.9 | 8.4 | 4.8 | 7.4 | 4.0 | D | 18.4 | 3,400 | 242 | 250 | 6,903 | 2.5 |
| Henry | 6.4 | D | 3.1 | 25.3 | D | 5.2 | 3.0 | D | 18.9 | 4,640 | 234 | 316 | 8,469 | 2.2 |
| Howard | 23.5 | 0.0 | 6.5 | 23.2 | 1.7 | 5.9 | 3.8 | D | 15.0 | 2,275 | 249 | 115 | 4,368 | 0.0 |
| Humboldt | 16.4 | D | 7.0 | 21.0 | 2.8 | 4.7 | 3.0 | D | 15.8 | 2,305 | 243 | 149 | 4,756 | 1.6 |
| Ida | 12.2 | 0.0 | 5.2 | 40.0 | 1.3 | 4.4 | 6.4 | D | 7.8 | 1,720 | 252 | 74 | 3,454 | 0.8 |
| Iowa | 8.1 | D | 5.6 | 47.5 | D | 5.6 | 1.8 | 4.2 | 10.3 | 3,750 | 232 | 167 | 7,376 | 1.6 |
| Jackson | 10.5 | D | 7.1 | 15.6 | 4.6 | 8.4 | 6.2 | D | 16.5 | 5,125 | 265 | 353 | 9,661 | 2.6 |
| Jasper | 7.2 | D | 8.3 | 22.3 | 4.9 | 6.7 | 3.9 | 9.0 | 18.4 | 8,785 | 236 | 564 | 16,446 | 1.6 |
| Jefferson | 3.1 | D | 4.7 | 13.4 | 7.8 | 6.9 | 18.9 | D | 16.6 | 4,415 | 243 | 350 | 7,745 | 2.0 |
| Johnson | 0.6 | 0.1 | 4.4 | 5.9 | 4.4 | 5.3 | 4.5 | 7.3 | 48.2 | 20,765 | 136 | 1,826 | 67,364 | 20.4 |
| Jones | 10.6 | D | 11.2 | 12.9 | 3.7 | 6.5 | 4.4 | 9.3 | 19.8 | 4,960 | 240 | 301 | 9,008 | 1.1 |
| Keokuk | 27.8 | D | 5.2 | 5.6 | 3.0 | 2.9 | 3.8 | D | 16.0 | 2,595 | 255 | 209 | 4,866 | -1.3 |
| Kossuth | 23.4 | 0.0 | 5.5 | 17.5 | 4.2 | 4.3 | 9.5 | D | 12.2 | 4,135 | 280 | 168 | 7,517 | 0.4 |
| Lee | 2.2 | D | 7.6 | 34.3 | 2.9 | 5.5 | 3.3 | 10.5 | 13.8 | 8,845 | 263 | 885 | 16,214 | 0.1 |
| Linn | 0.3 | 0.1 | 6.9 | 20.7 | 9.5 | 5.1 | 10.1 | 11.4 | 10.5 | 43,860 | 193 | 4,030 | 99,471 | 7.8 |
| Louisa | 21.7 | D | 2.6 | 32.4 | D | 1.8 | 2.6 | 4.2 | 14.3 | 2,480 | 224 | 168 | 5,073 | 1.4 |
| Lucas | 2.2 | 0.0 | D | 7.4 | 1.9 | 5.0 | D | D | 17.6 | 2,210 | 258 | 180 | 4,205 | -0.8 |
| Lyon | 46.1 | D | 5.2 | 8.0 | 5.7 | 2.4 | 3.3 | 3.8 | 7.7 | 2,260 | 191 | 57 | 5,126 | 5.7 |
| Madison | 5.2 | D | 15.6 | 4.1 | 5.5 | 7.0 | 6.3 | 6.3 | 24.9 | 3,400 | 209 | 143 | 7,287 | 11.2 |
| Mahaska | 10.9 | 0.0 | 4.8 | 26.7 | 3.9 | 6.0 | 3.0 | 5.5 | 17.6 | 4,880 | 218 | 429 | 9,909 | 1.4 |
| Marion | 1.3 | D | 4.0 | 50.2 | 3.3 | 4.0 | 3.6 | 10.2 | 8.9 | 7,540 | 227 | 425 | 14,469 | 4.0 |
| Marshall | 4.5 | D | 4.9 | 29.4 | D | 6.0 | 3.1 | 10.7 | 19.2 | 9,040 | 229 | 638 | 16,901 | 0.4 |
| Mills | 7.9 | D | 4.2 | 18.3 | D | 11.9 | 3.2 | D | 26.6 | 3,430 | 227 | 197 | 6,066 | -0.7 |
| Mitchell | 10.3 | D | 19.5 | 34.5 | D | 1.9 | 2.4 | D | 8.8 | 2,540 | 240 | 100 | 5,041 | 3.9 |
| Monona | 23.2 | 0.0 | 3.8 | 3.2 | 1.9 | 6.1 | 4.7 | 18.3 | 16.5 | 2,515 | 292 | 162 | 4,728 | 0.7 |
| Monroe | 5.3 | D | 7.4 | 41.9 | D | 2.7 | D | 3.5 | 14.9 | 1,885 | 245 | 110 | 4,031 | 3.8 |
| Montgomery | 7.2 | D | 7.2 | 16.1 | D | 5.5 | 4.8 | D | 24.7 | 2,885 | 291 | 276 | 5,226 | -0.2 |
| Muscatine | 1.7 | D | 3.7 | 40.3 | 5.4 | 4.0 | 2.9 | 6.3 | 11.0 | 9,170 | 215 | 858 | 18,296 | 2.0 |
| O'Brien | 27.2 | D | 3.4 | 9.5 | 4.0 | 4.7 | 5.1 | 11.9 | 11.7 | 3,425 | 248 | 197 | 6,718 | 1.0 |
| Osceola | 40.2 | 0.2 | 4.0 | 13.0 | 2.1 | 1.7 | D | 6.2 | 7.8 | 1,465 | 246 | 51 | 2,955 | -1.2 |
| Page | 10.7 | D | 4.1 | 16.8 | D | 6.1 | 3.9 | D | 23.4 | 4,240 | 280 | 351 | 7,183 | 0.0 |
| Palo Alto | 26.4 | 0.3 | 3.7 | 21.4 | 2.5 | 3.5 | 3.4 | D | 16.1 | 2,190 | 247 | 123 | 4,593 | -0.8 |
| Plymouth | 17.0 | D | 4.6 | 26.3 | D | 3.7 | 3.6 | D | 10.3 | 5,470 | 217 | 197 | 11,087 | 5.1 |
| Pocahontas | 31.0 | D | 19.4 | 9.2 | 2.1 | 3.3 | D | D | 10.7 | 1,865 | 281 | 102 | 3,741 | -1.4 |
| Polk | 0.1 | 0.1 | 7.4 | 5.7 | 11.2 | 5.3 | 19.9 | 10.7 | 12.9 | 79,300 | 161 | 8,074 | 211,332 | 16.0 |
| Pottawattamie | 2.9 | 0.2 | 11.4 | 15.0 | 5.1 | 7.1 | 3.3 | 12.2 | 15.0 | 20,265 | 217 | 2,166 | 40,213 | 2.3 |
| Poweshiek | 5.4 | D | 8.8 | 14.9 | 2.4 | 4.4 | 14.9 | D | 8.4 | 4,390 | 238 | 213 | 9,209 | 2.9 |
| Ringgold | 34.5 | D | 6.0 | D | D | 3.7 | 2.8 | 6.0 | 25.3 | 1,335 | 274 | 84 | 2,606 | -0.2 |
| Sac | 32.8 | D | 3.1 | 16.7 | 3.0 | 2.7 | D | 7.1 | 10.7 | 2,610 | 270 | 93 | 5,396 | -0.6 |
| Scott | 0.6 | D | 9.8 | 17.4 | 6.4 | 7.4 | 5.8 | 13.8 | 10.7 | 34,755 | 200 | 4,024 | 75,646 | 5.3 |
| Shelby | 17.2 | 0.0 | 3.0 | 10.3 | 6.8 | 4.2 | 6.1 | D | 16.0 | 3,160 | 275 | 189 | 5,636 | 1.7 |
| Sioux | 22.4 | 0.2 | 7.2 | 23.1 | 3.7 | 3.6 | 4.8 | 5.8 | 8.9 | 6,055 | 173 | 175 | 13,232 | 7.8 |
| Story | 1.0 | D | 6.1 | 11.4 | 9.1 | 4.8 | 4.3 | 7.9 | 38.6 | 13,620 | 140 | 716 | 41,845 | 13.7 |
| Tama | 14.3 | D | 5.5 | 20.4 | D | 4.1 | 3.7 | D | 29.2 | 4,135 | 245 | 200 | 7,782 | 0.2 |
| Taylor | 22.5 | 0.0 | 6.2 | 22.3 | D | 3.0 | 3.0 | 4.6 | 14.6 | 1,615 | 261 | 99 | 3,109 | 0.1 |
| Union | 4.3 | D | 5.3 | 20.7 | D | 6.8 | 4.0 | 6.4 | 27.4 | 3,075 | 252 | 258 | 5,939 | 0.0 |
| Van Buren | 12.5 | 0.6 | 9.7 | 23.8 | D | 3.3 | 3.5 | D | 21.9 | 1,940 | 275 | 129 | 3,650 | -0.5 |
| Wapello | 1.0 | D | 6.4 | 27.1 | 2.5 | 7.4 | 3.6 | 14.2 | 16.7 | 8,290 | 236 | 1,120 | 16,082 | -0.1 |
| Warren | 1.7 | 0.0 | 16.4 | 7.2 | 7.9 | 8.5 | 5.9 | 9.6 | 20.7 | 9,805 | 190 | 431 | 20,995 | 14.3 |
| Washington | 23.4 | D | 10.8 | 14.1 | 3.0 | 4.7 | 2.8 | 4.8 | 12.2 | 5,205 | 236 | 343 | 9,868 | 3.7 |
| Wayne | 8.9 | 0.0 | 16.9 | 20.1 | 3.0 | 4.2 | D | 4.3 | 25.4 | 1,675 | 261 | 129 | 3,191 | -0.7 |
| Webster | 6.8 | D | 9.7 | 19.6 | 3.2 | 6.1 | 3.1 | 12.7 | 15.1 | 8,575 | 238 | 800 | 17,069 | 0.2 |
| Winnebago | 13.1 | D | 6.9 | 14.7 | 5.4 | 8.4 | 5.8 | D | 15.9 | 2,660 | 257 | 131 | 5,191 | -0.1 |
| Winneshiek | 7.3 | D | 8.1 | 17.4 | D | 5.8 | 3.9 | D | 17.1 | 4,745 | 237 | 141 | 9,098 | 4.3 |
| Woodbury | 2.2 | D | 9.4 | 14.6 | 4.7 | 7.7 | 4.2 | 15.6 | 14.8 | 19,870 | 193 | 1,990 | 42,830 | 3.2 |
| Worth | 11.1 | 1.0 | 10.7 | 19.3 | D | 2.7 | D | 3.9 | 14.0 | 1,805 | 244 | 80 | 3,507 | -1.1 |
| Wright | 13.7 | D | 3.6 | 29.8 | 3.6 | 2.8 | 3.2 | 4.7 | 21.5 | 3,170 | 253 | 174 | 6,520 | -0.1 |

1. Per 1,000 resident population estimated as of July 1 of the year shown.

# Table B. States and Counties — Housing, Labor Force, and Employment

| STATE County | Housing units, 2015-2019 | | | | | | | | Civilian labor force, 2020 | | | | Civilian employment[6], 2015-2019 | | |
|---|---|---|---|---|---|---|---|---|---|---|---|---|---|---|---|
| | Occupied units | | | | | | | Sub-standard units[4] (percent) | Total | Percent change, 2019-2020 | Unemployment | | Total | Percent | |
| | | | Owner-occupied | | | Renter-occupied | | | | | | | | | Management, business, science, and arts | Construction, production, and maintenance occupations |
| | | | Median value[1] | Median owner cost as a percent of income | | Median rent[3] | Median rent as a percent of income[2] | | | | Total | Rate[5] | | | |
| | Total | Percent | | With a mortgage | Without a mortgage[2] | | | | | | | | | | |
| | 89 | 90 | 91 | 92 | 93 | 94 | 95 | 96 | 97 | 98 | 99 | 100 | 101 | 102 | 103 |
| IOWA—Cont'd | | | | | | | | | | | | | | | |
| Franklin | 4,192 | 74.7 | 94,100 | 17.1 | 10.0 | 699 | 20.9 | 1.3 | 5,597 | -2.9 | 212 | 3.8 | 4,965 | 32.1 | 32.6 |
| Fremont | 2,945 | 76.3 | 109,300 | 17.7 | 10.9 | 661 | 20.5 | 1.8 | 3,712 | -1.1 | 115 | 3.1 | 3,262 | 34.0 | 29.5 |
| Greene | 3,971 | 73.4 | 94,700 | 18.5 | 10.2 | 638 | 23.5 | 2.9 | 5,333 | -2.6 | 226 | 4.2 | 4,540 | 37.6 | 30.7 |
| Grundy | 5,155 | 81.6 | 130,200 | 17.2 | 10.0 | 633 | 22.2 | 1.1 | 6,273 | -5.3 | 274 | 4.4 | 6,219 | 37.3 | 28.2 |
| Guthrie | 4,452 | 79.6 | 117,900 | 19.5 | 11.9 | 682 | 25.5 | 2.3 | 5,409 | -4.1 | 269 | 5.0 | 5,326 | 33.1 | 29.3 |
| Hamilton | 6,259 | 72.4 | 106,100 | 16.9 | 10.6 | 709 | 22.7 | 1.2 | 6,779 | -2.9 | 299 | 4.4 | 7,348 | 32.8 | 31.9 |
| Hancock | 4,790 | 77.1 | 102,200 | 17.1 | 10.0 | 642 | 24.7 | 0.8 | 5,764 | -3.4 | 269 | 4.7 | 5,647 | 37.1 | 30.7 |
| Hardin | 7,148 | 74.9 | 94,300 | 16.9 | 11.8 | 639 | 24.3 | 1.6 | 7,767 | -5.3 | 360 | 4.6 | 8,491 | 33.4 | 32.3 |
| Harrison | 6,019 | 75.7 | 120,600 | 16.8 | 10.0 | 715 | 24.9 | 1.1 | 7,060 | -4.2 | 284 | 4.0 | 7,292 | 34.8 | 30.8 |
| Henry | 7,743 | 71.2 | 115,500 | 18.8 | 11.2 | 694 | 26.4 | 0.8 | 9,564 | -3.5 | 504 | 5.3 | 9,264 | 28.1 | 35.2 |
| Howard | 3,810 | 79.7 | 107,100 | 18.1 | 10.9 | 622 | 22.2 | 0.8 | 5,070 | -2.9 | 226 | 4.5 | 4,728 | 30.9 | 37.9 |
| Humboldt | 4,174 | 74.2 | 103,800 | 16.7 | 10.7 | 568 | 23.9 | 0.8 | 4,939 | -4.6 | 209 | 4.2 | 4,656 | 32.4 | 32.7 |
| Ida | 3,011 | 75.6 | 100,900 | 14.7 | 10.0 | 532 | 20.9 | 1.0 | 3,966 | -4.9 | 142 | 3.6 | 3,452 | 34.6 | 31.4 |
| Iowa | 6,780 | 80.4 | 152,500 | 19.9 | 12.2 | 580 | 21.8 | 1.3 | 10,009 | -6.0 | 444 | 4.4 | 8,516 | 31.5 | 32.0 |
| Jackson | 8,239 | 79.0 | 128,200 | 17.7 | 10.6 | 624 | 29.0 | 0.9 | 10,622 | -4.5 | 647 | 6.1 | 9,547 | 31.7 | 34.9 |
| Jasper | 14,574 | 74.5 | 134,200 | 18.6 | 12.3 | 730 | 24.7 | 1.7 | 19,177 | -4.1 | 1,020 | 5.3 | 18,027 | 28.3 | 31.0 |
| Jefferson | 6,754 | 66.3 | 118,300 | 19.6 | 12.2 | 685 | 27.4 | 0.4 | 9,487 | -6.6 | 539 | 5.7 | 9,052 | 35.2 | 28.1 |
| Johnson | 59,134 | 58.5 | 227,600 | 19.4 | 10.0 | 972 | 34.7 | 2.6 | 84,088 | -4.2 | 4,032 | 4.8 | 84,129 | 47.6 | 16.3 |
| Jones | 8,130 | 78.3 | 138,400 | 19.6 | 11.0 | 641 | 25.8 | 1.7 | 10,253 | -5.0 | 584 | 5.7 | 9,850 | 31.4 | 31.9 |
| Keokuk | 4,394 | 78.4 | 90,400 | 18.2 | 11.9 | 679 | 28.3 | 1.5 | 5,151 | -2.8 | 254 | 4.9 | 4,856 | 32.6 | 36.4 |
| Kossuth | 6,646 | 75.7 | 102,100 | 17.4 | 10.4 | 679 | 24.4 | 1.0 | 8,087 | -5.2 | 308 | 3.8 | 7,322 | 31.8 | 32.7 |
| Lee | 14,233 | 74.4 | 94,500 | 19.0 | 11.0 | 670 | 27.1 | 1.2 | 15,693 | -3.6 | 1,027 | 6.5 | 15,554 | 27.3 | 36.3 |
| Linn | 90,877 | 73.8 | 157,000 | 19.4 | 11.6 | 761 | 26.8 | 1.7 | 118,853 | -4.2 | 7,553 | 6.4 | 118,971 | 38.8 | 23.5 |
| Louisa | 4,335 | 76.9 | 108,800 | 18.7 | 11.7 | 637 | 18.4 | 2.9 | 5,916 | -3.2 | 293 | 5.0 | 5,621 | 25.4 | 41.4 |
| Lucas | 3,701 | 78.1 | 96,400 | 18.6 | 11.3 | 664 | 25.7 | 1.5 | 4,644 | -1.1 | 169 | 3.6 | 4,113 | 27.4 | 35.8 |
| Lyon | 4,548 | 85.4 | 146,900 | 19.5 | 10.0 | 696 | 21.0 | 1.0 | 6,823 | -4.4 | 171 | 2.5 | 6,076 | 37.9 | 26.5 |
| Madison | 6,462 | 78.4 | 177,500 | 21.2 | 13.5 | 789 | 21.6 | 0.3 | 8,347 | -3.9 | 479 | 5.7 | 8,446 | 37.4 | 29.0 |
| Mahaska | 8,927 | 69.0 | 115,300 | 18.3 | 10.4 | 652 | 22.7 | 1.5 | 11,798 | -1.8 | 561 | 4.8 | 11,050 | 34.0 | 30.4 |
| Marion | 13,365 | 70.8 | 157,300 | 18.7 | 11.7 | 721 | 22.6 | 2.2 | 17,933 | -3.3 | 751 | 4.2 | 16,911 | 36.1 | 29.4 |
| Marshall | 15,419 | 70.5 | 106,000 | 18.3 | 11.8 | 689 | 25.2 | 3.6 | 17,514 | -3.4 | 1,236 | 7.1 | 18,901 | 26.9 | 37.9 |
| Mills | 5,554 | 77.9 | 168,900 | 19.1 | 10.0 | 826 | 28.8 | 1.7 | 7,150 | -4.1 | 281 | 3.9 | 7,404 | 33.1 | 28.5 |
| Mitchell | 4,355 | 80.8 | 119,400 | 18.7 | 10.0 | 621 | 20.5 | 1.4 | 6,061 | -11.1 | 205 | 3.4 | 5,392 | 32.6 | 30.6 |
| Monona | 4,000 | 72.4 | 86,900 | 17.1 | 10.2 | 613 | 28.5 | 0.8 | 4,347 | -3.2 | 208 | 4.8 | 4,314 | 26.8 | 31.2 |
| Monroe | 3,294 | 79.9 | 100,200 | 17.1 | 12.4 | 566 | 27.7 | 1.2 | 3,938 | -3.1 | 204 | 5.2 | 3,785 | 31.0 | 34.2 |
| Montgomery | 4,554 | 70.2 | 85,700 | 18.4 | 10.5 | 644 | 24.1 | 1.3 | 4,847 | -3.9 | 226 | 4.7 | 4,867 | 31.8 | 30.1 |
| Muscatine | 16,660 | 73.4 | 130,500 | 19.0 | 11.9 | 831 | 24.3 | 2.5 | 21,053 | -4.6 | 1,257 | 6.0 | 21,555 | 29.7 | 38.7 |
| O'Brien | 6,169 | 71.7 | 107,000 | 18.5 | 10.0 | 585 | 25.4 | 1.4 | 8,058 | -3.3 | 272 | 3.4 | 7,224 | 28.5 | 36.0 |
| Osceola | 2,633 | 78.4 | 93,700 | 18.0 | 10.0 | 639 | 22.2 | 1.7 | 3,498 | -1.8 | 101 | 2.9 | 3,109 | 32.3 | 35.9 |
| Page | 6,280 | 71.5 | 94,400 | 17.3 | 11.4 | 663 | 25.0 | 1.2 | 6,186 | -4.1 | 308 | 5.0 | 6,982 | 34.3 | 31.2 |
| Palo Alto | 3,702 | 72.3 | 102,200 | 16.9 | 11.8 | 580 | 20.6 | 0.8 | 4,540 | -6.4 | 210 | 4.6 | 4,581 | 35.5 | 27.5 |
| Plymouth | 10,250 | 77.5 | 163,500 | 17.6 | 10.0 | 660 | 19.7 | 1.0 | 14,284 | -4.8 | 514 | 3.6 | 13,367 | 32.8 | 29.7 |
| Pocahontas | 3,215 | 77.3 | 76,000 | 14.6 | 10.6 | 597 | 21.8 | 0.6 | 4,220 | -7.2 | 148 | 3.5 | 3,322 | 30.1 | 37.0 |
| Polk | 187,798 | 67.1 | 179,900 | 19.3 | 11.6 | 910 | 27.2 | 2.8 | 264,807 | -2.8 | 15,876 | 6.0 | 257,032 | 40.9 | 19.8 |
| Pottawattamie | 36,799 | 68.4 | 139,800 | 19.0 | 11.6 | 842 | 27.7 | 2.4 | 47,249 | -2.8 | 2,478 | 5.2 | 47,210 | 32.2 | 28.1 |
| Poweshiek | 7,767 | 68.6 | 141,100 | 20.0 | 12.4 | 719 | 24.3 | 1.3 | 9,727 | -7.1 | 480 | 4.9 | 9,602 | 33.7 | 28.8 |
| Ringgold | 1,971 | 77.2 | 98,700 | 17.3 | 12.6 | 667 | 27.6 | 1.5 | 2,377 | -4.3 | 92 | 3.9 | 2,203 | 36.4 | 29.6 |
| Sac | 4,313 | 79.4 | 95,800 | 16.7 | 10.0 | 625 | 20.8 | 1.4 | 5,167 | -3.5 | 181 | 3.5 | 5,024 | 30.3 | 31.3 |
| Scott | 67,021 | 69.0 | 163,200 | 18.8 | 11.4 | 793 | 29.2 | 1.4 | 86,403 | -4.4 | 5,862 | 6.8 | 84,565 | 37.6 | 24.2 |
| Shelby | 5,030 | 77.5 | 121,200 | 18.6 | 10.7 | 643 | 27.1 | 0.9 | 6,318 | -6.2 | 228 | 3.6 | 6,197 | 37.8 | 27.6 |
| Sioux | 12,248 | 80.3 | 167,500 | 18.3 | 10.0 | 726 | 24.0 | 2.1 | 20,757 | -2.6 | 558 | 2.7 | 18,732 | 38.8 | 28.6 |
| Story | 37,811 | 53.5 | 185,800 | 18.6 | 10.6 | 905 | 33.3 | 0.9 | 56,458 | -4.1 | 2,036 | 3.6 | 51,453 | 46.9 | 17.1 |
| Tama | 6,767 | 75.8 | 110,400 | 19.5 | 11.8 | 722 | 24.2 | 2.6 | 9,225 | -4.0 | 472 | 5.1 | 8,459 | 26.3 | 36.6 |
| Taylor | 2,616 | 78.2 | 80,600 | 17.9 | 11.3 | 568 | 21.7 | 1.2 | 3,064 | -4.8 | 112 | 3.7 | 2,982 | 32.9 | 36.4 |
| Union | 5,176 | 70.4 | 104,000 | 17.3 | 13.6 | 609 | 25.9 | 3.0 | 6,147 | -3.3 | 361 | 5.9 | 6,143 | 30.2 | 33.4 |
| Van Buren | 2,894 | 81.3 | 92,800 | 21.5 | 11.5 | 571 | 24.8 | 4.9 | 3,667 | -1.1 | 170 | 4.6 | 3,051 | 30.4 | 36.6 |
| Wapello | 14,481 | 71.5 | 82,100 | 19.4 | 12.3 | 718 | 29.2 | 3.0 | 17,328 | -3.4 | 1,035 | 6.0 | 16,377 | 26.5 | 37.2 |
| Warren | 19,258 | 79.0 | 187,200 | 18.4 | 11.4 | 807 | 28.1 | 1.7 | 27,577 | -4.0 | 1,256 | 4.6 | 26,361 | 41.1 | 22.2 |
| Washington | 8,832 | 71.0 | 151,600 | 19.6 | 10.0 | 754 | 23.9 | 0.8 | 11,490 | -5.3 | 469 | 4.1 | 11,307 | 30.6 | 31.9 |
| Wayne | 2,689 | 81.1 | 81,100 | 23.0 | 10.1 | 524 | 22.0 | 4.9 | 2,783 | -2.3 | 125 | 4.5 | 2,799 | 29.4 | 41.8 |
| Webster | 15,306 | 68.0 | 99,100 | 18.8 | 11.7 | 642 | 25.0 | 1.5 | 18,615 | -5.2 | 977 | 5.2 | 16,772 | 30.9 | 29.1 |
| Winnebago | 4,577 | 76.0 | 94,800 | 18.2 | 10.5 | 620 | 24.3 | 0.9 | 5,025 | -2.9 | 322 | 6.4 | 5,477 | 31.2 | 37.6 |
| Winneshiek | 8,257 | 79.3 | 177,300 | 20.4 | 10.3 | 702 | 27.8 | 0.6 | 11,095 | -7.9 | 545 | 4.9 | 11,678 | 34.3 | 29.9 |
| Woodbury | 39,016 | 66.4 | 123,300 | 17.6 | 10.1 | 787 | 26.4 | 2.7 | 54,467 | -3.6 | 2,861 | 5.3 | 51,942 | 30.8 | 31.1 |
| Worth | 3,172 | 77.6 | 106,700 | 15.8 | 10.3 | 591 | 21.3 | 1.7 | 4,082 | -3.7 | 221 | 5.4 | 3,889 | 28.7 | 34.0 |
| Wright | 5,718 | 71.7 | 85,500 | 18.2 | 10.0 | 626 | 24.1 | 1.0 | 6,960 | -1.2 | 276 | 4.0 | 5,738 | 30.8 | 36.6 |

1. Specified owner-occupied units.   2. A value of 10.0 represents 10 percent or less; a value of 50.0 represents 50 percent or more.   3. Specified renter-occupied units.   4. Overcrowded or lacking complete plumbing facilities.   5. Percent of civilian labor force.   6. Civilian employed persons 16 years old and over.

| STATE County | Private nonfarm establishments, employment and payroll, 2019 | | | | | | | | | Agriculture, 2017 | | | |
|---|---|---|---|---|---|---|---|---|---|---|---|---|---|
| | Number of establish-ments | Employment | | | | | | Annual payroll | | Farms | | | Farm producers whose primary occupation is farming (percent) |
| | | Total | Health care and social assistance | Manufac-turing | Retail trade | Finance and insurance | Professional, scientific, and technical services | Total (mil dol) | Average per employee (dollars) | Number | Percent with: | | |
| | | | | | | | | | | | Fewer than 50 acres | 1000 acres or more | |
| | 104 | 105 | 106 | 107 | 108 | 109 | 110 | 111 | 112 | 113 | 114 | 115 | 116 |

**IOWA—Cont'd**

| STATE County | 104 | 105 | 106 | 107 | 108 | 109 | 110 | 111 | 112 | 113 | 114 | 115 | 116 |
|---|---|---|---|---|---|---|---|---|---|---|---|---|---|
| Franklin | 295 | 3,232 | 494 | 977 | 359 | 124 | 59 | 132 | 40,731 | 835 | 35.2 | 12.5 | 44.1 |
| Fremont | 190 | 1,650 | 284 | 216 | 575 | 56 | 41 | 64 | 38,945 | 527 | 25.4 | 17.6 | 45.4 |
| Greene | 259 | 3,046 | 654 | 457 | 488 | 141 | 216 | 140 | 46,098 | 700 | 25.7 | 18.3 | 50.8 |
| Grundy | 297 | 2,957 | 601 | 396 | 434 | 201 | 34 | 152 | 51,533 | 760 | 36.2 | 9.6 | 45.5 |
| Guthrie | 311 | 2,215 | 486 | 191 | 390 | 144 | 79 | 79 | 35,527 | 802 | 25.8 | 10.8 | 37.0 |
| Hamilton | 382 | 4,448 | 692 | 700 | 756 | 189 | 64 | 187 | 42,051 | 732 | 39.8 | 13.3 | 44.3 |
| Hancock | 319 | 4,082 | 759 | 1,243 | 401 | 99 | 578 | 164 | 40,068 | 801 | 30.5 | 13.5 | 48.0 |
| Hardin | 546 | 5,046 | 977 | 606 | 987 | 274 | 113 | 194 | 38,506 | 837 | 35.4 | 14.7 | 50.1 |
| Harrison | 355 | 2,990 | 703 | 307 | 464 | 145 | 55 | 115 | 38,412 | 794 | 29.5 | 16.4 | 48.8 |
| Henry | 483 | 7,745 | 1,065 | 2,485 | 931 | 150 | 159 | 300 | 38,785 | 908 | 30.1 | 5.9 | 34.6 |
| Howard | 262 | 3,080 | 531 | 1,155 | 447 | 128 | 43 | 106 | 34,401 | 879 | 32.5 | 7.7 | 46.2 |
| Humboldt | 291 | 3,409 | 406 | 911 | 538 | 101 | 72 | 141 | 41,259 | 572 | 31.1 | 14.3 | 51.4 |
| Ida | 256 | 3,337 | 440 | 1,456 | 311 | 186 | 32 | 161 | 48,177 | 525 | 31.2 | 17.1 | 49.8 |
| Iowa | 488 | 8,688 | 730 | 4,707 | 1,270 | 99 | 224 | 353 | 40,627 | 970 | 32.0 | 7.8 | 41.4 |
| Jackson | 530 | 4,976 | 669 | 929 | 959 | 239 | 113 | 170 | 34,171 | 1,107 | 30.3 | 5.7 | 45.4 |
| Jasper | 785 | 9,158 | 1,530 | 2,073 | 1,456 | 243 | 540 | 348 | 38,015 | 986 | 34.3 | 10.0 | 47.0 |
| Jefferson | 661 | 7,100 | 716 | 801 | 999 | 1,123 | 433 | 292 | 41,058 | 636 | 31.4 | 6.3 | 38.1 |
| Johnson | 3,337 | 66,359 | 18,818 | 4,568 | 9,797 | 2,772 | 2,781 | 2,982 | 44,932 | 1,257 | 38.0 | 5.3 | 46.5 |
| Jones | 514 | 4,839 | 861 | 1,205 | 706 | 196 | 124 | 167 | 34,450 | 1,110 | 34.5 | 8.6 | 45.1 |
| Keokuk | 234 | 1,643 | 321 | 152 | 214 | 96 | 51 | 63 | 38,226 | 927 | 25.2 | 9.6 | 47.3 |
| Kossuth | 566 | 5,820 | 675 | 1,589 | 913 | 520 | 247 | 248 | 42,580 | 1,347 | 25.2 | 13.4 | 53.4 |
| Lee | 853 | 14,524 | 2,151 | 3,877 | 2,071 | 357 | 1,916 | 630 | 43,381 | 1,374 | 44.0 | 5.9 | 38.2 |
| Linn | 5,688 | 120,891 | 16,292 | 16,891 | 15,678 | 8,570 | 8,677 | 6,435 | 53,226 | 1,374 | 32.5 | 11.5 | 40.7 |
| Louisa | 213 | 2,994 | 350 | 1,465 | 203 | 52 | 31 | 129 | 43,168 | 576 | 32.5 | 11.5 | 40.7 |
| Lucas | 189 | 4,081 | 476 | 141 | 580 | 90 | 40 | 161 | 39,356 | 567 | 25.7 | 7.9 | 38.0 |
| Lyon | 398 | 3,682 | 544 | 834 | 374 | 146 | 132 | 161 | 43,690 | 1,122 | 26.1 | 7.4 | 55.0 |
| Madison | 431 | 2,792 | 520 | 163 | 598 | 138 | 108 | 103 | 36,931 | 977 | 40.4 | 7.6 | 34.0 |
| Mahaska | 553 | 6,592 | 968 | 1,359 | 1,120 | 183 | 141 | 242 | 36,730 | 943 | 31.0 | 9.2 | 43.9 |
| Marion | 862 | 16,886 | 2,066 | 8,221 | 1,637 | 303 | 357 | 725 | 42,922 | 1,030 | 39.1 | 6.5 | 35.2 |
| Marshall | 784 | 12,631 | 1,809 | 4,148 | 1,979 | 351 | 275 | 563 | 44,565 | 886 | 38.6 | 10.7 | 43.6 |
| Mills | 300 | 2,303 | 549 | 57 | 361 | 90 | 87 | 81 | 35,172 | 520 | 38.8 | 13.1 | 41.6 |
| Mitchell | 314 | 3,281 | 613 | 986 | 420 | 119 | 77 | 127 | 38,804 | 789 | 29.0 | 10.6 | 52.6 |
| Monona | 221 | 1,914 | 588 | 85 | 367 | 112 | 28 | 65 | 34,134 | 619 | 20.8 | 17.6 | 53.3 |
| Monroe | 175 | 2,236 | 400 | 891 | 295 | 80 | 47 | 121 | 54,083 | 618 | 24.6 | 6.1 | 30.3 |
| Montgomery | 292 | 3,306 | 769 | 605 | 568 | 109 | 65 | 124 | 37,529 | 516 | 20.0 | 14.0 | 50.7 |
| Muscatine | 934 | 20,619 | 1,963 | 7,411 | 2,099 | 393 | 808 | 1,034 | 50,153 | 714 | 33.3 | 8.0 | 47.5 |
| O'Brien | 499 | 5,400 | 1,396 | 727 | 806 | 219 | 189 | 191 | 35,429 | 876 | 23.2 | 7.3 | 58.5 |
| Osceola | 191 | 1,607 | 244 | 254 | 136 | 89 | 26 | 65 | 40,459 | 591 | 26.7 | 11.0 | 55.7 |
| Page | 397 | 5,290 | 1,542 | 1,489 | 607 | 174 | 111 | 203 | 38,299 | 715 | 22.7 | 15.8 | 51.3 |
| Palo Alto | 292 | 2,987 | 768 | 680 | 375 | 123 | 36 | 108 | 36,140 | 785 | 28.9 | 13.6 | 45.1 |
| Plymouth | 676 | 10,482 | 1,178 | 3,222 | 1,092 | 343 | 292 | 514 | 49,021 | 1,219 | 24.6 | 10.7 | 54.3 |
| Pocahontas | 217 | 1,800 | 282 | 477 | 242 | 94 | 43 | 71 | 39,415 | 730 | 25.2 | 14.2 | 55.6 |
| Polk | 13,306 | 276,067 | 37,404 | 16,790 | 32,468 | 32,140 | 19,209 | 15,119 | 54,765 | 755 | 56.6 | 9.1 | 38.5 |
| Pottawattamie | 1,978 | 30,984 | 4,829 | 5,339 | 5,779 | 646 | 689 | 1,213 | 39,156 | 1,114 | 30.8 | 15.3 | 51.4 |
| Poweshiek | 565 | 9,372 | 1,108 | 1,289 | 1,437 | 949 | 136 | 398 | 42,476 | 852 | 29.5 | 10.3 | 47.8 |
| Ringgold | 143 | 1,098 | 346 | 37 | 228 | 37 | 53 | 36 | 32,447 | 675 | 12.9 | 11.0 | 41.9 |
| Sac | 334 | 2,212 | 511 | 435 | 339 | 119 | 37 | 84 | 38,049 | 889 | 31.3 | 12.9 | 47.5 |
| Scott | 4,505 | 84,229 | 12,963 | 12,295 | 12,543 | 2,866 | 3,077 | 3,786 | 44,949 | 684 | 35.4 | 8.8 | 46.6 |
| Shelby | 384 | 4,939 | 823 | 844 | 520 | 215 | 111 | 216 | 43,784 | 890 | 27.2 | 12.0 | 48.3 |
| Sioux | 1,323 | 18,918 | 2,692 | 5,508 | 1,617 | 644 | 724 | 806 | 42,609 | 1,724 | 35.6 | 4.9 | 48.6 |
| Story | 2,108 | 31,737 | 5,674 | 4,859 | 5,033 | 688 | 1,114 | 1,436 | 45,253 | 955 | 48.2 | 10.8 | 37.8 |
| Tama | 312 | 4,109 | 514 | 1,115 | 487 | 120 | 57 | 149 | 36,145 | 1,072 | 27.1 | 10.0 | 47.6 |
| Taylor | 140 | 1,803 | 286 | 797 | 153 | 28 | 53 | 82 | 45,204 | 667 | 21.9 | 12.0 | 41.2 |
| Union | 316 | 4,740 | 1,001 | 1,055 | 759 | 114 | 66 | 183 | 38,615 | 627 | 31.4 | 10.8 | 40.6 |
| Van Buren | 146 | 1,490 | 270 | 584 | 247 | 54 | 36 | 54 | 36,282 | 690 | 25.7 | 9.4 | 39.1 |
| Wapello | 737 | 13,912 | 2,551 | 3,081 | 2,276 | 373 | 217 | 549 | 39,465 | 715 | 33.7 | 6.9 | 38.1 |
| Warren | 917 | 8,997 | 1,212 | 724 | 1,837 | 427 | 263 | 332 | 36,859 | 1,214 | 45.3 | 5.7 | 31.0 |
| Washington | 670 | 6,537 | 1,340 | 581 | 1,126 | 219 | 176 | 229 | 34,978 | 1,129 | 31.0 | 4.3 | 46.2 |
| Wayne | 147 | 1,427 | 368 | 412 | 254 | 31 | 51 | 53 | 36,917 | 743 | 24.1 | 8.7 | 39.0 |
| Webster | 1,015 | 16,849 | 2,814 | 1,990 | 2,591 | 383 | 763 | 718 | 42,621 | 960 | 35.7 | 14.8 | 47.4 |
| Winnebago | 316 | 6,530 | 380 | 3,282 | 646 | 176 | 146 | 263 | 40,352 | 571 | 31.5 | 12.3 | 41.6 |
| Winneshiek | 620 | 9,266 | 1,460 | 1,551 | 1,159 | 269 | 234 | 350 | 37,819 | 1,458 | 30.1 | 5.5 | 46.7 |
| Woodbury | 2,626 | 48,232 | 9,873 | 7,975 | 7,388 | 1,184 | 1,094 | 1,969 | 40,821 | 1,037 | 30.9 | 13.1 | 42.5 |
| Worth | 188 | 1,831 | 207 | 484 | 170 | 56 | 15 | 63 | 34,532 | 582 | 33.5 | 15.6 | 48.3 |
| Wright | 366 | 4,249 | 829 | 1,511 | 509 | 135 | 147 | 196 | 46,199 | 735 | 35.9 | 16.9 | 50.6 |

# Table B. States and Counties — **Agriculture**

| STATE County | Acreage (1,000) [117] | Percent change, 2012-2017 [118] | Average size of farm [119] | Total irrigated (1,000) [120] | Total cropland (1,000) [121] | Average per farm [122] | Average per acre [123] | Value of machinery and equipment, average per farm (dollars) [124] | Total (mil dol) [125] | Average per farm (acres) [126] | Crops [127] | Livestock and poultry products [128] | Organic farms (number) [129] | Farms with internet access (per-cent) [130] | Total ($1,000) [131] | Percent of farms [132] |
|---|---|---|---|---|---|---|---|---|---|---|---|---|---|---|---|---|
| **IOWA—Cont'd** | | | | | | | | | | | | | | | | |
| Franklin | 349 | -1.7 | 418 | 0.2 | 330.0 | 3,267,910 | 7,814 | 267,635 | 382.7 | 458,354 | 49.9 | 50.1 | 2 | 79.9 | 9,716 | 85.4 |
| Fremont | 274 | -4.6 | 520 | 17.1 | 247.2 | 3,123,290 | 6,002 | 290,739 | 150.7 | 286,046 | 86.8 | 13.2 | 1 | 75.1 | 5,143 | 78.9 |
| Greene | 351 | -1.6 | 502 | 0.1 | 323.0 | 4,163,712 | 8,297 | 298,887 | 299.4 | 427,783 | 58.3 | 41.7 | 13 | 84.0 | 4,531 | 63.9 |
| Grundy | 308 | -3.1 | 405 | 0.0 | 296.3 | 3,682,729 | 9,086 | 287,687 | 266.7 | 350,933 | 70.8 | 29.2 | NA | 81.1 | 10,419 | 78.9 |
| Guthrie | 332 | 1.4 | 414 | D | 268.4 | 2,610,011 | 6,301 | 239,048 | 227.1 | 283,158 | 54.6 | 45.4 | 1 | 80.8 | 8,954 | 71.3 |
| Hamilton | 315 | -3.6 | 431 | D | 296.2 | 3,495,669 | 8,115 | 303,864 | 586.8 | 801,623 | 28.7 | 71.3 | 5 | 86.1 | 5,877 | 59.2 |
| Hancock | 345 | -2.2 | 431 | 3.6 | 334.9 | 3,286,656 | 7,621 | 331,122 | 422.6 | 527,586 | 44.7 | 55.3 | NA | 79.9 | 9,659 | 86.1 |
| Hardin | 337 | 1.3 | 402 | D | 310.6 | 3,142,054 | 7,813 | 299,034 | 484.6 | 578,983 | 39.0 | 61.0 | 5 | 82.2 | 4,740 | 79.3 |
| Harrison | 380 | -3.6 | 478 | 43.4 | 336.2 | 2,945,210 | 6,161 | 275,156 | 216.2 | 272,293 | 81.1 | 18.9 | NA | 81.6 | 5,349 | 78.6 |
| Henry | 262 | -2.9 | 288 | 0.0 | 213.8 | 1,807,194 | 6,267 | 167,085 | 177.8 | 195,843 | 58.8 | 41.2 | 7 | 74.2 | 6,564 | 73.2 |
| Howard | 299 | -0.3 | 340 | 0.5 | 273.7 | 2,537,364 | 7,455 | 246,741 | 285.0 | 324,225 | 52.1 | 47.9 | 27 | 73.7 | 6,702 | 75.4 |
| Humboldt | 240 | 2.0 | 419 | 0.3 | 230.8 | 3,387,096 | 8,088 | 295,137 | 180.9 | 316,248 | 69.4 | 30.6 | NA | 79.0 | 4,733 | 83.0 |
| Ida | 263 | 0.8 | 501 | D | 239.3 | 3,681,819 | 7,347 | 326,692 | 222.8 | 424,358 | 63.3 | 36.7 | 11 | 78.5 | 3,738 | 77.9 |
| Iowa | 347 | 3.1 | 357 | D | 283.4 | 2,291,958 | 6,415 | 209,704 | 227.9 | 234,951 | 61.6 | 38.4 | 11 | 81.2 | 7,532 | 73.5 |
| Jackson | 316 | 2.2 | 285 | 0.2 | 228.4 | 1,703,564 | 5,973 | 189,407 | 264.1 | 238,539 | 37.5 | 62.5 | 7 | 78.1 | 7,264 | 73.4 |
| Jasper | 378 | 1.2 | 384 | 0.6 | 320.0 | 2,497,555 | 6,512 | 256,238 | 252.8 | 256,395 | 70.7 | 29.3 | 3 | 84.0 | 3,560 | 51.8 |
| Jefferson | 206 | 3.9 | 324 | 0.1 | 170.4 | 1,792,949 | 5,538 | 161,501 | 108.6 | 170,704 | 62.6 | 37.4 | 25 | 71.9 | 5,626 | 70.0 |
| Johnson | 304 | -7.4 | 242 | 0.9 | 264.9 | 1,945,168 | 8,037 | 174,726 | 219.6 | 174,717 | 65.3 | 34.7 | 94 | 75.6 | 10,976 | 63.2 |
| Jones | 344 | 9.5 | 310 | 0.1 | 285.9 | 2,155,654 | 6,962 | 232,937 | 288.0 | 259,458 | 54.3 | 45.7 | 2 | 80.5 | 8,615 | 74.4 |
| Keokuk | 318 | 7.7 | 343 | 0.1 | 262.9 | 2,003,712 | 5,837 | 195,144 | 216.3 | 233,309 | 52.0 | 48.0 | 13 | 79.0 | 10,997 | 79.6 |
| Kossuth | 594 | -0.9 | 441 | 2.1 | 573.4 | 3,480,288 | 7,892 | 300,789 | 588.1 | 436,566 | 54.3 | 45.7 | 15 | 80.3 | 14,891 | 87.8 |
| Lee | 220 | -6.6 | 263 | 1.1 | 166.4 | 1,360,388 | 5,166 | 124,972 | 122.7 | 146,585 | 61.3 | 38.7 | 3 | 75.9 | 3,860 | 62.5 |
| Linn | 325 | -4.4 | 236 | 0.7 | 284.6 | 1,966,919 | 8,328 | 167,553 | 218.8 | 159,214 | 74.0 | 26.0 | 1 | 84.2 | 7,745 | 63.5 |
| Louisa | 190 | 12.7 | 330 | 16.9 | 165.3 | 2,187,445 | 6,631 | 230,894 | 205.9 | 357,387 | 45.4 | 54.6 | 2 | 79.7 | 4,825 | 80.7 |
| Lucas | 175 | -1.1 | 309 | 0.0 | 106.9 | 1,055,881 | 3,413 | 108,610 | 50.1 | 88,383 | 45.3 | 54.7 | NA | 74.3 | 4,943 | 65.8 |
| Lyon | 346 | -6.4 | 309 | 1.4 | 318.2 | 3,082,560 | 9,988 | 287,450 | 923.6 | 823,148 | 20.6 | 79.4 | 4 | 83.1 | 3,718 | 75.7 |
| Madison | 271 | -1.8 | 277 | 0.5 | 179.3 | 1,572,389 | 5,668 | 140,797 | 119.3 | 122,129 | 60.0 | 40.0 | 4 | 76.3 | 5,316 | 59.3 |
| Mahaska | 311 | -3.7 | 330 | 0.1 | 262.9 | 2,043,469 | 6,193 | 198,211 | 280.7 | 297,707 | 43.4 | 56.6 | 1 | 82.7 | 7,046 | 76.6 |
| Marion | 255 | -3.7 | 248 | 0.1 | 195.7 | 1,428,870 | 5,771 | 129,159 | 110.6 | 107,378 | 76.1 | 23.9 | 9 | 79.6 | 6,125 | 58.5 |
| Marshall | 316 | 1.3 | 357 | 0.0 | 289.6 | 2,714,523 | 7,600 | 259,796 | 266.0 | 300,265 | 65.9 | 34.1 | 4 | 84.2 | 4,640 | 49.8 |
| Mills | 208 | 0.6 | 399 | 1.9 | 188.9 | 2,391,040 | 5,992 | 221,104 | 98.8 | 189,944 | 97.9 | 2.1 | 2 | 82.5 | 5,150 | 73.1 |
| Mitchell | 289 | -2.4 | 367 | 1.8 | 272.0 | 2,883,359 | 7,865 | 275,666 | 315.9 | 400,393 | 47.7 | 52.3 | 9 | 80.1 | 5,099 | 77.1 |
| Monona | 334 | -1.3 | 539 | 60.4 | 293.2 | 3,107,691 | 5,762 | 317,189 | 192.6 | 311,189 | 80.1 | 19.9 | NA | 77.9 | 8,869 | 82.1 |
| Monroe | 193 | -1.0 | 312 | D | 113.9 | 1,314,106 | 4,206 | 119,256 | 61.6 | 99,710 | 48.0 | 52.0 | 1 | 75.2 | 6,442 | 65.7 |
| Montgomery | 245 | 0.0 | 475 | D | 213.2 | 2,695,892 | 5,680 | 266,552 | 155.7 | 301,806 | 71.6 | 28.4 | 3 | 83.3 | 4,661 | 82.8 |
| Muscatine | 219 | 2.0 | 307 | 10.4 | 187.4 | 2,063,703 | 6,720 | 204,328 | 185.4 | 259,639 | 55.8 | 44.2 | 4 | 77.0 | 3,374 | 70.0 |
| O'Brien | 314 | 3.2 | 359 | 0.8 | 294.0 | 3,489,359 | 9,727 | 273,559 | 501.2 | 572,092 | 35.5 | 64.5 | 7 | 83.7 | 6,056 | 81.4 |
| Osceola | 235 | -1.4 | 397 | 1.5 | 225.0 | 3,481,261 | 8,762 | 297,874 | 459.0 | 776,591 | 27.4 | 72.6 | NA | 85.3 | 5,537 | 83.9 |
| Page | 325 | 2.6 | 455 | 0.5 | 272.6 | 2,444,334 | 5,370 | 212,178 | 168.1 | 235,057 | 80.1 | 19.9 | 4 | 84.5 | 3,359 | 66.4 |
| Palo Alto | 342 | -4.7 | 436 | 6.8 | 329.4 | 3,491,702 | 8,014 | 322,625 | 468.3 | 596,596 | 38.9 | 61.1 | 15 | 81.4 | 11,659 | 87.4 |
| Plymouth | 503 | -7.1 | 413 | 2.8 | 452.6 | 3,535,476 | 8,561 | 285,527 | 738.2 | 605,578 | 33.9 | 66.1 | 5 | 80.3 | 7,240 | 79.2 |
| Pocahontas | 329 | -0.8 | 451 | 0.4 | 319.5 | 3,638,642 | 8,064 | 311,860 | 356.0 | 487,684 | 48.3 | 51.7 | 5 | 78.6 | 5,254 | 89.5 |
| Polk | 194 | -1.9 | 257 | 0.8 | 173.9 | 2,269,262 | 8,841 | 163,985 | 110.9 | 146,873 | 88.1 | 11.9 | 2 | 83.7 | 2,098 | 33.0 |
| Pottawattamie | 512 | -40.0 | 459 | 2.6 | 465.2 | 3,116,311 | 6,784 | 285,527 | 409.3 | 367,376 | 69.8 | 30.2 | 9 | 80.6 | 4,495 | 48.1 |
| Poweshiek | 340 | 1.5 | 399 | 0.1 | 293.7 | 2,673,805 | 6,708 | 240,513 | 314.5 | 369,148 | 47.9 | 52.1 | 2 | 77.8 | 4,962 | 74.4 |
| Ringgold | 303 | 12.4 | 449 | D | 208.9 | 1,731,086 | 3,854 | 142,521 | 121.4 | 179,914 | 41.3 | 58.7 | 9 | 74.7 | 10,630 | 75.9 |
| Sac | 353 | -1.1 | 397 | 0.7 | 330.7 | 3,155,661 | 7,943 | 295,984 | 459.5 | 516,830 | 39.9 | 60.1 | 11 | 81.2 | 5,280 | 64.0 |
| Scott | 220 | -0.3 | 322 | 1.9 | 201.7 | 3,083,078 | 9,588 | 263,956 | 223.8 | 327,151 | 58.0 | 42.0 | 6 | 82.7 | 3,084 | 73.4 |
| Shelby | 371 | -0.5 | 416 | D | 343.5 | 2,853,345 | 6,853 | 242,943 | 305.4 | 343,181 | 64.2 | 35.8 | 9 | 79.9 | 7,416 | 83.9 |
| Sioux | 484 | -0.2 | 280 | 9.6 | 453.5 | 2,918,016 | 10,405 | 297,115 | 1,696.1 | 983,814 | 16.4 | 83.6 | 5 | 86.9 | 4,371 | 46.1 |
| Story | 304 | -0.6 | 318 | 0.4 | 284.2 | 3,013,807 | 9,467 | 199,884 | 224.9 | 235,519 | 73.0 | 27.0 | 12 | 86.4 | 4,910 | 51.8 |
| Tama | 407 | 1.1 | 380 | 0.1 | 362.2 | 2,754,577 | 7,256 | 245,618 | 288.0 | 268,670 | 68.4 | 31.6 | 8 | 77.5 | 7,483 | 61.1 |
| Taylor | 289 | 3.6 | 433 | D | 225.3 | 2,081,546 | 4,812 | 169,533 | 128.0 | 191,942 | 65.8 | 34.2 | 2 | 72.4 | 10,007 | 77.4 |
| Union | 246 | 14.8 | 392 | D | 176.6 | 1,749,559 | 4,458 | 169,761 | 129.2 | 206,137 | 41.9 | 58.1 | 15 | 80.7 | 6,028 | 66.8 |
| Van Buren | 210 | -1.4 | 305 | 0.0 | 139.8 | 1,424,733 | 4,674 | 130,212 | 97.8 | 141,791 | 40.4 | 59.6 | 5 | 70.1 | 5,276 | 57.2 |
| Wapello | 199 | 5.6 | 279 | 0.0 | 153.0 | 1,359,960 | 4,876 | 148,496 | 78.1 | 109,175 | 72.7 | 27.3 | 1 | 78.3 | 4,434 | 61.8 |
| Warren | 247 | -6.2 | 204 | 0.0 | 181.7 | 1,147,348 | 5,636 | 105,972 | 79.0 | 65,045 | 85.0 | 15.0 | 3 | 78.3 | 6,028 | 48.2 |
| Washington | 310 | -1.3 | 275 | 0.3 | 263.7 | 1,894,242 | 6,889 | 209,805 | 671.9 | 595,126 | 20.5 | 79.5 | 38 | 75.6 | 9,990 | 74.7 |
| Wayne | 285 | 4.2 | 384 | 0.0 | 201.8 | 1,632,207 | 4,252 | 159,124 | 91.8 | 123,524 | 57.9 | 42.1 | 7 | 64.6 | 7,493 | 71.3 |
| Webster | 409 | 0.1 | 426 | 0.0 | 387.1 | 3,388,576 | 7,948 | 258,965 | 320.7 | 334,043 | 67.0 | 33.0 | 1 | 79.8 | 11,444 | 83.4 |
| Winnebago | 246 | 4.5 | 431 | 0.0 | 234.5 | 3,012,461 | 6,989 | 297,495 | 239.9 | 420,168 | 54.5 | 45.5 | 1 | 83.0 | 8,604 | 84.9 |
| Winneshiek | 391 | 4.0 | 268 | 0.1 | 322.6 | 1,718,306 | 6,401 | 199,801 | 337.4 | 231,388 | 41.9 | 58.1 | 33 | 82.7 | 8,883 | 74.7 |
| Woodbury | 451 | 1.1 | 435 | 4.6 | 395.6 | 3,205,600 | 7,375 | 255,301 | 368.8 | 355,602 | 54.1 | 45.9 | 1 | 81.6 | 12,463 | 74.3 |
| Worth | 239 | 1.6 | 410 | 1.5 | 223.1 | 2,854,365 | 6,956 | 308,236 | 144.4 | 248,129 | 86.1 | 13.9 | 6 | 79.7 | 7,896 | 84.2 |
| Wright | 356 | -0.9 | 485 | NA | 340.6 | 3,671,012 | 7,573 | 344,709 | 481.9 | 655,631 | 38.9 | 61.1 | NA | 81.0 | 9,844 | 84.5 |

| STATE County | Water use, 2015 | | Wholesale Trade[1], 2017 | | | | Retail Trade[2], 2017 | | | | Real estate and rental and leasing,[2] 2017 | | | |
|---|---|---|---|---|---|---|---|---|---|---|---|---|---|---|
| | Public supply water withdrawn (mil gal/day) | Public supply gallons withdrawn per person per day | Number of establishments | Number of employees | Sales (mil dol) | Average payroll (mil dol) | Number of establishments | Number of employees | Sales (mil dol) | Average payroll (mil dol) | Number of establishments | Number of employees | Sales (mil dol) | Average payroll (mil dol) |
| | 133 | 134 | 135 | 136 | 137 | 138 | 139 | 140 | 141 | 142 | 143 | 144 | 145 | 146 |
| IOWA—Cont'd | | | | | | | | | | | | | | |
| Franklin | 0.67 | 65.1 | 28 | 306 | 910.5 | 16.6 | 35 | 312 | 69.8 | 7.7 | 11 | 23 | 5.5 | 0.6 |
| Fremont | 0.50 | 72.4 | 15 | 133 | 140.8 | 6.9 | 30 | 580 | 160.6 | 14.1 | D | D | D | 0.1 |
| Greene | 0.66 | 73.1 | 16 | 375 | 1,211.6 | 19.0 | 40 | 463 | 96.6 | 10.1 | 6 | 8 | 1.3 | 0.2 |
| Grundy | 0.26 | 20.9 | 20 | 245 | 217.7 | 13.5 | 43 | 439 | 93.5 | 9.2 | D | D | D | D |
| Guthrie | 0.85 | 79.6 | 15 | 140 | 129.7 | 7.3 | 38 | 326 | 115.5 | 9.5 | D | D | D | D |
| Hamilton | 1.14 | 75.0 | 32 | 936 | 796.8 | 47.9 | 52 | 697 | 190.2 | 14.9 | 9 | 19 | 2.2 | 0.4 |
| Hancock | 0.58 | 52.9 | 24 | 185 | 232.8 | 9.7 | 44 | 427 | 228.4 | 11.6 | 8 | 41 | 5.1 | 2.0 |
| Hardin | 1.47 | 84.6 | D | D | D | D | 86 | 909 | 201.9 | 22.5 | 15 | 45 | 8.9 | 1.6 |
| Harrison | 0.92 | 64.5 | D | D | D | 26.7 | 44 | 463 | 273.2 | 14.1 | 7 | 14 | 2.9 | 0.3 |
| Henry | 2.04 | 102.3 | 24 | 245 | 153.2 | 9.3 | 60 | 885 | 244.5 | 21.7 | 11 | 32 | 4.3 | 0.6 |
| Howard | 0.54 | 57.4 | 17 | 123 | 167.7 | 6.6 | 47 | 444 | 116.9 | 10.1 | 4 | 3 | 1.9 | 0.2 |
| Humboldt | 0.94 | 98.4 | 27 | 276 | 321.4 | 15.7 | 42 | 524 | 102.3 | 10.8 | 4 | D | 0.9 | D |
| Ida | 0.42 | 59.8 | 23 | 339 | 307.8 | 18.6 | 33 | 326 | 58.9 | 6.6 | 8 | 15 | 2.1 | 0.2 |
| Iowa | 1.05 | 64.0 | 22 | 257 | 1,698.5 | 19.6 | 122 | 1,351 | 251.7 | 22.3 | D | D | D | 0.4 |
| Jackson | 1.10 | 56.6 | 32 | 223 | 163.8 | 10.7 | 73 | 923 | 281.7 | 24.0 | 8 | 11 | 1.8 | 0.3 |
| Jasper | 4.96 | 134.7 | D | D | D | D | 97 | 1,424 | 386.9 | 32.3 | 24 | 49 | 7.0 | 1.4 |
| Jefferson | 1.60 | 91.1 | D | D | D | 16.4 | 88 | 1,047 | 223.2 | 26.6 | 29 | 39 | 6.4 | 1.3 |
| Johnson | 10.04 | 69.6 | 99 | 1,343 | 1,391.7 | 75.3 | 470 | 9,142 | 2,264.7 | 236.6 | 149 | 702 | 147.9 | 31.4 |
| Jones | 0.89 | 43.5 | 25 | 252 | 183.6 | 10.7 | 72 | 834 | 257.0 | 25.2 | D | D | D | D |
| Keokuk | 1.24 | 122.0 | 28 | 200 | 149.1 | 8.0 | 35 | 246 | 53.5 | 4.1 | 7 | D | 1.8 | D |
| Kossuth | 1.25 | 82.4 | 36 | 404 | 289.7 | 23.6 | 84 | 890 | 237.8 | 21.9 | D | D | D | D |
| Lee | 12.38 | 352.8 | 35 | 451 | 395.3 | 24.4 | 126 | 1,881 | 440.3 | 43.7 | 26 | 72 | 9.1 | 1.9 |
| Linn | 41.89 | 190.5 | 305 | 5,149 | 3,401.3 | 327.9 | 679 | 15,106 | 6,405.8 | 369.6 | 251 | 1,123 | 256.0 | 47.3 |
| Louisa | 0.49 | 43.8 | 10 | 317 | 324.3 | 11.2 | 24 | 224 | 60.3 | 4.7 | 6 | D | 1.2 | D |
| Lucas | 0.47 | 54.1 | 6 | 31 | 10.6 | 0.8 | 34 | 517 | 95.1 | 11.3 | 4 | 12 | 1.5 | 0.5 |
| Lyon | 2.41 | 205.2 | 24 | 221 | 283.3 | 10.5 | 54 | 402 | 89.4 | 7.8 | 10 | 41 | 10.6 | 3.4 |
| Madison | 0.36 | 22.9 | 19 | 224 | 187.2 | 12.4 | 53 | 564 | 104.5 | 11.6 | 13 | 18 | 7.5 | 0.7 |
| Mahaska | 2.58 | 115.6 | 33 | 395 | 273.6 | 17.4 | 92 | 1,066 | 205.2 | 23.2 | 15 | 39 | 7.4 | 1.1 |
| Marion | 2.91 | 87.4 | 41 | D | 283.7 | D | 121 | 1,542 | 385.9 | 39.4 | 32 | 56 | 12.1 | 1.6 |
| Marshall | 5.01 | 123.0 | 43 | 376 | 311.2 | 20.7 | 135 | 2,024 | 493.8 | 46.7 | 27 | 478 | 100.1 | 17.1 |
| Mills | 0.87 | 58.6 | D | D | D | 6.1 | 35 | 329 | 85.4 | 7.7 | 14 | 34 | 5.3 | 0.7 |
| Mitchell | 0.75 | 69.2 | 20 | 274 | 333.3 | 14.8 | 52 | 450 | 72.7 | 8.3 | 4 | D | 1.5 | D |
| Monona | 0.82 | 91.3 | 18 | 159 | 205.0 | 7.4 | 38 | 354 | 102.7 | 7.5 | 4 | D | 0.6 | D |
| Monroe | 0.00 | 0.0 | 9 | 67 | 45.2 | 2.2 | 24 | 261 | 60.2 | 6.1 | 6 | 14 | 2.2 | 0.3 |
| Montgomery | 1.45 | 141.7 | 13 | 128 | 191.0 | 5.4 | 44 | 498 | 96.7 | 10.5 | 11 | 43 | 5.0 | 1.1 |
| Muscatine | 28.94 | 672.9 | D | D | D | D | 128 | 2,108 | 535.6 | 51.4 | 47 | 171 | 32.8 | 5.4 |
| O'Brien | 2.31 | 165.2 | 34 | 441 | 494.8 | 22.0 | 88 | 778 | 162.5 | 16.5 | 7 | 22 | 4.3 | 0.7 |
| Osceola | 4.30 | 698.7 | 11 | 115 | 207.6 | 7.0 | 28 | 153 | 118.9 | 4.0 | NA | NA | NA | NA |
| Page | 1.78 | 114.6 | 21 | 192 | 147.1 | 7.4 | 66 | 619 | 122.1 | 14.0 | 10 | 15 | 2.9 | 0.2 |
| Palo Alto | 0.78 | 85.4 | 18 | 174 | 236.3 | 8.7 | 39 | 348 | 78.1 | 7.8 | 6 | 13 | 1.1 | 0.3 |
| Plymouth | 3.34 | 134.7 | D | D | D | D | 94 | 1,019 | 327.2 | 25.0 | 27 | 62 | 12.0 | 1.7 |
| Pocahontas | 0.48 | 68.5 | 15 | 169 | 258.0 | 7.6 | 35 | 224 | 48.2 | 5.3 | NA | NA | NA | NA |
| Polk | 56.09 | 119.9 | 698 | 13,485 | 16,611.0 | 834.1 | 1,495 | 30,271 | 8,958.0 | 840.0 | 684 | 3,765 | 900.9 | 167.8 |
| Pottawattamie | 15.02 | 160.3 | 94 | 1,485 | 1,441.6 | 85.4 | 265 | 5,815 | 1,575.9 | 138.4 | 93 | 392 | 71.7 | 14.6 |
| Poweshiek | 1.33 | 71.7 | 22 | 160 | 136.1 | 7.7 | 78 | 1,605 | 584.4 | 61.3 | D | D | D | D |
| Ringgold | 0.00 | 0.0 | D | D | D | D | 23 | 211 | 48.9 | 4.6 | NA | NA | NA | NA |
| Sac | 1.14 | 113.8 | 31 | 306 | 259.6 | 16.1 | 52 | 306 | 95.7 | 7.7 | NA | NA | NA | NA |
| Scott | 17.82 | 103.5 | 253 | 3,698 | 2,221.0 | 195.7 | 633 | 11,830 | 3,209.8 | 308.2 | 199 | 821 | 189.7 | 31.4 |
| Shelby | 1.96 | 164.3 | 27 | 333 | 256.2 | 13.6 | 50 | 528 | 148.1 | 10.9 | 9 | 16 | 1.8 | 0.4 |
| Sioux | 9.13 | 261.3 | 85 | 1,632 | 1,309.9 | 76.2 | 159 | 1,599 | 423.2 | 37.8 | 33 | 157 | 35.9 | 7.3 |
| Story | 8.42 | 87.7 | 78 | 1,444 | 1,083.4 | 56.5 | 281 | 5,073 | 1,193.9 | 119.7 | 117 | 522 | 90.8 | 23.1 |
| Tama | 2.66 | 153.4 | 26 | 201 | 183.7 | 8.5 | 58 | 564 | 156.6 | 13.0 | 5 | 11 | 2.3 | 0.3 |
| Taylor | 0.00 | 0.0 | D | D | D | 2.9 | 15 | 152 | 40.0 | 2.5 | 5 | D | 1.2 | D |
| Union | 4.15 | 332.8 | 17 | 238 | 231.9 | 12.6 | 51 | 778 | 172.1 | 17.9 | 8 | D | 2.0 | D |
| Van Buren | 0.00 | 0.0 | D | D | D | 1.4 | 24 | 225 | 45.7 | 4.5 | 3 | 4 | 0.2 | 0.1 |
| Wapello | 7.94 | 225.7 | 30 | 220 | 267.3 | 10.8 | 136 | 2,315 | 592.7 | 58.4 | 26 | 124 | 24.8 | 3.8 |
| Warren | 1.00 | 20.6 | 42 | 388 | 197.7 | 18.4 | 103 | 1,739 | 409.7 | 43.3 | D | D | D | D |
| Washington | 1.81 | 81.4 | 45 | 639 | 365.1 | 33.3 | 87 | 1,006 | 201.1 | 24.0 | 9 | 13 | 1.8 | 0.3 |
| Wayne | 0.00 | 0.0 | D | D | D | 3.7 | 25 | 213 | 50.2 | 4.4 | NA | NA | NA | NA |
| Webster | 11.01 | 297.0 | 63 | 771 | 390.6 | 45.4 | 160 | 2,555 | 596.8 | 60.6 | 42 | 227 | 36.2 | 6.2 |
| Winnebago | 0.89 | 83.9 | 25 | 273 | 250.9 | 15.0 | 52 | 546 | 173.5 | 13.9 | 3 | 6 | 0.9 | 0.4 |
| Winneshiek | 1.33 | 64.2 | 41 | 350 | 249.3 | 15.6 | 116 | 1,201 | 300.3 | 29.5 | 14 | 23 | 2.3 | 0.4 |
| Woodbury | 13.97 | 135.9 | 163 | 2,368 | 2,167.6 | 123.5 | 407 | 7,373 | 2,009.5 | 176.1 | 95 | 497 | 81.0 | 15.6 |
| Worth | 0.45 | 59.5 | 13 | 106 | 131.8 | 4.9 | 26 | 189 | 37.6 | 3.9 | 3 | 2 | 0.3 | 0.0 |
| Wright | 1.24 | 97.1 | 19 | 319 | 556.9 | 18.0 | 54 | 533 | 89.9 | 11.2 | 8 | 12 | 1.9 | 0.3 |

1  Merchant wholesalers, except manufacturers' sales branches and offices.     2. Employer establishments.

# Table B. States and Counties — Professional Services, Manufacturing, and Accommodation and Food Services

| STATE County | Professional, scientific, and technical services, 2017 | | | | Manufacturing, 2017 | | | | Accommodation and food services, 2017 | | | |
|---|---|---|---|---|---|---|---|---|---|---|---|---|
| | Number of establishments | Number of employees | Sales (mil dol) | Average payroll (mil dol) | Number of establishments | Number of employees | Sales (mil dol) | Average payroll (mil dol) | Number of establishments | Number of employees | Sales (mil dol) | Annual payroll (mil dol) |
| | 147 | 148 | 149 | 150 | 151 | 152 | 153 | 154 | 155 | 156 | 157 | 158 |
| IOWA—Cont'd | | | | | | | | | | | | |
| Franklin | 14 | 59 | 5 | 1.6 | 25 | 860 | 249.3 | 35.6 | 16 | 172 | 6.6 | 1.7 |
| Fremont | 9 | 37 | 4 | 1.2 | 10 | 960 | 332.4 | 24.8 | 19 | 193 | 9.5 | 2.6 |
| Greene | 18 | 65 | 7 | 3.4 | 11 | 498 | 141.5 | 23.3 | 20 | 161 | 9.1 | 2.6 |
| Grundy | 8 | 33 | 5 | 1.0 | 13 | 418 | 141.8 | 21.1 | 20 | 161 | 5.8 | 1.3 |
| Guthrie | 20 | 75 | 14 | 3.7 | 8 | 495 | 350.0 | 17.2 | D | D | D | D |
| Hamilton | 22 | 82 | 17 | 5.4 | 22 | 779 | 303.4 | 35.5 | 26 | 238 | 11.2 | 2.9 |
| Hancock | 18 | 504 | 35 | 23.0 | 26 | 1,002 | 379.3 | 52.2 | 16 | 140 | 4.9 | 1.5 |
| Hardin | 33 | 131 | 13 | 4.2 | 28 | 567 | 974.0 | 30.9 | 31 | 316 | 13.7 | 3.7 |
| Harrison | 18 | 49 | 8 | 1.8 | 16 | 235 | 96.5 | 12.3 | 29 | 289 | 14.4 | 3.6 |
| Henry | 35 | 127 | 17 | 5.1 | 32 | 2,423 | 858.2 | 111.9 | D | D | D | D |
| Howard | D | D | D | D | 22 | 1,004 | 271.0 | 42.7 | D | D | D | D |
| Humboldt | 16 | 58 | 7 | 2.3 | 19 | 877 | 241.0 | 43.6 | 19 | 208 | 8.0 | 2.2 |
| Ida | 11 | 54 | 3 | 1.2 | 14 | 1,463 | 575.6 | 78.6 | 12 | 66 | 3.4 | 0.8 |
| Iowa | 25 | 194 | 35 | 21.5 | 24 | 4,376 | 1,777.8 | 185.9 | 36 | 455 | 19.7 | 6.4 |
| Jackson | D | D | D | D | 31 | 904 | 241.4 | 36.2 | 43 | 497 | 20.7 | 5.4 |
| Jasper | 50 | 480 | 132 | 25.0 | 36 | 1,945 | 518.8 | 90.0 | 60 | 779 | 40.8 | 10.9 |
| Jefferson | 132 | 502 | 88 | 29.2 | 31 | 864 | 203.3 | 46.5 | 41 | 482 | 24.6 | 7.3 |
| Johnson | 291 | 2,899 | 367 | 144.5 | 82 | 5,024 | 6,863.4 | 261.0 | D | D | D | D |
| Jones | 23 | 105 | 11 | 3.5 | 25 | 690 | 194.1 | 29.3 | 39 | 386 | 19.9 | 5.6 |
| Keokuk | D | D | 7 | D | 12 | 141 | 56.0 | 6.6 | 13 | D | 3.6 | D |
| Kossuth | 38 | 257 | 36 | 11.2 | 28 | 1,415 | 889.8 | 72.4 | 32 | 347 | 15.4 | 4.5 |
| Lee | 41 | 261 | 24 | 9.8 | 63 | 4,107 | 2,331.1 | 239.5 | 95 | 1,072 | 49.5 | 13.1 |
| Linn | 526 | 6,517 | 1,041 | 468.8 | 206 | 16,906 | 10,451.7 | 1,448.7 | 540 | 9,764 | 480.5 | 148.8 |
| Louisa | 12 | 32 | 3 | 0.8 | D | 1,489 | D | 65.5 | 16 | D | 3.6 | D |
| Lucas | 15 | 45 | 4 | 1.2 | 7 | 119 | 26.0 | 6.9 | D | D | D | D |
| Lyon | 24 | 131 | 23 | 13.8 | 27 | 623 | 133.5 | 32.9 | 20 | 602 | 70.6 | 12.5 |
| Madison | 29 | 95 | 10 | 3.5 | 19 | 159 | 38.7 | 7.6 | D | D | D | D |
| Mahaska | 27 | 132 | 14 | 5.7 | 29 | 1,234 | 756.6 | 62.0 | 38 | 540 | 23.8 | 6.3 |
| Marion | 57 | 403 | 65 | 18.3 | 41 | 6,668 | 2,143.2 | 375.5 | 71 | 1,099 | 45.1 | 11.9 |
| Marshall | 41 | 323 | 58 | 14.2 | 38 | 4,551 | 2,485.0 | 278.4 | 78 | 1,035 | 49.7 | 14.3 |
| Mills | 32 | 119 | 12 | 5.4 | 9 | 37 | 7.7 | 1.4 | D | D | D | D |
| Mitchell | 13 | 70 | 6 | 2.3 | 19 | 848 | 646.9 | 46.0 | 19 | 158 | 6.9 | 1.7 |
| Monona | 12 | 39 | 4 | 1.9 | 6 | 49 | 18.9 | 2.7 | 20 | 150 | 8.0 | 2.0 |
| Monroe | 13 | 42 | 7 | 1.5 | 12 | 763 | 618.4 | 45.0 | 21 | 168 | 6.6 | 1.7 |
| Montgomery | 17 | 61 | 10 | 2.8 | 12 | 543 | 238.5 | 27.5 | 22 | 270 | 10.3 | 3.2 |
| Muscatine | D | D | D | D | 70 | 8,104 | 4,472.0 | 452.0 | 103 | 1,221 | 60.4 | 15.3 |
| O'Brien | 29 | 174 | 27 | 7.1 | 25 | 589 | 685.9 | 31.5 | D | D | D | D |
| Osceola | 10 | 30 | 5 | 1.0 | 17 | 282 | 170.0 | 13.7 | 10 | 73 | 2.8 | 0.7 |
| Page | 20 | 119 | 7 | 3.0 | 16 | 1,246 | 346.1 | 70.9 | 26 | 289 | 13.2 | 4.1 |
| Palo Alto | 14 | 56 | 7 | 2.4 | 20 | 639 | 562.0 | 36.3 | 32 | 481 | 41.1 | 7.2 |
| Plymouth | 37 | 303 | 24 | 17.0 | 32 | 3,143 | 1,210.7 | 148.6 | D | D | D | D |
| Pocahontas | 15 | 42 | 3 | 1.2 | 12 | 400 | 90.7 | 18.3 | 13 | D | 3.2 | D |
| Polk | 1,602 | 16,639 | 3,137 | 1,175.8 | 343 | 16,918 | 8,337.4 | 921.3 | 1,205 | 24,155 | 1,485.0 | 426.6 |
| Pottawattamie | D | D | D | D | 61 | 4,966 | 3,442.5 | 224.7 | 209 | 5,738 | 601.5 | 116.2 |
| Poweshiek | 39 | 125 | 13 | 4.1 | 26 | 1,555 | 416.0 | 61.8 | 48 | 597 | 24.7 | 7.3 |
| Ringgold | 8 | 48 | 7 | 1.6 | 5 | D | 3.1 | D | 13 | D | 3.0 | D |
| Sac | 17 | 37 | 4 | 1.0 | 21 | 417 | 262.3 | 17.5 | D | D | D | 1.3 |
| Scott | D | D | D | D | 158 | 11,233 | 6,399.0 | 705.0 | 449 | 9,733 | 575.9 | 150.0 |
| Shelby | 21 | 171 | 20 | 7.1 | 17 | 739 | 323.5 | 35.6 | 26 | 236 | 8.9 | 2.3 |
| Sioux | 79 | 916 | 133 | 41.2 | 89 | 4,974 | 1,890.8 | 234.2 | 70 | 1,103 | 45.0 | 12.9 |
| Story | D | D | D | D | 76 | 4,599 | 3,279.3 | 269.0 | 248 | 5,077 | 244.7 | 69.8 |
| Tama | 17 | 53 | 7 | 1.8 | 18 | 1,190 | 746.0 | 57.0 | 27 | 1,062 | 118.1 | 24.3 |
| Taylor | D | D | 3 | D | 6 | 621 | 230.0 | 27.0 | 5 | 37 | 1.4 | 0.4 |
| Union | 18 | 67 | 7 | 2.5 | 12 | 1,474 | 364.7 | 49.5 | 29 | 429 | 16.7 | 4.6 |
| Van Buren | 7 | 32 | 3 | 1.2 | 12 | 580 | 129.4 | 27.7 | 12 | D | 2.8 | D |
| Wapello | D | D | D | D | 27 | 3,391 | 1,740.9 | 188.2 | D | D | D | D |
| Warren | 72 | 271 | 47 | 11.5 | 27 | 430 | 133.5 | 27.5 | 67 | 854 | 36.4 | 11.0 |
| Washington | 53 | 173 | 19 | 7.3 | 35 | 666 | 206.9 | 24.7 | D | D | D | 11.0 |
| Wayne | 9 | 45 | 5 | 1.5 | 5 | 344 | 61.0 | 14.2 | D | D | D | D |
| Webster | 67 | 1,253 | 34 | 76.7 | 46 | 1,863 | 1,793.2 | 117.3 | 95 | 1,390 | 71.2 | 20.0 |
| Winnebago | 18 | 137 | 11 | 3.5 | 19 | 3,169 | 1,007.6 | 152.3 | 21 | 191 | 6.1 | 1.8 |
| Winneshiek | D | D | D | D | 30 | 1,360 | 344.9 | 66.9 | 64 | 752 | 39.1 | 9.8 |
| Woodbury | D | D | D | D | 87 | 5,692 | 3,518.7 | 261.5 | 266 | 5,708 | 378.5 | 90.7 |
| Worth | 7 | 15 | 1 | 0.4 | 13 | 454 | 238.5 | 19.4 | D | D | D | D |
| Wright | 20 | 142 | 16 | 3.0 | 22 | 1,461 | 1,024.6 | 70.3 | 36 | 264 | 9.2 | 2.4 |

# Table B. States and Counties — Health Care and Social Assistance, Other Services, Nonemployer Businesses, and Residential Construction

| STATE County | Health care and social assistance, 2017 | | | | Other services, 2017 | | | | Nonemployer businesses, 2018 | | Value of residential construction authorized by building permits, 2020 | |
|---|---|---|---|---|---|---|---|---|---|---|---|---|
| | Number of establishments | Number of employees | Receipts (mil dol) | Annual payroll (mil dol) | Number of establishments | Number of employees | Receipts (mil dol) | Annual payroll (mil dol) | Number | Receipts (mil dol) | New construction ($1,000) | Number of housing units |
| | 159 | 160 | 161 | 162 | 163 | 164 | 165 | 166 | 167 | 168 | 169 | 170 |
| IOWA—Cont'd | | | | | | | | | | | | |
| Franklin | 27 | 465 | 35.6 | 12.9 | 25 | 74 | 8.3 | 2.0 | 779 | 35.5 | 0 | 0 |
| Fremont | 16 | 314 | 25.5 | 10.6 | D | D | D | 0.3 | 445 | 18.1 | 3,899 | 13 |
| Greene | 32 | 575 | 40.8 | 16.8 | 31 | 96 | 13.0 | 2.4 | 694 | 34.1 | 3,475 | 12 |
| Grundy | 26 | 504 | 41.2 | 16.4 | 15 | 49 | 4.7 | 1.0 | 918 | 40.3 | 4,856 | 20 |
| Guthrie | 27 | 347 | 21.6 | 10.2 | 29 | 93 | 11.6 | 2.7 | 1,037 | 40.5 | 11,776 | 32 |
| Hamilton | 36 | 733 | 57.8 | 31.7 | 31 | 114 | 22.3 | 3.9 | 1,039 | 43.9 | 1,630 | 7 |
| Hancock | 21 | 547 | 44.5 | 16.1 | 26 | 203 | 28.6 | 7.7 | 932 | 36.0 | 2,288 | 16 |
| Hardin | 50 | 974 | 70.1 | 30.3 | 46 | 155 | 17.9 | 4.5 | 1,233 | 55.7 | 4,571 | 15 |
| Harrison | 32 | 843 | 80.1 | 31.2 | 22 | 52 | 5.7 | 1.3 | 1,037 | 51.5 | 7,689 | 29 |
| Henry | 59 | 1,149 | 92.2 | 36.0 | 43 | 143 | 10.7 | 4.1 | 1,271 | 43.4 | 7,426 | 32 |
| Howard | 20 | 550 | 41.2 | 15.0 | 16 | 38 | 5.5 | 1.0 | 792 | 41.3 | 2,356 | 10 |
| Humboldt | 23 | 513 | 32.8 | 15.3 | 17 | 49 | 9.8 | 1.9 | 767 | 34.2 | 3,164 | 15 |
| Ida | 20 | 491 | 39.2 | 16.8 | D | D | D | 1.0 | 538 | 25.0 | 4,063 | 16 |
| Iowa | 38 | 736 | 59.2 | 25.5 | 28 | 88 | 11.2 | 3.0 | 1,196 | 49.6 | 4,252 | 21 |
| Jackson | 42 | 609 | 47.3 | 19.7 | 48 | 122 | 16.2 | 3.3 | 1,480 | 60.3 | 9,402 | 28 |
| Jasper | 72 | 1,519 | 108.9 | 46.7 | 63 | 214 | 22.7 | 5.7 | 2,287 | 88.8 | 23,889 | 108 |
| Jefferson | 48 | 849 | 79.5 | 28.7 | 45 | 193 | 30.8 | 7.1 | 1,689 | 64.7 | 2,144 | 14 |
| Johnson | 437 | 18,459 | 2,629.6 | 972.6 | 223 | 1,417 | 303.6 | 52.0 | 10,302 | 497.4 | 185,671 | 956 |
| Jones | 41 | 855 | 60.9 | 25.8 | D | D | D | 2.7 | 1,349 | 63.4 | 3,061 | 18 |
| Keokuk | 17 | 317 | 24.2 | 11.6 | D | D | D | 1.7 | 745 | 42.0 | 898 | 3 |
| Kossuth | 34 | 717 | 61.9 | 23.1 | 45 | 128 | 14.7 | 3.3 | 1,385 | 63.3 | 4,029 | 14 |
| Lee | 100 | 2,235 | 171.9 | 79.7 | 71 | 246 | 23.8 | 5.8 | 1,804 | 69.3 | 780 | 6 |
| Linn | 661 | 15,576 | 1,839.9 | 709.5 | 383 | 2,815 | 303.6 | 92.0 | 13,608 | 630.6 | 88,694 | 606 |
| Louisa | 26 | 415 | 17.3 | 9.4 | D | D | D | 0.8 | 638 | 28.9 | 3,613 | 18 |
| Lucas | 24 | 522 | 35.1 | 17.0 | 14 | 47 | 6.0 | 1.6 | 629 | 32.1 | 350 | 3 |
| Lyon | 24 | 520 | 46.2 | 13.9 | 27 | 98 | 16.1 | 3.0 | 1,036 | 50.0 | 6,558 | 22 |
| Madison | 30 | 531 | 42.5 | 18.3 | 32 | 100 | 13.9 | 3.0 | 1,418 | 67.1 | 29,922 | 103 |
| Mahaska | 48 | 1,014 | 96.5 | 46.0 | 42 | 144 | 16.0 | 4.2 | 1,448 | 55.2 | 5,898 | 37 |
| Marion | 94 | 2,117 | 207.0 | 87.9 | 59 | 197 | 17.0 | 4.9 | 2,250 | 90.4 | 27,809 | 220 |
| Marshall | 95 | 2,442 | 204.7 | 88.5 | 58 | 247 | 26.6 | 6.7 | 1,904 | 68.8 | 8,044 | 54 |
| Mills | 35 | 603 | 34.9 | 18.2 | 22 | 65 | 9.3 | 2.0 | 960 | 42.8 | 1,078 | 6 |
| Mitchell | 25 | 583 | 46.0 | 16.5 | 32 | 89 | 8.2 | 2.1 | 853 | 37.2 | 10,679 | 56 |
| Monona | 26 | 674 | 54.1 | 24.7 | 14 | 55 | 7.9 | 1.9 | 597 | 26.1 | 1,250 | 8 |
| Monroe | D | D | D | D | 11 | 25 | 3.9 | 0.7 | 527 | 20.0 | 1,701 | 9 |
| Montgomery | 29 | 796 | 68.4 | 30.4 | 26 | 94 | 11.9 | 2.9 | 734 | 30.3 | 1,902 | 9 |
| Muscatine | 99 | 2,127 | 159.9 | 71.6 | 64 | 444 | 75.8 | 14.7 | 2,192 | 94.5 | 3,070 | 17 |
| O'Brien | 48 | 1,410 | 92.6 | 35.9 | 45 | 149 | 18.7 | 4.1 | 992 | 49.4 | 8,123 | 34 |
| Osceola | 12 | 245 | 19.2 | 7.4 | D | D | D | D | 425 | 17.5 | 238 | 1 |
| Page | 60 | 1,490 | 134.8 | 59.4 | 23 | 63 | 6.5 | 1.4 | 901 | 30.8 | 2,399 | 10 |
| Palo Alto | 32 | 747 | 52.9 | 22.2 | D | D | D | D | 778 | 30.7 | 2,405 | 10 |
| Plymouth | 58 | 1,070 | 78.7 | 31.2 | 57 | 225 | 27.5 | 6.7 | 1,771 | 92.9 | 13,839 | 48 |
| Pocahontas | 18 | 266 | 20.9 | 9.9 | D | D | D | 1.8 | 544 | 27.8 | 285 | 1 |
| Polk | 1,304 | 35,596 | 4,322.6 | 1,857.5 | 1,036 | 7,898 | 1,239.2 | 328.1 | 33,975 | 1,674.1 | 1,049,665 | 4,580 |
| Pottawattamie | 235 | 5,267 | 563.5 | 215.0 | 139 | 664 | 105.4 | 20.3 | 5,123 | 223.9 | 41,122 | 195 |
| Poweshiek | 53 | 1,049 | 106.7 | 48.8 | 29 | 89 | 13.2 | 3.0 | 1,281 | 55.8 | 6,784 | 26 |
| Ringgold | 12 | 344 | 30.5 | 12.1 | 15 | 56 | 6.5 | 1.3 | 422 | 20.7 | 673 | 3 |
| Sac | 32 | 513 | 33.9 | 15.2 | 19 | 45 | 4.5 | 1.1 | 865 | 40.3 | 1,450 | 6 |
| Scott | 533 | 12,882 | 1,288.1 | 536.0 | 301 | 2,121 | 199.1 | 61.9 | 10,665 | 515.9 | 97,586 | 439 |
| Shelby | 38 | 813 | 74.5 | 27.0 | 29 | 94 | 10.4 | 2.4 | 963 | 44.5 | 3,508 | 13 |
| Sioux | 92 | 2,786 | 200.0 | 83.3 | 97 | 305 | 36.1 | 8.6 | 2,848 | 129.9 | 33,298 | 141 |
| Story | 217 | 6,102 | 701.7 | 285.0 | 160 | 1,164 | 274.7 | 36.7 | 5,610 | 242.6 | 62,434 | 220 |
| Tama | 28 | 476 | 25.7 | 12.5 | 19 | 62 | 6.0 | 1.5 | 1,100 | 44.2 | 2,029 | 18 |
| Taylor | D | D | D | D | D | D | 6.0 | D | 575 | 24.6 | 1,282 | 6 |
| Union | 42 | 982 | 89.1 | 36.7 | 24 | 152 | 31.8 | 6.1 | 775 | 30.5 | 733 | 4 |
| Van Buren | 12 | 276 | 20.7 | 9.5 | D | D | 2.3 | D | 652 | 33.3 | 350 | 1 |
| Wapello | 95 | 2,924 | 255.6 | 99.5 | 44 | 210 | 18.2 | 5.3 | 1,672 | 62.6 | 2,068 | 25 |
| Warren | 77 | 1,260 | 88.0 | 41.1 | 77 | 284 | 27.2 | 7.6 | 3,749 | 167.1 | 129,275 | 530 |
| Washington | 66 | 1,381 | 104.1 | 44.3 | 50 | 113 | 17.4 | 3.3 | 1,812 | 79.3 | 10,939 | 63 |
| Wayne | 19 | 365 | 37.0 | 14.9 | D | D | D | 0.6 | 618 | 47.0 | 0 | 0 |
| Webster | 122 | 2,921 | 289.2 | 123.6 | 67 | 386 | 46.7 | 15.7 | 2,020 | 84.7 | 18,435 | 154 |
| Winnebago | 30 | 452 | 23.8 | 10.0 | D | D | D | D | 862 | 35.2 | 925 | 4 |
| Winneshiek | 57 | 1,557 | 126.1 | 58.1 | 48 | 205 | 21.7 | 6.0 | 1,914 | 72.6 | 16,790 | 86 |
| Woodbury | 336 | 9,461 | 1,051.2 | 411.8 | 173 | 1,110 | 115.9 | 30.2 | 5,570 | 259.8 | 72,007 | 566 |
| Worth | 10 | 191 | 9.4 | 4.7 | D | D | D | D | 570 | 19.9 | 710 | 5 |
| Wright | 28 | 835 | 135.2 | 41.1 | 27 | 68 | 10.1 | 1.8 | 869 | 39.1 | 3,851 | 17 |

# Government Employment and Payroll, and Local Government Finances

| STATE County | Government employment and payroll, 2017 | | | | | | | | | Local government finances, 2017 | | | | |
| | | | March payroll (percent of total) | | | | | | | General revenue | | | | |
| | | | | | | | | | | | | Taxes | | |
| | Full-time equivalent employees | March payroll (dollars) | Administration, judicial, and legal | Police and corrections | Fire protection | Highways and transportation | Health and welfare | Natural resources and utilities | Education and libraries | Total (mil dol) | Inter-govern-mental (mil dol) | Total (mil dol) | Per capita[1] (dollars) | |
| | | | | | | | | | | | | | Total | Property |
| | 171 | 172 | 173 | 174 | 175 | 176 | 177 | 178 | 179 | 180 | 181 | 182 | 183 | 184 |

**IOWA—Cont'd**

| | | | | | | | | | | | | | | |
|---|---|---|---|---|---|---|---|---|---|---|---|---|---|---|
| Franklin | 730 | 2,473,290 | 5.0 | 4.0 | 0.0 | 6.0 | 26.0 | 2.6 | 54.1 | 58.5 | 25.8 | 24.1 | 2,375 | 2,081 |
| Fremont | 330 | 980,119 | 11.7 | 10.5 | 0.0 | 12.2 | 0.3 | 2.4 | 57.7 | 33.0 | 15.7 | 15.2 | 2,189 | 1,933 |
| Greene | 642 | 2,310,941 | 4.8 | 4.0 | 0.2 | 6.4 | 34.8 | 4.2 | 44.9 | 77.5 | 17.9 | 19.0 | 2,122 | 1,805 |
| Grundy | 568 | 2,237,450 | 6.0 | 4.3 | 0.0 | 2.9 | 33.5 | 4.6 | 48.2 | 87.4 | 22.5 | 21.8 | 1,773 | 1,504 |
| Guthrie | 732 | 2,681,463 | 4.8 | 3.0 | 0.0 | 5.3 | 26.4 | 3.5 | 56.6 | 77.1 | 24.3 | 29.0 | 2,723 | 2,430 |
| Hamilton | 937 | 3,558,349 | 5.5 | 6.1 | 0.9 | 5.2 | 35.5 | 4.4 | 41.5 | 101.6 | 32.2 | 31.3 | 2,079 | 1,814 |
| Hancock | 523 | 2,009,716 | 5.3 | 5.5 | 0.0 | 7.1 | 34.2 | 2.2 | 44.5 | 66.7 | 16.8 | 19.9 | 1,854 | 1,634 |
| Hardin | 785 | 2,773,954 | 6.1 | 8.6 | 0.6 | 6.7 | 2.6 | 5.8 | 69.4 | 111.5 | 33.9 | 36.4 | 2,131 | 1,839 |
| Harrison | 678 | 2,337,407 | 7.1 | 5.5 | 0.0 | 6.7 | 2.9 | 3.1 | 72.3 | 63.7 | 28.0 | 28.1 | 1,993 | 1,792 |
| Henry | 1,053 | 4,316,495 | 4.3 | 4.6 | 0.2 | 2.9 | 27.8 | 4.6 | 53.5 | 109.1 | 35.9 | 29.8 | 1,487 | 1,282 |
| Howard | 540 | 1,907,281 | 6.4 | 2.9 | 0.2 | 6.9 | 33.2 | 3.9 | 45.8 | 80.0 | 15.9 | 18.2 | 1,970 | 1,793 |
| Humboldt | 535 | 2,050,493 | 7.4 | 4.6 | 0.1 | 5.9 | 29.6 | 2.8 | 48.0 | 65.3 | 18.6 | 19.3 | 2,012 | 1,766 |
| Ida | 318 | 1,172,919 | 6.0 | 5.6 | 0.0 | 7.5 | 0.0 | 3.8 | 77.0 | 27.9 | 13.0 | 11.9 | 1,739 | 1,487 |
| Iowa | 847 | 3,089,125 | 4.2 | 5.6 | 0.0 | 6.3 | 25.7 | 2.3 | 55.2 | 95.8 | 26.8 | 29.9 | 1,855 | 1,506 |
| Jackson | 703 | 2,797,720 | 6.5 | 9.2 | 0.0 | 6.5 | 17.5 | 6.7 | 52.9 | 87.2 | 31.3 | 31.7 | 1,637 | 1,355 |
| Jasper | 1,190 | 4,811,184 | 6.7 | 7.7 | 3.0 | 6.0 | 1.8 | 4.5 | 68.5 | 133.7 | 62.4 | 54.0 | 1,458 | 1,290 |
| Jefferson | 764 | 3,055,439 | 4.6 | 7.0 | 0.6 | 4.6 | 46.0 | 4.7 | 32.2 | 101.9 | 19.0 | 25.8 | 1,414 | 1,242 |
| Johnson | 4,304 | 18,758,728 | 6.6 | 8.3 | 2.2 | 5.6 | 4.8 | 6.2 | 62.9 | 578.7 | 198.4 | 290.9 | 1,948 | 1,752 |
| Jones | 694 | 2,531,220 | 4.8 | 7.1 | 0.0 | 9.4 | 2.7 | 4.7 | 70.9 | 74.2 | 33.7 | 32.0 | 1,554 | 1,293 |
| Keokuk | 591 | 2,136,703 | 4.9 | 3.1 | 0.0 | 6.8 | 24.0 | 2.3 | 58.4 | 56.9 | 18.8 | 17.6 | 1,736 | 1,588 |
| Kossuth | 756 | 3,193,530 | 7.9 | 4.7 | 0.8 | 6.2 | 33.9 | 3.9 | 41.4 | 112.0 | 22.9 | 28.2 | 1,885 | 1,615 |
| Lee | 1,283 | 4,891,782 | 4.7 | 7.2 | 3.4 | 4.7 | 4.1 | 9.1 | 63.3 | 132.8 | 58.5 | 47.5 | 1,388 | 1,122 |
| Linn | 9,297 | 42,529,348 | 3.7 | 8.5 | 2.7 | 4.3 | 2.6 | 5.6 | 71.2 | 1,225.4 | 483.6 | 472.9 | 2,107 | 1,750 |
| Louisa | 528 | 1,907,077 | 5.4 | 7.4 | 0.0 | 4.7 | 1.9 | 2.1 | 77.8 | 57.5 | 26.1 | 20.8 | 1,861 | 1,527 |
| Lucas | 567 | 2,212,003 | 3.2 | 2.9 | 0.0 | 4.7 | 37.7 | 14.5 | 36.5 | 49.3 | 17.1 | 11.1 | 1,302 | 1,186 |
| Lyon | 456 | 1,705,855 | 5.7 | 7.4 | 0.0 | 8.0 | 2.3 | 4.8 | 69.1 | 64.6 | 26.9 | 23.7 | 2,014 | 1,736 |
| Madison | 750 | 2,843,194 | 5.6 | 4.1 | 0.0 | 5.2 | 20.6 | 7.1 | 57.3 | 90.5 | 34.3 | 28.7 | 1,794 | 1,602 |
| Mahaska | 929 | 4,381,532 | 1.0 | 2.0 | 0.9 | 1.0 | 56.4 | 1.2 | 36.6 | 123.5 | 32.8 | 30.6 | 1,376 | 1,161 |
| Marion | 1,294 | 4,521,779 | 6.4 | 7.7 | 0.0 | 5.8 | 3.3 | 8.6 | 66.7 | 132.3 | 56.7 | 53.8 | 1,627 | 1,428 |
| Marshall | 2,003 | 7,151,943 | 4.2 | 7.8 | 2.1 | 4.3 | 0.2 | 4.3 | 75.4 | 179.6 | 83.2 | 61.5 | 1,535 | 1,257 |
| Mills | 605 | 2,098,241 | 6.2 | 6.9 | 0.0 | 8.0 | 3.9 | 1.3 | 70.7 | 55.2 | 25.6 | 24.8 | 1,644 | 1,435 |
| Mitchell | 558 | 2,106,103 | 5.4 | 4.5 | 0.1 | 4.7 | 37.2 | 3.2 | 43.0 | 39.7 | 16.7 | 14.0 | 1,321 | 1,122 |
| Monona | 456 | 1,498,814 | 11.4 | 6.5 | 0.0 | 1.5 | 2.2 | 6.4 | 61.8 | 37.0 | 14.4 | 17.5 | 2,001 | 1,883 |
| Monroe | 420 | 1,624,442 | 5.6 | 4.4 | 0.2 | 6.0 | 41.6 | 1.3 | 40.3 | 31.0 | 15.3 | 13.9 | 1,781 | 1,582 |
| Montgomery | 694 | 2,965,707 | 3.5 | 3.4 | 1.5 | 4.4 | 53.2 | 3.2 | 29.3 | 86.3 | 20.4 | 21.1 | 2,091 | 1,845 |
| Muscatine | 1,830 | 8,225,672 | 4.1 | 7.2 | 2.6 | 3.3 | 1.6 | 24.0 | 53.1 | 205.2 | 83.8 | 72.0 | 1,680 | 1,448 |
| O'Brien | 738 | 2,672,396 | 4.5 | 7.0 | 0.2 | 6.1 | 0.8 | 5.6 | 74.9 | 88.4 | 33.6 | 33.8 | 2,453 | 2,119 |
| Osceola | 205 | 739,165 | 10.8 | 9.0 | 0.0 | 14.8 | 0.0 | 8.4 | 56.0 | 29.2 | 6.2 | 10.0 | 1,655 | 1,512 |
| Page | 744 | 2,874,743 | 5.4 | 5.3 | 0.7 | 5.0 | 26.7 | 3.1 | 52.6 | 92.7 | 27.9 | 25.7 | 1,689 | 1,452 |
| Palo Alto | 660 | 2,358,550 | 5.2 | 3.6 | 0.8 | 6.3 | 30.0 | 3.3 | 49.8 | 73.6 | 18.3 | 22.5 | 2,488 | 2,246 |
| Plymouth | 1,185 | 4,839,864 | 4.5 | 5.8 | 0.5 | 5.9 | 26.5 | 4.1 | 52.2 | 185.6 | 42.7 | 47.0 | 1,877 | 1,670 |
| Pocahontas | 353 | 1,372,747 | 9.0 | 7.1 | 0.0 | 8.4 | 28.0 | 7.9 | 38.9 | 37.8 | 13.9 | 12.7 | 2,594 | 2,435 |
| Polk | 18,292 | 91,866,993 | 5.2 | 8.5 | 3.7 | 4.4 | 9.3 | 5.8 | 61.6 | 2,662.3 | 963.7 | 1,059.4 | 2,205 | 2,004 |
| Pottawattamie | 4,139 | 17,992,820 | 4.7 | 11.8 | 3.7 | 3.2 | 1.9 | 2.8 | 71.0 | 509.9 | 221.8 | 209.6 | 2,244 | 1,841 |
| Poweshiek | 617 | 2,484,969 | 5.8 | 7.7 | 1.1 | 6.3 | 0.7 | 7.0 | 70.5 | 73.5 | 29.2 | 33.7 | 1,837 | 1,590 |
| Ringgold | 398 | 1,587,151 | 4.8 | 3.2 | 0.3 | 4.9 | 40.1 | 1.5 | 45.0 | 45.8 | 11.8 | 10.4 | 2,075 | 1,892 |
| Sac | 427 | 1,636,145 | 8.3 | 7.7 | 0.0 | 9.0 | 1.8 | 10.3 | 60.9 | 41.7 | 18.0 | 17.9 | 1,831 | 1,639 |
| Scott | 6,416 | 29,440,373 | 5.0 | 8.2 | 3.6 | 3.5 | 1.6 | 5.3 | 70.0 | 821.4 | 348.6 | 341.7 | 1,981 | 1,677 |
| Shelby | 747 | 2,290,978 | 6.2 | 4.7 | 0.2 | 6.8 | 59.6 | 6.3 | 14.2 | 78.1 | 16.1 | 16.8 | 1,450 | 1,286 |
| Sioux | 1,589 | 6,244,439 | 6.2 | 4.9 | 0.1 | 4.0 | 31.3 | 5.1 | 47.3 | 192.7 | 52.7 | 58.2 | 1,674 | 1,406 |
| Story | 2,794 | 12,489,459 | 6.8 | 7.5 | 2.7 | 7.8 | 7.4 | 10.3 | 54.3 | 525.8 | 110.4 | 157.1 | 1,615 | 1,375 |
| Tama | 888 | 3,064,417 | 4.9 | 6.6 | 0.0 | 7.2 | 4.2 | 4.8 | 70.1 | 79.6 | 40.3 | 31.7 | 1,859 | 1,630 |
| Taylor | 269 | 882,058 | 10.7 | 5.8 | 0.0 | 10.5 | 7.3 | 2.6 | 62.7 | 27.2 | 14.5 | 9.8 | 1,613 | 1,430 |
| Union | 995 | 4,230,257 | 2.4 | 3.0 | 0.6 | 2.8 | 45.7 | 1.3 | 43.9 | 140.1 | 39.1 | 24.8 | 1,988 | 1,760 |
| Van Buren | 370 | 1,423,594 | 4.8 | 3.1 | 0.0 | 5.2 | 47.1 | 3.1 | 36.3 | 36.4 | 9.6 | 9.8 | 1,365 | 1,210 |
| Wapello | 2,085 | 8,499,871 | 2.7 | 4.5 | 1.8 | 2.8 | 2.2 | 4.4 | 80.9 | 188.5 | 82.6 | 51.4 | 1,467 | 1,210 |
| Warren | 1,616 | 6,933,651 | 4.8 | 6.0 | 1.6 | 3.6 | 2.4 | 5.2 | 75.0 | 202.9 | 91.7 | 76.6 | 1,529 | 1,349 |
| Washington | 1,143 | 4,532,672 | 4.5 | 5.5 | 0.4 | 3.8 | 31.6 | 4.4 | 49.5 | 135.6 | 38.4 | 39.8 | 1,792 | 1,505 |
| Wayne | 461 | 1,828,631 | 3.8 | 4.0 | 0.0 | 4.9 | 57.2 | 2.3 | 27.5 | 71.7 | 8.9 | 9.5 | 1,466 | 1,296 |
| Webster | 2,010 | 8,073,006 | 4.5 | 5.0 | 2.0 | 4.2 | 2.7 | 4.5 | 76.1 | 214.8 | 76.5 | 66.0 | 1,802 | 1,466 |
| Winnebago | 592 | 2,060,290 | 6.5 | 5.5 | 0.3 | 5.1 | 4.9 | 5.3 | 70.9 | 58.9 | 23.5 | 25.6 | 2,419 | 2,167 |
| Winneshiek | 1,424 | 5,316,173 | 3.2 | 3.5 | 0.4 | 3.7 | 34.6 | 2.8 | 51.3 | 181.8 | 46.9 | 43.9 | 2,178 | 1,924 |
| Woodbury | 4,357 | 20,192,238 | 4.3 | 8.9 | 3.8 | 5.0 | 2.4 | 4.9 | 69.5 | 520.8 | 255.6 | 182.2 | 1,784 | 1,389 |
| Worth | 333 | 1,159,232 | 10.3 | 8.6 | 0.2 | 10.3 | 4.0 | 2.7 | 62.3 | 38.7 | 14.8 | 17.2 | 2,303 | 1,999 |
| Wright | 1,003 | 4,509,426 | 3.8 | 3.7 | 0.1 | 3.9 | 51.5 | 1.9 | 34.2 | 159.5 | 31.6 | 27.6 | 2,162 | 1,959 |

1. Based on the resident population estimated as of July 1 of the year shown.

# Table B. States and Counties — Local Government Finances, Government Employment, and Income Taxes

| STATE County | Direct general expenditure — Total (mil dol) | Per capita (dollars) | Education | Health and hospitals | Police protection | Public welfare | Highways | Debt outstanding — Total (mil dol) | Per capita (dollars) | Federal civilian | Federal military | State and local | Number of returns | Mean adjusted gross income | Mean income tax |
|---|---|---|---|---|---|---|---|---|---|---|---|---|---|---|---|
| | 185 | 186 | 187 | 188 | 189 | 190 | 191 | 192 | 193 | 194 | 195 | 196 | 197 | 198 | 199 |
| **IOWA—Cont'd** | | | | | | | | | | | | | | | |
| Franklin | 83 | 8,178 | 35.3 | 31.1 | 2.7 | 0.1 | 8.6 | 49.4 | 4,857 | 39 | 36 | 723 | 4,330 | 56,975 | 4,780 |
| Fremont | 40 | 5,703 | 45.2 | 5.6 | 4.8 | 0.1 | 22.3 | 10.2 | 1,468 | 28 | 25 | 807 | 3,040 | 58,237 | 4,723 |
| Greene | 69 | 7,649 | 28.7 | 43.3 | 2.9 | 0.1 | 9.0 | 33.5 | 3,744 | 30 | 32 | 737 | 4,120 | 54,198 | 4,518 |
| Grundy | 77 | 6,281 | 37.4 | 36.3 | 3.0 | 0.0 | 8.3 | 70.6 | 5,742 | 28 | 44 | 862 | 5,720 | 70,954 | 6,945 |
| Guthrie | 78 | 7,368 | 40.0 | 24.7 | 2.4 | 0.1 | 10.5 | 54.8 | 5,145 | 57 | 38 | 1,250 | 5,090 | 65,808 | 6,614 |
| Hamilton | 115 | 7,636 | 35.7 | 30.3 | 3.1 | 0.1 | 12.0 | 99.5 | 6,619 | 39 | 53 | 1,250 | 6,960 | 59,037 | 5,752 |
| Hancock | 69 | 6,453 | 36.6 | 32.9 | 3.8 | 0.1 | 11.2 | 29.9 | 2,782 | 39 | 38 | 659 | 5,040 | 58,415 | 5,408 |
| Hardin | 111 | 6,484 | 39.9 | 22.5 | 3.7 | 0.0 | 10.9 | 72.0 | 4,216 | 69 | 58 | 1,658 | 7,710 | 53,473 | 4,770 |
| Harrison | 64 | 4,557 | 55.7 | 6.4 | 4.2 | 0.4 | 13.3 | 24.3 | 1,726 | 67 | 50 | 860 | 6,540 | 59,535 | 5,237 |
| Henry | 118 | 5,891 | 37.6 | 32.1 | 2.7 | 0.1 | 6.2 | 297.7 | 14,853 | 32 | 33 | 765 | 8,750 | 54,661 | 4,572 |
| Howard | 68 | 7,367 | 34.7 | 32.2 | 2.8 | 0.0 | 9.3 | 37.3 | 4,044 | 32 | 33 | 765 | 4,480 | 53,771 | 4,634 |
| Humboldt | 63 | 6,561 | 32.7 | 34.0 | 3.6 | 0.2 | 10.4 | 29.4 | 3,071 | 44 | 34 | 744 | 4,450 | 64,008 | 6,108 |
| Ida | 31 | 4,546 | 53.5 | 1.7 | 8.3 | 1.0 | 16.5 | 17.8 | 2,598 | 29 | 24 | 379 | 3,260 | 69,473 | 7,829 |
| Iowa | 101 | 6,282 | 34.8 | 36.2 | 3.5 | 0.4 | 9.7 | 78.2 | 4,846 | 57 | 58 | 1,016 | 7,990 | 63,142 | 6,077 |
| Jackson | 88 | 4,524 | 43.0 | 16.9 | 4.1 | 0.1 | 8.8 | 27.7 | 1,430 | 79 | 70 | 1,008 | 9,380 | 52,854 | 4,456 |
| Jasper | 143 | 3,853 | 51.3 | 0.9 | 4.9 | 0.0 | 2.9 | 121.6 | 3,287 | 87 | 129 | 1,949 | 16,770 | 59,041 | 5,070 |
| Jefferson | 97 | 5,343 | 22.5 | 44.7 | 3.0 | 0.2 | 8.8 | 103.0 | 5,656 | 66 | 60 | 1,184 | 7,240 | 52,339 | 5,046 |
| Johnson | 615 | 4,117 | 47.2 | 2.6 | 5.6 | 0.5 | 8.1 | 732.8 | 4,907 | 2,117 | 562 | 37,166 | 66,220 | 80,462 | 10,418 |
| Jones | 82 | 3,962 | 56.2 | 6.3 | 4.4 | 0.1 | 12.9 | 64.9 | 3,150 | 52 | 71 | 1,310 | 9,020 | 59,861 | 5,618 |
| Keokuk | 52 | 5,122 | 46.6 | 21.4 | 2.6 | 0.1 | 12.5 | 24.3 | 2,401 | 43 | 37 | 526 | 4,490 | 49,506 | 3,690 |
| Kossuth | 107 | 7,183 | 27.6 | 34.5 | 2.4 | 0.1 | 11.0 | 56.8 | 3,803 | 65 | 53 | 1,054 | 7,300 | 60,335 | 5,510 |
| Lee | 127 | 3,709 | 47.1 | 5.5 | 5.6 | 0.6 | 7.8 | 114.0 | 3,331 | 97 | 136 | 1,941 | 15,200 | 54,810 | 5,078 |
| Linn | 1,263 | 5,630 | 51.1 | 1.5 | 5.2 | 0.8 | 3.5 | 1,327.9 | 5,918 | 1,059 | 806 | 12,781 | 108,750 | 73,291 | 8,282 |
| Louisa | 54 | 4,807 | 60.7 | 2.4 | 5.0 | 0.2 | 9.9 | 75.2 | 6,720 | 58 | 40 | 649 | 5,020 | 53,964 | 4,130 |
| Lucas | 48 | 5,634 | 30.5 | 43.5 | 2.4 | 0.3 | 9.8 | 68.8 | 8,058 | 38 | 31 | 606 | 3,890 | 50,431 | 4,019 |
| Lyon | 54 | 4,545 | 47.1 | 1.0 | 5.9 | 0.7 | 19.2 | 32.0 | 2,715 | 39 | 42 | 681 | 5,200 | 62,323 | 5,577 |
| Madison | 87 | 5,410 | 48.6 | 25.4 | 2.7 | 0.1 | 8.3 | 121.1 | 7,571 | 39 | 59 | 970 | 7,460 | 71,047 | 7,023 |
| Mahaska | 125 | 5,635 | 31.2 | 43.0 | 2.4 | 0.2 | 7.2 | 37.6 | 1,694 | 56 | 78 | 1,351 | 9,600 | 60,014 | 5,191 |
| Marion | 135 | 4,087 | 59.9 | 3.6 | 4.8 | 0.1 | 8.8 | 124.2 | 3,753 | 130 | 116 | 1,697 | 14,930 | 71,547 | 7,218 |
| Marshall | 187 | 4,659 | 64.6 | 1.8 | 5.1 | 0.2 | 8.7 | 120.2 | 3,000 | 103 | 138 | 3,093 | 17,850 | 57,366 | 4,882 |
| Mills | 56 | 3,715 | 53.9 | 2.7 | 6.2 | 0.8 | 13.5 | 32.4 | 2,152 | 38 | 53 | 1,475 | 6,650 | 70,667 | 7,294 |
| Mitchell | 50 | 4,775 | 43.3 | 2.1 | 5.4 | 1.4 | 14.1 | 69.1 | 6,546 | 34 | 38 | 708 | 4,830 | 57,091 | 4,917 |
| Monona | 39 | 4,503 | 54.2 | 0.8 | 4.9 | 0.0 | 17.8 | 14.7 | 1,681 | 40 | 31 | 525 | 4,030 | 53,245 | 4,707 |
| Monroe | 57 | 7,309 | 23.7 | 38.6 | 4.3 | 0.0 | 1.1 | 18.4 | 2,360 | 41 | 28 | 533 | 3,430 | 54,137 | 4,815 |
| Montgomery | 83 | 8,236 | 25.9 | 50.4 | 3.0 | 0.1 | 6.6 | 28.0 | 2,769 | 39 | 35 | 934 | 4,640 | 53,211 | 4,566 |
| Muscatine | 184 | 4,300 | 47.7 | 2.3 | 5.2 | 0.5 | 5.9 | 78.6 | 1,834 | 83 | 153 | 2,521 | 20,450 | 58,697 | 5,254 |
| O'Brien | 90 | 6,531 | 60.8 | 2.9 | 3.8 | 0.1 | 9.8 | 41.7 | 3,026 | 46 | 49 | 1,078 | 6,610 | 61,184 | 5,689 |
| Osceola | 32 | 5,348 | 28.1 | 0.6 | 6.6 | 0.7 | 13.1 | 8.1 | 1,343 | 25 | 21 | 309 | 2,870 | 50,319 | 4,231 |
| Page | 86 | 5,654 | 36.2 | 39.5 | 3.1 | 0.0 | 7.6 | 29.6 | 1,943 | 77 | 49 | 1,276 | 6,420 | 52,471 | 4,266 |
| Palo Alto | 73 | 8,043 | 32.0 | 35.0 | 2.8 | 0.2 | 10.4 | 34.1 | 3,763 | 37 | 31 | 996 | 4,150 | 52,836 | 4,673 |
| Plymouth | 201 | 8,032 | 26.7 | 46.7 | 2.5 | 0.1 | 6.5 | 129.3 | 5,158 | 83 | 90 | 1,505 | 12,230 | 66,406 | 6,615 |
| Pocahontas | 34 | 4,934 | 39.9 | 3.6 | 4.8 | 0.5 | 14.5 | 32.4 | 4,749 | 32 | 24 | 559 | 3,200 | 54,521 | 4,449 |
| Polk | 2,663 | 5,542 | 49.0 | 8.5 | 4.7 | 1.1 | 3.6 | 2,914.1 | 6,065 | 5,993 | 1,875 | 30,532 | 231,440 | 75,290 | 8,934 |
| Pottawattamie | 488 | 5,228 | 59.5 | 1.5 | 5.1 | 0.6 | 4.6 | 394.2 | 4,219 | 215 | 332 | 5,480 | 42,540 | 60,002 | 5,681 |
| Poweshiek | 79 | 4,301 | 45.9 | 0.9 | 5.6 | 0.0 | 25.1 | 55.5 | 3,028 | 61 | 61 | 869 | 8,190 | 60,868 | 5,732 |
| Ringgold | 48 | 9,583 | 26.9 | 41.9 | 2.3 | 0.1 | 12.3 | 40.5 | 8,065 | 28 | 17 | 458 | 2,050 | 43,782 | 4,428 |
| Sac | 44 | 4,527 | 50.2 | 7.8 | 5.2 | 0.2 | 16.5 | 9.4 | 957 | 42 | 35 | 601 | 4,780 | 55,851 | 4,674 |
| Scott | 824 | 4,777 | 58.0 | 0.8 | 5.5 | 0.5 | 3.7 | 602.0 | 3,491 | 624 | 618 | 8,569 | 82,080 | 72,545 | 8,489 |
| Shelby | 87 | 7,530 | 20.4 | 47.4 | 2.2 | 0.0 | 8.7 | 40.5 | 3,502 | 39 | 41 | 1,040 | 5,700 | 59,930 | 5,407 |
| Sioux | 191 | 5,503 | 32.2 | 31.7 | 3.3 | 0.1 | 6.9 | 146.2 | 4,202 | 91 | 119 | 2,346 | 14,520 | 71,663 | 7,752 |
| Story | 503 | 5,168 | 30.3 | 42.8 | 3.3 | 0.5 | 2.6 | 467.5 | 4,806 | 894 | 346 | 19,173 | 37,750 | 72,631 | 7,999 |
| Tama | 83 | 4,859 | 61.4 | 5.6 | 3.9 | 0.1 | 12.9 | 41.7 | 2,449 | 63 | 60 | 2,003 | 8,310 | 53,606 | 4,536 |
| Taylor | 31 | 5,145 | 43.4 | 5.2 | 4.5 | 0.2 | 28.4 | 14.0 | 2,291 | 40 | 22 | 383 | 2,750 | 44,132 | 3,419 |
| Union | 135 | 10,853 | 35.9 | 43.0 | 1.5 | 0.1 | 4.9 | 45.3 | 3,636 | 47 | 43 | 1,383 | 5,480 | 50,347 | 4,094 |
| Van Buren | 42 | 5,829 | 36.1 | 40.3 | 0.5 | 0.0 | 9.9 | 12.9 | 1,808 | 34 | 25 | 509 | 3,060 | 49,449 | 4,109 |
| Wapello | 223 | 6,379 | 69.2 | 0.7 | 3.7 | 0.5 | 4.1 | 80.6 | 2,301 | 110 | 124 | 2,367 | 15,410 | 49,697 | 3,955 |
| Warren | 187 | 3,731 | 58.7 | 2.8 | 6.1 | 0.4 | 8.5 | 248.6 | 4,964 | 89 | 181 | 2,438 | 23,880 | 74,888 | 7,735 |
| Washington | 131 | 5,895 | 40.9 | 31.0 | 3.2 | 0.2 | 7.3 | 139.4 | 6,278 | 56 | 79 | 1,504 | 10,300 | 60,604 | 5,191 |
| Wayne | 73 | 11,311 | 15.0 | 68.1 | 2.3 | 0.4 | 6.3 | 19.2 | 2,953 | 29 | 23 | 616 | 2,670 | 38,823 | 3,028 |
| Webster | 265 | 7,224 | 60.6 | 2.5 | 2.6 | 0.2 | 4.7 | 218.9 | 5,974 | 182 | 120 | 2,859 | 16,090 | 55,667 | 4,998 |
| Winnebago | 63 | 5,994 | 50.1 | 5.4 | 4.4 | 0.1 | 19.2 | 37.5 | 3,544 | 43 | 39 | 721 | 4,970 | 53,513 | 4,327 |
| Winneshiek | 182 | 9,023 | 45.4 | 33.3 | 2.0 | 0.1 | 6.1 | 88.0 | 4,374 | 67 | 65 | 1,987 | 9,290 | 59,447 | 5,826 |
| Woodbury | 530 | 5,193 | 56.4 | 1.8 | 5.0 | 0.7 | 3.9 | 434.4 | 4,253 | 640 | 367 | 6,095 | 47,460 | 57,544 | 5,487 |
| Worth | 38 | 5,049 | 46.8 | 2.8 | 7.1 | 0.1 | 15.4 | 25.8 | 3,460 | 27 | 27 | 370 | 3,560 | 53,901 | 4,197 |
| Wright | 126 | 9,829 | 27.9 | 45.3 | 2.3 | 0.3 | 6.3 | 71.1 | 5,572 | 71 | 45 | 1,412 | 5,670 | 54,332 | 4,330 |

1. Based on the resident population estimated as of July 1 of the year shown.

# Table B. States and Counties — Land Area and Population

| State / county code | CBSA code[1] | County Type code[2] | STATE County | Land area[3] (sq. mi) | Total persons 2019 | Rank | Per square mile | White | Black | American Indian, Alaska Native | Asian and Pacific Islander | Percent Hispanic or Latino[4] | Under 5 years | 5 to 14 years | 15 to 24 years | 25 to 34 years | 35 to 44 years | 45 to 54 years |
|---|---|---|---|---|---|---|---|---|---|---|---|---|---|---|---|---|---|---|
| | | | | 1 | 2 | 3 | 4 | 5 | 6 | 7 | 8 | 9 | 10 | 11 | 12 | 13 | 14 | 15 |
| 20000 | | 0 | KANSAS | 81,758.5 | 2,913,805 | X | 35.6 | 77.6 | 7.1 | 1.8 | 4.0 | 12.4 | 6.3 | 13.5 | 14.1 | 13.1 | 12.5 | 11.1 |
| 20001 | | 7 | Allen | 500.3 | 12,399 | 2,256 | 24.8 | 91.9 | 3.3 | 1.9 | 1.5 | 4.1 | 6.1 | 12.1 | 13.1 | 11.1 | 11.4 | 11.1 |
| 20003 | | 6 | Anderson | 579.6 | 7,949 | 2,595 | 13.7 | 95.5 | 1.4 | 1.7 | 1.0 | 2.3 | 6.8 | 14.6 | 11.8 | 10.0 | 10.6 | 10.9 |
| 20005 | 11860 | 6 | Atchison | 431.2 | 16,015 | 2,034 | 37.1 | 90.7 | 6.2 | 1.5 | 1.1 | 3.3 | 5.5 | 13.0 | 17.9 | 10.7 | 11.0 | 10.7 |
| 20007 | | 9 | Barber | 1,134.1 | 4,358 | 2,868 | 3.8 | 92.6 | 1.8 | 1.7 | 0.7 | 5.0 | 6.4 | 13.2 | 9.9 | 10.4 | 10.5 | 9.4 |
| 20009 | 24460 | 7 | Barton | 895.3 | 25,658 | 1,571 | 28.7 | 82.1 | 2.2 | 1.2 | 0.5 | 15.6 | 6.1 | 13.6 | 12.4 | 11.4 | 11.4 | 9.9 |
| 20011 | | 6 | Bourbon | 635.5 | 14,435 | 2,125 | 22.7 | 92.0 | 4.1 | 2.1 | 1.6 | 3.0 | 6.7 | 14.6 | 13.1 | 11.1 | 11.1 | 10.2 |
| 20013 | | 6 | Brown | 570.9 | 9,482 | 2,456 | 16.6 | 84.6 | 2.6 | 8.9 | 1.0 | 5.6 | 6.8 | 14.6 | 10.7 | 10.7 | 11.4 | 10.5 |
| 20015 | 48620 | 2 | Butler | 1,429.7 | 66,992 | 801 | 46.9 | 90.3 | 3.0 | 2.0 | 2.0 | 5.3 | 5.8 | 14.6 | 13.2 | 11.8 | 13.3 | 11.7 |
| 20017 | 21380 | 9 | Chase | 773.1 | 2,586 | 2,997 | 3.3 | 91.9 | 2.6 | 1.5 | 0.5 | 5.4 | 5.0 | 11.8 | 11.3 | 11.6 | 10.1 | 11.3 |
| 20019 | | 8 | Chautauqua | 638.9 | 3,230 | 2,950 | 5.1 | 88.9 | 2.8 | 8.0 | 0.8 | 4.6 | 5.7 | 11.7 | 9.9 | 10.2 | 10.7 | 9.7 |
| 20021 | | 6 | Cherokee | 587.6 | 19,681 | 1,841 | 33.5 | 90.8 | 1.7 | 7.1 | 1.3 | 3.1 | 6.0 | 12.5 | 12.0 | 10.8 | 11.6 | 12.3 |
| 20023 | | 9 | Cheyenne | 1,019.6 | 2,600 | 2,994 | 2.5 | 90.8 | 0.8 | 0.8 | 1.3 | 7.1 | 6.0 | 11.7 | 9.5 | 10.7 | 10.3 | 10.0 |
| 20025 | | 9 | Clark | 974.6 | 1,963 | 3,051 | 2.0 | 85.5 | 1.7 | 2.1 | 2.0 | 11.3 | 5.9 | 14.5 | 11.4 | 11.1 | 11.7 | 9.9 |
| 20027 | | 6 | Clay | 645.3 | 8,025 | 2,587 | 12.4 | 94.8 | 1.6 | 1.3 | 1.1 | 3.1 | 5.9 | 13.8 | 10.2 | 9.9 | 12.4 | 10.9 |
| 20029 | | 7 | Cloud | 715.3 | 8,642 | 2,524 | 12.1 | 93.5 | 1.9 | 1.2 | 1.4 | 3.7 | 6.1 | 13.1 | 13.4 | 11.1 | 10.7 | 10.5 |
| 20031 | | 6 | Coffey | 626.9 | 8,158 | 2,569 | 13.0 | 94.1 | 1.3 | 1.6 | 1.2 | 3.3 | 5.0 | 12.2 | 11.0 | 11.1 | 11.2 | 11.9 |
| 20033 | | 9 | Comanche | 788.3 | 1,690 | 3,069 | 2.1 | 90.8 | 1.4 | 1.0 | 0.8 | 7.9 | 4.5 | 15.1 | 11.4 | 8.0 | 9.8 | 10.5 |
| 20035 | 49060 | 4 | Cowley | 1,125.7 | 34,628 | 1,314 | 30.8 | 81.7 | 3.9 | 3.2 | 2.8 | 11.4 | 5.6 | 13.6 | 13.9 | 11.8 | 11.8 | 11.4 |
| 20037 | 38260 | 4 | Crawford | 589.8 | 38,730 | 1,206 | 65.7 | 88.2 | 3.2 | 2.1 | 2.7 | 6.4 | 6.0 | 12.1 | 21.5 | 11.5 | 11.3 | 10.1 |
| 20039 | | 9 | Decatur | 893.5 | 2,776 | 2,980 | 3.1 | 94.8 | 1.8 | 1.3 | 0.5 | 3.3 | 5.7 | 11.3 | 8.1 | 10.5 | 9.3 | 9.4 |
| 20041 | | 7 | Dickinson | 847.1 | 18,266 | 1,911 | 21.6 | 92.1 | 2.0 | 1.5 | 1.2 | 5.3 | 5.1 | 13.4 | 11.2 | 11.3 | 11.9 | 11.5 |
| 20043 | 41140 | 3 | Doniphan | 393.5 | 7,496 | 2,624 | 19.0 | 92.4 | 4.1 | 2.0 | 0.9 | 2.7 | 5.6 | 11.8 | 14.8 | 10.7 | 11.2 | 11.7 |
| 20045 | 29940 | 3 | Douglas | 455.8 | 122,530 | 517 | 268.8 | 81.8 | 6.1 | 3.5 | 6.1 | 6.6 | 4.7 | 10.1 | 26.0 | 14.2 | 12.2 | 9.5 |
| 20047 | | 9 | Edwards | 621.9 | 2,750 | 2,986 | 4.4 | 76.9 | 1.8 | 1.8 | 0.7 | 20.6 | 5.5 | 12.6 | 12.2 | 9.0 | 10.8 | 10.9 |
| 20049 | | 8 | Elk | 644.3 | 2,507 | 3,003 | 3.9 | 93.1 | 1.6 | 3.6 | 1.3 | 4.2 | 5.7 | 14.0 | 8.4 | 8.8 | 8.6 | 10.6 |
| 20051 | 25700 | 5 | Ellis | 899.9 | 28,671 | 1,471 | 31.9 | 90.5 | 2.1 | 0.9 | 1.9 | 6.4 | 5.4 | 12.0 | 20.5 | 12.8 | 11.6 | 9.6 |
| 20053 | | 7 | Ellsworth | 715.6 | 6,034 | 2,741 | 8.4 | 86.6 | 6.0 | 1.5 | 0.8 | 6.3 | 4.3 | 10.6 | 11.2 | 13.6 | 13.2 | 11.3 |
| 20055 | 23780 | 5 | Finney | 1,302.0 | 35,917 | 1,278 | 27.6 | 40.9 | 4.4 | 0.8 | 4.2 | 50.9 | 8.4 | 16.6 | 15.5 | 14.2 | 12.0 | 10.9 |
| 20057 | 19980 | 5 | Ford | 1,098.3 | 33,094 | 1,354 | 30.1 | 39.5 | 2.9 | 0.8 | 1.7 | 56.1 | 8.5 | 16.5 | 15.0 | 13.9 | 12.0 | 11.3 |
| 20059 | 36840 | 6 | Franklin | 571.8 | 25,703 | 1,569 | 45.0 | 92.8 | 2.3 | 1.9 | 0.9 | 4.6 | 6.0 | 13.2 | 12.7 | 12.2 | 11.8 | 11.7 |
| 20061 | 31740 | 4 | Geary | 384.7 | 32,218 | 1,374 | 83.7 | 60.9 | 19.6 | 2.1 | 6.4 | 17.0 | 12.0 | 16.1 | 17.7 | 21.9 | 11.7 | 6.0 |
| 20063 | | 9 | Gove | 1,071.7 | 2,621 | 2,991 | 2.4 | 94.5 | 1.0 | 0.5 | 1.4 | 3.8 | 6.7 | 14.2 | 10.4 | 9.1 | 11.1 | 9.6 |
| 20065 | | 9 | Graham | 898.5 | 2,389 | 3,010 | 2.7 | 89.9 | 4.5 | 1.8 | 2.4 | 4.2 | 4.1 | 13.3 | 9.5 | 8.9 | 11.1 | 9.7 |
| 20067 | | 7 | Grant | 574.8 | 7,077 | 2,652 | 12.3 | 49.4 | 0.9 | 1.1 | 0.8 | 49.0 | 7.9 | 17.1 | 14.4 | 11.3 | 11.9 | 10.7 |
| 20069 | | 9 | Gray | 868.9 | 5,954 | 2,747 | 6.9 | 81.8 | 1.1 | 0.8 | 0.8 | 16.4 | 7.0 | 16.7 | 13.7 | 10.9 | 12.2 | 10.6 |
| 20071 | | 9 | Greeley | 778.4 | 1,196 | 3,100 | 1.5 | 77.8 | 0.6 | 0.9 | 0.7 | 20.9 | 6.6 | 16.0 | 10.2 | 10.8 | 13.3 | 6.8 |
| 20073 | | 6 | Greenwood | 1,143.3 | 5,868 | 2,756 | 5.1 | 93.6 | 1.2 | 2.7 | 0.8 | 4.1 | 4.9 | 12.5 | 10.1 | 8.6 | 10.3 | 10.7 |
| 20075 | | 9 | Hamilton | 996.5 | 2,425 | 3,009 | 2.4 | 61.1 | 1.2 | 1.8 | 0.7 | 36.8 | 5.9 | 16.8 | 11.8 | 13.2 | 12.7 | 10.7 |
| 20077 | | 8 | Harper | 801.3 | 5,336 | 2,800 | 6.7 | 90.9 | 1.3 | 2.4 | 0.8 | 6.5 | 6.2 | 14.7 | 10.1 | 10.4 | 10.9 | 10.6 |
| 20079 | 48620 | 2 | Harvey | 539.8 | 34,291 | 1,324 | 63.5 | 84.0 | 2.7 | 1.6 | 1.5 | 12.5 | 5.5 | 14.0 | 14.0 | 11.1 | 12.1 | 10.2 |
| 20081 | | 9 | Haskell | 577.5 | 3,923 | 2,901 | 6.8 | 66.5 | 1.0 | 1.2 | 0.9 | 31.8 | 7.3 | 15.4 | 14.6 | 12.4 | 10.8 | 11.3 |
| 20083 | | 9 | Hodgeman | 860.0 | 1,779 | 3,067 | 2.1 | 86.9 | 2.2 | 0.8 | 4.0 | 7.4 | 5.7 | 13.4 | 10.7 | 10.8 | 10.2 | 9.7 |
| 20085 | 45820 | 3 | Jackson | 656.2 | 13,171 | 2,209 | 20.1 | 86.0 | 1.8 | 9.7 | 0.9 | 5.3 | 6.7 | 14.0 | 12.0 | 10.3 | 11.2 | 12.2 |
| 20087 | 45820 | 3 | Jefferson | 532.6 | 19,032 | 1,874 | 35.7 | 94.9 | 1.2 | 2.0 | 0.7 | 3.1 | 5.4 | 12.5 | 11.4 | 10.4 | 11.5 | 12.5 |
| 20089 | | 9 | Jewell | 910.0 | 2,833 | 2,975 | 3.1 | 96.0 | 0.8 | 1.2 | 0.7 | 2.6 | 5.4 | 11.5 | 9.7 | 9.0 | 8.9 | 8.6 |
| 20091 | 28140 | 1 | Johnson | 473.6 | 607,220 | 113 | 1,282.1 | 81.3 | 5.9 | 0.9 | 6.4 | 8.1 | 6.1 | 13.4 | 12.1 | 13.7 | 14.2 | 12.5 |
| 20093 | 23780 | 9 | Kearny | 870.5 | 3,745 | 2,917 | 4.3 | 65.3 | 1.9 | 2.1 | 0.9 | 31.9 | 7.5 | 17.4 | 12.4 | 12.9 | 11.4 | 10.1 |
| 20095 | | 2 | Kingman | 863.4 | 6,974 | 2,660 | 8.1 | 94.2 | 0.8 | 1.9 | 1.0 | 3.9 | 5.3 | 12.1 | 11.3 | 10.2 | 11.4 | 11.2 |
| 20097 | | 9 | Kiowa | 722.6 | 2,456 | 3,005 | 3.4 | 90.3 | 2.3 | 1.7 | 2.2 | 5.4 | 6.8 | 13.3 | 13.9 | 10.3 | 9.9 | 10.1 |
| 20099 | 37660 | 7 | Labette | 645.4 | 19,586 | 1,848 | 30.3 | 88.5 | 5.8 | 4.0 | 1.0 | 4.7 | 6.6 | 13.7 | 11.3 | 11.7 | 10.7 | 11.4 |
| 20101 | | 9 | Lane | 717.4 | 1,518 | 3,078 | 2.1 | 87.7 | 1.6 | 2.6 | 0.7 | 9.9 | 6.0 | 13.7 | 9.6 | 10.7 | 8.8 | 12.1 |
| 20103 | 28140 | 1 | Leavenworth | 463.4 | 82,246 | 692 | 177.5 | 81.3 | 10.3 | 1.6 | 2.8 | 7.3 | 6.4 | 13.3 | 11.6 | 13.8 | 14.6 | 12.2 |
| 20105 | | 9 | Lincoln | 719.4 | 2,986 | 2,959 | 4.2 | 94.5 | 1.1 | 1.2 | 0.8 | 3.8 | 5.2 | 12.8 | 11.1 | 7.7 | 11.6 | 10.3 |
| 20107 | 28140 | 1 | Linn | 594.1 | 9,654 | 2,441 | 16.2 | 94.9 | 1.4 | 2.2 | 0.8 | 3.0 | 4.9 | 12.3 | 11.1 | 9.2 | 11.4 | 12.1 |
| 20109 | | 9 | Logan | 1,073.0 | 2,732 | 2,987 | 2.5 | 91.4 | 1.6 | 1.3 | 1.4 | 6.4 | 8.2 | 13.0 | 10.0 | 12.6 | 11.2 | 8.9 |
| 20111 | 21380 | 4 | Lyon | 847.5 | 33,045 | 1,355 | 39.0 | 72.8 | 3.2 | 1.4 | 2.9 | 22.1 | 6.0 | 12.3 | 20.9 | 12.6 | 10.6 | 9.6 |
| 20113 | 32700 | 7 | McPherson | 898.3 | 28,448 | 1,481 | 31.7 | 92.5 | 2.1 | 1.1 | 1.4 | 4.7 | 5.5 | 12.8 | 12.9 | 11.6 | 12.2 | 10.2 |
| 20115 | | 6 | Marion | 944.3 | 11,652 | 2,313 | 12.3 | 93.6 | 1.4 | 1.7 | 1.0 | 4.1 | 4.7 | 12.2 | 13.9 | 9.4 | 10.4 | 10.4 |
| 20117 | | 6 | Marshall | 900.2 | 9,652 | 2,442 | 10.7 | 95.8 | 1.3 | 1.2 | 0.9 | 2.6 | 6.4 | 13.8 | 10.6 | 9.7 | 11.8 | 9.6 |
| 20119 | | 9 | Meade | 978.1 | 4,029 | 2,894 | 4.1 | 77.5 | 1.5 | 1.6 | 1.1 | 19.5 | 7.0 | 15.0 | 12.2 | 12.2 | 10.2 | 11.4 |
| 20121 | 28140 | 1 | Miami | 575.9 | 34,334 | 1,321 | 59.6 | 93.9 | 2.0 | 1.5 | 1.1 | 3.4 | 5.8 | 13.7 | 11.9 | 10.4 | 12.6 | 12.9 |
| 20123 | | 7 | Mitchell | 701.8 | 5,879 | 2,755 | 8.4 | 95.6 | 0.6 | 1.1 | 1.4 | 2.6 | 6.1 | 13.1 | 13.1 | 9.1 | 10.8 | 9.8 |
| 20125 | 17700 | 5 | Montgomery | 643.6 | 31,502 | 1,391 | 48.9 | 83.3 | 6.6 | 6.2 | 1.8 | 7.4 | 5.8 | 13.4 | 13.1 | 10.9 | 11.2 | 10.8 |
| 20127 | | 8 | Morris | 695.3 | 5,559 | 2,784 | 8.0 | 92.8 | 1.2 | 1.7 | 1.3 | 5.5 | 5.5 | 11.8 | 9.3 | 10.6 | 11.2 | 10.1 |
| 20129 | | 9 | Morton | 729.7 | 2,538 | 2,999 | 3.5 | 72.7 | 1.7 | 2.3 | 1.8 | 23.2 | 5.5 | 13.9 | 11.8 | 9.3 | 11.4 | 11.8 |
| 20131 | | 6 | Nemaha | 717.4 | 10,121 | 2,408 | 14.1 | 96.0 | 1.2 | 1.2 | 0.7 | 2.3 | 7.3 | 15.1 | 11.7 | 10.5 | 10.9 | 10.3 |

1. CBSA = Core Based Statistical Area. See Appendix A for explanation. See Appendix B for list of metropolitan areas with component counties. Service of USDA Rural-Urban Continuum Codes. See Appendix A for definition. 3. Dry land or land partially or temporarily covered by water.
2. County type code from the Economic Research
4. May be of any race.

Items 1—15

# Table B. States and Counties — Population and Households

| STATE County | Age (percent) (cont.) | | | | Population change, 2000-2020 | | | | | | | Households, 2015-2019 | | | | |
|---|---|---|---|---|---|---|---|---|---|---|---|---|---|---|---|---|
| | 55 to 64 years | 65 to 74 years | 75 years and over | Percent female | Total persons | | Percent change | | Components of change, 2010-2020 | | | | Persons per household | Family house-holds | Female family house-holder[1] | One person |
| | | | | | 2000 | 2010 | 2000-2010 | 2010-2020 | Births | Deaths | Net Migration | Number | | | | |
| | 16 | 17 | 18 | 19 | 20 | 21 | 22 | 23 | 24 | 25 | 26 | 27 | 28 | 29 | 30 | 31 |
| KANSAS | 12.6 | 9.7 | 7.1 | 50.2 | 2,688,418 | 2,853,120 | 6.1 | 2.1 | 391,375 | 267,583 | -63,016 | 1,129,227 | 2.51 | 64.7 | 9.7 | 29.1 |
| Allen | 13.6 | 11.9 | 9.7 | 50.9 | 14,385 | 13,372 | -7.0 | -7.3 | 1,517 | 1,771 | -722 | 5,372 | 2.26 | 60.7 | 11.8 | 31.4 |
| Anderson | 14.0 | 11.0 | 10.4 | 50.3 | 8,110 | 8,104 | -0.1 | -1.9 | 1,020 | 1,015 | -158 | 3,101 | 2.49 | 69.8 | 6.1 | 28.2 |
| Atchison | 13.2 | 9.7 | 8.3 | 51.6 | 16,774 | 16,921 | 0.9 | -5.4 | 2,017 | 1,855 | -1,075 | 5,958 | 2.48 | 64.8 | 10.3 | 31.7 |
| Barber | 15.6 | 14.0 | 10.5 | 49.7 | 5,307 | 4,864 | -8.3 | -10.4 | 610 | 602 | -516 | 1,923 | 2.38 | 61.8 | 8.5 | 32.8 |
| Barton | 14.7 | 11.4 | 9.1 | 50.7 | 28,205 | 27,672 | -1.9 | -7.3 | 3,504 | 3,076 | -2,439 | 10,624 | 2.45 | 63.6 | 8.7 | 30.7 |
| Bourbon | 12.8 | 11.6 | 8.8 | 50.6 | 15,379 | 15,173 | -1.3 | -4.9 | 2,080 | 1,942 | -875 | 5,636 | 2.52 | 67.3 | 9.5 | 27.9 |
| Brown | 14.2 | 12.3 | 8.8 | 50.5 | 10,724 | 9,984 | -6.9 | -5.0 | 1,354 | 1,319 | -538 | 3,872 | 2.46 | 66.6 | 7.6 | 29.2 |
| Butler | 13.4 | 9.6 | 6.6 | 49.6 | 59,482 | 65,884 | 10.8 | 1.7 | 7,628 | 6,557 | 41 | 24,870 | 2.57 | 71.9 | 8.7 | 24.1 |
| Chase | 13.8 | 13.7 | 11.4 | 48.8 | 3,030 | 2,790 | -7.9 | -7.3 | 255 | 350 | -107 | 1,051 | 2.42 | 62.2 | 7.5 | 34.5 |
| Chautauqua | 16.1 | 13.2 | 12.7 | 49.0 | 4,359 | 3,669 | -15.8 | -12.0 | 348 | 559 | -228 | 1,456 | 2.24 | 61.8 | 9.5 | 33.7 |
| Cherokee | 15.0 | 11.3 | 8.5 | 50.7 | 22,605 | 21,610 | -4.4 | -8.9 | 2,373 | 2,768 | -1,539 | 7,929 | 2.51 | 65.4 | 9.7 | 30.4 |
| Cheyenne | 14.3 | 14.0 | 13.6 | 49.7 | 3,165 | 2,726 | -13.9 | -4.6 | 299 | 391 | -32 | 1,221 | 2.15 | 55.4 | 7.1 | 41.1 |
| Clark | 13.2 | 10.7 | 11.5 | 51.7 | 2,390 | 2,215 | -7.3 | -11.4 | 244 | 320 | -178 | 873 | 2.25 | 63.2 | 8.0 | 30.6 |
| Clay | 13.2 | 12.6 | 11.3 | 50.3 | 8,822 | 8,542 | -3.2 | -6.1 | 1,019 | 1,073 | -462 | 3,514 | 2.26 | 64.7 | 5.7 | 33.2 |
| Cloud | 13.5 | 11.2 | 10.6 | 50.2 | 10,268 | 9,533 | -7.2 | -9.3 | 1,127 | 1,407 | -609 | 3,689 | 2.33 | 63.7 | 7.9 | 30.5 |
| Coffey | 15.0 | 12.8 | 9.9 | 50.7 | 8,865 | 8,598 | -3.0 | -5.1 | 859 | 1,006 | -290 | 3,585 | 2.27 | 69.2 | 8.7 | 23.9 |
| Comanche | 13.7 | 14.1 | 12.8 | 51.2 | 1,967 | 1,891 | -3.9 | -10.6 | 202 | 305 | -100 | 751 | 2.22 | 61.0 | 6.7 | 36.1 |
| Cowley | 12.9 | 10.4 | 8.5 | 50.0 | 36,291 | 36,309 | 0.0 | -4.6 | 4,400 | 4,436 | -1,649 | 13,499 | 2.45 | 65.8 | 10.7 | 29.6 |
| Crawford | 11.3 | 9.1 | 7.0 | 50.2 | 38,242 | 39,135 | 2.3 | -1.0 | 4,903 | 4,269 | -1,031 | 15,284 | 2.42 | 59.8 | 11.9 | 30.1 |
| Decatur | 16.2 | 15.0 | 14.4 | 50.0 | 3,472 | 2,965 | -14.6 | -6.4 | 325 | 473 | -41 | 1,442 | 1.94 | 54.2 | 6.2 | 40.1 |
| Dickinson | 14.8 | 11.1 | 9.6 | 49.9 | 19,344 | 19,747 | 2.1 | -7.5 | 2,199 | 2,382 | -1,297 | 7,911 | 2.34 | 63.1 | 6.2 | 33.1 |
| Doniphan | 14.0 | 11.5 | 8.7 | 49.7 | 8,249 | 7,948 | -3.6 | -5.7 | 844 | 820 | -473 | 3,046 | 2.28 | 62.3 | 6.7 | 31.0 |
| Douglas | 10.1 | 8.2 | 5.1 | 50.3 | 99,962 | 110,826 | 10.9 | 10.6 | 12,245 | 7,120 | 6,605 | 46,936 | 2.38 | 54.0 | 7.4 | 29.5 |
| Edwards | 16.5 | 12.1 | 10.4 | 49.2 | 3,449 | 3,037 | -11.9 | -9.5 | 324 | 369 | -248 | 1,308 | 2.12 | 64.2 | 7.0 | 33.4 |
| Elk | 15.4 | 15.0 | 13.6 | 50.5 | 3,261 | 2,882 | -11.6 | -13.0 | 273 | 431 | -221 | 1,128 | 2.21 | 62.4 | 6.2 | 34.1 |
| Ellis | 11.3 | 9.5 | 7.2 | 50.0 | 27,507 | 28,452 | 3.4 | 0.8 | 3,562 | 2,567 | -785 | 11,544 | 2.39 | 56.3 | 9.5 | 34.1 |
| Ellsworth | 14.2 | 12.0 | 9.6 | 43.1 | 6,525 | 6,499 | -0.4 | -7.2 | 600 | 853 | -211 | 2,388 | 2.24 | 63.8 | 6.3 | 32.4 |
| Finney | 10.7 | 6.9 | 4.7 | 48.6 | 40,523 | 36,785 | -9.2 | -2.4 | 6,901 | 2,202 | -5,611 | 12,519 | 2.89 | 70.9 | 11.8 | 24.2 |
| Ford | 10.8 | 6.8 | 5.0 | 48.1 | 32,458 | 33,844 | 4.3 | -2.2 | 6,512 | 2,474 | -4,831 | 11,344 | 2.95 | 72.1 | 11.0 | 22.8 |
| Franklin | 14.6 | 10.4 | 7.3 | 50.0 | 24,784 | 25,994 | 4.9 | -1.1 | 3,194 | 2,677 | -806 | 10,118 | 2.47 | 67.1 | 8.5 | 26.7 |
| Geary | 5.5 | 5.0 | 4.1 | 47.3 | 27,947 | 34,352 | 22.9 | -6.2 | 10,154 | 2,108 | -10,398 | 12,684 | 2.61 | 71.6 | 12.6 | 23.5 |
| Gove | 14.3 | 12.3 | 12.4 | 49.6 | 3,068 | 2,701 | -12.0 | -3.0 | 364 | 380 | -67 | 1,217 | 2.13 | 66.1 | 3.8 | 29.7 |
| Graham | 16.4 | 13.2 | 13.9 | 51.1 | 2,946 | 2,588 | -12.2 | -7.7 | 243 | 359 | -85 | 1,196 | 2.07 | 60.9 | 6.4 | 35.9 |
| Grant | 11.7 | 8.8 | 6.1 | 49.5 | 7,909 | 7,824 | -1.1 | -9.5 | 1,242 | 539 | -1,468 | 2,551 | 2.90 | 70.8 | 6.7 | 24.3 |
| Gray | 12.4 | 8.9 | 7.6 | 49.2 | 5,904 | 6,001 | 1.6 | -0.8 | 866 | 495 | -426 | 2,151 | 2.77 | 73.5 | 5.9 | 23.9 |
| Greeley | 16.4 | 10.1 | 9.9 | 50.1 | 1,534 | 1,248 | -18.6 | -4.2 | 182 | 195 | -41 | 519 | 2.22 | 65.1 | 6.0 | 31.4 |
| Greenwood | 16.1 | 15.0 | 11.7 | 49.6 | 7,673 | 6,689 | -12.8 | -12.3 | 621 | 959 | -483 | 2,769 | 2.17 | 63.1 | 8.5 | 34.0 |
| Hamilton | 13.7 | 9.1 | 6.1 | 48.0 | 2,670 | 2,687 | 0.6 | -9.8 | 373 | 251 | -386 | 890 | 2.91 | 60.8 | 6.9 | 36.2 |
| Harper | 13.2 | 12.6 | 11.4 | 49.9 | 6,536 | 6,034 | -7.7 | -11.6 | 727 | 929 | -493 | 2,265 | 2.40 | 65.6 | 6.8 | 30.6 |
| Harvey | 13.1 | 10.3 | 9.7 | 50.4 | 32,869 | 34,684 | 5.5 | -1.1 | 4,211 | 4,053 | -542 | 13,290 | 2.49 | 68.4 | 9.6 | 27.6 |
| Haskell | 12.6 | 9.5 | 6.2 | 49.9 | 4,307 | 4,256 | -1.2 | -7.8 | 559 | 276 | -624 | 1,320 | 3.01 | 75.7 | 7.9 | 20.2 |
| Hodgeman | 16.6 | 11.2 | 11.6 | 48.6 | 2,085 | 1,918 | -8.0 | -7.2 | 201 | 224 | -120 | 767 | 2.40 | 71.1 | 5.9 | 26.1 |
| Jackson | 14.1 | 11.2 | 8.3 | 49.8 | 12,657 | 13,460 | 6.3 | -2.1 | 1,717 | 1,412 | -598 | 5,465 | 2.40 | 69.3 | 9.7 | 26.0 |
| Jefferson | 16.2 | 11.8 | 8.4 | 49.3 | 18,426 | 19,108 | 3.7 | -0.4 | 1,928 | 1,856 | -141 | 7,575 | 2.46 | 71.4 | 7.6 | 22.8 |
| Jewell | 15.7 | 16.5 | 14.8 | 48.6 | 3,791 | 3,073 | -18.9 | -7.8 | 287 | 418 | -109 | 1,380 | 2.07 | 61.3 | 5.7 | 35.7 |
| Johnson | 12.3 | 9.3 | 6.2 | 50.9 | 451,086 | 544,181 | 20.6 | 11.6 | 74,925 | 39,136 | 27,622 | 228,592 | 2.57 | 68.3 | 8.5 | 26.0 |
| Kearny | 12.3 | 9.3 | 6.2 | 50.1 | 4,531 | 3,982 | -12.1 | -6.0 | 628 | 360 | -510 | 1,201 | 3.20 | 74.9 | 10.6 | 19.3 |
| Kingman | 11.6 | 8.9 | 7.8 | 49.9 | 8,673 | 7,860 | -9.4 | -11.3 | 779 | 1,260 | -407 | 3,133 | 2.30 | 68.1 | 8.1 | 29.6 |
| Kiowa | 16.5 | 11.6 | 10.4 | 50.9 | 3,278 | 2,553 | -22.1 | -3.8 | 360 | 273 | -188 | 985 | 2.35 | 65.4 | 4.0 | 30.8 |
| Labette | 13.2 | 12.9 | 9.6 | 50.0 | 22,835 | 21,609 | -5.4 | -9.4 | 2,743 | 2,698 | -2,081 | 8,176 | 2.39 | 68.2 | 12.1 | 28.8 |
| Lane | 14.5 | 11.2 | 9.0 | 50.3 | 2,155 | 1,749 | -18.8 | -13.2 | 193 | 268 | -156 | 757 | 2.06 | 53.1 | 4.2 | 42.8 |
| Leavenworth | 15.3 | 13.0 | 10.9 | 46.8 | 68,691 | 76,220 | 11.0 | 7.9 | 10,016 | 6,171 | 2,219 | 27,253 | 2.75 | 71.8 | 10.2 | 24.9 |
| Lincoln | 12.7 | 9.4 | 6.1 | 49.8 | 3,578 | 3,237 | -9.5 | -7.8 | 311 | 378 | -187 | 1,248 | 2.39 | 59.3 | 9.4 | 36.9 |
| Linn | 15.5 | 14.7 | 11.1 | 48.9 | 9,570 | 9,656 | 0.9 | 0.0 | 1,008 | 1,181 | 170 | 4,497 | 2.14 | 66.4 | 8.5 | 29.2 |
| Logan | 16.1 | 13.3 | 9.7 | 49.8 | 3,046 | 2,759 | -9.4 | -1.0 | 413 | 355 | -83 | 1,153 | 2.40 | 64.9 | 8.0 | 32.2 |
| Lyon | 14.6 | 11.4 | 10.1 | 51.2 | 35,935 | 33,692 | -6.2 | -1.9 | 4,217 | 2,933 | -1,964 | 13,569 | 2.34 | 58.9 | 8.0 | 29.3 |
| McPherson | 11.8 | 9.3 | 7.0 | 50.7 | 29,554 | 29,186 | -1.2 | -2.5 | 3,301 | 3,799 | -240 | 12,334 | 2.21 | 64.9 | 7.0 | 30.4 |
| Marion | 14.0 | 10.9 | 9.8 | 50.1 | 13,361 | 12,660 | -5.2 | -8.0 | 1,170 | 1,670 | -507 | 4,846 | 2.30 | 66.0 | 6.8 | 28.6 |
| Marshall | 15.0 | 12.3 | 11.7 | 50.2 | 10,965 | 10,117 | -7.7 | -4.6 | 1,228 | 1,303 | -384 | 4,085 | 2.35 | 63.0 | 5.3 | 32.5 |
| Meade | 15.1 | 12.1 | 10.9 | 48.9 | 4,631 | 4,575 | -1.2 | -11.9 | 570 | 494 | -633 | 1,689 | 2.40 | 69.6 | 9.3 | 27.8 |
| Miami | 12.9 | 9.9 | 9.3 | 50.0 | 28,351 | 32,781 | 15.6 | 4.7 | 3,744 | 2,946 | 755 | 12,835 | 2.55 | 71.2 | 8.4 | 24.9 |
| Mitchell | 14.7 | 10.1 | 7.8 | 49.2 | 6,932 | 6,373 | -8.1 | -7.8 | 785 | 923 | -359 | 2,570 | 2.31 | 60.5 | 9.8 | 32.1 |
| Montgomery | 13.9 | 12.6 | 11.5 | 50.3 | 36,252 | 35,468 | -2.2 | -11.2 | 4,226 | 4,483 | -3,741 | 13,576 | 2.33 | 61.1 | 10.4 | 33.8 |
| Morris | 13.9 | 11.5 | 9.3 | 50.1 | 6,104 | 5,923 | -3.0 | -6.1 | 643 | 772 | -236 | 2,280 | 2.41 | 70.3 | 7.1 | 25.0 |
| Morton | 16.1 | 13.9 | 11.6 | 50.4 | 3,496 | 3,233 | -7.5 | -21.5 | 341 | 352 | -690 | 1,038 | 2.58 | 59.9 | 5.0 | 32.6 |
| Nemaha | 14.7 | 10.7 | 11.0 | 49.0 | 10,717 | 10,178 | -5.0 | -0.6 | 1,455 | 1,296 | -207 | 4,007 | 2.45 | 68.3 | 4.4 | 27.7 |

1. No spouse present.

# Table B. States and Counties — Population, Vital Statistics, Health, and Crime

| STATE County | Persons in group quarters, 2020 | Daytime Population, 2015-2019 Number | Employment/residence ratio | Births, 2020 Total | Births, 2020 Rate[1] | Deaths, 2020 Number | Deaths, 2020 Rate[1] | Persons under 65 with no health insurance, 2019 Number | Percent | Medicare, 2020 Total beneficiaries | Enrolled in Original Medicare | Enrolled in Medicare Advantage | Crimes Violent | Crimes Property |
|---|---|---|---|---|---|---|---|---|---|---|---|---|---|---|
| | 32 | 33 | 34 | 35 | 36 | 37 | 38 | 39 | 40 | 41 | 42 | 43 | 44 | 45 |
| KANSAS | 79,138 | 2,927,674 | 1.01 | 35,467 | 12.2 | 27,169 | 9.3 | 254,180 | 10.7 | 546,213 | 429,673 | 116,539 | X | X |
| Allen | 382 | 12,790 | 1.04 | 157 | 12.7 | 166 | 13.4 | 927 | 9.8 | 3,114 | 2,723 | 390 | 14 | 51 |
| Anderson | 103 | 6,798 | 0.69 | 102 | 12.8 | 106 | 13.3 | 654 | 10.8 | 1,852 | 1,628 | 225 | 8 | 31 |
| Atchison | 1,386 | 15,614 | 0.91 | 180 | 11.2 | 178 | 11.1 | 1,041 | 8.6 | 3,405 | 2,843 | 563 | NA | NA |
| Barber | 40 | 4,335 | 0.86 | 51 | 11.7 | 51 | 11.7 | 473 | 14.1 | 1,200 | 1,149 | 51 | 6 | 19 |
| Barton | 672 | 26,418 | 1.00 | 309 | 12.0 | 317 | 12.4 | 2,597 | 12.9 | 5,667 | 5,506 | 161 | 20 | 69 |
| Bourbon | 409 | 14,913 | 1.05 | 195 | 13.5 | 160 | 11.1 | 1,204 | 10.8 | 3,252 | 2,415 | 838 | 10 | 33 |
| Brown | 116 | 10,108 | 1.11 | 134 | 14.1 | 128 | 13.5 | 819 | 10.9 | 2,350 | 2,233 | 117 | 5 | 16 |
| Butler | 2,298 | 56,090 | 0.65 | 690 | 10.3 | 693 | 10.3 | 4,801 | 8.8 | 12,450 | 9,312 | 3,138 | 49 | 301 |
| Chase | 145 | 2,422 | 0.80 | 25 | 9.7 | 18 | 7.0 | 211 | 11.5 | 663 | 618 | 45 | NA | NA |
| Chautauqua | 79 | 3,049 | 0.80 | 33 | 10.2 | 51 | 15.8 | 372 | 16.1 | 941 | 863 | 78 | NA | NA |
| Cherokee | 202 | 18,031 | 0.75 | 230 | 11.7 | 281 | 14.3 | 1,746 | 11.1 | 4,625 | 3,618 | 1,006 | NA | NA |
| Cheyenne | 49 | 2,579 | 0.93 | 22 | 8.5 | 49 | 18.8 | 299 | 15.7 | 752 | 688 | 64 | NA | NA |
| Clark | 57 | 1,948 | 0.92 | 24 | 12.2 | 20 | 10.2 | 178 | 11.5 | 471 | 453 | 18 | 3 | 14 |
| Clay | 140 | 7,362 | 0.81 | 88 | 11.0 | 109 | 13.6 | 580 | 9.6 | 2,083 | 1,970 | 113 | 2 | 5 |
| Cloud | 498 | 8,511 | 0.90 | 104 | 12.0 | 133 | 15.4 | 624 | 9.6 | 2,277 | 2,171 | 106 | NA | NA |
| Coffey | 130 | 8,321 | 1.02 | 69 | 8.5 | 88 | 10.8 | 525 | 8.2 | 2,065 | 1,864 | 201 | NA | NA |
| Comanche | 65 | 1,684 | 0.94 | 14 | 8.3 | 15 | 8.9 | 171 | 13.9 | 502 | 482 | 20 | NA | NA |
| Cowley | 2,014 | 34,455 | 0.94 | 372 | 10.7 | 447 | 12.9 | 2,825 | 10.6 | 7,422 | 6,367 | 1,055 | 30 | 125 |
| Crawford | 1,780 | 39,430 | 1.03 | 434 | 11.2 | 380 | 9.8 | 3,638 | 11.7 | 7,444 | 6,087 | 1,357 | 22 | 157 |
| Decatur | 81 | 2,533 | 0.76 | 29 | 10.4 | 40 | 14.4 | 267 | 13.3 | 808 | 745 | 63 | NA | NA |
| Dickinson | 307 | 17,440 | 0.84 | 179 | 9.8 | 208 | 11.4 | 1,600 | 11.0 | 4,386 | 4,075 | 311 | 23 | 106 |
| Doniphan | 427 | 6,843 | 0.78 | 88 | 11.7 | 74 | 9.9 | 574 | 10.1 | 1,680 | 1,539 | 142 | 1 | 8 |
| Douglas | 8,814 | 112,061 | 0.88 | 1,101 | 9.0 | 754 | 6.2 | 9,568 | 9.7 | 18,145 | 14,566 | 3,578 | NA | NA |
| Edwards | 35 | 2,487 | 0.73 | 30 | 10.9 | 25 | 9.1 | 355 | 16.4 | 718 | 698 | 20 | 0 | 15 |
| Elk | 40 | 2,348 | 0.82 | 23 | 9.2 | 34 | 13.6 | 288 | 16.4 | 787 | 713 | 75 | NA | NA |
| Ellis | 1,037 | 29,374 | 1.04 | 296 | 10.3 | 261 | 9.1 | 2,333 | 10.1 | 5,321 | 5,121 | 200 | 14 | 32 |
| Ellsworth | 978 | 6,244 | 1.00 | 48 | 8.0 | 75 | 12.4 | 369 | 9.4 | 1,439 | 1,383 | 57 | 8 | 47 |
| Finney | 547 | 36,905 | 1.01 | 608 | 16.9 | 223 | 6.2 | 5,443 | 17.4 | 4,759 | 4,563 | 196 | 28 | 135 |
| Ford | 708 | 35,006 | 1.05 | 562 | 17.0 | 240 | 7.3 | 5,301 | 18.4 | 4,201 | 4,044 | 157 | 13 | 59 |
| Franklin | 479 | 23,214 | 0.81 | 293 | 11.4 | 262 | 10.2 | 1,867 | 9.0 | 5,425 | 4,481 | 944 | 52 | 155 |
| Geary | 842 | 41,390 | 1.43 | 874 | 27.1 | 195 | 6.1 | 2,249 | 8.1 | 4,135 | 3,653 | 482 | 15 | 32 |
| Gove | 56 | 2,762 | 1.09 | 41 | 15.6 | 43 | 16.4 | 318 | 16.2 | 660 | 640 | 20 | 4 | 7 |
| Graham | 40 | 2,516 | 1.00 | 23 | 9.6 | 42 | 17.6 | 226 | 12.8 | 716 | 674 | 42 | NA | NA |
| Grant | 79 | 7,057 | 0.88 | 119 | 16.8 | 34 | 4.8 | 1,005 | 16.7 | 1,163 | 1,100 | 63 | 2 | 13 |
| Gray | 78 | 6,005 | 0.99 | 75 | 12.6 | 45 | 7.6 | 896 | 17.8 | 1,059 | 1,023 | 36 | 4 | 42 |
| Greeley | 26 | 1,280 | 1.16 | 12 | 10.0 | 11 | 9.2 | 133 | 13.6 | 284 | D | D | NA | NA |
| Greenwood | 90 | 5,237 | 0.71 | 54 | 9.2 | 71 | 12.1 | 495 | 11.3 | 1,802 | 1,651 | 152 | 25 | 72 |
| Hamilton | 0 | 2,566 | 0.98 | 27 | 11.1 | 15 | 6.2 | 455 | 21.4 | 410 | 394 | 16 | 9 | 21 |
| Harper | 129 | 5,854 | 1.11 | 66 | 12.4 | 94 | 17.6 | 634 | 15.5 | 1,400 | 1,291 | 109 | NA | NA |
| Harvey | 1,317 | 32,636 | 0.89 | 385 | 11.2 | 367 | 10.7 | 2,884 | 10.8 | 7,862 | 6,000 | 1,862 | 11 | 24 |
| Haskell | 33 | 4,159 | 1.08 | 64 | 16.3 | 26 | 6.6 | 691 | 20.9 | 613 | 589 | 24 | 0 | 22 |
| Hodgeman | 14 | 1,727 | 0.85 | 13 | 7.3 | 31 | 17.4 | 189 | 14.0 | 423 | 411 | 12 | NA | NA |
| Jackson | 133 | 11,339 | 0.68 | 164 | 12.5 | 142 | 10.8 | 1,209 | 11.5 | 2,885 | 2,288 | 597 | 15 | 122 |
| Jefferson | 235 | 13,647 | 0.43 | 194 | 10.2 | 156 | 8.2 | 1,432 | 9.3 | 4,167 | 3,283 | 884 | 33 | 224 |
| Jewell | 27 | 2,643 | 0.79 | 25 | 8.8 | 40 | 14.1 | 242 | 12.2 | 899 | 857 | 42 | 2 | 16 |
| Johnson | 5,161 | 619,283 | 1.09 | 7,020 | 11.6 | 4,438 | 7.3 | 36,800 | 7.1 | 99,855 | 66,483 | 33,372 | 50 | 220 |
| Kearny | 81 | 3,941 | 1.02 | 56 | 15.0 | 25 | 6.7 | 562 | 17.8 | 648 | 627 | 21 | 12 | 74 |
| Kingman | 190 | 6,512 | 0.76 | 65 | 9.3 | 94 | 13.5 | 548 | 9.9 | 1,899 | 1,759 | 140 | 3 | 30 |
| Kiowa | 149 | 2,686 | 1.15 | 38 | 15.5 | 24 | 9.8 | 272 | 15.5 | 575 | 560 | 15 | 3 | 18 |
| Labette | 480 | 20,549 | 1.05 | 250 | 12.8 | 235 | 12.0 | 1,961 | 12.6 | 4,858 | 4,045 | 812 | 6 | 55 |
| Lane | 3 | 1,540 | 0.97 | 18 | 11.9 | 17 | 11.2 | 147 | 12.7 | 440 | 427 | 14 | NA | NA |
| Leavenworth | 6,507 | 73,156 | 0.80 | 991 | 12.0 | 694 | 8.4 | 5,102 | 8.0 | 14,050 | 10,746 | 3,304 | NA | NA |
| Lincoln | 50 | 2,788 | 0.81 | 24 | 8.0 | 41 | 13.7 | 290 | 13.3 | 820 | 798 | 22 | NA | NA |
| Linn | 63 | 8,612 | 0.75 | 94 | 9.7 | 128 | 13.3 | 884 | 12.0 | 2,488 | 1,642 | 846 | 13 | 86 |
| Logan | 41 | 2,765 | 0.97 | 46 | 16.8 | 19 | 7.0 | 228 | 10.4 | 675 | 642 | 33 | 1 | 10 |
| Lyon | 1,452 | 33,435 | 1.01 | 401 | 12.1 | 299 | 9.0 | 3,679 | 13.8 | 6,077 | 5,425 | 652 | 17 | 57 |
| McPherson | 928 | 30,688 | 1.15 | 294 | 10.3 | 356 | 12.5 | 1,989 | 8.9 | 6,663 | 5,885 | 777 | 14 | 37 |
| Marion | 727 | 10,436 | 0.73 | 108 | 9.3 | 151 | 13.0 | 1,011 | 11.8 | 3,031 | 2,907 | 124 | 9 | 40 |
| Marshall | 164 | 9,958 | 1.04 | 114 | 11.8 | 110 | 11.4 | 668 | 8.9 | 2,513 | 2,313 | 199 | NA | NA |
| Meade | 119 | 3,981 | 0.91 | 53 | 13.2 | 38 | 9.4 | 542 | 17.1 | 804 | 765 | 39 | NA | NA |
| Miami | 644 | 26,240 | 0.57 | 374 | 10.9 | 320 | 9.3 | 2,087 | 7.4 | 6,850 | 4,769 | 2,080 | NA | NA |
| Mitchell | 246 | 6,797 | 1.25 | 72 | 12.2 | 104 | 17.7 | 442 | 10.1 | 1,587 | 1,527 | 60 | 2 | 29 |
| Montgomery | 1,113 | 33,489 | 1.07 | 344 | 10.9 | 371 | 11.8 | 3,220 | 13.3 | 7,646 | 6,354 | 1,292 | 27 | NA |
| Morris | 66 | 4,941 | 0.77 | 66 | 11.9 | 73 | 13.1 | 480 | 11.4 | 1,437 | 1,365 | 71 | 7 | 21 |
| Morton | 83 | 2,663 | 0.92 | 20 | 7.9 | 25 | 9.9 | 313 | 15.8 | 579 | 551 | 29 | NA | NA |
| Nemaha | 259 | 10,320 | 1.04 | 146 | 14.4 | 94 | 9.3 | 655 | 8.0 | 2,260 | 2,174 | 86 | 5 | 51 |

1. Per 1,000 estimated resident population.

Items 32—45

# Table B. States and Counties — Crime, Education, Money Income, and Poverty

| STATE County | County law enforcement employment, 2019 | | School enrollment and attainment, 2015-2019 | | | | Local government expenditures,[3] 2017-2018 | | Money income, 2015-2019 | | | | Income and poverty, 2019 | | | |
| | | | Enrollment[1] | | Attainment[2] (percent) | | | | | Households | | | | Percent below poverty level | | |
| | | | | | | | | | | | Percent | | | | | |
| | Officers | Civilians | Total | Percent private | High school graduate or less | Bachelor's degree or more | Total current spending (mil dol) | Current spending per student (dollars) | Per capita income[4] | Median income (dollars) | with income of less than $50,000 | with income of $200,000 or more | Median household income (dollars) | All persons | Children under 18 years | Children 5 to 17 years in families |
| | 46 | 47 | 48 | 49 | 50 | 51 | 52 | 53 | 54 | 55 | 56 | 57 | 58 | 59 | 60 | 61 |
| KANSAS | X | X | 772,562 | 13.7 | 35.0 | 33.4 | 5,806 | 11,680 | 31,814 | 59,597 | 42.2 | 5.3 | 62,028 | 11.3 | 14.3 | 12.9 |
| Allen | 12 | 18 | 3,017 | 5.1 | 41.2 | 20.3 | 38 | 16,044 | 23,734 | 45,333 | 53.5 | 2.0 | 46,690 | 17.3 | 24.0 | 19.9 |
| Anderson | NA | NA | 1,889 | 11.5 | 51.0 | 18.1 | 13 | 10,564 | 24,309 | 50,213 | 49.8 | 3.1 | 51,352 | 12.7 | 13.4 | 14.2 |
| Atchison | NA | NA | 5,097 | 44.0 | 45.6 | 22.0 | 26 | 11,626 | 24,237 | 50,439 | 49.8 | 1.7 | 51,860 | 13.8 | 16.6 | 15.2 |
| Barber | NA | NA | 943 | 16.8 | 42.6 | 20.7 | 10 | 12,989 | 27,648 | 50,174 | 49.9 | 2.9 | 50,660 | 13.6 | 20.0 | 19.1 |
| Barton | NA | NA | 6,222 | 9.4 | 42.5 | 20.9 | 53 | 12,370 | 27,495 | 49,723 | 50.3 | 1.9 | 50,754 | 13.4 | 17.0 | 15.0 |
| Bourbon | 13 | 3 | 3,695 | 10.2 | 40.2 | 21.7 | 25 | 10,632 | 23,102 | 43,917 | 56.8 | 2.2 | 47,452 | 15.6 | 22.2 | 20.4 |
| Brown | 9 | 17 | 2,257 | 10.7 | 44.2 | 21.2 | 21 | 13,998 | 25,208 | 48,333 | 52.0 | 1.9 | 47,558 | 15.4 | 20.5 | 17.8 |
| Butler | NA | NA | 18,274 | 12.5 | 33.4 | 29.8 | 166 | 9,017 | 29,986 | 64,782 | 39.0 | 4.5 | 66,199 | 8.5 | 10.1 | 7.8 |
| Chase | 5 | 5 | 625 | 3.4 | 37.3 | 25.5 | 5 | 13,098 | 23,055 | 45,353 | 55.6 | 0.8 | 51,022 | 11.4 | 15.6 | 14.3 |
| Chautauqua | 6 | 9 | 718 | 7.7 | 48.2 | 15.9 | 7 | 12,485 | 22,962 | 40,298 | 59.5 | 2.1 | 43,664 | 16.4 | 24.7 | 21.7 |
| Cherokee | 24 | 27 | 4,948 | 9.5 | 42.9 | 19.6 | 43 | 12,032 | 22,615 | 43,175 | 55.7 | 1.3 | 47,682 | 15.4 | 20.9 | 18.8 |
| Cheyenne | 5 | 1 | 564 | 1.6 | 37.3 | 28.9 | 6 | 13,988 | 29,041 | 47,639 | 52.1 | 2.6 | 47,323 | 11.6 | 18.3 | 18.6 |
| Clark | 5 | 5 | 444 | 9.2 | 32.4 | 26.1 | 7 | 14,703 | 28,227 | 53,348 | 47.2 | 1.9 | 59,085 | 12.4 | 18.3 | 17.1 |
| Clay | 7 | 6 | 1,901 | 12.7 | 42.4 | 26.7 | 19 | 14,401 | 29,235 | 53,929 | 48.6 | 3.4 | 52,781 | 10.0 | 15.1 | 14.3 |
| Cloud | 8 | 1 | 2,065 | 10.3 | 38.6 | 24.0 | 24 | 14,645 | 27,279 | 45,373 | 54.3 | 2.2 | 46,299 | 13.5 | 17.8 | 16.2 |
| Coffey | 12 | 29 | 1,685 | 6.7 | 39.4 | 24.5 | 23 | 14,912 | 33,371 | 59,583 | 41.3 | 3.9 | 62,715 | 11.2 | 12.1 | 11.4 |
| Comanche | NA | NA | 343 | 5.5 | 40.9 | 16.0 | 5 | 13,971 | 30,355 | 54,821 | 47.3 | 2.5 | 47,854 | 11.6 | 15.1 | 12.5 |
| Cowley | 24 | 3 | 9,101 | 13.0 | 39.3 | 21.8 | 76 | 12,668 | 24,726 | 50,102 | 49.9 | 2.0 | 51,537 | 13.5 | 18.3 | 16.6 |
| Crawford | 31 | 5 | 12,211 | 8.1 | 37.2 | 29.9 | 122 | 19,776 | 23,091 | 41,004 | 58.5 | 2.2 | 42,723 | 18.5 | 20.2 | 18.6 |
| Decatur | 3 | 0 | 490 | 8.6 | 41.4 | 22.5 | 5 | 14,470 | 30,905 | 48,125 | 51.9 | 2.1 | 46,878 | 13.2 | 22.8 | 22.3 |
| Dickinson | 21 | 3 | 4,474 | 10.0 | 41.7 | 22.9 | 41 | 11,559 | 27,219 | 49,991 | 50.0 | 2.7 | 58,066 | 9.8 | 13.7 | 12.8 |
| Doniphan | NA | NA | 2,042 | 2.3 | 46.4 | 21.1 | 17 | 12,494 | 25,775 | 49,703 | 50.2 | 2.5 | 52,596 | 11.4 | 14.4 | 12.9 |
| Douglas | 77 | 82 | 43,075 | 10.8 | 23.0 | 49.6 | 162 | 10,661 | 32,281 | 59,435 | 43.5 | 5.6 | 62,559 | 15.1 | 12.4 | 10.4 |
| Edwards | 6 | 4 | 550 | 4.4 | 42.8 | 18.1 | 6 | 13,734 | 27,431 | 50,902 | 48.3 | 2.6 | 49,047 | 11.7 | 17.4 | 15.3 |
| Elk | 5 | 5 | 517 | 6.0 | 54.1 | 15.6 | 10 | 20,311 | 23,875 | 38,750 | 60.9 | 2.4 | 41,930 | 19.4 | 29.9 | 28.0 |
| Ellis | 17 | 17 | 8,913 | 12.0 | 33.6 | 36.3 | 48 | 12,127 | 30,956 | 52,883 | 47.5 | 3.9 | 51,237 | 13.1 | 11.3 | 10.8 |
| Ellsworth | 7 | 13 | 1,230 | 4.8 | 42.7 | 19.5 | 14 | 12,029 | 26,443 | 54,902 | 45.2 | 1.3 | 56,988 | 9.6 | 12.1 | 11.0 |
| Finney | NA | NA | 11,173 | 4.4 | 51.2 | 17.1 | 90 | 10,408 | 24,152 | 60,798 | 41.0 | 2.5 | 57,412 | 13.0 | 17.0 | 15.3 |
| Ford | 26 | 9 | 9,018 | 3.6 | 51.7 | 17.7 | 83 | 10,677 | 23,293 | 51,711 | 48.3 | 2.6 | 54,604 | 13.8 | 19.5 | 18.4 |
| Franklin | 30 | 15 | 6,526 | 15.1 | 45.4 | 20.4 | 50 | 11,311 | 26,998 | 56,582 | 45.3 | 2.9 | 55,984 | 10.4 | 13.4 | 11.6 |
| Geary | 33 | 72 | 10,081 | 11.4 | 31.2 | 23.2 | 87 | 11,533 | 23,897 | 53,133 | 47.3 | 0.8 | 58,978 | 11.7 | 18.6 | 17.5 |
| Gove | 4 | 1 | 520 | 9.4 | 44.2 | 19.7 | 8 | 16,163 | 28,859 | 48,317 | 51.4 | 2.5 | 52,986 | 9.6 | 15.9 | 15.3 |
| Graham | 3 | 5 | 457 | 2.2 | 39.3 | 24.7 | 5 | 11,761 | 27,236 | 46,375 | 57.4 | 0.9 | 48,912 | 11.8 | 15.8 | 12.5 |
| Grant | 5 | 10 | 2,228 | 10.5 | 48.9 | 14.9 | 29 | 16,553 | 25,411 | 53,413 | 45.0 | 2.5 | 65,178 | 10.5 | 15.8 | 14.8 |
| Gray | 11 | 7 | 1,513 | 10.8 | 45.1 | 20.2 | 30 | 23,542 | 29,421 | 64,930 | 33.8 | 3.6 | 69,077 | 7.6 | 9.6 | 8.2 |
| Greeley | 4 | 4 | 287 | 0.7 | 39.9 | 27.2 | 3 | 11,455 | 28,982 | 54,112 | 45.3 | 1.9 | 58,492 | 10.9 | 16.3 | 15.5 |
| Greenwood | 13 | 9 | 1,242 | 4.7 | 43.9 | 19.9 | 13 | 13,759 | 27,500 | 41,982 | 55.7 | 1.7 | 44,876 | 14.9 | 21.5 | 19.4 |
| Hamilton | 6 | 0 | 691 | 7.1 | 59.0 | 18.7 | 6 | 10,405 | 22,003 | 46,944 | 51.8 | 0.6 | 55,857 | 13.1 | 18.9 | 17.1 |
| Harper | NA | NA | 1,189 | 7.9 | 49.5 | 16.2 | 13 | 12,736 | 24,800 | 49,865 | 50.2 | 2.3 | 47,540 | 14.2 | 19.6 | 17.7 |
| Harvey | 23 | 2 | 8,750 | 21.4 | 33.3 | 33.2 | 70 | 11,682 | 28,226 | 57,982 | 43.1 | 2.4 | 62,961 | 9.3 | 11.3 | 10.1 |
| Haskell | 10 | 5 | 966 | 12.6 | 46.9 | 17.1 | 12 | 16,029 | 26,814 | 55,064 | 44.5 | 5.0 | 62,477 | 9.7 | 14.5 | 12.8 |
| Hodgeman | NA | NA | 368 | 0.0 | 29.8 | 30.8 | 4 | 11,875 | 30,044 | 61,211 | 41.5 | 1.8 | 58,360 | 10.4 | 11.7 | 11.3 |
| Jackson | 22 | 21 | 3,233 | 8.1 | 44.4 | 20.2 | 32 | 13,215 | 28,127 | 57,914 | 42.6 | 3.5 | 60,615 | 10.9 | 15.5 | 14.9 |
| Jefferson | 23 | 16 | 4,134 | 10.5 | 46.2 | 23.2 | 61 | 16,952 | 30,655 | 64,864 | 34.6 | 2.6 | 68,549 | 7.8 | 10.5 | 9.0 |
| Jewell | 5 | 5 | 481 | 5.8 | 46.4 | 19.3 | 4 | 13,144 | 27,731 | 40,676 | 59.9 | 3.6 | 46,017 | 13.0 | 20.5 | 20.3 |
| Johnson | 475 | 160 | 154,952 | 19.3 | 18.5 | 56.0 | 994 | 10,581 | 46,517 | 89,087 | 25.6 | 12.8 | 91,913 | 5.3 | 5.5 | 5.0 |
| Kearny | 9 | 7 | 955 | 0.9 | 56.5 | 14.7 | 11 | 11,682 | 23,415 | 52,599 | 49.4 | 3.2 | 67,301 | 10.4 | 12.9 | 11.0 |
| Kingman | 8 | 12 | 1,549 | 13.0 | 39.4 | 24.4 | 14 | 11,551 | 34,578 | 60,469 | 42.4 | 6.0 | 56,907 | 11.3 | 17.4 | 16.5 |
| Kiowa | NA | NA | 614 | 19.4 | 38.9 | 28.1 | 7 | 12,365 | 26,321 | 53,274 | 46.7 | 1.5 | 49,588 | 12.1 | 17.4 | 16.1 |
| Labette | 18 | 2 | 4,109 | 8.1 | 42.0 | 18.7 | 43 | 10,869 | 24,572 | 47,643 | 52.2 | 2.7 | 44,170 | 15.8 | 21.5 | 20.0 |
| Lane | 5 | 5 | 303 | 1.0 | 37.6 | 23.1 | 5 | 14,396 | 28,435 | 52,125 | 48.9 | 2.2 | 60,315 | 10.3 | 16.8 | 15.4 |
| Leavenworth | 62 | 35 | 21,428 | 15.8 | 36.1 | 31.5 | 141 | 10,313 | 32,444 | 73,013 | 32.7 | 5.0 | 75,840 | 8.3 | 8.9 | 7.9 |
| Lincoln | 8 | 5 | 720 | 3.1 | 37.8 | 22.8 | 8 | 12,954 | 25,803 | 45,076 | 54.5 | 1.8 | 47,557 | 12.1 | 16.6 | 14.5 |
| Linn | 18 | 15 | 2,093 | 4.0 | 49.1 | 16.2 | 24 | 12,607 | 27,926 | 48,778 | 51.0 | 3.0 | 62,876 | 12.0 | 17.2 | 15.6 |
| Logan | 5 | 0 | 659 | 6.4 | 37.1 | 16.6 | 19 | 36,392 | 28,135 | 51,740 | 46.9 | 2.3 | 53,538 | 10.0 | 14.8 | 14.9 |
| Lyon | 25 | 5 | 10,537 | 5.8 | 41.1 | 27.5 | 72 | 12,763 | 26,904 | 46,338 | 53.0 | 2.7 | 49,840 | 15.1 | 16.0 | 13.7 |
| McPherson | 19 | 3 | 6,693 | 26.6 | 36.6 | 28.0 | 62 | 11,774 | 31,006 | 59,089 | 42.3 | 3.5 | 65,834 | 7.2 | 8.9 | 7.5 |
| Marion | 11 | 6 | 3,010 | 23.8 | 38.8 | 25.0 | 30 | 13,170 | 26,577 | 52,123 | 48.1 | 2.1 | 51,771 | 10.7 | 12.5 | 11.1 |
| Marshall | 11 | 15 | 2,175 | 13.6 | 50.0 | 18.9 | 23 | 12,238 | 28,469 | 51,895 | 47.1 | 3.1 | 55,858 | 9.4 | 12.1 | 11.7 |
| Meade | NA | NA | 890 | 9.4 | 45.7 | 20.3 | 7 | 12,599 | 28,945 | 59,316 | 40.4 | 3.4 | 64,640 | 9.7 | 13.6 | 12.0 |
| Miami | 32 | 29 | 8,296 | 10.7 | 35.8 | 29.8 | 103 | 11,501 | 35,659 | 71,995 | 31.0 | 6.5 | 74,428 | 6.8 | 8.7 | 7.7 |
| Mitchell | 8 | 1 | 1,508 | 17.2 | 35.5 | 26.1 | 18 | 16,077 | 25,801 | 46,203 | 53.8 | 2.5 | 51,979 | 11.1 | 16.9 | 16.7 |
| Montgomery | 25 | 14 | 7,905 | 12.2 | 40.3 | 19.5 | 72 | 12,677 | 24,647 | 45,157 | 53.9 | 2.1 | 44,240 | 15.7 | 20.7 | 19.4 |
| Morris | 8 | 4 | 1,133 | 5.1 | 44.9 | 21.4 | 13 | 11,912 | 29,558 | 55,658 | 45.5 | 4.6 | 54,775 | 10.5 | 14.4 | 14.0 |
| Morton | 6 | 6 | 568 | 0.0 | 44.0 | 17.3 | 12 | 8,797 | 23,447 | 47,750 | 51.7 | 2.2 | 56,175 | 11.7 | 17.0 | 14.7 |
| Nemaha | 9 | 9 | 2,477 | 16.3 | 45.6 | 25.4 | 21 | 12,153 | 30,344 | 63,216 | 39.2 | 3.2 | 63,373 | 8.8 | 10.7 | 10.4 |

1. All persons 3 years old and over enrolled in nursery school through college.  2. Persons 25 years old and over.  3. Elementary and secondary education expenditures.  4. Based on population estimated by the American Community Survey, 2014–2018.

# Table B. States and Counties — Personal Income

| STATE County | Personal income, 2019 | | | | | | | | | | Earnings, 2019 | | |
|---|---|---|---|---|---|---|---|---|---|---|---|---|---|
| | Total (mil dol) | Percent change 2018-2019 | Per capita[1] Dollars | Per capita[1] Rank | Wages and salaries (mil dol) | Supplements to wages and salaries, employer contributions (mil dol) Pension and insurance | Supplements to wages and salaries, employer contributions (mil dol) Government social insurance | Proprietors' income (mil dol) | Dividends, interest, and rent (mil dol) | Personal transfer receipts (mil dol) | Total (mil dol) | Contributions for government social insurance (mil dol) From employee and self-employed | Contributions for government social insurance (mil dol) From employer |
| | 62 | 63 | 64 | 65 | 66 | 67 | 68 | 69 | 70 | 71 | 72 | 73 | 74 |
| KANSAS ................ | 155,648 | 4.3 | 53,439 | X | 74,270 | 11,915 | 5,596 | 19,349 | 30,278 | 24,724 | 111,130 | 6,549 | 5,596 |
| Allen....................... | 534 | 7.4 | 43,163 | 1,591 | 233 | 46 | 18 | 54 | 94 | 146 | 351 | 23 | 18 |
| Anderson................. | 330 | 9.0 | 42,046 | 1,754 | 84 | 17 | 6 | 67 | 47 | 83 | 174 | 10 | 6 |
| Atchison.................. | 648 | 5.3 | 40,335 | 2,009 | 274 | 45 | 21 | 63 | 107 | 150 | 403 | 25 | 21 |
| Barber..................... | 215 | 9.7 | 48,489 | 973 | 66 | 13 | 5 | 37 | 46 | 54 | 121 | 8 | 5 |
| Barton..................... | 1,275 | 5.9 | 49,460 | 883 | 530 | 92 | 40 | 190 | 241 | 264 | 852 | 51 | 40 |
| Bourbon................... | 671 | 2.3 | 46,149 | 1,232 | 257 | 47 | 20 | 141 | 85 | 149 | 465 | 28 | 20 |
| Brown...................... | 458 | 4.3 | 47,930 | 1,025 | 235 | 41 | 17 | 69 | 79 | 107 | 362 | 21 | 17 |
| Butler...................... | 3,226 | 4.7 | 48,216 | 999 | 856 | 179 | 64 | 298 | 517 | 554 | 1,397 | 88 | 64 |
| Chase...................... | 143 | 5.4 | 53,882 | 549 | 41 | 9 | 3 | 28 | 41 | 23 | 81 | 4 | 3 |
| Chautauqua ............. | 141 | 4.8 | 43,327 | 1,565 | 27 | 6 | 2 | 13 | 32 | 42 | 48 | 4 | 2 |
| Cherokee ................ | 855 | 4.2 | 42,872 | 1,640 | 272 | 51 | 20 | 145 | 115 | 211 | 489 | 31 | 20 |
| Cheyenne................. | 146 | 16.3 | 54,890 | 500 | 38 | 7 | 3 | 46 | 23 | 32 | 93 | 4 | 3 |
| Clark....................... | 129 | 6.1 | 64,457 | 198 | 31 | 7 | 2 | 34 | 23 | 23 | 75 | 3 | 2 |
| Clay........................ | 375 | 10.0 | 46,888 | 1,143 | 108 | 22 | 8 | 64 | 81 | 86 | 202 | 12 | 8 |
| Cloud...................... | 368 | 8.8 | 41,892 | 1,785 | 119 | 22 | 9 | 65 | 69 | 98 | 215 | 13 | 9 |
| Coffey..................... | 451 | 3.3 | 55,119 | 488 | 242 | 58 | 18 | 57 | 65 | 94 | 375 | 21 | 18 |
| Comanche................ | 80 | 5.1 | 46,878 | 1,145 | 22 | 5 | 2 | 12 | 16 | 21 | 41 | 3 | 2 |
| Cowley.................... | 1,398 | 4.7 | 40,042 | 2,044 | 605 | 110 | 46 | 121 | 240 | 362 | 883 | 58 | 46 |
| Crawford ................. | 1,545 | 3.8 | 39,788 | 2,077 | 710 | 137 | 55 | 134 | 323 | 369 | 1,036 | 66 | 55 |
| Decatur ................... | 143 | 14.9 | 50,635 | 786 | 31 | 6 | 2 | 38 | 28 | 33 | 77 | 4 | 2 |
| Dickinson ................ | 814 | 5.7 | 44,101 | 1,474 | 259 | 53 | 21 | 88 | 151 | 194 | 421 | 27 | 21 |
| Doniphan................. | 338 | 12.1 | 44,486 | 1,418 | 96 | 20 | 8 | 55 | 50 | 75 | 179 | 10 | 8 |
| Douglas................... | 5,522 | 3.8 | 45,163 | 1,342 | 2,274 | 437 | 167 | 450 | 1,157 | 782 | 3,327 | 200 | 167 |
| Edwards.................. | 175 | 10.7 | 62,707 | 235 | 41 | 7 | 3 | 60 | 25 | 31 | 111 | 4 | 3 |
| Elk.......................... | 105 | 7.6 | 41,529 | 1,843 | 19 | 4 | 1 | 14 | 23 | 35 | 39 | 3 | 1 |
| Ellis........................ | 1,364 | 3.5 | 47,758 | 1,048 | 664 | 129 | 48 | 181 | 247 | 240 | 1,022 | 60 | 48 |
| Ellsworth................. | 295 | 8.8 | 48,380 | 989 | 100 | 20 | 8 | 39 | 59 | 63 | 166 | 10 | 8 |
| Finney..................... | 1,707 | 5.4 | 46,823 | 1,149 | 876 | 142 | 64 | 321 | 226 | 245 | 1,403 | 74 | 64 |
| Ford........................ | 1,391 | 4.7 | 41,373 | 1,864 | 800 | 128 | 57 | 192 | 186 | 209 | 1,176 | 66 | 57 |
| Franklin................... | 1,136 | 3.9 | 44,472 | 1,421 | 424 | 64 | 33 | 121 | 160 | 249 | 642 | 41 | 33 |
| Geary...................... | 1,628 | 1.4 | 51,410 | 728 | 1,591 | 466 | 149 | 53 | 313 | 263 | 2,259 | 97 | 149 |
| Gove....................... | 175 | 14.5 | 66,403 | 163 | 51 | 9 | 4 | 66 | 28 | 29 | 131 | 6 | 4 |
| Graham.................... | 141 | 14.9 | 56,940 | 407 | 36 | 7 | 3 | 41 | 21 | 32 | 86 | 4 | 3 |
| Grant....................... | 392 | 10.2 | 54,892 | 499 | 147 | 26 | 11 | 132 | 46 | 51 | 316 | 13 | 11 |
| Gray........................ | 455 | 11.5 | 75,961 | 82 | 132 | 22 | 11 | 193 | 58 | 41 | 357 | 12 | 11 |
| Greeley.................... | 122 | 30.4 | 98,916 | 24 | 26 | 5 | 2 | 73 | 11 | 13 | 105 | 2 | 2 |
| Greenwood............... | 267 | 7.3 | 44,689 | 1,396 | 67 | 14 | 5 | 26 | 50 | 76 | 113 | 8 | 5 |
| Hamilton.................. | 176 | 20.8 | 69,339 | 136 | 45 | 7 | 4 | 82 | 18 | 21 | 138 | 3 | 4 |
| Harper..................... | 298 | 14.0 | 54,894 | 498 | 117 | 21 | 9 | 61 | 53 | 62 | 208 | 12 | 9 |
| Harvey..................... | 1,529 | 6.0 | 44,416 | 1,429 | 639 | 107 | 53 | 155 | 234 | 340 | 955 | 62 | 53 |
| Haskell.................... | 309 | 20.6 | 77,955 | 72 | 82 | 14 | 7 | 155 | 35 | 28 | 258 | 6 | 7 |
| Hodgeman ............... | 118 | 12.0 | 65,688 | 176 | 23 | 5 | 2 | 47 | 18 | 17 | 77 | 3 | 2 |
| Jackson................... | 572 | 3.3 | 43,447 | 1,551 | 162 | 29 | 12 | 47 | 96 | 126 | 250 | 18 | 12 |
| Jefferson................. | 880 | 4.5 | 46,234 | 1,223 | 158 | 28 | 13 | 86 | 120 | 169 | 285 | 20 | 13 |
| Jewell..................... | 166 | 15.4 | 57,690 | 383 | 29 | 7 | 2 | 50 | 34 | 33 | 88 | 4 | 2 |
| Johnson................... | 45,907 | 3.1 | 76,206 | 81 | 22,912 | 2,809 | 1,639 | 5,778 | 9,343 | 4,117 | 33,137 | 1,942 | 1,639 |
| Kearny..................... | 227 | 12.5 | 59,187 | 331 | 56 | 11 | 5 | 81 | 36 | 31 | 152 | 5 | 5 |
| Kingman.................. | 347 | 10.2 | 48,512 | 970 | 90 | 17 | 7 | 49 | 61 | 76 | 164 | 10 | 7 |
| Kiowa ..................... | 118 | 10.7 | 47,726 | 1,051 | 40 | 8 | 3 | 31 | 22 | 25 | 83 | 4 | 3 |
| Labette.................... | 878 | 4.0 | 44,760 | 1,390 | 404 | 78 | 32 | 97 | 132 | 255 | 611 | 38 | 32 |
| Lane........................ | 130 | 17.0 | 85,007 | 46 | 31 | 6 | 2 | 50 | 22 | 19 | 89 | 3 | 2 |
| Leavenworth ............ | 3,603 | 3.7 | 44,075 | 1,475 | 1,483 | 372 | 126 | 175 | 658 | 674 | 2,156 | 124 | 126 |
| Lincoln.................... | 143 | 12.0 | 48,445 | 977 | 37 | 8 | 3 | 24 | 32 | 36 | 72 | 4 | 3 |
| Linn........................ | 398 | 5.9 | 41,013 | 1,918 | 99 | 24 | 8 | 42 | 58 | 102 | 173 | 12 | 8 |
| Logan...................... | 150 | 14.8 | 53,800 | 555 | 50 | 11 | 4 | 34 | 28 | 29 | 99 | 5 | 4 |
| Lyon........................ | 1,327 | 3.4 | 39,962 | 2,056 | 622 | 119 | 46 | 109 | 227 | 287 | 896 | 56 | 46 |
| McPherson............... | 1,616 | 9.6 | 56,617 | 416 | 824 | 166 | 61 | 259 | 269 | 286 | 1,309 | 76 | 61 |
| Marion..................... | 543 | 8.3 | 45,681 | 1,285 | 135 | 27 | 10 | 96 | 84 | 127 | 268 | 16 | 10 |
| Marshall................... | 452 | 6.1 | 46,580 | 1,179 | 217 | 38 | 19 | 51 | 102 | 105 | 324 | 21 | 19 |
| Meade..................... | 267 | 12.0 | 66,270 | 166 | 72 | 14 | 5 | 86 | 55 | 35 | 178 | 7 | 5 |
| Miami...................... | 1,719 | 5.1 | 50,210 | 813 | 382 | 63 | 30 | 153 | 271 | 287 | 628 | 42 | 30 |
| Mitchell................... | 375 | 7.2 | 62,780 | 233 | 126 | 24 | 9 | 145 | 52 | 69 | 304 | 16 | 9 |
| Montgomery............. | 1,243 | 3.7 | 39,041 | 2,184 | 595 | 123 | 46 | 127 | 194 | 360 | 892 | 58 | 46 |
| Morris...................... | 261 | 5.0 | 46,409 | 1,198 | 58 | 12 | 5 | 52 | 48 | 62 | 127 | 7 | 5 |
| Morton..................... | 138 | 14.7 | 53,296 | 590 | 47 | 8 | 4 | 47 | 22 | 26 | 105 | 4 | 4 |
| Nemaha ................... | 544 | 2.2 | 53,211 | 596 | 209 | 36 | 16 | 71 | 158 | 91 | 331 | 19 | 16 |

1. Based on the resident population estimated as of July 1 of the year shown.

| STATE County | Farm | Mining, quarrying, and extractions | Construction | Manu-facturing | Information; professional, scientific, technical services | Retail trade | Finance, insurance, real estate, and leasing | Health care and social assistance | Govern-ment | Number | Rate[1] | Supplemental Security Income recipients, 2019 | Total | Percent change, 2010-2020 |
|---|---|---|---|---|---|---|---|---|---|---|---|---|---|---|
| | 75 | 76 | 77 | 78 | 79 | 80 | 81 | 82 | 83 | 84 | 85 | 86 | 87 | 88 |
| KANSAS | 3.5 | 1.2 | 5.6 | 11.8 | 10.5 | 5.1 | 9.4 | 10.7 | 16.4 | 561,634 | 193 | 47,365 | 1,294,889 | 5.0 |
| Allen | 5.1 | D | 7.1 | 29.5 | 3.0 | 5.7 | 4.6 | D | 19.7 | 3,370 | 272 | 355 | 6,348 | 1.9 |
| Anderson | 20.7 | 3.5 | 10.3 | 4.6 | 2.0 | 10.1 | D | 11.4 | 14.8 | 1,965 | 249 | 123 | 3,767 | 1.2 |
| Atchison | 8.1 | D | 5.1 | 22.7 | D | 4.9 | 9.4 | D | 11.8 | 3,400 | 211 | 337 | 6,943 | -0.7 |
| Barber | 8.3 | 12.8 | 6.1 | 10.9 | D | 6.2 | D | D | 25.0 | 1,240 | 278 | 62 | 2,730 | -1.3 |
| Barton | 7.9 | 10.4 | 9.7 | 6.9 | 3.3 | 6.8 | 9.3 | D | 14.0 | 5,935 | 229 | 483 | 12,754 | 0.5 |
| Bourbon | 4.2 | 0.1 | 4.8 | 18.4 | D | 3.9 | 10.5 | 10.4 | 11.9 | 3,550 | 245 | 363 | 7,103 | -0.9 |
| Brown | 12.5 | D | 2.5 | 26.3 | 4.2 | 4.4 | 3.2 | 10.3 | 19.2 | 2,455 | 257 | 217 | 4,728 | -1.1 |
| Butler | 3.9 | 1.5 | 10.0 | 16.1 | D | 5.5 | 7.5 | 12.1 | 20.6 | 13,310 | 199 | 800 | 27,430 | 5.3 |
| Chase | 21.4 | D | 6.8 | 18.3 | D | 1.5 | 3.0 | D | 12.8 | 585 | 226 | 27 | 1,490 | -0.9 |
| Chautauqua | 14.0 | 5.6 | 5.4 | 7.3 | D | 4.1 | D | 14.3 | 22.9 | 1,065 | 333 | 110 | 2,137 | -0.6 |
| Cherokee | 6.1 | D | 8.8 | 23.2 | D | 4.2 | 14.7 | 10.7 | 13.2 | 5,075 | 254 | 606 | 9,896 | 0.0 |
| Cheyenne | 36.8 | 1.6 | 2.3 | 1.8 | D | 1.8 | D | 13.5 | 11.0 | 775 | 295 | 21 | 1,498 | -1.3 |
| Clark | 40.0 | 0.7 | D | D | D | 3.9 | D | D | 28.7 | 495 | 247 | 29 | 1,132 | -0.3 |
| Clay | 21.4 | 0.0 | 5.6 | 7.7 | 5.4 | 7.1 | 4.8 | 5.8 | 23.2 | 2,110 | 263 | 107 | 4,133 | 2.1 |
| Cloud | 13.7 | D | 3.7 | 7.5 | 6.1 | 6.9 | 6.6 | D | 17.5 | 2,340 | 267 | 185 | 4,585 | -1.6 |
| Coffey | 7.8 | 1.4 | 2.8 | 2.0 | 1.3 | 3.4 | 3.2 | D | 16.2 | 2,280 | 280 | 144 | 4,173 | 5.3 |
| Comanche | 3.5 | D | 4.0 | 6.1 | D | 6.5 | D | D | 26.8 | 490 | 288 | 40 | 1,026 | -1.7 |
| Cowley | 2.8 | 1.1 | 4.3 | 34.6 | 2.8 | 5.5 | 5.4 | D | 20.5 | 7,945 | 228 | 702 | 16,292 | 1.6 |
| Crawford | 2.1 | D | 3.5 | 14.3 | 4.4 | 6.4 | 5.6 | 12.7 | 23.3 | 7,910 | 204 | 1,031 | 18,297 | 2.8 |
| Decatur | 32.5 | D | 5.7 | 0.5 | D | 5.0 | D | 10.1 | 12.7 | 840 | 298 | 46 | 1,792 | -1.5 |
| Dickinson | 10.0 | D | 4.5 | 21.7 | 3.0 | 5.6 | 5.7 | 3.8 | 20.4 | 4,495 | 245 | 294 | 9,198 | 2.6 |
| Doniphan | 21.2 | D | 7.4 | 16.7 | D | 2.1 | 5.0 | 3.9 | 22.5 | 1,710 | 225 | 90 | 3,579 | 0.1 |
| Douglas | 0.7 | 0.0 | 5.2 | 9.6 | 9.8 | 6.2 | 8.0 | 7.7 | 31.7 | 18,235 | 149 | 1,349 | 52,181 | 11.7 |
| Edwards | 47.9 | D | 1.2 | 4.9 | D | 2.1 | 2.4 | 8.1 | 10.1 | 740 | 266 | 51 | 1,617 | -1.2 |
| Elk | 10.1 | 13.2 | D | D | 3.3 | 3.9 | D | D | 31.0 | 855 | 341 | 52 | 1,756 | -0.2 |
| Ellis | 1.9 | 6.3 | 5.3 | 5.6 | 5.4 | 7.2 | 9.3 | 18.5 | 19.7 | 5,515 | 192 | 316 | 13,349 | 3.7 |
| Ellsworth | 10.8 | D | 4.7 | 11.9 | 7.5 | 3.8 | D | 6.5 | 29.1 | 1,495 | 245 | 54 | 3,226 | -0.4 |
| Finney | 8.5 | 1.6 | 6.3 | 18.2 | 2.3 | 6.4 | 5.4 | 9.7 | 12.4 | 5,095 | 140 | 558 | 13,707 | 3.2 |
| Ford | 5.2 | 0.1 | 4.3 | 33.4 | 3.6 | 5.5 | 5.1 | D | 13.2 | 4,510 | 135 | 432 | 12,380 | 3.1 |
| Franklin | 5.7 | 0.7 | 7.0 | 8.9 | D | 7.9 | D | D | 18.2 | 5,765 | 225 | 484 | 11,388 | 2.2 |
| Geary | 0.5 | 0.0 | 1.1 | 2.1 | D | 1.9 | 1.8 | 1.2 | 79.6 | 4,505 | 141 | 565 | 15,342 | 5.7 |
| Gove | 15.9 | 1.5 | 3.1 | 6.6 | 1.1 | 14.7 | 9.9 | 7.4 | 14.9 | 680 | 255 | 17 | 1,511 | 9.9 |
| Graham | 21.9 | 10.0 | D | D | D | 4.7 | D | 3.9 | 17.4 | 790 | 324 | 38 | 1,478 | -0.3 |
| Grant | 31.2 | 9.9 | 5.0 | 3.5 | D | 2.8 | 4.1 | 4.8 | 9.3 | 1,195 | 167 | 72 | 2,974 | 1.0 |
| Gray | 43.9 | 0.0 | 8.1 | 1.5 | 2.1 | 2.7 | 5.0 | 1.9 | 10.9 | 1,035 | 173 | 28 | 2,452 | 4.9 |
| Greeley | 69.6 | D | 1.3 | 0.2 | D | 3.6 | 2.9 | D | 6.5 | 275 | 227 | 0 | 637 | 1.1 |
| Greenwood | 16.9 | 13.5 | 4.7 | 6.3 | D | 3.5 | D | D | 18.1 | 1,840 | 307 | 186 | 4,031 | -0.9 |
| Hamilton | 65.1 | D | D | D | D | 2.0 | 6.3 | 0.3 | 8.7 | 420 | 167 | 19 | 1,235 | -0.1 |
| Harper | 8.0 | 3.6 | 6.5 | 20.0 | D | 5.8 | D | D | 18.1 | 1,460 | 268 | 91 | 3,153 | 1.2 |
| Harvey | 3.9 | 0.8 | 6.8 | 26.9 | 3.5 | 5.2 | 6.1 | D | 10.9 | 7,880 | 229 | 495 | 14,904 | 2.6 |
| Haskell | 57.4 | 2.6 | 2.6 | 3.2 | D | 1.2 | D | D | 11.2 | 590 | 148 | 27 | 1,692 | 1.6 |
| Hodgeman | 50.5 | D | 1.3 | D | D | 2.5 | 3.5 | D | 17.0 | 430 | 244 | 0 | 967 | -0.7 |
| Jackson | 5.3 | D | 7.3 | 6.7 | 3.0 | 5.6 | 6.0 | 11.0 | 40.1 | 3,175 | 241 | 175 | 5,907 | 2.2 |
| Jefferson | 10.6 | D | 22.3 | 7.2 | D | 3.0 | 4.4 | 5.6 | 18.9 | 4,320 | 225 | 231 | 8,513 | 4.4 |
| Jewell | 46.6 | 0.6 | D | 0.8 | D | 3.7 | 3.9 | 0.4 | 18.5 | 920 | 322 | 45 | 2,014 | -0.8 |
| Johnson | 0.0 | 0.1 | 5.5 | 4.7 | 21.7 | 4.8 | 15.1 | 11.3 | 6.7 | 96,315 | 160 | 3,889 | 251,262 | 10.9 |
| Kearny | 39.9 | D | D | D | D | 2.4 | 6.2 | 0.5 | 23.3 | 685 | 180 | 46 | 1,577 | 1.3 |
| Kingman | 14.4 | 2.5 | 11.2 | 12.8 | D | 4.4 | 8.6 | 10.0 | 15.7 | 1,945 | 276 | 85 | 3,866 | 1.2 |
| Kiowa | 22.5 | 2.7 | D | D | D | 2.8 | 2.6 | D | 16.6 | 610 | 247 | 57 | 1,242 | 1.8 |
| Labette | 6.4 | D | 2.4 | 19.5 | 2.1 | 5.0 | 3.6 | 12.5 | 25.1 | 5,050 | 256 | 704 | 10,023 | -0.7 |
| Lane | 51.4 | D | D | 0.1 | D | D | 5.5 | 0.1 | 12.4 | 425 | 278 | 11 | 974 | -1.5 |
| Leavenworth | 0.4 | 0.0 | 5.4 | 3.7 | 5.2 | 3.5 | 6.0 | 4.9 | 59.3 | 14,345 | 175 | 884 | 30,464 | 6.2 |
| Lincoln | 22.5 | 0.3 | 6.2 | D | D | 2.5 | 3.9 | 4.3 | 27.1 | 830 | 280 | 41 | 1,841 | -1.2 |
| Linn | 9.7 | 2.4 | 7.6 | 5.0 | D | 5.5 | 5.5 | D | 21.3 | 2,605 | 268 | 195 | 5,735 | 5.3 |
| Logan | 18.0 | D | D | D | D | 4.3 | D | D | 34.9 | 665 | 240 | 28 | 1,444 | 0.1 |
| Lyon | 3.4 | D | 2.8 | 25.4 | D | 6.4 | 4.3 | 7.2 | 26.0 | 6,355 | 191 | 627 | 15,465 | 1.5 |
| McPherson | 2.3 | 0.5 | 7.2 | 37.3 | D | 3.0 | 3.3 | 7.8 | 16.1 | 6,895 | 241 | 297 | 13,470 | 5.9 |
| Marion | 16.9 | 0.4 | 4.8 | 12.1 | 3.7 | 5.3 | 3.9 | D | 16.1 | 3,080 | 260 | 145 | 6,015 | 1.2 |
| Marshall | 5.2 | D | 4.3 | 25.8 | 6.7 | 5.7 | 8.3 | D | 10.7 | 2,395 | 247 | 127 | 4,919 | 1.1 |
| Meade | 39.6 | 0.6 | 6.1 | 1.0 | D | D | 4.0 | D | 15.0 | 830 | 206 | 31 | 2,001 | 0.2 |
| Miami | 2.2 | 0.6 | 13.6 | 6.9 | 4.1 | 6.5 | 10.8 | 11.1 | 20.2 | 6,775 | 198 | 389 | 14,214 | 7.8 |
| Mitchell | 7.4 | D | 4.6 | 6.7 | D | 5.3 | D | D | 16.5 | 1,580 | 265 | 93 | 3,301 | 0.2 |
| Montgomery | 3.0 | 0.4 | 5.4 | 28.7 | 2.8 | 6.2 | 4.6 | D | 14.2 | 8,295 | 261 | 1,104 | 16,427 | -0.9 |
| Morris | 23.6 | D | 6.3 | 6.3 | 6.3 | 4.6 | D | D | 19.5 | 1,435 | 255 | 76 | 3,209 | 0.1 |
| Morton | 43.7 | 4.8 | D | 3.2 | 3.2 | 3.9 | D | | 18.2 | 615 | 243 | 34 | 1,452 | -1.0 |
| Nemaha | 13.8 | D | 5.4 | 18.1 | 3.2 | 4.9 | D | 12.7 | 10.8 | 2,260 | 222 | 76 | 4,630 | 1.5 |

1. Per 1,000 resident population estimated as of July 1 of the year shown.

# Table B. States and Counties — Housing, Labor Force, and Employment

Column groups:
- **Housing units, 2015-2019 — Occupied units:** Total (89), Percent (90); **Owner-occupied:** Median value¹ (91), Median owner cost as a percent of income — With a mortgage (92), Without a mortgage² (93); **Renter-occupied:** Median rent³ (94), Median rent as a percent of income² (95); Sub-standard units⁴ (percent) (96)
- **Civilian labor force, 2020:** Total (97), Percent change, 2019-2020 (98); **Unemployment:** Total (99), Rate⁵ (100)
- **Civilian employment⁶, 2015-2019:** Total (101); **Percent:** Management, business, science, and arts (102), Construction, production, and maintenance occupations (103)

| STATE County | 89 | 90 | 91 | 92 | 93 | 94 | 95 | 96 | 97 | 98 | 99 | 100 | 101 | 102 | 103 |
|---|---|---|---|---|---|---|---|---|---|---|---|---|---|---|---|
| KANSAS | 1,129,227 | 66.3 | 151,900 | 19.4 | 11.4 | 850 | 27.2 | 2.4 | 1,497,003 | 0.2 | 88,008 | 5.9 | 1,440,453 | 38.6 | 24.1 |
| Allen | 5,372 | 70.8 | 79,300 | 18.4 | 10.8 | 675 | 28.6 | 1.2 | 6,392 | -1.2 | 347 | 5.4 | 5,873 | 29.6 | 35.3 |
| Anderson | 3,101 | 74.4 | 95,700 | 19.9 | 12.8 | 647 | 18.4 | 3.2 | 4,250 | 1.3 | 175 | 4.1 | 3,368 | 30.6 | 33.0 |
| Atchison | 5,958 | 69.4 | 110,400 | 19.3 | 12.4 | 660 | 24.4 | 2.0 | 6,856 | -2.3 | 471 | 6.9 | 7,792 | 32.3 | 25.4 |
| Barber | 1,923 | 71.8 | 72,300 | 16.5 | 10.8 | 563 | 24.2 | 2.0 | 2,324 | 0.0 | 78 | 3.4 | 2,127 | 31.9 | 36.1 |
| Barton | 10,624 | 70.3 | 93,300 | 19.2 | 10.2 | 648 | 24.9 | 2.3 | 13,680 | 0.1 | 634 | 4.6 | 13,276 | 31.3 | 28.1 |
| Bourbon | 5,636 | 71.4 | 82,700 | 21.0 | 12.3 | 648 | 27.3 | 4.7 | 6,747 | 0.4 | 380 | 5.6 | 6,412 | 29.3 | 31.0 |
| Brown | 3,872 | 70.2 | 88,500 | 18.1 | 10.0 | 633 | 26.1 | 2.2 | 5,153 | -4.0 | 230 | 4.5 | 4,575 | 34.5 | 26.7 |
| Butler | 24,870 | 74.5 | 148,600 | 20.7 | 12.4 | 855 | 28.2 | 2.8 | 32,533 | -0.4 | 2,130 | 6.5 | 31,016 | 40.1 | 23.3 |
| Chase | 1,051 | 80.7 | 103,600 | 20.4 | 15.4 | 570 | 24.8 | 1.7 | 1,465 | -4.8 | 81 | 5.5 | 1,111 | 30.8 | 37.0 |
| Chautauqua | 1,456 | 77.7 | 49,500 | 18.7 | 13.2 | 601 | 22.4 | 3.7 | 1,482 | 3.1 | 90 | 6.1 | 1,364 | 34.2 | 33.1 |
| Cherokee | 7,929 | 72.7 | 82,500 | 19.7 | 12.2 | 613 | 27.0 | 2.5 | 10,016 | -1.4 | 522 | 5.2 | 8,566 | 33.0 | 33.2 |
| Cheyenne | 1,221 | 80.0 | 78,700 | 17.5 | 12.7 | 591 | 21.4 | 1.0 | 1,314 | 0.3 | 29 | 2.2 | 1,269 | 38.9 | 29.3 |
| Clark | 873 | 74.3 | 75,900 | 19.4 | 11.3 | 589 | 16.7 | 1.8 | 1,145 | 6.9 | 23 | 2.0 | 988 | 40.7 | 25.7 |
| Clay | 3,514 | 71.1 | 98,000 | 17.8 | 11.5 | 607 | 28.7 | 2.4 | 3,807 | 0.4 | 171 | 4.5 | 3,848 | 35.0 | 30.2 |
| Cloud | 3,689 | 72.9 | 75,800 | 20.2 | 10.6 | 631 | 25.8 | 1.3 | 3,865 | 1.8 | 157 | 4.1 | 4,377 | 37.3 | 27.9 |
| Coffey | 3,585 | 75.9 | 119,100 | 17.6 | 10.4 | 658 | 22.2 | 2.4 | 3,901 | -2.7 | 204 | 5.2 | 4,126 | 33.0 | 31.0 |
| Comanche | 751 | 79.6 | 70,300 | 17.0 | 12.1 | 443 | 18.4 | 2.1 | 860 | -1.1 | 30 | 3.5 | 925 | 35.8 | 29.5 |
| Cowley | 13,499 | 67.4 | 89,000 | 19.1 | 12.4 | 694 | 25.6 | 1.9 | 16,807 | 2.7 | 1,098 | 6.5 | 15,488 | 29.7 | 31.9 |
| Crawford | 15,284 | 59.4 | 93,900 | 19.1 | 12.2 | 717 | 32.9 | 3.9 | 18,947 | 0.4 | 1,046 | 5.5 | 18,488 | 36.4 | 25.8 |
| Decatur | 1,442 | 73.7 | 67,900 | 19.5 | 10.0 | 669 | 23.2 | 0.3 | 1,241 | 0.1 | 46 | 3.7 | 1,387 | 32.0 | 32.1 |
| Dickinson | 7,911 | 70.6 | 114,700 | 20.3 | 12.4 | 687 | 31.2 | 2.0 | 8,873 | -0.2 | 431 | 4.9 | 8,958 | 31.6 | 28.5 |
| Doniphan | 3,046 | 72.0 | 100,600 | 20.1 | 12.9 | 649 | 22.6 | 2.1 | 4,153 | 0.0 | 188 | 4.5 | 3,841 | 29.2 | 31.0 |
| Douglas | 46,936 | 51.3 | 199,400 | 19.2 | 10.6 | 926 | 30.0 | 2.0 | 64,225 | -2.1 | 3,888 | 6.1 | 68,190 | 44.3 | 16.7 |
| Edwards | 1,308 | 78.0 | 61,400 | 20.2 | 10.0 | 586 | 19.6 | 0.5 | 1,465 | -0.1 | 44 | 3.0 | 1,416 | 29.4 | 34.6 |
| Elk | 1,128 | 79.5 | 54,700 | 18.8 | 14.4 | 495 | 27.8 | 2.9 | 1,171 | 1.3 | 56 | 4.8 | 967 | 30.2 | 29.7 |
| Ellis | 11,544 | 61.2 | 169,100 | 19.5 | 10.7 | 711 | 28.8 | 0.8 | 16,920 | -0.9 | 592 | 3.5 | 16,115 | 34.2 | 21.5 |
| Ellsworth | 2,388 | 76.8 | 99,400 | 18.0 | 10.8 | 639 | 23.1 | 3.1 | 2,725 | -0.9 | 89 | 3.3 | 2,686 | 32.1 | 25.1 |
| Finney | 12,519 | 64.4 | 152,500 | 20.5 | 11.6 | 806 | 23.1 | 7.7 | 20,618 | -0.1 | 712 | 3.5 | 18,616 | 22.1 | 40.6 |
| Ford | 11,344 | 61.5 | 112,500 | 20.9 | 11.3 | 778 | 24.4 | 6.9 | 17,224 | 0.4 | 584 | 3.4 | 16,617 | 25.3 | 45.0 |
| Franklin | 10,118 | 72.3 | 128,000 | 20.0 | 12.5 | 793 | 25.0 | 2.2 | 14,228 | 1.0 | 769 | 5.4 | 12,732 | 28.6 | 35.3 |
| Geary | 12,684 | 37.9 | 148,800 | 22.1 | 13.0 | 1,057 | 27.1 | 3.4 | 11,652 | 2.8 | 800 | 6.9 | 12,001 | 32.2 | 27.8 |
| Gove | 1,217 | 75.8 | 89,600 | 18.0 | 12.4 | 693 | 23.8 | 1.6 | 1,465 |  | 42 | 2.9 | 1,288 | 36.5 | 29.4 |
| Graham | 1,196 | 82.3 | 72,500 | 23.2 | 11.1 | 554 | 29.2 | 0.0 | 1,126 | 0.9 | 52 | 4.6 | 1,229 | 33.4 | 27.2 |
| Grant | 2,551 | 74.5 | 116,100 | 18.1 | 10.0 | 647 | 27.3 | 1.6 | 3,097 | -3.6 | 127 | 4.1 | 3,438 | 27.6 | 37.9 |
| Gray | 2,151 | 78.1 | 135,500 | 18.5 | 10.0 | 663 | 18.0 | 3.3 | 3,244 | 0.1 | 72 | 2.2 | 3,158 | 32.5 | 31.5 |
| Greeley | 519 | 63.0 | 111,200 | 20.9 | 10.0 | 720 | 19.4 | 2.1 | 825 | 1.5 | 15 | 1.8 | 583 | 32.8 | 24.2 |
| Greenwood | 2,769 | 73.8 | 66,700 | 19.4 | 11.4 | 578 | 24.4 | 2.3 | 3,129 | -3.4 | 164 | 5.2 | 2,875 | 30.3 | 31.5 |
| Hamilton | 890 | 71.7 | 81,800 | 22.2 | 13.2 | 668 | 15.5 | 1.6 | 1,633 | 2.2 | 28 | 1.7 | 1,231 | 29.8 | 39.5 |
| Harper | 2,265 | 69.6 | 74,300 | 17.3 | 13.1 | 681 | 22.0 | 2.1 | 2,792 | -2.5 | 124 | 4.4 | 2,371 | 25.9 | 35.5 |
| Harvey | 13,290 | 71.6 | 126,100 | 20.3 | 11.5 | 733 | 27.2 | 1.8 | 16,930 | -1.3 | 900 | 5.3 | 17,037 | 39.1 | 27.7 |
| Haskell | 1,320 | 76.2 | 107,000 | 18.3 | 10.5 | 623 | 18.2 | 5.5 | 2,250 | 1.2 | 53 | 2.4 | 1,916 | 29.6 | 40.5 |
| Hodgeman | 767 | 81.5 | 84,800 | 19.4 | 10.9 | 626 | 22.8 | 2.0 | 1,008 | 1.6 | 27 | 2.7 | 989 | 43.6 | 27.4 |
| Jackson | 5,465 | 75.6 | 134,700 | 22.0 | 11.7 | 716 | 23.8 | 2.5 | 7,213 | 0.9 | 393 | 5.4 | 6,081 | 30.3 | 28.4 |
| Jefferson | 7,575 | 84.1 | 150,600 | 20.4 | 12.3 | 760 | 21.0 | 1.3 | 10,207 | 0.1 | 499 | 4.9 | 9,450 | 35.3 | 31.0 |
| Jewell | 1,380 | 81.3 | 56,100 | 19.0 | 10.6 | 508 | 36.5 | 1.4 | 1,249 | -4.2 | 39 | 3.1 | 1,220 | 44.6 | 25.0 |
| Johnson | 228,592 | 69.2 | 259,600 | 19.0 | 10.0 | 1,109 | 25.8 | 1.6 | 342,146 | 0.4 | 17,602 | 5.1 | 322,681 | 52.2 | 12.8 |
| Kearny | 1,201 | 76.1 | 116,700 | 20.2 | 12.9 | 645 | 20.9 | 1.6 | 2,035 | -0.2 | 55 | 2.7 | 1,659 | 38.7 | 38.0 |
| Kingman | 3,133 | 78.6 | 91,500 | 19.4 | 10.0 | 707 | 19.5 | 1.8 | 3,343 | -0.9 | 174 | 5.2 | 3,576 | 37.5 | 31.6 |
| Kiowa | 985 | 70.6 | 113,900 | 25.2 | 10.0 | 620 | 25.7 | 1.2 | 1,256 | -2.6 | 41 | 3.3 | 1,212 | 40.9 | 24.5 |
| Labette | 8,176 | 69.5 | 76,400 | 19.3 | 11.8 | 633 | 29.5 | 3.6 | 9,667 | -3.9 | 544 | 5.6 | 9,518 | 32.0 | 31.9 |
| Lane | 757 | 75.6 | 90,800 | 18.4 | 12.4 | 503 | 22.4 | 7.5 | 755 | -2.3 | 22 | 2.9 | 864 | 30.0 | 31.7 |
| Leavenworth | 27,253 | 67.4 | 182,900 | 20.3 | 11.7 | 987 | 24.8 | 1.9 | 36,746 | 0.4 | 2,135 | 5.8 | 35,647 | 38.3 | 23.1 |
| Lincoln | 1,248 | 77.9 | 72,700 | 19.3 | 11.5 | 530 | 25.6 | 2.3 | 1,705 | 2.0 | 50 | 2.9 | 1,440 | 35.6 | 28.5 |
| Linn | 4,497 | 77.5 | 110,100 | 19.6 | 15.8 | 667 | 24.3 | 4.4 | 4,379 | -0.8 | 306 | 7.0 | 4,302 | 24.8 | 38.1 |
| Logan | 1,153 | 69.2 | 86,600 | 19.2 | 11.0 | 681 | 23.8 | 1.5 | 1,614 | -0.7 | 41 | 2.5 | 1,425 | 34.0 | 26.2 |
| Lyon | 13,569 | 60.0 | 106,800 | 20.3 | 11.0 | 674 | 29.9 | 4.5 | 16,598 | -0.9 | 790 | 4.8 | 17,330 | 32.7 | 30.6 |
| McPherson | 12,334 | 69.5 | 151,800 | 19.2 | 11.1 | 746 | 23.1 | 1.2 | 17,211 | -0.2 | 599 | 3.5 | 14,832 | 35.5 | 29.0 |
| Marion | 4,846 | 78.5 | 90,800 | 19.0 | 12.3 | 585 | 19.9 | 1.3 | 5,913 | -2.1 | 245 | 4.1 | 5,726 | 31.7 | 33.9 |
| Marshall | 4,085 | 78.7 | 98,900 | 20.2 | 12.0 | 596 | 22.5 | 1.4 | 5,504 | -0.1 | 173 | 3.1 | 4,854 | 31.5 | 34.5 |
| Meade | 1,689 | 69.2 | 96,800 | 19.7 | 10.0 | 633 | 18.0 | 0.7 | 2,150 | -0.2 | 60 | 2.8 | 2,103 | 41.1 | 29.0 |
| Miami | 12,835 | 79.2 | 201,400 | 21.1 | 12.6 | 827 | 25.4 | 1.6 | 17,681 | -0.3 | 888 | 5.0 | 16,663 | 36.9 | 25.5 |
| Mitchell | 2,570 | 71.3 | 88,900 | 19.2 | 12.4 | 561 | 20.5 | 1.2 | 3,429 | 1.6 | 94 | 2.7 | 2,728 | 37.8 | 31.4 |
| Montgomery | 13,576 | 68.8 | 76,200 | 19.8 | 12.1 | 659 | 27.2 | 3.0 | 14,852 | 0.0 | 1,078 | 7.3 | 14,494 | 28.1 | 32.5 |
| Morris | 2,280 | 80.4 | 103,000 | 17.8 | 10.9 | 659 | 19.9 | 0.3 | 3,079 | 2.7 | 121 | 3.9 | 2,719 | 34.8 | 31.4 |
| Morton | 1,038 | 69.5 | 88,500 | 18.6 | 11.2 | 609 | 19.8 | 0.0 | 1,139 | 0.2 | 34 | 3.0 | 1,150 | 39.9 | 37.0 |
| Nemaha | 4,007 | 74.9 | 130,500 | 17.4 | 10.0 | 644 | 21.3 | 1.0 | 5,596 | 4.0 | 166 | 3.0 | 5,292 | 39.6 | 27.6 |

1. Specified owner-occupied units. lacking complete plumbing facilities.  2. A value of 10.0 represents 10 percent or less; a value of 50.0 represents 50 percent or more.  3. Specified renter-occupied units.  4. Overcrowded or  5. Percent of civilian labor force.  6. Civilian employed persons 16 years old and over.

| STATE County | Private nonfarm establishments, employment and payroll, 2019 | | | | | | | | | Agriculture, 2017 | | | |
| --- | --- | --- | --- | --- | --- | --- | --- | --- | --- | --- | --- | --- | --- |
| | | Employment | | | | | | Annual payroll | | Farms | | | Farm producers whose primary occupation is farming (percent) |
| | | | | | | | | | | | | Percent with: | |
| | Number of establishments | Total | Health care and social assistance | Manufacturing | Retail trade | Finance and insurance | Professional, scientific, and technical services | Total (mil dol) | Average per employee (dollars) | Number | Fewer than 50 acres | 1000 acres or more | |
| | 104 | 105 | 106 | 107 | 108 | 109 | 110 | 111 | 112 | 113 | 114 | 115 | 116 |
| KANSAS ..................... | 74,292 | 1,209,318 | 190,075 | 170,774 | 146,997 | 63,750 | 66,432 | 57,272 | 47,359 | 58,569 | 21.8 | 20.2 | 41.9 |
| Allen...................... | 376 | 4,255 | 539 | 1,704 | 603 | 118 | 160 | 161 | 37,773 | 505 | 24.0 | 13.5 | 41.3 |
| Anderson.................. | 196 | 1,553 | 356 | 100 | 457 | 102 | 33 | 54 | 34,867 | 611 | 18.7 | 16.5 | 43.7 |
| Atchison .................. | 362 | 5,701 | 834 | 1,233 | 539 | 123 | 84 | 209 | 36,741 | 595 | 17.5 | 9.1 | 44.4 |
| Barber .................... | 179 | 1,183 | 232 | 168 | 213 | 31 | 41 | 48 | 40,438 | 362 | 11.6 | 36.7 | 54.7 |
| Barton .................... | 928 | 9,991 | 2,348 | 1,155 | 1,419 | 558 | 423 | 389 | 38,896 | 628 | 19.9 | 24.8 | 46.0 |
| Bourbon ................... | 352 | 4,379 | 539 | 1,456 | 657 | 200 | 108 | 149 | 33,963 | 813 | 18.0 | 10.6 | 41.9 |
| Brown ..................... | 258 | 3,624 | 691 | 1,093 | 359 | 193 | 76 | 145 | 39,908 | 510 | 22.4 | 19.2 | 47.5 |
| Butler..................... | 1,313 | 13,923 | 2,689 | 1,748 | 2,107 | 475 | 685 | 537 | 38,601 | 1,471 | 38.0 | 12.8 | 33.4 |
| Chase..................... | 66 | 515 | NA | 171 | 51 | 11 | 28 | 19 | 37,821 | 238 | 13.4 | 33.6 | 51.4 |
| Chautauqua .............. | 64 | 466 | 135 | 63 | 89 | 37 | 8 | 13 | 27,524 | 351 | 16.2 | 21.4 | 45.8 |
| Cherokee.................. | 326 | 5,258 | 850 | 1,429 | 455 | 133 | 85 | 261 | 49,573 | 756 | 29.6 | 12.4 | 42.4 |
| Cheyenne................. | 109 | 612 | 263 | 25 | 80 | 35 | 17 | 22 | 35,191 | 384 | 11.2 | 31.8 | 40.6 |
| Clark...................... | 58 | 405 | 199 | NA | 38 | 34 | 22 | 18 | 44,346 | 230 | 2.2 | 33.5 | 41.7 |
| Clay ...................... | 257 | 2,106 | 425 | 333 | 400 | 98 | 54 | 68 | 32,201 | 547 | 19.6 | 22.5 | 42.7 |
| Cloud..................... | 282 | 2,666 | 649 | 275 | 479 | 89 | 82 | 84 | 31,480 | 412 | 18.7 | 23.3 | 42.9 |
| Coffey.................... | 218 | 2,720 | 458 | 146 | 356 | 103 | 28 | 195 | 71,640 | 699 | 16.5 | 15.2 | 38.3 |
| Comanche................ | 68 | 395 | 153 | 49 | 69 | 20 | 9 | 12 | 29,625 | 197 | 0.5 | 53.8 | 54.9 |
| Cowley ................... | 699 | 11,489 | 1,840 | 4,501 | 1,371 | 347 | 225 | 448 | 39,022 | 921 | 24.1 | 15.0 | 44.2 |
| Crawford ................. | 934 | 14,352 | 2,468 | 2,769 | 1,862 | 337 | 338 | 487 | 33,957 | 777 | 25.6 | 11.2 | 37.3 |
| Decatur................... | 98 | 588 | 152 | NA | 78 | 27 | 34 | 16 | 27,641 | 270 | 12.6 | 40.0 | 52.4 |
| Dickinson ................. | 451 | 5,029 | 832 | 1,437 | 776 | 188 | 130 | 179 | 35,627 | 919 | 22.0 | 18.2 | 43.4 |
| Doniphan.................. | 150 | 1,574 | 127 | 493 | 166 | 53 | 16 | 67 | 42,633 | 430 | 23.3 | 13.3 | 45.2 |
| Douglas................... | 2,757 | 40,811 | 6,486 | 4,134 | 6,576 | 1,282 | 3,635 | 1,478 | 36,214 | 998 | 44.7 | 5.6 | 33.5 |
| Edwards................... | 81 | 510 | 114 | 111 | 75 | 24 | 10 | 19 | 36,876 | 249 | 6.8 | 32.9 | 49.0 |
| Elk ....................... | 59 | 309 | 17 | NA | 47 | 21 | 9 | 7 | 24,104 | 318 | 12.6 | 22.3 | 48.3 |
| Ellis...................... | 1,098 | 12,517 | 2,845 | 1,085 | 2,017 | 448 | 319 | 465 | 37,152 | 603 | 16.1 | 20.6 | 37.3 |
| Ellsworth ................. | 168 | 1,763 | 287 | 363 | 233 | 93 | 73 | 89 | 50,680 | 384 | 10.7 | 28.6 | 45.5 |
| Finney.................... | 1,006 | 15,085 | 1,460 | 3,843 | 2,725 | 453 | 335 | 584 | 38,727 | 450 | 11.3 | 41.8 | 55.7 |
| Ford...................... | 786 | 14,840 | 1,115 | 6,248 | 1,562 | 298 | 497 | 570 | 38,440 | 505 | 11.9 | 32.9 | 44.1 |
| Franklin .................. | 537 | 8,286 | 1,530 | 668 | 1,184 | 142 | 132 | 316 | 38,117 | 1,020 | 37.5 | 9.1 | 37.4 |
| Geary..................... | 560 | 7,727 | 1,401 | 689 | 1,237 | 270 | 256 | 260 | 33,602 | 213 | 19.2 | 27.2 | 41.4 |
| Gove ..................... | 119 | 927 | 203 | 114 | 194 | 42 | 12 | 32 | 34,859 | 350 | 7.7 | 40.3 | 46.2 |
| Graham ................... | 87 | 560 | 188 | NA | 110 | 28 | 5 | 20 | 35,664 | 429 | 12.1 | 27.5 | 38.7 |
| Grant ..................... | 218 | 1,881 | 113 | 168 | 200 | 111 | 38 | 91 | 48,430 | 315 | 7.6 | 27.0 | 40.0 |
| Gray ...................... | 235 | 1,512 | 187 | 80 | 184 | 74 | 55 | 62 | 41,255 | 422 | 13.3 | 29.9 | 49.7 |
| Greeley ................... | 43 | 339 | 141 | NA | 84 | 27 | 21 | 14 | 40,147 | 227 | 4.0 | 45.8 | 52.8 |
| Greenwood................ | 169 | 1,201 | 300 | 91 | 184 | 56 | 44 | 44 | 36,487 | 540 | 16.3 | 26.1 | 48.1 |
| Hamilton.................. | 74 | 533 | 38 | NA | 71 | 63 | NA | 20 | 37,516 | 353 | 4.8 | 37.1 | 41.8 |
| Harper .................... | 196 | 1,862 | 406 | 529 | 260 | 105 | 48 | 75 | 40,165 | 477 | 13.2 | 29.8 | 45.3 |
| Harvey.................... | 756 | 13,451 | 2,894 | 4,064 | 1,349 | 369 | 256 | 506 | 37,631 | 752 | 38.2 | 14.2 | 40.5 |
| Haskell ................... | 116 | 825 | 198 | 101 | 79 | 34 | 23 | 35 | 42,295 | 207 | 8.2 | 43.0 | 70.9 |
| Hodgeman ................ | 48 | 235 | 71 | NA | 28 | 23 | 66 | NA | 36,855 | 351 | 4.0 | 38.7 | 50.6 |
| Jackson................... | 258 | 3,202 | 494 | 284 | 421 | 105 | 66 | 99 | 31,052 | 972 | 23.7 | 8.1 | 34.9 |
| Jefferson.................. | 299 | 2,109 | 328 | 274 | 292 | 63 | 86 | 82 | 39,079 | 1,012 | 33.1 | 6.1 | 33.7 |
| Jewell .................... | 77 | 432 | 87 | NA | 54 | 28 | 10 | 12 | 27,796 | 455 | 12.1 | 30.3 | 53.3 |
| Johnson .................. | 18,120 | 344,466 | 41,700 | 23,579 | 38,008 | 30,888 | 32,223 | 20,172 | 58,561 | 564 | 60.5 | 4.1 | 28.8 |
| Kearny.................... | 84 | 720 | 313 | NA | 91 | 57 | 9 | 27 | 37,368 | 299 | 6.4 | 34.1 | 45.4 |
| Kingman .................. | 208 | 1,650 | 337 | 371 | 248 | 88 | 33 | 61 | 36,915 | 740 | 16.4 | 19.9 | 40.2 |
| Kiowa .................... | 94 | 766 | 209 | NA | 85 | 27 | NA | 26 | 34,389 | 359 | 6.4 | 25.9 | 39.9 |
| Labette ................... | 443 | 7,594 | 2,436 | 2,250 | 839 | 251 | 85 | 284 | 37,359 | 997 | 26.5 | 11.4 | 38.0 |
| Lane...................... | 64 | 351 | 64 | NA | 25 | 32 | 14 | 15 | 41,655 | 242 | 8.3 | 39.3 | 48.4 |
| Leavenworth .............. | 1,252 | 14,248 | 2,692 | 1,217 | 2,254 | 674 | 1,131 | 608 | 42,670 | 1,213 | 44.8 | 2.5 | 29.5 |
| Lincoln.................... | 85 | 532 | 145 | NA | 50 | 47 | 10 | 20 | 50,892 | 864 | 21.6 | 7.2 | 33.4 |
| Linn...................... | 187 | 1,239 | 61 | 145 | 214 | 77 | 20 | 63 | 40,499 | 270 | 7.8 | 45.9 | 54.9 |
| Logan .................... | 110 | 698 | 170 | NA | 105 | 55 | 26 | 28 | 34,738 | 867 | 22.8 | 16.1 | 40.1 |
| Lyon ..................... | 831 | 12,325 | 1,988 | 3,627 | 1,724 | 356 | 185 | 428 | 44,162 | 988 | 25.0 | 17.8 | 40.2 |
| McPherson................ | 887 | 14,765 | 2,510 | 5,168 | 1,219 | 614 | 311 | 652 | 28,783 | 892 | 26.3 | 17.3 | 40.5 |
| Marion ................... | 278 | 2,601 | 584 | 319 | 255 | 88 | 95 | 75 | 39,316 | 802 | 15.1 | 21.6 | 42.0 |
| Marshall .................. | 362 | 4,033 | 522 | 1,321 | 603 | 190 | 123 | 159 | 36,320 | 407 | 11.1 | 32.9 | 48.1 |
| Meade .................... | 120 | 934 | 178 | NA | 132 | 81 | 27 | 34 | 36,320 | 407 | 11.1 | 32.9 | 48.1 |
| Miami .................... | 734 | 7,045 | 2,175 | 466 | 963 | 234 | 305 | 274 | 38,828 | 1,400 | 45.6 | 5.1 | 35.4 |
| Mitchell................... | 239 | 2,505 | 575 | 306 | 380 | 142 | 64 | 90 | 35,857 | 365 | 18.9 | 36.7 | 53.2 |
| Montgomery............... | 762 | 13,073 | 2,789 | 4,031 | 1,305 | 327 | 204 | 508 | 38,887 | 1,006 | 28.1 | 7.4 | 33.8 |
| Morris.................... | 132 | 1,212 | 296 | 212 | 148 | 47 | 42 | 43 | 35,828 | 430 | 17.4 | 23.0 | 52.4 |
| Morton ................... | 79 | 653 | 96 | NA | 65 | 41 | 10 | 28 | 42,286 | 323 | 4.3 | 28.8 | 42.7 |
| Nemaha .................. | 363 | 4,501 | 872 | 1,084 | 443 | 185 | 110 | 185 | 41,206 | 809 | 16.8 | 13.0 | 46.8 |

# Table B. States and Counties — **Agriculture**

| STATE County | Acreage (1,000) | Percent change, 2012-2017 | Average size of farm | Total irrigated (1,000) | Total cropland (1,000) | Average per farm | Average per acre | Value of machinery and equipment, average per farm (dollars) | Total (mil dol) | Average per farm (acres) | Crops | Livestock and poultry products | Organic farms (number) | Farms with internet access (percent) | Total ($1,000) | Percent of farms |
|---|---|---|---|---|---|---|---|---|---|---|---|---|---|---|---|---|
| | 117 | 118 | 119 | 120 | 121 | 122 | 123 | 124 | 125 | 126 | 127 | 128 | 129 | 130 | 131 | 132 |
| KANSAS | 45,759 | -0.8 | 781 | 2,503.4 | 29,125.5 | 1,443,891 | 1,848 | 180,725 | 18,782.7 | 320,694 | 34.4 | 65.6 | 117 | 76.5 | 509,205 | 61.7 |
| Allen | 240 | -2.2 | 475 | 0.5 | 133.0 | 964,120 | 2,029 | 104,166 | 47.9 | 94,921 | 65.1 | 34.9 | NA | 74.7 | 1,578 | 53.7 |
| Anderson | 365 | -0.5 | 597 | 2.9 | 242.1 | 1,216,033 | 2,038 | 170,102 | 108.8 | 178,031 | 74.3 | 25.7 | 2 | 68.4 | 2,651 | 65.5 |
| Atchison | 236 | 7.0 | 396 | 1.0 | 174.3 | 1,205,388 | 3,040 | 161,609 | 85.2 | 143,200 | 78.5 | 21.5 | NA | 73.4 | 2,051 | 58.8 |
| Barber | 632 | 6.9 | 1,745 | 14.2 | 236.5 | 2,506,422 | 1,436 | 218,751 | 93.6 | 258,472 | 40.8 | 59.2 | NA | 79.8 | 4,125 | 60.8 |
| Barton | 558 | -1.4 | 888 | 28.9 | 425.3 | 1,395,047 | 1,570 | 191,756 | 365.7 | 582,280 | 20.5 | 79.5 | NA | 72.1 | 7,953 | 75.8 |
| Bourbon | 336 | 0.5 | 413 | 0.6 | 132.0 | 830,177 | 2,008 | 97,866 | 78.9 | 97,090 | 31.6 | 68.4 | NA | 75.8 | 1,658 | 40.3 |
| Brown | 312 | 5.7 | 611 | 8.0 | 258.6 | 2,142,870 | 3,507 | 234,470 | 131.8 | 258,516 | 85.0 | 15.0 | NA | 81.0 | 5,985 | 75.3 |
| Butler | 798 | 3.9 | 543 | 2.7 | 323.5 | 1,231,054 | 2,268 | 125,659 | 266.2 | 180,942 | 29.7 | 70.3 | 4 | 78.9 | 3,696 | 31.2 |
| Chase | 360 | -8.3 | 1,513 | D | 65.3 | 2,891,131 | 1,911 | 199,983 | 85.4 | 358,954 | 21.9 | 78.1 | NA | 89.9 | 650 | 51.7 |
| Chautauqua | 288 | -7.1 | 822 | 0.0 | 41.6 | 1,340,024 | 1,631 | 97,412 | 31.3 | 89,268 | 32.9 | 67.1 | NA | 67.8 | 507 | 23.9 |
| Cherokee | 319 | 3.6 | 422 | 1.0 | 234.9 | 934,030 | 2,211 | 141,016 | 107.0 | 141,476 | 75.9 | 24.1 | NA | 77.2 | 2,126 | 48.0 |
| Cheyenne | 529 | -3.2 | 1,378 | 43.6 | 340.7 | 2,035,736 | 1,477 | 237,954 | 132.8 | 345,714 | 51.1 | 48.9 | 3 | 67.4 | 2,764 | 77.6 |
| Clark | 434 | -13.7 | 1,888 | 2.7 | 156.2 | 2,241,930 | 1,187 | 150,734 | 111.4 | 484,439 | 13.5 | 86.5 | NA | 59.6 | 6,014 | 87.8 |
| Clay | 386 | 6.5 | 706 | 30.9 | 259.7 | 1,914,252 | 2,712 | 226,844 | 121.2 | 221,528 | 67.5 | 32.5 | NA | 80.6 | 5,062 | 75.5 |
| Cloud | 322 | 0.0 | 782 | 16.9 | 203.2 | 1,799,344 | 2,302 | 219,484 | 77.5 | 188,070 | 73.8 | 26.2 | NA | 74.0 | 3,234 | 71.4 |
| Coffey | 386 | 17.3 | 553 | 1.1 | 219.0 | 1,008,529 | 1,825 | 128,316 | 71.7 | 102,564 | 65.4 | 34.6 | 1 | 76.5 | 3,260 | 68.5 |
| Comanche | 454 | -6.5 | 2,302 | 5.0 | 152.3 | 2,669,162 | 1,159 | 228,061 | 51.8 | 262,959 | 31.6 | 68.4 | NA | 76.1 | 3,306 | 79.2 |
| Cowley | 563 | -1.9 | 612 | 1.8 | 232.9 | 1,131,273 | 1,849 | 130,373 | 96.5 | 104,794 | 55.9 | 44.1 | 1 | 76.8 | 3,311 | 55.4 |
| Crawford | 335 | 3.7 | 431 | 1.4 | 208.4 | 868,757 | 2,014 | 127,884 | 85.9 | 110,605 | 72.6 | 27.4 | NA | 80.2 | 2,304 | 57.4 |
| Decatur | 420 | -9.2 | 1,556 | 6.6 | 254.3 | 2,267,881 | 1,458 | 261,822 | 233.4 | 864,559 | 20.3 | 79.7 | 4 | 80.7 | 2,069 | 61.9 |
| Dickinson | 519 | 1.8 | 565 | 4.4 | 364.4 | 1,209,027 | 2,140 | 176,779 | 149.5 | 162,723 | 51.0 | 49.0 | 2 | 75.7 | 7,399 | 76.8 |
| Doniphan | 177 | -1.1 | 413 | 1.8 | 144.9 | 1,416,197 | 3,431 | 214,601 | 81.2 | 188,902 | 94.3 | 5.7 | NA | 75.1 | 4,128 | 71.6 |
| Douglas | 230 | 9.3 | 231 | 3.5 | 159.3 | 939,826 | 4,072 | 94,196 | 65.9 | 65,999 | 76.7 | 23.3 | 11 | 78.3 | 1,316 | 38.8 |
| Edwards | 392 | -0.6 | 1,574 | 71.3 | 275.1 | 2,902,328 | 1,843 | 359,573 | 228.8 | 918,795 | 30.5 | 69.5 | NA | 85.9 | 7,290 | 91.2 |
| Elk | 247 | -22.0 | 777 | 0.0 | 56.9 | 1,199,924 | 1,545 | 92,972 | 37.7 | 118,428 | 21.6 | 78.4 | NA | 73.3 | 461 | 34.9 |
| Ellis | 502 | 1.0 | 832 | 1.6 | 270.4 | 1,122,301 | 1,349 | 133,106 | 65.0 | 107,813 | 48.9 | 51.1 | NA | 71.3 | 4,333 | 65.8 |
| Ellsworth | 390 | 2.3 | 1,016 | 0.5 | 200.0 | 1,632,908 | 1,608 | 200,669 | 48.3 | 125,828 | 55.6 | 44.4 | NA | 82.3 | 3,065 | 83.1 |
| Finney | 791 | -3.1 | 1,757 | 186.4 | 679.5 | 2,749,764 | 1,565 | 507,327 | 823.1 | 1,829,091 | 22.0 | 78.0 | NA | 75.8 | 14,410 | 74.9 |
| Ford | 670 | -4.3 | 1,326 | 67.1 | 529.2 | 2,051,886 | 1,547 | 293,262 | 515.3 | 1,020,301 | 21.1 | 78.9 | NA | 79.2 | 9,951 | 77.6 |
| Franklin | 355 | -1.8 | 348 | 4.3 | 222.5 | 837,366 | 2,403 | 110,830 | 140.9 | 138,123 | 53.8 | 46.2 | NA | 76.4 | 1,602 | 37.1 |
| Geary | 155 | 6.5 | 728 | 2.8 | 65.1 | 1,676,183 | 2,301 | 148,272 | 31.8 | 149,451 | 46.7 | 53.3 | NA | 72.3 | 934 | 63.4 |
| Gove | 567 | -1.9 | 1,621 | 13.4 | 362.3 | 2,108,777 | 1,301 | 243,452 | 201.5 | 575,757 | 29.7 | 70.3 | NA | 76.6 | 5,853 | 81.7 |
| Graham | 470 | -2.6 | 1,097 | 12.3 | 278.7 | 1,409,160 | 1,285 | 160,299 | 58.2 | 135,676 | 76.5 | 23.5 | NA | 72.5 | 5,961 | 83.7 |
| Grant | 359 | -1.3 | 1,139 | 83.2 | 304.6 | 1,794,738 | 1,576 | 356,803 | 814.1 | 2,584,575 | 9.9 | 90.1 | 1 | 78.4 | 5,423 | 86.3 |
| Gray | 556 | 1.6 | 1,318 | 116.9 | 439.4 | 2,103,461 | 1,596 | 317,212 | 990.7 | 2,347,519 | 12.0 | 88.0 | NA | 80.8 | 11,783 | 76.1 |
| Greeley | 475 | -4.5 | 2,092 | 19.8 | 437.2 | 2,962,429 | 1,416 | 335,255 | 251.3 | 1,107,084 | 23.9 | 76.1 | 1 | 74.4 | 6,405 | 78.9 |
| Greenwood | 616 | -12.1 | 1,141 | D | 112.4 | 1,901,459 | 1,667 | 128,315 | 105.5 | 195,311 | 16.4 | 83.6 | NA | 75.2 | 1,120 | 37.4 |
| Hamilton | 544 | -14.3 | 1,541 | 20.5 | 435.4 | 1,642,940 | 1,066 | 242,221 | 335.7 | 950,878 | 12.3 | 87.7 | NA | 67.7 | 9,079 | 87.3 |
| Harper | 489 | -3.3 | 1,026 | 3.3 | 337.1 | 1,684,429 | 1,642 | 198,557 | 93.1 | 195,279 | 50.3 | 49.7 | NA | 72.5 | 2,397 | 71.9 |
| Harvey | 344 | 1.3 | 457 | 40.4 | 297.9 | 1,448,397 | 3,167 | 188,621 | 140.0 | 186,137 | 57.4 | 42.6 | 11 | 82.0 | 7,020 | 59.8 |
| Haskell | 364 | 0.0 | 1,757 | 117.0 | 320.9 | 2,682,671 | 1,527 | 589,216 | 1,159.1 | 5,599,502 | 9.2 | 90.8 | NA | 84.1 | 8,597 | 90.8 |
| Hodgeman | 495 | -8.8 | 1,410 | 27.3 | 319.9 | 1,658,340 | 1,176 | 271,148 | 191.9 | 546,698 | 20.8 | 79.2 | NA | 78.6 | 8,878 | 86.9 |
| Jackson | 335 | 1.6 | 344 | 0.9 | 168.6 | 921,574 | 2,677 | 103,920 | 71.0 | 73,086 | 56.6 | 43.4 | NA | 79.8 | 2,468 | 44.7 |
| Jefferson | 255 | 4.8 | 252 | 4.1 | 153.3 | 732,914 | 2,904 | 86,460 | 75.7 | 74,833 | 59.3 | 40.7 | 5 | 72.9 | 1,968 | 37.0 |
| Jewell | 463 | -0.1 | 1,018 | 5.6 | 293.5 | 2,097,604 | 2,060 | 271,175 | 149.5 | 328,574 | 57.2 | 42.8 | NA | 79.8 | 5,204 | 80.7 |
| Johnson | 87 | -12.3 | 154 | 0.5 | 56.4 | 582,891 | 3,773 | 81,219 | D | D | D | D | NA | 82.1 | 390 | 22.0 |
| Kearny | 516 | -5.6 | 1,727 | 53.2 | 416.0 | 2,490,399 | 1,442 | 287,798 | 281.0 | 939,726 | 27.2 | 72.8 | NA | 83.6 | 8,766 | 86.3 |
| Kingman | 517 | -4.7 | 698 | 16.6 | 316.5 | 1,130,791 | 1,619 | 137,458 | 78.8 | 106,458 | 62.8 | 37.2 | NA | 72.4 | 6,379 | 76.6 |
| Kiowa | 443 | -2.7 | 1,234 | 56.8 | 256.3 | 1,938,782 | 1,571 | 193,895 | 72.3 | 201,340 | 74.2 | 25.8 | NA | 68.0 | 5,038 | 84.7 |
| Labette | 399 | 7.8 | 400 | 0.3 | 220.2 | 849,632 | 2,121 | 129,803 | 176.0 | 176,565 | 37.3 | 62.7 | 1 | 72.3 | 2,239 | 40.5 |
| Lane | 417 | -7.8 | 1,723 | 10.2 | 311.4 | 2,195,555 | 1,274 | 220,218 | D | D | D | D | NA | 61.2 | 6,680 | 93.0 |
| Leavenworth | 195 | 5.5 | 160 | 0.1 | 112.0 | 548,881 | 3,421 | 69,466 | 44.0 | 36,235 | 70.3 | 29.7 | 1 | 77.8 | 1,027 | 22.5 |
| Lincoln | 385 | -3.1 | 981 | 0.5 | 199.6 | 1,695,466 | 1,727 | 161,747 | 58.2 | 148,347 | 54.1 | 45.9 | NA | 76.3 | 3,203 | 72.7 |
| Linn | 302 | -14.8 | 350 | D | 156.9 | 892,889 | 2,554 | 91,016 | 60.3 | 69,764 | 68.3 | 31.7 | NA | 73.5 | 2,040 | 51.4 |
| Logan | 605 | 6.7 | 2,239 | 11.3 | 340.1 | 3,014,159 | 1,346 | 268,005 | 70.9 | 262,481 | 67.3 | 32.7 | NA | 84.4 | 4,466 | 75.2 |
| Lyon | 523 | -2.2 | 603 | D | 267.5 | 1,199,352 | 1,988 | 150,828 | 134.4 | 155,065 | 44.9 | 55.1 | NA | 76.2 | 3,709 | 58.7 |
| McPherson | 558 | -2.4 | 565 | 39.3 | 415.5 | 1,547,157 | 2,739 | 198,986 | 155.0 | 156,919 | 60.8 | 39.2 | 3 | 81.4 | 9,838 | 71.2 |
| Marion | 568 | -4.8 | 637 | 3.3 | 334.7 | 1,299,945 | 2,042 | 163,154 | 146.5 | 164,247 | 45.7 | 54.3 | 3 | 82.6 | 5,929 | 69.3 |
| Marshall | 500 | 14.0 | 623 | 5.0 | 361.5 | 1,928,648 | 3,094 | 224,692 | 125.4 | 156,353 | 74.1 | 25.9 | 1 | 80.5 | 6,159 | 74.1 |
| Meade | 588 | -4.9 | 1,445 | 93.8 | 331.6 | 2,131,999 | 1,476 | 304,190 | 233.4 | 573,428 | 38.7 | 61.3 | NA | 77.9 | 10,913 | 76.9 |
| Miami | 296 | 0.0 | 211 | 1.6 | 181.6 | 653,977 | 3,095 | 85,262 | 71.8 | 51,285 | 73.9 | 26.1 | 2 | 81.1 | 1,356 | 28.3 |
| Mitchell | 414 | -5.6 | 1,135 | 6.7 | 297.8 | 2,119,970 | 1,868 | 258,391 | 126.5 | 346,474 | 54.0 | 46.0 | NA | 79.5 | 5,379 | 79.5 |
| Montgomery | 366 | 8.9 | 364 | 2.8 | 202.9 | 730,827 | 2,010 | 112,280 | 95.3 | 94,683 | 58.6 | 41.4 | NA | 72.9 | 1,761 | 31.2 |
| Morris | 409 | 5.2 | 952 | 0.8 | 170.7 | 1,543,925 | 1,622 | 184,687 | 138.6 | 322,360 | 26.6 | 73.4 | 1 | 76.5 | 2,094 | 67.0 |
| Morton | 401 | -12.2 | 1,242 | 32.0 | 331.0 | 1,272,145 | 1,024 | 161,831 | 134.8 | 417,381 | 29.7 | 70.3 | NA | 70.9 | 8,627 | 83.0 |
| Nemaha | 400 | 4.6 | 495 | 1.0 | 286.1 | 1,487,190 | 3,006 | 225,584 | 197.4 | 244,049 | 38.6 | 61.4 | 4 | 78.1 | 5,377 | 69.8 |

# Table B. States and Counties — Water Use, Wholesale Trade, Retail Trade, and Real Estate

| STATE County | Water use, 2015 | | Wholesale Trade[1], 2017 | | | | Retail Trade[2], 2017 | | | | Real estate and rental and leasing,[2] 2017 | | | |
|---|---|---|---|---|---|---|---|---|---|---|---|---|---|---|
| | Public supply water withdrawn (mil gal/day) | Public supply gallons withdrawn per person per day | Number of establishments | Number of employees | Sales (mil dol) | Average payroll (mil dol) | Number of establishments | Number of employees | Sales (mil dol) | Average payroll (mil dol) | Number of establishments | Number of employees | Sales (mil dol) | Average payroll (mil dol) |
| | 133 | 134 | 135 | 136 | 137 | 138 | 139 | 140 | 141 | 142 | 143 | 144 | 145 | 146 |
| KANSAS | 351.15 | 120.6 | 3,755 | 52,611 | 61,889.4 | 3,159.9 | 10,095 | 149,845 | 39,337.5 | 3,784.9 | 3,415 | 14,532 | 3,942.5 | 586.4 |
| Allen | 1.78 | 140.0 | 22 | 149 | 79.7 | 8.1 | 51 | 588 | 146.7 | 13.2 | 31 | 52 | 3.2 | 0.6 |
| Anderson | 0.60 | 76.8 | 10 | 44 | 42.8 | 2.0 | 30 | 377 | 131.4 | 12.1 | 3 | D | 0.5 | D |
| Atchison | 4.23 | 258.0 | 18 | 360 | 247.3 | 15.3 | 46 | 577 | 142.7 | 14.1 | 8 | 27 | 4.1 | 0.8 |
| Barber | 0.63 | 130.6 | 10 | 80 | 53.5 | 4.2 | 29 | 199 | 74.3 | 5.1 | 5 | D | 4.5 | D |
| Barton | 2.42 | 89.3 | 62 | 552 | 319.9 | 26.1 | 122 | 1,577 | 382.9 | 38.7 | 36 | 115 | 15.8 | 4.1 |
| Bourbon | 2.08 | 141.4 | D | D | D | 13.5 | 52 | 607 | 152.7 | 15.5 | 8 | 26 | 2.9 | 0.6 |
| Brown | 0.94 | 96.2 | 13 | 178 | 315.2 | 10.5 | 41 | 371 | 84.1 | 8.7 | 5 | 14 | 0.9 | 0.2 |
| Butler | 9.84 | 147.4 | 51 | 543 | 406.6 | 32.4 | 168 | 2,227 | 654.1 | 55.9 | 67 | 168 | 24.9 | 5.4 |
| Chase | 0.17 | 63.5 | D | D | D | 0.7 | 7 | 55 | 13.2 | 0.9 | NA | NA | NA | NA |
| Chautauqua | 0.42 | 123.5 | NA | NA | NA | NA | 12 | 89 | 16.1 | 1.2 | NA | NA | NA | NA |
| Cherokee | 1.83 | 89.1 | 21 | 183 | 243.0 | 9.6 | 48 | 414 | 114.2 | 9.6 | 5 | 12 | 1.3 | 0.2 |
| Cheyenne | 0.47 | 175.4 | 10 | 98 | 93.0 | 5.0 | 20 | 100 | 18.5 | 1.8 | D | D | D | 0.0 |
| Clark | 0.30 | 143.1 | 3 | 18 | 2.9 | 0.8 | 5 | 38 | 7.9 | 0.9 | NA | NA | NA | NA |
| Clay | 0.88 | 105.4 | 17 | 130 | 98.2 | 5.5 | 39 | 409 | 87.8 | 8.8 | 4 | 10 | 2.2 | 0.3 |
| Cloud | 0.89 | 96.5 | 21 | 305 | 321.8 | 15.1 | 48 | 484 | 103.3 | 11.3 | D | D | D | 0.2 |
| Coffey | 0.52 | 62.0 | 13 | 103 | 96.7 | 4.3 | 42 | 384 | 92.9 | 8.2 | 5 | D | 0.3 | D |
| Comanche | 0.36 | 195.3 | NA | NA | NA | NA | 14 | 78 | 16.5 | 1.6 | NA | NA | NA | NA |
| Cowley | 5.40 | 150.9 | 19 | 178 | 83.1 | 9.0 | 116 | 1,438 | 323.8 | 36.2 | 26 | 61 | 8.7 | 1.4 |
| Crawford | 4.90 | 124.9 | 43 | 853 | 449.3 | 37.8 | 141 | 1,895 | 428.4 | 40.3 | 36 | 99 | 15.0 | 2.5 |
| Decatur | 0.37 | 126.2 | 12 | 95 | 94.2 | 3.7 | 16 | 75 | 17.7 | 1.7 | 4 | 10 | 1.8 | 0.2 |
| Dickinson | 1.93 | 100.0 | 17 | 315 | 312.7 | 15.8 | 70 | 703 | 175.7 | 17.0 | 9 | 15 | 1.4 | 0.2 |
| Doniphan | 0.21 | 26.9 | D | D | D | 4.9 | 21 | 144 | 47.1 | 3.5 | D | D | D | D |
| Douglas | 12.67 | 107.3 | 78 | 619 | 464.2 | 29.0 | 371 | 6,373 | 1,529.3 | 146.7 | 156 | 711 | 114.2 | 24.2 |
| Edwards | 0.25 | 84.2 | 9 | 88 | 251.7 | 5.1 | 10 | 76 | 13.6 | 1.3 | NA | NA | NA | NA |
| Elk | 0.09 | 34.5 | NA | NA | NA | NA | 12 | 57 | 14.9 | 1.0 | NA | NA | NA | NA |
| Ellis | 2.53 | 87.2 | 54 | 476 | 224.4 | 20.2 | 177 | 2,151 | 570.1 | 51.7 | 48 | 130 | 21.3 | 3.8 |
| Ellsworth | 1.40 | 220.7 | D | D | D | 3.3 | 31 | 210 | 45.8 | 3.8 | NA | NA | NA | NA |
| Finney | 7.62 | 205.3 | D | D | D | D | 175 | 2,855 | 717.0 | 66.1 | 38 | 140 | 25.9 | 4.5 |
| Ford | 6.79 | 196.6 | 64 | 895 | 541.9 | 43.1 | 118 | 1,762 | 476.3 | 41.8 | 31 | 87 | 27.4 | 3.3 |
| Franklin | 1.93 | 75.4 | 19 | 188 | 215.0 | 12.4 | 85 | 1,200 | 281.1 | 27.4 | 23 | 45 | 7.7 | 1.3 |
| Geary | 4.49 | 121.3 | D | D | D | D | 88 | 1,327 | 324.3 | 30.7 | 52 | 265 | 77.6 | 9.9 |
| Gove | 0.30 | 113.6 | 10 | 85 | 79.8 | 3.9 | 18 | 190 | 83.4 | 5.6 | NA | NA | NA | NA |
| Graham | 0.54 | 208.4 | 8 | D | 22.2 | D | 17 | 100 | 24.5 | 2.3 | NA | NA | NA | NA |
| Grant | 1.22 | 157.8 | 13 | 210 | 162.2 | 10.2 | 32 | 232 | 59.7 | 5.5 | D | D | D | D |
| Gray | 0.67 | 109.2 | 25 | 165 | 384.3 | 8.3 | 25 | 170 | 37.7 | 4.3 | 7 | D | 1.4 | D |
| Greeley | 0.25 | 188.0 | D | D | D | 1.3 | 8 | 68 | 14.8 | 1.9 | NA | NA | NA | NA |
| Greenwood | 0.63 | 100.9 | 9 | 78 | 13.0 | 2.0 | 32 | 198 | 34.6 | 3.5 | NA | NA | NA | NA |
| Hamilton | 0.70 | 282.9 | 6 | 56 | 45.4 | 3.3 | 11 | 75 | 25.1 | 1.6 | NA | NA | NA | NA |
| Harper | 0.86 | 147.8 | 12 | 81 | 75.8 | 3.6 | 39 | 244 | 50.2 | 5.7 | 3 | 7 | 0.7 | 0.1 |
| Harvey | 9.14 | 260.6 | 19 | 192 | 92.5 | 7.1 | 108 | 1,451 | 340.5 | 32.2 | 29 | 56 | 8.1 | 1.3 |
| Haskell | 0.54 | 132.9 | 17 | 126 | 201.5 | 6.8 | 11 | 70 | 17.5 | 1.1 | NA | NA | NA | NA |
| Hodgeman | 0.15 | 79.2 | 5 | 32 | 66.5 | 1.4 | 7 | 27 | 5.4 | 0.6 | NA | NA | NA | NA |
| Jackson | 1.06 | 79.5 | 14 | 107 | 56.9 | 5.2 | 40 | 370 | 86.9 | 8.8 | 8 | 31 | 4.5 | 1.0 |
| Jefferson | 1.20 | 63.4 | D | D | D | 2.0 | 51 | 291 | 69.4 | 5.5 | NA | NA | NA | NA |
| Jewell | 0.59 | 198.7 | 9 | D | 23.2 | D | 13 | 56 | 11.1 | 1.0 | NA | NA | NA | NA |
| Johnson | 11.53 | 19.9 | 905 | 16,313 | 25,385.5 | 1,173.2 | 1,846 | 37,452 | 10,360.8 | 1,046.5 | 1,088 | 5,204 | 2,000.9 | 264.8 |
| Kearny | 0.55 | 139.0 | NA | NA | NA | NA | 13 | 74 | 16.6 | 1.6 | 3 | D | 0.9 | D |
| Kingman | 0.85 | 110.6 | 22 | 123 | 153.9 | 5.1 | 27 | 210 | 49.4 | 4.7 | 7 | 13 | 1.2 | 0.3 |
| Kiowa | 0.33 | 128.7 | 6 | D | 18.8 | D | 12 | 50 | 11.3 | 1.3 | NA | NA | NA | NA |
| Labette | 2.62 | 125.9 | D | D | D | D | 84 | 812 | 184.2 | 19.0 | 14 | 47 | 4.3 | 0.9 |
| Lane | 0.27 | 161.7 | 11 | D | 58.0 | D | 9 | 31 | 6.7 | 0.7 | NA | NA | NA | NA |
| Leavenworth | 28.76 | 362.6 | 22 | 69 | 64.0 | 3.3 | 156 | 2,155 | 631.2 | 56.1 | 66 | 222 | 60.4 | 6.5 |
| Lincoln | 0.15 | 48.3 | 5 | D | 30.4 | D | 10 | 61 | 7.9 | 0.8 | NA | NA | NA | NA |
| Linn | 1.09 | 114.3 | D | D | D | 1.5 | 33 | 243 | 59.7 | 4.9 | D | D | D | D |
| Logan | 0.58 | 205.3 | 13 | 93 | 309.1 | 4.8 | 12 | 103 | 31.1 | 2.9 | NA | NA | NA | NA |
| Lyon | 4.03 | 120.9 | 23 | 433 | 301.6 | 17.6 | 141 | 1,731 | 459.4 | 38.2 | 44 | 174 | 23.1 | 4.1 |
| McPherson | 4.33 | 149.6 | 38 | 317 | 200.2 | 15.1 | 116 | 1,250 | 338.9 | 29.2 | 26 | 54 | 10.4 | 1.5 |
| Marion | 0.99 | 81.8 | 17 | 127 | 59.5 | 6.5 | 39 | 277 | 61.6 | 5.5 | 9 | 12 | 2.8 | 0.3 |
| Marshall | 1.11 | 111.7 | 24 | 220 | 183.4 | 9.5 | 68 | 679 | 229.9 | 16.0 | 5 | 3 | 0.4 | 0.1 |
| Meade | 0.65 | 150.1 | D | D | D | 8.7 | 17 | 130 | 22.5 | 2.3 | NA | NA | NA | NA |
| Miami | 5.67 | 174.2 | 17 | 55 | 85.8 | 2.4 | 78 | 1,030 | 294.4 | 26.7 | D | D | D | D |
| Mitchell | 1.22 | 194.2 | 23 | 200 | 499.6 | 10.1 | 46 | 405 | 128.2 | 10.6 | 6 | 36 | 3.8 | 0.6 |
| Montgomery | 5.17 | 155.2 | 32 | 293 | 135.9 | 12.7 | 121 | 1,359 | 300.3 | 32.5 | 29 | 76 | 9.5 | 1.7 |
| Morris | 1.62 | 287.0 | 5 | D | 2.9 | D | 23 | 160 | 35.7 | 3.3 | NA | NA | NA | NA |
| Morton | 0.62 | 206.2 | 10 | 64 | 50.6 | 3.2 | 11 | 77 | 15.3 | 2.1 | NA | NA | NA | NA |
| Nemaha | 1.33 | 130.0 | 28 | 274 | 210.2 | 13.7 | 55 | 468 | 137.7 | 12.2 | NA | NA | NA | NA |

1 Merchant wholesalers, except manufacturers' sales branches and offices.   2. Employer establishments.

# Professional Services, Manufacturing, and Accommodation and Food Services

| STATE County | Professional, scientific, and technical services, 2017 | | | | Manufacturing, 2017 | | | | Accommodation and food services, 2017 | | | |
|---|---|---|---|---|---|---|---|---|---|---|---|---|
| | Number of establish-ments | Number of employees | Sales (mil dol) | Average payroll (mil dol) | Number of establish-ments | Number of employees | Sales (mil dol) | Average payroll (mil dol) | Number of establis-hments | Number of employees | Sales (mil dol) | Annual payroll (mil dol) |
| | 147 | 148 | 149 | 150 | 151 | 152 | 153 | 154 | 155 | 156 | 157 | 158 |
| KANSAS ........................ | 7,162 | 65,238 | 10,824 | 4,151.0 | 2,760 | 155,968 | 83,418.1 | 9,032.0 | 6,253 | 118,905 | 5,907.5 | 1,723.7 |
| Allen............................ | 27 | 107 | 9 | 3.6 | 26 | 1,743 | 442.3 | 78.0 | 28 | 332 | 14.8 | 3.9 |
| Anderson...................... | 16 | 32 | 4 | 0.9 | 10 | 154 | 117.7 | 7.8 | D | D | D | D |
| Atchison....................... | 21 | 80 | 6 | 2.5 | 19 | 814 | 678.5 | 63.7 | 31 | 484 | 19.3 | 5.8 |
| Barber.......................... | D | D | D | 1.6 | D | 153 | D | 10.2 | 14 | D | 4.9 | D |
| Barton.......................... | 61 | 332 | 32 | 12.3 | 38 | 1,086 | 403.4 | 45.7 | 62 | 895 | 39.2 | 10.3 |
| Bourbon....................... | 26 | 96 | 11 | 4.4 | 24 | 1,201 | 202.6 | 47.6 | 33 | 365 | 17.4 | 4.5 |
| Brown.......................... | D | D | D | D | 19 | 843 | 411.7 | 52.0 | D | D | D | D |
| Butler.......................... | D | D | D | D | 44 | 1,399 | 4,649.7 | 100.2 | 114 | 1,773 | 77.2 | 21.6 |
| Chase.......................... | 5 | 13 | 3 | 0.7 | D | D | D | 7.1 | 7 | D | 2.7 | D |
| Chautauqua ................. | NA | NA | NA | NA | D | 67 | D | 2.1 | 3 | 33 | 1.1 | 0.3 |
| Cherokee ..................... | 23 | 84 | 8 | 2.0 | 29 | 1,400 | 439.2 | 64.7 | 32 | 357 | 16.7 | 4.5 |
| Cheyenne..................... | 7 | 19 | 2 | 0.6 | 4 | 24 | 3.1 | 0.6 | 8 | 56 | 1.5 | 0.3 |
| Clark............................ | 6 | 18 | 5 | 1.0 | NA | NA | NA | NA | D | D | D | 0.1 |
| Clay............................. | 15 | 58 | 4 | 2.0 | 11 | 265 | 59.6 | 13.1 | 18 | 205 | 7.4 | 2.0 |
| Cloud........................... | 12 | 68 | 7 | 2.6 | 10 | 294 | 72.5 | 16.8 | 25 | 432 | 16.4 | 4.5 |
| Coffey.......................... | 15 | 47 | 5 | 1.5 | 8 | 123 | 24.5 | 4.9 | 15 | 133 | 5.0 | 1.4 |
| Comanche..................... | 4 | 6 | 0 | 0.1 | 5 | 23 | 8.3 | 1.4 | 3 | 27 | 0.9 | 0.3 |
| Cowley......................... | D | D | D | D | 39 | 3,680 | 2,691.9 | 206.1 | 68 | 1,171 | 52.9 | 15.2 |
| Crawford ...................... | D | D | D | D | 50 | 2,435 | 1,050.4 | 103.2 | 80 | 1,985 | 93.8 | 24.3 |
| Decatur ........................ | 8 | 33 | 2 | 0.7 | NA | NA | NA | NA | 9 | 51 | 2.3 | 0.6 |
| Dickinson ..................... | 26 | 110 | 15 | 3.8 | 15 | 1,166 | 353.3 | 53.3 | 39 | 400 | 17.8 | 4.9 |
| Doniphan...................... | D | D | D | D | 10 | 370 | 94.6 | 19.9 | 7 | 46 | 2.1 | 0.6 |
| Douglas........................ | 304 | 4,993 | 332 | 131.3 | 65 | 3,723 | 1,499.1 | 183.9 | 327 | 7,255 | 326.0 | 96.0 |
| Edwards ....................... | 5 | 9 | 1 | 0.5 | 6 | 95 | 24.8 | 4.1 | 5 | D | 0.9 | D |
| Elk.............................. | 8 | 14 | 2 | 0.3 | NA | NA | NA | NA | 8 | D | 7.0 | D |
| Ellis............................ | D | D | D | D | 30 | 768 | 207.2 | 34.6 | 96 | 1,891 | 75.6 | 24.3 |
| Ellsworth ..................... | 10 | 71 | 10 | 4.3 | 7 | 259 | 46.5 | 10.1 | D | D | D | D |
| Finney.......................... | 57 | 384 | 44 | 15.8 | 30 | 4,240 | 2,506.2 | 168.8 | 90 | 1,615 | 89.5 | 23.6 |
| Ford............................ | D | D | D | D | 25 | 5,727 | 5,520.4 | 261.3 | 92 | 1,523 | 99.9 | 27.0 |
| Franklin ....................... | 28 | 144 | 15 | 5.5 | 24 | 658 | 323.8 | 34.6 | 52 | 762 | 35.6 | 9.4 |
| Geary........................... | D | D | D | D | 8 | 611 | 299.4 | 28.2 | 80 | 1,577 | 70.0 | 19.2 |
| Gove............................ | D | D | 1 | D | 7 | 101 | 104.6 | 4.8 | 7 | D | 2.5 | D |
| Graham........................ | D | D | D | 0.4 | NA | NA | NA | NA | 5 | D | 0.8 | D |
| Grant........................... | 7 | 35 | 3 | 0.9 | 9 | 151 | 93.4 | 8.7 | 22 | 247 | 9.8 | 2.7 |
| Gray............................ | 18 | 57 | 7 | 2.9 | 5 | 74 | 10.9 | 3.7 | D | D | D | 0.6 |
| Greeley ........................ | 4 | 13 | 1 | 0.4 | NA | NA | NA | NA | NA | NA | NA | 0.6 |
| Greenwood ................... | 15 | 33 | 3 | 1.0 | 6 | 57 | 41.6 | 4.7 | 9 | 76 | 3.0 | 0.9 |
| Hamilton....................... | NA | NA | NA | NA | NA | NA | NA | NA | 9 | 50 | 2.1 | 0.6 |
| Harper.......................... | D | D | 2 | D | 19 | 496 | 170.9 | 21.9 | 17 | 128 | 4.6 | 1.3 |
| Harvey......................... | 45 | 272 | 26 | 8.9 | 58 | 3,297 | 1,022.9 | 157.8 | 67 | 1,030 | 42.6 | 12.6 |
| Haskell ........................ | 9 | 26 | 3 | 0.9 | 6 | 79 | 79.7 | 4.7 | D | D | D | 0.3 |
| Hodgeman .................... | NA | NA | NA | NA | NA | NA | NA | NA | 3 | 12 | 0.4 | 0.1 |
| Jackson........................ | 17 | 68 | 6 | 1.3 | 7 | 236 | 92.1 | 11.3 | 24 | 972 | 144.8 | 26.4 |
| Jefferson ...................... | D | D | D | D | D | 292 | D | 13.3 | D | D | D | D |
| Jewell.......................... | 4 | 11 | 1 | 0.2 | NA | NA | NA | NA | D | D | D | 0.4 |
| Johnson........................ | 2,777 | 30,470 | 6,128 | 2,346.8 | 446 | 21,797 | 8,772.1 | 1,361.9 | 1,223 | 29,848 | 1,637.9 | 510.1 |
| Kearny......................... | 5 | 13 | 2 | 0.5 | NA | NA | NA | NA | 10 | 70 | 2.0 | 0.6 |
| Kingman....................... | 12 | 52 | 5 | 1.6 | 12 | 263 | 47.7 | 14.0 | 16 | 404 | 26.1 | 6.5 |
| Kiowa.......................... | NA | NA | NA | NA | NA | NA | NA | NA | D | D | D | 0.7 |
| Labette......................... | 22 | 87 | 12 | 3.3 | 39 | 2,106 | 473.5 | 91.4 | 40 | 547 | 20.7 | 5.6 |
| Lane............................ | 4 | 14 | 1 | 0.6 | NA | NA | NA | NA | NA | NA | NA | NA |
| Leavenworth ................. | D | D | D | D | 35 | 1,340 | 222.6 | 55.3 | 110 | 1,913 | 92.3 | 29.4 |
| Lincoln......................... | 8 | 13 | 1 | 0.6 | NA | NA | NA | NA | NA | NA | NA | NA |
| Linn............................ | D | D | D | D | 13 | D | 48.7 | D | 13 | 43 | 2.6 | 0.7 |
| Logan.......................... | 7 | 20 | 3 | 0.6 | NA | NA | NA | NA | 11 | 77 | 3.2 | 1.1 |
| Lyon............................ | D | D | D | D | 40 | 3,513 | 1,973.6 | 156.1 | 102 | 1,531 | 64.2 | 16.9 |
| McPherson.................... | 57 | 258 | 34 | 12.3 | 54 | 4,441 | 3,982.7 | 283.8 | 73 | 987 | 45.2 | 14.1 |
| Marion......................... | 15 | 82 | 18 | 3.6 | 24 | 378 | 255.8 | 19.2 | 24 | D | 7.8 | D |
| Marshall....................... | 26 | 108 | 15 | 4.3 | 17 | 1,182 | 320.0 | 62.8 | 23 | 211 | 12.4 | 2.8 |
| Meade.......................... | 7 | 23 | 5 | 0.7 | NA | NA | NA | NA | D | D | D | 0.4 |
| Miami.......................... | D | D | D | D | 23 | 465 | 110.8 | 26.4 | 48 | 689 | 32.3 | 8.8 |
| Mitchell........................ | 14 | 62 | 7 | 2.1 | 9 | 231 | 58.5 | 10.6 | 19 | 168 | 6.7 | 1.7 |
| Montgomery .................. | 46 | 243 | 23 | 8.1 | 50 | 3,180 | 6,658.4 | 181.6 | 72 | 1,026 | 43.5 | 11.9 |
| Morris.......................... | 10 | 56 | 5 | 1.6 | 6 | 161 | 97.8 | 9.1 | 13 | D | 4.3 | D |
| Morton......................... | 4 | 12 | 1 | 0.3 | NA | NA | NA | NA | 6 | D | 1.8 | D |
| Nemaha........................ | D | D | D | D | 21 | 1,112 | 443.8 | 57.7 | 22 | 228 | 8.3 | 2.4 |

# Table B. States and Counties — Health Care and Social Assistance, Other Services, Nonemployer Businesses, and Residential Construction

| STATE County | Health care and social assistance, 2017 | | | | Other services, 2017 | | | | Nonemployer businesses, 2018 | | Value of residential construction authorized by building permits, 2020 | |
|---|---|---|---|---|---|---|---|---|---|---|---|---|
| | Number of establish-ments | Number of employees | Receipts (mil dol) | Annual payroll (mil dol) | Number of establish-ments | Number of employees | Receipts (mil dol) | Annual payroll (mil dol) | Number | Receipts (mil dol) | New construction ($1,000) | Number of housing units |
| | 159 | 160 | 161 | 162 | 163 | 164 | 165 | 166 | 167 | 168 | 169 | 170 |
| KANSAS | 8,104 | 195,941 | 21,439.7 | 8,522.4 | 5,069 | 27,987 | 3,986.2 | 927.7 | 202,389 | 9,509.8 | 1,991,033 | 8,211 |
| Allen | 38 | 521 | 46.7 | 17.1 | 22 | 71 | 7.9 | 1.7 | 804 | 31.0 | 1,033 | 8 |
| Anderson | 18 | 357 | 32.3 | 12.5 | 9 | 22 | 3.0 | 0.5 | 622 | 26.2 | 2,532 | 18 |
| Atchison | 42 | 813 | 75.5 | 34.8 | 25 | 72 | 7.6 | 1.9 | 854 | 32.5 | 538 | 4 |
| Barber | 14 | 230 | 17.5 | 7.9 | 15 | 33 | 3.1 | 0.9 | 517 | 20.4 | 0 | 0 |
| Barton | 109 | 2,061 | 123.7 | 55.2 | 76 | 277 | 36.6 | 9.1 | 2,212 | 101.3 | 1,336 | 9 |
| Bourbon | 41 | 920 | 62.1 | 30.3 | 28 | 97 | 9.5 | 2.3 | 953 | 37.1 | 238 | 1 |
| Brown | 34 | 773 | 76.8 | 32.1 | 20 | 62 | 6.0 | 1.3 | 731 | 28.2 | 577 | 2 |
| Butler | 162 | 2,920 | 270.2 | 106.3 | 89 | 299 | 37.9 | 10.9 | 4,380 | 176.6 | 65,847 | 308 |
| Chase | D | D | D | D | D | D | 2.2 | D | 232 | 10.1 | 300 | 1 |
| Chautauqua | 8 | 146 | 10.1 | 4.5 | NA | NA | NA | NA | 279 | 12.4 | 0 | 0 |
| Cherokee | 36 | 942 | 80.0 | 29.0 | 19 | 47 | 4.6 | 1.1 | 1,097 | 45.8 | 1,815 | 19 |
| Cheyenne | 11 | 201 | 15.7 | 7.4 | D | D | D | 0.2 | 264 | 10.8 | 0 | 0 |
| Clark | D | D | D | D | NA | NA | NA | NA | 226 | 8.6 | 0 | 0 |
| Clay | 23 | 556 | 38.6 | 18.4 | 23 | 78 | 8.5 | 1.8 | 651 | 25.4 | 1,515 | 12 |
| Cloud | 32 | 700 | 45.1 | 18.6 | D | D | D | D | 664 | 25.2 | 0 | 0 |
| Coffey | 30 | 462 | 32.9 | 17.0 | 15 | 89 | 27.0 | 7.3 | 656 | 26.2 | 4,966 | 32 |
| Comanche | 4 | 147 | 8.8 | 4.8 | D | D | 0.8 | D | 176 | 8.5 | 0 | 0 |
| Cowley | 104 | 1,965 | 155.7 | 62.1 | 50 | 210 | 25.0 | 5.3 | 1,920 | 70.7 | 6,508 | 43 |
| Crawford | 121 | 2,800 | 241.9 | 101.3 | 61 | 232 | 20.3 | 5.8 | 2,100 | 79.4 | 16,928 | 89 |
| Decatur | 6 | D | 10.6 | D | D | D | D | 1.7 | 263 | 12.8 | 0 | 0 |
| Dickinson | 38 | 832 | 59.0 | 26.1 | 37 | 108 | 18.5 | 2.7 | 1,224 | 47.5 | 3,862 | 24 |
| Doniphan | D | D | D | D | D | D | 3.0 | D | 456 | 19.8 | 2,085 | 9 |
| Douglas | 316 | 6,533 | 649.4 | 260.4 | 197 | 1,251 | 226.5 | 42.9 | 8,650 | 366.2 | 109,517 | 414 |
| Edwards | 8 | D | 9.8 | D | D | D | 1.3 | D | 211 | 9.7 | 0 | 0 |
| Elk | NA | NA | NA | NA | D | D | D | 0.4 | 252 | 11.4 | NA | NA |
| Ellis | 126 | 3,016 | 307.7 | 130.6 | 80 | 403 | 62.0 | 10.5 | 2,898 | 132.5 | 12,480 | 53 |
| Ellsworth | 23 | 323 | 20.2 | 9.3 | D | D | 5.3 | D | 415 | 15.7 | 1,320 | 12 |
| Finney | 89 | 1,794 | 206.5 | 79.7 | 73 | 369 | 49.2 | 12.5 | 2,384 | 153.4 | 15,823 | 74 |
| Ford | 79 | 1,155 | 138.0 | 47.6 | 53 | 270 | 36.5 | 7.9 | 1,939 | 114.5 | 16,492 | 87 |
| Franklin | 72 | 1,503 | 112.4 | 52.6 | 50 | 240 | 21.9 | 7.3 | 1,688 | 64.9 | 16,325 | 85 |
| Geary | 42 | 1,481 | 169.1 | 71.3 | 48 | 278 | 20.0 | 7.0 | 1,409 | 45.8 | 4,142 | 16 |
| Gove | 8 | 212 | 16.3 | 6.3 | D | D | D | D | 312 | 17.5 | 80 | 1 |
| Graham | 9 | 189 | 12.4 | 5.8 | 10 | 20 | 2.2 | 0.5 | 292 | 11.0 | 383 | 2 |
| Grant | 14 | 191 | 22.7 | 8.2 | 15 | 43 | 5.0 | 1.2 | 504 | 36.5 | 320 | 2 |
| Gray | 10 | 175 | 8.2 | 4.1 | D | D | 6.9 | D | 625 | 38.1 | 1,733 | 9 |
| Greeley | NA | NA | NA | NA | D | D | D | D | 145 | 5.3 | 0 | 0 |
| Greenwood | 17 | 305 | 26.9 | 12.1 | D | D | 1.7 | D | 544 | 18.4 | 250 | 1 |
| Hamilton | 7 | D | 1.3 | D | 4 | 15 | 1.7 | 0.3 | 189 | 8.2 | 0 | 0 |
| Harper | 16 | 296 | 24.0 | 10.5 | D | D | D | 0.3 | 500 | 21.1 | 174 | 2 |
| Harvey | 103 | 2,779 | 254.9 | 110.2 | 54 | 270 | 28.1 | 8.3 | 2,382 | 83.6 | 11,282 | 50 |
| Haskell | D | D | D | D | D | D | D | 0.9 | 413 | 25.7 | 0 | 0 |
| Hodgeman | D | D | D | D | D | D | 0.8 | D | 169 | 9.3 | 0 | 0 |
| Jackson | 25 | 496 | 44.5 | 19.0 | 24 | 109 | 10.1 | 2.3 | 843 | 32.0 | 6,033 | 31 |
| Jefferson | 30 | 408 | 21.6 | 10.8 | D | D | D | D | 1,255 | 48.1 | 14,867 | 71 |
| Jewell | D | D | D | D | D | D | 1.6 | D | 270 | 10.7 | 0 | 0 |
| Johnson | 1,904 | 39,937 | 5,792.6 | 2,018.8 | 1,038 | 7,674 | 901.1 | 278.9 | 51,958 | 2,917.4 | 892,539 | 3,080 |
| Kearny | D | D | D | D | 5 | 18 | 2.0 | 0.5 | 330 | 16.4 | 360 | 3 |
| Kingman | 18 | 334 | 21.6 | 10.7 | 14 | 31 | 1.6 | 0.5 | 623 | 24.5 | 3,763 | 24 |
| Kiowa | 9 | 196 | 12.6 | 6.2 | D | D | D | 0.4 | 258 | 9.1 | 0 | 0 |
| Labette | 69 | 2,203 | 176.4 | 83.2 | 31 | 193 | 22.8 | 5.7 | 1,147 | 47.7 | 205 | 6 |
| Lane | D | D | D | D | D | D | 0.4 | D | 183 | 7.4 | 0 | 0 |
| Leavenworth | 138 | 3,355 | 347.9 | 143.0 | 104 | 445 | 38.8 | 11.9 | 3,971 | 152.1 | 69,259 | 356 |
| Lincoln | 6 | D | 10.4 | D | 7 | 20 | 1.5 | 0.3 | 233 | 7.4 | 0 | 0 |
| Linn | D | D | D | D | D | D | D | D | 696 | 26.7 | 8,608 | 65 |
| Logan | D | D | D | D | 15 | 27 | 3.5 | 0.7 | 276 | 13.1 | 1,189 | 5 |
| Lyon | 109 | 2,050 | 182.7 | 71.2 | 58 | 211 | 24.0 | 5.7 | 1,680 | 64.4 | 6,887 | 40 |
| McPherson | 87 | 2,406 | 150.5 | 69.4 | 73 | 289 | 44.1 | 9.0 | 2,419 | 107.7 | 23,072 | 115 |
| Marion | 30 | 548 | 43.5 | 16.7 | 22 | 95 | 23.0 | 3.1 | 961 | 28.5 | 3,428 | 16 |
| Marshall | 36 | 539 | 42.5 | 19.3 | 26 | 85 | 7.8 | 1.9 | 775 | 30.7 | 553 | 3 |
| Meade | 12 | D | 14.8 | D | D | D | D | 0.6 | 382 | 17.5 | 85 | 1 |
| Miami | 70 | 2,015 | 150.9 | 73.2 | D | D | D | D | 2,580 | 133.1 | 26,988 | 85 |
| Mitchell | 20 | 594 | 44.5 | 20.9 | 21 | 52 | 9.1 | 1.4 | 617 | 23.6 | 175 | 1 |
| Montgomery | 100 | 2,964 | 146.0 | 75.1 | D | D | D | D | 1,729 | 60.5 | 494 | 3 |
| Morris | 16 | 271 | 25.0 | 8.5 | D | D | 1.9 | D | 462 | 20.3 | 1,300 | 4 |
| Morton | D | D | D | D | 4 | 8 | 0.9 | 0.2 | 204 | 5.5 | 0 | 0 |
| Nemaha | 43 | 874 | 60.0 | 28.3 | 24 | 72 | 9.0 | 1.9 | 877 | 35.8 | 1,265 | 7 |

# Table B. States and Counties — Government Employment and Payroll, and Local Government Finances

| | Government employment and payroll, 2017 | | | | | | | | | Local government finances, 2017 | | | | |
| | | | March payroll (percent of total) | | | | | | | General revenue | | | | |
| | | | | | | | | | | | | Taxes | | |
| STATE County | Full-time equivalent employees | March payroll (dollars) | Administration, judicial, and legal | Police and corrections | Fire protection | Highways and transportation | Health and welfare | Natural resources and utilities | Education and libraries | Total (mil dol) | Intergovernmental (mil dol) | Total (mil dol) | Per capita[1] (dollars) Total | Per capita[1] (dollars) Property |
|---|---|---|---|---|---|---|---|---|---|---|---|---|---|---|
| | 171 | 172 | 173 | 174 | 175 | 176 | 177 | 178 | 179 | 180 | 181 | 182 | 183 | 184 |
| KANSAS | X | X | X | X | X | X | X | X | X | X | X | X | X | X |
| Allen | 1,206 | 3,386,282 | 6.4 | 5.7 | 4.0 | 3.5 | 1.1 | 8.7 | 68.7 | 72.3 | 33.7 | 26.6 | 2,120 | 1,732 |
| Anderson | 339 | 1,143,624 | 11.1 | 12.1 | 1.2 | 8.1 | 0.8 | 10.8 | 52.8 | 22.7 | 12.5 | 7.7 | 977 | 682 |
| Atchison | 689 | 2,141,245 | 7.6 | 8.8 | 3.1 | 4.1 | 11.7 | 6.3 | 55.8 | 56.3 | 23.5 | 23.6 | 1,448 | 1,004 |
| Barber | 443 | 1,439,745 | 7.6 | 3.1 | 0.1 | 7.3 | 48.3 | 3.7 | 28.6 | 45.4 | 7.2 | 18.2 | 3,978 | 3,138 |
| Barton | 1,621 | 5,594,991 | 5.1 | 6.4 | 2.4 | 3.4 | 6.3 | 5.1 | 71.0 | 134.9 | 58.1 | 41.5 | 1,573 | 1,088 |
| Bourbon | 825 | 2,577,083 | 7.0 | 4.6 | 3.0 | 4.8 | 0.9 | 4.5 | 73.8 | 68.0 | 30.4 | 25.7 | 1,756 | 1,407 |
| Brown | 504 | 1,480,659 | 8.7 | 10.8 | 0.7 | 4.1 | 1.8 | 5.1 | 67.9 | 69.8 | 15.7 | 48.5 | 5,050 | 4,889 |
| Butler | 3,820 | 13,594,071 | 5.2 | 6.4 | 1.9 | 2.6 | 1.8 | 3.9 | 76.3 | 344.1 | 149.9 | 119.6 | 1,790 | 1,532 |
| Chase | 133 | 400,846 | 19.1 | 17.4 | 1.1 | 8.5 | 5.6 | 1.4 | 46.0 | 13.9 | 5.6 | 5.5 | 2,085 | 1,915 |
| Chautauqua | 187 | 585,546 | 9.5 | 8.1 | 0.0 | 7.9 | 4.7 | 10.3 | 57.9 | 13.9 | 6.4 | 6.4 | 1,926 | 1,541 |
| Cherokee | 796 | 3,157,542 | 10.6 | 14.1 | 1.3 | 6.0 | 2.2 | 3.1 | 61.6 | 67.0 | 39.3 | 20.6 | 1,023 | 746 |
| Cheyenne | 159 | 418,712 | 11.2 | 5.0 | 0.0 | 7.9 | 2.0 | 6.5 | 64.9 | 13.4 | 4.8 | 6.8 | 2,519 | 2,092 |
| Clark | 324 | 1,163,469 | 4.6 | 3.5 | 0.0 | 3.3 | 62.1 | 3.3 | 23.0 | 31.0 | 4.5 | 7.4 | 3,699 | 3,481 |
| Clay | 814 | 2,573,844 | 4.2 | 3.5 | 1.0 | 4.4 | 42.0 | 7.0 | 36.2 | 54.3 | 13.9 | 14.3 | 1,783 | 1,412 |
| Cloud | 863 | 2,648,207 | 3.1 | 6.5 | 1.7 | 4.6 | 1.2 | 3.0 | 78.7 | 48.2 | 20.0 | 17.2 | 1,927 | 1,559 |
| Coffey | 721 | 2,935,460 | 6.3 | 5.5 | 0.0 | 7.3 | 38.8 | 4.8 | 35.4 | 72.2 | 14.4 | 30.2 | 3,671 | 3,518 |
| Comanche | 171 | 551,982 | 7.0 | 6.7 | 0.0 | 6.9 | 42.4 | 3.7 | 31.4 | 21.3 | 5.3 | 10.8 | 6,136 | 5,927 |
| Cowley | 2,518 | 8,522,490 | 5.0 | 5.5 | 2.3 | 1.8 | 21.8 | 6.3 | 56.2 | 172.5 | 77.3 | 47.0 | 1,330 | 1,037 |
| Crawford | 2,951 | 8,874,046 | 5.2 | 6.1 | 2.8 | 1.9 | 13.9 | 3.9 | 65.2 | 159.5 | 82.1 | 49.8 | 1,278 | 803 |
| Decatur | 129 | 411,896 | 16.1 | 6.0 | 0.3 | 11.4 | 5.4 | 6.8 | 47.1 | 18.3 | 9.9 | 6.5 | 2,287 | 1,992 |
| Dickinson | 1,316 | 4,765,557 | 5.2 | 5.5 | 1.1 | 2.9 | 35.8 | 3.2 | 45.5 | 131.5 | 40.4 | 33.0 | 1,753 | 1,383 |
| Doniphan | 705 | 2,334,745 | 3.7 | 3.4 | 0.1 | 3.3 | 2.1 | 4.0 | 82.9 | 48.5 | 25.0 | 12.9 | 1,682 | 1,460 |
| Douglas | 5,231 | 23,730,724 | 5.5 | 10.3 | 4.9 | 3.3 | 33.9 | 8.7 | 31.5 | 655.6 | 134.8 | 204.1 | 1,697 | 1,155 |
| Edwards | 171 | 531,403 | 18.9 | 6.3 | 0.6 | 15.7 | 3.2 | 5.6 | 49.2 | 15.9 | 5.5 | 8.1 | 2,817 | 2,523 |
| Elk | 248 | 634,742 | 9.0 | 4.1 | 0.4 | 6.7 | 4.2 | 2.8 | 71.2 | 13.6 | 7.2 | 4.5 | 1,796 | 1,552 |
| Ellis | 1,047 | 3,344,250 | 8.4 | 10.2 | 3.1 | 6.1 | 4.7 | 6.8 | 56.9 | 104.0 | 34.0 | 51.6 | 1,797 | 1,152 |
| Ellsworth | 359 | 1,095,132 | 16.1 | 3.1 | 0.0 | 7.5 | 5.1 | 8.2 | 59.4 | 26.8 | 10.2 | 13.3 | 2,120 | 1,799 |
| Finney | 2,241 | 7,951,552 | 6.0 | 10.9 | 2.1 | 1.9 | 2.5 | 6.6 | 68.4 | 225.6 | 100.7 | 71.4 | 1,943 | 1,425 |
| Ford | 2,552 | 7,227,791 | 5.1 | 7.8 | 3.3 | 2.6 | 2.6 | 3.3 | 73.2 | 210.6 | 104.6 | 59.3 | 1,733 | 1,155 |
| Franklin | 1,523 | 6,868,830 | 5.0 | 6.2 | 1.3 | 1.7 | 27.0 | 4.7 | 52.0 | 133.1 | 45.9 | 37.5 | 1,465 | 1,145 |
| Geary | 2,182 | 7,774,641 | 3.9 | 8.4 | 3.6 | 1.7 | 20.2 | 1.3 | 59.9 | 223.4 | 104.1 | 52.2 | 1,544 | 959 |
| Gove | 287 | 1,056,581 | 5.3 | 1.7 | 0.0 | 2.9 | 44.5 | 1.6 | 42.8 | 26.2 | 7.2 | 7.7 | 2,942 | 2,483 |
| Graham | 217 | 944,986 | 7.1 | 4.1 | 0.5 | 7.7 | 34.2 | 3.3 | 41.5 | 19.5 | 3.9 | 7.4 | 2,976 | 2,622 |
| Grant | 642 | 2,042,378 | 3.2 | 4.9 | 0.1 | 4.9 | 3.8 | 4.1 | 78.1 | 49.5 | 19.9 | 25.5 | 3,400 | 3,111 |
| Gray | 331 | 1,148,498 | 8.3 | 7.6 | 0.0 | 6.8 | 1.6 | 6.0 | 66.7 | 29.3 | 12.9 | 11.9 | 1,979 | 1,801 |
| Greeley | 89 | 275,669 | 10.9 | 7.8 | 0.1 | 8.9 | 7.2 | 7.5 | 54.5 | 4.3 | 2.3 | 1.6 | 1,330 | 1,293 |
| Greenwood | 348 | 1,033,163 | 9.3 | 7.5 | 1.0 | 6.7 | 3.8 | 4.3 | 66.6 | 28.1 | 11.8 | 14.7 | 2,428 | 2,402 |
| Hamilton | 115 | 383,001 | 16.0 | 9.0 | 1.4 | 8.7 | 3.3 | 9.3 | 41.5 | 14.9 | 5.5 | 8.6 | 3,287 | 3,068 |
| Harper | 557 | 1,854,247 | 5.7 | 4.5 | 0.0 | 6.1 | 49.8 | 5.8 | 26.3 | 57.4 | 13.6 | 17.4 | 3,121 | 2,895 |
| Harvey | 1,515 | 5,347,985 | 8.9 | 8.9 | 5.3 | 4.0 | 3.0 | 5.8 | 62.3 | 138.4 | 57.3 | 50.4 | 1,466 | 964 |
| Haskell | 390 | 1,604,338 | 6.9 | 6.3 | 0.0 | 7.3 | 33.2 | 1.2 | 44.7 | 37.3 | 8.8 | 16.7 | 4,152 | 3,877 |
| Hodgeman | 169 | 392,549 | 12.6 | 7.8 | 0.0 | 10.8 | 7.8 | 13.2 | 42.4 | 18.7 | 4.9 | 11.9 | 6,386 | 6,176 |
| Jackson | 1,193 | 3,448,832 | 3.9 | 5.1 | 0.5 | 3.3 | 0.5 | 4.8 | 80.5 | 48.9 | 26.8 | 17.9 | 1,339 | 1,125 |
| Jefferson | 1,064 | 3,187,375 | 4.2 | 6.2 | 0.1 | 3.7 | 4.7 | 3.6 | 75.2 | 81.3 | 38.1 | 24.6 | 1,299 | 1,209 |
| Jewell | 227 | 656,812 | 7.5 | 3.7 | 0.4 | 14.0 | 32.6 | 7.0 | 29.9 | 15.7 | 3.8 | 7.1 | 2,471 | 2,208 |
| Johnson | 22,476 | 95,446,190 | 7.0 | 9.7 | 3.9 | 2.4 | 5.0 | 7.4 | 63.0 | 2,929.7 | 836.6 | 1,369.1 | 2,315 | 1,502 |
| Kearny | 522 | 1,977,242 | 3.9 | 3.5 | 1.2 | 3.4 | 57.9 | 3.6 | 25.2 | 48.5 | 9.5 | 14.6 | 3,702 | 3,540 |
| Kingman | 313 | 1,057,325 | 10.2 | 8.9 | 0.5 | 9.6 | 4.0 | 5.7 | 57.6 | 32.3 | 12.6 | 15.5 | 2,125 | 1,876 |
| Kiowa | 175 | 522,866 | 13.9 | 11.1 | 0.3 | 13.4 | 2.5 | 8.3 | 43.7 | 14.9 | 4.4 | 8.5 | 3,409 | 3,206 |
| Labette | 1,562 | 5,650,058 | 3.1 | 4.6 | 1.2 | 2.3 | 39.0 | 3.1 | 44.5 | 145.7 | 42.1 | 26.0 | 1,294 | 874 |
| Lane | 191 | 585,401 | 9.4 | 5.7 | 0.4 | 7.8 | 35.8 | 3.3 | 36.9 | 13.8 | 3.0 | 9.8 | 6,318 | 5,994 |
| Leavenworth | 2,831 | 9,904,997 | 6.6 | 9.7 | 2.6 | 3.0 | 3.2 | 5.4 | 68.1 | 261.2 | 131.7 | 90.3 | 1,113 | 809 |
| Lincoln | 287 | 873,911 | 5.1 | 4.6 | 0.1 | 6.8 | 36.1 | 4.5 | 42.1 | 17.2 | 6.0 | 8.6 | 2,826 | 2,435 |
| Linn | 512 | 1,578,522 | 12.2 | 8.0 | 3.9 | 4.9 | 2.1 | 7.4 | 60.5 | 48.6 | 19.5 | 25.5 | 2,634 | 2,538 |
| Logan | 506 | 1,674,444 | 2.9 | 2.8 | 0.1 | 2.9 | 32.5 | 1.9 | 56.9 | 18.0 | 9.7 | 6.0 | 2,139 | 1,796 |
| Lyon | 2,381 | 8,466,398 | 4.0 | 7.5 | 2.8 | 3.7 | 29.1 | 3.3 | 47.6 | 205.0 | 66.5 | 50.6 | 1,522 | 1,126 |
| McPherson | 1,681 | 6,302,623 | 4.9 | 6.8 | 1.5 | 4.1 | 1.4 | 14.0 | 64.4 | 113.4 | 41.5 | 48.5 | 1,690 | 1,348 |
| Marion | 797 | 2,604,278 | 5.4 | 5.2 | 0.1 | 4.5 | 19.6 | 5.4 | 58.7 | 61.1 | 23.6 | 21.3 | 1,786 | 1,490 |
| Marshall | 521 | 1,631,580 | 7.8 | 6.8 | 0.3 | 8.2 | 2.2 | 4.2 | 67.9 | 45.2 | 16.6 | 22.2 | 2,293 | 2,053 |
| Meade | 415 | 1,383,508 | 5.3 | 3.8 | 1.0 | 5.1 | 54.6 | 4.3 | 24.8 | 39.9 | 9.4 | 13.7 | 3,248 | 2,923 |
| Miami | 1,389 | 4,586,183 | 7.1 | 10.7 | 0.8 | 4.2 | 3.7 | 4.0 | 66.6 | 121.8 | 50.8 | 49.2 | 1,473 | 1,088 |
| Mitchell | 593 | 1,586,312 | 6.7 | 6.5 | 0.0 | 6.4 | 2.9 | 12.0 | 61.8 | 44.4 | 15.2 | 18.6 | 3,028 | 2,154 |
| Montgomery | 2,150 | 6,554,386 | 5.0 | 7.5 | 4.1 | 3.1 | 1.1 | 7.9 | 70.3 | 176.0 | 89.8 | 54.8 | 1,693 | 1,230 |
| Morris | 241 | 681,932 | 15.0 | 9.4 | 0.1 | 12.0 | 0.0 | 7.6 | 54.4 | 19.8 | 7.1 | 10.4 | 1,906 | 1,557 |
| Morton | 254 | 932,368 | 7.1 | 4.1 | 0.3 | 6.3 | 31.0 | 3.5 | 42.5 | 27.4 | 12.9 | 12.3 | 4,461 | 4,133 |
| Nemaha | 466 | 1,441,764 | 8.3 | 6.7 | 0.1 | 7.1 | 1.7 | 10.0 | 65.3 | 35.9 | 14.0 | 17.0 | 1,692 | 1,397 |

1. Based on the resident population estimated as of July 1 of the year shown.

# Table B. States and Counties — Local Government Finances, Government Employment, and Income Taxes

| STATE County | Local government finances, 2017 (cont.) | | | | | | | Debt outstanding | | Government employment, 2019 | | | Individual income tax returns, 2016 | | |
|---|---|---|---|---|---|---|---|---|---|---|---|---|---|---|---|
| | Direct general expenditure | | | | | | | | | | | | | | |
| | | | Percent of total for: | | | | | | | | | | | Mean adjusted gross income | Mean income tax |
| | Total (mil dol) | Per capita[1] (dollars) | Education | Health and hospitals | Police protection | Public welfare | Highways | Total (mil dol) | Per capita[1] (dollars) | Federal civilian | Federal military | State and local | Number of returns | | |
| | 185 | 186 | 187 | 188 | 189 | 190 | 191 | 192 | 193 | 194 | 195 | 196 | 197 | 198 | 199 |
| KANSAS .......................... | X | X | X | X | X | X | X | X | X | 25,536 | 32,232 | 241,779 | 1,334,930 | 68,819 | 7,899 |
| Allen.............. | 78 | 6,207 | 62.4 | 2.0 | 3.4 | 0.7 | 3.3 | 63.8 | 5,086 | 52 | 45 | 1,512 | 5,830 | 43,501 | 3,203 |
| Anderson............... | 32 | 4,115 | 41.9 | 2.9 | 3.9 | 0.0 | 9.8 | 23.4 | 2,990 | 34 | 29 | 556 | 3,520 | 46,172 | 3,391 |
| Atchison.............. | 56 | 3,426 | 44.8 | 1.2 | 5.8 | 5.8 | 10.2 | 99.7 | 6,115 | 48 | 55 | 913 | 6,850 | 52,601 | 4,497 |
| Barber.............. | 46 | 10,141 | 19.3 | 44.1 | 2.5 | 0.0 | 11.7 | 30.7 | 6,726 | 28 | 16 | 641 | 2,080 | 44,424 | 4,092 |
| Barton.............. | 152 | 5,752 | 65.8 | 3.2 | 2.9 | 0.0 | 5.5 | 55.0 | 2,085 | 70 | 93 | 2,388 | 11,830 | 52,036 | 4,591 |
| Bourbon.............. | 95 | 6,493 | 69.0 | 0.3 | 2.2 | 0.0 | 4.9 | 80.5 | 5,512 | 77 | 53 | 1,168 | 6,180 | 41,737 | 2,895 |
| Brown.............. | 39 | 4,057 | 65.6 | 0.1 | 5.0 | 0.0 | 5.7 | 37.5 | 3,902 | 102 | 35 | 1,471 | 4,350 | 49,723 | 3,834 |
| Butler.............. | 329 | 4,918 | 69.5 | 1.2 | 3.1 | 0.0 | 5.4 | 875.2 | 13,093 | 135 | 240 | 5,532 | 29,410 | 69,285 | 7,332 |
| Chase.............. | 27 | 10,053 | 62.3 | 0.6 | 1.6 | 0.0 | 5.2 | 29.1 | 10,978 | 14 | 9 | 221 | 1,180 | 50,131 | 4,328 |
| Chautauqua .............. | 14 | 4,103 | 49.0 | 6.2 | 4.2 | 0.0 | 11.0 | 8.6 | 2,585 | 17 | 12 | 240 | 1,390 | 48,237 | 3,845 |
| Cherokee.............. | 71 | 3,533 | 63.7 | 2.1 | 4.4 | 0.0 | 8.0 | 35.8 | 1,779 | 60 | 73 | 1,308 | 8,430 | 49,815 | 3,869 |
| Cheyenne............... | 14 | 5,032 | 44.0 | 4.8 | 3.4 | 0.0 | 10.7 | 1.0 | 357 | 16 | 10 | 237 | 1,220 | 43,139 | 3,430 |
| Clark.............. | 34 | 16,979 | 18.7 | 61.9 | 2.2 | 0.0 | 4.5 | 5.8 | 2,912 | 11 | 7 | 388 | 940 | 49,411 | 3,870 |
| Clay.............. | 55 | 6,809 | 34.1 | 38.2 | 3.4 | 0.0 | 5.2 | 59.3 | 7,411 | 36 | 29 | 961 | 3,950 | 49,164 | 3,513 |
| Cloud.............. | 53 | 5,991 | 63.4 | 2.7 | 3.3 | 0.1 | 6.8 | 21.1 | 2,363 | 39 | 31 | 794 | 3,990 | 44,908 | 3,182 |
| Coffey.............. | 61 | 7,415 | 39.7 | 38.3 | 0.9 | 0.0 | 12.2 | 27.9 | 3,387 | 59 | 30 | 1,093 | 3,940 | 60,804 | 5,930 |
| Comanche............... | 14 | 7,716 | 31.8 | 36.1 | 0.9 | 0.0 | 1.6 | 7.2 | 4,084 | 6 | 6 | 249 | 790 | 43,235 | 3,133 |
| Cowley.............. | 172 | 4,868 | 56.4 | 9.8 | 4.4 | 0.0 | 5.6 | 211.7 | 5,997 | 99 | 122 | 3,401 | 15,000 | 52,018 | 4,238 |
| Crawford.............. | 201 | 5,161 | 58.8 | 12.0 | 4.6 | 0.0 | 3.2 | 187.7 | 4,823 | 88 | 142 | 5,193 | 15,970 | 56,082 | 5,703 |
| Decatur.............. | 14 | 4,761 | 32.0 | 3.1 | 3.9 | 0.0 | 13.9 | 5.1 | 1,787 | 18 | 10 | 259 | 1,400 | 40,118 | 3,059 |
| Dickinson.............. | 117 | 6,211 | 40.8 | 29.1 | 3.1 | 0.0 | 6.4 | 95.9 | 5,093 | 95 | 68 | 1,664 | 8,810 | 50,528 | 3,903 |
| Doniphan.............. | 46 | 5,990 | 75.0 | 2.1 | 3.0 | 0.0 | 6.4 | 11.2 | 1,463 | 31 | 27 | 937 | 3,270 | 52,139 | 4,157 |
| Douglas.............. | 638 | 5,305 | 25.4 | 35.8 | 4.3 | 0.0 | 3.0 | 783.8 | 6,517 | 430 | 456 | 17,860 | 52,010 | 68,522 | 7,791 |
| Edwards.............. | 14 | 4,973 | 42.7 | 4.2 | 5.1 | 0.0 | 13.7 | 7.1 | 2,456 | 19 | 10 | 223 | 1,390 | 48,832 | 3,896 |
| Elk.............. | 15 | 5,942 | 62.2 | 2.0 | 3.4 | 0.7 | 13.4 | 3.0 | 1,205 | 10 | 9 | 305 | 1,120 | 40,448 | 2,926 |
| Ellis.............. | 108 | 3,767 | 41.2 | 3.5 | 4.5 | 0.0 | 6.7 | 47.2 | 1,642 | 138 | 102 | 4,208 | 13,050 | 58,107 | 5,644 |
| Ellsworth.............. | 27 | 4,259 | 52.9 | 4.1 | 5.0 | 0.0 | 12.9 | 17.3 | 2,754 | 29 | 19 | 887 | 2,600 | 54,109 | 4,782 |
| Finney.............. | 209 | 5,692 | 51.9 | 1.9 | 7.1 | 0.0 | 3.7 | 168.8 | 4,598 | 121 | 134 | 3,103 | 17,590 | 52,450 | 4,676 |
| Ford.............. | 294 | 8,600 | 68.1 | 1.7 | 2.5 | 1.0 | 2.5 | 436.5 | 12,756 | 210 | 122 | 2,489 | 14,980 | 49,138 | 3,794 |
| Franklin .............. | 164 | 6,414 | 41.1 | 28.5 | 2.9 | 0.0 | 3.7 | 143.5 | 5,599 | 72 | 93 | 1,936 | 11,870 | 53,192 | 4,370 |
| Geary.............. | 203 | 6,021 | 43.8 | 22.2 | 6.7 | 0.0 | 3.3 | 296.9 | 8,791 | 3,221 | 15,342 | 2,490 | 16,860 | 44,468 | 2,691 |
| Gove.............. | 28 | 10,625 | 25.8 | 51.6 | 1.9 | 0.0 | 8.0 | 2.4 | 924 | 13 | 10 | 421 | 1,330 | 44,351 | 3,745 |
| Graham.............. | 22 | 8,929 | 21.2 | 38.3 | 2.9 | 0.0 | 8.8 | 12.1 | 4,851 | 22 | 9 | 300 | 1,180 | 40,747 | 3,040 |
| Grant.............. | 47 | 6,329 | 53.9 | 4.0 | 4.0 | 0.0 | 6.0 | 18.7 | 2,499 | 19 | 26 | 574 | 3,110 | 54,771 | 4,548 |
| Gray.............. | 29 | 4,877 | 52.7 | 2.9 | 0.8 | 0.0 | 13.1 | 15.2 | 2,531 | 22 | 22 | 887 | 2,710 | 59,249 | 5,745 |
| Greeley .............. | 4 | 3,226 | 77.3 | 0.0 | 1.2 | 0.0 | 4.3 | 3.6 | 2,955 | 10 | 4 | 180 | 590 | 52,517 | 4,078 |
| Greenwood .............. | 17 | 2,846 | 76.9 | 0.1 | 0.0 | 0.0 | 9.7 | 5.0 | 817 | 37 | 22 | 483 | 2,700 | 44,004 | 3,377 |
| Hamilton.............. | 18 | 6,982 | 67.0 | 1.4 | 4.4 | 0.0 | 5.8 | 7.7 | 2,950 | 10 | 9 | 266 | 1,010 | 32,784 | 3,184 |
| Harper.............. | 49 | 8,845 | 25.8 | 37.7 | 2.3 | 7.2 | 10.0 | 17.3 | 3,100 | 34 | 20 | 745 | 2,520 | 44,431 | 3,554 |
| Harvey.............. | 133 | 3,877 | 51.1 | 1.4 | 6.4 | 0.0 | 6.9 | 425.5 | 12,379 | 73 | 123 | 1,997 | 15,780 | 57,926 | 5,043 |
| Haskell.............. | 38 | 9,471 | 33.3 | 36.2 | 3.6 | 0.0 | 11.6 | 4.3 | 1,061 | 16 | 15 | 553 | 1,750 | 65,885 | 6,639 |
| Hodgeman .............. | 11 | 6,121 | 34.0 | 2.3 | 4.4 | 0.0 | 13.4 | 9.2 | 4,954 | 18 | 7 | 269 | 840 | 51,115 | 3,871 |
| Jackson.............. | 56 | 4,233 | 63.6 | 1.3 | 4.7 | 0.0 | 8.6 | 39.2 | 2,938 | 52 | 48 | 1,975 | 6,170 | 51,025 | 3,740 |
| Jefferson.............. | 85 | 4,495 | 70.1 | 2.6 | 4.5 | 0.0 | 9.8 | 46.4 | 2,453 | 74 | 70 | 1,073 | 8,690 | 59,099 | 5,120 |
| Jewell.............. | 15 | 5,306 | 25.2 | 21.5 | 2.5 | 0.0 | 17.8 | 0.8 | 286 | 30 | 11 | 352 | 1,420 | 43,823 | 3,047 |
| Johnson.............. | 2,699 | 4,564 | 44.1 | 2.2 | 7.8 | 0.9 | 9.6 | 5,985.2 | 10,122 | 2,697 | 2,256 | 29,681 | 294,410 | 107,800 | 16,101 |
| Kearny.............. | 45 | 11,503 | 22.9 | 51.4 | 2.8 | 0.0 | 4.9 | 6.1 | 1,550 | 13 | 14 | 667 | 1,750 | 56,513 | 4,883 |
| Kingman.............. | 22 | 2,984 | 68.1 | 0.5 | 0.3 | 0.0 | 9.2 | 15.5 | 2,125 | 36 | 30 | 564 | 3,490 | 53,216 | 4,589 |
| Kiowa.............. | 16 | 6,575 | 37.8 | 2.8 | 4.9 | 0.0 | 16.2 | 6.8 | 2,722 | 14 | 9 | 330 | 1,070 | 46,051 | 3,514 |
| Labette.............. | 146 | 7,255 | 35.4 | 43.4 | 2.8 | 0.0 | 3.0 | 91.2 | 4,530 | 81 | 71 | 2,885 | 8,920 | 44,689 | 3,398 |
| Lane.............. | 14 | 8,961 | 32.3 | 1.8 | 1.2 | 0.0 | 1.3 | 13.8 | 8,892 | 10 | 6 | 252 | 830 | 44,652 | 3,866 |
| Leavenworth .............. | 258 | 3,185 | 55.1 | 1.7 | 5.9 | 0.7 | 6.5 | 395.3 | 4,874 | 4,443 | 3,656 | 4,063 | 34,100 | 66,474 | 6,304 |
| Lincoln.............. | 18 | 5,737 | 43.8 | 7.4 | 2.8 | 0.0 | 13.1 | 1.0 | 337 | 29 | 11 | 377 | 1,390 | 40,063 | 3,004 |
| Linn.............. | 60 | 6,212 | 59.7 | 2.2 | 3.8 | 0.0 | 6.9 | 45.7 | 4,712 | 46 | 36 | 705 | 4,380 | 49,118 | 4,159 |
| Logan.............. | 22 | 7,797 | 82.5 | 0.0 | 3.4 | 0.0 | 2.8 | 3.5 | 1,233 | 20 | 10 | 714 | 1,360 | 53,692 | 4,808 |
| Lyon.............. | 196 | 5,889 | 37.4 | 33.9 | 3.8 | 0.4 | 4.5 | 94.8 | 2,849 | 91 | 118 | 4,734 | 14,610 | 50,049 | 4,000 |
| McPherson.............. | 112 | 3,897 | 52.1 | 1.9 | 4.6 | 0.4 | 11.1 | 191.4 | 6,672 | 90 | 103 | 1,966 | 13,820 | 63,427 | 6,097 |
| Marion.............. | 62 | 5,187 | 44.8 | 17.2 | 2.7 | 0.0 | 15.0 | 36.1 | 3,024 | 61 | 41 | 969 | 5,250 | 51,719 | 3,757 |
| Marshall.............. | 49 | 5,032 | 62.0 | 2.7 | 3.7 | 0.0 | 9.1 | 31.5 | 3,248 | 52 | 35 | 826 | 4,860 | 55,604 | 5,101 |
| Meade.............. | 40 | 9,389 | 18.1 | 40.8 | 2.8 | 3.9 | 7.4 | 11.4 | 2,692 | 19 | 15 | 548 | 1,870 | 57,706 | 4,650 |
| Miami.............. | 117 | 3,496 | 55.8 | 2.4 | 5.0 | 0.0 | 7.3 | 112.6 | 3,367 | 77 | 125 | 2,235 | 15,670 | 69,743 | 7,286 |
| Mitchell.............. | 45 | 7,273 | 61.9 | 4.9 | 3.9 | 0.0 | 9.6 | 33.6 | 5,467 | 31 | 21 | 1,022 | 2,880 | 48,402 | 4,067 |
| Montgomery.............. | 165 | 5,090 | 60.0 | 2.4 | 3.6 | 0.0 | 4.7 | 203.2 | 6,273 | 112 | 114 | 2,623 | 13,180 | 47,591 | 3,799 |
| Morris.............. | 19 | 3,543 | 44.9 | 8.2 | 3.5 | 0.0 | 15.2 | 13.8 | 2,521 | 33 | 21 | 501 | 2,550 | 51,538 | 4,256 |
| Morton.............. | 24 | 8,687 | 47.4 | 1.9 | 1.0 | 0.0 | 4.3 | 2.7 | 978 | 18 | 9 | 377 | 1,280 | 50,205 | 4,216 |
| Nemaha.............. | 32 | 3,212 | 59.7 | 1.3 | 2.7 | 0.0 | 11.6 | 25.7 | 2,550 | 52 | 37 | 759 | 5,000 | 66,910 | 6,584 |

1. Based on the resident population estimated as of July 1 of the year shown.

# Table B. States and Counties — Land Area and Population

| State / county code | CBSA code[1] | County Type code[2] | STATE County | Land area[3] (sq. mi) | Population, 2020 Total persons 2019 | Rank | Per square mile | White | Black | American Indian, Alaska Native | Asian and Pacific Islancer | Percent Hispanic or Latino[4] | Under 5 years | 5 to 14 years | 15 to 24 years | 25 to 34 years | 35 to 44 years | 45 to 54 years |
|---|---|---|---|---|---|---|---|---|---|---|---|---|---|---|---|---|---|---|
| | | | | 1 | 2 | 3 | 4 | 5 | 6 | 7 | 8 | 9 | 10 | 11 | 12 | 13 | 14 | 15 |
| | | | **KANSAS— Cont'd** | | | | | | | | | | | | | | | |
| 20133 | | 7 | Neosho | 571.5 | 15,929 | 2,039 | 27.9 | 90.7 | 1.9 | 2.0 | 1.2 | 6.4 | 6.2 | 14.2 | 12.5 | 11.2 | 11.2 | 10.8 |
| 20135 | | 9 | Ness | 1,074.7 | 2,768 | 2,982 | 2.6 | 87.2 | 1.0 | 1.1 | 0.4 | 11.4 | 5.8 | 12.2 | 10.8 | 9.8 | 9.2 | 9.2 |
| 20137 | | 7 | Norton | 878.0 | 5,328 | 2,802 | 6.1 | 89.5 | 4.1 | 1.1 | 1.7 | 5.5 | 5.4 | 10.4 | 11.1 | 14.1 | 12.0 | 12.9 |
| 20139 | 45820 | 3 | Osage | 705.5 | 15,770 | 2,047 | 22.4 | 94.6 | 1.4 | 1.5 | 0.8 | 3.6 | 5.6 | 12.9 | 11.4 | 10.2 | 11.6 | 11.5 |
| 20141 | | 9 | Osborne | 892.5 | 3,439 | 2,937 | 3.9 | 95.0 | 1.3 | 1.3 | 1.5 | 2.4 | 6.0 | 11.8 | 9.6 | 10.6 | 10.6 | 9.2 |
| 20143 | 41460 | 9 | Ottawa | 720.7 | 5,712 | 2,766 | 7.9 | 94.9 | 1.9 | 0.9 | 0.7 | 2.9 | 5.1 | 13.4 | 11.9 | 9.4 | 11.6 | 11.9 |
| 20145 | | 7 | Pawnee | 754.3 | 6,366 | 2,716 | 8.4 | 85.3 | 6.5 | 1.4 | 0.8 | 7.8 | 4.0 | 7.8 | 11.6 | 12.5 | 13.4 | 12.5 |
| 20147 | | 7 | Phillips | 885.9 | 5,181 | 2,814 | 5.8 | 94.4 | 0.9 | 1.0 | 1.4 | 3.5 | 6.0 | 12.7 | 10.8 | 9.9 | 10.2 | 10.2 |
| 20149 | 31740 | 3 | Pottawatomie | 840.7 | 24,722 | 1,619 | 29.4 | 91.6 | 2.0 | 1.4 | 1.8 | 5.4 | 7.6 | 16.4 | 12.7 | 12.2 | 13.7 | 10.3 |
| 20151 | | 7 | Pratt | 735.0 | 9,127 | 2,484 | 12.4 | 89.5 | 2.4 | 1.5 | 1.3 | 7.3 | 6.2 | 13.8 | 13.8 | 11.0 | 11.4 | 9.1 |
| 20153 | | 9 | Rawlins | 1,069.4 | 2,511 | 3,001 | 2.3 | 90.4 | 1.6 | 1.0 | 0.9 | 7.9 | 6.9 | 12.3 | 8.8 | 10.2 | 10.2 | 9.7 |
| 20155 | 26740 | 4 | Reno | 1,255.3 | 61,793 | 858 | 49.2 | 85.6 | 4.3 | 1.4 | 1.1 | 10.0 | 5.4 | 12.6 | 12.7 | 12.1 | 12.0 | 11.0 |
| 20157 | | 9 | Republic | 717.4 | 4,536 | 2,855 | 6.3 | 96.0 | 1.4 | 0.7 | 0.9 | 2.4 | 5.5 | 12.2 | 9.6 | 9.2 | 10.5 | 9.4 |
| 20159 | | 7 | Rice | 726.2 | 9,362 | 2,467 | 12.9 | 84.9 | 2.2 | 1.8 | 1.6 | 11.8 | 5.7 | 13.2 | 16.0 | 10.6 | 10.8 | 9.9 |
| 20161 | 31740 | 3 | Riley | 609.7 | 73,202 | 748 | 120.1 | 79.4 | 7.9 | 1.2 | 6.3 | 8.6 | 5.4 | 8.7 | 33.7 | 16.8 | 10.6 | 6.9 |
| 20163 | | 9 | Rooks | 890.5 | 4,827 | 2,839 | 5.4 | 95.5 | 1.5 | 0.9 | 0.9 | 2.4 | 6.4 | 12.4 | 10.5 | 10.6 | 10.6 | 10.6 |
| 20165 | | 9 | Rush | 717.8 | 2,947 | 2,965 | 4.1 | 93.1 | 1.0 | 1.5 | 0.7 | 5.0 | 4.4 | 11.3 | 10.2 | 9.4 | 10.7 | 11.8 |
| 20167 | | 7 | Russell | 886.3 | 6,804 | 2,678 | 7.7 | 93.0 | 2.4 | 1.6 | 0.9 | 4.2 | 5.8 | 13.0 | 9.6 | 10.2 | 10.6 | 10.4 |
| 20169 | 41460 | 5 | Saline | 720.2 | 53,926 | 946 | 74.9 | 82.0 | 4.9 | 1.1 | 3.0 | 12.0 | 6.0 | 13.0 | 12.8 | 12.6 | 11.6 | 11.4 |
| 20171 | | 7 | Scott | 717.6 | 4,790 | 2,843 | 6.7 | 78.6 | 1.0 | 1.0 | 1.3 | 19.1 | 6.3 | 15.3 | 12.2 | 12.1 | 10.6 | 10.7 |
| 20173 | 48620 | 2 | Sedgwick | 996.9 | 519,907 | 137 | 521.5 | 70.3 | 10.6 | 1.9 | 5.5 | 15.4 | 6.6 | 14.4 | 13.4 | 14.3 | 12.6 | 11.0 |
| 20175 | 30580 | 5 | Seward | 639.7 | 21,038 | 1,767 | 32.9 | 29.9 | 4.0 | 0.8 | 2.9 | 63.7 | 8.7 | 17.2 | 16.0 | 14.4 | 11.5 | 11.3 |
| 20177 | 45820 | 3 | Shawnee | 544.0 | 175,999 | 378 | 323.5 | 76.4 | 10.0 | 2.1 | 2.3 | 13.1 | 6.0 | 13.1 | 12.5 | 12.3 | 12.2 | 11.2 |
| 20179 | | 9 | Sheridan | 896.0 | 2,520 | 3,000 | 2.8 | 92.7 | 1.1 | 0.8 | 0.6 | 6.0 | 5.5 | 14.0 | 11.5 | 9.6 | 11.0 | 9.1 |
| 20181 | | 7 | Sherman | 1,056.1 | 5,777 | 2,762 | 5.5 | 85.5 | 2.0 | 0.9 | 0.8 | 12.7 | 6.3 | 13.9 | 12.0 | 11.1 | 11.8 | 10.1 |
| 20183 | | 9 | Smith | 895.5 | 3,544 | 2,931 | 4.0 | 95.6 | 1.4 | 1.8 | 0.8 | 2.8 | 5.6 | 11.0 | 9.8 | 8.2 | 10.5 | 9.1 |
| 20185 | | 7 | Stafford | 792.0 | 4,046 | 2,890 | 5.1 | 85.0 | 1.6 | 1.8 | 0.7 | 13.1 | 6.4 | 13.3 | 11.8 | 10.1 | 10.6 | 10.2 |
| 20187 | | 9 | Stanton | 680.4 | 1,969 | 3,050 | 2.9 | 56.7 | 2.7 | 2.3 | 1.4 | 40.0 | 6.3 | 17.3 | 13.4 | 10.1 | 12.0 | 10.6 |
| 20189 | | 7 | Stevens | 727.3 | 5,388 | 2,794 | 7.4 | 59.9 | 1.2 | 1.4 | 0.7 | 38.1 | 6.6 | 17.3 | 13.5 | 10.3 | 12.0 | 11.7 |
| 20191 | 48620 | 2 | Sumner | 1,181.7 | 22,578 | 1,701 | 19.1 | 91.2 | 2.1 | 2.6 | 0.8 | 5.9 | 5.8 | 13.9 | 11.8 | 11.0 | 11.7 | 11.1 |
| 20193 | | 7 | Thomas | 1,074.7 | 7,702 | 2,613 | 7.2 | 89.6 | 1.7 | 0.9 | 1.3 | 7.8 | 6.9 | 13.6 | 16.4 | 12.3 | 10.6 | 9.3 |
| 20195 | | 9 | Trego | 889.5 | 2,758 | 2,985 | 3.1 | 95.4 | 1.4 | 1.0 | 0.9 | 2.8 | 5.1 | 10.3 | 8.7 | 10.1 | 10.8 | 11.2 |
| 20197 | 45820 | 3 | Wabaunsee | 794.3 | 6,906 | 2,668 | 8.7 | 93.4 | 1.7 | 1.9 | 0.8 | 4.5 | 5.4 | 13.3 | 11.4 | 9.9 | 11.0 | 11.0 |
| 20199 | | 9 | Wallace | 913.7 | 1,536 | 3,077 | 1.7 | 90.8 | 1.6 | 1.0 | 0.3 | 7.9 | 8.9 | 13.9 | 10.9 | 10.3 | 10.0 | 8.3 |
| 20201 | | 9 | Washington | 894.8 | 5,427 | 2,792 | 6.1 | 93.6 | 1.1 | 0.7 | 1.0 | 4.8 | 7.2 | 12.8 | 10.7 | 9.9 | 9.7 | 10.5 |
| 20203 | | 9 | Wichita | 718.6 | 2,074 | 3,044 | 2.9 | 68.1 | 1.7 | 1.1 | 0.5 | 29.8 | 6.5 | 16.4 | 11.9 | 10.3 | 10.8 | 9.8 |
| 20205 | | 7 | Wilson | 570.4 | 8,362 | 2,548 | 14.7 | 94.4 | 1.2 | 2.7 | 1.0 | 3.4 | 6.4 | 12.8 | 11.2 | 10.2 | 11.2 | 10.6 |
| 20207 | | 9 | Woodson | 497.8 | 3,015 | 2,958 | 6.1 | 94.6 | 1.8 | 2.6 | 0.5 | 3.3 | 4.9 | 11.5 | 10.0 | 10.0 | 11.0 | 10.0 |
| 20209 | 28140 | 1 | Wyandotte | 151.6 | 165,265 | 402 | 1,090.1 | 42.1 | 23.1 | 1.4 | 6.1 | 30.1 | 7.7 | 15.4 | 13.0 | 14.9 | 13.0 | 11.1 |
| 21000 | | 0 | KENTUCKY | 39,491.4 | 4,477,251 | X | 113.4 | 85.7 | 9.5 | 0.7 | 2.2 | 4.0 | 6.0 | 12.5 | 13.0 | 13.2 | 12.3 | 12.4 |
| 21001 | | 7 | Adair | 405.3 | 19,555 | 1,850 | 48.2 | 93.9 | 3.8 | 0.7 | 0.6 | 2.4 | 5.1 | 10.6 | 16.4 | 11.5 | 10.3 | 12.5 |
| 21003 | 14540 | 3 | Allen | 344.3 | 21,303 | 1,753 | 61.9 | 95.9 | 1.8 | 0.8 | 0.4 | 2.3 | 6.0 | 12.9 | 11.1 | 12.9 | 11.5 | 13.3 |
| 21005 | 23180 | 6 | Anderson | 202.2 | 22,833 | 1,690 | 112.9 | 94.7 | 3.0 | 0.7 | 1.1 | 2.2 | 5.8 | 13.4 | 11.4 | 12.2 | 12.1 | 14.3 |
| 21007 | 37140 | 9 | Ballard | 246.9 | 7,769 | 2,608 | 31.5 | 94.2 | 4.8 | 1.0 | 0.8 | 1.5 | 5.2 | 11.6 | 11.5 | 11.9 | 11.8 | 12.9 |
| 21009 | 23980 | 6 | Barren | 487.6 | 44,300 | 1,091 | 90.9 | 91.6 | 4.9 | 0.6 | 1.0 | 3.5 | 6.5 | 13.0 | 11.4 | 12.1 | 12.2 | 12.6 |
| 21011 | 34460 | 8 | Bath | 278.8 | 12,481 | 2,248 | 44.8 | 96.4 | 2.1 | 0.6 | 0.6 | 1.7 | 7.1 | 14.0 | 11.9 | 11.4 | 11.4 | 13.0 |
| 21013 | 33180 | 7 | Bell | 359.1 | 25,482 | 1,578 | 71.0 | 95.6 | 3.2 | 1.1 | 0.6 | 1.2 | 5.9 | 11.4 | 11.5 | 12.4 | 11.3 | 13.0 |
| 21015 | 17140 | 1 | Boone | 246.3 | 135,396 | 480 | 549.7 | 88.6 | 5.1 | 0.6 | 3.4 | 4.5 | 6.5 | 14.6 | 12.7 | 12.3 | 13.6 | 13.5 |
| 21017 | 30460 | 2 | Bourbon | 289.7 | 19,901 | 1,829 | 68.7 | 86.8 | 7.1 | 0.6 | 0.7 | 6.9 | 5.8 | 12.9 | 11.6 | 12.0 | 10.9 | 12.7 |
| 21019 | 26580 | 2 | Boyd | 159.9 | 46,516 | 1,045 | 290.9 | 94.5 | 3.5 | 0.7 | 0.9 | 2.0 | 5.5 | 12.1 | 10.6 | 12.0 | 12.4 | 12.7 |
| 21021 | 19220 | 7 | Boyle | 180.4 | 30,367 | 1,421 | 168.3 | 87.4 | 9.4 | 0.7 | 1.4 | 3.4 | 5.4 | 11.4 | 15.0 | 12.3 | 11.3 | 12.1 |
| 21023 | 17140 | 1 | Bracken | 202.7 | 8,286 | 2,558 | 40.9 | 97.0 | 1.6 | 0.8 | 0.4 | 1.8 | 5.6 | 13.5 | 11.7 | 11.2 | 12.3 | 12.8 |
| 21025 | | 7 | Breathitt | 492.4 | 12,550 | 2,244 | 25.5 | 97.2 | 0.9 | 0.6 | 1.0 | 1.3 | 6.2 | 11.3 | 10.7 | 12.5 | 12.4 | 13.5 |
| 21027 | | 8 | Breckinridge | 569.8 | 20,537 | 1,793 | 36.0 | 95.1 | 3.2 | 0.9 | 0.5 | 2.0 | 6.0 | 12.2 | 11.5 | 10.7 | 11.6 | 12.4 |
| 21029 | 31140 | 1 | Bullitt | 297.1 | 82,182 | 693 | 276.6 | 95.0 | 2.0 | 0.9 | 1.1 | 2.5 | 5.2 | 12.1 | 11.4 | 12.3 | 13.1 | 13.8 |
| 21031 | 14540 | 3 | Butler | 426.1 | 12,703 | 2,234 | 29.8 | 95.4 | 1.2 | 0.7 | 0.5 | 3.5 | 6.2 | 12.3 | 11.0 | 11.9 | 12.7 | 12.3 |
| 21033 | | 7 | Caldwell | 344.8 | 12,687 | 2,237 | 36.8 | 91.8 | 6.5 | 0.7 | 0.7 | 1.8 | 6.0 | 12.9 | 10.8 | 11.8 | 11.0 | 12.5 |
| 21035 | 34660 | 7 | Calloway | 385.0 | 39,300 | 1,197 | 102.1 | 91.1 | 5.0 | 0.7 | 2.1 | 2.8 | 4.8 | 10.0 | 22.1 | 12.0 | 10.5 | 10.7 |
| 21037 | 17140 | 1 | Campbell | 151.3 | 94,020 | 633 | 621.4 | 93.5 | 4.0 | 0.5 | 1.4 | 2.2 | 5.7 | 11.6 | 13.0 | 14.5 | 12.5 | 11.8 |
| 21039 | | 9 | Carlisle | 189.4 | 4,692 | 2,846 | 24.8 | 95.2 | 3.1 | 1.2 | 0.8 | 2.4 | 6.7 | 12.5 | 10.5 | 11.0 | 11.8 | 12.7 |
| 21041 | | 6 | Carroll | 128.7 | 10,730 | 2,366 | 83.4 | 90.7 | 3.2 | 0.9 | 0.8 | 6.8 | 7.2 | 15.2 | 11.5 | 12.4 | 12.0 | 12.0 |
| 21043 | 26580 | 6 | Carter | 409.5 | 26,542 | 1,545 | 64.8 | 97.3 | 1.0 | 0.7 | 0.4 | 1.4 | 5.9 | 12.5 | 12.4 | 11.5 | 11.5 | 12.9 |
| 21045 | | 9 | Casey | 444.2 | 16,066 | 2,028 | 36.2 | 95.5 | 1.6 | 0.6 | 0.5 | 2.9 | 6.2 | 12.9 | 10.9 | 11.9 | 11.1 | 12.4 |
| 21047 | 17300 | 2 | Christian | 717.5 | 71,478 | 764 | 99.6 | 68.2 | 23.3 | 1.1 | 2.8 | 8.2 | 9.4 | 14.1 | 19.4 | 17.2 | 10.1 | 8.3 |
| 21049 | 30460 | 2 | Clark | 252.5 | 36,463 | 1,264 | 144.4 | 91.2 | 5.6 | 0.6 | 1.0 | 3.2 | 6.0 | 12.6 | 11.1 | 12.8 | 12.0 | 13.2 |

1. CBSA = Core Based Statistical Area. See Appendix A for explanation. See Appendix B for list of metropolitan areas with component counties. Service of USDA Rural-Urban Continuum Codes. See Appendix A for definition. 3. Dry land or land partially or temporarily covered by water. 2. County type code from the Economic Research 4. May be of any race.

| STATE County | 55 to 64 years | 65 to 74 years | 75 years and over | Percent female | 2000 | 2010 | 2000-2010 | 2010-2020 | Births | Deaths | Net Migration | Number | Persons per household | Family house-holds | Female family house-holder[1] | One person |
|---|---|---|---|---|---|---|---|---|---|---|---|---|---|---|---|---|
| | 16 | 17 | 18 | 19 | 20 | 21 | 22 | 23 | 24 | 25 | 26 | 27 | 28 | 29 | 30 | 31 |
| **KANSAS— Cont'd** | | | | | | | | | | | | | | | | |
| Neosho | 13.6 | 11.2 | 9.2 | 50.2 | 16,997 | 16,511 | -2.9 | -3.5 | 2,053 | 2,060 | -575 | 6,601 | 2.37 | 65.9 | 11.8 | 31.1 |
| Ness | 16.9 | 13.2 | 12.8 | 51.2 | 3,454 | 3,107 | -10.0 | -10.9 | 338 | 454 | -219 | 1,269 | 2.23 | 66.3 | 3.2 | 29.9 |
| Norton | 13.3 | 10.3 | 10.5 | 44.0 | 5,953 | 5,669 | -4.8 | -6.0 | 556 | 638 | -259 | 1,830 | 2.50 | 62.1 | 6.5 | 32.2 |
| Osage | 16.2 | 12.1 | 8.6 | 49.9 | 16,712 | 16,294 | -2.5 | -3.2 | 1,712 | 1,995 | -238 | 6,607 | 2.37 | 67.3 | 8.5 | 27.9 |
| Osborne | 16.1 | 12.5 | 13.5 | 49.8 | 4,452 | 3,861 | -13.3 | -10.9 | 441 | 587 | -273 | 1,648 | 2.08 | 58.1 | 5.4 | 36.3 |
| Ottawa | 15.5 | 11.9 | 9.3 | 48.1 | 6,163 | 6,091 | -1.2 | -6.2 | 590 | 703 | -263 | 2,446 | 2.35 | 69.9 | 7.8 | 25.2 |
| Pawnee | 15.7 | 12.5 | 10.0 | 43.9 | 7,233 | 6,971 | -3.6 | -8.7 | 655 | 824 | -435 | 2,448 | 2.27 | 57.8 | 8.1 | 39.3 |
| Phillips | 14.7 | 14.5 | 10.8 | 49.9 | 6,001 | 5,640 | -6.0 | -8.1 | 623 | 737 | -346 | 2,316 | 2.26 | 63.9 | 5.6 | 29.9 |
| Pottawatomie | 12.1 | 8.9 | 6.1 | 50.3 | 18,209 | 21,608 | 18.7 | 14.4 | 3,751 | 1,800 | 1,174 | 8,698 | 2.70 | 73.1 | 7.4 | 23.8 |
| Pratt | 13.5 | 11.2 | 9.9 | 50.4 | 9,647 | 9,653 | 0.1 | -5.4 | 1,312 | 1,190 | -645 | 3,652 | 2.47 | 65.8 | 8.4 | 27.7 |
| Rawlins | 14.1 | 14.5 | 13.4 | 49.1 | 2,966 | 2,519 | -15.1 | -0.3 | 307 | 384 | 69 | 1,176 | 2.10 | 63.9 | 3.1 | 31.5 |
| Reno | 13.7 | 11.1 | 9.4 | 49.7 | 64,790 | 64,511 | -0.4 | -4.2 | 7,189 | 7,634 | -2,241 | 25,014 | 2.40 | 62.1 | 9.4 | 31.3 |
| Republic | 15.5 | 14.4 | 13.7 | 50.5 | 5,835 | 4,980 | -14.7 | -8.9 | 477 | 880 | -40 | 2,218 | 2.06 | 60.6 | 6.0 | 37.6 |
| Rice | 13.7 | 10.9 | 9.3 | 49.4 | 10,761 | 10,082 | -6.3 | -7.1 | 1,199 | 1,235 | -681 | 3,917 | 2.29 | 65.8 | 7.0 | 30.2 |
| Riley | 7.6 | 6.1 | 4.2 | 47.7 | 62,843 | 71,134 | 13.2 | 2.9 | 10,266 | 3,624 | -4,855 | 26,490 | 2.46 | 52.1 | 6.6 | 31.6 |
| Rooks | 15.2 | 12.8 | 10.8 | 51.2 | 5,685 | 5,181 | -8.9 | -6.8 | 639 | 681 | -307 | 2,176 | 2.23 | 61.9 | 6.1 | 34.1 |
| Rush | 16.3 | 13.8 | 12.1 | 48.9 | 3,551 | 3,307 | -6.9 | -10.9 | 288 | 531 | -115 | 1,421 | 2.08 | 53.7 | 5.6 | 41.4 |
| Russell | 14.9 | 13.1 | 12.4 | 50.8 | 7,370 | 6,967 | -5.5 | -2.3 | 850 | 939 | -72 | 3,015 | 2.26 | 62.3 | 4.7 | 30.8 |
| Saline | 13.7 | 10.6 | 8.3 | 50.3 | 53,597 | 55,604 | 3.7 | -3.0 | 7,357 | 5,538 | -3,480 | 21,959 | 2.42 | 60.8 | 8.4 | 32.8 |
| Scott | 12.5 | 10.8 | 9.5 | 49.4 | 5,120 | 4,936 | -3.6 | -3.0 | 643 | 550 | -238 | 1,967 | 2.46 | 69.1 | 4.6 | 24.4 |
| Sedgwick | 12.3 | 9.3 | 6.1 | 50.6 | 452,869 | 498,356 | 10.0 | 4.3 | 74,904 | 44,911 | -8,270 | 197,229 | 2.57 | 63.4 | 12.0 | 30.9 |
| Seward | 9.9 | 6.2 | 4.8 | 48.9 | 22,510 | 22,950 | 2.0 | -8.3 | 4,355 | 1,325 | -4,974 | 7,321 | 2.98 | 71.5 | 13.7 | 23.3 |
| Shawnee | 13.4 | 11.2 | 8.0 | 51.7 | 169,871 | 177,943 | 4.8 | -1.1 | 23,328 | 18,831 | -6,378 | 72,267 | 2.39 | 61.2 | 9.9 | 33.1 |
| Sheridan | 14.7 | 12.4 | 12.2 | 49.1 | 2,813 | 2,544 | -9.6 | -0.9 | 272 | 299 | 6 | 1,134 | 2.18 | 65.6 | 7.1 | 30.3 |
| Sherman | 14.1 | 11.3 | 9.4 | 49.8 | 6,760 | 6,010 | -11.1 | -3.9 | 768 | 719 | -282 | 2,544 | 2.25 | 63.8 | 9.6 | 33.5 |
| Smith | 16.8 | 15.2 | 13.8 | 50.0 | 4,536 | 3,853 | -15.1 | -8.0 | 350 | 595 | -61 | 1,701 | 2.09 | 65.4 | 6.1 | 28.9 |
| Stafford | 15.4 | 11.3 | 10.9 | 49.5 | 4,789 | 4,439 | -7.3 | -8.9 | 486 | 590 | -291 | 1,771 | 2.29 | 67.4 | 6.7 | 28.3 |
| Stanton | 11.9 | 9.9 | 8.4 | 50.5 | 2,406 | 2,236 | -7.1 | -11.9 | 281 | 199 | -351 | 733 | 2.76 | 66.3 | 6.0 | 28.8 |
| Stevens | 12.8 | 8.1 | 7.7 | 50.7 | 5,463 | 5,726 | 4.8 | -5.9 | 792 | 464 | -674 | 1,814 | 3.03 | 73.6 | 10.1 | 21.2 |
| Sumner | 14.8 | 11.2 | 8.6 | 49.9 | 25,946 | 24,137 | -7.0 | -6.5 | 2,732 | 2,724 | -1,573 | 9,416 | 2.40 | 65.5 | 9.4 | 29.6 |
| Thomas | 12.7 | 9.8 | 8.5 | 50.9 | 8,180 | 7,902 | -3.4 | -2.5 | 1,138 | 767 | -580 | 3,375 | 2.20 | 67.0 | 8.4 | 25.2 |
| Trego | 17.5 | 15.0 | 11.2 | 49.5 | 3,319 | 3,006 | -9.4 | -8.3 | 321 | 460 | -106 | 1,379 | 2.00 | 63.8 | 5.8 | 34.1 |
| Wabaunsee | 16.8 | 12.4 | 8.7 | 48.9 | 6,885 | 7,055 | 2.5 | -2.1 | 810 | 662 | -298 | 2,744 | 2.48 | 68.3 | 8.6 | 28.6 |
| Wallace | 14.7 | 11.7 | 11.3 | 49.3 | 1,749 | 1,485 | -15.1 | 3.4 | 225 | 174 | -2 | 636 | 2.44 | 71.1 | 5.5 | 28.6 |
| Washington | 14.5 | 12.3 | 12.4 | 49.1 | 6,483 | 5,794 | -10.6 | -6.3 | 734 | 742 | -364 | 2,332 | 2.29 | 62.8 | 5.1 | 33.3 |
| Wichita | 13.1 | 11.1 | 10.2 | 47.7 | 2,531 | 2,234 | -11.7 | -7.2 | 262 | 247 | -175 | 934 | 2.25 | 67.2 | 6.5 | 30.1 |
| Wilson | 14.4 | 13.1 | 10.0 | 50.6 | 10,332 | 9,406 | -9.0 | -11.1 | 1,123 | 1,243 | -922 | 3,712 | 2.3 | 64.0 | 11.9 | 31.4 |
| Woodson | 17.0 | 15.0 | 10.6 | 49.2 | 3,788 | 3,309 | -12.6 | -8.9 | 314 | 501 | -111 | 1,434 | 2.2 | 60.4 | 9.7 | 33.0 |
| Wyandotte | 11.7 | 8.0 | 5.0 | 50.3 | 157,882 | 157,523 | -0.2 | 4.9 | 27,465 | 14,252 | -5,357 | 60,128 | 2.72 | 63.6 | 16.4 | 30.5 |
| **KENTUCKY** | 13.3 | 10.3 | 6.9 | 50.7 | 4,041,769 | 4,339,330 | 7.4 | 3.2 | 563,501 | 471,069 | 46,257 | 1,734,618 | 2.49 | 65.5 | 12.3 | 28.6 |
| Adair | 14.2 | 11.3 | 8.1 | 50.2 | 17,244 | 18,656 | 8.2 | 4.8 | 2,061 | 2,037 | 867 | 6,977 | 2.56 | 69.4 | 10.2 | 28.9 |
| Allen | 14.0 | 10.9 | 7.4 | 50.2 | 17,800 | 19,943 | 12.0 | 6.8 | 2,544 | 2,338 | 1,167 | 7,605 | 2.72 | 72.6 | 11.9 | 24.2 |
| Anderson | 14.1 | 10.2 | 6.5 | 50.8 | 19,111 | 21,469 | 12.3 | 6.4 | 2,606 | 2,222 | 991 | 8,694 | 2.57 | 68.3 | 12.0 | 25.8 |
| Ballard | 14.4 | 12.2 | 9.4 | 50.2 | 8,286 | 8,237 | -0.6 | -5.7 | 824 | 1,077 | -212 | 3,065 | 2.58 | 69.1 | 9.7 | 26.6 |
| Barren | 14.0 | 10.4 | 7.7 | 51.7 | 38,033 | 42,164 | 10.9 | 5.1 | 5,664 | 5,140 | 1,634 | 16,931 | 2.55 | 67.2 | 13.9 | 26.6 |
| Bath | 13.7 | 10.6 | 7.0 | 50.4 | 11,085 | 11,592 | 4.6 | 7.7 | 1,756 | 1,488 | 626 | 4,827 | 2.54 | 73.0 | 11.6 | 21.2 |
| Bell | 14.1 | 12.0 | 8.3 | 50.9 | 30,060 | 28,688 | -4.6 | -11.2 | 3,492 | 4,022 | -2,683 | 10,624 | 2.44 | 65.5 | 14.3 | 31.1 |
| Boone | 12.6 | 8.9 | 5.5 | 50.4 | 85,991 | 118,810 | 38.2 | 14.0 | 17,337 | 8,681 | 8,007 | 46,997 | 2.76 | 73.9 | 11.0 | 21.3 |
| Bourbon | 14.2 | 11.4 | 8.5 | 51.0 | 19,360 | 20,008 | 3.3 | -0.5 | 2,256 | 2,266 | -79 | 8,106 | 2.44 | 67.2 | 11.9 | 26.2 |
| Boyd | 14.1 | 11.9 | 8.7 | 50.6 | 49,752 | 49,533 | -0.4 | -6.1 | 5,752 | 6,326 | -2,425 | 18,210 | 2.52 | 67.3 | 13.0 | 27.6 |
| Boyle | 12.8 | 11.0 | 8.7 | 50.3 | 27,697 | 28,434 | 2.7 | 6.8 | 3,197 | 3,327 | 2,038 | 11,023 | 2.44 | 63.6 | 11.6 | 32.0 |
| Bracken | 14.8 | 11.3 | 6.7 | 50.1 | 8,279 | 8,485 | 2.5 | -2.3 | 1,046 | 964 | -281 | 3,318 | 2.49 | 64.5 | 8.6 | 29.5 |
| Breathitt | 15.2 | 11.7 | 6.4 | 50.2 | 16,100 | 13,875 | -13.8 | -9.5 | 1,723 | 2,010 | -1,040 | 5,358 | 2.35 | 67.0 | 12.8 | 29.7 |
| Breckinridge | 15.1 | 12.5 | 8.0 | 49.8 | 18,648 | 20,054 | 7.5 | 2.4 | 2,331 | 2,370 | 526 | 7,598 | 2.61 | 67.9 | 9.2 | 25.5 |
| Bullitt | 14.3 | 10.3 | 6.5 | 50.3 | 61,236 | 74,293 | 21.3 | 10.6 | 8,022 | 6,086 | 5,940 | 29,690 | 2.69 | 73.2 | 11.1 | 22.6 |
| Butler | 14.4 | 11.1 | 8.0 | 49.7 | 13,010 | 12,706 | -2.3 | 0.0 | 1,581 | 1,570 | -6 | 4,958 | 2.53 | 73.2 | 14.1 | 23.4 |
| Caldwell | 13.6 | 12.5 | 9.0 | 51.4 | 13,060 | 12,989 | -0.5 | -2.3 | 1,506 | 1,718 | -80 | 5,163 | 2.43 | 68.7 | 11.9 | 26.5 |
| Calloway | 12.0 | 10.3 | 7.6 | 51.3 | 34,177 | 37,199 | 8.8 | 5.6 | 3,869 | 4,091 | 2,319 | 14,996 | 2.34 | 59.6 | 8.8 | 32.6 |
| Campbell | 14.0 | 10.2 | 6.6 | 50.9 | 88,616 | 90,340 | 1.9 | 4.1 | 11,432 | 8,775 | 1,033 | 36,746 | 2.43 | 63.4 | 11.4 | 30.1 |
| Carlisle | 13.6 | 11.3 | 9.9 | 50.5 | 5,351 | 5,099 | -4.7 | -8.0 | 647 | 709 | -346 | 1,980 | 2.39 | 66.5 | 7.6 | 28.8 |
| Carroll | 13.7 | 9.6 | 6.5 | 49.9 | 10,155 | 10,807 | 6.4 | -0.7 | 1,587 | 1,347 | -328 | 4,101 | 2.52 | 62.8 | 11.6 | 30.8 |
| Carter | 13.7 | 11.5 | 8.2 | 50.7 | 26,889 | 27,716 | 3.1 | -4.2 | 3,529 | 3,318 | -1,372 | 9,606 | 2.75 | 68.7 | 12.7 | 28.7 |
| Casey | 13.7 | 12.2 | 8.6 | 51.2 | 15,447 | 15,963 | 3.3 | 0.6 | 1,989 | 1,984 | 110 | 6,099 | 2.53 | 64.3 | 10.8 | 30.6 |
| Christian | 8.7 | 7.2 | 5.7 | 46.8 | 72,265 | 73,943 | 2.3 | -3.3 | 15,231 | 6,231 | -11,531 | 25,721 | 2.52 | 67.5 | 14.1 | 27.4 |
| Clark | 13.7 | 11.0 | 7.5 | 51.4 | 33,144 | 35,599 | 7.4 | 2.4 | 4,356 | 4,250 | 774 | 14,509 | 2.45 | 68.5 | 14.4 | 25.0 |

1. No spouse present.

# Table B. States and Counties — Population, Vital Statistics, Health, and Crime

| STATE County | Persons in group quarters, 2020 | Daytime Population, 2015-2019 Number | Employment/residence ratio | Births, 2020 Total | Rate[1] | Deaths, 2020 Number | Rate[1] | Persons under 65 with no health insurance, 2019 Number | Percent | Medicare, 2020 Total beneficiaries | Enrolled in Original Medicare | Enrolled in Medicare Advantage | Crimes reported by county police or sheriff, 2019 Violent | Property |
|---|---|---|---|---|---|---|---|---|---|---|---|---|---|---|
| | 32 | 33 | 34 | 35 | 36 | 37 | 38 | 39 | 40 | 41 | 42 | 43 | 44 | 45 |
| **KANSAS— Cont'd** | | | | | | | | | | | | | | |
| Neosho | 451 | 15,964 | 0.98 | 161 | 10.1 | 184 | 11.6 | 1,186 | 9.6 | 3,631 | 3,279 | 353 | 8 | 30 |
| Ness | 70 | 2,860 | 0.99 | 35 | 12.6 | 38 | 13.7 | 326 | 16.5 | 779 | 749 | 29 | 8 | 18 |
| Norton | 818 | 5,462 | 1.01 | 56 | 10.5 | 68 | 12.8 | 391 | 11.3 | 1,149 | 1,094 | 54 | NA | NA |
| Osage | 194 | 11,767 | 0.43 | 166 | 10.5 | 230 | 14.6 | 1,247 | 10.0 | 3,981 | 3,325 | 656 | 14 | NA |
| Osborne | 100 | 3,312 | 0.87 | 41 | 11.9 | 40 | 11.6 | 327 | 13.0 | 968 | 935 | 34 | 1 | 26 |
| Ottawa | 98 | 4,509 | 0.54 | 59 | 10.3 | 66 | 11.6 | 430 | 9.7 | 1,390 | 1,344 | 46 | NA | NA |
| Pawnee | 1,023 | 6,916 | 1.11 | 51 | 8.0 | 81 | 12.7 | 386 | 9.5 | 1,485 | 1,431 | 54 | NA | NA |
| Phillips | 68 | 5,480 | 1.05 | 61 | 11.8 | 66 | 12.7 | 423 | 10.9 | 1,473 | 1,418 | 55 | 5 | 5 |
| Pottawatomie | 299 | 22,967 | 0.92 | 371 | 15.0 | 210 | 8.5 | 1,808 | 8.7 | 4,248 | 3,694 | 554 | 27 | 177 |
| Pratt | 381 | 9,897 | 1.10 | 111 | 12.2 | 91 | 10.0 | 799 | 11.6 | 2,090 | 2,031 | 59 | 5 | 20 |
| Rawlins | 40 | 2,490 | 0.99 | 37 | 14.7 | 23 | 9.2 | 265 | 14.7 | 724 | 677 | 47 | NA | NA |
| Reno | 3,113 | 61,631 | 0.96 | 620 | 10.0 | 749 | 12.1 | 5,664 | 12.2 | 14,325 | 12,410 | 1,915 | 32 | 178 |
| Republic | 103 | 4,466 | 0.91 | 41 | 9.0 | 65 | 14.3 | 390 | 11.8 | 1,384 | 1,322 | 61 | NA | NA |
| Rice | 651 | 9,071 | 0.87 | 108 | 11.5 | 98 | 10.5 | 822 | 11.6 | 2,122 | 2,032 | 90 | 4 | 28 |
| Riley | 9,232 | 71,882 | 0.92 | 852 | 11.6 | 389 | 5.3 | 5,883 | 10.3 | 8,057 | 7,524 | 533 | 263 | NA |
| Rooks | 72 | 5,044 | 1.00 | 64 | 13.3 | 82 | 17.0 | 433 | 11.6 | 1,290 | 1,230 | 60 | 3 | 16 |
| Rush | 66 | 2,916 | 0.92 | 21 | 7.1 | 29 | 9.8 | 254 | 11.3 | 877 | 840 | 37 | 11 | 31 |
| Russell | 89 | 6,648 | 0.90 | 81 | 11.9 | 98 | 14.4 | 607 | 12.0 | 1,813 | 1,759 | 54 | 5 | 40 |
| Saline | 1,470 | 57,099 | 1.09 | 627 | 11.6 | 526 | 9.8 | 4,599 | 10.7 | 11,671 | 10,600 | 1,071 | NA | NA |
| Scott | 105 | 4,912 | 1.00 | 66 | 13.8 | 63 | 13.2 | 530 | 13.9 | 1,042 | 999 | 43 | NA | NA |
| Sedgwick | 7,238 | 529,595 | 1.07 | 6,669 | 12.8 | 4,863 | 9.4 | 53,090 | 12.3 | 91,128 | 64,930 | 26,199 | 94 | 205 |
| Seward | 476 | 23,100 | 1.07 | 363 | 17.3 | 115 | 5.5 | 3,648 | 19.7 | 2,477 | 2,379 | 99 | 4 | 19 |
| Shawnee | 4,389 | 191,215 | 1.16 | 2,081 | 11.8 | 1,946 | 11.1 | 12,400 | 8.8 | 39,376 | 31,967 | 7,409 | 87 | 958 |
| Sheridan | 33 | 2,414 | 0.93 | 23 | 9.1 | 20 | 7.9 | 277 | 14.5 | 599 | 585 | 14 | NA | NA |
| Sherman | 102 | 5,833 | 0.96 | 65 | 11.3 | 77 | 13.3 | 423 | 9.1 | 1,310 | 1,261 | 49 | NA | NA |
| Smith | 56 | 3,582 | 0.97 | 26 | 7.3 | 41 | 11.6 | 298 | 11.7 | 1,167 | 1,115 | 53 | NA | NA |
| Stafford | 69 | 3,738 | 0.78 | 40 | 9.9 | 58 | 14.3 | 460 | 14.6 | 973 | 932 | 41 | NA | NA |
| Stanton | 47 | 2,058 | 1.01 | 18 | 9.1 | 17 | 8.6 | 258 | 15.9 | 346 | 334 | 12 | NA | NA |
| Stevens | 72 | 5,853 | 1.10 | 67 | 12.4 | 38 | 7.1 | 874 | 19.4 | 851 | 817 | 34 | NA | NA |
| Sumner | 409 | 20,626 | 0.76 | 249 | 11.0 | 276 | 12.2 | 1,807 | 9.9 | 5,093 | 3,971 | 1,122 | 12 | 156 |
| Thomas | 353 | 8,076 | 1.06 | 102 | 13.2 | 63 | 8.2 | 653 | 10.7 | 1,563 | 1,478 | 86 | 4 | 19 |
| Trego | 64 | 2,528 | 0.80 | 31 | 11.2 | 46 | 16.7 | 224 | 11.1 | 799 | 780 | 18 | NA | NA |
| Wabaunsee | 90 | 5,242 | 0.51 | 68 | 9.8 | 58 | 8.4 | 459 | 8.5 | 1,579 | 1,369 | 211 | 14 | 77 |
| Wallace | 19 | 1,553 | 0.97 | 31 | 20.2 | 10 | 6.5 | 122 | 10.7 | 343 | D | D | NA | NA |
| Washington | 114 | 5,195 | 0.89 | 76 | 14.0 | 88 | 16.2 | 520 | 12.8 | 1,463 | 1,403 | 60 | 3 | 5 |
| Wichita | 26 | 2,097 | 0.97 | 22 | 10.6 | 21 | 10.1 | 288 | 17.3 | 451 | D | D | 4 | 5 |
| Wilson | 121 | 8,992 | 1.08 | 94 | 11.2 | 114 | 13.6 | 747 | 11.5 | 2,338 | 2,108 | 230 | 10 | 47 |
| Woodson | 39 | 2,817 | 0.74 | 21 | 7.0 | 45 | 14.9 | 306 | 13.3 | 821 | 741 | 81 | 1 | 19 |
| Wyandotte | 1,335 | 182,956 | 1.24 | 2,572 | 15.6 | 1,523 | 9.2 | 23,968 | 17.0 | 24,605 | 13,215 | 11,390 | 21 | 92 |
| **KENTUCKY** | 131,670 | 4,466,644 | 1.01 | 52,831 | 11.8 | 49,879 | 11.1 | 275,158 | 7.6 | 942,915 | 578,633 | 364,282 | X | X |
| Adair | 1,487 | 16,954 | 0.71 | 182 | 9.3 | 233 | 11.9 | 1,213 | 8.5 | 4,444 | 3,112 | 1,332 | 1 | 5 |
| Allen | 197 | 18,582 | 0.72 | 247 | 11.6 | 265 | 12.4 | 1,454 | 8.5 | 4,538 | 2,876 | 1,661 | 8 | 86 |
| Anderson | 108 | 17,025 | 0.48 | 241 | 10.6 | 214 | 9.4 | 1,271 | 6.7 | 4,787 | 2,509 | 2,278 | 2 | 15 |
| Ballard | 127 | 6,521 | 0.54 | 74 | 9.5 | 107 | 13.8 | 438 | 7.2 | 2,089 | 1,455 | 633 | 9 | 62 |
| Barren | 702 | 42,974 | 0.95 | 558 | 12.6 | 563 | 12.7 | 2,866 | 8.1 | 10,249 | 6,711 | 3,538 | 30 | 84 |
| Bath | 103 | 9,877 | 0.46 | 183 | 14.7 | 167 | 13.4 | 885 | 8.7 | 2,909 | 1,610 | 1,299 | 4 | 33 |
| Bell | 919 | 27,661 | 1.12 | 317 | 12.4 | 403 | 15.8 | 1,765 | 8.9 | 6,794 | 4,557 | 2,237 | 18 | 137 |
| Boone | 834 | 152,470 | 1.33 | 1,668 | 12.3 | 1,081 | 8.0 | 6,474 | 5.6 | 21,582 | 11,933 | 9,649 | 66 | 817 |
| Bourbon | 236 | 18,408 | 0.82 | 209 | 10.5 | 250 | 12.6 | 1,465 | 9.4 | 4,616 | 2,579 | 2,036 | 3 | 55 |
| Boyd | 1,911 | 54,321 | 1.37 | 476 | 10.2 | 647 | 13.9 | 2,519 | 7.1 | 11,904 | 7,178 | 4,726 | 20 | 181 |
| Boyle | 3,193 | 33,620 | 1.31 | 309 | 10.2 | 360 | 11.9 | 1,385 | 6.6 | 7,048 | 4,377 | 2,672 | 6 | 34 |
| Bracken | 41 | 6,181 | 0.41 | 99 | 11.9 | 102 | 12.3 | 501 | 7.4 | 2,029 | 1,202 | 827 | 2 | 50 |
| Breathitt | 321 | 12,060 | 0.76 | 162 | 12.9 | 202 | 16.1 | 780 | 7.8 | 3,486 | 2,042 | 1,443 | 2 | 10 |
| Breckinridge | 299 | 16,861 | 0.55 | 235 | 11.4 | 219 | 10.7 | 1,427 | 8.9 | 4,975 | 3,478 | 1,497 | 7 | 40 |
| Bullitt | 340 | 65,945 | 0.64 | 775 | 9.4 | 731 | 8.9 | 4,141 | 6.1 | 16,286 | 9,732 | 6,553 | 20 | 347 |
| Butler | 203 | 11,356 | 0.72 | 160 | 12.6 | 167 | 13.1 | 992 | 9.7 | 2,939 | 1,748 | 1,191 | 2 | 32 |
| Caldwell | 148 | 11,715 | 0.81 | 143 | 11.3 | 183 | 14.4 | 785 | 7.9 | 3,330 | 2,298 | 1,031 | 2 | 22 |
| Calloway | 3,238 | 40,080 | 1.07 | 339 | 8.6 | 445 | 11.3 | 2,618 | 9.0 | 8,072 | 5,086 | 2,986 | 10 | 194 |
| Campbell | 3,201 | 78,436 | 0.69 | 1,088 | 11.6 | 962 | 10.2 | 4,324 | 5.7 | 17,378 | 9,540 | 7,838 | 13 | 141 |
| Carlisle | 60 | 3,894 | 0.51 | 59 | 12.6 | 73 | 15.6 | 350 | 9.5 | 1,257 | 827 | 429 | 2 | 19 |
| Carroll | 325 | 13,486 | 1.63 | 144 | 13.4 | 138 | 12.9 | 713 | 8.2 | 2,333 | 1,378 | 955 | 5 | 54 |
| Carter | 595 | 24,712 | 0.71 | 289 | 10.9 | 391 | 14.7 | 1,860 | 8.9 | 6,821 | 4,126 | 2,695 | 6 | 57 |
| Casey | 477 | 14,447 | 0.74 | 189 | 11.8 | 220 | 13.7 | 1,327 | 10.8 | 3,969 | 2,699 | 1,270 | 1 | 44 |
| Christian | 6,200 | 92,705 | 1.69 | 1,530 | 21.4 | 660 | 9.2 | 4,739 | 8.5 | 11,567 | 8,198 | 3,369 | 18 | 258 |
| Clark | 458 | 34,399 | 0.90 | 420 | 11.5 | 421 | 11.5 | 1,981 | 6.8 | 8,216 | 4,511 | 3,705 | 6 | 142 |

1. Per 1,000 estimated resident population.

# Table B. States and Counties — Crime, Education, Money Income, and Poverty

| | | Education | | | | | | Money income, 2015-2019 | | | | Income and poverty, 2019 | | | |
|---|---|---|---|---|---|---|---|---|---|---|---|---|---|---|---|
| | | School enrollment and attainment, 2015-2019 | | | | Local government expenditures,[3] 2017-2018 | | | Households | | | | Percent below poverty level | | |
| County law enforcement employment, 2019 | | Enrollment[1] | | Attainment[2] (percent) | | | | | | Percent | | | | | |
| STATE County | | | | High school graduate or less | Bachelor's degree or more | Total current spending (mil dol) | Current spending per student (dollars) | Per capita income[4] | Median income (dollars) | with income of less than $50,000 | with income of $200,000 or more | Median household income (dollars) | All persons | Children under 18 years | Children 5 to 17 years in families |
| | Officers | Civilians | Total | Percent private | | | | | | | | | | | |
| | 46 | 47 | 48 | 49 | 50 | 51 | 52 | 53 | 54 | 55 | 56 | 57 | 58 | 59 | 60 | 61 |

| STATE County | 46 | 47 | 48 | 49 | 50 | 51 | 52 | 53 | 54 | 55 | 56 | 57 | 58 | 59 | 60 | 61 |
|---|---|---|---|---|---|---|---|---|---|---|---|---|---|---|---|---|
| KANSAS— Cont'd | | | | | | | | | | | | | | | | |
| Neosho | 16 | 14 | 4,375 | 7.5 | 41.8 | 18.1 | 26 | 10,770 | 23,845 | 46,291 | 53.5 | 1.8 | 49,533 | 15.7 | 22.9 | 19.0 |
| Ness | 6 | 5 | 582 | 18.2 | 40.2 | 19.4 | 6 | 14,029 | 34,243 | 56,492 | 45.3 | 3.8 | 54,727 | 10.6 | 12.5 | 11.1 |
| Norton | 5 | 4 | 983 | 4.7 | 45.6 | 18.8 | 11 | 12,472 | 25,352 | 49,038 | 51.1 | 1.9 | 49,739 | 11.6 | 15.1 | 13.5 |
| Osage | NA | NA | 3,514 | 6.9 | 48.3 | 19.2 | 41 | 14,661 | 27,951 | 54,090 | 46.8 | 1.9 | 55,655 | 14.2 | 18.1 | 15.4 |
| Osborne | 7 | 7 | 652 | 6.6 | 40.7 | 23.6 | 7 | 16,130 | 30,843 | 51,286 | 48.9 | 2.4 | 46,351 | 13.5 | 21.8 | 20.2 |
| Ottawa | 5 | 13 | 1,262 | 8.0 | 40.4 | 24.5 | 18 | 14,737 | 29,230 | 54,784 | 44.4 | 2.2 | 58,729 | 9.9 | 13.2 | 11.6 |
| Pawnee | 11 | 1 | 1,209 | 5.5 | 42.5 | 19.6 | 16 | 15,146 | 25,131 | 49,917 | 50.0 | 1.1 | 65,141 | 14.0 | 18.1 | 16.6 |
| Phillips | 9 | 5 | 1,159 | 5.3 | 44.2 | 21.8 | 17 | 21,008 | 27,431 | 50,093 | 49.8 | 1.9 | 51,546 | 11.0 | 15.9 | 14.7 |
| Pottawatomie | 30 | 13 | 6,632 | 14.4 | 34.8 | 33.8 | 49 | 11,782 | 29,345 | 66,835 | 36.5 | 3.3 | 73,685 | 7.2 | 8.8 | 8.1 |
| Pratt | 10 | 7 | 2,521 | 8.9 | 32.8 | 30.3 | 30 | 17,202 | 27,442 | 52,327 | 46.7 | 3.9 | 53,980 | 11.2 | 15.0 | 13.6 |
| Rawlins | 3 | 1 | 510 | 6.3 | 34.0 | 24.1 | 4 | 12,439 | 28,590 | 53,207 | 47.4 | 1.4 | 50,768 | 14.2 | 20.0 | 20.5 |
| Reno | NA | NA | 15,114 | 13.7 | 39.2 | 20.9 | 125 | 13,027 | 27,073 | 49,936 | 50.1 | 3.1 | 53,371 | 13.4 | 16.9 | 14.3 |
| Republic | 9 | 6 | 968 | 8.4 | 38.6 | 25.5 | 9 | 11,694 | 27,424 | 47,976 | 52.3 | 1.8 | 47,082 | 12.2 | 17.9 | 18.5 |
| Rice | 9 | 0 | 2,513 | 21.4 | 37.7 | 22.8 | 27 | 14,715 | 26,737 | 53,012 | 48.7 | 1.9 | 56,350 | 11.8 | 15.8 | 14.8 |
| Riley | 108 | 98 | 29,529 | 5.1 | 23.3 | 45.6 | 81 | 10,664 | 27,272 | 51,208 | 49.1 | 4.1 | 51,996 | 20.9 | 13.3 | 12.5 |
| Rooks | 6 | 11 | 1,108 | 8.9 | 41.7 | 23.1 | 10 | 12,436 | 26,420 | 49,415 | 50.5 | 1.3 | 49,485 | 10.5 | 14.0 | 13.7 |
| Rush | 7 | 5 | 632 | 4.7 | 37.7 | 23.7 | 7 | 12,269 | 26,552 | 47,981 | 50.7 | 1.6 | 53,189 | 12.2 | 17.2 | 16.4 |
| Russell | 10 | 1 | 1,404 | 4.3 | 39.2 | 23.1 | 10 | 11,899 | 26,090 | 44,792 | 53.6 | 2.1 | 46,112 | 14.4 | 20.1 | 18.5 |
| Saline | 38 | 11 | 13,254 | 15.0 | 38.6 | 27.3 | 123 | 14,027 | 28,813 | 52,200 | 46.9 | 3.1 | 56,741 | 12.3 | 16.9 | 15.0 |
| Scott | 5 | 4 | 1,122 | 13.9 | 42.5 | 25.1 | 10 | 10,251 | 29,315 | 65,417 | 39.1 | 1.9 | 62,914 | 7.9 | 11.3 | 10.3 |
| Sedgwick | 177 | 339 | 139,434 | 16.6 | 36.7 | 30.9 | 952 | 11,191 | 29,530 | 56,524 | 44.4 | 4.4 | 60,014 | 12.6 | 16.5 | 15.4 |
| Seward | 15 | 22 | 6,595 | 4.0 | 62.4 | 11.0 | 62 | 10,914 | 20,754 | 49,291 | 50.7 | 2.0 | 55,727 | 14.2 | 18.4 | 16.6 |
| Shawnee | 112 | 70 | 42,084 | 12.2 | 39.2 | 30.9 | 313 | 11,097 | 30,974 | 56,762 | 44.7 | 3.7 | 59,451 | 9.8 | 13.3 | 11.8 |
| Sheridan | 4 | 0 | 499 | 3.4 | 38.0 | 22.0 | 8 | 11,640 | 35,465 | 62,885 | 41.4 | 6.3 | 54,627 | 11.8 | 16.1 | 14.6 |
| Sherman | 5 | 6 | 1,518 | 5.7 | 37.9 | 21.8 | 11 | 10,601 | 27,919 | 54,754 | 42.3 | 1.4 | 54,000 | 12.5 | 18.7 | 17.2 |
| Smith | 6 | 0 | 679 | 8.1 | 42.9 | 22.3 | 9 | 14,493 | 28,994 | 43,429 | 56.3 | 3.2 | 45,915 | 11.6 | 15.6 | 16.5 |
| Stafford | NA | NA | 889 | 6.3 | 38.7 | 25.2 | 13 | 14,646 | 28,060 | 49,375 | 50.3 | 2.8 | 51,007 | 13.7 | 19.3 | 17.9 |
| Stanton | 7 | 4 | 643 | 10.0 | 51.1 | 19.0 | 6 | 13,170 | 23,033 | 52,054 | 48.4 | 0.1 | 63,401 | 10.6 | 12.8 | 11.3 |
| Stevens | NA | NA | 1,488 | 6.9 | 47.6 | 13.7 | 15 | 12,285 | 24,652 | 57,806 | 46.6 | 2.4 | 61,264 | 9.7 | 13.7 | 12.1 |
| Sumner | 30 | 5 | 5,816 | 10.6 | 42.6 | 23.3 | 47 | 12,441 | 26,954 | 55,000 | 46.9 | 2.3 | 53,618 | 11.4 | 14.8 | 13.4 |
| Thomas | NA | NA | 1,878 | 17.1 | 33.1 | 22.6 | 11 | 10,835 | 30,442 | 60,124 | 40.8 | 4.1 | 58,652 | 9.1 | 9.7 | 9.2 |
| Trego | 4 | 2 | 518 | 5.4 | 37.9 | 24.8 | 5 | 11,536 | 34,230 | 57,966 | 41.6 | 1.7 | 52,597 | 10.8 | 14.6 | 14.0 |
| Wabaunsee | 10 | 13 | 1,594 | 10.8 | 39.5 | 24.2 | 13 | 14,163 | 27,876 | 61,178 | 39.3 | 1.6 | 66,168 | 6.9 | 9.2 | 8.4 |
| Wallace | 3 | 0 | 362 | 3.3 | 31.1 | 32.4 | 5 | 14,720 | 27,966 | 63,269 | 42.3 | 1.3 | 54,714 | 12.1 | 17.8 | 16.9 |
| Washington | 8 | 0 | 1,089 | 34.9 | 49.1 | 20.3 | 9 | 11,728 | 27,841 | 50,588 | 49.1 | 2.5 | 46,263 | 9.5 | 12.7 | 12.2 |
| Wichita | 4 | 5 | 462 | 2.4 | 40.9 | 20.6 | 5 | 12,889 | 29,126 | 57,978 | 43.7 | 1.4 | 54,009 | 11.0 | 16.2 | 14.1 |
| Wilson | NA | NA | 1,709 | 5.4 | 45.4 | 16.5 | 20 | 12,434 | 27,471 | 48,341 | 51.9 | 2.8 | 46,243 | 13.3 | 17.4 | 16.4 |
| Woodson | 7 | 4 | 635 | 1.7 | 44.5 | 17.2 | 6 | 12,040 | 22,672 | 39,643 | 59.8 | 1.0 | 43,751 | 14.3 | 22.4 | 20.0 |
| Wyandotte | NA | NA | 42,906 | 10.1 | 53.2 | 18.1 | 386 | 11,987 | 22,335 | 46,881 | 53.9 | 1.6 | 47,292 | 20.6 | 31.0 | 27.8 |
| KENTUCKY | X | X | 1,054,000 | 15.5 | 46.6 | 24.2 | 7,564 | 11,107 | 28,178 | 50,589 | 49.5 | 3.9 | 52,256 | 16.0 | 20.9 | 19.2 |
| Adair | 6 | 2 | 4,764 | 28.0 | 59.5 | 15.7 | 28 | 10,435 | 21,196 | 38,021 | 62.4 | 1.4 | 37,650 | 21.4 | 29.8 | 28.2 |
| Allen | 14 | 2 | 4,368 | 10.8 | 56.4 | 16.1 | 31 | 9,876 | 22,294 | 44,036 | 56.3 | 1.6 | 47,307 | 16.8 | 24.1 | 20.5 |
| Anderson | 19 | 2 | 5,172 | 12.2 | 46.7 | 20.2 | 36 | 9,697 | 27,250 | 55,334 | 45.1 | 3.2 | 61,308 | 10.2 | 14.1 | 12.2 |
| Ballard | 10 | 1 | 1,820 | 7.9 | 49.2 | 15.9 | 14 | 10,649 | 25,217 | 45,048 | 53.8 | 2.3 | 48,053 | 15.5 | 23.7 | 21.6 |
| Barren | 18 | 5 | 9,730 | 8.6 | 59.0 | 16.3 | 88 | 10,659 | 21,869 | 41,459 | 60.7 | 1.7 | 44,859 | 20.1 | 25.9 | 25.3 |
| Bath | 2 | 1 | 2,886 | 6.1 | 64.3 | 14.5 | 21 | 10,296 | 22,238 | 44,898 | 55.1 | 1.8 | 42,248 | 19.4 | 29.4 | 26.7 |
| Bell | NA | NA | 5,578 | 4.4 | 68.8 | 9.2 | 48 | 10,514 | 15,501 | 26,272 | 76.5 | 0.6 | 30,940 | 30.3 | 41.1 | 40.6 |
| Boone | NA | NA | 34,161 | 15.6 | 34.3 | 32.1 | 233 | 10,533 | 35,991 | 78,327 | 31.0 | 7.7 | 78,806 | 7.1 | 8.7 | 7.9 |
| Bourbon | 11 | 1 | 4,157 | 13.8 | 51.7 | 19.4 | 38 | 10,604 | 27,802 | 49,637 | 50.2 | 2.9 | 47,046 | 14.1 | 20.0 | 19.0 |
| Boyd | 34 | 12 | 10,020 | 9.1 | 45.7 | 19.4 | 82 | 11,310 | 26,193 | 48,308 | 51.7 | 2.9 | 50,017 | 20.6 | 32.1 | 29.7 |
| Boyle | 14 | 1 | 7,586 | 28.9 | 46.4 | 25.2 | 55 | 11,402 | 26,413 | 46,382 | 52.1 | 3.4 | 53,644 | 15.5 | 19.1 | 17.6 |
| Bracken | 5 | 1 | 1,692 | 10.0 | 60.6 | 13.2 | 16 | 10,136 | 24,124 | 49,158 | 51.0 | 0.9 | 53,289 | 14.1 | 19.2 | 16.9 |
| Breathitt | 2 | 0 | 2,414 | 12.6 | 61.1 | 15.2 | 26 | 11,223 | 19,053 | 27,344 | 67.4 | 2.0 | 33,407 | 29.2 | 37.7 | 38.1 |
| Breckinridge | 10 | 5 | 4,262 | 14.5 | 60.9 | 11.9 | 34 | 10,451 | 23,286 | 47,190 | 52.1 | 1.6 | 46,069 | 18.3 | 24.8 | 23.6 |
| Bullitt | 35 | 15 | 17,377 | 15.3 | 53.1 | 14.7 | 129 | 9,687 | 29,333 | 63,348 | 37.5 | 2.9 | 66,335 | 7.8 | 11.2 | 10.5 |
| Butler | 7 | 3 | 2,651 | 13.7 | 63.3 | 13.9 | 22 | 9,739 | 22,951 | 41,763 | 55.6 | 1.8 | 46,305 | 18.0 | 27.2 | 24.8 |
| Caldwell | 8 | 3 | 2,734 | 9.6 | 58.0 | 15.3 | 19 | 9,597 | 28,799 | 44,775 | 54.5 | 3.8 | 44,144 | 15.8 | 22.4 | 21.2 |
| Calloway | 25 | 9 | 12,587 | 5.8 | 42.6 | 28.5 | 52 | 10,904 | 23,219 | 42,273 | 56.6 | 2.3 | 45,574 | 16.5 | 19.2 | 17.4 |
| Campbell | NA | NA | 22,805 | 21.3 | 37.5 | 36.3 | 137 | 11,687 | 34,025 | 63,050 | 40.3 | 5.6 | 66,520 | 10.9 | 12.8 | 11.7 |
| Carlisle | 2 | 1 | 1,053 | 9.3 | 57.1 | 14.9 | 9 | 10,420 | 30,065 | 45,109 | 54.0 | 3.2 | 49,659 | 15.5 | 24.7 | 22.0 |
| Carroll | 5 | 2 | 2,313 | 7.3 | 62.0 | 8.2 | 27 | 12,679 | 25,217 | 43,524 | 55.2 | 4.8 | 51,672 | 15.2 | 22.3 | 21.8 |
| Carter | 7 | 4 | 5,774 | 13.5 | 61.1 | 12.6 | 45 | 10,119 | 18,778 | 34,736 | 63.3 | 0.7 | 43,492 | 20.0 | 27.7 | 26.3 |
| Casey | 10 | 7 | 3,087 | 14.0 | 64.5 | 12.8 | 25 | 10,848 | 18,457 | 34,819 | 65.7 | 0.9 | 34,751 | 25.2 | 32.9 | 32.2 |
| Christian | NA | NA | 17,509 | 16.2 | 45.5 | 18.1 | 90 | 10,134 | 23,021 | 43,919 | 56.8 | 2.5 | 46,390 | 18.9 | 24.5 | 26.2 |
| Clark | 12 | 4 | 7,564 | 15.0 | 48.1 | 20.1 | 56 | 10,268 | 28,802 | 54,953 | 45.5 | 2.9 | 55,428 | 13.0 | 18.2 | 16.9 |

1. All persons 3 years old and over enrolled in nursery school through college.   2. Persons 25 years old and over.   3. Elementary and secondary education expenditures.   4. Based on population estimated by the American Community Survey, 2014–2018.

# Table B. States and Counties — Personal Income

| STATE County | Personal income, 2019 Total (mil dol) | Percent change 2018-2019 | Per capita[1] Dollars | Per capita[1] Rank | Wages and salaries (mil dol) | Supplements to wages and salaries, employer contributions (mil dol) Pension and insurance | Government social insurance | Proprietors' income (mil dol) | Dividends, interest, and rent (mil dol) | Personal transfer receipts (mil dol) | Earnings, 2019 Total (mil dol) | Contributions for government social insurance (mil dol) From employee and self-employed | From employer |
|---|---|---|---|---|---|---|---|---|---|---|---|---|---|
| | 62 | 63 | 64 | 65 | 66 | 67 | 68 | 69 | 70 | 71 | 72 | 73 | 74 |
| KANSAS— Cont'd | | | | | | | | | | | | | |
| Neosho | 661 | 5.5 | 41,324 | 1,875 | 263 | 49 | 20 | 60 | 103 | 172 | 393 | 26 | 20 |
| Ness | 162 | 13.2 | 58,901 | 339 | 49 | 9 | 4 | 38 | 40 | 34 | 100 | 5 | 4 |
| Norton | 256 | 11.9 | 47,838 | 1,039 | 101 | 20 | 7 | 50 | 56 | 49 | 179 | 9 | 7 |
| Osage | 693 | 4.0 | 43,460 | 1,550 | 102 | 21 | 8 | 66 | 95 | 161 | 196 | 15 | 8 |
| Osborne | 181 | 8.4 | 52,889 | 623 | 60 | 10 | 4 | 34 | 35 | 38 | 110 | 6 | 4 |
| Ottawa | 259 | 7.7 | 45,417 | 1,312 | 49 | 10 | 4 | 35 | 38 | 56 | 97 | 6 | 4 |
| Pawnee | 296 | 11.8 | 46,226 | 1,224 | 116 | 23 | 9 | 76 | 49 | 65 | 224 | 11 | 9 |
| Phillips | 294 | 9.4 | 56,113 | 443 | 94 | 24 | 7 | 52 | 64 | 62 | 177 | 10 | 7 |
| Pottawatomie | 1,348 | 5.4 | 55,271 | 481 | 431 | 75 | 32 | 115 | 182 | 173 | 653 | 40 | 32 |
| Pratt | 494 | 10.5 | 53,891 | 548 | 200 | 33 | 15 | 109 | 96 | 90 | 357 | 19 | 15 |
| Rawlins | 157 | 22.1 | 62,147 | 245 | 40 | 7 | 3 | 46 | 31 | 34 | 97 | 5 | 3 |
| Reno | 2,611 | 4.4 | 42,115 | 1,749 | 1,146 | 200 | 84 | 247 | 498 | 623 | 1,677 | 110 | 84 |
| Republic | 232 | 12.9 | 50,019 | 831 | 65 | 13 | 5 | 57 | 43 | 54 | 139 | 7 | 5 |
| Rice | 451 | 9.7 | 47,243 | 1,104 | 177 | 37 | 13 | 78 | 70 | 90 | 305 | 16 | 13 |
| Riley | 3,157 | 3.0 | 42,528 | 1,683 | 1,340 | 286 | 98 | 207 | 759 | 405 | 1,931 | 111 | 98 |
| Rooks | 218 | 8.5 | 44,398 | 1,433 | 73 | 15 | 5 | 38 | 45 | 52 | 132 | 8 | 5 |
| Rush | 158 | 10.3 | 52,199 | 663 | 43 | 9 | 3 | 34 | 30 | 36 | 88 | 5 | 3 |
| Russell | 323 | 4.7 | 47,049 | 1,129 | 96 | 21 | 7 | 45 | 62 | 79 | 168 | 11 | 7 |
| Saline | 2,756 | 2.6 | 50,820 | 769 | 1,319 | 218 | 99 | 443 | 508 | 509 | 2,079 | 126 | 99 |
| Scott | 358 | 14.4 | 74,223 | 95 | 97 | 16 | 8 | 137 | 62 | 51 | 257 | 9 | 8 |
| Sedgwick | 27,648 | 3.5 | 53,577 | 565 | 13,869 | 2,209 | 1,061 | 3,118 | 6,816 | 4,218 | 20,257 | 1,217 | 1,061 |
| Seward | 891 | 5.0 | 41,567 | 1,835 | 502 | 87 | 36 | 161 | 105 | 136 | 785 | 42 | 36 |
| Shawnee | 8,488 | 2.0 | 47,991 | 1,015 | 5,078 | 810 | 382 | 678 | 1,513 | 1,866 | 6,948 | 442 | 382 |
| Sheridan | 157 | 34.7 | 62,156 | 244 | 46 | 8 | 4 | 60 | 28 | 24 | 118 | 5 | 4 |
| Sherman | 276 | 4.8 | 46,588 | 1,177 | 101 | 18 | 8 | 46 | 53 | 64 | 172 | 10 | 8 |
| Smith | 191 | 17.0 | 53,381 | 585 | 50 | 10 | 4 | 48 | 41 | 44 | 112 | 6 | 4 |
| Stafford | 231 | 16.6 | 55,629 | 466 | 50 | 10 | 4 | 74 | 39 | 44 | 137 | 6 | 4 |
| Stanton | 152 | 21.1 | 75,841 | 84 | 42 | 7 | 3 | 66 | 27 | 16 | 118 | 4 | 3 |
| Stevens | 272 | 16.1 | 49,540 | 874 | 83 | 15 | 7 | 95 | 41 | 38 | 200 | 8 | 7 |
| Sumner | 971 | 8.5 | 42,524 | 1,685 | 296 | 52 | 24 | 103 | 153 | 214 | 475 | 31 | 24 |
| Thomas | 409 | 10.6 | 52,540 | 643 | 169 | 28 | 13 | 98 | 68 | 67 | 307 | 16 | 13 |
| Trego | 150 | 10.3 | 53,593 | 564 | 49 | 11 | 4 | 32 | 27 | 36 | 96 | 5 | 4 |
| Wabaunsee | 365 | 4.4 | 52,706 | 637 | 53 | 10 | 4 | 33 | 90 | 65 | 100 | 7 | 4 |
| Wallace | 98 | 24.2 | 64,805 | 190 | 23 | 4 | 2 | 38 | 15 | 16 | 68 | 3 | 2 |
| Washington | 282 | 12.8 | 52,141 | 669 | 71 | 15 | 5 | 61 | 53 | 60 | 153 | 8 | 5 |
| Wichita | 183 | 19.4 | 86,215 | 41 | 52 | 8 | 4 | 80 | 27 | 22 | 144 | 4 | 4 |
| Wilson | 380 | 4.6 | 44,521 | 1,415 | 146 | 33 | 11 | 54 | 61 | 105 | 244 | 15 | 11 |
| Woodson | 133 | 9.9 | 42,276 | 1,724 | 26 | 6 | 2 | 20 | 24 | 35 | 54 | 4 | 2 |
| Wyandotte | 5,710 | 3.4 | 34,518 | 2,743 | 5,577 | 815 | 442 | 497 | 671 | 1,399 | 7,331 | 447 | 442 |
| KENTUCKY | 195,549 | 3.5 | 43,724 | X | 96,606 | 17,036 | 7,234 | 12,803 | 33,037 | 47,036 | 133,679 | 8,528 | 7,234 |
| Adair | 625 | 3.4 | 32,549 | 2,916 | 166 | 35 | 13 | 33 | 77 | 235 | 246 | 19 | 13 |
| Allen | 701 | 2.7 | 32,881 | 2,898 | 186 | 35 | 14 | 53 | 86 | 218 | 289 | 22 | 14 |
| Anderson | 923 | 3.9 | 40,581 | 1,977 | 205 | 41 | 16 | 29 | 128 | 201 | 290 | 23 | 16 |
| Ballard | 305 | 3.7 | 38,711 | 2,235 | 83 | 16 | 6 | 23 | 46 | 93 | 129 | 9 | 6 |
| Barren | 1,652 | 3.2 | 37,337 | 2,407 | 643 | 121 | 49 | 156 | 238 | 483 | 969 | 67 | 49 |
| Bath | 392 | 3.6 | 31,327 | 2,987 | 81 | 16 | 7 | 14 | 42 | 142 | 118 | 11 | 7 |
| Bell | 798 | 1.1 | 30,649 | 3,019 | 297 | 66 | 23 | 31 | 99 | 402 | 416 | 33 | 23 |
| Boone | 6,649 | 5.4 | 49,778 | 849 | 4,884 | 740 | 368 | 306 | 819 | 964 | 6,299 | 380 | 368 |
| Bourbon | 1,019 | 3.3 | 51,505 | 726 | 314 | 52 | 24 | 224 | 169 | 204 | 615 | 30 | 24 |
| Boyd | 1,890 | 1.2 | 40,453 | 1,993 | 1,242 | 246 | 91 | 77 | 257 | 618 | 1,656 | 107 | 91 |
| Boyle | 1,171 | 3.1 | 38,958 | 2,198 | 632 | 119 | 49 | 77 | 215 | 313 | 877 | 58 | 49 |
| Bracken | 321 | 4.8 | 38,692 | 2,239 | 51 | 11 | 4 | 16 | 39 | 90 | 82 | 7 | 4 |
| Breathitt | 432 | 1.2 | 34,232 | 2,778 | 105 | 25 | 8 | 9 | 43 | 227 | 147 | 13 | 8 |
| Breckinridge | 705 | 3.7 | 34,418 | 2,752 | 137 | 30 | 10 | 44 | 126 | 224 | 221 | 19 | 10 |
| Bullitt | 3,496 | 4.9 | 42,799 | 1,648 | 1,126 | 172 | 89 | 162 | 336 | 703 | 1,550 | 109 | 89 |
| Butler | 452 | 4.0 | 35,121 | 2,672 | 111 | 26 | 8 | 33 | 50 | 151 | 179 | 13 | 8 |
| Caldwell | 452 | 6.2 | 35,437 | 2,630 | 160 | 31 | 13 | 27 | 65 | 157 | 231 | 18 | 13 |
| Calloway | 1,399 | 2.3 | 35,878 | 2,583 | 637 | 150 | 47 | 80 | 246 | 362 | 914 | 57 | 47 |
| Campbell | 4,716 | 3.9 | 50,398 | 804 | 1,460 | 253 | 106 | 171 | 762 | 824 | 1,991 | 133 | 106 |
| Carlisle | 205 | 8.3 | 42,991 | 1,617 | 38 | 8 | 3 | 27 | 38 | 58 | 75 | 5 | 3 |
| Carroll | 402 | 4.1 | 37,806 | 2,344 | 434 | 70 | 33 | 12 | 48 | 117 | 549 | 34 | 33 |
| Carter | 850 | 2.6 | 31,713 | 2,959 | 198 | 42 | 15 | 39 | 97 | 338 | 295 | 26 | 15 |
| Casey | 512 | 4.4 | 31,688 | 2,961 | 137 | 32 | 11 | 24 | 59 | 207 | 204 | 17 | 11 |
| Christian | 2,784 | 3.7 | 39,504 | 2,115 | 3,201 | 887 | 298 | 233 | 520 | 684 | 4,619 | 208 | 298 |
| Clark | 1,466 | 3.7 | 40,413 | 1,999 | 675 | 120 | 51 | 56 | 225 | 378 | 903 | 62 | 51 |

1. Based on the resident population estimated as of July 1 of the year shown.

Items 62—74

# Table B. States and Counties — Earnings, Social Security, and Housing

| STATE County | Earnings, 2019 (cont.) | | | | | | | | | Social Security beneficiaries, December 2019 | | Supplemental Security Income recipients, 2019 | Housing units, 2020 | |
|---|---|---|---|---|---|---|---|---|---|---|---|---|---|---|
| | | | | | Percent by selected industries | | | | | | | | | |
| | Farm | Mining, quarrying, and extractions | Construction | Manu- facturing | Information; professional, scientific, technical services | Retail trade | Finance, insurance, real estate, and leasing | Health care and social assistance | Govern- ment | Number | Rate[1] | | Total | Percent change, 2010-2020 |
| | 75 | 76 | 77 | 78 | 79 | 80 | 81 | 82 | 83 | 84 | 85 | 86 | 87 | 88 |
| KANSAS— Cont'd | | | | | | | | | | | | | | |
| Neosho | 4.7 | 6.6 | 4.7 | 17.0 | D | 6.7 | 6.6 | D | 24.9 | 3,815 | 240 | 371 | 7,727 | 2.9 |
| Ness | 19.8 | 18.2 | 2.7 | 1.2 | D | 2.5 | D | D | 21.3 | 770 | 277 | 23 | 1,724 | -0.9 |
| Norton | 18.9 | D | 2.3 | 9.3 | 3.4 | 5.1 | 5.6 | 8.0 | 23.0 | 1,200 | 223 | 46 | 2,541 | 0.0 |
| Osage | 14.2 | D | 9.5 | 4.1 | D | 6.2 | 8.0 | D | 27.5 | 3,955 | 249 | 342 | 7,687 | 2.5 |
| Osborne | 16.3 | 2.7 | D | D | 3.9 | 4.1 | 7.0 | 7.2 | 15.2 | 1,010 | 294 | 45 | 2,173 | -1.5 |
| Ottawa | 19.7 | 0.1 | 2.1 | 9.9 | D | 2.3 | 10.0 | 9.4 | 20.6 | 1,450 | 254 | 78 | 2,792 | 0.5 |
| Pawnee | 24.9 | D | 5.3 | 1.3 | D | 4.3 | 5.6 | 8.0 | 35.9 | 1,520 | 238 | 78 | 3,142 | -0.3 |
| Phillips | 16.7 | 1.9 | 1.8 | 15.9 | 3.7 | 4.0 | D | 2.0 | 22.4 | 1,520 | 290 | 51 | 3,083 | 1.1 |
| Pottawatomie | 3.7 | 0.1 | 10.0 | 23.1 | 4.7 | 7.5 | 8.6 | D | 10.5 | 4,320 | 177 | 179 | 10,164 | 17.8 |
| Pratt | 17.6 | 5.1 | 2.7 | 2.5 | 4.1 | 9.3 | 7.4 | 14.3 | 15.7 | 2,125 | 233 | 120 | 4,459 | -1.2 |
| Rawlins | 17.4 | 0.4 | 4.3 | 3.3 | 5.3 | 14.1 | D | 4.2 | 15.1 | 740 | 297 | 28 | 1,458 | 0.0 |
| Reno | 3.2 | 0.6 | 5.6 | 11.7 | 5.6 | 6.4 | 8.2 | 14.7 | 16.5 | 15,200 | 245 | 1,252 | 28,535 | 0.9 |
| Republic | 33.2 | D | 2.8 | 7.2 | 3.6 | 5.0 | 5.3 | D | 14.4 | 1,435 | 314 | 69 | 2,870 | -0.2 |
| Rice | 20.3 | 4.7 | 3.1 | 11.7 | 3.7 | 2.8 | D | D | 15.5 | 2,195 | 231 | 130 | 4,599 | 1.1 |
| Riley | 0.8 | D | 5.5 | 1.9 | 7.8 | 6.4 | 8.6 | 11.4 | 37.9 | 8,190 | 113 | 597 | 31,507 | 11.7 |
| Rooks | 9.9 | 9.2 | D | D | 2.7 | 3.4 | 4.4 | 4.2 | 24.6 | 1,360 | 275 | 60 | 2,742 | -0.9 |
| Rush | 25.8 | D | 2.7 | 14.9 | 5.1 | 2.8 | 5.2 | 3.9 | 18.5 | 880 | 294 | 54 | 1,840 | -1.6 |
| Russell | 9.8 | 10.9 | 4.8 | 12.1 | D | 3.9 | 5.9 | 6.6 | 15.9 | 1,910 | 278 | 130 | 3,894 | -0.4 |
| Saline | 1.0 | D | 5.4 | 15.2 | 7.1 | 6.6 | 6.6 | 19.2 | 11.5 | 12,275 | 227 | 1,071 | 24,439 | 1.4 |
| Scott | 43.1 | 0.3 | 2.5 | 2.0 | D | 3.8 | D | D | 13.4 | 1,065 | 220 | 49 | 2,249 | 2.6 |
| Sedgwick | 0.2 | 3.1 | 6.1 | 21.8 | 7.3 | 5.8 | 8.4 | 11.7 | 12.2 | 95,990 | 186 | 10,957 | 223,763 | 5.8 |
| Seward | 7.7 | 3.0 | D | D | D | 6.1 | 5.9 | D | 18.0 | 2,660 | 124 | 308 | 8,300 | 3.0 |
| Shawnee | 0.2 | 0.1 | 5.2 | 7.3 | 8.8 | 4.9 | 12.2 | 15.8 | 21.6 | 40,000 | 226 | 4,768 | 80,317 | 1.5 |
| Sheridan | 39.1 | 0.9 | 2.6 | 1.1 | D | 2.8 | D | 0.7 | 17.2 | 595 | 236 | 14 | 1,249 | |
| Sherman | 18.7 | D | 2.3 | 2.9 | 2.5 | 8.1 | 7.4 | D | 23.9 | 1,360 | 229 | 112 | 3,106 | -1.3 |
| Smith | 33.5 | D | 3.3 | 1.0 | D | 5.6 | 3.1 | 10.8 | 14.4 | 1,210 | 339 | 53 | 2,228 | -0.2 |
| Stafford | 41.2 | 10.0 | 1.3 | 1.9 | D | 1.4 | D | 3.5 | 16.1 | 1,010 | 247 | 59 | 2,361 | 1.8 |
| Stanton | 44.2 | 0.9 | 1.9 | 1.3 | D | 2.3 | 5.4 | D | 14.2 | 335 | 169 | 20 | 1,000 | 0.9 |
| Stevens | 37.6 | 3.7 | 4.3 | 2.4 | D | 3.2 | 5.9 | D | 17.6 | 905 | 167 | 35 | 2,336 | 1.3 |
| Sumner | 12.1 | 0.7 | 4.2 | 12.7 | 3.8 | 4.4 | 5.9 | 7.5 | 20.2 | 5,295 | 232 | 349 | 11,066 | 1.8 |
| Thomas | 13.2 | 1.1 | 5.5 | 2.3 | 5.5 | 7.6 | 11.8 | D | 11.4 | 1,530 | 199 | 66 | 3,612 | 2.1 |
| Trego | 21.3 | 8.7 | 2.8 | 2.5 | D | 4.0 | D | D | 25.9 | 820 | 295 | 30 | 1,658 | -1.5 |
| Wabaunsee | 16.7 | 3.1 | 14.0 | 8.7 | D | 2.5 | D | D | 20.0 | 1,635 | 237 | 78 | 3,333 | 3.3 |
| Wallace | 36.4 | 0.6 | D | D | D | 3.8 | D | 0.3 | 12.3 | 365 | 240 | 22 | 773 | |
| Washington | 30.1 | D | 6.6 | 5.9 | D | 3.1 | D | 4.0 | 20.6 | 1,515 | 278 | 51 | 2,926 | -0.9 |
| Wichita | 54.7 | 0.2 | D | D | 0.1 | 1.5 | D | 0.5 | 10.0 | 455 | 215 | 16 | 1,052 | -0.2 |
| Wilson | 8.5 | 0.7 | 8.1 | 33.7 | 1.3 | 3.5 | 5.1 | 6.3 | 19.5 | 2,460 | 288 | 220 | 4,668 | -0.3 |
| Woodson | 12.0 | 11.8 | D | D | D | 6.0 | D | D | 22.0 | 830 | 268 | 57 | 2,026 | 0.2 |
| Wyandotte | 0.0 | D | 6.9 | 12.3 | 6.8 | 4.2 | 3.0 | 12.7 | 20.6 | 26,675 | 161 | 4,961 | 68,514 | 2.6 |
| KENTUCKY | 0.8 | 0.7 | 6.0 | 14.8 | 7.2 | 5.9 | 7.2 | 12.9 | 17.5 | 1,001,700 | 224 | 171,487 | 2,016,971 | 4.7 |
| Adair | 2.2 | D | 8.2 | 7.2 | D | 11.5 | 5.7 | D | 20.3 | 4,830 | 248 | 900 | 8,651 | 1.0 |
| Allen | 2.2 | 0.1 | 7.3 | 24.7 | D | 5.9 | 4.3 | D | 15.5 | 4,975 | 233 | 789 | 9,548 | 2.6 |
| Anderson | -2 | D | D | 29.5 | 4.8 | 9.2 | 4.0 | 5.3 | 16.7 | 2,155 | 276 | 269 | 9,601 | 5.0 |
| Ballard | 11.1 | 0.0 | 9.8 | 25.4 | D | 5.5 | 1.5 | 5.1 | 15.2 | 2,155 | 276 | 269 | 3,927 | 1.2 |
| Barren | 0.4 | D | 5.8 | 16.5 | 4.4 | 10.0 | 4.2 | 16.5 | 13.3 | 11,160 | 253 | 1,690 | 20,060 | 4.6 |
| Bath | -2.6 | 0.0 | 21.7 | D | D | 3.7 | 2.5 | D | 23.7 | 3,205 | 256 | 784 | 5,555 | 2.7 |
| Bell | -0.3 | 6.7 | 3.4 | 14.1 | 2.7 | 12.2 | 5.0 | D | 21.4 | 7,315 | 281 | 2,691 | 13,283 | 1.0 |
| Boone | 0.0 | D | 4.4 | 17.2 | 5.2 | 6.5 | 5.4 | 5.6 | 7.7 | 22,370 | 167 | 1,659 | 51,046 | 10.6 |
| Bourbon | 31.3 | D | 3.8 | 10.9 | D | 7.5 | 3.8 | 6.6 | 8.6 | 4,885 | 247 | 582 | 9,134 | 2.3 |
| Boyd | -0.1 | 0.9 | 7.6 | 15.1 | 7.0 | 7.3 | 3.3 | 23.8 | 12.1 | 12,665 | 270 | 2,389 | 21,619 | -0.8 |
| Boyle | -0.3 | D | 3.9 | 16.5 | D | 7.1 | 4.9 | D | 12.4 | 7,335 | 245 | 1,125 | 12,532 | 1.8 |
| Bracken | -0.8 | 0.0 | 8.2 | D | D | 2.7 | 2.5 | 7.1 | 26.1 | 2,170 | 261 | 315 | 3,878 | 1.0 |
| Breathitt | 0.0 | D | 1.2 | D | 1.7 | 9.7 | 5.1 | 32.0 | 33.6 | 3,830 | 304 | 1,790 | 6,433 | 3.3 |
| Breckinridge | 2.3 | D | 13.5 | 8.4 | D | 8.6 | 5.4 | 12.0 | 20.6 | 5,500 | 269 | 769 | 10,887 | 2.4 |
| Bullitt | -0.2 | D | 12.4 | 14.9 | 2.2 | 4.6 | 2.9 | 4.7 | 11.3 | 17,510 | 214 | 1,251 | 32,591 | 11.2 |
| Butler | 4.9 | D | 8.2 | 32.8 | 1.5 | 3.5 | 4.5 | D | 18.2 | 3,330 | 259 | 454 | 6,010 | 2.2 |
| Caldwell | -0.9 | D | 6.4 | 25.4 | D | 10.5 | 5.5 | 4.5 | 14.5 | 3,480 | 273 | 478 | 6,287 | -0.1 |
| Calloway | 1.7 | D | 5.4 | 18.0 | 4.9 | 7.1 | 4.7 | 12.7 | 29.6 | 8,490 | 217 | 884 | 19,069 | 5.6 |
| Campbell | 0.0 | D | D | 8.8 | 10.4 | 7.6 | 5.5 | 5.5 | 21.0 | 17,665 | 188 | 1,711 | 40,954 | 3.6 |
| Carlisle | 25.0 | 0.0 | 11.7 | 3.2 | D | 4.2 | D | D | 15.8 | 1,385 | 293 | 155 | 2,482 | 1.7 |
| Carroll | -0.2 | D | D | 51.4 | D | 4.2 | 1.1 | D | 7.8 | 2,655 | 249 | 494 | 4,752 | 1.2 |
| Carter | | 1.3 | 11.0 | 12.1 | 3.4 | 10.0 | 4.6 | D | 22.7 | 7,200 | 269 | 1,527 | 12,597 | 2.3 |
| Casey | -1.7 | D | 8.9 | 24.9 | 1.7 | 5.6 | D | 12.1 | 18.9 | 4,305 | 266 | 841 | 7,565 | 1.0 |
| Christian | 0.6 | 0.2 | 1.6 | 9.3 | 2.2 | 2.5 | 1.7 | 5.4 | 65.8 | 12,770 | 178 | 2,272 | 30,101 | 2.2 |
| Clark | -1.2 | D | 6.5 | 23.2 | D | 6.7 | 3.6 | 11.7 | 10.7 | 8,730 | 241 | 1,305 | 16,073 | 2.4 |

1. Per 1,000 resident population estimated as of July 1 of the year shown.

| STATE County | Housing units, 2015-2019 | | | | | | | | Civilian labor force, 2020 | | | | Civilian employment[6], 2015-2019 | | |
|---|---|---|---|---|---|---|---|---|---|---|---|---|---|---|---|
| | Occupied units | | | | | | | | | | Unemployment | | | Percent | |
| | Owner-occupied | | | | | Renter-occupied | | | | | | | | | |
| | | | | Median owner cost as a percent of income | | | Median rent as a percent of income[2] | | | | | | | | Construction, production, and maintenance occupations |
| | Total | Percent | Median value[1] | With a mort-gage | Without a mort-gage[2] | Median rent[3] | | Sub-standard units[4] (percent) | Total | Percent change, 2019-2020 | Total | Rate[5] | Total | Management, business, science, and arts | |
| | 89 | 90 | 91 | 92 | 93 | 94 | 95 | 96 | 97 | 98 | 99 | 100 | 101 | 102 | 103 |
| KANSAS— Cont'd | | | | | | | | | | | | | | | |
| Neosho | 6,601 | 70.4 | 82,000 | 18.2 | 12.6 | 595 | 25.7 | 4.1 | 6,246 | 1.8 | 401 | 6.4 | 7,207 | 28.9 | 33.0 |
| Ness | 1,269 | 83.4 | 73,300 | 18.3 | 10.6 | 627 | 13.7 | 0.3 | 1,334 | -0.2 | 46 | 3.4 | 1,389 | 33.9 | 35.3 |
| Norton | 1,830 | 80.2 | 82,900 | 18.4 | 10.0 | 629 | 22.8 | 0.5 | 2,644 | -1.9 | 71 | 2.7 | 2,350 | 30.5 | 23.0 |
| Osage | 6,607 | 74.7 | 115,700 | 21.4 | 13.1 | 716 | 28.6 | 2.2 | 8,009 | -0.2 | 409 | 5.1 | 7,406 | 32.1 | 29.2 |
| Osborne | 1,648 | 74.8 | 69,000 | 14.2 | 10.3 | 535 | 23.2 | 1.5 | 1,965 | -3.5 | 57 | 2.9 | 1,751 | 38.4 | 24.1 |
| Ottawa | 2,446 | 79.6 | 105,700 | 21.2 | 12.1 | 662 | 24.0 | 1.0 | 2,978 | -1.3 | 127 | 4.3 | 2,927 | 34.1 | 27.4 |
| Pawnee | 2,448 | 66.1 | 82,400 | 17.7 | 11.6 | 602 | 20.6 | 0.2 | 2,846 | -0.2 | 107 | 3.8 | 2,694 | 34.3 | 23.3 |
| Phillips | 2,316 | 75.7 | 79,600 | 18.5 | 10.4 | 616 | 25.7 | 2.6 | 2,651 | 0.2 | 91 | 3.4 | 2,621 | 34.8 | 30.9 |
| Pottawatomie | 8,698 | 78.7 | 183,100 | 20.5 | 11.4 | 854 | 27.3 | 2.0 | 11,981 | -3.6 | 520 | 4.3 | 11,216 | 40.5 | 26.2 |
| Pratt | 3,652 | 67.7 | 91,900 | 18.4 | 10.0 | 692 | 25.0 | 3.5 | 4,926 | 0.2 | 178 | 3.6 | 4,493 | 36.0 | 24.6 |
| Rawlins | 1,176 | 74.1 | 101,600 | 18.8 | 12.5 | 583 | 25.2 | 2.0 | 1,547 | 3.4 | 39 | 2.5 | 1,255 | 42.9 | 26.1 |
| Reno | 25,014 | 69.3 | 101,300 | 19.2 | 12.0 | 696 | 27.7 | 2.5 | 29,626 | -0.5 | 1,629 | 5.5 | 29,344 | 32.0 | 27.9 |
| Republic | 2,218 | 72.9 | 66,000 | 19.4 | 11.5 | 601 | 23.3 | 0.8 | 2,318 | 1.6 | 77 | 3.3 | 2,264 | 43.3 | 26.5 |
| Rice | 3,917 | 74.5 | 74,700 | 17.3 | 10.0 | 581 | 20.5 | 1.7 | 5,080 | -3.2 | 215 | 4.2 | 4,715 | 37.4 | 30.1 |
| Riley | 26,490 | 43.2 | 201,000 | 20.0 | 10.5 | 946 | 32.2 | 2.3 | 34,049 | -3.2 | 1,688 | 5.0 | 36,374 | 42.8 | 15.9 |
| Rooks | 2,176 | 77.0 | 78,800 | 20.0 | 12.2 | 550 | 21.3 | 0.6 | 2,509 | | 101 | 4.0 | 2,465 | 32.6 | 29.5 |
| Rush | 1,421 | 75.8 | 70,800 | 16.8 | 11.9 | 576 | 26.8 | 0.4 | 1,576 | -3.4 | 73 | 4.6 | 1,436 | 37.0 | 26.7 |
| Russell | 3,015 | 78.0 | 93,400 | 17.7 | 14.3 | 636 | 29.2 | 1.3 | 3,317 | 0.2 | 140 | 4.2 | 3,135 | 31.5 | 27.5 |
| Saline | 21,959 | 66.7 | 136,800 | 20.7 | 11.6 | 761 | 28.4 | 1.4 | 29,511 | -2.6 | 1,618 | 5.5 | 27,568 | 35.1 | 26.6 |
| Scott | 1,967 | 70.0 | 136,200 | 18.5 | 10.0 | 815 | 25.5 | 0.8 | 2,826 | 0.1 | 72 | 2.5 | 2,502 | 40.4 | 31.1 |
| Sedgwick | 197,229 | 62.9 | 140,700 | 19.1 | 11.0 | 824 | 28.0 | 2.6 | 257,217 | 1.6 | 22,448 | 8.7 | 249,591 | 35.8 | 25.5 |
| Seward | 7,321 | 66.7 | 105,300 | 24.3 | 11.9 | 787 | 23.5 | 4.9 | 9,816 | 2.7 | 430 | 4.4 | 10,870 | 21.5 | 45.8 |
| Shawnee | 72,267 | 64.4 | 132,500 | 18.8 | 11.5 | 825 | 28.2 | 2.8 | 91,343 | 0.8 | 5,433 | 5.9 | 85,691 | 38.8 | 23.3 |
| Sheridan | 1,134 | 76.7 | 113,900 | 19.1 | 10.0 | 640 | 20.9 | 1.0 | 1,462 | 3.4 | 33 | 2.3 | 1,308 | 35.9 | 26.8 |
| Sherman | 2,544 | 66.9 | 111,400 | 20.6 | 12.2 | 819 | 24.8 | 3.8 | 2,894 | 1.0 | 99 | 3.4 | 2,931 | 36.8 | 29.1 |
| Smith | 1,701 | 77.1 | 70,200 | 20.7 | 11.9 | 473 | 19.7 | 1.3 | 1,996 | 0.4 | 53 | 2.7 | 1,745 | 37.1 | 26.2 |
| Stafford | 1,771 | 81.2 | 65,900 | 17.1 | 10.0 | 617 | 19.7 | 3.1 | 2,043 | 2.2 | 76 | 3.7 | 2,017 | 33.2 | 26.0 |
| Stanton | 733 | 74.4 | 92,600 | 18.9 | 10.0 | 675 | 22.8 | 1.0 | 1,080 | 8.8 | 26 | 2.4 | 917 | 36.4 | 34.2 |
| Stevens | 1,814 | 70.2 | 114,900 | 19.3 | 10.3 | 671 | 25.3 | 2.3 | 2,807 | 8.8 | 80 | 2.9 | 2,493 | 26.1 | 42.9 |
| Sumner | 9,416 | 74.1 | 89,800 | 18.3 | 13.4 | 716 | 23.7 | 1.6 | 11,017 | 1.4 | 891 | 8.1 | 10,559 | 32.5 | 31.9 |
| Thomas | 3,375 | 67.6 | 129,700 | 18.2 | 11.8 | 638 | 23.8 | 0.8 | 4,230 | -2.1 | 122 | 2.9 | 4,232 | 30.5 | 22.9 |
| Trego | 1,379 | 74.9 | 99,300 | 20.7 | 10.0 | 580 | 29.0 | 0.0 | 1,470 | 4.0 | 51 | 3.5 | 1,603 | 29.9 | 30.1 |
| Wabaunsee | 2,744 | 85.5 | 146,700 | 21.4 | 12.6 | 669 | 19.9 | 1.6 | 3,677 | -0.2 | 155 | 4.2 | 3,379 | 36.3 | 29.7 |
| Wallace | 636 | 75.9 | 89,400 | 18.1 | 10.0 | 525 | 14.9 | 1.3 | 825 | 1.0 | 18 | 2.2 | 772 | 49.7 | 17.4 |
| Washington | 2,332 | 77.9 | 82,000 | 18.0 | 12.1 | 456 | 17.6 | 1.5 | 3,062 | 4.4 | 76 | 2.5 | 2,637 | 34.7 | 33.8 |
| Wichita | 934 | 73.3 | 88,900 | 18.0 | 13.8 | 707 | 27.2 | 3.0 | 1,258 | -9.0 | 26 | 2.1 | 1,119 | 38.0 | 28.2 |
| Wilson | 3,712 | 73.3 | 73,500 | 17.2 | 12.2 | 675 | 26.8 | 3.2 | 3,913 | -0.8 | 304 | 7.8 | 3,993 | 33.8 | 30.5 |
| Woodson | 1,434 | 77.1 | 73,600 | 18.7 | 14.1 | 599 | 24.3 | 2.9 | 1,567 | -3.3 | 93 | 5.9 | 1,340 | 36.0 | 29.2 |
| Wyandotte | 60,128 | 57.4 | 100,300 | 21.9 | 13.8 | 884 | 29.5 | 4.6 | 78,392 | 1.6 | 6,089 | 7.8 | 76,548 | 24.7 | 34.7 |
| KENTUCKY | 1,734,618 | 67.2 | 141,000 | 19.1 | 10.3 | 763 | 27.6 | 2.4 | 2,019,887 | -2.4 | 134,242 | 6.6 | 1,978,477 | 34.5 | 27.6 |
| Adair | 6,977 | 76.1 | 96,400 | 19.6 | 10.3 | 564 | 28.3 | 4.2 | 6,961 | -2.6 | 501 | 7.2 | 7,857 | 31.0 | 36.8 |
| Allen | 7,605 | 74.9 | 120,600 | 19.8 | 10.5 | 677 | 26.9 | 4.4 | 8,732 | -3.9 | 476 | 5.5 | 8,744 | 27.9 | 36.6 |
| Anderson | 8,694 | 76.1 | 149,800 | 18.7 | 10.0 | 708 | 30.1 | 0.7 | 11,752 | -2.3 | 718 | 6.1 | 10,461 | 35.2 | 29.6 |
| Ballard | 3,065 | 80.7 | 102,900 | 19.3 | 10.0 | 627 | 23.5 | 2.3 | 3,356 | -3.5 | 222 | 6.6 | 3,319 | 28.2 | 31.3 |
| Barren | 16,931 | 66.7 | 125,400 | 20.0 | 10.6 | 681 | 29.6 | 2.9 | 18,195 | -3.4 | 1,366 | 7.5 | 18,598 | 30.1 | 35.7 |
| Bath | 4,827 | 72.1 | 89,700 | 18.3 | 10.6 | 631 | 25.6 | 2.4 | 4,561 | -2.0 | 379 | 8.3 | 4,895 | 28.8 | 38.0 |
| Bell | 10,624 | 66.0 | 59,800 | 22.3 | 11.6 | 522 | 34.9 | 2.9 | 8,248 | -1.8 | 642 | 7.8 | 7,569 | 24.7 | 33.7 |
| Boone | 46,997 | 74.5 | 189,800 | 17.4 | 10.0 | 986 | 26.7 | 1.4 | 70,071 | -3.2 | 3,791 | 5.4 | 67,668 | 39.1 | 22.8 |
| Bourbon | 8,106 | 63.2 | 146,700 | 19.1 | 10.0 | 730 | 25.5 | 1.4 | 9,577 | -0.6 | 561 | 5.9 | 9,284 | 31.6 | 28.0 |
| Boyd | 18,210 | 68.8 | 111,200 | 19.0 | 10.4 | 699 | 28.2 | 1.2 | 17,551 | -1.5 | 1,465 | 8.3 | 18,345 | 38.6 | 23.0 |
| Boyle | 11,023 | 67.2 | 146,400 | 19.4 | 10.5 | 727 | 28.6 | 1.1 | 12,540 | -0.9 | 922 | 7.4 | 12,253 | 35.5 | 29.8 |
| Bracken | 3,318 | 74.0 | 103,100 | 18.3 | 11.1 | 617 | 26.1 | 2.5 | 3,744 | -3.2 | 233 | 6.2 | 3,665 | 27.6 | 38.1 |
| Breathitt | 5,358 | 71.0 | 54,000 | 19.0 | 10.9 | 488 | 31.8 | 4.1 | 3,285 | -5.4 | 336 | 10.2 | 3,911 | 32.7 | 25.2 |
| Breckinridge | 7,598 | 80.6 | 102,900 | 18.6 | 10.0 | 600 | 25.7 | 3.6 | 7,846 | -1.2 | 572 | 7.3 | 7,714 | 26.7 | 40.6 |
| Bullitt | 29,690 | 80.3 | 165,500 | 19.5 | 11.2 | 825 | 26.0 | 2.1 | 42,031 | -2.9 | 2,672 | 6.4 | 40,256 | 27.6 | 35.8 |
| Butler | 4,958 | 72.0 | 92,100 | 16.6 | 10.0 | 559 | 26.2 | 3.4 | 4,961 | -3.5 | 342 | 6.9 | 5,179 | 29.1 | 42.4 |
| Caldwell | 5,163 | 76.4 | 87,900 | 17.9 | 10.0 | 577 | 30.6 | 1.6 | 5,570 | 1.7 | 346 | 6.2 | 5,238 | 27.1 | 32.6 |
| Calloway | 14,996 | 62.4 | 137,300 | 19.6 | 10.5 | 705 | 30.8 | 1.5 | 17,865 | -4.6 | 1,113 | 6.2 | 18,318 | 31.6 | 25.4 |
| Campbell | 36,746 | 69.7 | 172,000 | 18.9 | 11.3 | 850 | 28.5 | 1.4 | 49,647 | -3.2 | 2,761 | 5.6 | 46,997 | 43.7 | 20.6 |
| Carlisle | 1,980 | 81.7 | 89,300 | 20.8 | 10.0 | 637 | 29.3 | 1.7 | 2,168 | -3.1 | 102 | 4.7 | 1,865 | 24.8 | 35.8 |
| Carroll | 4,101 | 64.3 | 115,900 | 16.9 | 10.6 | 650 | 29.1 | 5.3 | 5,289 | -0.8 | 311 | 5.9 | 4,458 | 19.9 | 42.2 |
| Carter | 9,606 | 77.4 | 95,000 | 20.0 | 11.8 | 579 | 28.3 | 2.9 | 9,586 | -1.5 | 923 | 9.6 | 8,723 | 25.9 | 31.8 |
| Casey | 6,099 | 77.3 | 87,600 | 21.2 | 10.0 | 507 | 28.3 | 3.1 | 6,666 | | 416 | 6.2 | 5,659 | 27.0 | 39.5 |
| Christian | 25,721 | 48.1 | 117,400 | 19.8 | 10.0 | 854 | 29.6 | 2.5 | 24,587 | 0.4 | 1,925 | 7.8 | 23,554 | 28.0 | 30.7 |
| Clark | 14,509 | 67.4 | 145,400 | 19.5 | 10.5 | 727 | 26.6 | 2.5 | 17,006 | -1.6 | 1,115 | 6.6 | 16,357 | 33.3 | 28.8 |

1. Specified owner-occupied units. lacking complete plumbing facilities.   2. A value of 10.0 represents 10 percent or less; a value of 50.0 represents 50 percent or more.   3. Specified renter-occupied units.   4. Overcrowded or
5. Percent of civilian labor force.   6. Civilian employed persons 16 years old and over.

| STATE County | Number of establish-ments | Employment Total | Health care and social assistance | Manufac-turing | Retail trade | Finance and insurance | Professional, scientific, and technical services | Annual payroll Total (mil dol) | Average per employee (dollars) | Farms Number | Percent with: Fewer than 50 acres | 1000 acres or more | Farm producers whose primary occupation is farming (percent) |
|---|---|---|---|---|---|---|---|---|---|---|---|---|---|
| | 104 | 105 | 106 | 107 | 108 | 109 | 110 | 111 | 112 | 113 | 114 | 115 | 116 |
| **KANSAS— Cont'd** | | | | | | | | | | | | | |
| Neosho | 425 | 4,746 | 1,043 | 989 | 789 | 218 | 129 | 174 | 36,590 | 687 | 25.5 | 9.3 | 36.3 |
| Ness | 130 | 922 | 259 | 24 | 72 | 44 | 22 | 35 | 37,973 | 523 | 8.2 | 33.8 | 39.8 |
| Norton | 173 | 1,760 | 489 | 245 | 183 | 94 | 31 | 69 | 39,124 | 328 | 14.0 | 39.6 | 52.1 |
| Osage | 242 | 1,592 | 402 | 149 | 339 | 114 | 78 | 42 | 26,564 | 1,042 | 24.8 | 12.7 | 36.1 |
| Osborne | 135 | 923 | 177 | 106 | 143 | 52 | 24 | 31 | 33,790 | 319 | 10.7 | 36.1 | 46.3 |
| Ottawa | 138 | 826 | 231 | 51 | 91 | 71 | 49 | 27 | 32,455 | 438 | 18.9 | 28.1 | 45.7 |
| Pawnee | 147 | 1,902 | 1,088 | 51 | 179 | 56 | 70 | 76 | 39,722 | 362 | 10.8 | 31.5 | 39.1 |
| Phillips | 226 | 1,585 | 272 | 187 | 263 | 114 | 111 | 60 | 37,681 | 415 | 11.8 | 33.3 | 50.1 |
| Pottawatomie | 619 | 9,038 | 1,494 | 1,614 | 1,848 | 263 | 233 | 384 | 42,510 | 774 | 27.4 | 14.6 | 34.4 |
| Pratt | 362 | 3,158 | 656 | 103 | 592 | 129 | 107 | 129 | 40,844 | 481 | 8.9 | 27.7 | 46.1 |
| Rawlins | 104 | 645 | 181 | 55 | 88 | 38 | 11 | 26 | 40,552 | 298 | 9.4 | 49.3 | 60.3 |
| Reno | 1,570 | 22,446 | 4,224 | 3,468 | 3,024 | 904 | 635 | 885 | 39,449 | 1,552 | 21.1 | 13.0 | 37.6 |
| Republic | 169 | 1,163 | 276 | 134 | 201 | 58 | 34 | 34 | 29,383 | 561 | 13.2 | 23.0 | 53.6 |
| Rice | 268 | 2,495 | 331 | 396 | 271 | 111 | 132 | 89 | 35,555 | 470 | 17.9 | 26.6 | 43.3 |
| Riley | 1,610 | 21,030 | 3,507 | 441 | 3,997 | 832 | 1,158 | 677 | 32,176 | 504 | 31.7 | 11.7 | 33.1 |
| Rooks | 181 | 1,334 | 281 | 108 | 192 | 64 | 48 | 46 | 34,280 | 412 | 10.9 | 32.3 | 45.5 |
| Rush | 93 | 1,012 | 153 | 320 | 70 | 47 | 15 | 38 | 37,426 | 488 | 10.5 | 25.4 | 38.9 |
| Russell | 245 | 1,945 | 344 | 251 | 257 | 63 | 57 | 67 | 34,380 | 500 | 15.0 | 23.0 | 38.1 |
| Saline | 1,465 | 25,929 | 4,611 | 5,004 | 3,849 | 823 | 1,131 | 995 | 38,386 | 609 | 25.1 | 17.1 | 41.5 |
| Scott | 199 | 1,510 | 480 | 52 | 212 | 103 | 51 | 56 | 36,878 | 236 | 17.8 | 39.4 | 58.9 |
| Sedgwick | 12,200 | 233,692 | 33,223 | 45,879 | 28,412 | 7,027 | 12,006 | 11,547 | 49,413 | 1,360 | 40.7 | 11.4 | 33.6 |
| Seward | 550 | 8,919 | 1,063 | 2,942 | 1,198 | 191 | 112 | 303 | 33,947 | 282 | 7.1 | 27.0 | 44.8 |
| Shawnee | 4,091 | 76,730 | 17,575 | 7,462 | 9,688 | 5,687 | 3,429 | 3,642 | 47,472 | 847 | 41.4 | 5.5 | 34.8 |
| Sheridan | 110 | 728 | 182 | NA | 98 | 48 | 19 | 28 | 38,720 | 318 | 2.8 | 49.1 | 65.0 |
| Sherman | 262 | 1,995 | 407 | 82 | 356 | 94 | 51 | 69 | 34,633 | 386 | 10.6 | 33.7 | 54.5 |
| Smith | 131 | 891 | 293 | 15 | 164 | 60 | 21 | 26 | 28,948 | 425 | 11.1 | 33.9 | 55.9 |
| Stafford | 126 | 599 | 180 | NA | 79 | 47 | 15 | 21 | 34,671 | 466 | 12.0 | 29.2 | 44.5 |
| Stanton | 63 | 495 | NA | NA | 65 | 49 | 16 | 22 | 45,022 | 220 | 2.7 | 42.3 | 51.2 |
| Stevens | 132 | 1,121 | 239 | 47 | 105 | 66 | 17 | 44 | 39,612 | 377 | 7.4 | 27.1 | 37.9 |
| Sumner | 451 | 4,261 | 944 | 973 | 569 | 192 | 83 | 155 | 36,352 | 953 | 21.9 | 24.3 | 43.3 |
| Thomas | 322 | 2,856 | 518 | 86 | 640 | 146 | 75 | 102 | 35,761 | 402 | 8.2 | 45.8 | 57.7 |
| Trego | 116 | 859 | 203 | 39 | 116 | 32 | 13 | 29 | 33,602 | 343 | 9.9 | 37.0 | 50.1 |
| Wabaunsee | 122 | 826 | 86 | 126 | 91 | 45 | NA | 26 | 30,915 | 638 | 21.0 | 16.1 | 35.3 |
| Wallace | 57 | 321 | 54 | NA | 47 | 17 | 16 | 11 | 34,146 | 281 | 5.0 | 39.9 | 47.0 |
| Washington | 197 | 1,516 | 384 | 180 | 161 | 68 | 154 | 52 | 34,142 | 694 | 13.0 | 23.2 | 50.0 |
| Wichita | 84 | 458 | 114 | 54 | 73 | 19 | 11 | 17 | 37,568 | 254 | 10.2 | 46.9 | 55.2 |
| Wilson | 208 | 3,832 | 624 | 1,142 | 243 | 72 | 42 | 172 | 44,800 | 420 | 14.3 | 21.7 | 49.6 |
| Woodson | 80 | 417 | 75 | NA | 60 | 25 | 5 | 12 | 28,724 | 289 | 16.3 | 23.5 | 55.1 |
| Wyandotte | 3,035 | 70,613 | 13,995 | 10,425 | 7,203 | 1,149 | 1,514 | 3,529 | 49,972 | 158 | 74.7 | 2.5 | 32.1 |
| **KENTUCKY** | 91,219 | 1,666,637 | 260,342 | 252,706 | 217,788 | 75,466 | 74,279 | 74,884 | 44,931 | 75,966 | 40.1 | 2.5 | 36.4 |
| Adair | 320 | 4,045 | 539 | 377 | 717 | 240 | 60 | 117 | 29,038 | 1,154 | 30.4 | 1.1 | 37.1 |
| Allen | 228 | 3,229 | 334 | 573 | 443 | 131 | 63 | 118 | 36,586 | 1,127 | 36.6 | 1.8 | 43.0 |
| Anderson | 331 | 3,768 | 366 | 1,219 | 725 | 113 | 82 | 147 | 38,982 | 774 | 45.7 | 0.6 | 31.4 |
| Ballard | 138 | 1,152 | 140 | 303 | 140 | 36 | 117 | 60 | 51,968 | 295 | 33.6 | 8.1 | 42.9 |
| Barren | 818 | 14,900 | 2,352 | 3,363 | 2,812 | 333 | 474 | 529 | 35,520 | 1,899 | 42.5 | 1.5 | 38.3 |
| Bath | 129 | 1,500 | 216 | 443 | 164 | 62 | 29 | 58 | 38,500 | 728 | 30.1 | 1.4 | 33.2 |
| Bell | 457 | 6,856 | 1,328 | 1,333 | 1,474 | 246 | 149 | 221 | 32,272 | 110 | 49.1 | 1.8 | 24.0 |
| Boone | 3,129 | 85,686 | 4,955 | 11,563 | 13,154 | 3,666 | 2,841 | 4,135 | 48,254 | 721 | 54.5 | 1.0 | 25.7 |
| Bourbon | 373 | 5,134 | 688 | 1,332 | 961 | 223 | 158 | 228 | 44,493 | 915 | 38.6 | 2.1 | 45.1 |
| Boyd | 1,230 | 21,906 | 6,613 | 1,958 | 3,427 | 561 | 741 | 1,004 | 45,824 | 203 | 46.8 | NA | 33.6 |
| Boyle | 745 | 13,575 | 3,259 | 2,832 | 1,756 | 334 | 336 | 514 | 37,897 | 602 | 45.0 | 2.3 | 39.5 |
| Bracken | 90 | 899 | 141 | 327 | 122 | 29 | 15 | 31 | 34,378 | 531 | 29.0 | 0.6 | 35.8 |
| Breathitt | 178 | 3,207 | 881 | 47 | 423 | 118 | 14 | 89 | 27,760 | 160 | 36.9 | 1.3 | 29.2 |
| Breckinridge | 277 | 2,475 | 564 | 284 | 498 | 160 | 66 | 85 | 34,274 | 1,357 | 31.9 | 2.8 | 35.8 |
| Bullitt | 1,171 | 21,995 | 1,515 | 3,910 | 5,390 | 324 | 269 | 866 | 39,390 | 486 | 58.4 | 1.0 | 31.7 |
| Butler | 177 | 1,691 | 250 | 572 | 208 | 78 | 25 | 63 | 37,122 | 642 | 29.3 | 4.5 | 25.3 |
| Caldwell | 291 | 3,315 | 527 | 830 | 716 | 116 | 55 | 113 | 34,100 | 475 | 36.4 | 5.3 | 33.5 |
| Calloway | 845 | 13,710 | 1,752 | 2,623 | 1,831 | 438 | 355 | 461 | 33,646 | 710 | 48.9 | 3.8 | 36.9 |
| Campbell | 1,689 | 25,316 | 3,411 | 2,286 | 4,016 | 473 | 1,133 | 1,034 | 40,841 | 577 | 54.8 | 0.3 | 32.3 |
| Carlisle | 93 | 646 | 118 | 48 | 94 | 64 | 15 | 20 | 31,070 | 273 | 40.3 | 7.0 | 41.7 |
| Carroll | 234 | 6,285 | 358 | 3,144 | 582 | 74 | 91 | 390 | 62,118 | 308 | 25.6 | 1.0 | 32.5 |
| Carter | 404 | 4,904 | 504 | 751 | 932 | 179 | 101 | 136 | 27,755 | 718 | 27.7 | 0.4 | 38.3 |
| Casey | 227 | 3,226 | 543 | 1,340 | 363 | 81 | 188 | 96 | 29,713 | 1,106 | 33.3 | 2.1 | 40.7 |
| Christian | 1,312 | 24,363 | 3,561 | 6,913 | 3,076 | 616 | 992 | 963 | 39,509 | 1,137 | 33.3 | 6.9 | 42.4 |
| Clark | 751 | 12,302 | 1,737 | 2,887 | 1,670 | 262 | 547 | 499 | 40,575 | 871 | 41.8 | 2.3 | 37.7 |

| STATE County | Land in farms | | | | | Value of land and buildings (dollars) | | Value of machinery and equiopmnet, average per farm (dollars) | Value of products sold: | | | | Organic farms (number) | Farms with internet access (per-cent) | Government payments | |
|---|---|---|---|---|---|---|---|---|---|---|---|---|---|---|---|---|
| | Acreage (1,000) | Percent change, 2012-2017 | Acres | | | Average per farm | Average per acre | | Total (mil dol) | Average per farm (acres) | Percent from: | | | | Total ($1,000) | Percent of farms |
| | | | Average size of farm | Total irrigated (1,000) | Total cropland (1,000) | | | | | | Crops | Livestock and poultry products | | | | |
| | 117 | 118 | 119 | 120 | 121 | 122 | 123 | 124 | 125 | 126 | 127 | 128 | 129 | 130 | 131 | 132 |
| **KANSAS— Cont'd** | | | | | | | | | | | | | | | | |
| Neosho | 323 | 4.8 | 470 | 0.8 | 169.9 | 923,397 | 1,963 | 93,313 | 81.9 | 119,167 | 56.0 | 44.0 | 2 | 72.9 | 1,845 | 49.5 |
| Ness | 668 | -1.4 | 1,278 | 3.7 | 404.6 | 1,385,852 | 1,084 | 159,980 | 60.8 | 116,216 | 61.4 | 38.6 | NA | 73.0 | 8,605 | 91.6 |
| Norton | 495 | -1.4 | 1,509 | 20.3 | 284.0 | 2,053,567 | 1,361 | 299,003 | 143.3 | 436,744 | 43.0 | 57.0 | NA | 69.8 | 2,110 | 74.1 |
| Osage | 440 | -0.6 | 422 | 1.3 | 252.6 | 840,108 | 1,992 | 126,651 | 92.4 | 88,677 | 72.4 | 27.6 | NA | 72.6 | 2,606 | 52.3 |
| Osborne | 437 | -0.7 | 1,370 | 5.6 | 228.6 | 1,932,419 | 1,410 | 206,187 | 62.5 | 195,922 | 71.3 | 28.7 | NA | 75.5 | 2,266 | 81.8 |
| Ottawa | 439 | 4.6 | 1,003 | 6.5 | 242.2 | 2,017,300 | 2,011 | 217,607 | 108.4 | 247,436 | 46.7 | 53.3 | 2 | 74.7 | 3,278 | 74.9 |
| Pawnee | 474 | -1.3 | 1,310 | 60.7 | 397.9 | 2,171,753 | 1,658 | 243,559 | 307.9 | 850,519 | 25.1 | 74.9 | 1 | 68.2 | 6,089 | 77.3 |
| Phillips | 497 | 0.5 | 1,198 | 6.4 | 254.1 | 1,728,534 | 1,442 | 208,663 | 107.6 | 259,294 | 54.5 | 45.5 | NA | 75.4 | 1,925 | 68.4 |
| Pottawatomie | 406 | -0.9 | 525 | 19.6 | 158.7 | 1,290,997 | 2,461 | 124,388 | 101.4 | 130,961 | 41.3 | 58.7 | NA | 80.2 | 2,520 | 52.2 |
| Pratt | 465 | 0.1 | 967 | 63.8 | 366.5 | 1,987,147 | 2,055 | 245,812 | 271.3 | 564,046 | 33.6 | 66.4 | NA | 72.8 | 9,324 | 83.4 |
| Rawlins | 604 | -0.9 | 2,025 | 14.3 | 356.0 | 3,123,663 | 1,542 | 323,136 | 100.4 | 336,748 | 66.6 | 33.4 | 1 | 84.2 | 3,971 | 79.5 |
| Reno | 789 | -0.1 | 508 | 54.1 | 590.7 | 1,087,423 | 2,139 | 154,602 | 216.7 | 139,645 | 53.5 | 46.5 | NA | 69.3 | 14,557 | 69.5 |
| Republic | 373 | 3.4 | 665 | 40.2 | 269.1 | 1,720,315 | 2,586 | 200,912 | 187.5 | 334,275 | 51.9 | 48.1 | 2 | 73.4 | 5,249 | 77.4 |
| Rice | 463 | 1.2 | 986 | 29.9 | 366.1 | 2,006,778 | 2,036 | 254,152 | 235.5 | 501,162 | 36.0 | 64.0 | 1 | 81.5 | 6,079 | 73.8 |
| Riley | 214 | -1.8 | 425 | 3.8 | 106.1 | 1,285,194 | 3,022 | 137,948 | 51.2 | 101,532 | 63.6 | 36.4 | 3 | 82.7 | 1,107 | 49.2 |
| Rooks | 559 | 1.3 | 1,356 | 5.7 | 320.2 | 1,696,528 | 1,251 | 189,770 | 76.6 | 185,934 | 64.1 | 35.9 | NA | 77.7 | 4,218 | 81.3 |
| Rush | 448 | -1.1 | 919 | 8.6 | 323.2 | 1,193,377 | 1,299 | 146,600 | 59.5 | 121,994 | 69.4 | 30.6 | NA | 68.4 | 8,957 | 82.0 |
| Russell | 492 | 13.1 | 985 | 0.5 | 263.3 | 1,318,319 | 1,339 | 125,089 | 50.1 | 100,106 | 65.9 | 34.1 | NA | 70.4 | 5,057 | 79.6 |
| Saline | 358 | -1.7 | 588 | 4.9 | 223.3 | 1,394,788 | 2,371 | 157,229 | 73.6 | 120,823 | 52.8 | 47.2 | 1 | 77.0 | 4,352 | 71.8 |
| Scott | 460 | 1.5 | 1,951 | 29.1 | 365.7 | 2,955,605 | 1,515 | 503,414 | 1,135 | 4,809,487 | 7.7 | 92.3 | 7 | 79.2 | 6,872 | 74.2 |
| Sedgwick | 497 | 2.0 | 365 | 40.4 | 408.9 | 1,264,948 | 3,464 | 135,560 | 118.9 | 87,440 | 79.9 | 20.1 | 1 | 79.9 | 7,283 | 52.6 |
| Seward | 361 | -10.2 | 1,279 | 95.5 | 263.7 | 1,803,833 | 1,410 | 305,603 | 424.7 | 1,506,021 | 18.9 | 81.1 | NA | 69.9 | 7,997 | 76.2 |
| Shawnee | 202 | 3.8 | 238 | 14.7 | 126.5 | 712,656 | 2,993 | 88,748 | 49.2 | 58,035 | 79.8 | 20.2 | 2 | 81.3 | 1,312 | 33.4 |
| Sheridan | 512 | -8.9 | 1,610 | 68.4 | 358.5 | 2,518,737 | 1,564 | 431,777 | 348.9 | 1,097,022 | 25.4 | 74.6 | 1 | 83.6 | 6,021 | 87.4 |
| Sherman | 618 | 4.0 | 1,602 | 97.1 | 491.6 | 2,821,024 | 1,761 | 358,514 | 139.2 | 360,567 | 72.8 | 27.2 | 4 | 74.6 | 8,255 | 78.2 |
| Smith | 541 | 8.2 | 1,274 | 7.4 | 363.2 | 2,431,582 | 1,909 | 275,597 | 129.3 | 304,144 | 75.9 | 24.1 | 5 | 84.0 | 4,089 | 86.4 |
| Stafford | 494 | | 1,059 | 76.4 | 393.0 | 1,909,862 | 1,803 | 300,164 | 198.6 | 426,120 | 41.5 | 58.5 | NA | 74.2 | 8,445 | 83.9 |
| Stanton | 435 | 1.4 | 1,978 | 54.3 | 396.1 | 2,105,893 | 1,064 | 364,126 | 133.5 | 606,786 | 54.5 | 45.5 | NA | 80.5 | 9,689 | 88.2 |
| Stevens | 455 | 0.0 | 1,208 | 138.4 | 370.0 | 1,675,641 | 1,387 | 330,668 | 340.6 | 903,358 | 32.3 | 67.7 | NA | 71.6 | 9,552 | 86.5 |
| Sumner | 758 | 5.3 | 795 | 25.7 | 630.3 | 1,461,530 | 1,838 | 204,872 | 155.7 | 163,348 | 87.8 | 12.2 | NA | 77.8 | 8,737 | 70.3 |
| Thomas | 670 | -0.8 | 1,667 | 81.4 | 569.5 | 2,911,418 | 1,747 | 417,933 | 251.1 | 624,515 | 52.6 | 47.4 | NA | 84.8 | 6,720 | 79.6 |
| Trego | 515 | 15.4 | 1,503 | 5.5 | 276.3 | 1,714,222 | 1,141 | 222,541 | 57.2 | 166,706 | 59.5 | 40.5 | NA | 73.2 | 4,819 | 82.2 |
| Wabaunsee | 379 | -4.4 | 594 | 8.7 | 114.2 | 1,144,196 | 1,927 | 113,289 | 63.1 | 98,975 | 40.0 | 60.0 | NA | 80.9 | 1,949 | 48.9 |
| Wallace | 446 | -8.6 | 1,587 | 33.9 | 303.1 | 2,144,593 | 1,352 | 252,476 | 81.8 | 291,050 | 68.3 | 31.7 | 2 | 74.0 | 5,388 | 81.5 |
| Washington | 526 | 7.3 | 757 | 13.9 | 336.7 | 1,837,799 | 2,426 | 218,194 | 182.0 | 262,218 | 47.9 | 52.1 | NA | 76.9 | 4,929 | 80.3 |
| Wichita | 438 | -5.6 | 1,724 | 40.9 | 367.9 | 2,416,112 | 1,401 | 385,925 | 559.3 | 2,202,157 | 12.8 | 87.2 | 16 | 76.0 | 6,749 | 84.3 |
| Wilson | 287 | 12.6 | 683 | 1.8 | 180.2 | 1,329,365 | 1,947 | 182,949 | 62.3 | 148,298 | 79.8 | 20.2 | NA | 71.2 | 1,941 | 51.0 |
| Woodson | 283 | -4.0 | 979 | D | 135.4 | 1,627,566 | 1,662 | 184,288 | 52.6 | 181,834 | 52.8 | 47.2 | 2 | 78.2 | 1,573 | 55.4 |
| Wyandotte | 12 | 3.2 | 78 | 0.6 | 8.8 | 528,919 | 6,742 | 73,098 | 5.3 | 33,380 | 90.5 | 9.5 | 2 | 82.9 | 52 | 4.4 |
| **KENTUCKY** | 12,962 | -0.7 | 171 | 83.9 | 6,630.4 | 643,019 | 3,769 | 82,740 | 5,737.9 | 75,533 | 44.3 | 55.7 | 227 | 72.4 | 126,697 | 22.2 |
| Adair | 172 | 1.1 | 149 | 0.0 | 76.8 | 448,542 | 3,008 | 72,779 | 69.4 | 60,101 | 26.2 | 73.8 | 1 | 68.5 | 2,151 | 33.6 |
| Allen | 169 | 15.8 | 150 | 0.2 | 74.2 | 518,199 | 3,462 | 70,602 | 88.6 | 78,613 | 26.4 | 73.6 | 8 | 68.1 | 2,489 | 24.2 |
| Anderson | 82 | 1.0 | 106 | 0.0 | 29.1 | 392,778 | 3,713 | 53,625 | 12.3 | 15,926 | 34.6 | 65.4 | 6 | 67.8 | 276 | 4.9 |
| Ballard | 94 | -12.0 | 320 | D | 75.3 | 1,155,670 | 3,614 | 205,371 | 70.6 | 239,366 | 54.0 | 46.0 | NA | 87.5 | 1,653 | 49.8 |
| Barren | 254 | 2.1 | 134 | 0.1 | 141.4 | 481,503 | 3,602 | 90,193 | 127.2 | 66,982 | 38.3 | 61.7 | NA | 77.6 | 2,218 | 26.1 |
| Bath | 127 | -10.6 | 175 | 0.0 | 46.7 | 412,199 | 2,360 | 61,331 | 18.3 | 25,161 | 30.8 | 69.2 | 4 | 76.2 | 217 | 6.2 |
| Bell | 15 | 87.3 | 137 | 0.0 | 3.4 | 273,172 | 1,990 | 43,893 | 0.5 | 4,591 | 48.5 | 51.5 | NA | 68.2 | 10 | 9.1 |
| Boone | 79 | 17.2 | 109 | 0.1 | 36.6 | 655,823 | 6,004 | 58,670 | 15.4 | 21,315 | 75.2 | 24.8 | 2 | 73.0 | 440 | 9.0 |
| Bourbon | 171 | -7.0 | 187 | 0.6 | 74.9 | 1,061,516 | 5,684 | 104,627 | 209.6 | 229,030 | 21.9 | 78.1 | 3 | 76.2 | 695 | 17.9 |
| Boyd | 19 | -10.7 | 96 | 0.0 | 3.5 | 230,681 | 2,405 | 57,311 | 1.2 | 5,842 | 38.0 | 62.0 | NA | 83.3 | 5 | 3.0 |
| Boyle | 89 | -12.7 | 147 | 0.0 | 36.8 | 547,715 | 3,720 | 64,909 | 31.3 | 52,056 | 20.6 | 79.4 | 7 | 71.9 | 251 | 12.5 |
| Bracken | 87 | 0.0 | 164 | 0.1 | 32.4 | 408,483 | 2,498 | 60,419 | 10.7 | 20,190 | 60.2 | 39.8 | 1 | 64.6 | 84 | 10.0 |
| Breathitt | 23 | 4.4 | 145 | 0.0 | 3.6 | 227,411 | 1,569 | 54,188 | 0.5 | 3,175 | 59.4 | 40.6 | NA | 72.5 | 21 | 13.8 |
| Breckinridge | 275 | 5.9 | 203 | 0.1 | 126.8 | 625,871 | 3,087 | 92,986 | 99.4 | 73,279 | 47.0 | 53.0 | 6 | 68.8 | 3,162 | 37.1 |
| Bullitt | 44 | -3.7 | 91 | D | 19.3 | 471,740 | 5,161 | 46,835 | 6.2 | 12,848 | 69.8 | 30.2 | NA | 78.8 | 41 | 2.5 |
| Butler | 147 | -3.8 | 229 | 0.2 | 75.6 | 605,594 | 2,648 | 78,210 | 48.9 | 76,221 | 55.8 | 44.2 | NA | 65.1 | 2,423 | 34.6 |
| Caldwell | 130 | -2.5 | 274 | 6.2 | 86.5 | 915,071 | 3,339 | 122,658 | 45.1 | 95,038 | 89.0 | 11.0 | NA | 71.8 | 2,555 | 45.8 |
| Calloway | 136 | -23.0 | 191 | 2.9 | 102.5 | 786,844 | 4,122 | 115,703 | 97.7 | 137,631 | 60.0 | 40.0 | NA | 77.9 | 2,938 | 57.3 |
| Campbell | 46 | 9.3 | 80 | 0.0 | 15.9 | 404,186 | 5,060 | 51,900 | 7.1 | 12,334 | 48.3 | 51.7 | NA | 77.8 | 28 | 3.6 |
| Carlisle | 88 | -10.8 | 322 | 1.7 | 73.1 | 1,224,443 | 3,798 | 170,980 | 67.6 | 247,681 | 56.3 | 43.7 | NA | 68.1 | 1,602 | 65.2 |
| Carroll | 51 | -5.0 | 165 | 0.0 | 17.9 | 560,794 | 3,393 | 55,839 | 5.8 | 18,867 | 57.6 | 42.4 | NA | 68.8 | 140 | 10.4 |
| Carter | 92 | -13.1 | 128 | 0.0 | 23.4 | 247,720 | 1,933 | 49,400 | 6.9 | 9,586 | 31.9 | 68.1 | 4 | 68.4 | 153 | 5.0 |
| Casey | 179 | 0.1 | 162 | 0.1 | 65.4 | 379,677 | 2,343 | 60,628 | 32.2 | 29,071 | 50.2 | 49.8 | 14 | 66.3 | 581 | 23.1 |
| Christian | 346 | -4.1 | 304 | 5.1 | 245.8 | 1,464,184 | 4,817 | 152,416 | 205.3 | 180,542 | 81.1 | 18.9 | 13 | 56.6 | 6,215 | 42.9 |
| Clark | 147 | 7.2 | 169 | 0.1 | 66.4 | 749,313 | 4,431 | 74,443 | 34.0 | 39,068 | 45.5 | 54.5 | 2 | 78.5 | 289 | 10.1 |

| STATE County | Water use, 2015 | | Wholesale Trade[1], 2017 | | | | Retail Trade[2], 2017 | | | | Real estate and rental and leasing,[2] 2017 | | | |
|---|---|---|---|---|---|---|---|---|---|---|---|---|---|---|
| | Public supply water withdrawn (mil gal/day) | Public supply gallons withdrawn per person per day | Number of establishments | Number of employees | Sales (mil dol) | Average payroll (mil dol) | Number of establishments | Number of employees | Sales (mil dol) | Average payroll (mil dol) | Number of establishments | Number of employees | Sales (mil dol) | Average payroll (mil dol) |
| | 133 | 134 | 135 | 136 | 137 | 138 | 139 | 140 | 141 | 142 | 143 | 144 | 145 | 146 |
| KANSAS— Cont'd | | | | | | | | | | | | | | |
| Neosho | 1.08 | 66.1 | 32 | 284 | 162.1 | 11.7 | 79 | 843 | 219.6 | 21.6 | 19 | 41 | 6.8 | 1.2 |
| Ness | 0.32 | 106.5 | 17 | 108 | 127.6 | 5.0 | 12 | 71 | 11.9 | 1.5 | NA | NA | NA | NA |
| Norton | 0.75 | 135.1 | 9 | 77 | 39.0 | 3.0 | 27 | 191 | 50.1 | 4.6 | NA | NA | NA | NA |
| Osage | 1.52 | 95.9 | 12 | 54 | 38.4 | 1.8 | 46 | 349 | 76.1 | 6.5 | 6 | 16 | 1.4 | 0.3 |
| Osborne | 0.45 | 122.2 | D | D | D | 5.5 | 22 | 160 | 30.6 | 2.9 | NA | NA | NA | NA |
| Ottawa | 0.53 | 88.7 | 8 | 143 | 64.1 | 7.1 | 14 | 91 | 13.8 | 1.2 | 5 | 7 | 2.1 | 0.2 |
| Pawnee | 0.94 | 137.5 | 7 | 63 | 69.5 | 3.9 | 24 | 191 | 46.4 | 3.8 | D | D | D | D |
| Phillips | 0.98 | 180.5 | 14 | 67 | 40.2 | 3.0 | 31 | 230 | 40.9 | 4.1 | NA | NA | NA | NA |
| Pottawatomie | 6.22 | 267.0 | 32 | 614 | 195.6 | 26.4 | 80 | 1,602 | 346.7 | 46.8 | 27 | 85 | 17.0 | 2.8 |
| Pratt | 1.67 | 172.3 | 21 | 243 | 356.2 | 14.1 | 51 | 645 | 166.3 | 16.8 | 12 | 26 | 3.3 | 0.6 |
| Rawlins | 0.30 | 119.7 | D | D | D | 8.3 | 17 | 103 | 22.0 | 2.0 | NA | NA | NA | NA |
| Reno | 7.61 | 119.4 | 81 | 1,036 | 1,149.4 | 43.8 | 240 | 3,085 | 764.3 | 73.1 | 77 | 221 | 35.3 | 7.0 |
| Republic | 0.65 | 137.6 | 16 | 128 | 91.9 | 4.5 | 30 | 207 | 55.0 | 4.4 | D | D | D | D |
| Rice | 1.16 | 116.3 | 23 | 109 | 125.0 | 5.5 | 40 | 287 | 49.3 | 4.9 | 5 | 13 | 2.0 | 0.3 |
| Riley | 3.03 | 40.3 | 32 | 280 | 152.4 | 12.2 | 246 | 4,206 | 904.5 | 93.2 | 125 | 619 | 80.6 | 19.9 |
| Rooks | 0.68 | 131.4 | 17 | 163 | 141.9 | 7.3 | 26 | 206 | 48.6 | 3.2 | 3 | 5 | 0.9 | 0.1 |
| Rush | 0.51 | 162.9 | 11 | D | 28.8 | D | 10 | 65 | 24.4 | 2.0 | NA | NA | NA | NA |
| Russell | 0.44 | 62.5 | 18 | 137 | 103.7 | 4.8 | 39 | 255 | 68.5 | 5.1 | 6 | D | 1.1 | D |
| Saline | 5.90 | 105.9 | 89 | 1,229 | 813.9 | 66.9 | 237 | 4,021 | 1,094.8 | 93.9 | 64 | 205 | 50.4 | 6.6 |
| Scott | 0.84 | 169.2 | 12 | 119 | 146.6 | 6.0 | 33 | 226 | 47.1 | 4.3 | NA | NA | NA | NA |
| Sedgwick | 51.98 | 101.6 | 589 | 8,732 | 8,030.5 | 527.5 | 1,718 | 29,929 | 8,082.9 | 773.3 | 617 | 3,195 | 764.6 | 122.8 |
| Seward | 4.87 | 210.3 | 38 | 279 | 185.5 | 13.3 | 93 | 1,282 | 335.5 | 31.5 | 21 | 60 | 12.3 | 2.2 |
| Shawnee | 14.34 | 80.2 | 154 | 2,070 | 1,972.0 | 116.7 | 600 | 9,622 | 2,395.4 | 227.0 | 222 | 952 | 175.4 | 32.1 |
| Sheridan | 0.48 | 191.1 | 11 | 119 | 147.4 | 7.9 | 17 | 86 | 17.1 | 1.7 | NA | NA | NA | NA |
| Sherman | 1.34 | 224.0 | 21 | 251 | 493.0 | 12.8 | 34 | 408 | 110.8 | 9.4 | 4 | 6 | 1.5 | 0.2 |
| Smith | 0.49 | 132.3 | 10 | 136 | 83.0 | 4.8 | 29 | 180 | 58.0 | 4.2 | 3 | 3 | 0.3 | 0.0 |
| Stafford | 0.35 | 82.6 | 8 | D | 31.6 | D | 20 | 86 | 28.3 | 2.0 | NA | NA | NA | NA |
| Stanton | 0.40 | 193.1 | D | D | D | 2.8 | 14 | 63 | 25.4 | 1.8 | NA | NA | NA | NA |
| Stevens | 1.37 | 236.0 | 13 | 113 | 126.2 | 5.9 | 17 | 120 | 35.6 | 3.1 | NA | NA | NA | NA |
| Sumner | 2.10 | 89.2 | 19 | 123 | 114.0 | 6.4 | 70 | 630 | 168.2 | 14.7 | 16 | 44 | 4.3 | 1.0 |
| Thomas | 1.27 | 160.7 | 30 | 339 | 458.7 | 17.4 | 64 | 634 | 197.2 | 16.2 | 9 | 31 | 7.9 | 0.7 |
| Trego | 0.68 | 232.3 | 8 | 65 | 80.3 | 2.6 | 20 | 130 | 41.7 | 2.6 | D | D | D | D |
| Wabaunsee | 0.54 | 77.7 | D | D | D | 1.4 | 20 | 94 | 29.0 | 2.0 | D | D | D | D |
| Wallace | 0.22 | 144.9 | D | D | D | D | 8 | 42 | 7.0 | 1.1 | NA | NA | NA | NA |
| Washington | 0.78 | 139.3 | 17 | 106 | 86.3 | 4.4 | 35 | 176 | 35.3 | 3.9 | 4 | 8 | 0.6 | 0.1 |
| Wichita | 0.29 | 134.4 | 11 | 71 | 231.7 | 3.2 | 12 | 60 | 14.8 | 1.5 | NA | NA | NA | NA |
| Wilson | 1.36 | 153.6 | D | D | D | 0.6 | 31 | 244 | 48.2 | 4.7 | 5 | D | 0.4 | D |
| Woodson | 0.30 | 96.3 | 5 | 84 | 54.6 | 2.4 | 13 | 73 | 20.3 | 1.7 | NA | NA | NA | NA |
| Wyandotte | 62.15 | 380.4 | 220 | 6,602 | 8,955.5 | 442.3 | 442 | 7,813 | 2,158.2 | 212.9 | 130 | 706 | 138.7 | 29.6 |
| KENTUCKY | 552.83 | 124.9 | 3,646 | 55,784 | 80,128.5 | 3,171.6 | 15,021 | 225,127 | 64,294.3 | 5,630.0 | 3,902 | 18,070 | 5,125.3 | 731.1 |
| Adair | 0.00 | 0.0 | 16 | 81 | 29.1 | 2.1 | 55 | 782 | 223.1 | 19.6 | D | D | D | D |
| Allen | 0.95 | 46.0 | 8 | 221 | 63.1 | 11.2 | 49 | 430 | 117.0 | 10.3 | 10 | 66 | 2.9 | 1.3 |
| Anderson | 2.21 | 100.6 | 9 | 67 | 23.5 | 2.0 | 59 | 843 | 237.8 | 20.1 | 16 | 66 | 6.8 | 1.2 |
| Ballard | 0.58 | 70.6 | D | D | D | 3.1 | 24 | 150 | 40.6 | 3.2 | D | D | D | D |
| Barren | 7.76 | 178.1 | D | D | D | D | 182 | 2,503 | 734.6 | 61.0 | 27 | 83 | 11.7 | 2.1 |
| Bath | 0.00 | 0.0 | 3 | 10 | 6.2 | 0.2 | 28 | 164 | 40.3 | 4.0 | D | D | D | D |
| Bell | 3.83 | 140.1 | 11 | 149 | 155.4 | 5.3 | 114 | 1,459 | 396.1 | 34.3 | 17 | 93 | 8.2 | 2.1 |
| Boone | 0.04 | 0.3 | 178 | 7,394 | 37,571.3 | 635.5 | 467 | 14,413 | 4,772.0 | 363.9 | 144 | 765 | 207.1 | 31.6 |
| Bourbon | 2.10 | 104.4 | 9 | 141 | 32.7 | 10.1 | 61 | 945 | 430.4 | 30.6 | 12 | 26 | 5.1 | 0.6 |
| Boyd | 11.09 | 229.5 | 63 | 682 | 583.7 | 30.4 | 223 | 3,505 | 855.4 | 74.9 | 43 | 257 | 56.7 | 11.8 |
| Boyle | 5.58 | 187.2 | 19 | 155 | 110.1 | 7.1 | 126 | 1,773 | 527.8 | 43.5 | 21 | 134 | 14.5 | 1.9 |
| Bracken | 0.65 | 78.1 | NA | NA | NA | NA | 19 | 135 | 28.1 | 2.4 | NA | NA | NA | NA |
| Breathitt | 1.41 | 104.6 | NA | NA | NA | NA | 40 | 437 | 112.3 | 10.5 | D | D | D | 0.8 |
| Breckinridge | 1.80 | 89.9 | 9 | D | 35.5 | D | 53 | 543 | 160.4 | 13.1 | D | D | D | D |
| Bullitt | 0.00 | 0.0 | 30 | 1,148 | 935.0 | 49.6 | 170 | 5,350 | 1,505.3 | 151.2 | 42 | 188 | 31.3 | 6.5 |
| Butler | 1.21 | 93.5 | D | D | D | D | 31 | 250 | 61.7 | 5.1 | 6 | D | 1.6 | D |
| Caldwell | 0.00 | 0.0 | D | D | D | D | 64 | 727 | 190.8 | 17.2 | 31 | 272 | 15.6 | 6.5 |
| Calloway | 3.47 | 90.5 | 29 | 328 | 193.1 | 13.9 | 160 | 1,978 | 610.5 | 47.3 | 83 | 613 | 144.1 | 46.4 |
| Campbell | 26.34 | 286.1 | 53 | 829 | 1,492.8 | 47.5 | 242 | 4,851 | 1,187.8 | 102.9 | 3 | 7 | 1.0 | 0.2 |
| Carlisle | 0.14 | 28.7 | D | D | D | 0.7 | 14 | 100 | 20.6 | 1.9 | | | | |
| Carroll | 1.87 | 174.8 | D | D | D | 1.7 | 41 | 608 | 177.1 | 15.3 | 6 | 31 | 9.3 | 1.3 |
| Carter | 4.09 | 150.6 | 9 | 280 | 46.0 | 7.2 | 102 | 939 | 256.1 | 19.2 | 18 | 32 | 5.6 | 1.0 |
| Casey | 1.01 | 63.9 | 13 | 96 | 64.3 | 3.1 | 57 | 396 | 104.0 | 8.5 | 3 | 8 | 0.9 | 0.1 |
| Christian | 12.19 | 166.3 | 65 | 743 | 804.0 | 35.5 | 247 | 3,183 | 927.2 | 78.3 | 62 | 334 | 57.4 | 15.3 |
| Clark | 6.90 | 193.0 | 36 | 627 | 823.1 | 30.7 | 121 | 1,582 | 480.1 | 40.6 | 30 | 109 | 24.4 | 3.7 |

1 Merchant wholesalers, except manufacturers' sales branches and offices.    2. Employer establishments.

# Table B. States and Counties — Professional Services, Manufacturing, and Accommodation and Food Services

| STATE County | Professional, scientific, and technical services, 2017 | | | | Manufacturing, 2017 | | | | Accommodation and food services, 2017 | | | |
|---|---|---|---|---|---|---|---|---|---|---|---|---|
| | Number of establishments | Number of employees | Sales (mil dol) | Average payroll (mil dol) | Number of establishments | Number of employees | Sales (mil dol) | Average payroll (mil dol) | Number of establishments | Number of employees | Sales (mil dol) | Annual payroll (mil dol) |
| | 147 | 148 | 149 | 150 | 151 | 152 | 153 | 154 | 155 | 156 | 157 | 158 |
| **KANSAS— Cont'd** | | | | | | | | | | | | |
| Neosho | 29 | 111 | 18 | 5.7 | 33 | 750 | 214.7 | 42.3 | D | D | D | D |
| Ness | D | D | D | 0.6 | 5 | 20 | 9.3 | 1.0 | 6 | D | 0.7 | D |
| Norton | 14 | 31 | 4 | 0.9 | 7 | 278 | 47.8 | 15.0 | 14 | 207 | 8.2 | 2.5 |
| Osage | 17 | 63 | 6 | 2.6 | D | D | D | D | 19 | 145 | 6.5 | 1.7 |
| Osborne | 7 | 27 | 2 | 0.7 | 5 | 113 | 21.1 | 4.8 | 9 | D | 1.3 | D |
| Ottawa | 13 | 37 | 7 | 1.9 | 5 | 36 | 5.0 | 1.6 | 10 | 37 | 1.4 | 0.4 |
| Pawnee | 16 | 79 | 7 | 2.6 | NA | NA | NA | NA | 13 | 161 | 9.2 | 2.4 |
| Phillips | D | D | 19 | D | 9 | 197 | 239.8 | 15.2 | 17 | 98 | 4.8 | 1.2 |
| Pottawatomie | D | D | D | D | 33 | 1,513 | 390.9 | 94.6 | 40 | 575 | 27.9 | 8.3 |
| Pratt | 30 | 115 | 13 | 4.6 | 8 | 99 | 117.6 | 5.3 | 40 | 457 | 18.9 | 4.9 |
| Rawlins | 9 | 14 | 1 | 0.3 | 6 | 25 | 8.2 | 1.6 | 5 | 34 | 1.0 | 0.3 |
| Reno | D | D | D | D | 89 | 3,499 | 1,152.4 | 169.4 | 132 | 2,578 | 110.4 | 33.4 |
| Republic | 9 | 38 | 5 | 1.1 | 8 | 153 | 46.3 | 5.8 | 7 | D | 2.8 | D |
| Rice | 15 | 143 | 13 | 2.8 | 16 | 399 | 298.3 | 22.6 | D | D | D | D |
| Riley | D | D | D | D | 26 | 490 | 135.6 | 19.2 | 177 | 4,427 | 177.6 | 54.4 |
| Rooks | 11 | 47 | 5 | 1.7 | 7 | 110 | 18.7 | 4.1 | 12 | D | 2.5 | D |
| Rush | D | D | D | D | 6 | 269 | 90.4 | 13.2 | 7 | D | 1.4 | D |
| Russell | 13 | 48 | 5 | 1.4 | D | 136 | D | 5.4 | 17 | 218 | 11.5 | 2.7 |
| Saline | D | D | D | D | 72 | 5,017 | 1,336.9 | 222.4 | 146 | 3,014 | 140.4 | 39.1 |
| Scott | 18 | 55 | 7 | 2.0 | 6 | 26 | 19.7 | 1.6 | 16 | 176 | 7.5 | 2.2 |
| Sedgwick | 1,189 | 11,563 | 1,920 | 716.3 | 501 | 41,132 | 16,434.6 | 2,920.7 | 1,177 | 24,300 | 1,166.6 | 346.1 |
| Seward | 28 | 110 | 22 | 4.5 | D | D | D | D | 53 | 891 | 43.8 | 12.5 |
| Shawnee | D | D | D | D | 112 | 6,691 | 3,344.3 | 353.8 | D | D | D | D |
| Sheridan | 4 | 24 | 3 | 0.6 | NA | NA | NA | NA | D | D | D | 0.3 |
| Sherman | 17 | 63 | 6 | 1.5 | 4 | 89 | 59.0 | 3.7 | 26 | 314 | 14.0 | 4.1 |
| Smith | 8 | 21 | 2 | 0.4 | 4 | 9 | 2.8 | 0.4 | D | D | D | 0.7 |
| Stafford | D | D | 2 | D | NA | NA | NA | NA | 9 | D | 1.9 | D |
| Stanton | 6 | 18 | 1 | 0.4 | NA | NA | NA | NA | 4 | 36 | 0.9 | 0.3 |
| Stevens | 8 | 27 | 3 | 0.8 | D | 62 | D | 2.9 | 15 | 128 | 6.0 | 1.5 |
| Sumner | 30 | 76 | 9 | 3.0 | 29 | 714 | 152.3 | 38.3 | 35 | 446 | 19.9 | 5.5 |
| Thomas | 24 | 56 | 6 | 2.1 | 11 | 73 | 27.0 | 3.2 | 30 | 472 | 21.5 | 5.4 |
| Trego | 7 | 11 | 1 | 0.3 | 7 | 34 | 4.9 | 1.3 | 12 | 79 | 2.9 | 0.7 |
| Wabaunsee | D | D | D | D | D | D | D | D | D | D | D | D |
| Wallace | D | D | D | D | NA | NA | NA | NA | NA | NA | NA | NA |
| Washington | 12 | 116 | 13 | 4.8 | 9 | 169 | 39.9 | 5.9 | D | D | D | D |
| Wichita | 4 | 7 | 0 | 0.1 | 4 | 54 | 21.2 | 2.2 | 3 | 11 | 0.6 | 0.2 |
| Wilson | 11 | 36 | 5 | 1.6 | 19 | 935 | 237.4 | 50.4 | D | D | D | D |
| Woodson | 5 | 8 | 1 | 0.1 | NA | NA | NA | NA | D | D | D | D |
| Wyandotte | D | D | D | D | 178 | 10,742 | 8,310.9 | 668.0 | 277 | 6,065 | 347.0 | 97.6 |
| **KENTUCKY** | 8,091 | 66,995 | 10,185 | 3,605.7 | 3,699 | 234,010 | 133,415.5 | 12,768.7 | 8,228 | 174,910 | 9,191.2 | 2,621.2 |
| Adair | 16 | 53 | 6 | 1.5 | 21 | 374 | 50.2 | 12.6 | D | D | D | D |
| Allen | 11 | 67 | 6 | 1.8 | 12 | D | 237.0 | D | 23 | 347 | 20.5 | 5.4 |
| Anderson | 29 | 106 | 10 | 3.2 | 23 | 1,029 | 825.3 | 58.6 | 32 | D | 24.5 | D |
| Ballard | 12 | 124 | 22 | 9.8 | D | 184 | D | 9.7 | D | D | D | 0.4 |
| Barren | 42 | 438 | 32 | 11.6 | 47 | 3,605 | 920.6 | 168.1 | 89 | 1,890 | 91.3 | 25.0 |
| Bath | D | D | D | D | D | D | D | D | D | D | D | D |
| Bell | 25 | 130 | 14 | 5.3 | 17 | 1,045 | 327.9 | 39.4 | 53 | 1,092 | 47.2 | 12.0 |
| Boone | 245 | 1,783 | 328 | 99.5 | 169 | 11,471 | 4,702.5 | 666.6 | 309 | 6,882 | 403.1 | 110.5 |
| Bourbon | 30 | 306 | 27 | 12.3 | 23 | 1,352 | 1,106.1 | 71.2 | D | D | D | D |
| Boyd | D | D | D | D | 35 | 1,825 | 6,621.9 | 157.1 | 129 | 3,117 | 147.7 | 41.4 |
| Boyle | 56 | 374 | 66 | 19.7 | 27 | 2,444 | 696.3 | 108.0 | 66 | 1,515 | 75.3 | 20.9 |
| Bracken | D | D | 3 | D | D | D | D | D | D | D | D | D |
| Breathitt | D | D | D | D | 4 | 19 | 3.7 | 0.9 | D | D | D | D |
| Breckinridge | 15 | 64 | 6 | 2.0 | 15 | 261 | 39.1 | 11.5 | 16 | 189 | 9.6 | 2.7 |
| Bullitt | 81 | 283 | 30 | 10.4 | 52 | 2,372 | 1,368.6 | 123.7 | 96 | 2,066 | 109.6 | 29.6 |
| Butler | 9 | 39 | 3 | 1.0 | D | 591 | D | 26.6 | D | D | D | D |
| Caldwell | 23 | 47 | 5 | 1.2 | 16 | 848 | 461.5 | 37.5 | 31 | 409 | 17.6 | 4.7 |
| Calloway | 63 | 335 | 36 | 12.1 | 25 | 2,457 | 1,138.8 | 106.3 | D | D | D | D |
| Campbell | D | D | D | D | 72 | 2,199 | 771.0 | 112.8 | 223 | 4,585 | 231.8 | 67.0 |
| Carlisle | 4 | 12 | 1 | 0.3 | 5 | 67 | 12.1 | 1.8 | 8 | 43 | 1.6 | 0.4 |
| Carroll | 15 | 88 | 9 | 4.7 | 17 | 2,612 | 3,854.4 | 212.1 | 26 | 515 | 30.9 | 7.8 |
| Carter | 25 | 111 | 9 | 2.9 | D | 620 | D | 23.9 | 41 | 726 | 32.7 | 9.2 |
| Casey | D | D | 10 | D | 30 | 954 | 257.9 | 36.5 | 15 | 217 | 9.1 | 3.3 |
| Christian | D | D | D | D | 82 | 6,271 | 2,447.7 | 298.4 | 121 | 2,460 | 120.7 | 34.9 |
| Clark | 48 | 1,188 | 174 | 30.2 | 41 | 2,854 | 1,172.4 | 140.5 | 64 | 1,685 | 64.9 | 18.4 |

# Table B. States and Counties — Health Care and Social Assistance, Other Services, Nonemployer Businesses, and Residential Construction

| STATE County | Health care and social assistance, 2017 | | | | Other services, 2017 | | | | Nonemployer businesses, 2018 | | Value of residential construction authorized by building permits, 2020 | |
|---|---|---|---|---|---|---|---|---|---|---|---|---|
| | Number of establishments | Number of employees | Receipts (mil dol) | Annual payroll (mil dol) | Number of establishments | Number of employees | Receipts (mil dol) | Annual payroll (mil dol) | Number | Receipts (mil dol) | New construction ($1,000) | Number of housing units |
| | 159 | 160 | 161 | 162 | 163 | 164 | 165 | 166 | 167 | 168 | 169 | 170 |
| **KANSAS— Cont'd** | | | | | | | | | | | | |
| Neosho | 50 | 1,151 | 112.7 | 42.3 | 26 | 76 | 6.3 | 1.3 | 1,110 | 39.0 | 1,292 | 10 |
| Ness | 10 | D | 20.4 | D | D | D | D | 1.5 | 300 | 14.9 | 0 | 0 |
| Norton | 22 | 469 | 31.1 | 15.6 | D | D | D | D | 422 | 17.9 | 0 | 0 |
| Osage | 28 | 420 | 23.0 | 10.6 | 16 | 44 | 5.7 | 1.2 | 1,032 | 43.9 | 5,730 | 46 |
| Osborne | 13 | 263 | 13.5 | 6.6 | D | D | 5.5 | D | 342 | 11.6 | 383 | 2 |
| Ottawa | 12 | 254 | 15.0 | 6.9 | 14 | 34 | 3.2 | 0.8 | 452 | 19.3 | 751 | 9 |
| Pawnee | 18 | 1,055 | 99.9 | 45.8 | 14 | 29 | 3.7 | 0.5 | 407 | 19.6 | 75 | 1 |
| Phillips | 16 | 274 | 21.7 | 9.0 | 26 | 72 | 47.3 | 2.1 | 1,854 | 81.0 | 238 | 1 |
| Pottawatomie | 58 | 1,534 | 102.6 | 39.5 | 50 | 159 | 17.6 | 4.2 | 817 | 37.8 | 46,418 | 164 |
| Pratt | 35 | 696 | 87.4 | 37.5 | 31 | 103 | 10.3 | 2.9 | 264 | 9.8 | 0 | 0 |
| Rawlins | 10 | 173 | 13.4 | 5.5 | D | D | 1.1 | D | 3,775 | 145.0 | 0 | 0 |
| Reno | 167 | 4,304 | 473.0 | 182.7 | 109 | 503 | 57.7 | 12.8 | 441 | 15.3 | 9,402 | 39 |
| Republic | 18 | 306 | 23.3 | 10.1 | 13 | 44 | 4.1 | 1.1 | 650 | 23.1 | 0 | 0 |
| Rice | 21 | 351 | 20.9 | 8.5 | 18 | 50 | 7.6 | 1.3 | 3,588 | 151.2 | 1,563 | 7 |
| Riley | 215 | 3,673 | 389.9 | 137.5 | 119 | 1,108 | 293.5 | 48.3 | 537 | 19.8 | 45,607 | 225 |
| Rooks | 13 | 270 | 22.6 | 9.5 | 13 | 35 | 5.4 | 0.8 | 254 | 9.5 | 175 | 1 |
| Rush | 6 | 156 | 8.9 | 4.8 | D | D | D | D | 836 | 41.7 | 0 | 0 |
| Russell | 15 | 368 | 27.4 | 13.2 | D | D | D | D | 3,446 | 165.1 | 460 | 4 |
| Saline | 156 | 4,993 | 563.9 | 230.0 | 108 | 581 | 105.3 | 16.9 | 452 | 16.9 | 10,677 | 61 |
| Scott | 16 | D | 35.7 | D | 18 | 47 | 9.5 | 1.7 | 33,067 | 1,580.7 | 350 | 2 |
| Sedgwick | 1,493 | 37,305 | 4,197.1 | 1,678.6 | 779 | 5,283 | 730.5 | 165.4 | 1,298 | 89.3 | 377,851 | 1,637 |
| Seward | 70 | 1,064 | 112.6 | 40.2 | 46 | 172 | 20.5 | 4.3 | 9,619 | 441.0 | 2,440 | 21 |
| Shawnee | 492 | 16,848 | 1,984.7 | 832.2 | 349 | 2,381 | 325.8 | 94.5 | 267 | 13.4 | 73,836 | 385 |
| Sheridan | D | D | D | D | 8 | 27 | 1.9 | 0.5 | 476 | 20.6 | NA | NA |
| Sherman | 32 | 417 | 34.7 | 15.1 | 22 | 78 | 8.5 | 2.7 | 329 | 11.1 | 0 | 0 |
| Smith | 8 | 228 | 21.0 | 7.6 | D | D | 2.3 | 0.7 | 392 | 16.4 | 0 | 0 |
| Stafford | 15 | 178 | 13.4 | 4.6 | 8 | 18 | D | 0.4 | 200 | 10.5 | 1,999 | 11 |
| Stanton | D | D | D | D | D | D | 4.8 | D | 384 | 19.4 | 0 | 0 |
| Stevens | D | D | D | D | 11 | 26 | 11.1 | 1.0 | 1,437 | 49.6 | 700 | 5 |
| Sumner | 54 | 959 | 64.1 | 26.9 | 38 | 108 | 7.1 | 2.9 | 793 | 29.8 | 3,607 | 29 |
| Thomas | 24 | 415 | 45.1 | 18.4 | 24 | 70 | 2.3 | 2.0 | 306 | 13.5 | 500 | 1 |
| Trego | D | D | D | D | 10 | 20 | D | 0.5 | 478 | 18.7 | 211 | 1 |
| Wabaunsee | 9 | 114 | 5.0 | 2.4 | D | D | 0.7 | D | 166 | 6.1 | 2,615 | 11 |
| Wallace | 4 | 59 | 2.6 | 1.1 | D | D | 6.1 | D | 460 | 20.7 | 100 | 1 |
| Washington | 22 | 334 | 19.1 | 9.2 | D | D | D | D | 183 | 7.4 | 238 | 1 |
| Wichita | D | D | D | D | D | D | 10.7 | 0.3 | 606 | 23.6 | 0 | 0 |
| Wilson | 35 | 678 | 50.6 | 21.9 | D | D | 0.5 | D | 261 | 13.1 | 2,406 | 13 |
| Woodson | 4 | D | 4.0 | D | 7 | 7 | 364.3 | 0.1 | 8,246 | 340.9 | 1,028 | 8 |
| Wyandotte | 297 | 16,184 | 1,915.6 | 843.2 | 206 | 1,447 | | 50.6 | | | 38,912 | 212 |
| **KENTUCKY** | 11,597 | 268,711 | 32,369.5 | 12,412.1 | 5,900 | 38,040 | 4,755.2 | 1,198.2 | 295,814 | 13,198.6 | 2,142,572 | 11,281 |
| Adair | 38 | 500 | 46.0 | 17.6 | D | D | D | D | 1,623 | 68.8 | 0 | 0 |
| Allen | D | D | D | D | 18 | 43 | 4.6 | 1.3 | 1,504 | 66.6 | 900 | 8 |
| Anderson | 35 | 359 | 24.3 | 10.4 | 29 | 100 | 15.5 | 2.9 | 1,526 | 52.5 | 14,814 | 92 |
| Ballard | D | D | D | D | D | D | D | D | 476 | 13.0 | NA | NA |
| Barren | 97 | 2,397 | 295.5 | 110.1 | 64 | 249 | 21.6 | 5.9 | 3,441 | 152.3 | 42,535 | 246 |
| Bath | 8 | 101 | 8.3 | 3.6 | D | D | D | D | 810 | 28.3 | 0 | 0 |
| Bell | 68 | 1,363 | 145.1 | 51.2 | 28 | 108 | 10.9 | 2.6 | 1,310 | 42.8 | 4,027 | 38 |
| Boone | 281 | 4,843 | 516.7 | 209.7 | 194 | 1,780 | 164.1 | 55.5 | 8,189 | 371.1 | 132,047 | 817 |
| Bourbon | 52 | 742 | 61.8 | 24.2 | 21 | 101 | 7.2 | 1.9 | 1,311 | 65.9 | 11,668 | 53 |
| Boyd | 195 | 7,103 | 812.7 | 348.7 | 83 | 659 | 60.5 | 24.6 | 2,390 | 100.3 | 519 | 3 |
| Boyle | 137 | 2,692 | 336.5 | 113.1 | 53 | 266 | 26.6 | 6.8 | 2,047 | 76.4 | 7,370 | 52 |
| Bracken | 7 | 129 | 11.3 | 4.6 | D | D | D | 0.8 | 554 | 17.7 | NA | NA |
| Breathitt | 47 | 737 | 82.6 | 28.8 | D | D | D | D | 583 | 17.1 | 209 | 1 |
| Breckinridge | 31 | 540 | 46.7 | 18.2 | D | D | D | D | 1,317 | 59.2 | 270 | 2 |
| Bullitt | 115 | 1,441 | 112.9 | 50.5 | 96 | 1,187 | 62.5 | 36.6 | 4,562 | 195.4 | 121,297 | 463 |
| Butler | D | D | D | D | 14 | 75 | 6.7 | 1.8 | 812 | 34.5 | 0 | 0 |
| Caldwell | 38 | 532 | 38.9 | 15.2 | D | D | D | D | 728 | 29.9 | 1,464 | 7 |
| Calloway | 99 | 1,730 | 185.4 | 78.1 | 50 | 214 | 17.5 | 5.1 | 2,478 | 107.5 | 5,385 | 51 |
| Campbell | 170 | 3,607 | 368.9 | 150.0 | 121 | 988 | 138.0 | 34.6 | 5,703 | 230.8 | 79,225 | 235 |
| Carlisle | 8 | 79 | 5.8 | 2.2 | 4 | 11 | 0.9 | 0.4 | 416 | 15.5 | NA | NA |
| Carroll | 24 | 411 | 37.4 | 16.3 | 15 | 70 | 7.6 | 1.9 | 498 | 21.0 | 285 | 9 |
| Carter | 34 | 592 | 42.4 | 16.6 | 30 | 106 | 13.4 | 2.3 | 1,650 | 65.8 | 400 | 7 |
| Casey | D | D | D | D | D | D | D | 0.7 | 1,234 | 57.9 | 0 | 0 |
| Christian | 153 | 3,457 | 373.9 | 137.6 | 93 | 472 | 44.3 | 12.8 | 3,690 | 157.9 | 9,074 | 88 |
| Clark | 118 | 1,642 | 182.5 | 69.5 | 49 | 216 | 20.6 | 5.5 | 2,240 | 94.8 | 23,528 | 139 |

# Table B. States and Counties — Government Employment and Payroll, and Local Government Finances

| STATE County | Full-time equivalent employees | March payroll (dollars) | Adminis-tration, judicial, and legal | Police and corrections | Fire protection | Highways and transpor-tation | Health and welfare | Natural resources and utilities | Education and libraries | Total (mil dol) | Inter-govern-mental (mil dol) | Total (mil dol) | Per capita[1] Total | Per capita[1] Property |
|---|---|---|---|---|---|---|---|---|---|---|---|---|---|---|
| | 171 | 172 | 173 | 174 | 175 | 176 | 177 | 178 | 179 | 180 | 181 | 182 | 183 | 184 |
| **KANSAS— Cont'd** | | | | | | | | | | | | | | |
| Neosho | 1,297 | 4,801,938 | 3.0 | 4.4 | 1.4 | 2.8 | 39.4 | 8.0 | 38.6 | 120.4 | 29.8 | 22.5 | 1,398 | 1,037 |
| Ness | 353 | 1,117,456 | 5.7 | 3.8 | 0.0 | 7.3 | 52.0 | 1.9 | 27.7 | 27.8 | 5.2 | 9.7 | 3,409 | 3,248 |
| Norton | 415 | 1,502,007 | 19.4 | 1.7 | 0.1 | 1.6 | 37.4 | 5.4 | 34.1 | 23.9 | 9.5 | 10.5 | 1,931 | 1,643 |
| Osage | 681 | 2,328,841 | 6.6 | 8.3 | 0.2 | 4.4 | 1.4 | 7.7 | 69.7 | 54.7 | 29.6 | 19.3 | 1,221 | 1,010 |
| Osborne | 190 | 579,543 | 9.9 | 8.7 | 0.0 | 9.0 | 5.6 | 14.1 | 49.4 | 13.5 | 3.7 | 7.3 | 2,062 | 1,718 |
| Ottawa | 360 | 1,026,767 | 7.3 | 5.9 | 0.4 | 7.2 | 4.4 | 6.2 | 65.2 | 25.8 | 13.4 | 10.8 | 1,860 | 1,781 |
| Pawnee | 516 | 1,671,513 | 7.5 | 7.9 | 0.1 | 5.2 | 3.4 | 6.3 | 67.1 | 31.4 | 12.1 | 15.7 | 2,350 | 1,938 |
| Phillips | 490 | 1,755,399 | 5.9 | 2.7 | 0.0 | 4.5 | 4.5 | 4.9 | 76.1 | 24.5 | 10.8 | 9.7 | 1,808 | 1,558 |
| Pottawatomie | 1,086 | 3,349,875 | 6.0 | 8.3 | 0.6 | 5.4 | 1.3 | 6.0 | 69.0 | 87.2 | 35.4 | 37.4 | 1,564 | 1,300 |
| Pratt | 1,145 | 2,943,117 | 4.5 | 6.0 | 0.7 | 4.6 | 3.1 | 7.3 | 72.8 | 57.5 | 20.2 | 26.6 | 2,797 | 2,290 |
| Rawlins | 141 | 412,477 | 14.2 | 8.8 | 5.6 | 13.6 | 4.3 | 5.8 | 47.7 | 10.7 | 4.0 | 5.3 | 2,140 | 1,677 |
| Reno | 3,405 | 11,880,654 | 4.4 | 9.9 | 3.6 | 4.2 | 2.2 | 4.6 | 69.6 | 296.9 | 118.3 | 123.3 | 1,966 | 1,387 |
| Republic | 265 | 785,802 | 8.5 | 9.5 | 0.1 | 12.9 | 6.3 | 14.0 | 44.1 | 25.5 | 9.6 | 12.5 | 2,677 | 2,391 |
| Rice | 713 | 2,394,086 | 5.6 | 6.4 | 1.8 | 4.6 | 26.0 | 3.9 | 51.0 | 60.3 | 22.9 | 18.6 | 1,940 | 1,739 |
| Riley | 2,444 | 8,630,897 | 9.2 | 12.1 | 6.1 | 4.6 | 2.9 | 8.6 | 55.7 | 251.1 | 74.8 | 132.7 | 1,794 | 1,119 |
| Rooks | 436 | 1,399,610 | 6.4 | 5.4 | 0.0 | 5.9 | 41.2 | 4.0 | 35.9 | 45.0 | 12.5 | 13.0 | 2,566 | 2,272 |
| Rush | 244 | 772,126 | 6.0 | 3.8 | 0.0 | 9.0 | 27.1 | 7.8 | 42.9 | 20.1 | 5.5 | 6.9 | 2,264 | 2,211 |
| Russell | 432 | 1,355,647 | 10.1 | 9.1 | 3.0 | 10.6 | 6.4 | 15.1 | 45.0 | 31.7 | 11.3 | 14.2 | 2,055 | 1,922 |
| Saline | 2,668 | 9,840,370 | 4.9 | 9.6 | 5.0 | 4.2 | 2.8 | 5.9 | 65.7 | 234.4 | 97.0 | 98.0 | 1,798 | 1,108 |
| Scott | 257 | 752,664 | 11.3 | 9.5 | 0.2 | 6.8 | 2.9 | 5.8 | 62.9 | 25.2 | 7.8 | 14.9 | 3,016 | 2,370 |
| Sedgwick | 19,778 | 74,886,985 | 6.5 | 11.9 | 4.8 | 3.1 | 4.6 | 5.1 | 62.0 | 2,058.2 | 954.7 | 660.8 | 1,288 | 924 |
| Seward | 1,967 | 6,874,718 | 4.4 | 5.1 | 1.6 | 2.1 | 27.3 | 5.1 | 52.2 | 184.4 | 73.7 | 35.4 | 1,590 | 1,087 |
| Shawnee | 8,734 | 32,886,735 | 5.1 | 11.3 | 4.8 | 2.8 | 2.4 | 5.2 | 67.6 | 788.7 | 323.9 | 316.0 | 1,775 | 1,198 |
| Sheridan | 269 | 807,205 | 4.8 | 5.0 | 0.0 | 4.8 | 59.3 | 3.4 | 22.2 | 24.4 | 4.7 | 7.9 | 3,111 | 2,610 |
| Sherman | 565 | 1,972,655 | 6.5 | 4.0 | 0.8 | 4.7 | 35.9 | 5.0 | 38.9 | 31.2 | 13.1 | 11.3 | 1,907 | 1,503 |
| Smith | 261 | 796,134 | 12.9 | 3.6 | 4.6 | 10.1 | 4.6 | 7.7 | 52.6 | 25.6 | 7.9 | 14.9 | 4,114 | 3,559 |
| Stafford | 275 | 1,012,071 | 11.0 | 5.7 | 0.6 | 9.8 | 3.4 | 4.8 | 61.9 | 29.2 | 9.8 | 12.2 | 2,918 | 2,649 |
| Stanton | 250 | 919,988 | 4.9 | 3.7 | 0.0 | 6.2 | 42.7 | 5.0 | 28.0 | 16.4 | 4.4 | 10.5 | 5,109 | 4,866 |
| Stevens | 516 | 1,791,369 | 4.3 | 5.4 | 0.3 | 5.7 | 39.0 | 5.0 | 38.1 | 57.8 | 12.3 | 22.6 | 4,075 | 3,789 |
| Sumner | 1,337 | 4,256,344 | 5.8 | 8.0 | 2.2 | 5.5 | 17.1 | 5.8 | 52.4 | 102.9 | 42.7 | 37.3 | 1,616 | 1,280 |
| Thomas | 574 | 1,857,168 | 17.9 | 3.7 | 1.2 | 2.5 | 1.0 | 5.1 | 66.2 | 52.2 | 14.5 | 22.7 | 2,906 | 2,208 |
| Trego | 348 | 1,157,339 | 7.9 | 4.7 | 0.2 | 5.2 | 59.5 | 2.8 | 18.1 | 13.4 | 3.6 | 7.9 | 2,780 | 2,419 |
| Wabaunsee | 273 | 775,032 | 17.0 | 9.0 | 0.0 | 6.6 | 2.3 | 7.5 | 55.7 | 24.7 | 10.1 | 11.3 | 1,652 | 1,561 |
| Wallace | 107 | 308,513 | 8.6 | 10.2 | 0.0 | 10.2 | 1.5 | 3.9 | 61.4 | 9.0 | 3.0 | 5.2 | 3,437 | 3,356 |
| Washington | 357 | 1,093,990 | 6.7 | 5.7 | 0.0 | 7.6 | 19.8 | 5.9 | 51.0 | 30.7 | 10.5 | 13.4 | 2,447 | 2,288 |
| Wichita | 254 | 742,160 | 4.0 | 3.4 | 0.1 | 6.1 | 47.2 | 4.4 | 34.5 | 19.2 | 3.8 | 7.7 | 3,612 | 3,096 |
| Wilson | 645 | 2,126,416 | 4.4 | 8.6 | 0.7 | 3.5 | 31.0 | 6.7 | 42.2 | 45.5 | 17.4 | 13.3 | 1,529 | 1,306 |
| Woodson | 165 | 496,876 | 11.0 | 8.9 | 0.3 | 9.4 | 5.4 | 8.1 | 50.4 | 12.9 | 5.4 | 6.4 | 2,036 | 1,771 |
| Wyandotte | 8,721 | 39,159,416 | 3.6 | 11.1 | 9.4 | 2.5 | 2.8 | 10.8 | 56.5 | 853.3 | 419.3 | 287.6 | 1,742 | 920 |
| **KENTUCKY** | X | X | X | X | X | X | X | X | X | X | X | X | X | X |
| Adair | 672 | 1,911,619 | 4.8 | 5.8 | 0.0 | 1.5 | 2.9 | 6.2 | 77.6 | 54.2 | 23.9 | 10.4 | 540 | 368 |
| Allen | 694 | 1,911,037 | 2.2 | 11.0 | 0.0 | 1.2 | 7.3 | 6.9 | 70.6 | 44.3 | 25.8 | 13.2 | 632 | 375 |
| Anderson | 708 | 2,270,158 | 6.2 | 6.3 | 0.2 | 2.1 | 4.7 | 4.9 | 74.8 | 51.4 | 23.5 | 20.9 | 926 | 700 |
| Ballard | 312 | 902,100 | 4.7 | 11.1 | 0.1 | 2.5 | 4.8 | 5.1 | 71.2 | 22.5 | 12.0 | 7.6 | 953 | 722 |
| Barren | 1,798 | 5,562,968 | 2.0 | 6.1 | 2.2 | 1.4 | 1.3 | 13.4 | 71.2 | 131.2 | 71.1 | 41.7 | 953 | 579 |
| Bath | 440 | 1,283,556 | 6.7 | 2.8 | 0.0 | 1.9 | 3.4 | 5.3 | 78.8 | 28.4 | 18.3 | 6.3 | 512 | 324 |
| Bell | 1,028 | 2,854,698 | 2.9 | 7.9 | 4.4 | 2.8 | 3.1 | 1.1 | 76.8 | 74.4 | 45.7 | 18.1 | 672 | 359 |
| Boone | 4,012 | 15,531,882 | 2.2 | 9.7 | 7.7 | 1.5 | 0.6 | 2.8 | 72.6 | 382.2 | 117.4 | 238.7 | 1,823 | 1,071 |
| Bourbon | 1,007 | 2,702,469 | 4.9 | 5.0 | 4.8 | 3.2 | 1.9 | 5.5 | 73.4 | 55.8 | 26.6 | 22.9 | 1,142 | 634 |
| Boyd | 2,240 | 7,334,452 | 3.2 | 9.5 | 7.5 | 4.4 | 4.9 | 10.1 | 55.5 | 157.3 | 71.4 | 63.2 | 1,322 | 685 |
| Boyle | 1,169 | 4,242,008 | 3.9 | 8.1 | 3.3 | 1.6 | 4.1 | 3.0 | 74.6 | 88.9 | 34.1 | 43.5 | 1,451 | 796 |
| Bracken | 360 | 1,060,350 | 8.3 | 8.2 | 0.0 | 2.4 | 5.1 | 6.1 | 67.1 | 21.0 | 12.3 | 6.5 | 785 | 558 |
| Breathitt | 517 | 1,741,469 | 4.8 | 4.6 | 1.3 | 5.3 | 10.1 | 2.6 | 70.0 | 39.1 | 25.6 | 8.0 | 617 | 288 |
| Breckinridge | 646 | 1,872,417 | 5.7 | 10.5 | 0.0 | 2.0 | 2.2 | 1.8 | 77.2 | 43.7 | 27.1 | 12.8 | 635 | 471 |
| Bullitt | 2,632 | 8,353,177 | 2.0 | 9.0 | 4.0 | 1.2 | 3.4 | 1.9 | 78.0 | 183.5 | 74.3 | 87.7 | 1,093 | 786 |
| Butler | 467 | 1,484,940 | 3.0 | 5.3 | 0.0 | 5.4 | 9.3 | 2.9 | 72.8 | 22.8 | 16.3 | 5.4 | 423 | 206 |
| Caldwell | 446 | 1,290,394 | 6.2 | 6.0 | 2.9 | 2.7 | 10.4 | 2.8 | 68.5 | 30.8 | 17.0 | 9.8 | 771 | 288 |
| Calloway | 2,192 | 7,589,227 | 2.2 | 4.3 | 2.0 | 1.6 | 52.2 | 1.9 | 35.3 | 186.4 | 37.3 | 28.7 | 739 | 533 |
| Campbell | 2,556 | 9,933,993 | 5.0 | 14.6 | 6.9 | 2.6 | 2.0 | 0.7 | 67.3 | 254.5 | 81.5 | 143.6 | 1,546 | 965 |
| Carlisle | 196 | 529,702 | 7.0 | 2.7 | 0.0 | 3.8 | 0.0 | 3.5 | 75.2 | 38.9 | 7.8 | 3.3 | 689 | 533 |
| Carroll | 489 | 1,542,741 | 6.6 | 11.6 | 0.9 | 1.2 | 2.7 | 4.8 | 71.7 | 46.3 | 16.1 | 13.4 | 1,254 | 592 |
| Carter | 1,032 | 2,802,794 | 2.1 | 4.3 | 0.0 | 1.9 | 10.7 | 4.2 | 74.9 | 55.4 | 39.3 | 11.6 | 425 | 218 |
| Casey | 766 | 2,410,352 | 2.4 | 5.7 | 0.0 | 1.1 | 37.6 | 2.6 | 50.4 | 54.0 | 25.0 | 7.3 | 462 | 323 |
| Christian | 1,508 | 5,687,041 | 3.4 | 11.0 | 6.8 | 1.4 | 4.6 | 12.3 | 51.9 | 147.9 | 73.5 | 61.5 | 866 | 415 |
| Clark | 1,224 | 4,097,901 | 3.7 | 9.0 | 9.0 | 1.9 | 5.8 | 4.5 | 63.3 | 105.5 | 39.2 | 43.9 | 1,224 | 674 |

1. Based on the resident population estimated as of July 1 of the year shown.

# Table B. States and Counties — Local Government Finances, Government Employment, and Income Taxes

| STATE County | Direct general expenditure — Total (mil dol) [185] | Per capita¹ (dollars) [186] | Percent of total for: Education [187] | Health and hospitals [188] | Police protection [189] | Public welfare [190] | Highways [191] | Debt outstanding — Total (mil dol) [192] | Per capita¹ (dollars) [193] | Gov't employment 2019 — Federal civilian [194] | Federal military [195] | State and local [196] | Income tax returns 2018 — Number of returns [197] | Mean adjusted gross income [198] | Mean income tax [199] |
|---|---|---|---|---|---|---|---|---|---|---|---|---|---|---|---|
| **KANSAS— Cont'd** | | | | | | | | | | | | | | | |
| Neosho | 136 | 8,431 | 30.4 | 44.7 | 2.1 | 0.0 | 2.9 | 174.8 | 10,859 | 55 | 58 | 1,778 | 6,960 | 47,792 | 3,747 |
| Ness | 28 | 9,748 | 22.0 | 55.5 | 2.5 | 0.0 | 8.0 | 1.2 | 436 | 28 | 10 | 425 | 1,350 | 50,771 | 4,290 |
| Norton | 32 | 5,808 | 54.4 | 4.8 | 3.6 | 0.0 | 5.7 | 17.9 | 3,305 | 23 | 17 | 765 | 2,250 | 49,741 | 4,484 |
| Osage | 53 | 3,319 | 59.3 | 1.2 | 5.9 | 0.0 | 10.5 | 43.2 | 2,726 | 74 | 59 | 1,133 | 7,400 | 51,298 | 3,776 |
| Osborne | 14 | 3,899 | 29.1 | 8.8 | 8.0 | 0.0 | 14.8 | 2.6 | 732 | 28 | 12 | 337 | 1,700 | 44,041 | 3,717 |
| Ottawa | 27 | 4,706 | 66.9 | 2.4 | 4.0 | 0.0 | 9.7 | 40.0 | 6,884 | 23 | 21 | 453 | 2,730 | 48,516 | 3,715 |
| Pawnee | 53 | 7,979 | 63.4 | 4.2 | 2.4 | 0.0 | 7.2 | 41.0 | 6,151 | 38 | 20 | 1,449 | 2,740 | 49,516 | 4,201 |
| Phillips | 31 | 5,698 | 54.1 | 3.3 | 3.3 | 0.0 | 11.1 | 9.3 | 1,737 | 36 | 19 | 878 | 2,530 | 44,989 | 3,424 |
| Pottawatomie | 87 | 3,618 | 58.9 | 7.0 | 3.7 | 0.6 | 8.6 | 147.3 | 6,159 | 58 | 94 | 1,353 | 11,060 | 65,287 | 5,758 |
| Pratt | 67 | 7,047 | 63.8 | 3.1 | 4.4 | 0.0 | 7.7 | 24.2 | 2,543 | 35 | 33 | 1,062 | 4,240 | 56,108 | 5,162 |
| Rawlins | 12 | 4,756 | 32.7 | 5.4 | 4.0 | 0.0 | 11.8 | 11.6 | 4,707 | 19 | 9 | 320 | 1,250 | 50,990 | 5,251 |
| Reno | 328 | 5,234 | 53.1 | 1.7 | 4.2 | 0.0 | 6.6 | 329.2 | 5,250 | 168 | 219 | 5,164 | 27,790 | 53,339 | 4,785 |
| Republic | 22 | 4,702 | 40.2 | 6.4 | 4.2 | 0.0 | 10.7 | 8.5 | 1,834 | 31 | 17 | 451 | 2,340 | 44,381 | 3,437 |
| Rice | 57 | 5,992 | 46.8 | 27.2 | 2.9 | 0.0 | 5.2 | 67.1 | 7,018 | 39 | 33 | 1,034 | 4,100 | 51,315 | 3,936 |
| Riley | 227 | 3,072 | 37.1 | 2.4 | 15.6 | 0.0 | 7.5 | 516.9 | 6,988 | 519 | 268 | 13,745 | 25,700 | 60,217 | 6,052 |
| Rooks | 46 | 9,020 | 22.2 | 34.9 | 3.0 | 0.0 | 6.1 | 23.1 | 4,574 | 32 | 18 | 672 | 2,370 | 42,694 | 3,564 |
| Rush | 20 | 6,524 | 33.5 | 32.7 | 0.0 | 0.0 | 10.4 | 8.1 | 2,646 | 23 | 11 | 353 | 1,470 | 45,083 | 3,286 |
| Russell | 36 | 5,216 | 31.8 | 5.9 | 3.0 | 0.0 | 13.5 | 15.4 | 2,222 | 31 | 25 | 635 | 3,150 | 45,582 | 3,603 |
| Saline | 237 | 4,349 | 53.0 | 1.8 | 5.7 | 0.0 | 6.6 | 274.4 | 5,034 | 230 | 196 | 4,214 | 26,310 | 57,214 | 5,502 |
| Scott | 24 | 4,813 | 41.2 | 4.0 | 6.7 | 2.7 | 8.8 | 54.4 | 11,043 | 25 | 18 | 611 | 2,310 | 69,827 | 7,939 |
| Sedgwick | 2,037 | 3,968 | 48.5 | 3.4 | 6.2 | 0.1 | 6.6 | 4,317.8 | 8,414 | 4,929 | 4,776 | 28,590 | 237,760 | 66,501 | 7,427 |
| Seward | 235 | 10,572 | 52.2 | 19.8 | 1.9 | 0.2 | 2.7 | 168.6 | 7,580 | 93 | 78 | 2,402 | 9,350 | 47,110 | 3,450 |
| Shawnee | 852 | 4,787 | 51.0 | 1.6 | 6.5 | 0.1 | 6.8 | 1,169.8 | 6,570 | 3,427 | 739 | 18,698 | 83,640 | 60,820 | 6,392 |
| Sheridan | 22 | 8,694 | 18.7 | 51.2 | 1.2 | 0.0 | 8.6 | 1.9 | 753 | 14 | 9 | 393 | 1,260 | 43,098 | 4,748 |
| Sherman | 34 | 5,713 | 54.8 | 5.5 | 3.6 | 0.1 | 6.1 | 24.6 | 4,136 | 51 | 22 | 720 | 2,660 | 49,026 | 3,976 |
| Smith | 21 | 5,932 | 40.3 | 2.7 | 3.5 | 0.0 | 13.0 | 8.2 | 2,282 | 34 | 13 | 354 | 1,800 | 42,086 | 3,213 |
| Stafford | 30 | 7,101 | 43.3 | 4.1 | 3.5 | 0.0 | 12.8 | 1.4 | 328 | 46 | 15 | 479 | 1,910 | 43,619 | 3,407 |
| Stanton | 15 | 7,042 | 39.6 | 1.4 | 1.6 | 0.0 | 5.7 | 14.4 | 6,991 | 8 | 7 | 334 | 940 | 66,205 | 6,618 |
| Stevens | 52 | 9,307 | 28.3 | 34.6 | 4.1 | 0.0 | 9.8 | 17.6 | 3,168 | 22 | 20 | 698 | 2,240 | 59,037 | 5,470 |
| Sumner | 110 | 4,757 | 52.6 | 11.7 | 3.6 | 0.0 | 7.1 | 101.6 | 4,394 | 73 | 83 | 1,901 | 10,170 | 54,877 | 4,613 |
| Thomas | 47 | 6,030 | 60.4 | 1.8 | 3.3 | 0.0 | 7.9 | 29.9 | 3,817 | 30 | 28 | 794 | 3,710 | 54,969 | 5,240 |
| Trego | 13 | 4,675 | 34.8 | 5.5 | 2.9 | 0.6 | 15.0 | 13.0 | 4,558 | 19 | 10 | 527 | 1,460 | 46,997 | 3,595 |
| Wabaunsee | 25 | 3,676 | 47.9 | 1.3 | 4.6 | 0.0 | 14.1 | 21.6 | 3,159 | 19 | 25 | 456 | 3,190 | 58,044 | 4,858 |
| Wallace | 10 | 6,800 | 39.6 | 2.9 | 3.8 | 2.4 | 13.6 | 4.6 | 3,036 | 11 | 6 | 209 | 720 | 49,622 | 3,931 |
| Washington | 30 | 5,408 | 44.0 | 18.2 | 2.3 | 0.0 | 10.8 | 11.0 | 2,007 | 46 | 20 | 709 | 2,760 | 46,482 | 3,751 |
| Wichita | 20 | 9,384 | 26.4 | 37.0 | 3.3 | 0.0 | 7.5 | 5.0 | 2,341 | 16 | 8 | 271 | 1,090 | 53,583 | 4,852 |
| Wilson | 47 | 5,384 | 42.2 | 27.4 | 5.6 | 0.0 | 5.2 | 50.4 | 5,790 | 36 | 31 | 945 | 3,750 | 46,246 | 3,437 |
| Woodson | 13 | 4,240 | 44.4 | 2.4 | 6.6 | 0.0 | 13.8 | 2.3 | 725 | 14 | 12 | 286 | 1,380 | 43,389 | 3,192 |
| Wyandotte | 891 | 5,399 | 48.2 | 2.6 | 6.3 | 0.0 | 1.6 | 2,776.8 | 16,820 | 1,293 | 610 | 14,681 | 71,640 | 43,445 | 3,160 |
| **KENTUCKY** | X | X | X | X | X | X | X | X | X | 34,918 | 45,355 | 273,713 | 1,921,210 | 59,764 | 6,200 |
| Adair | 49 | 2,538 | 46.6 | 31.5 | 1.9 | 0.0 | 4.4 | 92.7 | 4,807 | 47 | 54 | 872 | 7,160 | 37,602 | 2,472 |
| Allen | 49 | 2,361 | 56.6 | 8.7 | 5.5 | 0.1 | 3.9 | 81.3 | 3,889 | 45 | 64 | 834 | 7,910 | 45,376 | 3,370 |
| Anderson | 46 | 2,039 | 58.6 | 6.7 | 2.5 | 0.1 | 4.1 | 156.8 | 6,953 | 43 | 68 | 807 | 10,620 | 53,822 | 4,273 |
| Ballard | 23 | 2,805 | 62.0 | 6.2 | 1.7 | 0.0 | 6.0 | 17.9 | 2,231 | 25 | 23 | 373 | 3,330 | 48,846 | 3,849 |
| Barren | 125 | 2,858 | 58.5 | 11.9 | 3.7 | 0.5 | 3.3 | 184.9 | 4,227 | 107 | 131 | 2,184 | 18,470 | 46,217 | 3,729 |
| Bath | 26 | 2,077 | 69.1 | 4.9 | 1.4 | 0.0 | 6.5 | 89.0 | 7,186 | 27 | 37 | 505 | 4,610 | 39,802 | 2,562 |
| Bell | 71 | 2,637 | 58.8 | 7.7 | 3.3 | 0.0 | 3.6 | 68.0 | 2,532 | 118 | 75 | 1,550 | 8,490 | 36,722 | 2,368 |
| Boone | 320 | 2,445 | 62.5 | 0.3 | 5.3 | 0.0 | 4.6 | 1,384.5 | 10,572 | 1,235 | 400 | 5,570 | 63,030 | 72,222 | 7,865 |
| Bourbon | 63 | 3,138 | 58.8 | 3.6 | 4.5 | 0.1 | 2.6 | 34.5 | 1,717 | 38 | 59 | 975 | 8,820 | 52,156 | 5,358 |
| Boyd | 144 | 3,001 | 54.4 | 4.8 | 4.4 | 0.1 | 5.3 | 320.0 | 6,693 | 385 | 135 | 3,033 | 18,800 | 54,618 | 5,170 |
| Boyle | 87 | 2,886 | 60.0 | 3.6 | 4.3 | 0.3 | 3.5 | 380.6 | 12,697 | 66 | 81 | 1,858 | 12,110 | 55,186 | 5,201 |
| Bracken | 21 | 2,487 | 63.0 | 7.0 | 2.7 | 0.0 | 7.3 | 16.5 | 1,987 | 18 | 25 | 433 | 3,690 | 47,886 | 3,441 |
| Breathitt | 36 | 2,748 | 61.5 | 10.9 | 2.4 | 0.3 | 7.7 | 31.9 | 2,467 | 53 | 37 | 798 | 4,410 | 39,193 | 2,437 |
| Breckinridge | 54 | 2,692 | 71.6 | 2.9 | 1.3 | 0.1 | 4.5 | 321.5 | 15,973 | 59 | 61 | 840 | 8,190 | 46,731 | 3,534 |
| Bullitt | 184 | 2,286 | 62.0 | 3.1 | 4.8 | 0.0 | 4.9 | 230.7 | 2,874 | 75 | 245 | 2,859 | 38,710 | 56,986 | 4,833 |
| Butler | 22 | 1,696 | 78.4 | 4.8 | 1.6 | 0.0 | 1.8 | 41.3 | 3,230 | 32 | 38 | 618 | 5,010 | 44,144 | 2,924 |
| Caldwell | 33 | 2,607 | 44.9 | 14.5 | 5.5 | 0.0 | 5.7 | 15.6 | 1,232 | 34 | 38 | 629 | 5,440 | 45,913 | 3,438 |
| Calloway | 178 | 4,595 | 23.9 | 61.6 | 2.4 | 0.0 | 2.2 | 209.2 | 5,387 | 73 | 110 | 4,949 | 14,850 | 50,400 | 4,496 |
| Campbell | 238 | 2,564 | 54.4 | 0.6 | 6.8 | 0.1 | 4.3 | 277.7 | 2,990 | 296 | 272 | 6,115 | 43,760 | 71,584 | 8,265 |
| Carlisle | 18 | 3,836 | 67.1 | 6.5 | 0.8 | 0.0 | 4.9 | 24.3 | 5,069 | 14 | 14 | 235 | 2,010 | 51,250 | 3,812 |
| Carroll | 143 | 13,361 | 15.4 | 0.7 | 0.8 | 0.0 | 0.8 | 3,436.3 | 320,729 | 36 | 31 | 727 | 4,650 | 48,912 | 3,747 |
| Carter | 53 | 1,953 | 68.0 | 3.0 | 3.2 | 0.0 | 5.0 | 61.9 | 2,274 | 61 | 79 | 1,259 | 10,190 | 45,096 | 3,253 |
| Casey | 55 | 3,490 | 42.1 | 34.0 | 0.8 | 0.0 | 3.5 | 54.9 | 3,469 | 28 | 63 | 729 | 5,810 | 36,800 | 2,532 |
| Christian | 131 | 1,847 | 55.6 | 3.2 | 5.0 | 0.1 | 3.6 | 433.2 | 6,099 | 3,484 | 27,347 | 3,247 | 29,360 | 45,037 | 3,532 |
| Clark | 101 | 2,805 | 49.4 | 7.3 | 3.5 | 0.0 | 2.7 | 199.1 | 5,546 | 105 | 108 | 1,561 | 16,160 | 52,808 | 4,541 |

1. Based on the resident population estimated as of July 1 of the year shown.

| State / county code | CBSA code[1] | County Type code[2] | STATE County | Land area[3] (sq. mi) | Total persons 2019 | Rank | Per square mile | White | Black | American Indian, Alaska Native | Asian and Pacific Islancer | Percent Hispanic or Latino[4] | Under 5 years | 5 to 14 years | 15 to 24 years | 25 to 34 years | 35 to 44 years | 45 to 54 years |
|---|---|---|---|---|---|---|---|---|---|---|---|---|---|---|---|---|---|---|
| | | | | 1 | 2 | 3 | 4 | 5 | 6 | 7 | 8 | 9 | 10 | 11 | 12 | 13 | 14 | 15 |
| | | | KENTUCKY—Cont'd | | | | | | | | | | | | | | | |
| 21051 | 30940 | 7 | Clay | 469.3 | 19,631 | 1,845 | 41.8 | 93.8 | 4.2 | 0.7 | 0.4 | 2.1 | 5.8 | 11.3 | 10.7 | 15.4 | 13.5 | 13.6 |
| 21053 | | 9 | Clinton | 197.3 | 10,110 | 2,409 | 51.2 | 95.7 | 1.0 | 0.8 | 0.6 | 3.0 | 5.6 | 11.9 | 11.2 | 11.7 | 11.4 | 13.0 |
| 21055 | | 7 | Crittenden | 360.0 | 8,847 | 2,504 | 24.6 | 96.9 | 1.6 | 0.9 | 0.5 | 1.4 | 5.4 | 13.4 | 10.9 | 11.0 | 11.4 | 12.4 |
| 21057 | | 9 | Cumberland | 305.2 | 6,523 | 2,703 | 21.4 | 95.5 | 4.0 | 0.6 | 0.3 | 1.4 | 5.8 | 12.0 | 9.5 | 11.0 | 10.3 | 11.9 |
| 21059 | 36980 | 3 | Daviess | 458.4 | 101,978 | 597 | 222.5 | 89.5 | 6.5 | 0.4 | 2.4 | 3.4 | 6.6 | 13.7 | 12.2 | 13.0 | 11.9 | 11.7 |
| 21061 | 14540 | 3 | Edmonson | 302.9 | 12,235 | 2,270 | 40.4 | 95.6 | 2.5 | 0.9 | 0.7 | 1.6 | 4.4 | 9.9 | 11.7 | 12.2 | 11.9 | 13.3 |
| 21063 | | 9 | Elliott | 234.3 | 7,372 | 2,628 | 31.5 | 94.5 | 3.9 | 0.6 | 0.5 | 1.3 | 4.2 | 10.3 | 10.3 | 14.2 | 12.9 | 13.9 |
| 21065 | 40080 | 6 | Estill | 253.1 | 14,109 | 2,152 | 55.7 | 97.9 | 0.8 | 0.8 | 0.3 | 1.3 | 5.4 | 12.2 | 10.8 | 11.7 | 11.8 | 13.6 |
| 21067 | 30460 | 2 | Fayette | 283.6 | 324,735 | 221 | 1,145.0 | 72.9 | 17.0 | 0.7 | 5.1 | 7.3 | 5.9 | 11.5 | 17.3 | 15.2 | 13.0 | 11.4 |
| 21069 | | 7 | Fleming | 348.5 | 14,603 | 2,114 | 41.9 | 96.7 | 2.0 | 0.8 | 0.5 | 1.5 | 6.7 | 13.3 | 11.9 | 11.4 | 11.4 | 13.3 |
| 21071 | | 7 | Floyd | 393.3 | 34,974 | 1,306 | 88.9 | 97.5 | 1.4 | 0.5 | 0.4 | 1.0 | 6.0 | 12.3 | 11.2 | 11.5 | 12.3 | 13.0 |
| 21073 | 23180 | 4 | Franklin | 207.8 | 51,118 | 975 | 246.0 | 83.5 | 12.0 | 0.8 | 2.5 | 3.8 | 5.5 | 11.7 | 12.5 | 13.0 | 12.2 | 13.0 |
| 21075 | | 9 | Fulton | 205.9 | 5,952 | 2,748 | 28.9 | 72.3 | 25.8 | 1.0 | 1.0 | 2.7 | 5.6 | 12.3 | 10.7 | 13.8 | 11.3 | 10.7 |
| 21077 | 17140 | 1 | Gallatin | 98.4 | 8,779 | 2,512 | 89.2 | 92.7 | 2.4 | 0.7 | 0.6 | 5.2 | 6.0 | 13.3 | 12.1 | 13.3 | 11.6 | 14.1 |
| 21079 | | 6 | Garrard | 230.1 | 17,719 | 1,935 | 77.0 | 95.1 | 2.9 | 0.7 | 0.5 | 2.2 | 5.4 | 12.5 | 10.5 | 11.0 | 11.7 | 13.5 |
| 21081 | 17140 | 1 | Grant | 258.0 | 25,387 | 1,587 | 98.4 | 95.2 | 1.4 | 0.6 | 0.7 | 3.1 | 7.1 | 14.8 | 13.0 | 12.7 | 12.1 | 13.1 |
| 21083 | 32460 | 7 | Graves | 551.8 | 36,818 | 1,255 | 66.7 | 87.8 | 5.6 | 0.8 | 0.7 | 7.3 | 6.7 | 13.2 | 11.2 | 12.1 | 11.6 | 12.0 |
| 21085 | | 6 | Grayson | 499.9 | 26,480 | 1,549 | 53.0 | 96.8 | 1.7 | 0.8 | 0.5 | 1.4 | 6.2 | 12.9 | 11.7 | 12.5 | 11.8 | 12.3 |
| 21087 | 15820 | 8 | Green | 286.0 | 10,995 | 2,348 | 38.4 | 95.2 | 2.7 | 1.0 | 0.4 | 2.2 | 5.6 | 11.7 | 10.3 | 11.0 | 11.5 | 12.9 |
| 21089 | 26580 | 2 | Greenup | 344.5 | 34,865 | 1,307 | 101.2 | 97.1 | 1.3 | 0.9 | 0.7 | 1.1 | 5.3 | 12.1 | 10.8 | 11.2 | 11.8 | 12.9 |
| 21091 | 36980 | 3 | Hancock | 187.7 | 8,742 | 2,516 | 46.6 | 96.1 | 2.1 | 0.7 | 0.6 | 1.8 | 6.2 | 13.9 | 12.4 | 11.1 | 12.0 | 13.0 |
| 21093 | 21060 | 3 | Hardin | 623.4 | 111,309 | 551 | 178.6 | 79.2 | 14.0 | 1.1 | 3.8 | 5.9 | 6.4 | 13.8 | 13.1 | 13.5 | 13.2 | 12.2 |
| 21095 | | 7 | Harlan | 465.8 | 25,566 | 1,574 | 54.9 | 96.2 | 2.7 | 0.7 | 0.6 | 1.1 | 6.2 | 13.3 | 10.2 | 12.6 | 11.5 | 12.7 |
| 21097 | | 6 | Harrison | 306.4 | 18,920 | 1,878 | 61.7 | 94.9 | 2.8 | 0.7 | 0.5 | 2.6 | 6.3 | 12.3 | 11.7 | 12.1 | 11.2 | 13.3 |
| 21099 | | 8 | Hart | 412.6 | 19,013 | 1,875 | 46.1 | 92.8 | 5.4 | 0.7 | 0.7 | 2.1 | 6.9 | 13.2 | 11.9 | 12.2 | 11.0 | 12.7 |
| 21101 | 21780 | 2 | Henderson | 436.3 | 44,740 | 1,080 | 102.5 | 88.1 | 9.8 | 0.6 | 0.9 | 2.8 | 5.7 | 12.9 | 11.5 | 12.2 | 12.2 | 12.4 |
| 21103 | 31140 | 1 | Henry | 286.3 | 16,067 | 2,027 | 56.1 | 92.9 | 3.9 | 1.0 | 0.7 | 3.5 | 5.8 | 13.2 | 11.9 | 11.7 | 11.8 | 12.6 |
| 21105 | | 9 | Hickman | 242.3 | 4,364 | 2,867 | 18.0 | 87.6 | 10.3 | 1.0 | 0.7 | 2.3 | 4.7 | 10.8 | 10.4 | 10.2 | 10.1 | 13.1 |
| 21107 | 31580 | 5 | Hopkins | 542.1 | 44,662 | 1,082 | 82.4 | 90.3 | 8.3 | 0.7 | 1.0 | 2.2 | 6.0 | 12.9 | 11.1 | 12.5 | 12.0 | 12.3 |
| 21109 | | 9 | Jackson | 345.2 | 13,340 | 2,198 | 38.6 | 98.3 | 0.6 | 0.6 | 0.3 | 1.0 | 6.3 | 12.6 | 10.7 | 11.7 | 12.3 | 13.8 |
| 21111 | 31140 | 1 | Jefferson | 380.8 | 767,452 | 85 | 2,015.4 | 68.1 | 23.8 | 0.6 | 3.9 | 6.2 | 6.2 | 12.1 | 12.1 | 14.9 | 12.7 | 11.8 |
| 21113 | 30460 | 2 | Jessamine | 172.2 | 54,057 | 945 | 313.9 | 90.1 | 5.6 | 0.7 | 1.9 | 3.8 | 6.1 | 13.6 | 13.3 | 12.5 | 12.6 | 12.5 |
| 21115 | | 7 | Johnson | 262.0 | 22,002 | 1,718 | 84.0 | 98.0 | 0.6 | 0.7 | 0.7 | 0.9 | 5.5 | 12.6 | 11.7 | 11.2 | 12.6 | 13.1 |
| 21117 | 17140 | 1 | Kenton | 160.3 | 167,949 | 397 | 1,047.7 | 90.1 | 6.2 | 0.5 | 1.9 | 3.6 | 6.4 | 13.1 | 11.9 | 15.1 | 13.1 | 12.0 |
| 21119 | | 9 | Knott | 351.5 | 14,512 | 2,121 | 41.3 | 97.7 | 1.3 | 0.7 | 0.3 | 1.0 | 4.9 | 11.3 | 13.3 | 10.4 | 11.0 | 13.6 |
| 21121 | 30940 | 7 | Knox | 386.3 | 31,022 | 1,403 | 80.3 | 96.6 | 1.8 | 0.9 | 0.6 | 1.5 | 6.3 | 13.1 | 12.6 | 12.9 | 11.5 | 12.9 |
| 21123 | 21060 | 3 | Larue | 261.6 | 14,431 | 2,126 | 55.2 | 93.0 | 4.1 | 0.9 | 0.7 | 3.3 | 5.8 | 12.8 | 11.2 | 12.3 | 12.6 | 12.3 |
| 21125 | 30940 | 5 | Laurel | 434.0 | 61,238 | 862 | 141.1 | 96.7 | 1.3 | 0.9 | 0.8 | 1.6 | 6.1 | 12.9 | 11.7 | 13.0 | 12.7 | 13.1 |
| 21127 | | 6 | Lawrence | 415.6 | 15,436 | 2,065 | 37.1 | 97.7 | 1.0 | 0.8 | 0.5 | 1.5 | 6.6 | 13.6 | 10.7 | 11.6 | 12.1 | 12.9 |
| 21129 | | 9 | Lee | 208.9 | 7,268 | 2,637 | 34.8 | 95.0 | 3.2 | 0.9 | 0.4 | 1.5 | 5.2 | 9.8 | 10.1 | 14.6 | 12.7 | 14.0 |
| 21131 | | 9 | Leslie | 400.9 | 9,637 | 2,444 | 24.0 | 98.2 | 0.8 | 0.7 | 0.5 | 0.8 | 5.2 | 12.6 | 10.0 | 12.5 | 11.7 | 13.4 |
| 21133 | | 9 | Letcher | 337.9 | 21,213 | 1,760 | 62.8 | 98.0 | 0.9 | 0.5 | 0.4 | 0.9 | 5.6 | 12.4 | 10.1 | 11.2 | 12.3 | 12.6 |
| 21135 | | 8 | Lewis | 482.8 | 13,262 | 2,202 | 27.5 | 98.3 | 0.9 | 0.7 | 0.2 | 0.9 | 6.0 | 12.4 | 11.0 | 11.6 | 11.4 | 13.1 |
| 21137 | 19220 | 7 | Lincoln | 332.8 | 24,466 | 1,628 | 73.5 | 95.5 | 2.9 | 0.8 | 0.5 | 1.8 | 6.7 | 13.0 | 11.2 | 11.8 | 11.1 | 13.2 |
| 21139 | 37140 | 9 | Livingston | 313.3 | 9,041 | 2,494 | 28.9 | 96.1 | 1.1 | 1.1 | 0.7 | 2.4 | 4.9 | 11.5 | 9.8 | 10.9 | 10.4 | 13.3 |
| 21141 | | 6 | Logan | 552.2 | 27,416 | 1,516 | 49.6 | 90.3 | 7.4 | 0.7 | 0.6 | 3.0 | 6.6 | 13.0 | 11.6 | 12.0 | 11.9 | 12.1 |
| 21143 | | 9 | Lyon | 213.8 | 8,133 | 2,572 | 38.0 | 90.9 | 6.1 | 1.1 | 0.8 | 2.6 | 3.3 | 8.3 | 9.2 | 12.0 | 11.5 | 13.0 |
| 21145 | 37140 | 5 | McCracken | 248.7 | 65,644 | 812 | 263.9 | 85.0 | 12.3 | 0.8 | 1.4 | 2.8 | 5.9 | 12.4 | 11.0 | 12.2 | 11.9 | 12.0 |
| 21147 | | 9 | McCreary | 426.8 | 17,071 | 1,970 | 40.0 | 90.6 | 5.9 | 1.4 | 0.9 | 2.7 | 5.7 | 12.1 | 11.8 | 14.6 | 13.7 | 13.2 |
| 21149 | 36980 | 3 | McLean | 252.5 | 9,075 | 2,489 | 35.9 | 96.7 | 1.5 | 0.9 | 0.5 | 1.8 | 5.6 | 13.3 | 11.7 | 10.6 | 11.3 | 12.7 |
| 21151 | 40080 | 4 | Madison | 437.4 | 94,265 | 632 | 215.5 | 91.5 | 5.5 | 0.9 | 1.6 | 2.6 | 5.6 | 11.4 | 20.6 | 12.8 | 11.8 | 11.8 |
| 21153 | | 7 | Magoffin | 308.4 | 12,017 | 2,293 | 39.0 | 97.9 | 0.5 | 0.7 | 0.3 | 1.2 | 5.9 | 12.6 | 10.8 | 11.8 | 11.5 | 13.8 |
| 21155 | | 7 | Marion | 343.0 | 19,314 | 1,862 | 56.3 | 89.3 | 8.1 | 0.6 | 1.0 | 2.8 | 6.2 | 13.6 | 11.9 | 12.0 | 12.4 | 12.5 |
| 21157 | | 7 | Marshall | 302.2 | 31,163 | 1,399 | 103.1 | 97.1 | 0.8 | 0.7 | 0.7 | 1.8 | 5.0 | 11.4 | 10.5 | 11.2 | 11.5 | 12.7 |
| 21159 | | 9 | Martin | 229.6 | 11,031 | 2,345 | 48.0 | 89.1 | 7.1 | 0.8 | 0.4 | 3.5 | 4.7 | 11.0 | 10.9 | 15.9 | 13.9 | 12.9 |
| 21161 | 32500 | 6 | Mason | 240.1 | 17,035 | 1,972 | 70.9 | 90.5 | 7.8 | 0.8 | 1.2 | 2.2 | 6.2 | 13.2 | 11.3 | 12.7 | 10.6 | 13.2 |
| 21163 | 21060 | 6 | Meade | 305.4 | 28,616 | 1,474 | 93.7 | 90.7 | 4.5 | 1.2 | 1.6 | 4.1 | 5.1 | 13.0 | 12.1 | 13.1 | 13.9 | 13.2 |
| 21165 | 34460 | 9 | Menifee | 203.6 | 6,502 | 2,705 | 31.9 | 95.2 | 3.3 | 0.8 | 0.4 | 1.4 | 4.5 | 9.4 | 12.1 | 10.6 | 10.9 | 14.1 |
| 21167 | | 6 | Mercer | 249.1 | 21,889 | 1,725 | 87.9 | 92.4 | 4.6 | 0.8 | 1.0 | 3.1 | 6.0 | 12.2 | 11.0 | 12.1 | 11.2 | 12.9 |
| 21169 | 23980 | 8 | Metcalfe | 289.6 | 10,058 | 2,415 | 34.7 | 95.7 | 2.3 | 0.8 | 0.4 | 2.2 | 6.0 | 13.1 | 11.0 | 11.5 | 11.5 | 12.6 |
| 21171 | | 8 | Monroe | 329.4 | 10,549 | 2,386 | 32.0 | 94.3 | 2.8 | 0.7 | 0.4 | 3.1 | 6.7 | 12.6 | 11.5 | 11.0 | 11.4 | 12.7 |
| 21173 | 34460 | 6 | Montgomery | 197.4 | 28,186 | 1,490 | 142.8 | 94.0 | 3.2 | 0.6 | 0.6 | 2.9 | 6.3 | 12.9 | 11.5 | 13.2 | 12.5 | 13.8 |
| 21175 | | 6 | Morgan | 381.1 | 13,142 | 2,211 | 34.5 | 93.0 | 4.9 | 0.7 | 1.2 | 1.2 | 4.5 | 10.1 | 11.3 | 15.1 | 13.7 | 14.2 |
| 21177 | 16420 | 6 | Muhlenberg | 467.4 | 30,457 | 1,417 | 65.2 | 92.9 | 5.4 | 0.6 | 0.5 | 1.7 | 5.7 | 11.2 | 12.5 | 12.3 | 12.0 | 12.7 |
| 21179 | 12680 | 6 | Nelson | 417.5 | 46,450 | 1,046 | 111.3 | 92.0 | 6.1 | 0.6 | 0.8 | 2.3 | 6.0 | 13.1 | 11.7 | 13.1 | 12.4 | 12.9 |
| 21181 | | 8 | Nicholas | 195.2 | 7,234 | 2,639 | 37.1 | 96.6 | 1.2 | 0.5 | 0.3 | 2.3 | 6.1 | 13.8 | 12.0 | 11.5 | 11.7 | 13.6 |

1. CBSA = Core Based Statistical Area. See Appendix A for explanation. See Appendix B for list of metropolitan areas with component counties. 2. County type code from the Economic Research Service of USDA Rural-Urban Continuum Codes. See Appendix A for definition. 3. Dry land or land partially or temporarily covered by water. 4. May be of any race.

| STATE County | Age (percent) (cont.) 55 to 64 years | 65 to 74 years | 75 years and over | Percent female | Total persons 2000 | 2010 | Percent change 2000-2010 | 2010-2020 | Components of change, 2010-2020 Births | Deaths | Net Migration | Number | Persons per household | Family households | Female family householder[1] | One person |
|---|---|---|---|---|---|---|---|---|---|---|---|---|---|---|---|---|
| | 16 | 17 | 18 | 19 | 20 | 21 | 22 | 23 | 24 | 25 | 26 | 27 | 28 | 29 | 30 | 31 |
| **KENTUCKY—Cont'd** | | | | | | | | | | | | | | | | |
| Clay | 13.7 | 10.0 | 6.1 | 48.2 | 24,556 | 21,736 | -11.5 | -9.7 | 2,736 | 2,748 | -2,115 | 7,622 | 2.48 | 72.3 | 15.2 | 25.5 |
| Clinton | 14.7 | 11.9 | 8.5 | 51.1 | 9,634 | 10,279 | 6.7 | -1.6 | 1,174 | 1,359 | 22 | 3,985 | 2.52 | 62.6 | 9.1 | 35.1 |
| Crittenden | 14.2 | 12.3 | 9.1 | 49.6 | 9,384 | 9,315 | -0.7 | -5.0 | 1,015 | 1,309 | -162 | 3,567 | 2.46 | 67.1 | 10.7 | 29.2 |
| Cumberland | 16.3 | 13.4 | 9.9 | 51.3 | 7,147 | 6,845 | -4.2 | -4.7 | 787 | 1,040 | -64 | 2,616 | 2.53 | 63.0 | 10.6 | 33.5 |
| Daviess | 13.2 | 10.2 | 7.4 | 51.2 | 91,545 | 96,659 | 5.6 | 5.5 | 13,629 | 10,697 | 2,471 | 40,179 | 2.42 | 66.6 | 12.8 | 27.9 |
| Edmonson | 15.2 | 12.5 | 8.9 | 49.8 | 11,644 | 12,187 | 4.7 | 0.4 | 1,136 | 1,441 | 358 | 4,885 | 2.41 | 69.1 | 9.7 | 26.5 |
| Elliott | 13.0 | 12.0 | 9.1 | 43.3 | 6,748 | 7,852 | 16.4 | -6.1 | 686 | 775 | -396 | 2,523 | 2.37 | 76.3 | 11.7 | 21.5 |
| Estill | 14.9 | 11.7 | 7.9 | 50.7 | 15,307 | 14,686 | -4.1 | -3.9 | 1,564 | 1,989 | -146 | 5,499 | 2.56 | 65.4 | 17.3 | 31.0 |
| Fayette | 11.4 | 8.7 | 5.7 | 51.1 | 260,512 | 295,875 | 13.6 | 9.8 | 40,858 | 23,705 | 11,788 | 129,784 | 2.37 | 57.6 | 12.1 | 31.6 |
| Fleming | 13.4 | 11.3 | 7.3 | 50.9 | 13,792 | 14,344 | 4.0 | 1.8 | 2,007 | 1,791 | 50 | 5,841 | 2.48 | 71.9 | 10.6 | 24.1 |
| Floyd | 14.4 | 12.1 | 7.1 | 51.4 | 42,441 | 39,454 | -7.0 | -11.4 | 4,865 | 5,460 | -4,000 | 15,000 | 2.37 | 68.8 | 15.1 | 26.4 |
| Franklin | 13.5 | 11.4 | 7.2 | 51.8 | 47,687 | 49,262 | 3.3 | 3.8 | 5,896 | 5,519 | 1,520 | 21,076 | 2.31 | 60.0 | 13.4 | 34.3 |
| Fulton | 14.1 | 12.5 | 9.1 | 50.0 | 7,752 | 6,803 | -12.2 | -12.5 | 748 | 1,039 | -570 | 2,441 | 2.33 | 58.4 | 19.3 | 38.3 |
| Gallatin | 14.5 | 9.8 | 5.3 | 49.2 | 7,870 | 8,581 | 9.0 | 2.3 | 1,145 | 998 | 52 | 3,105 | 2.77 | 65.5 | 10.6 | 28.3 |
| Garrard | 16.3 | 11.5 | 7.6 | 50.6 | 14,792 | 16,905 | 14.3 | 4.8 | 1,908 | 1,780 | 691 | 6,706 | 2.58 | 73.7 | 12.2 | 21.3 |
| Grant | 12.8 | 8.8 | 5.7 | 50.1 | 22,384 | 24,660 | 10.2 | 2.9 | 3,730 | 2,403 | -602 | 9,141 | 2.69 | 73.9 | 15.0 | 21.2 |
| Graves | 13.7 | 11.2 | 8.2 | 51.1 | 37,028 | 37,142 | 0.3 | -0.9 | 5,129 | 4,661 | -773 | 14,316 | 2.56 | 62.7 | 9.0 | 33.2 |
| Grayson | 14.1 | 11.3 | 7.2 | 49.7 | 24,053 | 25,749 | 7.1 | 2.8 | 3,338 | 3,340 | 749 | 9,758 | 2.66 | 69.7 | 11.5 | 27.0 |
| Green | 15.8 | 12.4 | 8.9 | 50.4 | 11,518 | 11,285 | -2.0 | -2.6 | 1,147 | 1,455 | 20 | 4,413 | 2.46 | 66.1 | 10.3 | 30.3 |
| Greenup | 14.1 | 12.5 | 9.3 | 51.5 | 36,891 | 36,910 | 0.1 | -5.5 | 3,876 | 4,819 | -1,082 | 14,056 | 2.50 | 72.2 | 11.6 | 22.4 |
| Hancock | 13.2 | 10.6 | 7.6 | 48.6 | 8,392 | 8,565 | 2.1 | 2.1 | 1,123 | 871 | -69 | 3,311 | 2.60 | 75.4 | 9.2 | 20.8 |
| Hardin | 13.0 | 8.9 | 6.0 | 50.2 | 94,174 | 105,542 | 12.1 | 5.5 | 15,579 | 9,322 | -641 | 41,646 | 2.54 | 68.2 | 11.7 | 26.9 |
| Harlan | 14.0 | 12.3 | 7.2 | 52.3 | 33,202 | 29,278 | -11.8 | -12.7 | 3,636 | 4,364 | -3,007 | 11,134 | 2.34 | 66.9 | 13.2 | 29.3 |
| Harrison | 15.0 | 10.9 | 7.3 | 51.0 | 17,983 | 18,839 | 4.8 | 0.4 | 2,241 | 2,326 | 174 | 7,283 | 2.52 | 70.8 | 10.7 | 25.3 |
| Hart | 14.6 | 10.5 | 6.8 | 50.5 | 17,445 | 18,190 | 4.3 | 4.5 | 2,623 | 2,196 | 405 | 7,351 | 2.51 | 65.0 | 10.8 | 31.4 |
| Henderson | 14.3 | 11.2 | 7.7 | 51.5 | 44,829 | 46,249 | 3.2 | -3.3 | 5,539 | 5,231 | -1,784 | 18,643 | 2.38 | 65.9 | 13.3 | 29.1 |
| Henry | 14.9 | 11.2 | 6.9 | 50.6 | 15,060 | 15,411 | 2.3 | 4.3 | 1,805 | 1,828 | 686 | 6,044 | 2.62 | 70.5 | 13.6 | 23.9 |
| Hickman | 14.6 | 13.8 | 12.3 | 52.1 | 5,262 | 4,911 | -6.7 | -11.1 | 412 | 662 | -295 | 1,798 | 2.43 | 59.7 | 9.1 | 34.6 |
| Hopkins | 13.9 | 11.3 | 7.9 | 51.4 | 46,519 | 46,917 | 0.9 | -4.8 | 5,507 | 6,135 | -1,611 | 18,588 | 2.38 | 66.7 | 12.0 | 28.6 |
| Jackson | 14.3 | 11.3 | 7.0 | 50.2 | 13,495 | 13,486 | -0.1 | -1.1 | 1,669 | 1,711 | -104 | 5,484 | 2.42 | 70.4 | 13.3 | 26.5 |
| Jefferson | 13.2 | 10.2 | 6.8 | 51.6 | 693,604 | 741,095 | 6.8 | 3.6 | 101,330 | 79,480 | 5,163 | 312,679 | 2.40 | 59.5 | 13.9 | 33.5 |
| Jessamine | 13.3 | 9.6 | 6.5 | 51.4 | 39,041 | 48,579 | 24.4 | 11.3 | 6,724 | 4,573 | 3,329 | 18,821 | 2.72 | 72.9 | 15.0 | 21.5 |
| Johnson | 14.1 | 11.9 | 7.4 | 50.7 | 23,445 | 23,372 | -0.3 | -5.9 | 2,789 | 3,141 | -1,009 | 8,488 | 2.59 | 67.2 | 11.2 | 29.7 |
| Kenton | 13.2 | 9.5 | 5.8 | 50.5 | 151,464 | 159,728 | 5.5 | 5.1 | 23,471 | 15,116 | -9 | 63,966 | 2.55 | 62.9 | 12.0 | 29.9 |
| Knott | 15.4 | 12.5 | 7.7 | 50.8 | 17,649 | 16,360 | -7.3 | -11.3 | 1,693 | 2,093 | -1,448 | 6,388 | 2.28 | 67.5 | 11.1 | 28.0 |
| Knox | 12.9 | 10.2 | 7.7 | 51.1 | 31,795 | 31,882 | 0.3 | -2.7 | 4,124 | 4,141 | -835 | 11,961 | 2.56 | 65.9 | 12.7 | 30.5 |
| Larue | 14.7 | 10.7 | 7.7 | 50.5 | 13,373 | 14,190 | 6.1 | 1.7 | 1,610 | 1,728 | 365 | 5,741 | 2.42 | 67.6 | 12.9 | 25.7 |
| Laurel | 13.5 | 10.3 | 6.7 | 51.0 | 52,715 | 58,845 | 11.6 | 4.1 | 7,534 | 6,341 | 1,250 | 22,573 | 2.63 | 69.8 | 13.1 | 26.8 |
| Lawrence | 14.1 | 11.6 | 6.9 | 50.3 | 15,569 | 15,849 | 1.8 | -2.6 | 2,111 | 2,149 | -366 | 5,784 | 2.69 | 68.2 | 9.1 | 29.4 |
| Lee | 15.1 | 12.0 | 6.5 | 44.1 | 7,916 | 7,883 | -0.4 | -7.8 | 756 | 1,056 | -358 | 2,839 | 2.28 | 55.3 | 15.5 | 40.0 |
| Leslie | 15.6 | 11.6 | 7.3 | 50.3 | 12,401 | 11,310 | -8.8 | -14.8 | 1,290 | 1,686 | -1,284 | 4,076 | 2.45 | 69.4 | 13.9 | 27.6 |
| Letcher | 15.2 | 13.0 | 7.6 | 50.9 | 25,277 | 24,511 | -3.0 | -13.5 | 2,741 | 3,411 | -2,653 | 9,721 | 2.27 | 66.8 | 15.4 | 29.6 |
| Lewis | 15.0 | 11.6 | 7.8 | 50.1 | 14,092 | 13,882 | -1.5 | -4.5 | 1,630 | 1,718 | -529 | 5,268 | 2.52 | 67.6 | 13.8 | 29.5 |
| Lincoln | 13.8 | 11.0 | 8.1 | 51.0 | 23,361 | 24,741 | 5.9 | -1.1 | 3,356 | 2,999 | -623 | 9,697 | 2.5 | 69.9 | 10.8 | 25.2 |
| Livingston | 15.9 | 13.7 | 9.7 | 50.7 | 9,804 | 9,519 | -2.9 | -5.0 | 994 | 1,325 | -143 | 3,857 | 2.38 | 65.1 | 11.5 | 29.0 |
| Logan | 13.9 | 10.8 | 8.1 | 50.8 | 26,573 | 26,835 | 1.0 | 2.2 | 3,603 | 3,198 | 185 | 10,418 | 2.55 | 68.0 | 10.4 | 27.2 |
| Lyon | 16.1 | 15.7 | 10.9 | 44.6 | 8,080 | 8,319 | 3.0 | -2.2 | 542 | 1,174 | 442 | 3,333 | 2.10 | 60.7 | 6.5 | 33.5 |
| McCracken | 13.6 | 11.9 | 9.0 | 52.2 | 65,514 | 65,547 | 0.1 | 0.1 | 7,921 | 8,280 | 517 | 27,854 | 2.29 | 58.3 | 11.2 | 35.9 |
| McCreary | 12.4 | 10.1 | 6.6 | 45.6 | 17,080 | 18,306 | 7.2 | -6.7 | 2,245 | 2,172 | -1,316 | 6,052 | 2.58 | 70.8 | 9.9 | 24.6 |
| McLean | 14.4 | 11.5 | 8.8 | 50.6 | 9,938 | 9,527 | -4.1 | -4.7 | 1,102 | 1,243 | -307 | 3,809 | 2.41 | 72.3 | 8.0 | 25.6 |
| Madison | 11.4 | 8.8 | 5.7 | 51.5 | 70,872 | 82,910 | 17.0 | 13.7 | 10,259 | 7,692 | 8,708 | 33,359 | 2.52 | 63.5 | 12.0 | 28.3 |
| Magoffin | 15.0 | 11.6 | 7.0 | 50.1 | 13,332 | 13,333 | 0.0 | -9.9 | 1,551 | 1,644 | -1,227 | 5,020 | 2.46 | 70.0 | 12.3 | 28.8 |
| Marion | 14.1 | 10.2 | 7.1 | 50.3 | 18,212 | 19,844 | 9.0 | -2.7 | 2,450 | 2,162 | -856 | 7,405 | 2.48 | 60.0 | 13.1 | 33.7 |
| Marshall | 14.9 | 13.0 | 9.7 | 50.6 | 30,125 | 31,449 | 4.4 | -0.9 | 3,180 | 4,402 | 958 | 13,121 | 2.33 | 68.4 | 8.8 | 28.2 |
| Martin | 12.9 | 11.2 | 6.6 | 44.8 | 12,578 | 12,929 | 2.8 | -14.7 | 1,277 | 1,441 | -1,749 | 4,153 | 2.50 | 67.5 | 7.2 | 26.2 |
| Mason | 13.8 | 11.8 | 7.2 | 51.3 | 16,800 | 17,488 | 4.1 | -2.6 | 2,149 | 2,221 | -381 | 6,700 | 2.51 | 66.5 | 14.4 | 29.7 |
| Meade | 14.2 | 9.6 | 5.7 | 49.5 | 26,349 | 28,600 | 8.5 | 0.1 | 2,733 | 2,506 | -262 | 10,685 | 2.61 | 71.8 | 7.6 | 22.3 |
| Menifee | 16.6 | 13.3 | 8.4 | 49.7 | 6,556 | 6,306 | -3.8 | 3.1 | 710 | 807 | 289 | 2,651 | 2.39 | 64.5 | 5.6 | 31.9 |
| Mercer | 15.4 | 11.4 | 7.8 | 50.8 | 20,817 | 21,324 | 2.4 | 2.6 | 2,539 | 2,754 | 793 | 8,556 | 2.51 | 69.2 | 9.6 | 26.7 |
| Metcalfe | 14.2 | 11.2 | 8.7 | 50.7 | 10,037 | 10,105 | 0.7 | -0.5 | 1,306 | 1,340 | -10 | 4,074 | 2.43 | 71.8 | 9.8 | 24.6 |
| Monroe | 14.6 | 11.2 | 8.3 | 50.0 | 11,756 | 10,963 | -6.7 | -3.8 | 1,331 | 1,572 | -168 | 4,491 | 2.33 | 63.3 | 9.0 | 32.8 |
| Montgomery | 13.1 | 10.1 | 6.6 | 51.3 | 22,554 | 26,511 | 17.5 | 6.3 | 3,829 | 2,954 | 813 | 10,481 | 2.62 | 71.0 | 15.5 | 22.4 |
| Morgan | 13.9 | 10.9 | 6.5 | 42.9 | 13,948 | 13,923 | -0.2 | -5.6 | 1,307 | 1,515 | -600 | 4,865 | 2.39 | 67.2 | 10.8 | 27.5 |
| Muhlenberg | 13.6 | 11.4 | 8.6 | 48.9 | 31,839 | 31,501 | -1.1 | -3.3 | 3,589 | 3,959 | -669 | 11,351 | 2.65 | 70.9 | 9.8 | 25.1 |
| Nelson | 14.1 | 10.5 | 6.2 | 50.6 | 37,477 | 43,441 | 15.9 | 6.9 | 5,882 | 4,127 | 1,274 | 17,853 | 2.53 | 68.6 | 10.5 | 24.9 |
| Nicholas | 14.2 | 10.2 | 6.8 | 50.4 | 6,813 | 7,142 | 4.8 | 1.3 | 898 | 1,015 | 210 | 2,804 | 2.51 | 69.2 | 14.8 | 26.2 |

1. No spouse present.

| STATE County | Persons in group quarters, 2020 | Daytime Population, 2015-2019 Number | Daytime Population, 2015-2019 Employment/ residence ratio | Births, 2020 Total | Births, 2020 Rate[1] | Deaths, 2020 Number | Deaths, 2020 Rate[1] | Persons under 65 with no health insurance, 2019 Number | Persons under 65 with no health insurance, 2019 Percent | Medicare, 2020 Total beneficiaries | Medicare, 2020 Enrolled in Original Medicare | Medicare, 2020 Enrolled in Medicare Advantage | Crimes reported by county police or sheriff, 2019 Violent | Crimes reported by county police or sheriff, 2019 Property |
|---|---|---|---|---|---|---|---|---|---|---|---|---|---|---|
| | 32 | 33 | 34 | 35 | 36 | 37 | 38 | 39 | 40 | 41 | 42 | 43 | 44 | 45 |
| KENTUCKY—Cont'd | | | | | | | | | | | | | | |
| Clay | 1,621 | 19,007 | 0.75 | 217 | 11.1 | 258 | 13.1 | 1,382 | 9.3 | 4,847 | 2,399 | 2,448 | 4 | 95 |
| Clinton | 136 | 10,744 | 1.16 | 112 | 11.1 | 134 | 13.3 | 723 | 9.1 | 2,480 | 1,750 | 731 | 0 | 15 |
| Crittenden | 210 | 8,245 | 0.79 | 92 | 10.4 | 116 | 13.1 | 553 | 8.2 | 2,224 | 1,558 | 665 | 1 | 24 |
| Cumberland | 86 | 6,205 | 0.80 | 76 | 11.7 | 119 | 18.2 | 408 | 8.1 | 1,812 | 1,226 | 586 | 0 | 10 |
| Daviess | 2,741 | 102,552 | 1.05 | 1,308 | 12.8 | 1,143 | 11.2 | 6,033 | 7.4 | 22,216 | 14,714 | 7,502 | 20 | 374 |
| Edmonson | 354 | 9,610 | 0.43 | 114 | 9.3 | 163 | 13.3 | 888 | 9.4 | 2,982 | 1,864 | 1,118 | 9 | 22 |
| Elliott | 1,032 | 7,023 | 0.71 | 61 | 8.3 | 84 | 11.4 | 336 | 6.8 | 1,426 | 686 | 740 | 0 | 0 |
| Estill | 118 | 11,930 | 0.52 | 145 | 10.3 | 194 | 13.8 | 929 | 8.4 | 3,675 | 2,420 | 1,255 | 0 | 29 |
| Fayette | 13,048 | 352,594 | 1.20 | 3,839 | 11.8 | 2,545 | 7.8 | 23,429 | 8.8 | 50,444 | 28,573 | 21,871 | 0 | 0 |
| Fleming | 22 | 12,406 | 0.64 | 199 | 13.6 | 183 | 12.5 | 1,220 | 10.3 | 3,508 | 2,140 | 1,368 | 2 | 12 |
| Floyd | 668 | 35,839 | 0.94 | 415 | 11.9 | 589 | 16.8 | 2,465 | 8.7 | 10,336 | 6,497 | 3,839 | 5 | 50 |
| Franklin | 1,878 | 59,168 | 1.36 | 544 | 10.6 | 576 | 11.3 | 3,134 | 7.8 | 14,244 | 7,572 | 6,672 | 25 | 209 |
| Fulton | 444 | 6,466 | 1.17 | 61 | 10.2 | 102 | 17.1 | 290 | 6.8 | 1,776 | 1,129 | 647 | 0 | 13 |
| Gallatin | 85 | 7,331 | 0.63 | 100 | 11.4 | 82 | 9.3 | 696 | 9.2 | 1,627 | 861 | 767 | 2 | 17 |
| Garrard | 109 | 13,156 | 0.41 | 182 | 10.3 | 196 | 11.1 | 1,336 | 9.3 | 4,018 | 2,549 | 1,469 | 3 | 69 |
| Grant | 456 | 19,620 | 0.49 | 363 | 14.3 | 261 | 10.3 | 1,414 | 6.7 | 4,984 | 2,523 | 2,461 | 6 | 135 |
| Graves | 463 | 33,848 | 0.78 | 483 | 13.1 | 474 | 12.9 | 2,607 | 8.7 | 8,720 | 5,376 | 3,344 | 28 | 223 |
| Grayson | 766 | 23,814 | 0.77 | 314 | 11.9 | 326 | 12.3 | 1,850 | 8.9 | 6,346 | 4,271 | 2,075 | 4 | 84 |
| Green | 113 | 8,859 | 0.52 | 107 | 9.7 | 150 | 13.6 | 887 | 10.4 | 2,847 | 1,847 | 1,000 | 2 | 2 |
| Greenup | 454 | 30,288 | 0.62 | 354 | 10.2 | 522 | 15.0 | 1,802 | 6.6 | 9,539 | 5,896 | 3,643 | 3 | 34 |
| Hancock | 90 | 9,085 | 1.10 | 100 | 11.4 | 98 | 11.2 | 438 | 6.1 | 2,002 | 1,476 | 526 | 2 | 27 |
| Hardin | 3,190 | 113,130 | 1.09 | 1,377 | 12.4 | 1,041 | 9.4 | 5,818 | 6.3 | 20,664 | 14,512 | 6,152 | 16 | 140 |
| Harlan | 627 | 26,349 | 0.95 | 293 | 11.5 | 436 | 17.1 | 1,670 | 8.2 | 7,412 | 4,713 | 2,699 | 1 | 39 |
| Harrison | 275 | 16,625 | 0.74 | 232 | 12.3 | 243 | 12.8 | 1,177 | 7.7 | 4,257 | 2,719 | 1,539 | 7 | 73 |
| Hart | 227 | 17,896 | 0.88 | 254 | 13.4 | 218 | 11.5 | 1,376 | 8.9 | 4,224 | 2,759 | 1,465 | 3 | 79 |
| Henderson | 1,164 | 45,138 | 0.97 | 497 | 11.1 | 572 | 12.8 | 2,556 | 7.1 | 10,284 | 6,250 | 4,034 | 19 | 163 |
| Henry | 80 | 12,700 | 0.55 | 161 | 10.0 | 190 | 11.8 | 1,240 | 9.5 | 3,587 | 1,893 | 1,694 | 1 | 8 |
| Hickman | 212 | 4,204 | 0.81 | 37 | 8.5 | 70 | 16.0 | 271 | 8.8 | 1,272 | 855 | 417 | 0 | 13 |
| Hopkins | 1,084 | 45,251 | 1.00 | 523 | 11.7 | 598 | 13.4 | 2,419 | 6.8 | 10,843 | 7,186 | 3,658 | 18 | 204 |
| Jackson | 110 | 11,386 | 0.53 | 158 | 11.8 | 173 | 13.0 | 899 | 8.4 | 3,195 | 1,969 | 1,226 | 4 | 76 |
| Jefferson | 17,013 | 857,598 | 1.24 | 9,433 | 12.3 | 8,281 | 10.8 | 44,606 | 7.1 | 150,001 | 90,471 | 59,529 | 1 | 0 |
| Jessamine | 1,634 | 49,633 | 0.86 | 619 | 11.5 | 492 | 9.1 | 3,632 | 8.2 | 9,911 | 5,569 | 4,343 | 14 | 206 |
| Johnson | 528 | 21,495 | 0.83 | 223 | 10.1 | 321 | 14.6 | 1,359 | 7.8 | 6,079 | 3,822 | 2,257 | 4 | 45 |
| Kenton | 2,661 | 147,415 | 0.78 | 2,180 | 13.0 | 1,612 | 9.6 | 9,138 | 6.5 | 29,461 | 16,138 | 13,322 | 19 | 122 |
| Knott | 764 | 13,569 | 0.63 | 134 | 9.2 | 204 | 14.1 | 964 | 8.6 | 3,937 | 2,284 | 1,653 | 2 | 26 |
| Knox | 656 | 30,446 | 0.90 | 396 | 12.8 | 470 | 15.2 | 1,982 | 8.0 | 7,826 | 4,966 | 2,860 | 17 | 81 |
| Larue | 313 | 11,056 | 0.50 | 157 | 10.9 | 171 | 11.8 | 968 | 8.4 | 3,421 | 2,238 | 1,183 | 3 | 18 |
| Laurel | 695 | 63,150 | 1.12 | 713 | 11.6 | 737 | 12.0 | 4,115 | 8.3 | 14,055 | 8,708 | 5,348 | 25 | 460 |
| Lawrence | 108 | 14,095 | 0.65 | 214 | 13.9 | 223 | 14.4 | 1,077 | 8.7 | 3,979 | 2,565 | 1,415 | 3 | 42 |
| Lee | 990 | 6,546 | 0.83 | 75 | 10.3 | 120 | 16.5 | 371 | 7.4 | 1,820 | 1,154 | 666 | 1 | 5 |
| Leslie | 227 | 9,089 | 0.59 | 96 | 10.0 | 178 | 18.5 | 647 | 8.3 | 2,866 | 1,470 | 1,395 | 1 | 17 |
| Letcher | 243 | 20,910 | 0.80 | 228 | 10.7 | 342 | 16.1 | 1,355 | 8.0 | 6,066 | 3,654 | 2,412 | 1 | 13 |
| Lewis | 139 | 11,189 | 0.47 | 168 | 12.7 | 169 | 12.7 | 950 | 9.0 | 3,167 | 1,968 | 1,199 | 4 | 18 |
| Lincoln | 219 | 20,179 | 0.56 | 317 | 13.0 | 317 | 13.0 | 1,863 | 9.5 | 6,004 | 3,864 | 2,140 | 5 | 57 |
| Livingston | 69 | 7,918 | 0.65 | 82 | 9.1 | 134 | 14.8 | 552 | 7.8 | 2,622 | 1,852 | 770 | 2 | 52 |
| Logan | 277 | 25,111 | 0.84 | 377 | 13.8 | 318 | 11.6 | 1,811 | 8.3 | 6,218 | 4,191 | 2,027 | 26 | 80 |
| Lyon | 1,123 | 7,859 | 0.86 | 45 | 5.5 | 131 | 16.1 | 422 | 8.5 | 2,374 | 1,617 | 758 | 3 | 60 |
| McCracken | 1,190 | 77,086 | 1.41 | 735 | 11.2 | 863 | 13.1 | 3,612 | 7.0 | 16,434 | 11,212 | 5,222 | 38 | 472 |
| McCreary | 1,903 | 15,814 | 0.65 | 196 | 11.5 | 201 | 11.8 | 1,090 | 8.7 | 3,883 | 2,592 | 1,291 | 1 | 58 |
| McLean | 82 | 7,810 | 0.59 | 96 | 10.6 | 116 | 12.8 | 543 | 7.4 | 2,385 | 1,499 | 886 | 2 | 53 |
| Madison | 6,480 | 87,643 | 0.92 | 1,004 | 10.7 | 843 | 8.9 | 5,166 | 7.0 | 17,034 | 9,376 | 7,659 | 18 | 226 |
| Magoffin | 127 | 11,508 | 0.70 | 135 | 11.2 | 180 | 15.0 | 982 | 10.0 | 3,255 | 1,855 | 1,401 | 1 | 19 |
| Marion | 504 | 20,450 | 1.16 | 229 | 11.9 | 207 | 10.7 | 1,175 | 7.5 | 4,190 | 2,831 | 1,359 | 3 | 49 |
| Marshall | 464 | 30,547 | 0.95 | 278 | 8.9 | 460 | 14.8 | 1,723 | 7.2 | 8,633 | 5,872 | 2,761 | 19 | 204 |
| Martin | 1,376 | 11,485 | 0.95 | 102 | 9.2 | 151 | 13.7 | 617 | 7.9 | 2,788 | 1,763 | 1,025 | 4 | 13 |
| Mason | 298 | 18,954 | 1.26 | 210 | 12.3 | 221 | 13.0 | 1,114 | 8.2 | 4,026 | 2,495 | 1,531 | 4 | 73 |
| Meade | 212 | 21,476 | 0.47 | 260 | 9.1 | 259 | 9.1 | 1,548 | 6.4 | 5,494 | 3,726 | 1,768 | 14 | 122 |
| Menifee | 271 | 5,222 | 0.48 | 60 | 9.2 | 92 | 14.1 | 406 | 8.1 | 1,750 | 922 | 828 | 0 | 10 |
| Mercer | 128 | 19,370 | 0.76 | 241 | 11.0 | 300 | 13.7 | 1,401 | 7.9 | 5,350 | 3,252 | 2,098 | 2 | 49 |
| Metcalfe | 121 | 8,823 | 0.68 | 114 | 11.3 | 164 | 16.3 | 652 | 8.2 | 2,566 | 1,557 | 1,009 | 2 | 34 |
| Monroe | 143 | 9,820 | 0.81 | 129 | 12.2 | 155 | 14.7 | 867 | 10.3 | 2,709 | 1,886 | 823 | 0 | 9 |
| Montgomery | 357 | 28,860 | 1.08 | 345 | 12.2 | 319 | 11.3 | 1,738 | 7.5 | 6,045 | 3,386 | 2,660 | 10 | 285 |
| Morgan | 1,811 | 13,142 | 0.96 | 115 | 8.8 | 164 | 12.5 | 844 | 9.4 | 3,041 | 1,650 | 1,390 | 2 | 14 |
| Muhlenberg | 2,065 | 29,161 | 0.84 | 358 | 11.8 | 438 | 14.4 | 1,934 | 8.3 | 7,703 | 4,742 | 2,961 | 14 | 66 |
| Nelson | 496 | 40,977 | 0.78 | 551 | 11.9 | 453 | 9.8 | 2,420 | 6.3 | 10,040 | 6,778 | 3,262 | 22 | 213 |
| Nicholas | 95 | 5,740 | 0.48 | 81 | 11.2 | 109 | 15.1 | 610 | 10.3 | 1,742 | 1,130 | 612 | 0 | 2 |

1. Per 1,000 estimated resident population.

# Table B. States and Counties — Crime, Education, Money Income, and Poverty

| STATE County | County law enforcement employment, 2019 | | School enrollment and attainment, 2015-2019 | | | | Local government expenditures,[3] 2017-2018 | | Money income, 2015-2019 | | | | Income and poverty, 2019 | | | |
| | | | Enrollment[1] | | Attainment[2] (percent) | | | | | | Households | | | Percent below poverty level | | |
| | | | | | | | | | | | | Percent | | | | |
| | Officers | Civilians | Total | Percent private | High school graduate or less | Bachelor's degree or more | Total current spending (mil dol) | Current spending per student (dollars) | Per capita income[4] | Median income (dollars) | with income of less than $50,000 | with income of $200,000 or more | Median household income (dollars) | All persons | Children under 18 years | Children 5 to 17 years in families |
| | 46 | 47 | 48 | 49 | 50 | 51 | 52 | 53 | 54 | 55 | 56 | 57 | 58 | 59 | 60 | 61 |
| **KENTUCKY—Cont'd** | | | | | | | | | | | | | | | | |
| Clay | 10 | 6 | 3,913 | 7.6 | 70.6 | 9.8 | 36 | 10,920 | 15,391 | 26,840 | 71.9 | 0.9 | 30,175 | 32.6 | 42.6 | 42.5 |
| Clinton | NA | NA | 2,076 | 13.2 | 66.6 | 10.3 | 20 | 10,715 | 21,019 | 32,184 | 68.1 | 2.0 | 34,005 | 23.4 | 32.6 | 31.3 |
| Crittenden | 5 | 1 | 1,929 | 10.3 | 58.2 | 13.1 | 14 | 10,254 | 23,959 | 45,244 | 54.6 | 2.5 | 44,595 | 19.2 | 31.9 | 29.4 |
| Cumberland | 3 | 2 | 1,024 | 12.6 | 61.3 | 13.3 | 11 | 11,201 | 18,728 | 35,344 | 69.4 | 0.5 | 34,897 | 23.0 | 33.1 | 31.4 |
| Daviess | 35 | 31 | 24,333 | 15.3 | 44.1 | 22.9 | 179 | 10,606 | 28,806 | 51,673 | 48.4 | 3.5 | 52,620 | 15.4 | 19.5 | 17.5 |
| Edmonson | 6 | 3 | 2,089 | 6.5 | 61.2 | 10.3 | 20 | 10,022 | 21,802 | 43,401 | 57.0 | 1.1 | 47,482 | 16.0 | 21.9 | 21.2 |
| Elliott | NA | NA | 1,321 | 4.5 | 68.3 | 9.7 | 11 | 10,892 | 13,856 | 32,306 | 66.0 | 0.2 | 34,262 | 27.7 | 31.5 | 29.4 |
| Estill | 3 | 2 | 2,769 | 4.5 | 70.2 | 9.9 | 24 | 10,231 | 17,746 | 31,688 | 67.3 | 1.0 | 35,004 | 22.7 | 32.2 | 31.7 |
| Fayette | 62 | 27 | 90,758 | 14.8 | 28.4 | 43.6 | 535 | 12,855 | 34,442 | 57,291 | 43.9 | 6.2 | 58,981 | 15.4 | 16.7 | 14.6 |
| Fleming | 6 | 1 | 3,196 | 7.7 | 60.7 | 12.9 | 24 | 10,482 | 23,593 | 44,612 | 55.6 | 2.7 | 43,943 | 21.1 | 33.0 | 32.8 |
| Floyd | 10 | 10 | 7,685 | 10.6 | 62.5 | 11.3 | 64 | 10,691 | 19,471 | 32,730 | 67.9 | 1.2 | 35,828 | 27.4 | 34.2 | 34.1 |
| Franklin | 33 | 3 | 11,532 | 17.4 | 42.4 | 29.9 | 76 | 10,509 | 30,711 | 56,274 | 43.1 | 2.7 | 58,917 | 11.5 | 16.8 | 14.9 |
| Fulton | 4 | 1 | 1,293 | 4.3 | 58.4 | 13.4 | 11 | 11,390 | 18,247 | 30,114 | 71.2 | 0.6 | 34,376 | 25.6 | 37.8 | 35.3 |
| Gallatin | NA | NA | 1,988 | 6.1 | 63.0 | 11.6 | 17 | 10,439 | 23,638 | 52,167 | 47.0 | 0.2 | 55,273 | 13.3 | 17.3 | 16.6 |
| Garrard | 10 | 1 | 3,566 | 12.5 | 56.0 | 17.6 | 27 | 9,850 | 26,272 | 52,631 | 47.8 | 3.1 | 51,757 | 15.7 | 22.4 | 21.3 |
| Grant | 16 | 2 | 5,907 | 12.1 | 60.8 | 12.9 | 48 | 10,194 | 23,721 | 48,714 | 50.8 | 2.4 | 55,210 | 13.3 | 18.7 | 17.2 |
| Graves | 17 | 3 | 8,696 | 7.3 | 52.1 | 17.4 | 65 | 10,239 | 24,750 | 44,043 | 54.5 | 2.4 | 48,038 | 17.9 | 25.6 | 24.6 |
| Grayson | 9 | 7 | 5,716 | 7.2 | 61.5 | 10.5 | 43 | 10,002 | 21,231 | 38,612 | 61.6 | 1.1 | 43,771 | 20.3 | 29.6 | 23.8 |
| Green | 4 | 0 | 2,171 | 11.4 | 62.3 | 13.2 | 18 | 11,050 | 22,506 | 35,899 | 63.2 | 1.7 | 38,896 | 18.6 | 23.2 | 21.7 |
| Greenup | 14 | 3 | 7,917 | 4.8 | 47.0 | 17.1 | 64 | 10,311 | 27,722 | 51,655 | 48.4 | 3.9 | 52,179 | 15.9 | 21.9 | 19.1 |
| Hancock | NA | NA | 2,139 | 9.6 | 57.9 | 11.4 | 19 | 10,999 | 24,830 | 57,217 | 43.2 | 0.6 | 60,626 | 12.2 | 15.6 | 14.2 |
| Hardin | 40 | 16 | 28,416 | 13.8 | 39.9 | 22.0 | 177 | 10,233 | 28,606 | 54,367 | 45.7 | 2.9 | 57,711 | 10.7 | 13.3 | 12.3 |
| Harlan | 18 | 3 | 5,947 | 10.9 | 61.8 | 10.8 | 49 | 10,125 | 16,468 | 26,478 | 72.2 | 0.6 | 30,050 | 31.1 | 38.9 | 38.2 |
| Harrison | NA | NA | 3,717 | 9.5 | 58.3 | 14.6 | 30 | 10,202 | 24,154 | 48,438 | 50.9 | 2.4 | 51,475 | 15.3 | 20.5 | 19.5 |
| Hart | 8 | 10 | 3,840 | 9.1 | 64.1 | 11.4 | 27 | 11,342 | 20,742 | 38,396 | 62.0 | 1.6 | 42,718 | 20.1 | 27.0 | 28.4 |
| Henderson | 19 | 5 | 10,124 | 11.4 | 49.1 | 18.2 | 75 | 10,033 | 26,212 | 48,926 | 50.8 | 2.9 | 53,199 | 16.3 | 22.0 | 19.1 |
| Henry | NA | NA | 3,461 | 15.2 | 56.1 | 13.1 | 32 | 10,345 | 25,202 | 53,926 | 46.4 | 1.9 | 51,806 | 14.7 | 21.8 | 20.8 |
| Hickman | 3 | 2 | 911 | 6.7 | 61.6 | 15.8 | 9 | 11,288 | 28,114 | 42,929 | 57.5 | 4.9 | 52,331 | 16.9 | 27.8 | 25.1 |
| Hopkins | 32 | 8 | 9,909 | 10.8 | 53.7 | 16.5 | 77 | 10,303 | 25,042 | 47,170 | 53.1 | 2.2 | 48,769 | 18.1 | 24.7 | 22.6 |
| Jackson | 8 | 1 | 2,653 | 7.4 | 68.0 | 13.8 | 25 | 11,362 | 18,560 | 32,138 | 65.5 | 1.4 | 34,940 | 27.8 | 36.0 | 33.8 |
| Jefferson | 223 | 54 | 178,483 | 24.8 | 36.1 | 33.4 | 1,388 | 13,986 | 33,251 | 56,586 | 44.5 | 5.6 | 59,025 | 14.2 | 19.5 | 17.6 |
| Jessamine | NA | NA | 13,804 | 24.3 | 43.2 | 29.5 | 84 | 10,066 | 31,146 | 58,245 | 43.5 | 5.9 | 62,533 | 11.0 | 16.7 | 15.3 |
| Johnson | 8 | 2 | 5,107 | 10.7 | 55.7 | 16.3 | 48 | 10,793 | 21,974 | 37,055 | 62.6 | 2.0 | 31,202 | 25.8 | 31.7 | 26.7 |
| Kenton | 72 | 10 | 40,078 | 24.2 | 38.8 | 31.7 | 245 | 10,569 | 33,330 | 64,339 | 38.4 | 5.5 | 68,692 | 11.2 | 15.9 | 14.4 |
| Knott | 5 | 0 | 3,332 | 21.5 | 58.6 | 14.9 | 27 | 11,550 | 18,475 | 31,198 | 68.2 | 0.9 | 32,613 | 30.5 | 34.1 | 31.0 |
| Knox | NA | NA | 7,464 | 10.7 | 64.5 | 14.2 | 59 | 11,491 | 17,832 | 30,181 | 70.1 | 1.2 | 31,634 | 31.5 | 39.3 | 37.1 |
| Larue | 5 | 2 | 2,915 | 9.0 | 55.2 | 12.4 | 26 | 10,703 | 24,803 | 47,643 | 52.7 | 4.2 | 49,785 | 15.4 | 23.2 | 19.8 |
| Laurel | NA | NA | 13,511 | 9.8 | 58.0 | 14.4 | 94 | 9,639 | 22,398 | 41,526 | 58.4 | 2.4 | 44,033 | 21.4 | 27.2 | 26.1 |
| Lawrence | 11 | 2 | 3,088 | 9.6 | 67.4 | 9.8 | 27 | 10,289 | 18,124 | 32,798 | 67.0 | 0.8 | 38,055 | 23.4 | 28.6 | 28.2 |
| Lee | NA | NA | 1,403 | 7.9 | 68.2 | 8.3 | 10 | 10,881 | 17,386 | 25,275 | 74.2 | 1.3 | 30,905 | 34.9 | 44.3 | 43.6 |
| Leslie | 7 | 1 | 1,998 | 5.1 | 70.5 | 8.7 | 19 | 10,502 | 15,838 | 31,627 | 70.8 | 0.4 | 32,202 | 32.3 | 38.5 | 36.8 |
| Letcher | 5 | 4 | 4,599 | 9.2 | 61.3 | 12.1 | 40 | 11,358 | 20,641 | 29,886 | 69.2 | 1.3 | 32,435 | 28.9 | 34.7 | 31.9 |
| Lewis | 6 | 2 | 2,741 | 8.3 | 67.0 | 9.1 | 23 | 10,026 | 19,860 | 31,147 | 65.3 | 1.1 | 37,410 | 23.2 | 31.1 | 29.6 |
| Lincoln | 8 | 3 | 5,018 | 11.0 | 62.8 | 13.0 | 43 | 11,050 | 21,480 | 42,919 | 56.3 | 1.5 | 43,220 | 19.7 | 25.7 | 24.7 |
| Livingston | 9 | 1 | 1,779 | 18.4 | 59.8 | 11.8 | 15 | 12,441 | 27,728 | 50,839 | 49.3 | 1.9 | 50,521 | 12.2 | 20.6 | 19.3 |
| Logan | 23 | 3 | 5,895 | 4.9 | 59.5 | 16.4 | 48 | 10,357 | 24,221 | 48,014 | 52.2 | 2.2 | 48,902 | 15.3 | 21.5 | 20.8 |
| Lyon | 7 | 3 | 1,088 | 8.0 | 53.3 | 16.9 | 9 | 9,730 | 26,994 | 52,528 | 49.2 | 1.5 | 52,978 | 14.5 | 20.2 | 17.6 |
| McCracken | 49 | 5 | 14,026 | 9.2 | 40.9 | 24.2 | 109 | 10,699 | 32,357 | 46,080 | 52.7 | 4.2 | 48,219 | 15.5 | 22.7 | 21.6 |
| McCreary | NA | NA | 3,901 | 13.6 | 64.4 | 7.2 | 32 | 10,722 | 14,505 | 28,105 | 71.6 | 0.7 | 32,701 | 34.5 | 43.6 | 40.8 |
| McLean | NA | NA | 2,054 | 10.1 | 53.9 | 13.5 | 16 | 10,200 | 25,610 | 51,861 | 46.5 | 1.2 | 51,184 | 14.0 | 18.6 | 16.4 |
| Madison | 31 | 3 | 28,119 | 14.1 | 40.6 | 30.8 | 126 | 9,728 | 25,837 | 50,060 | 49.9 | 2.9 | 50,316 | 17.7 | 18.2 | 15.6 |
| Magoffin | NA | NA | 2,625 | 9.1 | 66.7 | 9.5 | 24 | 11,017 | 18,243 | 28,147 | 71.9 | 0.9 | 32,045 | 29.4 | 39.1 | 33.0 |
| Marion | 7 | 3 | 4,626 | 9.2 | 63.9 | 12.6 | 37 | 11,318 | 22,241 | 40,107 | 59.0 | 1.7 | 48,062 | 16.8 | 20.0 | 18.8 |
| Marshall | 31 | 11 | 6,214 | 7.3 | 49.6 | 19.2 | 50 | 10,376 | 29,792 | 55,113 | 45.6 | 3.4 | 57,945 | 11.8 | 16.9 | 16.0 |
| Martin | 1 | 3 | 2,114 | 8.0 | 64.6 | 11.7 | 22 | 11,242 | 18,017 | 41,013 | 61.3 | 0.4 | 34,291 | 34.4 | 38.3 | 34.3 |
| Mason | 11 | 8 | 3,639 | 11.0 | 48.4 | 19.5 | 29 | 10,281 | 28,411 | 45,377 | 53.6 | 3.4 | 47,861 | 15.7 | 23.2 | 20.8 |
| Meade | 16 | 2 | 6,480 | 9.2 | 46.8 | 19.4 | 45 | 9,075 | 27,861 | 56,603 | 41.0 | 1.9 | 54,946 | 11.4 | 13.1 | 11.0 |
| Menifee | 5 | 0 | 1,297 | 5.7 | 67.3 | 11.2 | 11 | 11,188 | 20,588 | 39,325 | 61.3 | 0.0 | 37,457 | 26.1 | 38.5 | 36.0 |
| Mercer | 9 | 3 | 4,710 | 7.4 | 52.4 | 18.5 | 37 | 11,010 | 27,061 | 55,093 | 45.9 | 2.4 | 55,267 | 13.8 | 19.8 | 18.6 |
| Metcalfe | 4 | 1 | 2,041 | 13.4 | 63.0 | 13.0 | 17 | 10,690 | 19,121 | 37,386 | 65.1 | 0.9 | 38,978 | 22.6 | 32.3 | 29.2 |
| Monroe | 5 | 1 | 2,136 | 9.5 | 64.5 | 14.4 | 23 | 12,007 | 23,481 | 34,879 | 61.4 | 2.5 | 36,512 | 21.7 | 30.8 | 30.9 |
| Montgomery | 16 | 2 | 6,239 | 11.0 | 55.5 | 15.7 | 48 | 10,112 | 22,381 | 45,516 | 52.0 | 1.5 | 52,903 | 15.8 | 22.0 | 19.6 |
| Morgan | 5 | 1 | 2,238 | 7.1 | 65.1 | 12.5 | 21 | 10,555 | 18,310 | 36,134 | 62.8 | 1.6 | 37,433 | 26.5 | 33.2 | 26.8 |
| Muhlenberg | 17 | 2 | 6,258 | 6.2 | 62.6 | 12.6 | 50 | 10,645 | 22,612 | 43,590 | 56.1 | 1.6 | 45,049 | 18.8 | 24.5 | 23.5 |
| Nelson | NA | NA | 10,252 | 20.7 | 47.5 | 18.0 | 80 | 10,945 | 32,735 | 60,127 | 40.5 | 3.7 | 64,795 | 10.5 | 15.1 | 14.0 |
| Nicholas | 2 | 1 | 1,446 | 18.9 | 66.3 | 8.9 | 11 | 10,059 | 22,212 | 38,149 | 61.3 | 2.0 | 38,084 | 17.3 | 26.7 | 25.7 |

1. All persons 3 years old and over enrolled in nursery school through college.   2. Persons 25 years old and over.   3. Elementary and secondary education expenditures.   4. Based on population estimated by the American Community Survey, 2014–2018.

# Table B. States and Counties — Personal Income

| STATE County | Personal income, 2019 | | | | | | | | | | Earnings, 2019 | | |
|---|---|---|---|---|---|---|---|---|---|---|---|---|---|
| | Total (mil dol) | Percent change 2018-2019 | Per capita[1] Dollars | Per capita[1] Rank | Wages and salaries (mil dol) | Supplements to wages and salaries, employer contributions (mil dol) Pension and insurance | Government social insurance | Proprietors' income (mil dol) | Dividends, interest, and rent (mil dol) | Personal transfer receipts (mil dol) | Total (mil dol) | Contributions for government social insurance (mil dol) From employee and self-employed | From employer |
| | 62 | 63 | 64 | 65 | 66 | 67 | 68 | 69 | 70 | 71 | 72 | 73 | 74 |

KENTUCKY—Cont'd

| | | | | | | | | | | | | | |
|---|---|---|---|---|---|---|---|---|---|---|---|---|---|
| Clay | 621 | 1.2 | 31,200 | 2,994 | 156 | 39 | 12 | 33 | 62 | 317 | 240 | 20 | 12 |
| Clinton | 323 | 1.2 | 31,601 | 2,969 | 128 | 29 | 11 | 20 | 36 | 143 | 187 | 13 | 11 |
| Crittenden | 320 | 3.8 | 36,354 | 2,529 | 68 | 17 | 5 | 23 | 46 | 110 | 113 | 9 | 5 |
| Cumberland | 241 | 4.9 | 36,467 | 2,513 | 78 | 16 | 6 | 15 | 27 | 108 | 116 | 9 | 6 |
| Daviess | 4,301 | 2.7 | 42,372 | 1,708 | 2,087 | 367 | 156 | 193 | 772 | 1,098 | 2,802 | 184 | 156 |
| Edmonson | 414 | 3.2 | 34,037 | 2,798 | 60 | 16 | 5 | 25 | 47 | 140 | 106 | 10 | 5 |
| Elliott | 172 | 2.6 | 22,828 | 3,107 | 32 | 9 | 2 | 6 | 19 | 86 | 49 | 5 | 2 |
| Estill | 464 | 2.7 | 32,928 | 2,896 | 20 | 20 | 7 | 11 | 47 | 186 | 132 | 13 | 7 |
| Fayette | 16,709 | 3.4 | 51,707 | 705 | 10,442 | 1,840 | 751 | 1,200 | 4,214 | 2,471 | 14,232 | 813 | 751 |
| Fleming | 506 | 3.1 | 34,732 | 2,724 | 128 | 28 | 10 | 45 | 61 | 162 | 211 | 16 | 10 |
| Floyd | 1,322 | 2.0 | 37,133 | 2,430 | 444 | 90 | 33 | 133 | 148 | 605 | 699 | 54 | 33 |
| Franklin | 2,206 | 3.6 | 43,271 | 1,579 | 1,516 | 333 | 103 | 91 | 412 | 595 | 2,043 | 121 | 103 |
| Fulton | 210 | 4.0 | 35,240 | 2,653 | 77 | 18 | 7 | 12 | 35 | 90 | 114 | 8 | 7 |
| Gallatin | 299 | 4.6 | 33,697 | 2,828 | 142 | 24 | 10 | 7 | 26 | 73 | 183 | 12 | 10 |
| Garrard | 623 | 3.4 | 35,251 | 2,651 | 89 | 19 | 7 | 28 | 84 | 174 | 143 | 14 | 7 |
| Grant | 930 | 4.0 | 37,083 | 2,437 | 218 | 39 | 16 | 42 | 96 | 253 | 315 | 25 | 16 |
| Graves | 1,434 | 1.9 | 38,485 | 2,262 | 479 | 93 | 37 | 135 | 207 | 441 | 743 | 50 | 37 |
| Grayson | 910 | 4.3 | 34,450 | 2,749 | 301 | 63 | 23 | 49 | 119 | 303 | 437 | 32 | 23 |
| Green | 383 | 1.7 | 34,997 | 2,687 | 63 | 14 | 5 | 22 | 50 | 145 | 104 | 10 | 5 |
| Greenup | 1,492 | 2.0 | 42,516 | 1,687 | 370 | 72 | 32 | 62 | 179 | 469 | 537 | 45 | 32 |
| Hancock | 338 | 3.5 | 38,722 | 2,233 | 294 | 46 | 21 | 22 | 40 | 88 | 383 | 25 | 21 |
| Hardin | 4,909 | 4.4 | 44,244 | 1,453 | 2,612 | 618 | 209 | 347 | 814 | 1,156 | 3,786 | 216 | 209 |
| Harlan | 828 | 1.8 | 31,824 | 2,956 | 234 | 53 | 17 | 22 | 97 | 451 | 327 | 29 | 17 |
| Harrison | 690 | 5.1 | 36,545 | 2,502 | 231 | 47 | 18 | 35 | 98 | 188 | 331 | 24 | 18 |
| Hart | 640 | 3.3 | 33,648 | 2,830 | 204 | 41 | 16 | 116 | 83 | 197 | 377 | 26 | 16 |
| Henderson | 1,866 | 3.9 | 41,279 | 1,883 | 884 | 159 | 67 | 90 | 270 | 502 | 1,199 | 79 | 67 |
| Henry | 632 | 4.2 | 39,175 | 2,166 | 119 | 27 | 9 | 30 | 85 | 163 | 185 | 15 | 9 |
| Hickman | 187 | -2.2 | 42,635 | 1,667 | 32 | 7 | 3 | 45 | 29 | 65 | 86 | 5 | 3 |
| Hopkins | 1,802 | 3.1 | 40,317 | 2,012 | 760 | 139 | 60 | 110 | 270 | 539 | 1,069 | 72 | 60 |
| Jackson | 398 | 3.0 | 29,848 | 3,042 | 60 | 17 | 4 | 13 | 44 | 159 | 95 | 10 | 4 |
| Jefferson | 41,524 | 3.2 | 54,155 | 536 | 28,636 | 4,157 | 2,088 | 3,246 | 8,230 | 7,567 | 38,127 | 2,339 | 2,088 |
| Jessamine | 2,503 | 3.8 | 46,250 | 1,220 | 752 | 127 | 58 | 301 | 407 | 429 | 1,238 | 76 | 58 |
| Johnson | 762 | 3.3 | 34,356 | 2,761 | 196 | 43 | 15 | 29 | 98 | 333 | 282 | 24 | 15 |
| Kenton | 9,744 | 3.7 | 58,349 | 361 | 4,178 | 661 | 296 | 693 | 2,389 | 1,432 | 5,828 | 364 | 296 |
| Knott | 473 | 3.1 | 31,929 | 2,947 | 86 | 20 | 7 | 19 | 47 | 238 | 131 | 13 | 7 |
| Knox | 938 | 2.6 | 30,105 | 3,032 | 284 | 64 | 22 | 51 | 100 | 436 | 421 | 35 | 22 |
| Larue | 552 | 5.6 | 38,316 | 2,289 | 94 | 20 | 7 | 22 | 81 | 160 | 144 | 13 | 7 |
| Laurel | 2,085 | 1.8 | 34,289 | 2,771 | 1,036 | 182 | 82 | 124 | 235 | 685 | 1,424 | 99 | 82 |
| Lawrence | 509 | 3.3 | 33,213 | 2,875 | 135 | 27 | 11 | 21 | 53 | 219 | 193 | 16 | 11 |
| Lee | 216 | 3.3 | 29,162 | 3,058 | 64 | 13 | 5 | 10 | 25 | 111 | 91 | 8 | 5 |
| Leslie | 337 | 1.9 | 34,079 | 2,792 | 62 | 15 | 5 | 12 | 31 | 177 | 93 | 9 | 5 |
| Letcher | 687 | 3.1 | 31,871 | 2,952 | 166 | 36 | 12 | 23 | 69 | 368 | 237 | 22 | 12 |
| Lewis | 427 | 3.7 | 32,147 | 2,937 | 71 | 16 | 5 | 24 | 42 | 161 | 117 | 11 | 5 |
| Lincoln | 807 | 3.6 | 32,881 | 2,898 | 153 | 34 | 11 | 44 | 108 | 280 | 242 | 22 | 11 |
| Livingston | 354 | 4.3 | 38,515 | 2,258 | 129 | 24 | 10 | 14 | 51 | 118 | 178 | 14 | 10 |
| Logan | 998 | 4.7 | 36,834 | 2,467 | 431 | 79 | 33 | 87 | 144 | 300 | 630 | 43 | 33 |
| Lyon | 306 | 4.6 | 37,276 | 2,419 | 90 | 21 | 6 | 15 | 53 | 103 | 132 | 10 | 6 |
| McCracken | 3,211 | 4.5 | 49,087 | 917 | 1,879 | 326 | 141 | 280 | 612 | 789 | 2,625 | 168 | 141 |
| McCreary | 458 | 2.5 | 26,600 | 3,095 | 107 | 30 | 8 | 19 | 54 | 243 | 164 | 15 | 8 |
| McLean | 361 | -2.1 | 39,163 | 2,168 | 82 | 17 | 6 | 41 | 46 | 111 | 145 | 10 | 6 |
| Madison | 3,378 | 4.0 | 36,331 | 2,531 | 1,529 | 309 | 115 | 112 | 470 | 818 | 2,065 | 133 | 115 |
| Magoffin | 401 | 4.2 | 32,995 | 2,890 | 62 | 15 | 5 | 18 | 43 | 197 | 99 | 10 | 5 |
| Marion | 720 | 3.2 | 37,340 | 2,406 | 377 | 72 | 29 | 38 | 97 | 204 | 516 | 34 | 29 |
| Marshall | 1,317 | 4.1 | 42,342 | 1,714 | 611 | 109 | 47 | 69 | 211 | 376 | 836 | 58 | 47 |
| Martin | 335 | 0.4 | 29,905 | 3,039 | 109 | 27 | 8 | 15 | 53 | 175 | 160 | 13 | 8 |
| Mason | 706 | 4.0 | 41,350 | 1,868 | 379 | 77 | 29 | 39 | 121 | 194 | 524 | 34 | 29 |
| Meade | 1,170 | 3.9 | 40,942 | 1,923 | 182 | 40 | 14 | 42 | 172 | 257 | 278 | 23 | 14 |
| Menifee | 206 | 5.6 | 31,681 | 2,962 | 35 | 9 | 3 | 6 | 22 | 88 | 53 | 5 | 3 |
| Mercer | 829 | 3.9 | 37,806 | 2,344 | 330 | 65 | 25 | 53 | 123 | 235 | 474 | 34 | 25 |
| Metcalfe | 321 | 3.1 | 31,848 | 2,953 | 75 | 17 | 6 | 15 | 37 | 127 | 114 | 10 | 6 |
| Monroe | 392 | -0.2 | 36,837 | 2,464 | 120 | 26 | 9 | 45 | 44 | 146 | 200 | 14 | 9 |
| Montgomery | 1,011 | 4.1 | 35,915 | 2,578 | 441 | 79 | 36 | 50 | 127 | 283 | 606 | 42 | 36 |
| Morgan | 372 | 3.7 | 27,978 | 3,081 | 115 | 31 | 9 | 16 | 45 | 154 | 170 | 13 | 9 |
| Muhlenberg | 1,076 | 4.2 | 35,138 | 2,671 | 368 | 75 | 29 | 53 | 141 | 369 | 525 | 39 | 29 |
| Nelson | 2,025 | 4.8 | 43,809 | 1,515 | 744 | 134 | 56 | 71 | 276 | 438 | 1,006 | 70 | 56 |
| Nicholas | 253 | 4.0 | 34,856 | 2,707 | 31 | 8 | 2 | 11 | 28 | 87 | 52 | 5 | 2 |

1. Based on the resident population estimated as of July 1 of the year shown.

Items 62—74

# Table B. States and Counties — Earnings, Social Security, and Housing

| STATE County | Earnings, 2019 (cont.) — Percent by selected industries | | | | | | | | | Social Security beneficiaries, December 2019 | | Supplemental Security Income recipients, 2019 | Housing units, 2020 | |
| | Farm | Mining, quarrying, and extractions | Construction | Manu-facturing | Information; professional, scientific, technical services | Retail trade | Finance, insurance, real estate, and leasing | Health care and social assistance | Govern-ment | Number | Rate[1] | | Total | Percent change, 2010-2020 |
| | 75 | 76 | 77 | 78 | 79 | 80 | 81 | 82 | 83 | 84 | 85 | 86 | 87 | 88 |
|---|---|---|---|---|---|---|---|---|---|---|---|---|---|---|
| **KENTUCKY—Cont'd** | | | | | | | | | | | | | | |
| Clay | 0.3 | 0.7 | 3.8 | 3.8 | D | 8.6 | 2.6 | D | 37.9 | 5,365 | 269 | 2,833 | 9,116 | 2.7 |
| Clinton | 2.3 | D | 4.2 | 34.6 | 1.8 | 6.0 | D | 15.5 | 16.1 | 2,725 | 267 | 787 | 5,327 | 0.2 |
| Crittenden | 3.2 | D | D | 22.3 | D | 6.7 | 5.1 | 15.5 | 18.2 | 2,455 | 279 | 258 | 4,595 | 0.6 |
| Cumberland | -0.4 | 0.1 | D | 8.6 | D | 4.4 | 3.8 | 40.2 | 15.1 | 1,965 | 297 | 413 | 3,684 | 0.0 |
| Daviess | 1.2 | 0.4 | 4.4 | 15.6 | 4.7 | 7.2 | 9.2 | 20.5 | 11.8 | 23,945 | 236 | 3,303 | 43,712 | 5.4 |
| Edmonson | 4.1 | 0.1 | 16.5 | D | D | 4.7 | D | D | 37.4 | 3,225 | 265 | 374 | 6,642 | 2.5 |
| Elliott | -1.5 | 0.1 | 4.7 | 0.4 | D | 3.6 | D | 11.2 | 60.1 | 1,770 | 236 | 432 | 3,495 | 3.7 |
| Estill | -1.8 | 2.0 | 5.8 | D | 2.6 | 6.9 | 4.3 | 19.8 | 25.6 | 3,745 | 265 | 1,185 | 6,946 | 1.1 |
| Fayette | 1.7 | 0.1 | 7.1 | 7.4 | 10.4 | 5.7 | 5.9 | D | 18.1 | 50,730 | 157 | 6,613 | 145,159 | 7.4 |
| Fleming | 0.6 | D | D | 15.4 | 2.5 | 7.9 | 5.3 | D | 17.1 | 4,010 | 275 | 662 | 6,780 | 2.4 |
| Floyd | 0.1 | 6.0 | 6.9 | 5.5 | 9.4 | 7.1 | 2.4 | 21.4 | 17.1 | 11,725 | 330 | 3,726 | 18,679 | 2.8 |
| Franklin | 0.1 | 0.2 | 3.8 | 10.2 | 6.6 | 4.2 | 4.3 | 8.5 | 43.9 | 14,635 | 288 | 1,495 | 23,335 | 0.8 |
| Fulton | 7.6 | D | 3.3 | 19.5 | D | 10.8 | D | D | 23.6 | 1,920 | 321 | 424 | 3,323 | -1.4 |
| Gallatin | 0.1 | 0.0 | 2.3 | D | D | 3.4 | D | 6.2 | 12.8 | 1,300 | 147 | 189 | 3,983 | 5.3 |
| Garrard | -4.9 | D | 26.6 | D | 5.1 | 4.4 | 4.3 | D | 24.3 | 4,370 | 247 | 601 | 7,509 | 0.6 |
| Grant | -2.3 | 0.0 | D | 11.7 | 8.0 | 11.1 | 3.1 | D | 20.7 | 5,835 | 232 | 812 | 10,403 | 4.6 |
| Graves | 8.3 | D | 5.3 | 15.1 | 4.6 | 7.8 | 4.7 | D | 15.4 | 9,435 | 254 | 1,437 | 16,887 | 0.6 |
| Grayson | 0.8 | D | 9.3 | 26.7 | D | 7.0 | 2.6 | D | 20.7 | 7,115 | 269 | 1,215 | 13,774 | 1.6 |
| Green | 0.5 | 0.2 | 6.6 | 3.9 | 6.1 | 8.6 | 5.7 | 20.2 | 23.6 | 3,115 | 285 | 512 | 5,340 | 0.0 |
| Greenup | -0.4 | D | D | 11.8 | D | 6.0 | 3.4 | 24.2 | 15.3 | 9,265 | 263 | 1,356 | 16,297 | -0.2 |
| Hancock | 0.9 | 0.0 | D | 75.2 | D | 1.4 | 0.9 | D | 6.4 | 2,280 | 262 | 279 | 3,767 | 0.9 |
| Hardin | -0.2 | D | 2.9 | 14.6 | 7.6 | 6.2 | 4.3 | 6.3 | 42.6 | 22,620 | 204 | 3,335 | 47,568 | 10.0 |
| Harlan | -0.1 | D | 2.4 | 1.1 | 6.1 | 7.7 | 2.3 | D | 27.2 | 8,325 | 321 | 2,556 | 13,619 | 0.8 |
| Harrison | -1.7 | D | 8.4 | 33.4 | 2.5 | 6.4 | 2.0 | D | 13.5 | 4,635 | 246 | 716 | 8,400 | 2.4 |
| Hart | 1.6 | D | D | 42.2 | D | 4.2 | 1.6 | D | 11.9 | 4,535 | 239 | 851 | 9,328 | 9.0 |
| Henderson | 1.4 | 0.2 | 5.3 | 32.4 | 2.9 | 5.7 | 3.5 | D | 12.1 | 11,100 | 245 | 1,556 | 20,623 | 1.5 |
| Henry | -0.3 | D | 8.5 | 17.8 | D | 4.2 | D | 7.1 | 24.7 | 3,935 | 245 | 491 | 6,847 | 3.2 |
| Hickman | 42.4 | 0.0 | D | D | D | 2.7 | D | 7.1 | 13.7 | 1,385 | 317 | 158 | 2,371 | 1.1 |
| Hopkins | 2.8 | D | 7.3 | 15.2 | D | 7.1 | 3.7 | 17.4 | 15.8 | 11,710 | 262 | 1,821 | 21,610 | 2.0 |
| Jackson | -3.3 | D | D | 4.4 | 5.8 | 5.1 | 3.8 | D | 35.5 | 3,470 | 261 | 1,172 | 6,676 | 2.4 |
| Jefferson | 0.0 | 0.0 | 5.0 | 13.1 | 10.2 | 4.7 | 12.8 | 13.2 | 10.1 | 154,065 | 201 | 24,618 | 351,734 | 4.2 |
| Jessamine | 8.6 | D | 9.3 | 14.6 | D | 10.9 | 3.0 | 6.3 | 11.2 | 10,370 | 192 | 1,286 | 20,963 | 8.5 |
| Johnson | -0.7 | D | 5.0 | 0.5 | 7.6 | 15.2 | 4.4 | D | 26.5 | 6,720 | 303 | 1,903 | 10,679 | 0.5 |
| Kenton | 0.0 | D | 9.0 | 8.9 | 7.6 | 5.2 | 12.3 | 16.5 | 12.3 | 30,095 | 180 | 3,627 | 70,001 | 1.5 |
| Knott | -0.5 | D | D | D | 9.5 | 5.3 | D | D | 27.1 | 4,165 | 280 | 1,407 | 7,706 | 3.2 |
| Knox | -0.4 | 0.5 | 3.6 | 14.9 | 7.6 | 10.5 | 3.6 | D | 20.1 | 8,170 | 262 | 2,966 | 14,879 | 2.7 |
| Larue | 0.8 | 0.0 | 11.7 | 16.1 | D | 4.7 | 5.8 | 11.6 | 23.4 | 3,710 | 258 | 487 | 6,472 | 4.8 |
| Laurel | -0.3 | 0.8 | 4.1 | 15.4 | 4.5 | 10.0 | 4.6 | 13.9 | 11.8 | 14,935 | 245 | 3,014 | 26,093 | 2.6 |
| Lawrence | -0.4 | 2.2 | 7.2 | 1.0 | D | 9.6 | 2.0 | 31.1 | 18.3 | 4,360 | 283 | 1,308 | 7,508 | 3.1 |
| Lee | -1.6 | 4.5 | D | D | 2.0 | 8.0 | 2.1 | 18.4 | 21.3 | 1,895 | 256 | 747 | 3,506 | 2.0 |
| Leslie | -0.1 | D | D | D | D | 7.1 | D | D | 29.0 | 3,205 | 324 | 997 | 5,459 | 3.4 |
| Letcher | -0.1 | D | 2.7 | 2.7 | D | 9.0 | 2.7 | 15.0 | 23.1 | 7,165 | 332 | 1,921 | 11,861 | 2.3 |
| Lewis | -0.7 | 0.0 | 12.4 | 11.3 | D | 3.6 | D | 15.0 | 26.1 | 3,260 | 245 | 916 | 6,606 | 1.9 |
| Lincoln | -0.5 | 0.2 | D | 11.1 | 9.9 | 7.3 | 4.3 | 12.0 | 21.1 | 6,450 | 263 | 1,476 | 11,136 | 3.0 |
| Livingston | 0.2 | D | 17.6 | 3.4 | D | 3.6 | D | 9.2 | 17.5 | 2,850 | 312 | 275 | 4,907 | 1.7 |
| Logan | 1.7 | D | 7.5 | 44.4 | 2.8 | 5.3 | 2.8 | 6.0 | 10.5 | 6,805 | 251 | 882 | 12,346 | 0.1 |
| Lyon | 0.3 | 0.1 | 6.9 | D | D | 5.5 | 2.2 | D | 40.5 | 2,550 | 309 | 183 | 4,933 | 3.0 |
| McCracken | 0.0 | D | 5.8 | 4.6 | 10.0 | 7.9 | 4.7 | 23.7 | 10.9 | 17,365 | 265 | 2,200 | 32,392 | 4.2 |
| McCreary | -1.7 | D | D | 7.5 | 2.0 | 6.9 | 3.5 | 8.1 | 53.2 | 4,225 | 245 | 1,657 | 7,561 | 0.7 |
| McLean | 21.4 | D | 4.3 | 6.3 | D | 5.1 | D | 4.7 | 16.6 | 2,615 | 285 | 288 | 4,293 | 0.7 |
| Madison | -0.2 | D | 6.3 | 18.1 | 11.5 | 7.6 | 3.3 | 9.9 | 23.8 | 18,100 | 194 | 2,951 | 37,606 | 7.3 |
| Magoffin | -1.9 | D | D | D | D | 7.1 | D | 14.9 | 29.6 | 3,585 | 294 | 1,418 | 6,168 | 3.7 |
| Marion | 0.0 | 0.0 | D | 52.2 | D | 4.3 | D | D | 9.9 | 4,475 | 233 | 872 | 8,390 | 2.4 |
| Marshall | 0.7 | D | 17.7 | 33.8 | 2.7 | 6.0 | 4.3 | D | 11.4 | 9,220 | 296 | 727 | 16,357 | 3.9 |
| Martin | -0.3 | D | D | D | D | 7.3 | D | 8.4 | 44.4 | 3,215 | 286 | 1,120 | 5,345 | 3.5 |
| Mason | 0.2 | 0.0 | 9.2 | 22.0 | D | 7.4 | 3.0 | 14.7 | 12.6 | 4,435 | 260 | 657 | 8,156 | 0.6 |
| Meade | -1.4 | 3.5 | 14.7 | 13.6 | 5.5 | 9.5 | 5.0 | 6.5 | 20.4 | 6,190 | 216 | 522 | 12,694 | 7.9 |
| Menifee | -2.2 | D | D | 13.4 | D | 4.3 | D | D | 35.8 | 1,975 | 305 | 506 | 3,822 | 2.1 |
| Mercer | -0.3 | D | 7.3 | 39.5 | D | 4.7 | 2.4 | D | 10.4 | 5,780 | 264 | 666 | 10,243 | 2.0 |
| Metcalfe | -2.3 | 0.2 | D | 34.8 | 2.6 | 4.2 | 2.8 | 6.4 | 21.4 | 2,885 | 286 | 565 | 5,275 | 1.4 |
| Monroe | 8.5 | 0.0 | 10.1 | 15.8 | 2.3 | 7.4 | 4.5 | D | 15.4 | 2,935 | 275 | 624 | 5,275 | 1.4 |
| Montgomery | -0.7 | D | 7.2 | 34.6 | D | 9.1 | 4.2 | D | 11.4 | 6,645 | 236 | 1,194 | 12,268 | 4.9 |
| Morgan | | D | 13.0 | | 5.5 | 6.3 | D | 9.7 | 32.5 | 3,280 | 247 | 889 | 6,095 | 4.5 |
| Muhlenberg | 3.0 | 6.8 | 11.4 | 9.6 | D | 7.3 | 2.4 | 14.6 | 9.9 | 8,360 | 272 | 1,202 | 13,881 | 1.3 |
| Nelson | -0.7 | D | 10.4 | 35.0 | 3.1 | 7.0 | 4.1 | 9.9 | 10.9 | 10,920 | 236 | 1,183 | 19,653 | 8.7 |
| Nicholas | 1.9 | 0.0 | 15.8 | 7.1 | D | 4.5 | D | 14.4 | 30.6 | 1,865 | 257 | 282 | 3,307 | 1.4 |

1. Per 1,000 resident population estimated as of July 1 of the year shown.

# Table B. States and Counties — **Housing, Labor Force, and Employment**

| STATE County | Total | Percent | Median value[1] | With a mortgage | Without a mortgage[2] | Median rent[3] | Median rent as a percent of income[2] | Sub-standard units[4] (percent) | Total | Percent change, 2019-2020 | Total | Rate[5] | Total | Management, business, science, and arts | Construction, production, and maintenance occupations |
|---|---|---|---|---|---|---|---|---|---|---|---|---|---|---|---|
| | 89 | 90 | 91 | 92 | 93 | 94 | 95 | 96 | 97 | 98 | 99 | 100 | 101 | 102 | 103 |
| **KENTUCKY—Cont'd** | | | | | | | | | | | | | | | |
| Clay | 7,622 | 67.9 | 54,100 | 21.6 | 11.6 | 533 | 28.9 | 4.1 | 5,199 | 0.5 | 448 | 8.6 | 5,509 | 22.1 | 36.6 |
| Clinton | 3,985 | 69.5 | 72,800 | 22.6 | 11.6 | 472 | 23.3 | 2.5 | 3,978 | -0.3 | 202 | 5.1 | 3,635 | 31.0 | 30.9 |
| Crittenden | 3,567 | 81.9 | 89,900 | 19.6 | 10.7 | 550 | 22.8 | 1.0 | 3,663 | -3.0 | 203 | 5.5 | 3,704 | 23.6 | 35.8 |
| Cumberland | 2,616 | 72.9 | 89,400 | 26.0 | 10.9 | 465 | 22.1 | 2.1 | 3,271 | 1.5 | 159 | 4.9 | 2,587 | 26.3 | 40.1 |
| Daviess | 40,179 | 67.9 | 133,600 | 19.4 | 10.2 | 769 | 27.8 | 2.6 | 45,710 | -3.2 | 2,799 | 6.1 | 45,985 | 31.7 | 28.3 |
| Edmonson | 4,885 | 80.4 | 87,300 | 19.6 | 11.6 | 660 | 24.9 | 4.2 | 4,599 | -3.4 | 352 | 7.7 | 4,615 | 27.9 | 34.4 |
| Elliott | 2,523 | 75.3 | 82,700 | 18.2 | 11.2 | 655 | 29.5 | 1.0 | 1,967 | -3.0 | 198 | 10.1 | 1,711 | 32.1 | 34.0 |
| Estill | 5,499 | 72.1 | 76,800 | 20.3 | 11.7 | 574 | 27.7 | 3.9 | 5,153 | -4.1 | 387 | 7.5 | 4,831 | 21.2 | 43.3 |
| Fayette | 129,784 | 54.4 | 189,800 | 18.7 | 10.0 | 896 | 29.2 | 2.2 | 172,284 | -2.0 | 9,879 | 5.7 | 167,423 | 43.5 | 17.3 |
| Fleming | 5,841 | 73.2 | 89,900 | 18.2 | 10.0 | 584 | 25.6 | 3.7 | 6,135 | -1.4 | 449 | 7.3 | 6,061 | 30.3 | 39.0 |
| Floyd | 15,000 | 69.9 | 74,400 | 22.3 | 11.8 | 646 | 32.6 | 2.6 | 10,688 | -3.6 | 983 | 9.2 | 10,603 | 30.4 | 25.7 |
| Franklin | 21,076 | 62.8 | 142,900 | 18.8 | 10.0 | 762 | 24.6 | 2.1 | 24,745 | -1.8 | 1,628 | 6.6 | 24,531 | 36.8 | 22.8 |
| Fulton | 2,441 | 61.3 | 66,100 | 22.7 | 11.9 | 579 | 33.2 | 0.2 | 1,983 | -2.6 | 129 | 6.5 | 2,005 | 28.0 | 28.4 |
| Gallatin | 3,105 | 72.0 | 128,100 | 19.5 | 11.2 | 701 | 21.5 | 1.6 | 4,012 | -2.6 | 241 | 6.0 | 3,863 | 19.9 | 40.5 |
| Garrard | 6,706 | 82.4 | 140,500 | 19.0 | 11.4 | 688 | 28.2 | 2.6 | 7,528 | -2.5 | 473 | 6.3 | 7,412 | 31.2 | 30.6 |
| Grant | 9,141 | 68.5 | 130,200 | 19.5 | 11.4 | 727 | 26.9 | 2.7 | 11,525 | -2.0 | 818 | 7.1 | 10,773 | 19.0 | 44.6 |
| Graves | 14,316 | 75.5 | 104,300 | 18.7 | 10.4 | 639 | 29.0 | 3.0 | 15,631 | -1.2 | 930 | 5.9 | 15,681 | 31.0 | 31.8 |
| Grayson | 9,758 | 72.6 | 113,900 | 21.3 | 10.7 | 589 | 30.2 | 3.1 | 10,704 | -2.7 | 831 | 7.8 | 10,697 | 26.4 | 42.0 |
| Green | 4,413 | 73.8 | 79,500 | 23.3 | 10.0 | 547 | 27.5 | 1.6 | 5,142 | 0.2 | 276 | 5.4 | 4,446 | 23.8 | 36.2 |
| Greenup | 14,056 | 77.6 | 116,700 | 18.8 | 11.1 | 704 | 28.6 | 1.2 | 13,238 | -1.8 | 1,174 | 8.9 | 14,322 | 38.2 | 25.0 |
| Hancock | 3,311 | 79.2 | 112,300 | 16.3 | 10.0 | 656 | 22.7 | 1.7 | 3,808 | -3.2 | 233 | 6.1 | 3,713 | 22.4 | 44.9 |
| Hardin | 41,646 | 61.5 | 156,600 | 19.1 | 10.0 | 827 | 25.2 | 2.6 | 47,581 | -3.4 | 3,385 | 7.1 | 47,906 | 31.5 | 27.3 |
| Harlan | 11,134 | 68.8 | 60,300 | 21.9 | 11.6 | 528 | 32.2 | 3.3 | 6,548 | -2.8 | 780 | 11.9 | 7,174 | 25.9 | 30.2 |
| Harrison | 7,283 | 69.0 | 140,000 | 18.3 | 11.6 | 617 | 27.0 | 3.2 | 8,712 | -1.4 | 535 | 6.1 | 8,134 | 28.1 | 39.2 |
| Hart | 7,351 | 74.1 | 94,400 | 18.1 | 10.9 | 530 | 27.8 | 3.3 | 7,694 | -3.1 | 496 | 6.4 | 7,248 | 24.5 | 41.5 |
| Henderson | 18,643 | 61.3 | 128,400 | 18.4 | 10.0 | 689 | 26.9 | 1.5 | 20,921 | -3.5 | 1,278 | 6.1 | 20,338 | 30.0 | 32.9 |
| Henry | 6,044 | 72.1 | 128,700 | 21.0 | 10.0 | 780 | 27.7 | 3.4 | 8,071 | -2.4 | 460 | 5.7 | 7,291 | 24.8 | 32.3 |
| Hickman | 1,798 | 82.8 | 80,600 | 20.6 | 11.6 | 516 | 20.3 | 2.6 | 1,720 | -2.6 | 95 | 5.5 | 1,598 | 34.0 | 32.2 |
| Hopkins | 18,588 | 69.5 | 108,700 | 17.6 | 10.0 | 715 | 24.5 | 1.5 | 18,149 | -1.6 | 1,334 | 7.4 | 18,846 | 30.4 | 32.8 |
| Jackson | 5,484 | 75.2 | 87,200 | 22.2 | 11.6 | 519 | 28.4 | 2.9 | 4,051 | -1.8 | 351 | 8.7 | 4,226 | 23.4 | 42.3 |
| Jefferson | 312,679 | 61.8 | 170,100 | 19.4 | 11.1 | 871 | 27.8 | 2.3 | 390,487 | -2.3 | 26,717 | 6.8 | 382,608 | 38.1 | 23.9 |
| Jessamine | 18,821 | 64.6 | 172,100 | 19.5 | 10.0 | 789 | 27.1 | 2.2 | 26,160 | -2.4 | 1,471 | 5.6 | 24,409 | 34.7 | 26.0 |
| Johnson | 8,488 | 71.9 | 101,300 | 20.2 | 12.1 | 616 | 31.2 | 2.7 | 6,589 | -2.4 | 573 | 8.7 | 6,782 | 38.8 | 20.1 |
| Kenton | 63,966 | 66.1 | 161,100 | 17.8 | 10.0 | 811 | 25.4 | 1.5 | 86,062 | -3.0 | 4,985 | 5.8 | 85,375 | 39.8 | 22.3 |
| Knott | 6,388 | 74.8 | 51,500 | 20.5 | 10.0 | 504 | 36.5 | 6.2 | 4,350 | -3.0 | 371 | 8.5 | 4,601 | 30.0 | 23.1 |
| Knox | 11,961 | 65.5 | 86,800 | 19.9 | 11.6 | 562 | 29.3 | 2.5 | 9,962 | 0.1 | 819 | 8.2 | 9,838 | 33.8 | 26.7 |
| Larue | 5,741 | 74.1 | 115,600 | 19.2 | 10.0 | 703 | 26.9 | 1.6 | 5,838 | -2.8 | 421 | 7.2 | 6,387 | 24.3 | 42.9 |
| Laurel | 22,573 | 69.9 | 112,000 | 19.6 | 11.0 | 651 | 29.6 | 3.1 | 24,857 | -0.2 | 1,727 | 6.9 | 23,932 | 29.6 | 29.5 |
| Lawrence | 5,784 | 74.9 | 79,400 | 22.8 | 11.3 | 568 | 35.2 | 3.9 | 5,217 | 0.5 | 433 | 8.3 | 4,685 | 25.9 | 33.8 |
| Lee | 2,839 | 68.9 | 63,000 | 21.6 | 13.6 | 548 | 38.3 | 5.7 | 2,047 | -2.8 | 143 | 7.0 | 2,002 | 27.6 | 35.0 |
| Leslie | 4,076 | 81.4 | 64,100 | 25.3 | 11.5 | 517 | 32.2 | 7.4 | 2,628 | -1.9 | 284 | 10.8 | 3,000 | 26.2 | 32.7 |
| Letcher | 9,721 | 74.2 | 58,600 | 23.8 | 12.4 | 559 | 29.8 | 3.8 | 6,111 | -2.8 | 623 | 10.2 | 7,050 | 36.1 | 26.0 |
| Lewis | 5,268 | 74.0 | 65,500 | 19.5 | 12.0 | 518 | 29.2 | 1.8 | 4,648 | -0.3 | 443 | 9.5 | 4,330 | 24.8 | 45.4 |
| Lincoln | 9,697 | 78.3 | 106,400 | 18.6 | 10.9 | 651 | 25.1 | 4.0 | 9,317 | -0.6 | 739 | 7.9 | 10,037 | 28.2 | 33.0 |
| Livingston | 3,857 | 79.3 | 102,400 | 20.4 | 11.2 | 676 | 23.9 | 2.4 | 3,534 | -3.0 | 266 | 7.5 | 3,888 | 21.7 | 40.4 |
| Logan | 10,418 | 69.7 | 111,000 | 20.4 | 11.5 | 661 | 27.0 | 1.5 | 12,136 | -3.6 | 670 | 5.5 | 11,162 | 27.7 | 39.3 |
| Lyon | 3,333 | 79.4 | 141,500 | 18.7 | 10.0 | 603 | 29.7 | 2.2 | 3,095 | -0.8 | 164 | 5.3 | 2,877 | 25.8 | 31.5 |
| McCracken | 27,854 | 65.1 | 143,400 | 18.8 | 10.0 | 717 | 27.6 | 2.5 | 28,486 | -2.6 | 2,051 | 7.2 | 28,954 | 34.8 | 21.5 |
| McCreary | 6,052 | 70.1 | 69,700 | 23.0 | 10.1 | 555 | 32.2 | 3.4 | 4,807 | 1.8 | 350 | 7.3 | 4,797 | 29.0 | 34.1 |
| McLean | 3,809 | 78.4 | 104,700 | 18.3 | 10.0 | 520 | 27.2 | 2.2 | 4,023 | -3.5 | 227 | 5.6 | 3,729 | 29.2 | 40.1 |
| Madison | 33,359 | 59.1 | 159,300 | 18.7 | 10.8 | 710 | 27.9 | 1.7 | 45,855 | -3.8 | 2,764 | 6.0 | 42,683 | 36.4 | 23.6 |
| Magoffin | 5,020 | 72.5 | 66,800 | 24.0 | 12.7 | 533 | 36.5 | 4.1 | 3,208 | 2.9 | 516 | 16.1 | 3,372 | 28.4 | 39.1 |
| Marion | 7,405 | 73.4 | 105,400 | 20.2 | 10.0 | 630 | 32.5 | 3.7 | 9,233 | -5.1 | 718 | 7.8 | 7,700 | 27.5 | 40.7 |
| Marshall | 13,121 | 81.0 | 131,100 | 18.1 | 10.0 | 690 | 25.4 | 1.8 | 14,831 | -1.9 | 912 | 6.1 | 13,466 | 29.8 | 33.0 |
| Martin | 4,153 | 70.5 | 84,300 | 24.1 | 10.1 | 560 | 26.5 | 4.7 | 2,498 | -6.7 | 290 | 11.6 | 3,276 | 33.1 | 25.1 |
| Mason | 6,700 | 67.6 | 132,800 | 18.3 | 10.4 | 661 | 26.5 | 2.1 | 6,909 | -2.9 | 551 | 8.0 | 7,255 | 34.1 | 30.3 |
| Meade | 10,685 | 72.6 | 153,500 | 18.1 | 10.2 | 856 | 23.1 | 3.4 | 11,749 | -3.7 | 836 | 7.1 | 12,001 | 29.7 | 36.1 |
| Menifee | 2,651 | 79.8 | 76,800 | 18.9 | 11.8 | 603 | 22.9 | 3.6 | 2,256 | -3.2 | 201 | 8.9 | 2,425 | 31.5 | 37.3 |
| Mercer | 8,556 | 72.3 | 147,300 | 18.4 | 10.0 | 624 | 27.0 | 1.9 | 9,907 | -1.6 | 720 | 7.3 | 9,681 | 31.6 | 34.9 |
| Metcalfe | 4,074 | 78.4 | 77,200 | 20.9 | 11.3 | 523 | 23.9 | 2.4 | 4,017 | -2.3 | 313 | 7.8 | 3,918 | 25.8 | 40.9 |
| Monroe | 4,491 | 69.7 | 80,500 | 17.8 | 10.8 | 523 | 32.7 | 2.6 | 4,564 | -2.3 | 246 | 5.4 | 4,394 | 31.3 | 39.5 |
| Montgomery | 10,481 | 67.9 | 116,700 | 20.0 | 10.4 | 710 | 23.9 | 3.2 | 11,355 | -2.5 | 905 | 8.0 | 11,657 | 24.8 | 36.2 |
| Morgan | 4,865 | 74.7 | 73,100 | 20.0 | 10.4 | 553 | 24.9 | 2.8 | 4,352 | -5.0 | 300 | 6.9 | 3,699 | 32.8 | 30.7 |
| Muhlenberg | 11,351 | 79.6 | 93,900 | 19.5 | 10.4 | 600 | 24.5 | 0.7 | 10,004 | -4.8 | 866 | 8.7 | 11,945 | 26.3 | 36.7 |
| Nelson | 17,853 | 76.0 | 154,700 | 18.4 | 10.0 | 761 | 23.6 | 2.5 | 22,755 | -2.1 | 1,570 | 6.9 | 21,654 | 29.4 | 39.5 |
| Nicholas | 2,804 | 68.9 | 85,600 | 18.0 | 11.5 | 593 | 24.9 | 5.8 | 3,275 | -1.2 | 222 | 6.8 | 2,821 | 21.8 | 39.3 |

1. Specified owner-occupied units.   2. A value of 10.0 represents 10 percent or less; a value of 50.0 represents 50 percent or more.   3. Specified renter-occupied units.   4. Overcrowded or lacking complete plumbing facilities.   5. Percent of civilian labor force.   6. Civilian employed persons 16 years old and over.

| STATE County | Number of establishments | Total | Health care and social assistance | Manufacturing | Retail trade | Finance and insurance | Professional, scientific, and technical services | Total (mil dol) | Average per employee (dollars) | Number | Fewer than 50 acres | 1000 acres or more | Farm producers whose primary occupation is farming (percent) |
|---|---|---|---|---|---|---|---|---|---|---|---|---|---|
| | 104 | 105 | 106 | 107 | 108 | 109 | 110 | 111 | 112 | 113 | 114 | 115 | 116 |
| KENTUCKY—Cont'd | | | | | | | | | | | | | |
| Clay........................ | 230 | 2,681 | 968 | 233 | 614 | 85 | 90 | 85 | 31,551 | 233 | 28.3 | 3.0 | 33.6 |
| Clinton..................... | 179 | 2,966 | 413 | 1,446 | 362 | 71 | 33 | 84 | 28,348 | 512 | 40.6 | 1.0 | 32.2 |
| Crittenden ............... | 163 | 1,662 | 344 | 402 | 240 | 44 | 35 | 49 | 29,524 | 575 | 23.7 | 6.1 | 33.0 |
| Cumberland ............ | 102 | 1,107 | 339 | 189 | 158 | 64 | 41 | 39 | 35,352 | 395 | 27.1 | 3.5 | 31.0 |
| Daviess ................... | 2,289 | 41,465 | 8,439 | 5,560 | 5,276 | 2,984 | 828 | 1,706 | 41,147 | 919 | 52.6 | 7.7 | 39.9 |
| Edmonson............... | 109 | 854 | 190 | NA | 155 | 57 | 11 | 25 | 29,359 | 578 | 37.0 | 0.5 | 33.1 |
| Elliott ...................... | 41 | 258 | 85 | NA | 75 | 14 | NA | 8 | 29,988 | 363 | 25.1 | NA | 41.8 |
| Estill ....................... | 174 | 1,590 | 383 | 203 | 272 | 78 | 42 | 47 | 29,665 | 622 | 51.8 | 3.2 | 40.2 |
| Fayette .................... | 8,782 | 166,086 | 33,169 | 8,244 | 21,867 | 5,077 | 11,861 | 7,632 | 45,953 | 1,013 | 32.0 | 1.8 | 39.8 |
| Fleming ................... | 224 | 2,231 | 445 | 538 | 447 | 110 | 36 | 76 | 34,134 | 136 | 58.8 | NA | 31.7 |
| Floyd ...................... | 693 | 7,884 | 1,990 | 195 | 1,438 | 236 | 164 | 310 | 39,265 | 599 | 40.7 | 0.3 | 34.9 |
| Franklin .................. | 1,183 | 16,516 | 2,336 | 2,737 | 2,587 | 983 | 822 | 682 | 41,268 | 146 | 41.1 | 19.9 | 53.6 |
| Fulton ..................... | 125 | 1,395 | 131 | 557 | 217 | 71 | 20 | 52 | 37,128 | 235 | 43.8 | 3.0 | 25.5 |
| Gallatin ................... | 83 | 866 | 217 | 41 | 209 | 23 | 16 | 29 | 33,828 | 793 | 37.6 | 1.9 | 42.4 |
| Garrard.................... | 221 | 1,490 | 261 | 234 | 161 | 58 | 33 | 50 | 33,793 | | | | |
| Grant ...................... | 360 | 4,222 | 511 | 1,175 | 909 | 72 | 69 | 163 | 38,574 | 811 | 36.1 | 0.9 | 34.3 |
| Graves .................... | 660 | 9,380 | 1,561 | 2,665 | 1,501 | 326 | 268 | 339 | 36,092 | 1,104 | 43.2 | 6.3 | 39.3 |
| Grayson .................. | 440 | 6,533 | 1,063 | 1,880 | 974 | 208 | 68 | 209 | 31,959 | 1,339 | 33.8 | 2.1 | 35.9 |
| Green ...................... | 164 | 1,210 | 521 | 22 | 205 | 53 | 36 | 37 | 30,473 | 1,004 | 33.4 | 1.5 | 28.8 |
| Greenup .................. | 454 | 4,950 | 953 | 585 | 1,005 | 223 | 151 | 175 | 35,328 | 584 | 37.0 | 0.9 | 34.4 |
| Hancock .................. | 121 | 3,538 | 106 | 2,613 | 128 | 81 | 45 | 217 | 61,245 | 321 | 35.8 | 0.9 | 26.5 |
| Hardin ..................... | 2,125 | 37,383 | 7,231 | 6,885 | 6,262 | 1,294 | 1,235 | 1,509 | 40,361 | 1,305 | 49.0 | 2.9 | 34.3 |
| Harlan ..................... | 384 | 4,589 | 1,254 | 68 | 862 | 99 | 134 | 171 | 37,216 | 39 | 51.3 | 5.1 | 21.1 |
| Harrison .................. | 273 | 4,325 | 1,130 | 1,400 | 564 | 85 | 58 | 187 | 43,336 | 1,138 | 36.7 | 1.1 | 39.0 |
| Hart ........................ | 266 | 4,170 | 335 | 2,561 | 472 | 70 | 51 | 149 | 35,799 | 1,287 | 40.9 | 1.2 | 36.7 |
| Henderson ............... | 998 | 16,450 | 2,119 | 5,606 | 1,948 | 369 | 558 | 706 | 42,892 | 458 | 46.7 | 12.4 | 39.4 |
| Henry ...................... | 209 | 1,962 | 246 | 389 | 317 | 81 | 56 | 68 | 34,856 | 771 | 35.3 | 2.2 | 41.7 |
| Hickman .................. | 70 | 875 | 173 | 106 | 218 | 51 | 8 | 32 | 36,245 | 246 | 34.6 | 10.2 | 54.8 |
| Hopkins ................... | 946 | 14,310 | 2,875 | 1,912 | 2,833 | 429 | 441 | 618 | 43,187 | 551 | 36.7 | 0.7 | 33.6 |
| Jackson................... | 91 | 993 | 237 | 52 | 165 | 48 | 10 | 36 | 36,049 | 343 | 70.0 | NA | 35.8 |
| Jefferson................. | 19,914 | 459,514 | 66,109 | 54,587 | 44,848 | 30,418 | 23,470 | 24,622 | 53,583 | 671 | 56.9 | 1.2 | 35.7 |
| Jessamine ............... | 1,161 | 16,792 | 1,197 | 2,250 | 2,647 | 232 | 1,335 | 651 | 38,794 | 224 | 34.8 | NA | 24.8 |
| Johnson................... | 357 | 3,821 | 607 | 58 | 1,265 | 121 | 141 | 134 | 35,065 | 506 | 54.3 | 0.2 | 24.4 |
| Kenton..................... | 3,157 | 60,453 | 11,746 | 5,627 | 5,169 | 5,354 | 3,289 | 3,167 | 52,390 | 62 | 38.7 | 3.2 | 50.0 |
| Knott....................... | 156 | 1,817 | 377 | NA | 282 | 46 | 48 | 63 | 34,858 | | | | |
| Knox........................ | 458 | 9,050 | 1,316 | 1,446 | 1,160 | 172 | 2,169 | 231 | 25,509 | 336 | 50.9 | 0.9 | 34.3 |
| Larue....................... | 197 | 1,984 | 318 | 562 | 198 | 363 | 39 | 73 | 36,924 | 718 | 49.7 | 2.1 | 38.0 |
| Laurel...................... | 1,197 | 23,288 | 2,760 | 3,863 | 3,328 | 967 | 1,855 | 812 | 34,859 | 955 | 48.6 | 0.3 | 33.0 |
| Lawrence ................ | 198 | 2,494 | 724 | NA | 568 | 57 | 68 | 90 | 36,254 | 284 | 25.4 | 2.1 | 32.8 |
| Lee.......................... | 95 | 1,206 | 290 | NA | 184 | 12 | 15 | 41 | 39,365 | 26 | 53.8 | NA | 12.5 |
| Leslie....................... | 99 | 1,142 | 319 | NA | 235 | 63 | 31 | 45 | 37,375 | 103 | 68.9 | NA | 27.3 |
| Letcher.................... | 291 | 3,145 | 992 | 153 | 603 | 99 | 86 | 118 | 33,136 | 550 | 21.5 | 1.5 | 37.9 |
| Lewis....................... | 119 | 1,206 | 247 | 352 | 264 | 77 | 13 | 40 | 33,092 | 1,090 | 42.8 | 2.0 | 42.8 |
| Lincoln..................... | 298 | 2,478 | 578 | 313 | 487 | 118 | 128 | 82 | 45,139 | 365 | 20.3 | 5.5 | 36.3 |
| Livingston................ | 140 | 1,626 | 360 | 76 | 166 | 29 | 33 | 73 | | | | | |
| Logan ...................... | 486 | 8,109 | 779 | 3,975 | 872 | 195 | 155 | 383 | 47,283 | 1,078 | 37.9 | 4.0 | 41.6 |
| Lyon........................ | 168 | 1,478 | 270 | 9 | 254 | 20 | 28 | 45 | 30,131 | 208 | 31.7 | 2.4 | 33.1 |
| McCracken............... | 2,093 | 34,544 | 6,893 | 2,242 | 5,743 | 1,233 | 1,180 | 1,480 | 42,834 | 173 | 49.1 | NA | 35.5 |
| McCreary ................. | 156 | 1,433 | 254 | 300 | 315 | 87 | 87 | 60 | 45,816 | 439 | 42.8 | 8.4 | 39.5 |
| McLean .................... | 167 | 1,299 | 140 | 144 | 207 | 66 | 18 | 37 | 38,738 | 1,187 | 37.3 | 1.9 | 38.6 |
| Madison................... | 1,751 | 25,276 | 4,258 | 5,139 | 4,199 | 566 | 1,233 | 979 | 33,252 | 335 | 36.7 | 0.9 | 30.4 |
| Magoffin .................. | 154 | 1,102 | 253 | NA | 198 | 48 | 137 | 109 | 37,510 | 954 | 35.6 | 1.5 | 38.0 |
| Marion..................... | 333 | 8,701 | 1,491 | 4,408 | 611 | 148 | 109 | 326 | 37,510 | 699 | 49.4 | 2.3 | 27.9 |
| Marshall .................. | 691 | 9,262 | 1,005 | 2,530 | 1,338 | 396 | 199 | 453 | 48,954 | 30 | 30.0 | 6.7 | 39.5 |
| Martin...................... | 133 | 1,094 | 208 | NA | 349 | 61 | 14 | 35 | 31,701 | | | | |
| Mason...................... | 427 | 6,756 | 1,329 | 922 | 1,287 | 174 | 91 | 265 | 39,270 | 680 | 32.5 | 2.6 | 36.6 |
| Meade...................... | 330 | 3,494 | 302 | 454 | 649 | 138 | 254 | 135 | 38,657 | 781 | 51.3 | 3.3 | 31.7 |
| Menifee ................... | 53 | 535 | 198 | 138 | 72 | NA | 6 | 15 | 28,942 | 283 | 36.7 | 0.4 | 27.8 |
| Mercer..................... | 335 | 4,594 | 497 | 1,927 | 566 | 107 | 73 | 201 | 43,857 | 1,108 | 43.4 | 1.2 | 33.0 |
| Metcalfe................... | 114 | 1,328 | 123 | 677 | 147 | 45 | 23 | 41 | 31,198 | 945 | 38.7 | 1.6 | 40.4 |
| Monroe.................... | 202 | 2,644 | 457 | 735 | 518 | 76 | 27 | 94 | 35,583 | 765 | 30.5 | 3.7 | 37.1 |
| Montgomery............. | 531 | 9,289 | 982 | 3,545 | 1,513 | 306 | 181 | 323 | 34,789 | 659 | 38.7 | 3.2 | 32.6 |
| Morgan..................... | 145 | 1,613 | 327 | 57 | 350 | 104 | 75 | 59 | 36,719 | 657 | 26.2 | 2.1 | 37.5 |
| Muhlenberg.............. | 541 | 6,437 | 1,222 | 772 | 1,134 | 177 | 217 | 235 | 36,488 | 549 | 32.8 | 4.0 | 37.6 |
| Nelson..................... | 949 | 15,262 | 1,342 | 5,749 | 1,874 | 355 | 227 | 682 | 44,693 | 1,434 | 54.0 | 2.0 | 38.6 |
| Nicholas .................. | 78 | 470 | 157 | NA | 69 | 25 | 12 | 14 | 29,957 | 556 | 26.3 | 0.9 | 44.7 |

| STATE County | Agriculture, 2017 (cont.) | | | | | | | | | | | | | | |
|---|---|---|---|---|---|---|---|---|---|---|---|---|---|---|---|
| | Land in farms | | | | | Value of land and buildings (dollars) | | Value of machinery and equiopmnet, average per farm (dollars) | Value of products sold: | | | | Organic farms (number) | Farms with internet access (percent) | Government payments | |
| | Acreage (1,000) | Percent change, 2012-2017 | Acres | | | | | | Total (mil dol) | Average per farm (acres) | Percent from: | | | | | |
| | | | Average size of farm | Total irrigated (1,000) | Total cropland (1,000) | Average per farm | Average per acre | | | | Crops | Livestock and poultry products | | | Total ($1,000) | Percent of farms |
| | 117 | 118 | 119 | 120 | 121 | 122 | 123 | 124 | 125 | 126 | 127 | 128 | 129 | 130 | 131 | 132 |
| KENTUCKY—Cont'd | | | | | | | | | | | | | | | | |
| Clay | 44 | 26.3 | 190 | 0.0 | 8.3 | 345,381 | 1,814 | 53,754 | 5.2 | 22,399 | 75.2 | 24.8 | 2 | 74.7 | 9 | 3.9 |
| Clinton | 65 | -12.0 | 127 | 0.0 | 23.8 | 354,745 | 2,788 | 72,223 | 40.6 | 79,334 | 7.5 | 92.5 | NA | 76.6 | 358 | 17.8 |
| Crittenden | 158 | 6.2 | 275 | D | 79.2 | 754,514 | 2,747 | 84,792 | 39.3 | 68,289 | 69.8 | 30.2 | 4 | 51.5 | 2,296 | 42.8 |
| Cumberland | 82 | 24.6 | 207 | 0.0 | 22.6 | 423,878 | 2,053 | 57,763 | 10.7 | 26,965 | 46.1 | 53.9 | 3 | 66.6 | 118 | 8.4 |
| Daviess | 238 | 0.3 | 259 | 10.9 | 190.8 | 1,291,082 | 4,986 | 163,634 | 185.9 | 202,262 | 64.4 | 35.6 | NA | 71.7 | 4,337 | 44.0 |
| Edmonson | 82 | -4.0 | 141 | 0.0 | 38.1 | 448,629 | 3,178 | 75,656 | 27.5 | 47,500 | 40.9 | 59.1 | NA | 74.7 | 1,188 | 26.8 |
| Elliott | 55 | -2.4 | 152 | 0.0 | 15.1 | 245,278 | 1,619 | 49,503 | 3.1 | 8,628 | 32.0 | 68.0 | NA | 79.3 | 72 | 9.9 |
| Estill | 53 | 1.6 | 145 | 0.0 | 14.2 | 347,874 | 2,404 | 48,838 | 4.1 | 11,183 | 35.4 | 64.6 | 1 | 75.5 | 126 | 16.1 |
| Fayette | 115 | -0.2 | 184 | 0.3 | 33.7 | 2,694,042 | 14,619 | 122,300 | 215.5 | 346,492 | 5.9 | 94.1 | 6 | 90.0 | 576 | 10.1 |
| Fleming | 171 | -6.4 | 169 | D | 77.3 | 447,405 | 2,643 | 71,101 | 48.8 | 48,124 | 38.3 | 61.7 | 2 | 68.4 | 783 | 46.5 |
| Floyd | 9 | 12.0 | 67 | D | 2.1 | 207,547 | 3,088 | 34,195 | 0.6 | 4,294 | 59.8 | 40.2 | NA | 88.2 | D | 1.5 |
| Franklin | 75 | -5.1 | 124 | 0.2 | 29.4 | 493,851 | 3,967 | 53,515 | 18.9 | 31,482 | 42.3 | 57.7 | 10 | 77.6 | 666 | 7.3 |
| Fulton | 98 | 17.1 | 669 | 7.0 | 86.6 | 2,456,571 | 3,674 | 304,536 | 62.1 | 425,521 | 71.5 | 28.5 | NA | 76.7 | 2,263 | 73.3 |
| Gallatin | 33 | 17.0 | 138 | 0.0 | 14.2 | 393,568 | 2,845 | 61,824 | 7.7 | 32,936 | 85.7 | 14.3 | NA | 70.2 | 6 | 4.7 |
| Garrard | 141 | 10.8 | 178 | 0.0 | 56.8 | 518,773 | 2,917 | 76,025 | 35.4 | 44,608 | 32.0 | 68.0 | 5 | 70.6 | 345 | 12.4 |
| Grant | 97 | -1.6 | 119 | D | 35.5 | 394,863 | 3,309 | 54,915 | 10.1 | 12,486 | 58.3 | 41.7 | 1 | 73.0 | 67 | 6.2 |
| Graves | 251 | -13.9 | 228 | 5.1 | 194.8 | 870,326 | 3,825 | 144,824 | 346.2 | 313,547 | 29.7 | 70.3 | NA | 70.4 | 5,088 | 50.5 |
| Grayson | 212 | 5.7 | 159 | 0.2 | 95.3 | 468,818 | 2,956 | 64,374 | 59.6 | 44,494 | 37.4 | 62.6 | 11 | 65.1 | 2,219 | 30.1 |
| Green | 156 | 2.4 | 156 | 0.1 | 76.1 | 403,864 | 2,595 | 76,652 | 53.2 | 53,021 | 59.4 | 40.6 | 2 | 65.9 | 3,100 | 39.4 |
| Greenup | 74 | -6.2 | 126 | 0.1 | 19.0 | 312,060 | 2,471 | 57,719 | 5.8 | 9,911 | 50.4 | 49.6 | 2 | 74.7 | 132 | 3.6 |
| Hancock | 48 | -8.6 | 149 | 0.0 | 21.4 | 442,673 | 2,977 | 78,019 | 11.6 | 36,131 | 71.9 | 28.1 | NA | 80.1 | 472 | 38.9 |
| Hardin | 199 | -1.9 | 153 | D | 109.6 | 701,236 | 4,594 | 81,201 | 59.2 | 45,396 | 70.1 | 29.9 | 1 | 73.9 | 1,777 | 22.1 |
| Harlan | 7 | 7.7 | 173 | D | 0.5 | 270,697 | 1,564 | 32,631 | 0.2 | 5,436 | 68.9 | 31.1 | NA | 79.5 | D | 5.1 |
| Harrison | 168 | 1.9 | 147 | 1.2 | 71.2 | 456,684 | 3,100 | 75,066 | 35.1 | 30,850 | 57.8 | 42.2 | 3 | 74.2 | 300 | 12.0 |
| Hart | 170 | -6.9 | 132 | 0.2 | 73.8 | 406,053 | 3,078 | 59,043 | 36.8 | 28,584 | 51.6 | 48.4 | 3 | 65.3 | 3,457 | 29.1 |
| Henderson | 181 | 2.7 | 394 | 9.2 | 156.8 | 2,010,156 | 5,096 | 193,308 | 97.7 | 213,312 | 88.9 | 11.1 | NA | 83.2 | 4,739 | 53.1 |
| Henry | 131 | 1.9 | 170 | 0.2 | 65.2 | 656,023 | 3,861 | 79,554 | 30.6 | 39,696 | 53.9 | 46.1 | 2 | 73.9 | 195 | 10.1 |
| Hickman | 118 | -16.1 | 482 | 9.5 | 103.9 | 1,933,385 | 4,014 | 317,655 | 159.6 | 648,817 | 35.7 | 64.3 | NA | 72.4 | 3,156 | 65.4 |
| Hopkins | 146 | -10.2 | 223 | 0.1 | 91.0 | 752,392 | 3,372 | 98,293 | 119.6 | 182,309 | 32.6 | 67.4 | 2 | 73.6 | 1,661 | 31.9 |
| Jackson | 75 | -3.3 | 136 | 0.0 | 26.1 | 321,560 | 2,356 | 50,666 | 5.7 | 10,430 | 35.2 | 64.8 | 4 | 73.7 | 55 | 6.4 |
| Jefferson | 20 | -12.4 | 59 | 0.1 | 7.6 | 725,049 | 12,296 | 43,708 | 6.4 | 18,551 | 78.5 | 21.5 | 3 | 80.5 | 99 | 2.9 |
| Jessamine | 76 | -8.8 | 114 | 0.0 | 32.4 | 754,661 | 6,640 | 57,978 | 79.9 | 119,028 | 7.4 | 92.6 | 1 | 85.5 | 387 | 7.7 |
| Johnson | 22 | -9.4 | 98 | 0.0 | 3.8 | 294,847 | 3,001 | 37,113 | 0.7 | 3,152 | 29.7 | 70.3 | NA | 82.1 | 10 | 3.6 |
| Kenton | 37 | -3.7 | 73 | 0.0 | 15.9 | 406,271 | 5,597 | 54,195 | 5.4 | 10,652 | 53.3 | 46.7 | NA | 77.7 | 38 | 5.3 |
| Knott | 13 | 89.1 | 206 | D | 0.6 | 263,055 | 1,279 | 38,779 | 0.4 | 6,581 | 8.6 | 91.4 | NA | 88.7 | NA | NA |
| Knox | 39 | 15.5 | 115 | 0.0 | 12.0 | 275,946 | 2,394 | 58,675 | 2.5 | 7,542 | 53.2 | 46.8 | 2 | 70.5 | 54 | 3.6 |
| Larue | 110 | -1.4 | 154 | 0.0 | 67.7 | 545,514 | 3,549 | 93,130 | 41.1 | 57,259 | 78.9 | 21.1 | NA | 81.2 | 986 | 20.6 |
| Laurel | 90 | -6.6 | 94 | 0.0 | 35.1 | 313,976 | 3,343 | 59,396 | 15.4 | 16,153 | 43.2 | 56.8 | NA | 69.0 | 147 | 7.0 |
| Lawrence | 51 | 22.9 | 180 | D | 7.9 | 288,449 | 1,599 | 47,327 | 1.3 | 4,423 | 40.8 | 59.2 | NA | 79.2 | 19 | 3.5 |
| Lee | 25 | 12.8 | 174 | NA | 8.5 | 356,475 | 2,047 | 49,063 | 1.4 | 9,667 | 25.2 | 74.8 | NA | 70.1 | 55 | 19.4 |
| Leslie | 2 | D | 63 | 0.0 | 0.2 | 242,308 | 3,820 | 23,981 | 0.0 | 1,269 | 69.7 | 30.3 | NA | 69.2 | NA | NA |
| Letcher | 6 | 95.5 | 55 | NA | 0.4 | 163,784 | 3,000 | 31,980 | 0.1 | 1,340 | 44.2 | 55.8 | NA | 82.5 | NA | NA |
| Lewis | 117 | -0.5 | 213 | D | 41.7 | 447,476 | 2,096 | 58,968 | 12.7 | 23,065 | 68.7 | 31.3 | NA | 71.6 | 338 | 13.3 |
| Lincoln | 163 | -9.7 | 150 | 0.0 | 69.0 | 486,123 | 3,246 | 73,186 | 59.0 | 54,136 | 31.5 | 68.5 | 20 | 70.1 | 939 | 21.8 |
| Livingston | 122 | -1.2 | 333 | 0.0 | 61.6 | 937,557 | 2,812 | 108,341 | 24.2 | 66,178 | 63.8 | 36.2 | NA | 67.9 | 1,309 | 51.2 |
| Logan | 276 | 0.2 | 256 | 4.8 | 195.2 | 1,237,795 | 4,829 | 133,156 | 152.2 | 141,154 | 68.8 | 31.2 | 5 | 71.1 | 6,869 | 46.0 |
| Lyon | 33 | -21.5 | 157 | D | 17.6 | 490,391 | 3,121 | 61,700 | 6.7 | 32,029 | 80.2 | 19.8 | NA | 64.4 | 391 | 28.8 |
| McCracken | 62 | -7.6 | 195 | 0.2 | 50.6 | 869,914 | 4,456 | 123,106 | 28.9 | 90,953 | 77.4 | 22.6 | NA | 84.6 | 821 | 33.0 |
| McCreary | 18 | 1.3 | 106 | D | 3.8 | 259,977 | 2,455 | 59,199 | 2.6 | 14,740 | 23.6 | 76.4 | 1 | 63.0 | 44 | 21.4 |
| McLean | 129 | 4.0 | 295 | D | 108.1 | 1,315,544 | 4,466 | 184,326 | 190.8 | 434,711 | 29.7 | 70.3 | NA | 83.6 | 2,369 | 58.8 |
| Madison | 230 | -1.3 | 194 | 0.1 | 74.8 | 671,408 | 3,468 | 71,336 | 50.6 | 42,592 | 15.3 | 84.7 | 6 | 72.4 | 361 | 14.7 |
| Magoffin | 44 | -1.5 | 131 | 0.0 | 7.1 | 270,577 | 2,070 | 54,356 | 1.4 | 4,269 | 55.7 | 44.3 | NA | 76.7 | 7 | 3.3 |
| Marion | 163 | -2.1 | 171 | 0.0 | 85.2 | 594,733 | 3,482 | 86,494 | 60.2 | 63,134 | 45.3 | 54.7 | 2 | 76.8 | 893 | 36.2 |
| Marshall | 85 | -10.8 | 121 | 0.1 | 52.9 | 459,882 | 3,796 | 82,985 | 51.4 | 73,504 | 33.2 | 66.8 | NA | 73.8 | 1,139 | 40.1 |
| Martin | 11 | D | 368 | NA | 2.3 | 1,270,003 | 3,451 | 72,023 | 0.2 | 7,133 | 10.7 | 89.3 | NA | 70.0 | NA | NA |
| Mason | 132 | 4.3 | 194 | 0.0 | 68.7 | 574,464 | 2,957 | 76,841 | 31.3 | 46,081 | 63.5 | 36.5 | 4 | 75.3 | 557 | 25.3 |
| Meade | 141 | 18.0 | 181 | 0.0 | 80.4 | 781,623 | 4,328 | 98,021 | 45.8 | 58,634 | 61.2 | 38.8 | NA | 83.1 | 1,798 | 31.6 |
| Menifee | 35 | -13.7 | 125 | 0.0 | 8.2 | 241,589 | 1,933 | 44,239 | 2.4 | 8,449 | 27.7 | 72.3 | NA | 73.5 | 6 | 1.8 |
| Mercer | 136 | -5.4 | 123 | 0.0 | 59.7 | 483,392 | 3,925 | 61,328 | 45.9 | 41,428 | 26.2 | 73.8 | 3 | 73.6 | 2,019 | 12.4 |
| Metcalfe | 138 | 10.5 | 147 | 0.0 | 58.6 | 392,272 | 2,677 | 74,723 | 43.1 | 45,654 | 47.4 | 52.6 | 2 | 77.6 | 1,290 | 24.2 |
| Monroe | 160 | -7.2 | 209 | 0.1 | 61.1 | 594,999 | 2,847 | 83,370 | 102.5 | 133,949 | 16.6 | 83.4 | 2 | 73.1 | 754 | 28.8 |
| Montgomery | 103 | 3.6 | 156 | 0.0 | 41.0 | 547,952 | 3,505 | 74,291 | 21.4 | 32,461 | 26.6 | 73.4 | NA | 71.9 | 114 | 13.1 |
| Morgan | 107 | -10.3 | 162 | 0.1 | 23.2 | 315,781 | 1,948 | 55,320 | 5.8 | 8,804 | 38.4 | 61.6 | NA | 68.9 | 139 | 7.0 |
| Muhlenberg | 128 | | 232 | 0.0 | 69.3 | 695,899 | 2,996 | 102,576 | 94.5 | 172,069 | 30.3 | 69.7 | NA | 70.1 | 1,142 | 28.8 |
| Nelson | 214 | 13.7 | 149 | 0.2 | 114.0 | 604,664 | 4,061 | 76,095 | 67.1 | 46,791 | 62.7 | 37.3 | NA | 73.4 | 927 | 10.0 |
| Nicholas | 93 | -9.1 | 167 | 0.0 | 39.4 | 387,800 | 2,322 | 66,211 | 25.9 | 46,653 | 49.3 | 50.7 | 1 | 60.1 | 139 | 8.6 |

Items 117—132

# Table B. States and Counties — Water Use, Wholesale Trade, Retail Trade, and Real Estate

| STATE County | Water use, 2015 Public supply water withdrawn (mil gal/day) | Public supply gallons withdrawn per person per day | Wholesale Trade[1], 2017 Number of establishments | Number of employees | Sales (mil dol) | Average payroll (mil dol) | Retail Trade[2], 2017 Number of establishments | Number of employees | Sales (mil dol) | Average payroll (mil dol) | Real estate and rental and leasing,[2] 2017 Number of establishments | Number of employees | Sales (mil dol) | Average payroll (mil dol) |
|---|---|---|---|---|---|---|---|---|---|---|---|---|---|---|
| | 133 | 134 | 135 | 136 | 137 | 138 | 139 | 140 | 141 | 142 | 143 | 144 | 145 | 146 |
| KENTUCKY—Cont'd | | | | | | | | | | | | | | |
| Clay | 5.20 | 247.5 | NA | NA | NA | NA | 58 | 635 | 204.7 | 15.1 | 8 | D | 2.0 | D |
| Clinton | 3.75 | 368.6 | 8 | D | 20.5 | D | 43 | 323 | 73.8 | 6.8 | 4 | 13 | 1.8 | 0.3 |
| Crittenden | 0.47 | 51.2 | D | D | D | D | 28 | 258 | 54.6 | 5.8 | 7 | D | 1.3 | D |
| Cumberland | 0.63 | 93.2 | NA | NA | NA | NA | 24 | 174 | 51.2 | 3.9 | 5 | D | 1.8 | D |
| Daviess | 13.36 | 134.6 | 97 | 1,267 | 1,008.3 | 65.5 | 400 | 5,513 | 1,505.5 | 132.2 | 97 | 549 | 90.2 | 15.7 |
| Edmonson | 1.67 | 139.1 | NA | NA | NA | NA | 23 | 154 | 44.9 | 3.8 | NA | NA | NA | NA |
| Elliott | 0.19 | 24.8 | NA | NA | NA | NA | 12 | 67 | 16.4 | 1.4 | NA | NA | NA | NA |
| Estill | 1.31 | 91.1 | D | D | D | 3.1 | 39 | 269 | 66.1 | 5.9 | D | D | D | 0.2 |
| Fayette | 35.21 | 112.0 | 328 | 4,645 | 3,402.9 | 258.8 | 1,211 | 23,102 | 6,341.8 | 597.9 | 486 | 2,223 | 580.8 | 86.3 |
| Fleming | 0.21 | 14.3 | 9 | 42 | 14.1 | 1.6 | 54 | 492 | 144.0 | 14.5 | 8 | D | 1.8 | D |
| Floyd | 4.52 | 119.7 | 33 | 366 | 237.8 | 14.1 | 148 | 1,462 | 473.5 | 37.6 | 27 | 113 | 26.6 | 5.5 |
| Franklin | 16.1 | 319.6 | 56 | 522 | 523.5 | 30.2 | 190 | 2,688 | 809.9 | 67.6 | 43 | D | D | 0.1 |
| Fulton | 1.18 | 189.2 | D | D | D | 3.0 | 32 | 205 | 45.0 | 3.8 | D | D | D | 0.1 |
| Gallatin | 0.60 | 69.5 | NA | NA | NA | NA | 17 | 152 | 67.7 | 3.1 | D | D | D | 0.2 |
| Garrard | 1.43 | 83.0 | D | D | D | 0.7 | 30 | 153 | 56.0 | 3.7 | 5 | D | 1.5 | D |
| Grant | 1.70 | 68.7 | 10 | 145 | 83.0 | 6.3 | 70 | 959 | 348.5 | 26.5 | 19 | 31 | 5.6 | 0.9 |
| Graves | 2.62 | 70.0 | D | D | D | D | 134 | 1,515 | 564.1 | 43.4 | 30 | 85 | 12.2 | 2.0 |
| Grayson | 2.64 | 100.7 | D | D | D | D | 100 | 992 | 279.5 | 23.2 | 15 | 211 | 9.8 | 5.4 |
| Green | 0.84 | 76.3 | D | D | D | 1.0 | 26 | 234 | 59.0 | 5.9 | D | D | D | 0.7 |
| Greenup | 3.80 | 105.4 | D | D | D | D | 85 | 946 | 271.8 | 20.5 | 13 | 31 | 7.0 | 0.7 |
| Hancock | 0.53 | 61.0 | NA | NA | NA | NA | 17 | 127 | 34.4 | 2.8 | D | D | D | D |
| Hardin | 14.33 | 134.6 | 64 | 594 | 477.2 | 27.1 | 391 | 6,326 | 1,744.6 | 157.5 | 116 | 556 | 86.8 | 15.3 |
| Harlan | 3.40 | 122.7 | D | D | D | 4.1 | 84 | 886 | 204.6 | 20.5 | 9 | 37 | 5.1 | 0.8 |
| Harrison | 2.78 | 148.2 | 9 | 46 | 28.5 | 1.3 | 49 | 590 | 145.8 | 13.5 | 9 | 37 | 5.1 | 0.8 |
| Hart | 4.10 | 222.2 | 7 | 29 | 18.9 | 1.3 | 63 | 495 | 147.8 | 8.7 | 7 | 15 | 1.9 | 0.3 |
| Henderson | 8.68 | 187.0 | 37 | 444 | 237.8 | 24.2 | 149 | 1,944 | 663.9 | 50.2 | 53 | 200 | 24.7 | 5.7 |
| Henry | 0.00 | 0.0 | D | D | D | D | 35 | 349 | 168.1 | 7.1 | 4 | 4 | 0.8 | 0.1 |
| Hickman | 0.13 | 28.2 | 8 | 62 | 65.8 | 2.0 | 17 | 239 | 60.4 | 6.0 | NA | NA | NA | NA |
| Hopkins | 9.63 | 208.3 | D | D | D | D | 164 | 2,583 | 955.8 | 70.5 | 31 | 105 | 15.4 | 3.1 |
| Jackson | 1.47 | 110.1 | NA | NA | NA | NA | 21 | 148 | 34.8 | 2.8 | 3 | 8 | 0.8 | 0.2 |
| Jefferson | 133.75 | 175.2 | 983 | 15,570 | 10,876.9 | 971.9 | 2,669 | 44,590 | 12,543.4 | 1,182.5 | 1,062 | 5,688 | 2,649.4 | 290.2 |
| Jessamine | 5.34 | 102.8 | 38 | 930 | 1,463.9 | 44.1 | 154 | 2,850 | 986.1 | 77.3 | 44 | 157 | 22.8 | 5.3 |
| Johnson | 2.48 | 107.0 | 10 | 72 | 36.6 | 2.6 | 84 | 1,350 | 385.0 | 33.4 | 16 | 50 | 7.9 | 1.4 |
| Kenton | 0.00 | 0.0 | 131 | 2,313 | 1,317.0 | 121.4 | 390 | 6,337 | 1,601.6 | 148.0 | 140 | 809 | 195.5 | 35.7 |
| Knott | 1.84 | 117.2 | NA | NA | NA | NA | 29 | 258 | 61.0 | 5.3 | NA | NA | NA | NA |
| Knox | 0.30 | 9.5 | 12 | 45 | 20.3 | 1.2 | 109 | 1,370 | 294.9 | 28.7 | 17 | 65 | 8.1 | 1.7 |
| Larue | 0.67 | 47.0 | 4 | 14 | 6.7 | 0.6 | 28 | 200 | 56.7 | 5.0 | NA | NA | NA | NA |
| Laurel | 8.90 | 148.1 | 55 | 680 | 505.0 | 31.4 | 246 | 3,376 | 1,086.6 | 88.4 | 48 | 265 | 39.3 | 6.8 |
| Lawrence | 1.49 | 94.6 | 7 | 189 | 113.2 | 8.7 | 48 | 711 | 161.3 | 14.4 | 9 | 21 | 4.0 | 0.4 |
| Lee | 0.95 | 140.7 | D | D | D | 0.8 | 19 | 171 | 36.9 | 3.9 | 3 | 5 | 0.3 | 0.0 |
| Leslie | 1.06 | 99.0 | NA | NA | NA | NA | 27 | 236 | 55.6 | 4.4 | NA | NA | NA | NA |
| Letcher | 1.61 | 69.6 | 10 | 151 | 247.8 | 6.1 | 57 | 635 | 138.8 | 16.1 | 10 | 33 | 6.2 | 1.1 |
| Lewis | 2.42 | 176.9 | D | D | D | 0.5 | 30 | 278 | 63.0 | 5.1 | 4 | 5 | 1.0 | 0.1 |
| Lincoln | 2.11 | 86.3 | 10 | 42 | 11.3 | 1.1 | 61 | 551 | 146.3 | 11.9 | 4 | 5 | 1.0 | 0.1 |
| Livingston | 1.66 | 178.2 | D | D | D | 2.8 | 27 | 206 | 45.5 | 4.2 | NA | NA | NA | NA |
| Logan | 0.00 | 0.0 | 26 | 208 | 147.7 | 9.4 | 98 | 961 | 276.1 | 24.8 | 13 | 31 | 5.3 | 1.0 |
| Lyon | 2.67 | 321.5 | D | D | D | D | 25 | 230 | 65.3 | 6.2 | 5 | 5 | 3.5 | 0.5 |
| McCracken | 8.21 | 126.3 | 103 | 1,377 | 4,324.5 | 63.1 | 399 | 6,077 | 1,715.6 | 155.7 | 68 | 393 | 90.5 | 11.9 |
| McCreary | 1.96 | 109.6 | NA | NA | NA | NA | 44 | 338 | 88.4 | 8.0 | D | D | D | 0.3 |
| McLean | 0.69 | 72.5 | D | D | D | D | 31 | 236 | 78.6 | 6.2 | NA | NA | NA | NA |
| Madison | 11.42 | 130.0 | 38 | 347 | 216.1 | 13.1 | 298 | 4,314 | 1,232.2 | 103.7 | 80 | 241 | 54.0 | 7.7 |
| Magoffin | 0.78 | 60.9 | 3 | 15 | 3.6 | 0.5 | 32 | 199 | 58.2 | 5.0 | NA | NA | NA | NA |
| Marion | 4.13 | 213.3 | 12 | 77 | 34.9 | 2.4 | 62 | 669 | 176.3 | 14.9 | 5 | 16 | 2.0 | 0.7 |
| Marshall | 4.10 | 131.8 | 25 | 193 | 112.3 | 9.3 | 125 | 1,395 | 453.1 | 37.2 | 22 | 89 | 30.6 | 5.5 |
| Martin | 4.45 | 361.6 | 5 | D | 49.5 | D | 38 | 394 | 106.0 | 7.9 | 3 | D | 1.4 | D |
| Mason | 2.73 | 159.7 | 15 | 253 | 86.4 | 12.3 | 97 | 1,479 | 383.3 | 32.3 | 12 | 77 | 9.8 | 2.0 |
| Meade | 0.75 | 26.9 | 9 | 73 | 121.6 | 3.8 | 61 | 660 | 225.8 | 16.3 | D | D | D | D |
| Menifee | 1.08 | 169.9 | NA | NA | NA | NA | 12 | 64 | 17.5 | 1.3 | NA | NA | NA | NA |
| Mercer | 2.94 | 137.3 | D | D | D | 1.2 | 52 | 644 | 172.0 | 16.2 | 7 | 14 | 1.8 | 0.4 |
| Metcalfe | 0.00 | 0.0 | NA | NA | NA | NA | 22 | 172 | 46.5 | 3.7 | 3 | 4 | 0.5 | 0.1 |
| Monroe | 1.41 | 132.2 | 6 | D | 24.2 | D | 46 | 485 | 118.7 | 10.9 | 7 | 35 | 2.2 | 0.9 |
| Montgomery | 3.00 | 108.7 | 18 | 249 | 65.9 | 6.8 | 124 | 1,534 | 467.2 | 36.9 | 19 | 40 | 6.0 | 1.0 |
| Morgan | 1.00 | 75.3 | NA | NA | NA | NA | 33 | 355 | 85.9 | 7.4 | NA | NA | NA | NA |
| Muhlenberg | 3.47 | 111.3 | D | D | D | D | 108 | 1,275 | 310.1 | 28.6 | 11 | 54 | 3.8 | 1.0 |
| Nelson | 5.50 | 121.9 | 34 | 414 | 251.0 | 20.1 | 161 | 2,036 | 590.3 | 53.8 | 34 | 98 | 15.9 | 2.1 |
| Nicholas | 1.71 | 239.8 | NA | NA | NA | NA | 12 | 76 | 16.2 | 1.3 | NA | NA | NA | NA |

1  Merchant wholesalers, except manufacturers' sales branches and offices.     2. Employer establishments.

| STATE County | Professional, scientific, and technical services, 2017 | | | | Manufacturing, 2017 | | | | Accommodation and food services, 2017 | | | |
|---|---|---|---|---|---|---|---|---|---|---|---|---|
| | Number of establish-ments | Number of employees | Sales (mil dol) | Average payroll (mil dol) | Number of establish-ments | Number of employees | Sales (mil dol) | Average payroll (mil dol) | Number of establis-hments | Number of employees | Sales (mil dol) | Annual payroll (mil dol) |
| | 147 | 148 | 149 | 150 | 151 | 152 | 153 | 154 | 155 | 156 | 157 | 158 |
| KENTUCKY—Cont'd | | | | | | | | | | | | |
| Clay | D | D | 12 | D | 6 | 133 | 28.0 | 6.2 | 19 | 328 | 15.5 | 4.5 |
| Clinton | 12 | 21 | 2 | 0.5 | D | 1,587 | D | 45.6 | D | D | D | D |
| Crittenden | 13 | 33 | 3 | 1.0 | 11 | 470 | 160.3 | 17.0 | 12 | 263 | 7.1 | 2.5 |
| Cumberland | D | D | D | 0.6 | 7 | 175 | 29.3 | 4.8 | 11 | 90 | 6.6 | 1.9 |
| Daviess | D | D | D | D | 97 | 5,070 | 3,078.5 | 285.5 | D | D | D | D |
| Edmonson | 6 | 16 | 2 | 0.4 | NA | NA | NA | NA | D | D | D | D |
| Elliott | NA | NA | NA | NA | NA | NA | NA | NA | D | D | D | 0.8 |
| Estill | 9 | 43 | 2 | 0.5 | 9 | 139 | 12.2 | 4.2 | 18 | 255 | 13.4 | 3.0 |
| Fayette | 1,158 | 11,056 | 2,001 | 698.3 | 219 | 7,509 | 2,739.9 | 385.9 | 849 | 20,035 | 1,112.4 | 325.4 |
| Fleming | 14 | 38 | 3 | 1.0 | 20 | 459 | 103.0 | 17.4 | D | D | D | D |
| Floyd | D | D | D | D | 12 | 172 | 33.4 | 6.1 | 58 | 849 | 40.9 | 10.8 |
| Franklin | 130 | 1,086 | 140 | 51.5 | 38 | 2,538 | 1,605.1 | 138.3 | 110 | D | 120.6 | D |
| Fulton | 4 | 23 | 2 | 0.6 | 7 | 450 | 198.4 | 19.2 | D | D | D | D |
| Gallatin | D | D | 5 | D | D | D | D | D | D | D | D | D |
| Garrard | 17 | 38 | 3 | 1.1 | 13 | 190 | 34.6 | 7.4 | D | D | D | D |
| Grant | 30 | 75 | 10 | 2.8 | D | 779 | D | 45.5 | D | D | D | D |
| Graves | 48 | 263 | 24 | 9.6 | 39 | 2,482 | 909.9 | 91.7 | 51 | 817 | 38.8 | 10.6 |
| Grayson | 18 | 57 | 7 | 2.3 | 28 | 1,682 | 434.6 | 68.1 | 31 | 551 | 26.0 | 6.8 |
| Green | 12 | 32 | 3 | 0.8 | 7 | 72 | 8.2 | 1.8 | 8 | 153 | 6.0 | 1.7 |
| Greenup | 26 | 152 | 14 | 5.3 | D | 546 | D | 29.9 | D | D | D | D |
| Hancock | 8 | 50 | 4 | 1.9 | 16 | 2,333 | 2,239.3 | 177.8 | D | D | D | D |
| Hardin | 162 | 1,044 | 185 | 66.5 | 66 | 6,183 | 2,438.9 | 331.5 | 194 | 4,763 | 232.5 | 66.6 |
| Harlan | 25 | 139 | 10 | 4.8 | 7 | 52 | 17.5 | 2.9 | 36 | 573 | 23.9 | 6.1 |
| Harrison | 16 | 54 | 6 | 1.2 | 20 | 1,236 | 756.2 | 72.9 | D | D | D | D |
| Hart | 18 | 53 | 5 | 1.6 | 21 | 2,560 | 876.9 | 102.0 | 18 | 309 | 14.2 | 3.7 |
| Henderson | 73 | 434 | 38 | 14.0 | 68 | 5,451 | 2,662.4 | 269.9 | 77 | 1,384 | 71.5 | 18.6 |
| Henry | 18 | 50 | 5 | 1.7 | 5 | 368 | 359.3 | 17.2 | D | D | D | D |
| Hickman | D | D | D | D | 4 | 143 | 24.3 | 3.2 | NA | NA | NA | NA |
| Hopkins | D | D | D | D | 44 | 1,719 | 755.2 | 98.4 | 65 | 1,274 | 58.9 | 16.6 |
| Jackson | D | D | 1 | D | 7 | 21 | 4.9 | 0.8 | 7 | 51 | 4.4 | 0.7 |
| Jefferson | 2,264 | 24,439 | 4,188 | 1,546.4 | 690 | 49,360 | 37,068.9 | 3,037.4 | 1,830 | 43,548 | 2,517.3 | 720.2 |
| Jessamine | 98 | 850 | 143 | 46.4 | 69 | 2,364 | 837.7 | 108.6 | 82 | 1,754 | 77.6 | 23.7 |
| Johnson | 21 | 125 | 24 | 5.7 | 8 | 44 | 7.4 | 1.5 | 34 | 576 | 25.9 | 7.3 |
| Kenton | D | D | D | D | 99 | 4,738 | 2,537.5 | 307.6 | 317 | 6,987 | 405.6 | 117.9 |
| Knott | 11 | 56 | 10 | 2.6 | NA | NA | NA | NA | 10 | 97 | 5.8 | 1.4 |
| Knox | D | D | D | D | 17 | 1,350 | 244.3 | 51.9 | 44 | 959 | 48.8 | 12.6 |
| Larue | 15 | 41 | 3 | 1.1 | D | 495 | D | D | 10 | 105 | 5.5 | 1.5 |
| Laurel | 93 | 1,513 | 146 | 33.3 | 51 | 3,831 | 1,026.3 | 163.4 | 101 | 2,587 | 128.2 | 36.2 |
| Lawrence | 12 | 70 | 6 | 2.7 | NA | NA | NA | NA | D | D | D | 4.7 |
| Lee | 5 | 14 | 2 | 0.3 | NA | NA | NA | NA | D | D | D | 0.6 |
| Leslie | D | D | D | D | NA | NA | NA | NA | 7 | 77 | 4.1 | 1.2 |
| Letcher | D | D | D | D | 7 | 69 | 26.6 | D | D | D | D | D |
| Lewis | D | D | D | 0.3 | 15 | 349 | 104.6 | 13.0 | 6 | 72 | 3.5 | 0.9 |
| Lincoln | 19 | 98 | 10 | 3.4 | 16 | 368 | 71.0 | 14.8 | 17 | 438 | 11.5 | 3.7 |
| Livingston | 8 | 37 | 6 | 2.1 | D | D | D | D | D | D | D | D |
| Logan | 29 | 170 | 16 | 7.3 | 37 | 2,913 | 2,129.4 | 166.5 | 32 | 572 | 25.8 | 6.7 |
| Lyon | 10 | 31 | 3 | 0.9 | NA | NA | NA | NA | 24 | 327 | 19.1 | 5.2 |
| McCracken | D | D | D | D | 51 | 2,269 | 547.4 | 108.4 | 220 | 4,650 | 218.2 | 64.2 |
| McCreary | 8 | 98 | 6 | 1.5 | 9 | 241 | 46.6 | 7.9 | 22 | 283 | 10.2 | 2.8 |
| McLean | 8 | 23 | 2 | 0.8 | 8 | 135 | 123.2 | 6.7 | 13 | 72 | 3.4 | 0.9 |
| Madison | D | D | D | D | 60 | 4,957 | 3,059.3 | 273.8 | 159 | 3,755 | 184.7 | 54.1 |
| Magoffin | 16 | 137 | 21 | 5.4 | NA | NA | NA | NA | 10 | 191 | 8.3 | 2.1 |
| Marion | 22 | 108 | 17 | 4.3 | 31 | 4,245 | 1,286.2 | 198.0 | 34 | 567 | 23.6 | 6.3 |
| Marshall | 44 | 241 | 39 | 14.5 | 40 | 2,349 | 2,840.1 | 204.8 | 80 | 1,425 | 58.6 | 16.0 |
| Martin | 8 | 19 | 3 | 0.5 | NA | NA | NA | NA | 10 | 148 | 6.2 | 1.6 |
| Mason | 22 | 99 | 11 | 3.9 | 15 | 1,223 | 773.4 | 60.9 | 41 | 856 | 39.3 | 11.8 |
| Meade | 16 | 276 | 11 | 4.0 | D | 522 | D | D | 31 | 554 | 23.3 | 6.2 |
| Menifee | NA | NA | NA | NA | D | D | D | D | NA | NA | NA | NA |
| Mercer | 25 | 56 | 5 | 1.7 | 11 | 1,854 | 1,270.7 | 112.1 | 31 | 458 | 23.5 | 6.6 |
| Metcalfe | 10 | 15 | 1 | 0.4 | 8 | 646 | 382.6 | 23.3 | 10 | 115 | 4.6 | 1.2 |
| Monroe | D | D | D | 0.7 | 17 | 667 | 104.7 | 25.0 | D | D | D | D |
| Montgomery | 35 | 165 | 19 | 5.2 | 29 | 3,423 | 1,012.8 | 153.0 | D | D | D | D |
| Morgan | 16 | 83 | 5 | 2.1 | 5 | 43 | 8.0 | 1.6 | 11 | 165 | 6.9 | 2.0 |
| Muhlenberg | 27 | 355 | 27 | 7.6 | 33 | D | 210.6 | D | 49 | 727 | 32.8 | 8.9 |
| Nelson | D | D | D | D | 60 | 5,104 | 2,152.7 | 292.2 | 71 | 1,264 | 65.3 | 17.9 |
| Nicholas | 7 | 13 | 1 | 0.5 | D | D | D | D | 4 | 16 | 0.9 | 0.1 |

# Health Care and Social Assistance, Other Services, Nonemployer Businesses, and Residential Construction

| STATE County | Health care and social assistance, 2017 | | | | Other services, 2017 | | | | Nonemployer businesses, 2018 | | Value of residential construction authorized by building permits, 2020 | |
|---|---|---|---|---|---|---|---|---|---|---|---|---|
| | Number of establishments | Number of employees | Receipts (mil dol) | Annual payroll (mil dol) | Number of establishments | Number of employees | Receipts (mil dol) | Annual payroll (mil dol) | Number | Receipts (mil dol) | New construction ($1,000) | Number of housing units |
| | 159 | 160 | 161 | 162 | 163 | 164 | 165 | 166 | 167 | 168 | 169 | 170 |
| KENTUCKY—Cont'd | | | | | | | | | | | | |
| Clay | 40 | 852 | 81.1 | 29.8 | D | D | D | 1.0 | 1,170 | 33.3 | 0 | 0 |
| Clinton | 26 | 484 | 49.2 | 16.7 | D | D | 1.5 | D | 757 | 28.8 | 0 | 0 |
| Crittenden | 16 | 401 | 25.9 | 11.7 | D | D | D | D | 559 | 21.6 | 150 | 1 |
| Cumberland | 13 | 382 | 28.2 | 13.1 | D | D | 2.7 | D | 559 | 21.6 | 0 | 0 |
| Daviess | 334 | 9,664 | 1,080.4 | 474.3 | 146 | 1,072 | 87.9 | 30.9 | 6,021 | 311.5 | 31,665 | 338 |
| Edmonson | D | D | D | D | 7 | 24 | 2.6 | 0.6 | 929 | 37.4 | NA | NA |
| Elliott | 11 | D | 12.4 | D | 3 | 7 | 0.4 | 0.1 | 379 | 11.3 | NA | NA |
| Estill | D | D | D | 18.0 | 12 | 34 | 4.1 | 0.8 | 787 | 19.4 | 0 | 0 |
| Fayette | 1,154 | 32,976 | 4,708.5 | 1,706.1 | 595 | 4,860 | 961.7 | 165.0 | 24,485 | 1,214.4 | 224,453 | 1,437 |
| Fleming | 17 | 495 | 40.6 | 16.6 | 19 | 55 | 6.0 | 1.5 | 1,276 | 66.7 | 0 | 0 |
| Floyd | 124 | 2,079 | 251.9 | 85.7 | 36 | 169 | 21.0 | 5.5 | 2,302 | 91.9 | 1,075 | 6 |
| Franklin | 160 | 2,301 | 333.6 | 101.9 | 121 | 659 | 100.6 | 27.1 | 3,472 | 128.1 | 20,889 | 97 |
| Fulton | D | D | D | D | D | D | D | D | 302 | 10.2 | 0 | 0 |
| Gallatin | 6 | 185 | 14.4 | 6.1 | D | D | 6.6 | D | 421 | 15.5 | 12,589 | 54 |
| Garrard | 25 | 317 | 17.0 | 8.8 | D | D | D | D | 1,314 | 51.8 | 0 | 0 |
| Grant | 39 | 505 | 56.9 | 23.4 | D | D | D | D | 1,351 | 62.4 | 30,462 | 51 |
| Graves | 76 | 1,466 | 153.5 | 56.8 | D | D | 35.3 | D | 2,630 | 116.1 | 304 | 2 |
| Grayson | 60 | 1,076 | 93.9 | 36.3 | D | D | D | D | 1,908 | 113.2 | 1,399 | 10 |
| Green | 23 | 543 | 41.8 | 18.5 | D | D | 6.8 | D | 925 | 36.5 | 0 | 0 |
| Greenup | 77 | 1,123 | 102.1 | 47.9 | D | D | D | D | 1,839 | 66.9 | 6,590 | 17 |
| Hancock | 16 | 131 | 9.9 | 3.2 | D | D | D | 0.3 | 468 | 17.4 | 1,782 | 10 |
| Hardin | 303 | 8,248 | 860.6 | 383.9 | 138 | 785 | 79.7 | 22.1 | 5,789 | 244.1 | 85,063 | 595 |
| Harlan | 45 | 1,318 | 145.3 | 48.5 | 20 | 64 | 6.9 | 1.6 | 1,286 | 39.2 | 0 | 0 |
| Harrison | 43 | 1,086 | 101.7 | 41.4 | D | D | D | D | 1,071 | 46.2 | 11,917 | 57 |
| Hart | 28 | 389 | 45.1 | 14.3 | 19 | 67 | 5.3 | 1.4 | 1,704 | 103.4 | 4,935 | 46 |
| Henderson | 145 | 2,521 | 226.7 | 91.6 | 62 | 414 | 83.3 | 17.9 | 2,382 | 92.7 | 10,921 | 63 |
| Henry | D | D | D | 6.9 | D | D | D | D | 1,031 | 49.8 | 11,426 | 61 |
| Hickman | 8 | D | 11.2 | D | 3 | 9 | 1.1 | 0.4 | 367 | 11.4 | NA | NA |
| Hopkins | 151 | 3,163 | 332.9 | 132.1 | 75 | 410 | 50.4 | 14.0 | 2,370 | 91.5 | 11,589 | 76 |
| Jackson | 19 | 219 | 16.3 | 6.9 | 4 | 12 | 1.4 | 0.2 | 891 | 28.6 | 0 | 0 |
| Jefferson | 2,472 | 69,939 | 9,367.8 | 3,580.8 | 1,339 | 11,113 | 1,497.6 | 382.1 | 57,630 | 2,757.4 | 259,621 | 1,264 |
| Jessamine | 124 | 1,368 | 101.6 | 47.2 | 85 | 313 | 27.8 | 7.6 | 4,454 | 215.3 | 72,970 | 239 |
| Johnson | 44 | 705 | 84.3 | 28.2 | D | D | D | 2.6 | 1,341 | 57.9 | 0 | 0 |
| Kenton | 372 | 13,524 | 1,964.5 | 736.7 | 220 | 1,785 | 150.9 | 47.2 | 10,116 | 462.2 | 86,641 | 311 |
| Knott | 24 | 205 | 20.0 | 6.3 | D | D | D | D | 824 | 27.8 | NA | NA |
| Knox | 70 | 1,222 | 103.6 | 41.2 | D | D | D | D | 1,954 | 75.2 | 0 | 0 |
| Larue | D | D | D | D | 18 | 75 | 6.0 | 1.7 | 1,037 | 39.6 | 8,652 | 37 |
| Laurel | 148 | 2,600 | 344.4 | 137.2 | D | D | D | D | 4,112 | 183.9 | 2,069 | 20 |
| Lawrence | 23 | 617 | 77.5 | 28.8 | D | D | D | D | 709 | 21.7 | NA | NA |
| Lee | 19 | 742 | 30.8 | 12.3 | NA | NA | NA | NA | 420 | 12.5 | NA | NA |
| Leslie | 15 | 309 | 26.5 | 10.2 | 3 | 17 | 1.1 | 0.4 | 490 | 21.2 | NA | NA |
| Letcher | 57 | 963 | 116.6 | 44.5 | D | D | D | D | 1,137 | 37.1 | 0 | 0 |
| Lewis | 12 | 453 | 50.6 | 22.6 | D | D | 1.8 | D | 873 | 33.4 | 0 | 0 |
| Lincoln | 37 | 564 | 39.7 | 15.8 | D | D | D | D | 1,619 | 66.8 | 5,159 | 55 |
| Livingston | 16 | 333 | 27.1 | 10.7 | D | D | 1.2 | D | 562 | 18.0 | NA | NA |
| Logan | 47 | 693 | 62.9 | 23.4 | 31 | 120 | 14.3 | 3.9 | 1,779 | 89.1 | 2,071 | 15 |
| Lyon | 20 | 289 | 20.4 | 8.3 | 6 | 14 | 1.2 | 0.3 | 499 | 17.1 | 1,947 | 18 |
| McCracken | 279 | 7,108 | 920.6 | 312.7 | 129 | 821 | 113.7 | 25.2 | 4,460 | 193.1 | 26,808 | 116 |
| McCreary | 27 | 266 | 21.4 | 8.4 | 9 | 22 | 2.6 | 0.5 | 868 | 34.0 | 0 | 0 |
| McLean | 12 | 136 | 10.3 | 4.2 | D | D | D | D | 524 | 15.3 | 0 | 0 |
| Madison | 253 | 3,714 | 386.4 | 147.2 | 104 | 602 | 72.0 | 17.2 | 5,889 | 243.6 | 58,696 | 404 |
| Magoffin | 21 | 285 | 31.4 | 11.3 | D | D | D | 0.5 | 606 | 25.5 | NA | NA |
| Marion | 39 | 1,077 | 98.3 | 36.1 | 23 | 96 | 14.3 | 3.1 | 1,219 | 52.9 | 7,118 | 58 |
| Marshall | D | D | D | D | 44 | 263 | 31.8 | 8.5 | 2,226 | 88.9 | 23,564 | 91 |
| Martin | 18 | 227 | 18.1 | 8.4 | D | D | D | D | 466 | 14.4 | NA | NA |
| Mason | 57 | 1,220 | 137.8 | 54.8 | 35 | 160 | 11.1 | 3.0 | 1,158 | 54.2 | 6,778 | 32 |
| Meade | D | D | D | D | 24 | 146 | 12.9 | 4.0 | 1,585 | 65.4 | 23,665 | 130 |
| Menifee | 9 | 152 | 11.2 | 5.1 | NA | NA | NA | NA | 403 | 12.7 | NA | NA |
| Mercer | 39 | 579 | 41.7 | 20.0 | 31 | 99 | 11.0 | 2.6 | 1,571 | 73.9 | 11,677 | 74 |
| Metcalfe | 16 | 90 | 7.3 | 2.7 | D | D | D | D | 838 | 28.2 | NA | NA |
| Monroe | 26 | 474 | 38.6 | 17.0 | 9 | 33 | 4.7 | 1.0 | 910 | 39.2 | NA | NA |
| Montgomery | 68 | 963 | 119.2 | 41.4 | 27 | 124 | 12.8 | 3.5 | 1,763 | 71.4 | 8,606 | 40 |
| Morgan | 18 | 359 | 43.8 | 15.9 | 12 | 66 | 9.1 | 2.3 | 784 | 28.0 | 0 | 0 |
| Muhlenberg | 80 | 1,291 | 97.5 | 48.4 | 38 | 136 | 14.6 | 4.1 | 1,586 | 56.0 | 1,064 | 9 |
| Nelson | 105 | 1,371 | 142.8 | 50.5 | 47 | 180 | 18.8 | 4.2 | 3,081 | 133.7 | 45,854 | 282 |
| Nicholas | 12 | 169 | 13.4 | 6.5 | D | D | 2.7 | D | 461 | 17.7 | 0 | 0 |

# Table B. States and Counties — Government Employment and Payroll, and Local Government Finances

| | Government employment and payroll, 2017 | | | | | | | | | Local government finances, 2017 | | | | |
| | | | March payroll (percent of total) | | | | | | | General revenue | | | | |
| | | | | | | | | | | | | | Taxes | |
| STATE County | Full-time equivalent employees | March payroll (dollars) | Adminis-tration, judicial, and legal | Police and corrections | Fire protection | Highways and transpor-tation | Health and welfare | Natural resources and utilities | Education and libraries | Total (mil dol) | Inter-govern-mental (mil dol) | Total (mil dol) | Per capita[1] (dollars) | |
| | | | | | | | | | | | | | Total | Property |
| | 171 | 172 | 173 | 174 | 175 | 176 | 177 | 178 | 179 | 180 | 181 | 182 | 183 | 184 |

| KENTUCKY—Cont'd | | | | | | | | | | | | | | |
| Clay | 859 | 2,375,911 | 5.0 | 5.6 | 0.0 | 2.3 | 4.5 | 4.4 | 77.8 | 44.3 | 31.9 | 7.6 | 373 | 215 |
| Clinton | 398 | 1,121,150 | 2.1 | 4.3 | 0.0 | 2.4 | 5.6 | 1.4 | 83.8 | 23.2 | 15.2 | 5.9 | 582 | 291 |
| Crittenden | 298 | 866,183 | 5.8 | 13.4 | 0.0 | 1.8 | 0.0 | 6.3 | 71.5 | 20.5 | 13.7 | 5.3 | 588 | 348 |
| Cumberland | 451 | 1,495,373 | 3.2 | 1.9 | 0.3 | 1.6 | 46.9 | 4.1 | 40.1 | 29.6 | 17.1 | 4.6 | 683 | 311 |
| Daviess | 4,180 | 14,060,220 | 3.8 | 7.2 | 4.1 | 3.8 | 4.8 | 13.0 | 59.1 | 336.4 | 122.9 | 120.0 | 1,195 | 713 |
| Edmonson | 423 | 1,188,212 | 3.3 | 4.5 | 0.4 | 2.0 | 3.9 | 8.6 | 77.1 | 22.9 | 14.9 | 6.1 | 498 | 384 |
| Elliott | 263 | 721,216 | 4.3 | 0.5 | 0.0 | 3.7 | 11.3 | 2.9 | 71.9 | 14.5 | 11.5 | 2.3 | 301 | 176 |
| Estill | 550 | 1,582,577 | 1.9 | 5.5 | 1.0 | 2.2 | 7.7 | 5.4 | 73.2 | 33.5 | 22.4 | 7.0 | 493 | 315 |
| Fayette | 9,865 | 41,635,220 | 4.8 | 14.0 | 9.4 | 2.7 | 3.3 | 6.4 | 56.0 | 1,138.7 | 245.8 | 687.2 | 2,134 | 1,001 |
| Fleming | 478 | 1,690,476 | 4.4 | 3.2 | 1.1 | 2.3 | 5.2 | 3.9 | 79.5 | 58.4 | 21.8 | 8.9 | 619 | 406 |
| Floyd | 1,297 | 3,808,363 | 2.3 | 5.6 | 2.8 | 1.8 | 5.2 | 4.6 | 76.6 | 82.8 | 53.6 | 18.1 | 498 | 381 |
| Franklin | 1,723 | 6,510,417 | 4.0 | 6.7 | 10.0 | 3.1 | 0.9 | 12.4 | 50.1 | 176.9 | 46.4 | 75.0 | 1,485 | 755 |
| Fulton | 412 | 1,101,709 | 9.6 | 16.5 | 5.3 | 8.6 | 3.1 | 8.8 | 46.2 | 28.1 | 16.0 | 5.6 | 917 | 552 |
| Gallatin | 329 | 988,148 | 7.6 | 2.4 | 0.0 | 2.3 | 2.3 | 4.1 | 81.1 | 24.3 | 13.4 | 8.6 | 989 | 598 |
| Garrard | 483 | 1,479,069 | 4.4 | 2.3 | 0.2 | 2.3 | 6.1 | 2.4 | 82.0 | 41.1 | 21.7 | 12.3 | 705 | 468 |
| Grant | 934 | 2,920,679 | 4.4 | 7.4 | 1.7 | 2.1 | 0.0 | 6.0 | 77.1 | 63.5 | 39.0 | 16.0 | 640 | 464 |
| Graves | 1,238 | 3,892,508 | 3.6 | 8.9 | 4.4 | 2.7 | 0.0 | 2.7 | 76.6 | 79.4 | 47.5 | 25.5 | 685 | 409 |
| Grayson | 978 | 2,948,785 | 4.3 | 17.1 | 0.3 | 2.1 | 0.1 | 4.9 | 68.3 | 71.5 | 43.9 | 19.5 | 741 | 393 |
| Green | 386 | 1,076,131 | 3.8 | 3.0 | 0.0 | 2.0 | 4.8 | 4.6 | 80.8 | 23.2 | 15.0 | 5.9 | 533 | 333 |
| Greenup | 1,163 | 3,780,181 | 4.1 | 7.1 | 0.7 | 2.4 | 0.2 | 4.9 | 79.0 | 85.7 | 43.5 | 33.5 | 943 | 750 |
| Hancock | 416 | 1,244,317 | 9.1 | 3.7 | 0.0 | 2.3 | 3.9 | 4.4 | 70.6 | 24.8 | 11.4 | 10.8 | 1,227 | 658 |
| Hardin | 5,558 | 22,988,050 | 1.3 | 3.3 | 1.6 | 1.1 | 51.5 | 3.8 | 36.5 | 586.1 | 125.3 | 105.6 | 977 | 545 |
| Harlan | 1,155 | 3,185,246 | 3.0 | 7.1 | 0.3 | 2.3 | 8.2 | 5.7 | 71.8 | 78.9 | 52.2 | 13.7 | 514 | 369 |
| Harrison | 644 | 1,959,248 | 6.6 | 10.6 | 3.7 | 2.9 | 0.1 | 3.4 | 70.9 | 41.4 | 21.3 | 16.1 | 858 | 387 |
| Hart | 508 | 1,727,761 | 5.1 | 7.0 | 0.1 | 2.1 | 5.1 | 10.1 | 68.9 | 40.6 | 23.1 | 11.7 | 626 | 318 |
| Henderson | 1,909 | 6,339,040 | 6.8 | 12.3 | 4.7 | 4.1 | 1.8 | 13.5 | 55.6 | 133.7 | 65.2 | 52.2 | 1,138 | 664 |
| Henry | 595 | 1,840,305 | 10.3 | 3.8 | 0.0 | 1.5 | 1.8 | 5.6 | 77.0 | 37.6 | 21.2 | 12.0 | 749 | 576 |
| Hickman | 175 | 523,679 | 5.9 | 5.5 | 0.0 | 5.6 | 0.2 | 0.8 | 78.6 | 11.6 | 7.3 | 3.4 | 761 | 567 |
| Hopkins | 1,826 | 5,753,345 | 4.9 | 9.9 | 4.7 | 3.6 | 3.1 | 9.8 | 62.9 | 125.6 | 61.0 | 46.7 | 1,032 | 515 |
| Jackson | 528 | 1,248,716 | 4.0 | 7.5 | 0.0 | 3.0 | 0.0 | 2.2 | 80.3 | 33.3 | 25.3 | 5.6 | 415 | 229 |
| Jefferson | 24,734 | 113,244,260 | 3.2 | 10.3 | 4.3 | 4.0 | 5.6 | 9.2 | 61.8 | 2,682.0 | 667.8 | 1,404.9 | 1,825 | 929 |
| Jessamine | 1,768 | 5,333,658 | 5.4 | 8.0 | 4.3 | 1.4 | 4.9 | 5.3 | 67.2 | 130.3 | 48.4 | 68.8 | 1,292 | 772 |
| Johnson | 925 | 2,854,090 | 3.6 | 2.5 | 0.9 | 2.0 | 7.5 | 1.3 | 81.1 | 63.4 | 35.7 | 15.8 | 701 | 416 |
| Kenton | 5,801 | 23,671,220 | 2.9 | 9.6 | 5.9 | 15.5 | 3.6 | 9.3 | 52.5 | 687.9 | 177.3 | 254.0 | 1,534 | 892 |
| Knott | 533 | 1,546,838 | 5.4 | 1.3 | 0.0 | 1.9 | 0.0 | 5.1 | 81.7 | 31.0 | 22.2 | 7.0 | 460 | 392 |
| Knox | 1,143 | 3,414,140 | 2.4 | 2.7 | 0.1 | 1.7 | 12.0 | 0.9 | 80.0 | 85.3 | 54.2 | 15.2 | 483 | 252 |
| Larue | 504 | 1,104,219 | 5.1 | 7.6 | 0.0 | 2.6 | 0.3 | 5.2 | 71.4 | 31.0 | 20.0 | 8.6 | 605 | 442 |
| Laurel | 1,952 | 5,552,219 | 3.1 | 7.6 | 0.7 | 2.7 | 2.9 | 5.7 | 76.2 | 134.3 | 74.5 | 46.0 | 763 | 351 |
| Lawrence | 567 | 1,592,040 | 5.9 | 3.3 | 1.1 | 2.3 | 5.0 | 3.1 | 78.9 | 33.5 | 22.1 | 8.6 | 543 | 415 |
| Lee | 243 | 764,630 | 10.4 | 3.5 | 0.0 | 4.3 | 1.4 | 6.9 | 73.4 | 15.6 | 9.9 | 3.9 | 586 | 413 |
| Leslie | 421 | 1,213,626 | 5.4 | 3.2 | 0.0 | 3.2 | 0.0 | 4.8 | 77.1 | 26.6 | 20.4 | 5.2 | 509 | 344 |
| Letcher | 732 | 2,218,820 | 1.6 | 2.7 | 0.6 | 0.8 | 0.0 | 5.7 | 88.2 | 53.2 | 35.2 | 11.8 | 531 | 340 |
| Lewis | 575 | 1,833,270 | 3.6 | 5.3 | 0.2 | 1.7 | 1.0 | 4.3 | 82.3 | 31.0 | 20.8 | 7.5 | 563 | 422 |
| Lincoln | 836 | 2,289,112 | 2.1 | 4.8 | 0.5 | 2.0 | 2.5 | 3.7 | 83.9 | 59.6 | 38.7 | 13.3 | 542 | 357 |
| Livingston | 321 | 886,221 | 4.4 | 4.8 | 2.2 | 3.9 | 7.1 | 5.1 | 69.1 | 24.0 | 13.0 | 8.7 | 943 | 630 |
| Logan | 1,001 | 2,855,484 | 4.5 | 7.7 | 1.4 | 1.8 | 0.0 | 5.0 | 77.6 | 65.8 | 35.9 | 25.3 | 937 | 489 |
| Lyon | 243 | 717,253 | 9.1 | 6.6 | 0.4 | 4.8 | 10.4 | 5.7 | 57.7 | 18.1 | 7.0 | 7.9 | 951 | 709 |
| McCracken | 2,315 | 10,412,476 | 3.0 | 7.4 | 3.6 | 3.7 | 6.8 | 10.2 | 62.4 | 205.9 | 78.8 | 84.0 | 1,285 | 662 |
| McCreary | 620 | 1,867,437 | 2.1 | 5.1 | 0.0 | 1.8 | 8.4 | 4.4 | 77.0 | 44.3 | 28.0 | 14.8 | 854 | 743 |
| McLean | 377 | 975,438 | 7.1 | 2.0 | 0.0 | 2.3 | 0.0 | 6.6 | 75.0 | 29.8 | 15.6 | 9.3 | 1,014 | 501 |
| Madison | 2,654 | 8,082,172 | 2.1 | 4.0 | 3.3 | 2.3 | 9.0 | 3.2 | 72.3 | 216.5 | 94.7 | 90.3 | 990 | 478 |
| Magoffin | 497 | 1,549,269 | 4.5 | 3.8 | 0.9 | 3.8 | 4.6 | 2.8 | 79.6 | 33.8 | 24.7 | 5.5 | 442 | 216 |
| Marion | 756 | 2,298,284 | 3.3 | 14.6 | 0.4 | 2.4 | 2.3 | 9.4 | 66.7 | 50.2 | 24.4 | 18.5 | 958 | 640 |
| Marshall | 1,244 | 5,690,803 | 2.0 | 3.2 | 0.4 | 1.7 | 16.5 | 4.9 | 71.3 | 109.3 | 39.7 | 42.3 | 1,352 | 722 |
| Martin | 490 | 1,622,222 | 7.1 | 2.9 | 0.0 | 2.3 | 2.7 | 4.4 | 80.2 | 30.3 | 20.2 | 6.6 | 571 | 375 |
| Mason | 912 | 2,313,200 | 5.6 | 11.3 | 4.5 | 2.9 | 5.1 | 9.5 | 59.5 | 50.4 | 22.1 | 20.2 | 1,173 | 558 |
| Meade | 893 | 2,746,337 | 3.8 | 6.0 | 0.0 | 1.6 | 3.4 | 5.0 | 79.4 | 61.6 | 36.5 | 18.7 | 666 | 514 |
| Menifee | 236 | 632,187 | 2.9 | 0.9 | 0.0 | 3.5 | 0.4 | 4.5 | 85.5 | 14.3 | 10.6 | 2.7 | 421 | 240 |
| Mercer | 705 | 2,044,636 | 6.4 | 7.6 | 3.3 | 2.5 | 0.0 | 4.4 | 72.0 | 52.0 | 23.1 | 22.8 | 1,057 | 650 |
| Metcalfe | 316 | 999,594 | 4.1 | 3.7 | 0.0 | 5.1 | 0.6 | 6.0 | 80.5 | 22.6 | 15.6 | 5.9 | 589 | 309 |
| Monroe | 549 | 1,359,308 | 6.7 | 7.5 | 0.0 | 1.8 | 2.3 | 10.3 | 71.0 | 39.8 | 17.7 | 7.7 | 723 | 396 |
| Montgomery | 902 | 3,035,358 | 3.3 | 6.6 | 6.5 | 3.0 | 6.6 | 0.9 | 73.1 | 69.2 | 39.1 | 23.5 | 841 | 426 |
| Morgan | 448 | 1,238,906 | 5.1 | 5.0 | 0.0 | 2.1 | 5.2 | 5.7 | 76.4 | 33.2 | 18.6 | 7.8 | 592 | 261 |
| Muhlenberg | 1,070 | 3,150,997 | 3.5 | 5.2 | 1.5 | 1.1 | 2.6 | 8.3 | 76.1 | 73.5 | 43.3 | 17.4 | 563 | 437 |
| Nelson | 1,438 | 4,943,456 | 2.2 | 6.2 | 1.6 | 1.4 | 2.5 | 8.0 | 76.3 | 123.4 | 45.5 | 44.8 | 983 | 750 |
| Nicholas | 210 | 598,380 | 8.6 | 4.7 | 0.1 | 3.1 | 0.0 | 9.8 | 71.8 | 16.9 | 12.1 | 2.6 | 362 | 270 |

1. Based on the resident population estimated as of July 1 of the year shown.

Items 171—184

# Table B. States and Counties — Local Government Finances, Government Employment, and Income Taxes

| STATE County | Local government finances, 2017 (cont.) | | | | | | | | | Government employment, 2019 | | | Individual income tax returns, 2018 | | |
|---|---|---|---|---|---|---|---|---|---|---|---|---|---|---|---|
| | Direct general expenditure | | | | | | | Debt outstanding | | | | | | | |
| | Total (mil dol) | Per capita¹ (dollars) | Percent of total for: | | | | | Total (mil dol) | Per capita¹ (dollars) | Federal civilian | Federal military | State and local | Number of returns | Mean adjusted gross income | Mean income tax |
| | | | Education | Health and hospitals | Police protection | Public welfare | Highways | | | | | | | | |
| | 185 | 186 | 187 | 188 | 189 | 190 | 191 | 192 | 193 | 194 | 195 | 196 | 197 | 198 | 199 |

KENTUCKY—Cont'd

| STATE County | 185 | 186 | 187 | 188 | 189 | 190 | 191 | 192 | 193 | 194 | 195 | 196 | 197 | 198 | 199 |
|---|---|---|---|---|---|---|---|---|---|---|---|---|---|---|---|
| Clay | 42 | 2,085 | 68.2 | 3.0 | 2.2 | 0.0 | 5.2 | 33.3 | 1,645 | 334 | 55 | 1,068 | 6,150 | 36,579 | 2,196 |
| Clinton | 23 | 2,287 | 68.1 | 4.7 | 2.4 | 0.0 | 5.3 | 28.1 | 2,756 | 45 | 30 | 575 | 3,730 | 34,759 | 1,938 |
| Crittenden | 20 | 2,194 | 53.9 | 0.9 | 3.1 | 0.1 | 11.0 | 30.1 | 3,340 | 18 | 26 | 423 | 3,500 | 45,227 | 3,327 |
| Cumberland | 31 | 4,638 | 29.5 | 51.5 | 2.7 | 0.0 | 4.4 | 15.8 | 2,357 | 10 | 20 | 327 | 2,520 | 37,357 | 2,477 |
| Daviess | 320 | 3,180 | 47.7 | 6.7 | 3.4 | 0.1 | 3.8 | 916.4 | 9,122 | 261 | 321 | 5,185 | 45,850 | 56,500 | 5,283 |
| Edmonson | 23 | 1,844 | 70.2 | 5.3 | 3.2 | 0.0 | 4.0 | 40.6 | 3,318 | 203 | 35 | 493 | 4,770 | 44,142 | 2,939 |
| Elliott | 14 | 1,906 | 66.5 | 0.9 | 1.3 | 0.0 | 10.7 | 17.8 | 2,381 | 8 | 19 | 569 | 2,120 | 40,080 | 2,392 |
| Estill | 41 | 2,857 | 67.2 | 7.1 | 1.3 | 0.0 | 3.0 | 43.0 | 3,030 | 20 | 42 | 653 | 5,380 | 40,872 | 2,652 |
| Fayette | 1,062 | 3,298 | 42.6 | 1.5 | 5.7 | 1.0 | 1.6 | 1,535.9 | 4,770 | 4,243 | 1,007 | 41,283 | 143,040 | 69,317 | 8,529 |
| Fleming | 70 | 4,854 | 39.8 | 38.6 | 0.8 | 0.0 | 2.6 | 211.0 | 14,616 | 45 | 44 | 650 | 6,020 | 41,605 | 2,842 |
| Floyd | 109 | 2,999 | 72.8 | 2.7 | 1.3 | 0.1 | 0.7 | 160.7 | 4,427 | 114 | 105 | 2,228 | 12,370 | 43,188 | 3,375 |
| Franklin | 167 | 3,298 | 38.2 | 4.0 | 5.5 | 0.2 | 3.3 | 146.6 | 2,904 | 438 | 165 | 13,231 | 24,930 | 58,912 | 5,515 |
| Fulton | 43 | 6,889 | 23.5 | 0.0 | 2.4 | 0.0 | 1.4 | 23.1 | 3,753 | 27 | 35 | 497 | 2,270 | 38,100 | 2,563 |
| Gallatin | 23 | 2,688 | 64.9 | 3.3 | 2.5 | 0.1 | 5.2 | 362.6 | 41,582 | 35 | 26 | 426 | 3,640 | 48,198 | 3,580 |
| Garrard | 38 | 2,193 | 55.0 | 20.0 | 2.2 | 0.0 | 3.6 | 51.8 | 2,971 | 21 | 53 | 605 | 6,950 | 49,439 | 3,857 |
| Grant | 65 | 2,602 | 63.9 | 0.4 | 3.4 | 0.0 | 2.8 | 101.8 | 4,075 | 48 | 74 | 1,166 | 10,720 | 50,040 | 3,983 |
| Graves | 78 | 2,088 | 70.0 | 3.4 | 2.2 | 0.3 | 6.4 | 125.2 | 3,365 | 193 | 111 | 1,706 | 15,120 | 47,871 | 3,807 |
| Grayson | 68 | 2,601 | 50.5 | 1.6 | 1.8 | 0.0 | 5.6 | 91.0 | 3,461 | 89 | 77 | 1,612 | 10,400 | 43,206 | 3,068 |
| Green | 26 | 2,355 | 67.5 | 5.5 | 1.7 | 0.0 | 6.3 | 44.6 | 4,039 | 22 | 33 | 470 | 4,330 | 39,422 | 2,607 |
| Greenup | 83 | 2,340 | 66.7 | 4.9 | 3.1 | 0.0 | 5.1 | 69.8 | 1,966 | 55 | 104 | 1,464 | 14,650 | 56,270 | 5,307 |
| Hancock | 26 | 2,907 | 58.7 | 3.0 | 2.2 | 0.4 | 6.1 | 1,321.1 | 150,506 | 21 | 26 | 431 | 3,850 | 51,527 | 3,870 |
| Hardin | 578 | 5,341 | 26.6 | 56.0 | 1.5 | 0.0 | 2.2 | 423.2 | 3,912 | 5,154 | 5,004 | 8,213 | 49,290 | 53,754 | 4,559 |
| Harlan | 75 | 2,795 | 54.1 | 15.2 | 2.9 | 0.6 | 3.8 | 51.0 | 1,913 | 61 | 76 | 1,860 | 8,500 | 35,846 | 2,584 |
| Harrison | 41 | 2,170 | 59.5 | 1.2 | 4.1 | 0.2 | 6.7 | 33.9 | 1,810 | 39 | 56 | 768 | 7,910 | 49,328 | 3,876 |
| Hart | 43 | 2,303 | 59.3 | 5.5 | 3.1 | 0.0 | 4.9 | 114.9 | 6,130 | 41 | 57 | 818 | 7,420 | 40,194 | 2,670 |
| Henderson | 125 | 2,715 | 51.0 | 0.2 | 5.7 | 0.3 | 3.8 | 965.7 | 21,054 | 99 | 134 | 2,424 | 19,780 | 53,886 | 4,786 |
| Henry | 37 | 2,291 | 71.2 | 2.2 | 2.7 | 0.0 | 3.5 | 44.5 | 2,785 | 102 | 48 | 720 | 7,050 | 50,144 | 3,998 |
| Hickman | 11 | 2,477 | 62.7 | 0.0 | 3.7 | 0.2 | 8.6 | 6.1 | 1,358 | 20 | 13 | 214 | 1,810 | 45,319 | 3,403 |
| Hopkins | 120 | 2,642 | 51.2 | 3.3 | 6.1 | 1.7 | 4.4 | 84.7 | 1,871 | 168 | 131 | 2,912 | 18,730 | 53,392 | 4,615 |
| Jackson | 34 | 2,563 | 60.7 | 15.7 | 0.4 | 0.0 | 3.5 | 30.0 | 2,239 | 34 | 40 | 690 | 4,530 | 37,769 | 2,119 |
| Jefferson | 2,872 | 3,730 | 42.0 | 4.2 | 7.1 | 0.6 | 3.2 | 5,585.4 | 7,255 | 6,917 | 2,529 | 41,088 | 374,950 | 66,922 | 7,911 |
| Jessamine | 116 | 2,186 | 56.3 | 4.8 | 5.3 | 0.1 | 3.2 | 204.2 | 3,834 | 76 | 158 | 2,298 | 22,700 | 67,410 | 7,836 |
| Johnson | 61 | 2,718 | 66.3 | 5.7 | 2.0 | 0.0 | 4.3 | 59.0 | 2,610 | 42 | 65 | 1,408 | 7,750 | 47,737 | 3,844 |
| Kenton | 628 | 3,790 | 37.0 | 3.1 | 4.9 | 0.0 | 6.3 | 1,835.7 | 11,085 | 3,005 | 495 | 6,816 | 79,470 | 111,070 | 16,134 |
| Knott | 29 | 1,928 | 75.7 | 0.2 | 0.5 | 0.1 | 6.0 | 22.6 | 1,484 | 34 | 42 | 658 | 4,700 | 46,749 | 4,149 |
| Knox | 82 | 2,605 | 55.8 | 10.5 | 1.2 | 0.0 | 2.7 | 121.0 | 3,849 | 122 | 92 | 1,333 | 10,870 | 38,469 | 2,461 |
| Larue | 35 | 2,471 | 70.6 | 1.9 | 1.9 | 0.0 | 4.1 | 45.6 | 3,214 | 43 | 42 | 567 | 6,250 | 45,715 | 3,334 |
| Laurel | 129 | 2,138 | 61.8 | 2.5 | 2.5 | 0.0 | 3.4 | 294.9 | 4,889 | 245 | 181 | 2,603 | 23,810 | 46,091 | 3,759 |
| Lawrence | 33 | 2,104 | 64.5 | 5.3 | 1.1 | 0.0 | 6.6 | 314.1 | 19,936 | 34 | 46 | 648 | 5,350 | 45,242 | 3,091 |
| Lee | 17 | 2,496 | 51.3 | 1.3 | 4.3 | 0.2 | 14.6 | 16.9 | 2,561 | 9 | 19 | 396 | 2,050 | 39,313 | 2,828 |
| Leslie | 28 | 2,757 | 57.0 | 0.3 | 0.3 | 0.0 | 7.3 | 25.1 | 2,433 | 12 | 29 | 530 | 3,260 | 40,635 | 2,492 |
| Letcher | 49 | 2,202 | 68.0 | 0.2 | 2.0 | 0.1 | 3.9 | 38.1 | 1,709 | 34 | 64 | 1,074 | 7,090 | 42,216 | 2,958 |
| Lewis | 29 | 2,157 | 65.2 | 4.9 | 2.2 | 0.0 | 5.2 | 38.2 | 2,860 | 23 | 40 | 617 | 4,820 | 40,501 | 2,535 |
| Lincoln | 63 | 2,557 | 63.0 | 18.1 | 1.2 | 0.0 | 3.1 | 44.9 | 1,834 | 55 | 73 | 942 | 9,440 | 42,369 | 3,066 |
| Livingston | 36 | 3,838 | 34.0 | 37.5 | 1.8 | 0.0 | 5.1 | 26.0 | 2,808 | 94 | 27 | 453 | 3,830 | 47,804 | 3,934 |
| Logan | 74 | 2,727 | 68.8 | 0.5 | 5.1 | 0.1 | 3.6 | 88.4 | 3,274 | 65 | 81 | 1,193 | 11,510 | 48,879 | 3,829 |
| Lyon | 19 | 2,280 | 39.0 | 4.7 | 4.9 | 0.0 | 7.9 | 17.5 | 2,117 | 37 | 21 | 896 | 3,230 | 55,032 | 4,928 |
| McCracken | 196 | 2,998 | 46.6 | 0.1 | 4.0 | 0.1 | 7.4 | 855.8 | 13,093 | 570 | 221 | 3,735 | 29,980 | 62,280 | 6,860 |
| McCreary | 47 | 2,675 | 75.6 | 1.0 | 0.4 | 0.0 | 2.1 | 37.5 | 2,160 | 475 | 46 | 754 | 4,800 | 34,235 | 1,765 |
| McLean | 30 | 3,248 | 43.5 | 4.1 | 0.2 | 0.8 | 6.6 | 36.5 | 3,961 | 32 | 27 | 486 | 3,870 | 49,218 | 3,698 |
| Madison | 188 | 2,059 | 54.2 | 10.1 | 4.5 | 0.0 | 3.6 | 389.7 | 4,274 | 1,171 | 268 | 6,988 | 36,990 | 54,013 | 4,691 |
| Magoffin | 35 | 2,792 | 64.9 | 3.7 | 2.1 | 0.1 | 4.2 | 46.9 | 3,743 | 12 | 36 | 567 | 3,890 | 38,829 | 2,439 |
| Marion | 56 | 2,888 | 49.9 | 1.2 | 2.3 | 0.0 | 4.2 | 221.3 | 11,481 | 45 | 56 | 925 | 8,440 | 47,391 | 3,642 |
| Marshall | 114 | 3,623 | 38.6 | 21.8 | 4.0 | 0.5 | 3.9 | 892.8 | 28,506 | 90 | 92 | 1,586 | 13,940 | 55,916 | 5,064 |
| Martin | 33 | 2,828 | 72.5 | 7.0 | 0.5 | 0.0 | 5.5 | 76.5 | 6,657 | 382 | 29 | 514 | 3,020 | 39,322 | 2,858 |
| Mason | 46 | 2,645 | 54.7 | 5.9 | 5.3 | 0.0 | 8.4 | 181.3 | 10,545 | 50 | 50 | 1,299 | 7,320 | 51,938 | 4,607 |
| Meade | 63 | 2,254 | 69.5 | 2.5 | 0.9 | 0.0 | 2.2 | 90.1 | 3,213 | 34 | 85 | 995 | 13,040 | 53,647 | 4,268 |
| Menifee | 14 | 2,120 | 69.1 | 0.3 | 1.5 | 0.0 | 6.3 | 8.9 | 1,383 | 44 | 19 | 332 | 2,270 | 37,376 | 2,138 |
| Mercer | 55 | 2,537 | 51.0 | 3.6 | 2.4 | 0.0 | 3.4 | 121.3 | 5,629 | 46 | 66 | 852 | 9,840 | 49,906 | 3,819 |
| Metcalfe | 24 | 2,375 | 66.4 | 1.7 | 2.2 | 0.0 | 4.7 | 33.8 | 3,354 | 21 | 30 | 453 | 3,990 | 36,380 | 2,288 |
| Monroe | 32 | 3,029 | 54.2 | 7.1 | 2.4 | 1.1 | 4.8 | 39.9 | 3,766 | 24 | 32 | 601 | 4,230 | 41,133 | 3,070 |
| Montgomery | 66 | 2,367 | 61.8 | 6.6 | 3.4 | 0.0 | 3.7 | 65.9 | 2,361 | 68 | 84 | 1,165 | 11,780 | 47,566 | 3,758 |
| Morgan | 38 | 2,832 | 75.3 | 0.5 | 1.7 | 0.0 | 3.2 | 93.3 | 7,050 | 32 | 34 | 1,010 | 4,100 | 41,601 | 2,711 |
| Muhlenberg | 71 | 2,303 | 55.8 | 8.1 | 3.5 | 0.1 | 4.1 | 102.0 | 3,298 | 316 | 86 | 1,713 | 11,820 | 47,275 | 3,446 |
| Nelson | 131 | 2,875 | 53.0 | 2.1 | 2.9 | 0.1 | 3.0 | 385.7 | 8,459 | 85 | 138 | 1,821 | 22,180 | 55,994 | 5,147 |
| Nicholas | 26 | 3,647 | 75.9 | 2.1 | 0.2 | 0.2 | 3.2 | 33.4 | 4,669 | 14 | 22 | 314 | 3,030 | 37,465 | 2,464 |

1. Based on the resident population estimated as of July 1 of the year shown.

| State / county code | CBSA code[1] | County Type code[2] | STATE County | Land area[3] (sq. mi) | Total persons 2019 | Rank | Per square mile | White | Black | American Indian, Alaska Native | Asian and Pacific Islander | Percent Hispanic or Latino[4] | Under 5 years | 5 to 14 years | 15 to 24 years | 25 to 34 years | 35 to 44 years | 45 to 54 years |
|---|---|---|---|---|---|---|---|---|---|---|---|---|---|---|---|---|---|---|
| | | | | 1 | 2 | 3 | 4 | 5 | 6 | 7 | 8 | 9 | 10 | 11 | 12 | 13 | 14 | 15 |
| | | | KENTUCKY—Cont'd | | | | | | | | | | | | | | | |
| 21183 | | 6 | Ohio | 587.3 | 23,899 | 1,647 | 40.7 | 95.0 | 1.5 | 0.6 | 0.4 | 3.4 | 6.0 | 13.6 | 11.6 | 11.8 | 12.5 | 12.4 |
| 21185 | 31140 | 1 | Oldham | 187.2 | 66,999 | 800 | 357.9 | 89.7 | 4.9 | 0.7 | 2.3 | 4.1 | 5.1 | 14.9 | 13.4 | 10.0 | 14.6 | 15.2 |
| 21187 | | 8 | Owen | 351.1 | 11,017 | 2,346 | 31.4 | 95.9 | 1.7 | 0.6 | 0.4 | 2.6 | 5.4 | 12.5 | 11.5 | 11.5 | 11.8 | 13.5 |
| 21189 | | 9 | Owsley | 197.4 | 4,331 | 2,872 | 21.9 | 97.2 | 0.9 | 0.9 | 0.3 | 1.6 | 6.3 | 12.2 | 10.3 | 11.5 | 11.9 | 12.8 |
| 21191 | 17140 | 1 | Pendleton | 277.2 | 14,586 | 2,115 | 52.6 | 97.0 | 1.5 | 0.8 | 0.6 | 1.5 | 6.1 | 13.0 | 11.2 | 12.5 | 11.1 | 13.6 |
| 21193 | | 7 | Perry | 339.7 | 25,456 | 1,579 | 74.9 | 95.8 | 2.3 | 0.6 | 1.3 | 1.1 | 6.5 | 12.7 | 10.6 | 12.5 | 12.3 | 13.4 |
| 21195 | | 7 | Pike | 786.7 | 57,057 | 906 | 72.5 | 97.5 | 1.0 | 0.4 | 0.7 | 1.0 | 5.2 | 11.7 | 11.6 | 11.4 | 12.1 | 13.5 |
| 21197 | | 6 | Powell | 179.0 | 12,218 | 2,273 | 68.3 | 96.9 | 1.4 | 0.7 | 0.6 | 1.6 | 6.4 | 13.5 | 11.5 | 12.7 | 12.7 | 12.9 |
| 21199 | 43700 | 5 | Pulaski | 658.4 | 65,530 | 814 | 99.5 | 95.0 | 1.6 | 0.8 | 1.1 | 2.7 | 5.8 | 12.2 | 11.1 | 12.0 | 11.9 | 13.2 |
| 21201 | | 8 | Robertson | 99.9 | 2,136 | 3,035 | 21.4 | 97.5 | 1.0 | 1.1 | 0.3 | 1.7 | 5.4 | 11.8 | 9.9 | 10.6 | 10.2 | 12.6 |
| 21203 | | 7 | Rockcastle | 316.5 | 16,750 | 1,987 | 52.9 | 97.9 | 0.7 | 1.0 | 0.4 | 1.1 | 5.2 | 11.9 | 11.1 | 12.0 | 11.1 | 14.3 |
| 21205 | | 7 | Rowan | 279.8 | 24,682 | 1,621 | 88.2 | 95.3 | 2.4 | 0.6 | 1.2 | 2.0 | 5.5 | 10.9 | 25.1 | 12.6 | 9.8 | 10.4 |
| 21207 | | 9 | Russell | 253.7 | 17,998 | 1,923 | 70.9 | 94.7 | 1.3 | 0.7 | 0.8 | 3.9 | 6.1 | 12.9 | 10.6 | 11.1 | 10.6 | 12.9 |
| 21209 | 30460 | 2 | Scott | 281.8 | 58,470 | 884 | 207.5 | 88.5 | 6.6 | 0.6 | 1.8 | 4.5 | 6.3 | 14.0 | 13.4 | 14.3 | 14.0 | 13.6 |
| 21211 | 31140 | 1 | Shelby | 379.8 | 49,611 | 998 | 130.6 | 82.2 | 8.2 | 0.8 | 1.5 | 9.6 | 5.9 | 12.3 | 12.4 | 12.7 | 12.8 | 13.6 |
| 21213 | | 6 | Simpson | 234.2 | 18,635 | 1,891 | 79.6 | 87.0 | 10.3 | 0.8 | 1.1 | 2.7 | 6.1 | 13.5 | 12.1 | 13.1 | 11.7 | 12.6 |
| 21215 | 31140 | 1 | Spencer | 186.7 | 19,585 | 1,849 | 104.9 | 95.0 | 2.4 | 0.7 | 0.7 | 2.6 | 5.5 | 12.8 | 11.0 | 11.1 | 13.4 | 15.2 |
| 21217 | 15820 | 7 | Taylor | 266.4 | 25,707 | 1,568 | 96.5 | 91.2 | 6.6 | 0.6 | 1.1 | 2.5 | 6.3 | 12.4 | 15.2 | 12.4 | 11.1 | 10.8 |
| 21219 | | 8 | Todd | 374.5 | 12,448 | 2,251 | 33.2 | 87.1 | 8.5 | 0.8 | 0.6 | 4.9 | 7.1 | 15.1 | 12.5 | 11.7 | 11.6 | 12.0 |
| 21221 | 17300 | 2 | Trigg | 441.5 | 14,776 | 2,102 | 33.5 | 89.9 | 8.3 | 0.9 | 0.8 | 2.3 | 5.3 | 12.1 | 10.7 | 9.8 | 10.1 | 12.6 |
| 21223 | | 1 | Trimble | 151.6 | 8,481 | 2,539 | 55.9 | 94.7 | 1.2 | 1.0 | 1.0 | 3.6 | 5.6 | 12.3 | 11.0 | 11.7 | 11.3 | 14.6 |
| 21225 | | 6 | Union | 342.8 | 14,443 | 2,124 | 42.1 | 83.7 | 14.2 | 0.7 | 0.9 | 2.2 | 4.5 | 9.5 | 19.2 | 11.7 | 11.6 | 11.8 |
| 21227 | 14540 | 3 | Warren | 541.7 | 134,510 | 483 | 248.3 | 79.2 | 10.9 | 0.6 | 6.0 | 5.6 | 6.4 | 12.8 | 19.3 | 13.6 | 12.1 | 11.2 |
| 21229 | | 9 | Washington | 297.0 | 12,147 | 2,282 | 40.9 | 89.7 | 6.6 | 0.5 | 1.0 | 4.1 | 6.4 | 12.8 | 11.6 | 11.7 | 11.2 | 12.8 |
| 21231 | | 7 | Wayne | 458.2 | 20,209 | 1,812 | 44.1 | 94.0 | 2.3 | 0.8 | 0.6 | 3.7 | 5.5 | 11.2 | 10.9 | 11.7 | 11.3 | 12.7 |
| 21233 | | 8 | Webster | 332.1 | 12,923 | 2,227 | 38.9 | 89.5 | 4.7 | 0.7 | 0.7 | 5.9 | 6.6 | 12.8 | 11.0 | 11.7 | 12.7 | 12.7 |
| 21235 | 30940 | 7 | Whitley | 437.8 | 36,451 | 1,265 | 83.3 | 96.8 | 1.4 | 0.9 | 0.7 | 1.5 | 7.2 | 14.0 | 15.3 | 12.4 | 11.1 | 11.6 |
| 21237 | | 9 | Wolfe | 222.2 | 7,106 | 2,650 | 32.0 | 97.9 | 0.8 | 0.8 | 0.3 | 1.2 | 6.7 | 11.7 | 11.5 | 10.7 | 12.2 | 12.0 |
| 21239 | 30460 | 2 | Woodford | 190.1 | 26,765 | 1,536 | 140.8 | 87.8 | 5.6 | 0.5 | 1.0 | 6.7 | 5.4 | 12.4 | 12.2 | 10.2 | 12.2 | 12.5 |
| 22000 | | 0 | LOUISIANA | 43,204.5 | 4,645,318 | X | 107.5 | 59.5 | 33.3 | 1.2 | 2.3 | 5.4 | 6.4 | 13.1 | 12.8 | 14.0 | 12.9 | 11.6 |
| 22001 | 29180 | 2 | Acadia | 655.2 | 61,918 | 856 | 94.5 | 78.5 | 19.0 | 0.6 | 0.5 | 3.0 | 6.8 | 14.3 | 12.7 | 13.1 | 12.2 | 11.5 |
| 22003 | | 6 | Allen | 762.1 | 25,440 | 1,582 | 33.4 | 72.1 | 23.4 | 2.8 | 1.0 | 2.6 | 5.9 | 12.4 | 11.5 | 16.0 | 14.0 | 12.8 |
| 22005 | 12940 | 2 | Ascension | 290.0 | 128,665 | 499 | 443.7 | 68.2 | 24.6 | 0.6 | 1.8 | 6.2 | 6.9 | 15.1 | 12.5 | 13.7 | 14.5 | 12.9 |
| 22007 | 12940 | 6 | Assumption | 344.0 | 21,621 | 1,738 | 62.9 | 66.2 | 29.8 | 0.9 | 0.6 | 3.4 | 5.2 | 11.9 | 11.2 | 12.7 | 11.8 | 12.5 |
| 22009 | | 6 | Avoyelles | 831.9 | 39,966 | 1,187 | 48.0 | 66.1 | 31.1 | 1.9 | 0.9 | 2.1 | 6.5 | 13.1 | 11.9 | 13.7 | 12.6 | 11.7 |
| 22011 | 19760 | 6 | Beauregard | 1,157.5 | 37,881 | 1,227 | 32.7 | 82.7 | 12.9 | 1.8 | 1.3 | 3.7 | 7.1 | 13.8 | 12.1 | 13.9 | 12.4 | 12.1 |
| 22013 | | 6 | Bienville | 811.3 | 12,983 | 2,221 | 16.0 | 55.5 | 42.4 | 0.9 | 0.7 | 2.0 | 6.0 | 12.6 | 10.6 | 11.7 | 11.0 | 11.6 |
| 22015 | 43340 | 2 | Bossier | 839.5 | 127,275 | 506 | 151.6 | 67.4 | 24.4 | 1.0 | 2.8 | 6.9 | 6.6 | 13.9 | 12.9 | 14.6 | 13.7 | 11.1 |
| 22017 | 43340 | 2 | Caddo | 879.5 | 237,479 | 290 | 270.0 | 45.1 | 50.7 | 1.0 | 1.9 | 3.0 | 6.3 | 13.2 | 12.1 | 13.4 | 12.3 | 11.2 |
| 22019 | 29340 | 3 | Calcasieu | 1,064.1 | 203,310 | 336 | 191.1 | 67.8 | 26.9 | 1.1 | 1.9 | 4.3 | 7.2 | 13.8 | 12.5 | 14.6 | 12.7 | 11.2 |
| 22021 | | 8 | Caldwell | 529.8 | 9,839 | 2,430 | 18.6 | 79.8 | 16.6 | 0.9 | 0.6 | 3.5 | 6.2 | 12.4 | 12.1 | 13.3 | 12.1 | 12.1 |
| 22023 | 29340 | 3 | Cameron | 1,284.6 | 7,003 | 2,656 | 5.5 | 91.2 | 4.6 | 1.1 | 0.7 | 3.9 | 5.9 | 12.3 | 11.6 | 12.6 | 11.8 | 11.6 |
| 22025 | | 9 | Catahoula | 708.0 | 9,226 | 2,479 | 13.0 | 66.0 | 32.3 | 1.0 | 0.3 | 1.8 | 5.6 | 11.8 | 11.7 | 13.8 | 13.1 | 10.8 |
| 22027 | | 6 | Claiborne | 754.8 | 15,508 | 2,057 | 20.5 | 45.4 | 52.7 | 0.9 | 0.8 | 1.5 | 4.5 | 10.2 | 10.5 | 15.7 | 12.6 | 12.3 |
| 22029 | 35020 | 7 | Concordia | 697.1 | 18,914 | 1,879 | 27.1 | 57.4 | 40.3 | 0.8 | 0.6 | 1.8 | 6.3 | 13.1 | 12.7 | 13.6 | 11.6 | 10.6 |
| 22031 | 43340 | 2 | De Soto | 876.4 | 27,650 | 1,509 | 31.5 | 60.7 | 35.5 | 1.4 | 0.6 | 3.2 | 5.9 | 13.8 | 11.3 | 12.2 | 12.7 | 12.0 |
| 22033 | 12940 | 2 | East Baton Rouge | 455.5 | 439,729 | 162 | 965.4 | 44.8 | 47.7 | 0.6 | 3.9 | 4.5 | 6.4 | 12.5 | 17.2 | 14.2 | 12.3 | 10.4 |
| 22035 | | 7 | East Carroll | 420.9 | 6,589 | 2,694 | 15.7 | 28.6 | 68.2 | 0.7 | 0.7 | 2.9 | 5.5 | 13.6 | 13.3 | 15.6 | 12.6 | 10.7 |
| 22037 | 12940 | 2 | East Feliciana | 453.3 | 18,882 | 1,882 | 41.7 | 55.4 | 42.8 | 0.9 | 0.5 | 1.8 | 4.8 | 9.8 | 10.9 | 12.9 | 12.8 | 13.6 |
| 22039 | | 6 | Evangeline | 662.4 | 33,276 | 1,350 | 50.2 | 67.5 | 28.2 | 0.7 | 0.7 | 4.1 | 7.0 | 13.8 | 13.1 | 13.9 | 12.0 | 11.4 |
| 22041 | | 7 | Franklin | 624.3 | 19,723 | 1,838 | 31.6 | 65.8 | 32.4 | 0.6 | 0.6 | 1.7 | 6.8 | 13.8 | 12.2 | 12.0 | 11.8 | 11.0 |
| 22043 | 10780 | 3 | Grant | 643.2 | 22,254 | 1,709 | 34.6 | 77.7 | 16.1 | 1.7 | 0.9 | 5.2 | 5.6 | 11.7 | 11.0 | 17.1 | 14.6 | 12.1 |
| 22045 | 29180 | 2 | Iberia | 573.7 | 68,991 | 783 | 120.3 | 59.9 | 33.7 | 0.8 | 3.1 | 4.4 | 6.7 | 14.4 | 12.6 | 13.0 | 11.8 | 11.4 |
| 22047 | 12940 | 2 | Iberville | 618.7 | 32,070 | 1,380 | 51.8 | 48.4 | 48.4 | 0.6 | 0.6 | 3.0 | 5.6 | 11.0 | 11.7 | 14.9 | 13.4 | 12.7 |
| 22049 | | 6 | Jackson | 569.4 | 15,574 | 2,054 | 27.4 | 69.1 | 28.9 | 0.9 | 0.9 | 1.9 | 4.8 | 12.5 | 11.6 | 12.8 | 12.2 | 12.0 |
| 22051 | 35380 | 1 | Jefferson | 300.9 | 432,346 | 163 | 1,436.8 | 52.7 | 28.0 | 0.8 | 5.0 | 15.0 | 6.4 | 12.2 | 10.8 | 14.0 | 12.8 | 11.8 |
| 22053 | 27660 | 6 | Jefferson Davis | 651.5 | 31,208 | 1,397 | 47.9 | 79.8 | 17.8 | 1.2 | 0.8 | 2.6 | 6.7 | 14.5 | 12.1 | 13.0 | 12.0 | 11.3 |
| 22055 | 29180 | 2 | Lafayette | 268.8 | 246,518 | 279 | 917.1 | 66.2 | 27.6 | 0.7 | 2.4 | 4.7 | 6.6 | 13.2 | 12.6 | 15.7 | 13.8 | 11.5 |
| 22057 | 26380 | 3 | Lafourche | 1,067.8 | 97,596 | 618 | 91.4 | 78.3 | 14.3 | 3.6 | 1.2 | 4.5 | 6.1 | 13.0 | 12.0 | 13.9 | 12.7 | 12.1 |
| 22059 | | 6 | La Salle | 624.9 | 15,021 | 2,092 | 24.0 | 83.7 | 12.1 | 1.7 | 0.5 | 3.1 | 6.1 | 12.6 | 12.3 | 14.4 | 13.3 | 11.7 |
| 22061 | 40820 | 4 | Lincoln | 471.6 | 46,552 | 1,044 | 98.7 | 54.7 | 40.7 | 0.8 | 1.9 | 3.2 | 5.4 | 10.9 | 28.5 | 11.8 | 10.4 | 8.8 |
| 22063 | 12940 | 2 | Livingston | 648.1 | 143,737 | 459 | 221.8 | 87.1 | 7.9 | 0.8 | 1.1 | 4.4 | 6.6 | 14.4 | 12.1 | 14.2 | 14.1 | 12.7 |
| 22065 | | 7 | Madison | 624.2 | 10,635 | 2,377 | 17.0 | 34.2 | 63.2 | 0.8 | 0.5 | 2.5 | 6.6 | 13.2 | 12.2 | 17.1 | 11.1 | 12.1 |
| 22067 | 33740 | 6 | Morehouse | 795.0 | 24,227 | 1,640 | 30.5 | 49.9 | 48.2 | 0.7 | 0.8 | 1.6 | 6.6 | 13.2 | 11.6 | 11.8 | 11.5 | 11.5 |
| 22069 | 35060 | 6 | Natchitoches | 1,253.3 | 37,655 | 1,232 | 30.0 | 54.2 | 42.6 | 1.9 | 1.1 | 2.5 | 6.1 | 12.7 | 20.6 | 11.5 | 10.2 | 10.2 |

1. CBSA = Core Based Statistical Area. See Appendix A for explanation. See Appendix B for list of metropolitan areas with component counties.
Service of USDA Rural-Urban Continuum Codes. See Appendix A for definition. 3. Dry land or land partially or temporarily covered by water.
2. County type code from the Economic Research
4. May be of any race.

Items 1—15

# Table B. States and Counties — Population and Households

| STATE County | Population, 2020 (cont.) Age (percent) (cont.) | | | | Population change, 2000-2020 Total persons | | Percent change | | Components of change, 2010-2020 | | | Households, 2015-2019 | | Percent | | |
|---|---|---|---|---|---|---|---|---|---|---|---|---|---|---|---|---|
| | 55 to 64 years | 65 to 74 years | 75 years and over | Percent female | 2000 | 2010 | 2000-2010 | 2010-2020 | Births | Deaths | Net Migration | Number | Persons per household | Family households | Female family householder[1] | One person |
| | 16 | 17 | 18 | 19 | 20 | 21 | 22 | 23 | 24 | 25 | 26 | 27 | 28 | 29 | 30 | 31 |
| **KENTUCKY—Cont'd** | | | | | | | | | | | | | | | | |
| Ohio | 13.3 | 11.0 | 7.8 | 50.0 | 22,916 | 23,849 | 4.1 | 0.2 | 2,974 | 2,898 | -3 | 9,122 | 2.61 | 71.5 | 11.6 | 23.0 |
| Oldham | 12.7 | 9.0 | 5.1 | 47.8 | 46,178 | 60,356 | 30.7 | 11.0 | 5,627 | 4,095 | 5,123 | 20,911 | 2.98 | 82.2 | 8.6 | 16.0 |
| Owen | 14.6 | 12.0 | 7.4 | 50.3 | 10,547 | 10,835 | 2.7 | 1.7 | 1,077 | 1,167 | 275 | 3,995 | 2.68 | 67.7 | 8.1 | 27.5 |
| Owsley | 14.2 | 11.7 | 9.1 | 50.8 | 4,858 | 4,757 | -2.1 | -9.0 | 544 | 788 | -188 | 1,704 | 2.56 | 68.0 | 18.8 | 28.8 |
| Pendleton | 15.8 | 10.4 | 6.3 | 48.9 | 14,390 | 14,874 | 3.4 | -1.9 | 1,773 | 1,602 | -458 | 5,253 | 2.73 | 71.9 | 10.0 | 25.5 |
| Perry | 14.1 | 11.3 | 6.7 | 50.6 | 29,390 | 28,697 | -2.4 | -11.3 | 3,843 | 4,406 | -2,702 | 11,226 | 2.31 | 68.7 | 19.0 | 27.0 |
| Pike | 14.6 | 12.2 | 7.6 | 51.1 | 68,736 | 65,029 | -5.4 | -12.3 | 6,871 | 8,409 | -6,483 | 25,702 | 2.27 | 68.4 | 14.0 | 28.0 |
| Powell | 13.6 | 10.6 | 6.0 | 50.2 | 13,237 | 12,613 | -4.7 | -3.1 | 1,643 | 1,744 | -294 | 4,770 | 2.52 | 72.7 | 15.9 | 24.5 |
| Pulaski | 14.1 | 11.7 | 8.0 | 51.2 | 56,217 | 63,062 | 12.2 | 3.9 | 7,683 | 8,281 | 3,121 | 25,301 | 2.51 | 66.3 | 12.5 | 27.7 |
| Robertson | 15.6 | 12.4 | 11.6 | 49.3 | 2,266 | 2,285 | 0.8 | -6.5 | 237 | 322 | -66 | 861 | 2.39 | 57.4 | 6.9 | 33.8 |
| Rockcastle | 15.0 | 11.5 | 7.8 | 51.0 | 16,582 | 17,060 | 2.9 | -1.8 | 1,798 | 2,095 | -9 | 6,544 | 2.52 | 70.5 | 11.8 | 27.1 |
| Rowan | 11.1 | 8.3 | 6.2 | 51.6 | 22,094 | 23,332 | 5.6 | 5.8 | 2,782 | 2,350 | 892 | 8,623 | 2.5 | 59.1 | 11.2 | 28.5 |
| Russell | 14.9 | 12.3 | 8.6 | 51.0 | 16,315 | 17,575 | 7.7 | 2.4 | 2,201 | 2,449 | 682 | 6,922 | 2.53 | 65.3 | 10.7 | 30.2 |
| Scott | 11.7 | 8.1 | 4.6 | 50.7 | 33,061 | 47,089 | 42.4 | 24.2 | 6,893 | 3,793 | 8,216 | 20,551 | 2.59 | 73.3 | 11.1 | 20.4 |
| Shelby | 13.8 | 10.2 | 6.2 | 51.3 | 33,337 | 42,004 | 26.0 | 18.1 | 6,024 | 3,697 | 5,263 | 16,624 | 2.75 | 72.4 | 10.7 | 22.7 |
| Simpson | 13.9 | 10.2 | 6.8 | 51.2 | 16,405 | 17,330 | 5.6 | 7.5 | 2,289 | 2,052 | 1,069 | 7,028 | 2.53 | 70.7 | 13.9 | 25.8 |
| Spencer | 16.0 | 9.8 | 5.3 | 49.6 | 11,766 | 17,108 | 45.4 | 14.5 | 1,959 | 1,435 | 1,954 | 6,763 | 2.74 | 80.8 | 8.0 | 14.7 |
| Taylor | 13.3 | 10.7 | 7.8 | 51.2 | 22,927 | 24,474 | 6.7 | 5.0 | 3,340 | 3,082 | 972 | 9,857 | 2.47 | 67.2 | 12.7 | 26.5 |
| Todd | 13.2 | 9.7 | 7.0 | 49.8 | 11,971 | 12,456 | 4.1 | -0.1 | 1,815 | 1,375 | -449 | 4,644 | 2.61 | 72.1 | 12.2 | 23.6 |
| Trigg | 15.7 | 14.0 | 9.7 | 50.3 | 12,597 | 14,327 | 13.7 | 3.1 | 1,564 | 1,772 | 661 | 5,892 | 2.44 | 66.6 | 9.6 | 28.1 |
| Trimble | 15.1 | 11.3 | 7.2 | 49.8 | 8,125 | 8,806 | 8.4 | -3.7 | 969 | 963 | -332 | 3,498 | 2.43 | 65.5 | 10.0 | 29.9 |
| Union | 13.5 | 11.7 | 6.4 | 47.6 | 15,637 | 15,006 | -4.0 | -3.8 | 1,670 | 1,742 | -528 | 5,374 | 2.43 | 70.8 | 10.1 | 26.2 |
| Warren | 10.9 | 8.2 | 5.4 | 50.9 | 92,522 | 113,782 | 23.0 | 18.2 | 16,593 | 10,073 | 14,169 | 48,224 | 2.52 | 63.4 | 12.4 | 27.4 |
| Washington | 14.6 | 10.7 | 8.2 | 50.8 | 10,916 | 11,701 | 7.2 | 3.8 | 1,461 | 1,433 | 412 | 4,581 | 2.52 | 70.1 | 13.1 | 24.0 |
| Wayne | 14.6 | 12.9 | 9.2 | 50.4 | 19,923 | 20,798 | 4.4 | -2.8 | 2,365 | 2,406 | -539 | 8,231 | 2.46 | 64.1 | 10.6 | 32.2 |
| Webster | 13.7 | 11.1 | 7.6 | 50.1 | 14,120 | 13,620 | -3.5 | -5.1 | 1,692 | 1,726 | -662 | 4,953 | 2.56 | 66.9 | 8.0 | 30.3 |
| Whitley | 12.2 | 9.4 | 6.8 | 50.8 | 35,865 | 35,634 | -0.6 | 2.3 | 5,615 | 4,952 | 164 | 12,620 | 2.71 | 71.6 | 13.1 | 23.2 |
| Wolfe | 15.3 | 11.9 | 7.9 | 50.9 | 7,065 | 7,358 | 4.1 | -3.4 | 924 | 1,088 | -92 | 2,898 | 2.44 | 62.2 | 17.1 | 34.1 |
| Woodford | 15.2 | 12.3 | 7.6 | 52.0 | 23,208 | 24,945 | 7.5 | 7.3 | 2,873 | 2,418 | 1,381 | 10,355 | 2.51 | 70.4 | 11.3 | 24.2 |
| **LOUISIANA** | 12.9 | 9.9 | 6.6 | 51.3 | 4,468,976 | 4,533,500 | 1.4 | 2.5 | 633,464 | 453,494 | -68,962 | 1,739,497 | 2.61 | 64.0 | 15.8 | 30.3 |
| Acadia | 13.2 | 9.6 | 6.7 | 51.0 | 58,861 | 61,788 | 5.0 | 0.2 | 8,915 | 6,752 | -1,994 | 22,236 | 2.76 | 69.8 | 14.8 | 25.3 |
| Allen | 12.0 | 8.9 | 6.6 | 42.9 | 25,440 | 25,750 | 1.2 | -1.2 | 3,258 | 2,666 | -890 | 7,925 | 2.69 | 70.2 | 12.4 | 27.6 |
| Ascension | 11.8 | 8.0 | 4.7 | 50.8 | 76,627 | 107,219 | 39.9 | 20.0 | 17,153 | 7,772 | 11,963 | 43,032 | 2.84 | 73.8 | 12.3 | 21.9 |
| Assumption | 15.0 | 11.6 | 8.2 | 51.4 | 23,388 | 23,417 | 0.1 | -7.7 | 2,605 | 2,291 | -2,113 | 8,552 | 2.6 | 64.1 | 15.1 | 32.1 |
| Avoyelles | 12.8 | 10.2 | 7.5 | 49.6 | 41,481 | 42,079 | 1.4 | -5.0 | 5,554 | 5,070 | -2,588 | 15,163 | 2.43 | 64.9 | 16.8 | 30.2 |
| Beauregard | 12.4 | 9.8 | 6.5 | 49.1 | 32,986 | 35,653 | 8.1 | 6.2 | 5,160 | 3,862 | 937 | 13,520 | 2.71 | 68.2 | 11.3 | 27.3 |
| Bienville | 14.7 | 11.6 | 10.1 | 52.3 | 15,752 | 14,355 | -8.9 | -9.6 | 1,744 | 2,084 | -1,032 | 5,812 | 2.28 | 61.7 | 19.4 | 35.0 |
| Bossier | 11.8 | 8.9 | 6.4 | 50.6 | 98,310 | 116,989 | 19.0 | 8.8 | 17,988 | 10,592 | 2,831 | 49,377 | 2.5 | 67.5 | 14.0 | 26.2 |
| Caddo | 13.1 | 10.7 | 7.6 | 52.7 | 252,161 | 254,969 | 1.1 | -6.9 | 35,766 | 28,294 | -25,035 | 95,864 | 2.51 | 61.5 | 18.7 | 33.5 |
| Calcasieu | 12.5 | 9.4 | 6.3 | 51.2 | 183,577 | 192,767 | 5.0 | 5.5 | 29,234 | 20,540 | 1,991 | 77,780 | 2.55 | 65.1 | 15.9 | 28.8 |
| Caldwell | 12.8 | 11.3 | 7.8 | 48.6 | 10,560 | 10,126 | -4.1 | -2.8 | 1,286 | 1,296 | -271 | 3,665 | 2.6 | 69.1 | 14.2 | 24.6 |
| Cameron | 16.3 | 10.4 | 7.7 | 49.9 | 9,991 | 6,868 | -31.3 | 2.0 | 710 | 516 | -64 | 2,734 | 2.52 | 77.2 | 14.8 | 18.1 |
| Catahoula | 14.4 | 11.5 | 7.3 | 46.8 | 10,920 | 10,409 | -4.7 | -11.4 | 1,216 | 1,365 | -1,028 | 3,364 | 2.52 | 72.0 | 11.6 | 27.1 |
| Claiborne | 13.7 | 11.5 | 8.9 | 43.3 | 16,851 | 17,195 | 2.0 | -9.8 | 1,594 | 1,952 | -1,335 | 5,917 | 2.34 | 62.8 | 15.9 | 34.1 |
| Concordia | 12.9 | 11.3 | 7.8 | 49.4 | 20,247 | 20,822 | 2.8 | -9.2 | 2,676 | 2,524 | -2,063 | 7,162 | 2.48 | 63.7 | 21.9 | 33.2 |
| De Soto | 13.6 | 10.8 | 7.7 | 51.8 | 25,494 | 26,656 | 4.6 | 3.7 | 3,446 | 3,058 | 624 | 10,821 | 2.5 | 67.0 | 18.8 | 29.2 |
| East Baton Rouge | 11.6 | 9.3 | 5.9 | 52.4 | 412,852 | 440,531 | 6.7 | -0.2 | 61,359 | 38,941 | -23,086 | 164,346 | 2.64 | 59.6 | 16.3 | 32.3 |
| East Carroll | 12.8 | 9.0 | 6.8 | 44.8 | 9,421 | 7,759 | -17.6 | -15.1 | 1,070 | 825 | -1,421 | 2,037 | 2.17 | 58.6 | 24.5 | 36.2 |
| East Feliciana | 16.0 | 11.9 | 7.3 | 45.4 | 21,360 | 20,254 | -5.2 | -6.8 | 2,090 | 2,474 | -983 | 6,959 | 2.24 | 68.6 | 17.8 | 27.5 |
| Evangeline | 12.8 | 9.3 | 6.7 | 49.0 | 35,434 | 33,986 | -4.1 | -2.1 | 4,969 | 3,892 | -1,785 | 12,172 | 2.63 | 63.6 | 17.0 | 32.8 |
| Franklin | 13.5 | 10.6 | 8.3 | 51.3 | 21,263 | 20,767 | -2.3 | -5.0 | 2,855 | 2,805 | -1,084 | 7,423 | 2.55 | 69.1 | 20.5 | 28.6 |
| Grant | 12.2 | 9.3 | 6.4 | 44.0 | 18,698 | 22,318 | 19.4 | -0.3 | 2,563 | 2,221 | -391 | 6,989 | 2.74 | 66.5 | 12.3 | 31.2 |
| Iberia | 13.9 | 9.7 | 6.4 | 51.1 | 73,266 | 72,856 | -0.6 | -5.3 | 10,511 | 7,558 | -6,812 | 26,184 | 2.71 | 70.5 | 17.4 | 24.7 |
| Iberville | 13.8 | 9.9 | 7.0 | 49.5 | 33,320 | 33,404 | 0.3 | -4.0 | 4,060 | 3,359 | -2,022 | 10,903 | 2.7 | 67.6 | 20.2 | 28.5 |
| Jackson | 13.5 | 11.3 | 9.5 | 48.6 | 15,397 | 16,274 | 5.7 | -4.3 | 1,726 | 1,977 | -439 | 5,971 | 2.45 | 67.2 | 13.7 | 30.0 |
| Jefferson | 13.8 | 10.8 | 7.4 | 51.7 | 455,466 | 432,576 | -5.0 | -0.1 | 58,872 | 43,035 | -15,900 | 169,452 | 2.54 | 63.2 | 15.6 | 31.3 |
| Jefferson Davis | 13.5 | 9.6 | 7.3 | 51.0 | 31,435 | 31,589 | 0.5 | -1.2 | 4,354 | 3,882 | -840 | 11,726 | 2.64 | 69.5 | 15.7 | 27.7 |
| Lafayette | 12.4 | 8.8 | 5.4 | 51.4 | 190,503 | 221,970 | 16.5 | 11.1 | 33,829 | 18,199 | 8,853 | 91,543 | 2.59 | 63.0 | 13.6 | 28.9 |
| Lafourche | 13.8 | 9.5 | 6.9 | 51.0 | 89,974 | 96,591 | 7.4 | 1.0 | 12,489 | 9,160 | -2,281 | 36,895 | 2.6 | 68.9 | 12.8 | 25.0 |
| La Salle | 12.2 | 10.2 | 7.2 | 48.0 | 14,282 | 14,888 | 4.2 | 0.9 | 1,812 | 1,736 | 67 | 4,814 | 2.81 | 75.7 | 15.1 | 23.3 |
| Lincoln | 9.8 | 8.3 | 6.2 | 51.3 | 42,509 | 46,740 | 10.0 | -0.4 | 5,731 | 3,931 | -1,994 | 17,712 | 2.49 | 58.8 | 18.0 | 28.9 |
| Livingston | 11.9 | 8.6 | 5.3 | 50.9 | 91,814 | 127,688 | 39.1 | 12.6 | 18,735 | 10,926 | 8,197 | 48,410 | 2.85 | 63.8 | 13.5 | 22.5 |
| Madison | 11.8 | 9.8 | 6.1 | 50.5 | 13,728 | 12,101 | -11.9 | -12.1 | 1,693 | 1,355 | -1,823 | 3,832 | 2.43 | 63.8 | 27.3 | 32.8 |
| Morehouse | 13.7 | 12.1 | 8.1 | 51.9 | 31,021 | 27,979 | -9.8 | -13.4 | 3,577 | 3,841 | -3,500 | 9,732 | 2.54 | 63.6 | 16.7 | 32.0 |
| Natchitoches | 11.1 | 10.1 | 7.5 | 52.1 | 39,080 | 39,563 | 1.2 | -4.8 | 5,203 | 4,109 | -2,985 | 14,659 | 2.58 | 51.4 | 12.6 | 39.4 |

1. No spouse present.

# Table B. States and Counties — Population, Vital Statistics, Health, and Crime

| STATE County | Persons in group quarters, 2020 | Daytime Population, 2015-2019 | | Births, 2020 | | Deaths, 2020 | | Persons under 65 with no health insurance, 2019 | | Medicare, 2020 | | | Crimes reported by county police or sheriff, 2019 | |
|---|---|---|---|---|---|---|---|---|---|---|---|---|---|---|
| | | Number | Employment/ residence ratio | Total | Rate[1] | Number | Rate[1] | Number | Percent | Total beneficiaries | Enrolled in Original Medicare | Enrolled in Medicare Advantage | Violent | Property |
| | 32 | 33 | 34 | 35 | 36 | 37 | 38 | 39 | 40 | 41 | 42 | 43 | 44 | 45 |
| KENTUCKY—Cont'd | | | | | | | | | | | | | | |
| Ohio | 307 | 23,195 | 0.91 | 269 | 11.3 | 284 | 11.9 | 1,501 | 7.8 | 5,705 | 3,515 | 2,190 | 19 | 116 |
| Oldham | 4,347 | 53,320 | 0.59 | 571 | 8.5 | 454 | 6.8 | 2,356 | 4.3 | 10,368 | 6,752 | 3,616 | 25 | 458 |
| Owen | 0 | 8,871 | 0.56 | 104 | 9.4 | 121 | 11.0 | 741 | 8.5 | 2,388 | 1,308 | 1,080 | 2 | 28 |
| Owsley | 90 | 4,035 | 0.68 | 51 | 11.8 | 78 | 18.0 | 253 | 7.3 | 1,155 | 757 | 399 | 0 | 16 |
| Pendleton | 215 | 11,151 | 0.46 | 180 | 12.3 | 177 | 12.1 | 962 | 8.0 | 3,152 | 1,733 | 1,419 | 0 | 16 |
| Perry | 611 | 29,012 | 1.28 | 313 | 12.3 | 470 | 18.5 | 1,657 | 8.0 | 6,932 | 4,277 | 2,655 | 4 | 60 |
| Pike | 1,443 | 62,137 | 1.13 | 573 | 10.0 | 869 | 15.2 | 4,051 | 9.0 | 16,661 | 10,434 | 6,227 | 8 | 51 |
| Powell | 188 | 10,558 | 0.63 | 162 | 13.3 | 170 | 13.9 | 754 | 7.5 | 3,154 | 1,768 | 1,387 | 2 | 33 |
| Pulaski | 962 | 66,173 | 1.07 | 748 | 11.4 | 851 | 13.0 | 4,845 | 9.4 | 17,037 | 10,438 | 6,599 | 0 | 8 |
| Robertson | 57 | 1,732 | 0.51 | 20 | 9.4 | 22 | 10.3 | 136 | 8.4 | 570 | 368 | 203 | 30 | 315 |
| Rockcastle | 341 | 15,515 | 0.78 | 160 | 9.6 | 260 | 15.5 | 1,004 | 7.6 | 4,092 | 2,652 | 1,440 | 0 | 8 |
| Rowan | 2,991 | 26,143 | 1.16 | 254 | 10.3 | 224 | 9.1 | 1,431 | 8.0 | 4,866 | 2,520 | 2,345 | 3 | 40 |
| Russell | 177 | 17,521 | 0.96 | 210 | 11.7 | 247 | 13.7 | 1,394 | 9.9 | 4,626 | 3,195 | 1,431 | 8 | 65 |
| Scott | 1,092 | 58,643 | 1.14 | 713 | 12.2 | 448 | 7.7 | 3,076 | 6.3 | 8,685 | 4,737 | 3,948 | 3 | 63 |
| Shelby | 1,759 | 42,980 | 0.81 | 570 | 11.5 | 389 | 7.8 | 3,759 | 9.4 | 9,033 | 5,532 | 3,501 | 8 | 161 |
| Simpson | 327 | 19,497 | 1.17 | 218 | 11.7 | 205 | 11.0 | 1,129 | 7.5 | 4,110 | 2,691 | 1,420 | 18 | 244 |
| Spencer | 113 | 11,862 | 0.28 | 189 | 9.7 | 174 | 8.9 | 1,029 | 6.2 | 3,562 | 2,194 | 1,369 | 2 | 40 |
| Taylor | 1,337 | 26,727 | 1.11 | 307 | 11.9 | 352 | 13.7 | 1,442 | 7.3 | 6,171 | 4,205 | 1,966 | 0 | 43 |
| Todd | 195 | 10,256 | 0.55 | 170 | 13.7 | 131 | 10.5 | 1,266 | 12.6 | 2,469 | 1,793 | 677 | 13 | 148 |
| | | | | | | | | | | | | | 1 | 45 |
| Trigg | 78 | 12,721 | 0.7 | 165 | 11.2 | 188 | 12.7 | 943 | 8.5 | 3,928 | 2,717 | 1,211 | 6 | 98 |
| Trimble | 41 | 6,710 | 0.5 | 92 | 10.8 | 109 | 12.9 | 495 | 7.2 | 1,986 | 1,228 | 757 | 0 | 11 |
| Union | 1,660 | 14,148 | 0.92 | 151 | 10.5 | 172 | 11.9 | 998 | 8.4 | 3,202 | 1,905 | 1,297 | 3 | 35 |
| Warren | 6,496 | 136,954 | 1.13 | 1,728 | 12.8 | 1,086 | 8.1 | 9,688 | 9.0 | 21,237 | 13,559 | 7,678 | 39 | 537 |
| Washington | 205 | 11,117 | 0.84 | 158 | 13.0 | 146 | 12.0 | 852 | 8.7 | 2,721 | 1,899 | 822 | 6 | 43 |
| Wayne | 331 | 19,487 | 0.86 | 213 | 10.5 | 253 | 12.5 | 1,298 | 8.4 | 5,173 | 3,489 | 1,684 | 8 | 35 |
| Webster | 393 | 11,677 | 0.71 | 175 | 13.5 | 148 | 11.5 | 1,057 | 10.3 | 3,109 | 1,933 | 1,176 | 0 | 21 |
| Whitley | 1,895 | 36,397 | 1.02 | 511 | 14.0 | 482 | 13.2 | 2,178 | 7.6 | 8,658 | 5,739 | 2,919 | 9 | 176 |
| Wolfe | 132 | 6,863 | 0.81 | 80 | 11.3 | 128 | 18.0 | 438 | 7.9 | 1,965 | 1,099 | 867 | 0 | 18 |
| Woodford | 519 | 24,937 | 0.89 | 266 | 9.9 | 266 | 9.9 | 1,652 | 7.8 | 5,922 | 3,197 | 2,725 | 0 | 14 |
| LOUISIANA | 129,720 | 4,678,393 | 1.01 | 57,616 | 12.4 | 50,108 | 10.8 | 391,933 | 10.3 | 883,811 | 512,547 | 371,264 | X | X |
| Acadia | 1,050 | 53,324 | 0.63 | 802 | 13.0 | 753 | 12.2 | 5,206 | 10.1 | 12,140 | 9,247 | 2,893 | 109 | 514 |
| Allen | 4,181 | 25,618 | 1.0 | 297 | 11.7 | 269 | 10.6 | 1,784 | 10.1 | 4,566 | 3,623 | 944 | 19 | NA |
| Ascension | 788 | 109,682 | 0.77 | 1,680 | 13.1 | 1,029 | 8.0 | 8,059 | 7.3 | 18,222 | 7,322 | 10,900 | 340 | 1,519 |
| Assumption | 199 | 17,940 | 0.5 | 231 | 10.7 | 240 | 11.1 | 1,771 | 10.1 | 4,956 | 2,780 | 2,177 | 70 | 215 |
| Avoyelles | 3,348 | 37,073 | 0.75 | 497 | 12.4 | 490 | 12.3 | 3,206 | 10.6 | 8,958 | 6,575 | 2,383 | NA | NA |
| Beauregard | 1,294 | 32,748 | 0.7 | 513 | 13.5 | 449 | 11.9 | 3,327 | 11.0 | 7,260 | 5,832 | 1,427 | 18 | 241 |
| Bienville | 319 | 12,632 | 0.8 | 155 | 11.9 | 213 | 16.4 | 971 | 9.5 | 3,408 | 2,372 | 1,036 | 83 | 196 |
| Bossier | 2,560 | 119,063 | 0.87 | 1,635 | 12.8 | 1,207 | 9.5 | 10,025 | 9.4 | 21,573 | 15,371 | 6,202 | 64 | 467 |
| Caddo | 6,246 | 260,940 | 1.15 | 2,918 | 12.3 | 2,925 | 12.3 | 18,842 | 9.8 | 50,479 | 33,333 | 17,146 | 8 | 309 |
| Calcasieu | 3,792 | 212,500 | 1.12 | 2,969 | 14.6 | 2,214 | 10.9 | 16,151 | 9.6 | 36,363 | 27,564 | 8,799 | 583 | 3,819 |
| Caldwell | 584 | 9,265 | 0.78 | 122 | 12.4 | 137 | 13.9 | 879 | 11.6 | 2,244 | 1,584 | 660 | 38 | 207 |
| Cameron | 20 | 10,476 | 2.13 | 74 | 10.6 | 77 | 11.0 | 591 | 10.2 | 1,201 | 946 | 255 | NA | NA |
| Catahoula | 866 | 8,574 | 0.63 | 99 | 10.7 | 127 | 13.8 | 708 | 10.2 | 2,286 | 1,677 | 609 | NA | NA |
| Claiborne | 2,923 | 14,905 | 0.78 | 118 | 7.6 | 209 | 13.5 | 1,040 | 10.7 | 3,387 | 2,481 | 906 | 16 | 51 |
| Concordia | 1,480 | 18,847 | 0.85 | 243 | 12.8 | 256 | 13.5 | 1,409 | 9.8 | 4,233 | 3,222 | 1,011 | 73 | 304 |
| De Soto | 215 | 24,162 | 0.71 | 316 | 11.4 | 316 | 11.4 | 2,383 | 10.7 | 6,127 | 4,334 | 1,792 | 72 | 358 |
| East Baton Rouge | 11,126 | 498,710 | 1.26 | 5,610 | 12.8 | 4,199 | 9.5 | 38,246 | 10.5 | 76,280 | 36,598 | 39,682 | 609 | 6,475 |
| East Carroll | 1,067 | 7,159 | 1.05 | 69 | 10.5 | 76 | 11.5 | 411 | 8.8 | 1,373 | 934 | 439 | 21 | 28 |
| East Feliciana | 2,417 | 18,394 | 0.85 | 172 | 9.1 | 246 | 13.0 | 1,222 | 9.2 | 4,240 | 2,285 | 1,956 | 70 | 131 |
| Evangeline | 1,379 | 30,391 | 0.72 | 480 | 14.4 | 422 | 12.7 | 2,885 | 10.8 | 7,133 | 5,573 | 1,560 | NA | NA |
| Franklin | 732 | 18,903 | 0.8 | 245 | 12.4 | 294 | 14.9 | 1,871 | 11.9 | 4,577 | 3,277 | 1,300 | 58 | 143 |
| Grant | 2,960 | 19,016 | 0.54 | 242 | 10.9 | 209 | 9.4 | 1,676 | 10.5 | 4,238 | 3,056 | 1,181 | NA | NA |
| Iberia | 944 | 70,379 | 0.95 | 888 | 12.9 | 836 | 12.1 | 6,592 | 11.4 | 14,429 | 9,594 | 4,835 | NA | NA |
| Iberville | 3,988 | 39,117 | 1.5 | 377 | 11.8 | 339 | 10.6 | 2,043 | 8.7 | 6,278 | 2,527 | 3,751 | 206 | 335 |
| Jackson | 1,037 | 14,276 | 0.7 | 133 | 8.5 | 242 | 15.5 | 1,134 | 9.8 | 3,629 | 2,569 | 1,060 | 6 | 52 |
| Jefferson | 3,311 | 427,170 | 0.96 | 5,467 | 12.6 | 4,679 | 10.8 | 45,498 | 12.9 | 88,616 | 32,654 | 55,962 | 1,240 | 9,618 |
| Jefferson Davis | 567 | 27,885 | 0.7 | 398 | 12.8 | 432 | 13.8 | 2,632 | 10.2 | 6,584 | 5,383 | 1,201 | NA | NA |
| Lafayette | 5,160 | 271,051 | 1.25 | 3,157 | 12.8 | 2,188 | 8.9 | 19,679 | 9.5 | 40,130 | 29,979 | 10,152 | NA | NA |
| Lafourche | 1,627 | 91,545 | 0.84 | 1,116 | 11.4 | 984 | 10.1 | 9,931 | 12.3 | 18,897 | 10,610 | 8,287 | 231 | 2,084 |
| La Salle | 1,240 | 14,291 | 0.88 | 176 | 11.7 | 191 | 12.7 | 1,299 | 11.6 | 3,077 | 2,484 | 594 | 30 | 66 |
| Lincoln | 4,904 | 47,616 | 1.02 | 502 | 10.8 | 452 | 9.7 | 4,094 | 11.5 | 7,443 | 5,044 | 2,399 | 79 | 197 |
| Livingston | 1,189 | 105,435 | 0.47 | 1,775 | 12.3 | 1,269 | 8.8 | 11,767 | 9.7 | 22,329 | 9,394 | 12,935 | 550 | 2,948 |
| Madison | 1,600 | 11,273 | 0.99 | 137 | 12.9 | 157 | 14.8 | 707 | 9.1 | 2,020 | 1,549 | 471 | 11 | 27 |
| Morehouse | 768 | 23,321 | 0.75 | 293 | 12.1 | 363 | 15.0 | 1,837 | 9.5 | 6,475 | 4,149 | 2,327 | 30 | 226 |
| Natchitoches | 2,196 | 39,453 | 1.06 | 440 | 11.7 | 424 | 11.3 | 3,042 | 10.3 | 7,666 | 5,793 | 1,872 | 79 | 371 |

1. Per 1,000 estimated resident population.

Items 32—45

# Table B. States and Counties — Crime, Education, Money Income, and Poverty

| | County law enforcement employment, 2019 | | Education | | | | | | Money income, 2015-2019 | | | | Income and poverty, 2019 | | | |
|---|---|---|---|---|---|---|---|---|---|---|---|---|---|---|---|---|
| | | | School enrollment and attainment, 2015-2019 | | | | Local government expenditures,[3] 2017-2018 | | | Households | | | | Percent below poverty level | | |
| | | | Enrollment[1] | | Attainment[2] (percent) | | | | | | Percent | | | | | |
| STATE County | Officers | Civilians | Total | Percent private | High school graduate or less | Bachelor's degree or more | Total current spending (mil dol) | Current spending per student (dollars) | Per capita income[4] | Median income (dollars) | with income of less than $50,000 | with income of $200,000 or more | Median household income (dollars) | All persons | Children under 18 years | Children 5 to 17 years in families |
| | 46 | 47 | 48 | 49 | 50 | 51 | 52 | 53 | 54 | 55 | 56 | 57 | 58 | 59 | 60 | 61 |
| KENTUCKY—Cont'd | | | | | | | | | | | | | | | | |
| Ohio | NA | NA | 5,482 | 10.5 | 59.7 | 13.6 | 41 | 9,451 | 24,297 | 45,564 | 55.3 | 1.9 | 47,542 | 15.6 | 20.4 | 19.2 |
| Oldham | 51 | 6 | 18,597 | 18.4 | 26.8 | 42.4 | 124 | 9,805 | 42,272 | 99,128 | 22.5 | 16.5 | 101,152 | 4.9 | 4.8 | 4.1 |
| Owen | 5 | 1 | 2,248 | 11.6 | 67.1 | 10.9 | 19 | 9,755 | 25,640 | 48,801 | 50.9 | 4.6 | 51,494 | 15.7 | 20.3 | 18.7 |
| Owsley | 3 | 1 | 784 | 1.9 | 68.0 | 16.2 | 11 | 13,584 | 18,940 | 30,284 | 66.4 | 0.4 | 28,727 | 35.5 | 43.6 | 43.3 |
| Pendleton | 8 | 1 | 3,224 | 11.8 | 60.0 | 13.2 | 24 | 9,984 | 24,120 | 54,375 | 47.0 | 1.2 | 58,962 | 12.4 | 19.9 | 18.6 |
| Perry | NA | NA | 5,644 | 8.0 | 56.1 | 14.4 | 53 | 10,385 | 23,135 | 33,640 | 63.7 | 1.9 | 38,000 | 24.2 | 32.0 | 31.6 |
| Pike | 23 | 13 | 12,971 | 10.5 | 62.4 | 13.0 | 109 | 11,086 | 22,437 | 34,856 | 65.4 | 1.9 | 36,875 | 24.0 | 30.0 | 25.7 |
| Powell | 3 | 2 | 2,964 | 10.0 | 59.7 | 15.9 | 24 | 10,156 | 20,841 | 37,469 | 57.4 | 1.0 | 41,760 | 21.5 | 31.7 | 30.5 |
| Pulaski | 39 | 8 | 13,335 | 9.6 | 52.8 | 16.5 | 109 | 10,321 | 24,646 | 39,998 | 57.9 | 2.2 | 43,568 | 22.5 | 30.8 | 26.5 |
| Robertson | 1 | 1 | 434 | 5.1 | 53.6 | 16.3 | 5 | 11,514 | 22,552 | 41,902 | 55.3 | 2.0 | 47,017 | 22.0 | 28.7 | 26.5 |
| Rockcastle | 6 | 2 | 3,600 | 4.4 | 65.0 | 11.4 | 30 | 10,536 | 20,892 | 35,720 | 59.0 | 1.4 | 40,790 | 21.0 | 31.1 | 25.6 |
| Rowan | 14 | 2 | 8,059 | 5.1 | 49.2 | 25.1 | 34 | 10,042 | 20,488 | 38,230 | 62.4 | 3.3 | 44,068 | 23.3 | 26.0 | 21.9 |
| Russell | NA | NA | 3,545 | 6.9 | 59.5 | 14.7 | 33 | 10,467 | 20,674 | 38,390 | 61.1 | 1.6 | 44,056 | 22.6 | 31.6 | 28.6 |
| Scott | NA | NA | 14,316 | 19.8 | 39.3 | 28.4 | 89 | 9,609 | 32,796 | 70,817 | 34.6 | 5.7 | 74,719 | 8.6 | 12.0 | 8.9 |
| Shelby | 22 | 3 | 10,602 | 21.2 | 43.8 | 24.9 | 77 | 10,637 | 32,468 | 67,056 | 37.5 | 7.5 | 71,737 | 9.4 | 12.8 | 11.6 |
| Simpson | 12 | 6 | 4,115 | 14.1 | 57.9 | 14.2 | 33 | 10,960 | 24,458 | 48,623 | 50.5 | 3.0 | 52,931 | 12.5 | 18.8 | 17.9 |
| Spencer | 5 | 6 | 4,138 | 12.3 | 45.9 | 20.3 | 29 | 9,649 | 33,661 | 80,166 | 29.1 | 5.6 | 79,863 | 7.3 | 9.5 | 8.3 |
| Taylor | 12 | 3 | 6,660 | 19.7 | 52.1 | 19.6 | 42 | 10,969 | 23,602 | 44,522 | 55.0 | 1.6 | 48,485 | 16.4 | 21.1 | 21.3 |
| Todd | 6 | 2 | 2,828 | 15.3 | 63.0 | 13.2 | 20 | 9,995 | 23,143 | 44,005 | 54.1 | 2.7 | 47,742 | 17.1 | 22.7 | 22.3 |
| Trigg | 7 | 3 | 2,877 | 20.1 | 45.2 | 16.6 | 22 | 10,543 | 28,264 | 50,536 | 49.6 | 2.7 | 49,955 | 14.6 | 22.8 | 20.3 |
| Trimble | 1 | 0 | 1,721 | 6.8 | 63.5 | 14.1 | 13 | 10,676 | 27,237 | 53,056 | 48.4 | 2.8 | 58,998 | 11.9 | 16.1 | 15.1 |
| Union | NA | NA | 2,839 | 17.1 | 57.8 | 11.1 | 24 | 11,138 | 23,097 | 46,673 | 54.7 | 1.7 | 49,870 | 17.3 | 21.2 | 18.2 |
| Warren | 50 | 39 | 39,135 | 8.6 | 38.5 | 32.2 | 196 | 9,839 | 28,558 | 52,270 | 48.1 | 4.6 | 53,238 | 15.4 | 19.1 | 17.1 |
| Washington | 8 | 6 | 2,691 | 23.6 | 58.4 | 18.8 | 19 | 10,892 | 26,035 | 53,743 | 45.3 | 3.8 | 51,286 | 14.4 | 19.8 | 18.8 |
| Wayne | 14 | 2 | 3,759 | 6.7 | 64.1 | 12.3 | 36 | 10,854 | 25,707 | 34,914 | 62.7 | 2.2 | 34,449 | 23.8 | 33.3 | 31.3 |
| Webster | 7 | 9 | 2,659 | 11.1 | 64.6 | 10.0 | 22 | 9,596 | 22,911 | 42,119 | 58.2 | 2.2 | 49,079 | 16.3 | 20.1 | 19.4 |
| Whitley | 14 | 5 | 9,558 | 20.3 | 55.5 | 20.4 | 87 | 10,488 | 20,547 | 39,005 | 58.5 | 2.2 | 38,191 | 22.6 | 26.1 | 27.0 |
| Wolfe | NA | NA | 1,387 | 12.2 | 65.7 | 8.7 | 16 | 12,484 | 14,923 | 24,623 | 75.9 | 0.2 | 32,773 | 30.1 | 44.0 | 42.7 |
| Woodford | NA | NA | 6,050 | 19.7 | 38.7 | 33.8 | 42 | 10,396 | 32,264 | 63,820 | 40.8 | 4.7 | 67,815 | 8.3 | 12.0 | 10.9 |
| LOUISIANA | X | X | 1,154,067 | 19.2 | 48.7 | 24.1 | 8,213 | 11,487 | 27,923 | 49,469 | 50.4 | 4.6 | 51,108 | 18.8 | 26.4 | 25.4 |
| Acadia | 61 | 43 | 14,863 | 17.7 | 62.4 | 13.3 | 91 | 9,277 | 23,122 | 43,396 | 55.0 | 2.8 | 44,728 | 20.3 | 26.5 | 24.6 |
| Allen | 28 | 75 | 5,597 | 12.8 | 64.6 | 11.3 | 47 | 10,802 | 20,964 | 46,446 | 53.6 | 1.6 | 47,111 | 21.6 | 26.1 | 23.8 |
| Ascension | 285 | 69 | 32,228 | 18.9 | 43.7 | 26.4 | 264 | 11,896 | 34,168 | 80,527 | 32.8 | 8.2 | 83,072 | 9.6 | 12.9 | 12.7 |
| Assumption | 59 | 24 | 4,877 | 12.4 | 70.7 | 9.6 | 41 | 11,689 | 25,752 | 43,759 | 53.3 | 4.2 | 51,763 | 18.8 | 25.1 | 23.4 |
| Avoyelles | NA | NA | 10,008 | 17.5 | 61.0 | 12.3 | 58 | 9,474 | 22,017 | 38,565 | 60.6 | 3.2 | 38,316 | 24.4 | 34.2 | 31.3 |
| Beauregard | 68 | 20 | 8,432 | 9.6 | 52.7 | 17.8 | 59 | 9,907 | 26,331 | 53,209 | 47.2 | 3.0 | 55,343 | 14.3 | 19.2 | 19.6 |
| Bienville | 40 | 24 | 3,055 | 12.6 | 60.4 | 12.8 | 36 | 15,954 | 21,491 | 30,272 | 64.7 | 1.8 | 37,198 | 24.4 | 32.9 | 31.1 |
| Bossier | 330 | 84 | 31,091 | 12.0 | 40.1 | 25.2 | 243 | 10,779 | 28,766 | 54,268 | 46.4 | 4.0 | 54,080 | 16.4 | 25.6 | 25.5 |
| Caddo | 414 | 201 | 59,330 | 13.2 | 46.7 | 23.5 | 459 | 11,672 | 26,545 | 41,797 | 56.2 | 4.3 | 45,613 | 24.1 | 35.8 | 34.2 |
| Calcasieu | 591 | 278 | 49,351 | 13.6 | 47.2 | 22.3 | 409 | 11,881 | 28,778 | 51,148 | 48.9 | 4.5 | 53,050 | 18.9 | 26.0 | 24.6 |
| Caldwell | NA | NA | 2,204 | 5.3 | 65.9 | 13.8 | 20 | 12,934 | 23,216 | 37,691 | 60.5 | 3.0 | 44,007 | 19.5 | 28.3 | 26.5 |
| Cameron | 71 | 11 | 1,549 | 13.9 | 55.6 | 12.7 | 24 | 17,803 | 28,358 | 53,423 | 48.2 | 4.1 | 60,317 | 13.7 | 18.1 | 17.3 |
| Catahoula | 22 | 108 | 2,002 | 9.0 | 65.1 | 13.8 | 16 | 12,221 | 22,412 | 40,129 | 55.6 | 3.2 | 36,794 | 26.4 | 34.3 | 31.5 |
| Claiborne | 31 | 39 | 2,767 | 9.7 | 64.0 | 12.8 | 19 | 10,991 | 16,770 | 26,776 | 76.0 | 0.8 | 36,497 | 32.5 | 39.4 | 37.8 |
| Concordia | 26 | 267 | 4,758 | 5.4 | 60.7 | 14.1 | 44 | 11,304 | 19,092 | 32,500 | 64.7 | 1.7 | 36,276 | 27.5 | 39.4 | 35.6 |
| De Soto | 101 | 45 | 6,231 | 5.0 | 60.0 | 14.8 | 79 | 15,273 | 26,767 | 46,006 | 52.5 | 4.3 | 47,347 | 18.0 | 24.3 | 20.5 |
| East Baton Rouge | 717 | 124 | 127,370 | 21.0 | 36.8 | 34.9 | 811 | 12,785 | 32,431 | 54,948 | 45.6 | 6.9 | 56,303 | 17.7 | 24.5 | 24.5 |
| East Carroll | 17 | 29 | 1,047 | 16.5 | 70.0 | 8.9 | 14 | 13,658 | 19,483 | 22,346 | 75.2 | 1.9 | 29,943 | 38.4 | 52.0 | 47.0 |
| East Feliciana | 53 | 0 | 3,732 | 22.0 | 63.4 | 12.7 | 23 | 11,983 | 23,612 | 51,803 | 48.6 | 3.3 | 50,041 | 18.7 | 23.8 | 22.5 |
| Evangeline | 18 | 28 | 8,068 | 14.5 | 65.8 | 11.6 | 59 | 9,857 | 18,770 | 31,965 | 67.0 | 2.0 | 35,146 | 28.6 | 33.0 | 30.9 |
| Franklin | 118 | 0 | 4,681 | 16.9 | 67.1 | 12.0 | 33 | 10,634 | 19,206 | 35,282 | 63.1 | 1.7 | 35,484 | 25.8 | 38.4 | 36.8 |
| Grant | NA | NA | 4,738 | 14.9 | 64.4 | 8.5 | 31 | 10,150 | 18,801 | 42,505 | 59.8 | 1.7 | 49,544 | 18.6 | 23.4 | 22.3 |
| Iberia | 112 | 64 | 17,772 | 12.5 | 63.8 | 13.6 | 132 | 10,069 | 23,290 | 46,861 | 52.5 | 1.8 | 47,731 | 21.9 | 30.8 | 29.3 |
| Iberville | 77 | 57 | 7,241 | 18.9 | 61.7 | 13.9 | 77 | 14,951 | 23,751 | 50,161 | 49.9 | 3.6 | 55,198 | 18.7 | 27.7 | 26.2 |
| Jackson | 41 | 224 | 3,360 | 9.9 | 61.9 | 14.0 | 25 | 11,143 | 21,594 | 39,139 | 59.9 | 2.5 | 43,232 | 19.2 | 27.2 | 25.5 |
| Jefferson | NA | NA | 100,134 | 31.0 | 45.1 | 26.8 | 569 | 11,503 | 30,374 | 54,032 | 46.3 | 4.6 | 55,909 | 14.6 | 23.3 | 22.3 |
| Jefferson Davis | NA | NA | 7,632 | 12.5 | 59.1 | 16.9 | 60 | 10,152 | 25,474 | 42,105 | 54.0 | 4.8 | 47,988 | 19.1 | 25.1 | 23.9 |
| Lafayette | NA | NA | 61,072 | 20.6 | 41.8 | 33.0 | 335 | 10,179 | 31,892 | 56,999 | 45.2 | 6.5 | 60,053 | 16.6 | 22.5 | 22.0 |
| Lafourche | 228 | 108 | 22,548 | 16.5 | 62.3 | 16.9 | 146 | 9,952 | 28,683 | 55,506 | 46.0 | 4.1 | 54,145 | 17.5 | 26.8 | 21.6 |
| La Salle | 30 | 17 | 3,243 | 13.2 | 64.9 | 13.6 | 29 | 11,099 | 21,654 | 42,104 | 56.4 | 2.9 | 51,014 | 16.9 | 21.4 | 20.1 |
| Lincoln | 58 | 19 | 17,456 | 7.2 | 38.3 | 35.6 | 77 | 11,714 | 22,863 | 35,467 | 60.6 | 3.9 | 41,608 | 29.5 | 37.0 | 38.6 |
| Livingston | 304 | 0 | 34,000 | 12.9 | 52.9 | 19.7 | 245 | 9,713 | 29,354 | 63,389 | 39.1 | 4.4 | 63,852 | 12.0 | 15.0 | 14.3 |
| Madison | 23 | 10 | 2,481 | 18.8 | 63.1 | 12.8 | 19 | 10,929 | 16,671 | 30,350 | 69.8 | 2.1 | 32,229 | 41.1 | 54.1 | 55.0 |
| Morehouse | NA | NA | 5,204 | 14.1 | 63.3 | 13.0 | 49 | 12,611 | 18,494 | 32,929 | 65.9 | 1.6 | 34,922 | 31.0 | 39.9 | 37.9 |
| Natchitoches | 90 | 78 | 12,531 | 16.6 | 46.7 | 19.0 | 74 | 11,284 | 19,198 | 28,567 | 69.6 | 2.0 | 49,536 | 19.6 | 23.9 | 22.2 |

1. All persons 3 years old and over enrolled in nursery school through college.   2. Persons 25 years old and over.   3. Elementary and secondary education expenditures.   4. Based on population estimated by the American Community Survey, 2014–2018.

# Table B. States and Counties — **Personal Income**

| STATE County | Personal income, 2019 | | | | | | | | | | Earnings, 2019 | | |
| | Total (mil dol) | Percent change 2018-2019 | Per capita¹ Dollars | Rank | Wages and salaries (mil dol) | Supplements to wages and salaries, employer contributions (mil dol) Pension and insurance | Government social insurance | Proprietors' income (mil dol) | Dividends, interest, and rent (mil dol) | Personal transfer receipts (mil dol) | Total (mil dol) | Contributions for government social insurance (mil dol) From employee and self-employed | From employer |
|---|---|---|---|---|---|---|---|---|---|---|---|---|---|
| | 62 | 63 | 64 | 65 | 66 | 67 | 68 | 69 | 70 | 71 | 72 | 73 | 74 |
| **KENTUCKY—Cont'd** | | | | | | | | | | | | | |
| Ohio | 799 | 1.2 | 33,313 | 2,864 | 303 | 63 | 24 | 47 | 98 | 281 | 436 | 31 | 24 |
| Oldham | 4,425 | 4.0 | 66,238 | 167 | 824 | 137 | 58 | 252 | 870 | 445 | 1,271 | 82 | 58 |
| Owen | 364 | 0.9 | 33,365 | 2,861 | 63 | 14 | 5 | 11 | 48 | 114 | 93 | 9 | 5 |
| Owsley | 143 | 2.3 | 32,461 | 2,921 | 22 | 6 | 2 | 5 | 14 | 82 | 35 | 4 | 2 |
| Pendleton | 657 | 3.6 | 45,039 | 1,356 | 103 | 22 | 8 | 28 | 85 | 147 | 160 | 13 | 8 |
| Perry | 1,012 | 2.4 | 39,308 | 2,148 | 474 | 95 | 35 | 48 | 124 | 456 | 652 | 45 | 35 |
| Pike | 2,170 | 1.8 | 37,490 | 2,390 | 959 | 176 | 73 | 92 | 272 | 890 | 1,299 | 97 | 73 |
| Powell | 402 | 2.8 | 32,532 | 2,917 | 90 | 21 | 7 | 15 | 39 | 161 | 133 | 12 | 7 |
| Pulaski | 2,501 | 4.9 | 38,492 | 2,261 | 1,050 | 205 | 81 | 158 | 303 | 947 | 1,494 | 106 | 81 |
| Robertson | 72 | 2.9 | 34,196 | 2,781 | 10 | 3 | 1 | 2 | 9 | 25 | 16 | 2 | 1 |
| Rockcastle | 539 | 3.1 | 32,294 | 2,926 | 148 | 37 | 11 | 15 | 53 | 213 | 210 | 17 | 11 |
| Rowan | 772 | 2.3 | 31,566 | 2,973 | 387 | 88 | 28 | 30 | 109 | 258 | 534 | 34 | 28 |
| Russell | 669 | 4.0 | 37,346 | 2,404 | 205 | 46 | 16 | 147 | 85 | 224 | 414 | 29 | 16 |
| Scott | 2,527 | 5.3 | 44,335 | 1,440 | 1,567 | 244 | 116 | 120 | 282 | 385 | 2,047 | 125 | 116 |
| Shelby | 2,287 | 5.4 | 46,657 | 1,164 | 738 | 132 | 56 | 136 | 393 | 376 | 1,061 | 70 | 56 |
| Simpson | 729 | 4.9 | 39,235 | 2,157 | 399 | 68 | 31 | 67 | 97 | 194 | 565 | 36 | 31 |
| Spencer | 875 | 5.6 | 45,208 | 1,338 | 81 | 17 | 6 | 32 | 81 | 157 | 137 | 12 | 6 |
| Taylor | 916 | 4.3 | 35,546 | 2,613 | 446 | 84 | 35 | 35 | 130 | 314 | 601 | 41 | 35 |
| Todd | 469 | 3.8 | 38,112 | 2,319 | 90 | 22 | 7 | 69 | 67 | 121 | 189 | 12 | 7 |
| Trigg | 551 | 4.1 | 37,633 | 2,369 | 110 | 24 | 9 | 33 | 89 | 169 | 175 | 15 | 9 |
| Trimble | 334 | 4.1 | 39,400 | 2,126 | 61 | 15 | 5 | 6 | 36 | 90 | 87 | 8 | 5 |
| Union | 523 | 5.7 | 36,348 | 2,530 | 251 | 45 | 19 | 32 | 83 | 169 | 346 | 23 | 19 |
| Warren | 5,086 | 4.2 | 38,268 | 2,302 | 2,989 | 533 | 220 | 375 | 767 | 1,118 | 4,118 | 255 | 220 |
| Washington | 462 | 3.0 | 38,210 | 2,304 | 147 | 29 | 11 | 25 | 78 | 126 | 211 | 16 | 11 |
| Wayne | 588 | 0.6 | 28,935 | 3,066 | 188 | 41 | 15 | 28 | 82 | 252 | 273 | 22 | 15 |
| Webster | 517 | 1.5 | 39,965 | 2,055 | 160 | 32 | 12 | 60 | 67 | 147 | 264 | 17 | 12 |
| Whitley | 1,257 | 3.5 | 34,663 | 2,729 | 515 | 99 | 40 | 62 | 170 | 535 | 717 | 52 | 40 |
| Wolfe | 227 | 1.6 | 31,658 | 2,965 | 41 | 10 | 3 | 8 | 22 | 118 | 62 | 6 | 3 |
| Woodford | 1,425 | 4.5 | 53,292 | 591 | 446 | 77 | 36 | 191 | 251 | 242 | 750 | 42 | 36 |
| **LOUISIANA** | 220,630 | 2.5 | 47,363 | X | 104,024 | 18,203 | 6,619 | 19,782 | 39,389 | 48,362 | 148,628 | 8,428 | 6,619 |
| Acadia | 2,344 | 1.7 | 37,786 | 2,346 | 609 | 121 | 39 | 153 | 344 | 651 | 922 | 60 | 39 |
| Allen | 888 | 3.7 | 34,644 | 2,732 | 348 | 94 | 19 | 48 | 108 | 241 | 509 | 27 | 19 |
| Ascension | 6,415 | 4.0 | 50,671 | 782 | 2,960 | 464 | 189 | 299 | 694 | 934 | 3,912 | 224 | 189 |
| Assumption | 1,050 | 0.6 | 47,947 | 1,019 | 197 | 38 | 12 | 74 | 131 | 236 | 322 | 23 | 12 |
| Avoyelles | 1,566 | 2.3 | 39,001 | 2,193 | 387 | 92 | 23 | 113 | 217 | 513 | 615 | 39 | 23 |
| Beauregard | 1,630 | 3.1 | 43,473 | 1,548 | 401 | 79 | 25 | 50 | 222 | 398 | 554 | 38 | 25 |
| Bienville | 518 | 1.0 | 39,111 | 2,175 | 186 | 41 | 11 | 21 | 75 | 196 | 260 | 17 | 11 |
| Bossier | 5,731 | 2.7 | 45,110 | 1,349 | 2,453 | 507 | 173 | 281 | 991 | 1,227 | 3,415 | 187 | 173 |
| Caddo | 12,176 | 1.8 | 50,690 | 779 | 5,709 | 1,014 | 375 | 1,192 | 2,824 | 2,882 | 8,291 | 472 | 375 |
| Calcasieu | 10,103 | 2.4 | 49,664 | 860 | 5,834 | 1,037 | 370 | 711 | 1,467 | 2,038 | 7,952 | 445 | 370 |
| Caldwell | 351 | 4.1 | 35,353 | 2,639 | 92 | 22 | 6 | 18 | 46 | 128 | 138 | 9 | 6 |
| Cameron | 342 | 1.8 | 49,000 | 923 | 835 | 93 | 51 | 10 | 65 | 52 | 988 | 55 | 51 |
| Catahoula | 333 | 4.8 | 35,037 | 2,681 | 78 | 19 | 5 | 22 | 44 | 120 | 124 | 9 | 5 |
| Claiborne | 548 | 1.1 | 34,954 | 2,693 | 153 | 39 | 9 | 56 | 90 | 174 | 257 | 15 | 9 |
| Concordia | 657 | 2.9 | 34,110 | 2,789 | 208 | 50 | 13 | 49 | 99 | 242 | 320 | 19 | 13 |
| De Soto | 1,162 | 2.6 | 42,298 | 1,719 | 398 | 79 | 23 | 26 | 212 | 309 | 527 | 34 | 23 |
| East Baton Rouge | 23,324 | 2.2 | 53,002 | 615 | 15,705 | 2,588 | 971 | 2,034 | 4,838 | 4,179 | 21,298 | 1,129 | 971 |
| East Carroll | 247 | 0.2 | 36,035 | 2,563 | 61 | 15 | 4 | 31 | 49 | 91 | 111 | 6 | 4 |
| East Feliciana | 810 | 2.3 | 42,353 | 1,712 | 250 | 74 | 12 | 30 | 124 | 230 | 365 | 19 | 12 |
| Evangeline | 1,118 | 0.9 | 33,474 | 2,848 | 294 | 66 | 19 | 78 | 161 | 406 | 457 | 31 | 19 |
| Franklin | 666 | 3.2 | 33,250 | 2,870 | 179 | 47 | 11 | 55 | 89 | 269 | 291 | 18 | 11 |
| Grant | 714 | 2.7 | 31,890 | 2,950 | 167 | 46 | 12 | 24 | 83 | 222 | 248 | 18 | 12 |
| Iberia | 2,821 | 1.6 | 40,403 | 2,001 | 1,411 | 232 | 89 | 152 | 502 | 823 | 1,884 | 114 | 89 |
| Iberville | 1,379 | 3.0 | 42,418 | 1,700 | 1,161 | 221 | 71 | 67 | 201 | 359 | 1,519 | 83 | 71 |
| Jackson | 549 | 1.8 | 34,843 | 2,708 | 158 | 38 | 9 | 33 | 71 | 190 | 237 | 16 | 9 |
| Jefferson | 22,608 | 2.4 | 52,274 | 659 | 10,945 | 1,539 | 702 | 2,221 | 4,578 | 4,453 | 15,407 | 914 | 702 |
| Jefferson Davis | 1,334 | 1.7 | 42,519 | 1,686 | 357 | 76 | 22 | 62 | 189 | 330 | 517 | 32 | 22 |
| Lafayette | 12,129 | 2.4 | 49,629 | 865 | 6,986 | 1,035 | 452 | 1,122 | 2,520 | 2,142 | 9,595 | 539 | 452 |
| Lafourche | 4,471 | 3.0 | 45,806 | 1,270 | 1,940 | 360 | 121 | 457 | 827 | 929 | 2,878 | 166 | 121 |
| La Salle | 506 | 6.5 | 33,992 | 2,805 | 211 | 47 | 13 | 20 | 66 | 158 | 290 | 17 | 13 |
| Lincoln | 1,895 | 0.3 | 40,541 | 1,981 | 790 | 167 | 48 | 194 | 415 | 440 | 1,199 | 63 | 48 |
| Livingston | 5,944 | 3.6 | 42,217 | 1,736 | 1,309 | 237 | 82 | 294 | 618 | 1,137 | 1,921 | 124 | 82 |
| Madison | 364 | 1.9 | 33,275 | 2,865 | 119 | 31 | 8 | 32 | 47 | 136 | 190 | 10 | 8 |
| Morehouse | 983 | 2.5 | 39,521 | 2,113 | 258 | 51 | 17 | 82 | 131 | 380 | 409 | 28 | 17 |
| Natchitoches | 1,597 | 2.9 | 41,846 | 1,790 | 545 | 132 | 32 | 227 | 241 | 439 | 936 | 51 | 32 |

1. Based on the resident population estimated as of July 1 of the year shown.

# Table B. States and Counties — Earnings, Social Security, and Housing

| STATE County | Earnings, 2019 (cont.) | | | | | | | | | Social Security beneficiaries, December 2019 | | | Housing units, 2020 | |
|---|---|---|---|---|---|---|---|---|---|---|---|---|---|---|
| | Percent by selected industries | | | | | | | | | | | Supple-mental Security Income recipients, 2019 | | |
| | Farm | Mining, quarrying, and extractions | Construction | Manu-facturing | Information; professional, scientific, technical services | Retail trade | Finance, insurance, real estate, and leasing | Health care and social assistance | Govern-ment | Number | Rate[1] | | Total | Percent change, 2010-2020 |
| | 75 | 76 | 77 | 78 | 79 | 80 | 81 | 82 | 83 | 84 | 85 | 86 | 87 | 88 |
| KENTUCKY—Cont'd | | | | | | | | | | | | | | |
| Ohio | 3.5 | D | 4.8 | 30.5 | D | 6.0 | 1.9 | 6.5 | 19.9 | 6,400 | 266 | 966 | 10,502 | 2.8 |
| Oldham | 0.9 | D | 10.1 | 5.9 | 12.0 | 5.0 | 17.3 | 12.8 | 16.4 | 10,385 | 155 | 496 | 22,742 | 9.8 |
| Owen | -3.7 | 0.0 | 5.1 | D | D | 5.7 | D | 8.9 | 26.5 | 2,755 | 253 | 401 | 5,790 | 2.9 |
| Owsley | -1.5 | 0.1 | D | D | D | D | D | 25.5 | 37.3 | 1,185 | 269 | 740 | 2,373 | 1.9 |
| Pendleton | -1.6 | D | D | 13.5 | 6.3 | 3.3 | D | 8.0 | 20.4 | 3,505 | 241 | 499 | 6,405 | 1.1 |
| Perry | 0.0 | D | 2.5 | 0.4 | D | 8.7 | 3.4 | D | 19.4 | 7,740 | 300 | 2,702 | 13,103 | 2.5 |
| Pike | 0.0 | 10.1 | 2.5 | 3.3 | 4.1 | 9.3 | 4.5 | D | 13.3 | 18,205 | 314 | 4,368 | 31,289 | 3.2 |
| Powell | -0.3 | D | D | 10.9 | D | 8.2 | 2.3 | D | 27.1 | 3,505 | 285 | 984 | 5,707 | 1.9 |
| Pulaski | -0.2 | D | 5.3 | 15.9 | 4.0 | 8.6 | 4.3 | 23.3 | 13.6 | 18,660 | 286 | 3,750 | 31,633 | 0.6 |
| Robertson | -1.4 | 0.1 | 10.3 | 0.2 | D | D | D | 17.9 | 40.8 | 520 | 242 | 103 | 1,105 | 0.8 |
| Rockcastle | -1.4 | D | D | 3.2 | D | 4.2 | 2.8 | D | 18.6 | 4,395 | 263 | 1,060 | 7,935 | 3.0 |
| Rowan | -0.3 | D | D | 11.3 | D | 8.5 | 3.9 | D | 29.7 | 5,185 | 211 | 1,175 | 10,456 | 3.5 |
| Russell | 0.4 | 1.9 | D | 34.0 | D | 6.7 | 3.0 | D | 13.3 | 4,900 | 273 | 1,059 | 10,230 | 2.3 |
| Scott | 0.8 | D | 4.7 | 50.7 | 3.0 | 3.0 | 1.9 | D | 6.7 | 9,090 | 159 | 1,053 | 23,447 | 21.6 |
| Shelby | 0.4 | 0.0 | 5.5 | 25.7 | 6.3 | 7.7 | 5.5 | 7.3 | 12.4 | 9,265 | 189 | 733 | 18,695 | 12.7 |
| Simpson | 2.9 | 0.0 | 7.0 | 39.8 | D | 9.6 | 2.5 | 4.8 | 8.7 | 4,385 | 237 | 520 | 8,182 | 10.0 |
| Spencer | 0.0 | 0.0 | 15.5 | D | 8.5 | 7.1 | 6.5 | D | 26.6 | 3,820 | 198 | 277 | 7,631 | 13.6 |
| Taylor | -0.4 | 0.0 | D | 11.4 | D | 8.3 | 4.3 | D | 19.6 | 6,725 | 261 | 1,227 | 10,960 | 1.0 |
| Todd | 15.4 | D | 9.5 | 12.5 | D | 8.6 | 4.1 | D | 15.6 | 2,695 | 219 | 364 | 5,383 | 1.9 |
| Trigg | 2.2 | D | 9.9 | 14.8 | D | 7.9 | 4.8 | D | 21.7 | 4,180 | 285 | 416 | 8,011 | 2.6 |
| Trimble | -2.3 | D | 7.2 | D | D | 2.4 | 6.5 | 5.5 | 19.9 | 2,175 | 255 | 248 | 4,033 | 2.6 |
| Union | 3.0 | D | 2.5 | 14.2 | D | 4.6 | 3.2 | D | 10.9 | 3,550 | 245 | 379 | 6,231 | 1.5 |
| Warren | 0.2 | 0.2 | 7.1 | 18.0 | 5.4 | 6.9 | 5.2 | 14.8 | 13.6 | 22,185 | 167 | 3,479 | 56,644 | 20.0 |
| Washington | -0.6 | 0.0 | 13.8 | 36.4 | D | 3.9 | D | 9.2 | 12.9 | 2,975 | 246 | 407 | 5,236 | 4.0 |
| Wayne | 3.0 | D | 4.6 | 18.7 | D | 8.2 | 4.5 | 11.0 | 19.2 | 5,575 | 275 | 1,679 | 10,961 | 0.2 |
| Webster | 11.5 | D | 24.0 | 6.8 | 1.4 | 3.0 | 2.7 | 5.9 | 12.8 | 3,435 | 265 | 493 | 5,993 | 0.9 |
| Whitley | -0.2 | D | 3.6 | 9.2 | D | 6.8 | 3.7 | D | 16.3 | 9,500 | 261 | 3,732 | 15,581 | 2.7 |
| Wolfe | -1.9 | D | 2.1 | 3.1 | D | 8.6 | 3.7 | D | 33.4 | 2,165 | 303 | 1,135 | 3,753 | 2.5 |
| Woodford | 18.4 | D | 5.2 | 18.5 | 6.5 | 4.1 | 2.7 | D | 11.8 | 6,160 | 231 | 385 | 11,460 | 7.0 |
| LOUISIANA | 0.4 | 5.0 | 8.9 | 9.7 | 8.5 | 6.2 | 6.3 | 12.2 | 17.0 | 922,223 | 198 | 173,485 | 2,103,997 | 7.1 |
| Acadia | 1.9 | 4.5 | 11.1 | 8.1 | 5.3 | 7.6 | 5.5 | 12.6 | 17.2 | 12,535 | 202 | 2,551 | 26,621 | 4.8 |
| Allen | 0.3 | 3.8 | 1.7 | 11.9 | D | 4.5 | 1.6 | 8.0 | 51.8 | 5,110 | 200 | 742 | 10,068 | 3.5 |
| Ascension | 0.0 | D | 21.4 | 26.8 | 4.3 | 5.8 | 4.5 | 5.1 | 9.4 | 19,340 | 153 | 2,085 | 50,355 | 23.5 |
| Assumption | 1.5 | D | 14.5 | 14.1 | D | 4.9 | 8.2 | D | 17.0 | 5,165 | 236 | 910 | 11,005 | 6.3 |
| Avoyelles | 4.3 | D | 11.8 | 2.4 | 4.6 | 8.9 | 6.3 | D | 27.8 | 9,655 | 241 | 2,419 | 18,776 | 4.1 |
| Beauregard | -1.1 | 0.5 | 6.7 | 18.5 | D | 7.6 | 8.1 | D | 18.2 | 7,845 | 208 | 1,100 | 16,432 | 9.3 |
| Bienville | 0.4 | 12.3 | 1.1 | 20.6 | D | 4.1 | 5.2 | D | 22.2 | 3,530 | 267 | 783 | 7,967 | 3.2 |
| Bossier | -0.1 | 4.5 | 4.4 | 5.4 | 6.2 | 8.5 | 4.4 | 8.8 | 34.1 | 22,035 | 173 | 3,405 | 58,855 | 19.2 |
| Caddo | -0.1 | 6.7 | 4.6 | 6.5 | 6.6 | 6.8 | 5.6 | 21.0 | 18.0 | 52,370 | 218 | 13,383 | 113,632 | 1.5 |
| Calcasieu | -0.1 | 0.4 | 19.4 | 19.1 | 6.3 | 6.8 | 3.7 | 11.2 | 12.1 | 40,400 | 198 | 6,048 | 94,567 | 15.2 |
| Caldwell | 2.2 | 2.9 | 3.6 | 2.3 | D | 4.7 | D | 23.6 | 25.1 | 2,350 | 236 | 493 | 5,328 | 6.8 |
| Cameron | 0.1 | 0.1 | 75.2 | 8.5 | D | 0.4 | D | 0.6 | 4.9 | 1,325 | 190 | 78 | 4,361 | 21.2 |
| Catahoula | 7.6 | 4.7 | D | D | 4.7 | 7.5 | D | D | 24.8 | 2,425 | 256 | 510 | 5,068 | 3.9 |
| Claiborne | 6.3 | 20.6 | 3.3 | 3.8 | D | 3.9 | D | D | 29.8 | 3,500 | 223 | 788 | 7,926 | 2.1 |
| Concordia | 6.0 | 2.4 | 2.5 | 3.3 | D | 10.4 | 5.2 | D | 28.9 | 4,575 | 237 | 1,161 | 9,789 | 4.3 |
| De Soto | -1.1 | 22.2 | 5.7 | 18.1 | D | 6.1 | 2.5 | D | 21.6 | 6,505 | 237 | 1,270 | 13,485 | 9.7 |
| East Baton Rouge | 0.0 | 0.4 | 14.1 | 6.8 | 12.9 | 5.2 | 7.8 | 13.0 | 17.3 | 75,930 | 172 | 14,253 | 197,023 | 5.1 |
| East Carroll | 22.5 | D | 0.3 | 2.9 | D | 6.1 | 4.6 | D | 28.0 | 1,375 | 201 | 565 | 2,902 | 0.0 |
| East Feliciana | -1.7 | 1.0 | 4.7 | 8.0 | D | 3.7 | 4.4 | 11.0 | 54.5 | 4,325 | 226 | 789 | 8,702 | 8.7 |
| Evangeline | 3.9 | 1.7 | 4.8 | 11.9 | 3.0 | 8.4 | 5.1 | D | 19.5 | 7,525 | 225 | 2,041 | 15,233 | 3.9 |
| Franklin | 10.4 | 0.8 | 3.5 | 3.1 | D | 10.7 | 5.9 | D | 30.3 | 4,775 | 239 | 1,200 | 9,431 | 4.4 |
| Grant | -1.3 | D | 5.7 | 8.9 | 1.2 | 3.6 | D | 5.9 | 52.6 | 4,615 | 206 | 736 | 9,569 | 7.7 |
| Iberia | 0.4 | 16.8 | 6.6 | 14.4 | 3.5 | 7.0 | 9.2 | 7.4 | 14.2 | 15,785 | 227 | 3,199 | 30,843 | 4.4 |
| Iberville | 0.3 | 0.3 | 14.8 | 43.2 | 1.9 | 2.4 | 3.9 | D | 14.1 | 6,785 | 208 | 1,269 | 13,758 | 8.3 |
| Jackson | 2.5 | D | D | D | D | 6.0 | D | 8.0 | 29.9 | 3,815 | 242 | 702 | 7,991 | 4.1 |
| Jefferson | 0.0 | 0.8 | 7.7 | 4.8 | 8.4 | 7.6 | 10.0 | 16.3 | 9.7 | 89,855 | 207 | 12,784 | 188,728 | -0.2 |
| Jefferson Davis | 3.8 | 3.4 | 6.2 | 6.0 | 2.4 | 8.3 | 6.9 | 18.0 | 23.7 | 6,525 | 208 | 1,040 | 14,284 | 7.4 |
| Lafayette | 0.0 | 10.3 | 5.2 | 7.2 | 11.4 | 6.9 | 7.6 | 17.3 | 10.6 | 40,955 | 167 | 6,103 | 106,189 | 13.2 |
| Lafourche | 0.6 | 7.8 | 6.5 | 8.8 | 3.1 | 4.5 | 2.7 | 6.6 | 16.6 | 20,115 | 206 | 3,081 | 42,202 | 9.1 |
| La Salle | -0.8 | 9.2 | 4.4 | 10.7 | D | 5.8 | 3.2 | D | 30.8 | 3,205 | 215 | 493 | 6,807 | 3.8 |
| Lincoln | 2.1 | 1.7 | 6.3 | 7.7 | 7.8 | 7.5 | 8.4 | 13.1 | 23.4 | 7,355 | 157 | 1,527 | 20,872 | 7.1 |
| Livingston | 0.4 | 0.5 | 17.3 | 9.6 | D | 9.1 | 10.9 | 7.0 | 18.7 | 23,730 | 168 | 2,627 | 59,577 | 19.0 |
| Madison | 8.5 | D | 0.9 | 10.1 | D | 7.6 | 3.7 | D | 26.3 | 2,130 | 194 | 699 | 5,318 | 10.7 |
| Morehouse | 10.5 | D | 8.3 | 10.3 | 2.5 | 6.9 | 4.1 | D | 14.9 | 7,075 | 285 | 1,769 | 12,746 | 2.6 |
| Natchitoches | 1.5 | 0.8 | 3.9 | 27.6 | D | 5.5 | 4.3 | 5.9 | 27.7 | 8,075 | 212 | 2,042 | 19,460 | 4.7 |

1. Per 1,000 resident population estimated as of July 1 of the year shown.

## Table B. States and Counties — Housing, Labor Force, and Employment

| STATE County | Housing units, 2015-2019 | | | | | | | | Civilian labor force, 2020 | | | | Civilian employment[6], 2015-2019 | | |
|---|---|---|---|---|---|---|---|---|---|---|---|---|---|---|---|
| | Occupied units | | | | | | | Sub-standard units[4] (percent) | | Percent change, 2019-2020 | Unemployment | | | Percent | |
| | Owner-occupied | | | | | Renter-occupied | | | | | | | | | |
| | | | | Median owner cost as a percent of income | | | Median rent as a percent of income[2] | | | | | | | Management, business, science, and arts | Construction, production, and maintenance occupations |
| | Total | Percent | Median value[1] | With a mortgage | Without a mortgage[2] | Median rent[3] | | | Total | | Total | Rate[5] | Total | | |
| | 89 | 90 | 91 | 92 | 93 | 94 | 95 | 96 | 97 | 98 | 99 | 100 | 101 | 102 | 103 |
| **KENTUCKY—Cont'd** | | | | | | | | | | | | | | | |
| Ohio | 9,122 | 76.1 | 92,500 | 19.7 | 10.0 | 627 | 25.6 | 2.0 | 9,289 | -4.8 | 669 | 7.2 | 9,677 | 25.2 | 38.1 |
| Oldham | 20,911 | 84.7 | 282,100 | 18.8 | 10.0 | 919 | 23.8 | 1.8 | 32,438 | -3.4 | 1,571 | 4.8 | 31,732 | 46.4 | 17.3 |
| Owen | 3,995 | 74.8 | 116,100 | 19.8 | 11.1 | 624 | 30.1 | 3.7 | 5,082 | -1.1 | 325 | 6.4 | 4,488 | 30.7 | 38.9 |
| Owsley | 1,704 | 68.4 | 70,200 | 18.8 | 12.3 | 316 | 29.1 | 2.2 | 1,061 | -2.7 | 79 | 7.4 | 1,262 | 39.4 | 21.0 |
| Pendleton | 5,253 | 75.3 | 116,400 | 21.8 | 11.2 | 756 | 30.8 | 2.7 | 6,803 | -3.4 | 356 | 5.2 | 6,497 | 26.0 | 37.6 |
| Perry | 11,226 | 75.9 | 75,400 | 19.1 | 12.1 | 649 | 23.7 | 3.4 | 8,181 | -1.9 | 728 | 8.9 | 8,855 | 31.8 | 24.6 |
| Pike | 25,702 | 72.9 | 78,600 | 22.8 | 11.0 | 665 | 30.9 | 1.9 | 19,196 | -4.3 | 1,622 | 8.4 | 20,047 | 32.8 | 27.7 |
| Powell | 4,770 | 69.0 | 89,800 | 17.1 | 12.1 | 669 | 28.9 | 4.3 | 5,070 | 0.5 | 372 | 7.3 | 4,936 | 25.5 | 36.8 |
| Pulaski | 25,301 | 69.0 | 116,200 | 19.7 | 10.5 | 682 | 29.7 | 1.7 | 25,795 | -3.3 | 1,888 | 7.3 | 25,852 | 29.9 | 30.4 |
| Robertson | 861 | 74.9 | 114,700 | 22.0 | 12.0 | 556 | 27.0 | 4.2 | 788 | -2.8 | 48 | 6.1 | 821 | 28.4 | 36.9 |
| Rockcastle | 6,544 | 76.2 | 81,900 | 18.7 | 11.7 | 575 | 28.1 | 3.4 | 6,428 | -3.5 | 457 | 7.1 | 6,124 | 29.5 | 36.6 |
| Rowan | 8,623 | 59.8 | 124,200 | 20.7 | 10.1 | 652 | 31.0 | 2.3 | 9,932 | -4.6 | 741 | 7.5 | 10,155 | 36.3 | 23.8 |
| Russell | 6,922 | 74.1 | 95,000 | 20.2 | 10.4 | 549 | 28.1 | 3.3 | 6,085 | -2.9 | 499 | 8.2 | 6,759 | 30.6 | 34.9 |
| Scott | 20,551 | 70.0 | 189,500 | 17.8 | 10.0 | 900 | 24.8 | 3.0 | 29,378 | -1.5 | 1,855 | 6.3 | 28,293 | 34.9 | 31.4 |
| Shelby | 16,624 | 70.2 | 197,300 | 19.0 | 10.0 | 849 | 26.1 | 3.2 | 25,191 | -2.6 | 1,374 | 5.5 | 23,414 | 35.9 | 29.7 |
| Simpson | 7,028 | 65.0 | 135,200 | 19.9 | 11.2 | 804 | 26.0 | 1.3 | 8,596 | -3.6 | 617 | 7.2 | 8,082 | 23.2 | 36.1 |
| Spencer | 6,763 | 85.1 | 213,300 | 18.5 | 10.0 | 649 | 22.9 | 2.8 | 10,254 | -2.7 | 592 | 5.8 | 9,737 | 35.6 | 31.7 |
| Taylor | 9,857 | 60.3 | 120,200 | 17.2 | 10.0 | 643 | 30.2 | 0.5 | 12,935 | 0.9 | 775 | 6.0 | 11,199 | 26.9 | 31.7 |
| Todd | 4,644 | 67.5 | 106,300 | 18.7 | 11.9 | 637 | 21.3 | 4.5 | 5,391 | -0.8 | 260 | 4.8 | 4,702 | 29.5 | 40.2 |
| Trigg | 5,892 | 81.0 | 129,400 | 19.5 | 10.0 | 564 | 24.5 | 1.4 | 6,030 | 0.7 | 419 | 6.9 | 5,685 | 32.0 | 34.2 |
| Trimble | 3,498 | 77.1 | 124,000 | 19.8 | 11.5 | 715 | 25.5 | 4.0 | 3,789 | -2.2 | 256 | 6.8 | 3,814 | 27.9 | 35.8 |
| Union | 5,374 | 70.8 | 92,100 | 16.6 | 10.0 | 597 | 25.1 | 0.7 | 5,939 | -2.5 | 377 | 6.3 | 6,064 | 22.4 | 39.9 |
| Warren | 48,224 | 58.7 | 167,600 | 19.2 | 10.0 | 781 | 27.8 | 3.3 | 63,256 | -3.8 | 4,148 | 6.6 | 64,595 | 33.8 | 26.7 |
| Washington | 4,581 | 75.0 | 128,300 | 18.1 | 10.0 | 608 | 19.5 | 1.9 | 6,099 | -2.4 | 381 | 6.2 | 5,775 | 30.1 | 39.4 |
| Wayne | 8,231 | 73.5 | 88,800 | 22.6 | 10.0 | 538 | 25.9 | 2.8 | 7,370 | -1.1 | 482 | 6.5 | 7,377 | 22.7 | 34.5 |
| Webster | 4,953 | 71.8 | 75,600 | 18.9 | 11.7 | 631 | 23.4 | 2.5 | 5,424 | -5.7 | 347 | 6.4 | 4,935 | 24.7 | 45.0 |
| Whitley | 12,620 | 68.8 | 92,500 | 17.9 | 10.0 | 604 | 26.2 | 2.6 | 13,784 | -0.3 | 983 | 7.1 | 12,811 | 33.1 | 26.8 |
| Wolfe | 2,898 | 63.4 | 64,400 | 22.9 | 13.5 | 497 | 36.5 | 2.2 | 2,059 | 4.7 | 158 | 7.7 | 1,774 | 16.2 | 33.7 |
| Woodford | 10,355 | 70.7 | 203,000 | 18.8 | 10.0 | 771 | 28.2 | 1.8 | 14,929 | | 747 | 5.0 | 12,870 | 41.7 | 23.8 |
| **LOUISIANA** | 1,739,497 | 65.6 | 163,100 | 19.6 | 10.0 | 866 | 32.5 | 2.7 | 2,076,643 | -2.0 | 171,405 | 8.3 | 2,033,758 | 34.1 | 24.7 |
| Acadia | 22,236 | 71.1 | 118,000 | 17.6 | 10.0 | 662 | 35.5 | 2.5 | 23,463 | -2.7 | 1,705 | 7.3 | 25,238 | 28.1 | 32.9 |
| Allen | 7,925 | 74.5 | 97,500 | 17.5 | 10.0 | 601 | 24.0 | 4.6 | 8,460 | -4.8 | 778 | 9.2 | 8,686 | 25.3 | 31.2 |
| Ascension | 43,032 | 81.5 | 207,400 | 17.3 | 10.0 | 982 | 28.0 | 2.5 | 64,136 | -2.2 | 4,329 | 6.7 | 59,423 | 36.1 | 26.2 |
| Assumption | 8,552 | 78.1 | 118,600 | 17.2 | 10.0 | 797 | 35.8 | 2.6 | 8,822 | -1.6 | 893 | 10.1 | 9,265 | 26.7 | 36.6 |
| Avoyelles | 15,163 | 69.5 | 104,200 | 19.1 | 10.0 | 663 | 33.8 | 3.0 | 15,048 | -1.7 | 1,202 | 8.0 | 14,953 | 30.1 | 27.0 |
| Beauregard | 13,520 | 77.5 | 127,500 | 17.4 | 10.0 | 741 | 26.1 | 3.0 | 14,405 | -2.4 | 961 | 6.7 | 14,342 | 29.8 | 34.8 |
| Bienville | 5,812 | 73.1 | 75,900 | 20.8 | 10.0 | 487 | 31.4 | 4.4 | 5,417 | -2.6 | 418 | 7.7 | 4,661 | 25.9 | 32.9 |
| Bossier | 49,377 | 63.4 | 172,400 | 20.2 | 10.0 | 992 | 30.6 | 1.9 | 56,750 | -3.0 | 3,612 | 6.4 | 54,094 | 36.1 | 21.6 |
| Caddo | 95,864 | 59.8 | 149,200 | 21.1 | 10.0 | 797 | 36.6 | 2.0 | 102,615 | -1.2 | 9,231 | 9.0 | 103,071 | 32.9 | 22.8 |
| Calcasieu | 77,780 | 67.5 | 156,800 | 18.1 | 10.0 | 832 | 29.5 | 2.8 | 98,110 | -8.7 | 8,874 | 9.0 | 89,962 | 31.0 | 27.4 |
| Caldwell | 3,665 | 76.5 | 77,300 | 18.1 | 10.7 | 611 | 21.8 | 3.8 | 3,670 | -3.7 | 267 | 7.3 | 3,355 | 26.9 | 31.2 |
| Cameron | 2,734 | 84.4 | 119,500 | 15.0 | 10.0 | 905 | 29.3 | 3.8 | 3,520 | -11.6 | 202 | 5.7 | 3,148 | 29.1 | 31.9 |
| Catahoula | 3,364 | 80.2 | 89,000 | 18.8 | 10.0 | 443 | 37.9 | 3.5 | 3,495 | -2.3 | 247 | 7.1 | 3,386 | 35.4 | 27.9 |
| Claiborne | 5,917 | 65.9 | 78,100 | 27.6 | 10.9 | 618 | 44.9 | 1.5 | 5,495 | -6.5 | 362 | 6.6 | 5,193 | 23.3 | 39.1 |
| Concordia | 7,162 | 66.1 | 80,600 | 20.4 | 11.4 | 636 | 35.8 | 2.2 | 6,949 | -1.4 | 564 | 8.1 | 6,541 | 26.7 | 24.7 |
| De Soto | 10,821 | 71.3 | 117,300 | 17.8 | 10.0 | 635 | 28.5 | 3.0 | 10,623 | -3.7 | 771 | 7.3 | 10,811 | 28.5 | 31.9 |
| East Baton Rouge | 164,346 | 59.8 | 194,000 | 19.6 | 10.0 | 933 | 32.5 | 2.9 | 228,974 | -1.4 | 17,777 | 7.8 | 217,043 | 39.0 | 20.3 |
| East Carroll | 2,037 | 51.6 | 76,800 | 20.4 | 11.9 | 485 | 34.1 | 6.8 | 1,824 | -4.3 | 214 | 11.7 | 1,359 | 34.4 | 17.8 |
| East Feliciana | 6,959 | 80.5 | 139,300 | 19.2 | 10.0 | 647 | 25.3 | 6.1 | 7,762 | -2.7 | 518 | 6.7 | 6,502 | 27.5 | 27.7 |
| Evangeline | 12,172 | 65.5 | 96,600 | 18.9 | 10.8 | 577 | 30.1 | 3.4 | 11,933 | -0.7 | 920 | 7.7 | 11,740 | 26.4 | 32.7 |
| Franklin | 7,423 | 72.7 | 88,000 | 19.2 | 10.0 | 587 | 40.1 | 3.7 | 7,380 | -0.8 | 597 | 8.1 | 6,796 | 28.7 | 30.8 |
| Grant | 6,989 | 68.7 | 103,400 | 19.3 | 10.0 | 714 | 28.6 | 2.1 | 8,168 | -1.2 | 533 | 6.5 | 7,499 | 22.8 | 34.4 |
| Iberia | 26,184 | 67.9 | 115,600 | 19.5 | 10.0 | 768 | 30.8 | 3.7 | 27,907 | -0.5 | 2,623 | 9.4 | 30,188 | 26.9 | 32.0 |
| Iberville | 10,903 | 73.4 | 143,700 | 19.3 | 10.0 | 755 | 30.4 | 2.8 | 13,975 | -2.1 | 1,374 | 9.8 | 12,795 | 30.7 | 25.9 |
| Jackson | 5,971 | 71.0 | 89,500 | 17.7 | 10.0 | 539 | 31.2 | 2.3 | 6,738 | -2.0 | 360 | 5.3 | 5,559 | 30.3 | 31.1 |
| Jefferson | 169,452 | 61.2 | 188,200 | 21.5 | 10.0 | 972 | 31.8 | 2.9 | 211,563 | -1.9 | 19,629 | 9.3 | 210,542 | 34.0 | 23.6 |
| Jefferson Davis | 11,726 | 72.9 | 114,000 | 16.4 | 10.0 | 635 | 32.1 | 4.3 | 12,827 | -4.4 | 964 | 7.5 | 12,292 | 27.5 | 31.8 |
| Lafayette | 91,543 | 64.8 | 185,300 | 18.8 | 10.0 | 874 | 28.7 | 2.6 | 113,811 | -1.8 | 8,069 | 7.1 | 119,065 | 37.4 | 21.4 |
| Lafourche | 36,895 | 76.5 | 158,000 | 18.6 | 10.0 | 793 | 27.8 | 2.1 | 41,475 | -2.1 | 2,689 | 6.5 | 43,281 | 29.5 | 33.1 |
| La Salle | 4,814 | 76.6 | 93,000 | 17.6 | 10.0 | 656 | 34.7 | 4.9 | 6,912 | 1.8 | 328 | 4.7 | 5,271 | 28.8 | 35.3 |
| Lincoln | 17,712 | 51.5 | 154,500 | 17.8 | 10.0 | 729 | 37.4 | 3.4 | 20,927 | -5.5 | 1,314 | 6.3 | 20,796 | 35.3 | 19.0 |
| Livingston | 48,410 | 81.8 | 167,100 | 18.5 | 10.0 | 934 | 24.7 | 2.7 | 68,265 | -2.5 | 4,391 | 6.4 | 63,662 | 31.9 | 29.0 |
| Madison | 3,832 | 53.2 | 78,100 | 21.6 | 10.0 | 616 | 32.8 | 6.6 | 3,573 | 0.3 | 301 | 8.4 | 3,414 | 26.2 | 25.0 |
| Morehouse | 9,732 | 66.8 | 91,100 | 19.9 | 10.0 | 562 | 35.7 | 2.5 | 10,096 | 0.4 | 994 | 9.8 | 9,217 | 23.6 | 33.4 |
| Natchitoches | 14,659 | 43.8 | 137,400 | 20.7 | 10.0 | 713 | 38.1 | 1.3 | 16,129 | -2.3 | 1,101 | 6.8 | 12,171 | 30.1 | 32.7 |

1. Specified owner-occupied units. lacking complete plumbing facilities.   2. A value of 10.0 represents 10 percent or less; a value of 50.0 represents 50 percent or more.   3. Specified renter-occupied units.   4. Overcrowded or
5. Percent of civilian labor force.   6. Civilian employed persons 16 years old and over.

## Table B. States and Counties — Nonfarm Employment and Agriculture

| STATE County | Private nonfarm establishments, employment and payroll, 2019 | | | | | | | | | Agriculture, 2017 | | | |
| | Number of establishments | Employment | | | | | | Annual payroll | | Farms | | | Farm producers whose primary occupation is farming (percent) |
| | | Total | Health care and social assistance | Manufacturing | Retail trade | Finance and insurance | Professional, scientific, and technical services | Total (mil dol) | Average per employee (dollars) | Number | Percent with: | | |
| | | | | | | | | | | | Fewer than 50 acres | 1000 acres or more | |
| | 104 | 105 | 106 | 107 | 108 | 109 | 110 | 111 | 112 | 113 | 114 | 115 | 116 |
| **KENTUCKY—Cont'd** | | | | | | | | | | | | | |
| Ohio | 361 | 7,043 | 1,112 | 2,149 | 640 | 126 | 107 | 285 | 40,434 | 813 | 35.1 | 2.8 | 39.3 |
| Oldham | 1,262 | 13,309 | 2,784 | 960 | 1,595 | 1,806 | 594 | 588 | 44,196 | 466 | 54.3 | 1.5 | 28.0 |
| Owen | 116 | 927 | 125 | 31 | 156 | 42 | 12 | 31 | 33,806 | 821 | 28.4 | 2.9 | 33.7 |
| Owsley | 43 | 413 | 241 | NA | 102 | NA | NA | 11 | 25,838 | 153 | 20.9 | 2.6 | 35.3 |
| Pendleton | 169 | 1,718 | 216 | 580 | 184 | 49 | 50 | 70 | 40,838 | 919 | 33.7 | 0.3 | 29.2 |
| Perry | 556 | 8,867 | 2,951 | 91 | 1,674 | 301 | 397 | 353 | 39,826 | 63 | 31.7 | 9.5 | 34.5 |
| Pike | 1,184 | 17,705 | 4,698 | 678 | 3,796 | 793 | 580 | 761 | 42,975 | 85 | 23.5 | 2.4 | 51.0 |
| Powell | 165 | 1,621 | 206 | 201 | 374 | 66 | 10 | 48 | 29,496 | 194 | 36.1 | 1.5 | 29.5 |
| Pulaski | 1,406 | 21,246 | 5,060 | 4,213 | 3,667 | 715 | 406 | 781 | 36,764 | 1,704 | 41.4 | 0.8 | 37.8 |
| Robertson | 19 | 172 | NA | NA | 18 | NA | NA | 4 | 23,384 | 245 | 20.4 | NA | 35.8 |
| Rockcastle | 194 | 2,912 | 1,112 | 116 | 264 | 87 | 159 | 93 | 32,066 | 681 | 37.7 | 0.9 | 26.8 |
| Rowan | 490 | 8,155 | 2,225 | 1,106 | 1,367 | 218 | 67 | 280 | 34,350 | 329 | 45.3 | 0.3 | 34.7 |
| Russell | 335 | 4,752 | 921 | 1,141 | 722 | 138 | 51 | 150 | 31,558 | 672 | 45.5 | 1.9 | 37.7 |
| Scott | 943 | 23,252 | 1,723 | 11,540 | 1,851 | 291 | 423 | 1,211 | 52,089 | 851 | 47.5 | 2.1 | 38.1 |
| Shelby | 1,051 | 18,935 | 1,323 | 3,993 | 2,764 | 368 | 350 | 692 | 36,537 | 1,548 | 55.1 | 2.3 | 37.3 |
| Simpson | 386 | 8,526 | 469 | 3,716 | 1,071 | 138 | 80 | 362 | 42,433 | 471 | 53.1 | 6.6 | 40.9 |
| Spencer | 226 | 1,328 | 236 | 20 | 345 | 49 | 33 | 35 | 26,681 | 606 | 49.3 | 1.0 | 34.3 |
| Taylor | 628 | 10,461 | 1,635 | 1,365 | 2,458 | 241 | 120 | 325 | 31,104 | 803 | 47.3 | 1.9 | 34.1 |
| Todd | 201 | 1,457 | 196 | 307 | 348 | 77 | 43 | 49 | 33,351 | 593 | 27.7 | 6.6 | 49.4 |
| Trigg | 226 | 2,306 | 375 | 560 | 403 | 87 | 45 | 65 | 28,019 | 405 | 34.1 | 4.7 | 40.1 |
| Trimble | 80 | 786 | 110 | 114 | 83 | 56 | 8 | 43 | 54,866 | 469 | 43.3 | 2.1 | 37.5 |
| Union | 256 | 4,039 | 842 | 495 | 520 | 98 | 54 | 188 | 46,528 | 284 | 41.2 | 16.2 | 45.3 |
| Warren | 2,864 | 56,153 | 8,542 | 9,803 | 7,859 | 1,414 | 3,222 | 2,403 | 42,799 | 1,755 | 49.9 | 2.1 | 30.9 |
| Washington | 240 | 3,002 | 589 | 1,116 | 305 | 78 | 43 | 118 | 39,352 | 1,102 | 34.1 | 0.5 | 33.4 |
| Wayne | 263 | 4,270 | 713 | 1,362 | 652 | 161 | 38 | 118 | 27,660 | 710 | 39.2 | 1.1 | 39.1 |
| Webster | 201 | 1,727 | 256 | 275 | 288 | 84 | 52 | 69 | 39,928 | 499 | 33.9 | 8.2 | 34.5 |
| Whitley | 593 | 11,230 | 2,682 | 983 | 1,419 | 241 | 819 | 401 | 35,728 | 548 | 45.1 | 0.5 | 33.1 |
| Wolfe | 89 | 705 | 265 | 69 | 130 | 17 | 5 | 21 | 29,237 | 294 | 32.3 | 0.3 | 33.8 |
| Woodford | 566 | 8,234 | 683 | 2,681 | 893 | 176 | 515 | 368 | 44,707 | 689 | 44.8 | 2.8 | 46.3 |
| **LOUISIANA** | 106,302 | 1,719,561 | 305,357 | 121,412 | 223,830 | 63,847 | 98,059 | 82,452 | 47,950 | 27,386 | 46.5 | 7.6 | 37.7 |
| Acadia | 1,120 | 12,391 | 2,243 | 1,237 | 2,285 | 406 | 315 | 438 | 35,363 | 964 | 50.0 | 8.8 | 40.2 |
| Allen | 303 | 3,415 | 769 | 785 | 639 | 108 | 60 | 126 | 36,978 | 420 | 47.4 | 5.2 | 37.0 |
| Ascension | 2,397 | 40,126 | 3,573 | 5,890 | 6,203 | 976 | 1,862 | 2,342 | 58,362 | 221 | 69.7 | 6.8 | 31.5 |
| Assumption | 243 | 2,718 | 279 | 457 | 463 | 206 | 75 | 121 | 44,401 | 103 | 35.9 | 36.9 | 48.4 |
| Avoyelles | 684 | 8,058 | 2,087 | 287 | 1,396 | 406 | 227 | 261 | 32,398 | 770 | 40.1 | 8.6 | 38.0 |
| Beauregard | 618 | 6,743 | 1,227 | 1,022 | 1,223 | 670 | 137 | 273 | 40,478 | 730 | 40.8 | 3.4 | 31.2 |
| Bienville | 220 | 3,268 | 529 | 983 | 268 | 137 | 14 | 146 | 44,797 | 221 | 41.6 | 1.8 | 41.2 |
| Bossier | 2,537 | 36,668 | 4,920 | 1,721 | 6,638 | 1,124 | 1,475 | 1,342 | 36,599 | 519 | 54.5 | 5.6 | 44.2 |
| Caddo | 6,142 | 101,831 | 29,905 | 5,444 | 12,674 | 2,810 | 4,455 | 4,408 | 43,292 | 706 | 54.1 | 8.9 | 34.7 |
| Calcasieu | 4,715 | 81,292 | 13,857 | 9,199 | 10,992 | 2,122 | 5,329 | 3,857 | 47,446 | 931 | 50.2 | 7.0 | 28.2 |
| Caldwell | 172 | 1,669 | 665 | 27 | 290 | 106 | 89 | 55 | 33,138 | 329 | 42.9 | 4.0 | 30.4 |
| Cameron | 132 | 2,115 | NA | NA | 112 | 18 | 167 | 227 | 107,521 | 294 | 24.8 | 17.0 | 31.5 |
| Catahoula | 169 | 1,437 | 396 | NA | 275 | 96 | 88 | 45 | 31,561 | 433 | 26.6 | 14.3 | 33.6 |
| Claiborne | 217 | 2,364 | 763 | 124 | 343 | 74 | 29 | 90 | 38,202 | 255 | 32.5 | 6.3 | 37.0 |
| Concordia | 347 | 3,805 | 998 | 142 | 767 | 222 | 87 | 134 | 35,127 | 371 | 24.0 | 22.6 | 36.3 |
| De Soto | 440 | 4,614 | 652 | 326 | 857 | 166 | 116 | 241 | 52,291 | 599 | 41.1 | 5.8 | 41.7 |
| East Baton Rouge | 12,271 | 251,568 | 37,518 | 10,514 | 26,619 | 11,700 | 20,477 | 13,172 | 52,358 | 449 | 58.4 | 1.3 | 40.5 |
| East Carroll | 113 | 1,165 | 290 | 58 | 101 | 24 | 19 | 45 | 38,591 | 209 | 18.2 | 35.9 | 62.4 |
| East Feliciana | 243 | 3,448 | 1,596 | 287 | 363 | 107 | 89 | 140 | 40,459 | 412 | 50.5 | 9.2 | 38.3 |
| Evangeline | 518 | 6,558 | 2,694 | 999 | 960 | 207 | 112 | 215 | 32,763 | 596 | 45.1 | 7.7 | 38.0 |
| Franklin | 396 | 3,671 | 1,059 | 223 | 754 | 233 | 109 | 96 | 26,206 | 797 | 31.0 | 8.7 | 36.7 |
| Grant | 170 | 1,808 | 272 | 382 | 288 | 58 | 7 | 63 | 34,755 | 207 | 35.7 | 8.2 | 28.4 |
| Iberia | 1,541 | 23,037 | 3,358 | 3,114 | 2,990 | 767 | 805 | 1,170 | 50,769 | 334 | 64.1 | 10.2 | 40.5 |
| Iberville | 554 | 11,966 | 823 | 4,116 | 1,129 | 296 | 401 | 827 | 69,085 | 151 | 31.8 | 27.2 | 55.6 |
| Jackson | 240 | 2,538 | 632 | 494 | 502 | 105 | 47 | 112 | 44,163 | 191 | 46.1 | 0.5 | 46.0 |
| Jefferson | 11,757 | 180,453 | 32,450 | 6,460 | 28,062 | 9,023 | 10,474 | 8,899 | 49,314 | 52 | 59.6 | 1.9 | 21.1 |
| Jefferson Davis | 613 | 6,259 | 1,429 | 410 | 1,356 | 253 | 372 | 240 | 38,315 | 703 | 37.8 | 12.4 | 37.7 |
| Lafayette | 8,616 | 130,868 | 26,587 | 7,262 | 17,246 | 4,037 | 8,758 | 6,467 | 49,416 | 549 | 73.4 | 2.9 | 33.3 |
| Lafourche | 1,700 | 26,367 | 4,311 | 1,829 | 3,884 | 789 | 864 | 1,381 | 52,359 | 379 | 53.0 | 6.9 | 27.1 |
| La Salle | 360 | 3,581 | 836 | 209 | 455 | 141 | 175 | 147 | 41,081 | 235 | 48.5 | 0.4 | 32.7 |
| Lincoln | 1,093 | 15,547 | 3,181 | 1,300 | 2,480 | 620 | 791 | 566 | 36,390 | 378 | 39.9 | 1.6 | 37.7 |
| Livingston | 1,850 | 22,334 | 2,267 | 1,619 | 5,045 | 700 | 1,143 | 865 | 38,730 | 436 | 74.5 | 0.2 | 38.0 |
| Madison | 198 | 2,501 | 1,044 | 175 | 331 | 68 | 16 | 81 | 32,332 | 250 | 16.4 | 32.4 | 50.2 |
| Morehouse | 435 | 5,517 | 1,948 | 612 | 828 | 292 | 84 | 177 | 32,097 | 409 | 27.1 | 18.6 | 49.0 |
| Natchitoches | 809 | 11,655 | 1,925 | 2,662 | 1,762 | 439 | 219 | 428 | 36,740 | 627 | 35.2 | 7.5 | 38.8 |

# Table B. States and Counties — **Agriculture**

| STATE County | Acreage (1,000) 117 | Percent change, 2012-2017 118 | Average size of farm 119 | Total irrigated (1,000) 120 | Total cropland (1,000) 121 | Average per farm 122 | Average per acre 123 | Value of machinery and equiopmnet, average per farm (dollars) 124 | Total (mil dol) 125 | Average per farm (acres) 126 | Crops 127 | Livestock and poultry products 128 | Organic farms (number) 129 | Farms with internet access (percent) 130 | Total ($1,000) 131 | Percent of farms 132 |
|---|---|---|---|---|---|---|---|---|---|---|---|---|---|---|---|---|
| KENTUCKY—Cont'd | | | | | | | | | | | | | | | | |
| Ohio | 158 | -0.4 | 194 | D | 86.4 | 662,879 | 3,418 | 96,947 | 135.8 | 166,998 | 27.8 | 72.2 | 3 | 67.2 | 1,810 | 29.5 |
| Oldham | 51 | -14.7 | 110 | 0.1 | 21.3 | 879,589 | 7,964 | 66,919 | 18.5 | 39,730 | 46.2 | 53.8 | 1 | 80.3 | 252 | 4.7 |
| Owen | 157 | 19.3 | 192 | 0.5 | 61.7 | 542,771 | 2,831 | 72,383 | 23.9 | 29,107 | 54.9 | 45.1 | NA | 69.7 | 158 | 5.7 |
| Owsley | 29 | 5.7 | 190 | D | 10.3 | 255,706 | 1,342 | 37,727 | 1.3 | 8,458 | 64.0 | 36.0 | NA | 74.5 | 41 | 5.2 |
| Pendleton | 111 | 9.9 | 121 | 0.5 | 38.2 | 344,033 | 2,840 | 55,326 | 9.4 | 10,282 | 64.2 | 35.8 | NA | 71.3 | 112 | 8.7 |
| Perry | 16 | 49.3 | 260 | 0.0 | 5.6 | 297,441 | 1,146 | 49,250 | 0.3 | 5,175 | 27.6 | 72.4 | NA | 77.8 | D | 1.6 |
| Pike | 19 | 45.5 | 229 | 0.0 | 3.4 | 324,062 | 1,414 | 56,010 | 0.7 | 8,753 | 49.6 | 50.4 | NA | 85.9 | NA | NA |
| Powell | 28 | -6.8 | 144 | 0.0 | 9.6 | 336,855 | 2,334 | 46,794 | 2.7 | 13,938 | 73.4 | 26.6 | NA | 64.9 | 32 | 8.2 |
| Pulaski | 226 | -0.9 | 133 | 0.2 | 100.3 | 429,851 | 3,238 | 67,220 | 55.5 | 32,570 | 45.2 | 54.8 | 2 | 69.2 | 632 | 21.0 |
| Robertson | 42 | 8.8 | 172 | 0.0 | 14.5 | 377,155 | 2,187 | 51,275 | 3.5 | 14,473 | 31.5 | 68.5 | 3 | 64.1 | 22 | 5.3 |
| Rockcastle | 81 | -11.2 | 119 | 0.0 | 27.8 | 258,213 | 2,179 | 41,923 | 7.3 | 10,742 | 39.5 | 60.5 | NA | 71.5 | 215 | 13.1 |
| Rowan | 40 | -6.4 | 120 | 0.0 | 13.1 | 333,138 | 2,770 | 56,262 | 4.1 | 12,605 | 40.1 | 59.9 | NA | 66.9 | 126 | 17.6 |
| Russell | 92 | 3.5 | 138 | 0.0 | 42.2 | 425,730 | 3,093 | 68,898 | 37.5 | 55,823 | 22.0 | 78.0 | NA | 69.8 | 456 | 20.7 |
| Scott | 131 | 2.4 | 153 | 0.3 | 56.1 | 858,492 | 5,595 | 78,301 | 51.7 | 60,734 | 30.2 | 69.8 | 1 | 78.7 | 224 | 6.8 |
| Shelby | 201 | 0.8 | 130 | 0.9 | 124.4 | 753,848 | 5,809 | 85,413 | 72.1 | 46,577 | 67.6 | 32.4 | 5 | 77.1 | 1,603 | 14.6 |
| Simpson | 111 | 9.2 | 235 | 0.3 | 88.3 | 1,335,785 | 5,675 | 146,265 | 79.0 | 167,658 | 64.5 | 35.5 | NA | 77.5 | 1,628 | 39.9 |
| Spencer | 74 | 7.3 | 122 | 0.1 | 34.8 | 556,079 | 4,542 | 62,736 | 18.3 | 30,163 | 75.1 | 24.9 | NA | 75.1 | 255 | 7.4 |
| Taylor | 112 | -2.5 | 139 | 0.0 | 61.1 | 422,767 | 3,038 | 82,616 | 46.1 | 57,365 | 49.3 | 50.7 | 2 | 71.9 | 1,440 | 35.4 |
| Todd | 168 | -7.2 | 283 | 0.8 | 127.5 | 1,468,492 | 5,185 | 159,843 | 175.6 | 296,152 | 47.2 | 52.8 | 13 | 50.3 | 2,136 | 43.2 |
| Trigg | 122 | -5.3 | 302 | 2.2 | 75.9 | 1,239,797 | 4,110 | 130,967 | 58.8 | 145,183 | 74.6 | 25.4 | 5 | 60.2 | 2,747 | 35.8 |
| Trimble | 66 | 18.6 | 141 | D | 30.4 | 456,362 | 3,245 | 57,517 | 12.1 | 25,823 | 80.0 | 20.0 | NA | 65.9 | 356 | 8.3 |
| Union | 195 | -0.1 | 685 | 3.8 | 172.8 | 3,453,444 | 5,039 | 337,598 | 108.9 | 383,285 | 92.5 | 7.5 | NA | 79.2 | 4,606 | 53.9 |
| Warren | 262 | 6.3 | 149 | 0.2 | 156.6 | 796,970 | 5,334 | 85,047 | 110.9 | 63,189 | 54.5 | 45.5 | NA | 79.2 | 5,481 | 21.8 |
| Washington | 161 | 14.5 | 147 | 0.1 | 67.5 | 470,574 | 3,212 | 62,124 | 34.4 | 31,208 | 34.9 | 65.1 | 2 | 75.6 | 459 | 18.2 |
| Wayne | 96 | -25.3 | 135 | 0.1 | 34.5 | 360,738 | 2,666 | 63,632 | 58.0 | 81,655 | 17.2 | 82.8 | NA | 68.1 | 371 | 15.1 |
| Webster | 163 | 6.9 | 327 | 0.6 | 123.0 | 1,202,901 | 3,683 | 145,475 | 141.9 | 284,333 | 37.5 | 62.5 | NA | 74.1 | 4,513 | 64.5 |
| Whitley | 59 | 1.0 | 108 | 0.0 | 20.2 | 272,502 | 2,532 | 52,085 | 6.1 | 11,057 | 34.2 | 65.8 | 1 | 73.9 | 206 | 13.7 |
| Wolfe | 34 | -20.6 | 115 | 0.0 | 10.2 | 231,135 | 2,016 | 53,278 | 1.5 | 5,044 | 56.4 | 43.6 | NA | 70.4 | 17 | 9.2 |
| Woodford | 112 | 0.2 | 163 | 0.0 | 42.0 | 1,320,705 | 8,111 | 92,879 | 132.6 | 192,441 | 9.7 | 90.3 | 2 | 88.2 | 403 | 9.3 |
| LOUISIANA | 7,998 | 1.2 | 292 | 1,235.8 | 4,345.8 | 889,146 | 3,045 | 121,758 | 3,173 | 115,861 | 65.0 | 35.0 | 29 | 69.9 | 177,399 | 28.4 |
| Acadia | 266 | 11.5 | 275 | 86.1 | 207.9 | 760,885 | 2,762 | 166,113 | 100.7 | 104,466 | 84.7 | 15.3 | NA | 62.1 | 13,935 | 59.1 |
| Allen | 95 | 7.9 | 226 | 18.9 | 46.4 | 580,390 | 2,566 | 68,258 | 19.8 | 47,055 | 71.2 | 28.8 | NA | 64.8 | 3,122 | 31.2 |
| Ascension | 38 | -23.9 | 174 | 0.0 | 26.3 | 629,741 | 3,626 | 104,541 | 13.1 | 59,217 | 89.4 | 10.6 | 3 | 72.4 | 198 | 10.0 |
| Assumption | 92 | 47.8 | 892 | D | 86.5 | 2,626,865 | 2,944 | 469,713 | 45.6 | 442,806 | 98.8 | 1.2 | NA | 83.5 | 427 | 27.2 |
| Avoyelles | 282 | -5.7 | 366 | 26.0 | 201.1 | 931,760 | 2,544 | 158,024 | 104.3 | 135,473 | 89.9 | 10.1 | NA | 69.5 | 7,776 | 39.0 |
| Beauregard | 148 | 0.9 | 202 | 3.3 | 39.1 | 614,319 | 3,039 | 69,645 | 15.8 | 21,705 | 45.4 | 54.6 | 1 | 76.7 | 2,855 | 12.9 |
| Bienville | 34 | -39 | 154 | D | 10.0 | 392,811 | 2,547 | 59,842 | 32.3 | 146,068 | 2.6 | 97.4 | NA | 69.7 | 116 | 7.7 |
| Bossier | 107 | 31.4 | 206 | 2.0 | 34.0 | 642,592 | 3,122 | 73,951 | 14.4 | 27,825 | 65.7 | 34.3 | NA | 76.7 | 398 | 13.7 |
| Caddo | 199 | 42.5 | 282 | 22.6 | 84.7 | 841,843 | 2,983 | 110,644 | 61.1 | 86,482 | 62.7 | 37.3 | NA | 71.2 | 2,145 | 10.8 |
| Calcasieu | 354 | 4.8 | 380 | 13.1 | 77.5 | 1,323,640 | 3,480 | 77,967 | 25.8 | 27,676 | 51.6 | 48.4 | NA | 75.8 | 3,924 | 20.4 |
| Caldwell | 68 | 8.8 | 206 | 3.6 | 21.8 | 515,954 | 2,507 | 60,462 | 18.7 | 56,714 | 38.6 | 61.4 | NA | 60.2 | 1,822 | 36.2 |
| Cameron | 188 | -20.1 | 639 | 5.9 | 50.9 | 1,584,171 | 2,478 | 87,142 | 12.0 | 40,667 | 46.1 | 53.9 | NA | 72.4 | 1,266 | 28.9 |
| Catahoula | 210 | -6.3 | 485 | 58.5 | 145.7 | 1,309,375 | 2,702 | 184,305 | 70.6 | 163,046 | 97.6 | 2.4 | NA | 62.8 | 6,969 | 66.1 |
| Claiborne | 63 | 10.8 | 249 | 2.2 | 12.9 | 594,447 | 2,389 | 78,967 | 77.9 | 305,533 | 2.5 | 97.5 | NA | 60.4 | 324 | 7.8 |
| Concordia | 281 | 17.1 | 759 | 46.3 | 206.5 | 2,125,125 | 2,802 | 219,661 | 100.3 | 270,415 | 97.7 | 2.3 | NA | 66.0 | 8,888 | 78.7 |
| De Soto | 137 | -16.9 | 228 | 0.2 | 28.3 | 676,324 | 2,966 | 97,326 | 20.6 | 34,347 | 12.6 | 87.4 | NA | 74.8 | 182 | 2.7 |
| East Baton Rouge | 58 | 1.3 | 130 | 0.5 | 13.4 | 1,071,013 | 8,251 | 89,008 | 12.6 | 28,058 | 22.3 | 77.7 | 1 | 83.1 | 132 | 5.8 |
| East Carroll | 221 | -11.8 | 1,059 | 128.4 | 193.4 | 3,932,394 | 3,713 | 491,558 | 116.5 | 557,191 | 99.9 | 0.1 | NA | 76.6 | 9,631 | 86.6 |
| East Feliciana | 131 | 16.4 | 318 | 0.0 | 24.7 | 889,716 | 2,799 | 75,308 | 9.2 | 22,303 | 18.2 | 81.8 | NA | 80.3 | 511 | 11.2 |
| Evangeline | 159 | -17.2 | 267 | 45.9 | 113.5 | 650,632 | 2,441 | 100,421 | 54.7 | 91,851 | 84.7 | 15.3 | NA | 74.2 | 6,872 | 41.4 |
| Franklin | 255 | 3.5 | 320 | 103.8 | 178.4 | 906,715 | 2,836 | 122,690 | 115.3 | 144,689 | 94.5 | 5.5 | NA | 64.0 | 8,085 | 64.9 |
| Grant | 51 | 6.0 | 246 | D | 17.0 | 672,574 | 2,738 | 75,464 | 7.7 | 37,411 | 74.8 | 25.2 | NA | 71.0 | 506 | 21.7 |
| Iberia | 115 | 7.0 | 344 | 2.6 | 88.2 | 985,830 | 2,868 | 215,874 | 57.9 | 173,263 | 96.5 | 3.5 | NA | 71.3 | 846 | 14.1 |
| Iberville | 182 | 11.2 | 1,203 | D | 76.8 | 2,159,687 | 1,796 | 484,293 | 48.3 | 320,126 | 95.3 | 4.7 | NA | 84.1 | 501 | 21.2 |
| Jackson | 21 | 10.5 | 108 | 0.2 | 3.2 | 431,104 | 4,003 | 76,776 | 42.8 | 223,874 | 1.1 | 98.9 | NA | 73.8 | 82 | 23.0 |
| Jefferson | 8 | 5.1 | 157 | 0.0 | 0.5 | 365,529 | 2,334 | 58,697 | 0.7 | 13,115 | 46.5 | 53.5 | NA | 75.0 | 11 | 9.6 |
| Jefferson Davis | 243 | -8.1 | 346 | 68.2 | 167.7 | 771,190 | 2,227 | 119,746 | 73.3 | 104,239 | 78.3 | 21.7 | NA | 69.4 | 13,380 | 48.6 |
| Lafayette | 50 | -10.8 | 90 | 5.1 | 29.9 | 504,748 | 5,581 | 86,348 | 21.4 | 38,998 | 79.0 | 21.0 | 2 | 71.8 | 1,039 | 8.2 |
| Lafourche | 157 | -0.7 | 414 | 0.3 | 58.6 | 1,200,913 | 2,899 | 125,407 | 39.0 | 102,821 | 66.8 | 33.2 | NA | 72.8 | 124 | 3.2 |
| La Salle | 24 | 20.9 | 100 | D | 4.5 | 319,962 | 3,185 | 67,387 | 1.6 | 6,817 | 27.3 | 72.7 | NA | 74.9 | 166 | 24.3 |
| Lincoln | 72 | 28.6 | 190 | 0.6 | 14.8 | 609,592 | 3,210 | 84,608 | 134.2 | 354,944 | 1.0 | 99.0 | NA | 75.7 | 282 | 6.9 |
| Livingston | 30 | 7.5 | 68 | 0.1 | 5.8 | 363,820 | 5,351 | 57,006 | 10.6 | 24,232 | 8.2 | 91.8 | NA | 67.4 | 239 | 5.7 |
| Madison | 245 | 9.2 | 981 | 83.4 | 190.9 | 2,697,499 | 2,749 | 404,533 | 106.8 | 427,108 | 99.4 | 0.6 | NA | 63.2 | 8,445 | 84.0 |
| Morehouse | 257 | -8.1 | 627 | 144.6 | 196.1 | 1,994,849 | 3,180 | 255,996 | 129.4 | 316,421 | 97.1 | 2.9 | NA | 71.4 | 10,044 | 63.6 |
| Natchitoches | 213 | 6.2 | 340 | 14.1 | 81.1 | 794,915 | 2,335 | 111,180 | 98.5 | 157,166 | 30.3 | 69.7 | NA | 67.5 | 2,910 | 37.8 |

| STATE County | Water use, 2015 Public supply water withdrawn (mil gal/day) | Public supply gallons withdrawn per person per day | Wholesale Trade[1], 2017 Number of establishments | Number of employees | Sales (mil dol) | Average payroll (mil dol) | Retail Trade[2], 2017 Number of establishments | Number of employees | Sales (mil dol) | Average payroll (mil dol) | Real estate and rental and leasing,[2] 2017 Number of establishments | Number of employees | Sales (mil dol) | Average payroll (mil dol) |
|---|---|---|---|---|---|---|---|---|---|---|---|---|---|---|
| | 133 | 134 | 135 | 136 | 137 | 138 | 139 | 140 | 141 | 142 | 143 | 144 | 145 | 146 |
| KENTUCKY—Cont'd | | | | | | | | | | | | | | |
| Ohio | 2.62 | 108.2 | D | D | D | 3.8 | 70 | 776 | 207.7 | 19.5 | 10 | 35 | 4.0 | 1.0 |
| Oldham | 3.93 | 60.6 | 47 | 319 | 131.8 | 16.7 | 100 | 1,467 | 493.0 | 41.5 | 61 | 141 | 41.1 | 6.8 |
| Owen | 0.00 | 0.0 | 5 | D | 18.9 | D | 23 | 167 | 63.6 | 4.3 | D | D | D | D |
| Owsley | 0.53 | 118.8 | NA | NA | NA | NA | 12 | 112 | 20.2 | 2.5 | 4 | 9 | 1.0 | 0.2 |
| Pendleton | 0.97 | 67.3 | D | D | D | 1.7 | 24 | 206 | 41.5 | 3.6 | D | D | D | D |
| Perry | 4.18 | 151.6 | 22 | 267 | 143.0 | 13.2 | 112 | 1,586 | 467.3 | 39.8 | 18 | 77 | 28.6 | 3.1 |
| Pike | 6.09 | 98.6 | 36 | 295 | 282.8 | 14.5 | 263 | 3,813 | 1,001.1 | 91.4 | 38 | 118 | 16.2 | 3.6 |
| Powell | 1.45 | 118.2 | D | D | D | 4.7 | 39 | 379 | 97.8 | 7.4 | 4 | 14 | 3.6 | 0.3 |
| Pulaski | 7.56 | 118.5 | 63 | 914 | 695.7 | 32.1 | 308 | 3,844 | 1,171.1 | 98.7 | 52 | 210 | 36.7 | 6.1 |
| Robertson | 0.00 | 0.0 | NA | NA | NA | NA | 3 | 24 | 2.7 | 0.3 | NA | NA | NA | NA |
| Rockcastle | 1.95 | 115.1 | 6 | 35 | 11.0 | 1.1 | 38 | 277 | 69.0 | 6.9 | 5 | 16 | 2.6 | 0.5 |
| Rowan | 6.03 | 252.4 | D | D | D | 6.2 | 98 | 1,428 | 380.4 | 32.5 | 22 | 101 | 13.9 | 2.5 |
| Russell | 2.01 | 113.8 | D | D | D | D | 86 | 741 | 206.0 | 16.1 | 3 | 8 | 0.6 | 0.2 |
| Scott | 2.89 | 55.1 | 30 | 692 | 1,771.2 | 49.9 | 115 | 1,835 | 597.9 | 43.8 | 52 | 164 | 35.2 | 4.9 |
| Shelby | 2.93 | 64.2 | 37 | 606 | 251.2 | 28.4 | 191 | 2,719 | 724.6 | 58.5 | 42 | 183 | 34.1 | 5.0 |
| Simpson | 1.65 | 91.6 | 16 | 215 | 281.5 | 10.1 | 68 | 1,133 | 365.5 | 26.9 | 14 | 40 | 6.0 | 0.9 |
| Spencer | 0.00 | 0.0 | NA | NA | NA | NA | 25 | 221 | 83.4 | 5.0 | D | D | D | D |
| Taylor | 5.20 | 204.6 | 18 | 157 | 185.5 | 6.2 | 119 | 2,581 | 732.5 | 70.0 | 25 | 104 | 15.5 | 2.6 |
| Todd | 0.00 | 0.0 | D | D | D | 2.1 | 40 | 350 | 92.8 | 7.9 | D | D | D | 0.1 |
| Trigg | 2.09 | 146.8 | 5 | 38 | 13.0 | 1.1 | 46 | 343 | 104.0 | 7.8 | 8 | 35 | 5.5 | 1.6 |
| Trimble | 2.75 | 313.6 | NA | NA | NA | NA | 13 | 65 | 20.4 | 1.1 | NA | NA | NA | NA |
| Union | 1.87 | 124.3 | 17 | 388 | 226.6 | 15.7 | 51 | 526 | 130.8 | 12.3 | 6 | D | 9.9 | D |
| Warren | 17.47 | 142.2 | 145 | 2,031 | 2,476.8 | 104.1 | 502 | 7,917 | 1,968.8 | 190.0 | 144 | 582 | 122.1 | 19.0 |
| Washington | 1.17 | 97.0 | 12 | 68 | 41.7 | 2.4 | 35 | 324 | 99.5 | 8.4 | 4 | D | 0.6 | D |
| Wayne | 2.76 | 134.9 | 11 | 97 | 35.5 | 3.2 | 59 | 700 | 162.2 | 15.9 | NA | NA | NA | NA |
| Webster | 2.02 | 153.4 | D | D | D | 4.9 | 39 | 273 | 62.2 | 6.3 | 3 | 6 | 0.6 | 0.1 |
| Whitley | 4.84 | 134.0 | 19 | 278 | 151.5 | 12.5 | 125 | 1,550 | 422.6 | 38.2 | 22 | 60 | 15.3 | 2.7 |
| Wolfe | 0.47 | 64.7 | NA | NA | NA | NA | 23 | 138 | 52.4 | 3.5 | D | D | D | D |
| Woodford | 3.12 | 121.0 | 15 | 304 | 134.5 | 15.1 | 73 | 894 | 271.8 | 21.7 | 19 | 30 | 7.7 | 1.2 |
| LOUISIANA | 708.92 | 151.8 | 4,673 | 63,503 | 59,523.7 | 3,493.7 | 16,564 | 233,385 | 65,000.8 | 5,988.4 | 5,121 | 29,345 | 7,300.6 | 1,386.0 |
| Acadia | 5.62 | 89.8 | 49 | 758 | 462.3 | 28.7 | 185 | 2,557 | 575.4 | 58.0 | 39 | 380 | 62.5 | 17.1 |
| Allen | 4.24 | 165.1 | D | D | D | D | 71 | 702 | 204.5 | 15.7 | D | D | D | 0.3 |
| Ascension | 2.65 | 22.2 | 133 | 2,105 | 1,208.0 | 123.3 | 373 | 6,371 | 2,031.2 | 167.5 | 138 | 847 | 302.5 | 50.7 |
| Assumption | 4.19 | 183.4 | 9 | 51 | 87.5 | 1.5 | 40 | 450 | 117.6 | 10.6 | 5 | D | 4.5 | D |
| Avoyelles | 4.05 | 98.5 | 17 | 308 | 454.7 | 13.2 | 149 | 1,408 | 378.7 | 33.7 | 22 | 56 | 9.7 | 1.6 |
| Beauregard | 4.42 | 121.2 | 11 | 63 | 41.3 | 3.4 | 105 | 1,273 | 337.3 | 29.8 | 24 | 52 | 10.4 | 2.1 |
| Bienville | 2.72 | 197.3 | D | D | D | D | 37 | 296 | 77.9 | 6.1 | 7 | 78 | 15.8 | 3.5 |
| Bossier | 12.88 | 102.9 | 106 | 1,682 | 1,047.4 | 85.3 | 440 | 7,643 | 2,182.4 | 187.7 | 137 | 600 | 145.1 | 22.8 |
| Caddo | 43.28 | 172.1 | 322 | 4,997 | 7,739.1 | 277.0 | 887 | 12,909 | 3,729.0 | 343.8 | 341 | 1,841 | 380.7 | 73.4 |
| Calcasieu | 27.51 | 138.4 | D | D | D | 113.3 | 790 | 11,232 | 3,622.5 | 288.6 | D | D | D | D |
| Caldwell | 1.03 | 103.1 | D | D | D | 1.3 | 26 | 296 | 76.9 | 6.8 | NA | NA | NA | NA |
| Cameron | 1.50 | 220.0 | D | D | D | 2.6 | 18 | 152 | 37.4 | 2.9 | D | D | D | D |
| Catahoula | 1.17 | 115.3 | 9 | 67 | 36.4 | 2.9 | 32 | 285 | 54.4 | 5.6 | NA | NA | NA | NA |
| Claiborne | 3.01 | 184.7 | 10 | 64 | 138.0 | 3.9 | 43 | 336 | 87.9 | 8.4 | 3 | 5 | 0.6 | 0.1 |
| Concordia | 2.63 | 130.6 | 18 | 270 | 512.3 | 14.5 | 77 | 822 | 228.6 | 22.3 | 12 | 112 | 10.3 | 3.3 |
| De Soto | 3.08 | 113.9 | 13 | 37 | 22.5 | 1.7 | 72 | 814 | 288.8 | 21.9 | 8 | 39 | 8.5 | 2.1 |
| East Baton Rouge | 72.12 | 161.4 | 571 | 8,111 | 5,706.2 | 496.8 | 1,834 | 28,939 | 8,006.5 | 779.5 | 632 | 3,270 | 823.8 | 148.5 |
| East Carroll | 0.91 | 124.5 | D | D | D | D | 20 | 120 | 25.5 | 2.6 | 6 | 8 | 1.5 | 0.1 |
| East Feliciana | 2.21 | 112.2 | 5 | 162 | 80.5 | 8.7 | 44 | 306 | 78.5 | 7.1 | 7 | 8 | 1.1 | 0.1 |
| Evangeline | 6.50 | 192.6 | 16 | 118 | 85.2 | 2.9 | 113 | 971 | 224.5 | 21.7 | 23 | 65 | 6.2 | 1.4 |
| Franklin | 1.95 | 95.5 | 19 | 137 | 214.3 | 6.7 | 74 | 866 | 258.0 | 20.6 | 11 | 30 | 4.1 | 0.9 |
| Grant | 4.54 | 203.2 | NA | NA | NA | NA | 29 | 269 | 89.0 | 7.7 | NA | NA | NA | NA |
| Iberia | 8.65 | 116.7 | 85 | 1,153 | 701.1 | 60.0 | 269 | 3,275 | 1,003.6 | 89.2 | 91 | 985 | 288.7 | 58.5 |
| Iberville | 1.98 | 59.8 | 26 | 269 | 147.6 | 12.1 | 81 | 1,104 | 291.9 | 27.7 | 25 | 179 | 73.1 | 10.7 |
| Jackson | 1.73 | 109.1 | D | D | D | 0.3 | 46 | 502 | 129.7 | 11.3 | 6 | 6 | 0.9 | 0.1 |
| Jefferson | 61.79 | 141.6 | 680 | 10,121 | 7,582.7 | 568.7 | 1,760 | 29,102 | 8,987.6 | 811.3 | 601 | 3,812 | 1,036.2 | 165.4 |
| Jefferson Davis | 3.82 | 121.5 | D | D | D | D | 122 | 1,404 | 380.8 | 32.6 | 18 | 176 | 35.5 | 5.6 |
| Lafayette | 25.42 | 105.9 | 459 | 6,489 | 3,658.9 | 357.9 | 1,090 | 17,143 | 4,471.0 | 447.5 | 503 | 3,007 | 831.1 | 169.2 |
| Lafourche | 25.66 | 261.0 | D | D | D | D | 300 | 4,022 | 1,042.2 | 95.5 | 64 | 210 | 43.6 | 8.9 |
| La Salle | 1.72 | 114.9 | 7 | 167 | 364.8 | 6.1 | 46 | 472 | 132.3 | 12.2 | 4 | 12 | 2.0 | 0.3 |
| Lincoln | 7.18 | 150.3 | 35 | 546 | 308.5 | 27.2 | 170 | 2,458 | 611.6 | 59.2 | 68 | 430 | 42.0 | 11.8 |
| Livingston | 11.54 | 83.8 | 41 | 554 | 541.1 | 32.4 | 319 | 5,322 | 1,493.2 | 125.8 | 79 | 283 | 46.1 | 12.0 |
| Madison | 1.65 | 143.3 | 14 | 140 | 212.2 | 6.8 | 28 | 363 | 108.9 | 7.6 | D | D | D | D |
| Morehouse | 3.47 | 131.5 | 19 | 257 | 111.0 | 10.8 | 73 | 900 | 196.6 | 20.8 | 25 | 47 | 9.2 | 1.3 |
| Natchitoches | 7.72 | 197.0 | 21 | 153 | 365.4 | 6.4 | 137 | 1,687 | 442.6 | 37.9 | 48 | 198 | 35.0 | 5.3 |

1 Merchant wholesalers, except manufacturers' sales branches and offices.  2. Employer establishments.

# Professional Services, Manufacturing, and Accommodation and Food Services

| STATE County | Professional, scientific, and technical services, 2017 | | | | Manufacturing, 2017 | | | | Accommodation and food services, 2017 | | | |
|---|---|---|---|---|---|---|---|---|---|---|---|---|
| | Number of establishments | Number of employees | Sales (mil dol) | Average payroll (mil dol) | Number of establishments | Number of employees | Sales (mil dol) | Average payroll (mil dol) | Number of establishments | Number of employees | Sales (mil dol) | Annual payroll (mil dol) |
| | 147 | 148 | 149 | 150 | 151 | 152 | 153 | 154 | 155 | 156 | 157 | 158 |
| **KENTUCKY—Cont'd** | | | | | | | | | | | | |
| Ohio | 24 | 111 | 12 | 3.7 | 20 | 2,368 | 732.3 | 82.1 | 31 | 395 | 19.3 | 5.5 |
| Oldham | 173 | 661 | 97 | 35.6 | 40 | 1,303 | 519.2 | 75.4 | 69 | 1,507 | 78.2 | 21.4 |
| Owen | 6 | 9 | 2 | 0.3 | D | D | D | D | 15 | 205 | 8.5 | 2.2 |
| Owsley | NA | NA | NA | NA | NA | NA | NA | NA | NA | NA | NA | NA |
| Pendleton | 11 | 53 | 4 | 1.7 | 13 | D | 158.9 | D | D | NA | NA | NA |
| Perry | D | D | D | D | 7 | 74 | 15.0 | 4.2 | 44 | 881 | 46.2 | 11.7 |
| Pike | 92 | 583 | 81 | 28.2 | 23 | 550 | 183.4 | 28.2 | 100 | 2,002 | 103.3 | 28.7 |
| Powell | 7 | 12 | 1 | 0.2 | 7 | 92 | 32.1 | 5.2 | 18 | 271 | 14.7 | 3.9 |
| Pulaski | D | D | D | D | 77 | 4,093 | 1,369.0 | 171.2 | 99 | 2,443 | 115.4 | 33.0 |
| Robertson | NA | NA | NA | NA | NA | NA | NA | NA | NA | NA | NA | NA |
| Rockcastle | 16 | 209 | 13 | 6.8 | 9 | 105 | 14.9 | 4.4 | 19 | 245 | 13.0 | 3.1 |
| Rowan | D | D | D | D | 21 | 1,157 | 381.6 | 45.6 | 50 | 1,259 | 57.5 | 16.2 |
| Russell | 21 | 48 | 4 | 1.1 | 24 | 1,251 | 350.4 | 38.3 | 32 | 404 | 22.0 | 5.9 |
| Scott | 88 | 382 | 53 | 19.0 | 41 | 11,077 | 10,376.6 | 797.5 | 107 | 2,252 | 127.1 | 35.8 |
| Shelby | 95 | 320 | 41 | 12.7 | 56 | 3,861 | 1,656.0 | 194.5 | D | D | D | D |
| Simpson | 15 | 64 | 8 | 2.2 | 35 | 2,695 | 1,414.5 | 160.5 | 50 | 994 | 49.5 | 13.6 |
| Spencer | D | D | 5 | D | D | D | D | D | D | D | D | D |
| Taylor | 56 | 110 | 14 | 4.0 | 29 | 1,208 | 288.6 | 43.4 | 47 | 933 | 47.0 | 11.9 |
| Todd | 11 | 41 | 4 | 1.7 | 19 | 298 | 111.1 | 12.0 | 10 | 113 | 4.0 | 1.1 |
| Trigg | 11 | 35 | 4 | 1.5 | 12 | 545 | 135.2 | 20.9 | 20 | 357 | 18.5 | 4.9 |
| Trimble | D | D | 0 | D | D | D | D | D | D | D | D | D |
| Union | 9 | 46 | 5 | 1.8 | 13 | 439 | 175.8 | 20.2 | D | D | D | D |
| Warren | D | D | D | D | 106 | 9,632 | 5,906.2 | 556.2 | 293 | 6,870 | 352.1 | 107.9 |
| Washington | 10 | 44 | 3 | 0.9 | 14 | 1,163 | 412.5 | 56.5 | 16 | 182 | 8.2 | 2.1 |
| Wayne | 20 | 43 | 3 | 1.2 | 23 | 1,224 | 138.3 | 40.3 | D | D | D | D |
| Webster | 10 | 32 | 4 | 1.1 | 11 | 224 | 140.6 | 8.4 | D | D | D | D |
| Whitley | 45 | 830 | 49 | 20.7 | 16 | 817 | 236.2 | 36.2 | 65 | 1,211 | 63.1 | 18.2 |
| Wolfe | 3 | 5 | 1 | 0.2 | 4 | 23 | 2.8 | 0.6 | 4 | 55 | 2.1 | 0.6 |
| Woodford | 80 | 437 | 78 | 29.5 | 33 | 2,839 | 788.4 | 149.1 | D | D | D | D |
| **LOUISIANA** | 12,010 | 94,813 | 15,936 | 6,130.5 | 3,190 | 117,910 | 187,439.9 | 8,257.3 | 9,877 | 215,048 | 14,553.0 | 3,808.8 |
| Acadia | 121 | 293 | 39 | 11.4 | 44 | 988 | 372.6 | 41.3 | 80 | 1,273 | 60.0 | 15.9 |
| Allen | D | D | D | D | 9 | 700 | 367.4 | 38.6 | 30 | 378 | 22.1 | 4.8 |
| Ascension | 204 | 2,272 | 280 | 158.4 | 100 | 5,970 | 9,047.0 | 603.6 | 230 | 4,051 | 217.9 | 61.0 |
| Assumption | 31 | 70 | 8 | 2.5 | 13 | 357 | 157.7 | 19.5 | 12 | 149 | 4.8 | 1.4 |
| Avoyelles | 57 | 214 | 35 | 8.5 | 20 | 253 | 41.0 | 12.8 | 43 | 1,663 | 127.3 | 32.8 |
| Beauregard | 54 | 215 | 22 | 7.8 | 18 | 905 | 919.0 | 70.1 | 46 | 710 | 39.6 | 10.1 |
| Bienville | 6 | 18 | 2 | 0.7 | 8 | 1,269 | 354.5 | 37.4 | D | D | D | D |
| Bossier | D | D | D | D | 63 | 1,444 | 654.2 | 73.9 | 263 | 8,989 | 777.1 | 168.1 |
| Caddo | D | D | D | D | 160 | 5,097 | 3,150.9 | 329.7 | 509 | 12,034 | 772.7 | 197.9 |
| Calcasieu | 471 | 4,473 | 661 | 299.8 | D | D | D | D | 451 | 14,129 | 1,374.8 | 296.1 |
| Caldwell | 30 | 75 | 9 | 2.3 | 3 | 36 | 5.5 | 1.1 | D | D | D | 1.7 |
| Cameron | 18 | 227 | 60 | 23.8 | NA | NA | NA | NA | 6 | 31 | 2.2 | 0.3 |
| Catahoula | 28 | 82 | 11 | 3.9 | NA | NA | NA | NA | D | D | D | D |
| Claiborne | 8 | 29 | 4 | 1.2 | 5 | 108 | 49.6 | 5.5 | D | D | D | 1.3 |
| Concordia | 27 | 81 | 12 | 3.4 | 10 | 118 | 56.5 | 7.1 | 26 | 326 | 17.5 | 4.2 |
| De Soto | 28 | 127 | 14 | 4.7 | D | D | D | D | 30 | 369 | 21.3 | 5.0 |
| East Baton Rouge | D | D | D | D | 325 | 10,327 | 31,999.7 | 779.2 | 1,066 | 26,112 | 1,553.1 | 419.5 |
| East Carroll | 4 | 16 | 1 | 0.5 | 5 | 65 | 25.6 | 1.7 | 8 | 67 | 2.9 | 0.6 |
| East Feliciana | D | D | 5 | D | D | D | D | D | 16 | 185 | 7.5 | 2.3 |
| Evangeline | 39 | 99 | 12 | 3.3 | 15 | 675 | 302.5 | 42.1 | D | D | D | D |
| Franklin | 34 | 121 | 12 | 3.4 | 8 | 392 | 57.0 | 6.5 | D | D | D | D |
| Grant | 6 | 3 | 2 | 0.4 | D | D | D | D | D | D | D | 0.5 |
| Iberia | 139 | 770 | 127 | 44.7 | 110 | 3,076 | 1,259.5 | 181.8 | 108 | 1,726 | 83.3 | 23.1 |
| Iberville | 57 | 473 | 72 | 30.6 | 36 | 4,629 | 8,228.6 | 463.8 | 39 | 633 | 38.2 | 9.4 |
| Jackson | 18 | 41 | 6 | 1.5 | D | D | D | D | D | D | D | D |
| Jefferson | D | D | D | D | 288 | 6,620 | 2,486.9 | 373.4 | 1,127 | 21,335 | 1,394.8 | 370.4 |
| Jefferson Davis | 67 | 254 | 36 | 11.2 | 21 | 431 | 100.6 | 18.3 | 46 | 670 | 33.7 | 9.1 |
| Lafayette | 1,286 | 8,329 | 1,518 | 534.8 | 268 | 6,356 | 1,812.9 | 333.7 | 759 | 16,127 | 1,027.0 | 265.6 |
| Lafourche | 172 | 1,298 | 134 | 67.9 | 55 | 2,111 | 678.3 | 114.9 | 156 | 2,419 | 128.8 | 32.9 |
| La Salle | D | D | D | D | 6 | 57 | 19.0 | 3.4 | 21 | 220 | 12.1 | 3.1 |
| Lincoln | 97 | 750 | 113 | 48.6 | 39 | 1,236 | 467.5 | 69.2 | 94 | 2,165 | 115.4 | 31.8 |
| Livingston | D | D | D | D | 55 | 1,795 | 559.7 | 90.5 | 170 | 3,316 | 164.5 | 43.2 |
| Madison | 9 | 22 | 2 | 0.5 | D | 166 | D | 9.3 | 18 | 211 | 12.2 | 2.7 |
| Morehouse | 20 | 84 | 9 | 3.1 | 18 | 626 | 183.6 | 18.1 | 30 | D | 24.0 | D |
| Natchitoches | 54 | 207 | 32 | 8.0 | 18 | 2,459 | 1,474.8 | 128.1 | 85 | 1,495 | 76.8 | 19.4 |

| STATE County | Health care and social assistance, 2017 | | | | Other services, 2017 | | | | Nonemployer businesses, 2018 | | Value of residential construction authorized by building permits, 2020 | |
|---|---|---|---|---|---|---|---|---|---|---|---|---|
| | Number of establishments | Number of employees | Receipts (mil dol) | Annual payroll (mil dol) | Number of establishments | Number of employees | Receipts (mil dol) | Annual payroll (mil dol) | Number | Receipts (mil dol) | New construction ($1,000) | Number of housing units |
| | 159 | 160 | 161 | 162 | 163 | 164 | 165 | 166 | 167 | 168 | 169 | 170 |
| KENTUCKY—Cont'd | | | | | | | | | | | | |
| Ohio | 46 | 1,161 | 79.7 | 39.4 | 35 | 93 | 10.4 | 2.7 | 1,071 | 40.7 | 1,135 | 23 |
| Oldham | 145 | 2,828 | 309.5 | 106.0 | 75 | 315 | 30.7 | 9.6 | 5,476 | 278.3 | 123,741 | 442 |
| Owen | 12 | 181 | 15.8 | 7.0 | D | D | 4.0 | D | 756 | 29.3 | 5,018 | 24 |
| Owsley | 11 | 202 | 14.0 | 5.6 | NA | NA | NA | NA | 279 | 9.7 | NA | NA |
| Pendleton | 16 | 235 | 16.7 | 7.4 | D | D | D | D | 737 | 31.5 | 0 | 0 |
| Perry | 113 | 2,905 | 351.7 | 128.9 | 24 | 100 | 10.2 | 2.4 | 1,335 | 45.7 | 0 | 0 |
| Pike | 183 | 5,226 | 763.6 | 291.8 | 70 | 302 | 27.1 | 8.1 | 2,824 | 118.3 | 1,771 | 11 |
| Powell | 16 | 203 | 16.5 | 7.1 | 6 | 12 | 1.1 | 0.3 | 798 | 30.4 | 209 | 1 |
| Pulaski | 228 | 4,985 | 551.0 | 223.8 | 62 | 307 | 34.9 | 8.4 | 4,529 | 191.9 | 9,150 | 116 |
| Robertson | NA | NA | NA | NA | NA | NA | NA | NA | 133 | 3.1 | NA | NA |
| Rockcastle | 25 | 990 | 83.8 | 35.8 | 6 | 25 | 2.6 | 0.6 | 897 | 29.9 | NA | NA |
| Rowan | 85 | 2,028 | 235.5 | 89.5 | 22 | 111 | 6.7 | 1.7 | 1,431 | 61.1 | 460 | 10 |
| Russell | 41 | 887 | 67.5 | 29.2 | D | D | D | D | 1,355 | 52.7 | 1,500 | 12 |
| Scott | 126 | 1,824 | 180.2 | 69.9 | 64 | 373 | 35.9 | 10.2 | 3,503 | 144.9 | 68,831 | 363 |
| Shelby | 98 | 1,532 | 151.6 | 57.1 | 66 | 549 | 53.0 | 17.2 | 3,564 | 175.5 | 75,077 | 328 |
| Simpson | 33 | 443 | 52.5 | 16.2 | 23 | 107 | 10.9 | 3.4 | 1,310 | 63.5 | 24,506 | 166 |
| Spencer | D | D | D | 7.3 | D | D | 4.1 | D | 1,354 | 59.2 | 39,770 | 161 |
| Taylor | 83 | 1,605 | 157.2 | 62.9 | D | D | D | D | 1,759 | 70.5 | 2,522 | 8 |
| Todd | 16 | 189 | 11.5 | 5.1 | D | D | D | 0.9 | 924 | 53.1 | 354 | 4 |
| Trigg | 27 | 357 | 31.6 | 12.8 | 16 | 97 | 6.3 | 1.9 | 1,004 | 42.1 | 8,259 | 81 |
| Trimble | 6 | 104 | 11.5 | 4.6 | D | D | 1.9 | D | 464 | 18.4 | NA | NA |
| Union | 33 | 864 | 69.7 | 28.6 | 11 | 42 | 7.4 | 1.4 | 732 | 26.1 | NA | NA |
| Warren | 400 | 9,097 | 1,196.9 | 438.1 | 217 | 1,273 | 116.1 | 32.7 | 10,027 | 596.3 | 179,092 | 970 |
| Washington | 27 | 430 | 32.0 | 13.3 | D | D | D | D | 839 | 34.2 | 476 | 12 |
| Wayne | 37 | 595 | 42.4 | 19.8 | D | D | D | D | 1,157 | 43.3 | 0 | 0 |
| Webster | 25 | 276 | 17.4 | 8.0 | D | D | D | 1.2 | 621 | 19.7 | 40 | 5 |
| Whitley | 95 | 2,940 | 333.9 | 124.5 | D | D | D | D | 2,289 | 96.3 | 1,487 | 18 |
| Wolfe | 12 | 245 | 18.1 | 6.4 | 3 | 16 | 1.3 | 0.4 | 432 | 17.8 | NA | NA |
| Woodford | 64 | 664 | 63.4 | 24.5 | 33 | 136 | 14.3 | 3.6 | 2,319 | 101.4 | 23,991 | 99 |
| LOUISIANA | 12,685 | 302,408 | 34,618.0 | 12,912.7 | 6,531 | 43,472 | 5,702.0 | 1,571.6 | 385,074 | 16,923.6 | 3,427,774 | 17,283 |
| Acadia | 125 | 2,475 | 206.8 | 76.3 | 60 | 259 | 27.7 | 6.2 | 4,533 | 194.8 | 21,045 | 124 |
| Allen | 41 | 770 | 66.7 | 25.2 | D | D | 12.9 | D | 1,169 | 37.9 | 5,368 | 41 |
| Ascension | 203 | 3,496 | 300.5 | 120.4 | 173 | 1,820 | 272.6 | 90.2 | 9,291 | 392.3 | 152,910 | 902 |
| Assumption | 16 | 436 | 19.6 | 8.9 | 20 | 55 | 13.4 | 1.9 | 1,627 | 57.0 | 8,912 | 25 |
| Avoyelles | 90 | 2,267 | 154.1 | 60.8 | 42 | 138 | 18.2 | 3.2 | 2,472 | 98.9 | 14,365 | 228 |
| Beauregard | 63 | 1,249 | 88.8 | 37.5 | 32 | 152 | 16.9 | 4.4 | 2,147 | 82.1 | 23,024 | 97 |
| Bienville | 20 | 526 | 45.2 | 14.8 | 9 | 16 | 2.2 | 0.4 | 796 | 31.0 | 2,509 | 12 |
| Bossier | 242 | 4,864 | 518.0 | 200.9 | 130 | 1,055 | 107.0 | 31.0 | 9,017 | 423.8 | 113,018 | 623 |
| Caddo | 876 | 28,411 | 4,122.4 | 1,490.0 | D | D | D | D | 20,754 | 889.6 | 62,662 | 255 |
| Calcasieu | 553 | 13,776 | 1,429.0 | 553.2 | D | D | D | D | 14,321 | 654.1 | 177,694 | 1,144 |
| Caldwell | 25 | 710 | 57.8 | 24.8 | 11 | 42 | 4.5 | 1.1 | 690 | 22.1 | 5,957 | 25 |
| Cameron | 3 | 117 | 8.4 | 4.1 | D | D | D | D | 688 | 30.5 | 12,592 | 86 |
| Catahoula | 17 | 365 | 21.1 | 9.6 | D | D | D | 0.8 | 615 | 21.8 | 25 | 1 |
| Claiborne | 24 | 770 | 48.7 | 20.8 | 9 | 38 | 2.5 | 0.6 | 855 | 31.5 | 258 | 3 |
| Concordia | 40 | 763 | 63.9 | 26.2 | 24 | 90 | 6.2 | 2.4 | 1,184 | 44.7 | 1,505 | 6 |
| De Soto | D | D | D | D | D | D | D | D | 2,094 | 78.7 | 28,616 | 122 |
| East Baton Rouge | 1,433 | 40,379 | 4,979.8 | 1,735.7 | 890 | 6,557 | 1,008.3 | 257.9 | 39,319 | 1,812.4 | 356,624 | 1,363 |
| East Carroll | 10 | 318 | 20.7 | 8.9 | D | D | D | D | 417 | 14.2 | 0 | 0 |
| East Feliciana | 31 | 1,521 | 120.8 | 65.1 | 18 | 60 | 7.8 | 2.2 | 1,428 | 52.8 | 11,248 | 54 |
| Evangeline | 101 | 2,521 | 188.8 | 74.8 | D | D | D | D | 1,718 | 66.2 | 12,732 | 95 |
| Franklin | 46 | 1,063 | 79.6 | 34.0 | 27 | 81 | 8.9 | 1.7 | 1,476 | 54.0 | 4,791 | 16 |
| Grant | 24 | 266 | 19.4 | 6.8 | 9 | 28 | 3.6 | 0.7 | 1,066 | 42.3 | 12,704 | 62 |
| Iberia | 200 | 3,804 | 322.8 | 116.6 | 110 | 668 | 92.7 | 27.7 | 6,071 | 234.3 | 15,704 | 72 |
| Iberville | 45 | 842 | 85.9 | 25.3 | 35 | 262 | 34.2 | 10.8 | 2,297 | 75.0 | 27,088 | 111 |
| Jackson | 29 | 639 | 53.4 | 19.8 | 15 | 87 | 8.1 | 2.7 | 878 | 31.3 | 4,292 | 14 |
| Jefferson | 1,389 | 31,801 | 4,091.2 | 1,643.8 | 782 | 5,061 | 657.3 | 180.7 | 45,909 | 2,122.4 | 128,378 | 594 |
| Jefferson Davis | 65 | 1,566 | 138.6 | 59.6 | 38 | 135 | 16.3 | 3.9 | 1,942 | 71.3 | 21,071 | 132 |
| Lafayette | 1,212 | 25,143 | 2,980.6 | 1,066.1 | 451 | 3,231 | 381.0 | 104.6 | 24,831 | 1,278.2 | 385,853 | 2,110 |
| Lafourche | 188 | 4,168 | 446.2 | 184.8 | 106 | 548 | 72.1 | 18.2 | 7,177 | 298.6 | 78,401 | 369 |
| La Salle | 22 | 835 | 61.6 | 26.0 | D | D | 48.3 | D | 943 | 41.1 | 1,227 | 6 |
| Lincoln | 129 | 2,965 | 253.2 | 95.5 | 53 | 243 | 20.3 | 5.5 | 3,123 | 154.0 | 72,136 | 281 |
| Livingston | 149 | 1,947 | 146.5 | 59.4 | 127 | 534 | 59.3 | 16.9 | 9,797 | 411.8 | 195,479 | 937 |
| Madison | 28 | 1,055 | 58.8 | 24.8 | D | D | D | D | 667 | 23.7 | 510 | 2 |
| Morehouse | 74 | 1,960 | 126.2 | 48.9 | 33 | 116 | 8.5 | 2.2 | 1,721 | 55.0 | 360 | 1 |
| Natchitoches | 95 | 2,073 | 192.4 | 63.8 | 42 | 135 | 14.9 | 3.3 | 2,456 | 103.7 | 18,610 | 75 |

# Table B. States and Counties — Government Employment and Payroll, and Local Government Finances

| | Government employment and payroll, 2017 | | | | | | | | | Local government finances, 2017 | | | | |
|---|---|---|---|---|---|---|---|---|---|---|---|---|---|---|
| | | | March payroll (percent of total) | | | | | | | General revenue | | | | |
| | | | | | | | | | | | | Taxes | | |
| | | | | | | | | | | | | | Per capita[1] (dollars) | |
| STATE County | Full-time equivalent employees | March payroll (dollars) | Administration, judicial, and legal | Police and corrections | Fire protection | Highways and transportation | Health and welfare | Natural resources and utilities | Education and libraries | Total (mil dol) | Intergovernmental (mil dol) | Total (mil dol) | Total | Property |
| | 171 | 172 | 173 | 174 | 175 | 176 | 177 | 178 | 179 | 180 | 181 | 182 | 183 | 184 |
| KENTUCKY—Cont'd | | | | | | | | | | | | | | |
| Ohio | 1,341 | 3,789,958 | 3.2 | 4.4 | 0.2 | 1.5 | 39.8 | 3.0 | 47.8 | 55.0 | 33.9 | 16.0 | 662 | 349 |
| Oldham | 1,966 | 6,901,073 | 3.1 | 3.6 | 1.2 | 1.0 | 3.4 | 1.5 | 85.5 | 160.9 | 64.6 | 77.5 | 1,164 | 877 |
| Owen | 344 | 892,449 | 8.5 | 4.4 | 0.0 | 3.0 | 7.9 | 0.7 | 75.3 | 23.5 | 14.3 | 7.0 | 651 | 490 |
| Owsley | 258 | 647,309 | 6.0 | 3.1 | 0.1 | 6.7 | 0.0 | 1.9 | 79.7 | 13.1 | 10.0 | 1.7 | 388 | 256 |
| Pendleton | 536 | 1,655,442 | 4.8 | 4.1 | 0.1 | 2.1 | 15.0 | 5.3 | 66.3 | 36.4 | 22.2 | 9.4 | 641 | 512 |
| Perry | 1,400 | 4,272,954 | 3.5 | 3.5 | 1.7 | 2.4 | 12.1 | 4.4 | 71.8 | 85.3 | 50.7 | 22.2 | 835 | 475 |
| Pike | 1,971 | 6,872,000 | 3.8 | 4.3 | 1.8 | 3.2 | 4.9 | 2.7 | 78.9 | 164.4 | 91.5 | 53.4 | 907 | 527 |
| Powell | 554 | 1,636,898 | 9.8 | 8.1 | 0.1 | 0.8 | 4.4 | 4.5 | 70.6 | 32.1 | 22.9 | 6.3 | 511 | 274 |
| Pulaski | 2,449 | 7,448,220 | 4.1 | 5.7 | 3.1 | 1.7 | 12.2 | 6.2 | 66.9 | 152.2 | 76.4 | 56.9 | 884 | 543 |
| Robertson | 84 | 240,471 | 9.8 | 5.0 | 0.0 | 5.1 | 0.0 | 1.4 | 78.2 | 8.1 | 5.8 | 1.4 | 655 | 415 |
| Rockcastle | 706 | 1,557,825 | 7.6 | 8.9 | 0.0 | 2.1 | 3.2 | 0.9 | 73.3 | 36.7 | 27.1 | 8.1 | 485 | 190 |
| Rowan | 790 | 2,523,845 | 3.3 | 7.9 | 0.4 | 3.7 | 10.2 | 3.6 | 69.7 | 63.0 | 26.2 | 25.1 | 1,024 | 441 |
| Russell | 908 | 2,789,361 | 3.5 | 5.7 | 0.0 | 1.0 | 33.4 | 3.4 | 52.0 | 111.1 | 24.3 | 14.6 | 824 | 491 |
| Scott | 1,731 | 5,777,032 | 3.5 | 9.1 | 7.1 | 1.9 | 8.9 | 0.9 | 67.3 | 159.7 | 53.0 | 81.9 | 1,495 | 587 |
| Shelby | 1,713 | 6,943,492 | 2.5 | 5.4 | 1.7 | 1.4 | 4.9 | 1.3 | 81.4 | 106.0 | 42.1 | 54.3 | 1,150 | 795 |
| Simpson | 715 | 2,207,301 | 8.7 | 11.3 | 1.4 | 2.4 | 6.1 | 3.7 | 65.5 | 56.9 | 26.6 | 20.2 | 1,121 | 578 |
| Spencer | 511 | 1,635,338 | 5.9 | 3.9 | 1.0 | 1.4 | 2.7 | 2.6 | 82.1 | 35.4 | 19.2 | 13.3 | 716 | 550 |
| Taylor | 1,428 | 5,528,404 | 2.0 | 2.6 | 0.9 | 1.0 | 57.1 | 0.9 | 34.7 | 147.5 | 63.8 | 23.7 | 928 | 493 |
| Todd | 443 | 1,285,695 | 3.1 | 6.5 | 1.7 | 1.8 | 5.8 | 8.5 | 72.5 | 32.0 | 19.9 | 8.2 | 671 | 352 |
| Trigg | 600 | 1,948,983 | 2.9 | 5.2 | 0.5 | 2.7 | 28.5 | 7.9 | 51.8 | 29.2 | 15.5 | 12.0 | 829 | 521 |
| Trimble | 294 | 914,679 | 6.1 | 3.0 | 2.6 | 2.2 | 0.0 | 6.8 | 78.6 | 17.0 | 8.3 | 7.5 | 878 | 593 |
| Union | 558 | 1,747,883 | 6.8 | 6.7 | 1.7 | 4.3 | 0.7 | 6.7 | 70.7 | 36.1 | 19.9 | 12.3 | 844 | 606 |
| Warren | 4,152 | 13,982,618 | 2.6 | 9.4 | 4.7 | 1.8 | 4.5 | 9.2 | 63.5 | 330.8 | 120.6 | 163.7 | 1,269 | 609 |
| Washington | 362 | 1,217,688 | 3.2 | 2.5 | 6.7 | 1.9 | 3.1 | 6.6 | 73.8 | 32.0 | 16.6 | 10.4 | 874 | 440 |
| Wayne | 686 | 1,957,101 | 3.2 | 7.1 | 0.2 | 2.0 | 4.0 | 1.4 | 79.9 | 43.9 | 29.5 | 11.0 | 532 | 324 |
| Webster | 516 | 1,504,866 | 5.3 | 11.0 | 3.2 | 2.8 | 0.7 | 9.0 | 64.5 | 33.6 | 21.6 | 8.8 | 673 | 517 |
| Whitley | 1,666 | 4,944,984 | 2.7 | 5.4 | 2.3 | 2.1 | 9.3 | 5.5 | 71.3 | 117.7 | 72.6 | 26.8 | 743 | 340 |
| Wolfe | 323 | 905,288 | 3.4 | 1.8 | 0.0 | 1.4 | 0.2 | 3.4 | 86.1 | 18.1 | 14.4 | 2.3 | 323 | 185 |
| Woodford | 855 | 2,924,644 | 3.9 | 12.7 | 3.0 | 3.1 | 4.3 | 4.7 | 67.1 | 69.8 | 23.0 | 38.2 | 1,448 | 824 |
| LOUISIANA | X | X | X | X | X | X | X | X | X | X | X | X | X | X |
| Acadia | 1,865 | 5,132,550 | 7.7 | 6.0 | 2.5 | 3.4 | 0.8 | 5.6 | 73.0 | 169.8 | 77.3 | 68.7 | 1,099 | 431 |
| Allen | 1,217 | 3,011,014 | 8.7 | 15.0 | 0.5 | 3.2 | 16.4 | 1.2 | 54.8 | 267.5 | 110.5 | 85.2 | 3,330 | 1,784 |
| Ascension | 4,015 | 15,039,268 | 7.3 | 9.2 | 2.6 | 2.5 | 3.0 | 4.0 | 69.2 | 465.0 | 161.7 | 259.1 | 2,104 | 858 |
| Assumption | 1,168 | 4,001,399 | 7.8 | 6.0 | 0.7 | 15.6 | 15.7 | 6.0 | 47.4 | 63.7 | 30.9 | 26.6 | 1,179 | 644 |
| Avoyelles | 1,505 | 3,666,589 | 8.2 | 23.1 | 1.7 | 1.0 | 13.8 | 5.7 | 46.2 | 109.9 | 45.5 | 22.0 | 540 | 132 |
| Beauregard | 1,523 | 4,885,820 | 3.1 | 2.7 | 1.6 | 1.8 | 35.8 | 1.4 | 52.8 | 143.8 | 44.1 | 53.8 | 1,461 | 689 |
| Bienville | 652 | 2,118,412 | 15.2 | 15.6 | 0.0 | 5.0 | 0.1 | 3.1 | 60.7 | 57.7 | 16.8 | 39.3 | 2,885 | 2,130 |
| Bossier | 4,800 | 16,718,972 | 8.1 | 14.2 | 5.7 | 3.3 | 0.3 | 2.9 | 64.6 | 474.7 | 161.6 | 239.7 | 1,888 | 815 |
| Caddo | 10,182 | 35,717,990 | 6.9 | 16.7 | 7.8 | 2.2 | 3.1 | 5.0 | 55.4 | 1,104.5 | 336.7 | 572.6 | 2,328 | 1,246 |
| Calcasieu | 9,753 | 32,272,710 | 5.9 | 12.9 | 3.8 | 5.4 | 9.4 | 5.0 | 54.9 | 1,094.7 | 277.6 | 593.2 | 2,930 | 1,002 |
| Caldwell | 351 | 883,483 | 7.9 | 0.7 | 0.1 | 4.0 | 0.0 | 2.7 | 83.8 | 35.3 | 17.9 | 16.2 | 1,630 | 1,319 |
| Cameron | 469 | 1,641,476 | 6.8 | 0.0 | 0.3 | 7.9 | 10.9 | 3.7 | 70.4 | 46.9 | 12.6 | 28.9 | 4,168 | 4,125 |
| Catahoula | 504 | 1,610,614 | 10.2 | 10.1 | 0.5 | 4.4 | 0.0 | 3.2 | 70.4 | 29.4 | 17.6 | 7.8 | 792 | 256 |
| Claiborne | 703 | 2,118,405 | 5.7 | 2.4 | 0.2 | 2.5 | 55.0 | 1.8 | 32.3 | 71.5 | 18.6 | 13.8 | 864 | 502 |
| Concordia | 1,215 | 3,846,133 | 5.2 | 22.1 | 2.4 | 2.9 | 18.8 | 5.7 | 42.7 | 108.1 | 31.6 | 43.6 | 2,197 | 585 |
| De Soto | 1,230 | 4,077,671 | 3.7 | 19.0 | 0.9 | 6.1 | 1.1 | 5.6 | 63.2 | 113.9 | 21.8 | 80.3 | 2,944 | 1,980 |
| East Baton Rouge | 14,509 | 57,542,743 | 7.2 | 13.3 | 6.7 | 5.4 | 8.4 | 5.0 | 53.9 | 1,882.7 | 474.8 | 1,033.1 | 2,324 | 998 |
| East Carroll | 326 | 915,955 | 5.6 | 6.1 | 0.9 | 1.7 | 42.3 | 2.8 | 40.1 | 47.2 | 23.4 | 8.3 | 1,164 | 583 |
| East Feliciana | 588 | 1,575,919 | 11.9 | 28.9 | 0.2 | 2.6 | 0.5 | 5.9 | 49.7 | 46.9 | 22.2 | 19.7 | 1,014 | 397 |
| Evangeline | 1,505 | 4,498,451 | 5.1 | 7.6 | 2.8 | 3.2 | 16.2 | 6.2 | 57.9 | 129.4 | 52.4 | 42.4 | 1,261 | 597 |
| Franklin | 1,063 | 3,246,498 | 5.1 | 13.9 | 1.1 | 1.7 | 35.0 | 1.9 | 41.3 | 97.0 | 34.3 | 26.0 | 1,283 | 480 |
| Grant | 536 | 1,408,063 | 9.7 | 2.7 | 0.0 | 2.1 | 3.1 | 5.9 | 76.0 | 47.9 | 27.2 | 14.6 | 655 | 353 |
| Iberia | 3,364 | 11,267,277 | 3.2 | 9.1 | 2.8 | 2.7 | 26.7 | 2.7 | 52.1 | 335.0 | 128.8 | 102.9 | 1,429 | 625 |
| Iberville | 1,420 | 4,860,218 | 7.6 | 15.9 | 1.3 | 5.5 | 1.5 | 5.5 | 61.4 | 172.8 | 32.8 | 124.9 | 3,797 | 1,710 |
| Jackson | 927 | 2,823,927 | 4.3 | 17.0 | 1.6 | 5.1 | 28.1 | 4.9 | 38.1 | 63.8 | 17.7 | 30.1 | 1,894 | 1,216 |
| Jefferson | 12,733 | 58,502,045 | 9.8 | 14.0 | 3.4 | 2.8 | 16.6 | 7.5 | 43.9 | 1,960.3 | 534.4 | 797.7 | 1,829 | 825 |
| Jefferson Davis | 1,389 | 3,649,957 | 4.4 | 11.4 | 2.1 | 2.3 | 1.2 | 2.8 | 75.3 | 126.0 | 48.3 | 55.6 | 1,766 | 640 |
| Lafayette | 8,689 | 26,026,218 | 7.5 | 14.2 | 4.9 | 3.7 | 1.2 | 9.4 | 58.1 | 814.0 | 210.7 | 478.2 | 1,978 | 829 |
| Lafourche | 4,703 | 18,572,811 | 6.0 | 10.7 | 0.0 | 3.8 | 35.6 | 3.2 | 38.8 | 639.0 | 210.9 | 204.4 | 2,083 | 1,459 |
| La Salle | 1,090 | 3,428,899 | 2.1 | 12.0 | 0.3 | 0.6 | 46.6 | 2.9 | 35.0 | 89.8 | 21.7 | 21.5 | 1,441 | 703 |
| Lincoln | 1,670 | 4,901,269 | 8.2 | 10.3 | 5.5 | 3.9 | 1.0 | 10.0 | 60.5 | 146.8 | 50.6 | 78.0 | 1,642 | 735 |
| Livingston | 4,347 | 31,808,462 | 58.0 | 4.0 | 0.4 | 0.8 | 0.2 | 1.7 | 34.6 | 424.4 | 223.5 | 156.6 | 1,137 | 366 |
| Madison | 571 | 1,239,152 | 10.5 | 36.2 | 0.0 | 5.0 | 0.3 | 1.7 | 45.6 | 58.8 | 20.3 | 19.7 | 1,733 | 1,052 |
| Morehouse | 1,147 | 3,631,339 | 4.9 | 13.6 | 4.3 | 2.1 | 26.2 | 1.3 | 46.5 | 108.4 | 42.5 | 31.4 | 1,223 | 546 |
| Natchitoches | 2,527 | 8,229,190 | 6.7 | 10.8 | 2.2 | 2.0 | 39.9 | 5.3 | 31.4 | 168.2 | 57.3 | 61.1 | 1,566 | 612 |

1. Based on the resident population estimated as of July 1 of the year shown.

# Table B. States and Counties — Local Government Finances, Government Employment, and Income Taxes

| | Local government finances, 2017 (cont.) | | | | | | | | | Government employment, 2019 | | | Individual income tax returns, 2018 | | |
|---|---|---|---|---|---|---|---|---|---|---|---|---|---|---|---|
| | Direct general expenditure | | | | | | | Debt outstanding | | | | | | Mean adjusted gross income | Mean income tax |
| STATE County | Total (mil dol) | Per capita[1] (dollars) | Education | Health and hospitals | Police protection | Public welfare | Highways | Total (mil dol) | Per capita[1] (dollars) | Federal civilian | Federal military | State and local | Number of returns | | |
| | | | | | Percent of total for: | | | | | | | | | | |
| | 185 | 186 | 187 | 188 | 189 | 190 | 191 | 192 | 193 | 194 | 195 | 196 | 197 | 198 | 199 |

| STATE County | 185 | 186 | 187 | 188 | 189 | 190 | 191 | 192 | 193 | 194 | 195 | 196 | 197 | 198 | 199 |
|---|---|---|---|---|---|---|---|---|---|---|---|---|---|---|---|
| KENTUCKY—Cont'd | | | | | | | | | | | | | | | |
| Ohio | 53 | 2,201 | 64.8 | 0.5 | 4.5 | 0.1 | 6.6 | 256.1 | 10,621 | 81 | 71 | 1,521 | 9,450 | 45,446 | 3,082 |
| Oldham | 163 | 2,442 | 68.7 | 1.7 | 4.3 | 0.1 | 2.8 | 709.0 | 10,652 | 71 | 189 | 3,230 | 28,850 | 103,751 | 14,374 |
| Owen | 23 | 2,127 | 74.7 | 4.8 | 1.2 | 2.5 | 5.5 | 207.1 | 19,183 | 19 | 33 | 432 | 4,400 | 48,683 | 3,595 |
| Owsley | 13 | 2,863 | 71.8 | 0.1 | 1.7 | 0.0 | 7.6 | 4.9 | 1,118 | 9 | 13 | 284 | 1,300 | 36,221 | 1,968 |
| Pendleton | 36 | 2,470 | 53.3 | 15.5 | 3.4 | 0.0 | 5.6 | 42.4 | 2,902 | 31 | 43 | 589 | 6,240 | 51,608 | 4,041 |
| Perry | 122 | 4,601 | 51.5 | 28.0 | 2.0 | 0.0 | 2.9 | 363.2 | 13,663 | 145 | 76 | 2,352 | 9,710 | 45,217 | 3,839 |
| Pike | 176 | 2,991 | 63.8 | 3.1 | 1.7 | 0.2 | 3.8 | 190.9 | 3,240 | 189 | 170 | 2,905 | 19,490 | 50,382 | 4,645 |
| Powell | 37 | 3,016 | 67.4 | 5.3 | 2.1 | 0.0 | 3.3 | 31.3 | 2,543 | 31 | 37 | 752 | 4,970 | 39,352 | 2,526 |
| Pulaski | 148 | 2,296 | 58.2 | 1.5 | 3.6 | 0.0 | 4.4 | 197.7 | 3,073 | 213 | 194 | 3,766 | 25,940 | 47,289 | 4,185 |
| Robertson | 7 | 3,151 | 52.5 | 1.8 | 2.9 | 0.1 | 5.2 | 22.3 | 10,497 | 1 | 6 | 150 | 840 | 41,863 | 2,642 |
| Rockcastle | 36 | 2,163 | 66.8 | 8.3 | 2.6 | 0.0 | 3.1 | 39.2 | 2,336 | 32 | 49 | 727 | 5,980 | 41,709 | 2,638 |
| Rowan | 68 | 2,792 | 41.0 | 10.9 | 2.8 | 0.4 | 4.1 | 92.3 | 3,764 | 90 | 68 | 3,088 | 8,660 | 47,561 | 3,807 |
| Russell | 62 | 3,498 | 41.3 | 35.6 | 2.0 | 0.0 | 3.0 | 72.0 | 4,064 | 61 | 53 | 1,011 | 7,070 | 41,105 | 2,902 |
| Scott | 140 | 2,551 | 51.0 | 8.4 | 5.0 | 0.1 | 2.7 | 615.6 | 11,237 | 61 | 168 | 2,428 | 25,290 | 65,757 | 6,454 |
| Shelby | 104 | 2,210 | 63.7 | 2.5 | 3.4 | 0.2 | 2.9 | 159.5 | 3,378 | 85 | 142 | 2,402 | 21,720 | 68,654 | 7,512 |
| Simpson | 48 | 2,649 | 54.6 | 4.0 | 6.6 | 0.1 | 4.8 | 76.6 | 4,246 | 37 | 55 | 846 | 8,450 | 46,089 | 3,631 |
| Spencer | 47 | 2,512 | 78.8 | 2.3 | 3.0 | 0.1 | 1.7 | 46.9 | 2,520 | 24 | 58 | 635 | 8,670 | 66,244 | 6,086 |
| Taylor | 161 | 6,320 | 32.9 | 52.6 | 1.0 | 0.0 | 1.4 | 194.0 | 7,596 | 82 | 74 | 1,829 | 10,530 | 43,856 | 3,247 |
| Todd | 29 | 2,371 | 57.8 | 5.5 | 3.2 | 0.0 | 5.3 | 127.1 | 10,441 | 35 | 36 | 619 | 4,800 | 48,255 | 3,616 |
| Trigg | 30 | 2,090 | 57.5 | 4.8 | 4.5 | 0.1 | 8.4 | 35.0 | 2,429 | 76 | 44 | 581 | 6,090 | 47,051 | 3,750 |
| Trimble | 24 | 2,849 | 45.9 | 2.6 | 1.0 | 0.0 | 4.1 | 557.0 | 65,374 | 17 | 25 | 330 | 3,710 | 50,252 | 3,842 |
| Union | 40 | 2,710 | 50.8 | 0.6 | 4.4 | 1.4 | 4.7 | 20.2 | 1,383 | 48 | 38 | 685 | 5,710 | 51,268 | 4,472 |
| Warren | 281 | 2,174 | 56.4 | 0.1 | 5.6 | 0.1 | 5.2 | 618.5 | 4,792 | 368 | 388 | 9,418 | 54,970 | 57,371 | 5,790 |
| Washington | 33 | 2,760 | 53.0 | 6.0 | 1.9 | 0.1 | 7.4 | 80.5 | 6,776 | 31 | 36 | 476 | 5,320 | 47,197 | 3,819 |
| Wayne | 42 | 2,024 | 65.8 | 3.3 | 2.6 | 0.1 | 4.8 | 45.9 | 2,223 | 39 | 60 | 1,003 | 7,440 | 38,287 | 2,483 |
| Webster | 35 | 2,684 | 55.0 | 3.4 | 3.0 | 0.1 | 6.2 | 120.2 | 9,241 | 41 | 38 | 648 | 5,230 | 50,853 | 3,778 |
| Whitley | 122 | 3,382 | 64.8 | 6.4 | 3.1 | 0.0 | 3.9 | 226.9 | 6,290 | 98 | 104 | 2,236 | 13,450 | 42,491 | 3,220 |
| Wolfe | 21 | 2,869 | 76.0 | 0.1 | 0.6 | 0.0 | 6.8 | 34.0 | 4,685 | 21 | 21 | 400 | 2,310 | 35,794 | 2,068 |
| Woodford | 63 | 2,405 | 55.7 | 4.6 | 9.7 | 0.0 | 3.1 | 76.2 | 2,887 | 44 | 79 | 1,499 | 12,620 | 63,410 | 7,012 |
| LOUISIANA | X | X | X | X | X | X | X | X | X | 31,692 | 33,774 | 293,551 | 1,962,550 | 61,210 | 6,945 |
| Acadia | 192 | 3,071 | 49.0 | 17.9 | 8.3 | 0.3 | 4.8 | 37.1 | 593 | 100 | 232 | 2,554 | 24,240 | 50,309 | 4,458 |
| Allen | 272 | 10,640 | 17.3 | 2.3 | 12.9 | 0.8 | 2.2 | 352.6 | 13,784 | 577 | 81 | 3,684 | 8,490 | 50,824 | 3,978 |
| Ascension | 615 | 4,996 | 52.9 | 1.0 | 1.4 | 0.0 | 25.0 | 737.2 | 5,985 | 173 | 479 | 5,028 | 54,080 | 73,718 | 8,049 |
| Assumption | 65 | 2,866 | 67.1 | 0.1 | 5.3 | 0.7 | 2.6 | 11.9 | 529 | 33 | 82 | 851 | 9,100 | 56,988 | 5,230 |
| Avoyelles | 100 | 2,448 | 53.8 | 11.9 | 14.5 | 0.0 | 3.3 | 14.2 | 348 | 84 | 140 | 3,153 | 15,520 | 45,624 | 3,941 |
| Beauregard | 140 | 3,791 | 44.7 | 27.3 | 8.2 | 0.0 | 1.4 | 85.2 | 2,313 | 73 | 138 | 1,698 | 14,130 | 59,844 | 5,298 |
| Bienville | 49 | 3,614 | 69.9 | 0.4 | 3.6 | 0.0 | 8.8 | 26.1 | 1,912 | 43 | 49 | 859 | 5,380 | 44,309 | 3,329 |
| Bossier | 444 | 3,499 | 60.3 | 1.3 | 5.3 | 0.2 | 2.6 | 854.8 | 6,733 | 1,945 | 5,664 | 6,405 | 54,380 | 60,595 | 5,939 |
| Caddo | 1,112 | 4,520 | 45.1 | 2.6 | 8.6 | 0.0 | 2.7 | 418.8 | 1,703 | 2,834 | 923 | 15,174 | 105,730 | 58,314 | 6,928 |
| Calcasieu | 998 | 4,931 | 39.1 | 8.6 | 7.2 | 0.4 | 7.5 | 1,108.5 | 5,475 | 560 | 841 | 13,405 | 88,560 | 64,743 | 7,201 |
| Caldwell | 26 | 2,624 | 74.6 | 0.0 | 0.9 | 0.0 | 4.3 | 14.0 | 1,405 | 28 | 35 | 599 | 3,650 | 54,127 | 4,664 |
| Cameron | 54 | 7,798 | 45.5 | 10.4 | 16.4 | 0.0 | 6.8 | 3.8 | 543 | 15 | 26 | 750 | 3,130 | 70,330 | 8,641 |
| Catahoula | 33 | 3,373 | 51.9 | 13.2 | 3.4 | 0.0 | 5.7 | 9.7 | 990 | 45 | 33 | 468 | 3,380 | 47,480 | 4,036 |
| Claiborne | 68 | 4,254 | 28.4 | 53.2 | 1.9 | 0.4 | 4.2 | 7.7 | 485 | 45 | 48 | 1,188 | 5,250 | 49,046 | 4,100 |
| Concordia | 87 | 4,366 | 45.0 | 19.9 | 5.0 | 0.0 | 2.2 | 41.6 | 2,098 | 61 | 68 | 1,472 | 6,990 | 45,566 | 3,697 |
| De Soto | 109 | 4,001 | 77.3 | 4.3 | 0.9 | 2.2 | 0.8 | 103.5 | 3,798 | 51 | 103 | 1,673 | 11,820 | 56,797 | 5,624 |
| East Baton Rouge | 1,989 | 4,474 | 38.0 | 7.3 | 8.2 | 0.3 | 2.5 | 1,561.7 | 3,513 | 2,291 | 1,835 | 46,278 | 193,420 | 72,133 | 9,501 |
| East Carroll | 62 | 8,683 | 20.6 | 17.9 | 2.8 | 0.0 | 3.7 | 35.2 | 4,954 | 19 | 22 | 489 | 2,360 | 39,738 | 3,353 |
| East Feliciana | 55 | 2,833 | 52.2 | 0.3 | 9.0 | 0.4 | 13.8 | 11.1 | 573 | 29 | 63 | 2,738 | 7,960 | 64,197 | 7,636 |
| Evangeline | 138 | 4,089 | 41.8 | 24.6 | 5.3 | 0.0 | 7.5 | 46.8 | 1,391 | 50 | 122 | 1,545 | 12,400 | 47,994 | 3,962 |
| Franklin | 99 | 4,868 | 33.1 | 34.6 | 7.2 | 0.0 | 3.3 | 49.7 | 2,452 | 56 | 73 | 1,441 | 7,570 | 40,065 | 3,285 |
| Grant | 44 | 1,977 | 70.1 | 0.1 | 6.5 | 0.0 | 2.5 | 18.5 | 832 | 695 | 82 | 818 | 7,710 | 51,115 | 3,800 |
| Iberia | 338 | 4,689 | 41.2 | 22.9 | 6.7 | 0.1 | 2.9 | 189.4 | 2,631 | 104 | 267 | 4,091 | 29,580 | 52,597 | 5,029 |
| Iberville | 157 | 4,774 | 54.3 | 0.5 | 7.6 | 1.3 | 5.5 | 131.8 | 4,005 | 88 | 108 | 2,903 | 13,170 | 55,337 | 5,347 |
| Jackson | 65 | 4,059 | 40.0 | 31.0 | 2.7 | 0.0 | 7.3 | 13.5 | 852 | 34 | 56 | 1,147 | 5,880 | 50,076 | 3,837 |
| Jefferson | 1,961 | 4,496 | 29.8 | 21.5 | 9.6 | 2.1 | 4.7 | 978.7 | 2,244 | 1,453 | 2,165 | 17,198 | 201,400 | 61,911 | 7,395 |
| Jefferson Davis | 118 | 3,747 | 54.6 | 3.5 | 5.1 | 0.0 | 5.3 | 162.8 | 5,171 | 91 | 117 | 1,933 | 13,160 | 52,331 | 4,654 |
| Lafayette | 837 | 3,462 | 46.0 | 0.2 | 12.4 | 0.1 | 6.9 | 1,034.4 | 4,279 | 1,012 | 956 | 12,979 | 107,570 | 69,595 | 8,621 |
| Lafourche | 462 | 4,712 | 34.9 | 35.5 | 1.6 | 1.1 | 2.9 | 180.4 | 1,838 | 143 | 386 | 6,614 | 39,370 | 47,180 | 6,177 |
| La Salle | 94 | 6,322 | 30.6 | 0.4 | 12.9 | 0.0 | 1.4 | 12.2 | 817 | 48 | 52 | 1,411 | 5,160 | 56,623 | 4,625 |
| Lincoln | 144 | 3,038 | 55.9 | 0.5 | 6.9 | 0.5 | 7.3 | 75.8 | 1,597 | 122 | 169 | 4,344 | 17,410 | 60,637 | 6,769 |
| Livingston | 400 | 2,903 | 75.1 | 0.4 | 4.0 | 0.0 | 4.5 | 278.6 | 2,023 | 156 | 530 | 5,725 | 57,770 | 59,149 | 5,439 |
| Madison | 52 | 4,613 | 33.2 | 30.3 | 8.3 | 0.0 | 5.8 | 57.2 | 5,039 | 35 | 36 | 760 | 3,790 | 33,332 | 2,443 |
| Morehouse | 100 | 3,898 | 50.2 | 19.1 | 2.9 | 0.1 | 2.9 | 88.1 | 3,435 | 54 | 92 | 1,012 | 10,340 | 42,275 | 3,329 |
| Natchitoches | 169 | 4,331 | 42.6 | 19.6 | 5.9 | 0.4 | 2.9 | 49.4 | 1,265 | 164 | 140 | 3,913 | 14,720 | 52,178 | 5,003 |

1. Based on the resident population estimated as of July 1 of the year shown.

# Table B. States and Counties — Land Area and Population

| State / county code | CBSA code[1] | County Type code[2] | STATE County | Land area[3] (sq. mi) | Total persons 2019 | Rank | Per square mile | White | Black | American Indian, Alaska Native | Asian and Pacific Islander | Percent Hispanic or Latino[4] | Under 5 years | 5 to 14 years | 15 to 24 years | 25 to 34 years | 35 to 44 years | 45 to 54 years |
|---|---|---|---|---|---|---|---|---|---|---|---|---|---|---|---|---|---|---|
| | | | | 1 | 2 | 3 | 4 | 5 | 6 | 7 | 8 | 9 | 10 | 11 | 12 | 13 | 14 | 15 |
| | | | LOUISIANA—Cont'd | | | | | | | | | | | | | | | |
| 22071 | 35380 | 1 | Orleans | 169.5 | 389,476 | 182 | 2,297.8 | 32.1 | 59.7 | 0.7 | 3.4 | 5.6 | 5.6 | 10.9 | 11.3 | 16.9 | 14.5 | 11.6 |
| 22073 | 33740 | 3 | Ouachita | 610.3 | 152,439 | 441 | 249.8 | 58.4 | 38.5 | 0.6 | 1.4 | 2.3 | 6.6 | 13.7 | 13.7 | 14.0 | 12.5 | 11.5 |
| 22075 | 35380 | 1 | Plaquemines | 780.2 | 23,113 | 1,677 | 29.6 | 64.9 | 22.0 | 2.5 | 5.1 | 8.1 | 6.5 | 14.7 | 12.9 | 13.0 | 13.3 | 12.6 |
| 22077 | 12940 | 2 | Pointe Coupee | 556.9 | 21,529 | 1,746 | 38.7 | 61.5 | 35.5 | 0.6 | 0.5 | 3.1 | 5.9 | 12.0 | 11.3 | 11.2 | 11.2 | 11.3 |
| 22079 | 10780 | 3 | Rapides | 1,320.4 | 128,567 | 500 | 97.4 | 62.2 | 32.9 | 1.5 | 1.9 | 3.5 | 6.6 | 13.7 | 12.7 | 13.0 | 12.3 | 11.5 |
| 22081 | | 8 | Red River | 389.0 | 8,286 | 2,558 | 21.3 | 56.4 | 40.2 | 0.9 | 0.4 | 3.0 | 5.9 | 13.5 | 11.5 | 11.8 | 10.8 | 12.2 |
| 22083 | | 6 | Richland | 555.6 | 20,014 | 1,823 | 36.0 | 61.7 | 35.8 | 0.6 | 0.6 | 2.4 | 6.7 | 13.0 | 11.9 | 13.2 | 12.4 | 11.1 |
| 22085 | | 6 | Sabine | 866.6 | 23,803 | 1,650 | 27.5 | 70.6 | 17.3 | 10.7 | 0.7 | 4.1 | 5.7 | 13.2 | 11.3 | 11.5 | 11.4 | 11.3 |
| 22087 | 35380 | 1 | St. Bernard | 377.5 | 47,647 | 1,028 | 126.2 | 62.6 | 24.7 | 1.3 | 2.8 | 10.6 | 6.6 | 15.2 | 11.6 | 15.6 | 14.4 | 11.3 |
| 22089 | 35380 | 1 | St. Charles | 277.7 | 52,987 | 955 | 190.8 | 66.0 | 26.7 | 0.8 | 1.6 | 6.5 | 5.9 | 14.0 | 12.4 | 12.8 | 13.4 | 12.5 |
| 22091 | 12940 | 2 | St. Helena | 408.5 | 10,081 | 2,412 | 24.7 | 45.0 | 52.5 | 0.9 | 0.5 | 2.3 | 5.7 | 11.5 | 11.1 | 13.5 | 11.1 | 11.7 |
| 22093 | 35380 | 1 | St. James | 237.9 | 20,727 | 1,781 | 87.1 | 49.3 | 48.6 | 0.5 | 0.5 | 1.8 | 5.7 | 12.7 | 11.7 | 13.2 | 12.0 | 11.4 |
| 22095 | 35380 | 1 | St. John the Baptist | 214.5 | 42,516 | 1,128 | 198.2 | 33.2 | 58.8 | 0.6 | 1.6 | 7.0 | 6.3 | 13.7 | 12.8 | 13.3 | 12.3 | 12.1 |
| 22097 | 36660 | 4 | St. Landry | 924.0 | 81,440 | 701 | 88.1 | 55.2 | 42.3 | 0.6 | 0.7 | 2.4 | 7.2 | 14.8 | 12.7 | 12.4 | 11.6 | 11.1 |
| 22099 | 29180 | 2 | St. Martin | 737.5 | 52,954 | 956 | 71.8 | 65.7 | 30.5 | 0.8 | 1.3 | 3.1 | 6.4 | 13.4 | 11.7 | 13.6 | 12.4 | 12.2 |
| 22101 | 34020 | 4 | St. Mary | 555.8 | 48,330 | 1,019 | 87.0 | 57.9 | 32.6 | 2.4 | 2.0 | 7.2 | 6.7 | 13.3 | 11.3 | 12.7 | 11.3 | 11.8 |
| 22103 | 35380 | 1 | St. Tammany | 845.3 | 263,446 | 265 | 311.7 | 79.0 | 13.8 | 1.0 | 2.0 | 6.0 | 5.9 | 13.4 | 11.8 | 11.3 | 13.0 | 12.6 |
| 22105 | 25220 | 3 | Tangipahoa | 791.2 | 136,765 | 475 | 172.9 | 64.0 | 31.0 | 0.8 | 1.1 | 4.6 | 7.0 | 13.5 | 13.4 | 14.7 | 12.7 | 11.1 |
| 22107 | | 6 | Tensas | 603.0 | 4,178 | 2,881 | 6.9 | 42.5 | 55.0 | 0.6 | 0.6 | 2.6 | 5.2 | 12.7 | 11.0 | 8.5 | 9.6 | 10.3 |
| 22109 | 26380 | 3 | Terrebonne | 1,229.9 | 109,859 | 557 | 89.3 | 68.5 | 20.5 | 7.0 | 1.7 | 5.3 | 6.7 | 14.0 | 12.3 | 13.6 | 12.9 | 11.8 |
| 22111 | 33740 | 3 | Union | 876.9 | 22,170 | 1,711 | 25.3 | 69.8 | 24.8 | 0.7 | 0.4 | 5.1 | 5.7 | 11.9 | 11.3 | 11.7 | 11.0 | 11.7 |
| 22113 | 29180 | 2 | Vermilion | 1,173.6 | 59,378 | 877 | 50.6 | 79.6 | 15.0 | 0.7 | 2.4 | 3.7 | 6.4 | 14.2 | 12.2 | 12.4 | 12.8 | 11.8 |
| 22115 | 22860 | 5 | Vernon | 1,326.7 | 47,894 | 1,024 | 36.1 | 72.9 | 15.3 | 2.1 | 3.6 | 9.7 | 7.8 | 13.1 | 17.8 | 17.5 | 12.3 | 9.4 |
| 22117 | 14220 | 6 | Washington | 669.6 | 45,773 | 1,059 | 68.4 | 66.4 | 31.0 | 0.9 | 0.6 | 2.7 | 6.3 | 13.1 | 12.1 | 12.2 | 12.1 | 11.5 |
| 22119 | 33380 | 2 | Webster | 593.3 | 37,943 | 1,226 | 64.0 | 62.7 | 34.8 | 1.0 | 0.7 | 2.2 | 6.0 | 12.6 | 11.6 | 12.1 | 11.5 | 11.9 |
| 22121 | 12940 | 2 | West Baton Rouge | 192.3 | 26,792 | 1,535 | 139.3 | 55.6 | 40.8 | 0.6 | 1.0 | 3.5 | 6.8 | 13.3 | 11.5 | 15.0 | 13.8 | 11.4 |
| 22123 | | 9 | West Carroll | 359.7 | 10,646 | 2,372 | 29.6 | 78.9 | 16.7 | 0.9 | 0.4 | 4.2 | 5.3 | 13.0 | 12.2 | 12.0 | 11.8 | 11.8 |
| 22125 | 12940 | 2 | West Feliciana | 403.3 | 15,465 | 2,064 | 38.3 | 53.1 | 44.7 | 0.5 | 1.0 | 1.7 | 4.1 | 9.4 | 8.3 | 13.6 | 18.3 | 16.9 |
| 22127 | | 6 | Winn | 950.0 | 13,839 | 2,168 | 14.6 | 66.0 | 31.7 | 1.4 | 0.7 | 2.0 | 5.6 | 11.4 | 11.5 | 13.5 | 12.8 | 12.9 |
| 23000 | | 0 | MAINE | 30,844.8 | 1,350,141 | X | 43.8 | 94.4 | 2.2 | 1.4 | 1.9 | 1.9 | 4.7 | 10.3 | 11.2 | 12.2 | 11.6 | 12.6 |
| 23001 | 30340 | 3 | Androscoggin | 468.0 | 108,547 | 562 | 231.9 | 91.9 | 5.8 | 1.1 | 1.6 | 2.0 | 5.7 | 12.1 | 12.1 | 12.5 | 12.0 | 12.6 |
| 23003 | | 7 | Aroostook | 6,671.1 | 66,804 | 803 | 10.0 | 95.0 | 1.5 | 2.7 | 0.9 | 1.5 | 4.7 | 10.2 | 10.6 | 10.1 | 10.3 | 12.5 |
| 23005 | 38860 | 2 | Cumberland | 836.2 | 298,111 | 232 | 356.5 | 91.7 | 3.9 | 0.9 | 3.3 | 2.3 | 4.7 | 10.2 | 11.7 | 13.9 | 12.7 | 12.8 |
| 23007 | | 6 | Franklin | 1,697.0 | 29,986 | 1,436 | 17.7 | 97.0 | 0.8 | 1.4 | 0.9 | 1.4 | 4.2 | 10.1 | 12.5 | 11.0 | 10.5 | 11.8 |
| 23009 | | 6 | Hancock | 1,587.1 | 55,088 | 931 | 34.7 | 95.6 | 1.3 | 1.1 | 1.6 | 1.7 | 4.1 | 9.4 | 9.4 | 11.2 | 10.6 | 12.3 |
| 23011 | 12300 | 4 | Kennebec | 867.5 | 122,955 | 516 | 141.7 | 96.0 | 1.3 | 1.4 | 1.5 | 1.7 | 4.9 | 10.7 | 11.6 | 12.0 | 11.5 | 12.8 |
| 23013 | | 7 | Knox | 365.1 | 39,951 | 1,188 | 109.4 | 96.6 | 1.1 | 1.3 | 1.0 | 1.6 | 4.0 | 10.0 | 9.7 | 10.3 | 11.1 | 12.2 |
| 23015 | | 8 | Lincoln | 455.9 | 34,775 | 1,310 | 76.3 | 96.8 | 0.9 | 1.1 | 1.2 | 1.4 | 4.4 | 9.3 | 8.9 | 10.2 | 9.8 | 11.8 |
| 23017 | | 6 | Oxford | 2,077.0 | 58,132 | 895 | 28.0 | 96.5 | 1.0 | 1.4 | 1.2 | 1.7 | 4.3 | 10.1 | 10.0 | 10.7 | 11.3 | 12.9 |
| 23019 | 12620 | 3 | Penobscot | 3,397.2 | 151,655 | 443 | 44.6 | 95.0 | 1.5 | 2.1 | 1.7 | 1.6 | 4.6 | 9.8 | 13.6 | 13.2 | 11.5 | 12.6 |
| 23021 | | 8 | Piscataquis | 3,960.9 | 16,996 | 1,974 | 4.3 | 94.8 | 1.0 | 1.4 | 1.5 | 2.8 | 5.4 | 8.8 | 9.3 | 8.8 | 10.0 | 12.5 |
| 23023 | 38860 | 2 | Sagadahoc | 254.0 | 36,044 | 1,274 | 141.9 | 95.8 | 1.5 | 1.0 | 1.4 | 2.0 | 4.5 | 10.5 | 9.7 | 11.4 | 11.7 | 12.6 |
| 23025 | | 6 | Somerset | 3,924.3 | 50,635 | 981 | 12.9 | 97.0 | 1.0 | 1.3 | 1.0 | 1.2 | 4.6 | 10.3 | 10.5 | 10.9 | 11.2 | 13.7 |
| 23027 | | 6 | Waldo | 730.0 | 39,923 | 1,190 | 54.7 | 96.6 | 0.9 | 1.3 | 0.9 | 1.7 | 4.6 | 10.4 | 10.2 | 10.7 | 11.8 | 12.4 |
| 23029 | | 7 | Washington | 2,562.7 | 31,473 | 1,392 | 12.3 | 91.0 | 1.0 | 6.4 | 0.8 | 2.9 | 4.8 | 10.9 | 10.2 | 9.7 | 10.6 | 11.8 |
| 23031 | 38860 | 2 | York | 991.2 | 209,066 | 324 | 210.9 | 95.3 | 1.5 | 1.0 | 1.9 | 2.0 | 4.6 | 10.2 | 10.6 | 12.5 | 11.7 | 12.7 |
| 24000 | | 0 | MARYLAND | 9,711.1 | 6,055,802 | X | 623.6 | 51.7 | 31.6 | 0.8 | 7.9 | 10.8 | 5.9 | 12.4 | 12.3 | 13.6 | 13.1 | 12.8 |
| 24001 | 19060 | 3 | Allegany | 422.2 | 70,057 | 776 | 165.9 | 88.3 | 9.7 | 0.5 | 1.5 | 2.0 | 4.6 | 9.6 | 15.1 | 13.0 | 11.6 | 12.0 |
| 24003 | 12580 | 1 | Anne Arundel | 414.8 | 582,777 | 117 | 1,405.0 | 68.5 | 19.6 | 0.8 | 5.7 | 8.6 | 6.1 | 12.6 | 12.1 | 14.1 | 13.7 | 12.7 |
| 24005 | 12580 | 1 | Baltimore | 598.4 | 826,017 | 75 | 1,380.4 | 56.8 | 31.7 | 0.8 | 7.3 | 6.0 | 5.9 | 12.1 | 12.3 | 13.5 | 12.5 | 12.7 |
| 24009 | 47900 | 1 | Calvert | 213.2 | 93,072 | 642 | 436.5 | 79.6 | 14.8 | 1.0 | 3.3 | 4.5 | 5.4 | 13.2 | 12.3 | 11.7 | 12.5 | 12.1 |
| 24011 | | 6 | Caroline | 319.4 | 33,492 | 1,340 | 104.9 | 76.9 | 15.2 | 0.7 | 1.5 | 8.1 | 6.2 | 13.3 | 11.5 | 12.8 | 12.6 | 13.5 |
| 24013 | 12580 | 1 | Carroll | 447.6 | 169,092 | 396 | 377.8 | 89.6 | 4.8 | 0.5 | 3.1 | 4.0 | 5.5 | 12.1 | 12.1 | 11.5 | 11.9 | 12.6 |
| 24015 | 37980 | 1 | Cecil | 346.3 | 103,419 | 587 | 298.6 | 86.3 | 8.4 | 0.8 | 2.2 | 4.8 | 5.7 | 12.5 | 11.8 | 13.0 | 12.2 | 13.6 |
| 24017 | 47900 | 1 | Charles | 457.8 | 164,436 | 405 | 359.2 | 38.6 | 52.7 | 1.6 | 4.7 | 6.6 | 5.9 | 13.6 | 12.7 | 12.9 | 13.1 | 13.5 |
| 24019 | 15700 | 6 | Dorchester | 540.8 | 31,853 | 1,384 | 58.9 | 64.0 | 29.7 | 0.9 | 1.6 | 6.1 | 5.6 | 11.7 | 10.3 | 12.3 | 10.4 | 14.6 |
| 24021 | 47900 | 1 | Frederick | 660.6 | 265,161 | 261 | 401.4 | 73.2 | 11.9 | 0.7 | 6.3 | 10.9 | 5.9 | 13.1 | 12.3 | 12.7 | 13.7 | 11.7 |
| 24023 | | 6 | Garrett | 649.1 | 28,852 | 1,465 | 44.4 | 97.0 | 1.5 | 0.5 | 0.7 | 1.2 | 4.7 | 9.8 | 10.5 | 11.2 | 11.0 | 13.5 |
| 24025 | 12580 | 1 | Harford | 437.1 | 256,805 | 272 | 587.5 | 76.9 | 16.2 | 0.8 | 4.2 | 4.9 | 5.5 | 12.7 | 11.9 | 12.5 | 12.9 | 13.1 |
| 24027 | 12580 | 1 | Howard | 251.0 | 328,200 | 216 | 1,307.6 | 52.3 | 21.6 | 0.8 | 21.6 | 7.5 | 5.7 | 14.0 | 12.2 | 12.1 | 14.4 | 13.8 |
| 24029 | | 6 | Kent | 277.0 | 19,192 | 1,869 | 69.3 | 79.5 | 15.2 | 0.7 | 1.8 | 4.7 | 4.0 | 8.5 | 14.3 | 10.5 | 9.1 | 10.3 |
| 24031 | 47900 | 1 | Montgomery | 493.1 | 1,051,816 | 45 | 2,133.1 | 44.9 | 20.2 | 0.6 | 17.3 | 20.0 | 6.0 | 12.9 | 11.7 | 12.4 | 13.8 | 13.5 |
| 24033 | 47900 | 1 | Prince George's | 482.7 | 909,612 | 62 | 1,884.4 | 13.4 | 63.3 | 0.9 | 4.9 | 19.8 | 6.5 | 12.2 | 12.8 | 14.5 | 13.4 | 13.2 |
| 24035 | 12580 | 1 | Queen Anne's | 371.7 | 51,167 | 973 | 137.7 | 87.9 | 7.0 | 0.7 | 1.9 | 4.5 | 5.2 | 11.9 | 11.2 | 10.5 | 11.2 | 13.7 |

1. CBSA = Core Based Statistical Area. See Appendix A for explanation. See Appendix B for list of metropolitan areas with component counties. Service of USDA Rural-Urban Continuum Codes. See Appendix A for definition.   2. County type code from the Economic Research Service.   3. Dry land or land partially or temporarily covered by water.   4. May be of any race.

| STATE County | 55 to 64 years (16) | 65 to 74 years (17) | 75 years and over (18) | Percent female (19) | 2000 (20) | 2010 (21) | 2000-2010 (22) | 2010-2020 (23) | Births (24) | Deaths (25) | Net Migration (26) | Number (27) | Persons per household (28) | Family households (29) | Female family householder[1] (30) | One person (31) |
|---|---|---|---|---|---|---|---|---|---|---|---|---|---|---|---|---|
| **LOUISIANA—Cont'd** | | | | | | | | | | | | | | | | |
| Orleans | 13.0 | 10.2 | 6.1 | 52.9 | 484,674 | 343,828 | -29.1 | 13.3 | 48,524 | 33,412 | 29,591 | 153,819 | 2.45 | 47.0 | 16.4 | 45.9 |
| Ouachita | 12.2 | 9.3 | 6.5 | 52.2 | 147,250 | 153,732 | 4.4 | -0.8 | 22,542 | 16,032 | -7,707 | 56,556 | 2.63 | 62.9 | 17.8 | 32.1 |
| Plaquemines | 12.5 | 8.6 | 6.0 | 50.0 | 26,757 | 23,037 | -13.9 | 0.3 | 2,964 | 1,861 | -1,042 | 8,919 | 2.57 | 75.3 | 17.2 | 21.5 |
| Pointe Coupee | 15.1 | 13.0 | 8.9 | 52.1 | 22,763 | 22,800 | 0.2 | -5.6 | 2,806 | 2,613 | -1,459 | 8,960 | 2.44 | 62.8 | 17.3 | 32.6 |
| Rapides | 13.1 | 9.9 | 7.2 | 51.6 | 126,337 | 131,592 | 4.2 | -2.3 | 18,231 | 14,777 | -6,421 | 48,488 | 2.61 | 67.1 | 18.7 | 28.4 |
| Red River | 14.5 | 11.4 | 8.5 | 51.9 | 9,622 | 9,091 | -5.5 | -8.9 | 1,176 | 1,132 | -850 | 3,372 | 2.49 | 58.8 | 12.3 | 37.7 |
| Richland | 13.4 | 10.9 | 7.4 | 51.9 | 20,981 | 20,722 | -1.2 | -3.4 | 2,796 | 2,442 | -1,055 | 7,459 | 2.59 | 66.7 | 18.8 | 30.1 |
| Sabine | 13.9 | 12.4 | 9.3 | 50.7 | 23,459 | 24,229 | 3.3 | -1.8 | 2,956 | 2,830 | -539 | 9,158 | 2.58 | 66.1 | 12.8 | 30.1 |
| St. Bernard | 12.7 | 8.1 | 4.5 | 50.9 | 67,229 | 35,897 | -46.6 | 32.7 | 6,388 | 3,610 | 8,773 | 15,005 | 3.06 | 68.4 | 18.7 | 26.0 |
| St. Charles | 14.5 | 9.2 | 5.3 | 51.1 | 48,072 | 52,888 | 10.0 | 0.2 | 6,411 | 4,360 | -1,973 | 19,212 | 2.72 | 75.3 | 14.8 | 20.2 |
| St. Helena | 14.3 | 12.5 | 8.7 | 51.8 | 10,525 | 11,213 | 6.5 | -10.1 | 1,181 | 1,216 | -1,106 | 3,857 | 2.63 | 71.7 | 16.9 | 26.7 |
| St. James | 14.7 | 11.1 | 7.5 | 51.6 | 21,216 | 22,105 | 4.2 | -6.2 | 2,692 | 2,101 | -1,971 | 7,719 | 2.73 | 74.3 | 20.7 | 22.9 |
| St. John the Baptist | 14.2 | 9.6 | 5.6 | 51.4 | 43,044 | 45,810 | 6.4 | -7.2 | 5,733 | 4,227 | -4,876 | 15,270 | 2.79 | 72.7 | 19.2 | 22.9 |
| St. Landry | 13.2 | 10.0 | 7.0 | 51.8 | 87,700 | 83,400 | -4.9 | -2.4 | 12,844 | 10,028 | -4,726 | 30,485 | 2.69 | 66.7 | 17.5 | 29.7 |
| St. Martin | 14.0 | 9.9 | 6.4 | 50.8 | 48,583 | 52,160 | 7.4 | 1.5 | 7,304 | 4,985 | -1,499 | 19,749 | 2.71 | 70.7 | 15.2 | 25.0 |
| St. Mary | 14.9 | 10.5 | 7.5 | 50.8 | 53,500 | 54,647 | 2.1 | -11.6 | 7,372 | 5,767 | -7,984 | 19,856 | 2.52 | 57.2 | 13.5 | 38.8 |
| St. Tammany | 13.9 | 11.2 | 6.9 | 51.4 | 191,268 | 233,760 | 22.2 | 12.7 | 29,404 | 22,486 | 22,818 | 92,962 | 2.72 | 70.3 | 13.0 | 24.7 |
| Tangipahoa | 12.3 | 9.6 | 5.6 | 51.6 | 100,588 | 121,083 | 20.4 | 13.0 | 19,674 | 12,629 | 8,635 | 47,597 | 2.7 | 66.8 | 17.6 | 27.6 |
| Tensas | 16.2 | 15.6 | 11.0 | 51.1 | 6,618 | 5,250 | -20.7 | -20.4 | 566 | 578 | -1,068 | 1,792 | 2.53 | 61.5 | 17.1 | 34.6 |
| Terrebonne | 13.3 | 9.3 | 6.1 | 50.8 | 104,503 | 111,589 | 6.8 | -1.6 | 16,435 | 10,758 | -7,387 | 39,972 | 2.77 | 67.9 | 16.4 | 25.3 |
| Union | 15.2 | 12.4 | 9.2 | 50.6 | 22,803 | 22,781 | -0.1 | -2.7 | 2,744 | 2,713 | -640 | 7,582 | 2.88 | 64.6 | 12.1 | 32.9 |
| Vermilion | 13.8 | 9.5 | 6.8 | 51.7 | 53,807 | 57,958 | 7.7 | 2.5 | 7,992 | 5,820 | -710 | 22,086 | 2.69 | 68.6 | 12.8 | 24.3 |
| Vernon | 9.2 | 7.6 | 5.4 | 46.8 | 52,531 | 52,357 | -0.3 | -8.5 | 10,086 | 4,079 | -10,596 | 17,696 | 2.68 | 69.9 | 11.0 | 26.9 |
| Washington | 13.7 | 11.3 | 7.8 | 50.8 | 43,926 | 47,134 | 7.3 | -2.9 | 5,891 | 6,086 | -1,143 | 17,613 | 2.53 | 67.7 | 19.1 | 28.3 |
| Webster | 13.8 | 11.5 | 9.1 | 51.3 | 41,831 | 41,206 | -1.5 | -7.9 | 5,006 | 5,451 | -2,820 | 16,551 | 2.31 | 62.2 | 14.5 | 32.8 |
| West Baton Rouge | 13.3 | 9.0 | 5.9 | 51.1 | 21,601 | 23,798 | 10.2 | 12.6 | 3,670 | 2,100 | 1,399 | 9,643 | 2.61 | 74.6 | 13.6 | 21.0 |
| West Carroll | 13.6 | 11.1 | 9.2 | 49.4 | 12,314 | 11,604 | -5.8 | -8.3 | 1,327 | 1,464 | -813 | 4,084 | 2.6 | 65.9 | 14.9 | 30.6 |
| West Feliciana | 13.3 | 9.8 | 6.3 | 34.8 | 15,111 | 15,620 | 3.4 | | 1,258 | 1,292 | -118 | 3,869 | 3.09 | 64.5 | 13.2 | 32.6 |
| Winn | 12.7 | 11.0 | 8.6 | 46.3 | 16,894 | 15,313 | -9.4 | -9.6 | 1,688 | 1,785 | -1,379 | 5,483 | 2.26 | 66.8 | 14.6 | 30.2 |
| **MAINE** | 15.6 | 13.0 | 8.8 | 51.0 | 1,274,923 | 1,328,354 | 4.2 | 1.6 | 129,051 | 143,362 | 36,945 | 559,921 | 2.32 | 62.0 | 9.0 | 29.6 |
| Androscoggin | 14.5 | 10.9 | 7.5 | 51.1 | 103,793 | 107,713 | 3.8 | 0.8 | 13,026 | 11,262 | -890 | 45,630 | 2.29 | 61.4 | 10.1 | 29.1 |
| Aroostook | 16.2 | 14.4 | 10.9 | 50.4 | 73,938 | 71,873 | -2.8 | -7.1 | 6,679 | 8,934 | -2,816 | 29,516 | 2.22 | 62.0 | 8.1 | 33.0 |
| Cumberland | 14.4 | 11.6 | 8.0 | 51.5 | 265,612 | 281,687 | 6.1 | 5.8 | 28,407 | 27,029 | 15,216 | 120,644 | 2.35 | 59.7 | 7.8 | 30.1 |
| Franklin | 16.7 | 14.1 | 9.2 | 50.9 | 29,467 | 30,767 | 4.4 | -2.5 | 2,565 | 3,386 | 50 | 11,848 | 2.43 | 61.2 | 7.7 | 31.6 |
| Hancock | 16.7 | 15.9 | 10.4 | 51.6 | 51,791 | 54,407 | 5.1 | 1.3 | 4,684 | 6,180 | 2,228 | 23,661 | 2.24 | 63.0 | 8.3 | 29.8 |
| Kennebec | 15.7 | 12.5 | 8.4 | 51.3 | 117,114 | 122,151 | 4.3 | 0.7 | 12,047 | 13,701 | 2,556 | 52,105 | 2.26 | 61.2 | 10.7 | 30.3 |
| Knox | 15.8 | 16.0 | 11.0 | 50.6 | 39,618 | 39,732 | 0.3 | 0.6 | 3,436 | 4,765 | 1,588 | 17,020 | 2.24 | 62.2 | 9.7 | 31.9 |
| Lincoln | 16.3 | 17.1 | 12.2 | 51.0 | 33,616 | 34,444 | 2.5 | 1.0 | 2,881 | 4,305 | 1,789 | 15,336 | 2.19 | 63.9 | 9.7 | 29.3 |
| Oxford | 17.7 | 14.0 | 9.0 | 50.3 | 54,755 | 57,829 | 5.6 | 0.5 | 5,140 | 6,757 | 1,963 | 21,338 | 2.66 | 65.6 | 9.0 | 28.1 |
| Penobscot | 15.2 | 11.7 | 7.8 | 50.4 | 144,919 | 153,932 | 6.2 | -1.5 | 14,727 | 16,299 | -615 | 62,156 | 2.33 | 60.7 | 9.6 | 29.7 |
| Piscataquis | 17.9 | 16.4 | 10.8 | 49.5 | 17,235 | 17,535 | 1.7 | -3.1 | 1,427 | 2,399 | 450 | 7,025 | 2.34 | 62.7 | 8.6 | 30.6 |
| Sagadahoc | 15.8 | 13.8 | 10.0 | 51.3 | 35,214 | 35,287 | 0.2 | 2.1 | 3,359 | 3,704 | 1,119 | 15,980 | 2.21 | 63.4 | 7.7 | 30.9 |
| Somerset | 16.5 | 13.5 | 8.8 | 50.2 | 50,888 | 52,222 | 2.6 | -3.0 | 4,909 | 6,009 | -460 | 21,321 | 2.33 | 64.2 | 10.8 | 27.8 |
| Waldo | 16.1 | 15.3 | 8.7 | 50.7 | 36,280 | 38,791 | 6.9 | 2.9 | 3,676 | 4,160 | 1,652 | 17,236 | 2.26 | 64.4 | 8.7 | 27.1 |
| Washington | 16.4 | 15.6 | 10.0 | 51.0 | 33,941 | 32,855 | -3.2 | -4.2 | 3,069 | 4,406 | -21 | 13,791 | 2.21 | 61.8 | 11.1 | 32.0 |
| York | 16.1 | 13.0 | 8.7 | 51.2 | 186,742 | 197,129 | 5.6 | 6.1 | 19,019 | 20,066 | 13,136 | 85,314 | 2.36 | 64.4 | 8.9 | 27.4 |
| **MARYLAND** | 13.5 | 9.6 | 6.8 | 51.6 | 5,296,486 | 5,773,787 | 9.0 | 4.9 | 741,516 | 489,918 | 33,437 | 2,205,204 | 2.67 | 66.6 | 13.8 | 27.4 |
| Allegany | 13.2 | 11.3 | 9.6 | 47.7 | 74,930 | 75,042 | 0.1 | -6.6 | 6,844 | 9,311 | -2,504 | 27,399 | 2.31 | 60.5 | 12.0 | 31.9 |
| Anne Arundel | 13.4 | 9.2 | 6.3 | 50.5 | 489,656 | 537,635 | 9.8 | 8.4 | 70,690 | 43,554 | 18,334 | 209,814 | 2.64 | 69.8 | 11.5 | 24.5 |
| Baltimore | 13.6 | 10.2 | 7.8 | 52.7 | 754,292 | 805,311 | 6.8 | 2.6 | 100,239 | 83,462 | 4,542 | 313,559 | 2.58 | 64.9 | 14.1 | 29.0 |
| Calvert | 15.4 | 9.4 | 6.6 | 50.5 | 74,563 | 88,741 | 19.0 | 4.9 | 9,331 | 6,952 | 1,990 | 31,973 | 2.71 | 71.2 | 13.2 | 21.8 |
| Caroline | 14.6 | 10.2 | 7.0 | 51.1 | 29,772 | 33,078 | 11.1 | 1.3 | 4,085 | 3,497 | -181 | 12,024 | 2.71 | 74.4 | 8.1 | 20.8 |
| Carroll | 15.2 | 10.3 | 7.5 | 50.5 | 150,897 | 167,155 | 10.8 | 1.2 | 16,827 | 15,737 | 925 | 60,758 | 2.7 | 74.4 | 8.1 | 20.8 |
| Cecil | 15.1 | 10.3 | 6.3 | 50.4 | 85,951 | 101,091 | 17.6 | 2.3 | 11,545 | 9,619 | 480 | 37,058 | 2.73 | 70.1 | 11.9 | 23.6 |
| Charles | 13.9 | 8.0 | 5.3 | 51.9 | 120,546 | 146,564 | 21.6 | 12.2 | 18,978 | 10,682 | 9,659 | 56,520 | 2.78 | 73.0 | 16.4 | 22.8 |
| Dorchester | 15.5 | 13.0 | 9.6 | 52.8 | 30,674 | 32,623 | 6.4 | -2.4 | 3,800 | 3,963 | -588 | 13,183 | 2.39 | 64.8 | 17.0 | 30.3 |
| Frederick | 13.5 | 9.0 | 6.3 | 50.7 | 195,277 | 233,415 | 19.5 | 13.6 | 29,021 | 18,076 | 20,993 | 92,526 | 2.67 | 71.6 | 9.9 | 22.2 |
| Garrett | 16.4 | 13.4 | 10.1 | 50.5 | 29,846 | 30,139 | 1.0 | -4.3 | 2,911 | 3,315 | -875 | 12,425 | 2.3 | 66.3 | 9.4 | 28.9 |
| Harford | 14.5 | 10.1 | 6.9 | 51.1 | 218,590 | 244,824 | 12.0 | 4.9 | 27,495 | 21,280 | 5,930 | 93,955 | 2.67 | 72.0 | 10.1 | 23.1 |
| Howard | 13.1 | 8.7 | 5.9 | 51.1 | 247,842 | 287,110 | 15.8 | 14.3 | 35,516 | 16,850 | 22,579 | 114,170 | 2.77 | 73.4 | 10.4 | 21.5 |
| Kent | 15.8 | 14.7 | 12.7 | 52.1 | 19,197 | 20,195 | 5.2 | -5.0 | 1,644 | 2,662 | 28 | 8,025 | 2.23 | 60.4 | 8.8 | 33.3 |
| Montgomery | 13.1 | 9.3 | 7.2 | 51.6 | 873,341 | 971,279 | 11.2 | 8.3 | 131,789 | 61,323 | 10,140 | 370,950 | 2.79 | 69.9 | 11.0 | 24.9 |
| Prince George's | 13.2 | 8.9 | 5.3 | 52.0 | 801,515 | 864,028 | 7.8 | 5.3 | 124,424 | 61,060 | -17,304 | 311,343 | 2.86 | 65.3 | 18.9 | 29.0 |
| Queen Anne's | 16.4 | 11.5 | 8.4 | 50.3 | 40,563 | 47,789 | 17.8 | 7.1 | 4,872 | 4,265 | 2,815 | 18,577 | 2.65 | 72.8 | 8.7 | 22.7 |

1. No spouse present.

# Table B. States and Counties — Population, Vital Statistics, Health, and Crime

| STATE County | Persons in group quarters, 2020 | Daytime Population, 2015-2019 Number | Employment/ residence ratio | Births, 2020 Total | Rate[1] | Deaths, 2020 Number | Rate[1] | Persons under 65 with no health insurance, 2019 Number | Percent | Medicare, 2020 Total beneficiaries | Enrolled in Original Medicare | Enrolled in Medicare Advantage | Crimes reported by county police or sheriff, 2019 Violent | Property |
|---|---|---|---|---|---|---|---|---|---|---|---|---|---|---|
| | 32 | 33 | 34 | 35 | 36 | 37 | 38 | 39 | 40 | 41 | 42 | 43 | 44 | 45 |
| LOUISIANA—Cont'd | | | | | | | | | | | | | | |
| Orleans | 13,521 | 441,140 | 1.28 | 4,445 | 11.4 | 3,961 | 10.2 | 32,121 | 10.2 | 64,567 | 27,804 | 36,763 | NA | NA |
| Ouachita | 5,546 | 162,466 | 1.11 | 1,985 | 13.0 | 1,777 | 11.7 | 11,677 | 9.4 | 28,829 | 18,779 | 10,050 | 425 | 2,512 |
| Plaquemines | 272 | 30,765 | 1.71 | 257 | 11.1 | 206 | 8.9 | 2,231 | 11.3 | 3,746 | 1,561 | 2,185 | 37 | 108 |
| Pointe Coupee | 116 | 19,126 | 0.67 | 254 | 11.8 | 270 | 12.5 | 1,721 | 10.2 | 5,084 | 2,366 | 2,717 | 154 | 286 |
| Rapides | 4,501 | 135,127 | 1.08 | 1,662 | 12.9 | 1,533 | 11.9 | 11,205 | 10.7 | 28,011 | 20,652 | 7,358 | 326 | 1,512 |
| Red River | 164 | 8,053 | 0.84 | 98 | 11.8 | 145 | 17.5 | 638 | 9.6 | 1,780 | 1,269 | 511 | 33 | 136 |
| Richland | 1,037 | 19,546 | 0.89 | 274 | 13.7 | 266 | 13.3 | 1,730 | 11.1 | 4,420 | 3,042 | 1,378 | 12 | 40 |
| Sabine | 438 | 22,821 | 0.85 | 269 | 11.3 | 257 | 10.8 | 2,353 | 12.7 | 5,560 | 4,262 | 1,298 | 27 | 194 |
| St. Bernard | 293 | 40,332 | 0.68 | 556 | 11.7 | 438 | 9.2 | 4,446 | 10.8 | 6,740 | 2,838 | 3,902 | NA | NA |
| St. Charles | 592 | 55,185 | 1.1 | 576 | 10.9 | 494 | 9.3 | 3,228 | 7.2 | 9,176 | 3,325 | 5,852 | 133 | 762 |
| St. Helena | 121 | 8,552 | 0.54 | 115 | 11.4 | 109 | 10.8 | 918 | 11.6 | 2,988 | 1,551 | 1,436 | 54 | 137 |
| St. James | 204 | 21,749 | 1.05 | 238 | 11.5 | 255 | 12.3 | 1,370 | 8.0 | 4,493 | 1,889 | 2,604 | 59 | 371 |
| St. John the Baptist | 471 | 41,404 | 0.9 | 493 | 11.6 | 449 | 10.6 | 3,243 | 9.0 | 8,325 | 3,212 | 5,113 | 93 | 965 |
| St. Landry | 1,105 | 76,162 | 0.77 | 1,119 | 13.7 | 1,083 | 13.3 | 7,016 | 10.4 | 18,524 | 13,347 | 5,177 | 159 | 746 |
| St. Martin | 701 | 46,585 | 0.68 | 649 | 12.3 | 584 | 11.0 | 4,678 | 10.6 | 10,797 | 7,541 | 3,257 | 156 | 627 |
| St. Mary | 868 | 54,216 | 1.17 | 602 | 12.5 | 595 | 12.3 | 4,450 | 11.1 | 10,696 | 6,445 | 4,251 | 121 | 563 |
| St. Tammany | 1,277 | 237,959 | 0.85 | 2,892 | 11.0 | 2,656 | 10.1 | 19,767 | 9.2 | 53,851 | 24,800 | 29,052 | 259 | 1,485 |
| Tangipahoa | 3,646 | 121,736 | 0.82 | 1,874 | 13.7 | 1,503 | 11.0 | 11,986 | 10.8 | 24,079 | 13,084 | 10,996 | 688 | 2,350 |
| Tensas | 16 | 4,357 | 0.84 | 38 | 9.1 | 67 | 16.0 | 333 | 10.5 | 1,157 | 841 | 316 | 6 | 18 |
| Terrebonne | 1,479 | 120,778 | 1.19 | 1,437 | 13.1 | 1,207 | 11.0 | 11,841 | 12.8 | 21,223 | 13,829 | 7,394 | NA | NA |
| Union | 476 | 19,592 | 0.68 | 232 | 10.5 | 266 | 12.0 | 1,750 | 10.3 | 5,303 | 3,639 | 1,664 | NA | NA |
| Vermilion | 537 | 50,495 | 0.61 | 732 | 12.3 | 653 | 11.0 | 5,413 | 10.9 | 11,373 | 8,455 | 2,918 | NA | NA |
| Vernon | 3,373 | 49,941 | 1.01 | 809 | 16.9 | 399 | 8.3 | 3,566 | 9.2 | 7,754 | 6,507 | 1,247 | 58 | 451 |
| Washington | 1,583 | 42,819 | 0.79 | 513 | 11.2 | 682 | 14.9 | 3,744 | 10.4 | 10,569 | 6,082 | 4,487 | 274 | 657 |
| Webster | 1,231 | 38,070 | 0.92 | 420 | 11.1 | 581 | 15.3 | 3,199 | 10.9 | 9,503 | 6,496 | 3,008 | 21 | 151 |
| West Baton Rouge | 556 | 27,309 | 1.1 | 351 | 13.1 | 240 | 9.0 | 1,769 | 8.0 | 4,588 | 1,921 | 2,667 | 85 | 545 |
| West Carroll | 434 | 10,151 | 0.76 | 115 | 10.8 | 159 | 14.9 | 998 | 12.0 | 2,623 | 1,774 | 849 | 33 | 57 |
| West Feliciana | 5,320 | 15,327 | 0.98 | 124 | 8.0 | 136 | 8.8 | 667 | 8.2 | 2,210 | 1,301 | 909 | 34 | 60 |
| Winn | 1,765 | 13,496 | 0.84 | 141 | 10.2 | 224 | 16.2 | 957 | 9.9 | 3,028 | 2,226 | 802 | 32 | 63 |
| MAINE | 36,221 | 1,319,385 | 0.98 | 12,209 | 9.0 | 15,562 | 11.5 | 104,963 | 10.2 | 347,452 | 202,526 | 144,926 | X | X |
| Androscoggin | 3,165 | 104,196 | 0.94 | 1,211 | 11.2 | 1,176 | 10.8 | 8,323 | 9.7 | 25,081 | 12,421 | 12,660 | 13 | 100 |
| Aroostook | 2,192 | 67,392 | 0.99 | 613 | 9.2 | 942 | 14.1 | 6,194 | 12.7 | 20,366 | 14,283 | 6,083 | 6 | 77 |
| Cumberland | 9,327 | 320,208 | 1.17 | 2,737 | 9.2 | 2,958 | 9.9 | 18,262 | 7.8 | 66,621 | 34,722 | 31,900 | 29 | 285 |
| Franklin | 1,024 | 29,040 | 0.93 | 232 | 7.7 | 364 | 12.1 | 2,685 | 12.1 | 8,148 | 4,016 | 4,132 | 6 | 38 |
| Hancock | 927 | 53,310 | 0.95 | 418 | 7.6 | 715 | 13.0 | 4,621 | 11.4 | 15,803 | 10,509 | 5,294 | 4 | 72 |
| Kennebec | 3,749 | 125,158 | 1.06 | 1,141 | 9.3 | 1,503 | 12.2 | 9,885 | 10.5 | 32,065 | 17,016 | 15,049 | 21 | 174 |
| Knox | 1,414 | 41,475 | 1.09 | 305 | 7.6 | 481 | 12.0 | 3,227 | 11.5 | 12,155 | 7,188 | 4,967 | 5 | 156 |
| Lincoln | 510 | 31,661 | 0.84 | 292 | 8.4 | 454 | 13.1 | 3,345 | 13.6 | 11,343 | 6,282 | 5,060 | 25 | 128 |
| Oxford | 938 | 51,927 | 0.78 | 471 | 8.1 | 775 | 13.3 | 5,035 | 11.4 | 16,046 | 9,103 | 6,943 | 25 | 223 |
| Penobscot | 7,264 | 155,117 | 1.05 | 1,350 | 8.9 | 1,728 | 11.4 | 13,631 | 11.7 | 37,389 | 23,162 | 14,228 | 5 | 226 |
| Piscataquis | 250 | 16,298 | 0.91 | 140 | 8.2 | 247 | 14.5 | 1,504 | 12.7 | 5,415 | 3,640 | 1,775 | 4 | 45 |
| Sagadahoc | 273 | 34,020 | 0.92 | 305 | 8.4 | 445 | 12.3 | 2,470 | 8.9 | 9,838 | 5,472 | 4,367 | 8 | 58 |
| Somerset | 722 | 47,480 | 0.86 | 475 | 9.4 | 585 | 11.6 | 4,684 | 12.1 | 14,231 | 9,094 | 5,137 | 33 | 296 |
| Waldo | 574 | 35,373 | 0.77 | 346 | 8.7 | 485 | 12.1 | 3,393 | 11.4 | 11,156 | 6,606 | 4,550 | 13 | 75 |
| Washington | 341 | 31,296 | 0.98 | 302 | 9.6 | 447 | 14.2 | 3,679 | 16.2 | 9,624 | 7,541 | 2,082 | 14 | 82 |
| York | 3,551 | 175,434 | 0.73 | 1,871 | 8.9 | 2,257 | 10.8 | 14,025 | 8.6 | 52,171 | 31,472 | 20,699 | 20 | 291 |
| MARYLAND | 140,092 | 5,752,593 | 0.91 | 69,753 | 11.5 | 55,635 | 9.2 | 345,782 | 7.0 | 1,057,571 | 919,440 | 138,132 | X | X |
| Allegany | 7,600 | 75,481 | 1.14 | 650 | 9.3 | 931 | 13.3 | 2,842 | 5.8 | 16,994 | 16,181 | 813 | 43 | 178 |
| Anne Arundel | 14,228 | 564,640 | 0.98 | 6,775 | 11.6 | 5,025 | 8.6 | 23,587 | 4.9 | 96,978 | 85,798 | 11,180 | 1,580 | 9,594 |
| Baltimore | 21,400 | 777,367 | 0.88 | 9,563 | 11.6 | 9,067 | 11.0 | 46,135 | 6.9 | 162,576 | 138,899 | 23,677 | 4,718 | 20,246 |
| Calvert | 642 | 68,981 | 0.52 | 888 | 9.5 | 830 | 8.9 | 3,299 | 4.2 | 16,423 | 15,307 | 1,116 | 111 | 760 |
| Caroline | 441 | 27,473 | 0.65 | 425 | 12.7 | 381 | 11.4 | 2,354 | 8.5 | 6,669 | 6,331 | 338 | 12 | 184 |
| Carroll | 3,367 | 139,147 | 0.67 | 1,713 | 10.1 | 1,704 | 10.1 | 6,153 | 4.5 | 33,653 | 30,313 | 3,340 | NA | NA |
| Cecil | 1,357 | 86,978 | 0.69 | 1,152 | 11.1 | 1,096 | 10.6 | 4,944 | 5.8 | 19,513 | 18,017 | 1,496 | 69 | 398 |
| Charles | 1,407 | 124,696 | 0.57 | 1,847 | 11.2 | 1,297 | 7.9 | 6,710 | 4.8 | 24,441 | 22,312 | 2,129 | 539 | 2,213 |
| Dorchester | 499 | 29,756 | 0.84 | 355 | 11.1 | 407 | 12.8 | 1,735 | 7.1 | 8,237 | 7,871 | 365 | 15 | 137 |
| Frederick | 4,367 | 230,194 | 0.84 | 2,994 | 11.3 | 2,013 | 7.6 | 12,034 | 5.5 | 43,691 | 39,132 | 4,558 | 196 | 1,183 |
| Garrett | 528 | 28,974 | 0.98 | 262 | 9.1 | 378 | 13.1 | 1,665 | 7.5 | 7,279 | 6,445 | 833 | 60 | 159 |
| Harford | 2,755 | 225,982 | 0.8 | 2,659 | 10.4 | 2,401 | 9.3 | 9,770 | 4.6 | 49,154 | 44,188 | 4,966 | 252 | 1,624 |
| Howard | 1,690 | 315,118 | 0.98 | 3,370 | 10.3 | 2,067 | 6.3 | 12,605 | 4.5 | 49,440 | 43,630 | 5,810 | 512 | 4,304 |
| Kent | 1,502 | 19,896 | 1.04 | 147 | 7.7 | 321 | 16.7 | 1,084 | 8.3 | 5,685 | 5,333 | 353 | 5 | 94 |
| Montgomery | 8,907 | 1,013,461 | 0.95 | 11,964 | 11.4 | 6,906 | 6.6 | 70,658 | 8.1 | 170,566 | 146,711 | 23,855 | 1,498 | 14,209 |
| Prince George's | 20,003 | 785,284 | 0.74 | 11,802 | 13.0 | 7,791 | 8.6 | 77,960 | 10.2 | 134,973 | 110,926 | 24,047 | 1,983 | 10,750 |
| Queen Anne's | 436 | 40,787 | 0.65 | 500 | 9.8 | 437 | 8.5 | 2,307 | 5.7 | 10,562 | 9,910 | 652 | 45 | 317 |

1. Per 1,000 estimated resident population.

# Table B. States and Counties — Crime, Education, Money Income, and Poverty

| | | | | | | | | | | | | | | | |
|---|---|---|---|---|---|---|---|---|---|---|---|---|---|---|---|
| | County law enforcement employment, 2019 | | Education | | | | | | Money income, 2015-2019 | | | | Income and poverty, 2019 | | |
| | | | School enrollment and attainment, 2015-2019 | | | | Local government expenditures,[3] 2017-2018 | | | | Households | | | Percent below poverty level | |
| | | | Enrollment[1] | | Attainment[2] (percent) | | | | | | Percent | | | | |
| STATE County | Officers | Civilians | Total | Percent private | High school graduate or less | Bachelor's degree or more | Total current spending (mil dol) | Current spending per student (dollars) | Per capita income[4] | Median income (dollars) | with income of less than $50,000 | with income of $200,000 or more | Median household income (dollars) | All persons | Children under 18 years | Children 5 to 17 years in families |
| | 46 | 47 | 48 | 49 | 50 | 51 | 52 | 53 | 54 | 55 | 56 | 57 | 58 | 59 | 60 | 61 |
| LOUISIANA—Cont'd | | | | | | | | | | | | | | | | |
| Orleans | NA | NA | 97,177 | 32.5 | 36.3 | 37.6 | 673 | 12,988 | 31,385 | 41,604 | 56.1 | 6.4 | 45,092 | 23.5 | 33.4 | 36.8 |
| Ouachita | 392 | 0 | 39,607 | 13.9 | 47.6 | 24.2 | 301 | 10,726 | 24,361 | 41,121 | 57.6 | 3.7 | 43,235 | 23.9 | 37.4 | 37.0 |
| Plaquemines | 257 | 0 | 6,629 | 23.4 | 51.7 | 19.1 | 75 | 14,919 | 29,258 | 57,204 | 44.5 | 3.8 | 60,795 | 15.1 | 18.9 | 17.7 |
| Pointe Coupee | 57 | 46 | 4,856 | 27.9 | 59.5 | 14.9 | 34 | 11,450 | 28,180 | 41,480 | 55.6 | 5.2 | 48,011 | 20.0 | 27.6 | 26.5 |
| Rapides | 379 | 95 | 32,877 | 17.4 | 51.2 | 20.9 | 244 | 10,438 | 25,984 | 47,269 | 52.3 | 3.4 | 49,687 | 18.3 | 26.1 | 23.1 |
| Red River | 42 | 28 | 2,029 | 14.4 | 62.4 | 13.9 | 23 | 15,656 | 22,924 | 33,816 | 62.2 | 6.1 | 39,806 | 23.9 | 33.7 | 30.6 |
| Richland | 115 | 30 | 4,347 | 7.5 | 62.6 | 13.2 | 40 | 10,557 | 20,107 | 34,029 | 61.2 | 2.0 | 39,586 | 25.1 | 37.7 | 34.7 |
| Sabine | 89 | 0 | 5,220 | 11.1 | 60.9 | 13.3 | 48 | 10,870 | 22,687 | 40,336 | 59.3 | 2.0 | 45,486 | 18.4 | 25.1 | 23.1 |
| St. Bernard | NA | NA | 12,692 | 14.8 | 52.4 | 11.6 | 86 | 11,209 | 20,763 | 44,661 | 54.8 | 1.4 | 45,806 | 19.2 | 27.8 | 26.7 |
| St. Charles | 233 | 178 | 13,307 | 12.7 | 44.5 | 25.3 | 154 | 16,522 | 32,935 | 69,019 | 35.3 | 6.1 | 71,579 | 11.1 | 16.0 | 14.3 |
| St. Helena | 29 | 18 | 2,421 | 15.0 | 61.0 | 16.7 | 14 | 11,423 | 24,380 | 43,886 | 57.3 | 1.7 | 43,376 | 19.6 | 29.6 | 29.1 |
| St. James | 59 | 44 | 4,835 | 15.4 | 55.8 | 16.3 | 64 | 16,324 | 26,739 | 51,603 | 48.2 | 4.0 | 57,250 | 16.6 | 23.8 | 21.2 |
| St. John the Baptist | 225 | 23 | 11,141 | 19.6 | 52.7 | 15.9 | 79 | 13,141 | 25,968 | 57,429 | 44.5 | 2.9 | 54,944 | 16.0 | 25.6 | 23.8 |
| St. Landry | 87 | 117 | 19,842 | 20.3 | 64.5 | 14.8 | 142 | 10,197 | 21,162 | 36,403 | 62.0 | 2.2 | 39,599 | 22.6 | 29.6 | 27.5 |
| St. Martin | 146 | 45 | 12,330 | 21.4 | 61.1 | 14.5 | 78 | 9,679 | 24,997 | 48,656 | 51.0 | 3.4 | 47,247 | 18.7 | 26.1 | 25.9 |
| St. Mary | 145 | 0 | 11,182 | 11.9 | 68.2 | 10.2 | 98 | 10,969 | 22,839 | 40,485 | 60.9 | 2.3 | 40,918 | 23.8 | 31.8 | 33.5 |
| St. Tammany | NA | NA | 63,698 | 25.0 | 36.1 | 33.8 | 460 | 12,056 | 34,658 | 68,905 | 37.3 | 8.1 | 71,526 | 11.5 | 15.7 | 13.6 |
| Tangipahoa | NA | NA | 34,172 | 16.7 | 53.5 | 20.0 | 199 | 10,123 | 25,163 | 47,832 | 51.5 | 3.1 | 47,860 | 21.7 | 31.6 | 30.6 |
| Tensas | 19 | 11 | 917 | 11.9 | 68.7 | 11.4 | 8 | 15,983 | 15,912 | 27,500 | 76.8 | 1.2 | 31,436 | 28.9 | 48.0 | 45.0 |
| Terrebonne | NA | NA | 27,167 | 20.6 | 60.0 | 15.6 | 180 | 10,150 | 25,924 | 48,747 | 51.1 | 3.8 | 51,719 | 19.0 | 27.3 | 24.6 |
| Union | 28 | 18 | 4,712 | 10.5 | 59.1 | 17.2 | 36 | 11,824 | 24,030 | 44,100 | 53.5 | 3.2 | 44,617 | 19.4 | 31.3 | 29.8 |
| Vermilion | 100 | 23 | 14,043 | 14.8 | 63.4 | 14.2 | 93 | 9,558 | 25,342 | 51,945 | 48.4 | 2.7 | 50,193 | 17.0 | 23.4 | 22.9 |
| Vernon | 64 | 54 | 11,971 | 9.2 | 53.4 | 17.5 | 92 | 10,398 | 24,554 | 49,141 | 51.1 | 2.2 | 51,652 | 15.3 | 21.1 | 20.6 |
| Washington | 65 | 10 | 10,670 | 15.2 | 65.2 | 11.6 | 83 | 11,398 | 20,177 | 37,570 | 63.1 | 1.6 | 35,505 | 24.9 | 33.6 | 31.9 |
| Webster | 122 | 28 | 8,390 | 10.0 | 58.1 | 15.4 | 64 | 10,226 | 18,962 | 28,951 | 72.1 | 1.3 | 38,386 | 29.6 | 37.3 | 32.2 |
| West Baton Rouge | 63 | 135 | 6,230 | 17.9 | 50.0 | 22.0 | 57 | 14,700 | 29,697 | 65,385 | 36.6 | 4.2 | 65,344 | 14.4 | 19.7 | 19.5 |
| West Carroll | 8 | 9 | 2,287 | 4.7 | 66.9 | 11.7 | 22 | 10,522 | 22,439 | 38,500 | 59.5 | 2.7 | 42,614 | 21.0 | 29.3 | 26.8 |
| West Feliciana | 67 | 0 | 3,430 | 7.2 | 49.8 | 24.2 | 31 | 14,194 | 26,945 | 59,637 | 38.9 | 6.8 | 65,296 | 22.1 | 19.3 | 17.7 |
| Winn | 37 | 0 | 2,935 | 22.4 | 63.9 | 13.9 | 25 | 11,057 | 22,107 | 38,353 | 61.4 | 2.7 | 43,166 | 23.4 | 28.8 | 27.8 |
| MAINE | X | X | 283,155 | 18.4 | 38.9 | 31.8 | 2,618 | 14,571 | 32,637 | 57,918 | 43.5 | 4.6 | 58,824 | 10.9 | 13.8 | 12.5 |
| Androscoggin | 22 | 2 | 25,016 | 22.0 | 45.3 | 22.8 | 228 | 13,260 | 28,956 | 53,509 | 47.1 | 2.7 | 58,958 | 10.4 | 15.3 | 12.7 |
| Aroostook | 19 | 7 | 13,154 | 12.0 | 48.7 | 19.2 | 137 | 14,333 | 25,477 | 41,123 | 57.5 | 2.1 | 40,675 | 15.5 | 20.2 | 19.5 |
| Cumberland | 62 | 12 | 65,779 | 21.0 | 27.0 | 47.6 | 588 | 15,602 | 40,527 | 73,072 | 33.7 | 8.5 | 75,254 | 7.8 | 9.2 | 8.5 |
| Franklin | 19 | 1 | 6,594 | 9.0 | 45.5 | 24.8 | 51 | 15,458 | 27,252 | 51,422 | 48.9 | 3.3 | 51,136 | 12.6 | 16.6 | 16.0 |
| Hancock | 20 | 2 | 10,316 | 15.8 | 36.7 | 34.1 | 112 | 16,724 | 34,361 | 57,178 | 43.4 | 5.4 | 58,926 | 10.6 | 14.2 | 12.6 |
| Kennebec | 24 | 3 | 26,360 | 20.7 | 40.2 | 28.1 | 208 | 13,491 | 30,685 | 55,365 | 45.8 | 3.6 | 55,358 | 11.5 | 13.9 | 12.6 |
| Knox | 21 | 1 | 7,362 | 18.4 | 39.4 | 33.5 | 100 | 16,982 | 33,126 | 57,751 | 42.1 | 4.0 | 55,910 | 10.9 | 17.5 | 14.6 |
| Lincoln | 23 | 2 | 5,598 | 21.2 | 41.3 | 33.5 | 58 | 15,321 | 33,595 | 57,720 | 43.3 | 4.2 | 58,619 | 11.0 | 16.4 | 15.0 |
| Oxford | 24 | 2 | 11,290 | 15.1 | 50.1 | 19.2 | 127 | 14,404 | 25,464 | 49,204 | 50.9 | 1.8 | 50,344 | 13.0 | 17.6 | 15.1 |
| Penobscot | 36 | 5 | 37,025 | 14.5 | 41.6 | 27.7 | 281 | 13,074 | 28,523 | 50,808 | 49.3 | 3.2 | 50,702 | 12.4 | 13.9 | 12.5 |
| Piscataquis | 9 | 1 | 3,160 | 21.2 | 50.5 | 18.4 | 19 | 10,458 | 24,074 | 40,890 | 59.8 | 1.7 | 43,509 | 16.5 | 23.8 | 21.8 |
| Sagadahoc | 19 | 2 | 6,911 | 13.0 | 36.0 | 36.4 | 70 | 15,828 | 34,675 | 63,694 | 39.6 | 3.0 | 65,841 | 8.6 | 11.4 | 10.4 |
| Somerset | 23 | 4 | 9,779 | 13.8 | 52.0 | 16.5 | 108 | 13,992 | 24,386 | 44,256 | 55.1 | 1.4 | 45,333 | 18.3 | 22.6 | 20.5 |
| Waldo | 21 | 2 | 7,914 | 22.3 | 40.0 | 31.4 | 62 | 17,287 | 29,674 | 51,931 | 48.0 | 3.2 | 51,073 | 13.7 | 18.6 | 17.8 |
| Washington | 16 | 2 | 6,337 | 12.3 | 48.7 | 22.0 | 68 | 16,379 | 25,157 | 41,347 | 57.9 | 1.5 | 39,068 | 19.6 | 24.6 | 23.0 |
| York | 30 | 3 | 40,560 | 20.9 | 36.7 | 32.5 | 400 | 14,237 | 36,093 | 67,830 | 36.2 | 5.6 | 66,803 | 7.9 | 9.9 | 9.0 |
| MARYLAND | X | X | 1,536,208 | 19.1 | 34.4 | 40.2 | 10,010 | 14,751 | 42,122 | 84,805 | 29.2 | 12.7 | 86,644 | 9.1 | 12.3 | 11.9 |
| Allegany | 32 | 2 | 16,607 | 6.4 | 51.6 | 18.9 | 121 | 14,063 | 23,607 | 45,893 | 53.8 | 2.0 | 48,170 | 16.0 | 20.8 | 18.0 |
| Anne Arundel | 851 | 250 | 143,308 | 19.4 | 31.3 | 41.7 | 1,151 | 13,896 | 46,629 | 100,798 | 21.3 | 15.1 | 100,916 | 5.8 | 8.1 | 7.7 |
| Baltimore | 1,942 | 243 | 207,022 | 22.3 | 34.3 | 39.4 | 1,600 | 14,122 | 40,105 | 76,866 | 31.2 | 9.3 | 76,972 | 8.9 | 11.0 | 11.2 |
| Calvert | 138 | 26 | 23,127 | 13.7 | 35.6 | 32.4 | 225 | 14,116 | 45,783 | 109,313 | 19.2 | 16.2 | 111,056 | 5.7 | 7.1 | 6.7 |
| Caroline | 39 | 4 | 7,813 | 6.8 | 55.6 | 18.4 | 80 | 13,786 | 29,624 | 58,638 | 41.4 | 4.4 | 60,143 | 12.1 | 19.6 | 17.9 |
| Carroll | 122 | 151 | 42,041 | 23.1 | 36.7 | 36.4 | 351 | 13,886 | 42,083 | 96,769 | 23.8 | 12.4 | 101,810 | 5.1 | 5.7 | 4.9 |
| Cecil | 95 | 18 | 24,934 | 17.8 | 46.7 | 23.9 | 211 | 13,739 | 34,810 | 76,887 | 32.5 | 6.8 | 75,307 | 10.3 | 13.6 | 11.1 |
| Charles | 288 | 141 | 42,106 | 13.6 | 38.5 | 28.9 | 384 | 14,287 | 41,717 | 100,003 | 21.5 | 12.5 | 102,510 | 6.4 | 9.2 | 9.2 |
| Dorchester | 39 | 5 | 6,636 | 6.8 | 50.6 | 21.2 | 73 | 15,215 | 30,293 | 52,917 | 47.9 | 4.4 | 48,709 | 16.4 | 24.2 | 23.9 |
| Frederick | 186 | 71 | 66,729 | 18.3 | 31.6 | 41.4 | 547 | 12,974 | 43,582 | 97,730 | 23.7 | 13.2 | 102,951 | 5.7 | 6.6 | 5.7 |
| Garrett | 39 | 15 | 5,740 | 12.9 | 53.5 | 20.9 | 54 | 14,098 | 30,617 | 52,617 | 47.3 | 3.6 | 59,253 | 12.8 | 17.7 | 17.4 |
| Harford | 302 | 87 | 62,114 | 17.7 | 33.3 | 36.7 | 500 | 13,237 | 41,147 | 89,147 | 26.5 | 10.9 | 91,492 | 6.7 | 8.8 | 8.0 |
| Howard | 530 | 225 | 90,049 | 17.0 | 18.0 | 62.6 | 886 | 15,595 | 54,628 | 121,160 | 16.6 | 24.2 | 121,329 | 5.0 | 5.6 | 5.3 |
| Kent | 24 | 5 | 4,440 | 36.4 | 41.2 | 35.1 | 31 | 15,292 | 36,813 | 58,598 | 43.9 | 7.7 | 65,615 | 12.4 | 18.2 | 17.3 |
| Montgomery | 1,566 | 600 | 272,487 | 21.4 | 22.4 | 58.9 | 2,586 | 16,005 | 54,510 | 108,820 | 21.3 | 22.8 | 110,012 | 7.3 | 9.3 | 8.5 |
| Prince George's | 1,551 | 239 | 239,482 | 16.9 | 39.1 | 33.1 | 2,047 | 15,378 | 37,191 | 84,920 | 26.5 | 9.9 | 85,357 | 8.7 | 13.0 | 13.6 |
| Queen Anne's | 62 | 8 | 11,566 | 16.5 | 35.7 | 36.5 | 103 | 13,247 | 44,754 | 97,034 | 24.3 | 12.6 | 101,350 | 6.0 | 7.4 | 6.9 |

1. All persons 3 years old and over enrolled in nursery school through college.   2. Persons 25 years old and over.   3. Elementary and secondary education expenditures.   4. Based on population estimated by the American Community Survey, 2014–2018.

| STATE County | Personal income, 2019 | | | | | | | | | | Earnings, 2019 | | |
|---|---|---|---|---|---|---|---|---|---|---|---|---|---|
| | Total (mil dol) | Percent change 2018-2019 | Per capita[1] Dollars | Rank | Wages and salaries (mil dol) | Supplements to wages and salaries, employer contributions (mil dol) Pension and insurance | Government social insurance | Proprietors' income (mil dol) | Dividends, interest, and rent (mil dol) | Personal transfer reecipts (mil dol) | Total (mil dol) | Contributions for government social insurance (mil dol) From employee and self-employed | From employer |
| | 62 | 63 | 64 | 65 | 66 | 67 | 68 | 69 | 70 | 71 | 72 | 73 | 74 |
| LOUISIANA—Cont'd | | | | | | | | | | | | | |
| Orleans | 21,038 | 2.5 | 53,923 | 545 | 12,162 | 1,940 | 798 | 2,204 | 5,123 | 3,843 | 17,104 | 931 | 798 |
| Ouachita | 6,499 | 1.7 | 42,398 | 1,705 | 3,163 | 564 | 203 | 523 | 1,116 | 1,724 | 4,453 | 255 | 203 |
| Plaquemines | 1,148 | 2.5 | 49,507 | 879 | 918 | 187 | 58 | 126 | 190 | 202 | 1,288 | 67 | 58 |
| Pointe Coupee | 1,023 | 2.3 | 47,083 | 1,123 | 276 | 55 | 19 | 77 | 154 | 246 | 427 | 28 | 19 |
| Rapides | 6,043 | 2.9 | 46,614 | 1,173 | 2,634 | 525 | 171 | 755 | 1,005 | 1,677 | 4,085 | 229 | 171 |
| Red River | 375 | -1.8 | 44,440 | 1,427 | 112 | 25 | 7 | 54 | 68 | 104 | 198 | 11 | 7 |
| Richland | 786 | 3.5 | 39,080 | 2,181 | 245 | 50 | 16 | 53 | 95 | 268 | 365 | 23 | 16 |
| Sabine | 860 | 1.1 | 36,025 | 2,568 | 225 | 49 | 14 | 91 | 135 | 284 | 379 | 24 | 14 |
| St. Bernard | 1,585 | 4.4 | 33,556 | 2,843 | 581 | 151 | 34 | 83 | 218 | 416 | 850 | 48 | 34 |
| St. Charles | 2,637 | 3.2 | 49,660 | 861 | 1,818 | 386 | 110 | 112 | 351 | 458 | 2,426 | 130 | 110 |
| St. Helena | 439 | 3.2 | 43,338 | 1,563 | 69 | 18 | 4 | 11 | 43 | 152 | 102 | 9 | 4 |
| St. James | 1,109 | 2.9 | 52,567 | 641 | 635 | 143 | 38 | 153 | 116 | 225 | 969 | 53 | 38 |
| St. John the Baptist | 1,772 | 2.7 | 41,368 | 1,865 | 928 | 200 | 57 | 85 | 196 | 460 | 1,271 | 73 | 57 |
| St. Landry | 3,443 | 2.4 | 41,922 | 1,782 | 1,016 | 213 | 62 | 198 | 521 | 1,107 | 1,489 | 97 | 62 |
| St. Martin | 2,087 | 2.4 | 39,053 | 2,183 | 576 | 101 | 37 | 117 | 282 | 546 | 830 | 57 | 37 |
| St. Mary | 2,011 | 1.9 | 40,746 | 1,957 | 1,166 | 207 | 72 | 110 | 368 | 593 | 1,556 | 91 | 72 |
| St. Tammany | 17,171 | 3.1 | 65,938 | 170 | 4,835 | 796 | 297 | 3,437 | 2,854 | 2,579 | 9,364 | 515 | 297 |
| Tangipahoa | 5,229 | 3.6 | 38,800 | 2,224 | 1,878 | 420 | 111 | 306 | 672 | 1,455 | 2,714 | 153 | 111 |
| Tensas | 179 | 2.4 | 41,323 | 1,876 | 35 | 8 | 2 | 32 | 41 | 61 | 78 | 4 | 2 |
| Terrebonne | 4,669 | 2.3 | 42,267 | 1,727 | 2,836 | 416 | 182 | 252 | 802 | 1,157 | 3,686 | 220 | 182 |
| Union | 894 | -0.7 | 40,441 | 1,996 | 188 | 40 | 12 | 59 | 117 | 279 | 299 | 21 | 12 |
| Vermilion | 2,395 | 1.9 | 40,237 | 2,021 | 595 | 120 | 36 | 166 | 377 | 598 | 917 | 57 | 36 |
| Vernon | 2,028 | 1.7 | 42,755 | 1,651 | 1,085 | 308 | 92 | 60 | 365 | 471 | 1,545 | 73 | 92 |
| Washington | 1,605 | 3.3 | 34,744 | 2,723 | 426 | 98 | 26 | 64 | 194 | 644 | 613 | 44 | 26 |
| Webster | 1,567 | 1.9 | 40,860 | 1,937 | 513 | 101 | 33 | 74 | 234 | 508 | 720 | 49 | 33 |
| West Baton Rouge | 1,274 | 2.7 | 48,144 | 1,006 | 768 | 135 | 51 | 94 | 138 | 236 | 1,049 | 58 | 51 |
| West Carroll | 358 | -7.5 | 33,073 | 2,883 | 78 | 21 | 4 | 22 | 50 | 138 | 125 | 9 | 4 |
| West Feliciana | 606 | 2.8 | 38,926 | 2,206 | 379 | 96 | 21 | 51 | 127 | 108 | 546 | 25 | 21 |
| Winn | 528 | 3.0 | 37,988 | 2,325 | 192 | 39 | 13 | 78 | 63 | 167 | 322 | 20 | 13 |
| MAINE | 68,062 | 4.2 | 50,575 | X | 31,483 | 5,546 | 2,280 | 5,202 | 12,651 | 15,069 | 44,511 | 2,985 | 2,280 |
| Androscoggin | 4,652 | 4.2 | 42,968 | 1,624 | 2,426 | 405 | 179 | 253 | 618 | 1,225 | 3,263 | 220 | 179 |
| Aroostook | 2,896 | 3.5 | 43,189 | 1,586 | 1,156 | 249 | 84 | 216 | 423 | 951 | 1,705 | 119 | 84 |
| Cumberland | 19,160 | 3.9 | 64,948 | 187 | 11,358 | 1,678 | 811 | 1,563 | 4,111 | 2,878 | 15,410 | 979 | 811 |
| Franklin | 1,214 | 4.9 | 40,189 | 2,030 | 447 | 92 | 32 | 80 | 224 | 349 | 651 | 47 | 32 |
| Hancock | 2,876 | 3.7 | 52,294 | 658 | 1,014 | 177 | 77 | 387 | 701 | 622 | 1,655 | 117 | 77 |
| Kennebec | 5,743 | 5.0 | 46,958 | 1,139 | 3,012 | 610 | 206 | 375 | 903 | 1,401 | 4,203 | 265 | 206 |
| Knox | 2,067 | 3.4 | 51,959 | 683 | 809 | 144 | 59 | 266 | 527 | 468 | 1,277 | 90 | 59 |
| Lincoln | 1,790 | 3.6 | 51,687 | 707 | 470 | 84 | 34 | 165 | 474 | 422 | 753 | 59 | 34 |
| Oxford | 2,275 | 4.0 | 39,240 | 2,156 | 735 | 138 | 53 | 155 | 377 | 690 | 1,080 | 84 | 53 |
| Penobscot | 6,563 | 3.9 | 43,136 | 1,594 | 3,415 | 651 | 243 | 307 | 981 | 1,718 | 4,616 | 305 | 243 |
| Piscataquis | 665 | 2.8 | 39,614 | 2,099 | 224 | 53 | 17 | 46 | 121 | 227 | 340 | 26 | 17 |
| Sagadahoc | 1,891 | 4.3 | 52,736 | 635 | 861 | 171 | 69 | 102 | 423 | 388 | 1,203 | 81 | 69 |
| Somerset | 1,983 | 4.0 | 39,282 | 2,151 | 719 | 136 | 53 | 145 | 292 | 666 | 1,052 | 79 | 53 |
| Waldo | 1,739 | 5.9 | 43,793 | 1,516 | 529 | 96 | 38 | 145 | 337 | 461 | 808 | 63 | 38 |
| Washington | 1,289 | 3.3 | 41,094 | 1,909 | 435 | 99 | 32 | 121 | 214 | 475 | 687 | 51 | 32 |
| York | 11,259 | 4.8 | 54,225 | 534 | 3,874 | 764 | 295 | 876 | 1,925 | 2,127 | 5,809 | 399 | 295 |
| MARYLAND | 390,792 | 3.3 | 64,541 | X | 186,162 | 30,878 | 13,456 | 32,090 | 76,429 | 54,797 | 262,586 | 15,372 | 13,456 |
| Allegany | 2,919 | 2.9 | 41,454 | 1,853 | 1,351 | 270 | 106 | 138 | 527 | 977 | 1,865 | 128 | 106 |
| Anne Arundel | 39,988 | 3.6 | 69,035 | 139 | 23,277 | 4,440 | 1,743 | 2,212 | 7,961 | 4,666 | 31,672 | 1,797 | 1,743 |
| Baltimore | 52,105 | 2.9 | 62,976 | 230 | 24,613 | 3,804 | 1,764 | 3,276 | 11,121 | 8,318 | 33,456 | 2,047 | 1,764 |
| Calvert | 5,919 | 3.8 | 63,976 | 208 | 1,272 | 230 | 90 | 251 | 1,046 | 772 | 1,843 | 121 | 90 |
| Caroline | 1,566 | 2.9 | 46,883 | 1,144 | 464 | 91 | 35 | 139 | 233 | 395 | 729 | 46 | 35 |
| Carroll | 10,829 | 3.5 | 64,288 | 201 | 2,950 | 470 | 215 | 607 | 1,737 | 1,486 | 4,242 | 274 | 215 |
| Cecil | 5,117 | 4.3 | 49,749 | 853 | 1,859 | 358 | 146 | 271 | 782 | 1,034 | 2,633 | 167 | 146 |
| Charles | 9,432 | 4.1 | 57,774 | 379 | 2,228 | 454 | 167 | 260 | 1,494 | 1,297 | 3,110 | 197 | 167 |
| Dorchester | 1,523 | 3.7 | 47,699 | 1,054 | 531 | 107 | 39 | 118 | 291 | 442 | 795 | 53 | 39 |
| Frederick | 16,649 | 4.8 | 64,147 | 204 | 6,353 | 1,026 | 463 | 1,109 | 2,640 | 1,984 | 8,951 | 533 | 463 |
| Garrett | 1,385 | 3.9 | 47,735 | 1,050 | 498 | 95 | 38 | 130 | 279 | 354 | 761 | 51 | 38 |
| Harford | 15,394 | 3.8 | 60,266 | 295 | 5,881 | 1,135 | 448 | 654 | 2,453 | 2,308 | 8,118 | 501 | 448 |
| Howard | 25,812 | 3.9 | 79,253 | 67 | 14,062 | 1,618 | 968 | 1,704 | 4,680 | 2,147 | 18,352 | 1,080 | 968 |
| Kent | 1,226 | 3.1 | 63,141 | 228 | 368 | 66 | 27 | 125 | 398 | 265 | 587 | 38 | 27 |
| Montgomery | 94,708 | 3.1 | 90,139 | 34 | 40,742 | 6,147 | 2,870 | 13,435 | 22,050 | 7,392 | 63,195 | 3,492 | 2,870 |
| Prince George's | 46,034 | 3.1 | 50,625 | 789 | 21,882 | 4,216 | 1,621 | 1,866 | 7,507 | 6,932 | 29,585 | 1,744 | 1,621 |
| Queen Anne's | 3,362 | 3.9 | 66,733 | 159 | 737 | 127 | 54 | 301 | 676 | 460 | 1,218 | 77 | 54 |

1. Based on the resident population estimated as of July 1 of the year shown.

# Table B. States and Counties — Earnings, Social Security, and Housing

| STATE County | Farm (75) | Mining, quarrying, and extractions (76) | Construction (77) | Manufacturing (78) | Information; professional, scientific, technical services (79) | Retail trade (80) | Finance, insurance, real estate, and leasing (81) | Health care and social assistance (82) | Government (83) | Social Security beneficiaries, Dec 2019 — Number (84) | Rate[1] (85) | Supplemental Security Income recipients, 2019 (86) | Housing units, 2020 — Total (87) | Percent change, 2010-2020 (88) |
|---|---|---|---|---|---|---|---|---|---|---|---|---|---|---|
| **LOUISIANA—Cont'd** | | | | | | | | | | | | | | |
| Orleans | 0.0 | 2.5 | 2.6 | 2.5 | 16.7 | 3.6 | 6.3 | 10.3 | 19.8 | 65,375 | 167 | 19,696 | 192,822 | 1.5 |
| Ouachita | 0.3 | 0.4 | 5.7 | 9.0 | 9.4 | 8.4 | 6.1 | 18.7 | 16.1 | 30,410 | 198 | 7,005 | 69,943 | 8.5 |
| Plaquemines | 0.4 | 13.6 | 3.6 | 16.7 | D | 2.3 | 3.4 | 1.5 | 16.0 | 3,895 | 167 | 596 | 10,428 | 8.7 |
| Pointe Coupee | 3.3 | 2.2 | 5.3 | 6.6 | 4.3 | 8.0 | 11.0 | D | 15.5 | 5,060 | 233 | 945 | 11,800 | 6 |
| Rapides | 1.4 | 1.4 | 6.6 | 7.6 | 4.7 | 10.7 | 4.8 | 19.7 | 21.5 | 29,380 | 226 | 6,700 | 59,079 | 6.1 |
| Red River | 2.8 | 25.0 | 3.6 | 11.3 | 2.7 | 3.7 | 3.0 | D | 19.9 | 1,730 | 206 | 450 | 4,253 | 3 |
| Richland | 6.0 | 1.1 | 5.3 | 11.5 | D | 7.7 | 4.8 | D | 16.1 | 4,685 | 233 | 1,121 | 9,098 | 5.5 |
| Sabine | 10.6 | 2.0 | 5.4 | 15.9 | 5.6 | 7.0 | 4.9 | D | 19.4 | 5,875 | 246 | 998 | 15,474 | 9.5 |
| St. Bernard | 0.2 | D | 8.6 | 32.1 | D | 6.1 | 1.8 | 6.6 | 17.6 | 7,320 | 155 | 1,700 | 17,446 | 3.9 |
| St. Charles | 0.0 | D | 11.3 | 33.0 | 5.6 | 3.3 | 2.6 | 3.1 | 10.6 | 9,735 | 183 | 1,109 | 21,069 | 5.6 |
| St. Helena | 1.4 | D | 5.3 | 20.5 | D | 4.7 | D | D | 35.4 | 3,385 | 334 | 948 | 5,466 | 6.1 |
| St. James | 0.3 | D | 2.1 | 50.6 | D | 2.2 | 2.6 | D | 11.1 | 4,785 | 227 | 705 | 9,035 | 6.9 |
| St. John the Baptist | 0.0 | D | 12.4 | 33.0 | 3.4 | 5.1 | 3.1 | 4.6 | 12.6 | 9,135 | 213 | 1,742 | 17,951 | 2.8 |
| St. Landry | 1.1 | 2.0 | 6.8 | 7.6 | 5.1 | 10.8 | 4.7 | 14.9 | 21.6 | 20,355 | 248 | 5,318 | 37,465 | 4.9 |
| St. Martin | 0.6 | 5.9 | 6.6 | 23.1 | 5.1 | 8.3 | 6.3 | 7.4 | 14.0 | 11,575 | 216 | 1,717 | 24,075 | 9.7 |
| St. Mary | 0.6 | 13.2 | 5.2 | 18.6 | D | 4.8 | 7.3 | 4.3 | 17.0 | 11,880 | 241 | 2,269 | 23,616 | 2.6 |
| St. Tammany | 0.0 | 27.9 | 6.6 | 2.5 | 7.4 | 5.7 | 5.9 | 10.3 | 11.8 | 54,455 | 208 | 4,956 | 107,200 | 12.3 |
| Tangipahoa | 0.1 | 0.8 | 5.8 | 7.4 | 4.6 | 10.5 | 5.4 | 11.1 | 28.7 | 24,925 | 185 | 4,952 | 58,725 | 17.3 |
| Tensas | 32.2 | 0.7 | 3.3 | 1.9 | D | 4.1 | D | D | 20.3 | 1,235 | 286 | 359 | 3,446 | 2.7 |
| Terrebonne | 0.2 | 14.3 | 6.9 | 10.1 | 6.4 | 6.9 | 6.5 | 12.4 | 9.6 | 23,525 | 213 | 4,702 | 45,906 | 4.8 |
| Union | 9.4 | D | 8.8 | D | D | 6.9 | 3.0 | 9.5 | 17.9 | 5,645 | 256 | 945 | 12,084 | 6.5 |
| Vermilion | 5.7 | 10.7 | 7.8 | 4.8 | 5.2 | 7.8 | 4.4 | D | 22.2 | 12,410 | 208 | 1,905 | 27,181 | 7.8 |
| Vernon | 0.3 | 0.1 | 2.5 | 0.8 | 9.5 | 3.1 | 1.9 | 5.1 | 67.1 | 8,145 | 168 | 1,190 | 22,469 | 4.8 |
| Washington | 0.3 | 0.4 | 8.2 | 15.3 | 2.5 | 7.2 | 3.8 | D | 26.9 | 11,675 | 253 | 2,830 | 21,869 | 4 |
| Webster | -0.2 | 7.6 | 8.6 | 14.5 | D | 9.7 | 4.4 | D | 17.6 | 10,140 | 265 | 1,905 | 20,042 | 3.7 |
| West Baton Rouge | 0.7 | D | 22.0 | 28.2 | D | 4.8 | D | D | 11.0 | 4,815 | 181 | 752 | 12,013 | 28.8 |
| West Carroll | 7.9 | D | D | D | 2.5 | 9.7 | D | D | 37.6 | 2,775 | 254 | 467 | 5,244 | 3.9 |
| West Feliciana | -0.2 | D | 9.0 | D | D | 3.2 | 5.3 | D | 30.7 | 2,115 | 136 | 288 | 5,599 | 9.9 |
| Winn | 0.0 | 2.2 | 5.1 | 15.3 | D | 4.2 | 10.1 | D | 15.2 | 3,170 | 227 | 570 | 7,329 | 1.3 |
| **MAINE** | 0.5 | 0.0 | 7.1 | 9.2 | 8.8 | 7.6 | 7.3 | 16.0 | 17.2 | 349,962 | 260 | 36,599 | 755,449 | 4.7 |
| Androscoggin | 0.8 | D | 7.9 | 10.5 | 7.0 | 8.0 | 5.8 | 20.5 | 11.4 | 26,475 | 244 | 4,131 | 50,515 | 2.9 |
| Aroostook | 4.1 | 0.1 | 4.6 | 11.7 | 3.1 | 8.7 | 4.8 | 17.9 | 23.9 | 20,965 | 312 | 2,561 | 40,147 | 1.5 |
| Cumberland | 0.1 | D | 5.7 | 6.5 | 13.8 | 6.2 | 12.7 | 16.5 | 10.6 | 63,930 | 216 | 5,252 | 148,788 | 7.3 |
| Franklin | 0.4 | 0.0 | 7.1 | 15.2 | 2.6 | 10.4 | 4.3 | D | 19.5 | 8,305 | 276 | 911 | 22,453 | 3.5 |
| Hancock | 0.4 | D | 10.6 | 3.6 | 13.7 | 9.1 | 4.4 | 12.3 | 13.0 | 15,380 | 279 | 1,020 | 41,871 | 4.2 |
| Kennebec | 0.3 | D | 5.7 | 5.1 | 6.0 | 8.0 | 3.8 | 16.0 | 30.0 | 32,390 | 265 | 4,011 | 63,525 | 4.2 |
| Knox | 0.4 | 0.0 | 9.3 | 9.0 | 6.5 | 9.0 | 5.0 | 14.5 | 13.9 | 11,975 | 301 | 806 | 24,670 | 4 |
| Lincoln | 1.2 | D | 11.8 | D | 7.8 | 10.2 | 5.2 | 13.6 | 13.8 | 11,045 | 318 | 668 | 24,254 | 3.3 |
| Oxford | 0.6 | 0.1 | 10.0 | 16.1 | 4.0 | 9.3 | 3.0 | 13.4 | 17.3 | 16,980 | 293 | 2,052 | 37,455 | 3.9 |
| Penobscot | 0.1 | D | 6.0 | 4.5 | 5.3 | 8.5 | 4.8 | 23.9 | 20.0 | 38,390 | 253 | 5,454 | 76,615 | 3.7 |
| Piscataquis | 0.3 | D | 7.2 | 24.0 | D | D | 2.3 | D | 26.0 | 5,440 | 322 | 683 | 15,546 | 1.3 |
| Sagadahoc | 0.1 | 0.0 | 8.6 | D | 6.9 | 5.7 | 2.8 | 5.2 | 12.8 | 9,735 | 270 | 635 | 19,146 | 4.7 |
| Somerset | 2.6 | D | 11.5 | 17.8 | 3.7 | 8.5 | 3.4 | 14.4 | 15.2 | 15,320 | 302 | 2,266 | 31,143 | 1.9 |
| Waldo | 0.6 | D | 10.6 | 8.4 | D | 8.5 | 7.9 | 14.6 | 11.9 | 11,380 | 286 | 1,231 | 22,899 | 6.2 |
| Washington | 2.6 | D | 5.5 | 10.8 | 3.0 | D | 3.7 | 14.3 | 24.1 | 9,880 | 316 | 1,340 | 23,456 | 2 |
| York | 0.1 | 0.1 | 8.9 | 11.6 | 6.5 | 7.6 | 4.3 | 10.9 | 27.2 | 52,370 | 252 | 3,578 | 112,966 | 6.8 |
| **MARYLAND** | 0.2 | 0.1 | 6.8 | 4.3 | 16.9 | 4.9 | 9.0 | 11.0 | 23.8 | 1,020,436 | 169 | 121,691 | 2,482,087 | 4.3 |
| Allegany | 0.0 | 0.1 | 4.6 | 8.1 | 3.9 | 7.7 | 4.8 | 22.5 | 25.5 | 17,655 | 250 | 2,361 | 32,678 | -1.8 |
| Anne Arundel | 0.0 | 0.0 | 6.3 | 5.5 | 14.7 | 4.5 | 4.2 | 7.7 | 35.5 | 94,885 | 164 | 7,692 | 230,031 | 8.2 |
| Baltimore | 0.1 | 0.0 | 7.5 | 4.6 | 12.6 | 6.4 | 13.1 | 13.5 | 17.5 | 161,075 | 194 | 16,626 | 338,658 | 0.9 |
| Calvert | -0.1 | 0.1 | 12.6 | 2.4 | 7.7 | 6.7 | 5.4 | 14.2 | 20.5 | 16,175 | 175 | 1,052 | 35,884 | 6.2 |
| Caroline | 6.0 | D | 12.0 | 10.9 | D | 8.5 | D | D | 18.2 | 7,385 | 220 | 811 | 13,693 | 1.5 |
| Carroll | 0.6 | D | 14.0 | 7.9 | 10.7 | 7.4 | 6.1 | 13.6 | 15.0 | 33,405 | 198 | 1,576 | 64,168 | 2.8 |
| Cecil | 1.4 | D | D | 21.1 | 4.4 | 6.0 | 2.3 | 9.2 | 24.1 | 20,535 | 199 | 1,869 | 42,832 | 4.2 |
| Charles | 0.0 | D | 11.0 | 1.5 | 6.8 | 11.5 | 4.1 | 11.2 | 33.6 | 23,980 | 146 | 2,370 | 62,506 | 13.7 |
| Dorchester | 2.7 | 0.2 | D | 20.5 | 4.2 | 4.9 | 3.7 | D | 23.7 | 8,695 | 273 | 1,303 | 16,779 | 1.4 |
| Frederick | 0.5 | D | D | 5.3 | D | 6.9 | D | D | 18.9 | 42,465 | 163 | 2,383 | 103,021 | 14.3 |
| Garrett | 2.1 | 3.5 | 10.7 | 7.8 | 5.5 | 9.8 | D | D | 15.4 | 7,660 | 264 | 605 | 19,510 | 3.4 |
| Harford | 0.2 | D | 7.1 | 5.3 | 13.7 | 6.6 | 4.8 | 10.3 | 32.7 | 49,090 | 192 | 3,433 | 102,314 | 7.1 |
| Howard | 0.0 | D | 7.3 | 5.3 | 31.9 | 4.3 | 7.5 | 7.3 | 9.0 | 43,780 | 134 | 3,613 | 123,173 | 12.7 |
| Kent | 4.6 | D | 7.6 | 11.6 | D | 6.9 | 9.1 | 10.4 | 13.4 | 5,705 | 294 | 392 | 10,685 | 1.3 |
| Montgomery | 0.0 | 0.0 | 4.8 | 3.3 | 23.9 | 3.8 | 14.2 | 9.2 | 21.0 | 144,995 | 138 | 14,084 | 393,194 | 4.8 |
| Prince George's | 0.0 | D | 11.3 | 2.2 | 11.3 | 5.8 | 3.6 | 8.0 | 37.4 | 124,985 | 137 | 14,502 | 337,402 | 2.6 |
| Queen Anne's | 2.6 | D | 10.9 | 8.6 | 9.3 | 8.2 | 4.5 | 5.7 | 16.9 | 10,470 | 207 | 350 | 21,829 | 8.4 |

1. Per 1,000 resident population estimated as of July 1 of the year shown.

# Table B. States and Counties — Housing, Labor Force, and Employment

| STATE County | Total [89] | Percent [90] | Median value[1] [91] | With a mortgage [92] | Without a mortgage[2] [93] | Median rent[3] [94] | Median rent as a percent of income[2] [95] | Substandard units[4] (percent) [96] | Total [97] | Percent change, 2019-2020 [98] | Total [99] | Rate[5] [100] | Total [101] | Management, business, science, and arts [102] | Construction, production, and maintenance occupations [103] |
|---|---|---|---|---|---|---|---|---|---|---|---|---|---|---|---|
| **LOUISIANA—Cont'd** | | | | | | | | | | | | | | | |
| Orleans | 153,819 | 48.3 | 231,500 | 25.1 | 13.6 | 998 | 36.5 | 1.9 | 181,868 | 0.6 | 22,134 | 12.2 | 181,973 | 43.1 | 14.8 |
| Ouachita | 56,556 | 58.4 | 146,700 | 19.5 | 10.0 | 768 | 33.4 | 2.7 | 69,823 | -1.4 | 5,160 | 7.4 | 63,687 | 34.9 | 21.4 |
| Plaquemines | 8,919 | 68.8 | 190,600 | 20.6 | 10.0 | 1,289 | 28.1 | 2.8 | 9,649 | -4.4 | 678 | 7.0 | 9,689 | 33.8 | 31.6 |
| Pointe Coupee | 8,960 | 76.3 | 142,600 | 20.1 | 10.4 | 833 | 31.4 | 3.4 | 9,638 | -2.3 | 750 | 7.8 | 8,952 | 25.8 | 29.9 |
| Rapides | 48,488 | 61.7 | 143,100 | 19.0 | 10.0 | 821 | 31.7 | 2.5 | 55,157 | -0.7 | 3,443 | 6.2 | 54,389 | 33.9 | 22.6 |
| Red River | 3,372 | 73.6 | 78,700 | 19.1 | 13.3 | 645 | 29.9 | 1.7 | 3,612 | -4.5 | 223 | 6.2 | 3,226 | 20.8 | 34.6 |
| Richland | 7,459 | 66.7 | 91,900 | 18.0 | 10.0 | 611 | 31.4 | 4.4 | 8,139 | -2.0 | 621 | 7.6 | 7,731 | 25.9 | 28.1 |
| Sabine | 9,158 | 68.2 | 94,300 | 17.2 | 10.0 | 583 | 32.7 | 2.4 | 9,236 | -0.8 | 506 | 5.5 | 8,112 | 29.9 | 34.1 |
| St. Bernard | 15,005 | 67.7 | 151,300 | 19.5 | 10.4 | 951 | 36.4 | 3.7 | 20,022 | -1.5 | 2,038 | 10.2 | 18,851 | 27.1 | 29.1 |
| St. Charles | 19,212 | 80.7 | 201,700 | 18.5 | 10.0 | 978 | 29.8 | 2.0 | 24,644 | -3.6 | 1,929 | 7.8 | 24,457 | 38.5 | 28.4 |
| St. Helena | 3,857 | 78.9 | 100,100 | 25.1 | 10.6 | 656 | 17.3 | 1.2 | 4,299 | 1.0 | 494 | 11.5 | 3,812 | 34.9 | 32.9 |
| St. James | 7,719 | 79.0 | 158,500 | 17.6 | 10.0 | 644 | 32.1 | 2.1 | 9,058 | -3.3 | 882 | 9.7 | 9,051 | 24.9 | 35.4 |
| St. John the Baptist | 15,270 | 78.5 | 153,900 | 19.7 | 10.0 | 933 | 28.1 | 2.9 | 19,473 | -1.3 | 2,179 | 11.2 | 18,972 | 30.8 | 31.2 |
| St. Landry | 30,485 | 68.3 | 108,400 | 18.4 | 10.2 | 652 | 30.6 | 4.0 | 32,418 | 0.6 | 2,807 | 8.7 | 31,300 | 29.0 | 29.6 |
| St. Martin | 19,749 | 78.8 | 124,900 | 18.6 | 10.0 | 658 | 29.9 | 5.3 | 22,057 | -1.6 | 1,789 | 8.1 | 22,634 | 25.0 | 34.2 |
| St. Mary | 19,856 | 60.7 | 106,900 | 20.0 | 11.4 | 781 | 29.0 | 2.3 | 19,516 | -2.2 | 1,794 | 9.2 | 19,256 | 24.5 | 34.7 |
| St. Tammany | 92,962 | 78.4 | 218,500 | 20.8 | 10.2 | 1,086 | 32.9 | 1.7 | 116,466 | -4.5 | 7,876 | 6.8 | 117,839 | 41.0 | 19.2 |
| Tangipahoa | 47,597 | 69.0 | 162,800 | 18.5 | 10.0 | 832 | 34.0 | 3.6 | 56,389 | 2.1 | 5,377 | 9.5 | 57,761 | 30.7 | 26.2 |
| Tensas | 1,792 | 67.5 | 66,400 | 25.8 | 12.4 | 561 | 28.9 | 6.4 | 1,359 | -3.2 | 114 | 8.4 | 1,332 | 28.7 | 30.0 |
| Terrebonne | 39,972 | 71.5 | 152,500 | 19.6 | 10.0 | 853 | 31.6 | 3.0 | 46,224 | | 3,561 | 7.7 | 46,657 | 28.8 | 30.6 |
| Union | 7,582 | 81.7 | 89,200 | 17.0 | 10.0 | 568 | 28.3 | 3.1 | 8,970 | -2.3 | 640 | 7.1 | 8,620 | 26.6 | 38.5 |
| Vermilion | 22,086 | 77.2 | 122,200 | 17.8 | 10.0 | 685 | 27.3 | 3.5 | 23,746 | -2.2 | 1,753 | 7.4 | 24,330 | 29.2 | 29.3 |
| Vernon | 17,696 | 53.0 | 119,300 | 17.6 | 10.0 | 951 | 26.0 | 1.7 | 15,795 | | 1,118 | 7.1 | 16,340 | 31.3 | 28.2 |
| Washington | 17,613 | 67.4 | 101,000 | 20.1 | 10.0 | 655 | 31.1 | 3.3 | 16,660 | -1.6 | 1,361 | 8.2 | 17,372 | 24.9 | 30.3 |
| Webster | 16,551 | 66.9 | 81,500 | 23.0 | 11.2 | 688 | 44.6 | 3.0 | 14,088 | -3.2 | 1,111 | 7.9 | 14,567 | 27.7 | 34.9 |
| West Baton Rouge | 9,643 | 73.1 | 200,000 | 16.8 | 10.0 | 884 | 29.9 | 2.4 | 13,502 | -2.1 | 972 | 7.2 | 12,525 | 35.2 | 27.1 |
| West Carroll | 4,084 | 76.9 | 90,300 | 16.8 | 10.0 | 547 | 28.3 | 2.0 | 3,637 | -0.8 | 367 | 10.1 | 3,767 | 35.3 | 30.2 |
| West Feliciana | 3,869 | 73.2 | 229,500 | 19.6 | 10.0 | 785 | 29.6 | 1.2 | 5,276 | -3.5 | 287 | 5.4 | 6,042 | 34.0 | 21.3 |
| Winn | 5,483 | 67.2 | 77,700 | 16.8 | 10.0 | 581 | 31.3 | 1.3 | 4,709 | -4.0 | 332 | 7.1 | 5,193 | 29.9 | 33.2 |
| **MAINE** | 559,921 | 72.3 | 190,400 | 21.3 | 12.5 | 853 | 28.8 | 2.1 | 676,547 | -2.8 | 36,788 | 5.4 | 670,417 | 38.1 | 22.5 |
| Androscoggin | 45,630 | 64.3 | 158,200 | 21.0 | 13.7 | 771 | 28.0 | 2.3 | 54,470 | -1.6 | 2,984 | 5.5 | 54,573 | 34.6 | 23.9 |
| Aroostook | 29,516 | 72.0 | 99,600 | 19.0 | 11.9 | 574 | 29.4 | 2.0 | 29,255 | -3.5 | 1,613 | 5.5 | 29,593 | 30.8 | 28.8 |
| Cumberland | 120,644 | 69.3 | 278,100 | 21.3 | 13.1 | 1,131 | 28.8 | 1.9 | 160,863 | -2.8 | 8,495 | 5.3 | 163,077 | 46.6 | 15.7 |
| Franklin | 11,848 | 79.4 | 139,800 | 19.6 | 10.8 | 635 | 26.5 | 2.5 | 13,795 | -4.4 | 839 | 6.1 | 14,474 | 30.3 | 28.3 |
| Hancock | 23,661 | 76.3 | 212,700 | 21.8 | 11.2 | 818 | 27.4 | 2.7 | 27,648 | -5.4 | 1,589 | 5.7 | 27,238 | 36.1 | 23.4 |
| Kennebec | 52,105 | 70.5 | 159,400 | 20.0 | 12.0 | 761 | 28.6 | 1.9 | 61,376 | -1.4 | 3,044 | 5.0 | 59,678 | 38.9 | 21.3 |
| Knox | 17,020 | 77.2 | 213,400 | 23.0 | 13.3 | 856 | 27.9 | 1.2 | 19,535 | -4.3 | 1,054 | 5.4 | 19,979 | 36.3 | 26.3 |
| Lincoln | 15,336 | 79.2 | 215,400 | 22.7 | 11.5 | 819 | 28.8 | 1.7 | 16,386 | -3.4 | 870 | 5.3 | 16,426 | 35.1 | 28.1 |
| Oxford | 21,338 | 80.8 | 144,100 | 21.4 | 12.2 | 713 | 31.9 | 2.3 | 25,815 | -2.9 | 1,718 | 6.7 | 26,073 | 30.7 | 29.2 |
| Penobscot | 62,156 | 69.4 | 144,700 | 19.9 | 12.3 | 799 | 29.3 | 2.5 | 74,717 | -2.8 | 4,007 | 5.4 | 72,962 | 37.0 | 20.6 |
| Piscataquis | 7,025 | 75.9 | 109,500 | 23.7 | 13.0 | 618 | 29.0 | 3.3 | 7,151 | -1.9 | 391 | 5.5 | 6,403 | 29.5 | 32.7 |
| Sagadahoc | 15,980 | 74.5 | 213,400 | 22.6 | 12.5 | 894 | 28.5 | 1.2 | 19,259 | -1.7 | 911 | 4.7 | 18,674 | 38.3 | 25.1 |
| Somerset | 21,321 | 76.1 | 115,700 | 20.6 | 13.1 | 728 | 33.1 | 3.7 | 22,024 | -2.0 | 1,461 | 6.6 | 21,939 | 27.2 | 32.3 |
| Waldo | 17,236 | 79.2 | 163,000 | 21.9 | 12.6 | 814 | 29.0 | 3.3 | 19,961 | -3.7 | 1,028 | 5.2 | 18,889 | 37.8 | 24.8 |
| Washington | 13,791 | 76.5 | 112,400 | 22.2 | 12.5 | 603 | 29.1 | 2.2 | 13,408 | -2.1 | 833 | 6.2 | 12,571 | 29.2 | 34.0 |
| York | 85,314 | 73.9 | 252,300 | 22.3 | 13.0 | 983 | 28.9 | 1.8 | 110,884 | -3.1 | 5,949 | 5.4 | 107,868 | 37.5 | 23.4 |
| **MARYLAND** | 2,205,204 | 66.9 | 314,800 | 21.6 | 10.6 | 1,392 | 29.9 | 2.5 | 3,172,796 | -3.0 | 214,509 | 6.8 | 3,073,886 | 46.3 | 16.7 |
| Allegany | 27,399 | 68.8 | 120,700 | 18.9 | 12.3 | 694 | 30.8 | 1.4 | 30,835 | -4.4 | 2,398 | 7.8 | 28,291 | 30.5 | 23.8 |
| Anne Arundel | 209,814 | 74.2 | 361,200 | 21.5 | 10.0 | 1,663 | 28.9 | 1.9 | 310,413 | -3.3 | 18,104 | 5.8 | 292,548 | 48.5 | 15.5 |
| Baltimore | 313,519 | 66.1 | 261,500 | 21.0 | 10.6 | 1,302 | 29.8 | 1.9 | 445,695 | -2.9 | 30,432 | 6.8 | 422,955 | 44.4 | 16.5 |
| Calvert | 31,973 | 84.7 | 358,800 | 21.3 | 10.0 | 1,520 | 29.7 | 0.8 | 48,543 | -4.4 | 2,537 | 5.2 | 46,981 | 43.1 | 20.9 |
| Caroline | 12,024 | 72.8 | 202,500 | 24.8 | 12.7 | 895 | 30.5 | 3.4 | 17,436 | -5.0 | 967 | 5.5 | 16,071 | 32.1 | 28.0 |
| Carroll | 60,758 | 82.0 | 339,600 | 21.4 | 10.0 | 1,132 | 29.0 | 1.2 | 93,073 | -4.0 | 4,740 | 5.1 | 87,632 | 45.7 | 18.7 |
| Cecil | 37,058 | 73.0 | 242,700 | 21.7 | 11.8 | 1,126 | 30.7 | 2.3 | 52,083 | -3.4 | 3,095 | 5.9 | 51,057 | 35.9 | 26.8 |
| Charles | 56,520 | 76.9 | 313,300 | 22.9 | 10.0 | 1,682 | 29.9 | 1.8 | 86,099 | -3.3 | 5,790 | 6.7 | 80,788 | 42.0 | 18.4 |
| Dorchester | 13,183 | 68.0 | 183,300 | 23.4 | 12.5 | 879 | 32.0 | 1.9 | 15,990 | | 1,067 | 6.7 | 14,860 | 33.3 | 25.5 |
| Frederick | 92,586 | 75.2 | 331,600 | 20.9 | 10.0 | 1,422 | 29.2 | 1.5 | 132,867 | -3.7 | 7,898 | 5.9 | 134,149 | 47.4 | 16.2 |
| Garrett | 12,425 | 78.6 | 173,900 | 21.5 | 11.0 | 614 | 24.7 | 1.4 | 15,274 | -4.4 | 1,013 | 6.6 | 13,870 | 35.9 | 30.1 |
| Harford | 93,955 | 78.7 | 293,400 | 20.7 | 10.8 | 1,257 | 29.0 | 1.6 | 138,337 | -3.6 | 8,001 | 5.8 | 130,345 | 44.5 | 17.7 |
| Howard | 114,170 | 73.2 | 455,700 | 20.9 | 10.0 | 1,716 | 27.4 | 2.3 | 184,474 | -3.7 | 9,558 | 5.2 | 170,318 | 62.1 | 9.4 |
| Kent | 8,025 | 69.2 | 249,900 | 23.6 | 14.1 | 1,007 | 33.7 | 0.6 | 9,664 | -5.9 | 638 | 6.6 | 9,304 | 37.9 | 20.5 |
| Montgomery | 370,950 | 65.4 | 484,900 | 21.2 | 10.0 | 1,768 | 30.3 | 3.4 | 548,398 | -3.0 | 34,735 | 6.3 | 560,950 | 56.3 | 11.4 |
| Prince George's | 311,343 | 62.1 | 302,800 | 23.3 | 10.7 | 1,475 | 30.7 | 4.3 | 504,276 | -1.8 | 41,397 | 8.2 | 483,867 | 39.8 | 19.8 |
| Queen Anne's | 18,577 | 81.0 | 353,100 | 22.8 | 12.1 | 1,503 | 28.9 | 1.2 | 27,202 | -3.8 | 1,490 | 5.5 | 25,558 | 44.3 | 18.7 |

1. Specified owner-occupied units. lacking complete plumbing facilities.  2. A value of 10.0 represents 10 percent or less; a value of 50.0 represents 50 percent or more.  3. Specified renter-occupied units.  4. Overcrowded or  5. Percent of civilian labor force.  6. Civilian employed persons 16 years old and over.

# Table B. States and Counties — Nonfarm Employment and Agriculture

| STATE County | Number of establishments | Employment — Total | Health care and social assistance | Manufacturing | Retail trade | Finance and insurance | Professional, scientific, and technical services | Annual payroll — Total (mil dol) | Average per employee (dollars) | Farms — Number | Percent with: Fewer than 50 acres | 1000 acres or more | Farm producers whose primary occupation is farming (percent) |
|---|---|---|---|---|---|---|---|---|---|---|---|---|---|
| (item) | 104 | 105 | 106 | 107 | 108 | 109 | 110 | 111 | 112 | 113 | 114 | 115 | 116 |
| **LOUISIANA—Cont'd** | | | | | | | | | | | | | |
| Orleans | 9,593 | 179,443 | 26,263 | 4,082 | 14,104 | 6,304 | 13,613 | 8,865 | 49,402 | 39 | 94.9 | NA | 55.6 |
| Ouachita | 4,185 | 61,398 | 13,456 | 4,465 | 9,088 | 3,449 | 2,640 | 2,462 | 40,105 | 488 | 54.1 | 3.7 | 35.1 |
| Plaquemines | 642 | 8,484 | 389 | 1,681 | 508 | 90 | 408 | 582 | 68,595 | 113 | 52.2 | 11.5 | 38.3 |
| Pointe Coupee | 330 | 4,084 | 769 | 279 | 864 | 189 | 101 | 188 | 46,094 | 482 | 45.4 | 10.2 | 43.2 |
| Rapides | 3,118 | 45,923 | 14,278 | 2,950 | 7,458 | 1,412 | 1,801 | 1,907 | 41,517 | 856 | 51.1 | 6.2 | 45.1 |
| Red River | 128 | 1,799 | 504 | 371 | 119 | 78 | 37 | 67 | 37,424 | 197 | 25.9 | 13.2 | 46.8 |
| Richland | 418 | 5,593 | 1,788 | 664 | 757 | 316 | 107 | 200 | 35,789 | 626 | 33.1 | 10.5 | 30.2 |
| Sabine | 486 | 4,419 | 695 | 1,142 | 795 | 259 | 150 | 171 | 38,775 | 442 | 39.1 | 1.4 | 47.2 |
| St. Bernard | 692 | 8,074 | 693 | 1,450 | 1,723 | 174 | 141 | 396 | 49,051 | 33 | 30.3 | 21.2 | 28.6 |
| St. Charles | 989 | 21,554 | 1,487 | 4,603 | 1,370 | 478 | 1,551 | 1,561 | 72,414 | 67 | 32.8 | NA | 47.3 |
| St. Helena | 118 | 1,143 | 437 | 240 | 168 | 28 | 28 | 46 | 40,226 | 348 | 45.1 | 1.1 | 42.0 |
| St. James | 308 | 6,949 | 548 | 2,615 | 544 | 213 | 130 | 531 | 76,347 | 56 | 50.0 | 32.1 | 77.4 |
| St. John the Baptist | 719 | 17,080 | 1,024 | 2,544 | 1,576 | 350 | 339 | 929 | 54,382 | 22 | 45.5 | 31.8 | 62.5 |
| St. Landry | 1,600 | 22,257 | 6,428 | 1,420 | 3,768 | 647 | 558 | 854 | 38,368 | 1,200 | 53.7 | 6.5 | 37.1 |
| St. Martin | 930 | 10,941 | 1,389 | 1,656 | 1,698 | 224 | 307 | 487 | 44,523 | 360 | 70.0 | 6.7 | 34.1 |
| St. Mary | 1,136 | 19,084 | 1,965 | 3,465 | 2,039 | 477 | 730 | 1,052 | 55,101 | 98 | 48.0 | 36.7 | 50.0 |
| St. Tammany | 6,576 | 80,199 | 17,313 | 2,807 | 13,457 | 3,427 | 4,763 | 3,596 | 44,834 | 994 | 80.7 | 0.4 | 30.7 |
| Tangipahoa | 2,492 | 37,041 | 8,959 | 2,528 | 6,785 | 1,694 | 1,062 | 1,377 | 37,163 | 967 | 55.0 | 1.1 | 37.4 |
| Tensas | 70 | 412 | 49 | 25 | 70 | 51 | NA | 14 | 34,677 | 231 | 16.9 | 24.7 | 42.5 |
| Terrebonne | 2,713 | 45,595 | 7,889 | 3,464 | 6,501 | 1,110 | 2,465 | 2,295 | 50,330 | 213 | 55.9 | 8.5 | 42.6 |
| Union | 348 | 4,456 | 809 | 1,405 | 596 | 106 | 70 | 139 | 31,250 | 426 | 36.6 | 3.1 | 36.4 |
| Vermilion | 970 | 9,177 | 1,584 | 475 | 2,222 | 342 | 316 | 384 | 41,879 | 1,304 | 41.9 | 8.5 | 34.6 |
| Vernon | 681 | 7,202 | 1,719 | 144 | 1,503 | 252 | 337 | 271 | 37,652 | 432 | 57.6 | 1.2 | 35.5 |
| Washington | 625 | 8,091 | 1,772 | 1,294 | 1,385 | 315 | 825 | 300 | 37,026 | 735 | 55.5 | 1.6 | 44.7 |
| Webster | 765 | 10,053 | 2,349 | 1,336 | 1,784 | 321 | 162 | 408 | 40,622 | 431 | 42.0 | 1.9 | 33.0 |
| West Baton Rouge | 558 | 12,203 | 477 | 2,545 | 1,208 | 122 | 302 | 598 | 48,986 | 111 | 61.3 | 11.7 | 44.3 |
| West Carroll | 167 | 1,525 | 557 | 73 | 328 | 63 | 19 | 45 | 29,566 | 548 | 27.7 | 8.2 | 34.9 |
| West Feliciana | 188 | 3,110 | 431 | 313 | 322 | 66 | 165 | 204 | 65,685 | 153 | 34.6 | 17.0 | 30.6 |
| Winn | 262 | 3,836 | 979 | 715 | 464 | 107 | 54 | 156 | 40,753 | 184 | 31.5 | 2.2 | 34.4 |
| **MAINE** | 41,843 | 522,191 | 112,725 | 50,526 | 83,728 | 28,320 | 22,925 | 24,129 | 46,208 | 7,600 | | | 61.0 |
| Androscoggin | 2,804 | 44,696 | 9,674 | 5,487 | 6,048 | 3,185 | 1,608 | 1,934 | 43,264 | 496 | 58.7 | 1.0 | 46.4 |
| Aroostook | 1,884 | 20,882 | 6,138 | 2,641 | 4,052 | 828 | 333 | 796 | 38,133 | 766 | 21.5 | 10.6 | 43.3 |
| Cumberland | 11,719 | 175,812 | 35,367 | 10,622 | 23,111 | 14,395 | 10,647 | 9,265 | 52,698 | 668 | 66.8 | 0.6 | 38.1 |
| Franklin | 865 | 9,636 | 1,841 | 1,221 | 1,825 | 626 | 119 | 325 | 33,730 | 354 | 49.4 | 2.0 | 44.0 |
| Hancock | 2,311 | 17,219 | 2,982 | 942 | 3,513 | 530 | 1,927 | 761 | 44,220 | 416 | 45.7 | 1.7 | 35.0 |
| Kennebec | 3,229 | 49,340 | 13,036 | 2,911 | 8,734 | 1,282 | 1,477 | 2,083 | 42,215 | 642 | 49.4 | 1.2 | 44.4 |
| Knox | 1,766 | 14,959 | 3,180 | 1,808 | 2,688 | 644 | 534 | 633 | 42,315 | 308 | 58.1 | NA | 45.5 |
| Lincoln | 1,395 | 8,910 | 1,821 | 667 | 1,766 | 304 | 299 | 360 | 40,388 | 309 | 53.4 | 0.3 | 36.2 |
| Oxford | 1,324 | 14,147 | 2,759 | 2,346 | 2,352 | 329 | 301 | 562 | 39,704 | 545 | 46.8 | 0.7 | 44.6 |
| Penobscot | 4,118 | 57,335 | 15,302 | 2,897 | 10,856 | 2,130 | 1,747 | 2,473 | 43,139 | 601 | 41.3 | 3.2 | 45.7 |
| Piscataquis | 448 | 4,716 | 1,305 | 1,181 | 887 | 76 | 67 | 169 | 35,913 | 188 | 29.8 | 3.7 | 52.1 |
| Sagadahoc | 976 | 13,797 | 1,130 | 5,731 | 1,951 | 307 | 784 | 645 | 46,778 | 209 | 49.3 | 1.0 | 43.2 |
| Somerset | 1,193 | 12,903 | 2,545 | 2,424 | 2,412 | 245 | 311 | 559 | 43,359 | 467 | 29.3 | 5.4 | 42.5 |
| Waldo | 990 | 9,261 | 1,822 | 986 | 1,667 | 384 | 255 | 456 | 49,232 | 517 | 49.1 | 0.6 | 33.8 |
| Washington | 821 | 7,315 | 1,710 | 1,139 | 1,699 | 312 | 149 | 282 | 38,512 | 379 | 44.6 | 1.6 | 43.0 |
| York | 5,671 | 56,440 | 11,767 | 7,520 | 10,163 | 1,704 | 1,709 | 2,529 | 44,810 | 735 | 59.0 | 0.8 | 42.1 |
| **MARYLAND** | 139,449 | 2,380,865 | 398,360 | 99,365 | 288,258 | 99,037 | 290,158 | 136,778 | 57,449 | 12,429 | 54.7 | 3.2 | 42.1 |
| Allegany | 1,502 | 23,807 | 5,955 | 2,755 | 3,951 | 862 | 579 | 860 | 36,143 | 290 | 43.8 | 0.3 | 35.5 |
| Anne Arundel | 14,382 | 247,074 | 29,866 | 11,326 | 32,817 | 6,537 | 34,897 | 14,620 | 59,172 | 390 | 67.7 | 0.3 | 38.2 |
| Baltimore | 19,957 | 335,413 | 63,471 | 14,572 | 46,625 | 19,920 | 28,459 | 17,825 | 53,143 | 708 | 67.1 | 1.7 | 40.0 |
| Calvert | 1,728 | 19,332 | 3,537 | 408 | 3,182 | 309 | 1,525 | 917 | 47,434 | 280 | 67.1 | 1.4 | 37.3 |
| Caroline | 611 | 7,426 | 815 | 1,283 | 1,163 | 166 | 252 | 303 | 40,751 | 588 | 48.3 | 4.9 | 50.7 |
| Carroll | 4,218 | 51,824 | 10,340 | 3,817 | 8,444 | 1,079 | 2,847 | 2,201 | 42,477 | 1,174 | 61.0 | 2.1 | 40.5 |
| Cecil | 1,761 | 27,213 | 4,342 | 4,523 | 3,847 | 351 | 650 | 1,379 | 50,667 | 533 | 55.7 | 1.7 | 42.3 |
| Charles | 2,616 | 33,750 | 5,145 | 413 | 8,332 | 706 | 1,703 | 1,351 | 40,015 | 385 | 58.7 | 1.3 | 44.3 |
| Dorchester | 679 | 9,872 | 1,693 | 2,922 | 1,295 | 202 | 279 | 390 | 39,482 | 371 | 39.1 | 11.1 | 53.1 |
| Frederick | 6,236 | 92,488 | 13,189 | 7,336 | 13,087 | 4,958 | 9,780 | 4,608 | 49,820 | 1,373 | 53.5 | 2.5 | 40.8 |
| Garrett | 906 | 10,643 | 1,805 | 998 | 1,617 | 332 | 465 | 373 | 35,032 | 707 | 37.3 | 0.8 | 32.7 |
| Harford | 5,539 | 75,948 | 12,369 | 5,195 | 12,670 | 1,785 | 9,230 | 3,566 | 46,947 | 628 | 64.0 | 1.6 | 38.4 |
| Howard | 9,548 | 178,425 | 18,659 | 5,904 | 15,942 | 7,285 | 41,381 | 12,460 | 69,835 | 321 | 72.9 | 1.6 | 35.9 |
| Kent | 588 | 6,623 | 1,113 | 798 | 849 | 184 | 213 | 244 | 36,823 | 346 | 32.9 | 11.6 | 48.2 |
| Montgomery | 27,376 | 444,313 | 77,582 | 6,043 | 45,739 | 20,269 | 81,652 | 30,287 | 68,167 | 558 | 69.7 | 2.3 | 39.2 |
| Prince George's | 15,447 | 273,474 | 35,031 | 7,605 | 37,666 | 5,374 | 30,340 | 13,856 | 50,667 | 367 | 71.9 | 2.2 | 31.5 |
| Queen Anne's | 1,394 | 12,760 | 1,253 | 1,602 | 2,454 | 272 | 712 | 522 | 40,880 | 483 | 40.2 | 8.9 | 47.3 |

# Table B. States and Counties — Agriculture

Agriculture, 2017 (cont.)

| STATE County | Land in farms Acreage (1,000) | Percent change, 2012-2017 | Average size of farm (acres) | Total irrigated (1,000) | Total cropland (1,000) | Value of land and buildings Average per farm (dollars) | Average per acre | Value of machinery and equipment, average per farm (dollars) | Value of products sold Total (mil dol) | Average per farm (acres) | Percent from: Crops | Livestock and poultry products | Organic farms (number) | Farms with internet access (percent) | Government payments Total ($1,000) | Percent of farms |
|---|---|---|---|---|---|---|---|---|---|---|---|---|---|---|---|---|
| | 117 | 118 | 119 | 120 | 121 | 122 | 123 | 124 | 125 | 126 | 127 | 128 | 129 | 130 | 131 | 132 |
| **LOUISIANA—Cont'd** | | | | | | | | | | | | | | | | |
| Orleans | 1 | 376.6 | 14 | 0.0 | 0.0 | 140,455 | 10,355 | 17,365 | 0.2 | 5,231 | 92.2 | 7.8 | NA | 87.2 | 8 | 10.3 |
| Ouachita | 93 | 0.1 | 191 | 16.1 | 49.0 | 813,699 | 4,261 | 89,169 | 45.3 | 92,887 | 52.5 | 47.5 | 3.0 | 81.6 | 2,315 | 22.7 |
| Plaquemines | 94 | 5.8 | 833 | 0.1 | 8.0 | 1,174,762 | 1,411 | 60,963 | 10.8 | 95,478 | 45.5 | 54.5 | 1.0 | 75.2 | NA | NA |
| Pointe Coupee | 188 | 3.0 | 389 | 5.9 | 132.7 | 1,203,401 | 3,091 | 191,613 | 80.2 | 166,309 | 89.8 | 10.2 | NA | 73.0 | 3,798 | 27.4 |
| Rapides | 203 | -3.7 | 237 | 26.3 | 131.8 | 889,536 | 3,747 | 134,402 | 147.5 | 172,346 | 91.3 | 8.7 | 1.0 | 73.6 | 3,482 | 23.8 |
| Red River | 95 | -29.5 | 485 | 2.4 | 26.8 | 1,084,557 | 2,238 | 150,566 | 17.6 | 89,589 | 38.6 | 61.4 | NA | 71.1 | 1,147 | 29.9 |
| Richland | 218 | -21.7 | 349 | 73.0 | 135.5 | 987,603 | 2,831 | 141,410 | 74.7 | 119,335 | 94.3 | 5.7 | NA | 62.9 | 9,359 | 61.7 |
| Sabine | 62 | 20.0 | 141 | 0.1 | 14.9 | 531,768 | 3,779 | 81,312 | 145.9 | 330,118 | 1.2 | 98.8 | NA | 64.3 | 1,411 | 25.3 |
| St. Bernard | 32 | 1.7 | 983 | NA | 2.8 | 1,722,736 | 1,752 | 116,121 | 4.0 | 121,939 | 1.8 | 98.2 | NA | 74.8 | D | 3.0 |
| St. Charles | 14 | -11.6 | 214 | 0.0 | 2.9 | 746,957 | 3,491 | 60,454 | 1.4 | 21,537 | 17.0 | 83.0 | NA | 78.8 | D | 3.0 |
| St. Helena | 37 | -30.5 | 106 | 0.2 | 12.2 | 441,244 | 4,159 | 50,934 | 20.6 | 59,075 | 5.3 | 94.7 | NA | 62.4 | D | 3.0 |
| St. James | 51 | 26.6 | 903 | D | 45.0 | 2,765,668 | 3,062 | 645,194 | 27.0 | 481,393 | 99.7 | 0.3 | NA | 76.8 | 102 | 12.1 |
| St. John the Baptist | 20 | 84.8 | 904 | 0.0 | 18.0 | 3,163,313 | 3,500 | 651,269 | 7.2 | 328,364 | 99.5 | 0.5 | NA | 59.1 | 49 | 7.1 |
| St. Landry | 267 | -11.1 | 223 | 24.3 | 195.2 | 650,514 | 2,919 | 110,986 | 93.1 | 77,605 | 85.4 | 14.6 | 2.0 | 65.9 | 6,348 | 28.0 |
| St. Martin | 83 | 10.0 | 232 | 3.9 | 63.7 | 743,147 | 3,205 | 182,286 | 39.6 | 109,994 | 88.3 | 11.7 | NA | 66.1 | 1,723 | 16.4 |
| St. Mary | 80 | 5.4 | 818 | D | 72.2 | 2,697,959 | 3,298 | 552,607 | 45.1 | 460,510 | 99.0 | 1.0 | NA | 73.5 | 368 | 15.3 |
| St. Tammany | 43 | 26.2 | 43 | 0.4 | 7.7 | 464,732 | 10,731 | 39,617 | 10.0 | 10,063 | 55.7 | 44.3 | 1.0 | 77.4 | NA | NA |
| Tangipahoa | 98 | -8.1 | 101 | 1.2 | 29.6 | 513,368 | 5,061 | 63,007 | 42.4 | 43,845 | 47.3 | 52.7 | 5.0 | 67.6 | 208 | 5.0 |
| Tensas | 202 | 2.7 | 874 | 55.5 | 166.0 | 2,450,887 | 2,803 | 324,337 | 93.5 | 404,827 | 99.7 | 0.3 | NA | 64.5 | 7,907 | 82.3 |
| Terrebonne | 79 | -15.2 | 370 | 0.1 | 35.8 | 1,523,117 | 4,118 | 175,237 | 30.1 | 141,479 | 56.8 | 43.2 | NA | 75.1 | 131 | 3.8 |
| Union | 75 | 20.2 | 177 | 0.0 | 19.5 | 538,032 | 3,044 | 85,109 | 121.1 | 284,319 | 1.5 | 98.5 | NA | 70.7 | 185 | 3.5 |
| Vermilion | 410 | 44.4 | 314 | 68.1 | 217.0 | 903,947 | 2,877 | 117,984 | 117.3 | 89,923 | 57.1 | 42.9 | NA | 61.8 | 13,719 | 53.5 |
| Vernon | 41 | -16.5 | 95 | 0.3 | 8.1 | 350,847 | 3,703 | 53,429 | 25.5 | 59,113 | 2.7 | 97.3 | NA | 70.4 | 184 | 15.3 |
| Washington | 81 | -0.9 | 110 | 0.5 | 30.8 | 396,447 | 3,612 | 80,648 | 32.4 | 44,020 | 36.4 | 63.6 | 8.0 | 69.5 | 255 | 4.6 |
| Webster | 58 | 11.0 | 135 | 0.2 | 11.7 | 395,957 | 2,939 | 68,771 | 9.4 | 21,717 | 13.6 | 86.4 | NA | 66.6 | 74 | 4.4 |
| West Baton Rouge | 34 | 12.5 | 307 | 0.1 | 29.8 | 865,851 | 2,820 | 203,536 | 25.6 | 230,991 | 63.0 | 37.0 | NA | 80.2 | 225 | 18.0 |
| West Carroll | 168 | 1.4 | 307 | 69.8 | 111.1 | 1,010,538 | 3,296 | 115,396 | 62.5 | 114,117 | 96.2 | 3.8 | NA | 54.9 | 5,591 | 79.4 |
| West Feliciana | 86 | -14.8 | 564 | 0.1 | 24.7 | 1,655,865 | 2,935 | 110,413 | 9.2 | 60,261 | 42.3 | 57.7 | 1.0 | 72.5 | 329 | 20.3 |
| Winn | 30 | 21.1 | 164 | 0.0 | 5.1 | 460,561 | 2,804 | 97,682 | 20.2 | 110,049 | 2.5 | 97.5 | NA | 76.1 | 282 | 11.4 |
| **MAINE** | 1,308 | -10.1 | 172 | 32.3 | 472.5 | 446,614 | 2,596 | 81,792 | 667.0 | 87,758 | 61.3 | 38.7 | 621.0 | 83.6 | 8,947 | 10.9 |
| Androscoggin | 56 | -6.4 | 112 | 1.0 | 24.6 | 395,349 | 3,526 | 68,417 | 40.5 | 81,726 | 37.6 | 62.4 | 18.0 | 85.1 | 481 | 10.5 |
| Aroostook | 317 | -9.6 | 414 | 12.7 | 174.0 | 720,511 | 1,741 | 194,251 | 202.0 | 263,674 | 92.9 | 7.1 | 49.0 | 79.6 | 2,817 | 28.6 |
| Cumberland | 50 | -20.2 | 75 | 0.8 | 15.3 | 511,437 | 6,830 | 59,232 | 25.6 | 38,389 | 60.8 | 39.2 | 38.0 | 90.6 | 539 | 5.1 |
| Franklin | 47 | -4.5 | 133 | 0.1 | 11.1 | 331,253 | 2,484 | 59,248 | D | D | D | D | 37.0 | 87.0 | 277 | 15.5 |
| Hancock | 65 | 22.2 | 157 | 0.2 | 15.2 | 438,042 | 2,794 | 65,383 | 18.4 | 44,163 | 57.0 | 43.0 | 65.0 | 80.8 | 247 | 7.2 |
| Kennebec | 82 | 5.2 | 128 | 0.2 | 36.5 | 386,694 | 3,023 | 77,600 | 49.0 | 76,333 | 57.0 | 43.0 | 48.0 | 83.5 | 422 | 10.6 |
| Knox | 26 | -12.9 | 83 | 0.3 | 9.9 | 392,714 | 4,725 | 49,648 | 9.1 | 29,597 | 69.9 | 30.1 | 25.0 | 78.9 | 46 | 3.6 |
| Lincoln | 25 | -19.3 | 82 | 0.1 | 6.5 | 382,809 | 4,644 | 50,811 | 12.9 | 41,689 | 35.2 | 64.8 | 32.0 | 86.7 | 177 | 5.5 |
| Oxford | 77 | 2.1 | 141 | 0.8 | 17.5 | 410,809 | 2,914 | 65,022 | 24.1 | 44,253 | 78.5 | 21.5 | 31.0 | 85.7 | 836 | 13.8 |
| Penobscot | 105 | -6.6 | 175 | 1.6 | 41.1 | 432,698 | 2,466 | 93,820 | 50.9 | 84,717 | 35.6 | 64.4 | 33.0 | 79.0 | 603 | 9.2 |
| Piscataquis | 51 | 9.4 | 272 | 0.0 | 8.7 | 382,344 | 1,408 | 84,800 | 9.1 | 48,447 | 47.2 | 52.8 | 24.0 | 79.8 | 107 | 16.0 |
| Sagadahoc | 18 | -12 | 85 | 0.1 | 5.2 | 349,864 | 4,134 | 52,958 | D | D | D | D | 31.0 | 85.2 | 379 | 8.1 |
| Somerset | 146 | 4.1 | 312 | 0.3 | 35.8 | 460,950 | 1,476 | 117,809 | 83.9 | 179,724 | 66.3 | 33.7 | 53.0 | 84.2 | 475 | 13.7 |
| Waldo | 57 | -56.7 | 109 | D | 21.5 | 327,769 | 2,998 | 52,008 | 23.0 | 44,400 | 36.9 | 63.1 | 81.0 | 85.9 | 908 | 9.5 |
| Washington | 125 | -16.2 | 329 | D | 31.2 | 505,172 | 1,534 | 65,445 | D | D | D | D | 19.0 | 78.6 | 55 | 3.2 |
| York | 61 | -5.4 | 83 | 1.6 | 18.3 | 425,179 | 5,120 | 60,944 | 28.6 | 38,846 | 82.5 | 17.5 | 37.0 | 83.9 | 578 | 5.4 |
| **MARYLAND** | 1,990 | -2 | 160 | 124.8 | 1,426.7 | 1,258,691 | 7,861 | 124,871 | 2,472.8 | 198,955 | 38.3 | 61.7 | 134.0 | 76.9 | 44,410 | 28.7 |
| Allegany | 35 | -2.7 | 122 | 0.0 | 13.2 | 680,294 | 5,592 | 53,733 | 4.2 | 14,362 | 73.0 | 27.0 | 1.0 | 61.4 | 245 | 22.8 |
| Anne Arundel | 27 | -3.9 | 69 | 0.2 | 14.6 | 713,958 | 10,312 | 65,583 | 18.2 | 46,549 | 70.7 | 29.3 | 1.0 | 85.9 | 322 | 7.2 |
| Baltimore | 76 | 8.1 | 108 | 0.7 | 50.5 | 1,594,001 | 14,825 | 103,714 | 67.5 | 95,367 | 86.9 | 13.1 | 10.0 | 82.5 | 1,327 | 14.8 |
| Calvert | 25 | -23.6 | 90 | 0.3 | 12.7 | 921,361 | 10,257 | 67,836 | 6.3 | 22,579 | 90.2 | 9.8 | NA | 76.4 | 147 | 9.6 |
| Caroline | 128 | -14.8 | 218 | 32.4 | 109.0 | 1,551,852 | 7,126 | 175,992 | 277.4 | 471,815 | 25.5 | 74.5 | 3.0 | 68.5 | 3,482 | 46.8 |
| Carroll | 147 | 10.7 | 125 | 1.6 | 109.4 | 1,022,466 | 8,178 | 117,957 | 110.4 | 94,078 | 65.6 | 34.4 | 5.0 | 81.9 | 3,501 | 33.4 |
| Cecil | 74 | -3.7 | 138 | 1.5 | 54.6 | 1,109,934 | 8,017 | 139,348 | 136.8 | 256,700 | 57.0 | 43.0 | 8.0 | 79.5 | 1,475 | 22.1 |
| Charles | 41 | -12.1 | 107 | 0.5 | 26.0 | 1,007,714 | 9,458 | 100,760 | 14.1 | 36,532 | 88.4 | 11.6 | 4.0 | 76.1 | 789 | 17.7 |
| Dorchester | 132 | 4.5 | 356 | 29.9 | 97.0 | 2,022,110 | 5,676 | 229,945 | 188.7 | 508,553 | 30.5 | 69.5 | NA | 70.4 | 4,484 | 70.6 |
| Frederick | 189 | 3.9 | 137 | 1.2 | 140.7 | 1,307,843 | 9,522 | 122,029 | 131.6 | 95,835 | 48.3 | 51.7 | 31.0 | 80.3 | 3,980 | 26.5 |
| Garrett | 90 | -5.1 | 128 | 0.1 | 42.7 | 582,887 | 4,561 | 82,214 | 29.0 | 41,068 | 42.4 | 57.6 | 11.0 | 63.6 | 260 | 7.4 |
| Harford | 74 | 13.4 | 118 | 0.6 | 51.7 | 1,289,892 | 10,906 | 107,318 | 45.9 | 84,919 | 74.3 | 25.7 | 2.0 | 84.7 | 1,411 | 20.7 |
| Howard | 32 | -13.4 | 101 | 0.4 | 17.8 | 925,165 | 9,156 | 86,214 | 27.3 | 84,919 | 86.3 | 13.7 | NA | 85.7 | 240 | 10.6 |
| Kent | 134 | 0.8 | 388 | 10.1 | 109.2 | 2,564,409 | 6,609 | 248,381 | 111.2 | 321,428 | 61.6 | 38.4 | 16.0 | 82.1 | 3,429 | 69.7 |
| Montgomery | 66 | 3.2 | 117 | 1.2 | 48.7 | 964,640 | 8,213 | 102,089 | 42.6 | 76,310 | 88.8 | 11.2 | 4.0 | 90.0 | 1,179 | 14.3 |
| Prince George's | 34 | 5.5 | 94 | 0.8 | 17.7 | 762,242 | 8,132 | 76,284 | 17.6 | 47,872 | 86.8 | 13.2 | NA | 80.4 | 135 | 8.4 |
| Queen Anne's | 163 | 3.9 | 337 | 16.7 | 135.8 | 2,476,771 | 7,339 | 191,888 | 180.6 | 373,822 | 50.7 | 49.3 | 15.0 | 84.7 | 4,235 | 59.2 |

| STATE County | Water use, 2015 | | Wholesale Trade[1], 2017 | | | | Retail Trade[2], 2017 | | | | Real estate and rental and leasing,[2] 2017 | | | |
|---|---|---|---|---|---|---|---|---|---|---|---|---|---|---|
| | Public supply water withdrawn (mil gal/day) | Public supply gallons withdrawn per person per day | Number of establishments | Number of employees | Sales (mil dol) | Average payroll (mil dol) | Number of establishments | Number of employees | Sales (mil dol) | Average payroll (mil dol) | Number of establishments | Number of employees | Sales (mil dol) | Average payroll (mil dol) |
| | 133 | 134 | 135 | 136 | 137 | 138 | 139 | 140 | 141 | 142 | 143 | 144 | 145 | 146 |
| LOUISIANA—Cont'd | | | | | | | | | | | | | | |
| Orleans | 140.9 | 361.6 | 255 | 3,124 | 2,506.4 | 187.2 | 1,327 | 14,795 | 3,499.1 | 387.0 | 470 | 2,327 | 573.6 | 99.7 |
| Ouachita | 24.18 | 154.2 | 182 | 2,148 | 1,700.6 | 102.8 | 681 | 9,452 | 2,410.5 | 229.6 | 216 | 1,317 | 267.9 | 49.9 |
| Plaquemines | 7.14 | 303.9 | 60 | 929 | 1,366.3 | 57.4 | 63 | 504 | 113.6 | 11.1 | 28 | 291 | 76.5 | 17.1 |
| Pointe Coupee | 3.54 | 159.1 | 11 | 157 | 182.4 | 7.9 | 76 | 909 | 231.8 | 22.1 | 11 | 37 | 5.9 | 1.4 |
| Rapides | 18.91 | 143.1 | D | D | D | D | 550 | 7,461 | 2,100.4 | 193.7 | D | D | D | D |
| Red River | 1.00 | 116.4 | 5 | 44 | 16.3 | 1.8 | 19 | 120 | 34.1 | 2.5 | 4 | 9 | 1.2 | 0.4 |
| Richland | 3.64 | 177.4 | 23 | 255 | 427.4 | 12.3 | 59 | 692 | 222.2 | 17.8 | D | D | D | D |
| Sabine | 2.34 | 96.8 | 12 | 110 | 62.7 | 4.2 | 80 | 941 | 247.7 | 22.6 | 11 | 23 | 2.6 | 0.6 |
| St. Bernard | 7.16 | 157.7 | 25 | 239 | 77.8 | 9.5 | 134 | 1,716 | 401.2 | 40.4 | 23 | 44 | 15.8 | 2.6 |
| St. Charles | 9.09 | 172.1 | 67 | 1,611 | 1,451.1 | 94.0 | 115 | 1,487 | 403.8 | 37.5 | 37 | 268 | 96.2 | 23.9 |
| St. Helena | 0.90 | 85.2 | NA | NA | NA | NA | 24 | 223 | 44.0 | 4.1 | 3 | 18 | 0.9 | 0.1 |
| St. James | 4.00 | 185.5 | 9 | 39 | 114.8 | 2.5 | 49 | 608 | 138.1 | 12.6 | 8 | 95 | 9.1 | 3.3 |
| St. John the Baptist | 7.45 | 170.8 | 31 | 568 | 307.5 | 48.4 | 115 | 1,698 | 412.3 | 38.0 | 38 | 510 | 85.9 | 25.1 |
| St. Landry | 10.31 | 123.0 | 56 | 738 | 940.4 | 33.0 | 311 | 3,921 | 1,087.5 | 100.8 | 52 | 174 | 26.0 | 5.6 |
| St. Martin | 4.67 | 86.7 | 58 | 749 | 355.3 | 33.8 | 150 | 1,643 | 581.0 | 37.4 | 42 | 695 | 222.1 | 51.6 |
| St. Mary | 9.32 | 176.5 | 65 | 795 | 412.4 | 42.7 | 183 | 2,203 | 529.5 | 55.2 | 78 | 595 | 119.6 | 31.7 |
| St. Tammany | 23.64 | 94.5 | 235 | 2,359 | 8,962.2 | 154.7 | 920 | 14,341 | 4,018.1 | 352.2 | 278 | 1,011 | 238.7 | 46.5 |
| Tangipahoa | 15.06 | 117.0 | 89 | 2,288 | 1,747.3 | 111.0 | 454 | 6,671 | 1,984.3 | 168.0 | 113 | 541 | 86.5 | 20.8 |
| Tensas | 1.16 | 244.7 | 11 | 85 | 137.3 | 3.9 | 16 | 93 | 23.8 | 1.9 | NA | NA | NA | NA |
| Terrebonne | 1.88 | 16.5 | 195 | 1,750 | 824.5 | 87.2 | 470 | 6,758 | 1,860.4 | 177.7 | 158 | 1,501 | 476.1 | 102.0 |
| Union | 4.40 | 195.8 | 9 | 53 | 109.3 | 2.8 | 59 | 553 | 175.2 | 15.3 | 13 | 16 | 2.0 | 0.4 |
| Vermilion | 7.07 | 118.1 | 33 | 299 | 213.8 | 14.7 | 188 | 2,325 | 566.4 | 54.3 | 31 | 191 | 22.8 | 7.7 |
| Vernon | 4.35 | 85.6 | D | D | D | D | 131 | 1,574 | 437.4 | 37.0 | 35 | 215 | 75.8 | 9.2 |
| Washington | 4.63 | 99.8 | 19 | 118 | 95.3 | 4.8 | 144 | 1,459 | 334.2 | 31.5 | 11 | 32 | 5.0 | 0.7 |
| Webster | 5.36 | 133.9 | 20 | 312 | 144.9 | 14.5 | 155 | 2,090 | 512.8 | 54.3 | 28 | 93 | 14.2 | 3.5 |
| West Baton Rouge | 7.21 | 282.9 | 38 | 699 | 1,030.8 | 38.5 | 92 | 1,183 | 340.0 | 27.5 | 14 | 182 | 47.2 | 10.1 |
| West Carroll | 1.39 | 123.1 | D | D | D | D | 25 | 333 | 80.9 | 7.9 | D | D | D | D |
| West Feliciana | 1.67 | 108.5 | D | D | D | D | 31 | 345 | 88.4 | 7.4 | 12 | 32 | 6.9 | 1.1 |
| Winn | 2.01 | 138.0 | 9 | 60 | 29.2 | 2.3 | 43 | 516 | 117.6 | 13.9 | 7 | 26 | 4.5 | 0.9 |
| MAINE | 84.97 | 63.9 | 1,309 | 15,527 | 13,951.8 | 803.4 | 6,250 | 81,733 | 23,878.7 | 2,249.6 | 1,821 | 7,138 | 1,484.2 | 287.4 |
| Androscoggin | 7.92 | 73.9 | 104 | 1,274 | 559.5 | 64.1 | 431 | 5,933 | 1,816.8 | 162.8 | 130 | 466 | 86.1 | 17.1 |
| Aroostook | 4.11 | 59.9 | 70 | 494 | 346.7 | 24.9 | 327 | 4,036 | 1,059.3 | 96.2 | 76 | 303 | 37.7 | 7.5 |
| Cumberland | 25.35 | 87.4 | 437 | 5,773 | 5,520.2 | 346.3 | 1,460 | 22,472 | 7,031.7 | 650.5 | 659 | 3,050 | 642.1 | 142.6 |
| Franklin | 1.36 | 45.3 | D | D | D | 2.3 | 175 | 1,779 | 428.6 | 43.8 | D | D | D | D |
| Hancock | 6.69 | 122.4 | 61 | 344 | 341.6 | 16.6 | 339 | 3,259 | 948.6 | 95.1 | 89 | 244 | 47.0 | 8.6 |
| Kennebec | 6.15 | 51.3 | 91 | 3,101 | 2,488.5 | 138.2 | 527 | 8,659 | 2,485.5 | 249.7 | 110 | 543 | 81.8 | 19.2 |
| Knox | 3.02 | 75.8 | 51 | 280 | 248.0 | 11.6 | 252 | 2,873 | 720.2 | 79.0 | 70 | 187 | 38.2 | 7.7 |
| Lincoln | 0.73 | 21.5 | 31 | 276 | 130.3 | 10.0 | 211 | 1,647 | 484.9 | 47.9 | 48 | 94 | 17.0 | 3.3 |
| Oxford | 2.85 | 49.8 | D | D | D | 7.8 | 234 | 2,361 | 698.0 | 63.6 | 49 | 185 | 22.8 | 6.7 |
| Penobscot | 4.79 | 31.4 | 156 | 1,816 | 1,056.4 | 93.9 | 697 | 10,573 | 3,261.5 | 276.5 | 184 | 924 | 270.0 | 29.3 |
| Piscataquis | 0.88 | 52.0 | 6 | 8 | 4.1 | 0.2 | 85 | 947 | 249.8 | 26.9 | D | D | D | D |
| Sagadahoc | 1.41 | 40.1 | 18 | 84 | 39.5 | 3.4 | 140 | 1,827 | 545.2 | 50.0 | 35 | 65 | 15.9 | 2.6 |
| Somerset | 1.82 | 35.6 | 29 | 164 | 77.5 | 7.1 | 208 | 2,270 | 649.0 | 59.9 | 34 | 208 | 55.8 | 9.8 |
| Waldo | 0.99 | 25.3 | 24 | 319 | 2,151.2 | 14.8 | 165 | 1,614 | 395.3 | 41.9 | D | D | D | D |
| Washington | 1.57 | 49.6 | D | D | D | 7.0 | 148 | 1,663 | 415.5 | 43.8 | D | D | D | D |
| York | 15.33 | 76.2 | 135 | 1,138 | 747.3 | 55.1 | 851 | 9,820 | 2,688.7 | 262.2 | 248 | 618 | 141.0 | 25.9 |
| MARYLAND | 749.52 | 124.8 | 4,598 | 73,388 | 62,762.8 | 4,872.3 | 17,911 | 291,814 | 84,966.2 | 8,240.9 | 6,811 | 49,157 | 18,087.4 | 2,851.5 |
| Allegany | 0.67 | 9.2 | 43 | 449 | 141.0 | 19.6 | 270 | 3,856 | 1,018.1 | 90.4 | 61 | 192 | 35.8 | 6.7 |
| Anne Arundel | 39.49 | 70.0 | 465 | 7,006 | 7,067.3 | 436.3 | 1,959 | 33,976 | 9,441.6 | 934.8 | 628 | 4,376 | 1,447.4 | 223.4 |
| Baltimore | 209.46 | 252.0 | 700 | 11,309 | 6,870.4 | 735.2 | 2,662 | 46,685 | 13,216.5 | 1,309.8 | 988 | 7,618 | 4,077.8 | 446.9 |
| Calvert | 2.49 | 27.5 | 32 | D | 109.6 | D | 210 | 3,182 | 938.1 | 89.2 | 87 | 278 | 66.2 | 11.3 |
| Caroline | 0.98 | 30.1 | 23 | 145 | 78.0 | 6.8 | 80 | 1,147 | 377.2 | 32.1 | 17 | 67 | 9.9 | 2.3 |
| Carroll | 7.05 | 42.1 | 141 | 1,197 | 735.1 | 67.9 | 506 | 8,205 | 2,387.9 | 216.2 | 139 | 488 | 100.4 | 20.8 |
| Cecil | 3.60 | 35.2 | D | D | D | D | 243 | 3,803 | 1,179.9 | 93.3 | 89 | 241 | 47.1 | 8.5 |
| Charles | 7.32 | 46.9 | 54 | 460 | 800.6 | 27.5 | 467 | 8,248 | 2,318.6 | 222.3 | 113 | 462 | 128.1 | 19.6 |
| Dorchester | 2.65 | 81.8 | 28 | 385 | 191.4 | 18.4 | 90 | 1,289 | 489.6 | 38.2 | 34 | 141 | 12.0 | 2.9 |
| Frederick | 13.79 | 56.2 | 232 | 2,703 | 1,672.7 | 163.1 | 732 | 12,831 | 3,906.5 | 363.8 | 280 | 1,024 | 255.5 | 49.4 |
| Garrett | 2.74 | 93.0 | 20 | 95 | 38.3 | 3.4 | 144 | 1,589 | 491.8 | 43.0 | 34 | 420 | 40.3 | 13.0 |
| Harford | 9.84 | 39.3 | 162 | 2,320 | 2,563.2 | 140.0 | 689 | 13,882 | 3,897.3 | 383.5 | 245 | 944 | 233.3 | 39.7 |
| Howard | 0.01 | 0.0 | 474 | 13,365 | 11,476.8 | 959.3 | 867 | 16,013 | 4,643.6 | 465.8 | 442 | 3,969 | 1,163.8 | 249.6 |
| Kent | 1.09 | 55.1 | 28 | 156 | 115.4 | 6.5 | 90 | 902 | 196.0 | 20.8 | 23 | 78 | 24.0 | 2.9 |
| Montgomery | 354.94 | 341.3 | 647 | 8,452 | 7,355.0 | 767.1 | 2,567 | 46,463 | 14,659.2 | 1,502.6 | 1,463 | 12,197 | 5,976.1 | 877.7 |
| Prince George's | 52.34 | 57.5 | 528 | 11,030 | 8,127.0 | 639.0 | 2,268 | 37,683 | 10,485.3 | 1,081.7 | 752 | 8,804 | 2,240.5 | 459.6 |
| Queen Anne's | 1.77 | 36.2 | 63 | 873 | 734.5 | 61.0 | 226 | 2,619 | 623.9 | 58.2 | 54 | 165 | 43.0 | 6.7 |

1  Merchant wholesalers, except manufacturers' sales branches and offices.     2. Employer establishments.

| STATE County | Professional, scientific, and technical services, 2017 | | | | Manufacturing, 2017 | | | | Accommodation and food services, 2017 | | | |
|---|---|---|---|---|---|---|---|---|---|---|---|---|
| | Number of establishments | Number of employees | Sales (mil dol) | Average payroll (mil dol) | Number of establishments | Number of employees | Sales (mil dol) | Average payroll (mil dol) | Number of establishments | Number of employees | Sales (mil dol) | Annual payroll (mil dol) |
| | 147 | 148 | 149 | 150 | 151 | 152 | 153 | 154 | 155 | 156 | 157 | 158 |
| LOUISIANA—Cont'd | | | | | | | | | | | | |
| Orleans | D | D | D | D | 180 | 3,395 | 2,462.0 | 211.8 | 1,510 | 42,732 | 3,768.7 | 1,026.2 |
| Ouachita | 428 | 2,681 | 411 | 139.6 | D | D | D | 271.8 | D | D | D | D |
| Plaquemines | D | D | D | D | D | 1,745 | D | 166.7 | 55 | 852 | 54.2 | 15.6 |
| Pointe Coupee | 27 | 87 | 15 | 4.8 | 6 | 319 | 176.4 | 14.9 | 30 | 323 | 18.4 | 4.0 |
| Rapides | 271 | 1,823 | 241 | 86.0 | D | D | D | D | 250 | 5,012 | 271.7 | 74.3 |
| Red River | 8 | 33 | 3 | 0.8 | 6 | 340 | 110.1 | 17.8 | 9 | 91 | 4.3 | 1.3 |
| Richland | 29 | 89 | 15 | 3.3 | D | D | D | D | 26 | 341 | 17.8 | 4.4 |
| Sabine | 46 | 110 | 14 | 4.9 | 11 | 825 | 201.1 | 40.6 | D | D | D | D |
| St. Bernard | 43 | 173 | 21 | 6.6 | 33 | 1,382 | 9,273.2 | 153.0 | 80 | 1,288 | 66.4 | 17.7 |
| St. Charles | 104 | 1,550 | 219 | 101.4 | 45 | 4,938 | 23,469.7 | 571.1 | 83 | 1,004 | 56.6 | 14.0 |
| St. Helena | D | D | 5 | D | D | 232 | D | 13.4 | D | D | D | 0.1 |
| St. James | D | D | D | D | 28 | 2,535 | 10,036.6 | 268.6 | 27 | 376 | 19.3 | 5.7 |
| St. John the Baptist | D | D | D | D | 24 | 2,375 | 17,844.4 | 247.0 | 79 | 1,213 | 62.2 | 16.0 |
| St. Landry | 151 | 702 | 100 | 34.5 | 62 | 1,184 | 2,078.5 | 61.2 | 102 | 1,641 | 96.0 | 23.4 |
| St. Martin | 79 | 263 | 39 | 12.3 | 71 | 1,552 | 426.5 | 76.3 | 70 | 1,167 | 55.7 | 14.9 |
| St. Mary | 98 | 665 | 101 | 39.8 | 76 | 3,228 | 1,381.4 | 191.7 | 104 | 2,256 | 129.5 | 41.6 |
| St. Tammany | D | D | D | D | D | 2,638 | D | 125.4 | 611 | 11,444 | 594.0 | 177.7 |
| Tangipahoa | 209 | 1,142 | 174 | 60.4 | 77 | 2,023 | 625.2 | 87.3 | 252 | 4,913 | 238.7 | 67.2 |
| Tensas | NA | NA | NA | NA | D | 22 | D | 1.0 | D | D | D | D |
| Terrebonne | 254 | 2,868 | 389 | 177.1 | 127 | 3,912 | 936.0 | 227.7 | 249 | 4,752 | 259.9 | 77.2 |
| Union | 15 | 48 | 6 | 2.0 | D | D | D | 33.5 | D | D | D | D |
| Vermilion | 110 | 311 | 35 | 13.3 | 36 | 543 | 220.7 | 31.3 | 76 | 970 | 45.3 | 12.4 |
| Vernon | 82 | 373 | 44 | 15.3 | 9 | 131 | 45.6 | 4.8 | 78 | 1,138 | 57.0 | 16.3 |
| Washington | 41 | 1,632 | 157 | 19.0 | 26 | 1,098 | 843.5 | 81.6 | 61 | 872 | 38.6 | 9.4 |
| Webster | 43 | 183 | 20 | 6.2 | D | D | D | D | 63 | 739 | 37.6 | 9.4 |
| West Baton Rouge | 35 | 310 | 55 | 21.3 | 42 | 2,798 | 5,682.6 | 172.4 | 61 | 1,063 | 56.5 | 13.5 |
| West Carroll | D | D | D | 0.4 | D | D | D | D | 61 | D | 56.5 | 13.5 |
| West Feliciana | 22 | 264 | 27 | 11.3 | D | D | D | D | 12 | D | 7.9 | D |
| Winn | 15 | 42 | 5 | 1.5 | 12 | 702 | 342.0 | 37.8 | 16 | 180 | 9.2 | 2.2 |
| MAINE | 3,506 | 21,683 | 3,485 | 1,361.9 | 1,691 | 48,620 | 15,089.2 | 2,598.7 | 4,257 | 55,746 | 4,017.7 | 1,167.1 |
| Androscoggin | D | D | D | D | 152 | 5,665 | 1,693.9 | 263.2 | 238 | 3,455 | 207.1 | 61.3 |
| Aroostook | D | D | D | D | 78 | 2,736 | 1,072.9 | 138.7 | 140 | 2,054 | 93.8 | 29.7 |
| Cumberland | D | D | D | D | 410 | 10,418 | 3,307.5 | 578.6 | 1,096 | 17,219 | 1,177.5 | 364.9 |
| Franklin | D | D | D | D | 31 | 1,203 | 461.4 | 49.4 | 111 | 1,187 | 58.8 | 17.0 |
| Hancock | D | D | D | D | 85 | 920 | 231.7 | 44.4 | 334 | 2,404 | 319.3 | 77.5 |
| Kennebec | 268 | 1,404 | 210 | 79.5 | 92 | 2,540 | 675.5 | 130.0 | 293 | 4,860 | 315.9 | 91.5 |
| Knox | D | D | D | D | 92 | 1,615 | 559.9 | 86.9 | 175 | 1,612 | 128.5 | 39.3 |
| Lincoln | D | D | D | D | 72 | 659 | 108.7 | 29.3 | 167 | 1,232 | 113.9 | 33.7 |
| Oxford | 75 | 347 | 38 | 16.7 | 58 | 2,095 | 787.3 | 121.5 | 149 | 3,012 | 222.3 | 54.2 |
| Penobscot | D | D | D | D | 147 | 2,881 | 686.3 | 142.8 | 332 | 6,001 | 365.7 | 102.4 |
| Piscataquis | 10 | 49 | 4 | 1.3 | 22 | 936 | 175.5 | 38.0 | 57 | 303 | 21.9 | 5.7 |
| Sagadahoc | D | D | D | D | D | D | D | D | 86 | 1,127 | 73.9 | 23.0 |
| Somerset | 50 | 524 | 61 | 31.8 | 66 | 2,478 | 1,076.9 | 148.0 | 97 | 851 | 54.3 | 15.6 |
| Waldo | 65 | 221 | 22 | 9.8 | 60 | 898 | 168.3 | 39.7 | 93 | 863 | 56.3 | 17.3 |
| Washington | D | D | D | D | 39 | 895 | 486.0 | 45.8 | 82 | 608 | 33.3 | 9.7 |
| York | 424 | 1,756 | 261 | 96.4 | D | D | D | D | 807 | 8,958 | 774.9 | 224.2 |
| MARYLAND | 20,776 | 270,959 | 52,616 | 22,323.3 | 2,967 | 97,992 | 41,776.9 | 6,343.2 | 12,139 | 237,730 | 16,930.7 | 4,624.5 |
| Allegany | D | D | D | D | 51 | 2,691 | 1,192.5 | 119.0 | D | D | D | 55.2 |
| Anne Arundel | 2,277 | 34,752 | 8,176 | 3,447.1 | 241 | 11,289 | 5,065.7 | 1,102.6 | 1,280 | 32,200 | 2,613.1 | 636.3 |
| Baltimore | D | D | D | D | 417 | 14,707 | 7,606.6 | 980.5 | 1,711 | 29,741 | 1,876.1 | 519.6 |
| Calvert | D | D | D | D | 42 | 372 | 79.5 | 17.6 | 153 | 3,177 | 192.8 | 51.9 |
| Caroline | 36 | 226 | 18 | 7.8 | 23 | 1,345 | 362.7 | 58.7 | D | D | D | 6.1 |
| Carroll | D | D | D | D | 120 | 3,977 | 1,283.2 | 234.7 | 278 | 6,440 | 318.7 | 94.7 |
| Cecil | D | D | D | D | 51 | 4,360 | 1,974.6 | 324.9 | 174 | 3,410 | 221.4 | 56.6 |
| Charles | D | D | D | D | 47 | 397 | 107.5 | 22.4 | 262 | 5,459 | 312.0 | 85.6 |
| Dorchester | D | D | D | D | 41 | 2,538 | 818.5 | 110.2 | 61 | 1,301 | 69.7 | 24.8 |
| Frederick | 865 | 8,819 | 1,308 | 604.6 | 166 | 6,991 | 3,175.9 | 448.4 | 489 | 9,826 | 569.5 | 165.9 |
| Garrett | 54 | 434 | 52 | 23.4 | 43 | 867 | 176.0 | 33.2 | 83 | 1,370 | 61.5 | 19.0 |
| Harford | D | D | D | D | 127 | 4,661 | 2,179.4 | 289.3 | 408 | 8,638 | 472.5 | 137.6 |
| Howard | D | D | D | D | 178 | 5,848 | 1,738.3 | 339.7 | 615 | 13,048 | 782.6 | 232.7 |
| Kent | 52 | 263 | 35 | 14.2 | 29 | 602 | 322.0 | 39.9 | D | D | D | 13.8 |
| Montgomery | 5,753 | 79,924 | 14,770 | 6,583.2 | 357 | 7,081 | 2,098.8 | 540.3 | 1,936 | 34,764 | 2,561.3 | 726.1 |
| Prince George's | 1,869 | 28,347 | 5,513 | 2,188.7 | 262 | 6,665 | 2,041.5 | 382.1 | 1,427 | 33,174 | 2,877.1 | 746.8 |
| Queen Anne's | 161 | 686 | 97 | 41.4 | 60 | 1,595 | 329.6 | 70.7 | 101 | 2,150 | 136.9 | 40.6 |

# Table B. States and Counties — Health Care and Social Assistance, Other Services, Nonemployer Businesses, and Residential Construction

| STATE County | Health care and social assistance, 2017 | | | | Other services, 2017 | | | | Nonemployer businesses, 2018 | | Value of residential construction authorized by building permits, 2020 | |
|---|---|---|---|---|---|---|---|---|---|---|---|---|
| | Number of establishments | Number of employees | Receipts (mil dol) | Annual payroll (mil dol) | Number of establishments | Number of employees | Receipts (mil dol) | Annual payroll (mil dol) | Number | Receipts (mil dol) | New construction ($1,000) | Number of housing units |
| | 159 | 160 | 161 | 162 | 163 | 164 | 165 | 166 | 167 | 168 | 169 | 170 |
| LOUISIANA—Cont'd | | | | | | | | | | | | |
| Orleans | 994 | 26,020 | 3,804.3 | 1,273.3 | 722 | 5,256 | 751.2 | 181.7 | 40,229 | 1,717.5 | 258,232 | 1,360 |
| Ouachita | 671 | 14,356 | 1,487.5 | 546.2 | D | D | D | D | 12,824 | 533.8 | 95,240 | 463 |
| Plaquemines | 22 | 426 | 38.0 | 17.3 | D | D | D | D | 2,387 | 123.3 | 36,259 | 112 |
| Pointe Coupee | 36 | 741 | 60.6 | 23.6 | D | D | D | D | 1,541 | 72.3 | 20,739 | 68 |
| Rapides | 537 | 13,681 | 1,712.3 | 627.9 | 180 | 1,040 | 125.2 | 33.3 | 8,234 | 397.9 | 54,900 | 251 |
| Red River | 14 | 454 | 43.5 | 15.9 | 8 | 55 | 3.0 | 1.4 | 530 | 20.3 | 1,250 | 5 |
| Richland | 73 | 1,693 | 121.5 | 50.5 | 13 | 52 | 4.5 | 1.2 | 1,549 | 55.5 | 8,138 | 30 |
| Sabine | 49 | 708 | 55.7 | 21.4 | 27 | 100 | 16.3 | 2.8 | 1,378 | 58.2 | 18,206 | 104 |
| St. Bernard | 61 | 966 | 73.0 | 32.5 | D | D | D | D | 3,809 | 148.8 | 59,901 | 340 |
| St. Charles | 78 | 1,469 | 127.5 | 54.9 | 51 | 423 | 70.4 | 19.4 | 4,164 | 168.1 | 36,651 | 138 |
| St. Helena | D | D | D | 11.7 | D | D | D | D | 919 | 24.1 | 6,781 | 38 |
| St. James | 26 | 500 | 53.0 | 19.6 | D | D | D | D | 1,259 | 37.3 | 14,592 | 53 |
| St. John the Baptist | 75 | 991 | 78.2 | 32.3 | D | D | D | D | 3,127 | 101.6 | 18,348 | 86 |
| St. Landry | 269 | 6,737 | 610.1 | 236.0 | 74 | 410 | 31.6 | 11.6 | 5,806 | 224.0 | 34,375 | 171 |
| St. Martin | 82 | 1,363 | 100.9 | 40.5 | 46 | 160 | 22.7 | 6.0 | 4,632 | 166.9 | 16,917 | 72 |
| St. Mary | 109 | 2,186 | 166.2 | 67.6 | 69 | 343 | 54.1 | 14.3 | 3,901 | 139.5 | 4,747 | 22 |
| St. Tammany | 848 | 16,034 | 1,881.0 | 739.9 | 396 | 2,113 | 235.7 | 66.9 | 25,992 | 1,386.8 | 380,207 | 1,665 |
| Tangipahoa | 321 | 8,978 | 880.7 | 357.3 | 160 | 1,207 | 127.0 | 37.2 | 10,518 | 406.7 | 231,959 | 1,577 |
| Tensas | 8 | 71 | 4.5 | 2.0 | NA | NA | NA | NA | 336 | 12.0 | 0 | 4 |
| Terrebonne | 276 | 7,323 | 773.6 | 319.0 | 158 | 1,263 | 192.8 | 66.2 | 8,397 | 361.1 | 43,368 | 178 |
| Union | 29 | 722 | 65.5 | 22.9 | D | D | D | D | 1,502 | 56.6 | 466 | 2 |
| Vermilion | 97 | 1,843 | 153.0 | 59.0 | 60 | 257 | 26.7 | 8.3 | 4,686 | 180.4 | 34,533 | 172 |
| Vernon | 65 | 1,679 | 200.3 | 78.3 | 37 | 163 | 15.3 | 3.6 | 2,104 | 82.9 | 6,539 | 44 |
| Washington | 82 | 1,908 | 153.7 | 64.0 | 29 | 130 | 16.1 | 3.2 | 3,078 | 116.2 | 4,540 | 22 |
| Webster | D | D | D | D | 39 | 174 | 20.0 | 4.9 | 2,565 | 112.2 | 6,524 | 24 |
| West Baton Rouge | 27 | 454 | 27.1 | 10.3 | 43 | 320 | 70.4 | 18.0 | 1,893 | 69.7 | 39,883 | 229 |
| West Carroll | 19 | 596 | 47.0 | 18.5 | D | D | 5.5 | D | 677 | 25.5 | 2,356 | 12 |
| West Feliciana | D | D | D | 16.1 | D | D | D | D | 840 | 50.8 | 14,459 | 53 |
| Winn | 38 | 913 | 73.0 | 31.2 | D | D | D | D | 708 | 40.5 | 0 | 0 |
| MAINE | 4,771 | 112,594 | 11,777.3 | 5,000.0 | 2,921 | 14,477 | 1,809.2 | 474.8 | 117,341 | 5,469.0 | 13,546 | 102 |
| Androscoggin | 380 | 10,347 | 1,145.9 | 488.6 | 212 | 1,029 | 103.3 | 30.8 | 6,622 | 306.0 | 46,088 | 275 |
| Aroostook | 223 | 5,842 | 549.2 | 246.7 | 126 | 444 | 55.7 | 11.5 | 3,935 | 171.0 | 8,688 | 56 |
| Cumberland | 1,372 | 33,663 | 4,053.5 | 1,622.2 | 792 | 4,782 | 638.1 | 169.6 | 30,346 | 1,609.5 | 406,000 | 1,624 |
| Franklin | 107 | 1,798 | 147.3 | 66.3 | 49 | 191 | 21.2 | 5.1 | 2,400 | 82.0 | 23,786 | 98 |
| Hancock | 168 | 3,225 | 303.9 | 138.4 | 149 | 767 | 123.9 | 27.7 | 7,799 | 363.2 | 73,199 | 254 |
| Kennebec | 469 | 14,228 | 1,410.9 | 632.1 | 287 | 1,463 | 177.2 | 48.8 | 8,484 | 353.6 | 73,826 | 348 |
| Knox | 167 | 2,977 | 260.9 | 113.8 | 145 | 671 | 82.6 | 24.0 | 5,901 | 298.8 | 30,887 | 136 |
| Lincoln | 108 | 1,852 | 186.5 | 64.0 | 90 | 450 | 54.3 | 16.1 | 4,657 | 189.3 | 39,046 | 152 |
| Oxford | 135 | 2,947 | 227.9 | 104.6 | 93 | 374 | 31.4 | 8.9 | 4,429 | 185.8 | 77,178 | 339 |
| Penobscot | 541 | 15,666 | 1,719.6 | 776.9 | 274 | 1,425 | 189.9 | 44.6 | 9,386 | 394.3 | 62,913 | 337 |
| Piscataquis | 41 | 1,419 | 112.1 | 51.9 | 21 | 58 | 7.9 | 1.4 | 1,186 | 43.2 | 6,663 | 31 |
| Sagadahoc | 115 | 1,130 | 78.2 | 33.2 | 72 | 369 | 50.5 | 13.4 | 3,379 | 128.4 | 23,748 | 123 |
| Somerset | 132 | 2,946 | 232.1 | 104.7 | 85 | 262 | 29.4 | 6.8 | 3,195 | 124.1 | 18,724 | 141 |
| Waldo | 115 | 1,766 | 174.8 | 76.3 | 82 | 353 | 35.7 | 11.0 | 3,981 | 151.7 | 20,974 | 127 |
| Washington | 103 | 1,816 | 148.5 | 66.7 | 52 | 196 | 33.6 | 5.8 | 3,719 | 177.6 | 7,131 | 53 |
| York | 595 | 10,972 | 1,025.8 | 413.6 | 392 | 1,643 | 174.6 | 49.3 | 17,922 | 890.3 | 261,741 | 1,108 |
| MARYLAND | 16,800 | 384,096 | 48,675.7 | 19,294.2 | 10,355 | 81,469 | 12,118.3 | 3,518.8 | 510,744 | 23,049.1 | 3,853,541 | 17,982 |
| Allegany | 235 | 5,814 | 665.8 | 250.6 | 133 | 749 | 66.4 | 19.7 | 2,937 | 110.4 | 4,798 | 19 |
| Anne Arundel | 1,390 | 28,993 | 3,448.7 | 1,433.7 | 1,151 | 8,972 | 1,075.8 | 331.5 | 44,895 | 2,255.9 | 350,092 | 1,976 |
| Baltimore | 2,723 | 62,797 | 7,069.8 | 2,887.2 | 1,439 | 10,460 | 1,348.4 | 380.1 | 70,590 | 3,292.0 | 326,528 | 1,447 |
| Calvert | 200 | 3,889 | 424.7 | 183.2 | 128 | 819 | 86.4 | 29.4 | 6,564 | 297.1 | 67,932 | 339 |
| Caroline | 51 | 843 | 70.4 | 31.5 | 44 | 264 | 33.2 | 10.3 | 2,608 | 122.7 | 13,192 | 75 |
| Carroll | 470 | 9,927 | 952.9 | 417.8 | 361 | 2,231 | 234.7 | 72.6 | 12,795 | 573.2 | 99,197 | 488 |
| Cecil | 204 | 5,314 | 555.2 | 254.8 | 170 | 1,005 | 98.7 | 28.9 | 5,609 | 263.8 | 51,606 | 234 |
| Charles | 323 | 5,020 | 544.5 | 223.6 | 222 | 1,442 | 149.1 | 46.7 | 11,889 | 408.2 | 223,327 | 735 |
| Dorchester | 74 | 1,835 | 170.3 | 72.4 | 57 | 316 | 22.9 | 5.5 | 2,442 | 95.7 | 12,998 | 58 |
| Frederick | 680 | 13,050 | 1,475.2 | 635.5 | 448 | 2,996 | 421.4 | 117.1 | 20,296 | 960.8 | 608,211 | 2,577 |
| Garrett | 79 | 1,959 | 144.5 | 65.0 | 57 | 301 | 26.7 | 7.5 | 2,339 | 105.1 | 48,380 | 118 |
| Harford | 621 | 11,750 | 1,330.8 | 543.0 | 436 | 2,526 | 232.0 | 83.3 | 17,530 | 779.0 | 292,456 | 1,198 |
| Howard | 1,048 | 17,547 | 2,137.6 | 894.2 | 552 | 4,423 | 611.3 | 186.3 | 29,187 | 1,475.5 | 194,827 | 1,062 |
| Kent | 74 | 1,165 | 121.4 | 39.7 | 43 | 357 | 33.9 | 8.6 | 1,825 | 90.0 | 10,807 | 46 |
| Montgomery | 3,726 | 71,254 | 9,203.0 | 3,955.6 | 1,947 | 20,029 | 4,498.3 | 1,259.9 | 118,612 | 6,239.1 | 326,527 | 1,486 |
| Prince George's | 1,967 | 33,146 | 3,958.5 | 1,549.8 | 1,184 | 9,453 | 1,140.9 | 382.0 | 82,444 | 2,598.7 | 586,356 | 2,721 |
| Queen Anne's | 102 | 1,116 | 117.5 | 47.3 | 107 | 498 | 53.5 | 14.3 | 5,037 | 284.0 | 59,607 | 302 |

# Table B. States and Counties — Government Employment and Payroll, and Local Government Finances

| | | | Government employment and payroll, 2017 | | | | | | | Local government finances, 2017 | | | | |
| | | | | | March payroll (percent of total) | | | | | General revenue | | | | |
| | | | | | | | | | | | | | Taxes | |
| STATE County | Full-time equivalent employees | March payroll (dollars) | Adminis-tration, judicial, and legal | Police and corrections | Fire protection | Highways and transpor-tation | Health and welfare | Natural resources and utilities | Education and libraries | Total (mil dol) | Inter-govern-mental (mil dol) | Total (mil dol) | Per capita[1] (dollars) | |
| | | | | | | | | | | | | | Total | Property |
| | 171 | 172 | 173 | 174 | 175 | 176 | 177 | 178 | 179 | 180 | 181 | 182 | 183 | 184 |

| LOUISIANA—Cont'd | | | | | | | | | | | | | | |
|---|---|---|---|---|---|---|---|---|---|---|---|---|---|---|
| Orleans | 7,488 | 35,682,619 | 11.9 | 38.7 | 8.7 | 3.0 | 7.4 | 16.4 | 7.5 | 1,989.3 | 464.0 | 925.8 | 2,365 | 1,080 |
| Ouachita | 6,979 | 21,934,354 | 7.7 | 10.6 | 6.3 | 2.0 | 4.1 | 5.0 | 63.5 | 637.1 | 248.3 | 308.1 | 1,977 | 711 |
| Plaquemines | 1,446 | 5,546,053 | 9.3 | 17.1 | 0.6 | 7.7 | 4.3 | 8.6 | 49.3 | 169.5 | 74.6 | 80.7 | 3,457 | 2,061 |
| Pointe Coupee | 945 | 3,003,522 | 8.0 | 14.2 | 0.7 | 1.0 | 25.1 | 5.5 | 45.4 | 87.5 | 26.5 | 36.1 | 1,630 | 950 |
| Rapides | 5,625 | 17,644,805 | 8.9 | 14.5 | 5.6 | 3.3 | 0.8 | 6.8 | 58.4 | 429.2 | 188.0 | 194.6 | 1,482 | 556 |
| Red River | 485 | 1,561,297 | 2.9 | 35.8 | 0.0 | 0.5 | 0.0 | 1.0 | 58.2 | 42.3 | 15.3 | 25.0 | 2,933 | 2,190 |
| Richland | 1,287 | 4,862,880 | 4.5 | 9.5 | 0.1 | 2.4 | 39.8 | 1.3 | 42.3 | 60.4 | 27.5 | 27.1 | 1,326 | 566 |
| Sabine | 932 | 2,481,071 | 6.8 | 15.6 | 0.0 | 4.3 | 3.3 | 8.6 | 49.2 | 83.9 | 45.3 | 33.0 | 1,381 | 636 |
| St. Bernard | 1,733 | 5,768,654 | 9.7 | 14.3 | 4.7 | 5.3 | 2.2 | 5.9 | 55.0 | 256.2 | 135.6 | 85.8 | 1,860 | 893 |
| St. Charles | 2,873 | 11,610,582 | 5.7 | 15.3 | 0.0 | 1.5 | 6.6 | 7.7 | 61.7 | 264.7 | 50.1 | 149.9 | 2,848 | 1,906 |
| St. Helena | 450 | 1,367,476 | 9.1 | 11.0 | 0.0 | 0.9 | 41.7 | 1.3 | 35.6 | 41.8 | 25.2 | 11.7 | 1,134 | 678 |
| St. James | 1,315 | 5,199,157 | 6.2 | 10.1 | 0.0 | 2.5 | 20.1 | 6.1 | 50.4 | 175.3 | 34.4 | 84.1 | 3,933 | 2,657 |
| St. John the Baptist | 1,521 | 6,556,079 | 8.8 | 14.4 | 0.0 | 9.9 | 1.0 | 2.5 | 62.1 | 188.5 | 54.4 | 97.1 | 2,241 | 1,052 |
| St. Landry | 4,353 | 13,939,901 | 7.5 | 9.2 | 4.3 | 2.7 | 28.7 | 4.0 | 42.6 | 386.9 | 151.4 | 110.0 | 1,317 | 459 |
| St. Martin | 1,699 | 5,351,403 | 8.4 | 14.9 | 0.1 | 2.3 | 1.3 | 3.1 | 69.0 | 140.1 | 62.7 | 62.6 | 1,158 | 590 |
| St. Mary | 3,036 | 9,397,015 | 6.2 | 10.1 | 1.6 | 2.8 | 11.4 | 10.8 | 55.3 | 236.8 | 84.5 | 104.1 | 2,053 | 1,182 |
| St. Tammany | 11,627 | 46,247,933 | 4.1 | 9.2 | 0.2 | 1.9 | 35.5 | 1.4 | 46.3 | 1,348.0 | 307.5 | 506.8 | 1,980 | 1,112 |
| Tangipahoa | 6,779 | 25,167,545 | 4.7 | 6.0 | 1.3 | 2.0 | 46.1 | 2.3 | 36.2 | 779.7 | 229.4 | 257.9 | 1,949 | 370 |
| Tensas | 280 | 767,261 | 12.5 | 18.0 | 0.0 | 3.8 | 0.5 | 6.6 | 57.9 | 22.6 | 6.8 | 9.7 | 2,119 | 1,299 |
| Terrebonne | 5,488 | 19,422,115 | 5.2 | 7.7 | 1.2 | 1.3 | 33.3 | 4.7 | 46.5 | 655.8 | 190.0 | 166.1 | 1,485 | 548 |
| Union | 687 | 2,277,187 | 5.4 | 18.1 | 0.4 | 3.0 | 14.8 | 5.9 | 51.4 | 78.6 | 40.4 | 25.8 | 1,151 | 515 |
| Vermilion | 2,457 | 8,494,425 | 5.1 | 8.5 | 1.8 | 3.0 | 24.7 | 8.0 | 48.9 | 221.7 | 85.4 | 79.5 | 1,326 | 616 |
| Vernon | 2,374 | 4,433,559 | 6.9 | 12.9 | 1.9 | 4.2 | 0.1 | 2.3 | 71.0 | 138.8 | 79.6 | 47.7 | 959 | 359 |
| Washington | 1,764 | 6,180,189 | 5.1 | 8.2 | 2.1 | 3.1 | 16.5 | 2.2 | 62.4 | 120.7 | 67.8 | 44.3 | 950 | 405 |
| Webster | 1,467 | 4,469,650 | 6.8 | 15.2 | 1.4 | 5.6 | 0.9 | 6.1 | 62.7 | 123.3 | 49.4 | 60.5 | 1,541 | 700 |
| West Baton Rouge | 1,289 | 4,372,750 | 5.8 | 3.1 | 0.8 | 21.0 | 0.7 | 11.4 | 53.2 | 116.4 | 30.9 | 67.9 | 2,594 | 1,197 |
| West Carroll | 374 | 1,208,152 | 2.4 | 1.9 | 0.0 | 3.5 | 0.2 | 5.4 | 86.3 | 34.8 | 18.3 | 12.5 | 1,139 | 354 |
| West Feliciana | 736 | 2,986,222 | 7.7 | 10.3 | 0.2 | 1.8 | 19.4 | 5.0 | 55.1 | 63.0 | 17.3 | 30.0 | 1,954 | 1,068 |
| Winn | 623 | 1,659,435 | 9.0 | 15.3 | 2.4 | 4.0 | 1.1 | 10.6 | 57.0 | 43.3 | 23.2 | 16.8 | 1,168 | 502 |
| MAINE | X | X | X | X | X | X | X | X | X | X | X | X | X | X |
| Androscoggin | 4,097 | 15,739,689 | 4.1 | 8.0 | 5.5 | 4.7 | 2.1 | 5.4 | 68.8 | 419.1 | 171.6 | 196.9 | 1,833 | 1,822 |
| Aroostook | 3,200 | 11,361,475 | 4.7 | 5.3 | 2.6 | 4.1 | 17.8 | 3.5 | 61.2 | 303.1 | 109.6 | 100.0 | 1,479 | 1,473 |
| Cumberland | 11,250 | 47,445,493 | 5.7 | 8.9 | 5.8 | 5.6 | 4.6 | 6.8 | 59.1 | 1,250.5 | 267.8 | 748.8 | 2,563 | 2,523 |
| Franklin | 1,047 | 4,032,002 | 6.4 | 7.6 | 1.0 | 4.7 | 0.1 | 5.0 | 73.2 | 102.3 | 32.2 | 61.2 | 2,053 | 2,048 |
| Hancock | 2,225 | 7,597,830 | 7.6 | 7.2 | 2.6 | 3.7 | 1.1 | 3.4 | 73.2 | 210.9 | 30.7 | 161.4 | 2,959 | 2,922 |
| Kennebec | 4,188 | 14,675,209 | 5.3 | 7.1 | 3.5 | 3.4 | 1.3 | 5.1 | 73.2 | 370.3 | 132.0 | 188.9 | 1,549 | 1,540 |
| Knox | 1,325 | 5,194,338 | 8.4 | 9.0 | 2.7 | 5.0 | 1.1 | 4.9 | 67.2 | 146.7 | 18.3 | 101.2 | 2,545 | 2,526 |
| Lincoln | 1,294 | 4,530,885 | 6.7 | 6.2 | 1.1 | 2.1 | 0.5 | 5.8 | 75.6 | 154.5 | 41.9 | 100.2 | 2,928 | 2,912 |
| Oxford | 2,266 | 7,726,655 | 5.2 | 5.4 | 2.1 | 5.3 | 0.1 | 3.1 | 77.8 | 210.2 | 71.0 | 120.2 | 2,090 | 2,068 |
| Penobscot | 4,862 | 18,469,574 | 6.5 | 8.5 | 5.4 | 7.2 | 2.8 | 5.2 | 62.9 | 493.2 | 177.7 | 234.2 | 1,544 | 1,531 |
| Piscataquis | 1,041 | 4,537,173 | 3.8 | 4.4 | 0.1 | 2.3 | 54.6 | 3.1 | 31.2 | 120.2 | 34.9 | 30.7 | 1,827 | 1,820 |
| Sagadahoc | 1,157 | 4,864,484 | 5.8 | 7.1 | 3.3 | 3.7 | 1.1 | 4.0 | 72.2 | 124.5 | 33.2 | 79.5 | 2,243 | 2,220 |
| Somerset | 2,578 | 8,814,246 | 4.1 | 5.6 | 1.1 | 2.8 | 0.1 | 2.0 | 83.0 | 220.2 | 93.3 | 111.9 | 2,223 | 2,218 |
| Waldo | 827 | 3,030,537 | 15.0 | 9.7 | 2.2 | 4.5 | 0.3 | 4.1 | 60.7 | 114.6 | 37.3 | 68.9 | 1,732 | 1,724 |
| Washington | 956 | 2,995,324 | 10.1 | 9.8 | 2.5 | 4.5 | 6.1 | 2.9 | 62.4 | 99.6 | 32.7 | 57.2 | 1,822 | 1,818 |
| York | 7,092 | 29,054,999 | 5.7 | 9.4 | 4.4 | 4.1 | 0.9 | 6.3 | 68.0 | 739.6 | 176.8 | 485.7 | 2,378 | 2,340 |
| MARYLAND | X | X | X | X | X | X | X | X | X | X | X | X | X | X |
| Allegany | 2,446 | 12,517,043 | 3.6 | 7.6 | 2.0 | 3.9 | 1.3 | 5.2 | 74.5 | 307.4 | 159.1 | 93.0 | 1,303 | 816 |
| Anne Arundel | 18,444 | 92,732,222 | 4.8 | 9.5 | 6.4 | 4.6 | 2.8 | 3.7 | 66.1 | 2,522.4 | 685.9 | 1,434.8 | 2,511 | 1,362 |
| Baltimore | 28,278 | 139,892,085 | 4.1 | 11.9 | 5.1 | 1.3 | 3.0 | 2.5 | 69.7 | 3,266.2 | 1,046.8 | 1,753.5 | 2,116 | 1,147 |
| Calvert | 3,174 | 15,760,271 | 6.0 | 8.9 | 0.2 | 2.1 | 2.4 | 5.8 | 70.5 | 415.7 | 119.5 | 246.1 | 2,691 | 1,687 |
| Caroline | 1,357 | 5,637,819 | 5.4 | 8.4 | 0.0 | 2.6 | 2.7 | 2.7 | 76.4 | 141.7 | 72.3 | 54.8 | 1,654 | 1,092 |
| Carroll | 5,634 | 25,017,473 | 6.1 | 6.8 | 0.0 | 2.7 | 0.8 | 3.6 | 76.4 | 662.3 | 194.3 | 400.5 | 2,391 | 1,329 |
| Cecil | 3,515 | 15,724,891 | 4.0 | 6.2 | 0.0 | 2.4 | 3.7 | 2.4 | 78.2 | 407.3 | 161.9 | 190.6 | 1,861 | 1,183 |
| Charles | 6,140 | 30,314,851 | 5.2 | 13.5 | 0.0 | 0.8 | 2.8 | 4.6 | 70.9 | 730.8 | 235.9 | 385.2 | 2,416 | 1,427 |
| Dorchester | 1,266 | 5,345,396 | 5.2 | 12.5 | 0.0 | 2.7 | 3.8 | 3.0 | 68.5 | 145.0 | 68.7 | 55.8 | 1,737 | 1,178 |
| Frederick | 9,651 | 48,627,596 | 5.0 | 7.3 | 6.0 | 2.5 | 4.5 | 4.1 | 69.0 | 1,239.6 | 355.8 | 687.1 | 2,742 | 1,683 |
| Garrett | 1,635 | 5,678,012 | 5.6 | 5.1 | 0.0 | 9.6 | 12.4 | 4.3 | 62.1 | 191.8 | 42.8 | 71.9 | 2,458 | 1,740 |
| Harford | 8,074 | 39,526,360 | 5.7 | 9.3 | 1.1 | 3.0 | 1.3 | 4.5 | 74.0 | 988.5 | 290.4 | 580.9 | 2,306 | 1,295 |
| Howard | 12,539 | 72,691,324 | 4.3 | 7.5 | 2.5 | 1.0 | 3.9 | 4.2 | 75.1 | 1,787.5 | 407.1 | 1,177.5 | 3,688 | 1,927 |
| Kent | 641 | 2,955,712 | 8.8 | 12.1 | 0.0 | 4.9 | 0.0 | 8.2 | 58.7 | 80.6 | 25.1 | 49.9 | 2,564 | 1,712 |
| Montgomery | 38,557 | 270,040,080 | 3.6 | 7.9 | 4.3 | 3.2 | 7.0 | 3.4 | 65.7 | 7,013.3 | 1,284.1 | 4,343.4 | 4,149 | 2,071 |
| Prince George's | 32,780 | 177,507,257 | 3.2 | 11.3 | 2.5 | 0.8 | 2.3 | 11.0 | 67.5 | 4,203.7 | 1,609.2 | 2,203.2 | 2,422 | 1,388 |
| Queen Anne's | 2,044 | 9,249,932 | 6.4 | 6.0 | 0.0 | 2.7 | 6.4 | 6.4 | 71.2 | 216.5 | 60.9 | 127.4 | 2,570 | 1,403 |

1. Based on the resident population estimated as of July 1 of the year shown.

# Table B. States and Counties — Local Government Finances, Government Employment, and Income Taxes

| STATE County | Local government finances, 2017 (cont.) | | | | | | | | | Government employment, 2019 | | | Individual income tax returns, 2018 | | |
| | Direct general expenditure | | | | | | | Debt outstanding | | | | | | | |
| | | | Percent of total for: | | | | | | | | | | | | |
| | Total (mil dol) | Per capita[1] (dollars) | Education | Health and hospitals | Police protection | Public welfare | Highways | Total (mil dol) | Per capita[1] (dollars) | Federal civilian | Federal military | State and local | Number of returns | Mean adjusted gross income | Mean income tax |
| | 185 | 186 | 187 | 188 | 189 | 190 | 191 | 192 | 193 | 194 | 195 | 196 | 197 | 198 | 199 |
| **LOUISIANA—Cont'd** | | | | | | | | | | | | | | | |
| Orleans | 1,776 | 4,536 | 23.3 | 2.6 | 4.9 | 0.0 | 3.2 | 2,802.7 | 7,159 | 10,258 | 3,589 | 25,017 | 159,990 | 67,862 | 9,557 |
| Ouachita | 614 | 3,941 | 55.1 | 0.8 | 7.4 | 0.3 | 3.3 | 614.2 | 3,941 | 438 | 562 | 10,150 | 64,480 | 56,089 | 5,879 |
| Plaquemines | 193 | 8,260 | 37.4 | 3.6 | 1.3 | 0.4 | 2.1 | 255.6 | 10,948 | 620 | 414 | 1,520 | 10,110 | 65,669 | 7,484 |
| Pointe Coupee | 79 | 3,552 | 41.9 | 19.4 | 6.5 | 0.4 | 2.0 | 20.6 | 931 | 63 | 82 | 1,011 | 9,590 | 58,107 | 6,115 |
| Rapides | 484 | 3,684 | 51.2 | 0.0 | 10.3 | 0.0 | 2.6 | 455.0 | 3,466 | 2,129 | 485 | 9,945 | 54,670 | 56,887 | 5,783 |
| Red River | 39 | 4,530 | 56.5 | 0.0 | 2.5 | 0.0 | 2.4 | 2.0 | 239 | 19 | 31 | 554 | 3,240 | 55,198 | 6,005 |
| Richland | 80 | 3,901 | 42.7 | 0.2 | 7.7 | 0.0 | 33.1 | 62.8 | 3,078 | 85 | 72 | 913 | 8,110 | 49,682 | 4,772 |
| Sabine | 81 | 3,405 | 65.3 | 0.2 | 7.0 | 0.0 | 6.5 | 30.2 | 1,262 | 40 | 89 | 1,284 | 8,690 | 56,021 | 5,132 |
| St. Bernard | 283 | 6,134 | 41.6 | 16.0 | 4.4 | 0.0 | 0.3 | 118.2 | 2,563 | 54 | 178 | 2,171 | 17,370 | 43,858 | 3,385 |
| St. Charles | 284 | 5,404 | 63.0 | 0.0 | 14.1 | 0.0 | 11.0 | 834.5 | 15,859 | 182 | 199 | 3,189 | 23,920 | 66,261 | 6,848 |
| St. Helena | 37 | 3,601 | 38.5 | 31.0 | 5.1 | 0.0 | 6.2 | 30.0 | 2,902 | 11 | 38 | 639 | 5,160 | 42,341 | 3,142 |
| St. James | 240 | 11,221 | 34.6 | 23.1 | 3.2 | 0.7 | 2.1 | 1,264.4 | 59,139 | 35 | 79 | 1,494 | 9,620 | 59,849 | 5,756 |
| St. John the Baptist | 234 | 5,396 | 40.1 | 1.0 | 12.2 | 0.0 | 4.9 | 1,335.2 | 30,816 | 98 | 161 | 2,090 | 19,220 | 49,037 | 4,102 |
| St. Landry | 390 | 4,669 | 36.6 | 31.9 | 5.0 | 1.1 | 2.6 | 161.3 | 1,931 | 172 | 308 | 4,925 | 34,970 | 49,177 | 4,447 |
| St. Martin | 138 | 2,548 | 67.5 | 0.7 | 7.0 | 0.2 | 4.6 | 199.5 | 3,689 | 64 | 200 | 1,815 | 22,820 | 50,461 | 4,439 |
| St. Mary | 257 | 5,066 | 40.8 | 10.8 | 6.4 | 0.1 | 3.7 | 100.4 | 1,980 | 136 | 271 | 4,146 | 20,690 | 48,811 | 4,352 |
| St. Tammany | 1,074 | 4,197 | 46.5 | 17.0 | 5.8 | 0.2 | 5.3 | 1,012.0 | 3,954 | 556 | 985 | 14,197 | 116,820 | 78,445 | 10,406 |
| Tangipahoa | 689 | 5,203 | 29.1 | 46.7 | 3.7 | 0.0 | 2.2 | 306.9 | 2,318 | 383 | 499 | 10,909 | 52,260 | 51,820 | 4,805 |
| Tensas | 24 | 5,196 | 37.6 | 2.3 | 10.5 | 0.0 | 5.3 | 2.7 | 592 | 21 | 16 | 266 | 1,610 | 45,842 | 4,425 |
| Terrebonne | 719 | 6,429 | 25.7 | 38.8 | 5.5 | 0.4 | 4.7 | 164.6 | 1,472 | 270 | 493 | 5,009 | 45,310 | 57,380 | 5,897 |
| Union | 102 | 4,558 | 52.7 | 24.8 | 5.2 | 0.0 | 3.1 | 64.0 | 2,858 | 93 | 82 | 768 | 8,840 | 48,786 | 3,975 |
| Vermilion | 230 | 3,837 | 41.4 | 20.3 | 5.7 | 2.1 | 4.8 | 35.6 | 594 | 154 | 236 | 3,050 | 24,000 | 54,573 | 5,078 |
| Vernon | 139 | 2,786 | 68.2 | 0.3 | 6.1 | 0.1 | 4.8 | 61.5 | 1,235 | 2,095 | 7,960 | 2,287 | 19,540 | 49,496 | 3,583 |
| Washington | 118 | 2,523 | 65.6 | 0.6 | 4.6 | 0.0 | 4.7 | 30.8 | 661 | 103 | 169 | 2,552 | 16,140 | 41,817 | 2,949 |
| Webster | 118 | 3,011 | 57.8 | 0.4 | 7.1 | 0.1 | 4.3 | 86.8 | 2,210 | 106 | 141 | 2,010 | 15,720 | 50,620 | 4,881 |
| West Baton Rouge | 130 | 4,973 | 40.5 | 1.0 | 8.5 | 0.2 | 4.6 | 173.1 | 6,612 | 96 | 98 | 1,547 | 11,780 | 61,880 | 6,273 |
| West Carroll | 34 | 3,071 | 66.8 | 2.0 | 6.7 | 0.0 | 8.1 | 3.1 | 281 | 30 | 39 | 732 | 4,130 | 48,460 | 3,988 |
| West Feliciana | 78 | 5,050 | 39.4 | 39.3 | 1.2 | 0.0 | 3.8 | 38.3 | 2,495 | 15 | 39 | 2,245 | 4,530 | 88,103 | 12,493 |
| Winn | 55 | 3,818 | 46.7 | 0.4 | 9.4 | 0.0 | 3.3 | 15.6 | 1,087 | 55 | 46 | 816 | 4,830 | 50,098 | 4,321 |
| **MAINE** | X | X | X | X | X | X | X | X | X | 16,090 | 6,807 | 85,870 | 666,460 | 61,652 | 6,417 |
| Androscoggin | 392 | 3,651 | 53.5 | 0.3 | 3.8 | 0.3 | 4.7 | 282.6 | 2,631 | 277 | 323 | 5,327 | 51,120 | 51,775 | 4,487 |
| Aroostook | 316 | 4,678 | 43.0 | 20.8 | 3.0 | 0.5 | 6.4 | 96.7 | 1,430 | 1,231 | 198 | 4,820 | 29,800 | 47,008 | 3,668 |
| Cumberland | 1,247 | 4,268 | 44.5 | 1.0 | 4.6 | 3.1 | 4.7 | 1,161.5 | 3,975 | 1,998 | 2,302 | 18,577 | 158,250 | 80,675 | 10,227 |
| Franklin | 88 | 2,950 | 50.4 | 0.9 | 5.2 | 0.3 | 11.5 | 85.7 | 2,877 | 126 | 89 | 2,047 | 13,320 | 49,606 | 4,111 |
| Hancock | 201 | 3,687 | 54.3 | 0.3 | 3.3 | 0.3 | 7.7 | 143.7 | 2,634 | 336 | 270 | 2,925 | 28,070 | 60,046 | 6,068 |
| Kennebec | 374 | 3,064 | 55.8 | 0.3 | 4.3 | 0.2 | 6.4 | 196.9 | 1,615 | 2,205 | 404 | 14,090 | 59,550 | 55,472 | 5,150 |
| Knox | 147 | 3,694 | 46.1 | 2.3 | 5.3 | 0.3 | 8.4 | 65.9 | 1,657 | 123 | 204 | 2,447 | 20,530 | 60,226 | 6,270 |
| Lincoln | 132 | 3,867 | 63.7 | 1.1 | 3.8 | 0.3 | 7.9 | 60.0 | 1,753 | 82 | 129 | 1,553 | 18,280 | 59,091 | 5,777 |
| Oxford | 217 | 3,775 | 62.9 | 0.5 | 3.5 | 0.3 | 8.0 | 96.1 | 1,670 | 128 | 174 | 3,005 | 26,410 | 47,362 | 3,837 |
| Penobscot | 508 | 3,352 | 50.2 | 1.4 | 5.1 | 0.3 | 5.1 | 383.0 | 2,526 | 1,131 | 457 | 13,257 | 69,070 | 54,375 | 5,139 |
| Piscataquis | 103 | 6,154 | 27.0 | 48.9 | 2.7 | 0.2 | 5.2 | 28.9 | 1,722 | 49 | 51 | 1,310 | 7,350 | 45,553 | 3,612 |
| Sagadahoc | 130 | 3,681 | 57.7 | 0.3 | 3.9 | 0.2 | 6.5 | 67.2 | 1,897 | 386 | 124 | 1,505 | 19,020 | 62,817 | 6,026 |
| Somerset | 197 | 3,919 | 64.2 | 0.2 | 3.1 | 0.2 | 7.2 | 87.5 | 1,738 | 194 | 152 | 2,371 | 22,540 | 47,559 | 3,921 |
| Waldo | 104 | 2,615 | 57.0 | 2.3 | 3.7 | 0.5 | 11.5 | 72.0 | 1,810 | 89 | 129 | 1,486 | 18,870 | 51,396 | 4,429 |
| Washington | 99 | 3,146 | 55.7 | 3.4 | 2.7 | 0.1 | 9.7 | 32.1 | 1,022 | 353 | 162 | 2,333 | 14,000 | 44,182 | 3,442 |
| York | 772 | 3,779 | 59.7 | 0.5 | 5.8 | 0.3 | 6.3 | 635.5 | 3,111 | 7,382 | 1,639 | 8,817 | 110,320 | 64,462 | 6,673 |
| **MARYLAND** | X | X | X | X | X | X | X | X | X | 174,698 | 50,680 | 347,487 | 3,004,260 | 82,114 | 10,673 |
| Allegany | 308 | 4,316 | 55.9 | 0.8 | 5.5 | 0.0 | 5.3 | 214.6 | 3,008 | 466 | 211 | 5,553 | 28,530 | 50,059 | 4,259 |
| Anne Arundel | 2,480 | 4,340 | 54.5 | 2.4 | 6.5 | 0.6 | 3.8 | 2,203.5 | 3,857 | 42,442 | 18,636 | 34,509 | 293,130 | 91,949 | 12,293 |
| Baltimore | 3,663 | 4,420 | 53.7 | 1.6 | 5.8 | 0.3 | 2.5 | 4,519.6 | 5,454 | 14,071 | 2,667 | 41,991 | 416,170 | 79,522 | 10,266 |
| Calvert | 397 | 4,341 | 58.5 | 0.9 | 6.2 | 0.0 | 1.9 | 211.7 | 2,316 | 156 | 332 | 4,315 | 45,500 | 87,793 | 10,409 |
| Caroline | 137 | 4,144 | 57.6 | 0.6 | 4.8 | 0.0 | 4.0 | 68.4 | 2,066 | 71 | 107 | 1,750 | 15,660 | 51,790 | 4,512 |
| Carroll | 658 | 3,924 | 57.9 | 0.9 | 4.0 | 0.0 | 5.0 | 435.3 | 2,598 | 346 | 542 | 7,862 | 84,850 | 84,350 | 9,847 |
| Cecil | 398 | 3,882 | 60.9 | 1.1 | 6.3 | 0.0 | 3.9 | 298.3 | 2,914 | 2,066 | 329 | 4,462 | 47,980 | 64,639 | 6,469 |
| Charles | 716 | 4,488 | 64.3 | 0.9 | 9.6 | 0.0 | 2.9 | 387.4 | 2,429 | 2,512 | 1,078 | 7,639 | 82,580 | 74,053 | 7,818 |
| Dorchester | 134 | 4,174 | 54.5 | 0.6 | 7.5 | 0.0 | 5.0 | 49.3 | 1,535 | 174 | 102 | 2,148 | 15,380 | 49,544 | 4,475 |
| Frederick | 1,171 | 4,672 | 57.4 | 3.0 | 5.3 | 0.0 | 5.3 | 1,093.6 | 4,364 | 3,730 | 1,921 | 12,276 | 130,660 | 82,714 | 9,892 |
| Garrett | 186 | 6,354 | 38.0 | 29.8 | 2.1 | 0.0 | 10.6 | 93.3 | 3,190 | 69 | 92 | 1,646 | 13,570 | 55,635 | 5,411 |
| Harford | 918 | 3,644 | 62.9 | 0.4 | 7.4 | 0.0 | 3.8 | 900.3 | 3,575 | 11,254 | 1,682 | 9,309 | 129,090 | 78,485 | 9,038 |
| Howard | 1,875 | 5,872 | 56.7 | 0.6 | 5.9 | 0.5 | 2.9 | 3,003.3 | 9,407 | 665 | 1,244 | 16,555 | 158,830 | 114,059 | 16,712 |
| Kent | 78 | 3,996 | 39.8 | 0.9 | 5.7 | 0.0 | 6.2 | 39.7 | 2,042 | 61 | 58 | 1,024 | 9,270 | 73,113 | 8,780 |
| Montgomery | 7,716 | 7,371 | 40.5 | 1.6 | 3.7 | 1.1 | 2.8 | 12,338.3 | 11,786 | 47,402 | 7,996 | 43,686 | 534,940 | 115,598 | 18,584 |
| Prince George's | 4,371 | 4,805 | 52.3 | 1.6 | 9.1 | 0.7 | 2.4 | 1,835.7 | 2,018 | 26,876 | 7,762 | 66,363 | 488,300 | 59,621 | 5,817 |
| Queen Anne's | 226 | 4,566 | 54.7 | 1.0 | 4.9 | 0.0 | 4.0 | 204.0 | 4,115 | 107 | 162 | 2,599 | 24,960 | 89,144 | 11,485 |

1. Based on the resident population estimated as of July 1 of the year shown.

# Table B. States and Counties — Land Area and Population

| State / county code | CBSA code[1] | County Type code[2] | STATE County | Land area[3] (sq. mi) | Total persons 2019 | Rank | Per square mile | White | Black | American Indian, Alaska Native | Asian and Pacific Islander | Percent Hispanic or Latino[4] | Under 5 years | 5 to 14 years | 15 to 24 years | 25 to 34 years | 35 to 44 years | 45 to 54 years |
|---|---|---|---|---|---|---|---|---|---|---|---|---|---|---|---|---|---|---|
| | | | | 1 | 2 | 3 | 4 | 5 | 6 | 7 | 8 | 9 | 10 | 11 | 12 | 13 | 14 | 15 |
| | | | **MARYLAND— Cont'd** | | | | | | | | | | | | | | | |
| 24037 | 15680 | 3 | St. Mary's | 358.6 | 114,687 | 542 | 319.8 | 76.3 | 16.5 | 0.9 | 4.3 | 5.6 | 6.1 | 13.6 | 13.3 | 14.3 | 12.8 | 12.8 |
| 24039 | 41540 | 2 | Somerset | 319.7 | 25,453 | 1,580 | 79.6 | 53.9 | 42.4 | 1.0 | 1.5 | 3.9 | 4.5 | 9.6 | 19.7 | 12.8 | 11.6 | 11.1 |
| 24041 | 20660 | 6 | Talbot | 268.6 | 36,972 | 1,249 | 137.6 | 78.8 | 13.3 | 0.5 | 1.9 | 7.1 | 4.6 | 10.2 | 9.2 | 9.8 | 9.4 | 11.2 |
| 24043 | 25180 | 2 | Washington | 457.8 | 151,146 | 444 | 330.2 | 79.4 | 14.3 | 0.6 | 2.6 | 6.1 | 5.6 | 12.1 | 12.0 | 13.1 | 12.4 | 13.3 |
| 24045 | 41540 | 2 | Wicomico | 374.4 | 103,990 | 584 | 277.8 | 63.9 | 28.6 | 0.7 | 3.9 | 5.7 | 6.1 | 12.0 | 18.4 | 12.2 | 11.0 | 11.0 |
| 24047 | 41540 | 2 | Worcester | 468.4 | 52,403 | 963 | 111.9 | 81.4 | 13.6 | 0.6 | 2.1 | 4.0 | 4.1 | 10.0 | 9.6 | 10.5 | 9.7 | 11.5 |
| 24510 | 12580 | 1 | Baltimore city | 80.9 | 586,131 | 115 | 7,245.1 | 29.1 | 63.1 | 0.9 | 3.4 | 5.8 | 6.1 | 11.0 | 12.3 | 18.9 | 13.2 | 10.9 |
| 25000 | | 0 | MASSACHUSETTS | 7,801.0 | 6,893,574 | X | 883.7 | 72.3 | 8.4 | 0.6 | 8.2 | 12.6 | 5.1 | 10.8 | 13.4 | 14.5 | 12.5 | 12.6 |
| 25001 | 12700 | 3 | Barnstable | 394.2 | 213,164 | 319 | 540.8 | 90.9 | 4.2 | 1.2 | 2.2 | 3.5 | 3.6 | 8.1 | 9.7 | 9.1 | 8.8 | 10.9 |
| 25003 | 38340 | 3 | Berkshire | 926.9 | 124,571 | 511 | 134.4 | 89.4 | 4.6 | 0.6 | 2.4 | 5.3 | 4.1 | 9.1 | 12.3 | 10.9 | 10.7 | 11.9 |
| 25005 | 39300 | 1 | Bristol | 553.1 | 566,765 | 120 | 1,024.7 | 82.8 | 6.5 | 0.7 | 2.4 | 8.9 | 5.2 | 11.5 | 12.5 | 13.1 | 12.3 | 13.4 |
| 25007 | 47240 | 7 | Dukes | 103.2 | 17,461 | 1,947 | 169.2 | 89.1 | 5.8 | 2.3 | 2.1 | 3.5 | 4.8 | 10.0 | 9.0 | 10.7 | 11.2 | 13.4 |
| 25009 | 14460 | 1 | Essex | 492.5 | 791,263 | 82 | 1,606.6 | 69.7 | 4.4 | 0.4 | 4.3 | 22.7 | 5.5 | 11.6 | 12.8 | 12.6 | 11.2 | 12.2 |
| 25011 | 44140 | 4 | Franklin | 699.2 | 70,267 | 772 | 100.5 | 91.8 | 2.2 | 1.0 | 2.6 | 4.5 | 3.9 | 9.7 | 9.9 | 11.7 | 12.1 | 12.9 |
| 25013 | 44140 | 2 | Hampden | 617.0 | 463,986 | 154 | 752.0 | 62.5 | 8.9 | 0.6 | 3.1 | 26.8 | 5.4 | 11.9 | 13.7 | 13.6 | 12.1 | 12.4 |
| 25015 | 44140 | 2 | Hampshire | 527.2 | 161,401 | 416 | 306.1 | 84.9 | 3.8 | 0.6 | 6.7 | 6.2 | 3.2 | 8.1 | 26.1 | 10.4 | 11.8 | 12.2 |
| 25017 | 14460 | 1 | Middlesex | 817.9 | 1,609,379 | 22 | 1,967.7 | 72.6 | 6.1 | 0.4 | 14.8 | 8.3 | 5.2 | 10.9 | 13.1 | 15.5 | 10.0 | 10.5 |
| 25019 | | 7 | Nantucket | 46.2 | 11,376 | 2,325 | 246.2 | 72.9 | 10.9 | 0.6 | 2.4 | 15.0 | 6.6 | 10.9 | 9.9 | 14.0 | 13.5 | 12.8 |
| 25021 | 14460 | 1 | Norfolk | 396.1 | 709,409 | 94 | 1,791.0 | 74.6 | 8.3 | 0.4 | 13.6 | 5.2 | 5.2 | 11.5 | 12.4 | 13.2 | 14.7 | 14.7 |
| 25023 | 14460 | 1 | Plymouth | 658.5 | 523,738 | 136 | 795.4 | 82.2 | 12.3 | 0.7 | 2.4 | 4.3 | 5.2 | 11.7 | 12.4 | 11.2 | 12.9 | 13.3 |
| 25025 | 14460 | 1 | Suffolk | 58.3 | 801,582 | 80 | 13,749.3 | 47.0 | 21.0 | 0.7 | 10.3 | 23.3 | 5.0 | 8.5 | 16.2 | 23.3 | 11.6 | 13.6 |
| 25027 | 49340 | 2 | Worcester | 1,510.7 | 829,212 | 74 | 548.9 | 76.9 | 5.9 | 0.6 | 6.1 | 12.4 | 5.3 | 11.6 | 13.2 | 13.2 | 13.2 | 13.4 |
| 26000 | | 0 | MICHIGAN | 56,605.9 | 9,966,555 | X | 176.1 | 76.7 | 15.0 | 1.3 | 4.1 | 5.4 | 5.6 | 11.9 | 13.1 | 13.2 | 12.3 | 13.4 |
| 26001 | | 9 | Alcona | 674.7 | 10,505 | 2,390 | 15.6 | 96.4 | 1.1 | 1.3 | 0.6 | 1.8 | 3.4 | 6.9 | 7.2 | 7.5 | 11.8 | 12.3 |
| 26003 | | 7 | Alger | 915.0 | 9,015 | 2,497 | 9.9 | 86.3 | 7.9 | 6.2 | 0.7 | 1.9 | 3.5 | 8.6 | 9.9 | 11.6 | 7.6 | 10.9 |
| 26005 | 26090 | 4 | Allegan | 825.3 | 118,927 | 529 | 144.1 | 89.6 | 2.1 | 1.2 | 1.2 | 7.7 | 6.1 | 13.5 | 11.8 | 12.1 | 11.2 | 11.3 |
| 26007 | 10980 | 9 | Alpena | 571.9 | 28,238 | 1,488 | 49.4 | 96.7 | 1.2 | 1.2 | 0.8 | 1.5 | 4.7 | 10.4 | 10.1 | 10.8 | 12.0 | 12.4 |
| 26009 | | 9 | Antrim | 475.7 | 23,449 | 1,661 | 49.3 | 95.9 | 0.8 | 1.9 | 0.6 | 2.3 | 4.0 | 9.8 | 9.6 | 9.5 | 11.0 | 11.4 |
| 26011 | | 8 | Arenac | 363.2 | 14,953 | 2,095 | 41.2 | 95.6 | 1.1 | 2.0 | 0.8 | 2.2 | 4.7 | 10.3 | 9.4 | 9.9 | 9.7 | 11.2 |
| 26013 | | 9 | Baraga | 898.4 | 8,164 | 2,568 | 9.1 | 76.5 | 9.0 | 16.3 | 1.0 | 1.8 | 4.0 | 9.5 | 11.4 | 12.0 | 10.2 | 11.3 |
| 26015 | | 2 | Barry | 553.1 | 62,061 | 855 | 112.2 | 95.2 | 1.3 | 1.1 | 0.8 | 3.2 | 5.4 | 12.4 | 11.6 | 11.6 | 11.6 | 12.7 |
| 26017 | 13020 | 3 | Bay | 442.4 | 102,387 | 593 | 231.4 | 91.4 | 2.8 | 1.1 | 0.9 | 5.7 | 4.9 | 11.1 | 11.3 | 12.5 | 11.6 | 12.5 |
| 26019 | 45900 | 2 | Benzie | 319.7 | 17,852 | 1,931 | 55.8 | 95.0 | 1.3 | 2.1 | 0.8 | 2.4 | 4.4 | 9.9 | 9.4 | 10.1 | 11.5 | 12.1 |
| 26021 | 35660 | 3 | Berrien | 567.8 | 153,025 | 436 | 269.5 | 76.9 | 15.7 | 1.2 | 2.9 | 5.9 | 5.5 | 12.2 | 11.8 | 11.7 | 10.7 | 10.9 |
| 26023 | 17740 | 6 | Branch | 506.4 | 43,424 | 1,106 | 85.8 | 91.3 | 2.7 | 1.0 | 1.3 | 5.4 | 6.2 | 12.9 | 11.4 | 11.9 | 11.6 | 12.1 |
| 26025 | 12980 | 2 | Calhoun | 706.3 | 133,580 | 485 | 189.1 | 80.1 | 13.0 | 1.4 | 3.4 | 5.7 | 5.9 | 12.7 | 12.6 | 12.5 | 12.0 | 12.0 |
| 26027 | 43780 | 2 | Cass | 490.1 | 51,584 | 970 | 105.3 | 89.0 | 6.6 | 2.1 | 1.2 | 4.1 | 4.7 | 11.8 | 11.3 | 10.3 | 11.8 | 12.1 |
| 26029 | | 7 | Charlevoix | 416.3 | 26,105 | 1,565 | 62.7 | 95.1 | 1.0 | 2.3 | 1.0 | 2.4 | 4.4 | 10.1 | 10.4 | 10.1 | 11.2 | 12.9 |
| 26031 | | 7 | Cheboygan | 715.3 | 25,365 | 1,589 | 35.5 | 94.2 | 1.3 | 4.8 | 0.8 | 1.6 | 3.8 | 8.7 | 9.7 | 9.6 | 10.4 | 11.4 |
| 26033 | 42300 | 7 | Chippewa | 1,558.5 | 36,958 | 1,250 | 23.7 | 74.0 | 7.5 | 20.3 | 1.5 | 2.1 | 4.5 | 10.4 | 14.3 | 13.4 | 9.9 | 12.3 |
| 26035 | | 6 | Clare | 564.4 | 30,771 | 1,412 | 54.5 | 95.9 | 1.2 | 1.6 | 0.5 | 2.3 | 5.2 | 11.2 | 9.9 | 10.6 | 12.8 | 11.8 |
| 26037 | 29620 | 2 | Clinton | 566.3 | 79,753 | 711 | 140.8 | 91.2 | 2.8 | 1.0 | 2.2 | 4.8 | 5.5 | 12.2 | 12.2 | 12.1 | 9.8 | 11.9 |
| 26039 | | 7 | Crawford | 556.4 | 13,981 | 2,159 | 25.1 | 95.5 | 1.2 | 1.4 | 1.0 | 2.3 | 4.3 | 9.9 | 9.0 | 9.7 | 12.7 | 12.9 |
| 26041 | 21540 | 5 | Delta | 1,171.1 | 35,612 | 1,289 | 30.4 | 94.8 | 1.0 | 4.3 | 1.0 | 1.6 | 4.8 | 11.0 | 10.5 | 9.8 | 9.9 | 11.7 |
| 26043 | 27020 | 7 | Dickinson | 760.9 | 25,112 | 1,601 | 33.0 | 96.0 | 0.9 | 1.6 | 1.1 | 1.7 | 5.3 | 10.9 | 10.5 | 10.6 | 10.9 | 11.7 |
| 26045 | 29620 | 2 | Eaton | 575.2 | 110,148 | 554 | 191.5 | 84.5 | 8.2 | 1.1 | 3.0 | 5.7 | 5.3 | 11.5 | 12.2 | 13.0 | 10.6 | 11.7 |
| 26047 | 22420 | 2 | Emmet | 467.5 | 33,342 | 1,348 | 71.3 | 92.9 | 1.2 | 4.9 | 1.1 | 2.1 | 4.5 | 10.5 | 11.0 | 11.3 | 12.2 | 12.0 |
| 26049 | 22420 | 2 | Genesee | 636.9 | 404,794 | 178 | 635.6 | 74.5 | 21.8 | 1.3 | 1.6 | 3.7 | 5.8 | 12.5 | 12.3 | 12.7 | 11.1 | 11.9 |
| 26051 | | 6 | Gladwin | 501.8 | 25,424 | 1,584 | 50.7 | 96.4 | 1.0 | 1.2 | 0.7 | 2.0 | 5.2 | 10.4 | 9.0 | 9.5 | 11.5 | 12.6 |
| 26053 | | 7 | Gogebic | 1,102.1 | 13,842 | 2,167 | 12.6 | 93.7 | 1.1 | 4.3 | 0.8 | 2.0 | 4.2 | 9.5 | 9.2 | 9.6 | 9.5 | 11.7 |
| 26055 | 45900 | 5 | Grand Traverse | 464.3 | 93,592 | 635 | 201.6 | 94.1 | 1.4 | 1.8 | 1.3 | 3.1 | 4.9 | 11.2 | 11.0 | 12.5 | 9.7 | 11.7 |
| 26057 | 10940 | 6 | Gratiot | 568.4 | 40,283 | 1,175 | 70.9 | 86.5 | 6.4 | 1.0 | 0.8 | 6.6 | 4.9 | 11.0 | 14.6 | 13.2 | 12.3 | 11.8 |
| 26059 | 25880 | 6 | Hillsdale | 598.2 | 45,658 | 1,062 | 76.3 | 95.7 | 1.2 | 1.3 | 0.8 | 2.5 | 5.6 | 11.8 | 13.5 | 10.7 | 12.8 | 12.3 |
| 26061 | 26340 | 5 | Houghton | 1,009.1 | 35,126 | 1,300 | 34.8 | 93.8 | 1.4 | 1.5 | 3.2 | 1.8 | 5.2 | 11.4 | 24.7 | 10.1 | 10.8 | 11.8 |
| 26063 | | 7 | Huron | 836.0 | 30,653 | 1,414 | 36.7 | 95.6 | 0.9 | 0.8 | 0.8 | 2.7 | 4.8 | 10.7 | 10.2 | 9.6 | 9.4 | 9.2 |
| 26065 | 29620 | 2 | Ingham | 556.1 | 290,609 | 241 | 522.6 | 72.8 | 14.2 | 1.3 | 7.7 | 8.2 | 5.4 | 10.9 | 21.9 | 14.4 | 10.0 | 11.6 |
| 26067 | 24340 | 4 | Ionia | 571.3 | 64,553 | 829 | 113.0 | 89.4 | 5.2 | 1.1 | 0.8 | 5.1 | 5.6 | 12.4 | 12.9 | 14.1 | 11.5 | 10.4 |
| 26069 | | 7 | Iosco | 549.1 | 25,140 | 1,600 | 45.8 | 94.9 | 1.4 | 1.8 | 1.2 | 2.6 | 4.7 | 9.2 | 8.5 | 9.4 | 13.0 | 13.1 |
| 26071 | | 7 | Iron | 1,166.0 | 11,066 | 2,341 | 9.5 | 95.3 | 1.0 | 2.2 | 0.8 | 2.4 | 4.2 | 9.8 | 8.4 | 8.3 | 9.2 | 10.5 |
| 26073 | 34380 | 4 | Isabella | 572.7 | 69,504 | 779 | 121.4 | 87.9 | 3.8 | 4.6 | 2.4 | 4.3 | 4.5 | 10.0 | 28.5 | 12.3 | 9.7 | 10.1 |
| 26075 | 27100 | 3 | Jackson | 701.9 | 156,920 | 428 | 223.6 | 87.1 | 9.8 | 1.0 | 1.3 | 3.8 | 5.6 | 11.9 | 12.1 | 13.0 | 10.3 | 9.5 |
| 26077 | 28020 | 2 | Kalamazoo | 562.0 | 265,988 | 258 | 473.3 | 79.8 | 13.6 | 1.2 | 3.6 | 5.4 | 5.8 | 12.0 | 18.6 | 13.8 | 11.6 | 12.6 |
| 26079 | 45900 | 7 | Kalkaska | 559.7 | 18,003 | 1,922 | 32.2 | 95.4 | 1.3 | 2.0 | 1.0 | 2.2 | 5.2 | 11.7 | 10.4 | 11.9 | 11.9 | 10.6 |
| 26081 | 24340 | 2 | Kent | 848.9 | 658,708 | 101 | 776.0 | 75.3 | 11.5 | 1.0 | 4.0 | 11.1 | 6.5 | 13.3 | 13.1 | 16.0 | 13.0 | 11.4 |

1. CBSA = Core Based Statistical Area. See Appendix A for explanation. See Appendix B for list of metropolitan areas with component counties.    2. County type code from the Economic Research Service of USDA Rural-Urban Continuum Codes. See Appendix A for definition.    3. Dry land or land partially or temporarily covered by water.    4. May be of any race.

# Table B. States and Counties — Population and Households

| STATE County | Age (percent) (cont.) 55 to 64 years | 65 to 74 years | 75 years and over | Percent female | Total persons 2000 | Total persons 2010 | Percent change 2000-2010 | Percent change 2010-2020 | Births | Deaths | Net Migration | Households Number | Persons per household | Family households | Female family householder[1] | One person |
|---|---|---|---|---|---|---|---|---|---|---|---|---|---|---|---|---|
| | 16 | 17 | 18 | 19 | 20 | 21 | 22 | 23 | 24 | 25 | 26 | 27 | 28 | 29 | 30 | 31 |
| **MARYLAND— Cont'd** | | | | | | | | | | | | | | | | |
| St. Mary's | 13.4 | 8.1 | 5.6 | 50.0 | 86,211 | 105,146 | 22.0 | 9.1 | 14,370 | 8,156 | 3,353 | 40,552 | 2.70 | 70.3 | 12.1 | 24.4 |
| Somerset | 13.2 | 10.4 | 7.2 | 46.2 | 24,747 | 26,470 | 7.0 | -3.8 | 2,549 | 2,792 | -794 | 8,574 | 2.30 | 64.0 | 17.7 | 31.3 |
| Talbot | 15.1 | 15.6 | 14.8 | 52.7 | 33,812 | 37,777 | 11.7 | -2.1 | 3,423 | 4,596 | 405 | 16,826 | 2.18 | 65.3 | 10.1 | 28.7 |
| Washington | 13.7 | 10.1 | 7.8 | 49.1 | 131,923 | 147,416 | 11.7 | 2.5 | 17,513 | 15,739 | 2,020 | 56,035 | 2.52 | 66.9 | 12.6 | 27.6 |
| Wicomico | 12.5 | 9.9 | 6.8 | 52.6 | 84,644 | 98,734 | 16.6 | 5.3 | 12,754 | 9,920 | 2,480 | 37,793 | 2.60 | 64.0 | 14.8 | 26.7 |
| Worcester | 16.2 | 15.9 | 12.6 | 51.5 | 46,543 | 51,448 | 10.5 | 1.9 | 4,377 | 6,511 | 3,138 | 22,089 | 2.31 | 62.9 | 9.6 | 30.3 |
| Baltimore city | 12.7 | 9.0 | 6.0 | 53.2 | 651,154 | 620,777 | -4.7 | -5.6 | 86,519 | 66,596 | -54,128 | 239,116 | 2.45 | 50.3 | 21.0 | 39.8 |
| **MASSACHUSETTS** | 13.7 | 10.1 | 7.3 | 51.4 | 6,349,097 | 6,547,788 | 3.1 | 5.3 | 729,892 | 581,114 | 201,059 | 2,617,497 | 2.52 | 63.4 | 12.1 | 28.5 |
| Barnstable | 17.7 | 18.3 | 13.8 | 52.2 | 222,230 | 215,883 | -2.9 | -1.3 | 15,862 | 29,924 | 11,568 | 94,323 | 2.23 | 62.6 | 8.7 | 30.9 |
| Berkshire | 16.3 | 14.1 | 10.6 | 51.6 | 134,953 | 131,274 | -2.7 | -5.1 | 10,857 | 15,234 | -2,256 | 54,813 | 2.18 | 58.5 | 11.2 | 34.1 |
| Bristol | 14.2 | 10.1 | 7.6 | 51.6 | 534,678 | 548,239 | 2.5 | 3.4 | 57,911 | 54,466 | 15,558 | 217,912 | 2.50 | 65.9 | 14.3 | 27.8 |
| Dukes | 16.4 | 16.3 | 9.5 | 50.9 | 14,987 | 16,535 | 10.3 | 5.6 | 1,681 | 1,483 | 729 | 6,765 | 2.52 | 59.0 | 9.4 | 34.0 |
| Essex | 14.3 | 10.5 | 7.5 | 51.8 | 723,419 | 743,082 | 2.7 | 6.5 | 86,525 | 68,356 | 30,576 | 294,202 | 2.60 | 66.8 | 13.6 | 27.5 |
| Franklin | 16.4 | 15.3 | 8.6 | 51.4 | 71,535 | 71,381 | -0.2 | -1.6 | 5,914 | 7,225 | 241 | 30,569 | 2.26 | 58.0 | 10.2 | 31.9 |
| Hampden | 13.6 | 10.3 | 7.5 | 51.7 | 456,228 | 463,620 | 1.6 | 0.1 | 52,849 | 45,653 | -6,785 | 179,423 | 2.53 | 63.6 | 17.7 | 30.4 |
| Hampshire | 13.0 | 11.4 | 7.2 | 53.4 | 152,251 | 158,058 | 3.8 | 2.1 | 10,685 | 12,941 | 5,520 | 58,838 | 2.35 | 59.0 | 10.3 | 30.1 |
| Middlesex | 13.0 | 9.1 | 6.8 | 50.9 | 1,465,396 | 1,503,107 | 2.6 | 7.1 | 176,352 | 117,275 | 48,292 | 604,384 | 2.55 | 64.7 | 9.4 | 26.1 |
| Nantucket | 13.4 | 9.7 | 6.1 | 48.8 | 9,520 | 10,168 | 6.8 | 11.9 | 1,507 | 671 | 360 | 3,713 | 2.93 | 62.4 | 7.9 | 30.2 |
| Norfolk | 14.0 | 9.9 | 7.7 | 51.8 | 650,308 | 670,951 | 3.2 | 5.7 | 73,746 | 59,432 | 24,614 | 265,300 | 2.57 | 66.3 | 9.3 | 26.7 |
| Plymouth | 15.0 | 11.3 | 7.9 | 51.4 | 472,822 | 494,933 | 4.7 | 5.8 | 52,496 | 46,601 | 23,347 | 187,460 | 2.69 | 70.2 | 12.1 | 24.9 |
| Suffolk | 10.5 | 7.3 | 5.3 | 51.8 | 689,807 | 722,172 | 4.7 | 11.0 | 95,611 | 49,524 | 33,667 | 309,844 | 2.41 | 49.7 | 15.3 | 35.5 |
| Worcester | 14.4 | 9.8 | 6.8 | 50.7 | 750,963 | 798,385 | 6.3 | 3.9 | 87,896 | 72,329 | 15,628 | 309,951 | 2.56 | 65.4 | 11.9 | 27.6 |
| **MICHIGAN** | 13.9 | 10.8 | 7.4 | 50.8 | 9,938,444 | 9,884,112 | -0.5 | 0.8 | 1,151,986 | 972,901 | -95,692 | 3,935,041 | 2.47 | 64.0 | 12.1 | 29.6 |
| Alcona | 19.9 | 20.1 | 16.5 | 49.8 | 11,719 | 10,943 | -6.6 | -4.0 | 651 | 1,892 | 808 | 4,988 | 2.05 | 61.5 | 6.9 | 32.8 |
| Alger | 17.2 | 16.1 | 10.6 | 45.0 | 9,862 | 9,601 | -2.6 | -6.1 | 603 | 1,158 | -24 | 3,007 | 2.70 | 64.5 | 5.0 | 30.9 |
| Allegan | 14.6 | 10.7 | 6.8 | 50.0 | 105,665 | 111,398 | 5.4 | 6.8 | 14,015 | 9,922 | 3,495 | 43,416 | 2.65 | 73.0 | 9.4 | 21.7 |
| Alpena | 17.0 | 14.0 | 10.8 | 50.5 | 31,314 | 29,594 | -5.5 | -4.6 | 2,688 | 3,823 | -195 | 12,752 | 2.19 | 60.1 | 9.5 | 32.6 |
| Antrim | 17.8 | 16.6 | 11.9 | 50.2 | 23,110 | 23,577 | 2.0 | -0.5 | 1,942 | 2,807 | 757 | 9,899 | 2.32 | 70.0 | 8.2 | 24.7 |
| Arenac | 17.9 | 15.9 | 10.5 | 49.1 | 17,269 | 15,903 | -7.9 | -6.0 | 1,297 | 2,149 | -80 | 6,571 | 2.26 | 63.2 | 8.5 | 30.8 |
| Baraga | 15.1 | 13.9 | 9.9 | 44.9 | 8,746 | 8,866 | 1.4 | -7.9 | 743 | 1,089 | -353 | 3,107 | 2.34 | 63.2 | 10.0 | 31.8 |
| Barry | 15.6 | 11.7 | 7.6 | 49.7 | 56,755 | 59,177 | 4.3 | 4.9 | 6,508 | 5,627 | 2,041 | 24,296 | 2.46 | 70.8 | 7.9 | 23.5 |
| Bay | 15.2 | 12.5 | 9.1 | 50.9 | 110,157 | 107,773 | -2.2 | -5.0 | 10,569 | 12,501 | -3,402 | 44,339 | 2.32 | 63.3 | 11.8 | 30.7 |
| Benzie | 17.1 | 15.9 | 11.5 | 50.2 | 15,998 | 17,519 | 9.5 | 1.9 | 1,555 | 2,154 | 940 | 6,792 | 2.55 | 66.6 | 6.7 | 27.7 |
| Berrien | 14.4 | 11.9 | 8.8 | 51.0 | 162,453 | 156,810 | -3.5 | -2.4 | 18,137 | 17,778 | -4,121 | 63,665 | 2.36 | 64.2 | 13.1 | 30.1 |
| Branch | 14.4 | 11.3 | 7.8 | 48.2 | 45,787 | 45,242 | -1.2 | -4.0 | 5,513 | 4,601 | -2,737 | 16,650 | 2.47 | 67.1 | 9.8 | 26.1 |
| Calhoun | 13.6 | 10.9 | 7.7 | 51.0 | 137,985 | 136,150 | -1.3 | -1.9 | 16,516 | 15,414 | -3,657 | 53,991 | 2.41 | 62.5 | 13.5 | 32.2 |
| Cass | 15.6 | 13.6 | 8.6 | 49.7 | 51,104 | 52,287 | 2.3 | -1.3 | 4,942 | 5,397 | -217 | 21,019 | 2.43 | 69.5 | 10.4 | 25.7 |
| Charlevoix | 17.3 | 15.3 | 10.5 | 50.5 | 26,090 | 25,955 | -0.5 | 0.6 | 2,396 | 2,912 | 690 | 11,503 | 2.25 | 64.8 | 7.5 | 29.7 |
| Cheboygan | 17.3 | 17.0 | 11.7 | 50.0 | 26,448 | 26,143 | -1.2 | -3.0 | 2,076 | 3,346 | 511 | 11,195 | 2.24 | 65.3 | 8.2 | 29.3 |
| Chippewa | 13.6 | 11.3 | 8.0 | 44.9 | 38,543 | 38,676 | 0.3 | -4.4 | 3,497 | 3,779 | -1,453 | 13,999 | 2.43 | 61.1 | 11.9 | 32.5 |
| Clare | 16.6 | 15.2 | 9.7 | 50.0 | 31,252 | 30,924 | | -0.5 | 3,332 | 4,273 | 805 | 12,199 | 2.48 | 64.6 | 9.4 | 30.5 |
| Clinton | 14.3 | 10.9 | 7.2 | 50.7 | 64,753 | 75,362 | 16.4 | 5.8 | 8,252 | 6,239 | 2,416 | 29,728 | 2.61 | 69.8 | 8.9 | 24.2 |
| Crawford | 18.5 | 15.6 | 11.2 | 48.9 | 14,273 | 14,081 | -1.3 | -0.7 | 1,215 | 1,804 | 499 | 6,141 | 2.23 | 65.8 | 8.1 | 27.1 |
| Delta | 15.8 | 15.1 | 10.5 | 50.2 | 38,520 | 37,069 | -3.8 | -3.9 | 3,652 | 4,540 | -544 | 16,234 | 2.18 | 64.7 | 9.6 | 29.5 |
| Dickinson | 16.3 | 13.3 | 10.5 | 49.6 | 27,472 | 26,168 | -4.7 | -4.0 | 2,549 | 3,169 | -423 | 11,231 | 2.23 | 64.5 | 10.2 | 30.6 |
| Eaton | 14.4 | 11.7 | 7.8 | 50.8 | 103,655 | 107,759 | 4.0 | 2.2 | 11,816 | 10,331 | 983 | 44,480 | 2.42 | 64.6 | 10.3 | 29.1 |
| Emmet | 15.7 | 14.5 | 9.5 | 50.5 | 31,437 | 32,696 | 4.0 | 2.0 | 3,024 | 3,604 | 1,255 | 14,463 | 2.25 | 63.1 | 8.6 | 30.8 |
| Genesee | 14.2 | 10.8 | 7.6 | 51.8 | 436,141 | 425,790 | -2.4 | -4.9 | 49,570 | 46,214 | -24,602 | 167,902 | 2.40 | 63.4 | 15.7 | 31.0 |
| Gladwin | 17.0 | 16.3 | 11.3 | 49.6 | 26,023 | 25,696 | -1.3 | -1.1 | 2,584 | 3,629 | 786 | 11,047 | 2.27 | 64.1 | 8.0 | 31.1 |
| Gogebic | 17.0 | 16.7 | 12.4 | 49.4 | 17,370 | 16,424 | -5.4 | -15.7 | 1,278 | 2,139 | -1,715 | 6,744 | 2.05 | 53.5 | 9.4 | 40.7 |
| Grand Traverse | 15.0 | 13.0 | 8.3 | 50.9 | 77,654 | 86,988 | 12.0 | 7.6 | 9,325 | 8,749 | 6,037 | 37,319 | 2.41 | 65.1 | 8.5 | 27.2 |
| Gratiot | 13.1 | 9.9 | 8.2 | 46.4 | 42,285 | 42,478 | 0.5 | -5.2 | 4,134 | 4,588 | -1,747 | 15,035 | 2.33 | 65.8 | 12.3 | 28.5 |
| Hillsdale | 15.1 | 12.4 | 8.4 | 50.3 | 46,527 | 46,691 | 0.4 | -2.2 | 5,448 | 4,952 | -1,517 | 18,107 | 2.44 | 66.5 | 9.3 | 28.1 |
| Houghton | 11.5 | 10.6 | 7.9 | 46.1 | 36,016 | 36,624 | 1.7 | -4.1 | 3,804 | 3,643 | -1,677 | 13,386 | 2.50 | 55.3 | 6.7 | 34.6 |
| Huron | 16.5 | 15.4 | 11.4 | 50.3 | 36,079 | 33,118 | -8.2 | -7.4 | 3,001 | 4,637 | -808 | 13,847 | 2.22 | 62.6 | 8.0 | 33.2 |
| Ingham | 11.1 | 8.9 | 5.6 | 51.3 | 279,320 | 280,895 | 0.6 | 3.5 | 33,036 | 21,655 | -1,809 | 112,840 | 2.41 | 53.9 | 11.1 | 34.1 |
| Ionia | 13.4 | 9.6 | 6.0 | 46.4 | 61,518 | 63,901 | 3.9 | 1.0 | 7,415 | 5,379 | -1,367 | 22,964 | 2.63 | 69.6 | 9.7 | 24.1 |
| Iosco | 17.7 | 17.7 | 13.1 | 50.1 | 27,339 | 25,895 | -5.3 | -2.9 | 2,394 | 4,147 | 1,006 | 11,669 | 2.13 | 60.6 | 8.3 | 33.0 |
| Iron | 18.2 | 17.8 | 13.5 | 50.5 | 13,138 | 11,817 | -10.1 | -6.4 | 933 | 1,932 | 262 | 5,225 | 2.06 | 57.2 | 8.9 | 38.6 |
| Isabella | 11.1 | 8.5 | 5.3 | 51.3 | 63,351 | 70,313 | 11.0 | -1.2 | 6,668 | 4,954 | -2,572 | 24,739 | 2.59 | 56.6 | 10.9 | 28.0 |
| Jackson | 14.4 | 11.1 | 7.7 | 49.2 | 158,422 | 160,233 | 1.1 | -2.1 | 18,145 | 17,154 | -4,218 | 61,805 | 2.41 | 65.1 | 12.2 | 29.7 |
| Kalamazoo | 11.4 | 9.3 | 6.5 | 51.1 | 238,603 | 250,327 | 4.9 | 6.3 | 31,878 | 21,796 | 5,744 | 103,445 | 2.46 | 59.2 | 10.7 | 30.5 |
| Kalkaska | 16.4 | 13.0 | 8.0 | 49.0 | 16,571 | 17,147 | 3.5 | 5.0 | 1,825 | 1,983 | 1,018 | 7,145 | 2.44 | 67.7 | 9.6 | 26.1 |
| Kent | 12.2 | 8.7 | 5.8 | 50.6 | 574,335 | 602,625 | 4.9 | 9.3 | 89,568 | 47,880 | 14,662 | 241,746 | 2.63 | 66.2 | 11.3 | 26.1 |

1. No spouse present.

| STATE County | Persons in group quarters, 2020 | Daytime Population, 2015-2019 | | Births, 2020 | | Deaths, 2020 | | Persons under 65 with no health insurance, 2019 | | Medicare, 2020 | | | Crimes reported by county police or sheriff, 2019 | |
|---|---|---|---|---|---|---|---|---|---|---|---|---|---|---|
| | | Number | Employment/ residence ratio | Total | Rate[1] | Number | Rate[1] | Number | Percent | Total beneficiaries | Enrolled in Original Medicare | Enrolled in Medicare Advantage | Violent | Property |
| | 32 | 33 | 34 | 35 | 36 | 37 | 38 | 39 | 40 | 41 | 42 | 43 | 44 | 45 |
| MARYLAND— Cont'd | | | | | | | | | | | | | | |
| St. Mary's................. | 2,789 | 107,717 | 0.92 | 1,344 | 11.7 | 949 | 8.3 | 4,997 | 5.2 | 17,047 | 16,589 | 458 | 202 | 1,663 |
| Somerset.................. | 5,318 | 24,473 | 0.85 | 223 | 8.8 | 295 | 11.6 | 1,209 | 7.5 | 5,242 | 4,803 | 440 | 6 | 40 |
| Talbot..................... | 371 | 40,991 | 1.22 | 323 | 8.7 | 538 | 14.6 | 2,074 | 8.0 | 11,443 | 10,808 | 635 | 19 | 111 |
| Washington.............. | 8,316 | 151,289 | 1.02 | 1,667 | 11.0 | 1,641 | 10.9 | 7,906 | 6.7 | 31,690 | 27,532 | 4,158 | 254 | 1,081 |
| Wicomico................. | 4,876 | 102,644 | 1.00 | 1,253 | 12.0 | 1,140 | 11.0 | 6,984 | 8.5 | 20,156 | 18,796 | 1,360 | 50 | 486 |
| Worcester................ | 723 | 54,766 | 1.13 | 409 | 7.8 | 754 | 14.4 | 2,747 | 7.4 | 15,307 | 13,967 | 1,340 | 15 | 248 |
| Baltimore city .......... | 26,570 | 716,498 | 1.39 | 7,468 | 12.7 | 7,266 | 12.4 | 34,023 | 7.0 | 99,854 | 79,643 | 20,211 | NA | NA |
| MASSACHUSETTS........ | 247,482 | 6,928,771 | 1.02 | 68,502 | 9.9 | 63,072 | 9.1 | 195,726 | 3.5 | 1,352,025 | 988,677 | 363,348 | X | X |
| Barnstable............... | 4,112 | 209,677 | 0.96 | 1,459 | 6.8 | 3,077 | 14.4 | 5,496 | 3.8 | 78,016 | 64,380 | 13,636 | NA | NA |
| Berkshire................. | 5,794 | 127,809 | 1.02 | 975 | 7.8 | 1,494 | 12.0 | 3,482 | 3.9 | 35,270 | 31,640 | 3,630 | NA | NA |
| Bristol.................... | 14,999 | 505,547 | 0.80 | 5,442 | 9.6 | 5,745 | 10.1 | 17,581 | 3.9 | 122,595 | 94,553 | 28,042 | NA | NA |
| Dukes..................... | 119 | 17,811 | 1.06 | 166 | 9.5 | 148 | 8.5 | 651 | 5.1 | 5,082 | 4,845 | 238 | NA | NA |
| Essex..................... | 19,420 | 714,671 | 0.83 | 8,291 | 10.5 | 7,321 | 9.3 | 24,645 | 3.9 | 161,477 | 119,160 | 42,318 | NA | NA |
| Franklin.................. | 1,496 | 63,427 | 0.80 | 461 | 6.6 | 774 | 11.0 | 1,783 | 3.9 | 19,091 | 14,746 | 4,345 | NA | NA |
| Hampden................. | 14,454 | 466,515 | 0.99 | 4,843 | 10.4 | 4,654 | 10.0 | 15,798 | 3.3 | 103,062 | 65,601 | 37,461 | NA | NA |
| Hampshire............... | 23,014 | 156,426 | 0.94 | 932 | 5.8 | 1,404 | 8.7 | 3,540 | 4.3 | 34,679 | 26,609 | 8,070 | NA | NA |
| Middlesex................ | 54,996 | 1,668,138 | 1.08 | 16,438 | 10.2 | 13,015 | 8.1 | 41,317 | 3.2 | 274,646 | 198,701 | 75,945 | NA | NA |
| Nantucket................ | 55 | 11,516 | 1.06 | 158 | 13.9 | 61 | 5.4 | 400 | 3.1 | 2,026 | 1,929 | 97 | NA | NA |
| Norfolk................... | 16,580 | 681,956 | 0.95 | 7,058 | 9.9 | 6,404 | 9.0 | 13,598 | 4.1 | 133,898 | 101,686 | 32,213 | NA | NA |
| Plymouth................. | 11,383 | 456,995 | 0.78 | 5,023 | 9.6 | 5,223 | 10.0 | 12,445 | 2.4 | 114,440 | 90,154 | 24,287 | NA | NA |
| Suffolk................... | 53,252 | 1,075,055 | 1.64 | 8,976 | 11.2 | 5,967 | 7.4 | 31,313 | 3.0 | 107,480 | 73,037 | 34,443 | NA | NA |
| Worcester................ | 27,808 | 773,228 | 0.88 | 8,280 | 10.0 | 7,785 | 9.4 | 23,677 | 4.8 | 160,262 | 101,638 | 58,624 | NA | NA |
| MICHIGAN ................. | 218,618 | 9,928,767 | 0.99 | 108,474 | 10.9 | 103,445 | 10.4 | 564,804 | 3.5 | 2,099,584 | 1,095,821 | 1,003,763 | X | X |
| Alcona.................... | 127 | 9,197 | 0.65 | 82 | 7.8 | 180 | 17.1 | 635 | 7.0 | 4,374 | 2,567 | 1,807 | 10 | 62 |
| Alger..................... | 1,021 | 8,842 | 0.90 | 57 | 6.3 | 114 | 12.6 | 484 | 9.7 | 2,676 | 1,568 | 1,108 | 3 | 2 |
| Allegan................... | 955 | 107,039 | 0.83 | 1,423 | 12.0 | 1,021 | 8.6 | 6,888 | 8.3 | 24,220 | 9,663 | 14,557 | 100 | 535 |
| Alpena.................... | 513 | 29,282 | 1.06 | 233 | 8.3 | 365 | 12.9 | 1,501 | 7.1 | 8,922 | 6,072 | 2,850 | 23 | 51 |
| Antrim.................... | 226 | 20,599 | 0.73 | 186 | 7.9 | 300 | 12.8 | 1,536 | 7.0 | 7,658 | 4,322 | 3,335 | 33 | 130 |
| Arenac.................... | 207 | 14,179 | 0.85 | 137 | 9.2 | 218 | 14.6 | 1,001 | 9.2 | 4,824 | 2,682 | 2,142 | 25 | 71 |
| Baraga.................... | 1,011 | 8,470 | 1.02 | 70 | 8.6 | 125 | 15.3 | 498 | 9.2 | 2,197 | 1,405 | 792 | 3 | 15 |
| Barry..................... | 601 | 47,354 | 0.54 | 658 | 10.6 | 628 | 10.1 | 3,186 | 6.4 | 13,851 | 6,147 | 7,705 | 83 | 304 |
| Bay....................... | 1,438 | 94,139 | 0.78 | 921 | 9.0 | 1,242 | 12.1 | 5,631 | 7.0 | 26,776 | 14,635 | 12,142 | 68 | 422 |
| Benzie.................... | 252 | 14,697 | 0.62 | 141 | 7.9 | 230 | 12.9 | 1,099 | 8.5 | 5,582 | 3,113 | 2,469 | 19 | 38 |
| Berrien................... | 3,512 | 153,110 | 0.99 | 1,644 | 10.7 | 1,818 | 11.9 | 10,145 | 8.5 | 36,249 | 21,594 | 14,654 | 117 | 689 |
| Branch.................... | 1,868 | 41,326 | 0.88 | 568 | 13.1 | 449 | 10.3 | 2,877 | 8.6 | 9,651 | 5,762 | 3,889 | 13 | 119 |
| Calhoun................... | 4,294 | 139,899 | 1.10 | 1,508 | 11.3 | 1,489 | 11.1 | 7,646 | 7.2 | 30,566 | 19,215 | 11,351 | NA | NA |
| Cass...................... | 491 | 42,029 | 0.59 | 463 | 9.0 | 582 | 11.3 | 3,207 | 8.0 | 12,968 | 7,922 | 5,046 | 0 | 0 |
| Charlevoix................ | 279 | 25,028 | 0.90 | 240 | 9.2 | 319 | 12.2 | 1,597 | 8.2 | 7,529 | 4,523 | 3,007 | 30 | 96 |
| Cheboygan................ | 391 | 22,865 | 0.75 | 178 | 7.0 | 393 | 15.5 | 1,626 | 9.1 | 8,292 | 4,952 | 3,340 | 22 | 30 |
| Chippewa................. | 4,619 | 37,650 | 1.00 | 313 | 8.5 | 402 | 10.9 | 2,568 | 10.0 | 8,271 | 4,648 | 3,623 | 6 | 11 |
| Clare..................... | 391 | 28,371 | 0.79 | 301 | 9.8 | 437 | 14.2 | 2,229 | 9.7 | 9,445 | 5,256 | 4,189 | NA | NA |
| Clinton................... | 628 | 62,156 | 0.57 | 819 | 10.3 | 634 | 7.9 | 3,811 | 5.8 | 15,681 | 6,949 | 8,732 | 14 | 97 |
| Crawford.................. | 197 | 13,388 | 0.91 | 104 | 7.4 | 187 | 13.4 | 802 | 7.8 | 4,092 | 2,364 | 1,727 | 21 | 95 |
| Delta..................... | 633 | 35,494 | 0.97 | 329 | 9.2 | 460 | 12.9 | 2,023 | 7.6 | 10,813 | 6,562 | 4,251 | 4 | 46 |
| Dickinson................. | 433 | 27,196 | 1.16 | 255 | 10.2 | 334 | 13.3 | 1,224 | 6.3 | 7,002 | 4,472 | 2,529 | 1 | 0 |
| Eaton..................... | 1,811 | 107,111 | 0.96 | 1,090 | 9.9 | 1,079 | 9.8 | 5,201 | 5.9 | 24,174 | 10,882 | 13,292 | 154 | 981 |
| Emmet..................... | 505 | 36,620 | 1.22 | 279 | 8.4 | 341 | 10.2 | 2,058 | 8.1 | 9,253 | 5,740 | 3,513 | 50 | 72 |
| Genesee................... | 5,677 | 389,917 | 0.89 | 4,486 | 11.1 | 4,916 | 12.1 | 21,718 | 6.6 | 90,711 | 40,826 | 49,885 | 155 | NA |
| Gladwin................... | 289 | 22,456 | 0.68 | 263 | 10.3 | 410 | 16.1 | 1,725 | 9.4 | 8,304 | 4,404 | 3,900 | 31 | 115 |
| Gogebic................... | 372 | 15,208 | 1.02 | 128 | 9.2 | 219 | 15.8 | 836 | 8.5 | 4,501 | 2,828 | 1,673 | 4 | 29 |
| Grand Traverse........... | 1,490 | 101,217 | 1.19 | 870 | 9.3 | 953 | 10.2 | 5,245 | 7.2 | 22,492 | 11,549 | 10,942 | 145 | 510 |
| Gratiot................... | 5,421 | 39,843 | 0.93 | 384 | 9.5 | 460 | 11.4 | 1,878 | 6.6 | 8,777 | 4,965 | 3,813 | 34 | 273 |
| Hillsdale................. | 1,681 | 41,914 | 0.80 | 502 | 11.0 | 521 | 11.4 | 2,843 | 8.1 | 10,845 | 6,511 | 4,334 | 70 | 133 |
| Houghton.................. | 2,768 | 35,912 | 0.99 | 372 | 10.6 | 372 | 10.6 | 2,243 | 8.4 | 7,562 | 4,321 | 3,241 | NA | NA |
| Huron..................... | 524 | 32,098 | 1.06 | 247 | 8.1 | 483 | 15.8 | 1,883 | 8.3 | 9,488 | 6,232 | 3,256 | 16 | 91 |
| Ingham.................... | 17,368 | 322,660 | 1.23 | 3,014 | 10.4 | 2,426 | 8.3 | 17,593 | 7.5 | 48,702 | 25,839 | 22,863 | 182 | 479 |
| Ionia..................... | 5,069 | 55,929 | 0.70 | 673 | 10.4 | 602 | 9.3 | 3,465 | 7.0 | 11,810 | 5,178 | 6,632 | 61 | 217 |
| Iosco..................... | 400 | 26,067 | 1.10 | 260 | 10.3 | 421 | 16.7 | 1,546 | 9.0 | 9,020 | 5,338 | 3,682 | 3 | 1 |
| Iron...................... | 355 | 10,839 | 0.93 | 79 | 7.1 | 189 | 17.1 | 629 | 8.4 | 3,916 | 2,457 | 1,459 | 0 | 0 |
| Isabella.................. | 4,730 | 73,917 | 1.10 | 597 | 8.6 | 503 | 7.2 | 5,189 | 9.3 | 11,273 | 6,634 | 4,639 | 47 | 256 |
| Jackson................... | 7,634 | 153,141 | 0.92 | 1,701 | 10.8 | 1,744 | 11.1 | 8,320 | 6.9 | 34,986 | 21,058 | 13,929 | NA | NA |
| Kalamazoo................. | 8,156 | 268,647 | 1.05 | 3,044 | 11.4 | 2,383 | 9.0 | 14,695 | 6.8 | 49,789 | 23,993 | 25,796 | 290 | 2,529 |
| Kalkaska.................. | 138 | 15,459 | 0.71 | 192 | 10.7 | 182 | 10.1 | 1,329 | 9.4 | 4,750 | 2,712 | 2,038 | NA | NA |
| Kent...................... | 11,305 | 700,971 | 1.16 | 8,596 | 13.0 | 5,226 | 7.9 | 41,382 | 7.4 | 109,847 | 42,754 | 67,094 | 390 | 2,325 |

1. Per 1,000 estimated resident population.

| STATE County | Officers | Civilians | Total | Percent private | High school graduate or less | Bachelor's degree or more | Total current spending (mil dol) | Current spending per student (dollars) | Per capita income[4] | Median income (dollars) | with income of less than $50,000 | with income of $200,000 or more | Median household income (dollars) | All persons | Children under 18 years | Children 5 to 17 years in families |
|---|---|---|---|---|---|---|---|---|---|---|---|---|---|---|---|---|
| | 46 | 47 | 48 | 49 | 50 | 51 | 52 | 53 | 54 | 55 | 56 | 57 | 58 | 59 | 60 | 61 |
| MARYLAND— Cont'd | | | | | | | | | | | | | | | | |
| St. Mary's.......................... | 145 | 71 | 30,622 | 17.3 | 40.5 | 31.9 | 239 | 13,223 | 40,354 | 89,845 | 25.7 | 10.7 | 89,123 | 7.7 | 10.6 | 10.0 |
| Somerset........................... | 25 | 4 | 7,148 | 8.6 | 58.4 | 14.4 | 50 | 17,137 | 18,772 | 37,803 | 57.1 | 1.6 | 38,731 | 23.6 | 32.7 | 29.2 |
| Talbot.............................. | 34 | 3 | 7,595 | 20.2 | 33.2 | 40.6 | 60 | 12,831 | 49,136 | 73,547 | 34.5 | 10.8 | 75,714 | 8.7 | 13.1 | 12.8 |
| Washington........................ | 106 | 140 | 33,757 | 12.9 | 49.3 | 21.9 | 309 | 13,684 | 30,464 | 60,860 | 41.8 | 4.7 | 59,785 | 12.3 | 17.0 | 15.6 |
| Wicomico.......................... | 92 | 21 | 30,196 | 10.7 | 45.1 | 27.2 | 211 | 14,096 | 28,080 | 56,956 | 44.6 | 4.0 | 54,351 | 16.0 | 23.1 | 21.7 |
| Worcester......................... | 52 | 8 | 9,828 | 16.4 | 40.5 | 29.0 | 119 | 17,807 | 38,080 | 63,499 | 40.7 | 6.5 | 65,821 | 9.9 | 15.5 | 14.0 |
| Baltimore city.................... | NA | NA | 150,861 | 24.1 | 43.9 | 31.9 | 1,273 | 15,715 | 31,271 | 50,379 | 49.7 | 5.9 | 49,780 | 20.4 | 30.6 | 31.7 |
| MASSACHUSETTS............ | X | X | 1,723,161 | 27.0 | 33.3 | 43.7 | 16,193 | 16,973 | 43,761 | 81,215 | 32.6 | 13.2 | 85,700 | 9.5 | 12.0 | 11.5 |
| Barnstable........................ | NA | NA | 37,831 | 16.3 | 28.3 | 43.4 | 472 | 19,116 | 44,505 | 74,336 | 33.1 | 9.0 | 82,686 | 6.5 | 10.1 | 9.3 |
| Berkshire.......................... | NA | NA | 27,061 | 23.5 | 38.0 | 33.6 | 279 | 17,620 | 35,616 | 59,230 | 43.3 | 5.8 | 58,805 | 11.2 | 16.3 | 15.3 |
| Bristol.............................. | NA | NA | 132,356 | 17.0 | 44.2 | 28.7 | 1,206 | 15,187 | 35,747 | 69,095 | 38.1 | 7.4 | 69,857 | 11.3 | 16.0 | 14.1 |
| Dukes.............................. | NA | NA | 3,371 | 16.6 | 28.8 | 44.9 | 68 | 29,424 | 45,990 | 71,811 | 35.2 | 11.7 | 76,036 | 7.3 | 11.6 | 11.3 |
| Essex............................... | NA | NA | 186,058 | 21.7 | 35.8 | 39.9 | 1,870 | 16,177 | 42,347 | 79,263 | 33.1 | 13.1 | 83,253 | 8.8 | 11.9 | 10.7 |
| Franklin............................ | NA | NA | 14,579 | 16.1 | 35.2 | 37.3 | 169 | 18,098 | 35,908 | 60,950 | 41.1 | 4.7 | 60,143 | 9.3 | 11.5 | 11.4 |
| Hampden........................... | NA | NA | 118,159 | 16.5 | 44.9 | 27.1 | 1,171 | 16,200 | 30,346 | 55,429 | 45.9 | 5.2 | 59,918 | 13.8 | 20.2 | 17.9 |
| Hampshire......................... | NA | NA | 56,031 | 23.0 | 27.9 | 48.3 | 304 | 16,839 | 34,838 | 70,876 | 36.5 | 7.7 | 74,142 | 11.1 | 9.9 | 9.0 |
| Middlesex.......................... | NA | NA | 413,763 | 32.2 | 25.6 | 56.3 | 3,858 | 17,655 | 52,228 | 102,603 | 24.9 | 19.5 | 106,543 | 6.9 | 7.1 | 6.7 |
| Nantucket.......................... | NA | NA | 2,292 | 25.3 | 24.2 | 52.8 | 37 | 23,043 | 55,398 | 107,717 | 23.4 | 18.4 | 96,030 | 5.5 | 5.8 | 5.5 |
| Norfolk............................. | NA | NA | 177,508 | 30.6 | 25.3 | 53.6 | 1,777 | 17,291 | 53,889 | 103,291 | 24.4 | 19.5 | 106,851 | 5.8 | 5.5 | 5.5 |
| Plymouth........................... | NA | NA | 127,640 | 17.0 | 34.9 | 37.6 | 1,290 | 15,445 | 43,412 | 89,489 | 28.3 | 13.3 | 90,584 | 7.4 | 9.3 | 8.9 |
| Suffolk.............................. | NA | NA | 217,051 | 45.8 | 35.8 | 46.1 | 1,760 | 21,295 | 42,765 | 69,669 | 39.2 | 12.0 | 75,843 | 16.7 | 23.9 | 26.6 |
| Worcester.......................... | NA | NA | 209,461 | 22.1 | 37.0 | 36.4 | 1,934 | 15,164 | 37,574 | 74,679 | 34.8 | 9.6 | 77,795 | 9.4 | 11.9 | 12.0 |
| MICHIGAN........................ | X | X | 2,451,348 | 13.0 | 38.1 | 29.1 | 17,706 | 12,027 | 31,713 | 57,144 | 44.0 | 5.2 | 59,522 | 12.9 | 17.5 | 16.3 |
| Alcona............................. | 11 | 18 | 1,331 | 8.4 | 50.0 | 16.9 | 8 | 10,554 | 25,636 | 40,484 | 60.2 | 0.9 | 41,491 | 18.0 | 26.7 | 23.7 |
| Alger............................... | 9 | 9 | 1,485 | 7.7 | 55.3 | 17.4 | 13 | 12,360 | 20,851 | 45,570 | 53.5 | 1.2 | 52,285 | 12.6 | 17.4 | 16.0 |
| Allegan............................ | 55 | 48 | 26,910 | 14.8 | 46.9 | 22.6 | 201 | 11,131 | 29,215 | 62,965 | 38.8 | 3.6 | 66,568 | 10.8 | 16.0 | 15.4 |
| Alpena............................. | 13 | 15 | 5,303 | 9.1 | 41.8 | 18.0 | 46 | 12,117 | 25,957 | 43,363 | 57.3 | 2.0 | 43,676 | 13.9 | 18.3 | 17.2 |
| Antrim.............................. | 21 | 30 | 4,107 | 7.8 | 39.9 | 28.8 | 35 | 10,391 | 31,851 | 56,165 | 44.2 | 4.1 | 58,317 | 9.6 | 16.9 | 15.6 |
| Arenac............................. | 14 | 12 | 2,578 | 7.1 | 53.4 | 12.6 | 20 | 9,933 | 24,328 | 42,290 | 58.7 | 1.8 | 44,017 | 17.2 | 25.6 | 26.8 |
| Baraga............................. | 6 | 7 | 1,506 | 9.8 | 51.4 | 15.8 | 12 | 11,963 | 22,920 | 46,065 | 55.2 | 2.6 | 48,371 | 13.7 | 18.2 | 16.4 |
| Barry............................... | 30 | 22 | 12,974 | 12.6 | 45.1 | 21.1 | 90 | 10,078 | 31,760 | 64,490 | 37.0 | 4.2 | 67,703 | 7.8 | 10.9 | 10.1 |
| Bay.................................. | 35 | 47 | 22,671 | 12.1 | 44.1 | 19.3 | 165 | 11,873 | 27,469 | 48,819 | 51.2 | 2.3 | 50,130 | 13.4 | 19.7 | 16.7 |
| Benzie............................. | 15 | 18 | 3,310 | 13.1 | 37.9 | 30.3 | 21 | 10,719 | 29,425 | 57,974 | 44.5 | 2.8 | 56,934 | 9.6 | 15.8 | 13.8 |
| Berrien............................. | 70 | 83 | 36,847 | 18.4 | 38.4 | 27.8 | 314 | 12,311 | 30,864 | 50,795 | 49.3 | 4.4 | 50,659 | 15.9 | 22.3 | 21.2 |
| Branch............................. | 14 | 21 | 9,240 | 13.5 | 52.4 | 14.4 | 81 | 12,383 | 25,279 | 51,947 | 47.8 | 2.1 | 52,255 | 13.2 | 17.6 | 16.3 |
| Calhoun............................ | 85 | 94 | 31,574 | 12.5 | 45.8 | 21.0 | 268 | 13,631 | 26,855 | 48,607 | 51.3 | 3.0 | 50,205 | 13.9 | 20.3 | 18.6 |
| Cass................................ | 32 | 37 | 11,028 | 13.0 | 45.5 | 19.2 | 75 | 11,295 | 31,295 | 55,107 | 44.5 | 4.3 | 55,029 | 14.8 | 19.9 | 16.2 |
| Charlevoix......................... | 19 | 11 | 5,149 | 9.9 | 35.7 | 31.1 | 62 | 17,204 | 33,620 | 55,760 | 45.0 | 5.0 | 58,831 | 9.7 | 13.1 | 10.7 |
| Cheboygan........................ | 22 | 16 | 4,428 | 10.2 | 47.9 | 20.7 | 38 | 14,253 | 26,393 | 48,044 | 51.8 | 1.9 | 47,876 | 14.9 | 23.8 | 21.8 |
| Chippewa.......................... | 16 | 18 | 8,152 | 15.2 | 44.1 | 21.2 | 67 | 14,639 | 25,086 | 46,486 | 53.2 | 2.0 | 49,610 | 18.3 | 24.6 | 20.9 |
| Clare............................... | 26 | 6 | 5,558 | 12.0 | 52.1 | 12.6 | 56 | 13,060 | 22,573 | 39,565 | 61.0 | 1.8 | 42,480 | 19.6 | 30.4 | 26.0 |
| Clinton............................. | 26 | 35 | 19,705 | 12.6 | 32.3 | 32.1 | 117 | 11,453 | 34,761 | 70,390 | 34.1 | 6.5 | 72,666 | 7.8 | 7.6 | 6.6 |
| Crawford........................... | 15 | 10 | 2,426 | 9.1 | 43.1 | 19.4 | 17 | 10,449 | 26,294 | 47,977 | 51.8 | 0.9 | 49,114 | 14.0 | 24.1 | 23.0 |
| Delta............................... | 10 | 0 | 7,015 | 9.7 | 39.7 | 21.0 | 57 | 12,639 | 27,632 | 47,434 | 51.9 | 1.9 | 48,880 | 13.8 | 18.8 | 18.0 |
| Dickinson.......................... | 12 | 17 | 5,188 | 12.5 | 38.2 | 25.7 | 46 | 12,614 | 29,656 | 51,645 | 48.6 | 2.8 | 53,751 | 9.9 | 14.0 | 13.2 |
| Eaton.............................. | 63 | 55 | 25,356 | 18.1 | 33.4 | 27.8 | 202 | 11,111 | 33,427 | 64,348 | 37.4 | 3.6 | 65,891 | 7.5 | 10.8 | 9.9 |
| Emmet.............................. | 27 | 22 | 6,514 | 11.6 | 30.8 | 33.4 | 55 | 11,737 | 34,074 | 55,829 | 44.6 | 4.8 | 59,456 | 9.2 | 11.8 | 10.0 |
| Genesee........................... | 113 | 134 | 98,032 | 10.0 | 41.0 | 21.2 | 762 | 11,787 | 27,295 | 48,588 | 51.3 | 3.2 | 50,554 | 16.6 | 22.2 | 20.2 |
| Gladwin............................ | 16 | 26 | 4,604 | 12.2 | 52.3 | 13.5 | 27 | 9,953 | 25,657 | 44,619 | 55.9 | 1.6 | 46,458 | 13.5 | 21.1 | 22.3 |
| Gogebic............................ | 16 | 8 | 2,475 | 12.3 | 43.6 | 19.4 | 19 | 11,492 | 25,555 | 38,839 | 60.9 | 2.0 | 39,030 | 17.1 | 24.2 | 22.5 |
| Grand Traverse.................. | 69 | 55 | 20,317 | 14.9 | 28.8 | 36.0 | 183 | 14,623 | 35,405 | 63,575 | 38.6 | 6.2 | 65,041 | 7.2 | 9.6 | 8.8 |
| Gratiot............................. | 26 | 16 | 9,478 | 23.7 | 49.3 | 15.4 | 83 | 13,891 | 23,616 | 47,848 | 52.2 | 2.5 | 51,046 | 14.3 | 19.2 | 18.2 |
| Hillsdale........................... | 22 | 18 | 10,278 | 28.0 | 50.7 | 17.3 | 70 | 12,189 | 25,621 | 49,622 | 50.4 | 2.0 | 50,015 | 15.3 | 23.6 | 21.3 |
| Houghton.......................... | 19 | 9 | 12,404 | 6.0 | 40.3 | 32.6 | 63 | 11,791 | 23,421 | 43,183 | 55.1 | 2.5 | 45,006 | 14.7 | 12.6 | 10.9 |
| Huron.............................. | 20 | 16 | 5,926 | 13.2 | 52.5 | 16.0 | 65 | 13,798 | 27,852 | 48,289 | 51.9 | 2.9 | 49,126 | 13.7 | 18.4 | 19.3 |
| Ingham............................. | 79 | 77 | 98,273 | 6.9 | 29.0 | 38.9 | 525 | 12,655 | 29,380 | 52,872 | 47.4 | 4.4 | 54,704 | 17.3 | 17.6 | 16.7 |
| Ionia............................... | 15 | 2 | 14,903 | 14.6 | 48.6 | 16.4 | 96 | 12,225 | 24,864 | 57,043 | 43.3 | 1.5 | 59,581 | 11.5 | 13.0 | 12.3 |
| Iosco............................... | 5 | 17 | 3,963 | 10.7 | 49.3 | 15.9 | 42 | 11,432 | 25,264 | 43,678 | 59.3 | 1.0 | 43,605 | 14.2 | 22.5 | 21.4 |
| Iron................................. | 10 | 9 | 1,700 | 6.9 | 50.8 | 18.5 | 14 | 10,813 | 25,944 | 41,599 | 57.4 | 1.2 | 46,814 | 13.7 | 21.5 | 19.9 |
| Isabella............................ | 19 | 29 | 27,673 | 7.7 | 40.3 | 29.1 | 69 | 10,616 | 23,888 | 45,116 | 53.9 | 3.4 | 46,677 | 22.9 | 20.6 | 17.0 |
| Jackson............................ | 52 | 79 | 36,856 | 15.1 | 41.8 | 22.2 | 296 | 12,821 | 28,364 | 53,658 | 46.8 | 3.2 | 56,001 | 13.7 | 17.5 | 16.7 |
| Kalamazoo......................... | 129 | 82 | 78,573 | 11.8 | 28.4 | 38.8 | 443 | 12,740 | 31,975 | 56,511 | 44.3 | 5.5 | 56,739 | 13.3 | 13.6 | 12.8 |
| Kalkaska........................... | 16 | 18 | 3,399 | 12.2 | 54.8 | 11.9 | 21 | 10,292 | 24,358 | 46,898 | 53.6 | 1.7 | 47,620 | 13.9 | 18.6 | 17.1 |
| Kent................................ | 207 | 341 | 166,483 | 20.1 | 33.4 | 35.7 | 1,275 | 11,905 | 32,524 | 63,053 | 39.5 | 5.7 | 66,560 | 11.2 | 14.7 | 14.3 |

1. All persons 3 years old and over enrolled in nursery school through college.　2. Persons 25 years old and over.　3. Elementary and secondary education expenditures.　4. Based on population estimated by the American Community Survey, 2014–2018.

| STATE County | Personal income, 2019 | | | | | | | | | | Earnings, 2019 | | |
|---|---|---|---|---|---|---|---|---|---|---|---|---|---|
| | Total (mil dol) | Percent change 2018-2019 | Per capita[1] Dollars | Per capita[1] Rank | Wages and salaries (mil dol) | Supplements to wages and salaries, employer contributions (mil dol) Pension and insurance | Supplements to wages and salaries, employer contributions (mil dol) Government social insurance | Proprietors' income (mil dol) | Dividends, interest, and rent (mil dol) | Personal transfer receipts (mil dol) | Total (mil dol) | Contributions for government social insurance (mil dol) From employee and self-employed | Contributions for government social insurance (mil dol) From employer |
| | 62 | 63 | 64 | 65 | 66 | 67 | 68 | 69 | 70 | 71 | 72 | 73 | 74 |
| **MARYLAND— Cont'd** | | | | | | | | | | | | | |
| St. Mary's | 6,650 | 3.8 | 58,582 | 351 | 3,628 | 784 | 281 | 267 | 1,172 | 915 | 4,960 | 282 | 281 |
| Somerset | 811 | 0.6 | 31,668 | 2,964 | 358 | 87 | 26 | 47 | 150 | 299 | 518 | 33 | 26 |
| Talbot | 2,778 | 3.3 | 74,711 | 92 | 913 | 141 | 66 | 209 | 924 | 483 | 1,329 | 88 | 66 |
| Washington | 7,349 | 3.8 | 48,650 | 950 | 3,247 | 539 | 242 | 555 | 1,211 | 1,592 | 4,583 | 290 | 242 |
| Wicomico | 4,408 | 2.8 | 42,547 | 1,679 | 2,246 | 422 | 163 | 339 | 768 | 1,133 | 3,171 | 194 | 163 |
| Worcester | 3,148 | 2.6 | 60,222 | 297 | 1,019 | 169 | 80 | 341 | 853 | 672 | 1,609 | 107 | 80 |
| Baltimore city | 31,679 | 2.0 | 53,378 | 586 | 25,681 | 4,081 | 1,805 | 3,736 | 5,474 | 8,476 | 35,303 | 2,032 | 1,805 |
| **MASSACHUSETTS** | 511,334 | 3.5 | 74,161 | X | 281,710 | 40,655 | 18,464 | 44,340 | 104,700 | 71,419 | 385,169 | 20,546 | 18,464 |
| Barnstable | 16,493 | 3.4 | 77,435 | 75 | 5,234 | 954 | 386 | 1,594 | 4,890 | 3,139 | 8,169 | 488 | 386 |
| Berkshire | 7,284 | 2.1 | 58,299 | 362 | 3,119 | 578 | 227 | 561 | 1,561 | 1,785 | 4,484 | 261 | 227 |
| Bristol | 31,493 | 3.7 | 55,718 | 462 | 12,674 | 2,328 | 911 | 2,483 | 4,210 | 6,706 | 18,396 | 1,025 | 911 |
| Dukes | 1,750 | 3.4 | 100,996 | 21 | 520 | 89 | 39 | 375 | 656 | 186 | 1,022 | 52 | 39 |
| Essex | 55,218 | 3.6 | 69,981 | 129 | 21,213 | 3,463 | 1,480 | 4,438 | 10,879 | 8,276 | 30,594 | 1,673 | 1,480 |
| Franklin | 3,986 | 3.0 | 56,793 | 414 | 1,228 | 277 | 88 | 320 | 777 | 972 | 1,912 | 112 | 88 |
| Hampden | 24,833 | 2.5 | 53,248 | 594 | 11,374 | 2,221 | 809 | 1,437 | 3,394 | 6,893 | 15,840 | 876 | 809 |
| Hampshire | 8,824 | 3.5 | 54,867 | 502 | 3,536 | 873 | 223 | 787 | 1,799 | 1,385 | 5,419 | 265 | 223 |
| Middlesex | 140,527 | 3.6 | 87,192 | 39 | 88,065 | 10,674 | 5,683 | 10,893 | 31,756 | 12,937 | 115,315 | 6,172 | 5,683 |
| Nantucket | 1,553 | 4.3 | 136,204 | 8 | 491 | 67 | 39 | 400 | 568 | 80 | 998 | 48 | 39 |
| Norfolk | 67,300 | 3.7 | 95,221 | 26 | 25,318 | 3,640 | 1,748 | 5,037 | 16,019 | 5,959 | 35,743 | 1,907 | 1,748 |
| Plymouth | 36,312 | 4.1 | 69,669 | 133 | 11,538 | 2,091 | 802 | 3,020 | 6,140 | 5,638 | 17,451 | 955 | 802 |
| Suffolk | 67,118 | 2.8 | 83,490 | 51 | 76,249 | 9,489 | 4,553 | 9,497 | 14,993 | 9,230 | 99,788 | 5,118 | 4,553 |
| Worcester | 48,643 | 3.8 | 58,563 | 352 | 21,150 | 3,912 | 1,476 | 3,499 | 7,057 | 8,233 | 30,037 | 1,595 | 1,476 |
| **MICHIGAN** | 491,632 | 3.1 | 49,238 | X | 246,831 | 39,833 | 17,999 | 32,662 | 87,940 | 103,673 | 337,325 | 21,733 | 17,999 |
| Alcona | 427 | 6.5 | 41,049 | 1,917 | 75 | 16 | 6 | 18 | 99 | 171 | 115 | 14 | 6 |
| Alger | 312 | 2.3 | 34,233 | 2,777 | 102 | 22 | 8 | 12 | 64 | 110 | 144 | 12 | 8 |
| Allegan | 5,424 | 3.6 | 45,937 | 1,254 | 2,106 | 376 | 156 | 329 | 890 | 1,044 | 2,967 | 194 | 156 |
| Alpena | 1,187 | 3.8 | 41,783 | 1,805 | 491 | 98 | 38 | 74 | 196 | 438 | 700 | 52 | 38 |
| Antrim | 1,116 | 3.5 | 47,855 | 1,037 | 210 | 44 | 17 | 74 | 307 | 322 | 344 | 30 | 17 |
| Arenac | 581 | 2.4 | 39,009 | 2,191 | 177 | 35 | 14 | 36 | 89 | 217 | 263 | 22 | 14 |
| Baraga | 279 | 3.5 | 34,044 | 2,796 | 119 | 31 | 9 | 8 | 51 | 97 | 167 | 12 | 9 |
| Barry | 2,819 | 3.7 | 45,804 | 1,271 | 585 | 130 | 43 | 195 | 463 | 557 | 953 | 70 | 43 |
| Bay | 4,502 | 2.5 | 43,657 | 1,529 | 1,633 | 308 | 124 | 197 | 721 | 1,340 | 2,261 | 166 | 124 |
| Benzie | 787 | 3.5 | 44,283 | 1,448 | 170 | 31 | 14 | 52 | 211 | 221 | 266 | 23 | 14 |
| Berrien | 7,400 | 1.4 | 48,237 | 997 | 3,283 | 629 | 247 | 397 | 1,399 | 1,745 | 4,556 | 299 | 247 |
| Branch | 1,619 | 2.0 | 37,202 | 2,424 | 649 | 122 | 49 | 76 | 262 | 459 | 896 | 64 | 49 |
| Calhoun | 5,401 | 2.9 | 40,257 | 2,019 | 3,093 | 538 | 228 | 208 | 867 | 1,537 | 4,068 | 272 | 228 |
| Cass | 2,352 | 1.8 | 45,409 | 1,313 | 442 | 88 | 33 | 104 | 409 | 560 | 666 | 57 | 33 |
| Charlevoix | 1,389 | 1.5 | 53,136 | 606 | 504 | 92 | 39 | 92 | 412 | 307 | 726 | 51 | 39 |
| Cheboygan | 1,021 | 3.1 | 40,405 | 2,000 | 249 | 47 | 21 | 52 | 238 | 347 | 368 | 34 | 21 |
| Chippewa | 1,308 | 3.5 | 35,034 | 2,683 | 520 | 136 | 40 | 60 | 246 | 410 | 756 | 52 | 40 |
| Clare | 1,072 | 3.1 | 34,637 | 2,734 | 307 | 65 | 23 | 73 | 169 | 429 | 469 | 41 | 23 |
| Clinton | 3,774 | 4.0 | 47,411 | 1,091 | 796 | 131 | 61 | 212 | 611 | 631 | 1,199 | 88 | 61 |
| Crawford | 497 | 4.2 | 35,407 | 2,636 | 193 | 37 | 15 | 28 | 90 | 177 | 273 | 21 | 15 |
| Delta | 1,475 | 3.7 | 41,210 | 1,890 | 629 | 118 | 50 | 61 | 245 | 468 | 858 | 65 | 50 |
| Dickinson | 1,241 | 0.8 | 49,179 | 909 | 718 | 140 | 55 | 26 | 252 | 333 | 939 | 62 | 55 |
| Eaton | 4,782 | 3.6 | 43,365 | 1,561 | 2,213 | 379 | 160 | 210 | 757 | 1,043 | 2,962 | 200 | 160 |
| Emmet | 1,956 | 4.7 | 58,528 | 355 | 856 | 141 | 67 | 143 | 580 | 416 | 1,206 | 80 | 67 |
| Genesee | 17,015 | 3.0 | 41,929 | 1,777 | 6,983 | 1,173 | 523 | 840 | 2,514 | 5,012 | 9,519 | 671 | 523 |
| Gladwin | 955 | 3.2 | 37,512 | 2,386 | 178 | 35 | 14 | 54 | 153 | 353 | 281 | 30 | 14 |
| Gogebic | 616 | 2.5 | 44,113 | 1,471 | 226 | 50 | 17 | 21 | 138 | 220 | 314 | 24 | 17 |
| Grand Traverse | 4,805 | 3.9 | 51,619 | 712 | 2,515 | 421 | 189 | 460 | 1,112 | 970 | 3,585 | 230 | 189 |
| Gratiot | 1,558 | 4.2 | 38,274 | 2,299 | 632 | 125 | 48 | 96 | 235 | 458 | 902 | 61 | 48 |
| Hillsdale | 1,643 | 2.1 | 36,033 | 2,565 | 626 | 119 | 47 | 73 | 260 | 467 | 865 | 63 | 47 |
| Houghton | 1,396 | 3.5 | 39,124 | 2,172 | 548 | 124 | 41 | 63 | 308 | 409 | 776 | 53 | 41 |
| Huron | 1,484 | 4.5 | 47,909 | 1,027 | 494 | 99 | 39 | 161 | 348 | 435 | 792 | 53 | 39 |
| Ingham | 12,072 | 3.2 | 41,286 | 1,882 | 8,411 | 1,591 | 605 | 648 | 2,203 | 2,614 | 11,256 | 680 | 605 |
| Ionia | 2,298 | 2.9 | 35,518 | 2,617 | 741 | 160 | 56 | 79 | 287 | 540 | 1,036 | 73 | 56 |
| Iosco | 1,001 | 6.5 | 39,832 | 2,071 | 383 | 72 | 28 | 40 | 194 | 399 | 523 | 43 | 28 |
| Iron | 503 | 3.0 | 45,463 | 1,304 | 155 | 31 | 12 | 14 | 90 | 196 | 213 | 18 | 12 |
| Isabella | 2,540 | 3.8 | 36,356 | 2,528 | 1,235 | 282 | 92 | 112 | 454 | 734 | 1,721 | 107 | 92 |
| Jackson | 6,402 | 3.1 | 40,387 | 2,004 | 3,009 | 579 | 224 | 339 | 991 | 1,703 | 4,151 | 279 | 224 |
| Kalamazoo | 13,119 | 2.7 | 49,493 | 880 | 7,062 | 1,218 | 513 | 867 | 2,611 | 2,398 | 9,661 | 597 | 513 |
| Kalkaska | 638 | 3.9 | 35,387 | 2,637 | 240 | 42 | 18 | 43 | 106 | 205 | 343 | 26 | 18 |
| Kent | 35,255 | 3.1 | 53,664 | 562 | 22,220 | 3,412 | 1,624 | 2,841 | 7,622 | 5,230 | 30,098 | 1,817 | 1,624 |

1. Based on the resident population estimated as of July 1 of the year shown.

# Table B. States and Counties — **Earnings, Social Security, and Housing**

| STATE County | Farm | Mining, quarrying, and extractions | Construction | Manufacturing | Information; professional, scientific, technical services | Retail trade | Finance, insurance, real estate, and leasing | Health care and social assistance | Government | Social Security beneficiaries, December 2019 — Number | Rate[1] | Supplemental Security Income recipients, 2019 | Housing units, 2020 — Total | Percent change, 2010-2020 |
|---|---|---|---|---|---|---|---|---|---|---|---|---|---|---|
| | 75 | 76 | 77 | 78 | 79 | 80 | 81 | 82 | 83 | 84 | 85 | 86 | 87 | 88 |
| **MARYLAND— Cont'd** | | | | | | | | | | | | | | |
| St. Mary's | 0.1 | D | 4.0 | 1.0 | 23.2 | 3.7 | 2.0 | 6.7 | 47.2 | 17,170 | 151 | 1,699 | 47,425 | 14.9 |
| Somerset | 4.6 | 0.0 | 4.6 | 4.1 | D | 3.5 | D | D | 45.2 | 5,525 | 216 | 936 | 11,473 | 3.1 |
| Talbot | 1.4 | D | 7.4 | 3.0 | 11.6 | 8.0 | 9.3 | 19.0 | 12.0 | 11,290 | 304 | 563 | 20,310 | 3.7 |
| Washington | 1.2 | D | 5.8 | 11.8 | 5.4 | 9.1 | 9.7 | 15.3 | 14.6 | 33,500 | 222 | 3,747 | 61,847 | 1.7 |
| Wicomico | 1.5 | D | 6.4 | 6.1 | 6.3 | 8.9 | 5.8 | 20.1 | 18.8 | 21,350 | 205 | 2,831 | 42,696 | 3.6 |
| Worcester | 1.6 | D | 6.8 | 3.3 | 6.6 | 9.5 | 8.9 | 9.4 | 17.6 | 15,955 | 305 | 846 | 56,842 | 2.0 |
| Baltimore city | 0.0 | D | 3.1 | 2.8 | 16.0 | 1.7 | 9.8 | 17.4 | 19.5 | 102,705 | 173 | 36,047 | 293,137 | -1.2 |
| **MASSACHUSETTS** | 0.0 | 0.0 | 6.1 | 7.4 | 21.0 | 4.5 | 11.4 | 12.6 | 11.6 | 1,287,830 | 187 | 182,701 | 2,944,574 | 4.9 |
| Barnstable | 0.0 | D | 12.4 | 2.5 | 8.7 | 9.3 | 6.5 | 15.7 | 17.6 | 73,505 | 345 | 3,107 | 165,101 | 3.0 |
| Berkshire | 0.0 | 0.1 | 8.1 | 7.8 | 10.4 | 7.2 | 5.8 | 19.3 | 14.5 | 34,930 | 279 | 3,884 | 69,466 | 1.4 |
| Bristol | 0.0 | D | 7.6 | 12.5 | 6.9 | 7.7 | 3.9 | 14.9 | 14.6 | 124,250 | 220 | 18,585 | 237,456 | 3.0 |
| Dukes | 0.1 | 0.0 | 20.4 | D | D | 8.6 | 5.1 | 8.8 | 13.8 | 4,580 | 262 | 118 | 18,276 | 6.3 |
| Essex | 0.0 | 0.0 | 7.9 | 15.8 | 12.7 | 5.7 | 5.2 | 15.2 | 12.6 | 153,370 | 194 | 21,877 | 315,705 | 3.0 |
| Franklin | 0.2 | D | 9.2 | 14.5 | 6.0 | 7.2 | 2.8 | 13.1 | 19.4 | 18,000 | 256 | 1,971 | 34,267 | 1.5 |
| Hampden | 0.0 | 0.1 | 5.9 | 9.7 | 5.5 | 6.1 | 9.4 | 21.0 | 18.0 | 103,415 | 222 | 30,335 | 194,049 | 1.0 |
| Hampshire | 0.1 | 0.1 | 5.3 | 5.4 | 6.8 | 6.3 | 3.7 | 12.3 | 31.1 | 32,440 | 201 | 2,961 | 64,798 | 3.5 |
| Middlesex | 0.1 | 0.0 | 5.3 | 9.1 | 33.0 | 3.3 | 5.8 | 8.2 | 8.3 | 251,985 | 156 | 24,604 | 650,460 | 6.3 |
| Nantucket | 0.0 | 0.0 | 22.0 | D | D | 7.8 | 6.0 | 4.8 | 9.2 | 1,795 | 158 | 36 | 12,823 | 10.4 |
| Norfolk | 0.0 | 0.0 | 8.8 | 7.0 | 16.4 | 6.1 | 12.2 | 11.4 | 9.9 | 124,555 | 176 | 10,694 | 283,720 | 4.9 |
| Plymouth | 0.0 | 0.0 | 11.7 | 6.5 | 8.3 | 7.5 | 6.8 | 13.1 | 17.3 | 110,550 | 212 | 9,214 | 210,538 | 5.2 |
| Suffolk | 0.0 | D | 2.9 | 0.9 | 24.7 | 2.6 | 24.6 | 14.2 | 9.8 | 98,250 | 122 | 33,011 | 348,410 | 10.4 |
| Worcester | 0.0 | 0.1 | 7.3 | 11.8 | 10.5 | 5.9 | 6.8 | 15.9 | 16.1 | 156,205 | 188 | 22,304 | 339,505 | 3.9 |
| **MICHIGAN** | 0.4 | 0.2 | 5.7 | 16.4 | 12.5 | 5.8 | 7.3 | 12.3 | 13.5 | 2,236,852 | 224 | 270,396 | 4,645,046 | 2.5 |
| Alcona | 1.3 | 0.1 | 5.8 | 11.7 | 3.5 | 9.5 | 3.7 | D | 16.2 | 4,755 | 455 | 340 | 11,262 | 1.7 |
| Alger | 0.0 | 0.0 | 4.1 | 24.3 | D | 5.2 | 4.6 | 7.4 | 29.4 | 3,005 | 331 | 164 | 6,729 | 2.7 |
| Allegan | 2.7 | 0.3 | 9.4 | 37.8 | 3.8 | 4.9 | 3.0 | 3.8 | 12.6 | 25,645 | 217 | 1,634 | 52,357 | 5.9 |
| Alpena | 1.0 | 1.4 | 5.4 | 17.7 | 4.5 | 10.2 | 4.1 | 9.6 | 26.8 | 10,015 | 353 | 1,119 | 16,085 | 0.2 |
| Antrim | 2.4 | 0.5 | 13.6 | 14.9 | 5.5 | 6.9 | 5.4 | 4.4 | 21.3 | 8,005 | 343 | 464 | 18,267 | 2.5 |
| Arenac | 1.9 | D | 4.8 | 21.4 | D | 9.3 | 4.9 | D | 14.3 | 5,360 | 359 | 587 | 9,904 | 1.0 |
| Baraga | 0.1 | D | 3.7 | 22.3 | D | 3.7 | D | D | 49.6 | 2,260 | 274 | 156 | 5,312 | 0.7 |
| Barry | 3.8 | 0.1 | 8.4 | 32.3 | D | 4.4 | 6.8 | 7.0 | 16.9 | 14,455 | 235 | 708 | 28,165 | 4.3 |
| Bay | -0.1 | 0.1 | 5.4 | 16.8 | 7.1 | 8.1 | 4.1 | 16.6 | 17.2 | 29,605 | 288 | 3,112 | 48,388 | 0.3 |
| Benzie | 0.8 | D | 16.0 | 8.3 | D | 7.7 | 7.5 | 10.0 | 16.9 | 5,810 | 327 | 276 | 12,700 | 4.2 |
| Berrien | 1.2 | 0.2 | 5.6 | 29.3 | 3.9 | 5.7 | 5.1 | 11.6 | 13.4 | 38,245 | 249 | 4,527 | 77,979 | 1.4 |
| Branch | -0.1 | 0.0 | 5.4 | 21.6 | 2.8 | 6.7 | 5.4 | 5.3 | 22.4 | 10,765 | 248 | 907 | 21,098 | 1.2 |
| Calhoun | 0.2 | 0.1 | 4.2 | 21.0 | 9.1 | 5.2 | 2.4 | 14.5 | 21.3 | 33,385 | 250 | 4,751 | 60,888 | -0.3 |
| Cass | 0.0 | D | 7.5 | 22.1 | 4.6 | 4.6 | 5.5 | D | 20.8 | 13,380 | 258 | 939 | 26,596 | 2.8 |
| Charlevoix | 0.4 | 0.0 | 10.0 | 26.8 | D | 4.9 | 4.7 | D | 17.0 | 7,920 | 302 | 450 | 17,940 | 4.0 |
| Cheboygan | 0.3 | D | 15.3 | 3.3 | 6.0 | 11.0 | 5.8 | D | 20.5 | 8,950 | 353 | 673 | 18,699 | 2.2 |
| Chippewa | 0.5 | D | 4.1 | 5.0 | 2.7 | 7.1 | 3.0 | 4.4 | 56.3 | 9,015 | 242 | 870 | 21,591 | 1.6 |
| Clare | 1.4 | D | 9.6 | 12.7 | D | 9.7 | 3.3 | 8.4 | 22.7 | 10,465 | 339 | 1,287 | 23,517 | 1.2 |
| Clinton | 3.3 | 0.4 | 14.8 | 8.9 | 7.0 | 8.7 | 8.6 | 7.3 | 13.9 | 16,405 | 206 | 668 | 32,463 | 5.8 |
| Crawford | 0.0 | D | 8.1 | 17.8 | 3.5 | 6.8 | 3.7 | D | 20.4 | 4,350 | 310 | 367 | 11,287 | 1.7 |
| Delta | 0.3 | D | 9.0 | 21.4 | D | 8.6 | 4.2 | 12.6 | 16.1 | 11,785 | 330 | 953 | 20,447 | 1.2 |
| Dickinson | 0.0 | 0.0 | 20.3 | 22.1 | 3.7 | 6.7 | 2.3 | 6.4 | 23.4 | 7,615 | 301 | 485 | 14,143 | 1.1 |
| Eaton | 0.1 | 0.1 | 5.9 | 17.3 | 5.1 | 7.0 | 18.1 | 5.5 | 14.2 | 26,110 | 237 | 1,797 | 47,855 | 1.7 |
| Emmet | 0.0 | D | 9.4 | 10.5 | D | 9.7 | 4.9 | 19.9 | 14.1 | 9,525 | 286 | 500 | 21,922 | 2.9 |
| Genesee | 0.1 | 0.0 | 6.3 | 12.1 | 7.2 | 8.4 | 6.0 | 18.3 | 15.4 | 101,555 | 250 | 16,406 | 193,269 | 0.6 |
| Gladwin | 0.1 | D | 14.8 | 19.2 | 2.0 | 11.9 | 4.7 | D | 19.1 | 9,110 | 358 | 857 | 18,035 | 2.0 |
| Gogebic | 0.0 | 0.1 | 4.9 | 16.1 | 3.1 | 7.2 | 2.2 | D | 31.9 | 4,925 | 353 | 470 | 10,788 | -0.1 |
| Grand Traverse | 0.3 | 0.8 | 7.9 | 10.0 | 8.3 | 9.3 | 10.1 | 21.1 | 12.1 | 23,735 | 255 | 1,415 | 45,628 | 9.7 |
| Gratiot | 3.9 | D | 7.1 | 17.8 | D | 5.3 | 3.9 | D | 18.6 | 9,705 | 239 | 1,086 | 16,386 | 0.3 |
| Hillsdale | 1.8 | D | 4.4 | 36.7 | 4.7 | 6.8 | 3.4 | D | 17.8 | 11,470 | 252 | 1,125 | 22,186 | 2.0 |
| Houghton | 0.1 | D | 6.6 | 6.0 | 7.5 | 7.2 | 3.5 | 14.2 | 40.1 | 8,275 | 232 | 559 | 18,972 | 1.8 |
| Huron | 11.5 | D | 5.9 | 14.4 | D | 5.8 | 5.2 | 11.9 | 14.8 | 10,085 | 326 | 723 | 21,334 | 0.6 |
| Ingham | 0.2 | 0.0 | 4.1 | 9.1 | 9.4 | 4.8 | 7.1 | 14.8 | 30.1 | 52,415 | 180 | 7,950 | 124,745 | 2.9 |
| Ionia | 2.0 | D | 7.9 | 26.0 | 2.6 | 6.2 | 5.5 | 6.4 | 22.6 | 13,075 | 202 | 1,243 | 25,304 | 2.1 |
| Iosco | 0.4 | D | 6.3 | 16.7 | 3.3 | 6.8 | 3.6 | D | 18.5 | 9,905 | 392 | 820 | 20,618 | 0.8 |
| Iron | 0.3 | D | 7.2 | 5.9 | 4.5 | 13.7 | 3.8 | 12.3 | 25.7 | 4,355 | 394 | 291 | 9,360 | 1.8 |
| Isabella | 0.4 | 2.0 | 8.7 | 10.8 | 3.8 | 6.3 | 5.2 | 8.0 | 36.4 | 12,065 | 173 | 1,245 | 29,377 | 3.5 |
| Jackson | 0.0 | 0.2 | 5.1 | 18.3 | 4.6 | 6.0 | 4.9 | 15.6 | 13.9 | 38,195 | 241 | 4,536 | 70,102 | 0.9 |
| Kalamazoo | 0.8 | D | 6.6 | 23.6 | 6.2 | 6.1 | 8.2 | 15.7 | 12.3 | 52,525 | 198 | 6,369 | 113,565 | 3.2 |
| Kalkaska | 0.5 | 12.1 | 18.7 | 5.5 | D | 5.4 | 2.5 | 2.2 | 21.1 | 5,160 | 287 | 372 | 12,511 | 2.8 |
| Kent | 0.2 | 0.3 | 6.1 | 19.3 | 8.7 | 5.6 | 8.1 | 15.8 | 7.1 | 114,115 | 174 | 13,209 | 261,651 | 6.0 |

1. Per 1,000 resident population estimated as of July 1 of the year shown.

| | Housing units, 2015-2019 | | | | | | | | Civilian labor force, 2020 | | | | Civilian employment[6], 2015-2019 | | |
| | Occupied units | | | | | | | | | | Unemployment | | | Percent | |
| | | Owner-occupied | | | | Renter-occupied | | | | | | | | | |
| STATE County | Total | Percent | Median value[1] | Median owner cost as a percent of income — With a mortgage | Median owner cost as a percent of income — Without a mortgage[2] | Median rent[3] | Median rent as a percent of income[2] | Sub-standard units[4] (percent) | Total | Percent change, 2019-2020 | Total | Rate[5] | Total | Management, business, science, and arts | Construction, production, and maintenance occupations |
|---|---|---|---|---|---|---|---|---|---|---|---|---|---|---|---|
| | 89 | 90 | 91 | 92 | 93 | 94 | 95 | 96 | 97 | 98 | 99 | 100 | 101 | 102 | 103 |
| **MARYLAND— Cont'd** | | | | | | | | | | | | | | | |
| St. Mary's | 40,552 | 71.0 | 301,300 | 21.1 | 10.0 | 1,398 | 27.1 | 1.8 | 57,281 | 0.0 | 2,768 | 4.8 | 55,220 | 45.8 | 21.5 |
| Somerset | 8,574 | 65.0 | 139,100 | 24.3 | 14.0 | 716 | 39.0 | 3.4 | 8,720 | -4.6 | 724 | 8.3 | 8,888 | 29.0 | 23.8 |
| Talbot | 16,826 | 70.3 | 336,800 | 22.8 | 10.7 | 1,116 | 31.2 | 1.1 | 17,776 | -7.8 | 1,069 | 6.0 | 17,732 | 45.0 | 17.1 |
| Washington | 56,035 | 65.9 | 210,300 | 20.3 | 10.8 | 921 | 27.9 | 2.0 | 72,703 | -2.9 | 4,904 | 6.7 | 68,511 | 33.6 | 25.7 |
| Wicomico | 37,793 | 57.7 | 175,700 | 20.2 | 11.4 | 1,060 | 32.6 | 3.4 | 49,403 | -3.8 | 3,681 | 7.5 | 48,959 | 33.0 | 21.5 |
| Worcester | 22,089 | 75.1 | 262,200 | 25.2 | 12.4 | 1,035 | 30.1 | 1.8 | 25,069 | -2.6 | 2,799 | 11.2 | 24,144 | 34.9 | 17.6 |
| Baltimore city | 239,116 | 47.5 | 160,100 | 21.9 | 13.5 | 1,073 | 30.7 | 2.6 | 281,187 | -2.3 | 24,705 | 8.8 | 280,888 | 43.7 | 16.9 |
| **MASSACHUSETTS** | 2,617,497 | 62.4 | 381,600 | 22.5 | 13.9 | 1,282 | 29.8 | 2.3 | 3,658,321 | -3.3 | 324,195 | 8.9 | 3,612,375 | 46.8 | 16.0 |
| Barnstable | 94,323 | 79.8 | 393,500 | 25.5 | 13.8 | 1,311 | 32.7 | 1.5 | 108,128 | -5.2 | 11,055 | 10.2 | 105,715 | 38.3 | 18.6 |
| Berkshire | 54,813 | 69.2 | 216,100 | 22.0 | 13.9 | 872 | 30.5 | 1.1 | 61,645 | -3.6 | 5,629 | 9.1 | 64,008 | 38.6 | 19.1 |
| Bristol | 217,912 | 62.6 | 299,800 | 22.2 | 13.8 | 901 | 28.3 | 1.7 | 293,532 | -2.6 | 30,076 | 10.2 | 284,670 | 37.5 | 22.7 |
| Dukes | 6,765 | 72.3 | 699,500 | 28.2 | 18.0 | 1,459 | 35.5 | 3.3 | 9,517 | -0.6 | 877 | 9.2 | 8,730 | 42.4 | 24.9 |
| Essex | 294,202 | 64.2 | 409,900 | 23.2 | 14.3 | 1,241 | 32.3 | 2.9 | 414,660 | -2.7 | 40,585 | 9.8 | 408,720 | 42.3 | 18.6 |
| Franklin | 30,569 | 68.6 | 232,100 | 23.5 | 14.0 | 976 | 31.1 | 1.6 | 39,501 | -3.1 | 2,936 | 7.4 | 36,728 | 41.9 | 21.3 |
| Hampden | 179,423 | 61.1 | 207,800 | 22.1 | 14.3 | 906 | 32.2 | 2.2 | 220,259 | -3.2 | 22,887 | 10.4 | 217,690 | 35.8 | 21.7 |
| Hampshire | 58,838 | 66.9 | 283,100 | 22.0 | 13.0 | 1,119 | 31.8 | 1.1 | 85,661 | -5.1 | 6,184 | 7.2 | 84,132 | 47.8 | 14.5 |
| Middlesex | 604,384 | 62.4 | 500,700 | 21.9 | 13.4 | 1,636 | 28.0 | 2.3 | 882,398 | -4.1 | 64,193 | 7.3 | 888,780 | 57.0 | 11.8 |
| Nantucket | 3,713 | 70.5 | 1,084,700 | 29.8 | 13.5 | 1,764 | 31.3 | 1.2 | 7,664 | 0.6 | 791 | 10.3 | 6,418 | 35.3 | 23.6 |
| Norfolk | 265,300 | 68.9 | 470,800 | 22.3 | 14.1 | 1,589 | 29.5 | 2.0 | 380,061 | -3.4 | 31,733 | 8.3 | 376,303 | 54.0 | 12.0 |
| Plymouth | 187,460 | 76.5 | 370,300 | 23.5 | 14.8 | 1,279 | 30.7 | 1.5 | 276,733 | -3.1 | 26,154 | 9.5 | 267,634 | 40.2 | 18.9 |
| Suffolk | 309,844 | 36.1 | 496,500 | 23.3 | 13.0 | 1,590 | 30.5 | 3.8 | 446,558 | -2.1 | 43,170 | 9.7 | 441,009 | 47.7 | 12.3 |
| Worcester | 309,951 | 65.1 | 280,600 | 21.6 | 14.0 | 1,060 | 29.2 | 2.2 | 432,005 | -3.1 | 37,924 | 8.8 | 421,838 | 42.3 | 19.4 |
| **MICHIGAN** | 3,935,041 | 71.2 | 154,900 | 19.4 | 12.0 | 871 | 29.5 | 2.0 | 4,840,843 | -2.2 | 478,115 | 9.9 | 4,654,930 | 36.6 | 24.8 |
| Alcona | 4,988 | 88.8 | 112,000 | 22.2 | 12.8 | 627 | 32.0 | 1.1 | 3,861 | -0.7 | 382 | 9.9 | 3,349 | 27.1 | 31.2 |
| Alger | 3,007 | 84.0 | 136,200 | 22.9 | 12.8 | 650 | 30.8 | 2.3 | 3,147 | -1.2 | 338 | 10.7 | 3,112 | 26.0 | 26.1 |
| Allegan | 43,416 | 82.6 | 165,400 | 19.6 | 10.9 | 814 | 25.3 | 2.5 | 62,253 | -0.4 | 4,543 | 7.3 | 55,812 | 30.5 | 35.9 |
| Alpena | 12,752 | 77.9 | 96,500 | 19.4 | 11.8 | 596 | 27.3 | 0.8 | 13,245 | 0.3 | 1,021 | 7.7 | 12,665 | 30.0 | 27.2 |
| Antrim | 9,899 | 87.1 | 160,500 | 21.8 | 11.8 | 752 | 29.4 | 1.4 | 10,207 | | 994 | 9.7 | 10,050 | 29.0 | 28.2 |
| Arenac | 6,571 | 83.8 | 94,400 | 22.0 | 12.1 | 604 | 28.8 | 1.4 | 5,909 | -2.3 | 675 | 11.4 | 5,906 | 24.4 | 33.2 |
| Baraga | 3,107 | 80.3 | 102,200 | 19.0 | 11.6 | 528 | 26.7 | 3.5 | 3,163 | -1.4 | 370 | 11.7 | 2,917 | 28.1 | 25.8 |
| Barry | 24,296 | 83.5 | 158,400 | 19.8 | 10.7 | 869 | 25.8 | 1.7 | 31,350 | -1.7 | 2,317 | 7.4 | 29,100 | 29.3 | 34.2 |
| Bay | 44,339 | 76.8 | 102,000 | 19.6 | 12.5 | 663 | 27.9 | 1.2 | 49,573 | -1.2 | 4,564 | 9.2 | 47,048 | 29.8 | 26.8 |
| Benzie | 6,792 | 89.7 | 185,500 | 21.5 | 11.5 | 702 | 31.9 | 1.7 | 8,655 | -2.1 | 817 | 9.4 | 7,830 | 32.1 | 26.7 |
| Berrien | 63,665 | 70.7 | 150,700 | 19.3 | 11.3 | 743 | 29.0 | 2.1 | 72,765 | -1.7 | 6,289 | 8.6 | 71,622 | 34.9 | 26.1 |
| Branch | 16,650 | 74.2 | 109,800 | 19.3 | 12.1 | 753 | 25.2 | 2.5 | 19,739 | 1.8 | 1,591 | 8.1 | 18,727 | 23.1 | 39.9 |
| Calhoun | 53,991 | 69.9 | 110,000 | 19.1 | 12.3 | 751 | 29.9 | 1.6 | 61,838 | -1.1 | 5,994 | 9.7 | 60,032 | 30.1 | 30.7 |
| Cass | 21,019 | 80.0 | 142,800 | 19.6 | 11.1 | 738 | 27.8 | 2.6 | 24,470 | -1.3 | 2,143 | 8.8 | 23,869 | 29.4 | 37.0 |
| Charlevoix | 11,503 | 80.9 | 166,300 | 20.6 | 11.9 | 694 | 28.4 | 1.7 | 12,649 | -2.0 | 1,244 | 9.8 | 12,378 | 31.9 | 27.9 |
| Cheboygan | 11,195 | 82.4 | 130,900 | 20.8 | 11.2 | 699 | 29.2 | 1.8 | 10,277 | -2.2 | 1,498 | 14.6 | 10,470 | 26.8 | 28.5 |
| Chippewa | 13,999 | 67.4 | 119,100 | 18.9 | 11.2 | 674 | 27.5 | 1.5 | 16,121 | 0.2 | 1,437 | 8.9 | 15,523 | 29.2 | 21.8 |
| Clare | 12,199 | 82.9 | 91,100 | 22.2 | 12.6 | 647 | 33.9 | 3.1 | 11,798 | 0.0 | 1,358 | 11.5 | 10,891 | 27.0 | 31.4 |
| Clinton | 29,728 | 80.0 | 172,800 | 18.8 | 11.2 | 839 | 26.4 | 1.2 | 40,089 | -3.0 | 2,626 | 6.6 | 38,589 | 40.3 | 23.0 |
| Crawford | 6,141 | 81.2 | 105,000 | 19.9 | 11.9 | 735 | 29.1 | 1.6 | 5,660 | 1.2 | 590 | 10.4 | 5,482 | 28.3 | 27.8 |
| Delta | 16,234 | 77.4 | 114,300 | 19.1 | 11.2 | 584 | 27.7 | 1.9 | 16,948 | -1.1 | 1,482 | 8.7 | 15,578 | 28.4 | 33.2 |
| Dickinson | 11,231 | 77.4 | 101,800 | 16.8 | 11.2 | 665 | 28.7 | 2.0 | 12,206 | -0.6 | 859 | 7.0 | 11,249 | 36.4 | 27.3 |
| Eaton | 44,480 | 72.3 | 155,000 | 18.8 | 10.9 | 858 | 25.5 | 1.8 | 56,112 | -2.5 | 4,309 | 7.7 | 53,457 | 36.0 | 26.0 |
| Emmet | 14,463 | 72.8 | 186,900 | 20.5 | 11.5 | 818 | 27.0 | 1.3 | 17,219 | -2.8 | 1,804 | 10.5 | 16,414 | 35.1 | 21.3 |
| Genesee | 167,902 | 70.0 | 111,100 | 20.0 | 12.9 | 767 | 30.9 | 1.8 | 179,933 | -0.8 | 20,074 | 11.2 | 172,390 | 32.1 | 27.1 |
| Gladwin | 11,047 | 84.9 | 110,000 | 21.2 | 12.2 | 617 | 28.8 | 4.0 | 9,876 | -0.4 | 961 | 9.7 | 8,954 | 25.7 | 36.0 |
| Gogebic | 6,744 | 77.5 | 67,100 | 18.3 | 13.5 | 509 | 29.4 | 0.5 | 5,852 | -2.6 | 450 | 7.7 | 5,978 | 28.0 | 32.4 |
| Grand Traverse | 37,319 | 76.3 | 212,500 | 20.6 | 10.0 | 951 | 28.8 | 2.0 | 48,974 | -1.7 | 4,105 | 8.4 | 46,754 | 38.0 | 19.7 |
| Gratiot | 15,035 | 74.3 | 95,400 | 19.2 | 12.3 | 687 | 30.5 | 1.2 | 17,927 | -0.7 | 1,430 | 8.0 | 16,358 | 29.8 | 29.2 |
| Hillsdale | 18,107 | 76.5 | 120,100 | 19.2 | 12.1 | 703 | 26.9 | 3.4 | 20,088 | -2.1 | 1,827 | 9.1 | 19,421 | 29.2 | 36.5 |
| Houghton | 13,386 | 66.9 | 108,700 | 18.7 | 10.9 | 640 | 33.0 | 2.7 | 15,945 | -1.9 | 1,156 | 7.2 | 15,275 | 38.1 | 19.6 |
| Huron | 13,847 | 80.9 | 103,900 | 19.7 | 11.5 | 609 | 28.9 | 1.3 | 15,554 | 0.8 | 1,323 | 8.5 | 13,883 | 29.0 | 33.7 |
| Ingham | 112,840 | 58.5 | 135,600 | 19.2 | 12.3 | 877 | 30.7 | 1.9 | 146,960 | -2.8 | 10,954 | 7.5 | 143,566 | 40.6 | 18.3 |
| Ionia | 22,964 | 76.6 | 131,800 | 19.2 | 11.5 | 768 | 25.0 | 2.4 | 30,414 | 0.1 | 2,414 | 7.9 | 28,978 | 26.3 | 36.1 |
| Iosco | 11,669 | 79.9 | 92,600 | 20.6 | 10.8 | 652 | 26.2 | 2.1 | 10,152 | -0.4 | 1,063 | 10.5 | 9,108 | 22.6 | 37.2 |
| Iron | 5,225 | 80.8 | 77,700 | 21.5 | 12.4 | 513 | 28.9 | 1.4 | 4,985 | -2.6 | 404 | 8.1 | 4,402 | 25.2 | 31.4 |
| Isabella | 24,739 | 62.1 | 134,900 | 19.9 | 11.2 | 736 | 34.9 | 1.9 | 33,122 | -4.6 | 2,768 | 8.4 | 33,830 | 31.8 | 22.8 |
| Jackson | 61,805 | 73.5 | 131,600 | 19.6 | 11.1 | 782 | 29.2 | 1.6 | 74,075 | -0.8 | 6,668 | 9.0 | 68,695 | 31.9 | 28.0 |
| Kalamazoo | 103,445 | 64.0 | 159,300 | 18.8 | 11.4 | 812 | 28.5 | 1.7 | 131,618 | -1.9 | 9,488 | 7.2 | 133,538 | 40.5 | 20.1 |
| Kalkaska | 7,145 | 82.6 | 115,900 | 21.0 | 11.2 | 702 | 28.7 | 2.5 | 7,904 | -0.9 | 875 | 11.1 | 7,355 | 20.4 | 37.1 |
| Kent | 241,746 | 69.8 | 173,700 | 18.6 | 10.4 | 899 | 28.7 | 2.6 | 357,123 | -1.3 | 27,091 | 7.6 | 335,479 | 36.9 | 25.7 |

1. Specified owner-occupied units lacking complete plumbing facilities.  2. A value of 10.0 represents 10 percent or less; a value of 50.0 represents 50 percent or more.  3. Specified renter-occupied units.  4. Overcrowded or
5. Percent of civilian labor force.  6. Civilian employed persons 16 years old and over.

# Table B. States and Counties — Nonfarm Employment and Agriculture

Private nonfarm establishments, employment and payroll, 2019 — Agriculture, 2017

| STATE County | Number of establishments | Employment — Total | Health care and social assistance | Manufacturing | Retail trade | Finance and insurance | Professional, scientific, and technical services | Annual payroll — Total (mil dol) | Average per employee (dollars) | Farms — Number | Percent with: Fewer than 50 acres | 1000 acres or more | Farm producers whose primary occupation is farming (percent) |
|---|---|---|---|---|---|---|---|---|---|---|---|---|---|
|  | 104 | 105 | 106 | 107 | 108 | 109 | 110 | 111 | 112 | 113 | 114 | 115 | 116 |
| **MARYLAND— Cont'd** | | | | | | | | | | | | | |
| St. Mary's | 2,014 | 32,995 | 4,627 | 452 | 5,066 | 434 | 10,453 | 1,822 | 55,210 | 615 | 57.2 | 1.5 | 43.7 |
| Somerset | 361 | 4,011 | 1,247 | 393 | 510 | 83 | 75 | 156 | 38,860 | 255 | 42.4 | 3.9 | 56.2 |
| Talbot | 1,453 | 17,318 | 3,859 | 647 | 2,660 | 490 | 1,141 | 740 | 42,724 | 317 | 44.5 | 10.4 | 43.7 |
| Washington | 3,450 | 60,314 | 9,926 | 6,296 | 9,629 | 4,995 | 1,747 | 2,553 | 42,327 | 877 | 50.3 | 1.1 | 45.1 |
| Wicomico | 2,481 | 36,283 | 8,966 | 2,256 | 6,686 | 1,020 | 1,264 | 1,508 | 41,566 | 494 | 53.0 | 4.9 | 48.6 |
| Worcester | 2,173 | 19,535 | 2,686 | 597 | 3,637 | 593 | 654 | 749 | 38,318 | 369 | 49.3 | 7.9 | 45.9 |
| Baltimore city | 12,356 | 302,102 | 79,541 | 11,224 | 20,355 | 16,417 | 21,007 | 19,514 | 64,594 | NA | NA | NA | NA |
| **MASSACHUSETTS** | 181,061 | 3,386,372 | 625,474 | 233,428 | 363,220 | 187,058 | 315,966 | 238,938 | 70,559 | 7,241 | 67.8 | 0.3 | 42.8 |
| Barnstable | 8,679 | 77,116 | 15,761 | 2,069 | 14,998 | 2,017 | 4,757 | 3,814 | 49,458 | 321 | 93.5 | 0.0 | 46.5 |
| Berkshire | 3,780 | 52,368 | 11,544 | 5,115 | 7,747 | 1,814 | 2,698 | 2,471 | 47,179 | 475 | 50.9 | 1.3 | 42.6 |
| Bristol | 12,965 | 203,446 | 41,948 | 24,323 | 34,118 | 4,116 | 5,593 | 9,852 | 48,423 | 688 | 75.9 | NA | 50.7 |
| Dukes | 1,130 | 5,709 | 832 | 123 | 1,007 | 219 | 246 | 367 | 64,355 | 108 | 90.7 | 0.9 | 45.5 |
| Essex | 19,328 | 300,360 | 68,014 | 42,411 | 39,175 | 10,623 | 14,567 | 16,760 | 55,801 | 419 | 78.0 | 0.5 | 46.9 |
| Franklin | 1,569 | 20,378 | 3,671 | 3,728 | 2,845 | 447 | 650 | 886 | 43,470 | 830 | 46.7 | 0.5 | 37.0 |
| Hampden | 9,475 | 175,265 | 42,812 | 19,313 | 21,517 | 8,239 | 7,022 | 8,325 | 47,497 | 523 | 64.1 | 0.4 | 41.2 |
| Hampshire | 3,573 | 50,922 | 10,040 | 2,897 | 7,534 | 1,637 | 1,734 | 2,044 | 40,149 | 692 | 59.2 | NA | 41.3 |
| Middlesex | 44,830 | 930,320 | 120,794 | 61,954 | 82,293 | 30,830 | 140,393 | 79,723 | 85,694 | 620 | 73.9 | NA | 59.7 |
| Nantucket | 1,129 | 5,009 | 562 | 63 | 887 | 96 | 236 | 377 | 75,234 | 21 | 90.5 | NA | 41.0 |
| Norfolk | 20,449 | 336,592 | 53,274 | 17,636 | 44,060 | 24,648 | 21,772 | 21,126 | 62,764 | 197 | 81.2 | 0.5 | 47.4 |
| Plymouth | 12,832 | 178,045 | 36,440 | 9,990 | 28,740 | 7,419 | 12,408 | 8,872 | 49,832 | 758 | 74.8 | 0.8 | 43.8 |
| Suffolk | 22,054 | 662,595 | 143,203 | 9,004 | 39,763 | 75,348 | 80,249 | 63,185 | 95,361 | 21 | 100.0 | NA | 38.6 |
| Worcester | 18,570 | 316,770 | 71,106 | 34,802 | 38,421 | 15,441 | 20,121 | 16,388 | 51,735 | 1,568 | 67.7 | 0.1 | 43.1 |
| **MICHIGAN** | 222,226 | 3,978,872 | 642,379 | 613,798 | 459,711 | 169,072 | 288,626 | 203,510 | 51,148 | 47,641 | 46.3 | 4.5 | 43.1 |
| Alcona | 183 | 1,171 | 195 | 176 | 308 | 42 | 33 | 38 | 32,238 | 223 | 37.7 | 1.3 | 51.4 |
| Alger | 242 | 1,736 | 213 | 450 | 262 | 74 | 18 | 73 | 41,794 | 126 | 48.4 | 4.8 | 39.4 |
| Allegan | 2,452 | 38,861 | 3,041 | 14,475 | 3,995 | 414 | 1,828 | 1,852 | 47,655 | 1,172 | 52.2 | 5.0 | 43.9 |
| Alpena | 801 | 10,166 | 2,735 | 1,599 | 1,784 | 268 | 249 | 383 | 37,662 | 415 | 38.6 | 2.4 | 38.1 |
| Antrim | 573 | 3,908 | 346 | 950 | 547 | 113 | 125 | 141 | 35,998 | 333 | 37.8 | 2.4 | 47.4 |
| Arenac | 318 | 3,491 | 557 | 1,042 | 508 | 74 | 125 | 142 | 40,679 | 350 | 34.0 | 5.7 | 36.7 |
| Baraga | 174 | 1,875 | 290 | 624 | 292 | 50 | 19 | 70 | 37,243 | 65 | 29.2 | 7.7 | 40.8 |
| Barry | 909 | 11,259 | 1,365 | 3,822 | 1,265 | 740 | 306 | 444 | 39,443 | 938 | 48.8 | 4.1 | 34.2 |
| Bay | 2,088 | 30,070 | 6,248 | 4,463 | 5,028 | 900 | 1,235 | 1,220 | 40,583 | 726 | 36.8 | 8.8 | 48.8 |
| Benzie | 466 | 3,359 | 349 | 488 | 514 | 160 | 69 | 123 | 36,715 | 197 | 59.9 | 0.5 | 38.7 |
| Berrien | 3,549 | 55,043 | 8,982 | 9,352 | 6,782 | 1,248 | 2,692 | 2,721 | 49,429 | 872 | 57.8 | 3.3 | 48.8 |
| Branch | 808 | 11,545 | 1,521 | 2,598 | 1,587 | 455 | 564 | 482 | 41,774 | 789 | 38.7 | 7.5 | 39.9 |
| Calhoun | 2,440 | 51,721 | 9,300 | 13,713 | 5,831 | 1,075 | 2,562 | 2,661 | 51,447 | 958 | 42.6 | 5.8 | 44.2 |
| Cass | 745 | 8,144 | 990 | 2,943 | 798 | 198 | 229 | 304 | 37,368 | 747 | 51.1 | 5.2 | 43.1 |
| Charlevoix | 826 | 8,219 | 1,209 | 1,895 | 974 | 178 | 245 | 363 | 44,222 | 271 | 45.8 | 1.1 | 39.0 |
| Cheboygan | 738 | 4,307 | 708 | 291 | 1,048 | 177 | 97 | 170 | 39,381 | 330 | 37.0 | 0.9 | 32.3 |
| Chippewa | 764 | 8,445 | 2,068 | 560 | 1,593 | 275 | 302 | 269 | 31,897 | 427 | 26.7 | 3.5 | 36.1 |
| Clare | 572 | 6,013 | 1,078 | 1,072 | 1,082 | 124 | 185 | 228 | 37,999 | 396 | 40.9 | 2.3 | 48.5 |
| Clinton | 1,350 | 16,585 | 1,945 | 2,380 | 2,283 | 1,057 | 1,132 | 690 | 41,588 | 1,017 | 46.7 | 5.9 | 41.5 |
| Crawford | 297 | 3,315 | 992 | 586 | 571 | 51 | 54 | 131 | 39,380 | 45 | 62.2 | NA | 19.2 |
| Delta | 995 | 11,920 | 1,954 | 2,355 | 2,167 | 685 | 379 | 453 | 37,973 | 253 | 29.2 | 5.5 | 41.6 |
| Dickinson | 829 | 12,376 | 2,488 | 2,398 | 1,605 | 318 | 504 | 600 | 48,511 | 158 | 32.9 | 1.3 | 34.9 |
| Eaton | 2,139 | 42,324 | 3,805 | 7,822 | 5,929 | 4,280 | 1,136 | 1,971 | 46,563 | 962 | 41.6 | 5.0 | 42.2 |
| Emmet | 1,553 | 15,789 | 3,251 | 1,494 | 2,716 | 351 | 495 | 668 | 42,279 | 324 | 43.8 | 0.9 | 33.5 |
| Genesee | 7,614 | 120,486 | 26,327 | 12,292 | 18,844 | 4,005 | 4,566 | 5,371 | 44,580 | 820 | 59.4 | 3.4 | 46.2 |
| Gladwin | 412 | 3,975 | 656 | 992 | 627 | 96 | 61 | 152 | 38,252 | 459 | 37.5 | 1.5 | 39.6 |
| Gogebic | 360 | 3,854 | 566 | 727 | 657 | 90 | 87 | 125 | 32,407 | 54 | 51.9 | 1.9 | 18.8 |
| Grand Traverse | 3,408 | 46,727 | 9,704 | 5,453 | 8,120 | 2,429 | 1,913 | 2,110 | 45,162 | 497 | 57.3 | 0.8 | 38.4 |
| Gratiot | 716 | 11,276 | 2,282 | 2,752 | 1,316 | 360 | 289 | 446 | 39,543 | 812 | 38.3 | 9.7 | 49.8 |
| Hillsdale | 755 | 10,917 | 1,490 | 3,958 | 1,341 | 285 | 187 | 422 | 38,655 | 1,205 | 44.3 | 5.8 | 40.8 |
| Houghton | 874 | 9,198 | 1,960 | 837 | 1,653 | 445 | 503 | 320 | 34,780 | 208 | 36.5 | 0.5 | 38.7 |
| Huron | 929 | 10,999 | 1,858 | 3,888 | 1,349 | 393 | 232 | 424 | 38,522 | 1,153 | 29.7 | 14.0 | 52.0 |
| Ingham | 6,086 | 114,656 | 23,461 | 8,872 | 13,231 | 9,761 | 6,828 | 5,495 | 47,923 | 912 | 60.2 | 4.4 | 39.6 |
| Ionia | 872 | 11,539 | 1,179 | 4,125 | 1,736 | 545 | 178 | 449 | 38,871 | 954 | 44.3 | 5.6 | 46.3 |
| Iosco | 600 | 6,411 | 931 | 1,095 | 1,229 | 245 | 170 | 239 | 37,303 | 244 | 42.2 | 0.8 | 37.0 |
| Iron | 342 | 2,637 | 335 | 393 | 471 | 118 | 84 | 90 | 34,127 | 133 | 32.3 | 2.3 | 49.0 |
| Isabella | 1,369 | 24,860 | 2,991 | 2,863 | 3,414 | 538 | 588 | 980 | 39,405 | 959 | 36.9 | 3.6 | 42.1 |
| Jackson | 2,966 | 51,345 | 10,320 | 9,168 | 6,941 | 1,544 | 2,934 | 2,562 | 49,894 | 923 | 52.1 | 3.9 | 41.5 |
| Kalamazoo | 5,598 | 113,025 | 20,557 | 19,000 | 13,969 | 5,214 | 4,317 | 5,878 | 52,010 | 707 | 61.8 | 5.1 | 43.2 |
| Kalkaska | 324 | 4,041 | 642 | 452 | 596 | 64 | 44 | 213 | 52,824 | 225 | 39.1 | 1.3 | 35.5 |
| Kent | 16,962 | 372,282 | 53,899 | 68,339 | 36,017 | 15,876 | 16,537 | 18,678 | 50,172 | 1,010 | 54.1 | 3.5 | 43.1 |

Items 104—116

# Table B. States and Counties — **Agriculture**

| STATE County | Land in farms | | | | | Value of land and buildings (dollars) | | Value of machinery and equiopmnet, average per farm (dollars) | Value of products sold: | | | | Organic farms (number) | Farms with internet access (percent) | Government payments | |
|---|---|---|---|---|---|---|---|---|---|---|---|---|---|---|---|---|
| | Acreage (1,000) | Percent change, 2012-2017 | Acres | | | Average per farm | Average per acre | | Total (mil dol) | Average per farm (acres) | Percent from: | | | | Total ($1,000) | Percent of farms |
| | | | Average size of farm | Total irrigated (1,000) | Total cropland (1,000) | | | | | | Crops | Livestock and poultry products | | | | |
| | 117 | 118 | 119 | 120 | 121 | 122 | 123 | 124 | 125 | 126 | 127 | 128 | 129 | 130 | 131 | 132 |
| MARYLAND— Cont'd | | | | | | | | | | | | | | | | |
| St. Mary's | 62 | -7.9 | 100 | 0.7 | 37.0 | 999,805 | 9,949 | 75,425 | 26.0 | 42,203 | 78.8 | 21.2 | 2 | 58.0 | 970 | 19.5 |
| Somerset | 59 | -8.9 | 233 | 0.3 | 38.0 | 1,334,676 | 5,726 | 173,390 | 262.2 | 1,028,239 | 8.4 | 91.6 | 2 | 76.9 | 1,875 | 51.8 |
| Talbot | 94 | -21.6 | 295 | 8.3 | 81.1 | 2,075,678 | 7,028 | 158,736 | 68.5 | 216,199 | 63.0 | 37.0 | 3 | 80.4 | 3,891 | 58.4 |
| Washington | 119 | -8.0 | 136 | 0.6 | 82.2 | 1,095,597 | 8,057 | 121,316 | 153.7 | 175,285 | 24.8 | 75.2 | 12 | 69.1 | 995 | 17.8 |
| Wicomico | 89 | 5.8 | 179 | 11.0 | 65.6 | 1,261,012 | 7,034 | 158,378 | 304.0 | 615,350 | 22.5 | 77.5 | 3 | 75.3 | 2,410 | 44.1 |
| Worcester | 99 | -0.1 | 269 | 5.9 | 71.6 | 1,425,500 | 5,300 | 184,865 | 249.1 | 675,154 | 15.1 | 84.9 | 1 | 72.1 | 3,630 | 51.8 |
| Baltimore city | NA | NA | NA | NA | NA | NA | NA | NA | NA | NA | NA | NA | NA | NA | NA | NA |
| MASSACHUSETTS | 492 | -6.1 | 68 | 23.9 | 171.5 | 739,711 | 10,894 | 65,382 | 475.2 | 65,624 | 76.5 | 23.5 | 208 | 84.1 | 4,004 | 7.3 |
| Barnstable | 7 | 40.4 | 20 | 1.1 | 1.6 | 624,807 | 30,555 | 60,762 | 23.1 | 72,019 | 40.4 | 59.6 | 11 | 90.0 | 197 | 4.7 |
| Berkshire | 59 | -4.9 | 123 | 0.3 | 19.1 | 943,835 | 7,644 | 72,649 | 23.5 | 49,453 | 42.8 | 57.2 | 16 | 85.5 | 447 | 7.8 |
| Bristol | 32 | -8.2 | 47 | 2.0 | 13.0 | 846,518 | 18,186 | 68,617 | 35.0 | 50,901 | 79.0 | 21.0 | 18 | 75.3 | 429 | 12.6 |
| Dukes | 8 | -39.3 | 71 | 0.3 | 0.8 | 816,456 | 11,429 | 88,694 | 5.4 | 49,917 | 61.5 | 38.5 | 5 | 91.7 | NA | NA |
| Essex | 21 | -7.5 | 49 | 1.3 | 11.0 | 863,169 | 17,450 | 69,643 | 32.9 | 78,439 | 86.6 | 13.4 | 6 | 89.7 | 54 | 2.1 |
| Franklin | 88 | -1.7 | 106 | 1.8 | 24.6 | 682,435 | 6,419 | 67,251 | 68.9 | 83,000 | 73.0 | 27.0 | 32 | 84.3 | 476 | 8.6 |
| Hampden | 36 | -7.0 | 69 | 1.0 | 12.0 | 711,793 | 10,343 | 55,643 | 25.9 | 49,507 | 83.3 | 16.7 | 7 | 79.3 | 362 | 6.3 |
| Hampshire | 51 | -6.1 | 73 | 0.8 | 20.2 | 560,780 | 7,662 | 75,215 | 46.0 | 66,512 | 76.8 | 23.2 | 28 | 85.7 | 473 | 9.0 |
| Middlesex | 27 | -3.2 | 44 | 1.2 | 13.0 | 702,873 | 15,944 | 61,303 | 63.4 | 102,177 | 88.1 | 11.9 | 26 | 84.5 | 172 | 4.8 |
| Nantucket | 1 | -37.6 | 37 | 0.3 | 0.5 | 1,432,672 | 39,124 | 53,243 | D | D | D | D | 2 | 42.9 | NA | NA |
| Norfolk | 8 | -19.3 | 39 | 0.5 | 3.1 | 663,973 | 17,150 | 57,310 | D | D | D | D | 6 | 85.8 | 64 | 5.1 |
| Plymouth | 60 | -6.2 | 79 | 11.9 | 18.2 | 758,347 | 9,575 | 73,670 | 71.9 | 94,900 | 86.4 | 13.6 | 20 | 89.3 | 316 | 5.7 |
| Suffolk | 0 | -12.5 | 1 | 0.0 | 0.0 | 229,899 | 229,899 | 12,756 | 0.5 | 24,810 | 96.7 | 3.3 | 5 | 95.2 | NA | NA |
| Worcester | 95 | -6.4 | 61 | 1.5 | 34.5 | 747,474 | 12,297 | 57,372 | 65.2 | 41,579 | 70.9 | 29.1 | 26 | 82.7 | 1,015 | 8.4 |
| MICHIGAN | 9,764 | -1.9 | 205 | 670.2 | 7,924.5 | 1,015,631 | 4,955 | 154,740 | 8,220.9 | 172,560 | 56.5 | 43.5 | 764 | 77.2 | 167,189 | 32.2 |
| Alcona | 36 | -4.9 | 163 | 0.1 | 24.1 | 423,042 | 2,590 | 106,156 | 11.5 | 51,543 | 43.6 | 56.4 | 8 | 65.9 | 268 | 21.5 |
| Alger | 21 | 17.9 | 166 | 0.0 | 9.9 | 339,285 | 2,040 | 69,429 | 4.2 | 33,532 | 29.9 | 70.1 | 3 | 81.0 | 12 | 6.3 |
| Allegan | 230 | -15.0 | 196 | 24.8 | 196.9 | 1,172,374 | 5,981 | 223,117 | 584.4 | 498,612 | 30.2 | 69.8 | 12 | 80.5 | 4,521 | 24.2 |
| Alpena | 65 | -5.6 | 158 | 0.0 | 42.2 | 401,054 | 2,545 | 95,420 | 25.9 | 62,400 | 29.5 | 70.5 | 8 | 74.9 | 861 | 23.1 |
| Antrim | 56 | -13.4 | 167 | 4.0 | 34.2 | 701,340 | 4,203 | 107,992 | 35.5 | 106,520 | 81.7 | 18.3 | 9 | 73.3 | 226 | 14.7 |
| Arenac | 87 | 6.7 | 249 | 0.2 | 73.0 | 835,592 | 3,355 | 186,941 | 43.0 | 122,860 | 67.5 | 32.5 | 12 | 74.3 | 2,256 | 71.4 |
| Baraga | 18 | -0.7 | 271 | 0.0 | 9.5 | 592,800 | 2,189 | 89,014 | 2.2 | 34,000 | 58.4 | 41.6 | 3 | 67.7 | 7 | 26.2 |
| Barry | 155 | -6.4 | 165 | 4.9 | 119.2 | 791,437 | 4,801 | 119,827 | 139.7 | 148,915 | 30.9 | 69.1 | 6 | 82.2 | 2,959 | 28.1 |
| Bay | 210 | 8.3 | 289 | 5.2 | 197.1 | 1,460,329 | 5,052 | 242,520 | 116.5 | 160,519 | 88.3 | 11.7 | NA | 69.3 | 4,501 | 67.8 |
| Benzie | 19 | -10.3 | 94 | 0.7 | 8.7 | 372,275 | 3,961 | 66,072 | 10.0 | 50,898 | 78.5 | 21.5 | 2 | 84.8 | 189 | 12.2 |
| Berrien | 145 | -7.6 | 166 | 20.0 | 123.5 | 1,068,559 | 6,445 | 142,166 | 171.3 | 196,501 | 91.4 | 8.6 | 15 | 76.4 | 2,594 | 20.6 |
| Branch | 239 | -2.0 | 303 | 55.7 | 204.3 | 1,285,432 | 4,236 | 207,759 | 162.3 | 205,697 | 65.9 | 34.1 | 6 | 73.9 | 5,703 | 45.5 |
| Calhoun | 214 | -4.9 | 223 | 13.3 | 174.6 | 1,185,546 | 5,309 | 139,254 | 113.9 | 118,861 | 61.1 | 38.9 | 20 | 71.5 | 4,522 | 36.7 |
| Cass | 199 | 5.2 | 266 | 71.2 | 163.7 | 1,228,314 | 4,622 | 191,044 | 153.4 | 205,339 | 65.1 | 34.9 | 10 | 79.8 | 4,103 | 42.6 |
| Charlevoix | 30 | -20.2 | 110 | 0.1 | 14.1 | 416,535 | 3,770 | 59,479 | 8.4 | 31,052 | 59.3 | 40.7 | 6 | 69.0 | 104 | 11.8 |
| Cheboygan | 44 | -3.4 | 133 | D | 22.1 | 341,794 | 2,562 | 72,788 | 7.1 | 21,655 | 58.8 | 41.2 | 7 | 69.1 | 112 | 10.9 |
| Chippewa | 89 | -4.3 | 209 | 0.0 | 53.2 | 451,403 | 2,164 | 75,495 | 10.7 | 25,000 | 37.7 | 62.3 | 1 | 79.9 | 447 | 23.7 |
| Clare | 55 | -12.6 | 138 | D | 30.4 | 415,261 | 3,011 | 76,135 | 22.1 | 55,705 | 21.9 | 78.1 | 13 | 56.6 | 384 | 21.7 |
| Clinton | 230 | -5.8 | 226 | 2.6 | 200.9 | 1,277,549 | 5,655 | 194,330 | 224.3 | 220,535 | 32.2 | 67.8 | 20 | 80.3 | 4,392 | 41.7 |
| Crawford | 3 | 6.8 | 65 | NA | 0.3 | 228,867 | 3,502 | 36,385 | 0.2 | 4,778 | 10.2 | 89.8 | NA | 71.1 | NA | NA |
| Delta | 59 | -17.0 | 232 | 0.8 | 29.0 | 414,631 | 1,785 | 89,494 | 10.8 | 42,688 | 59.9 | 40.1 | 4 | 75.5 | 133 | 24.1 |
| Dickinson | 22 | -23.0 | 140 | 0.3 | 8.7 | 380,435 | 2,727 | 84,219 | 4.4 | 27,671 | 54.4 | 45.6 | NA | 77.8 | 33 | 11.4 |
| Eaton | 210 | -5.9 | 218 | 0.7 | 177.1 | 912,095 | 4,176 | 132,342 | 83.3 | 86,595 | 89.3 | 10.7 | 20 | 74.1 | 4,192 | 37.6 |
| Emmet | 39 | -1.4 | 121 | 0.3 | 20.5 | 456,379 | 3,767 | 63,450 | 8.7 | 26,895 | 70.3 | 29.7 | 15 | 78.1 | 97 | 12.7 |
| Genesee | 124 | 0.5 | 151 | 1.5 | 104.3 | 795,230 | 5,262 | 103,908 | 70.4 | 85,871 | 86.5 | 13.5 | 11 | 84.8 | 2,561 | 25.0 |
| Gladwin | 59 | -12.8 | 128 | 0.1 | 36.5 | 498,882 | 3,909 | 74,943 | 15.8 | 34,333 | 56.1 | 43.9 | 3 | 63.6 | 680 | 34.6 |
| Gogebic | 6 | -8.9 | 103 | D | 1.8 | 266,642 | 2,601 | 41,304 | 0.7 | 13,833 | 29.0 | 71.0 | NA | 83.3 | D | 1.9 |
| Grand Traverse | 51 | -6.7 | 102 | 2.8 | 36.1 | 651,401 | 6,362 | 85,866 | 34.1 | 68,610 | 83.2 | 16.8 | 3 | 81.7 | 501 | 17.3 |
| Gratiot | 297 | 2.5 | 365 | 15.9 | 272.4 | 1,984,846 | 5,432 | 280,110 | 281.4 | 346,607 | 43.3 | 56.7 | 13 | 80.9 | 5,365 | 60.7 |
| Hillsdale | 254 | -3.1 | 211 | 8.8 | 212.3 | 909,201 | 4,309 | 149,441 | 165.1 | 137,021 | 51.4 | 48.6 | 18 | 73.0 | 7,375 | 50.0 |
| Houghton | 26 | -4.3 | 125 | 0.0 | 13.4 | 294,396 | 2,354 | 49,491 | 6.3 | 30,322 | 36.3 | 63.7 | 4 | 70.2 | 52 | 13.0 |
| Huron | 495 | 9.5 | 430 | 3.1 | 458.8 | 2,597,983 | 6,048 | 360,764 | 610.8 | 529,729 | 43.1 | 56.9 | 20 | 76.6 | 12,201 | 75.5 |
| Ingham | 178 | -11.2 | 195 | 1.8 | 152.2 | 1,040,491 | 5,325 | 143,901 | 113.8 | 124,780 | 64.1 | 35.9 | 12 | 85.1 | 3,921 | 19.7 |
| Ionia | 234 | -5.8 | 245 | 5.1 | 200.1 | 1,215,280 | 4,955 | 206,048 | 382.9 | 401,410 | 21.9 | 78.1 | 15 | 79.5 | 5,053 | 45.5 |
| Iosco | 34 | -11.0 | 139 | 0.2 | 22.9 | 437,110 | 3,153 | 103,020 | 14.9 | 61,037 | 27.3 | 72.7 | 6 | 69.7 | 503 | 33.2 |
| Iron | 23 | 2.2 | 176 | 0.5 | 10.6 | 348,316 | 1,976 | 64,714 | 3.7 | 27,564 | 88.6 | 11.4 | NA | 74.4 | 38 | 7.5 |
| Isabella | 212 | 12.4 | 221 | 2.9 | 174.2 | 905,013 | 4,099 | 156,666 | 116.8 | 121,844 | 52.8 | 47.2 | 27 | 75.2 | 3,450 | 41.8 |
| Jackson | 160 | -12.4 | 174 | 4.7 | 125.8 | 866,522 | 4,986 | 116,065 | 70.8 | 76,714 | 61.6 | 38.4 | 6 | 78.2 | 3,526 | 25.9 |
| Kalamazoo | 139 | -3.5 | 196 | 43.0 | 111.8 | 1,382,696 | 7,055 | 206,186 | 236.9 | 335,109 | 74.5 | 25.5 | 2 | 83.7 | 2,664 | 21.6 |
| Kalkaska | 27 | 5.1 | 121 | 1.6 | 16.3 | 320,046 | 2,654 | 88,846 | 8.3 | 37,000 | 92.3 | 7.7 | 4 | 83.1 | 170 | 20.0 |
| Kent | 157 | 0.0 | 156 | 15.3 | 124.8 | 1,150,595 | 7,380 | 134,311 | 262.8 | 260,213 | 77.0 | 23.0 | 12 | 85.0 | 2,199 | 18.7 |

| STATE County | Water use, 2015 | | Wholesale Trade[1], 2017 | | | | Retail Trade[2], 2017 | | | | Real estate and rental and leasing,[2] 2017 | | | |
|---|---|---|---|---|---|---|---|---|---|---|---|---|---|---|
| | Public supply water withdrawn (mil gal/day) | Public supply gallons withdrawn per person per day | Number of establishments | Number of employees | Sales (mil dol) | Average payroll (mil dol) | Number of establishments | Number of employees | Sales (mil dol) | Average payroll (mil dol) | Number of establishments | Number of employees | Sales (mil dol) | Average payroll (mil dol) |
| | 133 | 134 | 135 | 136 | 137 | 138 | 139 | 140 | 141 | 142 | 143 | 144 | 145 | 146 |
| **MARYLAND— Cont'd** | | | | | | | | | | | | | | |
| St. Mary's | 4.15 | 37.2 | 37 | D | 131.0 | D | 306 | 4,740 | 1,370.3 | 125.0 | 88 | 274 | 95.6 | 11.5 |
| Somerset | 1.09 | 42.3 | D | D | D | D | 56 | 482 | 121.9 | 10.5 | 15 | 31 | 2.9 | 0.7 |
| Talbot | 2.24 | 59.7 | 52 | 458 | 302.9 | 25.1 | 216 | 2,750 | 716.7 | 66.9 | 69 | 199 | 51.1 | 8.3 |
| Washington | 17.27 | 115.5 | 136 | 2,065 | 2,279.7 | 109.1 | 602 | 10,208 | 2,618.6 | 239.0 | 146 | 893 | 210.4 | 33.6 |
| Wicomico | 6.86 | 67.0 | 102 | 1,051 | 1,110.8 | 49.6 | 362 | 6,450 | 1,794.3 | 160.1 | 130 | 649 | 113.6 | 23.4 |
| Worcester | 7.68 | 149.0 | D | D | D | 9.5 | 387 | 3,747 | 897.9 | 93.3 | 166 | 540 | 105.6 | 18.8 |
| Baltimore city | 0.00 | 0.0 | 508 | 7,678 | 9,253.2 | 509.0 | 1,912 | 21,064 | 7,175.3 | 600.3 | 748 | 5,107 | 1,607.3 | 314.5 |
| **MASSACHUSETTS** | 648.06 | 95.4 | 6,324 | 120,957 | 137,010.9 | 10,129.9 | 23,928 | 364,204 | 110,194.5 | 10,911.5 | 7,584 | 52,315 | 17,912.1 | 3,369.5 |
| Barnstable | 31.84 | 148.6 | D | D | D | D | 1,450 | 16,168 | 4,537.5 | 500.4 | 386 | 1,533 | 361.5 | 66.0 |
| Berkshire | 14.33 | 112.1 | 98 | 1,276 | 491.8 | 66.1 | 653 | 8,151 | 2,049.7 | 221.0 | 116 | 496 | 92.7 | 18.7 |
| Bristol | 26.60 | 47.8 | 510 | 12,278 | 13,332.2 | 790.1 | 2,101 | 33,691 | 9,616.4 | 943.3 | 453 | 2,126 | 439.7 | 75.9 |
| Dukes | 3.07 | 177.5 | D | D | D | D | 203 | 1,258 | 430.4 | 56.3 | 73 | 193 | 54.5 | 10.1 |
| Essex | 77.15 | 99.4 | 677 | 10,236 | 16,291.1 | 882.6 | 2,642 | 39,713 | 11,318.8 | 1,173.2 | 689 | 2,962 | 784.5 | 147.4 |
| Franklin | 3.73 | 52.8 | D | D | D | D | 252 | 2,860 | 741.9 | 81.4 | 39 | 101 | 28.8 | 3.0 |
| Hampden | 45.38 | 96.4 | 358 | 6,524 | 5,771.8 | 389.8 | 1,552 | 22,417 | 6,014.7 | 612.5 | 400 | 2,699 | 561.2 | 104.6 |
| Hampshire | 18.09 | 112.2 | 92 | 1,598 | 3,281.8 | 84.5 | 493 | 7,543 | 1,877.8 | 206.5 | 117 | 432 | 95.8 | 15.8 |
| Middlesex | 70.26 | 44.3 | 1,694 | 38,971 | 49,659.9 | 3,834.7 | 5,044 | 81,645 | 24,988.1 | 2,509.4 | 1,845 | 14,522 | 6,350.3 | 977.6 |
| Nantucket | 2.00 | 183.1 | D | D | D | D | 161 | 1,034 | 379.5 | 49.2 | 66 | 148 | 74.9 | 10.2 |
| Norfolk | 35.88 | 51.6 | 834 | 15,675 | 16,604.8 | 1,118.6 | 2,512 | 45,268 | 13,773.2 | 1,385.8 | 937 | 7,751 | 2,323.8 | 547.2 |
| Plymouth | 68.09 | 133.4 | 466 | 6,905 | 8,658.6 | 509.7 | 1,883 | 28,193 | 8,039.7 | 842.9 | 439 | 1,643 | 453.5 | 87.2 |
| Suffolk | 0.00 | 0.0 | 623 | 13,681 | 13,536.5 | 1,594.5 | 2,456 | 36,618 | 13,173.4 | 1,208.1 | 1,337 | 14,358 | 5,477.9 | 1,139.2 |
| Worcester | 251.64 | 307.3 | 740 | 11,938 | 8,117.7 | 739.2 | 2,526 | 39,645 | 13,253.5 | 1,121.5 | 687 | 3,351 | 813.0 | 166.6 |
| **MICHIGAN** | 1,030.44 | 103.8 | 9,173 | 147,939 | 138,545.8 | 9,136.3 | 34,201 | 469,987 | 143,437.1 | 12,481.8 | 8,467 | 54,808 | 17,782.6 | 2,374.4 |
| Alcona | 0.06 | 5.8 | NA | NA | NA | NA | 32 | 291 | 62.9 | 6.8 | D | D | D | 0.2 |
| Alger | 0.41 | 43.7 | NA | NA | NA | NA | 45 | 269 | 65.4 | 6.4 | D | D | D | D |
| Allegan | 4.03 | 35.2 | D | D | D | D | 349 | 3,848 | 1,324.8 | 108.4 | 70 | 233 | 43.3 | 10.4 |
| Alpena | 2.02 | 70.1 | 29 | 524 | 310.1 | 26.2 | 140 | 1,887 | 498.9 | 49.3 | 21 | 124 | 9.8 | 2.6 |
| Antrim | 1.19 | 51.4 | 9 | 41 | 22.7 | 1.9 | 93 | 684 | 188.9 | 16.7 | 21 | 60 | 7.7 | 1.4 |
| Arenac | 42.16 | 2,762.6 | D | D | D | 5.2 | 67 | 473 | 192.0 | 12.0 | D | D | D | D |
| Baraga | 0.51 | 59.5 | NA | NA | NA | NA | 29 | 288 | 50.7 | 6.7 | 26 | 107 | 15.1 | 4.2 |
| Barry | 1.57 | 26.5 | 36 | 256 | 178.8 | 12.4 | 125 | 1,309 | 326.5 | 30.5 | 60 | 182 | 26.3 | 4.4 |
| Bay | 8.97 | 84.9 | 80 | 1,168 | 691.9 | 50.3 | 385 | 5,258 | 1,465.6 | 139.5 | 20 | 32 | 5.1 | 1.0 |
| Benzie | 0.68 | 39.0 | D | D | D | 78.1 | 553 | 7,053 | 1,781.5 | 171.4 | 156 | 656 | 277.4 | 23.3 |
| Berrien | 12.42 | 80.3 | 27 | 381 | 172.8 | 17.3 | 143 | 1,656 | 508.9 | 45.2 | 32 | 71 | 14.8 | 2.1 |
| Branch | 2.77 | 63.4 | 83 | 808 | 1,385.5 | 43.5 | 467 | 6,105 | 1,855.4 | 157.7 | 82 | 387 | 59.1 | 12.5 |
| Calhoun | 12.53 | 93.3 | 31 | 240 | 210.2 | 14.0 | 121 | 871 | 289.7 | 22.5 | 29 | 53 | 13.7 | 1.9 |
| Cass | 1.15 | 22.3 | 12 | 53 | 11.5 | 1.9 | 121 | 892 | 258.0 | 24.8 | 30 | 156 | 26.1 | 6.9 |
| Charlevoix | 2.66 | 101.4 | D | D | D | 3.1 | 129 | 1,206 | 337.3 | 31.7 | D | D | D | 3.2 |
| Cheboygan | 0.98 | 38.5 | 27 | 311 | 89.6 | 10.7 | 144 | 1,641 | 494.1 | 42.8 | 25 | 80 | 12.5 | 2.1 |
| Chippewa | 2.69 | 70.7 | D | D | D | 7.8 | 114 | 1,091 | 306.9 | 28.2 | D | D | D | 0.6 |
| Clare | 1.12 | 36.7 | 52 | 1,007 | 1,281.3 | 53.0 | 173 | 2,430 | 912.3 | 65.0 | 67 | 327 | 63.2 | 12.7 |
| Clinton | 1.16 | 15.0 | 5 | 41 | 21.5 | 1.9 | 45 | 543 | 182.9 | 13.4 | 15 | 29 | 5.2 | 1.2 |
| Crawford | 0.66 | 47.8 | 42 | 395 | 216.2 | 15.7 | 170 | 2,073 | 555.8 | 51.7 | 29 | 56 | 7.5 | 1.3 |
| Delta | 2.81 | 77.2 | D | D | D | D | 137 | 1,725 | 440.9 | 42.3 | 28 | 89 | 13.6 | 2.2 |
| Dickinson | 2.35 | 91.1 | 64 | 1,528 | 1,125.3 | 65.9 | 334 | 5,796 | 1,727.4 | 143.8 | 95 | 416 | 110.7 | 17.8 |
| Eaton | 2.63 | 24.2 | 32 | 268 | 140.3 | 12.9 | 280 | 2,837 | 792.1 | 79.0 | 50 | 210 | 35.7 | 5.8 |
| Emmet | 3.84 | 115.8 | 269 | 4,989 | 3,403.5 | 270.6 | 1,355 | 19,591 | 8,429.7 | 497.7 | 328 | 2,107 | 338.1 | 73.5 |
| Genesee | 4.22 | 10.3 | 11 | 47 | 14.7 | 1.9 | 89 | 677 | 174.2 | 17.2 | 16 | 98 | 9.6 | 3.5 |
| Gladwin | 0.45 | 17.9 | 13 | 63 | 58.8 | 2.5 | 64 | 735 | 164.2 | 16.2 | D | D | D | D |
| Gogebic | 1.48 | 95.9 | 127 | 1,249 | 601.9 | 64.1 | 572 | 8,460 | 2,311.7 | 240.9 | 163 | 591 | 121.8 | 21.6 |
| Grand Traverse | 7.33 | 80.0 | D | D | D | D | 128 | 1,419 | 407.0 | 34.2 | 19 | 53 | 7.1 | 1.4 |
| Gratiot | 1.60 | 38.5 | 30 | 253 | 252.9 | 12.2 | 124 | 1,497 | 415.7 | 37.6 | 21 | 33 | 7.3 | 0.9 |
| Hillsdale | 2.01 | 43.8 | D | D | D | D | 138 | 1,729 | 390.1 | 38.7 | D | D | D | D |
| Houghton | 3.87 | 106.4 | 34 | 398 | 398.4 | 21.6 | 157 | 1,483 | 398.8 | 36.1 | 15 | 24 | 4.1 | 0.6 |
| Huron | 2.58 | 80.9 | 208 | 2,957 | 8,573.0 | 157.5 | 923 | 13,908 | 3,576.0 | 340.5 | 265 | 2,133 | 292.7 | 92.0 |
| Ingham | 26.09 | 91.2 | 23 | 259 | 160.4 | 11.7 | 148 | 1,903 | 528.3 | 43.5 | 23 | 58 | 7.1 | 1.5 |
| Ionia | 4.99 | 77.7 | D | D | D | 0.3 | 111 | 1,296 | 334.9 | 33.1 | 22 | 59 | 7.6 | 1.8 |
| Iosco | 1.20 | 47.3 | 9 | 111 | 18.3 | 2.1 | 55 | 588 | 107.6 | 13.0 | 20 | 47 | 4.2 | 0.8 |
| Iron | 1.26 | 111.0 | 49 | 556 | 289.0 | 23.5 | 211 | 3,491 | 910.6 | 85.8 | 63 | 1,160 | 99.3 | 32.3 |
| Isabella | 3.01 | 42.6 | 128 | 1,327 | 1,023.5 | 78.3 | 497 | 7,174 | 2,031.9 | 187.5 | 109 | 573 | 104.9 | 19.7 |
| Jackson | 10.84 | 68.0 | 243 | 4,193 | 2,158.4 | 235.8 | 876 | 13,882 | 3,706.7 | 350.7 | 224 | 2,412 | 297.0 | 86.6 |
| Kalamazoo | 24.35 | 93.6 | 16 | 192 | 187.6 | 12.7 | 59 | 648 | 186.6 | 15.7 | 8 | 94 | 20.5 | 5.8 |
| Kalkaska | 0.41 | 23.8 | 983 | 24,306 | 19,309.7 | 1,429.8 | 2,214 | 35,681 | 11,021.5 | 981.6 | 688 | 4,153 | 967.0 | 183.1 |
| Kent | 9.01 | 14.2 | | | | | | | | | | | | |

1  Merchant wholesalers, except manufacturers' sales branches and offices.  2. Employer establishments.

## Table B. States and Counties — Professional Services, Manufacturing, and Accommodation and Food Services

| STATE County | Professional, scientific, and technical services, 2017 | | | | Manufacturing, 2017 | | | | Accommodation and food services, 2017 | | | |
|---|---|---|---|---|---|---|---|---|---|---|---|---|
| | Number of establishments | Number of employees | Sales (mil dol) | Average payroll (mil dol) | Number of establishments | Number of employees | Sales (mil dol) | Average payroll (mil dol) | Number of establishments | Number of employees | Sales (mil dol) | Annual payroll (mil dol) |
| | 147 | 148 | 149 | 150 | 151 | 152 | 153 | 154 | 155 | 156 | 157 | 158 |
| MARYLAND— Cont'd | | | | | | | | | | | | |
| St. Mary's | D | D | D | D | 27 | 409.0 | 117.5 | 24.1 | 190 | 3,797 | 215.6 | 60.3 |
| Somerset | 19 | 112 | 13 | 4.9 | 14 | 358.0 | 204.9 | 15.9 | 29 | 401 | 19.5 | 5.1 |
| Talbot | 141 | 1,128 | 212 | 66.1 | 35 | 671.0 | 164.5 | 34.2 | 143 | 2,652 | 158.4 | 49.5 |
| Washington | D | D | D | D | 131 | 5,748 | 3,910.0 | 346.6 | 319 | 5,807 | 314.0 | 90.2 |
| Wicomico | D | D | D | D | 71 | 2,662 | 1,085.1 | 144.0 | 212 | 4,168 | 215.5 | 58.6 |
| Worcester | 150 | 708 | 87 | 34.7 | 39 | 622.0 | 259.9 | 30.2 | 455 | 6,205 | 680.5 | 196.8 |
| Baltimore city | 1,663 | 24,412 | 5,489 | 2,262.5 | 395 | 11,536 | 5,482.7 | 634.0 | 1,542 | 25,190 | 1,980.5 | 550.7 |
| MASSACHUSETTS | 21,741 | 297,565 | 75,347 | 31,147.6 | 6,437 | 231,593 | 82,308.5 | 15,749.4 | 17,773 | 311,058 | 22,892.8 | 6,857.1 |
| Barnstable | D | D | D | D | 188 | 2,046 | 645.3 | 129.0 | 1,112 | 14,179 | 1,266.5 | 386.3 |
| Berkshire | D | D | D | D | 142 | 5,115 | 1,554.7 | 324.2 | 516 | 7,311 | 487.9 | 156.4 |
| Bristol | 1,039 | 5,992 | 859 | 338.7 | 617 | 26,859 | 8,909.8 | 1,643.0 | 1,266 | 21,685 | 1,198.4 | 370.2 |
| Dukes | 65 | 253 | 50 | 16.2 | 22 | 143.0 | 26.7 | 7.8 | 145 | 944 | 159.8 | 47.6 |
| Essex | 2,073 | 15,287 | 3,715 | 1,280.6 | 856 | 39,878 | 12,797.5 | 2,983.5 | 1,886 | 29,227 | 2,008.2 | 599.0 |
| Franklin | D | D | D | D | 107 | 3,534 | 1,683.4 | 199.4 | 149 | 1,856 | 96.1 | 31.6 |
| Hampden | 780 | 6,365 | 944 | 381.4 | 544 | 18,712 | 6,164.1 | 1,077.3 | 912 | 14,904 | 838.3 | 250.6 |
| Hampshire | 362 | 1,726 | 266 | 99.8 | 137 | 2,920 | 848.2 | 163.5 | 397 | 5,847 | 314.3 | 105.2 |
| Middlesex | 6,909 | 125,269 | 32,571 | 14,474.8 | 1,550 | 58,446 | 22,733.0 | 4,509.7 | 3,971 | 72,065 | 5,085.2 | 1,556.3 |
| Nantucket | D | D | D | D | 14 | 59.0 | 15.3 | 3.0 | 124 | 1,168 | 182.8 | 48.0 |
| Norfolk | 2,659 | 22,041 | 5,047 | 1,901.4 | 565 | 18,299 | 8,217.9 | 1,201.5 | 1,679 | 30,121 | 2,117.0 | 615.0 |
| Plymouth | 1,242 | 10,104 | 1,544 | 605.1 | 449 | 10,348 | 2,659.8 | 605.0 | 1,142 | 20,602 | 1,235.5 | 396.9 |
| Suffolk | 3,519 | 83,941 | 24,824 | 9,761.7 | 319 | 10,594 | 4,079.7 | 623.1 | 2,680 | 64,498 | 6,270.4 | 1,821.6 |
| Worcester | 1,833 | 18,742 | 4,101 | 1,706.7 | 927 | 34,640 | 11,973.1 | 2,279.4 | 1,794 | 26,651 | 1,632.5 | 472.5 |
| MICHIGAN | 21,713 | 280,247 | 39,125 | 19,529.8 | 12,418 | 582,365 | 262,495.4 | 33,349.3 | 20,696 | 399,032 | 23,056.4 | 6,562.7 |
| Alcona | D | D | 6 | D | 17 | 200.0 | 28.5 | 9.8 | 18 | 75 | 5.4 | 1.4 |
| Alger | 10 | 28 | 3 | 1.5 | D | 445.0 | D | | 46 | 309 | 28.7 | 6.8 |
| Allegan | D | D | D | D | 223 | 14,158 | 5,606.3 | 708.2 | 216 | 3,006 | 167.8 | 49.8 |
| Alpena | D | D | D | D | 50 | 1,304 | 416.8 | 75.1 | 74 | 958 | 46.0 | 13.4 |
| Antrim | 31 | 112 | 11 | 3.9 | 42 | 954.0 | 199.7 | 47.0 | 66 | 1,124 | 56.7 | 19.8 |
| Arenac | D | D | 21 | D | 30 | 910.0 | 200.1 | 44.1 | 46 | 417 | 25.2 | 7.1 |
| Baraga | 3 | 16 | 1 | 0.5 | 19 | 439.0 | 133.6 | 23.1 | 21 | 190 | 6.5 | 2.0 |
| Barry | 61 | 308 | 28 | 12.5 | 58 | 3,449 | 1,555.9 | 171.7 | 76 | 1,158 | 52.1 | 16.1 |
| Bay | 143 | 1,201 | 100 | 51.5 | 116 | 3,824 | 1,209.0 | 242.1 | 225 | 3,858 | 176.3 | 50.1 |
| Benzie | 29 | 69 | 7 | 2.4 | 22 | 418.0 | 115.3 | 18.3 | 63 | 1,136 | 74.4 | 24.0 |
| Berrien | D | D | D | D | 281 | 8,819 | 2,277.9 | 455.5 | 379 | 6,404 | 331.2 | 101.2 |
| Branch | 47 | 580 | 119 | 34.8 | 66 | 2,550 | 813.3 | 122.4 | 82 | 1,264 | 63.7 | 16.6 |
| Calhoun | D | D | D | D | 149 | 14,437 | 6,760.7 | 781.4 | 272 | 6,655 | 567.1 | 126.1 |
| Cass | 52 | 183 | 17 | 6.8 | 68 | 2,697 | 870.7 | 123.4 | 64 | 800 | 34.5 | 9.6 |
| Charlevoix | 60 | 222 | 26 | 9.2 | 42 | 2,247 | 550.4 | 118.4 | 73 | 1,665 | 85.1 | 28.8 |
| Cheboygan | 32 | 103 | 12 | 4.3 | 37 | 294.0 | 59.8 | 12.1 | 122 | 825 | 92.6 | 21.6 |
| Chippewa | D | D | D | D | 28 | 450.0 | 79.0 | 20.5 | 104 | 3,581 | 401.1 | 97.0 |
| Clare | 33 | 162 | 22 | 7.8 | 24 | 804.0 | 249.5 | 40.1 | 69 | 879 | 42.2 | 11.5 |
| Clinton | 123 | 940 | 121 | 41.7 | 49 | 2,150 | 706.4 | 105.2 | 113 | 1,645 | 76.2 | 21.1 |
| Crawford | 22 | 72 | 23 | 4.1 | 21 | 563.0 | 251.5 | 32.2 | 43 | 525 | 30.3 | 8.5 |
| Delta | D | D | D | D | 66 | 1,959 | 815.9 | 131.0 | 102 | 1,304 | 53.1 | 16.4 |
| Dickinson | 53 | 377 | 46 | 16.1 | 34 | 1,965 | 866.8 | 111.8 | 71 | 855 | 37.9 | 10.5 |
| Eaton | 172 | 1,051 | 160 | 62.3 | 95 | 7,369 | 8,953.9 | 414.3 | 220 | 4,479 | 228.3 | 67.0 |
| Emmet | 115 | 458 | 71 | 29.5 | 63 | 1,302 | 331.1 | 64.7 | 153 | 2,701 | 159.9 | 55.5 |
| Genesee | D | D | D | D | 279 | 11,516 | 13,080.7 | 738.1 | 730 | 14,666 | 707.3 | 209.7 |
| Gladwin | 17 | 49 | 5 | 1.5 | 26 | 1,027 | 266.9 | 53.7 | 39 | 461 | 20.9 | 5.8 |
| Gogebic | 17 | 94 | 7 | 3.2 | 21 | 661.0 | 103.0 | 28.2 | 51 | 960 | 51.2 | 13.8 |
| Grand Traverse | D | D | D | D | 187 | 4,933 | 1,197.0 | 237.4 | 282 | 6,108 | 441.9 | 125.7 |
| Gratiot | 27 | 247 | 22 | 10.1 | 51 | 2,378 | 695.6 | 129.2 | 67 | 1,013 | 45.8 | 13.9 |
| Hillsdale | D | D | D | D | 72 | 3,611 | 1,471.4 | 184.5 | 65 | 856 | 39.6 | 11.1 |
| Houghton | D | D | D | D | 42 | 746.0 | 215.4 | 27.5 | 106 | 1,630 | 63.7 | 18.3 |
| Huron | 45 | 259 | 41 | 20.8 | 60 | 3,993 | 1,091.9 | 174.6 | 98 | 880 | 42.5 | 11.5 |
| Ingham | 704 | 6,528 | 1,092 | 413.6 | 195 | 8,591 | 5,338.5 | 531.2 | 648 | 12,887 | 607.9 | 189.8 |
| Ionia | 46 | 162 | 17 | 7.1 | 70 | 3,746 | 1,252.0 | 176.7 | 79 | 1,217 | 51.5 | 15.4 |
| Iosco | 31 | 181 | 12 | 7.2 | 32 | 1,143 | 395.3 | 49.5 | 83 | 796 | 39.5 | 11.2 |
| Iron | 29 | 109 | 13 | 4.9 | 17 | 526.0 | 111.5 | 21.8 | 35 | 345 | 17.3 | 4.2 |
| Isabella | D | D | D | D | 55 | 2,814 | 773.0 | 112.5 | 140 | 6,331 | 444.1 | 114.9 |
| Jackson | D | D | D | D | 248 | 9,000 | 3,017.6 | 486.8 | 272 | 4,870 | 243.9 | 69.4 |
| Kalamazoo | 543 | 4,855 | 723 | 282.0 | 294 | 17,105 | 8,166.0 | 1,137.7 | 598 | 13,151 | 622.3 | 198.8 |
| Kalkaska | 27 | 74 | 15 | 3.9 | 13 | 436.0 | 86.0 | 20.7 | 31 | 482 | 24.6 | 6.8 |
| Kent | D | D | D | D | 1,106 | 65,939 | 18,311.2 | 3,627.5 | 1,326 | 30,415 | 1,540.6 | 488.4 |

# Table B. States and Counties — Health Care and Social Assistance, Other Services, Nonemployer Businesses, and Residential Construction

| STATE County | Health care and social assistance, 2017 | | | | Other services, 2017 | | | | Nonemployer businesses, 2018 | | Value of residential construction authorized by building permits, 2020 | |
|---|---|---|---|---|---|---|---|---|---|---|---|---|
| | Number of establishments | Number of employees | Receipts (mil dol) | Annual payroll (mil dol) | Number of establishments | Number of employees | Receipts (mil dol) | Annual payroll (mil dol) | Number | Receipts (mil dol) | New construction ($1,000) | Number of housing units |
| | 159 | 160 | 161 | 162 | 163 | 164 | 165 | 166 | 167 | 168 | 169 | 170 |
| MARYLAND— Cont'd | | | | | | | | | | | | |
| St. Mary's | 181 | 4,735 | 528.0 | 201.5 | 140 | 816 | 83.4 | 29.8 | 6,998 | 297.4 | 90,238 | 555 |
| Somerset | 43 | 1,156 | 75.4 | 35.1 | 29 | 90 | 18.3 | 2.1 | 1,449 | 49.1 | 3,740 | 20 |
| Talbot | 184 | 4,341 | 567.2 | 194.8 | 124 | 669 | 112.5 | 23.3 | 8,918 | 409.0 | 20,164 | 56 |
| Washington | 439 | 9,539 | 1,088.1 | 442.5 | 256 | 1,706 | 148.6 | 52.2 | 6,531 | 302.7 | 63,514 | 284 |
| Wicomico | 339 | 8,973 | 1,010.0 | 429.5 | 188 | 1,356 | 137.3 | 41.5 | 5,309 | 281.0 | 48,792 | 300 |
| Worcester | 153 | 2,425 | 259.8 | 108.7 | 143 | 808 | 89.7 | 26.6 | 39,609 | 1,527.7 | 63,808 | 265 |
| Baltimore city | 1,494 | 77,508 | 12,756.8 | 4,397.6 | 996 | 9,183 | 1,395.2 | 359.6 | 39,609 | 1,527.7 | 286,444 | 1,621 |
| MASSACHUSETTS | 19,349 | 635,012 | 74,024.0 | 32,038.2 | 14,810 | 100,088 | 13,093.7 | 3,605.7 | 573,754 | 30,907.1 | 4,761,486 | 17,025 |
| Barnstable | 795 | 17,032 | 2,064.8 | 858.8 | 627 | 3,498 | 395.6 | 125.5 | 27,281 | 1,451.8 | 274,179 | 592 |
| Berkshire | 436 | 11,773 | 1,277.1 | 569.5 | 267 | 1,591 | 166.8 | 44.2 | 36,012 | 1,720.6 | 61,893 | 211 |
| Bristol | 1,434 | 41,959 | 4,402.1 | 1,912.9 | 1,112 | 5,879 | 561.0 | 170.1 | 4,174 | 270.0 | 140,521 | 694 |
| Dukes | 57 | 879 | 132.8 | 55.7 | 72 | 288 | 42.6 | 11.7 | 69,827 | 3,607.8 | 89,936 | 124 |
| Essex | 2,195 | 63,832 | 6,148.5 | 2,781.5 | 1,625 | 9,004 | 868.2 | 260.1 | 6,331 | 231.2 | 315,104 | 1,359 |
| Franklin | 178 | 3,965 | 325.3 | 145.5 | 137 | 508 | 50.6 | 15.2 | 26,567 | 1,276.2 | 20,280 | 91 |
| Hampden | 1,202 | 43,673 | 4,805.2 | 2,045.0 | 759 | 5,779 | 676.4 | 173.4 | 13,605 | 576.5 | 83,499 | 320 |
| Hampshire | 452 | 9,667 | 965.6 | 454.2 | 336 | 1,803 | 212.2 | 64.9 | 148,625 | 8,418.1 | 76,684 | 349 |
| Middlesex | 4,791 | 121,315 | 13,898.7 | 6,000.2 | 3,649 | 24,713 | 3,487.9 | 1,002.5 | 2,796 | 224.8 | 820,959 | 4,443 |
| Nantucket | 31 | 521 | 101.1 | 30.6 | 47 | 235 | 36.8 | 10.9 | 63,886 | 3,979.0 | 116,971 | 198 |
| Norfolk | 2,423 | 78,250 | 6,218.9 | 2,910.3 | 1,629 | 11,232 | 1,496.5 | 414.1 | 63,886 | 3,979.0 | 451,192 | 1,924 |
| Plymouth | 1,284 | 35,925 | 3,535.6 | 1,545.9 | 1,069 | 6,594 | 635.7 | 200.5 | 41,058 | 2,195.2 | 317,992 | 1,343 |
| Suffolk | 1,943 | 137,719 | 21,688.3 | 9,447.6 | 2,020 | 20,551 | 3,556.6 | 849.9 | 65,274 | 3,663.9 | 1,638,916 | 3,696 |
| Worcester | 2,128 | 68,502 | 8,460.0 | 3,280.6 | 1,461 | 8,413 | 906.7 | 262.9 | 57,923 | 2,819.2 | 353,361 | 1,681 |
| MICHIGAN | 26,977 | 627,808 | 74,194.5 | 29,309.9 | 16,545 | 104,291 | 13,205.0 | 3,415.1 | 730,700 | 32,700.7 | 12,579 | 67 |
| Alcona | D | D | D | 6.8 | D | D | D | 0.4 | 638 | 24.1 | 2,914 | 29 |
| Alger | 18 | 251 | 22.2 | 10.6 | 15 | 35 | 6.4 | 1.2 | 570 | 22.7 | 4,847 | 19 |
| Allegan | 180 | 3,346 | 273.4 | 120.0 | 178 | 784 | 92.4 | 24.2 | 8,324 | 376.5 | 114,649 | 382 |
| Alpena | 83 | 2,673 | 277.0 | 107.7 | 71 | 389 | 39.6 | 8.3 | 1,875 | 75.9 | 5,076 | 28 |
| Antrim | 49 | 303 | 20.7 | 9.9 | 41 | 157 | 17.8 | 5.2 | 2,153 | 92.1 | 15,035 | 82 |
| Arenac | 35 | 589 | 48.8 | 19.6 | 22 | 67 | 5.8 | 1.9 | 936 | 45.9 | 2,853 | 12 |
| Baraga | 16 | 280 | 38.4 | 12.2 | 16 | 37 | 5.4 | 0.9 | 355 | 15.4 | 3,323 | 14 |
| Barry | 87 | 1,323 | 130.0 | 52.6 | 90 | 462 | 57.9 | 14.6 | 3,848 | 179.1 | 42,123 | 188 |
| Bay | 339 | 6,674 | 706.0 | 257.4 | 168 | 900 | 81.1 | 23.0 | 5,702 | 229.2 | 23,850 | 135 |
| Benzie | 27 | 367 | 36.0 | 15.7 | D | D | D | D | 1,751 | 73.1 | 20,921 | 88 |
| Berrien | 387 | 9,499 | 987.6 | 414.4 | 258 | 1,325 | 158.5 | 40.3 | 10,074 | 422.5 | 87,516 | 260 |
| Branch | 89 | 1,663 | 136.1 | 60.0 | 56 | 261 | 19.2 | 7.1 | 2,495 | 135.5 | 10,051 | 34 |
| Calhoun | 321 | 9,318 | 1,138.8 | 443.5 | 215 | 1,304 | 519.5 | 57.0 | 6,565 | 255.1 | 21,785 | 90 |
| Cass | 65 | 1,074 | 75.8 | 35.7 | 56 | 291 | 29.6 | 9.4 | 3,039 | 137.8 | 43,674 | 269 |
| Charlevoix | 69 | 1,322 | 134.2 | 62.0 | 60 | 211 | 31.5 | 7.1 | 2,645 | 125.7 | 25,196 | 106 |
| Cheboygan | 51 | 828 | 96.2 | 29.3 | 72 | 197 | 22.7 | 5.5 | 2,000 | 78.9 | 15,209 | 67 |
| Chippewa | 83 | 2,113 | 205.6 | 80.2 | 67 | 224 | 22.1 | 5.0 | 1,932 | 52.4 | 7,766 | 39 |
| Clare | 74 | 1,197 | 114.3 | 38.3 | 37 | 173 | 14.7 | 3.6 | 1,784 | 83.2 | 15,116 | 69 |
| Clinton | 113 | 1,926 | 163.4 | 66.0 | 101 | 598 | 50.7 | 16.1 | 5,626 | 252.1 | 64,602 | 341 |
| Crawford | 33 | 973 | 113.3 | 45.5 | 17 | 71 | 7.2 | 1.5 | 904 | 37.9 | 7,017 | 37 |
| Delta | 102 | 1,958 | 201.0 | 63.2 | 89 | 410 | 45.4 | 10.5 | 1,967 | 67.3 | 6,681 | 35 |
| Dickinson | 101 | 2,673 | 361.1 | 155.6 | 63 | 214 | 23.9 | 5.6 | 1,489 | 55.2 | 7,303 | 34 |
| Eaton | 234 | 3,775 | 342.5 | 153.0 | 170 | 1,186 | 154.3 | 49.0 | 7,159 | 284.5 | 33,187 | 115 |
| Emmet | 177 | 3,125 | 423.2 | 164.3 | 107 | 554 | 68.3 | 19.9 | 3,592 | 169.4 | 33,414 | 84 |
| Genesee | 1,286 | 27,955 | 3,165.7 | 1,254.4 | 569 | 3,886 | 606.1 | 116.9 | 27,859 | 1,060.7 | 112,570 | 485 |
| Gladwin | 38 | 699 | 62.5 | 19.8 | 28 | 154 | 12.2 | 3.2 | 1,589 | 84.5 | 18,544 | 78 |
| Gogebic | 34 | 645 | 86.6 | 31.6 | 37 | 111 | 11.1 | 2.8 | 892 | 31.7 | 3,491 | 20 |
| Grand Traverse | 417 | 9,529 | 1,290.2 | 490.1 | 234 | 1,446 | 179.5 | 50.4 | 9,954 | 477.7 | 84,152 | 467 |
| Gratiot | 107 | 2,362 | 252.5 | 90.3 | 46 | 201 | 42.8 | 7.2 | 2,206 | 97.3 | 7,080 | 33 |
| Hillsdale | 94 | 1,325 | 122.3 | 48.4 | 53 | 273 | 19.8 | 5.4 | 2,810 | 118.4 | 18,057 | 75 |
| Houghton | 91 | 1,935 | 214.6 | 85.3 | D | D | D | D | 2,001 | 65.7 | 9,468 | 53 |
| Huron | 94 | 1,911 | 199.5 | 70.0 | 53 | 182 | 21.1 | 4.9 | 2,244 | 102.2 | 11,051 | 35 |
| Ingham | 783 | 22,927 | 2,730.5 | 1,083.3 | 560 | 4,929 | 768.1 | 211.6 | 19,711 | 936.5 | 90,182 | 571 |
| Ionia | 98 | 1,294 | 127.5 | 52.0 | 67 | 304 | 29.0 | 7.8 | 3,273 | 127.0 | 18,174 | 76 |
| Iosco | 61 | 1,007 | 85.9 | 35.9 | 50 | 179 | 14.0 | 3.5 | 1,530 | 53.2 | 8,930 | 59 |
| Iron | 26 | 418 | 55.1 | 19.9 | 25 | 152 | 11.1 | 3.2 | 756 | 26.1 | 8,102 | 138 |
| Isabella | 218 | 3,103 | 263.8 | 95.6 | 105 | 562 | 59.7 | 16.2 | 3,691 | 165.5 | 13,460 | 105 |
| Jackson | 361 | 9,998 | 1,123.1 | 492.2 | 211 | 1,210 | 152.8 | 37.1 | 9,115 | 364.1 | 33,347 | 159 |
| Kalamazoo | 685 | 21,274 | 2,818.2 | 1,154.6 | 421 | 3,340 | 410.2 | 101.4 | 16,752 | 730.2 | 126,214 | 496 |
| Kalkaska | 21 | 551 | 54.1 | 22.9 | D | D | D | D | 1,221 | 46.7 | 6,942 | 38 |
| Kent | 1,650 | 55,229 | 6,984.5 | 2,609.0 | 1,204 | 9,907 | 1,004.5 | 293.0 | 49,636 | 2,475.7 | 445,802 | 1,692 |

# Government Employment and Payroll, and Local Government Finances

| STATE County | Government employment and payroll, 2017 | | | | | | | | | Local government finances, 2017 | | | | |
| | | | March payroll (percent of total) | | | | | | | General revenue | | | | |
| | | | | | | | | | | | | Taxes | | |
| | | | | | | | | | | | | | Per capita[1] (dollars) | |
| | Full-time equivalent employees | March payroll (dollars) | Administration, judicial, and legal | Police and corrections | Fire protection | Highways and transportation | Health and welfare | Natural resources and utilities | Education and libraries | Total (mil dol) | Inter-govern-mental | Total (mil dol) | Total | Property |
|---|---|---|---|---|---|---|---|---|---|---|---|---|---|---|
| | 171 | 172 | 173 | 174 | 175 | 176 | 177 | 178 | 179 | 180 | 181 | 182 | 183 | 184 |
| MARYLAND— Cont'd | | | | | | | | | | | | | | |
| St. Mary's | 3,228 | 15,729,729 | 6.4 | 10.5 | 0.0 | 2.2 | 1.2 | 4.4 | 72.5 | 423.1 | 139.9 | 222.2 | 1,975 | 1,009 |
| Somerset | 889 | 3,682,461 | 7.3 | 9.6 | 0.0 | 2.7 | 3.0 | 4.9 | 69.6 | 92.9 | 54.5 | 28.0 | 1,080 | 760 |
| Talbot | 1,296 | 6,122,919 | 6.6 | 12.4 | 0.0 | 3.5 | 4.0 | 9.0 | 52.5 | 172.1 | 31.5 | 98.6 | 2,662 | 1,398 |
| Washington | 5,565 | 24,234,522 | 4.1 | 6.2 | 1.9 | 3.1 | 2.4 | 5.3 | 74.2 | 591.0 | 251.1 | 256.8 | 1,710 | 1,016 |
| Wicomico | 4,036 | 17,547,154 | 4.4 | 11.8 | 1.9 | 1.8 | 0.5 | 4.2 | 73.8 | 428.6 | 210.3 | 155.3 | 1,519 | 894 |
| Worcester | 2,840 | 13,064,394 | 8.0 | 16.4 | 1.1 | 3.3 | 3.9 | 12.3 | 51.0 | 374.0 | 45.5 | 260.2 | 5,036 | 3,501 |
| Baltimore city | 26,369 | 141,757,363 | 6.5 | 17.3 | 8.5 | 3.0 | 6.2 | 8.5 | 48.1 | 3,818.8 | 1,751.2 | 1,488.2 | 2,438 | 1,405 |
| MASSACHUSETTS | X | X | X | X | X | X | X | X | X | X | X | X | X | X |
| Barnstable | 8,286 | 44,597,245 | 6.6 | 11.6 | 9.9 | 4.4 | 3.2 | 6.7 | 55.0 | 1,119.8 | 177.1 | 766.0 | 3,587 | 3,359 |
| Berkshire | 4,678 | 20,287,892 | 4.8 | 7.3 | 3.1 | 4.6 | 2.3 | 4.1 | 72.2 | 589.6 | 246.5 | 294.0 | 2,326 | 2,238 |
| Bristol | 18,103 | 90,434,250 | 3.1 | 10.3 | 6.8 | 2.2 | 3.5 | 3.8 | 68.5 | 2,292.9 | 968.1 | 1,022.4 | 1,823 | 1,745 |
| Dukes | 1,098 | 5,345,754 | 7.2 | 10.0 | 3.4 | 6.9 | 5.0 | 6.0 | 59.2 | 167.4 | 27.3 | 113.2 | 6,537 | 6,246 |
| Essex | 24,071 | 128,610,626 | 3.6 | 9.1 | 6.8 | 2.6 | 3.1 | 4.9 | 67.7 | 3,461.2 | 1,235.5 | 1,798.7 | 2,293 | 2,216 |
| Franklin | 2,970 | 13,630,074 | 4.9 | 6.3 | 3.2 | 4.7 | 2.3 | 3.6 | 64.2 | 338.8 | 122.6 | 164.9 | 2,336 | 2,290 |
| Hampden | 19,157 | 91,093,159 | 3.1 | 9.6 | 6.1 | 2.4 | 2.6 | 7.0 | 67.4 | 2,085.5 | 1,015.3 | 830.3 | 1,775 | 1,714 |
| Hampshire | 6,083 | 29,149,694 | 5.0 | 9.8 | 7.6 | 3.2 | 3.1 | 4.9 | 64.1 | 538.9 | 184.5 | 288.8 | 1,793 | 1,734 |
| Middlesex | 56,125 | 305,641,359 | 3.9 | 8.9 | 6.9 | 2.7 | 8.1 | 3.7 | 63.6 | 8,055.5 | 2,092.0 | 4,661.9 | 2,907 | 2,786 |
| Nantucket | 584 | 3,800,935 | 4.9 | 8.7 | 5.6 | 10.9 | 11.6 | 6.4 | 48.8 | 151.4 | 8.5 | 103.3 | 9,217 | 6,940 |
| Norfolk | 23,735 | 132,202,384 | 3.4 | 8.6 | 7.1 | 2.7 | 2.3 | 4.9 | 68.5 | 3,218.7 | 670.2 | 2,090.6 | 2,984 | 2,886 |
| Plymouth | 18,407 | 93,269,080 | 3.7 | 8.6 | 6.3 | 2.6 | 2.0 | 4.1 | 71.1 | 2,256.2 | 774.2 | 1,251.0 | 2,427 | 2,343 |
| Suffolk | 23,532 | 152,449,799 | 3.5 | 19.7 | 10.7 | 1.8 | 8.7 | 4.7 | 49.6 | 4,645.7 | 1,501.5 | 2,556.7 | 3,193 | 2,870 |
| Worcester | 24,532 | 121,122,757 | 4.7 | 10.4 | 6.6 | 3.5 | 2.8 | 4.1 | 66.5 | 3,385.5 | 1,371.0 | 1,610.5 | 1,950 | 1,895 |
| MICHIGAN | X | X | X | X | X | X | X | X | X | X | X | X | X | X |
| Alcona | 217 | 864,091 | 20.4 | 10.7 | 3.7 | 11.9 | 6.8 | 1.7 | 38.5 | 25.4 | 7.3 | 12.7 | 1,235 | 1,233 |
| Alger | 306 | 1,216,793 | 11.3 | 7.1 | 0.5 | 17.9 | 3.3 | 3.2 | 54.9 | 36.7 | 17.4 | 10.4 | 1,132 | 1,101 |
| Allegan | 3,057 | 11,712,420 | 9.5 | 7.1 | 1.1 | 3.0 | 8.4 | 1.9 | 68.3 | 367.8 | 194.7 | 125.2 | 1,076 | 1,055 |
| Alpena | 1,467 | 5,723,752 | 7.4 | 3.5 | 3.5 | 7.0 | 20.8 | 1.0 | 55.8 | 150.7 | 89.0 | 30.4 | 1,068 | 1,052 |
| Antrim | 742 | 2,734,139 | 16.9 | 7.3 | 2.7 | 7.8 | 3.5 | 3.7 | 55.3 | 99.8 | 25.6 | 45.0 | 1,935 | 1,900 |
| Arenac | 374 | 1,470,122 | 17.0 | 6.7 | 1.8 | 6.6 | 0.0 | 2.9 | 60.6 | 45.5 | 22.2 | 16.0 | 1,066 | 1,032 |
| Baraga | 389 | 1,704,033 | 10.1 | 3.6 | 0.0 | 8.2 | 41.8 | 7.9 | 27.9 | 58.7 | 19.2 | 9.1 | 1,083 | 1,081 |
| Barry | 1,420 | 5,620,614 | 8.6 | 7.3 | 1.3 | 4.7 | 18.0 | 2.0 | 55.8 | 181.7 | 83.4 | 51.6 | 850 | 842 |
| Bay | 4,219 | 17,395,958 | 6.6 | 5.0 | 2.0 | 4.1 | 15.6 | 3.4 | 61.1 | 526.0 | 268.5 | 126.8 | 1,219 | 1,193 |
| Benzie | 469 | 1,875,231 | 12.3 | 7.1 | 1.4 | 12.6 | 9.4 | 4.2 | 50.6 | 66.4 | 17.9 | 30.5 | 1,728 | 1,707 |
| Berrien | 4,916 | 20,205,038 | 9.5 | 10.0 | 1.7 | 2.9 | 2.8 | 4.4 | 66.9 | 642.6 | 301.4 | 220.9 | 1,433 | 1,406 |
| Branch | 1,900 | 7,881,188 | 5.8 | 4.4 | 1.1 | 3.6 | 31.6 | 3.8 | 47.3 | 254.2 | 102.6 | 44.6 | 1,028 | 1,012 |
| Calhoun | 3,902 | 17,403,464 | 10.1 | 10.3 | 3.5 | 2.2 | 6.9 | 4.6 | 59.9 | 615.2 | 302.7 | 179.7 | 1,339 | 1,163 |
| Cass | 1,414 | 5,442,148 | 10.0 | 6.8 | 0.6 | 3.7 | 1.8 | 1.9 | 73.7 | 186.7 | 94.4 | 53.6 | 1,041 | 1,023 |
| Charlevoix | 1,353 | 5,458,888 | 5.4 | 4.4 | 0.5 | 5.1 | 27.2 | 3.3 | 53.3 | 180.2 | 45.3 | 72.1 | 2,751 | 2,725 |
| Cheboygan | 697 | 2,909,603 | 14.7 | 8.9 | 0.6 | 7.8 | 0.3 | 2.0 | 64.5 | 91.8 | 36.3 | 40.4 | 1,587 | 1,582 |
| Chippewa | 1,232 | 4,686,385 | 10.5 | 7.1 | 2.4 | 13.2 | 5.6 | 4.4 | 55.0 | 152.7 | 72.6 | 38.6 | 1,023 | 1,019 |
| Clare | 1,153 | 4,629,928 | 8.0 | 5.4 | 1.5 | 4.4 | 0.6 | 1.4 | 72.7 | 141.7 | 76.8 | 32.4 | 1,061 | 1,048 |
| Clinton | 1,641 | 6,877,274 | 11.2 | 9.5 | 0.9 | 3.5 | 0.2 | 3.0 | 68.9 | 215.8 | 119.6 | 63.4 | 807 | 787 |
| Crawford | 385 | 1,534,085 | 16.4 | 10.7 | 2.3 | 18.1 | 4.6 | 1.4 | 45.3 | 48.3 | 23.6 | 15.5 | 1,117 | 1,061 |
| Delta | 1,271 | 5,378,133 | 8.5 | 7.7 | 0.2 | 5.4 | 3.5 | 6.3 | 67.0 | 145.1 | 78.3 | 40.9 | 1,139 | 1,131 |
| Dickinson | 1,731 | 7,856,461 | 4.0 | 4.6 | 0.7 | 2.6 | 57.2 | 1.7 | 28.2 | 210.6 | 44.2 | 36.4 | 1,431 | 1,416 |
| Eaton | 2,770 | 11,226,348 | 8.4 | 8.8 | 3.4 | 4.1 | 2.5 | 4.2 | 65.6 | 336.6 | 166.9 | 106.1 | 969 | 931 |
| Emmet | 1,178 | 5,264,795 | 10.8 | 7.1 | 0.7 | 4.9 | 19.9 | 5.2 | 48.0 | 200.9 | 72.4 | 75.4 | 2,280 | 2,243 |
| Genesee | 14,500 | 64,991,149 | 5.0 | 5.5 | 1.6 | 4.1 | 26.9 | 2.5 | 53.4 | 2,164.1 | 1,125.9 | 370.3 | 909 | 840 |
| Gladwin | 549 | 2,201,794 | 12.0 | 6.1 | 1.2 | 11.4 | 6.3 | 2.2 | 57.5 | 67.0 | 32.8 | 22.3 | 883 | 865 |
| Gogebic | 575 | 2,483,667 | 10.5 | 6.5 | 0.3 | 10.1 | 2.0 | 7.0 | 61.2 | 90.5 | 30.2 | 18.3 | 1,192 | 1,182 |
| Grand Traverse | 4,372 | 17,331,924 | 6.7 | 5.0 | 1.4 | 4.8 | 17.7 | 2.9 | 60.1 | 502.7 | 207.2 | 165.9 | 1,807 | 1,786 |
| Gratiot | 1,416 | 5,098,790 | 9.6 | 6.3 | 0.3 | 5.3 | 0.8 | 3.1 | 74.1 | 159.2 | 86.2 | 46.5 | 1,134 | 1,122 |
| Hillsdale | 1,043 | 4,049,058 | 9.8 | 9.6 | 1.6 | 5.7 | 0.1 | 3.2 | 68.1 | 140.9 | 70.0 | 30.6 | 668 | 660 |
| Houghton | 998 | 3,962,541 | 10.2 | 6.3 | 0.4 | 9.5 | 6.1 | 4.2 | 62.9 | 165.5 | 82.9 | 33.9 | 936 | 924 |
| Huron | 1,157 | 4,639,308 | 11.7 | 6.8 | 0.4 | 6.0 | 16.2 | 3.7 | 53.9 | 157.0 | 54.5 | 62.4 | 1,993 | 1,964 |
| Ingham | 10,811 | 51,980,691 | 7.4 | 6.7 | 3.2 | 4.5 | 11.9 | 9.4 | 51.2 | 1,387.5 | 599.7 | 470.9 | 1,612 | 1,451 |
| Ionia | 1,907 | 7,894,357 | 7.0 | 6.0 | 0.4 | 4.0 | 2.3 | 2.1 | 77.5 | 202.0 | 118.5 | 50.4 | 785 | 724 |
| Iosco | 1,235 | 4,090,829 | 9.1 | 3.9 | 0.6 | 4.5 | 35.2 | 1.5 | 43.3 | 129.2 | 59.8 | 34.6 | 1,377 | 1,364 |
| Iron | 348 | 1,361,315 | 16.8 | 8.6 | 0.0 | 10.2 | 11.0 | 10.3 | 42.9 | 73.5 | 40.3 | 17.4 | 1,565 | 1,552 |
| Isabella | 1,787 | 6,981,989 | 9.1 | 6.1 | 1.3 | 6.4 | 29.6 | 3.6 | 41.9 | 255.4 | 159.1 | 49.4 | 695 | 679 |
| Jackson | 4,580 | 19,841,345 | 6.9 | 7.2 | 1.5 | 4.1 | 4.5 | 2.4 | 72.5 | 658.3 | 378.1 | 159.5 | 1,006 | 930 |
| Kalamazoo | 7,505 | 31,550,062 | 7.4 | 11.7 | 2.4 | 2.8 | 2.4 | 2.9 | 69.1 | 1,031.0 | 525.6 | 328.5 | 1,249 | 1,210 |
| Kalkaska | 724 | 2,976,208 | 6.8 | 4.6 | 0.3 | 5.6 | 57.2 | 2.9 | 22.2 | 89.4 | 19.8 | 20.4 | 1,163 | 1,142 |
| Kent | 17,290 | 79,275,077 | 6.9 | 10.0 | 2.7 | 5.0 | 2.2 | 3.7 | 66.9 | 2,821.1 | 1,438.1 | 862.3 | 1,328 | 1,121 |

1. Based on the resident population estimated as of July 1 of the year shown.

# Table B. States and Counties — Local Government Finances, Government Employment, and Income Taxes

| STATE County | Total (mil dol) | Per capita[1] (dollars) | Education | Health and hospitals | Police protection | Public welfare | Highways | Debt Total (mil dol) | Debt Per capita[1] (dollars) | Federal civilian | Federal military | State and local | Number of returns | Mean adjusted gross income | Mean income tax |
|---|---|---|---|---|---|---|---|---|---|---|---|---|---|---|---|
| | 185 | 186 | 187 | 188 | 189 | 190 | 191 | 192 | 193 | 194 | 195 | 196 | 197 | 198 | 199 |
| **MARYLAND— Cont'd** | | | | | | | | | | | | | | | |
| St. Mary's | 392 | 3,481 | 59.9 | 2.0 | 7.1 | 0.0 | 4.1 | 366.3 | 3,256 | 10,334 | 2,504 | 4,928 | 53,080 | 81,417 | 9,381 |
| Somerset | 149 | 5,769 | 33.4 | 0.7 | 3.2 | 0.0 | 2.7 | 65.7 | 2,537 | 52 | 102 | 2,778 | 8,680 | 43,750 | 3,316 |
| Talbot | 143 | 3,865 | 39.6 | 2.0 | 8.1 | 0.0 | 7.4 | 118.3 | 3,194 | 195 | 139 | 1,709 | 19,290 | 101,404 | 14,466 |
| Washington | 584 | 3,886 | 57.2 | 0.7 | 4.8 | 0.0 | 4.8 | 307.7 | 3,009 | 310 | 327 | 7,785 | 46,720 | 52,535 | 5,038 |
| Wicomico | 447 | 4,375 | 55.5 | 1.0 | 6.2 | 1.9 | 4.3 | 247.2 | 4,785 | 189 | 205 | 3,440 | 29,660 | 64,637 | 7,606 |
| Worcester | 355 | 6,874 | 33.4 | 1.9 | 8.2 | 0.0 | 4.4 | 1,009.0 | 1,653 | 10,712 | 2,005 | 55,162 | 256,860 | 56,096 | 6,429 |
| Baltimore city | 3,906 | 6,398 | 35.4 | 3.0 | 12.7 | 0.0 | 2.5 | X | X | 46,226 | 19,363 | 402,690 | 3,488,070 | 99,097 | 15,064 |
| **MASSACHUSETTS** | X | X | X | X | X | X | X | X | X | | | | | | |
| Barnstable | 1,227 | 5,746 | 47.1 | 1.0 | 5.2 | 0.4 | 3.7 | 916.3 | 4,290 | 1,612 | 1,211 | 13,462 | 128,690 | 81,733 | 10,493 |
| Berkshire | 682 | 5,397 | 58.0 | 0.3 | 3.3 | 0.4 | 6.4 | 435.0 | 3,443 | 391 | 288 | 7,994 | 64,630 | 67,844 | 7,792 |
| Bristol | 2,297 | 4,096 | 59.2 | 0.8 | 5.4 | 0.8 | 3.1 | 1,270.0 | 2,264 | 1,177 | 1,332 | 28,192 | 281,700 | 67,538 | 7,568 |
| Dukes | 189 | 10,923 | 47.8 | 2.9 | 5.4 | 0.1 | 3.6 | 74.8 | 4,323 | 48 | 65 | 1,527 | 11,700 | 83,987 | 11,449 |
| Essex | 3,543 | 4,516 | 56.6 | 0.3 | 4.6 | 0.3 | 3.3 | 2,102.2 | 2,680 | 3,504 | 1,907 | 37,526 | 408,430 | 91,192 | 13,185 |
| Franklin | 379 | 5,371 | 59.1 | 0.8 | 2.8 | 0.3 | 5.2 | 115.4 | 1,634 | 209 | 162 | 5,021 | 35,800 | 60,737 | 5,908 |
| Hampden | 2,094 | 4,478 | 57.1 | 0.7 | 5.1 | 0.3 | 3.1 | 1,364.2 | 2,917 | 3,764 | 1,282 | 29,440 | 222,710 | 59,621 | 6,262 |
| Hampshire | 581 | 3,607 | 58.1 | 0.5 | 4.0 | 0.5 | 5.0 | 226.1 | 1,404 | 1,436 | 336 | 19,035 | 72,910 | 75,152 | 8,812 |
| Middlesex | 8,127 | 5,067 | 51.0 | 7.8 | 4.3 | 0.2 | 3.3 | 5,173.1 | 3,225 | 11,624 | 4,853 | 80,451 | 813,180 | 126,251 | 21,116 |
| Nantucket | 155 | 13,800 | 43.5 | 0.5 | 3.8 | 5.0 | 0.6 | 248.2 | 22,153 | 51 | 53 | 783 | 8,040 | 98,117 | 14,464 |
| Norfolk | 3,449 | 4,923 | 53.9 | 0.5 | 4.7 | 0.2 | 3.4 | 2,246.2 | 3,206 | 1,509 | 1,637 | 34,490 | 363,490 | 138,528 | 24,428 |
| Plymouth | 2,496 | 4,842 | 60.5 | 0.4 | 5.1 | 0.3 | 3.2 | 1,641.9 | 3,185 | 3,490 | 1,249 | 28,047 | 270,120 | 91,460 | 12,950 |
| Suffolk | 4,402 | 5,497 | 37.9 | 6.5 | 9.1 | 0.1 | 3.0 | 2,769.9 | 3,459 | 14,435 | 3,081 | 64,920 | 394,380 | 107,407 | 18,287 |
| Worcester | 3,673 | 4,448 | 60.9 | 0.6 | 4.3 | 0.3 | 4.1 | 2,567.2 | 3,109 | 2,976 | 1,907 | 51,802 | 412,370 | 76,864 | 9,488 |
| **MICHIGAN** | X | X | X | X | X | X | X | X | X | 52,384 | 17,693 | 548,392 | 4,785,990 | 66,439 | 7,792 |
| Alcona | 28 | 2,747 | 29.1 | 5.1 | 4.8 | 0.5 | 27.2 | 1.8 | 173 | 20 | 16 | 319 | 4,950 | 46,305 | 3,905 |
| Alger | 36 | 3,895 | 36.6 | 3.6 | 1.7 | 0.0 | 21.2 | 27.7 | 3,029 | 77 | 13 | 605 | 4,060 | 50,279 | 4,255 |
| Allegan | 396 | 3,405 | 53.5 | 8.4 | 3.4 | 3.3 | 10.7 | 521.6 | 4,482 | 181 | 187 | 5,602 | 57,100 | 64,045 | 6,532 |
| Alpena | 145 | 5,089 | 42.6 | 23.8 | 3.1 | 0.6 | 6.7 | 24.4 | 859 | 144 | 44 | 2,567 | 14,140 | 48,175 | 4,447 |
| Antrim | 100 | 4,287 | 37.8 | 2.9 | 4.3 | 18.5 | 11.4 | 57.9 | 2,488 | 60 | 37 | 1,100 | 11,860 | 60,392 | 6,459 |
| Arenac | 45 | 3,023 | 47.3 | 1.9 | 2.9 | 0.0 | 17.4 | 26.8 | 1,784 | 42 | 23 | 559 | 7,250 | 44,974 | 3,643 |
| Baraga | 55 | 6,533 | 22.4 | 44.8 | 2.0 | 0.2 | 8.6 | 20.1 | 2,379 | 28 | 11 | 1,441 | 3,450 | 47,613 | 3,780 |
| Barry | 206 | 3,401 | 41.9 | 6.0 | 3.4 | 17.3 | 9.2 | 215.1 | 3,545 | 259 | 231 | 5,586 | 53,160 | 52,172 | 4,875 |
| Bay | 486 | 4,671 | 50.5 | 10.2 | 3.8 | 5.7 | 5.8 | 292.8 | 2,814 | 35 | 29 | 702 | 9,380 | 57,257 | 5,626 |
| Benzie | 63 | 3,569 | 34.5 | 19.3 | 2.7 | 0.3 | 13.2 | 24.9 | 1,410 | 342 | 263 | 8,872 | 73,510 | 61,416 | 6,929 |
| Berrien | 712 | 4,622 | 50.9 | 8.7 | 3.9 | 1.3 | 5.1 | 534.7 | 3,469 | 85 | 66 | 2,739 | 19,120 | 50,078 | 4,136 |
| Branch | 283 | 6,530 | 30.6 | 24.0 | 1.6 | 5.2 | 6.0 | 163.2 | 3,762 | 2,991 | 266 | 7,905 | 61,210 | 52,794 | 4,693 |
| Calhoun | 634 | 4,727 | 46.9 | 2.5 | 5.8 | 5.4 | 12.5 | 651.6 | 4,856 | 77 | 82 | 2,084 | 23,970 | 63,615 | 7,331 |
| Cass | 197 | 3,833 | 51.1 | 7.2 | 3.1 | 4.8 | 7.0 | 207.6 | 4,033 | 52 | 66 | 1,780 | 13,900 | 62,716 | 7,580 |
| Charlevoix | 186 | 7,106 | 39.0 | 9.6 | 2.2 | 15.4 | 8.3 | 114.1 | 4,353 | 59 | 103 | 987 | 12,750 | 52,971 | 5,419 |
| Cheboygan | 89 | 3,494 | 51.8 | 0.9 | 4.7 | 1.1 | 13.6 | 32.2 | 1,264 | 491 | 235 | 5,742 | 15,660 | 46,394 | 3,785 |
| Chippewa | 138 | 3,670 | 45.0 | 3.7 | 3.5 | 0.4 | 11.0 | 107.7 | 2,859 | 91 | 49 | 1,511 | 13,240 | 42,993 | 3,243 |
| Clare | 143 | 4,663 | 63.7 | 0.5 | 3.1 | 0.4 | 7.8 | 46.4 | 1,520 | 223 | 131 | 2,033 | 38,230 | 70,936 | 7,891 |
| Clinton | 213 | 2,718 | 54.3 | 0.4 | 4.4 | 0.6 | 10.1 | 255.2 | 3,250 | 144 | 22 | 643 | 6,250 | 45,773 | 3,629 |
| Crawford | 62 | 4,491 | 37.0 | 2.2 | 2.8 | 0.8 | 14.1 | 37.3 | 2,684 | 199 | 56 | 1,815 | 17,590 | 52,122 | 4,544 |
| Delta | 168 | 4,687 | 51.7 | 2.1 | 1.6 | 0.8 | 9.9 | 89.6 | 2,496 | 775 | 60 | 1,941 | 12,980 | 54,837 | 5,127 |
| Dickinson | 213 | 8,390 | 22.3 | 53.5 | 2.6 | 0.0 | 5.0 | 103.5 | 4,071 | 188 | 207 | 5,537 | 54,580 | 59,617 | 5,751 |
| Eaton | 368 | 3,361 | 51.7 | 4.6 | 5.4 | 4.9 | 8.0 | 450.3 | 4,114 | 104 | 52 | 2,437 | 18,240 | 70,621 | 8,827 |
| Emmet | 228 | 6,882 | 32.9 | 20.9 | 1.8 | 6.4 | 6.5 | 113.3 | 3,424 | 1,089 | 640 | 19,174 | 192,980 | 53,034 | 5,197 |
| Genesee | 2,110 | 5,179 | 40.0 | 28.9 | 4.1 | 1.6 | 3.9 | 912.4 | 2,239 | 48 | 40 | 801 | 11,300 | 48,598 | 4,079 |
| Gladwin | 66 | 2,596 | 42.8 | 1.2 | 3.5 | 0.4 | 18.2 | 36.2 | 1,435 | 171 | 23 | 1,521 | 6,630 | 58,909 | 6,591 |
| Gogebic | 92 | 6,005 | 32.0 | 8.5 | 2.9 | 12.6 | 9.9 | 71.6 | 4,671 | 497 | 270 | 5,714 | 49,810 | 72,136 | 9,022 |
| Grand Traverse | 518 | 5,638 | 48.5 | 14.8 | 2.7 | 6.3 | 5.4 | 327.7 | 3,570 | 72 | 56 | 2,336 | 16,920 | 50,184 | 4,392 |
| Gratiot | 162 | 3,947 | 57.0 | 0.5 | 3.7 | 0.7 | 9.4 | 135.0 | 3,295 | 87 | 70 | 2,253 | 20,200 | 48,396 | 3,897 |
| Hillsdale | 153 | 3,333 | 46.1 | 1.1 | 2.8 | 13.4 | 7.5 | 103.6 | 2,262 | 167 | 90 | 3,603 | 15,050 | 51,153 | 4,337 |
| Houghton | 162 | 4,469 | 45.2 | 3.2 | 2.1 | 13.2 | 9.5 | 172.5 | 4,767 | 101 | 48 | 1,762 | 16,380 | 48,463 | 4,263 |
| Huron | 153 | 4,875 | 38.4 | 2.9 | 3.2 | 8.9 | 17.4 | 94.4 | 3,017 | 1,516 | 601 | 40,677 | 125,920 | 60,959 | 6,872 |
| Ingham | 1,496 | 5,122 | 46.5 | 11.4 | 6.0 | 2.7 | 4.2 | 1,640.5 | 5,618 | 118 | 94 | 3,142 | 27,970 | 52,211 | 4,179 |
| Ionia | 203 | 3,157 | 58.5 | 1.7 | 2.3 | 0.5 | 11.3 | 260.8 | 4,057 | 116 | 60 | 1,455 | 12,080 | 45,319 | 3,675 |
| Iosco | 130 | 5,172 | 38.9 | 19.4 | 2.5 | 11.8 | 6.0 | 52.1 | 2,073 | 31 | 17 | 914 | 5,570 | 45,669 | 3,886 |
| Iron | 74 | 6,700 | 18.1 | 5.3 | 2.1 | 32.4 | 12.6 | 45.8 | 4,121 | 157 | 109 | 10,214 | 26,950 | 53,996 | 5,105 |
| Isabella | 262 | 3,680 | 28.7 | 33.2 | 2.8 | 6.2 | 7.0 | 143.8 | 2,023 | 325 | 240 | 7,560 | 71,860 | 55,655 | 5,304 |
| Jackson | 669 | 4,217 | 50.5 | 11.8 | 4.6 | 3.8 | 6.7 | 543.5 | 3,428 | 697 | 420 | 14,343 | 122,790 | 69,477 | 8,300 |
| Kalamazoo | 1,031 | 3,919 | 55.8 | 9.5 | 3.4 | 0.4 | 5.5 | 1,217.5 | 4,629 | 25 | 28 | 1,096 | 8,480 | 47,574 | 4,223 |
| Kalkaska | 85 | 4,810 | 18.8 | 53.7 | 2.4 | 0.5 | 5.7 | 26.2 | 1,491 | 2,900 | 1,081 | 24,889 | 316,440 | 75,010 | 8,841 |
| Kent | 3,033 | 4,671 | 51.6 | 6.0 | 4.4 | 0.9 | 7.5 | 3,759.1 | 5,790 | | | | | | |

1. Based on the resident population estimated as of July 1 of the year shown.

| State / county code | CBSA code[1] | County Type code[2] | STATE County | Land area[3] (sq. mi) | Total persons 2019 | Rank | Per square mile | White | Black | American Indian, Alaska Native | Asian and Pacific Islander | Percent Hispanic or Latino[4] | Under 5 years | 5 to 14 years | 15 to 24 years | 25 to 34 years | 35 to 44 years | 45 to 54 years |
|---|---|---|---|---|---|---|---|---|---|---|---|---|---|---|---|---|---|---|
| | | | | 1 | 2 | 3 | 4 | 5 | 6 | 7 | 8 | 9 | 10 | 11 | 12 | 13 | 14 | 15 |
| | | | **MICHIGAN— Cont'd** | | | | | | | | | | | | | | | |
| 26083 | 26340 | 9 | Keweenaw | 540.1 | 2,119 | 3,038 | 3.9 | 97.5 | 1.0 | 0.9 | 0.4 | 1.5 | 4.2 | 8.4 | 7.6 | 7.8 | 7.8 | 9.5 |
| 26085 | | 9 | Lake | 567.6 | 11,587 | 2,316 | 20.4 | 88.8 | 8.8 | 1.9 | 0.9 | 2.7 | 4.5 | 9.0 | 7.9 | 7.5 | 8.9 | 12.0 |
| 26087 | 19820 | 1 | Lapeer | 647.0 | 87,635 | 666 | 135.4 | 92.8 | 1.6 | 1.0 | 1.0 | 5.0 | 4.8 | 11.3 | 11.7 | 11.3 | 11.2 | 13.7 |
| 26089 | 45900 | 9 | Leelanau | 347.2 | 21,743 | 1,730 | 62.6 | 91.7 | 1.0 | 3.7 | 1.0 | 4.3 | 3.8 | 8.7 | 8.7 | 8.5 | 9.5 | 9.9 |
| 26091 | 10300 | 4 | Lenawee | 749.6 | 97,808 | 617 | 130.5 | 88.1 | 3.5 | 1.1 | 0.9 | 8.5 | 5.3 | 11.6 | 12.8 | 11.8 | 11.7 | 12.6 |
| 26093 | 19820 | 1 | Livingston | 565.3 | 192,335 | 351 | 340.2 | 95.3 | 1.0 | 0.9 | 1.5 | 2.7 | 4.9 | 11.6 | 12.0 | 11.2 | 11.6 | 14.0 |
| 26095 | | 7 | Luce | 899.1 | 6,126 | 2,737 | 6.8 | 80.9 | 12.0 | 8.2 | 0.7 | 1.8 | 4.4 | 9.2 | 10.2 | 13.4 | 12.7 | 12.9 |
| 26097 | | 7 | Mackinac | 1,021.9 | 10,839 | 2,359 | 10.6 | 77.4 | 3.9 | 20.1 | 1.4 | 2.3 | 4.5 | 8.1 | 9.4 | 9.6 | 9.3 | 12.3 |
| 26099 | 19820 | 1 | Macomb | 479.3 | 870,791 | 66 | 1,816.8 | 79.5 | 14.0 | 0.9 | 5.4 | 2.8 | 5.4 | 11.5 | 11.6 | 13.9 | 11.9 | 13.3 |
| 26101 | | 7 | Manistee | 542.3 | 24,738 | 1,617 | 45.6 | 90.3 | 3.9 | 3.2 | 0.8 | 3.7 | 4.2 | 9.3 | 10.8 | 10.7 | 9.9 | 11.5 |
| 26103 | 32100 | 5 | Marquette | 1,809.1 | 65,834 | 809 | 36.4 | 94.0 | 2.1 | 3.0 | 1.2 | 1.7 | 4.6 | 10.2 | 17.8 | 11.7 | 11.5 | 10.5 |
| 26105 | 31220 | 7 | Mason | 495.0 | 29,164 | 1,456 | 58.9 | 92.8 | 1.6 | 1.6 | 1.2 | 4.7 | 5.0 | 11.5 | 10.0 | 10.6 | 10.7 | 11.0 |
| 26107 | 13660 | 6 | Mecosta | 555.2 | 43,907 | 1,098 | 79.1 | 93.2 | 3.7 | 1.6 | 1.3 | 2.6 | 5.0 | 10.1 | 21.4 | 10.9 | 9.5 | 10.1 |
| 26109 | 31940 | 3 | Menominee | 1,044.0 | 22,608 | 1,700 | 21.7 | 94.0 | 1.2 | 3.6 | 0.8 | 2.1 | 4.2 | 10.1 | 10.1 | 9.9 | 10.5 | 12.1 |
| 26111 | 33220 | 3 | Midland | 517.5 | 83,441 | 685 | 161.2 | 92.6 | 2.1 | 1.0 | 2.9 | 3.1 | 5.3 | 11.7 | 12.1 | 12.5 | 12.1 | 12.4 |
| 26113 | 15620 | 9 | Missaukee | 564.8 | 15,152 | 2,086 | 26.8 | 95.1 | 1.0 | 1.3 | 0.9 | 3.3 | 5.8 | 12.8 | 10.6 | 11.6 | 10.7 | 11.3 |
| 26115 | 33780 | 3 | Monroe | 549.4 | 150,568 | 448 | 274.1 | 92.5 | 3.5 | 0.9 | 1.1 | 3.8 | 5.2 | 11.9 | 11.6 | 12.0 | 11.7 | 13.2 |
| 26117 | 24340 | 2 | Montcalm | 705.3 | 63,476 | 841 | 90.0 | 93.0 | 2.9 | 1.3 | 0.8 | 3.7 | 5.6 | 12.3 | 11.5 | 12.8 | 12.2 | 12.7 |
| 26119 | | 9 | Montmorency | 546.7 | 9,337 | 2,470 | 17.1 | 96.7 | 1.1 | 1.7 | 0.6 | 1.5 | 4.2 | 8.4 | 7.9 | 7.6 | 8.6 | 10.8 |
| 26121 | 34740 | 3 | Muskegon | 503.9 | 173,883 | 379 | 345.1 | 78.9 | 15.4 | 1.7 | 1.2 | 6.0 | 6.0 | 12.8 | 11.9 | 13.3 | 12.2 | 11.8 |
| 26123 | | 6 | Newaygo | 838.9 | 49,348 | 1,001 | 58.8 | 91.8 | 1.7 | 1.4 | 0.8 | 6.0 | 5.7 | 12.5 | 11.3 | 11.6 | 11.0 | 11.8 |
| 26125 | 19820 | 1 | Oakland | 867.3 | 1,253,459 | 34 | 1,445.2 | 73.1 | 14.6 | 0.8 | 9.2 | 4.6 | 5.3 | 11.3 | 11.7 | 13.6 | 12.5 | 13.5 |
| 26127 | | 6 | Oceana | 538.1 | 26,819 | 1,534 | 49.8 | 82.7 | 1.4 | 1.6 | 0.9 | 15.3 | 5.2 | 12.5 | 11.5 | 10.9 | 11.0 | 11.3 |
| 26129 | | 9 | Ogemaw | 563.5 | 20,923 | 1,774 | 37.1 | 95.6 | 0.9 | 1.7 | 0.9 | 2.2 | 4.5 | 10.3 | 9.7 | 9.3 | 9.9 | 10.9 |
| 26131 | | 9 | Ontonagon | 1,311.0 | 5,656 | 2,777 | 4.3 | 96.2 | 0.7 | 2.0 | 1.0 | 1.5 | 2.9 | 6.3 | 7.6 | 5.9 | 7.3 | 11.2 |
| 26133 | | 9 | Osceola | 566.3 | 23,466 | 1,660 | 41.4 | 95.9 | 1.6 | 1.5 | 0.7 | 2.2 | 5.6 | 12.3 | 11.4 | 11.1 | 10.8 | 11.7 |
| 26135 | | 9 | Oscoda | 565.7 | 8,368 | 2,544 | 14.8 | 96.5 | 1.0 | 1.6 | 0.6 | 1.9 | 5.4 | 11.0 | 9.0 | 8.7 | 8.8 | 9.6 |
| 26137 | | 7 | Otsego | 515.0 | 24,765 | 1,615 | 48.1 | 95.9 | 1.2 | 1.7 | 1.1 | 1.9 | 5.2 | 12.0 | 10.9 | 11.5 | 10.8 | 12.2 |
| 26139 | 24340 | 2 | Ottawa | 563.5 | 294,635 | 236 | 522.9 | 84.9 | 2.4 | 0.7 | 3.5 | 10.3 | 5.9 | 13.5 | 17.1 | 12.4 | 12.0 | 11.2 |
| 26141 | | 7 | Presque Isle | 658.7 | 12,665 | 2,239 | 19.2 | 96.2 | 1.0 | 1.5 | 0.8 | 1.7 | 3.7 | 8.5 | 8.6 | 7.8 | 8.9 | 10.5 |
| 26143 | | 7 | Roscommon | 519.9 | 23,986 | 1,644 | 46.1 | 95.8 | 1.0 | 1.5 | 1.0 | 2.1 | 3.4 | 8.4 | 7.6 | 8.5 | 8.1 | 10.9 |
| 26145 | 40980 | 3 | Saginaw | 800.8 | 189,868 | 353 | 237.1 | 70.7 | 19.9 | 0.8 | 1.7 | 9.0 | 5.8 | 11.7 | 13.0 | 12.8 | 10.9 | 11.8 |
| 26147 | 19820 | 3 | St. Clair | 721.5 | 159,293 | 423 | 220.8 | 92.9 | 3.6 | 1.1 | 1.0 | 3.6 | 5.1 | 11.4 | 11.5 | 11.5 | 11.2 | 13.5 |
| 26149 | 44780 | 4 | St. Joseph | 500.6 | 60,848 | 866 | 121.6 | 88.0 | 3.7 | 1.0 | 1.1 | 8.5 | 6.3 | 13.7 | 12.0 | 12.1 | 11.6 | 11.5 |
| 26151 | | 6 | Sanilac | 962.6 | 40,747 | 1,163 | 42.3 | 94.5 | 1.0 | 1.0 | 0.6 | 4.1 | 5.0 | 11.9 | 11.3 | 10.4 | 10.5 | 12.0 |
| 26153 | | 7 | Schoolcraft | 1,171.9 | 8,104 | 2,574 | 6.9 | 88.9 | 1.2 | 11.6 | 0.6 | 1.4 | 4.5 | 9.8 | 9.1 | 8.5 | 9.1 | 12.2 |
| 26155 | 29620 | 4 | Shiawassee | 531.0 | 67,738 | 793 | 127.6 | 95.2 | 1.3 | 1.0 | 0.9 | 3.1 | 5.3 | 12.0 | 12.1 | 11.3 | 11.3 | 12.9 |
| 26157 | | 6 | Tuscola | 804.9 | 52,289 | 965 | 65.0 | 94.0 | 1.6 | 1.2 | 0.6 | 3.7 | 5.1 | 11.1 | 11.1 | 11.4 | 11.1 | 12.7 |
| 26159 | | 2 | Van Buren | 607.8 | 75,474 | 737 | 124.2 | 83.1 | 4.8 | 1.7 | 1.1 | 11.9 | 5.9 | 13.0 | 11.5 | 11.3 | 11.7 | 12.2 |
| 26161 | 11460 | 2 | Washtenaw | 706.0 | 366,473 | 198 | 519.1 | 73.2 | 13.7 | 1.0 | 10.7 | 5.1 | 4.8 | 10.2 | 21.2 | 14.5 | 11.6 | 11.2 |
| 26163 | 19820 | 1 | Wayne | 611.8 | 1,740,623 | 19 | 2,845.1 | 51.3 | 39.6 | 1.1 | 4.4 | 6.3 | 6.5 | 13.0 | 12.3 | 14.7 | 11.7 | 12.5 |
| 26165 | 15620 | 7 | Wexford | 564.9 | 33,743 | 1,333 | 59.7 | 95.6 | 1.3 | 1.4 | 1.1 | 2.3 | 6.0 | 13.0 | 11.0 | 11.7 | 11.5 | 11.8 |
| 27000 | | 0 | **MINNESOTA** | 79,625.9 | 5,657,342 | X | 71.0 | 80.8 | 8.2 | 1.7 | 6.1 | 5.7 | 6.1 | 13.0 | 12.7 | 13.4 | 13.1 | 11.7 |
| 27001 | | 8 | Aitkin | 1,821.8 | 15,848 | 2,040 | 8.7 | 95.4 | 1.1 | 3.1 | 0.7 | 1.3 | 3.7 | 9.3 | 8.3 | 7.5 | 8.6 | 9.9 |
| 27003 | 33460 | 3 | Anoka | 422.0 | 359,921 | 203 | 852.9 | 81.1 | 9.2 | 1.4 | 6.3 | 5.0 | 6.2 | 13.3 | 11.8 | 13.2 | 13.7 | 12.9 |
| 27005 | | 6 | Becker | 1,315.1 | 34,456 | 1,318 | 26.2 | 89.3 | 1.1 | 9.6 | 1.0 | 2.3 | 6.0 | 13.5 | 11.5 | 10.2 | 11.5 | 10.7 |
| 27007 | 13420 | 7 | Beltrami | 2,504.7 | 47,442 | 1,031 | 18.9 | 74.9 | 1.6 | 23.0 | 1.4 | 2.5 | 7.0 | 14.2 | 16.8 | 12.3 | 11.2 | 9.4 |
| 27009 | 41060 | 3 | Benton | 408.3 | 40,958 | 1,156 | 100.3 | 90.1 | 6.4 | 0.9 | 1.8 | 3.0 | 6.9 | 14.4 | 12.0 | 14.4 | 14.1 | 11.7 |
| 27011 | | 9 | Big Stone | 499.2 | 4,923 | 2,830 | 9.9 | 96.2 | 0.9 | 1.1 | 0.7 | 2.3 | 6.0 | 12.1 | 10.0 | 9.3 | 9.9 | 9.4 |
| 27013 | 31860 | 3 | Blue Earth | 747.8 | 68,241 | 786 | 91.3 | 88.5 | 5.6 | 0.7 | 3.3 | 4.1 | 5.3 | 11.2 | 24.4 | 13.4 | 11.5 | 9.1 |
| 27015 | 35580 | 3 | Brown | 611.1 | 24,846 | 1,608 | 40.7 | 93.8 | 1.0 | 0.4 | 1.1 | 4.7 | 5.6 | 12.4 | 13.4 | 10.3 | 11.3 | 10.3 |
| 27017 | 20260 | 2 | Carlton | 861.2 | 35,769 | 1,285 | 41.5 | 90.4 | 2.3 | 7.0 | 1.1 | 1.8 | 5.3 | 12.8 | 11.4 | 12.0 | 13.1 | 12.8 |
| 27019 | 33460 | 1 | Carver | 354.0 | 106,565 | 576 | 301.0 | 89.7 | 2.9 | 0.6 | 4.3 | 4.4 | 6.2 | 15.0 | 12.7 | 10.6 | 14.6 | 13.9 |
| 27021 | 14660 | 9 | Cass | 2,021.5 | 29,928 | 1,438 | 14.8 | 85.2 | 1.1 | 13.0 | 1.0 | 2.3 | 5.0 | 12.0 | 9.6 | 9.3 | 10.1 | 10.5 |
| 27023 | | 7 | Chippewa | 581.2 | 11,758 | 2,309 | 20.2 | 87.3 | 1.5 | 1.8 | 3.2 | 8.1 | 6.8 | 14.0 | 10.8 | 11.4 | 11.7 | 10.1 |
| 27025 | 33460 | 1 | Chisago | 414.9 | 56,794 | 910 | 136.9 | 94.0 | 2.0 | 1.2 | 1.9 | 2.5 | 5.7 | 12.7 | 11.4 | 12.0 | 13.1 | 13.4 |
| 27027 | 22020 | 3 | Clay | 1,045.2 | 64,690 | 827 | 61.9 | 87.7 | 5.6 | 2.1 | 2.2 | 4.8 | 6.9 | 14.2 | 17.9 | 13.5 | 13.2 | 10.1 |
| 27029 | | 8 | Clearwater | 998.8 | 9,017 | 2,495 | 9.0 | 87.9 | 1.2 | 11.1 | 1.1 | 2.4 | 6.3 | 14.4 | 11.3 | 10.1 | 11.2 | 10.9 |
| 27031 | | 9 | Cook | 1,452.6 | 5,417 | 2,793 | 3.7 | 87.0 | 2.0 | 9.7 | 1.7 | 2.6 | 3.9 | 8.7 | 8.1 | 10.0 | 10.5 | 11.4 |
| 27033 | | 7 | Cottonwood | 640.0 | 11,242 | 2,332 | 17.6 | 85.4 | 1.8 | 0.9 | 4.6 | 8.8 | 6.3 | 14.5 | 11.2 | 10.1 | 10.8 | 10.4 |
| 27035 | 14660 | 4 | Crow Wing | 998.4 | 65,644 | 812 | 65.7 | 96.1 | 1.5 | 1.5 | 0.9 | 1.6 | 5.2 | 12.0 | 10.4 | 10.9 | 11.5 | 10.8 |
| 27037 | 33460 | 1 | Dakota | 562.5 | 431,807 | 164 | 767.7 | 78.8 | 9.1 | 0.9 | 6.4 | 7.6 | 6.3 | 13.6 | 11.9 | 13.0 | 13.9 | 12.6 |
| 27039 | 40340 | 3 | Dodge | 439.3 | 20,987 | 1,771 | 47.8 | 93.1 | 1.3 | 0.7 | 1.5 | 5.0 | 6.1 | 14.6 | 12.5 | 11.8 | 13.6 | 12.7 |
| 27041 | 10820 | 6 | Douglas | 636.9 | 38,328 | 1,216 | 60.2 | 96.4 | 1.1 | 0.7 | 0.9 | 1.9 | 5.8 | 12.1 | 10.4 | 11.2 | 11.7 | 10.7 |
| 27043 | | 6 | Faribault | 712.5 | 13,601 | 2,185 | 19.1 | 90.8 | 1.3 | 0.8 | 0.8 | 7.6 | 5.7 | 12.3 | 10.7 | 9.8 | 11.7 | 10.4 |

1. CBSA = Core Based Statistical Area. See Appendix A for explanation. See Appendix B for list of metropolitan areas with component counties.    2. County type code from the Economic Research Service of USDA Rural-Urban Continuum Codes. See Appendix A for definition.    3. Dry land or land partially or temporarily covered by water.    4. May be of any race.

Items 1—15

# Table B. States and Counties — Population and Households

| STATE County | 55 to 64 years | 65 to 74 years | 75 years and over | Percent female | Total persons 2000 | 2010 | Percent change 2000-2010 | 2010-2020 | Births | Deaths | Net Migration | Number | Persons per household | Family households | Female family householder[1] | One person |
|---|---|---|---|---|---|---|---|---|---|---|---|---|---|---|---|---|
| | 16 | 17 | 18 | 19 | 20 | 21 | 22 | 23 | 24 | 25 | 26 | 27 | 28 | 29 | 30 | 31 |
| **MICHIGAN— Cont'd** | | | | | | | | | | | | | | | | |
| Keweenaw | 17.4 | 22.3 | 14.9 | 49.4 | 2,301 | 2,157 | -6.3 | -1.8 | 170 | 238 | 32 | 1,081 | 1.94 | 69.4 | 8.9 | 26.8 |
| Lake | 19.1 | 19.8 | 11.3 | 49.5 | 11,333 | 11,542 | 1.8 | 0.4 | 1,008 | 1,690 | 718 | 4,631 | 2.49 | 58.8 | 6.8 | 34.8 |
| Lapeer | 16.6 | 12.0 | 7.5 | 49.2 | 87,904 | 88,314 | 0.5 | -0.8 | 8,347 | 8,650 | -358 | 33,700 | 2.56 | 72.1 | 8.9 | 23.1 |
| Leelanau | 17.5 | 19.8 | 13.5 | 50.9 | 21,119 | 21,711 | 2.8 | 0.1 | 1,688 | 2,453 | 807 | 9,139 | 2.32 | 68.6 | 6.1 | 27.8 |
| Lenawee | 14.2 | 11.9 | 8.0 | 49.5 | 98,890 | 99,898 | 1.0 | -2.1 | 10,573 | 10,449 | -2,208 | 38,345 | 2.43 | 65.7 | 10.8 | 29.4 |
| Livingston | 16.1 | 11.5 | 7.2 | 49.9 | 156,951 | 180,962 | 15.3 | 6.3 | 18,142 | 14,874 | 8,245 | 72,000 | 2.62 | 72.4 | 6.8 | 22.5 |
| Luce | 14.9 | 12.3 | 10.0 | 41.3 | 7,024 | 6,631 | -5.6 | -7.6 | 538 | 756 | -288 | 2,231 | 2.35 | 64.0 | 6.9 | 31.5 |
| Mackinac | 16.9 | 17.8 | 12.1 | 49.3 | 11,943 | 11,110 | -7.0 | -2.4 | 877 | 1,456 | 313 | 5,269 | 1.99 | 60.2 | 7.8 | 31.6 |
| Macomb | 14.5 | 10.4 | 7.5 | 51.3 | 788,149 | 841,037 | 6.7 | 3.5 | 94,813 | 87,736 | 23,309 | 346,402 | 2.49 | 65.6 | 12.8 | 29.3 |
| Manistee | 16.4 | 16.2 | 11.0 | 48.6 | 24,527 | 24,747 | 0.9 | 0.0 | 2,022 | 3,262 | 1,241 | 9,426 | 2.44 | 63.2 | 7.9 | 32.6 |
| Marquette | 13.4 | 12.4 | 8.0 | 49.6 | 64,634 | 67,071 | 3.8 | -1.8 | 6,347 | 6,924 | -625 | 26,552 | 2.37 | 59.8 | 8.4 | 30.2 |
| Mason | 15.9 | 15.2 | 10.2 | 50.3 | 28,274 | 28,689 | 1.5 | 1.7 | 2,931 | 3,464 | 1,033 | 12,186 | 2.34 | 63.6 | 9.5 | 30.4 |
| Mecosta | 13.7 | 11.6 | 7.7 | 49.7 | 40,553 | 42,796 | 5.5 | 2.6 | 4,305 | 3,956 | 779 | 15,808 | 2.55 | 60.7 | 9.6 | 29.8 |
| Menominee | 16.8 | 15.2 | 11.1 | 48.9 | 25,326 | 24,029 | -5.1 | -5.9 | 2,041 | 2,828 | -617 | 10,627 | 2.14 | 59.7 | 6.7 | 35.4 |
| Midland | 14.3 | 10.9 | 8.7 | 50.5 | 82,874 | 83,623 | 0.9 | -0.2 | 8,810 | 7,532 | -1,455 | 34,199 | 2.40 | 67.6 | 8.6 | 26.8 |
| Missaukee | 15.7 | 12.7 | 8.8 | 49.6 | 14,478 | 14,851 | 2.6 | 2.0 | 1,806 | 1,686 | 187 | 6,055 | 2.45 | 68.8 | 8.1 | 25.6 |
| Monroe | 15.3 | 11.6 | 7.7 | 50.6 | 145,945 | 152,031 | 4.2 | | 15,571 | 14,929 | -2,097 | 59,909 | 2.47 | 68.4 | 10.3 | 27.2 |
| Montcalm | 14.4 | 11.1 | 7.5 | 48.3 | 61,266 | 63,342 | 3.4 | 0.2 | 7,357 | 6,286 | -935 | 23,913 | 2.54 | 69.1 | 10.0 | 25.4 |
| Montmorency | 19.2 | 20.6 | 12.8 | 49.0 | 10,315 | 9,760 | -5.4 | -4.3 | 707 | 1,642 | 516 | 4,452 | 2.05 | 62.7 | 7.1 | 32.1 |
| Muskegon | 14.0 | 10.9 | 7.1 | 50.5 | 170,200 | 172,202 | 1.2 | 1.0 | 21,441 | 17,713 | -2,042 | 65,939 | 2.53 | 66.7 | 14.6 | 26.8 |
| Newaygo | 15.9 | 12.3 | 8.0 | 49.6 | 47,874 | 48,459 | 1.2 | 1.8 | 5,541 | 5,155 | 519 | 19,161 | 2.48 | 68.3 | 7.9 | 26.4 |
| Oakland | 14.3 | 10.6 | 7.3 | 50.9 | 1,194,156 | 1,202,388 | 0.7 | 4.2 | 136,539 | 107,325 | 22,612 | 504,585 | 2.46 | 64.1 | 9.8 | 29.7 |
| Oceana | 15.7 | 13.2 | 8.7 | 49.0 | 26,873 | 26,570 | -1.1 | 0.9 | 3,013 | 2,748 | -7 | 10,156 | 2.53 | 71.4 | 10.6 | 24.7 |
| Ogemaw | 18.0 | 15.7 | 11.2 | 50.3 | 21,645 | 21,686 | 0.2 | -3.5 | 1,930 | 3,225 | 551 | 9,184 | 2.24 | 63.9 | 10.3 | 29.8 |
| Ontonagon | 20.6 | 22.1 | 16.2 | 49.0 | 7,818 | 6,780 | -13.3 | -16.6 | 303 | 1,032 | -396 | 2,793 | 2.07 | 58.0 | 5.9 | 36.2 |
| Osceola | 15.3 | 12.7 | 9.1 | 49.4 | 23,197 | 23,528 | 1.4 | -0.3 | 2,730 | 2,536 | -247 | 9,181 | 2.48 | 66.9 | 10.6 | 28.7 |
| Oscoda | 18.9 | 17.0 | 11.6 | 49.1 | 9,418 | 8,638 | -8.3 | -3.1 | 879 | 1,308 | 159 | 3,806 | 2.15 | 61.2 | 7.3 | 32.3 |
| Otsego | 15.3 | 13.3 | 8.9 | 50.2 | 23,301 | 24,167 | 3.7 | 2.5 | 2,535 | 2,734 | 815 | 9,944 | 2.42 | 68.8 | 9.6 | 25.1 |
| Ottawa | 12.0 | 9.3 | 6.7 | 50.6 | 238,314 | 263,787 | 10.7 | 11.7 | 33,909 | 19,284 | 16,355 | 102,610 | 2.70 | 71.5 | 8.1 | 21.9 |
| Presque Isle | 18.6 | 19.4 | 13.9 | 50.2 | 14,411 | 13,380 | -7.2 | -5.3 | 955 | 2,025 | 366 | 5,797 | 2.16 | 66.6 | 6.5 | 29.2 |
| Roscommon | 19.1 | 20.5 | 13.5 | 49.8 | 25,469 | 24,448 | -4.0 | -1.9 | 1,712 | 4,139 | 1,958 | 11,139 | 2.12 | 61.1 | 8.9 | 34.0 |
| Saginaw | 13.9 | 11.5 | 8.6 | 51.4 | 210,039 | 200,169 | -4.7 | -5.1 | 22,928 | 21,691 | -11,627 | 78,933 | 2.35 | 63.0 | 14.3 | 30.6 |
| St. Clair | 15.9 | 11.9 | 8.0 | 50.3 | 164,235 | 163,051 | -0.7 | -2.3 | 16,000 | 17,964 | -1,754 | 64,850 | 2.43 | 66.1 | 10.7 | 28.0 |
| St. Joseph | 14.0 | 11.0 | 7.7 | 49.9 | 62,422 | 61,303 | -1.8 | -0.7 | 8,112 | 6,546 | -2,006 | 24,150 | 2.49 | 67.9 | 11.0 | 26.3 |
| Sanilac | 16.0 | 13.3 | 9.5 | 50.0 | 44,547 | 43,106 | -3.2 | -5.5 | 4,423 | 4,947 | -1,834 | 17,499 | 2.32 | 65.8 | 9.5 | 28.7 |
| Schoolcraft | 18.7 | 17.0 | 11.1 | 50.3 | 8,903 | 8,485 | -4.7 | -4.5 | 686 | 1,157 | 94 | 3,468 | 2.28 | 62.4 | 7.2 | 32.5 |
| Shiawassee | 15.4 | 11.7 | 7.8 | 50.6 | 71,687 | 70,671 | -1.4 | -4.2 | 7,181 | 7,450 | -2,676 | 27,593 | 2.45 | 68.2 | 10.1 | 25.2 |
| Tuscola | 16.0 | 12.3 | 9.2 | 49.7 | 58,266 | 55,722 | -4.4 | -6.2 | 5,596 | 6,078 | -2,978 | 21,777 | 2.38 | 68.7 | 8.9 | 26.0 |
| Van Buren | 15.0 | 11.9 | 7.4 | 50.5 | 76,263 | 76,273 | 0.0 | | 9,042 | 7,800 | -2,059 | 29,411 | 2.51 | 69.3 | 11.0 | 26.0 |
| Washtenaw | 11.4 | 9.1 | 5.9 | 50.5 | 322,895 | 345,417 | 7.0 | 6.1 | 37,579 | 23,063 | 6,731 | 141,245 | 2.45 | 57.7 | 8.7 | 29.6 |
| Wayne | 13.1 | 9.7 | 6.6 | 51.9 | 2,061,162 | 1,820,216 | -11.7 | -4.4 | 238,334 | 186,785 | -132,905 | 682,282 | 2.54 | 60.6 | 18.6 | 34.0 |
| Wexford | 14.8 | 12.0 | 8.1 | 50.0 | 30,484 | 32,730 | 7.4 | 3.1 | 4,071 | 3,645 | 605 | 12,963 | 2.54 | 68.0 | 10.6 | 26.6 |
| **MINNESOTA** | 13.3 | 9.8 | 7.0 | 50.2 | 4,919,479 | 5,303,933 | 7.8 | 6.7 | 703,159 | 435,727 | 88,673 | 2,185,603 | 2.49 | 64.1 | 9.0 | 28.5 |
| Aitkin | 18.5 | 19.4 | 14.8 | 49.6 | 15,301 | 16,194 | 5.8 | -2.1 | 1,174 | 2,230 | 723 | 7,681 | 2.03 | 64.2 | 7.3 | 30.9 |
| Anoka | 14.1 | 9.1 | 5.8 | 50.0 | 298,084 | 330,858 | 11.0 | 8.8 | 43,006 | 20,947 | 7,270 | 128,198 | 2.71 | 71.1 | 10.2 | 22.8 |
| Becker | 14.7 | 13.0 | 8.9 | 50.0 | 30,000 | 32,511 | 8.4 | 6.0 | 4,276 | 3,567 | 1,267 | 13,617 | 2.46 | 68.5 | 9.8 | 27.3 |
| Beltrami | 12.3 | 10.0 | 6.9 | 50.1 | 39,650 | 44,442 | 12.1 | 6.8 | 6,997 | 4,170 | 197 | 17,372 | 2.55 | 62.8 | 14.2 | 29.6 |
| Benton | 12.0 | 8.0 | 6.4 | 49.9 | 34,226 | 38,451 | 12.3 | 6.5 | 5,776 | 3,583 | 323 | 16,452 | 2.39 | 64.0 | 9.9 | 27.8 |
| Big Stone | 16.3 | 13.8 | 13.3 | 50.2 | 5,820 | 5,268 | -9.5 | -6.5 | 608 | 764 | -182 | 2,298 | 2.10 | 61.2 | 6.0 | 32.1 |
| Blue Earth | 10.3 | 8.4 | 6.3 | 49.9 | 55,941 | 64,013 | 14.4 | 6.6 | 7,544 | 4,961 | 1,659 | 26,091 | 2.41 | 57.6 | 8.5 | 26.6 |
| Brown | 14.6 | 11.5 | 10.5 | 50.2 | 26,911 | 25,893 | -3.8 | -4.0 | 2,778 | 3,052 | -767 | 10,764 | 2.23 | 65.1 | 6.9 | 29.1 |
| Carlton | 14.4 | 10.5 | 7.6 | 47.5 | 31,671 | 35,384 | 11.7 | 1.1 | 3,756 | 3,848 | 501 | 13,613 | 2.45 | 67.9 | 10.0 | 25.7 |
| Carver | 13.6 | 8.2 | 5.2 | 50.4 | 70,205 | 91,014 | 29.6 | 17.1 | 11,860 | 5,087 | 8,818 | 36,754 | 2.75 | 75.0 | 6.6 | 20.1 |
| Cass | 16.8 | 16.0 | 10.7 | 48.9 | 27,150 | 28,567 | 5.2 | 4.8 | 3,268 | 3,240 | 1,347 | 13,164 | 2.19 | 66.6 | 9.6 | 27.8 |
| Chippewa | 13.6 | 11.5 | 10.2 | 49.9 | 13,088 | 12,436 | -5.0 | -5.5 | 1,626 | 1,569 | -741 | 5,209 | 2.24 | 62.3 | 7.9 | 33.0 |
| Chisago | 15.2 | 9.7 | 6.6 | 48.4 | 41,101 | 53,892 | 31.1 | 5.4 | 5,836 | 4,173 | 1,251 | 20,242 | 2.64 | 73.5 | 8.0 | 20.7 |
| Clay | 10.5 | 7.6 | 6.2 | 50.5 | 51,229 | 58,994 | 15.2 | 9.7 | 8,624 | 4,690 | 1,788 | 24,364 | 2.46 | 63.3 | 7.6 | 29.1 |
| Clearwater | 14.9 | 11.1 | 9.7 | 49.4 | 8,423 | 8,695 | 3.2 | 3.7 | 1,106 | 992 | 213 | 3,405 | 2.56 | 65.7 | 8.1 | 30.0 |
| Cook | 17.6 | 18.6 | 11.2 | 50.4 | 5,168 | 5,176 | 0.2 | 4.7 | 455 | 491 | 278 | 2,691 | 1.97 | 64.1 | 7.6 | 28.8 |
| Cottonwood | 13.4 | 11.9 | 11.6 | 49.7 | 12,167 | 11,687 | -3.9 | -3.8 | 1,519 | 1,404 | -557 | 4,846 | 2.27 | 64.5 | 9.1 | 32.5 |
| Crow Wing | 15.5 | 13.7 | 10.0 | 50.2 | 55,099 | 62,512 | 13.5 | 5.0 | 7,163 | 6,886 | 2,917 | 26,820 | 2.36 | 65.6 | 8.0 | 28.0 |
| Dakota | 13.5 | 9.1 | 6.1 | 50.7 | 355,904 | 398,582 | 12.0 | 8.3 | 53,431 | 25,440 | 5,396 | 161,488 | 2.59 | 69.3 | 10.2 | 24.5 |
| Dodge | 13.0 | 8.9 | 6.7 | 49.9 | 17,731 | 20,087 | 13.3 | 4.5 | 2,590 | 1,502 | -184 | 7,756 | 2.65 | 73.2 | 8.1 | 22.7 |
| Douglas | 14.7 | 13.0 | 10.6 | 49.8 | 32,821 | 36,011 | 9.7 | 6.4 | 4,256 | 4,271 | 2,363 | 16,663 | 2.22 | 65.2 | 5.8 | 29.1 |
| Faribault | 15.1 | 12.8 | 11.4 | 49.9 | 16,181 | 14,553 | -10.1 | -6.5 | 1,510 | 1,906 | -559 | 6,137 | 2.20 | 62.0 | 6.9 | 33.1 |

1. No spouse present.

| STATE County | Persons in group quarters, 2020 | Daytime Population, 2015-2019 Number | Employment/ residence ratio | Births, 2020 Total | Births, 2020 Rate[1] | Deaths, 2020 Number | Deaths, 2020 Rate[1] | Persons under 65 with no health insurance, 2019 Number | Percent | Medicare, 2020 Total beneficiaries | Enrolled in Original Medicare | Enrolled in Medicare Advantage | Crimes reported by county police or sheriff, 2019 Violent | Property |
|---|---|---|---|---|---|---|---|---|---|---|---|---|---|---|
| | 32 | 33 | 34 | 35 | 36 | 37 | 38 | 39 | 40 | 41 | 42 | 43 | 44 | 45 |
| MICHIGAN— Cont'd | | | | | | | | | | | | | | |
| Keweenaw | 10 | 1,986 | 0.84 | 12 | 5.7 | 26 | 12.3 | 78 | 6.0 | 794 | 503 | 290 | 1 | 13 |
| Lake | 110 | 10,481 | 0.64 | 87 | 7.5 | 177 | 15.3 | 734 | 9.2 | 3,971 | 2,229 | 1,742 | NA | NA |
| Lapeer | 1,487 | 74,665 | 0.66 | 815 | 9.3 | 968 | 11.0 | 5,380 | 7.7 | 19,892 | 9,958 | 9,934 | 55 | 180 |
| Leelanau | 284 | 19,940 | 0.82 | 143 | 6.6 | 240 | 11.0 | 1,315 | 8.9 | 7,614 | 3,947 | 3,667 | 7 | 39 |
| Lenawee | 4,956 | 87,449 | 0.75 | 1,012 | 10.3 | 1,093 | 11.2 | 5,492 | 7.3 | 23,247 | 13,107 | 10,140 | 45 | 169 |
| Livingston | 1,398 | 157,451 | 0.67 | 1,710 | 8.9 | 1,766 | 9.2 | 7,914 | 5.0 | 39,363 | 21,640 | 17,723 | 63 | 416 |
| Luce | 1,105 | 6,582 | 1.12 | 47 | 7.7 | 90 | 14.7 | 334 | 8.8 | 1,643 | 800 | 844 | 35 | 84 |
| Mackinac | 95 | 10,988 | 1.05 | 95 | 8.8 | 151 | 13.9 | 995 | 13.0 | 3,659 | 1,959 | 1,699 | 10 | 107 |
| Macomb | 7,392 | 800,697 | 0.83 | 9,123 | 10.5 | 9,161 | 10.5 | 52,226 | 7.3 | 179,076 | 97,487 | 81,589 | 272 | 1,046 |
| Manistee | 1,326 | 23,835 | 0.93 | 198 | 8.0 | 326 | 13.2 | 1,384 | 8.2 | 7,743 | 4,448 | 3,295 | 39 | 118 |
| Marquette | 3,876 | 67,145 | 1.02 | 571 | 8.7 | 727 | 11.0 | 3,194 | 6.4 | 15,543 | 9,211 | 6,332 | 15 | 87 |
| Mason | 442 | 28,975 | 1.00 | 244 | 8.4 | 333 | 11.4 | 1,801 | 8.2 | 8,480 | 4,634 | 3,846 | 36 | 138 |
| Mecosta | 3,344 | 43,120 | 0.99 | 413 | 9.4 | 384 | 8.7 | 2,583 | 8.1 | 9,642 | 4,881 | 4,761 | 99 | 378 |
| Menominee | 382 | 20,492 | 0.76 | 190 | 8.4 | 257 | 11.4 | 1,352 | 8.0 | 6,560 | 3,845 | 2,715 | 19 | 63 |
| Midland | 1,275 | 85,859 | 1.07 | 836 | 10.0 | 803 | 9.6 | 4,200 | 6.3 | 18,636 | 10,009 | 8,626 | 47 | 259 |
| Missaukee | 194 | 12,912 | 0.66 | 178 | 11.7 | 176 | 11.6 | 1,155 | 9.8 | 3,848 | 2,237 | 1,611 | 26 | 51 |
| Monroe | 1,463 | 125,702 | 0.65 | 1,440 | 9.6 | 1,611 | 10.7 | 6,706 | 5.5 | 34,146 | 18,656 | 15,490 | NA | NA |
| Montcalm | 2,923 | 57,288 | 0.77 | 711 | 11.2 | 670 | 10.6 | 4,111 | 8.3 | 14,604 | 7,290 | 7,313 | 18 | 109 |
| Montmorency | 156 | 8,983 | 0.91 | 82 | 8.8 | 181 | 19.4 | 550 | 8.9 | 3,932 | 2,313 | 1,620 | 9 | 17 |
| Muskegon | 4,681 | 164,861 | 0.89 | 2,027 | 11.7 | 1,909 | 11.0 | 9,617 | 6.9 | 39,501 | 16,484 | 23,017 | 68 | 421 |
| Newaygo | 557 | 43,403 | 0.75 | 499 | 10.1 | 576 | 11.7 | 3,402 | 8.8 | 12,322 | 5,023 | 7,299 | 40 | 241 |
| Oakland | 13,859 | 1,342,326 | 1.14 | 13,035 | 10.4 | 11,862 | 9.5 | 59,635 | 5.7 | 246,767 | 140,652 | 106,114 | 23 | 37 |
| Oceana | 303 | 23,714 | 0.75 | 268 | 10.0 | 280 | 10.4 | 2,305 | 11.3 | 6,890 | 3,561 | 3,328 | 46 | 186 |
| Ogemaw | 244 | 20,787 | 0.99 | 175 | 8.4 | 306 | 14.6 | 1,328 | 8.7 | 6,915 | 4,204 | 2,711 | 12 | 138 |
| Ontonagon | 83 | 5,436 | 0.78 | 23 | 4.1 | 118 | 20.9 | 289 | 8.2 | 2,429 | 1,614 | 816 | 8 | 14 |
| Osceola | 454 | 21,965 | 0.85 | 293 | 12.5 | 277 | 11.8 | 1,574 | 8.7 | 6,077 | 3,421 | 2,656 | 31 | 140 |
| Oscoda | 65 | 7,998 | 0.91 | 93 | 11.1 | 117 | 14.0 | 623 | 10.6 | 2,952 | 1,772 | 1,180 | 18 | 74 |
| Otsego | 347 | 25,753 | 1.12 | 229 | 9.2 | 315 | 12.7 | 1,457 | 7.7 | 6,537 | 4,029 | 2,508 | 0 | 2 |
| Ottawa | 8,957 | 276,396 | 0.93 | 3,119 | 10.6 | 2,141 | 7.3 | 14,136 | 5.9 | 52,775 | 17,335 | 35,441 | 571 | 2,081 |
| Presque Isle | 230 | 11,679 | 0.77 | 93 | 7.3 | 196 | 15.5 | 779 | 9.3 | 4,900 | 2,961 | 1,940 | 10 | 33 |
| Roscommon | 277 | 23,787 | 0.99 | 159 | 6.6 | 406 | 16.9 | 1,306 | 8.3 | 9,518 | 5,178 | 4,340 | NA | NA |
| Saginaw | 6,736 | 204,220 | 1.15 | 2,132 | 11.2 | 2,276 | 12.0 | 8,608 | 5.8 | 45,811 | 21,892 | 23,920 | 99 | 522 |
| St. Clair | 1,998 | 141,549 | 0.75 | 1,529 | 9.6 | 1,877 | 11.8 | 8,002 | 6.3 | 37,927 | 20,867 | 17,060 | 146 | 540 |
| St. Joseph | 764 | 58,624 | 0.92 | 755 | 12.4 | 625 | 10.3 | 4,767 | 9.7 | 13,108 | 7,944 | 5,164 | 67 | 224 |
| Sanilac | 564 | 38,028 | 0.81 | 384 | 9.4 | 525 | 12.9 | 2,996 | 9.5 | 10,722 | 6,648 | 4,074 | 34 | 87 |
| Schoolcraft | 140 | 7,914 | 0.95 | 70 | 8.6 | 120 | 14.8 | 612 | 10.6 | 2,625 | 1,560 | 1,065 | 1 | 4 |
| Shiawassee | 822 | 57,783 | 0.66 | 674 | 10.0 | 797 | 11.8 | 3,462 | 6.4 | 16,263 | 8,384 | 7,878 | 15 | 152 |
| Tuscola | 1,079 | 45,121 | 0.65 | 533 | 10.2 | 708 | 13.5 | 2,938 | 7.2 | 13,802 | 7,033 | 6,769 | 61 | 160 |
| Van Buren | 877 | 68,025 | 0.79 | 769 | 10.2 | 873 | 11.6 | 5,560 | 9.1 | 17,356 | 9,101 | 8,255 | 97 | 376 |
| Washtenaw | 20,772 | 407,698 | 1.22 | 3,455 | 9.4 | 2,644 | 7.2 | 15,815 | 5.3 | 61,158 | 37,272 | 23,886 | 574 | 1,655 |
| Wayne | 22,422 | 1,819,462 | 1.09 | 22,477 | 12.9 | 19,638 | 11.3 | 110,013 | 7.6 | 330,614 | 164,264 | 166,350 | 20 | 1 |
| Wexford | 389 | 35,452 | 1.16 | 369 | 10.9 | 362 | 10.7 | 2,084 | 7.8 | 8,388 | 4,852 | 3,536 | 25 | 260 |
| MINNESOTA | 129,373 | 5,578,030 | 1.01 | 66,492 | 11.8 | 46,882 | 8.3 | 268,829 | 5.8 | 1,045,928 | 539,499 | 506,429 | X | X |
| Aitkin | 274 | 14,528 | 0.79 | 114 | 7.2 | 248 | 15.6 | 810 | 7.9 | 5,765 | 2,139 | 3,626 | 11 | 167 |
| Anoka | 2,937 | 292,654 | 0.69 | 4,224 | 11.7 | 2,430 | 6.8 | 15,598 | 5.1 | 59,451 | 27,419 | 32,032 | 75 | 936 |
| Becker | 458 | 32,981 | 0.94 | 398 | 11.6 | 399 | 11.6 | 2,095 | 7.8 | 8,462 | 4,823 | 3,639 | 29 | 134 |
| Beltrami | 2,004 | 47,012 | 1.03 | 660 | 13.9 | 426 | 9.0 | 3,851 | 10.3 | 9,236 | 6,212 | 3,025 | 50 | 268 |
| Benton | 1,049 | 35,042 | 0.76 | 535 | 13.1 | 365 | 8.9 | 2,118 | 6.1 | 7,425 | 4,192 | 3,233 | 9 | 173 |
| Big Stone | 137 | 4,803 | 0.92 | 67 | 13.6 | 73 | 14.8 | 228 | 6.3 | 1,502 | 955 | 547 | 0 | 14 |
| Blue Earth | 4,150 | 71,487 | 1.13 | 724 | 10.6 | 497 | 7.3 | 3,206 | 5.8 | 11,811 | 7,724 | 4,087 | 24 | 105 |
| Brown | 1,111 | 26,958 | 1.13 | 257 | 10.3 | 309 | 12.4 | 971 | 5.2 | 6,280 | 4,210 | 2,069 | 5 | 31 |
| Carlton | 1,992 | 32,502 | 0.81 | 367 | 10.3 | 393 | 11.0 | 1,670 | 5.9 | 7,760 | 3,239 | 4,521 | 12 | 166 |
| Carver | 856 | 88,560 | 0.76 | 1,154 | 10.8 | 618 | 5.8 | 3,074 | 3.3 | 14,817 | 7,056 | 7,761 | 38 | 497 |
| Cass | 223 | 27,605 | 0.87 | 275 | 9.2 | 347 | 11.6 | 2,215 | 10.3 | 9,090 | 5,087 | 4,002 | 37 | 486 |
| Chippewa | 237 | 11,843 | 0.98 | 147 | 12.5 | 141 | 12.0 | 672 | 7.4 | 2,921 | 1,837 | 1,084 | 2 | 1 |
| Chisago | 1,676 | 44,306 | 0.62 | 593 | 10.4 | 456 | 8.0 | 2,071 | 4.4 | 10,505 | 4,934 | 5,571 | 12 | 152 |
| Clay | 2,917 | 52,227 | 0.67 | 839 | 13.0 | 420 | 6.5 | 2,724 | 5.1 | 10,239 | 6,690 | 3,549 | 7 | 58 |
| Clearwater | 115 | 8,003 | 0.78 | 106 | 11.8 | 87 | 9.6 | 683 | 9.9 | 2,062 | 1,139 | 923 | 10 | 47 |
| Cook | 51 | 5,506 | 1.05 | 37 | 6.8 | 49 | 9.0 | 378 | 9.9 | 1,754 | 721 | 1,033 | 10 | 47 |
| Cottonwood | 244 | 11,602 | 1.06 | 137 | 12.2 | 152 | 13.5 | 620 | 7.3 | 2,875 | 1,859 | 1,016 | 2 | 13 |
| Crow Wing | 736 | 65,922 | 1.06 | 639 | 9.7 | 743 | 11.3 | 3,190 | 6.5 | 17,798 | 9,537 | 8,261 | 15 | 210 |
| Dakota | 2,815 | 380,347 | 0.82 | 5,167 | 12.0 | 2,995 | 6.9 | 16,258 | 4.4 | 71,969 | 36,478 | 35,492 | 17 | 48 |
| Dodge | 149 | 16,606 | 0.64 | 256 | 12.2 | 159 | 7.6 | 860 | 4.9 | 3,497 | 2,194 | 1,302 | 11 | 104 |
| Douglas | 522 | 39,174 | 1.09 | 415 | 10.8 | 473 | 12.3 | 1,420 | 4.9 | 10,626 | 5,809 | 4,818 | 18 | 180 |
| Faribault | 336 | 12,720 | 0.84 | 143 | 10.5 | 160 | 11.8 | 711 | 7.0 | 3,696 | 2,369 | 1,327 | 12 | 27 |

1. Per 1,000 estimated resident population.

# Table B. States and Counties — Crime, Education, Money Income, and Poverty

| STATE County | County law enforcement employment, 2019 | | School enrollment and attainment, 2015-2019 | | | | Local government expenditures[3] 2017-2018 | | Money income, 2015-2019 | | | | Income and poverty, 2019 | | | |
| | | | Enrollment[1] | | Attainment[2] (percent) | | | | | Households | | | | Percent below poverty level | | |
| | Officers | Civilians | Total | Percent private | High school graduate or less | Bachelor's degree or more | Total current spending (mil dol) | Current spending per student (dollars) | Per capita income[4] | Median income (dollars) | Percent with income of less than $50,000 | Percent with income of $200,000 or more | Median household income (dollars) | All persons | Children under 18 years | Children 5 to 17 years in families |
| | 46 | 47 | 48 | 49 | 50 | 51 | 52 | 53 | 54 | 55 | 56 | 57 | 58 | 59 | 60 | 61 |
| **MICHIGAN— Cont'd** | | | | | | | | | | | | | | | | |
| Keweenaw | 6 | 0 | 297 | 2.0 | 39.0 | 32.0 | 0 | 62,250 | 34,608 | 50,292 | 49.7 | 2.9 | 47,318 | 10.3 | 15.0 | 13.7 |
| Lake | 15 | 21 | 1,915 | 12.2 | 58.1 | 11.9 | 8 | 15,684 | 20,461 | 37,320 | 64.1 | 1.1 | 38,044 | 20.2 | 33.2 | 33.6 |
| Lapeer | 49 | 31 | 18,676 | 9.7 | 44.9 | 18.3 | 128 | 10,868 | 30,344 | 62,641 | 39.4 | 3.5 | 66,415 | 10.1 | 12.8 | 11.4 |
| Leelanau | 20 | 1 | 3,724 | 18.6 | 26.0 | 44.7 | 30 | 15,040 | 39,073 | 65,249 | 37.6 | 6.9 | 72,888 | 7.1 | 9.7 | 8.9 |
| Lenawee | 42 | 64 | 22,685 | 17.4 | 46.0 | 21.0 | 187 | 12,343 | 27,850 | 55,450 | 45.0 | 2.8 | 55,466 | 11.8 | 15.1 | 13.7 |
| Livingston | 58 | 77 | 45,017 | 13.8 | 30.6 | 35.4 | 297 | 11,408 | 40,351 | 84,221 | 27.1 | 9.4 | 87,827 | 4.4 | 4.6 | 4.2 |
| Luce | 5 | 1 | 1,166 | 7.4 | 48.4 | 18.4 | 7 | 10,234 | 22,350 | 45,469 | 52.8 | 1.3 | 45,464 | 19.2 | 24.2 | 22.9 |
| Mackinac | 12 | 16 | 1,759 | 9.6 | 43.3 | 23.3 | 17 | 12,543 | 28,966 | 47,938 | 53.1 | 2.0 | 44,986 | 14.1 | 23.2 | 21.5 |
| Macomb | 272 | 273 | 200,948 | 11.5 | 39.8 | 24.9 | 1,442 | 11,742 | 32,238 | 62,855 | 39.7 | 4.4 | 65,051 | 8.6 | 11.7 | 10.3 |
| Manistee | 17 | 20 | 4,518 | 10.9 | 45.7 | 20.3 | 56 | 10,184 | 26,668 | 50,055 | 49.9 | 3.1 | 50,047 | 12.3 | 19.2 | 17.9 |
| Marquette | 24 | 40 | 17,884 | 8.8 | 35.9 | 32.9 | 102 | 12,232 | 27,979 | 53,970 | 46.3 | 2.8 | 54,913 | 11.8 | 12.9 | 11.3 |
| Mason | 20 | 20 | 5,901 | 11.9 | 42.4 | 23.1 | 61 | 15,244 | 29,549 | 51,725 | 48.4 | 3.4 | 49,869 | 13.9 | 23.4 | 23.0 |
| Mecosta | 20 | 29 | 13,422 | 9.3 | 45.8 | 22.7 | 77 | 13,415 | 23,431 | 45,018 | 54.5 | 2.1 | 47,705 | 19.9 | 23.5 | 20.3 |
| Menominee | 15 | 0 | 4,170 | 10.7 | 50.0 | 16.3 | 41 | 11,893 | 27,942 | 46,828 | 53.7 | 1.6 | 49,189 | 13.5 | 17.1 | 14.6 |
| Midland | 30 | 40 | 19,410 | 16.9 | 34.7 | 35.2 | 134 | 11,212 | 36,142 | 62,625 | 39.6 | 7.3 | 69,859 | 9.7 | 11.3 | 9.8 |
| Missaukee | 13 | 13 | 3,011 | 13.7 | 52.4 | 14.0 | 22 | 9,876 | 23,838 | 47,194 | 52.7 | 1.6 | 51,317 | 10.9 | 16.8 | 16.1 |
| Monroe | 74 | 80 | 34,382 | 13.4 | 43.8 | 20.5 | 265 | 11,760 | 31,481 | 62,203 | 39.2 | 3.5 | 64,341 | 10.2 | 14.3 | 11.3 |
| Montcalm | 17 | 25 | 13,676 | 13.1 | 51.8 | 13.3 | 114 | 12,186 | 24,200 | 49,448 | 50.7 | 2.1 | 52,089 | 12.4 | 18.6 | 16.5 |
| Montmorency | 13 | 0 | 1,386 | 11.4 | 56.0 | 13.7 | 8 | 11,250 | 23,958 | 41,772 | 59.9 | 0.8 | 40,126 | 14.4 | 24.7 | 23.5 |
| Muskegon | 46 | 61 | 39,551 | 11.4 | 42.7 | 19.7 | 322 | 11,629 | 25,435 | 50,854 | 49.2 | 2.6 | 50,730 | 13.5 | 17.3 | 16.7 |
| Newaygo | 27 | 36 | 10,582 | 13.4 | 52.3 | 16.6 | 91 | 12,265 | 25,451 | 50,326 | 49.7 | 2.2 | 51,772 | 15.7 | 24.0 | 22.7 |
| Oakland | 912 | 177 | 302,052 | 16.5 | 24.5 | 47.2 | 2,260 | 12,395 | 44,629 | 79,698 | 31.5 | 11.6 | 81,257 | 7.8 | 9.2 | 8.7 |
| Oceana | 21 | 14 | 5,610 | 12.6 | 49.1 | 19.5 | 35 | 10,933 | 24,345 | 50,104 | 49.9 | 1.8 | 52,319 | 14.8 | 23.4 | 20.7 |
| Ogemaw | 15 | 19 | 3,740 | 10.3 | 52.3 | 12.2 | 21 | 9,847 | 23,787 | 40,373 | 60.5 | 1.5 | 41,347 | 18.2 | 26.7 | 23.2 |
| Ontonagon | 6 | 4 | 780 | 10.3 | 50.9 | 16.5 | 15 | 30,770 | 25,022 | 41,546 | 59.0 | 0.6 | 45,531 | 13.3 | 23.0 | 20.1 |
| Osceola | 17 | 25 | 4,606 | 12.3 | 55.1 | 14.2 | 40 | 10,016 | 22,692 | 44,032 | 56.3 | 1.4 | 47,343 | 14.6 | 21.2 | 19.6 |
| Oscoda | 12 | 5 | 1,397 | 18.5 | 54.0 | 10.6 | 9 | 10,739 | 24,889 | 42,335 | 61.1 | 2.2 | 44,594 | 15.4 | 25.9 | 26.1 |
| Otsego | 10 | 18 | 5,153 | 16.7 | 41.1 | 23.6 | 40 | 10,256 | 27,234 | 54,332 | 46.6 | 2.2 | 52,392 | 10.0 | 14.8 | 13.5 |
| Ottawa | 140 | 99 | 83,021 | 18.9 | 35.3 | 34.1 | 542 | 11,907 | 31,872 | 69,314 | 33.8 | 5.3 | 69,827 | 7.8 | 7.2 | 6.5 |
| Presque Isle | 13 | 0 | 2,034 | 18.1 | 48.6 | 17.9 | 15 | 10,379 | 28,103 | 47,948 | 51.9 | 2.0 | 47,339 | 10.8 | 19.7 | 18.6 |
| Roscommon | 26 | 15 | 3,638 | 12.8 | 48.4 | 15.5 | 40 | 14,013 | 25,807 | 42,054 | 59.2 | 1.7 | 42,546 | 15.3 | 27.9 | 26.3 |
| Saginaw | 61 | 55 | 46,555 | 10.3 | 43.9 | 20.8 | 323 | 12,043 | 28,030 | 48,000 | 52.0 | 3.3 | 48,471 | 18.6 | 29.0 | 29.5 |
| St. Clair | 84 | 100 | 34,489 | 10.6 | 43.3 | 18.6 | 305 | 11,135 | 30,703 | 56,951 | 43.8 | 3.5 | 60,208 | 10.4 | 14.0 | 13.1 |
| St. Joseph | 26 | 34 | 13,680 | 9.4 | 49.6 | 16.2 | 122 | 11,646 | 26,248 | 52,086 | 48.2 | 2.2 | 54,024 | 12.9 | 16.1 | 15.4 |
| Sanilac | 27 | 30 | 8,371 | 11.3 | 52.8 | 14.2 | 69 | 11,118 | 25,871 | 47,672 | 52.5 | 1.9 | 49,917 | 13.6 | 21.6 | 19.3 |
| Schoolcraft | 3 | 7 | 1,283 | 9.4 | 53.5 | 15.6 | 8 | 9,948 | 24,647 | 45,500 | 54.6 | 1.4 | 50,256 | 14.3 | 22.9 | 21.6 |
| Shiawassee | 21 | 30 | 15,154 | 12.1 | 45.1 | 17.0 | 131 | 11,960 | 28,437 | 56,032 | 44.4 | 2.0 | 55,633 | 10.9 | 15.3 | 13.7 |
| Tuscola | 20 | 24 | 10,770 | 15.2 | 53.1 | 13.5 | 99 | 13,486 | 25,996 | 49,988 | 50.0 | 1.8 | 52,342 | 11.6 | 16.5 | 15.4 |
| Van Buren | 65 | 45 | 17,619 | 10.6 | 44.2 | 21.7 | 202 | 13,343 | 28,049 | 54,485 | 46.1 | 3.1 | 58,733 | 13.0 | 17.6 | 15.4 |
| Washtenaw | 125 | 177 | 126,009 | 9.3 | 19.3 | 55.9 | 611 | 13,774 | 41,399 | 72,586 | 35.2 | 11.1 | 76,728 | 13.0 | 12.1 | 11.0 |
| Wayne | 632 | 291 | 438,387 | 10.9 | 43.6 | 23.9 | 3,218 | 11,520 | 27,282 | 47,301 | 52.2 | 4.2 | 50,753 | 19.8 | 29.1 | 28.1 |
| Wexford | 24 | 29 | 7,228 | 11.9 | 46.8 | 18.2 | 79 | 13,801 | 25,149 | 47,193 | 53.3 | 2.1 | 53,146 | 12.4 | 18.7 | 17.3 |
| **MINNESOTA** | X | X | 1,395,300 | 15.4 | 31.5 | 36.1 | 11,376 | 12,937 | 37,625 | 71,306 | 34.8 | 7.8 | 74,529 | 8.9 | 11.0 | 10.4 |
| Aitkin | 18 | 32 | 2,568 | 7.3 | 45.6 | 18.1 | 26 | 13,183 | 29,275 | 49,351 | 50.7 | 2.4 | 51,936 | 12.5 | 19.7 | 17.6 |
| Anoka | 133 | 139 | 86,531 | 12.0 | 33.9 | 30.5 | 781 | 12,057 | 36,978 | 82,175 | 27.3 | 7.0 | 83,363 | 6.2 | 7.9 | 7.3 |
| Becker | 22 | 39 | 7,869 | 9.8 | 38.4 | 25.0 | 55 | 11,551 | 31,729 | 60,284 | 42.0 | 4.4 | 61,311 | 10.2 | 14.8 | 13.1 |
| Beltrami | 28 | 53 | 13,471 | 9.7 | 35.6 | 29.2 | 126 | 15,161 | 25,418 | 49,160 | 50.7 | 2.7 | 54,638 | 16.3 | 23.1 | 21.5 |
| Benton | 25 | 45 | 9,806 | 11.3 | 38.2 | 23.6 | 71 | 10,741 | 29,831 | 57,715 | 43.1 | 2.6 | 63,395 | 8.6 | 10.0 | 9.0 |
| Big Stone | 5 | 2 | 914 | 1.3 | 52.3 | 15.4 | 11 | 13,557 | 29,459 | 53,900 | 45.7 | 2.3 | 52,692 | 11.0 | 13.6 | 14.6 |
| Blue Earth | 34 | 55 | 22,341 | 8.6 | 31.9 | 34.8 | 131 | 11,507 | 29,677 | 57,429 | 43.6 | 3.5 | 60,408 | 15.2 | 12.6 | 10.0 |
| Brown | 12 | 25 | 6,132 | 30.1 | 42.6 | 22.2 | 47 | 13,614 | 31,237 | 61,120 | 39.9 | 2.7 | 62,904 | 7.4 | 9.3 | 8.6 |
| Carlton | 25 | 29 | 8,504 | 8.5 | 38.2 | 22.1 | 75 | 11,563 | 29,440 | 63,098 | 38.2 | 3.8 | 65,749 | 9.1 | 10.0 | 8.7 |
| Carver | 71 | 74 | 27,953 | 17.1 | 22.1 | 48.9 | 196 | 11,311 | 47,221 | 101,496 | 21.1 | 17.0 | 101,891 | 4.3 | 4.0 | 3.2 |
| Cass | 42 | 32 | 5,673 | 10.5 | 40.1 | 23.2 | 60 | 13,548 | 30,326 | 52,204 | 47.5 | 3.4 | 54,267 | 11.4 | 19.9 | 16.9 |
| Chippewa | 10 | 10 | 2,456 | 6.3 | 44.7 | 19.0 | 29 | 13,453 | 30,772 | 55,269 | 44.2 | 3.1 | 54,984 | 10.4 | 13.2 | 11.8 |
| Chisago | 44 | 55 | 12,779 | 10.9 | 39.1 | 21.5 | 89 | 12,036 | 35,021 | 83,464 | 26.1 | 5.9 | 84,811 | 6.2 | 6.4 | 5.7 |
| Clay | 34 | 61 | 19,600 | 18.6 | 28.8 | 34.2 | 119 | 11,253 | 30,768 | 65,269 | 37.6 | 3.8 | 64,874 | 11.9 | 12.4 | 12.0 |
| Clearwater | 10 | 10 | 2,093 | 8.5 | 48.8 | 17.1 | 18 | 11,661 | 27,178 | 50,386 | 49.6 | 2.1 | 54,896 | 11.5 | 15.0 | 13.8 |
| Cook | 12 | 8 | 855 | 8.2 | 26.9 | 40.4 | 9 | 13,279 | 33,194 | 57,432 | 42.9 | 3.2 | 58,664 | 8.9 | 11.7 | 10.7 |
| Cottonwood | 10 | 11 | 2,460 | 10.8 | 46.6 | 20.0 | 26 | 13,230 | 27,459 | 52,087 | 48.0 | 1.6 | 53,725 | 10.8 | 15.2 | 14.5 |
| Crow Wing | 41 | 83 | 12,995 | 9.6 | 35.8 | 24.7 | 116 | 11,841 | 32,368 | 56,549 | 43.8 | 3.8 | 60,384 | 10.5 | 14.1 | 12.4 |
| Dakota | 83 | 114 | 106,800 | 14.9 | 25.4 | 42.2 | 943 | 12,615 | 41,552 | 86,036 | 27.0 | 10.2 | 89,075 | 6.3 | 7.6 | 5.7 |
| Dodge | 24 | 14 | 5,378 | 9.0 | 38.4 | 25.7 | 39 | 9,887 | 34,575 | 74,575 | 29.4 | 4.7 | 86,436 | 5.6 | 6.3 | 5.7 |
| Douglas | 36 | 49 | 7,939 | 11.8 | 34.0 | 26.0 | 64 | 11,636 | 36,535 | 63,819 | 38.0 | 4.6 | 67,592 | 6.3 | 7.5 | 7.5 |
| Faribault | 14 | 16 | 2,771 | 7.1 | 45.1 | 18.6 | 22 | 12,132 | 29,991 | 53,156 | 46.1 | 2.4 | 54,193 | 12.2 | 16.8 | 16.2 |

1. All persons 3 years old and over enrolled in nursery school through college.   2. Persons 25 years old and over.   3. Elementary and secondary education expenditures.   4. Based on population estimated by the American Community Survey, 2014–2018.

# Table B. States and Counties — **Personal Income**

| STATE County | Personal income, 2019 | | | | | | | | | | Earnings, 2019 | | |
|---|---|---|---|---|---|---|---|---|---|---|---|---|---|
| | Total (mil dol) | Percent change 2018-2019 | Per capita¹ Dollars | Per capita¹ Rank | Wages and salaries (mil dol) | Supplements to wages and salaries, employer contributions (mil dol) Pension and insurance | Supplements to wages and salaries, employer contributions (mil dol) Government social insurance | Proprietors' income (mil dol) | Dividends, interest, and rent (mil dol) | Personal transfer reecipts (mil dol) | Total (mil dol) | Contributions for government social insurance (mil dol) From employee and self-employed | Contributions for government social insurance (mil dol) From employer |
| | 62 | 63 | 64 | 65 | 66 | 67 | 68 | 69 | 70 | 71 | 72 | 73 | 74 |
| MICHIGAN— Cont'd | | | | | | | | | | | | | |
| Keweenaw | 101 | 3.3 | 47,963 | 1,018 | 14 | 3 | 1 | 4 | 26 | 33 | 23 | 3 | 1 |
| Lake | 396 | 4.3 | 33,368 | 2,860 | 71 | 15 | 6 | 18 | 82 | 177 | 109 | 13 | 6 |
| Lapeer | 3,917 | 2.9 | 44,706 | 1,395 | 948 | 181 | 72 | 220 | 525 | 904 | 1,420 | 108 | 72 |
| Leelanau | 1,486 | 3.3 | 68,281 | 146 | 278 | 53 | 22 | 154 | 530 | 271 | 506 | 39 | 22 |
| Lenawee | 4,030 | 3.9 | 40,932 | 1,925 | 1,196 | 234 | 89 | 172 | 578 | 1,026 | 1,690 | 128 | 89 |
| Livingston | 11,305 | 3.4 | 58,879 | 341 | 3,163 | 532 | 234 | 606 | 1,691 | 1,619 | 4,535 | 308 | 234 |
| Luce | 200 | 2.0 | 32,174 | 2,935 | 71 | 18 | 5 | 9 | 41 | 81 | 103 | 8 | 5 |
| Mackinac | 487 | 4.1 | 45,138 | 1,345 | 174 | 33 | 16 | 25 | 122 | 152 | 248 | 19 | 16 |
| Macomb | 41,502 | 3.2 | 47,487 | 1,078 | 20,198 | 3,107 | 1,486 | 2,521 | 5,847 | 8,355 | 27,312 | 1,784 | 1,486 |
| Manistee | 967 | 3.7 | 39,389 | 2,128 | 325 | 72 | 24 | 42 | 205 | 354 | 463 | 37 | 24 |
| Marquette | 2,719 | 3.2 | 40,772 | 1,952 | 1,251 | 245 | 95 | 99 | 497 | 766 | 1,690 | 116 | 95 |
| Mason | 1,209 | 3.1 | 41,488 | 1,851 | 463 | 91 | 36 | 57 | 233 | 391 | 647 | 48 | 36 |
| Mecosta | 1,481 | 2.7 | 34,090 | 2,791 | 578 | 135 | 43 | 52 | 252 | 441 | 807 | 58 | 43 |
| Menominee | 1,005 | 2.6 | 44,114 | 1,470 | 290 | 65 | 22 | 52 | 202 | 272 | 430 | 33 | 22 |
| Midland | 4,654 | 2.9 | 55,972 | 446 | 2,495 | 397 | 170 | 342 | 894 | 849 | 3,405 | 220 | 170 |
| Missaukee | 558 | 6.9 | 36,922 | 2,454 | 145 | 28 | 12 | 73 | 85 | 164 | 257 | 18 | 12 |
| Monroe | 7,311 | 3.9 | 48,581 | 959 | 2,179 | 400 | 157 | 456 | 977 | 1,516 | 3,192 | 225 | 157 |
| Montcalm | 2,270 | 3.6 | 35,537 | 2,615 | 711 | 150 | 54 | 118 | 297 | 651 | 1,033 | 77 | 54 |
| Montmorency | 355 | 4.2 | 38,079 | 2,321 | 87 | 18 | 7 | 20 | 73 | 161 | 131 | 14 | 7 |
| Muskegon | 6,880 | 3.5 | 39,637 | 2,097 | 2,916 | 534 | 219 | 311 | 1,005 | 1,964 | 3,981 | 279 | 219 |
| Newaygo | 1,882 | 4.3 | 38,429 | 2,276 | 550 | 108 | 42 | 124 | 266 | 537 | 825 | 62 | 42 |
| Oakland | 92,145 | 2.8 | 73,271 | 100 | 51,512 | 6,667 | 3,690 | 8,757 | 19,227 | 11,282 | 70,626 | 4,308 | 3,690 |
| Oceana | 1,028 | 3.6 | 38,856 | 2,215 | 290 | 58 | 23 | 53 | 185 | 312 | 424 | 33 | 23 |
| Ogemaw | 730 | -0.2 | 34,770 | 2,720 | 218 | 42 | 17 | 39 | 126 | 305 | 316 | 28 | 17 |
| Ontonagon | 237 | 3.7 | 41,363 | 1,866 | 44 | 11 | 3 | 10 | 47 | 103 | 69 | 8 | 3 |
| Osceola | 883 | 5.5 | 37,636 | 2,368 | 370 | 64 | 29 | 57 | 121 | 267 | 519 | 38 | 29 |
| Oscoda | 285 | 3.7 | 34,549 | 2,740 | 62 | 13 | 5 | 26 | 54 | 121 | 106 | 10 | 5 |
| Otsego | 1,034 | 4.1 | 41,925 | 1,781 | 471 | 82 | 36 | 63 | 192 | 273 | 652 | 46 | 36 |
| Ottawa | 14,400 | 3.2 | 49,345 | 893 | 6,665 | 1,166 | 495 | 945 | 2,720 | 2,166 | 9,272 | 577 | 495 |
| Presque Isle | 523 | 3.3 | 41,560 | 1,841 | 119 | 27 | 9 | 20 | 111 | 190 | 175 | 17 | 9 |
| Roscommon | 901 | 3.1 | 37,507 | 2,387 | 211 | 45 | 16 | 39 | 196 | 409 | 310 | 33 | 16 |
| Saginaw | 7,723 | 2.4 | 40,533 | 1,984 | 4,167 | 733 | 312 | 408 | 1,163 | 2,362 | 5,620 | 380 | 312 |
| St. Clair | 7,266 | 3.7 | 45,662 | 1,289 | 2,264 | 435 | 171 | 365 | 1,073 | 1,795 | 3,235 | 235 | 171 |
| St. Joseph | 2,374 | 2.3 | 38,936 | 2,204 | 934 | 179 | 69 | 121 | 360 | 608 | 1,303 | 91 | 69 |
| Sanilac | 1,602 | 3.0 | 38,917 | 2,208 | 446 | 88 | 35 | 134 | 273 | 524 | 703 | 52 | 35 |
| Schoolcraft | 332 | 4.2 | 41,073 | 1,912 | 123 | 28 | 9 | 10 | 60 | 126 | 170 | 13 | 9 |
| Shiawassee | 2,761 | 3.6 | 40,535 | 1,982 | 719 | 145 | 54 | 147 | 367 | 750 | 1,066 | 83 | 54 |
| Tuscola | 2,021 | 3.6 | 38,676 | 2,240 | 521 | 106 | 40 | 110 | 296 | 659 | 777 | 64 | 40 |
| Van Buren | 3,161 | 3.0 | 41,773 | 1,807 | 1,033 | 210 | 78 | 158 | 488 | 819 | 1,479 | 104 | 78 |
| Washtenaw | 22,366 | 2.9 | 60,843 | 281 | 13,037 | 2,544 | 914 | 1,394 | 4,963 | 2,796 | 17,888 | 1,043 | 914 |
| Wayne | 77,867 | 3.0 | 44,512 | 1,417 | 49,126 | 7,361 | 3,529 | 4,768 | 12,055 | 21,011 | 64,784 | 4,109 | 3,529 |
| Wexford | 1,252 | 4.8 | 37,240 | 2,420 | 617 | 126 | 47 | 73 | 184 | 390 | 863 | 59 | 47 |
| MINNESOTA | 331,802 | 3.2 | 58,830 | X | 178,863 | 25,890 | 12,937 | 25,453 | 65,074 | 52,837 | 243,143 | 14,933 | 12,937 |
| Aitkin | 657 | 3.1 | 41,327 | 1,873 | 164 | 33 | 13 | 33 | 152 | 226 | 242 | 21 | 13 |
| Anoka | 18,649 | 3.3 | 52,250 | 661 | 7,518 | 1,098 | 571 | 1,000 | 2,538 | 2,918 | 10,187 | 655 | 571 |
| Becker | 1,693 | 4.4 | 49,190 | 907 | 634 | 112 | 50 | 162 | 307 | 429 | 958 | 64 | 50 |
| Beltrami | 1,992 | 3.4 | 42,212 | 1,737 | 925 | 173 | 71 | 100 | 357 | 561 | 1,270 | 83 | 71 |
| Benton | 1,859 | 4.1 | 45,458 | 1,307 | 824 | 129 | 65 | 147 | 260 | 369 | 1,165 | 74 | 65 |
| Big Stone | 277 | 3.4 | 55,491 | 473 | 73 | 15 | 6 | 29 | 62 | 76 | 122 | 8 | 6 |
| Blue Earth | 3,166 | 3.8 | 46,793 | 1,151 | 1,950 | 318 | 149 | 266 | 609 | 565 | 2,683 | 160 | 149 |
| Brown | 1,398 | 3.1 | 55,904 | 450 | 614 | 116 | 49 | 157 | 296 | 293 | 936 | 56 | 49 |
| Carlton | 1,592 | 3.9 | 44,391 | 1,434 | 637 | 120 | 49 | 54 | 233 | 417 | 859 | 59 | 49 |
| Carver | 7,656 | 3.8 | 72,852 | 103 | 2,424 | 385 | 178 | 570 | 1,262 | 647 | 3,557 | 216 | 178 |
| Cass | 1,429 | 3.2 | 47,981 | 1,016 | 375 | 79 | 30 | 95 | 343 | 461 | 579 | 47 | 30 |
| Chippewa | 605 | 4.5 | 51,230 | 738 | 252 | 48 | 20 | 55 | 123 | 142 | 375 | 23 | 20 |
| Chisago | 2,834 | 4.1 | 50,093 | 826 | 766 | 123 | 61 | 161 | 393 | 495 | 1,112 | 77 | 61 |
| Clay | 2,827 | 3.3 | 44,023 | 1,481 | 940 | 166 | 77 | 191 | 429 | 570 | 1,373 | 91 | 77 |
| Clearwater | 395 | 2.1 | 44,765 | 1,389 | 120 | 23 | 10 | 29 | 79 | 109 | 181 | 13 | 10 |
| Cook | 300 | 3.1 | 54,964 | 492 | 102 | 20 | 8 | 35 | 85 | 70 | 165 | 12 | 8 |
| Cottonwood | 577 | 5.4 | 51,543 | 720 | 211 | 42 | 17 | 105 | 115 | 140 | 375 | 21 | 17 |
| Crow Wing | 2,972 | 2.7 | 45,685 | 1,284 | 1,343 | 223 | 105 | 222 | 574 | 810 | 1,893 | 134 | 105 |
| Dakota | 26,562 | 3.3 | 61,914 | 252 | 12,008 | 1,747 | 863 | 1,836 | 4,337 | 3,386 | 16,454 | 1,018 | 863 |
| Dodge | 998 | 3.2 | 47,665 | 1,061 | 317 | 54 | 25 | 63 | 152 | 163 | 459 | 30 | 25 |
| Douglas | 2,010 | 1.7 | 52,691 | 638 | 898 | 155 | 69 | 172 | 422 | 454 | 1,294 | 88 | 69 |
| Faribault | 641 | 8.8 | 46,941 | 1,140 | 200 | 37 | 16 | 76 | 136 | 173 | 329 | 22 | 16 |

1. Based on the resident population estimated as of July 1 of the year shown.

**Earnings, Social Security, and Housing**

| STATE County | Earnings, 2019 (cont.) | | | | | | | | | Social Security beneficiaries, December 2019 | | | Housing units, 2020 | |
|---|---|---|---|---|---|---|---|---|---|---|---|---|---|---|
| | Percent by selected industries | | | | | | | | | | | Supplemental Security Income recipients, 2019 | | |
| | Farm | Mining, quarrying, and extractions | Construction | Manufacturing | Information; professional, scientific, technical services | Retail trade | Finance, insurance, real estate, and leasing | Health care and social assistance | Government | Number | Rate[1] | | Total | Percent change, 2010-2020 |
| | 75 | 76 | 77 | 78 | 79 | 80 | 81 | 82 | 83 | 84 | 85 | 86 | 87 | 88 |
| MICHIGAN— Cont'd | | | | | | | | | | | | | | |
| Keweenaw | 0.0 | D | 14.6 | D | D | 5.7 | D | 0.6 | 30.1 | 820 | 393 | 29 | 2,522 | 2.2 |
| Lake | 0.5 | 0.2 | 7.4 | D | D | 5.0 | 4.4 | 22.7 | 28.3 | 4,380 | 373 | 643 | 15,554 | 3.9 |
| Lapeer | 1.1 | 0.7 | 10.2 | 21.7 | 3.6 | 8.5 | 5.0 | 6.5 | 20.0 | 22,045 | 251 | 1,245 | 37,102 | 2.1 |
| Leelanau | 1.3 | D | 13.4 | 4.6 | D | 5.4 | 7.3 | 8.5 | 21.9 | 7,385 | 340 | 107 | 16,063 | 7.5 |
| Lenawee | 1.3 | D | 6.2 | 24.2 | 3.5 | 7.1 | 6.5 | 10.0 | 19.0 | 25,140 | 255 | 1,930 | 44,022 | 1.3 |
| Livingston | 0.3 | 0.1 | 10.8 | 18.5 | 9.9 | 8.3 | 6.2 | 10.2 | 10.4 | 41,320 | 215 | 1,216 | 78,386 | 7.7 |
| Luce | 0.7 | 0.0 | 4.1 | 4.1 | D | 8.2 | 5.4 | D | 51.7 | 1,825 | 291 | 197 | 4,393 | 1.2 |
| Mackinac | 1.2 | D | 6.3 | 2.2 | D | 9.6 | D | D | 23.3 | 3,860 | 357 | 215 | 11,336 | 3.0 |
| Macomb | 0.1 | 0.0 | 7.8 | 22.4 | 14.6 | 6.6 | 4.5 | 9.8 | 12.7 | 188,235 | 216 | 23,009 | 371,202 | 4.1 |
| Manistee | 0.1 | D | 6.6 | 18.8 | 2.9 | 8.1 | 2.7 | D | 37.4 | 8,155 | 331 | 651 | 15,931 | 1.5 |
| Marquette | 0.0 | 9.4 | 5.7 | 4.7 | 6.7 | 6.8 | 4.2 | 19.8 | 22.1 | 16,935 | 254 | 1,055 | 35,021 | 2.0 |
| Mason | 1.7 | 1.7 | 7.8 | 20.0 | D | 7.8 | 4.3 | 11.1 | 18.2 | 9,095 | 312 | 687 | 17,800 | 3.0 |
| Mecosta | 0.7 | 0.1 | 7.1 | 14.4 | D | 10.5 | 3.2 | 8.1 | 36.4 | 10,475 | 240 | 1,130 | 22,019 | 4.2 |
| Menominee | 3.5 | 0.4 | 3.5 | 33.5 | D | 4.4 | 3.1 | D | 24.4 | 7,125 | 313 | 459 | 14,385 | 1.1 |
| Midland | 0.0 | D | 5.9 | 27.8 | 4.1 | 3.8 | 3.0 | 11.0 | 6.6 | 20,100 | 241 | 1,522 | 37,306 | 3.8 |
| Missaukee | 17.2 | D | 8.9 | 17.1 | 3.1 | 7.6 | 3.6 | D | 13.4 | 4,160 | 276 | 330 | 9,339 | 2.4 |
| Monroe | 0.8 | D | 6.5 | 19.4 | 8.1 | 6.0 | 3.5 | 8.3 | 11.9 | 36,480 | 243 | 2,523 | 65,530 | 4.1 |
| Montcalm | 2.2 | D | 8.0 | 22.4 | D | 7.9 | 3.3 | D | 20.8 | 16,150 | 254 | 1,820 | 28,899 | 2.4 |
| Montmorency | 1.4 | D | 9.3 | 13.3 | 2.2 | 5.6 | D | 12.9 | 16.5 | 4,310 | 463 | 354 | 9,620 | 0.3 |
| Muskegon | 0.4 | 0.1 | 6.6 | 26.6 | 4.4 | 10.4 | 3.8 | 16.2 | 14.0 | 43,430 | 250 | 6,091 | 74,776 | 1.6 |
| Newaygo | 4.4 | D | 8.3 | 19.6 | 6.1 | 10.6 | 4.5 | 7.3 | 17.7 | 13,360 | 273 | 1,317 | 25,946 | 3.5 |
| Oakland | 0.0 | 0.1 | 4.7 | 10.8 | 22.8 | 5.7 | 12.5 | 11.2 | 5.7 | 248,720 | 198 | 20,970 | 546,788 | 3.7 |
| Oceana | 4.4 | 0.6 | 8.0 | 24.7 | 2.1 | 7.3 | 3.3 | 2.5 | 22.1 | 7,485 | 282 | 819 | 16,324 | 2.4 |
| Ogemaw | 4.5 | 0.6 | 9.0 | 4.4 | 3.3 | 15.4 | 3.9 | D | 14.9 | 7,605 | 363 | 745 | 16,273 | 1.4 |
| Ontonagon | 0.7 | 0.1 | 4.8 | 1.0 | D | 17.1 | D | 13.3 | 30.9 | 2,635 | 461 | 136 | 5,711 | 0.7 |
| Osceola | 2.2 | 0.6 | 14.8 | 24.5 | D | 6.1 | 1.5 | 8.9 | 12.5 | 6,835 | 292 | 787 | 13,898 | 2.0 |
| Oscoda | 1.8 | 0.2 | 11.3 | 24.7 | 3.8 | 8.5 | 3.5 | 1.8 | 20.3 | 3,145 | 378 | 263 | 9,295 | 1.9 |
| Otsego | 0.1 | 4.8 | 7.4 | 12.5 | 3.6 | 14.0 | 5.4 | D | 14.8 | 7,165 | 291 | 576 | 14,944 | 1.4 |
| Ottawa | 1.1 | 0.1 | 7.7 | 36.2 | 5.6 | 5.1 | 4.5 | 5.4 | 12.8 | 54,820 | 188 | 2,564 | 113,280 | 10.5 |
| Presque Isle | 2.5 | D | 6.1 | 4.5 | 3.1 | 7.4 | 4.5 | D | 21.9 | 5,370 | 426 | 377 | 10,512 | 0.8 |
| Roscommon | 0.0 | 2.0 | 9.3 | D | 2.8 | 17.0 | 4.4 | D | 26.9 | 10,605 | 442 | 948 | 24,627 | 0.7 |
| Saginaw | 0.1 | 0.3 | 4.8 | 18.9 | 6.3 | 7.6 | 6.3 | 19.2 | 13.7 | 50,150 | 263 | 8,483 | 88,162 | 1.5 |
| St. Clair | 0.2 | 0.1 | 9.0 | 18.5 | 4.2 | 7.3 | 4.4 | 14.3 | 16.7 | 41,200 | 259 | 3,585 | 72,860 | 1.4 |
| St. Joseph | 1.1 | 0.0 | 5.2 | 42.5 | 3.1 | 6.8 | 2.7 | 7.8 | 14.3 | 14,510 | 239 | 1,300 | 28,114 | 1.2 |
| Sanilac | 9.3 | 0.8 | 7.0 | 25.7 | 2.4 | 8.0 | 5.0 | D | 14.9 | 11,665 | 283 | 968 | 23,182 | 2.0 |
| Schoolcraft | 0.4 | 9.5 | 5.9 | 8.7 | 3.2 | 7.1 | 7.1 | 3.7 | 38.9 | 2,745 | 340 | 237 | 6,397 | 1.3 |
| Shiawassee | 1.1 | D | 7.6 | 13.6 | 6.3 | 9.1 | 3.7 | 14.8 | 19.6 | 17,820 | 262 | 1,659 | 30,292 | -0.1 |
| Tuscola | 1.8 | D | 8.1 | 14.6 | D | 6.7 | 3.8 | D | 24.9 | 15,325 | 293 | 1,418 | 24,553 | 0.4 |
| Van Buren | 3.1 | D | 6.1 | 12.7 | 11.9 | 5.6 | 2.5 | 6.8 | 22.1 | 18,955 | 251 | 2,160 | 37,697 | 2.5 |
| Washtenaw | 0.0 | D | 3.1 | 7.2 | 18.6 | 3.8 | 4.4 | 11.0 | 33.9 | 60,730 | 165 | 5,343 | 153,091 | 3.6 |
| Wayne | 0.0 | 0.1 | 3.9 | 14.4 | 13.2 | 4.3 | 6.4 | 13.0 | 11.7 | 357,785 | 204 | 82,982 | 815,330 | -0.7 |
| Wexford | 0.1 | 0.3 | 4.4 | 25.4 | D | 7.0 | 3.8 | 12.5 | 17.3 | 9,295 | 276 | 1,136 | 17,139 | 2.4 |
| MINNESOTA | 1.2 | 0.3 | 6.1 | 12.1 | 11.2 | 5.2 | 9.9 | 13.3 | 12.8 | 1,053,166 | 187 | 93,151 | 2,503,126 | 6.6 |
| Aitkin | -0.1 | 2.4 | 8.9 | 9.6 | D | 8.2 | 3.9 | D | 21.8 | 4,945 | 312 | 262 | 17,144 | 7.0 |
| Anoka | 0.4 | 0.0 | 10.1 | 23.6 | 5.2 | 6.3 | 4.2 | 11.9 | 12.1 | 59,620 | 167 | 4,606 | 135,831 | 7.2 |
| Becker | 4.4 | 0.4 | 9.1 | 16.1 | 3.2 | 7.6 | 5.0 | 11.8 | 23.1 | 9,130 | 265 | 651 | 20,304 | 8.1 |
| Beltrami | 0.4 | D | 8.9 | 4.4 | 4.2 | 7.9 | 3.4 | 22.9 | 30.4 | 9,715 | 206 | 1,172 | 22,114 | 7.7 |
| Benton | 2.9 | 0.0 | 18.7 | 20.6 | D | 6.2 | 2.9 | 7.9 | 10.4 | 7,695 | 188 | 924 | 17,729 | 9.9 |
| Big Stone | 10.4 | D | 14.9 | 0.8 | 1.3 | 4.1 | D | D | 30.3 | 1,515 | 305 | 107 | 3,175 | 1.9 |
| Blue Earth | 3.3 | D | 7.4 | 12.7 | 6.7 | 7.9 | 5.1 | 21.8 | 15.0 | 11,650 | 172 | 1,012 | 29,333 | 11.9 |
| Brown | 8.8 | D | 7.4 | 21.7 | 7.2 | 5.5 | 3.9 | 12.8 | 12.1 | 6,485 | 260 | 256 | 11,830 | 2.9 |
| Carlton | -0.2 | D | 9.1 | 13.6 | 3.0 | 5.4 | 5.4 | 11.5 | 36.7 | 8,215 | 229 | 573 | 16,398 | 4.7 |
| Carver | 0.4 | 0.0 | 9.5 | 24.8 | 6.9 | 4.4 | 6.3 | 11.3 | 10.4 | 14,595 | 139 | 572 | 40,299 | 16.7 |
| Cass | 1.1 | 0.0 | 9.5 | 6.2 | D | 7.4 | 4.5 | D | 37.9 | 9,990 | 337 | 631 | 26,508 | 6.4 |
| Chippewa | 9.2 | D | 5.2 | 19.5 | D | 5.4 | 6.1 | D | 21.7 | 2,735 | 233 | 132 | 5,708 | -0.2 |
| Chisago | 0.7 | D | 12.3 | 14.5 | 9.6 | 8.0 | 2.7 | 19.8 | 14.9 | 10,960 | 194 | 570 | 22,608 | 6.8 |
| Clay | 1.4 | D | 6.2 | 8.3 | 5.2 | 7.1 | 4.3 | D | 22.9 | 10,525 | 163 | 1,215 | 27,385 | 14.3 |
| Clearwater | 4.2 | D | 19.1 | 22.1 | D | 3.7 | D | 13.0 | 16.8 | 2,230 | 252 | 191 | 4,806 | 0.7 |
| Cook | 0.0 | 0.1 | 10.1 | D | 2.4 | 7.8 | 3.6 | 4.0 | 33.0 | 1,745 | 322 | 32 | 6,311 | 8.1 |
| Cottonwood | 18.7 | D | 5.7 | 22.1 | D | 4.7 | 3.9 | D | 14.4 | 3,105 | 276 | 217 | 5,435 | 0.4 |
| Crow Wing | 0.1 | D | 10.8 | 9.1 | 6.4 | 10.4 | 7.0 | 20.5 | 14.6 | 18,620 | 286 | 1,122 | 43,149 | 7.4 |
| Dakota | 0.2 | 0.1 | 7.8 | 12.2 | 11.3 | 7.2 | 12.2 | 7.9 | 11.3 | 70,825 | 165 | 4,616 | 172,028 | 7.8 |
| Dodge | 2.0 | D | 13.6 | 27.2 | 2.4 | 3.4 | D | D | 16.7 | 3,660 | 174 | 175 | 8,480 | 6.7 |
| Douglas | 0.5 | 0.1 | 9.9 | 20.2 | 4.5 | 8.9 | 5.7 | 10.3 | 19.4 | 10,875 | 286 | 462 | 21,702 | 9.0 |
| Faribault | 12.7 | 0.0 | 5.5 | 19.6 | 4.0 | 4.4 | 5.9 | 10.6 | 15.0 | 3,810 | 279 | 245 | 7,037 | -0.7 |

1. Per 1,000 resident population estimated as of July 1 of the year shown.

# Table B. States and Counties — Housing, Labor Force, and Employment

| STATE County | Owner-occupied Total | Percent | Median value[1] | Median owner cost, With a mortgage | Median owner cost, Without a mortgage[2] | Renter-occupied Median rent[3] | Median rent as a percent of income[2] | Substandard units[4] (percent) | Civilian labor force Total | Percent change, 2019-2020 | Unemployment Total | Rate[5] | Civilian employment Total | Management, business, science, and arts (percent) | Construction, production, and maintenance occupations (percent) |
|---|---|---|---|---|---|---|---|---|---|---|---|---|---|---|---|
| | 89 | 90 | 91 | 92 | 93 | 94 | 95 | 96 | 97 | 98 | 99 | 100 | 101 | 102 | 103 |
| **MICHIGAN— Cont'd** | | | | | | | | | | | | | | | |
| Keweenaw | 1,081 | 88.0 | 121,100 | 17.7 | 11.9 | 525 | 19.1 | 2.7 | 884 | -1.9 | 79 | 8.9 | 822 | 42.1 | 17.6 |
| Lake | 4,631 | 84.2 | 88,000 | 24.1 | 13.0 | 589 | 31.5 | 2.1 | 3,989 | 5.8 | 429 | 10.8 | 3,881 | 23.6 | 32.2 |
| Lapeer | 33,700 | 83.9 | 168,800 | 19.9 | 10.9 | 788 | 30.8 | 1.8 | 40,138 | -3.3 | 5,015 | 12.5 | 40,033 | 31.6 | 33.5 |
| Leelanau | 9,139 | 88.3 | 268,400 | 22.9 | 11.0 | 959 | 31.0 | 0.9 | 10,261 | -2.0 | 816 | 8.0 | 9,850 | 37.9 | 21.7 |
| Lenawee | 38,345 | 77.5 | 139,400 | 20.2 | 12.1 | 767 | 27.4 | 1.4 | 45,597 | -1.8 | 4,105 | 9.0 | 44,628 | 29.6 | 31.9 |
| Livingston | 72,000 | 85.4 | 247,100 | 19.1 | 10.5 | 1,053 | 28.2 | 1.0 | 99,598 | -5.2 | 8,779 | 8.8 | 97,775 | 41.5 | 20.9 |
| Luce | 2,231 | 79.5 | 87,200 | 19.1 | 10.3 | 717 | 27.1 | 0.9 | 2,257 | -0.5 | 194 | 8.6 | 2,085 | 31.9 | 24.3 |
| Mackinac | 5,269 | 71.9 | 136,600 | 22.0 | 11.6 | 602 | 27.1 | 2.4 | 5,003 | -3.6 | 668 | 13.4 | 4,519 | 28.3 | 25.8 |
| Macomb | 346,402 | 73.3 | 166,800 | 19.5 | 12.6 | 962 | 29.3 | 2.1 | 439,510 | -2.9 | 52,568 | 12.0 | 427,598 | 35.6 | 25.0 |
| Manistee | 9,426 | 83.4 | 125,400 | 21.2 | 11.8 | 716 | 28.3 | 1.4 | 10,270 | -1.4 | 1,106 | 10.8 | 9,518 | 30.0 | 24.5 |
| Marquette | 26,552 | 70.8 | 153,400 | 19.1 | 10.7 | 738 | 29.7 | 1.5 | 31,758 | -2.2 | 2,637 | 8.3 | 30,540 | 33.8 | 20.8 |
| Mason | 12,186 | 76.9 | 141,400 | 21.1 | 11.6 | 698 | 28.0 | 1.2 | 13,374 | -2.1 | 1,232 | 9.2 | 12,825 | 29.8 | 29.2 |
| Mecosta | 15,808 | 73.3 | 121,400 | 19.8 | 11.7 | 712 | 32.6 | 2.9 | 18,415 | -0.1 | 1,675 | 9.1 | 18,616 | 28.4 | 30.7 |
| Menominee | 10,627 | 77.7 | 101,600 | 18.5 | 11.7 | 578 | 26.7 | 1.9 | 10,694 | -2.9 | 686 | 6.4 | 10,898 | 23.5 | 39.8 |
| Midland | 34,199 | 76.6 | 141,700 | 18.4 | 11.1 | 777 | 28.0 | 1.4 | 39,706 | -1.2 | 2,964 | 7.5 | 38,139 | 41.4 | 22.1 |
| Missaukee | 6,055 | 80.1 | 118,700 | 21.1 | 11.7 | 736 | 28.4 | 3.8 | 6,923 | -0.2 | 632 | 9.1 | 6,322 | 24.2 | 41.9 |
| Monroe | 59,929 | 79.7 | 161,500 | 18.7 | 11.9 | 834 | 30.7 | 1.5 | 74,397 | -1.8 | 6,633 | 8.9 | 70,276 | 31.7 | 32.0 |
| Montcalm | 23,913 | 78.1 | 113,900 | 21.1 | 12.9 | 697 | 30.8 | 2.7 | 28,133 | -0.7 | 2,555 | 9.1 | 27,378 | 25.5 | 36.2 |
| Montmorency | 4,452 | 84.1 | 102,700 | 22.5 | 12.1 | 668 | 32.9 | 1.5 | 3,021 | 0.9 | 397 | 13.1 | 3,046 | 28.1 | 30.0 |
| Muskegon | 65,939 | 74.7 | 117,900 | 18.8 | 11.8 | 760 | 30.8 | 2.2 | 77,540 | 0.3 | 8,859 | 11.4 | 75,858 | 28.0 | 32.5 |
| Newaygo | 19,161 | 83.9 | 121,300 | 21.0 | 12.4 | 704 | 28.5 | 3.4 | 23,340 | -0.4 | 2,021 | 8.7 | 20,392 | 28.1 | 36.2 |
| Oakland | 504,585 | 71.0 | 242,700 | 19.0 | 11.4 | 1,080 | 26.8 | 1.3 | 648,742 | -4.9 | 60,170 | 9.3 | 646,682 | 50.0 | 14.9 |
| Oceana | 10,156 | 82.6 | 118,800 | 21.5 | 11.9 | 729 | 27.3 | 4.9 | 12,024 | -0.4 | 1,241 | 10.3 | 11,170 | 27.2 | 39.7 |
| Ogemaw | 9,184 | 81.4 | 98,200 | 22.7 | 12.4 | 701 | 36.7 | 2.5 | 8,048 | 0.1 | 881 | 10.9 | 7,599 | 25.4 | 29.9 |
| Ontonagon | 2,793 | 88.2 | 70,000 | 21.6 | 12.5 | 523 | 27.4 | 1.8 | 2,053 | -1.4 | 212 | 10.3 | 2,033 | 29.8 | 22.7 |
| Osceola | 9,181 | 80.7 | 98,500 | 21.2 | 12.4 | 650 | 28.0 | 2.9 | 11,606 | 3.0 | 965 | 8.3 | 8,934 | 24.1 | 40.9 |
| Oscoda | 3,806 | 85.3 | 90,500 | 22.1 | 11.4 | 750 | 25.2 | 3.9 | 2,910 | -2.6 | 347 | 11.9 | 2,828 | 24.9 | 35.2 |
| Otsego | 9,944 | 78.9 | 139,500 | 20.0 | 10.9 | 768 | 28.5 | 2.3 | 11,651 | 0.2 | 1,196 | 10.3 | 10,833 | 31.5 | 26.4 |
| Ottawa | 102,610 | 77.7 | 191,800 | 18.5 | 10.5 | 898 | 27.5 | 2.2 | 158,936 | -1.9 | 10,859 | 6.8 | 148,950 | 35.5 | 27.5 |
| Presque Isle | 5,797 | 88.8 | 105,400 | 19.8 | 10.8 | 542 | 27.3 | 1.6 | 4,951 | 0.1 | 549 | 11.1 | 4,678 | 30.8 | 30.4 |
| Roscommon | 11,139 | 82.0 | 108,200 | 22.2 | 11.1 | 684 | 31.4 | 1.9 | 7,835 | 0.9 | 950 | 12.1 | 7,944 | 26.4 | 26.2 |
| Saginaw | 78,933 | 71.3 | 101,300 | 19.0 | 12.3 | 780 | 31.8 | 1.4 | 84,807 | -2.3 | 8,410 | 9.9 | 82,783 | 31.6 | 25.1 |
| St. Clair | 64,850 | 77.3 | 154,300 | 19.8 | 11.7 | 814 | 29.8 | 1.4 | 73,897 | -3.4 | 8,855 | 12.0 | 73,161 | 30.3 | 31.8 |
| St. Joseph | 24,150 | 74.6 | 122,300 | 19.6 | 10.7 | 722 | 26.9 | 2.6 | 29,052 | 3.0 | 2,634 | 9.1 | 28,156 | 24.8 | 43.7 |
| Sanilac | 17,499 | 78.1 | 111,200 | 20.1 | 11.6 | 678 | 27.6 | 2.0 | 19,541 | -0.2 | 1,900 | 9.7 | 17,842 | 27.6 | 37.7 |
| Schoolcraft | 3,468 | 83.7 | 109,900 | 21.3 | 11.3 | 571 | 31.1 | 1.2 | 3,248 | -1.6 | 314 | 9.7 | 2,985 | 25.0 | 29.4 |
| Shiawassee | 27,593 | 75.6 | 122,300 | 19.0 | 10.5 | 732 | 27.3 | 1.6 | 32,859 | -1.8 | 3,020 | 9.2 | 31,878 | 27.7 | 32.5 |
| Tuscola | 21,777 | 82.4 | 104,000 | 19.6 | 12.2 | 715 | 26.9 | 2.1 | 23,739 | -0.7 | 2,395 | 10.1 | 22,922 | 26.2 | 34.9 |
| Van Buren | 29,411 | 77.0 | 139,400 | 19.4 | 11.7 | 705 | 29.4 | 3.1 | 34,835 | -1.5 | 2,903 | 8.3 | 34,755 | 28.9 | 32.6 |
| Washtenaw | 141,245 | 61.1 | 263,600 | 19.4 | 11.6 | 1,114 | 29.7 | 1.6 | 193,659 | -2.7 | 12,365 | 6.4 | 189,559 | 54.3 | 13.8 |
| Wayne | 682,282 | 62.0 | 113,000 | 20.0 | 13.5 | 875 | 32.6 | 2.7 | 796,974 | -1.6 | 110,086 | 13.8 | 744,194 | 33.0 | 25.9 |
| Wexford | 12,963 | 76.8 | 108,300 | 20.0 | 11.9 | 706 | 28.8 | 2.6 | 14,918 | 0.4 | 1,516 | 10.2 | 14,011 | 28.2 | 35.3 |
| **MINNESOTA** | 2,185,603 | 71.6 | 223,900 | 19.7 | 10.6 | 977 | 28.3 | 2.6 | 3,094,702 | 0.1 | 191,140 | 6.2 | 2,958,615 | 41.3 | 21.8 |
| Aitkin | 7,681 | 82.2 | 183,200 | 24.1 | 12.5 | 768 | 29.4 | 3.0 | 7,487 | 3.0 | 603 | 8.1 | 6,272 | 32.5 | 28.2 |
| Anoka | 128,198 | 80.1 | 232,400 | 19.9 | 10.0 | 1,118 | 28.2 | 2.1 | 198,304 | -0.2 | 12,590 | 6.3 | 191,712 | 37.5 | 24.8 |
| Becker | 13,617 | 78.5 | 198,400 | 21.7 | 11.2 | 731 | 26.7 | 2.9 | 18,817 | 0.2 | 1,173 | 6.2 | 16,304 | 34.4 | 30.1 |
| Beltrami | 17,372 | 67.1 | 162,600 | 20.5 | 11.4 | 765 | 31.3 | 2.5 | 24,865 | 2.1 | 1,559 | 6.3 | 21,064 | 38.0 | 21.0 |
| Benton | 16,452 | 67.4 | 174,900 | 21.3 | 10.7 | 760 | 27.7 | 3.2 | 22,062 | 0.6 | 1,465 | 6.6 | 21,302 | 32.5 | 28.4 |
| Big Stone | 2,298 | 72.9 | 97,900 | 16.7 | 11.4 | 606 | 25.4 | 1.2 | 2,517 | 1.3 | 119 | 4.7 | 2,354 | 34.4 | 28.3 |
| Blue Earth | 26,091 | 62.0 | 184,800 | 19.6 | 10.3 | 883 | 30.7 | 2.2 | 40,363 | -0.6 | 2,181 | 5.4 | 38,033 | 34.8 | 24.0 |
| Brown | 10,764 | 79.2 | 141,900 | 17.9 | 10.2 | 629 | 24.0 | 1.1 | 14,270 | -0.6 | 731 | 5.1 | 13,695 | 31.1 | 31.9 |
| Carlton | 13,613 | 79.8 | 171,400 | 20.5 | 11.0 | 776 | 26.5 | 2.8 | 17,715 | -0.9 | 1,280 | 7.2 | 16,825 | 33.9 | 26.1 |
| Carver | 36,754 | 81.9 | 313,200 | 18.9 | 10.9 | 1,146 | 27.0 | 1.4 | 58,673 | | 3,047 | 5.2 | 56,939 | 48.1 | 16.6 |
| Cass | 13,164 | 81.5 | 194,200 | 22.4 | 11.2 | 739 | 27.4 | 3.4 | 14,206 | -1.3 | 1,301 | 9.2 | 12,654 | 30.2 | 26.9 |
| Chippewa | 5,209 | 66.7 | 119,700 | 17.9 | 10.0 | 613 | 25.6 | 2.7 | 6,779 | -1.6 | 344 | 5.1 | 6,012 | 33.0 | 28.4 |
| Chisago | 20,242 | 86.4 | 231,500 | 20.5 | 12.3 | 884 | 27.5 | 1.3 | 30,083 | -0.5 | 1,980 | 6.6 | 29,175 | 35.6 | 28.4 |
| Clay | 24,364 | 67.5 | 197,100 | 19.6 | 11.6 | 865 | 32.4 | 1.7 | 36,323 | 0.7 | 1,561 | 4.3 | 34,306 | 38.6 | 21.0 |
| Clearwater | 3,405 | 81.3 | 137,200 | 21.5 | 12.5 | 680 | 30.1 | 3.5 | 4,507 | 1.1 | 423 | 9.4 | 3,871 | 30.1 | 32.8 |
| Cook | 2,691 | 76.9 | 240,100 | 25.0 | 10.0 | 651 | 23.4 | 8.4 | 2,948 | -5.4 | 249 | 8.4 | 2,634 | 36.0 | 19.7 |
| Cottonwood | 4,846 | 78.1 | 104,000 | 18.6 | 10.2 | 629 | 22.6 | 3.7 | 6,265 | 6.0 | 291 | 4.6 | 5,283 | 29.7 | 36.7 |
| Crow Wing | 26,820 | 76.5 | 198,300 | 21.3 | 10.9 | 806 | 29.7 | 1.5 | 32,291 | 0.2 | 2,337 | 7.2 | 30,841 | 33.9 | 23.7 |
| Dakota | 161,488 | 74.0 | 266,000 | 19.4 | 10.0 | 1,174 | 28.1 | 1.9 | 242,394 | -0.2 | 15,019 | 6.2 | 233,358 | 43.6 | 18.9 |
| Dodge | 7,756 | 83.9 | 183,900 | 19.4 | 10.1 | 653 | 23.5 | 1.7 | 12,042 | 1.4 | 647 | 5.4 | 11,378 | 34.6 | 29.8 |
| Douglas | 16,663 | 74.3 | 214,800 | 20.5 | 11.4 | 733 | 28.7 | 1.2 | 21,271 | 1.2 | 1,019 | 4.8 | 19,398 | 34.9 | 27.5 |
| Faribault | 6,137 | 77.5 | 89,500 | 18.1 | 10.0 | 590 | 25.2 | 1.0 | 7,087 | 1.0 | 436 | 6.2 | 7,016 | 30.7 | 34.7 |

1. Specified owner-occupied units.   2. A value of 10.0 represents 10 percent or less; a value of 50.0 represents 50 percent or more.   3. Specified renter-occupied units.   4. Overcrowded or lacking complete plumbing facilities.   5. Percent of civilian labor force.   6. Civilian employed persons 16 years old and over.

# Table B. States and Counties — Nonfarm Employment and Agriculture

| | Private nonfarm establishments, employment and payroll, 2019 | | | | | | | | | Agriculture, 2017 | | | |
| | | Employment | | | | | | Annual payroll | | Farms | | | Farm producers whose primary occupation is farming (percent) |
| STATE County | Number of establish-ments | Total | Health care and social assistance | Manufac-turing | Retail trade | Finance and insurance | Professional, scientific, and technical services | Total (mil dol) | Average per employee (dollars) | Number | Fewer than 50 acres | 1000 acres or more | |
| | 104 | 105 | 106 | 107 | 108 | 109 | 110 | 111 | 112 | 113 | 114 | 115 | 116 |
|---|---|---|---|---|---|---|---|---|---|---|---|---|---|
| **MICHIGAN— Cont'd** | | | | | | | | | | | | | |
| Keweenaw | 67 | 312 | NA | 27 | 27 | NA | NA | 10 | 31,974 | 9 | 100.0 | NA | NA |
| Lake | 151 | 1,016 | 252 | 86 | 214 | 69 | 43 | 32 | 31,722 | 168 | 37.5 | 1.2 | 32.5 |
| Lapeer | 1,703 | 19,971 | 2,972 | 5,675 | 3,359 | 451 | 606 | 725 | 36,325 | 1,013 | 52.0 | 3.4 | 46.5 |
| Leelanau | 761 | 4,749 | 531 | 458 | 702 | 238 | 184 | 204 | 42,899 | 470 | 50.6 | 0.9 | 46.9 |
| Lenawee | 1,785 | 23,766 | 3,005 | 6,060 | 3,316 | 822 | 465 | 928 | 39,061 | 1,361 | 42.7 | 7.6 | 41.9 |
| Livingston | 4,402 | 58,733 | 9,203 | 10,380 | 9,083 | 4,339 | 2,880 | 2,356 | 40,106 | 724 | 61.7 | 2.6 | 41.9 |
| Luce | 161 | 1,458 | 386 | 200 | 270 | 54 | 15 | 54 | 37,117 | 71 | 56.3 | 1.4 | 35.5 |
| Mackinac | 449 | 2,178 | 386 | 111 | 355 | 92 | 31 | 131 | 60,075 | 101 | 37.6 | 7.9 | 32.0 |
| Macomb | 19,101 | 315,000 | 42,116 | 75,293 | 41,215 | 5,642 | 33,942 | 16,494 | 52,362 | 404 | 55.7 | 4.7 | 53.3 |
| Manistee | 559 | 5,805 | 898 | 1,033 | 974 | 161 | 113 | 233 | 40,180 | 274 | 37.6 | 1.1 | 41.2 |
| Marquette | 1,615 | 21,239 | 5,327 | 939 | 3,605 | 923 | 675 | 883 | 41,551 | 179 | 49.7 | 3.4 | 41.6 |
| Mason | 713 | 9,097 | 1,285 | 2,001 | 1,465 | 236 | 241 | 361 | 39,681 | 472 | 41.5 | 2.8 | 38.7 |
| Mecosta | 707 | 9,619 | 1,551 | 2,409 | 2,044 | 206 | 214 | 324 | 33,709 | 694 | 31.7 | 2.2 | 45.6 |
| Menominee | 435 | 5,526 | 236 | 2,106 | 728 | 123 | 211 | 197 | 35,715 | 353 | 29.7 | 3.4 | 42.9 |
| Midland | 1,783 | 33,924 | 6,824 | 4,488 | 3,510 | 1,306 | 3,023 | 2,219 | 65,415 | 530 | 51.3 | 4.2 | 39.2 |
| Missaukee | 299 | 2,371 | 359 | 398 | 384 | 51 | 30 | 85 | 35,666 | 406 | 35.7 | 6.9 | 46.1 |
| Monroe | 2,311 | 35,888 | 4,557 | 6,864 | 5,068 | 780 | 1,122 | 1,672 | 46,589 | 1,085 | 57.7 | 4.1 | 40.5 |
| Montcalm | 995 | 13,829 | 2,969 | 3,271 | 2,569 | 344 | 190 | 477 | 34,482 | 962 | 45.2 | 4.6 | 40.1 |
| Montmorency | 205 | 1,742 | 259 | 494 | 259 | 35 | 20 | 58 | 33,022 | 178 | 34.8 | 1.7 | 38.1 |
| Muskegon | 3,089 | 52,799 | 9,992 | 14,465 | 7,749 | 1,062 | 1,542 | 2,322 | 43,982 | 476 | 61.1 | 2.3 | 44.0 |
| Newaygo | 813 | 10,509 | 1,703 | 2,394 | 1,588 | 786 | 319 | 435 | 41,371 | 850 | 46.2 | 3.4 | 42.8 |
| Oakland | 39,188 | 721,092 | 120,803 | 54,610 | 73,591 | 43,721 | 106,609 | 44,481 | 61,685 | 514 | 68.9 | 0.6 | 38.1 |
| Oceana | 484 | 4,959 | 591 | 1,707 | 592 | 115 | 101 | 194 | 39,081 | 545 | 36.7 | 5.0 | 47.3 |
| Ogemaw | 568 | 5,348 | 1,228 | 310 | 1,283 | 144 | 92 | 189 | 35,331 | 294 | 34.4 | 4.8 | 50.6 |
| Ontonagon | 166 | 869 | 189 | 13 | 265 | 51 | 14 | 26 | 29,557 | 114 | 11.4 | 1.8 | 33.7 |
| Osceola | 432 | 6,792 | 1,136 | 2,839 | 485 | 122 | 78 | 306 | 45,018 | 625 | 37.4 | 2.2 | 36.9 |
| Oscoda | 185 | 1,620 | 140 | 379 | 213 | 33 | 29 | 56 | 34,629 | 144 | 54.9 | 0.7 | 21.6 |
| Otsego | 807 | 9,544 | 1,619 | 1,381 | 2,101 | 281 | 167 | 382 | 40,007 | 193 | 43.5 | 2.6 | 36.9 |
| Ottawa | 6,324 | 115,670 | 11,375 | 40,433 | 11,150 | 2,326 | 4,814 | 5,236 | 45,267 | 1,130 | 56.8 | 3.0 | 50.4 |
| Presque Isle | 325 | 2,041 | 280 | 185 | 359 | 99 | 29 | 74 | 36,136 | 322 | 25.5 | 3.4 | 40.6 |
| Roscommon | 537 | 4,161 | 573 | 605 | 1,246 | 159 | 81 | 133 | 31,852 | 48 | 43.8 | 2.1 | 32.5 |
| Saginaw | 4,180 | 79,213 | 17,570 | 11,867 | 12,348 | 2,761 | 2,422 | 3,396 | 42,867 | 1,250 | 44.2 | 5.8 | 42.1 |
| St. Clair | 3,041 | 40,959 | 7,554 | 9,215 | 6,461 | 1,264 | 1,195 | 1,705 | 41,638 | 1,077 | 52.6 | 3.0 | 42.8 |
| St. Joseph | 1,132 | 19,520 | 1,949 | 9,244 | 2,552 | 420 | 577 | 884 | 45,302 | 896 | 45.5 | 8.3 | 39.7 |
| Sanilac | 806 | 8,583 | 1,281 | 2,997 | 1,442 | 369 | 160 | 317 | 36,988 | 1,315 | 33.7 | 8.1 | 58.9 |
| Schoolcraft | 216 | 1,781 | 377 | 172 | 325 | 149 | 38 | 76 | 42,689 | 62 | 37.1 | 6.5 | 24.5 |
| Shiawassee | 1,150 | 13,994 | 3,337 | 1,952 | 2,557 | 329 | 359 | 511 | 36,528 | 972 | 51.1 | 4.9 | 41.5 |
| Tuscola | 816 | 9,244 | 2,555 | 1,312 | 1,590 | 281 | 210 | 376 | 40,726 | 1,241 | 42.4 | 6.5 | 50.3 |
| Van Buren | 1,284 | 16,368 | 2,212 | 2,956 | 2,692 | 283 | 2,277 | 728 | 44,502 | 953 | 52.2 | 2.3 | 42.4 |
| Washtenaw | 8,227 | 159,215 | 40,021 | 15,905 | 17,742 | 3,397 | 15,347 | 9,480 | 59,543 | 1,245 | 59.7 | 2.7 | 39.8 |
| Wayne | 32,575 | 661,250 | 114,085 | 90,385 | 69,209 | 34,706 | 45,342 | 38,340 | 57,982 | 248 | 79.8 | NA | 47.5 |
| Wexford | 876 | 12,652 | 1,869 | 3,425 | 1,996 | 343 | 235 | 492 | 38,865 | 304 | 43.1 | 1.6 | 45.3 |
| **MINNESOTA** | 151,495 | 2,729,420 | 480,571 | 316,348 | 299,285 | 167,558 | 184,142 | 153,756 | 56,333 | 68,822 | 28.8 | 9.3 | 45.5 |
| Aitkin | 400 | 3,552 | 908 | 437 | 663 | 95 | 33 | 127 | 35,881 | 462 | 19.3 | 2.4 | 34.9 |
| Anoka | 8,012 | 124,985 | 17,390 | 22,186 | 16,178 | 2,257 | 4,947 | 6,287 | 50,302 | 360 | 60.0 | 1.7 | 37.4 |
| Becker | 924 | 10,798 | 2,318 | 2,701 | 1,735 | 267 | 204 | 436 | 40,411 | 943 | 20.5 | 7.7 | 39.9 |
| Beltrami | 1,212 | 15,425 | 4,123 | 900 | 3,185 | 386 | 440 | 584 | 37,881 | 583 | 25.6 | 4.3 | 32.8 |
| Benton | 928 | 15,991 | 2,536 | 3,952 | 1,797 | 253 | 257 | 720 | 45,018 | 816 | 32.2 | 4.0 | 41.3 |
| Big Stone | 186 | 1,432 | 578 | 10 | 255 | 64 | 29 | 57 | 40,129 | 438 | 17.8 | 19.4 | 47.9 |
| Blue Earth | 1,977 | 35,276 | 7,795 | 4,194 | 5,878 | 1,055 | 1,161 | 1,393 | 39,481 | 983 | 30.9 | 10.9 | 49.2 |
| Brown | 801 | 13,664 | 2,405 | 3,138 | 1,663 | 468 | 428 | 575 | 42,076 | 1,040 | 25.1 | 5.9 | 44.8 |
| Carlton | 673 | 7,391 | 1,981 | 1,354 | 968 | 340 | 153 | 336 | 45,397 | 529 | 25.7 | 0.9 | 35.4 |
| Carver | 2,527 | 40,024 | 5,382 | 9,686 | 3,762 | 898 | 2,995 | 2,208 | 55,157 | 689 | 38.3 | 5.2 | 45.0 |
| Cass | 841 | 6,750 | 1,392 | 324 | 1,032 | 173 | 176 | 238 | 35,193 | 432 | 20.8 | 4.4 | 45.8 |
| Chippewa | 387 | 5,091 | 1,183 | 1,243 | 637 | 230 | 89 | 194 | 38,109 | 623 | 26.6 | 15.4 | 52.3 |
| Chisago | 1,270 | 14,361 | 4,124 | 2,398 | 1,886 | 231 | 985 | 664 | 46,247 | 821 | 52.1 | 1.7 | 34.8 |
| Clay | 1,346 | 17,973 | 4,551 | 944 | 2,625 | 498 | 659 | 644 | 35,814 | 694 | 20.2 | 23.9 | 55.2 |
| Clearwater | 215 | 2,351 | 543 | 531 | 254 | 80 | 63 | 116 | 49,474 | 414 | 23.9 | 8.7 | 36.6 |
| Cook | 291 | 2,145 | 221 | 59 | 340 | 37 | 59 | 74 | 34,609 | 32 | 78.1 | NA | 20.4 |
| Cottonwood | 342 | 3,784 | 724 | 1,145 | 546 | 98 | 71 | 127 | 33,660 | 744 | 22.4 | 17.7 | 55.9 |
| Crow Wing | 2,205 | 27,612 | 6,061 | 2,835 | 4,833 | 1,271 | 1,091 | 1,091 | 39,511 | 494 | 29.8 | 2.6 | 34.0 |
| Dakota | 10,702 | 183,470 | 26,066 | 18,495 | 24,733 | 12,015 | 8,874 | 9,922 | 54,078 | 820 | 48.5 | 6.1 | 49.1 |
| Dodge | 413 | 4,668 | 306 | 1,695 | 486 | 107 | 94 | 212 | 45,517 | 611 | 40.9 | 13.3 | 47.9 |
| Douglas | 1,354 | 17,718 | 3,754 | 3,855 | 2,897 | 449 | 447 | 786 | 44,389 | 960 | 27.8 | 5.7 | 36.6 |
| Faribault | 407 | 3,713 | 704 | 1,059 | 507 | 187 | 86 | 140 | 37,639 | 822 | 28.6 | 14.4 | 55.0 |

# Table B. States and Counties — Agriculture

| STATE County | Acreage (1,000) | Percent change, 2012-2017 | Average size of farm | Total irrigated (1,000) | Total cropland (1,000) | Average per farm | Average per acre | Value of machinery and equipment, average per farm (dollars) | Total (mil dol) | Average per farm (acres) | Crops | Livestock and poultry products | Organic farms (number) | Farms with internet access (percent) | Total ($1,000) | Percent of farms |
|---|---|---|---|---|---|---|---|---|---|---|---|---|---|---|---|---|
| | 117 | 118 | 119 | 120 | 121 | 122 | 123 | 124 | 125 | 126 | 127 | 128 | 129 | 130 | 131 | 132 |
| **MICHIGAN— Cont'd** | | | | | | | | | | | | | | | | |
| Keweenaw | 0 | -24.5 | 27 | NA | 0.1 | 90,000 | 3,375 | 26,667 | 0.0 | 222 | 100.0 | NA | NA | 66.7 | NA | NA |
| Lake | 22 | -16.9 | 129 | 0.1 | 11.8 | 316,975 | 2,462 | 52,897 | 3.2 | 19,321 | 41.7 | 58.3 | NA | 69.6 | NA | 6.5 |
| Lapeer | 165 | -5.8 | 163 | 2.4 | 128.7 | 752,544 | 4,607 | 141,624 | 87.0 | 85,838 | 72.9 | 27.1 | 16 | 76.8 | 1,424 | 18.0 |
| Leelanau | 50 | -15.9 | 106 | 2.3 | 29.9 | 779,662 | 7,321 | 111,622 | 42.4 | 90,302 | 93.7 | 6.3 | 12 | 87.7 | 376 | 14.0 |
| Lenawee | 386 | 12.0 | 283 | 9.8 | 345.2 | 1,481,398 | 5,226 | 196,185 | 259.9 | 190,951 | 64.7 | 35.3 | 6 | 81.3 | 11,691 | 53.3 |
| Livingston | 89 | 3.7 | 123 | 1.1 | 67.2 | 703,733 | 5,701 | 85,997 | 48.5 | 66,965 | 62.5 | 37.5 | 7 | 84.0 | 1,444 | 13.3 |
| Luce | 10 | -14.7 | 139 | D | 4.0 | 351,299 | 2,521 | 64,881 | 3.8 | 53,113 | 88.4 | 11.6 | NA | 60.6 | D | 11.3 |
| Mackinac | 25 | 11.7 | 248 | 0.0 | 13.2 | 495,340 | 1,997 | 74,432 | 7.1 | 70,683 | 11.5 | 88.5 | 3 | 76.2 | 173 | 13.9 |
| Macomb | 74 | 8.4 | 182 | 4.3 | 66.6 | 928,906 | 5,095 | 149,403 | 78.8 | 195,111 | 87.2 | 12.8 | 1 | 81.9 | 800 | 25.0 |
| Manistee | 41 | -6.6 | 151 | 1.0 | 19.3 | 435,471 | 2,884 | 81,371 | 10.3 | 37,686 | 75.6 | 24.4 | 7 | 72.6 | 246 | 15.0 |
| Marquette | 30 | -1.4 | 169 | 0.1 | 11.3 | 372,759 | 2,206 | 85,310 | 3.7 | 20,536 | 30.8 | 69.2 | 8 | 83.2 | 46 | 9.5 |
| Mason | 85 | 8.1 | 181 | 4.2 | 63.0 | 563,138 | 3,111 | 132,221 | 56.7 | 120,097 | 58.1 | 41.9 | 21 | 75.8 | 943 | 25.8 |
| Mecosta | 115 | -6.4 | 166 | 17.6 | 81.5 | 581,429 | 3,505 | 145,703 | 180.1 | 259,458 | 21.5 | 78.5 | 6 | 64.3 | 909 | 27.2 |
| Menominee | 80 | -13.4 | 226 | 0.1 | 47.2 | 479,385 | 2,125 | 113,385 | 37.6 | 106,507 | 11.5 | 88.5 | 2 | 78.5 | 610 | 26.9 |
| Midland | 88 | -2.1 | 165 | 0.6 | 69.9 | 916,909 | 5,542 | 121,112 | 46.9 | 88,483 | 57.1 | 42.9 | 5 | 80.2 | 2,366 | 40.0 |
| Missaukee | 114 | 14.2 | 280 | 8.5 | 88.7 | 1,023,840 | 3,658 | 205,707 | 148.7 | 366,276 | 14.8 | 85.2 | 4 | 77.1 | 759 | 25.6 |
| Monroe | 210 | -2.2 | 193 | 4.3 | 194.1 | 1,190,516 | 6,156 | 150,143 | 174.5 | 160,833 | 96.2 | 3.8 | NA | 83.5 | 4,555 | 43.7 |
| Montcalm | 230 | -2.9 | 239 | 59.5 | 191.9 | 955,776 | 3,991 | 170,672 | 180.5 | 187,598 | 71.2 | 28.8 | 9 | 73.5 | 2,732 | 27.4 |
| Montmorency | 26 | 7.2 | 147 | 0.0 | 16.3 | 314,723 | 2,146 | 110,173 | 9.2 | 51,511 | 39.4 | 60.6 | NA | 71.9 | 177 | 21.9 |
| Muskegon | 63 | -14.9 | 133 | 3.6 | 45.8 | 852,881 | 6,425 | 112,539 | 74.7 | 156,884 | 58.9 | 41.1 | 5 | 82.8 | 173 | 11.1 |
| Newaygo | 136 | 8.4 | 160 | 12.0 | 96.7 | 623,318 | 3,889 | 126,490 | 128.3 | 150,958 | 37.6 | 62.4 | 16 | 78.4 | 296 | 13.3 |
| Oakland | 29 | -9.0 | 56 | 0.6 | 17.4 | 652,488 | 11,623 | 67,181 | 22.3 | 43,442 | 88.0 | 12.0 | 9 | 85.0 | 161 | 4.1 |
| Oceana | 127 | -0.5 | 233 | 13.1 | 93.6 | 969,750 | 4,159 | 183,666 | 124.7 | 228,791 | 66.2 | 33.8 | 9 | 81.3 | 699 | 14.9 |
| Ogemaw | 70 | 2.8 | 238 | D | 47.7 | 666,927 | 2,797 | 177,989 | 49.8 | 169,429 | 25.1 | 74.9 | NA | 63.9 | 677 | 41.8 |
| Ontonagon | 27 | -6.7 | 238 | D | 12.5 | 379,699 | 1,597 | 68,540 | 3.1 | 27,167 | 80.5 | 19.5 | NA | 77.2 | 14 | 15.8 |
| Osceola | 104 | -6.3 | 166 | 2.0 | 67.4 | 497,961 | 3,003 | 79,214 | 43.5 | 69,650 | 26.8 | 73.2 | 4 | 63.4 | 1,101 | 18.6 |
| Oscoda | 16 | -3.4 | 112 | D | 6.5 | 307,892 | 2,741 | 39,097 | 5.6 | 38,653 | 23.9 | 76.1 | 7 | 41.7 | 20 | 4.9 |
| Otsego | 33 | 3.0 | 172 | 1.1 | 17.5 | 479,329 | 2,780 | 80,181 | 6.2 | 32,311 | 80.9 | 19.1 | 6 | 80.8 | 64 | 18.7 |
| Ottawa | 172 | -7.7 | 152 | 23.0 | 145.2 | 1,258,057 | 8,271 | 167,813 | 506.7 | 448,373 | 55.7 | 44.3 | 6 | 85.2 | 2,675 | 19.6 |
| Presque Isle | 64 | -20.9 | 200 | 1.8 | 41.0 | 469,658 | 2,345 | 93,591 | 18.9 | 58,661 | 68.9 | 31.1 | 1 | 62.1 | 374 | 26.4 |
| Roscommon | 6 | -22.5 | 120 | 0.7 | 2.9 | 293,961 | 2,451 | 76,198 | 0.8 | 16,292 | 38.4 | 61.6 | 1 | 70.8 | 30 | 16.7 |
| Saginaw | 327 | 5.6 | 262 | 1.7 | 300.5 | 1,518,681 | 5,805 | 162,379 | 170.3 | 136,206 | 87.9 | 12.1 | 12 | 79.4 | 6,853 | 63.1 |
| St. Clair | 182 | 1.2 | 169 | 1.0 | 160.3 | 888,682 | 5,254 | 132,186 | 80.9 | 75,105 | 88.6 | 11.4 | 11 | 73.4 | 2,041 | 22.1 |
| St. Joseph | 245 | 10.5 | 273 | 123.1 | 212.3 | 1,562,992 | 5,717 | 235,762 | 230.7 | 257,450 | 68.8 | 31.2 | 16 | 64.6 | 5,839 | 39.7 |
| Sanilac | 437 | -4.5 | 332 | 1.7 | 401.0 | 1,636,584 | 4,930 | 274,204 | 357.7 | 271,993 | 54.3 | 45.7 | 52 | 72.5 | 7,801 | 49.4 |
| Schoolcraft | 15 | -22.6 | 242 | 0.0 | 5.9 | 441,401 | 1,820 | 65,397 | 2.1 | 34,016 | 48.7 | 51.3 | NA | 85.5 | 98 | 21.0 |
| Shiawassee | 210 | -5.8 | 217 | 1.5 | 187.1 | 976,860 | 4,511 | 151,599 | 97.5 | 100,284 | 67.4 | 32.6 | 21 | 80.9 | 3,670 | 45.8 |
| Tuscola | 330 | 1.4 | 266 | 8.7 | 298.5 | 1,386,278 | 5,217 | 226,605 | 231.0 | 186,103 | 69.4 | 30.6 | 79 | 79.7 | 7,168 | 52.8 |
| Van Buren | 152 | -13.3 | 159 | 35.6 | 118.1 | 921,952 | 5,789 | 150,630 | 205.5 | 215,686 | 81.5 | 18.5 | 7 | 74.5 | 1,602 | 14.4 |
| Washtenaw | 179 | 5.2 | 144 | 4.0 | 150.4 | 1,124,667 | 7,823 | 112,400 | 91.2 | 73,227 | 76.3 | 23.7 | 29 | 84.3 | 3,438 | 24.6 |
| Wayne | 10 | -36.3 | 40 | 0.6 | 7.8 | 467,878 | 11,561 | 70,633 | 23.1 | 93,274 | 96.8 | 3.2 | 2 | 88.3 | 56 | 4.8 |
| Wexford | 40 | -0.3 | 132 | 4.3 | 26.2 | 445,486 | 3,368 | 60,875 | 18.1 | 59,701 | 48.0 | 52.0 | 8 | 67.8 | 164 | 9.5 |
| **MINNESOTA** | 25,517 | -2.0 | 371 | 611.6 | 21,786.8 | 1,799,201 | 4,853 | 223,666 | 18,395.4 | 267,289 | 55.4 | 44.6 | 735 | 79.0 | 394,491 | 59.9 |
| Aitkin | 106 | -13.8 | 229 | 3.0 | 54.1 | 482,701 | 2,109 | 72,031 | 12.5 | 26,970 | 56.5 | 43.5 | 1 | 68.2 | 282 | 16.5 |
| Anoka | 39 | -12.9 | 108 | 3.0 | 29.4 | 823,272 | 7,590 | 122,561 | 67.8 | 188,222 | 51.9 | 48.1 | 5 | 85.6 | 239 | 9.2 |
| Becker | 368 | -15.4 | 390 | 13.4 | 256.0 | 1,133,301 | 2,906 | 151,842 | 174.5 | 185,081 | 59.4 | 40.6 | 4 | 73.6 | 5,253 | 58.4 |
| Beltrami | 169 | -6.6 | 289 | 3.5 | 79.8 | 568,690 | 1,966 | 67,182 | 23.8 | 40,823 | 55.8 | 44.2 | NA | 81.0 | 775 | 21.1 |
| Benton | 195 | 3.2 | 239 | 16.7 | 155.8 | 1,048,133 | 4,390 | 181,337 | 207.2 | 253,893 | 30.8 | 69.2 | 5 | 75.4 | 1,702 | 43.5 |
| Big Stone | 269 | 8.0 | 614 | 4.1 | 246.4 | 2,667,748 | 4,348 | 308,736 | 138.8 | 316,790 | 77.1 | 22.9 | 2 | 78.1 | 3,315 | 75.1 |
| Blue Earth | 383 | 1.7 | 389 | 4.3 | 355.5 | 2,804,263 | 7,202 | 284,239 | 483.5 | 491,860 | 41.9 | 58.1 | 7 | 82.3 | 12,899 | 76.4 |
| Brown | 356 | 9.1 | 342 | 4.1 | 330.8 | 2,248,890 | 6,574 | 241,577 | 381.5 | 366,838 | 45.8 | 54.2 | 3 | 80.1 | 6,242 | 82.2 |
| Carlton | 93 | 0.9 | 177 | D | 41.3 | 380,506 | 2,155 | 70,143 | 11.0 | 20,766 | 43.5 | 56.5 | 7 | 80.5 | 196 | 4.3 |
| Carver | 159 | 2.2 | 230 | 0.6 | 136.1 | 1,552,374 | 6,742 | 218,040 | 111.4 | 161,652 | 61.9 | 38.1 | 7 | 82.0 | 963 | 47.2 |
| Cass | 134 | -15.1 | 309 | 8.4 | 57.9 | 773,837 | 2,503 | 83,876 | 26.5 | 61,259 | 28.3 | 71.7 | 1 | 75.0 | 325 | 14.1 |
| Chippewa | 341 | 1.8 | 547 | 4.5 | 320.5 | 3,226,230 | 5,894 | 352,983 | 256.7 | 412,037 | 73.3 | 26.7 | 7 | 76.6 | 7,059 | 84.6 |
| Chisago | 116 | 1.5 | 141 | 4.0 | 82.7 | 643,553 | 4,575 | 94,546 | 52.8 | 64,358 | 74.6 | 25.4 | 6 | 83.3 | 1,379 | 31.5 |
| Clay | 577 | -5.6 | 831 | 2.9 | 538.1 | 3,280,220 | 3,948 | 367,559 | 277.8 | 400,218 | 86.5 | 13.5 | 12 | 84.7 | 5,759 | 73.5 |
| Clearwater | 156 | -6.7 | 376 | 4.9 | 76.1 | 761,920 | 2,026 | 113,597 | 30.1 | 72,609 | 65.7 | 34.3 | 3 | 77.1 | 417 | 27.1 |
| Cook | 1 | -38.7 | 44 | 0.0 | 0.2 | 258,025 | 5,927 | 19,944 | 0.4 | 11,906 | 95.8 | 4.2 | 2 | 84.4 | D | 3.1 |
| Cottonwood | 370 | -0.6 | 498 | 2.1 | 345.5 | 3,146,803 | 6,321 | 388,964 | 382.2 | 513,669 | 50.8 | 49.2 | 4 | 86.7 | 5,155 | 85.5 |
| Crow Wing | 89 | -10.8 | 181 | 4.5 | 41.8 | 485,403 | 2,688 | 93,161 | 19.1 | 38,571 | 48.6 | 51.4 | 6 | 73.5 | 159 | 21.1 |
| Dakota | 227 | 3.3 | 277 | 62.8 | 205.4 | 1,911,437 | 6,902 | 227,421 | 235.4 | 287,091 | 76.3 | 23.7 | 13 | 78.2 | 2,990 | 43.4 |
| Dodge | 248 | 10.0 | 406 | 4.9 | 231.9 | 2,797,971 | 6,892 | 283,496 | 238.4 | 390,185 | 58.0 | 42.0 | 6 | 83.6 | 6,605 | 64.6 |
| Douglas | 263 | -1.6 | 274 | 3.3 | 205.3 | 1,082,155 | 3,946 | 138,654 | 100.3 | 104,526 | 73.9 | 26.1 | 9 | 75.5 | 3,133 | 66.6 |
| Faribault | 408 | 4.5 | 496 | D | 388.9 | 3,372,928 | 6,799 | 352,244 | 337.7 | 410,869 | 68.4 | 31.6 | 5 | 82.8 | 5,464 | 81.8 |

| STATE County | Water use, 2015 — Public supply water withdrawn (mil gal/day) | Public supply gallons withdrawn per person per day | Wholesale Trade[1], 2017 — Number of establishments | Number of employees | Sales (mil dol) | Average payroll (mil dol) | Retail Trade[2], 2017 — Number of establishments | Number of employees | Sales (mil dol) | Average payroll (mil dol) | Real estate and rental and leasing,[2] 2017 — Number of establishments | Number of employees | Sales (mil dol) | Average payroll (mil dol) |
|---|---|---|---|---|---|---|---|---|---|---|---|---|---|---|
| | 133 | 134 | 135 | 136 | 137 | 138 | 139 | 140 | 141 | 142 | 143 | 144 | 145 | 146 |
| **MICHIGAN— Cont'd** | | | | | | | | | | | | | | |
| Keweenaw | 0.05 | 23.1 | NA | NA | NA | NA | 9 | 21 | 3.7 | 0.4 | NA | NA | NA | NA |
| Lake | 0.26 | 22.8 | NA | NA | NA | NA | 27 | 229 | 57.6 | 4.9 | D | D | D | 0.3 |
| Lapeer | 0.40 | 4.5 | 52 | 358 | 173.1 | 16.2 | 260 | 3,401 | 1,072.1 | 84.0 | 54 | 217 | 34.3 | 7.1 |
| Leelanau | 0.50 | 22.7 | D | D | D | D | 124 | 701 | 164.4 | 20.2 | 30 | 50 | 14.2 | 2.4 |
| Lenawee | 6.10 | 61.9 | D | D | D | D | 285 | 3,755 | 1,051.9 | 92.6 | 64 | 172 | 29.9 | 4.9 |
| Livingston | 6.92 | 36.9 | 175 | 1,649 | 1,921.2 | 132.9 | 603 | 9,428 | 2,808.7 | 240.5 | 156 | 540 | 171.5 | 23.5 |
| Luce | 0.50 | 77.9 | D | D | D | 2.2 | 29 | 274 | 78.5 | 7.2 | 8 | 69 | 8.3 | 1.6 |
| Mackinac | 1.29 | 118.5 | D | D | D | D | 99 | 440 | 130.2 | 12.3 | 12 | 37 | 6.5 | 1.2 |
| Macomb | 3.97 | 4.6 | 774 | 12,393 | 7,801.9 | 757.0 | 2,777 | 42,208 | 13,734.8 | 1,161.1 | 656 | 3,947 | 908.7 | 147.3 |
| Manistee | 1.54 | 63.0 | 15 | 103 | 173.4 | 6.2 | 104 | 1,126 | 303.9 | 25.3 | 21 | 46 | 7.7 | 1.4 |
| Marquette | 5.16 | 76.8 | 46 | 404 | 215.6 | 18.9 | 274 | 3,683 | 926.8 | 90.9 | 66 | 324 | 82.0 | 13.5 |
| Mason | 2.45 | 85.1 | 19 | 153 | 144.7 | 7.7 | 112 | 1,544 | 414.0 | 39.5 | 33 | 361 | 50.2 | 19.9 |
| Mecosta | 1.27 | 29.5 | D | D | D | D | 145 | 1,977 | 537.9 | 47.7 | 36 | 101 | 20.7 | 3.3 |
| Menominee | 1.07 | 45.4 | 17 | 369 | 189.7 | 15.5 | 68 | 639 | 134.4 | 12.9 | 11 | 55 | 5.4 | 0.9 |
| Midland | 0.14 | 1.7 | 50 | 735 | 2,781.2 | 71.8 | 270 | 3,902 | 1,116.0 | 98.0 | 64 | 338 | 70.3 | 17.3 |
| Missaukee | 0.38 | 25.5 | 19 | 158 | 87.9 | 6.5 | 35 | 424 | 147.9 | 13.3 | 5 | 5 | 1.2 | 0.1 |
| Monroe | 9.27 | 62.0 | D | D | D | D | 358 | 5,252 | 1,705.6 | 133.1 | 70 | 273 | 49.8 | 8.7 |
| Montcalm | 3.06 | 48.6 | 41 | 329 | 112.3 | 13.3 | 196 | 2,542 | 807.5 | 64.5 | 18 | 44 | 7.7 | 1.2 |
| Montmorency | 0.12 | 13.0 | NA | NA | NA | NA | 30 | 238 | 69.6 | 5.9 | D | D | D | D |
| Muskegon | 16.08 | 93.1 | D | D | D | 62.3 | 564 | 7,930 | 2,191.2 | 195.3 | 87 | 599 | 78.3 | 18.8 |
| Newaygo | 1.87 | 39.0 | 31 | 248 | 81.2 | 9.7 | 145 | 1,563 | 439.7 | 41.0 | 29 | 112 | 15.3 | 3.5 |
| Oakland | 19.11 | 15.4 | 2,006 | 33,497 | 35,641.4 | 2,461.8 | 4,781 | 75,952 | 24,638.8 | 2,243.9 | 1,826 | 16,580 | 3,291.9 | 846.6 |
| Oceana | 1.16 | 44.4 | 13 | 484 | 104.3 | 12.6 | 92 | 687 | 176.1 | 15.1 | 16 | 29 | 7.7 | 1.2 |
| Ogemaw | 0.38 | 18.1 | 18 | 487 | 147.9 | 21.3 | 112 | 1,412 | 401.8 | 36.0 | D | D | D | D |
| Ontonagon | 0.44 | 73.2 | 4 | 6 | 1.5 | 0.2 | 25 | 245 | 53.8 | 6.5 | 3 | 5 | 0.4 | 0.1 |
| Osceola | 1.37 | 59.4 | 10 | 97 | 86.5 | 4.3 | 65 | 552 | 149.2 | 12.4 | D | D | D | 0.5 |
| Oscoda | 0.07 | 8.5 | D | D | D | D | 34 | 252 | 58.4 | 5.1 | D | D | D | D |
| Otsego | 1.11 | 45.8 | 32 | 358 | 144.4 | 14.5 | 138 | 2,139 | 655.1 | 57.9 | 26 | 70 | 12.6 | 2.0 |
| Ottawa | 88.23 | 315.2 | 300 | 4,277 | 2,707.7 | 253.5 | 776 | 10,951 | 3,302.6 | 315.4 | 239 | 1,354 | 470.5 | 59.4 |
| Presque Isle | 0.46 | 35.8 | 3 | 3 | 0.8 | 0.1 | 61 | 371 | 110.6 | 9.0 | D | D | D | 0.3 |
| Roscommon | 0.28 | 11.7 | 7 | 23 | 5.9 | | 113 | 1,334 | 404.6 | 35.1 | D | D | D | 2.2 |
| Saginaw | 0.42 | 2.2 | 160 | 2,070 | 1,307.0 | 111.1 | 826 | 12,804 | 3,233.8 | 309.9 | 123 | 652 | 139.2 | 20.5 |
| St. Clair | 137.89 | 862.5 | 88 | 1,115 | 622.5 | 63.2 | 505 | 6,823 | 2,030.8 | 173.0 | 94 | 988 | 78.5 | 24.8 |
| St. Joseph | 3.25 | 53.3 | D | D | D | D | 185 | 2,580 | 763.9 | 65.6 | 32 | 103 | 18.9 | 2.9 |
| Sanilac | 1.56 | 37.6 | 26 | 293 | 227.4 | 15.1 | 143 | 1,606 | 447.5 | 40.5 | D | D | D | D |
| Schoolcraft | 0.35 | 42.8 | 4 | 5 | 3.0 | 0.1 | 44 | 382 | 99.9 | 9.2 | D | D | D | D |
| Shiawassee | 2.89 | 42.1 | 44 | 558 | 204.1 | 24.9 | 196 | 2,577 | 862.2 | 70.3 | 32 | 112 | 20.9 | 3.3 |
| Tuscola | 1.89 | 35.1 | 40 | 411 | 282.0 | 22.7 | 153 | 1,608 | 484.6 | 40.6 | D | D | D | 1.8 |
| Van Buren | 3.55 | 47.3 | 60 | 524 | 406.4 | 24.0 | 256 | 2,668 | 753.0 | 66.6 | 38 | 103 | 17.5 | 2.4 |
| Washtenaw | 18.36 | 51.2 | 270 | 3,870 | 5,384.1 | 267.6 | 1,100 | 17,534 | 5,121.3 | 473.5 | 373 | 2,686 | 1,101.3 | 140.5 |
| Wayne | 466.55 | 265.2 | 1,434 | 26,754 | 29,897.5 | 1,708.7 | 5,927 | 69,229 | 21,293.3 | 1,772.1 | 1,162 | 7,082 | 7,908.7 | 332.8 |
| Wexford | 2.65 | 80.3 | 28 | 434 | 211.7 | 26.0 | 167 | 2,096 | 643.2 | 54.5 | 41 | 207 | 21.5 | 4.6 |
| **MINNESOTA** | 515.24 | 93.9 | 6,397 | 108,895 | 106,476.5 | 7,091.2 | 18,827 | 302,886 | 91,993.6 | 8,117.5 | 7,218 | 38,077 | 10,432.9 | 1,915.0 |
| Aitkin | 0.27 | 17.2 | 12 | 173 | 156.7 | 6.5 | 69 | 582 | 140.4 | 13.1 | 10 | 8 | 2.3 | 0.4 |
| Anoka | 108.23 | 314.5 | 334 | 5,578 | 3,491.1 | 335.3 | 910 | 16,357 | 4,474.3 | 428.4 | 427 | 1,419 | 330.9 | 51.8 |
| Becker | 1.66 | 49.7 | 31 | 173 | 121.1 | 8.0 | 129 | 1,812 | 513.6 | 45.8 | 33 | 74 | 20.0 | 3.1 |
| Beltrami | 1.45 | 31.7 | 40 | 389 | 132.1 | 17.1 | 214 | 3,089 | 766.6 | 80.6 | D | D | D | D |
| Benton | 2.08 | 52.4 | 43 | 1,238 | 827.7 | 68.8 | 119 | 1,869 | 506.8 | 49.5 | 36 | 81 | 15.5 | 2.2 |
| Big Stone | 0.49 | 97.2 | 9 | D | 172.5 | D | 26 | 178 | 35.1 | 3.6 | D | D | D | 0.2 |
| Blue Earth | 6.23 | 94.7 | 89 | 1,397 | 1,802.5 | 81.5 | 311 | 6,187 | 1,464.4 | 143.9 | 20 | 184 | 10.6 | 3.6 |
| Brown | 1.07 | 42.3 | 29 | 389 | 853.2 | 18.6 | 104 | 1,667 | 395.0 | 39.7 | 13 | 86 | 8.9 | 1.4 |
| Carlton | 1.51 | 42.5 | 17 | 304 | 145.0 | 15.2 | 101 | 1,310 | 268.2 | 33.4 | 138 | 745 | 152.6 | 40.5 |
| Carver | 8.32 | 84.3 | 116 | 2,167 | 1,652.4 | 201.8 | 207 | 3,596 | 935.2 | 89.0 | D | D | D | D |
| Cass | 0.54 | 18.8 | 12 | 306 | 72.2 | 14.6 | 141 | 1,071 | 260.7 | 26.0 | 10 | 99 | 3.4 | 1.7 |
| Chippewa | 1.17 | 96.6 | 13 | 433 | 443.3 | 22.9 | 58 | 716 | 217.9 | 19.5 | 51 | 102 | 19.3 | 3.7 |
| Chisago | 2.00 | 36.8 | 39 | 358 | 174.3 | 15.1 | 168 | 1,806 | 495.9 | 47.5 | 44 | 161 | 24.8 | 4.6 |
| Clay | 4.92 | 78.9 | 64 | 1,115 | 1,110.4 | 61.9 | 165 | 2,910 | 738.7 | 73.6 | NA | NA | NA | NA |
| Clearwater | 0.22 | 25.0 | 4 | D | 7.1 | D | 38 | 284 | 61.6 | 5.9 | D | D | D | 1.7 |
| Cook | 0.29 | 55.8 | 5 | 18 | 3.5 | 0.6 | 50 | 319 | 64.9 | 9.2 | 8 | 20 | 2.8 | 1.0 |
| Cottonwood | 1.76 | 152.4 | 23 | 246 | 306.7 | 13.5 | 50 | 484 | 92.1 | 10.7 | D | D | D | D |
| Crow Wing | 3.40 | 53.6 | 65 | 522 | 248.6 | 21.2 | 379 | 4,656 | 1,378.7 | 131.9 | D | D | D | D |
| Dakota | 40.71 | 98.2 | 528 | 8,246 | 5,546.8 | 586.9 | 1,202 | 25,251 | 7,844.3 | 703.4 | 595 | 2,578 | 706.7 | 117.8 |
| Dodge | 0.92 | 45.2 | 17 | 549 | 507.4 | 30.8 | 48 | 455 | 128.7 | 9.8 | 8 | 10 | 1.7 | 0.3 |
| Douglas | 2.20 | 59.3 | 45 | 983 | 553.3 | 45.4 | 237 | 3,095 | 828.9 | 80.2 | 45 | 165 | 34.2 | 9.3 |
| Faribault | 0.92 | 65.5 | 23 | 144 | 156.7 | 6.1 | 63 | 510 | 109.4 | 11.3 | 6 | 9 | 1.3 | 0.2 |

1 Merchant wholesalers, except manufacturers' sales branches and offices.    2. Employer establishments.

# Table B. States and Counties — Professional Services, Manufacturing, and Accommodation and Food Services

| STATE County | Professional, scientific, and technical services, 2017 | | | | Manufacturing, 2017 | | | | Accommodation and food services, 2017 | | | |
|---|---|---|---|---|---|---|---|---|---|---|---|---|
| | Number of establishments | Number of employees | Sales (mil dol) | Average payroll (mil dol) | Number of establishments | Number of employees | Sales (mil dol) | Average payroll (mil dol) | Number of establishments | Number of employees | Sales (mil dol) | Annual payroll (mil dol) |
| | 147 | 148 | 149 | 150 | 151 | 152 | 153 | 154 | 155 | 156 | 157 | 158 |
| **MICHIGAN— Cont'd** | | | | | | | | | | | | |
| Keweenaw | D | D | D | D | | | | | 18 | 99 | 8.0 | 2.2 |
| Lake | D | D | 7 | D | 4 | 19 | 1.6 | 0.9 | 24 | D | 8.8 | D |
| Lapeer | D | D | D | D | 125 | 5,014 | 1,299.4 | 227.3 | 132 | 2,341 | 112.5 | 33.0 |
| Leelanau | 68 | 210 | 33 | 11.3 | 38 | 392 | 82.6 | 16.4 | 83 | 820 | 88.1 | 28.6 |
| Lenawee | 117 | 478 | 44 | 17.5 | 119 | 5,356 | 2,303.0 | 289.6 | 179 | 2,856 | 129.0 | 38.9 |
| Livingston | D | D | D | D | 246 | 9,377 | 4,053.5 | 527.4 | 299 | 6,151 | 301.9 | 93.6 |
| Luce | 8 | 12 | 1 | 0.4 | 5 | 160 | 111.7 | 10.1 | 17 | 193 | 10.4 | 2.6 |
| Mackinac | 18 | 32 | 4 | 1.3 | 19 | 132 | 27.5 | 5.1 | 102 | 763 | 160.6 | 48.9 |
| Macomb | D | D | D | D | 1,610 | 68,148 | 29,940.0 | 4,232.1 | 1,669 | 31,943 | 1,642.4 | 468.6 |
| Manistee | 37 | 106 | 13 | 3.8 | 27 | 971 | 489.3 | 58.4 | 65 | 1,467 | 138.2 | 34.3 |
| Marquette | D | D | D | D | 50 | 889 | 324.2 | 43.2 | 179 | 3,151 | 140.0 | 44.0 |
| Mason | 44 | 175 | 18 | 6.8 | 36 | 1,928 | 512.4 | 89.2 | 89 | 1,099 | 59.4 | 18.1 |
| Mecosta | 39 | 225 | 17 | 7.0 | 37 | 2,140 | 795.1 | 90.4 | 74 | 1,269 | 60.4 | 16.2 |
| Menominee | 30 | 218 | 18 | 6.4 | 45 | 1,814 | 486.1 | 96.6 | 43 | 594 | 23.8 | 6.6 |
| Midland | D | D | D | D | 65 | 4,271 | 2,280.2 | 298.0 | 152 | 3,306 | 170.6 | 51.1 |
| Missaukee | 11 | 40 | 3 | 1.0 | 24 | 321 | 115.9 | 13.0 | 24 | 204 | 13.5 | 3.5 |
| Monroe | D | D | D | D | 127 | 6,664 | 2,660.6 | 337.7 | 275 | 4,478 | 216.9 | 60.9 |
| Montcalm | 49 | 164 | 18 | 6.1 | 67 | 2,319 | 613.7 | 115.5 | 82 | 1,244 | 60.7 | 17.5 |
| Montmorency | D | D | 2 | D | 16 | 408 | 65.2 | 16.7 | 24 | 159 | 11.8 | 2.6 |
| Muskegon | D | D | D | D | 267 | 13,190 | 4,003.7 | 712.1 | 351 | 6,148 | 288.0 | 86.0 |
| Newaygo | 48 | 318 | 49 | 22.4 | 40 | 1,925 | 617.2 | 94.9 | 77 | 1,058 | 53.4 | 15.7 |
| Oakland | D | D | D | D | 1,647 | 53,627 | 19,516.2 | 3,336.5 | 3,032 | 59,174 | 3,360.0 | 974.0 |
| Oceana | 28 | 88 | 9 | 3.3 | 39 | 1,870 | 522.8 | 69.3 | 65 | 644 | 41.1 | 12.1 |
| Ogemaw | 26 | 89 | 9 | 2.8 | 34 | 262 | 83.1 | 11.3 | 59 | 826 | 39.3 | 11.6 |
| Ontonagon | D | D | D | 0.3 | D | D | D | 0.2 | D | D | D | D |
| Osceola | 27 | 77 | 9 | 3.4 | 41 | 2,085 | 784.5 | 90.5 | 39 | 601 | 31.3 | 10.5 |
| Oscoda | 13 | 25 | 3 | 1.2 | 17 | 371 | 107.0 | 15.2 | D | D | D | D |
| Otsego | D | D | D | D | 30 | 968 | 241.9 | 39.2 | 78 | 1,586 | 78.9 | 24.8 |
| Ottawa | D | D | D | D | 575 | 40,564 | 13,878.8 | 2,095.7 | 434 | 9,415 | 466.5 | 148.1 |
| Presque Isle | 15 | 41 | 3 | 1.0 | 13 | 196 | 38.9 | 8.7 | 38 | 255 | 11.9 | 3.6 |
| Roscommon | 30 | 72 | 7 | 2.5 | 8 | D | 119.4 | 21.6 | 76 | 1,401 | 66.1 | 20.5 |
| Saginaw | 298 | 2,414 | 375 | 142.5 | 194 | 12,288 | 4,461.3 | 799.8 | 400 | 9,735 | 493.3 | 137.5 |
| St. Clair | D | D | D | D | 231 | 8,755 | 3,666.0 | 420.5 | 269 | 5,025 | 218.8 | 66.4 |
| St. Joseph | 70 | 891 | 43 | 20.0 | 134 | 8,797 | 3,887.8 | 440.7 | 110 | 1,606 | 77.4 | 22.8 |
| Sanilac | 51 | 177 | 17 | 5.3 | 64 | 2,724 | 859.2 | 119.7 | 70 | 709 | 37.0 | 10.1 |
| Schoolcraft | 11 | 50 | 4 | 1.7 | 8 | 162 | 77.5 | 10.9 | 34 | 258 | 13.5 | 3.9 |
| Shiawassee | 66 | 355 | 42 | 16.8 | 66 | 2,055 | 672.9 | 88.9 | 111 | 1,703 | 78.5 | 21.7 |
| Tuscola | D | D | D | D | 43 | 1,319 | 394.4 | 66.6 | 62 | 727 | 32.1 | 9.5 |
| Van Buren | 74 | 2,018 | 306 | 98.9 | 81 | 2,735 | 968.5 | 142.6 | 155 | 2,008 | 108.3 | 32.3 |
| Washtenaw | 1,257 | 15,568 | 2,913 | 1,205.4 | 337 | 14,480 | 4,603.1 | 850.4 | 860 | 18,272 | 1,027.3 | 310.7 |
| Wayne | 2,870 | 51,817 | 8,008 | 3,606.7 | 1,454 | 88,033 | 67,028.2 | 5,655.0 | 3,354 | 71,006 | 5,250.8 | 1,354.7 |
| Wexford | 52 | 231 | 28 | 12.4 | 47 | 3,125 | 981.4 | 154.0 | 86 | 1,395 | 64.4 | 18.6 |
| **MINNESOTA** | D | D | D | D | 7,198 | 309,097 | 122,013.5 | 17,925.8 | 12,022 | 239,194 | 14,234.3 | 4,271.9 |
| Aitkin | D | D | D | D | 28 | 383 | 83.5 | 17.2 | 51 | 447 | 23.6 | 7.0 |
| Anoka | 736 | 4,325 | 855 | 290.8 | 589 | 21,661 | 7,087.7 | 1,301.6 | 529 | 10,983 | 558.5 | 173.9 |
| Becker | 53 | 188 | 24 | 9.6 | 41 | 2,243 | 492.4 | 109.3 | 86 | 1,266 | 66.6 | 19.7 |
| Beltrami | D | D | D | D | 35 | 902 | 346.6 | 36.0 | 114 | 2,193 | 115.9 | 36.1 |
| Benton | D | D | D | D | 64 | 3,475 | 824.0 | 167.9 | 61 | 1,099 | 50.1 | 15.5 |
| Big Stone | D | D | D | D | 5 | 13 | 0.9 | 0.3 | 18 | 121 | 4.9 | 1.0 |
| Blue Earth | D | D | D | D | 91 | 3,775 | 3,284.1 | 213.3 | 168 | 4,190 | 178.2 | 54.4 |
| Brown | 47 | 441 | 63 | 24.0 | 37 | 4,658 | 1,633.8 | 178.6 | 64 | 962 | 41.3 | 12.2 |
| Carlton | 40 | 148 | 13 | 4.7 | 28 | 1,359 | 813.7 | 93.4 | 61 | 815 | 34.6 | 10.7 |
| Carver | D | D | D | D | 153 | 9,317 | 3,173.2 | 547.8 | 181 | 3,066 | 164.9 | 50.1 |
| Cass | D | D | D | D | 38 | 271 | 50.8 | 12.4 | 126 | 1,678 | 174.5 | 45.8 |
| Chippewa | 20 | 78 | 9 | 3.0 | 27 | 1,015 | 301.2 | 43.1 | 24 | 310 | 12.2 | 4.1 |
| Chisago | D | D | 34 | D | 87 | 2,049 | 575.0 | 97.5 | D | D | D | 18.7 |
| Clay | 91 | 510 | 75 | 24.4 | 39 | 840 | 368.5 | 38.8 | 102 | 1,789 | 90.1 | 28.0 |
| Clearwater | D | D | D | D | D | 421 | D | 22.2 | 23 | 188 | 6.5 | 1.5 |
| Cook | D | D | D | D | 7 | 41 | 14.1 | 2.2 | 66 | 768 | 79.3 | 22.4 |
| Cottonwood | 18 | 75 | 9 | 3.0 | 18 | 1,033 | 455.4 | 38.5 | 20 | 295 | 11.1 | 3.3 |
| Crow Wing | D | D | D | D | 107 | 2,516 | 463.9 | 117.9 | 203 | 3,255 | 196.1 | 66.4 |
| Dakota | D | D | D | D | 434 | 18,463 | 12,835.4 | 1,087.9 | 733 | 16,485 | 901.3 | 270.1 |
| Dodge | D | D | 9 | D | 22 | 1,708 | 1,023.0 | 84.8 | 29 | 314 | 14.9 | 4.5 |
| Douglas | D | D | D | D | 91 | 3,632 | 1,103.6 | 210.5 | 113 | 1,915 | 94.3 | 30.4 |
| Faribault | D | D | D | D | 29 | 886 | 279.4 | 43.5 | 35 | 300 | 12.6 | 3.5 |

# Table B. States and Counties — Health Care and Social Assistance, Other Services, Nonemployer Businesses, and Residential Construction

| STATE County | Health care and social assistance, 2017 | | | | Other services, 2017 | | | | Nonemployer businesses, 2018 | | Value of residential construction authorized by building permits, 2020 | |
|---|---|---|---|---|---|---|---|---|---|---|---|---|
| | Number of establishments | Number of employees | Receipts (mil dol) | Annual payroll (mil dol) | Number of establishments | Number of employees | Receipts (mil dol) | Annual payroll (mil dol) | Number | Receipts (mil dol) | New construction ($1,000) | Number of housing units |
| | 159 | 160 | 161 | 162 | 163 | 164 | 165 | 166 | 167 | 168 | 169 | 170 |
| **MICHIGAN— Cont'd** | | | | | | | | | | | | |
| Keweenaw | NA | NA | NA | NA | NA | NA | NA | NA | 152 | 3.9 | 3,009 | 18 |
| Lake | 16 | 285 | 20.4 | 9.2 | D | D | 2.7 | D | 605 | 28.8 | 4,385 | 19 |
| Lapeer | 202 | 2,842 | 324.0 | 110.5 | 138 | 536 | 60.6 | 15.7 | 6,484 | 294.0 | 40,522 | 198 |
| Leelanau | 56 | 552 | 51.5 | 25.1 | 39 | 89 | 14.1 | 3.0 | 2,955 | 144.4 | 29,799 | 119 |
| Lenawee | 245 | 3,093 | 329.2 | 129.4 | 132 | 774 | 63.1 | 16.3 | 6,026 | 233.5 | 32,090 | 141 |
| Livingston | 389 | 6,075 | 502.8 | 216.1 | 317 | 1,795 | 166.2 | 55.6 | 15,408 | 770.7 | 171,113 | 677 |
| Luce | 17 | 406 | 43.5 | 17.7 | D | D | D | D | 330 | 10.8 | 502 | 3 |
| Mackinac | 17 | 420 | 57.7 | 19.7 | 16 | 41 | 7.8 | 1.5 | 852 | 32.2 | 4,636 | 25 |
| Macomb | 2,789 | 38,936 | 4,244.3 | 1,693.2 | 1,509 | 8,129 | 863.7 | 251.8 | 68,207 | 3,017.8 | 460,853 | 1,613 |
| Manistee | 65 | 1,069 | 114.3 | 45.1 | 47 | 175 | 26.3 | 5.1 | 1,686 | 61.3 | 22,711 | 96 |
| Marquette | 233 | 5,625 | 553.3 | 246.0 | 126 | 575 | 55.9 | 14.9 | 3,664 | 112.7 | 23,278 | 85 |
| Mason | 80 | 1,399 | 161.9 | 57.5 | 53 | 218 | 22.9 | 6.1 | 2,011 | 80.8 | 20,391 | 88 |
| Mecosta | 82 | 1,784 | 161.1 | 62.5 | 69 | 366 | 34.2 | 8.9 | 2,227 | 85.9 | 18,119 | 82 |
| Menominee | 32 | 251 | 15.9 | 7.3 | 30 | 103 | 15.6 | 3.0 | 1,251 | 64.7 | 4,453 | 27 |
| Midland | 282 | 6,998 | 883.2 | 294.3 | 142 | 980 | 158.5 | 29.7 | 5,185 | 198.6 | 26,149 | 123 |
| Missaukee | 33 | 434 | 21.1 | 9.2 | 18 | 58 | 5.2 | 1.4 | 1,149 | 50.6 | 3,687 | 23 |
| Monroe | 297 | 4,744 | 435.4 | 188.8 | 158 | 701 | 77.9 | 20.8 | 8,448 | 378.0 | 64,956 | 367 |
| Montcalm | 112 | 3,454 | 316.4 | 131.8 | 85 | 316 | 41.5 | 9.3 | 3,714 | 164.5 | 5,682 | 32 |
| Montmorency | 10 | 256 | 35.6 | 8.8 | D | D | 2.8 | D | 568 | 22.5 | 0 | 0 |
| Muskegon | 370 | 10,922 | 1,416.0 | 541.0 | 248 | 1,332 | 132.7 | 35.5 | 9,534 | 391.5 | 55,179 | 268 |
| Newaygo | 80 | 2,078 | 215.5 | 77.2 | 68 | 277 | 45.3 | 7.1 | 2,847 | 132.3 | 17,405 | 117 |
| Oakland | 5,176 | 107,384 | 12,796.2 | 5,033.8 | 2,537 | 19,132 | 2,438.7 | 648.2 | 118,561 | 7,027.0 | 638,506 | 2,475 |
| Oceana | 49 | 689 | 57.3 | 23.9 | 40 | 108 | 12.5 | 2.7 | 1,652 | 67.4 | 19,187 | 94 |
| Ogemaw | 77 | 1,256 | 132.5 | 45.7 | D | D | 13.6 | D | 1,306 | 52.1 | 8,565 | 20 |
| Ontonagon | 15 | 210 | 17.8 | 7.0 | 8 | 20 | 3.5 | 0.5 | 356 | 13.5 | 1,740 | 7 |
| Osceola | 46 | 1,202 | 114.7 | 42.2 | 34 | 165 | 19.1 | 5.9 | 1,401 | 58.8 | 7,311 | 43 |
| Oscoda | D | D | D | 3.6 | 15 | 41 | 5.0 | 0.9 | 586 | 25.4 | 3,792 | 13 |
| Otsego | 78 | 1,546 | 263.9 | 81.3 | 55 | 273 | 30.0 | 8.3 | 2,006 | 87.9 | 1,701 | 21 |
| Ottawa | 514 | 11,314 | 1,067.7 | 456.1 | 456 | 2,732 | 322.1 | 95.0 | 20,264 | 1,006.3 | 355,842 | 1,636 |
| Presque Isle | 24 | 290 | 23.5 | 10.0 | 22 | 54 | 5.7 | 1.2 | 893 | 31.0 | 8,321 | 35 |
| Roscommon | 48 | 570 | 48.1 | 19.0 | 53 | 135 | 14.9 | 3.7 | 1,510 | 60.4 | 15,349 | 62 |
| Saginaw | 597 | 17,800 | 2,040.7 | 811.3 | 325 | 1,989 | 183.1 | 49.7 | 10,640 | 436.5 | 44,779 | 245 |
| St. Clair | 400 | 7,724 | 782.4 | 315.4 | 206 | 837 | 105.7 | 25.5 | 10,500 | 453.9 | 46,943 | 206 |
| St. Joseph | 101 | 1,956 | 184.9 | 87.0 | 102 | 410 | 43.0 | 11.7 | 3,531 | 154.1 | 25,981 | 99 |
| Sanilac | 86 | 1,409 | 124.6 | 44.3 | 56 | 174 | 20.5 | 5.0 | 3,147 | 140.4 | 4,390 | 37 |
| Schoolcraft | 23 | 380 | 43.9 | 19.4 | 17 | 91 | 6.3 | 1.7 | 447 | 16.2 | 2,144 | 13 |
| Shiawassee | 130 | 2,830 | 263.2 | 109.3 | 100 | 415 | 53.8 | 11.8 | 4,249 | 173.8 | 12,457 | 73 |
| Tuscola | 107 | 2,728 | 229.0 | 110.5 | 46 | 155 | 14.8 | 3.3 | 3,339 | 131.6 | 7,567 | 43 |
| Van Buren | 111 | 2,268 | 197.3 | 81.7 | 93 | 369 | 33.7 | 7.6 | 4,841 | 211.1 | 38,675 | 176 |
| Washtenaw | 943 | 40,974 | 5,693.3 | 2,382.4 | 553 | 4,465 | 650.9 | 195.0 | 30,716 | 1,382.0 | 239,097 | 1,065 |
| Wayne | 4,118 | 111,962 | 14,321.6 | 5,577.6 | 2,725 | 18,581 | 2,525.3 | 643.2 | 132,583 | 4,544.5 | 435,269 | 1,996 |
| Wexford | 101 | 1,937 | 183.0 | 79.1 | 67 | 303 | 37.6 | 8.8 | 2,206 | 93.2 | 10,868 | 51 |
| **MINNESOTA** | 17,066 | 473,338 | 50,491.0 | 21,206.5 | 11,339 | 77,400 | 10,077.4 | 2,553.6 | 416,487 | 19,994.8 | 6,179,311 | 28,148 |
| Aitkin | 35 | 869 | 84.5 | 34.9 | 32 | 130 | 20.0 | 3.0 | 1,132 | 50.2 | 35,264 | 149 |
| Anoka | 774 | 18,582 | 2,155.7 | 866.1 | 614 | 4,087 | 461.9 | 131.9 | 23,918 | 1,081.9 | 386,727 | 1,670 |
| Becker | 83 | 2,186 | 219.3 | 89.5 | 80 | 351 | 33.9 | 8.3 | 2,858 | 155.6 | 48,570 | 199 |
| Beltrami | 169 | 3,157 | 368.1 | 131.1 | 86 | 408 | 61.5 | 14.4 | 2,600 | 113.3 | 24,295 | 195 |
| Benton | 94 | 2,735 | 159.6 | 64.3 | 68 | 370 | 36.9 | 10.5 | 2,853 | 122.4 | 29,906 | 117 |
| Big Stone | 19 | 561 | 54.0 | 17.8 | 17 | 39 | 5.0 | 0.9 | 402 | 21.7 | 1,746 | 13 |
| Blue Earth | 272 | 7,808 | 769.6 | 357.4 | 126 | 933 | 94.8 | 25.8 | 4,189 | 205.5 | 71,553 | 360 |
| Brown | 75 | 2,296 | 213.0 | 93.3 | 73 | 233 | 33.6 | 7.9 | 1,762 | 74.7 | 15,900 | 45 |
| Carlton | 92 | 2,148 | 181.6 | 83.3 | 64 | 282 | 28.0 | 7.5 | 1,941 | 75.8 | 19,269 | 101 |
| Carver | 246 | 6,261 | 647.3 | 286.9 | 178 | 1,008 | 107.0 | 30.5 | 8,356 | 441.0 | 209,040 | 690 |
| Cass | 70 | 1,275 | 92.8 | 45.4 | 57 | 226 | 22.5 | 4.6 | 2,648 | 131.1 | 61,464 | 263 |
| Chippewa | 44 | 1,205 | 83.7 | 37.3 | 33 | 200 | 23.1 | 5.4 | 823 | 35.7 | 1,905 | 12 |
| Chisago | 157 | 4,333 | 430.6 | 179.1 | 112 | 484 | 49.4 | 14.2 | 3,756 | 166.5 | 50,288 | 273 |
| Clay | 182 | 4,749 | 311.0 | 141.3 | 112 | 487 | 52.0 | 14.2 | 4,243 | 168.7 | 64,472 | 310 |
| Clearwater | 17 | 572 | 36.3 | 19.6 | 15 | 45 | 5.1 | 0.9 | 642 | 27.6 | 500 | 4 |
| Cook | 13 | 224 | 26.5 | 10.0 | 13 | 17 | 2.4 | 0.6 | 753 | 29.7 | 10,964 | 69 |
| Cottonwood | 31 | 762 | 55.4 | 22.4 | 26 | 101 | 10.3 | 2.8 | 820 | 36.2 | 1,945 | 8 |
| Crow Wing | 254 | 6,360 | 638.0 | 273.1 | 154 | 948 | 81.5 | 21.1 | 5,182 | 259.6 | 109,811 | 497 |
| Dakota | 1,146 | 23,039 | 2,049.6 | 884.9 | 769 | 5,962 | 688.3 | 209.6 | 30,500 | 1,443.8 | 469,486 | 2,165 |
| Dodge | D | D | D | 10.1 | 39 | 242 | 31.0 | 8.8 | 1,381 | 70.2 | 18,228 | 81 |
| Douglas | 147 | 3,662 | 362.4 | 137.2 | 106 | 456 | 53.1 | 10.9 | 3,404 | 172.1 | 63,026 | 309 |
| Faribault | 34 | 724 | 57.1 | 25.3 | 27 | 105 | 11.6 | 2.3 | 1,055 | 54.5 | 2,448 | 10 |

# Table B. States and Counties — Government Employment and Payroll, and Local Government Finances

| STATE County | Government employment and payroll, 2017 | | March payroll (percent of total) | | | | | | | Local government finances, 2017 — General revenue | | | | |
| | Full-time equivalent employees | March payroll (dollars) | Administration, judicial, and legal | Police and corrections | Fire protection | Highways and transportation | Health and welfare | Natural resources and utilities | Education and libraries | Total (mil dol) | Inter-governmental (mil dol) | Taxes Total (mil dol) | Taxes Per capita¹ (dollars) Total | Taxes Per capita¹ (dollars) Property |
| | 171 | 172 | 173 | 174 | 175 | 176 | 177 | 178 | 179 | 180 | 181 | 182 | 183 | 184 |
|---|---|---|---|---|---|---|---|---|---|---|---|---|---|---|
| **MICHIGAN— Cont'd** | | | | | | | | | | | | | | |
| Keweenaw | 53 | 176,536 | 29.4 | 14.4 | 0.5 | 34.1 | 0.0 | 9.6 | 5.4 | 8.7 | 3.8 | 2.4 | 1,162 | 1,127 |
| Lake | 258 | 1,122,787 | 23.5 | 24.8 | 0.7 | 12.0 | 1.6 | 0.3 | 31.6 | 49.5 | 25.8 | 15.9 | 1,331 | 1,330 |
| Lapeer | 2,111 | 8,342,424 | 9.9 | 7.1 | 1.0 | 4.3 | 6.3 | 3.0 | 66.6 | 294.4 | 156.7 | 66.3 | 752 | 699 |
| Leelanau | 555 | 2,222,787 | 15.1 | 8.8 | 6.3 | 6.8 | 0.6 | 2.3 | 55.9 | 73.0 | 16.9 | 42.7 | 1,974 | 1,934 |
| Lenawee | 3,305 | 13,215,049 | 7.3 | 7.0 | 1.3 | 2.6 | 2.3 | 3.6 | 74.3 | 385.6 | 192.3 | 103.9 | 1,055 | 1,015 |
| Livingston | 3,548 | 16,236,810 | 11.0 | 8.2 | 2.8 | 4.1 | 1.4 | 3.1 | 66.3 | 573.2 | 277.0 | 194.9 | 1,026 | 1,013 |
| Luce | 477 | 1,899,175 | 4.8 | 1.2 | 0.1 | 4.7 | 66.6 | 1.3 | 20.3 | 51.3 | 6.0 | 7.0 | 1,103 | 1,101 |
| Mackinac | 616 | 2,469,253 | 9.8 | 5.6 | 0.4 | 6.2 | 46.0 | 4.5 | 25.8 | 52.0 | 14.1 | 26.2 | 2,434 | 2,357 |
| Macomb | 20,652 | 100,313,575 | 6.2 | 9.5 | 4.3 | 2.0 | 3.8 | 3.1 | 69.1 | 3,390.0 | 1,593.4 | 968.7 | 1,112 | 1,074 |
| Manistee | 1,083 | 4,904,556 | 8.7 | 4.0 | 2.3 | 5.4 | 48.5 | 1.3 | 29.4 | 118.7 | 51.7 | 34.0 | 1,392 | 1,378 |
| Marquette | 2,499 | 9,820,143 | 8.6 | 6.8 | 1.8 | 7.5 | 15.7 | 11.4 | 44.0 | 346.1 | 160.6 | 91.5 | 1,376 | 1,342 |
| Mason | 1,065 | 4,415,112 | 8.3 | 6.3 | 0.0 | 7.2 | 0.6 | 3.5 | 72.8 | 149.5 | 47.6 | 65.3 | 2,252 | 2,233 |
| Mecosta | 1,812 | 7,267,045 | 7.9 | 4.7 | 1.3 | 3.8 | 35.7 | 1.4 | 44.2 | 150.4 | 77.2 | 44.5 | 1,029 | 953 |
| Menominee | 664 | 2,424,168 | 11.7 | 7.1 | 2.7 | 8.0 | 1.5 | 2.1 | 61.6 | 69.9 | 32.3 | 22.0 | 958 | 951 |
| Midland | 2,197 | 9,909,177 | 10.3 | 8.4 | 3.0 | 5.2 | 3.8 | 5.2 | 62.3 | 319.5 | 151.8 | 114.0 | 1,369 | 1,353 |
| Missaukee | 375 | 1,449,856 | 13.9 | 9.0 | 0.1 | 6.1 | 5.1 | 1.8 | 62.3 | 42.6 | 23.1 | 12.7 | 846 | 844 |
| Monroe | 4,140 | 17,254,732 | 6.2 | 6.0 | 1.8 | 3.3 | 2.1 | 3.4 | 75.6 | 493.3 | 232.5 | 176.4 | 1,181 | 1,139 |
| Montcalm | 1,818 | 7,488,261 | 6.7 | 4.7 | 0.7 | 3.6 | 7.0 | 1.8 | 73.9 | 223.6 | 124.6 | 61.6 | 969 | 964 |
| Montmorency | 206 | 737,287 | 21.0 | 8.5 | 1.6 | 12.0 | 4.3 | 7.2 | 39.6 | 23.0 | 8.6 | 11.6 | 1,253 | 1,234 |
| Muskegon | 5,624 | 23,739,095 | 7.6 | 7.6 | 2.2 | 3.3 | 9.2 | 3.3 | 65.9 | 734.8 | 410.0 | 175.1 | 1,008 | 917 |
| Newaygo | 1,547 | 6,890,797 | 9.6 | 5.8 | 0.5 | 2.8 | 11.4 | 1.1 | 67.0 | 167.8 | 97.9 | 46.8 | 969 | 959 |
| Oakland | 31,208 | 151,195,089 | 8.7 | 10.5 | 4.4 | 2.7 | 1.8 | 3.3 | 66.6 | 5,319.8 | 2,331.1 | 1,974.7 | 1,572 | 1,503 |
| Oceana | 792 | 3,172,763 | 14.7 | 6.7 | 0.0 | 10.2 | 5.3 | 3.8 | 58.2 | 97.1 | 32.3 | 33.1 | 1,254 | 1,233 |
| Ogemaw | 479 | 1,957,201 | 15.1 | 8.9 | 0.1 | 9.0 | 9.4 | 2.8 | 51.7 | 52.6 | 24.7 | 18.7 | 894 | 874 |
| Ontonagon | 253 | 974,226 | 13.2 | 4.1 | 0.3 | 27.2 | 1.5 | 2.0 | 50.6 | 38.3 | 19.0 | 12.3 | 2,090 | 2,085 |
| Osceola | 534 | 2,141,546 | 6.0 | 7.7 | 0.8 | 5.0 | 4.3 | 2.1 | 72.7 | 72.8 | 42.3 | 20.8 | 893 | 875 |
| Oscoda | 269 | 927,896 | 21.0 | 6.8 | 0.7 | 7.0 | 8.5 | 1.9 | 51.5 | 22.2 | 9.4 | 9.7 | 1,177 | 1,161 |
| Otsego | 668 | 2,754,676 | 13.5 | 4.9 | 0.4 | 10.2 | 0.4 | 2.1 | 64.4 | 85.7 | 34.2 | 36.4 | 1,484 | 1,480 |
| Ottawa | 6,943 | 30,282,853 | 7.0 | 5.8 | 1.5 | 3.1 | 3.5 | 3.7 | 72.1 | 980.2 | 490.6 | 339.3 | 1,184 | 1,153 |
| Presque Isle | 330 | 1,244,727 | 14.7 | 9.5 | 0.0 | 10.6 | 0.3 | 3.2 | 51.8 | 40.1 | 19.9 | 14.9 | 1,170 | 1,166 |
| Roscommon | 893 | 3,337,633 | 8.3 | 8.6 | 2.7 | 7.3 | 1.6 | 1.8 | 66.2 | 101.5 | 35.1 | 44.2 | 1,863 | 1,855 |
| Saginaw | 5,450 | 22,845,829 | 8.3 | 7.8 | 1.9 | 2.7 | 12.0 | 4.9 | 60.6 | 801.5 | 464.7 | 162.3 | 845 | 736 |
| St. Clair | 4,506 | 19,039,840 | 9.6 | 10.3 | 2.5 | 5.3 | 2.8 | 4.2 | 63.5 | 646.1 | 334.7 | 190.1 | 1,195 | 1,128 |
| St. Joseph | 2,396 | 9,388,918 | 7.8 | 5.5 | 1.3 | 3.5 | 18.8 | 3.7 | 57.7 | 296.0 | 130.3 | 68.9 | 1,135 | 1,119 |
| Sanilac | 1,234 | 4,735,671 | 9.8 | 7.5 | 0.3 | 4.7 | 2.5 | 3.6 | 67.2 | 186.4 | 105.6 | 44.6 | 1,081 | 1,064 |
| Schoolcraft | 422 | 2,006,463 | 7.0 | 4.0 | 0.3 | 8.9 | 61.8 | 2.1 | 15.7 | 74.9 | 14.2 | 8.5 | 1,061 | 1,060 |
| Shiawassee | 2,525 | 10,046,356 | 6.3 | 7.3 | 1.1 | 2.6 | 19.5 | 2.6 | 59.0 | 263.6 | 155.5 | 56.3 | 822 | 804 |
| Tuscola | 1,762 | 6,398,509 | 9.0 | 5.9 | 0.4 | 4.2 | 0.2 | 2.3 | 76.6 | 239.0 | 112.8 | 83.0 | 1,571 | 1,554 |
| Van Buren | 2,944 | 11,710,625 | 8.0 | 6.9 | 1.1 | 3.2 | 4.0 | 2.3 | 73.6 | 374.7 | 176.5 | 120.1 | 1,595 | 1,576 |
| Washtenaw | 10,291 | 50,170,923 | 7.5 | 9.9 | 2.6 | 5.9 | 4.7 | 5.2 | 61.5 | 1,623.7 | 629.7 | 665.3 | 1,804 | 1,745 |
| Wayne | 48,307 | 223,215,723 | 8.0 | 14.5 | 4.2 | 6.5 | 1.3 | 6.5 | 55.6 | 9,545.1 | 4,481.6 | 2,659.3 | 1,513 | 1,151 |
| Wexford | 921 | 3,817,311 | 9.5 | 7.4 | 2.2 | 7.4 | 0.9 | 3.0 | 68.2 | 135.7 | 72.8 | 38.4 | 1,155 | 1,143 |
| **MINNESOTA** | X | X | X | X | X | X | X | X | X | X | X | X | X | X |
| Aitkin | 638 | 2,441,866 | 16.1 | 10.9 | 0.7 | 7.1 | 11.3 | 3.9 | 46.9 | 68.4 | 35.2 | 20.4 | 1,293 | 1,275 |
| Anoka | 11,869 | 57,343,988 | 6.4 | 9.8 | 1.3 | 2.5 | 5.7 | 3.1 | 69.2 | 1,464.4 | 809.1 | 426.1 | 1,217 | 1,159 |
| Becker | 1,242 | 5,527,703 | 9.9 | 7.8 | 0.0 | 6.9 | 16.2 | 7.1 | 50.9 | 152.6 | 77.9 | 45.9 | 1,347 | 1,297 |
| Beltrami | 2,161 | 8,819,155 | 5.9 | 10.4 | 0.7 | 2.8 | 8.4 | 1.9 | 67.8 | 239.4 | 154.2 | 48.2 | 1,037 | 880 |
| Benton | 1,284 | 5,492,962 | 6.6 | 8.8 | 0.3 | 12.6 | 8.9 | 2.0 | 60.1 | 156.2 | 90.8 | 46.2 | 1,151 | 1,103 |
| Big Stone | 504 | 1,927,996 | 7.7 | 3.0 | 0.1 | 4.8 | 49.4 | 1.7 | 32.3 | 75.8 | 22.2 | 8.4 | 1,688 | 1,636 |
| Blue Earth | 2,477 | 11,010,929 | 8.4 | 9.1 | 1.3 | 4.8 | 7.1 | 4.6 | 62.2 | 366.3 | 177.4 | 101.6 | 1,518 | 1,279 |
| Brown | 1,171 | 5,188,028 | 10.3 | 8.5 | 0.2 | 4.7 | 21.6 | 11.3 | 41.4 | 134.9 | 57.8 | 35.1 | 1,395 | 1,321 |
| Carlton | 1,695 | 7,460,434 | 6.1 | 5.7 | 4.8 | 3.9 | 24.1 | 3.2 | 51.3 | 198.3 | 93.7 | 49.5 | 1,394 | 1,301 |
| Carver | 3,152 | 15,411,363 | 8.3 | 7.6 | 0.4 | 3.4 | 9.0 | 7.5 | 61.6 | 465.7 | 191.7 | 178.0 | 1,743 | 1,625 |
| Cass | 1,263 | 5,156,214 | 11.1 | 11.5 | 0.0 | 5.0 | 13.2 | 1.8 | 55.6 | 157.6 | 85.5 | 45.6 | 1,555 | 1,498 |
| Chippewa | 839 | 3,978,680 | 4.7 | 4.2 | 0.4 | 2.8 | 38.0 | 2.0 | 45.3 | 77.3 | 36.6 | 18.8 | 1,574 | 1,536 |
| Chisago | 1,560 | 6,808,243 | 11.0 | 9.0 | 0.2 | 4.8 | 8.0 | 3.2 | 62.4 | 209.8 | 100.8 | 75.7 | 1,370 | 1,286 |
| Clay | 2,420 | 10,374,791 | 7.4 | 10.1 | 2.4 | 3.5 | 11.3 | 8.4 | 55.3 | 323.2 | 181.7 | 64.5 | 1,012 | 974 |
| Clearwater | 418 | 1,645,740 | 10.1 | 13.6 | 0.0 | 4.7 | 12.2 | 6.2 | 50.3 | 59.6 | 22.9 | 12.8 | 1,446 | 1,440 |
| Cook | 348 | 1,508,292 | 11.9 | 6.8 | 0.0 | 7.0 | 39.1 | 5.9 | 25.5 | 50.9 | 18.2 | 12.5 | 2,311 | 1,888 |
| Cottonwood | 650 | 2,488,355 | 8.3 | 7.2 | 0.7 | 6.0 | 16.5 | 10.0 | 50.2 | 96.9 | 40.8 | 24.5 | 2,179 | 2,158 |
| Crow Wing | 3,471 | 14,319,009 | 6.9 | 8.6 | 0.7 | 3.3 | 30.9 | 1.4 | 48.0 | 299.8 | 141.0 | 104.7 | 1,629 | 1,526 |
| Dakota | 15,457 | 71,499,309 | 5.4 | 8.3 | 1.2 | 2.4 | 6.6 | 3.6 | 70.4 | 1,873.9 | 938.3 | 608.4 | 1,443 | 1,375 |
| Dodge | 804 | 3,401,794 | 8.0 | 7.8 | 0.2 | 4.9 | 8.5 | 4.2 | 64.7 | 104.2 | 52.2 | 28.8 | 1,392 | 1,363 |
| Douglas | 1,485 | 5,931,587 | 8.5 | 9.5 | 0.1 | 4.7 | 7.1 | 6.9 | 61.7 | 164.6 | 77.1 | 60.0 | 1,600 | 1,460 |
| Faribault | 588 | 2,199,089 | 9.5 | 10.2 | 0.4 | 6.1 | 3.4 | 11.5 | 55.0 | 71.7 | 34.0 | 22.2 | 1,615 | 1,578 |

1. Based on the resident population estimated as of July 1 of the year shown.

| STATE County | Local government finances, 2017 (cont.) | | | | | | | | | Government employment, 2019 | | | Individual income tax returns, 2018 | | |
| | Direct general expenditure | | | | | | | Debt outstanding | | | | | | | |
| | | | Percent of total for: | | | | | | | | | | | | |
| | Total (mil dol) | Per capita¹ (dollars) | Education | Health and hospitals | Police protection | Public welfare | Highways | Total (mil dol) | Per capita¹ (dollars) | Federal civilian | Federal military | State and local | Number of returns | Mean adjusted gross income | Mean income tax |
| | 185 | 186 | 187 | 188 | 189 | 190 | 191 | 192 | 193 | 194 | 195 | 196 | 197 | 198 | 199 |
| **MICHIGAN— Cont'd** | | | | | | | | | | | | | | | |
| Keweenaw | 8 | 4,037 | 2.6 | 0.8 | 9.4 | 0.4 | 35.8 | 3.9 | 1,861 | 26 | 3 | 117 | 1,040 | 55,688 | 4,835 |
| Lake | 37 | 3,066 | 22.8 | 1.4 | 6.5 | 1.1 | 21.5 | 18.1 | 1,509 | 47 | 18 | 406 | 4,570 | 40,320 | 3,001 |
| Lapeer | 279 | 3,170 | 46.4 | 8.2 | 5.0 | 8.8 | 8.4 | 220.2 | 2,499 | 143 | 137 | 4,181 | 42,990 | 59,426 | 5,874 |
| Leelanau | 79 | 3,640 | 40.0 | 2.6 | 6.2 | 0.0 | 9.2 | 40.7 | 1,880 | 132 | 34 | 1,699 | 12,100 | 89,571 | 12,533 |
| Lenawee | 386 | 3,923 | 54.0 | 5.6 | 3.2 | 4.4 | 7.6 | 268.6 | 2,729 | 196 | 148 | 4,583 | 46,160 | 54,138 | 4,839 |
| Livingston | 550 | 2,898 | 57.9 | 2.3 | 3.5 | 0.6 | 5.0 | 842.9 | 4,439 | 261 | 303 | 6,639 | 98,750 | 83,157 | 10,309 |
| Luce | 66 | 10,410 | 11.1 | 60.2 | 0.7 | 0.3 | 11.6 | 23.9 | 3,753 | 14 | 8 | 749 | 2,490 | 46,575 | 4,010 |
| Mackinac | 56 | 5,219 | 32.0 | 9.5 | 4.0 | 0.1 | 14.2 | 32.6 | 3,025 | 63 | 66 | 911 | 6,080 | 45,894 | 4,046 |
| Macomb | 3,408 | 3,911 | 50.6 | 9.2 | 6.5 | 1.2 | 5.6 | 4,181.8 | 4,799 | 8,328 | 1,769 | 29,686 | 453,160 | 58,805 | 6,041 |
| Manistee | 110 | 4,499 | 31.2 | 12.4 | 2.7 | 10.0 | 8.0 | 64.4 | 2,640 | 129 | 79 | 2,520 | 11,800 | 49,044 | 4,350 |
| Marquette | 331 | 4,982 | 35.2 | 10.9 | 4.3 | 10.4 | 9.0 | 339.8 | 5,110 | 274 | 126 | 5,285 | 30,640 | 56,706 | 5,201 |
| Mason | 147 | 5,059 | 54.0 | 0.3 | 2.7 | 10.5 | 8.6 | 47.1 | 1,623 | 69 | 47 | 1,798 | 14,490 | 52,205 | 4,997 |
| Mecosta | 147 | 3,392 | 50.8 | 2.2 | 4.2 | 0.5 | 15.8 | 43.0 | 993 | 96 | 65 | 4,653 | 17,360 | 50,268 | 4,541 |
| Menominee | 69 | 3,009 | 46.9 | 0.6 | 7.2 | 0.5 | 11.1 | 20.5 | 891 | 60 | 36 | 1,973 | 11,240 | 48,534 | 3,990 |
| Midland | 325 | 3,896 | 55.6 | 2.2 | 4.0 | 2.1 | 7.2 | 513.6 | 6,165 | 152 | 130 | 3,072 | 40,440 | 77,674 | 9,818 |
| Missaukee | 44 | 2,906 | 53.6 | 1.8 | 2.9 | 1.2 | 15.5 | 51.5 | 3,426 | 26 | 24 | 517 | 6,750 | 44,943 | 3,467 |
| Monroe | 544 | 3,638 | 53.6 | 6.9 | 2.8 | 0.2 | 7.7 | 514.5 | 3,443 | 246 | 237 | 5,471 | 75,290 | 62,405 | 6,208 |
| Montcalm | 233 | 3,668 | 62.3 | 8.6 | 1.7 | 0.5 | 7.5 | 163.8 | 2,579 | 136 | 97 | 3,065 | 28,220 | 46,452 | 3,594 |
| Montmorency | 23 | 2,504 | 37.8 | 2.0 | 5.3 | 1.0 | 15.6 | 11.2 | 1,209 | 18 | 15 | 350 | 4,510 | 40,923 | 3,008 |
| Muskegon | 780 | 4,489 | 51.7 | 8.7 | 3.7 | 2.8 | 4.5 | 744.5 | 4,287 | 356 | 281 | 7,333 | 79,760 | 51,293 | 4,634 |
| Newaygo | 192 | 3,973 | 52.9 | 6.8 | 2.7 | 7.0 | 8.7 | 200.0 | 4,140 | 71 | 77 | 2,225 | 22,000 | 49,108 | 4,061 |
| Oakland | 5,622 | 4,476 | 49.7 | 6.1 | 5.8 | 0.1 | 7.1 | 4,996.7 | 3,978 | 4,613 | 1,994 | 46,810 | 644,940 | 99,537 | 15,164 |
| Oceana | 105 | 3,991 | 41.8 | 3.7 | 3.4 | 16.2 | 10.2 | 46.2 | 1,748 | 172 | 42 | 1,231 | 12,400 | 48,696 | 4,307 |
| Ogemaw | 51 | 2,421 | 40.5 | 2.6 | 4.2 | 0.0 | 17.8 | 28.6 | 1,370 | 59 | 33 | 706 | 9,480 | 42,010 | 3,321 |
| Ontonagon | 36 | 6,054 | 49.7 | 1.4 | 1.4 | 0.1 | 24.2 | 24.1 | 4,100 | 30 | 9 | 337 | 2,700 | 46,353 | 3,662 |
| Osceola | 81 | 3,497 | 59.5 | 3.3 | 2.8 | 0.3 | 9.5 | 61.2 | 2,632 | 57 | 37 | 956 | 10,230 | 44,806 | 3,209 |
| Oscoda | 22 | 2,631 | 41.9 | 6.2 | 4.4 | 1.9 | 14.8 | 5.9 | 718 | 52 | 13 | 313 | 3,690 | 40,636 | 2,962 |
| Otsego | 80 | 3,240 | 50.8 | 2.7 | 3.1 | 1.0 | 12.2 | 23.0 | 939 | 136 | 41 | 1,094 | 12,640 | 51,578 | 4,579 |
| Ottawa | 1,185 | 4,137 | 54.7 | 4.2 | 3.1 | 0.7 | 10.7 | 1,517.8 | 5,296 | 434 | 524 | 16,057 | 137,590 | 73,462 | 8,186 |
| Presque Isle | 43 | 3,341 | 37.8 | 0.6 | 3.4 | 0.5 | 14.0 | 22.4 | 1,756 | 59 | 20 | 606 | 6,420 | 54,224 | 5,843 |
| Roscommon | 109 | 4,587 | 50.2 | 2.1 | 3.8 | 1.8 | 10.3 | 70.5 | 2,970 | 29 | 38 | 1,365 | 11,520 | 45,825 | 4,139 |
| Saginaw | 843 | 4,389 | 39.0 | 18.6 | 4.0 | 0.6 | 8.2 | 484.2 | 2,522 | 1,517 | 309 | 9,264 | 90,190 | 52,225 | 5,102 |
| St. Clair | 672 | 4,223 | 44.6 | 9.3 | 4.7 | 1.0 | 10.8 | 620.4 | 3,902 | 723 | 339 | 6,274 | 79,730 | 56,960 | 5,548 |
| St. Joseph | 318 | 5,236 | 43.3 | 23.5 | 2.7 | 0.3 | 4.1 | 196.4 | 3,236 | 106 | 96 | 2,759 | 28,020 | 51,301 | 4,455 |
| Sanilac | 189 | 4,577 | 38.7 | 7.9 | 3.7 | 1.1 | 12.6 | 137.6 | 3,335 | 103 | 65 | 1,498 | 19,450 | 46,969 | 3,932 |
| Schoolcraft | 69 | 8,612 | 12.4 | 40.1 | 0.1 | 11.3 | 7.8 | 59.0 | 7,358 | 36 | 13 | 989 | 3,870 | 49,676 | 4,217 |
| Shiawassee | 268 | 3,916 | 52.5 | 1.4 | 2.4 | 16.5 | 7.6 | 239.9 | 3,507 | 137 | 107 | 3,155 | 33,360 | 51,325 | 4,428 |
| Tuscola | 218 | 4,130 | 55.2 | 2.6 | 2.5 | 9.2 | 12.2 | 142.4 | 2,696 | 125 | 81 | 2,825 | 25,170 | 47,090 | 3,571 |
| Van Buren | 467 | 6,196 | 56.1 | 10.6 | 2.5 | 0.5 | 7.7 | 485.0 | 6,440 | 147 | 119 | 4,597 | 35,300 | 53,500 | 4,946 |
| Washtenaw | 1,638 | 4,442 | 49.2 | 6.2 | 7.2 | 0.6 | 3.7 | 1,748.4 | 4,741 | 4,066 | 620 | 73,942 | 168,360 | 90,653 | 12,815 |
| Wayne | 9,880 | 5,622 | 33.7 | 3.2 | 5.7 | 5.6 | 3.5 | 11,792.4 | 6,711 | 13,945 | 3,207 | 75,892 | 802,460 | 58,925 | 6,786 |
| Wexford | 123 | 3,707 | 60.0 | 1.4 | 3.8 | 0.4 | 8.8 | 61.3 | 1,844 | 124 | 53 | 2,087 | 15,640 | 47,462 | 3,885 |
| **MINNESOTA** | X | X | X | X | X | X | X | X | X | 32,187 | 19,829 | 377,623 | 2,795,760 | 77,556 | 9,548 |
| Aitkin | 83 | 5,283 | 34.6 | 2.2 | 4.3 | 7.6 | 24.2 | 14.1 | 893 | 44 | 54 | 867 | 7,400 | 52,900 | 4,471 |
| Anoka | 1,507 | 4,304 | 59.0 | 0.6 | 6.5 | 5.1 | 8.1 | 1,631.9 | 4,660 | 433 | 1,232 | 15,461 | 181,440 | 71,381 | 7,639 |
| Becker | 151 | 4,429 | 40.4 | 1.2 | 4.8 | 13.1 | 12.6 | 61.2 | 1,795 | 238 | 118 | 3,141 | 16,310 | 59,776 | 5,719 |
| Beltrami | 256 | 5,498 | 51.0 | 1.3 | 4.5 | 10.9 | 9.7 | 275.2 | 5,922 | 382 | 157 | 5,213 | 20,360 | 53,677 | 4,876 |
| Benton | 133 | 3,325 | 54.8 | 1.1 | 5.5 | 8.3 | 9.6 | 128.9 | 3,212 | 58 | 138 | 1,771 | 19,970 | 56,705 | 5,029 |
| Big Stone | 75 | 14,987 | 16.5 | 41.7 | 1.8 | 18.4 | 6.7 | 610.7 | 122,043 | 29 | 17 | 633 | 2,470 | 51,139 | 4,174 |
| Blue Earth | 393 | 5,869 | 41.2 | 1.9 | 4.1 | 5.1 | 12.8 | 381.5 | 5,702 | 266 | 251 | 5,284 | 31,660 | 58,128 | 5,785 |
| Brown | 160 | 6,344 | 37.3 | 12.8 | 8.3 | 6.4 | 8.9 | 113.7 | 4,519 | 75 | 83 | 1,686 | 13,110 | 60,123 | 5,491 |
| Carlton | 252 | 7,082 | 53.8 | 12.8 | 3.3 | 6.6 | 9.2 | 296.3 | 8,344 | 68 | 118 | 5,163 | 16,590 | 58,076 | 4,915 |
| Carver | 529 | 5,181 | 52.1 | 0.4 | 5.1 | 4.6 | 8.9 | 818.3 | 8,014 | 218 | 362 | 4,976 | 51,030 | 110,146 | 15,720 |
| Cass | 176 | 6,001 | 52.0 | 1.4 | 3.8 | 6.7 | 12.3 | 106.0 | 3,615 | 254 | 103 | 3,883 | 14,480 | 56,963 | 5,662 |
| Chippewa | 80 | 6,708 | 40.4 | 7.6 | 4.3 | 6.5 | 12.5 | 95.9 | 8,030 | 66 | 40 | 1,200 | 5,970 | 56,432 | 4,725 |
| Chisago | 216 | 3,906 | 45.9 | 0.8 | 5.2 | 5.2 | 12.5 | 200.3 | 3,624 | 99 | 191 | 2,342 | 27,910 | 69,546 | 7,095 |
| Clay | 376 | 5,896 | 44.6 | 2.9 | 4.5 | 5.4 | 18.2 | 710.7 | 11,146 | 110 | 213 | 4,452 | 28,820 | 59,241 | 5,070 |
| Clearwater | 55 | 6,241 | 35.5 | 33.4 | 3.4 | 0.0 | 11.9 | 23.6 | 2,672 | 33 | 30 | 498 | 3,820 | 55,069 | 5,119 |
| Cook | 64 | 11,897 | 11.3 | 44.1 | 4.8 | 4.0 | 10.6 | 66.6 | 12,304 | 134 | 19 | 753 | 3,080 | 57,188 | 5,358 |
| Cottonwood | 108 | 9,590 | 25.8 | 23.7 | 3.9 | 11.8 | 12.3 | 66.8 | 5,943 | 56 | 38 | 862 | 5,480 | 52,690 | 4,169 |
| Crow Wing | 281 | 4,374 | 43.0 | 0.8 | 6.2 | 8.1 | 12.3 | 266.1 | 4,139 | 221 | 250 | 3,694 | 32,470 | 59,413 | 5,909 |
| Dakota | 1,913 | 4,537 | 55.9 | 0.6 | 5.3 | 4.0 | 10.2 | 1,965.7 | 4,663 | 3,276 | 1,484 | 18,884 | 222,460 | 83,647 | 10,204 |
| Dodge | 111 | 5,359 | 49.7 | 6.4 | 6.3 | 4.2 | 11.4 | 133.3 | 6,441 | 40 | 72 | 1,267 | 10,290 | 65,144 | 6,057 |
| Douglas | 174 | 4,631 | 40.6 | 0.4 | 6.8 | 5.3 | 17.3 | 301.2 | 8,027 | 138 | 131 | 3,556 | 19,540 | 67,390 | 7,195 |
| Faribault | 75 | 5,480 | 33.1 | 0.7 | 5.0 | 2.6 | 22.2 | 91.2 | 6,625 | 55 | 46 | 836 | 6,760 | 50,124 | 3,920 |

1. Based on the resident population estimated as of July 1 of the year shown.

| State / county code | CBSA code[1] | County Type code[2] | STATE County | Land area[3] (sq. mi) | Total persons 2019 | Rank | Per square mile | White | Black | American Indian, Alaska Native | Asian and Pacific Islander | Percent Hispanic or Latino[4] | Under 5 years | 5 to 14 years | 15 to 24 years | 25 to 34 years | 35 to 44 years | 45 to 54 years |
|---|---|---|---|---|---|---|---|---|---|---|---|---|---|---|---|---|---|---|
| | | | | 1 | 2 | 3 | 4 | 5 | 6 | 7 | 8 | 9 | 10 | 11 | 12 | 13 | 14 | 15 |
| | | | MINNESOTA— Cont'd | | | | | | | | | | | | | | | |
| 27045 | 40340 | 3 | Fillmore | 861.3 | 21,135 | 1,765 | 24.5 | 96.8 | 0.9 | 0.5 | 0.9 | 1.8 | 6.2 | 14.3 | 10.7 | 10.0 | 12.0 | 10.7 |
| 27047 | 10660 | 7 | Freeborn | 707.2 | 30,364 | 1,423 | 42.9 | 84.3 | 1.9 | 0.7 | 3.9 | 10.6 | 5.4 | 12.4 | 11.0 | 11.0 | 11.2 | 10.8 |
| 27049 | 39860 | 4 | Goodhue | 756.7 | 46,318 | 1,050 | 61.2 | 92.9 | 2.2 | 1.8 | 1.2 | 3.6 | 5.6 | 12.4 | 10.9 | 11.1 | 12.2 | 11.6 |
| 27051 | | 9 | Grant | 547.8 | 6,026 | 2,743 | 11.0 | 96.1 | 1.3 | 1.2 | 0.7 | 2.3 | 5.4 | 13.5 | 9.7 | 10.7 | 11.4 | 10.3 |
| 27053 | 33460 | 1 | Hennepin | 554.0 | 1,268,408 | 33 | 2,289.5 | 70.8 | 15.5 | 1.4 | 8.6 | 7.0 | 6.2 | 12.0 | 11.8 | 16.8 | 14.2 | 11.6 |
| 27055 | 29100 | 3 | Houston | 552.0 | 18,632 | 1,892 | 33.8 | 96.6 | 1.5 | 0.7 | 1.3 | 1.4 | 5.7 | 12.3 | 10.2 | 10.0 | 11.9 | 11.3 |
| 27057 | | 7 | Hubbard | 926.0 | 21,783 | 1,727 | 23.5 | 94.1 | 1.1 | 3.4 | 0.9 | 2.5 | 5.0 | 12.6 | 9.3 | 9.0 | 10.8 | 10.6 |
| 27059 | 33460 | 1 | Isanti | 435.8 | 41,429 | 1,147 | 95.1 | 94.6 | 1.6 | 1.4 | 2.1 | 2.3 | 5.9 | 13.5 | 11.2 | 13.0 | 12.7 | 12.4 |
| 27061 | 24330 | 6 | Itasca | 2,667.7 | 45,268 | 1,072 | 17.0 | 93.7 | 1.1 | 5.1 | 0.8 | 1.7 | 5.1 | 11.5 | 10.5 | 9.7 | 11.3 | 11.2 |
| 27063 | | 7 | Jackson | 703.0 | 9,768 | 2,433 | 13.9 | 92.7 | 1.6 | 1.1 | 2.1 | 4.2 | 5.5 | 12.2 | 10.9 | 10.3 | 11.7 | 11.2 |
| 27065 | | 6 | Kanabec | 521.6 | 16,416 | 2,011 | 31.5 | 96.2 | 1.2 | 1.7 | 1.2 | 1.7 | 5.3 | 11.9 | 10.9 | 10.4 | 11.6 | 12.2 |
| 27067 | 48820 | 4 | Kandiyohi | 797.4 | 43,130 | 1,118 | 54.1 | 79.1 | 6.7 | 0.7 | 1.6 | 12.9 | 6.6 | 13.6 | 12.0 | 12.4 | 11.4 | 10.6 |
| 27069 | | 9 | Kittson | 1,099.0 | 4,214 | 2,879 | 3.8 | 95.8 | 1.2 | 0.9 | 0.9 | 2.4 | 5.3 | 12.9 | 10.6 | 8.9 | 9.6 | 11.0 |
| 27071 | | 6 | Koochiching | 3,104.6 | 12,059 | 2,292 | 3.9 | 94.8 | 1.2 | 3.5 | 1.0 | 1.4 | 4.0 | 10.1 | 9.9 | 9.5 | 10.1 | 11.6 |
| 27073 | | 9 | Lac qui Parle | 765.0 | 6,527 | 2,701 | 8.5 | 95.6 | 1.3 | 0.9 | 1.2 | 2.6 | 4.8 | 11.9 | 9.9 | 8.4 | 10.6 | 10.1 |
| 27075 | 20260 | 6 | Lake | 2,109.1 | 10,639 | 2,374 | 5.0 | 96.6 | 0.9 | 1.6 | 1.0 | 1.5 | 4.7 | 11.2 | 8.9 | 9.4 | 11.0 | 10.6 |
| 27077 | | 9 | Lake of the Woods | 1,297.8 | 3,754 | 2,916 | 2.9 | 94.8 | 1.4 | 2.3 | 2.3 | 1.9 | 5.3 | 10.1 | 8.8 | 9.2 | 10.9 | 11.1 |
| 27079 | 33460 | 1 | Le Sueur | 448.7 | 28,741 | 1,468 | 64.1 | 91.5 | 1.1 | 0.8 | 1.1 | 6.7 | 5.8 | 13.2 | 11.7 | 10.9 | 12.6 | 12.6 |
| 27081 | | 9 | Lincoln | 536.8 | 5,568 | 2,782 | 10.4 | 96.4 | 0.8 | 0.8 | 0.9 | 2.2 | 6.4 | 13.2 | 10.3 | 9.5 | 10.7 | 10.4 |
| 27083 | 32140 | 7 | Lyon | 714.4 | 25,271 | 1,597 | 35.4 | 83.8 | 3.9 | 0.8 | 5.4 | 7.5 | 7.3 | 14.5 | 13.2 | 12.2 | 12.3 | 10.6 |
| 27085 | 26780 | 6 | McLeod | 491.5 | 35,710 | 1,287 | 72.7 | 91.0 | 1.3 | 0.7 | 1.2 | 7.0 | 5.6 | 12.6 | 11.9 | 11.8 | 12.1 | 12.5 |
| 27087 | | 8 | Mahnomen | 557.9 | 5,473 | 2,790 | 9.8 | 53.3 | 1.3 | 48.7 | 0.8 | 4.8 | 8.7 | 17.9 | 12.9 | 10.0 | 10.3 | 9.6 |
| 27089 | | 8 | Marshall | 1,775.1 | 9,321 | 2,472 | 5.3 | 94.0 | 0.8 | 1.1 | 0.5 | 4.7 | 6.2 | 12.9 | 10.7 | 10.4 | 11.6 | 10.7 |
| 27091 | 21860 | 8 | Martin | 712.3 | 19,484 | 1,855 | 27.4 | 92.7 | 0.9 | 0.7 | 1.1 | 5.5 | 6.0 | 12.4 | 10.5 | 10.2 | 10.8 | 10.7 |
| 27093 | | 6 | Meeker | 608.0 | 23,341 | 1,664 | 38.4 | 94.5 | 1.0 | 0.6 | 0.7 | 4.2 | 6.1 | 13.6 | 11.9 | 10.3 | 11.3 | 11.2 |
| 27095 | 33460 | 1 | Mille Lacs | 572.3 | 26,146 | 1,563 | 45.7 | 90.2 | 1.2 | 6.7 | 1.1 | 2.7 | 6.0 | 13.3 | 11.1 | 11.7 | 12.2 | 11.9 |
| 27097 | | 6 | Morrison | 1,125.1 | 33,187 | 1,351 | 29.5 | 96.7 | 1.1 | 0.9 | 0.7 | 1.8 | 5.7 | 13.4 | 11.0 | 10.6 | 12.3 | 11.3 |
| 27099 | 12380 | 4 | Mower | 711.3 | 40,150 | 1,179 | 56.4 | 77.6 | 4.7 | 0.6 | 6.5 | 12.3 | 6.6 | 14.2 | 12.4 | 12.1 | 12.0 | 11.2 |
| 27101 | | 9 | Murray | 704.7 | 8,155 | 2,570 | 11.6 | 92.6 | 0.8 | 0.7 | 2.3 | 4.7 | 5.4 | 12.5 | 10.1 | 8.7 | 11.1 | 10.1 |
| 27103 | 31860 | 3 | Nicollet | 448.6 | 34,482 | 1,317 | 76.9 | 88.8 | 4.8 | 0.7 | 2.3 | 5.0 | 5.5 | 12.5 | 16.9 | 12.2 | 13.4 | 10.7 |
| 27105 | 49380 | 7 | Nobles | 715.1 | 21,400 | 1,752 | 29.9 | 57.7 | 5.0 | 0.8 | 7.6 | 30.1 | 8.6 | 15.1 | 12.5 | 11.7 | 11.6 | 11.0 |
| 27107 | | 3 | Norman | 872.8 | 6,338 | 2,719 | 7.3 | 90.8 | 1.4 | 3.8 | 1.1 | 5.6 | 6.2 | 13.3 | 11.4 | 9.7 | 11.1 | 11.6 |
| 27109 | 40340 | 3 | Olmsted | 653.5 | 159,298 | 422 | 243.8 | 80.6 | 8.1 | 0.7 | 7.7 | 5.4 | 6.6 | 13.6 | 11.8 | 14.5 | 13.7 | 11.0 |
| 27111 | 22260 | 6 | Otter Tail | 1,971.6 | 58,741 | 881 | 29.8 | 93.4 | 2.0 | 1.2 | 1.1 | 3.8 | 5.7 | 12.4 | 10.3 | 9.6 | 10.8 | 10.3 |
| 27113 | | 6 | Pennington | 616.6 | 13,874 | 2,165 | 22.5 | 91.9 | 1.7 | 2.5 | 1.5 | 4.4 | 6.1 | 12.7 | 11.3 | 12.9 | 12.3 | 11.6 |
| 27115 | | 6 | Pine | 1,411.3 | 29,359 | 1,450 | 20.8 | 90.9 | 2.8 | 4.0 | 1.3 | 3.1 | 4.5 | 11.1 | 9.9 | 11.0 | 12.6 | 12.1 |
| 27117 | | 6 | Pipestone | 465.1 | 9,121 | 2,485 | 19.6 | 88.8 | 1.9 | 2.3 | 1.4 | 7.8 | 7.1 | 14.7 | 11.2 | 10.4 | 10.8 | 10.6 |
| 27119 | 24220 | 3 | Polk | 1,971.0 | 30,900 | 1,406 | 15.7 | 88.0 | 3.4 | 2.5 | 1.4 | 6.9 | 6.7 | 13.7 | 12.5 | 11.6 | 11.9 | 10.7 |
| 27121 | | 8 | Pope | 669.5 | 11,277 | 2,331 | 16.8 | 96.7 | 0.9 | 0.8 | 1.0 | 1.9 | 5.6 | 12.5 | 9.4 | 10.7 | 11.5 | 10.1 |
| 27123 | 33460 | 1 | Ramsey | 152.2 | 547,903 | 127 | 3,599.9 | 63.3 | 14.7 | 1.4 | 16.6 | 7.6 | 6.7 | 12.9 | 13.0 | 16.6 | 13.0 | 10.7 |
| 27125 | | 8 | Red Lake | 432.4 | 4,046 | 2,890 | 9.4 | 93.2 | 1.7 | 2.5 | 0.6 | 4.0 | 6.0 | 13.6 | 10.6 | 9.6 | 12.5 | 10.8 |
| 27127 | | 7 | Redwood | 878.6 | 15,079 | 2,090 | 17.2 | 88.1 | 1.4 | 5.3 | 3.3 | 3.7 | 6.6 | 13.8 | 11.5 | 10.7 | 11.1 | 10.6 |
| 27129 | | 8 | Renville | 982.9 | 14,403 | 2,130 | 14.7 | 88.2 | 1.1 | 1.6 | 1.2 | 9.3 | 6.2 | 13.0 | 11.3 | 10.2 | 11.2 | 11.0 |
| 27131 | 22060 | 4 | Rice | 495.8 | 67,084 | 797 | 135.3 | 82.3 | 7.0 | 0.7 | 3.0 | 8.6 | 5.7 | 12.0 | 18.2 | 11.6 | 11.8 | 11.5 |
| 27133 | | 6 | Rock | 482.5 | 9,301 | 2,473 | 19.3 | 94.3 | 1.4 | 1.1 | 1.3 | 3.4 | 5.4 | 14.4 | 12.2 | 10.5 | 11.7 | 11.3 |
| 27135 | | 7 | Roseau | 1,671.7 | 15,117 | 2,087 | 9.0 | 92.8 | 1.3 | 2.8 | 3.5 | 1.7 | 5.5 | 14.0 | 11.9 | 10.2 | 11.2 | 12.7 |
| 27137 | 20260 | 2 | St. Louis | 6,247.6 | 198,538 | 340 | 31.8 | 93.1 | 2.6 | 3.3 | 1.7 | 1.9 | 4.8 | 10.6 | 15.6 | 11.6 | 11.6 | 10.7 |
| 27139 | 33460 | 1 | Scott | 356.3 | 150,689 | 446 | 422.9 | 81.3 | 6.6 | 1.4 | 7.7 | 5.5 | 6.3 | 15.4 | 12.8 | 11.9 | 14.7 | 14.3 |
| 27141 | 33460 | 1 | Sherburne | 432.9 | 98,811 | 611 | 228.3 | 91.7 | 4.3 | 1.0 | 2.2 | 2.8 | 6.6 | 14.7 | 12.8 | 13.1 | 14.3 | 13.6 |
| 27143 | | 1 | Sibley | 588.8 | 14,715 | 2,106 | 25.0 | 88.8 | 1.5 | 0.6 | 1.1 | 9.2 | 5.6 | 12.9 | 11.1 | 11.1 | 12.3 | 11.6 |
| 27145 | 41060 | 3 | Stearns | 1,342.8 | 162,038 | 413 | 120.7 | 85.2 | 9.3 | 0.7 | 3.0 | 3.8 | 6.5 | 12.9 | 18.3 | 12.1 | 11.5 | 10.6 |
| 27147 | 36940 | 5 | Steele | 429.6 | 36,596 | 1,259 | 85.2 | 86.9 | 4.3 | 0.6 | 1.4 | 8.2 | 5.8 | 14.3 | 11.9 | 11.4 | 12.6 | 12.0 |
| 27149 | | 7 | Stevens | 563.6 | 9,765 | 2,434 | 17.3 | 87.3 | 1.9 | 2.7 | 2.5 | 7.9 | 6.0 | 12.1 | 22.7 | 11.3 | 11.1 | 8.5 |
| 27151 | | 7 | Swift | 742.0 | 9,176 | 2,481 | 12.4 | 90.9 | 2.1 | 0.9 | 2.1 | 5.7 | 6.4 | 12.5 | 10.9 | 10.9 | 11.7 | 10.4 |
| 27153 | | 6 | Todd | 945.1 | 24,732 | 1,618 | 26.2 | 91.0 | 0.9 | 1.0 | 1.0 | 7.4 | 6.6 | 13.5 | 10.8 | 9.8 | 10.5 | 10.6 |
| 27155 | | 9 | Traverse | 573.9 | 3,218 | 2,951 | 5.6 | 88.1 | 1.9 | 6.5 | 1.1 | 4.9 | 5.6 | 12.0 | 10.4 | 9.4 | 11.0 | 10.1 |
| 27157 | 40340 | 3 | Wabasha | 522.9 | 21,642 | 1,736 | 41.4 | 95.3 | 1.1 | 0.7 | 1.0 | 3.2 | 5.7 | 12.3 | 10.1 | 10.3 | 11.1 | 11.9 |
| 27159 | | 7 | Wadena | 536.3 | 13,807 | 2,170 | 25.7 | 95.4 | 1.9 | 1.7 | 0.9 | 2.1 | 7.0 | 14.6 | 11.2 | 10.9 | 10.7 | 10.7 |
| 27161 | | 6 | Waseca | 423.4 | 18,550 | 1,895 | 43.8 | 89.7 | 3.0 | 1.0 | 1.2 | 6.5 | 5.7 | 13.1 | 12.1 | 12.0 | 13.4 | 11.6 |
| 27163 | 33460 | 1 | Washington | 384.7 | 265,476 | 259 | 690.1 | 83.0 | 6.3 | 0.9 | 7.8 | 4.6 | 5.7 | 14.0 | 12.2 | 11.5 | 13.6 | 13.1 |
| 27165 | | 6 | Watonwan | 434.9 | 10,792 | 2,363 | 24.8 | 70.9 | 1.1 | 0.5 | 1.2 | 26.9 | 7.3 | 14.3 | 11.8 | 10.6 | 11.2 | 10.7 |
| 27167 | 47420 | 4 | Wilkin | 751.0 | 6,161 | 2,733 | 8.2 | 94.0 | 1.2 | 2.1 | 0.9 | 3.4 | 5.8 | 12.8 | 11.3 | 10.5 | 11.8 | 11.8 |
| 27169 | 49100 | 4 | Winona | 626.2 | 50,485 | 983 | 80.6 | 92.0 | 2.4 | 0.7 | 3.0 | 3.2 | 4.6 | 9.9 | 22.9 | 11.1 | 10.5 | 10.0 |
| 27171 | 33460 | 1 | Wright | 661.2 | 140,249 | 466 | 212.1 | 93.0 | 2.8 | 0.7 | 2.1 | 3.3 | 6.7 | 15.9 | 12.3 | 11.6 | 14.4 | 13.2 |
| 27173 | | 9 | Yellow Medicine | 759.1 | 9,580 | 2,450 | 12.6 | 90.4 | 1.2 | 4.1 | 0.9 | 4.9 | 5.7 | 13.0 | 11.8 | 10.5 | 11.2 | 10.9 |

1. CBSA = Core Based Statistical Area. See Appendix A for explanation. See Appendix B for list of metropolitan areas with component counties. Service of USDA Rural-Urban Continuum Codes. See Appendix A for definition. 3. Dry land or land partially or temporarily covered by water. 2. County type code from the Economic Research 4. May be of any race.

# Table B. States and Counties — Population and Households

| STATE County | 55 to 64 years | 65 to 74 years | 75 years and over | Percent female | 2000 | 2010 | 2000-2010 | 2010-2020 | Births | Deaths | Net Migration | Number | Persons per household | Family households | Female family householder[1] | One person |
|---|---|---|---|---|---|---|---|---|---|---|---|---|---|---|---|---|
| | 16 | 17 | 18 | 19 | 20 | 21 | 22 | 23 | 24 | 25 | 26 | 27 | 28 | 29 | 30 | 31 |
| **MINNESOTA— Cont'd** | | | | | | | | | | | | | | | | |
| Fillmore | 14.4 | 11.6 | 10.1 | 49.9 | 21,122 | 20,870 | -1.2 | 1.3 | 2,603 | 2,396 | 68 | 8,616 | 2.38 | 68.2 | 5.7 | 26.5 |
| Freeborn | 15.0 | 12.3 | 10.9 | 50.1 | 32,584 | 31,254 | -4.1 | -2.8 | 3,424 | 3,728 | -573 | 13,009 | 2.30 | 65.2 | 10.2 | 28.7 |
| Goodhue | 15.4 | 11.8 | 9.0 | 50.4 | 44,127 | 46,194 | 4.7 | 0.3 | 5,262 | 4,997 | -114 | 19,452 | 2.33 | 65.8 | 9.9 | 27.8 |
| Grant | 14.9 | 12.6 | 11.4 | 49.8 | 6,289 | 6,018 | -4.3 | 0.1 | 677 | 712 | 45 | 2,612 | 2.24 | 66.4 | 7.9 | 29.8 |
| Hennepin | 12.5 | 8.9 | 6.0 | 50.5 | 1,116,200 | 1,152,487 | 3.3 | 10.1 | 166,544 | 86,869 | 36,936 | 508,030 | 2.40 | 57.4 | 9.0 | 33.0 |
| Houston | 15.8 | 12.8 | 10.1 | 50.0 | 19,718 | 19,020 | -3.5 | -2.0 | 1,964 | 1,777 | -572 | 8,253 | 2.23 | 65.5 | 7.6 | 29.0 |
| Hubbard | 16.6 | 15.4 | 10.7 | 49.2 | 18,376 | 20,428 | 11.2 | 6.6 | 2,193 | 2,107 | 1,282 | 8,688 | 2.39 | 68.9 | 7.7 | 26.2 |
| Isanti | 14.6 | 9.6 | 7.1 | 49.6 | 31,287 | 37,811 | 20.9 | 9.6 | 4,659 | 3,236 | 2,207 | 14,903 | 2.61 | 68.3 | 9.1 | 23.6 |
| Itasca | 15.7 | 14.8 | 10.3 | 49.6 | 43,992 | 45,055 | 2.4 | 0.5 | 4,704 | 5,344 | 901 | 19,506 | 2.25 | 65.3 | 7.8 | 28.4 |
| Jackson | 15.3 | 12.1 | 10.9 | 48.8 | 11,268 | 10,266 | -8.9 | -4.9 | 1,100 | 1,087 | -510 | 4,419 | 2.23 | 64.8 | 5.7 | 28.2 |
| Kanabec | 15.9 | 13.1 | 8.7 | 49.6 | 14,996 | 16,244 | 8.3 | 1.1 | 1,640 | 1,564 | 98 | 6,439 | 2.46 | 67.7 | 7.5 | 25.4 |
| Kandiyohi | 13.8 | 11.2 | 8.5 | 49.6 | 41,203 | 42,237 | 2.5 | 2.1 | 5,912 | 3,989 | -1,027 | 16,889 | 2.47 | 67.5 | 9.5 | 26.7 |
| Kittson | 16.1 | 13.5 | 12.1 | 50.1 | 5,285 | 4,552 | -13.9 | -7.4 | 495 | 656 | -172 | 1,860 | 2.17 | 63.1 | 6.7 | 32.2 |
| Koochiching | 17.1 | 16.0 | 11.7 | 49.9 | 14,355 | 13,319 | -7.2 | -9.5 | 1,060 | 1,557 | -759 | 5,538 | 2.20 | 62.2 | 8.6 | 32.9 |
| Lac qui Parle | 15.9 | 14.0 | 14.4 | 49.5 | 8,067 | 7,258 | -10.0 | -10.1 | 667 | 903 | -502 | 3,090 | 2.13 | 62.8 | 5.2 | 31.7 |
| Lake | 16.8 | 15.3 | 12.1 | 48.8 | 11,058 | 10,862 | -1.8 | -2.1 | 1,077 | 1,441 | 149 | 5,192 | 1.98 | 63.6 | 6.8 | 32.3 |
| Lake of the Woods | 18.4 | 16.5 | 9.7 | 47.9 | 4,522 | 4,045 | -10.5 | -7.2 | 356 | 450 | -196 | 1,470 | 2.53 | 62.2 | 7.7 | 35.0 |
| Le Sueur | 14.6 | 10.6 | 7.8 | 49.4 | 25,426 | 27,701 | 8.9 | 3.8 | 3,264 | 2,307 | 88 | 10,939 | 2.55 | 70.8 | 8.4 | 23.7 |
| Lincoln | 14.3 | 11.7 | 13.5 | 49.6 | 6,429 | 5,895 | -8.3 | -5.5 | 709 | 816 | -214 | 2,468 | 2.23 | 60.9 | 5.6 | 34.6 |
| Lyon | 12.7 | 9.2 | 7.9 | 50.3 | 25,425 | 25,858 | 1.7 | -2.3 | 3,710 | 2,256 | -2,069 | 10,018 | 2.44 | 65.7 | 9.6 | 28.1 |
| McLeod | 14.1 | 10.1 | 9.0 | 50.0 | 34,898 | 36,645 | 5.0 | -2.6 | 4,195 | 3,474 | -1,668 | 14,714 | 2.41 | 64.4 | 8.3 | 28.9 |
| Mahnomen | 12.8 | 10.1 | 7.8 | 49.4 | 5,190 | 5,413 | 4.3 | 1.1 | 1,039 | 641 | -334 | 1,946 | 2.80 | 64.9 | 15.5 | 30.4 |
| Marshall | 14.6 | 12.1 | 10.8 | 49.2 | 10,155 | 9,437 | -7.1 | -1.2 | 1,121 | 796 | -437 | 3,971 | 2.33 | 67.0 | 6.9 | 29.2 |
| Martin | 15.0 | 12.7 | 11.7 | 50.5 | 21,802 | 20,843 | -4.4 | -6.5 | 2,343 | 2,559 | -1,140 | 8,759 | 2.23 | 62.9 | 7.7 | 31.9 |
| Meeker | 14.6 | 11.7 | 9.2 | 49.1 | 22,644 | 23,302 | 2.9 | 0.2 | 2,832 | 2,303 | -484 | 9,209 | 2.47 | 67.4 | 7.1 | 27.7 |
| Mille Lacs | 14.9 | 10.7 | 8.3 | 49.6 | 22,330 | 26,094 | 16.9 | 0.2 | 3,258 | 3,026 | -175 | 10,249 | 2.47 | 64.5 | 10.2 | 29.8 |
| Morrison | 15.2 | 11.6 | 8.9 | 49.3 | 31,712 | 33,198 | 4.7 | 0.0 | 3,876 | 3,314 | -558 | 13,371 | 2.43 | 67.5 | 8.0 | 27.9 |
| Mower | 12.9 | 9.9 | 8.6 | 49.8 | 38,603 | 39,165 | 1.5 | 2.5 | 5,273 | 4,056 | -212 | 15,565 | 2.51 | 64.2 | 8.9 | 30.4 |
| Murray | 15.3 | 14.0 | 12.8 | 50.1 | 9,165 | 8,725 | -4.8 | -6.5 | 892 | 1,013 | -448 | 3,645 | 2.24 | 64.7 | 4.8 | 31.2 |
| Nicollet | 12.2 | 9.8 | 6.9 | 49.8 | 29,771 | 32,729 | 9.9 | 5.4 | 3,843 | 2,396 | 312 | 12,885 | 2.40 | 66.8 | 8.3 | 27.3 |
| Nobles | 12.2 | 9.1 | 8.1 | 48.2 | 20,832 | 21,378 | 2.6 | 0.1 | 3,797 | 1,828 | -1,971 | 8,033 | 2.65 | 69.2 | 9.3 | 27.1 |
| Norman | 14.8 | 11.3 | 10.7 | 49.6 | 7,442 | 6,845 | -8.0 | -7.4 | 751 | 981 | -280 | 2,746 | 2.32 | 62.2 | 6.7 | 31.7 |
| Olmsted | 12.6 | 8.9 | 7.3 | 51.2 | 124,277 | 144,268 | 16.1 | 10.4 | 21,833 | 10,540 | 3,807 | 62,108 | 2.45 | 65.1 | 8.8 | 28.0 |
| Otter Tail | 15.8 | 14.0 | 11.1 | 49.8 | 57,159 | 57,301 | 0.2 | 2.5 | 6,595 | 6,899 | 1,804 | 24,687 | 2.31 | 66.6 | 7.0 | 28.6 |
| Pennington | 13.6 | 10.3 | 9.1 | 50.4 | 13,584 | 13,932 | 2.6 | -0.4 | 1,801 | 1,376 | -469 | 5,995 | 2.32 | 59.5 | 10.2 | 33.5 |
| Pine | 16.8 | 13.0 | 9.1 | 46.6 | 26,530 | 29,747 | 12.1 | -1.3 | 2,780 | 2,912 | -255 | 10,760 | 2.56 | 63.2 | 8.6 | 30.9 |
| Pipestone | 13.6 | 10.4 | 11.0 | 50.8 | 9,895 | 9,597 | -3.0 | -5.0 | 1,266 | 1,142 | -605 | 3,973 | 2.27 | 61.6 | 6.4 | 33.1 |
| Polk | 13.8 | 10.5 | 8.6 | 49.7 | 31,369 | 31,600 | 0.7 | -2.2 | 4,215 | 3,470 | -1,434 | 12,584 | 2.40 | 62.0 | 7.7 | 33.7 |
| Pope | 15.5 | 13.8 | 10.9 | 48.8 | 11,236 | 10,990 | -2.2 | 2.6 | 1,275 | 1,313 | 332 | 4,981 | 2.18 | 64.8 | 5.5 | 29.5 |
| Ramsey | 11.9 | 9.0 | 6.2 | 51.1 | 511,035 | 508,611 | -0.5 | 7.7 | 78,250 | 42,615 | 4,035 | 209,043 | 2.53 | 58.5 | 11.6 | 33.0 |
| Red Lake | 14.7 | 12.1 | 10.1 | 48.5 | 4,299 | 4,089 | -4.9 | -1.1 | 498 | 366 | -177 | 1,669 | 2.39 | 66.6 | 8.1 | 28.9 |
| Redwood | 13.8 | 11.3 | 10.5 | 50.0 | 16,815 | 16,058 | -4.5 | -6.1 | 1,964 | 1,891 | -1,052 | 6,261 | 2.38 | 65.2 | 7.1 | 30.8 |
| Renville | 15.3 | 11.3 | 10.5 | 49.2 | 17,154 | 15,728 | -8.3 | -8.4 | 1,827 | 1,918 | -1,247 | 6,085 | 2.35 | 64.2 | 7.1 | 30.5 |
| Rice | 12.8 | 9.4 | 7.0 | 48.9 | 56,665 | 64,133 | 13.2 | 4.6 | 7,495 | 5,024 | 488 | 23,005 | 2.51 | 67.0 | 8.3 | 27.4 |
| Rock | 13.6 | 10.7 | 10.1 | 50.6 | 9,721 | 9,686 | -0.4 | -4.0 | 1,088 | 1,208 | -262 | 3,914 | 2.34 | 69.2 | 7.2 | 25.9 |
| Roseau | 15.7 | 10.7 | 8.1 | 48.4 | 16,338 | 15,629 | -4.3 | -3.3 | 1,894 | 1,401 | -1,016 | 5,980 | 2.53 | 67.0 | 5.6 | 27.2 |
| St. Louis | 14.3 | 12.4 | 8.4 | 50.0 | 200,528 | 200,229 | -0.1 | -0.8 | 20,352 | 21,501 | -357 | 85,807 | 2.22 | 56.5 | 8.2 | 34.0 |
| Scott | 12.8 | 7.2 | 4.6 | 50.2 | 89,498 | 129,916 | 45.2 | 16.0 | 18,956 | 6,611 | 8,484 | 49,479 | 2.90 | 76.8 | 9.0 | 18.3 |
| Sherburne | 12.7 | 7.6 | 4.7 | 48.8 | 64,417 | 88,492 | 37.4 | 11.7 | 12,225 | 5,267 | 3,389 | 32,206 | 2.87 | 75.1 | 8.7 | 17.6 |
| Sibley | 15.8 | 10.3 | 9.0 | 49.1 | 15,356 | 15,234 | -0.8 | -3.4 | 1,738 | 1,482 | -775 | 6,031 | 2.43 | 67.9 | 5.5 | 27.3 |
| Stearns | 12.3 | 9.1 | 6.9 | 49.7 | 133,166 | 150,646 | 13.1 | 7.6 | 20,366 | 10,432 | 1,485 | 59,479 | 2.53 | 64.2 | 8.8 | 27.1 |
| Steele | 13.3 | 10.3 | 8.5 | 50.1 | 33,680 | 36,581 | 8.6 | 0.0 | 4,577 | 3,161 | -1,408 | 14,692 | 2.46 | 66.7 | 8.6 | 28.6 |
| Stevens | 10.4 | 9.2 | 8.5 | 50.0 | 10,053 | 9,722 | -3.3 | 0.4 | 1,212 | 875 | -292 | 3,624 | 2.60 | 59.1 | 4.9 | 30.4 |
| Swift | 14.5 | 11.9 | 10.9 | 49.5 | 11,956 | 9,786 | -18.1 | -6.2 | 1,132 | 1,147 | -595 | 4,263 | 2.16 | 62.8 | 10.2 | 33.7 |
| Todd | 15.5 | 12.8 | 10.0 | 48.4 | 24,426 | 24,889 | 1.9 | -0.6 | 3,263 | 2,169 | -1,260 | 9,819 | 2.46 | 68.1 | 5.4 | 27.2 |
| Traverse | 15.3 | 12.5 | 13.6 | 49.5 | 4,134 | 3,558 | -13.9 | -9.6 | 341 | 547 | -138 | 1,608 | 2.00 | 63.2 | 5.1 | 33.1 |
| Wabasha | 15.6 | 12.9 | 9.9 | 49.9 | 21,610 | 21,665 | 0.3 | -0.1 | 2,401 | 1,942 | -473 | 9,020 | 2.36 | 67.2 | 6.2 | 26.1 |
| Wadena | 13.4 | 11.3 | 10.2 | 50.4 | 13,713 | 13,843 | 0.9 | -0.3 | 1,846 | 2,042 | 170 | 5,666 | 2.33 | 62.5 | 9.2 | 30.4 |
| Waseca | 13.5 | 10.9 | 7.8 | 51.6 | 19,526 | 19,138 | -2.0 | -3.1 | 2,163 | 1,736 | -1,020 | 7,425 | 2.38 | 66.1 | 9.5 | 29.9 |
| Washington | 13.9 | 9.6 | 6.4 | 50.5 | 201,130 | 238,109 | 18.4 | 11.5 | 28,789 | 15,943 | 14,735 | 94,431 | 2.67 | 72.0 | 8.7 | 22.7 |
| Watonwan | 13.9 | 10.9 | 10.3 | 50.3 | 11,876 | 11,211 | -5.6 | -3.7 | 1,586 | 1,232 | -776 | 4,310 | 2.51 | 66.5 | 9.3 | 29.4 |
| Wilkin | 15.9 | 11.0 | 9.1 | 47.5 | 7,138 | 6,576 | -7.9 | -6.3 | 731 | 791 | -355 | 2,818 | 2.18 | 65.0 | 8.3 | 32.5 |
| Winona | 12.6 | 10.5 | 7.8 | 50.4 | 49,985 | 51,461 | 3.0 | -1.9 | 4,822 | 4,258 | -1,550 | 19,442 | 2.41 | 58.3 | 6.8 | 31.3 |
| Wright | 12.5 | 8.0 | 5.4 | 49.4 | 89,986 | 124,695 | 38.6 | 12.5 | 18,323 | 7,992 | 5,318 | 48,242 | 2.76 | 74.0 | 7.4 | 20.4 |
| Yellow Medicine | 14.9 | 11.2 | 10.7 | 49.6 | 11,080 | 10,445 | -5.7 | -8.3 | 1,156 | 1,172 | -850 | 4,087 | 2.33 | 68.2 | 8.1 | 26.9 |

1. No spouse present.

# Table B. States and Counties — Population, Vital Statistics, Health, and Crime

| STATE County | Persons in group quarters, 2020 | Daytime Population, 2015-2019 Number | Daytime Population, 2015-2019 Employment/residence ratio | Births, 2020 Total | Births, 2020 Rate[1] | Deaths, 2020 Number | Deaths, 2020 Rate[1] | Persons under 65 with no health insurance, 2019 Number | Persons under 65 with no health insurance, 2019 Percent | Medicare, 2020 Total beneficiaries | Medicare, 2020 Enrolled in Original Medicare | Medicare, 2020 Enrolled in Medicare Advantage | Crimes reported by county police or sheriff, 2019 Violent | Crimes reported by county police or sheriff, 2019 Property |
|---|---|---|---|---|---|---|---|---|---|---|---|---|---|---|
| | 32 | 33 | 34 | 35 | 36 | 37 | 38 | 39 | 40 | 41 | 42 | 43 | 44 | 45 |
| **MINNESOTA— Cont'd** | | | | | | | | | | | | | | |
| Fillmore | 359 | 17,459 | 0.67 | 245 | 11.6 | 242 | 11.5 | 1,303 | 7.9 | 4,960 | 3,127 | 1,834 | 9 | 59 |
| Freeborn | 637 | 29,125 | 0.91 | 309 | 10.2 | 360 | 11.9 | 1,596 | 7.0 | 7,809 | 4,248 | 3,561 | 11 | 84 |
| Goodhue | 857 | 45,314 | 0.96 | 502 | 10.8 | 519 | 11.2 | 1,908 | 5.2 | 10,843 | 4,197 | 6,646 | 10 | 171 |
| Grant | 110 | 5,398 | 0.81 | 58 | 9.6 | 56 | 9.3 | 270 | 6.0 | 1,581 | 954 | 627 | 2 | 6 |
| Hennepin | 25,354 | 1,471,103 | 1.33 | 15,616 | 12.3 | 9,617 | 7.6 | 62,212 | 5.8 | 201,379 | 98,435 | 102,945 | 20 | 44 |
| Houston | 257 | 15,107 | 0.65 | 182 | 9.8 | 172 | 9.2 | 730 | 5.1 | 4,649 | 2,793 | 1,856 | 5 | 29 |
| Hubbard | 154 | 18,814 | 0.76 | 202 | 9.3 | 205 | 9.4 | 1,214 | 7.7 | 6,020 | 3,819 | 2,201 | 18 | 170 |
| Isanti | 465 | 31,370 | 0.60 | 466 | 11.2 | 357 | 8.6 | 1,956 | 5.8 | 7,940 | 3,883 | 4,057 | 15 | 232 |
| Itasca | 1,005 | 43,510 | 0.92 | 433 | 9.6 | 520 | 11.5 | 2,336 | 7.0 | 12,548 | 5,337 | 7,212 | 46 | 134 |
| Jackson | 110 | 10,001 | 1.01 | 96 | 9.8 | 110 | 11.3 | 445 | 6.0 | 2,443 | 1,750 | 693 | 6 | 30 |
| Kanabec | 246 | 13,123 | 0.61 | 169 | 10.3 | 190 | 11.6 | 830 | 6.5 | 3,991 | 1,786 | 2,205 | NA | NA |
| Kandiyohi | 1,047 | 43,862 | 1.05 | 556 | 12.9 | 387 | 9.0 | 2,787 | 8.1 | 9,640 | 5,403 | 4,237 | 25 | 170 |
| Kittson | 111 | 3,982 | 0.85 | 47 | 11.2 | 58 | 13.8 | 198 | 6.2 | 1,161 | 850 | 311 | 0 | 0 |
| Koochiching | 241 | 12,427 | 0.98 | 95 | 7.9 | 161 | 13.4 | 648 | 7.4 | 3,759 | 1,714 | 2,044 | 6 | 32 |
| Lac qui Parle | 154 | 6,331 | 0.88 | 65 | 10.0 | 93 | 14.2 | 331 | 7.0 | 1,899 | 1,227 | 672 | 0 | 20 |
| Lake | 224 | 10,681 | 1.03 | 95 | 8.9 | 126 | 11.8 | 409 | 5.3 | 3,153 | 1,360 | 1,793 | 0 | 19 |
| Lake of the Woods | 53 | 3,428 | 0.82 | 37 | 9.9 | 36 | 9.6 | 203 | 7.4 | 1,147 | 680 | 467 | 6 | 9 |
| Le Sueur | 270 | 23,066 | 0.65 | 317 | 11.0 | 265 | 9.2 | 1,543 | 6.5 | 5,758 | 2,238 | 3,520 | 4 | 60 |
| Lincoln | 135 | 4,916 | 0.72 | 70 | 12.6 | 82 | 14.7 | 277 | 6.6 | 1,511 | 1,062 | 450 | NA | NA |
| Lyon | 867 | 27,627 | 1.14 | 340 | 13.5 | 236 | 9.3 | 1,351 | 6.6 | 4,869 | 3,479 | 1,389 | 4 | 34 |
| McLeod | 476 | 35,027 | 0.96 | 390 | 10.9 | 376 | 10.5 | 1,526 | 5.3 | 7,844 | 2,679 | 5,165 | 11 | 55 |
| Mahnomen | 79 | 5,894 | 1.19 | 95 | 17.4 | 53 | 9.7 | 432 | 9.7 | 1,201 | 842 | 359 | 10 | 42 |
| Marshall | 76 | 7,895 | 0.69 | 102 | 10.9 | 85 | 9.1 | 478 | 6.7 | 2,219 | 1,358 | 861 | NA | NA |
| Martin | 348 | 20,046 | 1.02 | 247 | 12.7 | 235 | 12.1 | 953 | 6.5 | 5,368 | 3,612 | 1,755 | 0 | 14 |
| Meeker | 354 | 19,953 | 0.73 | 280 | 12.0 | 233 | 10.0 | 1,025 | 5.6 | 5,361 | 1,738 | 3,622 | 4 | 103 |
| Mille Lacs | 523 | 24,080 | 0.85 | 294 | 11.2 | 302 | 11.6 | 1,761 | 8.4 | 6,084 | 2,333 | 3,751 | 34 | 354 |
| Morrison | 521 | 29,790 | 0.80 | 363 | 10.9 | 382 | 11.5 | 1,762 | 6.7 | 7,922 | 4,223 | 3,699 | 10 | 195 |
| Mower | 632 | 38,260 | 0.92 | 535 | 13.3 | 416 | 10.4 | 2,483 | 7.8 | 8,516 | 5,257 | 3,259 | 9 | 68 |
| Murray | 163 | 7,482 | 0.80 | 86 | 10.5 | 89 | 10.9 | 422 | 7.0 | 2,325 | 1,679 | 646 | 2 | 6 |
| Nicollet | 2,870 | 32,954 | 0.95 | 339 | 9.8 | 296 | 8.6 | 1,280 | 4.9 | 6,195 | 3,973 | 2,221 | 7 | 52 |
| Nobles | 385 | 22,174 | 1.04 | 377 | 17.6 | 171 | 8.0 | 2,290 | 13.1 | 3,911 | 2,807 | 1,104 | 6 | 35 |
| Norman | 149 | 5,683 | 0.73 | 75 | 11.8 | 85 | 13.4 | 331 | 6.7 | 1,640 | 1,175 | 465 | 0 | 0 |
| Olmsted | 2,436 | 171,969 | 1.21 | 2,049 | 12.9 | 1,147 | 7.2 | 7,185 | 5.4 | 27,592 | 18,968 | 8,625 | 40 | 228 |
| Otter Tail | 1,186 | 54,608 | 0.87 | 615 | 10.5 | 686 | 11.7 | 2,979 | 6.8 | 16,458 | 9,141 | 7,317 | 30 | 214 |
| Pennington | 327 | 16,850 | 1.35 | 152 | 11.0 | 133 | 9.6 | 570 | 5.1 | 2,907 | 1,678 | 1,230 | 2 | 45 |
| Pine | 1,531 | 26,944 | 0.82 | 230 | 7.8 | 320 | 10.9 | 1,847 | 8.6 | 7,152 | 2,891 | 4,261 | 138 | 563 |
| Pipestone | 205 | 9,260 | 1.02 | 131 | 14.4 | 105 | 11.5 | 704 | 9.9 | 2,183 | 1,240 | 943 | 5 | 80 |
| Polk | 1,196 | 29,485 | 0.87 | 394 | 12.8 | 325 | 10.5 | 1,376 | 5.6 | 6,768 | 4,705 | 2,063 | 34 | 167 |
| Pope | 184 | 10,277 | 0.86 | 125 | 11.1 | 138 | 12.2 | 530 | 6.3 | 3,064 | 1,988 | 1,076 | 6 | 49 |
| Ramsey | 17,338 | 604,705 | 1.22 | 7,283 | 13.3 | 4,563 | 8.3 | 31,455 | 6.9 | 91,774 | 46,380 | 45,394 | 141 | 1,249 |
| Red Lake | 34 | 3,250 | 0.61 | 45 | 11.1 | 29 | 7.2 | 204 | 6.5 | 908 | 450 | 458 | 3 | 16 |
| Redwood | 373 | 15,245 | 1.00 | 197 | 13.1 | 170 | 11.3 | 973 | 8.3 | 3,511 | 2,255 | 1,256 | 6 | 44 |
| Renville | 342 | 13,618 | 0.86 | 168 | 11.7 | 188 | 13.1 | 837 | 7.4 | 3,540 | 2,328 | 1,212 | 5 | 81 |
| Rice | 7,328 | 62,764 | 0.90 | 731 | 10.9 | 528 | 7.9 | 3,647 | 7.4 | 11,816 | 4,585 | 7,231 | 18 | 142 |
| Rock | 260 | 8,373 | 0.78 | 89 | 9.6 | 115 | 12.4 | 403 | 5.6 | 2,173 | 1,065 | 1,107 | 7 | 57 |
| Roseau | 191 | 16,031 | 1.08 | 167 | 11.0 | 138 | 9.1 | 804 | 6.6 | 3,231 | 1,955 | 1,276 | 6 | 31 |
| St. Louis | 8,803 | 208,283 | 1.09 | 1,847 | 9.3 | 2,225 | 11.2 | 8,005 | 5.2 | 47,406 | 19,114 | 28,292 | 50 | 566 |
| Scott | 1,289 | 125,824 | 0.76 | 1,704 | 11.3 | 776 | 5.1 | 6,168 | 4.7 | 19,146 | 9,685 | 9,462 | 4 | 25 |
| Sherburne | 2,190 | 72,659 | 0.57 | 1,202 | 12.2 | 607 | 6.1 | 3,517 | 4.2 | 14,222 | 7,250 | 6,972 | 30 | 272 |
| Sibley | 230 | 12,044 | 0.64 | 168 | 11.4 | 137 | 9.3 | 922 | 7.7 | 3,112 | 1,190 | 1,922 | 2 | 1 |
| Stearns | 6,843 | 170,202 | 1.14 | 2,024 | 12.5 | 1,143 | 7.1 | 8,634 | 6.6 | 28,305 | 15,731 | 12,574 | 21 | 164 |
| Steele | 594 | 39,267 | 1.14 | 396 | 10.8 | 332 | 9.1 | 1,630 | 5.5 | 7,765 | 4,425 | 3,341 | 12 | 76 |
| Stevens | 927 | 11,000 | 1.23 | 123 | 12.6 | 92 | 9.4 | 420 | 5.8 | 1,805 | 1,051 | 755 | 2 | 33 |
| Swift | 150 | 8,941 | 0.91 | 113 | 12.3 | 103 | 11.2 | 538 | 7.6 | 2,313 | 1,507 | 805 | 2 | 2 |
| Todd | 346 | 21,249 | 0.70 | 308 | 12.5 | 225 | 9.1 | 1,669 | 8.9 | 5,736 | 2,866 | 2,870 | 17 | 140 |
| Traverse | 100 | 3,103 | 0.88 | 33 | 10.3 | 33 | 10.3 | 207 | 8.7 | 1,006 | 516 | 490 | 6 | 12 |
| Wabasha | 246 | 18,536 | 0.73 | 223 | 10.3 | 206 | 9.5 | 1,031 | 6.2 | 5,425 | 3,031 | 2,394 | 7 | 22 |
| Wadena | 482 | 14,509 | 1.14 | 178 | 12.9 | 183 | 13.3 | 705 | 6.7 | 3,661 | 2,118 | 1,543 | 10 | 10 |
| Waseca | 974 | 16,205 | 0.72 | 201 | 10.8 | 165 | 8.9 | 835 | 5.8 | 3,985 | 2,334 | 1,651 | 4 | 55 |
| Washington | 3,437 | 219,589 | 0.73 | 2,729 | 10.3 | 1,903 | 7.2 | 8,961 | 4.1 | 46,514 | 22,553 | 23,961 | 52 | 579 |
| Watonwan | 150 | 10,168 | 0.85 | 166 | 15.4 | 114 | 10.6 | 935 | 11.0 | 2,433 | 1,640 | 794 | 8 | 26 |
| Wilkin | 152 | 5,590 | 0.78 | 67 | 10.9 | 76 | 12.3 | 238 | 4.8 | 1,426 | 926 | 500 | 0 | 20 |
| Winona | 3,913 | 49,873 | 0.97 | 462 | 9.2 | 461 | 9.1 | 2,341 | 6.1 | 10,078 | 6,008 | 4,069 | 6 | 79 |
| Wright | 1,105 | 109,932 | 0.66 | 1,731 | 12.3 | 956 | 6.8 | 5,358 | 4.4 | 21,069 | 10,145 | 10,924 | 98 | 1,396 |
| Yellow Medicine | 290 | 9,407 | 0.92 | 104 | 10.9 | 105 | 11.0 | 488 | 6.4 | 2,261 | 1,144 | 1,117 | 3 | 35 |

1. Per 1,000 estimated resident population.

# Table B. States and Counties — Crime, Education, Money Income, and Poverty

| STATE County | County law enforcement employment, 2019 | | Education — School enrollment and attainment, 2015-2019 | | | | Local government expenditures[3] 2017-2018 | | Money income, 2015-2019 | | | | Income and poverty, 2019 | | | |
|---|---|---|---|---|---|---|---|---|---|---|---|---|---|---|---|---|
| | | | Enrollment[1] | | Attainment[2] (percent) | | | | Per capita income[4] | Households | | | Median household income (dollars) | Percent below poverty level | | |
| | Officers | Civilians | Total | Percent private | High school graduate or less | Bachelor's degree or more | Total current spending (mil dol) | Current spending per student (dollars) | | Median income (dollars) | Percent with income of less than $50,000 | Percent with income of $200,000 or more | | All persons | Children under 18 years | Children 5 to 17 years in families |
| | 46 | 47 | 48 | 49 | 50 | 51 | 52 | 53 | 54 | 55 | 56 | 57 | 58 | 59 | 60 | 61 |
| **MINNESOTA— Cont'd** | | | | | | | | | | | | | | | | |
| Fillmore | 20 | 11 | 4,631 | 11.7 | 43.1 | 22.2 | 28 | 11,211 | 30,392 | 61,207 | 41.4 | 3.4 | 62,716 | 9.4 | 13.3 | 12.0 |
| Freeborn | 26 | 53 | 6,501 | 5.4 | 46.2 | 17.1 | 57 | 13,247 | 29,316 | 53,631 | 46.4 | 2.2 | 56,716 | 9.6 | 13.3 | 12.3 |
| Goodhue | 41 | 67 | 10,388 | 11.3 | 38.4 | 26.0 | 84 | 12,754 | 34,803 | 66,800 | 37.1 | 4.6 | 71,497 | 7.1 | 8.7 | 8.2 |
| Grant | 11 | 4 | 1,226 | 6.1 | 40.4 | 18.8 | 14 | 12,380 | 32,871 | 55,466 | 45.2 | 4.1 | 55,247 | 8.8 | 13.7 | 12.6 |
| Hennepin | 326 | 504 | 309,608 | 16.5 | 23.1 | 50.1 | 2,413 | 14,231 | 45,768 | 78,167 | 32.2 | 12.2 | 82,323 | 9.7 | 11.9 | 11.7 |
| Houston | 13 | 18 | 3,942 | 16.9 | 39.1 | 24.8 | 47 | 11,041 | 33,032 | 60,382 | 39.1 | 3.7 | 64,920 | 7.8 | 9.3 | 8.4 |
| Hubbard | 19 | 29 | 4,293 | 10.8 | 38.2 | 27.4 | 29 | 10,955 | 29,896 | 56,709 | 43.6 | 3.4 | 56,064 | 11.2 | 14.8 | 14.3 |
| Isanti | 23 | 39 | 9,192 | 13.4 | 42.1 | 18.5 | 77 | 12,711 | 34,054 | 74,616 | 31.2 | 4.9 | 77,963 | 6.9 | 8.8 | 8.4 |
| Itasca | 33 | 44 | 9,613 | 11.5 | 36.0 | 23.1 | 97 | 14,098 | 30,286 | 55,139 | 45.6 | 2.7 | 58,660 | 11.2 | 15.2 | 13.8 |
| Jackson | 14 | 13 | 2,138 | 7.2 | 40.0 | 22.6 | 18 | 12,191 | 34,325 | 58,727 | 43.0 | 3.1 | 57,915 | 9.8 | 13.6 | 12.8 |
| Kanabec | 21 | 32 | 3,370 | 9.7 | 49.2 | 14.7 | 24 | 11,099 | 28,427 | 57,163 | 43.5 | 2.3 | 63,595 | 10.1 | 14.5 | 14.1 |
| Kandiyohi | 34 | 75 | 9,640 | 11.2 | 36.9 | 23.3 | 73 | 12,057 | 31,039 | 60,294 | 41.9 | 4.0 | 64,129 | 9.8 | 14.4 | 14.6 |
| Kittson | 6 | 5 | 806 | 6.9 | 41.7 | 25.9 | 9 | 14,850 | 30,539 | 59,643 | 40.6 | 2.3 | 62,762 | 9.2 | 11.5 | 10.8 |
| Koochiching | 10 | 10 | 2,586 | 7.0 | 43.8 | 15.5 | 23 | 13,645 | 29,834 | 50,870 | 49.3 | 1.6 | 52,545 | 13.0 | 19.1 | 17.5 |
| Lac qui Parle | 7 | 5 | 1,308 | 7.1 | 43.7 | 18.8 | 17 | 12,321 | 32,437 | 53,071 | 46.1 | 3.7 | 56,825 | 9.5 | 12.3 | 11.5 |
| Lake | 17 | 13 | 1,744 | 8.1 | 36.3 | 28.8 | 17 | 12,536 | 34,207 | 61,452 | 40.0 | 2.3 | 61,082 | 9.0 | 11.4 | 10.1 |
| Lake of the Woods | 7 | 7 | 692 | 2.0 | 45.6 | 16.7 | 6 | 13,942 | 25,640 | 49,967 | 50.1 | 1.1 | 56,244 | 9.2 | 12.2 | 11.8 |
| Le Sueur | 21 | 26 | 6,605 | 11.8 | 41.9 | 23.0 | 47 | 11,120 | 33,311 | 71,080 | 33.9 | 4.7 | 75,519 | 7.1 | 7.8 | 7.3 |
| Lincoln | 7 | 6 | 1,223 | 10.1 | 46.4 | 19.3 | 13 | 13,220 | 30,253 | 53,077 | 46.3 | 2.5 | 55,328 | 10.1 | 11.6 | 11.0 |
| Lyon | 18 | 35 | 7,269 | 10.5 | 39.7 | 26.1 | 88 | 18,570 | 29,634 | 57,730 | 43.8 | 3.4 | 63,849 | 10.6 | 13.3 | 13.4 |
| McLeod | 25 | 36 | 8,092 | 11.2 | 42.9 | 17.6 | 62 | 11,523 | 32,465 | 62,121 | 36.8 | 3.7 | 64,380 | 6.3 | 8.3 | 7.5 |
| Mahnomen | 13 | 6 | 1,511 | 6.1 | 49.8 | 13.2 | 20 | 14,026 | 21,279 | 44,688 | 54.5 | 1.7 | 46,702 | 21.3 | 33.6 | 31.4 |
| Marshall | 16 | 8 | 1,943 | 6.9 | 46.8 | 19.1 | 20 | 14,651 | 30,746 | 60,118 | 40.9 | 2.6 | 59,793 | 7.5 | 9.9 | 9.4 |
| Martin | 14 | 20 | 4,010 | 12.8 | 44.7 | 20.5 | 42 | 14,008 | 31,066 | 52,798 | 47.3 | 4.4 | 54,602 | 10.5 | 16.5 | 16.2 |
| Meeker | 20 | 29 | 5,273 | 11.0 | 42.7 | 19.5 | 38 | 11,231 | 32,368 | 63,452 | 37.4 | 3.9 | 63,812 | 7.9 | 8.9 | 8.5 |
| Mille Lacs | 36 | 44 | 5,535 | 11.6 | 48.8 | 14.7 | 73 | 11,533 | 27,758 | 56,135 | 44.0 | 3.0 | 59,262 | 11.4 | 14.9 | 13.5 |
| Morrison | 20 | 34 | 7,161 | 11.4 | 47.6 | 17.3 | 60 | 11,168 | 29,680 | 57,815 | 43.5 | 3.2 | 61,847 | 8.9 | 10.3 | 9.9 |
| Mower | 26 | 51 | 9,611 | 9.7 | 44.3 | 22.2 | 91 | 13,358 | 29,720 | 54,295 | 46.2 | 3.8 | 55,276 | 11.5 | 15.2 | 14.6 |
| Murray | 12 | 7 | 1,649 | 10.9 | 46.1 | 20.7 | 14 | 12,913 | 31,874 | 60,231 | 42.6 | 3.1 | 61,007 | 8.5 | 10.9 | 10.1 |
| Nicollet | 16 | 26 | 9,348 | 38.6 | 31.2 | 33.8 | 43 | 15,569 | 33,031 | 67,399 | 36.2 | 5.3 | 71,067 | 9.9 | 11.0 | 10.4 |
| Nobles | 14 | 22 | 5,355 | 5.7 | 55.1 | 15.3 | 51 | 11,171 | 25,939 | 55,304 | 45.9 | 2.1 | 57,300 | 11.0 | 14.2 | 14.8 |
| Norman | 7 | 5 | 1,471 | 8.8 | 45.0 | 19.1 | 14 | 12,532 | 30,422 | 55,085 | 45.4 | 3.4 | 53,961 | 10.4 | 15.8 | 13.9 |
| Olmsted | 72 | 107 | 39,218 | 17.8 | 25.3 | 44.9 | 298 | 12,091 | 41,066 | 76,951 | 31.8 | 9.2 | 80,631 | 6.7 | 8.0 | 7.6 |
| Otter Tail | 39 | 45 | 11,636 | 13.3 | 37.4 | 24.9 | 125 | 16,002 | 32,059 | 58,682 | 43.6 | 4.1 | 59,901 | 9.8 | 10.5 | 10.0 |
| Pennington | 10 | 25 | 3,069 | 6.8 | 39.9 | 19.3 | 73 | 32,160 | 31,159 | 60,122 | 42.0 | 2.5 | 55,870 | 11.6 | 15.9 | 14.5 |
| Pine | 32 | 50 | 5,895 | 11.1 | 50.3 | 14.9 | 46 | 12,101 | 26,407 | 53,422 | 47.4 | 2.8 | 55,684 | 11.1 | 16.2 | 15.1 |
| Pipestone | 14 | 11 | 2,072 | 15.0 | 45.8 | 21.3 | 19 | 12,153 | 30,540 | 52,917 | 47.3 | 3.9 | 55,405 | 11.1 | 16.2 | 15.1 |
| Polk | 32 | 20 | 7,593 | 8.0 | 38.8 | 25.3 | 62 | 11,732 | 30,078 | 59,343 | 43.0 | 3.2 | 60,582 | 10.8 | 14.0 | 12.1 |
| Pope | 8 | 6 | 2,169 | 10.4 | 36.4 | 22.9 | 18 | 13,443 | 34,166 | 61,275 | 41.1 | 3.5 | 59,442 | 8.9 | 10.9 | 10.7 |
| Ramsey | 248 | 211 | 144,787 | 22.6 | 30.6 | 42.3 | 1,363 | 14,857 | 35,013 | 64,660 | 38.6 | 6.9 | 68,652 | 12.4 | 17.3 | 17.8 |
| Red Lake | 9 | 2 | 898 | 5.7 | 44.8 | 16.2 | 11 | 14,159 | 30,037 | 58,576 | 44.2 | 2.3 | 58,384 | 9.9 | 10.8 | 10.0 |
| Redwood | 13 | 17 | 3,522 | 13.0 | 46.5 | 19.1 | 30 | 11,959 | 28,611 | 55,404 | 44.4 | 2.6 | 60,097 | 9.7 | 12.1 | 11.5 |
| Renville | 15 | 22 | 3,175 | 8.5 | 46.0 | 15.2 | 21 | 11,832 | 31,253 | 59,028 | 41.2 | 3.5 | 59,110 | 10.0 | 13.4 | 12.9 |
| Rice | 31 | 26 | 19,860 | 39.1 | 40.3 | 28.3 | 115 | 13,112 | 31,221 | 68,584 | 37.1 | 5.3 | 72,438 | 8.2 | 9.1 | 8.3 |
| Rock | 12 | 5 | 2,225 | 13.2 | 39.7 | 23.8 | 18 | 10,883 | 31,345 | 63,005 | 38.8 | 4.3 | 63,451 | 6.0 | 8.4 | 7.4 |
| Roseau | 9 | 8 | 3,420 | 9.4 | 42.2 | 20.3 | 32 | 11,766 | 30,193 | 62,202 | 38.7 | 2.9 | 62,924 | 8.7 | 10.5 | 9.4 |
| St. Louis | 108 | 147 | 49,065 | 13.3 | 32.7 | 29.3 | 316 | 12,431 | 31,537 | 55,646 | 45.4 | 3.8 | 60,228 | 12.8 | 12.8 | 10.8 |
| Scott | 45 | 95 | 40,201 | 14.8 | 27.6 | 39.9 | 282 | 11,150 | 42,155 | 102,152 | 22.4 | 14.0 | 108,718 | 4.2 | 4.9 | 4.2 |
| Sherburne | 81 | 214 | 24,925 | 9.0 | 33.9 | 26.6 | 220 | 10,932 | 35,439 | 89,250 | 23.9 | 7.2 | 95,323 | 5.5 | 6.0 | 5.6 |
| Sibley | 14 | 15 | 3,329 | 13.7 | 48.5 | 16.4 | 27 | 11,957 | 32,016 | 63,439 | 38.7 | 3.7 | 66,757 | 10.7 | 16.8 | 17.9 |
| Stearns | 71 | 136 | 46,675 | 18.3 | 36.3 | 27.5 | 308 | 12,338 | 30,994 | 62,789 | 39.6 | 4.3 | 66,449 | 11.2 | 14.0 | 13.1 |
| Steele | 24 | 6 | 9,088 | 11.8 | 41.1 | 26.0 | 75 | 11,306 | 32,477 | 64,903 | 38.7 | 4.0 | 68,603 | 6.3 | 8.9 | 8.0 |
| Stevens | 7 | 8 | 2,911 | 7.5 | 37.5 | 30.5 | 19 | 12,273 | 32,542 | 60,559 | 40.9 | 5.2 | 59,045 | 11.9 | 8.7 | 7.6 |
| Swift | 9 | 7 | 2,057 | 3.0 | 42.1 | 19.6 | 18 | 11,285 | 31,765 | 51,620 | 48.8 | 4.0 | 60,127 | 9.5 | 13.6 | 12.7 |
| Todd | 17 | 18 | 4,969 | 16.8 | 49.6 | 15.1 | 60 | 19,596 | 26,678 | 53,585 | 46.3 | 2.4 | 51,746 | 11.8 | 16.6 | 17.1 |
| Traverse | 6 | 8 | 553 | 5.1 | 42.1 | 17.5 | 7 | 12,945 | 32,548 | 51,957 | 48.9 | 4.1 | 57,602 | 11.9 | 20.2 | 19.7 |
| Wabasha | 18 | 31 | 4,578 | 12.2 | 39.6 | 23.7 | 46 | 10,335 | 34,442 | 65,226 | 36.4 | 4.6 | 63,435 | 8.4 | 10.2 | 10.3 |
| Wadena | 11 | 11 | 3,169 | 5.5 | 46.5 | 13.8 | 42 | 13,417 | 24,214 | 46,605 | 53.9 | 2.0 | 50,663 | 12.5 | 16.1 | 15.3 |
| Waseca | 14 | 14 | 4,320 | 12.0 | 43.4 | 21.4 | 42 | 11,509 | 29,929 | 56,762 | 43.9 | 4.7 | 63,907 | 8.9 | 11.5 | 11.1 |
| Washington | 116 | 133 | 66,333 | 15.6 | 24.9 | 44.3 | 470 | 11,554 | 45,822 | 96,671 | 21.5 | 13.4 | 100,857 | 4.6 | 5.1 | 4.6 |
| Watonwan | 8 | 14 | 2,538 | 8.7 | 56.8 | 15.5 | 23 | 12,545 | 28,341 | 54,065 | 46.7 | 2.6 | 57,290 | 10.1 | 13.7 | 13.4 |
| Wilkin | 7 | 12 | 1,405 | 15.1 | 33.8 | 23.8 | 13 | 11,681 | 34,543 | 60,595 | 41.1 | 2.9 | 69,890 | 9.1 | 11.1 | 10.8 |
| Winona | 21 | 42 | 15,800 | 18.2 | 33.7 | 30.1 | 74 | 14,201 | 30,013 | 59,329 | 42.6 | 3.4 | 64,219 | 11.0 | 11.3 | 10.7 |
| Wright | 146 | 118 | 36,008 | 12.9 | 34.5 | 29.6 | 323 | 11,611 | 36,260 | 84,974 | 26.6 | 6.4 | 91,374 | 5.2 | 5.3 | 4.7 |
| Yellow Medicine | 11 | 15 | 2,215 | 7.9 | 44.1 | 17.2 | 20 | 13,875 | 29,908 | 59,210 | 40.8 | 2.9 | 61,074 | 9.6 | 13.2 | 12.1 |

1. All persons 3 years old and over enrolled in nursery school through college. 2. Persons 25 years old and over. 3. Elementary and secondary education expenditures. 4. Based on population estimated by the American Community Survey, 2014–2018.

# Table B. States and Counties — **Personal Income**

| STATE County | Personal income, 2019 | | | | | | | | | | Earnings, 2019 | | |
|---|---|---|---|---|---|---|---|---|---|---|---|---|---|
| | Total (mil dol) | Percent change 2018-2019 | Per capita[1] Dollars | Rank | Wages and salaries (mil dol) | Supplements to wages and salaries, employer contributions (mil dol) Pension and insurance | Government social insurance | Proprietors' income (mil dol) | Dividends, interest, and rent (mil dol) | Personal transfer receipts (mil dol) | Total (mil dol) | Contributions for government social insurance (mil dol) From employee and self-employed | From employer |
| | 62 | 63 | 64 | 65 | 66 | 67 | 68 | 69 | 70 | 71 | 72 | 73 | 74 |
| **MINNESOTA— Cont'd** | | | | | | | | | | | | | |
| Fillmore | 991 | 4.1 | 47,061 | 1,127 | 235 | 47 | 19 | 126 | 176 | 219 | 427 | 30 | 19 |
| Freeborn | 1,434 | 3.6 | 47,356 | 1,095 | 547 | 93 | 42 | 117 | 268 | 355 | 798 | 55 | 42 |
| Goodhue | 2,587 | 3.7 | 55,816 | 454 | 1,123 | 194 | 86 | 331 | 435 | 471 | 1,733 | 108 | 86 |
| Grant | 301 | 4.8 | 50,366 | 806 | 85 | 15 | 7 | 38 | 62 | 77 | 146 | 10 | 7 |
| Hennepin | 96,903 | 2.6 | 76,552 | 78 | 73,662 | 9,346 | 5,096 | 8,118 | 23,350 | 11,224 | 96,222 | 5,666 | 5,096 |
| Houston | 994 | 3.7 | 53,446 | 579 | 209 | 43 | 17 | 83 | 183 | 197 | 353 | 25 | 17 |
| Hubbard | 973 | 4.5 | 45,257 | 1,328 | 256 | 46 | 20 | 107 | 191 | 267 | 429 | 33 | 20 |
| Isanti | 1,859 | 4.5 | 45,781 | 1,276 | 491 | 90 | 39 | 93 | 241 | 391 | 713 | 52 | 39 |
| Itasca | 2,077 | 3.7 | 46,020 | 1,244 | 738 | 137 | 59 | 180 | 393 | 610 | 1,114 | 82 | 59 |
| Jackson | 570 | 8.2 | 57,933 | 372 | 229 | 45 | 20 | 100 | 129 | 109 | 393 | 21 | 20 |
| Kanabec | 709 | 4.5 | 43,416 | 1,558 | 174 | 34 | 14 | 62 | 99 | 199 | 284 | 22 | 14 |
| Kandiyohi | 2,338 | 2.1 | 54,111 | 539 | 1,035 | 174 | 81 | 387 | 413 | 477 | 1,678 | 106 | 81 |
| Kittson | 218 | -0.9 | 50,628 | 788 | 66 | 13 | 5 | 20 | 52 | 55 | 104 | 7 | 5 |
| Koochiching | 551 | 5.3 | 45,054 | 1,354 | 224 | 41 | 19 | 27 | 92 | 184 | 311 | 24 | 19 |
| Lac qui Parle | 374 | 9.1 | 56,527 | 420 | 91 | 19 | 7 | 60 | 82 | 97 | 178 | 11 | 7 |
| Lake | 529 | 3.3 | 49,743 | 854 | 204 | 37 | 18 | 32 | 103 | 150 | 290 | 21 | 18 |
| Lake of the Woods | 214 | 1.0 | 57,121 | 400 | 68 | 14 | 5 | 35 | 39 | 50 | 122 | 8 | 5 |
| Le Sueur | 1,465 | 4.4 | 50,728 | 776 | 425 | 76 | 34 | 80 | 265 | 267 | 615 | 41 | 34 |
| Lincoln | 295 | 12.3 | 52,251 | 660 | 68 | 14 | 5 | 53 | 58 | 69 | 141 | 8 | 5 |
| Lyon | 1,378 | 5.5 | 54,108 | 540 | 672 | 121 | 49 | 216 | 265 | 257 | 1,059 | 60 | 49 |
| McLeod | 1,808 | 3.2 | 50,380 | 805 | 782 | 133 | 59 | 135 | 324 | 366 | 1,110 | 74 | 59 |
| Mahnomen | 201 | 3.3 | 36,427 | 2,517 | 76 | 17 | 6 | 10 | 38 | 72 | 109 | 7 | 6 |
| Marshall | 481 | -0.9 | 51,545 | 718 | 113 | 21 | 9 | 46 | 90 | 112 | 189 | 13 | 9 |
| Martin | 1,124 | 6.8 | 57,091 | 402 | 408 | 67 | 30 | 163 | 256 | 251 | 668 | 39 | 30 |
| Meeker | 1,070 | 3.4 | 46,073 | 1,240 | 329 | 60 | 26 | 82 | 186 | 252 | 497 | 35 | 26 |
| Mille Lacs | 1,131 | 3.7 | 43,058 | 1,609 | 383 | 74 | 29 | 66 | 161 | 315 | 552 | 40 | 29 |
| Morrison | 1,459 | 3.5 | 43,703 | 1,526 | 449 | 88 | 35 | 162 | 247 | 373 | 734 | 48 | 35 |
| Mower | 1,901 | 2.0 | 47,445 | 1,086 | 865 | 134 | 61 | 106 | 351 | 442 | 1,166 | 77 | 61 |
| Murray | 485 | 11.1 | 59,183 | 332 | 128 | 25 | 10 | 107 | 95 | 99 | 271 | 14 | 10 |
| Nicollet | 1,749 | 3.8 | 51,036 | 755 | 745 | 138 | 57 | 153 | 345 | 292 | 1,093 | 66 | 57 |
| Nobles | 1,065 | 7.5 | 49,217 | 903 | 490 | 83 | 36 | 195 | 181 | 195 | 805 | 44 | 36 |
| Norman | 265 | 4.4 | 41,584 | 1,833 | 72 | 14 | 6 | 16 | 64 | 82 | 108 | 8 | 6 |
| Olmsted | 9,367 | 2.9 | 59,176 | 333 | 6,741 | 878 | 482 | 494 | 1,574 | 1,326 | 8,595 | 521 | 482 |
| Otter Tail | 2,895 | 3.1 | 49,277 | 897 | 999 | 191 | 78 | 268 | 618 | 698 | 1,536 | 107 | 78 |
| Pennington | 810 | 1.3 | 57,337 | 392 | 521 | 84 | 39 | 35 | 208 | 150 | 679 | 41 | 39 |
| Pine | 1,123 | 3.8 | 37,971 | 2,327 | 309 | 67 | 25 | 64 | 189 | 347 | 465 | 36 | 25 |
| Pipestone | 564 | 9.8 | 61,830 | 255 | 194 | 36 | 16 | 158 | 94 | 100 | 405 | 20 | 16 |
| Polk | 1,522 | 3.2 | 48,513 | 969 | 551 | 100 | 45 | 98 | 272 | 373 | 793 | 52 | 45 |
| Pope | 587 | 3.4 | 52,217 | 662 | 221 | 39 | 17 | 51 | 135 | 133 | 328 | 22 | 17 |
| Ramsey | 30,589 | 2.5 | 55,583 | 468 | 23,121 | 3,237 | 1,615 | 1,661 | 6,626 | 5,521 | 29,635 | 1,788 | 1,615 |
| Red Lake | 204 | 1.1 | 50,221 | 811 | 41 | 9 | 3 | 28 | 30 | 42 | 83 | 5 | 3 |
| Redwood | 830 | 4.6 | 54,719 | 512 | 267 | 51 | 20 | 199 | 158 | 172 | 538 | 30 | 20 |
| Renville | 783 | 3.4 | 53,855 | 553 | 265 | 50 | 22 | 116 | 166 | 174 | 451 | 26 | 22 |
| Rice | 3,020 | 3.8 | 45,092 | 1,352 | 1,276 | 203 | 97 | 203 | 520 | 567 | 1,780 | 116 | 97 |
| Rock | 537 | 6.4 | 57,617 | 386 | 157 | 29 | 11 | 136 | 105 | 101 | 333 | 16 | 11 |
| Roseau | 748 | -0.6 | 49,328 | 895 | 392 | 76 | 32 | 28 | 170 | 144 | 528 | 33 | 32 |
| St. Louis | 9,698 | 3.2 | 48,718 | 948 | 5,081 | 857 | 398 | 495 | 1,741 | 2,454 | 6,831 | 451 | 398 |
| Scott | 9,263 | 3.6 | 62,164 | 243 | 3,027 | 435 | 232 | 770 | 1,349 | 902 | 4,464 | 274 | 232 |
| Sherburne | 4,639 | 3.6 | 47,712 | 1,052 | 1,313 | 227 | 103 | 262 | 555 | 688 | 1,904 | 125 | 103 |
| Sibley | 734 | 3.5 | 49,408 | 888 | 166 | 33 | 14 | 66 | 140 | 154 | 280 | 19 | 14 |
| Stearns | 7,722 | 3.4 | 47,938 | 1,020 | 4,504 | 753 | 345 | 638 | 1,383 | 1,469 | 6,240 | 376 | 345 |
| Steele | 1,780 | 3.8 | 48,580 | 960 | 1,092 | 171 | 80 | 108 | 301 | 353 | 1,450 | 90 | 80 |
| Stevens | 497 | 8.0 | 50,696 | 778 | 243 | 45 | 20 | 81 | 113 | 94 | 388 | 20 | 20 |
| Swift | 489 | 3.1 | 52,813 | 632 | 158 | 33 | 13 | 95 | 86 | 117 | 299 | 17 | 13 |
| Todd | 1,085 | 4.1 | 43,979 | 1,489 | 298 | 56 | 23 | 186 | 157 | 285 | 563 | 38 | 23 |
| Traverse | 197 | 5.0 | 60,535 | 286 | 54 | 11 | 4 | 38 | 50 | 49 | 107 | 6 | 4 |
| Wabasha | 1,082 | 3.4 | 50,042 | 830 | 282 | 55 | 23 | 104 | 199 | 221 | 463 | 31 | 23 |
| Wadena | 558 | 2.1 | 40,756 | 1,954 | 235 | 46 | 18 | 48 | 96 | 189 | 348 | 24 | 18 |
| Waseca | 824 | 3.6 | 44,262 | 1,450 | 277 | 54 | 22 | 74 | 147 | 186 | 427 | 28 | 22 |
| Washington | 17,984 | 3.4 | 68,525 | 142 | 4,638 | 725 | 342 | 1,063 | 3,542 | 2,013 | 6,769 | 443 | 342 |
| Watonwan | 531 | 7.4 | 48,684 | 949 | 178 | 34 | 14 | 69 | 89 | 115 | 295 | 17 | 14 |
| Wilkin | 354 | 2.9 | 57,087 | 403 | 95 | 16 | 8 | 68 | 65 | 74 | 187 | 12 | 8 |
| Winona | 2,495 | 2.3 | 49,412 | 887 | 1,122 | 220 | 86 | 136 | 594 | 470 | 1,564 | 97 | 86 |
| Wright | 7,142 | 3.8 | 51,609 | 714 | 2,146 | 365 | 169 | 449 | 955 | 977 | 3,130 | 202 | 169 |
| Yellow Medicine | 562 | 7.2 | 57,881 | 375 | 161 | 34 | 12 | 100 | 147 | 128 | 307 | 19 | 12 |

1. Based on the resident population estimated as of July 1 of the year shown.

# Table B. States and Counties — Earnings, Social Security, and Housing

| STATE County | Earnings, 2019 (cont.) Percent by selected industries | | | | | | | | | Social Security beneficiaries, December 2019 | | Supplemental Security Income recipients, 2019 | Housing units, 2020 | |
|---|---|---|---|---|---|---|---|---|---|---|---|---|---|---|
| | Farm | Mining, quarrying, and extractions | Construction | Manufacturing | Information; professional, scientific, technical services | Retail trade | Finance, insurance, real estate, and leasing | Health care and social assistance | Government | Number | Rate[1] | | Total | Percent change, 2010-2020 |
| | 75 | 76 | 77 | 78 | 79 | 80 | 81 | 82 | 83 | 84 | 85 | 86 | 87 | 88 |
| **MINNESOTA— Cont'd** | | | | | | | | | | | | | | |
| Fillmore | 7.7 | D | 7.9 | 19.3 | 2.8 | 7.2 | 5.2 | D | 16.4 | 5,150 | 244 | 177 | 10,133 | 4.1 |
| Freeborn | 3.7 | D | 6.3 | 19.7 | 3.4 | 9.8 | 7.3 | 16.0 | 12.7 | 8,115 | 267 | 568 | 14,355 | 0.9 |
| Goodhue | 3.2 | D | 6.0 | 27.3 | D | 5.5 | 3.1 | 10.8 | 15.7 | 11,010 | 237 | 513 | 20,963 | 3.1 |
| Grant | 8.6 | 0.0 | 11.2 | 6.9 | D | 4.2 | 6.8 | D | 17.6 | 1,745 | 292 | 59 | 3,297 | -0.8 |
| Hennepin | 0.0 | 0.2 | 4.1 | 8.0 | 18.4 | 3.9 | 15.0 | 10.2 | 9.1 | 194,595 | 154 | 26,349 | 552,023 | 8.3 |
| Houston | 7.3 | 0.0 | 12.4 | 8.5 | D | 4.8 | 4.1 | 10.6 | 18.4 | 4,845 | 260 | 202 | 8,823 | 2.6 |
| Hubbard | 2.0 | D | 9.7 | 23.8 | D | 7.7 | 4.4 | D | 15.8 | 6,265 | 290 | 373 | 14,971 | 2.4 |
| Isanti | -0.6 | 0.0 | 9.9 | 12.9 | D | 10.0 | 4.2 | 19.5 | 18.8 | 8,390 | 207 | 422 | 16,725 | 9.2 |
| Itasca | 0.0 | D | 7.5 | 13.5 | 3.1 | 7.8 | 4.4 | 16.3 | 19.9 | 13,525 | 299 | 855 | 27,926 | 3.2 |
| Jackson | 16.0 | 0.0 | 3.7 | 29.5 | 2.5 | 2.1 | 2.3 | D | 10.6 | 2,425 | 247 | 96 | 5,079 | 1.8 |
| Kanabec | 0.5 | D | 19.6 | 8.6 | 2.4 | 6.9 | 5.4 | 9.5 | 31.1 | 4,390 | 268 | 237 | 8,057 | 2.6 |
| Kandiyohi | 2.8 | D | 6.7 | 16.3 | 3.8 | 7.1 | 4.4 | D | 12.8 | 9,970 | 231 | 730 | 20,198 | 3.7 |
| Kittson | 10.2 | D | 5.2 | 11.0 | D | 5.2 | D | 15.0 | 19.7 | 1,200 | 280 | 47 | 2,611 | 0.2 |
| Koochiching | 0.3 | 0.0 | 12.3 | D | 1.9 | 6.9 | 3.7 | D | 20.7 | 4,015 | 329 | 271 | 7,964 | 0.8 |
| Lac qui Parle | 20.0 | 0.0 | 5.8 | 6.5 | 1.3 | 6.4 | 8.8 | 7.5 | 23.8 | 1,950 | 294 | 64 | 3,698 | 0.2 |
| Lake | 0.0 | D | D | 17.7 | 2.6 | 4.4 | 3.8 | D | 20.0 | 3,230 | 304 | 107 | 8,077 | 5.2 |
| Lake of the Woods | 1.0 | 0.0 | D | D | D | 5.6 | 2.1 | 11.3 | 15.7 | 1,160 | 310 | 33 | 3,926 | 6.9 |
| Le Sueur | 3.9 | D | 11.4 | 36.4 | 2.4 | 4.4 | 3.8 | 5.2 | 13.0 | 5,540 | 192 | 293 | 12,959 | 4.4 |
| Lincoln | 29.1 | 0.0 | 12.5 | 2.8 | D | 5.0 | D | D | 12.5 | 1,490 | 265 | 47 | 3,149 | 1.4 |
| Lyon | 9.9 | 0.0 | 4.2 | 13.7 | 3.3 | 6.0 | 11.3 | 7.6 | 20.9 | 5,180 | 203 | 416 | 11,417 | 2.9 |
| McLeod | 0.3 | D | 6.1 | 33.6 | 2.5 | 6.2 | 4.7 | 14.5 | 10.8 | 8,090 | 226 | 375 | 16,116 | 2.3 |
| Mahnomen | 0.1 | 0.0 | 5.1 | D | D | D | D | 4.4 | 56.4 | 975 | 176 | 133 | 2,782 | -0.1 |
| Marshall | 4.2 | D | 9.5 | 17.3 | 2.3 | 5.0 | D | D | 21.6 | 2,265 | 242 | 84 | 4,816 | 0.1 |
| Martin | 18.3 | 0.0 | 4.3 | 9.0 | 3.4 | 5.8 | 5.7 | 13.0 | 11.4 | 5,600 | 285 | 390 | 9,970 | -0.4 |
| Meeker | 4.9 | 0.2 | 11.6 | 21.4 | 4.3 | 5.9 | 4.5 | D | 16.4 | 5,505 | 238 | 228 | 11,082 | 3.8 |
| Mille Lacs | 0.3 | D | 9.6 | 7.1 | 5.1 | 6.7 | 4.3 | 35.3 | 22.0 | 6,805 | 260 | 467 | 13,123 | 2.9 |
| Morrison | 10.7 | 0.2 | 7.4 | 8.8 | 5.8 | 7.2 | 3.7 | D | 15.4 | 8,265 | 248 | 617 | 16,414 | 4.3 |
| Mower | 2.8 | D | 4.3 | 19.1 | D | 4.7 | 3.1 | 13.6 | | 8,905 | 222 | 757 | 17,171 | 0.8 |
| Murray | 29.2 | 0.6 | 6.0 | 9.7 | 2.3 | 6.1 | D | D | 13.9 | 2,240 | 273 | 87 | 4,672 | 2.5 |
| Nicollet | 4.3 | 0.0 | 3.8 | 23.8 | D | 4.1 | 4.0 | 9.0 | 21.0 | 6,160 | 180 | 325 | 13,724 | 6.6 |
| Nobles | 15.0 | 0.0 | 4.3 | 28.3 | 2.8 | 6.1 | 4.6 | D | 11.8 | 4,045 | 187 | 295 | 8,695 | 1.9 |
| Norman | 5.6 | 0.0 | 6.0 | 1.8 | 8.2 | 5.2 | D | 14.5 | 22.2 | 1,720 | 269 | 112 | 3,440 | 0.6 |
| Olmsted | 0.4 | 0.0 | 5.0 | 8.8 | 3.5 | 4.6 | 2.9 | 54.4 | 8.3 | 28,195 | 178 | 2,251 | 68,995 | 14.0 |
| Otter Tail | 3.8 | 0.1 | 9.4 | 18.3 | 4.9 | 6.6 | 3.8 | 13.4 | 16.6 | 16,845 | 287 | 739 | 36,434 | 2.4 |
| Pennington | 1.1 | 0.0 | 2.0 | 11.3 | 1.8 | 4.4 | 2.1 | D | 12.3 | 2,990 | 213 | 170 | 6,689 | 6.2 |
| Pine | 2.1 | D | 11.2 | 4.3 | D | 7.0 | 3.4 | 9.3 | 38.2 | 7,555 | 257 | 558 | 17,829 | 3.2 |
| Pipestone | 23.0 | D | 6.6 | 9.4 | D | 4.5 | D | D | 13.9 | 2,015 | 220 | 132 | 4,518 | 0.8 |
| Polk | 4.0 | 0.2 | 7.5 | 15.5 | 3.6 | 6.3 | 4.2 | D | 21.0 | 6,985 | 223 | 541 | 15,132 | 3.6 |
| Pope | 6.3 | D | 5.2 | 20.0 | 2.8 | 6.4 | 5.5 | D | 19.4 | 2,895 | 258 | 141 | 6,747 | 4.9 |
| Ramsey | 0.0 | 0.0 | 4.5 | 10.3 | 9.0 | 3.9 | 8.8 | 14.1 | 17.0 | 90,155 | 164 | 16,603 | 223,566 | 2.9 |
| Red Lake | 20.2 | 0.1 | 5.0 | D | D | 5.1 | D | D | 19.8 | 950 | 236 | 34 | 1,949 | 0.1 |
| Redwood | 14.5 | D | 3.4 | 25.5 | 2.5 | 4.3 | 6.6 | D | 12.5 | 3,590 | 237 | 199 | 7,353 | 1.1 |
| Renville | 19.0 | D | 4.4 | 15.6 | 3.5 | 2.9 | D | 6.0 | 19.9 | 3,560 | 245 | 215 | 7,349 | -0.1 |
| Rice | 1.8 | 0.2 | 7.3 | 22.9 | 3.0 | 5.6 | 4.0 | 9.3 | 14.4 | 12,345 | 184 | 736 | 25,445 | 4.1 |
| Rock | 32.2 | D | 4.9 | 3.9 | 3.5 | 3.7 | 10.3 | 11.7 | 12.7 | 2,345 | 251 | 87 | 4,351 | 2.1 |
| Roseau | 2.2 | 0.0 | 1.6 | 55.9 | 1.3 | 3.8 | 2.5 | 9.9 | 12.9 | 3,455 | 228 | 121 | 7,672 | 2.7 |
| St. Louis | 0.0 | 6.7 | 6.2 | 5.5 | 6.4 | 7.2 | 5.1 | 25.5 | 16.7 | 49,065 | 246 | 4,849 | 105,839 | 2.7 |
| Scott | 0.2 | 0.1 | 14.0 | 18.0 | 7.1 | 5.6 | 4.0 | 7.7 | 15.4 | 19,110 | 128 | 1,242 | 54,185 | 15.0 |
| Sherburne | 0.5 | D | 13.7 | 16.1 | 3.4 | 7.6 | 3.7 | 10.7 | 16.7 | 14,850 | 152 | 539 | 35,226 | 8.8 |
| Sibley | 11.6 | 0.0 | 9.0 | 17.5 | 2.2 | 2.7 | 2.7 | D | 15.2 | 3,270 | 220 | 154 | 6,676 | 1.4 |
| Stearns | 2.5 | 0.2 | 8.6 | 12.8 | 5.7 | 7.8 | 6.8 | 16.7 | 14.5 | 29,145 | 181 | 2,408 | 66,045 | 6.6 |
| Steele | 2.2 | D | 3.6 | 27.4 | D | 8.1 | 16.3 | 10.3 | 10.7 | 7,960 | 217 | 480 | 15,880 | 3.5 |
| Stevens | 17.5 | D | 6.0 | 18.5 | 4.4 | 5.5 | 3.2 | 10.9 | 20.0 | 1,750 | 178 | 111 | 4,270 | 2.7 |
| Swift | 15.6 | D | 5.3 | 18.8 | 2.9 | 4.6 | 3.4 | D | 18.5 | 2,315 | 249 | 154 | 4,839 | 0.1 |
| Todd | 7.0 | D | 4.8 | 33.5 | 2.3 | 3.8 | 3.8 | 15.0 | 15.5 | 6,120 | 248 | 363 | 13,455 | 4.2 |
| Traverse | 25.3 | 0.0 | D | 2.7 | D | 5.4 | 3.2 | 12.3 | 16.6 | 995 | 305 | 72 | 2,104 | 1.5 |
| Wabasha | 8.8 | D | 6.5 | 21.1 | 2.4 | 5.2 | 5.2 | D | 15.2 | 5,065 | 235 | 182 | 10,354 | 3.6 |
| Wadena | 3.1 | 0.0 | 7.7 | 9.6 | 3.2 | 6.1 | 3.2 | 19.9 | 23.4 | 3,985 | 292 | 387 | 7,224 | 4.7 |
| Waseca | 8.3 | D | 6.6 | 19.7 | 4.4 | 4.8 | 5.3 | 10.9 | 22.8 | 4,330 | 233 | 226 | 7,971 | 0.8 |
| Washington | 0.3 | D | 7.1 | 13.9 | 8.3 | 8.2 | 8.5 | 13.1 | 12.4 | 45,665 | 174 | 2,187 | 102,664 | 11.2 |
| Watonwan | 16.2 | D | 7.1 | 22.0 | 2.0 | 2.9 | 4.8 | 8.9 | 14.9 | 2,370 | 219 | 120 | 5,055 | 0.2 |
| Wilkin | 2.5 | D | 4.0 | 1.6 | D | D | 8.1 | 12.1 | 13.4 | 1,495 | 240 | 91 | 3,168 | 2.9 |
| Winona | 3.3 | D | 4.1 | 27.7 | D | 5.4 | 3.8 | 9.3 | 15.1 | 10,230 | 202 | 593 | 21,388 | 3.0 |
| Wright | 0.5 | D | 15.0 | 16.0 | 3.7 | 8.3 | 4.6 | 10.3 | 13.8 | 21,625 | 156 | 849 | 54,320 | 10.9 |
| Yellow Medicine | 12.0 | D | 7.2 | 10.8 | 4.4 | 4.1 | 5.9 | D | 24.8 | 2,530 | 260 | 145 | 4,764 | 0.1 |

1. Per 1,000 resident population estimated as of July 1 of the year shown.

# Table B. States and Counties — Housing, Labor Force, and Employment

| STATE County | Housing units, 2015-2019 — Occupied units | | Owner-occupied | Median owner cost as a percent of income | | Renter-occupied | | | Civilian labor force, 2020 | | Unemployment | | Civilian employment[6], 2015-2019 | Percent | |
|---|---|---|---|---|---|---|---|---|---|---|---|---|---|---|---|
| | Total | Percent | Median value[1] | With a mortgage | Without a mortgage[2] | Median rent[3] | Median rent as a percent of income[2] | Sub-standard units[4] (percent) | Total | Percent change, 2019-2020 | Total | Rate[5] | Total | Management, business, science, and arts | Construction, production, and maintenance occupations |
| | 89 | 90 | 91 | 92 | 93 | 94 | 95 | 96 | 97 | 98 | 99 | 100 | 101 | 102 | 103 |
| **MINNESOTA— Cont'd** | | | | | | | | | | | | | | | |
| Fillmore | 8,616 | 80.8 | 153,800 | 19.9 | 10.7 | 659 | 25.1 | 3.5 | 11,654 | 1.2 | 585 | 5.0 | 10,758 | 35.0 | 29.3 |
| Freeborn | 13,009 | 76.7 | 113,800 | 18.7 | 11.2 | 681 | 26.1 | 1.0 | 16,258 | 1.1 | 914 | 5.6 | 15,095 | 27.9 | 34.7 |
| Goodhue | 19,452 | 75.3 | 204,100 | 19.8 | 11.5 | 862 | 28.6 | 1.4 | 26,659 | -1.2 | 1,560 | 5.9 | 24,227 | 34.6 | 27.9 |
| Grant | 2,612 | 77.4 | 118,700 | 19.4 | 10.7 | 657 | 27.9 | 1.1 | 3,337 | 0.9 | 176 | 5.3 | 2,940 | 33.7 | 29.4 |
| Hennepin | 508,030 | 62.4 | 276,900 | 19.6 | 11.0 | 1,135 | 28.2 | 2.9 | 710,270 | 0.4 | 46,532 | 6.6 | 696,128 | 50.0 | 14.0 |
| Houston | 8,253 | 80.1 | 173,700 | 19.9 | 11.8 | 722 | 23.5 | 0.9 | 10,311 | -0.7 | 442 | 4.3 | 10,246 | 36.9 | 28.4 |
| Hubbard | 8,688 | 81.2 | 193,600 | 21.9 | 11.9 | 696 | 27.4 | 3.7 | 9,907 | -0.7 | 700 | 7.1 | 9,495 | 32.0 | 28.6 |
| Isanti | 14,903 | 82.5 | 196,800 | 20.9 | 11.4 | 1,019 | 28.8 | 2.0 | 21,430 | -0.6 | 1,454 | 6.8 | 20,397 | 30.8 | 32.6 |
| Itasca | 19,506 | 81.3 | 164,100 | 19.8 | 12.2 | 696 | 31.2 | 3.0 | 22,100 | 0.9 | 1,860 | 8.4 | 19,911 | 32.4 | 27.3 |
| Jackson | 4,419 | 81.0 | 124,900 | 17.6 | 10.0 | 710 | 22.0 | 1.6 | 5,772 | 1.2 | 279 | 4.8 | 5,251 | 32.3 | 33.7 |
| Kanabec | 6,439 | 84.8 | 163,900 | 22.2 | 13.3 | 816 | 28.8 | 3.5 | 9,109 | 0.6 | 729 | 8.0 | 7,836 | 28.5 | 34.2 |
| Kandiyohi | 16,889 | 74.0 | 168,900 | 20.0 | 11.9 | 776 | 27.6 | 3.0 | 24,696 | 1.7 | 1,184 | 4.8 | 21,982 | 33.4 | 29.5 |
| Kittson | 1,860 | 80.5 | 84,200 | 17.1 | 10.0 | 617 | 23.4 | 0.4 | 2,411 | 1.8 | 111 | 4.6 | 2,232 | 35.5 | 30.2 |
| Koochiching | 5,538 | 78.8 | 108,400 | 19.2 | 10.0 | 668 | 32.0 | 1.5 | 5,796 | -2.9 | 429 | 7.4 | 5,889 | 28.7 | 30.2 |
| Lac qui Parle | 3,090 | 80.9 | 88,600 | 18.5 | 10.0 | 541 | 21.1 | 1.3 | 3,529 | -0.3 | 166 | 4.7 | 3,356 | 36.8 | 28.4 |
| Lake | 5,192 | 82.0 | 174,700 | 21.6 | 12.1 | 662 | 26.2 | 2.3 | 5,273 | -4.0 | 384 | 7.3 | 4,814 | 38.9 | 27.4 |
| Lake of the Woods | 1,470 | 80.9 | 161,900 | 22.4 | 13.1 | 678 | 26.9 | 1.7 | 2,442 | -0.1 | 150 | 6.1 | 1,931 | 21.3 | 44.7 |
| Le Sueur | 10,939 | 81.6 | 207,400 | 20.7 | 10.5 | 773 | 27.6 | 1.9 | 16,186 | -0.9 | 1,094 | 6.8 | 15,086 | 32.4 | 33.4 |
| Lincoln | 2,468 | 78.8 | 103,100 | 18.3 | 11.5 | 616 | 26.4 | 1.3 | 3,125 | -3.8 | 139 | 4.4 | 2,786 | 35.8 | 27.4 |
| Lyon | 10,018 | 68.4 | 148,900 | 18.7 | 10.1 | 666 | 27.0 | 3.6 | 14,364 | -0.9 | 645 | 4.5 | 13,437 | 34.8 | 28.0 |
| McLeod | 14,714 | 77.7 | 164,700 | 20.5 | 11.1 | 781 | 25.2 | 1.1 | 19,236 | -0.4 | 1,091 | 5.7 | 19,062 | 28.4 | 34.6 |
| Mahnomen | 1,946 | 68.0 | 104,800 | 19.9 | 12.7 | 586 | 25.8 | 4.0 | 2,198 | -4.2 | 286 | 13.0 | 2,140 | 30.4 | 25.7 |
| Marshall | 3,971 | 82.6 | 117,000 | 18.1 | 10.4 | 636 | 24.0 | 1.5 | 5,451 | 1.0 | 353 | 6.5 | 4,827 | 34.2 | 33.0 |
| Martin | 8,759 | 73.8 | 118,300 | 17.4 | 10.7 | 620 | 25.7 | 1.7 | 9,994 | -0.5 | 520 | 5.2 | 9,910 | 34.0 | 28.0 |
| Meeker | 9,209 | 80.4 | 170,200 | 20.3 | 11.0 | 755 | 26.0 | 2.1 | 13,251 | -0.6 | 684 | 5.2 | 11,603 | 31.6 | 35.9 |
| Mille Lacs | 10,249 | 75.1 | 163,300 | 23.0 | 12.3 | 756 | 28.6 | 2.8 | 12,937 | -0.2 | 1,090 | 8.4 | 12,223 | 26.4 | 35.9 |
| Morrison | 13,371 | 77.2 | 170,100 | 21.0 | 11.6 | 742 | 25.4 | 1.8 | 17,936 | 2.2 | 1,212 | 6.8 | 16,637 | 27.8 | 34.3 |
| Mower | 15,565 | 73.0 | 123,900 | 18.9 | 10.1 | 716 | 27.3 | 3.4 | 20,929 | 2.6 | 1,040 | 5.0 | 19,254 | 29.8 | 34.5 |
| Murray | 3,645 | 81.2 | 120,100 | 17.8 | 10.1 | 572 | 21.3 | 0.7 | 4,834 | -0.1 | 245 | 5.1 | 4,202 | 35.6 | 31.2 |
| Nicollet | 12,885 | 73.9 | 191,900 | 19.6 | 10.8 | 837 | 26.1 | 0.5 | 20,752 | -0.8 | 997 | 4.8 | 19,230 | 37.5 | 23.1 |
| Nobles | 8,033 | 70.7 | 127,100 | 20.5 | 10.1 | 726 | 23.3 | 3.9 | 11,602 | 3.6 | 488 | 4.2 | 10,744 | 24.9 | 42.2 |
| Norman | 2,746 | 81.9 | 100,700 | 17.5 | 11.6 | 599 | 28.1 | 3.0 | 3,287 | 1.2 | 208 | 6.3 | 3,164 | 34.2 | 30.2 |
| Olmsted | 62,108 | 72.7 | 214,600 | 18.8 | 10.0 | 968 | 28.1 | 2.8 | 91,030 | 2.2 | 4,937 | 5.4 | 83,119 | 51.1 | 15.9 |
| Otter Tail | 24,687 | 78.3 | 189,100 | 20.0 | 11.0 | 678 | 27.0 | 2.2 | 31,488 | 0.4 | 1,698 | 5.4 | 28,566 | 34.0 | 29.6 |
| Pennington | 5,995 | 72.8 | 155,100 | 19.2 | 10.6 | 693 | 25.4 | 1.5 | 8,663 | -0.4 | 533 | 6.2 | 7,591 | 30.7 | 30.7 |
| Pine | 10,760 | 80.9 | 164,500 | 23.3 | 13.1 | 781 | 29.3 | 2.9 | 14,581 | -3.0 | 1,255 | 8.6 | 12,812 | 26.6 | 31.7 |
| Pipestone | 3,973 | 74.9 | 97,500 | 18.8 | 10.0 | 590 | 21.5 | 1.9 | 4,670 | -4.9 | 189 | 4.0 | 4,302 | 35.0 | 34.2 |
| Polk | 12,584 | 72.9 | 165,000 | 19.2 | 11.2 | 716 | 28.0 | 2.2 | 16,403 | -1.6 | 918 | 5.6 | 15,705 | 34.5 | 28.3 |
| Pope | 4,981 | 77.5 | 174,300 | 20.0 | 11.7 | 609 | 25.2 | 1.4 | 6,512 | -0.1 | 290 | 4.5 | 5,661 | 37.3 | 27.8 |
| Ramsey | 209,043 | 59.4 | 229,600 | 20.2 | 11.0 | 1,007 | 29.4 | 4.6 | 290,181 | 0.5 | 20,064 | 6.9 | 283,280 | 44.3 | 17.8 |
| Red Lake | 1,669 | 83.6 | 116,000 | 18.2 | 11.2 | 554 | 28.7 | 1.9 | 2,230 | 0.3 | 143 | 6.4 | 1,992 | 31.4 | 33.3 |
| Redwood | 6,261 | 78.5 | 106,300 | 18.5 | 10.0 | 641 | 23.6 | 0.9 | 7,782 | 2.4 | 430 | 5.5 | 7,519 | 33.2 | 30.0 |
| Renville | 6,085 | 78.0 | 106,300 | 18.3 | 10.0 | 635 | 28.9 | 1.1 | 8,598 | -0.3 | 476 | 5.5 | 7,189 | 32.1 | 34.3 |
| Rice | 23,005 | 74.0 | 208,600 | 21.3 | 10.0 | 843 | 29.6 | 2.0 | 37,193 | 0.0 | 1,951 | 5.2 | 34,116 | 35.6 | 27.6 |
| Rock | 3,914 | 75.5 | 153,000 | 18.7 | 10.0 | 678 | 22.0 | 1.9 | 5,808 | 2.9 | 176 | 3.0 | 4,801 | 37.1 | 27.4 |
| Roseau | 5,980 | 81.6 | 132,800 | 19.4 | 10.2 | 723 | 26.1 | 2.1 | 7,917 | 1.1 | 455 | 5.7 | 8,302 | 29.6 | 43.5 |
| St. Louis | 85,807 | 71.3 | 157,900 | 19.3 | 11.1 | 770 | 30.2 | 2.4 | 101,110 | -0.5 | 7,286 | 7.2 | 98,711 | 36.5 | 22.3 |
| Scott | 49,479 | 82.8 | 299,700 | 20.0 | 10.0 | 1,170 | 29.4 | 2.8 | 83,897 | -0.4 | 4,957 | 5.9 | 80,840 | 43.0 | 20.7 |
| Sherburne | 32,206 | 83.6 | 230,500 | 19.4 | 10.0 | 995 | 27.7 | 2.3 | 52,434 | -0.9 | 3,187 | 6.1 | 51,457 | 34.6 | 29.6 |
| Sibley | 6,031 | 79.1 | 159,200 | 19.6 | 11.3 | 719 | 24.0 | 1.7 | 8,373 | -1.1 | 461 | 5.5 | 7,935 | 29.7 | 39.2 |
| Stearns | 59,479 | 68.1 | 184,900 | 19.6 | 10.5 | 833 | 26.5 | 3.3 | 90,863 | 0.3 | 5,213 | 5.7 | 86,326 | 34.2 | 26.6 |
| Steele | 14,692 | 75.2 | 169,300 | 19.3 | 10.9 | 770 | 28.8 | 2.2 | 20,330 | -3.8 | 1,112 | 5.5 | 18,914 | 33.9 | 31.9 |
| Stevens | 3,624 | 68.5 | 160,100 | 16.5 | 10.0 | 753 | 32.0 | 1.1 | 5,446 | 0.1 | 177 | 3.3 | 5,383 | 36.4 | 27.5 |
| Swift | 4,263 | 70.3 | 104,100 | 18.2 | 10.6 | 656 | 25.3 | 1.4 | 4,881 | -0.3 | 282 | 5.8 | 4,766 | 33.2 | 31.4 |
| Todd | 9,819 | 82.6 | 153,400 | 21.0 | 12.2 | 649 | 25.5 | 5.8 | 13,786 | 1.9 | 729 | 5.3 | 11,089 | 28.0 | 38.5 |
| Traverse | 1,608 | 79.6 | 81,400 | 19.6 | 10.0 | 620 | 24.5 | 2.4 | 1,753 | -0.6 | 77 | 4.4 | 1,712 | 35.0 | 25.8 |
| Wabasha | 9,020 | 78.6 | 183,900 | 21.2 | 11.4 | 690 | 25.9 | 1.6 | 12,439 | 1.5 | 659 | 5.3 | 11,274 | 35.7 | 27.1 |
| Wadena | 5,666 | 76.5 | 132,000 | 20.6 | 13.0 | 668 | 29.5 | 5.6 | 6,024 | 1.7 | 460 | 7.6 | 5,998 | 27.0 | 37.5 |
| Waseca | 7,425 | 77.5 | 158,900 | 19.7 | 11.7 | 669 | 32.2 | 2.0 | 9,081 | 0.7 | 566 | 6.2 | 9,317 | 32.2 | 31.1 |
| Washington | 94,431 | 81.1 | 289,400 | 19.6 | 10.0 | 1,307 | 29.7 | 1.5 | 143,583 | -0.8 | 8,074 | 5.6 | 137,499 | 46.9 | 17.2 |
| Watonwan | 4,310 | 76.9 | 96,900 | 17.9 | 11.6 | 669 | 25.7 | 3.4 | 6,664 | 2.5 | 308 | 4.6 | 5,550 | 30.4 | 40.5 |
| Wilkin | 2,818 | 78.1 | 140,000 | 18.1 | 10.0 | 523 | 30.1 | 1.6 | 3,458 | -1.5 | 142 | 4.1 | 3,256 | 38.1 | 29.8 |
| Winona | 19,442 | 70.4 | 167,300 | 19.5 | 10.5 | 696 | 28.0 | 2.2 | 28,193 | -1.4 | 1,393 | 4.9 | 28,528 | 34.8 | 25.2 |
| Wright | 48,242 | 81.6 | 238,500 | 19.5 | 10.4 | 969 | 27.0 | 1.8 | 75,721 | -1.1 | 4,198 | 5.5 | 72,962 | 37.1 | 27.1 |
| Yellow Medicine | 4,087 | 79.4 | 110,000 | 18.6 | 10.6 | 584 | 26.0 | 1.2 | 5,314 | -0.7 | 245 | 4.6 | 4,924 | 30.6 | 31.2 |

1. Specified owner-occupied units. lacking complete plumbing facilities.   2. A value of 10.0 represents 10 percent or less; a value of 50.0 represents 50 percent or more.   3. Specified renter-occupied units.   4. Overcrowded or
5. Percent of civilian labor force.   6. Civilian employed persons 16 years old and over.

| STATE County | Private nonfarm establishments, employment and payroll, 2019 | | | | | | | | | Agriculture, 2017 | | | |
| | Number of establishments | Employment | | | | | | Annual payroll | | Farms | | | Farm producers whose primary occupation is farming (percent) |
| | | Total | Health care and social assistance | Manufacturing | Retail trade | Finance and insurance | Professional, scientific, and technical services | Total (mil dol) | Average per employee (dollars) | Number | Percent with: | | |
| | | | | | | | | | | | Fewer than 50 acres | 1000 acres or more | |
| | 104 | 105 | 106 | 107 | 108 | 109 | 110 | 111 | 112 | 113 | 114 | 115 | 116 |
|---|---|---|---|---|---|---|---|---|---|---|---|---|---|
| **MINNESOTA— Cont'd** | | | | | | | | | | | | | |
| Fillmore | 574 | 4,565 | 853 | 898 | 736 | 229 | 151 | 169 | 36,943 | 1,401 | 31.7 | 5.1 | 42.6 |
| Freeborn | 791 | 11,208 | 2,094 | 2,767 | 1,838 | 367 | 223 | 452 | 40,326 | 1,076 | 35.4 | 11.8 | 51.8 |
| Goodhue | 1,270 | 20,405 | 3,216 | 4,987 | 2,494 | 1,008 | 329 | 944 | 46,264 | 1,461 | 36.6 | 4.4 | 45.2 |
| Grant | 197 | 1,448 | 388 | 97 | 263 | 77 | 27 | 53 | 36,667 | 524 | 20.6 | 16.6 | 42.4 |
| Hennepin | 40,565 | 912,186 | 149,450 | 75,536 | 75,381 | 89,020 | 81,895 | 62,658 | 68,690 | 467 | 68.7 | 1.7 | 45.1 |
| Houston | 406 | 4,278 | 1,205 | 445 | 524 | 103 | 99 | 149 | 34,813 | 891 | 21.4 | 2.2 | 41.5 |
| Hubbard | 594 | 4,709 | 846 | 905 | 982 | 155 | 102 | 181 | 38,372 | 384 | 21.9 | 2.1 | 34.6 |
| Isanti | 900 | 9,717 | 2,301 | 1,239 | 2,089 | 225 | 251 | 362 | 37,239 | 805 | 46.0 | 3.4 | 34.7 |
| Itasca | 1,142 | 13,326 | 3,571 | 953 | 2,037 | 410 | 416 | 548 | 41,142 | 337 | 27.0 | 2.4 | 36.1 |
| Jackson | 297 | 4,151 | 417 | 1,386 | 293 | 78 | 78 | 172 | 41,513 | 799 | 25.4 | 13.0 | 55.1 |
| Kanabec | 309 | 3,178 | 1,169 | 485 | 501 | 96 | 63 | 135 | 42,349 | 624 | 33.2 | 1.6 | 41.2 |
| Kandiyohi | 1,400 | 19,123 | 5,330 | 2,674 | 2,825 | 537 | 611 | 761 | 39,795 | 1,220 | 32.3 | 9.3 | 39.9 |
| Kittson | 147 | 1,182 | 329 | 129 | 264 | 56 | 18 | 43 | 36,132 | 528 | 8.3 | 28.6 | 47.2 |
| Koochiching | 387 | 3,752 | 726 | 582 | 741 | 171 | 58 | 147 | 39,293 | 181 | 9.9 | 6.1 | 36.7 |
| Lac qui Parle | 204 | 1,779 | 583 | 224 | 224 | 89 | 11 | 58 | 32,805 | 853 | 20.2 | 14.8 | 44.9 |
| Lake | 309 | 3,283 | 547 | 673 | 429 | 107 | 43 | 132 | 40,100 | 42 | 54.8 | NA | 32.9 |
| Lake of the Woods | 155 | 1,407 | 162 | 215 | 155 | 23 | 22 | 55 | 38,911 | 134 | 14.2 | 14.2 | 35.2 |
| Le Sueur | 705 | 7,735 | 909 | 3,312 | 715 | 215 | 177 | 361 | 46,621 | 937 | 39.7 | 6.7 | 35.4 |
| Lincoln | 193 | 1,236 | 403 | 24 | 220 | 55 | 28 | 40 | 32,400 | 672 | 20.5 | 13.2 | 48.9 |
| Lyon | 808 | 12,617 | 2,430 | 1,892 | 1,920 | 1,146 | 383 | 529 | 41,904 | 893 | 19.9 | 12.1 | 50.6 |
| McLeod | 940 | 15,419 | 2,960 | 5,198 | 2,120 | 415 | 372 | 684 | 44,391 | 880 | 38.4 | 7.8 | 46.1 |
| Mahnomen | 101 | 1,759 | 357 | NA | 156 | 52 | 15 | 58 | 33,003 | 311 | 19.0 | 19.9 | 46.4 |
| Marshall | 249 | 1,512 | 222 | 343 | 233 | 124 | 30 | 65 | 42,708 | 1,086 | 7.2 | 22.4 | 42.9 |
| Martin | 618 | 7,582 | 1,562 | 1,404 | 1,374 | 356 | 170 | 333 | 43,897 | 911 | 23.6 | 16.4 | 56.7 |
| Meeker | 576 | 6,421 | 1,334 | 1,868 | 947 | 180 | 96 | 259 | 40,401 | 1,028 | 35.1 | 7.4 | 44.2 |
| Mille Lacs | 707 | 8,908 | 2,624 | 818 | 1,059 | 225 | 272 | 311 | 34,863 | 707 | 40.0 | 2.7 | 36.3 |
| Morrison | 888 | 7,916 | 1,592 | 1,118 | 1,494 | 235 | 157 | 296 | 37,351 | 1,760 | 25.0 | 2.4 | 46.7 |
| Mower | 817 | 13,809 | 2,600 | 3,500 | 1,832 | 280 | 216 | 703 | 50,912 | 1,068 | 34.9 | 12.5 | 47.7 |
| Murray | 283 | 2,622 | 434 | 676 | 337 | 174 | 69 | 90 | 34,510 | 864 | 24.1 | 14.0 | 53.2 |
| Nicollet | 663 | 14,111 | 2,757 | 4,031 | 1,187 | 213 | 416 | 650 | 46,035 | 689 | 22.5 | 7.7 | 51.9 |
| Nobles | 591 | 8,959 | 1,495 | 2,859 | 1,484 | 253 | 151 | 391 | 43,682 | 885 | 27.0 | 15.0 | 55.6 |
| Norman | 180 | 1,322 | 409 | NA | 247 | 97 | 27 | 58 | 43,595 | 505 | 13.5 | 35.8 | 55.1 |
| Olmsted | 3,594 | 100,499 | 21,031 | 4,345 | 11,289 | 1,680 | 35,918 | 5,898 | 58,686 | 1,139 | 43.2 | 6.2 | 43.2 |
| Otter Tail | 1,716 | 18,517 | 3,956 | 3,984 | 2,573 | 539 | 403 | 722 | 38,997 | 2,544 | 19.4 | 7.2 | 41.7 |
| Pennington | 375 | 9,538 | 1,164 | 1,459 | 1,090 | 158 | 108 | 429 | 44,939 | 409 | 15.4 | 20.8 | 44.6 |
| Pine | 590 | 6,927 | 1,145 | 379 | 1,118 | 177 | 196 | 209 | 30,219 | 823 | 26.6 | 1.8 | 43.7 |
| Pipestone | 328 | 3,127 | 696 | 516 | 519 | 96 | 94 | 114 | 36,322 | 595 | 32.9 | 8.9 | 49.9 |
| Polk | 741 | 9,215 | 2,327 | 1,792 | 1,396 | 249 | 190 | 356 | 38,600 | 1,258 | 14.2 | 25.4 | 56.8 |
| Pope | 373 | 4,196 | 714 | 1,005 | 476 | 123 | 248 | 181 | 43,193 | 837 | 25.9 | 12.1 | 42.7 |
| Ramsey | 13,549 | 304,348 | 63,668 | 23,920 | 25,740 | 18,953 | 14,849 | 18,634 | 61,226 | 55 | 96.4 | NA | 59.8 |
| Red Lake | 95 | 600 | 95 | NA | 125 | 45 | 21 | 20 | 33,687 | 263 | 11.0 | 19.8 | 48.1 |
| Redwood | 540 | 5,730 | 993 | 899 | 685 | 372 | 79 | 217 | 37,865 | 1,134 | 23.2 | 12.0 | 55.1 |
| Renville | 461 | 4,025 | 855 | 862 | 473 | 177 | 155 | 173 | 42,870 | 1,026 | 24.2 | 19.3 | 54.6 |
| Rice | 1,529 | 25,301 | 3,481 | 4,367 | 2,716 | 396 | 527 | 1,000 | 39,524 | 1,242 | 42.1 | 3.0 | 37.7 |
| Rock | 288 | 2,722 | 849 | 239 | 421 | 247 | 90 | 94 | 34,497 | 701 | 26.5 | 9.3 | 53.0 |
| Roseau | 410 | 6,816 | 781 | 3,558 | 780 | 179 | 71 | 297 | 43,545 | 842 | 14.4 | 16.2 | 39.0 |
| St. Louis | 5,395 | 88,309 | 24,236 | 4,729 | 12,388 | 3,517 | 3,303 | 4,087 | 46,286 | 779 | 27.3 | 1.8 | 36.1 |
| Scott | 3,426 | 50,693 | 5,283 | 7,912 | 6,187 | 663 | 1,966 | 2,564 | 50,574 | 740 | 50.0 | 2.7 | 32.4 |
| Sherburne | 2,170 | 23,074 | 3,538 | 4,101 | 3,364 | 447 | 737 | 1,097 | 47,532 | 501 | 47.7 | 5.8 | 40.7 |
| Sibley | 372 | 3,309 | 550 | 1,096 | 365 | 88 | 54 | 129 | 38,879 | 898 | 30.7 | 10.9 | 49.2 |
| Stearns | 4,410 | 85,227 | 18,005 | 11,990 | 11,487 | 4,043 | 2,387 | 4,027 | 47,254 | 2,951 | 24.4 | 2.3 | 49.8 |
| Steele | 1,002 | 20,521 | 2,672 | 5,733 | 2,673 | 2,304 | 319 | 985 | 48,002 | 746 | 39.9 | 9.9 | 47.4 |
| Stevens | 305 | 4,313 | 1,307 | 1,015 | 566 | 109 | 92 | 181 | 42,003 | 553 | 27.3 | 19.7 | 52.9 |
| Swift | 307 | 2,976 | 598 | 722 | 343 | 87 | 75 | 114 | 38,434 | 760 | 25.0 | 13.8 | 49.4 |
| Todd | 559 | 5,644 | 1,596 | 1,594 | 677 | 225 | 75 | 228 | 40,475 | 1,604 | 22.1 | 2.1 | 42.0 |
| Traverse | 117 | 738 | 228 | 39 | 189 | 44 | 8 | 31 | 41,636 | 411 | 22.4 | 30.9 | 59.1 |
| Wabasha | 555 | 5,296 | 970 | 1,266 | 783 | 169 | 121 | 197 | 37,114 | 809 | 22.9 | 4.8 | 50.5 |
| Wadena | 396 | 3,806 | 914 | 290 | 721 | 139 | 83 | 148 | 39,007 | 516 | 19.8 | 4.1 | 37.3 |
| Waseca | 458 | 4,478 | 998 | 1,027 | 595 | 176 | 195 | 183 | 40,798 | 729 | 33.2 | 9.1 | 48.8 |
| Washington | 6,032 | 81,632 | 12,602 | 9,706 | 14,064 | 3,711 | 4,011 | 3,636 | 44,536 | 612 | 64.4 | 3.4 | 44.4 |
| Watonwan | 296 | 3,686 | 508 | 1,264 | 394 | 153 | 54 | 131 | 35,543 | 497 | 27.8 | 15.1 | 52.0 |
| Wilkin | 154 | 1,657 | 372 | 44 | 190 | 65 | 466 | 65 | 39,308 | 391 | 16.9 | 37.9 | 64.5 |
| Winona | 1,155 | 23,291 | 3,231 | 5,265 | 2,748 | 546 | 321 | 939 | 40,310 | 1,034 | 25.0 | 4.5 | 49.5 |
| Wright | 3,413 | 39,381 | 5,684 | 6,572 | 7,648 | 702 | 1,178 | 1,753 | 44,519 | 1,338 | 45.1 | 3.3 | 37.2 |
| Yellow Medicine | 324 | 3,920 | 887 | 217 | 433 | 111 | 876 | 199 | 50,678 | 852 | 23.8 | 13.6 | 47.1 |

| STATE County | Land in farms | | | | | Value of land and buildings (dollars) | | Value of machinery and equipment, average per farm (dollars) | Value of products sold: | | | | Organic farms (number) | Farms with internet access (percent) | Government payments | |
|---|---|---|---|---|---|---|---|---|---|---|---|---|---|---|---|---|
| | | | Acres | | | | | | | | Percent from: | | | | | |
| | Acreage (1,000) | Percent change, 2012-2017 | Average size of farm | Total irrigated (1,000) | Total cropland (1,000) | Average per farm | Average per acre | | Total (mil dol) | Average per farm (acres) | Crops | Livestock and poultry products | | | Total ($1,000) | Percent of farms |
| | 117 | 118 | 119 | 120 | 121 | 122 | 123 | 124 | 125 | 126 | 127 | 128 | 129 | 130 | 131 | 132 |
| MINNESOTA— Cont'd | | | | | | | | | | | | | | | | |
| Fillmore | 376 | -11.1 | 268 | 0.7 | 287.6 | 1,488,701 | 5,554 | 185,396 | 291.7 | 208,242 | 48.4 | 51.6 | 44 | 73.7 | 8,723 | 67.1 |
| Freeborn | 394 | 3.1 | 366 | 1.7 | 374.9 | 2,208,230 | 6,030 | 250,401 | 364.0 | 338,289 | 60.7 | 39.3 | 12 | 84.3 | 11,605 | 74.9 |
| Goodhue | 385 | -3.4 | 263 | 6.0 | 325.8 | 1,633,399 | 6,204 | 219,974 | 348.6 | 238,596 | 49.9 | 50.1 | 41 | 83.4 | 11,779 | 59.3 |
| Grant | 324 | 7.0 | 619 | 5.9 | 303.0 | 2,724,030 | 4,403 | 298,988 | 190.3 | 363,141 | 75.6 | 24.4 | NA | 74.8 | 7,264 | 90.6 |
| Hennepin | 46 | -33.4 | 98 | 0.3 | 36.6 | 1,322,909 | 13,464 | 133,330 | 58.6 | 125,418 | 88.7 | 11.3 | 4 | 77.5 | 193 | 27.0 |
| Houston | 217 | -5.3 | 244 | 0.2 | 125.8 | 1,044,963 | 4,290 | 130,913 | 116.2 | 130,386 | 38.7 | 61.3 | 22 | 78.0 | 3,372 | 68.5 |
| Hubbard | 95 | -19.1 | 246 | 24.5 | 53.4 | 651,843 | 2,646 | 107,044 | 44.2 | 115,221 | 88.6 | 11.4 | 2 | 82.8 | 495 | 18.5 |
| Isanti | 132 | -7.0 | 164 | 4.0 | 96.3 | 639,469 | 3,887 | 100,672 | 48.7 | 60,463 | 74.2 | 25.8 | 10 | 81.7 | 635 | 12.5 |
| Itasca | 72 | -14.7 | 213 | D | 36.6 | 604,662 | 2,842 | 67,203 | 8.0 | 23,751 | 64.7 | 35.3 | NA | 77.7 | D | 5.9 |
| Jackson | 356 | -0.4 | 446 | D | 335.5 | 2,870,744 | 6,438 | 308,732 | 314.5 | 393,630 | 58.2 | 41.8 | 1 | 88.2 | 9,173 | 78.2 |
| Kanabec | 119 | -7.8 | 190 | 2.2 | 66.2 | 504,920 | 2,652 | 89,170 | 29.8 | 47,808 | 60.2 | 39.8 | 7 | 69.2 | 109 | 19.7 |
| Kandiyohi | 456 | 9.8 | 374 | 26.6 | 402.8 | 1,949,625 | 5,218 | 260,271 | 424.1 | 347,605 | 47.0 | 53.0 | 7 | 76.4 | 10,884 | 81.2 |
| Kittson | 479 | 1.9 | 908 | 2.4 | 408.6 | 2,041,371 | 2,249 | 319,681 | 128.3 | 243,081 | 95.6 | 4.4 | NA | 76.3 | 12,450 | 91.5 |
| Koochiching | 56 | 4.5 | 308 | D | 28.2 | 444,918 | 1,443 | 85,478 | 6.9 | 38,050 | 64.6 | 35.4 | NA | 72.4 | 320 | 14.4 |
| Lac qui Parle | 420 | -6.0 | 492 | 5.1 | 375.2 | 2,204,166 | 4,478 | 275,311 | 249.9 | 292,939 | 68.3 | 31.7 | 11 | 81.2 | 6,064 | 88.2 |
| Lake | 4 | -4.6 | 85 | 0.0 | 1.4 | 308,042 | 3,629 | 49,065 | 0.4 | 8,524 | 83.0 | 17.0 | 2 | 92.9 | 41 | 7.1 |
| Lake of the Woods | 91 | 1.1 | 681 | 1.2 | 67.8 | 1,156,249 | 1,697 | 172,314 | 17.3 | 128,940 | 95.9 | 4.1 | NA | 67.2 | 772 | 46.3 |
| Le Sueur | 249 | 3.1 | 266 | 1.8 | 224.2 | 1,643,885 | 6,175 | 180,636 | 181.4 | 193,551 | 64.0 | 36.0 | 8 | 75.5 | 4,652 | 72.8 |
| Lincoln | 298 | 2.4 | 443 | D | 269.8 | 2,143,527 | 4,836 | 261,768 | 186.0 | 276,847 | 60.5 | 39.5 | 3 | 75.3 | 7,124 | 80.4 |
| Lyon | 395 | -4.3 | 442 | 0.4 | 366.3 | 2,589,529 | 5,852 | 300,380 | 412.3 | 461,736 | 44.3 | 55.7 | 3 | 84.3 | 5,565 | 63.0 |
| McLeod | 269 | 1.8 | 305 | 0.1 | 248.9 | 1,747,482 | 5,724 | 226,601 | 185.6 | 210,928 | 69.3 | 30.7 | 9 | 78.5 | 4,093 | 66.4 |
| Mahnomen | 221 | 2.5 | 711 | D | 186.3 | 2,102,461 | 2,955 | 251,243 | 70.1 | 225,392 | 94.4 | 5.6 | 2 | 76.8 | 2,742 | 66.9 |
| Marshall | 902 | 10.0 | 831 | D | 824.1 | 2,056,182 | 2,474 | 274,618 | 261.5 | 240,750 | 94.2 | 5.8 | 2 | 71.3 | 18,205 | 85.7 |
| Martin | 449 | 4.8 | 493 | 0.5 | 434.3 | 3,308,478 | 6,712 | 386,978 | 635.5 | 697,610 | 41.8 | 58.2 | 4 | 86.2 | 8,613 | 79.7 |
| Meeker | 301 | -0.8 | 293 | 8.8 | 262.1 | 1,521,946 | 5,190 | 204,806 | 265.2 | 257,930 | 44.7 | 55.3 | 7 | 76.7 | 5,200 | 72.9 |
| Mille Lacs | 126 | -1.5 | 178 | 0.5 | 86.9 | 552,471 | 3,102 | 103,477 | 43.9 | 62,139 | 52.9 | 47.1 | 6 | 77.2 | 190 | 24.0 |
| Morrison | 382 | -12.4 | 217 | 30.3 | 232.8 | 719,062 | 3,310 | 156,473 | 394.7 | 224,273 | 17.3 | 82.7 | 10 | 75.7 | 2,301 | 39.8 |
| Mower | 447 | -0.6 | 419 | 5.0 | 420.9 | 2,852,833 | 6,813 | 279,549 | 413.2 | 386,915 | 58.7 | 41.3 | 15 | 84.5 | 10,064 | 74.3 |
| Murray | 395 | -3.1 | 457 | D | 362.6 | 2,819,738 | 6,166 | 316,318 | 337.8 | 391,005 | 54.5 | 45.5 | 10 | 82.1 | 5,522 | 65.5 |
| Nicollet | 265 | -3.4 | 384 | 0.0 | 249.5 | 2,691,522 | 7,002 | 299,514 | 339.3 | 492,462 | 40.6 | 59.4 | 1 | 83.5 | 2,277 | 47.8 |
| Nobles | 414 | 8.9 | 468 | 0.1 | 392.3 | 3,236,081 | 6,911 | 399,982 | 519.0 | 586,400 | 39.6 | 60.4 | 1 | 82.4 | 8,236 | 81.5 |
| Norman | 526 | -1.2 | 1,041 | 1.4 | 498.9 | 3,437,726 | 3,301 | 410,892 | 218.3 | 432,202 | 94.3 | 5.7 | 4 | 76.0 | 5,243 | 78.4 |
| Olmsted | 286 | 8.1 | 251 | 0.1 | 239.1 | 1,671,227 | 6,657 | 184,486 | 214.4 | 188,248 | 56.7 | 43.3 | 15 | 83.5 | 9,079 | 61.1 |
| Otter Tail | 794 | -10.0 | 312 | 75.5 | 576.2 | 927,172 | 2,969 | 150,686 | 349.9 | 137,547 | 57.4 | 42.6 | 33 | 72.9 | 8,675 | 64.2 |
| Pennington | 286 | 5.2 | 699 | D | 249.2 | 1,347,472 | 1,928 | 231,758 | 66.5 | 162,484 | 91.6 | 8.4 | NA | 77.0 | 5,293 | 75.3 |
| Pine | 160 | -21.3 | 195 | 0.4 | 81.7 | 460,529 | 2,364 | 73,400 | 39.0 | 47,361 | 38.6 | 61.4 | 20 | 75.8 | 174 | 18.1 |
| Pipestone | 240 | -0.8 | 403 | 2.2 | 210.6 | 2,352,844 | 5,833 | 343,986 | 326.1 | 547,988 | 30.0 | 70.0 | 5 | 88.1 | 1,582 | 49.2 |
| Polk | 1,023 | -6.6 | 813 | 5.6 | 938.5 | 2,667,646 | 3,280 | 391,409 | 429.8 | 341,630 | 93.8 | 6.2 | 18 | 76.1 | 17,113 | 77.8 |
| Pope | 333 | -0.3 | 398 | 32.0 | 280.3 | 1,648,542 | 4,144 | 254,858 | 199.3 | 238,106 | 63.8 | 36.2 | 8 | 75.7 | 4,173 | 74.8 |
| Ramsey | 1 | -10.8 | 12 | 0.1 | 0.4 | 312,488 | 26,646 | 45,670 | 3.0 | 53,655 | 74.7 | 25.3 | 8 | 94.5 | D | 1.8 |
| Red Lake | 209 | 5.1 | 794 | 1.1 | 189.3 | 1,881,249 | 2,370 | 305,925 | 65.6 | 249,426 | 89.0 | 11.0 | NA | 80.6 | 2,688 | 81.0 |
| Redwood | 524 | 0.5 | 462 | 0.0 | 494.9 | 2,882,387 | 6,239 | 315,623 | 453.2 | 399,613 | 58.2 | 41.8 | 4 | 80.2 | 6,220 | 66.6 |
| Renville | 624 | 0.4 | 608 | 1.2 | 598.2 | 3,990,664 | 6,560 | 376,170 | 609.2 | 593,752 | 61.1 | 38.9 | 1 | 80.9 | 5,570 | 84.6 |
| Rice | 226 | -4.3 | 182 | 2.8 | 192.6 | 1,249,054 | 6,857 | 145,442 | 205.0 | 165,042 | 49.6 | 50.4 | 13 | 76.9 | 6,228 | 62.0 |
| Rock | 288 | 2.6 | 411 | D | 266.4 | 3,133,759 | 7,631 | 322,110 | 419.1 | 597,825 | 34.2 | 65.8 | 1 | 89.7 | 2,105 | 73.9 |
| Roseau | 558 | 0.4 | 663 | 0.0 | 463.1 | 1,155,325 | 1,743 | 183,983 | 129.5 | 153,853 | 84.3 | 15.7 | 3 | 78.3 | 10,057 | 71.9 |
| St. Louis | 139 | 9.0 | 178 | D | 67.1 | 354,725 | 1,992 | 61,899 | 16.1 | 20,718 | 57.1 | 42.9 | 8 | 77.3 | 48 | 3.3 |
| Scott | 116 | -18.2 | 156 | 0.0 | 98.0 | 1,184,671 | 7,590 | 120,351 | 75.6 | 102,122 | 58.9 | 41.1 | 15 | 80.4 | 900 | 38.9 |
| Sherburne | 103 | -8.7 | 205 | 36.4 | 78.5 | 1,055,081 | 5,155 | 187,885 | 89.6 | 178,838 | 83.9 | 16.1 | 4 | 84.2 | 1,068 | 25.0 |
| Sibley | 350 | 1.1 | 390 | 0.5 | 326.6 | 2,629,735 | 6,746 | 283,199 | 318.7 | 354,924 | 57.2 | 42.8 | 8 | 82.4 | 4,299 | 62.0 |
| Stearns | 651 | -14.1 | 221 | 51.8 | 515.9 | 1,135,611 | 5,149 | 185,876 | 748.0 | 253,466 | 23.9 | 76.1 | 57 | 78.0 | 5,926 | 56.4 |
| Steele | 251 | 5.6 | 337 | 2.0 | 233.5 | 2,088,084 | 6,201 | 251,529 | 251.8 | 337,586 | 60.7 | 39.3 | 11 | 79.5 | 5,216 | 71.0 |
| Stevens | 330 | 3.4 | 597 | 21.3 | 308.0 | 3,032,762 | 5,077 | 289,508 | 327.4 | 592,118 | 42.6 | 57.4 | 4 | 85.7 | 3,945 | 75.9 |
| Swift | 345 | -4.4 | 454 | 26.8 | 316.4 | 2,289,023 | 5,043 | 305,866 | 284.2 | 373,895 | 53.0 | 47.0 | NA | 81.2 | 4,405 | 84.3 |
| Todd | 333 | -15.4 | 208 | 12.9 | 210.5 | 570,580 | 2,745 | 111,342 | 179.5 | 111,884 | 31.6 | 68.4 | 23 | 70.4 | 1,987 | 45.0 |
| Traverse | 365 | 4.6 | 887 | 0.8 | 350.9 | 4,415,585 | 4,979 | 484,191 | 210.5 | 512,088 | 78.3 | 21.7 | NA | 76.6 | 4,735 | 83.5 |
| Wabasha | 231 | -6.1 | 285 | 1.7 | 175.8 | 1,574,440 | 5,519 | 244,686 | 186.3 | 230,295 | 43.0 | 57.0 | 11 | 83.3 | 6,359 | 70.1 |
| Wadena | 128 | -13.9 | 249 | 15.7 | 75.3 | 546,823 | 2,200 | 88,995 | 52.8 | 102,411 | 56.5 | 43.5 | 15 | 71.1 | 865 | 41.3 |
| Waseca | 247 | 6.6 | 339 | D | 231.0 | 2,243,731 | 6,621 | 279,994 | 275.0 | 377,284 | 48.2 | 51.8 | 9 | 77.9 | 5,793 | 80.9 |
| Washington | 76 | -5.9 | 124 | 4.1 | 57.8 | 1,081,821 | 8,695 | 111,931 | 59.8 | 97,676 | 90.7 | 9.3 | 10 | 83.0 | 1,036 | 25.0 |
| Watonwan | 252 | 6.5 | 508 | 6.8 | 241.1 | 3,604,518 | 7,097 | 382,501 | 269.5 | 542,310 | 54.2 | 45.8 | NA | 82.3 | 9,171 | 87.3 |
| Wilkin | 428 | -3.6 | 1,095 | 1.2 | 414.6 | 4,239,436 | 3,872 | 442,751 | 185.6 | 474,673 | 99.5 | 0.5 | 3 | 79.0 | 4,555 | 85.9 |
| Winona | 269 | -3.1 | 260 | 0.0 | 180.8 | 1,494,722 | 5,753 | 181,789 | 228.2 | 220,662 | 30.6 | 69.4 | 58 | 78.9 | 3,952 | 60.3 |
| Wright | 241 | -16.5 | 180 | 6.8 | 203.7 | 1,158,403 | 6,441 | 150,834 | 196.5 | 146,867 | 57.5 | 42.5 | 20 | 79.8 | 1,301 | 38.1 |
| Yellow Medicine | 384 | -2.9 | 450 | 0.5 | 355.3 | 2,435,244 | 5,408 | 275,914 | 256.4 | 300,971 | 66.4 | 33.6 | 5 | 81.3 | 5,699 | 74.2 |

# Water Use, Wholesale Trade, Retail Trade, and Real Estate

| STATE County | Water use, 2015 | | Wholesale Trade[1], 2017 | | | | Retail Trade[2], 2017 | | | | Real estate and rental and leasing,[2] 2017 | | | |
|---|---|---|---|---|---|---|---|---|---|---|---|---|---|---|
| | Public supply water withdrawn (mil gal/day) | Public supply gallons withdrawn per person per day | Number of establishments | Number of employees | Sales (mil dol) | Average payroll (mil dol) | Number of establishments | Number of employees | Sales (mil dol) | Average payroll (mil dol) | Number of establishments | Number of employees | Sales (mil dol) | Average payroll (mil dol) |
| | 133 | 134 | 135 | 136 | 137 | 138 | 139 | 140 | 141 | 142 | 143 | 144 | 145 | 146 |
| MINNESOTA— Cont'd | | | | | | | | | | | | | | |
| Fillmore | 1.28 | 61.4 | 26 | 235 | 304.5 | 13.0 | 90 | 767 | 201.1 | 18.4 | 7 | 13 | 2.2 | 0.4 |
| Freeborn | 3.49 | 114.0 | 45 | 597 | 440.2 | 35.3 | 136 | 1,974 | 524.8 | 51.9 | 18 | 42 | 6.7 | 0.9 |
| Goodhue | 3.09 | 66.5 | D | D | D | D | 202 | 2,504 | 850.5 | 73.7 | 37 | 160 | 36.8 | 5.8 |
| Grant | 0.31 | 52.5 | 9 | 112 | 253.2 | 7.4 | 31 | 253 | 79.5 | 6.1 | NA | NA | NA | NA |
| Hennepin | 73.39 | 60.0 | 1,959 | 38,420 | 45,908.8 | 2,826.3 | 4,115 | 76,351 | 31,368.3 | 2,299.6 | 2,575 | 18,875 | 5,522.3 | 1,132.4 |
| Houston | 0.86 | 45.8 | 14 | 154 | 90.7 | 6.5 | 55 | 579 | 111.5 | 12.1 | 6 | 14 | 1.7 | 0.4 |
| Hubbard | 0.67 | 32.4 | D | D | D | D | 101 | 998 | 236.9 | 25.2 | D | D | D | 1.4 |
| Isanti | 1.32 | 34.3 | 22 | 127 | 29.1 | 4.7 | 108 | 1,905 | 577.9 | 49.7 | 33 | 120 | 17.1 | 2.8 |
| Itasca | 2.02 | 44.5 | D | D | D | D | 194 | 2,140 | 614.7 | 57.4 | 27 | 93 | 13.2 | 2.6 |
| Jackson | 0.56 | 55.6 | 13 | 282 | 171.0 | 11.5 | 43 | 356 | 82.3 | 7.3 | 4 | D | 1.1 | D |
| Kanabec | 0.38 | 24.0 | 6 | 24 | 8.0 | 0.9 | 47 | 530 | 135.0 | 11.8 | 3 | 53 | 2.9 | 1.7 |
| Kandiyohi | 4.56 | 107.2 | 68 | 922 | 1,156.7 | 55.6 | 199 | 2,985 | 865.8 | 81.2 | 45 | 95 | 18.5 | 3.4 |
| Kittson | 0.56 | 126.6 | 13 | 97 | 126.7 | 5.5 | 28 | 260 | 61.6 | 5.5 | NA | NA | NA | NA |
| Koochiching | 0.71 | 55.3 | 10 | 70 | 23.9 | 2.1 | 75 | 722 | 159.2 | 17.1 | 10 | 30 | 4.6 | 1.3 |
| Lac qui Parle | 0.37 | 54.0 | D | D | D | 9.2 | 36 | 240 | 48.9 | 5.1 | 3 | 7 | 0.5 | 0.1 |
| Lake | 0.78 | 73.4 | D | D | D | 1.8 | 46 | 416 | 149.9 | 11.0 | 6 | 19 | 1.3 | 0.5 |
| Lake of the Woods | 0.17 | 43.3 | 4 | 30 | 9.2 | 0.9 | 23 | 158 | 31.7 | 3.0 | 4 | 5 | 1.0 | 0.1 |
| Le Sueur | 2.32 | 83.9 | 25 | 262 | 123.9 | 12.8 | 82 | 735 | 138.2 | 12.6 | 23 | 68 | 8.4 | 1.6 |
| Lincoln | 0.99 | 171.5 | D | D | D | D | 32 | 216 | 50.9 | 4.5 | 4 | 9 | 0.3 | 0.1 |
| Lyon | 2.70 | 105.2 | 40 | 565 | 763.6 | 30.6 | 127 | 1,889 | 449.3 | 45.4 | 24 | 105 | 8.1 | 4.3 |
| McLeod | 2.74 | 76.3 | D | D | D | D | 141 | 2,154 | 513.6 | 49.2 | 28 | 93 | 13.7 | 2.9 |
| Mahnomen | 0.27 | 49.5 | D | D | D | 2.0 | 18 | 156 | 55.2 | 4.0 | NA | NA | NA | NA |
| Marshall | 0.45 | 47.8 | 19 | 141 | 302.2 | 8.0 | 29 | 263 | 110.7 | 8.4 | D | D | D | D |
| Martin | 1.61 | 80.4 | 37 | 668 | 1,165.7 | 60.1 | 92 | 1,341 | 296.5 | 30.5 | 16 | 35 | 3.3 | 1.2 |
| Meeker | 1.25 | 54.1 | 21 | 162 | 148.9 | 7.8 | 82 | 989 | 278.6 | 24.3 | D | D | D | 1.2 |
| Mille Lacs | 0.43 | 16.7 | 25 | 279 | 157.7 | 11.6 | 101 | 1,096 | 281.3 | 26.0 | 16 | 56 | 8.1 | 1.2 |
| Morrison | 1.95 | 59.5 | 18 | 228 | 81.9 | 9.6 | 125 | 1,547 | 344.6 | 38.1 | D | D | D | D |
| Mower | 5.87 | 150.1 | D | D | D | D | 116 | 1,705 | 352.0 | 38.0 | D | D | D | 0.7 |
| Murray | 0.55 | 65.4 | 13 | 178 | 299.4 | 8.6 | 44 | 276 | 57.0 | 5.3 | D | D | D | D |
| Nicollet | 4.22 | 126.5 | 31 | 385 | 268.7 | 21.0 | 81 | 1,103 | 327.3 | 33.8 | 12 | 43 | 4.5 | 0.6 |
| Nobles | 2.15 | 98.8 | D | D | D | D | 107 | 1,482 | 339.7 | 34.2 | D | D | D | 0.2 |
| Norman | 0.29 | 43.4 | 16 | 121 | 277.5 | 6.6 | 26 | 234 | 73.0 | 6.1 | D | D | D | D |
| Olmsted | 12.86 | 84.9 | 102 | 1,122 | 716.8 | 62.9 | 576 | 11,341 | 2,866.8 | 297.1 | 189 | 800 | 160.6 | 28.1 |
| Otter Tail | 3.44 | 59.6 | D | D | D | D | 253 | 2,742 | 793.1 | 69.3 | 44 | 84 | 29.3 | 3.0 |
| Pennington | 1.07 | 75.3 | D | D | D | D | 69 | 1,097 | 248.3 | 25.3 | 13 | 33 | 4.5 | 1.0 |
| Pine | 0.85 | 29.2 | 23 | 102 | 19.8 | 2.8 | 93 | 1,223 | 290.8 | 29.8 | D | D | D | D |
| Pipestone | 1.81 | 195.2 | 15 | 184 | 296.6 | 9.1 | 56 | 543 | 134.6 | 11.8 | 3 | D | 0.4 | D |
| Polk | 8.57 | 271.8 | 43 | 444 | 985.3 | 22.6 | 105 | 1,500 | 324.0 | 35.1 | 13 | 44 | 5.0 | 1.3 |
| Pope | 0.55 | 49.8 | 55 | 636 | 390.0 | 31.8 | 50 | 430 | 175.3 | 11.0 | 8 | 7 | 1.1 | 0.2 |
| Ramsey | 52.58 | 97.7 | 565 | 11,902 | 8,696.1 | 805.3 | 1,557 | 27,016 | 7,098.9 | 719.8 | 767 | 4,295 | 1,760.2 | 222.6 |
| Red Lake | 0.64 | 157.8 | 6 | D | 23.3 | D | 20 | 123 | 41.6 | 3.3 | NA | NA | NA | NA |
| Redwood | 0.97 | 62.7 | 28 | 405 | 439.1 | 24.7 | 55 | 713 | 178.0 | 17.1 | D | D | D | D |
| Renville | 1.02 | 68.5 | 30 | 300 | 509.2 | 18.5 | 62 | 472 | 123.5 | 10.3 | 4 | 3 | 0.6 | 0.1 |
| Rice | 5.54 | 84.7 | 55 | 1,237 | 2,046.3 | 73.5 | 194 | 2,636 | 712.2 | 69.0 | 63 | 224 | 64.8 | 8.4 |
| Rock | 1.51 | 157.3 | 17 | 173 | 274.6 | 9.5 | 36 | 397 | 112.5 | 9.6 | 9 | 21 | 2.2 | 0.7 |
| Roseau | 0.71 | 45.0 | 13 | 91 | 89.6 | 4.8 | 79 | 813 | 165.9 | 16.9 | D | D | D | D |
| St. Louis | 35.28 | 176.0 | 195 | 2,234 | 1,227.2 | 118.9 | 852 | 12,942 | 3,311.3 | 318.9 | 225 | 907 | 194.6 | 30.4 |
| Scott | 7.84 | 55.3 | 170 | 2,850 | 2,826.6 | 196.7 | 310 | 4,831 | 1,593.9 | 147.4 | 183 | 438 | 108.5 | 17.6 |
| Sherburne | 4.95 | 54.0 | 68 | 603 | 438.6 | 29.8 | 210 | 3,592 | 933.6 | 89.6 | 89 | 204 | 47.1 | 7.4 |
| Sibley | 1.45 | 97.5 | D | D | D | D | 50 | 332 | 62.5 | 5.9 | D | D | D | D |
| Stearns | 14.58 | 94.2 | 194 | 4,012 | 2,548.2 | 211.9 | 676 | 11,699 | 3,305.4 | 302.1 | 189 | 971 | 194.5 | 36.6 |
| Steele | 4.04 | 109.9 | 38 | 539 | 459.1 | 29.9 | 166 | 2,658 | 579.5 | 60.6 | 27 | 304 | 19.0 | 5.5 |
| Stevens | 0.79 | 80.6 | 19 | 139 | 277.5 | 7.3 | 50 | 564 | 234.9 | 16.6 | D | D | D | D |
| Swift | 0.79 | 84.6 | 16 | 206 | 427.2 | 13.3 | 40 | 340 | 75.9 | 7.4 | 6 | 64 | 3.4 | 1.6 |
| Todd | 1.18 | 48.6 | 16 | 123 | 134.7 | 5.0 | 100 | 729 | 162.5 | 14.7 | 18 | 30 | 3.3 | 0.5 |
| Traverse | 0.20 | 58.8 | D | D | D | D | 28 | 183 | 41.5 | 4.2 | NA | NA | NA | NA |
| Wabasha | 1.85 | 87.1 | 25 | 218 | 117.5 | 10.1 | 76 | 762 | 195.8 | 19.8 | 9 | 22 | 2.9 | 0.5 |
| Wadena | 0.93 | 67.0 | 17 | 546 | 290.1 | 22.7 | 71 | 725 | 189.1 | 18.7 | D | D | D | D |
| Waseca | 1.67 | 87.9 | 22 | 150 | 110.3 | 8.0 | 53 | 655 | 169.1 | 16.7 | 10 | 22 | 3.1 | 0.5 |
| Washington | 19.23 | 76.4 | 177 | 2,535 | 3,854.6 | 153.0 | 710 | 14,164 | 3,538.0 | 344.3 | 346 | 1,510 | 327.0 | 71.5 |
| Watonwan | 1.29 | 117.8 | 11 | 128 | 142.4 | 5.9 | 37 | 394 | 65.3 | 7.6 | 6 | 18 | 1.0 | 0.4 |
| Wilkin | 0.29 | 45.3 | D | D | D | D | 26 | 170 | 47.0 | 4.6 | 7 | 16 | 2.2 | 0.9 |
| Winona | 3.76 | 73.9 | 49 | 508 | 668.9 | 25.5 | 153 | 2,629 | 676.9 | 67.1 | 34 | 101 | 18.7 | 2.5 |
| Wright | 8.09 | 61.6 | 98 | 1,625 | 1,277.0 | 94.1 | 445 | 7,518 | 1,937.1 | 182.5 | 132 | 232 | 62.5 | 9.1 |
| Yellow Medicine | 2.24 | 226.8 | 12 | 170 | 731.6 | 8.7 | 48 | 425 | 120.0 | 10.8 | D | D | D | 0.1 |

1 Merchant wholesalers, except manufacturers' sales branches and offices.  2. Employer establishments.

| STATE County | Professional, scientific, and technical services, 2017 | | | | Manufacturing, 2017 | | | | Accommodation and food services, 2017 | | | |
|---|---|---|---|---|---|---|---|---|---|---|---|---|
| | Number of establish-ments | Number of employees | Sales (mil dol) | Average payroll (mil dol) | Number of establish-ments | Number of employees | Sales (mil dol) | Average payroll (mil dol) | Number of establis-hments | Number of employees | Sales (mil dol) | Annual payroll (mil dol) |
| | 147 | 148 | 149 | 150 | 151 | 152 | 153 | 154 | 155 | 156 | 157 | 158 |
| MINNESOTA— Cont'd | | | | | | | | | | | | |
| Fillmore | D | D | 14 | D | 43 | 764 | 248.0 | 35.8 | 68 | 474 | 20.3 | 5.7 |
| Freeborn | D | D | D | D | 53 | 2,679 | 962.3 | 123.4 | 68 | 940 | 41.7 | 12.3 |
| Goodhue | 77 | 836 | 149 | 68.4 | 85 | 4,918 | 1,589.3 | 256.5 | 106 | 2,870 | 373.1 | 81.6 |
| Grant | D | D | D | D | 9 | 102 | 20.4 | 3.9 | 10 | D | 2.6 | D |
| Hennepin | 6,988 | 86,259 | 17,796 | 7,221.5 | 1,614 | 76,502 | 23,595.6 | 5,056.8 | 3,025 | 73,216 | 4,655.1 | 1,492.8 |
| Houston | 19 | 93 | 10 | 3.6 | 28 | 378 | 83.9 | 17.3 | 34 | 266 | 9.9 | 2.4 |
| Hubbard | D | D | 8 | D | 29 | 896 | 369.4 | 38.3 | 81 | 550 | 36.3 | 10.5 |
| Isanti | D | D | D | D | 68 | 1,350 | 350.0 | 72.1 | D | D | D | 15.6 |
| Itasca | D | D | D | D | 42 | 1,172 | 564.6 | 72.1 | 109 | 1,319 | 70.0 | 20.7 |
| Jackson | 16 | 66 | 7 | 4.6 | 13 | 1,306 | 447.0 | 63.0 | 24 | 239 | 9.5 | 2.5 |
| Kanabec | 17 | 64 | 6 | 2.0 | 19 | 474 | 112.4 | 22.6 | 24 | 263 | 11.7 | 3.5 |
| Kandiyohi | D | D | D | D | 74 | 2,645 | 1,034.4 | 124.2 | 90 | 1,366 | 72.5 | 21.3 |
| Kittson | D | D | D | D | 11 | 204 | 356.6 | 11.3 | D | D | D | 0.7 |
| Koochiching | D | D | 6 | D | D | 671 | D | D | 43 | 544 | 26.7 | 7.7 |
| Lac qui Parle | D | D | D | D | 10 | 151 | 338.3 | 7.2 | 10 | 110 | 3.9 | 1.2 |
| Lake | D | D | D | D | 14 | 507 | 227.4 | 32.9 | 63 | 783 | 53.7 | 14.8 |
| Lake of the Woods | D | D | D | D | D | 204 | D | 15.8 | 39 | 459 | 32.8 | 8.8 |
| Le Sueur | D | D | 23 | D | 53 | 3,227 | 992.3 | 174.8 | D | D | D | 5.4 |
| Lincoln | D | D | D | D | 7 | 29 | 6.6 | 1.5 | 13 | D | 3.7 | D |
| Lyon | D | D | D | D | 29 | 1,847 | 1,082.0 | 91.0 | 65 | 1,070 | 44.1 | 13.6 |
| McLeod | D | D | D | D | 75 | 5,369 | 2,309.4 | 343.2 | 62 | 1,046 | 48.2 | 14.7 |
| Mahnomen | D | D | D | 0.3 | NA | NA | NA | NA | D | D | D | D |
| Marshall | D | D | 7 | D | 16 | 300 | 82.8 | 15.4 | D | D | D | 1.1 |
| Martin | 38 | 185 | 32 | 9.5 | 36 | 1,099 | 1,389.0 | 52.1 | 47 | 760 | 32.7 | 10.0 |
| Meeker | 26 | 110 | 14 | 4.6 | 53 | 1,661 | 933.2 | 83.7 | 41 | 436 | 16.8 | 4.9 |
| Mille Lacs | D | D | 39 | D | 42 | 764 | 155.0 | 33.7 | 73 | 1,815 | 197.2 | 38.0 |
| Morrison | 34 | 157 | 18 | 7.1 | 58 | 1,049 | 420.7 | 51.1 | 93 | 970 | 45.0 | 13.2 |
| Mower | D | D | D | D | 31 | 3,802 | 1,441.1 | 163.4 | 68 | 951 | 48.5 | 14.4 |
| Murray | D | D | D | D | D | D | D | D | 20 | 171 | 8.4 | 2.2 |
| Nicollet | 40 | 452 | 53 | 20.8 | 48 | 3,425 | 864.7 | 159.0 | 47 | 748 | 31.8 | 9.9 |
| Nobles | D | D | D | D | 27 | 2,538 | 2,082.0 | 164.7 | 40 | 565 | 28.3 | 8.0 |
| Norman | D | D | D | 1.1 | NA | NA | NA | NA | D | D | D | 0.6 |
| Olmsted | D | D | D | D | 95 | 4,029 | 2,082.4 | 218.5 | 348 | 8,096 | 496.6 | 152.7 |
| Otter Tail | 98 | 414 | 50 | 19.5 | 84 | 4,166 | 1,726.8 | 183.3 | 156 | 1,670 | 95.1 | 27.9 |
| Pennington | D | D | D | D | 17 | 1,977 | 918.2 | 89.2 | 33 | 877 | 56.9 | 14.0 |
| Pine | D | D | 16 | D | 26 | 325 | 57.7 | 13.8 | 66 | 2,313 | 268.9 | 48.9 |
| Pipestone | 12 | 102 | 79 | 4.5 | 13 | 473 | 120.6 | 17.9 | 21 | 310 | 14.1 | 3.7 |
| Polk | 41 | 187 | 25 | 10.2 | 36 | 1,545 | 1,207.3 | 73.8 | 71 | 1,066 | 44.0 | 14.0 |
| Pope | D | D | D | D | 25 | 726 | 195.3 | 44.9 | 29 | 238 | 13.2 | 3.6 |
| Ramsey | D | D | D | D | 567 | 23,768 | 6,540.7 | 1,464.6 | 1,187 | 24,210 | 1,343.3 | 427.7 |
| Red Lake | D | D | D | D | NA | NA | NA | NA | D | D | D | 0.6 |
| Redwood | D | D | D | D | 33 | 1,039 | 341.3 | 43.4 | 39 | 1,090 | 90.1 | 21.7 |
| Renville | D | D | D | D | 25 | 1,008 | 521.3 | 53.3 | 28 | 233 | 9.3 | 2.4 |
| Rice | 128 | 669 | 76 | 31.0 | 77 | 3,591 | 1,629.3 | 202.9 | 129 | 2,150 | 100.6 | 31.6 |
| Rock | D | D | 13 | D | 15 | 338 | 118.9 | 13.8 | 20 | 298 | 10.1 | 2.9 |
| Roseau | 23 | 73 | 7 | 2.4 | D | 3,477 | D | 164.5 | 33 | 545 | 35.2 | 8.9 |
| St. Louis | 372 | 3,177 | 459 | 189.6 | 200 | 4,084 | 1,340.3 | 249.0 | 550 | 10,353 | 544.3 | 168.1 |
| Scott | D | D | D | D | 177 | 7,532 | 3,252.0 | 499.5 | 220 | 7,164 | 681.3 | 167.3 |
| Sherburne | 173 | 732 | 82 | 32.3 | 158 | 3,732 | 1,106.4 | 215.8 | 125 | 2,379 | 112.9 | 33.8 |
| Sibley | D | D | D | D | 23 | 882 | 913.0 | 38.4 | 21 | 138 | 7.9 | 1.7 |
| Stearns | D | D | D | D | 249 | 11,838 | 4,314.8 | 609.9 | 374 | 6,985 | 319.0 | 94.4 |
| Steele | 51 | 205 | 36 | 7.0 | 60 | 4,836 | 1,526.4 | 253.9 | 83 | 1,568 | 66.3 | 20.8 |
| Stevens | 12 | 114 | 12 | 5.7 | 15 | 804 | 325.7 | 50.8 | 23 | 434 | 15.8 | 4.3 |
| Swift | 16 | 80 | 10 | 3.2 | 20 | 601 | 754.8 | 35.7 | D | D | D | D |
| Todd | D | D | 7 | D | 44 | 1,670 | 1,001.1 | 82.0 | 41 | 337 | 16.4 | 4.3 |
| Traverse | D | D | D | D | 5 | 27 | 5.6 | 1.3 | D | D | D | 0.2 |
| Wabasha | D | D | 14 | D | 25 | 1,093 | 511.8 | 56.0 | 65 | 572 | 25.5 | 7.2 |
| Wadena | 21 | 102 | 14 | 5.0 | 18 | 255 | 44.8 | 10.9 | 33 | 365 | 16.1 | 4.7 |
| Waseca | D | D | 19 | D | 25 | 1,674 | 914.9 | 103.8 | 37 | 373 | 16.0 | 4.4 |
| Washington | 843 | 3,933 | 640 | 263.1 | 203 | 9,111 | 6,677.1 | 606.6 | 443 | 9,926 | 507.1 | 160.3 |
| Watonwan | 8 | 43 | 5 | 1.8 | 15 | 1,401 | 502.1 | 51.2 | D | D | D | D |
| Wilkin | D | D | 20 | D | D | 49 | D | D | 8 | 90 | 3.2 | 0.8 |
| Winona | D | D | D | D | 102 | 5,393 | 1,565.7 | 297.3 | 128 | 2,240 | 88.7 | 25.3 |
| Wright | 319 | 1,096 | 167 | 52.0 | 207 | 6,179 | 1,650.5 | 349.1 | 209 | 3,640 | 161.7 | 51.6 |
| Yellow Medicine | D | D | 8 | D | 20 | 218 | 128.7 | 10.1 | 21 | 475 | 48.5 | 11.4 |

# Table B. States and Counties — Health Care and Social Assistance, Other Services, Nonemployer Businesses, and Residential Construction

| STATE County | Health care and social assistance, 2017 | | | | Other services, 2017 | | | | Nonemployer businesses, 2018 | | Value of residential construction authorized by building permits, 2020 | |
|---|---|---|---|---|---|---|---|---|---|---|---|---|
| | Number of establishments | Number of employees | Receipts (mil dol) | Annual payroll (mil dol) | Number of establishments | Number of employees | Receipts (mil dol) | Annual payroll (mil dol) | Number | Receipts (mil dol) | New construction ($1,000) | Number of housing units |
| | 159 | 160 | 161 | 162 | 163 | 164 | 165 | 166 | 167 | 168 | 169 | 170 |
| MINNESOTA— Cont'd | | | | | | | | | | | | |
| Fillmore | D | D | D | 24.4 | 49 | 142 | 16.5 | 3.4 | 1,709 | 74.9 | 13,719 | 76 |
| Freeborn | 72 | 2,190 | 202.7 | 104.0 | 65 | 316 | 23.2 | 5.9 | 1,861 | 90.4 | 2,698 | 13 |
| Goodhue | 152 | 3,462 | 334.1 | 141.8 | 102 | 498 | 53.8 | 13.5 | 3,097 | 154.3 | 32,405 | 176 |
| Grant | 17 | 368 | 32.0 | 11.8 | 16 | 49 | 2.5 | 0.7 | 547 | 24.4 | 388 | 2 |
| Hennepin | 4,470 | 143,909 | 17,502.6 | 7,313.5 | 2,834 | 25,569 | 3,790.7 | 976.0 | 110,717 | 5,836.2 | 1,761,533 | 7,694 |
| Houston | 46 | 1,416 | 67.0 | 37.5 | 40 | 107 | 15.2 | 2.9 | 1,384 | 64.2 | 11,127 | 49 |
| Hubbard | 49 | 836 | 101.2 | 41.0 | 36 | 150 | 17.3 | 3.5 | 1,676 | 68.0 | 6,730 | 31 |
| Isanti | 83 | 2,097 | 155.4 | 95.5 | 68 | 311 | 25.6 | 6.5 | 2,639 | 125.9 | 62,837 | 322 |
| Itasca | 167 | 3,477 | 307.0 | 134.3 | 79 | 420 | 66.4 | 13.0 | 2,921 | 113.7 | 34,386 | 204 |
| Jackson | 27 | 1,344 | 48.8 | 23.2 | 24 | 106 | 15.6 | 2.8 | 806 | 34.7 | 3,177 | 13 |
| Kanabec | 35 | 1,064 | 121.2 | 47.0 | 22 | 85 | 8.1 | 2.1 | 983 | 51.1 | 7,660 | 55 |
| Kandiyohi | 198 | 5,779 | 383.9 | 183.2 | 96 | 493 | 45.3 | 11.0 | 3,085 | 154.2 | 29,942 | 180 |
| Kittson | 10 | 349 | 21.4 | 11.0 | D | D | 3.7 | D | 341 | 14.1 | 212 | 1 |
| Koochiching | 50 | 721 | 63.0 | 26.1 | 20 | 72 | 4.8 | 1.3 | 841 | 30.6 | 3,305 | 24 |
| Lac qui Parle | 17 | 688 | 54.7 | 22.8 | 13 | 66 | 7.3 | 1.1 | 609 | 25.8 | 810 | 3 |
| Lake | 35 | 517 | 41.4 | 15.4 | 20 | 120 | 11.9 | 4.0 | 864 | 33.0 | 16,483 | 71 |
| Lake of the Woods | D | D | D | D | D | D | 6.4 | D | 382 | 15.4 | 6,799 | 30 |
| Le Sueur | 67 | 902 | 58.2 | 24.3 | 69 | 216 | 23.4 | 5.3 | 1,978 | 90.3 | 29,939 | 129 |
| Lincoln | 15 | 464 | 35.7 | 13.4 | D | D | D | 0.8 | 494 | 16.6 | 2,638 | 11 |
| Lyon | 95 | 2,057 | 177.6 | 77.8 | 59 | 234 | 23.2 | 5.7 | 1,839 | 88.9 | 8,120 | 32 |
| McLeod | 116 | 3,000 | 263.4 | 115.4 | 79 | 422 | 62.9 | 11.3 | 2,432 | 118.6 | 21,920 | 91 |
| Mahnomen | 9 | 399 | 16.7 | 9.1 | D | D | 4.7 | D | 319 | 14.3 | 0 | 0 |
| Marshall | 17 | 344 | 18.2 | 8.2 | 23 | 77 | 8.1 | 1.3 | 614 | 23.3 | 1,849 | 7 |
| Martin | 81 | 1,546 | 140.2 | 60.1 | 54 | 128 | 16.6 | 3.4 | 1,520 | 75.0 | 6,327 | 19 |
| Meeker | 59 | 1,519 | 110.9 | 46.0 | 45 | 174 | 13.1 | 2.8 | 1,717 | 73.0 | 14,712 | 87 |
| Mille Lacs | 77 | 2,646 | 233.7 | 92.8 | 49 | 230 | 19.3 | 4.1 | 1,671 | 71.5 | 20,896 | 104 |
| Morrison | 82 | 1,627 | 146.2 | 63.3 | 85 | 344 | 44.2 | 9.1 | 2,386 | 119.2 | 13,403 | 108 |
| Mower | 100 | 2,558 | 248.4 | 97.8 | 83 | 462 | 77.6 | 11.4 | 1,758 | 115.7 | 7,049 | 22 |
| Murray | 16 | 423 | 33.2 | 14.8 | 20 | 44 | 8.8 | 1.2 | 694 | 38.4 | 4,773 | 24 |
| Nicollet | 92 | 2,577 | 221.5 | 129.9 | 61 | 406 | 167.5 | 14.8 | 2,222 | 108.2 | 20,953 | 90 |
| Nobles | 62 | 1,288 | 87.4 | 41.8 | 52 | 307 | 31.6 | 8.4 | 1,226 | 68.6 | 3,180 | 12 |
| Norman | 16 | 406 | 32.5 | 14.0 | D | D | D | 0.7 | 477 | 22.9 | 1,991 | 6 |
| Olmsted | 479 | 21,541 | 2,776.3 | 947.8 | 269 | 2,055 | 196.6 | 57.9 | 10,219 | 486.6 | 153,090 | 673 |
| Otter Tail | 178 | 4,329 | 361.6 | 134.7 | 140 | 544 | 54.9 | 12.8 | 4,860 | 228.2 | 25,311 | 142 |
| Pennington | 43 | 1,193 | 172.7 | 51.3 | 39 | 238 | 16.2 | 4.1 | 879 | 32.8 | 8,548 | 97 |
| Pine | 75 | 1,329 | 83.1 | 38.3 | 42 | 217 | 20.6 | 4.7 | 1,721 | 75.7 | 22,435 | 118 |
| Pipestone | 24 | 703 | 45.4 | 20.9 | 22 | 65 | 6.8 | 1.2 | 728 | 33.3 | 3,203 | 13 |
| Polk | 93 | 2,370 | 222.6 | 94.5 | 41 | 246 | 19.6 | 4.6 | 2,259 | 97.6 | 14,776 | 68 |
| Pope | 25 | 676 | 54.0 | 25.7 | 29 | 83 | 10.4 | 2.1 | 1,029 | 45.6 | 9,791 | 40 |
| Ramsey | 2,077 | 64,912 | 7,010.9 | 2,999.2 | 1,147 | 10,406 | 1,731.1 | 400.3 | 39,771 | 1,704.3 | 384,185 | 2,929 |
| Red Lake | D | D | D | D | D | D | D | 0.3 | 246 | 10.0 | 225 | 1 |
| Redwood | 53 | 945 | 76.6 | 28.4 | 52 | 209 | 24.0 | 5.3 | 1,072 | 51.8 | 3,748 | 17 |
| Renville | 42 | 911 | 62.8 | 25.9 | 36 | 86 | 13.1 | 2.9 | 1,049 | 55.2 | 2,387 | 10 |
| Rice | 176 | 3,430 | 238.1 | 118.8 | 136 | 633 | 59.4 | 16.0 | 4,272 | 191.6 | 31,796 | 102 |
| Rock | 16 | 840 | 60.7 | 22.8 | 19 | 75 | 5.8 | 1.4 | 766 | 36.0 | 4,912 | 15 |
| Roseau | 33 | 785 | 71.5 | 30.3 | 35 | 127 | 15.2 | 2.8 | 1,166 | 44.4 | 4,488 | 25 |
| St. Louis | 805 | 25,628 | 2,868.0 | 1,231.6 | 382 | 2,431 | 251.1 | 67.2 | 12,014 | 479.8 | 52,137 | 297 |
| Scott | 266 | 5,664 | 566.2 | 226.3 | 235 | 1,855 | 134.2 | 49.3 | 11,234 | 578.7 | 214,364 | 614 |
| Sherburne | 178 | 3,802 | 233.2 | 104.9 | 151 | 764 | 90.2 | 21.5 | 6,717 | 312.1 | 115,650 | 570 |
| Sibley | 21 | 481 | 32.6 | 16.3 | 28 | 59 | 6.7 | 1.0 | 1,057 | 46.1 | 9,704 | 43 |
| Stearns | 477 | 17,582 | 2,185.9 | 923.7 | 375 | 2,214 | 249.5 | 68.7 | 11,153 | 593.5 | 123,928 | 498 |
| Steele | 134 | 2,639 | 251.6 | 110.5 | 87 | 528 | 69.3 | 15.1 | 2,437 | 103.4 | 29,171 | 161 |
| Stevens | 39 | 1,587 | 97.3 | 45.7 | 17 | 63 | 3.6 | 0.9 | 669 | 29.9 | 925 | 3 |
| Swift | 23 | 672 | 50.9 | 20.8 | D | D | D | 2.8 | 638 | 25.2 | 1,426 | 6 |
| Todd | 54 | 1,513 | 159.6 | 69.7 | 46 | 151 | 19.3 | 2.8 | 1,691 | 83.9 | 30,178 | 143 |
| Traverse | 12 | 299 | 28.1 | 10.2 | D | D | D | 0.4 | 281 | 11.2 | 1,225 | 5 |
| Wabasha | D | D | D | 34.8 | 40 | 156 | 13.6 | 3.5 | 1,550 | 77.7 | 18,621 | 76 |
| Wadena | 45 | 1,045 | 96.3 | 41.7 | 31 | 120 | 13.3 | 2.3 | 1,014 | 42.9 | 9,384 | 49 |
| Waseca | 56 | 998 | 64.9 | 26.7 | 34 | 111 | 15.6 | 2.7 | 1,149 | 56.8 | 6,112 | 24 |
| Washington | 685 | 10,930 | 1,319.5 | 550.7 | 456 | 2,995 | 254.8 | 81.7 | 19,068 | 896.6 | 635,359 | 2,684 |
| Watonwan | 27 | 567 | 52.3 | 20.5 | 27 | 137 | 17.0 | 2.9 | 648 | 27.6 | 3,124 | 15 |
| Wilkin | 22 | 465 | 49.1 | 20.1 | 9 | 24 | 4.1 | 0.8 | 486 | 43.4 | 1,682 | 7 |
| Winona | 133 | 3,084 | 232.7 | 105.5 | 76 | 359 | 46.3 | 9.6 | 2,930 | 124.9 | 18,956 | 79 |
| Wright | 301 | 5,727 | 488.7 | 223.9 | 252 | 1,302 | 151.3 | 35.4 | 9,887 | 443.1 | 387,459 | 1,388 |
| Yellow Medicine | 32 | 750 | 53.8 | 24.3 | 30 | 91 | 10.3 | 2.4 | 800 | 31.5 | 2,242 | 10 |

# Table B. States and Counties — Government Employment and Payroll, and Local Government Finances

| STATE County | Full-time equivalent employees | March payroll (dollars) | Administration, judicial, and legal | Police and corrections | Fire protection | Highways and transportation | Health and welfare | Natural resources and utilities | Education and libraries | Total (mil dol) | Intergovernmental (mil dol) | Taxes Total (mil dol) | Per capita Total (dollars) | Per capita Property (dollars) |
|---|---|---|---|---|---|---|---|---|---|---|---|---|---|---|
| | 171 | 172 | 173 | 174 | 175 | 176 | 177 | 178 | 179 | 180 | 181 | 182 | 183 | 184 |
| **MINNESOTA— Cont'd** | | | | | | | | | | | | | | |
| Fillmore | 728 | 3,091,737 | 12.3 | 10.2 | 1.1 | 8.5 | 11.1 | 6.9 | 49.2 | 90.0 | 46.2 | 27.9 | 1,331 | 1,235 |
| Freeborn | 1,078 | 4,540,042 | 9.0 | 11.3 | 2.5 | 6.0 | 12.5 | 7.8 | 49.9 | 149.1 | 78.1 | 46.9 | 1,534 | 1,326 |
| Goodhue | 1,835 | 8,024,257 | 9.3 | 11.4 | 2.1 | 4.1 | 8.2 | 5.3 | 57.4 | 249.5 | 110.4 | 87.5 | 1,889 | 1,870 |
| Grant | 314 | 1,187,787 | 14.0 | 6.1 | 0.0 | 8.9 | 8.7 | 4.9 | 57.1 | 33.9 | 19.1 | 9.5 | 1,598 | 1,512 |
| Hennepin | 43,507 | 233,875,110 | 8.5 | 12.1 | 2.2 | 3.1 | 10.5 | 6.3 | 54.9 | 7,525.3 | 2,620.1 | 2,562.6 | 2,054 | 1,799 |
| Houston | 825 | 3,164,378 | 10.2 | 8.8 | 0.2 | 4.1 | 7.5 | 3.3 | 64.0 | 95.9 | 56.5 | 26.7 | 1,427 | 1,395 |
| Hubbard | 683 | 2,507,742 | 11.1 | 11.3 | 0.4 | 5.6 | 8.9 | 3.5 | 57.4 | 86.4 | 40.4 | 30.3 | 1,445 | 1,418 |
| Isanti | 1,483 | 6,198,777 | 6.0 | 9.7 | 0.4 | 3.9 | 8.0 | 1.9 | 67.5 | 152.5 | 87.2 | 43.2 | 1,093 | 1,066 |
| Itasca | 2,318 | 9,160,293 | 7.6 | 7.8 | 0.4 | 9.1 | 16.9 | 3.5 | 53.6 | 324.9 | 119.6 | 71.4 | 1,583 | 1,547 |
| Jackson | 530 | 2,065,186 | 12.7 | 7.5 | 0.0 | 9.4 | 21.8 | 4.2 | 43.0 | 50.9 | 20.7 | 20.4 | 2,044 | 2,006 |
| Kanabec | 1,007 | 4,457,414 | 4.2 | 4.9 | 0.0 | 3.5 | 54.8 | 1.0 | 28.3 | 61.7 | 35.2 | 14.8 | 923 | 907 |
| Kandiyohi | 2,747 | 10,664,987 | 4.6 | 8.5 | 0.3 | 2.5 | 41.8 | 2.3 | 35.2 | 338.8 | 108.4 | 60.5 | 1,415 | 1,341 |
| Kittson | 264 | 995,915 | 12.6 | 5.9 | 0.0 | 9.7 | 5.5 | 10.4 | 52.2 | 31.7 | 15.7 | 8.4 | 1,981 | 1,963 |
| Koochiching | 531 | 2,050,764 | 7.7 | 8.8 | 1.4 | 7.2 | 12.3 | 9.1 | 50.8 | 55.3 | 30.4 | 16.1 | 1,286 | 1,187 |
| Lac qui Parle | 694 | 2,539,449 | 7.4 | 3.2 | 0.0 | 4.8 | 42.9 | 2.8 | 38.5 | 53.6 | 21.5 | 9.2 | 1,379 | 1,291 |
| Lake | 522 | 2,298,915 | 13.7 | 10.6 | 0.2 | 9.8 | 6.7 | 7.2 | 45.9 | 66.2 | 33.9 | 17.6 | 1,679 | 1,537 |
| Lake of the Woods | 213 | 792,289 | 15.8 | 10.2 | 0.0 | 9.0 | 8.4 | 7.2 | 44.9 | 27.5 | 17.3 | 5.4 | 1,458 | 1,425 |
| Le Sueur | 924 | 3,572,247 | 9.8 | 7.8 | 0.1 | 3.7 | 11.3 | 2.9 | 63.1 | 117.9 | 61.1 | 37.6 | 1,332 | 1,299 |
| Lincoln | 196 | 669,869 | 20.1 | 10.8 | 1.3 | 15.5 | 1.1 | 11.0 | 35.4 | 28.8 | 13.0 | 10.7 | 1,880 | 1,861 |
| Lyon | 1,419 | 5,698,091 | 5.5 | 6.8 | 0.3 | 4.9 | 1.0 | 9.6 | 69.3 | 171.8 | 87.9 | 39.2 | 1,513 | 1,404 |
| McLeod | 1,262 | 5,552,383 | 7.8 | 10.1 | 0.2 | 4.5 | 11.7 | 14.1 | 49.1 | 150.0 | 72.5 | 47.4 | 1,320 | 1,268 |
| Mahnomen | 419 | 1,553,453 | 8.8 | 6.5 | 0.2 | 3.3 | 23.9 | 1.5 | 54.4 | 45.9 | 28.2 | 6.9 | 1,237 | 1,233 |
| Marshall | 512 | 1,990,991 | 12.5 | 5.7 | 0.1 | 8.0 | 8.9 | 8.8 | 49.4 | 58.0 | 31.7 | 13.6 | 1,454 | 1,441 |
| Martin | 802 | 3,611,946 | 11.8 | 9.0 | 0.2 | 6.7 | 0.3 | 8.2 | 62.0 | 99.5 | 48.9 | 29.6 | 1,488 | 1,475 |
| Meeker | 1,166 | 4,241,950 | 6.5 | 7.5 | 0.1 | 4.6 | 7.5 | 3.6 | 68.6 | 157.4 | 72.7 | 33.0 | 1,434 | 1,421 |
| Mille Lacs | 1,212 | 5,029,294 | 5.6 | 11.0 | 0.2 | 3.0 | 7.3 | 2.9 | 67.3 | 127.9 | 79.6 | 36.2 | 1,399 | 1,364 |
| Morrison | 1,303 | 5,191,373 | 8.1 | 9.7 | 0.2 | 4.1 | 10.9 | 3.7 | 62.3 | 137.9 | 80.8 | 35.2 | 1,064 | 1,036 |
| Mower | 1,561 | 7,559,080 | 5.4 | 8.0 | 0.9 | 3.5 | 7.1 | 9.8 | 64.5 | 199.5 | 112.5 | 43.4 | 1,091 | 1,024 |
| Murray | 482 | 2,062,839 | 7.3 | 6.2 | 0.0 | 5.4 | 35.7 | 4.4 | 36.1 | 54.9 | 19.9 | 13.8 | 1,654 | 1,642 |
| Nicollet | 983 | 4,478,601 | 8.5 | 10.5 | 0.6 | 5.4 | 27.8 | 6.8 | 37.6 | 128.4 | 43.7 | 36.5 | 1,071 | 1,011 |
| Nobles | 1,185 | 4,298,914 | 7.7 | 8.5 | 0.2 | 4.6 | 7.4 | 5.9 | 61.5 | 127.4 | 75.8 | 29.1 | 1,343 | 1,230 |
| Norman | 436 | 1,582,516 | 10.5 | 4.2 | 0.3 | 9.3 | 5.7 | 7.7 | 60.3 | 40.5 | 25.0 | 10.9 | 1,661 | 1,627 |
| Olmsted | 5,602 | 27,791,782 | 9.6 | 10.8 | 2.9 | 3.3 | 10.6 | 9.4 | 52.2 | 853.5 | 359.8 | 265.0 | 1,713 | 1,447 |
| Otter Tail | 2,497 | 9,700,961 | 9.3 | 7.8 | 0.3 | 3.9 | 25.8 | 4.7 | 46.1 | 373.3 | 138.4 | 84.8 | 1,457 | 1,359 |
| Pennington | 739 | 2,999,461 | 6.6 | 8.3 | 1.5 | 3.8 | 9.4 | 10.0 | 54.8 | 130.2 | 37.1 | 19.6 | 1,383 | 1,332 |
| Pine | 951 | 3,821,837 | 10.3 | 12.4 | 0.3 | 5.3 | 9.3 | 0.5 | 60.4 | 118.1 | 66.8 | 32.8 | 1,126 | 1,099 |
| Pipestone | 654 | 2,685,480 | 7.4 | 4.4 | 0.1 | 4.8 | 36.8 | 2.6 | 43.4 | 52.3 | 32.0 | 12.6 | 1,381 | 1,277 |
| Polk | 1,559 | 6,278,987 | 7.4 | 13.3 | 1.7 | 3.6 | 11.7 | 8.9 | 48.3 | 193.8 | 103.4 | 46.2 | 1,462 | 1,381 |
| Pope | 686 | 3,078,501 | 4.2 | 5.0 | 0.0 | 7.2 | 52.4 | 2.0 | 28.0 | 109.5 | 28.1 | 17.7 | 1,614 | 1,587 |
| Ramsey | 21,340 | 125,504,794 | 7.1 | 9.9 | 3.4 | 6.0 | 7.7 | 5.0 | 59.1 | 3,345.0 | 1,468.3 | 870.8 | 1,598 | 1,453 |
| Red Lake | 280 | 1,002,315 | 11.4 | 5.6 | 0.0 | 4.7 | 5.7 | 2.7 | 69.7 | 27.6 | 16.0 | 6.1 | 1,532 | 1,525 |
| Redwood | 663 | 2,678,050 | 13.3 | 8.7 | 1.3 | 7.2 | 0.9 | 6.8 | 59.5 | 108.4 | 41.5 | 24.5 | 1,606 | 1,577 |
| Renville | 629 | 2,348,964 | 13.9 | 10.4 | 0.3 | 7.8 | 15.8 | 8.5 | 40.4 | 102.9 | 39.6 | 26.5 | 1,801 | 1,777 |
| Rice | 2,171 | 10,959,183 | 5.6 | 7.9 | 0.8 | 2.9 | 41.8 | 3.1 | 36.6 | 354.3 | 125.5 | 78.1 | 1,181 | 1,071 |
| Rock | 404 | 1,645,416 | 11.9 | 4.7 | 1.3 | 7.3 | 1.6 | 10.3 | 57.5 | 53.0 | 26.2 | 14.5 | 1,537 | 1,471 |
| Roseau | 707 | 2,928,463 | 8.4 | 5.7 | 0.2 | 5.9 | 6.7 | 4.0 | 67.7 | 64.1 | 38.4 | 17.2 | 1,125 | 1,116 |
| St. Louis | 9,040 | 40,619,148 | 7.9 | 13.5 | 3.2 | 7.3 | 10.0 | 8.9 | 45.2 | 1,139.6 | 562.0 | 300.0 | 1,502 | 1,293 |
| Scott | 4,646 | 23,159,356 | 8.9 | 8.6 | 0.5 | 3.1 | 6.7 | 4.4 | 65.9 | 650.4 | 316.1 | 231.1 | 1,588 | 1,458 |
| Sherburne | 3,328 | 16,214,184 | 5.7 | 12.1 | 0.5 | 2.7 | 5.2 | 2.9 | 69.4 | 430.4 | 228.3 | 138.3 | 1,464 | 1,430 |
| Sibley | 649 | 2,464,239 | 10.7 | 7.3 | 0.6 | 6.1 | 11.4 | 2.3 | 59.5 | 71.9 | 32.5 | 24.8 | 1,666 | 1,646 |
| Stearns | 5,585 | 25,416,178 | 9.4 | 11.9 | 1.9 | 3.4 | 8.5 | 4.0 | 57.4 | 715.5 | 368.4 | 214.6 | 1,355 | 1,191 |
| Steele | 1,417 | 6,167,793 | 8.4 | 12.7 | 0.8 | 4.3 | 2.2 | 11.0 | 58.7 | 176.8 | 94.2 | 53.7 | 1,459 | 1,432 |
| Stevens | 657 | 2,641,377 | 7.1 | 4.9 | 0.2 | 5.5 | 5.0 | 2.7 | 72.8 | 56.2 | 29.1 | 15.4 | 1,586 | 1,519 |
| Swift | 744 | 2,886,995 | 6.6 | 5.3 | 0.0 | 4.9 | 41.3 | 2.2 | 35.8 | 83.8 | 29.6 | 18.4 | 1,957 | 1,926 |
| Todd | 1,036 | 4,159,879 | 9.1 | 6.8 | 0.4 | 3.4 | 11.5 | 3.3 | 60.2 | 101.2 | 58.7 | 26.0 | 1,059 | 1,024 |
| Traverse | 111 | 369,011 | 6.6 | 4.2 | 1.7 | 4.2 | 0.0 | 5.6 | 77.7 | 32.3 | 16.7 | 9.5 | 2,890 | 2,884 |
| Wabasha | 753 | 3,138,464 | 10.0 | 13.9 | 0.2 | 5.6 | 10.9 | 6.2 | 52.7 | 95.5 | 45.8 | 31.2 | 1,445 | 1,405 |
| Wadena | 789 | 3,019,610 | 8.3 | 6.3 | 0.5 | 5.0 | 12.1 | 4.3 | 60.3 | 83.4 | 46.3 | 16.3 | 1,191 | 1,163 |
| Waseca | 737 | 2,944,545 | 7.3 | 10.0 | 0.7 | 6.4 | 3.5 | 5.1 | 65.6 | 103.0 | 53.0 | 29.9 | 1,600 | 1,516 |
| Washington | 6,926 | 34,001,217 | 7.1 | 9.5 | 1.6 | 2.8 | 6.1 | 3.7 | 67.8 | 976.1 | 432.2 | 373.0 | 1,460 | 1,344 |
| Watonwan | 595 | 2,176,624 | 8.9 | 7.0 | 0.4 | 6.9 | 10.5 | 6.3 | 57.1 | 54.2 | 29.7 | 13.8 | 1,261 | 1,168 |
| Wilkin | 341 | 1,324,740 | 10.1 | 9.3 | 0.0 | 8.3 | 10.4 | 5.9 | 54.7 | 45.4 | 20.3 | 13.6 | 2,154 | 2,145 |
| Winona | 1,414 | 6,227,334 | 8.1 | 9.9 | 2.2 | 4.1 | 9.2 | 6.0 | 58.1 | 176.7 | 95.1 | 51.6 | 1,017 | 970 |
| Wright | 4,348 | 20,124,567 | 6.7 | 7.7 | 0.4 | 3.1 | 5.5 | 3.0 | 69.9 | 581.1 | 293.6 | 179.1 | 1,334 | 1,279 |
| Yellow Medicine | 445 | 2,063,297 | 11.2 | 8.1 | 0.4 | 7.2 | 7.6 | 4.9 | 57.1 | 87.1 | 28.5 | 20.2 | 2,053 | 1,995 |

1. Based on the resident population estimated as of July 1 of the year shown.

# Table B. States and Counties — Local Government Finances, Government Employment, and Income Taxes

| STATE County | Direct general expenditure — Total (mil dol) | Per capita[1] (dollars) | Percent of total for: Education | Health and hospitals | Police protection | Public welfare | Highways | Debt outstanding — Total (mil dol) | Per capita[1] (dollars) | Government employment, 2019 — Federal civilian | Federal military | State and local | Individual income tax returns, 2018 — Number of returns | Mean adjusted gross income | Mean income tax |
|---|---|---|---|---|---|---|---|---|---|---|---|---|---|---|---|
| | 185 | 186 | 187 | 188 | 189 | 190 | 191 | 192 | 193 | 194 | 195 | 196 | 197 | 198 | 199 |
| **MINNESOTA— Cont'd** | | | | | | | | | | | | | | | |
| Fillmore | 111 | 5,312 | 48.0 | 1.9 | 3.0 | 3.0 | 12.8 | 155.0 | 7,387 | 67 | 72 | 1,224 | 9,950 | 55,575 | 4,696 |
| Freeborn | 153 | 5,003 | 42.0 | 1.5 | 7.5 | 7.6 | 14.7 | 90.8 | 2,968 | 77 | 103 | 1,453 | 14,790 | 54,927 | 4,960 |
| Goodhue | 282 | 6,090 | 45.4 | 1.8 | 5.6 | 4.5 | 14.0 | 288.2 | 6,224 | 126 | 158 | 4,243 | 23,830 | 67,906 | 6,974 |
| Grant | 33 | 5,515 | 47.6 | 1.8 | 2.4 | 2.9 | 9.9 | 30.4 | 5,137 | 22 | 20 | 397 | 2,930 | 55,490 | 4,939 |
| Hennepin | 8,491 | 6,804 | 30.8 | 17.6 | 5.5 | 3.0 | 7.7 | 9,298.7 | 7,452 | 13,570 | 4,697 | 87,808 | 650,220 | 99,393 | 14,993 |
| Houston | 99 | 5,301 | 53.7 | 1.7 | 3.6 | 4.3 | 12.7 | 84.2 | 4,508 | 68 | 64 | 1,059 | 9,440 | 61,095 | 5,856 |
| Hubbard | 85 | 4,028 | 39.4 | 0.0 | 6.6 | 9.6 | 16.6 | 94.7 | 4,518 | 42 | 74 | 1,134 | 9,880 | 56,749 | 5,066 |
| Isanti | 165 | 4,172 | 51.2 | 1.1 | 7.4 | 8.1 | 9.9 | 161.0 | 4,072 | 75 | 139 | 2,069 | 19,960 | 61,712 | 5,700 |
| Itasca | 322 | 7,141 | 32.1 | 21.4 | 3.7 | 11.1 | 13.0 | 255.6 | 5,669 | 162 | 153 | 3,329 | 21,090 | 58,088 | 5,355 |
| Jackson | 65 | 6,546 | 31.4 | 2.3 | 3.7 | 4.3 | 19.6 | 68.5 | 6,878 | 28 | 34 | 734 | 5,030 | 58,340 | 5,344 |
| Kanabec | 78 | 4,884 | 51.1 | 3.2 | 4.8 | 6.6 | 8.1 | 98.2 | 6,113 | 33 | 56 | 1,321 | 7,530 | 52,664 | 4,804 |
| Kandiyohi | 380 | 8,877 | 28.7 | 33.2 | 2.4 | 4.3 | 7.1 | 286.8 | 6,704 | 134 | 146 | 3,086 | 21,350 | 63,199 | 6,337 |
| Kittson | 33 | 7,664 | 33.0 | 0.3 | 3.8 | 4.8 | 25.1 | 24.0 | 5,633 | 38 | 15 | 289 | 2,130 | 57,175 | 5,099 |
| Koochiching | 56 | 4,479 | 45.5 | 11.2 | 4.2 | 3.6 | 11.4 | 52.6 | 4,208 | 181 | 42 | 765 | 5,990 | 50,661 | 4,222 |
| Lac qui Parle | 50 | 7,449 | 37.1 | 32.1 | 2.0 | 2.1 | 8.2 | 39.2 | 5,868 | 35 | 22 | 758 | 3,240 | 60,827 | 5,499 |
| Lake | 63 | 6,029 | 29.2 | 5.3 | 7.6 | 5.5 | 13.3 | 92.3 | 8,800 | 22 | 36 | 899 | 5,410 | 57,143 | 5,162 |
| Lake of the Woods | 28 | 7,461 | 27.4 | 0.4 | 6.4 | 6.0 | 30.2 | 226.8 | 60,847 | 30 | 13 | 270 | 2,020 | 51,650 | 4,280 |
| Le Sueur | 118 | 4,193 | 42.3 | 2.8 | 4.4 | 6.8 | 14.9 | 130.3 | 4,618 | 71 | 99 | 1,309 | 14,480 | 67,030 | 6,780 |
| Lincoln | 31 | 5,459 | 21.5 | 0.3 | 6.3 | 2.9 | 24.5 | 33.3 | 5,878 | 29 | 19 | 305 | 2,640 | 54,539 | 4,459 |
| Lyon | 183 | 7,040 | 55.0 | 2.7 | 7.2 | 1.4 | 10.8 | 216.3 | 8,342 | 117 | 85 | 3,033 | 11,870 | 60,887 | 5,547 |
| McLeod | 191 | 5,314 | 49.0 | 1.6 | 6.0 | 5.8 | 12.6 | 248.0 | 6,913 | 82 | 123 | 1,757 | 18,560 | 60,406 | 5,473 |
| Mahnomen | 48 | 8,677 | 45.7 | 20.6 | 6.0 | 5.7 | 8.9 | 29.4 | 5,286 | 23 | 19 | 1,260 | 2,320 | 44,400 | 3,255 |
| Marshall | 65 | 6,939 | 34.4 | 0.5 | 3.9 | 6.8 | 20.4 | 21.1 | 2,251 | 43 | 32 | 601 | 4,420 | 59,223 | 5,324 |
| Martin | 107 | 5,387 | 47.4 | 2.8 | 7.8 | 2.8 | 11.4 | 139.9 | 7,045 | 54 | 67 | 1,270 | 10,120 | 53,998 | 4,868 |
| Meeker | 157 | 6,820 | 42.7 | 19.9 | 4.8 | 5.6 | 8.6 | 113.9 | 4,943 | 66 | 79 | 1,256 | 11,260 | 61,050 | 5,640 |
| Mille Lacs | 146 | 5,655 | 62.5 | 0.7 | 5.1 | 7.7 | 6.8 | 151.6 | 5,863 | 67 | 89 | 3,450 | 12,520 | 52,266 | 4,447 |
| Morrison | 160 | 4,827 | 56.1 | 2.0 | 3.9 | 6.8 | 11.4 | 136.5 | 4,127 | 429 | 114 | 1,944 | 16,130 | 52,331 | 4,362 |
| Mower | 210 | 5,272 | 48.2 | 3.4 | 4.4 | 5.5 | 9.9 | 994.4 | 25,001 | 136 | 137 | 2,567 | 18,810 | 61,317 | 6,817 |
| Murray | 55 | 6,578 | 27.8 | 30.4 | 4.7 | 2.0 | 16.6 | 15.9 | 1,903 | 43 | 28 | 568 | 4,070 | 57,249 | 4,783 |
| Nicollet | 160 | 4,688 | 47.0 | 19.6 | 2.9 | 5.8 | 7.2 | 126.9 | 3,727 | 31 | 109 | 3,093 | 16,120 | 65,605 | 6,680 |
| Nobles | 123 | 5,667 | 45.2 | 3.0 | 5.1 | 5.9 | 14.0 | 83.6 | 3,858 | 84 | 74 | 1,477 | 10,080 | 52,126 | 4,129 |
| Norman | 39 | 5,862 | 39.2 | 2.1 | 4.1 | 5.2 | 21.9 | 15.7 | 2,377 | 29 | 22 | 431 | 3,040 | 56,044 | 4,906 |
| Olmsted | 939 | 6,071 | 36.2 | 1.4 | 4.8 | 6.7 | 10.4 | 3,233.2 | 20,896 | 766 | 547 | 8,208 | 80,190 | 81,670 | 10,264 |
| Otter Tail | 390 | 6,699 | 40.5 | 14.5 | 3.1 | 5.0 | 13.1 | 355.6 | 6,108 | 212 | 200 | 3,817 | 29,050 | 57,963 | 5,316 |
| Pennington | 130 | 9,200 | 71.0 | 0.8 | 4.4 | 3.9 | 4.5 | 121.2 | 8,548 | 55 | 48 | 1,443 | 6,950 | 57,654 | 5,343 |
| Pine | 126 | 4,307 | 43.3 | 8.8 | 4.3 | 6.0 | 13.6 | 206.2 | 7,071 | 284 | 97 | 2,903 | 13,110 | 50,614 | 4,133 |
| Pipestone | 53 | 5,836 | 54.2 | 0.7 | 2.8 | 2.7 | 11.8 | 32.8 | 3,581 | 47 | 31 | 907 | 4,440 | 51,677 | 4,135 |
| Polk | 219 | 6,935 | 37.0 | 4.8 | 4.2 | 7.3 | 10.3 | 169.4 | 5,354 | 98 | 105 | 2,604 | 14,560 | 58,546 | 5,115 |
| Pope | 63 | 5,766 | 35.7 | 18.4 | 4.1 | 6.5 | 12.3 | 54.1 | 4,919 | 43 | 38 | 877 | 5,710 | 61,170 | 6,051 |
| Ramsey | 3,624 | 6,650 | 37.9 | 1.9 | 5.9 | 5.7 | 5.2 | 6,544.0 | 12,010 | 2,379 | 1,940 | 55,356 | 270,980 | 71,614 | 8,493 |
| Red Lake | 31 | 7,637 | 58.7 | 0.2 | 4.2 | 4.2 | 15.2 | 12.6 | 3,131 | 14 | 14 | 278 | 1,880 | 54,248 | 4,161 |
| Redwood | 114 | 7,463 | 30.7 | 27.3 | 4.6 | 2.0 | 12.6 | 79.6 | 5,221 | 55 | 51 | 1,185 | 7,580 | 55,701 | 4,781 |
| Renville | 119 | 8,121 | 20.4 | 21.5 | 3.2 | 5.7 | 18.0 | 86.5 | 5,885 | 59 | 49 | 1,650 | 7,410 | 56,890 | 5,123 |
| Rice | 342 | 5,169 | 33.5 | 30.0 | 4.0 | 5.1 | 8.6 | 272.6 | 4,118 | 143 | 207 | 3,406 | 30,190 | 67,698 | 7,031 |
| Rock | 53 | 5,586 | 36.4 | 0.6 | 4.3 | 8.6 | 12.3 | 88.4 | 9,350 | 26 | 31 | 728 | 4,450 | 55,807 | 4,437 |
| Roseau | 62 | 4,020 | 66.1 | 0.8 | 2.2 | 0.0 | 7.7 | 62.9 | 4,112 | 95 | 52 | 957 | 7,630 | 67,218 | 7,705 |
| St. Louis | 1,203 | 6,022 | 28.0 | 1.2 | 4.8 | 7.5 | 14.8 | 1,581.6 | 7,917 | 1,388 | 798 | 14,724 | 94,550 | 63,338 | 6,532 |
| Scott | 753 | 5,171 | 51.8 | 0.7 | 4.3 | 3.1 | 11.4 | 1,073.2 | 7,374 | 130 | 513 | 9,863 | 72,750 | 94,032 | 12,704 |
| Sherburne | 469 | 4,962 | 61.4 | 0.5 | 3.7 | 3.6 | 5.1 | 600.0 | 6,350 | 141 | 330 | 4,173 | 46,140 | 70,531 | 7,037 |
| Sibley | 100 | 6,678 | 59.7 | 3.0 | 3.8 | 1.6 | 10.7 | 91.0 | 6,107 | 37 | 51 | 724 | 7,540 | 55,147 | 4,733 |
| Stearns | 787 | 4,971 | 48.5 | 0.7 | 5.2 | 5.2 | 9.7 | 1,464.0 | 9,248 | 2,411 | 543 | 9,392 | 75,400 | 63,923 | 6,804 |
| Steele | 197 | 5,355 | 56.6 | 1.4 | 4.3 | 5.0 | 10.0 | 176.5 | 4,797 | 75 | 125 | 2,225 | 18,580 | 61,183 | 5,690 |
| Stevens | 48 | 4,988 | 43.0 | 0.3 | 5.4 | 7.3 | 16.0 | 47.0 | 4,847 | 65 | 31 | 1,285 | 4,610 | 65,539 | 6,391 |
| Swift | 78 | 8,354 | 25.1 | 33.9 | 5.2 | 5.9 | 9.5 | 65.0 | 6,926 | 50 | 32 | 922 | 4,560 | 54,142 | 4,726 |
| Todd | 126 | 5,153 | 53.7 | 2.9 | 5.0 | 9.6 | 14.0 | 62.9 | 2,563 | 82 | 84 | 1,349 | 11,360 | 47,620 | 3,745 |
| Traverse | 38 | 11,543 | 29.1 | 0.5 | 4.2 | 5.5 | 16.5 | 13.9 | 4,220 | 24 | 11 | 323 | 1,680 | 49,373 | 4,781 |
| Wabasha | 95 | 4,397 | 42.2 | 3.3 | 6.9 | 5.2 | 11.2 | 72.7 | 3,369 | 56 | 74 | 1,057 | 11,010 | 61,417 | 5,749 |
| Wadena | 84 | 6,154 | 46.2 | 8.0 | 4.3 | 6.0 | 13.9 | 46.9 | 3,437 | 52 | 46 | 1,450 | 6,260 | 48,228 | 3,649 |
| Waseca | 106 | 5,677 | 46.5 | 7.5 | 5.1 | 2.2 | 16.7 | 78.5 | 4,196 | 244 | 61 | 1,155 | 9,040 | 54,562 | 4,451 |
| Washington | 1,061 | 4,151 | 58.5 | 1.1 | 5.0 | 0.8 | 10.8 | 1,569.8 | 6,145 | 385 | 901 | 10,867 | 133,320 | 96,286 | 12,891 |
| Watonwan | 74 | 6,734 | 58.8 | 2.6 | 2.6 | 2.3 | 10.7 | 82.7 | 7,565 | 52 | 37 | 755 | 5,180 | 52,713 | 4,227 |
| Wilkin | 38 | 6,064 | 43.1 | 3.7 | 5.8 | 6.7 | 16.9 | 42.1 | 6,662 | 20 | 21 | 410 | 3,170 | 56,769 | 4,699 |
| Winona | 182 | 3,588 | 43.1 | 2.6 | 5.9 | 6.8 | 13.2 | 62.2 | 1,226 | 149 | 163 | 3,296 | 22,570 | 59,624 | 6,895 |
| Wright | 644 | 4,798 | 57.4 | 1.2 | 4.4 | 3.5 | 11.0 | 1,100.7 | 8,201 | 201 | 478 | 6,325 | 66,620 | 75,979 | 8,172 |
| Yellow Medicine | 94 | 9,545 | 25.0 | 30.1 | 2.8 | 5.2 | 13.8 | 87.3 | 8,875 | 42 | 33 | 1,428 | 4,860 | 56,638 | 4,863 |

1. Based on the resident population estimated as of July 1 of the year shown.

| State / county code | CBSA code[1] | County Type code[2] | STATE County | Land area[3] (sq. mi) | Total persons 2019 | Rank | Per square mile | White | Black | American Indian, Alaska Native | Asian and Pacific Islander | Percent Hispanic or Latino[4] | Under 5 years | 5 to 14 years | 15 to 24 years | 25 to 34 years | 35 to 44 years | 45 to 54 years |
|---|---|---|---|---|---|---|---|---|---|---|---|---|---|---|---|---|---|---|
| | | | | 1 | 2 | 3 | 4 | 5 | 6 | 7 | 8 | 9 | 10 | 11 | 12 | 13 | 14 | 15 |
| 28000 | | 0 | MISSISSIPPI | 46,925.5 | 2,966,786 | X | 63.2 | 57.3 | 38.3 | 0.9 | 1.5 | 3.4 | 6.1 | 13.2 | 13.5 | 13.1 | 12.3 | 12.0 |
| 28001 | 35020 | 5 | Adams | 462.3 | 30,275 | 1,425 | 65.5 | 36.7 | 53.2 | 0.6 | 0.8 | 9.5 | 5.5 | 11.0 | 10.7 | 12.8 | 13.2 | 11.5 |
| 28003 | 18420 | 7 | Alcorn | 400.0 | 36,889 | 1,253 | 92.2 | 83.3 | 13.1 | 0.7 | 0.7 | 3.5 | 5.8 | 12.6 | 12.0 | 12.5 | 12.1 | 12.9 |
| 28005 | | 8 | Amite | 730.1 | 12,205 | 2,276 | 16.7 | 58.7 | 39.6 | 0.6 | 0.4 | 1.6 | 4.9 | 11.6 | 9.9 | 10.3 | 10.1 | 11.1 |
| 28007 | | 6 | Attala | 735.0 | 18,004 | 1,920 | 24.5 | 53.5 | 44.0 | 0.5 | 0.6 | 2.1 | 6.3 | 14.3 | 12.1 | 11.5 | 11.3 | 11.8 |
| 28009 | | 1 | Benton | 406.6 | 8,351 | 2,550 | 20.5 | 61.5 | 35.7 | 0.8 | 0.5 | 2.9 | 5.8 | 11.6 | 11.7 | 12.2 | 11.2 | 13.2 |
| 28011 | 17380 | 7 | Bolivar | 876.5 | 30,142 | 1,429 | 34.4 | 32.5 | 64.2 | 0.3 | 1.2 | 2.3 | 6.7 | 14.2 | 13.6 | 13.4 | 11.4 | 11.3 |
| 28013 | | 9 | Calhoun | 586.6 | 14,241 | 2,140 | 24.3 | 65.6 | 28.7 | 0.5 | 0.4 | 6.1 | 5.7 | 12.8 | 12.1 | 11.1 | 11.2 | 13.4 |
| 28015 | 24900 | 9 | Carroll | 628.4 | 9,732 | 2,435 | 15.5 | 65.0 | 33.5 | 0.6 | 0.4 | 1.6 | 4.3 | 10.2 | 10.7 | 10.9 | 11.3 | 12.3 |
| 28017 | | 7 | Chickasaw | 501.8 | 16,951 | 1,976 | 33.8 | 49.9 | 45.2 | 0.6 | 0.6 | 5.0 | 6.9 | 13.9 | 12.4 | 13.2 | 11.6 | 11.2 |
| 28019 | | 9 | Choctaw | 418.2 | 8,063 | 2,582 | 19.3 | 68.1 | 30.3 | 0.5 | 0.7 | 1.7 | 5.3 | 12.3 | 10.8 | 11.5 | 11.0 | 11.9 |
| 28021 | | 8 | Claiborne | 487.4 | 8,911 | 2,503 | 18.3 | 11.6 | 86.2 | 0.7 | 1.0 | 1.6 | 6.1 | 11.2 | 24.6 | 11.1 | 9.9 | 8.9 |
| 28023 | 32940 | 9 | Clarke | 691.6 | 15,299 | 2,071 | 22.1 | 63.7 | 35.0 | 0.6 | 0.3 | 1.6 | 6.1 | 11.9 | 11.5 | 11.7 | 10.8 | 12.1 |
| 28025 | 48500 | 7 | Clay | 410.1 | 19,352 | 1,858 | 47.2 | 38.9 | 59.2 | 0.5 | 0.6 | 1.2 | 5.8 | 12.6 | 12.4 | 13.1 | 11.5 | 11.7 |
| 28027 | 17260 | 7 | Coahoma | 553.1 | 21,564 | 1,744 | 39.0 | 20.3 | 77.7 | 0.3 | 0.6 | 1.6 | 7.7 | 15.0 | 14.1 | 12.5 | 10.7 | 10.8 |
| 28029 | 27140 | 2 | Copiah | 777.3 | 27,933 | 1,501 | 35.9 | 44.4 | 52.0 | 0.5 | 0.7 | 1.7 | 6.0 | 12.5 | 14.3 | 11.7 | 11.2 | 11.3 |
| 28031 | 25620 | 8 | Covington | 413.8 | 18,518 | 1,896 | 44.8 | 61.0 | 36.5 | 0.5 | 0.7 | 3.5 | 6.5 | 14.0 | 11.9 | 12.8 | 12.1 | 12.1 |
| 28033 | 32820 | 1 | DeSoto | 476.3 | 188,275 | 359 | 395.3 | 61.9 | 31.9 | 0.6 | 2.1 | 2.4 | 5.9 | 14.5 | 13.4 | 13.0 | 14.0 | 13.6 |
| 28035 | 25620 | 3 | Forrest | 466.0 | 75,009 | 739 | 161.0 | 58.1 | 38.1 | 0.6 | 1.5 | 5.1 | 6.7 | 13.0 | 17.5 | 14.9 | 11.9 | 10.6 |
| 28037 | | 9 | Franklin | 564.0 | 7,657 | 2,614 | 13.6 | 63.7 | 35.4 | 0.6 | 0.4 | 3.2 | 5.6 | 12.9 | 11.7 | 10.3 | 11.3 | 11.9 |
| 28039 | | 6 | George | 478.7 | 24,425 | 1,631 | 51.0 | 88.2 | 8.3 | 0.9 | 0.9 | 3.0 | 7.3 | 14.9 | 12.1 | 13.8 | 11.9 | 12.3 |
| 28041 | | 8 | Greene | 712.7 | 13,477 | 2,193 | 18.9 | 72.8 | 25.8 | 0.6 | 0.3 | 1.3 | 5.4 | 10.2 | 12.5 | 16.8 | 14.0 | 13.7 |
| 28043 | 24980 | 7 | Grenada | 422.1 | 20,610 | 1,790 | 48.8 | 54.7 | 43.4 | 0.6 | 0.7 | 1.6 | 6.1 | 13.7 | 11.9 | 12.9 | 11.3 | 12.3 |
| 28045 | 25060 | 2 | Hancock | 474.0 | 48,000 | 1,022 | 101.3 | 86.2 | 9.4 | 1.4 | 1.4 | 3.7 | 4.9 | 10.9 | 10.7 | 12.0 | 11.3 | 12.7 |
| 28047 | 25060 | 2 | Harrison | 573.6 | 208,801 | 326 | 364.0 | 65.1 | 27.2 | 1.0 | 3.9 | 5.5 | 6.4 | 13.4 | 13.1 | 13.9 | 12.6 | 11.7 |
| 28049 | 27140 | 2 | Hinds | 869.8 | 227,966 | 301 | 262.1 | 24.3 | 73.9 | 0.3 | 1.0 | 1.5 | 6.3 | 13.3 | 14.5 | 14.3 | 12.2 | 11.2 |
| 28051 | 27140 | 6 | Holmes | 756.7 | 16,726 | 1,988 | 22.1 | 15.5 | 83.3 | 0.4 | 0.4 | 1.2 | 6.3 | 15.2 | 15.3 | 12.9 | 11.2 | 10.5 |
| 28053 | | 8 | Humphreys | 418.5 | 7,827 | 2,602 | 18.7 | 20.6 | 75.5 | 0.4 | 0.7 | 3.6 | 5.8 | 15.4 | 12.6 | 11.1 | 11.7 | 10.9 |
| 28055 | | 8 | Issaquena | 413.0 | 1,220 | 3,098 | 3.0 | 35.6 | 61.9 | 0.5 | 0.6 | 1.7 | 2.5 | 4.4 | 16.1 | 16.7 | 12.7 | 12.0 |
| 28057 | 46180 | 7 | Itawamba | 532.8 | 23,261 | 1,671 | 43.7 | 90.5 | 7.6 | 0.6 | 0.6 | 1.7 | 5.5 | 12.2 | 13.7 | 13.0 | 11.2 | 13.4 |
| 28059 | 25060 | 2 | Jackson | 722.8 | 143,802 | 458 | 199.0 | 69.1 | 22.1 | 0.9 | 3.1 | 6.9 | 5.7 | 13.0 | 12.0 | 13.3 | 12.9 | 12.7 |
| 28061 | 29860 | 9 | Jasper | 676.3 | 16,332 | 2,016 | 24.1 | 45.0 | 53.5 | 0.6 | 0.6 | 1.5 | 5.9 | 13.2 | 11.2 | 11.8 | 11.2 | 11.8 |
| 28063 | | 8 | Jefferson | 519.9 | 6,997 | 2,658 | 13.5 | 13.9 | 85.0 | 0.5 | 0.4 | 0.9 | 6.1 | 12.7 | 11.9 | 13.1 | 12.3 | 11.2 |
| 28065 | | 8 | Jefferson Davis | 408.4 | 10,890 | 2,357 | 26.7 | 38.0 | 59.8 | 0.6 | 0.4 | 1.8 | 5.6 | 11.1 | 10.7 | 11.7 | 10.8 | 12.0 |
| 28067 | 29860 | 4 | Jones | 694.8 | 67,993 | 790 | 97.9 | 65.0 | 29.8 | 0.8 | 0.6 | 4.7 | 6.7 | 14.1 | 13.0 | 12.0 | 12.0 | 11.4 |
| 28069 | 32940 | 9 | Kemper | 766.2 | 9,521 | 2,452 | 12.4 | 34.1 | 61.2 | 4.0 | 0.3 | 1.1 | 4.9 | 10.4 | 14.3 | 12.4 | 10.8 | 12.1 |
| 28071 | 37060 | 9 | Lafayette | 631.7 | 54,408 | 939 | 86.1 | 71.2 | 24.0 | 0.6 | 2.9 | 2.6 | 4.9 | 10.4 | 23.9 | 13.8 | 12.2 | 10.4 |
| 28073 | 25620 | 3 | Lamar | 496.6 | 64,165 | 835 | 129.2 | 73.7 | 22.2 | 0.6 | 1.8 | 3.1 | 6.3 | 13.9 | 12.3 | 14.2 | 14.2 | 12.3 |
| 28075 | 32940 | 5 | Lauderdale | 703.7 | 73,751 | 746 | 104.8 | 52.2 | 44.9 | 0.5 | 1.1 | 2.3 | 6.1 | 13.0 | 12.7 | 13.2 | 11.8 | 11.9 |
| 28077 | | 8 | Lawrence | 430.6 | 12,480 | 2,249 | 29.0 | 64.8 | 32.7 | 0.5 | 0.6 | 2.2 | 5.9 | 14.2 | 11.3 | 11.6 | 11.9 | 11.8 |
| 28079 | | 6 | Leake | 582.9 | 22,741 | 1,695 | 39.0 | 47.0 | 42.7 | 5.8 | 0.6 | 4.8 | 6.6 | 13.1 | 13.8 | 12.5 | 12.2 | 12.0 |
| 28081 | 46180 | 5 | Lee | 450.0 | 85,466 | 676 | 189.9 | 65.1 | 31.5 | 0.5 | 1.2 | 3.0 | 6.6 | 14.0 | 12.5 | 13.4 | 12.7 | 12.6 |
| 28083 | 24900 | 5 | Leflore | 594.1 | 27,854 | 1,503 | 46.9 | 21.7 | 74.9 | 0.4 | 0.6 | 2.9 | 7.3 | 15.7 | 15.1 | 12.8 | 11.6 | 10.3 |
| 28085 | 15020 | 6 | Lincoln | 586.2 | 33,936 | 1,328 | 57.9 | 67.7 | 30.6 | 0.6 | 0.7 | 1.3 | 5.8 | 13.1 | 11.9 | 12.2 | 12.8 | 12.8 |
| 28087 | 18060 | 5 | Lowndes | 505.4 | 58,309 | 891 | 115.4 | 51.7 | 45.4 | 0.5 | 1.4 | 2.2 | 6.3 | 13.2 | 13.1 | 14.0 | 12.0 | 11.3 |
| 28089 | 27140 | 2 | Madison | 714.4 | 106,871 | 570 | 149.6 | 55.8 | 38.5 | 0.6 | 3.2 | 3.1 | 6.3 | 14.0 | 12.9 | 12.7 | 13.6 | 13.0 |
| 28091 | | 6 | Marion | 542.4 | 24,441 | 1,629 | 45.1 | 66.0 | 32.1 | 0.6 | 0.8 | 1.7 | 5.9 | 12.9 | 12.0 | 12.2 | 12.6 | 11.9 |
| 28093 | 32820 | 1 | Marshall | 706.2 | 35,301 | 1,296 | 50.0 | 48.7 | 47.0 | 0.6 | 0.5 | 4.2 | 5.5 | 11.7 | 12.7 | 12.8 | 11.3 | 12.6 |
| 28095 | | 7 | Monroe | 765.1 | 35,123 | 1,301 | 45.9 | 67.6 | 30.8 | 0.6 | 0.5 | 1.5 | 5.5 | 12.9 | 11.4 | 12.2 | 12.7 | 12.7 |
| 28097 | | 7 | Montgomery | 407.0 | 9,661 | 2,440 | 23.7 | 53.0 | 44.9 | 0.6 | 0.8 | 1.6 | 6.2 | 12.5 | 10.8 | 11.1 | 10.3 | 11.5 |
| 28099 | | 7 | Neshoba | 570.1 | 28,996 | 1,460 | 50.9 | 59.1 | 22.4 | 17.2 | 0.9 | 2.3 | 7.4 | 15.4 | 13.8 | 12.1 | 11.6 | 11.3 |
| 28101 | | 7 | Newton | 577.8 | 20,866 | 1,776 | 36.1 | 61.0 | 31.8 | 5.5 | 0.7 | 2.1 | 6.6 | 13.7 | 14.4 | 12.0 | 11.4 | 12.7 |
| 28103 | | 7 | Noxubee | 695.2 | 10,236 | 2,402 | 14.7 | 25.9 | 72.4 | 0.5 | 0.4 | 1.4 | 7.3 | 13.2 | 13.0 | 13.5 | 10.6 | 11.3 |
| 28105 | 44260 | 5 | Oktibbeha | 458.2 | 49,789 | 995 | 108.7 | 57.1 | 38.2 | 0.6 | 3.6 | 1.8 | 5.1 | 9.6 | 33.3 | 13.7 | 9.5 | 7.8 |
| 28107 | | 6 | Panola | 685.2 | 33,848 | 1,330 | 49.4 | 47.5 | 50.6 | 0.7 | 0.5 | 2.0 | 7.0 | 13.5 | 12.4 | 13.2 | 11.4 | 11.9 |
| 28109 | 38100 | 6 | Pearl River | 811.3 | 55,876 | 924 | 68.9 | 83.2 | 13.0 | 1.4 | 0.9 | 3.4 | 5.8 | 12.8 | 12.5 | 11.8 | 11.4 | 12.4 |
| 28111 | 25620 | 3 | Perry | 647.3 | 11,862 | 2,300 | 18.3 | 78.7 | 19.6 | 0.8 | 0.4 | 1.6 | 5.7 | 12.7 | 11.4 | 12.6 | 11.7 | 12.2 |
| 28113 | 32620 | 6 | Pike | 409.2 | 38,997 | 1,200 | 95.3 | 42.9 | 54.7 | 0.7 | 0.9 | 1.6 | 6.8 | 14.3 | 13.6 | 12.2 | 11.6 | 11.5 |
| 28115 | 46180 | 7 | Pontotoc | 497.8 | 32,461 | 1,370 | 65.2 | 77.1 | 15.8 | 0.7 | 0.6 | 7.3 | 7.1 | 14.5 | 12.9 | 12.8 | 12.5 | 12.1 |
| 28117 | 46180 | 7 | Prentiss | 415.0 | 25,013 | 1,604 | 60.3 | 83.5 | 15.3 | 0.5 | 0.5 | 1.5 | 6.1 | 12.7 | 13.1 | 12.7 | 11.5 | 12.3 |
| 28119 | | 6 | Quitman | 405.0 | 6,760 | 2,682 | 16.7 | 25.7 | 72.6 | 0.8 | 0.5 | 1.7 | 5.8 | 12.7 | 12.8 | 12.6 | 10.6 | 12.9 |
| 28121 | 27140 | 2 | Rankin | 775.5 | 155,975 | 430 | 201.1 | 74.0 | 22.3 | 0.5 | 1.7 | 2.7 | 5.5 | 12.7 | 12.0 | 13.8 | 14.3 | 13.1 |
| 28123 | | 6 | Scott | 609.2 | 28,061 | 1,496 | 46.1 | 49.9 | 38.3 | 0.6 | 0.7 | 11.7 | 8.2 | 14.7 | 12.1 | 12.5 | 11.9 | 12.2 |
| 28125 | | 8 | Sharkey | 431.7 | 4,160 | 2,883 | 9.6 | 25.7 | 71.3 | 0.2 | 1.0 | 2.4 | 6.1 | 13.4 | 12.0 | 10.2 | 10.7 | 11.0 |
| 28127 | 27140 | 2 | Simpson | 589.2 | 26,629 | 1,541 | 45.2 | 62.0 | 35.8 | 0.5 | 1.0 | 1.9 | 5.9 | 13.2 | 12.0 | 12.4 | 11.7 | 11.8 |
| 28129 | | 8 | Smith | 636.3 | 15,779 | 2,045 | 24.8 | 74.4 | 23.7 | 0.3 | 0.3 | 1.9 | 6.1 | 12.4 | 11.7 | 11.7 | 11.1 | 12.3 |
| 28131 | 25060 | 2 | Stone | 445.5 | 18,360 | 1,905 | 41.2 | 77.8 | 19.5 | 0.9 | 0.9 | 2.3 | 5.6 | 12.6 | 14.4 | 12.7 | 11.8 | 12.2 |

1. CBSA = Core Based Statistical Area. See Appendix A for explanation. See Appendix B for list of metropolitan areas with component counties. Service of USDA Rural-Urban Continuum Codes. See Appendix A for definition. 3. Dry land or land partially or temporarily covered by water.
2. County type code from the Economic Research
4. May be of any race.

Items 1—15

| STATE County | Population, 2020 (cont.) — Age (percent) (cont.) | | | | Population change, 2000-2020 — Total persons | | Percent change | | Components of change, 2010-2020 | | | Households, 2015-2019 | | | Percent | |
|---|---|---|---|---|---|---|---|---|---|---|---|---|---|---|---|---|
| | 55 to 64 years | 65 to 74 years | 75 years and over | Percent female | 2000 | 2010 | 2000-2010 | 2010-2020 | Births | Deaths | Net Migration | Number | Persons per household | Family households | Female family householder[1] | One person |
| | 16 | 17 | 18 | 19 | 20 | 21 | 22 | 23 | 24 | 25 | 26 | 27 | 28 | 29 | 30 | 31 |
| MISSISSIPPI | 12.9 | 10.1 | 6.8 | 51.6 | 2,844,658 | 2,968,129 | 4.3 | 0.0 | 389,874 | 319,113 | -72,673 | 1,104,394 | 2.62 | 66.5 | 17.3 | 29.2 |
| Adams | 15.1 | 12.0 | 8.2 | 48.2 | 34,340 | 32,304 | -5.9 | -6.3 | 3,763 | 4,244 | -1,588 | 11,237 | 2.56 | 59.2 | 21.6 | 37.5 |
| Alcorn | 13.1 | 10.8 | 8.3 | 51.4 | 34,558 | 37,062 | 7.2 | -0.5 | 4,445 | 4,756 | 169 | 14,649 | 2.49 | 65.5 | 14.9 | 30.2 |
| Amite | 16.8 | 14.3 | 10.9 | 51.8 | 13,599 | 13,135 | -3.4 | -7.1 | 1,344 | 1,562 | -719 | 5,218 | 2.36 | 65.2 | 15.8 | 32.4 |
| Attala | 13.0 | 11.0 | 8.7 | 52.5 | 19,661 | 19,562 | -0.5 | -8.0 | 2,501 | 2,802 | -1,257 | 6,941 | 2.60 | 68.2 | 21.2 | 29.4 |
| Benton | 14.8 | 11.2 | 8.2 | 50.5 | 8,026 | 8,730 | 8.8 | -4.3 | 966 | 1,064 | -291 | 3,149 | 2.59 | 66.1 | 15.9 | 32.4 |
| Bolivar | 12.5 | 10.8 | 6.3 | 53.4 | 40,633 | 34,151 | -16.0 | -11.7 | 4,953 | 4,203 | -4,768 | 12,111 | 2.51 | 63.7 | 24.9 | 31.9 |
| Calhoun | 13.7 | 11.3 | 8.8 | 51.8 | 15,069 | 14,957 | -0.7 | -4.8 | 1,778 | 1,845 | -639 | 5,846 | 2.44 | 64.2 | 15.4 | 32.7 |
| Carroll | 15.2 | 14.0 | 11.0 | 49.3 | 10,769 | 10,601 | -1.6 | -8.2 | 909 | 1,110 | -660 | 3,827 | 2.57 | 70.2 | 8.9 | 26.3 |
| Chickasaw | 12.6 | 10.3 | 8.0 | 50.8 | 19,440 | 17,389 | -10.6 | -2.5 | 2,510 | 1,910 | -1,036 | 6,476 | 2.58 | 66.1 | 18.1 | 32.1 |
| Choctaw | 14.2 | 12.5 | 10.5 | 51.3 | 9,758 | 8,552 | -12.4 | -5.7 | 895 | 998 | -381 | 3,228 | 2.51 | 62.3 | 11.6 | 36.8 |
| Claiborne | 11.1 | 10.4 | 6.7 | 52.9 | 11,831 | 9,595 | -18.9 | -7.1 | 1,210 | 1,045 | -878 | 2,908 | 2.86 | 65.2 | 27.5 | 31.9 |
| Clarke | 14.3 | 12.4 | 9.0 | 52.8 | 17,955 | 16,724 | -6.9 | -8.5 | 1,946 | 2,153 | -1,217 | 6,237 | 2.52 | 70.5 | 16.7 | 27.8 |
| Clay | 13.6 | 11.3 | 7.9 | 52.9 | 21,979 | 20,658 | -6.0 | -6.3 | 2,349 | 2,406 | -1,247 | 7,618 | 2.54 | 63.5 | 19.1 | 33.9 |
| Coahoma | 12.9 | 9.8 | 6.5 | 53.4 | 30,622 | 26,151 | -14.6 | -17.5 | 4,115 | 3,108 | -5,634 | 8,782 | 2.56 | 64.2 | 28.0 | 32.0 |
| Copiah | 14.3 | 11.1 | 7.6 | 51.9 | 28,757 | 29,451 | 2.4 | -5.2 | 3,736 | 3,339 | -1,909 | 9,414 | 2.93 | 70.2 | 20.7 | 28.0 |
| Covington | 13.4 | 9.9 | 7.3 | 51.2 | 19,407 | 19,573 | 0.9 | -5.4 | 2,648 | 2,704 | -993 | 6,759 | 2.77 | 73.5 | 18.4 | 24.4 |
| DeSoto | 11.9 | 8.3 | 5.3 | 52.0 | 107,199 | 161,267 | 50.4 | 16.7 | 21,562 | 13,528 | 19,002 | 62,890 | 2.83 | 72.8 | 14.3 | 22.2 |
| Forrest | 11.2 | 8.2 | 6.0 | 52.8 | 72,604 | 74,859 | 3.1 | 0.2 | 10,856 | 7,680 | -3,058 | 28,086 | 2.55 | 60.5 | 19.3 | 31.3 |
| Franklin | 14.7 | 12.3 | 9.2 | 50.8 | 8,448 | 8,114 | -4.0 | -5.6 | 883 | 1,035 | -308 | 2,928 | 2.61 | 71.6 | 11.4 | 25.9 |
| George | 12.6 | 9.4 | 5.6 | 49.4 | 19,144 | 22,582 | 18.0 | 8.2 | 3,693 | 2,724 | 881 | 7,592 | 3.08 | 69.9 | 7.9 | 28.3 |
| Greene | 11.6 | 9.1 | 6.7 | 42.5 | 13,299 | 14,387 | 8.2 | -6.3 | 1,339 | 1,336 | -931 | 3,925 | 2.50 | 66.9 | 10.0 | 26.9 |
| Grenada | 13.3 | 11.0 | 7.5 | 52.5 | 23,263 | 21,903 | -5.8 | -5.9 | 2,755 | 3,044 | -992 | 8,391 | 2.47 | 63.9 | 19.7 | 32.1 |
| Hancock | 16.1 | 12.9 | 8.4 | 51.3 | 42,967 | 44,037 | 2.5 | 9.0 | 4,887 | 4,959 | 4,026 | 20,036 | 2.33 | 65.3 | 14.2 | 29.4 |
| Harrison | 13.0 | 9.8 | 6.2 | 51.2 | 189,601 | 187,109 | -1.3 | 11.6 | 28,039 | 20,203 | 13,877 | 78,104 | 2.56 | 66.0 | 17.1 | 28.4 |
| Hinds | 12.7 | 9.4 | 6.1 | 53.4 | 250,800 | 245,360 | -2.2 | -7.1 | 33,056 | 22,012 | -28,566 | 88,611 | 2.59 | 62.3 | 23.4 | 32.0 |
| Holmes | 13.3 | 9.5 | 6.9 | 52.4 | 21,609 | 19,483 | -9.8 | -14.2 | 2,598 | 2,254 | -3,121 | 6,188 | 2.76 | 60.1 | 33.2 | 37.4 |
| Humphreys | 14.5 | 10.3 | 7.6 | 52.9 | 11,206 | 9,373 | -16.4 | -16.5 | 1,171 | 1,011 | -1,731 | 3,186 | 2.60 | 65.2 | 33.9 | 31.0 |
| Issaquena | 17.0 | 10.0 | 8.7 | 41.7 | 2,274 | 1,399 | -38.5 | -12.8 | 123 | 131 | -173 | 483 | 2.19 | 56.7 | 15.7 | 37.3 |
| Itawamba | 12.6 | 10.1 | 8.2 | 50.8 | 22,770 | 23,399 | 2.8 | -0.6 | 2,702 | 2,880 | 52 | 8,653 | 2.59 | 71.7 | 13.8 | 23.1 |
| Jackson | 13.7 | 10.2 | 6.5 | 51.1 | 131,420 | 139,669 | 6.3 | 3.0 | 16,590 | 13,987 | 1,579 | 52,423 | 2.69 | 66.2 | 16.1 | 29.3 |
| Jasper | 13.8 | 12.1 | 8.9 | 51.3 | 18,149 | 17,067 | -6.0 | -4.3 | 2,180 | 2,028 | -892 | 6,629 | 2.48 | 71.6 | 21.4 | 27.8 |
| Jefferson | 14.2 | 10.6 | 7.9 | 50.9 | 9,740 | 7,734 | -20.6 | -9.5 | 1,037 | 859 | -927 | 2,448 | 2.75 | 60.8 | 29.5 | 39.0 |
| Jefferson Davis | 14.5 | 13.3 | 10.3 | 52.8 | 13,962 | 12,475 | -10.7 | -12.7 | 1,366 | 1,615 | -1,333 | 4,713 | 2.38 | 63.7 | 17.0 | 33.0 |
| Jones | 12.9 | 10.5 | 7.4 | 51.5 | 64,958 | 67,778 | 4.3 | 0.3 | 9,789 | 7,801 | -1,746 | 24,858 | 2.68 | 68.3 | 16.7 | 27.7 |
| Kemper | 13.0 | 11.8 | 10.4 | 50.1 | 10,453 | 10,438 | -0.1 | -8.8 | 902 | 1,042 | -782 | 3,611 | 2.49 | 61.0 | 19.8 | 36.0 |
| Lafayette | 10.4 | 8.6 | 5.3 | 51.5 | 38,744 | 47,361 | 22.2 | 14.9 | 5,575 | 3,908 | 5,229 | 18,721 | 2.61 | 59.4 | 13.1 | 29.6 |
| Lamar | 11.9 | 8.9 | 5.9 | 52.2 | 39,070 | 55,732 | 42.6 | 15.1 | 8,012 | 4,337 | 4,713 | 22,116 | 2.79 | 70.5 | 13.0 | 24.4 |
| Lauderdale | 12.9 | 10.6 | 7.7 | 51.6 | 78,161 | 80,280 | 2.7 | -8.1 | 10,192 | 8,920 | -7,833 | 29,736 | 2.45 | 64.9 | 16.7 | 31.8 |
| Lawrence | 14.3 | 11.3 | 7.6 | 51.4 | 13,258 | 12,951 | -2.3 | -3.6 | 1,615 | 1,553 | -532 | 4,849 | 2.61 | 69.8 | 14.6 | 27.7 |
| Leake | 12.5 | 10.2 | 7.1 | 48.6 | 20,940 | 23,795 | 13.6 | -4.4 | 3,149 | 2,542 | -1,662 | 8,105 | 2.74 | 69.0 | 21.0 | 28.5 |
| Lee | 12.8 | 9.1 | 6.4 | 52.2 | 75,755 | 82,906 | 9.4 | 3.1 | 11,799 | 9,430 | 231 | 32,099 | 2.62 | 67.4 | 14.9 | 28.7 |
| Leflore | 11.9 | 9.0 | 6.2 | 53.4 | 37,947 | 32,382 | -14.7 | -14.0 | 4,770 | 3,771 | -5,607 | 9,962 | 2.79 | 60.7 | 27.5 | 36.9 |
| Lincoln | 13.3 | 10.7 | 7.2 | 52.2 | 33,166 | 34,867 | 5.1 | -2.7 | 4,446 | 4,269 | -1,090 | 12,699 | 2.65 | 71.3 | 16.2 | 26.2 |
| Lowndes | 13.0 | 9.9 | 6.8 | 52.4 | 61,586 | 59,779 | -2.9 | -2.5 | 8,139 | 6,323 | -3,273 | 22,436 | 2.56 | 65.3 | 16.0 | 30.9 |
| Madison | 13.2 | 9.4 | 5.0 | 52.1 | 74,674 | 95,208 | 27.5 | 12.3 | 13,349 | 10,353 | 8,634 | 39,635 | 2.59 | 71.1 | 15.0 | 25.6 |
| Marion | 13.3 | 11.2 | 8.2 | 51.4 | 25,595 | 27,075 | 5.8 | -9.7 | 3,206 | 3,512 | -2,339 | 9,483 | 2.56 | 61.8 | 12.7 | 35.8 |
| Marshall | 15.0 | 11.3 | 7.1 | 50.4 | 34,993 | 37,159 | 6.2 | -5.0 | 4,387 | 4,320 | -1,920 | 12,772 | 2.62 | 69.2 | 18.0 | 27.1 |
| Monroe | 13.6 | 11.5 | 8.7 | 52.3 | 38,014 | 36,971 | -2.7 | -5.0 | 4,237 | 4,426 | -1,655 | 13,966 | 2.52 | 65.3 | 15.3 | 32.0 |
| Montgomery | 14.6 | 13.0 | 10.0 | 51.9 | 12,189 | 10,929 | -10.3 | -11.6 | 1,307 | 1,511 | -1,067 | 4,539 | 2.19 | 67.5 | 21.3 | 29.3 |
| Neshoba | 11.9 | 9.8 | 6.7 | 52.4 | 28,684 | 29,694 | 3.5 | -2.4 | 4,445 | 3,859 | -1,277 | 10,667 | 2.71 | 67.4 | 19.9 | 29.4 |
| Newton | 12.0 | 10.0 | 7.3 | 52.1 | 21,838 | 21,717 | -0.6 | -3.9 | 2,914 | 2,757 | -999 | 8,037 | 2.57 | 70.9 | 15.6 | 27.4 |
| Noxubee | 13.7 | 10.1 | 7.4 | 52.7 | 12,548 | 11,546 | -8.0 | -11.3 | 1,617 | 1,263 | -1,674 | 3,986 | 2.63 | 65.0 | 23.0 | 33.9 |
| Oktibbeha | 9.0 | 6.8 | 5.3 | 50.0 | 42,902 | 47,657 | 11.1 | 4.5 | 5,684 | 3,191 | -402 | 17,798 | 2.51 | 52.0 | 12.1 | 31.9 |
| Panola | 13.8 | 10.3 | 6.5 | 52.2 | 34,274 | 34,701 | 1.2 | -2.5 | 5,138 | 4,128 | -1,847 | 12,488 | 2.72 | 68.2 | 18.8 | 28.9 |
| Pearl River | 13.6 | 11.8 | 7.9 | 50.9 | 48,621 | 55,726 | 14.6 | 0.3 | 6,500 | 6,629 | 297 | 21,020 | 2.56 | 71.7 | 13.8 | 25.3 |
| Perry | 14.2 | 11.3 | 8.2 | 51.1 | 12,138 | 12,252 | 0.9 | -3.2 | 1,455 | 1,387 | -458 | 4,623 | 2.57 | 72.4 | 16.9 | 25.3 |
| Pike | 12.3 | 10.6 | 7.0 | 52.7 | 38,940 | 40,402 | 3.8 | -3.5 | 5,622 | 5,241 | -1,777 | 14,404 | 2.67 | 59.2 | 16.3 | 38.8 |
| Pontotoc | 12.5 | 9.2 | 6.5 | 51.1 | 26,726 | 29,954 | 12.1 | 8.4 | 4,506 | 3,025 | 1,022 | 10,783 | 2.91 | 73.7 | 13.5 | 22.9 |
| Prentiss | 12.8 | 10.3 | 8.4 | 51.2 | 25,556 | 25,281 | -1.1 | -1.1 | 3,191 | 3,054 | -401 | 9,145 | 2.64 | 66.6 | 10.9 | 29.9 |
| Quitman | 13.8 | 11.0 | 7.9 | 53.2 | 10,117 | 8,216 | -18.8 | -17.7 | 949 | 1,030 | -1,391 | 3,027 | 2.32 | 60.6 | 27.9 | 35.0 |
| Rankin | 12.3 | 9.6 | 6.6 | 51.8 | 115,327 | 142,049 | 23.2 | 9.8 | 18,796 | 11,635 | 6,813 | 55,909 | 2.63 | 71.1 | 12.9 | 25.0 |
| Scott | 12.5 | 9.6 | 6.3 | 51.4 | 28,423 | 28,267 | -0.5 | -0.7 | 4,740 | 3,018 | -1,925 | 10,180 | 2.76 | 69.7 | 19.7 | 28.1 |
| Sharkey | 15.9 | 12.3 | 8.4 | 52.5 | 6,580 | 4,916 | -25.3 | -15.4 | 660 | 626 | -799 | 1,751 | 2.44 | 59.7 | 18.3 | 35.4 |
| Simpson | 14.0 | 11.3 | 7.7 | 51.8 | 27,639 | 27,500 | -0.5 | -3.2 | 3,323 | 3,125 | -1,052 | 9,486 | 2.77 | 69.8 | 18.4 | 27.0 |
| Smith | 13.8 | 11.7 | 9.0 | 51.9 | 16,182 | 16,484 | 1.9 | -4.3 | 1,958 | 1,734 | -930 | 5,820 | 2.73 | 70.6 | 11.4 | 27.5 |
| Stone | 13.8 | 11.0 | 6.0 | 49.8 | 13,622 | 17,790 | 30.6 | 3.2 | 2,193 | 1,869 | 223 | 6,334 | 2.77 | 72.0 | 13.7 | 26.2 |

1. No spouse present.

# Table B. States and Counties — Population, Vital Statistics, Health, and Crime

| STATE County | Persons in group quarters, 2020 | Daytime Population, 2015-2019 Number | Employment/residence ratio | Births, 2020 Total | Rate[1] | Deaths, 2020 Number | Rate[1] | Persons under 65 with no health insurance, 2019 Number | Percent | Medicare, 2020 Total beneficiaries | Enrolled in Original Medicare | Enrolled in Medicare Advantage | Crimes reported by county police or sheriff, 2019 Violent | Property |
|---|---|---|---|---|---|---|---|---|---|---|---|---|---|---|
| | 32 | 33 | 34 | 35 | 36 | 37 | 38 | 39 | 40 | 41 | 42 | 43 | 44 | 45 |
| MISSISSIPPI | 90,657 | 2,931,367 | 0.96 | 35,901 | 12.1 | 33,888 | 11.4 | 369,501 | 15.4 | 609,689 | 464,397 | 145,291 | X | X |
| Adams | 2,313 | 31,814 | 1.05 | 324 | 10.7 | 436 | 14.4 | 3,911 | 17.6 | 7,321 | 6,017 | 1,304 | NA | NA |
| Alcorn | 782 | 36,684 | 0.97 | 414 | 11.2 | 505 | 13.7 | 4,674 | 16.0 | 8,704 | 8,224 | 480 | NA | NA |
| Amite | 113 | 10,556 | 0.56 | 127 | 10.4 | 164 | 13.4 | 1,626 | 17.6 | 3,190 | 2,532 | 659 | NA | NA |
| Attala | 345 | 17,076 | 0.79 | 230 | 12.8 | 271 | 15.1 | 2,325 | 16.1 | 4,243 | 3,060 | 1,183 | NA | NA |
| Benton | 78 | 6,380 | 0.38 | 85 | 10.2 | 104 | 12.5 | 1,147 | 17.4 | 1,939 | 1,511 | 428 | NA | NA |
| Bolivar | 1,353 | 31,536 | 0.97 | 405 | 13.4 | 426 | 14.1 | 3,985 | 16.6 | 7,115 | 5,532 | 1,584 | NA | NA |
| Calhoun | 212 | 12,927 | 0.71 | 170 | 11.9 | 189 | 13.3 | 2,124 | 18.6 | 3,559 | 3,253 | 306 | NA | NA |
| Carroll | 365 | 8,038 | 0.40 | 88 | 9.0 | 120 | 12.3 | 1,164 | 16.2 | 2,681 | 2,247 | 434 | NA | NA |
| Chickasaw | 492 | 16,460 | 0.89 | 228 | 13.5 | 209 | 12.3 | 2,347 | 17.3 | 4,157 | 3,514 | 643 | NA | NA |
| Choctaw | 122 | 7,345 | 0.68 | 76 | 9.4 | 118 | 14.6 | 951 | 15.1 | 2,219 | 1,950 | 269 | 6 | 57 |
| Claiborne | 1,292 | 9,488 | 1.15 | 115 | 12.9 | 112 | 12.6 | 873 | 14.3 | 1,842 | 1,289 | 552 | 21 | 87 |
| Clarke | 52 | 12,728 | 0.50 | 189 | 12.4 | 212 | 13.9 | 1,984 | 16.3 | 4,130 | 3,118 | 1,012 | NA | NA |
| Clay | 303 | 17,981 | 0.76 | 237 | 12.2 | 241 | 12.5 | 2,423 | 15.8 | 4,662 | 4,099 | 563 | NA | NA |
| Coahoma | 758 | 23,396 | 1.02 | 329 | 15.3 | 310 | 14.4 | 2,693 | 15.2 | 5,000 | 3,673 | 1,328 | NA | NA |
| Copiah | 987 | 25,648 | 0.73 | 353 | 12.6 | 343 | 12.3 | 3,644 | 16.3 | 6,291 | 4,054 | 2,237 | NA | NA |
| Covington | 237 | 17,583 | 0.81 | 226 | 12.2 | 240 | 13.0 | 2,653 | 17.4 | 4,202 | 2,910 | 1,292 | NA | NA |
| DeSoto | 627 | 154,557 | 0.72 | 2,129 | 11.3 | 1,568 | 8.3 | 18,913 | 11.8 | 29,633 | 22,616 | 7,017 | NA | NA |
| Forrest | 2,823 | 84,512 | 1.29 | 1,040 | 13.9 | 791 | 10.5 | 9,468 | 15.3 | 13,486 | 9,361 | 4,125 | 35 | 359 |
| Franklin | 67 | 6,754 | 0.63 | 84 | 11.0 | 87 | 11.4 | 896 | 14.8 | 1,971 | 1,697 | 274 | NA | NA |
| George | 570 | 20,793 | 0.62 | 351 | 14.4 | 282 | 11.5 | 3,411 | 16.8 | 4,917 | 3,396 | 1,521 | 21 | 315 |
| Greene | 2,559 | 12,655 | 0.69 | 125 | 9.3 | 130 | 9.6 | 1,408 | 15.7 | 2,399 | 1,889 | 510 | NA | NA |
| Grenada | 226 | 23,192 | 1.26 | 254 | 12.3 | 271 | 13.1 | 2,608 | 15.6 | 5,478 | 4,631 | 847 | NA | NA |
| Hancock | 558 | 45,221 | 0.91 | 473 | 9.9 | 536 | 11.2 | 6,008 | 16.0 | 10,929 | 6,720 | 4,209 | 28 | 516 |
| Harrison | 5,136 | 217,156 | 1.14 | 2,624 | 12.6 | 2,201 | 10.5 | 28,528 | 16.7 | 40,535 | 28,796 | 11,740 | 60 | 869 |
| Hinds | 8,010 | 252,577 | 1.13 | 2,889 | 12.7 | 2,340 | 10.3 | 28,749 | 15.2 | 42,865 | 27,308 | 15,557 | 111 | 383 |
| Holmes | 844 | 16,069 | 0.66 | 211 | 12.6 | 250 | 14.9 | 2,059 | 15.5 | 4,047 | 2,730 | 1,317 | NA | NA |
| Humphreys | 83 | 7,950 | 0.84 | 92 | 11.8 | 126 | 16.1 | 1,033 | 16.0 | 1,929 | 1,504 | 424 | NA | NA |
| Issaquena | 280 | 1,354 | 0.98 | 10 | 8.2 | 16 | 13.1 | 136 | 17.7 | 257 | 195 | 62 | NA | NA |
| Itawamba | 884 | 20,275 | 0.68 | 261 | 11.2 | 299 | 12.9 | 2,964 | 16.2 | 5,414 | 5,019 | 396 | NA | NA |
| Jackson | 1,336 | 136,083 | 0.90 | 1,562 | 10.9 | 1,584 | 11.0 | 17,030 | 14.3 | 28,424 | 19,477 | 8,947 | 135 | 1,388 |
| Jasper | 87 | 14,354 | 0.66 | 195 | 11.9 | 212 | 13.0 | 2,130 | 16.7 | 4,113 | 2,861 | 1,252 | NA | NA |
| Jefferson | 365 | 6,735 | 0.74 | 81 | 11.6 | 78 | 11.1 | 880 | 16.4 | 1,753 | 1,362 | 391 | NA | NA |
| Jefferson Davis | 110 | 9,317 | 0.53 | 116 | 10.7 | 165 | 15.2 | 1,457 | 17.3 | 2,975 | 2,197 | 777 | NA | NA |
| Jones | 1,921 | 70,507 | 1.08 | 916 | 13.5 | 859 | 12.6 | 9,133 | 16.7 | 14,492 | 10,549 | 3,944 | NA | NA |
| Kemper | 832 | 9,547 | 0.88 | 85 | 8.9 | 111 | 11.7 | 1,265 | 19.0 | 2,173 | 1,609 | 564 | NA | NA |
| Lafayette | 4,349 | 55,485 | 1.08 | 520 | 9.6 | 428 | 7.9 | 6,590 | 15.6 | 8,568 | 7,665 | 903 | NA | NA |
| Lamar | 395 | 56,731 | 0.81 | 750 | 11.7 | 457 | 7.1 | 7,645 | 14.2 | 9,847 | 7,254 | 2,593 | 87 | 637 |
| Lauderdale | 3,727 | 81,296 | 1.16 | 880 | 11.9 | 906 | 12.3 | 8,366 | 14.5 | 15,283 | 11,856 | 3,428 | 27 | 410 |
| Lawrence | 0 | 11,072 | 0.66 | 139 | 11.1 | 159 | 12.7 | 1,546 | 15.1 | 2,985 | 2,306 | 679 | NA | NA |
| Leake | 1,688 | 20,091 | 0.67 | 320 | 14.1 | 290 | 12.8 | 3,749 | 21.8 | 4,503 | 3,208 | 1,295 | NA | NA |
| Lee | 1,060 | 98,299 | 1.35 | 1,076 | 12.6 | 1,050 | 12.3 | 11,132 | 15.5 | 17,318 | 15,863 | 1,455 | NA | NA |
| Leflore | 1,264 | 32,103 | 1.31 | 389 | 14.0 | 365 | 13.1 | 3,376 | 14.9 | 5,943 | 4,625 | 1,318 | NA | 250 |
| Lincoln | 648 | 33,680 | 0.95 | 393 | 11.6 | 440 | 13.0 | 4,531 | 16.3 | 7,496 | 6,045 | 1,452 | 31 | 299 |
| Lowndes | 1,232 | 63,641 | 1.18 | 794 | 13.6 | 678 | 11.6 | 6,397 | 13.4 | 11,970 | 10,622 | 1,348 | NA | NA |
| Madison | 1,771 | 111,391 | 1.13 | 1,343 | 12.6 | 1,174 | 11.0 | 9,831 | 10.9 | 19,020 | 14,451 | 4,568 | NA | 223 |
| Marion | 808 | 24,537 | 0.95 | 292 | 11.9 | 373 | 15.3 | 3,564 | 18.6 | 5,765 | 4,019 | 1,747 | NA | NA |
| Marshall | 1,786 | 30,678 | 0.65 | 406 | 11.5 | 486 | 13.8 | 4,185 | 15.5 | 7,867 | 5,621 | 2,246 | NA | NA |
| Monroe | 399 | 31,644 | 0.71 | 385 | 11.0 | 463 | 13.2 | 4,160 | 14.9 | 8,804 | 7,782 | 1,022 | NA | NA |
| Montgomery | 97 | 8,660 | 0.65 | 117 | 12.1 | 145 | 15.0 | 1,118 | 14.8 | 2,771 | 2,364 | 407 | NA | NA |
| Neshoba | 367 | 30,271 | 1.08 | 433 | 14.9 | 405 | 14.0 | 5,181 | 21.5 | 5,847 | 4,812 | 1,036 | NA | NA |
| Newton | 553 | 19,486 | 0.76 | 259 | 12.4 | 290 | 13.9 | 2,998 | 17.7 | 4,710 | 3,638 | 1,072 | NA | NA |
| Noxubee | 157 | 9,532 | 0.70 | 152 | 14.8 | 137 | 13.4 | 1,638 | 19.4 | 2,474 | 2,207 | 267 | NA | NA |
| Oktibbeha | 4,722 | 50,400 | 1.04 | 513 | 10.3 | 388 | 7.8 | 5,660 | 14.4 | 7,131 | 6,264 | 867 | NA | 464 |
| Panola | 316 | 32,713 | 0.88 | 485 | 14.3 | 456 | 13.5 | 4,612 | 16.3 | 7,480 | 5,417 | 2,063 | NA | NA |
| Pearl River | 1,236 | 47,103 | 0.62 | 616 | 11.0 | 713 | 12.8 | 7,202 | 16.5 | 12,960 | 8,528 | 4,432 | NA | NA |
| Perry | 102 | 10,603 | 0.69 | 140 | 11.8 | 136 | 11.5 | 1,699 | 17.6 | 2,825 | 2,069 | 756 | NA | NA |
| Pike | 945 | 41,330 | 1.13 | 521 | 13.4 | 553 | 14.2 | 5,000 | 16.0 | 8,933 | 6,536 | 2,397 | NA | NA |
| Pontotoc | 233 | 30,402 | 0.91 | 449 | 13.8 | 290 | 8.9 | 5,023 | 18.6 | 6,569 | 5,961 | 608 | NA | NA |
| Prentiss | 722 | 23,596 | 0.83 | 283 | 11.3 | 284 | 11.4 | 3,353 | 17.0 | 5,786 | 5,379 | 407 | NA | NA |
| Quitman | 142 | 6,357 | 0.63 | 70 | 10.4 | 107 | 15.8 | 868 | 16.2 | 1,639 | 1,179 | 460 | NA | NA |
| Rankin | 6,357 | 148,933 | 0.95 | 1,715 | 11.0 | 1,339 | 8.6 | 16,801 | 13.3 | 28,141 | 21,333 | 6,808 | NA | NA |
| Scott | 184 | 29,079 | 1.07 | 454 | 16.2 | 332 | 11.8 | 5,097 | 21.8 | 5,589 | 3,854 | 1,735 | NA | NA |
| Sharkey | 108 | 4,116 | 0.81 | 57 | 13.7 | 50 | 12.0 | 603 | 17.8 | 1,092 | 790 | 301 | NA | NA |
| Simpson | 545 | 23,371 | 0.67 | 324 | 12.2 | 324 | 12.2 | 3,713 | 17.3 | 6,113 | 4,460 | 1,653 | NA | NA |
| Smith | 118 | 13,928 | 0.66 | 188 | 11.9 | 197 | 12.5 | 2,208 | 17.6 | 3,400 | 2,581 | 820 | NA | NA |
| Stone | 1,276 | 16,276 | 0.74 | 218 | 11.9 | 199 | 10.8 | 2,417 | 17.2 | 3,820 | 2,600 | 1,220 | 47 | 174 |

1. Per 1,000 estimated resident population.

# Table B. States and Counties — Crime, Education, Money Income, and Poverty

| STATE County | County law enforcement employment, 2019 | | Education | | | | | | Money income, 2015-2019 | | | | Income and poverty, 2019 | | | |
|---|---|---|---|---|---|---|---|---|---|---|---|---|---|---|---|---|
| | | | School enrollment and attainment, 2015-2019 | | | | Local government expenditures,[3] 2017-2018 | | | | Households | | | Percent below poverty level | | |
| | | | Enrollment[1] | | Attainment[2] (percent) | | | | | | | Percent | | | | |
| | Officers | Civilians | Total | Percent private | High school graduate or less | Bachelor's degree or more | Total current spending (mil dol) | Current spending per student (dollars) | Per capita income[4] | Median income (dollars) | with income of less than $50,000 | with income of $200,000 or more | Median household income (dollars) | All persons | Children under 18 years | Children 5 to 17 years in families |
| | 46 | 47 | 48 | 49 | 50 | 51 | 52 | 53 | 54 | 55 | 56 | 57 | 58 | 59 | 60 | 61 |
| MISSISSIPPI | X | X | 769,205 | 13.4 | 45.9 | 22.0 | 4,282 | 8,954 | 24,369 | 45,081 | 54.2 | 3.0 | 45,928 | 19.5 | 27.6 | 26.4 |
| Adams | NA | NA | 6,628 | 14.4 | 55.1 | 17.4 | 38 | 11,533 | 18,386 | 29,936 | 69.0 | 2.0 | 34,583 | 27.9 | 40.8 | 37.4 |
| Alcorn | NA | NA | 8,242 | 10.0 | 50.9 | 17.1 | 49 | 8,280 | 23,909 | 42,086 | 58.4 | 2.5 | 43,496 | 17.3 | 23.7 | 18.4 |
| Amite | NA | NA | 2,620 | 23.2 | 57.4 | 12.6 | 13 | 13,021 | 21,568 | 33,981 | 63.9 | 1.7 | 41,575 | 20.9 | 29.1 | 28.5 |
| Attala | NA | NA | 4,406 | 6.2 | 54.8 | 13.9 | 31 | 8,872 | 21,276 | 33,767 | 64.5 | 2.7 | 36,905 | 24.1 | 35.0 | 34.1 |
| Benton | NA | NA | 1,848 | 4.2 | 60.8 | 11.5 | 11 | 9,884 | 20,643 | 36,258 | 63.9 | 1.9 | 37,875 | 20.7 | 31.1 | 28.5 |
| Bolivar | NA | NA | 9,597 | 7.6 | 47.0 | 23.6 | 56 | 9,726 | 18,843 | 29,854 | 65.9 | 2.0 | 30,309 | 36.6 | 49.0 | 48.7 |
| Calhoun | NA | NA | 3,251 | 5.2 | 59.8 | 11.8 | 20 | 7,991 | 19,229 | 37,263 | 63.5 | 0.3 | 38,648 | 17.2 | 25.8 | 23.6 |
| Carroll | NA | NA | 1,953 | 29.1 | 55.2 | 12.2 | 10 | 10,221 | 22,385 | 46,052 | 51.7 | 2.3 | 44,885 | 17.2 | 23.5 | 21.1 |
| Chickasaw | NA | NA | 4,162 | 6.0 | 62.9 | 12.4 | 26 | 9,295 | 19,926 | 36,548 | 64.4 | 1.3 | 37,533 | 22.7 | 32.8 | 32.9 |
| Choctaw | 8 | 6 | 2,033 | 14.8 | 55.4 | 19.1 | 16 | 11,800 | 22,257 | 36,777 | 60.1 | 2.1 | 41,146 | 20.9 | 31.8 | 29.7 |
| Claiborne | 13 | 16 | 3,084 | 5.0 | 47.6 | 24.0 | 15 | 10,204 | 14,320 | 29,338 | 72.0 | 0.8 | 31,221 | 37.5 | 48.9 | 45.6 |
| Clarke | NA | NA | 3,379 | 6.0 | 56.5 | 12.5 | 25 | 8,694 | 23,988 | 43,207 | 55.7 | 2.0 | 42,826 | 23.2 | 33.2 | 31.9 |
| Clay | NA | NA | 4,927 | 11.4 | 52.6 | 20.4 | 30 | 9,835 | 20,599 | 31,833 | 67.1 | 1.6 | 33,653 | 29.7 | 34.7 | 31.3 |
| Coahoma | NA | NA | 6,669 | 9.8 | 45.7 | 16.5 | 41 | 9,528 | 18,985 | 29,121 | 69.0 | 2.0 | 30,242 | 38.2 | 49.1 | 44.0 |
| Copiah | NA | NA | 6,881 | 12.1 | 54.2 | 14.5 | 35 | 8,166 | 20,093 | 42,151 | 57.5 | 1.7 | 42,299 | 22.1 | 33.0 | 32.1 |
| Covington | NA | NA | 4,101 | 14.0 | 53.4 | 14.7 | 26 | 9,048 | 19,180 | 38,178 | 61.9 | 0.8 | 42,076 | 21.3 | 33.3 | 30.0 |
| DeSoto | 134 | 140 | 48,832 | 13.8 | 38.8 | 25.0 | 250 | 7,368 | 30,195 | 67,038 | 36.8 | 4.0 | 67,450 | 9.4 | 13.4 | 12.2 |
| Forrest | NA | NA | 22,753 | 13.4 | 39.9 | 27.8 | 106 | 9,562 | 22,534 | 39,840 | 59.7 | 2.1 | 40,224 | 23.3 | 28.9 | 29.0 |
| Franklin | NA | NA | 2,007 | 6.7 | 50.8 | 12.0 | 13 | 10,087 | 23,067 | 40,219 | 57.7 | 2.8 | 41,420 | 19.6 | 27.9 | 25.9 |
| George | 17 | 34 | 5,937 | 15.2 | 52.3 | 13.8 | 32 | 7,814 | 22,732 | 47,292 | 51.4 | 1.5 | 51,858 | 16.6 | 23.9 | 23.9 |
| Greene | 7 | 15 | 2,551 | 13.4 | 57.5 | 9.7 | 17 | 8,780 | 16,006 | 49,677 | 50.2 | 1.7 | 48,641 | 21.8 | 23.9 | 22.2 |
| Grenada | NA | NA | 5,209 | 9.3 | 47.0 | 21.3 | 35 | 8,575 | 23,135 | 40,122 | 62.0 | 2.1 | 40,561 | 20.3 | 28.3 | 27.5 |
| Hancock | 72 | 74 | 10,173 | 15.0 | 40.7 | 22.0 | 56 | 8,678 | 27,462 | 48,119 | 51.5 | 2.8 | 50,401 | 16.7 | 26.1 | 25.2 |
| Harrison | 118 | 184 | 50,163 | 11.7 | 40.9 | 23.1 | 287 | 8,761 | 25,415 | 47,894 | 52.1 | 2.8 | 49,619 | 17.5 | 27.3 | 28.1 |
| Hinds | 80 | 258 | 69,012 | 18.6 | 37.4 | 29.0 | 348 | 9,214 | 23,734 | 44,625 | 55.0 | 2.9 | 44,773 | 20.1 | 29.9 | 28.2 |
| Holmes | NA | NA | 5,236 | 6.5 | 64.2 | 10.2 | 28 | 8,882 | 15,291 | 21,504 | 77.6 | 1.3 | 26,348 | 33.8 | 45.2 | 41.5 |
| Humphreys | NA | NA | 2,238 | 14.2 | 63.9 | 12.8 | 15 | 8,932 | 18,513 | 28,962 | 73.6 | 2.0 | 28,464 | 37.1 | 55.3 | 47.9 |
| Issaquena | NA | NA | 145 | 33.1 | 71.4 | 3.2 | NA | NA | 16,388 | 24,208 | 65.6 | 1.2 | 30,944 | 35.8 | 44.4 | 45.1 |
| Itawamba | NA | NA | 5,608 | 5.0 | 52.0 | 13.2 | 29 | 7,893 | 22,806 | 44,567 | 54.5 | 1.1 | 49,214 | 12.8 | 18.7 | 18.2 |
| Jackson | NA | NA | 35,985 | 10.5 | 42.5 | 21.1 | 236 | 9,785 | 27,141 | 51,657 | 48.5 | 3.5 | 53,414 | 15.1 | 21.5 | 20.7 |
| Jasper | NA | NA | 3,415 | 15.1 | 54.3 | 15.1 | 23 | 10,163 | 22,523 | 35,872 | 61.7 | 2.7 | 40,923 | 20.5 | 30.0 | 28.6 |
| Jefferson | NA | NA | 1,754 | 1.9 | 59.9 | 15.9 | 12 | 9,997 | 14,659 | 25,019 | 75.0 | 0.9 | 30,871 | 28.9 | 43.2 | 41.9 |
| Jefferson Davis | NA | NA | 2,217 | 20.1 | 54.2 | 14.1 | 16 | 11,367 | 20,638 | 32,116 | 65.0 | 1.7 | 32,077 | 24.3 | 36.4 | 36.0 |
| Jones | NA | NA | 16,202 | 15.4 | 45.8 | 20.8 | 94 | 7,932 | 23,163 | 41,775 | 57.2 | 2.3 | 41,733 | 23.8 | 38.1 | 37.6 |
| Kemper | 11 | 2 | 2,516 | 8.8 | 56.2 | 14.7 | 14 | 13,843 | 16,995 | 31,103 | 74.3 | 2.0 | 33,472 | 28.0 | 40.3 | 38.3 |
| Lafayette | NA | NA | 20,471 | 7.0 | 28.3 | 45.6 | 69 | 9,557 | 27,765 | 50,272 | 49.8 | 5.7 | 54,034 | 17.5 | 15.9 | 15.6 |
| Lamar | 46 | 7 | 17,264 | 17.5 | 34.8 | 34.1 | 89 | 8,467 | 30,062 | 60,328 | 41.8 | 4.9 | 62,606 | 16.0 | 22.1 | 20.9 |
| Lauderdale | 52 | 75 | 18,845 | 9.9 | 42.9 | 19.7 | 108 | 9,107 | 24,698 | 42,534 | 57.7 | 2.8 | 41,604 | 21.5 | 31.7 | 29.8 |
| Lawrence | NA | NA | 2,982 | 7.8 | 51.9 | 17.6 | 18 | 8,422 | 22,980 | 41,914 | 56.2 | 2.5 | 43,668 | 19.7 | 26.1 | 24.5 |
| Leake | NA | NA | 5,995 | 19.6 | 58.0 | 14.3 | 23 | 7,836 | 19,893 | 37,096 | 61.1 | 2.6 | 41,007 | 23.8 | 36.2 | 35.2 |
| Lee | NA | NA | 21,005 | 11.3 | 42.3 | 24.4 | 154 | 9,551 | 26,369 | 50,559 | 49.5 | 3.8 | 50,803 | 14.6 | 23.8 | 22.4 |
| Leflore | NA | NA | 8,830 | 11.0 | 56.5 | 18.5 | 47 | 9,182 | 18,419 | 26,735 | 70.0 | 2.3 | 29,687 | 35.7 | 50.8 | 50.4 |
| Lincoln | 24 | 28 | 8,006 | 11.5 | 51.3 | 16.2 | 51 | 8,265 | 23,541 | 42,606 | 54.9 | 3.1 | 45,062 | 19.6 | 24.2 | 23.4 |
| Lowndes | 49 | 59 | 14,720 | 12.8 | 45.9 | 23.4 | 91 | 9,582 | 26,438 | 50,441 | 49.7 | 2.8 | 49,251 | 17.2 | 25.2 | 25.4 |
| Madison | 81 | 4 | 29,118 | 26.5 | 25.3 | 48.4 | 154 | 9,198 | 38,901 | 71,824 | 35.4 | 9.6 | 70,824 | 10.1 | 15.0 | 14.7 |
| Marion | 15 | 6 | 5,652 | 20.6 | 57.8 | 12.3 | 35 | 9,316 | 19,545 | 32,090 | 63.9 | 1.6 | 38,402 | 23.8 | 30.8 | 29.6 |
| Marshall | NA | NA | 7,968 | 19.8 | 56.7 | 13.8 | 38 | 8,613 | 21,853 | 42,233 | 57.1 | 1.9 | 43,107 | 20.3 | 30.5 | 29.4 |
| Monroe | NA | NA | 7,670 | 6.9 | 55.0 | 16.7 | 45 | 8,736 | 23,844 | 42,354 | 56.5 | 2.1 | 42,464 | 17.9 | 23.9 | 22.9 |
| Montgomery | NA | NA | 2,184 | 14.5 | 55.4 | 19.3 | 15 | 11,098 | 24,054 | 39,840 | 63.5 | 1.8 | 38,475 | 23.7 | 37.0 | 37.0 |
| Neshoba | NA | NA | 7,715 | 7.1 | 55.4 | 11.6 | 33 | 7,755 | 19,540 | 37,987 | 61.7 | 1.6 | 41,795 | 21.7 | 30.0 | 27.9 |
| Newton | NA | NA | 6,006 | 8.7 | 45.7 | 16.4 | 34 | 8,932 | 20,787 | 35,958 | 61.5 | 1.9 | 42,962 | 20.3 | 27.4 | 25.2 |
| Noxubee | NA | NA | 2,429 | 11.2 | 58.7 | 15.6 | 17 | 10,438 | 18,525 | 33,784 | 65.1 | 0.8 | 32,916 | 29.2 | 41.3 | 40.6 |
| Oktibbeha | NA | NA | 22,009 | 8.0 | 31.0 | 42.6 | 49 | 9,655 | 22,512 | 40,453 | 57.1 | 4.0 | 41,631 | 31.1 | 27.5 | 25.8 |
| Panola | 43 | 40 | 8,350 | 12.8 | 56.0 | 14.9 | 56 | 9,582 | 19,892 | 38,304 | 62.0 | 0.9 | 40,655 | 22.8 | 32.6 | 30.8 |
| Pearl River | NA | NA | 13,204 | 13.8 | 46.6 | 16.7 | 79 | 9,062 | 23,920 | 46,901 | 51.4 | 2.5 | 45,874 | 18.1 | 24.8 | 23.5 |
| Perry | NA | NA | 2,848 | 8.1 | 57.4 | 10.7 | 17 | 9,890 | 23,166 | 42,469 | 57.5 | 1.3 | 43,106 | 19.9 | 28.8 | 27.1 |
| Pike | NA | NA | 9,837 | 10.5 | 52.7 | 16.6 | 60 | 8,903 | 17,621 | 31,784 | 68.8 | 1.5 | 35,044 | 26.2 | 36.5 | 35.9 |
| Pontotoc | 23 | 17 | 7,984 | 9.0 | 52.6 | 18.1 | 46 | 7,711 | 21,191 | 44,759 | 55.0 | 2.0 | 48,910 | 15.9 | 24.3 | 21.6 |
| Prentiss | NA | NA | 6,311 | 4.5 | 54.4 | 12.4 | 32 | 8,732 | 19,693 | 39,256 | 62.4 | 0.9 | 43,465 | 15.7 | 22.9 | 21.9 |
| Quitman | NA | NA | 1,658 | 17.2 | 59.7 | 10.3 | 13 | 12,355 | 15,026 | 25,283 | 72.6 | 0.3 | 29,318 | 35.0 | 51.8 | 48.5 |
| Rankin | NA | NA | 37,566 | 19.8 | 37.2 | 29.0 | 200 | 8,474 | 30,835 | 65,996 | 36.2 | 4.5 | 67,284 | 11.1 | 14.4 | 12.5 |
| Scott | NA | NA | 6,530 | 7.3 | 63.5 | 11.4 | 46 | 8,065 | 22,203 | 34,943 | 61.3 | 2.5 | 41,532 | 19.7 | 29.2 | 29.6 |
| Sharkey | NA | NA | 1,138 | 10.5 | 50.4 | 19.5 | 10 | 11,457 | 19,958 | 29,394 | 71.3 | 2.1 | 31,138 | 33.4 | 50.8 | 47.1 |
| Simpson | NA | NA | 6,944 | 18.3 | 50.4 | 16.2 | 30 | 8,308 | 23,520 | 43,850 | 56.3 | 4.3 | 47,664 | 17.8 | 28.3 | 26.7 |
| Smith | NA | NA | 3,539 | 6.8 | 60.0 | 12.2 | 23 | 8,409 | 23,698 | 43,105 | 55.2 | 2.7 | 46,296 | 18.9 | 27.4 | 24.8 |
| Stone | 22 | 4 | 4,204 | 11.1 | 53.6 | 15.8 | 22 | 8,591 | 23,820 | 45,483 | 53.7 | 3.0 | 47,187 | 18.5 | 28.7 | 25.0 |

1. All persons 3 years old and over enrolled in nursery school through college.　2. Persons 25 years old and over.　3. Elementary and secondary education expenditures.　4. Based on population estimated by the American Community Survey, 2014–2018.

# Table B. States and Counties — **Personal Income**

| STATE County | Total (mil dol) | Percent change 2018-2019 | Per capita[1] Dollars | Per capita[1] Rank | Wages and salaries (mil dol) | Pension and insurance | Government social insurance | Proprietors' income (mil dol) | Dividends, interest, and rent (mil dol) | Personal transfer receipts (mil dol) | Total (mil dol) | From employee and self-employed | From employer |
|---|---|---|---|---|---|---|---|---|---|---|---|---|---|
| | 62 | 63 | 64 | 65 | 66 | 67 | 68 | 69 | 70 | 71 | 72 | 73 | 74 |
| MISSISSIPPI | 115,814 | 2.6 | 38,887 | X | 51,152 | 8,700 | 3,808 | 8,146 | 18,591 | 30,683 | 71,805 | 5,062 | 3,808 |
| Adams | 1,092 | 2.8 | 35,563 | 2,608 | 438 | 68 | 32 | 92 | 206 | 378 | 630 | 49 | 32 |
| Alcorn | 1,352 | 1.6 | 36,592 | 2,497 | 589 | 96 | 43 | 74 | 193 | 422 | 803 | 62 | 43 |
| Amite | 432 | 1.0 | 35,151 | 2,667 | 89 | 15 | 6 | 34 | 52 | 147 | 144 | 12 | 6 |
| Attala | 629 | 3.7 | 34,583 | 2,737 | 176 | 32 | 13 | 38 | 83 | 218 | 259 | 22 | 13 |
| Benton | 259 | 3.4 | 31,411 | 2,983 | 42 | 9 | 3 | 11 | 27 | 93 | 65 | 7 | 3 |
| Bolivar | 1,200 | -1.3 | 39,179 | 2,165 | 449 | 84 | 34 | 80 | 208 | 415 | 647 | 47 | 34 |
| Calhoun | 480 | 3.8 | 33,450 | 2,850 | 118 | 22 | 9 | 35 | 62 | 174 | 184 | 16 | 9 |
| Carroll | 383 | 1.5 | 38,497 | 2,260 | 47 | 9 | 3 | 15 | 55 | 116 | 75 | 9 | 3 |
| Chickasaw | 633 | 3.6 | 36,994 | 2,449 | 204 | 35 | 15 | 90 | 105 | 194 | 344 | 24 | 15 |
| Choctaw | 287 | 5.0 | 34,999 | 2,686 | 111 | 23 | 8 | 17 | 35 | 93 | 160 | 12 | 8 |
| Claiborne | 283 | 2.2 | 31,501 | 2,978 | 213 | 52 | 15 | 4 | 36 | 131 | 284 | 18 | 15 |
| Clarke | 602 | 3.6 | 38,735 | 2,232 | 117 | 22 | 8 | 31 | 68 | 201 | 178 | 16 | 8 |
| Clay | 762 | 5.6 | 39,449 | 2,118 | 235 | 37 | 17 | 74 | 116 | 219 | 363 | 26 | 17 |
| Coahoma | 820 | 1.9 | 37,070 | 2,439 | 305 | 50 | 22 | 48 | 140 | 322 | 426 | 32 | 22 |
| Copiah | 952 | 1.9 | 33,932 | 2,808 | 256 | 52 | 20 | 46 | 106 | 317 | 374 | 31 | 20 |
| Covington | 689 | -1.5 | 36,985 | 2,451 | 224 | 41 | 17 | 85 | 89 | 213 | 366 | 25 | 17 |
| DeSoto | 7,769 | 4.0 | 42,007 | 1,763 | 2,656 | 370 | 198 | 533 | 797 | 1,283 | 3,757 | 269 | 198 |
| Forrest | 2,955 | 2.3 | 39,450 | 2,117 | 1,902 | 334 | 139 | 224 | 590 | 792 | 2,599 | 170 | 139 |
| Franklin | 273 | 4.7 | 35,353 | 2,639 | 71 | 14 | 5 | 13 | 35 | 99 | 104 | 9 | 5 |
| George | 823 | 3.3 | 33,595 | 2,837 | 193 | 37 | 14 | 23 | 104 | 213 | 267 | 23 | 14 |
| Greene | 389 | 1.1 | 28,617 | 3,072 | 76 | 16 | 6 | 8 | 41 | 122 | 105 | 10 | 6 |
| Grenada | 769 | 2.4 | 37,067 | 2,442 | 420 | 73 | 31 | 58 | 101 | 267 | 582 | 42 | 31 |
| Hancock | 1,740 | 3.3 | 36,533 | 2,504 | 827 | 158 | 66 | 74 | 371 | 468 | 1,124 | 79 | 66 |
| Harrison | 7,936 | 3.6 | 38,140 | 2,315 | 4,295 | 797 | 331 | 457 | 1,593 | 2,066 | 5,881 | 386 | 331 |
| Hinds | 9,338 | 1.6 | 40,279 | 2,018 | 6,477 | 1,097 | 466 | 772 | 1,918 | 2,351 | 8,812 | 576 | 466 |
| Holmes | 510 | 2.3 | 29,976 | 3,037 | 132 | 27 | 10 | 2 | 61 | 256 | 170 | 17 | 10 |
| Humphreys | 282 | 4.2 | 34,941 | 2,694 | 75 | 15 | 6 | 33 | 46 | 112 | 129 | 10 | 6 |
| Issaquena | 36 | 14.2 | 27,430 | 3,088 | 7 | 1 | 1 | 6 | 8 | 9 | 15 | 1 | 1 |
| Itawamba | 851 | 3.4 | 36,375 | 2,526 | 245 | 42 | 19 | 41 | 91 | 256 | 347 | 29 | 19 |
| Jackson | 5,496 | 2.9 | 38,269 | 2,301 | 2,732 | 542 | 210 | 262 | 903 | 1,325 | 3,747 | 256 | 210 |
| Jasper | 650 | 0.2 | 39,699 | 2,086 | 187 | 33 | 14 | 69 | 72 | 199 | 302 | 22 | 14 |
| Jefferson | 244 | 0.5 | 34,860 | 2,705 | 44 | 11 | 4 | 9 | 25 | 96 | 67 | 6 | 4 |
| Jefferson Davis | 339 | 0.6 | 30,429 | 3,024 | 68 | 13 | 5 | 19 | 41 | 138 | 106 | 10 | 5 |
| Jones | 2,638 | 2.4 | 38,740 | 2,231 | 1,278 | 245 | 96 | 156 | 415 | 824 | 1,774 | 121 | 96 |
| Kemper | 292 | 3.9 | 30,021 | 3,036 | 97 | 19 | 8 | 6 | 38 | 117 | 130 | 11 | 8 |
| Lafayette | 2,367 | 3.0 | 43,820 | 1,513 | 1,093 | 192 | 78 | 212 | 591 | 437 | 1,576 | 104 | 78 |
| Lamar | 2,547 | 2.9 | 40,207 | 2,028 | 756 | 113 | 55 | 202 | 419 | 452 | 1,126 | 82 | 55 |
| Lauderdale | 2,917 | 2.9 | 39,359 | 2,134 | 1,526 | 263 | 117 | 218 | 512 | 822 | 2,123 | 146 | 117 |
| Lawrence | 432 | 0.1 | 34,308 | 2,770 | 117 | 21 | 8 | 40 | 46 | 163 | 186 | 15 | 8 |
| Leake | 716 | -2.7 | 31,441 | 2,982 | 194 | 35 | 16 | 93 | 79 | 235 | 337 | 23 | 16 |
| Lee | 3,643 | 3.4 | 42,635 | 1,667 | 2,537 | 364 | 186 | 277 | 559 | 807 | 3,364 | 227 | 186 |
| Leflore | 1,098 | 1.8 | 38,974 | 2,197 | 573 | 105 | 43 | 86 | 224 | 362 | 808 | 54 | 43 |
| Lincoln | 1,347 | 2.3 | 39,441 | 2,119 | 535 | 81 | 39 | 100 | 166 | 378 | 755 | 56 | 39 |
| Lowndes | 2,386 | 4.7 | 40,717 | 1,959 | 1,360 | 236 | 107 | 123 | 423 | 598 | 1,826 | 121 | 107 |
| Madison | 6,987 | 3.0 | 65,746 | 173 | 2,836 | 380 | 207 | 683 | 1,588 | 854 | 4,106 | 277 | 207 |
| Marion | 951 | 2.2 | 38,701 | 2,237 | 348 | 55 | 25 | 191 | 122 | 307 | 618 | 46 | 25 |
| Marshall | 1,206 | 3.5 | 34,165 | 2,785 | 351 | 51 | 27 | 58 | 119 | 380 | 487 | 42 | 27 |
| Monroe | 1,279 | 3.2 | 36,295 | 2,538 | 456 | 79 | 35 | 76 | 191 | 398 | 646 | 51 | 35 |
| Montgomery | 351 | 2.8 | 35,910 | 2,579 | 85 | 15 | 6 | 22 | 53 | 136 | 128 | 11 | 6 |
| Neshoba | 1,071 | 1.4 | 36,785 | 2,474 | 506 | 85 | 36 | 114 | 152 | 312 | 740 | 49 | 36 |
| Newton | 747 | 0.7 | 35,563 | 2,608 | 222 | 43 | 17 | 51 | 98 | 248 | 333 | 25 | 17 |
| Noxubee | 356 | 5.9 | 34,204 | 2,780 | 98 | 19 | 7 | 31 | 43 | 135 | 155 | 12 | 7 |
| Oktibbeha | 1,758 | 3.3 | 35,457 | 2,624 | 904 | 176 | 65 | 76 | 365 | 390 | 1,220 | 79 | 65 |
| Panola | 1,146 | 1.9 | 33,528 | 2,847 | 434 | 73 | 32 | 86 | 153 | 374 | 625 | 48 | 32 |
| Pearl River | 2,065 | 3.5 | 37,178 | 2,427 | 423 | 80 | 31 | 95 | 258 | 625 | 629 | 59 | 31 |
| Perry | 397 | 3.1 | 33,188 | 2,879 | 107 | 18 | 8 | 19 | 40 | 144 | 151 | 13 | 8 |
| Pike | 1,239 | 2.6 | 31,536 | 2,975 | 552 | 98 | 40 | 59 | 172 | 459 | 750 | 57 | 40 |
| Pontotoc | 1,092 | 4.3 | 33,932 | 2,808 | 500 | 76 | 39 | 74 | 124 | 281 | 689 | 51 | 39 |
| Prentiss | 787 | 3.5 | 31,314 | 2,989 | 285 | 50 | 21 | 38 | 99 | 264 | 395 | 32 | 21 |
| Quitman | 205 | -0.7 | 30,185 | 3,030 | 39 | 8 | 3 | 9 | 31 | 98 | 58 | 6 | 3 |
| Rankin | 7,024 | 3.1 | 45,237 | 1,330 | 3,114 | 451 | 224 | 451 | 1,039 | 1,347 | 4,239 | 293 | 224 |
| Scott | 888 | -0.9 | 31,585 | 2,971 | 514 | 88 | 42 | 71 | 90 | 294 | 715 | 48 | 42 |
| Sharkey | 151 | 0.9 | 34,830 | 2,711 | 39 | 8 | 3 | 14 | 26 | 64 | 64 | 5 | 3 |
| Simpson | 975 | 0.2 | 36,587 | 2,498 | 238 | 47 | 18 | 76 | 106 | 337 | 379 | 30 | 18 |
| Smith | 564 | -1.6 | 35,445 | 2,627 | 124 | 25 | 9 | 84 | 58 | 172 | 242 | 17 | 9 |
| Stone | 612 | 4.0 | 33,369 | 2,859 | 180 | 33 | 13 | 28 | 76 | 209 | 253 | 21 | 13 |

1. Based on the resident population estimated as of July 1 of the year shown.

# Table B. States and Counties — Earnings, Social Security, and Housing

Earnings, 2019 (cont.) — Percent by selected industries. Social Security beneficiaries, December 2019. Housing units, 2020.

| STATE County | Farm | Mining, quarrying, and extractions | Construction | Manufacturing | Information; professional, scientific, technical services | Retail trade | Finance, insurance, real estate, and leasing | Health care and social assistance | Government | Number | Rate[1] | Supplemental Security Income recipients, 2019 | Total | Percent change, 2010-2020 |
|---|---|---|---|---|---|---|---|---|---|---|---|---|---|---|
| | 75 | 76 | 77 | 78 | 79 | 80 | 81 | 82 | 83 | 84 | 85 | 86 | 87 | 88 |
| MISSISSIPPI | 1.4 | 0.8 | 5.7 | 13.6 | 5.5 | 7.3 | 4.9 | 12.0 | 22.5 | 677,464 | 227 | 115,638 | 1,347,082 | 5.7 |
| Adams | 0.2 | 3.8 | 4.0 | 6.4 | 4.5 | 12.7 | 6.2 | D | 13.4 | 8,070 | 263 | 1,681 | 14,745 | 0.6 |
| Alcorn | 0.1 | 0.0 | 4.2 | 19.8 | 2.4 | 10.5 | 4.1 | 13.0 | 20.4 | 10,185 | 275 | 1,581 | 17,326 | 1.5 |
| Amite | 14.8 | D | 7.8 | 13.9 | D | 4.8 | D | D | 14.5 | 3,615 | 294 | 573 | 6,898 | 4.0 |
| Attala | 2.4 | 0.4 | 8.9 | 13.9 | D | 11.2 | 4.8 | D | 21.5 | 4,955 | 273 | 860 | 9,326 | 2.2 |
| Benton | 0.1 | 0.0 | D | 12.0 | 5.7 | D | 1.5 | 11.5 | 36.3 | 2,155 | 260 | 387 | 4,311 | 3.0 |
| Bolivar | 4.6 | 0.2 | 3.3 | 15.2 | D | 10.3 | 4.3 | 14.3 | 19.1 | 8,155 | 266 | 2,707 | 14,419 | 2.5 |
| Calhoun | 8.2 | 0.1 | 2.5 | 21.0 | 1.6 | 7.5 | 3.5 | D | 17.3 | 4,110 | 286 | 662 | 7,046 | 2.0 |
| Carroll | -4.2 | D | 19.6 | 12.3 | D | 4.4 | 3.1 | D | 23.1 | 3,055 | 308 | 413 | 5,238 | 3.7 |
| Chickasaw | 16.0 | 0.2 | 4.1 | 34.6 | 1.7 | 7.7 | 2.3 | D | 13.1 | 4,705 | 276 | 932 | 7,624 | 1.5 |
| Choctaw | 2.9 | D | 1.5 | 9.4 | D | 3.5 | D | D | 28.4 | 1,950 | 238 | 310 | 4,239 | 2.1 |
| Claiborne | -0.9 | 0.0 | D | 2.8 | D | 2.2 | D | D | 26.6 | 2,225 | 247 | 626 | 4,394 | 4.1 |
| Clarke | 7.5 | 7.6 | 11.7 | 11.5 | D | 5.4 | D | D | 21.1 | 4,625 | 298 | 700 | 8,072 | 2.5 |
| Clay | 12.0 | D | 4.0 | 17.1 | 3.1 | 9.3 | 3.1 | D | 12.7 | 5,210 | 269 | 982 | 9,387 | 2.2 |
| Coahoma | 3.6 | 0.1 | 2.7 | 6.7 | 5.0 | 6.3 | 9.6 | 17.1 | 22.0 | 5,660 | 256 | 1,933 | 10,697 | -0.9 |
| Copiah | 1.9 | 1.7 | 3.5 | 30.7 | 2.5 | 7.7 | 2.3 | 10.7 | 21.8 | 6,950 | 247 | 1,513 | 12,425 | 2.0 |
| Covington | 12.4 | D | 9.8 | 18.0 | D | 6.1 | 3.1 | 5.9 | 18.4 | 4,760 | 256 | 901 | 8,790 | 3.5 |
| DeSoto | 0.1 | D | 9.1 | 8.2 | 3.0 | 11.0 | 4.2 | 11.1 | 11.6 | 31,825 | 172 | 2,925 | 71,068 | 15.3 |
| Forrest | 0.0 | 0.1 | 4.6 | 9.2 | 3.7 | 6.3 | 4.5 | 17.5 | 29.0 | 15,400 | 205 | 2,608 | 33,289 | 3.2 |
| Franklin | 1.0 | D | D | D | D | 9.8 | D | D | 34.0 | 2,180 | 283 | 365 | 4,329 | 4.2 |
| George | -0.2 | D | 9.3 | 7.0 | 3.5 | 11.5 | 4.4 | D | 32.0 | 5,205 | 213 | 767 | 9,617 | 3.1 |
| Greene | -0.1 | 0.6 | D | D | D | 5.8 | D | D | 43.3 | 2,780 | 205 | 416 | 5,274 | 3.0 |
| Grenada | 0.7 | 0.0 | 2.8 | 27.7 | D | 11.2 | 3.8 | 7.3 | 19.2 | 6,120 | 295 | 1,243 | 10,247 | 1.0 |
| Hancock | -0.2 | D | 5.4 | 8.3 | 12.3 | 4.5 | 1.8 | D | 37.1 | 10,805 | 227 | 1,221 | 27,077 | 23.8 |
| Harrison | 0.0 | D | 5.6 | 3.9 | 5.3 | 7.0 | 4.2 | 8.5 | 35.3 | 44,235 | 213 | 6,503 | 95,689 | 12.3 |
| Hinds | 0.0 | 1.1 | 2.9 | 3.9 | 7.9 | 4.6 | 6.8 | 16.6 | 32.0 | 47,605 | 205 | 10,638 | 104,547 | 1.1 |
| Holmes | 0.0 | 0.0 | 3.8 | 18.9 | D | 7.4 | 5.7 | D | 39.5 | 4,600 | 270 | 1,764 | 8,677 | 3.1 |
| Humphreys | -7 | 0.0 | 3.0 | 19.5 | 3.5 | 4.5 | D | 9.6 | 17.8 | 2,195 | 272 | 970 | 3,812 | -1.1 |
| Issaquena | 11.8 | 1.2 | D | 0.0 | D | D | D | D | 24.2 | 180 | 136 | 55 | 570 | 2.3 |
| Itawamba | 46.9 | 0.0 | 11.6 | 31.9 | 1.7 | 6.4 | 2.8 | 7.7 | 17.5 | 6,350 | 272 | 807 | 10,370 | 2.4 |
| Jackson | 1.1 | 0.0 | 7.5 | 36.7 | 6.4 | 4.6 | 3.0 | 7.3 | 19.1 | 32,170 | 224 | 3,440 | 63,231 | 5.3 |
| Jasper | 14.3 | D | 13.0 | 29.2 | 3.7 | 3.9 | D | D | 16.5 | 4,595 | 281 | 732 | 8,500 | 3.4 |
| Jefferson | 5.4 | 0.0 | D | D | D | 4.0 | D | 10.6 | 39.9 | 1,975 | 281 | 633 | 3,830 | 4.2 |
| Jefferson Davis | 8.7 | D | 22.6 | D | D | 6.7 | D | D | 28.6 | 2,980 | 269 | 508 | 6,012 | 2.4 |
| Jones | 2.5 | 6.6 | 5.9 | 24.9 | 3.4 | 5.7 | 3.6 | 5.2 | 20.9 | 15,860 | 232 | 2,493 | 29,356 | 3.2 |
| Kemper | D | D | D | 11.8 | D | 2.6 | D | D | 25.5 | 2,310 | 236 | 471 | 4,795 | 1.7 |
| Lafayette | 0.0 | 0.1 | 4.3 | 7.1 | 10.6 | 6.4 | 5.2 | 15.8 | 34.5 | 8,820 | 163 | 865 | 26,747 | 17.6 |
| Lamar | 0.7 | 0.9 | 8.9 | 1.6 | 6.6 | 13.7 | 7.3 | 22.8 | 12.0 | 10,745 | 169 | 1,488 | 24,835 | 3.0 |
| Lauderdale | 0.0 | 0.2 | 3.5 | 6.8 | 3.8 | 8.6 | 4.6 | 21.9 | 20.7 | 17,095 | 229 | 2,983 | 35,435 | 2.1 |
| Lawrence | 11.1 | 0.8 | 4.1 | 33.3 | D | 5.0 | 2.9 | 5.4 | 18.0 | 3,720 | 295 | 607 | 6,261 | 3.8 |
| Leake | 19.1 | 0.2 | D | D | D | 8.1 | 3.0 | D | 12.1 | 5,155 | 226 | 934 | 9,662 | 2.7 |
| Lee | 0.2 | 0.0 | 3.4 | 19.4 | 7.4 | 8.5 | 6.3 | 20.4 | 9.4 | 19,800 | 231 | 2,893 | 37,330 | 4.1 |
| Leflore | 3.6 | 0.0 | 6.8 | 13.8 | 4.5 | 6.2 | 5.6 | 8.8 | 27.5 | 6,835 | 242 | 2,374 | 13,040 | -1.2 |
| Lincoln | 2.0 | 1.2 | 8.8 | 10.3 | 3.5 | 9.4 | 4.2 | 14.9 | 11.8 | 8,475 | 248 | 1,226 | 15,784 | 3.5 |
| Lowndes | 0.6 | D | 6.6 | 18.6 | 4.1 | 5.9 | 4.5 | 14.2 | 22.8 | 13,340 | 228 | 2,399 | 27,595 | 3.9 |
| Madison | 0.1 | 1.3 | 5.2 | 15.6 | 15.5 | 9.0 | 10.0 | 10.1 | 7.2 | 19,460 | 183 | 2,001 | 44,999 | 16.7 |
| Marion | 3.7 | 2.4 | 12.1 | 6.3 | 4.2 | 6.8 | 3.4 | D | 11.0 | 6,700 | 273 | 1,108 | 12,189 | 3.0 |
| Marshall | -0.2 | 0.0 | 11.3 | 17.9 | D | 6.9 | 4.8 | D | 14.3 | 8,755 | 248 | 1,710 | 16,069 | 7.9 |
| Monroe | 1.2 | 0.3 | 10.4 | 33.8 | D | 5.6 | 2.0 | D | 13.0 | 9,610 | 272 | 1,132 | 16,724 | 1.7 |
| Montgomery | 6.1 | 0.0 | 7.0 | D | 3.8 | 7.5 | D | D | 23.2 | 3,010 | 308 | 657 | 5,957 | 14.5 |
| Neshoba | 6.9 | D | 10.9 | 4.1 | 1.9 | 9.9 | 3.1 | 4.0 | 41.4 | 6,695 | 230 | 1,177 | 12,634 | 2.2 |
| Newton | 7.9 | D | 8.8 | 17.0 | D | 4.8 | 2.0 | D | 28.7 | 5,435 | 259 | 619 | 9,616 | 2.6 |
| Noxubee | 7.8 | 0.0 | 4.1 | 25.1 | 0.9 | 11.4 | D | 3.4 | 24.9 | 2,815 | 270 | 870 | 5,326 | 3.0 |
| Oktibbeha | -0.2 | D | 2.5 | 6.6 | 4.4 | 5.8 | 4.1 | 7.7 | 49.7 | 7,670 | 155 | 1,400 | 22,302 | 6.5 |
| Panola | 0.3 | 0.1 | 6.4 | 16.5 | D | 10.9 | 5.5 | D | 19.1 | 8,585 | 252 | 1,965 | 15,234 | 3.7 |
| Pearl River | -0.5 | D | 8.2 | 6.7 | 5.1 | 12.9 | 4.5 | D | 28.0 | 14,985 | 269 | 1,891 | 26,242 | 9.6 |
| Perry | 2.6 | 1.8 | 4.5 | 36.4 | D | 5.6 | D | 10.6 | 17.3 | 3,160 | 263 | 497 | 5,741 | 4.0 |
| Pike | 0.1 | 0.6 | 3.0 | 17.1 | 3.5 | 11.4 | 5.0 | 9.9 | 26.3 | 9,865 | 251 | 2,121 | 19,069 | 6.7 |
| Pontotoc | -0.2 | 0.1 | 3.3 | 45.9 | D | 6.1 | 2.4 | D | 10.3 | 7,210 | 223 | 837 | 13,188 | 6.0 |
| Prentiss | 0.8 | 0.0 | 7.9 | 21.9 | D | 7.2 | 3.9 | D | 20.9 | 6,310 | 252 | 655 | 11,297 | 2.2 |
| Quitman | 2.6 | D | D | D | D | 4.2 | 8.1 | D | 35.7 | 1,820 | 267 | 578 | 3,559 | -0.8 |
| Rankin | 0.7 | 1.8 | 9.4 | 7.2 | 4.8 | 9.3 | 7.5 | 12.2 | 12.5 | 30,230 | 195 | 2,508 | 62,174 | 10.1 |
| Scott | 5.0 | D | 6.2 | 49.2 | D | 6.0 | 1.9 | D | 11.4 | 6,360 | 226 | 1,226 | 11,851 | 3.3 |
| Sharkey | 15.6 | 0.3 | 1.2 | 0.1 | D | 6.6 | D | D | 35.3 | 1,300 | 301 | 368 | 2,155 | 2.6 |
| Simpson | 6.9 | 0.5 | 8.0 | 2.2 | 3.2 | 7.6 | 5.6 | D | 26.6 | 6,545 | 244 | 1,235 | 12,289 | 3.0 |
| Smith | 20.3 | D | 2.9 | 24.7 | D | 2.6 | 3.3 | 4.9 | 12.7 | 4,050 | 256 | 493 | 7,471 | 3.2 |
| Stone | 0.5 | D | 7.0 | 12.4 | 3.3 | 9.4 | 3.3 | D | 28.4 | 4,285 | 234 | 628 | 7,643 | 6.7 |

1. Per 1,000 resident population estimated as of July 1 of the year shown.

# Table B. States and Counties — Housing, Labor Force, and Employment

| STATE County | Housing units, 2015-2019 | | | | | | | | Civilian labor force, 2020 | | | | Civilian employment[6], 2015-2019 | | |
| | Occupied units | | | | | | | Sub-standard units[4] (percent) | Total | Percent change, 2019-2020 | Unemployment | | Total | Percent | |
| | Owner-occupied | | | | | Renter-occupied | | | | | | | | | Management, business, science, and arts | Construction, production, and maintenance occupations |
| | Total | Percent | Median value[1] | Median owner cost as a percent of income | | Median rent[3] | Median rent as a percent of income[2] | | | | Total | Rate[5] | | | |
| | | | | With a mortgage | Without a mortgage[2] | | | | | | | | | | |
| | 89 | 90 | 91 | 92 | 93 | 94 | 95 | 96 | 97 | 98 | 99 | 100 | 101 | 102 | 103 |
|---|---|---|---|---|---|---|---|---|---|---|---|---|---|---|---|
| MISSISSIPPI | 1,104,394 | 68.2 | 119,000 | 20.0 | 10.6 | 780 | 29.7 | 3.0 | 1,259,347 | -1.6 | 101,801 | 8.1 | 1,235,224 | 32.2 | 28.6 |
| Adams | 11,237 | 62.4 | 92,100 | 22.2 | 11.7 | 600 | 33.4 | 1.0 | 10,609 | -5.1 | 1,168 | 11.0 | 10,449 | 29.1 | 29.4 |
| Alcorn | 14,649 | 68.7 | 102,700 | 19.1 | 10.1 | 617 | 22.6 | 1.3 | 15,419 | -4.9 | 961 | 6.2 | 15,323 | 28.8 | 34.4 |
| Amite | 5,218 | 84.9 | 83,600 | 19.5 | 10.7 | 661 | 29.2 | 1.8 | 4,460 | 0.2 | 376 | 8.4 | 4,309 | 34.6 | 33.7 |
| Attala | 6,941 | 72.5 | 80,700 | 19.0 | 10.0 | 510 | 36.7 | 4.2 | 6,977 | -0.9 | 580 | 8.3 | 6,756 | 23.0 | 39.4 |
| Benton | 3,149 | 80.6 | 78,000 | 19.2 | 12.0 | 503 | 18.9 | 1.9 | 2,995 | -2.2 | 250 | 8.3 | 3,010 | 22.1 | 44.6 |
| Bolivar | 12,111 | 53.6 | 96,100 | 20.3 | 13.2 | 637 | 32.8 | 2.4 | 11,537 | -3.7 | 1,006 | 8.7 | 10,739 | 32.1 | 24.6 |
| Calhoun | 5,846 | 74.5 | 70,500 | 20.7 | 11.7 | 581 | 23.7 | 3.0 | 5,598 | 0.1 | 452 | 8.1 | 5,479 | 19.6 | 47.5 |
| Carroll | 3,827 | 81.2 | 99,000 | 17.8 | 12.4 | 474 | 37.8 | 2.9 | 3,462 | -2.2 | 267 | 7.7 | 3,458 | 37.0 | 33.0 |
| Chickasaw | 6,476 | 69.7 | 66,500 | 21.9 | 10.9 | 625 | 27.1 | 2.3 | 6,964 | 0.3 | 757 | 10.9 | 6,798 | 19.4 | 42.0 |
| Choctaw | 3,228 | 78.6 | 81,400 | 19.3 | 11.7 | 556 | 32.5 | 2.6 | 3,601 | -3.1 | 231 | 6.4 | 3,002 | 27.6 | 36.3 |
| Claiborne | 2,908 | 69.5 | 67,500 | 21.6 | 14.2 | 623 | 26.5 | 3.0 | 3,105 | 0.0 | 448 | 14.4 | 2,719 | 26.6 | 30.4 |
| Clarke | 6,237 | 83.3 | 84,900 | 16.6 | 10.4 | 669 | 36.7 | 3.4 | 5,846 | -2.1 | 491 | 8.4 | 6,367 | 27.2 | 34.2 |
| Clay | 7,618 | 71.4 | 89,900 | 23.4 | 13.9 | 750 | 33.0 | 0.7 | 7,946 | 0.9 | 925 | 11.6 | 7,037 | 27.4 | 36.9 |
| Coahoma | 8,782 | 50.6 | 68,300 | 19.5 | 11.5 | 579 | 30.7 | 2.7 | 8,362 | 0.1 | 1,048 | 12.5 | 8,024 | 29.8 | 24.9 |
| Copiah | 9,414 | 77.2 | 95,200 | 20.5 | 11.0 | 733 | 29.9 | 4.7 | 10,759 | -3.3 | 888 | 8.3 | 10,808 | 25.3 | 32.3 |
| Covington | 6,759 | 78.2 | 77,100 | 24.1 | 10.0 | 631 | 33.8 | 2.0 | 8,379 | 1.9 | 576 | 6.9 | 7,324 | 26.9 | 36.6 |
| DeSoto | 62,890 | 74.0 | 168,500 | 19.0 | 10.0 | 1,057 | 27.4 | 2.5 | 90,498 | -2.8 | 5,469 | 6.0 | 88,982 | 34.0 | 27.6 |
| Forrest | 28,086 | 54.5 | 122,700 | 19.5 | 11.0 | 773 | 33.6 | 2.7 | 32,983 | -0.7 | 2,655 | 8.0 | 32,568 | 34.7 | 22.9 |
| Franklin | 2,928 | 76.9 | 76,600 | 19.1 | 10.9 | 507 | 33.3 | 2.4 | 2,770 | -0.6 | 229 | 8.3 | 2,749 | 31.9 | 34.7 |
| George | 7,592 | 81.8 | 110,800 | 19.7 | 10.0 | 764 | 30.0 | 2.5 | 9,013 | 0.6 | 818 | 9.1 | 8,361 | 29.8 | 37.6 |
| Greene | 3,925 | 82.8 | 77,700 | 17.4 | 10.7 | 599 | 26.9 | 3.9 | 4,337 | -0.8 | 420 | 9.7 | 3,352 | 35.5 | 31.1 |
| Grenada | 8,391 | 64.0 | 105,000 | 21.6 | 10.0 | 682 | 25.4 | 2.2 | 9,305 | -3.5 | 667 | 7.2 | 8,317 | 38.2 | 24.0 |
| Hancock | 20,036 | 75.5 | 147,500 | 22.5 | 11.1 | 850 | 28.9 | 2.8 | 18,692 | -1.1 | 1,586 | 8.5 | 19,631 | 36.4 | 25.4 |
| Harrison | 78,104 | 56.6 | 150,500 | 21.3 | 10.0 | 895 | 31.3 | 3.2 | 87,410 | 0.4 | 8,262 | 9.5 | 84,893 | 31.2 | 23.6 |
| Hinds | 88,611 | 58.0 | 118,600 | 21.3 | 10.9 | 869 | 31.3 | 3.3 | 104,391 | | 9,755 | 9.3 | 104,928 | 32.9 | 23.5 |
| Holmes | 6,188 | 61.3 | 60,400 | 28.1 | 14.8 | 537 | 35.0 | 4.8 | 5,530 | -1.7 | 896 | 16.2 | 5,061 | 26.5 | 38.3 |
| Humphreys | 3,186 | 61.9 | 69,100 | 29.1 | 13.5 | 598 | 33.4 | 3.4 | 2,225 | 1.4 | 328 | 14.7 | 2,768 | 21.7 | 40.7 |
| Issaquena | 483 | 43.7 | 79,200 | 24.8 | 11.5 | 375 | 27.3 | 6.8 | 338 | -7.1 | 31 | 9.2 | 315 | 12.4 | 43.2 |
| Itawamba | 8,653 | 76.3 | 87,600 | 17.8 | 10.0 | 689 | 26.3 | 1.8 | 10,343 | -1.8 | 722 | 7.0 | 10,068 | 23.1 | 38.2 |
| Jackson | 52,423 | 69.8 | 135,900 | 20.0 | 10.0 | 879 | 29.4 | 1.6 | 58,435 | -0.5 | 5,567 | 9.5 | 61,808 | 33.0 | 27.4 |
| Jasper | 6,629 | 85.0 | 79,000 | 19.7 | 10.9 | 578 | 30.2 | 3.6 | 6,094 | -2.6 | 543 | 8.9 | 6,371 | 25.5 | 41.6 |
| Jefferson | 2,448 | 67.4 | 63,400 | 26.0 | 13.8 | 404 | 36.9 | 4.2 | 1,989 | -1.6 | 366 | 18.4 | 1,957 | 38.9 | 22.7 |
| Jefferson Davis | 4,713 | 75.6 | 94,600 | 24.3 | 11.4 | 605 | 38.4 | 3.1 | 4,022 | -2.2 | 386 | 9.6 | 4,303 | 20.5 | 40.7 |
| Jones | 24,858 | 72.6 | 98,600 | 20.4 | 10.0 | 690 | 32.1 | 3.7 | 25,621 | -3.2 | 1,808 | 7.1 | 28,482 | 30.0 | 32.9 |
| Kemper | 3,611 | 73.6 | 73,600 | 26.8 | 11.7 | 402 | 23.6 | 6.5 | 3,423 | -2.0 | 330 | 9.6 | 3,311 | 23.6 | 45.3 |
| Lafayette | 18,721 | 62.0 | 199,700 | 19.9 | 10.0 | 954 | 37.0 | 2.2 | 26,570 | -5.4 | 1,615 | 6.1 | 24,997 | 43.5 | 15.9 |
| Lamar | 22,116 | 67.5 | 177,700 | 18.7 | 10.0 | 927 | 26.0 | 2.7 | 30,544 | -1.9 | 1,835 | 6.0 | 27,710 | 41.0 | 19.5 |
| Lauderdale | 29,736 | 64.5 | 96,300 | 20.7 | 11.3 | 752 | 29.0 | 2.3 | 29,638 | -2.6 | 2,222 | 7.5 | 31,092 | 35.9 | 24.0 |
| Lawrence | 4,849 | 75.2 | 94,800 | 18.1 | 13.8 | 766 | 26.9 | 5.5 | 4,683 | 0.7 | 407 | 8.7 | 4,604 | 31.4 | 38.4 |
| Leake | 8,105 | 70.8 | 83,300 | 20.2 | 10.6 | 660 | 26.2 | 3.4 | 8,058 | 0.0 | 634 | 7.9 | 8,510 | 21.8 | 38.4 |
| Lee | 32,099 | 67.9 | 133,700 | 18.9 | 10.0 | 752 | 28.4 | 4.7 | 41,347 | -0.5 | 3,362 | 8.1 | 38,728 | 31.2 | 30.5 |
| Leflore | 9,962 | 50.6 | 81,200 | 21.7 | 12.8 | 585 | 33.6 | 3.8 | 9,873 | -0.7 | 1,057 | 10.7 | 9,306 | 26.7 | 29.4 |
| Lincoln | 12,699 | 75.3 | 99,500 | 18.8 | 10.5 | 713 | 26.8 | 2.5 | 14,553 | -0.8 | 1,069 | 7.3 | 13,394 | 32.6 | 29.8 |
| Lowndes | 22,436 | 63.5 | 134,200 | 19.8 | 10.0 | 805 | 28.9 | 1.4 | 25,123 | -2.6 | 2,085 | 8.3 | 24,392 | 30.4 | 32.2 |
| Madison | 39,635 | 71.7 | 219,700 | 18.6 | 10.0 | 940 | 28.8 | 1.8 | 52,376 | -3.0 | 3,449 | 6.6 | 52,872 | 47.8 | 18.5 |
| Marion | 9,483 | 82.3 | 89,000 | 26.4 | 12.4 | 605 | 37.1 | 3.1 | 10,178 | 1.5 | 719 | 7.1 | 8,517 | 27.0 | 36.5 |
| Marshall | 12,772 | 75.5 | 107,900 | 21.1 | 11.5 | 722 | 27.1 | 1.3 | 14,119 | -2.5 | 1,066 | 7.6 | 14,307 | 23.2 | 40.0 |
| Monroe | 13,966 | 74.3 | 91,700 | 19.9 | 10.9 | 618 | 23.9 | 2.1 | 15,692 | 0.2 | 1,395 | 8.9 | 14,104 | 27.4 | 39.0 |
| Montgomery | 4,539 | 69.2 | 84,800 | 19.7 | 12.7 | 554 | 20.1 | 3.3 | 3,926 | -0.8 | 316 | 8.0 | 4,038 | 26.7 | 40.3 |
| Neshoba | 10,657 | 71.5 | 83,000 | 21.3 | 10.7 | 666 | 27.9 | 7.5 | 10,382 | 0.1 | 967 | 9.3 | 11,463 | 30.1 | 23.9 |
| Newton | 8,037 | 77.2 | 85,600 | 18.8 | 11.7 | 658 | 28.8 | 5.1 | 8,010 | -4.6 | 636 | 7.9 | 8,047 | 29.3 | 38.2 |
| Noxubee | 3,986 | 72.9 | 59,700 | 22.3 | 11.8 | 532 | 31.5 | 5.1 | 3,888 | 2.8 | 429 | 11.0 | 3,969 | 21.1 | 42.8 |
| Oktibbeha | 17,798 | 51.7 | 162,900 | 19.9 | 10.0 | 788 | 38.6 | 2.0 | 22,096 | -3.1 | 1,682 | 7.6 | 20,699 | 45.5 | 16.3 |
| Panola | 12,488 | 69.6 | 83,300 | 21.8 | 12.6 | 679 | 31.0 | 4.0 | 12,739 | 0.3 | 1,251 | 9.8 | 12,728 | 29.4 | 36.4 |
| Pearl River | 21,020 | 77.8 | 138,200 | 21.0 | 10.7 | 765 | 27.1 | 2.6 | 23,200 | -2.5 | 1,698 | 7.3 | 21,710 | 31.7 | 31.3 |
| Perry | 4,623 | 82.2 | 86,700 | 20.7 | 10.5 | 608 | 28.8 | 4.2 | 4,245 | -1.5 | 369 | 8.7 | 4,600 | 25.3 | 39.2 |
| Pike | 14,404 | 69.0 | 95,500 | 23.7 | 12.2 | 717 | 31.7 | 2.8 | 14,674 | 0.7 | 1,308 | 8.9 | 13,966 | 22.5 | 37.5 |
| Pontotoc | 10,783 | 71.3 | 106,500 | 18.4 | 10.2 | 722 | 24.0 | 4.9 | 14,911 | -0.7 | 1,156 | 7.8 | 13,307 | 25.4 | 41.4 |
| Prentiss | 9,145 | 72.7 | 92,700 | 21.2 | 10.5 | 588 | 29.8 | 3.0 | 11,207 | -1.6 | 779 | 7.0 | 9,754 | 22.6 | 38.8 |
| Quitman | 3,027 | 58.7 | 50,900 | 22.7 | 14.3 | 568 | 30.6 | 3.5 | 2,355 | 0.9 | 256 | 10.9 | 2,296 | 25.3 | 31.7 |
| Rankin | 55,909 | 76.5 | 162,500 | 18.3 | 10.0 | 983 | 25.6 | 2.4 | 74,377 | -3.9 | 3,949 | 5.3 | 74,371 | 38.9 | 21.4 |
| Scott | 10,180 | 74.1 | 71,300 | 20.3 | 12.1 | 707 | 29.7 | 4.0 | 12,681 | -1.3 | 739 | 5.8 | 11,243 | 21.6 | 43.6 |
| Sharkey | 1,751 | 64.1 | 64,400 | 26.9 | 14.4 | 510 | 35.7 | 5.5 | 1,475 | -0.6 | 161 | 10.9 | 1,591 | 34.9 | 21.3 |
| Simpson | 9,486 | 79.7 | 87,900 | 19.4 | 11.6 | 660 | 33.2 | 4.6 | 10,445 | -3.3 | 715 | 6.8 | 10,756 | 31.7 | 31.3 |
| Smith | 5,820 | 83.0 | 102,600 | 19.2 | 10.0 | 587 | 23.1 | 2.1 | 6,631 | -2.9 | 383 | 5.8 | 6,257 | 28.9 | 37.9 |
| Stone | 6,334 | 74.5 | 112,800 | 22.9 | 10.0 | 680 | 25.6 | 4.1 | 7,012 | 1.2 | 529 | 7.5 | 7,886 | 25.3 | 36.1 |

1. Specified owner-occupied units. 2. A value of 10.0 represents 10 percent or less; a value of 50.0 represents 50 percent or more. 3. Specified renter-occupied units. 4. Overcrowded or lacking complete plumbing facilities. 5. Percent of civilian labor force. 6. Civilian employed persons 16 years old and over.

| STATE County | Private nonfarm establishments, employment and payroll, 2019 | | | | | | | | | Agriculture, 2017 | | | |
|---|---|---|---|---|---|---|---|---|---|---|---|---|---|
| | Number of establishments | Employment | | | | | | Annual payroll | | Farms | | | Farm producers whose primary occupation is farming (percent) |
| | | Total | Health care and social assistance | Manufacturing | Retail trade | Finance and insurance | Professional, scientific, and technical services | Total (mil dol) | Average per employee (dollars) | Number | Percent with: | | |
| | | | | | | | | | | | Fewer than 50 acres | 1000 acres or more | |
| | 104 | 105 | 106 | 107 | 108 | 109 | 110 | 111 | 112 | 113 | 114 | 115 | 116 |
| MISSISSIPPI | 59,130 | 958,126 | 176,031 | 148,901 | 137,257 | 32,351 | 31,394 | 37,731 | 39,379 | 34,988 | 31.6 | 6.4 | 37.8 |
| Adams | 764 | 10,394 | 2,097 | 657 | 1,809 | 346 | 201 | 393 | 37,767 | 171 | 40.4 | 7.6 | 41.7 |
| Alcorn | 788 | 13,016 | 2,510 | 2,742 | 2,209 | 287 | 786 | 502 | 38,566 | 457 | 39.4 | 3.1 | 29.5 |
| Amite | 181 | 1,528 | 242 | 388 | 198 | 33 | 19 | 59 | 38,476 | 484 | 26.9 | 2.7 | 43.1 |
| Attala | 352 | 4,532 | 587 | 701 | 705 | 138 | 77 | 168 | 36,965 | 468 | 28.0 | 4.7 | 33.5 |
| Benton | 58 | 1,070 | 184 | 182 | 107 | 13 | 30 | 36 | 33,939 | 285 | 24.2 | 3.9 | 29.9 |
| Bolivar | 687 | 8,689 | 1,757 | 1,813 | 1,464 | 206 | 215 | 340 | 39,112 | 412 | 21.4 | 27.9 | 63.1 |
| Calhoun | 255 | 2,069 | 190 | 565 | 404 | 66 | 24 | 63 | 30,680 | 518 | 26.4 | 6.8 | 31.4 |
| Carroll | 127 | 677 | 133 | 20 | 123 | 13 | 12 | 20 | 28,932 | 446 | 18.2 | 10.1 | 32.4 |
| Chickasaw | 325 | 5,520 | 370 | 3,369 | 654 | 104 | 62 | 163 | 29,491 | 506 | 22.7 | 7.5 | 33.7 |
| Choctaw | 124 | 1,462 | 417 | 290 | 139 | 16 | 11 | 77 | 52,386 | 227 | 17.6 | 4.8 | 33.7 |
| Claiborne | 105 | 2,156 | 567 | 96 | 129 | 31 | 147 | 144 | 66,626 | 224 | 17.0 | 8.0 | 35.9 |
| Clarke | 226 | 2,093 | 360 | 471 | 302 | 74 | 37 | 70 | 33,291 | 300 | 27.3 | 3.3 | 33.4 |
| Clay | 336 | 4,420 | 632 | 806 | 754 | 114 | 278 | 167 | 37,763 | 354 | 19.2 | 7.9 | 27.9 |
| Coahoma | 517 | 6,060 | 1,892 | 621 | 921 | 209 | 171 | 223 | 36,852 | 206 | 18.0 | 42.2 | 63.4 |
| Copiah | 437 | 6,068 | 835 | 2,230 | 799 | 131 | 84 | 187 | 30,768 | 478 | 28.0 | 3.6 | 40.4 |
| Covington | 328 | 4,100 | 535 | 1,396 | 599 | 112 | 51 | 141 | 34,292 | 523 | 33.7 | 2.1 | 43.6 |
| DeSoto | 3,012 | 56,215 | 6,685 | 4,262 | 9,820 | 1,120 | 1,015 | 1,954 | 34,758 | 398 | 51.0 | 9.0 | 39.1 |
| Forrest | 1,717 | 33,618 | 8,353 | 3,534 | 4,951 | 766 | 2,202 | 1,373 | 40,850 | 376 | 51.3 | 1.3 | 30.9 |
| Franklin | 118 | 1,566 | 357 | NA | 140 | 38 | 19 | 66 | 42,157 | 198 | 32.3 | 2.5 | 38.2 |
| George | 338 | 3,625 | 783 | 327 | 986 | 92 | 109 | 126 | 34,715 | 492 | 57.3 | 1.2 | 36.6 |
| Greene | 127 | 936 | 62 | 19 | 211 | 29 | 8 | 37 | 39,870 | 436 | 44.3 | 2.8 | 35.5 |
| Grenada | 554 | 8,388 | 1,368 | 2,690 | 1,290 | 210 | 230 | 307 | 36,546 | 245 | 24.1 | 6.1 | 28.0 |
| Hancock | 723 | 9,200 | 1,011 | 905 | 1,644 | 215 | 1,257 | 365 | 39,700 | 287 | 51.9 | 1.4 | 40.7 |
| Harrison | 4,215 | 75,069 | 15,183 | 2,978 | 11,453 | 2,617 | 2,083 | 2,951 | 39,312 | 322 | 72.0 | NA | 26.1 |
| Hinds | 5,118 | 99,108 | 32,498 | 3,384 | 10,142 | 4,706 | 4,511 | 4,769 | 48,123 | 872 | 39.8 | 4.8 | 34.1 |
| Holmes | 247 | 2,280 | 690 | 555 | 383 | 66 | 35 | 72 | 31,696 | 496 | 12.1 | 11.3 | 34.1 |
| Humphreys | 139 | 1,522 | 179 | 546 | 201 | 115 | 4 | 42 | 27,922 | 179 | 17.3 | 29.6 | 52.3 |
| Issaquena | 17 | 91 | NA | NA | NA | NA | NA | 3 | 33,330 | 119 | 16.8 | 35.3 | 46.7 |
| Itawamba | 369 | 5,309 | 605 | 2,288 | 553 | 104 | 31 | 191 | 35,987 | 364 | 28.0 | 2.2 | 30.1 |
| Jackson | 2,299 | 45,200 | 5,776 | 15,413 | 5,181 | 985 | 1,671 | 2,217 | 49,039 | 473 | 72.1 | 0.4 | 33.4 |
| Jasper | 205 | 4,298 | 269 | 1,539 | 339 | 69 | 57 | 178 | 41,452 | 507 | 26.4 | 3.0 | 46.8 |
| Jefferson | 57 | 547 | 221 | NA | 78 | NA | NA | 20 | 36,793 | 236 | 20.3 | 5.9 | 42.8 |
| Jefferson Davis | 130 | 2,778 | 280 | 81 | 235 | 33 | 9 | 108 | 38,754 | 355 | 32.1 | 1.7 | 47.7 |
| Jones | 1,302 | 22,910 | 3,246 | 7,163 | 3,094 | 556 | 467 | 889 | 38,807 | 882 | 40.4 | 1.5 | 40.0 |
| Kemper | 110 | 1,388 | 320 | 283 | 122 | 34 | 4 | 65 | 47,137 | 313 | 23.3 | 6.4 | 46.9 |
| Lafayette | 1,289 | 18,093 | 3,537 | 1,781 | 2,985 | 595 | 977 | 628 | 34,686 | 443 | 21.0 | 4.1 | 31.7 |
| Lamar | 1,442 | 18,834 | 3,606 | 108 | 4,821 | 936 | 857 | 619 | 32,890 | 491 | 51.5 | 1.8 | 36.5 |
| Lauderdale | 1,832 | 29,132 | 7,727 | 1,798 | 4,511 | 915 | 903 | 1,064 | 36,518 | 305 | 32.1 | 4.6 | 23.6 |
| Lawrence | 174 | 1,873 | 347 | 689 | 271 | 113 | 22 | 88 | 46,772 | 354 | 34.5 | 1.4 | 43.6 |
| Leake | 305 | 4,821 | 965 | 1,628 | 742 | 106 | 64 | 143 | 29,666 | 573 | 31.2 | 2.1 | 48.4 |
| Lee | 2,492 | 47,087 | 8,745 | 8,984 | 6,771 | 1,887 | 1,141 | 1,876 | 39,841 | 436 | 31.2 | 4.8 | 34.6 |
| Leflore | 670 | 12,060 | 2,791 | 2,470 | 1,598 | 297 | 251 | 444 | 36,792 | 257 | 11.3 | 35.0 | 51.7 |
| Lincoln | 809 | 11,293 | 1,842 | 1,636 | 1,713 | 328 | 223 | 422 | 37,336 | 611 | 32.6 | 1.3 | 35.8 |
| Lowndes | 1,449 | 21,746 | 3,749 | 4,366 | 3,200 | 520 | 427 | 926 | 42,595 | 444 | 34.5 | 6.3 | 27.6 |
| Madison | 3,201 | 52,439 | 5,067 | 8,410 | 7,048 | 3,310 | 3,312 | 2,240 | 42,713 | 524 | 29.0 | 8.4 | 34.7 |
| Marion | 551 | 7,391 | 996 | 797 | 1,060 | 277 | 219 | 296 | 40,058 | 511 | 39.1 | 2.0 | 41.5 |
| Marshall | 441 | 7,374 | 855 | 1,353 | 853 | 223 | 49 | 287 | 38,937 | 634 | 32.5 | 7.4 | 33.6 |
| Monroe | 588 | 7,689 | 1,554 | 2,237 | 1,006 | 182 | 71 | 320 | 41,680 | 644 | 34.6 | 5.6 | 28.7 |
| Montgomery | 203 | 2,062 | 678 | 183 | 356 | 99 | 31 | 65 | 31,502 | 270 | 26.7 | 3.7 | 31.5 |
| Neshoba | 509 | 10,056 | 982 | 422 | 1,167 | 269 | 551 | 454 | 45,190 | 652 | 30.8 | 1.8 | 44.9 |
| Newton | 315 | 3,392 | 653 | 907 | 589 | 116 | 48 | 122 | 36,006 | 527 | 28.3 | 3.2 | 41.4 |
| Noxubee | 185 | 1,641 | 304 | 531 | 263 | 62 | 11 | 57 | 34,628 | 517 | 19.5 | 8.5 | 39.1 |
| Oktibbeha | 913 | 13,385 | 2,048 | 1,034 | 2,274 | 404 | 499 | 405 | 30,244 | 412 | 27.2 | 3.2 | 28.5 |
| Panola | 583 | 8,582 | 1,723 | 1,728 | 1,594 | 347 | 72 | 311 | 36,291 | 627 | 18.7 | 9.7 | 32.4 |
| Pearl River | 806 | 8,069 | 1,615 | 645 | 2,001 | 343 | 413 | 258 | 31,966 | 717 | 47.4 | 2.1 | 37.9 |
| Perry | 138 | 1,836 | 340 | 656 | 278 | 27 | 17 | 89 | 48,702 | 306 | 43.1 | 1.3 | 41.9 |
| Pike | 913 | 12,218 | 2,032 | 2,714 | 2,535 | 385 | 285 | 391 | 32,016 | 508 | 38.8 | 0.2 | 45.2 |
| Pontotoc | 513 | 12,745 | 852 | 8,415 | 1,107 | 200 | 81 | 418 | 32,810 | 745 | 31.9 | 3.0 | 26.0 |
| Prentiss | 482 | 5,489 | 743 | 1,956 | 805 | 176 | 153 | 177 | 32,294 | 486 | 28.8 | 6.4 | 29.7 |
| Quitman | 97 | 591 | 200 | NA | 72 | 50 | 16 | 16 | 27,230 | 275 | 15.6 | 23.3 | 43.0 |
| Rankin | 3,736 | 55,810 | 8,712 | 3,678 | 8,738 | 2,696 | 1,645 | 2,421 | 43,371 | 577 | 38.1 | 3.8 | 37.2 |
| Scott | 488 | 10,695 | 736 | 6,252 | 1,262 | 202 | 72 | 400 | 37,368 | 660 | 34.5 | 2.1 | 39.6 |
| Sharkey | 106 | 752 | 258 | NA | 127 | 44 | 16 | 24 | 31,439 | 142 | 14.8 | 39.4 | 48.7 |
| Simpson | 418 | 5,679 | 2,137 | 488 | 941 | 283 | 112 | 166 | 29,172 | 498 | 25.7 | 2.6 | 47.2 |
| Smith | 153 | 1,968 | 235 | 870 | 227 | 49 | 58 | 91 | 46,151 | 540 | 25.6 | 1.1 | 48.4 |
| Stone | 269 | 3,191 | 573 | 500 | 681 | 107 | 99 | 106 | 33,127 | 323 | 46.7 | 1.2 | 44.6 |

| STATE County | Land in farms | | | | | Value of land and buildings (dollars) | | Value of machinery and equipmnet, average per farm (dollars) | Value of products sold: | | | | Organic farms (number) | Farms with internet access (percent) | Government payments | |
|---|---|---|---|---|---|---|---|---|---|---|---|---|---|---|---|---|
| | | | Acres | | | | | | | | Percent from: | | | | | |
| | Acreage (1,000) | Percent change, 2012-2017 | Average size of farm | Total irrigated (1,000) | Total cropland (1,000) | Average per farm | Average per acre | | Total (mil dol) | Average per farm (acres) | Crops | Livestock and poultry products | | | Total ($1,000) | Percent of farms |
| | 117 | 118 | 119 | 120 | 121 | 122 | 123 | 124 | 125 | 126 | 127 | 128 | 129 | 130 | 131 | 132 |
| MISSISSIPPI.................. | 10,415 | -4.7 | 298 | 1,814.5 | 4,960.6 | 817,041 | 2,745 | 109,875 | 6,196 | 177,088 | 37.0 | 63.0 | 37 | 66.0 | 213,785 | 40.8 |
| Adams............................ | 69 | 5.2 | 406 | D | 21.8 | 1,197,343 | 2,950 | 84,618 | 7.6 | 44,363 | 79.9 | 20.1 | NA | 74.9 | 464 | 34.5 |
| Alcorn.......................... | 83 | -11.8 | 181 | D | 39.4 | 385,117 | 2,131 | 65,906 | 19.1 | 41,755 | 78.7 | 21.3 | NA | 65.2 | 1,186 | 52.1 |
| Amite............................ | 92 | -23.9 | 190 | 0.2 | 19.3 | 607,952 | 3,192 | 69,786 | 85.2 | 176,112 | 2.4 | 97.6 | NA | 65.3 | 412 | 19.4 |
| Attala............................ | 118 | -5.9 | 252 | 0.3 | 22.1 | 465,417 | 1,846 | 47,727 | 14.3 | 30,585 | 30.4 | 69.6 | NA | 54.1 | 1,532 | 53.8 |
| Benton.......................... | 76 | -7.9 | 267 | D | 25.6 | 513,315 | 1,926 | 75,492 | 11.7 | 41,098 | 65.9 | 34.1 | NA | 67.0 | 510 | 35.8 |
| Bolivar.......................... | 409 | 4.8 | 993 | 275.0 | 374.6 | 3,004,756 | 3,025 | 475,613 | 208.0 | 504,847 | 99.9 | 0.1 | NA | 78.2 | 17,899 | 76.5 |
| Calhoun........................ | 148 | -15.6 | 286 | 0.8 | 73.8 | 507,660 | 1,776 | 142,990 | 61.8 | 119,373 | 90.9 | 9.1 | NA | 61.0 | 3,916 | 74.3 |
| Carroll.......................... | 177 | 5.0 | 398 | 24.5 | 71.9 | 850,650 | 2,138 | 112,567 | 39.4 | 88,312 | 83.1 | 16.9 | NA | 64.3 | 4,063 | 51.3 |
| Chickasaw.................... | 172 | 3.8 | 340 | 5.6 | 92.4 | 681,863 | 2,003 | 127,343 | 103.8 | 205,227 | 41.0 | 59.0 | 2 | 64.6 | 2,850 | 55.5 |
| Choctaw........................ | 64 | 1.5 | 284 | 1.8 | 10.5 | 522,082 | 1,841 | 53,109 | 12.7 | 55,859 | 43.5 | 56.5 | NA | 65.6 | 858 | 58.1 |
| Claiborne...................... | 72 | -13.0 | 323 | 2.2 | 18.8 | 874,076 | 2,708 | 100,254 | 8.6 | 38,438 | 73.9 | 26.1 | NA | 56.7 | 1,285 | 52.7 |
| Clarke.......................... | 103 | 82.8 | 344 | 0.4 | 9.5 | 768,676 | 2,237 | 65,877 | 29.5 | 98,260 | 4.8 | 95.2 | NA | 61.7 | 373 | 35.0 |
| Clay.............................. | 124 | -4.4 | 351 | 0.1 | 30.6 | 659,679 | 1,877 | 72,201 | 64.0 | 180,720 | 6.5 | 93.5 | NA | 59.0 | 1,977 | 58.2 |
| Coahoma ...................... | 267 | 2.2 | 1,294 | 165.8 | 237.5 | 4,279,639 | 3,307 | 617,813 | 156.3 | 758,699 | 91.3 | 8.7 | NA | 76.2 | 9,650 | 87.9 |
| Copiah.......................... | 122 | 5.3 | 255 | 0.8 | 22.5 | 654,533 | 2,563 | 76,441 | 92.1 | 192,770 | 3.3 | 96.7 | 1 | 63.0 | 1,412 | 38.1 |
| Covington...................... | 89 | -15.8 | 170 | 0.1 | 17.0 | 614,485 | 3,605 | 82,308 | 253.8 | 485,304 | 1.3 | 98.7 | NA | 67.5 | 691 | 22.8 |
| DeSoto.......................... | 121 | 1.9 | 304 | 10.7 | 74.2 | 914,718 | 3,009 | 116,103 | 39.4 | 98,912 | 85.9 | 14.1 | NA | 64.1 | 2,188 | 23.6 |
| Forrest.......................... | 48 | 12.5 | 128 | 0.4 | 10.7 | 543,054 | 4,244 | 65,447 | 12.4 | 33,016 | 33.5 | 66.5 | NA | 70.2 | 544 | 23.4 |
| Franklin........................ | 41 | -19.1 | 205 | 0.1 | 6.9 | 650,912 | 3,174 | 55,584 | 3.9 | 19,904 | 32.3 | 67.7 | NA | 68.7 | 240 | 18.2 |
| George.......................... | 55 | -10.0 | 111 | 1.1 | 24.5 | 415,726 | 3,731 | 68,803 | 18.2 | 36,988 | 80.1 | 19.9 | NA | 77.4 | 425 | 10.8 |
| Greene.......................... | 69 | 0.9 | 159 | 0.1 | 9.0 | 402,646 | 2,528 | 62,767 | 31.2 | 71,502 | 8.9 | 91.1 | NA | 66.7 | 376 | 31.7 |
| Grenada........................ | 73 | -17.2 | 299 | 1.0 | 21.2 | 657,052 | 2,200 | 89,278 | 11.7 | 47,833 | 76.0 | 24.0 | NA | 64.1 | 1,271 | 50.2 |
| Hancock........................ | 32 | 28.4 | 113 | 0.1 | 7.3 | 436,624 | 3,861 | 66,054 | 4.5 | 15,603 | 24.3 | 75.7 | NA | 79.8 | 66 | 5.2 |
| Harrison........................ | 16 | -34.3 | 49 | 0.1 | 4.1 | 412,695 | 8,374 | 52,978 | 3.4 | 10,519 | 70.0 | 30.0 | 3 | 72.4 | 39 | 4.3 |
| Hinds............................ | 213 | -15.1 | 244 | 0.5 | 54.1 | 804,312 | 3,292 | 61,962 | 57.4 | 65,807 | 29.1 | 70.9 | NA | 61.2 | 3,604 | 42.7 |
| Holmes.......................... | 241 | 1.6 | 487 | 58.4 | 118.6 | 1,180,015 | 2,425 | 120,527 | 66.3 | 133,647 | 96.1 | 3.9 | NA | 61.3 | 5,626 | 58.5 |
| Humphreys.................... | 164 | -15.3 | 917 | 72.1 | 134.1 | 2,998,513 | 3,270 | 419,098 | 95.9 | 535,721 | 75.7 | 24.3 | NA | 82.1 | 6,512 | 81.6 |
| Issaquena .................... | 131 | 5.5 | 1,103 | 26.3 | 87.2 | 3,018,677 | 2,737 | 610,300 | D | D | D | D | NA | 63.9 | 2,740 | 71.4 |
| Itawamba...................... | 75 | -20.5 | 207 | 0.1 | 25.7 | 354,780 | 1,712 | 59,971 | 7.9 | 21,791 | 82.0 | 18.0 | 1 | 69.0 | 950 | 51.9 |
| Jackson........................ | 36 | -3.0 | 77 | 0.2 | 8.4 | 386,270 | 5,012 | 64,869 | 7.6 | 16,163 | 47.1 | 52.9 | 3 | 75.7 | 97 | 7.0 |
| Jasper.......................... | 111 | 14.3 | 219 | 0.1 | 15.1 | 614,597 | 2,805 | 98,483 | 191.2 | 377,083 | 0.9 | 99.1 | NA | 64.9 | 1,301 | 43.6 |
| Jefferson ...................... | 60 | -30.4 | 253 | 0.1 | 11.4 | 636,903 | 2,520 | 51,954 | 30.5 | 129,186 | 9.8 | 90.2 | NA | 53.8 | 541 | 26.7 |
| Jefferson Davis ............ | 55 | -6.4 | 156 | 0.1 | 12.6 | 399,812 | 2,570 | 58,480 | 47.8 | 134,758 | 4.1 | 95.9 | NA | 71.3 | 158 | 14.1 |
| Jones............................ | 123 | -2.5 | 139 | 0.5 | 22.7 | 527,503 | 3,786 | 71,665 | 221.1 | 250,719 | 1.8 | 98.2 | NA | 68.6 | 1,299 | 17.7 |
| Kemper.......................... | 112 | -9.5 | 358 | D | 15.5 | 723,481 | 2,022 | 59,748 | 19.3 | 61,802 | 5.7 | 94.3 | NA | 60.1 | 1,199 | 46.3 |
| Lafayette ...................... | 105 | -3.9 | 236 | 0.6 | 29.0 | 646,938 | 2,740 | 57,637 | 9.4 | 21,176 | 72.0 | 28.0 | 2 | 69.1 | 1,031 | 42.7 |
| Lamar............................ | 73 | 14.3 | 149 | 0.2 | 11.0 | 529,531 | 3,548 | 59,665 | 53.2 | 108,360 | 5.4 | 94.6 | NA | 81.9 | 510 | 16.7 |
| Lauderdale.................... | 82 | 19.0 | 269 | 0.7 | 18.1 | 553,284 | 2,055 | 56,867 | 3.9 | 12,767 | 18.7 | 81.3 | NA | 73.4 | 312 | 20.7 |
| Lawrence...................... | 56 | -22.4 | 159 | 0.2 | 17.5 | 511,947 | 3,211 | 81,082 | 91.2 | 257,644 | 3.6 | 96.4 | NA | 68.9 | 490 | 11.0 |
| Leake............................ | 95 | -10.7 | 165 | 0.0 | 27.2 | 500,892 | 3,027 | 76,068 | 328.0 | 572,438 | 1.4 | 98.6 | NA | 64.2 | 793 | 43.6 |
| Lee................................ | 123 | -7.5 | 282 | 0.8 | 72.9 | 517,010 | 1,832 | 102,991 | 32.6 | 74,773 | 75.5 | 24.5 | NA | 70.6 | 2,112 | 37.2 |
| Leflore.......................... | 299 | 1.9 | 1,163 | 167.0 | 243.8 | 3,373,801 | 2,902 | 470,366 | 210.0 | 817,062 | 67.8 | 32.2 | NA | 67.7 | 13,507 | 85.6 |
| Lincoln.......................... | 101 | -7.2 | 165 | 0.1 | 14.9 | 506,676 | 3,071 | 60,954 | 72.2 | 118,187 | 2.5 | 97.5 | 7 | 66.6 | 239 | 12.6 |
| Lowndes........................ | 139 | 16.6 | 314 | 7.6 | 64.6 | 793,469 | 2,530 | 119,488 | 60.8 | 136,890 | 38.5 | 61.5 | NA | 75.9 | 2,939 | 50.7 |
| Madison........................ | 157 | -22.5 | 300 | D | 50.1 | 789,468 | 2,628 | 65,717 | 20.1 | 38,353 | 78.5 | 21.5 | NA | 66.2 | 2,653 | 38.7 |
| Marion.......................... | 81 | -0.5 | 159 | 0.1 | 18.8 | 458,585 | 2,881 | 75,413 | 91.2 | 178,462 | 2.5 | 97.5 | 3 | 67.1 | 1,174 | 27.6 |
| Marshall........................ | 212 | 4.4 | 335 | 2.3 | 72.4 | 804,766 | 2,403 | 72,214 | 24.7 | 38,882 | 76.2 | 23.8 | NA | 60.3 | 2,168 | 35.0 |
| Monroe.......................... | 186 | -18.6 | 288 | 0.3 | 90.4 | 633,479 | 2,198 | 97,367 | 39.7 | 61,691 | 71.7 | 28.3 | NA | 62.4 | 3,401 | 45.7 |
| Montgomery.................. | 68 | -30.4 | 251 | 0.5 | 20.9 | 501,369 | 1,999 | 67,717 | 14.7 | 54,567 | 55.2 | 44.8 | NA | 62.2 | 840 | 51.1 |
| Neshoba........................ | 112 | 10.8 | 172 | 0.1 | 25.9 | 459,112 | 2,664 | 77,883 | 248.7 | 381,434 | 1.5 | 98.5 | 2 | 71.0 | 741 | 29.0 |
| Newton.......................... | 105 | -3.7 | 199 | 0.8 | 25.1 | 472,560 | 2,371 | 79,938 | 116.6 | 221,264 | 4.6 | 95.4 | NA | 71.2 | 1,325 | 37.6 |
| Noxubee........................ | 203 | -4.7 | 393 | 24.4 | 98.1 | 995,194 | 2,533 | 173,217 | 124.7 | 241,246 | 39.2 | 60.8 | NA | 62.7 | 4,890 | 59.8 |
| Oktibbeha...................... | 95 | -9.1 | 231 | 0.3 | 21.0 | 645,315 | 2,796 | 69,080 | 17.1 | 41,398 | 17.5 | 82.5 | NA | 69.4 | 1,124 | 38.6 |
| Panola.......................... | 226 | -17.2 | 360 | 29.5 | 113.2 | 886,959 | 2,463 | 108,817 | 55.0 | 87,748 | 86.9 | 13.1 | NA | 61.7 | 7,916 | 55.2 |
| Pearl River.................... | 105 | -11.2 | 147 | 1.1 | 17.8 | 519,572 | 3,545 | 56,063 | 17.5 | 24,379 | 45.8 | 54.2 | 3 | 74.1 | 402 | 9.2 |
| Perry............................ | 44 | 1.3 | 144 | 0.4 | 10.6 | 462,067 | 3,210 | 67,901 | 33.7 | 110,196 | 11.6 | 88.4 | NA | 60.8 | 209 | 27.5 |
| Pike.............................. | 66 | -8.6 | 130 | 0.1 | 15.7 | 482,938 | 3,726 | 83,431 | 70.4 | 138,594 | 3.9 | 96.1 | 2 | 65.2 | 285 | 12.6 |
| Pontotoc........................ | 137 | -10.1 | 184 | D | 65.1 | 362,339 | 1,968 | 64,218 | 20.8 | 27,858 | 80.2 | 19.8 | NA | 52.1 | 2,165 | 51.9 |
| Prentiss........................ | 109 | 16.5 | 223 | 0.5 | 41.5 | 457,174 | 2,046 | 70,477 | 21.2 | 43,621 | 58.6 | 41.4 | NA | 52.7 | 1,606 | 62.1 |
| Quitman........................ | 199 | -4.9 | 722 | 80.1 | 150.4 | 2,065,226 | 2,861 | 210,343 | D | D | D | D | NA | 61.5 | 9,821 | 93.5 |
| Rankin.......................... | 121 | -3.6 | 210 | 0.1 | 37.3 | 693,646 | 3,300 | 90,066 | 114.0 | 197,615 | 7.7 | 92.3 | NA | 72.1 | 2,052 | 28.9 |
| Scott............................ | 120 | 4.2 | 182 | 0.2 | 30.7 | 462,657 | 2,546 | 95,592 | 272.7 | 413,121 | 2.1 | 97.9 | NA | 65.3 | 884 | 35.9 |
| Sharkey........................ | 172 | 10.9 | 1,214 | 70.6 | 140.8 | 4,016,682 | 3,310 | 476,427 | 86.5 | 608,972 | 93.3 | 6.7 | NA | 67.6 | 5,700 | 81.0 |
| Simpson........................ | 95 | -13.2 | 191 | 0.3 | 21.8 | 526,657 | 2,758 | 81,714 | 227.4 | 456,665 | 2.4 | 97.6 | NA | 63.9 | 535 | 29.7 |
| Smith............................ | 83 | -23.6 | 153 | 0.7 | 24.0 | 440,459 | 2,873 | 82,278 | 218.6 | 404,798 | 1.8 | 98.2 | NA | 61.5 | 1,135 | 31.5 |
| Stone............................ | 46 | -0.1 | 141 | 0.3 | 10.4 | 467,314 | 3,308 | 57,118 | 12.4 | 38,387 | 67.5 | 32.5 | NA | 74.3 | 408 | 11.8 |

# Water Use, Wholesale Trade, Retail Trade, and Real Estate

| STATE County | Water use, 2015 | | Wholesale Trade[1], 2017 | | | | Retail Trade[2], 2017 | | | | Real estate and rental and leasing,[2] 2017 | | | |
|---|---|---|---|---|---|---|---|---|---|---|---|---|---|---|
| | Public supply water withdrawn (mil gal/day) | Public supply gallons withdrawn per person per day | Number of establishments | Number of employees | Sales (mil dol) | Average payroll (mil dol) | Number of establishments | Number of employees | Sales (mil dol) | Average payroll (mil dol) | Number of establishments | Number of employees | Sales (mil dol) | Average payroll (mil dol) |
| | 133 | 134 | 135 | 136 | 137 | 138 | 139 | 140 | 141 | 142 | 143 | 144 | 145 | 146 |
| MISSISSIPPI.................. | 400.36 | 133.8 | 2,347 | 32,051 | 31,126.7 | 1,616.8 | 11,525 | 141,410 | 36,920.6 | 3,384.5 | 2,403 | 9,683 | 1,996.9 | 345.4 |
| Adams.................... | 4.92 | 157.4 | 42 | 289 | 196.9 | 10.7 | 141 | 1,895 | 472.7 | 47.9 | 41 | 128 | 23.8 | 4.0 |
| Alcorn................... | 4.75 | 127.0 | 36 | 546 | 263.8 | 25.7 | 178 | 2,254 | 548.8 | 56.1 | 20 | 206 | 21.9 | 6.5 |
| Amite..................... | 1.34 | 106.6 | 5 | 34 | 26.9 | 1.5 | 31 | 201 | 45.7 | 4.6 | 3 | 6 | 0.3 | 0.1 |
| Attala.................... | 2.24 | 117.6 | 12 | 56 | 18.1 | 2.1 | 72 | 714 | 188.0 | 16.8 | 12 | 32 | 6.2 | 1.0 |
| Benton................... | 0.49 | 59.9 | NA | NA | NA | NA | 15 | 113 | 27.4 | 2.5 | NA | NA | NA | NA |
| Bolivar.................. | 4.16 | 124.8 | 26 | 253 | 342.5 | 12.2 | 146 | 1,469 | 381.6 | 33.0 | 44 | 125 | 17.4 | 3.0 |
| Calhoun................. | 2.28 | 154.9 | 13 | 157 | 94.9 | 5.9 | 62 | 434 | 96.3 | 9.0 | 6 | D | 0.8 | D |
| Carroll .................. | 0.86 | 84.0 | D | D | D | D | 22 | 106 | 30.6 | 2.0 | 4 | 6 | 1.1 | 0.1 |
| Chickasaw............. | 1.81 | 104.5 | 12 | 78 | 67.6 | 3.1 | 79 | 679 | 153.2 | 13.4 | 5 | 63 | 13.4 | 3.0 |
| Choctaw................. | 0.96 | 115.7 | NA | NA | NA | NA | 25 | 165 | 30.4 | 3.0 | NA | NA | NA | NA |
| Claiborne.............. | 0.66 | 72.1 | NA | NA | NA | NA | 20 | 120 | 33.4 | 3.3 | NA | NA | NA | NA |
| Clarke................... | 2.00 | 125.0 | NA | NA | NA | NA | 49 | 296 | 63.4 | 6.0 | NA | NA | NA | NA |
| Clay...................... | 2.53 | 126.2 | 11 | 164 | 143.4 | 5.8 | 83 | 773 | 183.5 | 17.0 | 10 | 23 | 3.9 | 0.5 |
| Coahoma............... | 3.83 | 155.6 | 28 | 322 | 491.1 | 15.2 | 101 | 925 | 236.0 | 21.7 | 38 | 93 | 20.8 | 2.6 |
| Copiah................... | 3.79 | 131.7 | 11 | 102 | 50.1 | 3.2 | 88 | 855 | 199.1 | 20.0 | 11 | 32 | 5.4 | 0.8 |
| Covington.............. | 1.86 | 95.2 | 12 | 77 | 90.4 | 3.2 | 77 | 614 | 246.2 | 16.1 | 4 | 30 | 2.8 | 0.9 |
| DeSoto.................. | 19.51 | 112.6 | 121 | 3,258 | 4,857.6 | 172.5 | 602 | 9,783 | 2,862.1 | 254.0 | 119 | 409 | 146.1 | 17.6 |
| Forrest.................. | 12.10 | 159.3 | 76 | 836 | 664.8 | 38.7 | 342 | 4,033 | 1,155.7 | 106.7 | 95 | 444 | 84.6 | 16.7 |
| Franklin ................ | 0.98 | 126.6 | 4 | 45 | 23.9 | 2.8 | 20 | 114 | 32.6 | 2.4 | D | D | D | 0.2 |
| George .................. | 1.23 | 52.6 | 10 | 71 | 50.9 | 2.9 | 79 | 988 | 262.7 | 21.8 | 9 | 25 | 4.3 | 0.6 |
| Greene................... | 1.86 | 137.6 | NA | NA | NA | NA | 29 | 230 | 53.5 | 4.4 | NA | NA | NA | NA |
| Grenada................ | 3.96 | 183.5 | D | D | D | 6.2 | 118 | 1,390 | 455.1 | 37.7 | 24 | 77 | 20.4 | 2.9 |
| Hancock................ | 4.39 | 94.6 | 11 | 21 | 16.1 | 0.9 | 131 | 1,524 | 364.5 | 34.9 | 35 | 80 | 19.1 | 2.8 |
| Harrison................ | 20.84 | 103.5 | 147 | 1,363 | 716.9 | 66.6 | 807 | 11,916 | 3,124.5 | 286.3 | 221 | 1,007 | 244.8 | 34.8 |
| Hinds.................... | 50.54 | 208.1 | 247 | 3,059 | 2,998.3 | 181.2 | 795 | 10,782 | 3,330.6 | 307.1 | 263 | 1,039 | 259.9 | 48.5 |
| Holmes.................. | 2.20 | 120.0 | 9 | 53 | 62.3 | 2.5 | 66 | 422 | 87.2 | 8.7 | 14 | 33 | 4.0 | 0.5 |
| Humphreys............. | 0.92 | 106.1 | 6 | 36 | 59.5 | 1.6 | 27 | 220 | 48.9 | 4.7 | D | D | D | 0.2 |
| Issaquena............. | 0.12 | 89.8 | NA | NA | NA | NA | NA | NA | NA | NA | NA | NA | NA | NA |
| Itawamba............... | 10.39 | 440.1 | 10 | 216 | 53.0 | 5.7 | 70 | 682 | 164.4 | 15.8 | 7 | 18 | 1.9 | 0.3 |
| Jackson.................. | 14.50 | 102.5 | 61 | 375 | 271.3 | 18.9 | 417 | 5,034 | 1,336.2 | 123.8 | 100 | 450 | 73.5 | 15.0 |
| Jasper................... | 3.29 | 198.6 | 8 | 20 | 25.5 | 1.2 | 39 | 337 | 74.5 | 8.5 | NA | NA | NA | NA |
| Jefferson............... | 0.72 | 95.9 | NA | NA | NA | NA | 12 | 81 | 13.6 | 1.4 | NA | NA | NA | NA |
| Jefferson Davis...... | 1.45 | 124.3 | NA | NA | NA | NA | 36 | 239 | 61.2 | 5.4 | NA | NA | NA | NA |
| Jones.................... | 11.75 | 172.2 | 66 | 757 | 344.6 | 29.2 | 259 | 3,081 | 857.0 | 74.5 | D | D | D | D |
| Kemper.................. | 1.78 | 178.6 | NA | NA | NA | NA | 16 | 134 | 31.8 | 2.8 | D | D | D | D |
| Lafayette............... | 5.28 | 99.3 | 28 | 313 | 229.5 | 11.7 | 230 | 3,165 | 743.7 | 66.9 | 73 | 251 | 59.8 | 9.1 |
| Lamar................... | 5.92 | 97.7 | D | D | D | D | 311 | 5,000 | 1,213.1 | 110.4 | D | D | D | D |
| Lauderdale............. | 10.95 | 139.4 | 69 | 1,584 | 1,928.8 | 70.1 | 382 | 4,954 | 1,271.5 | 117.2 | 76 | 355 | 75.2 | 13.3 |
| Lawrence............... | 1.72 | 136.3 | NA | NA | NA | NA | 31 | 257 | 53.6 | 5.3 | NA | NA | NA | NA |
| Leake.................... | 2.21 | 97.1 | 3 | 44 | 12.9 | 1.8 | 85 | 767 | 179.4 | 17.6 | 6 | 17 | 1.3 | 0.3 |
| Lee....................... | 4.29 | 50.3 | 152 | 2,135 | 1,232.1 | 99.9 | 505 | 7,338 | 1,759.1 | 169.5 | 100 | 436 | 91.9 | 14.0 |
| Leflore.................. | 4.15 | 133.9 | D | D | D | D | 163 | 1,676 | 396.5 | 34.6 | 41 | 127 | 20.0 | 3.6 |
| Lincoln.................. | 4.04 | 116.6 | D | D | D | 31.8 | 159 | 1,939 | 553.7 | 53.0 | 22 | 76 | 10.7 | 2.3 |
| Lowndes................ | 8.37 | 140.2 | 70 | 1,287 | 1,005.6 | 75.1 | 306 | 3,582 | 862.1 | 82.0 | 60 | 211 | 43.8 | 7.5 |
| Madison................. | 16.52 | 159.7 | 134 | 2,414 | 3,481.4 | 142.8 | 522 | 7,094 | 1,798.9 | 173.9 | 170 | 915 | 169.9 | 39.7 |
| Marion.................. | 3.62 | 141.6 | 16 | 139 | 122.9 | 5.3 | 107 | 1,199 | 268.2 | 26.6 | 16 | 43 | 11.8 | 1.6 |
| Marshall................ | 2.53 | 70.4 | 20 | 504 | 210.4 | 19.2 | 108 | 838 | 214.9 | 19.9 | 7 | 15 | 2.6 | 0.3 |
| Monroe.................. | 3.63 | 101.3 | 23 | 243 | 130.8 | 12.4 | 112 | 1,017 | 254.7 | 23.4 | 15 | 39 | 6.4 | 1.3 |
| Montgomery............ | 1.30 | 128.1 | D | D | D | D | 44 | 499 | 123.7 | 9.1 | NA | NA | NA | NA |
| Neshoba................ | 4.47 | 151.7 | 16 | 324 | 328.8 | 15.2 | 111 | 1,213 | 281.0 | 28.0 | 10 | 26 | 2.3 | 0.8 |
| Newton.................. | 2.47 | 113.6 | 9 | 42 | 16.1 | 1.4 | 67 | 617 | 142.4 | 14.5 | D | D | D | D |
| Noxubee................ | 1.37 | 124.1 | 9 | 54 | 43.7 | 2.2 | 35 | 258 | 61.6 | 5.8 | NA | NA | NA | NA |
| Oktibbeha.............. | 7.58 | 152.2 | 16 | 828 | 559.6 | 33.4 | 176 | 2,391 | 529.4 | 51.7 | 58 | 271 | 55.8 | 7.8 |
| Panola.................. | 3.35 | 98.0 | 24 | 412 | 537.9 | 22.6 | 139 | 1,531 | 408.9 | 35.7 | 11 | 98 | 18.4 | 3.6 |
| Pearl River............ | 4.58 | 83.0 | 24 | 177 | 71.6 | 8.9 | 176 | 2,142 | 657.0 | 53.2 | 20 | 60 | 9.6 | 1.4 |
| Perry..................... | 1.23 | 100.2 | NA | NA | NA | NA | 32 | 359 | 83.8 | 8.3 | NA | NA | NA | NA |
| Pike...................... | 5.89 | 147.4 | 39 | 334 | 166.1 | 13.9 | 208 | 2,486 | 664.8 | 59.4 | 39 | 171 | 25.8 | 3.7 |
| Pontotoc............... | 3.45 | 111.6 | 18 | 130 | 74.2 | 4.4 | 95 | 1,071 | 326.8 | 25.3 | 9 | 30 | 4.4 | 0.5 |
| Prentiss................ | 3.28 | 128.8 | 8 | 41 | 5.5 | 0.9 | 97 | 832 | 185.8 | 17.4 | 16 | 44 | 5.4 | 1.3 |
| Quitman................ | 0.74 | 98.9 | 3 | 16 | 24.1 | 0.7 | 22 | 100 | 18.5 | 1.7 | 7 | D | 1.6 | D |
| Rankin.................. | 17.66 | 118.5 | 208 | 3,908 | 2,633.0 | 222.3 | 613 | 9,859 | 2,676.8 | 231.3 | 185 | 904 | 183.5 | 32.9 |
| Scott..................... | 8.02 | 283.7 | 14 | 46 | 81.3 | 1.5 | 124 | 1,195 | 296.1 | 28.1 | 10 | 44 | 5.1 | 1.0 |
| Sharkey................. | 0.52 | 113.4 | 13 | 78 | 106.2 | 4.5 | 19 | 150 | 30.3 | 3.0 | 7 | D | 2.5 | D |
| Simpson................ | 3.60 | 132.2 | 10 | 88 | 23.5 | 2.8 | 89 | 898 | 229.5 | 20.8 | 9 | 34 | 6.7 | 1.2 |
| Smith.................... | 1.18 | 73.5 | 4 | 39 | 19.5 | 1.5 | 29 | 204 | 52.0 | 4.4 | NA | NA | NA | NA |
| Stone.................... | 1.46 | 80.8 | 11 | 27 | 13.0 | 1.0 | 61 | 700 | 194.7 | 17.3 | 8 | D | 1.8 | D |

1 Merchant wholesalers, except manufacturers' sales branches and offices.   2. Employer establishments.

# Professional Services, Manufacturing, and Accommodation and Food Services

| STATE County | Professional, scientific, and technical services, 2017 | | | | Manufacturing, 2017 | | | | Accommodation and food services, 2017 | | | |
|---|---|---|---|---|---|---|---|---|---|---|---|---|
| | Number of establishments | Number of employees | Sales (mil dol) | Average payroll (mil dol) | Number of establishments | Number of employees | Sales (mil dol) | Average payroll (mil dol) | Number of establishments | Number of employees | Sales (mil dol) | Annual payroll (mil dol) |
| | 147 | 148 | 149 | 150 | 151 | 152 | 153 | 154 | 155 | 156 | 157 | 158 |
| MISSISSIPPI | 4,727 | 30,333 | 4,578 | 1,610.0 | 2,142 | 138,460 | 60,906.5 | 6,653.7 | 5,651 | 129,836 | 8,181.3 | 2,134.3 |
| Adams | 53 | 229 | 26 | 9.0 | 21 | 497 | 130.3 | 23.1 | 90 | 1,809 | 102.1 | 27.7 |
| Alcorn | 43 | 801 | 84 | 32.9 | 35 | 2,415 | 1,034.9 | 98.2 | 74 | 1,237 | 65.4 | 17.4 |
| Amite | 6 | 21 | 2 | 0.7 | 9 | 337 | 78.9 | 15.8 | 7 | 41 | 2.9 | 0.6 |
| Attala | 31 | 71 | 7 | 2.3 | 13 | 521 | 144.6 | 22.4 | 32 | 415 | 18.3 | 4.5 |
| Benton | D | D | D | D | 6 | 102 | 32.3 | 5.8 | NA | NA | NA | NA |
| Bolivar | 49 | 153 | 24 | 7.9 | 19 | 1,381 | 471.1 | 73.1 | 60 | 1,059 | 48.7 | 12.3 |
| Calhoun | 18 | 33 | 3 | 0.9 | 15 | 672 | 292.7 | 26.5 | 9 | 60 | 3.7 | 1.0 |
| Carroll | 6 | 8 | 2 | 0.4 | 6 | 20 | 2.7 | 0.7 | 4 | 10 | 0.7 | 0.1 |
| Chickasaw | 13 | 62 | 6 | 2.2 | 37 | 3,348 | 558.7 | 120.8 | D | D | D | D |
| Choctaw | 7 | 17 | 2 | 0.5 | 10 | 249 | 74.7 | 9.6 | D | D | D | D |
| Claiborne | D | D | D | D | 4 | 100 | 45.3 | 5.1 | 11 | 164 | 8.8 | 1.9 |
| Clarke | 8 | 38 | 2 | 0.5 | 10 | 373 | 125.9 | 14.5 | 14 | 195 | 9.5 | 2.2 |
| Clay | D | D | D | D | 16 | 720 | 384.5 | 47.0 | 39 | 642 | 26.5 | 7.6 |
| Coahoma | D | D | D | D | 15 | 569 | 178.0 | 25.2 | 43 | 1,027 | 76.6 | 17.2 |
| Copiah | 27 | 89 | 11 | 3.7 | 18 | 1,828 | 610.2 | 76.8 | 33 | 435 | 21.0 | 4.7 |
| Covington | 14 | 41 | 6 | 1.5 | 11 | 1,386 | 466.5 | 44.1 | D | D | D | 5.8 |
| DeSoto | 187 | 902 | 128 | 37.7 | 95 | 4,174 | 1,304.8 | 198.6 | 360 | 8,333 | 428.0 | 113.9 |
| Forrest | D | D | D | D | 68 | 3,493 | 1,003.6 | 151.6 | D | D | D | D |
| Franklin | 7 | 19 | 2 | 0.6 | D | D | D | D | 4 | D | 0.3 | D |
| George | 18 | 107 | 13 | 3.9 | 13 | 277 | 28.4 | 16.8 | 37 | 505 | 26.6 | 6.7 |
| Greene | 4 | 14 | 1 | 0.2 | 3 | 7 | 1.1 | 0.3 | 8 | 78 | 4.1 | 1.3 |
| Grenada | 34 | 191 | 25 | 8.3 | 25 | 2,803 | 896.8 | 123.7 | 60 | 1,055 | 46.6 | 12.6 |
| Hancock | 111 | 1,545 | 242 | 103.0 | 27 | 787 | 1,020.4 | 59.2 | 93 | 2,240 | 216.9 | 50.4 |
| Harrison | D | D | D | D | 101 | 2,546 | 1,360.8 | 162.7 | 510 | 21,999 | 1,974.3 | 495.4 |
| Hinds | 670 | 5,425 | 884 | 331.2 | 124 | 3,523 | 1,176.8 | 181.7 | 495 | 10,910 | 537.2 | 147.3 |
| Holmes | 12 | 40 | 10 | 1.8 | 5 | 505 | 207.7 | 14.9 | 18 | 147 | 6.2 | 1.6 |
| Humphreys | D | D | 2 | D | D | D | D | D | 5 | 131 | 4.3 | 1.1 |
| Issaquena | NA | NA | NA | NA | NA | NA | NA | NA | NA | NA | NA | NA |
| Itawamba | 11 | 30 | 8 | 2.1 | 34 | 2,204 | 706.7 | 87.0 | D | D | D | D |
| Jackson | 215 | 1,527 | 210 | 83.3 | 68 | 14,576 | 11,953.5 | 1,043.0 | 267 | 4,435 | 211.3 | 60.5 |
| Jasper | 15 | 49 | 5 | 1.5 | 17 | 1,235 | 369.9 | 53.7 | 11 | 121 | 6.0 | 1.3 |
| Jefferson | NA | NA | NA | NA | NA | NA | NA | NA | D | D | D | 0.2 |
| Jefferson Davis | 10 | 16 | 1 | 0.3 | D | 46 | D | D | 11 | 120 | 4.6 | 1.3 |
| Jones | 82 | 465 | 65 | 21.0 | 58 | 5,686 | 1,646.5 | 267.5 | 114 | 1,957 | 89.9 | 22.8 |
| Kemper | 5 | 14 | 1 | 0.1 | 7 | 210 | 26.8 | 6.6 | 12 | 88 | 3.5 | 0.9 |
| Lafayette | D | D | D | D | 23 | 1,796 | 259.8 | 81.4 | 199 | 4,000 | 217.2 | 68.0 |
| Lamar | 120 | 849 | 107 | 39.7 | 22 | 137 | 30.1 | 5.0 | D | D | D | D |
| Lauderdale | D | D | D | D | 56 | 1,925 | 659.0 | 81.1 | 175 | 3,990 | 190.1 | 51.6 |
| Lawrence | 11 | 13 | 2 | 0.5 | D | 674 | D | 51.0 | 10 | D | 3.7 | D |
| Leake | 19 | 52 | 6 | 1.5 | D | D | D | D | 24 | 385 | 16.1 | 4.3 |
| Lee | 184 | 1,071 | 134 | 54.1 | 133 | 8,457 | 3,026.1 | 358.0 | 234 | 4,717 | 235.5 | 65.5 |
| Leflore | 42 | 267 | 45 | 12.8 | 22 | 2,101 | 596.4 | 82.1 | 72 | 1,333 | 69.9 | 20.3 |
| Lincoln | 53 | 317 | 54 | 13.6 | 36 | 879 | 317.1 | 49.7 | 64 | 1,146 | 56.3 | 14.2 |
| Lowndes | 96 | 501 | 58 | 26.1 | 56 | 3,848 | 3,609.8 | 247.9 | 125 | 2,489 | 112.2 | 31.9 |
| Madison | 409 | 3,168 | 720 | 228.3 | 56 | 7,823 | 5,880.2 | 433.8 | 270 | 5,931 | 296.1 | 83.4 |
| Marion | 36 | 217 | 35 | 12.9 | 19 | 649 | 88.1 | 22.8 | 38 | 636 | 28.6 | 6.6 |
| Marshall | 16 | 61 | 7 | 1.9 | 29 | 748 | 317.0 | 49.4 | D | D | D | D |
| Monroe | 30 | 79 | 13 | 2.6 | 42 | 2,446 | 1,049.1 | 114.2 | D | D | D | D |
| Montgomery | 10 | 28 | 3 | 0.7 | 8 | 148 | 36.9 | 8.2 | D | D | D | D |
| Neshoba | 31 | 379 | 48 | 19.5 | 15 | 340 | 191.3 | 15.4 | 43 | 3,132 | 398.3 | 95.9 |
| Newton | 23 | 51 | 4 | 1.6 | 16 | 932 | 181.6 | 43.5 | 26 | 345 | 15.5 | 4.2 |
| Noxubee | 4 | 10 | 1 | 0.3 | 16 | 484 | 174.8 | 19.9 | D | D | D | D |
| Oktibbeha | 77 | 495 | 70 | 23.8 | 25 | 1,105 | 402.0 | 46.4 | 119 | 2,999 | 140.7 | 41.4 |
| Panola | 21 | 75 | 13 | 2.8 | 27 | 1,518 | 427.2 | 67.7 | 57 | 999 | 47.6 | 12.6 |
| Pearl River | 59 | 353 | 39 | 16.5 | 42 | 587 | 236.6 | 27.0 | 78 | 1,318 | 55.7 | 15.3 |
| Perry | 13 | 18 | 3 | 0.7 | 5 | 612 | 221.0 | 43.0 | 4 | 44 | 2.0 | 0.5 |
| Pike | D | D | D | D | 27 | 2,440 | 650.6 | 85.9 | 81 | 1,637 | 72.5 | 19.0 |
| Pontotoc | 22 | 91 | 11 | 4.0 | 61 | 7,262 | 1,634.9 | 267.3 | D | D | D | D |
| Prentiss | 31 | 162 | 28 | 5.7 | 40 | 1,714 | 873.8 | 67.8 | 35 | 573 | 26.4 | 6.3 |
| Quitman | D | D | D | D | NA | NA | NA | NA | 4 | 55 | 2.0 | 0.5 |
| Rankin | 310 | 1,522 | 219 | 74.8 | 120 | 3,691 | 1,553.5 | 202.1 | 326 | 7,005 | 386.2 | 100.4 |
| Scott | 30 | 83 | 7 | 2.1 | 22 | 6,221 | 1,703.9 | 222.9 | 46 | 731 | 33.3 | 8.3 |
| Sharkey | 8 | 16 | 1 | 0.4 | NA | NA | NA | NA | D | D | D | 0.6 |
| Simpson | 30 | 107 | 13 | 4.1 | 9 | 252 | 95.2 | 9.6 | 40 | 644 | 33.2 | 7.7 |
| Smith | D | D | 10 | D | 13 | 859 | 317.8 | 41.5 | D | D | D | 0.8 |
| Stone | 20 | 103 | 10 | 4.4 | 10 | 617 | 181.0 | 27.4 | 28 | 481 | 21.8 | 5.2 |

# Health Care and Social Assistance, Other Services, Nonemployer Businesses, and Residential Construction

| STATE County | Health care and social assistance, 2017 | | | | Other services, 2017 | | | | Nonemployer businesses, 2018 | | Value of residential construction authorized by building permits, 2020 | |
|---|---|---|---|---|---|---|---|---|---|---|---|---|
| | Number of establishments | Number of employees | Receipts (mil dol) | Annual payroll (mil dol) | Number of establishments | Number of employees | Receipts (mil dol) | Annual payroll (mil dol) | Number | Receipts (mil dol) | New construction ($1,000) | Number of housing units |
| | 159 | 160 | 161 | 162 | 163 | 164 | 165 | 166 | 167 | 168 | 169 | 170 |
| MISSISSIPPI | 6,391 | 169,010 | 18,752.3 | 7,533.3 | 3,417 | 18,345 | 2,278.6 | 603.2 | 222,159 | 9,164.2 | 1,428,531 | 7,810 |
| Adams | 94 | 1,871 | 198.8 | 66.1 | 37 | 166 | 16.2 | 4.5 | 2,404 | 96.4 | 1,308 | 6 |
| Alcorn | 105 | 2,378 | 288.9 | 85.7 | 47 | 152 | 17.1 | 4.1 | 2,466 | 133.7 | 2,807 | 16 |
| Amite | 13 | 210 | 13.1 | 5.3 | 7 | 26 | 2.3 | 0.6 | 815 | 35.2 | 0 | 0 |
| Attala | 29 | 559 | 57.5 | 18.7 | D | D | D | D | 1,302 | 49.9 | 0 | 0 |
| Benton | 5 | 139 | 12.8 | 4.4 | D | D | D | D | 588 | 23.1 | 0 | 0 |
| Bolivar | 83 | 1,811 | 172.2 | 65.8 | 50 | 198 | 17.1 | 4.2 | 2,084 | 72.2 | 2,917 | 20 |
| Calhoun | 25 | 204 | 16.7 | 7.1 | D | D | D | D | 947 | 36.7 | 0 | 0 |
| Carroll | 7 | 124 | 6.7 | 2.7 | D | D | D | D | 711 | 32.7 | NA | NA |
| Chickasaw | 23 | 429 | 24.9 | 13.6 | 18 | 44 | 3.5 | 0.8 | 1,127 | 43.5 | 276 | 3 |
| Choctaw | 8 | 377 | 33.5 | 14.4 | D | D | 1.9 | D | 678 | 25.5 | 0 | 0 |
| Claiborne | 12 | 623 | 36.1 | 16.7 | 6 | 40 | 2.4 | 0.6 | 395 | 12.2 | 0 | 0 |
| Clarke | 18 | 340 | 31.1 | 14.1 | D | D | 3.5 | D | 1,055 | 38.3 | 0 | 0 |
| Clay | 28 | 658 | 67.0 | 25.0 | 21 | 92 | 10.0 | 2.6 | 1,280 | 47.2 | 598 | 6 |
| Coahoma | 75 | 1,545 | 160.6 | 54.4 | D | D | D | D | 1,658 | 56.4 | 4,764 | 28 |
| Copiah | 52 | 763 | 55.7 | 21.1 | D | D | D | 1.9 | 2,040 | 64.2 | 841 | 4 |
| Covington | 24 | 507 | 47.2 | 19.4 | D | D | D | D | 1,511 | 57.6 | 0 | 0 |
| DeSoto | 290 | 5,553 | 727.1 | 254.4 | 163 | 1,003 | 99.3 | 28.0 | 14,396 | 645.4 | 300,505 | 1,694 |
| Forrest | 201 | 7,843 | 1,029.9 | 449.8 | 95 | 553 | 56.1 | 14.2 | 5,472 | 226.7 | 22,508 | 117 |
| Franklin | 11 | 388 | 30.7 | 12.8 | NA | NA | NA | NA | 576 | 14.3 | 0 | 0 |
| George | 32 | 791 | 80.7 | 34.1 | D | D | D | D | 1,467 | 51.3 | 660 | 4 |
| Greene | D | D | D | D | D | D | D | 1.9 | 690 | 24.2 | 836 | 4 |
| Grenada | 61 | 1,228 | 109.0 | 43.8 | D | D | D | D | 1,435 | 57.7 | 400 | 2 |
| Hancock | 50 | 986 | 92.5 | 36.5 | 33 | 235 | 22.7 | 6.3 | 3,603 | 150.3 | 113,284 | 717 |
| Harrison | 486 | 14,330 | 1,892.6 | 822.5 | 269 | 1,567 | 162.2 | 46.2 | 15,364 | 644.4 | 307,318 | 1,703 |
| Hinds | 719 | 31,975 | 3,912.0 | 1,773.7 | 384 | 2,465 | 334.8 | 99.1 | 20,626 | 777.8 | 38,911 | 196 |
| Holmes | 30 | 511 | 46.4 | 17.5 | 10 | 120 | 11.3 | 4.3 | 1,225 | 33.2 | 822 | 9 |
| Humphreys | 15 | 153 | 10.3 | 4.5 | D | D | D | 0.3 | 618 | 22.4 | 0 | 0 |
| Issaquena | NA | NA | NA | NA | NA | NA | NA | NA | 66 | 2.2 | NA | NA |
| Itawamba | 32 | 558 | 39.3 | 18.4 | D | D | D | D | 1,299 | 58.6 | 627 | 3 |
| Jackson | 307 | 5,720 | 653.5 | 280.6 | 134 | 703 | 97.4 | 19.7 | 9,626 | 374.8 | 103,192 | 546 |
| Jasper | 9 | 254 | 16.2 | 8.5 | D | D | D | D | 1,062 | 34.4 | 0 | 0 |
| Jefferson | 11 | 232 | 19.2 | 7.6 | D | D | 0.8 | D | 514 | 12.3 | 0 | 0 |
| Jefferson Davis | 12 | 265 | 22.7 | 10.0 | D | D | 1.5 | D | 867 | 22.8 | 0 | 0 |
| Jones | 111 | 3,244 | 290.5 | 140.2 | D | D | D | D | 4,783 | 196.4 | 12,079 | 93 |
| Kemper | 12 | 211 | 13.7 | 6.3 | D | D | 0.2 | D | 615 | 18.5 | NA | NA |
| Lafayette | 157 | 2,893 | 470.3 | 129.3 | 72 | 444 | 128.4 | 21.3 | 4,695 | 257.4 | 35,570 | 139 |
| Lamar | D | D | D | D | D | D | D | D | 4,990 | 261.4 | 4,089 | 26 |
| Lauderdale | 220 | 6,992 | 917.1 | 329.0 | 117 | 505 | 55.4 | 14.2 | 5,084 | 172.3 | 1,868 | 9 |
| Lawrence | 25 | 341 | 29.3 | 11.2 | D | D | D | D | 780 | 28.8 | 105 | 1 |
| Leake | 27 | 618 | 54.3 | 21.4 | 11 | 33 | 4.4 | 1.3 | 1,299 | 42.9 | 418 | 2 |
| Lee | 304 | 9,342 | 1,223.7 | 452.2 | 124 | 1,042 | 106.7 | 40.8 | 6,530 | 293.5 | 30,571 | 162 |
| Leflore | 70 | 2,817 | 203.2 | 96.2 | D | D | D | D | 1,622 | 68.8 | 3,104 | 8 |
| Lincoln | 78 | 1,791 | 204.0 | 73.0 | 48 | 303 | 27.8 | 10.2 | 2,718 | 115.9 | 6,513 | 67 |
| Lowndes | 166 | 3,253 | 417.8 | 129.7 | 89 | 418 | 46.7 | 12.2 | 4,042 | 143.7 | 13,709 | 128 |
| Madison | 298 | 8,737 | 539.7 | 223.8 | 159 | 1,269 | 192.8 | 57.5 | 11,866 | 709.7 | 170,620 | 633 |
| Marion | 45 | 952 | 63.7 | 29.3 | 32 | 311 | 34.7 | 9.6 | 1,957 | 77.7 | 0 | 0 |
| Marshall | 32 | 863 | 180.6 | 25.1 | D | D | D | D | 2,808 | 112.8 | 31,149 | 168 |
| Monroe | 73 | 1,408 | 144.4 | 49.0 | 38 | 162 | 15.6 | 3.6 | 2,006 | 79.8 | 905 | 7 |
| Montgomery | 20 | 738 | 53.0 | 23.9 | 17 | 61 | 5.7 | 1.4 | 646 | 20.5 | 2,465 | 93 |
| Neshoba | 37 | 992 | 92.4 | 38.5 | D | D | D | D | 1,880 | 62.4 | 0 | 0 |
| Newton | 31 | 513 | 44.3 | 15.2 | 18 | 53 | 5.1 | 1.3 | 1,316 | 51.9 | 0 | 0 |
| Noxubee | 13 | 307 | 25.4 | 9.0 | 10 | 15 | 2.0 | 0.4 | 926 | 30.8 | 0 | 0 |
| Oktibbeha | 91 | 2,026 | 186.6 | 79.8 | 62 | 374 | 123.0 | 14.4 | 3,200 | 115.8 | 10,000 | 52 |
| Panola | 65 | 1,694 | 125.5 | 48.5 | 22 | 57 | 6.9 | 1.7 | 2,745 | 95.7 | 1,415 | 11 |
| Pearl River | 98 | 1,516 | 125.2 | 52.6 | 54 | 220 | 20.0 | 5.2 | 4,062 | 170.2 | 38,311 | 328 |
| Perry | D | D | D | D | D | D | D | D | 720 | 26.6 | 250 | 1 |
| Pike | 114 | 2,612 | 233.6 | 100.0 | 51 | 245 | 24.1 | 6.5 | 3,307 | 118.0 | 2,280 | 45 |
| Pontotoc | 45 | 752 | 61.0 | 26.1 | D | D | D | D | 2,145 | 91.3 | 1,944 | 14 |
| Prentiss | 46 | 696 | 61.5 | 24.5 | D | D | D | D | 1,539 | 53.4 | 209 | 1 |
| Quitman | 13 | 159 | 8.6 | 4.3 | D | D | D | 0.2 | 522 | 14.2 |  |  |
| Rankin | 367 | 7,993 | 1,000.3 | 377.9 | 223 | 1,142 | 149.2 | 43.7 | 13,167 | 602.5 | 73,358 | 294 |
| Scott | 35 | 706 | 58.5 | 21.4 | 32 | 93 | 9.2 | 2.1 | 1,622 | 58.2 | 2,041 | 14 |
| Sharkey | 11 | 259 | 19.4 | 7.2 | D | D | D | 1.1 | 388 | 12.3 | 0 | 0 |
| Simpson | 51 | 2,201 | 154.7 | 60.2 | D | D | D | D | 1,884 | 81.1 | 8,119 | 42 |
| Smith | 7 | 224 | 15.2 | 6.3 | D | D | D | 0.8 | 993 | 37.1 | 263 | 3 |
| Stone | 25 | 524 | 47.8 | 18.6 | D | D | D | D | 1,284 | 54.0 | 8,860 | 55 |

Items 159—170

# Table B. States and Counties — Government Employment and Payroll, and Local Government Finances

| STATE County | Full-time equivalent employees | March payroll (dollars) | Administration, judicial, and legal | Police and corrections | Fire protection | Highways and transportation | Health and welfare | Natural resources and utilities | Education and libraries | Total (mil dol) | Inter-governmental (mil dol) | Taxes Total (mil dol) | Per capita[1] Total | Per capita[1] Property |
|---|---|---|---|---|---|---|---|---|---|---|---|---|---|---|
| | 171 | 172 | 173 | 174 | 175 | 176 | 177 | 178 | 179 | 180 | 181 | 182 | 183 | 184 |
| MISSISSIPPI | X | X | X | X | X | X | X | X | X | X | X | X | X | X |
| Adams | 1,154 | 3,640,747 | 9.8 | 19.4 | 4.1 | 8.0 | 2.0 | 7.8 | 48.6 | 104.3 | 46.1 | 38.5 | 1,225 | 1,077 |
| Alcorn | 2,738 | 9,492,949 | 2.6 | 3.6 | 1.4 | 0.9 | 63.9 | 1.4 | 26.2 | 304.7 | 71.1 | 26.8 | 722 | 693 |
| Amite | 303 | 826,040 | 13.7 | 8.8 | 0.0 | 5.6 | 0.0 | 4.4 | 65.2 | 22.5 | 11.2 | 7.3 | 587 | 582 |
| Attala | 904 | 2,596,804 | 5.9 | 4.4 | 3.4 | 4.4 | 22.7 | 1.8 | 56.7 | 68.0 | 27.6 | 18.4 | 997 | 969 |
| Benton | 306 | 790,316 | 11.4 | 7.6 | 0.6 | 4.6 | 0.0 | 3.4 | 71.0 | 21.0 | 15.5 | 4.4 | 528 | 507 |
| Bolivar | 1,507 | 4,354,776 | 6.3 | 13.4 | 0.4 | 4.9 | 1.1 | 3.6 | 69.1 | 126.3 | 62.6 | 40.4 | 1,275 | 1,213 |
| Calhoun | 491 | 1,342,068 | 7.2 | 6.2 | 0.5 | 3.3 | 0.0 | 5.2 | 75.5 | 54.6 | 22.2 | 10.1 | 698 | 680 |
| Carroll | 342 | 902,475 | 13.2 | 17.6 | 0.2 | 7.6 | 0.0 | 2.8 | 58.5 | 22.4 | 8.6 | 7.5 | 739 | 720 |
| Chickasaw | 837 | 2,241,340 | 7.3 | 10.5 | 1.8 | 3.3 | 8.8 | 6.2 | 61.2 | 53.1 | 27.3 | 13.8 | 802 | 767 |
| Choctaw | 368 | 1,024,805 | 14.0 | 5.2 | 0.3 | 3.1 | 0.0 | 2.6 | 74.8 | 31.7 | 13.1 | 10.7 | 1,294 | 1,281 |
| Claiborne | 406 | 1,271,692 | 7.2 | 7.6 | 2.2 | 4.6 | 23.6 | 2.8 | 50.9 | 29.7 | 19.7 | 6.9 | 766 | 742 |
| Clarke | 691 | 1,751,565 | 9.6 | 6.4 | 0.9 | 5.3 | 0.1 | 6.5 | 69.8 | 43.5 | 21.9 | 17.5 | 1,105 | 1,078 |
| Clay | 821 | 2,240,934 | 3.8 | 6.5 | 3.8 | 3.1 | 11.0 | 2.3 | 69.1 | 56.3 | 34.3 | 17.6 | 897 | 870 |
| Coahoma | 1,409 | 4,539,054 | 5.6 | 9.0 | 3.4 | 3.0 | 0.9 | 10.0 | 65.6 | 117.7 | 69.5 | 29.2 | 1,258 | 1,188 |
| Copiah | 1,373 | 4,392,339 | 5.5 | 5.2 | 0.9 | 2.2 | 10.6 | 0.9 | 74.2 | 101.2 | 52.7 | 22.5 | 788 | 762 |
| Covington | 869 | 2,781,519 | 6.4 | 5.7 | 0.1 | 2.5 | 33.6 | 2.9 | 47.7 | 48.2 | 23.0 | 22.1 | 1,161 | 1,148 |
| DeSoto | 5,975 | 19,064,532 | 5.9 | 13.1 | 7.1 | 2.0 | 0.7 | 4.0 | 66.1 | 522.3 | 229.3 | 217.7 | 1,216 | 1,102 |
| Forrest | 6,730 | 26,043,833 | 2.9 | 4.3 | 2.3 | 1.8 | 61.5 | 2.3 | 24.4 | 795.9 | 114.7 | 102.4 | 1,363 | 1,209 |
| Franklin | 313 | 888,977 | 12.4 | 4.4 | 0.0 | 4.9 | 0.1 | 3.1 | 74.7 | 23.8 | 12.3 | 6.1 | 791 | 772 |
| George | 817 | 2,059,025 | 5.9 | 10.2 | 0.8 | 4.5 | 0.1 | 2.0 | 75.0 | 101.2 | 33.6 | 15.9 | 663 | 639 |
| Greene | 500 | 1,389,648 | 7.4 | 4.1 | 0.1 | 6.3 | 13.5 | 3.8 | 64.7 | 37.8 | 19.8 | 10.6 | 782 | 764 |
| Grenada | 961 | 3,091,225 | 4.2 | 5.0 | 5.1 | 1.1 | 0.0 | 3.5 | 79.3 | 67.7 | 38.5 | 19.6 | 930 | 851 |
| Hancock | 1,751 | 5,914,245 | 7.1 | 9.4 | 4.5 | 3.1 | 18.2 | 5.9 | 49.5 | 228.8 | 95.0 | 55.2 | 1,176 | 1,121 |
| Harrison | 10,056 | 41,422,885 | 3.4 | 6.8 | 4.3 | 3.0 | 41.8 | 1.9 | 37.2 | 1,300.8 | 432.3 | 262.2 | 1,279 | 1,086 |
| Hinds | 10,432 | 34,791,313 | 5.4 | 9.8 | 4.2 | 1.9 | 2.0 | 3.7 | 71.3 | 895.4 | 423.4 | 297.4 | 1,239 | 1,172 |
| Holmes | 1,303 | 3,991,551 | 5.0 | 7.1 | 0.7 | 1.9 | 0.1 | 2.2 | 82.8 | 87.2 | 55.5 | 14.6 | 820 | 796 |
| Humphreys | 358 | 907,778 | 4.4 | 7.9 | 1.9 | 2.8 | 0.0 | 5.7 | 77.2 | 33.5 | 19.5 | 10.6 | 1,271 | 1,234 |
| Issaquena | 19 | 55,685 | 43.0 | 24.3 | 0.0 | 24.1 | 0.0 | 7.4 | 0.0 | 6.2 | 0.6 | 1.8 | 1,366 | 1,345 |
| Itawamba | 1,021 | 3,695,254 | 3.3 | 3.8 | 0.3 | 1.8 | 0.1 | 3.2 | 87.5 | 114.1 | 56.5 | 27.1 | 1,154 | 1,137 |
| Jackson | 7,086 | 27,480,316 | 3.9 | 5.8 | 2.8 | 3.5 | 41.1 | 3.2 | 38.9 | 855.5 | 198.9 | 217.4 | 1,529 | 1,466 |
| Jasper | 820 | 2,360,673 | 7.7 | 4.4 | 0.0 | 4.9 | 31.0 | 4.8 | 46.8 | 56.9 | 21.4 | 21.1 | 1,272 | 1,252 |
| Jefferson | 335 | 714,838 | 1.4 | 0.9 | 0.0 | 0.0 | 37.7 | 1.2 | 58.8 | 46.4 | 12.5 | 6.3 | 873 | 838 |
| Jefferson Davis | 383 | 1,028,403 | 15.1 | 4.2 | 0.9 | 5.8 | 0.0 | 3.2 | 70.1 | 28.3 | 13.6 | 11.2 | 995 | 962 |
| Jones | 5,005 | 13,746,901 | 2.9 | 4.1 | 2.5 | 2.2 | 34.9 | 2.2 | 49.7 | 384.5 | 134.0 | 63.9 | 935 | 889 |
| Kemper | 716 | 2,651,291 | 4.0 | 7.7 | 0.1 | 2.8 | 0.1 | 1.2 | 83.3 | 69.9 | 42.4 | 12.6 | 1,253 | 1,240 |
| Lafayette | 1,446 | 4,445,994 | 5.6 | 13.0 | 4.9 | 4.3 | 1.2 | 7.5 | 61.3 | 148.5 | 58.7 | 65.8 | 1,214 | 1,079 |
| Lamar | 2,022 | 5,076,981 | 6.5 | 7.2 | 0.8 | 5.3 | 0.0 | 1.7 | 76.1 | 140.3 | 69.3 | 60.6 | 985 | 961 |
| Lauderdale | 3,125 | 10,374,372 | 9.2 | 5.9 | 4.3 | 6.7 | 2.2 | 3.1 | 67.2 | 252.6 | 127.4 | 83.1 | 1,088 | 1,035 |
| Lawrence | 421 | 1,213,411 | 4.9 | 8.4 | 0.7 | 5.6 | 0.0 | 2.7 | 77.7 | 31.5 | 15.7 | 13.1 | 1,039 | 1,019 |
| Leake | 636 | 1,698,494 | 10.7 | 13.6 | 1.9 | 4.7 | 0.4 | 3.0 | 64.2 | 43.8 | 25.3 | 10.9 | 478 | 460 |
| Lee | 3,894 | 11,602,013 | 6.5 | 9.6 | 3.7 | 3.3 | 2.5 | 7.1 | 66.3 | 267.7 | 136.4 | 96.6 | 1,136 | 1,107 |
| Leflore | 2,143 | 8,712,788 | 3.1 | 4.4 | 2.0 | 1.8 | 57.4 | 3.1 | 27.6 | 228.2 | 54.8 | 31.8 | 1,086 | 1,038 |
| Lincoln | 1,197 | 3,479,462 | 7.9 | 8.4 | 3.8 | 3.8 | 0.5 | 2.7 | 72.4 | 94.4 | 53.0 | 30.4 | 884 | 839 |
| Lowndes | 2,221 | 7,680,900 | 8.6 | 11.6 | 3.6 | 4.4 | 3.2 | 7.0 | 60.1 | 193.5 | 100.8 | 60.1 | 1,016 | 948 |
| Madison | 3,126 | 11,817,063 | 7.2 | 10.8 | 4.4 | 2.6 | 0.3 | 3.0 | 69.5 | 341.6 | 133.2 | 161.0 | 1,539 | 1,460 |
| Marion | 885 | 2,847,886 | 6.3 | 11.9 | 2.5 | 4.8 | 0.0 | 1.8 | 71.2 | 81.7 | 35.0 | 20.5 | 816 | 785 |
| Marshall | 1,060 | 3,365,642 | 8.7 | 13.3 | 1.5 | 10.5 | 0.3 | 7.2 | 58.2 | 77.1 | 42.9 | 28.8 | 808 | 780 |
| Monroe | 1,199 | 3,461,119 | 5.8 | 10.3 | 2.6 | 5.3 | 1.0 | 8.8 | 65.0 | 93.5 | 48.0 | 32.6 | 910 | 874 |
| Montgomery | 499 | 1,627,928 | 7.8 | 5.3 | 1.6 | 3.5 | 33.4 | 1.7 | 44.8 | 26.5 | 14.9 | 7.9 | 782 | 744 |
| Neshoba | 1,352 | 4,328,328 | 2.8 | 5.5 | 2.7 | 3.0 | 44.9 | 3.0 | 38.0 | 62.8 | 37.4 | 17.6 | 597 | 578 |
| Newton | 1,081 | 3,275,267 | 4.6 | 3.7 | 0.6 | 3.1 | 3.5 | 2.6 | 80.9 | 88.4 | 61.7 | 15.0 | 702 | 670 |
| Noxubee | 577 | 2,460,355 | 5.7 | 2.2 | 0.1 | 2.1 | 32.7 | 0.5 | 56.2 | 31.7 | 16.5 | 9.5 | 891 | 848 |
| Oktibbeha | 1,807 | 6,409,671 | 1.2 | 5.3 | 3.3 | 1.8 | 43.4 | 7.8 | 37.2 | 197.7 | 57.5 | 46.1 | 928 | 908 |
| Panola | 1,359 | 4,142,779 | 7.1 | 13.5 | 2.4 | 5.0 | 1.0 | 4.6 | 65.5 | 100.8 | 52.1 | 35.4 | 1,037 | 961 |
| Pearl River | 2,501 | 7,674,433 | 5.9 | 5.5 | 1.9 | 2.3 | 11.7 | 1.9 | 70.2 | 193.9 | 95.1 | 44.4 | 804 | 763 |
| Perry | 470 | 1,179,800 | 9.8 | 5.1 | 0.4 | 5.6 | 0.0 | 3.1 | 75.7 | 36.6 | 19.4 | 10.4 | 867 | 836 |
| Pike | 2,663 | 9,831,060 | 3.5 | 4.2 | 1.2 | 1.6 | 41.9 | 0.9 | 46.3 | 281.9 | 83.9 | 35.6 | 900 | 866 |
| Pontotoc | 1,054 | 3,276,009 | 5.8 | 7.5 | 1.2 | 2.3 | 0.3 | 5.3 | 76.5 | 86.6 | 48.5 | 19.7 | 621 | 591 |
| Prentiss | 1,300 | 3,958,225 | 4.2 | 5.3 | 2.6 | 2.0 | 1.0 | 1.5 | 83.5 | 87.3 | 56.1 | 17.1 | 678 | 654 |
| Quitman | 518 | 1,904,312 | 5.9 | 4.5 | 0.1 | 4.7 | 42.8 | 1.2 | 40.9 | 23.1 | 10.9 | 10.3 | 1,430 | 1,395 |
| Rankin | 4,216 | 13,330,887 | 5.0 | 11.0 | 6.6 | 3.1 | 0.3 | 3.4 | 70.2 | 405.9 | 174.4 | 155.0 | 1,014 | 972 |
| Scott | 1,023 | 2,718,449 | 6.8 | 10.9 | 1.4 | 3.1 | 0.5 | 5.4 | 71.3 | 70.3 | 43.6 | 19.7 | 692 | 662 |
| Sharkey | 338 | 985,527 | 10.0 | 6.4 | 0.1 | 3.6 | 29.6 | 3.4 | 46.6 | 18.3 | 9.0 | 6.4 | 1,450 | 1,402 |
| Simpson | 924 | 2,483,290 | 7.0 | 10.8 | 0.1 | 6.4 | 0.0 | 2.4 | 69.9 | 57.5 | 31.7 | 19.7 | 730 | 701 |
| Smith | 549 | 1,526,484 | 10.2 | 9.5 | 0.0 | 4.3 | 0.1 | 3.8 | 72.1 | 39.3 | 21.0 | 13.5 | 837 | 810 |
| Stone | 1,356 | 4,896,983 | 3.3 | 2.7 | 0.7 | 0.9 | 0.1 | 0.9 | 91.3 | 135.2 | 76.2 | 14.3 | 771 | 723 |

1. Based on the resident population estimated as of July 1 of the year shown.

Items 171—184

## Table B. States and Counties — Local Government Finances, Government Employment, and Income Taxes

| STATE County | Total (mil dol) 185 | Per capita[1] (dollars) 186 | Education 187 | Health and hospitals 188 | Police protection 189 | Public welfare 190 | Highways 191 | Total (mil dol) 192 | Per capita[1] (dollars) 193 | Federal civilian 194 | Federal military 195 | State and local 196 | Number of returns 197 | Mean adjusted gross income 198 | Mean income tax 199 |
|---|---|---|---|---|---|---|---|---|---|---|---|---|---|---|---|
| MISSISSIPPI | X | X | X | X | X | X | X | X | X | 25,562 | 27,625 | 218,148 | 1,227,510 | 51,495 | 4,821 |
| Adams | 104 | 3,303 | 38.4 | 1.3 | 9.4 | 0.3 | 8.4 | 48.5 | 1,542 | 98 | 184 | 1,395 | 11,720 | 47,728 | 4,561 |
| Alcorn | 291 | 7,830 | 17.9 | 59.4 | 2.4 | 0.0 | 1.2 | 201.4 | 5,419 | 112 | 212 | 2,996 | 13,660 | 48,260 | 4,074 |
| Amite | 25 | 2,009 | 60.2 | 1.7 | 6.1 | 0.0 | 12.6 | 0.5 | 37 | 29 | 71 | 389 | 4,840 | 46,663 | 3,681 |
| Attala | 70 | 3,804 | 44.9 | 26.9 | 3.9 | 0.0 | 5.2 | 18.5 | 1,002 | 53 | 104 | 1,022 | 7,240 | 43,522 | 3,218 |
| Benton | 19 | 2,334 | 62.0 | 0.2 | 6.4 | 0.0 | 9.3 | 2.9 | 354 | 48 | 48 | 333 | 3,250 | 37,132 | 2,335 |
| Bolivar | 122 | 3,854 | 49.5 | 4.9 | 7.6 | 0.1 | 8.3 | 32.1 | 1,011 | 78 | 170 | 2,524 | 12,610 | 46,559 | 4,114 |
| Calhoun | 55 | 3,811 | 37.7 | 34.2 | 3.2 | 0.1 | 8.5 | 4.0 | 278 | 33 | 82 | 601 | 5,650 | 39,392 | 2,502 |
| Carroll | 23 | 2,263 | 43.2 | 0.4 | 4.8 | 0.0 | 13.1 | 8.9 | 879 | 21 | 56 | 348 | 4,140 | 49,440 | 4,107 |
| Chickasaw | 52 | 3,017 | 55.2 | 1.0 | 5.0 | 10.0 | 3.9 | 30.6 | 1,787 | 44 | 97 | 877 | 7,320 | 39,672 | 2,868 |
| Choctaw | 36 | 4,366 | 46.8 | 21.2 | 4.3 | 0.1 | 5.7 | 4.0 | 479 | 58 | 47 | 721 | 3,320 | 43,582 | 3,033 |
| Claiborne | 32 | 3,575 | 51.4 | 3.9 | 5.1 | 0.2 | 8.4 | 14.8 | 1,642 | 27 | 45 | 1,433 | 3,320 | 33,335 | 1,979 |
| Clarke | 43 | 2,718 | 60.3 | 0.9 | 6.7 | 0.0 | 10.2 | 10.5 | 665 | 33 | 90 | 764 | 6,350 | 46,016 | 3,467 |
| Clay | 56 | 2,867 | 58.3 | 0.6 | 7.3 | 0.1 | 7.4 | 33.3 | 1,695 | 52 | 111 | 843 | 8,250 | 42,789 | 3,635 |
| Coahoma | 126 | 5,421 | 59.8 | 0.7 | 4.4 | 0.0 | 4.0 | 164.5 | 7,086 | 59 | 125 | 1,711 | 8,890 | 38,129 | 3,253 |
| Copiah | 128 | 4,472 | 54.4 | 18.0 | 3.6 | 0.0 | 12.5 | 71.2 | 2,496 | 65 | 158 | 1,628 | 11,450 | 41,593 | 3,085 |
| Covington | 41 | 2,144 | 64.6 | 0.3 | 5.6 | 0.1 | 9.5 | 13.0 | 685 | 62 | 107 | 1,249 | 7,800 | 45,024 | 3,950 |
| DeSoto | 482 | 2,695 | 54.8 | 1.0 | 9.1 | 0.0 | 5.4 | 377.4 | 2,108 | 248 | 1,073 | 6,696 | 83,610 | 57,275 | 5,101 |
| Forrest | 760 | 10,107 | 14.2 | 66.6 | 2.5 | 0.0 | 2.1 | 403.2 | 5,365 | 699 | 810 | 11,128 | 30,890 | 55,521 | 6,117 |
| Franklin | 23 | 2,914 | 64.7 | 1.3 | 4.4 | 0.0 | 13.4 | 2.1 | 265 | 51 | 45 | 590 | 3,080 | 47,992 | 3,782 |
| George | 101 | 4,202 | 32.7 | 47.7 | 3.0 | 0.1 | 4.5 | 43.8 | 1,830 | 51 | 139 | 1,483 | 9,100 | 53,316 | 4,172 |
| Greene | 41 | 3,046 | 45.7 | 6.0 | 6.3 | 16.9 | 10.4 | 9.4 | 698 | 13 | 64 | 973 | 4,240 | 48,528 | 3,420 |
| Grenada | 70 | 3,323 | 51.7 | 0.1 | 10.6 | 0.0 | 4.6 | 22.9 | 1,089 | 210 | 120 | 1,591 | 9,350 | 45,209 | 3,732 |
| Hancock | 206 | 4,383 | 30.8 | 19.4 | 4.9 | 0.1 | 7.1 | 142.9 | 3,043 | 1,993 | 762 | 1,812 | 18,710 | 52,538 | 4,763 |
| Harrison | 1,329 | 6,483 | 26.3 | 36.6 | 4.1 | 0.2 | 3.5 | 922.8 | 4,502 | 5,883 | 8,377 | 13,638 | 87,460 | 50,140 | 4,743 |
| Hinds | 901 | 3,753 | 55.5 | 1.8 | 6.1 | 0.3 | 3.8 | 924.3 | 3,852 | 4,472 | 1,347 | 33,739 | 100,880 | 48,577 | 4,792 |
| Holmes | 92 | 5,178 | 81.3 | 0.4 | 3.9 | 0.0 | 4.0 | 33.7 | 1,887 | 49 | 94 | 1,251 | 6,780 | 30,659 | 1,607 |
| Humphreys | 34 | 4,077 | 47.2 | 0.9 | 5.7 | 0.1 | 21.0 | 4.1 | 486 | 21 | 46 | 495 | 3,150 | 33,650 | 2,536 |
| Issaquena | 7 | 5,042 | 0.7 | 5.4 | 10.6 | 0.0 | 13.5 | 1.0 | 772 | 1 | 6 | 91 | 390 | 42,333 | 2,918 |
| Itawamba | 106 | 4,520 | 81.8 | 0.5 | 2.1 | 0.1 | 4.3 | 32.1 | 1,368 | 49 | 130 | 1,048 | 9,290 | 47,498 | 3,501 |
| Jackson | 840 | 5,903 | 30.1 | 38.9 | 4.1 | 0.1 | 4.3 | 511.8 | 3,598 | 1,190 | 977 | 9,001 | 61,360 | 55,629 | 5,201 |
| Jasper | 53 | 3,204 | 46.9 | 14.7 | 5.1 | 0.1 | 12.5 | 15.0 | 905 | 48 | 95 | 926 | 6,950 | 44,909 | 3,566 |
| Jefferson | 39 | 5,356 | 32.7 | 30.5 | 4.4 | 7.3 | 3.3 | 2.4 | 335 | 21 | 44 | 546 | 3,040 | 33,005 | 1,975 |
| Jefferson Davis | 26 | 2,335 | 59.5 | 1.0 | 6.4 | 0.0 | 10.3 | 9.7 | 856 | 23 | 64 | 599 | 4,820 | 35,626 | 2,294 |
| Jones | 390 | 5,711 | 37.4 | 39.5 | 3.2 | 0.1 | 3.6 | 260.9 | 3,816 | 220 | 385 | 6,933 | 26,440 | 48,791 | 4,371 |
| Kemper | 83 | 8,218 | 81.3 | 0.6 | 1.4 | 0.1 | 3.2 | 58.3 | 5,776 | 29 | 51 | 611 | 3,770 | 35,475 | 2,157 |
| Lafayette | 132 | 2,440 | 52.4 | 0.5 | 8.2 | 0.0 | 7.4 | 110.1 | 2,032 | 334 | 309 | 8,358 | 20,860 | 70,295 | 8,669 |
| Lamar | 140 | 2,280 | 67.9 | 0.6 | 7.1 | 0.1 | 8.8 | 53.3 | 867 | 55 | 366 | 2,424 | 24,820 | 66,308 | 7,635 |
| Lauderdale | 247 | 3,233 | 61.3 | 4.3 | 6.3 | 0.0 | 5.1 | 134.7 | 1,766 | 789 | 1,374 | 4,925 | 30,170 | 53,189 | 5,455 |
| Lawrence | 30 | 2,382 | 62.8 | 0.9 | 9.6 | 0.0 | 9.1 | 7.5 | 595 | 41 | 73 | 658 | 4,910 | 45,938 | 3,254 |
| Leake | 46 | 1,998 | 52.5 | 0.3 | 4.9 | 0.0 | 7.7 | 10.8 | 472 | 59 | 123 | 732 | 8,410 | 41,824 | 2,961 |
| Lee | 281 | 3,306 | 56.4 | 0.7 | 9.6 | 0.2 | 7.2 | 305.5 | 3,591 | 446 | 522 | 4,956 | 37,140 | 57,789 | 6,115 |
| Leflore | 236 | 8,058 | 20.1 | 54.6 | 3.9 | 0.0 | 4.1 | 257.0 | 8,774 | 112 | 157 | 3,921 | 11,410 | 44,389 | 4,564 |
| Lincoln | 90 | 2,618 | 56.4 | 0.0 | 7.4 | 0.0 | 14.5 | 18.1 | 528 | 103 | 195 | 1,583 | 13,610 | 52,293 | 4,619 |
| Lowndes | 233 | 3,937 | 55.2 | 0.5 | 6.2 | 0.1 | 5.8 | 267.3 | 4,519 | 848 | 1,788 | 3,271 | 24,850 | 50,129 | 4,509 |
| Madison | 340 | 3,247 | 53.9 | 0.3 | 6.2 | 0.1 | 9.3 | 447.7 | 4,281 | 238 | 610 | 4,572 | 49,310 | 88,210 | 12,415 |
| Marion | 86 | 3,438 | 47.8 | 19.0 | 6.5 | 0.0 | 5.4 | 41.4 | 1,649 | 47 | 139 | 1,272 | 9,430 | 47,048 | 3,558 |
| Marshall | 80 | 2,252 | 50.8 | 1.1 | 7.5 | 0.0 | 11.2 | 46.3 | 1,302 | 84 | 195 | 1,159 | 15,240 | 40,271 | 2,965 |
| Monroe | 94 | 2,610 | 56.1 | 1.0 | 8.7 | 0.1 | 9.4 | 35.4 | 986 | 127 | 203 | 1,395 | 14,440 | 45,850 | 3,505 |
| Montgomery | 26 | 2,511 | 56.5 | 0.4 | 6.8 | 0.2 | 10.1 | 6.6 | 653 | 25 | 56 | 527 | 4,120 | 42,454 | 2,958 |
| Neshoba | 78 | 2,642 | 65.0 | 0.2 | 4.9 | 0.0 | 6.7 | 41.2 | 1,400 | 77 | 167 | 6,089 | 11,790 | 43,718 | 3,456 |
| Newton | 97 | 4,518 | 68.0 | 0.1 | 4.2 | 0.0 | 4.8 | 24.6 | 1,150 | 77 | 119 | 1,958 | 8,720 | 45,855 | 3,310 |
| Noxubee | 32 | 3,025 | 55.3 | 1.5 | 7.8 | 0.1 | 9.6 | 7.7 | 721 | 43 | 60 | 646 | 4,490 | 32,396 | 1,892 |
| Oktibbeha | 176 | 3,546 | 30.8 | 41.3 | 5.2 | 0.0 | 4.2 | 132.2 | 2,662 | 239 | 273 | 9,887 | 18,130 | 54,774 | 5,486 |
| Panola | 105 | 3,075 | 56.7 | 1.0 | 7.9 | 0.2 | 10.3 | 21.5 | 628 | 111 | 197 | 2,074 | 14,320 | 42,028 | 3,688 |
| Pearl River | 202 | 3,648 | 62.4 | 11.9 | 3.2 | 0.2 | 5.7 | 39.0 | 705 | 128 | 316 | 3,070 | 21,540 | 48,935 | 4,046 |
| Perry | 39 | 3,252 | 47.0 | 1.7 | 4.5 | 0.1 | 19.2 | 69.3 | 5,765 | 22 | 69 | 548 | 4,600 | 43,450 | 2,949 |
| Pike | 286 | 7,245 | 29.9 | 49.7 | 2.5 | 0.0 | 2.7 | 112.8 | 2,855 | 105 | 224 | 3,277 | 16,090 | 42,358 | 3,381 |
| Pontotoc | 79 | 2,488 | 63.6 | 1.0 | 5.2 | 0.1 | 7.2 | 26.0 | 821 | 52 | 186 | 1,235 | 12,930 | 45,162 | 3,133 |
| Prentiss | 95 | 3,772 | 79.4 | 0.8 | 3.9 | 0.0 | 3.4 | 26.6 | 1,053 | 46 | 142 | 1,510 | 9,620 | 41,934 | 2,764 |
| Quitman | 26 | 3,618 | 57.8 | 0.5 | 11.0 | 0.0 | 8.3 | 8.7 | 1,204 | 23 | 39 | 359 | 2,630 | 30,858 | 1,778 |
| Rankin | 414 | 2,707 | 56.8 | 0.6 | 7.4 | 0.1 | 5.9 | 590.2 | 3,860 | 599 | 869 | 8,587 | 66,400 | 63,484 | 6,120 |
| Scott | 76 | 2,666 | 66.1 | 0.4 | 5.3 | 0.0 | 7.7 | 36.6 | 1,286 | 194 | 163 | 1,265 | 11,200 | 40,408 | 2,649 |
| Sharkey | 18 | 4,039 | 54.9 | 2.5 | 6.1 | 0.0 | 10.7 | 4.3 | 968 | 20 | 25 | 467 | 1,760 | 38,165 | 2,765 |
| Simpson | 65 | 2,403 | 65.0 | 0.2 | 3.1 | 0.0 | 7.4 | 20.6 | 765 | 41 | 152 | 2,255 | 10,340 | 44,840 | 3,430 |
| Smith | 39 | 2,422 | 60.4 | 1.0 | 5.1 | 0.0 | 14.4 | 9.9 | 618 | 25 | 92 | 613 | 5,870 | 47,692 | 3,861 |
| Stone | 153 | 8,219 | 83.3 | 0.5 | 3.1 | 0.0 | 3.4 | 168.3 | 9,054 | 69 | 99 | 1,270 | 6,850 | 46,613 | 3,844 |

1. Based on the resident population estimated as of July 1 of the year shown.

# Table B. States and Counties — Land Area and Population

| State / county code | CBSA code[1] | County Type code[2] | STATE County | Land area[3] (sq. mi) | Population, 2020 Total persons 2019 | Rank | Per square mile | Population and population characteristics, 2020 — Race alone or in combination, not Hispanic or Latino (percent) White | Black | American Indian, Alaska Native | Asian and Pacific Islander | Percent Hispanic or Latino[4] | Age (percent) Under 5 years | 5 to 14 years | 15 to 24 years | 25 to 34 years | 35 to 44 years | 45 to 54 years |
|---|---|---|---|---|---|---|---|---|---|---|---|---|---|---|---|---|---|---|
| | | | | 1 | 2 | 3 | 4 | 5 | 6 | 7 | 8 | 9 | 10 | 11 | 12 | 13 | 14 | 15 |
| | | | **MISSISSIPPI— Cont'd** | | | | | | | | | | | | | | | |
| 28133 | 26940 | 7 | Sunflower | 697.8 | 24,740 | 1,616 | 35.5 | 23.9 | 73.7 | 0.5 | 0.6 | 1.9 | 5.6 | 12.3 | 13.9 | 16.3 | 13.1 | 12.3 |
| 28135 | | 7 | Tallahatchie | 645.2 | 13,707 | 2,176 | 21.2 | 35.7 | 56.2 | 0.7 | 1.4 | 7.2 | 5.7 | 10.6 | 13.0 | 17.7 | 12.8 | 12.2 |
| 28137 | 32820 | 1 | Tate | 404.8 | 28,539 | 1,479 | 70.5 | 65.7 | 31.6 | 0.6 | 0.6 | 2.8 | 6.0 | 12.6 | 14.8 | 12.5 | 11.4 | 12.2 |
| 28139 | | 6 | Tippah | 457.8 | 21,748 | 1,729 | 47.5 | 78.2 | 17.5 | 0.5 | 0.5 | 4.8 | 6.0 | 13.7 | 13.2 | 12.3 | 11.9 | 12.6 |
| 28141 | | 8 | Tishomingo | 424.3 | 19,275 | 1,863 | 45.4 | 93.5 | 3.1 | 0.6 | 0.4 | 3.4 | 5.4 | 12.2 | 11.4 | 12.2 | 11.1 | 12.6 |
| 28143 | 32820 | 1 | Tunica | 454.1 | 9,392 | 2,465 | 20.7 | 19.1 | 77.4 | 0.6 | 1.1 | 2.8 | 8.2 | 15.8 | 13.0 | 13.3 | 13.2 | 11.3 |
| 28145 | | 6 | Union | 415.6 | 28,866 | 1,464 | 69.5 | 78.7 | 16.6 | 0.6 | 1.4 | 4.4 | 6.5 | 14.1 | 12.4 | 12.9 | 12.3 | 12.8 |
| 28147 | | 9 | Walthall | 403.9 | 14,294 | 2,138 | 35.4 | 53.8 | 43.5 | 0.8 | 0.9 | 2.4 | 5.8 | 12.9 | 11.8 | 11.6 | 11.7 | 12.0 |
| 28149 | 46980 | 4 | Warren | 588.5 | 44,841 | 1,079 | 76.2 | 47.6 | 49.6 | 0.5 | 1.1 | 2.1 | 6.4 | 13.2 | 12.1 | 12.2 | 12.4 | 11.8 |
| 28151 | 24740 | 5 | Washington | 724.0 | 42,837 | 1,122 | 59.2 | 25.0 | 72.6 | 0.3 | 1.0 | 1.8 | 6.6 | 14.8 | 12.5 | 12.3 | 11.1 | 11.7 |
| 28153 | | 7 | Wayne | 810.7 | 20,317 | 1,804 | 25.1 | 57.8 | 40.6 | 0.6 | 0.5 | 1.7 | 6.5 | 14.0 | 12.1 | 12.6 | 11.6 | 11.5 |
| 28155 | 44260 | 9 | Webster | 421.2 | 9,676 | 2,438 | 23.0 | 79.6 | 19.0 | 0.7 | 0.3 | 1.5 | 6.2 | 13.3 | 11.5 | 12.1 | 12.1 | 12.2 |
| 28157 | | 8 | Wilkinson | 678.1 | 8,351 | 2,550 | 12.3 | 28.2 | 70.8 | 0.5 | 0.3 | 1.1 | 5.1 | 11.9 | 12.0 | 15.2 | 12.2 | 11.0 |
| 28159 | | 7 | Winston | 607.3 | 17,845 | 1,933 | 29.4 | 50.4 | 47.2 | 1.4 | 0.5 | 1.6 | 5.4 | 12.8 | 11.7 | 11.4 | 12.0 | 11.9 |
| 28161 | | 7 | Yalobusha | 467.2 | 11,982 | 2,295 | 25.6 | 58.9 | 39.5 | 0.6 | 0.5 | 1.8 | 5.9 | 11.8 | 11.3 | 11.4 | 11.3 | 12.2 |
| 28163 | 27140 | 2 | Yazoo | 922.3 | 26,982 | 1,526 | 29.3 | 35.2 | 57.4 | 0.6 | 0.8 | 6.9 | 5.6 | 12.8 | 11.9 | 17.0 | 15.3 | 11.4 |
| 29000 | | 0 | **MISSOURI** | 68,745.5 | 6,151,548 | X | 89.5 | 81.1 | 12.7 | 1.2 | 3.0 | 4.5 | 6.0 | 12.5 | 12.9 | 13.4 | 12.4 | 11.7 |
| 29001 | 28860 | 7 | Adair | 567.3 | 25,399 | 1,586 | 44.8 | 90.6 | 4.4 | 0.8 | 3.4 | 2.6 | 5.2 | 9.4 | 31.0 | 10.9 | 8.6 | 9.1 |
| 29003 | 41140 | 3 | Andrew | 432.6 | 17,586 | 1,940 | 40.7 | 95.1 | 1.8 | 0.9 | 1.0 | 2.7 | 5.6 | 13.0 | 11.1 | 11.3 | 12.4 | 12.4 |
| 29005 | | 9 | Atchison | 547.3 | 5,096 | 2,819 | 9.3 | 97.1 | 0.9 | 0.9 | 0.6 | 1.7 | 4.7 | 11.7 | 9.4 | 10.7 | 10.7 | 11.9 |
| 29007 | 33020 | 6 | Audrain | 692.3 | 24,835 | 1,611 | 35.9 | 89.3 | 7.5 | 0.9 | 1.0 | 3.3 | 6.1 | 12.3 | 12.0 | 12.8 | 12.4 | 11.9 |
| 29009 | | 6 | Barry | 778.1 | 35,818 | 1,283 | 46.0 | 86.4 | 0.9 | 2.0 | 2.6 | 9.7 | 5.9 | 12.6 | 11.1 | 10.8 | 10.8 | 11.7 |
| 29011 | | 6 | Barton | 592.0 | 11,592 | 2,315 | 19.6 | 94.6 | 1.2 | 2.7 | 0.9 | 3.0 | 5.7 | 13.1 | 12.0 | 10.6 | 11.5 | 11.1 |
| 29013 | 28140 | 3 | Bates | 836.7 | 16,242 | 2,019 | 19.4 | 95.1 | 1.8 | 1.6 | 0.7 | 2.6 | 5.8 | 12.8 | 11.8 | 11.6 | 11.6 | 11.5 |
| 29015 | | 7 | Benton | 704.0 | 19,627 | 1,846 | 27.9 | 95.9 | 1.0 | 1.8 | 0.8 | 2.2 | 4.5 | 9.6 | 8.3 | 8.2 | 8.4 | 11.3 |
| 29017 | 16020 | 3 | Bollinger | 617.9 | 12,111 | 2,286 | 19.6 | 97.0 | 0.9 | 1.3 | 0.6 | 1.6 | 5.6 | 12.1 | 10.2 | 11.0 | 11.2 | 12.6 |
| 29019 | 17860 | 3 | Boone | 685.6 | 182,991 | 364 | 266.9 | 81.1 | 11.6 | 1.0 | 6.0 | 3.6 | 5.7 | 11.3 | 21.6 | 15.0 | 12.2 | 10.2 |
| 29021 | 41140 | 3 | Buchanan | 408.2 | 86,530 | 669 | 212.0 | 85.1 | 7.3 | 1.1 | 2.4 | 6.8 | 6.1 | 12.5 | 12.8 | 13.7 | 12.7 | 11.6 |
| 29023 | 38740 | 5 | Butler | 694.8 | 42,178 | 1,136 | 60.7 | 90.4 | 7.0 | 1.5 | 1.4 | 2.3 | 6.3 | 12.9 | 11.8 | 12.3 | 11.8 | 11.9 |
| 29025 | 28140 | 1 | Caldwell | 426.4 | 9,051 | 2,492 | 21.2 | 95.5 | 1.5 | 1.3 | 1.0 | 2.8 | 5.8 | 13.1 | 12.2 | 10.4 | 11.5 | 12.2 |
| 29027 | 27620 | 3 | Callaway | 834.6 | 44,887 | 1,076 | 53.8 | 91.7 | 5.7 | 1.3 | 1.2 | 2.2 | 5.6 | 11.5 | 14.1 | 13.2 | 12.3 | 12.2 |
| 29029 | | 7 | Camden | 656.0 | 46,414 | 1,047 | 70.8 | 95.1 | 1.1 | 1.2 | 1.0 | 3.0 | 4.3 | 9.6 | 9.0 | 8.7 | 9.6 | 11.2 |
| 29031 | 16020 | 3 | Cape Girardeau | 578.5 | 79,512 | 713 | 137.4 | 87.6 | 9.1 | 0.7 | 2.4 | 2.5 | 5.8 | 11.9 | 17.5 | 12.7 | 11.5 | 10.7 |
| 29033 | | 6 | Carroll | 694.6 | 8,554 | 2,530 | 12.3 | 95.9 | 2.6 | 0.6 | 0.5 | 1.7 | 5.1 | 13.0 | 11.7 | 10.4 | 11.4 | 11.9 |
| 29035 | | 9 | Carter | 507.4 | 5,991 | 2,745 | 11.8 | 95.7 | 1.2 | 1.9 | 0.5 | 2.6 | 5.8 | 13.0 | 11.4 | 10.3 | 11.2 | 12.4 |
| 29037 | 28140 | 1 | Cass | 696.6 | 106,806 | 572 | 153.3 | 89.1 | 5.6 | 1.3 | 1.6 | 4.8 | 5.8 | 13.5 | 11.8 | 12.2 | 12.5 | 12.5 |
| 29039 | | 6 | Cedar | 474.5 | 14,322 | 2,136 | 30.2 | 95.9 | 0.9 | 1.8 | 0.8 | 2.4 | 6.7 | 13.3 | 11.1 | 10.0 | 10.1 | 11.4 |
| 29041 | | 9 | Chariton | 751.2 | 7,360 | 2,629 | 9.8 | 95.9 | 3.3 | 0.9 | 0.4 | 1.0 | 6.0 | 13.0 | 9.9 | 10.0 | 10.1 | 10.6 |
| 29043 | 44180 | 2 | Christian | 562.6 | 90,655 | 656 | 161.1 | 94.5 | 1.5 | 1.4 | 1.2 | 3.4 | 6.2 | 14.6 | 11.4 | 12.5 | 13.8 | 12.4 |
| 29045 | 22800 | 9 | Clark | 504.6 | 6,830 | 2,673 | 13.5 | 98.1 | 1.1 | 0.7 | 0.6 | 0.8 | 5.5 | 13.3 | 10.4 | 10.5 | 11.2 | 11.7 |
| 29047 | 28140 | 1 | Clay | 397.7 | 253,463 | 274 | 637.3 | 82.2 | 8.5 | 1.2 | 3.7 | 7.2 | 6.2 | 13.4 | 12.0 | 14.6 | 14.0 | 12.6 |
| 29049 | 28140 | 1 | Clinton | 418.9 | 20,553 | 1,792 | 49.1 | 95.0 | 2.3 | 1.3 | 0.9 | 2.5 | 5.8 | 13.1 | 11.4 | 11.7 | 11.9 | 12.3 |
| 29051 | 27620 | 3 | Cole | 391.5 | 76,191 | 731 | 194.6 | 83.5 | 13.0 | 0.8 | 2.0 | 2.9 | 5.8 | 12.6 | 12.5 | 13.2 | 12.8 | 12.3 |
| 29053 | 17860 | 6 | Cooper | 564.8 | 17,102 | 1,969 | 30.3 | 90.9 | 7.1 | 1.1 | 1.1 | 2.0 | 5.8 | 12.4 | 12.7 | 12.6 | 11.9 | 11.3 |
| 29055 | | 6 | Crawford | 742.5 | 23,739 | 1,652 | 32.0 | 96.3 | 1.1 | 1.4 | 0.6 | 2.2 | 5.5 | 13.0 | 11.2 | 11.4 | 11.5 | 11.8 |
| 29057 | | 8 | Dade | 490.0 | 7,568 | 2,622 | 15.4 | 95.6 | 1.2 | 2.6 | 0.9 | 2.3 | 4.9 | 12.4 | 10.5 | 9.2 | 11.6 | 11.6 |
| 29059 | 44180 | 2 | Dallas | 541.0 | 17,219 | 1,962 | 31.8 | 96.0 | 0.9 | 2.0 | 0.8 | 2.3 | 6.6 | 13.1 | 10.6 | 11.6 | 10.8 | 11.5 |
| 29061 | | 8 | Daviess | 563.3 | 8,283 | 2,560 | 14.7 | 96.9 | 1.2 | 1.0 | 0.6 | 1.7 | 7.2 | 13.1 | 12.9 | 10.5 | 10.4 | 11.0 |
| 29063 | 41140 | 3 | DeKalb | 421.4 | 10,944 | 2,353 | 26.0 | 88.9 | 7.7 | 1.0 | 0.8 | 2.7 | 4.8 | 11.2 | 11.1 | 14.1 | 13.4 | 13.3 |
| 29065 | | 7 | Dent | 752.8 | 15,481 | 2,061 | 20.6 | 95.6 | 1.0 | 1.9 | 1.2 | 1.9 | 5.8 | 12.5 | 10.6 | 10.3 | 11.1 | 11.4 |
| 29067 | | 6 | Douglas | 813.6 | 13,344 | 2,197 | 16.4 | 96.5 | 1.0 | 2.3 | 0.8 | 1.7 | 5.4 | 12.3 | 9.8 | 9.6 | 10.4 | 11.1 |
| 29069 | 28380 | 7 | Dunklin | 541.9 | 28,878 | 1,463 | 53.3 | 80.9 | 11.9 | 0.9 | 1.1 | 7.1 | 6.8 | 14.7 | 12.1 | 11.2 | 11.4 | 11.7 |
| 29071 | 41180 | 1 | Franklin | 922.6 | 104,469 | 583 | 113.2 | 96.2 | 1.6 | 0.9 | 0.9 | 1.9 | 5.9 | 12.7 | 11.3 | 12.2 | 11.6 | 12.3 |
| 29073 | | 6 | Gasconade | 519.0 | 14,566 | 2,116 | 28.1 | 96.8 | 1.1 | 1.0 | 0.7 | 1.7 | 5.4 | 11.4 | 10.5 | 10.7 | 10.4 | 11.7 |
| 29075 | | 8 | Gentry | 491.4 | 6,484 | 2,708 | 13.2 | 96.8 | 1.3 | 0.8 | 0.7 | 1.9 | 7.2 | 14.3 | 10.7 | 11.4 | 11.2 | 11.0 |
| 29077 | 44180 | 2 | Greene | 675.3 | 294,997 | 235 | 436.8 | 89.6 | 4.7 | 1.5 | 3.1 | 4.0 | 5.8 | 11.4 | 16.9 | 14.1 | 12.0 | 10.9 |
| 29079 | | 7 | Grundy | 435.3 | 9,595 | 2,447 | 22.0 | 94.9 | 1.5 | 1.2 | 1.4 | 2.5 | 6.9 | 13.8 | 11.1 | 12.2 | 9.9 | 11.4 |
| 29081 | | 7 | Harrison | 722.5 | 8,321 | 2,555 | 11.5 | 95.8 | 1.1 | 1.0 | 0.8 | 2.7 | 5.8 | 14.2 | 11.3 | 10.3 | 10.8 | 10.6 |
| 29083 | | 6 | Henry | 696.9 | 22,076 | 1,715 | 31.7 | 95.0 | 1.8 | 1.6 | 0.8 | 2.6 | 6.0 | 12.4 | 10.6 | 11.0 | 11.1 | 11.4 |
| 29085 | | 8 | Hickory | 398.8 | 9,586 | 2,448 | 24.0 | 95.9 | 1.0 | 2.1 | 0.5 | 2.1 | 4.1 | 9.3 | 8.1 | 7.5 | 8.3 | 11.3 |
| 29087 | | 8 | Holt | 462.7 | 4,232 | 2,877 | 9.1 | 96.4 | 1.0 | 1.9 | 0.6 | 1.5 | 5.5 | 10.6 | 9.9 | 9.9 | 9.5 | 12.2 |
| 29089 | 17860 | 6 | Howard | 463.8 | 10,001 | 2,418 | 21.6 | 92.1 | 6.1 | 1.4 | 0.8 | 1.9 | 5.4 | 12.7 | 15.9 | 10.6 | 10.7 | 10.6 |
| 29091 | 48460 | 7 | Howell | 927.2 | 40,262 | 1,176 | 43.4 | 95.5 | 0.9 | 1.7 | 1.1 | 2.4 | 6.3 | 13.8 | 11.5 | 11.9 | 11.3 | 11.6 |
| 29093 | | 7 | Iron | 550.3 | 10,098 | 2,410 | 18.3 | 95.7 | 2.0 | 1.7 | 0.4 | 1.9 | 5.3 | 11.5 | 11.5 | 12.0 | 11.8 | 12.8 |
| 29095 | 28140 | 1 | Jackson | 604.5 | 705,925 | 95 | 1,167.8 | 64.5 | 25.0 | 1.3 | 3.0 | 9.4 | 6.4 | 13.1 | 12.0 | 15.6 | 12.9 | 11.5 |

1. CBSA = Core Based Statistical Area. See Appendix A for explanation. See Appendix B for list of metropolitan areas with component counties. Service of USDA Rural-Urban Continuum Codes. See Appendix A for definition.
2. County type code from the Economic Research
3. Dry land or land partially or temporarily covered by water.
4. May be of any race.

Items 1—15

| STATE County | 55 to 64 years | 65 to 74 years | 75 years and over | Percent female | Total persons 2000 | 2010 | Percent change 2000-2010 | 2010-2020 | Births | Deaths | Net Migration | Number | Persons per household | Family households | Female family householder[1] | One person |
|---|---|---|---|---|---|---|---|---|---|---|---|---|---|---|---|---|
| | 16 | 17 | 18 | 19 | 20 | 21 | 22 | 23 | 24 | 25 | 26 | 27 | 28 | 29 | 30 | 31 |
| **MISSISSIPPI— Cont'd** | | | | | | | | | | | | | | | | |
| Sunflower | 11.7 | 9.1 | 5.7 | 46.6 | 34,369 | 29,383 | -14.5 | -15.8 | 3,242 | 3,147 | -4,822 | 8,322 | 2.63 | 63.6 | 27.2 | 31.0 |
| Tallahatchie | 11.9 | 9.3 | 7.0 | 44.4 | 14,903 | 15,385 | 3.2 | -10.9 | 1,647 | 1,578 | -1,762 | 4,369 | 2.70 | 63.5 | 22.3 | 31.5 |
| Tate | 13.2 | 10.4 | 6.8 | 51.8 | 25,370 | 28,863 | 13.8 | -1.1 | 3,483 | 2,954 | -865 | 10,324 | 2.57 | 73.4 | 18.8 | 21.3 |
| Tippah | 12.9 | 10.3 | 7.1 | 51.3 | 20,826 | 22,232 | 6.8 | -2.2 | 2,746 | 2,762 | -456 | 7,834 | 2.77 | 68.9 | 11.2 | 27.5 |
| Tishomingo | 14.1 | 12.3 | 8.9 | 51.3 | 19,163 | 19,598 | 2.3 | -1.6 | 2,052 | 2,948 | 584 | 7,635 | 2.51 | 65.9 | 9.6 | 29.5 |
| Tunica | 11.9 | 8.7 | 4.6 | 52.9 | 9,227 | 10,778 | 16.8 | -12.9 | 1,884 | 1,081 | -2,204 | 3,930 | 2.51 | 59.7 | 22.2 | 32.6 |
| Union | 11.9 | 10.0 | 7.1 | 51.1 | 25,362 | 27,131 | 7.0 | 6.4 | 3,730 | 3,006 | 1,022 | 9,808 | 2.87 | 70.6 | 10.5 | 27.9 |
| Walthall | 13.5 | 12.0 | 8.8 | 52.0 | 15,156 | 15,440 | 1.9 | -7.4 | 1,805 | 1,905 | -1,054 | 5,601 | 2.56 | 60.2 | 14.6 | 37.9 |
| Warren | 14.0 | 11.0 | 7.0 | 52.3 | 49,644 | 48,770 | -1.8 | -8.1 | 6,241 | 5,342 | -4,824 | 18,235 | 2.52 | 63.0 | 17.9 | 33.2 |
| Washington | 13.6 | 11.0 | 6.4 | 53.9 | 62,977 | 51,135 | -18.8 | -16.2 | 7,275 | 6,123 | -9,508 | 17,988 | 2.53 | 62.4 | 24.9 | 33.2 |
| Wayne | 14.0 | 10.1 | 7.6 | 52.0 | 21,216 | 20,754 | -2.2 | -2.1 | 2,900 | 2,275 | -1,057 | 7,683 | 2.63 | 70.2 | 17.8 | 26.8 |
| Webster | 14.6 | 10.5 | 7.4 | 51.5 | 10,294 | 10,263 | -0.3 | -5.7 | 1,242 | 1,435 | -394 | 3,782 | 2.56 | 71.0 | 14.3 | 26.4 |
| Wilkinson | 14.0 | 11.1 | 7.4 | 46.6 | 10,312 | 9,878 | -4.2 | -15.5 | 1,078 | 1,237 | -1,374 | 3,170 | 2.43 | 58.1 | 24.9 | 38.4 |
| Winston | 13.5 | 12.3 | 9.1 | 51.1 | 20,160 | 19,201 | -4.8 | -7.1 | 2,022 | 2,357 | -1,014 | 7,269 | 2.44 | 65.2 | 19.7 | 33.1 |
| Yalobusha | 14.5 | 13.2 | 8.6 | 51.9 | 13,051 | 12,678 | -2.9 | -5.5 | 1,561 | 1,815 | -436 | 5,181 | 2.35 | 64.7 | 17.2 | 31.9 |
| Yazoo | 11.4 | 8.4 | 6.1 | 43.8 | 28,149 | 28,065 | -0.3 | -3.9 | 3,496 | 3,024 | -1,570 | 8,542 | 2.78 | 60.9 | 25.3 | 35.7 |
| **MISSOURI** | 13.4 | 10.2 | 7.5 | 50.9 | 5,595,211 | 5,988,941 | 7.0 | 2.7 | 763,481 | 608,997 | 10,057 | 2,414,521 | 2.46 | 64.0 | 11.6 | 29.5 |
| Adair | 10.4 | 8.5 | 6.9 | 52.0 | 24,977 | 25,617 | 2.6 | -0.9 | 2,740 | 2,257 | -706 | 9,258 | 2.45 | 54.1 | 8.2 | 33.6 |
| Andrew | 14.3 | 11.4 | 8.5 | 50.5 | 16,492 | 17,296 | 4.9 | 1.7 | 1,930 | 1,698 | 68 | 6,763 | 2.56 | 70.1 | 8.3 | 26.0 |
| Atchison | 14.7 | 13.9 | 12.2 | 50.5 | 6,430 | 5,683 | -11.6 | -10.3 | 520 | 784 | -331 | 2,562 | 2.00 | 55.7 | 6.4 | 38.6 |
| Audrain | 13.8 | 10.5 | 8.2 | 53.3 | 25,853 | 25,531 | -1.2 | -2.7 | 3,444 | 3,092 | -1,039 | 9,286 | 2.56 | 65.1 | 11.8 | 28.0 |
| Barry | 15.1 | 12.6 | 9.3 | 49.8 | 34,010 | 35,601 | 4.7 | 0.6 | 4,256 | 4,295 | 285 | 13,645 | 2.58 | 71.7 | 9.5 | 24.3 |
| Barton | 14.7 | 11.5 | 9.7 | 50.5 | 12,541 | 12,396 | -1.2 | -6.5 | 1,406 | 1,442 | -776 | 4,895 | 2.37 | 65.2 | 9.0 | 26.9 |
| Bates | 14.7 | 11.3 | 9.0 | 50.2 | 16,653 | 17,047 | 2.4 | -4.7 | 1,902 | 2,152 | -562 | 6,547 | 2.44 | 70.5 | 8.0 | 26.3 |
| Benton | 17.6 | 18.6 | 13.3 | 49.6 | 17,180 | 19,070 | 11.0 | 2.9 | 1,665 | 3,039 | 1,935 | 7,872 | 2.40 | 63.7 | 6.2 | 30.4 |
| Bollinger | 15.8 | 12.4 | 9.0 | 49.9 | 12,029 | 12,365 | 2.8 | -2.1 | 1,380 | 1,397 | -234 | 4,593 | 2.62 | 68.1 | 8.3 | 28.9 |
| Boone | 10.7 | 8.2 | 5.1 | 51.6 | 135,454 | 162,652 | 20.1 | 12.5 | 21,470 | 11,040 | 9,899 | 70,619 | 2.39 | 57.2 | 9.2 | 30.3 |
| Buchanan | 13.5 | 9.9 | 7.4 | 49.5 | 85,998 | 89,191 | 3.7 | -3.0 | 11,755 | 9,810 | -4,592 | 33,501 | 2.52 | 61.3 | 13.0 | 31.7 |
| Butler | 13.5 | 11.0 | 8.7 | 51.5 | 40,867 | 42,792 | 4.7 | -1.4 | 5,589 | 5,869 | -291 | 16,390 | 2.54 | 67.1 | 14.3 | 27.5 |
| Caldwell | 14.5 | 11.1 | 9.2 | 49.2 | 8,969 | 9,418 | 5.0 | -3.9 | 972 | 1,052 | -294 | 3,639 | 2.43 | 64.5 | 8.2 | 29.8 |
| Callaway | 14.0 | 10.6 | 6.7 | 48.7 | 40,766 | 44,331 | 8.7 | 1.3 | 5,118 | 4,045 | -495 | 15,973 | 2.56 | 67.8 | 10.6 | 27.6 |
| Camden | 18.1 | 18.1 | 11.3 | 50.4 | 37,051 | 44,033 | 18.8 | 5.4 | 4,082 | 4,988 | 3,275 | 16,031 | 2.80 | 70.3 | 8.3 | 24.7 |
| Cape Girardeau | 12.3 | 9.9 | 7.7 | 51.7 | 68,693 | 75,676 | 10.2 | 5.1 | 9,295 | 7,938 | 2,527 | 29,594 | 2.54 | 63.8 | 10.5 | 26.7 |
| Carroll | 14.0 | 12.3 | 10.1 | 50.5 | 10,285 | 9,294 | -9.6 | -8.0 | 1,023 | 1,212 | -551 | 3,503 | 2.47 | 69.8 | 10.5 | 27.0 |
| Carter | 14.8 | 12.3 | 8.7 | 51.5 | 5,941 | 6,265 | 5.5 | -4.4 | 784 | 813 | -245 | 2,333 | 2.62 | 67.0 | 9.6 | 28.0 |
| Cass | 14.1 | 10.0 | 7.7 | 51.2 | 82,092 | 99,500 | 21.2 | 7.3 | 12,104 | 9,656 | 4,935 | 39,770 | 2.58 | 73.6 | 9.9 | 22.2 |
| Cedar | 14.0 | 12.9 | 10.4 | 49.8 | 13,733 | 13,982 | 1.8 | 2.4 | 1,826 | 2,117 | 640 | 5,842 | 2.37 | 65.8 | 10.5 | 28.6 |
| Chariton | 15.3 | 12.8 | 12.2 | 50.0 | 8,438 | 7,816 | -7.4 | -5.8 | 890 | 969 | -376 | 2,686 | 2.70 | 67.7 | 4.3 | 29.5 |
| Christian | 12.6 | 9.8 | 6.7 | 51.1 | 54,285 | 77,414 | 42.6 | 17.1 | 10,527 | 6,429 | 9,131 | 31,645 | 2.69 | 72.0 | 8.7 | 22.7 |
| Clark | 15.4 | 12.3 | 9.8 | 48.6 | 7,416 | 7,132 | -3.8 | -4.2 | 789 | 841 | -253 | 2,657 | 2.52 | 65.9 | 5.7 | 30.1 |
| Clay | 12.3 | 8.9 | 5.9 | 50.9 | 184,006 | 221,905 | 20.6 | 14.2 | 31,494 | 17,779 | 17,878 | 91,238 | 2.63 | 66.6 | 11.0 | 27.0 |
| Clinton | 12.3 | 8.9 | 5.9 | 50.9 | 18,979 | 20,742 | 9.3 | -0.9 | 2,343 | 2,557 | 22 | 8,177 | 2.45 | 69.7 | 9.0 | 23.9 |
| Cole | 15.0 | 10.6 | 8.1 | 49.9 | 71,397 | 75,970 | 6.4 | 0.3 | 9,318 | 6,707 | -2,360 | 30,154 | 2.38 | 62.6 | 10.2 | 31.8 |
| Cooper | 13.0 | 10.5 | 7.2 | 49.9 | 16,670 | 17,602 | 5.6 | -2.8 | 2,023 | 1,958 | -550 | 6,397 | 2.51 | 65.3 | 11.0 | 30.6 |
| Crawford | 13.9 | 10.8 | 8.6 | 49.6 | 22,804 | 24,720 | 8.4 | -4.0 | 2,826 | 3,088 | -710 | 9,578 | 2.47 | 64.6 | 9.3 | 29.6 |
| Dade | 15.3 | 11.8 | 8.4 | 50.3 | 7,923 | 7,879 | -0.6 | -3.9 | 704 | 1,088 | 70 | 3,068 | 2.41 | 68.9 | 12.8 | 24.6 |
| Dallas | 15.7 | 13.7 | 10.4 | 48.9 | 15,661 | 16,769 | 7.1 | 2.7 | 2,155 | 2,007 | 301 | 6,209 | 2.64 | 67.5 | 8.6 | 26.5 |
| | 15.0 | 12.0 | 8.9 | 50.6 | | | | | 1,206 | 907 | -437 | 3,035 | 2.67 | 70.4 | 6.9 | 25.0 |
| Daviess | 13.9 | 12.3 | 8.8 | 50.3 | 8,016 | 8,424 | 5.1 | -1.7 | 1,074 | 1,233 | -1,786 | 3,807 | 2.40 | 63.5 | 8.5 | 30.9 |
| DeKalb | 13.1 | 9.8 | 9.2 | 43.1 | 11,597 | 12,888 | 11.1 | -15.1 | 1,784 | 2,008 | 20 | 6,371 | 2.41 | 64.7 | 9.1 | 32.3 |
| Dent | 15.4 | 12.2 | 10.5 | 50.5 | 14,927 | 15,692 | 5.1 | -1.3 | 1,524 | 1,742 | -121 | 5,137 | 2.57 | 71.9 | 8.8 | 25.4 |
| Douglas | 15.8 | 14.7 | 10.8 | 50.1 | 13,084 | 13,686 | 4.6 | -2.5 | 4,338 | 4,502 | -2,933 | 12,342 | 2.37 | 62.3 | 16.5 | 31.8 |
| Dunklin | 13.3 | 10.5 | 8.5 | 52.1 | 33,155 | 31,959 | -3.6 | -9.6 | 12,617 | 10,493 | 948 | 40,943 | 2.50 | 68.5 | 10.2 | 25.7 |
| Franklin | 15.5 | 10.8 | 7.7 | 50.3 | 93,807 | 101,464 | 8.2 | 3.0 | 1,595 | 2,128 | -97 | 6,076 | 2.38 | 68.9 | 6.6 | 26.5 |
| Gasconade | 16.3 | 13.1 | 10.5 | 50.2 | 15,342 | 15,210 | -0.9 | -4.2 | 931 | 914 | -269 | 2,555 | 2.50 | 66.8 | 9.0 | 30.5 |
| Gentry | 13.9 | 10.2 | 10.0 | 52.0 | 6,861 | 6,740 | -1.8 | -3.8 | 35,778 | 28,568 | 12,668 | 125,201 | 2.22 | 58.1 | 9.7 | 32.1 |
| Greene | 11.8 | 9.6 | 7.5 | 51.4 | 240,391 | 275,179 | 14.5 | 7.2 | 1,497 | 1,363 | -794 | 3,936 | 2.46 | 60.7 | 9.8 | 32.9 |
| Grundy | 13.9 | 10.9 | 11.3 | 51.0 | 10,432 | 10,260 | -1.6 | -6.5 | | | | | | | | 29.7 |
| Harrison | 14.2 | 11.7 | 11.1 | 50.5 | 8,850 | 8,961 | 1.3 | -7.1 | 1,072 | 1,173 | -536 | 3,429 | 2.41 | 65.3 | 9.7 | 30.5 |
| Henry | 14.7 | 13.0 | 9.8 | 50.8 | 21,997 | 22,291 | 1.3 | -1.0 | 2,669 | 3,182 | 316 | 9,328 | 2.30 | 63.4 | 10.0 | 37.4 |
| Hickory | 18.0 | 18.5 | 14.9 | 50.6 | 8,940 | 9,629 | 7.7 | -0.4 | 770 | 1,526 | 715 | 3,976 | 2.34 | 59.4 | 7.5 | 32.6 |
| Holt | 15.7 | 14.1 | 12.7 | 51.3 | 5,351 | 4,912 | -8.2 | -13.8 | 505 | 633 | -554 | 2,022 | 2.14 | 61.8 | 7.8 | 26.8 |
| Howard | 14.0 | 11.6 | 8.6 | 50.1 | 10,212 | 10,142 | -0.7 | -1.4 | 1,196 | 1,048 | -292 | 3,471 | 2.68 | 68.7 | 11.1 | 28.3 |
| Howell | 13.3 | 11.5 | 8.7 | 51.4 | 37,238 | 40,378 | 8.4 | -0.3 | 5,216 | 5,241 | -64 | 15,878 | 2.49 | 67.2 | 10.3 | 27.8 |
| Iron | 15.5 | 12.7 | 9.6 | 50.0 | 10,697 | 10,615 | -0.8 | -4.9 | 1,072 | 1,513 | -76 | 4,074 | 2.40 | 66.5 | 10.7 | 34.1 |
| Jackson | 12.8 | 9.2 | 6.5 | 51.7 | 654,880 | 674,164 | 2.9 | 4.7 | 96,051 | 64,671 | 911 | 286,601 | 2.39 | 58.2 | 14.2 | |

1. No spouse present.

# Table B. States and Counties — Population, Vital Statistics, Health, and Crime

| STATE County | Persons in group quarters, 2020 | Daytime Population, 2015-2019 Number | Employment/ residence ratio | Births, 2020 Total | Rate[1] | Deaths, 2020 Number | Rate[1] | Persons under 65 with no health insurance, 2019 Number | Percent | Medicare, 2020 Total beneficiaries | Enrolled in Original Medicare | Enrolled in Medicare Advantage | Crimes reported by county police or sheriff, 2019 Violent | Property |
|---|---|---|---|---|---|---|---|---|---|---|---|---|---|---|
| | 32 | 33 | 34 | 35 | 36 | 37 | 38 | 39 | 40 | 41 | 42 | 43 | 44 | 45 |
| MISSISSIPPI— Cont'd | | | | | | | | | | | | | | |
| Sunflower | 4,053 | 25,901 | 0.97 | 271 | 11.0 | 322 | 13.0 | 2,659 | 15.3 | 5,140 | 4,101 | 1,039 | NA | NA |
| Tallahatchie | 2,462 | 12,966 | 0.72 | 155 | 11.3 | 157 | 11.5 | 1,474 | 16.0 | 2,722 | 2,158 | 565 | NA | NA |
| Tate | 1,066 | 22,831 | 0.53 | 352 | 12.3 | 318 | 11.1 | 3,473 | 15.5 | 5,965 | 4,392 | 1,573 | NA | NA |
| Tippah | 324 | 20,555 | 0.83 | 250 | 11.5 | 261 | 12.0 | 2,945 | 16.4 | 5,279 | 4,784 | 495 | NA | NA |
| Tishomingo | 283 | 18,501 | 0.88 | 190 | 9.9 | 300 | 15.6 | 2,626 | 17.3 | 5,159 | 4,847 | 312 | NA | NA |
| Tunica | 104 | 13,716 | 1.90 | 155 | 16.5 | 100 | 10.6 | 1,048 | 12.8 | 1,829 | 1,168 | 661 | 15 | 115 |
| Union | 277 | 28,977 | 1.04 | 348 | 12.1 | 297 | 10.3 | 4,106 | 17.3 | 6,192 | 5,557 | 635 | NA | 383 |
| Walthall | 102 | 12,346 | 0.59 | 160 | 11.2 | 190 | 13.3 | 2,072 | 18.4 | 3,412 | 2,530 | 882 | NA | NA |
| Warren | 522 | 48,217 | 1.09 | 555 | 12.4 | 534 | 11.9 | 5,590 | 15.1 | 9,874 | 7,536 | 2,338 | NA | NA |
| Washington | 595 | 46,369 | 1.02 | 573 | 13.4 | 671 | 15.7 | 5,120 | 14.4 | 10,399 | 7,474 | 2,925 | NA | NA |
| Wayne | 137 | 18,504 | 0.76 | 265 | 13.0 | 216 | 10.6 | 2,771 | 16.9 | 4,329 | 3,241 | 1,088 | NA | NA |
| Webster | 55 | 8,376 | 0.63 | 118 | 12.2 | 144 | 14.9 | 1,195 | 15.1 | 2,587 | 2,464 | 123 | NA | NA |
| Wilkinson | 1,061 | 8,304 | 0.76 | 94 | 11.3 | 132 | 15.8 | 994 | 16.7 | 1,968 | 1,507 | 461 | NA | NA |
| Winston | 459 | 16,524 | 0.75 | 192 | 10.8 | 248 | 13.9 | 2,341 | 17.0 | 4,554 | 3,721 | 833 | NA | NA |
| Yalobusha | 151 | 10,721 | 0.67 | 139 | 11.6 | 191 | 15.9 | 1,313 | 13.9 | 3,616 | 3,013 | 604 | NA | NA |
| Yazoo | 4,577 | 26,507 | 0.76 | 309 | 11.5 | 287 | 10.6 | 2,911 | 15.4 | 4,973 | 3,720 | 1,253 | NA | NA |
| MISSOURI | 166,664 | 6,144,939 | 1.01 | 71,982 | 11.7 | 65,045 | 10.6 | 589,071 | 11.9 | 1,247,990 | 747,869 | 500,121 | X | X |
| Adair | 2,916 | 25,994 | 1.06 | 288 | 11.3 | 229 | 9.0 | 2,890 | 15.4 | 4,772 | 4,078 | 694 | 5 | 97 |
| Andrew | 192 | 11,914 | 0.32 | 200 | 11.4 | 184 | 10.5 | 1,562 | 11.0 | 3,805 | 3,253 | 552 | 1 | 41 |
| Atchison | 98 | 4,910 | 0.88 | 44 | 8.6 | 84 | 16.5 | 482 | 12.8 | 1,470 | 1,341 | 129 | NA | NA |
| Audrain | 1,791 | 24,564 | 0.90 | 306 | 12.3 | 297 | 12.0 | 2,649 | 14.1 | 5,582 | 4,116 | 1,466 | NA | NA |
| Barry | 296 | 36,169 | 1.04 | 402 | 11.2 | 415 | 11.6 | 5,212 | 18.7 | 8,624 | 4,705 | 3,920 | NA | NA |
| Barton | 90 | 10,403 | 0.73 | 127 | 11.0 | 146 | 12.6 | 1,484 | 16.2 | 2,828 | 1,979 | 849 | NA | NA |
| Bates | 311 | 13,805 | 0.66 | 174 | 10.7 | 203 | 12.5 | 1,902 | 14.9 | 3,778 | 2,514 | 1,264 | 34 | 120 |
| Benton | 244 | 17,111 | 0.70 | 170 | 8.7 | 330 | 16.8 | 2,163 | 16.4 | 7,019 | 4,665 | 2,354 | 22 | 214 |
| Bollinger | 163 | 9,588 | 0.48 | 141 | 11.6 | 141 | 11.6 | 1,579 | 16.7 | 2,733 | 2,097 | 636 | NA | NA |
| Boone | 9,035 | 184,841 | 1.08 | 2,059 | 11.3 | 1,245 | 6.8 | 16,404 | 11.0 | 27,965 | 17,614 | 10,351 | NA | NA |
| Buchanan | 4,071 | 97,550 | 1.22 | 1,045 | 12.1 | 951 | 11.0 | 9,393 | 13.6 | 17,595 | 13,819 | 3,776 | 41 | 213 |
| Butler | 864 | 45,307 | 1.15 | 518 | 12.3 | 592 | 14.0 | 5,026 | 15.0 | 11,046 | 8,909 | 2,137 | NA | NA |
| Caldwell | 214 | 7,312 | 0.54 | 100 | 11.0 | 104 | 11.5 | 1,016 | 14.5 | 2,077 | 1,505 | 572 | 20 | 87 |
| Callaway | 3,622 | 39,779 | 0.74 | 491 | 10.9 | 454 | 10.1 | 3,855 | 11.4 | 9,191 | 5,630 | 3,561 | NA | NA |
| Camden | 774 | 45,610 | 1.01 | 360 | 7.8 | 582 | 12.5 | 5,113 | 15.7 | 13,290 | 9,490 | 3,800 | 89 | 413 |
| Cape Girardeau | 3,827 | 83,666 | 1.14 | 879 | 11.1 | 811 | 10.2 | 7,199 | 11.6 | 16,386 | 13,699 | 2,687 | NA | NA |
| Carroll | 88 | 8,109 | 0.81 | 86 | 10.1 | 115 | 13.4 | 891 | 13.3 | 2,332 | 1,792 | 541 | 6 | 32 |
| Carter | 40 | 5,467 | 0.73 | 72 | 12.0 | 92 | 15.4 | 713 | 15.3 | 1,559 | 1,228 | 331 | 21 | 54 |
| Cass | 1,045 | 81,225 | 0.56 | 1,155 | 10.8 | 1,079 | 10.1 | 8,734 | 10.0 | 21,240 | 11,938 | 9,302 | NA | NA |
| Cedar | 156 | 12,744 | 0.75 | 194 | 13.5 | 224 | 15.6 | 1,761 | 16.3 | 3,969 | 2,228 | 1,741 | 36 | 53 |
| Chariton | 213 | 6,549 | 0.69 | 90 | 12.2 | 103 | 14.0 | 746 | 13.5 | 2,014 | 1,690 | 324 | NA | NA |
| Christian | 582 | 65,031 | 0.49 | 1,062 | 11.7 | 759 | 8.4 | 8,870 | 12.0 | 17,020 | 8,409 | 8,611 | 64 | 250 |
| Clark | 90 | 5,507 | 0.59 | 73 | 10.7 | 73 | 10.7 | 804 | 15.2 | 1,570 | 1,385 | 185 | NA | NA |
| Clay | 2,657 | 220,148 | 0.82 | 3,118 | 12.3 | 2,000 | 7.9 | 19,670 | 9.3 | 42,171 | 27,235 | 14,936 | NA | NA |
| Clinton | 424 | 16,051 | 0.54 | 227 | 11.0 | 243 | 11.8 | 1,949 | 11.7 | 4,568 | 3,307 | 1,261 | 16 | 53 |
| Cole | 4,239 | 91,018 | 1.38 | 861 | 11.3 | 757 | 9.9 | 6,153 | 10.4 | 15,352 | 9,483 | 5,869 | 34 | 381 |
| Cooper | 982 | 15,194 | 0.68 | 204 | 11.9 | 198 | 11.6 | 1,577 | 12.1 | 3,793 | 2,533 | 1,259 | NA | NA |
| Crawford | 332 | 21,961 | 0.77 | 254 | 10.7 | 332 | 14.0 | 2,969 | 15.8 | 5,950 | 3,305 | 2,645 | NA | NA |
| Dade | 133 | 7,001 | 0.80 | 70 | 9.2 | 115 | 15.2 | 966 | 17.1 | 2,154 | 1,133 | 1,020 | 45 | 65 |
| Dallas | 199 | 13,502 | 0.50 | 233 | 13.5 | 231 | 13.4 | 2,267 | 17.2 | 4,480 | 2,042 | 2,437 | 5 | 23 |
| Daviess | 152 | 7,144 | 0.67 | 124 | 15.0 | 93 | 11.2 | 1,232 | 19.4 | 1,837 | 1,424 | 413 | 17 | 29 |
| DeKalb | 1,851 | 11,785 | 0.83 | 92 | 8.4 | 95 | 8.7 | 835 | 11.6 | 2,173 | 1,699 | 473 | 5 | 51 |
| Dent | 201 | 14,452 | 0.83 | 161 | 10.4 | 229 | 14.8 | 1,977 | 16.6 | 4,072 | 3,269 | 804 | 12 | 71 |
| Douglas | 123 | 11,818 | 0.69 | 134 | 10.0 | 179 | 13.4 | 1,725 | 17.7 | 2,917 | 1,299 | 1,618 | NA | NA |
| Dunklin | 645 | 28,951 | 0.91 | 365 | 12.6 | 452 | 15.7 | 3,694 | 16.1 | 7,180 | 5,524 | 1,656 | NA | NA |
| Franklin | 858 | 93,571 | 0.81 | 1,180 | 11.3 | 1,173 | 11.2 | 10,032 | 11.8 | 23,165 | 9,175 | 13,990 | 37 | NA |
| Gasconade | 250 | 13,702 | 0.85 | 151 | 10.4 | 227 | 15.6 | 1,589 | 14.1 | 3,893 | 2,285 | 1,608 | 49 | 178 |
| Gentry | 177 | 6,269 | 0.88 | 86 | 13.3 | 93 | 14.3 | 719 | 13.8 | 1,490 | 1,296 | 194 | 1 | 30 |
| Greene | 11,284 | 329,099 | 1.29 | 3,347 | 11.3 | 3,048 | 10.3 | 31,041 | 13.3 | 59,598 | 29,909 | 29,689 | NA | NA |
| Grundy | 350 | 10,014 | 1.01 | 129 | 13.4 | 129 | 13.4 | 1,145 | 15.4 | 2,457 | 2,174 | 283 | 2 | 15 |
| Harrison | 145 | 7,942 | 0.85 | 90 | 10.8 | 121 | 14.5 | 864 | 13.6 | 2,173 | 1,880 | 293 | 3 | 33 |
| Henry | 233 | 21,411 | 0.96 | 275 | 12.5 | 321 | 14.5 | 2,450 | 14.6 | 6,195 | 4,043 | 2,151 | 78 | 267 |
| Hickory | 75 | 8,259 | 0.63 | 82 | 8.6 | 155 | 16.2 | 1,098 | 17.5 | 2,775 | 1,388 | 1,387 | 1 | 81 |
| Holt | 101 | 4,011 | 0.79 | 47 | 11.1 | 69 | 16.3 | 462 | 14.4 | 1,215 | 1,054 | 161 | NA | NA |
| Howard | 720 | 8,155 | 0.59 | 106 | 10.6 | 90 | 9.0 | 950 | 12.9 | 2,154 | 1,486 | 668 | 17 | 35 |
| Howell | 608 | 41,676 | 1.10 | 485 | 12.0 | 555 | 13.8 | 5,057 | 16.1 | 10,580 | 6,973 | 3,608 | 70 | 349 |
| Iron | 276 | 9,691 | 0.87 | 104 | 10.3 | 174 | 17.2 | 1,064 | 13.9 | 2,963 | 2,211 | 752 | 4 | 63 |
| Jackson | 10,935 | 752,299 | 1.16 | 8,913 | 12.6 | 7,121 | 10.1 | 77,183 | 13.2 | 125,981 | 69,137 | 56,844 | NA | NA |

1. Per 1,000 estimated resident population.

Items 32—45

# Table B. States and Counties — Crime, Education, Money Income, and Poverty

| STATE County | County law enforcement employment, 2019 | | Education | | | | | | Money income, 2015-2019 | | | | Income and poverty, 2019 | | | |
| | | | School enrollment and attainment, 2015-2019 | | | | Local government expenditures,[3] 2017-2018 | | | Households | | | | Percent below poverty level | | |
| | | | Enrollment[1] | | Attainment[2] (percent) | | | | | | | Percent | | | | | |
| | Officers | Civilians | Total | Percent private | High school graduate or less | Bachelor's degree or more | Total current spending (mil dol) | Current spending per student (dollars) | Per capita income[4] | Median income (dollars) | with income of less than $50,000 | with income of $200,000 or more | Median household income (dollars) | All persons | Children under 18 years | Children 5 to 17 years in families |
| | 46 | 47 | 48 | 49 | 50 | 51 | 52 | 53 | 54 | 55 | 56 | 57 | 58 | 59 | 60 | 61 |
| **MISSISSIPPI— Cont'd** | | | | | | | | | | | | | | | | |
| Sunflower | NA | NA | 5,890 | 13.5 | 57.4 | 15.4 | 37 | 9,558 | 15,646 | 30,838 | 68.2 | 1.7 | 32,948 | 34.0 | 47.2 | 45.9 |
| Tallahatchie | NA | NA | 2,699 | 9.8 | 59.3 | 10.3 | 20 | 10,269 | 16,742 | 29,864 | 68.0 | 1.5 | 31,915 | 37.9 | 44.4 | 47.2 |
| Tate | NA | NA | 7,431 | 12.7 | 52.5 | 15.8 | 35 | 8,368 | 23,340 | 51,030 | 49.3 | 2.5 | 50,188 | 15.1 | 23.1 | 23.0 |
| Tippah | NA | NA | 5,443 | 10.9 | 56.3 | 15.2 | 35 | 8,751 | 19,501 | 39,246 | 62.1 | 1.3 | 44,696 | 16.2 | 22.0 | 19.6 |
| Tishomingo | 12 | 14 | 4,156 | 8.9 | 58.5 | 11.4 | 28 | 9,279 | 20,390 | 37,681 | 63.6 | 1.5 | 43,341 | 17.0 | 21.9 | 20.4 |
| Tunica | 45 | 55 | 2,943 | 8.3 | 49.9 | 16.9 | 24 | 11,536 | 20,579 | 39,370 | 62.6 | 0.7 | 33,993 | 28.1 | 43.2 | 41.6 |
| Union | NA | NA | 6,845 | 7.2 | 55.6 | 13.4 | 42 | 8,218 | 21,705 | 45,754 | 53.7 | 2.0 | 49,320 | 13.0 | 18.2 | 17.1 |
| Walthall | NA | NA | 3,213 | 24.0 | 62.1 | 14.4 | 17 | 8,987 | 21,123 | 30,961 | 67.6 | 2.3 | 35,465 | 21.7 | 33.2 | 31.9 |
| Warren | 39 | 3 | 11,126 | 13.7 | 43.4 | 22.4 | 76 | 9,344 | 25,661 | 45,113 | 54.8 | 2.3 | 46,606 | 20.1 | 33.3 | 30.7 |
| Washington | 36 | 14 | 12,232 | 14.2 | 53.5 | 17.9 | 79 | 9,416 | 20,174 | 29,705 | 67.9 | 2.4 | 31,018 | 33.7 | 48.0 | 49.0 |
| Wayne | NA | NA | 5,177 | 16.8 | 59.1 | 15.3 | 35 | 10,788 | 23,668 | 37,706 | 58.6 | 3.8 | 39,036 | 22.2 | 31.9 | 31.1 |
| Webster | NA | NA | 2,223 | 13.3 | 51.7 | 18.5 | 15 | 7,836 | 22,011 | 45,730 | 55.7 | 1.5 | 46,110 | 17.0 | 24.5 | 23.1 |
| Wilkinson | NA | NA | 1,647 | 11.7 | 65.8 | 13.7 | 11 | 9,441 | 15,475 | 27,313 | 74.8 | 1.5 | 32,612 | 31.2 | 38.9 | 37.6 |
| Winston | NA | NA | 4,196 | 13.6 | 54.1 | 14.7 | 28 | 9,623 | 25,263 | 34,724 | 65.9 | 2.4 | 37,596 | 25.4 | 37.4 | 34.7 |
| Yalobusha | NA | NA | 2,570 | 16.3 | 56.9 | 13.8 | 15 | 9,311 | 20,690 | 41,464 | 60.2 | 0.8 | 40,515 | 24.4 | 41.3 | 40.2 |
| Yazoo | NA | NA | 6,698 | 19.9 | 59.7 | 13.4 | 33 | 8,415 | 19,972 | 33,279 | 63.8 | 1.5 | 34,817 | 36.4 | 44.0 | 40.1 |
| **MISSOURI** | X | X | 1,480,841 | 18.8 | 40.7 | 29.2 | 10,016 | 10,941 | 30,810 | 55,461 | 45.2 | 4.8 | 57,375 | 12.9 | 17.0 | 15.9 |
| Adair | 9 | 15 | 9,765 | 9.2 | 44.3 | 31.0 | 30 | 10,097 | 22,847 | 41,929 | 58.1 | 1.9 | 43,044 | 19.2 | 18.9 | 18.0 |
| Andrew | 11 | 6 | 3,953 | 6.3 | 45.7 | 24.9 | 27 | 9,319 | 29,338 | 58,772 | 42.2 | 3.0 | 62,823 | 8.9 | 11.1 | 10.6 |
| Atchison | 5 | 5 | 1,004 | 3.1 | 47.5 | 23.1 | 11 | 13,165 | 31,248 | 50,236 | 49.7 | 2.7 | 50,516 | 12.9 | 17.5 | 16.5 |
| Audrain | 28 | 15 | 5,381 | 10.9 | 59.6 | 14.7 | 31 | 9,301 | 23,024 | 44,261 | 57.6 | 2.4 | 45,609 | 16.3 | 23.9 | 22.9 |
| Barry | 21 | 2 | 7,789 | 13.0 | 57.7 | 14.6 | 64 | 9,797 | 25,068 | 44,403 | 54.5 | 3.2 | 47,155 | 18.6 | 29.6 | 26.2 |
| Barton | NA | NA | 2,606 | 10.7 | 54.0 | 19.9 | 19 | 10,134 | 26,509 | 44,125 | 56.3 | 3.0 | 46,397 | 15.0 | 21.6 | 19.9 |
| Bates | 19 | 44 | 3,297 | 10.3 | 56.8 | 16.1 | 25 | 9,820 | 27,635 | 47,625 | 51.9 | 2.4 | 48,442 | 16.9 | 25.4 | 25.4 |
| Benton | 17 | 10 | 3,125 | 18.0 | 55.8 | 13.0 | 23 | 9,592 | 24,317 | 40,249 | 60.9 | 1.6 | 42,144 | 16.7 | 28.6 | 27.9 |
| Bollinger | 12 | 5 | 2,539 | 15.3 | 67.5 | 8.9 | 17 | 8,719 | 21,340 | 44,158 | 53.8 | 1.0 | 45,868 | 14.5 | 21.4 | 19.1 |
| Boone | 66 | 17 | 60,958 | 11.8 | 26.8 | 46.1 | 268 | 11,293 | 30,415 | 55,328 | 45.8 | 5.1 | 58,029 | 17.2 | 14.0 | 13.5 |
| Buchanan | 69 | 37 | 20,280 | 10.7 | 49.5 | 21.1 | 130 | 10,239 | 26,347 | 51,916 | 48.5 | 2.9 | 49,216 | 15.8 | 20.2 | 17.6 |
| Butler | 30 | 13 | 9,319 | 7.1 | 54.3 | 13.3 | 60 | 8,849 | 21,652 | 39,915 | 62.6 | 2.6 | 41,327 | 22.5 | 35.3 | 32.7 |
| Caldwell | 10 | 39 | 2,074 | 4.4 | 53.9 | 18.1 | 18 | 10,904 | 24,581 | 49,839 | 50.2 | 1.2 | 52,872 | 13.6 | 19.0 | 17.3 |
| Callaway | 30 | 3 | 10,516 | 24.4 | 49.9 | 23.1 | 46 | 9,240 | 25,849 | 56,938 | 43.3 | 2.6 | 58,720 | 11.6 | 16.4 | 14.5 |
| Camden | 54 | 39 | 7,962 | 12.8 | 47.5 | 21.1 | 55 | 10,582 | 28,275 | 53,478 | 46.4 | 4.7 | 51,460 | 13.5 | 21.2 | 19.2 |
| Cape Girardeau | 46 | 36 | 22,167 | 15.7 | 40.2 | 31.5 | 98 | 9,340 | 28,267 | 53,732 | 46.4 | 3.9 | 57,618 | 14.6 | 17.8 | 17.5 |
| Carroll | 8 | 1 | 1,697 | 9.9 | 58.4 | 16.7 | 17 | 12,417 | 25,715 | 50,830 | 48.7 | 3.5 | 50,628 | 13.8 | 21.1 | 18.9 |
| Carter | 8 | 6 | 1,517 | 4.2 | 58.4 | 15.5 | 13 | 10,413 | 22,104 | 39,530 | 58.4 | 1.5 | 36,383 | 19.8 | 29.4 | 31.5 |
| Cass | 90 | 21 | 24,037 | 13.6 | 41.4 | 26.3 | 179 | 9,806 | 32,868 | 69,433 | 33.9 | 4.8 | 73,882 | 7.9 | 11.6 | 9.6 |
| Cedar | 13 | 25 | 2,854 | 16.1 | 53.5 | 13.6 | 20 | 9,314 | 22,623 | 39,365 | 63.5 | 2.1 | 38,427 | 17.9 | 26.7 | 26.4 |
| Chariton | NA | NA | 1,603 | 19.5 | 58.7 | 17.2 | 13 | 12,099 | 23,903 | 46,738 | 54.2 | 2.0 | 50,589 | 12.5 | 16.3 | 15.3 |
| Christian | 58 | 35 | 22,333 | 12.9 | 37.4 | 28.6 | 138 | 8,889 | 28,216 | 60,645 | 41.6 | 3.6 | 63,400 | 10.4 | 12.8 | 11.5 |
| Clark | 5 | 7 | 1,356 | 15.3 | 57.2 | 12.8 | 10 | 9,213 | 26,490 | 48,909 | 51.0 | 2.1 | 46,001 | 14.2 | 22.2 | 21.3 |
| Clay | 132 | 64 | 60,191 | 15.7 | 35.0 | 33.2 | 444 | 10,580 | 34,560 | 70,510 | 34.1 | 5.6 | 70,705 | 9.1 | 12.5 | 11.4 |
| Clinton | 13 | 9 | 4,709 | 11.2 | 48.1 | 21.6 | 40 | 9,812 | 29,081 | 62,701 | 37.9 | 2.2 | 64,541 | 9.9 | 13.1 | 11.8 |
| Cole | 49 | 30 | 18,508 | 27.1 | 37.4 | 33.3 | 112 | 8,620 | 30,422 | 60,066 | 39.6 | 3.1 | 61,847 | 10.6 | 13.7 | 12.2 |
| Cooper | 9 | 1 | 3,939 | 17.9 | 48.8 | 22.1 | 25 | 10,188 | 23,961 | 52,735 | 47.4 | 1.8 | 56,957 | 11.8 | 16.5 | 15.8 |
| Crawford | 25 | 16 | 5,062 | 10.8 | 61.2 | 11.5 | 30 | 8,973 | 22,220 | 44,438 | 57.8 | 1.4 | 48,315 | 15.6 | 23.1 | 21.0 |
| Dade | 9 | 5 | 1,484 | 11.5 | 57.6 | 12.7 | 11 | 9,997 | 23,186 | 40,399 | 60.4 | 1.4 | 42,421 | 15.9 | 22.8 | 20.4 |
| Dallas | 15 | 14 | 3,436 | 18.2 | 61.1 | 12.9 | 15 | 9,083 | 23,647 | 43,542 | 56.6 | 3.3 | 42,389 | 17.8 | 28.0 | 27.4 |
| Daviess | 6 | 1 | 1,725 | 19.1 | 56.1 | 18.5 | 14 | 11,130 | 24,929 | 51,679 | 47.7 | 3.1 | 48,197 | 17.6 | 27.6 | 26.5 |
| DeKalb | 10 | 6 | 2,356 | 11.3 | 59.8 | 13.8 | 12 | 11,590 | 20,125 | 55,918 | 46.6 | 1.3 | 53,997 | 14.6 | 13.6 | 12.8 |
| Dent | 17 | 5 | 3,312 | 12.7 | 58.1 | 15.3 | 20 | 8,734 | 22,711 | 42,100 | 58.5 | 2.0 | 45,540 | 19.0 | 29.9 | 29.6 |
| Douglas | 7 | 6 | 2,658 | 16.8 | 61.9 | 10.8 | 14 | 8,790 | 21,083 | 37,425 | 61.6 | 2.1 | 39,773 | 18.7 | 26.8 | 25.3 |
| Dunklin | 14 | 10 | 6,935 | 4.3 | 61.9 | 13.8 | 51 | 9,299 | 21,801 | 36,380 | 64.4 | 2.1 | 38,880 | 26.7 | 35.3 | 29.7 |
| Franklin | 137 | 13 | 22,884 | 19.7 | 44.3 | 20.9 | 163 | 10,128 | 30,278 | 57,214 | 43.0 | 3.8 | 61,428 | 9.3 | 12.6 | 11.2 |
| Gasconade | 15 | 1 | 2,962 | 13.1 | 50.7 | 19.2 | 25 | 8,737 | 28,617 | 54,885 | 44.7 | 2.2 | 53,751 | 11.6 | 17.0 | 16.0 |
| Gentry | 5 | 0 | 1,372 | 7.5 | 53.3 | 19.1 | 13 | 11,513 | 24,269 | 47,790 | 51.9 | 1.4 | 53,663 | 11.5 | 18.0 | 17.7 |
| Greene | 141 | 257 | 77,946 | 16.4 | 36.2 | 30.5 | 367 | 9,542 | 27,525 | 46,086 | 53.4 | 2.9 | 45,313 | 14.4 | 13.3 | 13.1 |
| Grundy | 6 | 10 | 2,199 | 10.7 | 52.5 | 17.1 | 16 | 10,330 | 23,497 | 45,594 | 54.0 | 1.5 | 44,323 | 18.7 | 25.9 | 25.0 |
| Harrison | 4 | 7 | 1,844 | 10.1 | 57.7 | 16.5 | 15 | 10,934 | 22,032 | 42,917 | 56.9 | 0.6 | 43,016 | 15.4 | 23.1 | 20.2 |
| Henry | 25 | 14 | 4,612 | 13.1 | 52.8 | 15.9 | 29 | 10,389 | 20,736 | 45,795 | 53.8 | 2.2 | 46,995 | 17.3 | 27.5 | 30.0 |
| Hickory | 9 | 4 | 1,489 | 21.2 | 60.3 | 9.0 | 16 | 9,977 | 20,736 | 34,182 | 69.5 | 0.5 | 32,934 | 18.2 | 34.1 | 33.1 |
| Holt | 6 | 6 | 808 | 8.4 | 52.2 | 19.7 | 8 | 13,110 | 27,520 | 49,524 | 50.2 | 1.5 | 47,429 | 11.8 | 17.0 | 15.6 |
| Howard | 7 | 1 | 2,681 | 35.8 | 49.2 | 25.2 | 13 | 8,859 | 23,753 | 52,700 | 47.6 | 2.4 | 49,923 | 13.0 | 16.0 | 16.0 |
| Howell | 27 | 12 | 9,521 | 12.1 | 51.8 | 18.0 | 51 | 9,302 | 21,048 | 38,357 | 61.9 | 1.6 | 40,177 | 20.8 | 29.7 | 27.9 |
| Iron | 10 | 6 | 1,863 | 6.9 | 59.9 | 10.8 | 20 | 11,026 | 21,499 | 37,435 | 60.7 | 1.3 | 39,292 | 20.8 | 31.1 | 29.9 |
| Jackson | 96 | 39 | 165,004 | 15.4 | 37.7 | 31.6 | 1,272 | 11,534 | 31,480 | 55,134 | 45.6 | 4.3 | 57,907 | 13.7 | 19.6 | 18.4 |

1. All persons 3 years old and over enrolled in nursery school through college.  2. Persons 25 years old and over.  3. Elementary and secondary education expenditures.  4. Based on population estimated by the American Community Survey, 2014–2018.

| STATE County | Personal income, 2019 | | | | | | | | | | Earnings, 2019 | | |
|---|---|---|---|---|---|---|---|---|---|---|---|---|---|
| | Total (mil dol) | Percent change 2018-2019 | Per capita[1] Dollars | Rank | Wages and salaries (mil dol) | Supplements to wages and salaries, employer contributions (mil dol) Pension and insurance | Government social insurance | Proprietors' income (mil dol) | Dividends, interest, and rent (mil dol) | Personal transfer receipts (mil dol) | Total (mil dol) | Contributions for government social insurance (mil dol) From employee and self-employed | From employer |
| | 62 | 63 | 64 | 65 | 66 | 67 | 68 | 69 | 70 | 71 | 72 | 73 | 74 |
| MISSISSIPPI— Cont'd | | | | | | | | | | | | | |
| Sunflower | 800 | -1.4 | 31,846 | 2,954 | 307 | 58 | 23 | 71 | 128 | 295 | 459 | 33 | 23 |
| Tallahatchie | 420 | 0.8 | 30,451 | 3,023 | 119 | 20 | 9 | 50 | 57 | 147 | 198 | 14 | 9 |
| Tate | 1,043 | 3.4 | 36,836 | 2,465 | 221 | 40 | 16 | 62 | 115 | 289 | 340 | 29 | 16 |
| Tippah | 772 | 3.9 | 35,061 | 2,677 | 261 | 48 | 20 | 41 | 89 | 250 | 371 | 30 | 20 |
| Tishomingo | 676 | 3.3 | 34,871 | 2,703 | 254 | 47 | 20 | 33 | 82 | 229 | 354 | 29 | 20 |
| Tunica | 331 | 2.0 | 34,325 | 2,767 | 285 | 31 | 21 | 23 | 56 | 109 | 361 | 24 | 21 |
| Union | 952 | 4.0 | 33,027 | 2,886 | 527 | 80 | 41 | 61 | 123 | 270 | 709 | 52 | 41 |
| Walthall | 456 | -0.4 | 31,910 | 2,948 | 84 | 17 | 6 | 40 | 60 | 166 | 148 | 13 | 6 |
| Warren | 1,865 | 1.9 | 41,102 | 1,907 | 992 | 200 | 77 | 106 | 301 | 490 | 1,374 | 93 | 77 |
| Washington | 1,710 | 0.9 | 38,942 | 2,201 | 661 | 112 | 49 | 125 | 269 | 578 | 947 | 73 | 49 |
| Wayne | 706 | -0.4 | 34,961 | 2,692 | 215 | 39 | 16 | 95 | 107 | 217 | 364 | 25 | 16 |
| Webster | 386 | 4.4 | 39,794 | 2,076 | 83 | 15 | 6 | 19 | 47 | 136 | 124 | 12 | 6 |
| Wilkinson | 265 | 3.4 | 30,723 | 3,017 | 60 | 12 | 5 | 8 | 40 | 105 | 84 | 8 | 5 |
| Winston | 654 | 2.3 | 36,414 | 2,521 | 226 | 36 | 17 | 42 | 84 | 223 | 321 | 25 | 17 |
| Yalobusha | 466 | 1.7 | 38,478 | 2,265 | 119 | 23 | 9 | 27 | 58 | 174 | 178 | 16 | 9 |
| Yazoo | 821 | 4.2 | 27,666 | 3,084 | 297 | 65 | 23 | 47 | 122 | 276 | 432 | 30 | 23 |
| MISSOURI | 298,620 | 3.5 | 48,631 | X | 154,588 | 26,591 | 10,854 | 20,734 | 56,866 | 58,950 | 212,767 | 13,157 | 10,854 |
| Adair | 845 | 2.4 | 33,351 | 2,862 | 390 | 87 | 28 | 49 | 182 | 217 | 554 | 35 | 28 |
| Andrew | 808 | 5.4 | 45,594 | 1,293 | 103 | 24 | 7 | 50 | 125 | 145 | 185 | 14 | 7 |
| Atchison | 248 | 19.8 | 48,160 | 1,005 | 63 | 14 | 4 | 44 | 48 | 59 | 126 | 7 | 4 |
| Audrain | 993 | 5.3 | 39,118 | 2,174 | 369 | 83 | 26 | 137 | 163 | 234 | 615 | 38 | 26 |
| Barry | 1,253 | 0.0 | 35,003 | 2,685 | 654 | 123 | 47 | 80 | 238 | 345 | 904 | 61 | 47 |
| Barton | 424 | 3.1 | 36,035 | 2,563 | 124 | 26 | 9 | 52 | 68 | 117 | 211 | 15 | 9 |
| Bates | 652 | 5.9 | 40,325 | 2,011 | 143 | 34 | 10 | 68 | 104 | 176 | 255 | 17 | 10 |
| Benton | 749 | 4.5 | 38,536 | 2,255 | 128 | 30 | 9 | 84 | 125 | 270 | 252 | 23 | 9 |
| Bollinger | 408 | 3.7 | 33,632 | 2,833 | 66 | 15 | 5 | 25 | 54 | 137 | 111 | 10 | 5 |
| Boone | 8,757 | 5.0 | 48,525 | 967 | 4,969 | 1,072 | 327 | 428 | 1,739 | 1,363 | 6,795 | 377 | 327 |
| Buchanan | 3,508 | 2.6 | 40,156 | 2,033 | 2,410 | 434 | 176 | 194 | 518 | 915 | 3,214 | 197 | 176 |
| Butler | 1,555 | 2.1 | 36,606 | 2,495 | 724 | 157 | 53 | 100 | 252 | 553 | 1,035 | 69 | 53 |
| Caldwell | 351 | 4.1 | 38,939 | 2,203 | 71 | 16 | 5 | 33 | 49 | 89 | 125 | 9 | 5 |
| Callaway | 1,750 | 4.5 | 39,111 | 2,175 | 728 | 174 | 52 | 38 | 279 | 417 | 991 | 64 | 52 |
| Camden | 1,776 | 4.2 | 38,352 | 2,284 | 652 | 114 | 49 | 147 | 411 | 527 | 962 | 71 | 49 |
| Cape Girardeau | 3,688 | 3.9 | 46,754 | 1,154 | 1,925 | 360 | 137 | 394 | 675 | 733 | 2,816 | 172 | 137 |
| Carroll | 429 | 13.0 | 49,468 | 882 | 106 | 22 | 8 | 99 | 71 | 99 | 236 | 13 | 8 |
| Carter | 200 | 3.7 | 33,370 | 2,858 | 50 | 14 | 4 | 17 | 30 | 78 | 85 | 7 | 4 |
| Cass | 5,068 | 4.4 | 47,907 | 1,028 | 1,154 | 214 | 83 | 261 | 723 | 936 | 1,712 | 119 | 83 |
| Cedar | 448 | 2.9 | 31,208 | 2,993 | 101 | 25 | 7 | 39 | 78 | 177 | 172 | 15 | 7 |
| Chariton | 343 | 7.8 | 46,156 | 1,231 | 70 | 16 | 5 | 54 | 67 | 88 | 145 | 9 | 5 |
| Christian | 3,678 | 5.0 | 41,516 | 1,845 | 710 | 146 | 52 | 226 | 496 | 697 | 1,134 | 85 | 52 |
| Clark | 250 | 7.4 | 36,842 | 2,463 | 48 | 13 | 3 | 20 | 42 | 65 | 84 | 6 | 3 |
| Clay | 12,227 | 4.3 | 48,920 | 933 | 5,964 | 971 | 421 | 655 | 1,561 | 1,910 | 8,010 | 491 | 421 |
| Clinton | 881 | 3.8 | 43,200 | 1,585 | 174 | 36 | 12 | 27 | 136 | 190 | 249 | 19 | 12 |
| Cole | 3,705 | 4.6 | 48,273 | 995 | 2,557 | 585 | 169 | 223 | 687 | 710 | 3,535 | 195 | 169 |
| Cooper | 712 | 4.7 | 40,212 | 2,026 | 199 | 41 | 15 | 48 | 115 | 171 | 302 | 21 | 15 |
| Crawford | 867 | 2.9 | 36,242 | 2,544 | 284 | 59 | 21 | 57 | 135 | 274 | 421 | 31 | 21 |
| Dade | 268 | 4.1 | 35,409 | 2,635 | 63 | 15 | 5 | 27 | 45 | 81 | 110 | 8 | 5 |
| Dallas | 577 | 4.2 | 34,214 | 2,779 | 86 | 20 | 6 | 61 | 79 | 188 | 173 | 16 | 6 |
| Daviess | 293 | 7.4 | 35,453 | 2,625 | 54 | 14 | 4 | 46 | 55 | 81 | 117 | 8 | 4 |
| DeKalb | 372 | 5.1 | 29,649 | 3,048 | 122 | 30 | 8 | 22 | 52 | 99 | 182 | 12 | 8 |
| Dent | 531 | 4.1 | 34,112 | 2,788 | 146 | 34 | 10 | 42 | 82 | 187 | 232 | 18 | 10 |
| Douglas | 385 | 4.8 | 29,191 | 3,057 | 86 | 20 | 7 | 23 | 60 | 141 | 137 | 13 | 7 |
| Dunklin | 1,086 | 4.1 | 37,291 | 2,416 | 297 | 64 | 22 | 80 | 150 | 415 | 464 | 34 | 22 |
| Franklin | 4,748 | 3.9 | 45,672 | 1,287 | 1,884 | 359 | 136 | 208 | 791 | 1,028 | 2,587 | 173 | 136 |
| Gasconade | 602 | 4.5 | 40,932 | 1,925 | 191 | 44 | 14 | 43 | 115 | 172 | 292 | 21 | 14 |
| Gentry | 293 | 7.5 | 44,599 | 1,407 | 94 | 18 | 7 | 34 | 53 | 80 | 153 | 9 | 7 |
| Greene | 13,328 | 4.0 | 45,476 | 1,300 | 8,547 | 1,491 | 608 | 1,241 | 2,532 | 2,734 | 11,886 | 724 | 608 |
| Grundy | 356 | 2.9 | 36,146 | 2,550 | 117 | 29 | 9 | 45 | 61 | 123 | 200 | 13 | 9 |
| Harrison | 319 | 6.8 | 38,154 | 2,311 | 89 | 22 | 6 | 45 | 61 | 98 | 161 | 11 | 6 |
| Henry | 933 | 3.8 | 42,741 | 1,652 | 308 | 72 | 21 | 84 | 160 | 287 | 486 | 34 | 21 |
| Hickory | 277 | 2.5 | 29,033 | 3,062 | 44 | 11 | 3 | 22 | 48 | 127 | 80 | 9 | 3 |
| Holt | 237 | 19.0 | 53,880 | 550 | 54 | 12 | 4 | 66 | 38 | 48 | 136 | 7 | 4 |
| Howard | 415 | 5.8 | 41,534 | 1,842 | 95 | 22 | 7 | 28 | 70 | 110 | 152 | 11 | 7 |
| Howell | 1,415 | 4.0 | 35,261 | 2,648 | 606 | 131 | 45 | 110 | 227 | 478 | 890 | 62 | 45 |
| Iron | 350 | 1.9 | 34,554 | 2,739 | 152 | 33 | 12 | 9 | 47 | 146 | 206 | 15 | 12 |
| Jackson | 33,618 | 3.3 | 47,819 | 1,042 | 24,402 | 3,913 | 1,697 | 2,683 | 5,620 | 6,621 | 32,695 | 1,950 | 1,697 |

1.  Based on the resident population estimated as of July 1 of the year shown.

# Table B. States and Counties — Earnings, Social Security, and Housing

| STATE County | Earnings, 2019 (cont.) Percent by selected industries | | | | | | | | | Social Security beneficiaries, December 2019 | | Supplemental Security Income recipients, 2019 | Housing units, 2020 | |
|---|---|---|---|---|---|---|---|---|---|---|---|---|---|---|
| | Farm | Mining, quarrying, and extractions | Construction | Manufacturing | Information; professional, scientific, technical services | Retail trade | Finance, insurance, real estate, and leasing | Health care and social assistance | Government | Number | Rate[1] | | Total | Percent change, 2010-2020 |
| | 75 | 76 | 77 | 78 | 79 | 80 | 81 | 82 | 83 | 84 | 85 | 86 | 87 | 88 |
| MISSISSIPPI— Cont'd | | | | | | | | | | | | | | |
| Sunflower | 4.5 | 0.0 | 2.2 | 3.5 | 1.6 | 5.7 | 4.4 | 6.0 | 36.9 | 5,800 | 231 | 1,881 | 9,662 | -0.2 |
| Tallahatchie | 14.8 | 0.0 | 1.9 | 1.0 | D | 5.4 | 2.5 | 3.6 | 26.4 | 3,175 | 228 | 942 | 5,698 | 3.0 |
| Tate | 1.2 | D | 9.7 | 8.4 | 6.8 | 9.1 | 4.2 | 12.7 | 26.5 | 6,740 | 238 | 913 | 11,916 | 8.9 |
| Tippah | 0.7 | 1.1 | 6.4 | 27.3 | D | 7.3 | 3.2 | D | 17.3 | 6,060 | 275 | 998 | 9,999 | 3.1 |
| Tishomingo | -0.2 | 0.0 | 7.1 | 42.1 | 1.2 | 6.5 | 2.7 | D | 13.1 | 5,770 | 298 | 690 | 10,498 | 2.0 |
| Tunica | 4.3 | 0.0 | D | 6.0 | D | D | 2.8 | 2.7 | 11.5 | 2,145 | 223 | 700 | 5,118 | 6.6 |
| Union | -0.2 | 0.0 | D | 42.3 | D | 4.9 | 2.2 | D | 10.5 | 7,255 | 253 | 776 | 12,114 | 5.2 |
| Walthall | 10.3 | D | 10.3 | 11.8 | D | 8.5 | 2.8 | D | 21.0 | 3,740 | 261 | 604 | 7,361 | 3.2 |
| Warren | 0.5 | D | 2.9 | 13.8 | 6.1 | 6.0 | 3.2 | 11.1 | 31.9 | 10,765 | 237 | 1,915 | 22,167 | 1.3 |
| Washington | -0.7 | D | 3.4 | 8.5 | 5.0 | 8.3 | 4.5 | 11.8 | 24.8 | 11,840 | 269 | 4,133 | 21,592 | -0.5 |
| Wayne | 14.1 | 3.9 | 2.8 | 10.4 | 2.5 | 9.2 | 4.0 | D | 19.5 | 4,915 | 243 | 813 | 9,560 | 3.7 |
| Webster | 2.1 | 0.1 | 7.7 | D | D | 5.7 | D | D | 18.1 | 3,255 | 335 | 622 | 4,893 | 1.7 |
| Wilkinson | 0.3 | 0.3 | 2.0 | 8.9 | D | 9.7 | 5.0 | D | 34.1 | 2,165 | 251 | 590 | 5,251 | 4.2 |
| Winston | 6.8 | 0.1 | 3.9 | 28.1 | D | 7.3 | 1.0 | D | 12.9 | 5,160 | 287 | 832 | 8,825 | 0.9 |
| Yalobusha | 1.1 | D | 7.4 | 31.9 | D | 4.0 | 3.5 | D | 25.3 | 4,405 | 364 | 872 | 6,611 | 4.2 |
| Yazoo | 7.4 | D | 2.8 | 12.8 | 1.5 | 4.9 | 2.5 | D | 38.1 | 5,680 | 191 | 1,592 | 10,172 | 1.0 |
| MISSOURI | 0.6 | 0.2 | 6.2 | 11.1 | 11.7 | 5.8 | 8.6 | 12.7 | 14.8 | 1,312,639 | 214 | 136,094 | 2,833,720 | 4.5 |
| Adair | 1.3 | 0.0 | 4.0 | 13.0 | 3.0 | 9.3 | 3.7 | D | 23.8 | 4,970 | 195 | 633 | 11,674 | 3.6 |
| Andrew | 11.1 | D | 11.0 | 3.7 | D | 9.3 | 3.1 | D | 21.3 | 3,670 | 207 | 157 | 7,341 | 0.5 |
| Atchison | 23.5 | 0.0 | 3.9 | 0.6 | 5.7 | 5.4 | 6.8 | 12.4 | 15.5 | 1,500 | 294 | 84 | 2,947 | -1.3 |
| Audrain | 4.0 | 0.0 | 4.2 | 32.4 | D | 6.6 | 4.0 | D | 20.1 | 6,005 | 241 | 565 | 10,939 | 0.8 |
| Barry | 1.2 | 0.2 | 5.1 | 31.7 | 18.4 | 6.0 | 4.0 | D | 10.7 | 9,240 | 258 | 933 | 17,690 | 0.9 |
| Barton | 11.0 | 0.0 | 10.6 | 7.4 | D | 7.8 | 5.0 | 15.5 | 14.7 | 3,220 | 276 | 295 | 5,585 | -0.2 |
| Bates | 10.1 | D | 11.1 | 3.8 | 5.0 | 8.0 | D | D | 26.0 | 4,175 | 258 | 392 | 7,839 | 0.0 |
| Benton | 2.5 | D | 9.2 | 6.3 | 2.5 | 9.2 | 7.2 | D | 23.5 | 7,365 | 378 | 623 | 14,293 | 1.0 |
| Bollinger | 1.6 | D | 12.8 | 6.7 | 2.8 | 7.4 | D | D | 21.2 | 3,285 | 271 | 409 | 5,874 | -0.1 |
| Boone | 0.2 | D | 4.2 | 5.0 | 7.3 | 6.2 | 9.3 | 11.0 | 35.5 | 28,570 | 158 | 2,851 | 80,083 | 15.1 |
| Buchanan | 0.4 | D | 7.0 | 27.2 | 5.3 | 6.1 | 5.5 | 15.6 | 12.3 | 19,125 | 219 | 2,418 | 38,936 | 1.3 |
| Butler | 0.6 | D | 4.0 | 9.8 | 5.6 | 9.5 | 5.5 | D | 22.9 | 11,100 | 261 | 2,130 | 19,871 | 0.7 |
| Caldwell | 12.9 | D | 12.8 | 1.8 | D | 16.4 | D | D | 24.0 | 2,210 | 244 | 150 | 4,749 | 3.2 |
| Callaway | -0.9 | 0.3 | 6.2 | 15.1 | 3.5 | 5.9 | 2.3 | D | 21.4 | 10,045 | 224 | 882 | 19,122 | 3.2 |
| Camden | 0.1 | D | 11.6 | 4.0 | 4.7 | 12.4 | 6.8 | 20.7 | 11.0 | 13,505 | 292 | 689 | 42,306 | 2.7 |
| Cape Girardeau | 0.1 | 0.5 | 5.2 | 15.0 | 8.9 | 8.0 | 4.6 | 24.5 | 12.7 | 17,005 | 215 | 1,624 | 35,730 | 9.5 |
| Carroll | 20.0 | D | 13.7 | D | D | 4.0 | 4.5 | D | 13.0 | 2,355 | 271 | 210 | 4,647 | 0.4 |
| Carter | -0.2 | D | 10.4 | 12.9 | D | 6.6 | D | 9.6 | 28.0 | 1,850 | 309 | 315 | 3,263 | 0.5 |
| Cass | 2.1 | 0.3 | 12.1 | 6.9 | 4.9 | 9.9 | 4.7 | 10.0 | 18.1 | 21,490 | 203 | 1,034 | 43,340 | 8.2 |
| Cedar | -0.8 | D | 17.6 | 7.2 | 3.3 | 8.3 | 4.1 | D | 25.3 | 4,370 | 306 | 434 | 7,286 | 0.8 |
| Chariton | 25.1 | 0.0 | 6.6 | 3.6 | D | 8.4 | 6.7 | D | 15.7 | 1,985 | 266 | 155 | 4,153 | -0.2 |
| Christian | 0.2 | D | 15.3 | 7.4 | 6.9 | 9.8 | 6.1 | 6.5 | 17.8 | 18,005 | 203 | 1,081 | 35,989 | 14.0 |
| Clark | 12.7 | D | 6.4 | 8.8 | 2.7 | 10.5 | D | 4.4 | 26.7 | 1,620 | 238 | 115 | 3,503 | 0.9 |
| Clay | 0.0 | D | 6.6 | 17.1 | 12.9 | 7.3 | 4.6 | 8.0 | 14.9 | 43,965 | 175 | 2,698 | 100,352 | 6.9 |
| Clinton | -1.3 | 0.0 | 9.7 | 6.4 | D | 5.8 | 5.4 | D | 23.9 | 4,375 | 214 | 358 | 9,171 | 3.3 |
| Cole | 0.1 | D | 6.7 | 5.0 | 8.3 | 6.5 | 5.9 | 11.3 | 35.9 | 17,235 | 224 | 1,351 | 33,750 | 4.6 |
| Cooper | 5.9 | D | 10.7 | 6.3 | 3.5 | 9.6 | 6.1 | 11.0 | 20.2 | 4,000 | 228 | 343 | 7,497 | 0.5 |
| Crawford | -0.1 | 1.0 | 5.9 | 24.5 | D | 10.4 | 3.2 | 14.1 | 12.3 | 6,565 | 275 | 684 | 12,123 | 1.3 |
| Dade | 10.8 | D | 10.3 | 11.8 | D | 5.1 | D | 2.3 | 22.6 | 2,210 | 293 | 172 | 3,960 | -0.1 |
| Dallas | -1.6 | D | 20.4 | 6.1 | 2.7 | 11.1 | 5.6 | D | 20.9 | 4,675 | 275 | 534 | 7,717 | 0.8 |
| Daviess | 12.8 | D | 17.7 | 7.7 | D | 8.8 | 3.7 | D | 21.1 | 2,040 | 247 | 149 | 4,187 | -0.2 |
| DeKalb | 2.6 | 0.0 | 7.7 | 1.5 | 1.6 | 9.9 | 9.5 | 7.1 | 33.1 | 2,420 | 224 | 90 | 4,378 | 1.2 |
| Dent | 3.2 | D | 5.8 | 13.1 | 2.3 | 9.1 | 5.4 | D | 25.7 | 4,365 | 281 | 582 | 7,324 | 0.4 |
| Douglas | -0.4 | D | 8.6 | D | D | 9.2 | 2.6 | D | 19.0 | 3,865 | 292 | 336 | 6,523 | 0.0 |
| Dunklin | 9.2 | 0.0 | 5.1 | 4.8 | 3.0 | 9.9 | 5.0 | D | 19.0 | 8,005 | 274 | 1,989 | 14,467 | 0.3 |
| Franklin | -0.2 | D | 8.4 | 26.7 | 7.8 | 7.7 | 4.4 | 11.6 | 10.7 | 25,140 | 242 | 1,836 | 46,241 | 6.5 |
| Gasconade | 1.7 | D | 5.3 | 29.4 | D | 7.6 | 4.3 | D | 18.9 | 4,260 | 290 | 274 | 8,173 | -0.3 |
| Gentry | 13.4 | 0.1 | 4.4 | 12.2 | D | 6.4 | D | 22.0 | 14.9 | 1,650 | 251 | 137 | 3,204 | -0.2 |
| Greene | 0.0 | D | 4.6 | 9.3 | 10.7 | 7.0 | 5.8 | 19.6 | 11.7 | 61,515 | 210 | 6,559 | 136,320 | 8.7 |
| Grundy | 9.8 | D | 6.4 | 12.0 | 2.8 | 5.9 | 4.1 | 14.1 | 24.4 | 2,565 | 262 | 259 | 4,998 | -0.5 |
| Harrison | 7.7 | D | 4.4 | 2.9 | 2.7 | 22.5 | 5.3 | 5.7 | 27.7 | 2,260 | 271 | 216 | 4,373 | -0.8 |
| Henry | 2.4 | D | 6.4 | 16.3 | 3.1 | 10.2 | 5.0 | D | 25.0 | 6,770 | 308 | 677 | 10,983 | 0.9 |
| Hickory | 8.7 | D | 8.9 | 1.1 | D | 11.5 | D | 12.4 | 21.4 | 3,520 | 371 | 257 | 6,910 | 1.1 |
| Holt | 27.7 | D | 1.2 | 16.3 | D | 7.8 | 2.9 | 4.8 | 11.7 | 1,240 | 283 | 59 | 2,791 | -0.5 |
| Howard | 6.9 | D | 4.8 | 11.6 | D | 5.9 | 4.5 | D | 16.0 | 2,340 | 234 | 192 | 4,588 | 0.1 |
| Howell | 1.6 | 0.2 | 3.3 | 14.7 | 5.5 | 7.6 | 5.5 | 21.6 | 14.5 | 11,670 | 291 | 1,596 | 18,364 | 1.9 |
| Iron | -0.7 | D | 2.4 | 14.2 | D | 6.0 | 2.0 | 10.7 | 19.6 | 3,075 | 303 | 580 | 5,308 | -0.2 |
| Jackson | 0.0 | 0.0 | 6.2 | 7.0 | 19.8 | 4.4 | 11.4 | 11.8 | 15.1 | 132,005 | 187 | 17,231 | 331,034 | 6.1 |

1. Per 1,000 resident population estimated as of July 1 of the year shown.

# Table B. States and Counties — Housing, Labor Force, and Employment

| STATE County | Housing units, 2015-2019 | | | | | | | | Civilian labor force, 2020 | | | | Civilian employment[6], 2015-2019 | | |
| | Occupied units | | | | | | | | | | Unemployment | | | Percent | |
| | | Owner-occupied | | | | Renter-occupied | | | | | | | | | |
| | | | | Median owner cost as a percent of income | | | Median rent as a percent of income[2] | Substandard units[4] (percent) | | Percent change, 2019-2020 | | | | Management, business, science, and arts | Construction, production, and maintenance occupations |
| | Total | Percent | Median value[1] | With a mortgage | Without a mortgage[2] | Median rent[3] | | | Total | | Total | Rate[5] | Total | | |
| | 89 | 90 | 91 | 92 | 93 | 94 | 95 | 96 | 97 | 98 | 99 | 100 | 101 | 102 | 103 |
|---|---|---|---|---|---|---|---|---|---|---|---|---|---|---|---|
| **MISSISSIPPI— Cont'd** | | | | | | | | | | | | | | | |
| Sunflower | 8,322 | 53.6 | 78,400 | 21.1 | 12.8 | 642 | 31.9 | 4.5 | 7,491 | 0.1 | 840 | 11.2 | 8,212 | 29.3 | 29.8 |
| Tallahatchie | 4,369 | 72.0 | 66,400 | 23.0 | 13.3 | 545 | 27.2 | 2.9 | 5,197 | -3.4 | 387 | 7.4 | 4,340 | 30.1 | 33.7 |
| Tate | 10,324 | 74.8 | 119,200 | 21.6 | 10.7 | 751 | 30.9 | 5.0 | 11,990 | -2.0 | 920 | 7.7 | 12,190 | 27.6 | 34.0 |
| Tippah | 7,834 | 72.3 | 86,900 | 22.2 | 10.1 | 564 | 26.3 | 4.0 | 9,394 | -2.0 | 655 | 7.0 | 8,914 | 25.3 | 42.9 |
| Tishomingo | 7,635 | 75.5 | 88,100 | 20.8 | 10.1 | 564 | 25.1 | 3.7 | 8,399 | -3.5 | 550 | 6.5 | 7,695 | 25.1 | 37.4 |
| Tunica | 3,930 | 40.2 | 107,100 | 25.3 | 11.5 | 785 | 24.7 | 6.0 | 4,590 | 5.9 | 682 | 14.9 | 4,267 | 21.1 | 22.1 |
| Union | 9,808 | 73.1 | 100,600 | 18.6 | 10.0 | 696 | 22.7 | 3.7 | 14,156 | -1.7 | 1,006 | 7.1 | 12,282 | 26.2 | 36.4 |
| Walthall | 5,601 | 87.9 | 95,800 | 23.9 | 14.0 | 588 | 32.5 | 5.9 | 4,997 | 0.5 | 422 | 8.4 | 5,288 | 32.5 | 34.9 |
| Warren | 18,235 | 66.1 | 128,400 | 21.4 | 10.8 | 720 | 32.1 | 2.6 | 20,215 | -1.5 | 1,767 | 8.7 | 19,363 | 33.0 | 26.6 |
| Washington | 17,988 | 52.9 | 75,600 | 21.6 | 11.5 | 681 | 36.0 | 3.6 | 16,182 | -0.4 | 1,785 | 11.0 | 16,375 | 29.7 | 27.0 |
| Wayne | 7,683 | 83.5 | 85,800 | 21.3 | 12.4 | 557 | 24.0 | 2.4 | 7,494 | -0.7 | 651 | 8.7 | 7,893 | 29.2 | 36.4 |
| Webster | 3,782 | 76.3 | 84,000 | 18.7 | 10.0 | 528 | 36.4 | 2.2 | 3,891 | -2.1 | 288 | 7.4 | 3,902 | 29.1 | 30.2 |
| Wilkinson | 3,170 | 79.9 | 70,300 | 30.4 | 14.6 | 561 | 36.4 | 0.3 | 2,777 | -1.6 | 371 | 13.4 | 2,434 | 25.1 | 29.0 |
| Winston | 7,269 | 73.1 | 84,900 | 22.3 | 12.4 | 697 | 29.6 | 2.6 | 7,360 | -1.0 | 637 | 8.7 | 7,109 | 23.4 | 36.4 |
| Yalobusha | 5,181 | 70.0 | 79,700 | 21.9 | 11.8 | 665 | 22.3 | 3.5 | 4,900 | -1.9 | 387 | 7.9 | 5,043 | 22.7 | 41.7 |
| Yazoo | 8,542 | 63.2 | 88,900 | 23.3 | 11.9 | 636 | 32.6 | 5.5 | 9,877 | -0.9 | 960 | 9.7 | 8,876 | 27.4 | 33.2 |
| **MISSOURI** | 2,414,521 | 66.8 | 157,200 | 19.3 | 10.9 | 830 | 27.9 | 2.2 | 3,052,700 | -0.9 | 185,538 | 6.1 | 2,916,000 | 37.0 | 23.6 |
| Adair | 9,258 | 60.1 | 126,800 | 19.2 | 10.1 | 646 | 33.9 | 1.7 | 10,086 | -2.0 | 514 | 5.1 | 10,651 | 36.7 | 24.2 |
| Andrew | 6,763 | 79.9 | 147,900 | 18.9 | 10.9 | 801 | 24.3 | 1.4 | 9,749 | -1.0 | 394 | 4.0 | 8,417 | 34.3 | 29.6 |
| Atchison | 2,562 | 69.7 | 84,900 | 18.1 | 10.8 | 529 | 20.6 | 1.2 | 2,744 | 2.2 | 110 | 4.0 | 2,614 | 32.2 | 29.1 |
| Audrain | 9,286 | 68.5 | 97,900 | 18.4 | 11.3 | 649 | 25.5 | 2.0 | 10,339 | -2.7 | 611 | 5.9 | 10,920 | 26.6 | 35.1 |
| Barry | 13,645 | 73.7 | 120,600 | 19.8 | 11.1 | 666 | 29.7 | 5.7 | 15,692 | 1.2 | 1,019 | 6.5 | 14,677 | 28.0 | 38.0 |
| Barton | 4,895 | 70.1 | 97,500 | 20.3 | 10.5 | 557 | 28.8 | 3.3 | 5,187 | 0.6 | 256 | 4.9 | 5,135 | 34.2 | 30.6 |
| Bates | 6,547 | 72.7 | 114,600 | 19.2 | 11.2 | 654 | 26.0 | 4.4 | 7,662 | -1.3 | 407 | 5.3 | 7,527 | 26.9 | 36.5 |
| Benton | 7,872 | 81.6 | 117,500 | 20.0 | 11.9 | 638 | 28.5 | 2.6 | 7,235 | 1.1 | 526 | 7.3 | 6,710 | 29.8 | 29.3 |
| Bollinger | 4,593 | 80.8 | 110,800 | 19.1 | 12.1 | 647 | 26.3 | 2.6 | 5,409 | -0.6 | 304 | 5.6 | 5,257 | 25.6 | 39.1 |
| Boone | 70,619 | 55.6 | 185,700 | 19.1 | 10.2 | 876 | 31.0 | 2.0 | 96,381 | -1.9 | 4,070 | 4.2 | 93,541 | 45.7 | 15.5 |
| Buchanan | 33,501 | 62.4 | 122,200 | 18.7 | 10.7 | 779 | 26.6 | 2.8 | 43,827 | -0.9 | 2,160 | 4.9 | 41,340 | 29.2 | 32.1 |
| Butler | 16,390 | 63.7 | 113,500 | 21.2 | 11.8 | 675 | 29.5 | 1.8 | 18,183 | 1.5 | 1,151 | 6.3 | 17,421 | 26.1 | 32.1 |
| Caldwell | 3,639 | 76.1 | 109,300 | 18.7 | 12.6 | 631 | 24.2 | 5.3 | 4,279 | -0.2 | 229 | 5.4 | 3,854 | 28.4 | 35.2 |
| Callaway | 15,973 | 72.5 | 157,400 | 18.4 | 10.0 | 708 | 27.1 | 1.7 | 21,366 | 0.1 | 956 | 4.5 | 20,346 | 32.9 | 26.2 |
| Camden | 16,031 | 80.7 | 200,800 | 23.1 | 10.0 | 754 | 28.0 | 1.5 | 18,829 | 0.1 | 1,327 | 7.0 | 18,774 | 26.8 | 25.2 |
| Cape Girardeau | 29,594 | 65.0 | 160,500 | 18.7 | 10.8 | 808 | 27.6 | 1.9 | 40,333 | -0.9 | 2,079 | 5.2 | 38,028 | 36.2 | 23.1 |
| Carroll | 3,503 | 74.3 | 88,200 | 18.9 | 11.1 | 607 | 23.0 | 1.5 | 4,680 | -2.1 | 283 | 6.0 | 3,623 | 31.5 | 34.6 |
| Carter | 2,333 | 77.4 | 101,500 | 18.7 | 10.0 | 621 | 30.3 | 2.1 | 2,524 | 0.0 | 165 | 6.5 | 2,463 | 30.5 | 33.0 |
| Cass | 39,770 | 76.5 | 181,000 | 19.6 | 10.8 | 975 | 28.0 | 2.0 | 54,079 | -1.3 | 2,876 | 5.3 | 51,635 | 35.8 | 25.8 |
| Cedar | 5,842 | 70.7 | 98,900 | 20.7 | 11.5 | 612 | 31.2 | 2.0 | 5,734 | -0.3 | 256 | 4.5 | 5,164 | 33.3 | 30.8 |
| Chariton | 2,686 | 78.3 | 87,900 | 19.7 | 11.2 | 514 | 25.3 | 1.5 | 3,781 | 2.2 | 153 | 4.0 | 3,091 | 32.3 | 34.6 |
| Christian | 31,645 | 74.2 | 169,000 | 19.5 | 10.4 | 816 | 27.6 | 2.5 | 45,241 | -0.3 | 2,233 | 4.9 | 40,970 | 36.1 | 21.7 |
| Clark | 2,657 | 74.9 | 90,000 | 18.1 | 11.2 | 600 | 24.2 | 2.9 | 3,169 | -0.2 | 190 | 6.0 | 3,132 | 25.5 | 42.6 |
| Clay | 91,238 | 68.7 | 174,900 | 19.2 | 11.1 | 965 | 26.7 | 1.7 | 136,568 | -0.4 | 8,374 | 6.1 | 128,142 | 38.2 | 22.6 |
| Clinton | 8,177 | 76.4 | 159,700 | 19.3 | 11.3 | 855 | 23.5 | 1.5 | 10,564 | -0.3 | 628 | 5.9 | 9,658 | 31.8 | 32.2 |
| Cole | 30,154 | 66.3 | 162,300 | 18.6 | 10.0 | 661 | 23.0 | 1.3 | 39,151 | 0.2 | 1,719 | 4.4 | 37,598 | 39.6 | 21.9 |
| Cooper | 6,397 | 72.6 | 132,900 | 18.8 | 10.7 | 664 | 24.3 | 2.3 | 7,219 | -2.7 | 401 | 5.6 | 7,723 | 33.7 | 26.4 |
| Crawford | 9,578 | 71.4 | 124,700 | 21.8 | 11.4 | 624 | 28.8 | 2.4 | 10,852 | -0.8 | 695 | 6.4 | 9,530 | 26.3 | 32.7 |
| Dade | 3,068 | 76.8 | 83,500 | 21.5 | 10.6 | 630 | 28.1 | 3.0 | 3,560 | 0.8 | 155 | 4.4 | 3,045 | 24.4 | 40.5 |
| Dallas | 6,209 | 76.1 | 115,000 | 19.3 | 10.0 | 724 | 29.4 | 3.3 | 7,034 | -0.4 | 414 | 5.9 | 6,255 | 27.6 | 33.9 |
| Daviess | 3,035 | 79.3 | 110,400 | 19.6 | 11.0 | 614 | 27.7 | 2.9 | 4,123 | 1.1 | 175 | 4.2 | 3,540 | 31.4 | 32.1 |
| DeKalb | 3,807 | 69.1 | 120,100 | 17.8 | 11.2 | 656 | 24.0 | 2.3 | 4,791 | -0.5 | 206 | 4.3 | 4,387 | 29.9 | 34.7 |
| Dent | 6,371 | 73.5 | 102,600 | 21.4 | 10.2 | 537 | 28.9 | 3.3 | 6,351 | 1.3 | 306 | 4.8 | 6,378 | 30.3 | 32.3 |
| Douglas | 5,137 | 79.8 | 123,100 | 22.0 | 10.4 | 560 | 27.5 | 3.5 | 5,167 | 1.7 | 399 | 7.7 | 4,984 | 21.9 | 39.7 |
| Dunklin | 12,342 | 62.0 | 76,500 | 18.3 | 11.9 | 576 | 29.2 | 2.5 | 11,562 | 0.7 | 844 | 7.3 | 11,630 | 28.0 | 30.3 |
| Franklin | 40,943 | 74.5 | 165,700 | 19.2 | 10.8 | 746 | 24.6 | 1.7 | 52,661 | -1.6 | 3,265 | 6.2 | 50,296 | 31.0 | 33.5 |
| Gasconade | 6,076 | 78.4 | 134,200 | 19.2 | 10.0 | 611 | 22.5 | 1.7 | 7,606 | -1.5 | 438 | 5.8 | 7,019 | 28.5 | 37.1 |
| Gentry | 2,555 | 73.6 | 88,200 | 19.1 | 11.0 | 612 | 19.6 | 3.2 | 3,405 | -2.3 | 116 | 3.4 | 3,027 | 34.8 | 31.1 |
| Greene | 125,201 | 56.0 | 146,000 | 19.0 | 10.3 | 767 | 29.4 | 3.0 | 150,766 | 0.0 | 8,062 | 5.3 | 139,261 | 36.5 | 20.6 |
| Grundy | 3,936 | 68.2 | 98,400 | 18.6 | 11.6 | 534 | 19.4 | 3.4 | 4,333 | -0.7 | 199 | 4.6 | 4,413 | 31.1 | 32.8 |
| Harrison | 3,429 | 72.7 | 76,000 | 19.6 | 12.1 | 531 | 23.6 | 3.6 | 3,811 | 0.7 | 172 | 4.5 | 3,637 | 32.1 | 31.6 |
| Henry | 9,328 | 72.2 | 110,500 | 19.4 | 11.7 | 639 | 28.3 | 3.4 | 9,673 | -0.2 | 567 | 5.9 | 9,038 | 29.5 | 32.3 |
| Hickory | 3,976 | 82.2 | 83,400 | 25.5 | 11.5 | 608 | 28.0 | 3.1 | 3,879 | 1.3 | 196 | 5.1 | 3,183 | 23.2 | 36.4 |
| Holt | 2,022 | 76.2 | 88,600 | 16.8 | 12.4 | 510 | 17.8 | 0.6 | 2,578 | 0.1 | 87 | 3.4 | 2,017 | 29.3 | 31.6 |
| Howard | 3,471 | 78.2 | 130,100 | 19.9 | 10.0 | 601 | 24.3 | 1.9 | 4,822 | -0.9 | 200 | 4.1 | 4,648 | 36.8 | 26.4 |
| Howell | 15,878 | 68.4 | 112,200 | 20.5 | 10.6 | 598 | 26.5 | 3.3 | 16,201 | -1.3 | 1,083 | 6.7 | 16,431 | 32.2 | 28.0 |
| Iron | 4,074 | 70.8 | 83,600 | 17.8 | 10.7 | 597 | 28.6 | 3.7 | 3,500 | -0.3 | 246 | 7.0 | 3,826 | 20.4 | 32.4 |
| Jackson | 286,601 | 58.2 | 147,400 | 19.5 | 12.2 | 910 | 28.4 | 2.1 | 360,213 | -0.3 | 25,742 | 7.1 | 350,241 | 37.4 | 22.4 |

1. Specified owner-occupied units.  2. A value of 10.0 represents 10 percent or less; a value of 50.0 represents 50 percent or more.  3. Specified renter-occupied units.  4. Overcrowded or lacking complete plumbing facilities.  5. Percent of civilian labor force.  6. Civilian employed persons 16 years old and over.

# Table B. States and Counties — Nonfarm Employment and Agriculture

| STATE County | Private nonfarm establishments, employment and payroll, 2019 | | | | | | | | | Agriculture, 2017 | | | |
| | Employment | | | | | | | Annual payroll | | Farms | | | Farm producers whose primary occupation is farming (percent) |
| | Number of establishments | Total | Health care and social assistance | Manufacturing | Retail trade | Finance and insurance | Professional, scientific, and technical services | Total (mil dol) | Average per employee (dollars) | Number | Percent with: Fewer than 50 acres | 1000 acres or more | |
| | 104 | 105 | 106 | 107 | 108 | 109 | 110 | 111 | 112 | 113 | 114 | 115 | 116 |
|---|---|---|---|---|---|---|---|---|---|---|---|---|---|
| MISSISSIPPI— Cont'd | | | | | | | | | | | | | |
| Sunflower | 427 | 5,382 | 1,567 | 294 | 896 | 214 | 50 | 183 | 33,997 | 311 | 17.7 | 42.1 | 51.3 |
| Tallahatchie | 159 | 1,674 | 517 | NA | 213 | 24 | 36 | 59 | 35,364 | 436 | 14.7 | 17.7 | 40.3 |
| Tate | 386 | 3,754 | 616 | 402 | 806 | 204 | 84 | 125 | 33,202 | 593 | 33.6 | 5.9 | 40.1 |
| Tippah | 378 | 6,195 | 632 | 2,323 | 1,218 | 142 | 89 | 204 | 32,864 | 557 | 25.7 | 3.1 | 33.5 |
| Tishomingo | 369 | 4,905 | 584 | 2,213 | 634 | 132 | 48 | 167 | 34,079 | 274 | 25.2 | 2.9 | 28.2 |
| Tunica | 189 | 5,891 | 226 | 519 | 231 | 66 | 20 | 199 | 33,850 | 91 | 15.4 | 54.9 | 80.5 |
| Union | 504 | 10,485 | 1,280 | 4,314 | 1,129 | 221 | 89 | 390 | 37,213 | 618 | 33.8 | 3.2 | 27.2 |
| Walthall | 210 | 2,046 | 541 | 359 | 288 | 61 | 20 | 72 | 35,139 | 635 | 33.5 | 1.1 | 46.7 |
| Warren | 999 | 15,619 | 2,458 | 2,949 | 2,388 | 365 | 342 | 596 | 38,175 | 160 | 20.0 | 15.6 | 29.7 |
| Washington | 1,088 | 12,974 | 3,052 | 1,186 | 2,318 | 315 | 304 | 456 | 35,135 | 273 | 12.8 | 37.7 | 66.7 |
| Wayne | 370 | 4,268 | 668 | 892 | 815 | 201 | 90 | 154 | 36,123 | 562 | 40.9 | 2.5 | 41.4 |
| Webster | 154 | 1,706 | 480 | 235 | 302 | 62 | 65 | 61 | 35,704 | 292 | 21.6 | 5.5 | 30.8 |
| Wilkinson | 127 | 1,264 | 289 | 124 | 238 | 27 | 15 | 38 | 30,137 | 163 | 28.2 | 16.6 | 23.4 |
| Winston | 376 | 5,291 | 705 | 1,896 | 838 | 82 | 78 | 215 | 40,595 | 483 | 32.3 | 3.5 | 36.9 |
| Yalobusha | 155 | 2,547 | 403 | 990 | 268 | 75 | 11 | 85 | 33,566 | 348 | 21.0 | 3.7 | 33.7 |
| Yazoo | 365 | 4,087 | 1,034 | 603 | 834 | 119 | 136 | 145 | 35,360 | 574 | 17.2 | 14.6 | 32.7 |
| MISSOURI | 151,816 | 2,547,310 | 419,551 | 277,464 | 308,085 | 138,320 | 165,066 | 125,302 | 49,190 | 95,320 | 29.6 | 6.2 | 38.8 |
| Adair | 578 | 8,324 | 1,913 | 1,038 | 1,627 | 197 | 141 | 269 | 32,320 | 816 | 27.5 | 6.0 | 33.4 |
| Andrew | 284 | 1,677 | 303 | 46 | 322 | 56 | 61 | 57 | 33,735 | 706 | 32.6 | 7.8 | 40.8 |
| Atchison | 185 | 1,159 | 334 | 8 | 254 | 79 | 32 | 41 | 35,385 | 401 | 19.0 | 23.7 | 54.7 |
| Audrain | 508 | 6,806 | 990 | 1,774 | 968 | 226 | 141 | 258 | 37,969 | 911 | 24.6 | 12.4 | 47.1 |
| Barry | 721 | 12,589 | 1,028 | 5,287 | 1,622 | 297 | 1,151 | 543 | 43,115 | 1,392 | 34.2 | 2.9 | 41.0 |
| Barton | 235 | 2,371 | 449 | 227 | 475 | 133 | 95 | 77 | 32,357 | 865 | 25.7 | 10.3 | 44.5 |
| Bates | 355 | 2,937 | 828 | 128 | 581 | 150 | 109 | 95 | 32,383 | 1,160 | 30.0 | 9.6 | 43.1 |
| Benton | 383 | 2,180 | 364 | 93 | 651 | 119 | 78 | 57 | 26,213 | 749 | 21.5 | 5.5 | 47.2 |
| Bollinger | 210 | 1,483 | 320 | 191 | 300 | 50 | 29 | 44 | 29,970 | 756 | 23.3 | 4.0 | 39.6 |
| Boone | 4,659 | 78,049 | 18,809 | 4,564 | 11,274 | 6,910 | 4,172 | 3,383 | 43,350 | 1,184 | 47.1 | 3.0 | 30.5 |
| Buchanan | 2,102 | 44,179 | 8,029 | 12,124 | 5,678 | 1,587 | 1,286 | 2,067 | 46,780 | 797 | 40.4 | 6.0 | 35.8 |
| Butler | 1,032 | 14,097 | 4,257 | 1,991 | 2,700 | 562 | 320 | 516 | 36,578 | 441 | 30.8 | 20.6 | 47.1 |
| Caldwell | 156 | 1,317 | 123 | 25 | 637 | 56 | 18 | 45 | 34,259 | 924 | 32.5 | 6.0 | 34.0 |
| Callaway | 726 | 11,973 | 2,320 | 1,936 | 1,221 | 270 | 308 | 546 | 45,613 | 1,438 | 30.0 | 3.5 | 30.9 |
| Camden | 1,447 | 13,743 | 2,682 | 483 | 3,115 | 432 | 701 | 485 | 35,313 | 516 | 20.3 | 2.3 | 40.6 |
| Cape Girardeau | 2,307 | 37,774 | 10,721 | 3,449 | 5,938 | 989 | 1,251 | 1,510 | 39,986 | 1,111 | 32.3 | 5.1 | 41.8 |
| Carroll | 203 | 1,603 | 399 | 184 | 249 | 130 | 24 | 55 | 34,289 | 1,016 | 18.0 | 10.1 | 39.8 |
| Carter | 157 | 1,095 | 296 | 233 | 175 | 52 | 7 | 27 | 24,347 | 160 | 31.9 | 11.3 | 37.3 |
| Cass | 2,005 | 21,826 | 3,259 | 1,281 | 4,777 | 573 | 778 | 784 | 35,924 | 1,477 | 47.1 | 3.7 | 34.9 |
| Cedar | 278 | 2,402 | 518 | 363 | 449 | 94 | 67 | 70 | 29,019 | 854 | 26.3 | 2.9 | 44.0 |
| Chariton | 205 | 1,276 | 294 | 86 | 236 | 78 | 28 | 42 | 32,807 | 985 | 21.9 | 9.8 | 41.4 |
| Christian | 1,872 | 15,145 | 1,910 | 1,362 | 2,791 | 563 | 716 | 493 | 32,572 | 1,169 | 42.2 | 0.8 | 40.4 |
| Clark | 152 | 1,076 | 90 | 99 | 262 | 77 | 8 | 27 | 24,694 | 547 | 19.4 | 9.1 | 37.2 |
| Clay | 5,229 | 109,650 | 14,308 | 15,553 | 12,259 | 2,808 | 20,757 | 6,168 | 56,249 | 552 | 55.6 | 6.0 | 32.4 |
| Clinton | 355 | 2,715 | 843 | 153 | 487 | 140 | 91 | 100 | 36,972 | 684 | 41.1 | 8.8 | 36.4 |
| Cole | 2,207 | 36,498 | 6,649 | 2,766 | 5,206 | 1,872 | 1,543 | 1,531 | 41,945 | 1,169 | 30.2 | 0.8 | 32.1 |
| Cooper | 374 | 3,709 | 799 | 214 | 776 | 112 | 57 | 122 | 32,978 | 883 | 22.2 | 8.3 | 34.3 |
| Crawford | 487 | 5,563 | 857 | 1,916 | 675 | 133 | 125 | 212 | 38,020 | 628 | 21.8 | 3.7 | 34.6 |
| Dade | 138 | 1,177 | 25 | 317 | 180 | 40 | 27 | 35 | 29,616 | 699 | 22.2 | 9.7 | 50.9 |
| Dallas | 293 | 1,804 | 299 | 188 | 433 | 128 | 53 | 47 | 26,144 | 1,176 | 34.3 | 1.8 | 40.2 |
| Daviess | 153 | 999 | 133 | 201 | 212 | 53 | 33 | 29 | 29,397 | 1,015 | 20.8 | 6.1 | 31.5 |
| DeKalb | 216 | 2,180 | 455 | 46 | 562 | 180 | 25 | 81 | 37,171 | 708 | 28.2 | 5.8 | 33.2 |
| Dent | 325 | 3,515 | 803 | 648 | 478 | 177 | 45 | 130 | 36,984 | 694 | 27.4 | 5.8 | 37.7 |
| Douglas | 216 | 2,140 | 303 | 622 | 401 | 59 | 45 | 57 | 26,418 | 994 | 22.8 | 3.8 | 46.3 |
| Dunklin | 646 | 6,524 | 1,880 | 274 | 1,469 | 230 | 107 | 184 | 28,225 | 283 | 25.1 | 34.3 | 54.8 |
| Franklin | 2,627 | 36,834 | 4,518 | 10,347 | 4,880 | 1,059 | 1,052 | 1,537 | 41,738 | 1,818 | 42.2 | 2.3 | 30.5 |
| Gasconade | 399 | 5,386 | 917 | 1,910 | 611 | 151 | 81 | 148 | 27,463 | 823 | 17.4 | 2.3 | 36.0 |
| Gentry | 184 | 1,828 | 797 | 230 | 264 | 72 | 27 | 50 | 27,304 | 686 | 22.4 | 7.9 | 40.1 |
| Greene | 8,503 | 162,713 | 31,976 | 13,355 | 19,402 | 7,255 | 7,253 | 6,986 | 42,937 | 1,857 | 51.8 | 1.2 | 34.5 |
| Grundy | 224 | 1,962 | 573 | 185 | 404 | 65 | 68 | 56 | 28,617 | 662 | 24.5 | 8.5 | 36.3 |
| Harrison | 199 | 1,816 | 518 | 49 | 541 | 127 | 29 | 44 | 24,048 | 974 | 21.0 | 7.8 | 40.1 |
| Henry | 574 | 6,753 | 2,189 | 1,023 | 1,278 | 205 | 111 | 250 | 37,059 | 898 | 24.1 | 11.7 | 45.7 |
| Hickory | 134 | 852 | 220 | 14 | 321 | 39 | 22 | 22 | 26,112 | 529 | 17.4 | 4.2 | 46.4 |
| Holt | 129 | 902 | 184 | 182 | 108 | 41 | 13 | 33 | 36,551 | 380 | 24.2 | 14.2 | 50.5 |
| Howard | 181 | 2,493 | 592 | 299 | 259 | 51 | 64 | 65 | 26,100 | 690 | 16.8 | 6.1 | 36.4 |
| Howell | 1,108 | 12,607 | 3,411 | 2,402 | 2,097 | 380 | 323 | 439 | 34,822 | 1,451 | 33.7 | 2.4 | 39.8 |
| Iron | 218 | 1,876 | 638 | 74 | 309 | 64 | 13 | 65 | 34,727 | 270 | 24.8 | 4.1 | 35.1 |
| Jackson | 18,118 | 344,993 | 56,414 | 27,435 | 35,941 | 25,202 | 27,736 | 19,235 | 55,755 | 706 | 67.1 | 3.5 | 30.1 |

# Table B. States and Counties — **Agriculture**

| STATE County | Acreage (1,000) | Percent change, 2012-2017 | Average size of farm | Total irrigated (1,000) | Total cropland (1,000) | Value of land and buildings (dollars) Average per farm | Average per acre | Value of machinery and equipmnet, average per farm (dollars) | Total (mil dol) | Average per farm (acres) | Crops | Livestock and poultry products | Organic farms (number) | Farms with internet access (percent) | Government payments Total ($1,000) | Percent of farms |
|---|---|---|---|---|---|---|---|---|---|---|---|---|---|---|---|---|
| | 117 | 118 | 119 | 120 | 121 | 122 | 123 | 124 | 125 | 126 | 127 | 128 | 129 | 130 | 131 | 132 |
| MISSISSIPPI— Cont'd | | | | | | | | | | | | | | | | |
| Sunflower | 389 | 4.3 | 1,249 | 217.7 | 325.9 | 3,977,773 | 3,184 | 518,720 | 223.8 | 719,566 | 86.2 | 13.8 | NA | 72.3 | 14,647 | 80.7 |
| Tallahatchie | 310 | -9.1 | 710 | 130.4 | 234.5 | 2,002,926 | 2,820 | 212,704 | 134.9 | 309,404 | 99.0 | 1.0 | NA | 61.9 | 7,579 | 72.2 |
| Tate | 158 | 3.0 | 266 | 4.7 | 67.5 | 751,443 | 2,825 | 93,215 | 40.1 | 67,575 | 71.1 | 28.9 | NA | 72.2 | 2,548 | 37.1 |
| Tippah | 109 | -12.1 | 195 | 0.2 | 39.0 | 404,978 | 2,072 | 80,929 | 23.2 | 41,655 | 58.9 | 41.1 | NA | 62.3 | 1,086 | 64.3 |
| Tishomingo | 47 | -4.7 | 172 | 0.3 | 19.5 | 325,832 | 1,890 | 55,001 | 7.0 | 25,445 | 90.2 | 9.8 | NA | 61.3 | 484.0 | 51.5 |
| Tunica | 186 | -12.2 | 2,041 | 99.0 | 177.1 | 6,030,602 | 2,955 | 663,152 | D | D | D | D | NA | 64.8 | 5,159 | 86.8 |
| Union | 112 | -7.5 | 181 | 0.0 | 45.4 | 362,278 | 1,999 | 60,172 | 16.3 | 26,299 | 74.7 | 25.3 | NA | 65.4 | 1,906 | 57.8 |
| Walthall | 100 | -15.2 | 158 | 0.1 | 30.1 | 489,851 | 3,104 | 76,957 | 85.3 | 134,307 | 4.9 | 95.1 | 6 | 55.7 | 1,002 | 32.0 |
| Warren | 99 | -20.3 | 621 | 9.3 | 40.4 | 1,508,216 | 2,427 | 174,355 | 18.8 | 117,469 | 95.2 | 4.8 | NA | 68.8 | 2,283 | 59.4 |
| Washington | 371 | 8.2 | 1,357 | 238.6 | 349.2 | 5,033,096 | 3,708 | 634,141 | D | D | D | D | NA | 77.3 | 12,792 | 86.1 |
| Wayne | 97 | 4.2 | 173 | 0.6 | 27.8 | 536,506 | 3,096 | 84,892 | 242.6 | 431,692 | 3.1 | 96.9 | 2 | 69.2 | 759.0 | 35.8 |
| Webster | 74 | -8 | 254 | 0.4 | 29.6 | 464,975 | 1,833 | 90,512 | 19.7 | 67,548 | 76.6 | 23.4 | NA | 58.2 | 1,211 | 55.8 |
| Wilkinson | 87 | -15.8 | 532 | D | 11.7 | 1,278,484 | 2,401 | 86,434 | 4.6 | 28,258 | 48.4 | 51.6 | NA | 70.6 | 345.0 | 26.4 |
| Winston | 108 | 10.2 | 223 | 0.6 | 17.4 | 507,990 | 2,277 | 66,860 | 80.1 | 165,925 | 3.7 | 96.3 | NA | 64.0 | 804.0 | 41.4 |
| Yalobusha | 81 | -14.2 | 233 | 2.8 | 26.5 | 434,275 | 1,864 | 84,136 | 13.9 | 40,009 | 88.3 | 11.7 | NA | 56.6 | 741.0 | 42.5 |
| Yazoo | 308 | -12.3 | 536 | 62.8 | 167.5 | 1,495,962 | 2,789 | 170,781 | 105.2 | 183,207 | 84.9 | 15.1 | NA | 63.1 | 8,817 | 70.9 |
| MISSOURI | 27,782 | -1.7 | 291 | 1,529.2 | 15,599.4 | 986,481 | 3,385 | 104,066 | 10,525.9 | 110,427 | 52.0 | 48.0 | 415 | 72.5 | 323,801 | 32.8 |
| Adair | 268 | -1.9 | 328 | 0.1 | 141.9 | 916,811 | 2,792 | 94,641 | 52.8 | 64,721 | 65.3 | 34.7 | 1 | 68.3 | 2,179 | 36.8 |
| Andrew | 205 | 3.2 | 290 | 1.2 | 158.6 | 1,115,611 | 3,843 | 113,518 | 75.0 | 106,218 | 85.2 | 14.8 | NA | 82.3 | 3,325 | 55.4 |
| Atchison | 302 | 15.0 | 754 | 19.9 | 275.1 | 3,698,611 | 4,903 | 367,596 | 147.8 | 368,566 | 97.2 | 2.8 | NA | 81.3 | 5,332 | 78.1 |
| Audrain | 405 | -7.1 | 445 | 17.0 | 334.6 | 1,894,323 | 4,256 | 167,943 | 247.1 | 271,233 | 61.4 | 38.6 | 31 | 68.4 | 5,722 | 58.6 |
| Barry | 290 | 8.0 | 208 | 4.3 | 96.9 | 713,417 | 3,429 | 85,578 | 403.1 | 289,563 | 4.2 | 95.8 | 1 | 74.8 | 391.0 | 4.9 |
| Barton | 331 | -0.4 | 383 | 21.7 | 221.5 | 1,053,351 | 2,753 | 167,455 | 132.0 | 152,652 | 61.0 | 39.0 | NA | 74.1 | 6,312 | 48.6 |
| Bates | 460 | 2.5 | 396 | 3.6 | 281.7 | 1,224,243 | 3,090 | 137,790 | 159.8 | 137,752 | 63.3 | 36.7 | 3 | 71.6 | 4,732 | 49.1 |
| Benton | 224 | -7 | 299 | 0.7 | 77.4 | 817,792 | 2,737 | 78,088 | 82.7 | 110,463 | 19.5 | 80.5 | 2 | 72.9 | 785.0 | 21.6 |
| Bollinger | 180 | -10 | 238 | 13.1 | 77.9 | 597,725 | 2,507 | 75,632 | 32.1 | 42,450 | 56.9 | 43.1 | NA | 62.7 | 1,470 | 37.0 |
| Boone | 213 | -11.6 | 180 | 3.8 | 128.5 | 1,015,839 | 5,654 | 74,663 | 105.0 | 88,688 | 44.2 | 55.8 | 4 | 82.4 | 1,827 | 20.6 |
| Buchanan | 184 | -2.5 | 231 | 1.2 | 146.3 | 936,040 | 4,053 | 107,515 | 66.9 | 83,923 | 89.3 | 10.7 | NA | 70.3 | 2,939 | 45.5 |
| Butler | 242 | 3.3 | 548 | 154.3 | 207.0 | 2,573,693 | 4,695 | 233,342 | 112.8 | 255,841 | 98.6 | 1.4 | NA | 81.2 | 9,945 | 42.0 |
| Caldwell | 250 | 2.1 | 270 | D | 168.1 | 856,166 | 3,170 | 81,607 | 67.2 | 72,720 | 70.9 | 29.1 | 1 | 71.9 | 5,068 | 57.8 |
| Callaway | 297 | -6.2 | 206 | 8.8 | 166.6 | 816,689 | 3,960 | 97,888 | 124.5 | 86,602 | 47.6 | 52.4 | 4 | 81.1 | 3,457 | 27.6 |
| Camden | 123 | -11 | 239 | 0.0 | 25.9 | 527,680 | 2,208 | 59,654 | 15.1 | 29,250 | 11.3 | 88.7 | NA | 66.1 | 243.0 | 4.1 |
| Cape Girardeau | 290 | 14.6 | 261 | 36.8 | 208.5 | 1,132,758 | 4,343 | 120,048 | 99.3 | 89,351 | 75.8 | 24.2 | NA | 69.6 | 4,195 | 48.2 |
| Carroll | 426 | -1.4 | 419 | 3.9 | 333.2 | 1,500,209 | 3,580 | 151,887 | 144.7 | 142,399 | 87.4 | 12.6 | 4 | 72.1 | 10,429 | 79.1 |
| Carter | 72 | -2.7 | 448 | D | 9.1 | 854,925 | 1,909 | 68,650 | 3.8 | 23,844 | 11.2 | 88.8 | NA | 81.3 | 73.0 | 11.9 |
| Cass | 317 | -0.7 | 215 | 3.5 | 201.7 | 807,084 | 3,759 | 76,607 | 120.5 | 81,576 | 82.0 | 18.0 | 2 | 74.3 | 3,599 | 27.6 |
| Cedar | 207 | 9.4 | 243 | 0.8 | 62.2 | 604,805 | 2,489 | 65,465 | 49.5 | 58,000 | 13.4 | 86.6 | 13 | 69.7 | 665.0 | 13.8 |
| Chariton | 388 | -4.5 | 394 | 1.0 | 273.9 | 1,354,347 | 3,439 | 155,013 | 162.8 | 165,317 | 60.4 | 39.6 | 1 | 67.5 | 5,119 | 54.1 |
| Christian | 154 | -14.2 | 132 | 0.1 | 43.9 | 530,251 | 4,027 | 54,661 | 28.9 | 24,687 | 18.0 | 82.0 | NA | 72.3 | 105.0 | 2.5 |
| Clark | 256 | 6.2 | 468 | 4.3 | 193.8 | 1,673,148 | 3,575 | 177,466 | 99.8 | 182,404 | 77.0 | 23.0 | NA | 68.7 | 2,967 | 66.0 |
| Clay | 111 | 0.5 | 201 | 4.5 | 64.4 | 839,612 | 4,169 | 90,403 | 34.7 | 62,951 | 55.6 | 44.4 | NA | 81.3 | 1,272 | 18.7 |
| Clinton | 222 | 16.1 | 325 | 0.0 | 150.2 | 1,209,500 | 3,721 | 128,331 | 81.8 | 119,642 | 78.8 | 21.2 | NA | 77.3 | 2,035 | 33.3 |
| Cole | 186 | 5.3 | 159 | 0.4 | 70.5 | 578,816 | 3,646 | 69,979 | 36.8 | 31,506 | 36.0 | 64.0 | 6 | 73.7 | 919.0 | 26.3 |
| Cooper | 282 | -8.2 | 319 | 0.2 | 167.3 | 1,043,246 | 3,268 | 121,276 | 97.5 | 110,428 | 64.1 | 35.9 | 9 | 72.6 | 3,451 | 56.6 |
| Crawford | 160 | -17.6 | 255 | 0.4 | 37.7 | 658,502 | 2,583 | 58,391 | 14.8 | 23,556 | 25.2 | 74.8 | 1 | 73.1 | 261.0 | 8.9 |
| Dade | 266 | 8.2 | 380 | 3.7 | 113.5 | 1,067,218 | 2,807 | 117,152 | 70.2 | 100,418 | 42.2 | 57.8 | 3 | 66.2 | 1,928 | 25.0 |
| Dallas | 207 | -5.1 | 176 | 0.2 | 57.9 | 446,303 | 2,538 | 56,030 | 51.3 | 43,648 | 9.5 | 90.5 | 28 | 66.3 | 425.0 | 5.4 |
| Daviess | 307 | -2.7 | 302 | 0.6 | 192.8 | 985,987 | 3,265 | 98,261 | 131.1 | 129,122 | 36.7 | 63.3 | 9 | 64.5 | 7,048 | 65.2 |
| DeKalb | 202 | -17 | 285 | D | 133.5 | 932,047 | 3,273 | 89,075 | 64.8 | 91,479 | 65.4 | 34.6 | 1 | 76.4 | 3,612 | 54.5 |
| Dent | 190 | 0.8 | 273 | 0.3 | 27.3 | 571,102 | 2,091 | 62,969 | 21.8 | 31,438 | 11.8 | 88.2 | 7 | 79.0 | 204.0 | 9.7 |
| Douglas | 267 | 5.0 | 268 | 0.1 | 46.6 | 562,982 | 2,100 | 60,494 | 33.8 | 33,970 | 4.4 | 95.6 | 1 | 72.8 | 220.0 | 2.7 |
| Dunklin | 283 | 1.1 | 1,000 | 179.3 | 277.9 | 5,087,191 | 5,090 | 465,745 | 196.6 | 694,749 | 98.4 | 1.6 | NA | 75.3 | 7,431 | 68.9 |
| Franklin | 266 | -8.8 | 146 | 1.6 | 120.1 | 565,029 | 3,864 | 61,725 | 60.0 | 32,980 | 43.6 | 56.4 | 2 | 74.0 | 1,172 | 18.6 |
| Gasconade | 207 | -0.8 | 252 | 0.1 | 75.4 | 701,975 | 2,787 | 82,115 | 32.3 | 39,273 | 46.3 | 53.7 | 3 | 72.2 | 1,118 | 31.1 |
| Gentry | 239 | -5.8 | 348 | 0.0 | 152.0 | 1,077,559 | 3,098 | 118,270 | 120.2 | 175,265 | 31.3 | 68.7 | 22 | 75.7 | 4,896 | 69.2 |
| Greene | 223 | 6.0 | 120 | 0.5 | 78.3 | 570,292 | 4,745 | 53,364 | 38.7 | 20,854 | 21.2 | 78.8 | 2 | 74.3 | 807.0 | 6.1 |
| Grundy | 225 | 10.5 | 341 | 2.1 | 160.2 | 997,041 | 2,928 | 110,989 | 92.8 | 140,210 | 53.0 | 47.0 | 8 | 71.6 | 4,553 | 58.2 |
| Harrison | 392 | -2.2 | 403 | D | 259.3 | 1,114,054 | 2,768 | 101,414 | 92.7 | 95,181 | 74.1 | 25.9 | NA | 71.4 | 10,309 | 67.0 |
| Henry | 382 | 6.8 | 425 | 3.3 | 213.4 | 1,143,536 | 2,690 | 129,133 | 98.7 | 109,924 | 59.9 | 40.1 | 3 | 73.7 | 4,645 | 42.0 |
| Hickory | 164 | -9.7 | 310 | D | 50.9 | 654,352 | 2,108 | 79,880 | 29.9 | 56,609 | 18.8 | 81.2 | 2 | 67.9 | 301.0 | 14.6 |
| Holt | 209 | 4.2 | 550 | 31.9 | 188.3 | 2,688,737 | 4,893 | 253,959 | 108.8 | 286,195 | 96.4 | 3.6 | 4 | 76.3 | 4,386 | 75.0 |
| Howard | 219 | -10.2 | 317 | 4.6 | 133.3 | 957,809 | 3,023 | 115,515 | 55.9 | 81,077 | 76.1 | 23.9 | 2 | 71.4 | 2,818 | 59.1 |
| Howell | 333 | -5.8 | 230 | 0.6 | 54.7 | 492,355 | 2,144 | 62,506 | 56.9 | 39,193 | 7.8 | 92.2 | NA | 78.2 | 285.0 | 4.6 |
| Iron | 65 | -7.4 | 242 | D | 12.9 | 499,202 | 2,065 | 50,579 | 4.3 | 16,052 | 8.7 | 91.3 | NA | 75.2 | 32.0 | 4.1 |
| Jackson | 106 | -4.5 | 150 | 1.0 | 73.3 | 872,579 | 5,814 | 73,951 | 37.6 | 53,244 | 80.9 | 19.1 | 21 | 79.7 | 702.0 | 18.8 |

# Table B. States and Counties — Water Use, Wholesale Trade, Retail Trade, and Real Estate

| STATE County | Water use, 2015 | | Wholesale Trade[1], 2017 | | | | Retail Trade[2], 2017 | | | | Real estate and rental and leasing,[2] 2017 | | | |
|---|---|---|---|---|---|---|---|---|---|---|---|---|---|---|
| | Public supply water withdrawn (mil gal/day) | Public supply gallons withdrawn per person per day | Number of establishments | Number of employees | Sales (mil dol) | Average payroll (mil dol) | Number of establishments | Number of employees | Sales (mil dol) | Average payroll (mil dol) | Number of establishments | Number of employees | Sales (mil dol) | Average payroll (mil dol) |
| | 133 | 134 | 135 | 136 | 137 | 138 | 139 | 140 | 141 | 142 | 143 | 144 | 145 | 146 |
| MISSISSIPPI— Cont'd | | | | | | | | | | | | | | |
| Sunflower | 3.01 | 111.5 | 20 | 396 | 427.1 | 15.4 | 89 | 890 | 206.1 | 18.3 | 11 | 24 | 4.3 | 0.8 |
| Tallahatchie | 1.31 | 89.8 | 7 | D | 44.2 | D | 36 | 234 | 45.8 | 6.3 | 6 | D | 1.0 | D |
| Tate | 1.81 | 64.0 | 12 | 59 | 30.4 | 3.4 | 79 | 812 | 207.8 | 19.6 | 13 | 30 | 7.3 | 0.8 |
| Tippah | 2.69 | 121.5 | 18 | 134 | 287.5 | 9.3 | 74 | 629 | 143.7 | 14.0 | 6 | 47 | 2.6 | 0.7 |
| Tishomingo | 2.55 | 130.4 | 18 | 152 | 27.9 | 4.0 | 86 | 681 | 141.5 | 16.2 | 9 | D | 2.2 | D |
| Tunica | 3.10 | 299.7 | 9 | 95 | 159.8 | 4.9 | 47 | 253 | 60.8 | 5.6 | 9 | 21 | 3.5 | 0.5 |
| Union | 2.57 | 90.4 | 18 | 241 | 462.0 | 9.7 | 103 | 1,115 | 286.7 | 25.7 | 13 | 31 | 5.3 | 1.0 |
| Walthall | 2.64 | 180.4 | D | D | D | 1.6 | 40 | 325 | 73.0 | 7.1 | 5 | D | 2.6 | D |
| Warren | 7.99 | 168.3 | 37 | 303 | 354.3 | 14.4 | 189 | 2,486 | 672.1 | 57.1 | 35 | 137 | 30.2 | 4.6 |
| Washington | 11.55 | 240.0 | 57 | 692 | 780.8 | 36.6 | 219 | 2,612 | 591.6 | 55.1 | 41 | 130 | 22.7 | 4.5 |
| Wayne | 2.62 | 127.4 | 25 | 191 | 249.7 | 13.9 | 81 | 831 | 180.1 | 18.2 | 7 | 24 | 4.7 | 0.7 |
| Webster | 1.40 | 141.4 | D | D | D | D | 35 | 257 | 51.7 | 5.0 | NA | NA | NA | NA |
| Wilkinson | 1.02 | 111.8 | 7 | D | 27.2 | D | 23 | 245 | 51.1 | 5.4 | NA | NA | NA | NA |
| Winston | 2.00 | 109.2 | 11 | 140 | 125.6 | 7.0 | 84 | 859 | 200.7 | 20.2 | 23 | 81 | 13.0 | 2.3 |
| Yalobusha | 2.22 | 178.4 | NA | NA | NA | NA | 43 | 311 | 58.0 | 5.3 | NA | NA | NA | NA |
| Yazoo | 5.49 | 200.5 | 23 | 243 | 172.1 | 11.1 | 85 | 936 | 165.3 | 16.7 | 17 | 51 | 7.3 | 1.7 |
| MISSOURI | 797.09 | 131.0 | 6,293 | 95,156 | 102,651.6 | 5,631.7 | 20,694 | 312,616 | 100,394.0 | 8,147.5 | 6,644 | 37,144 | 8,975.9 | 1,625.7 |
| Adair | 2.50 | 98.5 | D | D | D | D | 102 | 1,503 | 307.4 | 31.0 | 18 | 51 | 12.2 | 1.9 |
| Andrew | 18.06 | 1,044.2 | D | D | D | 4.1 | 37 | 364 | 146.6 | 11.9 | 11 | 18 | 2.4 | 0.4 |
| Atchison | 0.49 | 92.3 | 12 | 139 | 171.2 | 4.9 | 34 | 374 | 73.4 | 11.0 | 5 | D | 0.8 | D |
| Audrain | 2.16 | 82.8 | 21 | 332 | 220.3 | 11.4 | 88 | 1,023 | 262.7 | 23.5 | 10 | 32 | 4.1 | 1.0 |
| Barry | 4.57 | 127.6 | 30 | 450 | 481.9 | 31.4 | 137 | 1,632 | 397.0 | 39.1 | 24 | 72 | 10.5 | 2.1 |
| Barton | 1.83 | 154.0 | 11 | 180 | 231.9 | 5.9 | 33 | 467 | 112.6 | 11.1 | D | D | D | D |
| Bates | 1.21 | 73.6 | 12 | 135 | 134.2 | 5.3 | 59 | 636 | 180.6 | 15.8 | 7 | 27 | 2.0 | 1.3 |
| Benton | 0.71 | 38.0 | 5 | D | 20.0 | D | 68 | 693 | 191.7 | 16.0 | 14 | 21 | 3.1 | 0.5 |
| Bollinger | 0.16 | 13.1 | D | D | D | 8.5 | 31 | 293 | 68.2 | 6.0 | NA | NA | NA | NA |
| Boone | 17.85 | 102.0 | 144 | 1,697 | 914.6 | 92.9 | 592 | 11,589 | 3,410.1 | 291.4 | 270 | 1,231 | 225.7 | 39.6 |
| Buchanan | 0.00 | 0.0 | 102 | 1,239 | 1,683.3 | 66.6 | 305 | 5,707 | 1,411.0 | 133.1 | 106 | 400 | 65.2 | 11.3 |
| Butler | 1.38 | 32.1 | 48 | 408 | 222.8 | 15.7 | 199 | 2,972 | 781.9 | 70.0 | 42 | 141 | 25.5 | 4.4 |
| Caldwell | 0.34 | 37.7 | D | D | D | 2.2 | 26 | 599 | 101.9 | 15.2 | D | D | D | D |
| Callaway | 3.60 | 80.3 | 22 | 323 | 151.9 | 14.9 | 120 | 1,244 | 340.0 | 27.2 | 27 | 105 | 17.7 | 3.2 |
| Camden | 3.87 | 87.5 | 42 | 297 | 142.0 | 11.6 | 289 | 3,300 | 791.9 | 81.1 | 89 | 335 | 63.4 | 10.0 |
| Cape Girardeau | 7.86 | 100.0 | 105 | 1,081 | 662.7 | 51.3 | 409 | 6,000 | 1,574.2 | 149.8 | 115 | 369 | 73.3 | 12.7 |
| Carroll | 0.81 | 90.1 | 12 | 72 | 120.0 | 3.3 | 37 | 245 | 52.9 | 5.2 | D | D | D | D |
| Carter | 0.46 | 73.4 | 5 | D | 17.2 | D | 27 | 163 | 28.7 | 2.8 | NA | NA | NA | NA |
| Cass | 1.55 | 15.3 | 59 | 545 | 533.3 | 32.6 | 259 | 4,481 | 1,344.4 | 114.0 | 80 | 244 | 59.7 | 9.3 |
| Cedar | 0.86 | 61.7 | 7 | 40 | 14.8 | 0.9 | 48 | 446 | 116.2 | 10.0 | 11 | 42 | 2.9 | 0.8 |
| Chariton | 0.37 | 48.8 | 15 | 227 | 389.5 | 10.1 | 35 | 220 | 57.4 | 5.2 | 6 | D | 0.6 | D |
| Christian | 6.55 | 78.7 | 73 | 700 | 339.4 | 29.9 | 247 | 2,863 | 752.6 | 70.2 | 80 | 191 | 43.5 | 5.8 |
| Clark | 0.69 | 101.5 | 12 | 143 | 116.6 | 6.1 | 30 | 252 | 80.9 | 4.6 | 3 | 5 | 0.4 | 0.2 |
| Clay | 124.21 | 527.1 | 287 | 4,540 | 4,081.9 | 273.4 | 612 | 12,795 | 3,927.9 | 350.5 | 255 | 1,060 | 284.6 | 47.5 |
| Clinton | 0.30 | 14.6 | 12 | 86 | 165.7 | 4.2 | 54 | 464 | 161.9 | 13.3 | D | D | D | D |
| Cole | 8.18 | 106.6 | 76 | 1,022 | 1,358.9 | 47.9 | 283 | 5,429 | 1,374.9 | 132.8 | 76 | 271 | 66.2 | 9.9 |
| Cooper | 1.33 | 75.4 | 13 | 141 | 99.5 | 5.0 | 58 | 756 | 277.4 | 18.8 | 12 | 35 | 4.0 | 0.7 |
| Crawford | 1.28 | 52.2 | 14 | 139 | 57.2 | 6.6 | 75 | 677 | 238.7 | 17.6 | 17 | 78 | 11.6 | 2.6 |
| Dade | 0.37 | 48.7 | 7 | 307 | 156.9 | 10.5 | 29 | 189 | 42.5 | 3.6 | 4 | D | 0.7 | D |
| Dallas | 0.47 | 28.7 | 8 | 87 | 80.0 | 2.6 | 47 | 435 | 115.3 | 10.2 | 10 | 21 | 1.7 | 0.5 |
| Daviess | 0.59 | 71.5 | 14 | 67 | 62.0 | 3.0 | 26 | 216 | 65.0 | 4.6 | 4 | 5 | 0.6 | 0.1 |
| DeKalb | 0.14 | 11.0 | D | D | D | 2.3 | 36 | 547 | 130.1 | 12.4 | D | D | D | D |
| Dent | 0.90 | 57.7 | D | D | D | D | 49 | 535 | 123.0 | 12.0 | 13 | 20 | 2.8 | 0.5 |
| Douglas | 0.56 | 41.9 | 11 | 36 | 17.1 | 1.1 | 33 | 448 | 101.5 | 10.1 | 5 | 5 | 0.7 | 0.1 |
| Dunklin | 3.06 | 99.0 | D | D | D | D | 131 | 1,462 | 428.1 | 35.5 | 24 | 601 | 24.0 | 13.5 |
| Franklin | 6.94 | 67.8 | 95 | 896 | 531.6 | 39.8 | 358 | 4,839 | 1,357.8 | 121.9 | 83 | 202 | 41.0 | 6.6 |
| Gasconade | 0.82 | 55.2 | 23 | 240 | 169.6 | 9.8 | 61 | 667 | 169.3 | 15.8 | 8 | 21 | 2.5 | 0.6 |
| Gentry | 0.62 | 92.6 | D | D | D | 4.6 | 31 | 244 | 54.3 | 6.0 | 4 | 47 | 1.1 | 0.7 |
| Greene | 33.20 | 115.2 | 386 | 8,467 | 6,884.5 | 426.4 | 1,138 | 19,066 | 5,495.5 | 515.1 | 477 | 3,066 | 501.5 | 103.0 |
| Grundy | 1.34 | 132.7 | 9 | 106 | 136.8 | 3.2 | 42 | 415 | 87.2 | 8.6 | 7 | 16 | 1.9 | 0.4 |
| Harrison | 0.91 | 105.6 | 11 | 136 | 99.6 | 3.6 | 38 | 560 | 206.9 | 14.8 | 5 | 4 | 1.5 | 0.1 |
| Henry | 2.00 | 92.0 | 21 | 165 | 89.1 | 6.5 | 112 | 1,207 | 347.8 | 29.1 | 27 | 81 | 9.6 | 1.7 |
| Hickory | 0.31 | 33.7 | 3 | 11 | 2.7 | 0.4 | 33 | 304 | 64.5 | 6.3 | 4 | D | 0.3 | D |
| Holt | 0.40 | 89.2 | 11 | 112 | 129.9 | 5.5 | 23 | 155 | 57.0 | 4.4 | NA | NA | NA | NA |
| Howard | 1.04 | 102.6 | D | D | D | D | 34 | 273 | 50.0 | 5.1 | 4 | D | 2.1 | D |
| Howell | 3.01 | 75.0 | 42 | 489 | 244.1 | 19.6 | 215 | 2,147 | 612.7 | 52.8 | 45 | 154 | 21.3 | 4.1 |
| Iron | 0.42 | 41.5 | D | D | D | 0.9 | 37 | 279 | 60.3 | 5.3 | 7 | D | 1.9 | D |
| Jackson | 26.90 | 39.1 | 804 | 12,327 | 12,797.4 | 739.9 | 2,118 | 35,602 | 10,337.7 | 949.9 | 860 | 5,490 | 1,842.2 | 305.3 |

1 Merchant wholesalers, except manufacturers' sales branches and offices.  2. Employer establishments.

# Table B. States and Counties — Professional Services, Manufacturing, and Accommodation and Food Services

| STATE County | Professional, scientific, and technical services, 2017 | | | | Manufacturing, 2017 | | | | Accommodation and food services, 2017 | | | |
|---|---|---|---|---|---|---|---|---|---|---|---|---|
| | Number of establishments | Number of employees | Sales (mil dol) | Average payroll (mil dol) | Number of establishments | Number of employees | Sales (mil dol) | Average payroll (mil dol) | Number of establishments | Number of employees | Sales (mil dol) | Annual payroll (mil dol) |
| | 147 | 148 | 149 | 150 | 151 | 152 | 153 | 154 | 155 | 156 | 157 | 158 |
| **MISSISSIPPI— Cont'd** | | | | | | | | | | | | |
| Sunflower | 16 | 58 | 7 | 1.9 | 18 | 315 | 183.7 | 14.4 | 36 | 562 | 23.1 | 6.4 |
| Tallahatchie | 8 | 43 | 4 | 1.2 | D | 42 | D | 1.1 | D | D | D | 0.7 |
| Tate | 22 | 79 | 8 | 3.0 | 11 | 465 | 203.6 | 20.4 | 36 | 571 | 25.9 | 6.5 |
| Tippah | 16 | 73 | 8 | 2.9 | 24 | 1,924 | 443.3 | 73.8 | D | D | D | D |
| Tishomingo | 16 | 51 | 5 | 1.3 | 38 | 2,034 | 537.1 | 71.7 | D | D | D | D |
| Tunica | D | D | D | D | 8 | 479 | 127.6 | 20.5 | 34 | 5,252 | 605.1 | 157.7 |
| Union | 28 | 102 | 12 | 4.1 | 32 | 4,367 | 3,050.5 | 186.9 | 45 | 787 | 43.8 | 11.0 |
| Walthall | 10 | 15 | 2 | 0.5 | 12 | 344 | 66.5 | 13.6 | 17 | 173 | 9.3 | 2.0 |
| Warren | D | D | D | D | 35 | 2,913 | 1,824.2 | 148.9 | 119 | 3,726 | 300.6 | 69.1 |
| Washington | D | D | D | D | 35 | 1,063 | 748.4 | 55.9 | 88 | 2,239 | 143.8 | 36.8 |
| Wayne | 42 | 99 | 11 | 3.1 | 10 | 1,112 | 250.8 | 50.1 | D | D | D | D |
| Webster | 10 | 77 | 6 | 1.8 | 8 | 188 | 55.0 | 8.8 | 13 | 132 | 5.6 | 1.3 |
| Wilkinson | 6 | 19 | 1 | 0.5 | 3 | 99 | 36.9 | 4.8 | D | D | D | D |
| Winston | 26 | 61 | 6 | 1.7 | 18 | 1,894 | 416.4 | 80.8 | 30 | 451 | 18.9 | 5.2 |
| Yalobusha | D | D | D | D | 6 | 1,201 | 308.2 | 36.4 | D | D | D | D |
| Yazoo | 23 | 181 | 13 | 5.0 | 10 | 586 | 307.7 | 34.6 | 32 | 385 | 18.3 | 4.4 |
| **MISSOURI** | 14,111 | 160,171 | 29,870 | 11,490.3 | 5,797 | 259,462 | 118,633.7 | 14,216.5 | 12,896 | 263,644 | 15,082.4 | 4,274.7 |
| Adair | D | D | D | D | 12 | D | 333.7 | 34.3 | 63 | 1,238 | 52.5 | 14.5 |
| Andrew | D | D | D | D | 11 | 31 | 5.7 | 1.3 | 9 | 118 | 4.8 | 1.4 |
| Atchison | D | D | D | D | 15 | D | D | 0.4 | D | D | D | D |
| Audrain | 27 | 140 | 12 | 4.9 | 36 | 1,838 | 902.1 | 91.5 | 41 | 579 | 22.0 | 5.7 |
| Barry | D | D | 297 | D | 42 | 4,814 | 1,510.1 | 197.2 | 55 | 761 | 35.7 | 10.4 |
| Barton | 14 | 112 | 11 | 5.8 | 19 | 325 | 96.7 | 13.7 | 19 | 289 | 11.0 | 3.0 |
| Bates | D | D | D | 2.8 | 16 | 74 | 26.5 | D | 29 | 368 | 14.9 | 4.2 |
| Benton | D | D | 6 | D | 17 | 254 | 18.8 | 5.4 | D | D | D | 3.8 |
| Bollinger | D | D | D | D | D | 160 | D | 5.2 | D | D | D | D |
| Boone | D | D | D | D | 96 | 4,129 | 2,175.4 | 205.9 | 460 | 10,333 | 485.2 | 141.5 |
| Buchanan | D | D | D | D | 85 | 11,918 | 7,943.0 | 583.8 | 201 | 4,071 | 198.3 | 58.0 |
| Butler | 47 | 277 | 31 | 10.4 | 40 | 2,198 | 530.6 | 83.7 | 84 | 1,689 | 74.9 | 21.7 |
| Caldwell | D | D | D | D | 7 | D | 9.9 | D | D | D | D | D |
| Callaway | 39 | 349 | 24 | 12.8 | 37 | 1,455 | 498.3 | 82.5 | 69 | 1,014 | 48.0 | 13.6 |
| Camden | D | D | D | 49.5 | 40 | 375 | 143.6 | 18.6 | 163 | 2,672 | 180.9 | 51.3 |
| Cape Girardeau | D | D | D | D | 92 | 3,682 | 2,892.5 | 200.1 | 187 | 4,214 | 197.1 | 59.8 |
| Carroll | D | D | 2 | D | 11 | 154 | 137.6 | 7.0 | D | D | D | D |
| Carter | D | D | 1 | D | 20 | 143 | 39.8 | 5.7 | 15 | 120 | 5.9 | 1.6 |
| Cass | 160 | 677 | 86 | 29.8 | 53 | 2,063 | 503.8 | 89.5 | 156 | 3,038 | 128.7 | 39.3 |
| Cedar | 18 | 69 | 5 | 1.5 | 21 | 339 | 162.9 | 15.6 | 28 | 333 | 11.6 | 3.4 |
| Chariton | D | D | D | D | 7 | 78 | 14.1 | 3.3 | D | D | D | 0.8 |
| Christian | 159 | 674 | 73 | 27.9 | 93 | 1,137 | 291.0 | 50.5 | 117 | 2,000 | 96.3 | 30.1 |
| Clark | 9 | 10 | 1 | 0.3 | 5 | 29 | 5.7 | 1.4 | 10 | 113 | 3.8 | 1.3 |
| Clay | D | D | D | D | 196 | 15,064 | 14,334.3 | 919.7 | 435 | 11,396 | 824.2 | 210.1 |
| Clinton | D | D | D | D | 14 | 118 | 21.9 | 4.8 | 27 | 337 | 9.7 | 3.4 |
| Cole | D | D | D | D | 56 | 2,420 | 748.7 | 117.8 | 169 | 3,374 | 163.4 | 47.0 |
| Cooper | D | D | D | D | 7 | 196 | 57.0 | 10.4 | D | D | D | 13.9 |
| Crawford | 27 | 98 | 8 | 2.6 | 48 | 1,938 | 403.1 | 68.3 | 55 | 619 | 31.4 | 9.0 |
| Dade | D | D | D | D | 12 | 209 | 94.2 | 9.0 | 13 | 91 | 3.5 | 0.9 |
| Dallas | 20 | 52 | 5 | 1.2 | 12 | 141 | 57.1 | 5.6 | 24 | 276 | 12.5 | 3.3 |
| Daviess | D | D | D | D | 9 | 170 | 52.2 | 6.0 | 8 | D | 4.0 | D |
| DeKalb | D | D | D | D | 6 | 25 | 7.8 | 1.4 | 18 | 246 | 9.5 | 2.7 |
| Dent | 14 | 46 | 3 | 1.2 | 25 | 556 | 291.3 | 31.9 | D | D | D | D |
| Douglas | 14 | 40 | 5 | 1.3 | 11 | D | 122.1 | 17.1 | 13 | 225 | 8.9 | 2.3 |
| Dunklin | 29 | 121 | 14 | 4.1 | 10 | 357 | 100.9 | 14.7 | 46 | 647 | 30.6 | 8.5 |
| Franklin | 186 | 1,019 | 123 | 45.0 | 207 | 9,398 | 2,677.6 | 469.7 | 205 | 3,601 | 166.1 | 49.3 |
| Gasconade | 23 | 114 | 9 | 3.0 | 41 | 1,402 | 216.8 | 49.0 | 46 | 497 | 24.0 | 7.0 |
| Gentry | D | D | D | D | 6 | 172 | 45.6 | 9.7 | D | D | D | 0.7 |
| Greene | D | D | D | D | 288 | 12,379 | 5,020.4 | 635.7 | 785 | 16,133 | 812.7 | 235.6 |
| Grundy | D | D | D | D | 10 | 527 | 400.1 | 27.3 | 15 | 236 | 6.6 | 2.4 |
| Harrison | D | D | 1 | D | D | 31 | D | 1.3 | 16 | 267 | 11.9 | 3.4 |
| Henry | 32 | 118 | 14 | 3.9 | 19 | 1,003 | 425.6 | 47.7 | 53 | 667 | 34.4 | 9.4 |
| Hickory | D | D | D | D | NA | NA | NA | NA | 14 | 78 | 4.3 | 1.2 |
| Holt | D | D | 2 | D | D | 152 | D | 11.2 | 7 | D | 3.8 | D |
| Howard | 14 | 58 | 5 | 1.8 | 10 | 324 | 71.8 | 14.4 | 15 | 158 | 4.9 | 1.5 |
| Howell | 70 | 306 | 30 | 12.4 | 62 | 2,465 | 606.8 | 91.1 | 72 | 1,172 | 56.0 | 15.4 |
| Iron | D | D | D | D | 12 | 46 | 7.7 | 1.9 | 17 | 147 | 5.9 | 1.8 |
| Jackson | D | D | D | D | 610 | 24,965 | 9,820.8 | 1,497.1 | 1,557 | 37,503 | 2,219.3 | 657.1 |

Table B. States and Counties — **Health Care and Social Assistance, Other Services, Nonemployer Businesses, and Residential Construction**

| STATE County | Health care and social assistance, 2017 | | | | Other services, 2017 | | | | Nonemployer businesses, 2018 | | Value of residential construction authorized by building permits, 2020 | |
|---|---|---|---|---|---|---|---|---|---|---|---|---|
| | Number of establishments | Number of employees | Receipts (mil dol) | Annual payroll (mil dol) | Number of establishments | Number of employees | Receipts (mil dol) | Annual payroll (mil dol) | Number | Receipts (mil dol) | New construction ($1,000) | Number of housing units |
| | 159 | 160 | 161 | 162 | 163 | 164 | 165 | 166 | 167 | 168 | 169 | 170 |
| MISSISSIPPI— Cont'd | | | | | | | | | | | | |
| Sunflower | 46 | 1,339 | 164.8 | 49.4 | 31 | 191 | 23.5 | 6.8 | 1,604 | 47.2 | 3,047 | 15 |
| Tallahatchie | 17 | 564 | 45.4 | 20.9 | 11 | 30 | 4.1 | 0.7 | 809 | 26.6 | 70 | 2 |
| Tate | 40 | 634 | 61.0 | 26.3 | D | D | D | D | 2,149 | 90.8 | 30,371 | 143 |
| Tippah | 33 | 658 | 44.9 | 18.4 | D | D | D | D | 1,486 | 67.3 | 2,300 | 11 |
| Tishomingo | 31 | 543 | 42.1 | 18.9 | 23 | 99 | 16.2 | 2.6 | 1,123 | 44.4 | 600 | 5 |
| Tunica | 20 | 232 | 18.9 | 6.6 | D | D | D | 0.3 | 645 | 26.8 | 21,179 | 94 |
| Union | 53 | 1,149 | 132.8 | 45.7 | D | D | D | D | 2,131 | 95.3 | 1,619 | 11 |
| Walthall | 17 | 469 | 50.2 | 20.8 | D | D | D | 0.4 | 1,139 | 39.8 | 0 | 0 |
| Warren | 125 | 2,394 | 293.7 | 105.6 | 56 | 264 | 24.4 | 6.2 | 2,937 | 110.1 | 1,405 | 7 |
| Washington | 149 | 3,480 | 295.8 | 122.9 | 76 | 462 | 45.4 | 15.3 | 3,298 | 121.1 | 2,191 | 10 |
| Wayne | 17 | 583 | 41.5 | 23.8 | D | D | D | D | 1,518 | 58.4 | 600 | 8 |
| Webster | 15 | 414 | 42.4 | 18.4 | D | D | D | 0.7 | 840 | 31.8 | 340 | 5 |
| Wilkinson | 17 | 296 | 27.1 | 9.4 | D | D | D | 1.2 | 581 | 16.5 | 0 | 0 |
| Winston | 22 | 555 | 54.5 | 19.1 | D | D | D | D | 1,289 | 45.1 | 1,054 | 5 |
| Yalobusha | 9 | 352 | 25.8 | 13.1 | 9 | 19 | 1.8 | 0.4 | 898 | 34.3 | 487 | 16 |
| Yazoo | 37 | 1,008 | 71.1 | 28.9 | D | D | D | 5.7 | 1,604 | 55.1 | 0 | 0 |
| MISSOURI | 19,097 | 423,057 | 48,192.5 | 18,682.2 | 10,513 | 68,206 | 8,693.9 | 2,297.4 | 426,915 | 19,463.4 | 4,021,896 | 19,839 |
| Adair | D | D | D | D | D | D | D | D | 1,638 | 58.0 | 8,930 | 30 |
| Andrew | 26 | 270 | 15.9 | 7.2 | 25 | 106 | 12.0 | 2.5 | 1,279 | 52.9 | 750 | 5 |
| Atchison | 20 | 345 | 30.5 | 15.1 | D | D | D | 1.4 | 394 | 16.4 | 518 | 3 |
| Audrain | 68 | 1,131 | 100.8 | 46.0 | 43 | 223 | 20.6 | 7.6 | 1,409 | 63.4 | 2,067 | 15 |
| Barry | 69 | 1,077 | 106.9 | 38.1 | 53 | 170 | 16.4 | 3.7 | 2,457 | 88.5 | 6,173 | 33 |
| Barton | 29 | 426 | 35.3 | 13.1 | D | D | D | D | 918 | 39.4 | 387 | 3 |
| Bates | 36 | 842 | 70.0 | 28.2 | D | D | D | D | 1,193 | 45.7 | 500 | 2 |
| Benton | 35 | 374 | 22.3 | 9.5 | 31 | 107 | 9.8 | 1.8 | 1,392 | 68.5 | 725 | 3 |
| Bollinger | D | D | D | D | D | D | D | D | 750 | 30.0 | 0 | 0 |
| Boone | 655 | 19,126 | 2,584.4 | 888.4 | 347 | 2,170 | 255.5 | 69.1 | 12,370 | 597.3 | 234,826 | 921 |
| Buchanan | 304 | 9,286 | 1,016.9 | 413.5 | D | D | 183.4 | D | 4,173 | 168.2 | 31,165 | 240 |
| Butler | 199 | 4,125 | 580.4 | 177.9 | 62 | 244 | 24.0 | 5.6 | 2,677 | 138.8 | 1,179 | 18 |
| Caldwell | D | D | D | D | D | D | D | D | 602 | 26.5 | 9,831 | 36 |
| Callaway | 71 | 2,543 | 167.3 | 86.6 | 64 | 256 | 26.0 | 6.1 | 2,465 | 94.8 | 13,034 | 61 |
| Camden | 140 | 2,537 | 295.5 | 108.8 | 101 | 359 | 35.8 | 9.5 | 3,942 | 207.3 | 53,802 | 202 |
| Cape Girardeau | 332 | 10,265 | 1,317.0 | 480.6 | D | D | D | D | 5,479 | 250.2 | 27,912 | 141 |
| Carroll | 21 | 444 | 40.2 | 15.0 | 14 | 52 | 5.2 | 1.2 | 604 | 23.0 | 3,003 | 13 |
| Carter | 28 | 241 | 11.5 | 6.6 | D | D | 2.5 | D | 512 | 25.5 | 0 | 0 |
| Cass | 200 | 3,233 | 323.4 | 128.0 | 148 | 617 | 55.4 | 16.3 | 7,397 | 374.5 | 199,318 | 1,108 |
| Cedar | 30 | 606 | 35.6 | 14.8 | 19 | 43 | 3.9 | 0.8 | 1,164 | 47.5 | 1,006 | 13 |
| Chariton | 16 | 283 | 16.0 | 6.4 | 18 | 49 | 9.0 | 1.4 | 658 | 23.1 | 400 | 1 |
| Christian | 141 | 1,554 | 119.5 | 47.5 | 135 | 496 | 41.0 | 12.2 | 7,243 | 326.7 | 137,428 | 638 |
| Clark | 15 | 130 | 7.8 | 3.0 | D | D | D | D | 451 | 17.7 | 1,634 | 9 |
| Clay | 608 | 13,243 | 1,470.4 | 648.9 | 350 | 2,307 | 267.5 | 78.7 | 16,567 | 685.5 | 90,345 | 367 |
| Clinton | 44 | 914 | 123.2 | 39.4 | D | D | D | D | 1,406 | 51.7 | 21,516 | 83 |
| Cole | 279 | 6,846 | 793.1 | 310.7 | 221 | 1,332 | 226.1 | 60.8 | 4,965 | 242.9 | 40,945 | 185 |
| Cooper | 51 | 835 | 57.8 | 24.0 | 36 | 104 | 11.1 | 2.6 | 1,092 | 46.8 | 1,854 | 8 |
| Crawford | 51 | 878 | 88.6 | 33.2 | 33 | 141 | 13.2 | 3.3 | 1,540 | 62.3 | 2,938 | 14 |
| Dade | D | D | D | 0.6 | D | D | D | 0.4 | 502 | 18.9 | 0 | 0 |
| Dallas | 36 | 886 | 25.2 | 8.9 | 23 | 52 | 5.2 | 1.0 | 1,441 | 65.1 | 1,664 | 7 |
| Daviess | 11 | 97 | 7.0 | 2.8 | D | D | 2.7 | D | 760 | 36.8 | 1,038 | 5 |
| DeKalb | D | D | D | D | 13 | 88 | 8.6 | 2.3 | 633 | 27.3 | 549 | 2 |
| Dent | 31 | 796 | 50.8 | 20.4 | 17 | 45 | 4.2 | 0.9 | 926 | 37.4 | 238 | 1 |
| Douglas | 17 | 301 | 20.9 | 7.8 | 15 | 43 | 5.0 | 1.1 | 1,120 | 42.6 | 1,189 | 5 |
| Dunklin | 105 | 2,207 | 163.8 | 57.3 | 40 | 105 | 12.2 | 2.6 | 1,794 | 109.4 | 3,960 | 23 |
| Franklin | 245 | 4,651 | 479.5 | 191.9 | 184 | 931 | 110.8 | 30.5 | 6,785 | 291.7 | 93,627 | 434 |
| Gasconade | 30 | 897 | 42.6 | 20.6 | 30 | 93 | 8.9 | 2.2 | 1,225 | 49.6 | 713 | 7 |
| Gentry | 41 | 778 | 42.6 | 18.2 | D | D | D | 2.9 | 548 | 21.5 | 130 | 3 |
| Greene | 894 | 29,480 | 3,798.3 | 1,417.7 | 592 | 4,716 | 492.7 | 147.9 | 21,787 | 1,144.9 | 285,380 | 1,327 |
| Grundy | 31 | 617 | 49.9 | 19.4 | 25 | 62 | 4.6 | 1.5 | 711 | 31.4 | 379 | 4 |
| Harrison | 28 | 341 | 18.4 | 7.8 | D | D | D | D | 623 | 27.3 | 0 | 0 |
| Henry | 58 | 2,021 | 176.9 | 81.0 | 41 | 169 | 11.6 | 3.2 | 1,459 | 57.3 | 1,711 | 11 |
| Hickory | 9 | 198 | 13.3 | 5.3 | D | D | D | 0.3 | 747 | 28.2 | NA | NA |
| Holt | 10 | 151 | 7.3 | 3.6 | D | D | 2.4 | D | 441 | 20.5 | 0 | 0 |
| Howard | 25 | 605 | 23.3 | 12.5 | D | D | D | D | 676 | 28.0 | 830 | 5 |
| Howell | 156 | 4,912 | 473.1 | 181.5 | 73 | 356 | 44.4 | 9.5 | 3,119 | 127.5 | 3,427 | 20 |
| Iron | 39 | 628 | 34.4 | 15.2 | 20 | 56 | 4.6 | 1.1 | 498 | 16.6 | 0 | 0 |
| Jackson | 2,436 | 57,154 | 7,797.2 | 2,920.4 | 1,269 | 9,916 | 1,962.3 | 361.5 | 47,652 | 2,117.9 | 748,811 | 4,767 |

# Table B. States and Counties — Government Employment and Payroll, and Local Government Finances

| STATE County | Full-time equivalent employees | March payroll (dollars) | March payroll (percent of total) | | | | | | | Local government finances, 2017 — General revenue | | | | |
| | | | Administration, judicial, and legal | Police and corrections | Fire protection | Highways and transportation | Health and welfare | Natural resources and utilities | Education and libraries | Total (mil dol) | Inter-governmental (mil dol) | Taxes Total (mil dol) | Taxes Per capita[1] (dollars) Total | Taxes Per capita[1] (dollars) Property |
| | 171 | 172 | 173 | 174 | 175 | 176 | 177 | 178 | 179 | 180 | 181 | 182 | 183 | 184 |
|---|---|---|---|---|---|---|---|---|---|---|---|---|---|---|
| **MISSISSIPPI— Cont'd** | | | | | | | | | | | | | | |
| Sunflower | 2,035 | 7,299,191 | 3.4 | 3.6 | 1.2 | 1.7 | 42.9 | 0.9 | 45.9 | 183.1 | 57.3 | 24.4 | 933 | 902 |
| Tallahatchie | 944 | 2,844,137 | 6.0 | 5.5 | 0.0 | 3.5 | 49.3 | 1.2 | 34.5 | 38.5 | 19.7 | 13.7 | 970 | 944 |
| Tate | 1,245 | 4,687,881 | 2.3 | 3.1 | 1.2 | 1.7 | 0.3 | 2.4 | 88.5 | 114.5 | 68.2 | 24.6 | 860 | 820 |
| Tippah | 788 | 2,243,776 | 2.9 | 2.0 | 0.1 | 0.9 | 22.9 | 3.9 | 67.3 | 69.0 | 36.1 | 13.2 | 600 | 563 |
| Tishomingo | 648 | 1,867,623 | 9.5 | 8.3 | 1.2 | 2.6 | 0.6 | 5.2 | 71.8 | 50.6 | 27.2 | 18.4 | 941 | 933 |
| Tunica | 613 | 1,660,838 | 1.7 | 18.9 | 0.0 | 4.2 | 5.2 | 14.6 | 55.0 | 77.7 | 42.0 | 23.0 | 2,300 | 1,836 |
| Union | 983 | 2,922,592 | 4.5 | 7.8 | 2.2 | 2.2 | 0.0 | 10.9 | 72.1 | 71.1 | 43.7 | 19.2 | 675 | 653 |
| Walthall | 397 | 1,089,504 | 5.9 | 4.7 | 0.1 | 4.9 | 0.0 | 6.5 | 77.3 | 39.6 | 16.8 | 10.4 | 716 | 700 |
| Warren | 1,841 | 5,441,149 | 4.4 | 5.4 | 4.5 | 2.4 | 4.4 | 5.9 | 71.0 | 172.9 | 71.1 | 70.9 | 1,520 | 1,336 |
| Washington | 3,096 | 10,743,574 | 4.5 | 7.4 | 2.4 | 2.9 | 44.3 | 2.5 | 34.8 | 300.2 | 88.8 | 63.2 | 1,367 | 1,263 |
| Wayne | 1,072 | 3,496,080 | 4.1 | 3.8 | 0.5 | 2.6 | 38.8 | 2.4 | 46.7 | 76.4 | 30.1 | 11.7 | 571 | 555 |
| Webster | 360 | 980,701 | 9.8 | 6.2 | 0.2 | 5.1 | 0.1 | 2.1 | 76.2 | 25.3 | 15.3 | 7.0 | 718 | 689 |
| Wilkinson | 337 | 922,832 | 18.9 | 9.7 | 0.3 | 6.1 | 0.0 | 4.1 | 60.6 | 54.6 | 12.6 | 6.4 | 718 | 698 |
| Winston | 593 | 1,520,585 | 8.8 | 15.4 | 2.3 | 3.4 | 1.5 | 2.8 | 64.8 | 51.8 | 28.2 | 12.7 | 695 | 665 |
| Yalobusha | 700 | 2,126,577 | 7.3 | 4.0 | 3.3 | 3.0 | 43.4 | 2.8 | 36.1 | 30.5 | 17.7 | 9.0 | 719 | 703 |
| Yazoo | 862 | 2,505,769 | 6.4 | 7.4 | 3.1 | 6.5 | 2.4 | 8.8 | 61.1 | 75.0 | 40.6 | 24.5 | 856 | 828 |
| **MISSOURI** | X | X | X | X | X | X | X | X | X | X | X | X | X | X |
| Adair | 864 | 2,743,554 | 7.4 | 5.9 | 3.6 | 3.4 | 14.4 | 5.3 | 58.6 | 83.3 | 35.1 | 27.3 | 1,072 | 505 |
| Andrew | 520 | 1,620,539 | 4.8 | 4.2 | 0.9 | 3.0 | 2.9 | 5.9 | 77.7 | 43.9 | 20.2 | 18.0 | 1,032 | 794 |
| Atchison | 271 | 779,674 | 9.8 | 6.3 | 0.1 | 9.7 | 10.0 | 4.2 | 58.9 | 23.9 | 6.5 | 13.4 | 2,547 | 1,906 |
| Audrain | 811 | 2,619,099 | 7.1 | 10.8 | 0.6 | 4.0 | 16.2 | 4.6 | 55.2 | 71.9 | 23.9 | 31.7 | 1,237 | 777 |
| Barry | 1,414 | 4,405,718 | 4.6 | 5.3 | 1.8 | 2.5 | 2.0 | 5.4 | 77.0 | 122.5 | 53.3 | 55.6 | 1,562 | 1,111 |
| Barton | 658 | 2,128,493 | 3.9 | 3.6 | 0.5 | 2.0 | 38.8 | 6.7 | 44.1 | 55.0 | 15.2 | 13.9 | 1,178 | 816 |
| Bates | 884 | 2,973,990 | 3.7 | 4.8 | 0.2 | 1.7 | 41.9 | 4.6 | 41.4 | 48.2 | 21.3 | 15.6 | 956 | 670 |
| Benton | 720 | 1,922,673 | 4.4 | 6.6 | 0.1 | 2.2 | 24.0 | 1.7 | 60.9 | 54.2 | 15.5 | 27.2 | 1,424 | 1,156 |
| Bollinger | 370 | 1,005,536 | 5.3 | 5.6 | 0.5 | 3.0 | 0.2 | 2.0 | 83.1 | 27.3 | 13.2 | 11.4 | 926 | 770 |
| Boone | 6,703 | 23,310,166 | 7.9 | 5.7 | 3.5 | 2.7 | 5.2 | 9.4 | 64.0 | 898.3 | 189.1 | 294.5 | 1,654 | 990 |
| Buchanan | 2,957 | 11,239,924 | 4.7 | 9.9 | 5.2 | 3.5 | 2.7 | 5.3 | 67.6 | 318.0 | 107.7 | 156.6 | 1,766 | 917 |
| Butler | 1,652 | 5,253,884 | 6.4 | 6.1 | 2.3 | 4.2 | 1.4 | 2.5 | 73.5 | 133.1 | 64.0 | 51.9 | 1,217 | 639 |
| Caldwell | 475 | 1,404,146 | 6.0 | 8.8 | 0.0 | 2.6 | 7.0 | 3.6 | 71.6 | 31.6 | 16.2 | 9.3 | 1,025 | 773 |
| Callaway | 1,257 | 3,411,112 | 8.6 | 7.5 | 2.9 | 4.3 | 5.6 | 8.2 | 59.2 | 115.3 | 38.4 | 44.6 | 992 | 689 |
| Camden | 1,489 | 4,730,188 | 10.2 | 8.8 | 5.3 | 5.7 | 2.5 | 3.0 | 62.6 | 138.6 | 35.1 | 78.9 | 1,734 | 1,143 |
| Cape Girardeau | 2,694 | 7,744,192 | 5.5 | 9.3 | 4.7 | 5.2 | 0.1 | 9.1 | 62.5 | 235.5 | 73.5 | 125.8 | 1,608 | 822 |
| Carroll | 377 | 1,180,272 | 4.7 | 5.0 | 1.4 | 4.3 | 4.9 | 3.0 | 76.4 | 32.2 | 11.6 | 16.0 | 1,828 | 1,562 |
| Carter | 285 | 715,720 | 7.0 | 8.6 | 0.1 | 2.6 | 6.2 | 1.3 | 73.2 | 17.3 | 9.8 | 5.8 | 940 | 712 |
| Cass | 4,148 | 15,759,622 | 5.2 | 6.9 | 4.5 | 2.5 | 12.9 | 4.5 | 62.6 | 422.1 | 129.7 | 158.7 | 1,533 | 984 |
| Cedar | 621 | 1,672,123 | 5.0 | 5.6 | 0.2 | 2.0 | 24.0 | 9.6 | 52.3 | 52.5 | 15.9 | 11.3 | 805 | 573 |
| Chariton | 312 | 878,644 | 9.7 | 6.1 | 0.2 | 2.9 | 5.4 | 6.9 | 68.2 | 23.1 | 8.0 | 12.2 | 1,639 | 1,357 |
| Christian | 2,299 | 7,714,341 | 5.3 | 7.0 | 3.1 | 2.0 | 0.1 | 4.4 | 76.7 | 220.7 | 101.0 | 90.9 | 1,064 | 856 |
| Clark | 366 | 956,972 | 6.2 | 6.0 | 0.0 | 3.5 | 20.5 | 6.7 | 53.1 | 21.2 | 9.4 | 7.6 | 1,133 | 885 |
| Clay | 11,912 | 49,984,164 | 2.4 | 4.4 | 2.0 | 0.6 | 40.2 | 1.4 | 48.4 | 1,499.3 | 269.1 | 411.9 | 1,699 | 1,183 |
| Clinton | 680 | 2,252,917 | 6.5 | 9.2 | 1.2 | 4.7 | 3.4 | 7.1 | 67.6 | 57.7 | 24.4 | 25.0 | 1,218 | 868 |
| Cole | 2,381 | 8,477,306 | 7.1 | 8.9 | 3.8 | 4.3 | 4.2 | 5.4 | 63.9 | 228.9 | 70.4 | 118.1 | 1,541 | 818 |
| Cooper | 716 | 2,385,908 | 5.3 | 6.6 | 1.5 | 2.5 | 27.8 | 4.5 | 49.8 | 52.7 | 16.8 | 26.0 | 1,474 | 735 |
| Crawford | 719 | 1,546,090 | 7.8 | 11.9 | 0.0 | 5.2 | 9.0 | 5.2 | 58.5 | 49.3 | 23.0 | 19.2 | 798 | 588 |
| Dade | 421 | 1,056,840 | 5.1 | 2.7 | 0.0 | 0.4 | 32.0 | 3.6 | 56.2 | 23.5 | 8.9 | 7.4 | 981 | 831 |
| Dallas | 365 | 976,127 | 7.9 | 6.8 | 0.4 | 5.1 | 1.0 | 3.0 | 75.9 | 24.5 | 13.1 | 9.0 | 536 | 325 |
| Daviess | 356 | 1,031,668 | 6.2 | 2.8 | 0.0 | 2.5 | 2.4 | 7.8 | 78.1 | 22.6 | 11.1 | 8.4 | 1,008 | 842 |
| DeKalb | 234 | 777,687 | 18.0 | 0.5 | 0.0 | 2.9 | 3.5 | 5.0 | 69.9 | 18.6 | 8.2 | 7.1 | 569 | 506 |
| Dent | 697 | 2,121,584 | 7.9 | 2.7 | 0.0 | 3.3 | 36.5 | 3.2 | 45.4 | 75.6 | 16.6 | 11.4 | 736 | 473 |
| Douglas | 473 | 1,321,306 | 5.3 | 8.7 | 0.0 | 4.9 | 3.8 | 4.6 | 71.2 | 23.7 | 13.2 | 7.0 | 528 | 326 |
| Dunklin | 1,240 | 3,617,815 | 4.4 | 7.5 | 1.7 | 2.9 | 4.4 | 9.3 | 65.5 | 87.9 | 47.6 | 28.8 | 955 | 582 |
| Franklin | 3,447 | 12,044,444 | 4.5 | 9.0 | 2.4 | 3.3 | 7.5 | 3.1 | 69.3 | 307.1 | 113.2 | 149.8 | 1,450 | 921 |
| Gasconade | 769 | 2,577,493 | 4.6 | 4.5 | 0.0 | 3.2 | 35.2 | 3.4 | 48.8 | 78.7 | 16.4 | 20.1 | 1,367 | 996 |
| Gentry | 331 | 926,821 | 11.4 | 4.9 | 0.4 | 6.3 | 7.9 | 7.2 | 61.2 | 24.6 | 9.3 | 11.7 | 1,764 | 1,613 |
| Greene | 10,598 | 41,070,291 | 5.2 | 8.9 | 4.1 | 4.8 | 1.5 | 12.8 | 54.1 | 1,021.2 | 349.5 | 481.6 | 1,664 | 860 |
| Grundy | 728 | 1,939,762 | 3.3 | 5.3 | 0.3 | 2.8 | 14.1 | 7.5 | 65.4 | 61.1 | 28.9 | 13.6 | 1,362 | 837 |
| Harrison | 657 | 2,013,855 | 3.9 | 4.5 | 0.2 | 2.5 | 48.2 | 3.1 | 37.1 | 57.8 | 12.6 | 12.4 | 1,461 | 893 |
| Henry | 1,448 | 5,199,624 | 3.0 | 4.0 | 0.9 | 1.1 | 62.9 | 2.3 | 25.6 | 160.4 | 51.4 | 28.0 | 1,290 | 772 |
| Hickory | 328 | 963,083 | 6.3 | 6.5 | 0.0 | 2.7 | 0.0 | 1.8 | 82.3 | 22.4 | 12.1 | 7.5 | 799 | 630 |
| Holt | 212 | 570,842 | 9.4 | 7.1 | 0.0 | 4.4 | 2.4 | 6.9 | 68.3 | 30.5 | 16.9 | 11.0 | 2,498 | 2,075 |
| Howard | 360 | 1,047,325 | 8.1 | 7.4 | 0.3 | 3.5 | 7.8 | 5.6 | 66.3 | 26.2 | 12.5 | 9.1 | 906 | 626 |
| Howell | 1,574 | 4,372,994 | 5.4 | 5.4 | 1.6 | 3.4 | 6.0 | 7.9 | 68.8 | 107.4 | 52.2 | 35.5 | 885 | 520 |
| Iron | 495 | 1,426,766 | 5.1 | 5.0 | 0.3 | 6.1 | 8.3 | 1.9 | 72.8 | 31.7 | 12.8 | 14.1 | 1,385 | 1,206 |
| Jackson | 26,895 | 111,806,751 | 5.7 | 10.2 | 10.4 | 6.5 | 3.3 | 9.5 | 53.7 | 3,867.8 | 951.9 | 1,775.4 | 2,543 | 1,159 |

1. Based on the resident population estimated as of July 1 of the year shown.

Table B. States and Counties — **Local Government Finances, Government Employment, and Income Taxes**

| STATE County | Direct general expenditure Total (mil dol) | Per capita[1] (dollars) | Education | Health and hospitals | Police protection | Public welfare | Highways | Debt outstanding Total (mil dol) | Per capita[1] (dollars) | Federal civilian | Federal military | State and local | Number of returns | Mean adjusted gross income | Mean income tax |
|---|---|---|---|---|---|---|---|---|---|---|---|---|---|---|---|
| | 185 | 186 | 187 | 188 | 189 | 190 | 191 | 192 | 193 | 194 | 195 | 196 | 197 | 198 | 199 |
| **MISSISSIPPI— Cont'd** | | | | | | | | | | 51 | 123 | 3,159 | 9,100 | 38,547 | 3,127 |
| Sunflower | 194 | 7,431 | 33.7 | 51.8 | 2.7 | 0.0 | 4.3 | 20.5 | 783 | 39 | 66 | 971 | 4,730 | 36,086 | 2,484 |
| Tallahatchie | 40 | 2,835 | 50.6 | 1.1 | 4.9 | 0.0 | 7.6 | 11.4 | 807 | 81 | 158 | 1,568 | 11,720 | 47,585 | 3,747 |
| Tate | 139 | 4,877 | 81.6 | 0.5 | 3.9 | 0.0 | 3.1 | 38.0 | 1,331 | 56 | 126 | 1,228 | 8,810 | 42,614 | 3,056 |
| Tippah | 74 | 3,349 | 49.1 | 23.7 | 3.1 | 0.3 | 9.1 | 9.8 | 448 | 61 | 111 | 859 | 7,360 | 46,156 | 3,289 |
| Tishomingo | 38 | 1,966 | 76.9 | 0.1 | 5.0 | 0.0 | 2.0 | 31.2 | 1,598 | 22 | 55 | 763 | 4,280 | 33,443 | 2,122 |
| Tunica | 80 | 7,980 | 30.8 | 3.2 | 11.3 | 0.5 | 8.5 | 35.0 | 3,504 | 42 | 166 | 1,233 | 11,550 | 46,993 | 3,444 |
| Union | 67 | 2,355 | 63.4 | 0.4 | 7.5 | 0.0 | 7.6 | 31.4 | 1,103 | 30 | 83 | 608 | 5,520 | 42,515 | 3,298 |
| Walthall | 37 | 2,570 | 47.7 | 30.1 | 5.1 | 0.3 | 4.9 | 0.6 | 39 | 2,503 | 304 | 2,198 | 20,000 | 52,784 | 4,940 |
| Warren | 176 | 3,759 | 45.7 | 0.8 | 6.8 | 0.2 | 8.5 | 254.0 | 5,440 | 386 | 274 | 3,455 | 18,550 | 41,331 | 3,420 |
| Washington | 290 | 6,282 | 28.5 | 41.9 | 4.6 | 0.1 | 4.1 | 113.0 | 2,447 | 40 | 117 | 1,318 | 8,110 | 43,523 | 3,307 |
| Wayne | 81 | 3,972 | 43.6 | 35.0 | 4.3 | 0.0 | 5.7 | 15.2 | 742 | 38 | 56 | 441 | 3,890 | 46,293 | 3,406 |
| Webster | 25 | 2,523 | 63.9 | 1.3 | 5.0 | 0.0 | 8.5 | 5.7 | 585 | 9 | 44 | 539 | 3,300 | 39,337 | 2,568 |
| Wilkinson | 57 | 6,398 | 20.6 | 32.8 | 2.0 | 0.1 | 3.5 | 25.3 | 2,854 | 45 | 102 | 747 | 7,320 | 45,235 | 3,807 |
| Winston | 53 | 2,879 | 52.1 | 1.3 | 7.0 | 0.1 | 7.7 | 3.4 | 189 | 59 | 70 | 803 | 5,210 | 39,500 | 2,638 |
| Yalobusha | 30 | 2,392 | 51.6 | 1.0 | 6.5 | 0.0 | 10.1 | 12.0 | 962 | 778 | 132 | 1,410 | 9,240 | 42,058 | 3,496 |
| Yazoo | 73 | 2,536 | 49.5 | 1.3 | 7.4 | 0.3 | 9.7 | 30.0 | 1,047 | | | | | | |
| **MISSOURI** | X | X | X | X | X | X | X | X | X | 59,184 | 35,892 | 377,551 | 2,809,420 | 64,986 | 7,409 |
| Adair | 72 | 2,846 | 46.5 | 4.6 | 5.9 | 7.5 | 6.0 | 53.7 | 2,110 | 80 | 79 | 2,407 | 9,830 | 47,622 | 3,957 |
| Andrew | 45 | 2,550 | 63.0 | 3.5 | 4.0 | 0.2 | 8.1 | 77.0 | 4,404 | 35 | 59 | 731 | 8,450 | 59,234 | 5,574 |
| Atchison | 23 | 4,360 | 49.8 | 0.3 | 5.1 | 0.0 | 17.7 | 16.2 | 3,089 | 32 | 17 | 355 | 2,460 | 54,790 | 4,652 |
| Audrain | 77 | 2,998 | 53.8 | 2.9 | 4.5 | 7.7 | 6.4 | 85.7 | 3,350 | 75 | 77 | 2,083 | 10,340 | 46,700 | 3,472 |
| Barry | 109 | 3,046 | 64.6 | 2.7 | 4.1 | 0.0 | 9.0 | 71.7 | 2,012 | 124 | 120 | 1,583 | 14,660 | 47,288 | 4,071 |
| Barton | 55 | 4,667 | 37.2 | 38.4 | 2.3 | 0.0 | 4.0 | 50.3 | 4,264 | 40 | 39 | 599 | 5,040 | 45,278 | 3,422 |
| Bates | 48 | 2,966 | 56.3 | 0.0 | 11.6 | 7.0 | 7.7 | 37.9 | 2,326 | 57 | 58 | 1,095 | 7,080 | 47,514 | 3,709 |
| Benton | 39 | 2,025 | 62.3 | 5.4 | 3.0 | 13.6 | 4.5 | 22.4 | 1,176 | 102 | 64 | 918 | 8,230 | 41,416 | 3,086 |
| Bollinger | 27 | 2,206 | 65.9 | 0.9 | 3.7 | 0.3 | 1.7 | 3.2 | 257 | 26 | 40 | 438 | 4,790 | 41,706 | 2,596 |
| Boone | 871 | 4,893 | 35.9 | 32.4 | 3.4 | 0.2 | 4.3 | 2,544.1 | 14,292 | 2,612 | 603 | 30,924 | 78,880 | 67,287 | 7,658 |
| Buchanan | 310 | 3,500 | 48.4 | 1.4 | 6.6 | 0.1 | 6.1 | 804.6 | 9,075 | 499 | 292 | 6,013 | 38,480 | 53,038 | 4,943 |
| Butler | 133 | 3,128 | 70.5 | 0.0 | 7.5 | 0.0 | 5.4 | 63.4 | 1,488 | 809 | 139 | 2,673 | 17,370 | 44,758 | 3,817 |
| Caldwell | 34 | 3,699 | 57.4 | 1.9 | 5.1 | 5.7 | 7.6 | 47.0 | 5,190 | 44 | 29 | 580 | 3,890 | 47,642 | 3,572 |
| Callaway | 92 | 2,044 | 55.9 | 4.2 | 5.6 | 0.0 | 5.7 | 52.7 | 1,173 | 119 | 137 | 3,683 | 19,700 | 51,478 | 4,238 |
| Camden | 134 | 2,940 | 45.7 | 3.2 | 7.8 | 0.0 | 10.8 | 141.7 | 3,117 | 80 | 152 | 1,792 | 20,660 | 57,665 | 6,028 |
| Cape Girardeau | 243 | 3,101 | 48.2 | 0.4 | 6.1 | 0.0 | 7.2 | 162.3 | 2,074 | 393 | 280 | 6,141 | 35,140 | 62,233 | 6,467 |
| Carroll | 33 | 3,729 | 56.2 | 3.0 | 3.7 | 0.0 | 8.2 | 22.8 | 2,597 | 41 | 29 | 534 | 3,970 | 48,946 | 3,807 |
| Carter | 18 | 2,868 | 72.8 | 1.2 | 2.4 | 0.0 | 0.6 | 3.2 | 510 | 91 | 20 | 344 | 2,310 | 36,470 | 2,253 |
| Cass | 450 | 4,350 | 51.9 | 13.7 | 3.3 | 0.0 | 6.1 | 519.7 | 5,023 | 286 | 351 | 4,610 | 50,210 | 66,890 | 6,858 |
| Cedar | 56 | 3,944 | 39.4 | 43.9 | 2.7 | 0.0 | 2.4 | 34.6 | 2,459 | 67 | 48 | 745 | 5,520 | 39,743 | 2,695 |
| Chariton | 23 | 3,107 | 60.9 | 0.0 | 5.5 | 0.0 | 12.7 | 7.2 | 971 | 40 | 24 | 430 | 3,380 | 50,086 | 4,186 |
| Christian | 227 | 2,655 | 69.5 | 0.7 | 5.0 | 0.0 | 5.0 | 233.0 | 2,729 | 141 | 295 | 3,244 | 38,990 | 61,180 | 5,802 |
| Clark | 21 | 3,050 | 50.6 | 1.2 | 4.2 | 14.8 | 5.2 | 16.4 | 2,441 | 34 | 22 | 433 | 2,850 | 43,991 | 3,203 |
| Clay | 1,516 | 6,251 | 30.9 | 48.6 | 2.1 | 0.0 | 2.1 | 1,044.1 | 4,305 | 1,339 | 896 | 14,844 | 121,230 | 66,732 | 6,970 |
| Clinton | 54 | 2,618 | 63.9 | 2.4 | 5.4 | 0.0 | 10.8 | 69.2 | 3,367 | 55 | 67 | 971 | 9,690 | 56,352 | 4,859 |
| Cole | 227 | 2,961 | 49.3 | 3.1 | 9.2 | 0.0 | 6.9 | 191.5 | 2,498 | 640 | 263 | 18,757 | 36,290 | 62,772 | 6,271 |
| Cooper | 56 | 3,154 | 46.9 | 2.3 | 4.8 | 4.9 | 6.8 | 57.4 | 3,252 | 55 | 54 | 1,069 | 7,620 | 47,122 | 3,527 |
| Crawford | 47 | 1,951 | 66.1 | 5.6 | 3.2 | 0.0 | 6.3 | 24.4 | 1,013 | 34 | 79 | 935 | 10,220 | 44,411 | 3,595 |
| Dade | 24 | 3,182 | 58.5 | 0.9 | 2.8 | 22.6 | 5.4 | 14.4 | 1,902 | 28 | 25 | 529 | 3,130 | 42,951 | 3,477 |
| Dallas | 20 | 1,171 | 86.6 | 0.0 | 2.2 | 0.0 | 2.1 | 3.3 | 200 | 35 | 56 | 585 | 6,850 | 39,717 | 2,622 |
| Daviess | 23 | 2,712 | 70.0 | 2.3 | 3.1 | 0.3 | 9.7 | 9.8 | 1,168 | 33 | 27 | 478 | 3,510 | 44,536 | 3,318 |
| DeKalb | 16 | 1,297 | 74.6 | 2.6 | 0.2 | 0.0 | 6.2 | 9.5 | 759 | 55 | 30 | 1,013 | 4,210 | 48,102 | 3,579 |
| Dent | 54 | 3,494 | 40.5 | 40.0 | 4.2 | 0.0 | 4.4 | 8.0 | 518 | 57 | 51 | 943 | 5,960 | 40,538 | 2,707 |
| Douglas | 23 | 1,723 | 64.7 | 0.4 | 5.5 | 0.0 | 11.6 | 4.0 | 298 | 57 | 44 | 392 | 5,070 | 35,011 | 2,075 |
| Dunklin | 83 | 2,747 | 64.3 | 1.0 | 6.6 | 0.0 | 6.1 | 38.4 | 1,276 | 91 | 95 | 1,472 | 10,710 | 42,530 | 3,360 |
| Franklin | 294 | 2,841 | 57.8 | 3.7 | 6.7 | 0.0 | 7.9 | 325.1 | 3,146 | 225 | 346 | 4,283 | 50,610 | 59,692 | 6,183 |
| Gasconade | 61 | 4,139 | 44.9 | 33.0 | 3.5 | 0.6 | 5.0 | 51.4 | 3,501 | 52 | 48 | 993 | 7,150 | 45,868 | 3,515 |
| Gentry | 22 | 3,350 | 61.2 | 5.7 | 2.3 | 0.0 | 8.1 | 14.8 | 2,233 | 39 | 21 | 413 | 2,760 | 45,423 | 3,188 |
| Greene | 969 | 3,345 | 51.7 | 1.1 | 9.0 | 0.2 | 6.9 | 1,429.1 | 4,936 | 2,189 | 960 | 19,344 | 133,770 | 59,067 | 6,789 |
| Grundy | 59 | 5,920 | 56.9 | 0.4 | 4.1 | 9.2 | 7.3 | 23.1 | 2,307 | 59 | 32 | 922 | 4,030 | 42,885 | 3,054 |
| Harrison | 57 | 6,701 | 27.8 | 48.0 | 1.5 | 0.0 | 6.9 | 18.5 | 2,170 | 41 | 27 | 778 | 3,630 | 37,956 | 2,673 |
| Henry | 146 | 6,706 | 21.9 | 60.1 | 2.6 | 0.0 | 4.3 | 95.0 | 4,377 | 71 | 72 | 1,708 | 9,680 | 47,025 | 3,748 |
| Hickory | 22 | 2,334 | 77.2 | 1.7 | 2.9 | 0.0 | 6.4 | 20.8 | 2,213 | 41 | 32 | 285 | 3,630 | 36,314 | 2,317 |
| Holt | 16 | 3,511 | 54.9 | 0.7 | 4.7 | 0.6 | 6.4 | 6.7 | 1,525 | 34 | 14 | 287 | 2,040 | 49,701 | 4,369 |
| Howard | 25 | 2,505 | 62.6 | 3.1 | 4.2 | 0.3 | 5.8 | 31.6 | 3,139 | 39 | 31 | 461 | 4,350 | 47,540 | 3,804 |
| Howell | 107 | 2,666 | 67.0 | 3.7 | 5.5 | 0.0 | 4.2 | 17.1 | 426 | 139 | 133 | 2,103 | 16,190 | 44,566 | 3,888 |
| Iron | 32 | 3,117 | 65.6 | 10.1 | 2.3 | 0.0 | 0.6 | 10.7 | 1,055 | 15 | 65 | 633 | 3,920 | 39,566 | 2,558 |
| Jackson | 3,940 | 5,645 | 34.2 | 2.3 | 9.0 | 0.3 | 3.7 | 6,617.9 | 9,480 | 17,287 | 2,489 | 42,665 | 332,910 | 61,244 | 6,714 |

1. Based on the resident population estimated as of July 1 of the year shown.

# Table B. States and Counties — Land Area and Population

| State / county code | CBSA code[1] | County Type code[2] | STATE County | Land area[3] (sq. mi) | Total persons 2019 | Rank | Per square mile | White | Black | American Indian, Alaska Native | Asian and Pacific Islander | Percent Hispanic or Latino[4] | Under 5 years | 5 to 14 years | 15 to 24 years | 25 to 34 years | 35 to 44 years | 45 to 54 years |
|---|---|---|---|---|---|---|---|---|---|---|---|---|---|---|---|---|---|---|
| | | | | 1 | 2 | 3 | 4 | 5 | 6 | 7 | 8 | 9 | 10 | 11 | 12 | 13 | 14 | 15 |
| | | | **MISSOURI— Cont'd** | | | | | | | | | | | | | | | |
| 29097 | 27900 | 3 | Jasper | 638.5 | 121648 | 520 | 190.5 | 86.3 | 3.3 | 3.0 | 2.1 | 8.6 | 6.5 | 13.9 | 13.3 | 13.9 | 12.7 | 11.5 |
| 29099 | 41180 | 1 | Jefferson | 656.4 | 226543 | 305 | 345.1 | 95.5 | 1.8 | 0.8 | 1.3 | 2.2 | 5.8 | 12.9 | 11.3 | 12.8 | 13.4 | 12.8 |
| 29101 | 47660 | 4 | Johnson | 829.3 | 54219 | 943 | 65.4 | 87.6 | 5.8 | 1.4 | 3.0 | 5.1 | 6.2 | 11.9 | 23.1 | 14.4 | 10.9 | 9.2 |
| 29103 | | 9 | Knox | 504.0 | 3,940 | 2,899 | 7.8 | 97.3 | 1.2 | 1.2 | 0.7 | 1.2 | 6.9 | 12.8 | 12.2 | 10.2 | 9.2 | 10.9 |
| 29105 | 30060 | 6 | Laclede | 764.6 | 35895 | 1,280 | 46.9 | 95.0 | 1.7 | 1.7 | 1.3 | 2.7 | 6.9 | 13.6 | 11.6 | 11.7 | 11.9 | 12.0 |
| 29107 | 28140 | 6 | Lafayette | 628.3 | 33006 | 1,356 | 52.5 | 93.3 | 3.1 | 1.3 | 1.1 | 3.5 | 6.1 | 13.0 | 11.5 | 11.9 | 11.5 | 12.5 |
| 29109 | | 6 | Lawrence | 611.7 | 38175 | 1,221 | 62.4 | 90.0 | 1.0 | 1.9 | 0.9 | 8.1 | 6.9 | 13.9 | 12.2 | 11.6 | 11.6 | 11.9 |
| 29111 | 39500 | 9 | Lewis | 505.0 | 9,810 | 2,431 | 19.4 | 94.2 | 4.0 | 1.0 | 0.7 | 1.9 | 6.0 | 12.3 | 16.1 | 10.6 | 10.9 | 10.9 |
| 29113 | 41180 | 1 | Lincoln | 626.6 | 60119 | 871 | 95.9 | 94.6 | 2.9 | 1.0 | 0.9 | 2.6 | 7.0 | 14.0 | 11.8 | 14.2 | 12.7 | 12.1 |
| 29115 | | 7 | Linn | 615.6 | 11830 | 2,304 | 19.2 | 95.8 | 1.6 | 0.7 | 0.5 | 2.7 | 5.7 | 13.4 | 11.4 | 11.1 | 11.2 | 10.7 |
| 29117 | | 6 | Livingston | 532.3 | 14413 | 2,129 | 27.1 | 94.0 | 3.6 | 0.9 | 1.1 | 2.0 | 5.8 | 12.5 | 11.2 | 12.9 | 12.4 | 11.7 |
| 29119 | | 2 | McDonald | 539.4 | 22900 | 1,686 | 42.5 | 79.7 | 2.7 | 4.7 | 5.1 | 11.3 | 7.3 | 13.6 | 12.5 | 11.7 | 11.8 | 13.0 |
| 29121 | | 7 | Macon | 801.2 | 15095 | 2,089 | 18.8 | 94.7 | 3.4 | 0.9 | 1.1 | 1.6 | 5.7 | 13.0 | 11.5 | 10.5 | 10.6 | 11.9 |
| 29123 | | 6 | Madison | 494.4 | 12113 | 2,285 | 24.5 | 95.7 | 1.0 | 1.2 | 1.1 | 2.4 | 5.5 | 13.3 | 11.2 | 11.5 | 11.7 | 12.2 |
| 29125 | | 8 | Maries | 527.0 | 8,795 | 2,510 | 16.7 | 96.5 | 1.3 | 1.6 | 0.8 | 1.6 | 4.8 | 11.5 | 10.9 | 10.9 | 11.0 | 12.7 |
| 29127 | 25300 | 5 | Marion | 436.9 | 28423 | 1,483 | 65.1 | 92.4 | 6.3 | 0.8 | 1.2 | 1.9 | 6.2 | 13.1 | 13.0 | 11.9 | 11.9 | 11.6 |
| 29129 | | 6 | Mercer | 453.8 | 3,558 | 2,929 | 7.8 | 95.2 | 0.6 | 1.3 | 1.0 | 3.0 | 6.5 | 13.2 | 11.8 | 9.4 | 9.9 | 10.8 |
| 29131 | | 6 | Miller | 592.6 | 25791 | 1,567 | 43.5 | 96.1 | 1.1 | 1.4 | 0.9 | 2.0 | 6.2 | 13.3 | 11.4 | 11.7 | 11.7 | 11.9 |
| 29133 | | 6 | Mississippi | 411.6 | 12691 | 2,236 | 30.8 | 72.9 | 24.9 | 0.8 | 0.5 | 2.5 | 5.7 | 13.3 | 11.4 | 11.7 | 11.7 | 11.9 |
| 29135 | 27620 | 3 | Moniteau | 415.0 | 15585 | 2,052 | 37.6 | 90.9 | 3.2 | 1.1 | 0.9 | 5.4 | 6.4 | 14.4 | 13.4 | 12.5 | 12.8 | 12.0 |
| 29137 | | 9 | Monroe | 647.6 | 8,672 | 2,523 | 13.4 | 94.7 | 3.9 | 1.1 | 0.8 | 1.6 | 5.7 | 12.9 | 12.1 | 13.4 | 13.2 | 12.0 |
| 29139 | | 6 | Montgomery | 535.0 | 11294 | 2,328 | 21.1 | 94.9 | 2.4 | 1.1 | 0.9 | 2.5 | 5.6 | 12.0 | 11.1 | 10.7 | 11.7 | 11.6 |
| 29141 | | 8 | Morgan | 597.6 | 20716 | 1,782 | 34.7 | 95.6 | 1.2 | 1.7 | 0.8 | 2.4 | 6.5 | 12.4 | 10.5 | 10.3 | 9.4 | 10.8 |
| 29143 | | 7 | New Madrid | 674.9 | 16693 | 1,994 | 24.7 | 81.3 | 17.0 | 0.8 | 0.7 | 2.1 | 6.3 | 13.1 | 10.8 | 11.5 | 11.2 | 12.6 |
| 29145 | 27900 | 3 | Newton | 624.8 | 58451 | 885 | 93.6 | 88.3 | 1.8 | 4.3 | 3.1 | 5.7 | 6.2 | 13.4 | 12.1 | 11.8 | 11.6 | 12.1 |
| 29147 | 32340 | 6 | Nodaway | 877.0 | 21743 | 1,730 | 24.8 | 93.6 | 3.1 | 0.6 | 1.9 | 1.8 | 4.5 | 8.3 | 31.8 | 10.4 | 9.4 | 8.7 |
| 29149 | | 9 | Oregon | 789.8 | 10411 | 2,393 | 13.2 | 95.9 | 0.9 | 2.8 | 0.7 | 1.9 | 5.6 | 12.9 | 10.6 | 10.5 | 9.7 | 11.4 |
| 29151 | 27620 | 9 | Osage | 606.6 | 13535 | 2,188 | 22.3 | 98.0 | 0.7 | 0.7 | 0.3 | 1.0 | 5.5 | 12.5 | 13.0 | 11.3 | 11.9 | 12.6 |
| 29153 | | 9 | Ozark | 745.0 | 9,083 | 2,488 | 12.2 | 96.8 | 0.7 | 1.9 | 0.6 | 1.9 | 4.3 | 10.7 | 8.7 | 8.2 | 9.9 | 11.1 |
| 29155 | | 7 | Pemiscot | 492.6 | 15600 | 2,051 | 31.7 | 70.2 | 27.5 | 0.9 | 0.7 | 2.9 | 7.6 | 14.2 | 11.9 | 11.8 | 11.6 | 11.3 |
| 29157 | | 6 | Perry | 474.4 | 19194 | 1,868 | 40.5 | 95.8 | 1.0 | 0.8 | 1.3 | 2.3 | 5.7 | 12.9 | 11.8 | 11.5 | 12.0 | 12.4 |
| 29159 | 42740 | 4 | Pettis | 682.2 | 42490 | 1,130 | 62.3 | 86.1 | 4.7 | 1.0 | 1.6 | 9.2 | 6.9 | 13.9 | 12.4 | 12.8 | 12.3 | 11.0 |
| 29161 | 40620 | 5 | Phelps | 671.8 | 44414 | 1,087 | 66.1 | 90.8 | 3.0 | 1.6 | 4.2 | 2.8 | 5.4 | 11.9 | 20.0 | 11.7 | 11.0 | 10.3 |
| 29163 | | 6 | Pike | 670.4 | 17552 | 1,942 | 26.2 | 91.1 | 6.8 | 0.7 | 0.8 | 2.3 | 5.9 | 13.0 | 11.4 | 13.1 | 12.1 | 12.0 |
| 29165 | 28140 | 1 | Platte | 419.8 | 106532 | 577 | 253.8 | 81.8 | 8.7 | 1.1 | 4.4 | 6.6 | 6.0 | 13.5 | 11.8 | 13.3 | 14.2 | 13.0 |
| 29167 | 44180 | 2 | Polk | 635.5 | 32490 | 1,368 | 51.1 | 94.8 | 1.4 | 1.6 | 1.3 | 2.7 | 6.3 | 12.6 | 16.5 | 11.5 | 10.8 | 11.1 |
| 29169 | 22780 | 5 | Pulaski | 547.1 | 52709 | 962 | 96.3 | 72.3 | 13.1 | 1.8 | 5.2 | 11.7 | 6.8 | 11.9 | 24.6 | 18.3 | 12.5 | 8.3 |
| 29171 | | 9 | Putnam | 517.3 | 4,688 | 2,847 | 9.1 | 96.1 | 0.7 | 0.9 | 0.9 | 2.8 | 6.5 | 12.3 | 10.3 | 10.3 | 10.3 | 11.2 |
| 29173 | 25300 | 9 | Ralls | 469.8 | 10299 | 2,400 | 21.9 | 96.4 | 1.9 | 0.8 | 0.8 | 1.5 | 5.1 | 11.5 | 10.3 | 10.3 | 11.5 | 12.7 |
| 29175 | 33620 | 6 | Randolph | 482.7 | 24409 | 1,632 | 50.6 | 91.0 | 7.1 | 1.1 | 1.1 | 2.2 | 6.0 | 11.8 | 12.8 | 14.1 | 12.7 | 12.6 |
| 29177 | 28140 | 1 | Ray | 569.0 | 22915 | 1,685 | 40.3 | 94.8 | 2.2 | 1.2 | 0.8 | 2.7 | 5.7 | 12.6 | 11.5 | 11.4 | 11.6 | 12.6 |
| 29179 | | 9 | Reynolds | 808.5 | 6,198 | 2,732 | 7.7 | 95.7 | 1.7 | 2.7 | 0.7 | 1.8 | 4.7 | 10.9 | 10.4 | 10.0 | 10.6 | 12.8 |
| 29181 | 38740 | 9 | Ripley | 629.5 | 13300 | 2,201 | 21.1 | 96.5 | 1.0 | 2.0 | 0.6 | 1.6 | 5.9 | 13.5 | 10.2 | 11.6 | 11.1 | 11.9 |
| 29183 | 41180 | 1 | St. Charles | 560.5 | 406204 | 176 | 724.7 | 88.1 | 6.3 | 0.6 | 3.5 | 3.5 | 5.7 | 13.0 | 12.3 | 12.7 | 13.8 | 12.5 |
| 29185 | | 8 | St. Clair | 675.0 | 9,689 | 2,437 | 14.4 | 96.0 | 1.0 | 1.8 | 0.7 | 2.3 | 5.4 | 11.2 | 9.8 | 9.9 | 9.5 | 11.6 |
| 29186 | | 6 | Ste. Genevieve | 499.2 | 17924 | 1,925 | 35.9 | 96.7 | 1.3 | 0.8 | 1.1 | 1.2 | 5.8 | 12.2 | 11.0 | 10.9 | 11.4 | 11.6 |
| 29187 | 22100 | 4 | St. Francois | 451.9 | 66485 | 805 | 147.1 | 93.3 | 4.6 | 0.9 | 0.8 | 1.7 | 5.4 | 11.9 | 11.9 | 14.3 | 13.5 | 12.7 |
| 29189 | 41180 | 1 | St. Louis | 507.9 | 994020 | 47 | 1,957.1 | 66.8 | 26.1 | 0.7 | 5.7 | 3.1 | 5.9 | 12.2 | 12.1 | 12.9 | 12.4 | 11.8 |
| 29195 | 32180 | 6 | Saline | 755.5 | 22858 | 1,688 | 30.3 | 80.7 | 6.4 | 1.0 | 2.8 | 11.4 | 5.5 | 12.8 | 15.1 | 11.6 | 11.4 | 11.3 |
| 29197 | 28860 | 9 | Schuyler | 307.3 | 4,534 | 2,856 | 14.8 | 97.8 | 1.0 | 0.7 | 0.6 | 1.4 | 8.0 | 14.0 | 11.0 | 12.5 | 9.7 | 10.8 |
| 29199 | | 9 | Scotland | 436.5 | 4,871 | 2,835 | 11.2 | 98.2 | 0.7 | 0.8 | 0.4 | 1.2 | 8.1 | 16.0 | 12.2 | 11.1 | 10.5 | 9.9 |
| 29201 | 43460 | 4 | Scott | 420.0 | 38288 | 1,218 | 91.2 | 84.5 | 13.0 | 0.9 | 0.7 | 2.7 | 6.4 | 13.4 | 11.5 | 12.7 | 11.7 | 12.0 |
| 29203 | | 9 | Shannon | 1,003.8 | 8,203 | 2,562 | 8.2 | 96.0 | 1.1 | 2.8 | 0.6 | 2.0 | 5.6 | 12.3 | 9.8 | 10.9 | 10.0 | 11.8 |
| 29205 | | 9 | Shelby | 500.9 | 5,919 | 2,751 | 11.8 | 95.7 | 2.1 | 0.8 | 0.4 | 2.5 | 6.4 | 13.8 | 10.3 | 10.2 | 12.1 | 10.0 |
| 29207 | | 6 | Stoddard | 823.1 | 29001 | 1,459 | 35.2 | 96.1 | 1.8 | 0.9 | 0.6 | 1.9 | 5.6 | 12.3 | 11.6 | 11.9 | 11.9 | 12.2 |
| 29209 | | 6 | Stone | 463.8 | 32465 | 1,369 | 70.0 | 95.9 | 0.9 | 1.8 | 0.7 | 2.4 | 4.1 | 9.0 | 8.9 | 8.2 | 9.0 | 11.2 |
| 29211 | | 9 | Sullivan | 648.0 | 6,033 | 2,742 | 9.3 | 77.6 | 3.2 | 1.0 | 0.5 | 18.8 | 6.9 | 12.1 | 11.6 | 11.3 | 10.4 | 13.2 |
| 29213 | 14700 | 4 | Taney | 632.3 | 56104 | 919 | 88.7 | 89.9 | 2.3 | 1.9 | 1.7 | 6.4 | 5.5 | 11.7 | 12.9 | 11.0 | 11.1 | 11.4 |
| 29215 | | 6 | Texas | 1,177.3 | 25112 | 1,601 | 21.3 | 93.1 | 3.9 | 2.0 | 0.7 | 2.4 | 5.5 | 12.0 | 11.0 | 11.4 | 11.7 | 11.6 |
| 29217 | | 7 | Vernon | 826.4 | 20388 | 1,799 | 24.7 | 95.3 | 1.4 | 1.7 | 1.0 | 2.5 | 6.2 | 13.0 | 12.8 | 10.9 | 11.0 | 11.6 |
| 29219 | 41180 | 1 | Warren | 428.6 | 36594 | 1,260 | 85.4 | 93.4 | 3.2 | 0.9 | 0.8 | 3.6 | 6.3 | 13.1 | 11.4 | 12.6 | 11.4 | 13.4 |
| 29221 | | 6 | Washington | 759.9 | 24604 | 1,624 | 32.4 | 95.2 | 2.9 | 1.2 | 0.5 | 1.6 | 5.8 | 12.6 | 11.5 | 12.2 | 11.9 | 13.4 |
| 29223 | | 9 | Wayne | 759.2 | 12769 | 2,230 | 16.8 | 96.4 | 1.4 | 1.9 | 0.5 | 2.0 | 4.8 | 11.6 | 9.6 | 9.8 | 10.3 | 12.2 |
| 29225 | 44180 | 2 | Webster | 592.6 | 39859 | 1,192 | 67.3 | 95.6 | 1.5 | 1.7 | 0.7 | 2.3 | 7.6 | 15.1 | 12.2 | 12.2 | 12.1 | 12.1 |
| 29227 | | 9 | Worth | 266.6 | 1,953 | 3,054 | 7.3 | 96.4 | 1.3 | 0.9 | 0.4 | 2.0 | 5.3 | 10.3 | 10.5 | 9.9 | 10.5 | 10.0 |
| 29229 | | 6 | Wright | 681.8 | 18325 | 1,908 | 26.9 | 95.7 | 1.0 | 1.6 | 0.7 | 2.4 | 7.1 | 14.1 | 11.9 | 10.9 | 10.8 | 11.5 |

1. CBSA = Core Based Statistical Area. See Appendix A for explanation. See Appendix B for list of metropolitan areas with component counties.
Service of USDA Rural-Urban Continuum Codes. See Appendix A for definition. 3. Dry land or land partially or temporarily covered by water.
2. County type code from the Economic Research
4. May be of any race.

# Table B. States and Counties — Population and Households

| STATE County | Age (percent): 55 to 64 years | 65 to 74 years | 75 years and over | Percent female | Total persons 2000 | Total persons 2010 | Percent change 2000-2010 | Percent change 2010-2020 | Births | Deaths | Net Migration | Households Number | Persons per household | Family households | Female family householder[1] | One person |
|---|---|---|---|---|---|---|---|---|---|---|---|---|---|---|---|---|
| | 16 | 17 | 18 | 19 | 20 | 21 | 22 | 23 | 24 | 25 | 26 | 27 | 28 | 29 | 30 | 31 |
| **MISSOURI— Cont'd** | | | | | | | | | | | | | | | | |
| Jasper | 12.1 | 9.3 | 6.9 | 51.3 | 104,686 | 117,395 | 12.1 | 3.6 | 16,948 | 12,274 | -364 | 45,759 | 2.57 | 65.9 | 12.4 | 27.9 |
| Jefferson | 14.8 | 10.1 | 6.0 | 50.3 | 198,099 | 218,722 | 10.4 | 3.6 | 26,665 | 20,602 | 1,929 | 84,444 | 2.63 | 73.0 | 11.3 | 21.5 |
| Johnson | 11.0 | 7.6 | 5.6 | 48.8 | 48,258 | 52,565 | 8.9 | 3.1 | 7,316 | 4,155 | -1,530 | 19,864 | 2.48 | 65.7 | 9.3 | 22.6 |
| Knox | 15.4 | 11.5 | 10.9 | 50.5 | 4,361 | 4,131 | -5.3 | -4.6 | 540 | 499 | -233 | 1,490 | 2.59 | 62.7 | 7.9 | 34.4 |
| Laclede | 13.9 | 10.6 | 7.7 | 50.4 | 32,513 | 35,594 | 9.5 | 0.8 | 4,883 | 4,185 | -395 | 14,112 | 2.49 | 72.8 | 9.9 | 22.7 |
| Lafayette | 14.2 | 10.4 | 8.8 | 50.2 | 32,960 | 33,369 | 1.2 | -1.1 | 3,863 | 3,898 | -320 | 13,050 | 2.45 | 68.5 | 10.2 | 25.2 |
| Lawrence | 13.4 | 10.1 | 8.4 | 50.1 | 35,204 | 38,647 | 9.8 | -1.2 | 5,120 | 4,748 | -832 | 14,787 | 2.54 | 69.8 | 10.7 | 25.9 |
| Lewis | 13.8 | 10.3 | 9.1 | 49.9 | 10,494 | 10,205 | -2.8 | -3.9 | 1,193 | 1,152 | -436 | 3,752 | 2.43 | 66.7 | 7.7 | 29.4 |
| Lincoln | 14.0 | 8.8 | 5.5 | 49.9 | 38,944 | 52,536 | 34.9 | 14.4 | 7,696 | 4,757 | 4,677 | 19,286 | 2.89 | 74.9 | 10.9 | 20.9 |
| Linn | 14.8 | 12.0 | 9.7 | 51.3 | 13,754 | 12,773 | -7.1 | -7.4 | 1,412 | 1,756 | -594 | 5,106 | 2.34 | 67.3 | 12.1 | 29.3 |
| Livingston | 12.8 | 10.7 | 10.1 | 54.5 | 14,558 | 15,196 | 4.4 | -5.2 | 1,643 | 1,947 | -484 | 5,915 | 2.32 | 65.0 | 9.5 | 28.4 |
| McDonald | 14.1 | 9.4 | 6.5 | 49.5 | 21,681 | 23,083 | 6.5 | -0.8 | 3,367 | 2,339 | -1,218 | 8,259 | 2.74 | 75.8 | 11.6 | 19.0 |
| Macon | 13.8 | 12.3 | 10.7 | 50.5 | 15,762 | 15,553 | -1.3 | -2.9 | 1,801 | 2,019 | -234 | 5,845 | 2.55 | 61.9 | 8.9 | 34.2 |
| Madison | 14.6 | 11.3 | 8.6 | 50.8 | 11,800 | 12,217 | 3.5 | -0.9 | 1,424 | 1,697 | 170 | 5,044 | 2.38 | 68.8 | 8.4 | 23.6 |
| Maries | 15.9 | 12.4 | 10.0 | 49.9 | 8,903 | 9,150 | 2.8 | -3.9 | 858 | 1,057 | -154 | 3,762 | 2.31 | 65.1 | 7.2 | 28.6 |
| Marion | 13.5 | 10.7 | 8.0 | 51.4 | 28,289 | 28,778 | 1.7 | -1.2 | 3,746 | 3,389 | -686 | 11,565 | 2.34 | 64.2 | 10.0 | 29.9 |
| Mercer | 15.1 | 12.4 | 11.0 | 50.1 | 3,757 | 3,785 | 0.7 | -6.0 | 456 | 450 | -231 | 1,271 | 2.81 | 61.5 | 5.8 | 34.1 |
| Miller | 14.1 | 11.6 | 8.2 | 47.3 | 23,564 | 24,733 | 5.0 | 4.3 | 3,116 | 2,895 | 861 | 10,111 | 2.46 | 69.6 | 9.9 | 26.6 |
| Mississippi | 12.9 | 10.5 | 8.3 | 48.3 | 13,427 | 14,377 | 7.1 | -11.7 | 1,632 | 1,855 | -1,458 | 5,026 | 2.36 | 72.4 | 17.6 | 22.8 |
| Moniteau | 12.6 | 9.5 | 7.2 | 49.1 | 14,827 | 15,610 | 5.3 | -0.2 | 2,084 | 1,518 | -580 | 5,436 | 2.70 | 70.5 | 7.4 | 25.6 |
| Monroe | 15.7 | 13.5 | 10.9 | 49.1 | 9,311 | 8,844 | -5.0 | -1.9 | 952 | 1,002 | -122 | 3,702 | 2.29 | 65.1 | 8.9 | 29.6 |
| Montgomery | 15.9 | 11.2 | 10.0 | 49.7 | 12,136 | 12,226 | 0.7 | -7.6 | 1,393 | 1,692 | -637 | 5,014 | 2.22 | 63.9 | 10.6 | 30.2 |
| Morgan | 15.8 | 13.9 | 10.4 | 49.5 | 19,309 | 20,567 | 6.5 | 0.7 | 2,646 | 3,048 | 558 | 7,588 | 2.62 | 65.0 | 8.7 | 30.4 |
| New Madrid | 14.7 | 11.4 | 8.6 | 52.4 | 19,760 | 18,932 | -4.2 | -11.8 | 2,334 | 2,499 | -2,079 | 7,371 | 2.34 | 65.0 | 15.8 | 29.9 |
| Newton | 13.8 | 10.9 | 8.0 | 50.1 | 52,636 | 58,114 | 10.4 | 0.6 | 7,403 | 6,503 | -536 | 22,202 | 2.57 | 69.9 | 8.8 | 25.1 |
| Nodaway | 10.8 | 8.5 | 7.6 | 50.4 | 21,912 | 23,373 | 6.7 | -7.0 | 2,271 | 1,976 | -1,940 | 8,395 | 2.29 | 57.7 | 8.4 | 28.8 |
| Oregon | 15.8 | 13.1 | 10.5 | 50.5 | 10,344 | 10,881 | 5.2 | -4.3 | 1,278 | 1,463 | -283 | 4,249 | 2.47 | 66.2 | 8.4 | 27.9 |
| Osage | 14.8 | 10.3 | 8.1 | 48.7 | 13,062 | 13,911 | 6.5 | -2.7 | 1,539 | 1,353 | -564 | 5,120 | 2.60 | 72.9 | 5.7 | 22.4 |
| Ozark | 17.3 | 17.2 | 12.7 | 49.4 | 9,542 | 9,728 | 1.9 | -6.6 | 860 | 1,294 | -202 | 3,958 | 2.30 | 60.4 | 4.5 | 35.0 |
| Pemiscot | 13.6 | 10.4 | 7.7 | 52.7 | 20,047 | 18,285 | -8.8 | -14.7 | 2,702 | 2,398 | -3,000 | 6,730 | 2.45 | 63.7 | 21.6 | 30.1 |
| Perry | 13.9 | 11.0 | 8.8 | 50.2 | 18,132 | 18,964 | 4.6 | 1.2 | 2,263 | 2,065 | 49 | 7,576 | 2.49 | 70.3 | 8.8 | 24.1 |
| Pettis | 13.2 | 9.9 | 7.6 | 50.3 | 39,403 | 42,201 | 7.1 | 0.7 | 6,000 | 4,510 | -1,178 | 16,033 | 2.58 | 65.9 | 11.0 | 29.4 |
| Phelps | 12.4 | 9.7 | 7.6 | 47.7 | 39,825 | 45,125 | 13.3 | -1.6 | 5,171 | 4,610 | -1,301 | 17,981 | 2.31 | 61.5 | 10.0 | 29.4 |
| Pike | 13.4 | 10.6 | 8.5 | 46.7 | 18,351 | 18,511 | 0.9 | -5.2 | 2,216 | 2,059 | -1,116 | 6,589 | 2.46 | 63.7 | 10.2 | 30.0 |
| Platte | 12.7 | 9.3 | 6.2 | 50.7 | 73,781 | 89,331 | 21.1 | 19.3 | 12,087 | 6,851 | 11,973 | 39,305 | 2.54 | 66.1 | 8.8 | 26.9 |
| Polk | 13.2 | 9.9 | 8.2 | 50.8 | 26,992 | 31,130 | 15.3 | 4.4 | 4,014 | 3,680 | 1,043 | 11,712 | 2.58 | 69.9 | 10.9 | 21.7 |
| Pulaski | 8.3 | 5.6 | 3.8 | 43.6 | 41,165 | 52,282 | 27.0 | 0.8 | 8,086 | 3,184 | -4,590 | 15,154 | 2.84 | 68.2 | 9.6 | 25.8 |
| Putnam | 14.5 | 13.6 | 11.1 | 49.4 | 5,223 | 4,979 | -4.7 | -5.8 | 579 | 680 | -186 | 1,744 | 2.68 | 60.8 | 10.3 | 35.0 |
| Ralls | 16.0 | 13.4 | 9.3 | 49.7 | 9,626 | 10,178 | 5.7 | 1.2 | 1,003 | 978 | 99 | 4,036 | 2.51 | 73.0 | 7.9 | 23.4 |
| Randolph | 12.9 | 9.8 | 7.4 | 47.9 | 24,663 | 25,415 | 3.0 | -4.0 | 2,939 | 2,851 | -1,094 | 8,626 | 2.63 | 65.7 | 10.0 | 30.2 |
| Ray | 15.3 | 10.8 | 8.4 | 50.1 | 23,354 | 23,516 | 0.7 | -2.6 | 2,596 | 2,628 | -572 | 8,868 | 2.55 | 70.7 | 8.4 | 23.9 |
| Reynolds | 16.5 | 13.2 | 10.9 | 48.3 | 6,689 | 6,690 | 0.0 | -7.4 | 630 | 901 | -216 | 2,596 | 2.34 | 69.5 | 10.2 | 26.2 |
| Ripley | 14.8 | 11.3 | 9.7 | 50.4 | 13,509 | 14,106 | 4.4 | -5.7 | 1,740 | 1,950 | -586 | 5,059 | 2.67 | 66.8 | 13.2 | 27.4 |
| St. Charles | 13.8 | 9.6 | 6.7 | 50.8 | 283,883 | 360,488 | 27.0 | 12.7 | 45,826 | 28,007 | 28,018 | 146,631 | 2.64 | 71.7 | 9.2 | 23.2 |
| St. Clair | 15.7 | 14.3 | 12.6 | 49.3 | 9,652 | 9,805 | 1.6 | -1.2 | 975 | 1,469 | 385 | 4,139 | 2.19 | 61.4 | 6.5 | 32.5 |
| Ste. Genevieve | 16.2 | 12.4 | 8.5 | 49.0 | 17,842 | 18,147 | 1.7 | -1.2 | 1,923 | 2,007 | -135 | 7,121 | 2.47 | 72.5 | 7.7 | 23.5 |
| St. Francois | 13.1 | 9.8 | 7.4 | 47.3 | 55,641 | 65,369 | 17.5 | 1.7 | 7,393 | 7,951 | 1,688 | 24,898 | 2.39 | 66.1 | 11.7 | 28.0 |
| St. Louis | 13.8 | 10.8 | 8.2 | 52.5 | 1,016,315 | 998,961 | -1.7 | -0.5 | 118,492 | 101,760 | -21,456 | 405,984 | 2.41 | 63.8 | 13.6 | 30.6 |
| Saline | 13.4 | 10.3 | 8.5 | 50.0 | 23,756 | 23,372 | -1.6 | -2.2 | 2,841 | 2,606 | -744 | 8,272 | 2.61 | 62.6 | 8.4 | 32.1 |
| Schuyler | 14.1 | 9.9 | 9.9 | 50.4 | 4,170 | 4,431 | 6.3 | 2.3 | 674 | 531 | -39 | 1,475 | 3.05 | 62.5 | 6.3 | 32.4 |
| Scotland | 12.7 | 10.1 | 9.5 | 50.2 | 4,983 | 4,854 | -2.6 | 0.4 | 809 | 568 | -224 | 1,768 | 2.72 | 67.1 | 6.9 | 30.6 |
| Scott | 13.2 | 10.9 | 8.1 | 51.6 | 40,422 | 39,204 | -3.0 | -2.3 | 5,159 | 4,500 | -1,554 | 15,230 | 2.50 | 68.0 | 14.5 | 27.6 |
| Shannon | 16.1 | 13.1 | 10.3 | 50.0 | 8,324 | 8,437 | 1.4 | -2.8 | 935 | 963 | -203 | 3,064 | 2.64 | 68.3 | 10.6 | 25.5 |
| Shelby | 14.8 | 11.9 | 10.4 | 50.0 | 6,799 | 6,372 | -6.3 | -7.1 | 757 | 837 | -371 | 2,432 | 2.36 | 65.5 | 8.4 | 31.5 |
| Stoddard | 13.9 | 11.3 | 9.4 | 50.7 | 29,705 | 29,968 | 0.9 | -3.2 | 3,586 | 3,998 | -527 | 11,530 | 2.51 | 65.7 | 10.8 | 29.5 |
| Stone | 17.5 | 19.4 | 12.6 | 50.8 | 28,658 | 32,211 | 12.4 | 0.8 | 2,610 | 4,032 | 1,714 | 12,783 | 2.44 | 71.2 | 5.7 | 24.1 |
| Sullivan | 14.0 | 11.6 | 8.9 | 48.5 | 7,219 | 6,714 | -7.0 | -10.1 | 890 | 765 | -813 | 2,128 | 2.83 | 62.7 | 8.1 | 33.9 |
| Taney | 13.6 | 13.1 | 9.8 | 51.5 | 39,703 | 51,672 | 30.1 | 8.6 | 6,469 | 5,782 | 3,750 | 22,272 | 2.40 | 68.4 | 10.5 | 26.6 |
| Texas | 14.7 | 12.2 | 9.9 | 47.8 | 23,003 | 26,018 | 13.1 | -3.5 | 2,945 | 3,116 | -718 | 9,728 | 2.49 | 67.0 | 7.8 | 28.1 |
| Vernon | 13.8 | 11.5 | 9.2 | 51.3 | 20,454 | 21,165 | 3.5 | -3.7 | 2,570 | 2,565 | -773 | 8,207 | 2.43 | 65.1 | 10.2 | 29.9 |
| Warren | 15.4 | 10.5 | 7.7 | 50.0 | 24,525 | 32,546 | 32.7 | 12.4 | 4,271 | 2,990 | 2,793 | 12,654 | 2.70 | 74.3 | 9.6 | 20.8 |
| Washington | 14.9 | 10.7 | 7.0 | 48.5 | 23,344 | 25,200 | 8.0 | -2.4 | 2,985 | 2,881 | -686 | 9,231 | 2.57 | 71.4 | 13.0 | 24.4 |
| Wayne | 16.7 | 13.7 | 11.3 | 50.7 | 13,259 | 13,535 | 2.1 | -5.7 | 1,422 | 1,994 | -189 | 5,426 | 2.41 | 60.2 | 10.0 | 32.0 |
| Webster | 12.9 | 9.1 | 6.8 | 49.3 | 31,045 | 36,264 | 16.8 | 9.9 | 5,702 | 3,468 | 1,377 | 13,575 | 2.78 | 72.8 | 9.6 | 22.0 |
| Worth | 16.9 | 13.5 | 13.0 | 51.0 | 2,382 | 2,169 | -8.9 | -10.0 | 235 | 309 | -143 | 859 | 2.31 | 69.0 | 5.2 | 29.0 |
| Wright | 13.6 | 11.4 | 8.8 | 50.8 | 17,955 | 18,740 | 4.4 | -2.2 | 2,496 | 2,449 | -457 | 7,063 | 2.55 | 67.2 | 11.7 | 28.0 |

1. No spouse present.

# Population, Vital Statistics, Health, and Crime

| STATE County | Persons in group quarters, 2020 | Daytime Population, 2015-2019 Number | Employment/ residence ratio | Births, 2020 Total | Rate[1] | Deaths, 2020 Number | Rate[1] | Persons under 65 with no health insurance, 2019 Number | Percent | Medicare, 2020 Total beneficiaries | Enrolled in Original Medicare | Enrolled in Medicare Advantage | Crimes reported by county police or sheriff, 2019 Violent | Property |
|---|---|---|---|---|---|---|---|---|---|---|---|---|---|---|
| | 32 | 33 | 34 | 35 | 36 | 37 | 38 | 39 | 40 | 41 | 42 | 43 | 44 | 45 |
| **MISSOURI— Cont'd** | | | | | | | | | | | | | | |
| Jasper | 2,488 | 124,767 | 1.09 | 1,548 | 12.7 | 1,290 | 10.6 | 16,611 | 16.7 | 23,697 | 15,432 | 8,265 | NA | NA |
| Jefferson | 1,984 | 168,561 | 0.50 | 2,496 | 11.0 | 2,347 | 10.4 | 21,129 | 11.2 | 44,649 | 20,342 | 24,307 | 386 | 2,722 |
| Johnson | 3,734 | 50,298 | 0.87 | 691 | 12.7 | 426 | 7.9 | 5,367 | 12.3 | 8,384 | 5,850 | 2,534 | NA | NA |
| Knox | 87 | 3,872 | 0.95 | 62 | 15.7 | 43 | 10.9 | 624 | 20.8 | 924 | 818 | 106 | NA | NA |
| Laclede | 345 | 35,792 | 1.02 | 481 | 13.4 | 432 | 12.0 | 4,117 | 14.2 | 8,376 | 3,850 | 4,525 | NA | NA |
| Lafayette | 724 | 28,352 | 0.72 | 398 | 12.1 | 383 | 11.6 | 2,981 | 11.4 | 7,372 | 4,641 | 2,731 | 49 | 333 |
| Lawrence | 534 | 33,365 | 0.71 | 509 | 13.3 | 486 | 12.7 | 5,462 | 17.7 | 8,399 | 4,106 | 4,292 | NA | NA |
| Lewis | 788 | 8,489 | 0.66 | 116 | 11.8 | 104 | 10.6 | 930 | 12.7 | 2,277 | 1,953 | 324 | 17 | 224 |
| Lincoln | 632 | 43,301 | 0.50 | 823 | 13.7 | 525 | 8.7 | 5,922 | 11.7 | 10,581 | 5,809 | 4,772 | NA | NA |
| Linn | 152 | 11,466 | 0.88 | 135 | 11.4 | 158 | 13.4 | 1,248 | 13.4 | 3,115 | 2,591 | 524 | 8 | 20 |
| Livingston | 1,257 | 16,017 | 1.14 | 152 | 10.5 | 193 | 13.4 | 1,370 | 13.1 | 3,485 | 2,864 | 621 | 2 | 26 |
| McDonald | 157 | 20,588 | 0.76 | 329 | 14.4 | 240 | 10.5 | 4,280 | 22.7 | 4,334 | 2,496 | 1,837 | NA | NA |
| Macon | 286 | 14,319 | 0.86 | 171 | 11.3 | 196 | 13.0 | 1,681 | 14.6 | 3,913 | 3,320 | 593 | 4 | 66 |
| Madison | 165 | 11,201 | 0.80 | 127 | 10.5 | 174 | 14.4 | 1,293 | 13.5 | 3,182 | 2,298 | 884 | 0 | 0 |
| Maries | 67 | 6,542 | 0.39 | 88 | 10.0 | 114 | 13.0 | 1,175 | 17.5 | 1,979 | 1,424 | 555 | NA | NA |
| Marion | 1,367 | 30,728 | 1.17 | 350 | 12.3 | 327 | 11.5 | 2,493 | 11.1 | 6,690 | 5,533 | 1,157 | NA | NA |
| Mercer | 51 | 3,613 | 0.98 | 43 | 12.1 | 34 | 9.6 | 478 | 17.2 | 809 | 697 | 112 | 0 | 2 |
| Miller | 271 | 22,045 | 0.72 | 323 | 12.5 | 296 | 11.5 | 3,203 | 15.7 | 5,578 | 3,674 | 1,904 | NA | NA |
| Mississippi | 1,427 | 12,735 | 0.81 | 150 | 11.8 | 198 | 15.6 | 1,299 | 14.5 | 2,959 | 2,421 | 538 | NA | NA |
| Moniteau | 883 | 13,378 | 0.61 | 189 | 12.1 | 167 | 10.7 | 2,073 | 16.9 | 3,020 | 1,999 | 1,022 | NA | NA |
| Monroe | 128 | 7,449 | 0.69 | 88 | 10.1 | 113 | 13.0 | 876 | 13.5 | 2,149 | 1,748 | 400 | NA | NA |
| Montgomery | 394 | 10,132 | 0.73 | 135 | 12.0 | 168 | 14.9 | 1,274 | 14.4 | 2,895 | 2,026 | 868 | NA | NA |
| Morgan | 302 | 19,498 | 0.90 | 254 | 12.3 | 303 | 14.6 | 3,243 | 21.0 | 5,686 | 4,013 | 1,673 | NA | NA |
| New Madrid | 332 | 17,198 | 0.95 | 215 | 12.9 | 251 | 15.0 | 1,870 | 13.9 | 4,170 | 3,302 | 868 | NA | NA |
| Newton | 880 | 57,074 | 0.96 | 723 | 12.4 | 652 | 11.2 | 7,327 | 15.7 | 12,708 | 8,451 | 4,257 | NA | NA |
| Nodaway | 3,444 | 22,129 | 0.98 | 206 | 9.5 | 183 | 8.4 | 2,043 | 13.3 | 4,038 | 3,613 | 425 | NA | NA |
| Oregon | 121 | 10,047 | 0.85 | 118 | 11.3 | 151 | 14.5 | 1,302 | 16.5 | 2,603 | 1,960 | 644 | 29 | 91 |
| Osage | 405 | 11,138 | 0.64 | 139 | 10.3 | 138 | 10.2 | 1,228 | 11.2 | 2,790 | 1,840 | 950 | NA | NA |
| Ozark | 123 | 7,961 | 0.58 | 70 | 7.7 | 130 | 14.3 | 1,237 | 19.4 | 2,966 | 1,746 | 1,220 | 20 | 45 |
| Pemiscot | 202 | 16,392 | 0.95 | 221 | 14.2 | 238 | 15.3 | 1,742 | 13.7 | 3,724 | 2,604 | 1,120 | 22 | 80 |
| Perry | 286 | 19,692 | 1.05 | 216 | 11.3 | 226 | 11.8 | 1,713 | 11.2 | 4,362 | 3,591 | 772 | 31 | 39 |
| Pettis | 851 | 44,399 | 1.11 | 587 | 13.8 | 438 | 10.3 | 5,348 | 15.6 | 8,713 | 6,190 | 2,524 | NA | NA |
| Phelps | 3,012 | 47,489 | 1.15 | 440 | 9.9 | 484 | 10.9 | 5,346 | 15.6 | 8,960 | 6,496 | 2,465 | NA | NA |
| Pike | 1,522 | 18,189 | 0.96 | 199 | 11.3 | 197 | 11.2 | 1,676 | 13.1 | 3,877 | 2,805 | 1,072 | NA | NA |
| Platte | 869 | 95,014 | 0.89 | 1,214 | 11.4 | 767 | 7.2 | 7,116 | 8.0 | 17,424 | 11,811 | 5,613 | 48 | 290 |
| Polk | 1,514 | 28,841 | 0.80 | 425 | 13.1 | 429 | 13.2 | 3,980 | 16.0 | 7,216 | 3,515 | 3,701 | 34 | 326 |
| Pulaski | 10,019 | 54,260 | 1.06 | 725 | 13.8 | 340 | 6.5 | 4,553 | 12.1 | 6,510 | 5,228 | 1,283 | 157 | 346 |
| Putnam | 57 | 4,260 | 0.74 | 53 | 11.3 | 72 | 15.4 | 552 | 15.8 | 1,301 | 1,105 | 196 | NA | NA |
| Ralls | 54 | 8,651 | 0.68 | 91 | 8.8 | 95 | 9.2 | 904 | 11.3 | 2,640 | 2,176 | 464 | 30 | 82 |
| Randolph | 2,279 | 25,060 | 1.02 | 287 | 11.8 | 269 | 11.0 | 2,330 | 12.7 | 5,381 | 3,934 | 1,446 | 11 | 101 |
| Ray | 311 | 17,665 | 0.51 | 256 | 11.2 | 287 | 12.5 | 2,259 | 12.2 | 5,067 | 3,245 | 1,821 | NA | NA |
| Reynolds | 101 | 6,208 | 0.97 | 67 | 10.8 | 99 | 16.0 | 745 | 16.0 | 1,621 | 1,195 | 426 | NA | NA |
| Ripley | 63 | 12,101 | 0.71 | 154 | 11.6 | 221 | 16.6 | 1,777 | 17.2 | 3,391 | 2,711 | 680 | NA | NA |
| St. Charles | 5,452 | 339,329 | 0.74 | 4,254 | 10.5 | 3,274 | 8.1 | 23,128 | 6.9 | 72,013 | 36,999 | 35,014 | 139 | 757 |
| St. Clair | 207 | 8,408 | 0.72 | 92 | 9.5 | 152 | 15.7 | 1,088 | 16.4 | 2,749 | 1,729 | 1,019 | 15 | 111 |
| Ste. Genevieve | 277 | 15,603 | 0.74 | 209 | 11.7 | 202 | 11.3 | 1,496 | 10.6 | 4,310 | 2,861 | 1,449 | 53 | 128 |
| St. Francois | 6,279 | 66,222 | 0.98 | 678 | 10.2 | 836 | 12.6 | 6,205 | 12.5 | 15,096 | 9,485 | 5,611 | NA | NA |
| St. Louis | 19,110 | 1,104,447 | 1.22 | 11,261 | 11.3 | 10,827 | 10.9 | 65,929 | 8.3 | 207,177 | 112,412 | 94,766 | 1,821 | 8,077 |
| Saline | 1,491 | 22,557 | 0.96 | 256 | 11.2 | 259 | 11.3 | 2,506 | 14.5 | 4,926 | 3,366 | 1,561 | NA | NA |
| Schuyler | 47 | 3,668 | 0.52 | 65 | 14.3 | 48 | 10.6 | 597 | 16.2 | 1,000 | 850 | 150 | 4 | 36 |
| Scotland | 76 | 4,676 | 0.89 | 72 | 14.8 | 45 | 9.2 | 850 | 21.6 | 917 | 827 | 90 | NA | NA |
| Scott | 559 | 39,480 | 1.05 | 488 | 12.7 | 495 | 12.9 | 4,177 | 13.6 | 9,183 | 7,236 | 1,947 | NA | NA |
| Shannon | 80 | 7,545 | 0.77 | 88 | 10.7 | 106 | 12.9 | 1,148 | 18.5 | 1,945 | 1,318 | 627 | 17 | 22 |
| Shelby | 198 | 5,346 | 0.77 | 73 | 12.3 | 59 | 10.0 | 667 | 14.7 | 1,558 | 1,309 | 250 | 3 | 8 |
| Stoddard | 689 | 28,368 | 0.92 | 334 | 11.5 | 409 | 14.1 | 3,400 | 14.9 | 7,543 | 5,555 | 1,988 | 52 | 214 |
| Stone | 300 | 27,775 | 0.68 | 260 | 8.0 | 432 | 13.3 | 3,610 | 16.7 | 10,585 | 5,625 | 4,960 | NA | NA |
| Sullivan | 108 | 6,596 | 1.13 | 93 | 15.4 | 71 | 11.8 | 726 | 15.2 | 1,431 | 1,217 | 215 | NA | NA |
| Taney | 1,756 | 58,922 | 1.15 | 627 | 11.2 | 602 | 10.7 | 7,520 | 18.0 | 14,525 | 7,687 | 6,838 | NA | NA |
| Texas | 1,698 | 25,984 | 1.04 | 263 | 10.5 | 321 | 12.8 | 3,492 | 19.4 | 6,402 | 4,932 | 1,470 | NA | NA |
| Vernon | 868 | 20,792 | 1.02 | 248 | 12.2 | 251 | 12.3 | 2,443 | 15.5 | 4,769 | 3,144 | 1,626 | NA | NA |
| Warren | 317 | 27,943 | 0.58 | 441 | 12.1 | 332 | 9.1 | 3,348 | 11.5 | 7,589 | 3,589 | 4,000 | NA | NA |
| Washington | 1,047 | 21,622 | 0.63 | 269 | 10.9 | 313 | 12.7 | 2,883 | 15.0 | 5,513 | 3,139 | 2,373 | NA | NA |
| Wayne | 133 | 12,273 | 0.78 | 123 | 9.6 | 190 | 14.9 | 1,617 | 16.9 | 3,153 | 2,271 | 881 | NA | NA |
| Webster | 729 | 32,408 | 0.59 | 616 | 15.5 | 410 | 10.3 | 4,792 | 14.8 | 7,818 | 3,396 | 4,423 | 9 | 178 |
| Worth | 52 | 1,676 | 0.61 | 21 | 10.8 | 23 | 11.8 | 211 | 14.3 | 534 | 458 | 76 | NA | NA |
| Wright | 195 | 16,873 | 0.80 | 243 | 13.3 | 243 | 13.3 | 2,729 | 19.0 | 4,935 | 2,608 | 2,327 | NA | NA |

1. Per 1,000 estimated resident population.

Items 32—45

# Table B. States and Counties — Crime, Education, Money Income, and Poverty

| STATE County | County law enforcement employment, 2019 | | School enrollment and attainment, 2015-2019 | | | | Local government expenditures, 2017-2018 | | Money income, 2015-2019 | | | | Income and poverty, 2019 | | | |
|---|---|---|---|---|---|---|---|---|---|---|---|---|---|---|---|---|
| | | | Enrollment[1] | | Attainment[2] (percent) | | | | | Households | | | | | Percent below poverty level | | |
| | Officers | Civilians | Total | Percent private | High school graduate or less | Bachelor's degree or more | Total current spending (mil dol) | Current spending per student (dollars) | Per capita income[4] | Median income (dollars) | Percent with income of less than $50,000 | Percent with income of $200,000 or more | Median household income (dollars) | All persons | Children under 18 years | Children 5 to 17 years in families |
| | 46 | 47 | 48 | 49 | 50 | 51 | 52 | 53 | 54 | 55 | 56 | 57 | 58 | 59 | 60 | 61 |
| **MISSOURI— Cont'd** | | | | | | | | | | | | | | | | |
| Jasper | 79 | 45 | 30,045 | 12.7 | 46.8 | 23.6 | 185 | 8,443 | 24,484 | 48,357 | 52.0 | 2.5 | 46,021 | 18.8 | 26.6 | 25.8 |
| Jefferson | 155 | 70 | 53,517 | 14.8 | 42.5 | 20.4 | 339 | 9,869 | 29,999 | 65,454 | 36.9 | 3.3 | 68,779 | 8.4 | 11.3 | 10.1 |
| Johnson | 33 | 29 | 16,564 | 10.1 | 38.2 | 28.9 | 76 | 10,178 | 25,380 | 55,273 | 44.8 | 1.7 | 58,396 | 11.4 | 12.0 | 11.7 |
| Knox | 2 | 0 | 925 | 14.2 | 56.3 | 17.2 | 5 | 10,501 | 22,326 | 40,000 | 61.2 | 1.3 | 41,755 | 16.7 | 27.7 | 28.0 |
| Laclede | 23 | 11 | 7,690 | 17.0 | 55.6 | 14.6 | 54 | 9,209 | 23,051 | 47,257 | 53.1 | 1.9 | 43,969 | 15.8 | 20.4 | 19.0 |
| Lafayette | 31 | 17 | 7,063 | 12.4 | 53.0 | 20.3 | 54 | 10,354 | 29,565 | 58,766 | 42.9 | 3.2 | 62,264 | 10.7 | 14.6 | 12.8 |
| Lawrence | 29 | 8 | 8,280 | 13.9 | 54.1 | 16.0 | 52 | 8,933 | 22,957 | 44,742 | 55.8 | 2.6 | 45,746 | 16.1 | 22.7 | 20.3 |
| Lewis | 7 | 6 | 2,497 | 26.2 | 57.4 | 13.6 | 14 | 9,735 | 22,739 | 45,988 | 53.7 | 1.5 | 40,782 | 16.6 | 20.3 | 19.1 |
| Lincoln | 55 | 40 | 13,366 | 15.9 | 52.8 | 16.8 | 87 | 9,664 | 26,848 | 64,196 | 40.1 | 3.1 | 69,178 | 9.4 | 11.9 | 11.7 |
| Linn | 9 | 1 | 2,588 | 8.8 | 56.3 | 16.9 | 23 | 10,178 | 25,130 | 45,930 | 54.4 | 1.6 | 45,774 | 18.9 | 30.7 | 25.5 |
| Livingston | 10 | 0 | 3,404 | 15.7 | 53.1 | 18.8 | 22 | 10,217 | 23,587 | 46,992 | 52.6 | 1.8 | 49,065 | 15.0 | 19.6 | 19.0 |
| McDonald | 20 | 9 | 4,969 | 9.8 | 63.8 | 11.5 | 34 | 9,135 | 20,468 | 41,643 | 57.2 | 1.6 | 39,615 | 18.5 | 28.8 | 27.5 |
| Macon | 11 | 10 | 3,351 | 13.7 | 56.9 | 16.3 | 23 | 10,420 | 22,136 | 42,746 | 57.0 | 1.0 | 45,717 | 12.5 | 18.4 | 15.4 |
| Madison | 9 | 1 | 2,623 | 10.1 | 52.1 | 15.6 | 19 | 9,262 | 24,418 | 43,636 | 54.7 | 1.5 | 43,919 | 16.7 | 24.2 | 23.0 |
| Maries | NA | NA | 1,802 | 22.3 | 53.5 | 15.8 | 12 | 9,817 | 26,226 | 47,569 | 51.9 | 2.6 | 49,712 | 14.3 | 19.9 | 17.9 |
| Marion | 16 | 22 | 6,854 | 19.2 | 49.0 | 22.4 | 46 | 9,430 | 25,069 | 48,784 | 50.9 | 2.1 | 50,803 | 13.9 | 18.3 | 17.1 |
| Mercer | 3 | 6 | 853 | 9.4 | 58.9 | 18.9 | 7 | 11,946 | 23,042 | 47,298 | 52.4 | 1.1 | 44,343 | 12.7 | 21.0 | 21.3 |
| Miller | 23 | 1 | 5,479 | 5.3 | 53.3 | 19.4 | 49 | 9,582 | 25,827 | 47,171 | 53.0 | 3.4 | 49,252 | 15.1 | 21.6 | 19.9 |
| Mississippi | 7 | 20 | 2,735 | 5.9 | 66.1 | 11.5 | 20 | 9,500 | 18,549 | 35,357 | 65.1 | 1.8 | 36,802 | 27.7 | 34.4 | 31.9 |
| Moniteau | NA | NA | 3,413 | 22.4 | 54.8 | 20.5 | 23 | 9,394 | 24,224 | 58,010 | 43.6 | 1.6 | 57,321 | 11.0 | 16.4 | 15.7 |
| Monroe | 8 | 1 | 1,598 | 17.4 | 59.1 | 12.4 | 15 | 10,302 | 24,697 | 43,966 | 57.9 | 0.9 | 48,399 | 13.4 | 20.0 | 18.8 |
| Montgomery | 18 | 2 | 2,242 | 9.7 | 59.5 | 14.0 | 16 | 10,083 | 24,682 | 46,757 | 53.1 | 1.8 | 50,282 | 13.3 | 19.7 | 18.6 |
| Morgan | 37 | 32 | 3,713 | 28.8 | 60.0 | 13.3 | 20 | 9,502 | 21,316 | 39,003 | 60.0 | 1.4 | 44,473 | 17.8 | 26.8 | 27.0 |
| New Madrid | 16 | 2 | 3,822 | 6.1 | 67.9 | 12.1 | 28 | 10,933 | 21,419 | 38,679 | 60.6 | 0.6 | 42,115 | 22.5 | 32.9 | 32.1 |
| Newton | NA | NA | 13,380 | 14.3 | 47.4 | 20.0 | 75 | 8,829 | 28,353 | 50,813 | 49.3 | 3.6 | 53,251 | 13.2 | 17.7 | 17.2 |
| Nodaway | 15 | 3 | 8,066 | 7.8 | 45.4 | 28.6 | 32 | 12,149 | 22,915 | 44,232 | 55.8 | 1.8 | 47,419 | 20.6 | 13.6 | 11.7 |
| Oregon | 7 | 3 | 2,198 | 8.3 | 59.9 | 14.6 | 18 | 9,370 | 18,398 | 33,601 | 70.1 | 0.5 | 35,356 | 20.6 | 31.8 | 30.4 |
| Osage | 11 | 7 | 3,015 | 21.6 | 53.8 | 18.9 | 16 | 9,789 | 29,061 | 61,687 | 38.8 | 2.3 | 60,728 | 9.1 | 10.3 | 9.2 |
| Ozark | 10 | 8 | 1,789 | 3.0 | 65.5 | 12.7 | 37 | 11,644 | 20,646 | 33,859 | 67.1 | 1.6 | 36,009 | 22.7 | 36.5 | 34.1 |
| Pemiscot | 21 | 28 | 4,049 | 3.0 | 65.5 | 12.7 | 37 | 11,644 | 20,646 | 33,859 | 67.1 | 1.6 | 34,280 | 26.9 | 40.7 | 37.2 |
| Perry | 22 | 12 | 4,565 | 30.1 | 55.6 | 17.7 | 23 | 9,711 | 26,609 | 55,863 | 43.0 | 2.1 | 57,032 | 9.1 | 14.0 | 13.2 |
| Pettis | 22 | 6 | 9,950 | 12.9 | 48.3 | 18.0 | 63 | 9,015 | 23,576 | 46,157 | 54.2 | 1.8 | 48,516 | 12.9 | 17.9 | 16.4 |
| Phelps | 35 | 39 | 14,351 | 8.8 | 42.2 | 29.1 | 64 | 9,499 | 25,202 | 44,154 | 54.5 | 3.2 | 46,573 | 17.0 | 19.4 | 17.0 |
| Pike | 11 | 12 | 3,415 | 11.1 | 63.4 | 15.6 | 26 | 10,155 | 22,297 | 46,385 | 54.0 | 2.3 | 49,089 | 17.6 | 21.8 | 21.1 |
| Platte | 99 | 31 | 25,021 | 16.8 | 26.3 | 43.1 | 200 | 11,580 | 40,567 | 80,393 | 30.7 | 9.1 | 84,456 | 5.3 | 6.5 | 5.8 |
| Polk | 25 | 15 | 8,224 | 28.7 | 50.6 | 19.9 | 50 | 9,751 | 22,773 | 45,660 | 54.0 | 2.4 | 44,837 | 19.6 | 24.4 | 22.9 |
| Pulaski | 26 | 14 | 14,791 | 10.5 | 34.7 | 28.3 | 89 | 10,053 | 23,651 | 53,492 | 46.6 | 1.2 | 55,620 | 14.6 | 14.8 | 14.0 |
| Putnam | 4 | 0 | 907 | 8.6 | 52.3 | 19.3 | 7 | 10,416 | 24,758 | 42,849 | 58.0 | 1.9 | 42,434 | 15.7 | 23.0 | 22.3 |
| Ralls | 12 | 1 | 1,990 | 15.3 | 59.3 | 14.9 | 6 | 8,177 | 27,023 | 55,377 | 44.4 | 1.4 | 63,419 | 10.0 | 13.6 | 11.7 |
| Randolph | 23 | 17 | 5,600 | 17.8 | 49.4 | 15.1 | 40 | 10,893 | 21,240 | 47,740 | 52.1 | 1.5 | 48,043 | 16.5 | 22.5 | 21.2 |
| Ray | 16 | 17 | 4,997 | 12.2 | 57.0 | 14.6 | 30 | 9,993 | 29,524 | 61,957 | 39.5 | 1.9 | 61,678 | 9.2 | 12.4 | 11.4 |
| Reynolds | NA | NA | 1,211 | 7.6 | 60.6 | 13.7 | 13 | 12,100 | 22,720 | 40,324 | 61.2 | 1.1 | 37,415 | 21.7 | 35.3 | 32.2 |
| Ripley | 11 | 2 | 2,750 | 6.3 | 62.3 | 10.2 | 19 | 8,758 | 18,119 | 34,971 | 64.6 | 0.4 | 35,788 | 19.7 | 30.2 | 30.0 |
| St. Charles | 177 | 52 | 102,808 | 26.6 | 29.8 | 38.9 | 680 | 11,385 | 38,943 | 84,978 | 26.3 | 8.7 | 89,236 | 4.6 | 5.4 | 4.8 |
| St. Clair | 19 | 47 | 1,694 | 8.6 | 56.6 | 13.9 | 13 | 9,743 | 22,520 | 38,870 | 60.2 | 0.9 | 39,319 | 19.3 | 27.5 | 26.3 |
| Ste. Genevieve | 47 | 27 | 3,838 | 21.9 | 49.9 | 19.7 | 24 | 12,824 | 29,742 | 60,129 | 41.5 | 3.1 | 61,662 | 9.7 | 12.9 | 11.8 |
| St. Francois | 61 | 14 | 14,384 | 10.6 | 52.4 | 14.3 | 98 | 9,189 | 22,578 | 46,466 | 54.0 | 1.6 | 43,886 | 17.0 | 19.8 | 18.0 |
| St. Louis | 949 | 282 | 247,428 | 28.8 | 27.7 | 43.6 | 1,994 | 13,975 | 41,426 | 67,420 | 36.8 | 10.1 | 70,161 | 9.3 | 12.3 | 11.2 |
| Saline | 16 | 10 | 5,953 | 22.7 | 53.8 | 19.4 | 36 | 9,834 | 22,942 | 44,720 | 54.8 | 1.9 | 47,324 | 15.7 | 21.3 | 21.4 |
| Schuyler | 6 | 3 | 1,124 | 14.9 | 58.7 | 12.6 | 6 | 9,609 | 19,862 | 42,694 | 58.8 | 1.6 | 43,183 | 20.8 | 20.5 | 22.2 |
| Scotland | 3 | 4 | 928 | 32.3 | 61.6 | 14.2 | 6 | 10,157 | 27,706 | 50,085 | 49.9 | 5.0 | 46,871 | 14.4 | 23.8 | 22.9 |
| Scott | NA | NA | 8,379 | 13.7 | 56.7 | 16.3 | 60 | 8,967 | 24,921 | 44,139 | 55.8 | 2.4 | 44,924 | 17.9 | 27.9 | 24.5 |
| Shannon | 3 | 5 | 1,525 | 8.5 | 64.9 | 14.9 | 17 | 8,776 | 17,387 | 34,265 | 67.8 | 0.1 | 36,477 | 22.6 | 35.2 | 33.6 |
| Shelby | 5 | 3 | 1,265 | 18.4 | 52.7 | 18.1 | 11 | 10,686 | 24,304 | 44,083 | 56.7 | 2.2 | 45,939 | 15.8 | 20.7 | 20.8 |
| Stoddard | 16 | 17 | 6,432 | 6.9 | 60.7 | 13.3 | 45 | 8,873 | 22,138 | 41,062 | 60.6 | 1.3 | 42,363 | 18.4 | 24.7 | 22.8 |
| Stone | 50 | 10 | 5,123 | 11.7 | 45.8 | 18.6 | 42 | 11,384 | 29,026 | 49,656 | 50.3 | 2.9 | 50,262 | 15.6 | 24.5 | 23.8 |
| Sullivan | 3 | 0 | 1,327 | 5.0 | 65.9 | 11.7 | 12 | 11,147 | 22,370 | 46,481 | 53.6 | 0.7 | 43,421 | 15.6 | 22.0 | 20.1 |
| Taney | 42 | 17 | 12,396 | 21.9 | 44.7 | 19.9 | 77 | 9,588 | 23,776 | 46,031 | 53.9 | 2.2 | 49,314 | 12.6 | 18.7 | 17.5 |
| Texas | 12 | 0 | 5,487 | 22.1 | 55.0 | 13.5 | 36 | 9,132 | 19,973 | 35,067 | 64.4 | 0.9 | 38,914 | 21.0 | 28.2 | 26.9 |
| Vernon | 17 | 29 | 4,971 | 22.3 | 52.1 | 18.9 | 29 | 9,618 | 24,312 | 43,276 | 55.9 | 2.5 | 46,182 | 17.1 | 26.6 | 24.9 |
| Warren | 36 | 22 | 7,604 | 18.4 | 48.9 | 19.1 | 49 | 10,270 | 28,641 | 60,125 | 43.1 | 4.8 | 63,726 | 10.0 | 14.1 | 12.5 |
| Washington | 21 | 15 | 5,242 | 3.5 | 60.4 | 9.9 | 36 | 9,954 | 20,581 | 41,483 | 57.9 | 1.6 | 41,198 | 22.4 | 30.3 | 28.2 |
| Wayne | 9 | 17 | 2,631 | 13.8 | 62.7 | 9.6 | 16 | 8,529 | 19,457 | 34,316 | 68.0 | 0.5 | 35,397 | 20.6 | 33.3 | 30.9 |
| Webster | 19 | 21 | 8,875 | 18.5 | 51.2 | 16.8 | 61 | 8,882 | 22,961 | 50,560 | 49.5 | 2.7 | 54,844 | 16.2 | 27.0 | 26.1 |
| Worth | 3 | 1 | 397 | 7.1 | 53.8 | 20.0 | 3 | 10,285 | 26,137 | 53,580 | 45.8 | 1.6 | 45,023 | 14.9 | 21.1 | 19.8 |
| Wright | 9 | 5 | 3,897 | 21.0 | 65.0 | 10.2 | 30 | 9,333 | 19,850 | 34,776 | 68.0 | 2.4 | 39,384 | 19.6 | 31.2 | 29.8 |

1. All persons 3 years old and over enrolled in nursery school through college.   2. Persons 25 years old and over.   3. Elementary and secondary education expenditures.   4. Based on population estimated by the American Community Survey, 2014–2018.

# Table B. States and Counties — **Personal Income**

| STATE County | Personal income, 2019 | | | | | | | | | | Earnings, 2019 | | |
|---|---|---|---|---|---|---|---|---|---|---|---|---|---|
| | Total (mil dol) | Percent change 2018-2019 | Per capita[1] Dollars | Per capita[1] Rank | Wages and salaries (mil dol) | Supplements to wages and salaries, employer contributions (mil dol) Pension and insurance | Supplements to wages and salaries, employer contributions (mil dol) Government social insurance | Proprietors' income (mil dol) | Dividends, interest, and rent (mil dol) | Personal transfer receipts (mil dol) | Total (mil dol) | Contributions for government social insurance (mil dol) From employee and self-employed | Contributions for government social insurance (mil dol) From employer |
| | 62 | 63 | 64 | 65 | 66 | 67 | 68 | 69 | 70 | 71 | 72 | 73 | 74 |
| **MISSOURI— Cont'd** | | | | | | | | | | | | | |
| Jasper | 5,138 | 3.6 | 42,348 | 1,713 | 2,625 | 497 | 192 | 561 | 873 | 1,153 | 3,875 | 237 | 192 |
| Jefferson | 9,737 | 3.6 | 43,259 | 1,582 | 2,183 | 407 | 161 | 393 | 1,126 | 2,043 | 3,143 | 235 | 161 |
| Johnson | 2,034 | 4.0 | 37,615 | 2,371 | 896 | 259 | 72 | 94 | 364 | 442 | 1,321 | 70 | 72 |
| Knox | 147 | 11.7 | 37,092 | 2,434 | 36 | 9 | 3 | 26 | 27 | 42 | 74 | 5 | 3 |
| Laclede | 1,330 | 3.1 | 37,228 | 2,422 | 543 | 125 | 42 | 156 | 189 | 386 | 867 | 57 | 42 |
| Lafayette | 1,401 | 3.6 | 42,819 | 1,647 | 339 | 76 | 25 | 90 | 214 | 373 | 530 | 38 | 25 |
| Lawrence | 1,322 | 1.3 | 34,465 | 2,748 | 379 | 79 | 28 | 75 | 197 | 382 | 560 | 42 | 28 |
| Lewis | 364 | 6.7 | 37,197 | 2,425 | 96 | 22 | 7 | 25 | 59 | 98 | 150 | 11 | 7 |
| Lincoln | 2,424 | 5.4 | 41,082 | 1,910 | 556 | 109 | 40 | 109 | 259 | 511 | 814 | 58 | 40 |
| Linn | 482 | 2.8 | 40,419 | 1,997 | 162 | 35 | 13 | 60 | 90 | 142 | 269 | 18 | 13 |
| Livingston | 598 | 3.8 | 39,305 | 2,149 | 247 | 53 | 18 | 73 | 111 | 159 | 390 | 24 | 18 |
| McDonald | 661 | 0.7 | 28,931 | 3,067 | 279 | 53 | 22 | 34 | 81 | 188 | 388 | 26 | 22 |
| Macon | 653 | 3.5 | 43,173 | 1,589 | 194 | 47 | 14 | 95 | 102 | 175 | 350 | 23 | 14 |
| Madison | 452 | 3.5 | 37,430 | 2,397 | 134 | 32 | 10 | 19 | 61 | 171 | 195 | 15 | 10 |
| Maries | 304 | 6.8 | 34,965 | 2,690 | 48 | 12 | 3 | 12 | 51 | 93 | 75 | 7 | 3 |
| Marion | 1,189 | 3.4 | 41,677 | 1,820 | 595 | 112 | 45 | 67 | 184 | 347 | 818 | 54 | 45 |
| Mercer | 124 | 6.9 | 34,283 | 2,772 | 38 | 8 | 3 | 20 | 21 | 38 | 69 | 4 | 3 |
| Miller | 935 | 5.0 | 36,501 | 2,507 | 270 | 58 | 20 | 59 | 143 | 256 | 407 | 29 | 20 |
| Mississippi | 438 | 6.6 | 33,213 | 2,875 | 129 | 30 | 9 | 49 | 72 | 152 | 218 | 14 | 9 |
| Moniteau | 592 | 3.0 | 36,675 | 2,486 | 169 | 38 | 12 | 43 | 102 | 132 | 263 | 18 | 12 |
| Monroe | 364 | 7.3 | 42,136 | 1,745 | 70 | 18 | 5 | 24 | 59 | 98 | 117 | 9 | 5 |
| Montgomery | 478 | 5.3 | 41,354 | 1,867 | 120 | 26 | 9 | 29 | 75 | 127 | 184 | 14 | 9 |
| Morgan | 844 | 1.4 | 40,924 | 1,929 | 153 | 34 | 11 | 79 | 286 | 254 | 277 | 24 | 11 |
| New Madrid | 654 | 9.0 | 38,281 | 2,297 | 311 | 63 | 23 | 56 | 88 | 205 | 453 | 28 | 23 |
| Newton | 2,300 | 2.6 | 39,495 | 2,116 | 941 | 183 | 68 | 134 | 354 | 544 | 1,327 | 88 | 68 |
| Nodaway | 739 | 6.0 | 33,466 | 2,849 | 316 | 79 | 22 | 79 | 147 | 173 | 496 | 30 | 22 |
| Oregon | 324 | 3.8 | 30,778 | 3,015 | 95 | 21 | 10 | 23 | 47 | 135 | 149 | 13 | 10 |
| Osage | 629 | 4.5 | 46,222 | 1,226 | 167 | 37 | 12 | 63 | 121 | 112 | 279 | 18 | 12 |
| Ozark | 289 | 4.0 | 31,526 | 2,976 | 47 | 12 | 3 | 19 | 52 | 118 | 81 | 9 | 3 |
| Pemiscot | 592 | 3.4 | 37,487 | 2,391 | 195 | 46 | 14 | 47 | 82 | 220 | 303 | 20 | 14 |
| Perry | 807 | 2.0 | 42,182 | 1,740 | 410 | 81 | 30 | 51 | 116 | 196 | 572 | 36 | 30 |
| Pettis | 1,632 | 3.1 | 38,545 | 2,254 | 822 | 164 | 61 | 124 | 258 | 449 | 1,170 | 73 | 61 |
| Phelps | 1,792 | 2.9 | 40,208 | 2,027 | 830 | 186 | 57 | 87 | 348 | 456 | 1,160 | 71 | 57 |
| Pike | 648 | 5.2 | 35,410 | 2,634 | 202 | 49 | 14 | 61 | 125 | 179 | 327 | 21 | 14 |
| Platte | 6,171 | 4.5 | 59,095 | 336 | 2,736 | 419 | 192 | 356 | 950 | 758 | 3,703 | 227 | 192 |
| Polk | 1,131 | 2.9 | 35,193 | 2,660 | 353 | 85 | 24 | 92 | 170 | 333 | 553 | 36 | 24 |
| Pulaski | 2,195 | 4.8 | 41,734 | 1,815 | 1,223 | 375 | 112 | 47 | 393 | 429 | 1,758 | 81 | 112 |
| Putnam | 175 | 4.6 | 37,336 | 2,408 | 39 | 10 | 3 | 28 | 36 | 53 | 80 | 5 | 3 |
| Ralls | 417 | 6.1 | 40,447 | 1,994 | 177 | 33 | 13 | 26 | 59 | 105 | 250 | 17 | 13 |
| Randolph | 1,011 | 3.8 | 40,856 | 1,938 | 417 | 93 | 31 | 106 | 141 | 277 | 648 | 40 | 31 |
| Ray | 1,003 | 4.6 | 43,569 | 1,538 | 171 | 44 | 12 | 70 | 125 | 237 | 297 | 21 | 12 |
| Reynolds | 222 | 4.7 | 35,467 | 2,621 | 79 | 18 | 6 | 4 | 42 | 92 | 107 | 9 | 6 |
| Ripley | 419 | 2.3 | 31,522 | 2,977 | 81 | 21 | 6 | 28 | 55 | 180 | 136 | 12 | 6 |
| St. Charles | 21,899 | 4.1 | 54,472 | 520 | 8,153 | 1,321 | 568 | 894 | 2,902 | 3,100 | 10,936 | 693 | 568 |
| St. Clair | 287 | 5.0 | 30,578 | 3,021 | 55 | 14 | 4 | 23 | 54 | 116 | 96 | 9 | 4 |
| Ste. Genevieve | 748 | 2.0 | 41,802 | 1,800 | 290 | 56 | 21 | 23 | 125 | 174 | 390 | 27 | 21 |
| St. Francois | 2,345 | 3.5 | 34,886 | 2,702 | 879 | 203 | 62 | 74 | 345 | 764 | 1,218 | 85 | 62 |
| St. Louis | 72,593 | 2.6 | 73,016 | 102 | 42,152 | 5,958 | 2,858 | 4,872 | 20,684 | 9,511 | 55,839 | 3,378 | 2,858 |
| Saline | 889 | 4.9 | 39,037 | 2,187 | 347 | 74 | 26 | 91 | 145 | 250 | 538 | 32 | 26 |
| Schuyler | 143 | 4.4 | 30,715 | 3,018 | 20 | 6 | 1 | 18 | 24 | 39 | 46 | 3 | 1 |
| Scotland | 176 | 7.4 | 36,001 | 2,570 | 45 | 13 | 3 | 35 | 33 | 40 | 95 | 5 | 3 |
| Scott | 1,593 | 2.9 | 41,626 | 1,827 | 686 | 130 | 52 | 122 | 257 | 481 | 990 | 67 | 52 |
| Shannon | 255 | 3.4 | 31,184 | 2,995 | 46 | 12 | 4 | 24 | 38 | 92 | 86 | 8 | 4 |
| Shelby | 246 | 6.1 | 41,446 | 1,857 | 62 | 15 | 4 | 25 | 46 | 65 | 107 | 8 | 4 |
| Stoddard | 1,085 | 3.4 | 37,365 | 2,403 | 417 | 83 | 32 | 41 | 165 | 352 | 574 | 44 | 32 |
| Stone | 1,253 | 4.1 | 39,225 | 2,158 | 247 | 45 | 20 | 74 | 257 | 378 | 387 | 36 | 20 |
| Sullivan | 254 | 7.2 | 41,690 | 1,819 | 124 | 23 | 10 | 35 | 34 | 68 | 191 | 10 | 10 |
| Taney | 2,058 | 3.9 | 36,794 | 2,472 | 1,087 | 179 | 83 | 257 | 299 | 581 | 1,606 | 108 | 83 |
| Texas | 752 | 4.2 | 29,618 | 3,049 | 200 | 55 | 14 | 60 | 128 | 261 | 330 | 24 | 14 |
| Vernon | 735 | 3.5 | 35,742 | 2,591 | 286 | 73 | 20 | 67 | 120 | 232 | 447 | 28 | 20 |
| Warren | 1,520 | 5.7 | 42,648 | 1,665 | 344 | 67 | 24 | 73 | 205 | 331 | 509 | 38 | 24 |
| Washington | 754 | 3.0 | 30,490 | 3,022 | 175 | 45 | 13 | 20 | 79 | 282 | 252 | 21 | 13 |
| Wayne | 398 | 3.5 | 30,930 | 3,010 | 77 | 21 | 6 | 16 | 70 | 172 | 120 | 12 | 6 |
| Webster | 1,322 | 4.7 | 33,387 | 2,855 | 305 | 69 | 23 | 96 | 172 | 340 | 492 | 38 | 23 |
| Worth | 81 | 12.4 | 40,013 | 2,048 | 12 | 4 | 1 | 15 | 15 | 22 | 31 | 2 | 1 |
| Wright | 608 | 3.5 | 33,243 | 2,871 | 147 | 36 | 11 | 48 | 91 | 215 | 241 | 20 | 11 |

1. Based on the resident population estimated as of July 1 of the year shown.

# Table B. States and Counties — Earnings, Social Security, and Housing

| STATE County | Earnings, 2019 (cont.) Percent by selected industries | | | | | | | | | Social Security beneficiaries, December 2019 | | Supplemental Security Income recipients, 2019 | Housing units, 2020 | |
|---|---|---|---|---|---|---|---|---|---|---|---|---|---|---|
| | Farm | Mining, quarrying, and extractions | Construction | Manu-facturing | Information; professional, scientific, technical services | Retail trade | Finance, insurance, real estate, and leasing | Health care and social assistance | Govern-ment | Number | Rate[1] | | Total | Percent change, 2010-2020 |
| | 75 | 76 | 77 | 78 | 79 | 80 | 81 | 82 | 83 | 84 | 85 | 86 | 87 | 88 |
| MISSOURI— Cont'd | | | | | | | | | | | | | | |
| Jasper | 0.3 | 0.2 | 4.2 | 27.8 | 4.3 | 7.3 | 3.6 | 10.8 | 10.2 | 25,870 | 213 | 3,622 | 52,862 | 4.3 |
| Jefferson | 0.0 | 0.4 | 12.8 | 10.4 | 4.9 | 8.7 | 4.9 | 12.3 | 17.7 | 48,105 | 213 | 2,846 | 92,783 | 5.9 |
| Johnson | 0.6 | D | 5.9 | 7.1 | 2.5 | 4.3 | 3.7 | 4.2 | 57.5 | 9,055 | 167 | 731 | 22,483 | 4.5 |
| Knox | 19.6 | D | 8.9 | 10.4 | D | 6.0 | 5.0 | D | 18.4 | 1,015 | 257 | 89 | 2,272 | -0.7 |
| Laclede | 0.5 | 0.1 | 4.3 | 43.8 | D | 8.7 | 3.5 | 9.6 | 10.2 | 9,265 | 258 | 1,149 | 16,114 | 2.0 |
| Lafayette | 2.1 | 0.0 | 10.8 | 10.7 | D | 7.5 | 4.8 | D | 22.0 | 7,730 | 236 | 587 | 14,916 | 1.4 |
| Lawrence | 3.6 | D | 9.8 | 16.9 | 5.0 | 8.7 | 4.1 | D | 18.0 | 9,610 | 251 | 893 | 16,725 | 0.4 |
| Lewis | 3.4 | D | 9.1 | 8.0 | 2.3 | 5.9 | 3.7 | D | 19.9 | 2,395 | 245 | 177 | 4,544 | 0.2 |
| Lincoln | 0.6 | 1.5 | 14.9 | 17.5 | 2.7 | 8.3 | 4.9 | 8.2 | 16.9 | 11,645 | 197 | 842 | 22,165 | 5.6 |
| Linn | 6.4 | 0.1 | 4.4 | 20.9 | 4.7 | 6.8 | 4.8 | D | 16.0 | 3,255 | 273 | 358 | 6,372 | -0.9 |
| Livingston | 6.3 | D | 7.0 | 8.9 | 2.5 | 11.2 | 4.7 | 12.4 | 20.0 | 3,680 | 246 | 316 | 6,871 | 2.1 |
| McDonald | 0.3 | 0.0 | 9.5 | 43.2 | D | 6.9 | 2.0 | 2.8 | 13.2 | 4,535 | 198 | 595 | 9,995 | 0.7 |
| Macon | 5.7 | 0.0 | 5.4 | 20.3 | 7.8 | 6.7 | 4.3 | D | 24.3 | 4,250 | 281 | 296 | 7,676 | 0.2 |
| Madison | 0.4 | D | 9.5 | 13.9 | D | 9.4 | 2.4 | 10.3 | 24.1 | 3,715 | 307 | 525 | 5,994 | 0.5 |
| Maries | 3.1 | 0.0 | D | 18.9 | D | 6.9 | D | 7.9 | 21.7 | 2,355 | 269 | 145 | 4,602 | 0.1 |
| Marion | 0.7 | D | 6.0 | 8.3 | 4.3 | 8.7 | 4.7 | D | 13.3 | 7,040 | 247 | 1,044 | 13,144 | 2.5 |
| Mercer | 28.5 | D | 3.3 | D | D | 4.1 | 3.1 | 5.5 | 19.8 | 965 | 267 | 66 | 2,120 | -0.7 |
| Miller | 1.5 | D | 12.9 | 8.3 | 3.2 | 14.9 | 4.2 | D | 19.7 | 6,200 | 242 | 564 | 12,972 | 1.7 |
| Mississippi | 17.3 | 0.0 | 2.1 | 1.6 | 1.2 | 6.8 | D | D | 25.0 | 3,225 | 244 | 666 | 5,729 | 0.2 |
| Moniteau | 2.1 | D | 20.0 | 17.5 | 2.3 | 6.0 | D | D | 21.2 | 3,305 | 209 | 199 | 6,172 | -0.1 |
| Monroe | 8.7 | D | 5.9 | 6.8 | D | 3.6 | D | D | 29.9 | 2,390 | 277 | 175 | 4,857 | 1.2 |
| Montgomery | 4.7 | 2.4 | 11.3 | 21.8 | 3.0 | 6.9 | D | 5.9 | 18.7 | 3,130 | 273 | 292 | 6,295 | 2.8 |
| Morgan | 0.5 | D | 15.9 | 9.6 | D | 16.5 | 4.9 | D | 18.6 | 6,440 | 311 | 640 | 15,618 | 0.6 |
| New Madrid | 8.9 | D | 1.7 | 17.6 | D | 10.3 | 2.3 | D | 11.2 | 4,125 | 242 | 783 | 8,618 | 1.1 |
| Newton | 1.0 | D | 6.9 | 13.8 | 3.0 | 7.3 | 3.1 | 29.4 | 11.3 | 13,355 | 229 | 939 | 24,909 | 2.5 |
| Nodaway | 3.2 | D | 4.7 | 19.6 | 3.5 | 7.2 | 3.9 | 12.5 | 27.0 | 4,125 | 187 | 289 | 9,856 | 3.5 |
| Oregon | 4.2 | D | 3.0 | 5.6 | D | 9.8 | 2.7 | 12.9 | 18.2 | 3,140 | 299 | 547 | 5,472 | -0.3 |
| Osage | 4.1 | 0.4 | 11.8 | 38.3 | D | 6.6 | D | 4.4 | 15.1 | 2,915 | 214 | 132 | 6,692 | 1.4 |
| Ozark | -1.6 | 0.2 | 10.1 | 8.2 | D | 7.0 | 7.1 | D | 26.7 | 3,270 | 357 | 283 | 5,706 | 0.9 |
| Pemiscot | 10.6 | 0.0 | D | D | D | 6.4 | 3.6 | D | 24.7 | 4,250 | 268 | 1,233 | 8,167 | 0.1 |
| Perry | 2.3 | 0.0 | 11.6 | 29.0 | 3.8 | 6.2 | 4.6 | 5.2 | 13.4 | 4,545 | 237 | 335 | 8,901 | 3.9 |
| Pettis | 0.4 | D | 5.2 | 26.3 | 5.7 | 7.6 | 3.3 | 7.7 | 17.9 | 9,380 | 221 | 1,309 | 18,225 | -0.1 |
| Phelps | -0.3 | D | 4.2 | 6.5 | 2.7 | 8.2 | 3.6 | 16.8 | 36.7 | 9,540 | 214 | 1,161 | 20,637 | 5.7 |
| Pike | 3.6 | D | 12.8 | 9.3 | 4.8 | 9.4 | 3.1 | 6.6 | 27.2 | 4,215 | 238 | 405 | 7,931 | 0.7 |
| Platte | 0.4 | D | 9.0 | 9.2 | 7.6 | 7.3 | 5.1 | 7.3 | 9.8 | 17,100 | 163 | 742 | 43,560 | 11.1 |
| Polk | 2.9 | D | 6.9 | 4.8 | D | 7.9 | 3.9 | 10.2 | 31.8 | 7,005 | 217 | 893 | 13,636 | 2.5 |
| Pulaski | 0.2 | D | 1.6 | 0.4 | D | 3.2 | 1.6 | 2.6 | 78.7 | 7,230 | 137 | 913 | 19,220 | 7.3 |
| Putnam | 23.0 | 0.0 | D | D | D | 6.8 | 6.1 | 1.5 | 27.2 | 1,315 | 281 | 121 | 2,978 | -0.1 |
| Ralls | 1.7 | 0.0 | 7.1 | 52.6 | D | 3.3 | 2.1 | D | 8.6 | 2,625 | 255 | 109 | 5,202 | 0.3 |
| Randolph | 2.7 | D | 5.0 | 11.6 | 2.4 | 11.0 | 6.7 | 9.3 | 17.8 | 5,425 | 218 | 797 | 10,847 | 1.2 |
| Ray | 8.5 | D | 9.8 | 14.0 | 4.3 | 8.3 | 3.1 | D | 25.8 | 5,480 | 238 | 331 | 10,176 | 1.8 |
| Reynolds | -3 | D | D | 14.2 | D | 3.8 | D | 9.2 | 17.9 | 2,165 | 344 | 234 | 4,030 | 0.0 |
| Ripley | 1.5 | D | 5.5 | 17.3 | 1.7 | 9.3 | 3.2 | 17.0 | 22.6 | 3,745 | 281 | 724 | 6,644 | 0.7 |
| St. Charles | 0.1 | D | 7.9 | 14.1 | 13.5 | 7.3 | 8.5 | 10.2 | 11.1 | 73,855 | 184 | 2,790 | 160,714 | 14.0 |
| St. Clair | 6.1 | D | 7.2 | 3.8 | 4.4 | 10.2 | 4.3 | 9.3 | 27.6 | 2,880 | 305 | 261 | 5,655 | 0.3 |
| Ste. Genevieve | 0.8 | 5.0 | 9.3 | 29.1 | D | 5.9 | 3.9 | D | 18.3 | 4,415 | 246 | 305 | 8,805 | 2.0 |
| St. Francois | -0.2 | 0.3 | 7.1 | 10.2 | 3.0 | 9.0 | 7.8 | 14.5 | 26.8 | 16,765 | 250 | 2,433 | 30,503 | 7.2 |
| St. Louis | 0.0 | 0.3 | 6.0 | 9.3 | 13.3 | 5.3 | 12.3 | 12.8 | 7.4 | 208,665 | 210 | 17,872 | 442,678 | 1.1 |
| Saline | 10.7 | 0.0 | 2.9 | 24.3 | 2.5 | 6.4 | 3.7 | D | 16.9 | 5,170 | 226 | 637 | 10,169 | 0.5 |
| Schuyler | 20.0 | 0.0 | 14.8 | D | D | 6.9 | 2.3 | 1.9 | 29.2 | 955 | 206 | 107 | 2,100 | -0.1 |
| Scotland | 12.6 | 0.0 | 6.0 | D | D | 8.7 | D | D | 33.4 | 960 | 194 | 60 | 2,371 | 0.0 |
| Scott | 2.0 | 0.1 | 5.7 | 16.7 | 4.7 | 4.8 | 4.7 | D | 13.8 | 10,425 | 271 | 1,658 | 17,371 | 2.2 |
| Shannon | 1.5 | D | 6.0 | 28.9 | D | 3.7 | 3.2 | D | 18.9 | 2,315 | 283 | 357 | 4,175 | 0.3 |
| Shelby | 1.5 | D | 6.8 | 13.6 | 4.6 | 9.3 | 6.0 | 4.3 | 25.5 | 1,570 | 265 | 130 | 3,202 | -0.1 |
| Stoddard | 3.2 | D | 9.6 | 30.8 | D | 7.7 | -2.9 | D | 14.1 | 8,210 | 282 | 1,061 | 13,763 | 1.1 |
| Stone | 1.1 | 0.2 | 14.5 | 1.6 | 4.3 | 9.6 | 4.1 | 5.7 | 15.8 | 10,455 | 327 | 546 | 21,538 | 5.7 |
| Sullivan | 20.5 | D | 2.7 | D | D | 3.3 | D | 2.8 | 13.4 | 1,530 | 252 | 178 | 3,332 | -0.8 |
| Taney | 0.2 | D | 2.9 | 2.0 | 3.5 | 9.8 | 5.0 | D | 9.6 | 15,230 | 272 | 1,187 | 31,452 | 7.5 |
| Texas | 5.9 | 0.3 | 6.0 | 14.7 | D | 7.2 | 3.4 | 5.2 | 31.9 | 6,340 | 250 | 757 | 11,742 | 0.5 |
| Vernon | 7.7 | D | 4.3 | 22.1 | 2.8 | 5.9 | 6.8 | D | 21.0 | 5,000 | 242 | 625 | 9,599 | 1.1 |
| Warren | 0.6 | D | 13.5 | 25.6 | D | 5.9 | 3.6 | D | 16.0 | 7,875 | 220 | 497 | 16,503 | 12.3 |
| Washington | | 1.7 | 5.6 | 12.3 | D | 7.4 | 4.4 | D | 36.5 | 6,085 | 246 | 1,094 | 11,091 | 0.7 |
| Wayne | 0.1 | D | 5.1 | 12.7 | D | 7.1 | 3.1 | D | 26.3 | 3,980 | 308 | 732 | 8,115 | 0.3 |
| Webster | -1.1 | 0.3 | 12.3 | 22.6 | D | 9.3 | 3.3 | 5.7 | 16.9 | 8,920 | 225 | 815 | 14,910 | 3.3 |
| Worth | 31.4 | D | 4.6 | D | D | 8.0 | 2.6 | 2.6 | 26.8 | 570 | 287 | 36 | 1,268 | -0.9 |
| Wright | 0.9 | D | 8.2 | 15.3 | D | 12.9 | 4.1 | D | 19.9 | 5,370 | 293 | 790 | 8,697 | 0.3 |

1. Per 1,000 resident population estimated as of July 1 of the year shown.

# Table B. States and Counties — Housing, Labor Force, and Employment

| STATE County | Housing units, 2015-2019 Occupied units — Total | Percent | Owner-occupied Median value[1] | Median owner cost as a percent of income — With a mortgage | Without a mortgage[2] | Renter-occupied — Median rent[3] | Median rent as a percent of income[2] | Sub-standard units[4] (percent) | Civilian labor force, 2020 — Total | Percent change, 2019-2020 | Unemployment — Total | Rate[5] | Civilian employment[6], 2015-2019 — Total | Percent Management, business, science, and arts | Construction, production, and maintenance occupations |
|---|---|---|---|---|---|---|---|---|---|---|---|---|---|---|---|
| | 89 | 90 | 91 | 92 | 93 | 94 | 95 | 96 | 97 | 98 | 99 | 100 | 101 | 102 | 103 |
| **MISSOURI— Cont'd** | | | | | | | | | | | | | | | |
| Jasper | 45,759 | 63.6 | 118,400 | 18.9 | 11.7 | 776 | 27.7 | 2.8 | 56,154 | -1.6 | 3,186 | 5.7 | 56,376 | 31.0 | 28.1 |
| Jefferson | 84,444 | 79.4 | 161,800 | 19.4 | 10.2 | 855 | 27.6 | 1.8 | 116,774 | -1.8 | 7,119 | 6.1 | 111,769 | 33.1 | 27.2 |
| Johnson | 19,864 | 61.0 | 156,500 | 19.1 | 10.0 | 806 | 28.5 | 2.9 | 23,108 | 0.3 | 1,312 | 5.7 | 23,931 | 32.2 | 26.0 |
| Knox | 1,490 | 82.6 | 75,500 | 22.1 | 13.8 | 556 | 20.8 | 2.7 | 1,757 | -0.4 | 71 | 4.0 | 1,563 | 25.7 | 33.7 |
| Laclede | 14,112 | 69.5 | 113,900 | 19.0 | 10.4 | 661 | 22.3 | 3.2 | 16,731 | -3.3 | 1,388 | 8.3 | 15,475 | 29.0 | 37.0 |
| Lafayette | 13,050 | 70.8 | 143,500 | 18.6 | 10.0 | 686 | 24.8 | 2.3 | 16,582 | -0.9 | 871 | 5.3 | 15,503 | 28.7 | 33.4 |
| Lawrence | 14,787 | 71.8 | 109,600 | 18.7 | 10.9 | 681 | 27.3 | 3.7 | 17,672 | -1.4 | 1,073 | 6.1 | 17,024 | 27.1 | 32.7 |
| Lewis | 3,752 | 76.1 | 87,900 | 17.8 | 11.1 | 569 | 22.7 | 3.1 | 4,966 | -0.2 | 243 | 4.9 | 4,372 | 21.9 | 37.8 |
| Lincoln | 19,286 | 78.5 | 163,300 | 19.5 | 10.6 | 854 | 27.7 | 2.1 | 28,723 | -1.5 | 1,797 | 6.3 | 26,776 | 29.2 | 35.7 |
| Linn | 5,106 | 72.9 | 87,200 | 16.8 | 11.7 | 558 | 23.8 | 2.2 | 5,093 | 0.4 | 312 | 6.1 | 5,612 | 27.5 | 30.5 |
| Livingston | 5,915 | 66.6 | 111,500 | 18.0 | 11.0 | 671 | 23.6 | 2.2 | 7,389 | 0.8 | 282 | 3.8 | 6,455 | 33.2 | 25.8 |
| McDonald | 8,259 | 68.6 | 110,500 | 19.2 | 11.5 | 631 | 24.8 | 9.4 | 10,578 | 1.6 | 526 | 5.0 | 9,381 | 21.8 | 41.3 |
| Macon | 5,845 | 74.6 | 97,200 | 18.9 | 12.8 | 549 | 24.3 | 2.6 | 7,435 | -0.6 | 356 | 4.8 | 6,354 | 30.0 | 35.7 |
| Madison | 5,044 | 71.8 | 106,100 | 19.3 | 10.2 | 607 | 29.9 | 5.2 | 5,475 | 0.6 | 358 | 6.5 | 5,005 | 24.1 | 35.5 |
| Maries | 3,762 | 74.9 | 133,200 | 18.5 | 10.0 | 559 | 29.2 | 4.6 | 3,909 | 1.2 | 189 | 4.8 | 3,758 | 30.1 | 37.9 |
| Marion | 11,565 | 65.4 | 123,400 | 19.0 | 10.4 | 665 | 25.1 | 1.3 | 14,299 | -0.9 | 687 | 4.8 | 13,038 | 30.5 | 28.5 |
| Mercer | 1,271 | 78.6 | 88,900 | 16.7 | 12.4 | 495 | 20.9 | 4.6 | 1,889 | 2.4 | 68 | 3.6 | 1,676 | 35.9 | 35.6 |
| Miller | 10,111 | 74.2 | 136,800 | 19.4 | 10.2 | 677 | 27.0 | 2.5 | 12,308 | 0.6 | 761 | 6.2 | 11,415 | 28.6 | 27.1 |
| Mississippi | 5,026 | 59.7 | 80,200 | 19.0 | 11.1 | 613 | 35.4 | 1.0 | 5,701 | -0.6 | 305 | 5.3 | 4,513 | 23.2 | 32.0 |
| Moniteau | 5,436 | 77.0 | 124,500 | 18.0 | 10.0 | 605 | 25.5 | 3.6 | 7,368 | 0.6 | 313 | 4.2 | 6,944 | 33.5 | 31.5 |
| Monroe | 3,702 | 74.5 | 105,300 | 18.8 | 11.1 | 561 | 19.9 | 2.5 | 3,789 | -0.9 | 225 | 5.9 | 3,846 | 24.7 | 34.6 |
| Montgomery | 5,014 | 70.0 | 108,100 | 20.2 | 11.0 | 670 | 24.5 | 2.4 | 5,750 | 0.6 | 286 | 5.0 | 5,105 | 25.2 | 39.1 |
| Morgan | 7,588 | 80.7 | 125,000 | 23.5 | 12.6 | 563 | 26.1 | 1.8 | 8,273 | 2.2 | 518 | 6.3 | 7,397 | 23.0 | 35.7 |
| New Madrid | 7,371 | 63.5 | 74,500 | 17.1 | 11.9 | 646 | 27.4 | 3.3 | 8,483 | 3.7 | 492 | 5.8 | 6,745 | 23.5 | 36.7 |
| Newton | 22,202 | 71.5 | 131,300 | 18.6 | 10.6 | 699 | 25.6 | 3.3 | 26,732 | -1.4 | 1,553 | 5.8 | 26,733 | 28.6 | 32.6 |
| Nodaway | 8,395 | 57.8 | 126,400 | 18.7 | 10.0 | 657 | 27.8 | 1.5 | 10,722 | -0.2 | 490 | 4.6 | 11,072 | 34.4 | 24.9 |
| Oregon | 4,249 | 73.8 | 93,600 | 21.8 | 10.3 | 541 | 34.0 | 3.2 | 3,938 | -0.4 | 265 | 6.7 | 3,977 | 25.1 | 36.0 |
| Osage | 5,120 | 84.6 | 151,900 | 18.3 | 10.0 | 535 | 23.2 | 2.7 | 7,078 | 0.2 | 254 | 3.6 | 6,968 | 31.7 | 33.2 |
| Ozark | 3,958 | 77.6 | 97,600 | 21.9 | 12.4 | 660 | 33.3 | 4.0 | 3,467 | 1.3 | 251 | 7.2 | 3,075 | 28.1 | 39.4 |
| Pemiscot | 6,730 | 54.1 | 82,100 | 18.9 | 12.4 | 617 | 29.3 | 4.3 | 6,271 | -0.6 | 501 | 8.0 | 5,898 | 25.1 | 33.5 |
| Perry | 7,576 | 75.0 | 147,700 | 19.2 | 10.0 | 694 | 28.5 | 1.3 | 9,826 | -1.7 | 562 | 5.7 | 9,395 | 30.3 | 36.4 |
| Pettis | 16,033 | 68.8 | 119,100 | 19.0 | 10.7 | 707 | 28.3 | 2.5 | 20,626 | | 1,408 | 6.8 | 18,681 | 27.9 | 34.8 |
| Phelps | 17,981 | 59.6 | 135,000 | 18.5 | 10.0 | 727 | 29.0 | 1.7 | 19,776 | 0.5 | 943 | 4.8 | 19,115 | 41.4 | 20.4 |
| Pike | 6,589 | 71.7 | 97,800 | 19.4 | 10.0 | 651 | 23.3 | 3.5 | 7,549 | 0.8 | 412 | 5.5 | 7,332 | 26.8 | 37.2 |
| Platte | 39,305 | 65.9 | 226,500 | 18.9 | 10.7 | 1,019 | 25.4 | 1.8 | 58,458 | -0.9 | 3,183 | 5.4 | 53,924 | 45.7 | 18.7 |
| Polk | 11,712 | 69.0 | 132,300 | 20.1 | 10.0 | 701 | 27.2 | 3.3 | 14,331 | -0.7 | 670 | 4.7 | 14,345 | 34.7 | 27.9 |
| Pulaski | 15,154 | 49.3 | 159,200 | 20.9 | 10.0 | 993 | 23.8 | 3.2 | 14,583 | -0.4 | 822 | 5.6 | 17,522 | 31.8 | 22.3 |
| Putnam | 1,744 | 77.2 | 96,600 | 23.5 | 12.1 | 572 | 25.2 | 2.5 | 2,275 | 0.4 | 100 | 4.4 | 2,060 | 27.2 | 38.1 |
| Ralls | 4,036 | 83.3 | 136,200 | 19.4 | 11.3 | 767 | 26.8 | 1.9 | 5,627 | -0.8 | 251 | 4.5 | 4,969 | 28.7 | 33.7 |
| Randolph | 8,626 | 71.5 | 105,000 | 18.2 | 10.6 | 622 | 29.0 | 1.2 | 10,210 | -0.8 | 605 | 5.9 | 10,370 | 27.8 | 29.2 |
| Ray | 8,868 | 78.5 | 133,700 | 18.3 | 10.4 | 717 | 25.3 | 1.8 | 10,966 | -0.2 | 753 | 6.9 | 10,953 | 24.9 | 38.4 |
| Reynolds | 2,596 | 78.7 | 96,000 | 17.0 | 11.1 | 600 | 28.0 | 2.0 | 2,922 | -3.1 | 166 | 5.7 | 2,601 | 24.3 | 42.0 |
| Ripley | 5,059 | 79.0 | 83,800 | 20.1 | 11.9 | 574 | 28.6 | 2.4 | 4,987 | -0.1 | 394 | 7.9 | 5,210 | 26.1 | 34.6 |
| St. Charles | 146,631 | 80.6 | 219,100 | 18.9 | 10.4 | 1,041 | 25.8 | 0.9 | 224,521 | -2.1 | 11,735 | 5.2 | 211,627 | 43.5 | 17.8 |
| St. Clair | 4,139 | 78.3 | 91,900 | 19.7 | 13.1 | 574 | 23.0 | 4.0 | 3,764 | 0.8 | 230 | 6.1 | 3,602 | 29.0 | 34.0 |
| Ste. Genevieve | 7,121 | 80.1 | 166,200 | 17.6 | 10.0 | 729 | 28.4 | 1.4 | 9,239 | 0.9 | 457 | 4.9 | 8,870 | 29.8 | 33.4 |
| St. Francois | 24,898 | 68.5 | 121,000 | 19.2 | 10.8 | 690 | 27.9 | 2.1 | 26,414 | -0.4 | 1,852 | 7.0 | 27,298 | 28.0 | 29.5 |
| St. Louis | 405,984 | 68.6 | 198,800 | 19.5 | 11.3 | 983 | 28.3 | 1.3 | 524,721 | -1.6 | 32,820 | 6.3 | 499,482 | 45.3 | 15.6 |
| Saline | 8,272 | 69.6 | 106,600 | 19.5 | 12.4 | 640 | 28.1 | 2.7 | 10,597 | -0.2 | 479 | 4.5 | 10,449 | 32.1 | 31.8 |
| Schuyler | 1,475 | 69.3 | 71,200 | 19.2 | 10.9 | 459 | 23.1 | 3.1 | 1,961 | 0.0 | 112 | 5.7 | 1,891 | 22.2 | 37.0 |
| Scotland | 1,768 | 77.5 | 113,000 | 22.4 | 12.6 | 511 | 23.5 | 3.3 | 2,393 | 1.5 | 78 | 3.3 | 2,010 | 36.3 | 35.4 |
| Scott | 15,230 | 66.7 | 109,200 | 18.3 | 10.1 | 687 | 26.9 | 2.9 | 19,872 | 0.1 | 1,058 | 5.3 | 16,745 | 29.1 | 29.0 |
| Shannon | 3,064 | 73.4 | 124,900 | 22.4 | 11.6 | 586 | 26.2 | 5.3 | 3,343 | 3.4 | 245 | 7.3 | 2,968 | 28.0 | 43.5 |
| Shelby | 2,432 | 73.8 | 72,200 | 19.6 | 11.3 | 564 | 19.8 | 0.7 | 3,047 | 0.6 | 136 | 4.5 | 2,905 | 29.5 | 32.9 |
| Stoddard | 11,530 | 68.0 | 104,200 | 19.8 | 10.6 | 621 | 28.0 | 2.3 | 13,238 | 1.6 | 788 | 6.0 | 12,370 | 29.1 | 34.0 |
| Stone | 12,783 | 81.7 | 172,600 | 21.9 | 11.2 | 695 | 25.7 | 2.9 | 13,120 | -5.3 | 1,284 | 9.8 | 12,358 | 27.0 | 29.9 |
| Sullivan | 2,128 | 74.5 | 77,000 | 18.5 | 12.4 | 622 | 20.4 | 3.1 | 2,785 | 4.8 | 121 | 4.3 | 2,691 | 27.0 | 44.1 |
| Taney | 22,272 | 61.7 | 139,800 | 21.6 | 11.9 | 816 | 25.2 | 4.3 | 26,259 | -2.7 | 3,393 | 12.9 | 25,289 | 30.4 | 17.9 |
| Texas | 9,728 | 72.4 | 111,700 | 22.0 | 10.8 | 591 | 34.0 | 3.4 | 8,953 | 0.5 | 519 | 5.8 | 8,906 | 27.3 | 33.6 |
| Vernon | 8,207 | 71.7 | 98,300 | 19.9 | 10.6 | 650 | 28.2 | 2.8 | 9,436 | 0.2 | 411 | 4.4 | 8,861 | 36.0 | 28.5 |
| Warren | 12,654 | 78.9 | 180,000 | 19.6 | 10.6 | 812 | 30.6 | 5.4 | 18,207 | -1.6 | 1,057 | 5.8 | 15,937 | 27.2 | 34.5 |
| Washington | 9,231 | 79.4 | 94,400 | 20.6 | 11.4 | 588 | 27.8 | 4.4 | 10,143 | -0.5 | 728 | 7.2 | 9,080 | 25.7 | 35.4 |
| Wayne | 5,426 | 73.5 | 85,100 | 21.6 | 11.4 | 566 | 27.0 | 3.8 | 5,085 | -2.9 | 310 | 6.1 | 4,322 | 25.9 | 33.6 |
| Webster | 13,575 | 75.0 | 136,200 | 19.2 | 10.0 | 661 | 25.1 | 5.5 | 17,322 | -0.2 | 926 | 5.3 | 15,538 | 27.0 | 36.5 |
| Worth | 859 | 76.4 | 63,800 | 16.1 | 10.0 | 569 | 27.5 | 0.9 | 1,180 | 2.2 | 49 | 4.2 | 907 | 30.7 | 31.5 |
| Wright | 7,063 | 70.0 | 103,600 | 21.5 | 10.6 | 557 | 29.9 | 4.4 | 7,369 | -0.6 | 518 | 7.0 | 6,629 | 21.5 | 37.1 |

1. Specified owner-occupied units. lacking complete plumbing facilities. 2. A value of 10.0 represents 10 percent or less; a value of 50.0 represents 50 percent or more. 3. Specified renter-occupied units. 4. Overcrowded or 5. Percent of civilian labor force. 6. Civilian employed persons 16 years old and over.

# Table B. States and Counties — Nonfarm Employment and Agriculture

| STATE County | Private nonfarm establishments, employment and payroll, 2019 | | | | | | | | | Agriculture, 2017 | | | |
|---|---|---|---|---|---|---|---|---|---|---|---|---|---|
| | Number of establishments | Employment | | | | | | Annual payroll | | Farms | | | Farm producers whose primary occupation is farming (percent) |
| | | Total | Health care and social assistance | Manufacturing | Retail trade | Finance and insurance | Professional, scientific, and technical services | Total (mil dol) | Average per employee (dollars) | Number | Percent with: | | |
| | | | | | | | | | | | Fewer than 50 acres | 1000 acres or more | |
| | 104 | 105 | 106 | 107 | 108 | 109 | 110 | 111 | 112 | 113 | 114 | 115 | 116 |
| **MISSOURI— Cont'd** | | | | | | | | | | | | | |
| Jasper | 2,788 | 51,908 | 6,782 | 10,597 | 7,999 | 1,118 | 1,036 | 2,069 | 39,849 | 1,315 | 42.1 | 3.3 | 38.1 |
| Jefferson | 3,905 | 40,761 | 5,531 | 4,270 | 7,245 | 1,152 | 1,197 | 1,545 | 37,902 | 721 | 43.7 | 1.1 | 37.2 |
| Johnson | 905 | 10,644 | 2,212 | 1,480 | 1,731 | 273 | 387 | 347 | 32,558 | 1,626 | 37.8 | 5.0 | 36.2 |
| Knox | 93 | 610 | 31 | 114 | 126 | 52 | 38 | 18 | 30,246 | 637 | 18.8 | 9.3 | 41.9 |
| Laclede | 747 | 11,345 | 1,354 | 4,586 | 1,762 | 288 | 156 | 390 | 34,360 | 1,304 | 27.7 | 2.3 | 36.5 |
| Lafayette | 694 | 6,568 | 1,172 | 1,162 | 858 | 228 | 166 | 211 | 32,134 | 1,175 | 38.5 | 7.7 | 39.2 |
| Lawrence | 712 | 6,634 | 1,077 | 1,165 | 1,326 | 183 | 152 | 228 | 34,318 | 1,697 | 41.8 | 3.2 | 40.0 |
| Lewis | 179 | 2,036 | 260 | 125 | 305 | 91 | 50 | 57 | 28,041 | 636 | 27.8 | 9.6 | 33.5 |
| Lincoln | 939 | 9,547 | 1,315 | 1,485 | 1,662 | 342 | 267 | 383 | 40,076 | 1,092 | 41.9 | 4.3 | 36.0 |
| Linn | 288 | 2,985 | 529 | 775 | 458 | 152 | 62 | 118 | 39,474 | 994 | 23.7 | 7.7 | 36.9 |
| Livingston | 415 | 5,020 | 959 | 613 | 1,118 | 188 | 127 | 167 | 33,282 | 784 | 22.1 | 9.3 | 35.9 |
| McDonald | 313 | 5,761 | 236 | 3,384 | 816 | 120 | 48 | 200 | 34,798 | 940 | 26.9 | 1.9 | 40.2 |
| Macon | 345 | 3,707 | 600 | 489 | 696 | 192 | 88 | 115 | 31,038 | 1,163 | 22.0 | 8.1 | 37.7 |
| Madison | 259 | 3,237 | 969 | 282 | 601 | 79 | 45 | 99 | 30,716 | 361 | 19.9 | 2.5 | 39.9 |
| Maries | 128 | 976 | 165 | 146 | 140 | 102 | 22 | 33 | 33,395 | 879 | 14.9 | 3.1 | 36.6 |
| Marion | 786 | 11,671 | 2,808 | 1,442 | 1,727 | 425 | 282 | 424 | 36,349 | 587 | 26.9 | 9.5 | 36.6 |
| Mercer | 70 | 386 | 60 | 21 | 92 | 33 | NA | 12 | 32,236 | 493 | 16.0 | 8.3 | 40.2 |
| Miller | 674 | 6,558 | 559 | 734 | 1,941 | 222 | 183 | 216 | 32,961 | 1,023 | 19.7 | 3.4 | 36.7 |
| Mississippi | 235 | 2,163 | 412 | 82 | 397 | 143 | 98 | 73 | 33,817 | 159 | 13.8 | 53.5 | 70.4 |
| Moniteau | 327 | 3,004 | 313 | 732 | 426 | 107 | 58 | 103 | 34,434 | 1,135 | 28.7 | 3.2 | 40.5 |
| Monroe | 180 | 1,345 | 229 | 358 | 255 | 68 | 19 | 41 | 30,577 | 978 | 23.6 | 7.3 | 35.1 |
| Montgomery | 244 | 2,206 | 346 | 531 | 383 | 109 | 24 | 80 | 36,111 | 698 | 26.8 | 9.5 | 38.8 |
| Morgan | 494 | 3,104 | 280 | 532 | 694 | 111 | 88 | 98 | 31,502 | 962 | 29.4 | 3.5 | 46.8 |
| New Madrid | 385 | 6,066 | 1,303 | 1,058 | 1,280 | 140 | 54 | 222 | 36,606 | 290 | 15.2 | 50.3 | 62.3 |
| Newton | 1,210 | 21,408 | 6,460 | 2,779 | 2,303 | 515 | 468 | 930 | 43,443 | 1,588 | 37.1 | 1.7 | 38.0 |
| Nodaway | 459 | 5,874 | 1,188 | 1,453 | 1,010 | 188 | 137 | 202 | 34,350 | 1,133 | 24.4 | 10.5 | 42.7 |
| Oregon | 188 | 1,663 | 509 | 127 | 463 | 42 | 32 | 40 | 23,753 | 564 | 19.5 | 8.0 | 44.6 |
| Osage | 268 | 4,292 | 298 | 2,499 | 475 | 132 | 27 | 164 | 38,244 | 1,277 | 15.8 | 3.4 | 33.2 |
| Ozark | 161 | 853 | 118 | 88 | 180 | 58 | 33 | 21 | 24,742 | 705 | 15.0 | 4.7 | 46.4 |
| Pemiscot | 330 | 3,951 | 1,520 | 441 | 614 | 139 | 37 | 123 | 31,199 | 184 | 14.1 | 49.5 | 63.6 |
| Perry | 516 | 9,584 | 1,091 | 3,697 | 1,022 | 321 | 162 | 370 | 38,657 | 921 | 27.3 | 4.3 | 37.4 |
| Pettis | 1,003 | 17,809 | 3,018 | 4,406 | 2,491 | 330 | 1,834 | 615 | 34,528 | 1,259 | 34.3 | 8.2 | 40.8 |
| Phelps | 1,053 | 14,334 | 3,606 | 1,249 | 2,567 | 394 | 354 | 494 | 34,475 | 728 | 32.1 | 2.6 | 32.0 |
| Pike | 391 | 4,163 | 748 | 658 | 635 | 148 | 96 | 149 | 35,734 | 926 | 26.0 | 9.1 | 36.0 |
| Platte | 2,459 | 43,759 | 4,615 | 3,354 | 6,514 | 2,346 | 1,734 | 2,030 | 46,389 | 490 | 41.6 | 8.0 | 30.9 |
| Polk | 588 | 7,539 | 2,278 | 450 | 1,094 | 320 | 823 | 317 | 42,067 | 1,562 | 32.8 | 2.9 | 42.0 |
| Pulaski | 665 | 8,500 | 1,618 | 154 | 1,751 | 355 | 346 | 282 | 33,204 | 502 | 25.7 | 1.6 | 36.8 |
| Putnam | 90 | 603 | 127 | 78 | 189 | 65 | 7 | 19 | 31,809 | 585 | 14.5 | 10.1 | 46.1 |
| Ralls | 215 | 2,680 | 186 | 1,449 | 208 | 43 | 26 | 136 | 50,601 | 672 | 25.0 | 10.1 | 37.3 |
| Randolph | 532 | 7,718 | 1,307 | 950 | 1,273 | 405 | 78 | 288 | 37,264 | 783 | 32.4 | 5.5 | 31.8 |
| Ray | 339 | 2,964 | 538 | 493 | 615 | 100 | 96 | 107 | 36,198 | 1,070 | 30.7 | 5.0 | 33.1 |
| Reynolds | 136 | 1,397 | 163 | 392 | 156 | 31 | 10 | 61 | 43,510 | 341 | 22.0 | 1.8 | 31.6 |
| Ripley | 316 | 2,315 | 908 | 443 | 444 | 76 | 18 | 47 | 20,452 | 438 | 25.1 | 5.3 | 37.4 |
| St. Charles | 8,664 | 137,412 | 18,048 | 14,986 | 20,087 | 10,446 | 6,506 | 6,390 | 46,499 | 604 | 45.0 | 8.1 | 39.1 |
| St. Clair | 166 | 1,252 | 374 | 70 | 301 | 83 | 32 | 34 | 27,335 | 734 | 18.1 | 7.4 | 43.8 |
| Ste. Genevieve | 387 | 5,305 | 1,049 | 1,605 | 446 | 158 | 78 | 230 | 43,371 | 660 | 26.4 | 3.6 | 39.4 |
| St. Francois | 1,419 | 18,752 | 5,030 | 1,922 | 3,270 | 1,304 | 375 | 643 | 34,273 | 688 | 34.9 | 1.9 | 31.7 |
| St. Louis | 31,044 | 617,583 | 90,005 | 44,419 | 64,281 | 39,299 | 51,732 | 37,306 | 60,407 | 184 | 67.4 | 3.8 | 33.7 |
| Saline | 474 | 7,687 | 1,432 | NA | 967 | 254 | 88 | 264 | 34,347 | 882 | 21.5 | 14.9 | 45.2 |
| Schuyler | 71 | 389 | 20 | NA | 106 | 17 | 8 | 14 | 36,203 | 541 | 23.7 | 6.3 | 42.4 |
| Scotland | 139 | 887 | 240 | 115 | 212 | 42 | 18 | 28 | 31,068 | 713 | 24.1 | 7.0 | 41.1 |
| Scott | 1,068 | 14,122 | 2,930 | 3,365 | 1,398 | 414 | 367 | 541 | 38,274 | 450 | 35.1 | 15.6 | 39.9 |
| Shannon | 163 | 1,239 | 198 | 594 | 136 | 43 | 7 | 31 | 25,229 | 435 | 25.3 | 6.0 | 37.0 |
| Shelby | 158 | 1,063 | 90 | 161 | 182 | 59 | 46 | 34 | 32,294 | 628 | 20.2 | 13.5 | 31.4 |
| Stoddard | 670 | 8,815 | 1,783 | 3,040 | 1,226 | 325 | 165 | 339 | 38,506 | 792 | 31.1 | 22.6 | 46.4 |
| Stone | 690 | 5,190 | 375 | 100 | 876 | 144 | 143 | 162 | 31,122 | 628 | 35.7 | 1.8 | 34.9 |
| Sullivan | 98 | 2,046 | 200 | 1,276 | 159 | 50 | 22 | 84 | 40,915 | 671 | 11.6 | 10.6 | 46.3 |
| Taney | 1,834 | 24,994 | 2,433 | 524 | 4,863 | 472 | 496 | 765 | 30,603 | 395 | 26.6 | 4.8 | 38.1 |
| Texas | 452 | 4,376 | 895 | 905 | 810 | 172 | 76 | 141 | 32,287 | 1,371 | 25.2 | 4.2 | 43.1 |
| Vernon | 466 | 5,742 | 1,400 | 1,254 | 775 | 393 | 92 | 202 | 35,240 | 1,265 | 24.5 | 8.3 | 39.8 |
| Warren | 617 | 6,823 | 719 | 1,402 | 970 | 214 | 124 | 263 | 38,559 | 568 | 37.7 | 4.8 | 32.7 |
| Washington | 350 | 3,207 | 623 | 464 | 607 | 133 | 60 | 97 | 30,136 | 502 | 27.5 | 3.0 | 41.8 |
| Wayne | 234 | 1,952 | 409 | 762 | 302 | 72 | 25 | 43 | 21,792 | 340 | 15.9 | 4.7 | 41.6 |
| Webster | 678 | 6,477 | 590 | 1,677 | 1,202 | 203 | 162 | 221 | 34,078 | 1,837 | 37.5 | 1.3 | 39.0 |
| Worth | 47 | 169 | 18 | NA | 55 | 13 | NA | 4 | 24,249 | 336 | 17.6 | 9.8 | 35.2 |
| Wright | 399 | 3,484 | 570 | 569 | 918 | 184 | 42 | 99 | 28,542 | 1,115 | 20.4 | 3.9 | 45.0 |

# Table B. States and Counties — Agriculture

| STATE County | Land in farms Acreage (1,000) | Percent change, 2012-2017 | Acres Average size of farm | Acres Total irrigated (1,000) | Acres Total cropland (1,000) | Value of land and buildings Average per farm | Value of land and buildings Average per acre | Value of machinery and equipment, average per farm (dollars) | Value of products sold: Total (mil dol) | Value of products sold: Average per farm (acres) | Percent from: Crops | Percent from: Livestock and poultry products | Organic farms (number) | Farms with internet access (percent) | Government payments Total ($1,000) | Government payments Percent of farms |
|---|---|---|---|---|---|---|---|---|---|---|---|---|---|---|---|---|
| | 117 | 118 | 119 | 120 | 121 | 122 | 123 | 124 | 125 | 126 | 127 | 128 | 129 | 130 | 131 | 132 |
| **MISSOURI— Cont'd** | | | | | | | | | | | | | | | | |
| Jasper | 265 | 7.2 | 201 | 5.3 | 141.2 | 647,953 | 3,221 | 94,672 | 97.2 | 73,947 | 48.1 | 51.9 | NA | 75.6 | 1,639 | 26.0 |
| Jefferson | 91 | -6.5 | 126 | 0.2 | 34.6 | 531,201 | 4,200 | 50,306 | 12.9 | 17,870 | 58.2 | 41.8 | NA | 76.1 | 263 | 10.0 |
| Johnson | 384 | -1.8 | 236 | 3.0 | 227.5 | 789,475 | 3,344 | 92,874 | 139.9 | 86,061 | 45.1 | 54.9 | 5 | 78.2 | 3,545 | 32.8 |
| Knox | 235 | -16.2 | 370 | 0.4 | 161.7 | 1,124,020 | 3,042 | 140,644 | 97.9 | 153,615 | 52.8 | 47.2 | 8 | 60.8 | 4,645 | 66.4 |
| Laclede | 298 | -6.8 | 229 | 0.3 | 75 | 575,705 | 2,517 | 64,847 | 45.5 | 34,913 | 11.6 | 88.4 | NA | 70.7 | 208 | 6.5 |
| Lafayette | 341 | 4.2 | 290 | 1.4 | 265.3 | 1,314,426 | 4,535 | 155,252 | 163.3 | 138,955 | 80.6 | 19.4 | 3 | 79.2 | 2,775 | 36.9 |
| Lawrence | 302 | -2.8 | 178 | 2.9 | 124.1 | 569,764 | 3,198 | 87,304 | 241.0 | 142,032 | 8.9 | 91.1 | 2 | 69.7 | 1,526 | 12.9 |
| Lewis | 214 | -24.8 | 336 | D | 153 | 1,112,669 | 3,312 | 120,524 | 83.6 | 131,465 | 72.2 | 27.8 | 13 | 68.6 | 3,114 | 58.2 |
| Lincoln | 227 | -19.1 | 208 | 1.3 | 157 | 908,336 | 4,361 | 98,978 | 86.3 | 79,056 | 69.5 | 30.5 | 3 | 75.9 | 2,891 | 42.1 |
| Linn | 331 | -1.5 | 333 | 1.3 | 206.7 | 958,778 | 2,883 | 107,802 | 76.4 | 76,814 | 59.7 | 40.3 | 12 | 68.8 | 7,019 | 56.0 |
| Livingston | 285 | 0.6 | 364 | D | 215.1 | 1,227,229 | 3,372 | 133,506 | 86.1 | 109,776 | 88.0 | 12.0 | 5 | 72.3 | 6,686 | 62.0 |
| McDonald | 191 | 2.6 | 204 | 0.1 | 44.8 | 546,226 | 2,682 | 77,040 | 197.5 | 210,061 | 2.3 | 97.7 | 7 | 75.7 | 409 | 7.3 |
| Macon | 393 | 1.7 | 338 | 0.6 | 236.9 | 1,078,661 | 3,194 | 101,249 | 116.9 | 100,521 | 59.9 | 40.1 | 6 | 69.1 | 4,513 | 50.6 |
| Madison | 94 | -12.1 | 261 | D | 22.3 | 547,402 | 2,099 | 56,907 | 16.3 | 45,169 | 4.9 | 95.1 | NA | 65.9 | 120 | 8.9 |
| Maries | 248 | 2.9 | 283 | 0.5 | 64.3 | 557,083 | 1,971 | 69,007 | 32.4 | 36,879 | 14.6 | 85.4 | NA | 70.0 | 513 | 13.5 |
| Marion | 233 | 5.0 | 396 | 4.8 | 182.5 | 1,571,407 | 3,966 | 156,078 | 99.1 | 168,853 | 76.7 | 23.3 | 1 | 68.3 | 2,268 | 59.6 |
| Mercer | 194 | -14.6 | 393 | 0.0 | 99.5 | 1,058,158 | 2,693 | 74,366 | 82.2 | 166,748 | 28.4 | 71.6 | 2 | 74.8 | 3,029 | 59.2 |
| Miller | 258 | 3.8 | 252 | 1.6 | 68.1 | 629,480 | 2,497 | 77,397 | 96.7 | 94,539 | 6.4 | 93.6 | NA | 72.5 | 256 | 10.0 |
| Mississippi | 251 | 2.3 | 1,576 | 116.8 | 243.1 | 9,200,783 | 5,837 | 631,538 | D | D | D | D | NA | 78.6 | 5,337 | 90.6 |
| Moniteau | 227 | -3.6 | 200 | 0.4 | 104.6 | 713,079 | 3,570 | 84,169 | 144.7 | 127,456 | 21.3 | 78.7 | 7 | 69.2 | 2,023 | 31.4 |
| Monroe | 340 | -4.4 | 348 | 1.2 | 226.8 | 1,237,626 | 3,559 | 106,640 | 133.8 | 136,833 | 59.6 | 40.4 | 4 | 69.1 | 5,907 | 61.8 |
| Montgomery | 221 | -20.8 | 317 | 3.3 | 144.7 | 1,194,722 | 3,771 | 123,954 | 93.7 | 134,219 | 63.5 | 36.5 | 7 | 71.8 | 2,612 | 50.1 |
| Morgan | 210 | 6.2 | 219 | 0.2 | 85.4 | 839,556 | 3,837 | 87,885 | 199.4 | 207,225 | 9.9 | 90.1 | 9 | 61.9 | 1,243 | 13.7 |
| New Madrid | 418 | 21.4 | 1,443 | 257.3 | 410.5 | 8,066,767 | 5,591 | 659,481 | 231.5 | 798,286 | 100.0 | 0.0 | NA | 78.6 | 11,682 | 91.4 |
| Newton | 261 | 5.5 | 165 | 0.5 | 91.3 | 598,888 | 3,639 | 75,574 | 246.0 | 154,909 | 6.2 | 93.8 | 1 | 73.0 | 776 | 12.1 |
| Nodaway | 440 | 3.8 | 388 | 1.6 | 325.4 | 1,557,349 | 4,012 | 144,528 | 152.2 | 134,308 | 84.0 | 16.0 | 6 | 76.8 | 5,833 | 54.8 |
| Oregon | 201 | -20.7 | 357 | 0.0 | 27.3 | 635,970 | 1,781 | 65,952 | 23.3 | 41,250 | 4.5 | 95.5 | 2 | 72.5 | 349 | 11.3 |
| Osage | 320 | 13.0 | 251 | 1.3 | 101.5 | 605,323 | 2,415 | 79,274 | 80.7 | 63,186 | 21.5 | 78.5 | NA | 68.8 | 1,009 | 21.1 |
| Ozark | 227 | -0.8 | 322 | 1.0 | 30.4 | 686,823 | 2,133 | 68,929 | 25.1 | 35,644 | 6.3 | 93.7 | 1 | 73.3 | 332 | 6.1 |
| Pemiscot | 296 | -2.9 | 1,610 | 148.1 | 286.7 | 8,426,555 | 5,235 | 614,890 | 159.2 | 865,217 | 100.0 | 0.0 | 1 | 77.2 | 6,158 | 89.1 |
| Perry | 219 | -3.4 | 237 | 1.4 | 124.6 | 776,958 | 3,273 | 100,770 | 62.8 | 68,172 | 57.0 | 43.0 | NA | 64.7 | 3,033 | 56.8 |
| Pettis | 389 | -7.2 | 309 | 0.7 | 240 | 1,082,193 | 3,500 | 118,802 | 239.1 | 189,934 | 35.5 | 64.5 | 17 | 69.7 | 5,057 | 42.2 |
| Phelps | 160 | 1.4 | 219 | 0.1 | 29.4 | 577,647 | 2,636 | 70,254 | 14.0 | 19,231 | 17.7 | 82.3 | 1 | 79.5 | 287 | 10.3 |
| Pike | 311 | -14.1 | 336 | 5.0 | 207.8 | 1,179,923 | 3,516 | 141,497 | 133.2 | 143,811 | 59.4 | 40.6 | 2 | 62.7 | 2,795 | 50.2 |
| Platte | 161 | 5.6 | 330 | 4.5 | 126.9 | 1,459,303 | 4,429 | 144,958 | 60.0 | 122,516 | 88.5 | 11.5 | 1 | 77.8 | 1,255 | 34.1 |
| Polk | 359 | 6.9 | 230 | 2.4 | 103.4 | 581,099 | 2,525 | 66,217 | 99.4 | 63,609 | 11.1 | 88.9 | 23 | 71.1 | 613 | 10.7 |
| Pulaski | 111 | -0.9 | 222 | 0.2 | 24.2 | 496,821 | 2,237 | 58,444 | 27.2 | 54,127 | 4.7 | 95.3 | NA | 73.7 | 60 | 4.8 |
| Putnam | 264 | -9.7 | 452 | D | 132.6 | 1,138,477 | 2,520 | 103,874 | 93.9 | 160,557 | 23.9 | 76.1 | NA | 73.8 | 1,964 | 52.6 |
| Ralls | 243 | -14.2 | 362 | D | 181.4 | 1,392,728 | 3,850 | 138,653 | 79.5 | 118,344 | 81.3 | 18.7 | NA | 73.4 | 3,524 | 68.2 |
| Randolph | 213 | 1.6 | 272 | 0.7 | 123.7 | 940,599 | 3,461 | 84,856 | 81.3 | 103,847 | 45.5 | 54.5 | 9 | 74.1 | 2,853 | 43.9 |
| Ray | 267 | -2.4 | 249 | 4.6 | 179.5 | 876,483 | 3,516 | 98,521 | 79.0 | 73,842 | 80.9 | 19.1 | 3 | 73.2 | 3,613 | 46.2 |
| Reynolds | 87 | -10.8 | 254 | D | 13.2 | 406,440 | 1,599 | 49,387 | D | D | D | D | NA | 70.7 | 105 | 8.5 |
| Ripley | 143 | 3.9 | 327 | 12.8 | 41.1 | 799,801 | 2,446 | 78,726 | 26.2 | 59,721 | 28.7 | 71.3 | NA | 73.3 | 1,303 | 20.8 |
| St. Charles | 156 | -1.7 | 258 | 1.7 | 124.2 | 1,257,799 | 4,885 | 137,288 | 61.8 | 102,308 | 87.1 | 12.9 | NA | 77.8 | 1,080 | 37.9 |
| St. Clair | 249 | 4.2 | 339 | 0.0 | 116 | 830,734 | 2,453 | 99,643 | 48.4 | 65,917 | 59.5 | 40.5 | 2 | 67.8 | 1,906 | 20.4 |
| Ste. Genevieve | 169 | 3.6 | 255 | 0.2 | 77.1 | 753,601 | 2,951 | 76,204 | 31.8 | 48,171 | 57.3 | 42.7 | NA | 65.3 | 2,166 | 39.2 |
| St. Francois | 125 | 7.4 | 182 | 0.3 | 37.6 | 532,762 | 2,933 | 62,812 | 14.6 | 21,150 | 53.7 | 46.3 | 3 | 68.9 | 203 | 7.4 |
| St. Louis | 46 | 54.3 | 249 | 0.9 | 13.6 | 1,082,987 | 4,348 | 60,124 | 20.5 | 111,560 | 95.4 | 4.6 | 3 | 84.8 | 215 | 15.2 |
| Saline | 441 | -4.3 | 500 | 7.2 | 355.8 | 2,007,141 | 4,010 | 214,564 | 257.4 | 291,841 | 70.4 | 29.6 | 3 | 74.9 | 4,231 | 68.8 |
| Schuyler | 167 | 4.7 | 309 | 0.0 | 80.8 | 796,550 | 2,581 | 77,623 | 38.4 | 71,002 | 38.8 | 61.2 | NA | 64.3 | 1,729 | 47.1 |
| Scotland | 250 | 2.5 | 351 | 0.0 | 181.5 | 1,182,524 | 3,370 | 135,426 | 156.7 | 219,724 | 36.7 | 63.3 | 19 | 63.5 | 4,814 | 56.1 |
| Scott | 223 | -0.1 | 495 | 91.5 | 203.4 | 2,518,197 | 5,087 | 274,600 | 184.8 | 410,682 | 55.0 | 45.0 | NA | 70.0 | 8,463 | 65.1 |
| Shannon | 130 | 4.7 | 298 | 0.0 | 29.1 | 585,434 | 1,963 | 69,516 | 11.6 | 26,632 | 7.8 | 92.2 | NA | 71.7 | 233 | 8.3 |
| Shelby | 278 | -7.2 | 442 | 1.4 | 204.8 | 1,648,469 | 3,729 | 163,179 | 105.5 | 168,002 | 72.4 | 27.6 | NA | 72.3 | 3,894 | 65.1 |
| Stoddard | 476 | 6.1 | 600 | 281.7 | 437.9 | 3,214,252 | 5,353 | 259,498 | 291.9 | 368,573 | 82.0 | 18.0 | NA | 69.6 | 17,568 | 63.9 |
| Stone | 106 | -10.0 | 169 | 0.0 | 20.9 | 489,224 | 2,894 | 47,276 | 35.2 | 55,989 | 4.3 | 95.7 | NA | 78.2 | 145 | 4.1 |
| Sullivan | 310 | -4.0 | 462 | D | 159.7 | 1,079,907 | 2,338 | 114,353 | 178.4 | 265,927 | 15.9 | 84.1 | NA | 70.5 | 5,031 | 57.8 |
| Taney | 108 | -6.6 | 274 | 0.1 | 18.8 | 612,082 | 2,235 | 50,546 | 13.1 | 33,225 | 9.3 | 90.7 | NA | 73.9 | 235 | 5.3 |
| Texas | 391 | -0.4 | 285 | 1.1 | 82.1 | 588,769 | 2,066 | 63,440 | 45.2 | 32,992 | 11.5 | 88.5 | 1 | 74.5 | 489 | 6.3 |
| Vernon | 436 | 4.1 | 344 | 12.2 | 272.2 | 978,555 | 2,841 | 126,119 | 215.6 | 170,442 | 38.2 | 61.8 | 10 | 69.6 | 4,835 | 37.4 |
| Warren | 128 | -6.1 | 225 | 0.9 | 81.8 | 911,284 | 4,048 | 105,482 | 46.5 | 81,917 | 77.9 | 22.1 | 3 | 80.6 | 1,442 | 36.3 |
| Washington | 104 | -16.3 | 207 | 0.1 | 20.4 | 490,458 | 2,372 | 50,335 | 10.0 | 19,859 | 16.3 | 83.7 | NA | 75.1 | 206 | 3.2 |
| Wayne | 98 | -16.2 | 287 | 0.5 | 33.2 | 675,333 | 2,350 | 60,230 | 12.6 | 37,156 | 61.9 | 38.1 | NA | 70.6 | 637 | 14.1 |
| Webster | 265 | -2.5 | 144 | 0.1 | 77 | 447,850 | 3,102 | 59,669 | 54.4 | 29,625 | 12.5 | 87.5 | 8 | 74.6 | 437 | 5.2 |
| Worth | 125 | 0.2 | 373 | NA | 72.8 | 1,061,053 | 2,844 | 106,629 | 41.9 | 124,705 | 37.9 | 62.1 | NA | 75.0 | 2,694 | 69.9 |
| Wright | 285 | -3.1 | 256 | 0.5 | 67.5 | 551,501 | 2,158 | 59,321 | 44.8 | 40,158 | 7.0 | 93.0 | 1 | 70.9 | 294 | 5.7 |

# Table B. States and Counties — Water Use, Wholesale Trade, Retail Trade, and Real Estate

| STATE County | Water use, 2015 | | Wholesale Trade[1], 2017 | | | | Retail Trade[2], 2017 | | | | Real estate and rental and leasing,[2] 2017 | | | |
|---|---|---|---|---|---|---|---|---|---|---|---|---|---|---|
| | Public supply water withdrawn (mil gal/day) | Public supply gallons withdrawn per person per day | Number of establishments | Number of employees | Sales (mil dol) | Average payroll (mil dol) | Number of establishments | Number of employees | Sales (mil dol) | Average payroll (mil dol) | Number of establishments | Number of employees | Sales (mil dol) | Average payroll (mil dol) |
| | 133 | 134 | 135 | 136 | 137 | 138 | 139 | 140 | 141 | 142 | 143 | 144 | 145 | 146 |
| MISSOURI— Cont'd | | | | | | | | | | | | | | |
| Jasper | 18.54 | 156.3 | 134 | 1,675 | 872.1 | 79.8 | 500 | 8,085 | 2,229.2 | 203.2 | 124 | 597 | 107.3 | 16.9 |
| Jefferson | 28.68 | 128.0 | 131 | 1,461 | 699.2 | 76.2 | 474 | 7,264 | 2,163.2 | 190.9 | 157 | 581 | 98.5 | 19.3 |
| Johnson | 5.14 | 95.3 | 30 | 116 | 59.4 | 5.3 | 129 | 1,687 | 427.7 | 39.8 | 33 | 83 | 22.1 | 2.5 |
| Knox | 0.00 | 0.0 | 6 | D | 113.3 | D | 14 | 120 | 30.0 | 2.4 | NA | NA | NA | NA |
| Laclede | 3.79 | 106.8 | 32 | 271 | 116.7 | 9.3 | 160 | 1,939 | 553.5 | 48.8 | 27 | 115 | 20.1 | 3.5 |
| Lafayette | 2.31 | 70.6 | 36 | 346 | 223.9 | 14.9 | 104 | 949 | 278.0 | 21.0 | 19 | 44 | 5.4 | 0.9 |
| Lawrence | 2.77 | 72.6 | 24 | 228 | 86.4 | 7.2 | 115 | 1,412 | 470.8 | 37.7 | 20 | 69 | 9.0 | 2.6 |
| Lewis | 0.69 | 67.6 | D | D | D | D | 27 | 321 | 79.1 | 6.6 | 3 | 5 | 0.6 | 0.1 |
| Lincoln | 2.51 | 45.9 | D | D | D | 9.8 | 141 | 1,675 | 604.5 | 47.4 | 31 | 71 | 17.8 | 2.1 |
| Linn | 2.10 | 170.6 | D | D | D | 5.6 | 52 | 515 | 137.3 | 12.7 | 6 | 12 | 0.9 | 0.2 |
| Livingston | 1.89 | 125.8 | 25 | 187 | 126.5 | 7.4 | 75 | 1,062 | 293.4 | 27.5 | 8 | 22 | 14.9 | 0.7 |
| McDonald | 1.86 | 82.1 | D | D | D | D | 60 | 957 | 243.2 | 19.6 | D | D | D | D |
| Macon | 1.86 | 121.3 | 14 | 158 | 41.4 | 4.1 | 60 | 623 | 150.2 | 13.4 | 9 | 33 | 2.8 | 0.6 |
| Madison | 0.62 | 50.0 | D | D | D | D | 43 | 533 | 141.7 | 13.3 | 5 | 18 | 2.2 | 0.4 |
| Maries | 0.30 | 33.5 | 4 | D | 5.1 | D | 17 | 144 | 41.5 | 2.6 | 3 | D | 1.1 | D |
| Marion | 3.66 | 126.7 | 26 | 449 | 252.2 | 19.8 | 131 | 1,817 | 496.2 | 41.5 | 27 | 69 | 9.3 | 2.2 |
| Mercer | 0.13 | 35.2 | NA | NA | NA | NA | 11 | 98 | 28.9 | 1.7 | 4 | 6 | 0.2 | 0.1 |
| Miller | 0.78 | 31.1 | 12 | 116 | 52.2 | 3.7 | 121 | 1,794 | 439.0 | 47.3 | 50 | 299 | 45.9 | 8.0 |
| Mississippi | 1.90 | 135.4 | 17 | 156 | 192.6 | 8.2 | 53 | 433 | 149.8 | 10.7 | NA | NA | NA | NA |
| Moniteau | 1.47 | 92.1 | 12 | 190 | 46.3 | 4.9 | 48 | 424 | 137.8 | 8.9 | D | D | D | D |
| Monroe | 4.71 | 548.8 | 6 | D | 76.0 | D | 31 | 270 | 53.8 | 5.5 | NA | NA | NA | NA |
| Montgomery | 0.44 | 37.6 | 13 | 126 | 216.3 | 6.0 | 42 | 348 | 100.9 | 9.1 | 4 | 8 | 1.1 | 0.1 |
| Morgan | 0.70 | 34.7 | 16 | 52 | 16.2 | 2.1 | 92 | 796 | 211.8 | 18.7 | 20 | 40 | 6.7 | 1.1 |
| New Madrid | 1.95 | 107.1 | 35 | 410 | 341.8 | 22.6 | 79 | 1,328 | 434.9 | 30.2 | 7 | 18 | 2.8 | 0.6 |
| Newton | 4.33 | 73.9 | 45 | 1,640 | 1,448.8 | 68.3 | 198 | 2,210 | 795.7 | 56.5 | 42 | 128 | 13.6 | 3.2 |
| Nodaway | 1.88 | 82.4 | 20 | 205 | 136.1 | 7.5 | 64 | 1,013 | 254.4 | 22.5 | 13 | 26 | 5.1 | D |
| Oregon | 0.88 | 80.3 | D | D | D | 1.8 | 48 | 461 | 103.8 | 9.8 | 6 | D | 1.5 | D |
| Osage | 0.55 | 40.4 | 5 | 47 | 23.7 | 1.5 | 51 | 479 | 152.6 | 11.9 | D | D | D | D |
| Ozark | 0.24 | 25.5 | D | D | D | D | 24 | 177 | 41.5 | 3.3 | D | D | D | 0.1 |
| Pemiscot | 2.13 | 121.8 | 21 | 182 | 319.1 | 9.9 | 55 | 569 | 179.1 | 11.6 | 10 | 33 | 7.4 | 1.6 |
| Perry | 1.48 | 77.2 | 11 | 301 | 119.5 | 14.9 | 80 | 1,040 | 279.3 | 26.7 | 15 | 34 | 6.9 | 0.8 |
| Pettis | 3.55 | 84.0 | 36 | 396 | 171.8 | 17.8 | 169 | 2,558 | 715.8 | 62.9 | 42 | 314 | 45.3 | 10.1 |
| Phelps | 2.83 | 63.2 | 37 | 301 | 108.8 | 12.3 | 184 | 2,464 | 726.4 | 61.9 | 39 | 161 | 25.8 | 5.1 |
| Pike | 1.13 | 61.6 | 26 | 362 | 247.2 | 14.4 | 58 | 683 | 161.5 | 16.4 | 7 | 27 | 1.7 | 0.3 |
| Platte | 2.42 | 25.2 | 100 | 1,937 | 8,079.7 | 132.8 | 319 | 6,788 | 3,520.2 | 170.7 | 151 | 1,222 | 308.3 | 62.8 |
| Polk | 1.31 | 41.9 | 18 | 528 | 59.4 | 8.6 | 100 | 1,141 | 311.5 | 28.7 | 16 | 62 | 7.1 | 1.6 |
| Pulaski | 5.85 | 109.9 | D | D | D | D | 139 | 1,760 | 475.7 | 41.4 | 42 | 170 | 47.1 | 4.7 |
| Putnam | 0.25 | 51.5 | 3 | 44 | 16.2 | 1.2 | 23 | 191 | 49.1 | 3.9 | 6 | D | 0.9 | D |
| Ralls | 0.00 | 0.0 | 15 | 166 | 125.1 | 8.0 | 30 | 221 | 73.9 | 5.3 | 4 | 6 | 0.2 | 0.0 |
| Randolph | 1.12 | 44.6 | 14 | D | 160.7 | D | 100 | 1,216 | 308.6 | 30.7 | 18 | 88 | 8.8 | 2.1 |
| Ray | 2.36 | 103.5 | 8 | 90 | 106.7 | 4.3 | 53 | 625 | 158.0 | 15.1 | 7 | 10 | 0.7 | 0.1 |
| Reynolds | 0.13 | 20.2 | D | D | D | 0.4 | 23 | 147 | 26.9 | 2.7 | D | D | D | 0.2 |
| Ripley | 0.63 | 45.6 | D | D | D | 1.6 | 43 | 446 | 112.8 | 8.9 | 6 | 7 | 0.9 | 0.2 |
| St. Charles | 17.94 | 46.5 | 344 | 4,375 | 8,155.1 | 255.6 | 1,124 | 20,236 | 5,882.9 | 536.7 | 415 | 1,932 | 572.9 | 82.8 |
| St. Clair | 0.32 | 33.9 | 3 | 6 | 1.1 | 0.2 | 35 | 333 | 80.3 | 6.4 | 3 | 4 | 0.3 | 0.1 |
| Ste. Genevieve | 1.43 | 79.8 | 17 | 148 | 125.4 | 7.6 | 48 | 479 | 115.3 | 9.4 | 7 | 31 | 3.9 | 0.6 |
| St. Francois | 5.07 | 76.2 | D | D | D | D | 241 | 3,344 | 842.1 | 80.1 | 62 | 210 | 29.5 | 5.6 |
| St. Louis | 226.20 | 225.4 | 1,443 | 27,555 | 35,553.3 | 2,077.5 | 3,611 | 66,288 | 29,641.9 | 1,909.3 | 1,471 | 11,174 | 3,039.4 | 582.9 |
| Saline | 2.88 | 123.8 | 34 | 356 | 394.0 | 17.0 | 87 | 941 | 251.0 | 22.5 | D | D | D | D |
| Schuyler | 0.00 | 0.0 | NA | NA | NA | NA | 18 | 110 | 30.5 | 2.0 | NA | NA | NA | NA |
| Scotland | 0.15 | 30.9 | 5 | 72 | 23.8 | 3.2 | 33 | 219 | 52.2 | 4.4 | NA | NA | NA | NA |
| Scott | 5.94 | 152.3 | 53 | 980 | 848.9 | 43.5 | 174 | 1,465 | 376.0 | 35.3 | 32 | 114 | 23.0 | 4.0 |
| Shannon | 0.44 | 53.3 | 4 | D | 36.5 | D | 31 | 150 | 32.4 | 2.6 | 8 | 6 | 1.3 | 0.3 |
| Shelby | 0.23 | 37.5 | 14 | D | 98.2 | D | 29 | 189 | 39.0 | 3.9 | 4 | D | 0.9 | D |
| Stoddard | 3.23 | 108.2 | 33 | 374 | 376.8 | 18.3 | 106 | 1,299 | 353.5 | 30.5 | 21 | 68 | 10.5 | 1.8 |
| Stone | 3.15 | 101.8 | D | D | D | 1.0 | 98 | 939 | 253.3 | 26.5 | 42 | 274 | 50.5 | 9.2 |
| Sullivan | 0.65 | 102.3 | 3 | 18 | 10.9 | 0.8 | 21 | 157 | 35.2 | 2.8 | 4 | D | 0.2 | D |
| Taney | 8.29 | 151.9 | D | D | D | 6.3 | 415 | 5,207 | 1,093.9 | 109.3 | 128 | 1,375 | 242.4 | 53.8 |
| Texas | 1.72 | 67.0 | 13 | 91 | 24.7 | 2.6 | 85 | 803 | 200.2 | 19.0 | 12 | 31 | 3.6 | 1.3 |
| Vernon | 2.02 | 97.0 | 26 | 202 | 105.0 | 8.0 | 81 | 936 | 238.6 | 20.5 | 13 | 40 | 5.5 | 1.1 |
| Warren | 1.63 | 48.6 | 27 | 475 | 531.3 | 29.7 | 73 | 915 | 297.6 | 22.2 | 18 | 93 | 15.0 | 2.4 |
| Washington | 0.66 | 26.6 | 11 | 79 | 51.6 | 1.8 | 54 | 569 | 142.9 | 12.8 | 7 | 16 | 2.0 | 0.5 |
| Wayne | 1.33 | 99.2 | 6 | D | 9.0 | D | 35 | 284 | 66.1 | 5.5 | NA | NA | NA | NA |
| Webster | 1.11 | 29.6 | 37 | 169 | 74.3 | 5.8 | 103 | 1,252 | 461.1 | 32.6 | 32 | 60 | 10.3 | 1.5 |
| Worth | 0.01 | 4.9 | NA | NA | NA | NA | 10 | 76 | 16.7 | 1.6 | NA | NA | NA | NA |
| Wright | 0.97 | 53.1 | 13 | 197 | 40.0 | 3.8 | 79 | 937 | 219.6 | 22.1 | 11 | 26 | 2.5 | 0.4 |

1  Merchant wholesalers, except manufacturers' sales branches and offices.  2. Employer establishments.

# Table B. States and Counties — Professional Services, Manufacturing, and Accommodation and Food Services

| STATE County | Professional, scientific, and technical services, 2017 | | | | Manufacturing, 2017 | | | | Accommodation and food services, 2017 | | | |
|---|---|---|---|---|---|---|---|---|---|---|---|---|
| | Number of establish-ments | Number of employees | Sales (mil dol) | Average payroll (mil dol) | Number of establish-ments | Number of employees | Sales (mil dol) | Average payroll (mil dol) | Number of establish-ments | Number of employees | Sales (mil dol) | Annual payroll (mil dol) |
| | 147 | 148 | 149 | 150 | 151 | 152 | 153 | 154 | 155 | 156 | 157 | 158 |
| **MISSOURI— Cont'd** | | | | | | | | | | | | |
| Jasper | 183 | 1,055 | 143 | 47.8 | 159 | 9,995 | 4,135.3 | 477.4 | 264 | 5,076 | 220.8 | 65.4 |
| Jefferson | 275 | 1,212 | 135 | 48.6 | 171 | 4,270 | 1,506.2 | 246.1 | 289 | 6,361 | 288.1 | 86.7 |
| Johnson | 74 | 412 | 86 | 28.2 | 37 | 1,534 | 350.5 | 60.5 | 104 | 1,987 | 86.4 | 23.2 |
| Knox | D | D | 2 | D | NA | NA | NA | NA | NA | NA | NA | NA |
| Laclede | D | D | D | D | 54 | 4,231 | 1,561.9 | 174.2 | 74 | 1,165 | 60.1 | 16.5 |
| Lafayette | D | D | 28 | D | 31 | 918 | 298.5 | 43.4 | D | D | D | D |
| Lawrence | 40 | 135 | 11 | 3.8 | 41 | 1,139 | 560.2 | 59.4 | 61 | 913 | 37.7 | 10.7 |
| Lewis | D | D | 4 | D | 5 | D | 42.4 | D | D | D | D | D |
| Lincoln | 57 | 234 | 34 | 11.2 | 50 | 1,312 | 510.5 | 71.6 | 58 | 1,128 | 53.6 | 13.7 |
| Linn | 18 | 57 | 7 | 1.5 | 15 | 869 | 169.0 | 30.6 | 23 | 249 | 9.5 | 2.5 |
| Livingston | D | D | 13 | D | 20 | 486 | 135.0 | 21.7 | 29 | 475 | 20.5 | 6.2 |
| McDonald | D | D | D | D | 24 | 3,304 | 656.9 | 125.5 | 32 | 315 | 17.4 | 4.3 |
| Macon | 20 | 78 | 8 | 3.0 | 11 | 464 | 210.4 | 21.6 | 31 | 458 | 20.3 | 6.0 |
| Madison | 9 | 58 | 6 | 1.4 | 15 | 321 | 60.9 | 11.7 | 22 | 290 | 9.5 | 2.8 |
| Maries | D | D | D | D | 7 | 147 | 74.6 | 8.0 | D | D | D | 0.5 |
| Marion | D | D | D | D | 35 | 1,374 | 1,860.2 | 73.8 | 79 | 1,221 | 58.2 | 16.9 |
| Mercer | NA | NA | NA | NA | NA | NA | NA | NA | 4 | 9 | 0.6 | 0.1 |
| Miller | 44 | 179 | 18 | 7.3 | 24 | 612 | 190.2 | 25.4 | 52 | 600 | 31.7 | 10.1 |
| Mississippi | 11 | 91 | 13 | 6.5 | 5 | 78 | 15.5 | 3.2 | 16 | 247 | 11.2 | 3.3 |
| Moniteau | 17 | 60 | 6 | 1.6 | D | 793 | D | 25.4 | D | D | D | D |
| Monroe | D | D | 2 | D | 11 | 393 | 85.9 | 16.6 | D | D | D | 1.5 |
| Montgomery | D | D | 1 | D | 17 | 462 | 212.3 | 20.0 | 20 | 163 | 8.2 | 1.9 |
| Morgan | 22 | 62 | 6 | 1.6 | 21 | 483 | 143.4 | 19.2 | 53 | 575 | 31.0 | 8.7 |
| New Madrid | 10 | 53 | 5 | 1.8 | 10 | 934 | 213.1 | 25.4 | 37 | 509 | 23.7 | 6.7 |
| Newton | 71 | 424 | 45 | 18.4 | 60 | 2,390 | 854.8 | 110.5 | 93 | 1,848 | 94.3 | 27.3 |
| Nodaway | 29 | 164 | 19 | 6.2 | 19 | 1,231 | 594.8 | 64.5 | 38 | 833 | 33.6 | 11.7 |
| Oregon | D | D | 2 | D | 14 | 106 | 28.7 | 3.9 | D | D | D | D |
| Osage | 12 | 22 | 3 | 0.8 | D | 1,271 | D | 69.2 | D | D | D | D |
| Ozark | D | D | 3 | D | 8 | 54 | 13.0 | 1.9 | 15 | 168 | 6.8 | 2.3 |
| Pemiscot | 11 | 41 | 3 | 1.0 | D | D | D | D | 33 | 303 | 16.2 | 3.8 |
| Perry | 24 | 153 | 14 | 6.7 | 38 | 3,530 | 1,239.6 | 138.1 | 37 | 546 | 22.9 | 7.2 |
| Pettis | 63 | 1,482 | 158 | 56.3 | 51 | 4,151 | 1,524.0 | 179.5 | 73 | 1,652 | 75.7 | 22.2 |
| Phelps | 80 | 359 | 38 | 13.7 | 47 | 1,050 | 911.5 | 57.7 | 131 | 2,426 | 110.1 | 29.2 |
| Pike | 18 | 89 | 9 | 2.4 | 14 | 700 | 193.4 | 38.9 | 31 | 352 | 18.8 | 4.9 |
| Platte | 267 | 1,378 | 255 | 81.4 | 51 | 3,569 | 2,317.9 | 175.3 | 216 | 5,990 | 503.7 | 123.4 |
| Polk | 46 | 1,424 | 297 | 142.8 | 23 | 333 | 101.3 | 12.8 | 45 | 536 | 28.0 | 7.1 |
| Pulaski | 60 | 374 | 39 | 15.0 | D | 151 | D | D | 102 | 2,418 | 113.5 | 48.1 |
| Putnam | D | D | 1 | D | D | 74 | D | 2.5 | 4 | 18 | 1.0 | 0.2 |
| Ralls | D | D | D | D | 13 | 1,091 | 1,277.8 | 71.5 | 14 | 80 | 3.9 | 1.1 |
| Randolph | 22 | 95 | 7 | 2.0 | 25 | 959 | 215.9 | 40.2 | 45 | 660 | 31.2 | 8.0 |
| Ray | 27 | 94 | 11 | 3.9 | 16 | 401 | 197.2 | 27.9 | D | D | D | D |
| Reynolds | D | D | D | D | 22 | 207 | 41.9 | 7.2 | D | D | D | 1.5 |
| Ripley | D | D | D | D | 26 | 404 | 53.1 | 10.4 | 18 | 173 | 7.5 | 2.0 |
| St. Charles | 844 | 7,456 | 863 | 348.0 | 241 | 15,165 | 12,017.9 | 927.0 | 759 | 19,028 | 1,136.7 | 311.5 |
| St. Clair | D | D | D | D | 11 | 33 | 8.0 | 2.1 | D | D | D | D |
| Ste. Genevieve | D | D | D | D | 30 | 1,127 | 440.6 | 68.1 | 29 | 406 | 15.6 | 4.5 |
| St. Francois | 86 | 359 | 34 | 11.1 | 43 | 1,695 | 333.3 | 75.3 | 120 | 2,240 | 98.3 | 28.2 |
| St. Louis | 3,599 | 44,516 | 8,643 | 3,240.7 | 867 | 42,197 | 12,718.8 | 3,067.8 | 2,263 | 50,715 | 2,880.1 | 850.9 |
| Saline | D | D | D | D | 22 | 1,862 | 927.7 | 79.1 | 43 | 499 | 21.9 | 6.0 |
| Schuyler | NA | NA | NA | NA | 3 | D | 14.9 | 2.6 | 3 | 16 | 0.6 | 0.1 |
| Scotland | D | D | D | D | 10 | 84 | 23.2 | 3.3 | D | D | D | D |
| Scott | 62 | 353 | 47 | 17.0 | 60 | 2,785 | 1,088.5 | 114.7 | 79 | 1,314 | 65.4 | 19.9 |
| Shannon | 4 | 6 | 1 | 0.2 | 28 | 644 | 76.0 | 15.1 | 20 | 100 | 7.6 | 1.8 |
| Shelby | D | D | 4 | D | D | 152 | D | 7.8 | D | D | D | 0.6 |
| Stoddard | 36 | 171 | 19 | 6.6 | 29 | 2,989 | 939.8 | 142.5 | 47 | 675 | 32.3 | 8.7 |
| Stone | 36 | 94 | 9 | 3.3 | 21 | 80 | 11.6 | 3.6 | 100 | 864 | 87.8 | 20.9 |
| Sullivan | D | D | 1 | D | D | D | D | D | 5 | D | 0.4 | D |
| Taney | 112 | 483 | 51 | 13.9 | 38 | 312 | 97.4 | 14.9 | 318 | 6,610 | 556.2 | 153.8 |
| Texas | 24 | 74 | 6 | 1.7 | 42 | 803 | 223.7 | 34.8 | 36 | 424 | 14.9 | 4.2 |
| Vernon | 31 | 109 | 12 | 3.4 | 20 | 1,064 | 1,266.7 | 60.6 | 41 | 481 | 27.8 | 6.8 |
| Warren | 28 | 124 | 11 | 4.3 | 30 | 1,297 | 449.6 | 65.8 | 48 | 711 | 33.9 | 9.1 |
| Washington | 17 | 48 | 4 | 1.2 | 21 | 521 | 166.5 | 19.6 | D | D | D | D |
| Wayne | D | D | 2 | D | 24 | 395 | 73.7 | 13.9 | 24 | 197 | 11.3 | 3.2 |
| Webster | 47 | 206 | 19 | 6.2 | 46 | 921 | 235.3 | 42.7 | 40 | 753 | 37.1 | 9.9 |
| Worth | NA | NA | NA | NA | NA | NA | NA | NA | NA | NA | NA | NA |
| Wright | 20 | 54 | 4 | 1.0 | 26 | 373 | 106.9 | 17.2 | 27 | 357 | 18.3 | 3.7 |

# Table B. States and Counties — Health Care and Social Assistance, Other Services, Nonemployer Businesses, and Residential Construction

| STATE County | Health care and social assistance, 2017 | | | | Other services, 2017 | | | | Nonemployer businesses, 2018 | | Value of residential construction authorized by building permits, 2020 | |
|---|---|---|---|---|---|---|---|---|---|---|---|---|
| | Number of establishments | Number of employees | Receipts (mil dol) | Annual payroll (mil dol) | Number of establishments | Number of employees | Receipts (mil dol) | Annual payroll (mil dol) | Number | Receipts (mil dol) | New construction ($1,000) | Number of housing units |
| | 159 | 160 | 161 | 162 | 163 | 164 | 165 | 166 | 167 | 168 | 169 | 170 |
| **MISSOURI— Cont'd** | | | | | | | | | | | | |
| Jasper | 316 | 6,386 | 596.0 | 217.4 | 207 | 1,186 | 105.9 | 32.1 | 7,000 | 292.3 | 111,587 | 930 |
| Jefferson | 384 | 5,685 | 512.1 | 216.8 | 334 | 1,701 | 185.6 | 54.5 | 13,662 | 564.3 | 202,433 | 833 |
| Johnson | 100 | 2,359 | 217.6 | 94.4 | 67 | 236 | 21.1 | 5.6 | 3,068 | 121.5 | 8,362 | 43 |
| Knox | 10 | 46 | 2.6 | 1.1 | 12 | 34 | 2.6 | 0.7 | 362 | 19.9 | 0 | 0 |
| Laclede | 74 | 1,215 | 87.4 | 35.2 | 49 | 154 | 20.5 | 3.7 | 2,706 | 138.8 | 6,849 | 43 |
| Lafayette | 67 | 1,260 | 84.3 | 38.4 | 45 | 171 | 15.7 | 5.4 | 2,069 | 88.6 | 19,379 | 84 |
| Lawrence | 76 | 1,035 | 87.4 | 29.6 | 48 | 152 | 18.9 | 4.5 | 2,655 | 108.8 | 1,938 | 11 |
| Lewis | D | D | D | 4.8 | D | D | D | 4.8 | 669 | 30.9 | 332 | 5 |
| Lincoln | 79 | 1,324 | 93.5 | 42.5 | D | D | D | 6.7 | 3,602 | 148.4 | 31,854 | 225 |
| Linn | 26 | 521 | 38.5 | 17.2 | 22 | 78 | 8.0 | 2.0 | 883 | 34.5 | 0 | 0 |
| Livingston | 38 | 995 | 94.1 | 37.6 | 25 | 208 | 15.3 | 4.2 | 1,036 | 42.9 | 4,023 | 16 |
| McDonald | 28 | 206 | 13.8 | 5.8 | D | D | D | 1.0 | 1,462 | 54.0 | 2,101 | 20 |
| Macon | 44 | 598 | 53.1 | 19.5 | D | D | D | D | 1,108 | 48.7 | 1,664 | 7 |
| Madison | 37 | 963 | 48.4 | 23.6 | 16 | 61 | 5.8 | 1.2 | 711 | 26.7 | 605 | 10 |
| Maries | 13 | 152 | 8.9 | 3.1 | D | D | 3.0 | D | 595 | 22.9 | 0 | 0 |
| Marion | 119 | 2,967 | 370.2 | 131.2 | D | D | D | D | 1,716 | 66.9 | 11,387 | 45 |
| Mercer | 6 | D | 3.7 | D | D | D | D | 0.4 | 268 | 14.7 | 0 | 0 |
| Miller | 44 | 571 | 38.7 | 15.8 | D | D | 14.4 | D | 1,716 | 84.4 | 7,913 | 27 |
| Mississippi | 31 | 454 | 26.1 | 11.4 | 17 | 43 | 3.6 | 0.9 | 579 | 28.2 | 120 | 1 |
| Moniteau | D | D | D | D | D | D | D | D | 1,052 | 48.5 | 560 | 2 |
| Monroe | 25 | 230 | 10.0 | 4.3 | D | D | 1.3 | D | 573 | 27.3 | 0 | 0 |
| Montgomery | 17 | 228 | 19.4 | 5.8 | 17 | 68 | 11.0 | 2.8 | 816 | 37.0 | 7,238 | 41 |
| Morgan | 40 | 267 | 15.0 | 6.2 | 26 | 79 | 8.8 | 2.3 | 1,713 | 77.3 | 490 | 22 |
| New Madrid | 53 | 1,052 | 48.3 | 22.8 | 19 | 48 | 4.5 | 0.9 | 814 | 26.5 | 746 | 7 |
| Newton | 148 | 6,542 | 724.6 | 315.4 | 64 | 302 | 30.2 | 8.7 | 3,855 | 177.2 | 5,936 | 43 |
| Nodaway | 48 | 1,131 | 80.0 | 41.1 | 36 | 121 | 11.6 | 3.1 | 1,406 | 53.7 | 7,981 | 37 |
| Oregon | 28 | 349 | 17.0 | 8.0 | 13 | 32 | 4.5 | 0.7 | 726 | 25.6 | 475 | 2 |
| Osage | D | D | D | D | D | D | 2.5 | D | 985 | 45.3 | 460 | 5 |
| Ozark | 11 | D | 6.2 | D | D | D | D | D | 796 | 25.2 | 238 | 1 |
| Pemiscot | 70 | 1,278 | 72.7 | 33.2 | 16 | 61 | 5.9 | 1.3 | 791 | 29.0 | 498 | 12 |
| Perry | 63 | 1,137 | 106.3 | 42.4 | 46 | 512 | 22.5 | 19.3 | 1,239 | 53.6 | 9,039 | 47 |
| Pettis | 144 | 3,481 | 261.5 | 108.8 | 89 | 319 | 27.8 | 8.6 | 2,733 | 140.3 | 1,253 | 10 |
| Phelps | 134 | 3,955 | 392.3 | 136.7 | 70 | 349 | 35.1 | 10.9 | 2,646 | 108.1 | 6,993 | 32 |
| Pike | 44 | 826 | 51.5 | 26.0 | 23 | 41 | 5.8 | 1.1 | 1,149 | 49.5 | 491 | 8 |
| Platte | 227 | 3,716 | 492.3 | 181.9 | 170 | 1,071 | 134.5 | 41.2 | 8,065 | 401.4 | 98,059 | 431 |
| Polk | 67 | 2,358 | 238.6 | 89.2 | 39 | 116 | 14.1 | 3.0 | 2,440 | 103.1 | 5,791 | 26 |
| Pulaski | 67 | 1,856 | 243.5 | 90.4 | 59 | 228 | 24.1 | 5.9 | 2,085 | 76.3 | 8,309 | 55 |
| Putnam | 12 | 176 | 18.7 | 5.7 | D | D | D | 0.5 | 399 | 18.4 | 0 | 0 |
| Ralls | 14 | 168 | 11.6 | 5.5 | D | D | D | D | 748 | 32.0 | 521 | 4 |
| Randolph | 70 | 1,293 | 117.4 | 46.3 | 38 | 156 | 11.4 | 3.2 | 1,262 | 54.8 | 6,820 | 43 |
| Ray | D | D | D | 18.2 | D | D | D | 1.8 | 1,339 | 53.5 | 14,919 | 52 |
| Reynolds | 14 | 152 | 6.0 | 2.7 | D | D | 1.9 | D | 509 | 22.6 | 0 | 0 |
| Ripley | 84 | 948 | 34.2 | 17.7 | D | D | 1.9 | D | 835 | 33.6 | 65 | 3 |
| St. Charles | 1,055 | 18,396 | 1,912.8 | 702.0 | 660 | 4,692 | 460.7 | 138.6 | 27,327 | 1,214.1 | 633,113 | 2,896 |
| St. Clair | 21 | 410 | 26.4 | 10.7 | D | D | 2.6 | D | 744 | 30.9 | 275 | 11 |
| Ste. Genevieve | 46 | 1,027 | 81.0 | 29.9 | 34 | 136 | 11.7 | 2.7 | 1,189 | 36.1 | 2,667 | 11 |
| St. Francois | 244 | 5,204 | 428.8 | 187.1 | 108 | 456 | 39.4 | 10.5 | 3,230 | 133.9 | 20,150 | 131 |
| St. Louis | 4,451 | 93,575 | 11,506.2 | 4,645.4 | 2,021 | 18,183 | 2,191.6 | 688.0 | 79,706 | 4,118.9 | 430,183 | 1,232 |
| Saline | 64 | 1,508 | 123.8 | 51.4 | 44 | 138 | 18.6 | 3.6 | 1,230 | 43.5 | 270 | 2 |
| Schuyler | D | D | D | D | D | D | D | D | 359 | 18.9 | 1,189 | 5 |
| Scotland | D | D | D | D | D | D | D | 0.8 | 508 | 31.6 | 122 | 1 |
| Scott | 146 | 2,801 | 247.1 | 117.6 | 58 | 366 | 38.2 | 10.9 | 2,328 | 113.1 | 9,781 | 68 |
| Shannon | 19 | 152 | 11.8 | 4.2 | 4 | 9 | 1.5 | 0.3 | 767 | 35.6 | 0 | 0 |
| Shelby | 9 | 94 | 5.1 | 2.1 | D | D | D | 1.2 | 434 | 22.7 | 500 | 2 |
| Stoddard | 96 | 1,667 | 110.1 | 45.5 | 38 | 106 | 10.7 | 2.8 | 2,016 | 109.9 | 135 | 2 |
| Stone | 47 | 435 | 34.6 | 12.9 | 53 | 440 | 27.5 | 7.7 | 2,783 | 128.1 | 42,786 | 196 |
| Sullivan | 11 | 213 | 17.5 | 6.3 | 7 | 17 | 2.6 | 0.5 | 365 | 16.4 | 0 | 0 |
| Taney | 125 | 2,094 | 309.5 | 96.3 | 90 | 518 | 56.2 | 14.5 | 4,619 | 199.8 | 53,551 | 293 |
| Texas | 53 | 836 | 61.9 | 27.4 | 27 | 59 | 6.1 | 1.1 | 1,752 | 70.8 | 340 | 3 |
| Vernon | 64 | 1,330 | 90.5 | 40.6 | 34 | 156 | 11.0 | 2.6 | 1,377 | 54.4 | 1,209 | 11 |
| Warren | 52 | 635 | 42.0 | 18.3 | 35 | 178 | 20.5 | 5.8 | 2,195 | 89.2 | 86,558 | 378 |
| Washington | 47 | 693 | 48.5 | 22.6 | 22 | 78 | 6.4 | 1.6 | 1,104 | 39.6 | 0 | 0 |
| Wayne | 54 | 392 | 19.1 | 8.1 | D | D | 3.0 | D | 632 | 25.5 | 0 | 0 |
| Webster | 57 | 615 | 38.6 | 15.5 | 49 | 122 | 14.1 | 3.5 | 3,198 | 136.1 | 33,054 | 215 |
| Worth | D | D | D | 0.3 | D | D | D | 0.3 | 191 | 7.0 | 0 | 0 |
| Wright | 47 | 562 | 33.0 | 14.0 | 24 | 75 | 7.3 | 1.4 | 1,468 | 67.7 | 2,113 | 13 |

# Table B. States and Counties — Government Employment and Payroll, and Local Government Finances

| STATE County | Government employment and payroll, 2017 | | | | | | | | | Local government finances, 2017 | | | | |
| | | | March payroll (percent of total) | | | | | | | General revenue | | | | |
| | | | | | | | | | | | | | Taxes | |
| | Full-time equivalent employees | March payroll (dollars) | Administration, judicial, and legal | Police and corrections | Fire protection | Highways and transportation | Health and welfare | Natural resources and utilities | Education and libraries | Total (mil dol) | Intergovernmental (mil dol) | Total (mil dol) | Per capita[1] (dollars) Total | Per capita[1] (dollars) Property |
| | 171 | 172 | 173 | 174 | 175 | 176 | 177 | 178 | 179 | 180 | 181 | 182 | 183 | 184 |
|---|---|---|---|---|---|---|---|---|---|---|---|---|---|---|
| **MISSOURI— Cont'd** | | | | | | | | | | | | | | |
| Jasper | 4,398 | 14,170,563 | 4.5 | 9.3 | 3.8 | 3.7 | 2.1 | 3.5 | 72.0 | 447.2 | 211.8 | 177.4 | 1,478 | 773 |
| Jefferson | 6,928 | 26,844,611 | 3.0 | 6.6 | 4.2 | 2.4 | 4.3 | 3.2 | 75.0 | 585.4 | 243.3 | 261.7 | 1,169 | 794 |
| Johnson | 2,338 | 9,601,545 | 3.0 | 3.4 | 1.9 | 2.1 | 42.0 | 2.2 | 45.2 | 347.2 | 52.5 | 62.4 | 1,160 | 701 |
| Knox | 214 | 535,366 | 8.7 | 3.6 | 0.1 | 6.1 | 21.3 | 5.6 | 54.0 | 11.0 | 3.0 | 4.7 | 1,180 | 864 |
| Laclede | 1,481 | 4,057,488 | 5.1 | 5.8 | 1.7 | 8.7 | 2.0 | 6.8 | 68.5 | 104.7 | 39.7 | 34.5 | 975 | 550 |
| Lafayette | 1,327 | 4,428,568 | 5.8 | 6.8 | 1.9 | 2.9 | 18.1 | 8.5 | 55.3 | 97.5 | 40.6 | 37.1 | 1,139 | 791 |
| Lawrence | 1,037 | 3,318,669 | 5.4 | 8.3 | 0.9 | 3.5 | 6.7 | 2.3 | 72.3 | 88.3 | 42.3 | 30.6 | 799 | 501 |
| Lewis | 449 | 1,201,239 | 7.8 | 5.2 | 0.0 | 3.7 | 21.0 | 4.7 | 57.1 | 28.6 | 12.7 | 11.5 | 1,152 | 844 |
| Lincoln | 1,819 | 5,994,734 | 5.2 | 7.4 | 1.4 | 2.6 | 3.9 | 1.2 | 76.6 | 191.5 | 76.6 | 57.3 | 1,023 | 731 |
| Linn | 601 | 1,786,369 | 6.8 | 6.1 | 0.8 | 3.8 | 6.2 | 7.2 | 68.4 | 47.2 | 22.0 | 18.2 | 1,492 | 1,043 |
| Livingston | 564 | 1,695,818 | 5.6 | 6.6 | 6.5 | 2.5 | 13.9 | 2.6 | 61.9 | 57.6 | 19.4 | 19.5 | 1,289 | 844 |
| McDonald | 773 | 2,253,514 | 5.8 | 6.2 | 0.0 | 3.8 | 3.1 | 3.7 | 77.1 | 60.6 | 30.0 | 24.6 | 1,085 | 922 |
| Macon | 1,075 | 3,112,412 | 4.1 | 3.9 | 1.1 | 4.6 | 42.0 | 8.5 | 35.6 | 47.1 | 17.5 | 18.4 | 1,207 | 770 |
| Madison | 657 | 2,048,596 | 3.0 | 2.5 | 0.2 | 2.0 | 42.6 | 3.5 | 45.3 | 29.9 | 15.9 | 10.2 | 836 | 554 |
| Maries | 324 | 876,182 | 6.2 | 6.2 | 0.1 | 7.0 | 8.3 | 3.6 | 67.0 | 19.5 | 8.8 | 8.2 | 935 | 694 |
| Marion | 1,388 | 4,194,523 | 5.8 | 8.1 | 3.7 | 2.6 | 11.7 | 5.9 | 57.6 | 108.5 | 33.5 | 49.1 | 1,718 | 982 |
| Mercer | 163 | 550,632 | 16.0 | 5.3 | 0.0 | 6.2 | 6.1 | 3.3 | 63.1 | 10.8 | 4.3 | 5.0 | 1,375 | 1,187 |
| Miller | 1,210 | 3,556,113 | 3.7 | 6.3 | 1.7 | 2.1 | 9.7 | 1.8 | 73.7 | 91.7 | 31.4 | 45.1 | 1,791 | 1,291 |
| Mississippi | 539 | 1,450,262 | 9.7 | 9.5 | 0.3 | 4.2 | 6.6 | 2.8 | 66.7 | 40.8 | 21.8 | 11.9 | 879 | 572 |
| Moniteau | 511 | 1,570,509 | 8.6 | 5.1 | 0.2 | 3.0 | 9.9 | 5.6 | 67.3 | 48.3 | 16.7 | 25.6 | 1,598 | 1,258 |
| Monroe | 484 | 1,338,886 | 5.2 | 5.3 | 0.5 | 4.7 | 22.5 | 8.8 | 52.2 | 39.1 | 11.5 | 11.6 | 1,347 | 1,045 |
| Montgomery | 416 | 1,286,249 | 8.4 | 10.9 | 0.0 | 5.2 | 6.3 | 5.4 | 60.3 | 30.0 | 9.1 | 15.8 | 1,393 | 982 |
| Morgan | 736 | 2,091,478 | 6.5 | 8.5 | 2.2 | 3.5 | 25.3 | 2.0 | 50.2 | 44.7 | 12.6 | 22.3 | 1,105 | 685 |
| New Madrid | 689 | 2,161,004 | 9.5 | 7.8 | 0.5 | 6.1 | 5.4 | 5.6 | 62.9 | 52.3 | 22.5 | 24.2 | 1,380 | 953 |
| Newton | 2,026 | 6,688,488 | 3.4 | 5.1 | 2.2 | 1.6 | 3.4 | 1.5 | 81.8 | 135.2 | 60.9 | 45.3 | 778 | 475 |
| Nodaway | 712 | 2,196,990 | 10.1 | 6.4 | 0.5 | 3.6 | 5.1 | 5.9 | 67.8 | 63.5 | 22.3 | 33.2 | 1,483 | 958 |
| Oregon | 437 | 1,150,265 | 5.4 | 4.1 | 0.0 | 3.7 | 5.6 | 3.1 | 77.6 | 24.5 | 13.3 | 7.2 | 686 | 456 |
| Osage | 395 | 968,305 | 7.4 | 5.2 | 0.0 | 4.2 | 9.0 | 5.9 | 67.8 | 29.4 | 10.7 | 14.9 | 1,091 | 836 |
| Ozark | 376 | 1,017,632 | 6.1 | 4.9 | 0.0 | 4.1 | 2.7 | 1.6 | 80.5 | 21.3 | 12.5 | 6.8 | 738 | 523 |
| Pemiscot | 816 | 2,452,299 | 4.9 | 7.9 | 0.9 | 2.0 | 5.1 | 7.8 | 70.7 | 56.9 | 30.5 | 17.7 | 1,056 | 642 |
| Perry | 584 | 2,058,654 | 3.6 | 9.8 | 0.0 | 3.2 | 6.4 | 9.4 | 66.6 | 45.5 | 14.9 | 23.5 | 1,220 | 642 |
| Pettis | 1,698 | 5,767,296 | 5.8 | 4.0 | 2.9 | 2.5 | 0.7 | 8.8 | 74.4 | 267.3 | 66.4 | 69.4 | 1,634 | 880 |
| Phelps | 2,566 | 10,173,000 | 2.7 | 3.6 | 1.2 | 1.4 | 54.5 | 4.7 | 30.9 | 315.0 | 57.5 | 44.2 | 992 | 545 |
| Pike | 642 | 2,385,096 | 2.8 | 2.2 | 0.2 | 0.7 | 29.7 | 2.4 | 61.3 | 43.8 | 17.7 | 19.4 | 1,046 | 681 |
| Platte | 2,890 | 11,780,285 | 5.3 | 7.8 | 3.0 | 1.7 | 2.6 | 2.2 | 77.1 | 343.0 | 102.8 | 180.3 | 1,782 | 1,326 |
| Polk | 2,111 | 8,047,769 | 1.7 | 2.2 | 0.5 | 1.6 | 61.1 | 1.5 | 31.0 | 192.9 | 38.3 | 22.5 | 708 | 485 |
| Pulaski | 1,770 | 5,316,782 | 3.6 | 4.2 | 1.3 | 2.5 | 8.1 | 5.1 | 74.2 | 148.8 | 87.8 | 36.6 | 705 | 470 |
| Putnam | 338 | 1,109,582 | 4.5 | 1.9 | 0.0 | 2.1 | 46.9 | 5.0 | 39.0 | 19.6 | 6.6 | 9.2 | 1,915 | 1,269 |
| Ralls | 179 | 551,668 | 13.1 | 9.4 | 0.7 | 7.6 | 3.5 | 8.4 | 56.0 | 26.7 | 6.3 | 7.5 | 735 | 481 |
| Randolph | 1,352 | 4,284,870 | 4.4 | 7.6 | 1.7 | 1.8 | 3.9 | 4.5 | 75.8 | 137.3 | 43.4 | 39.1 | 1,568 | 950 |
| Ray | 1,106 | 3,671,525 | 4.1 | 4.2 | 2.2 | 2.4 | 41.2 | 4.4 | 40.8 | 124.3 | 33.2 | 25.6 | 1,121 | 705 |
| Reynolds | 303 | 797,244 | 5.4 | 3.6 | 0.0 | 5.5 | 4.1 | 1.5 | 79.9 | 20.7 | 9.0 | 8.8 | 1,401 | 1,197 |
| Ripley | 589 | 1,595,247 | 4.3 | 3.3 | 1.0 | 2.0 | 27.1 | 2.5 | 59.9 | 44.8 | 22.6 | 16.3 | 1,198 | 828 |
| St. Charles | 13,553 | 52,183,204 | 5.5 | 9.1 | 4.6 | 3.1 | 2.7 | 4.8 | 67.6 | 1,402.6 | 398.5 | 775.6 | 1,963 | 1,245 |
| St. Clair | 433 | 1,258,079 | 4.9 | 10.0 | 0.0 | 2.7 | 24.9 | 1.4 | 52.8 | 34.6 | 11.9 | 8.0 | 851 | 733 |
| Ste. Genevieve | 491 | 1,927,342 | 7.2 | 15.6 | 0.1 | 2.9 | 4.6 | 6.6 | 62.4 | 45.4 | 16.4 | 22.8 | 1,280 | 789 |
| St. Francois | 2,394 | 8,376,327 | 5.2 | 6.7 | 0.6 | 2.6 | 4.1 | 4.3 | 75.5 | 209.1 | 98.5 | 75.7 | 1,135 | 744 |
| St. Louis | 34,807 | 159,175,124 | 4.4 | 10.4 | 7.2 | 2.5 | 1.7 | 2.7 | 69.6 | 4,109.3 | 1,101.0 | 2,519.9 | 2,529 | 1,593 |
| Saline | 772 | 2,331,088 | 7.0 | 11.6 | 3.9 | 5.1 | 6.4 | 17.5 | 41.0 | 72.4 | 29.3 | 26.6 | 1,162 | 824 |
| Schuyler | 213 | 500,136 | 7.2 | 1.9 | 0.0 | 4.4 | 28.1 | 4.8 | 53.4 | 13.9 | 7.7 | 4.3 | 953 | 708 |
| Scotland | 442 | 1,508,518 | 3.1 | 3.0 | 0.0 | 2.6 | 67.6 | 4.2 | 19.2 | 41.8 | 6.3 | 7.6 | 1,527 | 1,095 |
| Scott | 1,674 | 5,781,559 | 4.1 | 8.1 | 1.6 | 10.0 | 5.1 | 15.2 | 53.4 | 109.6 | 45.8 | 43.4 | 1,125 | 643 |
| Shannon | 233 | 602,275 | 8.6 | 8.3 | 0.0 | 6.4 | 6.9 | 5.2 | 64.3 | 12.2 | 7.9 | 3.1 | 379 | 257 |
| Shelby | 478 | 1,240,742 | 17.2 | 4.5 | 0.3 | 3.2 | 24.6 | 5.9 | 41.5 | 36.6 | 7.3 | 7.3 | 1,217 | 968 |
| Stoddard | 1,015 | 2,820,418 | 5.4 | 4.2 | 0.0 | 2.7 | 5.5 | 1.6 | 80.0 | 79.1 | 39.1 | 22.4 | 762 | 612 |
| Stone | 894 | 2,717,977 | 7.3 | 6.3 | 3.0 | 4.4 | 0.0 | 1.9 | 76.8 | 74.6 | 29.1 | 36.0 | 1,136 | 717 |
| Sullivan | 375 | 1,184,784 | 5.5 | 3.5 | 0.2 | 2.4 | 35.2 | 6.8 | 44.7 | 30.2 | 11.5 | 6.8 | 1,098 | 850 |
| Taney | 2,022 | 6,678,131 | 7.6 | 8.0 | 3.7 | 4.3 | 6.6 | 6.3 | 61.4 | 223.7 | 72.0 | 107.9 | 1,957 | 1,139 |
| Texas | 1,202 | 3,547,068 | 5.0 | 4.2 | 0.1 | 2.0 | 33.7 | 3.3 | 51.3 | 60.3 | 32.0 | 23.0 | 897 | 775 |
| Vernon | 1,130 | 3,728,284 | 3.7 | 4.9 | 1.3 | 1.5 | 34.6 | 3.8 | 48.6 | 50.4 | 22.4 | 20.8 | 1,012 | 576 |
| Warren | 1,085 | 3,585,264 | 5.2 | 9.2 | 4.2 | 1.5 | 5.7 | 3.3 | 69.1 | 86.9 | 36.8 | 38.5 | 1,117 | 787 |
| Washington | 955 | 3,010,693 | 3.5 | 3.1 | 0.3 | 3.0 | 32.9 | 1.0 | 55.7 | 75.2 | 30.8 | 15.0 | 600 | 434 |
| Wayne | 459 | 1,093,686 | 9.4 | 3.1 | 0.0 | 7.6 | 5.0 | 1.8 | 72.8 | 27.1 | 14.5 | 8.6 | 648 | 385 |
| Webster | 944 | 2,720,753 | 6.0 | 6.2 | 0.3 | 3.6 | 9.0 | 3.6 | 70.6 | 67.9 | 31.6 | 24.0 | 620 | 337 |
| Worth | 127 | 326,573 | 14.4 | 2.3 | 0.0 | 5.0 | 18.3 | 4.6 | 52.0 | 7.4 | 2.6 | 2.5 | 1,248 | 1,058 |
| Wright | 740 | 2,129,085 | 5.0 | 4.7 | 0.5 | 2.7 | 10.5 | 2.9 | 72.8 | 44.7 | 24.9 | 13.0 | 716 | 457 |

1. Based on the resident population estimated as of July 1 of the year shown.

# Table B. States and Counties — Local Government Finances, Government Employment, and Income Taxes

| STATE County | Local government finances, 2017 (cont.) | | | | | | | | | Government employment, 2019 | | | Individual income tax returns, 2018 | | |
| | Direct general expenditure | | | | | | | Debt outstanding | | Federal civilian | Federal military | State and local | Number of returns | Mean adjusted gross income | Mean income tax |
| | Total (mil dol) | Per capita[1] (dollars) | Education | Health and hospitals | Police protection | Public welfare | Highways | Total (mil dol) | Per capita[1] (dollars) | | | | | | |
| | | | Percent of total for: | | | | | | | | | | | | |
| | 185 | 186 | 187 | 188 | 189 | 190 | 191 | 192 | 193 | 194 | 195 | 196 | 197 | 198 | 199 |

| MISSOURI— Cont'd | | | | | | | | | | | | | | | |
| Jasper | 410 | 3,415 | 50.9 | 1.2 | 5.4 | 0.1 | 10.0 | 309.2 | 2,576 | 295 | 432 | 6,747 | 51,880 | 52,427 | 4,961 |
| Jefferson | 598 | 2,672 | 66.7 | 3.0 | 5.1 | 0.0 | 2.9 | 494.4 | 2,210 | 316 | 747 | 8,261 | 107,250 | 57,850 | 5,129 |
| Johnson | 213 | 3,956 | 39.8 | 37.7 | 3.1 | 0.0 | 5.4 | 296.1 | 5,510 | 1,313 | 3,980 | 5,821 | 22,570 | 52,875 | 4,344 |
| Knox | 11 | 2,730 | 49.6 | 1.3 | 2.8 | 20.1 | 10.6 | 8.5 | 2,147 | 27 | 13 | 280 | 1,680 | 37,463 | 2,270 |
| Laclede | 89 | 2,498 | 69.9 | 0.0 | 5.5 | 0.0 | 6.2 | 45.6 | 1,288 | 76 | 118 | 1,422 | 15,450 | 42,407 | 3,118 |
| Lafayette | 102 | 3,119 | 54.3 | 3.0 | 6.2 | 0.3 | 11.1 | 57.8 | 2,902 | 104 | 107 | 2,109 | 14,960 | 51,808 | 4,078 |
| Lawrence | 103 | 2,697 | 53.1 | 0.2 | 2.7 | 5.0 | 8.4 | 57.8 | 1,510 | 90 | 126 | 1,683 | 16,120 | 44,868 | 3,354 |
| Lewis | 30 | 2,988 | 50.2 | 1.9 | 4.5 | 1.5 | 11.9 | 9.1 | 919 | 48 | 30 | 526 | 4,100 | 47,061 | 3,498 |
| Lincoln | 220 | 3,931 | 43.8 | 19.8 | 3.7 | 0.0 | 4.5 | 929.3 | 16,581 | 116 | 195 | 2,107 | 26,970 | 56,002 | 4,706 |
| Linn | 47 | 3,827 | 51.0 | 1.8 | 5.3 | 0.1 | 16.2 | 34.5 | 2,834 | 57 | 39 | 810 | 5,390 | 45,455 | 3,300 |
| Livingston | 80 | 5,253 | 58.9 | 2.7 | 4.2 | 5.8 | 6.2 | 34.4 | 2,269 | 75 | 44 | 1,342 | 6,300 | 49,142 | 4,200 |
| McDonald | 46 | 2,036 | 79.4 | 3.5 | 3.0 | 0.0 | 1.1 | 35.0 | 1,539 | 81 | 76 | 798 | 9,100 | 40,516 | 2,619 |
| Macon | 44 | 2,852 | 55.4 | 1.6 | 3.9 | 11.3 | 4.6 | 27.1 | 1,775 | 68 | 50 | 1,428 | 7,060 | 44,074 | 3,123 |
| Madison | 28 | 2,293 | 70.2 | 3.0 | 6.1 | 0.0 | 3.8 | 6.9 | 563 | 40 | 40 | 820 | 5,030 | 43,884 | 2,813 |
| Maries | 20 | 2,287 | 59.6 | 6.1 | 3.9 | 0.0 | 10.3 | 7.4 | 844 | 11 | 29 | 324 | 3,790 | 41,384 | 2,813 |
| Marion | 104 | 3,619 | 45.0 | 4.2 | 6.5 | 6.5 | 4.2 | 127.2 | 4,450 | 109 | 91 | 1,738 | 12,690 | 53,039 | 4,813 |
| Mercer | 11 | 3,032 | 72.5 | 6.9 | 1.4 | 0.0 | 9.5 | 7.6 | 2,088 | 29 | 12 | 236 | 1,510 | 36,539 | 2,875 |
| Miller | 85 | 3,355 | 59.9 | 3.6 | 3.6 | 5.6 | 3.8 | 72.1 | 2,860 | 42 | 85 | 1,409 | 10,760 | 44,302 | 3,739 |
| Mississippi | 39 | 2,840 | 52.7 | 5.5 | 6.3 | 1.1 | 8.9 | 12.9 | 949 | 17 | 38 | 1,004 | 4,770 | 44,536 | 3,703 |
| Moniteau | 38 | 2,378 | 62.7 | 7.2 | 3.1 | 0.0 | 7.0 | 23.7 | 1,479 | 47 | 49 | 981 | 6,740 | 47,388 | 3,382 |
| Monroe | 32 | 3,703 | 49.8 | 2.4 | 4.0 | 17.2 | 11.0 | 48.6 | 5,656 | 72 | 29 | 594 | 3,900 | 45,564 | 3,327 |
| Montgomery | 32 | 2,818 | 51.2 | 4.0 | 5.3 | 0.0 | 7.1 | 25.2 | 2,218 | 46 | 37 | 602 | 5,130 | 46,764 | 3,351 |
| Morgan | 45 | 2,230 | 51.2 | 0.1 | 7.8 | 11.1 | 9.6 | 27.0 | 1,337 | 41 | 68 | 979 | 8,580 | 42,638 | 2,991 |
| New Madrid | 51 | 2,881 | 56.4 | 0.1 | 6.8 | 0.0 | 8.5 | 32.4 | 1,846 | 45 | 56 | 866 | 6,820 | 45,786 | 3,646 |
| Newton | 178 | 3,050 | 76.3 | 2.6 | 2.8 | 0.5 | 2.8 | 113.8 | 1,956 | 161 | 192 | 2,440 | 24,690 | 53,643 | 5,006 |
| Nodaway | 64 | 2,839 | 63.1 | 1.3 | 5.6 | 0.0 | 9.2 | 87.6 | 3,909 | 84 | 62 | 2,680 | 8,480 | 50,331 | 4,112 |
| Oregon | 24 | 2,249 | 75.0 | 5.2 | 2.0 | 0.0 | 4.4 | 3.4 | 320 | 34 | 35 | 525 | 3,830 | 36,398 | 2,490 |
| Osage | 29 | 2,136 | 58.1 | 11.2 | 3.5 | 0.0 | 6.2 | 13.5 | 986 | 27 | 44 | 911 | 6,220 | 51,267 | 3,698 |
| Ozark | 22 | 2,428 | 78.3 | 0.2 | 3.9 | 0.1 | 6.9 | 3.7 | 402 | 13 | 30 | 428 | 3,660 | 35,135 | 2,476 |
| Pemiscot | 60 | 3,543 | 64.2 | 0.4 | 5.2 | 0.0 | 3.3 | 13.0 | 776 | 53 | 52 | 1,320 | 6,020 | 42,200 | 3,372 |
| Perry | 41 | 2,151 | 57.6 | 0.3 | 9.0 | 0.2 | 4.5 | 3.6 | 185 | 48 | 63 | 1,211 | 9,000 | 51,270 | 3,983 |
| Pettis | 249 | 5,870 | 41.1 | 41.4 | 2.7 | 0.0 | 3.7 | 69.6 | 1,640 | 116 | 140 | 3,319 | 18,640 | 45,830 | 3,547 |
| Phelps | 307 | 6,881 | 24.1 | 58.1 | 2.1 | 0.3 | 3.1 | 80.4 | 1,804 | 338 | 251 | 6,209 | 18,050 | 50,783 | 4,573 |
| Pike | 47 | 2,543 | 63.3 | 0.9 | 6.6 | 0.6 | 10.9 | 32.3 | 1,741 | 68 | 53 | 1,548 | 7,120 | 47,388 | 3,817 |
| Platte | 308 | 3,049 | 70.7 | 1.1 | 4.9 | 0.0 | 7.5 | 263.9 | 2,609 | 689 | 368 | 4,056 | 50,390 | 88,520 | 11,605 |
| Polk | 201 | 6,318 | 27.9 | 64.3 | 1.2 | 0.0 | 1.0 | 61.5 | 1,939 | 69 | 103 | 2,509 | 12,700 | 45,512 | 3,525 |
| Pulaski | 150 | 2,885 | 67.4 | 1.7 | 3.1 | 0.0 | 4.6 | 41.4 | 798 | 3,508 | 11,098 | 2,084 | 18,590 | 46,406 | 3,027 |
| Putnam | 19 | 4,028 | 49.5 | 3.5 | 6.3 | 1.8 | 12.7 | 13.8 | 2,895 | 20 | 16 | 400 | 2,000 | 40,257 | 2,723 |
| Ralls | 21 | 2,033 | 37.9 | 4.6 | 4.1 | 0.0 | 25.3 | 20.5 | 2,006 | 29 | 34 | 328 | 4,670 | 54,273 | 4,677 |
| Randolph | 124 | 4,997 | 56.9 | 15.8 | 4.7 | 0.0 | 4.4 | 62.0 | 2,488 | 69 | 75 | 2,050 | 10,390 | 47,925 | 3,813 |
| Ray | 95 | 4,169 | 34.1 | 28.2 | 2.1 | 13.4 | 5.8 | 28.7 | 1,255 | 52 | 76 | 1,341 | 10,590 | 53,760 | 4,303 |
| Reynolds | 20 | 3,214 | 67.8 | 3.8 | 3.5 | 0.0 | 6.6 | 5.6 | 894 | 18 | 21 | 391 | 2,460 | 38,413 | 2,522 |
| Ripley | 41 | 3,049 | 50.9 | 2.5 | 5.9 | 0.7 | 14.9 | 11.9 | 876 | 59 | 44 | 481 | 4,840 | 35,461 | 2,270 |
| St. Charles | 1,357 | 3,435 | 55.0 | 2.6 | 6.6 | 0.2 | 6.3 | 1,640.9 | 4,153 | 798 | 1,332 | 16,795 | 197,980 | 77,929 | 8,902 |
| St. Clair | 34 | 3,661 | 38.4 | 24.7 | 2.3 | 0.0 | 13.7 | 12.9 | 1,374 | 41 | 31 | 464 | 3,740 | 40,218 | 2,787 |
| Ste. Genevieve | 43 | 2,410 | 58.5 | 1.1 | 5.8 | 0.8 | 6.7 | 32.9 | 1,850 | 27 | 59 | 1,079 | 8,360 | 58,658 | 5,150 |
| St. Francois | 220 | 3,300 | 60.2 | 4.4 | 5.3 | 0.0 | 7.9 | 174.5 | 2,617 | 146 | 201 | 5,816 | 26,350 | 47,356 | 3,813 |
| St. Louis | 3,942 | 3,957 | 54.9 | 1.3 | 7.8 | 1.6 | 5.0 | 4,876.3 | 4,895 | 5,894 | 3,325 | 47,939 | 498,640 | 98,692 | 14,896 |
| Saline | 74 | 3,245 | 53.5 | 3.3 | 7.4 | 0.0 | 11.0 | 9.8 | 429 | 78 | 71 | 1,600 | 9,690 | 47,204 | 3,653 |
| Schuyler | 13 | 2,892 | 48.0 | 25.0 | 2.0 | 0.7 | 13.9 | 9.5 | 2,113 | 27 | 15 | 277 | 1,740 | 35,248 | 2,264 |
| Scotland | 37 | 7,561 | 17.4 | 54.6 | 1.3 | 4.2 | 5.3 | 14.8 | 2,984 | 20 | 16 | 540 | 2,010 | 40,956 | 2,666 |
| Scott | 115 | 2,977 | 58.0 | 3.0 | 8.9 | 0.0 | 5.4 | 149.8 | 3,884 | 102 | 126 | 2,175 | 16,960 | 51,636 | 4,575 |
| Shannon | 11 | 1,349 | 69.2 | 1.0 | 3.9 | 0.0 | 11.5 | 0.1 | 6 | 21 | 27 | 303 | 3,070 | 35,325 | 2,459 |
| Shelby | 28 | 4,728 | 43.9 | 4.2 | 17.6 | 15.7 | 6.2 | 9.7 | 1,610 | 33 | 19 | 556 | 2,830 | 44,320 | 3,139 |
| Stoddard | 80 | 2,728 | 62.3 | 0.8 | 2.5 | 0.0 | 8.1 | 45.5 | 1,549 | 150 | 95 | 1,325 | 12,160 | 50,360 | 4,555 |
| Stone | 81 | 2,574 | 64.9 | 0.0 | 3.9 | 0.0 | 8.5 | 85.5 | 2,701 | 42 | 106 | 1,044 | 14,340 | 52,943 | 5,060 |
| Sullivan | 28 | 4,488 | 44.2 | 39.5 | 1.5 | 0.1 | 4.2 | 31.9 | 5,129 | 38 | 20 | 431 | 2,660 | 38,243 | 2,328 |
| Taney | 198 | 3,601 | 41.9 | 0.7 | 4.3 | 0.0 | 13.5 | 486.4 | 8,828 | 191 | 181 | 2,369 | 25,920 | 41,216 | 3,524 |
| Texas | 59 | 2,281 | 64.6 | 1.1 | 3.3 | 0.1 | 5.3 | 30.1 | 1,173 | 75 | 79 | 1,836 | 9,240 | 38,170 | 2,493 |
| Vernon | 51 | 2,498 | 57.7 | 2.6 | 4.3 | 0.2 | 7.7 | 49.3 | 2,402 | 94 | 66 | 1,604 | 8,310 | 42,457 | 3,067 |
| Warren | 87 | 2,518 | 59.1 | 7.7 | 5.6 | 0.0 | 6.6 | 68.7 | 1,994 | 44 | 119 | 1,272 | 16,560 | 58,461 | 5,348 |
| Washington | 72 | 2,893 | 52.5 | 31.0 | 2.7 | 0.0 | 3.4 | 15.9 | 637 | 62 | 79 | 1,492 | 8,810 | 39,074 | 2,325 |
| Wayne | 25 | 1,874 | 66.0 | 2.1 | 4.3 | 0.0 | 9.7 | 17.3 | 1,301 | 80 | 43 | 521 | 4,520 | 35,043 | 2,059 |
| Webster | 66 | 1,713 | 64.9 | 0.3 | 4.8 | 5.5 | 9.0 | 22.8 | 589 | 90 | 129 | 1,346 | 15,580 | 48,570 | 3,906 |
| Worth | 8 | 3,855 | 50.6 | 1.4 | 2.9 | 14.8 | 5.0 | 1.3 | 615 | 24 | 7 | 162 | 900 | 38,186 | 2,529 |
| Wright | 40 | 2,220 | 77.1 | 0.0 | 2.5 | 4.7 | 2.2 | 6.2 | 343 | 47 | 61 | 867 | 7,070 | 37,257 | 2,341 |

1. Based on the resident population estimated as of July 1 of the year shown.

| State / county code | CBSA code[1] | County Type code[2] | STATE County | Land area[3] (sq. mi) | Population, 2020 | | | Population and population characteristics, 2020 | | | | | | | | | | |
|---|---|---|---|---|---|---|---|---|---|---|---|---|---|---|---|---|---|---|
| | | | | | | | | Race alone or in combination, not Hispanic or Latino (percent) | | | | | Age (percent) | | | | | |
| | | | | | Total persons 2019 | Rank | Per square mile | White | Black | American Indian, Alaska Native | Asian and Pacific Islancer | Percent Hispanic or Latino[4] | Under 5 years | 5 to 14 years | 15 to 24 years | 25 to 34 years | 35 to 44 years | 45 to 54 years |
| | | | | 1 | 2 | 3 | 4 | 5 | 6 | 7 | 8 | 9 | 10 | 11 | 12 | 13 | 14 | 15 |
| | | | MISSOURI— Cont'd | | | | | | | | | | | | | | | |
| 29510 | 41180 | 1 | St. Louis city | 61.7 | 297,645 | 233 | 4,824.1 | 47.1 | 46.1 | 0.9 | 4.3 | 4.3 | 6.0 | 9.9 | 11.3 | 20.4 | 13.7 | 11.0 |
| 30000 | | 0 | MONTANA | 145,547.7 | 1,080,577 | X | 7.4 | 88.2 | 1.0 | 7.6 | 1.7 | 4.2 | 5.5 | 12.1 | 12.7 | 13.0 | 12.4 | 10.8 |
| 30001 | | 7 | Beaverhead | 5,542.7 | 9,483 | 2,455 | 1.7 | 91.4 | 0.9 | 2.7 | 1.6 | 5.3 | 4.1 | 9.3 | 17.7 | 10.7 | 10.9 | 9.5 |
| 30003 | | 6 | Big Horn | 4,997.8 | 13,063 | 2,216 | 2.6 | 29.2 | 0.7 | 65.8 | 1.1 | 6.0 | 8.3 | 19.6 | 13.6 | 11.9 | 11.0 | 9.9 |
| 30005 | | 9 | Blaine | 4,227.5 | 6,568 | 2,698 | 1.6 | 45.9 | 0.8 | 51.9 | 0.5 | 3.5 | 8.0 | 17.0 | 13.4 | 12.2 | 11.5 | 9.3 |
| 30007 | | 9 | Broadwater | 1,192.4 | 6,444 | 2,710 | 5.4 | 94.3 | 1.0 | 2.5 | 0.7 | 3.2 | 5.5 | 10.5 | 9.4 | 10.2 | 11.2 | 12.3 |
| 30009 | 13740 | 3 | Carbon | 2,047.8 | 10,921 | 2,356 | 5.3 | 95.2 | 1.0 | 1.9 | 0.7 | 2.6 | 4.1 | 10.1 | 8.6 | 8.3 | 11.9 | 12.0 |
| 30011 | | 9 | Carter | 3,340.5 | 1,235 | 3,097 | 0.4 | 97.4 | 1.3 | 1.5 | 0.4 | 1.3 | 7.6 | 10.9 | 8.1 | 10.9 | 10.3 | 7.5 |
| 30013 | 24500 | 3 | Cascade | 2,698.2 | 81,346 | 702 | 30.1 | 88.1 | 2.4 | 6.2 | 2.2 | 5.0 | 6.4 | 12.5 | 12.4 | 14.2 | 11.8 | 10.0 |
| 30015 | | 8 | Chouteau | 3,972.5 | 5,699 | 2,769 | 1.4 | 78.5 | 0.7 | 19.1 | 0.9 | 2.8 | 4.7 | 12.7 | 12.4 | 11.5 | 11.2 | 10.5 |
| 30017 | | 7 | Custer | 3,783.3 | 11,292 | 2,329 | 3.0 | 92.6 | 1.1 | 3.3 | 1.4 | 3.7 | 5.3 | 11.1 | 12.0 | 11.8 | 12.7 | 11.3 |
| 30019 | | 9 | Daniels | 1,426.3 | 1,638 | 3,074 | 1.1 | 93.7 | 0.9 | 4.0 | 0.6 | 3.3 | 4.8 | 12.7 | 9.8 | 9.9 | 11.2 | 8.4 |
| 30021 | | 7 | Dawson | 2,371.9 | 8,555 | 2,529 | 3.6 | 93.7 | 0.9 | 3.1 | 0.8 | 3.2 | 5.7 | 12.1 | 12.2 | 12.6 | 11.4 | 11.0 |
| 30023 | | 7 | Deer Lodge | 736.7 | 9,204 | 2,480 | 12.5 | 91.7 | 0.9 | 4.9 | 0.9 | 3.9 | 3.4 | 6.7 | 11.2 | 12.0 | 11.4 | 12.9 |
| 30025 | | 9 | Fallon | 1,620.6 | 2,826 | 2,978 | 1.7 | 96.1 | 0.7 | 1.7 | 1.3 | 1.7 | 6.4 | 15.7 | 10.5 | 12.4 | 11.9 | 10.2 |
| 30027 | | 7 | Fergus | 4,339.3 | 11,104 | 2,339 | 2.6 | 95.3 | 0.7 | 2.7 | 1.1 | 2.0 | 5.5 | 11.8 | 9.5 | 11.2 | 11.7 | 10.2 |
| 30029 | 28060 | 5 | Flathead | 5,087.2 | 105,851 | 579 | 20.8 | 94.5 | 0.6 | 2.4 | 1.5 | 3.1 | 5.5 | 12.3 | 10.3 | 12.1 | 12.9 | 11.6 |
| 30031 | 14580 | 5 | Gallatin | 2,605.4 | 116,806 | 534 | 44.8 | 92.8 | 0.8 | 1.7 | 2.6 | 4.2 | 5.1 | 11.2 | 18.8 | 16.6 | 14.0 | 10.3 |
| 30033 | | 9 | Garfield | 4,676.6 | 1,268 | 3,096 | 0.3 | 98.3 | 0.6 | 1.0 | 0.2 | 0.8 | 6.3 | 12.6 | 10.0 | 9.5 | 12.1 | 10.0 |
| 30035 | | 7 | Glacier | 2,995.1 | 13,594 | 2,186 | 4.5 | 32.9 | 0.5 | 65.6 | 0.7 | 3.4 | 8.4 | 17.7 | 14.2 | 13.1 | 11.7 | 9.8 |
| 30037 | | 3 | Golden Valley | 1,174.4 | 827 | 3,114 | 0.7 | 91.7 | 1.0 | 3.5 | 1.6 | 5.0 | 4.7 | 10.9 | 8.7 | 9.4 | 11.1 | 8.6 |
| 30039 | | 8 | Granite | 1,727.2 | 3,317 | 2,944 | 1.9 | 95.5 | 1.0 | 2.4 | 0.9 | 2.4 | 3.0 | 8.9 | 8.3 | 8.5 | 10.6 | 10.1 |
| 30041 | | 7 | Hill | 2,899.4 | 16,358 | 2,014 | 5.6 | 71.5 | 1.0 | 25.8 | 1.5 | 3.9 | 8.8 | 15.4 | 13.3 | 13.3 | 11.6 | 9.7 |
| 30043 | 25740 | 9 | Jefferson | 1,657.0 | 12,360 | 2,262 | 7.5 | 94.2 | 0.7 | 3.4 | 1.2 | 2.9 | 4.4 | 11.7 | 10.0 | 9.0 | 11.7 | 13.2 |
| 30045 | | 8 | Judith Basin | 1,869.7 | 1,994 | 3,046 | 1.1 | 95.8 | 0.1 | 1.9 | 0.7 | 2.5 | 5.7 | 10.6 | 8.9 | 9.1 | 9.7 | 10.0 |
| 30047 | | 7 | Lake | 1,490.5 | 30,986 | 1,404 | 20.8 | 71.6 | 0.7 | 28.4 | 1.4 | 4.5 | 5.5 | 13.4 | 11.3 | 10.7 | 10.9 | 10.3 |
| 30049 | 25740 | 5 | Lewis and Clark | 3,458.4 | 70,229 | 773 | 20.3 | 93.2 | 0.8 | 3.4 | 1.5 | 3.6 | 5.7 | 12.2 | 11.3 | 12.3 | 12.8 | 11.4 |
| 30051 | | 9 | Liberty | 1,430.0 | 2,369 | 3,011 | 1.7 | 97.6 | 0.9 | 1.9 | 0.8 | 0.8 | 5.7 | 12.2 | 10.0 | 11.9 | 11.8 | 10.2 |
| 30053 | | 7 | Lincoln | 3,612.6 | 20,343 | 1,802 | 5.6 | 94.4 | 0.7 | 2.9 | 1.1 | 3.4 | 4.2 | 10.3 | 8.4 | 8.3 | 10.0 | 10.8 |
| 30055 | | 9 | McCone | 2,642.2 | 1,648 | 3,073 | 0.6 | 96.0 | 1.2 | 2.2 | 1.0 | 1.9 | 5.2 | 12.0 | 9.4 | 11.3 | 9.6 | 9.6 |
| 30057 | | 9 | Madison | 3,588.2 | 8,959 | 2,502 | 2.5 | 93.8 | 0.6 | 1.8 | 0.8 | 4.5 | 4.1 | 8.5 | 8.3 | 10.2 | 10.8 | 10.6 |
| 30059 | | 9 | Meagher | 2,391.9 | 1,831 | 3,063 | 0.8 | 96.4 | 0.6 | 1.9 | 0.6 | 2.0 | 5.7 | 10.5 | 9.0 | 8.4 | 10.4 | 9.2 |
| 30061 | | 8 | Mineral | 1,219.6 | 4,544 | 2,854 | 3.7 | 93.9 | 0.9 | 3.5 | 1.2 | 3.1 | 5.3 | 10.4 | 7.7 | 9.4 | 10.4 | 10.2 |
| 30063 | 33540 | 8 | Missoula | 2,593.0 | 121,630 | 522 | 46.9 | 91.3 | 1.1 | 3.9 | 3.0 | 3.7 | 4.7 | 10.5 | 16.5 | 15.9 | 13.6 | 10.5 |
| 30065 | | 8 | Musselshell | 1,869.0 | 4,669 | 2,849 | 2.5 | 91.8 | 1.2 | 3.3 | 1.7 | 4.1 | 4.5 | 10.4 | 9.5 | 8.6 | 9.7 | 12.0 |
| 30067 | | 7 | Park | 2,802.5 | 16,760 | 1,986 | 6.0 | 94.8 | 0.7 | 2.2 | 1.1 | 3.2 | 4.7 | 9.4 | 8.4 | 11.7 | 12.8 | 12.0 |
| 30069 | | 9 | Petroleum | 1,655.6 | 500 | 3,136 | 0.3 | 94.0 | 0.4 | 2.0 | 0.4 | 3.6 | 3.4 | 9.6 | 8.4 | 10.4 | 8.0 | 10.6 |
| 30071 | | 9 | Phillips | 5,140.4 | 3,919 | 2,903 | 0.8 | 86.7 | 0.9 | 12.1 | 1.1 | 3.5 | 6.2 | 12.8 | 10.4 | 9.9 | 10.2 | 9.8 |
| 30073 | | 7 | Pondera | 1,624.6 | 5,782 | 2,760 | 3.6 | 82.6 | 0.8 | 15.5 | 1.1 | 2.4 | 6.8 | 13.2 | 11.4 | 11.3 | 10.7 | 10.1 |
| 30075 | | 9 | Powder River | 3,298.2 | 1,681 | 3,070 | 0.5 | 95.3 | 1.0 | 3.9 | 0.8 | 2.5 | 5.5 | 7.4 | 10.0 | 9.8 | 9.0 | 9.8 |
| 30077 | | 6 | Powell | 2,326.0 | 6,817 | 2,676 | 2.9 | 90.0 | 1.7 | 6.2 | 0.9 | 2.9 | 4.2 | 8.2 | 10.6 | 13.3 | 13.7 | 13.7 |
| 30079 | | 9 | Prairie | 1,736.6 | 1,067 | 3,106 | 0.6 | 91.9 | 1.3 | 3.8 | 1.3 | 5.2 | 6.8 | 10.9 | 6.4 | 8.2 | 9.2 | 9.5 |
| 30081 | | 6 | Ravalli | 2,391.0 | 45,002 | 1,075 | 18.8 | 94.3 | 0.6 | 2.2 | 1.2 | 3.6 | 4.5 | 10.5 | 9.8 | 9.3 | 11.1 | 11.3 |
| 30083 | | 7 | Richland | 2,084.6 | 11,043 | 2,343 | 5.3 | 90.8 | 1.2 | 3.7 | 1.1 | 5.7 | 6.6 | 14.8 | 11.3 | 13.8 | 12.7 | 11.4 |
| 30085 | | 7 | Roosevelt | 2,354.6 | 10,964 | 2,352 | 4.7 | 36.5 | 0.8 | 61.6 | 1.1 | 3.5 | 9.5 | 18.9 | 14.0 | 13.5 | 11.1 | 9.2 |
| 30087 | | 9 | Rosebud | 5,008.1 | 8,836 | 2,507 | 1.8 | 56.3 | 0.7 | 38.3 | 2.2 | 5.1 | 7.7 | 16.6 | 12.8 | 12.3 | 9.9 | 10.5 |
| 30089 | | 8 | Sanders | 2,760.4 | 12,157 | 2,280 | 4.4 | 92.2 | 0.8 | 5.5 | 0.8 | 3.5 | 4.2 | 9.7 | 8.3 | 8.1 | 9.1 | 10.7 |
| 30091 | | 9 | Sheridan | 1,676.9 | 3,261 | 2,949 | 1.9 | 92.6 | 1.3 | 3.5 | 1.2 | 3.9 | 5.7 | 11.5 | 9.7 | 9.5 | 12.2 | 10.2 |
| 30093 | 15580 | 5 | Silver Bow | 718.0 | 35,180 | 1,299 | 49.0 | 91.9 | 0.8 | 3.3 | 1.4 | 4.8 | 5.4 | 11.7 | 14.1 | 12.9 | 11.6 | 10.7 |
| 30095 | 13740 | 8 | Stillwater | 1,796.7 | 9,888 | 2,428 | 5.5 | 94.0 | 0.6 | 2.1 | 1.0 | 4.2 | 4.4 | 12.4 | 10.4 | 9.2 | 10.9 | 12.2 |
| 30097 | | 9 | Sweet Grass | 1,855.5 | 3,684 | 2,921 | 2.0 | 94.7 | 1.1 | 2.5 | 1.3 | 2.8 | 4.4 | 10.9 | 10.7 | 9.4 | 10.1 | 11.2 |
| 30099 | | 8 | Teton | 2,271.6 | 6,249 | 2,725 | 2.8 | 95.1 | 0.7 | 3.2 | 1.8 | 1.6 | 6.9 | 13.9 | 11.0 | 9.5 | 11.0 | 10.4 |
| 30101 | | 7 | Toole | 1,914.9 | 4,686 | 2,848 | 2.4 | 89.0 | 2.1 | 6.7 | 1.3 | 4.0 | 6.3 | 11.9 | 10.1 | 13.4 | 13.3 | 11.2 |
| 30103 | | 8 | Treasure | 977.8 | 695 | 3,130 | 0.7 | 90.5 | 0.6 | 4.3 | 1.2 | 5.5 | 10.4 | 11.2 | 7.5 | 10.5 | 8.6 | 6.5 |
| 30105 | | 7 | Valley | 4,926.1 | 7,359 | 2,630 | 1.5 | 87.0 | 0.9 | 10.2 | 1.5 | 3.0 | 5.8 | 12.8 | 10.8 | 10.3 | 11.1 | 11.5 |
| 30107 | | 9 | Wheatland | 1,422.5 | 2,157 | 3,030 | 1.5 | 93.2 | 1.4 | 2.7 | 0.9 | 4.4 | 5.7 | 14.0 | 10.7 | 9.1 | 9.8 | 11.5 |
| 30109 | | 9 | Wibaux | 889.6 | 939 | 3,108 | 1.1 | 93.4 | 1.1 | 2.1 | 1.4 | 3.6 | 4.4 | 12.1 | 10.3 | 7.8 | 11.4 | 11.1 |
| 30111 | 13740 | 3 | Yellowstone | 2,633.5 | 162,990 | 408 | 61.9 | 88.1 | 1.5 | 5.6 | 1.6 | 6.1 | 5.9 | 13.3 | 11.8 | 14.1 | 13.1 | 11.2 |
| 31000 | | 0 | NEBRASKA | 76,816.5 | 1,937,552 | X | 25.2 | 79.7 | 6.0 | 1.4 | 3.4 | 11.6 | 6.6 | 13.8 | 13.8 | 13.1 | 12.7 | 11.0 |
| 31001 | 25580 | 4 | Adams | 563.3 | 31,321 | 1,396 | 55.6 | 85.6 | 1.5 | 0.9 | 1.5 | 11.5 | 6.4 | 13.1 | 14.9 | 12.2 | 11.0 | 10.7 |
| 31003 | | 9 | Antelope | 857.2 | 6,264 | 2,723 | 7.3 | 95.4 | 0.7 | 0.5 | 0.6 | 3.5 | 6.8 | 13.4 | 10.1 | 9.4 | 9.1 | 9.1 |
| 31005 | | 9 | Arthur | 715.2 | 466 | 3,138 | 0.7 | 95.9 | 0.6 | 0.6 | 0.2 | 3.4 | 5.4 | 14.6 | 13.9 | 10.7 | 12.7 | 9.9 |
| 31007 | 42420 | 9 | Banner | 746.0 | 786 | 3,119 | 1.1 | 91.7 | 1.7 | 0.4 | 0.3 | 6.5 | 7.4 | 13.1 | 8.0 | 12.1 | 10.3 | 7.5 |

1. CBSA = Core Based Statistical Area. See Appendix A for explanation. See Appendix B for list of metropolitan areas with component counties. Service of USDA Rural-Urban Continuum Codes. See Appendix A for definition.  3. Dry land or land partially or temporarily covered by water.  2. County type code from the Economic Research  4. May be of any race.

# Table B. States and Counties — Population and Households

| STATE County | 55 to 64 years (16) | 65 to 74 years (17) | 75 years and over (18) | Percent female (19) | 2000 (20) | 2010 (21) | 2000-2010 (22) | 2010-2020 (23) | Births (24) | Deaths (25) | Net Migration (26) | Number (27) | Persons per household (28) | Family households (29) | Female family householder[1] (30) | One person (31) |
|---|---|---|---|---|---|---|---|---|---|---|---|---|---|---|---|---|
| **MISSOURI— Cont'd** | | | | | | | | | | | | | | | | |
| St. Louis city | 12.9 | 9.1 | 5.6 | 51.5 | 348,189 | 319,308 | -8.3 | -6.8 | 45,826 | 31,118 | -36,264 | 141,952 | 2.10 | 45.3 | 16.9 | 45.3 |
| **MONTANA** | 13.7 | 12.0 | 7.8 | 49.6 | 902,195 | 989,400 | 9.7 | 9.2 | 123,123 | 99,738 | 67,564 | 427,871 | 2.39 | 61.6 | 7.9 | 30.6 |
| Beaverhead | 14.0 | 13.8 | 9.9 | 49.1 | 9,202 | 9,246 | 0.5 | 2.6 | 890 | 888 | 234 | 3,992 | 2.22 | 57.3 | 4.5 | 37.8 |
| Big Horn | 11.7 | 8.8 | 5.2 | 50.6 | 12,671 | 12,863 | 1.5 | 1.6 | 2,626 | 1,311 | -1,113 | 3,609 | 3.67 | 75.6 | 16.3 | 19.7 |
| Blaine | 12.6 | 9.3 | 6.7 | 49.7 | 7,009 | 6,492 | -7.4 | 2.9 | 1,107 | 690 | -341 | 2,366 | 2.79 | 65.8 | 16.9 | 31.0 |
| Broadwater | 16.9 | 15.0 | 9.2 | 48.6 | 4,385 | 5,608 | 27.9 | 14.9 | 546 | 576 | 864 | 2,364 | 2.50 | 63.2 | 4.9 | 32.3 |
| Carbon | 18.4 | 16.6 | 10.0 | 48.8 | 9,552 | 10,078 | 5.5 | 8.4 | 806 | 1,018 | 1,054 | 4,524 | 2.32 | 60.5 | 5.4 | 34.4 |
| Carter | 17.8 | 15.2 | 11.6 | 49.0 | 1,360 | 1,160 | -14.7 | 6.5 | 160 | 148 | 61 | 628 | 2.07 | 64.0 | 3.8 | 32.5 |
| Cascade | 13.1 | 11.0 | 8.5 | 49.4 | 80,357 | 81,327 | 1.2 | 0.0 | 11,555 | 8,664 | -2,844 | 34,329 | 2.30 | 62.6 | 9.9 | 31.1 |
| Chouteau | 15.0 | 13.0 | 9.0 | 49.7 | 5,970 | 5,813 | -2.6 | -2.0 | 517 | 624 | -9 | 2,274 | 2.49 | 68.3 | 8.4 | 29.0 |
| Custer | 15.2 | 11.8 | 8.7 | 50.1 | 11,696 | 11,699 | 0.0 | -3.5 | 1,457 | 1,457 | -404 | 4,903 | 2.30 | 61.2 | 9.7 | 33.4 |
| Daniels | 16.2 | 13.2 | 13.7 | 48.5 | 2,017 | 1,751 | -13.2 | -6.5 | 181 | 238 | -57 | 857 | 1.99 | 58.7 | 5.0 | 37.6 |
| Dawson | 14.2 | 11.8 | 9.0 | 48.1 | 9,059 | 8,964 | -1.0 | -4.6 | 1,104 | 993 | -541 | 3,930 | 2.18 | 59.5 | 4.0 | 34.1 |
| Deer Lodge | 16.9 | 15.6 | 9.9 | 46.8 | 9,417 | 9,294 | -1.3 | -0.9 | 746 | 1,288 | 457 | 3,903 | 2.08 | 57.0 | 6.8 | 41.3 |
| Fallon | 13.9 | 11.4 | 7.6 | 49.9 | 2,837 | 2,890 | 1.9 | -2.2 | 449 | 312 | -204 | 1,334 | 2.12 | 57.8 | 7.6 | 35.8 |
| Fergus | 15.0 | 14.5 | 10.6 | 49.4 | 11,893 | 11,590 | -2.5 | -4.2 | 1,212 | 1,600 | -88 | 4,912 | 2.20 | 60.5 | 7.0 | 34.9 |
| Flathead | 14.5 | 13.2 | 7.6 | 50.1 | 74,471 | 90,922 | 22.1 | 16.4 | 11,423 | 8,967 | 12,441 | 38,773 | 2.55 | 65.2 | 7.1 | 29.6 |
| Gallatin | 10.7 | 8.6 | 4.7 | 48.1 | 67,831 | 89,515 | 32.0 | 30.5 | 11,805 | 5,684 | 20,934 | 43,777 | 2.37 | 57.5 | 5.7 | 27.0 |
| Garfield | 13.8 | 15.1 | 10.4 | 49.9 | 1,279 | 1,209 | -5.5 | 4.9 | 149 | 130 | 38 | 438 | 2.37 | 69.4 | 6.2 | 26.9 |
| Glacier | 12.4 | 7.8 | 4.8 | 51.6 | 13,247 | 13,399 | 1.1 | 1.5 | 2,434 | 1,483 | -758 | 4,163 | 3.11 | 61.5 | 17.3 | 34.1 |
| Golden Valley | 15.7 | 17.7 | 13.2 | 49.8 | 1,042 | 884 | -15.2 | -6.4 | 72 | 86 | -43 | 351 | 2.07 | 61.8 | 5.4 | 35.9 |
| Granite | 18.5 | 19.3 | 12.9 | 48.4 | 2,830 | 3,072 | 8.6 | 8.0 | 190 | 343 | 394 | 1,308 | 2.47 | 54.9 | 3.9 | 37.4 |
| Hill | 12.4 | 9.4 | 6.1 | 49.5 | 16,673 | 16,093 | -3.5 | 1.6 | 3,036 | 1,587 | -1,183 | 6,369 | 2.51 | 62.8 | 14.2 | 30.7 |
| Jefferson | 16.8 | 16.1 | 7.2 | 49.5 | 10,049 | 11,403 | 13.5 | 8.4 | 892 | 1,102 | 1,177 | 4,484 | 2.60 | 72.6 | 5.0 | 23.3 |
| Judith Basin | 17.7 | 16.8 | 11.6 | 48.9 | 2,329 | 2,072 | -11.0 | -3.8 | 167 | 195 | -48 | 906 | 2.15 | 59.9 | 5.7 | 36.8 |
| Lake | 14.6 | 13.8 | 9.5 | 51.0 | 26,507 | 28,748 | 8.5 | 7.8 | 3,548 | 3,071 | 1,783 | 11,802 | 2.49 | 67.1 | 12.1 | 25.6 |
| Lewis and Clark | 14.6 | 12.4 | 7.4 | 50.5 | 55,716 | 63,394 | 13.8 | 10.8 | 7,772 | 6,151 | 5,224 | 27,893 | 2.37 | 61.7 | 8.3 | 30.6 |
| Liberty | 14.4 | 11.4 | 12.4 | 51.9 | 2,158 | 2,339 | 8.4 | 1.3 | 250 | 223 | 1 | 895 | 2.60 | 62.1 | 4.1 | 32.1 |
| Lincoln | 17.6 | 18.7 | 11.6 | 49.8 | 18,837 | 19,681 | 4.5 | 3.4 | 1,792 | 2,548 | 1,431 | 8,122 | 2.38 | 62.9 | 5.0 | 32.2 |
| McCone | 15.5 | 15.2 | 12.1 | 48.5 | 1,977 | 1,734 | -12.3 | -5.0 | 161 | 192 | -52 | 723 | 2.45 | 66.9 | 3.3 | 30.2 |
| Madison | 16.6 | 19.5 | 11.4 | 47.6 | 6,851 | 7,691 | 12.3 | 16.5 | 654 | 777 | 1,389 | 3,570 | 2.24 | 58.8 | 3.1 | 35.3 |
| Meagher | 16.5 | 18.6 | 11.6 | 48.8 | 1,932 | 1,891 | -2.1 | -3.2 | 197 | 260 | 5 | 702 | 2.66 | 60.3 | 5.0 | 32.6 |
| Mineral | 16.6 | 18.1 | 11.8 | 48.7 | 3,884 | 4,230 | 8.9 | 7.4 | 445 | 539 | 409 | 1,786 | 2.34 | 60.4 | 6.9 | 37.5 |
| Missoula | 11.5 | 10.6 | 6.2 | 50.0 | 95,802 | 109,295 | 14.1 | 11.3 | 12,092 | 8,909 | 9,164 | 49,313 | 2.31 | 56.4 | 8.2 | 31.0 |
| Musselshell | 17.2 | 17.8 | 10.3 | 49.2 | 4,497 | 4,538 | 0.9 | 2.9 | 433 | 608 | 308 | 2,181 | 2.17 | 63.8 | 5.0 | 31.2 |
| Park | 16.7 | 15.1 | 9.1 | 49.7 | 15,694 | 15,635 | -0.4 | 7.2 | 1,525 | 1,573 | 1,166 | 7,782 | 2.07 | 57.6 | 4.9 | 33.8 |
| Petroleum | 20.8 | 14.0 | 14.8 | 46.4 | 493 | 488 | -1.0 | 2.5 | 26 | 40 | 27 | 204 | 2.22 | 59.3 | 2.0 | 27.0 |
| Phillips | 16.8 | 13.7 | 10.2 | 49.6 | 4,601 | 4,252 | -7.6 | -7.8 | 513 | 536 | -311 | 1,710 | 2.34 | 58.2 | 5.2 | 38.4 |
| Pondera | 15.0 | 11.6 | 10.0 | 50.8 | 6,424 | 6,155 | -4.2 | -6.1 | 736 | 717 | -392 | 2,103 | 2.75 | 67.2 | 10.0 | 30.1 |
| Powder River | 18.6 | 17.7 | 12.2 | 50.4 | 1,858 | 1,743 | -6.2 | -3.6 | 152 | 203 | -13 | 737 | 2.12 | 68.1 | 3.1 | 25.2 |
| Powell | 15.6 | 12.2 | 8.6 | 36.6 | 7,180 | 7,034 | -2.0 | -3.1 | 547 | 804 | 41 | 2,426 | 2.16 | 66.8 | 8.0 | 26.9 |
| Prairie | 15.1 | 20.3 | 13.7 | 48.0 | 1,199 | 1,179 | -1.7 | -9.5 | 142 | 182 | -75 | 518 | 2.37 | 66.8 | 1.7 | 28.2 |
| Ravalli | 16.3 | 16.6 | 10.7 | 50.1 | 36,070 | 40,210 | 11.5 | 11.9 | 3,969 | 4,579 | 5,422 | 17,354 | 2.43 | 68.8 | 7.3 | 26.9 |
| Richland | 13.8 | 9.3 | 6.2 | 48.0 | 9,667 | 9,746 | 0.8 | 13.3 | 1,579 | 941 | 603 | 4,535 | 2.45 | 60.1 | 6.6 | 35.5 |
| Roosevelt | 11.9 | 7.5 | 4.4 | 50.3 | 10,620 | 10,425 | -1.8 | 5.2 | 2,309 | 1,404 | -379 | 3,150 | 3.46 | 62.9 | 17.9 | 30.6 |
| Rosebud | 13.6 | 10.5 | 6.2 | 49.6 | 9,383 | 9,239 | -1.5 | -4.4 | 1,540 | 909 | -1,041 | 3,166 | 2.86 | 68.9 | 8.8 | 24.3 |
| Sanders | 17.8 | 19.3 | 12.8 | 48.8 | 10,227 | 11,413 | 11.6 | 6.5 | 1,080 | 1,449 | 1,124 | 5,036 | 2.27 | 60.4 | 5.1 | 32.3 |
| Sheridan | 16.6 | 13.8 | 10.8 | 48.6 | 4,105 | 3,384 | -17.6 | -3.6 | 400 | 542 | 7 | 1,639 | 2.07 | 55.1 | 4.9 | 34.8 |
| Silver Bow | 14.0 | 11.5 | 8.1 | 49.5 | 34,606 | 34,211 | -1.1 | 2.8 | 3,977 | 4,273 | 1,290 | 14,960 | 2.24 | 53.4 | 8.7 | 39.7 |
| Stillwater | 16.5 | 15.3 | 9.0 | 49.0 | 8,195 | 9,093 | 11.0 | 8.7 | 867 | 890 | 819 | 3,761 | 2.48 | 69.6 | 4.4 | 26.6 |
| Sweet Grass | 15.0 | 15.5 | 12.8 | 49.3 | 3,609 | 3,651 | 1.2 | 0.9 | 313 | 414 | 136 | 1,566 | 2.33 | 64.6 | 6.6 | 33.8 |
| Teton | 14.4 | 11.7 | 11.2 | 50.6 | 6,445 | 6,071 | -5.8 | 2.9 | 791 | 677 | 65 | 2,464 | 2.39 | 66.2 | 6.4 | 31.7 |
| Toole | 14.7 | 11.5 | 7.6 | 43.3 | 5,267 | 5,324 | 1.1 | -12.0 | 587 | 510 | -721 | 1,867 | 2.23 | 63.5 | 5.7 | 31.7 |
| Treasure | 15.7 | 15.4 | 14.2 | 50.1 | 861 | 714 | -17.1 | -2.7 | 109 | 86 | -41 | 351 | 1.90 | 62.4 | 0.6 | 35.3 |
| Valley | 14.9 | 13.1 | 11.1 | 49.9 | 7,675 | 7,369 | -4.0 | -0.1 | 880 | 838 | -50 | 3,386 | 2.17 | 64.7 | 7.3 | 28.8 |
| Wheatland | 13.6 | 15.2 | 10.6 | 48.8 | 2,259 | 2,168 | -4.0 | -0.5 | 250 | 255 | -7 | 792 | 2.66 | 61.7 | 8.0 | 33.8 |
| Wibaux | 16.7 | 14.0 | 12.2 | 49.1 | 1,068 | 1,017 | -4.8 | -7.7 | 108 | 148 | -42 | 498 | 2.21 | 70.1 | 4.8 | 29.3 |
| Yellowstone | 12.9 | 10.4 | 7.4 | 50.8 | 129,352 | 147,994 | 14.4 | 10.1 | 19,855 | 15,106 | 10,253 | 66,385 | 2.34 | 62.3 | 8.4 | 29.8 |
| **NEBRASKA** | 12.4 | 9.5 | 6.9 | 50.0 | 1,711,263 | 1,826,311 | 6.7 | 6.1 | 266,099 | 165,518 | 11,190 | 759,176 | 2.45 | 63.7 | 9.3 | 29.4 |
| Adams | 12.8 | 10.7 | 8.3 | 50.3 | 31,151 | 31,367 | 0.7 | -0.1 | 4,138 | 3,208 | -963 | 12,712 | 2.35 | 61.3 | 8.6 | 33.3 |
| Antelope | 15.2 | 13.4 | 11.7 | 49.8 | 7,452 | 6,685 | -10.3 | -6.3 | 851 | 733 | -541 | 2,714 | 2.31 | 65.2 | 6.6 | 31.2 |
| Arthur | 10.5 | 11.6 | 11.6 | 51.3 | 444 | 460 | 3.6 | 1.3 | 45 | 40 | 2 | 197 | 2.17 | 53.3 | 7.1 | 42.6 |
| Banner | 17.2 | 16.2 | 8.3 | 50.8 | 819 | 694 | -15.3 | 13.3 | 69 | 38 | 58 | 283 | 2.55 | 81.6 | 6.7 | 18.4 |

1. No spouse present.

# Table B. States and Counties — Population, Vital Statistics, Health, and Crime

| STATE County | Persons in group quarters, 2020 | Daytime Population, 2015-2019 Number | Employment/ residence ratio | Births, 2020 Total | Rate[1] | Deaths, 2020 Number | Rate[1] | Persons under 65 with no health insurance, 2019 Number | Percent | Medicare, 2020 Total beneficiaries | Enrolled in Original Medicare | Enrolled in Medicare Advantage | Crimes reported by county police or sheriff, 2019 Violent | Property |
|---|---|---|---|---|---|---|---|---|---|---|---|---|---|---|
| | 32 | 33 | 34 | 35 | 36 | 37 | 38 | 39 | 40 | 41 | 42 | 43 | 44 | 45 |
| **MISSOURI— Cont'd** | | | | | | | | | | | | | | |
| St. Louis city | 11,656 | 417,871 | 1.72 | 3,970 | 13.3 | 3,348 | 11.2 | 31,216 | 12.6 | 49,258 | 25,087 | 24,172 | NA | NA |
| **MONTANA** | 28,900 | 1,047,967 | 0.99 | 11,314 | 10.5 | 10,756 | 10.0 | 86,203 | 10.2 | 237,659 | 188,889 | 48,769 | X | X |
| Beaverhead | 442 | 9,407 | 1.00 | 80 | 8.4 | 105 | 11.1 | 741 | 10.7 | 2,433 | 2,130 | 303 | 3 | 4 |
| Big Horn | 137 | 13,914 | 1.11 | 211 | 16.2 | 143 | 10.9 | 1,633 | 14.8 | 1,866 | 1,598 | 269 | 53 | 46 |
| Blaine | 220 | 6,491 | 0.91 | 104 | 15.8 | 69 | 10.5 | 838 | 15.4 | 1,155 | 1,141 | 15 | NA | NA |
| Broadwater | 52 | 5,007 | 0.63 | 61 | 9.5 | 61 | 9.5 | 502 | 10.7 | 1,680 | 1,397 | 283 | 27 | 31 |
| Carbon | 53 | 9,141 | 0.72 | 84 | 7.7 | 93 | 8.5 | 915 | 11.7 | 2,923 | 2,239 | 684 | 21 | 43 |
| Carter | 11 | 1,294 | 0.94 | 20 | 16.2 | 15 | 12.1 | 121 | 13.3 | 340 | 323 | 17 | 0 | 0 |
| Cascade | 2,447 | 81,626 | 1.00 | 1,028 | 12.6 | 848 | 10.4 | 6,236 | 9.7 | 18,068 | 13,056 | 5,012 | NA | 248 |
| Chouteau | 135 | 5,152 | 0.77 | 51 | 8.9 | 69 | 12.1 | 615 | 14.1 | 1,325 | 1,078 | 247 | NA | NA |
| Custer | 429 | 11,580 | 0.98 | 109 | 9.7 | 157 | 13.9 | 830 | 9.5 | 2,685 | 2,481 | 204 | 6 | 29 |
| Daniels | 41 | 1,734 | 1.00 | 15 | 9.2 | 19 | 11.6 | 134 | 11.1 | 457 | 442 | 15 | NA | NA |
| Dawson | 498 | 9,068 | 1.01 | 91 | 10.6 | 92 | 10.8 | 482 | 7.4 | 1,971 | 1,865 | 105 | 10 | 12 |
| Deer Lodge | 852 | 9,144 | 1.01 | 73 | 7.9 | 129 | 14.0 | 639 | 10.2 | 2,805 | 2,357 | 448 | 13 | 268 |
| Fallon | 34 | 3,020 | 1.06 | 29 | 10.3 | 29 | 10.3 | 268 | 11.4 | 550 | 533 | 17 | NA | NA |
| Fergus | 465 | 11,357 | 1.03 | 109 | 9.8 | 143 | 12.9 | 917 | 11.4 | 3,120 | 2,416 | 704 | 13 | 32 |
| Flathead | 992 | 99,978 | 1.00 | 1,114 | 10.5 | 1,004 | 9.5 | 8,285 | 10.1 | 24,966 | 18,593 | 6,373 | 191 | 862 |
| Gallatin | 3,273 | 108,117 | 1.00 | 1,130 | 9.7 | 681 | 5.8 | 8,030 | 8.3 | 16,646 | 13,311 | 3,335 | 86 | 392 |
| Garfield | 0 | 1,045 | 1.02 | 15 | 11.8 | 12 | 9.5 | 155 | 16.6 | 303 | 289 | 14 | 0 | 0 |
| Glacier | 690 | 13,818 | 1.02 | 206 | 15.2 | 152 | 11.2 | 2,210 | 19.2 | 2,025 | 1,998 | 27 | NA | NA |
| Golden Valley | 89 | 654 | 0.79 | 5 | 6.0 | 2 | 2.4 | 76 | 13.0 | 263 | 201 | 62 | 2 | 7 |
| Granite | 40 | 3,366 | 1.05 | 16 | 4.8 | 35 | 10.6 | 307 | 13.4 | 930 | 800 | 130 | NA | NA |
| Hill | 593 | 17,026 | 1.08 | 288 | 17.6 | 176 | 10.8 | 1,723 | 12.7 | 2,934 | 2,904 | 30 | 22 | 146 |
| Jefferson | 230 | 9,246 | 0.50 | 82 | 6.6 | 152 | 12.3 | 843 | 8.9 | 3,159 | 2,599 | 560 | 23 | 83 |
| Judith Basin | 0 | 1,875 | 0.91 | 19 | 9.5 | 19 | 9.5 | 179 | 12.5 | 595 | 462 | 134 | NA | NA |
| Lake | 584 | 28,382 | 0.87 | 299 | 9.6 | 370 | 11.9 | 3,572 | 15.6 | 7,919 | 6,268 | 1,651 | 81 | 260 |
| Lewis and Clark | 1,958 | 71,029 | 1.09 | 727 | 10.4 | 711 | 10.1 | 4,363 | 7.9 | 15,792 | 12,271 | 3,521 | 63 | 330 |
| Liberty | 403 | 2,329 | 0.98 | 26 | 11.0 | 24 | 10.1 | 260 | 14.6 | 470 | 444 | 25 | NA | NA |
| Lincoln | 209 | 19,237 | 0.96 | 176 | 8.7 | 311 | 15.3 | 1,923 | 14.0 | 7,640 | 5,283 | 2,357 | 22 | 69 |
| McCone | 19 | 1,738 | 0.93 | 14 | 8.5 | 18 | 10.9 | 204 | 17.2 | 425 | D | D | 0 | 4 |
| Madison | 163 | 8,737 | 1.11 | 78 | 8.7 | 82 | 9.2 | 808 | 13.4 | 2,571 | 2,299 | 273 | 10 | 50 |
| Meagher | 174 | 1,877 | 0.99 | 19 | 10.4 | 23 | 12.6 | 125 | 10.0 | 561 | 532 | 28 | 6 | 8 |
| Mineral | 22 | 3,838 | 0.74 | 43 | 9.5 | 58 | 12.8 | 316 | 10.4 | 1,536 | 1,237 | 299 | NA | NA |
| Missoula | 3,644 | 120,755 | 1.05 | 1,057 | 8.7 | 973 | 8.0 | 8,936 | 9.2 | 23,220 | 18,409 | 4,811 | NA | NA |
| Musselshell | 52 | 4,569 | 0.89 | 45 | 9.6 | 73 | 15.6 | 441 | 13.4 | 1,523 | 1,228 | 295 | 11 | 28 |
| Park | 108 | 15,082 | 0.85 | 157 | 9.4 | 181 | 10.8 | 1,269 | 10.1 | 4,308 | 4,149 | 158 | 12 | 26 |
| Petroleum | 0 | 441 | 0.94 | 2 | 4.0 | 6 | 12.0 | 33 | 9.1 | 116 | 99 | 17 | NA | NA |
| Phillips | 127 | 3,905 | 0.90 | 44 | 11.2 | 45 | 11.5 | 481 | 16.2 | 1,035 | 980 | 55 | 15 | 52 |
| Pondera | 645 | 5,919 | 0.97 | 55 | 9.5 | 61 | 10.5 | 541 | 11.7 | 1,365 | 1,161 | 204 | NA | NA |
| Powder River | 35 | 1,663 | 1.06 | 18 | 10.7 | 16 | 9.5 | 156 | 13.0 | 405 | 372 | 33 | NA | NA |
| Powell | 1,699 | 6,863 | 1.00 | 52 | 7.6 | 91 | 13.3 | 435 | 11.5 | 1,597 | 1,400 | 196 | 12 | 19 |
| Prairie | 21 | 1,203 | 0.91 | 19 | 17.8 | 9 | 8.4 | 86 | 12.0 | 367 | 337 | 31 | 11 | 1 |
| Ravalli | 478 | 38,930 | 0.80 | 393 | 8.7 | 529 | 11.8 | 3,994 | 12.6 | 13,637 | 10,218 | 3,419 | 65 | 266 |
| Richland | 34 | 11,772 | 1.10 | 152 | 13.8 | 85 | 7.7 | 985 | 10.7 | 1,957 | 1,937 | 21 | 14 | 32 |
| Roosevelt | 186 | 11,201 | 1.01 | 205 | 18.7 | 179 | 16.3 | 1,875 | 20.2 | 1,645 | 1,598 | 47 | 60 | 65 |
| Rosebud | 65 | 9,948 | 1.21 | 118 | 13.4 | 107 | 12.1 | 946 | 12.9 | 1,698 | 1,521 | 177 | NA | NA |
| Sanders | 188 | 11,491 | 0.96 | 98 | 8.1 | 148 | 12.2 | 1,262 | 16.0 | 4,379 | 3,533 | 846 | 5 | 35 |
| Sheridan | 85 | 3,492 | 1.01 | 37 | 11.3 | 50 | 15.3 | 295 | 11.7 | 954 | 887 | 67 | NA | NA |
| Silver Bow | 999 | 34,594 | 0.99 | 369 | 10.5 | 448 | 12.7 | 2,514 | 9.2 | 8,114 | 6,697 | 1,416 | 120 | 1,372 |
| Stillwater | 112 | 8,963 | 0.89 | 92 | 9.3 | 89 | 9.0 | 586 | 7.9 | 2,477 | 1,878 | 599 | 18 | 52 |
| Sweet Grass | 45 | 3,856 | 1.12 | 27 | 7.3 | 40 | 10.9 | 293 | 10.9 | 998 | 860 | 138 | 20 | 46 |
| Teton | 461 | 5,782 | 0.88 | 73 | 11.7 | 70 | 11.2 | 543 | 11.7 | 1,590 | 1,100 | 490 | 9 | 28 |
| Toole | 668 | 4,960 | 1.03 | 55 | 11.7 | 50 | 10.7 | 432 | 13.3 | 972 | 941 | 32 | 30 | 51 |
| Treasure | 0 | 568 | 0.70 | 18 | 25.9 | 3 | 4.3 | 51 | 10.4 | 213 | 164 | 49 | NA | NA |
| Valley | 126 | 7,409 | 0.98 | 88 | 12.0 | 62 | 8.4 | 571 | 10.2 | 1,794 | 1,732 | 62 | 22 | 15 |
| Wheatland | 147 | 2,119 | 0.97 | 22 | 10.2 | 23 | 10.7 | 295 | 19.1 | 577 | 483 | 94 | 9 | 7 |
| Wibaux | 24 | 1,047 | 0.87 | 6 | 6.4 | 7 | 7.5 | 94 | 13.2 | 252 | D | D | NA | NA |
| Yellowstone | 3,696 | 161,208 | 1.03 | 1,790 | 11.0 | 1,609 | 9.9 | 10,839 | 8.3 | 32,355 | 23,991 | 8,364 | NA | NA |
| **NEBRASKA** | 50,039 | 1,934,811 | 1.02 | 25,061 | 12.9 | 16,824 | 8.7 | 150,342 | 9.5 | 353,641 | 282,431 | 71,210 | X | X |
| Adams | 1,261 | 31,589 | 1.00 | 391 | 12.5 | 330 | 10.5 | 2,320 | 9.6 | 6,739 | 5,819 | 920 | 4 | 72 |
| Antelope | 53 | 6,164 | 0.94 | 82 | 13.1 | 54 | 8.6 | 483 | 10.2 | 1,559 | 1,494 | 65 | NA | NA |
| Arthur | 0 | 415 | 0.94 | 6 | 12.9 | 1 | 2.1 | 44 | 12.4 | 102 | D | D | 0 | 0 |
| Banner | 0 | 651 | 0.82 | 9 | 11.5 | 1 | 1.3 | 39 | 7.0 | 103 | D | D | NA | NA |

1. Per 1,000 estimated resident population.

| STATE County | County law enforcement employment, 2019 | | School enrollment and attainment, 2015-2019 | | | | Local government expenditures,[3] 2017-2018 | | Money income, 2015-2019 | | | | Income and poverty, 2019 | | | |
|---|---|---|---|---|---|---|---|---|---|---|---|---|---|---|---|---|
| | | | Enrollment[1] | | Attainment[2] (percent) | | | | | | Households | | | Percent below poverty level | | |
| | | | | | | | | | | | Percent | | | | | |
| | Officers | Civilians | Total | Percent private | High school graduate or less | Bachelor's degree or more | Total current spending (mil dol) | Current spending per student (dollars) | Per capita income[4] | Median income (dollars) | with income of less than $50,000 | with income of $200,000 or more | Median household income (dollars) | All persons | Children under 18 years | Children 5 to 17 years in families |
| | 46 | 47 | 48 | 49 | 50 | 51 | 52 | 53 | 54 | 55 | 56 | 57 | 58 | 59 | 60 | 61 |
| MISSOURI— Cont'd | | | | | | | | | | | | | | | | |
| St. Louis city................... | NA | NA | 69,839 | 34.8 | 36.2 | 36.3 | 481 | 14,012 | 30,542 | 43,896 | 55.4 | 3.9 | 46,309 | 20.4 | 27.7 | 29.1 |
| MONTANA....................... | X | X | 238,987 | 13.3 | 35.3 | 32.0 | 1,712 | 11,675 | 31,151 | 54,970 | 45.7 | 4.2 | 57,248 | 12.6 | 15.4 | 14.2 |
| Beaverhead................... | 8 | 6 | 2,256 | 11.7 | 33.7 | 35.4 | 14 | 12,312 | 28,401 | 43,201 | 54.7 | 3.1 | 49,771 | 14.5 | 16.6 | 15.4 |
| Big Horn...................... | 17 | 16 | 4,012 | 12.5 | 45.7 | 18.1 | 38 | 14,390 | 18,934 | 49,859 | 50.1 | 2.5 | 43,718 | 26.1 | 32.3 | 30.2 |
| Blaine......................... | 7 | 6 | 1,882 | 6.5 | 37.5 | 22.9 | 21 | 15,900 | 19,258 | 41,279 | 58.5 | 1.6 | 44,090 | 21.3 | 27.2 | 25.8 |
| Broadwater.................. | 10 | 15 | 1,036 | 9.4 | 39.0 | 27.1 | 7 | 10,122 | 31,631 | 60,594 | 42.4 | 3.8 | 60,622 | 9.0 | 12.3 | 11.5 |
| Carbon....................... | 11 | 7 | 1,820 | 7.9 | 36.9 | 30.4 | 19 | 13,482 | 33,067 | 58,707 | 43.0 | 3.3 | 57,585 | 10.1 | 12.8 | 11.1 |
| Carter........................ | 4 | 1 | 222 | 45.9 | 40.2 | 20.2 | 3 | 19,695 | 28,577 | 48,000 | 50.6 | 3.7 | 49,057 | 14.2 | 20.7 | 24.0 |
| Cascade...................... | 39 | 101 | 17,773 | 12.6 | 41.0 | 25.7 | 118 | 10,149 | 30,110 | 49,913 | 50.1 | 4.0 | 51,732 | 14.2 | 17.8 | 14.5 |
| Chouteau .................... | 9 | 13 | 1,249 | 6.2 | 41.3 | 28.3 | 11 | 15,742 | 23,769 | 42,298 | 57.9 | 1.7 | 48,017 | 15.4 | 20.2 | 17.5 |
| Custer ....................... | 7 | 11 | 2,557 | 14.4 | 36.7 | 24.6 | 18 | 10,985 | 30,504 | 52,965 | 47.8 | 2.9 | 49,379 | 13.0 | 14.4 | 13.6 |
| Daniels ...................... | 3 | 3 | 333 | 6.6 | 41.7 | 24.0 | 4 | 12,380 | 35,133 | 57,440 | 39.1 | 4.7 | 50,001 | 11.0 | 11.6 | 10.2 |
| Dawson ...................... | 8 | 31 | 1,693 | 8.8 | 41.1 | 22.2 | 17 | 13,112 | 31,658 | 58,596 | 42.7 | 4.6 | 56,677 | 11.8 | 12.8 | 11.3 |
| Deer Lodge ................. | 21 | 0 | 1,418 | 13.0 | 47.0 | 20.4 | 13 | 11,302 | 26,241 | 41,820 | 61.1 | 1.9 | 41,780 | 15.9 | 19.9 | 19.0 |
| Fallon ........................ | 4 | 1 | 616 | 19.8 | 42.9 | 19.0 | 10 | 19,055 | 37,024 | 64,545 | 39.9 | 5.3 | 65,651 | 9.7 | 12.5 | 11.7 |
| Fergus ....................... | 9 | 11 | 2,346 | 5.4 | 35.8 | 29.1 | 22 | 14,128 | 30,529 | 50,540 | 49.5 | 3.8 | 48,703 | 13.2 | 15.7 | 14.8 |
| Flathead ..................... | 56 | 53 | 19,960 | 18.0 | 34.7 | 31.0 | 150 | 10,439 | 31,694 | 56,182 | 45.0 | 4.6 | 61,334 | 9.4 | 11.8 | 11.3 |
| Gallatin ...................... | 55 | 81 | 31,144 | 11.8 | 21.0 | 50.1 | 138 | 10,259 | 36,924 | 66,397 | 37.1 | 6.8 | 73,731 | 10.3 | 7.6 | 6.7 |
| Garfield ...................... | 3 | 1 | 183 | 3.8 | 46.8 | 14.3 | 3 | 16,541 | 28,541 | 42,000 | 57.8 | 1.1 | 43,106 | 16.4 | 27.2 | 25.2 |
| Glacier....................... | 12 | 13 | 3,690 | 3.5 | 49.7 | 21.4 | 40 | 14,284 | 18,043 | 36,045 | 66.9 | 2.6 | 41,808 | 25.7 | 31.9 | 32.5 |
| Golden Valley .............. | 2 | 0 | 91 | 23.1 | 48.7 | 22.7 | 3 | 20,778 | 30,790 | 43,875 | 54.1 | 6.3 | 40,469 | 19.7 | 32.9 | 33.0 |
| Granite ...................... | 7 | 7 | 712 | 10.7 | 39.3 | 28.9 | 5 | 15,058 | 30,752 | 49,881 | 50.2 | 4.9 | 49,646 | 12.5 | 19.0 | 16.9 |
| Hill............................ | 13 | 17 | 4,456 | 6.6 | 39.5 | 25.4 | 41 | 13,779 | 23,589 | 49,321 | 50.5 | 1.8 | 51,881 | 15.7 | 18.8 | 18.8 |
| Jefferson .................... | 14 | 14 | 2,687 | 10.9 | 32.2 | 35.0 | 18 | 10,416 | 32,773 | 69,646 | 35.1 | 3.8 | 70,929 | 8.0 | 9.6 | 8.5 |
| Judith Basin ................ | 5 | 1 | 372 | 5.1 | 37.8 | 26.6 | 5 | 19,573 | 28,811 | 43,661 | 54.0 | 3.6 | 49,401 | 14.5 | 21.6 | 20.1 |
| Lake .......................... | 23 | 22 | 6,639 | 14.1 | 36.7 | 29.6 | 52 | 11,891 | 25,388 | 48,829 | 51.3 | 3.4 | 49,019 | 17.2 | 23.0 | 21.0 |
| Lewis and Clark ........... | 48 | 59 | 14,844 | 23.7 | 27.9 | 40.1 | 102 | 10,557 | 34,504 | 65,791 | 38.3 | 4.1 | 66,075 | 9.4 | 11.9 | 10.5 |
| Liberty ....................... | 6 | 0 | 445 | 5.2 | 51.3 | 15.2 | 4 | 13,520 | 43,320 | 44,875 | 55.5 | 6.7 | 41,240 | 19.0 | 28.8 | 27.2 |
| Lincoln ....................... | 20 | 20 | 3,048 | 17.0 | 47.7 | 19.6 | 27 | 11,103 | 25,081 | 40,140 | 60.0 | 2.1 | 43,669 | 17.4 | 27.7 | 26.1 |
| McCone ...................... | 4 | 0 | 364 | 20.1 | 46.3 | 17.2 | 4 | 15,388 | 27,912 | 49,701 | 50.8 | 5.3 | 48,551 | 14.5 | 16.9 | 15.0 |
| Madison ..................... | 11 | 1 | 1,238 | 15.8 | 35.4 | 31.2 | 12 | 14,288 | 35,719 | 54,107 | 45.6 | 4.6 | 56,579 | 9.7 | 12.0 | 10.8 |
| Meagher ..................... | 4 | 5 | 398 | 9.5 | 45.1 | 29.3 | 3 | 14,271 | 24,723 | 46,607 | 55.6 | 0.7 | 43,946 | 15.7 | 24.2 | 23.1 |
| Mineral ...................... | 6 | 6 | 686 | 11.2 | 48.1 | 16.1 | 9 | 15,421 | 23,621 | 41,705 | 59.5 | 1.0 | 43,928 | 15.6 | 23.3 | 22.3 |
| Missoula ..................... | 58 | 14 | 31,460 | 11.1 | 26.0 | 43.0 | 156 | 11,196 | 31,897 | 54,062 | 46.5 | 4.3 | 57,279 | 11.8 | 11.3 | 10.7 |
| Musselshell ................ | 7 | 0 | 747 | 19.9 | 51.3 | 16.3 | 8 | 13,002 | 24,979 | 43,274 | 55.6 | 2.2 | 48,290 | 16.5 | 26.2 | 25.7 |
| Park........................... | 15 | 10 | 2,768 | 10.3 | 35.4 | 31.5 | 25 | 12,909 | 35,123 | 53,068 | 47.8 | 3.8 | 51,740 | 11.8 | 13.4 | 11.9 |
| Petroleum................... | 2 | 0 | 71 | 0.0 | 40.6 | 32.6 | 2 | 19,025 | 31,177 | 51,250 | 49.5 | 2.9 | 42,081 | 14.8 | 25.6 | 20.6 |
| Phillips ...................... | 6 | 4 | 744 | 3.8 | 48.8 | 17.2 | 11 | 16,800 | 25,605 | 46,212 | 56.3 | 3.0 | 47,679 | 14.3 | 20.1 | 19.5 |
| Pondera ..................... | 7 | 6 | 1,436 | 9.7 | 40.5 | 24.0 | 15 | 16,445 | 25,633 | 51,151 | 49.2 | 2.4 | 48,542 | 17.5 | 24.7 | 24.6 |
| Powder River ............... | 4 | 4 | 194 | 18.0 | 35.7 | 24.4 | 4 | 18,305 | 32,094 | 54,427 | 45.3 | 2.0 | 47,843 | 13.8 | 19.7 | 19.9 |
| Powell ....................... | 5 | 5 | 1,055 | 9.4 | 48.5 | 20.4 | 11 | 15,364 | 26,456 | 54,667 | 45.3 | 3.8 | 49,387 | 16.3 | 18.0 | 17.0 |
| Prairie ....................... | 3 | 1 | 247 | 30.8 | 42.7 | 22.2 | 2 | 15,664 | 27,377 | 43,625 | 55.4 | 2.5 | 46,663 | 13.1 | 19.1 | 19.4 |
| Ravalli ....................... | 32 | 42 | 8,214 | 14.0 | 38.8 | 27.8 | 58 | 10,366 | 29,761 | 53,054 | 47.2 | 3.6 | 57,209 | 11.2 | 16.0 | 14.8 |
| Richland ..................... | 10 | 22 | 2,598 | 4.5 | 42.9 | 18.9 | 26 | 13,372 | 33,109 | 67,205 | 35.9 | 3.3 | 69,095 | 8.8 | 10.3 | 9.3 |
| Roosevelt .................... | 15 | 27 | 3,070 | 3.4 | 44.0 | 17.4 | 43 | 17,136 | 18,657 | 43,194 | 56.4 | 1.4 | 38,409 | 24.3 | 27.6 | 27.6 |
| Rosebud ..................... | 11 | 13 | 2,207 | 7.9 | 40.7 | 17.8 | 28 | 17,459 | 24,922 | 57,992 | 44.0 | 2.9 | 58,139 | 16.3 | 21.8 | 21.3 |
| Sanders ..................... | 11 | 0 | 1,808 | 7.6 | 53.1 | 17.9 | 20 | 14,387 | 25,309 | 40,823 | 60.8 | 3.1 | 45,193 | 16.7 | 23.3 | 23.6 |
| Sheridan .................... | 8 | 2 | 551 | 12.5 | 40.2 | 23.9 | 9 | 18,219 | 36,091 | 55,521 | 45.1 | 4.8 | 51,111 | 12.3 | 12.9 | 12.0 |
| Silver Bow .................. | 52 | 45 | 8,297 | 13.0 | 39.5 | 27.2 | 47 | 10,764 | 27,706 | 45,718 | 53.9 | 2.9 | 49,249 | 15.3 | 17.7 | 15.1 |
| Stillwater .................... | 11 | 9 | 2,167 | 9.7 | 41.6 | 25.0 | 18 | 12,785 | 32,301 | 64,645 | 36.5 | 3.5 | 68,186 | 9.3 | 11.7 | 10.0 |
| Sweet Grass ................ | 7 | 6 | 672 | 17.0 | 42.0 | 25.4 | 6 | 12,036 | 30,050 | 49,886 | 50.2 | 2.3 | 55,259 | 10.3 | 13.9 | 12.8 |
| Teton ......................... | 10 | 3 | 1,321 | 21.3 | 39.8 | 24.9 | 14 | 13,539 | 28,022 | 55,000 | 46.0 | 2.9 | 54,458 | 13.5 | 17.6 | 16.4 |
| Toole ......................... | 13 | 8 | 746 | 11.0 | 51.3 | 15.2 | 9 | 13,903 | 27,528 | 50,721 | 49.6 | 2.4 | 49,448 | 15.8 | 19.0 | 18.1 |
| Treasure ..................... | 2 | 0 | 76 | 19.7 | 41.7 | 25.4 | 1 | 26,680 | 29,763 | 43,047 | 58.1 | 4.0 | 51,754 | 12.2 | 18.6 | 21.3 |
| Valley ........................ | 9 | 10 | 1,398 | 7.0 | 40.6 | 21.5 | 18 | 14,347 | 27,946 | 53,162 | 47.1 | 1.9 | 51,869 | 12.8 | 16.3 | 16.1 |
| Wheatland .................. | 5 | 6 | 450 | 0.2 | 52.3 | 19.6 | 5 | 15,594 | 21,619 | 39,655 | 59.2 | 1.6 | 42,059 | 18.3 | 29.6 | 28.3 |
| Wibaux....................... | 3 | 0 | 219 | 14.6 | 48.5 | 23.4 | 3 | 20,582 | 25,591 | 53,958 | 45.4 | 0.0 | 51,145 | 11.3 | 15.1 | 14.7 |
| Yellowstone................. | 56 | 126 | 36,301 | 15.9 | 36.1 | 32.1 | 248 | 10,520 | 34,451 | 61,264 | 40.7 | 5.3 | 61,435 | 11.3 | 14.5 | 13.5 |
| NEBRASKA ................. | X | X | 506,713 | 17.1 | 34.7 | 31.9 | 4,145 | 12,801 | 32,302 | 61,439 | 40.8 | 4.9 | 63,290 | 9.9 | 11.5 | 10.3 |
| Adams........................ | 18 | 3 | 8,559 | 25.5 | 38.1 | 24.3 | 70 | 13,578 | 28,729 | 53,023 | 47.6 | 3.0 | 55,182 | 10.6 | 12.9 | 12.1 |
| Antelope ..................... | 5 | 14 | 1,368 | 9.4 | 40.8 | 18.4 | 25 | 25,303 | 26,523 | 49,912 | 50.1 | 2.3 | 50,771 | 12.2 | 15.9 | 15.2 |
| Arthur ........................ | NA | NA | 123 | 13.0 | 32.2 | 26.2 | 3 | 22,189 | 25,906 | 42,813 | 59.4 | 3.0 | 46,417 | 13.6 | 19.8 | 17.6 |
| Banner ....................... | NA | NA | 113 | 4.4 | 28.6 | 23.3 | 3 | 22,373 | 27,552 | 51,750 | 46.3 | 3.5 | 54,706 | 11.3 | 18.9 | 19.2 |

1. All persons 3 years old and over enrolled in nursery school through college. 2. Persons 25 years old and over. 3. Elementary and secondary education expenditures. 4. Based on population estimated by the American Community Survey, 2014–2018.

| STATE County | Personal income, 2019 | | | | | Supplements to wages and salaries, employer contributions (mil dol) | | | | | Earnings, 2019 | Contributions for government social insurance (mil dol) | |
|---|---|---|---|---|---|---|---|---|---|---|---|---|---|
| | Total (mil dol) | Percent change 2018-2019 | Per capita[1] Dollars | Rank | Wages and salaries (mil dol) | Pension and insurance | Government social insurance | Proprietors' income (mil dol) | Dividends, interest, and rent (mil dol) | Personal transfer receipts (mil dol) | Total (mil dol) | From employee and self-employed | From employer |
| | 62 | 63 | 64 | 65 | 66 | 67 | 68 | 69 | 70 | 71 | 72 | 73 | 74 |
| MISSOURI— Cont'd | | | | | | | | | | | | | |
| St. Louis city | 14,488 | 2.0 | 48,202 | 1,001 | 16,189 | 2,671 | 1,143 | 1,465 | 2,610 | 3,134 | 21,467 | 1,237 | 1,143 |
| MONTANA | 53,168 | 4.0 | 49,684 | X | 22,502 | 3,726 | 1,973 | 4,973 | 13,358 | 10,411 | 33,174 | 2,216 | 1,973 |
| Beaverhead | 458 | 4.9 | 48,424 | 982 | 157 | 31 | 14 | 55 | 129 | 102 | 256 | 16 | 14 |
| Big Horn | 425 | 4.2 | 31,887 | 2,951 | 200 | 42 | 17 | 37 | 71 | 152 | 296 | 18 | 17 |
| Blaine | 224 | 6.9 | 33,560 | 2,842 | 63 | 17 | 5 | 24 | 54 | 69 | 109 | 7 | 5 |
| Broadwater | 282 | 7.7 | 45,211 | 1,337 | 56 | 10 | 5 | 34 | 59 | 61 | 105 | 7 | 5 |
| Carbon | 562 | 4.2 | 52,371 | 653 | 106 | 19 | 10 | 62 | 142 | 108 | 196 | 14 | 10 |
| Carter | 52 | -1.1 | 41,642 | 1,823 | 12 | 3 | 1 | 7 | 16 | 11 | 23 | 1 | 1 |
| Cascade | 4,052 | 3.6 | 49,803 | 842 | 1,849 | 342 | 164 | 243 | 919 | 862 | 2,598 | 170 | 164 |
| Chouteau | 233 | 1.9 | 41,312 | 1,878 | 46 | 10 | 4 | 44 | 70 | 49 | 104 | 5 | 4 |
| Custer | 539 | 3.4 | 47,268 | 1,102 | 219 | 39 | 19 | 67 | 115 | 114 | 343 | 23 | 19 |
| Daniels | 82 | -1.6 | 48,385 | 988 | 28 | 5 | 2 | 7 | 23 | 19 | 43 | 3 | 2 |
| Dawson | 407 | -0.8 | 47,273 | 1,101 | 179 | 31 | 17 | 18 | 72 | 87 | 244 | 18 | 17 |
| Deer Lodge | 379 | 4.4 | 41,454 | 1,853 | 132 | 25 | 11 | 15 | 81 | 119 | 184 | 15 | 11 |
| Fallon | 145 | 1.1 | 50,824 | 767 | 79 | 12 | 6 | 13 | 32 | 24 | 111 | 7 | 6 |
| Fergus | 515 | 8.3 | 46,646 | 1,166 | 210 | 37 | 19 | 43 | 134 | 116 | 308 | 23 | 19 |
| Flathead | 5,073 | 3.6 | 48,866 | 935 | 2,054 | 303 | 188 | 471 | 1,384 | 1,034 | 3,015 | 213 | 188 |
| Gallatin | 6,659 | 5.6 | 58,195 | 365 | 3,025 | 440 | 262 | 825 | 1,909 | 687 | 4,552 | 280 | 262 |
| Garfield | 43 | -3.1 | 33,995 | 2,804 | 12 | 3 | 1 | 3 | 16 | 10 | 18 | 1 | 1 |
| Glacier | 502 | 4.5 | 36,484 | 2,511 | 180 | 43 | 15 | 47 | 107 | 155 | 285 | 17 | 15 |
| Golden Valley | 43 | 4.0 | 52,862 | 627 | 7 | 2 | 1 | 7 | 15 | 10 | 16 | 1 | 1 |
| Granite | 151 | 2.8 | 44,717 | 1,392 | 35 | 6 | 3 | 8 | 49 | 36 | 53 | 5 | 3 |
| Hill | 776 | 2.7 | 47,082 | 1,124 | 339 | 63 | 35 | 59 | 198 | 173 | 496 | 32 | 35 |
| Jefferson | 608 | 4.8 | 49,780 | 848 | 112 | 19 | 10 | 39 | 131 | 122 | 180 | 15 | 10 |
| Judith Basin | 102 | 10.6 | 50,997 | 758 | 22 | 4 | 2 | 22 | 28 | 19 | 51 | 3 | 2 |
| Lake | 1,241 | 3.9 | 40,754 | 1,955 | 354 | 67 | 31 | 82 | 356 | 362 | 534 | 42 | 31 |
| Lewis and Clark | 3,513 | 4.4 | 50,600 | 792 | 1,853 | 334 | 155 | 232 | 853 | 672 | 2,574 | 168 | 155 |
| Liberty | 111 | 9.9 | 47,675 | 1,059 | 23 | 5 | 2 | 36 | 30 | 20 | 65 | 2 | 2 |
| Lincoln | 742 | 4.4 | 37,155 | 2,428 | 217 | 44 | 20 | 46 | 173 | 289 | 327 | 31 | 20 |
| McCone | 66 | 2.5 | 39,913 | 2,058 | 24 | 5 | 2 | 0 | 23 | 14 | 31 | 2 | 2 |
| Madison | 461 | 6.0 | 53,567 | 567 | 210 | 25 | 20 | 59 | 155 | 88 | 314 | 21 | 20 |
| Meagher | 87 | 4.6 | 46,461 | 1,190 | 21 | 4 | 2 | 12 | 28 | 24 | 38 | 3 | 2 |
| Mineral | 178 | 4.9 | 40,547 | 1,979 | 42 | 9 | 4 | 12 | 34 | 60 | 66 | 6 | 4 |
| Missoula | 6,110 | 3.4 | 51,090 | 749 | 2,936 | 464 | 258 | 531 | 1,710 | 1,034 | 4,190 | 275 | 258 |
| Musselshell | 213 | 3.5 | 45,902 | 1,260 | 63 | 11 | 5 | 17 | 53 | 63 | 96 | 8 | 5 |
| Park | 856 | 4.3 | 51,538 | 721 | 256 | 37 | 24 | 78 | 285 | 167 | 395 | 29 | 24 |
| Petroleum | 18 | 4.9 | 37,070 | 2,439 | 4 | 1 | 0 | 4 | 5 | 4 | 9 | 1 | 0 |
| Phillips | 166 | 0.6 | 42,055 | 1,753 | 50 | 11 | 4 | 13 | 48 | 43 | 78 | 6 | 4 |
| Pondera | 281 | 5.9 | 47,477 | 1,080 | 64 | 13 | 6 | 51 | 72 | 66 | 133 | 8 | 6 |
| Powder River | 59 | 1.1 | 35,063 | 2,676 | 18 | 4 | 2 | 7 | 19 | 13 | 30 | 2 | 2 |
| Powell | 304 | 3.2 | 44,111 | 1,472 | 102 | 23 | 9 | 29 | 94 | 73 | 163 | 11 | 9 |
| Prairie | 54 | -0.6 | 49,750 | 852 | 12 | 3 | 1 | 7 | 15 | 13 | 24 | 2 | 1 |
| Ravalli | 2,049 | 4.4 | 46,768 | 1,152 | 495 | 87 | 44 | 148 | 575 | 518 | 775 | 66 | 44 |
| Richland | 697 | 4.8 | 64,486 | 194 | 324 | 48 | 27 | 82 | 137 | 84 | 480 | 29 | 27 |
| Roosevelt | 397 | -1.2 | 36,067 | 2,559 | 159 | 33 | 13 | 12 | 73 | 136 | 216 | 15 | 13 |
| Rosebud | 410 | 2.8 | 45,929 | 1,256 | 214 | 47 | 19 | 23 | 69 | 98 | 302 | 18 | 19 |
| Sanders | 440 | 4.9 | 36,296 | 2,537 | 112 | 24 | 11 | 39 | 108 | 171 | 185 | 17 | 11 |
| Sheridan | 164 | 2.0 | 49,659 | 862 | 57 | 11 | 5 | 9 | 51 | 36 | 81 | 6 | 5 |
| Silver Bow | 1,721 | 2.7 | 49,285 | 896 | 714 | 127 | 62 | 271 | 332 | 378 | 1,174 | 78 | 62 |
| Stillwater | 520 | 4.4 | 53,913 | 546 | 240 | 36 | 18 | 35 | 135 | 95 | 329 | 22 | 18 |
| Sweet Grass | 182 | 5.4 | 48,591 | 955 | 85 | 13 | 7 | 13 | 65 | 37 | 118 | 8 | 7 |
| Teton | 306 | 5.2 | 49,840 | 841 | 66 | 14 | 6 | 62 | 93 | 61 | 149 | 9 | 6 |
| Toole | 266 | 9.6 | 56,222 | 440 | 98 | 19 | 9 | 69 | 62 | 42 | 196 | 11 | 9 |
| Treasure | 43 | 14.0 | 61,553 | 264 | 7 | 2 | 1 | 15 | 10 | 8 | 24 | 1 | 1 |
| Valley | 354 | 2.7 | 47,831 | 1,040 | 146 | 27 | 15 | 14 | 92 | 83 | 202 | 15 | 15 |
| Wheatland | 85 | 7.2 | 40,015 | 2,047 | 23 | 5 | 2 | 14 | 22 | 25 | 44 | 3 | 2 |
| Wibaux | 41 | 1.2 | 42,078 | 1,751 | 12 | 3 | 1 | 4 | 10 | 10 | 19 | 2 | 1 |
| Yellowstone | 8,721 | 3.5 | 54,069 | 542 | 4,381 | 666 | 378 | 798 | 1,820 | 1,486 | 6,223 | 406 | 378 |
| NEBRASKA | 105,454 | 3.5 | 54,567 | X | 52,330 | 9,328 | 3,913 | 11,861 | 21,283 | 16,383 | 77,432 | 4,594 | 3,913 |
| Adams | 1,581 | 3.4 | 50,399 | 803 | 707 | 136 | 52 | 146 | 346 | 306 | 1,041 | 65 | 52 |
| Antelope | 346 | 8.2 | 54,945 | 493 | 102 | 20 | 8 | 68 | 74 | 65 | 198 | 10 | 8 |
| Arthur | 23 | 6.7 | 48,955 | 927 | 6 | 1 | 1 | 7 | 5 | 4 | 14 | 1 | 1 |
| Banner | 41 | 8.1 | 55,695 | 465 | 9 | 1 | 1 | 17 | 6 | 7 | 27 | 1 | 1 |

1. Based on the resident population estimated as of July 1 of the year shown.

# Table B. States and Counties — **Earnings, Social Security, and Housing**

| STATE County | Earnings, 2019 (cont.) | | | | | | | | | Social Security beneficiaries, December 2019 | | Supplemental Security Income recipients, 2019 | Housing units, 2020 | |
|---|---|---|---|---|---|---|---|---|---|---|---|---|---|---|
| | Percent by selected industries | | | | | | | | | | | | | |
| | Farm | Mining, quarrying, and extractions | Construction | Manu-facturing | Information; professional, scientific, technical services | Retail trade | Finance, insurance, real estate, and leasing | Health care and social assistance | Govern-ment | Number | Rate[1] | | Total | Percent change, 2010-2020 |
| | 75 | 76 | 77 | 78 | 79 | 80 | 81 | 82 | 83 | 84 | 85 | 86 | 87 | 88 |
| MISSOURI— Cont'd | | | | | | | | | | | | | | |
| St. Louis city | 0.0 | D | D | 8.6 | 16.3 | 1.4 | 9.9 | 16.9 | 14.0 | 52,580 | 175 | 14,731 | 177,383 | 0.8 |
| MONTANA | 2.1 | 3.2 | 8.7 | 4.4 | 8.4 | 8.3 | 6.9 | 14.5 | 19.0 | 239,410 | 224 | 17,677 | 524,641 | 8.7 |
| Beaverhead | 13.9 | D | 6.1 | 1.1 | 3.3 | 10.3 | 6.1 | 13.0 | 24.7 | 2,440 | 258 | 134 | 5,378 | 2.0 |
| Big Horn | 9.3 | 21.0 | D | D | 2.2 | 5.1 | 2.0 | D | 37.8 | 2,320 | 175 | 329 | 4,728 | 0.7 |
| Blaine | 16.3 | 1.1 | 2.0 | 0.5 | D | 9.7 | D | D | 48.6 | 1,210 | 180 | 193 | 2,858 | 0.5 |
| Broadwater | 19.6 | D | 8.6 | 13.2 | D | 5.8 | 6.4 | D | 14.8 | 1,590 | 256 | 63 | 2,739 | 1.6 |
| Carbon | 13.3 | 2.9 | 11.8 | D | D | D | 5.0 | 8.1 | 18.1 | 2,870 | 268 | 108 | 6,599 | 2.5 |
| Carter | 28.7 | 0.1 | D | 0.1 | D | 6.1 | D | D | 24.0 | 320 | 257 | 0 | 836 | 3.2 |
| Cascade | 1.0 | 0.1 | 7.6 | 3.4 | 5.8 | 7.8 | 7.5 | 17.8 | 26.5 | 18,825 | 231 | 1,822 | 39,317 | 5.5 |
| Chouteau | 42.3 | D | 1.9 | 1.0 | D | 6.1 | 5.0 | 4.4 | 19.8 | 1,245 | 219 | 64 | 2,974 | 3.3 |
| Custer | 1.2 | 0.7 | 6.7 | 1.0 | D | 13.4 | 8.2 | D | 21.7 | 2,630 | 231 | 226 | 5,703 | 2.6 |
| Daniels | 9.4 | D | D | D | D | D | 7.1 | 14.0 | 14.7 | 435 | 264 | 11 | 1,131 | 1.8 |
| Dawson | 0.4 | 6.0 | 2.6 | 1.0 | 4.9 | 8.2 | 4.3 | D | 17.8 | 1,780 | 206 | 97 | 4,458 | 5.3 |
| Deer Lodge | 0.7 | 0.1 | 6.4 | 2.8 | 6.4 | 5.4 | 2.8 | 30.4 | 29.5 | 2,805 | 305 | 259 | 5,248 | 2.5 |
| Fallon | 0.5 | 28.7 | 11.2 | 0.5 | 2.7 | 5.1 | 2.9 | 6.4 | 15.7 | 565 | 196 | 0 | 1,774 | 20.7 |
| Fergus | -1.4 | D | 22.8 | 7.7 | 3.6 | 6.5 | 6.7 | 13.2 | 18.5 | 3,100 | 281 | 184 | 5,908 | 1.2 |
| Flathead | 0.2 | 0.7 | 10.2 | 6.5 | 7.9 | 8.9 | 9.1 | 18.9 | 11.8 | 25,100 | 242 | 1,289 | 50,092 | 6.7 |
| Gallatin | 1.2 | 0.5 | 13.9 | 5.2 | 13.9 | 12.3 | 7.3 | 9.2 | 14.1 | 16,200 | 142 | 528 | 54,240 | 28.3 |
| Garfield | 11.2 | 0.0 | 5.5 | 0.2 | D | D | D | 2.6 | 40.5 | 295 | 232 | 12 | 863 | 2.3 |
| Glacier | 11.3 | 2.3 | 1.8 | 0.3 | 1.6 | 6.5 | 1.1 | D | 53.6 | 2,210 | 161 | 603 | 5,399 | 1.0 |
| Golden Valley | 21.0 | D | 6.1 | D | D | D | 2.7 | D | 20.6 | 330 | 399 | 17 | 487 | 2.1 |
| Granite | 3.9 | 0.6 | 8.4 | 4.6 | D | 5.3 | D | 0.9 | 24.3 | 1,020 | 306 | 42 | 2,836 | 0.6 |
| Hill | 8.1 | 0.1 | 4.7 | 0.5 | 5.9 | 7.4 | 4.6 | D | 28.4 | 2,795 | 169 | 447 | 7,349 | 1.4 |
| Jefferson | 1.6 | 8.5 | 14.6 | 6.8 | D | 3.7 | 5.0 | 8.9 | 26.4 | 3,345 | 274 | 140 | 5,121 | 1.4 |
| Judith Basin | 38.9 | D | 14.5 | 0.0 | 3.2 | 3.6 | D | D | 17.5 | 565 | 281 | 22 | 1,357 | 1.6 |
| Lake | 1.6 | 0.7 | 8.7 | 5.1 | 6.8 | 8.1 | 4.0 | 14.3 | 34.6 | 7,780 | 254 | 698 | 16,971 | 2.3 |
| Lewis and Clark | 0.3 | 0.7 | 5.8 | 2.8 | 10.6 | 6.4 | 8.4 | 13.8 | 34.2 | 16,210 | 233 | 1,019 | 32,127 | 6.5 |
| Liberty | 54.4 | D | D | D | 1.1 | 2.8 | 1.7 | D | 10.8 | 435 | 182 | 46 | 1,048 | 0.5 |
| Lincoln | -0.6 | 0.6 | 12.6 | 2.6 | 3.7 | 9.5 | 3.1 | 17.7 | 27.5 | 7,450 | 371 | 569 | 11,612 | 1.8 |
| McCone | -4.6 | 0.0 | 4.8 | 0.1 | D | 4.9 | D | D | 22.3 | 375 | 227 | 0 | 1,026 | 1.7 |
| Madison | 7.0 | 5.4 | 9.4 | 1.8 | 4.0 | 4.0 | 3.2 | D | 9.4 | 2,500 | 286 | 73 | 7,012 | 1.1 |
| Meagher | 11.4 | D | D | D | 3.0 | 5.9 | D | D | 18.5 | 540 | 293 | 44 | 1,448 | 1.1 |
| Mineral | -1.4 | 0.5 | 11.5 | D | D | 11.8 | 1.3 | D | 27.4 | 1,595 | 361 | 136 | 2,510 | 2.5 |
| Missoula | 0.0 | 0.1 | 7.7 | 3.3 | 11.1 | 9.1 | 7.4 | 17.9 | 17.3 | 22,940 | 191 | 1,968 | 56,208 | 12.2 |
| Musselshell | 5.3 | 40.9 | 7.2 | 0.4 | D | 4.5 | D | 8.3 | 13.5 | 1,510 | 326 | 108 | 2,701 | 1.8 |
| Park | 1.4 | D | 11.5 | 8.0 | 6.7 | 7.2 | 5.9 | 13.1 | 11.8 | 4,000 | 241 | 227 | 9,749 | 4.0 |
| Petroleum | 34.8 | D | 2.3 | 0.1 | D | D | 1.2 | 1.5 | 28.4 | 115 | 231 | 0 | 332 | 3.1 |
| Phillips | 8.3 | D | 8.2 | 1.6 | D | 9.0 | 6.3 | 11.4 | 28.4 | 1,055 | 267 | 80 | 2,357 | 1.0 |
| Pondera | 22.3 | 1.4 | 7.1 | 1.8 | 3.4 | 13.1 | D | 10.8 | 15.7 | 1,320 | 225 | 197 | 2,680 | 0.8 |
| Powder River | 11.3 | D | D | D | D | 9.8 | D | D | 32.8 | 395 | 234 | 0 | 1,035 | 1.4 |
| Powell | 6.6 | 0.1 | D | D | 2.4 | 4.1 | 1.8 | 9.7 | 44.7 | 1,665 | 243 | 121 | 3,343 | 7.6 |
| Prairie | 20.4 | 0.6 | D | D | 5.3 | D | D | D | 48.7 | 355 | 334 | 16 | 681 | 1.2 |
| Ravalli | 0.9 | 0.4 | 12.6 | 5.1 | 9.8 | 7.9 | 5.9 | 14.0 | 17.7 | 13,740 | 311 | 731 | 19,876 | 1.5 |
| Richland | 6.5 | 20.3 | 8.8 | 5.0 | 3.1 | 6.0 | 5.8 | 10.2 | 9.9 | 1,965 | 180 | 93 | 5,297 | 16.4 |
| Roosevelt | -2.7 | 2.9 | 3.6 | 0.6 | D | 8.1 | 2.5 | D | 50.1 | 1,820 | 165 | 377 | 4,143 | 2.0 |
| Rosebud | 5.8 | 19.0 | 7.4 | 0.2 | D | 2.5 | 1.3 | D | 32.6 | 1,875 | 209 | 218 | 4,155 | 2.4 |
| Sanders | 1.3 | 2.7 | 11.1 | 5.7 | 3.7 | 7.3 | 4.1 | 14.6 | 21.9 | 4,515 | 378 | 299 | 6,768 | 1.3 |
| Sheridan | -3.2 | 4.0 | D | D | 4.2 | 11.8 | D | D | 25.3 | 955 | 288 | 37 | 2,158 | 3.3 |
| Silver Bow | 0.0 | D | 3.9 | 4.8 | 5.8 | 10.5 | 5.6 | 15.4 | 15.4 | 8,685 | 249 | 959 | 17,527 | 4.8 |
| Stillwater | 4.4 | D | 4.0 | 4.5 | D | 2.9 | 1.4 | D | 8.3 | 2,530 | 262 | 70 | 4,874 | 1.7 |
| Sweet Grass | 1.3 | D | 8.6 | 3.1 | 2.9 | 5.8 | 4.2 | 5.0 | 9.7 | 1,015 | 273 | 18 | 2,168 | 0.9 |
| Teton | 19.1 | D | 10.8 | 0.6 | D | 7.4 | 5.2 | 5.9 | 16.3 | 1,575 | 256 | 113 | 2,921 | 1.0 |
| Toole | 16.1 | 3.4 | D | D | D | 3.0 | 3.4 | 4.5 | 22.7 | 920 | 195 | 114 | 2,384 | 2.1 |
| Treasure | 64.8 | 0.0 | D | 0.0 | D | D | D | D | 10.0 | 210 | 301 | 0 | 431 | 2.4 |
| Valley | 0.6 | D | 4.7 | 0.7 | 5.3 | 5.7 | 6.0 | D | 24.1 | 1,795 | 243 | 129 | 4,906 | 0.6 |
| Wheatland | 28.6 | 0.3 | 6.6 | 5.8 | D | 6.1 | 3.8 | D | 18.5 | 550 | 262 | 63 | 1,209 | 1.1 |
| Wibaux | -9.9 | D | D | D | D | 6.9 | 4.5 | 2.6 | 40.6 | 255 | 267 | 0 | 581 | 8.0 |
| Yellowstone | 0.4 | 3.3 | 8.4 | 6.5 | 9.0 | 7.3 | 8.4 | 17.8 | 11.6 | 32,770 | 203 | 2,523 | 75,008 | 17.3 |
| NEBRASKA | 3.7 | 0.1 | 5.9 | 9.3 | 8.2 | 5.3 | 9.5 | 11.4 | 16.1 | 352,880 | 183 | 28,692 | 857,974 | 7.7 |
| Adams | 5.2 | 0.0 | 7.2 | 16.5 | 3.9 | 5.2 | 6.5 | D | 14.5 | 6,965 | 222 | 536 | 14,114 | 5.7 |
| Antelope | 24.3 | D | 11.6 | 3.7 | D | 3.7 | 7.3 | 8.8 | 13.8 | 1,560 | 247 | 84 | 3,351 | 2.0 |
| Arthur | 41.8 | 0.0 | D | 1.1 | D | D | D | D | 16.9 | 115 | 246 | 0 | 253 | -0.4 |
| Banner | 46.5 | 0.5 | D | 2.1 | 1.0 | 3.3 | D | 0.0 | 13.0 | 235 | 310 | 0 | 368 | -0.8 |

1. Per 1,000 resident population estimated as of July 1 of the year shown.

# Table B. States and Counties — Housing, Labor Force, and Employment

| STATE County | Total | Percent | Median value[1] | With a mortgage | Without a mortgage[2] | Median rent[3] | Median rent as a percent of income[2] | Sub-standard units[4] (percent) | Total | Percent change, 2019-2020 | Total | Rate[5] | Total | Management, business, science, and arts | Construction, production, and maintenance occupations |
|---|---|---|---|---|---|---|---|---|---|---|---|---|---|---|---|
| | 89 | 90 | 91 | 92 | 93 | 94 | 95 | 96 | 97 | 98 | 99 | 100 | 101 | 102 | 103 |
| MISSOURI— Cont'd | | | | | | | | | | | | | | | |
| St. Louis city | 141,952 | 43.7 | 138,700 | 19.8 | 12.5 | 828 | 29.7 | 2.2 | 152,292 | 0.0 | 12,958 | 8.5 | 155,835 | 43.0 | 16.3 |
| MONTANA | 427,871 | 68.1 | 230,600 | 22.1 | 10.6 | 810 | 27.7 | 2.7 | 539,883 | 0.5 | 31,788 | 5.9 | 512,329 | 37.3 | 22.9 |
| Beaverhead | 3,992 | 67.8 | 215,400 | 19.3 | 10.2 | 620 | 37.4 | 1.9 | 5,050 | -1.6 | 203 | 4.0 | 4,683 | 35.9 | 26.4 |
| Big Horn | 3,609 | 63.4 | 126,100 | 17.3 | 10.0 | 586 | 20.9 | 15.8 | 4,874 | -0.6 | 357 | 7.3 | 4,785 | 34.4 | 21.4 |
| Blaine | 2,366 | 58.0 | 94,300 | 22.4 | 11.3 | 466 | 18.2 | 6.5 | 2,278 | -0.7 | 109 | 4.8 | 2,502 | 41.5 | 21.9 |
| Broadwater | 2,364 | 82.5 | 222,200 | 23.8 | 10.0 | 752 | 22.9 | 0.2 | 2,605 | -0.4 | 163 | 6.3 | 2,595 | 37.2 | 29.9 |
| Carbon | 4,524 | 76.4 | 247,600 | 23.2 | 11.3 | 791 | 27.9 | 2.0 | 5,567 | 1.5 | 334 | 6.0 | 5,235 | 39.2 | 24.5 |
| Carter | 628 | 72.0 | 110,900 | 23.1 | 10.0 | 578 | 26.3 | 2.7 | 649 | -1.1 | 19 | 2.9 | 641 | 50.7 | 22.3 |
| Cascade | 34,329 | 64.4 | 179,900 | 21.2 | 11.1 | 770 | 28.7 | 2.2 | 37,797 | -0.8 | 2,236 | 5.9 | 36,897 | 35.4 | 21.8 |
| Chouteau | 2,274 | 63.8 | 143,200 | 21.2 | 11.4 | 384 | 18.1 | 2.6 | 2,454 | -0.2 | 95 | 3.9 | 2,569 | 39.3 | 26.3 |
| Custer | 4,903 | 68.8 | 170,400 | 19.8 | 10.0 | 758 | 25.5 | 1.5 | 6,230 | 2.6 | 300 | 4.8 | 6,170 | 33.4 | 21.5 |
| Daniels | 857 | 78.4 | 126,500 | 17.3 | 11.6 | 510 | 18.8 | 1.6 | 898 | 1.5 | 25 | 2.8 | 870 | 46.8 | 19.9 |
| Dawson | 3,930 | 68.3 | 160,600 | 19.4 | 10.0 | 784 | 24.3 | 2.5 | 4,668 | 1.1 | 260 | 5.6 | 4,707 | 32.8 | 30.9 |
| Deer Lodge | 3,903 | 67.8 | 131,300 | 19.7 | 10.0 | 557 | 26.0 | 1.7 | 5,035 | -0.9 | 256 | 5.1 | 3,705 | 33.0 | 20.8 |
| Fallon | 1,334 | 72.0 | 169,800 | 17.4 | 10.0 | 774 | 20.0 | 2.5 | 1,648 | -3.3 | 66 | 4.0 | 1,574 | 30.3 | 39.3 |
| Fergus | 4,912 | 72.8 | 143,000 | 21.6 | 10.2 | 759 | 24.2 | 2.2 | 5,844 | -3.2 | 266 | 4.6 | 5,476 | 34.6 | 23.6 |
| Flathead | 38,773 | 74.1 | 273,600 | 23.6 | 11.2 | 855 | 28.3 | 3.8 | 49,357 | 2.0 | 3,858 | 7.8 | 48,363 | 36.5 | 21.5 |
| Gallatin | 43,777 | 61.2 | 357,100 | 22.4 | 10.0 | 1,074 | 30.0 | 1.8 | 70,434 | 1.5 | 3,536 | 5.0 | 62,357 | 42.0 | 21.1 |
| Garfield | 438 | 73.1 | 124,100 | 16.2 | 10.0 | 570 | 19.5 | 5.5 | 741 | -0.3 | 20 | 2.7 | 550 | 47.8 | 24.0 |
| Glacier | 4,163 | 58.8 | 107,700 | 19.4 | 10.0 | 499 | 22.9 | 5.3 | 5,443 | -5.2 | 536 | 9.8 | 5,025 | 36.9 | 17.7 |
| Golden Valley | 351 | 76.9 | 106,100 | 27.5 | 10.0 | 818 | 16.5 | 2.3 | 363 | -0.5 | 17 | 4.7 | 365 | 41.4 | 30.1 |
| Granite | 1,308 | 79.1 | 224,000 | 25.2 | 10.0 | 515 | 18.4 | 1.8 | 1,606 | -6.0 | 131 | 8.2 | 1,208 | 35.9 | 24.5 |
| Hill | 6,369 | 61.8 | 143,600 | 21.0 | 10.0 | 615 | 27.8 | 1.8 | 7,608 | -1.2 | 379 | 5.0 | 7,119 | 37.0 | 27.2 |
| Jefferson | 4,484 | 84.3 | 282,700 | 21.4 | 10.0 | 788 | 23.2 | 2.5 | 5,789 | 0.7 | 297 | 5.1 | 5,424 | 44.8 | 21.9 |
| Judith Basin | 906 | 75.7 | 138,200 | 28.2 | 10.0 | 585 | 15.6 | 0.8 | 987 | 0.3 | 48 | 4.9 | 914 | 40.0 | 26.8 |
| Lake | 11,802 | 72.0 | 253,100 | 27.4 | 10.8 | 705 | 24.4 | 2.4 | 13,659 | 2.5 | 844 | 6.2 | 12,444 | 35.0 | 22.4 |
| Lewis and Clark | 27,893 | 69.4 | 243,000 | 20.0 | 10.0 | 833 | 26.4 | 1.4 | 36,479 | 0.9 | 1,813 | 5.0 | 34,814 | 45.5 | 17.4 |
| Liberty | 895 | 63.0 | 101,100 | 22.5 | 10.0 | 571 | 27.9 | 0.2 | 969 | 2.2 | 27 | 2.8 | 1,154 | 25.9 | 27.0 |
| Lincoln | 8,122 | 80.3 | 198,200 | 26.8 | 12.0 | 734 | 28.4 | 2.4 | 8,097 | -0.6 | 819 | 10.1 | 6,835 | 29.0 | 31.8 |
| McCone | 723 | 80.8 | 114,800 | 26.8 | 13.0 | 535 | 16.4 | 5.0 | 961 | 0.8 | 28 | 2.9 | 778 | 50.1 | 26.3 |
| Madison | 3,570 | 77.0 | 276,500 | 25.4 | 11.4 | 834 | 25.2 | 2.9 | 4,551 | -5.0 | 232 | 5.1 | 4,098 | 32.1 | 29.1 |
| Meagher | 702 | 72.5 | 154,100 | 25.1 | 15.6 | 647 | 24.1 | 0.0 | 966 | 0.5 | 45 | 4.7 | 947 | 37.5 | 28.8 |
| Mineral | 1,786 | 74.0 | 169,200 | 24.6 | 10.8 | 565 | 20.9 | 2.2 | 1,778 | 0.2 | 181 | 10.2 | 1,565 | 22.0 | 28.4 |
| Missoula | 49,313 | 58.9 | 284,000 | 22.8 | 12.0 | 863 | 29.7 | 2.9 | 64,123 | -0.2 | 4,017 | 6.3 | 65,170 | 38.9 | 18.7 |
| Musselshell | 2,181 | 72.0 | 172,700 | 24.8 | 12.8 | 756 | 29.4 | 3.9 | 2,328 | 2.0 | 143 | 6.1 | 1,841 | 28.1 | 33.6 |
| Park | 7,782 | 68.5 | 271,300 | 23.9 | 11.4 | 789 | 26.6 | 1.6 | 8,938 | -0.5 | 665 | 7.4 | 8,576 | 35.0 | 26.8 |
| Petroleum | 204 | 74.5 | 181,300 | 25.5 | 12.3 | 490 | 31.3 | 1.5 | 264 | -1.5 | 9 | 3.4 | 223 | 55.6 | 23.8 |
| Phillips | 1,710 | 77.1 | 129,200 | 22.3 | 10.0 | 528 | 20.3 | 1.9 | 1,891 | 0.4 | 93 | 4.9 | 1,790 | 29.6 | 33.1 |
| Pondera | 2,103 | 74.6 | 131,900 | 19.4 | 10.7 | 652 | 26.6 | 2.4 | 2,685 | 0.0 | 132 | 4.9 | 2,563 | 37.8 | 23.8 |
| Powder River | 737 | 74.6 | 142,600 | 18.0 | 10.1 | 657 | 19.0 | 0.0 | 976 | 1.9 | 30 | 3.1 | 877 | 29.5 | 37.2 |
| Powell | 2,426 | 67.1 | 143,100 | 20.2 | 10.5 | 688 | 24.2 | 5.0 | 2,956 | -0.4 | 137 | 4.6 | 2,577 | 42.9 | 24.8 |
| Prairie | 518 | 84.0 | 112,900 | 15.7 | 11.2 | 715 | 36.3 | 0.0 | 471 | -1.1 | 32 | 6.8 | 569 | 44.3 | 23.7 |
| Ravalli | 17,354 | 76.3 | 279,300 | 25.5 | 10.5 | 770 | 29.2 | 2.8 | 20,684 | 1.4 | 1,191 | 5.8 | 18,224 | 36.0 | 26.3 |
| Richland | 4,535 | 66.0 | 230,100 | 16.6 | 10.0 | 826 | 21.0 | 4.0 | 5,938 | 0.7 | 382 | 6.4 | 6,033 | 31.4 | 35.3 |
| Roosevelt | 3,150 | 62.9 | 112,400 | 15.7 | 10.0 | 438 | 18.1 | 5.7 | 4,451 | -1.7 | 266 | 6.0 | 3,865 | 33.0 | 23.8 |
| Rosebud | 3,166 | 69.2 | 114,900 | 17.7 | 10.0 | 605 | 15.0 | 6.1 | 3,742 | -2.6 | 212 | 5.7 | 3,922 | 30.6 | 34.7 |
| Sanders | 5,036 | 77.6 | 228,200 | 26.1 | 11.5 | 627 | 36.1 | 2.8 | 5,086 | 3.5 | 395 | 7.8 | 4,300 | 30.4 | 29.8 |
| Sheridan | 1,639 | 78.8 | 127,500 | 19.5 | 11.8 | 727 | 23.5 | 2.4 | 1,769 | 1.5 | 85 | 4.8 | 1,783 | 37.5 | 28.1 |
| Silver Bow | 14,960 | 68.9 | 149,800 | 20.8 | 10.5 | 638 | 29.9 | 1.8 | 17,295 | 0.0 | 1,142 | 6.6 | 16,043 | 35.4 | 22.8 |
| Stillwater | 3,761 | 80.7 | 255,500 | 21.0 | 10.2 | 791 | 20.5 | 1.3 | 5,332 | 1.7 | 270 | 5.1 | 4,493 | 29.2 | 36.3 |
| Sweet Grass | 1,566 | 72.3 | 255,100 | 18.9 | 13.0 | 708 | 18.6 | 0.6 | 1,861 | 2.5 | 67 | 3.6 | 1,687 | 32.7 | 32.4 |
| Teton | 2,464 | 69.9 | 172,700 | 23.2 | 10.6 | 777 | 27.9 | 4.0 | 2,765 | 0.2 | 121 | 4.4 | 2,640 | 38.4 | 27.3 |
| Toole | 1,867 | 59.4 | 137,900 | 17.9 | 10.0 | 547 | 22.8 | 1.9 | 2,141 | 0.8 | 107 | 5.0 | 2,189 | 33.5 | 24.6 |
| Treasure | 351 | 68.1 | 118,200 | 36.6 | 12.6 | 645 | 20.0 | 0.6 | 338 | -1.2 | 12 | 3.6 | 339 | 38.9 | 30.7 |
| Valley | 3,386 | 74.9 | 151,400 | 18.6 | 11.7 | 581 | 23.0 | 0.9 | 4,056 | -1.5 | 176 | 4.3 | 3,608 | 33.0 | 25.7 |
| Wheatland | 792 | 74.7 | 95,400 | 21.7 | 10.5 | 607 | 14.2 | 4.0 | 754 | -1.6 | 55 | 7.3 | 945 | 34.0 | 32.9 |
| Wibaux | 498 | 79.7 | 113,500 | 18.5 | 10.0 | 817 | 15.7 | 0.8 | 454 | 0.0 | 23 | 5.1 | 557 | 43.8 | 22.6 |
| Yellowstone | 66,385 | 68.4 | 226,000 | 21.6 | 10.9 | 892 | 28.0 | 2.8 | 83,203 | 1.1 | 4,527 | 5.4 | 81,146 | 35.4 | 23.7 |
| NEBRASKA | 759,176 | 66.1 | 155,800 | 19.4 | 11.4 | 833 | 26.4 | 2.2 | 1,035,175 | -0.5 | 43,787 | 4.2 | 999,212 | 37.8 | 24.4 |
| Adams | 12,712 | 66.5 | 128,500 | 18.5 | 12.0 | 722 | 27.3 | 2.3 | 16,839 | 0.0 | 670 | 4.0 | 15,862 | 32.7 | 29.6 |
| Antelope | 2,714 | 75.5 | 82,000 | 17.6 | 10.4 | 603 | 22.7 | 1.4 | 3,909 | 8.5 | 88 | 2.3 | 3,217 | 35.3 | 28.4 |
| Arthur | 197 | 66.5 | 111,300 | 30.4 | 11.4 | 600 | 28.3 | 2.5 | 231 | 1.3 | 12 | 5.2 | 199 | 45.7 | 30.7 |
| Banner | 283 | 69.6 | 139,900 | 25.6 | 10.0 | 444 | 33.3 | 2.1 | 380 | -0.3 | 10 | 2.6 | 397 | 58.4 | 26.2 |

1. Specified owner-occupied units. lacking complete plumbing facilities.  2. A value of 10.0 represents 10 percent or less; a value of 50.0 represents 50 percent or more.  3. Specified renter-occupied units.  4. Overcrowded or  5. Percent of civilian labor force.  6. Civilian employed persons 16 years old and over.

Items 89—103

| STATE County | Number of establishments | Total | Health care and social assistance | Manufacturing | Retail trade | Finance and insurance | Professional, scientific, and technical services | Total (mil dol) | Average per employee (dollars) | Number | Fewer than 50 acres | 1000 acres or more | Farm producers whose primary occupation is farming (percent) |
|---|---|---|---|---|---|---|---|---|---|---|---|---|---|
| | | | | | Employment | | | Annual payroll | | Farms | Percent with: | | |
| | 104 | 105 | 106 | 107 | 108 | 109 | 110 | 111 | 112 | 113 | 114 | 115 | 116 |
| **MISSOURI— Cont'd** | | | | | | | | | | | | | |
| St. Louis city | 9,861 | 204,516 | 37,471 | 16,994 | 9,848 | 10,752 | 18,130 | 11,629 | 56,861 | NA | NA | NA | NA |
| **MONTANA** | 38,959 | 375,176 | 72,131 | 20,393 | 58,214 | 16,555 | 19,247 | 16,051 | 42,784 | 27,048 | 30.9 | 31.7 | 48.5 |
| Beaverhead | 387 | 2,643 | 539 | 79 | 533 | 109 | 128 | 85 | 32,229 | 494 | 40.9 | 30.0 | 54.7 |
| Big Horn | 193 | 2,399 | 623 | NA | 352 | 58 | 38 | 131 | 54,705 | 353 | 19.0 | 46.2 | 55.8 |
| Blaine | 126 | 1,016 | 250 | 24 | 191 | 40 | 36 | 46 | 44,939 | 491 | 7.5 | 50.7 | 62.0 |
| Broadwater | 170 | 860 | 121 | 140 | 159 | 26 | 36 | 29 | 34,016 | 296 | 25.3 | 29.7 | 50.9 |
| Carbon | 414 | 2,179 | 338 | 53 | 232 | 63 | 112 | 66 | 30,189 | 725 | 28.0 | 23.7 | 52.8 |
| Carter | 31 | 125 | NA | NA | 19 | NA | 4 | 5 | 38,320 | 323 | 4.3 | 71.8 | 66.0 |
| Cascade | 2,447 | 30,486 | 7,342 | 1,249 | 4,806 | 1,460 | 1,125 | 1,253 | 41,107 | 1,027 | 37.7 | 22.6 | 42.1 |
| Chouteau | 149 | 729 | 223 | 24 | 121 | 55 | 11 | 23 | 31,985 | 633 | 4.3 | 61.5 | 66.0 |
| Custer | 426 | 3,806 | 796 | 91 | 784 | 290 | 97 | 147 | 38,683 | 441 | 24.3 | 39.0 | 48.1 |
| Daniels | 72 | 536 | 105 | NA | 86 | 37 | 9 | 24 | 45,336 | 277 | 4.0 | 54.5 | 50.7 |
| Dawson | 307 | 2,590 | 653 | 43 | 425 | 93 | 65 | 129 | 49,885 | 487 | 12.3 | 41.9 | 48.7 |
| Deer Lodge | 256 | 2,709 | 1,369 | 106 | 287 | 69 | 84 | 126 | 46,506 | 77 | 15.6 | 22.1 | 32.6 |
| Fallon | 133 | 997 | 106 | NA | 147 | 35 | 13 | 58 | 58,173 | 289 | 13.5 | 55.4 | 60.7 |
| Fergus | 426 | 3,408 | 757 | 414 | 481 | 140 | 94 | 127 | 37,169 | 845 | 18.6 | 48.3 | 56.3 |
| Flathead | 4,326 | 39,692 | 6,980 | 2,786 | 6,534 | 1,897 | 1,738 | 1,669 | 42,036 | 1,146 | 64.7 | 4.3 | 34.4 |
| Gallatin | 6,009 | 50,667 | 6,114 | 3,487 | 8,265 | 1,446 | 3,249 | 2,220 | 43,813 | 1,123 | 57.3 | 10.1 | 36.5 |
| Garfield | 25 | 130 | 8 | NA | 62 | NA | NA | 4 | 31,792 | 260 | 3.1 | 72.7 | 72.7 |
| Glacier | 219 | 1,805 | 472 | 2 | 434 | 51 | 35 | 84 | 46,802 | 637 | 16.0 | 31.1 | 45.1 |
| Golden Valley | 15 | 61 | NA | NA | NA | NA | NA | 1 | 23,049 | 157 | 10.8 | 52.2 | 58.2 |
| Granite | 109 | 651 | 153 | 13 | 105 | 11 | 12 | 22 | 33,069 | 151 | 17.2 | 41.7 | 50.8 |
| Hill | 518 | 4,891 | 1,338 | 76 | 950 | 221 | 141 | 175 | 35,711 | 698 | 9.3 | 51.0 | 50.3 |
| Jefferson | 291 | 1,810 | 272 | 246 | 169 | 41 | 53 | 71 | 39,490 | 370 | 37.0 | 16.8 | 32.1 |
| Judith Basin | 60 | 212 | 8 | NA | 41 | 26 | 16 | 8 | 36,481 | 357 | 13.7 | 53.5 | 59.6 |
| Lake | 825 | 6,036 | 1,247 | 261 | 1,125 | 219 | 375 | 231 | 38,339 | 1,170 | 55.7 | 6.6 | 43.6 |
| Lewis and Clark | 2,395 | 25,671 | 6,008 | 684 | 4,001 | 1,568 | 1,774 | 1,098 | 42,791 | 707 | 66.1 | 11.2 | 32.0 |
| Liberty | 66 | 339 | 121 | NA | 39 | 15 | 9 | 10 | 28,829 | 246 | 2.0 | 75.6 | 70.4 |
| Lincoln | 608 | 3,936 | 1,035 | 220 | 716 | 126 | 89 | 123 | 31,371 | 345 | 53.0 | 2.3 | 39.7 |
| McCone | 52 | 335 | 56 | NA | 66 | 34 | NA | 13 | 39,785 | 437 | 4.1 | 59.0 | 60.1 |
| Madison | 391 | 1,705 | 199 | 122 | 238 | 62 | 69 | 68 | 39,689 | 605 | 29.4 | 28.4 | 49.7 |
| Meagher | 75 | 300 | 72 | 15 | 62 | NA | 4 | 8 | 26,813 | 145 | 21.4 | 60.7 | 63.5 |
| Mineral | 124 | 785 | 133 | 148 | 240 | 9 | 13 | 25 | 31,455 | 93 | 44.1 | 4.3 | 19.9 |
| Missoula | 4,577 | 51,454 | 9,925 | 2,345 | 8,194 | 2,219 | 3,776 | 2,033 | 39,518 | 576 | 66.1 | 3.6 | 32.9 |
| Musselshell | 109 | 853 | 176 | 19 | 108 | 22 | 21 | 46 | 54,064 | 346 | 14.5 | 34.4 | 54.1 |
| Park | 852 | 5,135 | 858 | 533 | 754 | 164 | 193 | 198 | 38,463 | 575 | 39.8 | 22.6 | 46.2 |
| Petroleum | 8 | 18 | NA | NA | NA | NA | NA | 0 | 13,556 | 104 | 12.5 | 64.4 | 68.4 |
| Phillips | 136 | 857 | 238 | 3 | 191 | 46 | 31 | 26 | 30,426 | 445 | 11.5 | 54.2 | 58.8 |
| Pondera | 179 | 1,210 | 250 | 98 | 229 | 64 | 31 | 40 | 33,177 | 486 | 14.0 | 39.9 | 54.3 |
| Powder River | 70 | 288 | NA | NA | 83 | NA | 14 | 9 | 32,736 | 325 | 8.9 | 64.3 | 70.5 |
| Powell | 169 | 1,202 | 250 | 203 | 155 | 33 | 42 | 43 | 35,674 | 254 | 24.4 | 28.7 | 43.7 |
| Prairie | 38 | 154 | NA | NA | 26 | 16 | 4 | 4 | 29,013 | 179 | 4.5 | 65.9 | 67.0 |
| Ravalli | 1,525 | 9,558 | 1,825 | 921 | 1,639 | 355 | 309 | 338 | 35,358 | 1,576 | 74.6 | 3.0 | 39.1 |
| Richland | 505 | 4,703 | 648 | 441 | 583 | 131 | 131 | 255 | 54,188 | 527 | 13.5 | 50.7 | 53.2 |
| Roosevelt | 212 | 1,702 | 393 | 10 | 400 | 88 | 23 | 62 | 36,360 | 501 | 7.4 | 51.9 | 51.7 |
| Rosebud | 176 | 2,311 | 297 | 14 | 297 | 65 | 13 | 121 | 52,266 | 414 | 15.0 | 44.4 | 53.6 |
| Sanders | 373 | 2,271 | 583 | 232 | 338 | 52 | 52 | 71 | 31,258 | 521 | 37.2 | 9.6 | 44.3 |
| Sheridan | 155 | 827 | 231 | NA | 152 | 61 | 24 | 28 | 33,855 | 458 | 4.6 | 55.9 | 67.1 |
| Silver Bow | 1,143 | 13,014 | 3,101 | 749 | 2,007 | 291 | 494 | 531 | 40,789 | 142 | 45.8 | 11.3 | 24.9 |
| Stillwater | 253 | 2,674 | 198 | 536 | 269 | 48 | 41 | 199 | 74,474 | 562 | 22.8 | 28.8 | 42.0 |
| Sweet Grass | 175 | 1,221 | 109 | 64 | 146 | 38 | 23 | 73 | 59,878 | 301 | 19.3 | 36.9 | 45.6 |
| Teton | 200 | 1,057 | 229 | 9 | 166 | 78 | 32 | 37 | 34,807 | 686 | 23.6 | 26.2 | 46.5 |
| Toole | 164 | 1,357 | 241 | 17 | 182 | 57 | 28 | 52 | 38,082 | 362 | 4.4 | 62.7 | 62.4 |
| Treasure | 14 | 59 | NA | NA | NA | NA | 7 | 3 | 47,644 | 121 | 16.5 | 52.1 | 62.3 |
| Valley | 262 | 2,131 | 627 | 17 | 312 | 105 | 52 | 78 | 36,498 | 557 | 9.3 | 48.5 | 56.1 |
| Wheatland | 59 | 448 | 113 | 51 | 73 | 21 | 14 | 13 | 28,728 | 174 | 11.5 | 51.1 | 60.1 |
| Wibaux | 26 | 124 | NA | NA | 21 | NA | 4 | 4 | 33,298 | 137 | 10.2 | 56.9 | 54.5 |
| Yellowstone | 5,649 | 70,762 | 14,240 | 3,794 | 10,186 | 3,904 | 3,653 | 3,367 | 47,576 | 1,314 | 45.9 | 14.5 | 35.7 |
| **NEBRASKA** | 54,939 | 856,242 | 137,507 | 99,810 | 109,771 | 72,624 | 39,855 | 39,433 | 46,053 | 46,332 | 23.8 | 23.8 | 51.6 |
| Adams | 968 | 12,964 | 2,689 | 2,838 | 1,756 | 395 | 248 | 507 | 39,109 | 545 | 25.1 | 25.1 | 61.7 |
| Antelope | 225 | 1,580 | 301 | 181 | 253 | 92 | 138 | 63 | 39,615 | 704 | 20.7 | 22.2 | 58.5 |
| Arthur | 14 | 65 | NA | NA | NA | NA | NA | 2 | 30,400 | 95 | 4.2 | 63.2 | 62.9 |
| Banner | 6 | 32 | NA | NA | NA | NA | NA | 1 | 40,438 | 239 | 6.3 | 41.4 | 48.0 |

| STATE County | Land in farms | | | | | Value of land and buildings (dollars) | | Value of machinery and equiopmnet, average per farm (dollars) | Value of products sold: | | | | Organic farms (number) | Farms with internet access (per-cent) | Government payments | |
|---|---|---|---|---|---|---|---|---|---|---|---|---|---|---|---|---|
| | Acreage (1,000) | Percent change, 2012-2017 | Acres | | | Average per farm | Average per acre | | Total (mil dol) | Average per farm (acres) | Percent from: | | | | Total ($1,000) | Percent of farms |
| | | | Average size of farm | Total irrigated (1,000) | Total cropland (1,000) | | | | | | Crops | Livestock and poultry products | | | | |
| | 117 | 118 | 119 | 120 | 121 | 122 | 123 | 124 | 125 | 126 | 127 | 128 | 129 | 130 | 131 | 132 |
| MISSOURI— Cont'd | | | | | | | | | | | | | | | | |
| St. Louis city | NA | NA | NA | NA | NA | NA | NA | NA | NA | NA | NA | NA | NA | NA | NA | NA |
| MONTANA | 58,123 | -2.7 | 2,149 | 2,061.2 | 16,406.3 | 1,968,381 | 916 | 164,524 | 3,520.6 | 130,162 | 45.0 | 55.0 | 221 | 81.4 | 284,244 | 38.9 |
| Beaverhead | 1,234 | -10.7 | 2,498 | 263.8 | 162.8 | 3,290,191 | 1,317 | 201,987 | 118.2 | 239,277 | 31.5 | 68.5 | 2 | 85.6 | 577 | 8.5 |
| Big Horn | 3,188 | 1.2 | 9,032 | 44.1 | 223.8 | 3,546,651 | 393 | 218,911 | 83.6 | 236,833 | 39.5 | 60.5 | NA | 76.2 | 2,444 | 36.5 |
| Blaine | 2,040 | -7.4 | 4,155 | 44.2 | 599 | 2,460,670 | 592 | 215,403 | 90 | 183,253 | 44.0 | 56.0 | 29 | 79.0 | 10,435 | 63.5 |
| Broadwater | 467 | -2.1 | 1,577 | 47.1 | 112.5 | 1,844,915 | 1,170 | 166,279 | 39.7 | 134,081 | 56.7 | 43.3 | 1 | 77.4 | 2,336 | 44.9 |
| Carbon | 816 | 3.1 | 1,125 | 98.8 | 135 | 1,669,647 | 1,484 | 154,004 | 99 | 136,585 | 33.4 | 66.6 | NA | 82.3 | 1,388 | 32.6 |
| Carter | 1,768 | -0.6 | 5,473 | 2.3 | 247.5 | 2,943,872 | 538 | 270,868 | 70.6 | 218,663 | 8.5 | 91.5 | 1 | 79.6 | 7,229 | 68.7 |
| Cascade | 1,270 | 1.2 | 1,237 | 35.7 | 420 | 1,499,361 | 1,212 | 134,128 | 107.3 | 104,453 | 47.7 | 52.3 | 9 | 80.2 | 7,780 | 42.7 |
| Chouteau | 2,071 | -0.1 | 3,271 | 14.2 | 1,309.7 | 3,053,370 | 933 | 353,122 | 170.7 | 269,733 | 84.1 | 15.9 | 14 | 84.4 | 21,813 | 83.3 |
| Custer | 2,089 | -4.6 | 4,737 | 37.2 | 118.5 | 2,539,839 | 536 | 170,666 | 76.6 | 173,744 | 15.2 | 84.8 | 1 | 84.1 | 3,114 | 37.6 |
| Daniels | 770 | 0.2 | 2,779 | 0.6 | 533.7 | 1,762,211 | 634 | 297,586 | 42.2 | 152,300 | 77.3 | 22.7 | 1 | 78.7 | 7,683 | 85.6 |
| Dawson | 1,133 | -9.9 | 2,326 | 21.2 | 368.1 | 1,554,263 | 668 | 191,646 | 58.2 | 119,485 | 44.0 | 56.0 | NA | 74.5 | 7,569 | 64.9 |
| Deer Lodge | 74 | 11.3 | 962 | 13.1 | 11.8 | 1,869,813 | 1,943 | 145,035 | 6.5 | 84,247 | 11.3 | 88.7 | NA | 68.8 | 169 | 14.3 |
| Fallon | 902 | -7.9 | 3,121 | 1.4 | 191.8 | 1,883,893 | 604 | 196,591 | 45.3 | 156,706 | 15.0 | 85.0 | NA | 79.6 | 3,853 | 59.9 |
| Fergus | 2,188 | 11.6 | 2,589 | 16.3 | 623.9 | 2,833,620 | 1,094 | 184,680 | 133.6 | 158,135 | 26.9 | 73.1 | 2 | 84.3 | 12,281 | 51.8 |
| Flathead | 182 | 7.1 | 159 | 22.1 | 92.5 | 1,014,016 | 6,389 | 64,906 | 35.9 | 31,286 | 76.6 | 23.4 | 10 | 84.7 | 1,823 | 10.0 |
| Gallatin | 700 | -0.3 | 624 | 81.3 | 208.5 | 1,889,690 | 3,030 | 117,382 | 112.1 | 99,825 | 61.6 | 38.4 | 11 | 88.4 | 3,106 | 15.9 |
| Garfield | 2,215 | 1.1 | 8,519 | 2.8 | 294.2 | 3,441,672 | 404 | 239,724 | 54.5 | 209,765 | 16.6 | 83.4 | 1 | 79.2 | 5,608 | 64.6 |
| Glacier | 1,186 | -24.5 | 1,862 | 27.4 | 493.8 | 2,171,691 | 1,167 | 151,623 | 106.5 | 167,248 | 54.8 | 45.2 | 7 | 70.3 | 5,948 | 34.5 |
| Golden Valley | 683 | -3.5 | 4,351 | 7.3 | 108 | 2,914,246 | 670 | 142,583 | 18.6 | 118,478 | 27.9 | 72.1 | NA | 69.4 | 3,049 | 65.0 |
| Granite | 286 | 0.1 | 1,892 | 31.9 | 28.9 | 3,060,313 | 1,618 | 143,069 | 17.9 | 118,530 | 16.5 | 83.5 | NA | 90.1 | 268 | 32.5 |
| Hill | 1,616 | 1.1 | 2,315 | 4.1 | 1,222.3 | 1,992,745 | 861 | 259,353 | 130.8 | 187,322 | 81.9 | 18.1 | 12 | 81.5 | 20,360 | 75.9 |
| Jefferson | 352 | -5.1 | 952 | 33.5 | 57.9 | 1,619,472 | 1,702 | 88,418 | 20.2 | 54,497 | 20.1 | 79.9 | NA | 89.7 | 634 | 11.6 |
| Judith Basin | 860 | -16.8 | 2,409 | 13.7 | 278.9 | 2,237,680 | 929 | 226,807 | 88.9 | 249,022 | 28.2 | 71.8 | 4 | 86.0 | 3,687 | 58.3 |
| Lake | 641 | 15.4 | 548 | 100.4 | 93.4 | 880,495 | 1,606 | 72,903 | 64.8 | 55,381 | 34.0 | 66.0 | 32 | 82.8 | 1,442 | 13.4 |
| Lewis and Clark | 801 | -5.0 | 1,132 | 48.2 | 78.7 | 1,479,324 | 1,306 | 78,842 | 43.2 | 61,085 | 31.1 | 68.9 | 3 | 86.3 | 1,081 | 11.7 |
| Liberty | 914 | 1.7 | 3,714 | 7 | 698.8 | 3,329,506 | 897 | 424,336 | 88.3 | 358,911 | 78.2 | 21.8 | 6 | 83.3 | 10,873 | 90.7 |
| Lincoln | 48 | 1.1 | 139 | 4.9 | 12.5 | 727,103 | 5,250 | 54,620 | 3.1 | 9,017 | 32.5 | 67.5 | NA | 78.8 | 47 | 1.2 |
| McCone | 1,340 | -2.4 | 3,065 | 7.3 | 592 | 1,619,356 | 528 | 237,370 | 61.3 | 140,222 | 49.0 | 51.0 | NA | 76.4 | 13,400 | 80.5 |
| Madison | 923 | -14.9 | 1,526 | 130.3 | 151.2 | 2,610,305 | 1,710 | 149,921 | 83.6 | 138,238 | 27.6 | 72.4 | NA | 81.3 | 1,458 | 15.7 |
| Meagher | 882 | 8.6 | 6,084 | 52.1 | 85.2 | 5,524,417 | 908 | 270,602 | 36.8 | 253,586 | 12.1 | 87.9 | NA | 71.7 | 887 | 26.2 |
| Mineral | 18 | 8.0 | 198 | 0.6 | 5.4 | 989,976 | 5,002 | 42,783 | 0.6 | 6,495 | 37.7 | 62.3 | 2 | 73.1 | 82 | 12.9 |
| Missoula | 260 | 5.3 | 452 | 15.5 | 21.6 | 1,262,721 | 2,796 | 42,356 | 9.8 | 17,099 | 57.3 | 42.7 | 5 | 82.8 | 417 | 6.1 |
| Musselshell | 1,103 | 8.4 | 3,189 | 12.9 | 145.1 | 2,038,138 | 639 | 140,313 | 37.3 | 107,760 | 20.8 | 79.2 | NA | 78.6 | 2,231 | 30.1 |
| Park | 712 | -8.0 | 1,238 | 62 | 110.7 | 3,230,902 | 2,609 | 98,425 | 33.5 | 58,287 | 28.1 | 71.9 | 1 | 83.5 | 935 | 15.3 |
| Petroleum | 593 | -14.1 | 5,698 | 10.9 | 84.8 | 3,863,919 | 678 | 194,528 | 17.8 | 170,779 | 14.8 | 81.6 | NA | 86.5 | 861 | 51.9 |
| Phillips | 1,937 | -6.3 | 4,352 | 31.4 | 508.8 | 2,306,077 | 530 | 224,289 | 74.6 | 167,719 | 34.0 | 66.0 | 17 | 83.1 | 9,471 | 66.1 |
| Pondera | 805 | -15.9 | 1,656 | 69.8 | 551.1 | 1,998,397 | 1,207 | 272,432 | 111.5 | 229,477 | 68.2 | 31.8 | 1 | 80.7 | 10,629 | 69.5 |
| Powder River | 1,627 | 2.4 | 5,005 | 16 | 145.7 | 2,755,267 | 551 | 223,710 | 62.1 | 191,058 | 4.9 | 95.1 | 1 | 79.4 | 2,797 | 48.3 |
| Powell | 572 | -2.9 | 2,253 | 51.7 | 58 | 2,917,797 | 1,295 | 135,021 | 31.8 | 125,370 | 20.5 | 79.5 | NA | 79.5 | 377 | 23.2 |
| Prairie | 747 | -2.8 | 4,175 | 14.6 | 129.8 | 3,229,966 | 774 | 170,306 | 43.3 | 241,899 | 24.9 | 75.1 | 1 | 78.2 | 4,610 | 77.7 |
| Ravalli | 241 | 2.7 | 153 | 71 | 52.1 | 877,618 | 5,734 | 54,698 | 42.7 | 27,070 | 29.8 | 70.2 | 8 | 84.5 | 245 | 3.9 |
| Richland | 1,270 | -1.8 | 2,410 | 53.5 | 479.5 | 2,112,650 | 877 | 242,969 | 100.1 | 189,949 | 55.3 | 44.7 | 8 | 79.5 | 10,227 | 66.6 |
| Roosevelt | 1,307 | 5.5 | 2,610 | 24.3 | 757.3 | 1,883,016 | 722 | 268,384 | 64.1 | 128,004 | 69.4 | 30.6 | 2 | 76.6 | 13,696 | 76.2 |
| Rosebud | 2,732 | -13.0 | 6,600 | 38.7 | 169.8 | 2,835,966 | 430 | 161,371 | 86.6 | 209,271 | 19.6 | 80.4 | NA | 78.7 | 3,759 | 34.1 |
| Sanders | 643 | 89.7 | 1,233 | 21.5 | 43.3 | 1,219,226 | 988 | 65,694 | 16.9 | 32,457 | 30.6 | 69.4 | 8 | 76.4 | 570 | 8.6 |
| Sheridan | 1,064 | 2.1 | 2,323 | 4.6 | 786.1 | 1,611,743 | 694 | 443,294 | 70.6 | 154,083 | 82.8 | 17.2 | 3 | 86.2 | 11,702 | 80.6 |
| Silver Bow | 60 | -13.5 | 425 | 2.4 | 3.7 | 956,028 | 2,251 | 52,803 | 2.7 | 19,218 | 20.7 | 79.3 | NA | 78.9 | NA | NA |
| Stillwater | 763 | -5.8 | 1,357 | 23.2 | 175.7 | 1,823,923 | 1,344 | 103,840 | 51.5 | 91,557 | 30.4 | 69.6 | NA | 79.0 | 4,175 | 40.0 |
| Sweet Grass | 826 | -3.5 | 2,745 | 38.8 | 57.2 | 3,117,975 | 1,136 | 121,775 | 25.7 | 85,375 | 12.2 | 87.8 | NA | 84.7 | 617 | 19.6 |
| Teton | 887 | -9.0 | 1,294 | 93.8 | 460.1 | 1,533,701 | 1,186 | 159,196 | 107.2 | 156,277 | 64.1 | 35.9 | 8 | 82.4 | 8,367 | 57.0 |
| Toole | 1,095 | -3.0 | 3,025 | 5.3 | 734.5 | 2,653,458 | 877 | 280,722 | 84.2 | 232,547 | 79.5 | 20.5 | 4 | 77.1 | 11,549 | 82.9 |
| Treasure | 614 | -0.6 | 5,076 | 36.6 | 55.8 | 2,934,450 | 578 | 340,275 | 45 | 372,116 | 60.1 | 39.9 | NA | 77.7 | 1,197 | 47.1 |
| Valley | 1,630 | -0.3 | 2,926 | 45.1 | 779.2 | 1,883,822 | 644 | 274,232 | 96.6 | 173,345 | 55.2 | 44.8 | 3 | 74.0 | 13,469 | 74.9 |
| Wheatland | 860 | -1.6 | 4,944 | 23 | 138.8 | 3,425,165 | 693 | 199,401 | 43.1 | 247,718 | 20.3 | 79.7 | NA | 77.6 | 2,221 | 50.6 |
| Wibaux | 515 | -5.5 | 3,762 | 2.9 | 98.3 | 2,321,989 | 617 | 224,104 | 18 | 131,467 | 31.0 | 69.0 | NA | 77.4 | 2,083 | 83.2 |
| Yellowstone | 1,603 | -3.9 | 1,220 | 76.8 | 299 | 1,223,832 | 1,003 | 104,007 | 135.3 | 102,958 | 31.8 | 68.2 | 3 | 83.0 | 5,615 | 22.5 |
| NEBRASKA | 44,987 | -0.8 | 971 | 8,588.4 | 22,242.6 | 2,674,492 | 2,754 | 268,968 | 21,983.4 | 474,476 | 42.4 | 57.6 | 292 | 81.3 | 639,975 | 66.6 |
| Adams | 340 | -0.2 | 624 | 237 | 300.5 | 3,321,098 | 5,323 | 366,765 | 392.5 | 720,206 | 44.2 | 55.8 | 5 | 88.4 | 11,978 | 69.9 |
| Antelope | 492 | 3.6 | 699 | 255.3 | 364.4 | 3,205,709 | 4,588 | 385,363 | 529.5 | 752,134 | 36.3 | 63.7 | 4 | 86.5 | 10,289 | 72.4 |
| Arthur | 453 | 0.0 | 4,766 | 8.7 | 31.7 | 3,817,883 | 801 | 169,891 | 27.5 | 289,632 | 8.7 | 91.3 | NA | 89.5 | 506 | 25.3 |
| Banner | 423 | 0.2 | 1,770 | 26.1 | 191.2 | 1,543,874 | 872 | 205,818 | 100.5 | 420,540 | 20.7 | 79.3 | 12 | 83.7 | 4,845 | 80.3 |

Items 117—132

## Table B. States and Counties — Water Use, Wholesale Trade, Retail Trade, and Real Estate

| STATE County | Public supply water withdrawn (mil gal/day) [133] | Public supply gallons withdrawn per person per day [134] | Wholesale Trade[1] 2017 — Number of establishments [135] | Number of employees [136] | Sales (mil dol) [137] | Average payroll (mil dol) [138] | Retail Trade[2] 2017 — Number of establishments [139] | Number of employees [140] | Sales (mil dol) [141] | Average payroll (mil dol) [142] | Real estate and rental and leasing[2] 2017 — Number of establishments [143] | Number of employees [144] | Sales (mil dol) [145] | Average payroll (mil dol) [146] |
|---|---|---|---|---|---|---|---|---|---|---|---|---|---|---|
| **MISSOURI— Cont'd** | | | | | | | | | | | | | | |
| St. Louis city | 92.67 | 293.6 | 401 | 6,973 | 6,103.3 | 425.4 | 878 | 9,928 | 2,857.2 | 276.4 | 412 | 2,388 | 632.7 | 108.5 |
| **MONTANA** | 153.19 | 148.3 | 1,366 | 13,078 | 11,214.5 | 677.3 | 4,754 | 59,032 | 16,935.8 | 1,622.6 | 2,036 | 5,951 | 1,144.6 | 205.6 |
| Beaverhead | 1.82 | 195.7 | 13 | 74 | 41.2 | 2.9 | 48 | 512 | 146.0 | 10.8 | 19 | 122 | 14.0 | 4.9 |
| Big Horn | 1.21 | 91.4 | 8 | 35 | 20.3 | 1.6 | 40 | 372 | 110.3 | 9.1 | 8 | 15 | 1.6 | 0.4 |
| Blaine | 0.93 | 141.4 | 10 | D | 72.8 | D | 22 | 186 | 41.3 | 4.0 | NA | NA | NA | NA |
| Broadwater | 1.03 | 181.1 | 5 | D | 15.4 | D | 17 | 154 | 74.3 | 3.7 | D | D | D | 0.1 |
| Carbon | 1.76 | 169.1 | 15 | 64 | 22.7 | 2.1 | D | D | D | D | 16 | 23 | 2.2 | 0.5 |
| Carter | 0.04 | 33.9 | NA | NA | NA | NA | 5 | 33 | 17.2 | 0.9 | NA | NA | NA | NA |
| Cascade | 12.79 | 155.4 | 116 | 1,152 | 722.0 | 54.0 | 362 | 5,326 | 1,439.0 | 139.0 | 123 | 412 | 82.2 | 13.8 |
| Chouteau | 0.88 | 152.6 | 15 | 102 | 280.6 | 5.4 | 23 | 145 | 54.6 | 5.1 | 5 | D | 1.2 | D |
| Custer | 1.39 | 114.5 | D | D | 28.2 | D | 58 | 799 | 236.8 | 23.5 | 17 | 37 | 4.5 | 1.0 |
| Daniels | 0.21 | 119.3 | 5 | D | | D | 13 | 97 | 49.1 | 3.3 | NA | NA | NA | NA |
| Dawson | 1.75 | 181.8 | 20 | 185 | 184.4 | 10.5 | 41 | 454 | 122.9 | 11.9 | 12 | 41 | 5.9 | 1.7 |
| Deer Lodge | 2.30 | 251.7 | NA | NA | NA | NA | 30 | 276 | 73.8 | 6.4 | 7 | 16 | 1.6 | 0.3 |
| Fallon | 0.28 | 87.8 | 8 | 56 | 42.8 | 3.9 | 15 | 137 | 30.9 | 3.3 | D | D | D | 0.1 |
| Fergus | 1.06 | 92.8 | 18 | 162 | 1,033.7 | 9.6 | 59 | 519 | 157.6 | 12.4 | 19 | 61 | 9.1 | 1.4 |
| Flathead | 11.43 | 118.9 | 110 | 956 | 831.0 | 45.0 | 480 | 6,285 | 1,880.1 | 180.8 | 260 | 834 | 135.9 | 24.2 |
| Gallatin | 11.93 | 118.4 | 149 | 1,422 | 786.5 | 71.1 | 594 | 7,885 | 2,314.2 | 237.4 | 429 | 1,180 | 248.5 | 40.9 |
| Garfield | 0.04 | 30.4 | NA | NA | NA | NA | 5 | 58 | 13.5 | 1.5 | NA | NA | NA | NA |
| Glacier | 1.76 | 129.0 | 9 | 44 | 69.6 | 2.1 | 44 | 447 | 145.3 | 11.9 | D | D | D | 0.1 |
| Golden Valley | 0.04 | 48.4 | NA | NA | NA | NA | NA | NA | NA | NA | 4 | D | 0.6 | D |
| Granite | 0.05 | 15.4 | 4 | 20 | 6.4 | 0.8 | 17 | 104 | 34.9 | 2.4 | D | D | D | D |
| Hill | 2.15 | 129.7 | 28 | 248 | 261.2 | 11.3 | 80 | 1,042 | 266.8 | 26.5 | 27 | 85 | 8.6 | 1.9 |
| Jefferson | 1.28 | 109.9 | 9 | 21 | 5.1 | 0.4 | 24 | 168 | 40.1 | 3.1 | 9 | 19 | 1.3 | 0.4 |
| Judith Basin | 0.09 | 46.7 | 6 | D | 21.0 | D | 6 | 28 | 7.1 | 0.6 | NA | NA | NA | NA |
| Lake | 2.39 | 81.1 | 18 | 123 | 31.9 | 5.6 | 110 | 1,160 | 288.9 | 31.5 | D | D | D | 1.2 |
| Lewis and Clark | 7.93 | 119.4 | 56 | 459 | 330.6 | 19.8 | 278 | 3,879 | 1,100.4 | 107.2 | 122 | 291 | 69.5 | 10.9 |
| Liberty | 0.22 | 91.4 | 9 | D | 112.0 | D | 14 | 83 | 11.0 | 1.3 | NA | NA | NA | NA |
| Lincoln | 2.28 | 119.7 | D | D | D | 0.8 | 84 | 676 | 167.5 | 16.7 | 24 | 47 | 6.7 | 1.0 |
| McCone | 0.08 | 47.5 | D | D | D | D | 8 | 62 | 16.7 | 1.6 | NA | NA | NA | NA |
| Madison | 1.10 | 139.0 | 4 | 14 | 34.4 | 1.0 | 41 | 213 | 68.3 | 5.4 | D | D | D | D |
| Meagher | 0.24 | 131.1 | NA | NA | NA | NA | 12 | 64 | 13.4 | 1.0 | NA | NA | NA | NA |
| Mineral | 0.33 | 77.6 | NA | NA | NA | NA | 17 | 245 | 51.6 | 4.8 | 3 | 3 | 0.4 | 0.1 |
| Missoula | 30.41 | 266.3 | 152 | 1,774 | 1,059.9 | 93.6 | 569 | 8,373 | 2,360.2 | 228.3 | 252 | 902 | 178.0 | 31.0 |
| Musselshell | 0.25 | 54.6 | 7 | 29 | 11.2 | 0.9 | 18 | 114 | 40.7 | 2.7 | NA | NA | NA | NA |
| Park | 2.42 | 151.5 | 12 | 57 | 46.1 | 2.4 | 94 | 744 | 207.6 | 22.7 | 37 | 39 | 12.3 | 1.5 |
| Petroleum | 0.01 | 21.1 | NA | NA | NA | NA | D | D | D | 0.1 | NA | NA | NA | NA |
| Phillips | 0.47 | 112.7 | 3 | 50 | 32.5 | 1.7 | 23 | 202 | 43.9 | 4.6 | 5 | D | 1.1 | D |
| Pondera | 0.84 | 135.8 | 13 | 144 | 103.1 | 5.7 | 28 | 252 | 58.2 | 5.5 | NA | NA | NA | NA |
| Powder River | 0.08 | 45.1 | D | D | D | D | 10 | 88 | 19.9 | 2.1 | NA | NA | NA | NA |
| Powell | 1.37 | 200.3 | NA | NA | NA | NA | 13 | 155 | 53.7 | 3.8 | 8 | D | 1.5 | D |
| Prairie | 0.01 | 8.6 | NA | NA | NA | NA | 5 | D | 6.9 | D | NA | NA | NA | NA |
| Ravalli | 3.25 | 78.6 | 45 | 169 | 212.8 | 8.3 | 162 | 1,642 | 368.0 | 37.7 | 65 | 148 | 20.4 | 3.8 |
| Richland | 1.18 | 98.7 | 21 | 158 | 370.8 | 8.9 | 54 | 596 | 164.6 | 16.8 | 21 | 104 | 26.2 | 6.9 |
| Roosevelt | 1.72 | 149.9 | 9 | 79 | 185.3 | 5.0 | 39 | 439 | 120.7 | 11.0 | 5 | 13 | 1.2 | 0.4 |
| Rosebud | 1.52 | 161.7 | NA | NA | NA | NA | 27 | 306 | 65.1 | 5.2 | D | D | D | 0.7 |
| Sanders | 0.71 | 62.6 | 10 | 21 | 20.7 | 0.8 | 50 | 330 | 80.9 | 7.1 | 13 | D | 2.2 | D |
| Sheridan | 0.04 | 10.8 | 6 | D | 86.5 | D | 20 | 165 | 39.7 | 4.4 | 5 | D | 0.2 | D |
| Silver Bow | 7.73 | 223.3 | D | D | D | D | 162 | 2,175 | 573.3 | 54.5 | 57 | 156 | 20.4 | 3.9 |
| Stillwater | 0.38 | 40.1 | D | D | D | 1.7 | 28 | 240 | 80.6 | 5.3 | 6 | D | 1.8 | D |
| Sweet Grass | 0.51 | 140.3 | 3 | 10 | 2.5 | 0.2 | 24 | 163 | 43.9 | 4.0 | 8 | D | 1.9 | D |
| Teton | 1.13 | 185.1 | 16 | 91 | 80.5 | 3.5 | 28 | 174 | 50.7 | 5.0 | 5 | D | 1.0 | D |
| Toole | 1.21 | 237.9 | 13 | 60 | 153.4 | 4.4 | 27 | 186 | 67.8 | 4.1 | D | D | D | 0.3 |
| Treasure | 0.07 | 100.4 | NA | NA | NA | NA | D | D | D | D | NA | NA | NA | NA |
| Valley | 1.61 | 210.2 | 12 | 188 | 183.1 | 9.9 | 47 | 333 | 137.6 | 9.1 | 6 | 12 | 1.7 | 0.4 |
| Wheatland | 0.32 | 151.7 | NA | NA | NA | NA | 8 | 89 | 14.4 | 1.5 | NA | NA | NA | NA |
| Wibaux | 0.07 | 61.9 | NA | NA | NA | NA | 4 | 22 | 4.5 | 0.5 | NA | NA | NA | NA |
| Yellowstone | 25.09 | 159.8 | 322 | 4,277 | 3,090.8 | 247.3 | 715 | 10,550 | 3,296.5 | 312.8 | 334 | 1,149 | 251.1 | 48.0 |
| **NEBRASKA** | 275.18 | 145.1 | 2,782 | 35,143 | 39,908.6 | 1,964.7 | 7,154 | 109,729 | 31,214.7 | 2,853.4 | 2,354 | 11,293 | 2,291.9 | 486.8 |
| Adams | 6.50 | 205.8 | D | D | D | 36.7 | 141 | 1,936 | 499.2 | 50.7 | 48 | 111 | 24.1 | 3.2 |
| Antelope | 0.67 | 104.5 | 26 | 276 | 431.4 | 14.2 | 32 | 243 | 57.3 | 5.0 | 4 | D | 0.5 | D |
| Arthur | 0.00 | 0.0 | NA | NA | NA | NA | NA | NA | NA | NA | NA | NA | NA | NA |
| Banner | 0.02 | 25.4 | NA | NA | NA | NA | NA | NA | NA | NA | NA | NA | NA | NA |

1 Merchant wholesalers, except manufacturers' sales branches and offices.    2. Employer establishments.

Table B. States and Counties — **Professional Services, Manufacturing, and Accommodation and Food Services**

| STATE County | Professional, scientific, and technical services, 2017 | | | | Manufacturing, 2017 | | | | Accommodation and food services, 2017 | | | |
|---|---|---|---|---|---|---|---|---|---|---|---|---|
| | Number of establish-ments | Number of employees | Sales (mil dol) | Average payroll (mil dol) | Number of establish-ments | Number of employees | Sales (mil dol) | Average payroll (mil dol) | Number of establish-ments | Number of employees | Sales (mil dol) | Annual payroll (mil dol) |
| | 147 | 148 | 149 | 150 | 151 | 152 | 153 | 154 | 155 | 156 | 157 | 158 |
| MISSOURI— Cont'd | | | | | | | | | | | | |
| St. Louis city | 1,102 | 17,876 | 3,670 | 1,381.1 | 436 | 16,630 | 7,934.1 | 1,003.1 | 1,002 | 22,623 | 1,685.1 | 451.0 |
| MONTANA | 3,790 | 17,429 | 2,483 | 921.1 | 1,328 | 17,944 | 10,943.6 | 935.7 | 3,568 | 52,415 | 3,126.0 | 898.6 |
| Beaverhead | D | D | D | D | 14 | 63 | 10.3 | 2.1 | 51 | 404 | 22.0 | 6.2 |
| Big Horn | 15 | 59 | 5 | 2.0 | NA | NA | NA | NA | 24 | 196 | 16.4 | 3.8 |
| Blaine | 13 | 37 | 5 | 2.0 | 6 | 10 | 3.8 | 0.4 | 10 | D | 1.9 | D |
| Broadwater | 9 | 33 | 3 | 0.9 | D | D | D | D | 23 | 168 | 8.0 | 2.2 |
| Carbon | D | D | D | D | 12 | 45 | 7.6 | 1.5 | 60 | 531 | 35.5 | 9.6 |
| Carter | D | D | 0 | D | NA | NA | NA | NA | 5 | D | 0.9 | D |
| Cascade | D | D | D | D | 74 | 1,318 | 1,377.7 | 71.3 | 239 | 3,989 | 222.4 | 63.5 |
| Chouteau | 10 | 12 | 1 | 0.3 | 7 | 19 | 3.6 | 0.8 | 15 | D | 4.0 | D |
| Custer | 34 | 87 | 9 | 3.4 | 21 | 64 | 49.4 | 2.5 | D | D | D | D |
| Daniels | 4 | 13 | 1 | 0.3 | NA | NA | NA | NA | 8 | D | 1.9 | D |
| Dawson | 23 | 55 | 6 | 3.0 | 4 | 37 | 6.2 | 1.6 | 34 | 342 | 17.6 | 4.9 |
| Deer Lodge | 22 | 103 | 27 | 6.1 | D | 106 | D | 4.5 | 31 | 269 | 12.1 | 3.0 |
| Fallon | D | D | D | 0.7 | NA | NA | NA | NA | 16 | 79 | 5.9 | 1.3 |
| Fergus | 30 | 77 | 8 | 2.5 | 24 | 294 | 83.8 | 15.2 | 41 | 442 | 18.8 | 5.9 |
| Flathead | D | D | D | D | 173 | 2,688 | 726.9 | 145.2 | 380 | 5,398 | 352.9 | 97.6 |
| Gallatin | 765 | 2,962 | 473 | 171.5 | 237 | 3,101 | 728.1 | 136.7 | 470 | 8,695 | 564.7 | 163.7 |
| Garfield | NA | NA | NA | NA | NA | NA | NA | NA | 4 | D | 1.5 | D |
| Glacier | D | D | D | D | 3 | 5 | 1.1 | 0.2 | 43 | 638 | 67.2 | 17.7 |
| Golden Valley | NA | NA | NA | NA | NA | NA | NA | NA | 4 | 14 | 1.0 | 0.1 |
| Granite | 7 | 7 | 3 | 0.6 | 6 | 15 | 2.7 | 0.5 | 17 | 188 | 20.4 | 7.3 |
| Hill | D | D | D | D | 13 | 24 | 3.5 | 1.1 | 53 | 811 | 45.1 | 14.4 |
| Jefferson | 32 | 57 | 9 | 3.0 | 10 | 238 | 187.2 | 15.4 | 24 | 252 | 9.1 | 2.9 |
| Judith Basin | 4 | 11 | 2 | 0.4 | NA | NA | NA | NA | 7 | D | 1.7 | D |
| Lake | D | D | D | D | 37 | 330 | 60.5 | 11.9 | 87 | 796 | 46.9 | 14.2 |
| Lewis and Clark | 249 | 1,567 | 220 | 87.6 | 57 | 686 | 96.0 | 30.1 | 206 | 3,334 | 180.4 | 53.4 |
| Liberty | 4 | 7 | 0 | 0.2 | NA | NA | NA | NA | 4 | 34 | 1.0 | 0.3 |
| Lincoln | 36 | 92 | 9 | 2.8 | 23 | 195 | 27.9 | 5.8 | 68 | 575 | 35.5 | 10.1 |
| McCone | NA | NA | NA | NA | NA | NA | NA | NA | 5 | D | 2.3 | D |
| Madison | 32 | 54 | 12 | 2.7 | 13 | 78 | 9.4 | 3.0 | 60 | 281 | 28.4 | 7.2 |
| Meagher | D | D | D | 0.1 | NA | NA | NA | NA | 14 | D | 4.5 | D |
| Mineral | 6 | 17 | 1 | 0.3 | 7 | 290 | 78.2 | 12.4 | 18 | 124 | 7.7 | 1.9 |
| Missoula | 553 | 3,053 | 370 | 153.3 | 133 | 1,751 | 521.0 | 75.8 | 365 | 7,110 | 394.6 | 115.0 |
| Musselshell | 9 | 32 | 2 | 0.6 | 4 | 9 | 2.2 | 0.5 | 9 | D | 2.1 | D |
| Park | 74 | 176 | 19 | 7.1 | 29 | 441 | 76.3 | 19.5 | 123 | 1,269 | 107.6 | 30.9 |
| Petroleum | NA | NA | NA | NA | NA | NA | NA | NA | NA | NA | NA | NA |
| Phillips | 10 | 32 | 4 | 1.0 | NA | NA | NA | NA | 21 | 140 | 6.9 | 1.7 |
| Pondera | 12 | 31 | 3 | 1.2 | 10 | 65 | 63.9 | 3.0 | 15 | 104 | 3.8 | 0.9 |
| Powder River | D | D | 3 | D | NA | NA | NA | NA | 6 | D | 2.4 | D |
| Powell | 10 | 38 | 5 | 1.7 | D | 164 | D | D | 24 | 152 | 7.2 | 2.3 |
| Prairie | NA | NA | NA | NA | NA | NA | NA | NA | 4 | D | 0.4 | D |
| Ravalli | D | D | D | D | 88 | 872 | 155.1 | 42.7 | 89 | 943 | 49.6 | 15.3 |
| Richland | 37 | 133 | 21 | 8.3 | 13 | 335 | 183.9 | 19.8 | 47 | 580 | 27.6 | 9.4 |
| Roosevelt | 12 | 26 | 3 | 0.6 | 4 | 27 | 3.3 | 1.0 | D | D | D | D |
| Rosebud | 5 | 15 | 2 | 0.3 | NA | NA | NA | NA | 31 | 206 | 9.4 | 2.2 |
| Sanders | 25 | 39 | 3 | 1.2 | 21 | 180 | 44.8 | 8.9 | 45 | 395 | 21.3 | 6.1 |
| Sheridan | 9 | 30 | 3 | 1.2 | NA | NA | NA | NA | 18 | 117 | 6.3 | 1.6 |
| Silver Bow | D | D | D | D | 40 | 495 | 196.1 | 29.8 | 137 | 2,184 | 114.3 | 35.8 |
| Stillwater | 25 | 42 | 5 | 1.7 | D | 367 | D | 25.7 | 20 | 159 | 10.4 | 2.7 |
| Sweet Grass | 14 | 27 | 3 | 0.8 | 10 | 58 | 7.9 | 2.1 | 21 | 158 | 9.0 | 2.5 |
| Teton | D | D | D | 1.4 | D | 9 | D | 0.4 | D | D | D | 1.7 |
| Toole | 8 | 29 | 4 | 1.3 | D | 15 | D | D | 20 | 236 | 10.5 | 2.5 |
| Treasure | D | D | D | D | NA | NA | NA | NA | NA | NA | NA | NA |
| Valley | D | D | D | D | 6 | 36 | 3.6 | 0.9 | 31 | 372 | 18.8 | 5.5 |
| Wheatland | 4 | 15 | 2 | 0.4 | 4 | 16 | 0.8 | 0.3 | 9 | D | 1.8 | D |
| Wibaux | D | D | 1 | D | NA | NA | NA | NA | D | D | D | 0.5 |
| Yellowstone | D | D | D | 237.9 | 171 | 3,315 | 5,401.4 | 228.4 | 416 | 9,245 | 532.5 | 152.8 |
| NEBRASKA | 4,678 | 39,177 | 6,230 | 2,340.0 | 1,760 | 93,510 | 53,129.3 | 4,728.4 | 4,621 | 76,386 | 3,957.8 | 1,135.9 |
| Adams | 62 | 301 | 33 | 12.1 | 57 | 2,494 | 1,908.3 | 113.1 | 86 | 1,435 | 67.0 | 17.3 |
| Antelope | 11 | 25 | 4 | 0.8 | 10 | 209 | 59.9 | 9.5 | 11 | 49 | 1.9 | 0.5 |
| Arthur | NA | NA | NA | NA | NA | NA | NA | NA | NA | NA | NA | NA |
| Banner | NA | NA | NA | NA | NA | NA | NA | NA | NA | NA | NA | NA |

# Health Care and Social Assistance, Other Services, Nonemployer Businesses, and Residential Construction

| STATE County | Health care and social assistance, 2017 | | | | Other services, 2017 | | | | Nonemployer businesses, 2018 | | Value of residential construction authorized by building permits, 2020 | |
|---|---|---|---|---|---|---|---|---|---|---|---|---|
| | Number of establishments | Number of employees | Receipts (mil dol) | Annual payroll (mil dol) | Number of establishments | Number of employees | Receipts (mil dol) | Annual payroll (mil dol) | Number | Receipts (mil dol) | New construction ($1,000) | Number of housing units |
| | 159 | 160 | 161 | 162 | 163 | 164 | 165 | 166 | 167 | 168 | 169 | 170 |
| MISSOURI— Cont'd | | | | | | | | | | | | |
| St. Louis city | 1,554 | 39,629 | 4,647.7 | 1,753.1 | 625 | 5,377 | 827.8 | 199.1 | 22,366 | 889.9 | 80,644 | 422 |
| MONTANA | 3,716 | 73,254 | 8,447.9 | 3,303.9 | 2,409 | 12,077 | 1,608.6 | 403.0 | 93,753 | 4,414.9 | 0 | 0 |
| Beaverhead | 38 | 517 | 45.7 | 16.9 | 25 | 74 | 6.7 | 1.7 | 915 | 35.3 | 8,400 | 36 |
| Big Horn | 19 | 622 | 96.6 | 38.7 | D | D | D | D | 435 | 15.7 | 0 | 0 |
| Blaine | 16 | 273 | 50.0 | 20.0 | 10 | 41 | 4.6 | 0.8 | 327 | 8.6 | 297 | 1 |
| Broadwater | 9 | D | 10.0 | D | D | D | 3.9 | D | 558 | 28.6 | 0 | 0 |
| Carbon | D | D | D | D | D | D | D | NA | 1,241 | 55.1 | 8,921 | 38 |
| Carter | NA | NA | NA | NA | NA | NA | NA | NA | 134 | 7.3 | 317 | 5 |
| Cascade | 285 | 6,971 | 802.3 | 322.9 | 155 | 941 | 105.1 | 29.9 | 4,883 | 214.3 | 36,936 | 151 |
| Chouteau | 9 | D | 6.2 | D | D | D | 0.7 | D | 378 | 17.2 | 3,475 | 13 |
| Custer | 49 | 928 | 95.8 | 34.5 | 20 | 97 | 12.1 | 3.5 | 878 | 36.6 | 790 | 3 |
| Daniels | 7 | D | 8.5 | D | D | D | D | 0.7 | 126 | 7.2 | 0 | 0 |
| Dawson | 30 | 647 | 58.1 | 26.1 | 25 | 92 | 11.7 | 2.7 | 628 | 22.2 | 0 | 0 |
| Deer Lodge | 47 | 1,551 | 184.8 | 78.1 | 12 | 43 | 4.6 | 1.3 | 461 | 17.7 | 5,374 | 22 |
| Fallon | 4 | 121 | 13.6 | 5.2 | 8 | 23 | 2.7 | 0.5 | 320 | 15.5 | 3,589 | 35 |
| Fergus | 46 | 1,091 | 85.7 | 40.6 | 35 | 160 | 16.0 | 2.8 | 1,046 | 47.5 | 1,980 | 6 |
| Flathead | 380 | 7,578 | 870.8 | 379.2 | 257 | 1,108 | 117.8 | 31.8 | 11,359 | 566.7 | 180,755 | 799 |
| Gallatin | 428 | 5,578 | 631.9 | 230.6 | 315 | 1,681 | 258.4 | 62.9 | 13,841 | 710.9 | 386,981 | 2,038 |
| Garfield | 3 | 6 | 0.4 | 0.1 | NA | NA | NA | NA | 129 | 4.7 | 0 | 0 |
| Glacier | 18 | 455 | 65.9 | 27.2 | 14 | 67 | 8.5 | 2.4 | 745 | 20.2 | 0 | 0 |
| Golden Valley | NA | NA | NA | NA | NA | NA | NA | NA | 79 | 3.7 | NA | NA |
| Granite | D | D | D | D | D | D | 0.6 | D | 346 | 14.8 | NA | NA |
| Hill | 56 | 1,342 | 121.8 | 50.0 | 39 | 155 | 20.2 | 4.1 | 927 | 29.3 | 292 | 1 |
| Jefferson | 26 | 287 | 18.2 | 7.2 | 10 | 35 | 5.4 | 1.2 | 1,121 | 52.1 | 4,086 | 14 |
| Judith Basin | 3 | 3 | 0.1 | 0.0 | NA | NA | NA | NA | 201 | 8.5 | NA | NA |
| Lake | 86 | 1,338 | 128.9 | 51.1 | 50 | 159 | 14.8 | 3.9 | 2,502 | 108.0 | 9,321 | 44 |
| Lewis and Clark | 288 | 6,098 | 809.4 | 294.9 | 213 | 1,193 | 163.5 | 50.9 | 5,758 | 280.8 | 44,432 | 250 |
| Liberty | D | D | D | D | 3 | 5 | 0.6 | 0.1 | 137 | 4.9 | NA | NA |
| Lincoln | 56 | 1,207 | 92.6 | 38.3 | 37 | 107 | 10.7 | 2.5 | 1,715 | 67.3 | 2,192 | 23 |
| McCone | D | D | D | D | NA | NA | NA | NA | 165 | 6.9 | 0 | 0 |
| Madison | 24 | 208 | 16.9 | 8.0 | 21 | 79 | 11.4 | 2.2 | 1,132 | 54.3 | 3,015 | 11 |
| Meagher | D | D | D | D | 4 | 4 | 1.4 | 0.1 | 186 | 6.9 | NA | NA |
| Mineral | 11 | 106 | 7.8 | 3.5 | 4 | 5 | 0.8 | 0.1 | 350 | 12.9 | 1,523 | 11 |
| Missoula | 512 | 10,122 | 1,190.9 | 412.7 | 312 | 2,257 | 362.1 | 79.5 | 10,969 | 523.9 | 131,707 | 789 |
| Musselshell | 11 | D | 16.1 | D | D | D | 1.8 | D | 353 | 14.8 | 400 | 2 |
| Park | 58 | 864 | 90.8 | 38.7 | 50 | 196 | 20.7 | 5.6 | 2,296 | 91.0 | 6,022 | 31 |
| Petroleum | NA | NA | NA | NA | NA | NA | NA | NA | 34 | 1.5 | NA | NA |
| Phillips | 10 | 184 | 13.9 | 5.7 | 15 | 34 | 3.4 | 0.9 | 391 | 12.3 | 0 | 0 |
| Pondera | 20 | 276 | 22.3 | 9.1 | 8 | 19 | 2.1 | 0.4 | 435 | 16.4 | 0 | 0 |
| Powder River | NA | NA | NA | NA | D | D | D | 1.1 | 185 | 7.6 | NA | NA |
| Powell | 18 | 226 | 23.3 | 9.7 | D | D | D | D | 503 | 22.0 | 8,574 | 36 |
| Prairie | NA | NA | NA | NA | NA | NA | NA | NA | 88 | 3.4 | NA | NA |
| Ravalli | 137 | 1,674 | 147.4 | 64.1 | 82 | 278 | 26.8 | 7.3 | 4,585 | 214.9 | 10,297 | 56 |
| Richland | 38 | 620 | 70.9 | 27.9 | 32 | 116 | 23.4 | 4.0 | 945 | 52.4 | 1,755 | 13 |
| Roosevelt | 13 | D | 42.2 | D | 15 | 59 | 5.1 | 1.1 | 486 | 17.9 | 0 | 0 |
| Rosebud | 12 | 216 | 12.9 | 7.2 | D | D | 5.0 | D | 447 | 13.8 | 0 | 0 |
| Sanders | 42 | 498 | 38.2 | 15.5 | 15 | 62 | 7.2 | 1.9 | 1,108 | 52.6 | NA | NA |
| Sheridan | 12 | 271 | 15.3 | 7.1 | D | D | 4.3 | D | 308 | 16.1 | 0 | 0 |
| Silver Bow | 169 | 3,114 | 295.5 | 113.6 | 65 | 277 | 38.2 | 8.1 | 2,315 | 106.8 | 13,239 | 141 |
| Stillwater | 19 | 269 | 19.2 | 7.7 | D | D | 5.5 | D | 882 | 38.0 | 0 | 0 |
| Sweet Grass | 7 | D | 8.9 | D | D | D | 3.2 | D | 472 | 22.3 | 297 | 1 |
| Teton | 18 | 217 | 14.7 | 5.6 | 7 | 12 | 1.0 | 0.3 | 592 | 28.1 | 0 | 0 |
| Toole | 12 | 225 | 19.5 | 8.7 | 7 | 23 | 1.5 | 0.3 | 307 | 14.0 | 395 | 3 |
| Treasure | NA | NA | NA | NA | NA | NA | NA | NA | 40 | 1.7 | NA | NA |
| Valley | 28 | 548 | 52.1 | 22.7 | 17 | 57 | 5.9 | 1.6 | 528 | 18.9 | 1,033 | 9 |
| Wheatland | D | D | D | D | D | D | D | 0.1 | 134 | 5.8 | NA | NA |
| Wibaux | NA | NA | NA | NA | NA | NA | NA | NA | 74 | 3.6 | 300 | 1 |
| Yellowstone | 582 | 14,657 | 2,069.9 | 808.5 | 397 | 2,244 | 286.8 | 76.5 | 12,253 | 633.2 | 175,191 | 1,397 |
| NEBRASKA | 5,817 | 135,691 | 16,060.4 | 6,116.9 | 4,107 | 21,757 | 3,608.2 | 701.0 | 138,728 | 6,305.8 | 1,643,788 | 9,483 |
| Adams | 107 | 2,893 | 323.8 | 136.2 | 52 | 229 | 26.8 | 5.8 | 2,251 | 99.0 | 14,008 | 58 |
| Antelope | 17 | 335 | 25.6 | 11.2 | D | D | D | 0.8 | 703 | 31.0 | 6,723 | 27 |
| Arthur | NA | NA | NA | NA | NA | NA | NA | NA | 52 | 3.7 | NA | NA |
| Banner | NA | NA | NA | NA | NA | NA | NA | NA | 50 | 2.3 | NA | NA |

# Table B. States and Counties — Government Employment and Payroll, and Local Government Finances

| STATE County | Full-time equivalent employees | March payroll (dollars) | March payroll (percent of total) | | | | | | | Local government finances, 2017 General revenue | | | | |
|---|---|---|---|---|---|---|---|---|---|---|---|---|---|---|
| | | | Administration, judicial, and legal | Police and corrections | Fire protection | Highways and transportation | Health and welfare | Natural resources and utilities | Education and libraries | Total (mil dol) | Inter-governmental (mil dol) | Taxes Total (mil dol) | Taxes Per capita[1] (dollars) Total | Taxes Per capita[1] (dollars) Property |
| | 171 | 172 | 173 | 174 | 175 | 176 | 177 | 178 | 179 | 180 | 181 | 182 | 183 | 184 |
| MISSOURI— Cont'd | | | | | | | | | | | | | | |
| St. Louis city | 14,892 | 75,555,604 | 8.0 | 18.8 | 7.9 | 19.5 | 1.4 | 14.2 | 28.0 | 2,003.6 | 463.3 | 889.5 | 2,886 | 1,072 |
| MONTANA | X | X | X | X | X | X | X | X | X | X | X | X | X | X |
| Beaverhead | 298 | 1,044,238 | 13.3 | 8.3 | 0.7 | 4.7 | 1.1 | 5.9 | 64.6 | 65.9 | 13.5 | 11.6 | 1,227 | 1,182 |
| Big Horn | 658 | 2,554,319 | 5.8 | 5.0 | 0.0 | 3.6 | 6.2 | 2.6 | 75.4 | 68.3 | 41.6 | 17.9 | 1,328 | 1,320 |
| Blaine | 374 | 1,432,727 | 6.6 | 4.4 | 0.5 | 4.9 | 1.1 | 2.9 | 78.7 | 35.4 | 20.8 | 9.9 | 1,461 | 1,451 |
| Broadwater | 168 | 569,372 | 12.7 | 10.2 | 0.6 | 3.2 | 3.1 | 6.0 | 59.1 | 21.3 | 5.2 | 6.6 | 1,119 | 1,111 |
| Carbon | 405 | 1,434,878 | 11.1 | 9.4 | 1.2 | 6.2 | 0.5 | 3.6 | 65.6 | 43.9 | 14.4 | 19.4 | 1,815 | 1,718 |
| Carter | 29 | 101,518 | 12.1 | 13.9 | 0.0 | 25.2 | 2.5 | 4.7 | 9.7 | 10.5 | 3.0 | 6.8 | 5,540 | 5,536 |
| Cascade | 2,747 | 10,563,299 | 7.1 | 12.7 | 4.0 | 6.0 | 4.7 | 5.1 | 59.8 | 271.4 | 111.1 | 93.2 | 1,142 | 1,106 |
| Chouteau | 314 | 1,113,787 | 6.9 | 7.2 | 0.0 | 4.8 | 25.8 | 2.9 | 50.2 | 28.6 | 7.6 | 12.0 | 2,090 | 2,083 |
| Custer | 525 | 2,138,175 | 7.1 | 6.8 | 3.3 | 3.5 | 0.5 | 3.5 | 75.1 | 45.3 | 20.3 | 13.7 | 1,170 | 1,142 |
| Daniels | 110 | 385,999 | 15.9 | 5.2 | 0.0 | 7.1 | 3.6 | 7.1 | 55.3 | 14.0 | 3.5 | 4.0 | 2,348 | 2,327 |
| Dawson | 511 | 1,742,612 | 7.8 | 16.9 | 2.1 | 5.5 | 6.3 | 4.8 | 54.3 | 47.8 | 17.5 | 14.7 | 1,646 | 1,638 |
| Deer Lodge | 296 | 1,144,779 | 10.6 | 17.9 | 3.7 | 3.5 | 5.4 | 3.0 | 53.8 | 36.1 | 16.0 | 13.6 | 1,494 | 1,484 |
| Fallon | 208 | 918,812 | 13.3 | 7.4 | 0.0 | 10.9 | 4.6 | 8.3 | 50.2 | 25.8 | 11.2 | 8.0 | 2,663 | 2,629 |
| Fergus | 514 | 1,719,722 | 7.8 | 9.1 | 2.4 | 4.8 | 2.2 | 3.7 | 68.7 | 45.2 | 19.7 | 18.5 | 1,638 | 1,624 |
| Flathead | 2,793 | 11,502,242 | 6.7 | 8.3 | 3.1 | 4.9 | 4.6 | 5.0 | 66.1 | 328.7 | 128.2 | 127.4 | 1,273 | 1,213 |
| Gallatin | 2,808 | 11,950,725 | 7.4 | 9.3 | 4.1 | 2.2 | 3.3 | 5.7 | 64.1 | 344.5 | 100.5 | 161.4 | 1,486 | 1,374 |
| Garfield | 97 | 242,697 | 18.2 | 4.7 | 0.5 | 7.8 | 20.4 | 2.7 | 44.2 | 7.9 | 2.8 | 2.5 | 1,980 | 1,980 |
| Glacier | 609 | 2,218,362 | 4.3 | 4.5 | 0.1 | 5.7 | 0.9 | 3.9 | 78.0 | 61.8 | 38.7 | 16.7 | 1,213 | 1,204 |
| Golden Valley | 62 | 190,453 | 13.6 | 2.8 | 0.2 | 0.5 | 0.0 | 0.2 | 81.9 | 4.4 | 1.8 | 2.0 | 2,471 | 2,471 |
| Granite | 167 | 611,424 | 10.2 | 4.8 | 0.0 | 4.7 | 32.0 | 4.5 | 41.8 | 11.4 | 4.6 | 5.0 | 1,479 | 1,477 |
| Hill | 748 | 2,806,106 | 4.5 | 7.7 | 2.9 | 4.2 | 2.1 | 5.5 | 72.1 | 80.4 | 46.3 | 24.0 | 1,459 | 1,446 |
| Jefferson | 306 | 1,127,486 | 13.8 | 9.7 | 0.2 | 4.3 | 2.6 | 3.7 | 62.4 | 37.0 | 15.1 | 17.2 | 1,440 | 1,430 |
| Judith Basin | 127 | 400,014 | 7.8 | 4.1 | 0.1 | 15.6 | 0.1 | 0.9 | 69.4 | 10.4 | 3.5 | 5.1 | 2,620 | 2,620 |
| Lake | 1,024 | 3,728,543 | 6.0 | 7.2 | 0.4 | 2.1 | 1.7 | 4.3 | 75.2 | 89.9 | 44.1 | 34.1 | 1,130 | 1,103 |
| Lewis and Clark | 1,948 | 8,463,467 | 8.6 | 10.1 | 2.7 | 3.8 | 6.3 | 5.1 | 60.7 | 226.4 | 92.7 | 82.2 | 1,211 | 1,178 |
| Liberty | 115 | 395,871 | 11.5 | 8.8 | 0.2 | 6.8 | 4.1 | 6.3 | 58.9 | 11.9 | 3.5 | 6.0 | 2,503 | 2,445 |
| Lincoln | 531 | 1,926,072 | 11.0 | 7.2 | 0.6 | 3.6 | 2.6 | 4.9 | 66.1 | 48.0 | 22.6 | 17.5 | 893 | 883 |
| McCone | 168 | 381,253 | 15.9 | 5.1 | 0.1 | 1.3 | 2.0 | 1.5 | 42.2 | 10.1 | 4.2 | 4.8 | 2,852 | 2,839 |
| Madison | 471 | 1,526,402 | 5.8 | 4.4 | 0.5 | 3.1 | 43.8 | 3.4 | 38.5 | 39.9 | 11.0 | 19.7 | 2,377 | 2,355 |
| Meagher | 85 | 260,367 | 14.8 | 10.8 | 1.5 | 5.0 | 0.5 | 9.6 | 53.5 | 12.1 | 2.8 | 7.1 | 3,826 | 3,816 |
| Mineral | 266 | 899,835 | 7.1 | 5.3 | 0.2 | 2.5 | 33.2 | 0.7 | 48.6 | 22.4 | 7.1 | 6.5 | 1,535 | 1,480 |
| Missoula | 3,480 | 14,645,116 | 8.3 | 11.5 | 6.5 | 6.6 | 8.7 | 4.2 | 51.1 | 389.1 | 154.0 | 168.3 | 1,428 | 1,378 |
| Musselshell | 194 | 610,621 | 8.8 | 9.8 | 0.0 | 5.6 | 1.6 | 3.1 | 67.6 | 15.7 | 7.2 | 6.8 | 1,458 | 1,457 |
| Park | 536 | 1,963,411 | 9.2 | 9.4 | 5.2 | 3.0 | 3.1 | 6.1 | 63.7 | 54.5 | 21.7 | 21.1 | 1,288 | 1,231 |
| Petroleum | 44 | 170,919 | 6.1 | 2.4 | 0.0 | 3.9 | 0.0 | 7.4 | 80.2 | 3.5 | 1.6 | 1.0 | 1,922 | 1,920 |
| Phillips | 229 | 720,323 | 8.7 | 5.9 | 0.1 | 7.5 | 1.3 | 8.7 | 65.4 | 16.2 | 8.3 | 5.7 | 1,395 | 1,381 |
| Pondera | 226 | 800,158 | 8.9 | 8.9 | 0.0 | 5.7 | 0.9 | 5.8 | 67.5 | 43.4 | 13.4 | 10.3 | 1,722 | 1,566 |
| Powder River | 167 | 522,249 | 10.5 | 5.0 | 0.0 | 11.9 | 34.1 | 2.8 | 31.1 | 15.0 | 7.1 | 3.9 | 2,239 | 2,238 |
| Powell | 235 | 795,317 | 10.1 | 7.0 | 0.4 | 5.1 | 0.2 | 5.3 | 69.4 | 24.6 | 12.6 | 8.1 | 1,189 | 1,179 |
| Prairie | 104 | 286,187 | 7.4 | 3.7 | 0.0 | 6.5 | 29.1 | 4.8 | 42.8 | 9.5 | 2.5 | 2.4 | 2,145 | 2,126 |
| Ravalli | 1,163 | 3,979,716 | 7.6 | 8.6 | 0.1 | 1.4 | 1.1 | 2.6 | 76.7 | 99.6 | 49.5 | 37.5 | 882 | 868 |
| Richland | 626 | 2,170,542 | 9.0 | 10.8 | 0.2 | 9.1 | 5.9 | 9.4 | 53.0 | 67.9 | 32.8 | 14.5 | 1,314 | 1,308 |
| Roosevelt | 689 | 2,517,688 | 5.6 | 5.9 | 0.0 | 3.4 | 10.9 | 3.5 | 68.8 | 71.6 | 48.4 | 14.2 | 1,275 | 1,265 |
| Rosebud | 654 | 2,223,967 | 7.1 | 8.0 | 0.1 | 6.4 | 3.7 | 5.9 | 67.7 | 68.5 | 28.9 | 22.8 | 2,471 | 2,468 |
| Sanders | 391 | 1,371,035 | 8.2 | 8.8 | 0.1 | 5.0 | 0.7 | 4.7 | 70.6 | 35.3 | 14.6 | 16.1 | 1,377 | 1,362 |
| Sheridan | 255 | 1,363,756 | 7.0 | 4.1 | 0.3 | 6.0 | 31.4 | 1.6 | 46.4 | 20.7 | 8.3 | 8.0 | 2,313 | 2,309 |
| Silver Bow | 1,023 | 4,334,062 | 8.0 | 12.4 | 9.6 | 8.2 | 2.6 | 11.0 | 46.5 | 117.4 | 45.0 | 44.7 | 1,285 | 1,255 |
| Stillwater | 356 | 1,292,795 | 8.5 | 9.0 | 0.7 | 5.7 | 0.6 | 5.2 | 69.3 | 34.5 | 15.1 | 15.0 | 1,594 | 1,583 |
| Sweet Grass | 150 | 557,685 | 17.8 | 12.6 | 0.1 | 7.8 | 0.2 | 2.5 | 58.4 | 24.1 | 5.7 | 7.2 | 1,963 | 1,961 |
| Teton | 371 | 1,234,432 | 7.8 | 4.2 | 0.2 | 4.1 | 9.7 | 9.2 | 63.9 | 37.4 | 11.4 | 10.3 | 1,697 | 1,680 |
| Toole | 327 | 1,163,779 | 9.9 | 6.7 | 0.1 | 5.3 | 30.1 | 3.9 | 42.4 | 37.3 | 7.8 | 9.0 | 1,843 | 1,837 |
| Treasure | 36 | 115,623 | 14.2 | 3.5 | 0.0 | 5.7 | 0.5 | 7.7 | 63.6 | 3.6 | 1.4 | 1.9 | 2,890 | 2,813 |
| Valley | 410 | 1,411,240 | 6.8 | 9.9 | 0.0 | 7.6 | 3.3 | 6.3 | 64.5 | 39.9 | 15.1 | 16.9 | 2,279 | 2,196 |
| Wheatland | 149 | 476,915 | 22.0 | 6.6 | 0.0 | 14.6 | 3.5 | 5.1 | 44.5 | 12.7 | 5.5 | 5.5 | 2,538 | 2,536 |
| Wibaux | 88 | 333,672 | 15.7 | 5.3 | 0.0 | 13.0 | 4.6 | 1.0 | 59.6 | 9.8 | 2.9 | 6.1 | 5,968 | 5,828 |
| Yellowstone | 4,728 | 21,645,773 | 5.8 | 9.5 | 4.6 | 4.1 | 8.5 | 6.5 | 56.2 | 569.1 | 199.2 | 204.1 | 1,282 | 1,191 |
| NEBRASKA | X | X | X | X | X | X | X | X | X | X | X | X | X | X |
| Adams | 2,059 | 9,251,437 | 3.5 | 4.4 | 1.6 | 2.6 | 0.0 | 16.2 | 65.1 | 371.8 | 61.4 | 104.8 | 3,294 | 2,886 |
| Antelope | 446 | 1,495,755 | 4.0 | 6.2 | 0.0 | 5.4 | 0.6 | 3.0 | 80.3 | 38.4 | 10.2 | 24.9 | 3,929 | 3,553 |
| Arthur | 51 | 187,013 | 5.1 | 1.8 | 0.0 | 3.3 | 0.0 | 25.0 | 64.6 | 9.6 | 1.1 | 8.3 | 18,273 | 11,123 |
| Banner | 70 | 220,174 | 9.5 | 1.8 | 0.0 | 8.0 | 0.0 | 0.0 | 79.2 | 5.8 | 1.6 | 3.9 | 5,427 | 5,222 |

1. Based on the resident population estimated as of July 1 of the year shown.

| STATE County | Total (mil dol) | Per capita¹ (dollars) | Education | Health and hospitals | Police protection | Public welfare | Highways | Total (mil dol) | Per capita¹ (dollars) | Federal civilian | Federal military | State and local | Number of returns | Mean adjusted gross income | Mean income tax |
|---|---|---|---|---|---|---|---|---|---|---|---|---|---|---|---|
| | 185 | 186 | 187 | 188 | 189 | 190 | 191 | 192 | 193 | 194 | 195 | 196 | 197 | 198 | 199 |
| **MISSOURI— Cont'd** | | | | | | | | | | | | | | | |
| St. Louis city | 2,186 | 7,092 | 23.5 | 2.6 | 8.7 | 0.0 | 1.4 | 4,818.8 | 15,634 | 13,903 | 1,506 | 19,402 | 140,160 | 53,585 | 6,309 |
| **MONTANA** | X | X | X | X | X | X | X | X | X | 13,368 | 7,867 | 74,900 | 511,340 | 61,858 | 6,666 |
| Beaverhead | 64 | 6,719 | 25.0 | 48.2 | 4.2 | 0.4 | 3.9 | 84.4 | 8,934 | 193 | 39 | 816 | 4,420 | 56,858 | 5,557 |
| Big Horn | 70 | 5,180 | 62.9 | 1.2 | 5.4 | 0.7 | 3.8 | 80.8 | 5,990 | 432 | 58 | 1,314 | 4,220 | 40,786 | 2,556 |
| Blaine | 39 | 5,705 | 62.1 | 1.8 | 4.1 | 0.0 | 5.2 | 2.9 | 434 | 183 | 28 | 671 | 2,510 | 39,457 | 2,775 |
| Broadwater | 22 | 3,764 | 35.0 | 3.0 | 10.0 | 0.2 | 4.4 | 3.5 | 589 | 41 | 27 | 210 | 2,810 | 57,045 | 5,006 |
| Carbon | 42 | 3,959 | 48.8 | 2.4 | 7.2 | 0.1 | 9.3 | 28.4 | 2,661 | 72 | 47 | 509 | 5,130 | 59,169 | 6,227 |
| Carter | 11 | 8,904 | 30.1 | 7.7 | 4.7 | 0.0 | 31.3 | 4.4 | 3,594 | 14 | 5 | 104 | 580 | 50,697 | 4,091 |
| Cascade | 277 | 3,392 | 49.0 | 2.5 | 10.1 | 1.9 | 4.5 | 235.4 | 2,883 | 1,686 | 3,557 | 3,872 | 39,540 | 56,044 | 5,339 |
| Chouteau | 26 | 4,585 | 40.1 | 22.4 | 4.1 | 0.0 | 9.6 | 7.3 | 1,269 | 34 | 24 | 426 | 2,170 | 50,048 | 4,274 |
| Custer | 54 | 4,562 | 61.0 | 3.4 | 8.1 | 0.8 | 6.3 | 16.9 | 1,445 | 192 | 48 | 884 | 5,360 | 54,057 | 5,031 |
| Daniels | 13 | 7,806 | 28.8 | 6.4 | 3.5 | 0.0 | 5.4 | 4.0 | 2,336 | 14 | 7 | 109 | 820 | 56,821 | 5,807 |
| Dawson | 46 | 5,124 | 54.2 | 3.0 | 5.7 | 0.0 | 4.5 | 39.0 | 4,357 | 34 | 35 | 747 | 4,080 | 58,195 | 5,277 |
| Deer Lodge | 37 | 4,039 | 35.8 | 4.9 | 8.2 | 0.2 | 6.7 | 9.9 | 1,085 | 85 | 36 | 929 | 4,250 | 48,505 | 4,008 |
| Fallon | 32 | 10,644 | 40.2 | 2.2 | 8.0 | 0.6 | 22.4 | 1.7 | 552 | 13 | 12 | 279 | 1,330 | 67,006 | 7,014 |
| Fergus | 47 | 4,201 | 56.5 | 2.4 | 9.7 | 0.1 | 9.8 | 8.1 | 718 | 126 | 46 | 850 | 5,460 | 52,112 | 4,452 |
| Flathead | 368 | 3,675 | 58.5 | 3.6 | 6.3 | 0.1 | 4.6 | 332.4 | 3,321 | 764 | 450 | 4,281 | 51,840 | 64,150 | 7,108 |
| Gallatin | 361 | 3,323 | 52.3 | 1.2 | 8.4 | 4.2 | 4.3 | 424.5 | 3,910 | 587 | 494 | 9,667 | 58,890 | 79,849 | 10,141 |
| Garfield | 12 | 9,084 | 28.4 | 1.4 | 3.5 | 22.4 | 6.7 | 0.6 | 495 | 25 | 6 | 129 | 530 | 34,357 | 2,287 |
| Glacier | 65 | 4,697 | 74.1 | 3.1 | 4.0 | 0.1 | 3.0 | 14.8 | 1,076 | 458 | 57 | 2,066 | 5,100 | 37,494 | 2,543 |
| Golden Valley | 5 | 5,682 | 64.9 | 0.9 | 3.7 | 0.6 | 8.3 | 0.5 | 625 | 5 | 3 | 74 | 430 | 33,677 | 2,365 |
| Granite | 12 | 3,457 | 51.2 | 2.2 | 8.5 | 0.0 | 7.8 | 0.9 | 256 | 36 | 15 | 211 | 1,460 | 55,445 | 5,708 |
| Hill | 71 | 4,305 | 66.6 | 2.1 | 5.6 | 0.0 | 3.8 | 35.0 | 2,126 | 159 | 70 | 2,119 | 7,390 | 51,694 | 4,324 |
| Jefferson | 38 | 3,197 | 49.5 | 1.9 | 7.3 | 0.0 | 4.5 | 18.3 | 1,533 | 36 | 52 | 559 | 5,750 | 69,621 | 7,183 |
| Judith Basin | 11 | 5,598 | 55.9 | 0.6 | 3.3 | 0.1 | 9.7 | 2.9 | 1,491 | 33 | 9 | 146 | 950 | 41,781 | 3,243 |
| Lake | 91 | 3,016 | 62.8 | 1.1 | 5.1 | 0.4 | 2.8 | 30.6 | 1,012 | 105 | 131 | 3,061 | 13,100 | 50,004 | 4,416 |
| Lewis and Clark | 229 | 3,379 | 50.3 | 4.8 | 9.3 | 0.5 | 8.4 | 147.3 | 2,171 | 1,971 | 310 | 8,816 | 34,540 | 63,919 | 6,316 |
| Liberty | 12 | 4,759 | 34.0 | 3.6 | 7.5 | 6.0 | 11.8 | 0.8 | 343 | 21 | 8 | 129 | 900 | 43,248 | 3,016 |
| Lincoln | 54 | 2,751 | 58.7 | 1.8 | 7.7 | 0.3 | 6.7 | 25.9 | 1,326 | 431 | 86 | 753 | 8,390 | 47,256 | 3,864 |
| McCone | 10 | 6,058 | 48.6 | 4.2 | 3.3 | 2.3 | 9.5 | 2.9 | 1,688 | 15 | 7 | 130 | 830 | 34,692 | 2,661 |
| Madison | 40 | 4,876 | 34.2 | 3.6 | 6.8 | 17.7 | 10.1 | 12.4 | 1,500 | 59 | 37 | 472 | 4,200 | 66,135 | 7,231 |
| Meagher | 18 | 9,843 | 53.0 | 1.2 | 2.1 | 0.1 | 4.2 | 14.8 | 7,964 | 30 | 7 | 103 | 900 | 42,664 | 3,637 |
| Mineral | 23 | 5,387 | 44.7 | 29.9 | 6.0 | 0.4 | 3.1 | 3.7 | 864 | 58 | 19 | 277 | 1,990 | 47,987 | 4,143 |
| Missoula | 459 | 3,895 | 45.7 | 7.4 | 6.3 | 0.6 | 3.3 | 347.1 | 2,946 | 1,450 | 538 | 9,151 | 58,930 | 64,394 | 8,318 |
| Musselshell | 16 | 3,494 | 74.0 | 2.4 | 5.0 | 0.0 | 4.3 | 12.6 | 2,710 | 13 | 20 | 237 | 2,000 | 46,562 | 4,144 |
| Park | 56 | 3,402 | 47.9 | 2.0 | 7.3 | 0.0 | 6.3 | 19.7 | 1,201 | 79 | 72 | 633 | 9,090 | 58,857 | 6,223 |
| Petroleum | 3 | 6,725 | 49.0 | 0.9 | 2.8 | 0.0 | 12.5 | 0.4 | 738 | 3 | 2 | 55 | 210 | 36,967 | 2,624 |
| Phillips | 22 | 5,289 | 47.8 | 2.0 | 6.0 | 0.0 | 10.4 | 6.7 | 1,624 | 69 | 17 | 297 | 1,930 | 39,764 | 3,273 |
| Pondera | 44 | 7,339 | 37.5 | 35.4 | 3.3 | 0.0 | 4.4 | 7.8 | 1,302 | 34 | 23 | 364 | 2,720 | 43,164 | 3,066 |
| Powder River | 14 | 8,086 | 34.3 | 1.2 | 5.9 | 28.2 | 8.5 | 0.4 | 225 | 12 | 7 | 186 | 820 | 44,665 | 3,066 |
| Powell | 27 | 4,033 | 61.4 | 0.9 | 5.0 | 0.5 | 8.1 | 25.5 | 3,759 | 80 | 23 | 971 | 2,670 | 50,851 | 4,675 |
| Prairie | 10 | 9,389 | 23.3 | 2.2 | 3.6 | 45.5 | 7.1 | 1.6 | 1,475 | 35 | 5 | 143 | 530 | 46,370 | 3,513 |
| Ravalli | 99 | 2,336 | 66.0 | 1.0 | 6.6 | 0.5 | 5.6 | 37.8 | 888 | 529 | 190 | 1,431 | 20,850 | 58,782 | 5,995 |
| Richland | 78 | 7,069 | 39.5 | 1.8 | 5.7 | 1.0 | 8.3 | 8.1 | 732 | 71 | 47 | 670 | 5,620 | 69,850 | 7,898 |
| Roosevelt | 74 | 6,625 | 64.3 | 9.4 | 6.5 | 0.1 | 5.6 | 28.7 | 2,567 | 168 | 47 | 1,646 | 3,890 | 43,800 | 3,629 |
| Rosebud | 67 | 7,227 | 49.8 | 5.3 | 5.5 | 0.2 | 4.0 | 177.6 | 19,289 | 210 | 39 | 1,472 | 3,740 | 55,321 | 4,745 |
| Sanders | 37 | 3,141 | 53.5 | 1.1 | 7.9 | 1.1 | 8.8 | 5.8 | 494 | 145 | 52 | 515 | 5,200 | 45,777 | 3,768 |
| Sheridan | 23 | 6,702 | 46.1 | 5.5 | 7.8 | 0.0 | 10.9 | 6.8 | 1,959 | 59 | 14 | 253 | 1,720 | 54,573 | 5,078 |
| Silver Bow | 143 | 4,111 | 34.8 | 3.3 | 6.1 | 0.2 | 4.6 | 43.9 | 1,261 | 217 | 161 | 2,209 | 16,450 | 56,503 | 5,455 |
| Stillwater | 35 | 3,681 | 57.5 | 0.6 | 5.9 | 0.0 | 12.4 | 8.5 | 907 | 34 | 42 | 451 | 4,510 | 65,545 | 6,405 |
| Sweet Grass | 26 | 6,949 | 29.3 | 0.3 | 5.2 | 38.6 | 8.3 | 0.9 | 243 | 23 | 16 | 193 | 1,780 | 57,570 | 5,652 |
| Teton | 40 | 6,640 | 39.9 | 21.1 | 0.8 | 6.1 | 4.1 | 17.9 | 2,933 | 49 | 25 | 458 | 3,010 | 47,588 | 3,883 |
| Toole | 41 | 8,360 | 24.1 | 47.4 | 6.5 | 1.3 | 5.5 | 17.8 | 3,659 | 173 | 18 | 435 | 2,060 | 49,008 | 4,166 |
| Treasure | 3 | 5,105 | 49.9 | 1.2 | 4.4 | 0.0 | 13.6 | 4.1 | 6,146 | 6 | 3 | 55 | 340 | 49,941 | 3,865 |
| Valley | 38 | 5,068 | 54.7 | 1.8 | 5.5 | 0.8 | 6.3 | 25.3 | 3,414 | 160 | 32 | 606 | 3,580 | 54,384 | 5,254 |
| Wheatland | 11 | 5,273 | 58.6 | 4.8 | 6.7 | 0.1 | 5.8 | 3.9 | 1,810 | 21 | 9 | 139 | 830 | 42,481 | 3,140 |
| Wibaux | 8 | 8,215 | 36.4 | 3.4 | 4.6 | 3.4 | 26.0 | 1.1 | 1,098 | 7 | 4 | 139 | 430 | 47,919 | 3,705 |
| Yellowstone | 645 | 4,048 | 45.6 | 8.1 | 5.6 | 0.3 | 7.2 | 557.9 | 3,503 | 1,809 | 731 | 7,478 | 78,720 | 66,755 | 7,402 |
| **NEBRASKA** | X | X | X | X | X | X | X | X | X | 17,013 | 12,677 | 146,014 | 909,530 | 66,890 | 7,268 |
| Adams | 385 | 12,106 | 40.0 | 43.9 | 1.7 | 0.1 | 2.9 | 192.8 | 6,062 | 99 | 103 | 2,215 | 14,290 | 59,906 | 5,791 |
| Antelope | 38 | 6,029 | 67.7 | 0.0 | 0.9 | 0.0 | 5.4 | 17.4 | 2,748 | 29 | 21 | 481 | 2,980 | 50,230 | 4,227 |
| Arthur | 9 | 20,191 | 29.2 | 0.0 | 0.4 | 0.2 | 5.0 | 22.0 | 48,303 | 2 | 2 | 46 | 200 | 26,120 | 1,390 |
| Banner | 6 | 7,563 | 67.2 | 0.1 | 1.0 | 0.0 | 17.7 | 0.0 | 0 | 1 | 3 | 65 | 280 | 43,582 | 2,850 |

1. Based on the resident population estimated as of July 1 of the year shown.

| State / county code | CBSA code[1] | County Type code[2] | STATE County | Land area[3] (sq. mi) | Total persons 2019 | Rank | Per square mile | White | Black | American Indian, Alaska Native | Asian and Pacific Islander | Percent Hispanic or Latino[4] | Under 5 years | 5 to 14 years | 15 to 24 years | 25 to 34 years | 35 to 44 years | 45 to 54 years |
|---|---|---|---|---|---|---|---|---|---|---|---|---|---|---|---|---|---|---|
| | | | | 1 | 2 | 3 | 4 | 5 | 6 | 7 | 8 | 9 | 10 | 11 | 12 | 13 | 14 | 15 |
| | | | NEBRASKA— Cont'd | | | | | | | | | | | | | | | |
| 31009 | | 9 | Blaine | 710.7 | 457 | 3,139 | 0.6 | 97.6 | 1.1 | 0.7 | 0.0 | 2.0 | 6.3 | 9.8 | 10.3 | 11.2 | 8.8 | 10.7 |
| 31011 | | 9 | Boone | 686.5 | 5,096 | 2,819 | 7.4 | 96.4 | 0.9 | 0.4 | 0.3 | 2.5 | 6.5 | 13.1 | 10.6 | 9.9 | 10.2 | 10.2 |
| 31013 | | 7 | Box Butte | 1,075.4 | 10,696 | 2,368 | 9.9 | 82.5 | 1.7 | 3.6 | 1.2 | 13.2 | 6.3 | 14.9 | 11.0 | 11.7 | 12.0 | 10.3 |
| 31015 | | 9 | Boyd | 539.9 | 1,860 | 3,061 | 3.4 | 94.8 | 0.3 | 1.3 | 1.5 | 2.5 | 4.0 | 10.9 | 10.3 | 8.6 | 9.2 | 9.2 |
| 31017 | | 9 | Brown | 1,221.4 | 2,981 | 2,963 | 2.4 | 94.0 | 1.0 | 1.6 | 0.5 | 4.1 | 5.5 | 11.6 | 11.1 | 9.9 | 9.5 | 10.7 |
| 31019 | 28260 | 4 | Buffalo | 968.2 | 50,114 | 990 | 51.8 | 87.5 | 1.8 | 0.7 | 1.8 | 9.6 | 6.4 | 12.7 | 18.4 | 12.9 | 12.4 | 10.0 |
| 31021 | | 8 | Burt | 491.6 | 6,477 | 2,709 | 13.2 | 93.5 | 1.0 | 2.6 | 1.0 | 3.6 | 5.5 | 13.0 | 10.5 | 10.1 | 10.2 | 10.8 |
| 31023 | | 6 | Butler | 584.9 | 7,960 | 2,592 | 13.6 | 93.6 | 1.1 | 0.7 | 0.7 | 5.0 | 5.9 | 13.2 | 12.0 | 10.2 | 11.0 | 11.1 |
| 31025 | 36540 | 2 | Cass | 557.3 | 26,232 | 1,560 | 47.1 | 94.6 | 1.2 | 1.0 | 1.1 | 3.7 | 5.2 | 13.7 | 11.5 | 10.3 | 12.6 | 12.9 |
| 31027 | | 9 | Cedar | 740.2 | 8,414 | 2,541 | 11.4 | 96.6 | 0.7 | 0.8 | 0.6 | 2.2 | 6.5 | 14.6 | 11.6 | 9.3 | 10.2 | 10.1 |
| 31029 | | 9 | Chase | 894.4 | 3,840 | 2,905 | 4.3 | 84.2 | 0.9 | 0.4 | 0.5 | 14.9 | 6.2 | 14.5 | 11.0 | 9.7 | 11.8 | 11.1 |
| 31031 | | 7 | Cherry | 5,960.2 | 5,781 | 2,761 | 1.0 | 89.2 | 1.3 | 7.6 | 1.3 | 3.9 | 6.7 | 13.1 | 10.2 | 10.4 | 11.9 | 10.8 |
| 31033 | | 7 | Cheyenne | 1,196.0 | 9,111 | 2,486 | 7.6 | 89.1 | 1.2 | 1.4 | 1.6 | 8.3 | 6.2 | 13.1 | 10.6 | 11.0 | 11.8 | 10.9 |
| 31035 | | 8 | Clay | 572.3 | 6,216 | 2,731 | 10.9 | 89.9 | 0.8 | 0.7 | 0.4 | 9.0 | 6.3 | 13.6 | 11.4 | 11.6 | 10.5 | 10.9 |
| 31037 | | 7 | Colfax | 411.6 | 10,587 | 2,382 | 25.7 | 48.1 | 5.2 | 0.6 | 0.9 | 45.9 | 8.9 | 16.5 | 12.9 | 11.4 | 13.5 | 10.7 |
| 31039 | | 7 | Cuming | 570.5 | 8,798 | 2,509 | 15.4 | 88.5 | 0.9 | 0.8 | 0.6 | 10.2 | 6.5 | 13.6 | 12.3 | 10.2 | 10.5 | 10.7 |
| 31041 | | 7 | Custer | 2,575.6 | 10,626 | 2,380 | 4.1 | 94.9 | 1.2 | 0.9 | 0.5 | 3.6 | 6.0 | 13.7 | 10.7 | 10.9 | 10.8 | 10.3 |
| 31043 | 43580 | 3 | Dakota | 264.3 | 20,070 | 1,819 | 75.9 | 46.7 | 7.9 | 3.1 | 4.6 | 39.4 | 8.7 | 16.1 | 13.6 | 13.9 | 11.5 | 10.7 |
| 31045 | | 7 | Dawes | 1,396.4 | 8,361 | 2,549 | 6.0 | 86.6 | 2.6 | 4.4 | 2.5 | 5.9 | 4.6 | 9.7 | 24.7 | 10.1 | 9.9 | 9.0 |
| 31047 | 30420 | 7 | Dawson | 1,013.1 | 23,510 | 1,656 | 23.2 | 57.1 | 7.4 | 0.7 | 1.6 | 33.9 | 8.2 | 14.5 | 12.7 | 12.9 | 11.7 | 11.1 |
| 31049 | | 9 | Deuel | 439.9 | 1,793 | 3,066 | 4.1 | 91.4 | 0.3 | 1.4 | 0.6 | 7.4 | 4.7 | 12.2 | 9.6 | 9.0 | 8.9 | 12.3 |
| 31051 | 43580 | 3 | Dixon | 476.1 | 5,596 | 2,780 | 11.8 | 83.8 | 1.2 | 0.9 | 0.5 | 14.8 | 6.6 | 14.4 | 11.6 | 9.8 | 10.7 | 11.4 |
| 31053 | 23340 | 4 | Dodge | 529.1 | 36,222 | 1,270 | 68.5 | 83.0 | 1.5 | 0.9 | 1.0 | 14.7 | 6.5 | 13.6 | 13.5 | 11.3 | 11.5 | 11.0 |
| 31055 | 36540 | 2 | Douglas | 326.4 | 574,332 | 118 | 1,759.6 | 70.6 | 12.6 | 1.1 | 5.1 | 13.2 | 7.2 | 14.1 | 13.1 | 15.2 | 13.8 | 11.3 |
| 31057 | | 9 | Dundy | 919.7 | 1,671 | 3,072 | 1.8 | 89.8 | 1.4 | 1.6 | 0.7 | 8.1 | 5.1 | 10.8 | 11.6 | 8.7 | 10.2 | 11.4 |
| 31059 | | 8 | Fillmore | 575.4 | 5,519 | 2,786 | 9.6 | 94.0 | 1.1 | 0.9 | 0.9 | 4.0 | 5.3 | 10.7 | 10.1 | 11.7 | 11.3 | 11.0 |
| 31061 | | 9 | Franklin | 575.8 | 2,940 | 2,966 | 5.1 | 96.4 | 1.1 | 0.9 | 0.6 | 2.6 | 5.8 | 11.3 | 9.4 | 9.3 | 10.6 | 9.7 |
| 31063 | | 9 | Frontier | 974.6 | 2,587 | 2,996 | 2.7 | 95.7 | 1.0 | 0.9 | 0.5 | 3.0 | 4.9 | 11.0 | 13.6 | 9.5 | 9.7 | 10.8 |
| 31065 | | 9 | Furnas | 719.1 | 4,653 | 2,850 | 6.5 | 93.6 | 1.0 | 1.1 | 0.7 | 4.7 | 6.2 | 11.5 | 11.4 | 9.5 | 10.3 | 11.0 |
| 31067 | 13100 | 6 | Gage | 851.5 | 21,431 | 1,749 | 25.2 | 95.0 | 1.3 | 1.2 | 1.0 | 3.0 | 5.6 | 13.1 | 10.9 | 10.6 | 11.7 | 11.4 |
| 31069 | | 9 | Garden | 1,705.4 | 1,847 | 3,062 | 1.1 | 91.5 | 0.8 | 1.8 | 0.5 | 6.8 | 4.2 | 11.2 | 10.0 | 9.9 | 9.3 | 10.6 |
| 31071 | | 9 | Garfield | 569.3 | 1,956 | 3,052 | 3.4 | 97.9 | 0.4 | 0.3 | 0.2 | 1.6 | 5.6 | 9.0 | 10.6 | 9.9 | 9.2 | 9.3 |
| 31073 | 30420 | 9 | Gosper | 458.2 | 1,986 | 3,047 | 4.3 | 92.1 | 1.5 | 1.2 | 0.8 | 5.6 | 5.6 | 11.8 | 9.9 | 10.2 | 11.2 | 9.9 |
| 31075 | | 9 | Grant | 777.0 | 630 | 3,135 | 0.8 | 97.5 | 0.8 | 0.5 | 0.6 | 1.9 | 8.3 | 14.0 | 7.9 | 11.1 | 11.0 | 6.7 |
| 31077 | | 8 | Greeley | 569.8 | 2,319 | 3,014 | 4.1 | 96.2 | 0.9 | 0.5 | 0.3 | 2.6 | 5.8 | 12.8 | 10.9 | 9.6 | 10.1 | 9.1 |
| 31079 | 24260 | 3 | Hall | 546.4 | 61,028 | 863 | 111.7 | 65.2 | 3.5 | 0.8 | 1.7 | 30.0 | 7.8 | 15.3 | 12.7 | 12.9 | 12.6 | 11.4 |
| 31081 | | 3 | Hamilton | 542.1 | 9,237 | 2,477 | 17.0 | 95.2 | 0.8 | 0.7 | 0.6 | 3.8 | 6.5 | 13.1 | 11.6 | 10.8 | 11.7 | 11.5 |
| 31083 | | 9 | Harlan | 553.5 | 3,311 | 2,945 | 6.0 | 96.8 | 0.7 | 0.8 | 0.8 | 2.1 | 5.1 | 12.0 | 9.6 | 8.5 | 9.3 | 10.8 |
| 31085 | | 9 | Hayes | 713.1 | 916 | 3,110 | 1.3 | 93.6 | 1.2 | 0.9 | 1.1 | 4.7 | 8.1 | 12.3 | 8.6 | 9.4 | 9.4 | 5.8 |
| 31087 | | 9 | Hitchcock | 709.9 | 2,773 | 2,981 | 3.9 | 95.4 | 0.6 | 1.1 | 0.4 | 3.5 | 5.2 | 13.1 | 10.2 | 8.7 | 12.1 | 9.8 |
| 31089 | | 7 | Holt | 2,412.4 | 9,956 | 2,422 | 4.1 | 93.3 | 0.8 | 0.7 | 0.7 | 5.3 | 6.8 | 14.4 | 10.8 | 9.6 | 10.6 | 9.5 |
| 31091 | | 9 | Hooker | 721.2 | 647 | 3,132 | 0.9 | 96.0 | 0.3 | 1.1 | 0.6 | 3.2 | 4.8 | 12.5 | 10.5 | 8.0 | 11.3 | 8.5 |
| 31093 | 24260 | 3 | Howard | 569.3 | 6,488 | 2,707 | 11.4 | 95.9 | 0.8 | 0.6 | 0.7 | 2.8 | 6.4 | 13.7 | 10.3 | 11.2 | 11.3 | 11.6 |
| 31095 | | 7 | Jefferson | 570.2 | 7,099 | 2,651 | 12.5 | 93.5 | 0.8 | 0.9 | 0.7 | 5.1 | 6.0 | 12.4 | 10.2 | 10.3 | 10.9 | 10.8 |
| 31097 | | 8 | Johnson | 376.1 | 5,057 | 2,824 | 13.4 | 80.6 | 6.7 | 1.7 | 1.5 | 10.4 | 4.4 | 10.9 | 10.8 | 14.7 | 13.5 | 12.6 |
| 31099 | 28260 | 7 | Kearney | 516.2 | 6,652 | 2,691 | 12.9 | 91.4 | 0.5 | 0.6 | 0.7 | 7.7 | 6.3 | 14.2 | 11.3 | 11.5 | 11.9 | 10.7 |
| 31101 | | 7 | Keith | 1,061.7 | 7,983 | 2,590 | 7.5 | 89.9 | 1.2 | 1.2 | 1.2 | 8.0 | 5.7 | 11.2 | 9.7 | 9.6 | 10.3 | 10.4 |
| 31103 | | 9 | Keya Paha | 773.1 | 759 | 3,122 | 1.0 | 97.6 | 0.0 | 1.1 | 0.7 | 1.1 | 3.4 | 9.7 | 10.8 | 7.6 | 8.3 | 9.9 |
| 31105 | | 8 | Kimball | 951.9 | 3,495 | 2,935 | 3.7 | 88.2 | 0.9 | 2.5 | 1.7 | 8.9 | 5.5 | 11.1 | 10.4 | 10.1 | 9.8 | 11.3 |
| 31107 | | 9 | Knox | 1,108.4 | 8,304 | 2,556 | 7.5 | 86.3 | 0.9 | 10.5 | 0.8 | 3.5 | 6.2 | 13.7 | 11.5 | 8.5 | 10.0 | 9.7 |
| 31109 | 30700 | 2 | Lancaster | 837.6 | 320,650 | 223 | 382.8 | 82.9 | 5.6 | 1.2 | 5.5 | 7.6 | 6.0 | 12.7 | 18.6 | 13.8 | 12.7 | 10.5 |
| 31111 | 35820 | 5 | Lincoln | 2,564.1 | 34,347 | 1,320 | 13.4 | 87.9 | 1.6 | 0.9 | 1.2 | 9.6 | 6.0 | 13.2 | 11.3 | 11.0 | 12.5 | 11.7 |
| 31113 | 35820 | 9 | Logan | 570.7 | 747 | 3,124 | 1.3 | 92.9 | 0.4 | 2.0 | 0.5 | 5.6 | 5.5 | 12.6 | 11.5 | 9.2 | 13.7 | 9.9 |
| 31115 | | 9 | Loup | 563.5 | 650 | 3,131 | 1.2 | 94.8 | 1.1 | 0.2 | 0.2 | 3.8 | 4.9 | 12.0 | 9.7 | 9.8 | 6.3 | 10.3 |
| 31117 | 35820 | 9 | McPherson | 859.3 | 474 | 3,137 | 0.6 | 97.5 | 1.5 | 0.4 | 0.2 | 1.9 | 3.4 | 9.3 | 12.9 | 7.8 | 8.4 | 13.1 |
| 31119 | 35740 | 5 | Madison | 572.6 | 34,813 | 1,309 | 60.8 | 79.8 | 1.9 | 1.6 | 2.2 | 16.0 | 7.2 | 14.1 | 13.3 | 12.4 | 12.3 | 10.2 |
| 31121 | 24260 | 3 | Merrick | 487.3 | 7,809 | 2,605 | 16.0 | 92.5 | 0.8 | 1.0 | 1.4 | 5.4 | 5.7 | 12.1 | 11.4 | 11.1 | 11.6 | 11.7 |
| 31123 | | 9 | Morrill | 1,424.0 | 4,625 | 2,851 | 3.2 | 82.3 | 0.9 | 1.5 | 0.8 | 15.9 | 5.1 | 12.5 | 12.3 | 11.1 | 11.2 | 11.3 |
| 31125 | | 8 | Nance | 441.6 | 3,532 | 2,932 | 8.0 | 95.6 | 0.8 | 1.2 | 0.4 | 3.1 | 6.1 | 12.9 | 11.3 | 10.0 | 11.6 | 9.9 |
| 31127 | | 7 | Nemaha | 407.4 | 7,044 | 2,655 | 17.3 | 94.6 | 2.0 | 1.0 | 0.8 | 3.0 | 6.1 | 12.6 | 15.7 | 11.7 | 11.1 | 9.4 |
| 31129 | | 9 | Nuckolls | 575.2 | 4,134 | 2,885 | 7.2 | 94.8 | 0.6 | 1.1 | 1.7 | 3.1 | 5.2 | 10.7 | 10.5 | 9.3 | 10.1 | 9.9 |
| 31131 | | 6 | Otoe | 615.7 | 15,965 | 2,037 | 25.9 | 89.2 | 1.4 | 0.9 | 1.2 | 8.6 | 6.1 | 13.7 | 11.2 | 10.8 | 11.5 | 11.7 |
| 31133 | | 9 | Pawnee | 431.1 | 2,601 | 2,993 | 6.0 | 96.7 | 1.5 | 1.1 | 0.9 | 2.0 | 7.3 | 11.6 | 9.3 | 8.7 | 9.1 | 10.6 |
| 31135 | | 9 | Perkins | 883.3 | 2,867 | 2,973 | 3.2 | 94.2 | 0.8 | 0.6 | 0.6 | 4.7 | 6.1 | 13.8 | 11.2 | 9.1 | 12.5 | 10.0 |
| 31137 | | 7 | Phelps | 539.8 | 9,006 | 2,499 | 16.7 | 92.3 | 0.8 | 0.9 | 0.8 | 6.2 | 6.5 | 13.6 | 11.4 | 11.7 | 11.1 | 10.9 |
| 31139 | 35740 | 9 | Pierce | 573.2 | 7,184 | 2,642 | 12.5 | 96.5 | 0.9 | 0.7 | 0.6 | 2.2 | 6.4 | 14.3 | 11.5 | 10.2 | 12.0 | 10.9 |

1. CBSA = Core Based Statistical Area. See Appendix A for explanation. See Appendix B for list of metropolitan areas with component counties. Service of USDA Rural-Urban Continuum Codes. See Appendix A for definition. 3. Dry land or land partially or temporarily covered by water.

2. County type code from the Economic Research Service of USDA Rural-Urban Continuum Codes. 4. May be of any race.

# Table B. States and Counties — Population and Households

| STATE County | 55 to 64 years | 65 to 74 years | 75 years and over | Percent female | Total persons 2000 | Total persons 2010 | Percent change 2000-2010 | Percent change 2010-2020 | Births | Deaths | Net Migration | Number | Persons per household | Family households | Female family householder[1] | One person |
|---|---|---|---|---|---|---|---|---|---|---|---|---|---|---|---|---|
| | 16 | 17 | 18 | 19 | 20 | 21 | 22 | 23 | 24 | 25 | 26 | 27 | 28 | 29 | 30 | 31 |
| **NEBRASKA— Cont'd** | | | | | | | | | | | | | | | | |
| Blaine | 19.7 | 13.3 | 9.8 | 48.8 | 583 | 478 | -18.0 | -4.4 | 59 | 47 | -35 | 213 | 2.24 | 67.6 | 2.3 | 31.0 |
| Boone | 15.7 | 11.7 | 12.1 | 50.3 | 6,259 | 5,505 | -12.0 | -7.4 | 650 | 664 | -392 | 2,311 | 2.24 | 64.7 | 5.3 | 29.6 |
| Box Butte | 13.9 | 12.1 | 7.9 | 50.1 | 12,158 | 11,308 | -7.0 | -5.4 | 1,493 | 1,194 | -913 | 4,743 | 2.27 | 61.6 | 8.3 | 34.3 |
| Boyd | 16.8 | 16.7 | 14.2 | 50.4 | 2,438 | 2,099 | -13.9 | -11.4 | 171 | 321 | -86 | 904 | 2.11 | 64.0 | 2.3 | 33.3 |
| Brown | 15.2 | 13.4 | 13.0 | 50.3 | 3,525 | 3,143 | -10.8 | -5.2 | 298 | 397 | -64 | 1,387 | 2.14 | 58.2 | 5.9 | 37.2 |
| Buffalo | 11.1 | 9.0 | 7.0 | 50.0 | 42,259 | 46,099 | 9.1 | 8.7 | 6,945 | 3,645 | 724 | 19,062 | 2.47 | 61.7 | 8.8 | 29.1 |
| Burt | 14.7 | 13.3 | 11.8 | 49.9 | 7,791 | 6,858 | -12.0 | -5.6 | 702 | 987 | -97 | 2,856 | 2.24 | 66.2 | 8.8 | 29.5 |
| Butler | 15.4 | 11.7 | 9.5 | 49.4 | 8,767 | 8,395 | -4.2 | -5.2 | 938 | 1,058 | -312 | 3,363 | 2.34 | 64.6 | 5.6 | 31.6 |
| Cass | 15.2 | 11.4 | 7.2 | 49.3 | 24,334 | 25,241 | 3.7 | 3.9 | 2,824 | 2,428 | 608 | 9,945 | 2.57 | 70.2 | 6.3 | 24.5 |
| Cedar | 15.3 | 11.4 | 11.1 | 49.2 | 9,615 | 8,852 | -7.9 | -4.9 | 1,094 | 1,043 | -493 | 3,506 | 2.38 | 67.7 | 4.6 | 27.8 |
| Chase | 14.1 | 10.7 | 11.0 | 50.8 | 4,068 | 3,966 | -2.5 | -3.2 | 491 | 533 | -80 | 1,679 | 2.21 | 68.6 | 6.4 | 28.2 |
| Cherry | 14.0 | 11.6 | 11.4 | 48.9 | 6,148 | 5,713 | -7.1 | 1.2 | 757 | 636 | -50 | 2,537 | 2.23 | 66.1 | 7.1 | 29.9 |
| Cheyenne | 14.5 | 12.6 | 9.2 | 50.0 | 9,830 | 9,998 | 1.7 | -8.9 | 1,236 | 979 | -1,149 | 4,395 | 2.16 | 55.7 | 6.4 | 40.4 |
| Clay | 14.7 | 12.1 | 8.9 | 49.2 | 7,039 | 6,539 | -7.1 | -4.9 | 801 | 729 | -395 | 2,539 | 2.41 | 66.0 | 6.1 | 29.3 |
| Colfax | 12.3 | 7.7 | 6.1 | 46.3 | 10,441 | 10,515 | 0.7 | 0.7 | 1,918 | 805 | -1,051 | 3,731 | 2.85 | 64.6 | 5.6 | 24.1 |
| Cuming | 14.1 | 11.0 | 11.2 | 49.9 | 10,203 | 9,139 | -10.4 | -3.7 | 1,148 | 1,013 | -476 | 3,727 | 2.36 | 64.8 | 5.3 | 32.2 |
| Custer | 14.5 | 12.4 | 10.7 | 50.2 | 11,793 | 10,939 | -7.2 | -2.9 | 1,302 | 1,326 | -282 | 4,862 | 2.20 | 65.4 | 8.4 | 31.1 |
| Dakota | 11.5 | 8.1 | 5.9 | 49.5 | 20,253 | 21,006 | 3.7 | -4.5 | 3,721 | 1,639 | -3,050 | 7,404 | 2.69 | 72.7 | 14.4 | 18.5 |
| Dawes | 12.5 | 10.1 | 9.4 | 50.9 | 9,060 | 9,182 | 1.3 | -8.9 | 934 | 875 | -888 | 3,525 | 2.21 | 57.0 | 4.2 | 32.3 |
| Dawson | 11.6 | 9.7 | 7.5 | 48.0 | 24,365 | 24,316 | -0.2 | -3.3 | 4,047 | 2,165 | -2,718 | 8,965 | 2.61 | 70.2 | 11.4 | 23.9 |
| Deuel | 17.6 | 14.4 | 11.3 | 49.7 | 2,098 | 1,932 | -7.9 | -7.2 | 161 | 225 | -76 | 830 | 2.19 | 66.5 | 8.6 | 30.6 |
| Dixon | 14.0 | 12.5 | 9.0 | 49.1 | 6,339 | 6,003 | -5.3 | -6.8 | 761 | 637 | -536 | 2,352 | 2.40 | 66.7 | 5.8 | 30.5 |
| Dodge | 13.3 | 10.3 | 9.1 | 50.2 | 36,160 | 36,683 | 1.4 | -1.3 | 4,914 | 4,513 | -848 | 15,261 | 2.33 | 64.3 | 9.5 | 28.8 |
| Douglas | 11.5 | 8.4 | 5.4 | 50.6 | 463,585 | 517,120 | 11.5 | 11.1 | 86,359 | 41,130 | 12,363 | 218,061 | 2.51 | 61.8 | 11.3 | 31.0 |
| Dundy | 16.8 | 13.7 | 11.5 | 49.8 | 2,292 | 2,008 | -12.4 | -16.8 | 176 | 284 | -230 | 872 | 2.14 | 54.7 | 7.2 | 41.5 |
| Fillmore | 16.1 | 12.6 | 11.1 | 50.0 | 6,634 | 5,890 | -11.2 | -6.3 | 601 | 863 | -107 | 2,510 | 2.12 | 60.5 | 5.3 | 34.9 |
| Franklin | 15.6 | 15.1 | 13.2 | 49.8 | 3,574 | 3,225 | -9.8 | -8.8 | 312 | 396 | -199 | 1,377 | 2.11 | 63.8 | 6.0 | 31.5 |
| Frontier | 15.0 | 13.3 | 12.0 | 48.6 | 3,099 | 2,756 | -11.1 | -6.1 | 260 | 193 | -234 | 1,135 | 2.23 | 67.0 | 3.8 | 30.9 |
| Furnas | 14.2 | 14.3 | 11.6 | 50.1 | 5,324 | 4,959 | -6.9 | -6.2 | 549 | 681 | -170 | 2,128 | 2.19 | 63.8 | 5.1 | 31.3 |
| Gage | 15.2 | 11.5 | 10.1 | 50.0 | 22,993 | 22,311 | -3.0 | -3.9 | 2,401 | 2,833 | -444 | 8,979 | 2.34 | 60.9 | 6.1 | 34.3 |
| Garden | 17.1 | 13.0 | 14.7 | 47.6 | 2,292 | 2,057 | -10.3 | -10.2 | 186 | 299 | -100 | 868 | 2.10 | 60.0 | 2.9 | 35.1 |
| Garfield | 16.0 | 14.5 | 16.0 | 50.3 | 1,902 | 2,047 | 7.6 | -4.4 | 176 | 351 | 85 | 884 | 2.20 | 68.1 | 6.0 | 27.9 |
| Gosper | 17.1 | 12.7 | 11.6 | 48.8 | 2,143 | 2,044 | -4.6 | -2.8 | 224 | 243 | -38 | 942 | 2.10 | 64.3 | 4.1 | 30.0 |
| Grant | 16.8 | 14.8 | 9.5 | 47.5 | 747 | 614 | -17.8 | 2.6 | 94 | 62 | -17 | 288 | 2.35 | 68.8 | 2.4 | 28.8 |
| Greeley | 15.2 | 12.8 | 13.8 | 50.0 | 2,714 | 2,538 | -6.5 | -8.6 | 289 | 295 | -216 | 1,019 | 2.31 | 65.8 | 7.0 | 32.3 |
| Hall | 12.2 | 8.7 | 6.5 | 49.5 | 53,534 | 58,621 | 9.5 | 4.1 | 9,747 | 5,571 | -1,739 | 23,096 | 2.61 | 68.0 | 14.5 | 25.9 |
| Hamilton | 14.0 | 11.3 | 9.5 | 49.4 | 9,403 | 9,114 | -3.1 | 1.3 | 1,060 | 842 | -92 | 3,713 | 2.45 | 73.3 | 6.7 | 23.6 |
| Harlan | 16.4 | 15.5 | 12.7 | 48.5 | 3,786 | 3,417 | -9.7 | -3.1 | 378 | 425 | -58 | 1,521 | 2.21 | 64.8 | 4.6 | 33.5 |
| Hayes | 18.0 | 14.5 | 13.9 | 49.8 | 1,068 | 960 | -10.1 | -4.6 | 140 | 55 | -132 | 403 | 2.22 | 64.8 | 1.2 | 32.5 |
| Hitchcock | 14.5 | 15.1 | 11.2 | 49.8 | 3,111 | 2,908 | -6.5 | -4.6 | 302 | 393 | -45 | 1,243 | 2.23 | 58.6 | 6.3 | 37.2 |
| Holt | 15.5 | 12.4 | 10.3 | 49.6 | 11,551 | 10,435 | -9.7 | -4.6 | 1,410 | 1,265 | -621 | 4,487 | 2.22 | 62.5 | 5.5 | 31.0 |
| Hooker | 13.8 | 14.1 | 16.5 | 51.2 | 783 | 736 | -6.0 | -12.1 | 68 | 128 | -31 | 318 | 1.97 | 57.2 | 7.2 | 39.6 |
| Howard | 14.2 | 11.7 | 9.6 | 49.4 | 6,567 | 6,274 | -4.5 | 3.4 | 777 | 691 | 134 | 2,720 | 2.36 | 67.2 | 6.3 | 27.6 |
| Jefferson | 15.0 | 13.5 | 11.0 | 50.8 | 8,333 | 7,547 | -9.4 | -5.9 | 811 | 1,099 | -159 | 3,302 | 2.13 | 55.4 | 7.8 | 40.4 |
| Johnson | 14.4 | 10.1 | 8.6 | 41.0 | 4,488 | 5,217 | 16.2 | -3.1 | 436 | 636 | 34 | 1,809 | 2.28 | 67.6 | 8.2 | 29.2 |
| Kearney | 14.0 | 10.8 | 9.4 | 50.1 | 6,882 | 6,489 | -5.7 | 2.5 | 825 | 747 | 87 | 2,666 | 2.42 | 63.8 | 5.4 | 32.4 |
| Keith | 15.8 | 15.2 | 12.0 | 49.8 | 8,875 | 8,371 | -5.7 | -4.6 | 834 | 967 | -250 | 3,882 | 2.06 | 59.4 | 6.6 | 35.6 |
| Keya Paha | 17.7 | 15.5 | 17.0 | 48.9 | 983 | 824 | -16.2 | -7.9 | 69 | 72 | -63 | 304 | 2.50 | 65.1 | 5.3 | 30.6 |
| Kimball | 15.3 | 13.6 | 12.9 | 50.5 | 4,089 | 3,821 | -6.6 | -8.5 | 403 | 525 | -202 | 1,577 | 2.27 | 64.8 | 10.3 | 28.3 |
| Knox | 14.4 | 13.5 | 12.5 | 50.8 | 9,374 | 8,701 | -7.2 | -4.6 | 1,042 | 1,164 | -269 | 3,586 | 2.29 | 65.4 | 5.4 | 30.8 |
| Lancaster | 10.9 | 9.0 | 5.8 | 49.8 | 250,291 | 285,407 | 14.0 | 12.3 | 40,935 | 21,669 | 16,065 | 124,324 | 2.41 | 59.3 | 8.8 | 30.3 |
| Lincoln | 13.7 | 11.8 | 8.8 | 50.4 | 34,632 | 36,288 | 4.8 | -5.3 | 4,345 | 3,776 | -2,511 | 14,856 | 2.32 | 62.5 | 9.4 | 30.3 |
| Logan | 13.9 | 13.0 | 10.7 | 48.6 | 774 | 763 | -1.4 | -2.1 | 97 | 59 | -56 | 347 | 2.67 | 67.4 | 2.9 | 25.4 |
| Loup | 19.7 | 16.9 | 10.3 | 49.5 | 712 | 628 | -11.8 | 3.5 | 62 | 71 | 31 | 294 | 2.06 | 67.3 | 4.4 | 29.3 |
| McPherson | 19.8 | 13.9 | 11.4 | 50.0 | 533 | 539 | 1.1 | -12.1 | 29 | 39 | -57 | 191 | 2.07 | 62.9 | 1.0 | 39.3 |
| Madison | 13.7 | 9.3 | 7.3 | 50.4 | 35,226 | 34,876 | -1.0 | -0.2 | 5,295 | 3,710 | -1,641 | 14,205 | 2.39 | 62.9 | 9.1 | 30.2 |
| Merrick | 15.2 | 11.5 | 9.8 | 49.8 | 8,204 | 7,859 | -4.2 | -0.6 | 927 | 880 | -94 | 3,373 | 2.26 | 66.6 | 8.0 | 28.4 |
| Morrill | 14.0 | 12.0 | 10.5 | 48.7 | 5,440 | 5,042 | -7.3 | -8.3 | 533 | 597 | -354 | 1,954 | 2.36 | 60.7 | 8.3 | 34.6 |
| Nance | 16.6 | 12.6 | 9.0 | 50.3 | 4,038 | 3,735 | -7.5 | -5.4 | 420 | 508 | -117 | 1,546 | 2.20 | 63.5 | 7.0 | 31.4 |
| Nemaha | 12.8 | 11.6 | 8.9 | 50.1 | 7,576 | 7,248 | -4.3 | -2.8 | 804 | 860 | -147 | 2,977 | 2.16 | 61.3 | 7.9 | 33.2 |
| Nuckolls | 16.7 | 13.8 | 13.8 | 50.4 | 5,057 | 4,500 | -11.0 | -8.1 | 434 | 630 | -164 | 1,867 | 2.24 | 59.8 | 5.5 | 37.3 |
| Otoe | 14.6 | 11.0 | 9.4 | 50.3 | 15,396 | 15,740 | 2.2 | 1.4 | 1,985 | 1,889 | 139 | 6,510 | 2.42 | 65.1 | 9.9 | 29.1 |
| Pawnee | 13.9 | 14.9 | 14.7 | 50.5 | 3,087 | 2,773 | -10.2 | -6.2 | 323 | 373 | -119 | 1,222 | 2.14 | 63.0 | 9.8 | 31.8 |
| Perkins | 13.3 | 13.2 | 10.8 | 48.9 | 3,200 | 2,967 | -7.3 | -3.4 | 373 | 335 | -140 | 1,228 | 2.34 | 68.4 | 4.2 | 30.0 |
| Phelps | 13.8 | 10.3 | 10.7 | 49.8 | 9,747 | 9,188 | -5.7 | -2.0 | 1,173 | 1,159 | -188 | 3,905 | 2.26 | 62.7 | 4.1 | 30.6 |
| Pierce | 14.6 | 10.5 | 9.6 | 49.5 | 7,857 | 7,266 | -7.5 | -1.1 | 881 | 721 | -240 | 3,022 | 2.32 | 66.2 | 5.5 | 28.4 |

1. No spouse present.

# Table B. States and Counties — Population, Vital Statistics, Health, and Crime

| STATE County | Persons in group quarters, 2020 | Daytime Population, 2015-2019 | | Births, 2020 | | Deaths, 2020 | | Persons under 65 with no health insurance, 2019 | | Medicare, 2020 | | | Crimes reported by county police or sheriff, 2019 | |
| | | Number | Employment/ residence ratio | Total | Rate[1] | Number | Rate[1] | Number | Percent | Total beneficiaries | Enrolled in Original Medicare | Enrolled in Medicare Advantage | Violent | Property |
| | 32 | 33 | 34 | 35 | 36 | 37 | 38 | 39 | 40 | 41 | 42 | 43 | 44 | 45 |
| NEBRASKA— Cont'd | | | | | | | | | | | | | | |
| Blaine | 0 | 441 | 0.85 | 6 | 13.1 | 9 | 19.7 | 67 | 20.1 | 119 | D | D | NA | NA |
| Boone | 98 | 5,210 | 0.98 | 70 | 13.7 | 64 | 12.6 | 381 | 9.6 | 1,296 | 1,174 | 121 | NA | NA |
| Box Butte | 126 | 11,385 | 1.08 | 129 | 12.1 | 109 | 10.2 | 887 | 10.3 | 2,408 | 2,279 | 129 | 2 | 15 |
| Boyd | 28 | 1,828 | 0.89 | 12 | 6.5 | 28 | 15.1 | 165 | 12.6 | 615 | 596 | 19 | 4 | 14 |
| Brown | 26 | 3,007 | 0.99 | 33 | 11.1 | 33 | 11.1 | 282 | 13.2 | 813 | D | D | NA | NA |
| Buffalo | 2,156 | 50,513 | 1.05 | 609 | 12.2 | 413 | 8.2 | 3,849 | 9.5 | 8,475 | 7,163 | 1,312 | 13 | 87 |
| Burt | 127 | 5,752 | 0.74 | 63 | 9.7 | 100 | 15.4 | 541 | 11.2 | 1,838 | 1,504 | 334 | 3 | 20 |
| Butler | 166 | 6,779 | 0.70 | 100 | 12.6 | 99 | 12.4 | 500 | 8.0 | 1,868 | 1,636 | 232 | 6 | 28 |
| Cass | 253 | 18,664 | 0.45 | 251 | 9.6 | 243 | 9.3 | 1,416 | 6.6 | 5,422 | 4,167 | 1,255 | 27 | 138 |
| Cedar | 137 | 7,515 | 0.78 | 112 | 13.3 | 101 | 12.0 | 617 | 9.5 | 1,958 | 1,611 | 347 | 3 | 27 |
| Chase | 69 | 3,959 | 1.09 | 49 | 12.8 | 56 | 14.6 | 422 | 13.8 | 884 | D | D | 2 | 6 |
| Cherry | 38 | 5,970 | 1.06 | 76 | 13.1 | 57 | 9.9 | 700 | 16.0 | 1,359 | 1,346 | 14 | 2 | 9 |
| Cheyenne | 70 | 9,962 | 1.07 | 106 | 11.6 | 85 | 9.3 | 524 | 7.4 | 2,236 | 2,033 | 203 | 7 | 14 |
| Clay | 72 | 5,525 | 0.77 | 76 | 12.2 | 73 | 11.7 | 533 | 10.9 | 1,502 | 1,418 | 83 | NA | NA |
| Colfax | 51 | 10,193 | 0.90 | 190 | 17.9 | 74 | 7.0 | 1,612 | 17.8 | 1,565 | 1,438 | 127 | 9 | 27 |
| Cuming | 93 | 8,602 | 0.92 | 116 | 13.2 | 110 | 12.5 | 802 | 11.8 | 2,044 | 1,834 | 209 | 6 | 3 |
| Custer | 117 | 10,947 | 1.02 | 125 | 11.8 | 114 | 10.7 | 916 | 11.2 | 2,623 | 2,466 | 157 | 6 | 35 |
| Dakota | 253 | 21,819 | 1.16 | 342 | 17.0 | 136 | 6.8 | 2,606 | 15.5 | 3,196 | 2,195 | 1,001 | NA | NA |
| Dawes | 978 | 8,602 | 0.95 | 78 | 9.3 | 86 | 10.3 | 660 | 10.6 | 1,731 | 1,531 | 201 | NA | NA |
| Dawson | 296 | 23,441 | 0.98 | 396 | 16.8 | 217 | 9.2 | 2,873 | 15.1 | 4,320 | 3,911 | 409 | 18 | 36 |
| Deuel | 0 | 1,740 | 0.90 | 18 | 10.0 | 15 | 8.4 | 145 | 10.7 | 511 | 497 | 14 | 4 | 5 |
| Dixon | 74 | 4,466 | 0.57 | 68 | 12.2 | 53 | 9.5 | 468 | 10.6 | 1,304 | 999 | 305 | 2 | 30 |
| Dodge | 1,068 | 36,049 | 0.97 | 461 | 12.7 | 428 | 11.8 | 2,875 | 10.1 | 8,037 | 6,200 | 1,837 | 3 | 92 |
| Douglas | 12,040 | 612,047 | 1.18 | 8,194 | 14.3 | 4,329 | 7.5 | 49,720 | 10.3 | 89,674 | 59,732 | 29,943 | 71 | 605 |
| Dundy | 27 | 1,865 | 0.95 | 20 | 12.0 | 25 | 15.0 | 249 | 20.0 | 472 | D | D | NA | NA |
| Fillmore | 194 | 5,326 | 0.92 | 59 | 10.7 | 80 | 14.5 | 354 | 8.7 | 1,466 | 1,359 | 107 | 5 | 16 |
| Franklin | 23 | 2,518 | 0.68 | 36 | 12.2 | 36 | 12.2 | 215 | 10.2 | 838 | 784 | 54 | 0 | 10 |
| Frontier | 113 | 2,471 | 0.87 | 27 | 10.4 | 16 | 6.2 | 207 | 11.1 | 548 | 533 | 15 | 3 | 24 |
| Furnas | 60 | 4,824 | 1.03 | 62 | 13.3 | 48 | 10.3 | 423 | 12.3 | 1,365 | 1,292 | 72 | 7 | 57 |
| Gage | 536 | 20,227 | 0.87 | 209 | 9.8 | 278 | 13.0 | 1,477 | 8.8 | 5,411 | 4,366 | 1,045 | 11 | 77 |
| Garden | 21 | 1,896 | 1.04 | 12 | 6.5 | 29 | 15.7 | 171 | 13.2 | 577 | 563 | 14 | NA | NA |
| Garfield | 60 | 1,987 | 0.99 | 23 | 11.8 | 27 | 13.8 | 133 | 9.7 | 494 | 464 | 30 | NA | NA |
| Gosper | 40 | 1,616 | 0.65 | 23 | 11.6 | 23 | 11.6 | 129 | 8.7 | 504 | 445 | 59 | 2 | 9 |
| Grant | 3 | 699 | 0.93 | 6 | 9.5 | 8 | 12.7 | 54 | 11.3 | 166 | 166 | 0 | NA | NA |
| Greeley | 57 | 2,254 | 0.89 | 26 | 11.2 | 32 | 13.8 | 225 | 12.9 | 592 | 566 | 26 | 0 | 0 |
| Hall | 820 | 64,915 | 1.12 | 938 | 15.4 | 550 | 9.0 | 6,520 | 12.8 | 10,615 | 8,723 | 1,892 | 35 | 113 |
| Hamilton | 126 | 8,487 | 0.85 | 106 | 11.5 | 86 | 9.3 | 517 | 7.0 | 2,075 | 1,888 | 187 | 2 | 20 |
| Harlan | 45 | 3,105 | 0.81 | 34 | 10.3 | 35 | 10.6 | 269 | 11.1 | 931 | 878 | 53 | NA | NA |
| Hayes | 0 | 771 | 0.74 | 10 | 10.9 | 7 | 7.6 | 126 | 19.4 | 205 | 205 | 0 | NA | NA |
| Hitchcock | 25 | 2,472 | 0.74 | 25 | 9.0 | 42 | 15.1 | 231 | 11.5 | 785 | D | D | NA | NA |
| Holt | 99 | 10,203 | 1.00 | 128 | 12.9 | 110 | 11.0 | 773 | 10.0 | 2,522 | 2,378 | 144 | 1 | 2 |
| Hooker | 23 | 715 | 1.08 | 5 | 7.7 | 9 | 13.9 | 53 | 11.3 | 232 | D | D | 0 | 0 |
| Howard | 66 | 5,261 | 0.65 | 78 | 12.0 | 79 | 12.2 | 469 | 9.4 | 1,474 | 1,354 | 120 | NA | NA |
| Jefferson | 99 | 7,513 | 1.11 | 84 | 11.8 | 95 | 13.4 | 487 | 9.3 | 1,984 | 1,681 | 303 | 14 | 95 |
| Johnson | 1,056 | 4,868 | 0.86 | 35 | 6.9 | 58 | 11.5 | 323 | 10.4 | 1,030 | 935 | 95 | 5 | 39 |
| Kearney | 73 | 5,862 | 0.80 | 79 | 11.9 | 62 | 9.3 | 380 | 7.4 | 1,427 | 1,280 | 147 | 6 | 40 |
| Keith | 30 | 7,996 | 0.98 | 82 | 10.3 | 92 | 11.5 | 593 | 10.2 | 2,236 | 2,016 | 220 | 3 | 17 |
| Keya Paha | 0 | 723 | 0.90 | 5 | 6.6 | 8 | 10.5 | 101 | 18.8 | 203 | D | D | 2 | 0 |
| Kimball | 50 | 3,627 | 1.00 | 34 | 9.7 | 52 | 14.9 | 330 | 12.3 | 1,002 | 973 | 29 | NA | NA |
| Knox | 204 | 7,973 | 0.89 | 101 | 12.2 | 99 | 11.9 | 767 | 12.5 | 2,299 | 1,929 | 369 | 0 | 29 |
| Lancaster | 15,272 | 324,178 | 1.06 | 3,796 | 11.8 | 2,313 | 7.2 | 21,201 | 8.2 | 51,847 | 41,029 | 10,818 | 20 | 146 |
| Lincoln | 508 | 36,009 | 1.04 | 400 | 11.6 | 404 | 11.8 | 2,331 | 8.5 | 7,943 | 6,809 | 1,134 | 16 | 66 |
| Logan | 0 | 795 | 0.71 | 7 | 9.4 | 4 | 5.4 | 64 | 11.1 | 165 | 152 | 13 | NA | NA |
| Loup | 0 | 526 | 0.74 | 7 | 10.8 | 10 | 15.4 | 53 | 11.0 | 173 | 160 | 13 | NA | NA |
| McPherson | 0 | 343 | 0.75 | 2 | 4.2 | 4 | 8.4 | 48 | 12.6 | 85 | 72 | 13 | NA | NA |
| Madison | 1,218 | 38,588 | 1.20 | 495 | 14.2 | 375 | 10.8 | 3,176 | 11.2 | 6,707 | 5,388 | 1,319 | 10 | 19 |
| Merrick | 198 | 6,865 | 0.76 | 86 | 11.0 | 92 | 11.8 | 608 | 10.1 | 1,832 | 1,645 | 187 | 3 | 3 |
| Morrill | 82 | 4,510 | 0.88 | 50 | 10.8 | 43 | 9.3 | 472 | 13.2 | 1,148 | 1,066 | 82 | NA | NA |
| Nance | 142 | 3,286 | 0.85 | 41 | 11.6 | 37 | 10.5 | 256 | 9.5 | 856 | 798 | 58 | NA | NA |
| Nemaha | 502 | 7,060 | 1.03 | 78 | 11.1 | 74 | 10.5 | 398 | 7.7 | 1,553 | 1,443 | 111 | 10 | 17 |
| Nuckolls | 46 | 4,004 | 0.89 | 44 | 10.6 | 68 | 16.4 | 320 | 10.7 | 1,259 | 1,247 | 12 | NA | NA |
| Otoe | 240 | 15,127 | 0.90 | 179 | 11.2 | 172 | 10.8 | 1,212 | 9.5 | 3,482 | 2,834 | 648 | NA | NA |
| Pawnee | 31 | 2,613 | 0.97 | 37 | 14.2 | 33 | 12.7 | 216 | 11.8 | 762 | 710 | 52 | NA | NA |
| Perkins | 30 | 2,933 | 1.02 | 29 | 10.1 | 27 | 9.4 | 251 | 11.4 | 684 | 650 | 34 | 0 | 24 |
| Phelps | 244 | 9,900 | 1.17 | 115 | 12.8 | 100 | 11.1 | 562 | 8.0 | 2,119 | 2,061 | 59 | NA | NA |
| Pierce | 108 | 5,901 | 0.66 | 92 | 12.8 | 70 | 9.7 | 482 | 8.4 | 1,502 | 1,388 | 114 | NA | NA |

1. Per 1,000 estimated resident population.

# Table B. States and Counties — Crime, Education, Money Income, and Poverty

| STATE County | County law enforcement employment, 2019 | | Education | | | | | | Money income, 2015-2019 | | | | Income and poverty, 2019 | | | |
|---|---|---|---|---|---|---|---|---|---|---|---|---|---|---|---|---|
| | | | School enrollment and attainment, 2015-2019 | | | | Local government expenditures,[3] 2017-2018 | | | Households | | | | Percent below poverty level | | |
| | | | Enrollment[1] | | Attainment[2] (percent) | | | | | | Percent | | | | | |
| | Officers | Civilians | Total | Percent private | High school graduate or less | Bachelor's degree or more | Total current spending (mil dol) | Current spending per student (dollars) | Per capita income[4] | Median income (dollars) | with income of less than $50,000 | with income of $200,000 or more | Median household income (dollars) | All persons | Children under 18 years | Children 5 to 17 years in families |
| | 46 | 47 | 48 | 49 | 50 | 51 | 52 | 53 | 54 | 55 | 56 | 57 | 58 | 59 | 60 | 61 |
| NEBRASKA— Cont'd | | | | | | | | | | | | | | | | |
| Blaine................ | NA | NA | 92 | 7.6 | 36.0 | 23.0 | 3 | 28,089 | 25,097 | 51,094 | 48.4 | 0.0 | 45,411 | 15.9 | 28.2 | 27.1 |
| Boone................ | 5 | 9 | 1,111 | 17.6 | 42.6 | 21.1 | 13 | 16,777 | 33,027 | 54,979 | 45.0 | 5.1 | 58,204 | 10.0 | 12.1 | 11.3 |
| Box Butte........... | 6 | 1 | 2,425 | 13.3 | 42.3 | 17.1 | 22 | 12,069 | 28,156 | 54,004 | 47.8 | 1.3 | 61,522 | 10.3 | 13.7 | 12.4 |
| Boyd................. | NA | NA | 359 | 5.8 | 47.9 | 16.7 | 6 | 18,234 | 29,871 | 47,778 | 51.8 | 2.1 | 46,548 | 14.1 | 21.9 | 20.5 |
| Brown................ | 5 | 3 | 592 | 9.5 | 45.7 | 20.7 | 11 | 25,837 | 30,674 | 43,098 | 54.9 | 2.2 | 48,843 | 11.6 | 16.9 | 14.9 |
| Buffalo.............. | 29 | 18 | 14,381 | 9.8 | 32.5 | 34.6 | 101 | 12,214 | 31,043 | 59,431 | 42.1 | 4.0 | 64,759 | 11.6 | 10.1 | 9.1 |
| Burt................. | 5 | 5 | 1,399 | 9.6 | 44.4 | 21.3 | 19 | 15,100 | 31,635 | 54,203 | 46.2 | 3.5 | 54,554 | 10.1 | 12.9 | 11.5 |
| Butler............... | 11 | 10 | 1,810 | 26.9 | 44.5 | 21.4 | 17 | 18,201 | 30,095 | 58,979 | 42.2 | 2.2 | 63,310 | 8.2 | 9.2 | 7.8 |
| Cass................. | 26 | 15 | 6,333 | 13.0 | 35.5 | 27.7 | 47 | 12,245 | 33,851 | 71,846 | 33.0 | 4.8 | 72,915 | 5.3 | 7.2 | 6.3 |
| Cedar................ | 7 | 0 | 2,034 | 18.3 | 42.9 | 21.0 | 21 | 16,812 | 30,163 | 61,869 | 40.8 | 3.3 | 60,400 | 8.6 | 10.2 | 9.5 |
| Chase................ | 4 | 4 | 785 | 3.2 | 37.7 | 21.0 | 12 | 13,171 | 30,080 | 57,009 | 41.5 | 3.3 | 57,656 | 9.1 | 10.8 | 9.7 |
| Cherry............... | 4 | 1 | 1,209 | 11.7 | 36.6 | 22.9 | 13 | 17,615 | 29,415 | 56,651 | 44.2 | 2.7 | 50,677 | 10.7 | 17.2 | 16.4 |
| Cheyenne............ | 10 | 10 | 2,163 | 9.8 | 34.0 | 25.4 | 23 | 13,498 | 31,294 | 53,871 | 45.1 | 4.3 | 56,229 | 10.1 | 12.8 | 11.7 |
| Clay................. | 7 | 4 | 1,427 | 5.0 | 42.9 | 19.9 | 22 | 15,863 | 27,539 | 57,173 | 43.9 | 2.5 | 63,829 | 10.6 | 13.3 | 12.8 |
| Colfax............... | 10 | 8 | 2,584 | 5.2 | 59.7 | 15.4 | 33 | 12,061 | 23,603 | 58,872 | 43.0 | 4.2 | 62,472 | 8.2 | 11.4 | 11.3 |
| Cuming............... | 5 | 1 | 2,071 | 23.6 | 46.7 | 23.8 | 22 | 14,982 | 29,765 | 56,768 | 45.3 | 2.7 | 59,505 | 8.9 | 10.4 | 9.4 |
| Custer............... | 8 | 1 | 2,140 | 5.1 | 36.6 | 24.0 | 29 | 14,994 | 33,561 | 52,184 | 48.1 | 4.1 | 51,004 | 11.8 | 15.0 | 13.9 |
| Dakota............... | 17 | 2 | 5,357 | 9.4 | 58.8 | 13.0 | 54 | 12,631 | 25,269 | 59,231 | 38.7 | 3.5 | 61,010 | 10.6 | 14.0 | 13.2 |
| Dawes................ | 7 | 10 | 2,852 | 4.8 | 29.6 | 36.9 | 15 | 12,081 | 25,383 | 50,750 | 49.1 | 1.3 | 46,377 | 16.4 | 18.0 | 14.3 |
| Dawson............... | 24 | 42 | 5,660 | 4.2 | 53.6 | 15.8 | 69 | 12,607 | 25,956 | 53,721 | 46.2 | 1.8 | 56,839 | 11.9 | 15.9 | 12.6 |
| Deuel................ | 6 | 1 | 336 | 6.3 | 44.4 | 15.9 | 8 | 19,484 | 26,351 | 47,287 | 53.9 | 1.8 | 48,793 | 11.2 | 21.9 | 19.1 |
| Dixon................ | 8 | 6 | 1,363 | 8.3 | 47.4 | 21.2 | 14 | 15,679 | 27,634 | 56,905 | 44.3 | 2.9 | 57,324 | 8.7 | 10.8 | 10.0 |
| Dodge................ | 23 | 3 | 8,897 | 18.3 | 46.3 | 19.0 | 81 | 13,144 | 28,019 | 54,085 | 46.7 | 2.2 | 60,853 | 9.5 | 11.2 | 9.8 |
| Douglas.............. | 132 | 78 | 151,965 | 22.9 | 30.8 | 39.7 | 1,192 | 12,000 | 35,307 | 64,629 | 38.6 | 7.1 | 67,264 | 10.2 | 12.0 | 10.9 |
| Dundy................ | 4 | 4 | 437 | 3.4 | 41.1 | 23.3 | 6 | 18,586 | 25,479 | 41,716 | 58.0 | 0.7 | 45,446 | 13.4 | 18.2 | 17.3 |
| Fillmore............. | 7 | 7 | 1,077 | 11.1 | 41.4 | 21.4 | 16 | 17,001 | 34,483 | 60,313 | 42.0 | 3.6 | 60,440 | 8.5 | 9.9 | 9.2 |
| Franklin............. | 4 | 4 | 512 | 2.3 | 42.8 | 17.3 | 5 | 16,084 | 28,365 | 49,282 | 51.1 | 2.4 | 49,659 | 14.3 | 20.2 | 19.0 |
| Frontier............. | 5 | 4 | 664 | 4.5 | 33.2 | 21.5 | 11 | 19,227 | 28,899 | 54,659 | 45.9 | 2.0 | 55,712 | 12.9 | 15.3 | 14.2 |
| Furnas............... | 6 | 6 | 991 | 2.7 | 44.3 | 17.4 | 17 | 15,352 | 27,021 | 48,838 | 52.2 | 2.5 | 50,705 | 12.2 | 17.4 | 16.3 |
| Gage................. | 15 | 3 | 4,653 | 13.3 | 45.7 | 20.8 | 46 | 14,116 | 29,565 | 53,110 | 47.2 | 2.7 | 56,817 | 11.2 | 13.7 | 11.9 |
| Garden............... | 5 | 8 | 402 | 14.4 | 37.8 | 22.5 | 5 | 19,028 | 35,050 | 43,750 | 55.8 | 2.3 | 41,158 | 15.0 | 24.1 | 22.2 |
| Garfield............. | 2 | 0 | 334 | 14.4 | 35.0 | 25.5 | 5 | 15,654 | 29,146 | 53,000 | 45.2 | 4.8 | 52,707 | 12.1 | 15.3 | 15.3 |
| Gosper............... | 5 | 1 | 334 | 1.8 | 34.6 | 29.5 | 4 | 16,299 | 37,565 | 64,053 | 36.1 | 3.3 | 64,870 | 8.9 | 14.0 | 13.7 |
| Grant................ | 2 | NA | 165 | 18.2 | 38.8 | 26.8 | 3 | 18,663 | 23,264 | 46,500 | 51.4 | 1.7 | 60,773 | 9.6 | 15.8 | 16.2 |
| Greeley.............. | 2 | NA | 522 | 21.8 | 40.4 | 16.3 | 11 | 21,024 | 26,553 | 47,869 | 53.7 | 1.9 | 48,178 | 12.5 | 18.6 | 18.8 |
| Hall................. | 30 | 10 | 14,891 | 9.6 | 46.7 | 21.5 | 142 | 11,407 | 28,359 | 57,104 | 44.1 | 3.4 | 57,371 | 10.2 | 13.6 | 11.7 |
| Hamilton............. | 9 | 13 | 2,155 | 11.2 | 34.9 | 25.0 | 23 | 14,434 | 32,131 | 64,210 | 36.2 | 4.0 | 68,236 | 6.5 | 7.5 | 6.9 |
| Harlan............... | 4 | 4 | 743 | 6.1 | 34.1 | 22.4 | 5 | 15,560 | 28,919 | 49,835 | 50.2 | 1.8 | 53,362 | 11.0 | 14.9 | 13.1 |
| Hayes................ | 1 | 0 | 155 | 6.5 | 36.0 | 18.3 | 3 | 25,209 | 31,286 | 51,726 | 46.4 | 6.2 | 46,399 | 15.0 | 19.9 | 21.5 |
| Hitchcock............ | 5 | 3 | 612 | 8.8 | 41.4 | 17.6 | 7 | 23,582 | 30,037 | 49,962 | 50.0 | 4.0 | 49,451 | 11.3 | 15.9 | 14.3 |
| Holt................. | 6 | 2 | 2,041 | 24.5 | 39.5 | 22.4 | 27 | 15,879 | 31,785 | 60,387 | 40.7 | 2.9 | 54,474 | 11.1 | 14.7 | 13.6 |
| Hooker............... | 1 | 0 | 143 | 11.9 | 41.7 | 26.4 | 4 | 22,682 | 27,714 | 41,125 | 57.2 | 3.5 | 50,093 | 8.2 | 10.6 | 10.6 |
| Howard............... | 5 | 7 | 1,480 | 10.3 | 41.3 | 21.1 | 19 | 14,427 | 27,489 | 59,348 | 43.7 | 1.3 | 56,881 | 10.1 | 11.7 | 10.9 |
| Jefferson............ | 17 | 10 | 1,354 | 3.8 | 46.1 | 17.8 | 22 | 14,248 | 27,461 | 44,510 | 55.1 | 2.3 | 56,740 | 9.5 | 13.5 | 11.8 |
| Johnson.............. | 7 | 6 | 993 | 6.6 | 56.5 | 17.5 | 11 | 15,245 | 24,072 | 54,712 | 48.2 | 2.8 | 54,567 | 11.2 | 12.8 | 12.2 |
| Kearney.............. | 6 | 5 | 1,519 | 8.2 | 32.7 | 26.7 | 20 | 14,931 | 31,094 | 60,266 | 43.4 | 2.6 | 64,461 | 7.7 | 9.6 | 8.7 |
| Keith................ | 9 | 8 | 1,525 | 13.4 | 40.2 | 20.5 | 22 | 21,154 | 29,823 | 50,755 | 49.1 | 2.4 | 51,478 | 11.1 | 17.8 | 15.8 |
| Keya Paha............ | 1 | 0 | 183 | 20.8 | 37.1 | 25.0 | 2 | 27,186 | 28,853 | 56,500 | 45.4 | 4.6 | 46,010 | 18.9 | 24.1 | 20.7 |
| Kimball.............. | 4 | 5 | 717 | 6.8 | 47.5 | 16.1 | 7 | 16,600 | 27,478 | 53,403 | 47.6 | 2.2 | 52,284 | 12.2 | 16.5 | 15.3 |
| Knox................. | 4 | 10 | 1,859 | 13.0 | 41.7 | 21.6 | 28 | 18,211 | 28,409 | 52,332 | 47.3 | 3.1 | 53,676 | 12.2 | 17.1 | 15.0 |
| Lancaster............ | 89 | 20 | 94,971 | 16.7 | 27.9 | 39.5 | 549 | 11,563 | 32,378 | 60,527 | 41.4 | 5.0 | 61,383 | 10.7 | 10.6 | 9.5 |
| Lincoln.............. | 24 | 44 | 8,082 | 12.6 | 35.0 | 23.2 | 68 | 11,696 | 31,567 | 59,795 | 42.1 | 3.4 | 60,924 | 10.8 | 13.1 | 10.8 |
| Logan................ | 1 | 0 | 247 | 20.2 | 36.1 | 19.6 | 3 | 15,397 | 24,355 | 52,708 | 48.7 | 1.4 | 53,108 | 8.9 | 8.3 | 7.3 |
| Loup................. | NA | NA | 59 | 5.1 | 31.4 | 25.3 | 2 | 32,029 | 27,726 | 51,000 | 49.3 | 0.0 | 51,559 | 15.2 | 27.4 | 25.5 |
| McPherson............ | NA | NA | 67 | 7.5 | 34.7 | 22.0 | 2 | 26,096 | 30,063 | 48,125 | 55.5 | 4.7 | 52,033 | 12.6 | 26.4 | 23.3 |
| Madison.............. | 33 | 29 | 9,076 | 18.3 | 40.0 | 23.0 | 71 | 12,028 | 28,179 | 53,188 | 46.9 | 2.9 | 58,275 | 10.2 | 11.7 | 10.5 |
| Merrick.............. | 9 | 7 | 1,630 | 14.6 | 41.1 | 17.8 | 16 | 14,837 | 28,795 | 55,649 | 44.1 | 2.2 | 53,411 | 11.0 | 14.0 | 12.6 |
| Morrill.............. | 9 | 5 | 1,120 | 10.6 | 44.8 | 19.5 | 14 | 16,042 | 26,951 | 46,194 | 54.2 | 1.6 | 49,982 | 13.2 | 18.6 | 17.1 |
| Nance................ | 6 | 1 | 763 | 10.6 | 44.5 | 18.3 | 13 | 16,720 | 29,454 | 49,032 | 50.8 | 2.6 | 55,853 | 10.3 | 13.0 | 11.6 |
| Nemaha............... | 11 | 0 | 1,933 | 10.0 | 38.1 | 27.8 | 22 | 17,612 | 29,574 | 51,828 | 48.0 | 3.5 | 55,343 | 11.8 | 12.1 | 11.8 |
| Nuckolls............. | 3 | 5 | 805 | 1.5 | 42.1 | 20.9 | 7 | 16,171 | 30,078 | 43,388 | 54.9 | 3.2 | 48,624 | 11.2 | 14.8 | 11.8 |
| Otoe................. | 16 | 15 | 3,449 | 12.4 | 42.5 | 23.4 | 33 | 11,978 | 31,450 | 59,167 | 42.7 | 4.1 | 66,293 | 8.3 | 10.5 | 9.0 |
| Pawnee............... | 4 | 1 | 463 | 6.0 | 53.0 | 17.1 | 8 | 16,303 | 27,162 | 46,452 | 53.7 | 2.2 | 47,006 | 13.1 | 21.0 | 20.6 |
| Perkins.............. | 5 | 2 | 564 | 16.3 | 37.8 | 28.1 | 7 | 15,883 | 35,765 | 65,543 | 38.4 | 4.8 | 63,443 | 9.5 | 12.1 | 11.1 |
| Phelps............... | 7 | 21 | 2,008 | 3.5 | 36.6 | 23.5 | 26 | 15,803 | 32,864 | 59,040 | 40.5 | 3.3 | 61,420 | 10.0 | 13.2 | 12.3 |
| Pierce............... | NA | NA | 1,632 | 14.1 | 37.8 | 23.2 | 17 | 14,321 | 32,040 | 64,511 | 40.5 | 3.9 | 67,465 | 9.3 | 10.7 | 9.6 |

1. All persons 3 years old and over enrolled in nursery school through college.　2. Persons 25 years old and over.　3. Elementary and secondary education expenditures.　4. Based on population estimated by the American Community Survey, 2014–2018.

# Table B. States and Counties — **Personal Income**

| STATE County | Personal income, 2019 | | | | | | | | | | Earnings, 2019 | | |
|---|---|---|---|---|---|---|---|---|---|---|---|---|---|
| | | | Per capita[1] | | | Supplements to wages and salaries, employer contributions (mil dol) | | | | | | Contributions for government social insurance (mil dol) | |
| | Total (mil dol) | Percent change 2018-2019 | Dollars | Rank | Wages and salaries (mil dol) | Pension and insurance | Government social insurance | Proprietors' income (mil dol) | Dividends, interest, and rent (mil dol) | Personal transfer receipts (mil dol) | Total (mil dol) | From employee and self-employed | From employer |
| | 62 | 63 | 64 | 65 | 66 | 67 | 68 | 69 | 70 | 71 | 72 | 73 | 74 |
| **NEBRASKA— Cont'd** | | | | | | | | | | | | | |
| Blaine | 28 | 1.8 | 60,120 | 299 | 6 | 2 | 1 | 8 | 9 | 4 | 16 | 1 | 1 |
| Boone | 301 | 8.3 | 58,042 | 367 | 100 | 21 | 8 | 59 | 76 | 53 | 188 | 9 | 8 |
| Box Butte | 545 | 3.8 | 50,518 | 796 | 289 | 51 | 37 | 58 | 87 | 131 | 435 | 29 | 37 |
| Boyd | 111 | 8.5 | 57,949 | 370 | 24 | 6 | 2 | 34 | 21 | 26 | 66 | 3 | 2 |
| Brown | 170 | 6.3 | 57,563 | 387 | 51 | 11 | 4 | 46 | 35 | 33 | 112 | 5 | 4 |
| Buffalo | 2,853 | 2.7 | 57,453 | 389 | 1,247 | 236 | 90 | 324 | 830 | 385 | 1,897 | 112 | 90 |
| Burt | 329 | 7.6 | 50,943 | 762 | 78 | 17 | 6 | 60 | 54 | 79 | 161 | 9 | 6 |
| Butler | 401 | 4.9 | 50,050 | 828 | 117 | 25 | 9 | 39 | 81 | 80 | 190 | 12 | 9 |
| Cass | 1,402 | 4.4 | 53,422 | 582 | 267 | 53 | 20 | 87 | 251 | 236 | 427 | 30 | 20 |
| Cedar | 428 | 2.6 | 50,992 | 759 | 114 | 24 | 9 | 65 | 92 | 73 | 212 | 13 | 9 |
| Chase | 241 | 12.1 | 61,516 | 265 | 80 | 16 | 6 | 68 | 50 | 41 | 170 | 7 | 6 |
| Cherry | 265 | 3.3 | 46,632 | 1,168 | 91 | 19 | 7 | 46 | 72 | 51 | 164 | 9 | 7 |
| Cheyenne | 440 | -4.6 | 49,384 | 890 | 194 | 37 | 15 | 41 | 112 | 96 | 287 | 18 | 15 |
| Clay | 319 | 3.8 | 51,362 | 730 | 116 | 25 | 9 | 42 | 61 | 61 | 192 | 11 | 9 |
| Colfax | 480 | 4.6 | 44,785 | 1,387 | 249 | 43 | 18 | 72 | 83 | 71 | 382 | 20 | 18 |
| Cuming | 601 | 4.8 | 67,977 | 148 | 173 | 32 | 13 | 213 | 104 | 84 | 431 | 17 | 13 |
| Custer | 529 | 3.0 | 49,091 | 916 | 192 | 42 | 15 | 78 | 114 | 110 | 328 | 19 | 15 |
| Dakota | 879 | 3.8 | 43,872 | 1,504 | 677 | 114 | 48 | 92 | 99 | 159 | 931 | 56 | 48 |
| Dawes | 331 | 2.1 | 38,564 | 2,251 | 122 | 31 | 9 | 24 | 67 | 81 | 186 | 12 | 9 |
| Dawson | 1,091 | 4.6 | 46,256 | 1,218 | 519 | 101 | 38 | 151 | 173 | 202 | 808 | 45 | 38 |
| Deuel | 77 | -5.8 | 42,899 | 1,634 | 21 | 5 | 2 | 11 | 17 | 22 | 38 | 3 | 2 |
| Dixon | 248 | 4.4 | 43,920 | 1,498 | 77 | 16 | 6 | 22 | 46 | 46 | 121 | 8 | 6 |
| Dodge | 1,760 | 2.6 | 48,123 | 1,007 | 805 | 143 | 59 | 198 | 339 | 365 | 1,206 | 76 | 59 |
| Douglas | 36,481 | 3.1 | 63,853 | 212 | 21,216 | 3,295 | 1,558 | 5,175 | 7,802 | 4,396 | 31,245 | 1,840 | 1,558 |
| Dundy | 127 | 5.6 | 74,921 | 91 | 30 | 6 | 2 | 42 | 28 | 22 | 81 | 3 | 2 |
| Fillmore | 328 | 4.0 | 59,979 | 304 | 108 | 23 | 8 | 59 | 80 | 64 | 197 | 11 | 8 |
| Franklin | 148 | 4.8 | 49,619 | 866 | 29 | 7 | 2 | 10 | 48 | 35 | 48 | 4 | 2 |
| Frontier | 102 | 2.2 | 38,860 | 2,214 | 35 | 8 | 3 | 7 | 24 | 22 | 53 | 4 | 3 |
| Furnas | 215 | 6.7 | 46,012 | 1,247 | 83 | 18 | 6 | 10 | 47 | 59 | 117 | 8 | 6 |
| Gage | 1,055 | 3.0 | 49,057 | 919 | 383 | 82 | 29 | 59 | 191 | 269 | 553 | 38 | 29 |
| Garden | 89 | 4.0 | 48,213 | 1,000 | 23 | 5 | 2 | 15 | 17 | 27 | 44 | 3 | 2 |
| Garfield | 81 | 5.0 | 41,192 | 1,894 | 30 | 7 | 2 | 10 | 21 | 22 | 49 | 3 | 2 |
| Gosper | 100 | 4.2 | 50,478 | 797 | 23 | 4 | 2 | 5 | 27 | 21 | 35 | 3 | 2 |
| Grant | 32 | 6.1 | 51,082 | 753 | 7 | 2 | 1 | 7 | 9 | 7 | 16 | 1 | 1 |
| Greeley | 111 | 3.4 | 46,979 | 1,134 | 27 | 6 | 2 | 18 | 30 | 23 | 53 | 3 | 2 |
| Hall | 2,768 | 3.1 | 45,112 | 1,347 | 1,614 | 301 | 118 | 236 | 537 | 508 | 2,269 | 138 | 118 |
| Hamilton | 526 | 4.0 | 56,372 | 428 | 176 | 32 | 13 | 75 | 114 | 84 | 296 | 17 | 13 |
| Harlan | 180 | 6.0 | 53,210 | 597 | 35 | 8 | 3 | 45 | 32 | 39 | 90 | 4 | 3 |
| Hayes | 65 | 7.3 | 70,008 | 128 | 10 | 2 | 1 | 29 | 11 | 7 | 42 | 1 | 1 |
| Hitchcock | 110 | 5.0 | 39,690 | 2,089 | 31 | 8 | 2 | 4 | 25 | 35 | 46 | 4 | 2 |
| Holt | 552 | 5.4 | 54,796 | 507 | 193 | 38 | 15 | 139 | 100 | 108 | 386 | 20 | 15 |
| Hooker | 35 | 3.0 | 51,849 | 693 | 12 | 3 | 1 | 10 | 7 | 8 | 25 | 2 | 1 |
| Howard | 300 | 6.0 | 46,471 | 1,187 | 67 | 15 | 5 | 39 | 52 | 58 | 126 | 7 | 5 |
| Jefferson | 353 | 3.1 | 50,079 | 827 | 139 | 28 | 11 | 42 | 83 | 80 | 220 | 14 | 11 |
| Johnson | 184 | 5.1 | 36,262 | 2,541 | 69 | 18 | 5 | 21 | 34 | 40 | 114 | 7 | 5 |
| Kearney | 365 | 4.6 | 56,230 | 439 | 100 | 20 | 8 | 60 | 80 | 67 | 188 | 9 | 8 |
| Keith | 390 | 3.6 | 48,533 | 966 | 140 | 27 | 11 | 58 | 87 | 88 | 236 | 15 | 11 |
| Keya Paha | 43 | 4.6 | 52,917 | 622 | 7 | 1 | 1 | 13 | 10 | 8 | 22 | 1 | 1 |
| Kimball | 162 | 0.0 | 44,484 | 1,419 | 65 | 15 | 5 | 9 | 43 | 42 | 94 | 6 | 5 |
| Knox | 395 | 1.8 | 47,418 | 1,089 | 117 | 26 | 9 | 54 | 82 | 91 | 206 | 13 | 9 |
| Lancaster | 16,192 | 3.4 | 50,743 | 773 | 9,129 | 1,745 | 677 | 911 | 3,382 | 2,430 | 12,462 | 756 | 677 |
| Lincoln | 1,713 | 2.0 | 49,052 | 920 | 816 | 148 | 82 | 191 | 275 | 399 | 1,237 | 81 | 82 |
| Logan | 40 | 1.5 | 53,451 | 578 | 7 | 2 | 1 | 14 | 5 | 7 | 24 | 1 | 1 |
| Loup | 35 | 6.5 | 52,977 | 616 | 5 | 1 | 0 | 14 | 6 | 6 | 20 | 1 | 0 |
| McPherson | 24 | 7.3 | 47,684 | 1,057 | 4 | 1 | 0 | 7 | 5 | 4 | 12 | 1 | 0 |
| Madison | 1,818 | 3.3 | 51,785 | 696 | 1,028 | 192 | 74 | 214 | 340 | 319 | 1,507 | 90 | 74 |
| Merrick | 369 | 8.0 | 47,637 | 1,064 | 107 | 20 | 8 | 55 | 70 | 78 | 191 | 11 | 8 |
| Morrill | 272 | 5.2 | 58,632 | 348 | 72 | 15 | 6 | 80 | 47 | 50 | 173 | 7 | 6 |
| Nance | 144 | 7.4 | 40,934 | 1,924 | 48 | 11 | 4 | 10 | 28 | 37 | 73 | 4 | 4 |
| Nemaha | 343 | 5.3 | 49,201 | 904 | 170 | 39 | 12 | 35 | 70 | 74 | 255 | 15 | 12 |
| Nuckolls | 220 | 5.3 | 53,144 | 602 | 60 | 13 | 4 | 39 | 45 | 53 | 117 | 7 | 4 |
| Otoe | 885 | 4.0 | 55,284 | 480 | 283 | 58 | 21 | 68 | 222 | 157 | 430 | 27 | 21 |
| Pawnee | 118 | 2.0 | 45,191 | 1,341 | 34 | 8 | 3 | 20 | 27 | 27 | 64 | 4 | 3 |
| Perkins | 189 | 17.4 | 65,325 | 179 | 60 | 13 | 4 | 63 | 33 | 29 | 141 | 6 | 4 |
| Phelps | 490 | 4.1 | 54,249 | 533 | 233 | 48 | 18 | 83 | 94 | 94 | 382 | 20 | 18 |
| Pierce | 402 | 3.7 | 56,236 | 438 | 84 | 18 | 6 | 89 | 69 | 61 | 198 | 10 | 6 |

1.  Based on the resident population estimated as of July 1 of the year shown.

Items 62—74

# Table B. States and Counties — Earnings, Social Security, and Housing

| STATE County | Earnings, 2019 (cont.) — Percent by selected industries — Farm | Mining, quarrying, and extractions | Construction | Manufacturing | Information; professional, scientific, technical services | Retail trade | Finance, insurance, real estate, and leasing | Health care and social assistance | Government | Social Security beneficiaries, December 2019 — Number | Rate[1] | Supplemental Security Income recipients, 2019 | Housing units, 2020 — Total | Percent change, 2010-2020 |
|---|---|---|---|---|---|---|---|---|---|---|---|---|---|---|
| | 75 | 76 | 77 | 78 | 79 | 80 | 81 | 82 | 83 | 84 | 85 | 86 | 87 | 88 |
| **NEBRASKA— Cont'd** | | | | | | | | | | | | | | |
| Blaine | 47.2 | 0.0 | 2.0 | 0.4 | 0.2 | D | D | D | 28.1 | 110 | 242 | 0 | 324 | -0.6 |
| Boone | 25.8 | 0.0 | 5.2 | 6.6 | D | 6.0 | D | 5.2 | 21.6 | 1,260 | 243 | 46 | 2,633 | -0.6 |
| Box Butte | 9.5 | 0.0 | 3.0 | 4.7 | 2.7 | 3.5 | 2.8 | D | 15.4 | 1,920 | 178 | 183 | 5,426 | -0.9 |
| Boyd | 37.8 | D | 3.9 | 1.9 | D | 3.0 | 4.6 | 4.6 | 16.3 | 620 | 327 | 24 | 1,370 | -1.4 |
| Brown | 33.6 | 0.1 | 5.7 | 2.9 | D | 5.8 | D | 3.6 | 23.7 | 800 | 270 | 41 | 1,842 | -1.1 |
| Buffalo | 2.4 | 0.2 | 6.3 | 18.5 | D | 6.7 | 5.5 | 17.2 | 14.7 | 8,565 | 172 | 470 | 20,931 | 9.8 |
| Burt | 32.5 | 0.0 | 5.8 | 4.7 | D | 3.5 | 5.5 | 4.5 | 19.8 | 1,875 | 289 | 125 | 3,467 | 0.0 |
| Butler | 12.2 | D | 5.2 | 18.4 | D | 2.9 | 4.7 | D | 22.8 | 1,875 | 235 | 81 | 4,084 | 0.8 |
| Cass | 5.6 | D | 9.3 | 10.8 | 3.7 | 6.5 | 6.3 | 4.6 | 19.8 | 5,400 | 206 | 268 | 11,808 | 6.2 |
| Cedar | 13.9 | D | 8.7 | 6.4 | 3.2 | 4.4 | 12.4 | 2.0 | 20.1 | 1,915 | 227 | 64 | 4,169 | 0.5 |
| Chase | 31.3 | 0.0 | 3.0 | 1.5 | 3.4 | 6.7 | 6.7 | D | 17.4 | 870 | 223 | 33 | 1,975 | 1.5 |
| Cherry | 23.4 | 0.0 | 7.5 | 1.9 | 4.4 | 7.0 | 2.8 | 6.3 | 22.5 | 1,290 | 224 | 68 | 3,290 | 4.2 |
| Cheyenne | 6.8 | D | 3.5 | 6.0 | 2.5 | 10.1 | 3.8 | 14.0 | 16.2 | 2,265 | 251 | 139 | 5,020 | 2.7 |
| Clay | 17.0 | D | 7.8 | 10.9 | D | 3.6 | D | 4.3 | 25.5 | 1,535 | 247 | 74 | 3,026 | 0.9 |
| Colfax | 13.6 | D | D | D | D | 2.3 | 2.9 | D | 12.0 | 1,515 | 142 | 76 | 4,249 | 3.7 |
| Cuming | 42.8 | D | 4.2 | 6.4 | 3.6 | 2.6 | 6.8 | D | 9.7 | 2,150 | 244 | 74 | 4,272 | 1.6 |
| Custer | 16.8 | 0.0 | 7.3 | 17.2 | 3.6 | 5.3 | 6.7 | D | 16.4 | 2,525 | 234 | 138 | 5,674 | 1.7 |
| Dakota | 1.1 | 0.2 | 5.3 | 40.9 | D | 3.6 | 8.9 | 3.1 | 8.7 | 3,610 | 180 | 253 | 7,920 | 3.8 |
| Dawes | 1.8 | D | 5.0 | D | 3.2 | 12.8 | 4.5 | 12.7 | 38.2 | 1,720 | 201 | 102 | 4,247 | -0.1 |
| Dawson | 13.5 | D | 3.4 | 27.0 | 2.7 | 5.1 | 4.4 | D | 19.1 | 4,475 | 190 | 308 | 10,382 | 2.6 |
| Deuel | 10.0 | 0.2 | D | 0.6 | D | 8.5 | D | 2.4 | 27.2 | 510 | 281 | 25 | 1,022 | -1.9 |
| Dixon | 12.5 | 0.0 | 7.5 | D | D | 1.4 | 3.4 | D | 17.0 | 1,185 | 210 | 45 | 2,716 | 1.0 |
| Dodge | 5.1 | 0.1 | 7.4 | 22.0 | 2.8 | 8.1 | 5.0 | D | 12.3 | 8,535 | 234 | 563 | 17,022 | 2.7 |
| Douglas | 0.1 | 0.0 | 4.5 | 5.2 | 11.8 | 4.6 | 12.6 | 13.1 | 10.8 | 89,450 | 157 | 10,694 | 241,565 | 10.0 |
| Dundy | 34.0 | 0.4 | 5.5 | 1.9 | D | 4.3 | 4.2 | 2.6 | 17.7 | 470 | 279 | 24 | 1,119 | -0.5 |
| Fillmore | 17.0 | D | 10.5 | 7.5 | 1.8 | 4.0 | D | 4.6 | 21.6 | 1,545 | 282 | 51 | 2,934 | 0.7 |
| Franklin | 13.5 | 0.2 | D | D | D | 8.1 | 7.9 | 4.8 | 34.8 | 890 | 302 | 52 | 1,716 | |
| Frontier | 3.5 | 0.0 | 9.3 | 8.5 | D | 5.2 | 10.7 | 0.9 | 33.3 | 580 | 220 | 31 | 1,582 | 0.5 |
| Furnas | 1.4 | 0.0 | 5.8 | 6.8 | 6.8 | 5.6 | D | 12.8 | 23.2 | 1,400 | 299 | 87 | 2,711 | -0.4 |
| Gage | 3.2 | D | 5.4 | 20.1 | 2.9 | 7.1 | 4.0 | D | 21.4 | 5,565 | 258 | 422 | 10,530 | 0.8 |
| Garden | 29.1 | 0.2 | D | D | D | 4.3 | D | 11.6 | 20.1 | 590 | 324 | 39 | 1,305 | -0.7 |
| Garfield | 7.1 | D | 5.4 | 13.9 | D | 8.3 | D | 10.0 | 18.8 | 500 | 254 | 18 | 1,221 | 3.7 |
| Gosper | 4.0 | 0.0 | D | D | D | 8.1 | D | 19.1 | 22.0 | 515 | 260 | 18 | 1,311 | 3.5 |
| Grant | 40.7 | 0.0 | 3.5 | 1.3 | D | 4.0 | 8.2 | 1.2 | 25.0 | 180 | 285 | 0 | 388 | -0.8 |
| Greeley | 20.7 | 0.0 | 5.0 | 2.3 | D | 4.0 | 8.1 | 7.3 | 16.3 | 585 | 247 | 38 | 1,314 | 1.2 |
| Hall | 1.9 | 0.1 | 7.2 | 20.4 | 3.4 | 8.1 | 7.3 | 11.2 | 16.3 | 11,005 | 179 | 981 | 25,261 | 7.3 |
| Hamilton | 17.3 | 0.0 | 5.4 | 11.2 | D | 6.3 | 4.7 | D | 10.2 | 2,195 | 236 | 60 | 4,203 | 6.0 |
| Harlan | 40.9 | D | 1.9 | 1.2 | D | 3.4 | D | D | 19.8 | 910 | 270 | 36 | 2,420 | 1.9 |
| Hayes | 56.2 | D | 2.7 | 1.3 | D | D | D | 1.0 | 11.1 | 185 | 203 | 0 | 510 | 0.4 |
| Hitchcock | -5.6 | 9.4 | 11.2 | 15.1 | D | 4.0 | 1.7 | 1.0 | 34.3 | 800 | 290 | 47 | 1,739 | -1.4 |
| Holt | 16.7 | 0.5 | 6.5 | 2.8 | 2.5 | 11.1 | 7.8 | D | 12.5 | 2,510 | 249 | 141 | 5,270 | 1.1 |
| Hooker | 1.9 | 0.0 | D | D | D | 7.0 | D | D | 17.9 | 225 | 337 | 0 | 451 | 4.6 |
| Howard | 24.6 | D | 5.4 | 2.1 | D | 5.5 | D | 5.7 | 31.0 | 1,505 | 234 | 50 | 3,152 | 6.8 |
| Jefferson | 13.8 | D | 7.9 | 18.5 | D | 6.3 | 3.3 | 10.4 | 14.8 | 1,955 | 278 | 155 | 3,914 | -0.1 |
| Johnson | 9.0 | 0.1 | 6.2 | D | D | 3.9 | D | 3.9 | 46.7 | 915 | 181 | 51 | 2,168 | |
| Kearney | 26.0 | 0.0 | 5.6 | 14.5 | D | 2.6 | D | D | 14.4 | 1,235 | 187 | 71 | 2,998 | 3.9 |
| Keith | 11.5 | D | 6.6 | 5.8 | 3.7 | 9.2 | 6.4 | 9.0 | 13.2 | 2,250 | 281 | 99 | 5,524 | 1.8 |
| Keya Paha | 27.8 | 0.7 | D | D | D | D | 5.2 | D | 15.2 | 260 | 332 | 0 | 545 | -0.7 |
| Kimball | 2.8 | 8.8 | 3.0 | 12.4 | D | 5.3 | D | 1.1 | 26.6 | 1,015 | 282 | 59 | 1,945 | -0.9 |
| Knox | 18.6 | 0.0 | 6.4 | 4.2 | D | 5.7 | 5.4 | D | 29.3 | 2,315 | 278 | 126 | 4,987 | 4.2 |
| Lancaster | 0.1 | D | 6.6 | 7.6 | 10.3 | 5.6 | 8.8 | 13.5 | 21.7 | 50,745 | 159 | 4,955 | 136,694 | 13.1 |
| Lincoln | 6.1 | D | 4.7 | 2.2 | 3.2 | 6.2 | 5.4 | D | 16.9 | 6,740 | 193 | 703 | 16,912 | 2.0 |
| Logan | 15.1 | 0.0 | D | D | D | 8.5 | 5.9 | D | 18.7 | 155 | 207 | 0 | 396 | 0.3 |
| Loup | 50.8 | 0.1 | 1.9 | 3.1 | 0.6 | D | D | D | 16.3 | 160 | 243 | 0 | 475 | 11.8 |
| McPherson | 35.2 | 0.2 | 2.6 | D | D | D | D | 0.0 | 18.2 | 70 | 142 | 0 | 281 | -0.7 |
| Madison | 3.0 | D | 6.0 | 17.1 | 3.5 | 7.2 | 9.7 | 15.6 | 16.2 | 7,235 | 206 | 528 | 15,675 | 4.4 |
| Merrick | 25.1 | D | 8.7 | 8.0 | 2.6 | 3.6 | 6.4 | D | 14.3 | 1,890 | 242 | 118 | 3,866 | 4.4 |
| Morrill | 37.2 | D | 0.9 | 4.5 | D | 3.6 | 4.4 | 8.2 | 32.3 | 980 | 212 | 93 | 2,429 | -0.5 |
| Nance | 13.6 | D | D | D | D | 4.1 | D | 5.2 | 58.5 | 735 | 211 | 61 | 1,872 | 3.9 |
| Nemaha | 8.8 | 0.0 | 3.8 | 5.6 | 2.9 | 2.9 | 5.3 | 17.2 | 16.7 | 1,630 | 235 | 114 | 3,493 | -0.1 |
| Nuckolls | 21.3 | 0.1 | 4.1 | D | D | 6.0 | 4.9 | 8.2 | 24.7 | 1,200 | 289 | 58 | 2,448 | -0.7 |
| Otoe | 6.3 | D | 7.6 | 21.4 | D | 4.4 | D | 3.8 | 24.0 | 3,560 | 223 | 197 | 7,249 | 3.2 |
| Pawnee | 13.3 | D | D | 15.1 | D | 2.6 | 4.4 | 2.4 | 18.4 | 660 | 250 | 39 | 1,646 | 3.7 |
| Perkins | 33.0 | 0.0 | 8.6 | 3.3 | D | 5.6 | 6.3 | D | 12.8 | 695 | 241 | 17 | 1,444 | -0.3 |
| Phelps | 18.0 | D | D | D | 2.6 | 5.6 | 6.3 | 7.5 | 11.4 | 2,145 | 239 | 118 | 4,302 | 3.0 |
| Pierce | 20.3 | 1.3 | 8.5 | 12.8 | 3.6 | 3.6 | 6.9 | 7.5 | 11.4 | 1,445 | 202 | 61 | 3,295 | 2.3 |

1. Per 1,000 resident population estimated as of July 1 of the year shown.

# Table B. States and Counties — Housing, Labor Force, and Employment

| STATE County | Housing units, 2015-2019 — Occupied units — Owner-occupied Total [89] | Percent [90] | Median value[1] [91] | Median owner cost as a percent of income — With a mortgage [92] | Without a mortgage[2] [93] | Renter-occupied Median rent[3] [94] | Median rent as a percent of income[2] [95] | Sub-standard units[4] (percent) [96] | Civilian labor force, 2020 Total [97] | Percent change, 2019-2020 [98] | Unemployment Total [99] | Rate[5] [100] | Civilian employment[6], 2015-2019 Total [101] | Percent — Management, business, science, and arts [102] | Construction, production, and maintenance occupations [103] |
|---|---|---|---|---|---|---|---|---|---|---|---|---|---|---|---|

| STATE County | 89 | 90 | 91 | 92 | 93 | 94 | 95 | 96 | 97 | 98 | 99 | 100 | 101 | 102 | 103 |
|---|---|---|---|---|---|---|---|---|---|---|---|---|---|---|---|
| **NEBRASKA— Cont'd** | | | | | | | | | | | | | | | |
| Blaine | 213 | 73.7 | 75,700 | 26.8 | 10.8 | 567 | 12.5 | 3.3 | 255 | -0.4 | 9 | 3.5 | 255 | 42.0 | 35.7 |
| Boone | 2,311 | 76.2 | 118,400 | 17.9 | 10.0 | 572 | 17.8 | 1.5 | 2,937 | -1.1 | 73 | 2.5 | 2,904 | 34.0 | 31.5 |
| Box Butte | 4,743 | 68.6 | 113,600 | 19.2 | 11.8 | 668 | 23.7 | 3.1 | 5,405 | -0.5 | 257 | 4.8 | 5,094 | 31.4 | 39.5 |
| Boyd | 904 | 81.5 | 73,000 | 16.5 | 11.7 | 508 | 26.8 | 1.9 | 1,057 | -2.2 | 30 | 2.8 | 985 | 40.6 | 29.7 |
| Brown | 1,387 | 75.4 | 79,700 | 15.1 | 13.6 | 529 | 16.3 | 2.5 | 1,397 | -0.4 | 38 | 2.7 | 1,554 | 37.2 | 22.3 |
| Buffalo | 19,062 | 65.0 | 174,800 | 19.9 | 10.7 | 779 | 27.6 | 2.0 | 27,519 | -1.5 | 1,089 | 4.0 | 27,371 | 34.2 | 25.1 |
| Burt | 2,856 | 76.6 | 100,400 | 19.0 | 11.1 | 640 | 23.6 | 2.6 | 3,556 | 1.3 | 137 | 3.9 | 2,982 | 34.1 | 28.6 |
| Butler | 3,363 | 77.9 | 117,300 | 17.7 | 11.5 | 734 | 20.5 | 2.5 | 4,589 | -1.0 | 156 | 3.4 | 4,222 | 31.0 | 36.5 |
| Cass | 9,945 | 81.7 | 173,800 | 20.1 | 12.9 | 820 | 25.3 | 1.3 | 13,432 | -1.1 | 581 | 4.3 | 13,180 | 37.2 | 25.0 |
| Cedar | 3,506 | 80.5 | 121,300 | 17.5 | 11.0 | 655 | 20.7 | 0.5 | 4,506 | -0.7 | 122 | 2.7 | 4,584 | 34.9 | 29.7 |
| Chase | 1,679 | 75.6 | 133,300 | 20.9 | 11.1 | 687 | 19.3 | 0.4 | 2,285 | 0.1 | 41 | 1.8 | 1,934 | 34.3 | 29.7 |
| Cherry | 2,537 | 61.3 | 116,000 | 21.1 | 10.1 | 712 | 24.7 | 1.4 | 3,334 | -1.4 | 81 | 2.4 | 3,278 | 35.9 | 27.6 |
| Cheyenne | 4,395 | 66.6 | 103,000 | 18.5 | 12.0 | 792 | 23.2 | 3.4 | 4,224 | -4.0 | 186 | 4.4 | 5,067 | 36.1 | 24.9 |
| Clay | 2,539 | 79.8 | 91,300 | 19.3 | 10.0 | 607 | 24.5 | 1.7 | 3,256 | -1.5 | 104 | 3.2 | 3,016 | 31.9 | 37.5 |
| Colfax | 3,731 | 72.3 | 101,100 | 21.2 | 10.4 | 696 | 18.2 | 7.7 | 5,620 | 0.2 | 163 | 2.9 | 5,435 | 25.6 | 50.0 |
| Cuming | 3,727 | 69.7 | 122,200 | 18.1 | 10.5 | 683 | 22.6 | 2.0 | 4,830 | 1.4 | 122 | 2.5 | 4,610 | 31.7 | 35.7 |
| Custer | 4,862 | 69.6 | 108,100 | 21.1 | 11.8 | 666 | 26.6 | 3.2 | 6,348 | -1.4 | 174 | 2.7 | 5,503 | 32.5 | 29.8 |
| Dakota | 7,404 | 65.2 | 119,600 | 18.2 | 10.4 | 813 | 23.2 | 7.0 | 10,598 | -1.6 | 509 | 4.8 | 10,294 | 21.3 | 45.3 |
| Dawes | 3,525 | 64.6 | 125,200 | 21.5 | 12.5 | 713 | 25.3 | 1.1 | 4,952 | -1.1 | 140 | 2.8 | 4,661 | 36.3 | 23.2 |
| Dawson | 8,965 | 67.3 | 102,600 | 17.8 | 11.5 | 730 | 24.8 | 5.1 | 13,302 | 1.3 | 456 | 3.4 | 12,024 | 25.3 | 41.9 |
| Deuel | 830 | 76.3 | 87,300 | 21.2 | 12.5 | 631 | 21.5 | 1.2 | 956 | -0.2 | 29 | 3.0 | 900 | 30.9 | 28.2 |
| Dixon | 2,352 | 78.5 | 94,700 | 17.0 | 10.1 | 713 | 28.3 | 2.4 | 2,945 | -1.9 | 95 | 3.2 | 2,897 | 28.4 | 37.3 |
| Dodge | 15,261 | 62.1 | 130,200 | 19.5 | 11.7 | 760 | 25.4 | 2.1 | 20,100 | 2.0 | 709 | 3.5 | 18,444 | 30.6 | 30.2 |
| Douglas | 218,061 | 61.7 | 169,800 | 19.7 | 12.3 | 934 | 28.3 | 2.3 | 300,140 | -0.3 | 15,294 | 5.1 | 292,805 | 41.9 | 19.2 |
| Dundy | 872 | 72.1 | 75,900 | 19.3 | 13.0 | 492 | 25.9 | 0.8 | 1,158 | 0.2 | 25 | 2.2 | 923 | 40.1 | 26.0 |
| Fillmore | 2,510 | 77.5 | 80,100 | 16.6 | 10.2 | 561 | 22.6 | 0.6 | 3,154 | -0.8 | 87 | 2.8 | 2,876 | 35.7 | 29.2 |
| Franklin | 1,377 | 80.6 | 71,300 | 19.6 | 11.1 | 473 | 19.9 | 1.8 | 1,473 | -0.9 | 51 | 3.5 | 1,505 | 34.0 | 31.2 |
| Frontier | 1,135 | 70.2 | 101,200 | 17.1 | 11.1 | 596 | 23.3 | 1.2 | 1,539 | -4.9 | 45 | 2.9 | 1,294 | 35.9 | 29.4 |
| Furnas | 2,128 | 74.6 | 73,000 | 19.7 | 11.4 | 628 | 26.4 | 3.1 | 2,593 | 0.1 | 72 | 2.8 | 2,336 | 37.0 | 28.3 |
| Gage | 8,979 | 69.6 | 116,200 | 19.1 | 10.4 | 640 | 25.9 | 0.9 | 10,872 | -1.6 | 434 | 4.0 | 11,059 | 33.8 | 30.5 |
| Garden | 868 | 73.7 | 81,800 | 20.8 | 11.7 | 668 | 21.5 | 0.1 | 1,061 | -3.4 | 37 | 3.5 | 876 | 48.9 | 27.1 |
| Garfield | 884 | 77.8 | 105,100 | 22.1 | 10.6 | 443 | 24.5 | 1.1 | 1,147 | -0.5 | 23 | 2.0 | 1,030 | 38.7 | 29.4 |
| Gosper | 942 | 70.5 | 127,000 | 19.8 | 12.5 | 639 | 16.1 | 0.0 | 1,094 | -0.9 | 28 | 2.6 | 1,125 | 41.7 | 31.5 |
| Grant | 288 | 70.5 | 58,300 | 29.6 | 12.3 | 425 | 20.7 | 2.4 | 409 | -0.5 | 8 | 2.0 | 348 | 37.9 | 28.2 |
| Greeley | 1,019 | 81.0 | 72,700 | 22.2 | 10.5 | 536 | 19.9 | 2.8 | 1,232 | 0.2 | 36 | 2.9 | 1,172 | 31.9 | 33.2 |
| Hall | 23,096 | 62.3 | 146,000 | 19.5 | 10.8 | 768 | 26.0 | 3.5 | 32,030 | 0.0 | 1,851 | 5.8 | 31,540 | 27.5 | 35.5 |
| Hamilton | 3,713 | 77.0 | 154,300 | 19.8 | 10.3 | 739 | 23.4 | 0.9 | 4,760 | -1.4 | 175 | 3.7 | 4,909 | 37.5 | 23.2 |
| Harlan | 1,521 | 79.4 | 102,700 | 21.2 | 11.3 | 634 | 22.6 | 0.9 | 1,731 | -3.1 | 50 | 2.9 | 1,670 | 33.8 | 29.2 |
| Hayes | 403 | 75.2 | 80,900 | 14.3 | 12.2 | 478 | 11.8 | 1.0 | 630 | 1.6 | 14 | 2.2 | 473 | 46.1 | 24.9 |
| Hitchcock | 1,243 | 73.9 | 66,700 | 15.6 | 10.2 | 628 | 17.2 | 2.6 | 1,304 | 3.2 | 45 | 3.5 | 1,386 | 32.1 | 28.4 |
| Holt | 4,487 | 72.2 | 114,100 | 17.7 | 10.0 | 642 | 22.8 | 1.0 | 5,730 | -0.4 | 140 | 2.4 | 5,685 | 39.8 | 25.9 |
| Hooker | 318 | 62.6 | 92,200 | 22.8 | 10.6 | 573 | 17.4 | 1.3 | 412 | -1.9 | 11 | 2.7 | 287 | 32.4 | 25.1 |
| Howard | 2,720 | 76.5 | 148,800 | 20.7 | 12.6 | 685 | 27.5 | 2.0 | 3,340 | -1.2 | 147 | 4.4 | 3,323 | 37.5 | 26.4 |
| Jefferson | 3,302 | 71.2 | 84,000 | 21.4 | 12.6 | 650 | 23.9 | 0.8 | 4,193 | 0.8 | 115 | 2.7 | 3,438 | 32.7 | 32.5 |
| Johnson | 1,809 | 73.4 | 94,400 | 17.9 | 10.0 | 646 | 24.7 | 2.2 | 2,058 | -0.5 | 81 | 3.9 | 2,126 | 32.1 | 30.6 |
| Kearney | 2,666 | 73.1 | 149,700 | 18.3 | 10.0 | 695 | 23.2 | 2.8 | 3,776 | -0.7 | 121 | 3.2 | 3,388 | 36.9 | 29.9 |
| Keith | 3,882 | 68.8 | 128,300 | 21.4 | 11.9 | 633 | 26.9 | 1.4 | 4,547 | -1.7 | 147 | 3.2 | 4,076 | 31.9 | 26.5 |
| Keya Paha | 304 | 72.7 | 78,100 | 16.0 | 11.6 | 725 | 18.0 | 0.3 | 597 | 1.4 | 12 | 2.0 | 385 | 46.0 | 20.5 |
| Kimball | 1,577 | 71.4 | 82,600 | 21.5 | 12.7 | 741 | 22.9 | 1.4 | 1,975 | -2.9 | 70 | 3.5 | 1,848 | 21.6 | 34.8 |
| Knox | 3,586 | 74.1 | 95,900 | 17.6 | 11.8 | 517 | 20.1 | 3.2 | 4,655 | 0.7 | 138 | 3.0 | 4,239 | 37.4 | 27.8 |
| Lancaster | 124,324 | 59.8 | 177,400 | 19.6 | 10.5 | 852 | 27.7 | 2.3 | 177,006 | -0.6 | 7,499 | 4.2 | 171,824 | 41.0 | 19.2 |
| Lincoln | 14,856 | 67.0 | 139,400 | 19.0 | 11.7 | 753 | 26.0 | 1.1 | 18,367 | -0.1 | 722 | 3.9 | 17,538 | 32.5 | 32.8 |
| Logan | 347 | 73.8 | 110,300 | 27.5 | 11.3 | 614 | 14.2 | 2.0 | 471 | -0.6 | 12 | 2.5 | 478 | 29.3 | 43.9 |
| Loup | 294 | 76.5 | 112,500 | 26.8 | 13.8 | 533 | 10.0 | 0.0 | 409 | -1.9 | 12 | 2.9 | 317 | 50.5 | 26.5 |
| McPherson | 191 | 72.8 | 102,700 | 18.4 | 10.0 | 630 | 20.7 | 1.0 | 443 | -0.9 | 12 | 2.7 | 204 | 35.3 | 22.1 |
| Madison | 14,205 | 65.6 | 146,400 | 18.9 | 11.8 | 690 | 25.4 | 2.5 | 19,665 | -1.2 | 767 | 3.9 | 18,271 | 29.9 | 33.8 |
| Merrick | 3,373 | 74.5 | 115,400 | 19.7 | 10.0 | 689 | 23.1 | 2.5 | 4,004 | -1.3 | 159 | 4.0 | 4,023 | 33.8 | 32.6 |
| Morrill | 1,954 | 71.5 | 95,600 | 19.6 | 13.2 | 707 | 24.7 | 1.8 | 2,605 | -0.3 | 93 | 3.6 | 2,287 | 33.8 | 33.5 |
| Nance | 1,546 | 80.5 | 78,300 | 19.9 | 10.0 | 586 | 21.5 | 1.2 | 1,946 | -2.3 | 63 | 3.2 | 1,778 | 31.8 | 30.3 |
| Nemaha | 2,977 | 72.9 | 106,300 | 18.4 | 12.6 | 563 | 19.9 | 1.8 | 3,581 | -1.6 | 128 | 3.6 | 3,349 | 37.1 | 29.9 |
| Nuckolls | 1,867 | 75.8 | 68,300 | 17.2 | 11.0 | 495 | 27.7 | 1.3 | 2,454 | 2.0 | 66 | 2.7 | 2,206 | 32.0 | 24.1 |
| Otoe | 6,510 | 69.8 | 143,900 | 19.1 | 12.1 | 692 | 21.7 | 2.0 | 8,369 | -0.9 | 297 | 3.5 | 8,290 | 37.2 | 29.0 |
| Pawnee | 1,222 | 83.4 | 71,100 | 15.8 | 12.8 | 551 | 28.8 | 3.7 | 1,486 | 0.3 | 41 | 2.8 | 1,241 | 32.4 | 32.8 |
| Perkins | 1,228 | 77.9 | 118,400 | 17.5 | 10.9 | 680 | 26.5 | 0.6 | 1,834 | 0.5 | 33 | 1.8 | 1,516 | 31.7 | 28.3 |
| Phelps | 3,905 | 71.8 | 134,000 | 18.4 | 11.3 | 672 | 23.1 | 0.4 | 4,953 | -0.8 | 148 | 3.0 | 4,740 | 40.0 | 31.3 |
| Pierce | 3,022 | 75.9 | 132,700 | 17.3 | 10.7 | 636 | 20.0 | 0.5 | 4,125 | 0.2 | 136 | 3.3 | 3,799 | 34.8 | 36.2 |

1. Specified owner-occupied units, lacking complete plumbing facilities.  2. A value of 10.0 represents 10 percent or less; a value of 50.0 represents 50 percent or more.  3. Specified renter-occupied units.  4. Overcrowded or  5. Percent of civilian labor force.  6. Civilian employed persons 16 years old and over.

| STATE County | Private nonfarm establishments, employment and payroll, 2019 | | | | | | | | | Agriculture, 2017 | | | |
|---|---|---|---|---|---|---|---|---|---|---|---|---|---|
| | | Employment | | | | | | Annual payroll | | Farms | | | Farm producers whose primary occupation is farming (percent) |
| | | | | | | | Professional, scientific, and technical services | | | | Percent with: | | |
| | Number of establishments | Total | Health care and social assistance | Manufacturing | Retail trade | Finance and insurance | | Total (mil dol) | Average per employee (dollars) | Number | Fewer than 50 acres | 1000 acres or more | |
| | 104 | 105 | 106 | 107 | 108 | 109 | 110 | 111 | 112 | 113 | 114 | 115 | 116 |

| STATE County | 104 | 105 | 106 | 107 | 108 | 109 | 110 | 111 | 112 | 113 | 114 | 115 | 116 |
|---|---|---|---|---|---|---|---|---|---|---|---|---|---|
| **NEBRASKA— Cont'd** | | | | | | | | | | | | | |
| Blaine | 12 | 23 | NA | NA | NA | NA | NA | 1 | 35,957 | 101 | 5.9 | 72.3 | 70.0 |
| Boone | 198 | 1,462 | 282 | 94 | 253 | 91 | 29 | 56 | 38,207 | 524 | 15.6 | 23.5 | 57.3 |
| Box Butte | 292 | 2,705 | 473 | 327 | 454 | 129 | 94 | 101 | 37,392 | 431 | 13.7 | 33.4 | 54.1 |
| Boyd | 70 | 454 | 119 | 14 | 59 | 33 | NA | 14 | 30,485 | 286 | 15.7 | 34.3 | 64.6 |
| Brown | 139 | 897 | 161 | 36 | 295 | 49 | 29 | 31 | 34,957 | 268 | 21.6 | 40.7 | 51.2 |
| Buffalo | 1,716 | 22,962 | 4,098 | 3,723 | 3,952 | 710 | 744 | 897 | 39,084 | 953 | 27.7 | 18.4 | 48.0 |
| Burt | 197 | 1,090 | 208 | 38 | 194 | 63 | 67 | 40 | 36,916 | 521 | 28.0 | 17.5 | 49.4 |
| Butler | 211 | 1,869 | 391 | 528 | 217 | 125 | 39 | 78 | 41,525 | 723 | 23.9 | 14.9 | 50.1 |
| Cass | 596 | 4,413 | 424 | 418 | 921 | 208 | 165 | 166 | 37,512 | 766 | 37.9 | 15.7 | 43.2 |
| Cedar | 298 | 1,902 | 156 | 248 | 335 | 130 | 68 | 62 | 32,764 | 784 | 19.5 | 16.3 | 54.0 |
| Chase | 163 | 1,054 | 141 | 10 | 300 | 74 | 30 | 39 | 36,867 | 325 | 13.8 | 41.5 | 59.8 |
| Cherry | 231 | 1,511 | 267 | 19 | 353 | 51 | 103 | 46 | 30,260 | 567 | 10.8 | 61.9 | 63.9 |
| Cheyenne | 277 | 4,304 | 552 | 301 | 794 | 137 | 78 | 78 | 51,030 | 572 | 8.6 | 37.6 | 49.2 |
| Clay | 194 | 1,122 | 97 | 40 | 219 | 79 | 27 | 42 | 37,662 | 441 | 25.4 | 24.0 | 58.5 |
| Colfax | 253 | 4,480 | 213 | NA | 305 | 103 | 53 | 205 | 45,760 | 516 | 22.5 | 16.7 | 59.1 |
| Cuming | 342 | 2,771 | 406 | 471 | 397 | 215 | 136 | 107 | 38,751 | 804 | 23.9 | 13.2 | 53.0 |
| Custer | 372 | 3,127 | 692 | 652 | 609 | 213 | 101 | 113 | 36,201 | 1,108 | 18.1 | 36.2 | 55.9 |
| Dakota | 429 | 11,891 | 493 | 5,208 | 967 | 733 | 104 | 536 | 45,049 | 267 | 28.5 | 16.9 | 43.4 |
| Dawes | 247 | 1,945 | 411 | NA | 609 | 84 | 65 | 59 | 30,215 | 491 | 16.7 | 35.0 | 50.4 |
| Dawson | 688 | 9,006 | 1,227 | 3,435 | 1,115 | 250 | 174 | 362 | 40,168 | 686 | 26.4 | 25.1 | 56.1 |
| Deuel | 54 | 275 | 15 | NA | 91 | 21 | NA | 9 | 31,785 | 225 | 12.0 | 35.1 | 50.8 |
| Dixon | 110 | 1,082 | 71 | NA | 93 | 38 | NA | 43 | 39,767 | 567 | 28.7 | 14.3 | 47.1 |
| Dodge | 1,003 | 16,610 | 2,507 | 3,613 | 2,509 | 457 | 198 | 628 | 37,792 | 676 | 28.4 | 14.3 | 52.5 |
| Douglas | 15,941 | 322,661 | 55,184 | 22,793 | 35,390 | 37,352 | 19,055 | 16,924 | 52,451 | 367 | 61.3 | 6.8 | 32.5 |
| Dundy | 56 | 348 | 128 | 17 | 32 | 16 | 15 | 13 | 38,540 | 268 | 6.0 | 51.1 | 68.3 |
| Fillmore | 221 | 1,700 | 327 | 255 | 226 | 134 | 25 | 73 | 42,696 | 439 | 14.4 | 27.1 | 65.7 |
| Franklin | 75 | 376 | 99 | NA | 102 | 30 | 24 | 12 | 31,561 | 317 | 16.1 | 30.0 | 61.5 |
| Frontier | 69 | 455 | 58 | NA | 91 | 36 | 18 | 15 | 32,303 | 371 | 22.4 | 34.5 | 54.8 |
| Furnas | 153 | 1,176 | 320 | 137 | 199 | 59 | 22 | 45 | 38,562 | 377 | 12.7 | 34.2 | 54.1 |
| Gage | 655 | 6,940 | 1,625 | 1,565 | 1,115 | 213 | 149 | 261 | 37,591 | 1,188 | 33.2 | 14.8 | 42.3 |
| Garden | 48 | 231 | NA | NA | 55 | 19 | NA | 5 | 23,208 | 221 | 15.4 | 46.2 | 54.6 |
| Garfield | 100 | 563 | 123 | 108 | 111 | 21 | 22 | 19 | 32,982 | 202 | 27.7 | 31.7 | 54.4 |
| Gosper | 57 | 171 | 11 | NA | 27 | 26 | NA | 6 | 35,228 | 287 | 8.7 | 28.9 | 56.4 |
| Grant | 26 | 78 | NA | NA | 13 | NA | NA | 2 | 21,897 | 64 | 1.6 | 64.1 | 80.0 |
| Greeley | 74 | 323 | 23 | 17 | 84 | 32 | 7 | 10 | 30,765 | 369 | 8.4 | 28.5 | 59.8 |
| Hall | 1,889 | 31,828 | 4,388 | 7,422 | 4,846 | 1,262 | 600 | 1,268 | 39,829 | 582 | 28.9 | 18.6 | 54.7 |
| Hamilton | 312 | 2,990 | 418 | 606 | 345 | 191 | 111 | 127 | 42,602 | 586 | 31.2 | 17.9 | 55.4 |
| Harlan | 100 | 632 | 188 | NA | 107 | 46 | 29 | 21 | 33,816 | 281 | 23.5 | 38.4 | 54.9 |
| Hayes | 19 | 45 | NA | NA | 17 | NA | NA | 1 | 31,578 | 220 | 9.5 | 43.2 | 53.9 |
| Hitchcock | 68 | 406 | NA | NA | 94 | 48 | 20 | NA | 40,374 | 288 | 9.0 | 38.9 | 51.7 |
| Holt | 421 | 3,219 | 787 | 122 | 552 | 179 | 83 | 112 | 34,803 | 1,142 | 15.7 | 33.3 | 55.5 |
| Hooker | 31 | 108 | NA | NA | 26 | NA | NA | 5 | 43,139 | 97 | 7.2 | 67.0 | 62.3 |
| Howard | 180 | 1,135 | 411 | 27 | 225 | 102 | 37 | 39 | 34,391 | 617 | 25.9 | 13.0 | 49.6 |
| Jefferson | 228 | 2,616 | 439 | 781 | 373 | 79 | 43 | 95 | 34,826 | 590 | 20.2 | 20.0 | 50.8 |
| Johnson | 112 | 863 | 218 | 164 | 136 | 55 | 31 | 30 | 34,826 | 502 | 24.1 | 11.2 | 41.9 |
| Kearney | 172 | 1,736 | 524 | 387 | 130 | 77 | 33 | 69 | 39,681 | 342 | 14.3 | 34.8 | 68.6 |
| Keith | 330 | 2,426 | 238 | 243 | 620 | 120 | 111 | 80 | 33,063 | 318 | 19.8 | 32.7 | 45.4 |
| Keya Paha | 17 | 41 | NA | NA | 16 | NA | 3 | 2 | 38,732 | 237 | 6.3 | 46.0 | 68.9 |
| Kimball | 126 | 973 | 115 | 186 | 153 | 51 | 29 | 44 | 45,014 | 443 | 8.1 | 32.1 | 42.5 |
| Knox | 250 | 1,759 | 385 | 66 | 322 | 153 | 62 | 53 | 29,951 | 956 | 18.0 | 19.2 | 50.2 |
| Lancaster | 8,684 | 145,299 | 26,706 | 12,532 | 19,150 | 14,684 | 9,002 | 6,371 | 43,848 | 1,786 | 54.9 | 7.3 | 34.1 |
| Lincoln | 1,036 | 11,520 | 2,815 | 367 | 1,932 | 551 | 358 | 449 | 38,946 | 1,040 | 26.3 | 29.5 | 45.6 |
| Logan | 21 | 69 | NA | NA | NA | NA | NA | 2 | 29,130 | 117 | 24.8 | 33.3 | 61.6 |
| Loup | 16 | 43 | NA | NA | 11 | NA | NA | 1 | 17,279 | 130 | 19.2 | 37.7 | 62.9 |
| McPherson | 6 | 19 | NA | NA | NA | NA | NA | 0 | 24,000 | 109 | 8.3 | 63.3 | 62.1 |
| Madison | 1,271 | 18,264 | 3,724 | 3,408 | 3,002 | 635 | 538 | 743 | 40,693 | 659 | 24.6 | 16.4 | 52.3 |
| Merrick | 236 | 1,711 | 320 | 236 | 197 | 103 | 43 | 71 | 41,642 | 483 | 28.0 | 16.6 | 52.7 |
| Morrill | 108 | 764 | 161 | NA | 201 | 53 | 9 | 29 | 37,992 | 426 | 20.0 | 31.5 | 55.7 |
| Nance | 91 | 436 | 130 | NA | 111 | 50 | 15 | 13 | 28,856 | 375 | 21.1 | 25.1 | 58.7 |
| Nemaha | 187 | 1,465 | 367 | 206 | 253 | 97 | 37 | 48 | 32,608 | 410 | 24.6 | 21.5 | 49.2 |
| Nuckolls | 169 | 1,093 | 403 | 6 | 200 | 74 | 35 | 35 | 32,229 | 431 | 9.3 | 27.1 | 59.3 |
| Otoe | 470 | 4,752 | 728 | 1,489 | 692 | 159 | 76 | 173 | 36,472 | 815 | 28.6 | 16.6 | 42.5 |
| Pawnee | 66 | 615 | 111 | 276 | 74 | 39 | 10 | 23 | 36,896 | 460 | 15.2 | 18.7 | 47.0 |
| Perkins | 128 | 902 | 254 | 43 | 135 | 37 | 21 | 39 | 42,854 | 418 | 10.0 | 36.1 | 55.7 |
| Phelps | 343 | 4,167 | 702 | 903 | 460 | 164 | 222 | 175 | 41,902 | 371 | 12.4 | 33.7 | 66.6 |
| Pierce | 236 | 1,531 | 272 | 109 | 177 | 104 | 67 | 55 | 35,757 | 625 | 23.0 | 18.2 | 51.4 |

# Table B. States and Counties — Agriculture

| STATE County | Agriculture, 2017 (cont.) | | | | | | | | | | | | | | | |
|---|---|---|---|---|---|---|---|---|---|---|---|---|---|---|---|---|
| | Land in farms | | | | | Value of land and buildings (dollars) | | Value of machinery and equipment, average per farm (dollars) | Value of products sold: | | | | Organic farms (number) | Farms with internet access (per-cent) | Government payments | |
| | | | Acres | | | | | | | | Percent from: | | | | | |
| | Acreage (1,000) | Percent change, 2012-2017 | Average size of farm | Total irrigated (1,000) | Total cropland (1,000) | Average per farm | Average per acre | | Total (mil dol) | Average per farm (acres) | Crops | Livestock and poultry products | | | Total ($1,000) | Percent of farms |
| | 117 | 118 | 119 | 120 | 121 | 122 | 123 | 124 | 125 | 126 | 127 | 128 | 129 | 130 | 131 | 132 |
| **NEBRASKA— Cont'd** | | | | | | | | | | | | | | | | |
| Blaine | 367 | -8.9 | 3,630 | 8.5 | 28.8 | 3,603,549 | 993 | 173,445 | 32.1 | 317,376 | 6.6 | 93.4 | NA | 89.1 | 979 | 43.6 |
| Boone | 432 | -0.5 | 825 | 169.8 | 319.2 | 3,669,838 | 4,449 | 399,375 | 473.8 | 904,158 | 34.1 | 65.9 | 1 | 79.6 | 9,212 | 73.3 |
| Box Butte | 677 | 0.3 | 1,571 | 138.5 | 346.6 | 2,096,463 | 1,334 | 311,694 | 176.9 | 410,517 | 60.9 | 39.1 | 2 | 81.0 | 7,657 | 72.2 |
| Boyd | 323 | 11 | 1,129 | 13.9 | 135.6 | 2,290,609 | 2,028 | 258,985 | 104.3 | 364,577 | 38.7 | 61.3 | 2 | 76.6 | 2,160 | 72.4 |
| Brown | 615 | -15.2 | 2,295 | 40.1 | 108.1 | 2,689,620 | 1,172 | 259,836 | 290.7 | 1,084,873 | 7.8 | 92.2 | NA | 85.8 | 809 | 39.6 |
| Buffalo | 528 | -9 | 554 | 214.1 | 324.5 | 2,435,697 | 4,393 | 241,065 | 332.7 | 349,120 | 52.1 | 47.9 | NA | 80.9 | 8,917 | 62.4 |
| Burt | 298 | -3.8 | 572 | 47.5 | 275.2 | 3,524,417 | 6,160 | 318,618 | 263.7 | 506,226 | 55.1 | 44.9 | 6 | 80.4 | 8,438 | 73.5 |
| Butler | 374 | 1.1 | 517 | 102.3 | 319.1 | 2,946,211 | 5,693 | 273,973 | 259.8 | 359,288 | 62.2 | 37.8 | 10 | 79.4 | 11,674 | 77.3 |
| Cass | 346 | 0.4 | 452 | 3.9 | 306.4 | 2,534,769 | 5,607 | 202,195 | 164.2 | 214,405 | 92.5 | 7.5 | 9 | 82.4 | 5,164 | 62.9 |
| Cedar | 474 | 1.5 | 604 | 163.2 | 393.2 | 3,129,654 | 5,182 | 295,573 | 423.1 | 539,617 | 44.0 | 56.0 | 2 | 80.1 | 14,856 | 75.8 |
| Chase | 569 | 5.1 | 1,750 | 168.9 | 323.0 | 3,647,226 | 2,085 | 480,540 | 440.1 | 1,354,191 | 34.6 | 65.4 | NA | 88.3 | 10,653 | 72.9 |
| Cherry | 3,563 | -5.2 | 6,284 | 53.0 | 383.7 | 5,862,309 | 933 | 240,960 | 230.9 | 407,279 | 14.6 | 85.4 | 7 | 85.4 | 3,302 | 22.6 |
| Cheyenne | 759 | 8 | 1,328 | 52.6 | 528.8 | 1,354,420 | 1,020 | 193,548 | 163.9 | 286,594 | 37.2 | 62.8 | 15 | 73.6 | 13,034 | 83.6 |
| Clay | 319 | -3.5 | 723 | 190.6 | 259.5 | 3,508,217 | 4,850 | 345,579 | 356.1 | 807,372 | 44.2 | 55.8 | 2 | 90.0 | 10,804 | 68.3 |
| Colfax | 262 | 1.8 | 508 | 81.0 | 240.4 | 3,149,351 | 6,194 | 354,733 | 364.5 | 706,298 | 34.6 | 65.4 | 6 | 85.5 | 7,546 | 76.2 |
| Cuming | 364 | 0.2 | 452 | 53.1 | 330.1 | 2,752,249 | 6,087 | 334,881 | 1,132 | 1,407,956 | 15.7 | 84.3 | 3 | 76.5 | 9,249 | 76.9 |
| Custer | 1,505 | 0.1 | 1,358 | 258.0 | 481.9 | 3,361,912 | 2,475 | 269,342 | 781.2 | 705,014 | 23.5 | 76.5 | NA | 81.9 | 14,411 | 48.5 |
| Dakota | 167 | 5.4 | 624 | 29.2 | 151.1 | 3,527,505 | 5,656 | 268,570 | 85 | 318,180 | 91.7 | 8.3 | NA | 73.0 | 4,211 | 69.7 |
| Dawes | 750 | -9 | 1,528 | 17.8 | 174.5 | 1,358,907 | 889 | 115,299 | 60.9 | 124,100 | 22.9 | 77.1 | 1 | 84.1 | 2,924 | 63.5 |
| Dawson | 610 | -3.2 | 889 | 235.0 | 303.7 | 3,034,516 | 3,412 | 336,300 | 748.4 | 1,091,000 | 22.9 | 77.1 | 16 | 84.4 | 15,112 | 60.5 |
| Deuel | 276 | -0.2 | 1,227 | 17.4 | 226.8 | 1,601,034 | 1,305 | 277,172 | 71.3 | 316,960 | 49.5 | 50.5 | NA | 74.7 | 4,297 | 76.9 |
| Dixon | 279 | -6.6 | 492 | 28.1 | 221.8 | 2,303,075 | 4,676 | 232,241 | 271.6 | 478,968 | 38.0 | 62.0 | 2 | 83.8 | 8,379 | 68.1 |
| Dodge | 337 | 2.2 | 499 | 132.6 | 312.5 | 3,199,731 | 6,412 | 286,042 | 270.5 | 400,151 | 65.3 | 34.7 | 9 | 76.3 | 9,943 | 74.1 |
| Douglas | 91 | 5.4 | 247 | 24.6 | 81.6 | 1,891,447 | 7,644 | 169,943 | 55.5 | 151,324 | 93.7 | 6.3 | NA | 88.8 | 2,205 | 35.7 |
| Dundy | 540 | 3.7 | 2,016 | 78.4 | 210.5 | 3,082,953 | 1,530 | 404,375 | 161.1 | 601,254 | 37.9 | 62.1 | NA | 88.1 | 5,558 | 76.1 |
| Fillmore | 329 | 0.3 | 750 | 219.9 | 305.3 | 4,088,900 | 5,448 | 460,510 | 240.9 | 548,850 | 74.9 | 25.1 | NA | 84.5 | 12,680 | 79.5 |
| Franklin | 316 | 10 | 998 | 97.5 | 187.0 | 3,456,445 | 3,462 | 273,977 | 106.9 | 337,088 | 84.8 | 15.2 | NA | 82.0 | 5,786 | 75.7 |
| Frontier | 484 | 7.1 | 1,305 | 50.2 | 203.8 | 2,452,811 | 1,879 | 262,770 | 121.4 | 327,332 | 50.0 | 50.0 | NA | 84.6 | 4,408 | 57.7 |
| Furnas | 450 | 3.3 | 1,194 | 64.1 | 291.5 | 2,946,660 | 2,467 | 338,139 | 240.4 | 637,637 | 42.2 | 57.8 | NA | 75.9 | 5,050 | 79.3 |
| Gage | 539 | 0.9 | 454 | 90.8 | 449.4 | 2,008,444 | 4,427 | 195,339 | 280.2 | 235,836 | 64.4 | 35.6 | NA | 83.0 | 14,883 | 70.6 |
| Garden | 1,018 | -0.8 | 4,608 | 34.6 | 166.3 | 3,992,706 | 866 | 240,357 | 81.2 | 367,416 | 40.8 | 59.2 | 3 | 86.9 | 1,939 | 61.1 |
| Garfield | 342 | | 1,696 | 15.3 | 66.4 | 2,513,876 | 1,483 | 142,181 | 54.7 | 270,891 | 19.7 | 80.3 | NA | 83.7 | 671 | 36.6 |
| Gosper | 282 | -2.7 | 983 | 79.2 | 150.2 | 2,763,817 | 2,812 | 361,583 | 105.7 | 368,397 | 69.4 | 30.6 | 1 | 83.3 | 5,134 | 61.3 |
| Grant | 495 | 0.4 | 7,736 | 1.7 | 50.6 | 6,476,796 | 837 | 236,936 | D | D | D | D | NA | 90.6 | D | 3.1 |
| Greeley | 339 | 0.3 | 919 | 91.9 | 156.5 | 2,726,640 | 2,965 | 274,794 | 193.3 | 523,957 | 34.2 | 65.8 | NA | 82.1 | 4,895 | 77.8 |
| Hall | 328 | -0.4 | 564 | 230.8 | 272.0 | 2,866,675 | 5,083 | 354,102 | 302.4 | 519,589 | 57.7 | 42.3 | 1 | 85.2 | 13,806 | 69.8 |
| Hamilton | 312 | 2.6 | 533 | 248.1 | 286.7 | 3,379,235 | 6,341 | 333,006 | 275.7 | 470,561 | 69.3 | 30.7 | 9 | 85.0 | 13,043 | 69.1 |
| Harlan | 334 | 6.7 | 1,188 | 86.6 | 220.6 | 3,417,433 | 2,878 | 370,997 | 160.3 | 570,374 | 58.6 | 41.4 | 2 | 83.6 | 5,397 | 69.4 |
| Hayes | 437 | 13.4 | 1,985 | 55.8 | 195.7 | 2,809,923 | 1,415 | 293,526 | 167.2 | 760,155 | 35.8 | 64.2 | NA | 84.1 | 2,946 | 83.2 |
| Hitchcock | 393 | -1.7 | 1,363 | 24.3 | 228.3 | 2,145,874 | 1,574 | 230,905 | 59.6 | 207,024 | 76.5 | 23.5 | 1 | 71.9 | 3,038 | 75.3 |
| Holt | 1,393 | -1.5 | 1,220 | 303.7 | 608.0 | 2,904,705 | 2,380 | 268,044 | 453.5 | 397,144 | 49.4 | 50.6 | 5 | 75.7 | 11,202 | 49.0 |
| Hooker | 427 | -2.2 | 4,402 | 2.5 | 6.8 | 3,237,188 | 735 | 109,284 | D | D | D | D | NA | 72.2 | 451 | 24.7 |
| Howard | 281 | -10.1 | 455 | 115.6 | 178.0 | 1,587,631 | 3,491 | 215,883 | 235.2 | 381,172 | 35.2 | 64.8 | NA | 78.3 | 7,045 | 63.5 |
| Jefferson | 359 | 1.9 | 608 | 105.4 | 283.7 | 2,525,053 | 4,151 | 262,657 | 219.6 | 372,158 | 52.0 | 48.0 | NA | 77.6 | 8,991 | 76.1 |
| Johnson | 197 | -0.1 | 393 | 19.4 | 137.7 | 1,427,986 | 3,631 | 151,986 | 83.1 | 165,602 | 59.2 | 40.8 | 1 | 72.1 | 4,348 | 80.1 |
| Kearney | 291 | -0.8 | 852 | 189.5 | 244.9 | 4,456,076 | 5,232 | 482,408 | 369.7 | 1,081,094 | 42.4 | 57.6 | 1 | 93.3 | 10,195 | 72.8 |
| Keith | 491 | -9.2 | 1,546 | 79.9 | 226.7 | 2,513,223 | 1,626 | 294,853 | 161.9 | 508,972 | 48.5 | 51.5 | 9 | 88.4 | 4,957 | 60.4 |
| Keya Paha | 423 | -9.2 | 1,784 | 25.0 | 95.6 | 2,527,270 | 1,416 | 200,328 | 52.3 | 220,810 | 34.8 | 65.2 | NA | 84.0 | 1,305 | 38.0 |
| Kimball | 603 | 1 | 1,362 | 33.8 | 410.7 | 1,464,107 | 1,075 | 141,161 | 40 | 90,237 | 75.9 | 24.1 | 26 | 65.9 | 9,488 | 81.0 |
| Knox | 601 | -4.3 | 628 | 70.8 | 323.6 | 2,102,065 | 3,345 | 223,437 | 288.5 | 301,768 | 37.5 | 62.5 | 11 | 76.8 | 12,233 | 72.6 |
| Lancaster | 423 | -13.5 | 237 | 21.3 | 362.9 | 1,325,789 | 5,598 | 125,910 | 188.8 | 105,730 | 82.3 | 17.7 | 11 | 85.8 | 9,485 | 53.1 |
| Lincoln | 1,357 | -4.7 | 1,305 | 242.8 | 421.6 | 2,184,451 | 1,674 | 221,829 | 755.2 | 726,188 | 24.5 | 75.5 | 7 | 83.4 | 13,167 | 44.2 |
| Logan | 298 | -9.7 | 2,547 | 16.2 | 41.9 | 3,055,685 | 1,200 | 227,088 | 28.6 | 244,564 | 35.6 | 64.4 | NA | 68.4 | 1,227 | 41.9 |
| Loup | 280 | -1.1 | 2,152 | 9.4 | 24.0 | 2,502,612 | 1,163 | 167,861 | 30.8 | 236,946 | 14.3 | 85.7 | NA | 85.4 | 765 | 46.2 |
| McPherson | 489 | 3.9 | 4,486 | 8.1 | 22.7 | 3,675,404 | 819 | 168,099 | 28.4 | 260,541 | 7.5 | 92.5 | NA | 73.4 | 471 | 28.4 |
| Madison | 353 | 0.5 | 536 | 130.7 | 312.1 | 3,120,513 | 5,819 | 324,775 | 276.1 | 418,948 | 57.0 | 43.0 | 2 | 79.5 | 7,543 | 69.7 |
| Merrick | 243 | 3.3 | 503 | 163.8 | 201.5 | 2,551,386 | 5,074 | 278,229 | 240.3 | 497,573 | 47.4 | 52.6 | 5 | 80.7 | 8,619 | 67.1 |
| Morrill | 829 | 3.7 | 1,945 | 121.8 | 242.5 | 2,226,414 | 1,145 | 237,293 | 319.7 | 750,451 | 25.1 | 74.9 | 1 | 73.0 | 5,034 | 67.4 |
| Nance | 220 | 5.7 | 587 | 75.6 | 158.8 | 2,480,621 | 4,227 | 240,847 | 155.3 | 414,141 | 48.6 | 51.4 | 6 | 77.3 | 6,113 | 73.9 |
| Nemaha | 261 | 2.9 | 636 | 15.4 | 230.0 | 2,893,077 | 4,548 | 234,596 | 114.4 | 279,090 | 93.3 | 6.7 | 6 | 74.6 | 6,373 | 79.5 |
| Nuckolls | 357 | 2.2 | 829 | 76.7 | 248.7 | 2,975,636 | 3,588 | 321,294 | 147.5 | 342,276 | 79.0 | 21.0 | 9 | 82.6 | 8,084 | 81.9 |
| Otoe | 390 | 0.6 | 479 | 12.8 | 331.0 | 2,372,053 | 4,957 | 203,937 | 170.5 | 209,232 | 85.3 | 14.7 | 9 | 83.1 | 5,740 | 73.3 |
| Pawnee | 273 | 1.4 | 593 | 11.4 | 183.7 | 1,921,924 | 3,244 | 194,408 | 78.9 | 171,454 | 72.0 | 28.0 | NA | 68.7 | 3,568 | 76.5 |
| Perkins | 556 | -0.1 | 1,330 | 131.9 | 432.1 | 2,847,060 | 2,140 | 350,487 | 196.8 | 470,794 | 74.6 | 25.4 | 2 | 84.9 | 11,277 | 80.4 |
| Phelps | 342 | 3.1 | 921 | 225.4 | 277.1 | 4,707,223 | 5,114 | 603,438 | 578.2 | 1,558,601 | 29.8 | 70.2 | NA | 85.7 | 13,067 | 81.7 |
| Pierce | 344 | 4.4 | 550 | 127.1 | 275.2 | 2,679,363 | 4,872 | 300,420 | 255.5 | 408,749 | 48.8 | 51.2 | 2 | 72.6 | 10,199 | 71.8 |

| STATE County | Water use, 2015 Public supply water withdrawn (mil gal/day) | Public supply gallons withdrawn per person per day | Wholesale Trade[1], 2017 Number of establishments | Number of employees | Sales (mil dol) | Average payroll (mil dol) | Retail Trade[2], 2017 Number of establishments | Number of employees | Sales (mil dol) | Average payroll (mil dol) | Real estate and rental and leasing,[2] 2017 Number of establishments | Number of employees | Sales (mil dol) | Average payroll (mil dol) |
|---|---|---|---|---|---|---|---|---|---|---|---|---|---|---|
| | 133 | 134 | 135 | 136 | 137 | 138 | 139 | 140 | 141 | 142 | 143 | 144 | 145 | 146 |
| **NEBRASKA— Cont'd** | | | | | | | | | | | | | | |
| Blaine | 0.02 | 41.1 | NA | NA | NA | NA | NA | NA | NA | NA | D | D | D | D |
| Boone | 0.63 | 118.5 | 16 | 205 | 342.9 | 9.2 | 36 | 272 | 75.8 | 5.4 | D | D | D | D |
| Box Butte | 1.58 | 139.4 | 20 | 240 | 164.8 | 10.5 | 39 | 422 | 101.2 | 9.2 | NA | NA | NA | NA |
| Boyd | 0.15 | 74.8 | D | D | D | D | 12 | 64 | 13.0 | 1.3 | NA | NA | NA | NA |
| Brown | 0.49 | 166.3 | 6 | 98 | 81.8 | 5.3 | 31 | 287 | 86.1 | 6.9 | NA | NA | NA | NA |
| Buffalo | 6.45 | 132.0 | 81 | 933 | 1,024.1 | 48.2 | 242 | 4,132 | 1,018.7 | 98.9 | 67 | 169 | 49.2 | 5.9 |
| Burt | 0.85 | 129.1 | 18 | 216 | 182.4 | 9.6 | 33 | 221 | 36.7 | 3.4 | 5 | D | 0.4 | D |
| Butler | 0.68 | 83.8 | 15 | 150 | 155.2 | 8.3 | 27 | 205 | 33.7 | 4.1 | NA | NA | NA | NA |
| Cass | 13.53 | 530.3 | D | D | D | 6.2 | 63 | 931 | 314.2 | 22.2 | 29 | 41 | 8.3 | 1.6 |
| Cedar | 0.65 | 75.9 | 27 | 185 | 146.5 | 8.6 | 45 | 339 | 89.8 | 7.6 | 7 | D | 1.3 | D |
| Chase | 0.91 | 230.0 | 22 | 204 | 458.9 | 10.2 | 31 | 358 | 113.4 | 10.1 | 4 | 4 | 0.2 | 0.1 |
| Cherry | 0.61 | 104.3 | 8 | 42 | 137.6 | 1.7 | 41 | 332 | 93.4 | 7.8 | 12 | 17 | 2.1 | 0.4 |
| Cheyenne | 2.16 | 212.5 | 16 | 114 | 68.5 | 6.2 | 48 | 823 | 646.4 | 24.8 | 8 | 21 | 2.0 | 0.6 |
| Clay | 1.25 | 198.1 | 18 | 139 | 153.3 | 6.8 | 27 | 208 | 71.7 | 5.5 | 3 | 3 | 0.4 | 0.1 |
| Colfax | 1.11 | 105.5 | 18 | 231 | 212.6 | 12.7 | 39 | 317 | 55.9 | 7.7 | 5 | D | 0.8 | D |
| Cuming | 1.61 | 176.4 | 23 | 163 | 139.7 | 8.0 | 48 | 403 | 128.1 | 9.0 | 4 | 6 | 0.4 | 0.1 |
| Custer | 1.18 | 109.2 | 11 | 72 | 36.9 | 3.3 | 76 | 629 | 169.7 | 15.1 | 5 | D | 0.4 | D |
| Dakota | 2.02 | 97.2 | D | D | D | 8.8 | 69 | 1,023 | 212.2 | 22.4 | D | D | D | D |
| Dawes | 1.18 | 130.3 | 5 | 71 | 11.3 | 1.5 | 46 | 661 | 174.4 | 17.6 | 5 | D | 0.8 | D |
| Dawson | 5.97 | 249.9 | 41 | 449 | 415.2 | 22.4 | 110 | 1,123 | 397.1 | 31.5 | NA | NA | NA | NA |
| Deuel | 0.50 | 260.3 | 4 | D | 20.5 | D | 14 | 104 | 79.9 | 2.4 | NA | NA | NA | NA |
| Dixon | 0.75 | 129.4 | D | D | D | D | 10 | 73 | 14.2 | 1.3 | NA | NA | NA | NA |
| Dodge | 0.70 | 19.1 | 59 | 874 | 1,121.8 | 49.3 | 144 | 2,415 | 751.5 | 65.4 | 44 | 174 | 31.5 | 5.6 |
| Douglas | 56.63 | 103.0 | 730 | 10,638 | 14,090.4 | 700.7 | 1,678 | 34,438 | 9,991.2 | 939.7 | 927 | 6,264 | 1,320.6 | 299.5 |
| Dundy | 0.28 | 155.6 | D | D | D | D | 9 | 29 | 14.9 | 0.8 | NA | NA | NA | NA |
| Fillmore | 0.78 | 138.8 | 23 | 161 | 165.7 | 7.5 | 32 | 215 | 54.3 | 4.9 | 4 | 5 | 0.3 | 0.1 |
| Franklin | 0.67 | 224.5 | 8 | D | 55.6 | D | 13 | 100 | 19.3 | 2.5 | NA | NA | NA | NA |
| Frontier | 0.27 | 102.9 | D | D | D | D | 13 | 77 | 17.0 | 1.7 | NA | NA | NA | NA |
| Furnas | 0.58 | 119.3 | 7 | 106 | 163.0 | 7.4 | 26 | 198 | 39.0 | 4.1 | NA | NA | NA | NA |
| Gage | 3.60 | 164.4 | D | D | D | 18.2 | 109 | 1,104 | 277.0 | 26.3 | 20 | 33 | 5.8 | 0.9 |
| Garden | 0.22 | 114.7 | NA | NA | NA | NA | 11 | 59 | 12.1 | 1.1 | NA | NA | NA | NA |
| Garfield | 0.18 | 88.8 | 3 | 60 | 9.5 | 1.0 | 20 | 142 | 29.8 | 2.7 | NA | NA | NA | NA |
| Gosper | 0.28 | 141.9 | 6 | 23 | 28.8 | 1.1 | 4 | 29 | 4.1 | 0.4 | NA | NA | NA | NA |
| Grant | 0.05 | 78.0 | NA | NA | NA | NA | 5 | 27 | 4.6 | 0.3 | NA | NA | NA | NA |
| Greeley | 0.23 | 94.7 | 7 | 72 | 47.5 | 2.9 | 14 | 90 | 36.3 | 2.1 | NA | NA | NA | NA |
| Hall | 11.94 | 193.6 | 95 | 1,251 | 859.6 | 71.2 | 306 | 5,070 | 1,276.3 | 129.7 | 75 | 313 | 58.1 | 10.5 |
| Hamilton | 1.24 | 134.9 | 23 | 372 | 651.6 | 25.4 | 40 | 344 | 131.9 | 8.6 | 9 | 22 | 1.9 | 0.4 |
| Harlan | 0.45 | 130.4 | 11 | D | 80.1 | D | 17 | 113 | 45.3 | 2.6 | NA | NA | NA | NA |
| Hayes | 0.09 | 96.6 | NA | NA | NA | NA | 4 | 18 | 1.4 | 0.4 | NA | NA | NA | NA |
| Hitchcock | 0.37 | 128.3 | 9 | D | 76.4 | D | 12 | 63 | 11.6 | 1.2 | NA | NA | NA | NA |
| Holt | 1.15 | 111.5 | 37 | 415 | 305.7 | 16.7 | 71 | 575 | 157.6 | 12.4 | 9 | 13 | 1.9 | 0.4 |
| Hooker | 0.18 | 245.9 | NA | NA | NA | NA | 6 | 27 | 6.1 | 0.6 | NA | NA | NA | NA |
| Howard | 0.56 | 87.4 | 8 | 55 | 49.3 | 2.4 | 24 | 241 | 51.8 | 5.3 | D | D | D | D |
| Jefferson | 1.11 | 152.8 | 19 | 197 | 212.9 | 9.0 | 27 | 439 | 161.5 | 10.1 | 7 | 11 | 1.0 | 0.3 |
| Johnson | 1.00 | 193.3 | D | D | D | 1.2 | 24 | 133 | 36.7 | 3.3 | 3 | D | 3.2 | D |
| Kearney | 0.86 | 130.6 | 15 | 158 | 257.0 | 8.6 | 19 | 133 | 31.4 | 3.3 | NA | NA | NA | NA |
| Keith | 1.08 | 133.9 | 16 | 163 | 143.7 | 6.5 | 61 | 690 | 222.1 | 17.7 | 15 | 30 | 3.0 | 0.7 |
| Keya Paha | 0.05 | 62.2 | NA | NA | NA | NA | 4 | 21 | 3.4 | 0.3 | NA | NA | NA | NA |
| Kimball | 0.52 | 141.0 | D | D | D | D | 22 | 171 | 41.2 | 4.1 | NA | NA | NA | NA |
| Knox | 0.97 | 113.5 | 14 | 137 | 123.0 | 7.0 | 42 | 342 | 71.8 | 6.4 | 9 | D | 2.0 | D |
| Lancaster | 1.52 | 5.0 | 286 | 4,065 | 4,341.6 | 223.6 | 1,010 | 18,415 | 4,810.3 | 477.3 | 415 | 1,999 | 367.7 | 81.7 |
| Lincoln | 5.63 | 157.9 | D | D | D | D | 190 | 2,257 | 704.5 | 54.4 | 47 | 158 | 30.9 | 6.0 |
| Logan | 0.05 | 64.4 | NA | NA | NA | NA | NA | NA | NA | NA | NA | NA | NA | NA |
| Loup | 0.00 | 0.0 | NA | NA | NA | NA | NA | NA | NA | NA | NA | NA | NA | NA |
| McPherson | 0.00 | 0.0 | NA | NA | NA | NA | NA | NA | NA | NA | NA | NA | NA | NA |
| Madison | 4.94 | 141.0 | 58 | 1,828 | 1,959.1 | 78.9 | 209 | 3,008 | 767.5 | 72.6 | 68 | 203 | 41.7 | 8.7 |
| Merrick | 0.80 | 102.7 | 21 | 162 | 155.2 | 9.1 | 27 | 208 | 66.7 | 4.8 | D | D | D | D |
| Morrill | 0.43 | 88.6 | 9 | D | 242.6 | D | 20 | 172 | 34.1 | 3.6 | 3 | D | 0.7 | D |
| Nance | 0.37 | 102.9 | D | D | D | 1.7 | 18 | 116 | 26.2 | 2.4 | NA | NA | NA | NA |
| Nemaha | 0.89 | 126.3 | 10 | 57 | 40.9 | 2.2 | 28 | 261 | 64.5 | 5.7 | 6 | D | 0.9 | D |
| Nuckolls | 0.55 | 127.1 | 15 | 122 | 106.5 | 5.7 | 25 | 210 | 61.3 | 5.0 | NA | NA | NA | NA |
| Otoe | 2.54 | 158.9 | 23 | 140 | 190.7 | 7.4 | 66 | 764 | 194.6 | 17.6 | 20 | 56 | 6.7 | 1.7 |
| Pawnee | 0.37 | 139.2 | NA | NA | NA | NA | 12 | 78 | 25.0 | 1.6 | NA | NA | NA | NA |
| Perkins | 0.68 | 231.0 | 15 | 106 | 238.6 | 6.1 | 18 | 138 | 71.3 | 4.3 | NA | NA | NA | NA |
| Phelps | 1.62 | 174.3 | 28 | 329 | 509.2 | 17.6 | 51 | 474 | 141.1 | 12.0 | 6 | D | 1.2 | D |
| Pierce | 0.69 | 95.7 | D | D | D | D | 30 | 209 | 46.8 | 3.8 | D | D | D | D |

1  Merchant wholesalers, except manufacturers' sales branches and offices.   2. Employer establishments.

# Table B. States and Counties — Professional Services, Manufacturing, and Accommodation and Food Services

| STATE County | Professional, scientific, and technical services, 2017 | | | | Manufacturing, 2017 | | | | Accommodation and food services, 2017 | | | |
|---|---|---|---|---|---|---|---|---|---|---|---|---|
| | Number of establishments | Number of employees | Sales (mil dol) | Average payroll (mil dol) | Number of establishments | Number of employees | Sales (mil dol) | Average payroll (mil dol) | Number of establishments | Number of employees | Sales (mil dol) | Annual payroll (mil dol) |
| | 147 | 148 | 149 | 150 | 151 | 152 | 153 | 154 | 155 | 156 | 157 | 158 |
| NEBRASKA— Cont'd | | | | | | | | | | | | |
| Blaine | NA | NA | NA | NA | NA | NA | NA | NA | NA | NA | NA | NA |
| Boone | 7 | 28 | 6 | 1.0 | D | 93 | D | 6.2 | 11 | 83 | 4.0 | 1.0 |
| Box Butte | 19 | 90 | 9 | 2.8 | 7 | 277 | 97.5 | 14.3 | 30 | 338 | 17.0 | 5.0 |
| Boyd | NA | NA | NA | NA | 4 | 14 | 1.7 | 0.4 | 3 | 19 | 0.6 | 0.2 |
| Brown | 8 | 22 | 3 | 0.7 | D | D | D | D | 11 | D | 4.1 | D |
| Buffalo | 115 | 1,088 | 112 | 48.1 | 58 | 3,146 | 1,503.2 | 159.0 | 161 | 3,133 | 154.2 | 45.4 |
| Burt | 9 | 57 | 14 | 3.6 | 6 | 34 | 21.1 | 1.7 | 13 | D | 3.2 | D |
| Butler | 11 | 40 | 4 | 1.8 | 8 | 490 | 166.6 | 22.0 | D | D | D | D |
| Cass | 38 | 126 | 17 | 6.5 | 22 | 376 | 184.8 | 21.0 | 37 | 401 | 17.1 | 4.8 |
| Cedar | 12 | 75 | 11 | 2.8 | 9 | 195 | 74.0 | 12.6 | 16 | D | 4.7 | D |
| Chase | 13 | 24 | 4 | 0.9 | 5 | 13 | 0.8 | 0.2 | 9 | D | 2.6 | D |
| Cherry | 23 | 92 | 10 | 2.3 | 7 | 19 | 4.6 | 1.1 | 27 | 341 | 22.7 | 7.4 |
| Cheyenne | 15 | 62 | 9 | 3.1 | 11 | 389 | 118.0 | 17.2 | 36 | 443 | 23.6 | 7.6 |
| Clay | 9 | 36 | 4 | 1.2 | 5 | 43 | 21.5 | 2.0 | 11 | D | 1.7 | D |
| Colfax | 12 | 46 | 6 | 1.6 | D | D | D | D | D | D | D | D |
| Cuming | D | D | D | D | 21 | 360 | 442.2 | 18.3 | 24 | 240 | 9.5 | 2.7 |
| Custer | 28 | 95 | 12 | 3.2 | D | 556 | D | D | 32 | 307 | 12.3 | 3.7 |
| Dakota | D | D | D | D | 36 | 5,462 | 3,633.6 | 234.6 | 43 | 643 | 37.4 | 10.8 |
| Dawes | 17 | 66 | 5 | 2.1 | 3 | 8 | 1.6 | 0.3 | 37 | 459 | 20.2 | 5.7 |
| Dawson | D | D | D | D | D | D | D | D | 61 | 651 | 42.4 | 9.2 |
| Deuel | NA | NA | NA | NA | NA | NA | NA | NA | D | D | D | 0.6 |
| Dixon | NA | NA | NA | NA | NA | NA | NA | NA | D | D | D | D |
| Dodge | 45 | 217 | 27 | 9.1 | 51 | 3,553 | 1,600.2 | 170.3 | 87 | 1,358 | 61.5 | 17.5 |
| Douglas | 1,845 | 19,367 | 3,258 | 1,289.7 | 390 | 20,969 | 11,619.9 | 1,047.6 | 1,415 | 28,267 | 1,578.4 | 461.6 |
| Dundy | 5 | 16 | 2 | 0.4 | 3 | 8 | 2.7 | 0.5 | 4 | D | 1.2 | D |
| Fillmore | 10 | 27 | 3 | 0.9 | 13 | 230 | 259.0 | 11.9 | 12 | 42 | 2.2 | 0.5 |
| Franklin | 4 | 22 | 3 | 0.9 | NA | NA | NA | NA | 6 | D | 1.1 | D |
| Frontier | 3 | 15 | 2 | 0.6 | NA | NA | NA | NA | 6 | D | 1.6 | D |
| Furnas | 7 | 23 | 3 | 0.8 | 8 | 129 | 115.3 | 5.5 | 13 | 68 | 2.7 | 0.6 |
| Gage | 35 | 146 | 14 | 5.6 | 38 | 1,304 | 669.7 | 71.4 | 49 | 643 | 29.4 | 7.6 |
| Garden | NA | NA | NA | NA | NA | NA | NA | NA | 6 | D | 0.9 | D |
| Garfield | 10 | 18 | 3 | 0.6 | 5 | 92 | 32.9 | 4.0 | 10 | D | 1.5 | D |
| Gosper | NA | NA | NA | NA | NA | NA | NA | NA | NA | NA | NA | NA |
| Grant | NA | NA | NA | NA | NA | NA | NA | NA | NA | NA | NA | NA |
| Greeley | NA | NA | NA | NA | 3 | 14 | 1.6 | 0.4 | D | D | D | 0.1 |
| Hall | D | D | D | D | 64 | 6,799 | 4,629.0 | 332.7 | 165 | 2,662 | 147.2 | 42.9 |
| Hamilton | 23 | 100 | 15 | 5.2 | 19 | 582 | 664.9 | 23.9 | 19 | 163 | 6.5 | 2.2 |
| Harlan | 8 | 30 | 3 | 0.8 | 3 | 31 | 7.1 | D | 17 | 91 | 5.7 | 1.0 |
| Hayes | NA | NA | NA | NA | NA | NA | NA | NA | NA | NA | NA | NA |
| Hitchcock | NA | NA | NA | NA | 3 | 82 | 127.9 | 6.6 | 4 | 19 | 0.6 | 0.1 |
| Holt | 24 | 88 | 12 | 4.4 | 19 | 164 | 91.0 | 7.2 | 29 | 290 | 12.6 | 3.2 |
| Hooker | NA | NA | NA | NA | NA | NA | NA | NA | NA | NA | NA | NA |
| Howard | 9 | 36 | 5 | 1.4 | 5 | 23 | 3.6 | 0.7 | D | D | D | D |
| Jefferson | 13 | 52 | 7 | 1.8 | 18 | 642 | 188.7 | 29.9 | 20 | 162 | 7.6 | 1.8 |
| Johnson | 6 | 29 | 2 | 0.6 | D | D | D | D | D | D | D | 0.5 |
| Kearney | 9 | 23 | 3 | 1.0 | 11 | 301 | 238.0 | 15.2 | 14 | 95 | 3.9 | 1.2 |
| Keith | 32 | 117 | 14 | 4.5 | 13 | 205 | 22.5 | 6.8 | 48 | 527 | 26.7 | 7.8 |
| Keya Paha | 4 | 4 | 0 | 0.1 | NA | NA | NA | NA | NA | NA | NA | NA |
| Kimball | 8 | 15 | 1 | 0.4 | 7 | 185 | 33.1 | 7.4 | 12 | 46 | 3.5 | 0.7 |
| Knox | 17 | 64 | 7 | 1.9 | 7 | 29 | 9.7 | 1.4 | 20 | 197 | 14.2 | 3.4 |
| Lancaster | 891 | 8,517 | 1,468 | 495.8 | 229 | 12,008 | 4,863.0 | 683.8 | 777 | 15,006 | 761.1 | 216.1 |
| Lincoln | D | D | D | D | D | D | D | D | 100 | 1,848 | 92.3 | 25.8 |
| Logan | NA | NA | NA | NA | NA | NA | NA | NA | NA | NA | NA | NA |
| Loup | NA | NA | NA | NA | NA | NA | NA | NA | D | D | D | 0.1 |
| McPherson | NA | NA | NA | NA | NA | NA | NA | NA | NA | NA | NA | NA |
| Madison | D | D | D | D | D | 3,056 | D | D | 93 | 1,634 | 77.8 | 21.5 |
| Merrick | 8 | 39 | 5 | 1.5 | 14 | 212 | 279.3 | 10.3 | D | D | D | D |
| Morrill | 4 | 15 | 2 | 0.5 | NA | NA | NA | NA | 8 | 53 | 2.3 | 0.7 |
| Nance | D | D | D | D | NA | NA | NA | NA | 10 | D | 2.0 | D |
| Nemaha | 11 | 45 | 3 | 1.1 | D | D | D | D | 21 | 249 | 9.0 | 3.1 |
| Nuckolls | 11 | 28 | 3 | 0.8 | D | 7 | D | 0.2 | 11 | D | 2.9 | D |
| Otoe | 28 | 79 | 9 | 2.5 | 16 | 1,474 | 562.1 | 59.7 | 40 | 434 | 17.1 | 4.7 |
| Pawnee | D | D | D | D | D | D | D | D | 6 | D | 0.8 | D |
| Perkins | 7 | 18 | 1 | 0.5 | D | D | D | D | 5 | D | 0.8 | D |
| Phelps | 23 | 103 | 13 | 4.2 | D | 45 | D | 3.0 | 18 | 264 | 9.1 | 2.5 |
| Pierce | 11 | 53 | 4 | 1.3 | D | D | D | D | 11 | 82 | 2.5 | 0.6 |

# Health Care and Social Assistance, Other Services, Nonemployer Businesses, and Residential Construction

| STATE County | Health care and social assistance, 2017 | | | | Other services, 2017 | | | | Nonemployer businesses, 2018 | | Value of residential construction authorized by building permits, 2020 | |
|---|---|---|---|---|---|---|---|---|---|---|---|---|
| | Number of establishments | Number of employees | Receipts (mil dol) | Annual payroll (mil dol) | Number of establishments | Number of employees | Receipts (mil dol) | Annual payroll (mil dol) | Number | Receipts (mil dol) | New construction ($1,000) | Number of housing units |
| | 159 | 160 | 161 | 162 | 163 | 164 | 165 | 166 | 167 | 168 | 169 | 170 |
| **NEBRASKA— Cont'd** | | | | | | | | | | | | |
| Blaine | NA | NA | NA | NA | NA | NA | NA | NA | 41 | 0.9 | NA | NA |
| Boone | 18 | 277 | 19.6 | 7.3 | 15 | 30 | 4.2 | 1.1 | 540 | 22.9 | 605 | 2 |
| Box Butte | 27 | 548 | 47.9 | 21.0 | 28 | 83 | 10.8 | 2.5 | 716 | 27.4 | 0 | 0 |
| Boyd | 7 | D | 7.0 | D | 3 | 6 | 0.9 | 0.1 | 193 | 7.9 | 0 | 0 |
| Brown | 12 | 158 | 15.8 | 7.0 | D | D | 3.6 | D | 302 | 12.4 | 0 | 0 |
| Buffalo | 206 | 4,103 | 678.2 | 174.1 | 124 | 655 | 91.9 | 17.9 | 3,923 | 184.0 | 56,557 | 217 |
| Burt | 16 | 243 | 25.5 | 9.0 | D | D | 8.5 | D | 554 | 27.5 | 3,070 | 15 |
| Butler | 16 | 378 | 33.6 | 14.1 | D | D | D | D | 675 | 30.9 | 1,165 | 4 |
| Cass | 31 | 411 | 22.9 | 9.6 | 41 | 122 | 11.8 | 3.4 | 1,962 | 89.7 | 26,673 | 135 |
| Cedar | 14 | 166 | 8.3 | 4.1 | D | D | D | D | 773 | 37.4 | 2,095 | 11 |
| Chase | D | D | D | 4.7 | D | D | 4.9 | D | 457 | 23.2 | 920 | 3 |
| Cherry | 16 | 328 | 38.0 | 12.9 | 15 | 42 | 5.0 | 0.9 | 590 | 25.7 | 3,132 | 13 |
| Cheyenne | 20 | 565 | 75.1 | 25.4 | 21 | 61 | 7.2 | 1.8 | 692 | 32.1 | 0 | 0 |
| Clay | 14 | D | 5.5 | D | D | D | 3.6 | D | 540 | 23.7 | 352 | 8 |
| Colfax | 16 | 257 | 28.2 | 8.5 | 26 | 79 | 7.3 | 2.3 | 579 | 31.4 | 4,124 | 15 |
| Cuming | 23 | 379 | 57.4 | 17.8 | 35 | 136 | 22.9 | 4.8 | 753 | 44.0 | 4,793 | 33 |
| Custer | 42 | 648 | 51.0 | 20.8 | 28 | 50 | 9.3 | 1.5 | 1,169 | 47.0 | 5,217 | 22 |
| Dakota | 33 | 520 | 43.4 | 16.1 | 36 | 213 | 54.5 | 13.9 | 1,074 | 62.7 | 7,354 | 28 |
| Dawes | 34 | 414 | 45.7 | 16.5 | 20 | 37 | 3.3 | 0.7 | 626 | 22.9 | 2,220 | 10 |
| Dawson | 70 | 1,316 | 111.8 | 45.9 | D | D | D | D | 1,598 | 72.7 | 14,210 | 81 |
| Deuel | 4 | D | 2.5 | D | NA | NA | NA | NA | 178 | 7.4 | 250 | 2 |
| Dixon | 6 | 83 | 3.3 | 1.6 | D | D | D | D | 499 | 18.9 | 1,395 | 7 |
| Dodge | 122 | 2,949 | 268.9 | 109.4 | 87 | 349 | 38.2 | 9.5 | 2,202 | 109.1 | 44,985 | 384 |
| Douglas | 1,958 | 54,665 | 7,151.1 | 2,807.5 | 1,139 | 7,736 | 1,699.7 | 263.2 | 39,062 | 1,982.0 | 406,540 | 3,240 |
| Dundy | 4 | D | 11.9 | D | NA | NA | NA | NA | 172 | 6.6 | 543 | 4 |
| Fillmore | 14 | 314 | 29.5 | 11.2 | D | D | D | 2.0 | 521 | 26.1 | 2,219 | 8 |
| Franklin | 9 | 95 | 6.4 | 2.3 | D | D | 1.3 | D | 263 | 9.6 | 1,328 | 4 |
| Frontier | 8 | 51 | 1.3 | 0.8 | D | D | 2.0 | D | 215 | 9.6 | 185 | 2 |
| Furnas | 18 | 330 | 38.2 | 13.4 | 10 | 25 | 3.5 | 0.7 | 441 | 20.8 | 10,680 | 6 |
| Gage | 57 | 1,739 | 137.3 | 62.7 | 65 | 285 | 19.5 | 5.1 | 1,539 | 62.9 | 7,296 | 33 |
| Garden | NA | NA | NA | NA | D | D | 0.3 | D | 182 | 7.8 | 595 | 4 |
| Garfield | 8 | D | 6.1 | D | 13 | 33 | 2.9 | 1.0 | 216 | 8.5 | 1,462 | 8 |
| Gosper | 7 | 9 | 0.5 | 0.2 | NA | NA | NA | NA | 218 | 11.9 | 2,806 | 10 |
| Grant | NA | NA | NA | NA | NA | NA | NA | NA | 93 | 3.5 | 0 | 0 |
| Greeley | 3 | 21 | 0.4 | 0.2 | 3 | 10 | 1.6 | 0.3 | 270 | 15.1 | 1,946 | 10 |
| Hall | 192 | 4,518 | 514.4 | 196.8 | 152 | 1,067 | 98.0 | 27.3 | 3,908 | 186.1 | 34,404 | 221 |
| Hamilton | D | D | D | D | 24 | 68 | 9.2 | 2.5 | 818 | 29.5 | 8,345 | 27 |
| Harlan | 13 | D | 17.6 | D | 3 | 9 | 1.9 | 0.3 | 320 | 13.1 | 2,046 | 12 |
| Hayes | NA | NA | NA | NA | NA | NA | NA | NA | 84 | 4.3 | 0 | 0 |
| Hitchcock | NA | NA | NA | NA | NA | NA | NA | NA | 221 | 9.1 | 0 | 0 |
| Holt | 36 | 783 | 81.5 | 28.8 | 34 | 92 | 16.3 | 2.9 | 1,297 | 73.8 | 1,880 | 8 |
| Hooker | NA | NA | NA | NA | D | D | D | 0.1 | 107 | 3.6 | 1,850 | 3 |
| Howard | D | D | D | D | 14 | 42 | 3.5 | 0.9 | 486 | 19.4 | 4,942 | 18 |
| Jefferson | 19 | 436 | 33.5 | 14.9 | 15 | 60 | 6.5 | 1.5 | 469 | 20.4 | 1,525 | 15 |
| Johnson | 11 | 278 | 22.3 | 9.9 | D | D | 1.8 | D | 295 | 10.7 | 0 | 0 |
| Kearney | 10 | 477 | 31.6 | 16.9 | 14 | 25 | 3.4 | 0.7 | 538 | 23.6 | 6,315 | 21 |
| Keith | 22 | 284 | 31.1 | 11.9 | 24 | 80 | 10.3 | 2.5 | 814 | 41.1 | 6,235 | 33 |
| Keya Paha | NA | NA | NA | NA | 6 | 9 | 1.2 | 0.2 | 118 | 4.3 | 0 | 0 |
| Kimball | 6 | 95 | 11.1 | 5.4 | 10 | 25 | 9.1 | 0.8 | 259 | 8.7 | 238 | 1 |
| Knox | 19 | 274 | 19.8 | 8.4 | D | D | 3.3 | D | 693 | 25.3 | 3,285 | 16 |
| Lancaster | 1,050 | 24,911 | 2,890.7 | 1,102.1 | 727 | 4,334 | 779.5 | 157.9 | 22,126 | 918.8 | 362,510 | 1,968 |
| Lincoln | D | D | D | D | D | D | D | D | 2,234 | 92.5 | 7,066 | 33 |
| Logan | NA | NA | NA | NA | NA | NA | NA | NA | 63 | 2.7 | NA | NA |
| Loup | NA | NA | NA | NA | NA | NA | NA | NA | 80 | 2.3 | 693 | 3 |
| McPherson | NA | NA | NA | NA | NA | NA | NA | NA | 51 | 2.1 | 58 | 1 |
| Madison | 173 | 3,507 | 463.7 | 158.4 | 96 | 482 | 46.5 | 14.2 | 2,620 | 116.3 | 35,291 | 359 |
| Merrick | 17 | 311 | 30.3 | 10.7 | 16 | 64 | 11.5 | 2.2 | 686 | 28.3 | 6,617 | 32 |
| Morrill | 5 | D | 16.4 | D | D | D | 2.3 | D | 405 | 15.4 | 0 | 0 |
| Nance | D | D | D | D | D | D | D | 0.3 | 277 | 10.9 | 400 | 2 |
| Nemaha | 23 | 352 | 32.3 | 14.2 | 14 | 46 | 3.7 | 0.9 | 481 | 16.9 | 842 | 4 |
| Nuckolls | 17 | 397 | 43.1 | 13.7 | D | D | 3.3 | D | 314 | 14.3 | 600 | 4 |
| Otoe | 42 | 806 | 66.2 | 27.3 | 31 | 98 | 9.2 | 3.0 | 1,252 | 49.8 | 4,469 | 18 |
| Pawnee | D | D | D | 4.3 | D | D | D | 0.8 | 271 | 11.6 | 758 | 6 |
| Perkins | 7 | 254 | 23.8 | 9.0 | D | D | D | 0.7 | 312 | 16.1 | 2,050 | 6 |
| Phelps | 23 | 762 | 69.1 | 27.6 | 28 | 85 | 15.0 | 2.6 | 875 | 38.4 | 2,525 | 14 |
| Pierce | D | D | D | D | D | D | 2.5 | D | 674 | 28.7 | 3,574 | 15 |

# Table B. States and Counties — Government Employment and Payroll, and Local Government Finances

| | Government employment and payroll, 2017 | | | | | | | | | Local government finances, 2017 | | | | |
| | | | March payroll (percent of total) | | | | | | | General revenue | | | | |
| | | | | | | | | | | | | Taxes | | |
| | | | | | | | | | | | | | Per capita[1] (dollars) | |
| STATE County | Full-time equivalent employees | March payroll (dollars) | Administration, judicial, and legal | Police and corrections | Fire protection | Highways and transportation | Health and welfare | Natural resources and utilities | Education and libraries | Total (mil dol) | Intergovernmental (mil dol) | Total (mil dol) | Total | Property |
| | 171 | 172 | 173 | 174 | 175 | 176 | 177 | 178 | 179 | 180 | 181 | 182 | 183 | 184 |
| **NEBRASKA— Cont'd** | | | | | | | | | | | | | | |
| Blaine | 41 | 137,107 | 18.7 | 0.0 | 0.0 | 5.8 | 0.0 | 0.0 | 75.4 | 4.6 | 0.9 | 3.5 | 7,228 | 6,902 |
| Boone | 523 | 2,183,498 | 3.9 | 2.2 | 0.0 | 2.5 | 53.8 | 0.8 | 36.3 | 59.4 | 8.0 | 19.4 | 3,645 | 3,232 |
| Box Butte | 716 | 2,924,913 | 4.1 | 6.3 | 0.8 | 3.3 | 43.9 | 4.7 | 36.0 | 90.8 | 13.8 | 26.9 | 2,481 | 2,039 |
| Boyd | 132 | 462,199 | 12.1 | 1.6 | 0.0 | 18.9 | 0.8 | 4.6 | 61.1 | 11.3 | 2.6 | 6.2 | 3,154 | 2,951 |
| Brown | 284 | 1,161,041 | 4.2 | 2.5 | 0.0 | 3.9 | 34.1 | 13.1 | 41.7 | 28.6 | 5.0 | 10.2 | 3,415 | 3,091 |
| Buffalo | 1,824 | 7,821,914 | 6.4 | 9.3 | 1.3 | 3.5 | 0.9 | 7.1 | 69.3 | 186.1 | 42.4 | 110.0 | 2,230 | 1,747 |
| Burt | 393 | 1,395,814 | 7.7 | 5.6 | 0.0 | 7.6 | 0.7 | 13.2 | 63.7 | 41.3 | 6.6 | 27.1 | 4,150 | 3,805 |
| Butler | 494 | 1,871,938 | 5.2 | 3.6 | 0.0 | 7.7 | 30.0 | 16.4 | 36.9 | 64.3 | 17.4 | 28.7 | 3,577 | 3,263 |
| Cass | 913 | 3,916,978 | 7.5 | 8.9 | 0.1 | 4.2 | 1.0 | 2.7 | 72.4 | 108.9 | 30.9 | 61.5 | 2,370 | 2,118 |
| Cedar | 375 | 1,508,658 | 9.1 | 3.1 | 0.0 | 5.4 | 0.5 | 15.1 | 66.3 | 39.0 | 6.8 | 22.8 | 2,682 | 2,561 |
| Chase | 374 | 1,346,214 | 4.5 | 2.4 | 0.0 | 4.7 | 32.3 | 7.7 | 43.8 | 43.6 | 5.0 | 17.1 | 4,345 | 3,891 |
| Cherry | 356 | 1,368,079 | 5.4 | 5.3 | 0.0 | 5.7 | 37.6 | 5.0 | 39.9 | 52.0 | 7.7 | 19.9 | 3,433 | 2,691 |
| Cheyenne | 520 | 1,921,015 | 5.9 | 7.2 | 0.1 | 5.9 | 0.7 | 16.3 | 62.2 | 78.5 | 18.2 | 37.0 | 3,841 | 3,242 |
| Clay | 465 | 1,531,436 | 7.3 | 3.6 | 0.0 | 5.3 | 10.3 | 4.6 | 67.2 | 41.2 | 9.4 | 24.7 | 3,992 | 3,690 |
| Colfax | 523 | 2,018,305 | 5.5 | 5.4 | 0.2 | 3.4 | 0.7 | 6.5 | 76.7 | 70.0 | 16.7 | 35.7 | 3,333 | 3,090 |
| Cuming | 438 | 1,759,278 | 8.8 | 5.2 | 0.0 | 9.1 | 0.6 | 10.9 | 64.1 | 53.0 | 10.7 | 30.5 | 3,411 | 3,005 |
| Custer | 684 | 2,528,214 | 7.7 | 3.4 | 0.2 | 5.8 | 11.8 | 16.9 | 52.2 | 71.7 | 14.2 | 37.9 | 3,484 | 3,136 |
| Dakota | 879 | 3,512,617 | 6.8 | 10.0 | 1.2 | 2.6 | 1.1 | 7.0 | 70.7 | 91.5 | 44.5 | 30.7 | 1,534 | 1,197 |
| Dawes | 380 | 1,261,755 | 10.4 | 8.6 | 0.0 | 7.3 | 1.7 | 10.4 | 59.9 | 42.5 | 13.2 | 16.1 | 1,808 | 1,480 |
| Dawson | 1,784 | 7,529,079 | 3.1 | 4.5 | 0.0 | 2.3 | 31.6 | 11.1 | 46.4 | 235.4 | 43.1 | 105.2 | 4,449 | 3,676 |
| Deuel | 139 | 539,672 | 11.1 | 6.0 | 0.0 | 9.1 | 1.0 | 3.4 | 67.3 | 18.9 | 1.7 | 15.7 | 8,419 | 7,722 |
| Dixon | 439 | 1,573,949 | 7.5 | 4.4 | 0.0 | 4.0 | 0.0 | 1.2 | 81.9 | 40.5 | 10.4 | 22.1 | 3,849 | 3,635 |
| Dodge | 2,252 | 10,124,969 | 4.5 | 4.5 | 1.4 | 2.6 | 35.9 | 8.4 | 42.2 | 281.5 | 47.0 | 80.6 | 2,191 | 1,696 |
| Douglas | 23,705 | 112,709,866 | 3.4 | 9.3 | 4.7 | 3.4 | 5.5 | 21.8 | 50.8 | 2,804.7 | 924.4 | 1,317.4 | 2,349 | 1,684 |
| Dundy | 217 | 848,201 | 7.7 | 2.6 | 0.0 | 3.3 | 50.0 | 4.1 | 32.0 | 20.7 | 2.3 | 7.9 | 4,437 | 4,059 |
| Fillmore | 444 | 1,702,509 | 7.7 | 2.9 | 0.1 | 5.3 | 38.7 | 2.4 | 42.0 | 60.6 | 8.1 | 21.8 | 3,924 | 3,472 |
| Franklin | 198 | 688,947 | 18.6 | 3.8 | 0.0 | 6.8 | 28.0 | 3.0 | 37.7 | 20.3 | 3.0 | 10.0 | 3,375 | 3,219 |
| Frontier | 188 | 723,299 | 8.9 | 4.0 | 0.2 | 5.5 | 1.0 | 2.2 | 71.1 | 22.0 | 4.4 | 15.2 | 5,777 | 4,396 |
| Furnas | 428 | 1,519,779 | 6.6 | 2.8 | 0.6 | 3.6 | 19.2 | 15.5 | 51.0 | 35.8 | 9.5 | 19.6 | 4,125 | 3,530 |
| Gage | 954 | 3,899,069 | 7.0 | 7.1 | 4.0 | 5.7 | 0.5 | 20.2 | 55.1 | 99.5 | 25.3 | 55.8 | 2,579 | 2,058 |
| Garden | 125 | 382,534 | 13.3 | 10.0 | 1.8 | 9.2 | 0.8 | 0.4 | 61.5 | 23.8 | 3.5 | 6.5 | 3,428 | 3,124 |
| Garfield | 107 | 365,812 | 13.6 | 3.7 | 0.0 | 5.9 | 4.3 | 3.1 | 67.9 | 12.8 | 2.7 | 8.8 | 4,411 | 4,076 |
| Gosper | 86 | 342,689 | 12.4 | 6.9 | 0.0 | 15.4 | 0.0 | 15.4 | 46.5 | 15.9 | 2.0 | 9.7 | 4,762 | 4,132 |
| Grant | 63 | 174,460 | 8.4 | 5.0 | 0.0 | 7.4 | 0.0 | 1.3 | 77.4 | 3.8 | 0.6 | 2.9 | 4,516 | 4,276 |
| Greeley | 111 | 412,595 | 11.1 | 4.4 | 0.0 | 5.5 | 0.7 | 6.7 | 71.2 | 16.9 | 2.5 | 9.8 | 4,155 | 3,638 |
| Hall | 2,859 | 12,652,757 | 5.4 | 8.4 | 3.9 | 3.6 | 0.7 | 17.3 | 59.6 | 309.0 | 108.8 | 130.3 | 2,128 | 1,608 |
| Hamilton | 409 | 1,507,031 | 6.9 | 9.5 | 0.0 | 4.3 | 2.8 | 4.2 | 70.6 | 46.7 | 8.6 | 26.0 | 2,830 | 2,569 |
| Harlan | 250 | 843,465 | 7.5 | 3.9 | 0.0 | 4.8 | 38.7 | 7.5 | 36.5 | 21.9 | 1.8 | 10.3 | 3,027 | 2,431 |
| Hayes | 56 | 163,322 | 7.5 | 0.0 | 0.0 | 18.5 | 0.0 | 1.4 | 72.6 | 5.1 | 1.4 | 3.6 | 3,993 | 3,896 |
| Hitchcock | 163 | 655,498 | 14.0 | 4.4 | 0.0 | 3.0 | 0.3 | 27.0 | 51.3 | 21.3 | 8.6 | 8.2 | 2,921 | 2,557 |
| Holt | 498 | 1,700,633 | 8.5 | 5.0 | 0.0 | 10.1 | 0.8 | 4.4 | 70.5 | 61.8 | 12.9 | 36.2 | 3,556 | 3,033 |
| Hooker | 71 | 221,421 | 9.9 | 3.4 | 0.0 | 4.1 | 29.1 | 0.0 | 52.4 | 7.8 | 1.1 | 3.8 | 5,775 | 5,477 |
| Howard | 484 | 1,946,017 | 3.0 | 2.0 | 0.0 | 2.8 | 34.5 | 14.2 | 41.8 | 51.9 | 9.7 | 17.2 | 2,682 | 2,448 |
| Jefferson | 520 | 1,742,755 | 5.1 | 4.0 | 0.0 | 4.9 | 3.2 | 7.9 | 72.9 | 58.8 | 10.8 | 29.1 | 4,050 | 3,621 |
| Johnson | 315 | 1,301,494 | 3.7 | 2.6 | 0.0 | 3.6 | 43.0 | 7.0 | 40.0 | 21.2 | 4.5 | 14.5 | 2,800 | 2,472 |
| Kearney | 539 | 1,816,137 | 4.9 | 3.4 | 0.1 | 3.3 | 25.7 | 2.0 | 59.9 | 42.5 | 5.6 | 23.6 | 3,641 | 3,301 |
| Keith | 407 | 1,465,239 | 7.3 | 6.4 | 0.0 | 4.0 | 1.0 | 3.4 | 73.7 | 46.0 | 11.6 | 25.5 | 3,159 | 2,549 |
| Keya Paha | 43 | 135,802 | 8.4 | 2.4 | 0.0 | 9.4 | 0.0 | 0.0 | 78.0 | 4.4 | 1.0 | 3.0 | 3,852 | 3,651 |
| Kimball | 290 | 1,120,828 | 8.1 | 4.5 | 0.0 | 4.3 | 40.5 | 6.7 | 34.6 | 19.7 | 3.4 | 11.6 | 3,265 | 2,749 |
| Knox | 460 | 1,758,386 | 8.4 | 5.4 | 0.0 | 6.1 | 2.5 | 9.5 | 67.8 | 56.1 | 22.7 | 23.2 | 2,751 | 2,534 |
| Lancaster | 11,659 | 52,706,646 | 4.9 | 8.8 | 3.7 | 4.6 | 3.6 | 10.8 | 62.3 | 1,244.1 | 388.0 | 607.8 | 1,937 | 1,467 |
| Lincoln | 1,799 | 7,044,197 | 5.0 | 8.0 | 3.6 | 3.9 | 0.8 | 9.0 | 68.0 | 178.8 | 56.1 | 94.9 | 2,692 | 2,208 |
| Logan | 68 | 230,737 | 8.2 | 3.1 | 0.0 | 5.9 | 0.0 | 3.0 | 79.4 | 5.2 | 1.5 | 3.3 | 4,310 | 4,005 |
| Loup | 39 | 134,840 | 14.3 | 4.5 | 0.0 | 13.2 | 0.0 | 1.4 | 65.6 | 3.8 | 0.8 | 2.4 | 3,909 | 3,593 |
| McPherson | 54 | 130,557 | 33.7 | 3.6 | 0.0 | 2.6 | 0.0 | 0.0 | 60.0 | 2.8 | 0.6 | 2.1 | 4,207 | 4,030 |
| Madison | 1,901 | 8,282,646 | 4.7 | 6.6 | 2.2 | 2.8 | 0.5 | 7.2 | 75.7 | 203.7 | 47.9 | 113.0 | 3,216 | 2,626 |
| Merrick | 405 | 1,422,479 | 7.1 | 1.5 | 0.0 | 1.6 | 45.6 | 3.0 | 40.2 | 53.1 | 15.4 | 19.3 | 2,457 | 2,109 |
| Morrill | 433 | 1,761,898 | 4.7 | 3.9 | 0.1 | 4.2 | 30.9 | 9.3 | 45.8 | 44.4 | 16.8 | 15.8 | 3,299 | 2,839 |
| Nance | 268 | 991,657 | 10.6 | 5.1 | 0.0 | 6.0 | 24.9 | 1.6 | 50.7 | 31.2 | 4.3 | 14.1 | 3,962 | 3,630 |
| Nemaha | 448 | 1,888,679 | 5.1 | 2.3 | 0.0 | 3.8 | 23.9 | 6.9 | 56.2 | 56.1 | 13.2 | 16.6 | 2,381 | 2,171 |
| Nuckolls | 192 | 763,054 | 9.0 | 5.4 | 0.0 | 7.4 | 0.1 | 30.2 | 46.9 | 25.1 | 5.1 | 10.9 | 2,548 | 2,111 |
| Otoe | 813 | 3,087,094 | 5.9 | 5.8 | 1.2 | 3.8 | 19.0 | 11.1 | 49.7 | 89.8 | 17.6 | 39.7 | 2,486 | 2,042 |
| Pawnee | 201 | 759,976 | 7.1 | 2.5 | 0.0 | 4.3 | 36.7 | 4.0 | 44.9 | 26.6 | 11.1 | 9.0 | 3,419 | 3,176 |
| Perkins | 347 | 1,365,589 | 5.9 | 4.3 | 0.0 | 8.3 | 53.0 | 1.7 | 26.7 | 36.6 | 4.3 | 11.1 | 3,859 | 3,512 |
| Phelps | 424 | 1,712,705 | 5.9 | 7.4 | 0.0 | 6.4 | 0.9 | 7.6 | 71.4 | 51.6 | 9.2 | 31.8 | 3,510 | 2,976 |
| Pierce | 305 | 1,028,959 | 8.1 | 5.8 | 0.0 | 6.5 | 0.1 | 4.0 | 74.9 | 33.2 | 5.8 | 21.1 | 2,976 | 2,708 |

1. Based on the resident population estimated as of July 1 of the year shown.

| STATE County | Local government finances, 2017 (cont.) | | | | | | | | | Government employment, 2019 | | | Individual income tax returns, 2018 | | |
|---|---|---|---|---|---|---|---|---|---|---|---|---|---|---|---|
| | Direct general expenditure | | | | | | | Debt outstanding | | | | | | | |
| | Total (mil dol) | Per capita[1] (dollars) | Percent of total for: | | | | | Total (mil dol) | Per capita[1] (dollars) | Federal civilian | Federal military | State and local | Number of returns | Mean adjusted gross income | Mean income tax |
| | | | Education | Health and hospitals | Police protection | Public welfare | Highways | | | | | | | | |
| | 185 | 186 | 187 | 188 | 189 | 190 | 191 | 192 | 193 | 194 | 195 | 196 | 197 | 198 | 199 |
| NEBRASKA— Cont'd | | | | | | | | | | | | | | | |
| Blaine | 5 | 9,359 | 58.2 | 0.0 | 1.3 | 0.1 | 14.7 | 0.9 | 1,967 | 28 | 2 | 52 | 230 | -4,352 | 2,165 |
| Boone | 75 | 14,158 | 37.9 | 42.7 | 1.1 | 0.1 | 7.1 | 17.1 | 3,216 | 28 | 17 | 674 | 2,560 | 59,019 | 5,470 |
| Box Butte | 94 | 8,700 | 23.5 | 37.8 | 2.4 | 2.0 | 8.2 | 14.4 | 1,333 | 33 | 36 | 1,034 | 5,080 | 55,359 | 4,948 |
| Boyd | 14 | 7,110 | 47.2 | 0.5 | 1.4 | 0.2 | 25.3 | 4.5 | 2,309 | 15 | 6 | 219 | 870 | 39,764 | 3,064 |
| Brown | 33 | 11,063 | 33.8 | 30.0 | 2.2 | 0.2 | 7.3 | 29.6 | 9,864 | 19 | 10 | 387 | 1,440 | 39,381 | 3,413 |
| Buffalo | 211 | 4,283 | 55.1 | 0.1 | 4.2 | 0.2 | 11.3 | 270.8 | 5,487 | 133 | 164 | 3,984 | 22,360 | 65,952 | 6,958 |
| Burt | 43 | 6,633 | 50.3 | 0.1 | 2.5 | 8.2 | 17.5 | 46.9 | 7,179 | 35 | 22 | 543 | 3,100 | 53,563 | 4,386 |
| Butler | 66 | 8,231 | 28.2 | 22.4 | 1.7 | 0.3 | 15.2 | 45.3 | 5,639 | 45 | 27 | 638 | 3,920 | 56,766 | 4,875 |
| Cass | 105 | 4,044 | 48.0 | 0.1 | 3.8 | 4.5 | 12.0 | 96.9 | 3,737 | 70 | 89 | 1,343 | 12,780 | 70,046 | 7,064 |
| Cedar | 39 | 4,520 | 65.5 | 0.1 | 1.1 | 6.6 | 6.1 | 15.7 | 1,843 | 98 | 28 | 600 | 3,950 | 53,820 | 4,646 |
| Chase | 46 | 11,740 | 34.2 | 40.7 | 1.8 | 0.5 | 8.2 | 18.5 | 4,701 | 26 | 13 | 494 | 1,740 | 55,876 | 5,205 |
| Cherry | 47 | 8,126 | 29.2 | 37.9 | 1.4 | 0.2 | 11.4 | 6.6 | 1,134 | 51 | 19 | 563 | 2,750 | 41,717 | 3,471 |
| Cheyenne | 57 | 5,878 | 42.8 | 0.4 | 3.3 | 0.6 | 12.9 | 94.0 | 9,745 | 35 | 30 | 746 | 4,450 | 56,233 | 5,232 |
| Clay | 44 | 7,104 | 56.7 | 0.8 | 2.9 | 3.4 | 11.0 | 12.0 | 1,941 | 147 | 21 | 586 | 2,820 | 58,416 | 5,327 |
| Colfax | 58 | 5,451 | 62.4 | 0.0 | 2.4 | 0.6 | 14.4 | 20.9 | 1,950 | 61 | 36 | 738 | 5,000 | 50,438 | 3,520 |
| Cuming | 58 | 6,472 | 46.6 | 0.3 | 2.4 | 4.9 | 15.4 | 26.8 | 2,991 | 34 | 30 | 712 | 4,380 | 60,944 | 5,622 |
| Custer | 64 | 5,869 | 48.1 | 3.5 | 1.7 | 0.2 | 21.4 | 26.8 | 2,465 | 39 | 36 | 876 | 5,140 | 50,928 | 3,923 |
| Dakota | 94 | 4,719 | 58.0 | 0.4 | 8.2 | 0.2 | 9.3 | 64.5 | 3,222 | 74 | 68 | 1,093 | 10,190 | 47,931 | 3,573 |
| Dawes | 48 | 5,408 | 36.7 | 1.4 | 3.4 | 0.4 | 10.5 | 31.9 | 3,592 | 132 | 26 | 979 | 3,480 | 46,413 | 3,754 |
| Dawson | 190 | 8,027 | 37.5 | 36.0 | 1.3 | 0.1 | 7.5 | 196.8 | 8,321 | 98 | 80 | 2,228 | 11,510 | 49,452 | 3,566 |
| Deuel | 15 | 8,222 | 80.3 | 0.0 | 0.7 | 0.1 | 1.9 | 9.9 | 5,341 | 5 | 6 | 213 | 800 | 48,060 | 3,519 |
| Dixon | 51 | 8,867 | 49.6 | 0.4 | 2.7 | 6.1 | 12.3 | 11.5 | 2,003 | 35 | 19 | 394 | 2,700 | 49,458 | 4,024 |
| Dodge | 289 | 7,865 | 28.4 | 40.5 | 3.0 | 0.0 | 8.2 | 151.1 | 4,108 | 112 | 121 | 2,101 | 17,290 | 56,679 | 5,081 |
| Douglas | 2,745 | 4,896 | 56.3 | 2.6 | 5.7 | 0.5 | 3.9 | 6,476.4 | 11,550 | 6,064 | 2,191 | 35,203 | 271,390 | 78,887 | 10,085 |
| Dundy | 22 | 12,097 | 28.0 | 42.5 | 2.0 | 0.1 | 8.0 | 4.0 | 2,223 | 11 | 6 | 226 | 800 | 44,279 | 4,161 |
| Fillmore | 61 | 10,972 | 33.0 | 28.5 | 2.0 | 0.6 | 10.5 | 42.7 | 7,685 | 28 | 18 | 719 | 2,720 | 62,627 | 5,978 |
| Franklin | 20 | 6,544 | 27.3 | 30.6 | 2.3 | 0.1 | 5.3 | 3.9 | 1,322 | 17 | 10 | 299 | 1,380 | 52,006 | 4,932 |
| Frontier | 23 | 8,921 | 48.2 | 0.1 | 2.2 | 0.1 | 11.2 | 1.9 | 712 | 16 | 9 | 297 | 1,120 | 37,048 | 3,222 |
| Furnas | 41 | 8,677 | 47.2 | 0.2 | 3.6 | 0.2 | 7.7 | 65.5 | 13,760 | 33 | 16 | 478 | 2,230 | 45,911 | 3,581 |
| Gage | 112 | 5,175 | 43.9 | 0.5 | 4.9 | 0.9 | 14.9 | 47.0 | 2,174 | 80 | 72 | 1,820 | 10,240 | 54,601 | 4,528 |
| Garden | 25 | 13,231 | 21.7 | 41.7 | 1.6 | 0.1 | 6.2 | 4.8 | 2,566 | 15 | 6 | 149 | 860 | 38,967 | 2,742 |
| Garfield | 15 | 7,769 | 48.1 | 0.0 | 1.7 | 0.1 | 13.7 | 12.8 | 6,454 | 10 | 7 | 158 | 820 | 38,967 | 2,813 |
| Gosper | 13 | 6,514 | 33.4 | 0.0 | 3.1 | 21.5 | 15.7 | 12.7 | 6,238 | 6 | 7 | 144 | 940 | 60,536 | 6,322 |
| Grant | 7 | 10,008 | 95.6 | 0.0 | 0.1 | 0.0 | 1.7 | 5.9 | 9,108 | 5 | 2 | 71 | 300 | 46,950 | 3,340 |
| Greeley | 17 | 7,139 | 45.8 | 0.4 | 1.6 | 11.0 | 10.2 | 5.5 | 2,314 | 7 | 8 | 250 | 1,070 | 31,350 | 2,754 |
| Hall | 348 | 5,678 | 51.4 | 0.2 | 5.4 | 0.3 | 3.5 | 300.2 | 4,905 | 651 | 220 | 4,463 | 29,880 | 55,328 | 5,122 |
| Hamilton | 50 | 5,459 | 54.2 | 1.7 | 2.9 | 0.3 | 19.9 | 34.2 | 3,724 | 23 | 31 | 506 | 4,520 | 68,666 | 6,600 |
| Harlan | 20 | 5,761 | 31.1 | 47.1 | 0.7 | 0.0 | 5.3 | 10.0 | 2,920 | 31 | 11 | 282 | 1,480 | 46,309 | 3,674 |
| Hayes | 6 | 6,215 | 54.6 | 0.0 | 1.5 | 0.0 | 19.5 | 1.8 | 2,059 | 10 | 3 | 84 | 410 | 17,034 | 2,578 |
| Hitchcock | 14 | 5,051 | 53.1 | 0.0 | 2.3 | 0.1 | 12.5 | 18.9 | 6,734 | 10 | 9 | 283 | 1,210 | 43,369 | 3,027 |
| Holt | 61 | 5,990 | 47.4 | 0.1 | 1.9 | 0.3 | 21.4 | 17.5 | 1,723 | 51 | 34 | 789 | 4,950 | 44,520 | 3,937 |
| Hooker | 8 | 12,393 | 44.5 | 0.1 | 4.2 | 20.4 | 7.7 | 1.5 | 2,300 | 3 | 2 | 108 | 350 | 42,491 | 2,489 |
| Howard | 52 | 8,177 | 37.4 | 33.7 | 1.2 | 0.2 | 8.8 | 36.5 | 5,703 | 31 | 22 | 588 | 3,020 | 48,451 | 3,933 |
| Jefferson | 53 | 7,416 | 48.5 | 1.4 | 3.2 | 0.1 | 15.4 | 22.4 | 3,115 | 31 | 24 | 579 | 3,300 | 54,929 | 4,954 |
| Johnson | 22 | 4,242 | 50.7 | 0.0 | 2.7 | 0.0 | 20.0 | 18.4 | 3,552 | 32 | 14 | 696 | 1,980 | 47,330 | 3,261 |
| Kearney | 52 | 7,943 | 63.5 | 21.9 | 0.7 | 0.2 | 4.0 | 60.2 | 9,269 | 20 | 22 | 449 | 3,130 | 62,033 | 5,865 |
| Keith | 46 | 5,658 | 49.1 | 0.5 | 3.2 | 0.7 | 11.1 | 39.1 | 4,851 | 33 | 27 | 501 | 3,890 | 52,062 | 4,838 |
| Keya Paha | 4 | 5,605 | 54.5 | 0.1 | 1.6 | 0.3 | 23.0 | 0.0 | 51 | 1 | 3 | 66 | 390 | 30,267 | 2,287 |
| Kimball | 19 | 5,434 | 39.1 | 1.5 | 3.2 | 0.2 | 10.0 | 11.6 | 3,256 | 18 | 12 | 435 | 1,610 | 49,242 | 4,809 |
| Knox | 59 | 6,999 | 56.4 | 0.1 | 1.7 | 0.1 | 17.5 | 22.1 | 2,613 | 49 | 28 | 1,088 | 3,890 | 47,677 | 3,663 |
| Lancaster | 1,319 | 4,202 | 52.8 | 2.9 | 4.1 | 1.2 | 7.2 | 2,458.0 | 7,834 | 3,443 | 1,075 | 30,693 | 147,210 | 69,130 | 7,639 |
| Lincoln | 212 | 6,003 | 53.0 | 1.0 | 4.3 | 0.2 | 5.4 | 119.7 | 3,395 | 258 | 117 | 2,639 | 16,370 | 56,874 | 5,330 |
| Logan | 7 | 8,729 | 69.2 | 0.1 | 1.9 | 0.1 | 11.5 | 0.1 | 71 | 6 | 3 | 76 | 340 | 38,265 | 2,256 |
| Loup | 4 | 5,748 | 59.9 | 0.6 | 2.0 | 0.1 | 8.4 | 0.4 | 707 | 1 | 2 | 65 | 300 | 31,293 | 2,137 |
| McPherson | 3 | 6,249 | 66.6 | 0.0 | 2.1 | 0.0 | 16.3 | 1.1 | 2,318 | 2 | 2 | 40 | 220 | 38,514 | 2,027 |
| Madison | 194 | 5,519 | 72.0 | 0.2 | 3.1 | 6.7 | 0.9 | 176.8 | 5,032 | 176 | 117 | 3,634 | 16,680 | 58,867 | 5,616 |
| Merrick | 49 | 6,211 | 34.4 | 27.1 | 2.1 | 0.2 | 12.6 | 14.8 | 1,886 | 24 | 26 | 486 | 3,570 | 53,814 | 4,780 |
| Morrill | 46 | 9,599 | 31.2 | 24.6 | 3.9 | 17.3 | 6.6 | 35.1 | 7,322 | 20 | 16 | 609 | 2,110 | 52,158 | 5,884 |
| Nance | 24 | 6,594 | 57.0 | 0.0 | 5.7 | 0.1 | 12.2 | 2.2 | 607 | 16 | 12 | 387 | 1,600 | 43,204 | 3,376 |
| Nemaha | 51 | 7,309 | 44.0 | 27.5 | 2.3 | 0.1 | 7.4 | 23.9 | 3,431 | 31 | 22 | 1,624 | 2,990 | 58,923 | 5,246 |
| Nuckolls | 19 | 4,323 | 37.2 | 0.3 | 1.8 | 0.5 | 8.5 | 7.2 | 1,685 | 22 | 14 | 369 | 1,960 | 49,185 | 3,699 |
| Otoe | 101 | 6,356 | 41.3 | 14.9 | 2.6 | 0.2 | 19.0 | 91.7 | 5,747 | 58 | 54 | 1,347 | 7,560 | 59,985 | 5,355 |
| Pawnee | 27 | 10,119 | 33.5 | 30.2 | 1.1 | 13.2 | 9.6 | 7.3 | 2,792 | 22 | 9 | 248 | 1,210 | 44,467 | 3,296 |
| Perkins | 41 | 14,292 | 17.0 | 55.0 | 1.6 | 0.3 | 6.1 | 10.2 | 3,545 | 16 | 10 | 416 | 1,370 | 51,023 | 4,769 |
| Phelps | 51 | 5,618 | 55.9 | 0.1 | 2.6 | 0.4 | 13.9 | 37.0 | 4,087 | 57 | 30 | 719 | 4,370 | 58,477 | 5,553 |
| Pierce | 35 | 4,960 | 51.4 | 0.0 | 2.1 | 13.6 | 12.3 | 11.0 | 1,554 | 27 | 24 | 381 | 3,360 | 58,201 | 5,477 |

1. Based on the resident population estimated as of July 1 of the year shown.

# Table B. States and Counties — Land Area and Population

| State / county code | CBSA code[1] | County Type code[2] | STATE County | Land area[3] (sq. mi) | Population, 2020 | | | Population and population characteristics, 2020 | | | | | | | | | | |
|---|---|---|---|---|---|---|---|---|---|---|---|---|---|---|---|---|---|---|
| | | | | | | | | Race alone or in combination, not Hispanic or Latino (percent) | | | | | Age (percent) | | | | | |
| | | | | | Total persons 2019 | Rank | Per square mile | White | Black | American Indian, Alaska Native | Asian and Pacific Islancer[4] | Percent Hispanic or Latino[4] | Under 5 years | 5 to 14 years | 15 to 24 years | 25 to 34 years | 35 to 44 years | 45 to 54 years |
| | | | | 1 | 2 | 3 | 4 | 5 | 6 | 7 | 8 | 9 | 10 | 11 | 12 | 13 | 14 | 15 |
| | | | NEBRASKA— Cont'd | | | | | | | | | | | | | | | |
| 31141 | 18100 | 5 | Platte | 674.1 | 33,364 | 1,347 | 49.5 | 76.4 | 1.2 | 0.6 | 1.3 | 21.2 | 7.1 | 15.0 | 12.3 | 12.1 | 12.0 | 10.6 |
| 31143 | | 9 | Polk | 438.7 | 5,201 | 2,809 | 11.9 | 92.8 | 0.7 | 1.0 | 0.5 | 6.2 | 5.8 | 12.6 | 11.5 | 9.7 | 10.4 | 11.9 |
| 31145 | | 7 | Red Willow | 717.0 | 10,627 | 2,379 | 14.8 | 91.8 | 1.7 | 0.9 | 0.6 | 6.1 | 6.3 | 11.8 | 13.1 | 11.2 | 11.6 | 10.0 |
| 31147 | | 7 | Richardson | 551.8 | 7,791 | 2,606 | 14.1 | 93.6 | 1.3 | 4.6 | 1.1 | 2.1 | 5.4 | 12.4 | 10.6 | 9.4 | 10.2 | 10.8 |
| 31149 | | 9 | Rock | 1,008.3 | 1,377 | 3,089 | 1.4 | 97.2 | 0.3 | 0.8 | 0.4 | 1.7 | 6.2 | 10.7 | 10.4 | 9.4 | 11.7 | 7.2 |
| 31151 | | 6 | Saline | 574.1 | 13,987 | 2,158 | 24.4 | 68.8 | 1.5 | 0.8 | 3.5 | 26.6 | 6.6 | 15.0 | 16.5 | 10.3 | 12.3 | 11.9 |
| 31153 | 36540 | 2 | Sarpy | 238.1 | 188,856 | 357 | 793.2 | 82.3 | 5.4 | 0.9 | 4.1 | 10.4 | 6.9 | 15.4 | 13.0 | 14.2 | 14.6 | 12.1 |
| 31155 | 36540 | 2 | Saunders | 748.9 | 21,927 | 1,724 | 29.3 | 96.1 | 1.1 | 0.8 | 1.0 | 2.4 | 6.3 | 14.1 | 11.1 | 10.7 | 12.3 | 11.4 |
| 31157 | 42420 | 2 | Scotts Bluff | 739.6 | 35,299 | 1,298 | 47.7 | 72.1 | 1.1 | 1.9 | 1.2 | 24.8 | 6.6 | 14.0 | 12.2 | 11.9 | 11.7 | 10.6 |
| 31159 | 30700 | 2 | Seward | 571.4 | 17,186 | 1,964 | 30.1 | 95.6 | 1.3 | 0.9 | 0.9 | 2.8 | 5.8 | 13.6 | 17.3 | 10.2 | 11.8 | 10.6 |
| 31161 | | 9 | Sheridan | 2,440.9 | 5,150 | 2,817 | 2.1 | 83.6 | 1.1 | 11.0 | 1.6 | 6.2 | 6.1 | 11.8 | 11.1 | 9.3 | 10.4 | 10.8 |
| 31163 | | 8 | Sherman | 565.9 | 2,986 | 2,959 | 5.3 | 96.2 | 0.9 | 0.8 | 0.6 | 2.5 | 5.6 | 11.7 | 10.0 | 9.3 | 8.9 | 10.9 |
| 31165 | 42420 | 9 | Sioux | 2,066.7 | 1,200 | 3,099 | 0.6 | 92.8 | 0.8 | 1.8 | 0.3 | 5.6 | 4.3 | 8.8 | 11.2 | 7.0 | 10.2 | 10.0 |
| 31167 | 35740 | 9 | Stanton | 427.6 | 5,880 | 2,754 | 13.8 | 91.9 | 1.3 | 0.9 | 0.6 | 6.4 | 6.5 | 13.8 | 11.3 | 11.0 | 11.8 | 11.2 |
| 31169 | | 9 | Thayer | 573.8 | 4,887 | 2,832 | 8.5 | 95.4 | 1.3 | 0.9 | 0.5 | 3.3 | 5.8 | 13.5 | 10.7 | 9.4 | 10.2 | 10.9 |
| 31171 | | 9 | Thomas | 712.6 | 739 | 3,126 | 1.0 | 93.5 | 1.6 | 2.2 | 0.5 | 3.2 | 4.7 | 14.5 | 10.1 | 10.1 | 8.7 | 11.1 |
| 31173 | | 8 | Thurston | 393.6 | 7,220 | 2,641 | 18.3 | 36.6 | 1.1 | 56.1 | 1.2 | 7.7 | 10.0 | 20.0 | 15.6 | 12.9 | 10.1 | 8.7 |
| 31175 | | 9 | Valley | 568.1 | 4,103 | 2,888 | 7.2 | 95.5 | 0.7 | 0.6 | 0.7 | 3.4 | 5.8 | 13.4 | 10.4 | 9.9 | 10.5 | 10.9 |
| 31177 | 36540 | 2 | Washington | 390.0 | 20,901 | 1,775 | 53.6 | 95.5 | 1.3 | 0.8 | 0.9 | 2.8 | 5.5 | 14.2 | 11.7 | 9.7 | 13.1 | 12.1 |
| 31179 | | 6 | Wayne | 442.9 | 9,492 | 2,454 | 21.4 | 90.3 | 2.2 | 0.9 | 1.1 | 6.7 | 5.1 | 10.9 | 26.2 | 10.5 | 9.8 | 9.3 |
| 31181 | | 9 | Webster | 574.9 | 3,419 | 2,938 | 5.9 | 92.8 | 1.3 | 1.2 | 1.1 | 5.3 | 5.4 | 13.1 | 10.6 | 11.1 | 9.7 | 11.5 |
| 31183 | | 9 | Wheeler | 575.2 | 790 | 3,117 | 1.4 | 96.8 | 0.8 | 1.0 | 0.5 | 2.0 | 7.3 | 11.4 | 7.1 | 8.9 | 9.0 | 9.6 |
| 31185 | | 6 | York | 572.5 | 13,511 | 2,190 | 23.6 | 91.8 | 2.1 | 1.1 | 1.2 | 5.1 | 6.2 | 13.1 | 12.7 | 12.0 | 11.1 | 10.0 |
| 32000 | | 0 | NEVADA | 109,860.4 | 3,138,259 | X | 28.6 | 50.6 | 10.8 | 1.5 | 11.7 | 29.5 | 5.9 | 12.6 | 11.8 | 14.7 | 13.4 | 12.6 |
| 32001 | 21980 | 6 | Churchill | 4,950.2 | 25,363 | 1,591 | 5.1 | 75.0 | 3.6 | 5.1 | 5.1 | 15.1 | 6.7 | 12.7 | 11.2 | 14.5 | 11.4 | 10.4 |
| 32003 | 29820 | 1 | Clark | 7,891.7 | 2,315,963 | 11 | 293.5 | 44.2 | 13.6 | 1.1 | 13.7 | 31.8 | 6.0 | 12.9 | 11.9 | 14.9 | 13.9 | 13.0 |
| 32005 | 23820 | 4 | Douglas | 710.5 | 49,088 | 1,006 | 69.1 | 82.3 | 1.3 | 2.6 | 3.3 | 13.1 | 3.3 | 9.1 | 8.6 | 9.5 | 10.0 | 10.6 |
| 32007 | 21220 | 5 | Elko | 17,173.4 | 53,006 | 953 | 3.1 | 68.0 | 1.4 | 5.5 | 2.3 | 25.0 | 7.0 | 15.3 | 12.7 | 15.0 | 13.4 | 11.7 |
| 32009 | | 9 | Esmeralda | 3,581.9 | 886 | 3,112 | 0.2 | 69.8 | 4.2 | 7.6 | 2.6 | 19.6 | 2.5 | 9.8 | 6.2 | 8.8 | 11.6 | 11.6 |
| 32011 | 21220 | 9 | Eureka | 4,175.7 | 2,065 | 3,045 | 0.5 | 80.4 | 1.5 | 3.2 | 2.3 | 14.2 | 3.5 | 15.7 | 8.7 | 11.5 | 11.6 | 12.9 |
| 32013 | 49080 | 7 | Humboldt | 9,640.8 | 16,962 | 1,975 | 1.8 | 66.5 | 1.7 | 4.6 | 2.0 | 27.7 | 7.1 | 15.8 | 11.8 | 13.7 | 12.5 | 11.2 |
| 32015 | | 7 | Lander | 5,519.2 | 5,514 | 2,787 | 1.0 | 71.2 | 1.3 | 4.8 | 1.5 | 23.0 | 7.2 | 14.7 | 11.3 | 12.8 | 12.6 | 11.6 |
| 32017 | | 8 | Lincoln | 10,633.4 | 5,159 | 2,816 | 0.5 | 86.9 | 3.2 | 2.5 | 1.9 | 7.6 | 5.0 | 10.6 | 11.0 | 11.1 | 11.8 | 11.1 |
| 32019 | 22280 | 4 | Lyon | 2,003.1 | 58,319 | 890 | 29.1 | 76.1 | 2.1 | 3.5 | 3.0 | 18.6 | 5.3 | 11.8 | 10.3 | 13.2 | 11.8 | 11.2 |
| 32021 | | 7 | Mineral | 3,751.3 | 4,518 | 2,858 | 1.2 | 63.7 | 5.2 | 16.5 | 4.4 | 14.0 | 5.4 | 10.5 | 9.5 | 11.4 | 11.1 | 10.0 |
| 32023 | 37220 | 4 | Nye | 18,181.9 | 48,054 | 1,020 | 2.6 | 77.0 | 4.2 | 2.5 | 3.6 | 15.7 | 4.4 | 9.2 | 9.0 | 9.8 | 9.1 | 10.4 |
| 32027 | | 9 | Pershing | 6,036.6 | 6,573 | 2,697 | 1.1 | 66.6 | 4.7 | 4.3 | 2.4 | 24.7 | 4.6 | 9.7 | 10.1 | 15.5 | 14.3 | 15.2 |
| 32029 | 39900 | 2 | Storey | 263.8 | 4,207 | 2,880 | 15.9 | 85.9 | 2.2 | 2.7 | 3.4 | 8.2 | 2.4 | 7.0 | 7.4 | 9.0 | 9.2 | 12.0 |
| 32031 | 39900 | 2 | Washoe | 6,315.9 | 477,082 | 148 | 75.5 | 64.5 | 3.3 | 2.0 | 8.2 | 25.4 | 5.7 | 11.9 | 12.2 | 15.3 | 12.7 | 11.9 |
| 32033 | | 7 | White Pine | 8,886.7 | 9,466 | 2,459 | 1.1 | 73.8 | 5.0 | 5.0 | 1.8 | 16.7 | 5.5 | 11.0 | 11.0 | 14.6 | 13.0 | 11.3 |
| 32510 | 16180 | 1 | Carson City | 144.5 | 56,034 | 921 | 387.8 | 68.3 | 2.5 | 2.9 | 3.9 | 24.9 | 5.6 | 11.3 | 10.7 | 13.5 | 11.5 | 11.8 |
| 33000 | | 0 | NEW HAMPSHIRE | 8,953.4 | 1,366,275 | X | 152.6 | 91.0 | 2.1 | 0.8 | 3.7 | 4.2 | 4.6 | 10.4 | 12.4 | 12.8 | 11.7 | 13.0 |
| 33001 | 29060 | 4 | Belknap | 401.8 | 61,551 | 860 | 153.2 | 96.1 | 1.0 | 0.8 | 1.6 | 1.9 | 4.1 | 10.3 | 10.2 | 10.8 | 11.1 | 12.6 |
| 33003 | | 6 | Carroll | 931.9 | 49,167 | 1,004 | 52.8 | 96.6 | 1.0 | 1.0 | 1.0 | 1.6 | 3.5 | 8.8 | 8.7 | 9.1 | 9.4 | 11.9 |
| 33005 | 28300 | 4 | Cheshire | 706.7 | 76,228 | 729 | 107.9 | 95.2 | 1.2 | 0.9 | 2.2 | 2.1 | 4.5 | 10.0 | 13.6 | 12.3 | 10.9 | 11.8 |
| 33007 | 13620 | 7 | Coos | 1,794.6 | 31,174 | 1,398 | 17.4 | 96.0 | 1.2 | 1.3 | 1.0 | 2.0 | 4.0 | 9.1 | 9.8 | 10.9 | 11.4 | 13.1 |
| 33009 | 30100 | 5 | Grafton | 1,708.6 | 90,691 | 655 | 53.1 | 91.8 | 1.7 | 1.0 | 4.6 | 2.6 | 3.8 | 9.0 | 15.8 | 12.4 | 10.4 | 11.3 |
| 33011 | 31700 | 2 | Hillsborough | 876.5 | 418,735 | 172 | 477.7 | 85.0 | 3.4 | 0.7 | 5.3 | 7.6 | 5.2 | 11.1 | 12.0 | 14.3 | 12.4 | 13.4 |
| 33013 | 18180 | 4 | Merrimack | 932.9 | 152,622 | 440 | 163.6 | 93.3 | 2.1 | 0.8 | 2.8 | 2.4 | 4.6 | 10.5 | 12.0 | 12.7 | 12.0 | 12.9 |
| 33015 | 14460 | 1 | Rockingham | 695.4 | 311,307 | 227 | 447.7 | 93.2 | 1.3 | 0.6 | 2.9 | 3.4 | 4.6 | 10.5 | 10.9 | 12.2 | 11.8 | 14.0 |
| 33017 | 14460 | 1 | Strafford | 367.5 | 131,533 | 487 | 357.9 | 91.7 | 2.0 | 0.7 | 4.7 | 2.9 | 4.5 | 10.1 | 18.5 | 13.8 | 11.4 | 11.9 |
| 33019 | 30100 | 7 | Sullivan | 537.5 | 43,267 | 1,112 | 80.4 | 95.9 | 1.2 | 1.2 | 1.5 | 1.9 | 4.4 | 10.3 | 10.0 | 11.6 | 11.1 | 13.2 |
| 34000 | | 0 | NEW JERSEY | 7,354.8 | 8,882,371 | X | 1,207.7 | 55.4 | 13.8 | 0.5 | 10.9 | 21.1 | 5.8 | 12.2 | 12.2 | 13.0 | 12.8 | 13.2 |
| 34001 | 12100 | 2 | Atlantic | 555.5 | 262,945 | 266 | 473.3 | 57.3 | 15.7 | 0.6 | 8.9 | 19.7 | 5.3 | 11.9 | 12.5 | 12.1 | 11.0 | 12.8 |
| 34003 | 35620 | 1 | Bergen | 232.8 | 930,394 | 57 | 3,996.5 | 55.6 | 6.1 | 0.3 | 18.0 | 21.5 | 5.2 | 11.9 | 11.9 | 11.5 | 13.2 | 14.0 |
| 34005 | 37980 | 1 | Burlington | 799.3 | 446,596 | 159 | 558.7 | 67.9 | 18.6 | 0.7 | 6.7 | 8.9 | 5.1 | 11.6 | 11.9 | 12.7 | 12.6 | 13.4 |
| 34007 | 37980 | 1 | Camden | 221.4 | 506,809 | 139 | 2,289.1 | 56.7 | 20.0 | 0.6 | 6.8 | 18.1 | 6.1 | 12.6 | 11.9 | 14.0 | 13.0 | 12.7 |
| 34009 | 36140 | 3 | Cape May | 251.5 | 91,546 | 649 | 364.0 | 86.3 | 5.0 | 0.5 | 1.5 | 8.3 | 4.4 | 9.9 | 10.0 | 10.5 | 9.4 | 10.9 |
| 34011 | 47220 | 3 | Cumberland | 483.4 | 147,008 | 456 | 304.1 | 46.4 | 20.2 | 1.4 | 1.8 | 32.5 | 6.3 | 13.9 | 12.1 | 13.9 | 12.9 | 12.5 |
| 34013 | 35620 | 1 | Essex | 126.1 | 800,501 | 81 | 6,348.1 | 31.0 | 39.5 | 0.5 | 6.6 | 24.0 | 6.6 | 13.3 | 12.4 | 13.8 | 13.8 | 13.6 |
| 34015 | 37980 | 1 | Gloucester | 322.0 | 293,245 | 239 | 910.7 | 78.9 | 11.8 | 0.5 | 3.9 | 7.0 | 5.1 | 12.3 | 12.7 | 12.6 | 12.6 | 13.6 |
| 34017 | 35620 | 1 | Hudson | 46.2 | 671,666 | 100 | 14,538.2 | 30.1 | 11.6 | 0.5 | 17.3 | 42.1 | 6.9 | 10.6 | 10.4 | 21.2 | 16.0 | 11.8 |
| 34019 | 35620 | 1 | Hunterdon | 427.8 | 124,797 | 510 | 291.7 | 85.1 | 3.1 | 0.4 | 5.4 | 7.4 | 4.2 | 10.5 | 12.6 | 9.4 | 11.0 | 14.5 |

1. CBSA = Core Based Statistical Area. See Appendix A for explanation. See Appendix B for list of metropolitan areas with component counties. Service of USDA Rural-Urban Continuum Codes. See Appendix A for definition.  2. County type code from the Economic Research  3. Dry land or land partially or temporarily covered by water.  4. May be of any race.

# Table B. States and Counties — Population and Households

| STATE County | Age (percent) (cont.) | | | | Population change, 2000-2020 | | | | | | | Households, 2015-2019 | | | | |
|---|---|---|---|---|---|---|---|---|---|---|---|---|---|---|---|---|
| | 55 to 64 years | 65 to 74 years | 75 years and over | Percent female | Total persons | | Percent change | | Components of change, 2010-2020 | | | | Persons per household | Family house-holds | Female family house-holder[1] | One person |
| | | | | | 2000 | 2010 | 2000-2010 | 2010-2020 | Births | Deaths | Net Migration | Number | | | | |
| | 16 | 17 | 18 | 19 | 20 | 21 | 22 | 23 | 24 | 25 | 26 | 27 | 28 | 29 | 30 | 31 |

| STATE County | 16 | 17 | 18 | 19 | 20 | 21 | 22 | 23 | 24 | 25 | 26 | 27 | 28 | 29 | 30 | 31 |
|---|---|---|---|---|---|---|---|---|---|---|---|---|---|---|---|---|
| **NEBRASKA— Cont'd** | | | | | | | | | | | | | | | | |
| | 13.0 | 10.0 | 8.0 | 49.4 | 31,662 | 32,237 | 1.8 | 3.5 | 4,900 | 2,898 | -873 | 12,947 | 2.53 | 65.9 | 6.5 | 29.5 |
| Platte | 14.7 | 13.3 | 10.1 | 49.5 | 5,639 | 5,402 | -4.2 | -3.7 | 554 | 646 | -105 | 2,052 | 2.51 | 70.6 | 5.4 | 26.1 |
| Polk | 14.4 | 11.5 | 10.0 | 50.0 | 11,448 | 11,055 | -3.4 | -3.9 | 1,341 | 1,293 | -472 | 4,519 | 2.29 | 62.4 | 7.5 | 31.5 |
| Red Willow | 16.5 | 12.7 | 11.9 | 50.0 | 9,531 | 8,363 | -12.3 | -6.8 | 922 | 1,191 | -299 | 3,704 | 2.11 | 60.0 | 8.6 | 35.8 |
| Richardson | 16.0 | 17.4 | 10.9 | 50.9 | 1,756 | 1,528 | -13.0 | -9.9 | 152 | 227 | -77 | 623 | 2.17 | 66.9 | 5.5 | 28.6 |
| Rock | 12.0 | 8.7 | 6.6 | 48.7 | 13,843 | 14,200 | 2.6 | -1.5 | 1,946 | 1,450 | -714 | 5,157 | 2.59 | 68.5 | 6.8 | 27.8 |
| Saline | 11.3 | 7.6 | 4.9 | 50.0 | 122,595 | 158,837 | 29.6 | 18.9 | 26,027 | 9,689 | 13,711 | 66,260 | 2.72 | 72.5 | 9.6 | 21.8 |
| Sarpy | 15.2 | 10.8 | 8.0 | 49.3 | 19,830 | 20,778 | 4.8 | 5.5 | 2,517 | 2,109 | 751 | 8,501 | 2.46 | 68.8 | 6.5 | 26.8 |
| Saunders | 13.1 | 11.0 | 8.8 | 51.3 | 36,951 | 36,972 | 0.1 | -4.5 | 4,904 | 4,311 | -2,265 | 14,732 | 2.40 | 63.2 | 12.3 | 30.7 |
| Scotts Bluff | 13.0 | 10.0 | 7.8 | 48.8 | 16,496 | 16,750 | 1.5 | 2.6 | 1,961 | 1,742 | 223 | 6,600 | 2.38 | 67.5 | 6.0 | 26.9 |
| Seward | | | | | | | | | | | | | | | | |
| Sheridan | 13.4 | 14.6 | 12.5 | 50.4 | 6,198 | 5,469 | -11.8 | -5.8 | 606 | 765 | -157 | 2,273 | 2.24 | 60.4 | 7.2 | 34.8 |
| Sherman | 15.5 | 15.1 | 12.9 | 50.6 | 3,318 | 3,152 | -5.0 | -5.3 | 330 | 397 | -100 | 1,368 | 2.17 | 63.0 | 5.9 | 30.8 |
| Sioux | 17.2 | 16.6 | 14.8 | 47.0 | 1,475 | 1,307 | -11.4 | -8.2 | 96 | 71 | -133 | 544 | 2.24 | 68.2 | 5.7 | 26.5 |
| Stanton | 15.1 | 11.1 | 8.3 | 49.2 | 6,455 | 6,128 | -5.1 | -4.0 | 765 | 449 | -568 | 2,418 | 2.44 | 72.0 | 9.3 | 22.4 |
| Thayer | 14.3 | 13.7 | 11.4 | 50.4 | 6,055 | 5,228 | -13.7 | -6.5 | 549 | 801 | -88 | 2,244 | 2.19 | 63.5 | 4.5 | 32.7 |
| Thomas | 11.6 | 16.8 | 12.3 | 49.0 | 729 | 647 | -11.2 | 14.2 | 78 | 45 | 60 | 277 | 2.33 | 69.0 | 5.8 | 24.2 |
| Thurston | 10.7 | 7.2 | 4.9 | 50.5 | 7,171 | 6,939 | -3.2 | 4.0 | 1,529 | 815 | -436 | 2,176 | 3.28 | 73.6 | 22.6 | 23.0 |
| Valley | 13.9 | 13.5 | 11.7 | 50.3 | 4,647 | 4,262 | -8.3 | -3.7 | 512 | 561 | -111 | 1,865 | 2.22 | 65.3 | 6.0 | 31.0 |
| Washington | 14.9 | 11.5 | 7.3 | 50.0 | 18,780 | 20,232 | 7.7 | 3.3 | 2,157 | 1,820 | 295 | 8,185 | 2.45 | 70.2 | 7.1 | 24.8 |
| Wayne | 11.5 | 8.8 | 8.0 | 49.6 | 9,851 | 9,592 | -2.6 | | 1,014 | 656 | -466 | 3,708 | 2.18 | 60.8 | 3.6 | 27.6 |
| Webster | 15.6 | 12.1 | 10.9 | 50.1 | 4,061 | 3,812 | -6.1 | -10.3 | 375 | 604 | -164 | 1,507 | 2.30 | 63.8 | 6.4 | 33.6 |
| Wheeler | 18.6 | 15.8 | 12.3 | 49.2 | 886 | 818 | -7.7 | -3.4 | 93 | 52 | -69 | 346 | 2.26 | 63.9 | 8.1 | 33.8 |
| York | 13.7 | 11.3 | 9.9 | 50.7 | 14,598 | 13,665 | -6.4 | -1.1 | 1,758 | 1,562 | -344 | 5,656 | 2.25 | 65.0 | 5.2 | 30.5 |
| **NEVADA** | 12.4 | 10.1 | 6.5 | 49.9 | 1,998,257 | 2,700,683 | 35.2 | 16.2 | 365,815 | 236,627 | 307,049 | 1,098,602 | 2.67 | 63.5 | 12.9 | 28.5 |
| Churchill | 13.6 | 11.4 | 8.0 | 49.3 | 23,982 | 24,875 | 3.7 | 2.0 | 3,311 | 2,825 | -20 | 9,940 | 2.38 | 66.1 | 10.0 | 25.7 |
| Clark | 11.9 | 9.4 | 6.1 | 50.2 | 1,375,765 | 1,951,278 | 41.8 | 18.7 | 274,679 | 161,068 | 250,044 | 783,524 | 2.76 | 63.6 | 13.9 | 28.5 |
| Douglas | 17.8 | 18.5 | 12.6 | 49.7 | 41,259 | 47,000 | 13.9 | 4.4 | 3,386 | 5,034 | 3,771 | 20,741 | 2.31 | 68.5 | 8.9 | 24.7 |
| Elko | 12.4 | 7.8 | 4.7 | 48.1 | 45,291 | 48,942 | 8.1 | 8.3 | 7,206 | 2,804 | -432 | 18,065 | 2.85 | 70.1 | 8.1 | 25.1 |
| Esmeralda | 15.6 | 17.7 | 16.1 | 44.0 | 971 | 784 | -19.3 | 13.0 | 55 | 98 | 139 | 491 | 1.96 | 41.1 | 7.5 | 51.5 |
| Eureka | 17.5 | 11.2 | 7.3 | 45.6 | 1,651 | 1,987 | 20.4 | 3.9 | 175 | 134 | 30 | 774 | 2.40 | 72.1 | 3.1 | 23.1 |
| Humboldt | 12.8 | 9.9 | 5.2 | 48.1 | 16,106 | 16,519 | 2.6 | 2.7 | 2,594 | 1,295 | -888 | 6,500 | 2.56 | 66.2 | 6.0 | 27.8 |
| Lander | 13.2 | 9.5 | 7.1 | 48.5 | 5,794 | 5,775 | -0.3 | -4.5 | 867 | 439 | -705 | 2,198 | 2.54 | 71.3 | 9.7 | 25.5 |
| Lincoln | 13.9 | 14.5 | 10.9 | 46.6 | 4,165 | 5,339 | 28.2 | -3.4 | 390 | 489 | -89 | 2,018 | 2.28 | 60.8 | 7.0 | 35.3 |
| Lyon | 14.5 | 13.7 | 8.1 | 48.7 | 34,501 | 51,985 | 50.7 | 12.2 | 5,938 | 6,052 | 6,437 | 21,189 | 2.56 | 65.5 | 9.0 | 26.6 |
| Mineral | 15.7 | 14.9 | 11.5 | 50.3 | 5,071 | 4,770 | -5.9 | -5.3 | 500 | 754 | -4 | 2,048 | 2.14 | 49.5 | 11.0 | 47.1 |
| Nye | 17.0 | 18.5 | 12.6 | 49.0 | 32,485 | 43,944 | 35.3 | 9.4 | 3,882 | 7,389 | 7,620 | 19,088 | 2.30 | 58.4 | 8.7 | 33.8 |
| Pershing | 13.5 | 9.5 | 7.5 | 35.8 | 6,693 | 6,748 | 0.8 | -2.6 | 565 | 529 | -226 | 1,948 | 2.41 | 67.0 | 9.3 | 26.8 |
| Storey | 19.7 | 21.2 | 12.1 | 48.8 | 3,399 | 4,013 | 18.1 | 4.8 | 182 | 316 | 324 | 1,627 | 2.43 | 63.7 | 5.0 | 29.1 |
| Washoe | 13.0 | 10.8 | 6.5 | 49.5 | 339,486 | 421,429 | 24.1 | 13.2 | 55,040 | 39,171 | 39,684 | 182,180 | 2.47 | 62.3 | 10.8 | 28.3 |
| White Pine | 13.9 | 10.8 | 8.8 | 42.6 | 9,181 | 10,026 | 9.2 | -5.6 | 1,026 | 908 | -680 | 3,516 | 2.33 | 60.7 | 9.4 | 33.2 |
| Carson City | 14.3 | 12.3 | 8.9 | 49.0 | 52,457 | 55,269 | 5.4 | 1.4 | 6,019 | 7,322 | 2,044 | 22,755 | 2.31 | 62.2 | 12.6 | 32.3 |
| **NEW HAMPSHIRE** | 15.7 | 11.6 | 7.6 | 50.4 | 1,235,786 | 1,316,457 | 6.5 | 3.8 | 126,460 | 120,351 | 44,426 | 532,037 | 2.46 | 65.7 | 9.0 | 26.3 |
| Belknap | 17.4 | 14.4 | 9.1 | 50.6 | 56,325 | 60,075 | 6.7 | 2.5 | 5,291 | 7,132 | 3,370 | 25,052 | 2.40 | 66.2 | 10.5 | 26.8 |
| Carroll | 19.0 | 18.4 | 11.2 | 50.4 | 43,666 | 47,823 | 9.5 | 2.8 | 3,543 | 5,514 | 3,362 | 21,456 | 2.22 | 64.1 | 8.1 | 28.6 |
| Cheshire | 15.5 | 13.1 | 8.4 | 51.1 | 73,825 | 77,123 | 4.5 | -1.2 | 6,878 | 7,398 | -341 | 30,141 | 2.37 | 60.9 | 8.5 | 30.7 |
| Coos | 17.0 | 14.9 | 10.2 | 47.6 | 33,111 | 33,052 | -0.2 | -5.7 | 2,524 | 4,491 | 35 | 13,768 | 2.14 | 59.9 | 8.1 | 33.5 |
| Grafton | 15.4 | 13.0 | 9.1 | 50.7 | 81,743 | 89,135 | 9.0 | 1.7 | 7,490 | 8,466 | 2,589 | 34,727 | 2.39 | 60.6 | 8.1 | 32.5 |
| Hillsborough | 15.0 | 9.9 | 6.8 | 50.2 | 380,841 | 400,706 | 5.2 | 4.5 | 43,670 | 33,432 | 7,970 | 161,086 | 2.51 | 65.9 | 9.8 | 25.6 |
| Merrimack | 15.6 | 11.7 | 7.6 | 50.8 | 136,225 | 146,452 | 7.5 | 4.2 | 13,641 | 13,959 | 6,601 | 58,452 | 2.45 | 66.3 | 8.7 | 25.9 |
| Rockingham | 16.8 | 11.7 | 7.4 | 50.5 | 277,359 | 295,207 | 6.4 | 5.5 | 26,952 | 24,543 | 13,944 | 121,045 | 2.51 | 69.5 | 8.2 | 23.2 |
| Strafford | 13.8 | 9.6 | 6.4 | 50.9 | 112,233 | 123,146 | 9.7 | 6.8 | 12,430 | 10,862 | 6,825 | 48,982 | 2.45 | 68.7 | 8.2 | 26.1 |
| Sullivan | 16.5 | 13.9 | 9.0 | 50.4 | 40,458 | 43,738 | 8.1 | -1.1 | 4,041 | 4,554 | 71 | 17,328 | 2.44 | 65.5 | 11.7 | 28.2 |
| **NEW JERSEY** | 13.8 | 9.7 | 7.3 | 51.1 | 8,414,350 | 8,791,959 | 4.5 | 1.0 | 1,054,140 | 751,791 | -212,010 | 3,231,874 | 2.69 | 68.9 | 12.8 | 26.0 |
| Atlantic | 15.1 | 11.2 | 8.0 | 51.6 | 252,552 | 274,532 | 8.7 | -4.2 | 31,062 | 27,578 | -15,198 | 99,850 | 2.61 | 66.1 | 14.9 | 28.3 |
| Bergen | 14.3 | 10.0 | 8.0 | 51.4 | 884,118 | 905,107 | 2.4 | 2.8 | 94,809 | 73,868 | 4,937 | 339,856 | 2.71 | 71.9 | 10.6 | 24.0 |
| Burlington | 14.9 | 10.1 | 7.7 | 50.8 | 423,394 | 448,730 | 6.0 | -0.5 | 46,182 | 41,274 | -6,874 | 166,391 | 2.61 | 69.8 | 12.2 | 25.3 |
| Camden | 13.4 | 9.5 | 6.8 | 51.8 | 508,932 | 513,541 | 0.9 | -1.3 | 63,974 | 48,782 | -22,116 | 187,383 | 2.67 | 66.8 | 16.7 | 28.0 |
| Cape May | 16.7 | 16.4 | 11.9 | 51.4 | 102,326 | 97,261 | -4.9 | -5.9 | 8,901 | 13,502 | -1,017 | 40,171 | 2.25 | 65.6 | 9.3 | 28.7 |
| Cumberland | 12.4 | 9.3 | 6.7 | 49.3 | 146,438 | 156,610 | 6.9 | -6.1 | 20,374 | 15,718 | -14,342 | 50,729 | 2.74 | 68.0 | 18.6 | 26.6 |
| Essex | 12.2 | 8.2 | 6.1 | 51.8 | 793,633 | 783,855 | -1.2 | 2.1 | 106,241 | 61,264 | -28,699 | 285,908 | 2.71 | 64.4 | 18.8 | 30.9 |
| Gloucester | 14.4 | 10.1 | 6.7 | 51.4 | 254,673 | 288,763 | 13.4 | 1.6 | 30,216 | 26,795 | 1,231 | 104,908 | 2.73 | 70.3 | 11.7 | 24.3 |
| Hudson | 10.6 | 7.2 | 5.3 | 50.2 | 608,975 | 634,281 | 4.2 | 5.9 | 105,124 | 39,404 | -28,508 | 258,591 | 2.56 | 62.0 | 15.0 | 28.0 |
| Hunterdon | 17.7 | 11.9 | 8.2 | 50.6 | 121,989 | 127,364 | 4.4 | -2.0 | 9,583 | 9,221 | -2,948 | 47,175 | 2.56 | 73.2 | 7.2 | 22.2 |

1. No spouse present.

# Table B. States and Counties — **Population, Vital Statistics, Health, and Crime**

| STATE County | Persons in group quarters, 2020 | Daytime Population, 2015-2019 Number | Daytime Population, 2015-2019 Employment/ residence ratio | Births, 2020 Total | Births, 2020 Rate[1] | Deaths, 2020 Number | Deaths, 2020 Rate[1] | Persons under 65 with no health insurance, 2019 Number | Persons under 65 with no health insurance, 2019 Percent | Medicare, 2020 Total beneficiaries | Medicare, 2020 Enrolled in Original Medicare | Medicare, 2020 Enrolled in Medicare Advantage | Crimes reported by county police or sheriff, 2019 Violent | Crimes reported by county police or sheriff, 2019 Property |
|---|---|---|---|---|---|---|---|---|---|---|---|---|---|---|
| | 32 | 33 | 34 | 35 | 36 | 37 | 38 | 39 | 40 | 41 | 42 | 43 | 44 | 45 |
| **NEBRASKA— Cont'd** | | | | | | | | | | | | | | |
| Platte | 439 | 36,002 | 1.17 | 459 | 13.8 | 298 | 8.9 | 2,749 | 10.1 | 6,485 | 5,995 | 490 | 8 | 69 |
| Polk | 96 | 4,307 | 0.65 | 57 | 11.0 | 61 | 11.7 | 418 | 10.6 | 1,294 | 1,235 | 59 | 0 | 20 |
| Red Willow | 423 | 11,180 | 1.07 | 133 | 12.5 | 134 | 12.6 | 794 | 9.8 | 2,510 | 2,470 | 41 | NA | NA |
| Richardson | 129 | 7,259 | 0.81 | 86 | 11.0 | 116 | 14.9 | 570 | 9.8 | 2,175 | 2,136 | 39 | NA | NA |
| Rock | 22 | 1,393 | 0.97 | 17 | 12.3 | 14 | 10.2 | 121 | 12.3 | 385 | 367 | 18 | NA | NA |
| Saline | 846 | 14,129 | 0.98 | 188 | 13.4 | 132 | 9.4 | 1,399 | 12.5 | 2,495 | 2,212 | 283 | 7 | 46 |
| Sarpy | 1,101 | 157,265 | 0.75 | 2,458 | 13.0 | 1,186 | 6.3 | 9,378 | 5.7 | 26,689 | 19,810 | 6,879 | 35 | 798 |
| Saunders | 345 | 16,871 | 0.61 | 248 | 11.3 | 216 | 9.9 | 1,234 | 7.1 | 4,477 | 3,496 | 980 | 8 | 65 |
| Scotts Bluff | 764 | 36,490 | 1.02 | 419 | 11.9 | 449 | 12.7 | 3,215 | 11.5 | 8,283 | 7,522 | 762 | 10 | 54 |
| Seward | 1,234 | 15,390 | 0.80 | 187 | 10.9 | 159 | 9.3 | 886 | 6.7 | 3,424 | 2,971 | 453 | 3 | 16 |
| Sheridan | 144 | 4,941 | 0.89 | 63 | 12.2 | 67 | 13.0 | 624 | 16.5 | 1,384 | 1,251 | 133 | NA | NA |
| Sherman | 55 | 2,779 | 0.82 | 35 | 11.7 | 43 | 14.4 | 267 | 12.5 | 812 | 711 | 102 | 5 | 23 |
| Sioux | 0 | 1,074 | 0.77 | 7 | 5.8 | 2 | 1.7 | 114 | 14.3 | 289 | D | D | NA | NA |
| Stanton | 0 | 4,383 | 0.50 | 79 | 13.4 | 41 | 7.0 | 374 | 7.8 | 1,105 | 956 | 149 | 2 | 29 |
| Thayer | 66 | 5,243 | 1.07 | 49 | 10.0 | 80 | 16.4 | 340 | 9.2 | 1,424 | 1,336 | 88 | 4 | 25 |
| Thomas | 2 | 634 | 0.97 | 6 | 8.1 | 1 | 1.4 | 78 | 14.6 | 209 | 195 | 14 | NA | NA |
| Thurston | 52 | 7,790 | 1.22 | 138 | 19.1 | 93 | 12.9 | 984 | 15.9 | 1,078 | 1,031 | 48 | 7 | 13 |
| Valley | 38 | 4,233 | 1.01 | 49 | 11.9 | 57 | 13.9 | 342 | 11.2 | 1,102 | 1,027 | 75 | 5 | 4 |
| Washington | 137 | 19,063 | 0.87 | 204 | 9.8 | 191 | 9.1 | 1,134 | 6.7 | 4,326 | 3,136 | 1,190 | 2 | 79 |
| Wayne | 1,105 | 9,743 | 1.08 | 90 | 9.5 | 55 | 5.8 | 530 | 7.8 | 1,546 | 1,354 | 192 | 6 | 10 |
| Webster | 156 | 3,261 | 0.84 | 42 | 12.3 | 59 | 17.3 | 235 | 9.1 | 889 | 827 | 62 | 0 | 1 |
| Wheeler | 0 | 766 | 0.96 | 14 | 17.7 | 4 | 5.1 | 72 | 12.9 | 202 | D | D | 3 | 21 |
| York | 764 | 14,568 | 1.12 | 160 | 11.8 | 136 | 10.1 | 811 | 8.0 | 3,193 | 2,982 | 211 | NA | NA |
| **NEVADA** | 36,323 | 2,975,311 | 1.00 | 35,906 | 11.4 | 28,170 | 9.0 | 342,719 | 13.5 | 549,064 | 328,982 | 220,082 | X | X |
| Churchill | 327 | 23,890 | 0.96 | 348 | 13.7 | 302 | 11.9 | 2,662 | 13.4 | 5,885 | 5,011 | 873 | 20 | 186 |
| Clark | 22,331 | 2,187,099 | 1.01 | 26,886 | 11.6 | 19,618 | 8.5 | 260,442 | 13.7 | 371,680 | 203,111 | 168,568 | NA | NA |
| Douglas | 230 | 48,154 | 1.00 | 329 | 6.7 | 599 | 12.2 | 3,903 | 11.4 | 16,251 | 13,364 | 2,886 | NA | NA |
| Elko | 772 | 51,128 | 0.95 | 725 | 13.7 | 339 | 6.4 | 6,002 | 13.0 | 6,807 | 6,591 | 216 | 57 | 176 |
| Esmeralda | 1 | 1,040 | 1.19 | 4 | 4.5 | 4 | 4.5 | 103 | 18.2 | 258 | 244 | 13 | 6 | 6 |
| Eureka | 1 | 4,493 | 3.83 | 16 | 7.7 | 12 | 5.8 | 123 | 7.2 | 356 | D | D | 22 | 27 |
| Humboldt | 154 | 17,242 | 1.05 | 251 | 14.8 | 123 | 7.3 | 2,150 | 15.1 | 2,932 | 2,852 | 80 | 26 | 76 |
| Lander | 16 | 5,512 | 0.95 | 71 | 12.9 | 46 | 8.3 | 579 | 12.5 | 924 | 910 | 14 | 62 | 56 |
| Lincoln | 239 | 5,083 | 0.95 | 55 | 10.7 | 61 | 11.8 | 424 | 11.8 | 1,083 | 1,015 | 68 | NA | NA |
| Lyon | 361 | 44,521 | 0.56 | 614 | 10.5 | 702 | 12.0 | 5,867 | 13.1 | 14,635 | 10,909 | 3,726 | 179 | 522 |
| Mineral | 52 | 4,484 | 1.02 | 40 | 8.9 | 96 | 21.2 | 417 | 12.6 | 1,357 | 1,162 | 195 | 4 | 48 |
| Nye | 1,047 | 43,788 | 0.96 | 417 | 8.7 | 861 | 17.9 | 4,048 | 12.9 | 16,999 | 8,104 | 8,895 | 113 | 715 |
| Pershing | 1,698 | 6,751 | 1.07 | 50 | 7.6 | 58 | 8.8 | 545 | 14.1 | 1,035 | 1,002 | 32 | 26 | 39 |
| Storey | 5 | 4,569 | 1.35 | 22 | 5.2 | 36 | 8.6 | 252 | 9.0 | 1,232 | 919 | 313 | 31 | 74 |
| Washoe | 5,290 | 457,101 | 1.00 | 5,388 | 11.3 | 4,440 | 9.3 | 48,387 | 12.5 | 91,461 | 60,273 | 31,188 | 223 | 616 |
| White Pine | 1,110 | 10,072 | 1.11 | 93 | 9.8 | 87 | 9.2 | 684 | 10.4 | 1,921 | 1,848 | 73 | NA | NA |
| Carson City | 2,689 | 60,384 | 1.22 | 597 | 10.7 | 786 | 14.0 | 6,131 | 14.7 | 14,250 | 11,319 | 2,931 | NA | NA |
| **NEW HAMPSHIRE** | 40,463 | 1,310,091 | 0.95 | 12,016 | 8.8 | 13,114 | 9.6 | 82,733 | 7.7 | 307,453 | 240,053 | 67,400 | X | X |
| Belknap | 1,032 | 57,589 | 0.89 | 466 | 7.6 | 841 | 13.7 | 3,906 | 8.3 | 18,208 | 13,991 | 4,217 | NA | NA |
| Carroll | 399 | 46,999 | 0.95 | 323 | 6.6 | 624 | 12.7 | 3,792 | 11.0 | 16,153 | 12,842 | 3,311 | 4 | 49 |
| Cheshire | 3,924 | 73,698 | 0.94 | 643 | 8.4 | 755 | 9.9 | 4,263 | 7.5 | 19,031 | 15,127 | 3,904 | 0 | 7 |
| Coos | 2,134 | 31,146 | 0.96 | 240 | 7.7 | 450 | 14.4 | 1,928 | 8.7 | 10,184 | 8,414 | 1,771 | NA | NA |
| Grafton | 7,056 | 100,800 | 1.24 | 675 | 7.4 | 909 | 10.0 | 5,994 | 9.3 | 21,979 | 17,746 | 4,233 | 4 | 4 |
| Hillsborough | 7,678 | 394,600 | 0.92 | 4,196 | 10.0 | 3,708 | 8.9 | 27,581 | 8.0 | 81,114 | 61,914 | 19,200 | NA | NA |
| Merrimack | 6,505 | 152,282 | 1.03 | 1,273 | 8.3 | 1,437 | 9.4 | 8,643 | 7.3 | 35,120 | 26,020 | 9,100 | 1 | 1 |
| Rockingham | 2,412 | 296,800 | 0.94 | 2,674 | 8.6 | 2,754 | 8.8 | 16,057 | 6.4 | 68,220 | 53,985 | 14,235 | 0 | 3 |
| Strafford | 8,602 | 118,789 | 0.85 | 1,156 | 8.8 | 1,155 | 8.8 | 7,990 | 7.8 | 26,066 | 20,402 | 5,663 | 3 | 4 |
| Sullivan | 721 | 37,388 | 0.72 | 370 | 8.6 | 481 | 11.1 | 2,579 | 7.7 | 11,378 | 9,612 | 1,766 | NA | NA |
| **NEW JERSEY** | 180,001 | 8,538,209 | 0.92 | 99,603 | 11.2 | 81,630 | 9.2 | 675,782 | 9.3 | 1,636,000 | 1,108,201 | 527,799 | X | X |
| Atlantic | 5,995 | 266,957 | 1.01 | 2,761 | 10.5 | 3,055 | 11.6 | 22,420 | 10.7 | 58,468 | 40,777 | 17,692 | 0 | 0 |
| Bergen | 10,435 | 894,122 | 0.92 | 8,901 | 9.6 | 7,939 | 8.5 | 64,563 | 8.4 | 172,299 | 125,042 | 47,258 | 2 | 46 |
| Burlington | 11,774 | 427,115 | 0.92 | 4,331 | 9.7 | 4,404 | 9.9 | 21,430 | 6.0 | 90,581 | 59,972 | 30,609 | 0 | 0 |
| Camden | 7,501 | 468,420 | 0.84 | 6,100 | 12.0 | 5,251 | 10.4 | 34,449 | 8.2 | 97,175 | 64,494 | 32,681 | 2 | 44 |
| Cape May | 2,711 | 93,067 | 1.00 | 813 | 8.9 | 1,393 | 15.2 | 6,002 | 9.2 | 28,067 | 21,291 | 6,776 | 0 | 0 |
| Cumberland | 8,760 | 152,948 | 1.02 | 1,850 | 12.6 | 1,630 | 11.1 | 14,106 | 12.1 | 28,789 | 18,381 | 10,408 | 0 | 1 |
| Essex | 23,148 | 793,123 | 0.99 | 10,253 | 12.8 | 6,750 | 8.4 | 80,196 | 12.0 | 119,129 | 70,649 | 48,480 | 131 | 135 |
| Gloucester | 4,238 | 256,378 | 0.76 | 2,719 | 9.3 | 2,937 | 10.0 | 13,841 | 5.7 | 57,059 | 38,621 | 18,438 | 0 | 0 |
| Hudson | 9,488 | 609,024 | 0.83 | 10,142 | 15.1 | 4,459 | 6.6 | 78,318 | 13.5 | 82,089 | 48,000 | 34,089 | 0 | 0 |
| Hunterdon | 3,842 | 114,551 | 0.84 | 973 | 7.8 | 1,044 | 8.4 | 5,098 | 5.2 | 26,441 | 18,813 | 7,628 | 0 | 0 |

1. Per 1,000 estimated resident population.

# Table B. States and Counties — Crime, Education, Money Income, and Poverty

| STATE County | County law enforcement employment, 2019 — Officers | Civilians | Enrollment[1] Total | Percent private | Attainment[2] High school graduate or less | Attainment[2] Bachelor's degree or more | Local govt expenditures[3] 2017-2018 Total current spending (mil dol) | Current spending per student (dollars) | Per capita income[4] | Median income (dollars) | Households Percent with income of less than $50,000 | Percent with income of $200,000 or more | Median household income (dollars) | Percent below poverty — All persons | Children under 18 years | Children 5 to 17 years in families |
|---|---|---|---|---|---|---|---|---|---|---|---|---|---|---|---|---|
| | 46 | 47 | 48 | 49 | 50 | 51 | 52 | 53 | 54 | 55 | 56 | 57 | 58 | 59 | 60 | 61 |
| **NEBRASKA— Cont'd** | | | | | | | | | | | | | | | | |
| Platte | 22 | 43 | 8,368 | 24.3 | 40.8 | 21.8 | 71 | 14,035 | 30,331 | 62,305 | 40.6 | 4.0 | 62,804 | 9.0 | 11.5 | 10.2 |
| Polk | NA | NA | 1,299 | 10.3 | 39.6 | 21.8 | 22 | 17,527 | 30,234 | 67,719 | 34.4 | 2.2 | 66,462 | 6.9 | 8.1 | 6.9 |
| Red Willow | 5 | 16 | 2,586 | 11.6 | 40.1 | 19.1 | 22 | 12,405 | 27,392 | 48,808 | 51.4 | 2.3 | 50,260 | 12.0 | 13.8 | 12.6 |
| Richardson | 9 | 19 | 1,628 | 12.9 | 44.0 | 19.0 | 19 | 14,863 | 28,309 | 47,917 | 52.4 | 2.2 | 50,207 | 12.3 | 15.7 | 13.0 |
| Rock | 3 | 4 | 222 | 5.0 | 37.3 | 23.2 | 4 | 16,214 | 36,140 | 56,250 | 45.6 | 8.0 | 53,905 | 13.9 | 21.1 | 20.1 |
| Saline | 18 | 0 | 4,278 | 22.2 | 49.8 | 17.8 | 40 | 12,907 | 24,627 | 51,502 | 48.7 | 2.7 | 59,937 | 9.2 | 9.9 | 9.3 |
| Sarpy | 115 | 105 | 51,967 | 15.6 | 26.2 | 39.8 | 314 | 11,180 | 36,058 | 82,032 | 27.3 | 7.0 | 84,016 | 5.9 | 6.5 | 5.2 |
| Saunders | 13 | 12 | 5,167 | 20.1 | 37.8 | 25.4 | 40 | 13,141 | 33,851 | 68,682 | 35.3 | 3.8 | 71,086 | 6.9 | 6.8 | 5.8 |
| Scotts Bluff | 17 | 7 | 8,865 | 11.2 | 40.7 | 22.9 | 89 | 13,257 | 27,978 | 49,745 | 50.3 | 3.3 | 50,263 | 13.0 | 16.9 | 14.6 |
| Seward | 16 | 2 | 4,956 | 33.7 | 34.6 | 28.9 | 42 | 15,747 | 32,040 | 70,389 | 34.0 | 5.0 | 72,291 | 7.2 | 6.3 | 5.6 |
| Sheridan | 7 | 10 | 1,150 | 6.8 | 42.3 | 22.8 | 12 | 15,084 | 29,477 | 45,371 | 54.1 | 2.1 | 46,159 | 14.5 | 21.9 | 20.6 |
| Sherman | 5 | 1 | 613 | 15.0 | 43.3 | 20.2 | 9 | 17,842 | 28,448 | 50,781 | 49.3 | 1.0 | 48,735 | 12.4 | 19.8 | 17.6 |
| Sioux | NA | NA | 193 | 3.6 | 36.2 | 30.6 | 3 | 29,714 | 28,646 | 48,269 | 51.7 | 1.8 | 47,983 | 15.9 | 25.6 | 24.3 |
| Stanton | 8 | 1 | 1,520 | 8.6 | 38.0 | 17.8 | 6 | 14,750 | 29,117 | 63,986 | 37.3 | 1.6 | 62,045 | 7.0 | 9.3 | 9.1 |
| Thayer | 8 | 6 | 1,033 | 4.8 | 43.3 | 18.8 | 16 | 17,488 | 29,632 | 51,821 | 47.7 | 3.7 | 52,439 | 10.8 | 14.3 | 13.1 |
| Thomas | NA | NA | 140 | 17.1 | 29.0 | 24.7 | 2 | 20,461 | 36,572 | 59,432 | 42.6 | 5.4 | 51,401 | 13.9 | 22.9 | 21.0 |
| Thurston | 7 | 7 | 2,422 | 7.6 | 43.8 | 16.4 | 37 | 19,804 | 20,141 | 51,034 | 48.9 | 2.4 | 48,891 | 24.9 | 30.5 | 27.8 |
| Valley | 4 | 5 | 874 | 10.6 | 41.9 | 24.2 | 11 | 15,441 | 29,378 | 55,324 | 47.4 | 3.5 | 47,981 | 10.4 | 13.8 | 12.4 |
| Washington | 31 | 21 | 4,851 | 12.6 | 36.1 | 29.9 | 40 | 10,631 | 35,145 | 71,430 | 35.2 | 4.9 | 75,591 | 6.8 | 8.0 | 6.3 |
| Wayne | 6 | 0 | 3,578 | 4.3 | 31.5 | 32.4 | 32 | 19,280 | 27,671 | 56,456 | 44.8 | 2.7 | 57,655 | 11.8 | 9.7 | 8.8 |
| Webster | 6 | 4 | 766 | 9.7 | 41.7 | 23.0 | 10 | 17,088 | 29,969 | 46,188 | 52.1 | 3.6 | 49,518 | 12.3 | 15.5 | 14.1 |
| Wheeler | 2 | 0 | 107 | 28.0 | 36.9 | 19.4 | 3 | 25,768 | 29,037 | 51,591 | 46.0 | 2.9 | 50,207 | 12.3 | 21.9 | 21.4 |
| York | 12 | 11 | 3,346 | 22.2 | 34.3 | 28.4 | 26 | 13,302 | 32,975 | 60,298 | 42.0 | 5.4 | 59,749 | 9.4 | 11.6 | 10.7 |
| **NEVADA** | X | X | 699,523 | 11.5 | 41.4 | 24.7 | 4,409 | 9,075 | 31,557 | 60,365 | 41.6 | 5.4 | 63,268 | 12.7 | 17.6 | 16.5 |
| Churchill | 38 | 8 | 4,871 | 11.6 | 40.0 | 18.8 | 37 | 11,066 | 28,462 | 57,824 | 44.8 | 3.6 | 56,335 | 11.7 | 17.4 | 16.2 |
| Clark | NA | NA | 518,601 | 12.0 | 42.4 | 24.5 | 2,956 | 8,943 | 30,704 | 59,340 | 42.5 | 5.3 | 62,131 | 13.3 | 18.7 | 17.6 |
| Douglas | 108 | 16 | 8,919 | 9.8 | 31.2 | 29.4 | 61 | 10,513 | 38,814 | 66,810 | 34.4 | 6.3 | 76,296 | 8.1 | 11.9 | 10.4 |
| Elko | 62 | 13 | 13,464 | 5.5 | 46.4 | 16.6 | 113 | 11,405 | 33,875 | 81,232 | 29.7 | 4.4 | 83,287 | 11.0 | 14.3 | 13.4 |
| Esmeralda | 8 | 4 | 144 | 0.0 | 52.3 | 16.1 | 2 | 31,630 | 25,849 | 37,375 | 60.9 | 0.4 | 50,675 | 13.8 | 13.8 | 13.8 |
| Eureka | NA | NA | 374 | 20.6 | 52.8 | 8.8 | 8 | 26,271 | 34,249 | 67,105 | 31.1 | 6.6 | 71,541 | 9.9 | 12.6 | 10.5 |
| Humboldt | 35 | 21 | 4,058 | 7.3 | 48.4 | 16.8 | 43 | 12,113 | 33,362 | 67,756 | 38.7 | 4.5 | 73,106 | 11.0 | 15.2 | 13.5 |
| Lander | 16 | 15 | 1,432 | 7.8 | 49.7 | 12.4 | 12 | 11,377 | 34,000 | 88,030 | 36.9 | 5.1 | 81,239 | 10.9 | 14.9 | 14.6 |
| Lincoln | NA | NA | 1,091 | 10.2 | 48.1 | 19.6 | 14 | 12,846 | 26,432 | 58,462 | 40.4 | 3.1 | 58,782 | 13.8 | 17.4 | 17.1 |
| Lyon | 82 | 35 | 11,002 | 7.4 | 46.0 | 14.4 | 92 | 10,300 | 27,855 | 56,875 | 42.8 | 2.8 | 59,406 | 10.2 | 16.6 | 15.5 |
| Mineral | 18 | 6 | 778 | 15.6 | 46.7 | 14.9 | 9 | 15,131 | 22,921 | 35,806 | 64.7 | 1.1 | 43,771 | 15.9 | 23.4 | 23.0 |
| Nye | 110 | 37 | 6,894 | 10.6 | 52.8 | 10.7 | 63 | 11,782 | 25,558 | 47,300 | 52.1 | 0.7 | 48,490 | 16.1 | 26.9 | 26.3 |
| Pershing | 13 | 7 | 1,670 | 1.8 | 56.3 | 11.1 | 11 | 15,891 | 19,891 | 50,491 | 49.4 | 2.7 | 58,562 | 17.6 | 18.7 | 18.0 |
| Storey | 27 | 2 | 740 | 18.8 | 34.1 | 31.4 | 7 | 14,738 | 37,941 | 66,292 | 33.5 | 5.0 | 88,003 | 7.0 | 11.9 | 10.3 |
| Washoe | NA | NA | 111,804 | 11.4 | 34.9 | 30.8 | 627 | 9,330 | 36,071 | 64,791 | 37.9 | 6.8 | 71,733 | 10.7 | 12.4 | 11.2 |
| White Pine | NA | NA | 1,976 | 5.1 | 53.5 | 13.2 | 17 | 8,729 | 25,675 | 60,827 | 41.4 | 2.3 | 60,086 | 13.7 | 17.4 | 15.6 |
| Carson City | 100 | 44 | 11,705 | 7.6 | 39.0 | 22.2 | 338 | 7,507 | 31,549 | 55,718 | 44.0 | 4.2 | 58,126 | 12.2 | 17.4 | 15.6 |
| **NEW HAMPSHIRE** | X | X | 306,551 | 21.6 | 34.3 | 37.0 | 2,902 | 16,172 | 40,003 | 76,768 | 31.9 | 9.2 | 78,571 | 7.5 | 8.1 | 7.5 |
| Belknap | NA | NA | 12,839 | 14.6 | 36.3 | 32.6 | 152 | 17,130 | 37,430 | 69,447 | 36.5 | 6.2 | 68,752 | 8.6 | 11.0 | 9.9 |
| Carroll | 13 | 14 | 7,984 | 13.8 | 35.3 | 34.3 | 119 | 20,795 | 37,490 | 63,153 | 38.9 | 5.3 | 64,495 | 9.0 | 13.4 | 12.6 |
| Cheshire | 9 | 13 | 18,748 | 21.8 | 38.0 | 33.1 | 146 | 17,633 | 33,946 | 64,751 | 38.7 | 5.8 | 65,326 | 9.7 | 10.9 | 8.9 |
| Coos | NA | NA | 5,572 | 14.7 | 51.1 | 18.2 | 69 | 18,416 | 27,393 | 47,117 | 52.9 | 2.4 | 49,156 | 13.9 | 18.1 | 16.7 |
| Grafton | 10 | 15 | 21,508 | 33.8 | 33.6 | 40.6 | 229 | 20,201 | 37,750 | 63,389 | 39.7 | 8.5 | 62,468 | 9.2 | 9.9 | 9.0 |
| Hillsborough | NA | NA | 95,952 | 24.8 | 33.7 | 38.1 | 825 | 14,523 | 40,955 | 81,460 | 29.8 | 10.2 | 82,862 | 7.5 | 7.9 | 7.6 |
| Merrimack | 16 | 13 | 34,485 | 25.5 | 34.5 | 35.6 | 355 | 16,553 | 37,367 | 75,737 | 32.2 | 7.9 | 75,800 | 6.4 | 7.1 | 6.6 |
| Rockingham | 25 | 21 | 65,924 | 19.2 | 30.2 | 41.4 | 669 | 15,632 | 47,222 | 93,756 | 24.0 | 13.6 | 91,416 | 5.1 | 4.8 | 4.6 |
| Strafford | 28 | 12 | 35,449 | 12.6 | 34.8 | 35.8 | 229 | 15,421 | 35,601 | 72,960 | 33.4 | 6.4 | 71,486 | 8.8 | 8.3 | 7.9 |
| Sullivan | NA | NA | 8,090 | 17.9 | 45.2 | 28.1 | 108 | 19,579 | 33,042 | 61,312 | 40.6 | 4.2 | 63,518 | 10.1 | 12.7 | 11.4 |
| **NEW JERSEY** | X | X | 2,206,536 | 19.0 | 37.4 | 39.7 | 30,299 | 18,893 | 42,745 | 82,545 | 31.4 | 14.0 | 85,786 | 9.1 | 12.2 | 11.5 |
| Atlantic | 106 | 25 | 65,348 | 12.2 | 45.0 | 28.1 | 879 | 19,665 | 33,284 | 62,110 | 41.3 | 6.8 | 62,678 | 11.3 | 15.4 | 13.6 |
| Bergen | 496 | 85 | 225,951 | 22.2 | 30.2 | 49.3 | 2,855 | 20,892 | 51,214 | 101,144 | 25.5 | 19.6 | 107,971 | 5.7 | 5.8 | 5.5 |
| Burlington | 71 | 17 | 107,173 | 17.6 | 34.3 | 38.0 | 1,389 | 20,078 | 43,187 | 87,416 | 26.3 | 12.2 | 88,443 | 5.7 | 7.3 | 6.6 |
| Camden | 171 | 30 | 126,100 | 17.2 | 40.8 | 32.5 | 1,547 | 18,928 | 35,958 | 70,451 | 37.2 | 8.9 | 73,168 | 10.7 | 15.3 | 14.4 |
| Cape May | 136 | 27 | 17,400 | 15.2 | 39.5 | 32.8 | 275 | 22,081 | 40,389 | 67,074 | 37.0 | 8.0 | 66,565 | 9.8 | 14.9 | 14.4 |
| Cumberland | 63 | 0 | 35,808 | 7.4 | 60.0 | 15.6 | 524 | 17,840 | 25,694 | 54,179 | 45.8 | 3.5 | 54,522 | 15.9 | 22.9 | 21.0 |
| Essex | 367 | 72 | 209,452 | 17.0 | 42.0 | 35.5 | 2,793 | 19,519 | 38,722 | 61,510 | 42.7 | 12.5 | 64,522 | 13.8 | 18.4 | 17.1 |
| Gloucester | 95 | 10 | 74,271 | 14.7 | 39.1 | 33.0 | 882 | 18,562 | 39,337 | 87,283 | 28.5 | 10.9 | 87,996 | 7.5 | 7.7 | 6.6 |
| Hudson | 279 | 130 | 149,916 | 19.7 | 40.3 | 42.3 | 1,717 | 18,063 | 40,740 | 71,189 | 37.1 | 12.4 | 77,738 | 13.7 | 20.3 | 22.6 |
| Hunterdon | 38 | 5 | 29,329 | 17.1 | 26.0 | 52.7 | 437 | 23,134 | 57,098 | 115,379 | 19.1 | 23.7 | 117,275 | 4.0 | 3.6 | 3.2 |

1. All persons 3 years old and over enrolled in nursery school through college.  2. Persons 25 years old and over.  3. Elementary and secondary education expenditures.  4. Based on population estimated by the American Community Survey, 2014–2018.

# Table B. States and Counties — **Personal Income**

| STATE County | Personal income, 2019 | | | | | Supplements to wages and salaries, employer contributions (mil dol) | | | | | Earnings, 2019 | Contributions for government social insurance (mil dol) | |
|---|---|---|---|---|---|---|---|---|---|---|---|---|---|
| | Total (mil dol) | Percent change 2018-2019 | Per capita[1] Dollars | Per capita[1] Rank | Wages and salaries (mil dol) | Pension and insurance | Government social insurance | Proprietors' income (mil dol) | Dividends, interest, and rent (mil dol) | Personal transfer receipts (mil dol) | Total (mil dol) | From employee and self-employed | From employer |
| | 62 | 63 | 64 | 65 | 66 | 67 | 68 | 69 | 70 | 71 | 72 | 73 | 74 |
| **NEBRASKA— Cont'd** | | | | | | | | | | | | | |
| Platte | 1,610 | 3.2 | 48,102 | 1,008 | 922 | 195 | 69 | 122 | 315 | 271 | 1,308 | 78 | 69 |
| Polk | 279 | 2.7 | 53,462 | 576 | 61 | 14 | 5 | 56 | 52 | 51 | 136 | 7 | 5 |
| Red Willow | 504 | 3.9 | 47,027 | 1,131 | 227 | 46 | 18 | 41 | 120 | 115 | 332 | 21 | 18 |
| Richardson | 401 | 4.8 | 50,988 | 760 | 102 | 24 | 8 | 55 | 78 | 95 | 188 | 12 | 8 |
| Rock | 84 | 3.2 | 62,018 | 248 | 22 | 5 | 2 | 24 | 20 | 15 | 53 | 3 | 2 |
| Saline | 631 | 3.8 | 44,351 | 1,437 | 340 | 64 | 24 | 61 | 105 | 120 | 490 | 29 | 24 |
| Sarpy | 9,770 | 4.8 | 52,190 | 665 | 4,451 | 853 | 339 | 364 | 1,487 | 1,328 | 6,007 | 358 | 339 |
| Saunders | 1,153 | 4.7 | 53,427 | 580 | 253 | 53 | 19 | 99 | 195 | 193 | 424 | 27 | 19 |
| Scotts Bluff | 1,638 | 1.1 | 45,992 | 1,249 | 765 | 145 | 61 | 169 | 269 | 370 | 1,140 | 73 | 61 |
| Seward | 907 | 3.2 | 52,487 | 644 | 296 | 56 | 22 | 97 | 172 | 154 | 471 | 28 | 22 |
| Sheridan | 255 | 4.4 | 48,644 | 951 | 68 | 15 | 5 | 47 | 50 | 58 | 135 | 7 | 5 |
| Sherman | 129 | 3.1 | 42,984 | 1,619 | 30 | 6 | 2 | 15 | 26 | 33 | 54 | 4 | 2 |
| Sioux | 61 | 0.4 | 52,159 | 667 | 10 | 2 | 1 | 21 | 13 | 8 | 34 | 1 | 1 |
| Stanton | 272 | 4.0 | 45,869 | 1,266 | 91 | 16 | 6 | 30 | 52 | 40 | 142 | 9 | 6 |
| Thayer | 281 | 4.2 | 56,265 | 436 | 103 | 22 | 8 | 47 | 69 | 61 | 179 | 10 | 8 |
| Thomas | 40 | 4.4 | 55,960 | 448 | 11 | 2 | 1 | 12 | 8 | 7 | 27 | 1 | 1 |
| Thurston | 287 | 6.3 | 39,667 | 2,091 | 148 | 35 | 11 | 52 | 47 | 68 | 245 | 12 | 11 |
| Valley | 207 | 4.7 | 49,708 | 857 | 76 | 18 | 6 | 29 | 50 | 45 | 128 | 7 | 6 |
| Washington | 1,201 | 4.4 | 57,962 | 369 | 434 | 77 | 31 | 82 | 220 | 175 | 625 | 39 | 31 |
| Wayne | 421 | 4.0 | 44,879 | 1,377 | 174 | 41 | 13 | 55 | 91 | 71 | 282 | 15 | 13 |
| Webster | 184 | 6.7 | 52,857 | 629 | 39 | 9 | 3 | 48 | 31 | 40 | 100 | 5 | 3 |
| Wheeler | 65 | 10.4 | 82,976 | 54 | 12 | 2 | 1 | 36 | 11 | 6 | 51 | 1 | 1 |
| York | 705 | 4.1 | 51,573 | 717 | 354 | 65 | 27 | 75 | 158 | 136 | 520 | 31 | 27 |
| **NEVADA** | 157,584 | 4.1 | 50,985 | X | 76,530 | 11,927 | 5,656 | 10,808 | 38,545 | 26,263 | 104,920 | 6,297 | 5,656 |
| Churchill | 1,126 | 3.5 | 45,219 | 1,335 | 459 | 101 | 36 | 98 | 202 | 279 | 694 | 40 | 36 |
| Clark | 110,628 | 4.1 | 48,806 | 942 | 55,589 | 8,275 | 4,161 | 7,293 | 25,860 | 18,672 | 75,318 | 4,531 | 4,161 |
| Douglas | 3,699 | 3.5 | 75,635 | 87 | 1,022 | 154 | 74 | 295 | 1,474 | 560 | 1,545 | 107 | 74 |
| Elko | 2,564 | 4.1 | 48,584 | 958 | 1,276 | 203 | 86 | 102 | 335 | 337 | 1,667 | 95 | 86 |
| Esmeralda | 36 | 12.1 | 41,641 | 1,824 | 15 | 4 | 1 | | 7 | 9 | 18 | 1 | 1 |
| Eureka | 87 | 12.0 | 42,932 | 1,626 | 474 | 69 | 30 | 9 | 14 | 13 | 582 | 33 | 30 |
| Humboldt | 829 | 6.1 | 49,228 | 902 | 473 | 90 | 32 | 39 | 119 | 134 | 634 | 35 | 32 |
| Lander | 383 | 5.1 | 69,293 | 137 | 297 | 50 | 19 | 11 | 36 | 48 | 377 | 21 | 19 |
| Lincoln | 204 | 5.1 | 39,362 | 2,133 | 71 | 20 | 4 | 11 | 43 | 54 | 107 | 6 | 4 |
| Lyon | 2,280 | 6.5 | 39,650 | 2,094 | 585 | 117 | 43 | 98 | 328 | 568 | 844 | 64 | 43 |
| Mineral | 188 | 7.7 | 41,634 | 1,825 | 94 | 18 | 6 | 8 | 35 | 68 | 127 | 8 | 6 |
| Nye | 1,834 | 6.5 | 39,431 | 2,122 | 673 | 115 | 48 | 106 | 286 | 671 | 943 | 74 | 48 |
| Pershing | 221 | 2.5 | 32,859 | 2,901 | 115 | 29 | 7 | 5 | 36 | 54 | 155 | 8 | 7 |
| Storey | 246 | 6.6 | 59,628 | 319 | 1,070 | 240 | 94 | 14 | 44 | 38 | 1,417 | 76 | 94 |
| Washoe | 29,875 | 3.9 | 46,761 | 221 | 12,405 | 1,987 | 900 | 2,346 | 9,035 | 4,008 | 17,638 | 1,053 | 900 |
| White Pine | 448 | 4.7 | 46,761 | 1,153 | 256 | 59 | 15 | 25 | 67 | 102 | 355 | 18 | 15 |
| Carson City | 2,934 | 4.6 | 52,470 | 648 | 1,654 | 398 | 99 | 351 | 623 | 648 | 2,502 | 127 | 99 |
| **NEW HAMPSHIRE** | 86,345 | 3.8 | 63,452 | X | 40,200 | 6,187 | 2,702 | 8,791 | 14,834 | 12,968 | 57,879 | 3,586 | 2,702 |
| Belknap | 3,881 | 3.6 | 63,308 | 223 | 1,257 | 225 | 87 | 395 | 829 | 719 | 1,963 | 132 | 87 |
| Carroll | 2,942 | 3.1 | 60,157 | 298 | 853 | 149 | 59 | 370 | 852 | 630 | 1,431 | 101 | 59 |
| Cheshire | 4,043 | 2.6 | 53,141 | 603 | 1,630 | 293 | 112 | 269 | 882 | 799 | 2,304 | 161 | 112 |
| Coos | 1,401 | 3.0 | 44,399 | 1,432 | 515 | 115 | 36 | 92 | 243 | 460 | 758 | 55 | 36 |
| Grafton | 5,691 | 2.7 | 63,319 | 222 | 3,467 | 531 | 242 | 627 | 1,405 | 931 | 4,868 | 295 | 242 |
| Hillsborough | 26,364 | 4.0 | 63,218 | 226 | 14,443 | 2,051 | 957 | 2,399 | 3,891 | 3,596 | 19,849 | 1,203 | 957 |
| Merrimack | 9,022 | 4.0 | 59,593 | 321 | 4,602 | 811 | 309 | 866 | 1,458 | 1,596 | 6,588 | 399 | 309 |
| Rockingham | 23,822 | 4.1 | 76,902 | 77 | 9,852 | 1,351 | 658 | 3,015 | 3,818 | 2,656 | 14,876 | 909 | 658 |
| Strafford | 6,843 | 4.1 | 52,387 | 652 | 2,932 | 532 | 195 | 542 | 1,027 | 1,105 | 4,201 | 260 | 195 |
| Sullivan | 2,335 | 2.6 | 54,115 | 537 | 650 | 129 | 46 | 217 | 430 | 477 | 1,041 | 71 | 46 |
| **NEW JERSEY** | 625,938 | 3.8 | 70,399 | X | 284,970 | 43,912 | 20,276 | 63,001 | 112,465 | 87,021 | 412,159 | 24,558 | 20,276 |
| Atlantic | 13,350 | 4.2 | 50,631 | 787 | 6,852 | 1,204 | 546 | 1,098 | 2,153 | 3,124 | 9,700 | 613 | 546 |
| Bergen | 83,391 | 3.5 | 89,456 | 36 | 32,662 | 4,726 | 2,361 | 9,851 | 18,968 | 8,089 | 49,600 | 2,890 | 2,361 |
| Burlington | 28,327 | 4.0 | 63,605 | 217 | 13,550 | 2,250 | 1,023 | 1,953 | 4,473 | 4,414 | 18,776 | 1,147 | 1,023 |
| Camden | 28,251 | 3.8 | 55,780 | 457 | 12,601 | 2,109 | 948 | 1,984 | 4,213 | 5,689 | 17,642 | 1,103 | 948 |
| Cape May | 5,774 | 3.6 | 62,734 | 234 | 1,892 | 385 | 159 | 669 | 1,369 | 1,353 | 3,106 | 207 | 159 |
| Cumberland | 6,166 | 3.4 | 41,237 | 1,888 | 3,090 | 632 | 246 | 498 | 854 | 1,744 | 4,466 | 279 | 246 |
| Essex | 54,056 | 3.8 | 67,657 | 151 | 27,101 | 4,516 | 1,875 | 3,957 | 10,666 | 8,616 | 37,450 | 2,190 | 1,875 |
| Gloucester | 16,477 | 4.0 | 56,499 | 422 | 5,875 | 1,112 | 460 | 1,101 | 2,106 | 2,793 | 8,548 | 543 | 460 |
| Hudson | 46,668 | 4.1 | 69,406 | 134 | 23,146 | 3,333 | 1,539 | 6,886 | 5,451 | 5,790 | 34,905 | 1,972 | 1,539 |
| Hunterdon | 11,478 | 3.8 | 92,291 | 31 | 3,586 | 535 | 249 | 1,056 | 2,220 | 1,127 | 5,426 | 328 | 249 |

1. Based on the resident population estimated as of July 1 of the year shown.

# Table B. States and Counties — Earnings, Social Security, and Housing

| STATE County | Earnings, 2019 (cont.) — Percent by selected industries | | | | | | | | | Social Security beneficiaries, December 2019 | | Supplemental Security Income recipients, 2019 | Housing units, 2020 | |
|---|---|---|---|---|---|---|---|---|---|---|---|---|---|---|
| | Farm | Mining, quarrying, and extractions | Construction | Manufacturing | Information; professional, scientific, technical services | Retail trade | Finance, insurance, real estate, and leasing | Health care and social assistance | Government | Number | Rate[1] | | Total | Percent change, 2010-2020 |
| | 75 | 76 | 77 | 78 | 79 | 80 | 81 | 82 | 83 | 84 | 85 | 86 | 87 | 88 |
| **NEBRASKA— Cont'd** | | | | | | | | | | | | | | |
| Platte | 3.9 | D | 6.5 | 34.5 | 2.6 | 6.1 | 5.3 | 8.7 | 14.7 | 6,645 | 199 | 316 | 14,204 | 6.2 |
| Polk | 31.7 | D | 5.5 | 1.3 | D | 5.2 | D | 6.4 | 23.4 | 1,315 | 255 | 42 | 2,738 | 0.3 |
| Red Willow | 6.0 | 0.9 | 4.5 | 9.4 | 3.6 | 7.7 | 6.5 | 13.9 | 20.0 | 2,540 | 237 | 142 | 5,314 | 0.9 |
| Richardson | 17.8 | D | 4.5 | 9.2 | 3.7 | 5.4 | 4.1 | D | 18.9 | 2,215 | 282 | 170 | 4,396 | 0.1 |
| Rock | 28.2 | 0.2 | D | D | D | 5.0 | 6.1 | 1.1 | 23.2 | 350 | 255 | 0 | 921 | 0.8 |
| Saline | 5.1 | 0.0 | 3.7 | 40.4 | D | 4.2 | 3.3 | D | 17.7 | 2,565 | 181 | 171 | 5,924 | 2.8 |
| Sarpy | 0.1 | 0.0 | 10.3 | 4.2 | 8.6 | 5.7 | 10.4 | 5.5 | 24.1 | 26,680 | 143 | 1,457 | 72,387 | 16.9 |
| Saunders | 11.9 | D | 13.2 | 6.3 | 5.6 | 6.1 | 6.3 | 4.7 | 22.7 | 4,335 | 201 | 192 | 10,034 | 8.8 |
| Scotts Bluff | 5.0 | D | 6.8 | 5.2 | 4.4 | 6.6 | 5.4 | 17.2 | 18.7 | 8,455 | 238 | 803 | 16,361 | -0.3 |
| Seward | 10.8 | 0.0 | 8.8 | 17.3 | 3.8 | 4.4 | 5.9 | D | 15.9 | 3,465 | 200 | 127 | 7,463 | 8.6 |
| Sheridan | 27.4 | D | 2.4 | 7.9 | 2.1 | 5.4 | D | 4.3 | 25.6 | 1,375 | 263 | 72 | 2,912 | -0.8 |
| Sherman | 17.0 | 0.0 | D | D | D | 7.3 | 5.5 | 10.8 | 21.1 | 840 | 282 | 32 | 1,967 | 1.3 |
| Sioux | 53.7 | 0.4 | 1.7 | 1.7 | D | 3.3 | D | D | 12.9 | 260 | 225 | 0 | 827 | 1.7 |
| Stanton | 7.5 | D | 5.2 | 49.3 | D | 1.0 | 4.3 | 1.1 | 12.1 | 1,045 | 176 | 26 | 2,707 | 2.8 |
| Thayer | 21.7 | 0.0 | 4.2 | 16.3 | 1.1 | 2.9 | 6.8 | 3.3 | 20.8 | 1,360 | 273 | 69 | 2,761 | 1.1 |
| Thomas | 14.9 | D | D | D | D | 9.5 | D | D | 17.7 | 170 | 235 | 0 | 405 | 0.7 |
| Thurston | 13.5 | 0.0 | 2.1 | 4.9 | D | 4.3 | 2.6 | 6.9 | 48.8 | 1,145 | 158 | 246 | 2,452 | 1.9 |
| Valley | 16.1 | D | 6.1 | 5.4 | D | 7.0 | D | D | 27.0 | 1,075 | 261 | 55 | 2,309 | 1.5 |
| Washington | 5.5 | D | 10.9 | 19.8 | 5.5 | 9.6 | 6.0 | 5.8 | 15.5 | 4,260 | 206 | 171 | 8,753 | 5.5 |
| Wayne | 11.1 | 0.1 | 3.5 | 17.6 | 2.0 | 4.4 | 9.8 | 9.5 | 25.3 | 1,520 | 161 | 70 | 3,997 | 5.9 |
| Webster | 32.2 | D | D | D | D | 4.1 | D | 5.1 | 17.7 | 940 | 271 | 78 | 1,905 | -0.4 |
| Wheeler | 69.4 | 0.1 | D | D | D | D | D | 0.0 | 7.5 | 160 | 205 | 0 | 585 | 1.4 |
| York | 7.8 | D | 5.2 | 13.9 | D | 5.5 | 7.8 | D | 14.9 | 3,170 | 233 | 151 | 6,437 | 3.3 |
| **NEVADA** | 0.2 | 1.8 | 8.1 | 4.5 | 9.0 | 6.7 | 6.9 | 9.3 | 14.8 | 552,219 | 179 | 56,627 | 1,305,899 | 11.3 |
| Churchill | 6.0 | 0.4 | 9.0 | 7.3 | 3.5 | 5.5 | 4.2 | 8.6 | 29.6 | 5,965 | 238 | 495 | 11,161 | 3.1 |
| Clark | 0.0 | 0.1 | 7.6 | 2.6 | 8.9 | 7.1 | 7.6 | 9.3 | 14.2 | 376,685 | 166 | 44,355 | 938,258 | 11.7 |
| Douglas | 0.2 | 0.1 | 9.1 | 12.9 | 11.9 | 5.5 | 7.8 | 7.3 | 12.0 | 15,445 | 315 | 390 | 24,930 | 5.3 |
| Elko | 1.3 | 16.6 | 9.4 | 0.8 | 3.6 | 6.2 | 2.6 | 6.5 | 17.3 | 6,975 | 132 | 530 | 22,140 | 13.2 |
| Esmeralda | -6.7 | D | D | D | D | D | 0.2 | D | 0.0 | 270 | 310 | 23 | 865 | 1.5 |
| Eureka | 1.2 | 91.2 | D | D | D | 0.2 | D | D | 2.8 | 355 | 171 | 22 | 1,092 | 1.5 |
| Humboldt | 1.5 | 37.5 | 5.8 | 3.1 | D | 5.9 | 1.2 | D | 19.4 | 2,955 | 175 | 257 | 7,666 | 7.7 |
| Lander | 1.5 | 73.3 | D | D | 0.3 | 2.2 | D | 0.2 | 10.8 | 945 | 171 | 90 | 2,765 | 7.4 |
| Lincoln | 6.4 | D | D | D | D | D | D | D | 43.6 | 1,095 | 211 | 68 | 2,836 | 4.0 |
| Lyon | 1.9 | 5.8 | 11.1 | 18.5 | 5.1 | 6.8 | 3.2 | 4.2 | 19.3 | 15,200 | 264 | 943 | 24,354 | 8.0 |
| Mineral | 2.1 | 10.7 | 1.7 | 0.5 | 2.1 | 3.4 | D | 0.8 | 30.6 | 1,345 | 298 | 146 | 2,847 | 0.7 |
| Nye | 4.0 | 13.6 | 5.4 | 1.8 | 19.8 | 6.5 | 2.2 | D | 14.7 | 17,425 | 372 | 1,165 | 22,557 | 0.9 |
| Pershing | 5.0 | 45.8 | D | D | D | 3.4 | 0.9 | D | 35.9 | 1,030 | 155 | 89 | 2,515 | 2.2 |
| Storey | 0.0 | D | 5.4 | 70.9 | D | D | D | D | 1.5 | 1,190 | 284 | 11 | 2,087 | 4.8 |
| Washoe | 0.0 | 0.3 | 10.9 | 6.2 | D | 6.1 | 6.6 | 11.2 | 14.4 | 89,590 | 190 | 6,858 | 210,665 | 14.0 |
| White Pine | 2.5 | 41.1 | D | D | D | 3.1 | 1.5 | D | 31.4 | 1,910 | 202 | 161 | 4,549 | 1.2 |
| Carson City | 0.0 | D | 5.8 | 8.7 | 7.7 | 7.0 | 5.7 | 14.0 | 34.3 | 13,840 | 248 | 1,024 | 24,612 | 4.6 |
| **NEW HAMPSHIRE** | 0.1 | 0.1 | 7.5 | 11.6 | 12.8 | 7.8 | 8.6 | 12.5 | 11.9 | 312,043 | 229 | 18,328 | 646,867 | 5.2 |
| Belknap | 0.0 | 0.3 | 12.0 | 9.4 | 6.7 | 11.9 | 4.5 | 12.7 | 14.7 | 18,610 | 303 | 958 | 39,171 | 4.8 |
| Carroll | 0.1 | D | 13.9 | D | 7.7 | 11.1 | 5.3 | 11.3 | 13.8 | 16,200 | 332 | 623 | 41,949 | 5.3 |
| Cheshire | 0.1 | D | 10.2 | 16.0 | 6.6 | 9.9 | 7.1 | 11.7 | 14.7 | 19,850 | 262 | 1,217 | 35,979 | 3.5 |
| Coos | 0.2 | D | 6.3 | D | 2.9 | 8.9 | 3.4 | 18.1 | 26.7 | 10,950 | 348 | 941 | 21,795 | 2.2 |
| Grafton | 0.2 | 0.1 | 4.4 | 8.3 | 11.6 | 5.9 | 5.0 | 25.4 | 10.3 | 21,950 | 243 | 983 | 53,769 | 5.2 |
| Hillsborough | 0.0 | D | 6.6 | 14.6 | 15.2 | 7.3 | 10.9 | 11.5 | 9.5 | 83,030 | 199 | 6,909 | 174,506 | 5.1 |
| Merrimack | 0.2 | 0.3 | 7.4 | 7.7 | 8.9 | 6.9 | 9.2 | 13.7 | 20.4 | 35,815 | 236 | 1,906 | 65,908 | 3.7 |
| Rockingham | 0.0 | 0.1 | 8.3 | 10.5 | 15.0 | 8.1 | 8.0 | 8.9 | 7.7 | 67,480 | 218 | 2,158 | 135,402 | 6.9 |
| Strafford | 0.1 | D | 6.3 | 10.0 | 13.1 | 7.9 | 9.2 | 13.9 | 20.0 | 26,685 | 204 | 1,873 | 55,574 | 7.5 |
| Sullivan | 0.4 | D | 9.8 | 23.8 | 6.4 | 10.0 | 4.4 | 8.2 | 15.1 | 11,470 | 265 | 760 | 22,814 | 2.1 |
| **NEW JERSEY** | 0.1 | 0.3 | 5.3 | 8.1 | 15.1 | 6.0 | 10.4 | 11.4 | 13.9 | 1,646,792 | 185 | 178,009 | 3,655,018 | 2.9 |
| Atlantic | 0.6 | D | 6.9 | D | 6.5 | 6.7 | 4.5 | 16.0 | 22.6 | 60,700 | 230 | 7,257 | 128,943 | 1.8 |
| Bergen | 0.0 | 0.0 | 5.5 | 7.3 | 14.4 | 7.3 | 9.5 | 14.3 | 9.5 | 166,225 | 178 | 12,220 | 362,016 | 2.7 |
| Burlington | 0.2 | D | 5.5 | 8.8 | 11.3 | 7.4 | 12.0 | 13.5 | 17.3 | 92,600 | 208 | 6,167 | 181,020 | 3.1 |
| Camden | 0.1 | D | 5.9 | 7.8 | 9.4 | 6.6 | 5.4 | 18.8 | 17.1 | 100,065 | 198 | 16,521 | 207,028 | 1.1 |
| Cape May | 0.2 | D | 11.0 | D | 5.0 | 9.9 | 7.8 | 9.4 | 25.8 | 29,050 | 315 | 1,689 | 99,496 | 1.2 |
| Cumberland | 2.0 | 0.3 | 6.8 | 14.6 | D | 7.4 | 2.8 | 16.4 | 25.3 | 31,065 | 208 | 5,888 | 56,454 | 1.1 |
| Essex | 0.0 | D | 4.1 | 4.2 | 12.4 | 3.6 | 13.2 | 11.0 | 19.9 | 120,310 | 150 | 27,502 | 320,761 | 2.5 |
| Gloucester | 0.5 | D | 8.9 | 9.4 | 5.7 | 11.0 | 3.9 | 11.2 | 19.7 | 59,850 | 205 | 4,477 | 114,808 | 4.4 |
| Hudson | 0.0 | D | 2.0 | 2.2 | 15.4 | 5.0 | 27.9 | 6.3 | 11.5 | 82,910 | 123 | 20,149 | 290,593 | 7.5 |
| Hunterdon | 0.3 | D | 9.9 | 5.0 | 16.8 | 7.4 | 13.7 | 9.9 | 13.5 | 25,550 | 204 | 829 | 50,750 | 2.5 |

1. Per 1,000 resident population estimated as of July 1 of the year shown.

| STATE County | Housing units, 2015-2019 | | | | | | | | Civilian labor force, 2020 | | | | Civilian employment[6], 2015-2019 | | |
|---|---|---|---|---|---|---|---|---|---|---|---|---|---|---|---|
| | Occupied units | | | | | | | Sub-standard units[4] (percent) | | Percent change, 2019-2020 | Unemployment | | | Percent | |
| | Total | Percent | Owner-occupied | | | Renter-occupied | | | Total | | Total | Rate[5] | Total | Management, business, science, and arts | Construction, production, and maintenance occupations |
| | | | Median value[1] | Median owner cost as a percent of income | | Median rent[3] | Median rent as a percent of income[2] | | | | | | | | |
| | | | | With a mortgage | Without a mortgage[2] | | | | | | | | | | |
| | 89 | 90 | 91 | 92 | 93 | 94 | 95 | 96 | 97 | 98 | 99 | 100 | 101 | 102 | 103 |
| **NEBRASKA— Cont'd** | | | | | | | | | | | | | | | |
| Platte | 12,947 | 72.5 | 149,700 | 18.8 | 10.7 | 769 | 22.1 | 3.0 | 17,741 | -0.1 | 651 | 3.7 | 17,399 | 31.0 | 35.4 |
| Polk | 2,052 | 83.1 | 107,300 | 16.7 | 10.0 | 584 | 18.2 | 2.3 | 2,912 | -0.7 | 88 | 3.0 | 2,639 | 36.0 | 29.1 |
| Red Willow | 4,519 | 72.9 | 109,000 | 19.2 | 11.3 | 623 | 23.6 | 0.2 | 5,854 | -0.3 | 191 | 3.3 | 5,715 | 26.5 | 30.2 |
| Richardson | 3,704 | 75.9 | 69,200 | 15.2 | 11.7 | 531 | 18.3 | 1.3 | 4,197 | -1.4 | 135 | 3.2 | 3,823 | 33.7 | 33.4 |
| Rock | 623 | 79.8 | 96,200 | 17.6 | 10.0 | 488 | 18.4 | 1.0 | 873 | -0.1 | 18 | 2.1 | 754 | 39.5 | 25.1 |
| Saline | 5,157 | 70.6 | 99,300 | 19.5 | 12.1 | 761 | 31.6 | 4.9 | 7,310 | -1.4 | 263 | 3.6 | 7,140 | 25.4 | 39.2 |
| Sarpy | 66,260 | 68.9 | 195,300 | 19.2 | 11.2 | 1,006 | 25.3 | 2.1 | 96,776 | -1.0 | 3,943 | 4.1 | 94,810 | 43.9 | 18.4 |
| Saunders | 8,501 | 78.1 | 165,600 | 20.1 | 11.5 | 745 | 22.2 | 2.1 | 11,155 | -1.2 | 409 | 3.7 | 10,968 | 36.0 | 27.9 |
| Scotts Bluff | 14,732 | 66.9 | 128,100 | 20.9 | 12.6 | 761 | 28.1 | 1.6 | 18,160 | -0.9 | 758 | 4.2 | 17,175 | 30.4 | 28.7 |
| Seward | 6,600 | 71.7 | 168,200 | 18.5 | 10.0 | 724 | 23.0 | 1.2 | 8,852 | -0.7 | 363 | 4.1 | 9,129 | 36.5 | 28.6 |
| Sheridan | 2,273 | 68.8 | 72,300 | 18.4 | 12.9 | 684 | 22.3 | 2.6 | 2,708 | 0.8 | 65 | 2.4 | 2,714 | 35.2 | 29.2 |
| Sherman | 1,368 | 77.2 | 88,700 | 24.0 | 10.8 | 534 | 17.6 | 0.7 | 1,667 | -0.9 | 53 | 3.2 | 1,516 | 34.8 | 33.4 |
| Sioux | 544 | 72.1 | 117,700 | 25.9 | 13.1 | 592 | 18.2 | 4.8 | 747 | 0.3 | 17 | 2.3 | 629 | 50.6 | 21.9 |
| Stanton | 2,418 | 82.1 | 119,500 | 17.6 | 10.2 | 690 | 18.4 | 1.7 | 3,531 | 0.2 | 119 | 3.4 | 3,229 | 27.9 | 32.0 |
| Thayer | 2,244 | 78.6 | 71,400 | 17.7 | 10.0 | 591 | 18.5 | 1.8 | 2,781 | -0.4 | 76 | 2.7 | 2,568 | 35.5 | 30.0 |
| Thomas | 277 | 76.5 | 119,300 | 17.5 | 10.0 | 578 | 14.4 | 0.0 | 420 | -1.9 | 17 | 4.0 | 365 | 34.2 | 29.6 |
| Thurston | 2,176 | 60.8 | 79,800 | 16.5 | 10.0 | 611 | 21.5 | 11.9 | 3,070 | 0.4 | 161 | 5.2 | 2,785 | 36.9 | 23.4 |
| Valley | 1,865 | 74.3 | 110,400 | 17.3 | 12.2 | 579 | 24.6 | 1.9 | 2,124 | -0.7 | 60 | 2.8 | 2,183 | 39.2 | 25.1 |
| Washington | 8,185 | 78.3 | 196,100 | 20.5 | 11.9 | 739 | 24.5 | 0.5 | 11,136 | -1.3 | 407 | 3.7 | 10,468 | 36.0 | 25.2 |
| Wayne | 3,708 | 64.1 | 137,200 | 16.4 | 11.2 | 655 | 27.4 | 0.9 | 5,610 | -3.0 | 156 | 2.8 | 5,334 | 34.2 | 23.0 |
| Webster | 1,507 | 79.6 | 78,400 | 16.3 | 10.8 | 458 | 21.8 | 1.6 | 1,653 | -1.4 | 62 | 3.8 | 1,748 | 38.4 | 30.2 |
| Wheeler | 346 | 67.1 | 97,800 | 30.1 | 13.0 | 508 | 16.7 | 1.7 | 550 | -3.3 | 10 | 1.8 | 400 | 35.3 | 38.5 |
| York | 5,656 | 71.6 | 135,200 | 18.1 | 10.6 | 713 | 23.5 | 1.6 | 7,265 | -0.8 | 244 | 3.4 | 6,869 | 37.3 | 25.1 |
| **NEVADA** | 1,098,602 | 56.3 | 267,900 | 22.1 | 10.0 | 1,107 | 29.8 | 4.5 | 1,530,872 | -2.3 | 196,456 | 12.8 | 1,406,568 | 29.6 | 21.1 |
| Churchill | 9,940 | 65.4 | 191,500 | 19.7 | 10.0 | 821 | 25.4 | 3.2 | 11,074 | -1.3 | 623 | 5.6 | 9,620 | 33.4 | 27.0 |
| Clark | 783,524 | 53.8 | 262,700 | 22.3 | 10.0 | 1,132 | 30.8 | 4.6 | 1,123,582 | -2.3 | 165,513 | 14.7 | 1,031,774 | 28.7 | 19.4 |
| Douglas | 20,741 | 71.4 | 378,800 | 24.3 | 11.0 | 1,102 | 26.2 | 2.9 | 22,465 | -4.5 | 1,934 | 8.6 | 21,421 | 35.9 | 20.2 |
| Elko | 18,065 | 71.2 | 212,500 | 18.2 | 10.0 | 952 | 21.6 | 3.4 | 26,791 | -3.3 | 1,527 | 5.7 | 26,168 | 27.4 | 33.9 |
| Esmeralda | 491 | 58.2 | 65,000 | 15.8 | 10.2 | 551 | 25.4 | 0.8 | 406 | -7.9 | 21 | 5.2 | 398 | 36.2 | 37.2 |
| Eureka | 774 | 73.8 | 120,100 | 16.3 | 10.0 | 0 | 24.9 | 2.1 | 1,040 | -6.7 | 36 | 3.5 | 930 | 25.2 | 46.5 |
| Humboldt | 6,500 | 71.4 | 180,600 | 19.1 | 10.0 | 841 | 24.5 | 4.6 | 8,036 | -4.3 | 395 | 4.9 | 8,491 | 23.7 | 41.2 |
| Lander | 2,198 | 73.4 | 179,900 | 16.6 | 10.0 | 842 | 35.1 | 1.3 | 3,149 | -3.3 | 150 | 4.8 | 2,586 | 23.1 | 45.7 |
| Lincoln | 2,018 | 74.8 | 140,900 | 20.7 | 10.0 | 606 | 13.5 | 2.8 | 2,013 | -5.0 | 103 | 5.1 | 1,803 | 33.6 | 24.5 |
| Lyon | 21,189 | 70.7 | 214,900 | 21.4 | 10.3 | 1,033 | 27.3 | 3.0 | 23,045 | -2.0 | 1,968 | 8.5 | 22,593 | 26.4 | 34.1 |
| Mineral | 2,048 | 70.6 | 98,900 | 19.6 | 10.0 | 610 | 29.1 | 4.4 | 2,093 | 1.2 | 117 | 5.6 | 1,534 | 26.2 | 31.8 |
| Nye | 19,088 | 70.8 | 163,300 | 22.4 | 10.3 | 850 | 28.3 | 4.0 | 17,596 | 0.4 | 1,707 | 9.7 | 14,619 | 23.5 | 32.5 |
| Pershing | 1,948 | 66.8 | 113,000 | 17.5 | 10.0 | 678 | 25.3 | 5.3 | 2,530 | -3.0 | 123 | 4.9 | 2,072 | 30.5 | 38.0 |
| Storey | 1,627 | 90.4 | 240,700 | 23.4 | 10.0 | 669 | 21.0 | 0.2 | 2,050 | -1.8 | 171 | 8.3 | 1,638 | 32.5 | 26.1 |
| Washoe | 182,180 | 58.3 | 334,100 | 22.2 | 10.0 | 1,074 | 28.5 | 4.7 | 254,278 | -2.0 | 19,734 | 7.8 | 231,583 | 34.1 | 23.0 |
| White Pine | 3,516 | 75.6 | 154,100 | 14.8 | 10.0 | 699 | 20.6 | 3.6 | 4,595 | -2.1 | 192 | 4.2 | 3,633 | 31.0 | 30.6 |
| Carson City | 22,755 | 56.8 | 273,800 | 23.3 | 10.0 | 940 | 27.3 | 4.4 | 26,130 | -1.8 | 2,143 | 8.2 | 25,705 | 29.7 | 27.0 |
| **NEW HAMPSHIRE** | 532,037 | 71.1 | 261,700 | 22.4 | 14.7 | 1,111 | 28.6 | 2.0 | 761,732 | -1.5 | 50,915 | 6.7 | 729,701 | 41.3 | 20.9 |
| Belknap | 25,052 | 75.5 | 226,500 | 22.6 | 15.5 | 999 | 29.0 | 1.6 | 31,128 | -1.4 | 2,157 | 6.9 | 31,344 | 38.2 | 21.2 |
| Carroll | 21,456 | 79.3 | 242,800 | 23.3 | 13.4 | 948 | 29.9 | 2.2 | 23,134 | -3.6 | 1,776 | 7.7 | 24,125 | 33.4 | 26.3 |
| Cheshire | 30,141 | 69.4 | 195,300 | 22.8 | 16.3 | 949 | 28.5 | 2.2 | 39,851 | -2.1 | 2,444 | 6.1 | 39,891 | 39.7 | 22.8 |
| Coos | 13,768 | 70.7 | 128,000 | 21.2 | 15.7 | 745 | 27.3 | 2.0 | 14,399 | -2.2 | 1,136 | 7.9 | 14,358 | 32.5 | 24.7 |
| Grafton | 34,727 | 69.3 | 220,300 | 22.8 | 14.9 | 953 | 29.3 | 1.8 | 48,154 | -2.9 | 2,813 | 5.8 | 46,609 | 42.3 | 19.2 |
| Hillsborough | 161,086 | 66 | 274,800 | 22.1 | 14.0 | 1,191 | 28.7 | 2.4 | 240,135 | -1.1 | 16,792 | 7.0 | 227,110 | 42.4 | 20.0 |
| Merrimack | 58,452 | 71.8 | 237,600 | 22.3 | 14.9 | 1,042 | 28.3 | 2.0 | 82,622 | -1.4 | 4,800 | 5.8 | 79,851 | 40.2 | 21.8 |
| Rockingham | 121,045 | 77.7 | 326,300 | 22.4 | 14.7 | 1,228 | 28.4 | 1.5 | 185,979 | -1.4 | 13,147 | 7.1 | 175,088 | 44.4 | 19.3 |
| Strafford | 48,982 | 66 | 235,900 | 22.5 | 15.1 | 1,091 | 28.0 | 2.3 | 73,223 | -1.7 | 4,610 | 6.3 | 70,486 | 38.6 | 21.5 |
| Sullivan | 17,328 | 73.3 | 173,600 | 23.7 | 15.9 | 990 | 29.7 | 1.4 | 23,105 | -0.3 | 1,239 | 5.4 | 20,839 | 35.6 | 28.3 |
| **NEW JERSEY** | 3,231,874 | 63.9 | 335,600 | 24.3 | 16.3 | 1,334 | 30.8 | 3.5 | 4,495,166 | -0.6 | 439,906 | 9.8 | 4,422,491 | 43.1 | 18.6 |
| Atlantic | 99,850 | 67.1 | 217,900 | 26.3 | 18.3 | 1,120 | 34.3 | 3.0 | 121,037 | -0.6 | 21,491 | 17.8 | 127,399 | 32.2 | 18.0 |
| Bergen | 339,856 | 64.8 | 469,500 | 24.9 | 16.9 | 1,506 | 29.2 | 2.6 | 477,892 | -1.4 | 45,703 | 9.6 | 480,028 | 49.4 | 14.7 |
| Burlington | 166,391 | 75.4 | 251,200 | 23.1 | 15.1 | 1,346 | 29.6 | 1.5 | 230,784 | -0.6 | 18,996 | 8.2 | 223,483 | 44.7 | 17.2 |
| Camden | 187,383 | 66.8 | 197,800 | 23.6 | 16.9 | 1,079 | 32.3 | 2.5 | 256,048 | 0.6 | 25,761 | 10.1 | 249,732 | 40.3 | 19.4 |
| Cape May | 40,171 | 77.6 | 300,500 | 26.1 | 15.5 | 1,169 | 37.0 | 1.5 | 45,492 | -1.2 | 6,286 | 13.8 | 41,841 | 36.1 | 18.6 |
| Cumberland | 50,729 | 65.2 | 162,500 | 24.0 | 15.4 | 1,069 | 37.5 | 4.0 | 66,277 | 1.2 | 7,129 | 10.8 | 61,596 | 27.2 | 33.1 |
| Essex | 285,908 | 44 | 386,000 | 26.0 | 18.1 | 1,178 | 32.7 | 4.9 | 373,943 | 0.5 | 43,674 | 11.7 | 378,224 | 39.0 | 19.9 |
| Gloucester | 104,908 | 80.1 | 219,700 | 22.9 | 17.0 | 1,225 | 32.2 | 1.2 | 151,080 | 0.1 | 14,028 | 9.3 | 149,352 | 43.3 | 19.3 |
| Hudson | 258,591 | 31.7 | 378,000 | 26.1 | 17.3 | 1,391 | 28.1 | 8.0 | 360,498 | -0.6 | 38,221 | 10.6 | 357,147 | 42.4 | 20.4 |
| Hunterdon | 47,175 | 82.8 | 415,100 | 22.6 | 14.8 | 1,446 | 30.3 | 1.0 | 63,295 | -2.7 | 4,565 | 7.2 | 66,913 | 52.1 | 13.0 |

1. Specified owner-occupied units lacking complete plumbing facilities.    2. A value of 10.0 represents 10 percent or less; a value of 50.0 represents 50 percent or more.    3. Specified renter-occupied units.    4. Overcrowded or
5. Percent of civilian labor force.    6. Civilian employed persons 16 years old and over.

# Table B. States and Counties — Nonfarm Employment and Agriculture

| STATE County | Number of establishments | Total | Health care and social assistance | Manufacturing | Retail trade | Finance and insurance | Professional, scientific, and technical services | Total (mil dol) | Average per employee (dollars) | Number | Fewer than 50 acres | 1000 acres or more | Farm producers whose primary occupation is farming (percent) |
|---|---|---|---|---|---|---|---|---|---|---|---|---|---|
| | 104 | 105 | 106 | 107 | 108 | 109 | 110 | 111 | 112 | 113 | 114 | 115 | 116 |
| **NEBRASKA— Cont'd** | | | | | | | | | | | | | |
| Platte | 1,031 | 16,796 | 1,883 | 5,925 | 2,379 | 523 | 390 | 715 | 42,577 | 836 | 26.0 | 12.3 | 52.7 |
| Polk | 140 | 895 | 249 | 42 | 134 | 50 | 18 | 30 | 33,522 | 432 | 18.5 | 17.8 | 60.7 |
| Red Willow | 404 | 3,498 | 607 | 296 | 800 | 194 | 124 | 123 | 35,148 | 333 | 34.5 | 32.7 | 48.3 |
| Richardson | 258 | 1,763 | 388 | 271 | 274 | 64 | 69 | 55 | 31,467 | 708 | 22.2 | 15.7 | 44.0 |
| Rock | 47 | 348 | 126 | NA | 42 | 17 | NA | 13 | 36,368 | 220 | 7.7 | 52.3 | 70.5 |
| Saline | 316 | 6,056 | 546 | 2,932 | 518 | 156 | 48 | 269 | 44,478 | 717 | 19.1 | 19.7 | 52.3 |
| Sarpy | 3,827 | 58,654 | 5,790 | 3,095 | 9,680 | 4,314 | 3,174 | 2,740 | 46,713 | 417 | 60.7 | 8.4 | 34.5 |
| Saunders | 557 | 4,002 | 687 | 550 | 646 | 216 | 135 | 147 | 36,619 | 1,118 | 35.1 | 12.6 | 42.7 |
| Scotts Bluff | 1,048 | 12,213 | 2,734 | 973 | 2,081 | 608 | 354 | 479 | 39,197 | 760 | 29.7 | 13.2 | 48.0 |
| Seward | 471 | 5,783 | 811 | 1,204 | 505 | 210 | 129 | 196 | 33,968 | 944 | 39.4 | 12.7 | 40.8 |
| Sheridan | 154 | 952 | 235 | 9 | 214 | 100 | 33 | 29 | 29,972 | 525 | 11.2 | 45.3 | 60.0 |
| Sherman | 88 | 593 | 229 | 18 | 127 | 16 | 14 | 17 | 28,759 | 384 | 16.4 | 22.4 | 48.2 |
| Sioux | 16 | 32 | NA | NA | 8 | 8 | NA | 1 | 43,469 | 307 | 8.1 | 53.7 | 62.5 |
| Stanton | 109 | 981 | 27 | NA | 100 | 26 | 5 | 80 | 81,510 | 571 | 22.1 | 12.3 | 48.9 |
| Thayer | 208 | 1,854 | 361 | 576 | 190 | 92 | 24 | 69 | 37,004 | 414 | 20.3 | 29.0 | 54.6 |
| Thomas | 26 | 264 | NA | NA | 54 | NA | NA | 11 | 42,330 | 90 | 12.2 | 63.3 | 55.6 |
| Thurston | 130 | 1,571 | 304 | 244 | 159 | 70 | 331 | 89 | 56,350 | 309 | 21.7 | 26.5 | 59.7 |
| Valley | 176 | 1,276 | 352 | 81 | 316 | 65 | 71 | 46 | 35,773 | 362 | 18.0 | 29.0 | 53.1 |
| Washington | 598 | 6,253 | 639 | 1,024 | 1,059 | 253 | 252 | 313 | 49,990 | 747 | 44.4 | 11.5 | 41.2 |
| Wayne | 242 | 3,305 | 429 | 1,060 | 409 | 317 | 92 | 107 | 32,297 | 485 | 22.9 | 16.1 | 55.0 |
| Webster | 89 | 616 | 188 | NA | 123 | 25 | 7 | 21 | 34,208 | 406 | 16.7 | 26.1 | 49.8 |
| Wheeler | 20 | 108 | NA | NA | NA | NA | NA | 2 | 19,407 | 215 | 10.7 | 32.6 | 57.0 |
| York | 515 | 6,588 | 943 | 920 | 913 | 363 | 131 | 256 | 38,820 | 521 | 16.3 | 22.1 | 61.4 |
| **NEVADA** | 68,567 | 1,261,577 | 137,402 | 48,919 | 149,441 | 40,786 | 62,359 | 58,122 | 46,071 | 3,423 | 51.7 | 13.7 | 50.4 |
| Churchill | 469 | 5,250 | 694 | 403 | 837 | 142 | 307 | 235 | 44,669 | 504 | 59.3 | 5.4 | 53.1 |
| Clark | 47,435 | 915,555 | 97,560 | 23,191 | 109,793 | 30,964 | 43,129 | 40,977 | 44,756 | 179 | 77.1 | 0.6 | 30.5 |
| Douglas | 1,632 | 17,104 | 1,553 | 1,808 | 2,051 | 347 | 865 | 735 | 43,000 | 239 | 69.5 | 7.9 | 41.0 |
| Elko | 1,122 | 21,159 | 1,770 | 340 | 2,499 | 292 | 451 | 1,253 | 59,195 | 526 | 42.0 | 21.3 | 49.1 |
| Esmeralda | 9 | 145 | NA | NA | NA | NA | NA | 8 | 58,586 | 24 | 37.5 | 29.2 | 55.3 |
| Eureka | 36 | 1,824 | NA | NA | 32 | NA | NA | 172 | 94,357 | 86 | 14.0 | 40.7 | 65.6 |
| Humboldt | 412 | 6,275 | 617 | 233 | 900 | 80 | 88 | 378 | 60,279 | 298 | 33.9 | 24.5 | 59.3 |
| Lander | 96 | 1,050 | 188 | NA | 207 | 11 | 7 | 42 | 40,388 | 117 | 41.0 | 21.4 | 60.3 |
| Lincoln | 77 | 537 | 98 | NA | 201 | 30 | 16 | 16 | 30,682 | 166 | 36.1 | 9.6 | 48.4 |
| Lyon | 826 | 10,096 | 732 | 2,011 | 1,390 | 132 | 345 | 417 | 41,268 | 312 | 56.1 | 8.7 | 54.1 |
| Mineral | 63 | 1,145 | 156 | NA | 99 | NA | 8 | 53 | 46,582 | 59 | 71.2 | 1.7 | 51.2 |
| Nye | 730 | 10,124 | 939 | 78 | 1,553 | 134 | 1,453 | 484 | 47,804 | 211 | 53.6 | 12.8 | 53.3 |
| Pershing | 60 | 1,412 | 136 | 439 | 155 | 13 | 5 | 91 | 64,685 | 154 | 32.5 | 24.0 | 55.3 |
| Storey | 80 | 541 | NA | 91 | 55 | NA | 28 | 19 | 35,296 | 2 | 100.0 | NA | 100.0 |
| Washoe | 12,972 | 204,608 | 27,879 | 17,661 | 25,990 | 5,500 | 11,637 | 10,069 | 49,209 | 353 | 69.7 | 7.4 | 43.3 |
| White Pine | 182 | 2,856 | 300 | NA | 408 | 39 | 22 | 173 | 60,450 | 176 | 42.0 | 19.9 | 46.4 |
| Carson City | 1,890 | 22,261 | 4,274 | 2,529 | 3,155 | 1,030 | 1,035 | 1,033 | 46,403 | 17 | 70.6 | NA | 53.3 |
| **NEW HAMPSHIRE** | 38,494 | 620,164 | 96,722 | 67,926 | 98,230 | 28,929 | 35,315 | 33,048 | 53,289 | 4,123 | 57.1 | 0.8 | 38.6 |
| Belknap | 1,835 | 21,759 | 4,222 | 2,329 | 5,221 | 580 | 627 | 934 | 42,908 | 256 | 50.8 | 0.4 | 41.5 |
| Carroll | 1,862 | 17,130 | 2,586 | 793 | 3,826 | 361 | 514 | 625 | 36,491 | 285 | 51.6 | 1.1 | 41.5 |
| Cheshire | 1,932 | 28,383 | 3,943 | 4,694 | 5,542 | 1,205 | 665 | 1,317 | 46,414 | 420 | 49.8 | 0.7 | 32.1 |
| Coos | 843 | 8,903 | 2,160 | 654 | 1,730 | 250 | 107 | 338 | 37,945 | 272 | 40.4 | 2.2 | 36.8 |
| Grafton | 2,933 | 51,976 | 12,572 | 5,504 | 7,278 | 895 | 1,699 | 2,830 | 54,444 | 462 | 39.8 | 1.7 | 42.5 |
| Hillsborough | 11,049 | 191,526 | 32,464 | 22,802 | 28,831 | 9,356 | 15,703 | 11,051 | 57,699 | 605 | 63.3 | 0.2 | 42.0 |
| Merrimack | 4,123 | 63,450 | 12,848 | 5,948 | 9,989 | 3,172 | 2,788 | 3,119 | 49,152 | 545 | 58.9 | 0.6 | 38.5 |
| Rockingham | 9,918 | 147,801 | 16,895 | 16,457 | 26,290 | 7,275 | 9,940 | 8,126 | 54,980 | 618 | 73.6 | 0.2 | 42.6 |
| Strafford | 2,648 | 40,040 | 7,635 | 5,288 | 7,209 | 3,719 | 1,330 | 1,938 | 48,391 | 310 | 65.5 | 0.3 | 32.1 |
| Sullivan | 911 | 11,551 | 1,213 | 3,457 | 2,277 | 397 | 256 | 530 | 45,898 | 350 | 61.1 | 1.1 | 32.7 |
| **NEW JERSEY** | 233,888 | 3,805,357 | 624,050 | 223,226 | 451,038 | 197,485 | 335,558 | 239,862 | 63,033 | 9,883 | 75.2 | 1.1 | 39.6 |
| Atlantic | 6,187 | 110,453 | 18,860 | 2,237 | 15,662 | 2,458 | 4,007 | 4,638 | 41,991 | 450 | 75.3 | 0.9 | 52.1 |
| Bergen | 31,826 | 446,836 | 79,108 | 31,428 | 51,593 | 15,797 | 37,842 | 28,807 | 64,469 | 74 | 98.6 | NA | 50.4 |
| Burlington | 10,501 | 189,086 | 29,930 | 12,981 | 24,850 | 17,336 | 15,497 | 9,744 | 52,090 | 915 | 72.3 | 2.0 | 47.3 |
| Camden | 11,356 | 187,058 | 45,994 | 13,070 | 23,881 | 4,613 | 12,004 | 10,599 | 56,052 | 197 | 81.2 | NA | 38.8 |
| Cape May | 3,807 | 27,578 | 4,322 | 626 | 5,978 | 1,138 | 1,031 | 1,218 | 44,180 | 164 | 73.8 | NA | 48.5 |
| Cumberland | 2,835 | 47,003 | 10,299 | 7,946 | 6,772 | 1,030 | 1,021 | 1,930 | 41,066 | 560 | 63.6 | 2.0 | 48.0 |
| Essex | 18,913 | 298,755 | 60,828 | 17,353 | 26,657 | 20,801 | 21,663 | 19,320 | 64,669 | 22 | 100.0 | NA | 28.9 |
| Gloucester | 6,127 | 97,974 | 15,491 | 9,694 | 17,048 | 1,779 | 3,768 | 18,655 | 76,734 | 580 | 72.6 | 1.2 | 46.5 |
| Hudson | 13,858 | 243,114 | 30,792 | 8,365 | 24,884 | 39,645 | 13,444 | 4,359 | 44,490 | 4 | 100.0 | NA | 50.0 |
| Hunterdon | 3,726 | 46,480 | 9,655 | 3,578 | 6,655 | 3,509 | 4,110 | 2,917 | 62,759 | 1,604 | 75.3 | 0.7 | 34.7 |

# Table B. States and Counties — **Agriculture**

| STATE County | Land in farms Acreage (1,000) | Percent change, 2012-2017 | Acres Average size of farm | Total irrigated (1,000) | Total cropland (1,000) | Value of land and buildings (dollars) Average per farm | Average per acre | Value of machinery and equipment, average per farm (dollars) | Value of products sold: Total (mil dol) | Average per farm (acres) | Percent from: Crops | Livestock and poultry products | Organic farms (number) | Farms with internet access (percent) | Government payments Total ($1,000) | Percent of farms |
|---|---|---|---|---|---|---|---|---|---|---|---|---|---|---|---|---|
| | 117 | 118 | 119 | 120 | 121 | 122 | 123 | 124 | 125 | 126 | 127 | 128 | 129 | 130 | 131 | 132 |
| NEBRASKA— Cont'd | | | | | | | | | | | | | | | | |
| Platte | 384 | -10.0 | 459 | 191.8 | 336.4 | 2,903,960 | 6,328 | 345,260 | 688.6 | 823,639 | 27.8 | 72.2 | 6 | 79.7 | 12,585 | 75.1 |
| Polk | 251 | 2.3 | 581 | 164.2 | 225.0 | 3,434,576 | 5,911 | 369,062 | 330.7 | 765,505 | 39.4 | 60.6 | 1 | 85.6 | 7,235 | 83.1 |
| Red Willow | 439 | 4.7 | 1,319 | 52.7 | 248.2 | 2,568,525 | 1,947 | 307,287 | 188.2 | 565,150 | 38.2 | 61.8 | NA | 82.6 | 4,049 | 57.1 |
| Richardson | 342 | 7.1 | 483 | 11.2 | 272.4 | 2,061,048 | 4,267 | 202,966 | 149.3 | 210,944 | 78.5 | 21.5 | 3 | 79.4 | 8,915 | 75.3 |
| Rock | 584 | -9.4 | 2,655 | 35.2 | 121.4 | 3,019,965 | 1,138 | 221,861 | 108.1 | 491,364 | 20.6 | 79.4 | 1 | 87.3 | 1,581 | 41.4 |
| Saline | 360 | -0.4 | 503 | 135.7 | 305.0 | 2,370,146 | 4,716 | 285,163 | 206.9 | 288,576 | 69.9 | 30.1 | 1 | 77.3 | 9,762 | 80.6 |
| Sarpy | 99 | 8.5 | 239 | 11.5 | 93.0 | 1,774,545 | 7,439 | 127,332 | 54.9 | 131,707 | 91.8 | 8.2 | 9 | 83.2 | 1,930 | 47.0 |
| Saunders | 480 | 2.2 | 429 | 135.6 | 436.2 | 2,420,851 | 5,641 | 235,724 | 360.5 | 322,419 | 62.6 | 37.4 | 7 | 79.2 | 12,010 | 68.5 |
| Scotts Bluff | 442 | -0.8 | 581 | 170.7 | 218.1 | 1,184,110 | 2,038 | 231,435 | 322.7 | 424,641 | 30.6 | 69.4 | 3 | 84.1 | 5,405 | 54.2 |
| Seward | 363 | 2.4 | 385 | 138.7 | 310.5 | 2,198,064 | 5,710 | 252,569 | 251.0 | 265,841 | 61.5 | 38.5 | NA | 79.3 | 10,627 | 71.6 |
| Sheridan | 1,562 | 1.8 | 2,974 | 74.1 | 301.5 | 2,884,505 | 970 | 210,422 | 150.6 | 286,863 | 38.2 | 61.8 | 2 | 80.8 | 5,522 | 62.9 |
| Sherman | 311 | 10.5 | 809 | 93.3 | 159.0 | 2,165,691 | 2,676 | 223,342 | 139.3 | 362,885 | 58.1 | 41.9 | NA | 82.0 | 3,109 | 62.5 |
| Sioux | 1,230 | 0.4 | 4,006 | 41.8 | 97.6 | 3,607,478 | 901 | 175,807 | 133.3 | 434,283 | 13.2 | 86.8 | NA | 86.3 | 2,690 | 49.8 |
| Stanton | 266 | 4.6 | 466 | 34.5 | 203.7 | 2,327,041 | 4,991 | 214,703 | 208.4 | 364,956 | 45.4 | 54.6 | 2 | 79.3 | 7,411 | 79.5 |
| Thayer | 326 | -0.2 | 787 | 128.4 | 251.0 | 3,083,516 | 3,920 | 360,804 | 227.7 | 550,041 | 56.5 | 43.5 | 2 | 77.5 | 8,364 | 79.0 |
| Thomas | 388 | 5.6 | 4,313 | 2.6 | 7.4 | 3,364,667 | 780 | 159,890 | 24.4 | 271,067 | 8.3 | 91.7 | NA | 94.4 | D | 20.0 |
| Thurston | 232 | -6.3 | 751 | 14.9 | 206.8 | 3,625,111 | 4,827 | 377,357 | 207.3 | 670,731 | 49.1 | 50.9 | 1 | 76.7 | 6,000 | 79.3 |
| Valley | 351 | 0.4 | 969 | 111.3 | 172.6 | 2,788,775 | 2,878 | 303,059 | 223.9 | 618,483 | 34.7 | 65.3 | 4 | 82.6 | 4,664 | 68.8 |
| Washington | 248 | -0.1 | 332 | 22.4 | 215.9 | 2,163,878 | 6,519 | 197,031 | 150.4 | 201,325 | 71.8 | 28.2 | 1 | 86.7 | 2,839 | 58.0 |
| Wayne | 281 | 0.5 | 580 | 61.6 | 253.6 | 3,196,693 | 5,512 | 310,983 | 223.8 | 461,466 | 57.7 | 42.3 | 1 | 85.8 | 8,593 | 70.9 |
| Webster | 329 | 8.8 | 810 | 68.1 | 206.2 | 2,219,552 | 2,739 | 254,116 | 347.9 | 856,778 | 22.3 | 77.7 | 2 | 78.3 | 5,547 | 71.9 |
| Wheeler | 357 | 0.0 | 1,662 | 36.9 | 87.8 | 2,539,799 | 1,528 | 258,222 | 283.1 | 1,316,963 | 7.4 | 92.6 | 2 | 88.4 | 2,418 | 54.9 |
| York | 347 | 2.1 | 665 | 275.4 | 327.3 | 4,575,694 | 6,878 | 498,031 | 340.9 | 654,309 | 59.8 | 40.2 | NA | 84.6 | 14,518 | 77.2 |
| NEVADA | 6,128 | 3.6 | 1,790 | 790.4 | 794.7 | 1,627,858 | 909 | 155,033 | 665.8 | 194,495 | 41.5 | 58.5 | 51 | 82.9 | 5,049 | 9.1 |
| Churchill | 250 | 26.7 | 496 | 45.0 | 47.4 | 949,440 | 1,915 | 106,008 | 90.7 | 179,938 | 21.9 | 78.1 | 3 | 85.1 | 427 | 11.9 |
| Clark | D | D | D | 3.7 | 4.0 | 1,439,603 | 3,109 | 76,922 | 12.7 | 70,676 | 90.2 | 9.8 | 1 | 74.9 | 16 | 2.2 |
| Douglas | 118 | 17.2 | 495 | 34.7 | 22.0 | 1,017,364 | 2,055 | 108,770 | 23.5 | 98,464 | 25.6 | 74.4 | NA | 89.5 | 130 | 4.6 |
| Elko | 2,180 | 2.5 | 4,145 | 199.9 | 215.0 | 2,276,052 | 549 | 136,223 | 72.2 | 137,213 | 15.4 | 84.6 | 1 | 85.0 | 351 | 3.6 |
| Esmeralda | D | D | D | 15.0 | 14.4 | 2,138,151 | 2,083 | 517,623 | 12.0 | 498,083 | 96.2 | 3.8 | NA | 83.3 | 98 | 45.8 |
| Eureka | 579 | -9.4 | 6,729 | 50.4 | 59.8 | 3,362,856 | 500 | 423,423 | 40.4 | 470,128 | 73.2 | 26.8 | 1 | 84.9 | 1,066 | 11.6 |
| Humboldt | 990 | 22.4 | 3,323 | 162.6 | 158.3 | 2,945,636 | 887 | 309,287 | 105.7 | 354,534 | 71.2 | 28.8 | 5 | 89.6 | 747 | 19.5 |
| Lander | 329 | 4.9 | 2,815 | 37.7 | 46.7 | 2,951,893 | 1,049 | 227,081 | 31.8 | 272,043 | 67.1 | 32.9 | 7 | 74.4 | 144 | 12.8 |
| Lincoln | 66 | D | 399 | 24.2 | 20.6 | 1,028,189 | 2,576 | 237,470 | 22.0 | 132,404 | 58.6 | 41.4 | 2 | 73.5 | 10 | 3.0 |
| Lyon | 181 | -50.5 | 581 | 55.8 | 65.1 | 1,328,718 | 2,286 | 147,374 | 102.7 | 329,179 | 28.3 | 71.7 | 13 | 85.3 | 839 | 9.9 |
| Mineral | D | D | D | 2.0 | 2.2 | 1,168,722 | 293 | 66,181 | D | D | D | D | NA | 57.6 | 150 | 27.1 |
| Nye | 93 | 43.4 | 442 | 21.9 | 22.4 | 766,261 | 1,732 | 128,095 | 65.0 | 307,924 | 8.0 | 92.0 | 3 | 71.6 | 30 | 1.9 |
| Pershing | 330 | 10.4 | 2,145 | 42.0 | 56.8 | 1,717,125 | 801 | 189,476 | 33.8 | 219,610 | 58.3 | 41.7 | 5 | 58.0 | 689 | 26.0 |
| Storey | D | D | D | NA | NA | D | D | D | D | D | NA | D | NA | 83.8 | | |
| Washoe | 501 | 13.2 | 1,420 | 43.6 | 22.2 | 1,457,532 | 1,026 | 62,049 | 19.9 | 56,484 | 45.8 | 54.2 | 14 | 100.0 | NA | NA |
| White Pine | 165 | -14.5 | 939 | 51.5 | 37.2 | 1,245,500 | 1,326 | 168,506 | 30.0 | 170,682 | 38.8 | 61.2 | NA | 86.4 | 174 | 5.1 |
| Carson City | 1 | D | 57 | 0.4 | 0.6 | D | D | D | D | D | D | D | 1 | 79.5 | 177 | 5.7 |
| | | | | | | | | | | | | | | 100.0 | NA | NA |
| NEW HAMPSHIRE | 425 | -10.3 | 103 | 2.2 | 108.0 | 539,732 | 5,231 | 68,629 | 187.8 | 45,548 | 57.4 | 42.6 | 156 | 87.2 | 3,494 | 7.5 |
| Belknap | 25 | 6.4 | 99 | 0.1 | 5.2 | 487,697 | 4,914 | 62,719 | 7.6 | 29,754 | 56.7 | 43.3 | 5 | 85.9 | 148 | 9.0 |
| Carroll | 32 | 10.6 | 114 | 0.3 | 4.4 | 447,525 | 3,929 | 47,048 | 5.5 | 19,446 | 67.5 | 32.5 | 15 | 82.8 | 295 | 8.4 |
| Cheshire | 54 | -15.3 | 128 | 0.1 | 10.7 | 484,018 | 3,791 | 72,235 | 14.6 | 34,683 | 47.6 | 52.4 | 23 | 91.7 | 473 | 7.6 |
| Coos | 47 | -17.4 | 172 | 0.1 | 15.4 | 442,111 | 2,564 | 70,608 | 16.4 | 60,301 | 32.5 | 67.5 | 9 | 81.6 | 420 | 7.0 |
| Grafton | 74 | -10.7 | 159 | 0.1 | 16.3 | 622,137 | 3,907 | 74,022 | 23.4 | 50,587 | 30.3 | 69.7 | 10 | 84.6 | 426 | 9.5 |
| Hillsborough | 44 | -7.3 | 73 | 0.6 | 11.7 | 568,208 | 7,775 | 67,925 | 18.8 | 31,030 | 77.2 | 22.8 | 18 | 93.1 | 281 | 7.8 |
| Merrimack | 54 | -16.4 | 100 | 0.4 | 17.5 | 577,576 | 5,798 | 99,615 | 49.3 | 90,541 | 74.1 | 25.9 | 29 | 87.3 | 492 | 6.6 |
| Rockingham | 32 | -10.5 | 52 | 0.3 | 10.3 | 588,366 | 11,281 | 53,046 | 22.4 | 36,243 | 72.6 | 27.4 | 21 | 86.2 | 621 | 6.0 |
| Strafford | 23 | -24.4 | 75 | 0.1 | 5.6 | 485,607 | 6,493 | 53,461 | 10.5 | 33,803 | 61.6 | 38.4 | 13 | 91.3 | 171 | 5.8 |
| Sullivan | 39 | 1.2 | 113 | 0.1 | 10.9 | 540,738 | 4,792 | 71,638 | 19.3 | 55,146 | 34.4 | 65.6 | 13 | 81.4 | 168 | 8.0 |
| NEW JERSEY | 734 | 2.7 | 74 | 86.8 | 463.0 | 1,000,464 | 13,469 | 86,532 | 1,098 | 111,095 | 89.7 | 10.3 | 122 | 80.9 | 7,503 | 7.5 |
| Atlantic | 29 | 306.2 | 64 | 11.6 | 17.8 | 823,031 | 12,764 | 135,393 | 120.7 | 268,162 | 98.7 | 1.3 | 10 | 85.3 | 198 | 7.8 |
| Bergen | 1 | -26.6 | 14 | 0.1 | 0.3 | 1,412,811 | 99,475 | 48,545 | D | D | 100.0 | D | 1 | 79.7 | D | 1.4 |
| Burlington | 96 | 0.4 | 105 | 12.4 | 49.7 | 1,057,462 | 10,052 | 92,669 | 98.6 | 107,738 | 92.4 | 7.6 | 6 | 82.2 | 828 | 7.4 |
| Camden | 9 | 26.5 | 47 | 2.3 | 5.0 | 774,929 | 16,419 | 81,602 | 22.9 | 116,208 | 99.6 | 0.4 | 5 | 83.2 | D | 1.5 |
| Cape May | 8 | -87.4 | 50 | 1.4 | 3.8 | 722,281 | 14,561 | 59,087 | 9.8 | 59,988 | 89.2 | 10.8 | NA | 81.1 | D | 1.2 |
| Cumberland | 66 | 53.1 | 118 | 20.0 | 49.6 | 1,159,637 | 9,801 | 150,219 | 212.6 | 379,730 | 97.5 | 2.5 | 8 | 77.5 | 665 | 9.6 |
| Essex | 0 | 49.2 | 9 | 0.0 | 0.1 | 733,010 | 84,431 | 108,943 | D | D | D | D | NA | 59.1 | NA | NA |
| Gloucester | 49 | -51.5 | 85 | 8.7 | 35.6 | 1,079,229 | 12,676 | 107,289 | 102.5 | 176,645 | 92.6 | 7.4 | 1 | 78.6 | 1,196 | 12.6 |
| Hudson | 0 | -100.0 | 7 | NA | D | 327,000 | 50,308 | 35,385 | D | D | D | D | NA | 100.0 | NA | NA |
| Hunterdon | 101 | 600.6 | 63 | 1.8 | 65.6 | 986,211 | 15,617 | 66,403 | 92.2 | 57,510 | 85.5 | 14.5 | 31 | 82.6 | 524 | 5.9 |

# Table B. States and Counties — Water Use, Wholesale Trade, Retail Trade, and Real Estate

| STATE County | Water use, 2015 | | Wholesale Trade[1], 2017 | | | | Retail Trade[2], 2017 | | | | Real estate and rental and leasing,[2] 2017 | | | |
|---|---|---|---|---|---|---|---|---|---|---|---|---|---|---|
| | Public supply water withdrawn (mil gal/day) | Public supply gallons withdrawn per person per day | Number of establishments | Number of employees | Sales (mil dol) | Average payroll (mil dol) | Number of establishments | Number of employees | Sales (mil dol) | Average payroll (mil dol) | Number of establishments | Number of employees | Sales (mil dol) | Average payroll (mil dol) |
| | 133 | 134 | 135 | 136 | 137 | 138 | 139 | 140 | 141 | 142 | 143 | 144 | 145 | 146 |
| NEBRASKA— Cont'd | | | | | | | | | | | | | | |
| Platte | 4.96 | 151.0 | 55 | 584 | 615.8 | 34.1 | 164 | 2,238 | 557.5 | 56.2 | 39 | 147 | 21.5 | 4.9 |
| Polk | 0.39 | 75.0 | 10 | 105 | 89.0 | 5.1 | 20 | 137 | 48.6 | 3.1 | NA | NA | NA | NA |
| Red Willow | 2.34 | 216.1 | 23 | 223 | 183.1 | 9.2 | 79 | 890 | 217.1 | 21.9 | 10 | 35 | 3.7 | 0.4 |
| Richardson | 0.74 | 91.4 | 28 | 140 | 279.6 | 5.7 | 42 | 296 | 68.7 | 7.0 | 4 | 9 | 1.3 | 0.2 |
| Rock | 0.16 | 115.9 | D | D | D | D | 8 | 47 | 10.2 | 0.8 | NA | NA | NA | NA |
| Saline | 1.36 | 95.2 | 23 | 196 | 258.1 | 10.9 | 46 | 533 | 157.8 | 14.2 | 7 | 16 | 3.4 | 0.4 |
| Sarpy | 30.65 | 174.5 | 195 | 3,329 | 3,359.0 | 195.2 | 432 | 9,545 | 2,762.4 | 263.9 | 187 | 742 | 204.8 | 30.3 |
| Saunders | 49.56 | 2,358.2 | 29 | 242 | 288.7 | 12.9 | 70 | 650 | 213.1 | 17.1 | 17 | 25 | 3.1 | 0.6 |
| Scotts Bluff | 5.66 | 156.1 | D | D | D | D | 163 | D | 571.3 | D | 45 | 139 | 23.8 | 4.6 |
| Seward | 1.39 | 81.2 | 33 | 284 | 309.3 | 14.4 | 48 | 516 | 117.7 | 12.3 | 11 | 16 | 1.9 | 0.5 |
| Sheridan | 0.59 | 113.0 | 13 | 161 | 51.2 | 4.2 | 45 | 227 | 55.3 | 4.3 | NA | NA | NA | NA |
| Sherman | 0.23 | 74.4 | D | D | D | 1.4 | 21 | 138 | 43.1 | 3.5 | NA | NA | NA | NA |
| Sioux | 0.08 | 63.5 | NA | NA | NA | NA | 3 | D | 2.6 | D | NA | NA | NA | NA |
| Stanton | 0.36 | 60.6 | NA | NA | NA | NA | 13 | 127 | 18.7 | 2.2 | NA | NA | NA | NA |
| Thayer | 0.81 | 156.9 | 25 | 242 | 348.7 | 13.0 | 35 | 188 | 35.1 | 3.3 | D | D | D | D |
| Thomas | 0.11 | 160.8 | NA | NA | NA | NA | D | D | D | D | NA | NA | NA | NA |
| Thurston | 0.60 | 84.9 | 8 | 85 | 177.1 | 5.3 | 20 | 179 | 56.9 | 6.1 | 4 | 10 | 2.9 | 0.2 |
| Valley | 1.25 | 300.9 | 6 | D | 169.4 | D | 29 | 314 | 98.6 | 8.1 | NA | NA | NA | NA |
| Washington | 12.18 | 601.5 | D | D | D | 8.1 | 61 | 1,081 | 771.1 | 41.1 | 22 | 31 | 4.6 | 0.8 |
| Wayne | 0.87 | 92.9 | 14 | 157 | 116.4 | 8.0 | 36 | 418 | 71.3 | 7.5 | 6 | 16 | 1.3 | 0.5 |
| Webster | 0.39 | 107.6 | D | D | D | D | 20 | 130 | 27.9 | 2.6 | NA | NA | NA | NA |
| Wheeler | 0.04 | 53.3 | 3 | 54 | 10.1 | 0.9 | NA | NA | NA | NA | NA | NA | NA | NA |
| York | 1.87 | 135.4 | 38 | 323 | 306.9 | 15.0 | 62 | 919 | 269.9 | 24.4 | D | D | D | D |
| NEVADA | 558.26 | 193.1 | 2,665 | 32,879 | 25,395.7 | 1,915.2 | 8,745 | 145,773 | 45,110.7 | 4,220.2 | 4,684 | 30,562 | 7,359.0 | 1,312.2 |
| Churchill | 3.38 | 139.7 | 24 | 115 | 86.3 | 3.9 | 58 | 844 | 249.7 | 25.2 | 28 | 100 | 15.4 | 3.0 |
| Clark | 432.47 | 204.5 | 1,743 | 19,687 | 15,088.0 | 1,212.5 | 6,267 | 107,067 | 32,047.7 | 3,038.7 | 3,454 | 24,634 | 6,064.6 | 1,085.2 |
| Douglas | 14.08 | 295.1 | 51 | 345 | 180.2 | 19.8 | 152 | 2,101 | 632.4 | 56.0 | 113 | 655 | 104.7 | 25.9 |
| Elko | 12.61 | 242.8 | 65 | 800 | 1,060.9 | 53.8 | 163 | 2,547 | 864.2 | 73.9 | 53 | 296 | 53.3 | 9.7 |
| Esmeralda | 0.11 | 132.7 | NA | NA | NA | NA | NA | NA | NA | NA | NA | NA | NA | NA |
| Eureka | 0.32 | 158.7 | NA | NA | NA | NA | 5 | 42 | 4.7 | 0.6 | NA | NA | NA | NA |
| Humboldt | 3.68 | 216.2 | D | D | D | D | 66 | 919 | 314.2 | 24.3 | D | D | D | D |
| Lander | 1.56 | 264.3 | 3 | 26 | 62.6 | 1.3 | 23 | 204 | 48.1 | 4.5 | NA | NA | NA | NA |
| Lincoln | 1.16 | 230.3 | NA | NA | NA | NA | D | D | D | D | NA | NA | NA | NA |
| Lyon | 9.39 | 178.6 | 48 | 981 | 364.2 | 30.3 | 105 | 1,306 | 555.1 | 37.3 | D | D | D | D |
| Mineral | 1.05 | 234.5 | NA | NA | NA | NA | 15 | 110 | 27.5 | 3.1 | NA | NA | NA | NA |
| Nye | 6.29 | 148.1 | 15 | 189 | 308.3 | 6.8 | 116 | 1,527 | 483.6 | 40.9 | 42 | 100 | 13.6 | 2.8 |
| Pershing | 1.27 | 191.4 | NA | NA | NA | NA | 14 | 131 | 48.1 | 3.1 | NA | NA | NA | NA |
| Storey | 0.53 | 132.9 | NA | NA | NA | NA | 18 | 48 | 10.8 | 1.4 | NA | NA | NA | NA |
| Washoe | 57.68 | 129.1 | D | D | D | D | 1,477 | 25,264 | 8,507.8 | 795.2 | D | D | D | D |
| White Pine | 1.18 | 120.3 | 6 | 54 | 32.9 | 3.5 | 30 | 345 | 114.9 | 10.4 | 6 | D | 1.5 | D |
| Carson City | 11.50 | 210.9 | 81 | 620 | 335.2 | 29.4 | 220 | 3,141 | 1,170.8 | 102.3 | 118 | 361 | 94.1 | 13.0 |
| NEW HAMPSHIRE | 95.52 | 71.8 | 1,509 | 22,314 | 20,328.4 | 1,577.3 | 6,032 | 96,591 | 30,039.4 | 2,804.6 | 1,523 | 7,920 | 2,000.3 | 381.4 |
| Belknap | 3.12 | 51.5 | D | D | D | D | 321 | 4,991 | 1,706.5 | 155.3 | 80 | 261 | 57.9 | 10.5 |
| Carroll | 3.45 | 73.0 | D | D | D | D | 372 | 3,768 | 948.1 | 103.9 | 93 | 328 | 53.2 | 12.0 |
| Cheshire | 4.12 | 54.3 | D | D | D | D | 346 | 5,486 | 1,869.5 | 166.7 | 72 | 247 | 50.7 | 9.6 |
| Coos | 4.20 | 134.6 | D | D | D | D | 164 | 1,756 | 545.1 | 46.9 | 27 | 76 | 15.0 | 3.9 |
| Grafton | 7.57 | 84.8 | 78 | 700 | 806.8 | 46.1 | 495 | 7,320 | 2,251.8 | 220.0 | 154 | 544 | 96.1 | 21.1 |
| Hillsborough | 38.74 | 95.3 | 531 | 6,897 | 5,718.9 | 498.3 | 1,577 | 28,064 | 9,169.1 | 836.6 | 454 | 2,938 | 616.4 | 142.0 |
| Merrimack | 8.26 | 55.8 | 150 | 3,601 | 4,215.3 | 205.6 | 613 | 9,896 | 3,161.9 | 292.3 | 144 | 950 | 436.7 | 60.0 |
| Rockingham | 15.35 | 50.9 | 476 | 7,641 | 7,418.9 | 616.7 | 1,600 | 26,342 | 7,783.6 | 713.6 | 360 | 1,945 | 522.4 | 98.9 |
| Strafford | 8.00 | 63.1 | 76 | 1,120 | 418.7 | 68.5 | 387 | 6,695 | 1,959.3 | 205.1 | 104 | 332 | 73.1 | 12.1 |
| Sullivan | 2.71 | 63.1 | 35 | 559 | 539.5 | 52.1 | 157 | 2,273 | 644.5 | 64.4 | 35 | 299 | 78.9 | 11.4 |
| NEW JERSEY | 1,175.42 | 131.2 | 12,289 | 230,006 | 312,405.1 | 18,970.9 | 31,200 | 469,615 | 149,171.3 | 13,453.1 | 9,622 | 61,052 | 21,184.3 | 3,435.2 |
| Atlantic | 30.20 | 110.1 | 172 | 2,049 | 1,304.6 | 115.8 | 1,078 | 16,535 | 4,631.1 | 417.9 | 233 | 1,358 | 356.2 | 56.3 |
| Bergen | 127.42 | 135.8 | 2,627 | 40,918 | 96,148.0 | 3,090.6 | 3,703 | 53,862 | 19,846.4 | 1,686.1 | 1,485 | 8,053 | 2,890.4 | 471.0 |
| Burlington | 54.20 | 120.4 | 480 | 11,758 | 13,279.4 | 753.7 | 1,344 | 26,240 | 8,223.1 | 698.2 | 428 | 4,210 | 1,168.0 | 261.4 |
| Camden | 42.83 | 83.8 | 513 | 8,577 | 6,671.1 | 527.9 | 1,694 | 24,257 | 6,810.6 | 637.2 | 425 | 2,450 | 782.0 | 123.8 |
| Cape May | 13.61 | 143.7 | D | D | D | D | 639 | 7,307 | 1,935.5 | 182.6 | 202 | 779 | 217.0 | 32.5 |
| Cumberland | 15.72 | 100.9 | 153 | 4,405 | 2,833.3 | 200.5 | 492 | 7,027 | 2,011.0 | 182.1 | 102 | 486 | 108.7 | 18.9 |
| Essex | 31.22 | 39.2 | 953 | 15,257 | 15,695.0 | 1,097.4 | 2,667 | 28,700 | 9,070.9 | 921.6 | 915 | 6,233 | 1,820.3 | 274.8 |
| Gloucester | 17.52 | 60.1 | 299 | 7,933 | 11,213.8 | 512.1 | 912 | 17,236 | 4,813.2 | 461.8 | 175 | 1,374 | 472.5 | 80.4 |
| Hudson | 0.00 | 0.0 | 667 | 17,076 | 19,699.7 | 1,121.9 | 2,120 | 27,508 | 7,964.9 | 720.1 | 706 | 4,069 | 1,509.0 | 250.7 |
| Hunterdon | 119.9 | 955.5 | 141 | 1,318 | 907.6 | 83.9 | 458 | 7,917 | 2,478.7 | 230.0 | 127 | 417 | 146.8 | 21.5 |

1   Merchant wholesalers, except manufacturers' sales branches and offices.   2. Employer establishments.

# Table B. States and Counties — Professional Services, Manufacturing, and Accommodation and Food Services

| STATE County | Prof., sci., tech. services 2017 — No. of establishments (147) | No. of employees (148) | Sales (mil dol) (149) | Average payroll (mil dol) (150) | Manufacturing 2017 — No. of establishments (151) | No. of employees (152) | Sales (mil dol) (153) | Average payroll (mil dol) (154) | Accommodation and food services 2017 — No. of establishments (155) | No. of employees (156) | Sales (mil dol) (157) | Annual payroll (mil dol) (158) |
|---|---|---|---|---|---|---|---|---|---|---|---|---|
| **NEBRASKA— Cont'd** | | | | | | | | | | | | |
| Platte | 63 | 390 | 42 | 18.4 | 71 | 5,681 | 3,621.0 | 313.2 | 75 | 1,159 | 54.6 | 14.8 |
| Polk | 14 | 23 | 2 | 0.5 | 6 | 46 | 17.1 | 2.1 | 8 | 47 | 1.9 | 0.4 |
| Red Willow | 26 | 118 | 14 | 4.9 | 10 | 260 | 105.8 | 13.4 | 33 | 527 | 23.6 | 7.3 |
| Richardson | 11 | 65 | 4 | 1.7 | 14 | 264 | 53.9 | 12.1 | 22 | 178 | 6.6 | 2.0 |
| Rock | NA | NA | NA | NA | NA | NA | NA | NA | NA | NA | NA | NA |
| Saline | 17 | 44 | 5 | 1.4 | 14 | 2,820 | 1,879.5 | 151.4 | 25 | 286 | 11.3 | 3.0 |
| Sarpy | D | D | D | D | 91 | 2,991 | 1,176.3 | 178.9 | 279 | 5,693 | 301.5 | 89.4 |
| Saunders | 46 | 129 | 15 | 6.7 | 19 | 331 | 81.9 | 14.6 | D | D | D | D |
| Scotts Bluff | D | D | D | D | 34 | 959 | 280.5 | 39.8 | 107 | 1,511 | 72.8 | 21.3 |
| Seward | 28 | 118 | 21 | 4.2 | 22 | 1,129 | 397.7 | 55.8 | 31 | 457 | 19.9 | 6.2 |
| Sheridan | 9 | 20 | 3 | 0.6 | 3 | 5 | 2.9 | 0.3 | 20 | 84 | 5.7 | 1.6 |
| Sherman | 5 | 8 | 1 | 0.2 | 3 | 15 | 1.9 | 0.5 | 10 | 88 | 4.1 | 1.3 |
| Sioux | NA | NA | NA | NA | NA | NA | NA | NA | 10 | 88 | 4.1 | 1.3 |
| Stanton | 4 | 7 | 1 | 0.2 | NA | NA | NA | NA | NA | NA | NA | NA |
| Thayer | 10 | 28 | 2 | 0.7 | D | D | D | D | 4 | 22 | 1.4 | 0.4 |
| Thomas | NA | NA | NA | NA | NA | NA | NA | NA | 16 | D | 2.7 | D |
| Thurston | 12 | 192 | 29 | 11.0 | 7 | 295 | 60.6 | 15.7 | 3 | 6 | 0.3 | 0.1 |
| Valley | 15 | 65 | 8 | 2.6 | 7 | 70 | 103.5 | 3.7 | 4 | 26 | 0.9 | 0.2 |
| Washington | 56 | 249 | 36 | 12.2 | 26 | 1,302 | 1,549.8 | 86.0 | 12 | D | 3.4 | D |
| Wayne | 15 | 76 | 13 | 4.3 | 13 | 981 | 240.6 | 34.2 | 39 | 424 | 16.1 | 5.0 |
| Webster | 3 | 4 | 0 | 0.1 | NA | NA | NA | NA | 25 | 350 | 13.3 | 3.3 |
| Wheeler | NA | NA | NA | NA | NA | NA | NA | NA | 6 | D | 1.3 | D |
| York | 25 | 141 | 17 | 5.6 | 26 | 862 | 333.5 | 42.8 | 7 | D | 0.8 | D |
| **NEVADA** | 8,991 | 59,549 | 10,763 | 3,922.7 | 1,828 | 44,182 | 16,408.1 | 2,516.7 | 6,810 | 319,584 | 33,979.9 | 9,578.4 |
| Churchill | 37 | 311 | 48 | 18.7 | 16 | 373 | 223.2 | 27.0 | D | D | D | D |
| Clark | 6,477 | 45,018 | 8,261 | 2,991.7 | 992 | 19,985 | 6,795.1 | 974.2 | 4,839 | 268,757 | 30,017.1 | 8,468.5 |
| Douglas | 243 | 817 | 133 | 43.9 | 81 | 1,822 | 960.0 | 143.2 | 136 | 5,833 | 556.2 | 160.4 |
| Elko | D | D | D | D | D | D | D | D | 154 | 5,310 | 440.0 | 124.8 |
| Esmeralda | NA | NA | NA | NA | NA | NA | NA | NA | NA | NA | NA | NA |
| Eureka | NA | NA | NA | NA | NA | NA | NA | NA | 5 | 24 | 1.7 | 0.4 |
| Humboldt | D | D | D | D | 11 | 215 | 216.2 | 12.1 | 61 | 1,136 | 87.1 | 20.4 |
| Lander | D | D | D | 0.2 | NA | NA | NA | NA | 16 | 139 | 7.0 | 2.0 |
| Lincoln | D | D | D | 0.6 | NA | NA | NA | NA | D | D | D | 1.3 |
| Lyon | D | D | D | D | 79 | 1,830 | 1,004.6 | 100.3 | 70 | 846 | 56.3 | 13.3 |
| Mineral | D | D | D | D | NA | NA | NA | NA | 8 | 142 | 9.1 | 2.3 |
| Nye | D | D | 234 | D | 15 | 55 | 34.6 | 2.3 | 82 | 1,500 | 108.3 | 26.1 |
| Pershing | NA | NA | NA | NA | D | D | D | D | 9 | 99 | 4.3 | 1.4 |
| Storey | 13 | 21 | 4 | 1.4 | 5 | 69 | 31.3 | 3.6 | 18 | 118 | 7.4 | 2.4 |
| Washoe | D | D | D | D | 486 | 16,584 | 6,200.8 | 1,064.5 | 1,141 | 31,445 | 2,414.0 | 682.0 |
| White Pine | D | D | D | D | D | 11 | D | D | 36 | 459 | 31.3 | 9.2 |
| Carson City | D | D | D | D | 103 | 2,704 | 714.3 | 159.4 | 170 | 3,036 | 193.0 | 54.1 |
| **NEW HAMPSHIRE** | 3,638 | 32,034 | 5,537 | 2,361.0 | 1,790 | 65,211 | 20,304.1 | 3,875.2 | 3,784 | 59,531 | 3,722.0 | 1,132.7 |
| Belknap | D | D | D | D | 77 | 2,653 | 725.7 | 175.7 | 248 | 3,082 | 212.4 | 65.9 |
| Carroll | D | D | D | D | 69 | 689 | 181.3 | 37.2 | 296 | 4,103 | 302.2 | 96.3 |
| Cheshire | D | D | D | D | 124 | 4,316 | 1,135.8 | 231.9 | 185 | 2,547 | 151.1 | 47.0 |
| Coos | D | D | D | D | 39 | 590 | 190.1 | 27.2 | 112 | 2,041 | 95.4 | 37.7 |
| Grafton | D | D | D | D | 100 | 5,254 | 1,793.6 | 294.3 | 421 | 5,317 | 370.7 | 109.7 |
| Hillsborough | 1,334 | 12,550 | 2,565 | 1,158.2 | 518 | 21,931 | 7,038.6 | 1,432.2 | 922 | 16,755 | 973.2 | 292.3 |
| Merrimack | D | D | D | D | 203 | 5,681 | 1,752.7 | 327.3 | 325 | 5,425 | 329.7 | 101.3 |
| Rockingham | D | D | D | D | 429 | 15,793 | 5,506.0 | 891.7 | 902 | 15,041 | 1,001.4 | 296.5 |
| Strafford | D | D | D | D | 145 | 5,183 | 1,293.6 | 283.1 | 300 | 4,400 | 239.3 | 71.2 |
| Sullivan | D | D | D | D | 86 | 3,121 | 686.6 | 174.7 | 73 | 820 | 46.3 | 14.7 |
| **NEW JERSEY** | 28,863 | 323,560 | 66,882 | 28,572.7 | 7,332 | 219,835 | 95,483.0 | 13,908.6 | 21,495 | 318,734 | 23,785.1 | 6,431.5 |
| Atlantic | D | D | D | D | 108 | 2,013 | 474.1 | 98.6 | 830 | 35,488 | 3,550.7 | 1,003.8 |
| Bergen | 3,989 | 35,448 | 8,287 | 2,887.9 | 1,020 | 29,416 | 11,519.6 | 1,870.1 | 2,560 | 34,688 | 2,559.8 | 675.9 |
| Burlington | 1,297 | 13,902 | 2,507 | 963.0 | 319 | 14,429 | 5,853.6 | 1,060.9 | 924 | 15,270 | 902.3 | 251.6 |
| Camden | 1,358 | 13,321 | 2,709 | 980.5 | 389 | 11,295 | 3,701.2 | 671.3 | 1,055 | 15,647 | 913.3 | 251.8 |
| Cape May | D | D | D | D | 76 | 592 | 110.1 | 26.9 | 931 | 6,065 | 757.0 | 206.6 |
| Cumberland | D | D | D | D | 158 | 8,509 | 2,612.5 | 408.6 | 243 | 3,541 | 184.5 | 49.1 |
| Essex | 2,227 | 26,109 | 5,501 | 2,364.9 | 648 | 16,233 | 6,122.4 | 933.4 | 1,687 | 23,991 | 1,883.3 | 509.1 |
| Gloucester | 497 | 3,848 | 507 | 229.2 | 243 | 9,518 | 8,256.3 | 576.8 | 544 | 10,177 | 562.2 | 150.9 |
| Hudson | D | D | D | D | 366 | 8,600 | 2,850.2 | 405.5 | 1,564 | 19,766 | 1,579.9 | 413.6 |
| Hunterdon | D | D | D | D | 129 | 3,550 | 1,262.4 | 214.7 | 330 | 4,215 | 256.5 | 71.5 |

## Table B. States and Counties — Health Care and Social Assistance, Other Services, Nonemployer Businesses, and Residential Construction

| STATE County | Health care and social assistance, 2017 | | | | Other services, 2017 | | | | Nonemployer businesses, 2018 | | Value of residential construction authorized by building permits, 2020 | |
|---|---|---|---|---|---|---|---|---|---|---|---|---|
| | Number of establishments | Number of employees | Receipts (mil dol) | Annual payroll (mil dol) | Number of establishments | Number of employees | Receipts (mil dol) | Annual payroll (mil dol) | Number | Receipts (mil dol) | New construction ($1,000) | Number of housing units |
| | 159 | 160 | 161 | 162 | 163 | 164 | 165 | 166 | 167 | 168 | 169 | 170 |
| **NEBRASKA— Cont'd** | | | | | | | | | | | | |
| Platte | 97 | 2,099 | 295.5 | 92.2 | 78 | 393 | 37.7 | 10.5 | 2,381 | 111.0 | 26,465 | 111 |
| Polk | 9 | 238 | 16.1 | 7.5 | D | D | 3.6 | D | 470 | 19.3 | 1,013 | 5 |
| Red Willow | 43 | 650 | 72.7 | 28.0 | 27 | 80 | 8.3 | 2.1 | 933 | 38.0 | 960 | 5 |
| Richardson | 23 | 429 | 41.0 | 15.2 | 23 | 68 | 6.3 | 1.2 | 585 | 18.9 | 375 | 2 |
| Rock | D | D | D | D | D | D | 0.5 | D | 217 | 11.7 | 0 | 0 |
| Saline | 27 | 524 | 47.9 | 19.3 | 27 | 75 | 10.5 | 1.8 | 829 | 38.6 | 13,121 | 78 |
| Sarpy | 357 | 5,406 | 495.4 | 190.5 | 250 | 1,633 | 187.5 | 51.4 | 11,261 | 444.8 | 367,602 | 1,650 |
| Saunders | 30 | 622 | 51.7 | 21.0 | 30 | 77 | 10.1 | 2.1 | 1,872 | 80.2 | 34,131 | 147 |
| Scotts Bluff | 133 | 3,031 | 385.7 | 135.7 | D | D | D | D | 2,559 | 115.2 | 2,001 | 9 |
| Seward | 44 | 869 | 68.3 | 28.2 | 41 | 141 | 17.0 | 4.2 | 1,349 | 46.9 | 21,570 | 70 |
| Sheridan | 17 | 244 | 17.1 | 8.8 | 13 | 41 | 3.1 | 0.6 | 428 | 16.8 | 212 | 3 |
| Sherman | 6 | 147 | 15.0 | 5.1 | D | D | 1.3 | D | 260 | 12.6 | 1,245 | 5 |
| Sioux | NA | NA | NA | NA | NA | NA | 3.0 | D | 107 | 3.1 | 0 | 0 |
| Stanton | D | D | D | D | D | D | D | 1.5 | 509 | 25.4 | 4,306 | 14 |
| Thayer | 11 | 412 | 32.7 | 14.6 | D | D | D | NA | 421 | 20.8 | 1,687 | 7 |
| Thomas | NA | NA | NA | NA | 8 | 34 | 3.4 | 0.6 | 90 | 3.2 | 713 | 3 |
| Thurston | 12 | 325 | 39.3 | 17.2 | D | D | D | 1.4 | 293 | 14.8 | 1,426 | 6 |
| Valley | 18 | 387 | 28.7 | 13.1 | D | D | D | D | 399 | 22.1 | 1,110 | 5 |
| Washington | 34 | 794 | 57.3 | 26.0 | 43 | 215 | 15.2 | 6.0 | 1,628 | 74.3 | 21,055 | 71 |
| Wayne | 24 | 475 | 36.5 | 15.5 | D | D | D | D | 644 | 27.7 | 2,200 | 17 |
| Webster | 9 | 173 | 12.6 | 4.9 | 4 | 17 | 2.2 | 0.3 | 277 | 9.3 | 452 | 2 |
| Wheeler | NA | NA | NA | NA | NA | NA | NA | NA | 84 | 2.9 | 1,189 | 5 |
| York | 39 | 948 | 103.1 | 38.4 | 55 | 261 | 26.1 | 6.9 | 1,100 | 47.8 | 6,988 | 24 |
| **NEVADA** | 7,372 | 132,093 | 18,111.8 | 6,620.4 | 4,032 | 28,825 | 3,281.5 | 914.2 | 245,381 | 12,646.4 | 4,057,456 | 19,716 |
| Churchill | 43 | 811 | 80.3 | 34.3 | 41 | 181 | 13.9 | 4.1 | 1,265 | 58.6 | 20,777 | 99 |
| Clark | 5,349 | 94,935 | 13,080.9 | 4,731.3 | 2,784 | 20,781 | 2,256.4 | 638.8 | 189,386 | 9,286.9 | 2,824,227 | 14,100 |
| Douglas | 119 | 1,721 | 201.5 | 70.2 | 89 | 488 | 53.6 | 15.2 | 5,382 | 356.4 | 91,111 | 200 |
| Elko | 124 | 1,639 | 178.2 | 67.9 | D | D | D | D | 2,567 | 120.3 | 28,573 | 180 |
| Esmeralda | NA | NA | NA | NA | NA | NA | NA | NA | 58 | 1.5 | NA | NA |
| Eureka | NA | NA | NA | NA | NA | NA | NA | NA | 129 | 7.1 | NA | NA |
| Humboldt | 36 | 664 | 64.8 | 24.7 | 25 | 108 | 18.3 | 4.3 | 880 | 33.0 | 2,313 | 15 |
| Lander | D | D | D | D | 5 | 19 | 3.4 | 0.7 | 253 | 9.2 | 719 | 6 |
| Lincoln | 7 | D | 11.3 | D | NA | NA | NA | NA | 312 | 9.9 | 2,082 | 12 |
| Lyon | 56 | 597 | 53.7 | 22.1 | 38 | 196 | 20.3 | 6.1 | 2,788 | 137.3 | 79,963 | 329 |
| Mineral | D | D | D | D | D | D | D | D | 160 | 6.5 | 0 | 0 |
| Nye | 71 | 916 | 93.0 | 31.1 | 55 | 305 | 19.0 | 5.9 | 2,462 | 98.8 | NA | NA |
| Pershing | D | D | D | D | NA | NA | NA | NA | 238 | 7.6 | 0 | 0 |
| Storey | NA | NA | NA | NA | D | D | D | D | 323 | 15.3 | 1,407 | 15 |
| Washoe | D | D | D | D | D | D | D | D | 33,976 | 2,089.6 | 947,203 | 4,489 |
| White Pine | 15 | 285 | 45.4 | 17.7 | D | D | D | D | 450 | 15.5 | 1,492 | 8 |
| Carson City | 215 | 3,865 | 623.8 | 214.4 | 132 | 618 | 100.6 | 22.3 | 4,752 | 392.9 | 57,589 | 263 |
| **NEW HAMPSHIRE** | 3,737 | 94,594 | 11,931.6 | 4,741.7 | 3,011 | 17,864 | 2,096.2 | 619.2 | 108,327 | 6,435.5 | 1,059,391 | 4,320 |
| Belknap | 159 | 4,153 | 457.7 | 188.2 | 151 | 554 | 57.5 | 15.9 | 5,767 | 366.6 | 79,666 | 340 |
| Carroll | 152 | 2,498 | 254.0 | 103.8 | 125 | 505 | 56.3 | 15.8 | 5,807 | 325.2 | 128,744 | 291 |
| Cheshire | 164 | 4,174 | 471.7 | 185.5 | 141 | 979 | 96.9 | 28.4 | 5,889 | 332.3 | 29,489 | 130 |
| Coos | 102 | 2,317 | 224.3 | 97.0 | 59 | 298 | 29.9 | 10.0 | 2,271 | 99.2 | 11,487 | 73 |
| Grafton | 283 | 10,452 | 1,995.6 | 617.9 | 186 | 1,059 | 120.6 | 33.3 | 8,120 | 477.7 | 143,151 | 582 |
| Hillsborough | 1,167 | 32,354 | 3,938.1 | 1,643.7 | 883 | 6,272 | 746.1 | 231.2 | 29,810 | 1,759.0 | 181,799 | 868 |
| Merrimack | 415 | 13,052 | 1,428.6 | 640.8 | 406 | 2,245 | 333.0 | 91.0 | 11,453 | 641.7 | 119,127 | 477 |
| Rockingham | 896 | 16,733 | 2,047.9 | 830.7 | 760 | 4,285 | 466.8 | 139.0 | 27,736 | 1,841.6 | 274,350 | 1,166 |
| Strafford | 309 | 7,661 | 1,004.1 | 386.4 | 219 | 1,368 | 135.9 | 43.4 | 8,313 | 438.7 | 66,794 | 317 |
| Sullivan | 90 | 1,200 | 109.5 | 47.7 | 81 | 299 | 53.1 | 11.2 | 3,161 | 153.6 | 24,783 | 76 |
| **NEW JERSEY** | 28,005 | 613,406 | 74,723.4 | 30,134.1 | 19,162 | 113,846 | 14,025.6 | 3,679.3 | 745,483 | 44,074.1 | 4,796,438 | 36,146 |
| Atlantic | 801 | 18,708 | 2,202.5 | 951.0 | 540 | 3,437 | 291.1 | 83.9 | 17,834 | 887.3 | 155,745 | 667 |
| Bergen | 3,959 | 78,752 | 10,992.7 | 4,254.3 | 2,618 | 14,352 | 1,624.8 | 440.1 | 101,881 | 7,092.7 | 529,852 | 3,380 |
| Burlington | 1,283 | 28,853 | 3,281.1 | 1,296.8 | 781 | 4,731 | 413.8 | 134.5 | 30,302 | 1,759.9 | 163,524 | 1,749 |
| Camden | 1,559 | 44,197 | 5,497.5 | 2,316.9 | 954 | 6,282 | 641.4 | 190.4 | 32,429 | 1,677.7 | 77,956 | 895 |
| Cape May | 244 | 4,578 | 467.7 | 198.8 | 308 | 1,268 | 123.7 | 35.5 | 8,493 | 536.4 | 243,196 | 675 |
| Cumberland | 405 | 11,257 | 1,121.9 | 531.6 | 240 | 1,246 | 120.9 | 31.9 | 6,139 | 287.3 | 46,103 | 179 |
| Essex | 2,535 | 58,553 | 8,099.0 | 3,070.3 | 1,717 | 11,204 | 1,366.0 | 376.7 | 73,493 | 3,911.8 | 333,175 | 2,705 |
| Gloucester | 775 | 15,590 | 1,682.4 | 719.3 | 537 | 3,331 | 279.5 | 90.8 | 17,422 | 854.3 | 89,665 | 709 |
| Hudson | 1,480 | 28,379 | 3,091.1 | 1,210.4 | 1,187 | 5,368 | 608.1 | 150.5 | 61,548 | 2,953.1 | 950,776 | 6,950 |
| Hunterdon | 363 | 9,579 | 1,005.4 | 433.2 | 296 | 1,442 | 138.3 | 42.2 | 12,208 | 855.0 | 40,136 | 319 |

# Table B. States and Counties — Government Employment and Payroll, and Local Government Finances

| STATE County | Government employment and payroll, 2017 | | | | | | | | | Local government finances, 2017 | | | | |
| | | | March payroll (percent of total) | | | | | | | General revenue | | | | |
| | | | | | | | | | | | | Taxes | | |
| | | | | | | | | | | | | | Per capita[1] (dollars) | |
| | Full-time equivalent employees | March payroll (dollars) | Administration, judicial, and legal | Police and corrections | Fire protection | Highways and transportation | Health and welfare | Natural resources and utilities | Education and libraries | Total (mil dol) | Intergovernmental (mil dol) | Total (mil dol) | Total | Property |
| | 171 | 172 | 173 | 174 | 175 | 176 | 177 | 178 | 179 | 180 | 181 | 182 | 183 | 184 |
|---|---|---|---|---|---|---|---|---|---|---|---|---|---|---|
| **NEBRASKA— Cont'd** | | | | | | | | | | | | | | |
| Platte | 3,283 | 23,192,574 | 1.1 | 2.1 | 0.4 | 1.0 | 0.2 | 81.4 | 13.7 | 155.4 | 32.1 | 72.9 | 2,193 | 1,563 |
| Polk | 404 | 1,568,770 | 5.3 | 1.7 | 0.0 | 3.8 | 15.2 | 12.9 | 61.1 | 45.5 | 9.1 | 25.0 | 4,709 | 4,265 |
| Red Willow | 475 | 1,797,139 | 6.3 | 9.7 | 1.5 | 3.3 | 2.9 | 15.2 | 60.5 | 54.7 | 16.7 | 23.2 | 2,160 | 1,694 |
| Richardson | 441 | 1,401,615 | 8.0 | 7.5 | 0.7 | 8.0 | 1.5 | 8.9 | 63.4 | 44.6 | 11.2 | 29.1 | 3,649 | 3,276 |
| Rock | 140 | 475,568 | 6.5 | 7.0 | 0.0 | 4.1 | 43.9 | 0.7 | 36.3 | 8.3 | 2.0 | 5.1 | 3,596 | 3,292 |
| Saline | 666 | 2,570,074 | 6.8 | 10.1 | 0.3 | 4.9 | 1.4 | 5.2 | 70.6 | 93.1 | 22.7 | 41.0 | 2,853 | 2,475 |
| Sarpy | 5,171 | 21,553,098 | 7.3 | 11.7 | 3.4 | 2.3 | 0.8 | 2.1 | 69.9 | 600.0 | 186.0 | 294.0 | 1,621 | 1,307 |
| Saunders | 910 | 3,350,353 | 8.1 | 10.1 | 0.4 | 4.8 | 22.4 | 4.5 | 48.3 | 121.0 | 20.6 | 51.6 | 2,452 | 2,149 |
| Scotts Bluff | 2,351 | 8,529,773 | 4.6 | 7.2 | 1.1 | 2.8 | 1.3 | 9.8 | 71.4 | 218.0 | 82.0 | 77.7 | 2,150 | 1,749 |
| Seward | 755 | 2,861,348 | 7.0 | 6.5 | 0.0 | 4.6 | 1.1 | 8.7 | 70.1 | 75.6 | 16.1 | 44.6 | 2,592 | 2,253 |
| Sheridan | 517 | 1,874,447 | 14.1 | 4.2 | 0.0 | 5.9 | 21.6 | 11.2 | 42.7 | 45.5 | 6.3 | 17.0 | 3,224 | 2,930 |
| Sherman | 163 | 516,043 | 10.1 | 9.1 | 0.0 | 9.7 | 2.3 | 2.0 | 65.5 | 22.9 | 4.0 | 17.7 | 5,757 | 5,464 |
| Sioux | 54 | 188,007 | 16.7 | 3.9 | 0.0 | 15.8 | 0.0 | 0.7 | 60.6 | 5.6 | 1.6 | 3.6 | 2,983 | 2,751 |
| Stanton | 143 | 596,334 | 11.3 | 7.5 | 0.0 | 13.2 | 1.1 | 19.2 | 47.3 | 24.6 | 5.4 | 9.4 | 1,575 | 1,451 |
| Thayer | 367 | 1,508,260 | 6.2 | 3.2 | 0.1 | 5.1 | 47.1 | 2.5 | 35.7 | 74.8 | 9.3 | 36.7 | 7,281 | 6,635 |
| Thomas | 55 | 117,868 | 27.0 | 4.0 | 0.2 | 6.1 | 0.0 | 14.1 | 43.7 | 4.7 | 1.3 | 3.1 | 4,316 | 4,170 |
| Thurston | 686 | 2,767,589 | 4.3 | 1.8 | 0.0 | 2.3 | 27.4 | 0.9 | 63.0 | 88.0 | 42.0 | 13.6 | 1,892 | 1,538 |
| Valley | 465 | 1,847,702 | 4.6 | 2.6 | 0.0 | 2.8 | 41.4 | 18.8 | 29.6 | 51.3 | 5.7 | 16.2 | 3,839 | 3,029 |
| Washington | 685 | 2,820,809 | 6.9 | 10.5 | 0.0 | 5.5 | 0.6 | 2.8 | 71.0 | 97.6 | 21.0 | 52.2 | 2,567 | 2,250 |
| Wayne | 354 | 1,441,303 | 8.7 | 5.3 | 0.0 | 11.1 | 1.0 | 6.5 | 66.7 | 41.6 | 9.7 | 26.2 | 2,848 | 2,511 |
| Webster | 220 | 736,498 | 10.1 | 4.4 | 0.0 | 6.3 | 25.1 | 7.6 | 45.7 | 23.7 | 4.3 | 11.8 | 3,375 | 3,081 |
| Wheeler | 49 | 157,709 | 14.3 | 0.0 | 0.0 | 15.3 | 0.0 | 0.1 | 70.0 | 4.8 | 1.1 | 3.5 | 4,240 | 4,045 |
| York | 572 | 2,345,805 | 6.0 | 7.1 | 2.7 | 7.9 | 1.9 | 20.4 | 52.2 | 85.2 | 22.1 | 42.6 | 3,092 | 2,375 |
| **NEVADA** | X | X | X | X | X | X | X | X | X | X | X | X | X | X |
| Churchill | 833 | 3,650,301 | 12.0 | 14.2 | 0.7 | 2.2 | 1.1 | 12.7 | 45.5 | 108.9 | 45.8 | 25.8 | 1,074 | 810 |
| Clark | 61,549 | 328,902,480 | 8.5 | 16.4 | 5.5 | 3.3 | 10.2 | 7.9 | 46.5 | 10,404.2 | 4,197.7 | 3,614.1 | 1,657 | 853 |
| Douglas | 1,529 | 6,567,049 | 11.1 | 16.1 | 7.1 | 2.1 | 1.8 | 11.3 | 45.6 | 208.9 | 82.1 | 89.9 | 1,872 | 1,491 |
| Elko | 1,864 | 8,958,548 | 10.3 | 10.4 | 3.2 | 2.5 | 1.9 | 5.6 | 63.0 | 248.3 | 142.4 | 69.9 | 1,336 | 969 |
| Esmeralda | 88 | 249,417 | 22.1 | 19.3 | 0.6 | 11.0 | 1.5 | 4.1 | 39.6 | 23.2 | 18.9 | 2.9 | 3,400 | 3,356 |
| Eureka | 146 | 809,801 | 15.9 | 10.5 | 0.0 | 8.0 | 5.2 | 10.5 | 48.4 | 32.8 | 9.5 | 21.1 | 10,845 | 10,791 |
| Humboldt | 846 | 4,090,004 | 9.3 | 12.5 | 0.2 | 2.8 | 28.7 | 3.7 | 41.0 | 124.2 | 56.8 | 31.0 | 1,852 | 1,636 |
| Lander | 369 | 1,632,378 | 9.8 | 11.5 | 0.0 | 4.7 | 36.0 | 3.3 | 34.2 | 72.3 | 14.9 | 38.4 | 6,882 | 6,615 |
| Lincoln | 303 | 1,433,875 | 8.6 | 12.3 | 0.1 | 3.0 | 19.1 | 12.1 | 42.2 | 29.0 | 18.4 | 5.7 | 1,099 | 1,024 |
| Lyon | 1,686 | 6,408,283 | 8.3 | 7.8 | 5.1 | 1.6 | 3.2 | 5.0 | 65.9 | 174.5 | 104.0 | 47.4 | 880 | 684 |
| Mineral | 299 | 1,163,326 | 7.1 | 6.2 | 0.9 | 2.1 | 42.1 | 7.2 | 33.4 | 31.2 | 11.1 | 5.5 | 1,231 | 1,112 |
| Nye | 1,264 | 5,334,954 | 13.8 | 13.8 | 4.4 | 3.3 | 1.1 | 1.5 | 58.9 | 147.7 | 85.1 | 49.6 | 1,127 | 915 |
| Pershing | 339 | 1,413,302 | 11.8 | 11.5 | 0.1 | 3.7 | 25.9 | 5.2 | 38.5 | 34.9 | 14.5 | 8.0 | 1,231 | 1,160 |
| Storey | 179 | 955,773 | 19.4 | 15.9 | 16.3 | 4.0 | 1.6 | 2.7 | 33.3 | 35.5 | 8.9 | 18.0 | 4,519 | 4,384 |
| Washoe | 13,349 | 63,720,984 | 10.1 | 13.5 | 5.8 | 4.3 | 4.5 | 8.7 | 50.6 | 2,074.6 | 935.8 | 702.1 | 1,538 | 979 |
| White Pine | 441 | 2,113,814 | 9.9 | 9.3 | 2.9 | 3.5 | 38.6 | 4.6 | 29.4 | 81.9 | 27.2 | 16.7 | 1,730 | 1,348 |
| Carson City | 1,652 | 7,251,836 | 10.3 | 13.8 | 7.8 | 4.2 | 2.8 | 7.3 | 52.0 | 235.3 | 121.5 | 63.2 | 1,159 | 789 |
| **NEW HAMPSHIRE** | X | X | X | X | X | X | X | X | X | X | X | X | X | X |
| Belknap | 2,697 | 10,844,945 | 5.6 | 10.6 | 5.3 | 4.6 | 5.1 | 2.9 | 65.3 | 367.3 | 76.0 | 240.6 | 3,953 | 3,913 |
| Carroll | 2,052 | 7,917,481 | 6.0 | 9.6 | 3.1 | 4.6 | 4.0 | 5.2 | 65.7 | 314.9 | 72.7 | 218.1 | 4,531 | 4,476 |
| Cheshire | 3,237 | 12,506,344 | 6.3 | 8.7 | 2.3 | 4.1 | 7.5 | 3.0 | 66.5 | 410.2 | 99.0 | 265.5 | 3,495 | 3,452 |
| Coos | 1,570 | 5,804,393 | 5.0 | 7.6 | 3.3 | 4.3 | 16.7 | 5.1 | 57.4 | 180.8 | 60.2 | 94.9 | 3,022 | 3,010 |
| Grafton | 4,302 | 16,931,264 | 6.1 | 9.6 | 4.2 | 5.6 | 4.7 | 3.7 | 65.4 | 503.2 | 112.4 | 343.8 | 3,826 | 3,764 |
| Hillsborough | 13,924 | 65,255,845 | 4.2 | 10.8 | 6.1 | 5.1 | 3.7 | 4.7 | 63.9 | 1,758.2 | 438.5 | 1,096.8 | 2,654 | 2,624 |
| Merrimack | 5,953 | 24,716,272 | 5.8 | 10.4 | 4.1 | 3.8 | 7.4 | 3.7 | 62.8 | 665.2 | 175.8 | 416.1 | 2,777 | 2,762 |
| Rockingham | 12,488 | 50,374,277 | 4.8 | 9.9 | 5.9 | 2.3 | 2.4 | 2.8 | 69.8 | 1,401.9 | 293.6 | 977.1 | 3,186 | 3,149 |
| Strafford | 4,373 | 18,041,005 | 5.8 | 10.9 | 4.8 | 3.3 | 7.5 | 3.4 | 61.8 | 569.7 | 151.1 | 323.8 | 2,494 | 2,423 |
| Sullivan | 1,799 | 7,006,104 | 5.7 | 8.6 | 2.3 | 4.8 | 10.4 | 3.8 | 63.1 | 204.5 | 55.2 | 133.8 | 3,110 | 3,068 |
| **NEW JERSEY** | X | X | X | X | X | X | X | X | X | X | X | X | X | X |
| Atlantic | 13,889 | 75,732,427 | 5.6 | 14.7 | 5.0 | 2.0 | 4.6 | 4.0 | 62.7 | 1,644.3 | 578.1 | 930.0 | 3,503 | 3,445 |
| Bergen | 33,707 | 213,265,581 | 3.6 | 14.1 | 1.8 | 2.1 | 2.4 | 3.7 | 71.0 | 5,499.1 | 681.6 | 4,069.0 | 4,364 | 4,300 |
| Burlington | 16,622 | 96,494,090 | 4.4 | 8.2 | 1.8 | 3.1 | 1.0 | 3.2 | 76.7 | 2,196.7 | 589.7 | 1,322.5 | 2,968 | 2,929 |
| Camden | 21,189 | 121,311,072 | 4.8 | 14.5 | 2.6 | 4.2 | 3.8 | 2.9 | 62.8 | 3,377.4 | 1,071.2 | 1,407.7 | 2,781 | 2,753 |
| Cape May | 6,404 | 30,726,844 | 9.5 | 11.4 | 3.5 | 4.4 | 5.6 | 8.4 | 56.0 | 972.3 | 143.8 | 573.0 | 6,152 | 5,996 |
| Cumberland | 7,181 | 38,072,311 | 3.2 | 10.2 | 0.7 | 1.1 | 5.7 | 5.9 | 71.6 | 897.5 | 508.3 | 274.3 | 1,812 | 1,776 |
| Essex | 28,131 | 169,331,200 | 5.6 | 20.5 | 6.6 | 1.4 | 6.8 | 5.7 | 51.9 | 4,358.2 | 1,295.3 | 2,465.0 | 3,095 | 2,947 |
| Gloucester | 12,646 | 66,611,744 | 3.4 | 9.5 | 0.7 | 2.0 | 5.3 | 4.7 | 72.2 | 1,513.6 | 427.0 | 878.7 | 3,020 | 2,971 |
| Hudson | 19,609 | 113,757,554 | 5.5 | 23.4 | 8.6 | 1.9 | 6.1 | 4.0 | 47.1 | 3,179.1 | 1,130.6 | 1,338.1 | 1,989 | 1,898 |
| Hunterdon | 4,818 | 27,005,860 | 5.0 | 6.0 | 0.3 | 4.2 | 1.0 | 2.2 | 80.1 | 746.8 | 83.6 | 492.1 | 3,946 | 3,905 |

1. Based on the resident population estimated as of July 1 of the year shown.

# Table B. States and Counties — Local Government Finances, Government Employment, and Income Taxes

| STATE County | Total (mil dol) 185 | Per capita¹ (dollars) 186 | Education 187 | Health and hospitals 188 | Police protection 189 | Public welfare 190 | Highways 191 | Total (mil dol) 192 | Per capita¹ (dollars) 193 | Federal civilian 194 | Federal military 195 | State and local 196 | Number of returns 197 | Mean adjusted gross income 198 | Mean income tax 199 |
|---|---|---|---|---|---|---|---|---|---|---|---|---|---|---|---|
| | | | | | | Percent of total for: | | Debt outstanding | | Government employment, 2019 | | | Individual income tax returns, 2018 | | |
| **NEBRASKA— Cont'd** | | | | | | | | | | | | | | | |
| Platte | 147 | 4,417 | 57.2 | 0.1 | 4.2 | 0.0 | 7.0 | 2,579.0 | 77,607 | 93 | 113 | 2,688 | 16,750 | 59,210 | 5,271 |
| Polk | 49 | 9,162 | 59.4 | 13.9 | 1.3 | 0.4 | 10.3 | 51.3 | 9,656 | 23 | 17 | 527 | 2,570 | 57,241 | 4,684 |
| Red Willow | 62 | 5,794 | 35.9 | 0.6 | 3.8 | 13.0 | 6.8 | 35.0 | 3,252 | 80 | 35 | 1,045 | 4,860 | 52,448 | 5,039 |
| Richardson | 38 | 4,763 | 52.2 | 0.5 | 5.1 | 0.8 | 13.6 | 24.5 | 3,070 | 37 | 26 | 628 | 3,660 | 52,549 | 4,236 |
| Rock | 10 | 7,154 | 50.4 | 0.2 | 2.8 | 0.3 | 14.6 | 1.3 | 935 | 3 | 5 | 237 | 680 | 30,128 | 3,088 |
| Saline | 97 | 6,716 | 44.9 | 9.4 | 4.1 | 0.0 | 10.6 | 88.0 | 6,129 | 57 | 46 | 1,444 | 6,490 | 53,122 | 4,004 |
| Sarpy | 602 | 3,317 | 56.5 | 0.3 | 6.6 | 0.4 | 5.6 | 956.1 | 5,271 | 3,032 | 6,540 | 7,181 | 88,800 | 74,040 | 7,633 |
| Saunders | 114 | 5,405 | 35.6 | 28.3 | 2.4 | 0.1 | 16.2 | 71.1 | 3,379 | 98 | 73 | 1,574 | 10,270 | 69,967 | 6,951 |
| Scotts Bluff | 254 | 7,015 | 53.2 | 0.3 | 3.9 | 1.6 | 5.1 | 159.9 | 4,423 | 151 | 119 | 3,173 | 16,370 | 52,575 | 4,692 |
| Seward | 85 | 4,914 | 50.0 | 0.3 | 5.0 | 0.9 | 10.2 | 81.8 | 4,758 | 48 | 55 | 1,095 | 7,670 | 68,288 | 6,493 |
| Sheridan | 37 | 7,090 | 34.2 | 28.9 | 2.6 | 0.7 | 7.8 | 20.9 | 3,958 | 26 | 17 | 605 | 2,300 | 38,225 | 3,077 |
| Sherman | 12 | 3,895 | 68.4 | 0.4 | 1.1 | 0.5 | 6.2 | 2.0 | 656 | 15 | 10 | 205 | 1,420 | 36,441 | 2,780 |
| Sioux | 6 | 4,769 | 54.9 | 0.2 | 3.2 | 0.2 | 19.0 | 1.2 | 963 | 9 | 4 | 73 | 510 | 32,314 | 2,610 |
| Stanton | 24 | 4,060 | 25.8 | 0.0 | 7.2 | 0.3 | 41.5 | 18.0 | 3,008 | 18 | 20 | 273 | 2,730 | 55,468 | 4,767 |
| Thayer | 58 | 11,577 | 41.8 | 30.9 | 2.1 | 2.6 | 6.8 | 15.3 | 3,044 | 30 | 17 | 640 | 2,370 | 56,069 | 4,973 |
| Thomas | 4 | 6,097 | 58.5 | 0.0 | 2.4 | 0.0 | 6.5 | 2.2 | 3,112 | 10 | 2 | 89 | 330 | 31,597 | 2,661 |
| Thurston | 88 | 12,219 | 52.6 | 31.3 | 0.9 | 0.1 | 4.5 | 58.2 | 8,085 | 138 | 25 | 1,812 | 2,700 | 42,451 | 3,580 |
| Valley | 48 | 11,449 | 21.8 | 42.3 | 1.3 | 0.1 | 10.3 | 62.8 | 14,854 | 30 | 14 | 571 | 1,920 | 37,375 | 3,864 |
| Washington | 91 | 4,469 | 54.3 | 0.0 | 3.6 | 0.1 | 15.1 | 194.4 | 9,564 | 52 | 70 | 1,171 | 9,910 | 81,147 | 9,388 |
| Wayne | 47 | 5,114 | 48.3 | 0.3 | 2.9 | 0.7 | 18.3 | 34.9 | 3,791 | 33 | 29 | 1,048 | 3,760 | 58,992 | 5,411 |
| Webster | 22 | 6,300 | 40.7 | 23.8 | 3.0 | 0.2 | 12.2 | 9.7 | 2,766 | 25 | 11 | 305 | 1,560 | 44,683 | 3,097 |
| Wheeler | 5 | 6,368 | 52.0 | 0.0 | 2.1 | 0.0 | 24.6 | 0.4 | 551 | 3 | 3 | 69 | 360 | 33,847 | 3,325 |
| York | 78 | 5,642 | 34.4 | 0.3 | 5.0 | 0.6 | 19.8 | 66.9 | 4,863 | 53 | 44 | 1,093 | 6,510 | 61,012 | 5,650 |
| **NEVADA** | X | X | X | X | X | X | X | X | X | 19,777 | 18,932 | 141,472 | 1,449,500 | 73,350 | 9,640 |
| Churchill | 120 | 4,984 | 33.8 | 0.6 | 8.8 | 5.0 | 2.6 | 63.0 | 2,624 | 580 | 542 | 1,331 | 11,750 | 53,981 | 5,040 |
| Clark | 9,856 | 4,517 | 33.6 | 7.7 | 10.6 | 2.6 | 8.8 | 23,841.2 | 10,928 | 13,733 | 16,332 | 91,703 | 1,049,220 | 71,163 | 9,346 |
| Douglas | 205 | 4,261 | 31.9 | 1.1 | 9.3 | 2.3 | 4.8 | 132.4 | 2,756 | 87 | 126 | 2,234 | 26,920 | 95,499 | 14,702 |
| Elko | 246 | 4,689 | 54.6 | 1.7 | 7.5 | 1.5 | 4.4 | 46.0 | 878 | 347 | 135 | 3,451 | 23,730 | 71,568 | 7,367 |
| Esmeralda | 21 | 25,216 | 13.1 | 1.2 | 6.7 | 0.7 | 58.2 | 0.9 | 1,114 | 4 | 2 | 103 | 370 | 47,608 | 4,276 |
| Eureka | 22 | 11,528 | 38.6 | 5.0 | 7.7 | 0.2 | 10.0 | 0.0 | 0 | 7 | 5 | 205 | 780 | 60,637 | 6,367 |
| Humboldt | 142 | 8,472 | 31.6 | 39.2 | 7.1 | 1.7 | 4.7 | 2.6 | 158 | 144 | 43 | 1,423 | 7,860 | 64,308 | 6,337 |
| Lander | 53 | 9,574 | 25.3 | 31.9 | 6.3 | 1.5 | 3.7 | 1.6 | 294 | 69 | 14 | 487 | 2,430 | 70,188 | 7,614 |
| Lincoln | 30 | 5,682 | 48.5 | 1.0 | 7.2 | 2.9 | 8.2 | 7.2 | 1,395 | 52 | 13 | 601 | 1,690 | 53,511 | 4,156 |
| Lyon | 163 | 3,014 | 58.9 | 0.7 | 8.6 | 2.6 | 4.2 | 180.6 | 3,351 | 66 | 148 | 2,297 | 28,460 | 50,239 | 4,364 |
| Mineral | 33 | 7,391 | 24.7 | 39.2 | 6.8 | 0.6 | 3.6 | 6.1 | 1,359 | 54 | 13 | 485 | 1,930 | 46,683 | 3,552 |
| Nye | 139 | 3,155 | 45.1 | 3.5 | 11.5 | 1.5 | 7.5 | 100.6 | 2,287 | 120 | 119 | 1,727 | 20,010 | 53,252 | 4,860 |
| Pershing | 31 | 4,787 | 40.1 | 27.3 | 2.1 | 1.6 | 1.5 | 17.9 | 2,767 | 17 | 13 | 717 | 2,060 | 51,287 | 4,429 |
| Storey | 28 | 6,940 | 22.6 | 0.2 | 16.3 | 0.0 | 5.3 | 58.1 | 14,620 | 3 | 11 | 253 | 2,070 | 69,620 | 8,306 |
| Washoe | 1,963 | 4,300 | 34.4 | 1.4 | 7.5 | 4.8 | 8.6 | 3,086.8 | 6,760 | 3,767 | 1,253 | 24,226 | 237,680 | 88,988 | 12,604 |
| White Pine | 85 | 8,792 | 28.9 | 35.6 | 5.3 | 0.7 | 4.6 | 13.2 | 1,368 | 158 | 21 | 1,128 | 3,940 | 58,065 | 5,342 |
| Carson City | 226 | 4,152 | 42.3 | 4.1 | 8.4 | 2.9 | 6.4 | 397.7 | 7,292 | 569 | 142 | 9,101 | 28,630 | 59,571 | 6,443 |
| **NEW HAMPSHIRE** | X | X | X | X | X | X | X | X | X | 7,953 | 4,560 | 81,952 | 712,070 | 80,937 | 10,340 |
| Belknap | 328 | 5,390 | 55.5 | 0.2 | 6.2 | 3.4 | 6.7 | 101.8 | 1,672 | 141 | 198 | 4,027 | 33,760 | 74,352 | 9,145 |
| Carroll | 251 | 5,213 | 52.5 | 0.8 | 6.3 | 7.0 | 8.0 | 167.4 | 3,478 | 145 | 159 | 2,846 | 26,560 | 78,608 | 10,364 |
| Cheshire | 350 | 4,608 | 55.7 | 0.4 | 6.5 | 7.7 | 5.6 | 229.8 | 3,026 | 168 | 237 | 5,137 | 37,850 | 68,540 | 7,761 |
| Coos | 184 | 5,848 | 42.6 | 0.6 | 4.2 | 17.9 | 5.3 | 40.4 | 1,286 | 368 | 96 | 2,582 | 15,220 | 51,281 | 4,583 |
| Grafton | 452 | 5,027 | 56.9 | 0.4 | 5.6 | 5.5 | 6.8 | 260.7 | 2,902 | 553 | 274 | 6,705 | 44,790 | 88,551 | 12,168 |
| Hillsborough | 1,550 | 3,751 | 54.3 | 0.4 | 6.9 | 4.3 | 5.0 | 894.8 | 2,165 | 4,205 | 1,415 | 17,479 | 219,230 | 78,654 | 9,841 |
| Merrimack | 568 | 3,794 | 59.9 | 0.3 | 4.2 | 5.6 | 4.0 | 278.1 | 1,856 | 854 | 481 | 16,005 | 77,500 | 74,328 | 8,666 |
| Rockingham | 1,295 | 4,223 | 59.6 | 0.2 | 7.1 | 3.4 | 4.0 | 526.5 | 1,717 | 1,114 | 1,148 | 13,464 | 171,230 | 97,776 | 13,868 |
| Strafford | 503 | 3,873 | 56.6 | 2.1 | 5.7 | 5.4 | 5.7 | 326.5 | 2,514 | 314 | 413 | 11,370 | 64,220 | 69,016 | 7,658 |
| Sullivan | 174 | 4,049 | 51.5 | 0.8 | 4.7 | 12.1 | 6.5 | 73.1 | 1,700 | 91 | 139 | 2,337 | 21,740 | 69,732 | 8,029 |
| **NEW JERSEY** | X | X | X | X | X | X | X | X | X | 48,566 | 25,097 | 539,901 | 4,462,440 | 91,138 | 13,423 |
| Atlantic | 1,641 | 6,179 | 58.1 | 1.2 | 6.3 | 1.4 | 4.6 | 847.7 | 3,193 | 2,477 | 841 | 19,912 | 134,970 | 58,260 | 6,138 |
| Bergen | 5,309 | 5,694 | 57.7 | 5.9 | 6.5 | 1.0 | 2.9 | 2,273.6 | 2,438 | 2,483 | 1,908 | 46,768 | 471,350 | 119,361 | 20,351 |
| Burlington | 2,177 | 4,886 | 68.8 | 0.6 | 4.7 | 1.5 | 2.2 | 1,673.3 | 3,755 | 4,833 | 6,337 | 24,364 | 228,420 | 83,695 | 10,751 |
| Camden | 3,214 | 6,349 | 53.2 | 2.0 | 6.0 | 3.0 | 4.1 | 3,185.3 | 6,292 | 2,225 | 1,039 | 29,584 | 248,590 | 69,514 | 8,242 |
| Cape May | 1,326 | 14,234 | 23.2 | 1.0 | 3.2 | 2.4 | 6.9 | 959.9 | 10,307 | 440 | 1,081 | 8,632 | 49,920 | 68,351 | 8,023 |
| Cumberland | 984 | 6,499 | 63.7 | 2.4 | 3.2 | 1.9 | 4.6 | 340.3 | 2,247 | 621 | 280 | 11,801 | 65,950 | 50,790 | 4,546 |
| Essex | 4,558 | 5,724 | 42.1 | 2.9 | 8.1 | 2.4 | 1.6 | 3,386.7 | 4,253 | 9,032 | 1,572 | 62,081 | 382,140 | 89,391 | 14,279 |
| Gloucester | 1,616 | 5,553 | 60.0 | 0.7 | 4.4 | 1.7 | 1.5 | 958.9 | 3,296 | 505 | 576 | 19,700 | 143,360 | 75,593 | 8,613 |
| Hudson | 3,159 | 4,695 | 37.0 | 1.8 | 8.3 | 2.6 | 1.7 | 2,636.2 | 3,918 | 5,254 | 1,455 | 36,476 | 342,990 | 77,709 | 11,153 |
| Hunterdon | 827 | 6,632 | 53.2 | 1.0 | 2.9 | 1.3 | 4.2 | 380.9 | 3,054 | 269 | 246 | 7,819 | 65,410 | 126,731 | 20,692 |

1. Based on the resident population estimated as of July 1 of the year shown.

# Table B. States and Counties — Land Area and Population

| State / county code | CBSA code[1] | County Type code[2] | STATE County | Land area[3] (sq. mi) | Total persons 2019 | Rank | Per square mile | White | Black | American Indian, Alaska Native | Asian and Pacific Islancer | Percent Hispanic or Latino[4] | Under 5 years | 5 to 14 years | 15 to 24 years | 25 to 34 years | 35 to 44 years | 45 to 54 years |
|---|---|---|---|---|---|---|---|---|---|---|---|---|---|---|---|---|---|---|
| | | | | 1 | 2 | 3 | 4 | 5 | 6 | 7 | 8 | 9 | 10 | 11 | 12 | 13 | 14 | 15 |
| | | | **NEW JERSEY— Cont'd** | | | | | | | | | | | | | | | |
| 34021 | 45940 | 2 | Mercer | 224.4 | 367,239 | 197 | 1,636.5 | 48.8 | 20.5 | 0.5 | 13.3 | 18.9 | 5.6 | 11.9 | 14.8 | 12.5 | 12.6 | 13.4 |
| 34023 | 35620 | 1 | Middlesex | 309.2 | 822,736 | 76 | 2,660.9 | 42.0 | 10.7 | 0.5 | 26.1 | 22.2 | 5.7 | 12.2 | 13.0 | 13.2 | 13.8 | 13.3 |
| 34025 | 35620 | 1 | Monmouth | 468.2 | 618,381 | 109 | 1,320.8 | 76.3 | 7.4 | 0.4 | 6.4 | 11.2 | 4.9 | 11.8 | 12.2 | 11.0 | 11.5 | 13.8 |
| 34027 | 35620 | 1 | Morris | 461.0 | 491,087 | 144 | 1,065.3 | 71.3 | 3.9 | 0.3 | 12.2 | 14.0 | 4.9 | 11.6 | 12.5 | 11.2 | 12.4 | 14.3 |
| 34029 | 35620 | 1 | Ocean | 628.3 | 614,237 | 111 | 977.6 | 85.1 | 3.7 | 0.3 | 2.5 | 9.6 | 7.2 | 13.3 | 11.1 | 11.1 | 10.1 | 10.8 |
| 34031 | 35620 | 1 | Passaic | 186.0 | 500,382 | 140 | 2,690.2 | 41.0 | 10.6 | 0.4 | 6.1 | 43.2 | 6.6 | 13.1 | 13.4 | 13.6 | 12.5 | 12.7 |
| 34033 | 37980 | 1 | Salem | 331.9 | 62,451 | 851 | 188.2 | 74.6 | 15.0 | 0.7 | 1.6 | 10.2 | 5.3 | 12.1 | 11.6 | 12.0 | 12.0 | 12.6 |
| 34035 | 35620 | 1 | Somerset | 301.9 | 329,331 | 215 | 1,090.9 | 55.2 | 10.3 | 0.4 | 20.5 | 15.4 | 5.0 | 12.1 | 12.3 | 11.3 | 12.9 | 14.6 |
| 34037 | 35620 | 1 | Sussex | 518.7 | 140,002 | 467 | 269.9 | 85.8 | 2.8 | 0.4 | 2.8 | 9.6 | 4.6 | 10.7 | 11.8 | 11.4 | 11.5 | 14.4 |
| 34039 | 35620 | 1 | Union | 102.8 | 555,394 | 124 | 5,402.7 | 39.8 | 21.8 | 0.4 | 6.4 | 33.2 | 6.3 | 13.1 | 12.2 | 12.7 | 13.8 | 13.8 |
| 34041 | 10900 | 2 | Warren | 356.5 | 105,624 | 580 | 296.3 | 80.7 | 6.1 | 0.4 | 3.5 | 10.8 | 4.8 | 10.7 | 12.0 | 11.5 | 11.4 | 14.0 |
| 35000 | | 0 | **NEW MEXICO** | 121,312.2 | 2,106,319 | X | 17.4 | 37.8 | 2.5 | 9.4 | 2.3 | 49.7 | 5.6 | 12.8 | 13.3 | 13.5 | 12.3 | 11.1 |
| 35001 | 10740 | 2 | Bernalillo | 1,161.2 | 681,666 | 99 | 587.0 | 39.4 | 3.4 | 4.9 | 3.7 | 50.6 | 5.3 | 12.0 | 12.7 | 14.8 | 13.3 | 11.6 |
| 35003 | | 9 | Catron | 6,924.2 | 3,623 | 2,924 | 0.5 | 77.1 | 1.4 | 4.5 | 0.7 | 18.6 | 2.0 | 7.3 | 7.3 | 6.8 | 6.7 | 7.8 |
| 35005 | 40740 | 5 | Chaves | 6,067.2 | 64,711 | 825 | 10.7 | 38.0 | 1.8 | 1.2 | 1.3 | 58.7 | 6.5 | 14.5 | 14.6 | 13.0 | 12.5 | 10.7 |
| 35006 | 24380 | 6 | Cibola | 4,540.0 | 26,354 | 1,555 | 5.8 | 19.6 | 1.6 | 40.5 | 0.9 | 38.8 | 5.6 | 13.7 | 12.4 | 14.4 | 13.0 | 10.9 |
| 35007 | | 7 | Colfax | 3,758.0 | 11,927 | 2,297 | 3.2 | 47.9 | 0.9 | 2.1 | 1.0 | 49.3 | 4.2 | 10.0 | 9.9 | 10.8 | 9.7 | 10.7 |
| 35009 | 17580 | 5 | Curry | 1,405.5 | 48,793 | 1,012 | 34.7 | 47.5 | 6.7 | 1.4 | 2.7 | 44.2 | 7.8 | 14.4 | 16.2 | 17.1 | 11.9 | 9.1 |
| 35011 | | 9 | De Baca | 2,323.1 | 1,673 | 3,071 | 0.7 | 52.6 | 0.9 | 2.1 | 0.8 | 45.3 | 3.8 | 12.9 | 10.8 | 7.8 | 10.6 | 9.2 |
| 35013 | 29740 | 3 | Dona Ana | 3,808.2 | 221,262 | 312 | 58.1 | 27.5 | 2.0 | 1.2 | 1.6 | 68.8 | 6.1 | 13.6 | 18.5 | 13.2 | 11.2 | 9.9 |
| 35015 | 16100 | 5 | Eddy | 4,176.3 | 58,418 | 886 | 14.0 | 45.2 | 1.8 | 1.6 | 1.1 | 51.6 | 7.3 | 15.0 | 13.0 | 14.5 | 13.2 | 10.7 |
| 35017 | 43500 | 7 | Grant | 3,961.2 | 27,007 | 1,524 | 6.8 | 47.6 | 1.1 | 1.7 | 1.2 | 49.8 | 4.7 | 11.3 | 11.0 | 9.5 | 10.4 | 9.8 |
| 35019 | | 7 | Guadalupe | 3,029.8 | 4,275 | 2,875 | 1.4 | 16.9 | 1.9 | 2.0 | 1.6 | 78.6 | 5.0 | 11.6 | 11.9 | 14.7 | 13.6 | 9.7 |
| 35021 | | 9 | Harding | 2,125.5 | 638 | 3,133 | 0.3 | 54.4 | 0.5 | 0.5 | 0.3 | 44.8 | 3.0 | 8.2 | 4.9 | 8.0 | 10.7 | 6.4 |
| 35023 | | 9 | Hidalgo | 3,438.5 | 4,106 | 2,887 | 1.2 | 38.4 | 1.5 | 1.1 | 0.8 | 59.3 | 5.9 | 12.7 | 11.3 | 11.5 | 9.9 | 11.2 |
| 35025 | 26020 | 5 | Lea | 4,391.6 | 71,830 | 760 | 16.4 | 33.8 | 3.7 | 1.2 | 0.7 | 61.5 | 8.0 | 17.2 | 14.8 | 14.4 | 13.3 | 10.6 |
| 35027 | 40760 | 5 | Lincoln | 4,831.1 | 19,939 | 1,827 | 4.1 | 61.4 | 1.0 | 3.7 | 0.9 | 34.3 | 4.6 | 10.2 | 9.2 | 9.0 | 10.0 | 9.9 |
| 35028 | 31060 | 6 | Los Alamos | 109.1 | 19,462 | 1,856 | 178.4 | 72.8 | 1.6 | 1.5 | 7.8 | 18.1 | 5.0 | 12.5 | 10.9 | 13.5 | 13.2 | 12.4 |
| 35029 | 19700 | 6 | Luna | 2,965.2 | 23,905 | 1,646 | 8.1 | 29.1 | 1.2 | 1.2 | 1.4 | 68.2 | 7.5 | 14.3 | 13.1 | 12.4 | 10.4 | 9.8 |
| 35031 | 23700 | 5 | McKinley | 5,451.1 | 70,824 | 769 | 13.0 | 9.5 | 0.9 | 75.4 | 1.4 | 14.6 | 6.3 | 16.8 | 14.0 | 15.0 | 12.1 | 10.7 |
| 35033 | 29780 | 9 | Mora | 1,926.2 | 4,478 | 2,860 | 2.3 | 17.8 | 0.5 | 0.9 | 0.6 | 80.8 | 4.4 | 9.1 | 10.0 | 9.1 | 10.4 | 10.7 |
| 35035 | 10460 | 4 | Otero | 6,612.6 | 67,967 | 791 | 10.3 | 49.7 | 4.2 | 6.8 | 2.5 | 39.1 | 6.5 | 12.5 | 13.9 | 15.4 | 11.8 | 9.8 |
| 35037 | | 7 | Quay | 2,874.0 | 8,197 | 2,564 | 2.9 | 49.6 | 1.9 | 2.0 | 1.5 | 46.6 | 5.6 | 11.8 | 10.4 | 10.2 | 10.1 | 10.7 |
| 35039 | 21580 | 4 | Rio Arriba | 5,860.9 | 38,521 | 1,210 | 6.6 | 13.4 | 0.6 | 14.6 | 0.8 | 71.3 | 5.6 | 13.2 | 11.7 | 11.7 | 11.0 | 11.5 |
| 35041 | 38780 | 7 | Roosevelt | 2,446.1 | 18,350 | 1,906 | 7.5 | 51.5 | 2.8 | 1.9 | 1.8 | 43.8 | 6.7 | 13.4 | 21.1 | 13.4 | 10.6 | 9.8 |
| 35043 | 10740 | 2 | Sandoval | 3,710.2 | 148,904 | 451 | 40.1 | 43.8 | 2.8 | 12.6 | 2.5 | 40.4 | 5.2 | 13.2 | 11.8 | 12.3 | 12.9 | 12.0 |
| 35045 | 22140 | 3 | San Juan | 5,517.2 | 123,312 | 515 | 22.4 | 38.5 | 1.0 | 40.1 | 1.0 | 21.6 | 6.0 | 14.9 | 13.1 | 13.4 | 12.9 | 10.7 |
| 35047 | 29780 | 6 | San Miguel | 4,721.5 | 27,144 | 1,521 | 5.7 | 18.4 | 1.8 | 1.5 | 1.3 | 77.8 | 4.4 | 9.5 | 13.0 | 11.8 | 10.6 | 11.4 |
| 35049 | 42140 | 3 | Santa Fe | 1,910.4 | 151,946 | 442 | 79.5 | 44.5 | 1.1 | 3.2 | 1.9 | 50.6 | 4.0 | 9.9 | 10.3 | 11.1 | 11.5 | 11.8 |
| 35051 | | 6 | Sierra | 4,181.2 | 10,867 | 2,358 | 2.6 | 65.2 | 1.1 | 2.8 | 1.2 | 31.7 | 4.4 | 8.8 | 7.7 | 8.5 | 7.8 | 9.8 |
| 35053 | | 6 | Socorro | 6,646.4 | 16,541 | 2,003 | 2.5 | 34.3 | 1.4 | 12.8 | 2.7 | 50.6 | 5.9 | 12.5 | 14.7 | 11.6 | 10.4 | 10.6 |
| 35055 | 45340 | 7 | Taos | 2,202.4 | 32,593 | 1,365 | 14.8 | 37.1 | 0.7 | 6.0 | 1.1 | 56.3 | 4.2 | 9.6 | 9.6 | 10.0 | 11.0 | 11.4 |
| 35057 | 10740 | 2 | Torrance | 3,345.2 | 15,486 | 2,060 | 4.6 | 51.0 | 2.1 | 3.2 | 1.1 | 44.6 | 5.2 | 11.5 | 11.9 | 11.7 | 11.4 | 11.4 |
| 35059 | | 9 | Union | 3,825.1 | 4,026 | 2,895 | 1.1 | 51.6 | 2.7 | 2.3 | 1.0 | 43.7 | 5.2 | 11.0 | 10.5 | 14.5 | 13.3 | 10.9 |
| 35061 | 10740 | 2 | Valencia | 1,066.7 | 77,574 | 722 | 72.7 | 32.5 | 1.7 | 4.8 | 1.0 | 61.3 | 5.6 | 13.4 | 12.7 | 12.6 | 11.9 | 11.7 |
| 36000 | | 0 | **NEW YORK** | 47,123.8 | 19,336,776 | X | 410.3 | 56.6 | 15.5 | 0.7 | 10.0 | 19.3 | 5.7 | 11.4 | 12.4 | 14.7 | 12.6 | 12.4 |
| 36001 | 10580 | 2 | Albany | 522.9 | 303,654 | 231 | 580.7 | 73.2 | 14.4 | 0.5 | 8.1 | 6.5 | 4.9 | 10.2 | 17.4 | 13.4 | 11.8 | 11.4 |
| 36003 | | 7 | Allegany | 1,029.4 | 45,587 | 1,066 | 44.3 | 95.3 | 1.8 | 0.5 | 1.8 | 1.8 | 5.2 | 11.2 | 18.2 | 10.2 | 10.3 | 10.9 |
| 36005 | 35620 | 1 | Bronx | 42.2 | 1,401,142 | 29 | 33,202.4 | 9.6 | 29.9 | 0.6 | 4.5 | 56.4 | 7.0 | 13.6 | 13.4 | 15.9 | 12.7 | 12.0 |
| 36007 | 13780 | 2 | Broome | 705.6 | 189,420 | 354 | 268.5 | 84.8 | 7.3 | 0.7 | 5.4 | 4.7 | 5.0 | 10.8 | 17.5 | 11.4 | 10.6 | 10.7 |
| 36009 | 36460 | 4 | Cattaraugus | 1,308.2 | 75,863 | 734 | 58.0 | 92.1 | 2.5 | 4.0 | 1.2 | 2.3 | 5.6 | 12.5 | 12.8 | 10.9 | 11.0 | 11.9 |
| 36011 | 12180 | 4 | Cayuga | 691.6 | 76,029 | 732 | 109.9 | 91.6 | 5.5 | 0.8 | 1.1 | 3.2 | 4.9 | 10.9 | 11.5 | 12.4 | 11.9 | 12.5 |
| 36013 | 27460 | 4 | Chautauqua | 1,060.4 | 126,032 | 508 | 118.9 | 88.6 | 3.5 | 0.9 | 1.0 | 8.0 | 5.3 | 11.3 | 13.0 | 11.6 | 10.8 | 11.8 |
| 36015 | 21300 | 3 | Chemung | 407.3 | 82,622 | 690 | 202.9 | 88.8 | 8.2 | 0.8 | 2.0 | 3.5 | 5.4 | 12.0 | 11.7 | 12.5 | 11.9 | 12.1 |
| 36017 | | 6 | Chenango | 893.6 | 46,730 | 1,040 | 52.3 | 95.8 | 1.5 | 0.8 | 0.9 | 2.4 | 5.1 | 11.8 | 10.7 | 11.3 | 11.1 | 12.2 |
| 36019 | 38460 | 5 | Clinton | 1,037.8 | 79,778 | 710 | 76.9 | 91.3 | 4.5 | 0.8 | 1.9 | 3.0 | 4.7 | 10.2 | 15.4 | 12.5 | 12.2 | 12.2 |
| 36021 | 26460 | 6 | Columbia | 634.7 | 59,534 | 875 | 93.8 | 87.7 | 5.8 | 0.6 | 2.6 | 5.3 | 4.0 | 9.1 | 9.9 | 11.1 | 10.7 | 12.9 |
| 36023 | 18660 | 4 | Cortland | 498.8 | 47,173 | 1,035 | 94.6 | 93.7 | 2.7 | 0.8 | 1.7 | 2.9 | 5.0 | 10.5 | 21.3 | 11.4 | 10.5 | 11.0 |
| 36025 | | 6 | Delaware | 1,442.6 | 43,938 | 1,097 | 30.5 | 92.6 | 2.3 | 0.6 | 1.5 | 4.2 | 4.0 | 9.1 | 12.2 | 10.2 | 9.9 | 12.0 |
| 36027 | 39100 | 1 | Dutchess | 795.6 | 293,293 | 238 | 368.6 | 71.9 | 11.9 | 0.6 | 4.5 | 13.4 | 4.5 | 10.2 | 14.1 | 12.1 | 11.7 | 13.3 |
| 36029 | 15380 | 2 | Erie | 1,042.7 | 917,241 | 60 | 879.7 | 76.2 | 14.2 | 0.9 | 4.7 | 6.0 | 5.4 | 11.2 | 12.5 | 14.3 | 11.7 | 11.9 |
| 36031 | | 6 | Essex | 1,794.1 | 36,891 | 1,252 | 20.6 | 93.2 | 2.9 | 0.9 | 1.1 | 3.2 | 3.9 | 8.5 | 9.8 | 12.0 | 11.3 | 12.6 |
| 36033 | 31660 | 7 | Franklin | 1,629.3 | 49,965 | 991 | 30.7 | 82.9 | 6.1 | 7.8 | 0.9 | 3.6 | 4.8 | 11.1 | 12.7 | 14.3 | 12.6 | 12.5 |
| 36035 | 24100 | 4 | Fulton | 495.5 | 52,812 | 960 | 106.6 | 93.3 | 2.8 | 0.6 | 1.1 | 3.7 | 5.0 | 11.5 | 10.8 | 11.9 | 11.9 | 13.0 |

1. CBSA = Core Based Statistical Area. See Appendix A for explanation. See Appendix B for list of metropolitan areas with component counties. Service of USDA Rural-Urban Continuum Codes. See Appendix A for definition.
2. County type code from the Economic Research
3. Dry land or land partially or temporarily covered by water.
4. May be of any race.

| STATE County | Age (percent) 55 to 64 years | 65 to 74 years | 75 years and over | Percent female | Total persons 2000 | Total persons 2010 | Percent change 2000-2010 | Percent change 2010-2020 | Births | Deaths | Net Migration | Number | Persons per household | Family house-holds | Female family house-holder[1] | One person |
|---|---|---|---|---|---|---|---|---|---|---|---|---|---|---|---|---|
| | 16 | 17 | 18 | 19 | 20 | 21 | 22 | 23 | 24 | 25 | 26 | 27 | 28 | 29 | 30 | 31 |
| **NEW JERSEY— Cont'd** | | | | | | | | | | | | | | | | |
| Mercer | 13.2 | 9.1 | 6.9 | 51.0 | 350,761 | 367,485 | 4.8 | -0.1 | 42,239 | 30,039 | -12,588 | 129,936 | 2.7 | 67.3 | 13.3 | 27.6 |
| Middlesex | 13.1 | 9.2 | 6.7 | 50.6 | 750,162 | 810,099 | 8.0 | 1.6 | 96,611 | 60,728 | -23,620 | 285,005 | 2.79 | 73.4 | 11.5 | 22.2 |
| Monmouth | 16.0 | 10.9 | 7.9 | 51.3 | 615,301 | 630,347 | 2.4 | -1.9 | 60,904 | 56,505 | -16,350 | 235,362 | 2.61 | 68.8 | 9.7 | 26.3 |
| Morris | 15.1 | 9.9 | 8.1 | 50.8 | 470,212 | 492,281 | 4.7 | -0.2 | 47,629 | 39,193 | -9,581 | 181,884 | 2.66 | 71.5 | 8.9 | 23.8 |
| Ocean | 13.5 | 12.2 | 10.7 | 51.7 | 510,916 | 576,554 | 12.8 | 6.5 | 86,895 | 74,398 | 25,804 | 226,160 | 2.61 | 66.2 | 9.3 | 29.1 |
| Passaic | 12.8 | 8.8 | 6.5 | 51.1 | 489,049 | 501,595 | 2.6 | -0.2 | 70,094 | 38,086 | -33,561 | 165,429 | 2.99 | 72.4 | 17.8 | 23.4 |
| Salem | 14.8 | 11.1 | 8.2 | 51.1 | 64,285 | 66,072 | 2.8 | -5.5 | 6,910 | 7,568 | -2,993 | 23,933 | 2.57 | 65.9 | 12.8 | 29.3 |
| Somerset | 15.0 | 9.3 | 7.4 | 51.1 | 297,490 | 323,476 | 8.7 | 1.8 | 34,160 | 24,386 | -3,863 | 118,193 | 2.75 | 72.2 | 9.0 | 23.8 |
| Sussex | 16.9 | 11.6 | 7.2 | 50.2 | 144,166 | 148,941 | 3.3 | -6.0 | 12,655 | 12,337 | -9,378 | 53,322 | 2.61 | 71.8 | 8.4 | 22.9 |
| Union | 13.2 | 8.6 | 6.4 | 51.0 | 522,541 | 536,426 | 2.7 | 3.5 | 69,722 | 40,932 | -9,699 | 190,101 | 2.88 | 72.1 | 15.3 | 24.1 |
| Warren | 16.4 | 10.9 | 8.3 | 50.9 | 102,437 | 108,639 | 6.1 | -2.8 | 9,855 | 10,213 | -2,647 | 41,587 | 2.5 | 66.7 | 9.4 | 28.0 |
| **NEW MEXICO** | 12.8 | 11.0 | 7.5 | 50.5 | 1,819,046 | 2,059,199 | 13.2 | 2.3 | 258,988 | 182,459 | -29,366 | 780,249 | 2.63 | 63.2 | 13.8 | 30.8 |
| Bernalillo | 12.8 | 10.4 | 7.0 | 51.0 | 556,678 | 662,469 | 19.0 | 2.9 | 79,605 | 56,600 | -3,360 | 267,699 | 2.5 | 59.3 | 13.6 | 33.5 |
| Catron | 18.9 | 25.4 | 17.9 | 47.7 | 3,543 | 3,727 | 5.2 | -2.8 | 186 | 401 | 111.0 | 1,325 | 2.57 | 59.7 | 8.9 | 33.0 |
| Chaves | 12.0 | 9.3 | 6.9 | 50.2 | 61,382 | 65,648 | 6.9 | -1.4 | 9,148 | 6,798 | -3,269 | 23,284 | 2.7 | 67.6 | 15.1 | 29.0 |
| Cibola | 12.6 | 10.3 | 7.1 | 48.9 | 25,595 | 27,218 | 6.3 | -3.2 | 3,647 | 2,690 | -1,832 | 8,708 | 2.92 | 68.9 | 18.1 | 26.2 |
| Colfax | 16.0 | 16.6 | 12.0 | 49.1 | 14,189 | 13,751 | -3.1 | -13.3 | 1,252 | 1,501 | -1,595 | 5,853 | 1.98 | 56.0 | 11.7 | 37.7 |
| Curry | 10.4 | 7.4 | 5.7 | 48.4 | 45,044 | 48,376 | 7.4 | 0.9 | 9,060 | 4,055 | -4,653 | 18,548 | 2.63 | 64.4 | 14.4 | 30.7 |
| De Baca | 15.3 | 16.3 | 13.3 | 49.9 | 2,240 | 2,024 | -9.6 | -17.3 | 152 | 276 | -231.0 | 672 | 3.0 | 49.4 | 7.1 | 46.4 |
| Dona Ana | 10.8 | 9.6 | 7.1 | 50.8 | 174,682 | 209,212 | 19.8 | 5.8 | 29,575 | 15,857 | -1,709 | 77,842 | 2.7 | 67.9 | 16.0 | 25.3 |
| Eddy | 11.8 | 8.5 | 6.1 | 49.4 | 51,658 | 53,828 | 4.2 | 8.5 | 8,489 | 5,802 | 1,874 | 21,251 | 2.69 | 71.0 | 13.6 | 24.9 |
| Grant | 14.2 | 16.6 | 12.5 | 50.7 | 31,002 | 29,512 | -4.8 | -8.5 | 2,971 | 3,545 | -1,917 | 11,851 | 2.27 | 56.8 | 12.2 | 35.6 |
| Guadalupe | 13.2 | 10.6 | 9.7 | 43.3 | 4,680 | 4,688 | 0.2 | -8.8 | 430 | 463 | -386.0 | 1,384 | 2.58 | 43.4 | 10.9 | 54.2 |
| Harding | 16.9 | 22.7 | 19.3 | 48.0 | 810 | 695 | -14.2 | -8.2 | 40 | 81 | -16.0 | 211 | 2.09 | 49.8 | 0.9 | 48.8 |
| Hidalgo | 14.1 | 13.1 | 10.2 | 49.9 | 5,932 | 4,898 | -17.4 | -16.2 | 538 | 504 | -837.0 | 1,679 | 2.49 | 63.8 | 12.4 | 31.6 |
| Lea | 10.3 | 6.7 | 4.6 | 48.7 | 55,511 | 64,726 | 16.6 | 11.0 | 11,435 | 5,556 | 1,146 | 22,523 | 3.0 | 73.6 | 13.3 | 23.0 |
| Lincoln | 16.7 | 17.9 | 12.5 | 51.4 | 19,411 | 20,493 | 5.6 | -2.7 | 1,863 | 2,169 | -250.0 | 7,566 | 2.54 | 64.1 | 7.1 | 31.5 |
| Los Alamos | 14.1 | 10.4 | 8.0 | 48.8 | 18,343 | 17,950 | -2.1 | 8.4 | 1,744 | 1,267 | 1,007 | 7,931 | 2.34 | 62.5 | 5.4 | 33.0 |
| Luna | 11.4 | 11.2 | 9.7 | 49.6 | 25,016 | 25,095 | 0.3 | -4.7 | 3,894 | 3,070 | -2,017 | 8,904 | 2.65 | 60.1 | 12.6 | 35.9 |
| McKinley | 11.6 | 8.1 | 5.3 | 51.9 | 74,798 | 71,485 | -4.4 | -0.9 | 10,844 | 6,225 | -5,300 | 20,942 | 3.42 | 71.0 | 26.5 | 25.3 |
| Mora | 16.5 | 17.3 | 12.5 | 48.8 | 5,180 | 4,881 | -5.8 | -8.3 | 420 | 466 | -360.0 | 1,713 | 2.64 | 48.9 | 6.2 | 36.8 |
| Otero | 12.4 | 10.0 | 7.7 | 48.3 | 62,298 | 63,836 | 2.5 | 6.5 | 9,249 | 6,127 | 925.0 | 23,634 | 2.64 | 61.6 | 10.9 | 33.3 |
| Quay | 14.6 | 15.0 | 11.5 | 51.1 | 10,155 | 9,040 | -11.0 | -9.3 | 999 | 1,219 | -636.0 | 3,040 | 2.71 | 52.0 | 7.8 | 45.0 |
| Rio Arriba | 14.4 | 12.1 | 8.7 | 50.9 | 41,190 | 40,220 | -2.4 | -4.2 | 5,174 | 4,197 | -2,680 | 12,730 | 3.03 | 59.5 | 13.8 | 35.4 |
| Roosevelt | 10.0 | 8.3 | 6.7 | 49.9 | 18,018 | 19,840 | 10.1 | -7.5 | 2,843 | 1,598 | -2,790 | 6,814 | 2.61 | 60.5 | 12.6 | 30.4 |
| Sandoval | 13.5 | 11.9 | 7.2 | 50.9 | 89,908 | 131,623 | 46.4 | 13.1 | 14,845 | 10,871 | 13,345 | 51,001 | 2.78 | 70.6 | 11.9 | 23.2 |
| San Juan | 12.8 | 9.6 | 6.6 | 50.6 | 113,801 | 130,045 | 14.3 | -5.2 | 17,580 | 10,400 | -13,988 | 43,387 | 2.88 | 70.6 | 16.4 | 24.8 |
| San Miguel | 15.7 | 13.9 | 9.6 | 50.2 | 30,126 | 29,375 | -2.5 | -7.6 | 2,982 | 2,901 | -2,343 | 11,609 | 2.25 | 55.4 | 17.1 | 40.2 |
| Santa Fe | 15.1 | 16.4 | 9.9 | 51.7 | 129,292 | 144,237 | 11.6 | 5.3 | 13,243 | 11,800 | 6,407 | 61,921 | 2.36 | 59.4 | 12.1 | 34.0 |
| Sierra | 15.0 | 20.3 | 17.6 | 49.8 | 13,270 | 11,996 | -9.6 | -9.4 | 1,017 | 2,450 | 308.0 | 5,555 | 1.94 | 50.5 | 11.0 | 46.2 |
| Socorro | 13.7 | 12.1 | 8.5 | 49.7 | 18,078 | 17,861 | -1.2 | -7.4 | 2,113 | 1,775 | -1,685 | 4,520 | 3.62 | 53.2 | 11.0 | 40.3 |
| Taos | 15.5 | 17.2 | 11.5 | 51.1 | 29,979 | 32,933 | 9.9 | | 2,980 | 3,065 | -231.0 | 12,103 | 2.66 | 55.9 | 11.2 | 39.5 |
| Torrance | 14.9 | 13.8 | 8.2 | 47.2 | 16,911 | 16,375 | -3.2 | -5.4 | 1,605 | 1,577 | -945.0 | 5,644 | 2.69 | 64.6 | 10.9 | 32.2 |
| Union | 12.7 | 11.7 | 10.2 | 43.4 | 4,174 | 4,553 | 9.1 | -11.6 | 418 | 469 | -479.0 | 1,395 | 2.5 | 53.3 | 9.9 | 41.9 |
| Valencia | 13.5 | 11.3 | 7.4 | 49.7 | 66,152 | 76,589 | 15.8 | 1.3 | 8,647 | 6,684 | -980.0 | 27,010 | 2.76 | 68.1 | 13.8 | 26.5 |
| **NEW YORK** | 13.4 | 9.9 | 7.5 | 51.4 | 18,976,457 | 19,378,117 | 2.1 | -0.2 | 2,400,759 | 1,590,348 | -849,381 | 7,343,234 | 2.59 | 63.1 | 14.0 | 29.9 |
| Albany | 13.1 | 10.4 | 7.5 | 51.7 | 294,565 | 304,199 | 3.3 | -0.2 | 31,521 | 28,058 | -3,853 | 126,540 | 2.29 | 55.0 | 11.4 | 35.7 |
| Allegany | 13.9 | 12.0 | 8.1 | 49.3 | 49,927 | 48,923 | -2.0 | -6.8 | 4,978 | 4,909 | -3,436 | 17,948 | 2.33 | 64.0 | 9.8 | 29.8 |
| Bronx | 11.6 | 7.7 | 6.1 | 52.8 | 1,332,650 | 1,384,224 | 3.9 | 1.2 | 214,582 | 99,466 | -98,945 | 503,829 | 2.76 | 64.6 | 29.0 | 31.1 |
| Broome | 14.1 | 10.8 | 9.0 | 50.7 | 200,536 | 200,670 | 0.1 | -5.6 | 20,449 | 21,690 | -10,059 | 78,549 | 2.33 | 58.6 | 11.8 | 32.8 |
| Cattaraugus | 15.0 | 12.0 | 8.2 | 50.2 | 83,955 | 80,337 | -4.3 | -5.6 | 9,107 | 8,675 | -4,920 | 31,779 | 2.34 | 62.1 | 11.6 | 31.4 |
| Cayuga | 15.5 | 12.0 | 8.5 | 48.7 | 81,963 | 80,022 | -2.4 | -5.0 | 7,805 | 7,844 | -3,951 | 31,221 | 2.34 | 62.6 | 12.2 | 29.4 |
| Chautauqua | 15.0 | 12.1 | 9.1 | 50.6 | 139,750 | 134,907 | -3.5 | -6.6 | 13,986 | 15,082 | -7,827 | 53,075 | 2.31 | 60.9 | 11.4 | 32.6 |
| Chemung | 14.5 | 11.5 | 8.4 | 50.5 | 91,070 | 88,846 | -2.4 | -7.0 | 9,747 | 9,870 | -6,117 | 33,989 | 2.36 | 62.6 | 12.2 | 31.1 |
| Chenango | 16.2 | 12.4 | 9.1 | 49.9 | 51,401 | 50,516 | -1.7 | -7.5 | 5,123 | 5,798 | -3,108 | 20,697 | 2.28 | 63.9 | 10.5 | 28.9 |
| Clinton | 14.7 | 10.8 | 7.5 | 48.6 | 79,894 | 82,131 | 2.8 | -2.9 | 7,772 | 7,353 | -2,768 | 31,301 | 2.33 | 61.4 | 11.0 | 29.4 |
| Columbia | 16.8 | 14.7 | 10.8 | 50.0 | 63,094 | 63,063 | 0.0 | -5.6 | 5,249 | 6,954 | -1,786 | 24,966 | 2.32 | 63.0 | 10.7 | 29.6 |
| Cortland | 13.0 | 10.1 | 7.2 | 50.9 | 48,599 | 49,287 | 1.4 | -4.3 | 4,827 | 4,559 | -2,387 | 17,745 | 2.5 | 64.2 | 10.8 | 25.7 |
| Delaware | 16.5 | 14.7 | 11.4 | 49.6 | 48,055 | 47,962 | -0.2 | -8.4 | 3,996 | 5,580 | -2,425 | 18,968 | 2.25 | 60.0 | 8.6 | 33.0 |
| Dutchess | 15.3 | 10.6 | 8.0 | 50.3 | 280,150 | 297,444 | 6.2 | -1.4 | 26,993 | 25,568 | -5,594 | 108,413 | 2.54 | 67.1 | 10.9 | 27.0 |
| Erie | 14.2 | 10.8 | 8.0 | 51.6 | 950,265 | 919,129 | -3.3 | -0.2 | 101,743 | 100,550 | -2,544 | 389,585 | 2.29 | 58.8 | 13.0 | 34.1 |
| Essex | 16.8 | 14.5 | 10.6 | 48.4 | 38,851 | 39,371 | 1.3 | -6.3 | 3,172 | 4,201 | -1,448 | 15,790 | 2.19 | 63.4 | 10.6 | 30.8 |
| Franklin | 14.3 | 10.5 | 7.3 | 45.1 | 51,134 | 51,597 | 0.9 | -3.2 | 5,020 | 4,778 | -1,904 | 19,015 | 2.35 | 63.0 | 9.4 | 30.9 |
| Fulton | 15.3 | 12.0 | 8.7 | 50.4 | 55,073 | 55,509 | 0.8 | -4.9 | 5,429 | 6,218 | -1,891 | 22,557 | 2.33 | 63.7 | 12.9 | 29.4 |

1. No spouse present.

| STATE County | Persons in group quarters, 2020 | Daytime Population, 2015-2019 Number | Daytime Population Employment/ residence ratio | Births, 2020 Total | Births, 2020 Rate[1] | Deaths, 2020 Number | Deaths, 2020 Rate[1] | Persons under 65 with no health insurance, 2019 Number | Persons under 65... Percent | Medicare, 2020 Total beneficiaries | Medicare Enrolled in Original Medicare | Medicare Enrolled in Medicare Advantage | Crimes reported... Violent | Crimes reported... Property |
|---|---|---|---|---|---|---|---|---|---|---|---|---|---|---|
| | 32 | 33 | 34 | 35 | 36 | 37 | 38 | 39 | 40 | 41 | 42 | 43 | 44 | 45 |
| **NEW JERSEY— Cont'd** | | | | | | | | | | | | | | |
| Mercer | 19,977 | 423,248 | 1.31 | 3,920 | 10.7 | 3,219 | 8.8 | 28,211 | 9.6 | 65,853 | 40,457 | 25,396 | 2 | 14 |
| Middlesex | 23,006 | 812,782 | 0.97 | 8,607 | 10.5 | 6,707 | 8.2 | 61,727 | 9.1 | 138,668 | 94,802 | 43,866 | 0 | 0 |
| Monmouth | 6,654 | 583,588 | 0.88 | 5,799 | 9.4 | 6,241 | 10.1 | 35,682 | 7.1 | 127,486 | 95,653 | 31,833 | 7 | 1 |
| Morris | 8,587 | 535,823 | 1.17 | 4,419 | 9.0 | 4,204 | 8.6 | 22,979 | 5.7 | 93,105 | 68,701 | 24,404 | 0 | 0 |
| Ocean | 7,240 | 524,694 | 0.72 | 9,040 | 14.7 | 7,815 | 12.7 | 33,231 | 7.2 | 155,624 | 106,807 | 48,817 | 0 | 8 |
| Passaic | 10,155 | 449,152 | 0.77 | 6,378 | 12.7 | 4,130 | 8.3 | 58,912 | 14.2 | 83,228 | 50,967 | 32,261 | 0 | 0 |
| Salem | 1,258 | 57,144 | 0.80 | 597 | 9.6 | 803 | 12.9 | 4,229 | 8.5 | 14,278 | 10,216 | 4,062 | 0 | 0 |
| Somerset | 5,436 | 352,684 | 1.13 | 3,139 | 9.5 | 2,783 | 8.5 | 18,141 | 6.6 | 57,474 | 41,657 | 15,817 | 0 | 0 |
| Sussex | 1,662 | 111,522 | 0.60 | 1,210 | 8.6 | 1,378 | 9.8 | 6,509 | 5.7 | 29,805 | 21,927 | 7,878 | 1 | 1 |
| Union | 6,158 | 523,131 | 0.89 | 6,643 | 12.0 | 4,423 | 8.0 | 59,289 | 12.6 | 87,512 | 54,469 | 33,043 | 0 | 0 |
| Warren | 1,976 | 88,736 | 0.68 | 1,008 | 9.5 | 1,065 | 10.1 | 6,449 | 7.6 | 22,869 | 16,507 | 6,362 | 0 | 0 |
| **NEW MEXICO** | 42,967 | 2,085,177 | 0.99 | 22,671 | 10.8 | 19,912 | 9.5 | 202,767 | 12.1 | 432,330 | 261,857 | 170,473 | X | X |
| Bernalillo | 11,882 | 707,760 | 1.09 | 6,931 | 10.2 | 6,443 | 9.5 | 60,480 | 10.9 | 130,981 | 60,947 | 70,035 | NA | NA |
| Catron | 106 | 3,539 | 1.01 | 11 | 3.0 | 30 | 8.3 | 181 | 9.1 | 1,480 | 1,163 | 317 | 1 | 15 |
| Chaves | 1,776 | 62,713 | 0.91 | 798 | 12.3 | 677 | 10.5 | 6,736 | 12.9 | 11,991 | 8,845 | 3,145 | 54 | 223 |
| Cibola | 2,569 | 26,787 | 0.99 | 297 | 11.3 | 298 | 11.3 | 2,446 | 12.5 | 4,732 | 3,361 | 1,371 | 12 | 37 |
| Colfax | 408 | 12,202 | 1.01 | 109 | 9.1 | 150 | 12.6 | 841 | 10.1 | 3,693 | 2,720 | 973 | 3 | 8 |
| Curry | 1,503 | 50,326 | 1.03 | 802 | 16.4 | 399 | 8.2 | 4,520 | 11.0 | 7,576 | 6,141 | 1,436 | NA | NA |
| De Baca | 9 | 2,003 | 0.94 | 10 | 6.0 | 22 | 13.2 | 151 | 12.3 | 535 | 506 | 29 | 4 | 15 |
| Dona Ana | 4,412 | 207,435 | 0.90 | 2,598 | 11.7 | 1,777 | 8.0 | 23,882 | 13.5 | 42,625 | 24,542 | 18,084 | NA | NA |
| Eddy | 1,000 | 61,592 | 1.15 | 833 | 14.3 | 584 | 10.0 | 4,891 | 9.8 | 9,748 | 8,621 | 1,127 | NA | NA |
| Grant | 587 | 27,973 | 1.03 | 233 | 8.6 | 368 | 13.6 | 1,645 | 8.8 | 8,777 | 5,914 | 2,863 | 24 | 112 |
| Guadalupe | 574 | 4,652 | 1.24 | 37 | 8.7 | 42 | 9.8 | 217 | 7.7 | 1,002 | 879 | 123 | 0 | 6 |
| Harding | 0 | 409 | 0.81 | 2 | 3.1 | 4 | 6.3 | 37 | 9.7 | 231 | 208 | 23 | NA | NA |
| Hidalgo | 64 | 4,364 | 1.04 | 39 | 9.5 | 50 | 12.2 | 334 | 10.5 | 1,069 | 688 | 381 | 10 | 17 |
| Lea | 2,068 | 71,306 | 1.04 | 1,130 | 15.7 | 550 | 7.7 | 8,145 | 13.3 | 9,111 | 8,156 | 955 | 60 | 342 |
| Lincoln | 116 | 19,580 | 1.02 | 170 | 8.5 | 219 | 11.0 | 1,786 | 13.1 | 6,255 | 4,696 | 1,559 | 43 | 118 |
| Los Alamos | 94 | 26,573 | 1.82 | 176 | 9.0 | 110 | 5.7 | 458 | 2.8 | 3,520 | 3,151 | 370 | NA | NA |
| Luna | 548 | 24,086 | 1.00 | 344 | 14.4 | 287 | 12.0 | 2,524 | 14.2 | 6,642 | 3,512 | 3,130 | 63 | 222 |
| McKinley | 774 | 72,267 | 0.99 | 856 | 12.1 | 677 | 9.6 | 12,033 | 19.9 | 11,290 | 9,675 | 1,615 | 183 | 222 |
| Mora | 8 | 3,893 | 0.56 | 38 | 8.5 | 48 | 10.7 | 281 | 8.9 | 1,449 | 1,183 | 266 | NA | NA |
| Otero | 2,374 | 65,340 | 0.97 | 862 | 12.7 | 647 | 9.5 | 6,008 | 11.3 | 13,402 | 9,711 | 3,690 | 149 | 272 |
| Quay | 22 | 8,288 | 0.99 | 100 | 12.2 | 107 | 13.1 | 638 | 10.6 | 2,525 | 2,168 | 357 | 1 | 13 |
| Rio Arriba | 425 | 35,898 | 0.77 | 397 | 10.3 | 474 | 12.3 | 3,780 | 12.4 | 9,146 | 5,522 | 3,624 | NA | NA |
| Roosevelt | 1,078 | 18,240 | 0.92 | 253 | 13.8 | 177 | 9.6 | 1,916 | 13.1 | 3,082 | 2,713 | 369 | NA | NA |
| Sandoval | 760 | 120,697 | 0.64 | 1,365 | 9.2 | 1,252 | 8.4 | 10,765 | 9.0 | 31,816 | 16,065 | 15,752 | 144 | 102 |
| San Juan | 1,739 | 125,670 | 0.98 | 1,382 | 11.2 | 1,199 | 9.7 | 15,208 | 14.8 | 21,722 | 17,863 | 3,859 | NA | NA |
| San Miguel | 1,294 | 26,623 | 0.88 | 244 | 9.0 | 301 | 11.1 | 1,996 | 10.0 | 7,129 | 4,361 | 2,769 | 0 | 2 |
| Santa Fe | 2,634 | 150,540 | 1.02 | 1,163 | 7.7 | 1,336 | 8.8 | 15,703 | 14.2 | 41,546 | 26,525 | 15,021 | 103 | 414 |
| Sierra | 265 | 11,281 | 1.07 | 86 | 7.9 | 221 | 20.3 | 781 | 11.9 | 4,565 | 2,782 | 1,783 | 43 | 71 |
| Socorro | 581 | 16,158 | 0.87 | 185 | 11.2 | 163 | 9.9 | 1,548 | 12.3 | 3,791 | 2,484 | 1,307 | 15 | 61 |
| Taos | 470 | 32,912 | 1.01 | 244 | 7.5 | 323 | 9.9 | 3,425 | 14.7 | 9,904 | 6,713 | 3,191 | 28 | 93 |
| Torrance | 611 | 13,555 | 0.64 | 155 | 10.0 | 167 | 10.8 | 1,389 | 12.2 | 3,601 | 1,723 | 1,878 | 65 | 105 |
| Union | 667 | 4,099 | 0.98 | 38 | 9.4 | 43 | 10.7 | 330 | 13.1 | 966 | 914 | 52 | 2 | 15 |
| Valencia | 1,549 | 66,416 | 0.67 | 783 | 10.1 | 767 | 9.9 | 7,692 | 12.6 | 16,429 | 7,408 | 9,021 | 376 | 907 |
| **NEW YORK** | 564,192 | 19,993,630 | 1.05 | 220,375 | 11.4 | 177,602 | 9.2 | 973,636 | 6.2 | 3,670,152 | 2,067,099 | 1,603,054 | X | X |
| Albany | 16,698 | 373,668 | 1.42 | 2,831 | 9.3 | 2,964 | 9.8 | 9,639 | 4.0 | 61,754 | 31,905 | 29,848 | 25 | 121 |
| Allegany | 4,096 | 43,406 | 0.83 | 446 | 9.8 | 494 | 10.8 | 1,824 | 5.5 | 10,524 | 5,943 | 4,581 | 3 | 7 |
| Bronx | 44,149 | 1,239,855 | 0.66 | 19,182 | 13.7 | 12,017 | 8.6 | 100,053 | 8.5 | 209,336 | 81,615 | 127,720 | NA | NA |
| Broome | 11,052 | 198,908 | 1.07 | 1,823 | 9.6 | 2,249 | 11.9 | 7,520 | 5.2 | 44,220 | 23,108 | 21,112 | 51 | 486 |
| Cattaraugus | 2,571 | 74,274 | 0.91 | 834 | 11.0 | 914 | 12.0 | 4,088 | 6.9 | 18,473 | 8,901 | 9,571 | 48 | 202 |
| Cayuga | 3,762 | 69,372 | 0.77 | 691 | 9.1 | 759 | 10.0 | 2,770 | 4.8 | 17,100 | 10,074 | 7,025 | NA | NA |
| Chautauqua | 5,731 | 128,594 | 1.00 | 1,263 | 10.0 | 1,505 | 11.9 | 5,210 | 5.4 | 31,238 | 15,458 | 15,780 | 33 | 428 |
| Chemung | 3,606 | 85,648 | 1.02 | 846 | 10.2 | 1,014 | 12.3 | 2,636 | 4.1 | 20,120 | 10,557 | 9,563 | 9 | 221 |
| Chenango | 728 | 45,768 | 0.90 | 445 | 9.5 | 571 | 12.2 | 1,703 | 4.6 | 12,130 | 6,458 | 5,672 | 32 | 274 |
| Clinton | 6,102 | 80,597 | 1.00 | 707 | 8.9 | 806 | 10.1 | 3,018 | 5.0 | 18,510 | 13,287 | 5,224 | 11 | 34 |
| Columbia | 1,863 | 55,695 | 0.84 | 449 | 7.5 | 710 | 11.9 | 2,524 | 5.7 | 16,166 | 10,516 | 5,649 | 25 | 251 |
| Cortland | 3,689 | 46,767 | 0.95 | 444 | 9.4 | 458 | 9.7 | 1,506 | 4.2 | 9,441 | 6,028 | 3,413 | NA | NA |
| Delaware | 2,349 | 45,493 | 1.03 | 368 | 8.4 | 622 | 14.2 | 1,789 | 5.7 | 12,157 | 7,691 | 4,466 | 22 | 151 |
| Dutchess | 19,505 | 269,630 | 0.83 | 2,549 | 8.7 | 3,006 | 10.2 | 12,124 | 5.3 | 61,133 | 44,672 | 16,461 | 38 | 353 |
| Erie | 27,094 | 949,598 | 1.07 | 9,679 | 10.6 | 10,469 | 11.4 | 29,797 | 4.1 | 200,998 | 74,335 | 126,664 | 50 | 505 |
| Essex | 2,114 | 37,078 | 0.98 | 285 | 7.7 | 450 | 12.2 | 1,308 | 4.9 | 9,905 | 6,965 | 2,940 | 2 | 4 |
| Franklin | 6,139 | 50,622 | 1.01 | 464 | 9.3 | 567 | 11.3 | 2,492 | 7.0 | 11,298 | 7,991 | 3,308 | 0 | 0 |
| Fulton | 1,361 | 47,953 | 0.76 | 492 | 9.3 | 612 | 11.6 | 2,014 | 4.8 | 13,714 | 6,333 | 7,381 | 22 | 288 |

1. Per 1,000 estimated resident population.

# Table B. States and Counties — Crime, Education, Money Income, and Poverty

| | County law enforcement employment, 2019 | | Education | | | | | | Money income, 2015-2019 | | | | Income and poverty, 2019 | | |
| STATE County | | | School enrollment and attainment, 2015-2019 | | | | Local government expenditures,[3] 2017-2018 | | | Households | | | | Percent below poverty level | | |
| | | | Enrollment[1] | | Attainment[2] (percent) | | | | | | | Percent | | | | | |
| | Officers | Civilians | Total | Percent private | High school graduate or less | Bachelor's degree or more | Total current spending (mil dol) | Current spending per student (dollars) | Per capita income[4] | Median income (dollars) | with income of less than $50,000 | with income of $200,000 or more | Median household income (dollars) | All persons | Children under 18 years | Children 5 to 17 years in families |
| | 46 | 47 | 48 | 49 | 50 | 51 | 52 | 53 | 54 | 55 | 56 | 57 | 58 | 59 | 60 | 61 |
| **NEW JERSEY— Cont'd** | | | | | | | | | | | | | | | | |
| Mercer | 147 | 44 | 97,811 | 23.5 | 36.1 | 42.6 | 1,175 | 18,518 | 43,086 | 81,057 | 32.3 | 14.9 | 79,475 | 11.9 | 13.4 | 12.8 |
| Middlesex | 189 | 44 | 219,151 | 14.2 | 35.5 | 43.6 | 2,331 | 18,143 | 39,599 | 89,533 | 27.5 | 13.4 | 92,770 | 8.5 | 10.8 | 10.8 |
| Monmouth | 435 | 177 | 152,904 | 21.3 | 30.0 | 46.0 | 2,007 | 20,366 | 51,700 | 99,733 | 26.0 | 19.0 | 102,579 | 6.2 | 6.9 | 6.1 |
| Morris | 91 | 30 | 123,086 | 21.0 | 25.6 | 54.1 | 1,557 | 20,478 | 58,109 | 115,527 | 20.1 | 23.8 | 116,328 | 5.3 | 5.4 | 5.3 |
| Ocean | 157 | 121 | 146,281 | 37.4 | 42.0 | 30.0 | 1,323 | 18,772 | 36,100 | 70,909 | 35.8 | 8.0 | 75,627 | 9.0 | 14.0 | 12.9 |
| Passaic | 540 | 137 | 131,469 | 15.1 | 50.8 | 28.2 | 1,675 | 18,790 | 32,064 | 69,688 | 37.9 | 9.4 | 76,039 | 13.3 | 20.4 | 18.8 |
| Salem | 190 | 34 | 14,303 | 12.5 | 49.6 | 21.0 | 213 | 19,661 | 34,047 | 66,842 | 37.9 | 6.1 | 68,146 | 11.4 | 16.8 | 15.6 |
| Somerset | 177 | 33 | 83,243 | 18.4 | 25.8 | 54.7 | 1,057 | 19,660 | 55,828 | 113,611 | 20.3 | 24.4 | 112,722 | 5.4 | 6.2 | 5.3 |
| Sussex | 119 | 22 | 33,020 | 16.5 | 35.7 | 36.1 | 459 | 22,649 | 44,728 | 94,520 | 22.4 | 12.5 | 100,281 | 5.2 | 6.1 | 5.5 |
| Union | 213 | 47 | 139,721 | 14.8 | 42.3 | 36.0 | 1,856 | 18,715 | 41,603 | 80,198 | 31.6 | 14.3 | 79,953 | 9.1 | 11.9 | 11.3 |
| Warren | 20 | 4 | 24,799 | 17.0 | 39.1 | 33.0 | 325 | 19,632 | 39,802 | 81,307 | 30.6 | 9.6 | 83,998 | 7.1 | 9.2 | 8.5 |
| **NEW MEXICO** | X | X | 532,619 | 9.7 | 40.9 | 27.3 | 3,195 | 9,557 | 27,230 | 49,754 | 50.2 | 4.1 | 52,021 | 17.5 | 23.5 | 22.4 |
| Bernalillo | NA | NA | 173,685 | 12.4 | 34.3 | 34.4 | 893 | 9,085 | 30,734 | 53,329 | 47.1 | 5.0 | 56,148 | 15.5 | 20.4 | 18.6 |
| Catron | NA | NA | 543 | 3.3 | 53.0 | 15.7 | 6 | 21,342 | 21,989 | 41,910 | 59.2 | 2.1 | 37,190 | 20.6 | 38.3 | 36.9 |
| Chaves | NA | NA | 17,274 | 15.2 | 49.5 | 18.0 | 109 | 9,188 | 22,041 | 43,359 | 56.3 | 2.2 | 43,687 | 18.1 | 23.7 | 22.0 |
| Cibola | NA | NA | 6,582 | 11.2 | 49.2 | 14.2 | 37 | 10,372 | 20,018 | 39,413 | 58.3 | 0.8 | 42,754 | 25.5 | 32.5 | 31.7 |
| Colfax | 12 | 2 | 2,005 | 5.5 | 41.1 | 21.7 | 21 | 12,765 | 24,035 | 36,302 | 61.9 | 0.8 | 45,623 | 17.0 | 27.0 | 26.5 |
| Curry | NA | NA | 13,642 | 6.4 | 44.8 | 19.2 | 83 | 9,202 | 23,021 | 45,092 | 54.9 | 1.8 | 34,746 | 18.7 | 25.6 | 28.0 |
| De Baca | NA | NA | 700 | 6.9 | 52.0 | 13.0 | 4 | 13,685 | 17,376 | 31,625 | 64.9 | 0.0 | 43,618 | 18.6 | 29.8 | 31.6 |
| Dona Ana | NA | NA | 67,375 | 5.5 | 43.3 | 27.1 | 372 | 9,157 | 22,154 | 40,973 | 57.7 | 2.7 | 63,671 | 23.8 | 31.3 | 14.0 |
| Eddy | NA | NA | 13,578 | 5.2 | 52.2 | 16.2 | 113 | 9,718 | 30,246 | 65,328 | 40.2 | 4.6 | 43,175 | 11.2 | 14.5 | 25.2 |
| Grant | NA | NA | 5,777 | 8.2 | 39.0 | 28.8 | 45 | 11,047 | 25,590 | 37,843 | 61.6 | 2.6 | 36,554 | 18.1 | 27.0 | 26.6 |
| Guadalupe | NA | NA | 827 | 11.0 | 64.8 | 7.6 | 10 | 13,901 | 17,420 | 24,798 | 79.2 | 1.0 | 38,162 | 23.5 | 28.8 | 15.7 |
| Harding | NA | NA | 69 | 10.1 | 41.1 | 21.7 | 3 | 36,841 | 34,721 | 29,375 | 69.7 | 5.7 | 37,322 | 15.4 | 15.9 | 31.4 |
| Hidalgo | NA | NA | 711 | 7.3 | 52.8 | 14.6 | 9 | 13,673 | 22,009 | 42,526 | 58.0 | 1.0 | 63,012 | 22.9 | 32.4 | 16.8 |
| Lea | NA | NA | 18,966 | 5.3 | 57.2 | 13.6 | 131 | 8,567 | 25,585 | 60,546 | 41.3 | 3.8 | 47,254 | 15.1 | 18.1 | 25.7 |
| Lincoln | 25 | 12 | 3,469 | 8.9 | 40.4 | 25.8 | 32 | 11,304 | 27,454 | 46,216 | 52.8 | 3.4 | 122,001 | 13.7 | 25.9 | 2.8 |
| Los Alamos | NA | NA | 4,740 | 9.6 | 11.2 | 67.4 | 40 | 10,653 | 60,746 | 121,324 | 17.5 | 20.4 | 32,544 | 3.5 | 3.2 | 34.1 |
| Luna | NA | NA | 5,239 | 3.3 | 64.6 | 12.5 | 53 | 9,701 | 17,168 | 29,360 | 67.9 | 0.9 | 36,404 | 23.8 | 33.2 | 39.3 |
| McKinley | 34 | 9 | 20,477 | 5.5 | 57.1 | 11.4 | 147 | 11,377 | 15,931 | 33,834 | 63.0 | 2.1 | 36,968 | 30.1 | 39.0 | 28.3 |
| Mora | NA | NA | 678 | 1.9 | 44.8 | 11.0 | 10 | 20,175 | 19,864 | 28,446 | 68.8 | 0.8 | 43,135 | 21.2 | 29.7 | 26.7 |
| Otero | NA | NA | 15,058 | 9.5 | 43.8 | 17.5 | 67 | 9,248 | 23,170 | 41,988 | 57.8 | 1.4 | 34,596 | 20.1 | 27.7 | 33.6 |
| Quay | NA | NA | 1,931 | 2.1 | 53.8 | 17.1 | 18 | 12,003 | 19,784 | 29,035 | 68.6 | 0.7 | 44,579 | 21.7 | 34.2 | 25.8 |
| Rio Arriba | NA | NA | 8,998 | 6.8 | 45.8 | 19.3 | 65 | 12,634 | 22,911 | 39,952 | 59.5 | 2.9 | 44,928 | 22.2 | 27.0 | 24.8 |
| Roosevelt | NA | NA | 6,077 | 1.9 | 48.1 | 23.9 | 36 | 10,473 | 20,123 | 42,702 | 57.3 | 1.1 | 71,118 | 18.8 | 25.2 | 13.1 |
| Sandoval | NA | NA | 36,473 | 11.7 | 35.6 | 30.5 | 195 | 8,835 | 29,255 | 63,802 | 38.8 | 5.0 | 46,201 | 10.0 | 14.2 | 25.5 |
| San Juan | NA | NA | 34,222 | 9.0 | 45.6 | 14.9 | 219 | 9,218 | 23,235 | 50,518 | 49.5 | 2.4 | 36,986 | 19.9 | 26.8 | 30.4 |
| San Miguel | 8 | 3 | 6,951 | 8.7 | 44.4 | 21.4 | 36 | 11,526 | 20,762 | 30,946 | 68.2 | 1.3 | 61,791 | 23.8 | 31.7 | 14.8 |
| Santa Fe | 14 | 2 | 30,520 | 13.0 | 33.0 | 41.0 | 187 | 9,679 | 38,520 | 61,200 | 41.4 | 7.8 | 33,661 | 12.4 | 16.4 | 42.7 |
| Sierra | NA | NA | 1,595 | 7.5 | 44.7 | 21.1 | 14 | 10,689 | 25,344 | 29,755 | 71.2 | 2.3 | 39,582 | 27.6 | 39.7 | 35.9 |
| Socorro | NA | NA | 4,762 | 3.5 | 55.6 | 22.0 | 24 | 11,801 | 18,948 | 42,083 | 61.1 | 1.4 | 43,032 | 26.0 | 37.6 | 24.4 |
| Taos | NA | NA | 6,713 | 7.3 | 38.1 | 28.7 | 47 | 11,104 | 25,646 | 38,329 | 60.5 | 3.2 | 39,532 | 18.2 | 24.9 | 28.4 |
| Torrance | NA | NA | 3,358 | 9.9 | 53.1 | 14.8 | 45 | 10,305 | 20,876 | 36,120 | 61.0 | 2.0 | 38,221 | 21.6 | 30.2 | 26.4 |
| Union | NA | NA | 721 | 9.8 | 64.9 | 11.7 | 8 | 13,853 | 22,116 | 35,884 | 64.3 | 1.0 | 54,867 | 20.3 | 27.0 | 21.5 |
| Valencia | NA | NA | 18,903 | 11.1 | 48.6 | 18.5 | 118 | 9,147 | 23,675 | 48,945 | 51.5 | 2.2 | 72,038 | 17.6 | 23.2 | 17.2 |
| **NEW YORK** | X | X | 4,723,073 | 23.8 | 39.2 | 36.6 | 62,238 | 22,845 | 39,326 | 68,486 | 38.3 | 10.7 | 69,408 | 13.1 | 18.2 | 13.8 |
| Albany | 146 | 63 | 82,502 | 21.1 | 30.8 | 41.8 | 720 | 17,245 | 37,635 | 66,252 | 38.3 | 7.6 | 49,411 | 11.7 | 16.0 | 25.6 |
| Allegany | NA | NA | 13,345 | 24.5 | 45.9 | 20.9 | 144 | 21,590 | 24,857 | 48,412 | 51.7 | 1.6 | 41,470 | 17.9 | 26.8 | 35.6 |
| Bronx | NA | NA | 396,807 | 16.5 | 55.2 | 20.1 | (7) | (7) | 21,778 | 40,088 | 58.2 | 3.0 | 52,179 | 26.2 | 36.5 | 22.8 |
| Broome | 50 | 9 | 52,105 | 9.3 | 40.8 | 28.4 | 537 | 20,126 | 28,699 | 52,226 | 48.4 | 3.5 | 50,783 | 17.8 | 24.9 | 20.2 |
| Cattaraugus | 50 | 26 | 17,571 | 18.3 | 50.3 | 19.1 | 271 | 20,504 | 25,822 | 48,703 | 51.0 | 2.2 | 58,055 | 14.7 | 21.5 | 19.0 |
| Cayuga | 29 | 7 | 16,097 | 14.6 | 45.5 | 22.6 | 179 | 18,327 | 30,509 | 58,377 | 43.3 | 3.2 | 50,143 | 13.8 | 20.5 | 23.6 |
| Chautauqua | 71 | 43 | 28,561 | 8.1 | 46.7 | 22.2 | 373 | 19,123 | 26,165 | 46,820 | 52.8 | 1.9 | 58,400 | 16.3 | 24.4 | 19.5 |
| Chemung | 43 | 5 | 17,874 | 17.2 | 45.0 | 23.3 | 205 | 17,448 | 28,778 | 54,940 | 45.7 | 3.1 | 51,894 | 14.1 | 20.3 | 16.9 |
| Chenango | 22 | 0 | 9,527 | 10.2 | 50.9 | 18.7 | 170 | 22,466 | 27,708 | 52,002 | 48.5 | 2.5 | 59,069 | 13.7 | 20.1 | 15.6 |
| Clinton | 26 | 13 | 17,935 | 7.2 | 50.0 | 22.9 | 237 | 21,526 | 27,761 | 56,365 | 44.2 | 2.5 | 71,563 | 14.8 | 16.8 | 13.8 |
| Columbia | 59 | 16 | 10,373 | 17.8 | 39.7 | 32.7 | 170 | 24,106 | 38,268 | 66,787 | 37.0 | 6.8 | 55,143 | 10.1 | 15.4 | 18.4 |
| Cortland | 33 | 3 | 14,363 | 10.2 | 41.4 | 28.0 | 125 | 19,906 | 28,229 | 56,023 | 45.0 | 3.9 | 51,038 | 17.3 | 19.3 | 18.4 |
| Delaware | 18 | 4 | 9,497 | 7.4 | 48.6 | 22.6 | 149 | 24,602 | 27,701 | 49,544 | 50.4 | 2.7 | 84,564 | 12.6 | 19.1 | 9.2 |
| Dutchess | 105 | 22 | 72,053 | 27.0 | 35.7 | 35.0 | 901 | 22,482 | 40,093 | 81,219 | 31.4 | 11.0 | 60,620 | 8.6 | 9.8 | 18.8 |
| Erie | NA | NA | 217,704 | 17.1 | 35.8 | 34.4 | 2,214 | 17,371 | 33,598 | 58,121 | 43.6 | 5.0 | 55,091 | 13.3 | 19.3 | 16.2 |
| Essex | 16 | 1 | 6,387 | 13.4 | 43.0 | 26.8 | 99 | 25,600 | 31,667 | 56,763 | 43.4 | 3.6 | 48,943 | 12.4 | 17.1 | 23.6 |
| Franklin | 5 | 2 | 9,513 | 17.5 | 50.3 | 19.1 | 165 | 22,284 | 25,406 | 50,407 | 49.7 | 2.9 | 50,061 | 18.1 | 25.1 | 22.6 |
| Fulton | 28 | 4 | 10,691 | 6.6 | 48.4 | 18.2 | 149 | 18,865 | 27,443 | 50,482 | 49.6 | 2.4 | | 16.4 | 26.4 | |

1. All persons 3 years old and over enrolled in nursery school through college.    2. Persons 25 years old and over.    3. Elementary and secondary education expenditures.    4. Based on population estimated by the American Community Survey, 2014–2018.    7. Bronx, Kings, Queens, and Richmond counties are included with New York county.

# Table B. States and Counties — **Personal Income**

| STATE County | Personal income, 2019 | | | | | | | | | | Earnings, 2019 | | |
|---|---|---|---|---|---|---|---|---|---|---|---|---|---|
| | Total (mil dol) | Percent change 2018-2019 | Per capita[1] Dollars | Per capita[1] Rank | Wages and salaries (mil dol) | Supplements to wages and salaries, employer contributions (mil dol) Pension and insurance | Supplements — Government social insurance | Proprietors' income (mil dol) | Dividends, interest, and rent (mil dol) | Personal transfer receipts (mil dol) | Total (mil dol) | Contributions for government social insurance (mil dol) From employee and self-employed | From employer |
| | 62 | 63 | 64 | 65 | 66 | 67 | 68 | 69 | 70 | 71 | 72 | 73 | 74 |
| **NEW JERSEY— Cont'd** | | | | | | | | | | | | | |
| Mercer | 26,378 | 3.7 | 71,790 | 111 | 18,574 | 2,835 | 1,280 | 2,014 | 5,479 | 3,626 | 24,704 | 1,443 | 1,280 |
| Middlesex | 52,356 | 3.9 | 63,457 | 219 | 31,040 | 4,652 | 2,258 | 3,912 | 8,506 | 6,972 | 41,863 | 2,483 | 2,258 |
| Monmouth | 51,082 | 3.6 | 82,551 | 55 | 16,600 | 2,672 | 1,240 | 4,393 | 10,034 | 6,284 | 24,906 | 1,521 | 1,240 |
| Morris | 49,994 | 3.7 | 101,646 | 19 | 28,094 | 3,527 | 1,846 | 6,091 | 10,160 | 4,137 | 39,559 | 2,281 | 1,846 |
| Ocean | 32,265 | 4.5 | 53,139 | 605 | 8,682 | 1,661 | 684 | 2,731 | 5,590 | 7,458 | 13,758 | 984 | 684 |
| Passaic | 26,160 | 3.8 | 52,129 | 672 | 9,827 | 1,769 | 746 | 2,314 | 3,890 | 5,159 | 14,656 | 908 | 746 |
| Salem | 3,132 | 3.8 | 50,203 | 814 | 1,296 | 282 | 99 | 175 | 441 | 783 | 1,852 | 119 | 99 |
| Somerset | 36,155 | 4.1 | 109,916 | 17 | 18,822 | 2,324 | 1,221 | 7,246 | 6,610 | 2,606 | 29,612 | 1,641 | 1,221 |
| Sussex | 8,961 | 4.1 | 63,784 | 213 | 2,034 | 390 | 156 | 715 | 1,334 | 1,326 | 3,295 | 219 | 156 |
| Union | 39,425 | 3.7 | 70,865 | 123 | 17,785 | 2,654 | 1,199 | 3,945 | 7,019 | 4,855 | 25,584 | 1,504 | 1,199 |
| Warren | 6,090 | 4.1 | 57,854 | 376 | 1,862 | 342 | 139 | 411 | 928 | 1,078 | 2,754 | 182 | 139 |
| **NEW MEXICO** | 90,847 | 4.2 | 43,268 | X | 42,803 | 7,161 | 3,229 | 5,799 | 17,335 | 21,597 | 58,992 | 3,931 | 3,229 |
| Bernalillo | 30,853 | 3.6 | 45,431 | 1,310 | 18,397 | 2,920 | 1,382 | 1,518 | 6,022 | 6,439 | 24,218 | 1,572 | 1,382 |
| Catron | 116 | -0.7 | 32,866 | 2,900 | 24 | 7 | 2 | 10 | 30 | 48 | 42 | 4 | 2 |
| Chaves | 2,697 | 5.8 | 41,746 | 1,813 | 861 | 151 | 66 | 379 | 454 | 715 | 1,457 | 92 | 66 |
| Cibola | 784 | 4.0 | 29,384 | 3,054 | 308 | 64 | 23 | 29 | 111 | 287 | 423 | 30 | 23 |
| Colfax | 507 | 2.2 | 42,497 | 1,691 | 168 | 33 | 13 | 44 | 111 | 168 | 257 | 19 | 13 |
| Curry | 2,199 | 5.6 | 44,914 | 1,371 | 1,036 | 231 | 95 | 234 | 362 | 501 | 1,597 | 82 | 95 |
| De Baca | 91 | 5.0 | 52,191 | 664 | 18 | 4 | 2 | 23 | 21 | 28 | 48 | 3 | 2 |
| Dona Ana | 8,238 | 4.8 | 37,756 | 2,350 | 3,236 | 626 | 250 | 839 | 1,320 | 2,285 | 4,951 | 334 | 250 |
| Eddy | 3,488 | 9.9 | 59,661 | 317 | 2,355 | 328 | 165 | 213 | 609 | 555 | 3,062 | 190 | 165 |
| Grant | 1,160 | 3.1 | 42,959 | 1,625 | 431 | 90 | 31 | 53 | 217 | 421 | 605 | 47 | 31 |
| Guadalupe | 153 | 2.7 | 35,516 | 2,618 | 52 | 10 | 4 | 8 | 25 | 59 | 74 | 5 | 4 |
| Harding | 28 | 4.6 | 44,883 | 1,376 | 7 | 2 | 1 | 2 | 9 | 8 | 11 | 1 | 1 |
| Hidalgo | 188 | 3.9 | 44,716 | 1,393 | 71 | 18 | 6 | 19 | 29 | 61 | 113 | 7 | 6 |
| Lea | 3,485 | 6.1 | 49,039 | 921 | 2,048 | 277 | 143 | 425 | 366 | 589 | 2,893 | 178 | 143 |
| Lincoln | 840 | 4.6 | 42,900 | 1,633 | 247 | 43 | 18 | 65 | 222 | 261 | 373 | 31 | 18 |
| Los Alamos | 1,430 | 5.2 | 73,821 | 98 | 1,601 | 151 | 121 | 57 | 257 | 119 | 1,930 | 120 | 121 |
| Luna | 804 | 4.4 | 33,927 | 2,810 | 304 | 66 | 25 | 75 | 120 | 308 | 470 | 34 | 25 |
| McKinley | 2,121 | 5.2 | 29,714 | 3,045 | 873 | 212 | 69 | 91 | 303 | 755 | 1,245 | 83 | 69 |
| Mora | 181 | 5.2 | 39,987 | 2,052 | 26 | 7 | 2 | 17 | 33 | 72 | 52 | 4 | 2 |
| Otero | 2,422 | 4.1 | 35,889 | 2,581 | 1,037 | 251 | 87 | 103 | 478 | 655 | 1,477 | 93 | 87 |
| Quay | 343 | 4.6 | 41,569 | 1,834 | 112 | 21 | 8 | 32 | 60 | 136 | 174 | 14 | 8 |
| Rio Arriba | 1,405 | 3.6 | 36,100 | 2,556 | 416 | 84 | 31 | 47 | 199 | 508 | 577 | 46 | 31 |
| Roosevelt | 802 | 7.0 | 43,330 | 1,564 | 224 | 50 | 17 | 130 | 115 | 202 | 421 | 21 | 17 |
| Sandoval | 6,329 | 5.0 | 43,125 | 1,596 | 1,534 | 237 | 113 | 293 | 975 | 1,334 | 2,177 | 173 | 113 |
| San Juan | 4,462 | 0.8 | 35,999 | 2,571 | 2,373 | 409 | 176 | 134 | 665 | 1,160 | 3,091 | 205 | 176 |
| San Miguel | 985 | 2.9 | 36,105 | 2,555 | 293 | 71 | 22 | 33 | 179 | 427 | 419 | 33 | 22 |
| Santa Fe | 9,063 | 3.1 | 60,276 | 294 | 3,152 | 496 | 228 | 594 | 3,066 | 1,570 | 4,470 | 318 | 228 |
| Sierra | 456 | 5.3 | 42,255 | 1,730 | 122 | 25 | 10 | 30 | 98 | 216 | 187 | 17 | 10 |
| Socorro | 597 | 3.6 | 35,867 | 2,585 | 223 | 53 | 17 | 42 | 101 | 206 | 334 | 22 | 17 |
| Taos | 1,270 | 1.9 | 38,810 | 2,222 | 403 | 70 | 32 | 82 | 324 | 423 | 587 | 49 | 32 |
| Torrance | 517 | 4.6 | 33,446 | 2,851 | 134 | 27 | 11 | 46 | 70 | 189 | 217 | 16 | 11 |
| Union | 153 | 1.2 | 37,763 | 2,348 | 57 | 10 | 4 | 22 | 27 | 46 | 94 | 6 | 4 |
| Valencia | 2,681 | 4.4 | 34,964 | 2,691 | 663 | 117 | 55 | 109 | 360 | 843 | 945 | 79 | 55 |
| **NEW YORK** | 1,395,147 | 4.0 | 71,682 | X | 740,281 | 120,798 | 50,264 | 141,598 | 302,178 | 229,559 | 1,052,941 | 56,356 | 50,264 |
| Albany | 19,258 | 3.6 | 63,037 | 229 | 14,928 | 3,525 | 1,143 | 1,408 | 3,933 | 3,448 | 21,004 | 1,086 | 1,143 |
| Allegany | 1,667 | 3.4 | 36,164 | 2,548 | 558 | 181 | 46 | 92 | 253 | 507 | 877 | 56 | 46 |
| Bronx | 56,319 | 5.0 | 39,711 | 2,084 | 19,881 | 4,619 | 1,555 | 3,266 | 6,801 | 18,110 | 29,322 | 1,719 | 1,555 |
| Broome | 8,831 | 3.8 | 46,361 | 1,205 | 4,214 | 1,063 | 330 | 530 | 1,502 | 2,189 | 6,137 | 358 | 330 |
| Cattaraugus | 3,180 | 4.1 | 41,774 | 1,806 | 1,225 | 373 | 98 | 213 | 484 | 918 | 1,909 | 115 | 98 |
| Cayuga | 3,409 | 4.7 | 44,520 | 1,416 | 1,204 | 314 | 98 | 216 | 540 | 821 | 1,833 | 109 | 98 |
| Chautauqua | 5,375 | 4.2 | 42,355 | 1,711 | 2,087 | 545 | 172 | 350 | 800 | 1,585 | 3,154 | 194 | 172 |
| Chemung | 3,874 | 3.7 | 46,421 | 1,196 | 1,760 | 414 | 140 | 192 | 571 | 1,028 | 2,507 | 150 | 140 |
| Chenango | 2,084 | 4.5 | 44,146 | 1,464 | 868 | 229 | 68 | 125 | 322 | 556 | 1,290 | 78 | 68 |
| Clinton | 3,631 | 3.6 | 45,112 | 1,347 | 1,653 | 449 | 135 | 191 | 532 | 924 | 2,428 | 142 | 135 |
| Columbia | 3,456 | 3.9 | 58,123 | 366 | 945 | 241 | 77 | 248 | 787 | 749 | 1,511 | 93 | 77 |
| Cortland | 2,010 | 3.2 | 42,252 | 1,732 | 802 | 228 | 64 | 125 | 330 | 480 | 1,220 | 70 | 64 |
| Delaware | 1,863 | 5.7 | 42,220 | 1,735 | 740 | 219 | 58 | 121 | 354 | 532 | 1,138 | 70 | 58 |
| Dutchess | 17,205 | 4.7 | 58,478 | 357 | 6,597 | 1,363 | 514 | 987 | 3,104 | 3,102 | 9,461 | 543 | 514 |
| Erie | 49,148 | 4.0 | 53,498 | 573 | 25,772 | 5,457 | 1,996 | 3,558 | 8,240 | 10,409 | 36,783 | 2,051 | 1,996 |
| Essex | 1,777 | 3.8 | 48,181 | 1,003 | 665 | 188 | 53 | 106 | 364 | 456 | 1,013 | 62 | 53 |
| Franklin | 1,983 | 5.5 | 39,649 | 2,095 | 855 | 288 | 68 | 106 | 319 | 558 | 1,317 | 77 | 68 |
| Fulton | 2,427 | 4.2 | 45,459 | 1,305 | 731 | 189 | 59 | 175 | 351 | 690 | 1,153 | 74 | 59 |

1. Based on the resident population estimated as of July 1 of the year shown.

## Table B. States and Counties — Earnings, Social Security, and Housing

| STATE County | Farm | Mining, quarrying, and extractions | Construction | Manufacturing | Information; professional, scientific, technical services | Retail trade | Finance, insurance, real estate, and leasing | Health care and social assistance | Government | Social Security beneficiaries, Dec. 2019 — Number | Rate[1] | Supplemental Security Income recipients, 2019 | Housing units, 2020 — Total | Percent change, 2010-2020 |
|---|---|---|---|---|---|---|---|---|---|---|---|---|---|---|
| | 75 | 76 | 77 | 78 | 79 | 80 | 81 | 82 | 83 | 84 | 85 | 86 | 87 | 88 |
| **NEW JERSEY— Cont'd** | | | | | | | | | | | | | | |
| Mercer | 0.0 | 0.9 | 2.9 | 7.4 | 19.3 | 3.4 | 12.6 | 8.7 | 16.7 | 66,675 | 181 | 9,557 | 145,739 | 1.8 |
| Middlesex | 0.0 | D | 4.7 | 8.0 | 18.2 | 5.1 | 7.0 | 9.4 | 13.3 | 136,335 | 165 | 13,706 | 305,038 | 3.4 |
| Monmouth | 0.1 | D | 9.1 | 3.4 | 17.1 | 7.6 | 8.7 | 16.4 | 13.3 | 127,610 | 206 | 7,686 | 263,305 | 1.9 |
| Morris | 0.0 | D | 4.1 | 8.6 | 22.7 | 5.0 | 11.1 | 9.2 | 8.4 | 88,285 | 180 | 4,332 | 196,813 | 3.7 |
| Ocean | 0.1 | 0.6 | 10.3 | 3.9 | 8.7 | 10.9 | 4.6 | 17.6 | 19.4 | 159,720 | 263 | 7,461 | 287,044 | 3.2 |
| Passaic | 0.0 | 0.2 | 7.9 | 11.0 | 7.7 | 8.6 | 6.0 | 14.0 | 19.5 | 86,255 | 172 | 13,964 | 177,453 | 0.8 |
| Salem | 1.8 | 0.1 | D | 11.0 | D | 4.8 | 2.5 | 10.4 | 19.3 | 15,750 | 253 | 1,723 | 27,594 | 0.7 |
| Somerset | 0.0 | 0.2 | 3.5 | 25.0 | 19.5 | 4.2 | 7.6 | 5.9 | 6.1 | 54,935 | 166 | 2,917 | 127,868 | 3.8 |
| Sussex | 0.0 | 0.6 | 10.2 | 7.6 | 8.4 | 8.4 | 4.8 | 13.0 | 20.2 | 30,865 | 219 | 1,650 | 62,694 | 1.2 |
| Union | 0.0 | D | 5.6 | 10.0 | 17.6 | 5.1 | 6.7 | 10.2 | 12.9 | 88,480 | 159 | 10,754 | 203,877 | 2.2 |
| Warren | 1.0 | D | D | 12.2 | 9.6 | 9.2 | 2.3 | 12.4 | 17.3 | 23,560 | 223 | 1,560 | 45,724 | 1.8 |
| **NEW MEXICO** | 1.8 | 4.8 | 6.6 | 3.7 | 12.7 | 6.2 | 5.2 | 11.9 | 25.6 | 445,742 | 212 | 62,064 | 955,996 | 6.1 |
| Bernalillo | 0.0 | 0.1 | 6.4 | 3.5 | 17.3 | 6.1 | 7.0 | 13.3 | 25.0 | 132,820 | 195 | 17,412 | 297,669 | 4.7 |
| Catron | 11.1 | D | 5.4 | 3.5 | D | D | D | D | 42.6 | 1,510 | 431 | 79 | 3,424 | 4.1 |
| Chaves | 11.7 | 8.3 | 4.2 | 4.7 | 4.8 | 9.0 | 4.5 | 12.9 | 20.4 | 13,095 | 203 | 2,088 | 27,562 | 3.2 |
| Cibola | 1.0 | D | 2.2 | 1.1 | 1.2 | 6.6 | 1.6 | D | 39.8 | 5,365 | 201 | 930 | 11,484 | 3.4 |
| Colfax | 7.9 | 2.6 | 4.8 | 2.4 | 2.2 | 8.5 | 4.6 | 6.1 | 32.3 | 3,915 | 324 | 423 | 10,366 | 3.4 |
| Curry | 12.1 | 0.1 | 3.1 | 2.8 | 2.5 | 5.1 | 2.7 | 10.0 | 41.3 | 7,710 | 157 | 1,492 | 21,658 | 8.0 |
| De Baca | 35.6 | 1.1 | D | D | D | 6.8 | D | 5.1 | 19.1 | 555 | 319 | 73 | 1,378 | 2.5 |
| Dona Ana | 3.0 | 0.1 | 6.6 | 4.0 | 7.7 | 5.5 | 4.8 | 15.7 | 29.3 | 44,265 | 202 | 8,308 | 91,091 | 11.8 |
| Eddy | 1.5 | 35.3 | 11.0 | 4.0 | 4.5 | 4.5 | 3.7 | 5.7 | 10.7 | 10,480 | 179 | 1,139 | 25,602 | 13.4 |
| Grant | 1.4 | D | 3.7 | D | 3.0 | 6.3 | 2.8 | 8.4 | 32.1 | 9,090 | 337 | 884 | 15,193 | 3.4 |
| Guadalupe | 8.1 | 0.1 | D | D | D | 12.4 | D | 12.9 | 30.4 | 1,115 | 261 | 219 | 2,474 | 3.0 |
| Harding | 23.0 | 0.0 | D | D | D | D | 0.1 | D | 45.1 | 210 | 330 | 0 | 544 | 3.0 |
| Hidalgo | 10.0 | D | D | D | 2.3 | 6.8 | D | D | 50.5 | 1,165 | 277 | 169 | 2,492 | 4.1 |
| Lea | 3.1 | 28.8 | 11.2 | 4.3 | 3.0 | 5.7 | 4.6 | 4.7 | 9.1 | 9,905 | 139 | 1,467 | 27,403 | 10.0 |
| Lincoln | 2.0 | D | 7.3 | 1.4 | D | 12.8 | 8.2 | 12.9 | 21.2 | 6,420 | 325 | 383 | 18,478 | 5.5 |
| Los Alamos | 0.0 | 0.0 | 1.4 | D | D | 0.9 | 1.9 | 3.2 | 8.1 | 3,240 | 167 | 86 | 8,442 | 1.1 |
| Luna | 8.8 | 0.2 | 6.0 | 9.2 | D | 6.4 | 2.0 | D | 34.9 | 7,170 | 302 | 1,507 | 11,399 | 3.6 |
| McKinley | -0.1 | 0.3 | 2.7 | 6.9 | 1.9 | 9.3 | 3.2 | 12.0 | 43.3 | 11,910 | 167 | 4,520 | 26,450 | 2.5 |
| Mora | 25.3 | D | D | D | D | 4.7 | D | | 30.6 | 1,585 | 353 | 272 | 3,342 | 3.4 |
| Otero | 0.6 | 0.4 | 4.9 | 0.3 | 4.4 | 4.9 | 2.5 | 12.1 | 55.0 | 13,870 | 205 | 1,447 | 32,013 | 3.3 |
| Quay | 5.2 | D | 5.3 | D | 2.0 | 8.2 | 4.3 | 14.0 | 26.7 | 2,685 | 326 | 376 | 5,732 | 2.9 |
| Rio Arriba | 1.7 | 3.9 | 5.1 | 0.9 | D | 6.1 | 1.8 | 16.5 | 41.2 | 10,365 | 267 | 1,676 | 20,389 | 3.9 |
| Roosevelt | 28.0 | 0.3 | 2.5 | 5.4 | D | 5.4 | 2.7 | D | 22.7 | 3,275 | 177 | 543 | 8,602 | 5.4 |
| Sandoval | 0.2 | 0.6 | 8.1 | 16.9 | 6.3 | 5.9 | 4.1 | 9.2 | 22.7 | 32,240 | 219 | 2,645 | 59,322 | 13.4 |
| San Juan | 0.8 | 16.3 | 9.8 | 2.8 | D | 6.5 | 2.9 | 12.9 | 24.2 | 23,740 | 191 | 3,932 | 51,670 | 4.7 |
| San Miguel | 2.0 | 0.1 | 4.7 | 1.0 | 2.6 | 6.2 | 3.9 | D | 49.3 | 7,480 | 274 | 1,742 | 16,128 | 3.5 |
| Santa Fe | 0.0 | 0.2 | D | D | 12.6 | 8.5 | 6.4 | 13.6 | 26.6 | 40,045 | 265 | 2,516 | 75,053 | 5.3 |
| Sierra | 8.4 | D | 7.0 | 2.3 | 3.2 | 7.9 | 2.8 | D | 31.0 | 4,680 | 430 | 563 | 8,625 | 3.2 |
| Socorro | 10.3 | D | 0.8 | 1.7 | 9.1 | 4.0 | 1.7 | D | 46.7 | 3,990 | 240 | 970 | 8,315 | 3.2 |
| Taos | 0.3 | D | 6.3 | 1.9 | 7.2 | 9.5 | 4.5 | 16.3 | 21.3 | 10,015 | 306 | 1,116 | 21,319 | 5.2 |
| Torrance | 13.0 | D | 4.2 | 2.4 | 3.4 | 9.7 | D | D | 25.3 | 3,895 | 252 | 523 | 8,138 | 4.4 |
| Union | 15.2 | 1.6 | D | D | D | 7.4 | 6.0 | D | 20.6 | 990 | 245 | 107 | 2,364 | 2.5 |
| Valencia | 1.3 | D | 17.3 | 5.8 | 5.1 | 9.1 | 2.8 | 8.1 | 25.6 | 16,945 | 220 | 2,449 | 31,875 | 5.9 |
| **NEW YORK** | 0.2 | 0.3 | 4.5 | 3.9 | 17.9 | 4.5 | 17.1 | 11.2 | 15.7 | 3,667,022 | 188 | 621,220 | 8,444,231 | 4.1 |
| Albany | 0.1 | D | 4.7 | 3.3 | 13.0 | 4.7 | 10.1 | 11.5 | 32.1 | 62,510 | 205 | 7,107 | 143,314 | 4.1 |
| Allegany | 1.4 | 0.3 | 6.8 | 14.2 | 3.0 | 4.8 | 1.8 | 8.3 | 37.4 | 11,430 | 249 | 1,323 | 26,459 | 1.3 |
| Bronx | 0.0 | D | 5.3 | 1.3 | 2.5 | 5.3 | 3.6 | 24.6 | 31.4 | 207,515 | 146 | 103,699 | 541,001 | 5.8 |
| Broome | 0.2 | 0.1 | 5.7 | 10.5 | 6.4 | 6.5 | 5.7 | D | 25.8 | 47,770 | 250 | 7,149 | 91,650 | 1.2 |
| Cattaraugus | 1.5 | 0.6 | 4.3 | 12.5 | 3.3 | 8.0 | 3.7 | D | 37.2 | 20,090 | 264 | 2,426 | 41,804 | 1.7 |
| Cayuga | 4.1 | D | 6.5 | 13.6 | 4.6 | 7.0 | 2.9 | 12.9 | 28.8 | 18,030 | 236 | 1,824 | 37,349 | 2.4 |
| Chautauqua | 1.7 | 0.3 | 5.2 | 22.2 | 3.5 | 7.6 | 2.8 | 13.3 | 25.6 | 33,790 | 267 | 4,453 | 67,829 | 1.4 |
| Chemung | 0.1 | 0.9 | 5.5 | 16.6 | 4.8 | 6.8 | 4.6 | 16.5 | 24.2 | 21,825 | 262 | 3,019 | 39,271 | 2.3 |
| Chenango | 2.0 | D | 4.5 | 31.8 | 3.6 | 5.7 | 8.9 | 5.6 | 24.7 | 13,205 | 280 | 1,652 | 26,147 | 5.8 |
| Clinton | 1.3 | D | 7.3 | 10.1 | 3.9 | 8.5 | 2.1 | 16.1 | 32.7 | 20,705 | 259 | 2,810 | 36,986 | 3.1 |
| Columbia | 1.6 | D | 9.2 | 6.8 | 7.4 | 7.3 | 2.6 | 16.5 | 25.4 | 16,085 | 270 | 1,349 | 33,761 | 3.0 |
| Cortland | 1.5 | 0.5 | 4.7 | 13.2 | 5.4 | 7.1 | 4.0 | 13.1 | 30.7 | 10,160 | 214 | 1,088 | 20,872 | 1.5 |
| Delaware | 0.9 | 2.0 | 5.3 | 27.5 | 4.1 | 4.6 | 3.1 | 8.0 | 30.8 | 12,240 | 276 | 1,044 | 31,691 | 1.5 |
| Dutchess | 0.3 | 0.3 | 6.4 | 10.2 | 7.4 | 6.8 | 4.7 | 16.6 | 22.2 | 62,250 | 212 | 5,216 | 122,019 | 2.9 |
| Erie | 0.1 | 0.5 | 4.3 | 10.7 | 9.5 | 5.8 | 9.4 | 13.3 | 20.6 | 209,005 | 227 | 26,962 | 432,026 | 2.9 |
| Essex | 0.2 | 0.9 | 6.6 | 7.8 | 3.8 | 7.2 | 4.1 | 10.9 | 35.5 | 10,140 | 274 | 767 | 26,583 | 3.8 |
| Franklin | 1.8 | 0.1 | 3.3 | 2.2 | 3.4 | 5.7 | 1.9 | 20.1 | 49.8 | 12,355 | 247 | 1,799 | 26,059 | 3.0 |
| Fulton | 0.4 | 0.5 | 5.2 | 9.2 | 3.8 | 13.6 | 3.1 | 16.4 | 25.5 | 14,535 | 273 | 2,012 | 29,231 | 2.4 |

1. Per 1,000 resident population estimated as of July 1 of the year shown.

# Table B. States and Counties — Housing, Labor Force, and Employment

| STATE County | Housing units, 2015-2019 Occupied units — Owner-occupied Total | Percent | Median value[1] | Median owner cost as a percent of income — With a mortgage | Without a mortgage[2] | Renter-occupied Median rent[3] | Median rent as a percent of income[2] | Sub-standard units[4] (percent) | Civilian labor force, 2020 Total | Percent change, 2019-2020 | Unemployment Total | Rate[5] | Civilian employment[6], 2015-2019 Total | Percent Management, business, science, and arts | Construction, production, and maintenance occupations |
|---|---|---|---|---|---|---|---|---|---|---|---|---|---|---|---|
| | 89 | 90 | 91 | 92 | 93 | 94 | 95 | 96 | 97 | 98 | 99 | 100 | 101 | 102 | 103 |
| **NEW JERSEY— Cont'd** | | | | | | | | | | | | | | | |
| Mercer | 129,936 | 63.1 | 291,100 | 23.4 | 15.1 | 1,266 | 31.1 | 2.4 | 205,405 | 0.5 | 15,255 | 7.4 | 180,235 | 45.6 | 16.9 |
| Middlesex | 285,005 | 63.5 | 344,100 | 24.1 | 15.6 | 1,469 | 29.0 | 4.5 | 436,329 | -0.9 | 38,031 | 8.7 | 406,921 | 46.4 | 19.6 |
| Monmouth | 235,362 | 73.8 | 421,900 | 24.1 | 15.3 | 1,399 | 32.9 | 1.6 | 327,712 | -0.7 | 28,964 | 8.8 | 318,706 | 46.3 | 15.2 |
| Morris | 181,884 | 73.9 | 456,200 | 23.0 | 14.6 | 1,534 | 27.2 | 1.7 | 252,588 | -2.2 | 19,630 | 7.8 | 261,352 | 52.7 | 12.5 |
| Ocean | 226,160 | 80.0 | 279,000 | 25.4 | 16.8 | 1,428 | 35.7 | 2.6 | 278,413 | -0.5 | 26,397 | 9.5 | 260,353 | 37.3 | 18.4 |
| Passaic | 165,429 | 52.7 | 342,600 | 27.4 | 19.1 | 1,293 | 34.0 | 7.0 | 245,187 | 0.6 | 30,894 | 12.6 | 243,337 | 32.9 | 27.8 |
| Salem | 23,933 | 71.1 | 184,600 | 22.9 | 15.5 | 1,019 | 33.9 | 1.8 | 29,532 | 0.6 | 2,817 | 9.5 | 29,326 | 33.7 | 29.0 |
| Somerset | 118,193 | 75.8 | 430,700 | 22.9 | 14.2 | 1,594 | 28.8 | 1.8 | 167,938 | -2.4 | 13,065 | 7.8 | 173,721 | 54.3 | 12.6 |
| Sussex | 53,322 | 83.1 | 267,500 | 24.2 | 15.7 | 1,314 | 29.5 | 1.0 | 73,836 | | 6,936 | 9.4 | 75,773 | 40.7 | 19.8 |
| Union | 190,101 | 58.6 | 367,200 | 25.0 | 17.7 | 1,290 | 31.2 | 5.0 | 276,211 | -0.8 | 27,340 | 9.9 | 281,698 | 38.0 | 22.8 |
| Warren | 41,587 | 71.7 | 264,200 | 23.1 | 16.9 | 1,132 | 29.9 | 1.3 | 55,670 | -0.6 | 4,723 | 8.5 | 55,354 | 37.8 | 21.8 |
| **NEW MEXICO** | 780,249 | 67.7 | 171,400 | 21.6 | 10.0 | 844 | 29.3 | 4.4 | 943,287 | -1.8 | 79,413 | 8.4 | 888,646 | 36.5 | 20.8 |
| Bernalillo | 267,699 | 63.0 | 199,300 | 21.9 | 10.0 | 874 | 30.7 | 3.0 | 330,645 | -1.5 | 26,825 | 8.1 | 318,839 | 41.4 | 16.4 |
| Catron | 1,325 | 87.7 | 175,400 | 28.6 | 10.0 | 725 | 15.7 | 1.4 | 1,159 | 1.4 | 84 | 7.2 | 888 | 20.9 | 37.8 |
| Chaves | 23,284 | 68.9 | 108,700 | 20.6 | 10.4 | 771 | 25.7 | 4.3 | 28,106 | 2.2 | 2,274 | 8.1 | 26,889 | 28.0 | 27.2 |
| Cibola | 8,708 | 68.7 | 84,400 | 19.5 | 13.0 | 660 | 23.4 | 11.9 | 9,120 | 0.8 | 842 | 9.2 | 9,202 | 28.8 | 22.5 |
| Colfax | 5,853 | 71.0 | 104,800 | 25.1 | 13.0 | 676 | 27.9 | 2.4 | 5,191 | -6.3 | 367 | 7.1 | 4,900 | 35.6 | 25.8 |
| Curry | 18,548 | 57.0 | 125,000 | 23.1 | 10.0 | 843 | 27.8 | 3.8 | 22,292 | 2.3 | 1,224 | 5.5 | 20,253 | 27.9 | 31.9 |
| De Baca | 672 | 62.6 | 106,300 | 26.1 | 11.2 | 676 | 29.5 | 1.0 | 710 | -10.7 | 35 | 4.9 | 627 | 39.9 | 33.0 |
| Dona Ana | 77,842 | 63.1 | 147,400 | 21.9 | 10.0 | 754 | 30.9 | 4.9 | 97,153 | -1.5 | 8,073 | 8.3 | 88,305 | 33.2 | 19.8 |
| Eddy | 21,251 | 69.5 | 155,900 | 16.9 | 10.0 | 946 | 21.7 | 4.4 | 33,700 | -2.9 | 2,316 | 6.9 | 26,438 | 27.4 | 34.0 |
| Grant | 11,851 | 68.1 | 125,100 | 21.2 | 10.0 | 671 | 31.3 | 3.7 | 11,893 | -3.2 | 1,074 | 9.0 | 10,098 | 33.9 | 21.0 |
| Guadalupe | 1,384 | 62.6 | 85,000 | 18.6 | 13.3 | 395 | 27.8 | 2.5 | 1,531 | -2.8 | 103 | 6.7 | 1,262 | 29.7 | 10.7 |
| Harding | 211 | 65.4 | 102,400 | 17.5 | 13.5 | 543 | 31.4 | 0.9 | 276 | 2.2 | 12 | 4.3 | 172 | 45.9 | 18.0 |
| Hidalgo | 1,679 | 70.6 | 86,000 | 28.3 | 10.0 | 501 | 27.1 | 3.5 | 1,956 | -6.8 | 114 | 5.8 | 1,527 | 20.8 | 28.7 |
| Lea | 22,523 | 66.8 | 133,100 | 17.9 | 10.0 | 895 | 22.9 | 5.7 | 29,545 | -6.0 | 3,264 | 11.0 | 28,876 | 25.1 | 39.2 |
| Lincoln | 7,566 | 80.6 | 193,900 | 23.8 | 10.1 | 697 | 26.7 | 1.5 | 8,943 | -1.3 | 789 | 8.8 | 7,565 | 28.1 | 19.8 |
| Los Alamos | 7,931 | 74.1 | 302,800 | 14.8 | 10.0 | 1,062 | 18.6 | 0.5 | 9,974 | 3.7 | 371 | 3.7 | 9,792 | 71.4 | 7.4 |
| Luna | 8,904 | 60.9 | 86,900 | 23.6 | 10.0 | 562 | 30.3 | 2.0 | 10,372 | -0.7 | 1,645 | 15.9 | 7,587 | 22.6 | 33.4 |
| McKinley | 20,942 | 70.9 | 64,800 | 19.7 | 10.0 | 672 | 23.3 | 21.2 | 24,035 | -1.4 | 2,478 | 10.3 | 23,730 | 30.4 | 22.3 |
| Mora | 1,713 | 85.5 | 112,300 | 22.0 | 14.2 | 703 | 33.9 | 4.7 | 2,064 | -7.5 | 166 | 8.0 | 1,445 | 23.9 | 26.0 |
| Otero | 23,634 | 64.2 | 112,400 | 20.9 | 10.0 | 792 | 29.3 | 2.9 | 25,696 | 2.6 | 2,140 | 8.3 | 23,336 | 27.9 | 23.9 |
| Quay | 3,040 | 61.5 | 72,700 | 23.1 | 10.0 | 563 | 26.6 | 0.8 | 3,099 | -4.2 | 229 | 7.4 | 2,988 | 39.6 | 18.4 |
| Rio Arriba | 12,730 | 76.9 | 167,300 | 20.8 | 10.0 | 623 | 28.4 | 3.2 | 16,816 | -0.3 | 1,361 | 8.1 | 14,103 | 38.1 | 16.9 |
| Roosevelt | 6,814 | 58.5 | 118,200 | 19.3 | 10.0 | 778 | 31.1 | 5.4 | 7,934 | 0.3 | 491 | 6.2 | 7,901 | 38.1 | 16.9 |
| Sandoval | 51,001 | 78.6 | 200,900 | 21.8 | 10.0 | 1,117 | 31.8 | 4.0 | 66,325 | -1.7 | 5,698 | 8.6 | 62,530 | 32.2 | 31.3 |
| San Juan | 43,387 | 71.0 | 151,200 | 20.3 | 10.0 | 817 | 27.0 | 8.3 | 50,367 | -3.7 | 5,046 | 10.0 | 49,924 | 38.3 | 17.7 |
| San Miguel | 11,609 | 70.3 | 135,000 | 23.4 | 14.9 | 658 | 31.9 | 4.0 | 10,516 | -3.0 | 870 | 8.3 | 9,626 | 28.6 | 29.8 |
| Santa Fe | 61,921 | 70.9 | 291,800 | 23.4 | 10.0 | 1,048 | 28.6 | 3.6 | 71,346 | -3.8 | 5,891 | 8.3 | 70,383 | 36.0 | 18.9 |
| Sierra | 5,555 | 73.9 | 117,400 | 25.8 | 11.4 | 537 | 31.5 | 2.7 | 4,127 | 4.7 | 378 | 9.2 | 3,422 | 44.0 | 16.2 |
| Socorro | 4,520 | 73.4 | 124,100 | 20.7 | 10.0 | 715 | 31.2 | 3.5 | 6,248 | -3.8 | 443 | 7.1 | 5,488 | 37.3 | 21.4 |
| Taos | 12,103 | 76.4 | 239,500 | 23.1 | 11.7 | 784 | 34.5 | 3.2 | 14,967 | -2.2 | 1,729 | 11.6 | 13,868 | 44.0 | 17.9 |
| Torrance | 5,644 | 83.2 | 114,300 | 24.4 | 11.3 | 754 | 33.3 | 3.0 | 5,365 | -5.7 | 511 | 9.5 | 5,504 | 31.3 | 16.9 |
| Union | 1,395 | 64.9 | 88,000 | 19.4 | 10.0 | 557 | 21.0 | 1.1 | 1,653 | -11.5 | 84 | 5.1 | 1,496 | 28.1 | 28.5 |
| Valencia | 27,010 | 81.5 | 142,600 | 23.5 | 10.0 | 876 | 28.6 | 3.3 | 30,167 | -2.5 | 2,486 | 8.2 | 29,682 | 31.9 | 29.6 |
| **NEW YORK** | 7,343,234 | 53.9 | 313,700 | 23.0 | 13.5 | 1,280 | 31.2 | 5.5 | 9,289,171 | -2.3 | 928,165 | 10.0 | 9,498,320 | 31.2 | 26.4 |
| Albany | 126,540 | 56.4 | 222,500 | 19.8 | 10.7 | 1,022 | 29.1 | 1.5 | 157,326 | 0.3 | 11,299 | 7.2 | 160,761 | 41.5 | 17.1 |
| Allegany | 17,948 | 76.2 | 76,400 | 18.8 | 11.9 | 658 | 30.4 | 2.5 | 19,038 | -0.8 | 1,536 | 8.1 | 19,996 | 45.9 | 13.3 |
| Bronx | 503,829 | 19.7 | 404,700 | 28.3 | 11.5 | 1,212 | 35.4 | 13.1 | 589,087 | -1.2 | 94,436 | 16.0 | 601,341 | 32.6 | 28.3 |
| Broome | 78,549 | 65.2 | 117,000 | 19.4 | 12.0 | 776 | 32.5 | 1.6 | 81,897 | -1.5 | 7,099 | 8.7 | 86,505 | 25.4 | 19.0 |
| Cattaraugus | 31,779 | 71.7 | 88,100 | 18.7 | 12.3 | 655 | 28.7 | 3.2 | 33,142 | -1.3 | 3,120 | 9.4 | 33,684 | 38.6 | 19.5 |
| Cayuga | 31,221 | 71.3 | 128,000 | 19.7 | 12.4 | 763 | 26.6 | 1.9 | 35,417 | -0.9 | 2,799 | 7.9 | 36,570 | 31.7 | 27.6 |
| Chautauqua | 53,075 | 69.3 | 88,000 | 18.2 | 11.7 | 659 | 31.8 | 1.9 | 52,729 | -2.4 | 4,702 | 8.9 | 55,951 | 34.4 | 25.7 |
| Chemung | 33,989 | 68.3 | 105,800 | 18.8 | 11.3 | 813 | 30.0 | 1.1 | 35,430 | 1.1 | 3,102 | 8.8 | 37,333 | 31.0 | 28.9 |
| Chenango | 20,697 | 74.5 | 98,400 | 18.9 | 11.9 | 678 | 29.7 | 2.5 | 21,760 | -0.4 | 1,499 | 6.9 | 21,649 | 35.0 | 22.6 |
| Clinton | 31,301 | 66.5 | 134,100 | 20.0 | 10.7 | 810 | 27.2 | 1.8 | 35,447 | -1.7 | 2,824 | 8.0 | 35,759 | 33.2 | 29.7 |
| Columbia | 24,966 | 73.7 | 233,600 | 22.7 | 13.4 | 928 | 27.7 | 1.4 | 30,309 | -2.4 | 1,878 | 6.2 | 29,694 | 32.8 | 25.2 |
| Cortland | 17,745 | 65.6 | 123,800 | 18.6 | 13.0 | 778 | 28.5 | 2.1 | 22,453 | -1.7 | 1,785 | 7.9 | 23,241 | 39.1 | 21.1 |
| Delaware | 18,968 | 73.9 | 139,300 | 21.1 | 13.2 | 691 | 31.1 | 2.5 | 18,614 | -1.7 | 1,340 | 7.2 | 19,952 | 36.4 | 20.8 |
| Dutchess | 108,413 | 68.8 | 282,000 | 23.8 | 14.0 | 1,220 | 32.7 | 3.7 | 142,061 | -1.6 | 10,888 | 7.7 | 145,110 | 32.2 | 28.2 |
| Erie | 389,585 | 64.6 | 153,400 | 18.6 | 12.1 | 829 | 29.7 | 1.7 | 439,291 | -0.3 | 41,755 | 9.5 | 451,762 | 40.1 | 18.6 |
| Essex | 15,790 | 75.4 | 153,600 | 19.9 | 12.6 | 812 | 29.1 | 2.1 | 16,518 | -2.5 | 1,344 | 8.1 | 17,336 | 40.0 | 18.3 |
| Franklin | 19,015 | 71.4 | 108,700 | 19.4 | 12.2 | 703 | 28.7 | 2.7 | 19,197 | -1.2 | 1,575 | 8.2 | 19,380 | 32.1 | 24.1 |
| Fulton | 22,557 | 70.7 | 109,600 | 19.0 | 12.8 | 779 | 30.8 | 1.0 | 22,509 | 0.2 | 1,940 | 8.6 | 24,632 | 31.8 | 21.1 |
| | | | | | | | | | | | | | | 29.9 | 28.2 |

1. Specified owner-occupied units. lacking complete plumbing facilities.  2. A value of 10.0 represents 10 percent or less; a value of 50.0 represents 50 percent or more.  3. Specified renter-occupied units.  4. Overcrowded or  5. Percent of civilian labor force.  6. Civilian employed persons 16 years old and over.

# Table B. States and Counties — Nonfarm Employment and Agriculture

| STATE County | Number of establishments | Total | Health care and social assistance | Manufacturing | Retail trade | Finance and insurance | Professional, scientific, and technical services | Total (mil dol) | Average per employee (dollars) | Number | Fewer than 50 acres | 1000 acres or more | Farm producers whose primary occupation is farming (percent) |
|---|---|---|---|---|---|---|---|---|---|---|---|---|---|
| | 104 | 105 | 106 | 107 | 108 | 109 | 110 | 111 | 112 | 113 | 114 | 115 | 116 |
| **NEW JERSEY— Cont'd** | | | | | | | | | | | | | |
| Mercer | 9,694 | 199,859 | 30,193 | 6,738 | 24,701 | 14,247 | 24,806 | 14,129 | 70,692 | 323 | 72.1 | 0.6 | 35.1 |
| Middlesex | 21,913 | 412,429 | 56,558 | 26,999 | 44,009 | 14,373 | 62,394 | 26,713 | 64,770 | 217 | 82.0 | 0.9 | 38.4 |
| Monmouth | 19,308 | 248,962 | 49,428 | 8,616 | 39,209 | 11,584 | 20,302 | 13,224 | 53,117 | 838 | 85.3 | 0.7 | 46.4 |
| Morris | 16,700 | 309,210 | 39,957 | 15,350 | 29,133 | 17,597 | 47,878 | 26,417 | 85,435 | 418 | 81.6 | NA | 29.9 |
| Ocean | 13,639 | 152,419 | 40,069 | 5,031 | 28,492 | 3,371 | 7,997 | 6,167 | 40,460 | 260 | 87.3 | NA | 38.4 |
| Passaic | 12,069 | 145,363 | 28,037 | 16,699 | 22,826 | 4,335 | 6,974 | 7,403 | 50,925 | 89 | 88.8 | NA | 39.0 |
| Salem | 1,146 | 16,930 | 3,323 | 2,225 | 1,750 | 391 | 457 | 937 | 55,335 | 781 | 67.3 | 3.1 | 39.9 |
| Somerset | 9,878 | 192,626 | 24,146 | 14,282 | 18,187 | 9,623 | 24,046 | 16,544 | 85,888 | 452 | 76.8 | 2.0 | 31.5 |
| Sussex | 3,152 | 32,062 | 5,858 | 2,792 | 5,631 | 893 | 1,533 | 1,433 | 44,705 | 1,008 | 75.6 | 0.7 | 32.1 |
| Union | 13,976 | 204,662 | 34,423 | 13,545 | 27,004 | 5,764 | 15,198 | 13,739 | 67,129 | 9 | 100.0 | NA | 69.2 |
| Warren | 2,388 | 28,556 | 5,074 | 3,671 | 5,346 | 495 | 1,006 | 1,328 | 46,516 | 918 | 70.7 | 1.3 | 35.6 |
| **NEW MEXICO** | 43,804 | 644,537 | 126,174 | 26,567 | 93,316 | 24,766 | 57,910 | 28,177 | 43,717 | 25,044 | 52.2 | 18.2 | 42.1 |
| Bernalillo | 15,930 | 273,440 | 52,384 | 12,477 | 35,096 | 12,998 | 32,334 | 12,386 | 45,295 | 1,248 | 89.8 | 1.8 | 32.2 |
| Catron | 67 | 484 | 148 | 31 | 74 | NA | 75 | 16 | 32,202 | 341 | 22.9 | 29.0 | 41.3 |
| Chaves | 1,356 | 16,723 | 3,954 | 882 | 3,266 | 529 | 702 | 613 | 36,671 | 560 | 41.3 | 25.7 | 44.5 |
| Cibola | 298 | 5,464 | 1,638 | 77 | 861 | 123 | 54 | 198 | 36,226 | 640 | 47.5 | 18.9 | 44.9 |
| Colfax | 379 | 3,194 | 425 | 148 | 633 | 114 | 67 | 89 | 27,820 | 304 | 17.8 | 34.9 | 38.9 |
| Curry | 1,003 | 12,518 | 2,990 | 791 | 2,301 | 413 | 491 | 411 | 32,850 | 641 | 21.2 | 32.4 | 41.6 |
| De Baca | 43 | 224 | 44 | NA | 64 | NA | NA | 7 | 29,616 | 226 | 44.2 | 33.2 | 53.5 |
| Dona Ana | 3,642 | 53,532 | 15,225 | 2,380 | 7,911 | 1,809 | 3,825 | 1,873 | 34,981 | 1,946 | 86.7 | 1.8 | 33.1 |
| Eddy | 1,513 | 26,347 | 2,823 | 1,207 | 3,144 | 534 | 795 | 1,568 | 59,515 | 507 | 47.1 | 18.9 | 41.7 |
| Grant | 575 | 6,761 | 1,618 | 140 | 1,066 | 162 | 138 | 270 | 39,923 | 404 | 30.4 | 25.2 | 44.4 |
| Guadalupe | 89 | 1,055 | 115 | NA | 258 | 14 | NA | 31 | 29,096 | 297 | 20.2 | 46.5 | 54.7 |
| Harding | 14 | 42 | NA | NA | 15 | NA | NA | 1 | 26,881 | 184 | 4.9 | 54.9 | 49.0 |
| Hidalgo | 97 | 801 | 188 | NA | 257 | NA | 27 | 19 | 23,347 | 151 | 11.3 | 45.7 | 51.5 |
| Lea | 1,720 | 26,264 | 2,300 | 600 | 3,439 | 458 | 580 | 1,436 | 54,692 | 555 | 36.6 | 28.6 | 42.9 |
| Lincoln | 669 | 5,261 | 544 | 146 | 1,163 | 269 | 142 | 167 | 31,815 | 454 | 29.7 | 30.2 | 36.0 |
| Los Alamos | 376 | 15,074 | 964 | 80 | 466 | 257 | 11,120 | 1,363 | 90,434 | 2 | 100.0 | NA | 20.0 |
| Luna | 369 | 4,455 | 1,256 | 447 | 966 | 111 | 87 | 129 | 28,945 | 211 | 39.8 | 25.1 | 53.0 |
| McKinley | 954 | 16,111 | 4,495 | 601 | 3,389 | 473 | 206 | 535 | 33,189 | 2,441 | 42.4 | 22.6 | 53.1 |
| Mora | 41 | 282 | 138 | NA | 50 | 7 | NA | 8 | 27,426 | 700 | 40.4 | 9.1 | 36.0 |
| Otero | 953 | 11,947 | 2,927 | 89 | 2,337 | 291 | 336 | 391 | 32,765 | 473 | 56.0 | 15.4 | 43.7 |
| Quay | 214 | 1,840 | 355 | 49 | 409 | 84 | 53 | 50 | 27,198 | 613 | 12.6 | 36.2 | 40.7 |
| Rio Arriba | 560 | 5,855 | 1,721 | 99 | 1,147 | 148 | 80 | 187 | 31,858 | 1,439 | 61.8 | 11.3 | 36.0 |
| Roosevelt | 323 | 3,359 | 682 | 252 | 501 | 103 | 64 | 112 | 33,344 | 742 | 18.3 | 31.3 | 40.7 |
| Sandoval | 1,764 | 24,998 | 4,377 | 2,136 | 3,295 | 968 | 750 | 1,046 | 41,827 | 1,007 | 69.0 | 10.9 | 33.8 |
| San Juan | 2,527 | 36,497 | 7,634 | 1,586 | 5,752 | 852 | 1,213 | 1,622 | 44,447 | 2,965 | 56.5 | 21.0 | 54.9 |
| San Miguel | 435 | 6,307 | 2,985 | 58 | 900 | 187 | 93 | 198 | 31,386 | 1,170 | 35.9 | 18.6 | 38.0 |
| Santa Fe | 4,806 | 47,235 | 8,651 | 747 | 8,992 | 1,525 | 2,488 | 2,050 | 43,396 | 639 | 67.9 | 9.7 | 35.7 |
| Sierra | 210 | 2,225 | 655 | 102 | 406 | 64 | 33 | 66 | 29,515 | 257 | 38.5 | 25.3 | 56.7 |
| Socorro | 236 | 2,988 | 924 | 112 | 438 | 75 | 223 | 91 | 30,434 | 658 | 69.6 | 12.0 | 45.7 |
| Taos | 1,056 | 9,492 | 1,846 | 292 | 1,561 | 162 | 341 | 274 | 28,883 | 824 | 78.4 | 4.1 | 34.5 |
| Torrance | 205 | 1,499 | 169 | 83 | 514 | 35 | 55 | 44 | 29,384 | 716 | 18.7 | 22.6 | 43.2 |
| Union | 90 | 913 | 120 | NA | 164 | 76 | 10 | 29 | 32,298 | 369 | 1.1 | 59.9 | 47.8 |
| Valencia | 929 | 10,785 | 1,785 | 933 | 2,479 | 283 | 301 | 356 | 33,005 | 1,360 | 91.7 | 1.5 | 32.0 |
| **NEW YORK** | 547,351 | 8,597,216 | 1,734,587 | 415,886 | 914,248 | 554,574 | 657,125 | 603,078 | 70,148 | 33,438 | 36.7 | 3.3 | 48.1 |
| Albany | 9,526 | 183,540 | 38,118 | 8,096 | 22,351 | 12,916 | 18,755 | 9,733 | 53,031 | 440 | 37.7 | 0.9 | 42.6 |
| Allegany | 753 | 11,066 | 1,699 | 2,040 | 1,233 | 183 | 246 | 381 | 34,434 | 789 | 29.4 | 2.5 | 42.7 |
| Bronx | 18,451 | 281,091 | 112,917 | 6,037 | 33,255 | 4,049 | 4,985 | 13,803 | 49,106 | NA | NA | NA | NA |
| Broome | 4,141 | 71,656 | 16,295 | 7,314 | 10,390 | 2,054 | 3,653 | 3,028 | 42,255 | 494 | 37.2 | 1.0 | 41.5 |
| Cattaraugus | 1,576 | 20,814 | 3,212 | 2,902 | 3,365 | 688 | 515 | 805 | 38,696 | 956 | 31.1 | 1.5 | 42.9 |
| Cayuga | 1,552 | 19,261 | 4,108 | 3,135 | 3,299 | 317 | 417 | 767 | 39,844 | 842 | 38.7 | 6.7 | 53.5 |
| Chautauqua | 2,764 | 41,002 | 7,956 | 8,409 | 5,925 | 801 | 839 | 1,634 | 39,853 | 1,228 | 34.6 | 2.4 | 45.3 |
| Chemung | 1,721 | 31,018 | 7,122 | 5,101 | 4,936 | 883 | 749 | 1,315 | 42,385 | 398 | 29.1 | 1.3 | 37.8 |
| Chenango | 874 | 13,398 | 1,701 | 4,185 | 1,668 | 1,266 | 224 | 658 | 49,119 | 770 | 28.4 | 2.5 | 50.2 |
| Clinton | 1,789 | 25,619 | 5,648 | 4,542 | 4,786 | 424 | 577 | 1,020 | 39,822 | 588 | 27.7 | 5.8 | 46.6 |
| Columbia | 1,771 | 17,028 | 4,768 | 1,513 | 2,944 | 337 | 684 | 675 | 39,661 | 518 | 45.4 | 3.1 | 46.7 |
| Cortland | 1,040 | 14,976 | 3,190 | 2,335 | 2,281 | 386 | 866 | 545 | 36,421 | 536 | 33.2 | 2.8 | 43.6 |
| Delaware | 1,049 | 10,201 | 1,659 | 3,206 | 1,578 | 370 | 189 | 467 | 45,816 | 689 | 28.6 | 1.6 | 46.7 |
| Dutchess | 7,599 | 101,218 | 20,168 | 8,391 | 14,610 | 2,765 | 7,365 | 5,102 | 50,406 | 620 | 53.1 | 3.5 | 45.5 |
| Erie | 22,762 | 427,343 | 79,503 | 44,229 | 51,627 | 31,631 | 27,919 | 20,633 | 48,282 | 940 | 51.9 | 2.7 | 48.7 |
| Essex | 1,174 | 10,093 | 2,180 | 781 | 1,966 | 190 | 203 | 440 | 43,629 | 285 | 40.7 | 5.6 | 37.5 |
| Franklin | 960 | 10,805 | 3,637 | 491 | 1,802 | 197 | 328 | 402 | 37,210 | 636 | 25.8 | 3.3 | 52.0 |
| Fulton | 1,139 | 14,050 | 3,379 | 2,026 | 2,310 | 243 | 290 | 526 | 37,406 | 207 | 39.6 | NA | 39.1 |

| STATE County | Acreage (1,000) 117 | Percent change, 2012-2017 118 | Average size of farm 119 | Total irrigated (1,000) 120 | Total cropland (1,000) 121 | Average per farm 122 | Average per acre 123 | Value of machinery and equiopmnet, average per farm (dollars) 124 | Total (mil dol) 125 | Average per farm (acres) 126 | Crops 127 | Livestock and poultry products 128 | Organic farms (number) 129 | Farms with internet access (per-cent) 130 | Total ($1,000) 131 | Percent of farms 132 |
|---|---|---|---|---|---|---|---|---|---|---|---|---|---|---|---|---|
| NEW JERSEY— Cont'd | | | | | | | | | | | | | | | | |
| Mercer | 25 | 27.8 | 78 | 1.0 | 15.8 | 1,414,874 | 18,114 | 83,438 | 25.0 | 77,344 | 80.1 | 19.9 | 12 | 73.4 | 149 | 8.0 |
| Middlesex | 16 | -7.2 | 74 | 2.0 | 11.2 | 1,607,661 | 21,773 | 112,644 | 38.4 | 176,770 | 98.0 | 2.0 | 2 | 77.9 | 92 | 7.8 |
| Monmouth | 39 | 0.6 | 47 | 3.6 | 23.8 | 981,430 | 20,982 | 79,157 | 80.6 | 96,221 | 83.6 | 16.4 | 3 | 80.5 | 366 | 3.3 |
| Morris | 15 | 898.2 | 35 | 1.0 | 6.7 | 743,975 | 21,426 | 68,480 | 24.8 | 59,388 | 93.1 | 6.9 | 11 | 86.4 | 60 | 3.1 |
| Ocean | 9 | 6.8 | 33 | 0.8 | 4.4 | 622,892 | 19,031 | 60,677 | 24.6 | 94,769 | 81.1 | 18.9 | 3 | 81.2 | 59 | 3.1 |
| Passaic | 2 | -94.6 | 21 | 0.1 | 0.3 | 679,638 | 31,953 | 47,779 | 2.9 | 32,169 | 95.0 | 5.0 | NA | 85.4 | 8 | 3.4 |
| Salem | 98 | -95.0 | 126 | 17.1 | 80.9 | 1,059,096 | 8,420 | 139,528 | 102.3 | 131,040 | 88.3 | 11.7 | 4 | 79.1 | 1,869 | 19.3 |
| Somerset | 36 | -41.2 | 79 | 0.9 | 19.9 | 1,569,021 | 19,776 | 65,277 | 20.1 | 44,509 | 71.5 | 28.5 | 13 | 81.0 | 148 | 7.3 |
| Sussex | 60 | 62,156.3 | 59 | 0.4 | 25.7 | 683,936 | 11,535 | 47,928 | 18.2 | 18,081 | 59.4 | 40.6 | 9 | 80.0 | 310 | 3.9 |
| Union | 0 | -99.9 | 8 | 0.0 | D | 1,303,684 | 156,442 | 118,929 | D | D | D | D | NA | 100.0 | NA | NA |
| Warren | 74 | 150.6 | 80 | 1.4 | 46.8 | 980,498 | 12,184 | 76,777 | 93.2 | 101,542 | 72.0 | 28.0 | 3 | 80.7 | 952 | 10.6 |
| | | | | | | | | | | | | | | | | |
| NEW MEXICO | 40,660 | -5.9 | 1,624 | 626.0 | 1,825.8 | 845,740 | 521 | 63,619 | 2,582.3 | 103,112 | 25.2 | 74.8 | 183 | 60.5 | 63,660 | 13.8 |
| | | | | | | | | | | | | | | | | |
| Bernalillo | 221 | -85.8 | 177 | 4.3 | 6.5 | 489,688 | 2,759 | 34,855 | 9.3 | 7,466 | 51.8 | 48.2 | 23 | 70.5 | 2 | 0.4 |
| Catron | 1,261 | 18.4 | 3,697 | 1.8 | 5.1 | 1,551,601 | 420 | 52,857 | 9.5 | 27,774 | 1.4 | 98.6 | 5 | 68.0 | 970 | 9.1 |
| Chaves | 2,318 | 251.2 | 4,140 | 42.8 | 62.2 | 1,957,270 | 473 | 164,614 | 404.5 | 722,348 | 10.6 | 89.4 | NA | 83.0 | 2,236 | 17.0 |
| Cibola | 1,594 | 9,374,482.4 | 2,490 | 0.9 | 5.4 | 789,275 | 317 | 36,619 | D | D | 100.0 | D | 4 | 46.6 | 476 | 9.2 |
| Colfax | 2,073 | 135.4 | 6,819 | 16.7 | 19.2 | 3,442,286 | 505 | 77,901 | 25.1 | 82,543 | 3.0 | 97.0 | 5 | 73.4 | 237 | 7.9 |
| Curry | 902 | -15.5 | 1,407 | 58.8 | 428.4 | 977,762 | 695 | 163,697 | 480.6 | 749,735 | 6.6 | 93.4 | 5 | 78.5 | 14,607 | 69.4 |
| De Baca | 1,182 | -28.1 | 5,231 | 7.6 | 10.1 | 1,962,324 | 375 | 85,317 | 28.1 | 124,487 | 19.5 | 80.5 | NA | 87.6 | 1,989 | 36.7 |
| Dona Ana | 528 | -53.7 | 271 | 73.7 | 93.1 | 674,302 | 2,484 | 91,093 | 370.3 | 190,284 | 61.8 | 38.2 | 23 | 73.7 | 578 | 4.0 |
| Eddy | 1,088 | -45.1 | 2,146 | 31.5 | 49.3 | 1,268,345 | 591 | 145,419 | 97.3 | 191,870 | 49.5 | 50.5 | NA | 82.4 | 385 | 9.5 |
| Grant | 894 | -3.9 | 2,213 | 2.0 | 6.5 | 1,077,324 | 487 | 62,151 | 14.7 | 36,507 | 4.2 | 95.8 | NA | 77.7 | 945 | 8.4 |
| | | | | | | | | | | | | | | | | |
| Guadalupe | 1,444 | 39.7 | 4,862 | 4.2 | 13.3 | 1,512,859 | 311 | 59,332 | 13.3 | 44,643 | 2.3 | 97.7 | NA | 65.0 | 1,851 | 25.3 |
| Harding | 938 | 20.6 | 5,100 | 0.9 | 14.8 | 1,868,529 | 366 | D | 13.4 | 73,054 | 0.3 | 99.7 | NA | 70.7 | 951 | 33.2 |
| Hidalgo | 849 | 54.3 | 5,622 | 18.5 | 25.5 | D | D | 157,031 | 23.4 | 155,099 | 55.3 | 44.7 | NA | 79.5 | 1,565 | 40.4 |
| Lea | 1,938 | 24.8 | 3,492 | 42.9 | 111.9 | 1,439,038 | 412 | 115,880 | 192.2 | 346,382 | 17.3 | 82.7 | 4 | 71.2 | 2,204 | 27.7 |
| Lincoln | 1,466 | 19.8 | 3,230 | 2.1 | 6.7 | 1,607,027 | 498 | 56,708 | 15.9 | 34,930 | 2.5 | 97.5 | 1 | 74.9 | 1,665 | 17.4 |
| Los Alamos | D | D | D | NA | NA | D | D | D | D | D | NA | D | NA | 100.0 | NA | NA |
| Luna | 576 | -53.9 | 2,729 | 20.1 | 30.0 | 1,404,258 | 515 | 168,893 | 79.3 | 375,725 | 61.1 | 38.9 | 1 | 83.4 | 2,313 | 43.1 |
| McKinley | 2,570 | 79.3 | 1,053 | 2.5 | 38.7 | 450,977 | 428 | 26,248 | 8.1 | 3,334 | 7.8 | 92.2 | NA | 19.4 | 600 | 12.2 |
| Mora | 931 | -38.7 | 1,329 | 11.7 | 21.2 | 734,372 | 552 | 49,430 | 18.2 | 25,931 | 12.7 | 87.3 | 7 | 52.7 | 548 | 4.6 |
| Otero | 1,019 | -5.4 | 2,155 | 4.7 | 8.6 | 946,804 | 439 | 52,864 | 18.2 | 38,395 | 51.8 | 48.2 | 5 | 79.3 | 783 | 8.5 |
| | | | | | | | | | | | | | | | | |
| Quay | 1,548 | 14.8 | 2,526 | 13.1 | 199.6 | 889,432 | 352 | 83,453 | 39.5 | 64,388 | 16.2 | 83.8 | 1 | 71.3 | 7,352 | 61.3 |
| Rio Arriba | 1,362 | -47.2 | 947 | 26.1 | 41.7 | 812,334 | 858 | 51,633 | 14.6 | 10,172 | 35.7 | 64.3 | 32 | 60.7 | 1,114 | 7.9 |
| Roosevelt | 1,500 | -36.2 | 2,021 | 44.5 | 317.1 | 923,240 | 457 | 118,420 | 290.6 | 391,699 | 10.4 | 89.6 | 12 | 74.5 | 10,555 | 56.5 |
| Sandoval | 784 | 9.2 | 778 | 6.9 | 17.5 | 415,220 | 534 | 33,172 | 12.4 | 12,286 | 63.8 | 36.2 | 6 | 52.2 | 487 | 4.1 |
| San Juan | 2,551 | 168.5 | 861 | 73.6 | 107.9 | 345,734 | 402 | 39,425 | 74.1 | 24,998 | 91.5 | 8.5 | 1 | 44.9 | 1,383 | 8.3 |
| San Miguel | 2,270 | 21.7 | 1,940 | 13.3 | 22.4 | 870,177 | 449 | 37,470 | 19.1 | 16,323 | 3.7 | 96.3 | 7 | 50.0 | 1,385 | 9.2 |
| Santa Fe | D | D | D | 15.6 | 23.7 | 932,062 | 881 | 65,568 | 25.4 | 39,798 | 54.5 | 45.5 | 14 | 77.0 | 367 | 5.5 |
| Sierra | 1,012 | -20.4 | 3,939 | 5.8 | 8.5 | 1,422,045 | 361 | 87,711 | 31.9 | 124,093 | 30.8 | 69.2 | 3 | 75.1 | 316 | 9.7 |
| Socorro | 912 | 282.7 | 1,387 | 16.2 | 17.7 | 870,620 | 628 | 65,960 | 65.1 | 99,009 | 14.1 | 85.9 | 1 | 63.1 | 265 | 2.3 |
| Taos | 285 | -57.4 | 346 | 15.9 | 16.6 | 565,648 | 1,635 | 45,157 | 7.6 | 9,266 | 55.0 | 45.0 | 17 | 66.3 | 270 | 3.6 |
| | | | | | | | | | | | | | | | | |
| Torrance | 1,561 | -20.7 | 2,180 | 16.1 | 29.8 | 1,038,912 | 477 | 55,855 | 45.9 | 64,096 | 24.7 | 75.3 | NA | 63.5 | 2,311 | 18.6 |
| Union | 1,887 | 438.1 | 5,114 | 15.4 | 48.9 | 2,110,381 | 413 | 89,284 | 83.1 | 225,295 | 9.8 | 90.2 | NA | 66.9 | 2,896 | 27.4 |
| Valencia | 518 | -79.1 | 381 | 16.0 | 18.0 | 442,410 | 1,162 | 47,173 | 46.1 | 33,886 | 10.3 | 89.7 | 6 | 70.1 | 52 | 1.1 |
| | | | | | | | | | | | | | | | | |
| NEW YORK | 6,866 | -4.4 | 205 | 53.3 | 4,291.4 | 663,082 | 3,229 | 135,626 | 5,369.2 | 160,572 | 39.3 | 60.7 | 1,497 | 77.1 | 59,106 | 19.3 |
| | | | | | | | | | | | | | | | | |
| Albany | 60 | 86.9 | 135 | 1.1 | 34.5 | 601,883 | 4,446 | 78,697 | 47.3 | 107,564 | 76.6 | 23.4 | 10 | 79.1 | 250 | 16.1 |
| Allegany | 162 | 7.5 | 205 | 0.4 | 88.7 | 412,662 | 2,013 | 85,322 | 69.3 | 87,853 | 37.1 | 62.9 | 45 | 64.6 | 1,280 | 28.4 |
| Bronx | NA | NA | NA | NA | NA | NA | NA | NA | NA | NA | NA | NA | NA | NA | NA | NA |
| Broome | 62 | -21.6 | 126 | 0.5 | 33.5 | 390,574 | 3,089 | 72,467 | 32.1 | 64,953 | 32.4 | 67.6 | 10 | 73.9 | 346 | 14.8 |
| Cattaraugus | 166 | -15.7 | 174 | 0.2 | 88.7 | 398,333 | 2,291 | 102,151 | 93.4 | 97,711 | 21.3 | 78.7 | 11 | 65.1 | 650 | 24.0 |
| Cayuga | 225 | 34.7 | 267 | 0.3 | 177.9 | 1,057,750 | 3,955 | 211,196 | 287.9 | 341,868 | 24.3 | 75.7 | 27 | 76.5 | 3,092 | 28.7 |
| Chautauqua | 224 | -5.5 | 182 | 1.0 | 126.9 | 481,615 | 2,645 | 128,191 | 161.0 | 131,081 | 45.5 | 54.5 | 42 | 78.1 | 1,621 | 11.8 |
| Chemung | 67 | 15.1 | 168 | 0.1 | 35.0 | 440,268 | 2,619 | 91,732 | 19.0 | 47,771 | 50.6 | 49.4 | 5 | 78.6 | 384 | 19.1 |
| Chenango | 149 | 29.5 | 193 | 0.3 | 77.1 | 397,806 | 2,056 | 104,336 | 67.9 | 88,212 | 26.3 | 73.7 | 42 | 76.2 | 539 | 20.6 |
| Clinton | 162 | 9.8 | 275 | 0.8 | 85.1 | 727,023 | 2,645 | 179,004 | 167.8 | 285,355 | 26.6 | 73.4 | 8 | 80.4 | 127 | 7.0 |
| | | | | | | | | | | | | | | | | |
| Columbia | 99 | 4.0 | 191 | 2.1 | 64.5 | 801,814 | 4,188 | 123,888 | 88.4 | 170,718 | 55.9 | 44.1 | 33 | 83.0 | 913 | 12.9 |
| Cortland | 114 | -19.1 | 212 | 0.2 | 58.5 | 491,363 | 2,320 | 128,057 | 69.5 | 129,675 | 19.7 | 80.3 | 46 | 80.6 | 592 | 31.0 |
| Delaware | 140 | -3.7 | 204 | 0.2 | 67.9 | 586,225 | 2,880 | 89,482 | 45.7 | 66,335 | 29.1 | 70.9 | 38 | 83.3 | 871 | 26.1 |
| Dutchess | 102 | -9.4 | 164 | 0.8 | 40.1 | 1,486,434 | 9,040 | 84,120 | 43.9 | 70,818 | 64.8 | 35.2 | 15 | 88.5 | 242 | 8.4 |
| Erie | 143 | 0.3 | 152 | 2.0 | 96.4 | 696,510 | 4,576 | 137,967 | 131.0 | 139,333 | 47.3 | 52.7 | 28 | 81.6 | 1,405 | 19.1 |
| Essex | 58 | 5.1 | 202 | 0.1 | 22.7 | 481,641 | 2,382 | 84,723 | 13.2 | 46,239 | 51.4 | 48.6 | 12 | 87.0 | 209 | 12.3 |
| Franklin | 141 | -3.0 | 221 | 0.5 | 74.8 | 550,272 | 2,487 | 114,437 | 86.4 | 135,822 | 18.6 | 81.4 | 49 | 68.9 | 446 | 15.9 |
| Fulton | 22 | -83.1 | 107 | 0.1 | 12.8 | 299,090 | 2,791 | 84,953 | 10.3 | 49,609 | 42.2 | 57.8 | 4 | 78.7 | 107 | 11.1 |

# Table B. States and Counties — Water Use, Wholesale Trade, Retail Trade, and Real Estate

| STATE County | Water use, 2015 | | Wholesale Trade[1], 2017 | | | | Retail Trade[2], 2017 | | | | Real estate and rental and leasing,[2] 2017 | | | |
|---|---|---|---|---|---|---|---|---|---|---|---|---|---|---|
| | Public supply water withdrawn (mil gal/day) | Public supply gallons withdrawn per person per day | Number of establishments | Number of employees | Sales (mil dol) | Average payroll (mil dol) | Number of establishments | Number of employees | Sales (mil dol) | Average payroll (mil dol) | Number of establishments | Number of employees | Sales (mil dol) | Average payroll (mil dol) |
| | 133 | 134 | 135 | 136 | 137 | 138 | 139 | 140 | 141 | 142 | 143 | 144 | 145 | 146 |
| NEW JERSEY— Cont'd | | | | | | | | | | | | | | |
| Mercer | 37.62 | 101.3 | 352 | D | 10,156.9 | D | 1,249 | 25,101 | 7,573.9 | 702.8 | 355 | 2,142 | 721.0 | 109.9 |
| Middlesex | 39.08 | 46.5 | 1,508 | 38,219 | 47,076.2 | 3,107.7 | 2,578 | 44,440 | 14,856.0 | 1,271.5 | 782 | 7,053 | 2,843.9 | 435.0 |
| Monmouth | 71.17 | 113.2 | 759 | 6,431 | 5,475.2 | 404.8 | 2,628 | 40,226 | 12,274.6 | 1,122.6 | 709 | 4,059 | 1,221.7 | 217.1 |
| Morris | 90.03 | 180.2 | 903 | 20,136 | 24,702.2 | 2,628.2 | 1,810 | 29,662 | 10,433.2 | 935.9 | 719 | 6,579 | 3,419.9 | 522.0 |
| Ocean | 52.45 | 89.1 | 446 | 3,997 | 1,839.2 | 200.0 | 2,022 | 28,464 | 8,720.1 | 822.7 | 719 | 2,973 | 813.3 | 128.9 |
| Passaic | 270.78 | 530.0 | 744 | 9,561 | 6,109.2 | 524.3 | 1,857 | 26,191 | 8,086.8 | 739.8 | 450 | 2,461 | 671.7 | 114.2 |
| Salem | 4.24 | 66.1 | 36 | D | 1,693.4 | D | 168 | 1,917 | 575.0 | 46.7 | 43 | 285 | 60.5 | 12.8 |
| Somerset | 2.71 | 8.1 | 479 | 14,123 | 24,434.3 | 1,286.6 | 1,059 | 18,630 | 6,750.8 | 597.5 | 350 | 1,905 | 543.5 | 103.7 |
| Sussex | 5.94 | 41.3 | 116 | D | 521.0 | D | 412 | 5,962 | 1,832.8 | 160.8 | 86 | 265 | 79.0 | 13.1 |
| Union | 141.52 | 254.6 | 788 | 17,565 | 21,882.8 | 2,576.7 | 1,936 | 26,666 | 8,666.6 | 761.5 | 548 | 3,675 | 1,275.3 | 177.5 |
| Warren | 7.26 | 67.9 | D | D | D | D | 374 | 5,767 | 1,616.1 | 153.8 | 63 | 226 | 63.8 | 9.6 |
| NEW MEXICO | 254.10 | 121.9 | 1,507 | 16,914 | 11,936.9 | 834.9 | 6,335 | 92,557 | 26,404.3 | 2,511.3 | 2,408 | 9,229 | 2,185.6 | 375.8 |
| Bernalillo | 87.60 | 129.5 | 723 | 9,958 | 6,130.8 | 510.5 | 1,991 | 34,299 | 10,171.3 | 984.3 | 999 | 4,223 | 1,032.8 | 167.9 |
| Catron | 0.15 | 43.4 | NA | NA | NA | NA | 13 | 75 | 13.1 | 1.4 | NA | NA | NA | NA |
| Chaves | 11.57 | 175.9 | 48 | 427 | 212.4 | 18.0 | 211 | 3,114 | 866.1 | 78.5 | 73 | 200 | 41.9 | 7.4 |
| Cibola | 2.38 | 87.1 | D | D | D | D | 58 | 826 | 218.6 | 19.7 | D | D | D | D |
| Colfax | 1.80 | 145.0 | D | D | D | D | 73 | 661 | 139.8 | 13.2 | D | D | D | 1.4 |
| Curry | 5.20 | 103.2 | 34 | 378 | 252.4 | 13.8 | 182 | 2,361 | 633.6 | 60.0 | 61 | 198 | 32.7 | 5.2 |
| De Baca | 0.24 | 131.3 | NA | NA | NA | NA | 11 | 56 | 16.2 | 1.2 | NA | NA | NA | NA |
| Dona Ana | 33.99 | 158.6 | 112 | 1,069 | 621.0 | 48.3 | 498 | 8,070 | 2,098.3 | 191.8 | 218 | 686 | 197.3 | 19.8 |
| Eddy | 13.45 | 233.6 | 61 | 538 | 2,063.5 | 32.9 | 183 | 2,764 | 883.8 | 84.6 | 68 | 369 | 138.0 | 20.2 |
| Grant | 2.49 | 87.0 | D | D | D | D | 94 | 1,141 | 260.0 | 28.1 | 34 | 80 | 10.7 | 2.2 |
| Guadalupe | 0.58 | 132.7 | NA | NA | NA | NA | 18 | 254 | 123.0 | 6.0 | NA | NA | NA | NA |
| Harding | 0.08 | 114.6 | NA | NA | NA | NA | 4 | 10 | 5.0 | 0.2 | NA | NA | NA | NA |
| Hidalgo | 1.73 | 391.1 | NA | NA | NA | NA | 27 | 251 | 122.0 | 5.1 | NA | NA | NA | NA |
| Lea | 10.19 | 143.2 | 85 | 1,126 | 575.2 | 57.8 | 218 | 3,264 | 1,076.0 | 95.0 | 88 | 613 | 161.9 | 37.4 |
| Lincoln | 4.39 | 226.1 | 7 | 28 | 10.5 | 0.8 | 135 | 1,202 | 291.8 | 32.1 | 54 | 138 | 20.5 | 3.5 |
| Los Alamos | 3.06 | 172.1 | 3 | 29 | 21.6 | 1.7 | 30 | 508 | 146.2 | 14.2 | D | D | D | D |
| Luna | 2.99 | 122.0 | D | D | D | D | 64 | 906 | 251.8 | 21.3 | 19 | 45 | 7.0 | 1.3 |
| McKinley | 3.40 | 44.3 | D | D | D | D | 216 | 3,438 | 922.8 | 82.9 | 43 | 211 | 35.0 | 6.3 |
| Mora | 0.33 | 71.8 | NA | NA | NA | NA | 12 | 43 | 13.9 | 1.0 | NA | NA | NA | NA |
| Otero | 6.69 | 103.9 | 15 | 73 | 17.4 | 1.8 | 167 | 2,376 | 585.3 | 57.9 | 44 | 157 | 23.5 | 4.0 |
| Quay | 1.26 | 149.0 | D | D | D | D | 33 | 414 | 185.2 | 8.9 | 11 | 12 | 0.9 | 0.2 |
| Rio Arriba | 1.83 | 46.4 | 6 | 12 | 8.1 | 0.4 | 86 | 1,119 | 303.4 | 28.1 | D | D | D | D |
| Roosevelt | 2.72 | 142.3 | 12 | 82 | 65.2 | 2.9 | 43 | 616 | 135.2 | 14.9 | D | D | D | 0.9 |
| Sandoval | 11.24 | 80.6 | 39 | 259 | 128.8 | 12.7 | 191 | 3,450 | 954.9 | 91.0 | 87 | 270 | 52.4 | 10.5 |
| San Juan | 18.83 | 158.6 | 129 | 1,053 | 502.9 | 57.5 | 413 | 5,940 | 1,744.6 | 163.8 | 100 | 560 | 141.0 | 30.8 |
| San Miguel | 2.19 | 78.3 | 6 | 25 | 8.4 | 1.0 | 79 | 942 | 219.4 | 21.9 | 17 | 42 | 7.0 | 1.1 |
| Santa Fe | 10.57 | 71.1 | 99 | 856 | 923.5 | 44.1 | 797 | 9,053 | 2,588.1 | 269.2 | 278 | 815 | 200.7 | 41.4 |
| Sierra | 1.53 | 135.6 | NA | NA | NA | NA | 43 | 443 | 109.6 | 10.2 | 8 | 49 | 2.9 | 0.6 |
| Socorro | 1.99 | 115.3 | NA | NA | NA | NA | 33 | 400 | 106.3 | 9.4 | D | D | D | 0.4 |
| Taos | 2.41 | 73.2 | D | D | D | D | 206 | 1,637 | 352.5 | 40.6 | 58 | 198 | 20.0 | 5.3 |
| Torrance | 1.76 | 113.7 | 8 | 30 | 23.0 | 1.0 | 43 | 490 | 191.5 | 11.8 | 3 | 4 | 0.4 | 0.2 |
| Union | 0.52 | 123.8 | NA | NA | NA | NA | 16 | 124 | 39.1 | 3.1 | NA | NA | NA | NA |
| Valencia | 4.94 | 65.2 | 14 | 61 | 40.3 | 2.3 | 147 | 2,310 | 636.0 | 59.9 | 41 | 71 | 10.5 | 1.9 |
| NEW YORK | 2,424.65 | 122.5 | 26,900 | 330,990 | 367,972.4 | 22,510.3 | 78,260 | 945,360 | 291,724.9 | 27,814.8 | 34,076 | 193,442 | 70,693.0 | 11,356.8 |
| Albany | 45.78 | 148.0 | 395 | 5,782 | 6,199.4 | 355.0 | 1,296 | 21,679 | 6,617.7 | 619.1 | 441 | 2,823 | 763.1 | 125.4 |
| Allegany | 3.12 | 65.7 | 18 | 141 | 81.5 | 5.3 | 132 | 1,251 | 247.6 | 26.0 | 18 | 37 | 5.4 | 1.3 |
| Bronx | 0.00 | 0.0 | 666 | 11,040 | 10,255.6 | 663.8 | 4,143 | 34,642 | 9,150.3 | 874.3 | 2,216 | 9,499 | 2,401.4 | 379.4 |
| Broome | 18.73 | 95.3 | 173 | 3,505 | 2,579.6 | 160.3 | 686 | 10,488 | 2,736.0 | 258.0 | 152 | 901 | 188.4 | 32.5 |
| Cattaraugus | 9.45 | 121.3 | 51 | 725 | 621.7 | 25.1 | 309 | 3,749 | 1,096.0 | 94.8 | 50 | 209 | 25.7 | 5.4 |
| Cayuga | 5.43 | 69.4 | 59 | 909 | 502.5 | 45.1 | 233 | 3,320 | 886.8 | 85.4 | 55 | 160 | 37.4 | 4.6 |
| Chautauqua | 12.58 | 96.2 | 103 | 1,299 | 609.7 | 58.0 | 469 | 6,221 | 1,508.9 | 150.9 | 90 | 479 | 103.1 | 17.7 |
| Chemung | 8.37 | 96.1 | 85 | 1,216 | 608.6 | 56.7 | 316 | 5,032 | 1,378.8 | 126.2 | 89 | 365 | 98.5 | 13.6 |
| Chenango | 2.22 | 45.5 | 23 | 230 | 72.7 | 8.4 | 157 | 1,719 | 478.7 | 42.9 | 25 | 97 | 13.6 | 3.4 |
| Clinton | 4.93 | 60.7 | 73 | 837 | 639.1 | 39.8 | 346 | 4,750 | 1,409.4 | 121.0 | 83 | 311 | 68.4 | 11.7 |
| Columbia | 2.97 | 48.3 | 47 | 750 | 345.8 | 37.8 | 275 | 2,915 | 784.6 | 82.7 | 62 | 168 | 30.2 | 5.9 |
| Cortland | 3.80 | 78.4 | D | D | D | D | 176 | 2,287 | 723.5 | 58.5 | 38 | 139 | 22.3 | 3.6 |
| Delaware | 405.54 | 8,805.9 | 28 | 303 | 133.7 | 13.7 | 189 | 1,568 | 462.3 | 39.2 | 35 | 113 | 14.0 | 2.6 |
| Dutchess | 18.97 | 64.1 | 201 | 1,839 | 1,146.6 | 120.7 | 1,079 | 15,373 | 3,909.6 | 394.9 | 325 | 1,240 | 330.7 | 49.3 |
| Erie | 171.43 | 185.8 | 950 | 17,608 | 20,991.3 | 1,026.6 | 3,269 | 52,962 | 13,619.2 | 1,354.7 | 895 | 6,725 | 1,363.4 | 260.1 |
| Essex | 5.45 | 141.6 | D | D | D | D | 207 | 1,861 | 502.0 | 49.8 | 39 | 100 | 20.6 | 2.6 |
| Franklin | 3.83 | 75.6 | 23 | 179 | 139.6 | 7.1 | 167 | 1,858 | 551.1 | 52.4 | 35 | 115 | 16.3 | 3.3 |
| Fulton | 5.46 | 101.1 | 41 | 404 | 174.9 | 18.7 | 202 | 2,291 | 752.5 | 66.0 | 29 | 155 | 18.9 | 4.2 |

1 Merchant wholesalers, except manufacturers' sales branches and offices.    2. Employer establishments.

# Table B. States and Counties — Professional Services, Manufacturing, and Accommodation and Food Services

| STATE County | Professional, scientific, and technical services, 2017 | | | | Manufacturing, 2017 | | | | Accommodation and food services, 2017 | | | |
|---|---|---|---|---|---|---|---|---|---|---|---|---|
| | Number of establishments | Number of employees | Sales (mil dol) | Average payroll (mil dol) | Number of establishments | Number of employees | Sales (mil dol) | Average payroll (mil dol) | Number of establishments | Number of employees | Sales (mil dol) | Annual payroll (mil dol) |
| | 147 | 148 | 149 | 150 | 151 | 152 | 153 | 154 | 155 | 156 | 157 | 158 |
| **NEW JERSEY— Cont'd** | | | | | | | | | | | | |
| Mercer | 1,614 | 24,223 | 5,682 | 2,299.6 | 231 | 6,743 | 2,217.8 | 423.3 | 867 | 14,034 | 912.8 | 260.6 |
| Middlesex | 3,964 | 59,695 | 11,906 | 5,156.7 | 677 | 27,134 | 12,418.8 | 1,771.6 | 1,791 | 24,008 | 1,790.5 | 448.9 |
| Monmouth | 2,550 | 21,526 | 4,029 | 1,561.7 | 404 | 7,829 | 2,558.7 | 461.6 | 1,832 | 26,814 | 1,798.7 | 493.1 |
| Morris | 2,531 | 45,725 | 10,267 | 4,443.1 | 516 | 15,003 | 7,076.0 | 1,038.3 | 1,343 | 21,118 | 1,574.7 | 438.9 |
| Ocean | D | D | D | D | 289 | 5,060 | 1,100.4 | 259.8 | 1,240 | 14,722 | 1,114.8 | 288.1 |
| Passaic | D | D | D | D | 678 | 16,493 | 4,908.8 | 971.3 | 1,039 | 11,792 | 816.9 | 210.1 |
| Salem | 86 | 577 | 84 | 30.9 | 37 | 2,175 | 723.8 | 157.1 | 100 | 1,565 | 83.8 | 23.5 |
| Somerset | 1,739 | 23,308 | 5,524 | 2,797.4 | 263 | 16,258 | 7,139.3 | 1,344.2 | 849 | 12,469 | 941.7 | 259.4 |
| Sussex | D | D | D | D | 109 | 1,839 | 504.8 | 125.7 | 291 | 4,567 | 274.2 | 77.3 |
| Union | 1,367 | 14,942 | 2,057 | 1,834.3 | 565 | 13,670 | 12,052.0 | 877.4 | 1,221 | 15,848 | 1,158.1 | 301.8 |
| Warren | D | D | D | D | 107 | 3,476 | 2,020.2 | 211.6 | 254 | 2,949 | 169.0 | 46.0 |
| **NEW MEXICO** | D | D | 9,895 | D | 1,332 | 23,235 | 13,723.6 | 1,286.8 | 4,392 | 91,601 | 5,526.0 | 1,583.2 |
| Bernalillo | D | D | D | D | 555 | 11,441 | 3,654.6 | 613.9 | 1,512 | 35,165 | 2,064.5 | 616.6 |
| Catron | NA | NA | NA | NA | 4 | 15 | 4.4 | 1.0 | 7 | D | 1.1 | D |
| Chaves | D | D | D | D | 32 | 774 | 629.2 | 38.4 | 127 | 2,398 | 124.5 | 35.3 |
| Cibola | D | D | D | D | 8 | 62 | 5.2 | 1.8 | 40 | 1,124 | 85.3 | 20.5 |
| Colfax | D | D | D | D | 12 | 103 | 12.0 | 4.0 | 63 | 1,160 | 78.3 | 25.7 |
| Curry | D | D | D | D | 25 | 636 | 979.4 | 30.4 | 82 | 1,796 | 89.7 | 24.7 |
| De Baca | NA | NA | NA | NA | NA | NA | NA | NA | 5 | D | 1.4 | D |
| Dona Ana | D | D | D | D | 139 | 2,065 | 922.8 | 92.7 | 335 | 7,168 | 361.2 | 106.7 |
| Eddy | D | D | D | D | D | 1,161 | D | 104.7 | 129 | 2,455 | 163.8 | 40.9 |
| Grant | D | D | D | D | 16 | 124 | 12.1 | 4.1 | 70 | 994 | 38.1 | 10.5 |
| Guadalupe | NA | NA | NA | NA | NA | NA | NA | NA | 21 | 359 | 21.1 | 5.1 |
| Harding | NA | NA | NA | NA | NA | NA | NA | NA | NA | NA | NA | NA |
| Hidalgo | D | D | D | D | NA | NA | NA | NA | 17 | 193 | 10.9 | 3.5 |
| Lea | D | D | D | D | 33 | 702 | 615.9 | 54.7 | 150 | 2,768 | 229.0 | 47.3 |
| Lincoln | D | D | D | D | 14 | 67 | 9.6 | 2.6 | 85 | 1,149 | 67.7 | 19.5 |
| Los Alamos | D | D | D | D | 12 | 78 | 14.9 | 3.2 | 31 | 513 | 30.6 | 9.8 |
| Luna | D | D | D | D | 9 | 402 | 83.4 | 15.0 | 63 | 803 | 35.5 | 10.1 |
| McKinley | D | D | D | D | D | 402 | D | 30.1 | 149 | 2,905 | 151.4 | 39.5 |
| Mora | NA | NA | NA | NA | NA | NA | NA | NA | D | D | D | 0.1 |
| Otero | D | D | D | D | 30 | 74 | 12.2 | 2.5 | 96 | 2,258 | 179.5 | 49.0 |
| Quay | D | D | D | D | D | 20 | D | 1.3 | 36 | 480 | 25.0 | 6.9 |
| Rio Arriba | D | D | D | D | 14 | 94 | 19.6 | 3.7 | 71 | 1,533 | 118.7 | 31.0 |
| Roosevelt | D | D | D | D | 9 | 218 | 420.3 | 11.9 | 34 | 559 | 23.6 | 6.6 |
| Sandoval | D | D | D | D | 59 | 2,006 | 1,327.7 | 141.8 | 168 | 4,574 | 309.5 | 82.3 |
| San Juan | D | D | D | D | 69 | 776 | 154.0 | 43.1 | 240 | 4,900 | 237.5 | 68.0 |
| San Miguel | D | D | D | D | 6 | 38 | 5.4 | 1.5 | 71 | 976 | 49.6 | 12.1 |
| Santa Fe | 654 | 2,391 | 387 | 146.3 | 120 | 725 | 165.9 | 32.3 | 429 | 9,749 | 743.9 | 226.0 |
| Sierra | D | D | D | D | 5 | 95 | 23.6 | 4.3 | 43 | 486 | 22.2 | 6.7 |
| Socorro | D | D | D | D | 9 | 67 | 19.4 | 3.5 | 35 | 533 | 26.5 | 7.0 |
| Taos | D | D | D | D | 35 | 196 | 42.0 | 6.6 | 160 | 2,479 | 140.4 | 44.6 |
| Torrance | 12 | 45 | 3 | 1.1 | 14 | 101 | 35.1 | 4.0 | 19 | 174 | 9.4 | 2.3 |
| Union | D | D | D | D | NA | NA | NA | NA | 15 | 134 | 7.1 | 2.0 |
| Valencia | 72 | 336 | 38 | 13.5 | 35 | 772 | 168.1 | 32.8 | 85 | 1,743 | 78.6 | 22.0 |
| **NEW YORK** | 61,306 | 633,235 | 167,191 | 58,010.8 | 15,499 | 411,100 | 155,571.9 | 23,751.3 | 54,797 | 824,806 | 66,963.5 | 19,792.9 |
| Albany | 1,213 | 14,825 | 2,673 | 1,101.2 | 234 | 8,444 | 3,864.8 | 504.8 | 1,057 | 16,422 | 1,001.2 | 310.1 |
| Allegany | D | D | D | D | 36 | 2,047 | 810.3 | 109.8 | 94 | 1,433 | 69.3 | 17.6 |
| Bronx | D | D | D | D | 312 | 6,037 | 1,600.0 | 249.1 | 1,954 | 18,882 | 1,418.7 | 380.3 |
| Broome | D | D | D | D | 165 | 7,110 | 2,511.7 | 438.5 | 539 | 8,809 | 448.5 | 139.1 |
| Cattaraugus | D | D | D | D | 72 | 2,790 | 718.0 | 154.3 | 209 | 4,044 | 337.8 | 75.8 |
| Cayuga | 85 | 432 | 48 | 19.5 | 82 | 2,957 | 1,166.0 | 157.6 | 174 | 1,939 | 112.0 | 30.8 |
| Chautauqua | D | D | D | 41.1 | 186 | 9,055 | 3,675.5 | 446.5 | 361 | 5,209 | 228.1 | 73.0 |
| Chemung | D | D | D | D | 82 | 5,229 | 1,296.9 | 286.3 | 210 | 3,073 | 157.5 | 50.4 |
| Chenango | D | D | D | D | 64 | 4,336 | 2,290.0 | 247.3 | 100 | 842 | 44.8 | 14.8 |
| Clinton | 108 | 447 | 56 | 20.7 | 91 | 4,135 | 1,585.0 | 214.2 | 200 | 2,797 | 173.6 | 50.7 |
| Columbia | D | D | D | D | 81 | 1,396 | 381.5 | 62.2 | 197 | 1,726 | 107.8 | 34.2 |
| Cortland | D | D | D | D | 56 | 2,125 | 555.4 | 104.5 | 146 | 3,064 | 150.9 | 44.5 |
| Delaware | 71 | 190 | 20 | 6.2 | 53 | 3,319 | 3,299.5 | 192.1 | 118 | 999 | 43.2 | 12.8 |
| Dutchess | D | D | D | D | 186 | 5,837 | 2,038.1 | 453.8 | 846 | 10,846 | 655.2 | 196.8 |
| Erie | D | D | D | D | 963 | 44,061 | 18,569.3 | 2,749.0 | 2,377 | 45,506 | 2,373.7 | 738.0 |
| Essex | D | D | D | D | 30 | 833 | 339.5 | 57.0 | 219 | 2,455 | 187.7 | 60.7 |
| Franklin | D | D | D | D | 26 | 437 | 322.5 | 19.3 | 110 | 1,774 | 194.4 | 42.6 |
| Fulton | 66 | 282 | 28 | 9.4 | 71 | 1,951 | 997.0 | 84.9 | 132 | 1,171 | 70.0 | 20.5 |

## Table B. States and Counties — Health Care and Social Assistance, Other Services, Nonemployer Businesses, and Residential Construction

| STATE County | Health care and social assistance, 2017 | | | | Other services, 2017 | | | | Nonemployer businesses, 2018 | | Value of residential construction authorized by building permits, 2020 | |
|---|---|---|---|---|---|---|---|---|---|---|---|---|
| | Number of establishments | Number of employees | Receipts (mil dol) | Annual payroll (mil dol) | Number of establishments | Number of employees | Receipts (mil dol) | Annual payroll (mil dol) | Number | Receipts (mil dol) | New construction ($1,000) | Number of housing units |
| | 159 | 160 | 161 | 162 | 163 | 164 | 165 | 166 | 167 | 168 | 169 | 170 |
| **NEW JERSEY— Cont'd** | | | | | | | | | | | | |
| Mercer | 1,192 | 30,342 | 3,364.8 | 1,424.9 | 854 | 6,011 | 1,423.0 | 276.7 | 27,402 | 1,492.9 | 115,256 | 832 |
| Middlesex | 2,349 | 53,954 | 6,618.6 | 2,656.7 | 1,600 | 10,288 | 1,910.1 | 413.5 | 62,950 | 3,622.4 | 317,014 | 2,757 |
| Monmouth | 2,449 | 48,971 | 6,222.5 | 2,420.3 | 1,509 | 9,977 | 944.3 | 287.6 | 58,473 | 4,008.7 | 453,841 | 2,638 |
| Morris | 1,773 | 41,323 | 5,330.0 | 2,230.4 | 1,303 | 8,512 | 1,164.8 | 300.0 | 45,781 | 3,268.6 | 190,387 | 1,630 |
| Ocean | 1,693 | 38,798 | 4,219.5 | 1,700.7 | 1,120 | 5,885 | 640.5 | 159.4 | 45,228 | 2,738.3 | 490,382 | 3,625 |
| Passaic | 1,539 | 27,424 | 3,127.3 | 1,259.2 | 1,022 | 5,147 | 556.9 | 148.7 | 45,144 | 2,299.0 | 106,880 | 1,314 |
| Salem | 159 | 3,203 | 300.5 | 125.5 | 101 | 421 | 35.7 | 9.8 | 2,924 | 154.5 | 6,886 | 35 |
| Somerset | 1,235 | 23,730 | 2,963.2 | 1,216.8 | 728 | 4,733 | 703.4 | 172.5 | 28,681 | 1,902.7 | 202,531 | 1,913 |
| Sussex | 379 | 7,037 | 662.9 | 274.1 | 302 | 1,311 | 142.9 | 37.4 | 11,404 | 614.3 | 36,123 | 220 |
| Union | 1,526 | 35,061 | 3,951.0 | 1,634.5 | 1,214 | 7,816 | 775.5 | 261.8 | 48,345 | 2,769.9 | 213,741 | 2,650 |
| Warren | 307 | 5,117 | 521.9 | 208.3 | 231 | 1,084 | 120.9 | 35.4 | 7,402 | 387.4 | 33,272 | 304 |
| **NEW MEXICO** | 5,134 | 127,808 | 13,602.4 | 5,405.5 | 2,963 | 17,547 | 2,085.4 | 558.2 | 125,925 | 5,271.5 | 107,347 | 396 |
| Bernalillo | 1,967 | 52,986 | 6,259.9 | 2,466.1 | 1,116 | 7,479 | 868.5 | 246.4 | 42,174 | 1,846.2 | 302,969 | 1,084 |
| Catron | D | D | D | 2.4 | 3 | 10 | 2.0 | 0.3 | 370 | 12.8 | NA | NA |
| Chaves | 186 | 4,109 | 415.1 | 182.2 | 79 | 523 | 45.6 | 11.3 | 3,038 | 151.1 | 14,871 | 69 |
| Cibola | 52 | 1,666 | 150.9 | 64.3 | 17 | 58 | 7.0 | 1.9 | 1,111 | 34.1 | NA | NA |
| Colfax | 26 | 473 | 42.9 | 18.2 | 29 | 89 | 11.7 | 2.2 | 794 | 28.9 | NA | NA |
| Curry | 114 | 2,763 | 249.7 | 102.0 | 80 | 461 | 52.4 | 11.6 | 1,921 | 81.4 | 19,197 | 111 |
| De Baca | D | D | D | D | NA | NA | NA | NA | 115 | 5.8 | NA | NA |
| Dona Ana | 553 | 14,627 | 1,251.4 | 527.1 | 238 | 1,142 | 93.0 | 29.4 | 13,241 | 533.7 | 270,286 | 1,109 |
| Eddy | 113 | 2,814 | 291.0 | 122.4 | 85 | 651 | 93.7 | 21.6 | 2,854 | 164.1 | 75,923 | 448 |
| Grant | 72 | 1,411 | 134.7 | 56.1 | 43 | 215 | 14.7 | 4.4 | 1,542 | 46.7 | 1,621 | 10 |
| Guadalupe | 7 | 119 | 12.3 | 4.7 | D | D | D | D | 179 | 4.8 | NA | NA |
| Harding | NA | NA | NA | NA | D | D | D | 0.3 | 50 | 2.1 | NA | NA |
| Hidalgo | 14 | 185 | 10.0 | 4.6 | D | D | D | D | 205 | 6.0 | NA | NA |
| Lea | 116 | 2,422 | 276.1 | 101.7 | 94 | 643 | 117.5 | 28.4 | 3,854 | 231.2 | 38,496 | 147 |
| Lincoln | 43 | 566 | 62.7 | 26.2 | 37 | 151 | 14.6 | 4.5 | 1,895 | 82.0 | 22,909 | 63 |
| Los Alamos | 69 | 986 | 98.2 | 40.6 | 36 | 215 | 14.5 | 3.9 | 1,100 | 42.4 | 6,320 | 24 |
| Luna | 47 | 1,126 | 96.9 | 35.3 | 25 | 91 | 7.0 | 1.7 | 978 | 36.8 | 1,524 | 7 |
| McKinley | 121 | 5,228 | 518.2 | 196.9 | 68 | 403 | 38.7 | 10.5 | 3,127 | 72.8 | 1,941 | 13 |
| Mora | 6 | D | 6.1 | D | D | D | D | D | 305 | 9.8 | 891 | 3 |
| Otero | 112 | 3,083 | 361.9 | 125.4 | 68 | 334 | 23.7 | 6.8 | 3,414 | 119.9 | NA | NA |
| Quay | 20 | 369 | 34.8 | 11.9 | D | D | D | 2.2 | 384 | 13.5 | NA | NA |
| Rio Arriba | 80 | 1,736 | 169.7 | 71.0 | 27 | 179 | 18.7 | 5.6 | 1,767 | 56.6 | NA | NA |
| Roosevelt | 34 | 638 | 57.5 | 26.4 | D | D | D | D | 823 | 33.7 | 3,213 | 21 |
| Sandoval | 234 | 3,996 | 540.2 | 170.3 | 108 | 555 | 49.1 | 15.3 | 8,567 | 317.4 | 170,548 | 834 |
| San Juan | 282 | 7,628 | 874.3 | 352.8 | 202 | 1,305 | 207.0 | 44.1 | 5,045 | 201.7 | 23,144 | 98 |
| San Miguel | 73 | 3,090 | 187.3 | 90.0 | 28 | 74 | 6.2 | 1.6 | 1,265 | 39.3 | 664 | 3 |
| Santa Fe | 510 | 9,791 | 1,131.9 | 426.1 | 351 | 2,038 | 294.6 | 82.2 | 16,612 | 772.0 | 117,591 | 577 |
| Sierra | 20 | 720 | 47.9 | 25.2 | D | D | D | 2.8 | 733 | 27.6 | 5,782 | 23 |
| Socorro | 35 | 1,092 | 56.4 | 30.2 | D | D | D | D | 725 | 21.9 | NA | NA |
| Taos | 97 | 1,723 | 136.5 | 65.9 | 62 | 257 | 23.2 | 7.0 | 3,304 | 109.5 | 19,229 | 83 |
| Torrance | 15 | 168 | 8.5 | 4.2 | 7 | 13 | 1.5 | 0.4 | 804 | 29.1 | NA | NA |
| Union | 6 | 171 | 13.9 | 6.6 | 10 | 21 | 2.2 | 0.5 | 255 | 6.4 | NA | NA |
| Valencia | 101 | 1,748 | 97.5 | 42.8 | 64 | 250 | 22.8 | 6.2 | 3,374 | 130.0 | 21,516 | 96 |
| **NEW YORK** | 58,902 | 1,654,593 | 193,507.6 | 80,687.9 | 48,436 | 299,209 | 51,749.9 | 11,845.9 | 1,804,188 | 98,252.1 | 6,406,410 | 37,330 |
| Albany | 964 | 35,623 | 4,402.0 | 1,717.5 | 843 | 6,034 | 847.2 | 249.8 | 19,516 | 1,022.4 | 116,686 | 651 |
| Allegany | 94 | 1,882 | 133.3 | 63.4 | 73 | 282 | 23.7 | 5.5 | 2,526 | 93.6 | 3,876 | 48 |
| Bronx | 2,471 | 108,072 | 13,367.5 | 5,550.1 | 1,922 | 8,817 | 1,043.7 | 274.8 | 120,174 | 4,001.5 | 517,626 | 4,461 |
| Broome | 481 | 16,077 | 1,758.5 | 716.0 | 345 | 1,819 | 171.5 | 46.1 | 9,908 | 412.7 | 37,264 | 313 |
| Cattaraugus | 165 | 3,370 | 375.6 | 135.6 | 130 | 660 | 67.1 | 15.8 | 3,778 | 148.6 | 11,850 | 78 |
| Cayuga | 197 | 4,023 | 361.0 | 180.3 | 124 | 465 | 57.5 | 13.0 | 4,211 | 172.5 | 12,748 | 86 |
| Chautauqua | 269 | 7,795 | 653.3 | 278.3 | 267 | 1,417 | 117.1 | 27.6 | 7,046 | 281.5 | 24,340 | 109 |
| Chemung | 210 | 6,480 | 650.8 | 310.8 | 132 | 632 | 67.9 | 16.9 | 3,867 | 141.5 | 29,734 | 184 |
| Chenango | 112 | 1,890 | 172.0 | 68.0 | 81 | 271 | 29.2 | 7.9 | 2,700 | 109.2 | 9,122 | 175 |
| Clinton | 238 | 5,480 | 582.6 | 282.0 | 114 | 578 | 52.2 | 17.0 | 4,020 | 174.1 | 26,196 | 159 |
| Columbia | 173 | 4,319 | 378.7 | 172.1 | 105 | 517 | 56.2 | 14.3 | 6,383 | 292.0 | 23,978 | 71 |
| Cortland | 124 | 3,428 | 273.6 | 115.1 | 88 | 419 | 39.1 | 10.8 | 2,395 | 101.3 | 5,441 | 36 |
| Delaware | 121 | 2,325 | 179.0 | 75.4 | 85 | 365 | 59.0 | 11.8 | 3,623 | 148.1 | 11,822 | 63 |
| Dutchess | 888 | 20,003 | 2,503.7 | 977.5 | 636 | 2,897 | 317.4 | 86.2 | 22,767 | 1,111.1 | 140,112 | 590 |
| Erie | 2,725 | 79,702 | 9,058.2 | 3,756.3 | 1,865 | 12,072 | 1,197.5 | 353.9 | 52,711 | 2,383.0 | 340,877 | 1,269 |
| Essex | 157 | 2,046 | 175.2 | 79.4 | 66 | 284 | 40.6 | 9.9 | 3,075 | 126.3 | 31,619 | 101 |
| Franklin | 169 | 3,841 | 355.3 | 170.6 | 62 | 207 | 20.6 | 5.0 | 2,775 | 107.7 | 15,990 | 79 |
| Fulton | 185 | 4,117 | 316.0 | 147.2 | 85 | 561 | 49.3 | 15.4 | 2,732 | 109.3 | 8,609 | 49 |

## Table B. States and Counties — Government Employment and Payroll, and Local Government Finances

| STATE County | Full-time equivalent employees 171 | March payroll (dollars) 172 | Administration, judicial, and legal 173 | Police and corrections 174 | Fire protection 175 | Highways and transportation 176 | Health and welfare 177 | Natural resources and utilities 178 | Education and libraries 179 | Total (mil dol) 180 | Intergovernmental (mil dol) 181 | Taxes Total (mil dol) 182 | Taxes Per capita Total 183 | Per capita Property 184 |
|---|---|---|---|---|---|---|---|---|---|---|---|---|---|---|
| **NEW JERSEY— Cont'd** | | | | | | | | | | | | | | |
| Mercer | 14,557 | 89,476,151 | 6.3 | 11.2 | 3.5 | 2.2 | 5.1 | 4.7 | 64.1 | 2,186.7 | 669.9 | 1,243.6 | 3,378 | 3,332 |
| Middlesex | 28,682 | 174,474,309 | 3.9 | 12.8 | 3.1 | 1.5 | 4.1 | 4.0 | 67.9 | 4,052.3 | 1,017.5 | 2,417.5 | 2,922 | 2,859 |
| Monmouth | 25,418 | 150,306,285 | 5.1 | 13.0 | 0.8 | 3.0 | 1.9 | 4.3 | 70.3 | 3,544.9 | 749.5 | 2,385.0 | 3,835 | 3,757 |
| Morris | 20,249 | 117,120,799 | 4.5 | 10.0 | 0.7 | 2.8 | 2.9 | 4.7 | 72.3 | 2,941.5 | 383.6 | 2,069.1 | 4,189 | 4,143 |
| Ocean | 19,831 | 110,436,008 | 5.3 | 14.9 | 0.6 | 2.7 | 3.7 | 6.9 | 63.3 | 2,680.0 | 623.3 | 1,810.6 | 3,041 | 2,999 |
| Passaic | 16,364 | 100,071,382 | 4.9 | 16.6 | 4.9 | 1.6 | 7.6 | 4.5 | 58.5 | 2,409.2 | 721.9 | 1,497.3 | 2,969 | 2,935 |
| Salem | 3,265 | 17,256,254 | 5.4 | 13.2 | 0.1 | 2.6 | 3.8 | 2.7 | 70.7 | 374.1 | 142.8 | 179.1 | 2,848 | 2,819 |
| Somerset | 12,816 | 76,918,330 | 4.3 | 11.0 | 0.6 | 3.1 | 1.9 | 3.4 | 72.7 | 1,760.0 | 223.3 | 1,327.1 | 4,015 | 3,945 |
| Sussex | 5,872 | 32,546,214 | 6.3 | 9.5 | 0.2 | 4.2 | 1.6 | 1.6 | 75.0 | 809.3 | 168.7 | 564.0 | 3,994 | 3,976 |
| Union | 23,177 | 138,986,930 | 4.5 | 15.1 | 5.5 | 2.7 | 4.6 | 2.8 | 61.9 | 3,548.8 | 1,149.2 | 2,086.1 | 3,761 | 3,669 |
| Warren | 4,148 | 22,171,732 | 7.0 | 11.6 | 0.0 | 4.6 | 3.0 | 2.0 | 69.5 | 590.3 | 166.2 | 370.7 | 3,506 | 3,477 |
| **NEW MEXICO** | X | X | X | X | X | X | X | X | X | X | X | X | X | X |
| Bernalillo | 22,858 | 98,449,958 | 6.5 | 17.0 | 7.0 | 7.0 | 4.0 | 6.8 | 49.6 | 2,642.9 | 1,232.0 | 981.7 | 1,448 | 821 |
| Catron | 135 | 474,259 | 10.1 | 7.7 | 0.1 | 6.7 | 0.0 | 5.1 | 69.6 | 14.5 | 9.0 | 2.5 | 701 | 595 |
| Chaves | 2,412 | 8,352,745 | 4.6 | 13.6 | 5.0 | 4.0 | 0.5 | 7.3 | 63.8 | 230.8 | 170.6 | 37.7 | 579 | 502 |
| Cibola | 995 | 3,157,600 | 10.2 | 11.2 | 1.7 | 3.1 | 3.9 | 3.7 | 61.3 | 105.6 | 46.3 | 23.4 | 869 | 352 |
| Colfax | 675 | 2,163,084 | 8.0 | 10.4 | 4.2 | 4.0 | 9.1 | 12.5 | 45.4 | 67.0 | 29.8 | 23.4 | 1,925 | 1,114 |
| Curry | 2,069 | 7,073,175 | 3.3 | 8.6 | 5.0 | 2.7 | 1.0 | 3.6 | 74.1 | 190.0 | 106.4 | 58.3 | 1,170 | 379 |
| De Baca | 127 | 391,523 | 9.7 | 10.1 | 6.0 | 6.6 | 3.4 | 14.3 | 65.2 | 13.9 | 6.9 | 2.5 | 1,390 | 984 |
| Dona Ana | 8,033 | 28,795,832 | 4.8 | 11.2 | 3.2 | 1.7 | 1.5 | 6.9 | 49.9 | 760.4 | 405.6 | 260.8 | 1,206 | 537 |
| Eddy | 2,423 | 10,400,192 | 6.9 | 15.4 | 6.1 | 5.1 | 0.7 | 8.8 | 53.4 | 355.2 | 123.0 | 186.3 | 3,263 | 1,658 |
| Grant | 1,550 | 5,780,581 | 5.0 | 7.4 | 1.7 | 3.3 | 42.0 | 3.2 | 35.8 | 186.2 | 72.9 | 34.7 | 1,254 | 629 |
| Guadalupe | 265 | 783,103 | 12.6 | 8.6 | 0.0 | 5.8 | 1.5 | 7.4 | 60.5 | 23.8 | 14.4 | 6.9 | 1,573 | 779 |
| Harding | 69 | 216,806 | 19.0 | 3.5 | 0.0 | 11.8 | 4.0 | 3.2 | 56.0 | 9.5 | 5.3 | 2.6 | 3,857 | 3,437 |
| Hidalgo | 234 | 697,509 | 9.8 | 23.7 | 0.0 | 4.0 | 5.1 | 5.8 | 48.7 | 26.4 | 13.0 | 7.3 | 1,702 | 1,001 |
| Lea | 3,711 | 14,993,130 | 3.8 | 12.1 | 3.6 | 2.3 | 19.3 | 4.9 | 51.0 | 525.6 | 234.4 | 167.6 | 2,427 | 1,425 |
| Lincoln | 830 | 2,957,511 | 11.5 | 11.4 | 3.2 | 4.9 | 1.5 | 13.3 | 51.8 | 102.6 | 44.8 | 39.6 | 2,033 | 1,234 |
| Los Alamos | 1,188 | 4,973,128 | 7.7 | 6.7 | 16.3 | 4.1 | 1.5 | 14.0 | 44.1 | 133.5 | 74.3 | 39.6 | 2,112 | 830 |
| Luna | 1,175 | 3,345,403 | 6.3 | 15.3 | 2.2 | 2.9 | 1.2 | 4.4 | 63.7 | 108.0 | 69.2 | 23.7 | 983 | 523 |
| McKinley | 3,490 | 10,015,812 | 4.5 | 8.3 | 3.2 | 1.7 | 1.2 | 4.6 | 75.9 | 279.2 | 175.9 | 73.1 | 1,009 | 314 |
| Mora | 195 | 575,840 | 10.2 | | 0.0 | 3.4 | 3.1 | 3.6 | 72.8 | 26.5 | 10.3 | 8.6 | 1,898 | 1,650 |
| Otero | 1,684 | 5,230,317 | 7.5 | 14.4 | 1.7 | 2.1 | 1.6 | 5.4 | 64.2 | 161.8 | 92.2 | 50.4 | 762 | 358 |
| Quay | 576 | 1,668,229 | 5.4 | 8.5 | 0.2 | 3.0 | 5.2 | 9.2 | 64.8 | 57.4 | 36.8 | 11.8 | 1,421 | 707 |
| Rio Arriba | 1,409 | 4,355,659 | 9.9 | 7.3 | 1.8 | 2.2 | 4.6 | 6.2 | 64.2 | 138.1 | 68.9 | 47.6 | 1,212 | 790 |
| Roosevelt | 732 | 2,561,527 | 6.9 | 10.6 | 3.8 | 2.9 | 0.3 | 3.9 | 70.5 | 71.1 | 41.8 | 16.6 | 879 | 638 |
| Sandoval | 3,840 | 14,507,749 | 7.3 | 12.9 | 5.7 | 2.8 | 1.2 | 3.6 | 64.9 | 441.1 | 223.4 | 150.0 | 1,051 | 647 |
| San Juan | 5,621 | 20,270,322 | 4.8 | 12.1 | 2.9 | 2.2 | 1.4 | 10.3 | 64.6 | 518.5 | 323.9 | 118.0 | 930 | 655 |
| San Miguel | 1,122 | 3,368,535 | 8.8 | 7.8 | 3.9 | 2.6 | 1.2 | 8.0 | 67.5 | 101.2 | 58.7 | 31.0 | 1,120 | 540 |
| Santa Fe | 5,045 | 19,287,175 | 8.7 | 13.7 | 6.2 | 5.6 | 3.8 | 12.0 | 47.2 | 639.4 | 266.4 | 264.5 | 1,769 | 1,075 |
| Sierra | 670 | 2,240,852 | 7.0 | 6.5 | 0.2 | 2.3 | 45.9 | 4.7 | 30.1 | 56.6 | 20.9 | 11.1 | 1,004 | 593 |
| Socorro | 649 | 1,830,078 | 8.6 | 12.5 | 4.1 | 3.3 | 3.1 | 11.4 | 56.1 | 56.1 | 35.8 | 12.6 | 751 | 383 |
| Taos | 1,259 | 3,729,730 | 12.2 | 10.7 | 1.4 | 2.3 | 2.3 | 7.3 | 56.9 | 139.1 | 64.1 | 54.2 | 1,654 | 827 |
| Torrance | 744 | 2,293,228 | 6.7 | 8.0 | 0.8 | 2.8 | 1.2 | 2.4 | 76.9 | 82.9 | 42.7 | 18.9 | 1,217 | 842 |
| Union | 192 | 631,110 | 13.8 | 11.4 | 3.7 | 9.3 | 2.8 | 4.5 | 50.9 | 21.9 | 11.2 | 5.9 | 1,403 | 528 |
| Valencia | 2,566 | 8,587,011 | 4.5 | 7.4 | 1.2 | 1.1 | 1.8 | 3.3 | 79.9 | 209.0 | 121.7 | 70.0 | 923 | 558 |
| **NEW YORK** | X | X | X | X | X | X | X | X | X | X | X | X | X | X |
| Albany | 12,758 | 68,164,060 | 5.4 | 14.9 | 2.8 | 3.5 | 7.9 | 5.2 | 58.9 | 1,876.3 | 544.6 | 1,096.3 | 3,563 | 2,264 |
| Allegany | 2,356 | 9,274,123 | 7.5 | 7.8 | 0.0 | 7.6 | 7.6 | 3.5 | 63.5 | 306.4 | 173.2 | 107.9 | 2,314 | 1,861 |
| Bronx | (2) | (2) | (2) | (2) | (2) | (2) | (2) | (2) | (2) | (2) | (2) | (2) | (2) | (2) |
| Broome | 10,984 | 43,183,824 | 4.5 | 8.2 | 2.7 | 4.7 | 8.2 | 3.3 | 66.7 | 1,399.3 | 665.2 | 568.2 | 2,943 | 1,998 |
| Cattaraugus | 5,125 | 20,920,176 | 4.4 | 6.2 | 1.4 | 6.2 | 10.5 | 2.6 | 67.2 | 585.6 | 318.5 | 198.5 | 2,572 | 1,880 |
| Cayuga | 3,718 | 16,218,981 | 6.1 | 8.2 | 2.6 | 4.9 | 6.6 | 2.7 | 68.3 | 455.1 | 195.1 | 195.0 | 2,517 | 1,641 |
| Chautauqua | 6,354 | 28,121,337 | 4.9 | 7.5 | 2.0 | 6.6 | 8.3 | 2.9 | 66.6 | 836.7 | 417.4 | 317.9 | 2,477 | 1,721 |
| Chemung | 4,253 | 19,683,127 | 4.4 | 7.8 | 2.9 | 4.7 | 9.4 | 3.1 | 66.7 | 507.5 | 228.8 | 195.1 | 2,303 | 1,450 |
| Chenango | 3,098 | 12,262,824 | 6.1 | 5.9 | 1.1 | 6.9 | 5.9 | 1.7 | 71.5 | 326.2 | 173.8 | 116.7 | 2,442 | 1,844 |
| Clinton | 3,990 | 18,137,505 | 5.8 | 5.8 | 1.7 | 6.5 | 8.6 | 3.5 | 66.9 | 511.2 | 222.1 | 222.8 | 2,766 | 1,838 |
| Columbia | 2,773 | 12,572,682 | 8.9 | 8.7 | 0.3 | 7.7 | 9.6 | 1.9 | 60.9 | 399.3 | 132.8 | 223.3 | 3,702 | 2,760 |
| Cortland | 2,079 | 8,544,216 | 7.5 | 10.6 | 3.1 | 5.5 | 7.7 | 3.3 | 57.9 | 297.7 | 132.7 | 133.0 | 2,782 | 1,883 |
| Delaware | 2,416 | 9,750,539 | 7.6 | 5.4 | 0.1 | 11.6 | 7.4 | 2.4 | 64.3 | 315.4 | 130.7 | 151.3 | 3,361 | 2,826 |
| Dutchess | 11,879 | 65,652,723 | 5.5 | 9.3 | 2.8 | 3.6 | 4.9 | 1.7 | 71.1 | 1,807.5 | 556.7 | 1,088.9 | 3,710 | 2,909 |
| Erie | 38,244 | 195,946,904 | 3.5 | 10.3 | 2.6 | 3.0 | 13.9 | 4.7 | 60.7 | 5,952.6 | 2,418.3 | 2,549.9 | 2,775 | 1,656 |
| Essex | 1,873 | 7,617,938 | 11.6 | 6.3 | 0.3 | 11.2 | 8.9 | 5.6 | 54.0 | 273.4 | 84.1 | 156.6 | 4,176 | 3,231 |
| Franklin | 2,753 | 10,599,267 | 6.3 | 5.3 | 1.0 | 6.5 | 6.7 | 3.2 | 69.9 | 343.4 | 188.3 | 114.0 | 2,259 | 1,777 |
| Fulton | 2,755 | 10,604,219 | 5.2 | 7.9 | 2.9 | 5.3 | 5.1 | 2.1 | 70.7 | 323.1 | 159.4 | 132.3 | 2,459 | 1,786 |

1. Based on the resident population estimated as of July 1 of the year shown.  2. Bronx, Kings, Queens, and Richmond counties are included with New York county.

# Table B. States and Counties — Local Government Finances, Government Employment, and Income Taxes

| STATE County | Total (mil dol) | Per capita[1] (dollars) | Education | Health and hospitals | Police protection | Public welfare | Highways | Debt Total (mil dol) | Debt Per capita[1] (dollars) | Federal civilian | Federal military | State and local | Number of returns | Mean adjusted gross income | Mean income tax |
|---|---|---|---|---|---|---|---|---|---|---|---|---|---|---|---|
| | 185 | 186 | 187 | 188 | 189 | 190 | 191 | 192 | 193 | 194 | 195 | 196 | 197 | 198 | 199 |
| **NEW JERSEY— Cont'd** | | | | | | | | | | | | | | | |
| Mercer | 2,257 | 6,130 | 55.2 | 1.0 | 5.6 | 3.2 | 1.4 | 1,839.7 | 4,997 | 2,259 | 720 | 38,019 | 178,080 | 98,318 | 14,989 |
| Middlesex | 4,151 | 5,018 | 58.9 | 0.9 | 6.1 | 2.2 | 1.5 | 3,453.9 | 4,175 | 2,404 | 1,733 | 54,022 | 412,700 | 83,223 | 10,582 |
| Monmouth | 3,914 | 6,292 | 54.7 | 1.1 | 6.0 | 1.6 | 5.0 | 1,993.6 | 3,205 | 2,036 | 1,431 | 33,807 | 325,420 | 112,088 | 18,174 |
| Morris | 2,762 | 5,593 | 59.0 | 2.0 | 5.9 | 1.8 | 3.1 | 1,620.6 | 3,281 | 5,446 | 1,089 | 26,494 | 256,440 | 136,592 | 23,701 |
| Ocean | 2,624 | 4,407 | 55.0 | 0.4 | 7.2 | 3.0 | 3.7 | 1,565.9 | 2,630 | 3,306 | 1,435 | 25,411 | 293,030 | 67,705 | 7,906 |
| Passaic | 2,544 | 5,043 | 48.3 | 2.2 | 6.5 | 3.9 | 1.7 | 2,460.6 | 4,878 | 1,042 | 986 | 28,427 | 254,430 | 62,121 | 7,007 |
| Salem | 415 | 6,594 | 64.0 | 0.9 | 4.0 | 1.2 | 1.8 | 181.2 | 2,882 | 164 | 123 | 4,175 | 30,300 | 63,424 | 6,364 |
| Somerset | 1,852 | 5,601 | 61.7 | 3.4 | 4.4 | 0.2 | 1.7 | 1,194.1 | 3,612 | 1,806 | 649 | 16,983 | 171,460 | 131,346 | 22,744 |
| Sussex | 769 | 5,449 | 63.6 | 1.4 | 3.8 | 2.1 | 4.2 | 334.2 | 2,367 | 331 | 280 | 7,421 | 73,370 | 83,520 | 10,377 |
| Union | 3,483 | 6,279 | 57.0 | 1.9 | 6.0 | 1.5 | 2.1 | 2,156.3 | 3,887 | 1,403 | 1,109 | 32,687 | 279,650 | 92,623 | 14,221 |
| Warren | 613 | 5,802 | 59.8 | 2.3 | 4.1 | 2.3 | 3.4 | 178.2 | 1,686 | 230 | 207 | 5,318 | 54,530 | 73,306 | 8,185 |
| **NEW MEXICO** | X | X | X | X | X | X | X | X | X | 29,082 | 17,874 | 160,992 | 929,070 | 56,079 | 5,785 |
| Bernalillo | 2,383 | 3,513 | 49.8 | 1.6 | 9.2 | 1.7 | 2.3 | 3,424.6 | 5,050 | 14,034 | 5,278 | 55,972 | 315,520 | 60,168 | 6,468 |
| Catron | 19 | 5,247 | 36.7 | 1.0 | 5.1 | 0.0 | 24.1 | 4.5 | 1,261 | 88 | 9 | 195 | 1,380 | 40,917 | 3,748 |
| Chaves | 230 | 3,537 | 54.3 | 0.6 | 7.7 | 1.4 | 5.0 | 75.5 | 1,161 | 241 | 168 | 4,149 | 25,530 | 48,908 | 5,050 |
| Cibola | 111 | 4,105 | 37.4 | 27.2 | 4.3 | 0.7 | 5.6 | 38.2 | 1,420 | 252 | 62 | 2,803 | 9,300 | 39,345 | 2,745 |
| Colfax | 71 | 5,848 | 35.6 | 6.1 | 7.0 | 0.7 | 6.5 | 83.0 | 6,833 | 46 | 29 | 1,186 | 5,270 | 48,171 | 4,411 |
| Curry | 190 | 3,812 | 57.8 | 0.0 | 5.1 | 0.8 | 7.3 | 99.8 | 2,004 | 898 | 4,786 | 2,425 | 20,350 | 42,786 | 3,420 |
| De Baca | 15 | 8,416 | 30.9 | 23.8 | 4.9 | 0.3 | 9.5 | 4.3 | 2,374 | 11 | 4 | 161 | 730 | 41,804 | 3,138 |
| Dona Ana | 709 | 3,279 | 60.6 | 1.6 | 7.6 | 2.3 | 2.1 | 452.3 | 2,092 | 3,329 | 557 | 16,887 | 94,870 | 46,148 | 4,113 |
| Eddy | 341 | 5,977 | 45.0 | 1.5 | 7.8 | 1.5 | 6.3 | 163.4 | 2,862 | 683 | 147 | 3,230 | 24,920 | 82,562 | 10,899 |
| Grant | 186 | 6,725 | 25.4 | 46.6 | 4.9 | 1.7 | 2.3 | 94.7 | 3,425 | 192 | 67 | 3,043 | 12,090 | 47,963 | 4,075 |
| Guadalupe | 28 | 6,395 | 39.1 | 0.5 | 4.9 | 0.6 | 3.1 | 15.0 | 3,405 | 25 | 10 | 391 | 1,780 | 32,941 | 1,935 |
| Harding | 8 | 11,769 | 42.3 | 0.7 | 1.4 | 0.9 | 21.9 | 6.9 | 10,140 | 12 | 2 | 75 | 300 | 32,363 | 2,403 |
| Hidalgo | 26 | 6,105 | 41.5 | 2.6 | 17.8 | 0.5 | 5.7 | 9.2 | 2,133 | 268 | 11 | 357 | 1,740 | 38,366 | 2,680 |
| Lea | 498 | 7,219 | 44.3 | 16.0 | 6.4 | 1.3 | 3.1 | 311.0 | 4,504 | 88 | 176 | 3,772 | 28,590 | 65,878 | 7,299 |
| Lincoln | 97 | 4,955 | 38.6 | 2.9 | 12.9 | 1.1 | 3.6 | 98.5 | 5,057 | 92 | 50 | 1,118 | 9,060 | 51,576 | 5,392 |
| Los Alamos | 155 | 8,245 | 29.4 | 0.0 | 5.9 | 2.5 | 4.9 | 133.1 | 7,099 | 241 | 52 | 1,615 | 9,640 | 112,358 | 15,044 |
| Luna | 118 | 4,892 | 58.7 | 0.9 | 7.4 | 1.6 | 3.1 | 78.7 | 3,266 | 447 | 59 | 1,566 | 10,370 | 34,688 | 2,234 |
| McKinley | 268 | 3,700 | 61.8 | 1.8 | 4.1 | 1.2 | 3.4 | 90.6 | 1,251 | 2,394 | 180 | 4,635 | 30,050 | 35,267 | 2,397 |
| Mora | 25 | 5,503 | 41.9 | 0.9 | 2.2 | 0.0 | 0.1 | 7.3 | 1,620 | 33 | 12 | 257 | 2,010 | 35,383 | 2,407 |
| Otero | 149 | 2,258 | 50.8 | 1.2 | 9.4 | 1.4 | 4.5 | 207.2 | 3,134 | 1,727 | 4,526 | 4,592 | 27,200 | 44,547 | 3,473 |
| Quay | 53 | 6,349 | 56.2 | 4.8 | 4.0 | 2.3 | 5.3 | 23.8 | 2,872 | 29 | 21 | 793 | 3,570 | 34,040 | 2,378 |
| Rio Arriba | 125 | 3,172 | 54.4 | 2.1 | 4.1 | 0.1 | 2.9 | 83.6 | 2,130 | 270 | 98 | 4,270 | 16,570 | 43,684 | 3,416 |
| Roosevelt | 69 | 3,647 | 58.4 | 0.1 | 6.3 | 2.4 | 4.8 | 43.9 | 2,321 | 39 | 44 | 2,121 | 7,350 | 38,185 | 2,884 |
| Sandoval | 402 | 2,814 | 56.4 | 0.1 | 6.3 | 1.6 | 4.9 | 609.9 | 4,275 | 358 | 373 | 7,698 | 67,340 | 61,578 | 6,036 |
| San Juan | 565 | 4,449 | 58.2 | 1.5 | 6.0 | 1.7 | 3.7 | 2,061.0 | 16,239 | 1,483 | 314 | 9,703 | 48,350 | 52,504 | 4,832 |
| San Miguel | 107 | 3,865 | 59.1 | 0.8 | 4.9 | 0.7 | 4.0 | 90.4 | 3,260 | 127 | 66 | 3,307 | 11,530 | 37,363 | 2,782 |
| Santa Fe | 553 | 3,699 | 43.8 | 0.0 | 7.4 | 1.4 | 2.9 | 844.6 | 5,650 | 953 | 384 | 14,577 | 78,170 | 75,175 | 9,788 |
| Sierra | 50 | 4,507 | 35.1 | 36.7 | 3.0 | 0.0 | 3.4 | 29.5 | 2,667 | 82 | 27 | 801 | 4,810 | 34,365 | 2,705 |
| Socorro | 60 | 3,538 | 49.7 | 0.4 | 6.1 | 0.0 | 6.0 | 33.2 | 1,973 | 175 | 41 | 2,494 | 6,690 | 39,877 | 3,078 |
| Taos | 137 | 4,166 | 40.0 | 2.2 | 3.8 | 1.3 | 4.5 | 126.7 | 3,862 | 265 | 82 | 1,804 | 15,570 | 44,796 | 4,313 |
| Torrance | 78 | 5,002 | 59.9 | 3.4 | 3.7 | 0.0 | 2.6 | 239.4 | 15,408 | 64 | 38 | 901 | 5,840 | 39,138 | 2,947 |
| Union | 24 | 5,647 | 37.2 | 6.5 | 7.1 | 0.0 | 8.0 | 15.7 | 3,755 | 39 | 9 | 302 | 1,540 | 38,906 | 3,545 |
| Valencia | 175 | 2,309 | 78.4 | 0.0 | 5.2 | 0.0 | 1.2 | 180.8 | 2,382 | 97 | 192 | 3,792 | 31,160 | 47,123 | 3,690 |
| **NEW YORK** | X | X | X | X | X | X | X | X | X | 115,237 | 55,853 | 1,333,389 | 9,742,110 | 89,397 | 14,072 |
| Albany | 1,840 | 5,980 | 45.1 | 4.7 | 5.9 | 11.0 | 3.1 | 1,378.3 | 4,479 | 5,477 | 700 | 56,748 | 153,480 | 81,412 | 10,736 |
| Allegany | 299 | 6,399 | 52.7 | 2.5 | 1.4 | 9.4 | 9.1 | 222.8 | 4,777 | 110 | 66 | 3,939 | 18,640 | 51,002 | 4,431 |
| Bronx | (2) | (2) | (2) | (2) | (2) | (2) | (2) | (2) | (2) | 4,032 | 2,296 | 72,583 | 665,660 | 40,638 | 3,253 |
| Broome | 1,444 | 7,478 | 45.7 | 3.2 | 2.5 | 14.0 | 4.1 | 970.7 | 5,027 | 515 | 284 | 17,431 | 88,030 | 57,270 | 5,780 |
| Cattaraugus | 607 | 7,868 | 45.0 | 6.7 | 2.9 | 12.4 | 9.1 | 354.5 | 4,594 | 264 | 127 | 9,479 | 33,930 | 49,931 | 4,290 |
| Cayuga | 447 | 5,770 | 50.3 | 2.8 | 2.8 | 8.9 | 6.7 | 323.2 | 4,172 | 150 | 115 | 5,680 | 35,150 | 56,071 | 5,283 |
| Chautauqua | 850 | 6,618 | 52.5 | 2.7 | 3.1 | 10.1 | 6.5 | 596.2 | 4,644 | 303 | 192 | 9,075 | 56,540 | 49,920 | 4,275 |
| Chemung | 500 | 5,899 | 45.2 | 2.9 | 3.2 | 18.4 | 5.7 | 196.0 | 2,313 | 211 | 127 | 6,301 | 38,820 | 57,619 | 5,681 |
| Chenango | 323 | 6,746 | 57.5 | 3.0 | 1.4 | 7.3 | 7.4 | 154.1 | 3,223 | 91 | 73 | 4,151 | 22,380 | 49,577 | 4,121 |
| Clinton | 513 | 6,365 | 50.0 | 5.6 | 1.9 | 11.2 | 5.5 | 255.5 | 3,172 | 707 | 118 | 7,409 | 36,840 | 56,145 | 5,130 |
| Columbia | 400 | 6,624 | 49.2 | 4.4 | 2.2 | 11.6 | 8.4 | 214.4 | 3,554 | 155 | 91 | 4,402 | 30,320 | 71,285 | 9,001 |
| Cortland | 323 | 6,748 | 46.6 | 3.8 | 3.5 | 9.2 | 7.6 | 205.1 | 4,289 | 111 | 69 | 4,333 | 20,530 | 53,980 | 4,949 |
| Delaware | 318 | 7,065 | 48.9 | 2.6 | 1.3 | 8.5 | 11.5 | 127.5 | 2,830 | 135 | 100 | 4,263 | 19,660 | 52,071 | 4,725 |
| Dutchess | 1,777 | 6,055 | 56.4 | 3.1 | 3.4 | 7.5 | 4.9 | 1,010.3 | 3,442 | 1,321 | 434 | 18,347 | 144,440 | 79,438 | 10,015 |
| Erie | 5,952 | 6,477 | 45.4 | 12.2 | 4.1 | 10.2 | 3.5 | 5,083.2 | 5,531 | 8,445 | 1,618 | 65,166 | 453,510 | 65,825 | 7,596 |
| Essex | 263 | 7,026 | 38.8 | 5.2 | 1.3 | 7.6 | 11.6 | 183.8 | 4,904 | 284 | 55 | 4,008 | 17,960 | 56,351 | 5,482 |
| Franklin | 358 | 7,087 | 56.4 | 3.5 | 1.6 | 9.0 | 6.3 | 213.7 | 4,234 | 164 | 69 | 7,504 | 20,390 | 50,151 | 4,298 |
| Fulton | 337 | 6,265 | 55.1 | 1.7 | 2.4 | 13.2 | 4.6 | 134.4 | 2,499 | 84 | 83 | 3,444 | 24,930 | 52,210 | 4,594 |

1. Based on the resident population estimated as of July 1 of the year shown.    2. Bronx, Kings, Queens, and Richmond counties are included with New York county.

| State / county code | CBSA code[1] | County Type code[2] | STATE County | Land area[3] (sq. mi) | Total persons 2019 | Rank | Per square mile | White | Black | American Indian, Alaska Native | Asian and Pacific Islancer | Percent Hispanic or Latino[4] | Under 5 years | 5 to 14 years | 15 to 24 years | 25 to 34 years | 35 to 44 years | 45 to 54 years |
|---|---|---|---|---|---|---|---|---|---|---|---|---|---|---|---|---|---|---|
| | | | | 1 | 2 | 3 | 4 | 5 | 6 | 7 | 8 | 9 | 10 | 11 | 12 | 13 | 14 | 15 |
| | | | **NEW YORK— Cont'd** | | | | | | | | | | | | | | | |
| 36037 | 12860 | 4 | Genesee | 492.9 | 56,994 | 908 | 115.6 | 91.9 | 4.0 | 1.5 | 1.1 | 3.5 | 5.1 | 11.5 | 11.4 | 12.6 | 11.4 | 12.7 |
| 36039 | | 6 | Greene | 647.2 | 47,177 | 1,033 | 72.9 | 86.3 | 6.6 | 0.8 | 1.7 | 6.6 | 4.4 | 8.9 | 11.5 | 12.1 | 10.9 | 12.9 |
| 36041 | | 8 | Hamilton | 1,717.4 | 4,345 | 2,870 | 2.5 | 95.7 | 1.7 | 1.0 | 1.2 | 1.9 | 2.7 | 7.2 | 8.5 | 7.3 | 8.5 | 11.7 |
| 36043 | 46540 | 2 | Herkimer | 1,411.5 | 60,945 | 864 | 43.2 | 95.2 | 2.1 | 0.6 | 1.0 | 2.5 | 5.0 | 11.6 | 11.8 | 11.5 | 10.8 | 12.1 |
| 36045 | 48060 | 3 | Jefferson | 1,268.7 | 108,095 | 565 | 85.2 | 83.9 | 7.2 | 1.1 | 2.9 | 7.8 | 7.5 | 13.0 | 15.6 | 16.6 | 11.9 | 9.9 |
| 36047 | 35620 | 1 | Kings | 69.4 | 2,538,934 | 9 | 36,584.1 | 38.4 | 30.5 | 0.6 | 13.9 | 18.8 | 7.0 | 12.2 | 11.6 | 18.3 | 14.1 | 11.4 |
| 36049 | | 6 | Lewis | 1,274.6 | 26,187 | 1,562 | 20.5 | 96.9 | 1.1 | 0.7 | 0.8 | 1.6 | 6.2 | 12.7 | 11.1 | 11.5 | 11.3 | 11.8 |
| 36051 | 40380 | 1 | Livingston | 631.8 | 62,398 | 853 | 98.8 | 91.7 | 3.2 | 0.7 | 2.0 | 4.1 | 4.2 | 9.9 | 18.2 | 10.7 | 11.0 | 11.7 |
| 36053 | 45060 | 2 | Madison | 654.9 | 70,478 | 771 | 107.6 | 94.2 | 2.5 | 1.0 | 1.5 | 2.5 | 4.9 | 10.5 | 16.0 | 10.5 | 10.7 | 12.1 |
| 36055 | 40380 | 1 | Monroe | 656.9 | 740,900 | 90 | 1,127.9 | 71.7 | 16.1 | 0.6 | 4.5 | 9.5 | 5.4 | 11.5 | 13.2 | 14.2 | 11.8 | 11.8 |
| 36057 | 11220 | 4 | Montgomery | 403.1 | 49,170 | 1,003 | 122.0 | 81.0 | 3.3 | 0.6 | 1.3 | 15.4 | 6.3 | 12.5 | 11.7 | 13.2 | 12.0 | 11.9 |
| 36059 | 35620 | 1 | Nassau | 284.5 | 1,351,334 | 30 | 4,749.9 | 58.9 | 12.4 | 0.4 | 12.4 | 17.5 | 5.4 | 12.1 | 12.3 | 11.8 | 12.1 | 13.2 |
| 36061 | 35620 | 1 | New York | 22.7 | 1,611,989 | 21 | 71,012.7 | 48.8 | 13.4 | 0.5 | 13.9 | 25.6 | 4.6 | 7.6 | 10.8 | 21.9 | 14.4 | 11.9 |
| 36063 | 15380 | 1 | Niagara | 522.4 | 208,396 | 327 | 398.9 | 87.0 | 8.8 | 1.6 | 1.7 | 3.5 | 5.1 | 11.1 | 11.5 | 12.5 | 11.4 | 12.4 |
| 36065 | 46540 | 2 | Oneida | 1,212.3 | 227,346 | 303 | 187.5 | 82.9 | 7.4 | 0.6 | 4.8 | 6.4 | 5.5 | 12.0 | 12.9 | 12.5 | 11.4 | 12.1 |
| 36067 | 45060 | 2 | Onondaga | 778.4 | 459,214 | 155 | 589.9 | 78.8 | 12.9 | 1.4 | 4.7 | 5.4 | 5.7 | 11.8 | 13.6 | 12.5 | 11.4 | 12.1 |
| 36069 | 40380 | 1 | Ontario | 644.1 | 110,091 | 555 | 170.9 | 90.8 | 3.2 | 0.6 | 2.0 | 5.2 | 4.8 | 11.2 | 12.4 | 11.4 | 11.4 | 12.3 |
| 36071 | 39100 | 1 | Orange | 812.3 | 385,234 | 185 | 474.3 | 63.6 | 12.1 | 0.7 | 3.7 | 22.2 | 6.7 | 14.2 | 14.7 | 12.0 | 12.0 | 12.9 |
| 36073 | 40380 | 1 | Orleans | 391.3 | 39,978 | 1,186 | 102.2 | 87.9 | 6.8 | 1.0 | 1.0 | 5.2 | 4.7 | 10.5 | 11.9 | 13.0 | 11.3 | 13.0 |
| 36075 | 45060 | 2 | Oswego | 951.6 | 116,346 | 536 | 122.3 | 95.0 | 1.6 | 0.9 | 1.2 | 2.8 | 5.4 | 11.7 | 13.7 | 12.4 | 11.2 | 12.6 |
| 36077 | 36580 | 7 | Otsego | 1,001.7 | 58,701 | 882 | 58.6 | 92.0 | 2.8 | 0.7 | 2.0 | 4.0 | 4.1 | 8.6 | 19.3 | 10.1 | 9.8 | 11.0 |
| 36079 | 35620 | 1 | Putnam | 230.2 | 98,532 | 613 | 428.0 | 77.2 | 3.7 | 0.5 | 3.0 | 17.1 | 4.5 | 10.9 | 12.1 | 11.2 | 12.0 | 14.6 |
| 36081 | 35620 | 1 | Queens | 108.7 | 2,225,821 | 13 | 20,476.7 | 25.8 | 19.0 | 0.8 | 28.6 | 27.9 | 6.0 | 10.8 | 10.5 | 15.8 | 13.5 | 13.1 |
| 36083 | 10580 | 2 | Rensselaer | 652.5 | 158,108 | 426 | 242.3 | 84.3 | 8.5 | 0.6 | 3.6 | 5.4 | 5.1 | 10.8 | 13.0 | 14.2 | 12.3 | 13.3 |
| 36085 | 35620 | 1 | Richmond | 57.5 | 475,327 | 151 | 8,266.6 | 60.0 | 10.2 | 0.5 | 12.4 | 18.7 | 5.7 | 12.1 | 12.0 | 13.5 | 12.4 | 13.3 |
| 36087 | 35620 | 1 | Rockland | 173.4 | 326,225 | 219 | 1,881.3 | 63.8 | 12.0 | 0.4 | 6.9 | 18.5 | 8.4 | 15.6 | 13.8 | 11.6 | 11.1 | 11.4 |
| 36089 | 36300 | 4 | St. Lawrence | 2,679.5 | 107,185 | 569 | 40.0 | 93.2 | 2.7 | 1.5 | 1.6 | 2.4 | 5.0 | 11.4 | 16.8 | 11.6 | 11.2 | 11.7 |
| 36091 | 10580 | 2 | Saratoga | 810.0 | 230,298 | 298 | 284.3 | 91.6 | 2.6 | 0.5 | 3.7 | 3.4 | 4.8 | 11.1 | 11.4 | 12.2 | 12.5 | 13.7 |
| 36093 | 10580 | 2 | Schenectady | 204.6 | 155,358 | 433 | 759.3 | 74.7 | 13.5 | 1.1 | 6.6 | 7.7 | 6.0 | 11.9 | 12.4 | 13.4 | 12.5 | 12.2 |
| 36095 | 10580 | 2 | Schoharie | 621.8 | 31,132 | 1,400 | 50.1 | 93.9 | 2.0 | 0.8 | 1.2 | 3.6 | 4.4 | 9.7 | 13.3 | 10.7 | 10.6 | 12.2 |
| 36097 | | 6 | Schuyler | 328.3 | 17,685 | 1,936 | 53.9 | 95.9 | 1.7 | 0.9 | 1.2 | 1.9 | 4.9 | 10.6 | 9.7 | 11.2 | 11.3 | 13.1 |
| 36099 | 42900 | 6 | Seneca | 323.7 | 33,991 | 1,327 | 105.0 | 89.9 | 5.8 | 0.9 | 1.2 | 3.9 | 5.3 | 11.3 | 11.7 | 13.5 | 11.5 | 11.8 |
| 36101 | 18500 | 4 | Steuben | 1,390.5 | 94,657 | 630 | 68.1 | 94.7 | 2.4 | 0.7 | 2.0 | 1.8 | 5.5 | 12.0 | 11.1 | 11.9 | 11.5 | 12.5 |
| 36103 | 35620 | 1 | Suffolk | 911.2 | 1,474,273 | 27 | 1,617.9 | 67.2 | 8.4 | 0.5 | 4.9 | 20.5 | 5.3 | 11.5 | 12.7 | 12.2 | 11.7 | 13.8 |
| 36105 | | 4 | Sullivan | 968.1 | 75,802 | 735 | 78.3 | 72.2 | 9.6 | 0.8 | 2.6 | 17.1 | 6.0 | 11.6 | 11.5 | 12.1 | 11.6 | 12.8 |
| 36107 | 13780 | 2 | Tioga | 518.8 | 47,904 | 1,023 | 92.3 | 95.8 | 1.5 | 0.6 | 1.2 | 2.2 | 4.7 | 11.8 | 11.2 | 10.9 | 11.5 | 12.2 |
| 36109 | 27060 | 3 | Tompkins | 474.6 | 101,058 | 600 | 212.9 | 79.9 | 5.4 | 0.9 | 11.8 | 5.4 | 3.7 | 8.1 | 28.6 | 12.6 | 10.7 | 9.8 |
| 36111 | 28740 | 3 | Ulster | 1,124.2 | 177,716 | 375 | 158.1 | 80.4 | 7.4 | 0.8 | 2.9 | 11.0 | 4.4 | 9.7 | 12.0 | 12.6 | 11.9 | 12.9 |
| 36113 | 24020 | 3 | Warren | 867.2 | 63,756 | 839 | 73.5 | 94.6 | 1.9 | 0.7 | 1.4 | 2.9 | 4.4 | 10.1 | 10.4 | 11.9 | 10.9 | 13.4 |
| 36115 | 24020 | 3 | Washington | 831.2 | 60,606 | 867 | 72.9 | 93.1 | 3.6 | 0.7 | 1.0 | 3.0 | 4.5 | 10.4 | 10.9 | 12.5 | 12.0 | 13.4 |
| 36117 | 40380 | 1 | Wayne | 603.8 | 89,339 | 660 | 148.0 | 91.4 | 4.1 | 0.6 | 1.2 | 4.7 | 5.5 | 11.9 | 10.7 | 11.7 | 11.2 | 13.0 |
| 36119 | 35620 | 1 | Westchester | 430.7 | 965,802 | 51 | 2,242.4 | 53.6 | 14.7 | 0.4 | 7.3 | 25.8 | 5.4 | 12.1 | 12.7 | 11.6 | 12.8 | 13.8 |
| 36121 | | 6 | Wyoming | 592.8 | 39,465 | 1,195 | 66.6 | 90.6 | 5.5 | 0.6 | 0.8 | 3.5 | 4.6 | 10.4 | 10.6 | 14.0 | 12.7 | 13.5 |
| 36123 | 40380 | 1 | Yates | 338.1 | 24,780 | 1,613 | 73.3 | 95.7 | 1.5 | 0.5 | 1.1 | 2.3 | 6.5 | 12.1 | 14.6 | 10.8 | 9.7 | 10.0 |
| 37000 | | 0 | **NORTH CAROLINA** | 48,620.3 | 10,600,823 | X | 218.0 | 64.1 | 22.6 | 1.7 | 3.9 | 9.9 | 5.7 | 12.3 | 13.2 | 13.6 | 12.4 | 12.8 |
| 37001 | 15500 | 3 | Alamance | 423.5 | 171,346 | 385 | 404.6 | 63.7 | 21.9 | 0.8 | 2.3 | 13.2 | 5.8 | 12.3 | 14.3 | 12.8 | 11.5 | 12.8 |
| 37003 | 25860 | 2 | Alexander | 260.0 | 37,441 | 1,240 | 144.0 | 88.0 | 6.7 | 0.6 | 1.3 | 4.9 | 4.9 | 11.1 | 10.9 | 12.3 | 11.4 | 13.8 |
| 37005 | | 9 | Alleghany | 234.8 | 11,194 | 2,333 | 47.7 | 87.7 | 2.0 | 0.8 | 0.9 | 9.8 | 4.4 | 9.5 | 9.2 | 9.6 | 9.7 | 12.9 |
| 37007 | 16740 | 6 | Anson | 531.5 | 24,097 | 1,643 | 45.3 | 46.0 | 48.5 | 1.0 | 1.7 | 4.4 | 5.4 | 11.0 | 11.3 | 14.2 | 11.7 | 12.6 |
| 37009 | | 7 | Ashe | 426.3 | 27,166 | 1,520 | 63.7 | 93.2 | 1.2 | 0.8 | 0.7 | 5.1 | 3.8 | 9.9 | 9.6 | 9.8 | 10.9 | 12.7 |
| 37011 | | 8 | Avery | 247.3 | 17,571 | 1,941 | 71.1 | 89.4 | 4.9 | 0.9 | 0.9 | 5.0 | 4.1 | 8.1 | 11.8 | 12.7 | 12.2 | 13.8 |
| 37013 | 47820 | 6 | Beaufort | 832.7 | 47,073 | 1,036 | 56.5 | 67.3 | 24.3 | 0.6 | 0.7 | 8.4 | 4.8 | 11.0 | 10.7 | 10.1 | 10.5 | 12.4 |
| 37015 | | 7 | Bertie | 699.2 | 18,712 | 1,887 | 26.8 | 35.3 | 61.7 | 1.0 | 0.8 | 2.3 | 4.4 | 9.5 | 11.0 | 13.1 | 10.1 | 11.5 |
| 37017 | | 6 | Bladen | 875.0 | 32,911 | 1,359 | 37.6 | 56.3 | 33.9 | 2.9 | 0.6 | 7.8 | 5.2 | 11.3 | 11.1 | 10.8 | 10.7 | 12.5 |
| 37019 | 34820 | 2 | Brunswick | 850.1 | 149,039 | 450 | 175.3 | 84.1 | 10.3 | 1.2 | 1.1 | 5.0 | 3.6 | 8.2 | 7.7 | 9.0 | 9.1 | 10.6 |
| 37021 | 11700 | 2 | Buncombe | 656.5 | 263,477 | 264 | 401.3 | 85.1 | 7.0 | 1.0 | 2.0 | 7.0 | 4.7 | 10.1 | 10.7 | 13.9 | 13.5 | 12.5 |
| 37023 | 25860 | 2 | Burke | 506.2 | 90,418 | 657 | 178.6 | 82.7 | 7.5 | 0.8 | 4.1 | 6.6 | 4.8 | 9.9 | 11.4 | 12.4 | 10.8 | 13.9 |
| 37025 | 16740 | 1 | Cabarrus | 361.2 | 221,479 | 310 | 613.2 | 63.9 | 20.6 | 0.8 | 5.7 | 11.3 | 6.3 | 14.4 | 12.5 | 12.8 | 14.3 | 14.2 |
| 37027 | 25860 | 2 | Caldwell | 471.9 | 82,100 | 696 | 174.0 | 88.0 | 5.9 | 0.7 | 1.0 | 6.1 | 5.1 | 11.0 | 11.1 | 11.8 | 10.9 | 14.0 |
| 37029 | 47260 | 8 | Camden | 240.3 | 10,984 | 2,350 | 45.7 | 83.7 | 12.2 | 1.2 | 2.5 | 3.3 | 5.0 | 13.0 | 11.0 | 11.5 | 13.7 | 14.0 |
| 37031 | 33980 | 4 | Carteret | 507.6 | 69,558 | 778 | 137.0 | 88.4 | 6.2 | 1.1 | 1.9 | 4.5 | 3.8 | 9.7 | 9.4 | 10.0 | 10.9 | 12.5 |
| 37033 | | 8 | Caswell | 425.4 | 22,443 | 1,705 | 52.8 | 62.3 | 32.9 | 1.0 | 0.9 | 4.7 | 5.0 | 10.0 | 10.3 | 11.9 | 10.6 | 13.3 |
| 37035 | 25860 | 2 | Catawba | 401.4 | 160,307 | 419 | 399.4 | 76.2 | 9.7 | 0.6 | 5.1 | 10.3 | 5.5 | 12.3 | 12.2 | 12.2 | 11.4 | 13.5 |
| 37037 | 20500 | 2 | Chatham | 681.7 | 75,748 | 736 | 111.1 | 73.8 | 12.5 | 0.8 | 2.7 | 12.0 | 4.6 | 11.2 | 10.0 | 9.0 | 11.2 | 13.5 |
| 37039 | | 2 | Cherokee | 455.5 | 29,073 | 1,457 | 63.8 | 92.8 | 2.1 | 2.9 | 0.9 | 3.6 | 4.2 | 9.0 | 8.9 | 8.8 | 9.3 | 12.0 |

1. CBSA = Core Based Statistical Area. See Appendix A for explanation. See Appendix B for list of metropolitan areas with component counties. 2. County type code from the Economic Research Service of USDA Rural-Urban Continuum Codes. See Appendix A for definition. 3. Dry land or land partially or temporarily covered by water. 4. May be of any race.

# Table B. States and Counties — Population and Households

| STATE County | Population, 2020 (cont.) — Age (percent) (cont.) | | | | Population change, 2000-2020 — Total persons | | Percent change | | Components of change, 2010-2020 | | | Households, 2015-2019 | | | | |
|---|---|---|---|---|---|---|---|---|---|---|---|---|---|---|---|---|
| | 55 to 64 years | 65 to 74 years | 75 years and over | Percent female | 2000 | 2010 | 2000-2010 | 2010-2020 | Births | Deaths | Net Migration | Number | Persons per household | Family house-holds | Female family house-holder[1] | One person |
| | 16 | 17 | 18 | 19 | 20 | 21 | 22 | 23 | 24 | 25 | 26 | 27 | 28 | 29 | 30 | 31 |
| **NEW YORK— Cont'd** | | | | | | | | | | | | | | | | |
| Genesee | 15.5 | 11.0 | 8.7 | 50.2 | 60,370 | 59,937 | -0.7 | -4.9 | 6,066 | 6,427 | -2,587 | 23,759 | 2.40 | 62.7 | 9.6 | 30.6 |
| Greene | 16.2 | 13.3 | 9.8 | 47.9 | 48,195 | 49,212 | 2.1 | -4.1 | 4,249 | 5,396 | -866.0 | 17,100 | 2.58 | 59.7 | 8.4 | 34.4 |
| Hamilton | 20.7 | 21.2 | 12.2 | 49.6 | 5,379 | 4,842 | -10.0 | -10.3 | 281 | 621 | -154.0 | 1,157 | 3.79 | 67.9 | 7.3 | 27.9 |
| Herkimer | 15.4 | 12.6 | 9.1 | 50.5 | 64,427 | 64,469 | 0.1 | -5.5 | 6,423 | 7,029 | -2,909 | 24,524 | 2.48 | 63.7 | 8.8 | 30.3 |
| Jefferson | 11.2 | 8.6 | 6.1 | 47.5 | 111,738 | 116,234 | 4.0 | -7.0 | 20,811 | 9,304 | -19,763 | 42,734 | 2.49 | 66.0 | 10.7 | 27.8 |
| Kings | 11.1 | 8.5 | 6.3 | 52.7 | 2,465,326 | 2,504,856 | 1.6 | 1.4 | 416,347 | 167,730 | -213,664 | 958,567 | 2.66 | 61.4 | 17.9 | 29.1 |
| Lewis | 15.7 | 11.5 | 8.2 | 49.3 | 26,944 | 27,087 | 0.5 | -3.3 | 3,363 | 2,584 | -1,682 | 10,247 | 2.56 | 68.0 | 8.7 | 25.0 |
| Livingston | 15.1 | 11.3 | 7.9 | 49.7 | 64,328 | 65,194 | 1.3 | -4.3 | 5,376 | 5,646 | -2,510 | 24,176 | 2.37 | 64.0 | 11.1 | 27.9 |
| Madison | 15.7 | 11.5 | 8.0 | 50.5 | 69,441 | 73,457 | 5.8 | -4.1 | 6,804 | 6,361 | -3,424 | 25,877 | 2.55 | 64.3 | 8.6 | 29.2 |
| Monroe | 13.8 | 10.5 | 7.8 | 51.8 | 735,343 | 744,396 | 1.2 | -0.5 | 83,630 | 68,299 | -18,820 | 301,948 | 2.37 | 60.3 | 13.9 | 32.0 |
| Montgomery | 13.8 | 11.2 | 8.3 | 50.7 | 49,708 | 50,268 | 1.1 | -2.2 | 6,158 | 5,960 | -1,285 | 19,660 | 2.45 | 63.1 | 13.8 | 30.1 |
| Nassau | 14.4 | 10.4 | 8.2 | 51.2 | 1,334,544 | 1,339,854 | 0.4 | 0.9 | 145,334 | 113,411 | -19,771 | 446,977 | 2.99 | 76.7 | 11.2 | 20.0 |
| New York | 11.3 | 9.4 | 8.1 | 52.7 | 1,537,195 | 1,586,609 | 3.2 | 1.6 | 185,498 | 106,877 | -52,552 | 759,460 | 2.07 | 42.4 | 10.9 | 46.1 |
| Niagara | 15.8 | 11.8 | 8.4 | 51.2 | 219,846 | 216,479 | -1.5 | -3.7 | 21,941 | 24,510 | -5,405 | 88,519 | 2.34 | 62.0 | 12.1 | 31.9 |
| Oneida | 14.0 | 10.9 | 8.6 | 50.2 | 235,469 | 234,851 | -0.3 | -3.2 | 26,260 | 26,088 | -7,668 | 89,729 | 2.43 | 62.5 | 12.9 | 31.6 |
| Onondaga | 14.0 | 10.4 | 7.7 | 51.7 | 458,336 | 467,068 | 1.9 | -1.7 | 53,964 | 43,840 | -18,014 | 185,324 | 2.39 | 60.5 | 12.6 | 31.3 |
| Ontario | 15.2 | 12.6 | 8.8 | 51.1 | 100,224 | 108,106 | 7.9 | 1.8 | 10,406 | 11,061 | 2,738 | 44,787 | 2.36 | 63.4 | 10.1 | 29.9 |
| Orange | 12.8 | 8.6 | 6.0 | 49.9 | 341,367 | 372,823 | 9.2 | 3.3 | 50,596 | 27,744 | -10,522 | 128,016 | 2.89 | 70.1 | 11.9 | 24.8 |
| Orleans | 16.0 | 11.5 | 7.7 | 50.0 | 44,171 | 42,891 | -2.9 | -6.8 | 4,095 | 4,367 | -2,653 | 16,563 | 2.28 | 66.1 | 12.0 | 27.5 |
| Oswego | 15.4 | 10.6 | 6.9 | 50.0 | 122,377 | 122,109 | -0.2 | -4.7 | 13,106 | 11,092 | -7,824 | 46,079 | 2.46 | 66.1 | 12.3 | 25.8 |
| Otsego | 14.8 | 12.6 | 9.6 | 51.5 | 61,676 | 62,278 | 1.0 | -5.7 | 5,276 | 6,308 | -2,532 | 23,409 | 2.32 | 61.9 | 8.9 | 30.1 |
| Putnam | 16.1 | 10.9 | 7.7 | 50.0 | 95,745 | 99,633 | 4.1 | -1.1 | 8,574 | 7,271 | -2,401 | 34,854 | 2.76 | 73.1 | 8.9 | 22.3 |
| Queens | 13.4 | 9.6 | 7.3 | 51.5 | 2,229,379 | 2,230,511 | 0.1 | -0.2 | 303,380 | 151,261 | -156,967 | 778,932 | 2.90 | 67.8 | 15.8 | 25.6 |
| Rensselaer | 14.3 | 10.9 | 7.2 | 50.6 | 152,538 | 159,435 | 4.5 | -0.8 | 16,903 | 15,612 | -2,547 | 64,906 | 2.36 | 60.4 | 12.8 | 31.2 |
| Richmond | 13.9 | 10.1 | 7.0 | 51.4 | 443,728 | 468,730 | 5.6 | 1.4 | 54,884 | 37,825 | -10,349 | 166,246 | 2.82 | 73.9 | 14.1 | 22.7 |
| Rockland | 12.0 | 8.7 | 7.4 | 50.9 | 286,753 | 311,691 | 8.7 | 4.7 | 52,484 | 22,260 | -15,768 | 100,438 | 3.17 | 74.6 | 10.2 | 21.4 |
| St. Lawrence | 13.9 | 10.8 | 7.6 | 48.9 | 111,931 | 111,941 | 0.0 | -4.2 | 11,807 | 10,413 | -6,179 | 41,933 | 2.32 | 63.1 | 11.5 | 29.4 |
| Saratoga | 14.9 | 11.5 | 7.9 | 50.7 | 200,635 | 219,598 | 9.5 | 4.9 | 22,195 | 18,828 | 7,464 | 94,406 | 2.38 | 66.4 | 8.6 | 26.7 |
| Schenectady | 13.7 | 10.2 | 7.7 | 51.3 | 146,555 | 154,758 | 5.6 | 0.4 | 18,506 | 15,884 | -1,938 | 54,302 | 2.77 | 60.2 | 11.6 | 34.5 |
| Schoharie | 15.6 | 13.8 | 9.7 | 50.0 | 31,582 | 32,723 | 3.6 | -4.9 | 2,712 | 2,991 | -1,324 | 12,559 | 2.38 | 63.0 | 10.5 | 29.6 |
| Schuyler | 16.6 | 13.8 | 8.9 | 50.1 | 19,224 | 18,369 | -4.4 | -3.7 | 1,745 | 2,044 | -383.0 | 7,324 | 2.42 | 63.1 | 8.5 | 28.5 |
| Seneca | 14.8 | 12.1 | 8.1 | 47.5 | 33,342 | 35,253 | 5.7 | -3.6 | 3,844 | 3,556 | -1,543 | 13,564 | 2.31 | 62.9 | 9.2 | 28.9 |
| Steuben | 15.1 | 11.8 | 8.6 | 50.2 | 98,726 | 98,946 | 0.2 | -4.3 | 10,960 | 10,277 | -4,958 | 39,861 | 2.37 | 64.1 | 10.3 | 29.8 |
| Suffolk | 15.0 | 10.0 | 7.8 | 50.8 | 1,419,369 | 1,493,116 | 5.2 | -1.3 | 160,899 | 124,888 | -54,950 | 489,301 | 2.97 | 72.8 | 11.1 | 22.6 |
| Sullivan | 14.9 | 11.7 | 7.8 | 48.6 | 73,966 | 77,504 | 4.8 | -2.2 | 8,637 | 7,577 | -2,793 | 28,184 | 2.53 | 63.2 | 11.2 | 30.6 |
| Tioga | 16.3 | 12.0 | 9.4 | 50.3 | 51,784 | 51,052 | -1.4 | -6.2 | 4,826 | 4,654 | -3,343 | 20,030 | 2.40 | 68.1 | 10.4 | 26.8 |
| Tompkins | 10.9 | 9.6 | 6.0 | 50.9 | 96,501 | 101,581 | 5.3 | -0.5 | 8,432 | 6,996 | -2,022 | 39,618 | 2.25 | 51.6 | 8.1 | 32.1 |
| Ulster | 15.6 | 12.2 | 8.7 | 50.4 | 177,749 | 182,522 | 2.7 | -2.6 | 16,083 | 17,416 | -3,429 | 69,320 | 2.41 | 61.3 | 10.7 | 30.8 |
| Warren | 16.2 | 14.0 | 9.6 | 51.0 | 63,303 | 65,698 | 3.8 | -3.0 | 5,874 | 7,148 | -626.0 | 28,015 | 2.26 | 60.1 | 10.2 | 32.6 |
| Washington | 15.6 | 12.1 | 8.5 | 48.2 | 61,042 | 63,254 | 3.6 | -4.2 | 5,976 | 6,475 | -2,123 | 24,014 | 2.43 | 65.4 | 11.2 | 27.0 |
| Wayne | 16.0 | 11.9 | 8.1 | 50.3 | 93,765 | 93,751 | 0.0 | -4.7 | 9,992 | 8,955 | -5,470 | 36,720 | 2.42 | 66.9 | 10.3 | 26.8 |
| Westchester | 13.8 | 9.6 | 8.2 | 51.6 | 923,459 | 949,376 | 2.8 | 1.7 | 107,463 | 73,684 | -17,043 | 349,292 | 2.70 | 68.7 | 12.4 | 27.0 |
| Wyoming | 15.2 | 11.7 | 7.4 | 45.8 | 43,424 | 42,154 | -2.9 | -6.4 | 3,859 | 3,855 | -2,705 | 15,917 | 2.37 | 66.0 | 9.3 | 27.7 |
| Yates | 14.6 | 13.1 | 8.7 | 51.4 | 24,621 | 25,364 | 3.0 | -2.3 | 3,241 | 2,670 | -1,154 | 8,919 | 2.66 | 65.4 | 7.7 | 27.7 |
| **NORTH CAROLINA** | 13.0 | 10.2 | 6.9 | 51.4 | 8,049,313 | 9,535,762 | 18.5 | 11.2 | 1,229,915 | 907,793 | 737,659 | 3,965,482 | 2.52 | 65.4 | 13.0 | 28.5 |
| Alamance | 13.2 | 9.8 | 7.6 | 52.6 | 130,800 | 151,160 | 15.6 | 13.4 | 18,788 | 16,815 | 18,248 | 64,439 | 2.46 | 64.8 | 14.3 | 29.4 |
| Alexander | 14.8 | 12.0 | 8.9 | 49.2 | 33,603 | 37,182 | 10.7 | 0.7 | 3,626 | 3,909 | 568.0 | 14,169 | 2.53 | 72.2 | 11.6 | 24.5 |
| Alleghany | 16.1 | 17.0 | 11.8 | 50.6 | 10,677 | 11,150 | 4.4 | 0.4 | 950 | 1,395 | 497.0 | 4,920 | 2.21 | 68.9 | 9.3 | 27.3 |
| Anson | 13.9 | 11.8 | 8.1 | 49.1 | 25,275 | 26,929 | 6.5 | -10.5 | 2,625 | 3,173 | -2,312 | 9,521 | 2.42 | 61.0 | 18.2 | 32.9 |
| Ashe | 16.2 | 15.8 | 11.4 | 50.8 | 24,384 | 27,239 | 11.7 | -0.3 | 2,281 | 3,505 | 1,179 | 11,938 | 2.21 | 66.1 | 10.0 | 28.8 |
| Avery | 14.4 | 13.5 | 9.5 | 45.8 | 17,167 | 17,805 | 3.7 | -1.3 | 1,459 | 2,061 | 374.0 | 6,551 | 2.14 | 65.9 | 8.7 | 29.5 |
| Beaufort | 15.3 | 14.9 | 10.2 | 52.4 | 44,958 | 47,784 | 6.3 | -1.5 | 4,794 | 6,063 | 597.0 | 19,701 | 2.37 | 64.1 | 13.7 | 31.0 |
| Bertie | 15.8 | 13.8 | 10.7 | 49.3 | 19,773 | 21,273 | 7.6 | -12.0 | 1,826 | 2,578 | -1,821 | 7,909 | 2.28 | 59.8 | 20.8 | 36.4 |
| Bladen | 14.8 | 14.1 | 9.5 | 52.6 | 32,278 | 35,165 | 8.9 | -6.4 | 3,608 | 4,270 | -1,587 | 13,636 | 2.42 | 63.7 | 16.8 | 32.8 |
| Brunswick | 18.2 | 22.2 | 11.3 | 52.2 | 73,143 | 107,429 | 46.9 | 38.7 | 10,531 | 13,962 | 44,585 | 56,056 | 2.34 | 68.3 | 7.4 | 26.9 |
| Buncombe | 13.5 | 12.5 | 8.6 | 52.1 | 206,330 | 238,315 | 15.5 | 10.6 | 26,080 | 26,300 | 25,139 | 107,479 | 2.33 | 58.3 | 8.6 | 33.4 |
| Burke | 15.3 | 12.2 | 9.1 | 50.1 | 89,148 | 90,824 | 1.9 | -0.4 | 8,979 | 10,637 | 1,300 | 35,156 | 2.48 | 67.1 | 11.2 | 28.5 |
| Cabarrus | 11.8 | 8.2 | 5.4 | 51.3 | 131,063 | 178,116 | 35.9 | 24.3 | 24,847 | 15,830 | 34,192 | 72,238 | 2.84 | 73.5 | 13.0 | 22.2 |
| Caldwell | 15.0 | 12.4 | 8.6 | 50.6 | 77,415 | 83,059 | 7.3 | -1.2 | 8,262 | 9,807 | 651.0 | 32,864 | 2.46 | 68.2 | 13.1 | 27.4 |
| Camden | 14.8 | 10.0 | 7.1 | 49.8 | 6,885 | 9,980 | 45.0 | 10.1 | 945 | 798 | 863.0 | 3,933 | 2.67 | 80.2 | 11.3 | 18.9 |
| Carteret | 17.2 | 15.9 | 10.5 | 51.1 | 59,383 | 66,463 | 11.9 | 4.7 | 5,919 | 8,271 | 5,457 | 29,755 | 2.28 | 64.7 | 9.9 | 29.3 |
| Caswell | 15.6 | 13.7 | 9.6 | 49.4 | 23,501 | 23,727 | 1.0 | -5.4 | 2,116 | 2,773 | -635.0 | 9,145 | 2.36 | 71.8 | 13.4 | 24.6 |
| Catawba | 14.1 | 11.0 | 7.6 | 51.2 | 141,685 | 154,742 | 9.2 | 3.6 | 17,727 | 16,881 | 4,831 | 62,319 | 2.49 | 67.5 | 11.4 | 27.1 |
| Chatham | 15.0 | 14.2 | 11.3 | 52.0 | 49,329 | 63,486 | 28.7 | 19.3 | 6,512 | 6,940 | 12,618 | 29,033 | 2.43 | 69.5 | 9.5 | 26.2 |
| Cherokee | 16.6 | 18.4 | 12.8 | 51.1 | 24,298 | 27,447 | 13.0 | 5.9 | 2,366 | 3,785 | 3,052 | 12,471 | 2.21 | 67.9 | 9.8 | 26.7 |

1. No spouse present.

# Table B. States and Counties — Population, Vital Statistics, Health, and Crime

| STATE County | Persons in group quarters, 2020 | Daytime Population, 2015-2019 Number | Employment/residence ratio | Births, 2020 Total | Rate[1] | Deaths, 2020 Number | Rate[1] | Persons under 65 with no health insurance, 2019 Number | Percent | Medicare, 2020 Total beneficiaries | Enrolled in Original Medicare | Enrolled in Medicare Advantage | Crimes reported by county police or sheriff, 2019 Violent | Property |
|---|---|---|---|---|---|---|---|---|---|---|---|---|---|---|
| | 32 | 33 | 34 | 35 | 36 | 37 | 38 | 39 | 40 | 41 | 42 | 43 | 44 | 45 |
| **NEW YORK— Cont'd** | | | | | | | | | | | | | | |
| Genesee | 1,985 | 53,361 | 0.85 | 579 | 10.2 | 676 | 11.9 | 1,985 | 4.4 | 13,294 | 5,340 | 7,954 | NA | NA |
| Greene | 3,055 | 43,703 | 0.81 | 436 | 9.2 | 589 | 12.5 | 1,687 | 5.0 | 11,959 | 7,078 | 4,881 | 13 | 26 |
| Hamilton | 81 | 4,394 | 0.93 | 18 | 4.1 | 67 | 15.4 | 218 | 7.3 | 1,715 | 1,034 | 682 | 1 | 6 |
| Herkimer | 1,443 | 52,657 | 0.66 | 565 | 9.3 | 701 | 11.5 | 2,237 | 4.7 | 15,069 | 8,342 | 6,727 | 0 | 1 |
| Jefferson | 6,015 | 115,957 | 1.06 | 1,750 | 16.2 | 988 | 9.1 | 4,191 | 4.8 | 20,410 | 12,884 | 7,525 | 28 | 227 |
| Kings | 35,297 | 2,264,110 | 0.73 | 37,399 | 14.7 | 19,573 | 7.7 | 155,196 | 7.2 | 377,134 | 201,743 | 175,392 | NA | NA |
| Lewis | 313 | 23,596 | 0.75 | 315 | 12.0 | 295 | 11.3 | 1,047 | 5.0 | 6,084 | 3,745 | 2,339 | NA | NA |
| Livingston | 5,738 | 57,707 | 0.79 | 502 | 8.0 | 592 | 9.5 | 2,114 | 4.6 | 13,849 | 5,058 | 8,791 | 12 | 54 |
| Madison | 4,599 | 61,560 | 0.71 | 640 | 9.1 | 686 | 9.7 | 2,311 | 4.3 | 14,953 | 8,121 | 6,832 | 32 | 283 |
| Monroe | 25,887 | 773,993 | 1.09 | 7,653 | 10.3 | 7,288 | 9.8 | 24,849 | 4.2 | 160,018 | 48,782 | 111,236 | 20 | 85 |
| Montgomery | 922 | 48,021 | 0.94 | 608 | 12.4 | 569 | 11.6 | 2,044 | 5.2 | 11,567 | 5,483 | 6,084 | 209 | 2,341 |
| Nassau | 19,453 | 1,258,680 | 0.86 | 13,892 | 10.3 | 12,111 | 9.0 | 53,385 | 4.8 | 270,825 | 200,571 | 70,255 | 6 | 212 |
| New York | 68,180 | 3,350,054 | 2.94 | 16,643 | 10.3 | 13,234 | 8.2 | 73,874 | 5.6 | 286,566 | 169,801 | 116,765 | 1,179 | 9,655 |
| Niagara | 3,735 | 187,833 | 0.77 | 2,041 | 9.8 | 2,530 | 12.1 | 7,153 | 4.3 | 50,445 | 18,796 | 31,649 | NA | NA |
| Oneida | 12,765 | 237,768 | 1.08 | 2,443 | 10.7 | 2,651 | 11.7 | 9,299 | 5.3 | 52,712 | 29,403 | 23,309 | 56 | 475 |
| Onondaga | 16,666 | 493,528 | 1.14 | 5,148 | 11.2 | 4,546 | 9.9 | 17,751 | 4.8 | 96,258 | 47,026 | 49,232 | 70 | 309 |
| Ontario | 3,203 | 109,546 | 1.0 | 953 | 8.7 | 1,186 | 10.8 | 3,475 | 4.1 | 26,659 | 9,831 | 16,828 | 191 | 1,518 |
| Orange | 11,057 | 355,887 | 0.86 | 5,072 | 13.2 | 3,131 | 8.1 | 15,775 | 4.9 | 65,044 | 47,907 | 17,137 | 69 | 669 |
| Orleans | 2,522 | 37,161 | 0.79 | 336 | 8.4 | 448 | 11.2 | 1,817 | 5.9 | 8,968 | 3,539 | 5,429 | NA | NA |
| Oswego | 4,249 | 102,525 | 0.7 | 1,189 | 10.2 | 1,174 | 10.1 | 4,904 | 5.3 | 25,357 | 13,311 | 12,046 | 15 | 115 |
| Otsego | 5,237 | 61,125 | 1.04 | 489 | 8.3 | 694 | 11.8 | 2,160 | 5.1 | 14,364 | 10,037 | 4,327 | 33 | 268 |
| Putnam | 2,578 | 77,677 | 0.58 | 830 | 8.4 | 856 | 8.7 | 3,369 | 4.2 | 19,467 | 14,307 | 5,160 | 13 | 29 |
| | | | | | | | | | | | | | 18 | 127 |
| Queens | 27,742 | 1,923,743 | 0.67 | 26,781 | 12.0 | 17,672 | 7.9 | 191,502 | 10.3 | 369,532 | 185,080 | 184,452 | NA | NA |
| Rensselaer | 5,690 | 138,274 | 0.74 | 1,568 | 9.9 | 1,724 | 10.9 | 5,339 | 4.2 | 32,469 | 16,904 | 15,565 | 18 | 230 |
| Richmond | 7,017 | 394,651 | 0.62 | 5,086 | 10.7 | 4,496 | 9.5 | 19,502 | 4.9 | 90,445 | 53,617 | 36,828 | NA | NA |
| Rockland | 6,497 | 299,794 | 0.83 | 5,617 | 17.2 | 2,631 | 8.1 | 15,161 | 5.6 | 56,999 | 42,833 | 14,166 | NA | NA |
| St. Lawrence | 10,455 | 106,663 | 0.95 | 1,045 | 9.7 | 1,180 | 11.0 | 4,451 | 5.7 | 23,789 | 16,157 | 7,632 | 11 | 25 |
| Saratoga | 3,405 | 206,477 | 0.82 | 1,989 | 8.6 | 2,244 | 9.7 | 6,612 | 3.6 | 50,456 | 25,423 | 25,034 | NA | NA |
| Schenectady | 4,538 | 149,891 | 0.93 | 1,828 | 11.8 | 1,575 | 10.1 | 6,032 | 4.8 | 32,026 | 15,060 | 16,965 | NA | NA |
| Schoharie | 1,236 | 27,862 | 0.76 | 277 | 8.9 | 338 | 10.9 | 1,160 | 5.1 | 7,604 | 4,898 | 2,705 | 5 | 18 |
| Schuyler | 203 | 15,433 | 0.69 | 172 | 9.7 | 197 | 11.1 | 665 | 4.8 | 4,760 | 2,488 | 2,272 | 5 | 10 |
| Seneca | 2,670 | 33,877 | 0.97 | 381 | 11.2 | 339 | 10.0 | 1,362 | 5.5 | 7,818 | 3,976 | 3,842 | 2 | 43 |
| Steuben | 1,656 | 95,732 | 0.98 | 1,018 | 10.8 | 1,097 | 11.6 | 3,881 | 5.2 | 22,461 | 11,670 | 10,792 | 9 | 47 |
| Suffolk | 29,392 | 1,392,021 | 0.88 | 15,140 | 10.3 | 14,189 | 9.6 | 62,909 | 5.2 | 296,308 | 230,315 | 65,993 | 1,217 | 14,828 |
| Sullivan | 3,773 | 70,367 | 0.85 | 859 | 11.3 | 837 | 11.0 | 3,685 | 6.3 | 17,033 | 12,876 | 4,157 | 27 | 273 |
| Tioga | 480 | 41,053 | 0.66 | 427 | 8.9 | 530 | 11.1 | 1,621 | 4.3 | 11,447 | 6,045 | 5,393 | 7 | 140 |
| Tompkins | 12,890 | 114,525 | 1.24 | 721 | 7.1 | 686 | 6.8 | 3,475 | 4.7 | 17,166 | 11,646 | 5,520 | NA | NA |
| Ulster | 11,664 | 162,747 | 0.81 | 1,517 | 8.5 | 1,868 | 10.5 | 7,976 | 5.9 | 41,372 | 28,985 | 12,387 | 21 | 145 |
| Warren | 608 | 70,058 | 1.18 | 547 | 8.6 | 727 | 11.4 | 2,128 | 4.3 | 17,940 | 9,306 | 8,634 | NA | NA |
| Washington | 3,054 | 51,889 | 0.66 | 531 | 8.8 | 732 | 12.1 | 2,344 | 5.1 | 14,650 | 7,423 | 7,228 | 30 | 67 |
| Wayne | 1,060 | 78,015 | 0.71 | 962 | 10.8 | 953 | 10.7 | 3,067 | 4.3 | 22,067 | 8,359 | 13,709 | 63 | 333 |
| Westchester | 27,871 | 945,950 | 0.95 | 9,924 | 10.3 | 8,141 | 8.4 | 45,099 | 5.7 | 178,092 | 123,927 | 54,165 | 30 | 199 |
| Wyoming | 3,333 | 37,296 | 0.83 | 355 | 9.0 | 384 | 9.7 | 1,289 | 4.4 | 8,898 | 3,677 | 5,221 | 17 | 96 |
| Yates | 1,369 | 23,543 | 0.87 | 331 | 13.4 | 290 | 11.7 | 1,520 | 8.3 | 5,898 | 2,461 | 3,437 | 12 | 98 |
| **NORTH CAROLINA** | 279,336 | 10,264,330 | 1.0 | 118,855 | 11.2 | 102,921 | 9.7 | 1,139,623 | 13.4 | 2,034,911 | 1,210,686 | 824,225 | X | X |
| Alamance | 6,238 | 153,884 | 0.88 | 1,942 | 11.3 | 1,879 | 11.0 | 19,822 | 14.8 | 33,912 | 13,974 | 19,938 | 118 | 649 |
| Alexander | 1,450 | 30,669 | 0.6 | 344 | 9.2 | 431 | 11.5 | 3,867 | 13.6 | 8,542 | 4,683 | 3,859 | 86 | 429 |
| Alleghany | 102 | 10,543 | 0.9 | 84 | 7.5 | 154 | 13.8 | 1,535 | 19.5 | 3,368 | 2,021 | 1,347 | 19 | 72 |
| Anson | 1,425 | 22,426 | 0.76 | 256 | 10.6 | 312 | 12.9 | 2,372 | 13.3 | 5,546 | 3,362 | 2,184 | 26 | 181 |
| Ashe | 324 | 24,773 | 0.83 | 203 | 7.5 | 391 | 14.4 | 3,261 | 16.5 | 8,061 | 5,496 | 2,565 | 57 | 151 |
| Avery | 2,449 | 18,857 | 1.21 | 148 | 8.4 | 242 | 13.8 | 2,135 | 19.2 | 4,707 | 3,091 | 1,615 | 18 | 179 |
| Beaufort | 512 | 46,763 | 0.98 | 424 | 9.0 | 648 | 13.8 | 5,040 | 14.4 | 13,261 | 9,807 | 3,454 | NA | NA |
| Bertie | 1,547 | 17,996 | 0.81 | 165 | 8.8 | 280 | 15.0 | 1,639 | 12.6 | 5,333 | 3,948 | 1,385 | NA | NA |
| Bladen | 296 | 34,571 | 1.09 | 345 | 10.5 | 490 | 14.9 | 3,994 | 16.1 | 8,036 | 4,849 | 3,187 | 65 | 481 |
| Brunswick | 820 | 122,063 | 0.81 | 1,035 | 6.9 | 1,821 | 12.2 | 12,873 | 13.5 | 50,856 | 36,672 | 14,184 | NA | NA |
| Buncombe | 7,955 | 278,730 | 1.17 | 2,465 | 9.4 | 2,857 | 10.8 | 29,493 | 14.6 | 61,723 | 40,038 | 21,685 | 219 | 1,830 |
| Burke | 3,114 | 83,837 | 0.84 | 863 | 9.5 | 1,133 | 12.5 | 10,688 | 15.5 | 21,837 | 11,744 | 10,093 | 110 | 1,045 |
| Cabarrus | 1,567 | 186,520 | 0.8 | 2,574 | 11.6 | 1,821 | 8.2 | 20,893 | 11.2 | 34,338 | 18,559 | 15,779 | 38 | 640 |
| Caldwell | 935 | 73,348 | 0.76 | 828 | 10.1 | 1,100 | 13.4 | 9,855 | 15.3 | 20,002 | 10,348 | 9,654 | 75 | 1,199 |
| Camden | 17 | 7,333 | 0.3 | 101 | 9.2 | 74 | 6.7 | 1,016 | 11.2 | 1,979 | 1,581 | 398 | 11 | 74 |
| Carteret | 975 | 66,774 | 0.93 | 504 | 7.2 | 964 | 13.9 | 6,861 | 13.4 | 19,437 | 15,482 | 3,955 | 69 | 640 |
| Caswell | 1,190 | 16,996 | 0.4 | 210 | 9.4 | 271 | 12.1 | 2,162 | 13.3 | 5,987 | 2,757 | 3,230 | NA | NA |
| Catawba | 2,544 | 173,718 | 1.22 | 1,688 | 10.5 | 1,819 | 11.3 | 19,832 | 15.4 | 34,663 | 19,456 | 15,207 | NA | NA |
| Chatham | 764 | 59,015 | 0.61 | 629 | 8.3 | 845 | 11.2 | 7,803 | 14.0 | 20,175 | 11,560 | 8,616 | 60 | 478 |
| Cherokee | 433 | 27,313 | 0.94 | 248 | 8.5 | 379 | 13.0 | 3,261 | 16.6 | 9,852 | 6,488 | 3,364 | 68 | 763 |

1.  Per 1,000 estimated resident population.

# Table B. States and Counties — Crime, Education, Money Income, and Poverty

| STATE County | County law enforcement employment, 2019 | | Education | | | | | | Money income, 2015-2019 | | | | Income and poverty, 2019 | | | |
| | | | School enrollment and attainment, 2015-2019 | | | | Local government expenditures,[3] 2017-2018 | | | Households | | | | Percent below poverty level | | |
| | | | Enrollment[1] | | Attainment[2] (percent) | | | | | | | Percent | | | | | |
| | Officers | Civilians | Total | Percent private | High school graduate or less | Bachelor's degree or more | Total current spending (mil dol) | Current spending per student[4] (dollars) | Per capita income[4] | Median income (dollars) | with income of less than $50,000 | with income of $200,000 or more | Median household income (dollars) | All persons | Children under 18 years | Children 5 to 17 years in families |
| | 46 | 47 | 48 | 49 | 50 | 51 | 52 | 53 | 54 | 55 | 56 | 57 | 58 | 59 | 60 | 61 |
| **NEW YORK— Cont'd** | | | | | | | | | | | | | | | | |
| Genesee | NA | NA | 12,258 | 11.6 | 43.7 | 22.3 | 163 | 19,783 | 30,511 | 60,524 | 41.3 | 3.1 | 62,570 | 10.5 | 13.6 | 11.8 |
| Greene | 32 | 2 | 7,505 | 11.5 | 52.0 | 22.1 | 137 | 22,974 | 28,433 | 53,601 | 47.1 | 3.8 | 54,746 | 12.6 | 18.5 | 17.8 |
| Hamilton | 7 | 1 | 674 | 15.1 | 49.7 | 16.8 | 19 | 42,025 | 24,851 | 58,675 | 42.6 | 3.0 | 60,148 | 9.8 | 17.7 | 15.4 |
| Herkimer | 6 | 10 | 13,339 | 11.6 | 47.1 | 21.5 | 178 | 18,873 | 27,850 | 54,646 | 46.0 | 2.2 | 55,778 | 13.7 | 19.0 | 18.0 |
| Jefferson | 44 | 14 | 25,528 | 11.6 | 43.7 | 22.2 | 310 | 17,235 | 26,194 | 52,685 | 47.4 | 2.3 | 53,829 | 14.6 | 20.6 | 21.0 |
| Kings | NA | NA | 638,429 | 30.1 | 43.4 | 37.5 | (5) | (5) | 34,173 | 60,231 | 43.6 | 9.6 | 66,501 | 17.7 | 24.7 | 24.3 |
| Lewis | NA | NA | 5,637 | 12.7 | 54.5 | 18.4 | 81 | 19,230 | 26,708 | 54,524 | 45.0 | 2.2 | 56,599 | 12.4 | 18.6 | 17.8 |
| Livingston | 49 | 26 | 16,901 | 11.8 | 41.3 | 26.5 | 152 | 19,349 | 28,701 | 59,510 | 42.4 | 2.9 | 60,268 | 13.8 | 16.4 | 15.0 |
| Madison | 40 | 12 | 17,338 | 27.5 | 42.8 | 26.2 | 185 | 19,361 | 30,469 | 61,633 | 39.7 | 4.7 | 61,713 | 9.8 | 13.2 | 11.6 |
| Monroe | 277 | 53 | 185,791 | 26.2 | 33.0 | 38.6 | 2,180 | 19,806 | 33,840 | 60,075 | 42.5 | 5.7 | 62,159 | 12.7 | 18.9 | 18.0 |
| Montgomery | 27 | 21 | 10,765 | 15.2 | 48.6 | 17.8 | 150 | 19,136 | 26,521 | 49,462 | 50.4 | 1.8 | 52,092 | 15.9 | 25.5 | 24.8 |
| Nassau | 2,357 | 921 | 338,939 | 25.8 | 31.3 | 46.0 | 5,458 | 26,492 | 51,422 | 116,100 | 21.0 | 23.2 | 117,767 | 5.6 | 6.5 | 5.8 |
| New York | NA | NA | 322,582 | 45.1 | 25.1 | 61.3 | (5) 25,970 | (5) 25,791 | 76,592 | 86,553 | 33.7 | 22.7 | 93,007 | 14.1 | 17.3 | 17.1 |
| Niagara | 109 | 53 | 46,124 | 17.8 | 42.0 | 24.6 | 517 | 17,878 | 30,971 | 55,522 | 45.8 | 3.4 | 56,371 | 12.4 | 16.9 | 16.5 |
| Oneida | 102 | 39 | 52,867 | 15.3 | 43.4 | 25.7 | 611 | 17,532 | 29,687 | 56,027 | 44.9 | 4.0 | 56,516 | 14.6 | 19.3 | 18.0 |
| Onondaga | 212 | 43 | 118,977 | 24.9 | 34.4 | 35.9 | 1,331 | 18,776 | 34,002 | 61,359 | 41.1 | 5.7 | 61,597 | 13.8 | 18.7 | 17.3 |
| Ontario | 71 | 57 | 25,879 | 20.5 | 32.5 | 36.3 | 307 | 19,020 | 36,835 | 64,845 | 37.6 | 6.3 | 66,754 | 8.1 | 10.4 | 9.9 |
| Orange | NA | NA | 104,547 | 25.9 | 39.5 | 30.3 | 1,371 | 23,018 | 34,959 | 79,944 | 32.5 | 10.1 | 83,188 | 12.3 | 18.0 | 16.2 |
| Orleans | 27 | 14 | 7,933 | 12.1 | 50.6 | 16.2 | 109 | 18,629 | 26,298 | 52,620 | 47.1 | 2.2 | 53,115 | 13.2 | 19.2 | 18.3 |
| Oswego | NA | NA | 28,882 | 6.2 | 49.3 | 19.0 | 394 | 20,158 | 28,587 | 55,067 | 44.7 | 3.4 | 57,640 | 15.6 | 25.0 | 20.1 |
| Otsego | 20 | 2 | 15,651 | 17.3 | 39.6 | 31.1 | 153 | 21,843 | 28,849 | 54,028 | 46.1 | 2.9 | 52,414 | 16.2 | 18.6 | 16.6 |
| Putnam | 87 | 21 | 22,642 | 18.4 | 34.1 | 39.6 | 392 | 27,605 | 47,448 | 104,486 | 21.3 | 17.0 | 105,600 | 5.2 | 5.6 | 5.1 |
| Queens | NA | NA | 525,047 | 19.9 | 45.5 | 32.2 | (5) | (5) | 31,930 | 68,666 | 37.2 | 7.6 | 72,975 | 11.0 | 14.4 | 13.5 |
| Rensselaer | 37 | 6 | 37,493 | 30.1 | 37.1 | 32.0 | 393 | 18,987 | 35,903 | 68,991 | 36.9 | 5.8 | 70,688 | 11.5 | 15.1 | 14.4 |
| Richmond | NA | NA | 116,219 | 24.1 | 41.6 | 33.9 | (5) | (5) | 36,907 | 82,783 | 32.0 | 11.7 | 86,624 | 9.1 | 13.9 | 14.5 |
| Rockland | NA | NA | 96,521 | 47.4 | 34.3 | 41.1 | 1,063 | 26,190 | 39,286 | 93,024 | 28.7 | 17.3 | 99,831 | 12.5 | 21.6 | 19.9 |
| St. Lawrence | 31 | 2 | 29,455 | 23.0 | 48.2 | 23.8 | 316 | 20,627 | 25,378 | 50,940 | 49.2 | 2.7 | 51,742 | 18.9 | 27.1 | 23.1 |
| Saratoga | NA | NA | 50,470 | 19.9 | 30.8 | 41.4 | 534 | 17,515 | 43,065 | 84,291 | 28.3 | 9.3 | 89,071 | 6.0 | 7.2 | 6.9 |
| Schenectady | NA | NA | 36,539 | 19.5 | 38.8 | 32.1 | 459 | 18,388 | 32,417 | 65,499 | 38.9 | 5.6 | 64,960 | 12.1 | 19.6 | 19.8 |
| Schoharie | 20 | 15 | 7,067 | 8.0 | 46.1 | 22.4 | 98 | 23,452 | 30,397 | 57,714 | 43.6 | 3.6 | 58,927 | 12.1 | 16.6 | 15.7 |
| Schuyler | 18 | 3 | 3,397 | 12.4 | 43.1 | 23.5 | 44 | 20,539 | 27,678 | 52,327 | 47.6 | 2.3 | 52,215 | 15.7 | 22.3 | 22.4 |
| Seneca | NA | NA | 6,666 | 18.4 | 50.4 | 21.1 | 87 | 21,341 | 28,245 | 54,545 | 45.0 | 3.5 | 51,064 | 13.7 | 19.4 | 19.6 |
| Steuben | 35 | 13 | 19,990 | 11.2 | 46.2 | 23.6 | 304 | 21,045 | 29,886 | 53,663 | 46.8 | 3.4 | 53,167 | 15.0 | 21.8 | 18.4 |
| Suffolk | 2,789 | 658 | 364,996 | 15.1 | 36.8 | 36.3 | 6,081 | 25,319 | 44,465 | 101,031 | 23.8 | 16.7 | 105,241 | 6.8 | 9.3 | 8.0 |
| Sullivan | 57 | 0 | 15,621 | 17.1 | 47.1 | 24.8 | 265 | 27,431 | 30,446 | 57,426 | 45.4 | 4.7 | 58,851 | 16.1 | 26.2 | 25.8 |
| Tioga | NA | NA | 10,338 | 13.9 | 44.8 | 26.0 | 142 | 18,962 | 31,822 | 62,999 | 40.1 | 3.7 | 60,699 | 9.4 | 13.8 | 12.7 |
| Tompkins | 40 | 4 | 40,554 | 56.3 | 24.5 | 53.5 | 244 | 21,684 | 33,075 | 60,240 | 42.1 | 7.8 | 59,176 | 16.9 | 14.3 | 12.9 |
| Ulster | 60 | 23 | 38,728 | 15.4 | 39.0 | 32.5 | 575 | 25,125 | 34,834 | 64,304 | 39.2 | 6.6 | 64,090 | 12.0 | 16.1 | 14.1 |
| Warren | NA | NA | 12,677 | 11.7 | 38.7 | 31.3 | 180 | 20,314 | 35,271 | 61,024 | 40.3 | 4.8 | 63,240 | 10.1 | 15.0 | 13.7 |
| Washington | 40 | 5 | 11,735 | 17.1 | 50.8 | 20.5 | 172 | 19,671 | 28,530 | 57,258 | 43.0 | 2.5 | 56,897 | 11.9 | 17.2 | 15.9 |
| Wayne | 72 | 8 | 19,445 | 12.0 | 44.5 | 22.6 | 292 | 20,552 | 31,043 | 59,449 | 41.5 | 3.4 | 61,989 | 11.4 | 15.8 | 14.5 |
| Westchester | 293 | 68 | 245,368 | 26.4 | 31.2 | 48.9 | 3,889 | 25,952 | 57,049 | 96,610 | 28.5 | 22.4 | 101,741 | 8.4 | 10.1 | 8.9 |
| Wyoming | 31 | 0 | 7,744 | 14.4 | 50.1 | 17.7 | 79 | 19,854 | 27,947 | 58,052 | 41.5 | 1.5 | 58,214 | 9.9 | 12.6 | 11.9 |
| Yates | 26 | 4 | 4,910 | 38.1 | 46.3 | 25.2 | 46 | 21,563 | 28,270 | 56,563 | 45.4 | 3.7 | 60,612 | 13.9 | 26.2 | 25.1 |
| **NORTH CAROLINA** | X | X | 2,518,907 | 15.2 | 37.9 | 31.3 | 14,424 | 9,285 | 30,783 | 54,602 | 45.9 | 5.4 | 57,388 | 13.6 | 19.3 | 18.1 |
| Alamance | 136 | 160 | 41,266 | 22.3 | 41.3 | 24.8 | 215 | 8,682 | 27,312 | 49,688 | 50.3 | 2.9 | 57,963 | 14.6 | 21.8 | 21.4 |
| Alexander | 41 | 40 | 7,460 | 14.6 | 56.3 | 14.2 | 47 | 9,450 | 26,688 | 48,756 | 51.5 | 3.7 | 54,960 | 11.7 | 16.7 | 14.4 |
| Alleghany | 24 | 17 | 2,005 | 4.8 | 51.5 | 17.3 | 17 | 11,918 | 23,267 | 37,830 | 63.5 | 2.2 | 41,420 | 16.9 | 26.3 | 24.9 |
| Anson | 33 | 28 | 5,166 | 11.0 | 59.5 | 11.3 | 35 | 10,321 | 21,635 | 40,213 | 59.8 | 1.2 | 40,826 | 21.4 | 31.1 | 31.2 |
| Ashe | 32 | 43 | 4,844 | 14.0 | 46.7 | 19.7 | 33 | 10,721 | 24,814 | 40,962 | 58.8 | 1.6 | 41,542 | 14.6 | 22.6 | 19.9 |
| Avery | 28 | 24 | 3,744 | 29.3 | 46.6 | 20.7 | 26 | 11,819 | 21,167 | 40,340 | 62.2 | 3.7 | 45,823 | 16.4 | 24.4 | 22.5 |
| Beaufort | 56 | 40 | 9,477 | 7.5 | 44.8 | 20.6 | 69 | 9,655 | 27,504 | 45,212 | 53.2 | 2.8 | 49,410 | 17.6 | 29.5 | 27.1 |
| Bertie | 26 | 17 | 3,800 | 15.2 | 59.7 | 13.6 | 28 | 11,357 | 22,947 | 35,527 | 64.3 | 2.4 | 37,899 | 24.2 | 32.4 | 30.3 |
| Bladen | 53 | 44 | 6,825 | 12.2 | 52.0 | 15.8 | 51 | 10,171 | 23,440 | 36,173 | 62.4 | 1.7 | 42,260 | 21.2 | 29.9 | 28.0 |
| Brunswick | 179 | 123 | 20,868 | 12.2 | 36.6 | 29.9 | 139 | 9,979 | 33,312 | 58,236 | 43.0 | 4.2 | 63,712 | 10.2 | 21.0 | 20.3 |
| Buncombe | 234 | 151 | 52,958 | 17.8 | 32.2 | 40.1 | 318 | 10,287 | 32,426 | 52,207 | 47.8 | 5.0 | 55,448 | 12.2 | 17.3 | 17.5 |
| Burke | 93 | 32 | 17,416 | 14.7 | 49.8 | 16.4 | 113 | 8,810 | 24,481 | 44,557 | 55.4 | 2.6 | 47,890 | 18.4 | 30.5 | 30.0 |
| Cabarrus | 202 | 135 | 55,533 | 13.4 | 35.2 | 32.3 | 351 | 8,587 | 32,255 | 67,328 | 36.0 | 7.1 | 72,071 | 7.9 | 10.6 | 9.5 |
| Caldwell | 83 | 9 | 16,559 | 10.6 | 52.5 | 15.7 | 114 | 9,721 | 23,869 | 44,511 | 55.2 | 1.4 | 48,512 | 12.0 | 17.8 | 16.7 |
| Camden | 17 | 3 | 2,556 | 4.7 | 42.9 | 17.1 | 18 | 9,521 | 28,221 | 64,572 | 41.4 | 1.8 | 69,610 | 7.6 | 9.4 | 8.1 |
| Carteret | 67 | 33 | 13,670 | 12.4 | 34.0 | 29.1 | 83 | 9,670 | 33,722 | 57,194 | 44.1 | 4.7 | 60,058 | 10.4 | 17.0 | 15.8 |
| Caswell | 36 | 16 | 4,414 | 19.6 | 53.2 | 15.9 | 27 | 10,226 | 23,901 | 45,753 | 53.7 | 2.0 | 51,240 | 16.2 | 25.1 | 23.6 |
| Catawba | 144 | 59 | 36,123 | 15.9 | 44.1 | 22.6 | 206 | 8,697 | 28,224 | 52,056 | 48.0 | 4.1 | 53,688 | 13.3 | 18.6 | 17.7 |
| Chatham | 100 | 38 | 14,661 | 15.4 | 32.3 | 42.4 | 102 | 9,949 | 40,967 | 67,031 | 37.7 | 11.0 | 70,258 | 8.7 | 12.4 | 10.5 |
| Cherokee | 31 | 43 | 4,552 | 15.1 | 46.6 | 19.0 | 38 | 10,809 | 23,902 | 41,438 | 57.8 | 1.1 | 42,764 | 17.7 | 29.3 | 25.8 |

1. All persons 3 years old and over enrolled in nursery school through college.   2. Persons 25 years old and over.   3. Elementary and secondary education expenditures.   4. Based on population estimated by the American Community Survey, 2014–2018.   5. Bronx, Kings, Queens, and Richmond counties are included with New York county.

# Table B. States and Counties — **Personal Income**

| STATE County | Personal income, 2019 Total (mil dol) | Percent change 2018-2019 | Per capita[1] Dollars | Per capita Rank | Wages and salaries (mil dol) | Supplements to wages and salaries, employer contributions (mil dol) Pension and insurance | Government social insurance | Proprietors' income (mil dol) | Dividends, interest, and rent (mil dol) | Personal transfer receipts (mil dol) | Earnings, 2019 Total (mil dol) | Contributions for government social insurance (mil dol) From employee and self-employed | From employer |
|---|---|---|---|---|---|---|---|---|---|---|---|---|---|
| | 62 | 63 | 64 | 65 | 66 | 67 | 68 | 69 | 70 | 71 | 72 | 73 | 74 |
| **NEW YORK— Cont'd** | | | | | | | | | | | | | |
| Genesee | 2,645 | 4.4 | 46,181 | 1,230 | 1,030 | 275 | 86 | 184 | 397 | 607 | 1,575 | 90 | 86 |
| Greene | 2,317 | 4.1 | 49,100 | 914 | 685 | 204 | 55 | 167 | 410 | 561 | 1,111 | 70 | 55 |
| Hamilton | 245 | 3.0 | 55,578 | 469 | 67 | 28 | 5 | 13 | 66 | 61 | 113 | 8 | 5 |
| Herkimer | 2,666 | 4.5 | 43,479 | 1,547 | 711 | 201 | 58 | 131 | 395 | 736 | 1,101 | 74 | 58 |
| Jefferson | 5,422 | 3.6 | 49,365 | 891 | 2,776 | 830 | 249 | 235 | 954 | 1,152 | 4,089 | 201 | 249 |
| Kings | 143,558 | 4.4 | 56,080 | 445 | 42,082 | 9,135 | 3,322 | 10,330 | 21,290 | 31,706 | 64,869 | 3,610 | 3,322 |
| Lewis | 1,248 | 5.5 | 47,471 | 1,082 | 279 | 97 | 23 | 111 | 177 | 267 | 511 | 29 | 23 |
| Livingston | 3,037 | 4.5 | 48,268 | 996 | 900 | 281 | 74 | 470 | 418 | 643 | 1,724 | 95 | 74 |
| Madison | 3,249 | 4.3 | 45,794 | 1,273 | 979 | 249 | 80 | 230 | 518 | 686 | 1,537 | 92 | 80 |
| Monroe | 40,823 | 3.8 | 55,034 | 490 | 21,768 | 4,151 | 1,696 | 2,894 | 7,339 | 8,471 | 30,508 | 1,703 | 1,696 |
| Montgomery | 2,130 | 4.2 | 43,272 | 1,578 | 860 | 190 | 72 | 122 | 290 | 641 | 1,244 | 78 | 72 |
| Nassau | 126,522 | 3.9 | 93,241 | 29 | 42,444 | 7,381 | 3,233 | 10,939 | 31,582 | 14,640 | 63,997 | 3,429 | 3,233 |
| New York | 322,234 | 3.6 | 197,847 | 2 | 335,642 | 35,975 | 19,007 | 63,594 | 104,628 | 21,817 | 454,218 | 22,998 | 19,007 |
| Niagara | 9,880 | 3.9 | 47,207 | 1,108 | 3,335 | 787 | 273 | 510 | 1,392 | 2,483 | 4,906 | 310 | 273 |
| Oneida | 10,717 | 4.5 | 46,867 | 1,148 | 5,042 | 1,304 | 397 | 679 | 1,746 | 2,743 | 7,422 | 429 | 397 |
| Onondaga | 25,589 | 3.8 | 55,564 | 470 | 13,931 | 2,972 | 1,083 | 2,397 | 4,183 | 5,092 | 20,383 | 1,103 | 1,083 |
| Ontario | 6,097 | 4.2 | 55,541 | 471 | 2,691 | 577 | 218 | 337 | 1,090 | 1,202 | 3,822 | 220 | 218 |
| Orange | 20,654 | 5.3 | 53,656 | 563 | 7,975 | 1,814 | 636 | 1,204 | 2,981 | 3,885 | 11,630 | 640 | 636 |
| Orleans | 1,656 | 4.2 | 41,050 | 1,916 | 546 | 177 | 46 | 149 | 215 | 438 | 919 | 54 | 46 |
| Oswego | 4,922 | 4.3 | 42,025 | 1,758 | 1,668 | 498 | 131 | 159 | 601 | 1,314 | 2,457 | 152 | 131 |
| Otsego | 2,723 | 4.1 | 45,774 | 1,278 | 1,116 | 278 | 89 | 239 | 449 | 659 | 1,722 | 101 | 89 |
| Putnam | 6,820 | 3.9 | 69,365 | 135 | 1,517 | 323 | 119 | 381 | 1,202 | 935 | 2,339 | 139 | 119 |
| Queens | 119,472 | 4.5 | 53,008 | 613 | 43,810 | 9,396 | 3,428 | 9,701 | 16,804 | 28,447 | 66,334 | 3,678 | 3,428 |
| Rensselaer | 8,151 | 3.9 | 51,356 | 731 | 3,137 | 744 | 247 | 346 | 1,257 | 1,671 | 4,474 | 261 | 247 |
| Richmond | 28,040 | 4.8 | 58,890 | 340 | 7,292 | 1,630 | 581 | 1,465 | 4,117 | 6,538 | 10,968 | 656 | 581 |
| Rockland | 20,579 | 4.4 | 63,167 | 227 | 7,353 | 1,448 | 569 | 1,377 | 3,917 | 3,572 | 10,747 | 597 | 569 |
| St. Lawrence | 4,277 | 4.2 | 39,697 | 2,087 | 1,750 | 520 | 143 | 194 | 634 | 1,209 | 2,606 | 156 | 143 |
| Saratoga | 16,724 | 3.6 | 72,757 | 104 | 5,169 | 1,016 | 404 | 892 | 4,010 | 2,134 | 7,481 | 427 | 404 |
| Schenectady | 8,028 | 4.2 | 51,697 | 706 | 3,785 | 732 | 289 | 353 | 1,334 | 1,719 | 5,159 | 300 | 289 |
| Schoharie | 1,336 | 4.7 | 43,113 | 1,598 | 418 | 121 | 33 | 98 | 192 | 333 | 671 | 42 | 33 |
| Schuyler | 770 | 3.8 | 43,262 | 1,581 | 208 | 56 | 18 | 49 | 113 | 221 | 331 | 22 | 18 |
| Seneca | 1,379 | 4.6 | 40,534 | 1,983 | 655 | 166 | 51 | 102 | 216 | 361 | 973 | 57 | 51 |
| Steuben | 4,503 | 3.8 | 47,212 | 1,107 | 2,328 | 456 | 171 | 222 | 723 | 1,082 | 3,177 | 188 | 171 |
| Suffolk | 106,184 | 4.1 | 71,911 | 110 | 43,475 | 8,334 | 3,256 | 7,270 | 20,076 | 16,591 | 62,335 | 3,419 | 3,256 |
| Sullivan | 3,678 | 6.0 | 48,753 | 944 | 1,331 | 328 | 107 | 157 | 625 | 1,018 | 1,923 | 118 | 107 |
| Tioga | 2,264 | 4.3 | 46,969 | 1,137 | 744 | 162 | 56 | 147 | 324 | 517 | 1,109 | 70 | 56 |
| Tompkins | 4,751 | 2.8 | 46,496 | 1,186 | 2,913 | 541 | 236 | 355 | 966 | 795 | 4,045 | 217 | 236 |
| Ulster | 9,412 | 4.3 | 53,006 | 614 | 2,970 | 749 | 235 | 703 | 1,793 | 2,091 | 4,657 | 279 | 235 |
| Warren | 3,528 | 4.0 | 55,181 | 487 | 1,825 | 367 | 144 | 286 | 677 | 783 | 2,623 | 152 | 144 |
| Washington | 2,582 | 4.9 | 42,181 | 1,741 | 735 | 228 | 60 | 138 | 371 | 656 | 1,161 | 74 | 60 |
| Wayne | 4,306 | 4.2 | 47,886 | 1,032 | 1,313 | 382 | 106 | 407 | 539 | 1,028 | 2,209 | 134 | 106 |
| Westchester | 109,790 | 3.7 | 113,477 | 15 | 34,632 | 5,542 | 2,425 | 9,568 | 31,139 | 10,331 | 52,167 | 2,751 | 2,425 |
| Wyoming | 1,721 | 7.2 | 43,176 | 1,588 | 632 | 193 | 54 | 163 | 248 | 398 | 1,041 | 57 | 54 |
| Yates | 1,018 | 5.2 | 40,877 | 1,934 | 277 | 77 | 23 | 119 | 193 | 267 | 496 | 31 | 23 |
| **NORTH CAROLINA** | 500,974 | 4.4 | 47,706 | X | 257,909 | 39,818 | 18,116 | 35,812 | 93,303 | 97,751 | 351,656 | 22,114 | 18,116 |
| Alamance | 6,983 | 5.0 | 41,193 | 1,892 | 2,965 | 443 | 210 | 329 | 1,153 | 1,584 | 3,946 | 271 | 210 |
| Alexander | 1,411 | 2.8 | 37,622 | 2,370 | 363 | 69 | 27 | 95 | 203 | 384 | 554 | 43 | 27 |
| Alleghany | 429 | 4.2 | 38,514 | 2,259 | 116 | 24 | 9 | 34 | 110 | 138 | 183 | 15 | 9 |
| Anson | 888 | 2.3 | 36,307 | 2,536 | 288 | 63 | 21 | 82 | 113 | 285 | 455 | 30 | 21 |
| Ashe | 991 | 5.1 | 36,426 | 2,519 | 311 | 54 | 23 | 46 | 206 | 323 | 435 | 37 | 23 |
| Avery | 645 | 3.5 | 36,748 | 2,478 | 259 | 47 | 19 | 35 | 162 | 186 | 361 | 27 | 19 |
| Beaufort | 2,056 | 4.7 | 43,743 | 1,522 | 666 | 131 | 47 | 134 | 381 | 604 | 977 | 73 | 47 |
| Bertie | 709 | 4.2 | 37,424 | 2,398 | 222 | 48 | 17 | 55 | 99 | 251 | 341 | 24 | 17 |
| Bladen | 1,188 | 2.7 | 36,312 | 2,534 | 539 | 97 | 41 | 99 | 165 | 406 | 776 | 51 | 41 |
| Brunswick | 6,303 | 6.5 | 44,129 | 1,468 | 1,527 | 268 | 110 | 408 | 1,460 | 1,940 | 2,313 | 205 | 110 |
| Buncombe | 13,240 | 4.4 | 50,690 | 779 | 6,699 | 1,035 | 483 | 1,202 | 3,346 | 2,547 | 9,420 | 608 | 483 |
| Burke | 3,277 | 3.8 | 36,212 | 2,545 | 1,253 | 246 | 91 | 148 | 544 | 948 | 1,738 | 127 | 91 |
| Cabarrus | 10,090 | 5.5 | 46,615 | 1,171 | 3,673 | 596 | 257 | 578 | 1,339 | 1,695 | 5,104 | 330 | 257 |
| Caldwell | 3,008 | 3.7 | 36,609 | 2,493 | 1,147 | 199 | 81 | 151 | 461 | 896 | 1,578 | 117 | 81 |
| Camden | 504 | 4.9 | 46,366 | 1,203 | 54 | 12 | 4 | 21 | 82 | 98 | 91 | 8 | 4 |
| Carteret | 3,584 | 4.4 | 51,582 | 716 | 992 | 179 | 71 | 221 | 897 | 837 | 1,463 | 109 | 71 |
| Caswell | 817 | 4.6 | 36,153 | 2,549 | 128 | 29 | 9 | 37 | 114 | 267 | 203 | 21 | 9 |
| Catawba | 7,398 | 3.7 | 46,367 | 1,202 | 4,346 | 681 | 314 | 496 | 1,309 | 1,641 | 5,836 | 375 | 314 |
| Chatham | 4,826 | 4.5 | 64,807 | 189 | 667 | 113 | 48 | 183 | 1,295 | 773 | 1,011 | 87 | 48 |
| Cherokee | 956 | 5.3 | 33,402 | 2,853 | 313 | 59 | 23 | 57 | 168 | 396 | 452 | 40 | 23 |

1. Based on the resident population estimated as of July 1 of the year shown.

Items 62—74

# Table B. States and Counties — Earnings, Social Security, and Housing

| STATE County | Farm | Mining, quarrying, and extractions | Construction | Manufacturing | Information; professional, scientific, technical services | Retail trade | Finance, insurance, real estate, and leasing | Health care and social assistance | Government | Number | Rate[1] | Supplemental Security Income recipients, 2019 | Total | Percent change, 2010-2020 |
|---|---|---|---|---|---|---|---|---|---|---|---|---|---|---|
| | 75 | 76 | 77 | 78 | 79 | 80 | 81 | 82 | 83 | 84 | 85 | 86 | 87 | 88 |
| **NEW YORK— Cont'd** | | | | | | | | | | | | | | |
| Genesee | 4.7 | 0.7 | 5.9 | 14.7 | 2.8 | 7.2 | 3.3 | 10.0 | 28.9 | 14,225 | 248 | 1,094 | 25,938 | 1.6 |
| Greene | 0.3 | 0.2 | 5.2 | 7.1 | 4.9 | 7.6 | 3.2 | 5.4 | 35.8 | 12,740 | 270 | 1,297 | 29,864 | 2.3 |
| Hamilton | 0.0 | 0.0 | 7.1 | D | D | 6.2 | D | D | 55.9 | 1,695 | 382 | 60 | 9,026 | 3.8 |
| Herkimer | 1.7 | 1.1 | 7.8 | 14.1 | 2.8 | 8.8 | 2.5 | 9.5 | 30.4 | 15,930 | 260 | 1,553 | 33,955 | 1.8 |
| Jefferson | 1.5 | 0.4 | 4.1 | 3.4 | 2.6 | 6.0 | 2.3 | 11.0 | 56.9 | 22,640 | 206 | 2,713 | 60,327 | 4.1 |
| Kings | 0.0 | 0.0 | 5.5 | 1.9 | 9.4 | 6.4 | 5.0 | 22.0 | 23.6 | 349,585 | 136 | 122,399 | 1,076,939 | 7.7 |
| Lewis | 11.1 | D | 6.3 | 14.9 | 2.5 | 5.3 | 2.2 | D | 37.0 | 6,035 | 229 | 541 | 15,869 | 5.0 |
| Livingston | 3.3 | D | 5.6 | 8.0 | D | 5.2 | 1.9 | 7.2 | 31.0 | 14,720 | 234 | 1,256 | 27,772 | 2.8 |
| Madison | 3.0 | D | 6.4 | 13.3 | 5.1 | 8.5 | 3.0 | D | 23.1 | 15,420 | 218 | 1,252 | 32,503 | 2.3 |
| Monroe | 0.1 | 0.0 | 5.3 | 11.6 | 12.6 | 5.2 | 6.5 | 15.3 | 14.3 | 165,260 | 223 | 26,107 | 331,484 | 3.4 |
| Montgomery | 2.0 | 0.4 | 5.0 | 15.7 | 2.9 | 9.3 | 2.1 | 22.2 | 18.5 | 13,425 | 272 | 1,880 | 23,584 | 2.2 |
| Nassau | 0.0 | D | 6.4 | D | 11.7 | 7.2 | 10.9 | 20.3 | 15.7 | 263,910 | 195 | 16,376 | 475,426 | 1.5 |
| New York | 0.0 | 0.3 | 1.6 | 0.6 | 29.5 | 2.5 | 29.9 | 4.9 | 6.8 | 262,985 | 161 | 65,565 | 895,417 | 5.7 |
| Niagara | 0.6 | D | 5.7 | 17.7 | 5.2 | 7.6 | 2.8 | 12.9 | 24.7 | 54,040 | 258 | 5,764 | 100,826 | 1.7 |
| Oneida | 0.4 | 0.4 | 3.7 | 8.8 | 6.6 | 5.9 | 7.9 | 17.3 | 30.6 | 56,135 | 246 | 7,906 | 105,824 | 1.6 |
| Onondaga | 0.2 | 0.1 | 4.6 | 9.6 | 10.7 | 5.4 | 7.8 | 13.6 | 20.1 | 100,060 | 217 | 14,251 | 210,132 | 3.8 |
| Ontario | 1.2 | 0.2 | 10.5 | 14.2 | 5.9 | 7.6 | 3.9 | 12.4 | 20.1 | 27,495 | 250 | 1,824 | 51,932 | 7.5 |
| Orange | 0.3 | 0.2 | 6.9 | 5.9 | 8.1 | 7.9 | 4.0 | 15.0 | 28.0 | 68,030 | 177 | 6,858 | 146,539 | 6.9 |
| Orleans | 4.5 | 4.3 | 5.3 | 16.7 | 2.9 | 5.6 | 2.5 | D | 39.0 | 9,705 | 241 | 910 | 18,649 | 1.2 |
| Oswego | 0.6 | 0.2 | 6.5 | 11.7 | D | 7.4 | 2.7 | 10.3 | 31.9 | 28,010 | 240 | 3,293 | 54,862 | 2.4 |
| Otsego | 1.2 | D | 4.0 | 4.6 | 3.4 | 7.9 | 7.1 | 27.4 | 21.6 | 15,025 | 253 | 1,338 | 31,357 | 1.8 |
| Putnam | 0.1 | D | 12.5 | 5.6 | 9.3 | 5.6 | 4.5 | 15.3 | 23.5 | 19,520 | 198 | 805 | 38,802 | 1.6 |
| Queens | 0.0 | 0.0 | 10.1 | 2.1 | 4.9 | 4.9 | 6.4 | 14.8 | 24.1 | 348,140 | 154 | 68,198 | 875,036 | 4.8 |
| Rensselaer | 0.4 | 0.4 | 6.9 | 13.4 | 8.1 | 5.6 | 3.7 | 11.2 | 23.9 | 35,045 | 220 | 4,167 | 73,568 | 2.9 |
| Richmond | 0.0 | 0.0 | 11.1 | D | 5.3 | 6.9 | 4.5 | 21.4 | 26.6 | 92,255 | 194 | 14,834 | 182,060 | 3.1 |
| Rockland | 0.0 | D | 8.2 | 8.0 | 11.2 | 6.8 | 4.4 | 16.2 | 19.4 | 56,515 | 173 | 5,007 | 107,349 | 3.2 |
| St. Lawrence | 2.6 | 0.3 | 4.5 | 7.6 | 3.2 | 6.5 | 2.5 | 15.5 | 37.9 | 26,325 | 244 | 3,569 | 53,608 | 2.8 |
| Saratoga | 0.4 | D | 8.0 | 12.9 | 11.0 | 6.3 | 9.6 | 10.9 | 16.7 | 51,650 | 225 | 2,746 | 109,233 | 10.7 |
| Schenectady | 0.0 | D | 4.9 | 10.9 | 18.9 | 5.7 | 5.2 | 14.8 | 19.1 | 33,865 | 218 | 5,268 | 70,255 | 3.0 |
| Schoharie | 3.2 | 0.5 | 11.6 | 2.9 | 3.4 | 6.7 | 5.0 | 8.6 | 34.2 | 8,040 | 259 | 657 | 17,629 | 2.4 |
| Schuyler | 4.9 | D | 8.7 | 14.8 | 2.6 | 7.3 | D | 12.2 | 25.4 | 5,065 | 284 | 385 | 9,938 | 5.0 |
| Seneca | 2.9 | D | 3.6 | 24.7 | D | 7.8 | 2.1 | D | 28.8 | 8,375 | 245 | 753 | 16,470 | 2.6 |
| Steuben | 1.5 | 0.2 | 2.3 | 14.9 | 20.0 | 5.0 | 3.6 | 9.7 | 20.1 | 24,020 | 252 | 2,816 | 49,967 | 2.3 |
| Suffolk | 0.2 | 0.1 | 9.3 | 7.9 | 9.5 | 6.3 | 9.6 | 12.4 | 20.8 | 301,385 | 204 | 19,320 | 578,014 | 1.4 |
| Sullivan | -0.2 | 0.4 | 5.8 | 4.4 | 3.8 | 5.8 | 3.1 | 19.4 | 31.2 | 18,290 | 242 | 2,634 | 51,495 | 4.7 |
| Tioga | 1.0 | 1.8 | 4.6 | 36.5 | D | 5.5 | 2.3 | 5.2 | 18.8 | 12,640 | 262 | 1,158 | 22,639 | 2.1 |
| Tompkins | 0.5 | 0.8 | 2.4 | 6.2 | 8.9 | 4.6 | 3.3 | D | 13.2 | 16,890 | 166 | 1,443 | 45,392 | 8.9 |
| Ulster | 0.4 | 0.2 | 6.9 | 6.7 | 6.5 | 7.5 | 4.9 | 13.2 | 29.0 | 42,585 | 239 | 3,961 | 85,957 | 2.8 |
| Warren | 0.1 | 0.8 | 6.8 | 11.7 | 8.3 | 8.8 | 6.4 | 16.5 | 15.0 | 19,130 | 299 | 1,481 | 40,416 | 4.4 |
| Washington | 3.1 | 0.7 | 7.6 | 18.2 | 3.0 | 7.1 | 2.0 | 7.5 | 37.5 | 15,170 | 248 | 1,599 | 29,719 | 3.0 |
| Wayne | 0.9 | D | 7.2 | 28.2 | 3.4 | 6.1 | 2.7 | 6.7 | 26.4 | 23,610 | 263 | 2,214 | 41,981 | 2.3 |
| Westchester | 0.0 | 0.0 | 6.4 | 4.2 | 15.7 | 4.6 | 14.2 | 13.8 | 15.8 | 171,955 | 178 | 17,152 | 379,732 | 2.4 |
| Wyoming | 9.2 | D | 4.5 | 12.0 | D | 6.3 | 2.2 | D | 38.2 | 9,570 | 240 | 621 | 18,624 | 3.7 |
| Yates | 4.5 | D | 10.3 | 15.5 | D | 7.5 | 4.2 | D | 19.4 | 6,270 | 252 | 496 | 14,070 | 4.3 |
| **NORTH CAROLINA** | 0.6 | 0.1 | 6.4 | 10.7 | 11.5 | 5.9 | 9.1 | 10.0 | 18.3 | 2,144,804 | 204 | 228,518 | 4,813,611 | 11.2 |
| Alamance | -0.1 | D | 7.2 | 15.3 | 4.8 | 8.6 | 5.9 | 19.9 | D | 36,205 | 214 | 3,566 | 74,058 | 11.2 |
| Alexander | 4.0 | D | 5.5 | 33.2 | 2.0 | 4.5 | 3.5 | D | 21.6 | 9,595 | 256 | 623 | 16,695 | 3.2 |
| Alleghany | 2.2 | D | 9.6 | 17.9 | 3.1 | 7.4 | 4.7 | 11.5 | 18.6 | 3,570 | 321 | 291 | 8,257 | 2.4 |
| Anson | 11.7 | D | 4.5 | 18.9 | D | 4.6 | 1.6 | 4.3 | 26.1 | 6,145 | 261 | 1,018 | 11,698 | 1.1 |
| Ashe | -1.5 | D | 15.8 | 10.9 | D | 8.5 | 5.1 | 11.8 | 15.6 | 8,520 | 313 | 808 | 17,969 | 3.9 |
| Avery | 0.0 | D | 10.1 | 1.5 | D | 7.6 | 4.0 | D | 24.7 | 4,680 | 267 | 392 | 14,504 | 4.4 |
| Beaufort | 3.1 | D | 5.5 | 18.9 | 2.9 | 6.9 | 4.9 | 6.5 | 20.5 | 14,775 | 314 | 1,778 | 26,538 | 7.4 |
| Bertie | 15.1 | D | D | D | D | 2.4 | D | D | 26.0 | 5,975 | 315 | 1,096 | 9,854 | 0.3 |
| Bladen | 9.8 | 0.0 | 3.2 | 41.9 | D | 3.7 | 1.6 | 3.4 | 17.6 | 8,835 | 269 | 1,464 | 18,074 | 2.0 |
| Brunswick | 0.3 | D | 12.0 | 4.9 | D | 8.6 | 5.7 | 11.5 | 16.2 | 50,590 | 354 | 2,198 | 98,438 | 27.0 |
| Buncombe | 0.3 | D | 7.3 | 10.5 | 8.6 | 7.4 | 6.3 | 20.4 | 13.5 | 62,350 | 238 | 5,391 | 129,905 | 14.6 |
| Burke | 0.5 | D | 3.8 | 25.6 | 2.9 | 5.6 | 3.3 | 17.3 | 25.1 | 22,885 | 253 | 2,024 | 41,822 | 2.4 |
| Cabarrus | 0.3 | D | 7.5 | 8.2 | 5.9 | 8.4 | 4.3 | 6.8 | 22.1 | 36,720 | 170 | 3,251 | 84,954 | 18.1 |
| Caldwell | 0.8 | D | 4.9 | 26.1 | 4.9 | 6.2 | 2.6 | 11.2 | 15.4 | 21,390 | 260 | 1,962 | 38,295 | 1.6 |
| Camden | 3.7 | 0.1 | 10.6 | 2.4 | D | 5.7 | D | D | 32.1 | 2,190 | 203 | 114 | 4,459 | 8.7 |
| Carteret | 0.2 | D | 8.2 | 3.8 | 6.0 | 12.2 | 8.0 | 9.9 | 24.0 | 20,190 | 290 | 1,166 | 51,411 | 6.7 |
| Caswell | 3.1 | D | 14.0 | 7.0 | D | 4.0 | D | D | 35.3 | 6,650 | 294 | 671 | 10,887 | 2.5 |
| Catawba | 0.2 | D | 4.4 | 27.2 | 5.2 | 8.2 | 3.9 | 9.5 | 11.6 | 39,195 | 246 | 3,190 | 70,628 | 3.8 |
| Chatham | 0.6 | D | 7.7 | 9.4 | 11.9 | 7.4 | 3.2 | 12.9 | 17.1 | 19,615 | 264 | 912 | 34,627 | 20.4 |
| Cherokee | 1.3 | D | 8.6 | 9.8 | 9.1 | 10.3 | 5.1 | D | 21.9 | 10,670 | 372 | 770 | 18,791 | 7.3 |

1. Per 1,000 resident population estimated as of July 1 of the year shown.

| STATE County | Housing units, 2015-2019 | | | | | | | | Civilian labor force, 2020 | | | | Civilian employment[6], 2015-2019 | | |
|---|---|---|---|---|---|---|---|---|---|---|---|---|---|---|---|
| | Occupied units | | | | | | | | | | Unemployment | | | Percent | |
| | | | Owner-occupied | | | Renter-occupied | | | | | | | | | |
| | | | | Median owner cost as a percent of income | | | Median rent as a percent of income[2] | Sub-standard units[4] (percent) | | Percent change, 2019-2020 | | | | Management, business, science, and arts | Construction, production, and maintenance occupations |
| | Total | Percent | Median value[1] | With a mortgage | Without a mortgage[2] | Median rent[3] | | | Total | | Total | Rate[5] | Total | | |
| | 89 | 90 | 91 | 92 | 93 | 94 | 95 | 96 | 97 | 98 | 99 | 100 | 101 | 102 | 103 |
| **NEW YORK— Cont'd** | | | | | | | | | | | | | | | |
| Genesee | 23,759 | 72.3 | 119,800 | 18.3 | 11.6 | 747 | 27.1 | 1.9 | 28,962 | -2.3 | 2,129 | 7.4 | 29,570 | 30.9 | 29.5 |
| Greene | 17,100 | 74.8 | 176,300 | 23.6 | 13.5 | 908 | 38.1 | 2.3 | 20,281 | 0.6 | 1,683 | 8.3 | 20,451 | 33.3 | 23.4 |
| Hamilton | 1,157 | 86.7 | 166,000 | 22.7 | 10.0 | 783 | 24.7 | 3.0 | 2,141 | -3.0 | 169 | 7.9 | 1,787 | 26.5 | 29.6 |
| Herkimer | 24,524 | 73.6 | 100,000 | 19.0 | 11.8 | 675 | 24.1 | 1.8 | 27,477 | -1.2 | 2,250 | 8.2 | 28,769 | 32.4 | 26.7 |
| Jefferson | 42,734 | 56.5 | 149,900 | 19.9 | 12.1 | 987 | 29.4 | 2.1 | 43,295 | | 3,753 | 8.7 | 43,241 | 32.5 | 24.0 |
| Kings | 958,567 | 30.1 | 706,000 | 28.8 | 13.7 | 1,426 | 31.9 | 10.5 | 1,151,130 | -3.8 | 144,278 | 12.5 | 1,227,030 | 43.1 | 14.9 |
| Lewis | 10,247 | 80.8 | 120,200 | 20.1 | 12.1 | 768 | 29.1 | 1.3 | 11,389 | -1.3 | 874 | 7.7 | 11,917 | 31.6 | 32.6 |
| Livingston | 24,176 | 74.7 | 129,400 | 19.6 | 13.4 | 798 | 32.8 | 1.1 | 29,831 | | 2,104 | 7.1 | 29,283 | 35.7 | 26.0 |
| Madison | 25,877 | 77.4 | 136,800 | 19.6 | 12.4 | 783 | 26.4 | 1.3 | 32,437 | -2.5 | 2,529 | 7.8 | 34,103 | 36.2 | 23.1 |
| Monroe | 301,948 | 63.5 | 148,400 | 19.7 | 12.4 | 927 | 31.9 | 1.6 | 359,299 | -0.5 | 30,774 | 8.6 | 368,088 | 44.3 | 16.9 |
| Montgomery | 19,660 | 68.4 | 105,400 | 20.6 | 13.7 | 783 | 32.3 | 2.3 | 22,032 | -0.7 | 1,934 | 8.8 | 21,636 | 30.4 | 29.1 |
| Nassau | 446,977 | 80.7 | 493,500 | 26.0 | 17.1 | 1,772 | 32.7 | 2.9 | 698,942 | -1.3 | 58,907 | 8.4 | 687,898 | 46.3 | 14.3 |
| New York | 759,460 | 24.1 | 987,700 | 18.2 | 10.0 | 1,740 | 27.6 | 6.3 | 859,618 | -6.4 | 82,035 | 9.5 | 905,475 | 60.9 | 6.5 |
| Niagara | 88,519 | 70.7 | 125,600 | 19.0 | 12.6 | 703 | 28.5 | 1.3 | 98,105 | -0.1 | 10,199 | 10.4 | 100,670 | 34.2 | 23.7 |
| Oneida | 89,729 | 67.5 | 127,700 | 19.2 | 12.2 | 766 | 29.0 | 2.2 | 100,648 | -0.6 | 8,175 | 8.1 | 102,302 | 36.8 | 20.4 |
| Onondaga | 185,324 | 64.7 | 145,400 | 19.0 | 11.8 | 872 | 29.0 | 2.2 | 220,483 | 0.5 | 18,446 | 8.4 | 222,177 | 42.3 | 16.8 |
| Ontario | 44,787 | 73.8 | 161,800 | 19.3 | 12.6 | 896 | 28.7 | 1.4 | 54,401 | -1.7 | 3,954 | 7.3 | 54,822 | 43.2 | 19.2 |
| Orange | 128,016 | 66.9 | 271,200 | 24.6 | 15.6 | 1,259 | 33.6 | 3.9 | 183,265 | -1.7 | 15,308 | 8.4 | 177,356 | 36.8 | 20.7 |
| Orleans | 16,563 | 75.6 | 98,400 | 21.3 | 13.6 | 728 | 28.7 | 1.9 | 17,146 | -1.8 | 1,433 | 8.4 | 17,833 | 30.2 | 31.5 |
| Oswego | 46,079 | 73.7 | 104,600 | 19.0 | 12.3 | 777 | 30.7 | 3.0 | 52,209 | -0.5 | 4,660 | 8.9 | 54,134 | 30.3 | 28.7 |
| Otsego | 23,409 | 72.2 | 146,400 | 20.1 | 11.8 | 833 | 30.0 | 2.5 | 26,607 | -4.4 | 1,921 | 7.2 | 28,014 | 36.2 | 21.1 |
| Putnam | 34,854 | 81.8 | 358,500 | 25.2 | 15.7 | 1,453 | 31.6 | 1.5 | 50,079 | -1.7 | 3,785 | 7.6 | 51,264 | 43.1 | 17.6 |
| Queens | 778,932 | 44.8 | 543,800 | 28.7 | 14.1 | 1,583 | 32.1 | 9.7 | 1,098,937 | -3.2 | 137,723 | 12.5 | 1,134,877 | 34.6 | 19.7 |
| Rensselaer | 64,906 | 62.8 | 188,700 | 20.1 | 12.7 | 973 | 27.7 | 1.8 | 80,795 | -0.1 | 5,640 | 7.0 | 81,515 | 41.6 | 19.4 |
| Richmond | 166,246 | 69.3 | 504,800 | 26.0 | 14.7 | 1,319 | 32.1 | 4.6 | 211,064 | -5.8 | 22,456 | 10.6 | 214,492 | 41.5 | 17.6 |
| Rockland | 100,438 | 68.3 | 443,400 | 27.3 | 17.9 | 1,504 | 35.8 | 6.5 | 152,903 | -1.6 | 12,356 | 8.1 | 147,612 | 45.7 | 13.9 |
| St. Lawrence | 41,933 | 73.0 | 93,600 | 18.7 | 12.2 | 741 | 29.1 | 3.1 | 42,854 | -0.2 | 3,501 | 8.2 | 43,722 | 36.1 | 22.5 |
| Saratoga | 94,406 | 72.2 | 258,300 | 19.5 | 11.1 | 1,138 | 25.5 | 0.9 | 119,009 | 0.0 | 7,980 | 6.7 | 119,896 | 46.6 | 16.8 |
| Schenectady | 54,302 | 65.7 | 169,600 | 20.9 | 11.6 | 945 | 29.6 | 1.4 | 76,925 | 1.0 | 6,202 | 8.1 | 72,938 | 39.2 | 18.0 |
| Schoharie | 12,559 | 75.8 | 144,800 | 21.5 | 13.7 | 790 | 27.6 | 1.7 | 14,440 | -1.2 | 1,028 | 7.1 | 14,429 | 35.8 | 26.8 |
| Schuyler | 7,324 | 75.2 | 133,100 | 20.7 | 14.2 | 742 | 28.3 | 1.8 | 8,008 | -0.4 | 689 | 8.6 | 8,226 | 37.9 | 25.1 |
| Seneca | 13,564 | 73.9 | 107,100 | 19.0 | 12.9 | 781 | 28.1 | 1.5 | 15,214 | -4.3 | 1,244 | 8.2 | 15,280 | 34.2 | 26.1 |
| Steuben | 39,861 | 72.7 | 99,600 | 18.7 | 12.2 | 695 | 26.6 | 1.6 | 42,022 | -0.8 | 3,430 | 8.2 | 42,836 | 35.4 | 28.1 |
| Suffolk | 489,301 | 80.6 | 397,400 | 26.3 | 17.7 | 1,742 | 35.0 | 2.9 | 764,564 | -1.6 | 64,951 | 8.5 | 752,016 | 40.0 | 19.0 |
| Sullivan | 28,184 | 68.3 | 172,800 | 23.0 | 14.7 | 890 | 29.8 | 2.5 | 35,758 | -2.0 | 3,157 | 8.8 | 33,008 | 33.4 | 25.2 |
| Tioga | 20,030 | 76.9 | 119,700 | 18.8 | 11.4 | 751 | 29.6 | 1.4 | 22,142 | -2.0 | 1,732 | 7.8 | 22,805 | 36.6 | 24.1 |
| Tompkins | 39,618 | 54.9 | 204,600 | 20.9 | 11.3 | 1,164 | 36.5 | 1.8 | 47,929 | -2.4 | 2,962 | 6.2 | 50,882 | 52.9 | 11.9 |
| Ulster | 69,320 | 68.3 | 230,500 | 23.7 | 14.8 | 1,112 | 34.2 | 2.2 | 86,186 | -1.5 | 6,901 | 8.0 | 87,223 | 38.6 | 19.4 |
| Warren | 28,015 | 70.7 | 196,500 | 21.5 | 11.8 | 929 | 30.9 | 1.2 | 30,901 | -0.6 | 2,593 | 8.4 | 31,988 | 36.7 | 20.1 |
| Washington | 24,014 | 74.0 | 149,000 | 20.7 | 13.4 | 855 | 29.3 | 2.1 | 27,302 | -1.7 | 1,954 | 7.2 | 29,139 | 30.4 | 29.3 |
| Wayne | 36,720 | 76.4 | 123,800 | 20.2 | 12.7 | 779 | 28.7 | 1.5 | 42,921 | -1.9 | 3,172 | 7.4 | 44,700 | 33.1 | 27.8 |
| Westchester | 349,292 | 61.4 | 540,600 | 24.1 | 16.1 | 1,537 | 32.6 | 4.3 | 477,959 | -1.4 | 40,187 | 8.4 | 482,359 | 48.3 | 13.6 |
| Wyoming | 15,917 | 76.9 | 115,900 | 19.3 | 11.7 | 659 | 25.6 | 1.3 | 17,793 | -1.4 | 1,336 | 7.5 | 18,497 | 30.9 | 29.6 |
| Yates | 8,919 | 80.0 | 136,300 | 20.8 | 13.0 | 775 | 30.9 | 2.9 | 11,546 | -2.9 | 752 | 6.5 | 11,404 | 32.8 | 30.2 |
| **NORTH CAROLINA** | 3,965,482 | 65.2 | 172,500 | 19.9 | 10.4 | 907 | 28.9 | 2.6 | 4,950,859 | -2.5 | 363,452 | 7.3 | 4,764,135 | 38.0 | 23.9 |
| Alamance | 64,439 | 65.2 | 154,800 | 19.1 | 10.2 | 813 | 29.5 | 2.5 | 80,457 | -3.1 | 5,832 | 7.2 | 77,665 | 33.2 | 27.6 |
| Alexander | 14,169 | 75.7 | 136,500 | 17.7 | 10.0 | 654 | 23.4 | 3.3 | 17,694 | -2.2 | 1,317 | 7.4 | 16,913 | 27.0 | 42.3 |
| Alleghany | 4,920 | 76.3 | 139,800 | 23.8 | 10.2 | 622 | 30.6 | 2.6 | 4,210 | -4.7 | 307 | 7.3 | 4,644 | 30.0 | 30.4 |
| Anson | 9,521 | 66.1 | 91,200 | 20.0 | 13.5 | 754 | 31.8 | 3.4 | 10,422 | -3.4 | 805 | 7.7 | 10,383 | 25.3 | 41.3 |
| Ashe | 11,938 | 75.9 | 153,200 | 21.0 | 10.6 | 634 | 26.0 | 1.7 | 12,664 | -2.3 | 786 | 6.2 | 12,512 | 26.9 | 31.7 |
| Avery | 6,551 | 75.3 | 144,000 | 23.0 | 10.0 | 777 | 35.6 | 2.6 | 7,287 | -3.8 | 462 | 6.3 | 6,545 | 30.3 | 22.9 |
| Beaufort | 19,701 | 69.9 | 132,800 | 21.1 | 12.1 | 740 | 30.7 | 1.5 | 19,387 | -1.5 | 1,228 | 6.3 | 19,383 | 31.3 | 31.1 |
| Bertie | 7,909 | 74.4 | 82,200 | 22.1 | 15.8 | 638 | 29.3 | 3.7 | 7,618 | -3.1 | 486 | 6.4 | 7,338 | 26.6 | 38.0 |
| Bladen | 13,636 | 71.4 | 95,300 | 24.6 | 13.1 | 655 | 28.6 | 1.8 | 14,485 | -1.2 | 1,046 | 7.2 | 13,159 | 27.7 | 33.8 |
| Brunswick | 56,056 | 80.7 | 212,200 | 22.8 | 11.5 | 960 | 30.2 | 2.0 | 52,246 | -5.0 | 4,605 | 8.8 | 52,589 | 30.7 | 24.0 |
| Buncombe | 107,479 | 63.4 | 238,200 | 21.7 | 10.8 | 975 | 31.2 | 3.0 | 136,480 | -3.7 | 11,366 | 8.3 | 127,147 | 40.1 | 19.7 |
| Burke | 35,156 | 73.9 | 120,700 | 19.6 | 10.0 | 648 | 27.5 | 3.9 | 39,888 | -2.8 | 2,869 | 7.2 | 39,033 | 29.0 | 32.5 |
| Cabarrus | 72,238 | 71.5 | 203,100 | 19.1 | 10.0 | 925 | 26.9 | 4.0 | 108,444 | -2.1 | 7,571 | 7.0 | 103,251 | 39.7 | 22.2 |
| Caldwell | 32,864 | 72.5 | 118,100 | 19.6 | 10.3 | 645 | 27.2 | 3.0 | 35,754 | -2.4 | 2,823 | 7.9 | 36,848 | 25.3 | 38.7 |
| Camden | 3,933 | 81.9 | 208,800 | 23.0 | 11.5 | 953 | 30.6 | 1.2 | 4,574 | -3.5 | 251 | 5.5 | 4,466 | 35.7 | 21.2 |
| Carteret | 29,755 | 72.6 | 214,100 | 22.4 | 11.2 | 902 | 26.6 | 1.4 | 31,320 | -3.1 | 1,947 | 6.2 | 30,680 | 35.6 | 22.4 |
| Caswell | 9,145 | 76.4 | 112,800 | 19.6 | 10.0 | 558 | 33.9 | 1.6 | 9,621 | -3.1 | 752 | 7.8 | 9,519 | 30.7 | 31.5 |
| Catawba | 62,319 | 69.4 | 143,600 | 18.8 | 10.0 | 738 | 24.0 | 3.4 | 77,852 | -1.9 | 6,220 | 8.0 | 73,813 | 30.8 | 33.2 |
| Chatham | 29,033 | 76.4 | 281,700 | 19.2 | 10.4 | 845 | 27.2 | 2.2 | 35,248 | -3.3 | 1,910 | 5.4 | 32,344 | 44.2 | 21.2 |
| Cherokee | 12,471 | 79.0 | 159,100 | 21.7 | 11.1 | 724 | 29.5 | 1.9 | 10,952 | -3.0 | 868 | 7.9 | 10,844 | 29.9 | 28.0 |

1. Specified owner-occupied units lacking complete plumbing facilities.    2. A value of 10.0 represents 10 percent or less; a value of 50.0 represents 50 percent or more.    3. Specified renter-occupied units.    4. Overcrowded or
5. Percent of civilian labor force.    6. Civilian employed persons 16 years old and over.

# Table B. States and Counties — Nonfarm Employment and Agriculture

| STATE County | Private nonfarm establishments, employment and payroll, 2019 | | | | | | | | | Agriculture, 2017 | | | |
| | Number of establishments | Employment Total | Health care and social assistance | Manufacturing | Retail trade | Finance and insurance | Professional, scientific, and technical services | Annual payroll Total (mil dol) | Average per employee (dollars) | Farms Number | Percent with Fewer than 50 acres | Percent with 1000 acres or more | Farm producers whose primary occupation is farming (percent) |
| | 104 | 105 | 106 | 107 | 108 | 109 | 110 | 111 | 112 | 113 | 114 | 115 | 116 |
| **NEW YORK— Cont'd** | | | | | | | | | | | | | |
| Genesee | 1,259 | 17,042 | 2,844 | 3,262 | 2,765 | 358 | 310 | 643 | 37,740 | 485 | 38.8 | 7.8 | 51.5 |
| Greene | 1,167 | 10,660 | 1,217 | 718 | 2,278 | 327 | 288 | 387 | 36,322 | 206 | 35.9 | 2.9 | 50.3 |
| Hamilton | 190 | 938 | 35 | NA | 150 | 76 | 3 | 34 | 35,817 | 14 | 35.7 | NA | 33.3 |
| Herkimer | 1,108 | 12,669 | 2,123 | 2,904 | 1,897 | 298 | 175 | 470 | 37,073 | 596 | 24.3 | 1.2 | 47.4 |
| Jefferson | 2,420 | 30,055 | 6,871 | 2,388 | 6,388 | 748 | 863 | 1,183 | 39,345 | 792 | 26.5 | 5.6 | 51.5 |
| Kings | 60,707 | 677,323 | 254,232 | 18,039 | 75,621 | 18,071 | 24,157 | 29,561 | 43,644 | 19 | 100.0 | NA | 15.4 |
| Lewis | 518 | 4,711 | 962 | 1,114 | 845 | 78 | 115 | 194 | 41,123 | 625 | 23.8 | 5.0 | 56.6 |
| Livingston | 1,213 | 13,657 | 2,073 | 1,879 | 2,563 | 221 | 381 | 510 | 37,368 | 661 | 43.9 | 8.9 | 47.4 |
| Madison | 1,334 | 17,481 | 3,254 | 2,899 | 2,583 | 361 | 760 | 668 | 38,234 | 691 | 27.4 | 4.5 | 54.0 |
| Monroe | 17,565 | 361,787 | 72,971 | 36,814 | 40,637 | 12,820 | 26,765 | 17,458 | 48,255 | 527 | 56.4 | 6.1 | 51.8 |
| Montgomery | 1,053 | 15,934 | 4,098 | 3,026 | 2,975 | 294 | 323 | 638 | 40,048 | 564 | 28.4 | 2.8 | 62.4 |
| Nassau | 48,915 | 580,276 | 136,778 | 15,041 | 76,872 | 34,083 | 40,184 | 32,240 | 55,559 | 32 | 81.3 | NA | 23.1 |
| New York | 101,883 | 2,327,677 | 265,613 | 14,412 | 144,797 | 301,547 | 331,795 | 275,586 | 118,395 | 7 | 100.0 | NA | 50.0 |
| Niagara | 4,590 | 63,307 | 12,174 | 9,064 | 11,075 | 1,441 | 1,713 | 2,392 | 37,778 | 690 | 50.1 | 5.2 | 43.0 |
| Oneida | 4,874 | 89,008 | 21,588 | 9,401 | 11,797 | 6,290 | 3,969 | 3,718 | 41,771 | 967 | 34.1 | 2.5 | 45.5 |
| Onondaga | 11,581 | 220,326 | 43,134 | 16,779 | 28,114 | 10,178 | 14,343 | 10,582 | 48,027 | 623 | 44.6 | 5.1 | 48.9 |
| Ontario | 2,930 | 48,567 | 8,811 | 7,667 | 9,084 | 950 | 1,682 | 2,398 | 49,373 | 833 | 40.3 | 5.2 | 49.2 |
| Orange | 9,752 | 122,988 | 25,539 | 8,997 | 22,777 | 2,730 | 4,528 | 5,397 | 43,885 | 621 | 43.6 | 1.8 | 53.1 |
| Orleans | 661 | 7,684 | 1,439 | 1,927 | 1,050 | 376 | 170 | 283 | 36,824 | 498 | 43.4 | 5.8 | 56.8 |
| Oswego | 2,111 | 24,021 | 4,877 | 3,004 | 4,327 | 665 | 861 | 1,087 | 45,242 | 612 | 40.2 | 1.1 | 43.4 |
| Otsego | 1,337 | 19,241 | 6,397 | 970 | 3,021 | 1,122 | 485 | 828 | 43,017 | 880 | 27.0 | 1.8 | 45.1 |
| Putnam | 2,877 | 22,076 | 4,834 | 1,364 | 3,145 | 536 | 1,350 | 1,034 | 46,834 | 89 | 64.0 | NA | 32.8 |
| Queens | 51,163 | 602,904 | 159,021 | 18,460 | 61,197 | 20,183 | 16,677 | 31,283 | 51,888 | 4 | 100.0 | NA | 20.0 |
| Rensselaer | 3,063 | 45,372 | 8,595 | 4,840 | 5,794 | 1,144 | 3,212 | 2,302 | 50,737 | 470 | 37.0 | 1.9 | 44.6 |
| Richmond | 9,539 | 109,566 | 34,418 | 1,081 | 16,167 | 3,156 | 3,835 | 5,023 | 45,848 | 6 | 100.0 | NA | 6.1 |
| Rockland | 10,400 | 118,341 | 31,147 | 7,373 | 14,074 | 3,196 | 6,354 | 5,810 | 49,093 | 14 | 71.4 | NA | 59.5 |
| St. Lawrence | 1,893 | 26,986 | 6,498 | 2,245 | 4,778 | 687 | 630 | 1,126 | 41,721 | 1,253 | 16.7 | 3.9 | 49.0 |
| Saratoga | 5,406 | 74,465 | 12,218 | 6,163 | 11,166 | 3,088 | 5,025 | 3,783 | 50,804 | 591 | 54.0 | 2.0 | 47.3 |
| Schenectady | 3,004 | 55,660 | 12,246 | 5,533 | 6,980 | 2,174 | 5,872 | 3,058 | 54,947 | 185 | 43.8 | NA | 41.7 |
| Schoharie | 575 | 5,453 | 1,016 | 351 | 1,051 | 193 | 187 | 219 | 40,152 | 541 | 27.7 | 1.3 | 51.2 |
| Schuyler | 383 | 3,662 | 810 | 622 | 635 | 57 | 35 | 155 | 42,257 | 408 | 38.5 | 2.2 | 46.4 |
| Seneca | 716 | 10,346 | 1,088 | 1,789 | 2,142 | 196 | 170 | 405 | 39,177 | 516 | 39.7 | 3.5 | 55.6 |
| Steuben | 1,786 | 25,619 | 6,833 | 3,608 | 4,223 | 986 | 1,032 | 1,540 | 60,112 | 1,542 | 21.8 | 4.5 | 44.9 |
| Suffolk | 49,932 | 594,392 | 110,973 | 53,452 | 82,122 | 22,896 | 41,265 | 33,772 | 56,818 | 560 | 74.1 | 0.4 | 58.1 |
| Sullivan | 1,998 | 22,592 | 6,876 | 1,105 | 2,589 | 597 | 860 | 871 | 38,563 | 366 | 50.3 | 2.2 | 45.0 |
| Tioga | 767 | 8,058 | 1,111 | 1,192 | 1,267 | 212 | 240 | 317 | 39,382 | 535 | 29.0 | 2.2 | 46.4 |
| Tompkins | 2,348 | 50,689 | 6,363 | 2,666 | 4,794 | 1,201 | 2,552 | 2,193 | 43,255 | 523 | 47.6 | 2.9 | 39.6 |
| Ulster | 4,856 | 46,888 | 9,491 | 3,435 | 8,810 | 2,038 | 1,727 | 1,789 | 38,160 | 421 | 51.8 | 0.7 | 51.0 |
| Warren | 2,240 | 31,608 | 6,695 | 3,677 | 5,649 | 1,130 | 950 | 1,352 | 42,779 | 80 | 58.8 | NA | 40.8 |
| Washington | 1,070 | 10,567 | 1,770 | 2,769 | 1,974 | 211 | 283 | 455 | 43,078 | 915 | 37.2 | 2.8 | 43.6 |
| Wayne | 1,713 | 20,186 | 2,816 | 6,164 | 3,550 | 465 | 594 | 925 | 45,831 | 829 | 43.4 | 4.0 | 53.0 |
| Westchester | 31,580 | 392,824 | 84,498 | 10,190 | 47,729 | 21,843 | 24,087 | 27,460 | 69,905 | 115 | 81.7 | 1.7 | 49.5 |
| Wyoming | 776 | 8,720 | 1,224 | 1,765 | 1,321 | 360 | 296 | 348 | 39,854 | 729 | 41.0 | 6.4 | 51.9 |
| Yates | 528 | 5,276 | 874 | 998 | 970 | 86 | 108 | 179 | 33,966 | 867 | 30.8 | 1.4 | 60.4 |
| **NORTH CAROLINA** | 238,015 | 3,932,620 | 617,700 | 451,253 | 505,536 | 193,388 | 237,492 | 194,776 | 49,528 | 46,418 | 47.9 | 3.8 | 42.7 |
| Alamance | 3,346 | 61,650 | 12,055 | 8,839 | 9,868 | 1,351 | 1,461 | 2,454 | 39,803 | 720 | 45.8 | 0.7 | 40.6 |
| Alexander | 566 | 7,948 | 525 | 4,355 | 804 | 163 | 118 | 278 | 34,955 | 544 | 50.0 | 0.9 | 45.4 |
| Alleghany | 249 | 2,298 | 379 | 612 | 311 | 83 | 54 | 78 | 33,759 | 448 | 37.3 | 3.1 | 38.2 |
| Anson | 388 | 5,167 | 790 | 1,625 | 770 | 77 | 93 | 165 | 31,995 | 412 | 32.8 | 4.4 | 41.7 |
| Ashe | 543 | 6,060 | 1,102 | 1,224 | 1,157 | 185 | 77 | 221 | 36,418 | 864 | 42.0 | 1.9 | 43.5 |
| Avery | 484 | 4,741 | 774 | 124 | 869 | 66 | 77 | 149 | 31,384 | 351 | 50.4 | 1.7 | 44.7 |
| Beaufort | 1,087 | 12,997 | 2,192 | 2,543 | 2,315 | 365 | 262 | 474 | 36,457 | 310 | 38.7 | 15.5 | 46.1 |
| Bertie | 289 | 5,333 | 907 | 2,363 | 453 | 65 | 31 | 151 | 28,397 | 323 | 26.9 | 10.8 | 65.4 |
| Bladen | 531 | 10,869 | 981 | 5,316 | 871 | 162 | 131 | 358 | 32,944 | 512 | 35.9 | 7.4 | 48.6 |
| Brunswick | 2,590 | 28,026 | 3,983 | 1,392 | 5,893 | 600 | 760 | 1,047 | 37,352 | 231 | 42.0 | 5.2 | 43.9 |
| Buncombe | 8,892 | 123,369 | 25,704 | 13,546 | 18,923 | 2,776 | 5,210 | 5,081 | 41,188 | 1,073 | 67.3 | 0.5 | 38.0 |
| Burke | 1,467 | 24,409 | 5,677 | 7,698 | 3,263 | 352 | 378 | 876 | 35,896 | 508 | 56.1 | 0.6 | 38.9 |
| Cabarrus | 4,598 | 68,150 | 9,861 | 6,539 | 13,482 | 945 | 1,873 | 2,755 | 40,422 | 629 | 56.6 | 1.6 | 34.8 |
| Caldwell | 1,357 | 21,248 | 3,357 | 6,405 | 2,807 | 348 | 421 | 822 | 38,684 | 411 | 57.2 | 0.5 | 39.6 |
| Camden | 111 | 557 | 35 | NA | 80 | 15 | 18 | 16 | 29,289 | 81 | 48.1 | 24.7 | 59.1 |
| Carteret | 2,036 | 19,996 | 3,382 | 978 | 4,756 | 565 | 594 | 696 | 34,812 | 158 | 78.5 | 2.5 | 29.8 |
| Caswell | 237 | 1,759 | 389 | 95 | 317 | 43 | 42 | 55 | 31,354 | 493 | 28.6 | 3.4 | 44.9 |
| Catawba | 4,202 | 84,721 | 12,829 | 23,559 | 10,070 | 1,788 | 2,394 | 3,603 | 42,528 | 638 | 53.8 | 1.1 | 36.7 |
| Chatham | 1,506 | 14,183 | 2,765 | 1,643 | 2,246 | 311 | 660 | 519 | 36,600 | 1,116 | 48.7 | 0.4 | 37.8 |
| Cherokee | 600 | 7,404 | 1,154 | 1,338 | 1,455 | 185 | 106 | 232 | 31,385 | 277 | 57.4 | 0.7 | 39.0 |

# Table B. States and Counties — Agriculture

| STATE County | Land in farms | | Acres | | | Value of land and buildings (dollars) | | Value of machinery and equipment, average per farm (dollars) | Value of products sold: | | | | Organic farms (number) | Farms with internet access (percent) | Government payments | |
|---|---|---|---|---|---|---|---|---|---|---|---|---|---|---|---|---|
| | Acreage (1,000) | Percent change, 2012-2017 | Average size of farm | Total irrigated (1,000) | Total cropland (1,000) | Average per farm | Average per acre | | Total (mil dol) | Average per farm (acres) | Crops | Livestock and poultry products | | | Total ($1,000) | Percent of farms |
| | 117 | 118 | 119 | 120 | 121 | 122 | 123 | 124 | 125 | 126 | 127 | 128 | 129 | 130 | 131 | 132 |
| NEW YORK— Cont'd | | | | | | | | | | | | | | | | |
| Genesee | 177 | -5.5 | 365 | 3.3 | 148.6 | 1,220,292 | 3,345 | 312,434 | 234.9 | 484,402 | 34.4 | 65.6 | 13 | 77.5 | 2,668 | 44.7 |
| Greene | 35 | -18.6 | 170 | 0.3 | 13.7 | 535,865 | 3,156 | 74,288 | 19.8 | 95,927 | 46.7 | 53.3 | 7 | 72.3 | 138 | 14.1 |
| Hamilton | 1 | -55.1 | 67 | D | 0.1 | 222,415 | 3,341 | 32,734 | D | D | D | NA | NA | 92.9 | D | 14.3 |
| Herkimer | 118 | -37.2 | 198 | 0.2 | 69.5 | 452,284 | 2,289 | 111,551 | 58.0 | 97,277 | 23.4 | 76.6 | 28 | 67.1 | 467 | 18.3 |
| Jefferson | 247 | -14.9 | 312 | 0.4 | 168.8 | 836,501 | 2,677 | 170,553 | 165.1 | 208,404 | 21.6 | 78.4 | 31 | 74.7 | 1,986 | 22.3 |
| Kings | 0 | D | 1 | 0.0 | 0.0 | 536,994 | 443,604 | 12,004 | D | D | D | D | D | 73.7 | NA | NA |
| Lewis | 182 | 0.4 | 292 | 0.1 | 105.0 | 586,534 | 2,009 | 171,441 | 153.1 | 244,917 | 13.8 | 86.2 | 12 | 72.8 | 416 | 22.9 |
| Livingston | 189 | -2.8 | 287 | 0.2 | 145.9 | 1,064,151 | 3,712 | 196,665 | 183.7 | 277,905 | 30.1 | 69.9 | 5 | 82.8 | 4,427 | 30.7 |
| Madison | 172 | -16.2 | 249 | 0.5 | 105.5 | 652,012 | 2,621 | 142,460 | 113.6 | 164,444 | 24.7 | 75.3 | 64 | 79.7 | 1,418 | 22.0 |
| Monroe | 107 | 8.2 | 203 | 0.6 | 85.7 | 872,523 | 4,306 | 155,561 | 76.6 | 145,433 | 86.9 | 13.1 | 20 | 88.4 | 2,627 | 17.6 |
| Montgomery | 115 | 29.5 | 204 | 0.1 | 84.5 | 506,591 | 2,485 | 129,396 | 75.0 | 132,906 | 23.8 | 76.2 | 55 | 70.7 | 404 | 14.9 |
| Nassau | 1 | -66.1 | 28 | 0.1 | 0.2 | 493,741 | 17,362 | 39,261 | D | D | D | D | 2 | 56.3 | NA | NA |
| New York | 0 | D | 2 | D | 0.0 | D | D | 4,336 | 0.0 | 6,429 | 100.0 | NA | NA | 100.0 | NA | NA |
| Niagara | 140 | -1.8 | 203 | 1.5 | 116.2 | 624,058 | 3,070 | 161,032 | 118.6 | 171,909 | 63.6 | 36.4 | 19 | 82.3 | 2,095 | 21.3 |
| Oneida | 193 | 28.3 | 199 | 0.7 | 119.2 | 530,131 | 2,659 | 118,065 | 100.5 | 103,883 | 29.8 | 70.2 | 15 | 77.4 | 1,066 | 19.6 |
| Onondaga | 161 | 70.6 | 258 | 1.3 | 113.1 | 1,063,488 | 4,122 | 196,839 | 178.4 | 286,371 | 22.2 | 77.8 | 20 | 86.2 | 1,815 | 25.8 |
| Ontario | 200 | 3.9 | 240 | 0.8 | 163.8 | 967,405 | 4,027 | 189,335 | 205.2 | 246,291 | 30.8 | 69.2 | 53 | 72.6 | 3,712 | 29.7 |
| Orange | 81 | -7.8 | 131 | 2.5 | 49.1 | 906,875 | 6,936 | 113,299 | 87.9 | 141,572 | 75.7 | 24.3 | 17 | 81.3 | 463 | 11.0 |
| Orleans | 130 | -4.1 | 260 | 3.2 | 107.7 | 825,545 | 3,173 | 225,733 | 155.3 | 311,811 | 85.8 | 14.2 | 23 | 71.7 | 3,323 | 39.0 |
| Oswego | 86 | -52.3 | 141 | 0.8 | 40.3 | 311,560 | 2,213 | 87,168 | 41.2 | 67,369 | 50.7 | 49.3 | 11 | 80.1 | 126 | 14.4 |
| Otsego | 155 | 143.9 | 176 | 0.4 | 81.5 | 468,529 | 2,666 | 79,321 | 56.2 | 63,840 | 34.5 | 65.5 | 62 | 75.2 | 619 | 13.6 |
| Putnam | 7 | 26.5 | 84 | 0.1 | 2.6 | 503,785 | 6,001 | 55,525 | 3.1 | 35,337 | 91.1 | 8.9 | 3 | 97.8 | NA | NA |
| Queens | D | D | D | D | D | 679,231 | 48,517 | 6,091 | 0.1 | 23,500 | 61.7 | 38.3 | NA | 75.0 | NA | NA |
| Rensselaer | 83 | 5.0 | 176 | 1.0 | 46.8 | 618,709 | 3,513 | 107,314 | 41.0 | 87,255 | 59.2 | 40.8 | 26 | 81.9 | 1,040 | 22.8 |
| Richmond | D | D | D | 0.0 | D | D | D | 7,145 | D | D | D | NA | 2 | 66.7 | NA | NA |
| Rockland | 1 | 9.5 | 41 | 0.1 | 0.3 | 1,183,762 | 28,772 | 115,774 | 2.1 | 153,071 | 95.9 | 4.1 | 2 | 100.0 | NA | NA |
| St. Lawrence | 343 | -4.0 | 273 | 0.4 | 176.7 | 500,868 | 1,832 | 111,812 | 191.1 | 152,496 | 17.6 | 82.4 | 100 | 77.5 | 1,130 | 12.1 |
| Saratoga | 72 | 260.4 | 121 | 0.7 | 41.6 | 713,614 | 5,890 | 125,743 | 76.8 | 129,968 | 24.8 | 75.2 | 5 | 88.0 | 369 | 5.1 |
| Schenectady | 17 | -82.4 | 94 | 0.2 | 9.2 | 402,340 | 4,288 | 59,097 | 5.5 | 29,519 | 64.3 | 35.7 | NA | 78.4 | 21 | 8.1 |
| Schoharie | 100 | -47.3 | 185 | 0.1 | 61.1 | 449,208 | 2,435 | 108,434 | 47.9 | 88,590 | 42.0 | 58.0 | 28 | 76.0 | 259 | 21.6 |
| Schuyler | 79 | 13.8 | 193 | 0.3 | 48.4 | 631,614 | 3,270 | 127,031 | 45.8 | 112,140 | 33.1 | 66.9 | 23 | 73.5 | 203 | 13.5 |
| Seneca | 119 | -9.0 | 230 | 0.2 | 96.7 | 862,800 | 3,756 | 170,204 | 90.8 | 176,052 | 51.9 | 48.1 | 42 | 68.6 | 2,029 | 20.2 |
| Steuben | 397 | -2.1 | 258 | 0.9 | 240.4 | 540,734 | 2,099 | 128,101 | 196.0 | 127,075 | 35.5 | 64.5 | 59 | 76.6 | 3,739 | 29.4 |
| Suffolk | 30 | -16.5 | 54 | 12.1 | 23.1 | 655,303 | 12,219 | 153,042 | 225.6 | 402,818 | 90.4 | 9.6 | 30 | 86.1 | 76 | 2.1 |
| Sullivan | 60 | 11.3 | 164 | 0.2 | 26.2 | 689,944 | 4,213 | 78,765 | 28.4 | 77,549 | 19.7 | 80.3 | 15 | 82.0 | 152 | 7.7 |
| Tioga | 113 | 4.9 | 212 | 0.4 | 54.8 | 483,358 | 2,285 | 99,481 | 40.9 | 76,368 | 29.9 | 70.1 | 16 | 83.0 | 971 | 25.2 |
| Tompkins | 91 | 0.6 | 175 | 0.4 | 62.1 | 661,192 | 3,789 | 138,310 | 64.7 | 123,715 | 24.6 | 75.4 | 45 | 85.9 | 1,575 | 22.6 |
| Ulster | 59 | -17.3 | 140 | 3.7 | 24.1 | 965,681 | 10,646 | 95,712 | 54.3 | 129,088 | 88.3 | 11.7 | 24 | 86.9 | 164 | 8.1 |
| Warren | 10 | 5.9 | 126 | 0.0 | 0.9 | 526,197 | 4,174 | 60,826 | D | D | D | D | 1 | 85.0 | D | 2.5 |
| Washington | 185 | D | 203 | 1.0 | 103.0 | 565,739 | 2,794 | 133,639 | 135.8 | 148,428 | 18.0 | 82.0 | 18 | 83.9 | 1,418 | 18.7 |
| Wayne | 159 | -11.2 | 192 | 1.5 | 115.5 | 659,498 | 3,437 | 190,441 | 221.3 | 266,942 | 70.3 | 29.7 | 41 | 79.5 | 1,462 | 15.2 |
| Westchester | 7 | -10.0 | 61 | 0.1 | 4.1 | 645,915 | 10,646 | 72,483 | 7.0 | 60,487 | 72.6 | 27.4 | 2 | 89.6 | D | 1.7 |
| Wyoming | 235 | 4.0 | 322 | 1.7 | 167.3 | 1,115,529 | 3,463 | 283,473 | 307.5 | 421,838 | 21.1 | 78.9 | 14 | 73.8 | 2,817 | 30.0 |
| Yates | 115 | -9.5 | 133 | 0.6 | 83.0 | 692,852 | 5,227 | 138,824 | 114.7 | 132,246 | 41.2 | 58.8 | 119 | 40.0 | 825 | 9.5 |
| NORTH CAROLINA | 8,431 | 0.2 | 182 | 143.4 | 5,000.7 | 843,154 | 4,642 | 112,477 | 12,900.7 | 277,924 | 29.0 | 71.0 | 465 | 75.2 | 107,565 | 21.6 |
| Alamance | 80 | -4.2 | 111 | 0.7 | 33.8 | 667,068 | 6,000 | 64,708 | 41.8 | 57,986 | 35.3 | 64.7 | 23 | 76.4 | 147 | 8.6 |
| Alexander | 54 | -7.7 | 100 | 0.3 | 22.7 | 560,533 | 5,632 | 101,446 | 176.3 | 324,168 | 4.8 | 95.2 | NA | 71.1 | 170 | 5.5 |
| Alleghany | 71 | -22.4 | 158 | D | 26.2 | 777,876 | 4,937 | 87,835 | 31.4 | 70,092 | 57.9 | 42.1 | 6 | 72.1 | 32 | 5.6 |
| Anson | 85 | 2.1 | 207 | 0.3 | 33.6 | 846,181 | 4,086 | 116,392 | 303.7 | 737,248 | 4.3 | 95.7 | NA | 69.7 | 817 | 27.2 |
| Ashe | 110 | -2.4 | 127 | 0.1 | 36.4 | 668,101 | 5,258 | 71,937 | 57.1 | 66,135 | 76.9 | 23.1 | 6 | 74.1 | 76 | 1.9 |
| Avery | 29 | 1.6 | 82 | 0.0 | 11.9 | 530,636 | 6,494 | 63,220 | 20.1 | 57,222 | 96.1 | 3.9 | 6 | 74.1 | 37 | 3.1 |
| Beaufort | 139 | -5.9 | 450 | 2.0 | 118.6 | 1,538,781 | 3,420 | 265,851 | 112.0 | 361,413 | 65.4 | 34.6 | NA | 73.5 | 4,556 | 58.4 |
| Bertie | 148 | 0.9 | 459 | 3.2 | 112.3 | 1,390,119 | 3,032 | 238,427 | 260.5 | 806,412 | 30.9 | 69.1 | NA | 69.0 | 4,249 | 63.2 |
| Bladen | 180 | 53.7 | 352 | 6.0 | 76.8 | 1,174,444 | 3,334 | 225,098 | 446.4 | 871,873 | 15.8 | 84.2 | 1 | 77.7 | 3,178 | 35.0 |
| Brunswick | 45 | -1.6 | 193 | 0.5 | 22.8 | 758,353 | 3,920 | 114,405 | 46.1 | 199,437 | 34.2 | 65.8 | NA | 83.1 | 338 | 15.2 |
| Buncombe | 72 | 1.1 | 67 | 0.8 | 17.8 | 670,235 | 9,949 | 55,669 | 48.0 | 44,747 | 77.4 | 22.6 | 16 | 78.5 | 357 | 12.3 |
| Burke | 39 | 12.4 | 76 | 1.8 | 17.5 | 433,663 | 5,701 | 76,589 | 81.5 | 160,366 | 24.3 | 75.7 | 3 | 75.4 | 167 | 5.1 |
| Cabarrus | 64 | -3.9 | 101 | D | 32.8 | 881,469 | 8,709 | 83,381 | 57.8 | 91,967 | 30.7 | 69.3 | NA | 75.2 | 304 | 10.8 |
| Caldwell | 38 | 18.5 | 92 | 0.6 | 14.4 | 419,469 | 4,538 | 59,705 | 48.1 | 116,922 | 22.9 | 77.1 | 3 | 77.4 | 152 | 5.6 |
| Camden | 59 | 20.1 | 731 | D | 53.8 | 2,327,456 | 3,182 | 421,303 | 39.9 | 492,988 | 98.3 | 1.7 | NA | 64.2 | 1,262 | 45.7 |
| Carteret | 63 | -0.1 | 397 | 0.1 | 44.7 | 1,801,801 | 4,536 | 94,673 | 23.8 | 150,506 | 97.9 | 2.1 | NA | 75.3 | 137 | 10.1 |
| Caswell | 105 | 8.1 | 213 | 1.4 | 32.3 | 689,530 | 3,241 | 65,751 | 37.9 | 76,929 | 55.7 | 44.3 | 13 | 72.8 | 164 | 19.3 |
| Catawba | 64 | -5.3 | 100 | 0.7 | 34.2 | 651,547 | 6,543 | 87,564 | 77.3 | 121,130 | 28.2 | 71.8 | 1 | 75.4 | 620 | 8.5 |
| Chatham | 106 | -5.2 | 95 | 0.3 | 30.9 | 522,890 | 5,505 | 66,667 | 171.2 | 153,360 | 6.5 | 93.5 | 19 | 78.7 | 103 | 8.7 |
| Cherokee | 26 | 22.3 | 95 | 0.1 | 7.6 | 464,663 | 4,906 | 79,138 | D | D | D | D | 11 | 81.6 | 303 | 18.1 |

# Table B. States and Counties — Water Use, Wholesale Trade, Retail Trade, and Real Estate

| STATE County | Water use, 2015 | | Wholesale Trade[1], 2017 | | | | Retail Trade[2], 2017 | | | | Real estate and rental and leasing,[2] 2017 | | | |
|---|---|---|---|---|---|---|---|---|---|---|---|---|---|---|
| | Public supply water withdrawn (mil gal/ day) | Public supply gallons withdrawn per person per day | Number of establishments | Number of employees | Sales (mil dol) | Average payroll (mil dol) | Number of establishments | Number of employees | Sales (mil dol) | Average payroll (mil dol) | Number of establishments | Number of employees | Sales (mil dol) | Average payroll (mil dol) |
| | 133 | 134 | 135 | 136 | 137 | 138 | 139 | 140 | 141 | 142 | 143 | 144 | 145 | 146 |
| NEW YORK— Cont'd | | | | | | | | | | | | | | |
| Genesee | 3.74 | 63.5 | D | D | D | D | 214 | 2,766 | 889.6 | 69.3 | 34 | 158 | 29.3 | 6.6 |
| Greene | 3.07 | 64.5 | 24 | 770 | 866.2 | 42.0 | 204 | 2,197 | 714.0 | 57.3 | 43 | 145 | 30.0 | 6.0 |
| Hamilton | 0.75 | 159.2 | NA | NA | NA | NA | 33 | 156 | 37.8 | 4.8 | D | D | D | 0.2 |
| Herkimer | 28.84 | 457.1 | 33 | 629 | 193.7 | 32.2 | 168 | 2,007 | 555.1 | 50.9 | 36 | 139 | 21.1 | 3.3 |
| Jefferson | 9.29 | 79.0 | 56 | 678 | 285.3 | 33.0 | 484 | 6,485 | 1,911.8 | 171.6 | 128 | 624 | 102.3 | 24.3 |
| Kings | 0.00 | 0.0 | 3,202 | 27,329 | 15,914.1 | 1,213.4 | 10,799 | 76,885 | 28,734.8 | 2,187.0 | 4,935 | 18,848 | 5,339.0 | 798.8 |
| Lewis | 1.81 | 67.1 | D | D | D | D | 79 | 874 | 285.5 | 23.6 | D | D | D | D |
| Livingston | 3.78 | 58.4 | 37 | 366 | 360.1 | 26.4 | 209 | 2,648 | 700.0 | 64.9 | 36 | 193 | 25.4 | 5.6 |
| Madison | 1.22 | 17.0 | 36 | 399 | 172.3 | 18.0 | 202 | 2,488 | 834.5 | 73.7 | 48 | 165 | 16.9 | 4.3 |
| Monroe | 53.54 | 71.4 | 753 | 12,126 | 8,925.0 | 717.2 | 2,304 | 40,485 | 10,123.1 | 1,018.4 | 865 | 6,191 | 1,319.9 | 278.3 |
| Montgomery | 2.61 | 52.6 | 35 | 343 | 158.9 | 18.7 | 186 | 2,636 | 705.9 | 72.4 | 20 | 73 | 12.0 | 2.4 |
| Nassau | 194.47 | 142.9 | 2,707 | 28,344 | 27,806.7 | 1,869.2 | 6,033 | 79,455 | 26,841.2 | 2,506.6 | 2,634 | 11,791 | 4,187.6 | 686.3 |
| New York | 0.00 | 0.0 | 6,819 | 86,176 | 141,457.7 | 7,023.0 | 11,203 | 161,413 | 54,824.9 | 6,029.1 | 9,795 | 84,318 | 40,972.3 | 6,325.9 |
| Niagara | 49.46 | 232.6 | 167 | 1,978 | 3,544.7 | 101.2 | 747 | 11,361 | 2,721.4 | 259.6 | 158 | 704 | 118.9 | 23.9 |
| Oneida | 32.82 | 141.2 | 178 | 2,415 | 1,294.4 | 110.3 | 790 | 11,703 | 3,166.5 | 298.1 | 161 | 610 | 126.0 | 22.1 |
| Onondaga | 58.04 | 123.9 | 550 | 11,142 | 14,738.1 | 646.2 | 1,651 | 28,181 | 7,365.6 | 719.8 | 561 | 3,555 | 739.8 | 162.0 |
| Ontario | 46.53 | 424.7 | 128 | 1,621 | 847.3 | 89.1 | 537 | 9,457 | 2,309.9 | 234.5 | 113 | 500 | 96.0 | 15.4 |
| Orange | 33.04 | 87.5 | 460 | 7,225 | 8,476.8 | 363.2 | 1,582 | 23,477 | 7,023.7 | 619.0 | 387 | 1,775 | 440.8 | 76.1 |
| Orleans | 1.64 | 39.4 | 17 | 303 | 170.2 | 12.5 | 87 | 1,092 | 325.4 | 27.3 | 19 | 59 | 8.0 | 1.1 |
| Oswego | 34.87 | 290.2 | 49 | 800 | 841.5 | 42.1 | 360 | 4,357 | 1,417.3 | 120.1 | 75 | 232 | 45.7 | 7.9 |
| Otsego | 3.10 | 51.1 | 37 | 370 | 143.6 | 15.7 | 272 | 3,008 | 907.9 | 82.9 | 44 | 179 | 39.3 | 7.2 |
| Putnam | 2.82 | 28.5 | 118 | 1,124 | 588.5 | 67.9 | 323 | 2,921 | 909.9 | 82.6 | 88 | 243 | 57.9 | 10.7 |
| Queens | 0.00 | 0.0 | 2,788 | 24,120 | 20,233.8 | 1,293.4 | 7,940 | 63,862 | 19,297.1 | 1,800.3 | 3,134 | 13,605 | 4,164.0 | 672.1 |
| Rensselaer | 21.61 | 134.8 | 93 | 1,056 | 1,547.3 | 66.7 | 410 | 5,420 | 1,558.4 | 142.5 | 94 | 452 | 102.3 | 21.8 |
| Richmond | 0.00 | 0.0 | 338 | 1,630 | 2,725.1 | 90.9 | 1,403 | 16,292 | 4,302.7 | 404.0 | 355 | 1,250 | 372.0 | 50.2 |
| Rockland | 33.18 | 101.8 | 476 | 4,592 | 4,966.3 | 261.5 | 1,335 | 14,543 | 4,732.5 | 425.4 | 596 | 2,527 | 497.7 | 103.8 |
| St. Lawrence | 6.95 | 62.6 | D | D | D | D | 370 | 4,818 | 1,448.8 | 124.1 | 57 | 169 | 25.3 | 4.9 |
| Saratoga | 26.12 | 115.4 | 178 | 3,698 | 2,952.2 | 228.8 | 739 | 10,803 | 3,289.0 | 299.7 | 240 | 1,138 | 335.4 | 47.6 |
| Schenectady | 23.51 | 152.1 | 80 | 675 | 508.2 | 40.1 | 444 | 6,692 | 1,938.2 | 184.5 | 121 | 519 | 129.0 | 19.5 |
| Schoharie | 173.82 | 5,548 | 14 | 112 | 83.1 | 7.2 | 100 | 1,138 | 287.0 | 26.9 | 15 | 39 | 9.5 | 2.1 |
| Schuyler | 0.83 | 45.6 | NA | NA | NA | NA | 48 | 618 | 152.7 | 16.9 | 11 | D | 3.1 | D |
| Seneca | 4.41 | 126.6 | 32 | 303 | 190.0 | 12.4 | 163 | 2,350 | 505.7 | 47.9 | 17 | 56 | 18.6 | 1.6 |
| Steuben | 7.5 | 76.8 | 43 | 319 | 135.4 | 12.5 | 319 | 4,101 | 1,229.7 | 107.5 | 54 | 204 | 29.8 | 7.5 |
| Suffolk | 246.12 | 163.9 | 2,720 | 40,301 | 34,850.1 | 2,609.6 | 6,528 | 82,839 | 27,801.1 | 2,544.4 | 1,821 | 7,080 | 2,309.2 | 388.2 |
| Sullivan | 89.77 | 1,198.9 | 38 | 781 | 300.6 | 30.0 | 283 | 2,602 | 872.2 | 76.3 | 109 | 331 | 48.8 | 10.2 |
| Tioga | 3.45 | 69.8 | 28 | 460 | 315.9 | 23.0 | 128 | 1,241 | 350.3 | 32.5 | 11 | 25 | 2.7 | 0.6 |
| Tompkins | 7.87 | 75.0 | D | D | D | D | 346 | 5,033 | 1,194.1 | 121.5 | 129 | 712 | 171.4 | 26.2 |
| Ulster | 394.23 | 2,188.4 | 153 | 1,565 | 893.5 | 82.5 | 726 | 8,896 | 2,570.5 | 242.2 | 187 | 611 | 107.7 | 21.8 |
| Warren | 11.56 | 178.7 | 52 | 566 | 292.7 | 25.6 | 402 | 5,683 | 1,590.4 | 152.4 | 76 | 256 | 53.4 | 10.9 |
| Washington | 2.59 | 41.6 | 27 | 210 | 110.1 | 10.8 | 183 | 1,944 | 603.9 | 52.6 | 22 | 48 | 7.3 | 1.5 |
| Wayne | 7.84 | 85.7 | 63 | 616 | 303.4 | 29.1 | 267 | 3,512 | 997.1 | 97.4 | 61 | 209 | 35.8 | 5.7 |
| Westchester | 65.43 | 67.0 | 1,205 | 16,216 | 23,537.3 | 2,457.1 | 3,804 | 50,878 | 16,295.6 | 1,589.2 | 2,050 | 9,804 | 2,976.5 | 556.3 |
| Wyoming | 3.26 | 79.5 | 21 | 258 | 188.1 | 13.4 | 126 | 1,378 | 392.4 | 38.3 | 19 | 111 | 16.7 | 3.6 |
| Yates | 1.10 | 43.9 | 12 | 74 | 53.7 | 2.7 | 78 | 699 | 194.6 | 18.1 | 21 | 90 | 12.6 | 4.1 |
| NORTH CAROLINA | 938.01 | 93.4 | 9,831 | 154,877 | 132,342.1 | 10,273.4 | 34,926 | 496,081 | 141,134.3 | 12,673.3 | 12,450 | 56,360 | 14,647.0 | 2,638.6 |
| Alamance | 15.35 | 97.0 | 129 | 2,065 | 814.5 | 115.5 | 642 | 9,115 | 2,313.5 | 206.7 | 134 | 849 | 213.1 | 38.7 |
| Alexander | 0.03 | 0.8 | 13 | 80 | 50.2 | 4.1 | 78 | 789 | 197.9 | 19.1 | D | D | D | D |
| Alleghany | 0.31 | 28.6 | 3 | 7 | 1.4 | 0.4 | 38 | 317 | 85.8 | 7.1 | D | D | D | 0.4 |
| Anson | 8.80 | 341.6 | 14 | 345 | 146.0 | 13.5 | 72 | 750 | 181.2 | 16.2 | 8 | 96 | 3.1 | 1.6 |
| Ashe | 0.67 | 24.8 | 17 | 83 | 79.8 | 4.5 | 106 | 1,071 | 291.0 | 25.7 | 27 | 47 | 10.4 | 1.5 |
| Avery | 1.57 | 88.8 | 9 | D | 39.2 | D | 86 | 938 | 186.6 | 20.8 | D | D | D | 4.2 |
| Beaufort | 4.31 | 90.4 | 51 | 515 | 357.1 | 21.6 | 182 | 2,362 | 637.3 | 57.6 | 39 | 155 | 25.2 | 4.1 |
| Bertie | 1.68 | 83.2 | 14 | 111 | 72.2 | 4.9 | 45 | 406 | 86.0 | 8.0 | 4 | D | 0.2 | D |
| Bladen | 37.35 | 1,088.4 | 22 | 136 | 132.2 | 5.4 | 87 | 851 | 210.3 | 17.3 | D | D | D | D |
| Brunswick | 2.80 | 22.8 | 47 | 515 | 252.8 | 24.9 | 421 | 5,221 | 1,414.2 | 133.4 | 169 | 793 | 123.0 | 27.2 |
| Buncombe | 21.12 | 83.4 | 281 | 3,566 | 1,919.2 | 174.0 | 1,276 | 18,500 | 4,594.4 | 465.1 | 550 | 1,936 | 383.0 | 77.2 |
| Burke | 23.0 | 258.9 | 46 | 840 | 301.2 | 31.9 | 255 | 3,346 | 836.6 | 75.3 | D | D | D | D |
| Cabarrus | 17.54 | 89.1 | 186 | 2,705 | 2,442.7 | 166.8 | 729 | 13,155 | 3,302.5 | 312.9 | 248 | 837 | 201.7 | 30.4 |
| Caldwell | 7.16 | 88.1 | 62 | 1,991 | 1,925.1 | 102.0 | 234 | 2,816 | 758.1 | 64.6 | 54 | 128 | 21.8 | 3.7 |
| Camden | 0.71 | 68.9 | NA | NA | NA | NA | 16 | 108 | 31.6 | 1.8 | NA | NA | NA | NA |
| Carteret | 6.97 | 101.2 | 52 | 305 | 120.6 | 12.4 | 376 | 4,371 | 1,158.4 | 109.1 | 133 | 650 | 79.5 | 18.5 |
| Caswell | 0.43 | 18.7 | D | D | D | 2.1 | 50 | 321 | 69.8 | 6.5 | NA | NA | NA | NA |
| Catawba | 4.29 | 27.7 | 233 | 4,158 | 2,525.5 | 222.7 | 675 | 9,939 | 3,015.5 | 261.7 | 203 | 725 | 179.7 | 28.3 |
| Chatham | 25.42 | 358.4 | 57 | 451 | 245.1 | 21.7 | 176 | 2,501 | 624.7 | 54.4 | 54 | 137 | 46.5 | 5.3 |
| Cherokee | 1.73 | 63.7 | 12 | 118 | 20.9 | 4.4 | 118 | 1,445 | 450.5 | 36.2 | 29 | 67 | 17.6 | 2.2 |

1 Merchant wholesalers, except manufacturers' sales branches and offices. 2. Employer establishments.

# Table B. States and Counties — Professional Services, Manufacturing, and Accommodation and Food Services

| STATE County | Professional, scientific, and technical services, 2017 | | | | Manufacturing, 2017 | | | | Accommodation and food services, 2017 | | | |
|---|---|---|---|---|---|---|---|---|---|---|---|---|
| | Number of establishments | Number of employees | Sales (mil dol) | Average payroll (mil dol) | Number of establishments | Number of employees | Sales (mil dol) | Average payroll (mil dol) | Number of establishments | Number of employees | Sales (mil dol) | Annual payroll (mil dol) |
| | 147 | 148 | 149 | 150 | 151 | 152 | 153 | 154 | 155 | 156 | 157 | 158 |
| NEW YORK— Cont'd | | | | | | | | | | | | |
| Genesee | 69 | 286 | 30 | 11.8 | 84 | 2,903 | 1,171.9 | 146.8 | 137 | 2,036 | 103.5 | 32.8 |
| Greene | 78 | 282 | 39 | 12.9 | 33 | 910 | 345.5 | 63.7 | 200 | 2,298 | 130.8 | 38.0 |
| Hamilton | NA | NA | NA | NA | NA | NA | NA | NA | 61 | 301 | 28.3 | 7.6 |
| Herkimer | 54 | 170 | 20 | 6.6 | 60 | 2,829 | 705.5 | 150.1 | 171 | 1,445 | 84.4 | 24.3 |
| Jefferson | D | D | D | D | 68 | 2,334 | 906.1 | 113.5 | 333 | 4,376 | 249.2 | 81.6 |
| Kings | 5,449 | 20,614 | 3,696 | 1,142.8 | 1,716 | 17,332 | 4,043.7 | 801.8 | 6,092 | 54,847 | 4,361.7 | 1,197.6 |
| Lewis | 20 | 101 | 15 | 4.1 | 21 | 1,254 | 629.0 | 65.5 | 62 | 547 | 26.2 | 7.8 |
| Livingston | D | D | D | D | 61 | 2,195 | 635.1 | 101.2 | 159 | 2,395 | 102.5 | 32.1 |
| Madison | D | D | D | D | 56 | 2,771 | 983.1 | 141.0 | 167 | 2,159 | 110.0 | 32.6 |
| Monroe | 2,013 | 26,170 | 4,339 | 1,682.9 | 848 | 35,328 | 12,354.4 | 2,124.2 | 1,753 | 28,857 | 1,593.4 | 492.3 |
| Montgomery | D | D | D | D | 57 | 3,272 | 776.0 | 140.7 | 107 | 1,010 | 57.1 | 15.9 |
| Nassau | 7,086 | 40,370 | 7,940 | 2,875.1 | 949 | 16,079 | 4,541.1 | 893.3 | 3,808 | 54,397 | 4,066.7 | 1,203.9 |
| New York | 17,512 | 345,282 | 116,750 | 39,074.8 | 1,721 | 17,642 | 4,450.8 | 885.4 | 10,453 | 258,047 | 27,902.3 | 8,646.6 |
| Niagara | D | D | D | D | 273 | 8,189 | 3,040.1 | 478.3 | 548 | 9,696 | 834.4 | 191.3 |
| Oneida | 423 | 3,911 | 656 | 224.9 | 229 | 9,037 | 3,009.3 | 475.2 | 602 | 12,179 | 1,007.8 | 258.1 |
| Onondaga | 1,191 | 13,590 | 3,097 | 946.2 | 417 | 17,753 | 7,228.3 | 1,071.4 | 1,187 | 20,088 | 1,137.1 | 354.4 |
| Ontario | D | D | D | D | 168 | 6,878 | 2,451.3 | 386.8 | 339 | 5,281 | 283.2 | 94.7 |
| Orange | D | D | D | D | 304 | 7,836 | 2,940.3 | 408.9 | 937 | 10,678 | 715.0 | 200.7 |
| Orleans | 40 | 148 | 12 | 4.0 | 44 | 2,069 | 688.5 | 104.7 | 67 | 671 | 35.3 | 11.1 |
| Oswego | D | D | D | D | 86 | 3,211 | 2,318.3 | 218.3 | 283 | 3,941 | 181.9 | 53.8 |
| Otsego | D | D | D | D | 63 | 960 | 233.0 | 41.9 | 217 | 2,293 | 165.5 | 44.5 |
| Putnam | D | D | D | D | 73 | 1,472 | 334.4 | 78.5 | 223 | 2,091 | 146.3 | 42.4 |
| Queens | 3,819 | 17,802 | 2,445 | 930.4 | 1,207 | 19,083 | 4,531.7 | 960.3 | 5,496 | 55,803 | 4,593.7 | 1,193.1 |
| Rensselaer | D | D | D | D | 98 | 4,115 | 1,072.5 | 375.8 | 351 | 4,733 | 274.2 | 83.1 |
| Richmond | D | D | D | D | 144 | 1,035 | 398.0 | 59.9 | 876 | 9,328 | 665.4 | 161.3 |
| Rockland | 1,345 | 5,933 | 1,767 | 398.5 | 230 | 7,442 | 12,887.5 | 529.9 | 830 | 9,332 | 647.4 | 188.2 |
| St. Lawrence | D | D | D | D | 69 | 2,159 | 993.0 | 133.1 | 244 | 3,149 | 197.4 | 46.9 |
| Saratoga | D | D | D | D | 126 | 6,899 | 2,336.4 | 562.3 | 571 | 9,672 | 689.0 | 210.3 |
| Schenectady | D | D | D | D | 101 | 4,134 | 2,483.8 | 261.0 | 337 | 4,487 | 261.4 | 77.2 |
| Schoharie | D | D | D | D | 29 | 303 | 46.1 | 12.4 | 65 | 529 | 34.9 | 8.8 |
| Schuyler | 14 | 46 | 5 | 1.7 | D | D | D | D | 80 | 659 | 45.3 | 13.1 |
| Seneca | 31 | 162 | 20 | 7.0 | 54 | 1,395 | 564.6 | 83.5 | 79 | 1,942 | 204.9 | 50.1 |
| Steuben | 125 | 1,067 | 98 | 70.2 | 78 | 3,853 | 1,159.2 | 212.7 | 235 | 2,468 | 139.2 | 40.7 |
| Suffolk | 5,668 | 36,027 | 6,471 | 2,426.0 | 1,893 | 49,115 | 17,177.1 | 2,935.6 | 3,910 | 55,011 | 3,936.7 | 1,129.8 |
| Sullivan | 157 | 549 | 60 | 21.8 | 54 | 1,017 | 451.2 | 42.7 | 254 | 1,997 | 207.2 | 51.5 |
| Tioga | D | D | D | D | 35 | 1,225 | 411.2 | 61.4 | 87 | 885 | 44.6 | 12.9 |
| Tompkins | 284 | 2,037 | 319 | 123.2 | 88 | 2,873 | 939.4 | 150.9 | 334 | 4,972 | 279.4 | 89.9 |
| Ulster | D | D | D | D | 187 | 3,298 | 806.4 | 177.2 | 591 | 7,299 | 479.1 | 155.9 |
| Warren | 147 | 886 | 118 | 44.3 | 78 | 3,481 | 1,005.5 | 181.6 | 430 | 5,120 | 420.9 | 116.8 |
| Washington | 64 | 282 | 50 | 16.9 | 88 | 2,795 | 1,137.0 | 162.6 | 116 | 694 | 42.3 | 11.4 |
| Wayne | 89 | 613 | 101 | 27.4 | 130 | 5,511 | 1,834.8 | 278.5 | 161 | 1,663 | 78.4 | 24.0 |
| Westchester | 4,088 | 24,562 | 5,441 | 2,131.8 | 551 | 13,928 | 4,639.6 | 1,002.5 | 2,667 | 33,245 | 2,561.6 | 752.5 |
| Wyoming | D | D | D | D | 47 | 1,728 | 451.0 | 80.4 | 79 | 728 | 37.1 | 11.3 |
| Yates | 26 | 108 | 12 | 5.5 | 45 | 893 | 245.2 | 40.4 | 55 | 466 | 26.1 | 7.3 |
| NORTH CAROLINA | 24,650 | 215,951 | 39,509 | 15,555.9 | 8,834 | 422,891 | 200,380.8 | 21,480.2 | 21,437 | 429,125 | 24,913.0 | 6,865.6 |
| Alamance | D | D | D | D | 193 | 8,100 | 2,841.6 | 380.2 | 326 | 7,220 | 365.7 | 103.2 |
| Alexander | D | D | D | D | 69 | 3,834 | 748.6 | 157.1 | 37 | 578 | 25.6 | 7.2 |
| Alleghany | 16 | 48 | 4 | 1.5 | 15 | 436 | 138.2 | 17.7 | 25 | 241 | 15.8 | 4.2 |
| Anson | 24 | 109 | 10 | 3.1 | 20 | 1,343 | 366.6 | 53.3 | 32 | 456 | 20.6 | 5.3 |
| Ashe | 23 | 85 | 6 | 2.2 | 22 | 814 | 213.6 | 31.3 | D | D | D | D |
| Avery | 33 | 68 | 9 | 2.8 | 17 | 124 | 11.8 | 3.6 | 58 | 691 | 46.8 | 16.0 |
| Beaufort | D | D | D | D | 53 | 2,631 | 1,075.9 | 147.8 | 92 | 1,332 | 65.8 | 17.7 |
| Bertie | 10 | 40 | 4 | 1.2 | 9 | 2,233 | 915.2 | 71.3 | D | D | D | D |
| Bladen | 34 | 145 | 21 | 5.0 | 26 | 5,321 | 1,725.8 | 203.1 | 49 | 535 | 31.9 | 8.0 |
| Brunswick | D | D | D | D | 71 | 1,232 | 393.9 | 75.4 | 295 | 4,144 | 259.6 | 69.1 |
| Buncombe | 1,050 | 4,618 | 634 | 254.9 | 318 | 12,371 | 3,995.3 | 658.7 | 846 | 18,664 | 1,177.3 | 349.6 |
| Burke | D | D | D | D | 121 | 7,288 | 2,515.1 | 312.0 | 127 | 2,788 | 138.6 | 36.9 |
| Cabarrus | D | D | D | D | 165 | 6,298 | 2,274.0 | 308.9 | 401 | 10,502 | 569.6 | 154.2 |
| Caldwell | 83 | 367 | 65 | 14.1 | 117 | 6,997 | 1,497.3 | 279.9 | 120 | 2,095 | 97.7 | 26.1 |
| Camden | 11 | 30 | 3 | 1.4 | NA | NA | NA | NA | 7 | 86 | 4.2 | 1.2 |
| Carteret | 134 | 616 | 90 | 29.9 | 63 | 753 | 221.7 | 28.6 | 245 | 3,728 | 211.9 | 65.1 |
| Caswell | 12 | 39 | 4 | 1.5 | 9 | 211 | 33.5 | 7.6 | D | D | D | D |
| Catawba | 334 | 2,219 | 779 | 119.6 | 368 | 22,637 | 6,606.2 | 1,008.6 | 361 | 7,310 | 378.0 | 109.6 |
| Chatham | D | D | D | D | 78 | 1,969 | 583.7 | 85.7 | 95 | 1,539 | 86.3 | 25.4 |
| Cherokee | 32 | 89 | 9 | 3.1 | 25 | 1,278 | 362.7 | 56.4 | 65 | 1,867 | 230.4 | 45.3 |

# Health Care and Social Assistance, Other Services, Nonemployer Businesses, and Residential Construction

| STATE County | Health care and social assistance, 2017 | | | | Other services, 2017 | | | | Nonemployer businesses, 2018 | | Value of residential construction authorized by building permits, 2020 | |
| --- | --- | --- | --- | --- | --- | --- | --- | --- | --- | --- | --- | --- |
| | Number of establish-ments | Number of employees | Receipts (mil dol) | Annual payroll (mil dol) | Number of establish-ments | Number of employees | Receipts (mil dol) | Annual payroll (mil dol) | Number | Receipts (mil dol) | New construction ($1,000) | Number of housing units |
| | 159 | 160 | 161 | 162 | 163 | 164 | 165 | 166 | 167 | 168 | 169 | 170 |
| NEW YORK— Cont'd | | | | | | | | | | | | |
| Genesee | 138 | 3,141 | 270.2 | 114.9 | 93 | 572 | 54.3 | 14.8 | 2,958 | 130.6 | 12,129 | 89 |
| Greene | 88 | 1,210 | 100.1 | 41.2 | 82 | 329 | 36.4 | 9.4 | 3,681 | 160.3 | 15,526 | 68 |
| Hamilton | 6 | 27 | 1.6 | 1.0 | D | D | D | D | 506 | 20.5 | 14,118 | 48 |
| Herkimer | 120 | 2,244 | 188.5 | 71.0 | 99 | 702 | 40.7 | 12.5 | 3,263 | 142.7 | 31,476 | 77 |
| Jefferson | 273 | 6,535 | 660.4 | 304.3 | 183 | 866 | 102.3 | 22.8 | 5,293 | 213.4 | 16,547 | 129 |
| Kings | 6,877 | 229,944 | 20,989.5 | 9,197.5 | 5,125 | 21,666 | 2,169.1 | 573.3 | 288,852 | 14,340.1 | 753,530 | 6,802 |
| Lewis | 45 | 922 | 109.6 | 45.0 | 46 | 186 | 28.9 | 6.3 | 1,677 | 76.0 | 10,540 | 86 |
| Livingston | 132 | 2,090 | 164.8 | 74.4 | 97 | 375 | 40.7 | 10.2 | 3,574 | 152.6 | 11,938 | 50 |
| Madison | 176 | 3,151 | 312.4 | 128.0 | 105 | 374 | 38.0 | 9.6 | 4,350 | 181.9 | 18,358 | 80 |
| Monroe | 1,992 | 72,084 | 7,179.0 | 3,025.4 | 1,297 | 7,911 | 974.8 | 254.5 | 47,781 | 2,194.7 | 284,177 | 1,645 |
| Montgomery | 186 | 4,059 | 414.4 | 173.8 | 88 | 600 | 63.8 | 17.7 | 2,466 | 103.6 | 8,968 | 43 |
| Nassau | 5,953 | 119,939 | 15,609.6 | 6,621.3 | 4,315 | 24,423 | 2,756.9 | 755.5 | 150,025 | 10,716.4 | 430,676 | 1,338 |
| New York | 8,474 | 284,981 | 44,469.8 | 17,081.4 | 9,976 | 102,738 | 29,591.3 | 5,909.8 | 232,503 | 19,976.9 | 224,779 | 1,896 |
| Niagara | 509 | 10,709 | 873.7 | 402.3 | 331 | 1,393 | 126.1 | 34.0 | 10,336 | 405.2 | 37,863 | 175 |
| Oneida | 599 | 20,938 | 1,934.4 | 903.0 | 424 | 4,958 | 308.9 | 106.9 | 12,400 | 520.0 | 54,382 | 337 |
| Onondaga | 1,288 | 41,752 | 5,179.4 | 2,129.3 | 918 | 5,994 | 675.9 | 189.7 | 29,848 | 1,424.1 | 75,037 | 352 |
| Ontario | 266 | 8,686 | 755.8 | 387.5 | 230 | 1,348 | 121.7 | 37.8 | 7,248 | 322.1 | 76,298 | 322 |
| Orange | 987 | 24,486 | 2,906.3 | 1,182.6 | 799 | 4,832 | 639.9 | 160.0 | 26,842 | 1,310.0 | 216,507 | 1,106 |
| Orleans | 75 | 1,313 | 92.8 | 41.5 | 61 | 240 | 25.7 | 5.8 | 1,910 | 74.9 | 6,248 | 33 |
| Oswego | 214 | 5,054 | 387.5 | 184.3 | 196 | 705 | 75.4 | 18.9 | 5,464 | 208.5 | 68,374 | 345 |
| Otsego | 154 | 5,925 | 774.1 | 341.8 | 101 | 484 | 97.6 | 11.9 | 4,274 | 174.4 | 23,210 | 100 |
| Putnam | 252 | 5,250 | 609.9 | 257.5 | 257 | 1,147 | 126.7 | 36.8 | 9,661 | 481.0 | 10,625 | 40 |
| Queens | 5,592 | 155,364 | 15,735.8 | 6,645.8 | 5,184 | 21,641 | 2,442.3 | 626.7 | 266,193 | 11,028.9 | 715,879 | 5,840 |
| Rensselaer | 358 | 8,633 | 870.3 | 361.9 | 245 | 1,379 | 135.8 | 52.4 | 9,244 | 370.8 | 33,427 | 164 |
| Richmond | 1,345 | 33,206 | 3,540.7 | 1,528.2 | 912 | 4,484 | 441.4 | 108.4 | 38,651 | 1,882.7 | 70,364 | 408 |
| Rockland | 1,201 | 28,807 | 2,706.3 | 1,207.7 | 781 | 3,325 | 358.9 | 102.6 | 29,017 | 1,776.5 | 65,864 | 406 |
| St. Lawrence | 234 | 6,667 | 708.9 | 315.1 | 171 | 713 | 73.1 | 16.3 | 4,821 | 180.4 | 20,651 | 166 |
| Saratoga | 620 | 10,603 | 1,149.2 | 456.5 | 358 | 2,056 | 246.5 | 68.1 | 17,073 | 872.1 | 238,664 | 1,000 |
| Schenectady | 429 | 12,036 | 1,179.6 | 535.1 | D | D | 214.2 | D | 8,859 | 357.2 | 36,649 | 192 |
| Schoharie | 60 | 887 | 71.9 | 30.2 | D | D | 13.3 | D | 1,865 | 71.4 | 8,689 | 39 |
| Schuyler | 27 | 685 | 55.6 | 26.8 | D | D | D | D | 1,252 | 45.4 | 9,474 | 60 |
| Seneca | 64 | 1,362 | 85.2 | 40.4 | 56 | 241 | 27.5 | 8.7 | 1,901 | 91.9 | 9,143 | 56 |
| Steuben | 237 | 6,707 | 572.6 | 243.0 | 160 | 831 | 80.7 | 18.6 | 5,305 | 209.0 | 29,234 | 219 |
| Suffolk | 4,915 | 108,385 | 13,050.8 | 5,573.8 | 4,251 | 21,465 | 2,416.5 | 664.5 | 137,044 | 7,904.0 | 501,651 | 590 |
| Sullivan | 233 | 5,033 | 425.8 | 204.1 | 149 | 456 | 64.1 | 14.0 | 5,670 | 246.3 | 81,963 | 407 |
| Tioga | 70 | 1,024 | 64.8 | 30.7 | 59 | 168 | 15.5 | 3.6 | 2,764 | 112.5 | 21,300 | 133 |
| Tompkins | 284 | 6,258 | 643.2 | 273.4 | 164 | 962 | 197.3 | 31.3 | 7,936 | 307.9 | 47,602 | 329 |
| Ulster | 525 | 10,035 | 841.2 | 383.2 | 363 | 1,619 | 149.2 | 40.3 | 16,953 | 718.6 | 69,731 | 265 |
| Warren | 279 | 6,302 | 774.3 | 305.2 | 150 | 864 | 101.6 | 30.0 | 4,924 | 237.7 | 55,384 | 213 |
| Washington | 101 | 1,713 | 105.6 | 56.1 | 75 | 265 | 32.0 | 8.3 | 3,729 | 149.9 | 22,350 | 93 |
| Wayne | 140 | 3,229 | 228.1 | 106.7 | 131 | 396 | 38.7 | 10.2 | 4,818 | 203.2 | 20,899 | 114 |
| Westchester | 3,532 | 80,551 | 10,819.9 | 4,784.1 | 2,967 | 16,068 | 2,279.4 | 591.9 | 102,917 | 7,030.4 | 553,914 | 2,834 |
| Wyoming | 65 | 1,289 | 130.5 | 45.4 | 68 | 260 | 28.5 | 7.2 | 2,101 | 85.6 | 6,423 | 41 |
| Yates | 44 | 904 | 67.7 | 29.7 | 41 | 136 | 13.5 | 3.2 | 2,032 | 103.2 | 17,964 | 78 |
| NORTH CAROLINA | 24,080 | 602,444 | 72,732.3 | 27,627.2 | 15,118 | 93,642 | 13,022.2 | 3,152.9 | 787,883 | 34,400.9 | 16,111,594 | 80,474 |
| Alamance | 366 | 12,073 | 1,705.6 | 651.3 | 199 | 1,292 | 125.7 | 36.1 | 10,616 | 408.9 | 267,601 | 1,784 |
| Alexander | 47 | 648 | 42.5 | 18.3 | 35 | 159 | 11.8 | 2.9 | 2,371 | 98.2 | 20,082 | 117 |
| Alleghany | 27 | 400 | 27.0 | 12.0 | D | D | D | 1.2 | 954 | 40.3 | 12,978 | 49 |
| Anson | 49 | 814 | 54.0 | 22.0 | 24 | 53 | 8.2 | 1.2 | 1,292 | 43.5 | 8,243 | 55 |
| Ashe | 48 | 1,167 | 84.1 | 37.9 | 33 | 134 | 11.3 | 3.1 | 2,507 | 105.1 | 33,799 | 118 |
| Avery | 41 | 798 | 58.3 | 23.0 | 23 | 140 | 18.1 | 4.8 | 1,720 | 66.8 | 72,901 | 152 |
| Beaufort | 119 | 2,049 | 218.4 | 76.3 | 75 | 364 | 33.2 | 10.2 | 3,337 | 134.7 | 29,456 | 168 |
| Bertie | 43 | 955 | 64.9 | 29.0 | D | D | D | D | 959 | 29.7 | 4,295 | 23 |
| Bladen | 55 | 936 | 47.7 | 23.2 | 25 | 122 | 11.4 | 3.2 | 1,778 | 64.0 | 17,843 | 82 |
| Brunswick | 220 | 3,902 | 400.5 | 153.0 | 145 | 654 | 69.2 | 19.1 | 11,453 | 509.0 | 809,396 | 3,720 |
| Buncombe | 1,003 | 28,143 | 3,678.0 | 1,482.0 | 517 | 2,958 | 326.5 | 96.9 | 28,970 | 1,292.7 | 550,368 | 2,546 |
| Burke | 167 | 5,422 | 644.9 | 259.5 | 90 | 528 | 58.0 | 14.5 | 5,230 | 219.4 | 50,778 | 302 |
| Cabarrus | 392 | 9,447 | 1,119.5 | 482.2 | 296 | 1,725 | 176.8 | 47.9 | 16,583 | 631.6 | 418,862 | 3,134 |
| Caldwell | 131 | 3,226 | 271.5 | 104.2 | 88 | 466 | 49.1 | 13.8 | 4,984 | 205.1 | 53,205 | 231 |
| Camden | D | D | D | 1.0 | D | D | 3.9 | D | 730 | 22.0 | 28,620 | 101 |
| Carteret | 198 | 3,364 | 372.3 | 148.9 | 142 | 687 | 70.3 | 18.8 | 6,772 | 313.1 | 140,275 | 580 |
| Caswell | 30 | 384 | 32.0 | 12.1 | D | D | D | 0.5 | 1,149 | 38.7 | 6,996 | 30 |
| Catawba | 419 | 12,285 | 1,390.3 | 534.5 | 247 | 1,556 | 133.6 | 41.1 | 10,790 | 514.4 | 156,521 | 917 |
| Chatham | 137 | 2,702 | 215.8 | 79.5 | 100 | 411 | 39.9 | 12.6 | 6,553 | 295.3 | 201,673 | 601 |
| Cherokee | 71 | 1,587 | 143.1 | 57.1 | 36 | 131 | 10.5 | 3.2 | 2,292 | 93.5 | 39,417 | 247 |

# Table B. States and Counties — Government Employment and Payroll, and Local Government Finances

| | Government employment and payroll, 2017 | | | | | | | | | Local government finances, 2017 | | | | |
| | | | March payroll (percent of total) | | | | | | | General revenue | | | | |
| | | | | | | | | | | | | Taxes | | |
| | | | | | | | | | | | | | Per capita[1] (dollars) | |
| STATE County | Full-time equivalent employees | March payroll (dollars) | Administration, judicial, and legal | Police and corrections | Fire protection | Highways and transportation | Health and welfare | Natural resources and utilities | Education and libraries | Total (mil dol) | Inter-govern-mental (mil dol) | Total (mil dol) | Total | Property |
| | 171 | 172 | 173 | 174 | 175 | 176 | 177 | 178 | 179 | 180 | 181 | 182 | 183 | 184 |
| NEW YORK— Cont'd | | | | | | | | | | | | | | |
| Genesee | 4,222 | 19,296,881 | 3.9 | 4.9 | 1.2 | 3.4 | 3.5 | 1.1 | 49.5 | 405.7 | 182.5 | 156.1 | 2,700 | 1,663 |
| Greene | 2,298 | 10,115,224 | 7.5 | 6.4 | 0.0 | 7.3 | 10.3 | 2.2 | 65.1 | 316.4 | 105.8 | 172.3 | 3,632 | 2,887 |
| Hamilton | 491 | 2,080,685 | 13.7 | 4.2 | 0.7 | 18.1 | 7.9 | 8.2 | 46.0 | 62.5 | 12.4 | 46.3 | 10,356 | 9,408 |
| Herkimer | 2,865 | 11,466,437 | 6.7 | 6.1 | 2.3 | 7.7 | 6.3 | 3.6 | 65.4 | 385.8 | 191.0 | 155.9 | 2,507 | 1,851 |
| Jefferson | 5,641 | 24,270,939 | 5.8 | 5.5 | 1.8 | 7.2 | 5.7 | 3.2 | 67.8 | 739.1 | 370.0 | 279.4 | 2,469 | 1,419 |
| Kings | (2) | (2) | (2) | (2) | (2) | (2) | (2) | (2) | (2) | (2) | (2) | (2) | (2) | (2) |
| Lewis | 1,898 | 7,686,343 | 5.5 | 2.9 | 0.7 | 6.9 | 35.7 | 0.7 | 47.2 | 268.6 | 100.0 | 62.4 | 2,345 | 1,865 |
| Livingston | 3,484 | 13,651,324 | 6.4 | 7.3 | 0.0 | 5.5 | 15.4 | 1.5 | 62.7 | 406.5 | 171.2 | 146.6 | 2,310 | 1,765 |
| Madison | 2,857 | 12,223,460 | 5.5 | 7.3 | 1.1 | 8.0 | 6.4 | 2.9 | 66.0 | 381.6 | 167.1 | 177.9 | 2,507 | 1,913 |
| Monroe | 34,542 | 172,090,209 | 3.6 | 10.0 | 3.6 | 2.5 | 5.0 | 4.0 | 69.4 | 4,816.6 | 2,030.9 | 2,185.7 | 2,943 | 1,917 |
| Montgomery | 2,178 | 9,914,847 | 7.3 | 8.6 | 1.5 | 5.5 | 4.4 | 2.8 | 67.4 | 305.6 | 147.2 | 120.3 | 2,446 | 1,580 |
| Nassau | 61,582 | 417,805,788 | 3.4 | 12.5 | 0.8 | 2.1 | 6.7 | 4.4 | 67.9 | 11,881.8 | 2,704.9 | 7,683.7 | 5,661 | 4,662 |
| New York | (2) 442,294 | (2) 80,474,536 | (2) 3.4 | (2) 16.2 | (2) 4.8 | (2) 16.0 | (2) 17.3 | (2) 4.8 | (2) 33.6 | (2) 111,321.9 | (2) 37,041.5 | (2) 55,325.6 | (2) 6,557 | (2) 2,932 |
| Niagara | 8,142 | 40,535,142 | 5.1 | 9.4 | 3.1 | 4.3 | 6.2 | 3.8 | 66.6 | 1,306.8 | 582.1 | 541.3 | 2,567 | 1,656 |
| Oneida | 10,587 | 47,343,125 | 5.5 | 8.1 | 2.7 | 4.7 | 5.1 | 4.6 | 68.2 | 1,401.0 | 695.4 | 542.2 | 2,357 | 1,488 |
| Onondaga | 21,209 | 104,071,695 | 2.9 | 9.0 | 2.4 | 4.1 | 5.1 | 4.9 | 70.8 | 2,901.5 | 1,289.4 | 1,280.2 | 2,772 | 1,798 |
| Ontario | 6,340 | 27,805,597 | 6.5 | 7.4 | 1.0 | 4.2 | 3.9 | 2.8 | 73.0 | 690.4 | 243.5 | 361.6 | 3,301 | 2,172 |
| Orange | 16,171 | 90,101,768 | 4.4 | 9.8 | 1.4 | 3.8 | 6.2 | 2.4 | 70.9 | 2,752.3 | 998.8 | 1,462.4 | 3,851 | 2,871 |
| Orleans | 2,190 | 9,766,804 | 4.9 | 6.6 | 0.8 | 4.2 | 5.2 | 1.4 | 75.4 | 230.7 | 125.9 | 80.1 | 1,965 | 1,495 |
| Oswego | 6,356 | 26,849,806 | 3.9 | 5.3 | 2.1 | 5.1 | 4.8 | 2.6 | 73.7 | 798.7 | 378.4 | 284.3 | 2,400 | 1,786 |
| Otsego | 2,680 | 10,255,860 | 6.3 | 4.6 | 2.2 | 9.1 | 12.9 | 1.2 | 61.7 | 335.9 | 145.4 | 153.2 | 2,557 | 1,734 |
| Putnam | 3,769 | 22,766,685 | 6.7 | 8.5 | 0.0 | 6.3 | 5.8 | 1.9 | 68.8 | 687.0 | 170.1 | 479.8 | 4,854 | 4,138 |
| Queens | (2) | (2) | (2) | (2) | (2) | (2) | (2) | (2) | (2) | (2) | (2) | (2) | (2) | (2) |
| Rensselaer | 7,542 | 36,126,286 | 4.8 | 7.7 | 1.8 | 2.7 | 10.2 | 2.2 | 68.6 | 1,032.5 | 404.8 | 444.2 | 2,790 | 2,005 |
| Richmond | (2) | (2) | (2) | (2) | (2) | (2) | (2) | (2) | (2) | (2) | (2) | (2) | (2) | (2) |
| Rockland | 11,549 | 73,992,763 | 6.3 | 11.6 | 0.1 | 3.4 | 4.6 | 3.6 | 69.8 | 2,326.2 | 571.1 | 1,551.6 | 4,780 | 3,970 |
| St. Lawrence | 5,416 | 22,595,829 | 4.9 | 5.5 | 1.0 | 7.2 | 16.8 | 2.7 | 60.3 | 720.6 | 338.2 | 261.5 | 2,405 | 1,632 |
| Saratoga | 8,979 | 40,877,513 | 4.9 | 5.8 | 1.2 | 5.8 | 4.0 | 2.1 | 74.3 | 1,200.7 | 368.1 | 680.9 | 2,970 | 2,065 |
| Schenectady | 6,037 | 31,374,949 | 4.2 | 10.7 | 3.6 | 3.7 | 8.6 | 2.0 | 65.1 | 945.6 | 365.1 | 456.5 | 2,951 | 2,096 |
| Schoharie | 1,516 | 7,162,811 | 7.2 | 4.6 | 0.1 | 8.4 | 7.2 | 1.9 | 69.5 | 215.1 | 103.2 | 94.1 | 3,011 | 2,453 |
| Schuyler | 735 | 2,945,335 | 10.5 | 8.4 | 0.7 | 8.7 | 14.5 | 2.7 | 51.1 | 111.3 | 48.0 | 49.5 | 2,761 | 1,968 |
| Seneca | 1,446 | 5,681,942 | 7.9 | 12.0 | 0.0 | 5.7 | 10.9 | 3.0 | 57.3 | 208.6 | 96.4 | 84.9 | 2,480 | 1,633 |
| Steuben | 4,733 | 22,321,354 | 5.8 | 5.4 | 1.5 | 6.8 | 5.6 | 2.7 | 70.6 | 673.6 | 336.1 | 260.5 | 2,703 | 1,914 |
| Suffolk | 65,773 | 423,548,830 | 4.0 | 11.6 | 1.0 | 2.6 | 3.9 | 2.8 | 72.6 | 11,679.6 | 3,371.4 | 7,263.3 | 4,897 | 3,824 |
| Sullivan | 4,046 | 20,010,366 | 6.5 | 7.6 | 0.2 | 7.4 | 9.0 | 3.2 | 64.4 | 648.9 | 213.8 | 319.0 | 4,254 | 3,630 |
| Tioga | 2,181 | 8,320,257 | 4.4 | 6.8 | 0.1 | 5.4 | 8.9 | 1.3 | 71.1 | 285.2 | 150.5 | 106.5 | 2,191 | 1,651 |
| Tompkins | 4,615 | 22,463,397 | 5.1 | 5.9 | 2.8 | 4.9 | 6.6 | 3.9 | 68.1 | 599.8 | 194.0 | 318.4 | 3,102 | 2,262 |
| Ulster | 7,944 | 40,510,439 | 5.8 | 8.2 | 1.2 | 5.7 | 4.1 | 2.2 | 71.0 | 1,172.0 | 385.2 | 704.9 | 3,946 | 3,148 |
| Warren | 3,011 | 13,531,587 | 7.4 | 10.1 | 4.0 | 7.9 | 6.2 | 5.6 | 57.7 | 450.3 | 129.8 | 277.7 | 4,314 | 2,860 |
| Washington | 3,587 | 14,776,257 | 4.7 | 5.1 | 0.0 | 5.2 | 4.8 | 1.1 | 77.0 | 381.4 | 174.1 | 145.2 | 2,359 | 1,937 |
| Wayne | 4,441 | 18,062,784 | 5.4 | 5.5 | 0.1 | 4.7 | 9.8 | 1.9 | 71.2 | 565.1 | 261.1 | 218.6 | 2,417 | 1,816 |
| Westchester | 45,049 | 315,904,961 | 4.5 | 12.1 | 4.1 | 2.1 | 12.5 | 4.7 | 58.0 | 9,763.8 | 2,045.4 | 5,422.8 | 5,592 | 4,506 |
| Wyoming | 1,898 | 7,449,778 | 5.8 | 7.5 | 0.0 | 7.5 | 30.6 | 2.8 | 44.8 | 251.8 | 96.9 | 81.0 | 2,011 | 1,534 |
| Yates | 888 | 3,891,464 | 9.4 | 12.2 | 0.2 | 8.8 | 5.9 | 3.5 | 58.4 | 125.9 | 49.2 | 66.4 | 2,655 | 2,101 |
| NORTH CAROLINA | X | X | X | X | X | X | X | X | X | X | X | X | X | X |
| Alamance | 5,370 | 20,215,032 | 4.4 | 10.8 | 3.3 | 1.7 | 8.5 | 5.8 | 62.4 | 467.6 | 226.6 | 184.2 | 1,128 | 793 |
| Alexander | 1,064 | 3,474,220 | 3.9 | 8.7 | 0.4 | 0.0 | 13.8 | 2.3 | 68.4 | 85.9 | 47.0 | 30.6 | 824 | 595 |
| Alleghany | 370 | 1,198,979 | 4.8 | 10.0 | 0.0 | 1.9 | 10.5 | 3.2 | 65.9 | 33.0 | 16.8 | 12.4 | 1,130 | 847 |
| Anson | 1,234 | 4,221,291 | 2.9 | 6.4 | 2.6 | 0.6 | 7.3 | 4.9 | 73.7 | 61.3 | 48.9 | 4.5 | 180 | 114 |
| Ashe | 713 | 2,611,599 | 5.9 | 8.7 | 0.0 | 1.1 | 5.5 | 6.8 | 69.6 | 71.1 | 37.0 | 27.4 | 1,023 | 722 |
| Avery | 699 | 2,156,864 | 6.3 | 10.1 | 0.0 | 1.3 | 11.8 | 5.1 | 60.9 | 54.8 | 23.7 | 26.2 | 1,498 | 1,131 |
| Beaufort | 2,154 | 6,621,286 | 5.8 | 8.5 | 1.9 | 0.7 | 11.6 | 7.8 | 61.3 | 162.7 | 85.3 | 54.7 | 1,162 | 888 |
| Bertie | 662 | 2,275,555 | 4.1 | 7.4 | 0.2 | 1.1 | 13.7 | 5.1 | 65.4 | 56.1 | 32.5 | 15.6 | 812 | 619 |
| Bladen | 1,401 | 4,540,093 | 3.8 | 8.3 | 1.0 | 0.5 | 16.7 | 2.9 | 65.2 | 117.1 | 64.5 | 35.1 | 1,050 | 811 |
| Brunswick | 3,843 | 15,105,174 | 6.7 | 10.1 | 1.5 | 1.3 | 15.6 | 6.6 | 53.6 | 467.2 | 138.6 | 205.5 | 1,571 | 1,174 |
| Buncombe | 9,182 | 35,550,616 | 5.5 | 8.0 | 3.7 | 1.6 | 17.3 | 4.3 | 54.9 | 1,308.2 | 735.4 | 417.9 | 1,626 | 1,061 |
| Burke | 2,789 | 9,414,475 | 6.7 | 3.9 | 1.0 | 0.9 | 6.6 | 6.9 | 72.1 | 253.3 | 129.2 | 82.4 | 914 | 668 |
| Cabarrus | 8,191 | 32,543,046 | 3.8 | 7.0 | 4.4 | 1.3 | 19.4 | 5.3 | 55.8 | 737.9 | 293.6 | 313.8 | 1,516 | 1,139 |
| Caldwell | 3,080 | 10,733,278 | 3.7 | 6.6 | 2.1 | 0.6 | 9.5 | 3.8 | 72.2 | 238.7 | 137.0 | 77.1 | 941 | 718 |
| Camden | 377 | 1,104,342 | 26.9 | 0.0 | 0.0 | 0.0 | 0.0 | 0.3 | 71.7 | 29.5 | 17.2 | 10.5 | 995 | 799 |
| Carteret | 2,545 | 9,164,720 | 6.1 | 8.8 | 6.0 | 1.8 | 8.0 | 4.1 | 61.2 | 386.4 | 89.4 | 115.4 | 1,674 | 1,187 |
| Caswell | 726 | 2,304,482 | 3.8 | 7.9 | 0.0 | 0.0 | 12.6 | 2.4 | 63.0 | 50.7 | 29.5 | 16.1 | 712 | 516 |
| Catawba | 7,861 | 31,418,018 | 3.4 | 5.3 | 2.9 | 1.4 | 40.0 | 3.7 | 41.6 | 784.2 | 236.7 | 202.3 | 1,281 | 912 |
| Chatham | 1,925 | 6,848,499 | 4.6 | 9.1 | 0.3 | 0.5 | 8.5 | 5.2 | 66.7 | 198.1 | 76.4 | 96.9 | 1,361 | 1,090 |
| Cherokee | 1,138 | 3,665,143 | 3.9 | 7.6 | 0.2 | 0.8 | 8.0 | 8.4 | 63.9 | 91.8 | 51.7 | 29.6 | 1,058 | 695 |

1. Based on the resident population estimated as of July 1 of the year shown.　　2. Bronx, Kings, Queens, and Richmond counties are included with New York county.

Items 171—184

| STATE County | Local government finances, 2017 (cont.) | | | | | | | | | Government employment, 2019 | | | Individual income tax returns, 2018 | | |
|---|---|---|---|---|---|---|---|---|---|---|---|---|---|---|---|
| | Direct general expenditure | | | | | | | Debt outstanding | | | | | | | |
| | | | Percent of total for: | | | | | | | | | | | | |
| | Total (mil dol) | Per capita[1] (dollars) | Education | Health and hospitals | Police protection | Public welfare | Highways | Total (mil dol) | Per capita[1] (dollars) | Federal civilian | Federal military | State and local | Number of returns | Mean adjusted gross income | Mean income tax |
| | 185 | 186 | 187 | 188 | 189 | 190 | 191 | 192 | 193 | 194 | 195 | 196 | 197 | 198 | 199 |
| NEW YORK— Cont'd | | | | | | | | | | | | | | | |
| Genesee | 433 | 7,488 | 57.4 | 2.7 | 2.8 | 10.8 | 5.8 | 219.8 | 3,804 | 576 | 88 | 4,881 | 28,050 | 53,795 | 4,639 |
| Greene | 313 | 6,586 | 47.9 | 6.0 | 2.3 | 8.6 | 8.3 | 163.9 | 3,454 | 90 | 70 | 4,190 | 22,540 | 57,448 | 5,705 |
| Hamilton | 62 | 13,781 | 33.2 | 5.1 | 2.1 | 2.4 | 16.4 | 13.4 | 2,996 | 15 | 7 | 831 | 2,560 | 55,510 | 4,995 |
| Herkimer | 397 | 6,384 | 60.9 | 2.1 | 1.8 | 8.2 | 8.4 | 165.2 | 2,658 | 111 | 95 | 4,171 | 27,980 | 52,521 | 4,647 |
| Jefferson | 719 | 6,352 | 52.7 | 3.0 | 2.3 | 8.1 | 6.8 | 408.3 | 3,608 | 3,240 | 15,314 | 8,295 | 50,080 | 51,658 | 4,348 |
| Kings | (2) | (2) | (2) | (2) | (2) | (2) | (2) | (2) | (2) | 7,815 | 4,196 | 116,790 | 1,256,080 | 65,709 | 8,989 |
| Lewis | 326 | 12,263 | 28.5 | 42.4 | 1.0 | 4.9 | 5.6 | 105.0 | 3,947 | 60 | 41 | 2,359 | 11,670 | 50,965 | 4,041 |
| Livingston | 397 | 6,250 | 44.7 | 6.1 | 4.1 | 16.4 | 7.1 | 214.1 | 3,372 | 130 | 90 | 6,058 | 28,080 | 59,184 | 5,565 |
| Madison | 378 | 5,328 | 52.4 | 3.0 | 2.8 | 7.2 | 8.3 | 213.9 | 3,015 | 139 | 104 | 4,207 | 31,280 | 63,292 | 6,937 |
| Monroe | 5,085 | 6,846 | 50.9 | 4.2 | 4.3 | 8.9 | 3.0 | 3,146.9 | 4,237 | 2,921 | 1,236 | 45,617 | 366,870 | 66,769 | 7,558 |
| Montgomery | 313 | 6,356 | 50.3 | 2.2 | 2.3 | 8.2 | 6.7 | 208.4 | 4,238 | 107 | 76 | 2,652 | 22,820 | 49,390 | 4,271 |
| Nassau | 12,275 | 9,043 | 49.2 | 6.4 | 8.7 | 5.1 | 3.4 | 8,617.9 | 6,349 | 5,076 | 2,563 | 75,172 | 724,700 | 121,278 | 21,129 |
| New York | (2)105,453 | (2)12,498 | (2)30.2 | (2)9.6 | (2)5.4 | (2)13.4 | (2)2.8 | (2)181,077.3 | (2)21,461 | 22,116 | 2,692 | 215,127 | 894,310 | 228,933 | 51,360 |
| Niagara | 1,283 | 6,084 | 48.3 | 3.0 | 4.6 | 9.2 | 6.1 | 758.8 | 3,599 | 1,052 | 366 | 12,266 | 105,420 | 55,713 | 5,219 |
| Oneida | 1,378 | 5,990 | 52.7 | 2.2 | 3.2 | 11.1 | 4.5 | 1,087.6 | 4,728 | 2,348 | 409 | 24,098 | 105,520 | 55,969 | 5,429 |
| Onondaga | 3,170 | 6,864 | 48.3 | 3.1 | 3.4 | 9.2 | 4.6 | 2,019.1 | 4,372 | 4,638 | 827 | 36,811 | 224,800 | 68,487 | 7,967 |
| Ontario | 685 | 6,250 | 56.0 | 2.1 | 2.8 | 7.0 | 6.4 | 460.0 | 4,199 | 1,664 | 170 | 7,168 | 55,670 | 70,563 | 8,226 |
| Orange | 2,852 | 7,510 | 53.5 | 5.1 | 3.7 | 9.6 | 3.3 | 1,448.1 | 3,813 | 4,461 | 6,818 | 22,145 | 185,690 | 72,513 | 8,479 |
| Orleans | 230 | 5,645 | 55.2 | 3.3 | 2.3 | 9.3 | 6.8 | 168.1 | 4,121 | 79 | 60 | 3,773 | 18,440 | 48,639 | 3,837 |
| Oswego | 740 | 6,247 | 56.0 | 2.5 | 2.2 | 8.8 | 6.0 | 495.9 | 4,188 | 269 | 212 | 9,066 | 52,880 | 53,883 | 4,721 |
| Otsego | 321 | 5,350 | 51.5 | 3.4 | 1.7 | 8.2 | 9.0 | 174.7 | 2,915 | 132 | 86 | 4,395 | 26,500 | 55,631 | 5,358 |
| Putnam | 683 | 6,905 | 61.8 | 3.1 | 4.7 | 4.0 | 5.0 | 279.1 | 2,824 | 151 | 151 | 4,541 | 50,630 | 96,524 | 13,362 |
| Queens | (2) | (2) | (2) | (2) | (2) | (2) | (2) | (2) | (2) | 13,989 | 3,591 | 112,715 | 1,181,690 | 54,741 | 5,833 |
| Rensselaer | 1,044 | 6,557 | 50.3 | 4.8 | 3.1 | 13.0 | 3.7 | 826.7 | 5,193 | 375 | 257 | 11,328 | 79,010 | 64,383 | 6,704 |
| Richmond | (2) | (2) | (2) | (2) | (2) | (2) | (2) | (2) | (2) | 933 | 1,175 | 22,871 | 228,610 | 73,528 | 8,819 |
| Rockland | 2,466 | 7,597 | 46.4 | 9.1 | 4.7 | 6.8 | 3.8 | 1,690.8 | 5,209 | 489 | 505 | 17,428 | 149,670 | 94,641 | 13,741 |
| St. Lawrence | 706 | 6,497 | 50.1 | 9.6 | 2.0 | 8.6 | 7.5 | 454.1 | 4,178 | 598 | 261 | 10,012 | 43,160 | 53,963 | 4,794 |
| Saratoga | 1,119 | 4,882 | 57.3 | 3.1 | 3.4 | 5.5 | 5.8 | 637.4 | 2,780 | 427 | 2,130 | 12,035 | 119,950 | 103,568 | 15,937 |
| Schenectady | 1,003 | 6,483 | 49.7 | 5.2 | 3.7 | 13.1 | 3.5 | 639.6 | 4,134 | 584 | 247 | 9,645 | 77,430 | 63,277 | 6,547 |
| Schoharie | 203 | 6,504 | 49.6 | 3.0 | 1.2 | 8.8 | 9.6 | 106.8 | 3,418 | 71 | 47 | 2,635 | 13,980 | 53,678 | 4,850 |
| Schuyler | 101 | 5,655 | 38.5 | 5.1 | 2.0 | 10.2 | 10.3 | 118.6 | 6,620 | 42 | 28 | 1,090 | 8,720 | 52,973 | 4,742 |
| Seneca | 201 | 5,866 | 48.3 | 4.1 | 3.3 | 8.9 | 6.1 | 139.1 | 4,062 | 80 | 49 | 2,879 | 15,470 | 51,853 | 4,548 |
| Steuben | 661 | 6,863 | 57.3 | 3.4 | 1.7 | 8.8 | 8.9 | 424.3 | 4,403 | 1,038 | 170 | 6,668 | 43,540 | 59,526 | 6,226 |
| Suffolk | 11,755 | 7,925 | 56.1 | 2.6 | 5.6 | 5.2 | 2.7 | 8,810.0 | 5,939 | 10,855 | 2,558 | 97,551 | 793,640 | 92,182 | 13,646 |
| Sullivan | 657 | 8,759 | 44.5 | 5.6 | 2.3 | 11.2 | 8.5 | 458.7 | 6,116 | 184 | 113 | 5,860 | 34,430 | 59,523 | 6,397 |
| Tioga | 298 | 6,132 | 60.1 | 3.1 | 2.6 | 7.3 | 6.4 | 196.4 | 4,040 | 149 | 80 | 2,459 | 23,200 | 57,530 | 5,238 |
| Tompkins | 619 | 6,032 | 49.2 | 4.1 | 2.9 | 6.9 | 6.0 | 621.0 | 6,049 | 261 | 157 | 5,898 | 41,040 | 72,135 | 8,503 |
| Ulster | 1,210 | 6,773 | 55.7 | 1.7 | 3.2 | 9.5 | 6.0 | 668.9 | 3,745 | 416 | 284 | 12,830 | 87,620 | 66,561 | 7,565 |
| Warren | 421 | 6,537 | 46.2 | 4.0 | 3.8 | 7.9 | 7.4 | 212.6 | 3,303 | 189 | 102 | 4,564 | 34,220 | 63,269 | 7,026 |
| Washington | 373 | 6,052 | 60.0 | 3.2 | 2.2 | 8.0 | 7.5 | 156.5 | 2,542 | 127 | 92 | 4,972 | 28,470 | 52,020 | 4,489 |
| Wayne | 561 | 6,207 | 56.9 | 3.8 | 2.3 | 10.6 | 5.3 | 248.2 | 2,744 | 171 | 140 | 7,234 | 44,170 | 56,266 | 5,036 |
| Westchester | 10,234 | 10,554 | 40.9 | 15.6 | 4.2 | 5.4 | 1.7 | 6,415.1 | 6,616 | 4,359 | 1,484 | 58,506 | 491,430 | 162,316 | 32,426 |
| Wyoming | 252 | 6,244 | 32.7 | 23.9 | 2.6 | 7.7 | 11.0 | 124.0 | 3,077 | 90 | 58 | 4,150 | 18,200 | 54,677 | 4,628 |
| Yates | 120 | 4,817 | 43.7 | 3.7 | 4.2 | 7.8 | 11.7 | 85.9 | 3,436 | 61 | 37 | 1,183 | 10,950 | 53,953 | 4,895 |
| NORTH CAROLINA | X | X | X | X | X | X | X | X | X | 74,376 | 127,121 | 663,871 | 4,637,380 | 66,251 | 7,685 |
| Alamance | 439 | 2,691 | 53.1 | 4.0 | 9.9 | 6.0 | 1.8 | 147.6 | 904 | 255 | 358 | 7,004 | 75,020 | 55,405 | 5,198 |
| Alexander | 81 | 2,185 | 58.3 | 6.4 | 6.5 | 8.7 | 0.6 | 15.6 | 419 | 60 | 79 | 1,880 | 15,490 | 50,640 | 4,234 |
| Alleghany | 30 | 2,760 | 56.2 | 3.9 | 6.7 | 6.8 | 0.5 | 5.3 | 486 | 43 | 24 | 567 | 4,610 | 51,503 | 5,038 |
| Anson | 69 | 2,781 | 82.5 | 0.0 | 3.4 | 0.0 | 1.8 | 12.7 | 513 | 49 | 50 | 1,940 | 9,590 | 38,711 | 2,531 |
| Ashe | 68 | 2,542 | 49.7 | 2.5 | 6.8 | 14.9 | 1.1 | 10.5 | 393 | 65 | 59 | 1,067 | 11,290 | 48,240 | 4,027 |
| Avery | 54 | 3,108 | 46.8 | 4.6 | 9.5 | 8.6 | 4.9 | 9.3 | 533 | 43 | 33 | 1,442 | 7,170 | 48,622 | 4,193 |
| Beaufort | 158 | 3,356 | 53.5 | 4.6 | 8.1 | 8.8 | 0.7 | 88.4 | 1,879 | 125 | 102 | 3,181 | 20,550 | 53,391 | 4,890 |
| Bertie | 57 | 2,940 | 48.6 | 6.1 | 7.1 | 8.8 | 1.4 | 40.4 | 2,100 | 88 | 38 | 1,324 | 7,680 | 39,403 | 2,724 |
| Bladen | 123 | 3,665 | 48.5 | 5.8 | 6.2 | 7.9 | 0.8 | 37.9 | 1,133 | 113 | 71 | 2,149 | 12,560 | 43,136 | 3,333 |
| Brunswick | 434 | 3,319 | 33.7 | 12.6 | 10.2 | 4.8 | 2.3 | 433.6 | 3,314 | 508 | 387 | 4,976 | 66,210 | 67,556 | 7,343 |
| Buncombe | 1,366 | 5,314 | 27.1 | 31.0 | 6.1 | 6.9 | 1.5 | 691.4 | 2,690 | 3,377 | 624 | 13,049 | 129,700 | 66,068 | 7,720 |
| Burke | 221 | 2,450 | 51.5 | 3.6 | 8.3 | 8.4 | 1.6 | 118.9 | 1,319 | 157 | 191 | 6,912 | 38,210 | 47,071 | 3,732 |
| Cabarrus | 680 | 3,283 | 55.2 | 1.3 | 7.0 | 4.1 | 1.7 | 553.6 | 2,674 | 330 | 472 | 14,309 | 97,910 | 67,329 | 7,156 |
| Caldwell | 246 | 2,998 | 57.3 | 2.2 | 7.9 | 7.1 | 1.3 | 67.9 | 828 | 147 | 178 | 4,008 | 34,490 | 46,251 | 3,739 |
| Camden | 30 | 2,827 | 60.2 | 0.4 | 9.3 | 4.2 | 0.0 | 3.2 | 301 | 18 | 26 | 461 | 4,690 | 61,082 | 5,034 |
| Carteret | 371 | 5,384 | 27.4 | 41.5 | 5.5 | 3.5 | 1.5 | 94.8 | 1,376 | 301 | 420 | 4,730 | 33,120 | 64,680 | 7,038 |
| Caswell | 51 | 2,234 | 54.0 | 9.3 | 7.6 | 10.0 | 0.0 | 12.2 | 538 | 42 | 47 | 1,209 | 9,220 | 45,941 | 3,522 |
| Catawba | 771 | 4,881 | 34.1 | 37.0 | 4.5 | 5.1 | 0.9 | 438.2 | 2,776 | 334 | 344 | 9,483 | 73,740 | 60,081 | 6,492 |
| Chatham | 208 | 2,920 | 45.9 | 4.0 | 7.4 | 5.4 | 0.3 | 122.6 | 1,722 | 120 | 161 | 2,466 | 35,160 | 94,255 | 12,857 |
| Cherokee | 87 | 3,115 | 54.4 | 6.9 | 6.3 | 7.4 | 0.5 | 11.7 | 417 | 109 | 62 | 1,495 | 11,890 | 44,377 | 3,455 |

1. Based on the resident population estimated as of July 1 of the year shown. 2. Bronx, Kings, Queens, and Richmond counties are included with New York county.

# Table B. States and Counties — **Land Area and Population**

| State / county code | CBSA code[1] | County Type code[2] | STATE County | Land area[3] (sq. mi) | Total persons 2019 | Rank | Per square mile | White | Black | American Indian, Alaska Native | Asian and Pacific Islander | Percent Hispanic or Latino[4] | Under 5 years | 5 to 14 years | 15 to 24 years | 25 to 34 years | 35 to 44 years | 45 to 54 years |
|---|---|---|---|---|---|---|---|---|---|---|---|---|---|---|---|---|---|---|
| | | | | 1 | 2 | 3 | 4 | 5 | 6 | 7 | 8 | 9 | 10 | 11 | 12 | 13 | 14 | 15 |
| | | | NORTH CAROLINA— Cont'd | | | | | | | | | | | | | | | |
| 37041 | | 6 | Chowan | 172.7 | 13,815 | 2,169 | 80.0 | 61.4 | 34.5 | 0.8 | 1.1 | 3.6 | 4.8 | 11.2 | 10.4 | 10.3 | 10.5 | 10.8 |
| 37043 | | 9 | Clay | 215.0 | 11,505 | 2,319 | 53.5 | 94.2 | 1.9 | 1.2 | 0.7 | 3.6 | 4.5 | 8.5 | 9.0 | 8.4 | 9.1 | 11.3 |
| 37045 | 43140 | 4 | Cleveland | 464.2 | 99,035 | 610 | 213.3 | 73.8 | 22.0 | 0.6 | 1.4 | 4.0 | 5.9 | 12.4 | 12.7 | 12.4 | 10.6 | 13.1 |
| 37047 | | 6 | Columbus | 938.1 | 54,754 | 936 | 58.4 | 60.5 | 30.6 | 4.1 | 0.7 | 5.8 | 5.2 | 11.5 | 12.1 | 12.5 | 11.4 | 12.3 |
| 37049 | 35100 | 3 | Craven | 706.6 | 101,233 | 598 | 143.3 | 67.7 | 22.2 | 1.1 | 4.1 | 7.8 | 6.2 | 11.8 | 14.8 | 13.4 | 10.9 | 9.9 |
| 37051 | 22180 | 2 | Cumberland | 652.6 | 336,364 | 212 | 515.4 | 45.0 | 40.2 | 2.6 | 4.5 | 12.4 | 7.5 | 13.6 | 16.5 | 17.0 | 12.1 | 10.1 |
| 37053 | 47260 | 1 | Currituck | 261.9 | 29,052 | 1,458 | 110.9 | 88.3 | 6.4 | 1.2 | 1.5 | 4.6 | 5.4 | 12.6 | 10.1 | 12.5 | 12.3 | 13.2 |
| 37055 | 28620 | 4 | Dare | 383.2 | 37,547 | 1,236 | 98.0 | 88.7 | 3.4 | 0.9 | 1.3 | 7.5 | 4.3 | 10.5 | 9.7 | 9.8 | 11.9 | 13.1 |
| 37057 | 49180 | 2 | Davidson | 553.2 | 169,234 | 395 | 305.9 | 80.3 | 10.7 | 0.9 | 2.0 | 7.7 | 5.4 | 12.2 | 11.5 | 12.0 | 11.4 | 14.0 |
| 37059 | 49180 | 2 | Davie | 263.7 | 43,286 | 1,110 | 164.1 | 85.2 | 7.5 | 0.7 | 1.1 | 7.3 | 5.0 | 11.9 | 11.2 | 10.1 | 11.0 | 13.8 |
| 37061 | | 6 | Duplin | 814.7 | 58,794 | 879 | 72.2 | 51.8 | 24.8 | 0.8 | 0.7 | 23.1 | 6.1 | 13.5 | 12.4 | 11.2 | 11.1 | 12.2 |
| 37063 | 20500 | 2 | Durham | 286.5 | 327,306 | 217 | 1,142.4 | 45.0 | 36.4 | 1.0 | 6.4 | 13.7 | 6.2 | 10.9 | 13.2 | 18.5 | 14.0 | 11.8 |
| 37065 | 40580 | 3 | Edgecombe | 505.4 | 50,829 | 978 | 100.6 | 36.8 | 57.7 | 0.8 | 0.7 | 5.3 | 5.8 | 12.6 | 11.9 | 12.1 | 10.7 | 11.3 |
| 37067 | 49180 | 2 | Forsyth | 407.9 | 383,843 | 186 | 941.0 | 57.4 | 27.1 | 0.8 | 3.1 | 13.5 | 5.8 | 12.8 | 13.7 | 13.4 | 11.9 | 12.5 |
| 37069 | 39580 | 1 | Franklin | 491.8 | 71,859 | 758 | 146.1 | 64.7 | 25.9 | 1.1 | 1.2 | 9.1 | 5.5 | 12.2 | 11.8 | 12.2 | 11.9 | 13.9 |
| 37071 | 16740 | 1 | Gaston | 355.8 | 226,568 | 304 | 636.8 | 72.2 | 19.0 | 0.9 | 2.1 | 7.9 | 5.8 | 12.6 | 11.8 | 13.6 | 12.3 | 14.0 |
| 37073 | 47260 | 6 | Gates | 340.6 | 11,464 | 2,321 | 33.7 | 65.4 | 32.1 | 1.5 | 1.0 | 2.5 | 4.5 | 11.4 | 10.3 | 11.2 | 9.9 | 13.5 |
| 37075 | | 9 | Graham | 292.0 | 8,474 | 2,540 | 29.0 | 87.6 | 1.1 | 8.8 | 0.8 | 3.7 | 5.0 | 10.9 | 11.1 | 10.7 | 10.6 | 11.9 |
| 37077 | 20500 | 4 | Granville | 532.0 | 60,486 | 868 | 113.7 | 58.8 | 32.0 | 1.0 | 1.1 | 8.9 | 5.1 | 11.4 | 11.9 | 12.2 | 12.0 | 14.4 |
| 37079 | | 8 | Greene | 266.7 | 20,928 | 1,773 | 78.5 | 47.7 | 35.9 | 1.0 | 0.7 | 15.8 | 5.1 | 11.4 | 12.0 | 13.9 | 13.5 | 12.5 |
| 37081 | 24660 | 2 | Guilford | 645.9 | 540,521 | 129 | 836.8 | 50.5 | 36.2 | 1.1 | 6.1 | 8.6 | 5.7 | 12.4 | 14.7 | 13.8 | 12.2 | 12.7 |
| 37083 | 40260 | 4 | Halifax | 723.7 | 49,479 | 1,000 | 68.4 | 39.1 | 53.8 | 4.1 | 1.1 | 3.2 | 5.5 | 12.0 | 11.0 | 11.8 | 10.7 | 11.7 |
| 37085 | 22180 | 4 | Harnett | 594.9 | 137,058 | 474 | 230.4 | 62.8 | 22.8 | 1.8 | 2.3 | 13.6 | 7.1 | 14.5 | 12.6 | 15.5 | 13.7 | 12.1 |
| 37087 | 11700 | 4 | Haywood | 553.6 | 62,972 | 850 | 113.8 | 93.0 | 1.6 | 1.2 | 1.0 | 4.5 | 4.7 | 10.0 | 9.5 | 11.6 | 10.8 | 12.6 |
| 37089 | 11700 | 2 | Henderson | 372.9 | 118,445 | 532 | 317.6 | 84.5 | 4.0 | 1.0 | 1.9 | 10.4 | 4.6 | 10.6 | 9.5 | 10.5 | 11.1 | 12.3 |
| 37091 | | 6 | Hertford | 353.2 | 23,108 | 1,678 | 65.4 | 33.9 | 60.8 | 1.7 | 1.1 | 3.8 | 4.4 | 11.5 | 13.9 | 11.9 | 11.2 | 11.4 |
| 37093 | 22180 | 4 | Hoke | 390.2 | 55,830 | 925 | 143.1 | 41.4 | 36.4 | 9.3 | 2.6 | 14.6 | 7.9 | 15.0 | 11.8 | 17.4 | 14.2 | 11.5 |
| 37095 | | 9 | Hyde | 612.3 | 4,843 | 2,836 | 9.7 | 62.3 | 27.1 | 0.9 | 1.0 | 10.1 | 3.5 | 9.9 | 9.6 | 12.3 | 12.7 | 13.1 |
| 37097 | 16740 | 1 | Iredell | 574.4 | 185,770 | 360 | 323.4 | 76.9 | 13.0 | 0.8 | 3.2 | 8.2 | 5.4 | 12.9 | 12.2 | 12.5 | 12.5 | 14.4 |
| 37099 | 19000 | 6 | Jackson | 491.1 | 44,033 | 1,094 | 89.7 | 82.5 | 2.8 | 9.0 | 1.5 | 6.3 | 4.3 | 9.3 | 21.2 | 11.1 | 10.1 | 10.6 |
| 37101 | 39580 | 1 | Johnston | 792.0 | 216,246 | 316 | 273.0 | 67.8 | 17.5 | 1.0 | 1.4 | 14.4 | 6.2 | 14.4 | 12.2 | 12.7 | 13.6 | 14.5 |
| 37103 | 35100 | 3 | Jones | 471.4 | 9,250 | 2,476 | 19.6 | 64.2 | 29.5 | 1.4 | 1.0 | 5.7 | 4.7 | 10.1 | 9.6 | 11.5 | 10.5 | 12.3 |
| 37105 | 41820 | 4 | Lee | 255.1 | 62,353 | 854 | 244.4 | 59.3 | 19.9 | 1.1 | 1.9 | 20.0 | 6.3 | 13.6 | 11.9 | 13.0 | 12.0 | 12.6 |
| 37107 | 28820 | 4 | Lenoir | 399.1 | 55,720 | 926 | 139.6 | 49.8 | 41.8 | 0.8 | 1.1 | 8.1 | 5.6 | 12.8 | 11.6 | 11.7 | 10.6 | 12.0 |
| 37109 | 16740 | 1 | Lincoln | 295.8 | 88,097 | 663 | 297.8 | 86.1 | 6.1 | 0.8 | 1.1 | 7.3 | 5.1 | 11.7 | 10.9 | 11.2 | 11.9 | 14.9 |
| 37111 | 32000 | 6 | McDowell | 440.0 | 45,782 | 1,058 | 104.1 | 88.4 | 4.5 | 0.9 | 1.2 | 6.3 | 4.9 | 11.2 | 10.8 | 12.1 | 11.1 | 13.7 |
| 37113 | | 7 | Macon | 515.6 | 35,994 | 1,275 | 69.8 | 89.3 | 1.9 | 1.1 | 1.3 | 7.6 | 4.7 | 10.2 | 9.6 | 9.9 | 9.8 | 11.0 |
| 37115 | 11700 | 2 | Madison | 449.6 | 21,740 | 1,732 | 48.4 | 95.0 | 1.8 | 1.2 | 0.9 | 2.6 | 4.5 | 9.5 | 12.8 | 11.1 | 11.5 | 12.7 |
| 37117 | | 6 | Martin | 456.4 | 22,178 | 1,710 | 48.6 | 53.1 | 42.2 | 0.7 | 0.9 | 4.4 | 5.3 | 11.4 | 10.3 | 10.9 | 9.4 | 12.2 |
| 37119 | 16740 | 1 | Mecklenburg | 523.6 | 1,128,945 | 41 | 2,156.1 | 47.5 | 32.9 | 0.9 | 7.2 | 13.8 | 6.5 | 12.7 | 12.7 | 17.4 | 14.6 | 13.1 |
| 37121 | | 7 | Mitchell | 221.3 | 14,881 | 2,098 | 67.2 | 92.4 | 1.1 | 1.0 | 0.8 | 5.9 | 4.7 | 10.0 | 9.8 | 11.3 | 10.5 | 12.8 |
| 37123 | | 6 | Montgomery | 491.5 | 27,238 | 1,519 | 55.4 | 64.2 | 18.8 | 0.9 | 1.8 | 15.7 | 5.2 | 12.1 | 12.3 | 11.3 | 10.7 | 12.4 |
| 37125 | 38240 | 4 | Moore | 697.7 | 103,352 | 588 | 148.1 | 78.7 | 12.5 | 1.4 | 2.2 | 7.2 | 5.9 | 12.1 | 9.5 | 11.8 | 12.1 | 11.1 |
| 37127 | 40580 | 3 | Nash | 540.4 | 94,859 | 629 | 175.5 | 49.4 | 42.4 | 1.2 | 1.3 | 7.4 | 5.6 | 12.2 | 12.7 | 12.2 | 11.1 | 12.9 |
| 37129 | 48900 | 2 | New Hanover | 192.3 | 236,613 | 292 | 1,230.4 | 79.2 | 13.8 | 1.0 | 2.2 | 6.0 | 4.6 | 10.2 | 15.4 | 13.0 | 12.5 | 12.2 |
| 37131 | 40260 | 9 | Northampton | 536.7 | 19,088 | 1,873 | 35.6 | 40.0 | 57.2 | 1.0 | 0.5 | 2.6 | 4.9 | 9.5 | 9.4 | 10.5 | 9.0 | 11.4 |
| 37133 | 27340 | 3 | Onslow | 762.1 | 203,943 | 334 | 267.6 | 69.2 | 16.5 | 1.4 | 4.0 | 13.2 | 8.2 | 12.8 | 25.1 | 17.8 | 10.9 | 7.5 |
| 37135 | 20500 | 2 | Orange | 397.6 | 149,077 | 449 | 374.9 | 71.7 | 12.2 | 1.0 | 9.1 | 8.6 | 4.2 | 11.1 | 22.0 | 12.0 | 11.4 | 12.2 |
| 37137 | 35100 | 3 | Pamlico | 336.5 | 12,715 | 2,231 | 37.8 | 75.9 | 19.3 | 1.2 | 1.0 | 4.2 | 3.5 | 8.4 | 8.7 | 10.2 | 10.0 | 11.1 |
| 37139 | 21020 | 4 | Pasquotank | 226.9 | 40,372 | 1,171 | 177.9 | 56.2 | 37.0 | 1.0 | 2.1 | 6.2 | 6.0 | 12.1 | 13.2 | 13.6 | 12.4 | 11.2 |
| 37141 | 48900 | 2 | Pender | 871.3 | 64,671 | 828 | 74.2 | 77.3 | 14.9 | 1.2 | 1.0 | 7.7 | 5.4 | 12.8 | 10.7 | 11.1 | 12.8 | 13.7 |
| 37143 | 21020 | 8 | Perquimans | 247.2 | 13,667 | 2,180 | 55.3 | 74.2 | 22.8 | 0.9 | 0.7 | 3.0 | 4.6 | 10.8 | 9.4 | 9.9 | 9.8 | 11.2 |
| 37145 | 20500 | 2 | Person | 392.3 | 39,925 | 1,189 | 101.8 | 67.7 | 27.7 | 1.1 | 0.7 | 4.8 | 5.2 | 11.4 | 11.0 | 11.7 | 10.8 | 13.3 |
| 37147 | 24780 | 3 | Pitt | 652.4 | 182,924 | 365 | 280.4 | 55.4 | 36.7 | 0.8 | 2.7 | 6.6 | 5.6 | 11.8 | 20.8 | 13.9 | 11.7 | 10.9 |
| 37149 | | 8 | Polk | 237.7 | 21,030 | 1,769 | 88.5 | 89.0 | 4.6 | 0.9 | 1.1 | 5.9 | 3.7 | 8.6 | 9.4 | 8.3 | 8.9 | 12.4 |
| 37151 | 24660 | 2 | Randolph | 782.3 | 144,557 | 457 | 184.8 | 79.4 | 7.2 | 1.0 | 1.8 | 12.3 | 5.5 | 12.5 | 12.1 | 12.2 | 11.2 | 13.9 |
| 37153 | 40460 | 5 | Richmond | 473.7 | 44,332 | 1,090 | 93.6 | 57.8 | 32.6 | 3.5 | 1.4 | 7.0 | 6.2 | 13.0 | 12.5 | 12.5 | 11.0 | 12.8 |
| 37155 | 31300 | 4 | Robeson | 947.3 | 129,999 | 496 | 137.2 | 26.3 | 24.4 | 41.7 | 1.0 | 9.2 | 6.5 | 14.0 | 14.2 | 12.7 | 11.5 | 12.3 |
| 37157 | 24660 | 2 | Rockingham | 565.6 | 91,285 | 650 | 161.4 | 73.9 | 19.9 | 1.0 | 1.1 | 6.4 | 5.0 | 11.4 | 10.9 | 11.6 | 10.6 | 13.5 |
| 37159 | 16740 | 1 | Rowan | 511.6 | 142,495 | 461 | 278.5 | 72.6 | 17.2 | 0.8 | 1.5 | 9.6 | 5.7 | 12.2 | 12.7 | 12.5 | 11.9 | 13.0 |
| 37161 | 22580 | 4 | Rutherford | 565.4 | 67,076 | 798 | 118.6 | 84.4 | 11.0 | 0.8 | 1.0 | 4.9 | 5.0 | 11.3 | 11.1 | 11.0 | 10.6 | 13.5 |
| 37163 | | 6 | Sampson | 945.9 | 63,382 | 843 | 67.0 | 51.5 | 25.9 | 2.5 | 0.9 | 21.0 | 6.5 | 13.5 | 12.6 | 11.5 | 11.3 | 12.8 |
| 37165 | 29900 | 6 | Scotland | 319.1 | 34,481 | 1,313 | 108.5 | 44.5 | 39.8 | 13.6 | 1.5 | 3.5 | 6.6 | 13.0 | 12.6 | 12.8 | 11.2 | 11.9 |
| 37167 | 10620 | 6 | Stanly | 395.1 | 63,239 | 846 | 160.1 | 81.9 | 12.0 | 0.7 | 2.2 | 4.6 | 5.6 | 12.2 | 11.8 | 13.0 | 11.2 | 12.9 |
| 37169 | 49180 | 2 | Stokes | 449.3 | 45,743 | 1,060 | 101.8 | 91.8 | 4.6 | 0.8 | 0.6 | 3.5 | 4.6 | 10.4 | 10.5 | 11.2 | 10.6 | 14.2 |
| 37171 | 34340 | 4 | Surry | 532.6 | 71,683 | 761 | 134.6 | 83.9 | 4.3 | 0.7 | 0.9 | 11.4 | 5.3 | 11.8 | 11.6 | 11.4 | 10.7 | 13.7 |

1. CBSA = Core Based Statistical Area. See Appendix A for explanation. See Appendix B for list of metropolitan areas with component counties. Service of USDA Rural-Urban Continuum Codes. See Appendix A for definition. 3. Dry land or land partially or temporarily covered by water. 2. County type code from the Economic Research 4. May be of any race.

# Table B. States and Counties — **Population and Households**

| STATE County | Population, 2020 (cont.) — Age (percent) (cont.) | | | | Population change, 2000-2020 — Total persons | | Percent change | | Components of change, 2010-2020 | | | Households, 2015-2019 | | Percent | | |
|---|---|---|---|---|---|---|---|---|---|---|---|---|---|---|---|---|
| | 55 to 64 years | 65 to 74 years | 75 years and over | Percent female | 2000 | 2010 | 2000-2010 | 2010-2020 | Births | Deaths | Net Migration | Number | Persons per household | Family households | Female family householder[1] | One person |
| | 16 | 17 | 18 | 19 | 20 | 21 | 22 | 23 | 24 | 25 | 26 | 27 | 28 | 29 | 30 | 31 |
| **NORTH CAROLINA— Cont'd** | | | | | | | | | | | | | | | | |
| Chowan | 15.9 | 15.1 | 11.1 | 52.5 | 14,526 | 14,793 | 1.8 | -6.6 | 1,447 | 1,949 | -474 | 6,133 | 2.25 | 65.0 | 15.6 | 31.8 |
| Clay | 16.6 | 19.4 | 13.2 | 51.0 | 8,775 | 10,591 | 20.7 | 8.6 | 867 | 1,454 | 1,511 | 4,996 | 2.18 | 68.5 | 8.5 | 28.4 |
| Cleveland | 13.9 | 11.4 | 7.6 | 52.0 | 96,287 | 98,034 | 1.8 | 1.0 | 11,203 | 12,320 | 2,201 | 35,638 | 2.68 | 67.7 | 14.6 | 28.3 |
| Columbus | 14.0 | 12.1 | 9.0 | 50.6 | 54,749 | 58,129 | 6.2 | -5.8 | 6,184 | 7,219 | -2,319 | 21,580 | 2.46 | 66.0 | 17.3 | 31.4 |
| Craven | 12.8 | 11.7 | 8.6 | 49.9 | 91,436 | 103,498 | 13.2 | -2.2 | 15,153 | 10,869 | -6,582 | 41,226 | 2.36 | 67.4 | 10.3 | 28.0 |
| Cumberland | 10.6 | 7.5 | 5.1 | 50.6 | 302,963 | 319,455 | 5.4 | 5.3 | 57,121 | 25,609 | -15,768 | 125,427 | 2.53 | 62.3 | 16.9 | 33.0 |
| Currituck | 16.5 | 11.2 | 6.2 | 50.1 | 18,190 | 23,547 | 29.5 | 23.4 | 2,578 | 2,216 | 5,128 | 10,522 | 2.49 | 71.0 | 7.5 | 20.8 |
| Dare | 17.5 | 15.3 | 8.0 | 50.9 | 29,967 | 33,920 | 13.2 | 10.7 | 3,525 | 3,356 | 3,455 | 15,529 | 2.32 | 66.2 | 10.6 | 27.7 |
| Davidson | 14.5 | 11.0 | 7.8 | 51.1 | 147,246 | 162,824 | 10.6 | 3.9 | 17,877 | 18,096 | 6,785 | 66,653 | 2.45 | 69.2 | 12.1 | 25.4 |
| Davie | 15.0 | 12.5 | 9.5 | 51.3 | 34,835 | 41,218 | 18.3 | 5.0 | 3,968 | 4,634 | 2,756 | 16,405 | 2.55 | 70.6 | 9.3 | 24.6 |
| Duplin | 13.6 | 11.6 | 8.3 | 51.0 | 49,063 | 58,412 | 19.1 | 0.7 | 7,511 | 5,503 | -1,624 | 21,466 | 2.73 | 67.6 | 14.6 | 27.6 |
| Durham | 11.5 | 8.8 | 5.4 | 52.4 | 223,314 | 269,998 | 20.9 | 21.2 | 43,673 | 20,205 | 33,500 | 125,953 | 2.36 | 57.8 | 14.4 | 33.2 |
| Edgecombe | 14.5 | 12.7 | 8.5 | 53.9 | 55,606 | 56,511 | 1.6 | -10.1 | 6,392 | 6,610 | -5,498 | 21,151 | 2.46 | 68.1 | 24.6 | 28.0 |
| Forsyth | 13.1 | 9.9 | 6.9 | 52.7 | 306,067 | 350,635 | 14.6 | 9.5 | 45,918 | 33,775 | 21,176 | 146,816 | 2.48 | 62.2 | 13.6 | 32.5 |
| Franklin | 14.7 | 11.0 | 6.7 | 50.5 | 47,260 | 60,564 | 28.2 | 18.6 | 7,139 | 5,935 | 10,078 | 25,381 | 2.55 | 72.6 | 14.4 | 23.2 |
| Gaston | 13.3 | 10.0 | 6.6 | 51.8 | 190,365 | 206,100 | 8.3 | 9.9 | 26,037 | 23,577 | 18,070 | 83,735 | 2.58 | 67.4 | 13.7 | 26.8 |
| Gates | 17.2 | 12.7 | 9.3 | 51.0 | 10,516 | 12,185 | 15.9 | -5.9 | 1,057 | 1,234 | -547 | 4,638 | 2.48 | 69.1 | 13.7 | 27.2 |
| Graham | 14.5 | 13.9 | 11.4 | 50.0 | 7,993 | 8,858 | 10.8 | -4.3 | 880 | 1,106 | -157 | 3,393 | 2.47 | 64.2 | 8.9 | 31.5 |
| Granville | 14.9 | 10.8 | 7.3 | 49.5 | 48,498 | 57,550 | 18.7 | 5.1 | 5,873 | 5,423 | 2,516 | 21,400 | 2.6 | 70.9 | 14.2 | 24.9 |
| Greene | 13.8 | 11.0 | 6.8 | 45.5 | 18,974 | 21,351 | 12.5 | -2.0 | 2,207 | 2,014 | -642 | 7,164 | 2.59 | 66.3 | 18.2 | 27.8 |
| Guilford | 12.5 | 9.4 | 6.5 | 52.7 | 421,048 | 488,455 | 16.0 | 10.7 | 62,777 | 44,013 | 33,533 | 203,987 | 2.49 | 63.4 | 15.7 | 29.9 |
| Halifax | 15.1 | 13.0 | 9.1 | 52.2 | 57,370 | 54,631 | -4.8 | -9.4 | 5,835 | 6,941 | -4,053 | 21,017 | 2.37 | 65.1 | 21.6 | 31.2 |
| Harnett | 11.2 | 8.2 | 5.2 | 50.5 | 91,025 | 114,693 | 26.0 | 19.5 | 18,866 | 9,985 | 13,323 | 45,416 | 2.84 | 70.1 | 13.3 | 25.6 |
| Haywood | 15.3 | 14.6 | 11.0 | 51.7 | 54,033 | 59,030 | 9.2 | 6.7 | 5,799 | 7,851 | 6,002 | 26,653 | 2.27 | 64.4 | 10.8 | 30.9 |
| Henderson | 14.6 | 15.0 | 11.7 | 52.1 | 89,173 | 106,716 | 19.7 | 11.0 | 10,959 | 14,100 | 14,776 | 49,221 | 2.31 | 65.1 | 9.0 | 29.7 |
| Hertford | 15.1 | 12.5 | 9.1 | 50.6 | 22,601 | 24,684 | 9.2 | -6.4 | 2,347 | 2,781 | -1,136 | 8,845 | 2.46 | 61.3 | 20.2 | 34.5 |
| Hoke | 11.2 | 7.3 | 3.6 | 50.9 | 33,646 | 46,889 | 39.4 | 19.1 | 9,373 | 3,418 | 2,923 | 17,799 | 2.95 | 69.1 | 15.2 | 27.7 |
| Hyde | 15.5 | 14.4 | 9.0 | 46.3 | 5,826 | 5,817 | -0.2 | -16.7 | 444 | 613 | -813 | 1,947 | 2.36 | 70.8 | 17.4 | 23.7 |
| Iredell | 14.0 | 9.9 | 6.5 | 50.7 | 122,660 | 159,465 | 30.0 | 16.5 | 18,835 | 15,745 | 23,233 | 66,369 | 2.63 | 71.5 | 11.4 | 22.6 |
| Jackson | 12.7 | 12.5 | 8.1 | 51.0 | 33,121 | 40,273 | 21.6 | 9.3 | 3,947 | 3,765 | 3,566 | 16,773 | 2.33 | 59.4 | 10.7 | 29.9 |
| Johnston | 12.4 | 8.9 | 5.2 | 51.1 | 121,965 | 168,879 | 38.5 | 28.0 | 23,594 | 14,447 | 38,061 | 68,968 | 2.83 | 73.2 | 13.9 | 22.6 |
| Jones | 16.9 | 14.8 | 9.5 | 51.3 | 10,381 | 10,163 | -2.1 | -9.0 | 974 | 1,310 | -577 | 4,045 | 2.34 | 65.4 | 14.6 | 30.0 |
| Lee | 13.3 | 10.3 | 7.0 | 51.1 | 49,040 | 57,849 | 18.0 | 7.8 | 8,129 | 5,861 | 2,250 | 21,894 | 2.72 | 65.4 | 14.5 | 26.1 |
| Lenoir | 14.7 | 12.1 | 8.8 | 52.5 | 59,648 | 59,511 | -0.2 | -6.4 | 6,631 | 7,408 | -3,009 | 23,148 | 2.39 | 64.2 | 17.5 | 31.5 |
| Lincoln | 15.4 | 11.6 | 7.4 | 50.5 | 63,780 | 78,001 | 22.3 | 12.9 | 8,272 | 8,022 | 9,876 | 32,894 | 2.5 | 72.0 | 11.7 | 22.7 |
| McDowell | 15.4 | 12.6 | 8.8 | 49.9 | 42,151 | 45,002 | 6.8 | 1.7 | 4,624 | 5,281 | 1,481 | 18,173 | 2.43 | 71.9 | 11.1 | 24.0 |
| Macon | 15.4 | 16.8 | 12.7 | 51.8 | 29,811 | 33,925 | 13.8 | 6.1 | 3,414 | 4,620 | 3,304 | 15,921 | 2.16 | 64.4 | 9.2 | 29.9 |
| Madison | 14.6 | 14.0 | 9.4 | 50.8 | 19,635 | 20,803 | 5.9 | 4.5 | 1,994 | 2,509 | 1,451 | 8,403 | 2.42 | 64.9 | 8.4 | 28.2 |
| Martin | 15.6 | 14.7 | 10.2 | 53.0 | 25,593 | 24,515 | -4.2 | -9.5 | 2,484 | 3,277 | -1,541 | 9,378 | 2.42 | 66.1 | 17.7 | 28.3 |
| Mecklenburg | 11.2 | 7.2 | 4.6 | 51.9 | 695,454 | 919,675 | 32.2 | 22.8 | 147,138 | 60,901 | 121,969 | 411,097 | 2.58 | 61.1 | 13.1 | 31.2 |
| Mitchell | 14.9 | 14.6 | 11.4 | 50.9 | 15,687 | 15,581 | -0.7 | -4.5 | 1,436 | 2,280 | 156 | 6,344 | 2.31 | 63.5 | 9.3 | 31.6 |
| Montgomery | 14.4 | 13.1 | 8.5 | 51.4 | 26,822 | 27,782 | 3.6 | -2.0 | 3,134 | 3,054 | -614 | 10,195 | 2.56 | 66.6 | 12.1 | 30.1 |
| Moore | 13.2 | 12.9 | 11.4 | 51.5 | 74,769 | 88,250 | 18.0 | 17.1 | 10,874 | 11,355 | 15,512 | 39,821 | 2.41 | 65.0 | 10.3 | 30.3 |
| Nash | 14.1 | 11.7 | 7.5 | 52.2 | 87,420 | 95,858 | 9.7 | -0.9 | 11,038 | 10,783 | -1,247 | 37,011 | 2.48 | 66.8 | 15.9 | 29.0 |
| New Hanover | 13.1 | 11.4 | 7.6 | 52.5 | 160,307 | 202,683 | 26.4 | 16.7 | 22,854 | 19,341 | 30,088 | 95,638 | 2.31 | 56.6 | 10.4 | 33.5 |
| Northampton | 17.0 | 15.9 | 12.3 | 51.7 | 22,086 | 22,104 | 0.1 | -13.6 | 1,868 | 2,804 | -2,090 | 8,547 | 2.24 | 65.6 | 19.1 | 29.3 |
| Onslow | 8.0 | 5.9 | 3.9 | 44.1 | 150,355 | 177,801 | 18.3 | 14.7 | 42,537 | 11,014 | -6,585 | 64,386 | 2.72 | 71.0 | 13.0 | 22.1 |
| Orange | 11.9 | 9.6 | 5.7 | 52.6 | 118,227 | 133,692 | 13.1 | 11.5 | 12,291 | 8,180 | 11,243 | 53,376 | 2.49 | 59.9 | 8.5 | 29.7 |
| Pamlico | 17.4 | 18.1 | 12.5 | 49.1 | 12,934 | 13,143 | 1.6 | -3.3 | 888 | 1,683 | 370 | 5,416 | 2.21 | 66.3 | 8.9 | 30.8 |
| Pasquotank | 13.6 | 10.6 | 7.3 | 51.3 | 34,897 | 40,661 | 16.5 | -0.7 | 5,002 | 4,294 | -1,029 | 14,697 | 2.52 | 66.9 | 14.0 | 29.4 |
| Pender | 14.7 | 11.6 | 7.2 | 50.2 | 41,082 | 52,195 | 27.1 | 23.9 | 6,315 | 5,598 | 11,682 | 21,740 | 2.72 | 67.5 | 9.3 | 28.6 |
| Perquimans | 16.0 | 16.0 | 12.3 | 52.5 | 11,368 | 13,452 | 18.3 | 1.6 | 1,285 | 1,604 | 544 | 5,936 | 2.25 | 67.5 | 11.1 | 28.0 |
| Person | 15.9 | 12.3 | 8.3 | 51.8 | 35,623 | 39,479 | 10.8 | 1.1 | 4,243 | 4,533 | 760 | 15,896 | 2.44 | 67.3 | 15.5 | 28.3 |
| Pitt | 11.2 | 8.7 | 5.6 | 53.1 | 133,798 | 168,173 | 25.7 | 8.8 | 21,557 | 12,967 | 6,181 | 69,799 | 2.46 | 58.9 | 16.0 | 29.8 |
| Polk | 16.4 | 17.0 | 15.5 | 52.3 | 18,324 | 20,514 | 12.0 | 2.5 | 1,480 | 2,926 | 1,979 | 9,071 | 2.23 | 61.2 | 6.8 | 33.9 |
| Randolph | 14.2 | 10.8 | 7.6 | 50.7 | 130,454 | 141,825 | 8.7 | 1.9 | 16,187 | 15,248 | 1,929 | 56,541 | 2.51 | 68.1 | 12.2 | 27.1 |
| Richmond | 13.4 | 11.2 | 7.5 | 51.3 | 46,564 | 46,647 | 0.2 | -5.0 | 5,644 | 5,743 | -2,208 | 18,380 | 2.37 | 64.1 | 19.8 | 31.0 |
| Robeson | 12.6 | 10.0 | 6.2 | 51.9 | 123,339 | 134,205 | 8.8 | -3.1 | 18,581 | 13,843 | -8,915 | 45,927 | 2.81 | 66.2 | 20.3 | 29.2 |
| Rockingham | 15.6 | 12.5 | 8.9 | 51.8 | 91,928 | 93,663 | 1.9 | -2.5 | 9,449 | 11,717 | -41 | 37,388 | 2.4 | 66.2 | 13.9 | 30.4 |
| Rowan | 14.1 | 10.7 | 7.4 | 50.8 | 130,340 | 138,493 | 6.3 | 2.9 | 16,093 | 16,458 | 4,455 | 52,843 | 2.57 | 67.8 | 14.1 | 26.7 |
| Rutherford | 14.9 | 13.3 | 9.2 | 51.7 | 62,899 | 67,819 | 7.8 | -1.1 | 6,919 | 8,775 | 1,182 | 26,683 | 2.45 | 63.6 | 12.3 | 31.8 |
| Sampson | 13.0 | 10.9 | 7.8 | 50.9 | 60,161 | 63,471 | 5.5 | -0.1 | 8,521 | 6,929 | -1,658 | 23,416 | 2.66 | 67.1 | 15.2 | 29.7 |
| Scotland | 12.8 | 11.2 | 7.7 | 50.4 | 35,998 | 36,160 | 0.5 | -4.2 | 4,623 | 4,139 | -2,016 | 12,922 | 2.53 | 71.0 | 22.2 | 29.9 |
| Stanly | 13.8 | 11.5 | 8.1 | 50.2 | 58,100 | 60,585 | 4.3 | 4.4 | 6,919 | 7,159 | 2,949 | 23,332 | 2.55 | 68.2 | 11.5 | 24.9 |
| Stokes | 16.2 | 12.8 | 9.7 | 51.1 | 44,711 | 47,400 | 6.0 | -3.5 | 4,056 | 5,540 | -155 | 19,578 | 2.31 | 68.1 | 11.3 | 28.0 |
| Surry | 14.4 | 12.2 | 9.0 | 51.3 | 71,219 | 73,752 | 3.6 | -2.8 | 7,759 | 9,277 | -508 | 29,153 | 2.43 | 68.1 | 10.2 | 28.8 |

1. No spouse present.

— **Population, Vital Statistics, Health, and Crime**

| STATE County | Persons in group quarters, 2020 | Daytime Population, 2015-2019 | | Births, 2020 | | Deaths, 2020 | | Persons under 65 with no health insurance, 2019 | | Medicare, 2020 | | | Crimes reported by county police or sheriff, 2019 | |
|---|---|---|---|---|---|---|---|---|---|---|---|---|---|---|
| | | Number | Employment/ residence ratio | Total | Rate[1] | Number | Rate[1] | Number | Percent | Total beneficiaries | Enrolled in Original Medicare | Enrolled in Medicare Advantage | Violent | Property |
| | 32 | 33 | 34 | 35 | 36 | 37 | 38 | 39 | 40 | 41 | 42 | 43 | 44 | 45 |
| NORTH CAROLINA— Cont'd | | | | | | | | | | | | | | |
| Chowan | 253 | 13,553 | 0.90 | 126 | 9.1 | 216 | 15.6 | 1,207 | 11.8 | 4,093 | 3,250 | 843 | 21 | 95 |
| Clay | 117 | 10,073 | 0.79 | 84 | 7.3 | 144 | 12.5 | 1,302 | 17.2 | 3,982 | 2,800 | 1,182 | NA | NA |
| Cleveland | 1,945 | 92,410 | 0.88 | 1,151 | 11.6 | 1,358 | 13.7 | 10,527 | 13.6 | 23,526 | 14,070 | 9,455 | NA | NA |
| Columbus | 2,811 | 52,520 | 0.82 | 558 | 10.2 | 822 | 15.0 | 6,779 | 16.6 | 13,244 | 8,783 | 4,461 | NA | NA |
| Craven | 4,727 | 105,652 | 1.07 | 1,304 | 12.9 | 1,146 | 11.3 | 9,938 | 12.9 | 24,197 | 20,028 | 4,169 | 119 | 919 |
| Cumberland | 16,266 | 356,386 | 1.15 | 5,379 | 16.0 | 2,912 | 8.7 | 34,470 | 12.5 | 54,413 | 35,747 | 18,665 | 458 | 1,964 |
| Currituck | 114 | 20,643 | 0.55 | 271 | 9.3 | 205 | 7.1 | 2,944 | 12.7 | 5,557 | 4,583 | 975 | 63 | 268 |
| Dare | 159 | 37,995 | 1.10 | 312 | 8.3 | 355 | 9.5 | 3,918 | 13.6 | 9,375 | 7,972 | 1,404 | 12 | 239 |
| Davidson | 1,580 | 140,202 | 0.66 | 1,768 | 10.4 | 2,002 | 11.8 | 19,706 | 14.6 | 37,652 | 14,883 | 22,768 | NA | NA |
| Davie | 298 | 37,107 | 0.73 | 401 | 9.3 | 490 | 11.3 | 4,369 | 13.0 | 10,603 | 4,852 | 5,750 | NA | NA |
| Duplin | 363 | 56,302 | 0.89 | 700 | 11.9 | 625 | 10.6 | 10,019 | 21.6 | 10,876 | 7,618 | 3,258 | NA | NA |
| Durham | 14,454 | 361,444 | 1.31 | 4,188 | 12.8 | 2,429 | 7.4 | 36,915 | 14.0 | 48,534 | 30,189 | 18,345 | NA | NA |
| Edgecombe | 912 | 49,271 | 0.84 | 572 | 11.3 | 730 | 14.4 | 4,790 | 12.1 | 12,603 | 8,550 | 4,053 | NA | NA |
| Forsyth | 10,832 | 398,035 | 1.13 | 4,367 | 11.4 | 3,714 | 9.7 | 43,632 | 14.2 | 73,690 | 31,607 | 42,083 | NA | NA |
| Franklin | 1,461 | 51,436 | 0.51 | 752 | 10.5 | 714 | 9.9 | 7,749 | 13.8 | 13,717 | 8,227 | 5,490 | 77 | 616 |
| Gaston | 3,318 | 197,907 | 0.79 | 2,561 | 11.3 | 2,646 | 11.7 | 25,668 | 13.9 | 45,263 | 23,800 | 21,462 | 17 | 5 |
| Gates | 61 | 8,338 | 0.38 | 94 | 8.2 | 131 | 11.4 | 1,068 | 11.8 | 2,555 | 1,933 | 622 | NA | NA |
| Graham | 93 | 7,525 | 0.71 | 83 | 9.8 | 109 | 12.9 | 1,200 | 19.0 | 2,264 | 1,452 | 812 | NA | NA |
| Granville | 3,567 | 54,840 | 0.83 | 582 | 9.6 | 632 | 10.4 | 6,156 | 13.2 | 12,012 | 7,209 | 4,803 | 57 | 316 |
| Greene | 2,485 | 17,405 | 0.56 | 215 | 10.3 | 237 | 11.3 | 3,062 | 20.7 | 3,964 | 2,823 | 1,141 | 60 | 304 |
| Guilford | 20,099 | 574,640 | 1.19 | 6,074 | 11.2 | 5,065 | 9.4 | 57,058 | 13.1 | 97,416 | 40,477 | 56,939 | 206 | 1,063 |
| Halifax | 1,216 | 49,604 | 0.91 | 516 | 10.4 | 770 | 15.6 | 5,633 | 15.0 | 13,461 | 8,838 | 4,623 | NA | NA |
| Harnett | 3,134 | 107,153 | 0.55 | 1,850 | 13.5 | 1,175 | 8.6 | 17,293 | 15.1 | 21,203 | 13,598 | 7,605 | NA | NA |
| Haywood | 688 | 55,591 | 0.79 | 531 | 8.4 | 850 | 13.5 | 5,782 | 12.4 | 18,400 | 11,217 | 7,183 | 118 | 835 |
| Henderson | 1,208 | 108,753 | 0.88 | 1,055 | 8.9 | 1,553 | 13.1 | 12,599 | 14.7 | 33,926 | 22,029 | 11,897 | NA | NA |
| Hertford | 2,453 | 24,883 | 1.09 | 206 | 8.9 | 317 | 13.7 | 2,196 | 13.6 | 5,363 | 3,880 | 1,483 | 28 | 184 |
| Hoke | 862 | 43,393 | 0.51 | 846 | 15.2 | 422 | 7.6 | 7,934 | 16.5 | 7,385 | 4,564 | 2,821 | NA | NA |
| Hyde | 569 | 5,252 | 1.02 | 34 | 7.0 | 56 | 11.6 | 562 | 16.8 | 1,186 | 829 | 357 | NA | NA |
| Iredell | 1,159 | 173,693 | 0.98 | 1,921 | 10.3 | 1,814 | 9.8 | 19,485 | 12.8 | 35,362 | 19,921 | 15,441 | 121 | 780 |
| Jackson | 4,208 | 44,532 | 1.09 | 365 | 8.3 | 405 | 9.2 | 5,921 | 19.2 | 8,866 | 5,891 | 2,975 | NA | NA |
| Johnston | 1,387 | 163,041 | 0.64 | 2,481 | 11.5 | 1,794 | 8.3 | 25,998 | 14.5 | 35,078 | 21,248 | 13,830 | 152 | 1,463 |
| Jones | 87 | 7,676 | 0.47 | 84 | 9.1 | 126 | 13.6 | 1,226 | 17.5 | 2,674 | 1,991 | 683 | NA | NA |
| Lee | 1,029 | 63,699 | 1.12 | 791 | 12.7 | 624 | 10.0 | 8,032 | 15.9 | 12,568 | 8,186 | 4,383 | NA | NA |
| Lenoir | 1,107 | 59,895 | 1.14 | 620 | 11.1 | 805 | 14.4 | 6,522 | 14.9 | 14,217 | 10,608 | 3,609 | NA | NA |
| Lincoln | 675 | 70,175 | 0.68 | 874 | 9.9 | 965 | 11.0 | 9,115 | 13.0 | 19,031 | 11,367 | 7,663 | 131 | 943 |
| McDowell | 1,574 | 44,758 | 0.97 | 423 | 9.2 | 554 | 12.1 | 5,200 | 14.8 | 11,879 | 6,841 | 5,038 | 59 | 881 |
| Macon | 311 | 34,318 | 0.96 | 325 | 9.0 | 467 | 13.0 | 4,592 | 18.2 | 11,923 | 8,502 | 3,422 | 25 | 476 |
| Madison | 997 | 18,310 | 0.64 | 197 | 9.1 | 275 | 12.6 | 2,121 | 13.4 | 5,843 | 3,468 | 2,375 | 14 | 178 |
| Martin | 146 | 21,035 | 0.81 | 228 | 10.3 | 320 | 14.4 | 2,180 | 11.8 | 6,396 | 4,751 | 1,645 | NA | NA |
| Mecklenburg | 17,303 | 1,226,799 | 1.27 | 14,706 | 13.0 | 7,302 | 6.5 | 127,730 | 13.2 | 142,184 | 82,874 | 59,311 | NA | NA |
| Mitchell | 187 | 14,588 | 0.93 | 135 | 9.1 | 245 | 16.5 | 1,604 | 14.6 | 4,465 | 2,744 | 1,722 | 2 | 55 |
| Montgomery | 1,007 | 26,600 | 0.94 | 280 | 10.3 | 333 | 12.2 | 3,536 | 17.3 | 6,225 | 3,744 | 2,481 | NA | NA |
| Moore | 784 | 96,415 | 0.98 | 1,178 | 11.4 | 1,266 | 12.2 | 9,196 | 12.0 | 26,513 | 17,057 | 9,455 | 62 | 580 |
| Nash | 2,203 | 93,839 | 1.00 | 1,061 | 11.2 | 1,174 | 12.4 | 9,344 | 12.6 | 21,835 | 14,831 | 7,003 | NA | NA |
| New Hanover | 8,179 | 249,158 | 1.19 | 2,137 | 9.0 | 2,276 | 9.6 | 22,453 | 12.2 | 47,592 | 34,589 | 13,003 | 179 | 1,593 |
| Northampton | 441 | 18,563 | 0.80 | 164 | 8.6 | 309 | 16.2 | 1,740 | 13.0 | 5,640 | 3,922 | 1,718 | NA | NA |
| Onslow | 25,397 | 195,642 | 1.01 | 3,802 | 18.6 | 1,269 | 6.2 | 16,678 | 10.6 | 26,277 | 21,220 | 5,057 | 214 | 1,773 |
| Orange | 10,473 | 154,836 | 1.14 | 1,127 | 7.6 | 925 | 6.2 | 12,277 | 10.5 | 23,654 | 13,795 | 9,860 | NA | NA |
| Pamlico | 710 | 11,562 | 0.76 | 76 | 6.0 | 188 | 14.8 | 1,191 | 14.6 | 4,100 | 3,332 | 768 | 25 | 206 |
| Pasquotank | 2,031 | 39,503 | 1.00 | 455 | 11.3 | 408 | 10.1 | 3,756 | 12.2 | 8,661 | 6,533 | 2,127 | 64 | 233 |
| Pender | 973 | 49,002 | 0.56 | 619 | 9.6 | 650 | 10.1 | 6,755 | 13.4 | 13,227 | 9,100 | 4,127 | NA | NA |
| Perquimans | 72 | 11,162 | 0.57 | 119 | 8.7 | 148 | 10.8 | 1,242 | 12.8 | 3,969 | 3,106 | 863 | NA | NA |
| Person | 444 | 33,785 | 0.70 | 414 | 10.4 | 492 | 12.3 | 3,972 | 12.7 | 9,612 | 5,305 | 4,306 | 56 | 357 |
| Pitt | 6,612 | 180,595 | 1.03 | 2,062 | 11.3 | 1,548 | 8.5 | 17,828 | 12.0 | 30,816 | 21,944 | 8,872 | 138 | 915 |
| Polk | 339 | 18,841 | 0.79 | 153 | 7.3 | 296 | 14.1 | 1,969 | 14.1 | 6,575 | 4,661 | 1,914 | NA | NA |
| Randolph | 1,218 | 129,064 | 0.78 | 1,531 | 10.6 | 1,720 | 11.9 | 20,225 | 17.4 | 31,179 | 11,749 | 19,430 | NA | NA |
| Richmond | 866 | 42,559 | 0.86 | 552 | 12.5 | 621 | 14.0 | 5,159 | 14.7 | 10,484 | 6,802 | 3,682 | 178 | 1,007 |
| Robeson | 4,253 | 129,546 | 0.94 | 1,661 | 12.8 | 1,525 | 11.7 | 19,377 | 18.5 | 24,791 | 14,486 | 10,306 | 717 | 2,525 |
| Rockingham | 1,033 | 81,518 | 0.75 | 898 | 9.8 | 1,320 | 14.5 | 9,758 | 13.7 | 23,750 | 9,181 | 14,569 | 70 | 700 |
| Rowan | 4,173 | 131,919 | 0.86 | 1,573 | 11.0 | 1,723 | 12.1 | 16,376 | 14.5 | 31,830 | 15,948 | 15,882 | 149 | 999 |
| Rutherford | 1,181 | 60,369 | 0.76 | 663 | 9.9 | 928 | 13.8 | 7,051 | 13.8 | 16,714 | 10,619 | 6,095 | NA | NA |
| Sampson | 872 | 58,590 | 0.82 | 780 | 12.3 | 752 | 11.9 | 10,177 | 20.2 | 13,214 | 7,980 | 5,234 | 195 | 877 |
| Scotland | 2,496 | 35,728 | 1.06 | 435 | 12.6 | 451 | 13.0 | 3,511 | 13.7 | 8,229 | 5,013 | 3,217 | 59 | 325 |
| Stanly | 2,140 | 55,288 | 0.78 | 675 | 10.7 | 795 | 12.6 | 7,153 | 14.6 | 14,288 | 8,377 | 5,912 | 24 | 481 |
| Stokes | 455 | 35,335 | 0.50 | 384 | 8.4 | 599 | 13.1 | 4,374 | 12.3 | 11,511 | 4,211 | 7,300 | NA | NA |
| Surry | 824 | 71,772 | 0.99 | 759 | 10.6 | 1,006 | 14.0 | 9,396 | 16.7 | 18,378 | 8,327 | 10,051 | NA | NA |

1. Per 1,000 estimated resident population.

| STATE County | County law enforcement employment, 2019 | | Education | | | | | | Money income, 2015-2019 | | | | Income and poverty, 2019 | | | |
|---|---|---|---|---|---|---|---|---|---|---|---|---|---|---|---|---|
| | | | School enrollment and attainment, 2015-2019 | | | | Local government expenditures,[3] 2017-2018 | | | Households | | | Percent below poverty level | | | |
| | | | Enrollment[1] | | Attainment[2] (percent) | | | | | | Percent | | | | | |
| | Officers | Civilians | Total | Percent private | High school graduate or less | Bachelor's degree or more | Total current spending (mil dol) | Current spending per student (dollars) | Per capita income[4] | Median income (dollars) | with income of less than $50,000 | with income of $200,000 or more | Median household income (dollars) | All persons | Children under 18 years | Children 5 to 17 years in families |
| | 46 | 47 | 48 | 49 | 50 | 51 | 52 | 53 | 54 | 55 | 56 | 57 | 58 | 59 | 60 | 61 |
| NORTH CAROLINA—Cont'd | | | | | | | | | | | | | | | | |
| Chowan | 18 | 10 | 3,108 | 14.9 | 46.6 | 21.7 | 24 | 11,887 | 26,256 | 46,519 | 55.6 | 2.3 | 47,126 | 18.5 | 28.5 | 27.4 |
| Clay | 17 | 20 | 1,954 | 5.8 | 45.2 | 23.7 | 14 | 10,679 | 28,467 | 42,222 | 59.4 | 3.4 | 47,116 | 14.0 | 25.5 | 23.2 |
| Cleveland | 98 | 84 | 22,412 | 11.7 | 50.2 | 17.6 | 150 | 9,544 | 22,589 | 42,247 | 58.5 | 2.1 | 46,012 | 19.0 | 30.5 | 29.0 |
| Columbus | 83 | 40 | 11,199 | 7.2 | 53.4 | 13.8 | 88 | 9,766 | 22,608 | 37,628 | 62.1 | 2.2 | 39,531 | 22.3 | 32.5 | 30.4 |
| Craven | 79 | 7 | 23,429 | 15.5 | 37.2 | 25.3 | 122 | 8,761 | 29,521 | 52,687 | 47.5 | 3.7 | 53,372 | 13.8 | 20.0 | 18.6 |
| Cumberland | 281 | 260 | 92,212 | 18.1 | 35.0 | 25.5 | 464 | 8,917 | 24,936 | 46,875 | 53.2 | 2.3 | 46,599 | 18.0 | 25.7 | 24.5 |
| Currituck | 62 | 36 | 6,021 | 18.8 | 38.3 | 23.5 | 40 | 9,742 | 31,849 | 69,964 | 33.3 | 4.7 | 70,699 | 8.8 | 13.7 | 12.9 |
| Dare | 68 | 79 | 7,384 | 14.0 | 28.6 | 34.7 | 60 | 11,389 | 34,578 | 59,381 | 40.9 | 5.9 | 63,033 | 8.9 | 13.8 | 13.1 |
| Davidson | 123 | 54 | 36,714 | 14.4 | 48.9 | 18.5 | 217 | 8,780 | 25,988 | 49,546 | 50.4 | 2.5 | 53,924 | 15.2 | 22.6 | 23.2 |
| Davie | 60 | 36 | 9,106 | 10.7 | 44.2 | 24.0 | 59 | 9,573 | 31,173 | 60,434 | 42.1 | 4.7 | 63,828 | 10.9 | 15.1 | 13.7 |
| Duplin | 62 | 29 | 14,409 | 7.3 | 55.9 | 12.1 | 90 | 9,072 | 20,254 | 41,764 | 57.9 | 1.0 | 44,929 | 17.7 | 28.9 | 26.1 |
| Durham | 190 | 269 | 80,848 | 27.0 | 29.4 | 48.2 | 451 | 10,216 | 35,398 | 60,958 | 41.1 | 7.3 | 65,541 | 14.0 | 20.6 | 19.8 |
| Edgecombe | 53 | 74 | 11,277 | 6.0 | 56.3 | 13.6 | 71 | 10,083 | 20,626 | 36,866 | 62.6 | 1.2 | 40,784 | 21.0 | 33.3 | 33.2 |
| Forsyth | 220 | 341 | 95,874 | 18.7 | 36.5 | 34.0 | 550 | 9,431 | 30,769 | 51,569 | 48.6 | 5.4 | 53,054 | 15.2 | 22.5 | 24.5 |
| Franklin | 84 | 36 | 14,791 | 18.8 | 43.6 | 22.0 | 81 | 9,263 | 27,294 | 55,193 | 45.0 | 2.9 | 57,371 | 11.6 | 17.0 | 15.9 |
| Gaston | 244 | 198 | 49,243 | 16.3 | 45.9 | 21.2 | 296 | 8,518 | 27,352 | 52,835 | 47.5 | 3.1 | 56,542 | 11.6 | 16.5 | 15.2 |
| Gates | 14 | 1 | 2,574 | 17.2 | 48.9 | 14.3 | 19 | 11,039 | 25,846 | 50,750 | 49.7 | 1.5 | 54,204 | 14.7 | 21.1 | 19.1 |
| Graham | 18 | 17 | 1,765 | 5.6 | 54.4 | 12.9 | 14 | 11,750 | 20,323 | 39,571 | 62.5 | 1.5 | 45,813 | 16.8 | 23.2 | 21.8 |
| Granville | 57 | 57 | 13,096 | 10.7 | 46.3 | 22.5 | 82 | 9,075 | 27,131 | 55,856 | 44.6 | 4.0 | 54,300 | 14.6 | 22.5 | 22.2 |
| Greene | 22 | 8 | 4,547 | 9.3 | 57.9 | 11.0 | 34 | 11,160 | 19,168 | 39,837 | 57.9 | 1.3 | 44,648 | 20.2 | 30.4 | 28.3 |
| Guilford | 232 | 340 | 142,377 | 15.9 | 34.2 | 36.0 | 756 | 9,466 | 30,767 | 53,261 | 47.2 | 5.4 | 55,820 | 16.0 | 23.1 | 22.0 |
| Halifax | 71 | 31 | 11,069 | 13.5 | 58.6 | 14.4 | 78 | 11,148 | 21,848 | 35,502 | 64.5 | 1.4 | 38,727 | 23.8 | 37.5 | 34.8 |
| Harnett | 155 | 99 | 37,989 | 20.4 | 42.4 | 21.9 | 176 | 8,421 | 23,767 | 53,554 | 46.4 | 2.3 | 55,619 | 15.6 | 20.6 | 19.9 |
| Haywood | 65 | 61 | 11,377 | 12.2 | 38.4 | 26.0 | 70 | 9,108 | 30,490 | 51,659 | 48.2 | 3.5 | 51,612 | 10.6 | 16.7 | 15.9 |
| Henderson | 155 | 76 | 22,536 | 18.3 | 34.2 | 31.8 | 122 | 8,774 | 31,564 | 55,945 | 44.5 | 4.3 | 61,651 | 10.6 | 18.1 | 15.8 |
| Hertford | 24 | 32 | 5,670 | 22.0 | 51.9 | 14.9 | 33 | 11,255 | 20,119 | 41,028 | 61.9 | 1.0 | 42,374 | 23.0 | 28.3 | 24.4 |
| Hoke | 69 | 53 | 15,055 | 14.2 | 41.6 | 17.7 | 80 | 8,941 | 20,991 | 48,072 | 52.8 | 1.1 | 48,420 | 16.9 | 22.2 | 22.6 |
| Hyde | 13 | 2 | 1,102 | 5.5 | 57.9 | 9.2 | 11 | 17,154 | 18,245 | 39,663 | 57.0 | 0.0 | 43,112 | 19.2 | 28.1 | 26.0 |
| Iredell | 182 | 76 | 41,424 | 13.9 | 39.0 | 28.4 | 260 | 8,312 | 33,194 | 60,955 | 40.7 | 6.6 | 68,308 | 8.2 | 11.5 | 10.5 |
| Jackson | 55 | 24 | 13,035 | 8.2 | 37.5 | 30.4 | 40 | 9,885 | 25,347 | 47,252 | 52.1 | 3.3 | 47,759 | 19.3 | 24.8 | 24.9 |
| Johnston | 126 | 67 | 51,696 | 12.5 | 40.9 | 22.9 | 325 | 8,744 | 27,924 | 59,865 | 41.5 | 4.1 | 62,835 | 12.5 | 18.1 | 17.0 |
| Jones | 18 | 10 | 1,866 | 17.4 | 53.3 | 12.2 | 16 | 13,562 | 21,928 | 38,158 | 60.2 | 0.8 | 46,275 | 18.8 | 31.8 | 27.0 |
| Lee | 40 | 44 | 15,078 | 13.2 | 43.4 | 21.5 | 92 | 9,070 | 24,842 | 49,994 | 50.0 | 3.1 | 53,114 | 14.2 | 21.1 | 19.7 |
| Lenoir | 61 | 45 | 13,622 | 13.2 | 52.2 | 14.9 | 83 | 9,288 | 23,865 | 39,402 | 60.0 | 1.8 | 39,947 | 23.1 | 37.1 | 33.2 |
| Lincoln | 120 | 55 | 17,263 | 11.3 | 43.2 | 23.0 | 108 | 7,974 | 31,771 | 57,536 | 42.6 | 6.1 | 70,479 | 9.0 | 13.0 | 11.9 |
| McDowell | 50 | 19 | 8,889 | 11.7 | 48.5 | 17.4 | 61 | 10,091 | 24,281 | 43,646 | 54.9 | 1.8 | 46,370 | 13.6 | 19.8 | 17.0 |
| Macon | 57 | 22 | 6,006 | 12.5 | 41.8 | 23.8 | 43 | 9,768 | 28,936 | 45,507 | 54.0 | 3.1 | 46,279 | 14.3 | 22.9 | 21.7 |
| Madison | 19 | 29 | 4,707 | 35.7 | 44.3 | 28.5 | 24 | 10,450 | 25,956 | 45,873 | 54.1 | 2.8 | 50,062 | 14.6 | 19.7 | 19.9 |
| Martin | 38 | 4 | 4,555 | 10.9 | 50.5 | 16.4 | 40 | 10,684 | 23,575 | 40,090 | 62.4 | 2.4 | 39,413 | 20.6 | 33.5 | 32.9 |
| Mecklenburg | 282 | 801 | 274,784 | 17.4 | 26.7 | 45.4 | 1,509 | 9,246 | 38,819 | 66,641 | 37.3 | 9.7 | 69,455 | 10.3 | 13.8 | 13.0 |
| Mitchell | 19 | 1 | 3,036 | 21.2 | 47.8 | 20.5 | 21 | 11,044 | 25,009 | 46,103 | 53.2 | 0.9 | 47,675 | 14.8 | 22.1 | 21.2 |
| Montgomery | 31 | 26 | 5,660 | 10.6 | 55.1 | 15.7 | 43 | 10,487 | 24,636 | 44,146 | 54.9 | 3.3 | 46,497 | 16.1 | 24.2 | 21.9 |
| Moore | 80 | 79 | 21,150 | 16.9 | 30.9 | 37.4 | 126 | 9,244 | 34,606 | 59,963 | 41.8 | 6.1 | 63,942 | 11.3 | 17.1 | 16.2 |
| Nash | 84 | 48 | 21,666 | 16.9 | 48.1 | 20.8 | 153 | 9,221 | 27,406 | 49,537 | 50.5 | 3.6 | 50,902 | 16.4 | 22.1 | 19.4 |
| New Hanover | 377 | 170 | 56,287 | 12.6 | 28.1 | 40.6 | 277 | 9,885 | 34,288 | 54,891 | 45.6 | 6.4 | 57,252 | 13.0 | 16.2 | 16.9 |
| Northampton | 31 | 24 | 3,599 | 13.2 | 58.1 | 13.8 | 36 | 12,180 | 22,002 | 37,146 | 63.0 | 2.1 | 39,777 | 21.6 | 37.2 | 34.2 |
| Onslow | 137 | 130 | 47,761 | 13.0 | 38.5 | 22.9 | 233 | 8,665 | 24,413 | 50,278 | 49.7 | 1.5 | 50,645 | 12.5 | 17.9 | 18.9 |
| Orange | 97 | 55 | 52,111 | 9.7 | 20.9 | 59.7 | 245 | 11,914 | 42,231 | 71,723 | 36.7 | 14.5 | 74,314 | 13.4 | 13.3 | 12.0 |
| Pamlico | 22 | 27 | 2,061 | 5.8 | 43.0 | 18.5 | 22 | 11,998 | 29,092 | 46,728 | 52.9 | 3.6 | 52,522 | 15.9 | 28.4 | 26.4 |
| Pasquotank | 46 | 6 | 9,666 | 11.5 | 45.0 | 22.1 | 58 | 9,633 | 24,664 | 50,558 | 49.3 | 2.1 | 51,245 | 14.3 | 24.1 | 23.1 |
| Pender | 70 | 46 | 13,466 | 13.2 | 39.9 | 27.9 | 84 | 8,850 | 29,266 | 57,240 | 44.5 | 5.1 | 60,405 | 11.5 | 17.1 | 15.5 |
| Perquimans | 17 | 2 | 2,679 | 11.8 | 41.9 | 20.2 | 18 | 11,120 | 27,569 | 47,162 | 52.2 | 2.4 | 50,804 | 15.0 | 26.3 | 24.8 |
| Person | 51 | 37 | 8,211 | 10.3 | 50.9 | 15.6 | 55 | 9,831 | 27,329 | 51,020 | 49.0 | 1.3 | 54,553 | 15.4 | 26.8 | 27.5 |
| Pitt | 120 | 187 | 59,971 | 9.0 | 34.5 | 32.7 | 225 | 9,038 | 27,155 | 47,437 | 51.4 | 3.5 | 52,961 | 19.2 | 24.1 | 21.9 |
| Polk | 33 | 27 | 3,804 | 11.1 | 34.4 | 29.1 | 26 | 11,832 | 30,756 | 48,787 | 51.1 | 3.8 | 53,405 | 12.1 | 20.7 | 19.3 |
| Randolph | 184 | 60 | 31,969 | 12.6 | 52.1 | 15.7 | 198 | 8,783 | 24,397 | 47,288 | 52.2 | 1.6 | 50,129 | 14.1 | 18.9 | 18.3 |
| Richmond | 57 | 31 | 10,143 | 6.6 | 51.3 | 15.7 | 70 | 9,474 | 21,228 | 36,148 | 63.7 | 1.2 | 40,518 | 25.8 | 42.7 | 35.8 |
| Robeson | 131 | 15 | 33,637 | 5.7 | 56.1 | 13.7 | 219 | 9,445 | 18,709 | 34,976 | 64.1 | 1.4 | 36,366 | 31.5 | 48.1 | 45.7 |
| Rockingham | 96 | 48 | 18,545 | 10.6 | 52.2 | 15.1 | 112 | 8,870 | 24,209 | 43,579 | 55.9 | 1.8 | 44,686 | 18.4 | 25.5 | 21.4 |
| Rowan | 124 | 71 | 33,160 | 18.0 | 46.9 | 18.5 | 179 | 9,234 | 25,630 | 49,842 | 50.1 | 2.5 | 52,051 | 13.9 | 19.8 | 17.8 |
| Rutherford | 75 | 56 | 13,356 | 12.6 | 48.7 | 17.8 | 98 | 9,824 | 23,978 | 42,608 | 58.5 | 3.3 | 44,547 | 18.5 | 28.1 | 24.7 |
| Sampson | 93 | 41 | 15,337 | 7.7 | 54.9 | 13.7 | 104 | 9,031 | 22,875 | 42,151 | 58.1 | 1.9 | 45,997 | 16.8 | 26.1 | 24.3 |
| Scotland | 41 | 24 | 8,144 | 6.4 | 53.7 | 15.5 | 66 | 11,252 | 19,052 | 37,238 | 62.4 | 0.8 | 39,197 | 28.5 | 46.3 | 46.5 |
| Stanly | 53 | 43 | 13,944 | 18.8 | 48.0 | 17.3 | 80 | 8,772 | 25,863 | 52,623 | 47.3 | 2.7 | 58,303 | 10.7 | 16.7 | 16.3 |
| Stokes | 52 | 26 | 8,919 | 7.3 | 56.2 | 14.1 | 59 | 9,798 | 26,279 | 48,054 | 52.0 | 1.2 | 52,356 | 13.0 | 17.0 | 14.8 |
| Surry | 83 | 51 | 14,841 | 8.1 | 49.4 | 18.0 | 111 | 9,468 | 25,500 | 43,597 | 54.9 | 2.9 | 48,637 | 16.0 | 22.6 | 19.4 |

1. All persons 3 years old and over enrolled in nursery school through college.    2. Persons 25 years old and over.    3. Elementary and secondary education expenditures.    4. Based on population estimated by the American Community Survey, 2014–2018.

# Table B. States and Counties — **Personal Income**

| STATE County | Personal income, 2019 | | | | | | | | | | Earnings, 2019 | | |
|---|---|---|---|---|---|---|---|---|---|---|---|---|---|
| | | | Per capita[1] | | | Supplements to wages and salaries, employer contributions (mil dol) | | | | | | Contributions for government social insurance (mil dol) | |
| | Total (mil dol) | Percent change 2018-2019 | Dollars | Rank | Wages and salaries (mil dol) | Pension and insurance | Government social insurance | Proprietors' income (mil dol) | Dividends, interest, and rent (mil dol) | Personal transfer reecipts (mil dol) | Total (mil dol) | From employee and self-employed | From employer |
| | 62 | 63 | 64 | 65 | 66 | 67 | 68 | 69 | 70 | 71 | 72 | 73 | 74 |
| NORTH CAROLINA— Cont'd | | | | | | | | | | | | | |
| Chowan | 590 | 3.3 | 42,314 | 1,717 | 198 | 36 | 14 | 37 | 139 | 175 | 286 | 22 | 14 |
| Clay | 386 | 4.7 | 34,391 | 2,759 | 76 | 16 | 6 | 11 | 91 | 150 | 108 | 13 | 6 |
| Cleveland | 3,785 | 3.9 | 38,644 | 2,244 | 1,635 | 282 | 119 | 177 | 572 | 1,175 | 2,214 | 158 | 119 |
| Columbus | 1,873 | 4.0 | 33,743 | 2,826 | 626 | 124 | 45 | 96 | 288 | 682 | 891 | 68 | 45 |
| Craven | 4,744 | 4.7 | 46,446 | 1,192 | 2,428 | 587 | 193 | 200 | 1,045 | 1,171 | 3,408 | 204 | 193 |
| Cumberland | 13,498 | 4.5 | 40,233 | 2,022 | 9,356 | 2,391 | 796 | 445 | 2,823 | 3,817 | 12,989 | 672 | 796 |
| Currituck | 1,326 | 6.0 | 47,767 | 1,047 | 298 | 49 | 22 | 73 | 202 | 246 | 442 | 32 | 22 |
| Dare | 2,135 | 4.1 | 57,688 | 384 | 779 | 127 | 59 | 276 | 525 | 357 | 1,242 | 81 | 59 |
| Davidson | 6,851 | 4.2 | 40,876 | 1,935 | 2,090 | 337 | 152 | 365 | 980 | 1,702 | 2,943 | 222 | 152 |
| Davie | 2,121 | 3.8 | 49,508 | 878 | 562 | 86 | 42 | 108 | 397 | 449 | 798 | 61 | 42 |
| Duplin | 2,068 | 3.2 | 35,208 | 2,658 | 747 | 138 | 57 | 221 | 340 | 558 | 1,162 | 70 | 57 |
| Durham | 16,625 | 4.6 | 51,713 | 704 | 16,886 | 2,166 | 1,144 | 1,296 | 3,212 | 2,453 | 21,492 | 1,263 | 1,144 |
| Edgecombe | 1,895 | 3.1 | 36,814 | 2,469 | 671 | 141 | 48 | 80 | 366 | 642 | 941 | 67 | 48 |
| Forsyth | 18,827 | 3.8 | 49,247 | 901 | 11,141 | 1,543 | 778 | 1,404 | 3,768 | 3,559 | 14,866 | 933 | 778 |
| Franklin | 2,585 | 6.2 | 37,089 | 2,435 | 608 | 109 | 43 | 127 | 320 | 606 | 887 | 67 | 43 |
| Gaston | 9,319 | 4.2 | 41,506 | 1,847 | 3,495 | 566 | 252 | 447 | 1,308 | 2,248 | 4,760 | 332 | 252 |
| Gates | 449 | 4.5 | 38,851 | 2,216 | 63 | 13 | 5 | 26 | 65 | 116 | 107 | 9 | 5 |
| Graham | 296 | 4.8 | 35,040 | 2,679 | 88 | 17 | 7 | 24 | 46 | 103 | 135 | 11 | 7 |
| Granville | 2,394 | 4.5 | 39,610 | 2,100 | 1,041 | 250 | 76 | 104 | 348 | 536 | 1,471 | 96 | 76 |
| Greene | 693 | 4.0 | 32,904 | 2,897 | 180 | 42 | 13 | 56 | 87 | 184 | 291 | 19 | 13 |
| Guilford | 25,540 | 4.1 | 47,546 | 1,073 | 16,019 | 2,291 | 1,111 | 1,807 | 5,062 | 4,750 | 21,228 | 1,320 | 1,111 |
| Halifax | 1,893 | 3.2 | 37,854 | 2,338 | 628 | 125 | 46 | 78 | 304 | 683 | 876 | 69 | 46 |
| Harnett | 4,825 | 4.3 | 35,484 | 2,620 | 1,093 | 202 | 80 | 234 | 746 | 1,189 | 1,609 | 117 | 80 |
| Haywood | 2,605 | 4.7 | 41,798 | 1,801 | 748 | 132 | 54 | 174 | 510 | 778 | 1,107 | 90 | 54 |
| Henderson | 5,324 | 4.3 | 45,342 | 1,318 | 1,841 | 292 | 131 | 388 | 1,299 | 1,335 | 2,652 | 198 | 131 |
| Hertford | 787 | 3.5 | 33,251 | 2,869 | 399 | 73 | 29 | 25 | 117 | 270 | 526 | 37 | 29 |
| Hoke | 1,799 | 4.8 | 32,574 | 2,915 | 362 | 79 | 27 | 51 | 256 | 516 | 518 | 38 | 27 |
| Hyde | 198 | 5.6 | 40,044 | 2,043 | 77 | 17 | 6 | 11 | 43 | 47 | 112 | 7 | 6 |
| Iredell | 9,531 | 5.0 | 52,423 | 650 | 4,164 | 594 | 285 | 654 | 1,518 | 1,579 | 5,698 | 367 | 285 |
| Jackson | 1,532 | 5.4 | 34,860 | 2,705 | 761 | 160 | 54 | 73 | 367 | 406 | 1,047 | 69 | 54 |
| Johnston | 8,651 | 6.4 | 41,327 | 1,873 | 2,388 | 424 | 169 | 543 | 1,012 | 1,727 | 3,524 | 242 | 169 |
| Jones | 396 | 6.7 | 42,024 | 1,759 | 70 | 14 | 5 | 36 | 62 | 127 | 125 | 9 | 5 |
| Lee | 2,596 | 4.0 | 42,021 | 1,761 | 1,206 | 203 | 86 | 120 | 463 | 668 | 1,615 | 107 | 86 |
| Lenoir | 2,336 | 3.3 | 41,752 | 1,810 | 1,254 | 238 | 92 | 115 | 404 | 714 | 1,699 | 112 | 92 |
| Lincoln | 4,063 | 5.8 | 47,178 | 1,111 | 1,165 | 187 | 83 | 175 | 602 | 805 | 1,610 | 116 | 83 |
| McDowell | 1,612 | 3.9 | 35,227 | 2,655 | 662 | 140 | 48 | 70 | 221 | 529 | 920 | 67 | 48 |
| Macon | 1,465 | 4.9 | 40,848 | 1,940 | 467 | 78 | 34 | 106 | 399 | 447 | 684 | 56 | 34 |
| Madison | 777 | 4.9 | 35,729 | 2,593 | 162 | 33 | 12 | 55 | 133 | 240 | 262 | 24 | 12 |
| Martin | 798 | 3.7 | 35,541 | 2,614 | 282 | 54 | 20 | 25 | 131 | 298 | 380 | 30 | 20 |
| Mecklenburg | 69,830 | 4.5 | 62,890 | 231 | 55,004 | 6,522 | 3,639 | 9,326 | 12,434 | 7,536 | 74,491 | 4,344 | 3,639 |
| Mitchell | 549 | 2.5 | 36,684 | 2,485 | 205 | 39 | 15 | 23 | 96 | 180 | 283 | 23 | 15 |
| Montgomery | 983 | 2.8 | 36,181 | 2,547 | 402 | 72 | 31 | 59 | 187 | 279 | 564 | 38 | 31 |
| Moore | 5,237 | 4.2 | 51,913 | 686 | 1,700 | 253 | 120 | 350 | 1,431 | 1,183 | 2,422 | 175 | 120 |
| Nash | 4,146 | 3.9 | 43,971 | 1,492 | 1,909 | 334 | 136 | 216 | 740 | 1,047 | 2,595 | 173 | 136 |
| New Hanover | 11,262 | 4.0 | 48,029 | 1,011 | 6,168 | 957 | 427 | 708 | 2,964 | 2,198 | 8,260 | 526 | 427 |
| Northampton | 713 | 5.5 | 36,597 | 2,496 | 223 | 39 | 17 | 39 | 118 | 260 | 318 | 25 | 17 |
| Onslow | 9,407 | 3.0 | 47,525 | 1,075 | 4,772 | 1,362 | 436 | 312 | 1,717 | 1,791 | 6,883 | 321 | 436 |
| Orange | 9,678 | 3.7 | 65,185 | 183 | 4,782 | 1,028 | 321 | 609 | 2,742 | 1,015 | 6,740 | 385 | 321 |
| Pamlico | 551 | 5.7 | 43,290 | 1,572 | 118 | 25 | 9 | 24 | 126 | 168 | 175 | 16 | 9 |
| Pasquotank | 1,608 | 4.5 | 40,388 | 2,003 | 733 | 156 | 55 | 78 | 275 | 430 | 1,023 | 66 | 55 |
| Pender | 2,418 | 5.1 | 38,338 | 2,285 | 546 | 96 | 39 | 156 | 402 | 636 | 837 | 64 | 39 |
| Perquimans | 563 | 5.1 | 41,804 | 1,799 | 89 | 18 | 6 | 28 | 107 | 169 | 141 | 14 | 6 |
| Person | 1,545 | 4.1 | 39,119 | 2,173 | 453 | 84 | 33 | 55 | 222 | 437 | 624 | 49 | 33 |
| Pitt | 7,831 | 4.2 | 43,325 | 1,566 | 3,990 | 803 | 279 | 535 | 1,454 | 1,643 | 5,607 | 341 | 279 |
| Polk | 942 | 3.6 | 45,459 | 1,305 | 192 | 36 | 14 | 49 | 298 | 253 | 291 | 26 | 14 |
| Randolph | 5,587 | 3.8 | 38,892 | 2,210 | 1,907 | 326 | 139 | 449 | 789 | 1,441 | 2,821 | 202 | 139 |
| Richmond | 1,597 | 3.0 | 35,630 | 2,601 | 568 | 113 | 44 | 92 | 210 | 571 | 817 | 61 | 44 |
| Robeson | 3,974 | 3.1 | 30,423 | 3,025 | 1,578 | 304 | 116 | 249 | 513 | 1,505 | 2,247 | 157 | 116 |
| Rockingham | 3,497 | 3.8 | 38,419 | 2,277 | 1,025 | 183 | 74 | 163 | 527 | 1,061 | 1,446 | 117 | 74 |
| Rowan | 5,714 | 4.6 | 40,213 | 2,025 | 2,468 | 437 | 182 | 305 | 923 | 1,538 | 3,392 | 232 | 182 |
| Rutherford | 2,280 | 4.6 | 34,021 | 2,799 | 773 | 144 | 56 | 109 | 369 | 761 | 1,082 | 86 | 56 |
| Sampson | 2,279 | 3.9 | 35,871 | 2,584 | 782 | 139 | 57 | 184 | 358 | 674 | 1,161 | 76 | 57 |
| Scotland | 1,182 | 3.1 | 33,938 | 2,807 | 507 | 95 | 37 | 77 | 181 | 431 | 715 | 51 | 37 |
| Stanly | 2,522 | 4.5 | 40,157 | 2,032 | 800 | 149 | 57 | 196 | 396 | 655 | 1,202 | 87 | 57 |
| Stokes | 1,771 | 3.4 | 38,835 | 2,219 | 294 | 57 | 21 | 76 | 235 | 506 | 448 | 43 | 21 |
| Surry | 2,837 | 3.0 | 39,525 | 2,110 | 1,233 | 217 | 88 | 175 | 485 | 806 | 1,712 | 120 | 88 |

1. Based on the resident population estimated as of July 1 of the year shown.

Items 62—74

## Table B. States and Counties — Earnings, Social Security, and Housing

| STATE County | Farm | Mining, quarrying, and extractions | Construction | Manufacturing | Information; professional, scientific, technical services | Retail trade | Finance, insurance, real estate and leasing | Health care and social assistance | Government | Number | Rate[1] | Supplemental Security Income recipients, 2019 | Total | Percent change, 2010-2020 |
|---|---|---|---|---|---|---|---|---|---|---|---|---|---|---|
| | 75 | 76 | 77 | 78 | 79 | 80 | 81 | 82 | 83 | 84 | 85 | 86 | 87 | 88 |
| **NORTH CAROLINA—Cont'd** | | | | | | | | | | | | | | |
| Chowan | 4.1 | 0.1 | 4.2 | 11.8 | 7.4 | 7.3 | 4.8 | D | 16.7 | 4,425 | 318 | 515 | 7,349 | 0.8 |
| Clay | -11.1 | 0.0 | 16.0 | D | D | 12.5 | 6.0 | 13.6 | 28.1 | 4,255 | 377 | 270 | 7,487 | 4.8 |
| Cleveland | 0.8 | D | 6.5 | 24.3 | 5.3 | 6.2 | 3.5 | 12.5 | 15.9 | 26,770 | 273 | 3,339 | 43,889 | 1.2 |
| Columbus | 4.6 | 0.0 | 4.6 | 16.2 | D | 7.9 | 4.9 | 13.2 | 23.7 | 14,795 | 267 | 2,832 | 26,452 | 1.5 |
| Craven | 0.5 | 0.1 | 3.4 | 8.2 | 4.7 | 5.4 | 3.4 | 8.6 | 50.9 | 25,315 | 248 | 2,509 | 48,059 | 6.8 |
| Cumberland | 0.2 | 0.0 | 4.2 | 4.1 | 4.3 | 4.5 | 2.5 | 5.3 | 62.8 | 60,515 | 180 | 10,789 | 148,087 | 9.3 |
| Currituck | -0.1 | D | 13.4 | 0.9 | D | 11.6 | 9.7 | D | 18.6 | 5,720 | 205 | 325 | 16,873 | 16.7 |
| Dare | 0.0 | D | 11.7 | 2.4 | D | 11.8 | 13.4 | 3.1 | 20.4 | 9,515 | 257 | 280 | 35,582 | 6.2 |
| Davidson | 0.2 | D | 7.6 | 21.4 | 3.1 | 6.0 | 3.1 | 9.0 | 13.4 | 40,315 | 240 | 3,325 | 76,672 | 5.6 |
| Davie | 0.2 | D | 8.5 | 26.3 | 4.3 | 7.2 | 5.2 | 9.4 | 12.5 | 11,125 | 261 | 644 | 19,523 | 7.1 |
| Duplin | 15.1 | 0.0 | 5.5 | 25.9 | D | 5.0 | 1.3 | 4.2 | 17.3 | 11,875 | 202 | 1,519 | 26,074 | 1.5 |
| Durham | 0.0 | D | 2.6 | 16.2 | 20.0 | 2.6 | 8.0 | 14.7 | 9.4 | 48,725 | 151 | 6,218 | 145,711 | 21.2 |
| Edgecombe | 5.2 | D | 3.9 | 17.5 | D | 10.2 | D | D | 30.0 | 13,325 | 259 | 2,819 | 25,063 | 1.0 |
| Forsyth | 0.1 | 0.1 | 5.0 | 9.0 | 8.2 | 7.3 | 11.7 | 20.2 | 9.1 | 76,590 | 200 | 8,658 | 170,696 | 8.8 |
| Franklin | 2.6 | D | 12.4 | 25.1 | 5.2 | 5.8 | 3.1 | 5.5 | 16.9 | 13,730 | 197 | 1,363 | 30,635 | 15.4 |
| Gaston | 0.1 | D | 6.7 | 23.4 | 4.3 | 7.7 | 4.3 | 15.8 | 13.9 | 49,125 | 219 | 5,438 | 96,201 | 8.5 |
| Gates | 15.7 | 0.1 | D | D | D | 4.9 | D | D | 27.0 | 2,740 | 238 | 314 | 5,519 | 6.1 |
| Graham | 0.4 | D | 30.1 | 2.4 | D | 5.5 | 4.8 | 4.4 | 22.4 | 2,435 | 287 | 244 | 6,079 | 2.5 |
| Granville | 0.5 | D | 5.2 | 24.9 | D | 4.2 | 1.6 | D | 42.7 | 12,980 | 215 | 1,324 | 25,144 | 10.1 |
| Greene | 14.5 | 0.0 | 7.8 | 5.2 | 2.1 | 3.0 | 3.2 | 8.9 | 38.4 | 4,105 | 196 | 541 | 8,405 | 2.4 |
| Guilford | 0.0 | 0.0 | 5.9 | 13.0 | 9.2 | 6.0 | 10.4 | 12.1 | 11.2 | 101,345 | 188 | 11,671 | 233,159 | 6.9 |
| Halifax | 2.8 | 0.0 | 4.9 | 15.7 | 2.2 | 8.3 | 4.0 | 10.8 | 27.6 | 15,465 | 309 | 3,519 | 26,054 | 1.1 |
| Harnett | 2.1 | D | 11.9 | 6.9 | 3.9 | 10.4 | 4.4 | 8.1 | 23.0 | 22,700 | 167 | 2,787 | 53,645 | 14.8 |
| Haywood | 0.2 | D | 7.0 | 22.0 | 4.7 | 11.3 | 4.7 | 13.7 | 15.7 | 19,365 | 310 | 1,572 | 36,119 | 3.4 |
| Henderson | 1.0 | D | 8.2 | 14.4 | D | 9.5 | 4.5 | 13.5 | 15.1 | 34,520 | 294 | 1,692 | 59,397 | 8.6 |
| Hertford | 1.8 | 0.0 | 6.2 | 18.2 | D | 8.9 | 3.6 | D | 22.8 | 5,890 | 249 | 1,140 | 10,637 | 0.0 |
| Hoke | 3.9 | D | 7.2 | 19.6 | D | 4.2 | 1.4 | 10.5 | 31.9 | 8,355 | 151 | 1,179 | 21,600 | 18.7 |
| Hyde | 8.6 | 0.0 | 8.2 | 9.8 | D | 4.8 | 3.3 | D | 34.2 | 1,195 | 242 | 143 | 3,407 | 1.7 |
| Iredell | 0.4 | D | 7.6 | 14.7 | D | 7.5 | 4.1 | 8.5 | 10.4 | 37,265 | 205 | 2,620 | 78,641 | 13.9 |
| Jackson | 0.0 | D | 4.7 | 2.0 | D | 6.4 | 3.3 | 12.0 | 48.3 | 9,180 | 210 | 664 | 28,561 | 10.2 |
| Johnston | 1.0 | 0.2 | 11.8 | 18.3 | 4.1 | 8.3 | 4.7 | 6.0 | 18.4 | 37,545 | 179 | 3,612 | 82,457 | 21.8 |
| Jones | 20.9 | 0.0 | 10.0 | 1.2 | D | 3.1 | D | D | 26.1 | 2,715 | 291 | 347 | 5,042 | 4.1 |
| Lee | 0.3 | D | 6.6 | 32.5 | 4.5 | 7.7 | 2.9 | 9.8 | 13.0 | 13,600 | 220 | 1,610 | 24,871 | 3.1 |
| Lenoir | 1.8 | D | 8.1 | 27.4 | 3.5 | 5.5 | 4.5 | 7.8 | 20.8 | 15,655 | 279 | 2,859 | 27,783 | 1.3 |
| Lincoln | 0.1 | D | 12.9 | 20.9 | 6.4 | 7.0 | 4.9 | 5.1 | 16.0 | 18,680 | 216 | 1,446 | 38,229 | 14.0 |
| McDowell | 1.1 | 0.5 | 4.5 | 41.2 | D | 7.8 | 2.6 | 7.6 | 16.4 | 12,815 | 280 | 1,364 | 21,714 | 4.4 |
| Macon | 0.4 | 0.1 | 13.4 | 3.6 | 11.6 | 11.6 | 5.6 | 9.8 | 16.0 | 12,340 | 345 | 739 | 25,692 | 1.7 |
| Madison | -0.9 | D | 11.8 | 9.6 | D | 5.9 | 2.7 | D | 20.7 | 6,160 | 285 | 629 | 11,244 | 6.0 |
| Martin | 3.1 | 0.0 | 4.4 | 21.9 | 4.2 | 8.8 | 3.9 | D | 22.1 | 6,960 | 310 | 1,013 | 11,572 | -1.1 |
| Mecklenburg | 0.1 | 0.0 | 5.8 | 4.5 | 15.1 | 4.6 | 19.6 | 6.0 | 9.4 | 146,095 | 131 | 17,339 | 478,629 | 20.1 |
| Mitchell | -0.5 | D | 5.8 | D | D | 7.6 | 4.2 | 14.9 | 18.2 | 4,780 | 320 | 452 | 8,865 | 1.6 |
| Montgomery | 3.4 | 0.0 | 7.1 | 32.0 | D | 7.8 | D | D | 17.2 | 6,255 | 230 | 692 | 16,371 | 2.9 |
| Moore | 1.1 | D | 7.6 | 4.2 | 7.7 | 6.9 | 6.3 | 28.1 | 12.0 | 27,245 | 269 | 1,465 | 49,439 | 12.5 |
| Nash | 1.2 | D | 7.8 | 22.2 | 4.8 | 6.9 | 6.0 | | 15.3 | 23,115 | 245 | 3,491 | 43,486 | 2.8 |
| New Hanover | 0.2 | D | 8.9 | 5.9 | 15.4 | 8.2 | 7.6 | 10.4 | 19.8 | 48,405 | 207 | 3,982 | 117,638 | 16.0 |
| Northampton | 8.1 | 0.1 | 5.7 | 9.0 | D | 3.0 | D | D | 19.9 | 6,050 | 310 | 1,053 | 11,673 | -0.1 |
| Onslow | 0.7 | D | 2.8 | 0.8 | 2.5 | 4.3 | 2.4 | 3.1 | 72.7 | 29,400 | 145 | 3,280 | 82,375 | 20.7 |
| Orange | 0.2 | 0.0 | 3.6 | 1.8 | 11.1 | 4.3 | 4.6 | 6.7 | 54.7 | 22,250 | 150 | 1,461 | 60,725 | 9.4 |
| Pamlico | 2.4 | 0.0 | 6.0 | 5.4 | 3.8 | 11.2 | 3.6 | D | 30.0 | 4,280 | 338 | 231 | 7,851 | 4.2 |
| Pasquotank | 0.5 | 0.0 | 4.6 | 3.9 | 6.3 | 8.3 | 4.1 | 14.6 | 36.6 | 9,390 | 235 | 1,243 | 17,475 | 3.8 |
| Pender | 5.7 | D | 12.6 | 5.3 | 6.3 | 7.8 | 4.6 | 8.5 | 21.8 | 14,075 | 223 | 1,105 | 30,540 | 14.3 |
| Perquimans | 6.1 | 0.0 | 7.7 | 1.5 | 4.6 | 3.8 | 2.6 | D | 26.8 | 4,175 | 307 | 395 | 7,234 | 3.6 |
| Person | 0.9 | 0.0 | 8.2 | 17.3 | 4.7 | 7.9 | 3.9 | 11.3 | 17.5 | 10,380 | 262 | 1,085 | 18,719 | 2.8 |
| Pitt | 0.6 | D | 6.1 | 10.3 | 4.2 | 6.5 | 5.4 | 10.3 | 37.2 | 32,280 | 178 | 5,767 | 82,201 | 9.6 |
| Polk | 1.8 | 0.0 | 3.4 | 5.0 | 7.5 | 5.5 | 3.7 | 24.4 | 19.7 | 6,630 | 320 | 302 | 11,913 | 4.2 |
| Randolph | 1.2 | 0.1 | 8.6 | 31.6 | 3.1 | 5.9 | 3.9 | 8.3 | 13.1 | 34,620 | 241 | 3,437 | 63,187 | 3.5 |
| Richmond | 4.2 | 1.3 | 4.4 | 20.9 | 3.0 | 8.1 | 3.5 | 10.9 | 21.3 | 11,595 | 259 | 2,124 | 22,192 | 7.0 |
| Robeson | 2.8 | 0.0 | 6.9 | 17.8 | 2.4 | 6.8 | 3.5 | 16.7 | 23.0 | 28,345 | 217 | 7,247 | 53,807 | 2.0 |
| Rockingham | 0.8 | D | 8.0 | 21.6 | D | 8.6 | 3.8 | 12.0 | 16.0 | 25,885 | 284 | 3,059 | 44,526 | 1.9 |
| Rowan | 0.9 | 0.5 | 7.8 | 16.2 | 3.4 | 7.3 | 3.1 | 9.6 | 20.6 | 34,100 | 240 | 2,821 | 63,019 | 4.6 |
| Rutherford | 0.8 | D | 9.6 | 16.1 | 8.0 | 8.5 | 2.7 | 11.7 | 18.4 | 18,395 | 274 | 1,946 | 34,680 | 2.4 |
| Sampson | 12.1 | 0.0 | 7.4 | 16.4 | 2.3 | 7.2 | 3.0 | 7.1 | 21.2 | 14,395 | 227 | 2,032 | 28,123 | 3.2 |
| Scotland | 3.0 | 0.0 | 4.0 | 20.3 | D | 8.3 | 2.6 | D | 20.2 | 9,245 | 266 | 1,668 | 15,339 | 0.9 |
| Stanly | 1.4 | D | 10.3 | 19.9 | 3.2 | 10.0 | 3.6 | 6.2 | 22.7 | 15,460 | 247 | 1,355 | 28,498 | 5.1 |
| Stokes | 1.6 | 0.0 | 12.6 | 10.3 | D | 9.5 | 4.8 | 10.5 | 23.7 | 12,850 | 282 | 950 | 22,420 | 2.2 |
| Surry | 1.2 | D | 18.1 | 12.9 | 4.3 | 8.3 | 3.8 | D | 16.7 | 19,065 | 266 | 1,998 | 34,467 | 2.2 |

1. Per 1,000 resident population estimated as of July 1 of the year shown.

# Table B. States and Counties — Housing, Labor Force, and Employment

| STATE County | Housing units, 2015-2019 | | | | | | | | Civilian labor force, 2020 | | | | Civilian employment[6], 2015-2019 | | |
| | Occupied units | | | | | | | | | | Unemployment | | | Percent | |
| | | | Owner-occupied | | | Renter-occupied | | | | | | | | | |
| | | | | Median owner cost as a percent of income | | | Median rent as a percent of income[2] | Sub-standard units[4] (percent) | | Percent change, 2019-2020 | | | | Management, business, science, and arts | Construction, production, and maintenance occupations |
| | Total | Percent | Median value[1] | With a mortgage | Without a mortgage[2] | Median rent[3] | | | Total | | Total | Rate[5] | Total | | |
| | 89 | 90 | 91 | 92 | 93 | 94 | 95 | 96 | 97 | 98 | 99 | 100 | 101 | 102 | 103 |
|---|---|---|---|---|---|---|---|---|---|---|---|---|---|---|---|
| **NORTH CAROLINA— Cont'd** | | | | | | | | | | | | | | | |
| Chowan | 6,133 | 66.7 | 131,100 | 24.8 | 15.0 | 798 | 23.9 | 1.5 | 5,476 | -2.0 | 336 | 6.1 | 5,586 | 31.6 | 23.3 |
| Clay | 4,996 | 78.0 | 180,300 | 22.2 | 11.9 | 736 | 25.9 | 2.3 | 4,079 | -2.4 | 301 | 7.4 | 4,217 | 38.3 | 21.7 |
| Cleveland | 35,638 | 67.8 | 120,100 | 20.3 | 10.9 | 696 | 29.4 | 2.9 | 48,435 | -1.1 | 3,801 | 7.8 | 40,163 | 30.6 | 33.4 |
| Columbus | 21,580 | 72.5 | 88,900 | 22.6 | 12.9 | 633 | 33.8 | 2.7 | 22,501 | -1.9 | 1,839 | 8.2 | 19,767 | 28.8 | 31.5 |
| Craven | 41,226 | 62.9 | 157,900 | 20.8 | 10.2 | 910 | 28.7 | 2.7 | 40,820 | -2.7 | 2,699 | 6.6 | 40,394 | 32.5 | 26.6 |
| Cumberland | 125,427 | 50.7 | 135,300 | 21.9 | 11.8 | 941 | 31.3 | 2.3 | 126,387 | -1.9 | 12,001 | 9.5 | 127,212 | 33.3 | 22.4 |
| Currituck | 10,522 | 82.0 | 257,200 | 24.1 | 10.0 | 1,151 | 30.0 | 2.1 | 13,851 | -4.0 | 875 | 6.3 | 12,681 | 36.9 | 24.5 |
| Dare | 15,529 | 74.4 | 289,000 | 25.9 | 12.4 | 1,089 | 30.8 | 3.1 | 19,548 | -3.3 | 1,912 | 9.8 | 18,416 | 35.2 | 20.4 |
| Davidson | 66,653 | 69.8 | 142,100 | 19.5 | 10.0 | 720 | 28.5 | 2.3 | 78,916 | -3.3 | 5,605 | 7.1 | 75,087 | 29.4 | 33.7 |
| Davie | 16,405 | 79.8 | 172,600 | 19.6 | 10.0 | 740 | 26.2 | 2.9 | 20,079 | -3.4 | 1,370 | 6.8 | 19,509 | 33.0 | 31.7 |
| Duplin | 21,466 | 70.4 | 90,200 | 22.1 | 12.5 | 665 | 29.0 | 5.3 | 24,812 | -2.9 | 1,392 | 5.6 | 24,773 | 24.6 | 39.6 |
| Durham | 125,953 | 54.4 | 223,000 | 19.1 | 10.0 | 1,067 | 29.4 | 3.1 | 170,422 | -2.4 | 11,118 | 6.5 | 162,945 | 50.1 | 16.2 |
| Edgecombe | 21,151 | 58.8 | 85,700 | 22.4 | 14.4 | 695 | 28.7 | 3.6 | 20,702 | -1.8 | 2,191 | 10.6 | 21,563 | 25.2 | 32.6 |
| Forsyth | 146,816 | 61.6 | 159,300 | 19.3 | 10.0 | 812 | 29.5 | 2.3 | 184,149 | -3.1 | 13,700 | 7.4 | 172,626 | 40.3 | 21.2 |
| Franklin | 25,381 | 73.8 | 150,400 | 20.4 | 10.6 | 829 | 26.3 | 2.6 | 31,029 | -2.3 | 2,120 | 6.8 | 30,998 | 32.5 | 26.6 |
| Gaston | 83,735 | 65.3 | 144,200 | 19.2 | 10.8 | 832 | 28.0 | 3.1 | 109,526 | -1.3 | 8,782 | 8.0 | 102,010 | 31.4 | 29.9 |
| Gates | 4,638 | 77.5 | 147,700 | 23.8 | 12.8 | 860 | 21.3 | 2.1 | 5,069 | -5.0 | 290 | 5.7 | 5,258 | 30.6 | 37.9 |
| Graham | 3,393 | 83.0 | 122,300 | 17.5 | 10.3 | 499 | 22.9 | 2.9 | 3,077 | -5.5 | 305 | 9.9 | 3,388 | 24.1 | 27.9 |
| Granville | 21,400 | 72.1 | 156,000 | 19.4 | 10.7 | 831 | 30.2 | 1.4 | 29,566 | -2.8 | 1,734 | 5.9 | 26,927 | 34.9 | 27.6 |
| Greene | 7,164 | 70.7 | 88,600 | 22.4 | 13.3 | 695 | 32.1 | 3.4 | 9,551 | | 539 | 5.6 | 8,346 | 26.3 | 41.3 |
| Guilford | 203,987 | 58.8 | 167,000 | 19.9 | 10.0 | 878 | 29.4 | 3.0 | 256,997 | -2.9 | 22,064 | 8.6 | 253,250 | 39.0 | 21.9 |
| Halifax | 21,017 | 62.6 | 86,300 | 23.5 | 14.2 | 717 | 32.7 | 2.3 | 20,081 | -2.3 | 1,844 | 9.2 | 19,287 | 25.6 | 33.5 |
| Harnett | 45,416 | 65.4 | 154,000 | 20.9 | 11.1 | 893 | 28.9 | 2.2 | 52,873 | -1.6 | 3,957 | 7.5 | 52,441 | 34.6 | 27.2 |
| Haywood | 26,653 | 73.0 | 179,700 | 20.7 | 10.0 | 785 | 28.9 | 2.4 | 28,663 | -4.9 | 2,160 | 7.5 | 27,218 | 33.0 | 25.3 |
| Henderson | 49,221 | 73.3 | 214,000 | 20.3 | 10.0 | 853 | 28.6 | 2.5 | 52,454 | -5.4 | 3,618 | 6.9 | 53,229 | 34.0 | 27.2 |
| Hertford | 8,845 | 65.8 | 88,600 | 21.4 | 12.2 | 756 | 29.8 | 2.2 | 9,016 | -2.8 | 627 | 7.0 | 9,541 | 29.6 | 34.9 |
| Hoke | 17,799 | 67.1 | 139,300 | 23.5 | 12.0 | 929 | 30.7 | 2.2 | 19,759 | -2.6 | 1,710 | 8.7 | 19,173 | 30.3 | 30.5 |
| Hyde | 1,947 | 67.8 | 141,500 | 28.2 | 26.4 | 775 | 43.4 | 3.9 | 1,797 | -7.1 | 183 | 10.2 | 1,956 | 24.2 | 38.8 |
| Iredell | 66,369 | 72.3 | 190,300 | 19.8 | 10.0 | 868 | 26.7 | 2.3 | 88,855 | -1.7 | 6,519 | 7.3 | 86,063 | 35.1 | 28.1 |
| Jackson | 16,773 | 64.3 | 196,100 | 19.7 | 10.0 | 739 | 31.2 | 2.9 | 19,785 | -1.5 | 1,499 | 7.6 | 18,967 | 33.8 | 19.1 |
| Johnston | 68,968 | 73.0 | 165,100 | 19.7 | 10.0 | 857 | 29.3 | 2.2 | 98,297 | -2.4 | 6,241 | 6.3 | 94,439 | 34.5 | 25.8 |
| Jones | 4,045 | 73.3 | 89,000 | 21.9 | 12.7 | 634 | 34.4 | 2.5 | 4,199 | -3.3 | 242 | 5.8 | 3,701 | 23.2 | 41.0 |
| Lee | 21,894 | 66.7 | 144,200 | 20.7 | 11.9 | 775 | 27.5 | 3.0 | 25,965 | -1.4 | 2,071 | 8.0 | 26,643 | 29.1 | 33.4 |
| Lenoir | 23,148 | 59.3 | 95,500 | 22.2 | 12.9 | 717 | 28.7 | 2.5 | 27,600 | -2.1 | 1,708 | 6.2 | 22,856 | 28.5 | 32.4 |
| Lincoln | 32,894 | 76.4 | 172,000 | 19.2 | 10.0 | 727 | 26.9 | 2.2 | 43,262 | -2.3 | 2,852 | 6.6 | 39,889 | 32.1 | 30.3 |
| McDowell | 18,173 | 72.7 | 119,200 | 19.1 | 10.0 | 645 | 24.3 | 3.9 | 20,555 | -2.3 | 1,432 | 7.0 | 18,707 | 26.5 | 37.6 |
| Macon | 15,921 | 72.9 | 165,600 | 22.3 | 10.0 | 756 | 26.7 | 2.0 | 15,449 | -1.9 | 1,033 | 6.7 | 14,178 | 29.6 | 24.6 |
| Madison | 8,403 | 74.1 | 194,500 | 22.3 | 10.0 | 688 | 29.8 | 2.7 | 9,587 | -5.8 | 661 | 6.9 | 9,162 | 35.9 | 29.3 |
| Martin | 9,378 | 68.0 | 92,500 | 22.9 | 12.6 | 664 | 33.4 | 1.3 | 8,884 | -2.6 | 617 | 6.9 | 9,424 | 27.5 | 27.7 |
| Mecklenburg | 411,097 | 56.4 | 238,000 | 19.3 | 10.5 | 1,146 | 28.4 | 2.5 | 617,543 | -1.3 | 47,927 | 7.8 | 571,063 | 44.1 | 17.9 |
| Mitchell | 6,344 | 78.0 | 157,400 | 20.1 | 10.7 | 611 | 23.1 | 1.2 | 5,886 | -3.6 | 459 | 7.8 | 6,397 | 32.6 | 32.3 |
| Montgomery | 10,195 | 74.8 | 117,700 | 19.8 | 10.0 | 576 | 25.9 | 3.7 | 11,328 | -2.8 | 755 | 6.7 | 10,797 | 26.4 | 42.7 |
| Moore | 39,821 | 75.7 | 216,500 | 20.5 | 10.3 | 879 | 25.7 | 1.8 | 40,734 | -4.0 | 2,938 | 7.2 | 40,851 | 39.7 | 20.2 |
| Nash | 37,011 | 65.3 | 128,200 | 19.9 | 11.6 | 799 | 29.7 | 3.2 | 42,129 | -3.1 | 3,539 | 8.4 | 43,322 | 31.7 | 29.4 |
| New Hanover | 95,638 | 58.1 | 243,600 | 22.3 | 11.8 | 1,031 | 31.8 | 1.3 | 120,839 | -2.3 | 8,761 | 7.3 | 111,708 | 41.3 | 17.3 |
| Northampton | 8,547 | 68.4 | 83,300 | 23.3 | 13.9 | 717 | 33.4 | 1.3 | 7,528 | -3.6 | 550 | 7.3 | 7,357 | 28.5 | 34.8 |
| Onslow | 64,386 | 53.1 | 156,900 | 23.1 | 10.0 | 997 | 28.8 | 2.0 | 63,899 | -1.7 | 4,623 | 7.2 | 64,064 | 31.3 | 22.9 |
| Orange | 53,376 | 62.6 | 308,800 | 19.5 | 10.0 | 1,093 | 30.7 | 2.3 | 77,808 | -3.5 | 4,172 | 5.4 | 74,179 | 55.9 | 11.2 |
| Pamlico | 5,416 | 77.7 | 156,000 | 23.4 | 12.1 | 767 | 29.0 | 2.0 | 5,242 | -3.4 | 295 | 5.6 | 4,825 | 28.8 | 35.7 |
| Pasquotank | 14,697 | 59.6 | 163,100 | 21.7 | 11.9 | 909 | 30.2 | 1.4 | 16,517 | -2.4 | 1,211 | 7.3 | 16,462 | 31.9 | 26.0 |
| Pender | 21,740 | 81.2 | 198,700 | 22.5 | 12.7 | 919 | 29.8 | 2.3 | 28,001 | -3.4 | 1,844 | 6.6 | 26,243 | 33.0 | 26.4 |
| Perquimans | 5,936 | 75.4 | 172,300 | 22.3 | 13.2 | 914 | 46.7 | 2.1 | 4,845 | -3.3 | 324 | 6.7 | 5,338 | 32.0 | 27.9 |
| Person | 15,896 | 76.4 | 123,100 | 19.8 | 10.1 | 694 | 30.6 | 2.4 | 18,098 | -2.3 | 1,332 | 7.4 | 18,668 | 32.2 | 31.9 |
| Pitt | 69,799 | 52.1 | 148,600 | 19.3 | 10.5 | 794 | 32.3 | 2.6 | 87,450 | -3.1 | 5,946 | 6.8 | 84,097 | 38.2 | 20.2 |
| Polk | 9,071 | 74.0 | 225,700 | 22.1 | 11.5 | 851 | 27.5 | 2.4 | 8,948 | -2.4 | 556 | 6.2 | 8,359 | 34.0 | 27.1 |
| Randolph | 56,541 | 71.8 | 124,100 | 19.3 | 10.0 | 703 | 27.4 | 2.9 | 63,993 | -3.8 | 4,725 | 7.4 | 66,144 | 27.0 | 36.0 |
| Richmond | 18,380 | 66.2 | 83,200 | 20.7 | 12.3 | 664 | 35.0 | 2.7 | 16,733 | -0.3 | 1,533 | 9.2 | 17,762 | 27.5 | 35.2 |
| Robeson | 45,927 | 65.6 | 75,600 | 21.9 | 12.8 | 645 | 29.5 | 4.6 | 49,951 | -1.3 | 4,420 | 8.8 | 48,621 | 26.3 | 37.0 |
| Rockingham | 37,388 | 69.6 | 112,800 | 19.5 | 10.4 | 672 | 28.6 | 2.5 | 38,983 | -3.4 | 3,260 | 8.4 | 39,300 | 25.4 | 33.3 |
| Rowan | 52,843 | 69.3 | 136,400 | 18.9 | 10.0 | 772 | 26.8 | 2.6 | 65,083 | -1.5 | 5,174 | 7.9 | 62,623 | 28.8 | 33.9 |
| Rutherford | 26,683 | 71.5 | 118,300 | 19.3 | 10.3 | 636 | 27.2 | 3.6 | 24,588 | -3.4 | 2,281 | 9.3 | 26,817 | 27.5 | 35.6 |
| Sampson | 23,416 | 69.3 | 89,300 | 20.4 | 11.3 | 648 | 29.0 | 3.4 | 28,498 | -3.8 | 1,661 | 5.8 | 27,250 | 27.4 | 38.8 |
| Scotland | 12,922 | 60.5 | 85,500 | 20.6 | 12.7 | 674 | 31.4 | 5.3 | 11,407 | -1.5 | 1,373 | 12.0 | 12,113 | 27.6 | 33.7 |
| Stanly | 23,332 | 74.4 | 138,900 | 19.4 | 10.0 | 708 | 25.1 | 3.9 | 29,630 | -3.1 | 1,873 | 6.3 | 28,750 | 29.4 | 32.2 |
| Stokes | 19,578 | 77.6 | 137,000 | 20.5 | 10.3 | 687 | 28.5 | 2.4 | 20,991 | -3.9 | 1,353 | 6.4 | 20,997 | 30.6 | 32.0 |
| Surry | 29,153 | 73.5 | 126,900 | 19.0 | 10.3 | 628 | 28.1 | 3.0 | 32,640 | -4.7 | 2,236 | 6.9 | 31,291 | 32.6 | 31.0 |

1. Specified owner-occupied units. lacking complete plumbing facilities.   2. A value of 10.0 represents 10 percent or less; a value of 50.0 represents 50 percent or more.   3. Specified renter-occupied units.   4. Overcrowded or 5. Percent of civilian labor force.   6. Civilian employed persons 16 years old and over.

# Table B. States and Counties — Nonfarm Employment and Agriculture

| STATE County | Private nonfarm establishments, employment and payroll, 2019 | | | | | | | | | Agriculture, 2017 | | | |
| --- | --- | --- | --- | --- | --- | --- | --- | --- | --- | --- | --- | --- | --- |
| | Employment | | | | | | | Annual payroll | | Farms | | | Farm producers whose primary occupation is farming (percent) |
| | Number of establish-ments | Total | Health care and social assistance | Manufac-turing | Retail trade | Finance and insurance | Professional, scientific, and technical services | Total (mil dol) | Average per employee (dollars) | Number | Percent with: | | |
| | | | | | | | | | | | Fewer than 50 acres | 1000 acres or more | |
| | 104 | 105 | 106 | 107 | 108 | 109 | 110 | 111 | 112 | 113 | 114 | 115 | 116 |
| **NORTH CAROLINA— Cont'd** | | | | | | | | | | | | | |
| Chowan | 362 | 4,201 | 1,043 | 1,000 | 449 | 78 | 133 | 156 | 37,136 | 97 | 35.1 | 23.7 | 64.2 |
| Clay | 224 | 1,833 | 367 | 310 | 409 | 40 | 32 | 57 | 31,312 | 164 | 56.7 | 0.6 | 51.1 |
| Cleveland | 1,951 | 30,928 | 5,185 | 7,302 | 4,302 | 458 | 996 | 1,202 | 38,871 | 1,005 | 43.0 | 0.9 | 36.8 |
| Columbus | 963 | 11,713 | 2,594 | 1,922 | 2,456 | 811 | 201 | 415 | 35,425 | 514 | 33.9 | 8.4 | 47.7 |
| Craven | 2,158 | 29,044 | 6,673 | 3,612 | 4,781 | 613 | 1,725 | 1,172 | 40,358 | 245 | 36.3 | 8.6 | 48.1 |
| Cumberland | 5,726 | 93,688 | 18,910 | 6,731 | 17,557 | 1,804 | 6,042 | 3,509 | 37,453 | 336 | 48.5 | 4.5 | 37.7 |
| Currituck | 670 | 5,130 | 287 | 63 | 994 | 66 | 138 | 189 | 36,874 | 89 | 48.3 | 11.2 | 43.8 |
| Dare | 1,989 | 15,244 | 1,008 | 470 | 3,723 | 407 | 452 | 553 | 36,245 | 32 | 65.6 | 9.4 | 37.1 |
| Davidson | 2,826 | 38,091 | 4,060 | 9,631 | 5,299 | 613 | 868 | 1,501 | 39,409 | 1,003 | 57.2 | 0.9 | 39.0 |
| Davie | 838 | 10,160 | 1,515 | 1,896 | 1,588 | 174 | 297 | 387 | 38,083 | 591 | 46.9 | 2.2 | 41.2 |
| Duplin | 865 | 13,797 | 1,633 | 5,937 | 1,780 | 218 | 148 | 444 | 32,184 | 820 | 35.1 | 6.6 | 61.3 |
| Durham | 7,689 | 200,346 | 30,254 | 13,889 | 15,462 | 9,825 | 31,742 | 14,266 | 71,206 | 241 | 68.0 | 0.8 | 44.9 |
| Edgecombe | 721 | 11,571 | 2,505 | 2,484 | 1,709 | 164 | 144 | 388 | 33,538 | 249 | 29.3 | 16.1 | 40.9 |
| Forsyth | 8,661 | 185,736 | 42,798 | 17,276 | 22,361 | 10,162 | 6,816 | 9,431 | 50,779 | 557 | 68.6 | 0.4 | 37.1 |
| Franklin | 1,097 | 11,167 | 896 | 3,493 | 1,489 | 182 | 253 | 519 | 46,453 | 538 | 43.9 | 4.5 | 38.0 |
| Gaston | 4,258 | 69,160 | 12,453 | 15,752 | 9,751 | 1,104 | 1,394 | 2,821 | 40,789 | 522 | 58.6 | 0.4 | 32.4 |
| Gates | 117 | 845 | 107 | 154 | 170 | 34 | 34 | 29 | 34,709 | 141 | 40.4 | 11.3 | 46.6 |
| Graham | 163 | 1,964 | 244 | 73 | 351 | 57 | 12 | 74 | 37,627 | 123 | 65.9 | NA | 39.2 |
| Granville | 907 | 15,711 | 4,300 | 4,402 | 1,373 | 246 | 256 | 653 | 41,568 | 557 | 32.5 | 3.4 | 34.9 |
| Greene | 261 | 2,203 | 567 | 183 | 345 | 52 | 50 | 65 | 29,478 | 207 | 34.8 | 12.6 | 54.4 |
| Guilford | 13,660 | 262,198 | 34,226 | 31,432 | 27,918 | 11,225 | 11,601 | 12,526 | 47,772 | 854 | 58.0 | 0.6 | 39.9 |
| Halifax | 949 | 13,103 | 2,615 | 2,589 | 2,373 | 302 | 198 | 475 | 36,215 | 336 | 23.5 | 18.5 | 42.0 |
| Harnett | 1,744 | 23,344 | 3,702 | 1,902 | 4,304 | 548 | 521 | 770 | 32,991 | 643 | 54.9 | 3.7 | 40.8 |
| Haywood | 1,401 | 15,652 | 3,115 | 1,971 | 3,317 | 395 | 566 | 570 | 36,430 | 541 | 60.3 | 1.3 | 33.8 |
| Henderson | 2,800 | 35,339 | 7,019 | 6,307 | 5,872 | 656 | 915 | 1,478 | 41,820 | 455 | 70.1 | 1.1 | 43.6 |
| Hertford | 475 | 7,477 | 1,928 | 1,079 | 1,146 | 167 | 101 | 303 | 40,555 | 126 | 38.1 | 19.8 | 58.7 |
| Hoke | 448 | 5,774 | 1,447 | 1,742 | 795 | 69 | 102 | 178 | 30,829 | 189 | 45.0 | 7.4 | 32.2 |
| Hyde | 164 | 734 | 87 | 32 | 147 | 27 | 19 | 27 | 37,366 | 138 | 26.1 | 26.8 | 50.4 |
| Iredell | 4,926 | 72,738 | 9,651 | 12,675 | 9,440 | 1,242 | 3,438 | 3,170 | 43,577 | 1,055 | 47.3 | 1.7 | 42.9 |
| Jackson | 1,006 | 12,189 | 1,814 | 246 | 2,039 | 210 | 256 | 424 | 34,818 | 215 | 56.7 | 0.5 | 52.1 |
| Johnston | 3,537 | 44,097 | 6,183 | 7,694 | 8,060 | 826 | 1,151 | 1,708 | 38,736 | 1,063 | 51.3 | 3.7 | 41.2 |
| Jones | 132 | 800 | 93 | 24 | 125 | 13 | 27 | 26 | 31,910 | 177 | 39.5 | 9.6 | 51.0 |
| Lee | 1,336 | 24,077 | 2,922 | 8,935 | 3,269 | 344 | 346 | 1,080 | 44,869 | 250 | 49.6 | 3.2 | 38.3 |
| Lenoir | 1,173 | 24,138 | 4,053 | 6,107 | 3,135 | 491 | 598 | 895 | 37,068 | 386 | 37.8 | 8.3 | 53.6 |
| Lincoln | 1,782 | 20,220 | 2,862 | 3,678 | 3,514 | 394 | 573 | 801 | 39,601 | 614 | 58.6 | 0.8 | 31.9 |
| McDowell | 770 | 14,497 | 1,597 | 6,922 | 1,768 | 142 | 175 | 495 | 34,139 | 333 | 63.1 | NA | 41.6 |
| Macon | 1,146 | 9,864 | 1,461 | 483 | 1,941 | 364 | 247 | 343 | 34,734 | 340 | 69.7 | NA | 34.4 |
| Madison | 343 | 3,049 | 609 | 437 | 487 | 37 | 37 | 100 | 32,765 | 639 | 50.1 | 1.3 | 40.5 |
| Martin | 447 | 4,978 | 1,059 | 752 | 970 | 139 | 93 | 149 | 29,963 | 332 | 27.4 | 16.6 | 54.3 |
| Mecklenburg | 33,186 | 655,533 | 76,951 | 26,362 | 61,432 | 76,575 | 52,278 | 43,176 | 65,864 | 216 | 74.5 | NA | 36.7 |
| Mitchell | 360 | 4,131 | 889 | 444 | 657 | 95 | 66 | 164 | 39,635 | 250 | 63.6 | NA | 29.2 |
| Montgomery | 493 | 8,164 | 1,063 | 3,360 | 800 | 179 | 68 | 295 | 36,128 | 240 | 40.4 | 0.4 | 51.0 |
| Moore | 2,319 | 32,880 | 9,188 | 1,666 | 4,896 | 960 | 1,443 | 1,368 | 41,612 | 733 | 51.7 | 2.6 | 42.9 |
| Nash | 2,083 | 37,739 | 5,608 | 7,793 | 4,888 | 1,536 | 815 | 1,530 | 40,551 | 425 | 45.4 | 7.8 | 51.9 |
| New Hanover | 7,830 | 102,660 | 17,947 | 5,753 | 16,056 | 3,684 | 5,936 | 4,645 | 45,245 | 59 | 93.2 | NA | 43.9 |
| Northampton | 250 | 4,551 | 622 | 377 | 387 | 20 | 23 | 167 | 36,764 | 272 | 21.7 | 19.9 | 51.5 |
| Onslow | 2,904 | 37,064 | 5,389 | 1,095 | 8,891 | 1,025 | 1,676 | 1,106 | 29,844 | 340 | 50.3 | 3.5 | 55.9 |
| Orange | 3,328 | 49,242 | 19,210 | 1,658 | 5,591 | 1,036 | 2,698 | 2,259 | 45,874 | 686 | 51.7 | 0.6 | 44.1 |
| Pamlico | 237 | 2,285 | 290 | 126 | 597 | 30 | 31 | 65 | 28,379 | 100 | 44.0 | 12.0 | 39.3 |
| Pasquotank | 910 | 11,068 | 2,473 | 534 | 2,431 | 458 | 374 | 425 | 38,418 | 126 | 31.7 | 25.4 | 55.9 |
| Pender | 1,153 | 9,528 | 1,435 | 943 | 2,102 | 124 | 278 | 303 | 31,849 | 336 | 52.1 | 3.9 | 48.3 |
| Perquimans | 191 | 1,391 | 136 | 49 | 255 | 39 | 23 | 49 | 35,051 | 149 | 33.6 | 20.1 | 77.6 |
| Person | 699 | 8,492 | 1,231 | 1,650 | 1,491 | 192 | 110 | 336 | 39,528 | 393 | 41.7 | 4.8 | 50.9 |
| Pitt | 3,659 | 61,820 | 16,635 | 6,556 | 9,361 | 1,727 | 2,244 | 2,532 | 40,964 | 478 | 35.8 | 12.6 | 55.0 |
| Polk | 478 | 3,970 | 1,526 | 333 | 415 | 110 | 136 | 129 | 32,431 | 281 | 57.7 | 0.4 | 44.6 |
| Randolph | 2,494 | 40,211 | 4,435 | 15,236 | 4,922 | 792 | 745 | 1,441 | 35,825 | 1,368 | 45.5 | 0.7 | 41.6 |
| Richmond | 810 | 11,495 | 1,942 | 3,279 | 2,018 | 209 | 158 | 397 | 34,507 | 237 | 27.4 | 3.0 | 44.4 |
| Robeson | 1,777 | 33,444 | 6,817 | 10,304 | 4,715 | 1,102 | 630 | 1,116 | 33,368 | 722 | 43.4 | 11.5 | 46.0 |
| Rockingham | 1,614 | 21,114 | 2,439 | 5,469 | 3,550 | 430 | 387 | 707 | 33,486 | 844 | 37.3 | 2.5 | 40.1 |
| Rowan | 2,682 | 48,004 | 9,801 | 9,645 | 4,877 | 593 | 855 | 2,195 | 45,722 | 925 | 51.1 | 1.6 | 37.3 |
| Rutherford | 1,233 | 15,719 | 2,666 | 3,247 | 2,558 | 269 | 319 | 546 | 34,726 | 620 | 46.6 | 0.3 | 44.2 |
| Sampson | 971 | 12,852 | 2,270 | 3,035 | 2,193 | 211 | 220 | 489 | 38,062 | 960 | 33.2 | 7.1 | 52.2 |
| Scotland | 582 | 9,578 | 2,514 | 1,644 | 1,508 | 183 | 95 | 338 | 35,295 | 108 | 26.9 | 11.1 | 55.1 |
| Stanly | 1,340 | 16,698 | 2,824 | 3,630 | 2,948 | 389 | 289 | 593 | 35,533 | 672 | 54.0 | 2.8 | 32.6 |
| Stokes | 632 | 6,213 | 1,271 | 914 | 1,311 | 118 | 152 | 180 | 28,940 | 856 | 39.6 | 0.9 | 40.1 |
| Surry | 1,610 | 30,051 | 4,260 | 3,869 | 4,269 | 511 | 423 | 1,236 | 41,134 | 1,064 | 41.8 | 1.6 | 42.3 |

# Table B. States and Counties — Agriculture

| STATE County | Acreage (1,000) | Percent change, 2012-2017 | Average size of farm | Total irrigated (1,000) | Total cropland (1,000) | Average per farm | Average per acre | Value of machinery and equipment, average per farm (dollars) | Total (mil dol) | Average per farm (acres) | Crops | Livestock and poultry products | Organic farms (number) | Farms with internet access (percent) | Total ($1,000) | Percent of farms |
|---|---|---|---|---|---|---|---|---|---|---|---|---|---|---|---|---|
| | 117 | 118 | 119 | 120 | 121 | 122 | 123 | 124 | 125 | 126 | 127 | 128 | 129 | 130 | 131 | 132 |
| NORTH CAROLINA— Cont'd | | | | | | | | | | | | | | | | |
| Chowan | 54 | -7.9 | 552 | 4.2 | 43.9 | 1,928,100 | 3,494 | 396,748 | 46.6 | 480,258 | 82.7 | 17.3 | 3 | 79.4 | 2,814 | 66.0 |
| Clay | 13 | 6.6 | 76 | 0.2 | 4.7 | 479,404 | 6,277 | 42,034 | 2.9 | 17,860 | 69.8 | 30.2 | NA | 65.2 | 129 | 14.0 |
| Cleveland | 113 | -2.8 | 113 | 0.4 | 45.6 | 526,049 | 4,664 | 66,770 | 133.8 | 133,133 | 11.9 | 88.1 | 6 | 75.4 | 1,076 | 29.3 |
| Columbus | 141 | -11.4 | 274 | 2.1 | 113.0 | 895,784 | 3,264 | 156,073 | 162.0 | 315,191 | 36.1 | 63.9 | 1 | 66.9 | 2,708 | 45.3 |
| Craven | 81 | 15.2 | 332 | 0.9 | 61.9 | 1,307,554 | 3,937 | 175,519 | 71.6 | 292,273 | 51.4 | 48.6 | 2 | 77.1 | 1,461 | 47.8 |
| Cumberland | 66 | -19.8 | 196 | 1.2 | 35.5 | 1,014,731 | 5,166 | 123,059 | 95.8 | 285,116 | 26.5 | 73.5 | NA | 79.8 | 355 | 25.6 |
| Currituck | 45 | 26.8 | 504 | 0.1 | 36.4 | 2,471,600 | 4,906 | 200,095 | 18.2 | 204,730 | 99.6 | 0.4 | NA | 87.6 | 1,247 | 42.7 |
| Dare | 5 | D | 167 | 0.1 | 4.7 | 702,379 | 4,207 | 83,082 | 1.6 | 48,969 | 64.6 | 35.4 | NA | 78.1 | 225 | 9.4 |
| Davidson | 92 | 5.8 | 92 | 0.4 | 45.5 | 533,839 | 5,797 | 62,547 | 47.1 | 46,929 | 34.5 | 65.5 | 5 | 76.8 | 523 | 9.3 |
| Davie | 77 | 29.0 | 130 | 0.3 | 37.5 | 662,441 | 5,089 | 71,003 | 26.9 | 45,569 | 46.0 | 54.0 | NA | 72.1 | 279 | 9.8 |
| Duplin | 243 | 5.3 | 296 | 8.0 | 165.1 | 1,422,426 | 4,798 | 220,716 | 1,261.7 | 1,538,648 | 7.8 | 92.2 | 9 | 79.5 | 3,021 | 30.4 |
| Durham | 19 | -10.8 | 77 | 0.1 | 6.8 | 822,548 | 10,656 | 57,642 | 10.1 | 41,921 | 89.2 | 10.8 | 1 | 77.6 | 27 | 9.1 |
| Edgecombe | 149 | 17.6 | 598 | 2.5 | 106.8 | 1,823,586 | 3,049 | 277,862 | 176.2 | 707,614 | 52.5 | 47.5 | 3 | 61.4 | 3,812 | 61.4 |
| Forsyth | 35 | -14.0 | 62 | 0.1 | 15.7 | 581,509 | 9,307 | 46,871 | 10.9 | 19,575 | 85.1 | 14.9 | NA | 70.7 | 70 | 5.7 |
| Franklin | 108 | -7.6 | 201 | 1.8 | 42.8 | 713,579 | 3,556 | 101,007 | 58.5 | 108,669 | 59.4 | 40.6 | 7 | 72.1 | 575 | 27.0 |
| Gaston | 38 | -10.1 | 72 | 0.2 | 15.4 | 443,484 | 6,141 | 50,169 | 23.2 | 44,375 | 24.4 | 75.6 | 8 | 77.8 | 271 | 14.8 |
| Gates | 58 | -8.4 | 411 | 3.6 | 45.5 | 1,490,928 | 3,625 | 231,965 | 72.9 | 516,901 | 32.7 | 67.3 | NA | 82.3 | 1,997 | 61.7 |
| Graham | 11 | 60.9 | 89 | 0.0 | 3.0 | 428,116 | 4,788 | 78,974 | 1.4 | 11,089 | 28.5 | 71.5 | NA | 65.0 | 66 | 11.4 |
| Granville | 125 | 23.8 | 224 | 1.8 | 42.7 | 780,700 | 3,484 | 76,280 | D | D | D | D | 15 | 79.2 | 274 | 28.7 |
| Greene | 83 | -17.7 | 403 | 1.6 | 69.4 | 1,588,126 | 3,945 | 235,131 | 242.5 | 1,171,415 | 28.9 | 71.1 | NA | 77.3 | 1,055 | 63.3 |
| Guilford | 76 | -15.9 | 89 | 1.3 | 40.9 | 731,162 | 8,178 | 69,898 | 52.2 | 61,108 | 62.9 | 37.1 | 7 | 75.3 | 331 | 11.0 |
| Halifax | 209 | 6.7 | 622 | 2.5 | 138.1 | 1,596,584 | 2,566 | 245,400 | 133.2 | 396,411 | 62.4 | 37.6 | 10 | 68.8 | 5,417 | 64.3 |
| Harnett | 106 | -11.3 | 165 | 2.8 | 69.7 | 921,279 | 5,575 | 128,788 | 204.6 | 318,138 | 35.0 | 65.0 | 5 | 76.5 | 536 | 25.7 |
| Haywood | 52 | 6.7 | 97 | 0.3 | 11.0 | 623,971 | 6,461 | 61,440 | 18.2 | 33,567 | 43.1 | 56.9 | 6 | 71.7 | 525 | 22.2 |
| Henderson | 41 | 15.0 | 90 | 2.5 | 25.0 | 700,622 | 7,756 | 96,176 | 67.2 | 147,780 | 96.1 | 3.9 | 3 | 71.4 | 385 | 11.9 |
| Hertford | 81 | -2.6 | 642 | 2.1 | 60.0 | 2,053,706 | 3,199 | 275,035 | 139.3 | 1,105,683 | 32.9 | 67.1 | NA | 75.4 | 1,790 | 62.7 |
| Hoke | 54 | -8.4 | 284 | 2.2 | 32.9 | 1,249,580 | 4,402 | 131,728 | 76.8 | 406,328 | 16.5 | 83.5 | NA | 79.9 | 339 | 31.7 |
| Hyde | 125 | 16.1 | 905 | D | 93.1 | 2,690,902 | 2,974 | 326,488 | D | D | D | D | 1 | 64.5 | 2,515 | 71.0 |
| Iredell | 133 | -12.5 | 126 | 0.3 | 68.8 | 687,281 | 5,438 | 80,370 | 112.9 | 106,988 | 16.7 | 83.3 | 6 | 70.4 | 1,627 | 7.1 |
| Jackson | 16 | -2.9 | 73 | 0.1 | 5.3 | 599,496 | 8,195 | 62,010 | 11.6 | 54,126 | 96.2 | 3.8 | NA | 75.3 | 86 | 15.8 |
| Johnston | 183 | -5.9 | 172 | 2.1 | 129.8 | 817,375 | 4,741 | 117,962 | 267.8 | 251,888 | 55.0 | 45.0 | 12 | 77.0 | 1,938 | 28.2 |
| Jones | 66 | 10.6 | 371 | 0.5 | 49.2 | 1,503,580 | 4,054 | 228,303 | 213.6 | 1,206,910 | 13.8 | 86.2 | 2 | 87.0 | 1,039 | 45.8 |
| Lee | 35 | -10.0 | 141 | 1.4 | 19.5 | 810,854 | 5,764 | 108,487 | 54.4 | 217,568 | 34.5 | 65.5 | 9 | 80.4 | 124 | 12.8 |
| Lenoir | 114 | -6.9 | 295 | 2.7 | 85.2 | 1,172,512 | 3,980 | 256,044 | 311.4 | 806,666 | 23.5 | 76.5 | 1 | 77.2 | 1,589 | 50.3 |
| Lincoln | 54 | -2.7 | 88 | 0.2 | 30.1 | 504,795 | 5,731 | 60,532 | 53.8 | 87,577 | 15.7 | 84.3 | NA | 75.2 | 518 | 16.8 |
| McDowell | 23 | -7.7 | 69 | 0.3 | 6.2 | 363,086 | 5,258 | 63,862 | 24.6 | 73,880 | 56.6 | 43.4 | 3 | 73.3 | 73 | 14.7 |
| Macon | 20 | -12.8 | 58 | 0.0 | 5.2 | 463,306 | 7,966 | 42,489 | 7.8 | 23,065 | 21.4 | 78.6 | NA | 68.2 | 132 | 20.0 |
| Madison | 57 | 0.9 | 89 | 0.1 | 11.1 | 440,456 | 4,959 | 34,914 | D | D | D | D | 12 | 71.0 | 194 | 17.2 |
| Martin | 141 | 10.8 | 425 | D | 105.4 | 1,137,286 | 2,678 | 206,716 | 92.7 | 279,238 | 84.5 | 15.5 | NA | 69.6 | 5,810 | 84.9 |
| Mecklenburg | 12 | -24.4 | 54 | 0.5 | 5.6 | 1,853,146 | 34,288 | 183,007 | D | D | D | 100.0 | NA | 77.3 | D | 6.5 |
| Mitchell | 15 | -23.4 | 59 | 0.0 | 4.2 | 344,314 | 5,815 | 50,043 | 2.4 | 9,400 | 74.9 | 25.1 | NA | 77.2 | 38 | 4.0 |
| Montgomery | 34 | -4.4 | 140 | 1.5 | 9.4 | 662,440 | 4,743 | 110,080 | 143.3 | 597,100 | 5.7 | 94.3 | 2 | 75.0 | 86 | 8.3 |
| Moore | 89 | 8.4 | 122 | 2.5 | 29.0 | 621,977 | 5,101 | 82,084 | 150.3 | 205,115 | 12.2 | 87.8 | 11 | 79.1 | 685 | 9.4 |
| Nash | 129 | -7.9 | 305 | 2.2 | 96.8 | 1,409,586 | 4,627 | 247,776 | 191.7 | 450,993 | 65.8 | 34.2 | 14 | 71.8 | 1,092 | 34.8 |
| New Hanover | 1 | -69.5 | 15 | 0.1 | 0.2 | 507,127 | 34,039 | 106,406 | D | D | D | 100.0 | NA | 81.4 | D | 3.4 |
| Northampton | 170 | 4.6 | 626 | 2.9 | 110.6 | 1,749,629 | 2,797 | 235,146 | 114.4 | 420,706 | 51.4 | 48.6 | NA | 73.5 | 4,683 | 79.4 |
| Onslow | 52 | -9.0 | 154 | 0.4 | 35.6 | 969,753 | 6,284 | 93,973 | 171.6 | 504,629 | 13.7 | 86.3 | NA | 75.0 | 862 | 25.6 |
| Orange | 70 | 23.4 | 102 | 0.9 | 32.1 | 681,306 | 6,686 | 68,490 | 37.7 | 54,974 | 67.3 | 32.7 | 28 | 88.3 | 428 | 18.7 |
| Pamlico | 43 | -7.5 | 433 | 3.4 | 37.4 | 1,579,956 | 3,652 | 266,792 | 23.4 | 233,730 | 98.3 | 1.7 | NA | 86.0 | 1,079 | 64.0 |
| Pasquotank | 72 | -0.1 | 573 | 0.3 | 68.8 | 2,282,057 | 3,984 | 346,522 | 48.8 | 387,452 | 99.1 | 0.9 | 1 | 76.2 | 2,408 | 59.5 |
| Pender | 64 | 15.6 | 192 | 1.8 | 41.2 | 1,026,573 | 5,349 | 98,925 | 200.3 | 596,140 | 19.2 | 80.8 | 9 | 78.9 | 583 | 20.2 |
| Perquimans | 80 | 0.3 | 539 | D | 74.5 | 1,919,310 | 3,560 | 300,322 | 70.6 | 473,671 | 57.9 | 42.1 | 3 | 78.5 | 2,646 | 69.8 |
| Person | 82 | -13.8 | 209 | 1.6 | 45.6 | 790,145 | 3,778 | 107,022 | 39.3 | 100,000 | 87.6 | 12.4 | 10 | 72.8 | 345 | 22.4 |
| Pitt | 186 | 8.5 | 390 | 2.8 | 149.8 | 1,499,480 | 3,845 | 183,107 | 242.5 | 507,234 | 41.6 | 58.4 | 4 | 81.8 | 3,291 | 45.6 |
| Polk | 29 | 20.8 | 104 | 0.1 | 7.9 | 667,163 | 6,442 | 50,460 | 6.6 | 23,491 | 74.8 | 25.2 | NA | 85.4 | 104 | 13.9 |
| Randolph | 148 | -5.8 | 108 | 1.2 | 61.7 | 545,745 | 5,052 | 77,013 | 281.9 | 206,035 | 9.9 | 90.1 | 20 | 75.8 | 360 | 9.3 |
| Richmond | 59 | 24.4 | 250 | 0.5 | 18.9 | 1,089,872 | 4,365 | 114,969 | 189.2 | 798,186 | 3.8 | 96.2 | NA | 72.6 | 473 | 23.6 |
| Robeson | 264 | -0.7 | 365 | 11.5 | 212.8 | 1,211,105 | 3,315 | 182,661 | 385.8 | 534,294 | 27.0 | 73.0 | 1 | 72.6 | 3,535 | 43.4 |
| Rockingham | 125 | 11.1 | 148 | 3.0 | 51.3 | 571,868 | 3,873 | 69,782 | 39.1 | 46,294 | 75.1 | 24.9 | 13 | 71.4 | 580 | 15.5 |
| Rowan | 119 | -1.8 | 129 | 0.8 | 67.9 | 752,022 | 5,850 | 103,875 | 81.8 | 88,412 | 58.8 | 41.2 | NA | 73.9 | 1,162 | 14.3 |
| Rutherford | 60 | 0.6 | 97 | 0.1 | 18.6 | 462,120 | 4,782 | 47,498 | 45.4 | 73,289 | 8.5 | 91.5 | 1 | 76.6 | 392 | 16.0 |
| Sampson | 301 | 3.3 | 314 | 17.9 | 209.6 | 1,534,742 | 4,891 | 269,485 | 1,249.1 | 1,301,188 | 16.3 | 83.7 | 18 | 78.1 | 2,424 | 32.5 |
| Scotland | 55 | -20.4 | 508 | D | 28.6 | 1,828,369 | 3,600 | 196,309 | 112.2 | 1,038,500 | 10.3 | 89.7 | 1 | 72.2 | 500 | 33.3 |
| Stanly | 96 | 3.0 | 143 | 0.1 | 54.2 | 656,070 | 4,599 | 91,390 | 90.3 | 134,391 | 24.9 | 75.1 | NA | 72.9 | 909 | 20.8 |
| Stokes | 93 | 1.5 | 109 | 0.4 | 34.4 | 443,265 | 4,084 | 60,719 | 42.4 | 49,558 | 29.3 | 70.7 | 3 | 75.5 | 148 | 12.0 |
| Surry | 152 | 20.1 | 143 | 0.6 | 73.0 | 640,586 | 4,471 | 96,475 | 230.1 | 216,264 | 21.5 | 78.5 | 8 | 75.5 | 483 | 17.7 |

# Table B. States and Counties — Water Use, Wholesale Trade, Retail Trade, and Real Estate

| STATE County | Water use, 2015 | | Wholesale Trade[1], 2017 | | | | Retail Trade[2], 2017 | | | | Real estate and rental and leasing,[2] 2017 | | | |
|---|---|---|---|---|---|---|---|---|---|---|---|---|---|---|
| | Public supply water withdrawn (mil gal/day) | Public supply gallons withdrawn per person per day | Number of establishments | Number of employees | Sales (mil dol) | Average payroll (mil dol) | Number of establishments | Number of employees | Sales (mil dol) | Average payroll (mil dol) | Number of establishments | Number of employees | Sales (mil dol) | Average payroll (mil dol) |
| | 133 | 134 | 135 | 136 | 137 | 138 | 139 | 140 | 141 | 142 | 143 | 144 | 145 | 146 |
| **NORTH CAROLINA— Cont'd** | | | | | | | | | | | | | | |
| Chowan | 1.39 | 96.6 | 19 | 211 | 190.4 | 9.2 | 53 | 383 | 94.2 | 8.1 | D | D | D | D |
| Clay | 0.21 | 19.6 | D | D | D | 1.6 | 36 | 399 | 107.0 | 9.1 | D | D | D | 0.2 |
| Cleveland | 12.26 | 126.5 | 65 | 1,108 | 1,365.6 | 43.2 | 368 | 4,054 | 1,028.4 | 93.8 | 84 | 342 | 61.3 | 9.0 |
| Columbus | 2.02 | 35.6 | 24 | 306 | 191.2 | 13.7 | 201 | 2,280 | 570.6 | 57.4 | 31 | 88 | 16.5 | 2.6 |
| Craven | 10.76 | 104.0 | 58 | 610 | 933.4 | 30.0 | 372 | 4,539 | 1,285.0 | 114.0 | 118 | 335 | 55.4 | 10.6 |
| Cumberland | 35.85 | 110.7 | 147 | 2,143 | 952.6 | 79.2 | 1,092 | 17,305 | 4,707.6 | 431.1 | 310 | 1,722 | 414.4 | 63.6 |
| Currituck | 3.29 | 130.2 | 20 | D | 62.7 | D | 131 | 979 | 310.0 | 28.4 | 43 | 437 | 66.2 | 17.7 |
| Dare | 7.28 | 204.1 | D | D | D | D | 421 | 3,681 | 967.3 | 101.3 | D | D | D | D |
| Davidson | 14.9 | 90.5 | 112 | 1,721 | 1,038.0 | 75.0 | 461 | 5,063 | 1,437.9 | 128.7 | 116 | 425 | 104.5 | 17.6 |
| Davie | 3.58 | 85.7 | 37 | 330 | 459.2 | 20.6 | 118 | 1,527 | 396.4 | 37.3 | D | D | D | 3.3 |
| Duplin | 4.76 | 80.5 | 32 | 578 | 826.6 | 22.0 | 181 | 1,720 | 443.7 | 39.9 | 26 | 54 | 4.7 | 1.0 |
| Durham | 27.74 | 92.2 | 262 | 13,956 | 10,669.0 | 1,691.7 | 936 | 15,667 | 4,117.6 | 403.7 | 375 | 2,235 | 531.8 | 117.5 |
| Edgecombe | 3.29 | 60.8 | 18 | 279 | 104.7 | 11.1 | 117 | 1,415 | 329.9 | 30.6 | 31 | 105 | 30.6 | 3.5 |
| Forsyth | 48.46 | 131.3 | 367 | 5,651 | 5,370.8 | 308.5 | 1,380 | 22,503 | 6,259.9 | 599.1 | 444 | 1,871 | 709.3 | 84.0 |
| Franklin | 0.95 | 14.9 | 47 | 538 | 292.0 | 26.9 | 143 | 1,394 | 351.9 | 32.7 | 28 | 37 | 6.7 | 1.2 |
| Gaston | 25.08 | 117.5 | 197 | 2,697 | 1,811.3 | 120.6 | 637 | 9,596 | 2,511.7 | 236.2 | 190 | 740 | 187.6 | 24.4 |
| Gates | 0.85 | 74.4 | D | D | D | D | 19 | 174 | 30.6 | 3.4 | NA | NA | NA | NA |
| Graham | 0.67 | 77.8 | NA | NA | NA | NA | 31 | 244 | 54.3 | 6.0 | 6 | D | 1.0 | D |
| Granville | 2.99 | 51.0 | D | D | D | D | 143 | 1,453 | 389.3 | 33.6 | 40 | 108 | 27.2 | 3.4 |
| Greene | 1.07 | 50.6 | D | D | D | 1.1 | 41 | 283 | 68.6 | 6.5 | 7 | 11 | 1.9 | 0.3 |
| Guilford | 48.12 | 93.0 | 974 | 15,679 | 16,113.7 | 1,203.9 | 1,846 | 28,166 | 8,015.9 | 742.4 | 719 | 4,900 | 875.3 | 238.3 |
| Halifax | 7.60 | 144.9 | D | D | D | D | 217 | 2,322 | 627.8 | 55.1 | 32 | 118 | 15.4 | 2.4 |
| Harnett | 20.57 | 160.5 | D | D | D | D | 272 | 3,500 | 998.2 | 78.5 | 77 | 301 | 35.2 | 8.0 |
| Haywood | 5.92 | 98.9 | 41 | 427 | 198.2 | 14.9 | 236 | 3,214 | 929.4 | 78.5 | 66 | 201 | 29.6 | 5.7 |
| Henderson | 11.05 | 98.1 | 116 | 1,229 | 666.7 | 58.1 | 399 | 5,626 | 1,584.0 | 144.4 | 145 | 399 | 72.2 | 11.9 |
| Hertford | 2.09 | 86.4 | 18 | 152 | 59.3 | 5.6 | 98 | 1,134 | 261.5 | 25.9 | D | D | D | 0.9 |
| Hoke | 4.42 | 83.9 | 11 | 80 | 46.8 | 3.6 | 81 | 781 | 188.8 | 15.5 | 19 | 47 | 11.6 | 1.7 |
| Hyde | 1.08 | 195.4 | 11 | 95 | 44.4 | 3.2 | 40 | 170 | 29.4 | 3.4 | 9 | 60 | 6.4 | 3.5 |
| Iredell | 11.54 | 67.9 | 264 | 2,534 | 1,573.9 | 147.0 | 655 | 9,275 | 2,770.1 | 235.8 | 243 | 676 | 148.5 | 25.4 |
| Jackson | 1.94 | 47.0 | D | D | D | D | 163 | 1,903 | 449.3 | 47.2 | D | D | D | D |
| Johnston | 15.77 | 84.9 | 111 | 1,291 | 1,100.3 | 66.8 | 558 | 8,051 | 2,247.5 | 182.4 | 120 | 457 | 134.5 | 18.2 |
| Jones | 0.72 | 71.9 | 7 | 123 | 58.1 | 4.6 | 29 | 140 | 47.8 | 3.7 | NA | NA | NA | NA |
| Lee | 7.41 | 124.2 | 50 | 914 | 693.9 | 44.7 | 251 | 3,025 | 889.8 | 75.5 | 43 | 128 | 28.5 | 5.3 |
| Lenoir | 8.81 | 151.6 | 60 | 622 | 274.4 | 28.5 | 218 | 2,944 | 760.2 | 79.8 | 45 | 204 | 52.4 | 15.5 |
| Lincoln | 6.63 | 81.8 | 67 | 1,268 | 732.7 | 60.0 | 238 | 3,189 | 867.2 | 78.1 | 73 | 129 | 25.4 | 4.6 |
| McDowell | 1.95 | 43.3 | 32 | 277 | 80.6 | 10.6 | 127 | 1,634 | 473.8 | 39.9 | 23 | 53 | 8.6 | 1.4 |
| Macon | 2.07 | 60.5 | 21 | 137 | 82.5 | 8.0 | 218 | 1,962 | 474.9 | 51.7 | 61 | 116 | 25.8 | 4.3 |
| Madison | 0.93 | 44.0 | 6 | 19 | 9.5 | 0.6 | 40 | 451 | 98.2 | 8.5 | 18 | 20 | 4.1 | 0.6 |
| Martin | 0.56 | 24.0 | 23 | 186 | 122.0 | 7.6 | 77 | 966 | 222.4 | 21.3 | 13 | 35 | 5.7 | 1.0 |
| Mecklenburg | 112.24 | 108.5 | 1,889 | 31,210 | 25,783.5 | 2,104.7 | 3,608 | 61,937 | 18,367.7 | 1,714.1 | 2,389 | 12,728 | 4,500.6 | 766.5 |
| Mitchell | 1.07 | 70.2 | 8 | 23 | 19.4 | 1.0 | 58 | 743 | 179.7 | 16.9 | 14 | 79 | 7.2 | 2.0 |
| Montgomery | 2.25 | 81.7 | 22 | 249 | 109.5 | 10.5 | 76 | 736 | 202.0 | 17.3 | 15 | 20 | 3.1 | 0.6 |
| Moore | 6.78 | 71.9 | 60 | 440 | 245.7 | 21.0 | 361 | 4,797 | 1,439.6 | 116.0 | 113 | 254 | 57.3 | 9.5 |
| Nash | 9.13 | 97.2 | 112 | 2,462 | 3,704.2 | 115.6 | 393 | 4,931 | 1,309.3 | 115.4 | 81 | 415 | 61.7 | 14.6 |
| New Hanover | 6.25 | 28.4 | 306 | 3,030 | 1,613.6 | 171.5 | 1,082 | 16,087 | 4,908.8 | 447.4 | 501 | 1,820 | 459.9 | 77.9 |
| Northampton | 1.11 | 54.3 | D | D | D | D | 48 | 405 | 130.3 | 12.1 | 5 | 13 | 0.9 | 0.2 |
| Onslow | 17.87 | 95.9 | 55 | 320 | 144.1 | 13.4 | 547 | 8,416 | 2,334.6 | 206.5 | 200 | 744 | 146.7 | 24.8 |
| Orange | 8.78 | 62.1 | 80 | 852 | 580.6 | 53.0 | 346 | 5,825 | 1,618.6 | 177.3 | 188 | 768 | 146.3 | 29.6 |
| Pamlico | 1.25 | 97.8 | 11 | 75 | 46.4 | 2.5 | 40 | 622 | 131.4 | 15.2 | 13 | 38 | 3.2 | 0.8 |
| Pasquotank | 2.67 | 67.0 | 29 | 456 | 292.2 | 20.9 | 182 | 2,469 | 734.2 | 63.0 | 31 | 151 | 29.9 | 4.9 |
| Pender | 2.77 | 48.1 | 33 | 301 | 187.4 | 11.6 | 157 | 1,812 | 435.4 | 42.2 | 62 | 333 | 71.1 | 15.3 |
| Perquimans | 0.75 | 55.8 | D | D | D | D | 28 | 214 | 53.0 | 4.3 | D | D | D | D |
| Person | 2.53 | 64.4 | 21 | 206 | 68.9 | 10.0 | 128 | 1,500 | 384.0 | 35.6 | 25 | 60 | 13.5 | 2.0 |
| Pitt | 13.64 | 77.6 | 131 | 1,514 | 1,021.3 | 73.6 | 625 | 9,154 | 2,556.1 | 229.1 | 184 | 717 | 170.0 | 25.7 |
| Polk | 0.91 | 44.7 | 17 | 38 | 11.7 | 1.9 | 60 | 452 | 112.5 | 9.5 | D | D | D | D |
| Randolph | 16.76 | 117.4 | 131 | 1,902 | 858.9 | 98.8 | 395 | 4,659 | 1,269.8 | 115.2 | 77 | 267 | 43.8 | 9.6 |
| Richmond | 9.87 | 217.2 | 16 | 266 | 223.0 | 8.6 | 174 | 1,999 | 481.9 | 46.3 | 39 | 274 | 26.0 | 5.3 |
| Robeson | 21.28 | 158.6 | 65 | 690 | 630.7 | 30.1 | 404 | 4,898 | 1,414.4 | 118.5 | 59 | 169 | 34.2 | 4.8 |
| Rockingham | 11.41 | 124.3 | 51 | 1,168 | 607.8 | 42.3 | 299 | 3,611 | 892.2 | 86.1 | 61 | 154 | 22.9 | 4.0 |
| Rowan | 9.64 | 69.3 | 117 | 1,522 | 846.3 | 67.8 | 410 | 4,591 | 1,357.9 | 114.0 | 98 | 286 | 54.4 | 10.7 |
| Rutherford | 7.68 | 115.7 | D | D | D | D | 202 | 2,481 | 646.4 | 60.6 | 51 | 110 | 31.3 | 3.4 |
| Sampson | 3.67 | 57.6 | 39 | 675 | 369.8 | 31.0 | 184 | 2,246 | 621.8 | 50.5 | 29 | 67 | 14.8 | 2.1 |
| Scotland | 3.10 | 87.3 | D | D | D | D | 127 | 1,511 | 337.1 | 32.8 | 25 | 56 | 14.7 | 2.1 |
| Stanly | 6.40 | 105.4 | D | D | D | D | 228 | 2,768 | 733.6 | 62.4 | 42 | 186 | 28.7 | 5.2 |
| Stokes | 0.30 | 6.5 | D | D | D | 2.0 | 106 | 1,247 | 285.5 | 26.8 | D | D | D | D |
| Surry | 4.41 | 60.6 | D | D | D | D | 335 | 4,344 | 1,131.6 | 108.9 | 62 | 211 | 30.3 | 5.6 |

1   Merchant wholesalers, except manufacturers' sales branches and offices.   2. Employer establishments.

| STATE County | Professional, scientific, and technical services, 2017 | | | | Manufacturing, 2017 | | | | Accommodation and food services, 2017 | | | |
|---|---|---|---|---|---|---|---|---|---|---|---|---|
| | Number of establishments | Number of employees | Sales (mil dol) | Average payroll (mil dol) | Number of establishments | Number of employees | Sales (mil dol) | Average payroll (mil dol) | Number of establishments | Number of employees | Sales (mil dol) | Annual payroll (mil dol) |
| | 147 | 148 | 149 | 150 | 151 | 152 | 153 | 154 | 155 | 156 | 157 | 158 |
| NORTH CAROLINA— Cont'd | | | | | | | | | | | | |
| Chowan | D | D | D | D | 14 | 587 | 512.6 | 29.6 | 27 | 400 | 20.4 | 5.0 |
| Clay | 15 | 30 | 3 | 1.0 | D | 198 | D | 8.5 | D | D | D | 3.1 |
| Cleveland | 124 | 957 | 108 | 42.4 | 121 | 6,047 | 2,412.7 | 328.5 | 158 | 2,904 | 147.3 | 37.6 |
| Columbus | 62 | 208 | 28 | 7.6 | 33 | 1,923 | 761.7 | 117.2 | 87 | 1,128 | 61.9 | 15.4 |
| Craven | D | D | D | D | 68 | 3,408 | 1,408.6 | 179.4 | 204 | 3,647 | 199.8 | 52.9 |
| Cumberland | 531 | 6,492 | 796 | 325.9 | 95 | 5,745 | 2,570.0 | 329.9 | 703 | 15,036 | 760.9 | 208.5 |
| Currituck | 50 | 149 | 18 | 7.6 | 13 | 80 | 14.0 | 3.8 | 83 | D | 53.1 | D |
| Dare | D | D | D | D | 35 | 454 | 106.2 | 23.0 | D | D | D | D |
| Davidson | 186 | 821 | 84 | 29.1 | 234 | 9,287 | 2,697.4 | 405.4 | 237 | 3,901 | 192.5 | 53.3 |
| Davie | D | D | D | D | 42 | 1,402 | 695.7 | 65.9 | 72 | 1,162 | 56.8 | 14.6 |
| Duplin | 50 | 148 | 17 | 5.7 | 41 | 6,209 | 1,954.9 | 222.1 | 71 | 1,137 | 54.0 | 13.9 |
| Durham | D | D | D | D | 180 | 13,308 | 9,017.6 | 966.5 | 771 | 16,566 | 1,061.2 | 309.0 |
| Edgecombe | 39 | 118 | 14 | 4.5 | 33 | 2,529 | 779.7 | 100.8 | 57 | 986 | 45.8 | 12.3 |
| Forsyth | 963 | 6,564 | 1,059 | 393.8 | 304 | 16,032 | 26,160.4 | 914.9 | 802 | 17,489 | 934.5 | 265.8 |
| Franklin | 84 | 218 | 25 | 8.3 | 54 | 3,298 | 1,311.7 | 210.1 | 60 | 879 | 40.3 | 10.8 |
| Gaston | D | D | D | D | 271 | 12,792 | 5,397.7 | 666.0 | 373 | 7,221 | 394.5 | 105.4 |
| Gates | D | D | D | 1.0 | 4 | 150 | 45.6 | 6.6 | D | D | D | D |
| Graham | D | D | D | 0.6 | 7 | 42 | 8.5 | 1.4 | D | D | D | D |
| Granville | 68 | 273 | 29 | 10.9 | 45 | 4,467 | 4,335.3 | 229.8 | 66 | 1,128 | 54.3 | 6.3 |
| Greene | 8 | 56 | 6 | 2.4 | 13 | 116 | 37.6 | 5.8 | 19 | 235 | 11.2 | 3.1 |
| Guilford | D | D | D | D | 602 | 31,362 | 17,742.3 | 1,916.0 | 1,231 | 27,375 | 1,433.5 | 412.2 |
| Halifax | 42 | 195 | 16 | 5.2 | 33 | 1,667 | 599.2 | 74.5 | D | D | D | D |
| Harnett | 119 | 520 | 63 | 22.3 | 66 | 1,666 | 387.4 | 77.0 | 167 | 3,020 | 146.0 | 40.1 |
| Haywood | 115 | 442 | 52 | 18.8 | 43 | 1,896 | 993.6 | 116.3 | 152 | 2,383 | 129.6 | 38.1 |
| Henderson | 253 | 954 | 113 | 38.8 | 112 | 5,831 | 2,413.0 | 317.6 | 225 | 3,858 | 229.5 | 65.3 |
| Hertford | D | D | D | D | 16 | 1,030 | 1,161.7 | 66.0 | 43 | 685 | 30.6 | 8.2 |
| Hoke | 35 | 122 | 9 | 3.1 | 21 | 1,484 | 613.0 | 57.0 | 32 | 514 | 24.2 | 6.4 |
| Hyde | D | D | D | 0.6 | 4 | 58 | 12.7 | 2.3 | 30 | 189 | 17.6 | 4.8 |
| Iredell | 437 | 3,111 | 461 | 168.9 | 299 | 11,356 | 4,071.5 | 571.4 | 380 | 7,339 | 393.4 | 106.4 |
| Jackson | D | D | D | D | 17 | 270 | 71.9 | 9.8 | 122 | 4,558 | 715.1 | 131.6 |
| Johnston | 250 | 1,072 | 124 | 47.1 | 103 | 6,182 | 2,431.3 | 373.3 | 290 | 5,881 | 313.9 | 81.3 |
| Jones | 6 | 20 | 2 | 0.7 | 8 | 22 | 5.3 | 1.2 | 8 | 85 | 3.0 | 0.9 |
| Lee | D | D | D | D | 80 | 7,887 | 3,278.2 | 430.1 | 115 | 2,011 | 102.5 | 27.9 |
| Lenoir | 65 | 579 | 136 | 31.3 | 51 | 6,013 | 1,918.8 | 236.9 | 102 | 1,971 | 94.2 | 25.2 |
| Lincoln | D | D | D | D | 106 | 4,111 | 2,000.3 | 173.1 | 119 | 2,119 | 110.9 | 31.2 |
| McDowell | 38 | 196 | 16 | 6.8 | 47 | 4,651 | 1,638.2 | 224.3 | 80 | 1,277 | 58.1 | 15.5 |
| Macon | 74 | 214 | 23 | 8.5 | 29 | 385 | 117.5 | 18.8 | 109 | 1,538 | 122.5 | 32.1 |
| Madison | 24 | 32 | 4 | 1.2 | 14 | 284 | 107.2 | 14.9 | 38 | 388 | 21.7 | 6.3 |
| Martin | 18 | 80 | 7 | 2.3 | 18 | 897 | 641.0 | 38.1 | D | D | D | D |
| Mecklenburg | D | D | D | D | 760 | 25,616 | 10,022.7 | 1,423.5 | 2,758 | 62,105 | 4,016.2 | 1,101.5 |
| Mitchell | 17 | 69 | 6 | 2.0 | 23 | 450 | 76.3 | 21.9 | 38 | 456 | 23.6 | 6.1 |
| Montgomery | 17 | 64 | 6 | 2.2 | 63 | 2,907 | 1,035.8 | 130.2 | D | D | D | D |
| Moore | 220 | 1,375 | 245 | 84.3 | 73 | 1,415 | 425.6 | 68.8 | 224 | 5,154 | 309.4 | 91.9 |
| Nash | 135 | 897 | 124 | 37.9 | 88 | 7,157 | 4,144.5 | 339.2 | 188 | 3,846 | 194.3 | 51.4 |
| New Hanover | D | D | D | D | 185 | 5,596 | 2,092.4 | 449.5 | 771 | 15,249 | 856.2 | 243.6 |
| Northampton | 6 | 16 | 2 | 0.5 | 8 | 408 | 271.5 | 20.8 | D | D | D | D |
| Onslow | D | D | D | D | 47 | 1,040 | 391.8 | 42.3 | 399 | 7,870 | 435.4 | 117.0 |
| Orange | D | D | D | D | 75 | 2,233 | 724.5 | 120.7 | 328 | 6,267 | 349.4 | 110.4 |
| Pamlico | 17 | 44 | 5 | 1.7 | 8 | 96 | 34.6 | 3.7 | 27 | 526 | 32.8 | 9.2 |
| Pasquotank | 70 | 374 | 50 | 18.4 | 22 | 563 | 176.4 | 27.1 | 95 | 1,786 | 89.2 | 24.3 |
| Pender | D | D | D | D | 40 | 919 | 222.8 | 35.9 | 110 | 1,445 | 78.6 | 22.9 |
| Perquimans | 15 | 35 | 4 | 1.7 | D | D | D | D | 13 | 237 | 13.4 | 4.6 |
| Person | D | D | D | 3.1 | 29 | 1,368 | 949.5 | 75.2 | 57 | 940 | 49.7 | 11.9 |
| Pitt | D | D | D | D | 93 | 6,726 | 2,690.2 | 375.9 | 369 | 8,445 | 417.1 | 111.3 |
| Polk | 40 | 173 | 23 | 7.2 | 20 | 308 | 71.5 | 10.8 | D | D | D | 6.1 |
| Randolph | 139 | 658 | 70 | 22.8 | 281 | 15,975 | 3,568.9 | 581.9 | 207 | 3,883 | 193.9 | 53.6 |
| Richmond | 45 | 146 | 16 | 4.7 | 35 | 2,875 | 895.6 | 106.5 | 63 | 1,039 | 49.5 | 12.0 |
| Robeson | D | D | D | D | 57 | 7,537 | 3,487.2 | 301.0 | 184 | 3,678 | 190.5 | 46.5 |
| Rockingham | D | D | D | D | 82 | 5,998 | 2,177.2 | 270.5 | 146 | 2,264 | 109.4 | 30.0 |
| Rowan | 178 | 799 | 112 | 34.0 | 181 | 8,381 | 4,026.0 | 403.6 | 227 | 4,081 | 218.8 | 60.0 |
| Rutherford | D | D | D | D | 67 | 3,073 | 1,130.4 | 138.5 | 120 | 1,636 | 87.5 | 23.6 |
| Sampson | 51 | 268 | 32 | 11.6 | 41 | 3,287 | 955.4 | 132.2 | 88 | 1,323 | 61.3 | 15.7 |
| Scotland | 24 | 89 | 8 | 3.0 | 27 | 1,904 | 963.9 | 90.9 | 57 | 1,069 | 55.4 | 14.9 |
| Stanly | 63 | 290 | 36 | 11.5 | 91 | 2,713 | 909.3 | 132.3 | 115 | 1,877 | 88.0 | 23.5 |
| Stokes | 39 | 154 | 13 | 5.1 | 33 | 948 | 405.1 | 39.4 | 59 | 933 | 41.1 | 11.3 |
| Surry | 90 | 422 | 41 | 14.8 | 100 | 3,336 | 1,012.1 | 136.6 | 161 | 2,768 | 127.8 | 36.2 |

Items 147—158

# Health Care and Social Assistance, Other Services, Nonemployer Businesses, and Residential Construction

| STATE County | Health care and social assistance, 2017 | | | | Other services, 2017 | | | | Nonemployer businesses, 2018 | | Value of residential construction authorized by building permits, 2020 | |
|---|---|---|---|---|---|---|---|---|---|---|---|---|
| | Number of establishments | Number of employees | Receipts (mil dol) | Annual payroll (mil dol) | Number of establishments | Number of employees | Receipts (mil dol) | Annual payroll (mil dol) | Number | Receipts (mil dol) | New construction ($1,000) | Number of housing units |
| | 159 | 160 | 161 | 162 | 163 | 164 | 165 | 166 | 167 | 168 | 169 | 170 |
| NORTH CAROLINA— Cont'd | | | | | | | | | | | | |
| Chowan | D | D | D | D | 17 | 76 | 6.1 | 1.9 | 1,010 | 41.4 | 1,686 | 29 |
| Clay | 24 | 316 | 22.2 | 10.4 | D | D | D | 1.2 | 1,088 | 44.1 | 17,727 | 86 |
| Cleveland | 239 | 5,249 | 636.6 | 227.5 | 119 | 649 | 67.7 | 19.7 | 5,721 | 185.9 | 46,631 | 248 |
| Columbus | 171 | 2,863 | 262.7 | 98.1 | D | D | D | 4.4 | 3,506 | 137.6 | 14,144 | 78 |
| Craven | 264 | 7,776 | 825.0 | 342.0 | 147 | 603 | 61.9 | 16.8 | 6,270 | 246.1 | 115,626 | 602 |
| Cumberland | 722 | 20,168 | 2,318.6 | 927.1 | 415 | 2,560 | 254.4 | 74.0 | 18,982 | 717.0 | 135,474 | 710 |
| Currituck | 29 | 254 | 20.4 | 9.3 | 47 | 316 | 25.8 | 6.5 | 2,226 | 110.3 | 116,404 | 560 |
| Dare | 95 | 1,074 | 145.5 | 55.1 | D | D | D | D | 5,470 | 299.0 | 161,782 | 450 |
| Davidson | 194 | 4,938 | 398.2 | 165.5 | 197 | 757 | 82.4 | 22.3 | 11,596 | 472.0 | 215,817 | 768 |
| Davie | 58 | 1,442 | 133.9 | 50.1 | 66 | 222 | 22.5 | 5.1 | 3,238 | 149.1 | 48,727 | 256 |
| Duplin | 95 | 1,606 | 149.1 | 58.2 | D | D | D | D | 3,276 | 146.5 | 0 | 0 |
| Durham | 875 | 28,074 | 4,835.1 | 1,520.1 | 490 | 5,905 | 1,719.8 | 344.2 | 26,142 | 966.7 | 626,441 | 3,956 |
| Edgecombe | 98 | 2,565 | 231.9 | 77.8 | 40 | 144 | 12.3 | 3.3 | 2,650 | 82.2 | 6,465 | 42 |
| Forsyth | 830 | 35,474 | 4,304.1 | 1,490.1 | 611 | 3,625 | 631.1 | 116.0 | 27,723 | 1,190.2 | 561,583 | 2,918 |
| Franklin | 77 | 922 | 66.8 | 27.8 | 74 | 262 | 32.4 | 6.8 | 4,701 | 192.1 | 455,710 | 913 |
| Gaston | 473 | 11,894 | 1,321.5 | 524.5 | 308 | 1,520 | 156.6 | 40.6 | 14,291 | 562.9 | 455,710 | 1,728 |
| Gates | 12 | 99 | 6.9 | 2.8 | D | D | D | D | 549 | 18.6 | 5,228 | 27 |
| Graham | 10 | 264 | 14.8 | 6.7 | D | D | D | 0.6 | 715 | 25.3 | 3,000 | 14 |
| Granville | 103 | 4,059 | 299.4 | 156.1 | D | D | D | D | 3,480 | 129.6 | 65,277 | 315 |
| Greene | 38 | 632 | 39.7 | 17.6 | D | D | D | 1.9 | 1,075 | 38.8 | 5,183 | 29 |
| Guilford | 1,294 | 35,445 | 4,243.5 | 1,645.2 | 870 | 6,280 | 1,265.9 | 210.8 | 42,996 | 1,957.4 | 503,315 | 2,566 |
| Halifax | 122 | 2,808 | 222.9 | 91.4 | D | D | D | D | 2,582 | 83.6 | 18,401 | 64 |
| Harnett | 178 | 4,596 | 461.9 | 182.1 | 111 | 437 | 53.1 | 12.4 | 7,511 | 303.4 | 132,584 | 911 |
| Haywood | 138 | 3,171 | 303.1 | 123.8 | 96 | 467 | 51.6 | 14.4 | 5,045 | 193.9 | 58,666 | 219 |
| Henderson | 278 | 6,208 | 659.3 | 259.2 | 200 | 1,000 | 113.5 | 29.2 | 9,657 | 408.3 | 167,752 | 572 |
| Hertford | 78 | 2,037 | 177.1 | 71.3 | D | D | D | D | 1,067 | 34.4 | 160 | 1 |
| Hoke | 72 | 1,681 | 117.6 | 47.6 | 30 | 134 | 9.5 | 2.9 | 2,706 | 76.7 | 79,407 | 339 |
| Hyde | D | D | D | D | D | D | D | D | 593 | 22.7 | 1,093 | 10 |
| Iredell | 493 | 9,223 | 964.9 | 393.9 | 302 | 1,790 | 187.2 | 52.2 | 14,434 | 705.2 | 422,399 | 1,905 |
| Jackson | 91 | 1,772 | 219.2 | 81.6 | 59 | 219 | 26.6 | 6.4 | 3,200 | 133.5 | 166,767 | 832 |
| Johnston | 330 | 6,077 | 605.9 | 251.9 | 195 | 833 | 87.8 | 22.5 | 14,299 | 653.0 | 587,243 | 3,447 |
| Jones | 13 | 249 | 18.7 | 8.6 | D | D | D | D | 583 | 21.4 | 2,909 | 28 |
| Lee | 160 | 2,867 | 275.5 | 106.8 | 89 | 417 | 33.3 | 10.4 | 4,005 | 190.4 | 44,834 | 245 |
| Lenoir | 159 | 4,193 | 323.8 | 148.3 | 79 | 487 | 47.5 | 14.4 | 6,357 | 292.3 | 225,734 | 107 |
| Lincoln | 156 | 2,544 | 316.1 | 104.8 | 138 | 546 | 51.2 | 14.8 | 6,357 | 292.3 | 225,734 | 1,336 |
| McDowell | 79 | 1,741 | 157.0 | 57.9 | 39 | 204 | 24.8 | 6.8 | 2,722 | 99.7 | 40,927 | 128 |
| Macon | 94 | 1,505 | 165.8 | 60.9 | 81 | 355 | 48.2 | 10.7 | 3,313 | 136.0 | 34,508 | 109 |
| Madison | 33 | 544 | 42.2 | 18.2 | 17 | 62 | 7.4 | 1.9 | 2,037 | 82.3 | 22,147 | 95 |
| Martin | 55 | 1,094 | 82.4 | 32.2 | D | D | D | D | 1,298 | 45.4 | 0 | 0 |
| Mecklenburg | 2,950 | 73,197 | 11,317.4 | 4,008.5 | 2,043 | 16,834 | 2,629.8 | 581.9 | 103,292 | 4,974.7 | 2,024,696 | 11,067 |
| Mitchell | 39 | 817 | 80.2 | 29.5 | 30 | 81 | 6.6 | 1.7 | 1,162 | 41.0 | 7,167 | 36 |
| Montgomery | 48 | 1,281 | 67.2 | 32.8 | 34 | 342 | 23.7 | 10.3 | 1,508 | 55.2 | 15,276 | 85 |
| Moore | 293 | 9,678 | 1,243.4 | 482.7 | 162 | 880 | 80.0 | 25.5 | 8,133 | 407.9 | 229,017 | 1,021 |
| Nash | 227 | 5,927 | 515.3 | 229.8 | 102 | 562 | 171.1 | 16.6 | 21,300 | 1,070.5 | 44,844 | 297 |
| New Hanover | 825 | 18,772 | 2,310.5 | 901.7 | 514 | 3,163 | 320.7 | 94.6 | 21,300 | 1,070.5 | 697,907 | 3,095 |
| Northampton | 29 | 529 | 31.9 | 13.8 | D | D | D | D | 915 | 32.8 | 9,155 | 30 |
| Onslow | 263 | 5,331 | 642.5 | 218.9 | 211 | 1,071 | 104.4 | 28.1 | 10,712 | 387.6 | 215,040 | 1,435 |
| Orange | 406 | 16,410 | 2,030.8 | 876.2 | 216 | 1,930 | 369.8 | 81.9 | 13,826 | 692.0 | 126,891 | 465 |
| Pamlico | 26 | 333 | 21.4 | 10.1 | D | D | D | D | 993 | 42.3 | 13,125 | 105 |
| Pasquotank | 138 | 2,414 | 299.0 | 114.7 | 74 | 468 | 33.1 | 10.3 | 2,664 | 92.6 | 29,669 | 162 |
| Pender | 111 | 1,577 | 122.4 | 49.9 | 72 | 286 | 26.1 | 8.8 | 4,672 | 209.2 | 134,717 | 762 |
| Perquimans | D | D | D | 4.3 | D | D | 5.1 | D | 958 | 33.0 | 9,249 | 45 |
| Person | 85 | 1,287 | 106.2 | 40.9 | 47 | 209 | 24.1 | 8.2 | 2,262 | 83.3 | 40,069 | 168 |
| Pitt | 499 | 16,338 | 2,158.5 | 773.9 | 220 | 1,287 | 123.7 | 36.0 | 10,993 | 433.2 | 162,847 | 719 |
| Polk | 48 | 1,485 | 123.5 | 51.2 | 26 | 111 | 14.6 | 3.2 | 1,994 | 79.8 | 25,483 | 94 |
| Randolph | 229 | 4,612 | 383.6 | 165.3 | 157 | 830 | 91.5 | 24.3 | 9,940 | 418.8 | 83,676 | 372 |
| Richmond | 92 | 1,922 | 187.8 | 77.7 | D | D | D | D | 2,037 | 80.0 | 10,851 | 277 |
| Robeson | 267 | 6,823 | 588.2 | 259.1 | 89 | 330 | 30.2 | 7.6 | 7,458 | 265.9 | 34,925 | 189 |
| Rockingham | 158 | 3,038 | 326.4 | 109.4 | 110 | 439 | 40.7 | 12.5 | 4,980 | 193.4 | 58,319 | 209 |
| Rowan | 248 | 9,245 | 1,079.7 | 518.1 | 163 | 659 | 73.0 | 21.3 | 9,488 | 371.6 | 224,461 | 874 |
| Rutherford | 126 | 2,720 | 234.6 | 88.1 | 79 | 469 | 44.0 | 13.7 | 4,124 | 173.2 | 31,888 | 163 |
| Sampson | 121 | 2,485 | 183.5 | 79.0 | 50 | 305 | 25.2 | 7.5 | 3,670 | 151.5 | 21,413 | 107 |
| Scotland | 100 | 2,358 | 252.2 | 98.8 | 31 | 110 | 10.4 | 2.8 | 1,860 | 57.9 | 6,803 | 33 |
| Stanly | 165 | 2,798 | 282.2 | 108.8 | 88 | 440 | 38.9 | 12.1 | 4,130 | 180.1 | 50,515 | 300 |
| Stokes | 50 | 1,430 | 107.8 | 43.5 | 44 | 157 | 21.5 | 5.3 | 2,861 | 111.3 | 28,563 | 112 |
| Surry | 166 | 4,067 | 408.6 | 167.5 | 90 | 476 | 45.5 | 13.4 | 4,883 | 223.9 | 18,377 | 107 |

# Table B. States and Counties — Government Employment and Payroll, and Local Government Finances

| | Government employment and payroll, 2017 | | | | | | | | | Local government finances, 2017 | | | | |
| STATE County | | | March payroll (percent of total) | | | | | | | General revenue | | | | |
| | Full-time equivalent employees | March payroll (dollars) | Administration, judicial, and legal | Police and corrections | Fire protection | Highways and transportation | Health and welfare | Natural resources and utilities | Education and libraries | Total (mil dol) | Intergovernmental (mil dol) | Taxes Total (mil dol) | Per capita[1] Total (dollars) | Per capita[1] Property (dollars) |
| | 171 | 172 | 173 | 174 | 175 | 176 | 177 | 178 | 179 | 180 | 181 | 182 | 183 | 184 |
|---|---|---|---|---|---|---|---|---|---|---|---|---|---|---|
| **NORTH CAROLINA— Cont'd** | | | | | | | | | | | | | | |
| Chowan | 576 | 1,815,508 | 5.8 | 10.4 | 1.7 | 1.1 | 8.0 | 8.4 | 60.0 | 49.3 | 24.8 | 16.5 | 1,176 | 972 |
| Clay | 406 | 1,175,292 | 5.2 | 11.4 | 0.0 | 0.0 | 22.6 | 3.3 | 52.3 | 32.0 | 17.2 | 11.3 | 1,031 | 783 |
| Cleveland | 3,804 | 13,068,896 | 4.1 | 7.9 | 1.7 | 0.9 | 14.1 | 6.5 | 63.0 | 321.2 | 177.3 | 102.0 | 1,050 | 804 |
| Columbus | 2,426 | 7,265,072 | 4.9 | 6.3 | 0.8 | 0.5 | 11.7 | 2.5 | 67.6 | 119.9 | 88.1 | 16.0 | 286 | 252 |
| Craven | 5,659 | 23,933,631 | 3.2 | 4.3 | 1.7 | 1.0 | 53.9 | 4.6 | 29.7 | 629.0 | 146.8 | 84.1 | 821 | 581 |
| Cumberland | 19,164 | 73,834,537 | 2.4 | 7.0 | 2.0 | 1.0 | 45.1 | 4.1 | 35.1 | 1,032.1 | 550.1 | 382.2 | 1,155 | 831 |
| Currituck | 913 | 3,283,009 | 5.3 | 11.2 | 1.6 | 0.3 | 15.3 | 6.0 | 55.9 | 100.2 | 32.3 | 57.8 | 2,198 | 1,182 |
| Dare | 2,036 | 7,931,376 | 10.5 | 14.0 | 4.0 | 1.3 | 15.2 | 10.3 | 40.1 | 238.0 | 56.5 | 160.4 | 4,431 | 2,505 |
| Davidson | 5,250 | 17,412,155 | 3.5 | 7.6 | 2.7 | 1.0 | 6.4 | 4.9 | 67.7 | 423.5 | 233.7 | 149.6 | 906 | 655 |
| Davie | 1,287 | 5,002,042 | 4.8 | 7.8 | 0.3 | 0.0 | 13.0 | 2.2 | 69.7 | 117.0 | 54.6 | 48.2 | 1,140 | 915 |
| Duplin | 2,751 | 9,217,556 | 2.6 | 5.4 | 3.6 | 0.6 | 35.9 | 3.4 | 47.1 | 481.5 | 409.3 | 49.9 | 847 | 601 |
| Durham | 11,409 | 46,922,955 | 5.5 | 11.9 | 5.2 | 1.4 | 14.8 | 6.7 | 49.6 | 1,163.6 | 402.7 | 572.6 | 1,834 | 1,377 |
| Edgecombe | 1,866 | 6,236,459 | 1.8 | 8.6 | 1.2 | 1.1 | 14.6 | 6.6 | 63.7 | 141.1 | 84.1 | 39.7 | 753 | 615 |
| Forsyth | 14,332 | 48,791,745 | 5.1 | 10.2 | 5.5 | 1.6 | 7.5 | 4.8 | 64.2 | 1,248.9 | 549.4 | 525.0 | 1,397 | 1,059 |
| Franklin | 1,901 | 5,975,975 | 4.8 | 9.8 | 0.4 | 0.7 | 15.1 | 4.7 | 60.3 | 91.3 | 62.6 | 21.1 | 319 | 288 |
| Gaston | 8,063 | 29,522,445 | 6.9 | 6.5 | 2.7 | 2.4 | 17.8 | 6.2 | 55.6 | 1,010.3 | 618.6 | 259.9 | 1,183 | 892 |
| Gates | 410 | 1,369,610 | 3.9 | 2.9 | 0.0 | 1.2 | 6.1 | 2.3 | 81.7 | 28.9 | 17.8 | 9.2 | 795 | 604 |
| Graham | 393 | 1,257,751 | 4.8 | 6.4 | 0.0 | 2.1 | 15.5 | 2.2 | 62.1 | 31.5 | 19.2 | 9.5 | 1,119 | 845 |
| Granville | 1,572 | 5,369,066 | 5.3 | 10.7 | 0.7 | 1.1 | 6.8 | 4.6 | 66.7 | 163.2 | 77.9 | 58.7 | 989 | 798 |
| Greene | 729 | 2,325,290 | 3.3 | 4.8 | 2.9 | 0.2 | 7.1 | 4.5 | 75.9 | 37.1 | 30.7 | 3.4 | 162 | 158 |
| Guilford | 18,494 | 73,703,685 | 5.2 | 12.0 | 5.0 | 2.6 | 7.8 | 7.1 | 57.5 | 1,905.5 | 821.7 | 783.7 | 1,484 | 1,118 |
| Halifax | 2,203 | 7,410,028 | 5.5 | 6.6 | 1.7 | 1.7 | 13.8 | 6.6 | 60.6 | 161.4 | 92.3 | 46.6 | 909 | 684 |
| Harnett | 4,945 | 15,502,659 | 4.3 | 7.0 | 0.1 | 0.7 | 28.1 | 4.4 | 54.8 | 202.1 | 151.6 | 35.5 | 268 | 248 |
| Haywood | 2,174 | 7,275,501 | 5.1 | 10.0 | 1.4 | 1.0 | 13.6 | 5.6 | 58.1 | 189.8 | 81.7 | 79.3 | 1,300 | 962 |
| Henderson | 4,476 | 17,881,311 | 3.7 | 6.4 | 0.8 | 0.4 | 44.1 | 2.7 | 39.8 | 788.6 | 132.8 | 133.4 | 1,158 | 884 |
| Hertford | 1,040 | 3,454,715 | 5.9 | 8.8 | 0.7 | 0.9 | 11.7 | 5.4 | 62.6 | 70.9 | 43.0 | 20.5 | 858 | 630 |
| Hoke | 1,515 | 4,737,322 | 2.7 | 8.5 | 0.5 | 0.0 | 4.0 | 4.8 | 79.2 | 87.5 | 72.4 | 9.7 | 178 | 160 |
| Hyde | 301 | 980,614 | 4.8 | 7.7 | 0.0 | 0.3 | 18.8 | 9.5 | 55.5 | 25.0 | 11.6 | 9.8 | 1,864 | 1,476 |
| Iredell | 5,860 | 19,989,935 | 5.2 | 8.7 | 4.1 | 0.7 | 9.4 | 6.0 | 63.2 | 559.3 | 232.7 | 227.9 | 1,297 | 1,024 |
| Jackson | 1,467 | 4,605,328 | 8.5 | 2.5 | 0.0 | 1.0 | 10.6 | 3.4 | 65.2 | 120.3 | 55.7 | 48.9 | 1,131 | 838 |
| Johnston | 7,829 | 29,315,952 | 2.6 | 4.5 | 1.7 | 0.2 | 28.4 | 3.7 | 57.4 | 860.3 | 309.0 | 204.2 | 1,040 | 792 |
| Jones | 388 | 1,184,828 | 4.7 | 8.2 | 0.0 | 0.0 | 9.6 | 3.4 | 67.2 | 27.8 | 16.5 | 8.8 | 921 | 725 |
| Lee | 2,729 | 9,721,756 | 4.3 | 8.9 | 2.6 | 1.7 | 6.2 | 4.3 | 68.9 | 230.2 | 121.1 | 80.7 | 1,337 | 997 |
| Lenoir | 2,440 | 8,798,545 | 3.3 | 7.7 | 2.2 | 1.9 | 8.1 | 8.8 | 65.6 | 215.8 | 117.4 | 63.5 | 1,122 | 858 |
| Lincoln | 2,379 | 7,944,638 | 3.9 | 9.2 | 1.3 | 1.1 | 14.0 | 6.6 | 59.8 | 209.1 | 92.4 | 90.1 | 1,091 | 792 |
| McDowell | 1,469 | 4,927,665 | 1.4 | 2.0 | 0.4 | 0.7 | 5.9 | 2.6 | 85.0 | 122.6 | 68.9 | 39.0 | 865 | 583 |
| Macon | 1,314 | 4,399,214 | 5.8 | 9.8 | 0.8 | 1.3 | 14.1 | 7.3 | 55.8 | 198.0 | 59.7 | 105.7 | 3,056 | 2,360 |
| Madison | 754 | 2,186,041 | 5.3 | 7.0 | 0.7 | 0.1 | 18.0 | 5.4 | 63.3 | 61.0 | 33.7 | 19.9 | 921 | 675 |
| Martin | 1,039 | 3,593,479 | 3.7 | 6.7 | 1.9 | 1.2 | 12.6 | 3.9 | 67.3 | 76.5 | 48.7 | 20.4 | 896 | 670 |
| Mecklenburg | 70,012 | 366,856,715 | 2.5 | 4.8 | 2.0 | 1.9 | 61.7 | 2.8 | 23.3 | 10,817.9 | 1,742.4 | 2,294.2 | 2,128 | 1,440 |
| Mitchell | 666 | 2,302,636 | 2.4 | 5.3 | 0.1 | 1.4 | 11.3 | 2.5 | 74.0 | 56.9 | 34.0 | 15.2 | 1,015 | 724 |
| Montgomery | 1,203 | 3,707,129 | 6.3 | 5.9 | 0.0 | 1.2 | 7.2 | 2.7 | 73.6 | 120.0 | 55.2 | 47.4 | 1,738 | 1,518 |
| Moore | 3,723 | 14,117,302 | 5.7 | 7.8 | 2.6 | 1.4 | 15.7 | 4.2 | 60.4 | 584.5 | 421.1 | 121.7 | 1,250 | 948 |
| Nash | 5,745 | 23,157,319 | 4.0 | 6.6 | 2.9 | 1.9 | 38.6 | 6.3 | 38.1 | 344.9 | 176.7 | 111.9 | 1,191 | 872 |
| New Hanover | 13,503 | 56,814,312 | 2.7 | 6.8 | 2.6 | 1.1 | 49.2 | 3.0 | 30.3 | 2,039.2 | 295.9 | 399.2 | 1,745 | 1,168 |
| Northampton | 786 | 2,459,636 | 5.4 | 9.5 | 0.0 | 1.0 | 19.7 | 3.5 | 46.8 | 37.6 | 28.4 | 6.1 | 306 | 282 |
| Onslow | 5,784 | 20,744,428 | 4.2 | 8.1 | 2.3 | 1.0 | 8.3 | 4.8 | 68.0 | 507.1 | 241.2 | 182.5 | 935 | 624 |
| Orange | 5,470 | 22,525,778 | 6.4 | 8.3 | 3.0 | 3.3 | 5.9 | 6.6 | 60.0 | 408.0 | 156.1 | 212.8 | 1,481 | 1,211 |
| Pamlico | 505 | 1,720,040 | 5.0 | 7.4 | 0.2 | 0.5 | 10.5 | 5.5 | 68.2 | 55.6 | 21.6 | 15.5 | 1,224 | 958 |
| Pasquotank | 1,988 | 6,479,973 | 2.8 | 8.3 | 2.4 | 2.4 | 17.0 | 8.3 | 56.5 | 137.9 | 70.4 | 45.3 | 1,150 | 808 |
| Pender | 1,717 | 6,154,687 | 5.5 | 8.1 | 1.1 | 0.4 | 9.6 | 2.4 | 70.5 | 187.7 | 84.5 | 82.2 | 1,354 | 1,065 |
| Perquimans | 417 | 1,644,255 | 5.1 | 7.2 | 0.0 | 0.0 | 11.4 | 3.5 | 70.2 | 35.2 | 19.8 | 11.8 | 875 | 677 |
| Person | 1,447 | 5,177,815 | 3.5 | 9.3 | 2.1 | 0.0 | 10.4 | 5.8 | 62.4 | 111.5 | 60.0 | 40.4 | 1,026 | 815 |
| Pitt | 7,258 | 24,893,380 | 7.1 | 10.4 | 2.2 | 2.4 | 12.6 | 10.0 | 49.6 | 931.3 | 283.9 | 183.1 | 1,025 | 720 |
| Polk | 687 | 2,236,878 | 10.2 | 8.8 | 0.5 | 2.0 | 8.1 | 4.5 | 62.9 | 53.0 | 24.2 | 22.8 | 1,108 | 869 |
| Randolph | 4,153 | 15,512,894 | 3.9 | 8.1 | 2.1 | 1.2 | 6.9 | 3.4 | 71.9 | 387.1 | 212.4 | 132.9 | 929 | 684 |
| Richmond | 2,035 | 6,946,800 | 8.8 | 3.2 | 1.7 | 1.4 | 9.3 | 5.8 | 67.7 | 159.7 | 98.3 | 42.4 | 945 | 725 |
| Robeson | 4,797 | 16,250,632 | 3.1 | 9.1 | 1.1 | 1.6 | 12.8 | 4.8 | 66.4 | 301.0 | 227.9 | 40.6 | 306 | 243 |
| Rockingham | 3,057 | 10,462,110 | 6.2 | 8.7 | 2.2 | 1.3 | 9.9 | 5.1 | 62.4 | 244.2 | 129.4 | 81.3 | 895 | 696 |
| Rowan | 5,068 | 16,833,762 | 3.8 | 7.1 | 2.5 | 1.3 | 9.6 | 4.1 | 67.9 | 427.9 | 211.7 | 150.7 | 1,074 | 793 |
| Rutherford | 2,441 | 8,181,622 | 4.3 | 7.4 | 1.7 | 1.9 | 10.3 | 4.9 | 66.1 | 187.0 | 95.9 | 67.5 | 1,014 | 744 |
| Sampson | 2,554 | 7,843,044 | 3.5 | 6.3 | 1.2 | 0.4 | 11.7 | 2.9 | 72.0 | 132.8 | 99.1 | 21.9 | 347 | 292 |
| Scotland | 1,491 | 4,808,909 | 4.0 | 7.1 | 0.5 | 1.9 | 13.4 | 4.7 | 64.9 | 114.7 | 67.5 | 35.4 | 1,005 | 763 |
| Stanly | 2,490 | 8,648,656 | 4.6 | 7.4 | 2.7 | 1.8 | 12.4 | 6.0 | 63.3 | 181.7 | 96.6 | 58.3 | 947 | 698 |
| Stokes | 1,540 | 4,795,661 | 5.8 | 6.6 | 2.7 | 0.4 | 12.6 | 2.4 | 67.9 | 108.6 | 58.9 | 39.6 | 866 | 652 |
| Surry | 3,903 | 13,457,582 | 2.7 | 4.5 | 0.8 | 0.2 | 37.8 | 3.5 | 48.8 | 312.3 | 123.3 | 69.5 | 964 | 589 |

1. Based on the resident population estimated as of July 1 of the year shown.

| STATE County | Total (mil dol) [185] | Per capita¹ (dollars) [186] | Education [187] | Health and hospitals [188] | Police protection [189] | Public welfare [190] | Highways [191] | Total (mil dol) [192] | Per capita¹ (dollars) [193] | Federal civilian [194] | Federal military [195] | State and local [196] | Number of returns [197] | Mean adjusted gross income [198] | Mean income tax [199] |
|---|---|---|---|---|---|---|---|---|---|---|---|---|---|---|---|
| **NORTH CAROLINA— Cont'd** | | | | | | | | | | | | | | | |
| Chowan | 49 | 3,480 | 49.3 | 0.3 | 12.8 | 5.8 | 1.0 | 15.1 | 1,076 | 32 | 30 | 809 | 6,240 | 56,077 | 5,393 |
| Clay | 30 | 2,697 | 48.1 | 7.7 | 11.7 | 8.4 | 0.1 | 8.6 | 784 | 21 | 24 | 495 | 4,700 | 50,436 | 4,740 |
| Cleveland | 324 | 3,337 | 55.9 | 3.8 | 9.7 | 6.8 | 1.2 | 180.7 | 1,860 | 198 | 210 | 5,212 | 42,500 | 48,600 | 4,000 |
| Columbus | 199 | 3,547 | 52.1 | 40.6 | 1.5 | 0.0 | 0.4 | 84.5 | 1,507 | 135 | 115 | 3,351 | 21,500 | 43,202 | 3,717 |
| Craven | 611 | 5,958 | 23.8 | 59.5 | 2.1 | 3.7 | 0.2 | 152.1 | 1,484 | 6,067 | 7,182 | 7,235 | 46,740 | 58,309 | 5,756 |
| Cumberland | 1,025 | 3,095 | 53.0 | 2.6 | 9.7 | 7.7 | 1.7 | 455.3 | 1,375 | 15,802 | 47,426 | 22,706 | 143,060 | 48,071 | 4,081 |
| Currituck | 96 | 3,653 | 41.6 | 7.2 | 9.0 | 5.0 | 0.0 | 24.6 | 935 | 38 | 60 | 1,263 | 12,880 | 62,315 | 5,891 |
| Dare | 247 | 6,827 | 23.9 | 8.4 | 14.8 | 3.7 | 1.4 | 231.0 | 6,382 | 237 | 192 | 3,295 | 20,410 | 63,956 | 7,203 |
| Davidson | 448 | 2,710 | 63.5 | 3.2 | 6.4 | 5.5 | 1.0 | 179.6 | 1,087 | 199 | 363 | 6,185 | 74,820 | 53,196 | 4,719 |
| Davie | 138 | 3,256 | 60.4 | 5.2 | 7.7 | 4.8 | 1.1 | 71.0 | 1,679 | 70 | 93 | 1,551 | 19,490 | 67,631 | 7,483 |
| Duplin | 533 | 9,036 | 28.6 | 60.1 | 2.0 | 2.1 | 0.3 | 78.7 | 1,335 | 134 | 128 | 3,207 | 21,610 | 40,614 | 2,975 |
| Durham | 1,056 | 3,382 | 39.1 | 4.6 | 11.1 | 5.9 | 2.1 | 1,049.0 | 3,361 | 6,474 | 817 | 15,827 | 147,580 | 71,222 | 8,589 |
| Edgecombe | 134 | 2,544 | 62.1 | 4.7 | 5.9 | 11.0 | 0.0 | 29.0 | 550 | 197 | 112 | 4,359 | 21,260 | 40,260 | 3,187 |
| Forsyth | 1,373 | 3,654 | 43.4 | 2.6 | 9.1 | 3.9 | 3.0 | 1,609.1 | 4,281 | 1,558 | 858 | 18,348 | 171,970 | 66,137 | 7,724 |
| Franklin | 91 | 1,369 | 84.4 | 0.0 | 3.9 | 0.0 | 1.7 | 14.0 | 211 | 89 | 149 | 2,130 | 28,880 | 53,882 | 4,606 |
| Gaston | 1,030 | 4,690 | 35.4 | 33.8 | 7.0 | 4.2 | 2.0 | 342.0 | 1,557 | 383 | 486 | 9,576 | 99,890 | 55,823 | 5,275 |
| Gates | 27 | 2,377 | 69.9 | 0.5 | 7.5 | 6.9 | 0.1 | 3.3 | 287 | 18 | 25 | 502 | 4,610 | 49,381 | 3,723 |
| Graham | 29 | 3,376 | 51.1 | 9.1 | 7.4 | 8.6 | 0.7 | 11.4 | 1,332 | 37 | 18 | 486 | 3,320 | 42,475 | 2,790 |
| Granville | 143 | 2,402 | 53.5 | 1.1 | 12.3 | 6.5 | 2.6 | 120.9 | 2,036 | 1,362 | 124 | 6,575 | 26,010 | 55,742 | 4,963 |
| Greene | 37 | 1,780 | 90.6 | 0.0 | 0.9 | 0.0 | 0.8 | 23.3 | 1,114 | 38 | 40 | 1,742 | 7,540 | 42,113 | 3,088 |
| Guilford | 1,923 | 3,640 | 46.0 | 3.2 | 7.9 | 4.9 | 3.3 | 2,117.4 | 4,008 | 4,005 | 1,183 | 29,192 | 241,810 | 65,814 | 7,883 |
| Halifax | 161 | 3,137 | 54.1 | 7.0 | 5.3 | 10.1 | 0.5 | 132.9 | 2,590 | 127 | 106 | 3,826 | 20,690 | 41,647 | 3,308 |
| Harnett | 208 | 1,568 | 84.6 | 0.0 | 3.1 | 0.0 | 1.4 | 246.4 | 1,861 | 141 | 299 | 5,462 | 52,600 | 51,887 | 4,218 |
| Haywood | 182 | 2,983 | 47.9 | 6.8 | 9.0 | 8.8 | 2.8 | 61.4 | 1,007 | 125 | 135 | 2,806 | 28,700 | 51,423 | 4,502 |
| Henderson | 472 | 4,096 | 32.5 | 39.7 | 4.9 | 4.3 | 0.9 | 223.1 | 1,937 | 240 | 254 | 5,555 | 54,770 | 60,542 | 5,957 |
| Hertford | 74 | 3,099 | 55.8 | 7.3 | 6.0 | 8.4 | 0.5 | 105.5 | 4,409 | 68 | 46 | 1,932 | 8,240 | 43,145 | 3,367 |
| Hoke | 86 | 1,595 | 91.3 | 0.0 | 1.6 | 0.0 | 1.1 | 57.3 | 1,059 | 68 | 119 | 2,504 | 21,000 | 44,430 | 2,877 |
| Hyde | 25 | 4,698 | 43.3 | 12.5 | 9.3 | 7.9 | 0.0 | 14.5 | 2,761 | 37 | 10 | 589 | 2,080 | 43,423 | 3,507 |
| Iredell | 533 | 3,034 | 51.4 | 1.8 | 9.3 | 4.6 | 1.6 | 506.4 | 2,883 | 292 | 396 | 8,781 | 84,640 | 74,331 | 9,684 |
| Jackson | 117 | 2,707 | 51.2 | 6.1 | 6.0 | 8.8 | 0.6 | 51.5 | 1,192 | 57 | 88 | 8,631 | 16,330 | 51,839 | 4,960 |
| Johnston | 848 | 4,318 | 43.5 | 33.2 | 3.6 | 4.1 | 0.6 | 607.7 | 3,095 | 264 | 454 | 9,759 | 90,530 | 60,689 | 5,748 |
| Jones | 28 | 2,937 | 55.3 | 6.5 | 6.7 | 10.1 | 0.2 | 4.5 | 469 | 24 | 20 | 518 | 4,130 | 42,521 | 3,061 |
| Lee | 230 | 3,799 | 61.9 | 1.4 | 7.8 | 4.4 | 1.4 | 168.5 | 2,789 | 165 | 133 | 3,292 | 27,210 | 54,338 | 4,998 |
| Lenoir | 216 | 3,808 | 50.8 | 3.4 | 7.5 | 6.0 | 0.7 | 161.5 | 2,852 | 192 | 120 | 5,657 | 23,940 | 45,916 | 4,099 |
| Lincoln | 197 | 2,387 | 48.9 | 5.9 | 8.0 | 7.0 | 0.8 | 130.1 | 1,576 | 124 | 187 | 3,678 | 38,840 | 68,746 | 7,518 |
| McDowell | 123 | 2,723 | 60.2 | 2.7 | 6.9 | 9.6 | 1.2 | 14.7 | 326 | 86 | 97 | 2,449 | 18,960 | 44,791 | 3,254 |
| Macon | 176 | 5,094 | 35.0 | 8.6 | 11.9 | 12.1 | 0.4 | 119.3 | 3,450 | 172 | 78 | 1,735 | 16,150 | 55,381 | 5,705 |
| Madison | 58 | 2,699 | 44.2 | 6.9 | 8.9 | 11.1 | 1.1 | 13.0 | 603 | 67 | 45 | 882 | 9,180 | 47,508 | 3,829 |
| Martin | 81 | 3,575 | 56.5 | 8.3 | 5.4 | 7.7 | 0.1 | 42.6 | 1,870 | 49 | 49 | 1,478 | 9,960 | 41,590 | 3,074 |
| Mecklenburg | 10,082 | 9,352 | 17.1 | 55.0 | 4.3 | 2.1 | 1.4 | 10,115.2 | 9,382 | 5,989 | 2,601 | 73,173 | 517,980 | 88,240 | 13,040 |
| Mitchell | 58 | 3,897 | 64.1 | 2.1 | 5.2 | 11.6 | 2.0 | 1.7 | 111 | 41 | 32 | 927 | 6,100 | 45,923 | 3,374 |
| Montgomery | 109 | 4,003 | 46.5 | 3.1 | 4.2 | 5.0 | 0.5 | 55.9 | 2,048 | 62 | 57 | 1,610 | 11,120 | 47,918 | 4,223 |
| Moore | 586 | 6,014 | 25.8 | 54.5 | 3.9 | 2.0 | 1.2 | 152.1 | 1,562 | 201 | 239 | 4,557 | 46,020 | 72,816 | 8,914 |
| Nash | 350 | 3,720 | 51.4 | 4.3 | 8.5 | 5.2 | 1.6 | 290.3 | 3,088 | 183 | 202 | 5,996 | 43,360 | 51,974 | 5,238 |
| New Hanover | 1,949 | 8,521 | 20.3 | 56.7 | 4.2 | 2.2 | 0.6 | 1,558.7 | 6,814 | 1,046 | 704 | 20,541 | 109,060 | 76,607 | 10,177 |
| Northampton | 38 | 1,898 | 68.0 | 0.1 | 2.0 | 0.0 | 1.8 | 26.9 | 1,351 | 47 | 41 | 1,041 | 7,700 | 42,485 | 3,143 |
| Onslow | 520 | 2,664 | 54.1 | 3.9 | 6.9 | 9.7 | 1.1 | 462.0 | 2,368 | 6,521 | 43,835 | 8,077 | 81,540 | 46,790 | 3,436 |
| Orange | 418 | 2,912 | 59.9 | 3.8 | 6.3 | 6.5 | 0.4 | 365.0 | 2,541 | 289 | 350 | 41,958 | 62,300 | 105,988 | 16,074 |
| Pamlico | 44 | 3,471 | 53.4 | 2.2 | 8.1 | 8.3 | 0.3 | 10.5 | 832 | 31 | 51 | 836 | 5,490 | 57,339 | 5,410 |
| Pasquotank | 134 | 3,408 | 57.1 | 3.8 | 9.7 | 6.4 | 0.9 | 49.7 | 1,263 | 743 | 969 | 3,578 | 17,660 | 47,744 | 3,807 |
| Pender | 197 | 3,244 | 57.1 | 2.6 | 5.2 | 5.1 | 0.3 | 257.9 | 4,247 | 101 | 173 | 2,543 | 26,500 | 60,221 | 5,796 |
| Perquimans | 36 | 2,657 | 52.4 | 4.1 | 8.3 | 6.7 | 1.3 | 29.1 | 2,166 | 35 | 29 | 638 | 5,810 | 53,288 | 4,464 |
| Person | 140 | 3,559 | 45.8 | 7.1 | 5.0 | 8.6 | 0.0 | 200.8 | 5,104 | 58 | 85 | 1,747 | 17,320 | 50,211 | 4,070 |
| Pitt | 993 | 5,557 | 28.8 | 44.4 | 6.0 | 0.0 | 1.4 | 605.9 | 3,392 | 711 | 426 | 26,816 | 72,810 | 58,788 | 6,429 |
| Polk | 54 | 2,641 | 48.8 | 3.2 | 8.7 | 6.8 | 0.5 | 39.9 | 1,939 | 51 | 69 | 881 | 9,050 | 60,032 | 6,004 |
| Randolph | 377 | 2,635 | 59.9 | 2.4 | 9.7 | 5.9 | 1.1 | 100.4 | 702 | 216 | 312 | 5,757 | 63,110 | 48,557 | 3,990 |
| Richmond | 160 | 3,578 | 58.1 | 3.8 | 7.8 | 6.6 | 0.8 | 46.2 | 1,031 | 108 | 96 | 2,906 | 18,030 | 39,982 | 2,810 |
| Robeson | 306 | 2,306 | 79.9 | 0.0 | 4.7 | 0.0 | 1.3 | 53.5 | 403 | 301 | 279 | 8,106 | 47,710 | 38,436 | 2,677 |
| Rockingham | 235 | 2,588 | 53.8 | 2.7 | 9.8 | 9.0 | 0.9 | 119.7 | 1,319 | 159 | 197 | 3,695 | 39,380 | 48,499 | 3,991 |
| Rowan | 408 | 2,906 | 55.6 | 2.9 | 6.9 | 5.6 | 1.6 | 201.9 | 1,439 | 2,726 | 301 | 6,563 | 63,750 | 50,879 | 4,515 |
| Rutherford | 185 | 2,779 | 55.8 | 2.3 | 7.2 | 7.8 | 1.7 | 71.5 | 1,074 | 116 | 144 | 3,354 | 26,770 | 45,847 | 3,654 |
| Sampson | 133 | 2,097 | 89.3 | 0.1 | 2.3 | 0.0 | 1.1 | 111.2 | 1,758 | 119 | 137 | 3,847 | 26,040 | 44,858 | 3,886 |
| Scotland | 114 | 3,226 | 57.7 | 4.2 | 6.8 | 8.1 | 1.1 | 10.9 | 311 | 48 | 71 | 2,253 | 13,520 | 41,499 | 3,015 |
| Stanly | 191 | 3,098 | 54.7 | 4.8 | 7.9 | 5.6 | 1.2 | 65.6 | 1,066 | 150 | 133 | 3,924 | 26,720 | 53,915 | 4,830 |
| Stokes | 105 | 2,302 | 57.9 | 5.2 | 7.0 | 7.7 | 0.5 | 42.5 | 929 | 74 | 99 | 1,769 | 19,680 | 50,588 | 4,087 |
| Surry | 306 | 4,242 | 43.9 | 33.1 | 4.0 | 3.8 | 0.6 | 73.3 | 1,016 | 179 | 155 | 4,711 | 30,520 | 49,440 | 4,285 |

1. Based on the resident population estimated as of July 1 of the year shown.

| State / county code | CBSA code[1] | County Type code[2] | STATE County | Land area[3] (sq. mi) | Total persons 2019 | Rank | Per square mile | White | Black | American Indian, Alaska Native | Asian and Pacific Islancer | Percent Hispanic or Latino[4] | Under 5 years | 5 to 14 years | 15 to 24 years | 25 to 34 years | 35 to 44 years | 45 to 54 years |
|---|---|---|---|---|---|---|---|---|---|---|---|---|---|---|---|---|---|---|
| | | | | 1 | 2 | 3 | 4 | 5 | 6 | 7 | 8 | 9 | 10 | 11 | 12 | 13 | 14 | 15 |
| | | | **NORTH CAROLINA—Cont'd** | | | | | | | | | | | | | | | |
| 37173 | 19000 | 8 | Swain | 527.7 | 14,179 | 2,147 | 26.9 | 64.3 | 2.0 | 30.5 | 1.2 | 6.5 | 5.7 | 12.8 | 11.5 | 12.8 | 11.3 | 12.3 |
| 37175 | 14820 | 6 | Transylvania | 378.4 | 34,498 | 1,316 | 91.2 | 91.9 | 4.5 | 1.1 | 1.0 | 3.4 | 3.8 | 8.8 | 10.4 | 9.4 | 9.7 | 10.9 |
| 37177 | | 9 | Tyrrell | 390.8 | 3,774 | 2,915 | 9.7 | 53.5 | 35.7 | 1.0 | 1.9 | 10.3 | 4.8 | 11.1 | 10.0 | 13.4 | 12.1 | 11.4 |
| 37179 | 16740 | 1 | Union | 632.7 | 244,562 | 281 | 386.5 | 72.2 | 12.8 | 0.8 | 4.4 | 11.7 | 5.5 | 15.2 | 14.6 | 9.8 | 13.3 | 15.9 |
| 37181 | 25780 | 4 | Vance | 252.4 | 44,718 | 1,081 | 177.2 | 39.6 | 51.7 | 0.6 | 1.1 | 8.4 | 6.3 | 13.2 | 12.4 | 12.4 | 10.5 | 12.2 |
| 37183 | 39580 | 1 | Wake | 834.6 | 1,132,271 | 40 | 1,356.7 | 61.3 | 21.0 | 0.8 | 9.0 | 10.3 | 5.9 | 13.3 | 13.1 | 14.8 | 14.8 | 14.0 |
| 37185 | | 8 | Warren | 429.4 | 19,522 | 1,851 | 45.5 | 39.7 | 53.5 | 6.1 | 0.7 | 4.1 | 4.2 | 10.1 | 9.9 | 11.0 | 10.3 | 10.9 |
| 37187 | | 7 | Washington | 346.5 | 11,485 | 2,320 | 33.1 | 45.4 | 49.0 | 0.7 | 0.6 | 6.0 | 5.4 | 11.0 | 10.3 | 10.3 | 8.9 | 11.5 |
| 37189 | 14380 | 8 | Watauga | 312.4 | 56,441 | 914 | 180.7 | 92.7 | 2.2 | 0.9 | 1.7 | 4.1 | 3.1 | 7.1 | 29.4 | 11.1 | 9.7 | 10.2 |
| 37191 | 24140 | 3 | Wayne | 553.9 | 123,967 | 514 | 223.8 | 54.1 | 32.6 | 0.9 | 2.1 | 12.8 | 6.5 | 13.3 | 13.5 | 13.6 | 11.4 | 11.6 |
| 37193 | 35900 | 6 | Wilkes | 753.7 | 68,043 | 788 | 90.3 | 88.0 | 5.1 | 0.6 | 0.8 | 6.9 | 5.0 | 11.3 | 11.2 | 11.1 | 10.7 | 13.1 |
| 37195 | 48980 | 6 | Wilson | 367.6 | 81,979 | 697 | 223.0 | 47.9 | 40.5 | 0.7 | 1.6 | 11.0 | 6.0 | 12.8 | 12.7 | 12.1 | 11.4 | 12.2 |
| 37197 | 49180 | 2 | Yadkin | 334.9 | 37,625 | 1,234 | 112.3 | 84.8 | 3.8 | 0.6 | 0.5 | 11.5 | 5.4 | 11.5 | 11.3 | 11.5 | 10.4 | 14.0 |
| 37199 | | 8 | Yancey | 312.6 | 18,099 | 1,916 | 57.9 | 93.0 | 1.4 | 0.8 | 0.5 | 5.3 | 4.9 | 9.8 | 10.0 | 10.7 | 10.4 | 12.8 |
| 38000 | | 0 | NORTH DAKOTA | 68,994.8 | 765,309 | X | 11.1 | 85.3 | 4.0 | 6.2 | 2.3 | 4.3 | 6.9 | 13.2 | 14.4 | 14.8 | 12.4 | 10.1 |
| 38001 | | 9 | Adams | 987.5 | 2,188 | 3,027 | 2.2 | 92.7 | 1.4 | 1.5 | 4.0 | 2.1 | 5.7 | 10.5 | 9.4 | 10.1 | 9.9 | 9.3 |
| 38003 | | 6 | Barnes | 1,491.6 | 10,402 | 2,394 | 7.0 | 93.8 | 2.0 | 2.1 | 1.4 | 2.3 | 4.8 | 10.9 | 13.7 | 10.9 | 10.6 | 11.0 |
| 38005 | | 9 | Benson | 1,388.6 | 6,762 | 2,680 | 4.9 | 42.3 | 0.7 | 54.3 | 0.5 | 4.0 | 10.8 | 19.2 | 13.3 | 11.5 | 9.4 | 9.2 |
| 38007 | 19860 | 9 | Billings | 1,148.5 | 890 | 3,111 | 0.8 | 92.1 | 0.4 | 1.0 | 4.6 | 2.5 | 6.5 | 10.4 | 8.2 | 11.9 | 13.1 | 10.3 |
| 38009 | | 9 | Bottineau | 1,668.9 | 6,287 | 2,722 | 3.8 | 92.5 | 1.5 | 4.9 | 1.1 | 2.5 | 5.9 | 12.0 | 11.7 | 10.5 | 9.7 | 10.3 |
| 38011 | | 9 | Bowman | 1,161.8 | 2,986 | 2,959 | 2.6 | 93.4 | 0.7 | 1.7 | 0.3 | 4.9 | 5.9 | 14.9 | 10.1 | 10.2 | 11.9 | 9.3 |
| 38013 | | 9 | Burke | 1,103.6 | 2,118 | 3,039 | 1.9 | 93.2 | 1.4 | 2.3 | 1.2 | 3.4 | 7.7 | 13.8 | 8.9 | 11.2 | 11.3 | 9.8 |
| 38015 | 13900 | 3 | Burleigh | 1,632.7 | 96,212 | 626 | 58.9 | 89.9 | 2.8 | 5.0 | 1.5 | 2.7 | 6.3 | 13.3 | 12.6 | 13.7 | 13.4 | 10.8 |
| 38017 | 22020 | 3 | Cass | 1,764.9 | 183,904 | 362 | 104.2 | 86.1 | 7.1 | 2.1 | 4.1 | 2.9 | 6.7 | 12.4 | 17.0 | 17.1 | 13.6 | 10.1 |
| 38019 | | 9 | Cavalier | 1,489.1 | 3,713 | 2,919 | 2.5 | 95.5 | 0.8 | 2.4 | 0.9 | 1.9 | 7.1 | 11.5 | 9.9 | 9.5 | 9.5 | 9.8 |
| 38021 | | 9 | Dickey | 1,131.5 | 4,705 | 2,845 | 4.2 | 94.6 | 1.1 | 1.4 | 1.1 | 3.4 | 6.2 | 13.4 | 13.2 | 9.6 | 11.0 | 10.5 |
| 38023 | | 9 | Divide | 1,261.1 | 2,277 | 3,018 | 1.8 | 91.8 | 2.8 | 2.2 | 2.9 | 3.1 | 6.9 | 13.2 | 8.9 | 10.5 | 10.5 | 8.6 |
| 38025 | | 9 | Dunn | 2,008.5 | 4,465 | 2,861 | 2.2 | 82.8 | 1.9 | 9.5 | 1.9 | 5.8 | 7.5 | 14.2 | 9.7 | 15.1 | 11.6 | 10.4 |
| 38027 | | 9 | Eddy | 630.2 | 2,179 | 3,028 | 3.5 | 90.4 | 0.6 | 5.1 | 0.9 | 4.4 | 7.3 | 12.9 | 10.0 | 10.5 | 10.2 | 8.9 |
| 38029 | | 9 | Emmons | 1,510.0 | 3,187 | 2,952 | 2.1 | 96.2 | 1.1 | 1.6 | 0.9 | 1.5 | 4.9 | 11.1 | 10.5 | 8.7 | 8.3 | 10.4 |
| 38031 | | 9 | Foster | 634.1 | 3,165 | 2,953 | 5.0 | 95.6 | 1.0 | 1.8 | 0.6 | 2.1 | 5.9 | 12.3 | 10.3 | 10.9 | 10.7 | 11.2 |
| 38033 | | 9 | Golden Valley | 1,000.9 | 1,737 | 3,068 | 1.7 | 94.5 | 1.0 | 2.2 | 0.4 | 3.4 | 5.1 | 10.3 | 10.4 | 9.6 | 11.9 | 9.3 |
| 38035 | 24220 | 3 | Grand Forks | 1,436.2 | 69,481 | 780 | 48.4 | 85.4 | 4.9 | 3.7 | 4.0 | 4.9 | 6.7 | 11.6 | 22.1 | 16.0 | 11.0 | 8.6 |
| 38037 | | 8 | Grant | 1,659.2 | 2,215 | 3,023 | 1.3 | 96.9 | 0.9 | 2.3 | 0.8 | 1.1 | 5.0 | 12.1 | 7.7 | 7.7 | 9.7 | 10.3 |
| 38039 | | 9 | Griggs | 708.7 | 2,207 | 3,024 | 3.1 | 97.4 | 0.8 | 0.6 | 0.1 | 1.3 | 4.8 | 12.6 | 9.2 | 8.4 | 9.7 | 9.8 |
| 38041 | | 9 | Hettinger | 1,132.2 | 2,438 | 3,008 | 2.2 | 93.7 | 1.7 | 3.8 | 0.7 | 2.2 | 7.3 | 13.5 | 9.9 | 11.2 | 10.0 | 9.7 |
| 38043 | | 8 | Kidder | 1,351.1 | 2,458 | 3,004 | 1.8 | 94.1 | 0.6 | 0.7 | 1.0 | 4.4 | 6.2 | 12.0 | 8.7 | 10.3 | 10.7 | 10.6 |
| 38045 | | 9 | LaMoure | 1,146.0 | 4,033 | 2,892 | 3.5 | 96.8 | 1.0 | 0.8 | 0.3 | 2.1 | 5.9 | 11.9 | 9.8 | 8.8 | 9.4 | 9.6 |
| 38047 | | 9 | Logan | 992.6 | 1,880 | 3,058 | 1.9 | 96.0 | 0.9 | 1.9 | 1.0 | 2.2 | 6.6 | 12.8 | 10.7 | 8.8 | 8.7 | 10.3 |
| 38049 | 33500 | 9 | McHenry | 1,873.9 | 5,695 | 2,772 | 3.0 | 96.2 | 0.8 | 1.4 | 0.5 | 2.1 | 6.0 | 13.6 | 9.6 | 9.8 | 12.9 | 11.3 |
| 38051 | | 9 | McIntosh | 974.6 | 2,440 | 3,007 | 2.5 | 94.6 | 1.4 | 1.1 | 0.9 | 2.7 | 4.9 | 10.9 | 8.4 | 8.8 | 8.8 | 9.6 |
| 38053 | | 9 | McKenzie | 2,760.1 | 15,242 | 2,076 | 5.5 | 78.1 | 2.6 | 10.0 | 1.3 | 10.9 | 9.4 | 18.2 | 12.7 | 17.5 | 13.4 | 9.7 |
| 38055 | | 8 | McLean | 2,110.3 | 9,416 | 2,464 | 4.5 | 89.7 | 0.8 | 7.9 | 0.5 | 3.0 | 5.9 | 12.4 | 9.7 | 10.0 | 10.9 | 10.9 |
| 38057 | | 7 | Mercer | 1,042.8 | 8,174 | 2,567 | 7.8 | 93.7 | 1.3 | 3.1 | 0.8 | 2.6 | 5.9 | 13.3 | 10.0 | 10.4 | 12.0 | 10.3 |
| 38059 | 13900 | 3 | Morton | 1,925.9 | 31,503 | 1,390 | 16.4 | 90.2 | 1.9 | 4.6 | 1.0 | 4.3 | 7.0 | 12.9 | 10.5 | 14.9 | 13.3 | 11.2 |
| 38061 | | 9 | Mountrail | 1,825.2 | 10,502 | 2,391 | 5.8 | 61.1 | 1.9 | 29.9 | 1.0 | 9.2 | 8.6 | 15.1 | 13.1 | 15.9 | 13.0 | 10.8 |
| 38063 | | 8 | Nelson | 981.8 | 2,789 | 2,979 | 2.8 | 93.0 | 1.3 | 2.6 | 0.5 | 4.4 | 4.7 | 11.5 | 9.3 | 10.1 | 10.2 | 9.3 |
| 38065 | 13900 | 3 | Oliver | 722.4 | 1,926 | 3,055 | 2.7 | 94.5 | 0.6 | 3.2 | 0.6 | 2.0 | 6.0 | 14.8 | 9.2 | 8.6 | 11.1 | 10.0 |
| 38067 | | 9 | Pembina | 1,118.5 | 6,658 | 2,690 | 6.0 | 93.3 | 0.9 | 3.5 | 0.8 | 3.7 | 5.6 | 11.7 | 9.3 | 10.1 | 12.3 | 9.9 |
| 38069 | | 7 | Pierce | 1,018.3 | 3,928 | 2,900 | 3.9 | 91.3 | 1.2 | 5.8 | 1.3 | 1.8 | 5.0 | 13.1 | 10.6 | 9.5 | 11.5 | 10.6 |
| 38071 | | 7 | Ramsey | 1,185.4 | 11,388 | 2,324 | 9.6 | 84.7 | 1.3 | 12.0 | 1.4 | 3.5 | 7.0 | 13.3 | 11.7 | 12.2 | 10.4 | 10.0 |
| 38073 | | 8 | Ransom | 862.5 | 5,173 | 2,815 | 6.0 | 94.6 | 1.4 | 1.5 | 1.3 | 2.7 | 5.3 | 13.4 | 11.3 | 10.1 | 11.3 | 11.9 |
| 38075 | 33500 | 9 | Renville | 877.2 | 2,283 | 3,017 | 2.6 | 96.1 | 1.5 | 1.5 | 0.8 | 2.2 | 5.7 | 12.6 | 10.8 | 10.7 | 10.4 | 11.2 |
| 38077 | 47420 | 6 | Richland | 1,435.7 | 16,156 | 2,023 | 11.3 | 92.1 | 1.5 | 3.5 | 1.2 | 3.6 | 6.1 | 12.2 | 16.9 | 11.6 | 10.7 | 9.5 |
| 38079 | | 9 | Rolette | 903.0 | 14,165 | 2,149 | 15.7 | 20.6 | 0.8 | 78.7 | 0.4 | 2.3 | 8.2 | 19.9 | 13.7 | 11.9 | 11.3 | 9.9 |
| 38081 | | 9 | Sargent | 858.5 | 3,913 | 2,904 | 4.6 | 92.4 | 2.7 | 2.1 | 1.5 | 3.2 | 6.1 | 12.1 | 9.8 | 12.1 | 10.9 | 11.2 |
| 38083 | | 9 | Sheridan | 972.2 | 1,275 | 3,095 | 1.3 | 95.8 | 0.5 | 2.8 | 1.0 | 1.7 | 5.2 | 9.8 | 7.6 | 7.8 | 9.3 | 10.1 |
| 38085 | | 3 | Sioux | 1,094.1 | 4,173 | 2,882 | 3.8 | 15.2 | 1.2 | 81.0 | 0.8 | 5.6 | 9.4 | 20.9 | 15.7 | 11.8 | 11.0 | 11.4 |
| 38087 | | 9 | Slope | 1,215.3 | 747 | 3,124 | 0.6 | 92.9 | 1.2 | 2.1 | 0.0 | 4.4 | 5.9 | 10.0 | 8.3 | 11.5 | 8.0 | 9.9 |
| 38089 | 19860 | 7 | Stark | 1,334.9 | 32,107 | 1,378 | 24.1 | 87.8 | 3.4 | 2.3 | 2.1 | 6.2 | 8.6 | 15.2 | 11.6 | 16.9 | 13.1 | 9.9 |
| 38091 | | 8 | Steele | 712.1 | 1,890 | 3,057 | 2.7 | 95.9 | 0.5 | 1.7 | 0.5 | 2.4 | 6.7 | 11.7 | 9.6 | 9.4 | 10.0 | 10.3 |
| 38093 | 27420 | 7 | Stutsman | 2,221.9 | 20,498 | 1,794 | 9.2 | 92.8 | 2.1 | 2.4 | 1.4 | 2.9 | 5.4 | 11.4 | 13.6 | 12.3 | 12.0 | 10.8 |
| 38095 | | 9 | Towner | 1,025.1 | 2,108 | 3,040 | 2.1 | 91.5 | 0.9 | 5.2 | 0.9 | 3.0 | 4.7 | 13.0 | 8.8 | 9.3 | 9.4 | 10.4 |
| 38097 | | 8 | Traill | 861.9 | 7,959 | 2,593 | 9.2 | 84.5 | 1.5 | 2.0 | 1.1 | 3.6 | 6.0 | 13.2 | 12.5 | 11.9 | 10.6 | 11.2 |
| 38099 | | 6 | Walsh | 1,281.6 | 10,437 | 2,392 | 8.1 | 84.5 | 1.1 | 2.6 | 1.2 | 12.1 | 6.4 | 13.1 | 10.8 | 10.3 | 10.9 | 11.1 |

1. CBSA = Core Based Statistical Area. See Appendix A for explanation. See Appendix B for list of metropolitan areas with component counties.
2. County type code from the Economic Research Service of USDA Rural-Urban Continuum Codes. See Appendix A for definition.
3. Dry land or land partially or temporarily covered by water.
4. May be of any race.

Items 1—15

# Table B. States and Counties — Population and Households

| STATE County | 55 to 64 years (16) | 65 to 74 years (17) | 75 years and over (18) | Percent female (19) | Total persons 2000 (20) | Total persons 2010 (21) | Percent change 2000-2010 (22) | Percent change 2010-2020 (23) | Births (24) | Deaths (25) | Net Migration (26) | Number (27) | Persons per household (28) | Family households (29) | Female family householder[1] (30) | One person (31) |
|---|---|---|---|---|---|---|---|---|---|---|---|---|---|---|---|---|
| **NORTH CAROLINA—Cont'd** | | | | | | | | | | | | | | | | |
| Swain | 13.7 | 11.7 | 8.1 | 52.2 | 12,968 | 13,978 | 7.8 | 1.4 | 1,864 | 1,995 | 336.0 | 5,620 | 2.49 | 64.3 | 13.0 | 32.9 |
| Transylvania | 15.4 | 17.1 | 14.5 | 51.6 | 29,334 | 33,089 | 12.8 | 4.3 | 2,741 | 4,234 | 2,913 | 14,567 | 2.23 | 68.5 | 10.1 | 26.7 |
| Tyrrell | 14.8 | 12.5 | 9.9 | 45.5 | 4,149 | 4,407 | 6.2 | -14.4 | 403 | 443 | -606.0 | 1,594 | 2.1 | 64.9 | 10.4 | 30.6 |
| Union | 12.4 | 8.1 | 5.2 | 50.7 | 123,677 | 201,328 | 62.8 | 21.5 | 24,352 | 14,176 | 32,964 | 76,346 | 3.0 | 80.1 | 9.9 | 16.1 |
| Vance | 13.6 | 11.5 | 7.9 | 53.4 | 42,954 | 45,412 | 5.7 | -1.5 | 5,795 | 5,283 | -1,210 | 16,875 | 2.59 | 66.2 | 21.9 | 28.5 |
| Wake | 11.6 | 7.7 | 4.8 | 51.3 | 627,846 | 901,036 | 43.5 | 25.7 | 129,514 | 54,593 | 155,019 | 400,172 | 2.62 | 67.0 | 11.2 | 25.6 |
| Warren | 16.5 | 15.6 | 11.5 | 50.6 | 19,972 | 21,016 | 5.2 | -7.1 | 1,806 | 2,523 | -767.0 | 7,786 | 2.43 | 58.9 | 16.5 | 37.8 |
| Washington | 15.8 | 15.5 | 11.0 | 53.1 | 13,723 | 13,193 | -3.9 | -12.9 | 1,307 | 1,642 | -1,381 | 4,977 | 2.37 | 53.3 | 18.7 | 42.1 |
| Watauga | 12.2 | 10.5 | 6.7 | 50.2 | 42,695 | 51,061 | 19.6 | 10.5 | 3,632 | 3,570 | 5,276 | 21,077 | 2.32 | 54.3 | 6.2 | 27.5 |
| Wayne | 12.9 | 10.2 | 7.1 | 51.1 | 113,329 | 122,665 | 8.2 | 1.1 | 17,091 | 12,485 | -3,298 | 48,343 | 2.49 | 66.4 | 17.9 | 28.4 |
| Wilkes | 14.9 | 13.0 | 9.8 | 50.9 | 65,632 | 69,311 | 5.6 | -1.8 | 6,970 | 8,268 | 97.0 | 28,753 | 2.35 | 65.8 | 10.2 | 29.2 |
| Wilson | 13.7 | 11.3 | 7.8 | 52.7 | 73,814 | 81,225 | 10.0 | 0.9 | 9,878 | 8,909 | -160.0 | 32,014 | 2.5 | 63.6 | 18.3 | 31.6 |
| Yadkin | 15.1 | 11.8 | 9.1 | 50.4 | 36,348 | 38,412 | 5.7 | -2.0 | 3,971 | 4,397 | -338.0 | 15,425 | 2.42 | 69.9 | 10.8 | 26.9 |
| Yancey | 14.6 | 15.4 | 11.4 | 51.0 | 17,774 | 17,817 | 0.2 | 1.6 | 1,753 | 2,326 | 870.0 | 7,510 | 2.35 | 67.7 | 8.3 | 27.6 |
| **NORTH DAKOTA** | 12.0 | 9.1 | 7.0 | 48.8 | 642,200 | 672,575 | 4.7 | 13.8 | 108,282 | 64,102 | 46,920 | 318,322 | 2.3 | 59.5 | 7.6 | 31.7 |
| Adams | 16.0 | 15.7 | 13.4 | 50.5 | 2,593 | 2,343 | -9.6 | -6.6 | 240 | 306 | -87.0 | 1,035 | 2.16 | 60.7 | 1.4 | 36.7 |
| Barnes | 14.5 | 13.0 | 10.6 | 49.6 | 11,775 | 11,064 | -6.0 | -6.0 | 1,101 | 1,352 | -408.0 | 5,074 | 2.0 | 57.3 | 4.4 | 35.8 |
| Benson | 11.8 | 9.3 | 5.4 | 49.3 | 6,964 | 6,660 | -4.4 | 1.5 | 1,544 | 747 | -695.0 | 2,248 | 3.05 | 71.3 | 19.4 | 24.0 |
| Billings | 15.4 | 12.7 | 11.3 | 47.4 | 888 | 784 | -11.7 | 13.5 | 126 | 47 | 24.0 | 383 | 2.33 | 70.8 | 5.0 | 25.8 |
| Bottineau | 15.0 | 14.5 | 10.4 | 48.5 | 7,149 | 6,429 | -10.1 | -2.2 | 703 | 820 | -34.0 | 2,992 | 2.08 | 61.2 | 6.3 | 33.0 |
| Bowman | 14.5 | 12.3 | 11.0 | 49.5 | 3,242 | 3,151 | -2.8 | -5.2 | 404 | 416 | -154.0 | 1,350 | 2.27 | 64.4 | 6.0 | 29.6 |
| Burke | 15.4 | 12.5 | 9.3 | 47.6 | 2,242 | 1,968 | -12.2 | 7.6 | 335 | 212 | 18.0 | 905 | 2.41 | 62.9 | 5.4 | 33.8 |
| Burleigh | 12.6 | 9.7 | 7.5 | 50.1 | 69,416 | 81,308 | 17.1 | 18.3 | 12,719 | 7,271 | 9,265 | 39,507 | 2.32 | 62.2 | 8.6 | 31.0 |
| Cass | 10.2 | 7.7 | 5.2 | 49.4 | 123,138 | 149,778 | 21.6 | 22.8 | 25,191 | 10,719 | 19,381 | 75,843 | 2.27 | 55.9 | 8.1 | 31.9 |
| Cavalier | 14.8 | 14.1 | 13.9 | 48.1 | 4,831 | 3,994 | -17.3 | -7.0 | 473 | 500 | -251.0 | 1,737 | 2.13 | 64.3 | 3.1 | 29.5 |
| Dickey | 14.0 | 10.4 | 11.5 | 50.0 | 5,757 | 5,287 | -8.2 | -11.0 | 604 | 712 | -476.0 | 2,119 | 2.23 | 62.9 | 4.2 | 31.3 |
| Divide | 16.2 | 13.4 | 11.9 | 48.0 | 2,283 | 2,071 | -9.3 | 9.9 | 272 | 278 | 202.0 | 1,061 | 2.1 | 61.3 | 6.5 | 36.9 |
| Dunn | 14.8 | 10.0 | 6.8 | 46.0 | 3,600 | 3,536 | -1.8 | 26.3 | 635 | 349 | 615.0 | 1,692 | 2.53 | 65.9 | 4.1 | 30.6 |
| Eddy | 16.4 | 13.1 | 10.7 | 49.8 | 2,757 | 2,385 | -13.5 | -8.6 | 291 | 454 | -43.0 | 1,062 | 2.08 | 60.8 | 8.1 | 33.6 |
| Emmons | 17.4 | 14.5 | 14.2 | 48.5 | 4,331 | 3,545 | -18.1 | -10.1 | 300 | 498 | -160.0 | 1,585 | 2.05 | 61.1 | 2.3 | 34.3 |
| Foster | 15.7 | 12.2 | 10.8 | 50.0 | 3,759 | 3,337 | -11.2 | -5.2 | 372 | 428 | -116.0 | 1,423 | 2.22 | 60.2 | 3.7 | 35.6 |
| Golden Valley | 16.2 | 15.3 | 11.9 | 50.1 | 1,924 | 1,680 | -12.7 | 3.4 | 229 | 160 | -16.0 | 842 | 2.06 | 54.5 | 2.6 | 38.5 |
| Grand Forks | 10.3 | 8.0 | 5.7 | 48.6 | 66,109 | 66,864 | 1.1 | 3.9 | 10,114 | 5,059 | -2,454 | 30,502 | 2.16 | 53.0 | 8.1 | 34.1 |
| Grant | 16.7 | 16.3 | 14.5 | 49.3 | 2,841 | 2,394 | -15.7 | -7.5 | 268 | 296 | -152.0 | 1,074 | 2.14 | 59.7 | 2.9 | 36.0 |
| Griggs | 15.0 | 16.5 | 14.0 | 49.4 | 2,754 | 2,421 | -12.1 | -8.8 | 209 | 392 | -30.0 | 1,089 | 2.12 | 61.6 | 4.0 | 37.2 |
| Hettinger | 14.7 | 12.1 | 11.6 | 51.1 | 2,715 | 2,478 | -8.7 | -1.6 | 337 | 331 | -48.0 | 1,073 | 2.21 | 63.9 | 5.8 | 33.4 |
| Kidder | 14.5 | 14.7 | 12.4 | 47.8 | 2,753 | 2,435 | -11.6 | 0.9 | 292 | 186 | -81.0 | 1,092 | 2.26 | 68.5 | 4.5 | 28.1 |
| LaMoure | 16.5 | 14.1 | 13.9 | 48.9 | 4,701 | 4,141 | -11.9 | -2.6 | 456 | 443 | -116.0 | 1,822 | 2.22 | 64.0 | 2.7 | 29.8 |
| Logan | 15.1 | 12.3 | 14.7 | 47.8 | 2,308 | 1,988 | -13.9 | -5.4 | 224 | 268 | -62.0 | 849 | 2.11 | 61.8 | 4.2 | 32.9 |
| McHenry | 14.9 | 12.4 | 9.5 | 47.9 | 5,987 | 5,392 | -9.9 | 5.6 | 695 | 626 | 221.0 | 2,594 | 2.24 | 66.0 | 6.4 | 28.9 |
| McIntosh | 17.0 | 13.1 | 18.4 | 50.8 | 3,390 | 2,813 | -17.0 | -13.3 | 254 | 518 | -113.0 | 1,291 | 1.93 | 64.2 | 6.6 | 35.7 |
| McKenzie | 10.1 | 5.6 | 3.3 | 46.8 | 5,737 | 6,359 | 10.8 | 139.7 | 2,063 | 649 | 7,288 | 4,302 | 3.05 | 60.9 | 10.2 | 27.1 |
| McLean | 15.7 | 14.6 | 10.0 | 49.1 | 9,311 | 8,962 | -3.7 | 5.1 | 1,166 | 1,111 | 388.0 | 4,334 | 2.18 | 66.7 | 5.6 | 29.4 |
| Mercer | 17.2 | 11.9 | 9.1 | 48.4 | 8,644 | 8,424 | -2.5 | -3.0 | 1,012 | 829 | -439.0 | 3,709 | 2.24 | 68.5 | 5.1 | 27.2 |
| Morton | 13.1 | 9.8 | 7.5 | 49.7 | 25,303 | 27,469 | 8.6 | 14.7 | 4,562 | 2,938 | 2,352 | 13,565 | 2.23 | 63.9 | 7.4 | 25.4 |
| Mountrail | 11.7 | 7.7 | 4.2 | 45.8 | 6,631 | 7,663 | 15.6 | 37.0 | 1,723 | 885 | 1,927 | 3,360 | 3.04 | 64.8 | 9.0 | 30.7 |
| Nelson | 17.9 | 15.0 | 12.1 | 48.8 | 3,715 | 3,126 | -15.9 | -10.8 | 296 | 588 | -41.0 | 1,467 | 1.93 | 55.4 | 3.9 | 40.2 |
| Oliver | 16.3 | 14.9 | 9.0 | 47.3 | 2,065 | 1,848 | -10.5 | 4.2 | 225 | 121 | -27.0 | 738 | 2.5 | 79.0 | 5.4 | 20.5 |
| Pembina | 16.3 | 13.9 | 11.0 | 47.1 | 8,585 | 7,403 | -13.8 | -10.1 | 783 | 859 | -671.0 | 3,158 | 2.14 | 62.7 | 6.2 | 32.5 |
| Pierce | 14.8 | 12.0 | 12.8 | 49.3 | 4,675 | 4,359 | -6.8 | -9.9 | 444 | 604 | -272.0 | 1,908 | 2.08 | 56.7 | 7.5 | 34.7 |
| Ramsey | 14.7 | 11.3 | 9.5 | 49.0 | 12,066 | 11,451 | -5.1 | -0.6 | 1,671 | 1,387 | -339.0 | 4,828 | 2.3 | 56.9 | 6.4 | 36.4 |
| Ransom | 14.7 | 11.5 | 10.4 | 47.8 | 5,890 | 5,457 | -7.4 | -5.2 | 599 | 810 | -74.0 | 2,371 | 2.18 | 60.5 | 4.6 | 34.2 |
| Renville | 17.0 | 12.5 | 9.2 | 47.7 | 2,610 | 2,470 | -5.4 | -7.6 | 288 | 287 | -192.0 | 950 | 2.52 | 66.1 | 6.7 | 27.6 |
| Richland | 13.6 | 10.7 | 8.6 | 48.5 | 17,998 | 16,321 | -9.3 | 1.6 | 1,875 | 1,443 | -590.0 | 6,777 | 2.22 | 63.2 | 8.6 | 28.6 |
| Rolette | 12.3 | 7.9 | 4.8 | 50.2 | 13,674 | 13,939 | 1.9 | 1.6 | 2,812 | 1,462 | -1,124 | 4,608 | 3.11 | 68.7 | 22.7 | 25.6 |
| Sargent | 14.5 | 13.3 | 9.9 | 46.2 | 4,366 | 3,829 | -12.3 | 2.2 | 422 | 352 | 12.0 | 1,864 | 2.06 | 62.3 | 4.0 | 31.9 |
| Sheridan | 16.1 | 14.9 | 19.1 | 48.5 | 1,710 | 1,321 | -22.7 | -3.5 | 129 | 106 | -70.0 | 731 | 1.93 | 56.9 | 5.3 | 39.3 |
| Sioux | 10.8 | 5.3 | 3.7 | 50.2 | 4,044 | 4,154 | 2.7 | 0.5 | 936 | 486 | -430.0 | 1,112 | 3.88 | 73.7 | 30.9 | 20.6 |
| Slope | 18.9 | 15.0 | 11.5 | 47.4 | 767 | 727 | -5.2 | 2.8 | 89 | 39 | -31.0 | 326 | 2.28 | 71.2 | 6.7 | 27.3 |
| Stark | 11.3 | 7.1 | 6.4 | 48.1 | 22,636 | 24,198 | 6.9 | 32.7 | 4,941 | 2,394 | 5,161 | 12,524 | 2.42 | 59.4 | 7.2 | 32.4 |
| Steele | 15.7 | 15.3 | 11.3 | 47.5 | 2,258 | 1,975 | -12.5 | -4.3 | 225 | 176 | -135.0 | 777 | 2.24 | 63.6 | 0.9 | 33.2 |
| Stutsman | 14.6 | 11.2 | 8.8 | 48.9 | 21,908 | 21,100 | -3.7 | -2.9 | 2,367 | 2,464 | -492.0 | 8,908 | 2.17 | 56.5 | 7.5 | 36.9 |
| Towner | 18.7 | 13.9 | 11.8 | 48.1 | 2,876 | 2,246 | -21.9 | -6.1 | 238 | 286 | -92.0 | 1,058 | 2.07 | 57.4 | 4.1 | 36.1 |
| Traill | 14.2 | 10.8 | 9.5 | 49.0 | 8,477 | 8,121 | -4.2 | -2.0 | 962 | 943 | -183.0 | 3,383 | 2.25 | 62.5 | 6.5 | 33.3 |
| Walsh | 15.7 | 11.6 | 10.1 | 48.9 | 12,389 | 11,126 | -10.2 | -6.2 | 1,391 | 1,454 | -621.0 | 4,780 | 2.17 | 61.7 | 5.1 | 32.9 |

1. No spouse present.

# Table B. States and Counties — Population, Vital Statistics, Health, and Crime

| STATE County | Persons in group quarters, 2020 | Daytime Population, 2015-2019 Number | Daytime Population Employment/ residence ratio | Births, 2020 Total | Births, 2020 Rate[1] | Deaths, 2020 Number | Deaths, 2020 Rate[1] | Persons under 65 with no health insurance, 2019 Number | Percent | Medicare, 2020 Total beneficiaries | Enrolled in Original Medicare | Enrolled in Medicare Advantage | Crimes reported by county police or sheriff, 2019 Violent | Property |
|---|---|---|---|---|---|---|---|---|---|---|---|---|---|---|
| | 32 | 33 | 34 | 35 | 36 | 37 | 38 | 39 | 40 | 41 | 42 | 43 | 44 | 45 |
| **NORTH CAROLINA— Cont'd** | | | | | | | | | | | | | | |
| Swain | 246 | 15,118 | 1.16 | 161 | 11.4 | 250 | 17.6 | 2,529 | 22.3 | 3,821 | 2,795 | 1,026 | 27 | 242 |
| Transylvania | 1,007 | 32,360 | 0.90 | 268 | 7.8 | 448 | 13.0 | 3,625 | 15.9 | 10,444 | 6,704 | 3,741 | 38 | 257 |
| Tyrrell | 402 | 3,894 | 0.87 | 27 | 7.2 | 41 | 10.9 | 416 | 16.3 | 858 | 631 | 227 | 4 | 23 |
| Union | 2,849 | 198,651 | 0.71 | 2,374 | 9.7 | 1,673 | 6.8 | 23,543 | 11.3 | 34,660 | 21,099 | 13,562 | 235 | 1,987 |
| Vance | 802 | 43,339 | 0.94 | 558 | 12.5 | 567 | 12.7 | 5,118 | 14.6 | 10,427 | 5,897 | 4,531 | 125 | 546 |
| Wake | 21,362 | 1,114,138 | 1.08 | 12,716 | 11.2 | 6,757 | 6.0 | 97,433 | 10.1 | 150,059 | 91,443 | 58,616 | 215 | 1,376 |
| Warren | 912 | 17,477 | 0.66 | 148 | 7.6 | 272 | 13.9 | 2,322 | 17.0 | 5,407 | 3,300 | 2,107 | 40 | 216 |
| Washington | 149 | 11,766 | 0.96 | 128 | 11.1 | 179 | 15.6 | 1,196 | 14.1 | 3,443 | 2,457 | 986 | NA | NA |
| Watauga | 5,759 | 58,109 | 1.13 | 328 | 5.8 | 424 | 7.5 | 5,399 | 13.1 | 9,485 | 6,256 | 3,230 | 30 | 266 |
| Wayne | 2,896 | 120,668 | 0.95 | 1,609 | 13.0 | 1,521 | 12.3 | 15,268 | 15.4 | 24,427 | 16,536 | 7,891 | 206 | 1,181 |
| Wilkes | 927 | 65,171 | 0.89 | 679 | 10.0 | 906 | 13.3 | 8,605 | 16.4 | 17,601 | 8,230 | 9,372 | 132 | 676 |
| Wilson | 1,541 | 86,498 | 1.15 | 973 | 11.9 | 1,011 | 12.3 | 10,451 | 16.1 | 18,033 | 12,458 | 5,575 | NA | NA |
| Yadkin | 283 | 33,230 | 0.72 | 364 | 9.7 | 469 | 12.5 | 4,551 | 15.3 | 9,007 | 3,500 | 5,507 | 66 | 451 |
| Yancey | 153 | 15,759 | 0.72 | 178 | 9.8 | 249 | 13.8 | 2,069 | 15.7 | 5,615 | 3,436 | 2,179 | 5 | 47 |
| **NORTH DAKOTA** | 24,634 | 787,650 | 1.08 | 10,509 | 13.7 | 6,738 | 8.8 | 45,186 | 7.2 | 134,282 | 107,530 | 26,753 | X | X |
| Adams | 48 | 2,270 | 0.97 | 23 | 10.5 | 30 | 13.7 | 149 | 9.5 | 682 | 588 | 94 | 2 | 9 |
| Barnes | 532 | 10,461 | 0.96 | 89 | 8.6 | 130 | 12.5 | 526 | 6.9 | 2,673 | 2,154 | 519 | 2 | 18 |
| Benson | 16 | 6,883 | 1.00 | 141 | 20.9 | 81 | 12.0 | 635 | 11.1 | 1,088 | 1,025 | 63 | 1 | 13 |
| Billings | 9 | 903 | 0.96 | 10 | 11.2 | 7 | 7.9 | 64 | 9.0 | 175 | 153 | 21 | 0 | 6 |
| Bottineau | 219 | 6,306 | 0.94 | 64 | 10.2 | 84 | 13.4 | 383 | 8.4 | 1,672 | 1,659 | 13 | 4 | 55 |
| Bowman | 60 | 3,253 | 1.07 | 33 | 11.1 | 32 | 10.7 | 236 | 10.0 | 743 | 659 | 84 | 1 | 10 |
| Burke | 2 | 2,033 | 0.86 | 24 | 11.3 | 15 | 7.1 | 113 | 6.8 | 501 | D | D | 0 | 2 |
| Burleigh | 3,846 | 98,595 | 1.08 | 1,160 | 12.1 | 822 | 8.5 | 4,333 | 5.6 | 17,722 | 12,997 | 4,725 | 12 | 97 |
| Cass | 5,373 | 193,381 | 1.16 | 2,446 | 13.3 | 1,278 | 6.9 | 9,575 | 6.2 | 25,974 | 19,411 | 6,563 | 18 | 141 |
| Cavalier | 74 | 3,683 | 0.93 | 45 | 12.1 | 44 | 11.9 | 173 | 6.3 | 1,052 | 935 | 117 | 2 | 11 |
| Dickey | 186 | 4,757 | 0.93 | 55 | 11.7 | 56 | 11.9 | 295 | 7.9 | 1,135 | 950 | 185 | 0 | 8 |
| Divide | 62 | 2,420 | 1.08 | 27 | 11.9 | 23 | 10.1 | 170 | 10.1 | 520 | D | D | 2 | 6 |
| Dunn | 131 | 4,963 | 1.28 | 72 | 16.1 | 23 | 5.2 | 426 | 11.5 | 746 | 641 | 104 | 5 | 14 |
| Eddy | 69 | 2,149 | 0.87 | 28 | 12.8 | 39 | 17.9 | 144 | 8.3 | 631 | 546 | 85 | 1 | 20 |
| Emmons | 49 | 3,266 | 0.97 | 29 | 9.1 | 49 | 15.4 | 268 | 11.6 | 948 | 710 | 238 | 3 | 28 |
| Foster | 55 | 3,390 | 1.07 | 38 | 12.0 | 45 | 14.2 | 158 | 6.4 | 750 | 605 | 145 | 0 | 10 |
| Golden Valley | 52 | 1,765 | 0.92 | 21 | 12.1 | 7 | 4.0 | 115 | 9.2 | 418 | D | D | 0 | 0 |
| Grand Forks | 2,964 | 73,845 | 1.09 | 926 | 13.3 | 543 | 7.8 | 3,661 | 6.4 | 10,827 | 8,664 | 2,163 | 15 | 89 |
| Grant | 6 | 2,360 | 1.01 | 22 | 9.9 | 29 | 13.1 | 210 | 13.5 | 682 | 536 | 146 | 1 | 5 |
| Griggs | 38 | 2,453 | 1.04 | 21 | 9.5 | 34 | 15.4 | 118 | 7.7 | 689 | 559 | 130 | 3 | 6 |
| Hettinger | 132 | 2,385 | 0.86 | 32 | 13.1 | 24 | 9.8 | 213 | 11.7 | 610 | 515 | 94 | 1 | 22 |
| Kidder | 0 | 2,302 | 0.86 | 27 | 11.0 | 18 | 7.3 | 189 | 10.3 | 653 | 436 | 217 | 0 | 7 |
| LaMoure | 58 | 3,887 | 0.88 | 49 | 12.1 | 41 | 10.2 | 317 | 10.9 | 1,086 | 808 | 278 | 3 | 4 |
| Logan | 53 | 1,887 | 1.02 | 26 | 13.8 | 18 | 9.6 | 182 | 13.7 | 522 | 401 | 121 | 2 | 0 |
| McHenry | 44 | 4,815 | 0.62 | 62 | 10.9 | 56 | 9.8 | 385 | 8.5 | 1,337 | 1,123 | 214 | 6 | 34 |
| McIntosh | 93 | 2,662 | 1.05 | 26 | 10.7 | 47 | 19.3 | 171 | 10.2 | 857 | 672 | 186 | 0 | 0 |
| McKenzie | 149 | 15,832 | 1.38 | 288 | 18.9 | 79 | 5.2 | 1,220 | 8.9 | 1,178 | 1,148 | 30 | 15 | 167 |
| McLean | 147 | 9,819 | 1.05 | 104 | 11.0 | 107 | 11.4 | 651 | 9.1 | 2,430 | 2,003 | 427 | 12 | 79 |
| Mercer | 125 | 9,273 | 1.19 | 88 | 10.8 | 85 | 10.4 | 335 | 5.1 | 1,852 | 1,453 | 399 | 1 | 16 |
| Morton | 797 | 25,900 | 0.72 | 441 | 14.0 | 309 | 9.8 | 1,635 | 6.3 | 6,146 | 4,248 | 1,898 | 3 | 113 |
| Mountrail | 542 | 11,825 | 1.32 | 193 | 18.4 | 91 | 8.7 | 1,337 | 14.4 | 1,461 | 1,443 | 19 | 2 | 58 |
| Nelson | 79 | 2,665 | 0.84 | 24 | 8.6 | 48 | 17.2 | 169 | 8.1 | 931 | 841 | 89 | 5 | 15 |
| Oliver | 2 | 1,788 | 0.93 | 23 | 11.9 | 5 | 2.6 | 118 | 7.8 | 424 | 326 | 98 | 0 | 3 |
| Pembina | 146 | 7,426 | 1.14 | 61 | 9.2 | 89 | 13.4 | 426 | 8.3 | 1,794 | 1,560 | 234 | 6 | 36 |
| Pierce | 238 | 4,203 | 1.04 | 35 | 8.9 | 50 | 12.7 | 192 | 6.7 | 1,063 | 872 | 191 | 0 | 11 |
| Ramsey | 425 | 11,785 | 1.03 | 160 | 14.0 | 134 | 11.8 | 647 | 7.3 | 2,658 | 2,429 | 230 | 3 | 18 |
| Ransom | 139 | 5,142 | 0.94 | 51 | 9.9 | 79 | 15.3 | 269 | 6.6 | 1,316 | 1,138 | 179 | 3 | 26 |
| Renville | 54 | 2,200 | 0.79 | 21 | 9.2 | 24 | 10.5 | 123 | 6.7 | 550 | D | D | 2 | 24 |
| Richland | 909 | 16,688 | 1.05 | 189 | 11.7 | 151 | 9.3 | 857 | 6.9 | 3,335 | 2,428 | 907 | 8 | 94 |
| Rolette | 123 | 14,417 | 0.98 | 218 | 15.4 | 187 | 13.2 | 1,657 | 13.6 | 2,273 | 2,226 | 47 | 4 | 38 |
| Sargent | 38 | 4,854 | 1.48 | 49 | 12.5 | 39 | 10.0 | 215 | 7.2 | 936 | 839 | 97 | 7 | 21 |
| Sheridan | 3 | 1,274 | 0.76 | 12 | 9.4 | 8 | 6.3 | 107 | 12.4 | 422 | 338 | 83 | 1 | 4 |
| Sioux | 52 | 4,883 | 1.41 | 71 | 17.0 | 73 | 17.5 | 460 | 12.3 | 422 | 397 | 25 | 0 | 0 |
| Slope | 0 | 683 | 0.81 | 8 | 10.7 | 6 | 8.0 | 64 | 11.3 | 179 | 154 | 25 | 0 | 1 |
| Stark | 542 | 32,065 | 1.06 | 544 | 16.9 | 251 | 7.8 | 1,759 | 6.5 | 4,777 | 4,115 | 662 | 5 | 55 |
| Steele | 0 | 1,617 | 0.86 | 20 | 10.6 | 10 | 5.3 | 66 | 4.7 | 474 | 395 | 79 | 0 | 1 |
| Stutsman | 1,646 | 21,462 | 1.05 | 224 | 10.9 | 248 | 12.1 | 1,014 | 6.7 | 4,842 | 3,192 | 1,650 | 2 | 33 |
| Towner | 32 | 2,251 | 1.03 | 23 | 10.9 | 22 | 10.4 | 173 | 10.7 | 587 | 547 | 40 | 0 | 16 |
| Traill | 301 | 7,570 | 0.89 | 89 | 11.2 | 96 | 12.1 | 374 | 6.0 | 1,706 | 1,442 | 263 | 3 | 54 |
| Walsh | 329 | 10,858 | 1.02 | 125 | 12.0 | 132 | 12.6 | 794 | 9.6 | 2,574 | 2,211 | 363 | 7 | 96 |

1. Per 1,000 estimated resident population.

# Table B. States and Counties — Crime, Education, Money Income, and Poverty

| STATE County | County law enforcement employment, 2019 — Officers | Civilians | Education — Enrollment[1] Total | Percent private | Attainment[2] (percent) High school graduate or less | Bachelor's degree or more | Local government expenditures[3] 2017-2018 Total current spending (mil dol) | Current spending per student (dollars) | Money income, 2015-2019 Per capita income[4] | Households Median income (dollars) | Percent with income of less than $50,000 | Percent with income of $200,000 or more | Income and poverty, 2019 Median household income (dollars) | Percent below poverty level All persons | Children under 18 years | Children 5 to 17 years in families |
|---|---|---|---|---|---|---|---|---|---|---|---|---|---|---|---|---|
| | 46 | 47 | 48 | 49 | 50 | 51 | 52 | 53 | 54 | 55 | 56 | 57 | 58 | 59 | 60 | 61 |
| **NORTH CAROLINA—Cont'd** | | | | | | | | | | | | | | | | |
| Swain | 28 | 20 | 3,022 | 7.3 | 50.8 | 18.2 | 23 | 10,382 | 23,139 | 39,317 | 58.8 | 2.0 | 46,075 | 16.1 | 27.4 | 24.8 |
| Transylvania | 58 | 26 | 6,094 | 20.5 | 37.7 | 32.6 | 41 | 10,710 | 29,549 | 50,212 | 49.8 | 3.8 | 55,628 | 13.1 | 20.3 | 18.4 |
| Tyrrell | 12 | 2 | 687 | 11.9 | 62.4 | 10.6 | 9 | 14,956 | 19,743 | 35,300 | 61.4 | 0.8 | 37,680 | 25.4 | 34.9 | 33.3 |
| Union | 219 | 92 | 66,435 | 17.3 | 34.9 | 35.4 | 381 | 8,551 | 36,362 | 80,033 | 29.1 | 11.1 | 86,138 | 7.3 | 8.8 | 7.8 |
| Vance | 40 | 48 | 10,211 | 11.0 | 56.6 | 14.8 | 79 | 9,873 | 23,049 | 40,529 | 59.9 | 1.9 | 40,450 | 18.5 | 31.0 | 28.3 |
| Wake | 368 | 553 | 291,186 | 16.9 | 22.1 | 52.8 | 1,543 | 8,838 | 40,982 | 80,591 | 30.0 | 11.1 | 84,377 | 8.0 | 9.4 | 8.1 |
| Warren | 33 | 16 | 3,559 | 12.9 | 55.1 | 15.8 | 28 | 12,424 | 23,432 | 36,466 | 62.4 | 2.7 | 37,027 | 21.7 | 32.7 | 31.3 |
| Washington | 17 | 1 | 2,644 | 8.3 | 53.7 | 11.5 | 19 | 12,531 | 23,431 | 35,979 | 64.8 | 1.4 | 40,157 | 21.3 | 36.4 | 34.9 |
| Watauga | 49 | 33 | 21,261 | 7.9 | 30.0 | 42.2 | 48 | 9,996 | 26,882 | 47,526 | 52.4 | 4.0 | 51,630 | 21.4 | 14.3 | 13.6 |
| Wayne | 128 | 7 | 30,525 | 13.5 | 46.0 | 19.0 | 173 | 8,856 | 25,277 | 44,416 | 55.5 | 2.4 | 45,634 | 18.6 | 30.2 | 25.8 |
| Wilkes | 76 | 60 | 13,665 | 8.5 | 51.6 | 15.8 | 90 | 9,033 | 25,202 | 44,107 | 55.7 | 2.4 | 45,250 | 15.2 | 21.1 | 18.9 |
| Wilson | 92 | 61 | 19,996 | 13.1 | 51.2 | 19.6 | 119 | 8,968 | 24,790 | 43,877 | 55.1 | 2.6 | 42,414 | 21.5 | 32.2 | 27.7 |
| Yadkin | 40 | 36 | 7,359 | 10.8 | 51.5 | 11.6 | 49 | 9,055 | 25,218 | 44,682 | 55.1 | 2.6 | 50,929 | 13.9 | 23.8 | 21.8 |
| Yancey | 23 | 20 | 3,180 | 10.5 | 48.7 | 20.3 | 24 | 10,719 | 24,487 | 42,222 | 55.8 | 1.6 | 47,664 | 14.2 | 22.0 | 21.0 |
| **NORTH DAKOTA** | X | X | 184,783 | 11.0 | 33.8 | 30.0 | 1,539 | 13,752 | 36,062 | 64,894 | 38.5 | 5.9 | 67,402 | 10.5 | 10.9 | 10.2 |
| Adams | 5 | 1 | 459 | 9.4 | 39.8 | 25.6 | 4 | 13,252 | 33,458 | 56,681 | 44.9 | 3.4 | 55,880 | 10.7 | 15.0 | 14.4 |
| Barnes | 8 | 1 | 2,260 | 7.4 | 39.2 | 29.4 | 21 | 15,332 | 35,491 | 58,365 | 45.2 | 4.3 | 60,689 | 9.9 | 10.0 | 9.0 |
| Benson | 4 | 0 | 1,993 | 1.2 | 43.2 | 16.2 | 18 | 17,910 | 21,707 | 47,667 | 51.9 | 3.2 | 45,709 | 23.3 | 33.9 | 33.3 |
| Billings | 6 | 0 | 152 | 0.0 | 43.5 | 26.3 | 3 | 34,329 | 42,071 | 75,208 | 26.9 | 6.0 | 69,528 | 9.7 | 9.5 | 10.7 |
| Bottineau | 11 | 10 | 1,423 | 8.9 | 36.0 | 24.0 | 15 | 16,715 | 35,655 | 60,381 | 39.2 | 5.5 | 62,904 | 9.7 | 12.4 | 11.0 |
| Bowman | 4 | 1 | 540 | 10.9 | 45.6 | 24.2 | 10 | 15,222 | 34,880 | 62,442 | 37.3 | 4.9 | 65,411 | 8.0 | 9.9 | 9.4 |
| Burke | 6 | 1 | 486 | 6.6 | 35.2 | 22.8 | 7 | 19,642 | 39,350 | 77,232 | 31.3 | 7.2 | 67,545 | 8.8 | 8.8 | 8.8 |
| Burleigh | 52 | 84 | 21,953 | 23.4 | 28.4 | 35.8 | 165 | 12,434 | 38,804 | 71,524 | 34.7 | 6.4 | 73,565 | 6.9 | 6.7 | 5.7 |
| Cass | 110 | 57 | 49,019 | 10.3 | 25.2 | 40.2 | 321 | 12,940 | 37,620 | 64,482 | 38.5 | 7.1 | 63,977 | 11.4 | 9.2 | 8.6 |
| Cavalier | 6 | 6 | 672 | 3.1 | 39.6 | 19.5 | 8 | 15,980 | 43,457 | 64,798 | 35.2 | 8.1 | 63,484 | 9.1 | 12.0 | 12.2 |
| Dickey | 4 | 1 | 1,129 | 15.3 | 36.5 | 28.8 | 10 | 11,377 | 33,539 | 65,492 | 37.0 | 4.2 | 57,490 | 10.6 | 13.2 | 12.1 |
| Divide | 6 | 1 | 461 | 0.2 | 40.6 | 18.5 | 6 | 15,829 | 40,184 | 62,865 | 42.6 | 11.0 | 61,506 | 10.8 | 13.9 | 13.5 |
| Dunn | 16 | 2 | 1,015 | 16.4 | 42.6 | 21.3 | 10 | 15,961 | 42,691 | 76,719 | 34.2 | 10.1 | 75,723 | 9.8 | 11.9 | 10.9 |
| Eddy | 5 | 1 | 357 | 2.8 | 39.8 | 26.2 | 6 | 20,420 | 35,639 | 54,868 | 47.5 | 4.9 | 54,465 | 10.0 | 13.6 | 13.2 |
| Emmons | 5 | 1 | 529 | 4.7 | 48.0 | 18.2 | 9 | 17,025 | 33,088 | 55,902 | 43.8 | 4.1 | 53,073 | 13.7 | 17.7 | 16.5 |
| Foster | 3 | 1 | 712 | 3.4 | 35.6 | 26.7 | 6 | 12,508 | 35,299 | 61,425 | 38.6 | 5.6 | 54,917 | 8.8 | 10.6 | 10.2 |
| Golden Valley | 3 | 1 | 325 | 16.9 | 32.5 | 26.9 | 6 | 16,817 | 33,111 | 58,690 | 42.6 | 4.3 | 58,370 | 12.1 | 19.6 | 19.2 |
| Grand Forks | 34 | 6 | 21,918 | 5.9 | 27.1 | 35.6 | 121 | 12,971 | 32,153 | 54,051 | 46.3 | 3.9 | 57,388 | 13.2 | 11.2 | 10.5 |
| Grant | 5 | 0 | 452 | 8.0 | 46.0 | 18.1 | 4 | 18,448 | 34,786 | 50,938 | 49.4 | 4.7 | 41,841 | 14.9 | 22.0 | 20.5 |
| Griggs | 2 | 0 | 409 | 1.7 | 44.1 | 25.8 | 7 | 15,358 | 34,882 | 53,565 | 45.1 | 4.9 | 56,527 | 9.5 | 12.3 | 11.1 |
| Hettinger | 6 | 1 | 448 | 12.3 | 46.8 | 13.0 | 7 | 14,642 | 34,181 | 60,164 | 43.6 | 5.7 | 54,600 | 12.5 | 16.4 | 16.7 |
| Kidder | 3 | 1 | 458 | 5.5 | 43.0 | 17.7 | 6 | 14,212 | 31,000 | 54,643 | 45.7 | 4.3 | 50,570 | 14.1 | 17.5 | 17.2 |
| LaMoure | 4 | 1 | 802 | 11.1 | 42.8 | 24.6 | 13 | 15,701 | 37,220 | 60,806 | 40.1 | 6.6 | 56,059 | 12.2 | 14.8 | 14.4 |
| Logan | 3 | 0 | 300 | 4.7 | 45.2 | 19.6 | 6 | 16,871 | 32,619 | 59,375 | 43.3 | 3.3 | 51,496 | 13.2 | 18.7 | 18.2 |
| McHenry | 8 | 0 | 1,196 | 5.5 | 42.8 | 21.4 | 15 | 14,922 | 36,980 | 64,179 | 39.2 | 5.0 | 59,661 | 12.1 | 14.0 | 12.5 |
| McIntosh | 3 | 0 | 360 | 1.1 | 51.2 | 15.6 | 6 | 16,556 | 32,131 | 52,587 | 47.9 | 4.2 | 47,349 | 13.9 | 16.1 | 15.9 |
| McKenzie | 29 | 15 | 3,392 | 13.0 | 37.7 | 25.8 | 29 | 13,622 | 36,690 | 77,845 | 31.5 | 9.8 | 86,890 | 7.4 | 8.7 | 8.7 |
| McLean | 26 | 15 | 1,856 | 6.5 | 41.3 | 20.8 | 24 | 13,641 | 36,428 | 68,529 | 36.3 | 4.2 | 67,453 | 9.4 | 11.7 | 10.4 |
| Mercer | 16 | 14 | 1,772 | 12.4 | 35.6 | 21.7 | 17 | 13,139 | 39,025 | 82,181 | 30.1 | 5.0 | 83,341 | 7.7 | 8.1 | 7.4 |
| Morton | 36 | 8 | 6,011 | 9.8 | 37.5 | 26.9 | 62 | 13,049 | 39,384 | 70,556 | 34.1 | 4.7 | 68,051 | 7.9 | 9.7 | 9.5 |
| Mountrail | 9 | 13 | 2,521 | 3.1 | 41.5 | 20.4 | 28 | 14,443 | 36,364 | 72,147 | 35.1 | 9.3 | 69,040 | 9.5 | 13.3 | 13.3 |
| Nelson | 5 | 1 | 457 | 5.0 | 37.4 | 25.0 | 8 | 17,299 | 34,491 | 52,039 | 48.9 | 4.4 | 56,098 | 11.2 | 14.7 | 13.3 |
| Oliver | 4 | 1 | 415 | 7.5 | 39.5 | 21.1 | 4 | 15,262 | 35,565 | 78,929 | 33.2 | 3.5 | 72,919 | 10.6 | 17.4 | 15.9 |
| Pembina | 12 | 8 | 1,291 | 10.8 | 42.3 | 20.6 | 19 | 15,347 | 36,251 | 64,549 | 37.9 | 4.7 | 59,970 | 8.8 | 9.6 | 8.8 |
| Pierce | 4 | 34 | 679 | 8.2 | 41.9 | 21.0 | 9 | 12,944 | 30,756 | 55,660 | 46.2 | 3.4 | 56,753 | 11.6 | 13.6 | 11.3 |
| Ramsey | 8 | 1 | 2,711 | 7.9 | 36.9 | 26.4 | 29 | 15,762 | 34,041 | 58,910 | 43.5 | 4.3 | 60,710 | 12.3 | 17.0 | 17.5 |
| Ransom | 5 | 2 | 981 | 6.4 | 44.4 | 19.1 | 12 | 12,586 | 34,665 | 63,903 | 38.5 | 2.9 | 66,422 | 8.7 | 9.6 | 8.5 |
| Renville | 6 | 1 | 491 | 3.7 | 38.8 | 20.1 | 9 | 14,634 | 34,452 | 73,182 | 30.6 | 3.4 | 84,986 | 8.9 | 9.8 | 8.6 |
| Richland | 18 | 20 | 4,223 | 5.7 | 36.3 | 21.3 | 34 | 15,139 | 31,346 | 61,371 | 43.5 | 3.2 | 62,837 | 12.4 | 13.5 | 10.7 |
| Rolette | 8 | 14 | 3,896 | 1.6 | 38.3 | 19.6 | 48 | 15,742 | 20,484 | 43,158 | 54.3 | 2.8 | 45,296 | 25.9 | 30.1 | 26.0 |
| Sargent | 4 | 1 | 661 | 4.1 | 44.5 | 18.5 | 10 | 16,506 | 37,105 | 63,073 | 33.8 | 3.7 | 67,551 | 7.5 | 9.9 | 9.8 |
| Sheridan | 3 | 0 | 168 | 6.5 | 50.0 | 14.6 | 2 | 18,758 | 32,627 | 51,055 | 48.2 | 3.3 | 58,725 | 16.7 | 23.5 | 22.5 |
| Sioux | 1 | 0 | 1,393 | 5.4 | 48.7 | 15.0 | 11 | 21,900 | 15,649 | 38,939 | 59.4 | 3.2 | 37,133 | 32.1 | 35.4 | 31.9 |
| Slope | 1 | 0 | 127 | 1.6 | 38.4 | 25.9 | 0 | 26,000 | 33,798 | 63,611 | 38.0 | 3.7 | 52,695 | 13.5 | 19.1 | 19.4 |
| Stark | 24 | 5 | 7,223 | 17.3 | 41.6 | 22.4 | 58 | 12,468 | 38,103 | 72,045 | 33.7 | 7.4 | 68,759 | 8.5 | 8.5 | 8.5 |
| Steele | 3 | 0 | 301 | 0.0 | 29.0 | 30.9 | 4 | 26,215 | 39,011 | 70,724 | 30.0 | 4.8 | 67,661 | 7.4 | 10.5 | 10.1 |
| Stutsman | 13 | 2 | 4,416 | 27.4 | 42.9 | 23.7 | 39 | 14,748 | 31,991 | 57,674 | 43.5 | 2.9 | 60,297 | 10.8 | 10.5 | 9.7 |
| Towner | 6 | 1 | 409 | 2.9 | 42.4 | 17.0 | 4 | 12,217 | 37,239 | 52,300 | 47.8 | 6.6 | 53,024 | 10.7 | 15.5 | 13.1 |
| Traill | 10 | 5 | 1,955 | 6.1 | 32.5 | 28.9 | 20 | 14,565 | 33,321 | 64,453 | 36.0 | 3.6 | 69,443 | 8.0 | 8.3 | 7.6 |
| Walsh | 10 | 5 | 2,120 | 5.3 | 47.6 | 17.2 | 26 | 15,533 | 34,194 | 55,700 | 45.6 | 5.2 | 59,153 | 11.7 | 12.0 | 11.4 |

1. All persons 3 years old and over enrolled in nursery school through college.   2. Persons 25 years old and over.   3. Elementary and secondary education expenditures.   4. Based on population estimated by the American Community Survey, 2014–2018.

# Table B. States and Counties — **Personal Income**

| STATE County | Personal income, 2019 | | | | | Supplements to wages and salaries, employer contributions (mil dol) | | | | | Earnings, 2019 | Contributions for government social insurance (mil dol) | |
|---|---|---|---|---|---|---|---|---|---|---|---|---|---|
| | Total (mil dol) | Percent change 2018-2019 | Per capita¹ Dollars | Per capita¹ Rank | Wages and salaries (mil dol) | Pension and insurance | Government social insurance | Proprietors' income (mil dol) | Dividends, interest, and rent (mil dol) | Personal transfer receipts (mil dol) | Total (mil dol) | From employee and self-employed | From employer |
| | 62 | 63 | 64 | 65 | 66 | 67 | 68 | 69 | 70 | 71 | 72 | 73 | 74 |
| NORTH CAROLINA—Cont'd | | | | | | | | | | | | | |
| Swain | 583 | 5.4 | 40,819 | 1,947 | 265 | 47 | 20 | 33 | 99 | 187 | 365 | 26 | 20 |
| Transylvania | 1,462 | 3.6 | 42,515 | 1,688 | 385 | 65 | 28 | 113 | 433 | 403 | 591 | 49 | 28 |
| Tyrrell | 131 | 5.2 | 32,712 | 2,908 | 40 | 10 | 3 | 16 | 20 | 38 | 68 | 4 | 3 |
| Union | 12,813 | 4.9 | 53,417 | 584 | 3,355 | 514 | 240 | 867 | 1,740 | 1,639 | 4,976 | 320 | 240 |
| Vance | 1,571 | 3.6 | 35,278 | 2,644 | 577 | 100 | 42 | 58 | 269 | 559 | 778 | 60 | 42 |
| Wake | 69,223 | 4.9 | 62,264 | 241 | 40,067 | 5,095 | 2,700 | 4,356 | 13,249 | 7,057 | 52,217 | 3,117 | 2,700 |
| Warren | 599 | 3.5 | 30,377 | 3,026 | 117 | 28 | 8 | 12 | 104 | 219 | 165 | 17 | 8 |
| Washington | 434 | 3.8 | 37,481 | 2,393 | 110 | 23 | 8 | 11 | 74 | 166 | 153 | 14 | 8 |
| Watauga | 2,175 | 4.9 | 38,720 | 2,234 | 1,023 | 191 | 72 | 202 | 577 | 422 | 1,487 | 94 | 72 |
| Wayne | 4,912 | 3.7 | 39,894 | 2,065 | 2,091 | 439 | 161 | 272 | 872 | 1,293 | 2,963 | 185 | 161 |
| Wilkes | 2,553 | 1.4 | 37,313 | 2,412 | 912 | 160 | 66 | 189 | 494 | 791 | 1,326 | 100 | 66 |
| Wilson | 3,382 | 3.9 | 41,339 | 1,870 | 1,895 | 301 | 132 | 202 | 542 | 939 | 2,530 | 168 | 132 |
| Yadkin | 1,462 | 4.0 | 38,816 | 2,220 | 413 | 69 | 29 | 76 | 213 | 406 | 587 | 46 | 29 |
| Yancey | 683 | 4.9 | 37,812 | 2,343 | 194 | 37 | 14 | 40 | 138 | 223 | 284 | 25 | 14 |
| NORTH DAKOTA | 43,614 | 3.1 | 57,108 | X | 24,331 | 3,690 | 1,904 | 4,284 | 9,446 | 6,308 | 34,209 | 2,052 | 1,904 |
| Adams | 111 | 0.9 | 50,172 | 818 | 44 | 7 | 4 | 3 | 23 | 30 | 58 | 5 | 4 |
| Barnes | 581 | 4.8 | 55,764 | 458 | 198 | 36 | 17 | 86 | 139 | 122 | 336 | 20 | 17 |
| Benson | 247 | -4.4 | 36,139 | 2,552 | 89 | 20 | 7 | 30 | 50 | 71 | 147 | 8 | 7 |
| Billings | 74 | 6.2 | 79,550 | 66 | 25 | 5 | 2 | 12 | 27 | 6 | 44 | 2 | 2 |
| Bottineau | 384 | -3.7 | 61,071 | 275 | 119 | 22 | 10 | 61 | 93 | 73 | 212 | 13 | 10 |
| Bowman | 201 | 10.3 | 66,566 | 162 | 81 | 13 | 7 | 26 | 49 | 33 | 126 | 8 | 7 |
| Burke | 157 | 3.2 | 74,224 | 94 | 44 | 9 | 3 | 31 | 37 | 20 | 87 | 5 | 3 |
| Burleigh | 5,749 | 3.1 | 60,117 | 300 | 3,101 | 463 | 241 | 491 | 1,196 | 781 | 4,296 | 267 | 241 |
| Cass | 10,694 | 3.9 | 58,783 | 345 | 6,598 | 935 | 511 | 927 | 2,525 | 1,193 | 8,972 | 537 | 511 |
| Cavalier | 246 | -14.2 | 65,298 | 181 | 67 | 12 | 6 | 70 | 57 | 43 | 154 | 7 | 6 |
| Dickey | 295 | 9.7 | 60,512 | 288 | 76 | 13 | 6 | 68 | 78 | 56 | 163 | 8 | 6 |
| Divide | 119 | 3.7 | 52,476 | 646 | 45 | 8 | 4 | 10 | 41 | 24 | 66 | 4 | 4 |
| Dunn | 280 | 7.6 | 63,237 | 225 | 169 | 24 | 12 | 29 | 75 | 34 | 235 | 14 | 12 |
| Eddy | 137 | -0.3 | 59,794 | 313 | 29 | 5 | 3 | 33 | 25 | 29 | 70 | 3 | 3 |
| Emmons | 159 | 0.6 | 49,101 | 913 | 38 | 8 | 3 | 31 | 41 | 39 | 81 | 6 | 3 |
| Foster | 191 | -2.6 | 59,570 | 323 | 77 | 12 | 6 | 56 | 38 | 32 | 150 | 10 | 6 |
| Golden Valley | 66 | 2.0 | 37,688 | 2,361 | 25 | 4 | 2 | 3 | 25 | 14 | 35 | 3 | 2 |
| Grand Forks | 3,716 | 2.1 | 53,502 | 571 | 2,124 | 385 | 174 | 314 | 793 | 532 | 2,996 | 175 | 174 |
| Grant | 103 | 4.4 | 45,370 | 1,315 | 24 | 5 | 2 | 19 | 26 | 29 | 50 | 3 | 2 |
| Griggs | 127 | -1.2 | 56,996 | 405 | 37 | 6 | 3 | 29 | 31 | 30 | 76 | 4 | 3 |
| Hettinger | 124 | 3.8 | 49,775 | 850 | 37 | 7 | 3 | 9 | 34 | 27 | 56 | 4 | 3 |
| Kidder | 131 | 6.5 | 52,877 | 624 | 29 | 5 | 2 | 30 | 30 | 28 | 67 | 4 | 2 |
| LaMoure | 234 | 8.7 | 57,831 | 378 | 58 | 11 | 5 | 56 | 65 | 42 | 129 | 6 | 5 |
| Logan | 96 | -3.1 | 51,670 | 709 | 22 | 4 | 2 | 22 | 25 | 23 | 50 | 3 | 2 |
| McHenry | 299 | -1.7 | 52,038 | 679 | 67 | 12 | 6 | 42 | 51 | 58 | 127 | 8 | 6 |
| McIntosh | 145 | 4.3 | 58,245 | 363 | 41 | 8 | 3 | 27 | 36 | 41 | 79 | 5 | 3 |
| McKenzie | 979 | 10.8 | 65,130 | 185 | 1,066 | 124 | 79 | 79 | 262 | 63 | 1,348 | 78 | 79 |
| McLean | 526 | -0.6 | 55,710 | 464 | 221 | 42 | 17 | 51 | 123 | 112 | 330 | 21 | 17 |
| Mercer | 474 | | 57,936 | 371 | 308 | 64 | 24 | 21 | 84 | 83 | 418 | 25 | 24 |
| Morton | 1,677 | 3.7 | 53,454 | 577 | 617 | 105 | 52 | 117 | 283 | 274 | 891 | 60 | 52 |
| Mountrail | 649 | 4.6 | 61,569 | 263 | 440 | 57 | 32 | 60 | 189 | 77 | 589 | 36 | 32 |
| Nelson | 176 | -0.5 | 60,964 | 277 | 37 | 7 | 3 | 31 | 41 | 46 | 78 | 5 | 3 |
| Oliver | 86 | 0.9 | 44,140 | 1,466 | 60 | 12 | 5 | 3 | 17 | 17 | 80 | 5 | 5 |
| Pembina | 412 | 2.3 | 60,512 | 288 | 181 | 31 | 16 | 69 | 97 | 76 | 296 | 16 | 16 |
| Pierce | 197 | -5.5 | 49,600 | 868 | 78 | 13 | 6 | 32 | 38 | 47 | 130 | 9 | 6 |
| Ramsey | 576 | 0.1 | 50,047 | 829 | 248 | 45 | 20 | 66 | 125 | 131 | 379 | 24 | 20 |
| Ransom | 302 | 5.3 | 57,933 | 372 | 90 | 16 | 8 | 53 | 57 | 64 | 167 | 10 | 8 |
| Renville | 164 | -7.2 | 70,486 | 125 | 33 | 6 | 3 | 40 | 33 | 26 | 82 | 4 | 3 |
| Richland | 826 | 1.7 | 51,045 | 754 | 350 | 60 | 29 | 80 | 190 | 146 | 519 | 33 | 29 |
| Rolette | 523 | 1.0 | 36,875 | 2,458 | 187 | 52 | 15 | 37 | 79 | 181 | 292 | 17 | 15 |
| Sargent | 238 | 10.0 | 61,102 | 274 | 171 | 31 | 14 | 44 | 60 | 36 | 260 | 14 | 14 |
| Sheridan | 62 | -5.7 | 46,962 | 1,138 | 9 | 2 | 1 | 18 | 14 | 16 | 30 | 2 | 1 |
| Sioux | 121 | 0.3 | 28,524 | 3,073 | 72 | 19 | 6 | 9 | 21 | 46 | 106 | 6 | 6 |
| Slope | 35 | 14.0 | 46,328 | 1,209 | 12 | 2 | 1 | 7 | 12 | 6 | 21 | 2 | 1 |
| Stark | 2,013 | 6.0 | 63,942 | 209 | 1,329 | 166 | 99 | 174 | 332 | 219 | 1,769 | 107 | 99 |
| Steele | 109 | 3.9 | 57,770 | 380 | 36 | 6 | 3 | 17 | 33 | 18 | 62 | 3 | 3 |
| Stutsman | 1,099 | 1.6 | 53,065 | 610 | 495 | 84 | 40 | 136 | 228 | 214 | 755 | 47 | 40 |
| Towner | 137 | -12.6 | 62,530 | 238 | 32 | 6 | 3 | 43 | 27 | 29 | 83 | 4 | 3 |
| Traill | 427 | 6.7 | 53,140 | 604 | 162 | 29 | 13 | 64 | 86 | 82 | 268 | 16 | 13 |
| Walsh | 553 | 1.9 | 51,995 | 681 | 218 | 37 | 18 | 80 | 119 | 123 | 353 | 21 | 18 |

1. Based on the resident population estimated as of July 1 of the year shown.

Items 62—74

| STATE County | Earnings, 2019 (cont.) | | | | | | | | | Social Security beneficiaries, December 2019 | | Supplemental Security Income recipients, 2019 | Housing units, 2020 | |
|---|---|---|---|---|---|---|---|---|---|---|---|---|---|---|
| | Percent by selected industries | | | | | | | | | | | | | |
| | Farm | Mining, quarrying, and extractions | Construction | Manu-facturing | Information; professional, scientific, technical services | Retail trade | Finance, insurance, real estate, and leasing | Health care and social assistance | Govern-ment | Number | Rate[1] | | Total | Percent change, 2010-2020 |
| | 75 | 76 | 77 | 78 | 79 | 80 | 81 | 82 | 83 | 84 | 85 | 86 | 87 | 88 |
| **NORTH CAROLINA—Cont'd** | | | | | | | | | | | | | | |
| Swain | 0.4 | 0.0 | 4.7 | D | 1.5 | 5.5 | 1.6 | D | 49.5 | 4,430 | 310 | 346 | 9,207 | 5.6 |
| Transylvania | 0.8 | 0.1 | 11.2 | 7.6 | 8.0 | 8.2 | 6.8 | 13.7 | 16.0 | 10,550 | 308 | 588 | 19,920 | 3.9 |
| Tyrrell | 18.4 | 0.2 | 5.6 | 6.0 | D | 5.3 | D | D | 36.7 | 925 | 243 | 124 | 2,094 | 1.3 |
| Union | 1.3 | D | 17.6 | 18.2 | 6.9 | 6.7 | 3.2 | 4.8 | 13.4 | 35,840 | 149 | 2,379 | 85,517 | 17.3 |
| Vance | -0.6 | D | 4.1 | 13.2 | 4.6 | 9.5 | 5.0 | 14.2 | 18.3 | 11,720 | 262 | 2,243 | 20,175 | 0.5 |
| Wake | 0.1 | 0.0 | 6.9 | 6.9 | 24.0 | 5.4 | 8.8 | 9.4 | 12.8 | 149,555 | 134 | 11,967 | 462,507 | 24.4 |
| Warren | -2.6 | 0.0 | 4.8 | 10.5 | 3.3 | 6.2 | 4.0 | 4.9 | 43.0 | 5,030 | 255 | 825 | 12,136 | 2.7 |
| Washington | 1.6 | 0.0 | D | 14.5 | D | 6.6 | 3.9 | D | 27.3 | 3,730 | 321 | 649 | 6,445 | -0.5 |
| Watauga | 0.0 | 0.1 | 5.8 | 1.8 | 5.5 | 9.2 | 6.9 | 16.7 | 28.4 | 9,800 | 174 | 570 | 34,702 | 8.1 |
| Wayne | 2.7 | D | 5.5 | 11.2 | 2.9 | 7.0 | 5.0 | 11.1 | 34.2 | 26,595 | 215 | 4,047 | 55,010 | 3.9 |
| Wilkes | 2.4 | D | 5.6 | 17.7 | D | 7.7 | 4.7 | 11.3 | 15.9 | 19,870 | 291 | 1,930 | 33,830 | 2.3 |
| Wilson | 1.4 | D | 10.2 | 23.4 | 6.2 | 6.4 | 6.5 | 8.4 | 12.1 | 21,320 | 261 | 2,992 | 36,639 | 3.2 |
| Yadkin | 1.9 | 0.0 | 12.6 | 25.0 | D | 5.3 | 3.1 | D | 15.9 | 9,740 | 259 | 773 | 17,495 | 0.9 |
| Yancey | 0.9 | D | 11.7 | 28.0 | 3.2 | 7.9 | 2.3 | D | 19.7 | 6,070 | 336 | 548 | 11,354 | 2.8 |
| **NORTH DAKOTA** | 3.2 | 8.3 | 8.2 | 5.8 | 6.5 | 5.8 | 7.4 | 12.6 | 17.0 | 136,520 | 179 | 8,301 | 382,354 | 20.4 |
| Adams | -13.9 | D | 8.1 | 7.6 | 4.2 | 10.1 | 5.5 | 32.3 | 12.2 | 665 | 299 | 21 | 1,448 | 5.2 |
| Barnes | 17.4 | D | 7.1 | 11.3 | 3.6 | 4.8 | 4.4 | 12.6 | 19.2 | 2,765 | 266 | 120 | 6,028 | 5.7 |
| Benson | 15.3 | 0.2 | 6.6 | D | D | 1.2 | D | 1.5 | 54.6 | 1,225 | 179 | 171 | 3,085 | 4.6 |
| Billings | 11.6 | D | D | D | D | D | 6.3 | D | 24.3 | 165 | 176 | 0 | 563 | 16.1 |
| Bottineau | 16.3 | 13.1 | 6.3 | 3.7 | 3.5 | 5.7 | D | 10.8 | 11.5 | 1,670 | 265 | 33 | 4,513 | 4.0 |
| Bowman | 8.5 | 3.4 | 5.9 | 2.5 | 7.2 | 6.1 | D | 0.7 | 25.8 | 730 | 240 | 24 | 1,775 | 5.5 |
| Burke | 23.0 | D | 2.7 | 2.0 | D | 4.4 | D | 0.7 | 25.8 | 510 | 239 | 0 | 1,457 | 8.7 |
| Burleigh | 0.7 | 1.6 | 6.5 | 2.2 | 9.2 | 7.2 | 7.6 | 20.1 | 20.6 | 17,990 | 188 | 886 | 43,813 | 22.5 |
| Cass | 0.6 | 0.1 | 7.8 | 7.3 | 10.7 | 6.8 | 11.8 | 17.6 | 10.4 | 26,010 | 143 | 2,033 | 86,542 | 27.4 |
| Cavalier | 37.3 | D | 6.4 | D | 4.0 | 5.0 | D | 6.3 | 10.1 | 1,075 | 288 | 22 | 2,363 | 2.3 |
| Dickey | 33.1 | 0.0 | 2.8 | 6.4 | D | 4.7 | 5.2 | 10.1 | 10.1 | 1,140 | 235 | 54 | 2,648 | 0.5 |
| Divide | 3.9 | D | D | D | D | 3.4 | D | 2.3 | 22.0 | 525 | 234 | 14 | 1,533 | 15.8 |
| Dunn | 0.8 | 35.2 | 13.2 | 2.9 | D | 3.0 | 1.3 | 12.6 | 10.1 | 760 | 172 | 26 | 2,576 | 20.9 |
| Eddy | 38.5 | D | 5.3 | D | D | 3.9 | D | 11.4 | 12.7 | 615 | 274 | 34 | 1,353 | 2.3 |
| Emmons | 15.0 | 0.0 | D | D | D | 6.8 | D | 10.8 | 17.0 | 990 | 306 | 33 | 2,164 | 3.9 |
| Foster | 2.4 | 1.3 | 6.0 | D | D | 6.6 | 5.8 | 17.5 | 11.4 | 775 | 241 | 26 | 1,826 | 1.5 |
| Golden Valley | -18.4 | D | D | D | D | 8.9 | 6.9 | 17.1 | 23.3 | 430 | 244 | 12 | 1,054 | 9.0 |
| Grand Forks | 1.7 | 0.5 | 7.2 | 6.2 | 5.4 | 7.4 | 6.4 | 15.5 | 17.1 | 11,120 | 160 | 754 | 33,664 | 14.7 |
| Grant | 29.3 | 0.6 | 3.3 | 2.9 | D | 3.1 | 5.4 | D | 15.5 | 680 | 299 | 25 | 1,756 | 4.1 |
| Griggs | 25.6 | 0.0 | 5.0 | 7.6 | D | 5.7 | D | D | 11.8 | 670 | 302 | 19 | 1,459 | -0.2 |
| Hettinger | 5.0 | 0.4 | 9.8 | D | D | 3.6 | 6.2 | 5.9 | 20.2 | 660 | 262 | 26 | 1,434 | 1.3 |
| Kidder | 33.6 | D | D | D | D | 4.5 | D | 2.4 | 12.5 | 660 | 266 | 22 | 1,740 | 3.9 |
| LaMoure | 35.1 | 0.2 | 2.9 | 2.7 | D | 2.2 | D | 8.8 | 15.8 | 1,050 | 259 | 59 | 2,259 | 0.9 |
| Logan | 32.3 | 0.0 | 7.3 | D | D | 4.6 | D | 3.4 | 16.0 | 520 | 281 | 25 | 1,162 | 1.7 |
| McHenry | 14.9 | 4.5 | D | D | D | 3.8 | 4.3 | 19.6 | 12.3 | 1,310 | 228 | 37 | 3,179 | 7.9 |
| McIntosh | 18.8 | 0.0 | 4.9 | 3.3 | 3.3 | 1.9 | 6.0 | D | 12.8 | 895 | 362 | 25 | 1,856 | -0.2 |
| McKenzie | -0.1 | 20.3 | 22.7 | D | D | 2.3 | 5.2 | 5.3 | 16.0 | 1,210 | 80 | 49 | 6,962 | 125.3 |
| McLean | 5.4 | D | 4.7 | 3.1 | 2.9 | 3.0 | 2.3 | D | 16.0 | 2,575 | 273 | 85 | 6,330 | 13.2 |
| Mercer | 0.9 | 20.5 | 9.6 | 0.4 | D | 6.7 | D | 8.6 | 7.3 | 1,985 | 241 | 41 | 4,830 | 8.6 |
| Morton | -0.3 | 2.0 | 10.6 | 11.7 | D | 6.7 | D | D | 12.6 | 6,205 | 198 | 296 | 15,622 | 29.3 |
| Mountrail | -1.4 | 30.7 | D | D | 5.1 | 4.0 | 3.1 | D | 9.9 | 1,480 | 141 | 63 | 5,244 | 27.4 |
| Nelson | 27.9 | 0.0 | 8.8 | 0.8 | D | 2.7 | D | 13.4 | 16.2 | 955 | 334 | 35 | 1,952 | 1.2 |
| Oliver | 2.0 | D | 9.8 | D | D | 3.3 | D | 4.8 | 20.3 | 475 | 243 | 0 | 979 | 8.1 |
| Pembina | 18.4 | 0.3 | 7.1 | 17.0 | D | 5.8 | D | 16.8 | 11.7 | 1,830 | 271 | 51 | 3,862 | 0.2 |
| Pierce | 7.0 | D | 11.9 | D | 2.4 | 9.4 | 8.5 | D | 24.4 | 1,080 | 270 | 37 | 2,259 | 2.7 |
| Ramsey | 4.2 | D | 4.8 | 4.6 | 4.2 | 4.8 | 6.2 | D | 16.3 | 2,720 | 237 | 193 | 5,860 | 4.4 |
| Ransom | 17.1 | D | 3.0 | 9.6 | D | 4.9 | D | 4.6 | 15.8 | 1,285 | 246 | 44 | 2,694 | 1.4 |
| Renville | 42.5 | 3.5 | D | 24.6 | 4.1 | 6.6 | 6.3 | D | 20.6 | 565 | 243 | 0 | 1,438 | 3.8 |
| Richland | 4.1 | D | 6.6 | D | 1.1 | 4.9 | D | D | 64.5 | 3,400 | 210 | 152 | 7,809 | 4.1 |
| Rolette | 5.6 | D | 4.7 | D | D | 1.9 | D | 1.2 | 6.0 | 2,465 | 173 | 767 | 5,586 | 4.0 |
| Sargent | 14.2 | D | 4.7 | D | D | D | D | D | 16.3 | 980 | 248 | 32 | 2,203 | 9.9 |
| Sheridan | 47.1 | 0.5 | 5.1 | D | D | D | D | D | 84.9 | 400 | 309 | 25 | 936 | 4.7 |
| Sioux | 5.2 | 0.0 | 0.4 | 0.0 | D | D | 1.3 | 1.5 | 5.2 | 530 | 124 | 167 | 1,348 | 2.7 |
| Slope | -2 | 45.8 | D | 0.7 | 2.1 | D | 4.9 | 6.4 | 9.3 | 155 | 206 | 0 | 454 | 4.1 |
| Stark | 0.0 | 24.2 | 10.9 | 8.5 | 4.4 | 5.3 | D | D | 10.8 | 4,770 | 152 | 260 | 14,942 | 39.2 |
| Steele | 24.8 | 0.0 | 13.3 | 9.2 | D | 5.7 | D | 12.3 | 17.1 | 485 | 255 | 0 | 1,202 | 2.6 |
| Stutsman | 7.9 | 0.0 | 4.8 | 12.5 | 4.3 | 6.5 | 7.5 | D | 9.5 | 4,985 | 241 | 350 | 10,242 | 3.9 |
| Towner | 42.3 | 0.3 | 1.5 | 2.3 | D | 2.8 | 5.4 | D | 17.3 | 635 | 293 | 8 | 1,450 | 0.1 |
| Traill | 9.2 | D | 5.7 | 14.7 | 2.6 | 3.6 | 4.9 | D | 18.8 | 1,745 | 217 | 49 | 3,865 | 2.2 |
| Walsh | 15.7 | D | 4.3 | 11.2 | 4.0 | 4.0 | 5.3 | D | D | 2,610 | 247 | 117 | 5,616 | 2.1 |

1. Per 1,000 resident population estimated as of July 1 of the year shown.

# Table B. States and Counties — Housing, Labor Force, and Employment

| STATE County | Housing units, 2015-2019 | | | | | | | | Civilian labor force, 2020 | | | | Civilian employment[6], 2015-2019 | | |
|---|---|---|---|---|---|---|---|---|---|---|---|---|---|---|---|
| | Occupied units | | | | | | | Sub-standard units[4] (percent) | Total | Percent change, 2019-2020 | Unemployment | | Total | Percent | |
| | Owner-occupied | | | | | Renter-occupied | | | | | Total | Rate[5] | | Management, business, science, and arts | Construction, production, and maintenance occupations |
| | Total | Percent | Median value[1] | Median owner cost as a percent of income | | Median rent[3] | Median rent as a percent of income[2] | | | | | | | | |
| | | | | With a mortgage | Without a mortgage[2] | | | | | | | | | | |
| | 89 | 90 | 91 | 92 | 93 | 94 | 95 | 96 | 97 | 98 | 99 | 100 | 101 | 102 | 103 |
| **NORTH CAROLINA—Cont'd** | | | | | | | | | | | | | | | |
| Swain | 5,620 | 71.5 | 139,100 | 22.4 | 10.0 | 642 | 24.3 | 4.0 | 6,939 | -2.1 | 588 | 8.5 | 5,642 | 30.6 | 22.7 |
| Transylvania | 14,567 | 76.0 | 221,900 | 21.5 | 10.0 | 756 | 30.5 | 3.1 | 14,146 | -4.4 | 868 | 6.1 | 14,055 | 33.2 | 23.8 |
| Tyrrell | 1,594 | 68.3 | 114,800 | 23.1 | 13.2 | 760 | 27.2 | 2.2 | 1,359 | -5.2 | 121 | 8.9 | 1,536 | 21.3 | 28.8 |
| Union | 76,346 | 81.3 | 241,400 | 19.4 | 10.0 | 1,030 | 27.9 | 2.7 | 121,173 | -3.0 | 7,247 | 6.0 | 113,699 | 40.6 | 22.1 |
| Vance | 16,875 | 56.5 | 101,400 | 22.4 | 11.7 | 699 | 28.0 | 3.3 | 17,119 | -2.2 | 1,731 | 10.1 | 19,153 | 24.2 | 35.7 |
| Wake | 400,172 | 63.9 | 281,700 | 18.7 | 10.0 | 1,150 | 27.3 | 2.2 | 584,088 | -2.0 | 37,598 | 6.4 | 565,245 | 52.0 | 13.7 |
| Warren | 7,786 | 73.2 | 94,600 | 24.6 | 14.3 | 619 | 27.7 | 1.9 | 6,560 | -0.9 | 641 | 9.8 | 7,402 | 28.7 | 32.9 |
| Washington | 4,977 | 66.5 | 85,400 | 22.3 | 14.8 | 607 | 43.1 | 2.9 | 4,392 | -2.4 | 358 | 8.2 | 4,318 | 17.8 | 37.9 |
| Watauga | 21,077 | 60.1 | 238,000 | 21.1 | 10.0 | 870 | 38.9 | 2.5 | 27,997 | -5.7 | 1,736 | 6.2 | 25,748 | 39.0 | 15.6 |
| Wayne | 48,343 | 62.1 | 123,500 | 19.9 | 11.5 | 777 | 29.3 | 2.6 | 50,708 | -3.0 | 3,278 | 6.5 | 52,287 | 28.6 | 31.6 |
| Wilkes | 28,753 | 75.7 | 126,300 | 19.3 | 11.0 | 632 | 28.7 | 1.6 | 28,483 | -3.7 | 1,939 | 6.8 | 29,454 | 28.0 | 33.6 |
| Wilson | 32,014 | 59.1 | 123,600 | 20.7 | 12.7 | 753 | 29.5 | 3.7 | 34,442 | -2.3 | 3,062 | 8.9 | 35,593 | 32.2 | 31.0 |
| Yadkin | 15,425 | 76.2 | 133,100 | 19.8 | 10.0 | 635 | 29.3 | 3.9 | 17,305 | -3.1 | 1,231 | 7.1 | 16,233 | 28.9 | 36.8 |
| Yancey | 7,510 | 73.3 | 157,100 | 19.4 | 10.0 | 634 | 26.7 | 2.1 | 8,189 | -3.7 | 540 | 6.6 | 7,306 | 31.8 | 32.4 |
| **NORTH DAKOTA** | 318,322 | 62.4 | 193,900 | 18.4 | 10.0 | 826 | 24.4 | 2.4 | 406,839 | -0.6 | 20,833 | 5.1 | 402,322 | 37.3 | 25.4 |
| Adams | 1,035 | 73.5 | 121,200 | 19.7 | 10.0 | 604 | 22.6 | 1.3 | 999 | -4.6 | 38 | 3.8 | 1,191 | 37.4 | 31.2 |
| Barnes | 5,074 | 69.3 | 141,400 | 16.2 | 10.0 | 679 | 23.7 | 0.5 | 5,165 | -0.6 | 200 | 3.9 | 5,863 | 35.1 | 29.6 |
| Benson | 2,248 | 67.0 | 78,100 | 14.7 | 10.0 | 447 | 20.4 | 9.4 | 2,322 | 1.2 | 157 | 6.8 | 2,440 | 37.5 | 24.5 |
| Billings | 383 | 79.1 | 199,400 | 19.5 | 10.0 | 870 | 12.7 | 0.8 | 438 | -0.5 | 16 | 3.7 | 447 | 42.1 | 28.4 |
| Bottineau | 2,992 | 79.4 | 150,100 | 17.9 | 10.0 | 648 | 19.7 | 1.1 | 2,959 | -0.2 | 159 | 5.4 | 3,215 | 36.8 | 26.1 |
| Bowman | 1,350 | 75.3 | 148,600 | 16.2 | 10.0 | 640 | 25.3 | 2.6 | 1,639 | -4.5 | 55 | 3.4 | 1,623 | 34.9 | 28.2 |
| Burke | 905 | 78.9 | 114,400 | 12.8 | 10.0 | 716 | 15.4 | 1.7 | 1,022 | -4.3 | 54 | 5.3 | 1,053 | 38.8 | 29.8 |
| Burleigh | 39,507 | 68.7 | 263,600 | 19.2 | 10.0 | 871 | 26.1 | 1.8 | 50,588 | 1.4 | 2,202 | 4.4 | 51,291 | 41.2 | 20.8 |
| Cass | 75,843 | 52.1 | 223,800 | 18.7 | 10.0 | 833 | 25.0 | 2.1 | 104,926 | 1.0 | 4,592 | 4.4 | 103,886 | 40.8 | 21.2 |
| Cavalier | 1,737 | 80.7 | 104,500 | 17.3 | 10.0 | 571 | 21.5 | 1.4 | 1,911 | 1.0 | 66 | 3.5 | 1,894 | 37.9 | 29.8 |
| Dickey | 2,119 | 74.3 | 118,100 | 15.7 | 10.0 | 650 | 20.3 | 0.7 | 2,318 | 0.3 | 77 | 3.3 | 2,632 | 33.4 | 30.4 |
| Divide | 1,061 | 72.4 | 151,200 | 20.6 | 10.0 | 711 | 20.0 | 0.0 | 1,418 | -2.7 | 45 | 3.2 | 1,111 | 35.7 | 31.8 |
| Dunn | 1,692 | 77.0 | 192,500 | 18.2 | 10.0 | 1,031 | 22.0 | 3.1 | 3,097 | -4.1 | 141 | 4.6 | 2,048 | 36.2 | 35.7 |
| Eddy | 1,062 | 71.0 | 87,700 | 14.5 | 10.0 | 617 | 31.3 | 1.2 | 1,206 | -1.4 | 57 | 4.7 | 1,175 | 45.7 | 21.4 |
| Emmons | 1,585 | 85.9 | 97,000 | 14.7 | 10.0 | 484 | 23.1 | 1.1 | 1,493 | 3.8 | 74 | 5.0 | 1,535 | 41.8 | 25.1 |
| Foster | 1,423 | 74.2 | 143,500 | 18.6 | 10.0 | 595 | 24.3 | 1.8 | 1,555 | 2.4 | 54 | 3.5 | 1,753 | 42.2 | 27.0 |
| Golden Valley | 842 | 70.0 | 116,200 | 17.8 | 10.0 | 686 | 24.1 | 4.0 | 853 | 1.4 | 27 | 3.2 | 979 | 33.4 | 26.1 |
| Grand Forks | 30,502 | 48.5 | 206,500 | 19.2 | 10.0 | 824 | 29.1 | 2.8 | 36,730 | -2.1 | 1,699 | 4.6 | 38,382 | 36.6 | 21.0 |
| Grant | 1,074 | 86.7 | 81,800 | 19.4 | 10.0 | 545 | 19.6 | 2.2 | 1,189 | 0.5 | 33 | 2.8 | 1,078 | 50.7 | 23.3 |
| Griggs | 1,089 | 76.1 | 97,600 | 18.1 | 10.0 | 447 | 17.1 | 0.4 | 1,014 | -1.4 | 31 | 3.1 | 1,176 | 40.8 | 28.7 |
| Hettinger | 1,073 | 80.7 | 111,200 | 20.0 | 10.0 | 672 | 30.2 | 2.1 | 1,392 | 0.7 | 55 | 4.0 | 1,141 | 35.1 | 30.1 |
| Kidder | 1,092 | 75.3 | 114,400 | 16.3 | 10.0 | 523 | 19.0 | 3.8 | 1,284 | 1.1 | 66 | 5.1 | 1,155 | 34.9 | 27.9 |
| LaMoure | 1,822 | 76.9 | 107,700 | 15.9 | 10.0 | 501 | 18.9 | 0.7 | 2,055 | -0.3 | 54 | 2.6 | 2,051 | 41.9 | 26.5 |
| Logan | 849 | 83.9 | 86,800 | 18.5 | 10.0 | 738 | 23.6 | 1.5 | 854 | -0.9 | 27 | 3.2 | 948 | 44.6 | 28.2 |
| McHenry | 2,594 | 82.9 | 117,100 | 18.4 | 10.0 | 565 | 20.7 | 0.8 | 3,088 | 0.6 | 169 | 5.5 | 2,871 | 33.9 | 33.1 |
| McIntosh | 1,291 | 73.4 | 77,100 | 17.8 | 10.0 | 545 | 23.2 | 0.6 | 1,097 | -0.9 | 36 | 3.3 | 1,401 | 37.8 | 32.4 |
| McKenzie | 4,302 | 54.0 | 238,600 | 19.7 | 10.0 | 1,098 | 20.1 | 8.9 | 9,367 | -10.5 | 708 | 7.6 | 6,725 | 33.7 | 35.0 |
| McLean | 4,334 | 81.2 | 174,400 | 18.0 | 10.0 | 701 | 22.6 | 1.2 | 4,789 | 1.4 | 216 | 4.5 | 4,552 | 37.4 | 33.0 |
| Mercer | 3,709 | 82.7 | 177,800 | 15.5 | 10.0 | 769 | 21.0 | 2.5 | 3,616 | -0.6 | 208 | 5.8 | 4,311 | 28.4 | 37.1 |
| Morton | 13,565 | 70.7 | 208,600 | 19.0 | 10.0 | 861 | 24.0 | 1.3 | 16,746 | 1.5 | 850 | 5.1 | 18,045 | 35.6 | 27.5 |
| Mountrail | 3,360 | 68.1 | 165,700 | 17.4 | 10.0 | 797 | 14.6 | 6.1 | 6,880 | -7.3 | 354 | 5.1 | 4,719 | 36.3 | 28.1 |
| Nelson | 1,467 | 72.3 | 84,300 | 17.0 | 10.0 | 522 | 19.9 | 2.2 | 1,398 | 0.6 | 63 | 4.5 | 1,451 | 39.8 | 25.8 |
| Oliver | 738 | 88.6 | 190,600 | 18.1 | 10.0 | 545 | 50.0 | 1.6 | 882 | 1.0 | 46 | 5.2 | 802 | 39.5 | 31.4 |
| Pembina | 3,158 | 77.0 | 93,300 | 14.8 | 10.0 | 593 | 20.0 | 0.6 | 3,386 | 1.5 | 196 | 5.8 | 3,376 | 34.4 | 29.4 |
| Pierce | 1,908 | 73.6 | 128,600 | 18.5 | 10.0 | 680 | 35.8 | 0.5 | 1,710 | -0.5 | 73 | 4.3 | 1,999 | 47.5 | 20.2 |
| Ramsey | 4,828 | 60.3 | 164,700 | 15.3 | 10.0 | 565 | 20.9 | 2.3 | 5,703 | 0.8 | 256 | 4.5 | 6,114 | 33.7 | 18.5 |
| Ransom | 2,371 | 70.9 | 121,700 | 17.4 | 10.0 | 669 | 22.0 | 0.3 | 2,867 | 1.6 | 118 | 4.1 | 2,788 | 30.8 | 37.3 |
| Renville | 950 | 82.6 | 134,700 | 18.2 | 10.0 | 731 | 23.7 | 1.9 | 1,225 | 1.2 | 60 | 4.9 | 1,191 | 34.8 | 30.2 |
| Richland | 6,777 | 70.9 | 127,900 | 17.5 | 10.0 | 664 | 27.4 | 2.8 | 8,580 | -0.9 | 332 | 3.9 | 8,405 | 33.2 | 33.7 |
| Rolette | 4,608 | 71.3 | 84,000 | 15.1 | 10.0 | 397 | 20.0 | 6.7 | 4,743 | 2.6 | 637 | 13.4 | 5,299 | 34.6 | 23.9 |
| Sargent | 1,864 | 72.8 | 110,200 | 15.5 | 10.0 | 711 | 18.4 | 1.7 | 2,770 | 3.1 | 121 | 4.4 | 2,091 | 32.3 | 45.3 |
| Sheridan | 731 | 79.8 | 84,400 | 17.4 | 10.4 | 521 | 47.7 | 2.7 | 685 | 1.6 | 41 | 6.0 | 567 | 43.0 | 26.8 |
| Sioux | 1,112 | 43.4 | 83,800 | 22.8 | 10.0 | 450 | 14.1 | 13.0 | 1,205 | -0.4 | 47 | 3.9 | 1,258 | 39.9 | 15.1 |
| Slope | 326 | 82.2 | 101,800 | 14.4 | 10.0 | 633 | 17.5 | 0.3 | 391 | -0.5 | 12 | 3.1 | 325 | 59.7 | 18.8 |
| Stark | 12,524 | 61.1 | 239,000 | 19.0 | 10.0 | 917 | 21.6 | 3.2 | 18,332 | -2.4 | 1,308 | 7.1 | 16,014 | 32.1 | 34.5 |
| Steele | 777 | 84.7 | 85,300 | 14.0 | 10.0 | 555 | 14.3 | 1.3 | 949 | -3.6 | 28 | 3.0 | 927 | 45.3 | 28.6 |
| Stutsman | 8,908 | 65.6 | 155,500 | 18.3 | 10.0 | 683 | 24.3 | 1.6 | 10,495 | 1.0 | 364 | 3.5 | 10,663 | 35.0 | 23.3 |
| Towner | 1,058 | 73.1 | 70,200 | 17.9 | 10.4 | 550 | 22.2 | 1.9 | 1,157 | 1.0 | 32 | 2.8 | 1,038 | 33.7 | 28.4 |
| Traill | 3,383 | 71.5 | 150,600 | 17.6 | 10.0 | 624 | 22.5 | 0.4 | 4,458 | 1.0 | 162 | 3.6 | 4,059 | 40.1 | 26.0 |
| Walsh | 4,780 | 76.4 | 91,100 | 17.7 | 10.0 | 697 | 23.5 | 1.5 | 5,135 | -0.7 | 233 | 4.5 | 5,401 | 34.5 | 34.0 |

1. Specified owner-occupied units. lacking complete plumbing facilities.　2. A value of 10.0 represents 10 percent or less; a value of 50.0 represents 50 percent or more.　3. Specified renter-occupied units.　4. Overcrowded or
5. Percent of civilian labor force.　6. Civilian employed persons 16 years old and over.

| STATE County | Private nonfarm establishments, employment and payroll, 2019 | | | | | | | | | Agriculture, 2017 | | | |
|---|---|---|---|---|---|---|---|---|---|---|---|---|---|
| | Number of establishments | Employment | | | | | | Annual payroll | | Farms | | | Farm producers whose primary occupation is farming (percent) |
| | | Total | Health care and social assistance | Manufacturing | Retail trade | Finance and insurance | Professional, scientific, and technical services | Total (mil dol) | Average per employee (dollars) | Number | Fewer than 50 acres | 1000 acres or more | |
| | 104 | 105 | 106 | 107 | 108 | 109 | 110 | 111 | 112 | 113 | 114 | 115 | 116 |
| **NORTH CAROLINA— Cont'd** | | | | | | | | | | | | | |
| Swain | 375 | 3,658 | 1,141 | 129 | 609 | 61 | 73 | 121 | 32,990 | 99 | 71.7 | 1.0 | 51.4 |
| Transylvania | 875 | 7,973 | 1,384 | 729 | 1,449 | 230 | 296 | 270 | 33,819 | 215 | 69.8 | 0.9 | 39.7 |
| Tyrrell | 72 | 552 | 83 | 59 | 186 | 22 | NA | 15 | 26,409 | 68 | 25.0 | 22.1 | 46.5 |
| Union | 5,006 | 59,403 | 5,751 | 11,324 | 8,111 | 1,026 | 1,937 | 2,514 | 42,326 | 957 | 56.9 | 4.1 | 44.7 |
| Vance | 841 | 13,221 | 2,220 | 1,652 | 2,289 | 225 | 470 | 458 | 34,678 | 238 | 30.7 | 6.7 | 40.1 |
| Wake | 30,266 | 497,013 | 66,300 | 14,514 | 61,530 | 27,661 | 60,348 | 29,621 | 59,597 | 691 | 64.5 | 1.9 | 34.5 |
| Warren | 262 | 2,037 | 451 | 269 | 412 | 50 | 45 | 61 | 30,005 | 267 | 41.9 | 3.7 | 38.8 |
| Washington | 222 | 2,597 | 622 | 687 | 396 | 40 | 56 | 109 | 42,085 | 141 | 24.8 | 19.1 | 60.4 |
| Watauga | 1,668 | 19,081 | 4,425 | 695 | 3,774 | 369 | 553 | 665 | 34,860 | 520 | 58.1 | 1.0 | 35.3 |
| Wayne | 2,148 | 33,298 | 7,049 | 5,053 | 5,845 | 1,114 | 681 | 1,207 | 36,259 | 551 | 39.7 | 8.5 | 55.4 |
| Wilkes | 1,160 | 18,835 | 2,340 | 4,589 | 2,438 | 319 | 703 | 837 | 44,451 | 932 | 45.0 | 1.0 | 47.2 |
| Wilson | 1,767 | 31,696 | 4,525 | 6,770 | 4,014 | 2,816 | 655 | 1,365 | 43,069 | 276 | 43.8 | 12.3 | 51.2 |
| Yadkin | 613 | 8,516 | 935 | 2,757 | 862 | 158 | 195 | 297 | 34,864 | 818 | 55.4 | 1.8 | 42.5 |
| Yancey | 324 | 3,693 | 471 | 1,236 | 566 | 73 | 109 | 137 | 37,113 | 369 | 51.5 | 0.3 | 39.3 |
| **NORTH DAKOTA** | 24,654 | 353,333 | 65,064 | 26,627 | 47,661 | 18,547 | 16,731 | 18,292 | 51,770 | 26,364 | 11.7 | 40.0 | 54.3 |
| Adams | 97 | 717 | 294 | NA | 131 | 34 | 20 | 28 | 39,530 | 380 | 9.5 | 36.3 | 49.6 |
| Barnes | 356 | 3,509 | 1,070 | 494 | 503 | 127 | 63 | 128 | 36,509 | 749 | 14.4 | 34.3 | 54.3 |
| Benson | 72 | 906 | 42 | NA | 37 | 63 | 5 | 31 | 33,929 | 473 | 7.6 | 41.0 | 58.1 |
| Billings | 57 | 264 | NA | NA | 14 | NA | NA | 18 | 68,216 | 204 | 7.4 | 53.9 | 58.0 |
| Bottineau | 252 | 1,679 | 277 | 122 | 282 | 112 | 84 | 73 | 43,373 | 684 | 8.6 | 39.5 | 55.6 |
| Bowman | 156 | 1,093 | 256 | 20 | 190 | 79 | 57 | 51 | 46,964 | 341 | 7.0 | 42.5 | 46.9 |
| Burke | 70 | 325 | 15 | NA | 96 | 29 | 14 | 14 | 43,908 | 356 | 8.7 | 38.5 | 47.5 |
| Burleigh | 3,025 | 47,861 | 11,382 | 1,967 | 7,150 | 1,800 | 3,146 | 2,277 | 47,579 | 785 | 29.9 | 23.7 | 34.4 |
| Cass | 5,603 | 108,178 | 20,602 | 9,195 | 12,874 | 9,339 | 5,839 | 5,583 | 51,609 | 784 | 20.0 | 46.9 | 61.9 |
| Cavalier | 141 | 1,172 | 227 | 16 | 197 | 81 | 16 | 49 | 42,104 | 523 | 5.9 | 54.7 | 64.6 |
| Dickey | 201 | 1,482 | 346 | 207 | 277 | 55 | 21 | 53 | 36,068 | 419 | 6.0 | 45.4 | 52.0 |
| Divide | 89 | 483 | 117 | NA | 32 | 28 | 10 | 25 | 51,342 | 416 | 3.4 | 51.9 | 64.0 |
| Dunn | 189 | 2,773 | 108 | 479 | 183 | 18 | 93 | 201 | 72,623 | 524 | 3.1 | 40.2 | 46.3 |
| Eddy | 79 | 518 | 200 | NA | 66 | 25 | 13 | 18 | 34,591 | 291 | 6.8 | 43.8 | 53.5 |
| Emmons | 128 | 679 | 194 | NA | 133 | 59 | 17 | 24 | 34,742 | 516 | 14.3 | 56.1 | 61.9 |
| Foster | 139 | 1,353 | 243 | 310 | 218 | 67 | 10 | 63 | 46,554 | 230 | 10.8 | 42.9 | 51.5 |
| Golden Valley | 66 | 374 | 125 | NA | 89 | 29 | 18 | 13 | 35,198 | 287 | 14.5 | 24.5 | 48.5 |
| Grand Forks | 1,892 | 34,893 | 7,693 | 3,007 | 5,833 | 953 | 1,450 | 1,505 | 43,143 | 889 | 7.5 | 51.7 | 52.4 |
| Grant | 73 | 395 | 172 | 18 | 60 | 36 | 8 | 13 | 31,944 | 412 | 8.1 | 36.6 | 55.4 |
| Griggs | 91 | 555 | 126 | 99 | 46 | 19 | 16 | 21 | 36,982 | 393 | 8.9 | 36.7 | 44.1 |
| Hettinger | 93 | 453 | 72 | NA | 89 | 42 | 30 | 20 | 44,143 | 482 | 8.2 | 47.5 | 49.8 |
| Kidder | 74 | 569 | 90 | NA | 103 | 36 | 35 | 20 | 35,051 | 476 | 8.8 | 41.5 | 66.2 |
| LaMoure | 146 | 999 | 169 | 95 | 138 | 129 | 7 | 37 | 37,496 | 571 | 10.3 | 52.1 | 59.2 |
| Logan | 69 | 442 | 91 | NA | 60 | 29 | NA | 13 | 29,724 | 351 | 11.2 | 33.9 | 48.5 |
| McHenry | 123 | 795 | 127 | 98 | 111 | 45 | 18 | 38 | 47,341 | 750 | 13.5 | 36.1 | 53.7 |
| McIntosh | 114 | 793 | 310 | 46 | 119 | 48 | 20 | 28 | 35,366 | 363 | 11.3 | 51.6 | 68.0 |
| McKenzie | 576 | 8,004 | 346 | NA | 572 | 121 | 335 | 667 | 83,363 | 539 | 9.8 | 35.7 | 50.1 |
| McLean | 259 | 2,569 | 388 | 115 | 289 | 144 | 43 | 171 | 66,378 | 762 | 12.0 | 40.7 | 49.3 |
| Mercer | 240 | 3,770 | 506 | 734 | 450 | 122 | 46 | 265 | 70,312 | 317 | 12.0 | 41.5 | 54.7 |
| Morton | 869 | 10,463 | 1,745 | 940 | 1,417 | 387 | 138 | 500 | 47,777 | 781 | 6.8 | 49.0 | 54.7 |
| Mountrail | 408 | 4,104 | 260 | 106 | 601 | 93 | 138 | 270 | 65,789 | 584 | 8.4 | 30.7 | 48.8 |
| Nelson | 125 | 684 | 237 | NA | 95 | 55 | 8 | 25 | 35,931 | 489 | 14.5 | 33.3 | 52.1 |
| Oliver | 46 | 557 | 16 | 16 | 21 | 12 | 8 | 45 | 81,654 | 234 | 10.6 | 38.3 | 66.4 |
| Pembina | 260 | 2,469 | 392 | 743 | 294 | 114 | 33 | 104 | 42,048 | 481 | 13.2 | 37.9 | 55.6 |
| Pierce | 167 | 1,411 | 340 | 108 | 234 | 89 | 58 | 58 | 40,944 | 446 | 13.0 | 35.9 | 47.5 |
| Ramsey | 452 | 4,351 | 847 | 129 | 974 | 285 | 78 | 170 | 38,979 | 515 | 12.4 | 30.6 | 53.0 |
| Ransom | 201 | 1,488 | 371 | 249 | 200 | 69 | 59 | 54 | 36,557 | 507 | 12.6 | 60.2 | 71.7 |
| Renville | 111 | 492 | 87 | NA | 124 | 31 | 10 | 18 | 37,378 | 269 | 18.8 | 33.7 | 60.5 |
| Richland | 520 | 6,130 | 635 | 1,829 | 753 | 212 | 127 | 263 | 42,924 | 846 | 8.6 | 35.8 | 58.7 |
| Rolette | 194 | 2,171 | 611 | 63 | 459 | 111 | 24 | 87 | 40,075 | 453 | 12.6 | 33.7 | 56.9 |
| Sargent | 116 | 1,916 | 86 | 1,273 | 146 | 37 | 11 | 102 | 53,120 | 260 | 2.3 | 54.6 | 60.7 |
| Sheridan | 40 | 129 | 24 | NA | 17 | 7 | 13 | 4 | 32,388 | 187 | 16.0 | 53.5 | 49.5 |
| Sioux | 31 | 563 | 53 | NA | 54 | NA | 11 | 16 | 28,304 | 215 | 0.9 | 54.0 | 65.7 |
| Slope | 15 | 66 | NA | NA | NA | NA | 4 | NA | 63,455 | 678 | 19.8 | 29.6 | 43.4 |
| Stark | 1,177 | 15,241 | 2,094 | 1,256 | 2,008 | 382 | 493 | 909 | 59,634 | 358 | 16.8 | 37.2 | 56.1 |
| Steele | 63 | 526 | NA | 143 | 70 | 47 | 6 | 25 | 48,365 | 939 | 12.6 | 37.7 | 56.7 |
| Stutsman | 664 | 9,100 | 2,694 | 793 | 1,345 | 484 | 230 | 384 | 42,244 | 454 | 6.8 | 44.3 | 51.9 |
| Towner | 83 | 461 | 152 | NA | 64 | 47 | 20 | 19 | 40,729 | 415 | 18.3 | 45.1 | 53.3 |
| Traill | 285 | 2,635 | 595 | 429 | 365 | 135 | 47 | 119 | 45,228 | 763 | 12.1 | 32.2 | 47.6 |
| Walsh | 384 | 3,342 | 713 | 558 | 515 | 175 | 94 | 132 | 39,505 | 501 | | | |

# Table B. States and Counties — **Agriculture**

| STATE County | Land in farms — Acreage (1,000) | Percent change, 2012-2017 | Acres — Average size of farm | Total irrigated (1,000) | Total cropland (1,000) | Value of land and buildings (dollars) — Average per farm | Average per acre | Value of machinery and equipment, average per farm (dollars) | Value of products sold: Total (mil dol) | Average per farm (acres) | Percent from: Crops | Livestock and poultry products | Organic farms (number) | Farms with internet access (percent) | Government payments Total ($1,000) | Percent of farms |
|---|---|---|---|---|---|---|---|---|---|---|---|---|---|---|---|---|
| | 117 | 118 | 119 | 120 | 121 | 122 | 123 | 124 | 125 | 126 | 127 | 128 | 129 | 130 | 131 | 132 |
| **NORTH CAROLINA—Cont'd** | | | | | | | | | | | | | | | | |
| Swain | 10 | D | 102 | 0.2 | 1.1 | 374,350 | 3,658 | 43,804 | 2.2 | 21,717 | 58.7 | 41.3 | NA | 72.7 | 42 | 12.1 |
| Transylvania | 15 | -18.1 | 68 | 0.6 | 4.5 | 528,315 | 7,735 | 58,576 | 9.8 | 45,460 | 68.8 | 31.2 | 5 | 74.4 | 74 | 10.2 |
| Tyrrell | 53 | -18.0 | 779 | D | 51.0 | 3,229,747 | 4,148 | 436,706 | D | D | D | D | NA | 60.3 | 1,312 | 75.0 |
| Union | 187 | -7.5 | 195 | 0.2 | 142.3 | 1,040,504 | 5,336 | 130,984 | 482.0 | 503,637 | 17.9 | 82.1 | 4 | 76.0 | 2,529 | 12.6 |
| Vance | 66 | 20.5 | 278 | 1.2 | 20.6 | 827,874 | 2,978 | 114,092 | 17.2 | 72,315 | 97.6 | 2.4 | 9 | 78.6 | 164 | 34.5 |
| Wake | 77 | -8.6 | 111 | 1.2 | 46.3 | 1,299,196 | 11,658 | 82,844 | 63.7 | 92,122 | 88.8 | 11.2 | 2 | 82.6 | 557 | 13.5 |
| Warren | 61 | -7.5 | 228 | 1.2 | 29.2 | 662,147 | 2,909 | 78,681 | 40.1 | 150,176 | 30.7 | 69.3 | 2 | 69.7 | 428 | 33.0 |
| Washington | 80 | -12.8 | 565 | 1.9 | 71.3 | 2,072,945 | 3,668 | 358,266 | 49.0 | 347,844 | 88.5 | 11.5 | NA | 78.7 | 3,085 | 68.8 |
| Watauga | 50 | -11.0 | 95 | 0.0 | 14.4 | 614,311 | 6,439 | 58,217 | 16.7 | 32,162 | 53.4 | 46.6 | 16 | 76.2 | 351 | 13.3 |
| Wayne | 165 | -13.5 | 300 | 4.8 | 130.5 | 1,497,288 | 4,990 | 222,201 | 592.1 | 1,074,539 | 18.1 | 81.9 | 13 | 79.9 | 3,611 | 46.1 |
| Wilkes | 107 | -4.0 | 114 | 0.1 | 41.4 | 586,674 | 5,124 | 106,608 | 335.1 | 359,575 | 4.4 | 95.6 | 8 | 69.6 | 81 | 4.2 |
| Wilson | 123 | 10.4 | 445 | 1.1 | 100.3 | 1,685,619 | 3,784 | 304,394 | 210.7 | 763,399 | 77.2 | 22.8 | 6 | 76.4 | 1,129 | 49.6 |
| Yadkin | 88 | -12.9 | 107 | 0.9 | 48.3 | 585,403 | 5,471 | 74,938 | 139.7 | 170,722 | 19.5 | 80.5 | 7 | 70.8 | 489 | 13.7 |
| Yancey | 31 | -0.6 | 84 | 0.1 | 10.2 | 454,909 | 5,446 | 40,731 | 6.8 | 18,491 | 83.1 | 16.9 | 1 | 66.7 | 157 | 27.1 |
| **NORTH DAKOTA** | 39,342 | 0.2 | 1,492 | 263.9 | 27,951.7 | 2,546,783 | 1,707 | 375,872 | 8,234.1 | 312,324 | 81.1 | 18.9 | 129 | 78.9 | 467,034 | 77.8 |
| Adams | 600 | -0.3 | 1,578 | NA | 388.0 | 1,682,952 | 1,067 | 297,071 | 60.4 | 158,937 | 50.4 | 49.6 | NA | 73.2 | 8,748 | 82.1 |
| Barnes | 952 | 1.6 | 1,271 | 2.5 | 819.8 | 2,937,290 | 2,311 | 395,596 | 278.7 | 372,143 | 92.4 | 7.6 | 3 | 80.0 | 8,226 | 78.2 |
| Benson | 751 | -6.4 | 1,587 | 2.0 | 586.7 | 2,172,369 | 1,368 | 384,343 | 164.3 | 347,393 | 89.9 | 10.1 | 1 | 79.5 | 9,813 | 78.9 |
| Billings | 735 | 1.7 | 3,602 | D | 125.7 | 3,308,786 | 919 | 237,105 | 28.8 | 140,951 | 19.7 | 80.3 | 6 | 71.1 | 2,338 | 69.6 |
| Bottineau | 958 | 6.6 | 1,401 | 0.0 | 865.6 | 2,166,705 | 1,546 | 354,320 | 215.5 | 315,058 | 92.5 | 7.5 | 1 | 82.5 | 18,721 | 81.4 |
| Bowman | 711 | -2.6 | 2,086 | 0.9 | 334.9 | 2,144,932 | 1,028 | 254,275 | 74.1 | 217,240 | 26.4 | 73.6 | NA | 79.2 | 5,354 | 80.1 |
| Burke | 519 | -12.7 | 1,459 | NA | 393.4 | 1,537,171 | 1,054 | 303,368 | 71.6 | 201,062 | 85.4 | 14.6 | NA | 67.7 | 7,985 | 88.2 |
| Burleigh | 824 | -13.3 | 1,050 | 4.0 | 423.4 | 1,994,601 | 1,900 | 200,722 | 134.1 | 170,866 | 48.8 | 51.2 | NA | 80.5 | 6,674 | 52.9 |
| Cass | 1,126 | 1.7 | 1,436 | 13.9 | 1,085.4 | 5,339,580 | 3,718 | 557,103 | 439.5 | 560,528 | 94.9 | 5.1 | 7 | 85.1 | 8,418 | 71.3 |
| Cavalier | 928 | -1.3 | 1,775 | 0.3 | 877.2 | 3,691,559 | 2,080 | 546,736 | 280.6 | 536,507 | 98.5 | 1.5 | 2 | 82.8 | 25,654 | 90.4 |
| Dickey | 679 | 7.2 | 1,620 | 14.8 | 541.9 | 3,847,791 | 2,375 | 510,138 | 225.1 | 537,212 | 80.8 | 19.2 | 6 | 79.5 | 6,323 | 81.1 |
| Divide | 704 | 24.6 | 1,693 | 2.3 | 538.7 | 1,808,767 | 1,069 | 358,232 | 71.5 | 171,755 | 86.1 | 13.9 | 7 | 75.0 | 9,797 | 91.3 |
| Dunn | 1,017 | -1.4 | 1,941 | 0.8 | 417.8 | 2,225,966 | 1,147 | 297,360 | 77.6 | 148,101 | 35.7 | 64.3 | 5 | 84.0 | 6,276 | 66.0 |
| Eddy | 377 | -4.7 | 1,296 | NA | 285.6 | 1,886,332 | 1,455 | 312,599 | 82.2 | 282,584 | 77.3 | 22.7 | NA | 89.7 | 4,439 | 89.0 |
| Emmons | 812 | 9.1 | 1,573 | 10.1 | 500.4 | 2,493,059 | 1,585 | 357,337 | 162.1 | 314,151 | 69.0 | 31.0 | 3 | 79.7 | 7,581 | 81.8 |
| Foster | 394 | 5.3 | 1,713 | 2.8 | 342.9 | 3,438,653 | 2,008 | 609,701 | 118.7 | 516,230 | 76.5 | 23.5 | 2 | 89.1 | 5,532 | 85.7 |
| Golden Valley | 626 | 11.4 | 2,182 | 1.1 | 274.1 | 2,296,429 | 1,052 | 313,468 | 45.2 | 157,390 | 46.2 | 53.8 | NA | 79.1 | 5,890 | 76.0 |
| Grand Forks | 798 | -2.2 | 898 | 27.5 | 744.1 | 3,064,078 | 3,411 | 359,812 | 318.4 | 358,112 | 92.6 | 7.4 | 1 | 74.6 | 14,911 | 87.5 |
| Grant | 959 | -8.6 | 2,329 | 2.1 | 429.0 | 2,494,937 | 1,071 | 264,216 | 83.0 | 201,502 | 35.0 | 65.0 | NA | 76.9 | 8,094 | 87.4 |
| Griggs | 454 | 1.8 | 1,154 | 1.5 | 364.8 | 2,051,720 | 1,777 | 339,861 | 114.6 | 291,659 | 91.7 | 8.3 | NA | 75.1 | 7,670 | 88.3 |
| Hettinger | 705 | -1.5 | 1,462 | NA | 586.0 | 2,055,386 | 1,405 | 264,427 | 66.6 | 138,077 | 80.1 | 19.9 | NA | 76.8 | 13,029 | 86.7 |
| Kidder | 748 | -4.1 | 1,572 | 23.7 | 423.9 | 1,978,460 | 1,258 | 264,631 | 113.6 | 238,626 | 64.1 | 35.9 | 28 | 77.5 | 6,882 | 82.6 |
| LaMoure | 727 | 0.1 | 1,273 | 5.5 | 632.9 | 3,230,939 | 2,539 | 417,666 | 226.0 | 395,781 | 85.2 | 14.8 | 9 | 79.0 | 4,860 | 78.5 |
| Logan | 632 | 10.5 | 1,800 | 2.4 | 347.7 | 2,144,519 | 1,191 | 326,130 | 155.4 | 442,675 | 35.6 | 64.4 | 7 | 76.6 | 4,218 | 84.9 |
| McHenry | 1,040 | -2.0 | 1,387 | 6.2 | 629.4 | 1,510,122 | 1,089 | 241,112 | 146.3 | 195,129 | 56.2 | 43.8 | 3 | 78.3 | 10,874 | 77.6 |
| McIntosh | 488 | -17.3 | 1,343 | 0.7 | 298.3 | 1,832,362 | 1,364 | 275,666 | 94.8 | 261,157 | 56.6 | 43.4 | 6 | 79.6 | 3,200 | 76.6 |
| McKenzie | 1,119 | 5.2 | 2,077 | 26.7 | 433.3 | 2,130,024 | 1,026 | 372,024 | 104.7 | 194,230 | 47.9 | 52.1 | NA | 82.7 | 7,119 | 63.8 |
| McLean | 1,045 | -6.0 | 1,372 | 8.1 | 807.3 | 2,448,520 | 1,785 | 327,473 | 176.9 | 232,180 | 85.2 | 14.8 | NA | 82.8 | 17,822 | 81.4 |
| Mercer | 518 | 3.0 | 1,635 | 2.4 | 232.8 | 2,236,333 | 1,368 | 291,683 | 57.2 | 180,341 | 43.8 | 56.2 | 2 | 77.6 | 4,684 | 65.3 |
| Morton | 1,226 | 0.5 | 1,570 | 4.7 | 562.8 | 2,203,330 | 1,404 | 275,670 | 146.0 | 186,936 | 43.5 | 56.5 | 3 | 81.8 | 7,594 | 65.7 |
| Mountrail | 1,081 | 12.2 | 1,852 | 0.0 | 697.2 | 2,367,371 | 1,278 | 415,245 | 135.7 | 232,435 | 78.0 | 22.0 | NA | 80.8 | 14,849 | 76.5 |
| Nelson | 553 | -1.4 | 1,131 | 2.9 | 475.4 | 1,537,488 | 1,360 | 341,693 | 120.9 | 247,213 | 93.1 | 6.9 | NA | 67.5 | 9,416 | 93.5 |
| Oliver | 314 | -20.5 | 1,340 | 2.7 | 148.5 | 1,932,910 | 1,443 | 292,694 | 47.3 | 202,248 | 46.3 | 53.7 | NA | 76.1 | 2,639 | 71.4 |
| Pembina | 691 | -0.1 | 1,438 | 1.4 | 640.8 | 3,962,790 | 2,757 | 514,306 | 282.7 | 587,653 | 97.4 | 2.6 | NA | 78.4 | 11,608 | 78.0 |
| Pierce | 540 | -9.7 | 1,212 | 0.8 | 435.3 | 1,707,184 | 1,409 | 337,405 | 98.8 | 221,466 | 84.3 | 15.7 | 1 | 80.0 | 8,501 | 86.3 |
| Ramsey | 697 | -0.3 | 1,353 | D | 626.4 | 2,446,528 | 1,809 | 474,152 | 190.2 | 369,297 | 86.2 | 13.8 | NA | 77.3 | 12,149 | 84.9 |
| Ransom | 549 | 9.5 | 1,083 | 28.9 | 389.1 | 2,179,084 | 2,012 | 345,733 | 173.6 | 342,454 | 79.3 | 20.7 | 3 | 77.5 | 6,098 | 74.0 |
| Renville | 554 | 10.8 | 2,060 | 0.0 | 523.4 | 3,306,195 | 1,605 | 672,748 | 125.5 | 466,498 | 97.7 | 2.3 | NA | 86.2 | 10,660 | 84.0 |
| Richland | 875 | 0.8 | 1,035 | 6.1 | 823.0 | 3,300,158 | 3,189 | 467,825 | 390.8 | 461,935 | 90.8 | 9.2 | 3 | 78.3 | 6,568 | 77.9 |
| Rolette | 512 | -4.1 | 1,131 | 1.0 | 373.1 | 1,447,711 | 1,280 | 327,186 | 91.5 | 202,079 | 84.3 | 15.7 | NA | 72.6 | 9,446 | 70.6 |
| Sargent | 548 | 6.9 | 1,094 | 16.8 | 489.2 | 3,049,460 | 2,787 | 399,455 | 214.1 | 427,433 | 88.1 | 11.9 | NA | 76.0 | 5,981 | 79.0 |
| Sheridan | 551 | 7.4 | 2,121 | NA | 397.9 | 2,328,439 | 1,098 | 314,736 | 92.0 | 354,015 | 80.2 | 19.8 | 7 | 78.8 | 5,366 | 80.4 |
| Sioux | 588 | 2.6 | 3,145 | D | 226.1 | 3,909,369 | 1,243 | 390,741 | 61.9 | 331,048 | 35.1 | 64.9 | NA | 77.0 | 2,912 | 68.4 |
| Slope | 728 | 7.9 | 3,385 | NA | 286.5 | 3,579,166 | 1,057 | 418,177 | 47.9 | 222,809 | 41.3 | 58.7 | NA | 77.7 | 4,683 | 83.7 |
| Stark | 736 | -11.3 | 1,085 | 0.5 | 493.6 | 1,444,094 | 1,331 | 222,178 | 68.5 | 100,997 | 63.6 | 36.4 | NA | 77.6 | 7,576 | 59.6 |
| Steele | 421 | -1.1 | 1,177 | 6.6 | 398.6 | 2,584,785 | 2,197 | 525,062 | 149.9 | 418,782 | 97.3 | 2.7 | NA | 77.7 | 5,812 | 80.4 |
| Stutsman | 1,316 | 1.0 | 1,401 | 4.2 | 1,048.2 | 2,808,671 | 2,005 | 370,108 | 335.6 | 357,428 | 85.2 | 14.8 | 2 | 79.2 | 10,869 | 74.8 |
| Towner | 631 | -2.3 | 1,389 | NA | 552.4 | 1,923,869 | 1,385 | 399,244 | 166.1 | 365,965 | 84.8 | 15.2 | NA | 74.2 | 14,647 | 85.2 |
| Traill | 539 | -1.6 | 1,298 | D | 526.1 | 4,004,597 | 3,084 | 592,272 | 225.9 | 544,410 | 99.0 | 1.0 | NA | 81.2 | 4,053 | 80.0 |
| Walsh | 805 | 0.3 | 1,054 | 1.7 | 720.3 | 2,649,975 | 2,513 | 446,123 | 335.6 | 439,840 | 96.4 | 3.6 | 5 | 76.5 | 13,679 | 86.1 |

Items 117—132

# Table B. States and Counties — Water Use, Wholesale Trade, Retail Trade, and Real Estate

| STATE County | Water use, 2015 — Public supply water withdrawn (mil gal/day) [133] | Water use, 2015 — Public supply gallons withdrawn per person per day [134] | Wholesale Trade[1], 2017 — Number of establishments [135] | Wholesale Trade[1], 2017 — Number of employees [136] | Wholesale Trade[1], 2017 — Sales (mil dol) [137] | Wholesale Trade[1], 2017 — Average payroll (mil dol) [138] | Retail Trade[2], 2017 — Number of establishments [139] | Retail Trade[2], 2017 — Number of employees [140] | Retail Trade[2], 2017 — Sales (mil dol) [141] | Retail Trade[2], 2017 — Average payroll (mil dol) [142] | Real estate and rental and leasing,[2] 2017 — Number of establishments [143] | Real estate and rental and leasing,[2] 2017 — Number of employees [144] | Real estate and rental and leasing,[2] 2017 — Sales (mil dol) [145] | Real estate and rental and leasing,[2] 2017 — Average payroll (mil dol) [146] |
|---|---|---|---|---|---|---|---|---|---|---|---|---|---|---|
| **NORTH CAROLINA— Cont'd** | | | | | | | | | | | | | | |
| Swain | 0.50 | 34.6 | NA | NA | NA | NA | 81 | 594 | 124.5 | 12.3 | 10 | 52 | 9.4 | 1.7 |
| Transylvania | 1.81 | 54.5 | D | D | D | D | 119 | 1,448 | 327.9 | 33.1 | D | D | D | D |
| Tyrrell | 0.48 | 117.9 | NA | NA | NA | NA | 21 | 157 | 35.2 | 2.6 | D | D | D | D |
| Union | 6.57 | 29.5 | 266 | 2,863 | 1,661.6 | 163.7 | 535 | 7,999 | 2,655.9 | 213.0 | 226 | 457 | 113.7 | 19.9 |
| Vance | 5.95 | 133.5 | D | D | D | D | 184 | 2,212 | 587.4 | 52.3 | 45 | 185 | 37.3 | 6.1 |
| Wake | 56.66 | 55.3 | 1,160 | 22,292 | 30,140.0 | 1,790.4 | 3,493 | 59,325 | 18,099.7 | 1,613.2 | 1,740 | 9,061 | 2,566.3 | 509.3 |
| Warren | 0.05 | 2.5 | D | D | D | 1.3 | 46 | 397 | 93.7 | 9.2 | 12 | 25 | 3.2 | 0.6 |
| Washington | 0.79 | 63.8 | 7 | 61 | 35.0 | 3.1 | 42 | 368 | 97.2 | 7.9 | D | D | D | 0.5 |
| Watauga | 3.30 | 62.4 | 32 | 290 | 169.7 | 15.9 | 323 | 3,658 | 931.8 | 87.1 | 123 | 435 | 74.0 | 13.5 |
| Wayne | 11.27 | 90.8 | 98 | 1,920 | 1,385.3 | 75.9 | 461 | 6,214 | 1,619.3 | 144.9 | 67 | 335 | 53.7 | 10.5 |
| Wilkes | 6.61 | 96.5 | D | D | D | D | 220 | 2,665 | 3,243.8 | 62.3 | D | D | D | 8.5 |
| Wilson | 9.62 | 117.7 | 104 | 1,120 | 636.4 | 55.5 | 325 | 3,775 | 1,017.7 | 92.4 | 83 | 276 | 47.4 | 8.0 |
| Yadkin | 1.36 | 36.2 | D | D | D | 6.9 | 108 | 839 | 251.6 | 18.0 | D | D | D | D |
| Yancey | 0.70 | 39.8 | 10 | 50 | 7.1 | 0.9 | 59 | 541 | 134.9 | 12.7 | D | D | D | 0.9 |
| **NORTH DAKOTA** | 84.18 | 111.2 | 1,563 | 20,075 | 23,276.8 | 1,203.6 | 3,277 | 49,579 | 19,251.1 | 1,453.9 | 1,100 | 5,440 | 1,279.6 | 242.3 |
| Adams | 0.00 | 0.0 | 9 | 70 | 108.0 | 3.0 | 15 | 126 | 40.3 | 4.2 | D | D | D | D |
| Barnes | 1.15 | 103.6 | 21 | 159 | 298.9 | 8.3 | 47 | 457 | 102.1 | 12.1 | NA | NA | NA | NA |
| Benson | 0.07 | 10.4 | 12 | 75 | 165.7 | 4.5 | 8 | 29 | 8.9 | 0.8 | NA | NA | NA | NA |
| Billings | 0.00 | 0.0 | NA | NA | NA | NA | 10 | D | 3.9 | D | 4 | D | 0.3 | D |
| Bottineau | 0.48 | 71.5 | 12 | 77 | 255.9 | 4.8 | 26 | 431 | 104.2 | 10.6 | NA | NA | NA | NA |
| Bowman | 0.22 | 66.8 | 12 | 140 | 101.8 | 7.0 | 18 | 221 | 69.2 | 5.4 | NA | NA | NA | NA |
| Burke | 0.26 | 112.7 | 5 | 41 | 295.4 | 2.7 | 8 | 116 | 31.1 | 3.5 | NA | NA | NA | NA |
| Burleigh | 11.9 | 128.0 | 140 | 2,507 | 1,943.9 | 158.0 | 402 | 7,654 | 2,109.9 | 216.4 | 163 | 507 | 125.8 | 19.5 |
| Cass | 14.11 | 82.3 | 353 | 6,399 | 6,183.0 | 386.2 | 681 | 13,234 | 8,502.8 | 410.6 | 331 | 1,843 | 316.6 | 71.2 |
| Cavalier | 0.50 | 130.6 | 13 | 170 | 592.0 | 11.5 | 26 | 183 | 91.5 | 5.5 | NA | NA | NA | NA |
| Dickey | 0.83 | 162.6 | D | D | D | 8.4 | 40 | 319 | 82.5 | 7.7 | NA | NA | NA | NA |
| Divide | 0.00 | 0.0 | D | D | D | D | 11 | 57 | 9.2 | 1.2 | 3 | 3 | 0.4 | 0.1 |
| Dunn | 0.00 | 0.0 | 13 | 95 | 115.8 | 7.7 | 15 | 128 | 39.7 | 4.3 | D | D | D | D |
| Eddy | 0.98 | 414.4 | 5 | D | 49.9 | D | 10 | 62 | 13.6 | 1.9 | NA | NA | NA | NA |
| Emmons | 0.43 | 126.4 | 13 | 70 | 125.1 | 3.1 | 23 | 133 | 27.9 | 3.0 | NA | NA | NA | D |
| Foster | 0.43 | 128.1 | 19 | 138 | 332.3 | 9.0 | 24 | 237 | 66.8 | 6.1 | 5 | D | 1.4 | D |
| Golden Valley | 0.0 | 0.0 | 5 | 59 | 77.4 | 3.1 | 13 | 104 | 43.5 | 3.2 | NA | NA | NA | NA |
| Grand Forks | 8.78 | 123.8 | 91 | 1,242 | 1,161.2 | 70.5 | 317 | 6,060 | 1,598.4 | 154.6 | 86 | 589 | 91.4 | 20.7 |
| Grant | 0.00 | 0.0 | 7 | 39 | 52.3 | 1.7 | 10 | 47 | 8.0 | 1.0 | NA | NA | NA | NA |
| Griggs | 0.38 | 164.2 | 6 | 42 | 29.2 | 1.9 | 11 | 68 | 18.8 | 1.3 | NA | NA | NA | NA |
| Hettinger | 0.00 | 0.0 | 7 | D | 51.0 | D | 15 | 99 | 52.1 | 3.3 | NA | NA | NA | NA |
| Kidder | 0.09 | 37.2 | 7 | D | 14.3 | D | 11 | 37 | 10.0 | 1.3 | D | D | D | 0.0 |
| LaMoure | 0.01 | 2.4 | 26 | 197 | 262.1 | 10.8 | 18 | 109 | 24.5 | 2.9 | D | D | D | D |
| Logan | 0.09 | 46.5 | 6 | D | 170.1 | D | 10 | 82 | 21.4 | 2.2 | NA | NA | NA | NA |
| McHenry | 0.97 | 162.5 | 8 | 184 | 236.1 | 11.1 | 17 | 148 | 32.1 | 3.4 | NA | NA | NA | NA |
| McIntosh | 0.26 | 94.2 | 13 | 116 | 169.2 | 4.7 | 17 | 126 | 48.8 | 3.8 | NA | NA | NA | NA |
| McKenzie | 0.79 | 61.6 | 23 | 290 | 292.3 | 21.2 | 41 | 479 | 159.9 | 15.3 | 28 | 231 | 82.2 | 14.3 |
| McLean | 0.61 | 62.6 | 19 | 161 | 368.0 | 10.4 | 40 | 326 | 90.0 | 9.6 | 4 | D | 0.5 | D |
| Mercer | 6.67 | 753.4 | 9 | 50 | 52.4 | 2.9 | 45 | 466 | 102.8 | 12.1 | NA | NA | NA | NA |
| Morton | 2.80 | 92.4 | 45 | 344 | 504.9 | 17.5 | 97 | 1,527 | 587.5 | 57.3 | 36 | 122 | 39.1 | 5.1 |
| Mountrail | 0.52 | 50.3 | 17 | 164 | 282.4 | 10.9 | 50 | 616 | 183.6 | 18.9 | 8 | D | 2.8 | D |
| Nelson | 1.33 | 448.1 | 12 | 119 | 271.8 | 7.9 | 18 | 99 | 22.6 | 1.7 | 3 | D | 0.3 | D |
| Oliver | 0.00 | 0.0 | NA | NA | NA | NA | D | D | D | 0.9 | NA | NA | NA | NA |
| Pembina | 1.10 | 155.1 | 36 | 322 | 527.8 | 14.7 | 40 | 323 | 73.1 | 7.5 | NA | NA | NA | NA |
| Pierce | 0.75 | 173.9 | 16 | 137 | 192.7 | 8.4 | 20 | 205 | 78.5 | 6.2 | 8 | D | 10.9 | D |
| Ramsey | 0.00 | 0.0 | 31 | 318 | 499.2 | 17.9 | 82 | 1,020 | 335.7 | 32.3 | 10 | D | 1.4 | D |
| Ransom | 1.02 | 187.2 | 16 | 128 | 140.0 | 7.3 | 25 | 206 | 52.9 | 4.7 | NA | NA | NA | NA |
| Renville | 0.00 | 0.0 | 14 | 102 | 211.1 | 5.8 | 16 | 113 | 35.3 | 3.4 | NA | NA | NA | NA |
| Richland | 2.08 | 126.8 | D | D | D | D | 69 | 715 | 215.6 | 18.4 | 11 | 39 | 5.3 | 0.9 |
| Rolette | 1.59 | 108.5 | 12 | 47 | 68.5 | 2.4 | 39 | 480 | 111.8 | 10.3 | NA | NA | NA | NA |
| Sargent | 0.28 | 72.2 | 12 | 108 | 73.0 | 5.3 | 14 | 160 | 31.0 | 4.1 | 3 | 6 | 0.2 | 0.0 |
| Sheridan | 0.01 | 7.6 | NA | NA | NA | NA | D | D | D | 0.2 | NA | NA | NA | NA |
| Sioux | 0.02 | 4.6 | NA | NA | NA | NA | D | D | D | 1.1 | NA | NA | NA | NA |
| Slope | 0.03 | 39.1 | NA | NA | NA | NA | NA | NA | NA | NA | NA | NA | NA | NA |
| Stark | 0.00 | 0.0 | 71 | 736 | 528.2 | 48.2 | 157 | 2,191 | 743.5 | 73.0 | 53 | 256 | 99.4 | 16.1 |
| Steele | 0.67 | 342.5 | D | D | D | 3.9 | 6 | 58 | 37.9 | 2.2 | NA | NA | NA | NA |
| Stutsman | 3.32 | 157.3 | 38 | 387 | 488.7 | 22.2 | 102 | 1,512 | 576.9 | 46.0 | 28 | 63 | 11.3 | 1.8 |
| Towner | 0.19 | 83.6 | 9 | 102 | 178.6 | 5.7 | 13 | 47 | 32.5 | 1.5 | NA | NA | NA | NA |
| Traill | 0.92 | 114.8 | 34 | 323 | 371.8 | 16.4 | 46 | 349 | 104.8 | 8.6 | 7 | 14 | 0.5 | 0.2 |
| Walsh | 1.14 | 104.6 | 38 | 371 | 535.6 | 19.7 | 52 | 465 | 94.6 | 10.6 | 10 | 27 | 1.4 | 0.4 |

1 Merchant wholesalers, except manufacturers' sales branches and offices.    2. Employer establishments.

— **Professional Services, Manufacturing, and Accommodation and Food Services**

| STATE County | Professional, scientific, and technical services, 2017 | | | | Manufacturing, 2017 | | | | Accommodation and food services, 2017 | | | |
|---|---|---|---|---|---|---|---|---|---|---|---|---|
| | Number of establishments | Number of employees | Sales (mil dol) | Average payroll (mil dol) | Number of establishments | Number of employees | Sales (mil dol) | Average payroll (mil dol) | Number of establishments | Number of employees | Sales (mil dol) | Annual payroll (mil dol) |
| | 147 | 148 | 149 | 150 | 151 | 152 | 153 | 154 | 155 | 156 | 157 | 158 |
| NORTH CAROLINA— Cont'd | | | | | | | | | | | | |
| Swain | 17 | 50 | 5 | 1.5 | D | 390 | D | 18.8 | 89 | 835 | 64.9 | 14.5 |
| Transylvania | D | D | D | D | 31 | 602 | 117.5 | 30.0 | 78 | 1,305 | 91.1 | 27.9 |
| Tyrrell | NA | NA | NA | NA | 4 | 46 | 8.5 | 3.3 | D | D | D | D |
| Union | 500 | 1,917 | 296 | 99.1 | 239 | 11,168 | 4,033.2 | 602.6 | 315 | 5,582 | 302.0 | 80.4 |
| Vance | 51 | 289 | 41 | 16.8 | 37 | 1,564 | 653.4 | 71.4 | 70 | 1,326 | 68.6 | 17.5 |
| Wake | 4,934 | 59,018 | 12,407 | 4,942.2 | 610 | 12,947 | 13,115.7 | 779.1 | 2,424 | 53,773 | 3,206.7 | 903.5 |
| Warren | 11 | 39 | 3 | 0.8 | 9 | 377 | 163.4 | 15.2 | D | D | D | 3.2 |
| Washington | 7 | 56 | 7 | 1.8 | 10 | 753 | 455.2 | 53.3 | 22 | 326 | 15.4 | 4.1 |
| Watauga | 154 | 479 | 50 | 19.1 | 34 | 488 | 78.1 | 17.3 | 184 | 3,943 | 190.4 | 55.1 |
| Wayne | 130 | 679 | 75 | 29.3 | 56 | 4,957 | 1,351.3 | 226.3 | 208 | 4,152 | 202.4 | 56.5 |
| Wilkes | 73 | 566 | 59 | 21.4 | 65 | 4,645 | 1,323.3 | 169.0 | 101 | 1,715 | 88.5 | 22.8 |
| Wilson | D | D | D | D | 87 | 7,525 | 7,725.8 | 419.7 | D | D | D | D |
| Yadkin | 34 | 193 | 22 | 9.3 | 47 | 2,480 | 962.7 | 110.4 | 55 | 855 | 41.1 | 10.7 |
| Yancey | 20 | 105 | 7 | 2.4 | 13 | 1,037 | 216.2 | 59.5 | 20 | D | 13.6 | D |
| NORTH DAKOTA | 1,804 | 14,780 | 2,412 | 942.2 | 701 | 24,214 | 13,604.9 | 1,266.4 | 2,080 | 36,648 | 2,118.0 | 619.9 |
| Adams | 4 | 23 | 3 | 0.9 | NA | NA | NA | NA | 8 | D | 2.4 | D |
| Barnes | 13 | 70 | 10 | 3.2 | 12 | 365 | 178.6 | 19.6 | 33 | 346 | 15.3 | 4.2 |
| Benson | NA | NA | NA | NA | D | 15 | D | D | D | D | D | D |
| Billings | NA | NA | NA | NA | NA | NA | NA | NA | 10 | 85 | 20.2 | 6.5 |
| Bottineau | 21 | 75 | 8 | 3.2 | 9 | 90 | 28.5 | 4.5 | 25 | 193 | 9.5 | 2.6 |
| Bowman | 9 | 50 | 12 | 3.9 | D | 27 | D | 1.2 | 17 | 173 | 5.4 | 1.6 |
| Burke | 4 | 12 | 2 | 0.4 | NA | NA | NA | NA | 8 | 37 | 4.0 | 0.5 |
| Burleigh | D | D | D | D | 60 | 1,058 | 375.9 | 64.2 | 210 | 5,516 | 296.9 | 96.8 |
| Cass | 525 | 5,938 | 925 | 382.2 | 186 | 8,334 | 2,870.8 | 412.9 | 457 | 10,766 | 506.0 | 168.8 |
| Cavalier | 7 | 15 | 2 | 0.7 | 6 | 10 | 3.9 | 0.6 | 10 | D | 3.6 | D |
| Dickey | 12 | 35 | 3 | 1.0 | 11 | 210 | 44.7 | 8.0 | 18 | 149 | 4.8 | 1.3 |
| Divide | 4 | 9 | 1 | 0.3 | 3 | D | 0.3 | 0.1 | 10 | 55 | 3.4 | 0.8 |
| Dunn | D | D | D | D | D | D | D | D | 9 | 118 | 27.8 | 4.2 |
| Eddy | D | D | 1 | D | D | D | D | D | 6 | D | 1.1 | D |
| Emmons | 8 | 12 | 2 | 0.5 | NA | NA | NA | NA | D | D | D | 0.7 |
| Foster | 4 | 5 | 1 | 0.1 | D | D | D | D | 9 | 120 | 4.6 | 1.7 |
| Golden Valley | D | D | 1 | D | NA | NA | NA | NA | 4 | 20 | 0.8 | 0.3 |
| Grand Forks | 130 | 1,376 | 185 | 82.0 | 51 | 2,784 | 797.7 | 136.5 | 202 | 4,267 | 192.2 | 62.0 |
| Grant | NA | NA | NA | NA | 4 | 11 | 1.5 | 0.4 | 8 | D | 1.1 | D |
| Griggs | 6 | 38 | 2 | 1.0 | 5 | 121 | 22.7 | 5.0 | D | D | D | 0.5 |
| Hettinger | 6 | 28 | 4 | 2.5 | 3 | 9 | 2.2 | 0.4 | D | D | D | 0.4 |
| Kidder | 7 | 22 | 6 | 0.9 | NA | NA | NA | NA | 9 | D | 2.3 | D |
| LaMoure | NA | NA | NA | NA | 6 | 81 | 21.8 | 3.0 | D | D | D | 0.6 |
| Logan | NA | NA | NA | NA | NA | NA | NA | NA | 4 | 45 | 1.5 | 0.3 |
| McHenry | D | D | 4 | D | D | D | D | D | 10 | D | 1.8 | D |
| McIntosh | 7 | 21 | 2 | 0.8 | 5 | 44 | 4.9 | 1.3 | 10 | D | 1.5 | D |
| McKenzie | 38 | 205 | 52 | 17.5 | NA | NA | NA | NA | 39 | 413 | 44.7 | 10.9 |
| McLean | 8 | 30 | 3 | 1.2 | D | 100 | D | D | 28 | 238 | 13.9 | 3.1 |
| Mercer | 10 | 51 | 7 | 2.3 | D | D | D | D | 33 | 287 | 15.2 | 3.9 |
| Morton | D | D | D | D | 31 | 856 | 1,559.9 | 63.8 | D | D | D | D |
| Mountrail | D | D | D | D | 4 | D | 22.2 | 5.3 | 41 | 337 | 37.5 | 7.8 |
| Nelson | 4 | 9 | 1 | 0.5 | NA | NA | NA | NA | D | D | D | 0.7 |
| Oliver | NA | NA | NA | NA | D | D | D | D | NA | NA | NA | NA |
| Pembina | 13 | 33 | 4 | 1.2 | 15 | 670 | 489.8 | 31.9 | 14 | D | 4.5 | D |
| Pierce | 12 | 68 | 10 | 2.8 | D | D | D | 2.9 | 16 | 126 | 5.7 | 1.7 |
| Ramsey | 19 | 80 | 8 | 3.3 | 7 | 171 | 40.2 | 8.6 | 50 | 742 | 41.3 | 10.5 |
| Ransom | 17 | 56 | 6 | 1.8 | D | 284 | D | 16.1 | 21 | 130 | 5.0 | 1.1 |
| Renville | D | D | D | 0.2 | NA | NA | NA | NA | 14 | D | 2.7 | D |
| Richland | D | D | 12 | D | 37 | 1,831 | 1,049.5 | D | 39 | 833 | 91.4 | 19.5 |
| Rolette | D | D | D | 0.3 | D | 63 | D | 1.7 | 21 | 569 | 51.9 | 11.7 |
| Sargent | 6 | 17 | 2 | 0.8 | D | D | D | D | 12 | 36 | 3.6 | 0.6 |
| Sheridan | D | D | D | D | NA | NA | NA | NA | NA | NA | NA | NA |
| Sioux | NA | NA | NA | NA | NA | NA | NA | NA | D | D | D | D |
| Slope | NA | NA | NA | NA | NA | NA | NA | NA | NA | NA | NA | NA |
| Stark | 83 | 552 | 80 | 29.9 | 31 | 1,111 | 1,158.8 | 66.8 | 89 | 1,432 | 88.4 | 28.7 |
| Steele | NA | NA | NA | NA | 8 | 99 | 18.4 | 3.7 | D | D | D | 0.2 |
| Stutsman | 35 | 242 | 26 | 10.1 | 21 | 811 | 406.0 | 44.5 | 62 | 909 | 43.4 | 14.7 |
| Towner | 6 | 21 | 2 | 0.8 | 3 | D | 1.3 | 0.5 | 8 | 34 | 2.2 | 0.4 |
| Traill | 15 | 44 | 4 | 1.7 | D | 371 | D | 16.3 | 23 | 157 | 7.4 | 1.8 |
| Walsh | 25 | 95 | 13 | 4.0 | 12 | 544 | 105.1 | 19.7 | 30 | 145 | 7.5 | 1.8 |

# Table B. States and Counties — Health Care and Social Assistance, Other Services, Nonemployer Businesses, and Residential Construction

| STATE County | Health care and social assistance, 2017 | | | | Other services, 2017 | | | | Nonemployer businesses, 2018 | | Value of residential construction authorized by building permits, 2020 | |
|---|---|---|---|---|---|---|---|---|---|---|---|---|
| | Number of establishments | Number of employees | Receipts (mil dol) | Annual payroll (mil dol) | Number of establishments | Number of employees | Receipts (mil dol) | Annual payroll (mil dol) | Number | Receipts (mil dol) | New construction ($1,000) | Number of housing units |
| | 159 | 160 | 161 | 162 | 163 | 164 | 165 | 166 | 167 | 168 | 169 | 170 |
| NORTH CAROLINA—Cont'd | | | | | | | | | | | | |
| Swain | 27 | 884 | 109.0 | 38.0 | D | D | D | 1.1 | 1,330 | 48.9 | 17,342 | 83 |
| Transylvania | 88 | 1,686 | 182.9 | 61.7 | 52 | 351 | 31.5 | 9.1 | 3,254 | 137.7 | 0 | 0 |
| Tyrrell | 7 | 23 | 1.3 | 0.8 | D | D | D | D | 266 | 10.3 | 800 | 4 |
| Union | 333 | 5,911 | 629.4 | 247.1 | 345 | 1,678 | 161.5 | 50.5 | 19,542 | 960.2 | 469,193 | 2,354 |
| Vance | 93 | 2,538 | 222.2 | 95.2 | D | D | D | D | 2,408 | 104.0 | 15,602 | 87 |
| Wake | 3,040 | 63,396 | 7,715.6 | 3,097.5 | 2,004 | 13,651 | 1,716.5 | 518.0 | 95,345 | 4,481.6 | 2,665,544 | 12,598 |
| Warren | 19 | 395 | 23.0 | 11.0 | D | D | D | 1.2 | 1,027 | 33.4 | 20,400 | 75 |
| Washington | 35 | 644 | 39.9 | 18.5 | 8 | 21 | 4.1 | 1.0 | 603 | 17.9 | 994 | 5 |
| Watauga | 138 | 3,952 | 1,073.3 | 193.3 | 92 | 448 | 37.6 | 11.7 | 4,985 | 216.8 | 105,890 | 217 |
| Wayne | 259 | 7,604 | 781.9 | 310.8 | 152 | 810 | 71.4 | 21.2 | 6,621 | 264.2 | 75,705 | 414 |
| Wilkes | 152 | 3,137 | 239.1 | 86.5 | 62 | 255 | 29.0 | 7.3 | 4,441 | 203.2 | 25,544 | 110 |
| Wilson | 217 | 4,930 | 414.4 | 165.6 | 117 | 685 | 65.1 | 18.7 | 4,834 | 193.5 | 50,835 | 385 |
| Yadkin | 48 | 896 | 69.7 | 27.4 | 28 | 90 | 11.5 | 2.9 | 2,342 | 101.8 | 37,767 | 162 |
| Yancey | 31 | 481 | 33.8 | 13.4 | 25 | 108 | 12.1 | 3.6 | 1,642 | 60.4 | 12,308 | 51 |
| NORTH DAKOTA | 2,057 | 62,455 | 7,297.8 | 2,913.3 | 1,763 | 9,706 | 1,246.0 | 318.3 | 56,523 | 2,948.3 | 726,755 | 3,493 |
| Adams | 16 | 346 | 31.0 | 14.7 | D | D | D | 0.4 | 181 | 10.1 | 0 | 0 |
| Barnes | 41 | 1,043 | 69.8 | 32.0 | 24 | 146 | 11.5 | 3.1 | 869 | 44.3 | 3,378 | 10 |
| Benson | 5 | D | 0.9 | D | NA | NA | NA | NA | 290 | 11.1 | 0 | 0 |
| Billings | NA | NA | NA | NA | 4 | 14 | 3.7 | 0.8 | 112 | 5.9 | 829 | 4 |
| Bottineau | 13 | 269 | 18.9 | 8.7 | D | D | D | 0.5 | 629 | 27.9 | 2,930 | 14 |
| Bowman | 18 | 225 | 18.5 | 9.7 | D | D | D | 1.0 | 253 | 10.6 | 50 | 1 |
| Burke | D | D | D | 0.2 | NA | NA | NA | NA | 193 | 8.6 | 135 | 2 |
| Burleigh | 300 | 10,358 | 1,256.8 | 484.9 | 265 | 1,711 | 248.0 | 70.0 | 7,773 | 425.9 | 114,467 | 531 |
| Cass | 520 | 20,346 | 2,941.8 | 1,011.4 | 403 | 3,015 | 354.9 | 88.3 | 13,264 | 822.9 | 349,367 | 1,666 |
| Cavalier | 8 | 227 | 17.5 | 6.5 | D | D | 2.7 | D | 392 | 14.1 | 1,312 | 5 |
| Dickey | 22 | 371 | 34.6 | 13.4 | 16 | 49 | 7.3 | 1.8 | 444 | 17.8 | 1,658 | 9 |
| Divide | D | D | D | D | 6 | 18 | 1.5 | 0.2 | 207 | 8.4 | 400 | 2 |
| Dunn | D | D | D | D | D | D | D | D | 387 | 28.5 | 2,614 | 12 |
| Eddy | 9 | 190 | 11.6 | 6.2 | D | D | 2.0 | D | 197 | 7.9 | 550 | 1 |
| Emmons | 10 | 191 | 17.6 | 7.3 | D | D | D | 0.1 | 277 | 9.0 | 3,872 | 11 |
| Foster | 12 | 236 | 22.3 | 9.8 | D | D | 1.5 | D | 313 | 16.4 | 1,454 | 9 |
| Golden Valley | 7 | D | 8.2 | D | D | D | D | 0.2 | 189 | 7.8 | 1,171 | 5 |
| Grand Forks | 163 | 7,478 | 828.8 | 381.5 | 138 | 782 | 93.2 | 23.8 | 4,294 | 197.7 | 61,663 | 420 |
| Grant | 8 | 187 | 13.2 | 5.4 | D | D | 2.4 | D | 210 | 8.2 | 2,635 | 8 |
| Griggs | 5 | 128 | 9.2 | 3.9 | D | D | D | 0.2 | 220 | 9.4 | 0 | 0 |
| Hettinger | 8 | 97 | 5.9 | 3.1 | D | D | D | 0.3 | 211 | 6.4 | 0 | 0 |
| Kidder | 5 | D | 2.1 | D | 3 | 18 | 1.0 | 0.2 | 241 | 11.9 | 395 | 1 |
| LaMoure | 14 | D | 7.8 | D | D | D | 3.1 | D | 350 | 13.4 | 238 | 1 |
| Logan | 8 | 116 | 5.1 | 2.5 | D | D | D | 0.9 | 183 | 8.5 | 190 | 1 |
| McHenry | D | D | D | D | D | D | 2.7 | D | 443 | 21.3 | 1,139 | 6 |
| McIntosh | 10 | 316 | 21.5 | 10.6 | D | D | D | 0.7 | 229 | 9.8 | 0 | 0 |
| McKenzie | 14 | 454 | 28.7 | 13.5 | 25 | 130 | 13.4 | 4.5 | 1,018 | 59.1 | 18,815 | 73 |
| McLean | 19 | 389 | 24.4 | 11.6 | D | D | D | 1.3 | 715 | 28.0 | 7,430 | 33 |
| Mercer | 22 | 509 | 40.5 | 19.2 | D | D | D | D | 600 | 15.6 | 450 | 2 |
| Morton | 71 | 1,480 | 109.3 | 51.0 | 71 | 317 | 45.8 | 12.3 | 2,491 | 143.1 | 32,834 | 184 |
| Mountrail | 14 | 269 | 14.8 | 9.2 | 21 | 306 | 27.6 | 10.9 | 742 | 44.8 | 7,055 | 45 |
| Nelson | 12 | 277 | 15.6 | 7.8 | D | D | 2.5 | D | 269 | 9.3 | 0 | 0 |
| Oliver | D | D | D | 0.5 | D | D | D | 0.1 | 142 | 5.9 | 0 | 0 |
| Pembina | 17 | 338 | 24.3 | 9.4 | 14 | 30 | 3.6 | 0.8 | 552 | 20.9 | 1,664 | 7 |
| Pierce | 9 | D | 30.8 | D | D | D | D | D | 326 | 12.1 | 880 | 3 |
| Ramsey | 52 | 815 | 73.4 | 34.1 | 35 | 127 | 11.7 | 2.7 | 871 | 38.7 | 4,602 | 28 |
| Ransom | 30 | 386 | 34.5 | 14.1 | 16 | 68 | 5.4 | 1.2 | 386 | 16.5 | 930 | 5 |
| Renville | D | D | D | D | D | D | 2.1 | D | 181 | 8.2 | 0 | 0 |
| Richland | 44 | 551 | 49.9 | 21.3 | 33 | 122 | 10.2 | 2.7 | 1,253 | 67.3 | 7,492 | 33 |
| Rolette | D | D | D | 42.0 | D | D | 2.7 | D | 626 | 18.7 | 0 | 0 |
| Sargent | 8 | 71 | 5.2 | 2.5 | D | D | D | D | 286 | 13.1 | 350 | 2 |
| Sheridan | 3 | 23 | 1.0 | 0.6 | 5 | 8 | 0.9 | 0.2 | 107 | 4.7 | 975 | 4 |
| Sioux | D | D | D | 1.2 | D | D | D | 0.1 | 112 | 3.6 | 0 | 0 |
| Slope | NA | NA | NA | NA | NA | NA | NA | NA | 59 | 3.3 | 300 | 2 |
| Stark | 96 | 1,867 | 205.6 | 75.6 | 89 | 404 | 62.0 | 14.3 | 2,668 | 147.8 | 22,159 | 68 |
| Steele | NA | NA | NA | NA | 3 | 5 | 1.3 | 0.3 | 151 | 9.5 | 1,060 | 6 |
| Stutsman | 70 | 2,641 | 225.0 | 122.8 | 56 | 267 | 24.3 | 8.5 | 1,428 | 59.4 | 4,076 | 17 |
| Towner | NA | NA | NA | NA | 5 | 29 | 1.4 | 0.2 | 209 | 6.5 | 0 | 0 |
| Traill | 19 | 546 | 42.4 | 19.2 | 24 | 66 | 13.0 | 2.1 | 621 | 25.7 | 4,473 | 16 |
| Walsh | 36 | 692 | 56.7 | 26.5 | 28 | 71 | 6.2 | 1.9 | 800 | 35.1 | 238 | 1 |

# Table B. States and Counties — Government Employment and Payroll, and Local Government Finances

| STATE County | Full-time equivalent employees | March payroll (dollars) | March payroll (percent of total) Administration, judicial, and legal | Police and corrections | Fire protection | Highways and transportation | Health and welfare | Natural resources and utilities | Education and libraries | General revenue Total (mil dol) | Intergovernmental (mil dol) | Taxes Total (mil dol) | Per capita[1] Total (dollars) | Per capita[1] Property (dollars) |
|---|---|---|---|---|---|---|---|---|---|---|---|---|---|---|
| | 171 | 172 | 173 | 174 | 175 | 176 | 177 | 178 | 179 | 180 | 181 | 182 | 183 | 184 |
| **NORTH CAROLINA— Cont'd** | | | | | | | | | | | | | | |
| Swain | 572 | 1,870,157 | 4.2 | 9.9 | 0.0 | 0.6 | 14.5 | 6.1 | 59.2 | 47.2 | 29.2 | 11.6 | 815 | 461 |
| Transylvania | 1,126 | 3,434,915 | 5.9 | 13.1 | 1.2 | 1.0 | 13.6 | 6.1 | 53.3 | 94.8 | 35.7 | 48.2 | 1,428 | 1,112 |
| Tyrrell | 174 | 666,490 | 7.2 | 4.9 | 0.0 | 0.4 | 8.3 | 7.2 | 65.4 | 17.2 | 10.0 | 5.0 | 1,185 | 964 |
| Union | 7,152 | 25,616,934 | 4.2 | 7.2 | 1.5 | 0.7 | 4.5 | 5.3 | 74.2 | 676.5 | 302.7 | 297.0 | 1,284 | 1,005 |
| Vance | 1,972 | 7,360,563 | 2.8 | 6.5 | 2.4 | 2.5 | 10.2 | 2.6 | 70.6 | 157.5 | 92.5 | 42.4 | 958 | 703 |
| Wake | 33,845 | 151,028,554 | 3.7 | 9.2 | 3.9 | 4.1 | 5.6 | 8.0 | 61.0 | 4,021.1 | 1,405.6 | 1,881.3 | 1,755 | 1,250 |
| Warren | 558 | 2,079,861 | 3.4 | 9.3 | 0.0 | 0.4 | 11.1 | 4.1 | 69.7 | 57.9 | 27.5 | 23.9 | 1,203 | 1,031 |
| Washington | 542 | 1,589,913 | 7.6 | 11.1 | 0.0 | 0.7 | 16.4 | 4.7 | 57.0 | 42.2 | 23.5 | 12.7 | 1,060 | 752 |
| Watauga | 1,419 | 4,949,287 | 7.1 | 10.9 | 3.0 | 2.2 | 12.8 | 7.6 | 48.1 | 154.2 | 60.9 | 69.9 | 1,268 | 856 |
| Wayne | 4,543 | 14,430,137 | 3.0 | 6.6 | 2.3 | 1.5 | 7.8 | 6.9 | 69.4 | 355.9 | 205.6 | 110.1 | 895 | 614 |
| Wilkes | 3,313 | 11,432,319 | 2.1 | 4.8 | 0.3 | 1.2 | 33.7 | 2.3 | 53.3 | 190.7 | 106.8 | 58.6 | 856 | 619 |
| Wilson | 3,251 | 11,938,713 | 6.9 | 9.7 | 3.6 | 1.1 | 11.6 | 12.9 | 50.6 | 297.5 | 135.9 | 98.2 | 1,206 | 913 |
| Yadkin | 1,177 | 3,864,174 | 4.5 | 9.0 | 0.1 | 0.5 | 11.2 | 3.3 | 69.9 | 90.5 | 49.6 | 31.5 | 838 | 630 |
| Yancey | 753 | 2,341,687 | 2.6 | 6.5 | 0.3 | 1.4 | 7.4 | 4.9 | 64.8 | 66.7 | 25.1 | 36.5 | 2,063 | 1,620 |
| **NORTH DAKOTA** | X | X | X | X | X | X | X | X | X | X | X | X | X | X |
| Adams | 82 | 268,388 | 14.2 | 10.9 | 0.0 | 8.8 | 7.6 | 5.2 | 52.2 | 14.8 | 8.6 | 4.8 | 2,046 | 1,679 |
| Barnes | 386 | 1,620,424 | 9.9 | 11.1 | 0.5 | 6.9 | 5.9 | 14.8 | 49.9 | 67.4 | 31.8 | 24.2 | 2,262 | 1,953 |
| Benson | 296 | 1,067,230 | 8.8 | 1.6 | 0.6 | 6.7 | 7.6 | 1.5 | 73.1 | 34.6 | 23.8 | 6.3 | 915 | 899 |
| Billings | 78 | 365,632 | 15.9 | 12.5 | 1.0 | 27.8 | 8.0 | 2.8 | 29.0 | 23.7 | 18.4 | 4.1 | 4,478 | 3,951 |
| Bottineau | 292 | 1,136,510 | 12.0 | 5.6 | 0.4 | 8.1 | 4.3 | 8.7 | 58.7 | 70.5 | 46.1 | 16.6 | 2,545 | 2,304 |
| Bowman | 177 | 691,854 | 10.4 | 5.5 | 0.0 | 7.0 | 5.3 | 7.9 | 62.5 | 30.1 | 18.0 | 4.6 | 1,458 | 1,286 |
| Burke | 134 | 537,457 | 15.0 | 7.0 | 0.1 | 9.0 | 2.7 | 3.7 | 58.5 | 23.4 | 10.7 | 6.5 | 3,064 | 3,023 |
| Burleigh | 2,989 | 13,881,814 | 4.6 | 10.4 | 3.5 | 3.5 | 5.4 | 9.7 | 59.3 | 402.2 | 182.0 | 133.7 | 1,403 | 1,033 |
| Cass | 5,640 | 25,246,394 | 5.1 | 9.1 | 2.7 | 5.4 | 8.6 | 6.5 | 61.7 | 904.5 | 382.3 | 313.2 | 1,762 | 1,216 |
| Cavalier | 150 | 603,757 | 10.7 | 8.5 | 0.1 | 8.5 | 3.7 | 6.8 | 61.5 | 23.5 | 9.9 | 10.0 | 2,645 | 2,632 |
| Dickey | 186 | 725,275 | 10.7 | 6.6 | 0.0 | 5.8 | 9.7 | 5.4 | 58.7 | 39.5 | 15.5 | 15.7 | 3,244 | 3,123 |
| Divide | 124 | 536,408 | 2.6 | 8.4 | 0.0 | 20.0 | 4.9 | 4.7 | 48.5 | 31.8 | 15.2 | 9.6 | 4,192 | 3,941 |
| Dunn | 189 | 749,819 | 19.8 | 5.8 | 0.6 | 11.1 | 4.9 | 6.2 | 51.6 | 58.5 | 15.7 | 30.9 | 7,233 | 2,207 |
| Eddy | 123 | 378,514 | 7.6 | 5.1 | 0.0 | 5.2 | 2.2 | 5.6 | 73.7 | 12.2 | 6.7 | 3.7 | 1,596 | 1,487 |
| Emmons | 179 | 628,299 | 12.4 | 3.6 | 0.0 | 7.7 | 2.9 | 5.7 | 67.7 | 45.9 | 20.6 | 14.6 | 4,431 | 4,018 |
| Foster | 109 | 418,777 | 15.4 | 2.3 | 0.1 | 6.6 | 7.3 | 1.5 | 65.0 | 21.9 | 7.9 | 9.4 | 2,898 | 2,637 |
| Golden Valley | 110 | 361,391 | 6.1 | 2.6 | 0.0 | 4.0 | 3.0 | 3.4 | 80.3 | 21.0 | 11.9 | 6.6 | 3,689 | 3,581 |
| Grand Forks | 2,657 | 11,301,030 | 5.8 | 9.6 | 3.9 | 5.5 | 8.1 | 9.2 | 56.2 | 351.0 | 156.7 | 106.6 | 1,512 | 1,170 |
| Grant | 88 | 309,918 | 12.1 | 7.5 | 0.0 | 8.7 | 7.5 | 4.0 | 58.2 | 12.7 | 6.2 | 4.0 | 1,675 | 1,618 |
| Griggs | 111 | 405,400 | 8.5 | 2.7 | 0.0 | 8.0 | 8.5 | 3.2 | 68.5 | 18.4 | 6.3 | 10.0 | 4,478 | 4,390 |
| Hettinger | 144 | 556,770 | 12.0 | 4.3 | 0.0 | 6.7 | 4.9 | 3.6 | 68.3 | 20.2 | 10.1 | 6.3 | 2,560 | 2,384 |
| Kidder | 93 | 346,221 | 8.3 | 6.2 | 0.3 | 1.6 | 6.7 | 5.6 | 71.1 | 15.0 | 6.2 | 4.4 | 1,795 | 1,701 |
| LaMoure | 242 | 931,413 | 8.2 | 3.6 | 0.0 | 6.6 | 2.1 | 17.8 | 61.3 | 48.9 | 27.9 | 15.5 | 3,787 | 3,681 |
| Logan | 90 | 339,766 | 4.2 | 2.9 | 0.1 | 3.6 | 5.9 | 4.1 | 78.2 | 13.9 | 7.3 | 5.0 | 2,596 | 2,444 |
| McHenry | 252 | 947,289 | 11.5 | 3.1 | 0.1 | 5.6 | 3.3 | 3.2 | 71.2 | 30.3 | 15.9 | 10.9 | 1,846 | 1,803 |
| McIntosh | 140 | 501,588 | 11.6 | 4.2 | 0.1 | 12.8 | 4.7 | 7.7 | 58.1 | 18.8 | 7.2 | 8.4 | 3,229 | 3,105 |
| McKenzie | 562 | 2,637,964 | 10.0 | 21.9 | 0.0 | 11.4 | 4.4 | 5.8 | 45.8 | 151.2 | 82.1 | 39.7 | 3,126 | 2,613 |
| McLean | 539 | 2,091,520 | 8.1 | 8.9 | 0.0 | 5.5 | 13.1 | 3.0 | 61.0 | 64.4 | 32.2 | 21.7 | 2,250 | 2,117 |
| Mercer | 399 | 1,409,489 | 9.7 | 10.8 | 0.1 | 9.1 | 0.8 | 8.3 | 58.8 | 36.0 | 16.5 | 11.2 | 1,320 | 960 |
| Morton | 1,159 | 4,948,844 | 5.8 | 10.0 | 1.0 | 5.2 | 7.4 | 4.9 | 64.6 | 150.5 | 69.9 | 60.0 | 1,938 | 1,701 |
| Mountrail | 498 | 2,068,080 | 8.8 | 8.7 | 0.0 | 6.6 | 11.7 | 3.8 | 60.2 | 153.0 | 104.5 | 32.2 | 3,132 | 2,805 |
| Nelson | 167 | 597,962 | 16.5 | 3.2 | 0.0 | 9.5 | 5.2 | 7.9 | 57.6 | 22.9 | 12.5 | 7.2 | 2,479 | 2,336 |
| Oliver | 64 | 231,582 | 15.3 | 8.2 | 0.0 | 10.8 | 0.0 | 4.2 | 60.1 | 9.2 | 4.3 | 3.0 | 1,561 | 1,156 |
| Pembina | 335 | 1,313,469 | 12.7 | 7.2 | 0.0 | 4.7 | 7.6 | 7.6 | 59.7 | 43.2 | 15.3 | 16.4 | 2,361 | 2,162 |
| Pierce | 164 | 573,855 | 3.0 | 3.5 | 0.0 | 1.2 | 16.0 | 4.4 | 65.0 | 23.8 | 10.5 | 7.1 | 1,736 | 1,693 |
| Ramsey | 534 | 2,087,368 | 7.9 | 5.4 | 1.4 | 4.4 | 9.5 | 6.5 | 64.1 | 63.6 | 35.6 | 17.4 | 1,500 | 1,184 |
| Ransom | 237 | 824,513 | 8.5 | 4.1 | 0.7 | 4.6 | 5.6 | 4.8 | 70.1 | 29.1 | 16.6 | 8.4 | 1,589 | 1,434 |
| Renville | 147 | 551,394 | 10.1 | 4.8 | 0.0 | 8.7 | 3.1 | 1.8 | 70.5 | 20.6 | 10.8 | 6.3 | 2,571 | 2,376 |
| Richland | 608 | 2,389,542 | 6.1 | 7.8 | 0.1 | 4.7 | 8.5 | 6.6 | 61.8 | 102.7 | 40.0 | 42.1 | 2,585 | 2,368 |
| Rolette | 709 | 3,167,800 | 1.8 | 2.9 | 0.2 | 1.7 | 5.5 | 1.0 | 86.7 | 79.6 | 57.6 | 13.6 | 933 | 896 |
| Sargent | 179 | 634,102 | 8.5 | 3.5 | 0.1 | 3.2 | 4.2 | 4.5 | 72.3 | 31.5 | 12.0 | 12.3 | 3,192 | 3,094 |
| Sheridan | 63 | 208,074 | 11.4 | 5.7 | 0.1 | 10.8 | 0.0 | 1.8 | 63.7 | 54.4 | 40.0 | 13.3 | 9,861 | 9,784 |
| Sioux | 181 | 703,697 | 3.9 | 0.7 | 0.1 | 3.6 | 0.2 | 0.7 | 90.4 | 70.7 | 11.2 | 58.6 | 13,238 | 13,223 |
| Slope | 39 | 142,259 | 24.3 | 7.8 | 0.0 | 14.3 | 0.0 | 16.0 | 31.8 | 8.3 | 6.3 | 1.0 | 1,304 | 1,199 |
| Stark | 1,197 | 5,293,019 | 4.2 | 17.4 | 0.9 | 4.0 | 6.5 | 13.2 | 52.9 | 196.8 | 96.9 | 59.9 | 1,979 | 1,597 |
| Steele | 76 | 267,742 | 18.4 | 4.0 | 0.2 | 9.4 | 8.0 | 1.4 | 56.6 | 24.7 | 17.0 | 5.5 | 2,879 | 2,754 |
| Stutsman | 761 | 3,193,820 | 6.2 | 11.9 | 1.1 | 5.2 | 8.3 | 10.8 | 54.1 | 105.0 | 57.4 | 30.1 | 1,424 | 1,217 |
| Towner | 111 | 389,904 | 15.0 | 6.6 | 0.0 | 14.1 | 4.0 | 4.5 | 53.0 | 13.1 | 5.1 | 4.9 | 2,196 | 1,827 |
| Traill | 315 | 1,293,677 | 7.3 | 4.3 | 0.0 | 6.2 | 5.3 | 2.8 | 73.2 | 43.1 | 22.8 | 13.3 | 1,663 | 1,559 |
| Walsh | 566 | 2,218,535 | 6.0 | 4.8 | 0.1 | 3.9 | 4.6 | 5.4 | 72.4 | 68.3 | 40.9 | 18.3 | 1,690 | 1,512 |

1. Based on the resident population estimated as of July 1 of the year shown.

# Table B. States and Counties — Local Government Finances, Government Employment, and Income Taxes

| | Local government finances, 2017 (cont.) | | | | | | | | | Government employment, 2019 | | | Individual income tax returns, 2018 | | |
| | Direct general expenditure | | | | | | | Debt outstanding | | | | | | | |
| STATE County | Total (mil dol) | Per capita[1] (dollars) | Education | Health and hospitals | Police protection | Public welfare | Highways | Total (mil dol) | Per capita[1] (dollars) | Federal civilian | Federal military | State and local | Number of returns | Mean adjusted gross income | Mean income tax |
|---|---|---|---|---|---|---|---|---|---|---|---|---|---|---|---|
| | 185 | 186 | 187 | 188 | 189 | 190 | 191 | 192 | 193 | 194 | 195 | 196 | 197 | 198 | 199 |
| **NORTH CAROLINA—Cont'd** | | | | | | | | | | | | | | | |
| Swain | 50 | 3,475 | 45.6 | 6.2 | 8.2 | 9.6 | 0.4 | 12.6 | 887 | 136 | 31 | 2,710 | 8,000 | 42,520 | 3,100 |
| Transylvania | 91 | 2,691 | 42.6 | 5.6 | 11.0 | 7.6 | 1.3 | 13.8 | 408 | 142 | 73 | 1,382 | 14,910 | 61,448 | 6,150 |
| Tyrrell | 17 | 4,049 | 55.6 | 3.2 | 8.2 | 8.3 | 0.6 | 9.5 | 2,280 | 20 | 7 | 417 | 1,620 | 36,235 | 2,320 |
| Union | 666 | 2,877 | 55.9 | 1.9 | 8.4 | 4.8 | 1.6 | 531.9 | 2,299 | 296 | 519 | 9,307 | 102,270 | 88,513 | 11,966 |
| Vance | 158 | 3,573 | 62.8 | 5.9 | 7.4 | 7.9 | 0.7 | 41.5 | 937 | 91 | 96 | 2,418 | 18,740 | 41,774 | 3,003 |
| Wake | 3,798 | 3,544 | 50.6 | 3.1 | 6.2 | 3.4 | 2.8 | 8,157.5 | 7,612 | 5,898 | 3,103 | 81,521 | 512,360 | 92,888 | 12,608 |
| Warren | 55 | 2,773 | 49.1 | 5.7 | 14.8 | 11.3 | 0.5 | 17.7 | 894 | 32 | 41 | 1,155 | 7,630 | 41,383 | 3,084 |
| Washington | 41 | 3,397 | 48.1 | 5.1 | 7.1 | 10.7 | 0.9 | 5.8 | 490 | 31 | 25 | 747 | 4,970 | 41,489 | 3,116 |
| Watauga | 145 | 2,622 | 33.0 | 7.4 | 8.4 | 4.5 | 4.8 | 124.6 | 2,258 | 114 | 115 | 6,565 | 20,350 | 60,333 | 6,469 |
| Wayne | 362 | 2,941 | 53.6 | 5.6 | 8.1 | 6.3 | 0.9 | 253.1 | 2,057 | 1,213 | 4,882 | 8,118 | 51,600 | 47,597 | 4,002 |
| Wilkes | 187 | 2,723 | 63.1 | 5.0 | 5.2 | 9.1 | 0.4 | 63.0 | 920 | 172 | 148 | 3,437 | 27,730 | 48,193 | 4,004 |
| Wilson | 280 | 3,433 | 46.4 | 5.6 | 8.7 | 9.2 | 1.5 | 126.3 | 1,550 | 127 | 176 | 4,758 | 35,420 | 49,770 | 4,544 |
| Yadkin | 87 | 2,300 | 59.5 | 6.1 | 7.1 | 8.7 | 0.7 | 33.8 | 900 | 76 | 82 | 1,457 | 16,470 | 50,984 | 4,099 |
| Yancey | 50 | 2,806 | 51.9 | 11.8 | 6.6 | 8.3 | 0.6 | 15.7 | 890 | 47 | 39 | 843 | 7,590 | 46,031 | 3,448 |
| **NORTH DAKOTA** | X | X | X | X | X | X | X | X | X | 9,320 | 11,872 | 67,818 | 363,730 | 72,904 | 8,513 |
| Adams | 16 | 6,984 | 27.1 | 0.0 | 4.7 | 3.8 | 17.0 | 1.2 | 524 | 15 | 13 | 124 | 1,060 | 53,420 | 5,395 |
| Barnes | 79 | 7,375 | 29.3 | 3.3 | 2.6 | 1.7 | 14.9 | 40.8 | 3,819 | 65 | 60 | 1,062 | 5,100 | 63,433 | 6,008 |
| Benson | 42 | 6,137 | 45.8 | 0.8 | 1.8 | 3.7 | 28.7 | 7.9 | 1,135 | 102 | 42 | 1,379 | 2,370 | 49,519 | 4,014 |
| Billings | 24 | 25,963 | 13.0 | 3.9 | 5.0 | 0.0 | 56.0 | 0.6 | 682 | 37 | 6 | 125 | 460 | 69,180 | 8,315 |
| Bottineau | 53 | 8,130 | 32.1 | 0.0 | 3.7 | 1.9 | 20.7 | 24.9 | 3,813 | 68 | 37 | 603 | 3,050 | 67,611 | 7,465 |
| Bowman | 33 | 10,504 | 32.3 | 0.0 | 2.6 | 0.2 | 46.4 | 0.5 | 150 | 20 | 18 | 243 | 1,490 | 66,166 | 7,581 |
| Burke | 23 | 10,870 | 33.8 | 0.0 | 2.3 | 1.3 | 35.9 | 4.3 | 2,039 | 85 | 13 | 173 | 1,090 | 73,700 | 8,128 |
| Burleigh | 458 | 4,812 | 39.9 | 0.8 | 5.0 | 1.5 | 11.2 | 423.7 | 4,448 | 1,117 | 563 | 10,767 | 46,320 | 80,723 | 9,874 |
| Cass | 1,072 | 6,033 | 37.0 | 1.1 | 3.5 | 1.2 | 15.9 | 2,222.3 | 12,503 | 2,500 | 1,108 | 12,575 | 88,250 | 75,815 | 9,166 |
| Cavalier | 25 | 6,448 | 33.2 | 1.6 | 6.2 | 3.1 | 27.5 | 6.3 | 1,662 | 36 | 23 | 218 | 1,820 | 75,292 | 7,845 |
| Dickey | 41 | 8,417 | 26.6 | 3.9 | 3.7 | 0.0 | 20.2 | 34.1 | 7,034 | 25 | 28 | 301 | 2,380 | 58,027 | 5,492 |
| Divide | 31 | 13,490 | 44.8 | 4.9 | 2.8 | 1.6 | 24.5 | 11.8 | 5,153 | 34 | 13 | 158 | 1,080 | 59,546 | 6,431 |
| Dunn | 73 | 17,127 | 14.7 | 0.0 | 2.7 | 0.9 | 48.4 | 21.4 | 5,011 | 13 | 26 | 343 | 2,120 | 104,013 | 18,510 |
| Eddy | 14 | 6,061 | 49.5 | 2.7 | 4.0 | 0.0 | 21.7 | 2.0 | 873 | 21 | 14 | 166 | 1,110 | 48,929 | 4,483 |
| Emmons | 46 | 13,819 | 28.5 | 0.9 | 1.6 | 0.5 | 18.4 | 9.9 | 2,999 | 23 | 19 | 226 | 1,520 | 51,395 | 5,273 |
| Foster | 24 | 7,482 | 39.4 | 1.8 | 2.9 | 3.6 | 15.5 | 31.7 | 9,750 | 19 | 19 | 205 | 1,650 | 70,168 | 8,057 |
| Golden Valley | 28 | 15,430 | 21.5 | 5.9 | 3.5 | 2.7 | 8.6 | 0.0 | 14 | 9 | 10 | 152 | 810 | 59,330 | 5,774 |
| Grand Forks | 305 | 4,333 | 45.5 | 0.7 | 7.3 | 2.6 | 30.7 | 630.1 | 8,940 | 1,014 | 2,068 | 9,050 | 32,780 | 69,962 | 7,967 |
| Grant | 14 | 5,972 | 36.1 | 2.2 | 3.3 | 2.8 | 7.4 | 2.9 | 1,222 | 21 | 14 | 135 | 1,030 | 34,810 | 3,733 |
| Griggs | 14 | 6,064 | 51.7 | 7.1 | 1.9 | 0.0 | 7.4 | 5.9 | 2,638 | 20 | 13 | 164 | 1,090 | 59,281 | 5,294 |
| Hettinger | 19 | 7,577 | 55.2 | 0.0 | 0.4 | 0.0 | 24.9 | 5.8 | 2,352 | 21 | 14 | 191 | 1,150 | 56,647 | 6,141 |
| Kidder | 17 | 7,048 | 31.0 | 27.0 | 2.6 | 0.0 | 12.4 | 3.6 | 1,472 | 18 | 15 | 148 | 1,130 | 44,873 | 4,748 |
| LaMoure | 35 | 8,520 | 45.7 | 2.2 | 2.1 | 1.7 | 11.5 | 14.3 | 3,497 | 38 | 24 | 306 | 1,860 | 61,325 | 6,369 |
| Logan | 17 | 8,982 | 42.1 | 0.0 | 2.0 | 0.0 | 11.6 | 3.7 | 1,935 | 16 | 11 | 128 | 850 | 34,658 | 3,846 |
| McHenry | 41 | 6,924 | 38.4 | 0.0 | 1.6 | 1.2 | 14.7 | 6.5 | 1,108 | 41 | 35 | 311 | 2,650 | 56,215 | 5,383 |
| McIntosh | 15 | 5,760 | 43.7 | 1.6 | 2.8 | 1.8 | 11.1 | 2.1 | 787 | 16 | 15 | 179 | 1,270 | 45,113 | 4,346 |
| McKenzie | 281 | 22,154 | 12.0 | 0.5 | 3.3 | 0.6 | 31.8 | 274.4 | 21,630 | 77 | 91 | 2,506 | 5,610 | 128,872 | 21,906 |
| McLean | 74 | 7,657 | 37.7 | 1.1 | 5.0 | 4.7 | 19.6 | 59.0 | 6,122 | 118 | 57 | 751 | 4,550 | 67,252 | 7,155 |
| Mercer | 50 | 5,872 | 38.2 | 0.5 | 6.9 | 1.4 | 23.1 | 43.4 | 5,136 | 39 | 49 | 537 | 4,000 | 75,926 | 8,401 |
| Morton | 160 | 5,175 | 45.4 | 3.0 | 5.4 | 2.9 | 10.3 | 200.3 | 6,473 | 111 | 187 | 1,698 | 15,610 | 69,132 | 7,466 |
| Mountrail | 148 | 14,378 | 31.1 | 1.2 | 3.4 | 1.8 | 40.1 | 50.6 | 4,928 | 41 | 61 | 800 | 4,550 | 92,635 | 12,866 |
| Nelson | 23 | 7,909 | 35.8 | 6.7 | 2.2 | 0.0 | 28.3 | 10.6 | 3,663 | 19 | 17 | 216 | 1,510 | 55,423 | 5,405 |
| Oliver | 9 | 4,601 | 41.8 | 0.0 | 1.1 | 0.0 | 1.0 | 7.2 | 3,747 | 6 | 12 | 112 | 870 | 61,720 | 5,385 |
| Pembina | 50 | 7,228 | 48.5 | 0.6 | 5.2 | 2.6 | 25.2 | 22.8 | 3,277 | 218 | 72 | 521 | 3,380 | 68,822 | 6,975 |
| Pierce | 23 | 5,635 | 41.2 | 1.5 | 2.9 | 2.7 | 18.5 | 13.9 | 3,387 | 18 | 23 | 236 | 1,930 | 57,737 | 5,193 |
| Ramsey | 74 | 6,368 | 44.3 | 5.3 | 5.0 | 4.5 | 15.2 | 19.5 | 1,685 | 125 | 68 | 1,411 | 5,520 | 60,130 | 6,025 |
| Ransom | 29 | 5,512 | 47.9 | 1.9 | 2.9 | 1.5 | 13.4 | 19.3 | 3,640 | 37 | 31 | 477 | 2,730 | 60,333 | 5,328 |
| Renville | 20 | 8,297 | 49.8 | 1.1 | 2.2 | 1.7 | 16.9 | 4.2 | 1,704 | 22 | 14 | 207 | 1,180 | 63,547 | 5,872 |
| Richland | 106 | 6,479 | 40.6 | 3.8 | 4.3 | 1.5 | 8.9 | 87.4 | 5,368 | 64 | 97 | 1,850 | 7,790 | 66,665 | 6,604 |
| Rolette | 81 | 5,529 | 61.7 | 1.4 | 2.3 | 2.9 | 10.0 | 20.7 | 1,414 | 926 | 86 | 1,907 | 4,900 | 46,913 | 3,463 |
| Sargent | 33 | 8,631 | 32.9 | 0.0 | 3.7 | 0.0 | 14.7 | 20.2 | 5,243 | 30 | 24 | 248 | 2,020 | 63,273 | 6,111 |
| Sheridan | 6 | 4,599 | 39.8 | 1.7 | 3.4 | 0.0 | 13.0 | 1.6 | 1,187 | 10 | 8 | 107 | 590 | 47,203 | 3,856 |
| Sioux | 15 | 3,484 | 74.7 | 0.1 | 0.2 | 0.0 | 15.2 | 3.7 | 845 | 232 | 26 | 1,186 | 1,170 | 32,208 | 2,301 |
| Slope | 8 | 9,735 | 7.4 | 0.1 | 4.2 | 2.7 | 39.2 | 0.5 | 647 | 1 | 5 | 24 | 310 | 56,352 | 6,184 |
| Stark | 216 | 7,140 | 47.5 | 1.8 | 3.8 | 1.6 | 5.8 | 217.8 | 7,190 | 208 | 189 | 2,278 | 15,660 | 79,771 | 9,451 |
| Steele | 15 | 7,947 | 30.5 | 0.8 | 3.4 | 3.3 | 19.5 | 4.0 | 2,083 | 10 | 12 | 104 | 960 | 69,879 | 6,790 |
| Stutsman | 104 | 4,907 | 43.1 | 3.0 | 5.4 | 2.7 | 14.4 | 87.2 | 4,125 | 148 | 116 | 1,876 | 9,910 | 62,361 | 6,350 |
| Towner | 15 | 6,475 | 27.1 | 4.5 | 4.1 | 2.7 | 20.4 | 4.3 | 1,917 | 16 | 13 | 118 | 1,110 | 55,122 | 4,697 |
| Traill | 1,947 | 243,457 | 1.2 | 0.0 | 0.1 | 0.1 | 0.6 | 37.5 | 4,690 | 37 | 47 | 875 | 3,750 | 70,249 | 7,089 |
| Walsh | 76 | 7,000 | 48.3 | 1.7 | 4.3 | 3.3 | 16.1 | 41.7 | 3,865 | 52 | 63 | 1,042 | 5,140 | 61,367 | 5,775 |

1. Based on the resident population estimated as of July 1 of the year shown.

# Table B. States and Counties — Land Area and Population

| State / county code | CBSA code[1] | County Type code[2] | STATE County | Land area[3] (sq. mi) | Population, 2020 Total persons 2019 | Rank | Per square mile | White | Black | American Indian, Alaska Native | Asian and Pacific Islander | Percent Hispanic or Latino[4] | Under 5 years | 5 to 14 years | 15 to 24 years | 25 to 34 years | 35 to 44 years | 45 to 54 years |
|---|---|---|---|---|---|---|---|---|---|---|---|---|---|---|---|---|---|---|
| | | | | 1 | 2 | 3 | 4 | 5 | 6 | 7 | 8 | 9 | 10 | 11 | 12 | 13 | 14 | 15 |
| | | | **NORTH DAKOTA— Cont'd** | | | | | | | | | | | | | | | |
| 38101 | 33500 | 5 | Ward | 2,013.0 | 68,466 | 785 | 34.0 | 84.2 | 6.0 | 3.2 | 2.9 | 6.7 | 7.4 | 12.9 | 16.5 | 17.4 | 12.6 | 9.3 |
| 38103 | | 9 | Wells | 1,270.5 | 3,709 | 2,920 | 2.9 | 97.2 | 0.7 | 1.6 | 0.5 | 1.3 | 5.5 | 11.9 | 9.0 | 8.6 | 8.9 | 10.8 |
| 38105 | 48780 | 7 | Williams | 2,077.6 | 38,700 | 1,208 | 18.6 | 79.9 | 5.9 | 6.2 | 1.8 | 9.9 | 10.1 | 16.1 | 12.1 | 19.4 | 13.2 | 9.9 |
| 39000 | | 0 | OHIO | 40,858.8 | 11,693,217 | X | 286.2 | 80.1 | 14.2 | 0.7 | 3.2 | 4.2 | 5.9 | 12.3 | 12.8 | 13.4 | 12.1 | 12.1 |
| 39001 | | 6 | Adams | 583.9 | 27,531 | 1,511 | 47.2 | 97.5 | 1.0 | 1.2 | 0.6 | 1.1 | 6.4 | 13.4 | 11.5 | 11.4 | 11.1 | 13.2 |
| 39003 | 30620 | 5 | Allen | 402.5 | 101,980 | 596 | 253.4 | 82.9 | 14.7 | 0.7 | 1.3 | 3.5 | 6.1 | 12.8 | 13.8 | 12.4 | 11.6 | 11.4 |
| 39005 | 11740 | 4 | Ashland | 423.0 | 53,362 | 950 | 126.2 | 96.5 | 1.5 | 0.6 | 1.1 | 1.6 | 5.7 | 12.5 | 14.1 | 11.4 | 11.2 | 11.7 |
| 39007 | 11780 | 4 | Ashtabula | 702.1 | 96,513 | 624 | 137.5 | 91.2 | 4.8 | 0.8 | 0.8 | 4.6 | 5.7 | 12.3 | 12.3 | 11.0 | 11.0 | 12.7 |
| 39009 | 11900 | 4 | Athens | 503.6 | 65,481 | 815 | 130.0 | 91.9 | 3.8 | 1.1 | 3.7 | 1.9 | 3.7 | 7.9 | 30.6 | 12.0 | 10.4 | 9.9 |
| 39011 | 47540 | 4 | Auglaize | 401.4 | 45,680 | 1,061 | 113.8 | 96.7 | 1.2 | 0.5 | 0.9 | 1.9 | 6.3 | 13.5 | 11.9 | 11.9 | 11.4 | 11.8 |
| 39013 | 48540 | 3 | Belmont | 532.1 | 65,932 | 808 | 123.9 | 94.1 | 5.1 | 0.6 | 0.8 | 1.1 | 4.8 | 10.7 | 10.7 | 12.6 | 11.6 | 12.5 |
| 39015 | 17140 | 1 | Brown | 489.5 | 43,414 | 1,107 | 88.7 | 97.3 | 1.5 | 0.8 | 0.6 | 1.2 | 5.7 | 12.7 | 11.5 | 11.5 | 11.7 | 12.9 |
| 39017 | 17140 | 1 | Butler | 466.5 | 385,648 | 183 | 826.7 | 81.4 | 10.6 | 0.6 | 4.8 | 5.1 | 5.9 | 13.2 | 16.3 | 11.9 | 12.1 | 12.1 |
| 39019 | 15940 | 2 | Carroll | 394.6 | 26,897 | 1,531 | 68.2 | 97.3 | 1.3 | 0.8 | 0.5 | 1.4 | 5.3 | 11.6 | 11.0 | 10.3 | 10.7 | 12.6 |
| 39021 | 46500 | 6 | Champaign | 429.0 | 38,960 | 1,201 | 90.8 | 95.2 | 3.4 | 1.0 | 0.9 | 1.6 | 5.5 | 12.2 | 12.3 | 12.2 | 11.2 | 13.1 |
| 39023 | 44220 | 3 | Clark | 396.9 | 133,593 | 484 | 336.6 | 86.4 | 10.8 | 0.9 | 0.9 | 3.7 | 5.9 | 12.5 | 12.6 | 11.9 | 11.0 | 12.1 |
| 39025 | 17140 | 1 | Clermont | 452.6 | 207,449 | 328 | 458.3 | 94.4 | 2.5 | 0.7 | 1.9 | 2.2 | 5.6 | 12.7 | 11.7 | 12.7 | 12.4 | 12.8 |
| 39027 | 48940 | 6 | Clinton | 408.7 | 41,921 | 1,141 | 102.6 | 95.0 | 3.5 | 0.8 | 0.9 | 1.9 | 5.8 | 12.6 | 13.6 | 11.7 | 11.7 | 12.2 |
| 39029 | 41400 | 4 | Columbiana | 531.9 | 101,118 | 599 | 190.1 | 95.1 | 3.3 | 0.7 | 0.6 | 1.2 | 5.1 | 11.4 | 10.7 | 11.6 | 11.7 | 12.7 |
| 39031 | 18740 | 6 | Coshocton | 564.0 | 36,449 | 1,266 | 64.6 | 96.9 | 2.2 | 0.7 | 0.6 | 1.2 | 6.5 | 13.2 | 11.4 | 11.6 | 11.2 | 12.0 |
| 39033 | 15340 | 4 | Crawford | 401.8 | 41,338 | 1,150 | 102.9 | 96.1 | 1.8 | 0.6 | 0.9 | 1.8 | 5.7 | 12.1 | 11.4 | 11.4 | 11.1 | 12.7 |
| 39035 | 17460 | 1 | Cuyahoga | 457.2 | 1,227,883 | 35 | 2,685.7 | 60.1 | 31.1 | 0.7 | 4.0 | 6.5 | 5.6 | 11.3 | 12.0 | 14.5 | 11.8 | 11.7 |
| 39037 | 24820 | 6 | Darke | 598.1 | 51,205 | 972 | 85.6 | 96.8 | 1.4 | 0.6 | 0.8 | 1.7 | 6.2 | 13.2 | 11.8 | 11.0 | 11.1 | 12.0 |
| 39039 | 19580 | 4 | Defiance | 411.5 | 37,778 | 1,228 | 91.8 | 86.9 | 2.5 | 0.6 | 0.7 | 10.6 | 5.6 | 12.8 | 13.1 | 11.5 | 12.1 | 11.7 |
| 39041 | 18140 | 1 | Delaware | 443.1 | 213,554 | 318 | 482.0 | 84.7 | 4.9 | 0.5 | 9.1 | 2.9 | 5.8 | 14.8 | 12.7 | 9.7 | 15.1 | 15.0 |
| 39043 | 41780 | 4 | Erie | 251.3 | 73,719 | 747 | 293.4 | 85.5 | 11.0 | 0.8 | 1.1 | 4.7 | 5.2 | 11.1 | 11.3 | 11.7 | 10.6 | 11.9 |
| 39045 | 18140 | 1 | Fairfield | 504.4 | 159,709 | 421 | 316.6 | 86.0 | 10.2 | 0.8 | 2.8 | 2.5 | 5.9 | 13.7 | 12.4 | 12.1 | 13.0 | 13.3 |
| 39047 | 47920 | 6 | Fayette | 406.4 | 28,579 | 1,477 | 70.3 | 93.9 | 3.8 | 0.7 | 1.4 | 2.4 | 5.9 | 13.3 | 11.9 | 12.0 | 11.8 | 12.9 |
| 39049 | 18140 | 1 | Franklin | 532.4 | 1,324,624 | 31 | 2,488.0 | 64.2 | 25.6 | 0.9 | 6.8 | 6.0 | 6.8 | 12.8 | 13.1 | 18.2 | 13.7 | 11.5 |
| 39051 | 45780 | 2 | Fulton | 405.4 | 41,889 | 1,142 | 103.3 | 89.6 | 1.2 | 0.6 | 0.7 | 9.1 | 5.9 | 13.3 | 11.9 | 11.4 | 11.8 | 12.4 |
| 39053 | 38580 | 6 | Gallia | 466.5 | 29,802 | 1,441 | 63.9 | 94.9 | 3.3 | 1.0 | 1.2 | 1.4 | 6.1 | 12.9 | 12.0 | 11.6 | 11.4 | 12.0 |
| 39055 | 17460 | 1 | Geauga | 400.3 | 93,271 | 639 | 233.0 | 96.2 | 1.7 | 0.4 | 1.0 | 1.7 | 5.3 | 12.6 | 12.6 | 9.0 | 10.4 | 13.1 |
| 39057 | 19430 | 2 | Greene | 413.6 | 170,122 | 392 | 411.3 | 86.2 | 8.7 | 1.0 | 4.3 | 3.2 | 5.5 | 11.5 | 15.3 | 13.4 | 11.7 | 11.3 |
| 39059 | 15740 | 6 | Guernsey | 522.3 | 38,779 | 1,204 | 74.2 | 96.4 | 2.8 | 1.0 | 0.8 | 1.2 | 5.7 | 12.2 | 11.4 | 12.2 | 11.0 | 12.5 |
| 39061 | 17140 | 1 | Hamilton | 405.4 | 817,985 | 77 | 2,017.7 | 66.7 | 27.9 | 0.7 | 3.7 | 3.7 | 6.4 | 12.7 | 13.0 | 15.3 | 12.2 | 11.2 |
| 39063 | 22300 | 4 | Hancock | 531.3 | 75,407 | 738 | 141.9 | 89.8 | 2.8 | 0.6 | 2.5 | 6.0 | 5.8 | 12.3 | 12.5 | 13.5 | 11.8 | 12.1 |
| 39065 | | 6 | Hardin | 470.3 | 31,469 | 1,393 | 66.9 | 95.8 | 1.8 | 0.8 | 1.2 | 2.1 | 6.1 | 13.0 | 19.2 | 10.4 | 10.7 | 11.5 |
| 39067 | | 6 | Harrison | 402.3 | 15,014 | 2,093 | 37.3 | 96.2 | 3.1 | 0.8 | 0.4 | 1.3 | 4.8 | 11.6 | 10.7 | 10.6 | 10.9 | 12.4 |
| 39069 | | 6 | Henry | 416.0 | 26,904 | 1,530 | 64.7 | 90.7 | 1.1 | 0.7 | 0.7 | 7.8 | 5.9 | 13.1 | 11.7 | 11.5 | 11.8 | 12.2 |
| 39071 | | 6 | Highland | 553.1 | 43,304 | 1,109 | 78.3 | 96.5 | 2.5 | 0.8 | 0.7 | 1.3 | 6.4 | 13.0 | 11.8 | 11.6 | 11.6 | 12.8 |
| 39073 | 18140 | 1 | Hocking | 421.3 | 28,095 | 1,494 | 66.7 | 97.4 | 1.0 | 1.0 | 0.7 | 1.1 | 5.5 | 12.3 | 11.2 | 11.3 | 11.4 | 13.2 |
| 39075 | | 7 | Holmes | 422.6 | 44,004 | 1,095 | 104.1 | 98.2 | 0.6 | 0.3 | 0.4 | 1.0 | 8.3 | 16.7 | 15.4 | 13.0 | 10.9 | 10.4 |
| 39077 | 35940 | 4 | Huron | 492.2 | 57,979 | 896 | 117.8 | 91.3 | 0.7 | 0.7 | 0.6 | 6.9 | 6.2 | 13.3 | 12.1 | 12.1 | 11.6 | 12.4 |
| 39079 | 27160 | 7 | Jackson | 420.3 | 32,493 | 1,367 | 77.3 | 97.2 | 1.3 | 1.2 | 0.6 | 1.2 | 6.3 | 13.3 | 11.7 | 12.3 | 12.3 | 12.8 |
| 39081 | 48260 | 3 | Jefferson | 408.1 | 64,939 | 819 | 159.1 | 92.0 | 7.0 | 0.7 | 0.9 | 1.7 | 5.1 | 10.7 | 12.6 | 11.5 | 10.3 | 12.1 |
| 39083 | 34540 | 4 | Knox | 525.5 | 62,423 | 852 | 118.8 | 96.4 | 1.6 | 0.6 | 1.0 | 1.7 | 6.1 | 12.7 | 15.3 | 11.0 | 11.0 | 11.4 |
| 39085 | 17460 | 2 | Lake | 229.3 | 229,569 | 299 | 1,001.2 | 88.7 | 5.8 | 0.5 | 2.0 | 4.8 | 4.9 | 10.9 | 11.3 | 12.3 | 11.5 | 12.7 |
| 39087 | 26580 | 2 | Lawrence | 453.4 | 59,091 | 878 | 130.3 | 96.1 | 3.1 | 0.7 | 0.7 | 1.0 | 5.3 | 12.3 | 11.2 | 12.1 | 12.1 | 13.4 |
| 39089 | 18140 | 1 | Licking | 682.4 | 178,100 | 373 | 261.0 | 90.5 | 5.5 | 0.8 | 3.3 | 2.2 | 5.9 | 13.0 | 12.6 | 12.0 | 12.4 | 12.9 |
| 39091 | 13340 | 6 | Logan | 458.5 | 45,326 | 1,070 | 98.9 | 94.9 | 3.3 | 0.8 | 1.3 | 2.0 | 5.9 | 12.8 | 11.7 | 11.5 | 11.5 | 12.8 |
| 39093 | 17460 | 2 | Lorain | 490.5 | 312,172 | 226 | 636.4 | 80.1 | 9.6 | 0.8 | 1.8 | 10.5 | 5.5 | 12.3 | 12.2 | 11.6 | 12.0 | 12.8 |
| 39095 | 45780 | 2 | Lucas | 339.7 | 428,294 | 166 | 1,260.8 | 70.5 | 21.8 | 0.8 | 2.3 | 7.7 | 6.3 | 12.8 | 12.4 | 14.4 | 11.6 | 11.9 |
| 39097 | 18140 | 1 | Madison | 465.8 | 44,559 | 1,085 | 95.7 | 89.6 | 7.0 | 0.8 | 1.7 | 2.6 | 5.1 | 11.5 | 11.4 | 13.7 | 13.9 | 14.2 |
| 39099 | 49660 | 2 | Mahoning | 411.5 | 226,075 | 306 | 549.4 | 77.0 | 16.4 | 0.7 | 1.3 | 6.8 | 5.4 | 11.0 | 11.8 | 12.3 | 11.2 | 11.8 |
| 39101 | 32020 | 4 | Marion | 403.8 | 64,820 | 821 | 160.5 | 89.9 | 7.5 | 0.7 | 0.9 | 3.0 | 5.8 | 12.1 | 11.4 | 13.5 | 12.3 | 12.8 |
| 39103 | 17460 | 1 | Medina | 421.5 | 180,912 | 367 | 429.2 | 94.5 | 2.1 | 0.6 | 1.7 | 2.5 | 5.1 | 12.5 | 11.7 | 10.9 | 12.5 | 13.7 |
| 39105 | | 6 | Meigs | 430.1 | 22,678 | 1,698 | 52.7 | 97.8 | 1.4 | 0.9 | 0.5 | 0.8 | 5.1 | 11.4 | 11.1 | 10.7 | 12.2 | 13.3 |
| 39107 | 16380 | 7 | Mercer | 462.4 | 41,274 | 1,151 | 89.3 | 95.9 | 1.0 | 0.6 | 1.8 | 1.9 | 7.8 | 14.1 | 11.9 | 11.7 | 10.9 | 10.8 |
| 39109 | 19430 | 2 | Miami | 406.5 | 107,516 | 568 | 264.5 | 93.5 | 3.9 | 0.7 | 2.2 | 2.0 | 5.8 | 13.0 | 11.2 | 11.9 | 12.2 | 12.7 |
| 39111 | | 8 | Monroe | 455.7 | 13,586 | 2,187 | 29.8 | 98.2 | 1.1 | 1.0 | 0.5 | 0.8 | 4.8 | 11.7 | 10.3 | 10.2 | 11.0 | 12.5 |
| 39113 | 19430 | 2 | Montgomery | 461.4 | 531,610 | 132 | 1,152.2 | 72.4 | 23.2 | 0.9 | 3.2 | 3.4 | 6.1 | 12.1 | 12.8 | 13.9 | 11.7 | 11.5 |
| 39115 | | 6 | Morgan | 416.4 | 14,305 | 2,137 | 34.4 | 95.3 | 5.3 | 1.7 | 0.9 | 1.0 | 4.9 | 11.5 | 11.0 | 10.5 | 11.4 | 12.6 |
| 39117 | 18140 | 1 | Morrow | 406.1 | 35,411 | 1,292 | 87.2 | 96.9 | 1.3 | 0.9 | 0.7 | 1.6 | 5.5 | 12.3 | 11.6 | 11.2 | 12.0 | 13.7 |
| 39119 | 49780 | 4 | Muskingum | 664.5 | 86,020 | 672 | 129.5 | 94.0 | 6.0 | 1.0 | 1.0 | 1.3 | 5.9 | 12.7 | 12.4 | 12.5 | 11.6 | 12.5 |
| 39121 | | 7 | Noble | 398.0 | 14,364 | 2,133 | 36.1 | 95.9 | 3.2 | 0.8 | 0.5 | 0.8 | 5.0 | 10.6 | 8.7 | 9.1 | 8.3 | 10.5 |

1. CBSA = Core Based Statistical Area. See Appendix A for explanation. See Appendix B for list of metropolitan areas with component counties. 2. County type code from the Economic Research Service of USDA Rural-Urban Continuum Codes. See Appendix A for definition. 3. Dry land or land partially or temporarily covered by water. 4. May be of any race.

# Table B. States and Counties — Population and Households

| STATE County | 55 to 64 years (16) | 65 to 74 years (17) | 75 years and over (18) | Percent female (19) | 2000 (20) | 2010 (21) | 2000-2010 (22) | 2010-2020 (23) | Births (24) | Deaths (25) | Net Migration (26) | Number (27) | Persons per household (28) | Family households (29) | Female family householder[1] (30) | One person (31) |
|---|---|---|---|---|---|---|---|---|---|---|---|---|---|---|---|---|
| **NORTH DAKOTA—Cont'd** | | | | | | | | | | | | | | | | |
| Ward | 10.3 | 7.6 | 6.0 | 47.7 | 58,795 | 61,675 | 4.9 | 11.0 | 11,808 | 4,982 | -222 | 28,277 | 2.36 | 59.5 | 6.5 | 31.7 |
| Wells | 17.1 | 14.0 | 14.3 | 49.7 | 5,102 | 4,207 | -17.5 | -11.8 | 425 | 689 | -234 | 1,943 | 2.0 | 58.3 | 6.3 | 37.5 |
| Williams | 10.1 | 5.4 | 3.7 | 46.2 | 19,761 | 22,399 | 13.3 | 72.8 | 6,442 | 2,370 | 11,841 | 14,320 | 2.42 | 60.5 | 6.0 | 30.6 |
| **OHIO** | 13.6 | 10.5 | 7.4 | 51.0 | 11,353,140 | 11,536,763 | 1.6 | 1.4 | 1,408,338 | 1,204,444 | -43,576 | 4,676,358 | 2.43 | 63.2 | 12.5 | 30.5 |
| Adams | 14.4 | 11.2 | 7.5 | 50.4 | 27,330 | 28,532 | 4.4 | -3.5 | 3,484 | 3,534 | -942 | 10,673 | 2.57 | 69.4 | 13.7 | 25.9 |
| Allen | 13.3 | 10.8 | 7.7 | 49.5 | 108,473 | 106,295 | -2.0 | -4.1 | 12,966 | 11,420 | -5,865 | 40,650 | 2.44 | 64.7 | 14.1 | 30.1 |
| Ashland | 13.8 | 11.1 | 8.6 | 50.9 | 52,523 | 53,136 | 1.2 | 0.4 | 6,248 | 5,811 | -202 | 20,417 | 2.51 | 65.9 | 9.7 | 27.5 |
| Ashtabula | 14.8 | 11.9 | 8.3 | 49.7 | 102,728 | 101,497 | -1.2 | -4.9 | 11,394 | 12,204 | -4,174 | 37,832 | 2.49 | 63.5 | 12.1 | 30.5 |
| Athens | 11.3 | 8.9 | 5.2 | 50.3 | 62,223 | 64,763 | 4.1 | 1.1 | 5,382 | 5,037 | 263 | 22,557 | 2.49 | 54.9 | 9.3 | 31.2 |
| Auglaize | 14.0 | 10.9 | 8.3 | 50.2 | 46,611 | 45,933 | -1.5 | -0.6 | 5,630 | 5,007 | -867 | 18,888 | 2.39 | 69.3 | 9.2 | 26.1 |
| Belmont | 15.2 | 12.9 | 8.8 | 49.1 | 70,226 | 70,402 | 0.3 | -6.3 | 6,926 | 9,209 | -2,120 | 25,919 | 2.47 | 63.0 | 10.2 | 33.0 |
| Brown | 15.0 | 11.1 | 7.9 | 50.4 | 42,285 | 44,831 | 6.0 | -3.2 | 5,101 | 5,202 | -1,311 | 17,829 | 2.41 | 69.2 | 10.6 | 26.0 |
| Butler | 12.9 | 9.4 | 6.2 | 51.0 | 332,807 | 368,138 | 10.6 | 4.8 | 46,236 | 33,608 | 5,079 | 139,113 | 2.65 | 68.4 | 11.9 | 25.3 |
| Carroll | 16.2 | 13.1 | 9.3 | 49.6 | 28,836 | 28,847 | 0.0 | -6.8 | 2,836 | 3,266 | -1,521 | 11,298 | 2.38 | 67.5 | 7.8 | 26.9 |
| Champaign | 14.5 | 10.8 | 8.1 | 50.1 | 38,890 | 40,101 | 3.1 | -2.8 | 4,130 | 4,110 | -1,164 | 15,159 | 2.51 | 68.4 | 10.6 | 25.8 |
| Clark | 13.7 | 11.7 | 8.5 | 51.4 | 144,742 | 138,339 | -4.4 | -3.4 | 16,273 | 17,744 | -3,203 | 54,696 | 2.41 | 64.8 | 14.3 | 28.9 |
| Clermont | 14.4 | 10.7 | 6.9 | 50.7 | 177,977 | 197,370 | 10.9 | 5.1 | 23,868 | 17,820 | 4,197 | 78,009 | 2.59 | 69.8 | 10.0 | 24.8 |
| Clinton | 14.2 | 10.7 | 7.4 | 50.6 | 40,543 | 42,046 | 3.7 | -0.3 | 5,067 | 4,597 | -585 | 16,528 | 2.46 | 67.2 | 12.5 | 27.6 |
| Columbiana | 15.2 | 12.8 | 8.8 | 49.4 | 112,075 | 107,850 | -3.8 | -6.2 | 10,888 | 12,831 | -4,765 | 41,537 | 2.39 | 65.8 | 11.6 | 29.5 |
| Coshocton | 14.0 | 11.7 | 8.5 | 50.4 | 36,655 | 36,900 | 0.7 | -1.2 | 4,656 | 4,206 | -885 | 14,476 | 2.5 | 68.2 | 10.5 | 27.7 |
| Crawford | 14.0 | 12.1 | 9.5 | 51.0 | 46,966 | 43,783 | -6.8 | -5.6 | 4,838 | 5,621 | -1,656 | 17,782 | 2.32 | 63.1 | 11.7 | 31.1 |
| Cuyahoga | 14.0 | 10.9 | 8.2 | 52.3 | 1,393,978 | 1,280,119 | -8.2 | -4.1 | 149,786 | 141,721 | -60,134 | 540,965 | 2.25 | 55.8 | 15.3 | 37.9 |
| Darke | 14.3 | 11.2 | 9.1 | 50.3 | 53,309 | 52,968 | -0.6 | -3.3 | 6,372 | 6,170 | -1,954 | 20,861 | 2.44 | 68.7 | 8.5 | 25.9 |
| Defiance | 13.6 | 11.5 | 8.2 | 50.2 | 39,500 | 39,026 | -1.2 | -3.2 | 4,447 | 4,076 | -1,627 | 15,297 | 2.46 | 69.8 | 11.5 | 24.1 |
| Delaware | 12.1 | 9.1 | 5.6 | 50.3 | 109,989 | 174,169 | 58.4 | 22.6 | 22,018 | 11,110 | 28,414 | 69,985 | 2.84 | 76.8 | 7.6 | 19.4 |
| Erie | 15.2 | 13.3 | 9.8 | 51.3 | 79,551 | 77,063 | -3.1 | -4.3 | 8,037 | 9,666 | -1,671 | 31,183 | 2.36 | 64.0 | 12.6 | 30.1 |
| Fairfield | 13.2 | 9.8 | 6.7 | 50.4 | 122,759 | 146,180 | 19.1 | 9.3 | 17,424 | 13,205 | 9,423 | 56,339 | 2.69 | 71.6 | 10.5 | 23.4 |
| Fayette | 13.7 | 10.9 | 7.6 | 51.0 | 28,433 | 29,034 | 2.1 | -1.6 | 3,515 | 3,701 | -252 | 11,773 | 2.4 | 68.5 | 14.9 | 26.4 |
| Franklin | 11.2 | 7.8 | 4.8 | 51.2 | 1,068,978 | 1,163,473 | 8.8 | 13.9 | 189,785 | 96,659 | 68,162 | 511,447 | 2.47 | 58.7 | 13.9 | 32.1 |
| Fulton | 14.2 | 11.2 | 7.9 | 50.4 | 42,084 | 42,698 | 1.5 | -1.9 | 4,964 | 4,332 | -1,438 | 16,506 | 2.53 | 71.1 | 9.5 | 23.9 |
| Gallia | 14.4 | 11.1 | 8.4 | 50.5 | 31,069 | 30,937 | -0.4 | -3.7 | 3,811 | 3,869 | -1,089 | 11,588 | 2.52 | 65.3 | 10.2 | 29.3 |
| Geauga | 15.4 | 12.3 | 9.3 | 50.3 | 90,895 | 93,397 | 2.8 | -0.1 | 9,430 | 8,478 | -1,046 | 35,298 | 2.63 | 73.4 | 5.8 | 21.5 |
| Greene | 13.4 | 10.6 | 7.4 | 50.6 | 147,886 | 161,578 | 9.3 | 5.3 | 18,155 | 14,799 | 5,225 | 65,604 | 2.4 | 65.1 | 10.1 | 27.9 |
| Guernsey | 14.5 | 12.1 | 8.4 | 50.1 | 40,792 | 40,092 | -1.7 | -3.3 | 4,644 | 4,798 | -1,139 | 16,220 | 2.38 | 63.9 | 12.9 | 29.1 |
| Hamilton | 13.0 | 9.6 | 6.6 | 51.7 | 845,303 | 802,371 | -5.1 | 1.9 | 111,144 | 81,189 | -13,967 | 341,873 | 2.33 | 57.7 | 14.5 | 34.9 |
| Hancock | 13.7 | 10.6 | 7.7 | 50.4 | 71,295 | 74,790 | 4.9 | 0.8 | 9,234 | 7,637 | -955 | 31,937 | 2.31 | 62.4 | 8.4 | 29.7 |
| Hardin | 12.5 | 9.7 | 6.9 | 50.2 | 31,945 | 32,063 | 0.4 | -1.9 | 3,883 | 3,289 | -1,186 | 11,678 | 2.49 | 67.2 | 9.7 | 26.3 |
| Harrison | 16.4 | 13.3 | 9.2 | 50.4 | 15,856 | 15,852 | 0.0 | -5.3 | 1,578 | 2,081 | -322 | 6,223 | 2.4 | 67.7 | 7.0 | 28.6 |
| Henry | 14.3 | 11.0 | 8.5 | 50.6 | 29,210 | 28,215 | -3.4 | -4.6 | 3,249 | 2,904 | -1,666 | 10,992 | 2.43 | 69.7 | 11.6 | 25.6 |
| Highland | 13.8 | 11.1 | 8.0 | 50.9 | 40,875 | 43,605 | 6.7 | -0.7 | 5,566 | 5,006 | -850 | 16,772 | 2.53 | 66.9 | 13.1 | 27.3 |
| Hocking | 15.2 | 11.8 | 8.2 | 50.5 | 28,241 | 29,373 | 4.0 | -4.4 | 3,169 | 3,205 | -1,245 | 11,286 | 2.46 | 70.3 | 10.6 | 24.7 |
| Holmes | 10.9 | 8.2 | 6.2 | 49.9 | 38,943 | 42,364 | 8.8 | 3.9 | 7,861 | 3,227 | -2,988 | 12,342 | 3.48 | 79.4 | 6.3 | 17.3 |
| Huron | 14.1 | 10.9 | 7.3 | 50.4 | 59,487 | 59,625 | 0.2 | -2.8 | 7,449 | 6,132 | -2,980 | 22,935 | 2.5 | 66.7 | 11.8 | 26.9 |
| Jackson | 13.3 | 11.1 | 6.9 | 50.8 | 32,641 | 33,219 | 1.8 | -2.2 | 4,244 | 4,258 | -707 | 12,780 | 2.51 | 72.1 | 13.5 | 24.9 |
| Jefferson | 15.2 | 13.1 | 9.3 | 51.3 | 73,894 | 69,718 | -5.7 | -6.9 | 6,773 | 9,772 | -1,721 | 27,428 | 2.33 | 62.6 | 11.8 | 31.7 |
| Knox | 13.7 | 11.1 | 7.8 | 51.0 | 54,500 | 60,926 | 11.8 | 2.5 | 7,523 | 6,335 | 333 | 23,248 | 2.49 | 68.8 | 8.7 | 25.1 |
| Lake | 15.3 | 12.1 | 9.0 | 51.2 | 227,511 | 230,051 | 1.1 | -0.2 | 23,012 | 25,278 | 1,973 | 96,361 | 2.35 | 63.8 | 10.6 | 30.0 |
| Lawrence | 14.0 | 11.4 | 8.2 | 51.3 | 62,319 | 62,451 | 0.2 | -5.4 | 6,889 | 8,052 | -2,171 | 23,221 | 2.56 | 65.7 | 13.3 | 29.5 |
| Licking | 14.0 | 10.3 | 6.9 | 50.8 | 145,491 | 166,504 | 14.4 | 7.0 | 20,239 | 16,500 | 8,009 | 64,035 | 2.65 | 70.4 | 11.9 | 23.5 |
| Logan | 14.6 | 11.7 | 7.5 | 50.4 | 46,005 | 45,848 | -0.3 | -1.1 | 5,582 | 5,064 | -1,017 | 18,677 | 2.4 | 68.0 | 9.0 | 26.7 |
| Lorain | 14.2 | 11.4 | 7.9 | 50.7 | 284,664 | 301,374 | 5.9 | 3.6 | 34,052 | 31,388 | 8,383 | 120,281 | 2.48 | 66.1 | 12.9 | 28.6 |
| Lucas | 13.5 | 10.3 | 6.9 | 51.7 | 455,054 | 441,813 | -2.9 | -3.1 | 56,915 | 46,166 | -24,407 | 179,930 | 2.34 | 59.5 | 15.4 | 33.5 |
| Madison | 13.9 | 9.6 | 6.8 | 46.0 | 40,213 | 43,438 | 8.0 | 2.6 | 4,382 | 4,240 | 966 | 15,013 | 2.58 | 69.4 | 11.7 | 23.0 |
| Mahoning | 14.6 | 12.7 | 9.4 | 50.9 | 257,555 | 238,788 | -7.3 | -5.3 | 24,555 | 31,303 | -5,870 | 98,472 | 2.27 | 60.5 | 14.5 | 34.5 |
| Marion | 13.6 | 11.0 | 7.6 | 47.3 | 66,217 | 66,505 | 0.4 | -2.5 | 7,746 | 7,587 | -1,824 | 24,737 | 2.41 | 64.0 | 12.9 | 30.3 |
| Medina | 14.6 | 11.4 | 7.7 | 50.6 | 151,095 | 172,329 | 14.1 | 5.0 | 17,878 | 15,048 | 5,912 | 69,007 | 2.56 | 70.4 | 8.3 | 25.6 |
| Meigs | 15.0 | 12.7 | 8.6 | 50.5 | 23,072 | 23,767 | 3.0 | -4.6 | 2,462 | 2,943 | -596 | 9,045 | 2.53 | 67.6 | 11.5 | 27.7 |
| Mercer | 13.8 | 11.0 | 8.0 | 49.6 | 40,924 | 40,818 | -0.3 | 1.1 | 6,017 | 4,203 | -1,350 | 16,234 | 2.49 | 70.6 | 7.1 | 25.3 |
| Miami | 13.8 | 11.2 | 8.2 | 50.7 | 98,868 | 102,503 | 3.7 | 4.9 | 12,095 | 11,002 | 4,023 | 41,043 | 2.54 | 64.7 | 8.3 | 28.8 |
| Monroe | 15.0 | 13.5 | 10.9 | 49.6 | 15,180 | 14,631 | -3.6 | -7.1 | 1,456 | 1,808 | -689 | 5,745 | 2.4 | 65.7 | 9.1 | 31.7 |
| Montgomery | 13.3 | 10.6 | 8.0 | 51.9 | 559,062 | 535,199 | -4.3 | -0.7 | 67,812 | 62,126 | -9,039 | 224,328 | 2.29 | 59.9 | 14.9 | 33.6 |
| Morgan | 15.5 | 12.8 | 9.8 | 50.3 | 14,897 | 15,043 | 1.0 | -4.9 | 1,464 | 1,792 | -405 | 6,108 | 2.37 | 66.1 | 9.3 | 29.2 |
| Morrow | 15.2 | 11.1 | 7.5 | 49.8 | 31,628 | 34,825 | 10.1 | 1.7 | 3,897 | 3,319 | 21 | 12,922 | 2.69 | 73.4 | 10.3 | 21.7 |
| Muskingum | 13.8 | 10.7 | 7.8 | 51.4 | 84,585 | 86,086 | 1.8 | -0.1 | 10,486 | 10,090 | -413 | 33,878 | 2.48 | 65.1 | 12.4 | 29.8 |
| Noble | 19.0 | 17.2 | 11.6 | 41.4 | 14,058 | 14,655 | 4.2 | -2.0 | 1,458 | 1,375 | -366 | 5,067 | 2.38 | 62.8 | 5.8 | 32.6 |

1. No spouse present.

# Table B. States and Counties — Population, Vital Statistics, Health, and Crime

| STATE County | Persons in group quarters, 2020 | Daytime Population, 2015-2019 | | Births, 2020 | | Deaths, 2020 | | Persons under 65 with no health insurance, 2019 | | Medicare, 2020 | | | Crimes reported by county police or sheriff, 2019 | |
|---|---|---|---|---|---|---|---|---|---|---|---|---|---|---|
| | | Number | Employment/ residence ratio | Total | Rate[1] | Number | Rate[1] | Number | Percent | Total beneficiaries | Enrolled in Original Medicare | Enrolled in Medicare Advantage | Violent | Property |
| | 32 | 33 | 34 | 35 | 36 | 37 | 38 | 39 | 40 | 41 | 42 | 43 | 44 | 45 |
| **NORTH DAKOTA—Cont'd** | | | | | | | | | | | | | | |
| Ward | 3,151 | 69,195 | 1.01 | 1,070 | 15.6 | 530 | 7.7 | 4,236 | 7.5 | 10,242 | 8,551 | 1,691 | 38 | 182 |
| Wells | 71 | 4,177 | 1.11 | 37 | 10.0 | 58 | 15.6 | 194 | 7.2 | 1,177 | 1,059 | 117 | 3 | 7 |
| Williams | 423 | 40,724 | 1.30 | 845 | 21.8 | 252 | 6.5 | 2,885 | 8.5 | 3,845 | 3,468 | 377 | 15 | 166 |
| **OHIO** | 309,717 | 11,674,365 | 1.00 | 133,153 | 11.4 | 126,835 | 10.8 | 741,409 | 7.9 | 2,380,143 | 1,278,219 | 1,101,925 | X | X |
| Adams | 338 | 25,036 | 0.73 | 346 | 12.6 | 389 | 14.1 | 2,132 | 9.6 | 6,518 | 4,021 | 2,497 | 8 | 186 |
| Allen | 5,743 | 108,105 | 1.10 | 1,237 | 12.1 | 1,168 | 11.5 | 5,966 | 7.6 | 22,007 | 13,676 | 8,331 | 38 | 582 |
| Ashland | 2,024 | 48,639 | 0.80 | 596 | 11.2 | 620 | 11.6 | 4,419 | 10.7 | 11,594 | 6,708 | 4,886 | 37 | 103 |
| Ashtabula | 3,212 | 90,482 | 0.81 | 1,073 | 11.1 | 1,234 | 12.8 | 7,386 | 9.8 | 23,626 | 15,395 | 8,231 | 32 | 457 |
| Athens | 9,598 | 66,066 | 1.01 | 473 | 7.2 | 536 | 8.2 | 4,271 | 9.1 | 10,618 | 6,907 | 3,711 | NA | NA |
| Auglaize | 526 | 43,772 | 0.92 | 530 | 11.6 | 534 | 11.7 | 2,180 | 5.9 | 9,984 | 7,036 | 2,949 | 3 | 45 |
| Belmont | 3,552 | 63,731 | 0.85 | 639 | 9.7 | 907 | 13.8 | 3,694 | 7.5 | 16,098 | 8,298 | 7,801 | 42 | 205 |
| Brown | 575 | 33,330 | 0.47 | 486 | 11.2 | 541 | 12.5 | 2,887 | 8.2 | 10,083 | 5,441 | 4,642 | NA | NA |
| Butler | 12,209 | 354,947 | 0.86 | 4,426 | 11.5 | 3,747 | 9.7 | 24,275 | 7.7 | 68,346 | 36,592 | 31,754 | NA | NA |
| Carroll | 405 | 23,389 | 0.68 | 269 | 10.0 | 340 | 12.6 | 1,763 | 8.4 | 6,729 | 3,160 | 3,570 | 2 | 71 |
| Champaign | 672 | 32,981 | 0.67 | 388 | 10.0 | 429 | 11.0 | 2,106 | 6.8 | 8,546 | 4,631 | 3,915 | 25 | 153 |
| Clark | 3,293 | 124,155 | 0.82 | 1,530 | 11.5 | 1,785 | 13.4 | 8,896 | 8.5 | 31,048 | 14,935 | 16,114 | NA | NA |
| Clermont | 1,716 | 168,813 | 0.65 | 2,258 | 10.9 | 1,968 | 9.5 | 12,006 | 7.0 | 41,528 | 20,853 | 20,675 | 111 | 897 |
| Clinton | 1,123 | 43,067 | 1.06 | 455 | 10.9 | 502 | 12.0 | 2,382 | 7.1 | 8,766 | 5,004 | 3,761 | NA | NA |
| Columbiana | 3,943 | 91,186 | 0.73 | 976 | 9.7 | 1,326 | 13.1 | 6,206 | 8.1 | 24,838 | 12,493 | 12,346 | 37 | 315 |
| Coshocton | 428 | 33,639 | 0.81 | 448 | 12.3 | 442 | 12.1 | 3,060 | 10.6 | 8,329 | 5,366 | 2,963 | 40 | 345 |
| Crawford | 579 | 37,579 | 0.77 | 450 | 10.9 | 540 | 13.1 | 2,476 | 7.7 | 10,679 | 7,201 | 3,478 | 3 | 92 |
| Cuyahoga | 29,818 | 1,377,804 | 1.22 | 13,702 | 11.2 | 14,348 | 11.7 | 71,497 | 7.3 | 259,986 | 133,649 | 126,338 | NA | NA |
| Darke | 607 | 46,560 | 0.80 | 612 | 12.0 | 646 | 12.6 | 3,321 | 8.2 | 11,825 | 7,889 | 3,936 | NA | NA |
| Defiance | 724 | 35,882 | 0.88 | 404 | 10.7 | 438 | 11.6 | 2,059 | 6.8 | 8,932 | 5,831 | 3,101 | 14 | 117 |
| Delaware | 1,896 | 189,507 | 0.89 | 2,118 | 9.9 | 1,372 | 6.4 | 8,122 | 4.5 | 33,303 | 19,375 | 13,927 | NA | NA |
| Erie | 1,679 | 75,867 | 1.03 | 758 | 10.3 | 991 | 13.4 | 4,258 | 7.5 | 19,557 | 12,498 | 7,059 | 1 | 189 |
| Fairfield | 2,705 | 126,286 | 0.61 | 1,720 | 10.8 | 1,530 | 9.6 | 8,893 | 6.8 | 30,051 | 14,622 | 15,429 | 52 | 1,039 |
| Fayette | 593 | 28,461 | 0.99 | 321 | 11.2 | 344 | 12.0 | 1,957 | 8.5 | 6,587 | 3,852 | 2,735 | 19 | 307 |
| Franklin | 30,651 | 1,397,577 | 1.16 | 18,195 | 13.7 | 10,851 | 8.2 | 100,068 | 8.9 | 185,527 | 94,943 | 90,584 | NA | NA |
| Fulton | 392 | 40,254 | 0.90 | 466 | 11.1 | 459 | 11.0 | 2,540 | 7.4 | 9,347 | 5,850 | 3,497 | 14 | 167 |
| Gallia | 883 | 29,589 | 0.96 | 349 | 11.7 | 384 | 12.9 | 2,250 | 9.6 | 6,891 | 4,547 | 2,344 | 32 | 294 |
| Geauga | 864 | 86,161 | 0.84 | 912 | 9.8 | 919 | 9.9 | 7,271 | 9.8 | 21,267 | 12,511 | 8,756 | 10 | 134 |
| Greene | 8,406 | 174,108 | 1.10 | 1,713 | 10.1 | 1,683 | 9.9 | 9,227 | 7.0 | 32,851 | 19,181 | 13,670 | 36 | 348 |
| Guernsey | 510 | 39,753 | 1.04 | 436 | 11.2 | 515 | 13.3 | 2,717 | 8.8 | 9,437 | 6,059 | 3,378 | NA | NA |
| Hamilton | 23,395 | 935,754 | 1.31 | 10,619 | 13.0 | 8,412 | 10.3 | 51,666 | 7.7 | 149,597 | 78,669 | 70,928 | 176 | 2,025 |
| Hancock | 1,337 | 85,211 | 1.25 | 848 | 11.2 | 762 | 10.1 | 3,916 | 6.4 | 15,373 | 10,344 | 5,029 | 14 | 127 |
| Hardin | 1,856 | 27,574 | 0.72 | 368 | 11.7 | 349 | 11.1 | 1,905 | 7.8 | 6,220 | 4,277 | 1,943 | 13 | 68 |
| Harrison | 232 | 14,259 | 0.84 | 135 | 9.0 | 209 | 13.9 | 1,015 | 8.7 | 3,800 | 2,272 | 1,527 | NA | NA |
| Henry | 346 | 25,130 | 0.84 | 300 | 11.2 | 273 | 10.1 | 1,517 | 7.0 | 6,204 | 4,352 | 1,852 | 13 | 109 |
| Highland | 484 | 38,496 | 0.74 | 543 | 12.5 | 518 | 12.0 | 3,374 | 9.8 | 9,842 | 5,948 | 3,894 | 38 | 305 |
| Hocking | 308 | 25,428 | 0.76 | 291 | 10.4 | 349 | 12.4 | 1,721 | 7.6 | 6,439 | 4,095 | 2,344 | 28 | 346 |
| Holmes | 765 | 47,459 | 1.18 | 809 | 18.4 | 366 | 8.3 | 8,847 | 23.6 | 5,125 | 3,125 | 2,001 | NA | NA |
| Huron | 578 | 54,505 | 0.86 | 686 | 11.8 | 658 | 11.3 | 4,213 | 8.9 | 12,704 | 8,749 | 3,955 | NA | NA |
| Jackson | 325 | 30,462 | 0.85 | 376 | 11.6 | 419 | 12.9 | 2,218 | 8.4 | 7,486 | 4,970 | 2,517 | 26 | 229 |
| Jefferson | 2,353 | 62,049 | 0.84 | 659 | 10.1 | 990 | 15.2 | 3,833 | 7.9 | 16,722 | 10,048 | 6,674 | 10 | 44 |
| Knox | 3,559 | 56,401 | 0.82 | 738 | 11.8 | 646 | 10.3 | 4,173 | 8.8 | 13,335 | 8,254 | 5,081 | 56 | 330 |
| Lake | 2,811 | 211,050 | 0.84 | 2,185 | 9.5 | 2,694 | 11.7 | 13,520 | 7.4 | 53,891 | 29,328 | 24,564 | NA | NA |
| Lawrence | 669 | 50,405 | 0.60 | 612 | 10.4 | 829 | 14.0 | 3,684 | 7.8 | 14,381 | 9,989 | 4,392 | 65 | 446 |
| Licking | 3,485 | 160,292 | 0.84 | 1,980 | 11.1 | 1,826 | 10.3 | 10,388 | 7.2 | 35,522 | 18,709 | 16,813 | 53 | 793 |
| Logan | 450 | 45,107 | 0.99 | 542 | 12.0 | 544 | 12.0 | 2,677 | 7.3 | 10,198 | 7,031 | 3,166 | 31 | 171 |
| Lorain | 8,447 | 275,452 | 0.77 | 3,204 | 10.3 | 3,323 | 10.6 | 18,632 | 7.6 | 68,250 | 35,937 | 32,312 | 60 | 509 |
| Lucas | 9,462 | 447,989 | 1.09 | 5,238 | 12.2 | 4,808 | 11.2 | 28,110 | 8.1 | 86,525 | 43,798 | 42,727 | 93 | 801 |
| Madison | 4,598 | 43,048 | 0.94 | 449 | 10.1 | 442 | 9.9 | 2,604 | 7.9 | 8,020 | 4,025 | 3,995 | 20 | 264 |
| Mahoning | 8,974 | 230,746 | 1.01 | 2,350 | 10.4 | 3,116 | 13.8 | 13,311 | 7.8 | 57,420 | 25,109 | 32,311 | 1 | 57 |
| Marion | 5,466 | 65,557 | 1.01 | 722 | 11.1 | 808 | 12.5 | 3,632 | 7.6 | 14,378 | 8,306 | 6,072 | 38 | 935 |
| Medina | 1,199 | 153,040 | 0.73 | 1,664 | 9.2 | 1,613 | 8.9 | 8,465 | 5.8 | 37,968 | 19,693 | 18,275 | 11 | 166 |
| Meigs | 212 | 18,859 | 0.51 | 201 | 8.9 | 290 | 12.8 | 1,591 | 8.8 | 5,542 | 3,420 | 2,122 | 16 | 230 |
| Mercer | 439 | 39,682 | 0.94 | 637 | 15.4 | 408 | 9.9 | 2,411 | 7.2 | 8,743 | 6,417 | 2,326 | NA | NA |
| Miami | 1,056 | 97,605 | 0.85 | 1,192 | 11.1 | 1,153 | 10.7 | 6,516 | 7.5 | 23,530 | 13,028 | 10,503 | 38 | 280 |
| Monroe | 165 | 12,764 | 0.77 | 127 | 9.3 | 195 | 14.4 | 922 | 9.0 | 3,646 | 1,881 | 1,765 | 12 | 64 |
| Montgomery | 15,268 | 552,264 | 1.09 | 6,367 | 12.0 | 6,441 | 12.1 | 37,048 | 8.8 | 112,242 | 53,320 | 58,922 | 279 | 1,051 |
| Morgan | 188 | 12,333 | 0.59 | 125 | 8.7 | 177 | 12.4 | 1,046 | 9.3 | 3,483 | 2,018 | 1,466 | 29 | 80 |
| Morrow | 366 | 25,793 | 0.46 | 381 | 10.8 | 358 | 10.1 | 2,483 | 8.6 | 7,264 | 4,345 | 2,919 | NA | NA |
| Muskingum | 1,571 | 85,062 | 0.97 | 970 | 11.3 | 1,019 | 11.8 | 5,482 | 7.9 | 20,338 | 12,071 | 8,267 | NA | NA |
| Noble | 2,655 | 13,329 | 0.72 | 145 | 10.1 | 161 | 11.2 | 703 | 8.0 | 2,645 | 1,738 | 907 | 5 | 36 |

1. Per 1,000 estimated resident population.

Items 32—45

# Table B. States and Counties — Crime, Education, Money Income, and Poverty

| STATE County | County law enforcement employment, 2019 | | Education | | | | | | Money income, 2015-2019 | | | | Income and poverty, 2019 | | | |
| | | | School enrollment and attainment, 2015-2019 | | | | Local government expenditures, 2017-2018 | | Households | | | | | Percent below poverty level | | |
| | | | Enrollment[1] | | Attainment[2] (percent) | | | | | | Percent | | | | | |
| | Officers | Civilians | Total | Percent private | High school graduate or less | Bachelor's degree or more | Total current spending (mil dol) | Current spending per student (dollars) | Per capita income[4] | Median income (dollars) | with income of less than $50,000 | with income of $200,000 or more | Median household income (dollars) | All persons | Children under 18 years | Children 5 to 17 years in families |
| | 46 | 47 | 48 | 49 | 50 | 51 | 52 | 53 | 54 | 55 | 56 | 57 | 58 | 59 | 60 | 61 |
| **NORTH DAKOTA—Cont'd** | | | | | | | | | | | | | | | | |
| Ward | 45 | 61 | 16,859 | 9.5 | 36.3 | 29.3 | 140 | 13,461 | 34,923 | 68,871 | 35.6 | 5.2 | 68,634 | 9.0 | 8.9 | 8.8 |
| Wells | 4 | 1 | 682 | 8.2 | 45.9 | 22.2 | 9 | 14,632 | 35,134 | 57,582 | 44.4 | 3.1 | 56,167 | 12.4 | 11.8 | 11.0 |
| Williams | 38 | 45 | 7,945 | 11.1 | 36.6 | 23.6 | 80 | 13,815 | 44,235 | 87,161 | 26.7 | 10.5 | 84,606 | 6.4 | 7.9 | 7.8 |
| **OHIO** | X | X | 2,819,381 | 18.2 | 42.7 | 28.3 | 21,734 | 12,752 | 31,552 | 56,602 | 44.5 | 4.9 | 58,704 | 13.0 | 18.1 | 16.6 |
| Adams | NA | NA | 5,874 | 13.7 | 64.3 | 13.2 | 56 | 11,727 | 22,366 | 39,079 | 60.1 | 2.0 | 43,145 | 19.1 | 29.5 | 28.9 |
| Allen | 65 | 82 | 26,028 | 20.8 | 48.5 | 18.9 | 177 | 12,161 | 26,761 | 53,131 | 47.8 | 2.5 | 58,271 | 12.9 | 19.6 | 18.3 |
| Ashland | 49 | 38 | 13,090 | 28.4 | 53.1 | 20.9 | 85 | 11,728 | 26,017 | 52,823 | 47.0 | 2.5 | 52,554 | 12.6 | 19.1 | 19.4 |
| Ashtabula | 34 | 45 | 19,828 | 14.6 | 58.5 | 14.3 | 162 | 12,345 | 24,265 | 46,700 | 53.2 | 1.8 | 47,388 | 18.7 | 26.5 | 23.2 |
| Athens | 31 | 5 | 26,574 | 2.7 | 45.1 | 30.2 | 106 | 14,765 | 22,040 | 40,905 | 56.6 | 2.6 | 43,621 | 26.6 | 25.3 | 22.7 |
| Auglaize | NA | NA | 10,418 | 8.7 | 50.8 | 18.9 | 90 | 11,439 | 31,198 | 64,074 | 37.8 | 3.1 | 64,227 | 7.3 | 9.2 | 8.8 |
| Belmont | NA | NA | 13,441 | 12.8 | 52.2 | 16.7 | 99 | 11,187 | 27,580 | 50,904 | 49.0 | 3.0 | 50,166 | 11.6 | 15.7 | 13.8 |
| Brown | 29 | 20 | 8,853 | 8.1 | 57.7 | 14.9 | 86 | 12,399 | 26,696 | 54,575 | 45.9 | 1.9 | 58,441 | 12.1 | 18.0 | 16.4 |
| Butler | NA | NA | 104,600 | 13.9 | 42.3 | 30.2 | 666 | 11,595 | 31,921 | 66,117 | 37.9 | 6.1 | 68,611 | 11.7 | 14.1 | 13.0 |
| Carroll | 27 | 17 | 5,269 | 17.5 | 60.2 | 13.3 | 37 | 12,298 | 29,518 | 55,267 | 44.8 | 2.9 | 56,999 | 11.3 | 18.2 | 15.5 |
| Champaign | 26 | 4 | 8,627 | 13.6 | 55.0 | 16.1 | 89 | 13,083 | 27,722 | 60,112 | 39.8 | 1.8 | 62,077 | 7.7 | 11.1 | 10.6 |
| Clark | NA | NA | 31,237 | 19.6 | 49.2 | 18.7 | 257 | 12,400 | 27,066 | 50,873 | 49.2 | 2.4 | 50,832 | 14.6 | 21.0 | 18.5 |
| Clermont | NA | NA | 47,081 | 18.4 | 42.1 | 28.6 | 301 | 11,130 | 34,390 | 66,968 | 36.2 | 6.5 | 68,724 | 8.2 | 11.7 | 10.7 |
| Clinton | 35 | 35 | 9,797 | 19.5 | 50.8 | 18.9 | 80 | 10,696 | 26,690 | 52,815 | 47.2 | 2.4 | 52,870 | 14.8 | 22.2 | 19.3 |
| Columbiana | NA | NA | 20,743 | 10.2 | 56.2 | 14.9 | 180 | 12,405 | 26,469 | 48,345 | 51.6 | 2.1 | 52,693 | 13.2 | 18.1 | 15.8 |
| Coshocton | NA | NA | 7,717 | 17.2 | 61.9 | 13.2 | 57 | 12,014 | 23,432 | 46,606 | 54.2 | 1.7 | 49,679 | 12.5 | 18.8 | 18.6 |
| Crawford | NA | NA | 8,446 | 12.1 | 56.7 | 14.4 | 75 | 11,795 | 25,744 | 44,971 | 54.8 | 1.7 | 49,484 | 14.7 | 23.0 | 18.6 |
| Cuyahoga | NA | NA | 296,767 | 25.6 | 37.9 | 32.5 | 2,641 | 15,670 | 33,114 | 50,366 | 49.7 | 5.5 | 52,503 | 16.2 | 23.6 | 22.6 |
| Darke | NA | NA | 11,700 | 9.1 | 56.0 | 15.8 | 94 | 11,056 | 27,855 | 55,620 | 45.2 | 2.2 | 56,824 | 12.1 | 17.1 | 16.0 |
| Defiance | NA | NA | 9,123 | 16.3 | 51.0 | 18.0 | 72 | 11,512 | 29,816 | 59,931 | 41.2 | 3.2 | 60,918 | 8.8 | 13.6 | 11.7 |
| Delaware | 107 | 123 | 56,720 | 17.8 | 21.9 | 54.3 | 368 | 11,333 | 48,117 | 106,908 | 21.0 | 18.5 | 110,252 | 4.8 | 5.0 | 4.1 |
| Erie | NA | NA | 15,601 | 15.6 | 45.5 | 23.2 | 183 | 14,985 | 32,790 | 54,226 | 46.3 | 3.9 | 60,790 | 11.6 | 19.2 | 17.1 |
| Fairfield | 110 | 40 | 37,809 | 14.7 | 41.4 | 27.6 | 292 | 11,721 | 32,021 | 67,609 | 36.4 | 5.6 | 71,782 | 8.1 | 11.5 | 10.0 |
| Fayette | 23 | 18 | 6,055 | 9.2 | 60.2 | 16.3 | 50 | 10,330 | 26,079 | 47,308 | 52.0 | 2.8 | 51,023 | 13.6 | 21.0 | 19.6 |
| Franklin | NA | NA | 339,468 | 16.7 | 33.4 | 40.0 | 2,625 | 12,361 | 33,998 | 61,305 | 40.8 | 5.9 | 64,648 | 13.5 | 18.6 | 17.6 |
| Fulton | 21 | 2 | 9,798 | 10.2 | 50.4 | 17.6 | 111 | 14,603 | 29,600 | 63,092 | 39.2 | 3.3 | 64,334 | 6.9 | 9.8 | 9.1 |
| Gallia | NA | NA | 6,573 | 13.4 | 57.2 | 15.9 | 65 | 15,097 | 24,130 | 44,858 | 53.7 | 1.7 | 48,094 | 16.5 | 24.9 | 23.7 |
| Geauga | 49 | 68 | 20,669 | 27.8 | 35.0 | 38.0 | 144 | 14,363 | 42,958 | 82,303 | 28.9 | 11.2 | 79,865 | 5.5 | 6.2 | 5.3 |
| Greene | 68 | 93 | 47,178 | 21.6 | 30.7 | 39.2 | 275 | 12,555 | 35,833 | 68,720 | 36.2 | 7.0 | 69,709 | 9.9 | 12.9 | 12.1 |
| Guernsey | 28 | 31 | 8,722 | 14.0 | 57.2 | 14.6 | 63 | 13,918 | 24,742 | 45,917 | 54.1 | 2.5 | 48,283 | 15.5 | 22.2 | 19.6 |
| Hamilton | NA | NA | 206,075 | 23.4 | 34.9 | 37.9 | 1,480 | 13,380 | 35,570 | 57,212 | 44.5 | 6.9 | 60,251 | 14.6 | 20.8 | 18.6 |
| Hancock | 41 | 56 | 18,002 | 24.5 | 43.5 | 27.1 | 141 | 11,582 | 32,797 | 58,450 | 42.5 | 4.5 | 62,720 | 9.3 | 11.2 | 10.6 |
| Hardin | NA | NA | 8,751 | 31.2 | 58.6 | 16.2 | 49 | 11,793 | 23,383 | 50,506 | 49.5 | 1.8 | 51,155 | 13.9 | 18.2 | 16.6 |
| Harrison | NA | NA | 2,768 | 15.4 | 61.1 | 11.6 | 18 | 11,234 | 24,940 | 49,689 | 50.4 | 1.7 | 50,137 | 14.5 | 21.7 | 19.7 |
| Henry | NA | NA | 6,293 | 13.6 | 51.0 | 17.2 | 69 | 15,939 | 28,830 | 59,695 | 41.3 | 2.1 | 61,351 | 7.5 | 10.4 | 9.5 |
| Highland | NA | NA | 9,274 | 11.6 | 60.3 | 13.4 | 81 | 11,493 | 23,543 | 44,169 | 54.9 | 1.7 | 47,126 | 15.6 | 22.7 | 19.4 |
| Hocking | NA | NA | 6,208 | 8.5 | 56.5 | 14.4 | 42 | 11,111 | 25,588 | 52,363 | 47.7 | 1.8 | 51,016 | 15.0 | 21.3 | 20.3 |
| Holmes | NA | NA | 9,007 | 43.1 | 76.0 | 9.7 | 42 | 11,196 | 23,404 | 63,753 | 37.7 | 4.1 | 62,596 | 9.2 | 13.0 | 12.5 |
| Huron | NA | NA | 13,246 | 12.6 | 57.6 | 14.4 | 118 | 11,526 | 26,133 | 52,560 | 47.0 | 1.8 | 56,704 | 10.0 | 15.3 | 14.2 |
| Jackson | NA | NA | 7,130 | 10.3 | 58.6 | 14.5 | 57 | 11,375 | 23,254 | 47,550 | 53.1 | 1.5 | 50,671 | 17.2 | 26.7 | 25.7 |
| Jefferson | NA | NA | 14,146 | 25.0 | 53.7 | 15.8 | 101 | 11,541 | 26,391 | 46,581 | 53.5 | 2.1 | 47,652 | 17.1 | 25.4 | 23.0 |
| Knox | 49 | 24 | 15,909 | 29.2 | 48.5 | 23.0 | 97 | 12,672 | 26,390 | 57,749 | 43.1 | 2.3 | 59,957 | 11.3 | 16.7 | 15.2 |
| Lake | NA | NA | 50,855 | 19.1 | 39.5 | 27.4 | 396 | 12,682 | 34,409 | 64,466 | 38.0 | 4.8 | 65,040 | 8.3 | 11.2 | 10.0 |
| Lawrence | NA | NA | 13,197 | 8.6 | 56.4 | 14.9 | 127 | 13,741 | 24,562 | 45,118 | 53.9 | 1.7 | 46,433 | 16.2 | 22.6 | 21.4 |
| Licking | 96 | 133 | 42,894 | 19.9 | 42.8 | 26.1 | 311 | 11,312 | 31,617 | 64,589 | 38.8 | 5.3 | 66,321 | 9.3 | 12.4 | 11.0 |
| Logan | NA | NA | 9,679 | 10.8 | 55.8 | 16.5 | 93 | 14,832 | 29,424 | 56,754 | 43.3 | 2.4 | 60,624 | 10.5 | 14.6 | 12.7 |
| Lorain | NA | NA | 73,894 | 20.2 | 41.3 | 24.9 | 516 | 11,859 | 30,928 | 58,427 | 42.5 | 4.8 | 59,201 | 13.9 | 20.3 | 18.8 |
| Lucas | NA | NA | 106,609 | 17.0 | 40.4 | 26.3 | 863 | 12,165 | 29,226 | 48,736 | 51.0 | 3.7 | 49,924 | 18.0 | 26.3 | 24.6 |
| Madison | 29 | 14 | 9,737 | 14.9 | 54.6 | 17.5 | 85 | 12,658 | 29,164 | 68,022 | 34.2 | 4.9 | 65,696 | 9.6 | 12.4 | 11.5 |
| Mahoning | NA | NA | 51,453 | 13.1 | 46.7 | 24.2 | 432 | 14,106 | 28,378 | 46,042 | 53.5 | 3.1 | 48,018 | 18.4 | 27.4 | 24.1 |
| Marion | 33 | 26 | 14,239 | 6.3 | 56.6 | 12.5 | 130 | 11,459 | 24,484 | 47,498 | 52.1 | 2.7 | 51,479 | 14.8 | 20.3 | 18.7 |
| Medina | NA | NA | 41,945 | 17.1 | 35.9 | 33.9 | 298 | 11,304 | 37,788 | 76,600 | 31.2 | 7.3 | 78,540 | 5.5 | 7.0 | 5.6 |
| Meigs | NA | NA | 4,818 | 7.7 | 58.2 | 13.1 | 40 | 11,917 | 24,899 | 44,899 | 55.8 | 1.8 | 43,754 | 15.8 | 23.7 | 22.3 |
| Mercer | 30 | 37 | 9,860 | 6.0 | 53.2 | 18.3 | 97 | 12,222 | 29,765 | 62,952 | 38.3 | 2.5 | 67,075 | 6.7 | 8.8 | 8.4 |
| Miami | 57 | 62 | 24,608 | 13.4 | 45.4 | 23.3 | 199 | 12,811 | 31,254 | 61,041 | 41.3 | 4.0 | 63,524 | 8.8 | 11.9 | 10.8 |
| Monroe | NA | NA | 2,670 | 14.6 | 62.7 | 12.5 | 34 | 15,842 | 26,476 | 45,289 | 54.4 | 1.8 | 48,231 | 14.0 | 20.2 | 18.0 |
| Montgomery | NA | NA | 131,847 | 24.9 | 37.7 | 27.9 | 1,002 | 13,435 | 30,034 | 51,542 | 48.7 | 3.6 | 54,704 | 15.3 | 22.5 | 20.3 |
| Morgan | 10 | 7 | 2,635 | 11.7 | 58.6 | 10.2 | 23 | 12,639 | 22,937 | 42,341 | 60.6 | 1.5 | 46,883 | 15.7 | 21.4 | 19.6 |
| Morrow | 33 | 27 | 7,393 | 9.3 | 56.2 | 13.8 | 54 | 10,325 | 28,257 | 59,452 | 40.2 | 2.9 | 59,498 | 8.5 | 13.6 | 12.7 |
| Muskingum | NA | NA | 19,944 | 14.2 | 53.8 | 16.1 | 194 | 13,518 | 25,148 | 47,254 | 52.6 | 2.1 | 52,105 | 15.3 | 21.4 | 19.9 |
| Noble | 12 | 17 | 2,658 | 15.5 | 63.2 | 9.9 | 22 | 13,243 | 24,440 | 46,897 | 52.2 | 2.3 | 50,788 | 14.2 | 16.0 | 15.0 |

1. All persons 3 years old and over enrolled in nursery school through college. 2. Persons 25 years old and over. 3. Elementary and secondary education expenditures. 4. Based on population estimated by the American Community Survey, 2014–2018.

# Table B. States and Counties — **Personal Income**

| STATE County | Personal income, 2019 | | | | | | | | | | Earnings, 2019 | | |
|---|---|---|---|---|---|---|---|---|---|---|---|---|---|
| | Total (mil dol) | Percent change 2018-2019 | Per capita[1] Dollars | Per capita[1] Rank | Wages and salaries (mil dol) | Supplements to wages and salaries, employer contributions (mil dol) Pension and insurance | Government social insurance | Proprietors' income (mil dol) | Dividends, interest, and rent (mil dol) | Personal transfer receipts (mil dol) | Total (mil dol) | Contributions for government social insurance (mil dol) From employee and self-employed | From employer |
| | 62 | 63 | 64 | 65 | 66 | 67 | 68 | 69 | 70 | 71 | 72 | 73 | 74 |
| **NORTH DAKOTA—Cont'd** | | | | | | | | | | | | | |
| Ward | 3,794 | 3.5 | 56,088 | 444 | 2,101 | 369 | 180 | 195 | 733 | 515 | 2,845 | 165 | 180 |
| Wells | 227 | -3.4 | 59,292 | 328 | 70 | 11 | 7 | 50 | 53 | 54 | 137 | 8 | 7 |
| Williams | 2,567 | 4.6 | 68,291 | 145 | 2,446 | 253 | 168 | 191 | 499 | 199 | 3,058 | 181 | 168 |
| **OHIO** | 586,784 | 3.1 | 50,167 | X | 302,847 | 50,494 | 21,145 | 44,020 | 102,341 | 114,048 | 418,505 | 24,537 | 21,145 |
| Adams | 977 | 2.0 | 35,269 | 2,646 | 266 | 58 | 18 | 85 | 114 | 343 | 427 | 30 | 18 |
| Allen | 4,482 | 2.5 | 43,788 | 1,518 | 2,596 | 473 | 186 | 310 | 673 | 1,091 | 3,565 | 213 | 186 |
| Ashland | 2,138 | 2.8 | 39,977 | 2,054 | 838 | 160 | 61 | 147 | 319 | 506 | 1,206 | 77 | 61 |
| Ashtabula | 3,867 | 2.5 | 39,767 | 2,080 | 1,316 | 267 | 98 | 268 | 511 | 1,184 | 1,949 | 132 | 98 |
| Athens | 2,273 | 4.3 | 34,796 | 2,716 | 1,039 | 285 | 53 | 137 | 422 | 570 | 1,514 | 67 | 53 |
| Auglaize | 2,273 | 2.3 | 49,789 | 847 | 1,098 | 189 | 79 | 143 | 389 | 417 | 1,509 | 91 | 79 |
| Belmont | 2,880 | -0.8 | 42,987 | 1,618 | 1,038 | 184 | 73 | 117 | 488 | 747 | 1,413 | 95 | 73 |
| Brown | 1,668 | 3.5 | 38,401 | 2,279 | 309 | 73 | 20 | 104 | 190 | 475 | 506 | 38 | 20 |
| Butler | 18,829 | 4.4 | 49,146 | 911 | 8,938 | 1,429 | 622 | 1,748 | 2,749 | 3,329 | 12,737 | 744 | 622 |
| Carroll | 1,099 | 2.9 | 40,833 | 1,945 | 279 | 54 | 21 | 83 | 157 | 286 | 438 | 32 | 21 |
| Champaign | 1,650 | 2.8 | 42,432 | 1,698 | 511 | 99 | 36 | 91 | 218 | 376 | 737 | 48 | 36 |
| Clark | 5,606 | 2.7 | 41,811 | 1,795 | 2,224 | 402 | 158 | 265 | 849 | 1,538 | 3,049 | 200 | 158 |
| Clermont | 11,527 | 4.7 | 55,842 | 453 | 3,076 | 494 | 219 | 1,824 | 1,434 | 1,862 | 5,612 | 338 | 219 |
| Clinton | 1,956 | 2.8 | 46,599 | 1,174 | 980 | 188 | 69 | 328 | 280 | 420 | 1,565 | 91 | 69 |
| Columbiana | 4,005 | 1.8 | 39,311 | 2,147 | 1,295 | 258 | 94 | 219 | 583 | 1,173 | 1,866 | 131 | 94 |
| Coshocton | 1,316 | 2.6 | 35,962 | 2,575 | 451 | 96 | 33 | 118 | 179 | 413 | 698 | 47 | 33 |
| Crawford | 1,653 | 2.0 | 39,844 | 2,070 | 578 | 111 | 41 | 77 | 258 | 479 | 807 | 57 | 41 |
| Cuyahoga | 69,784 | 2.6 | 56,502 | 421 | 47,623 | 7,200 | 3,325 | 5,098 | 15,071 | 13,315 | 63,245 | 3,651 | 3,325 |
| Darke | 2,264 | -0.3 | 44,297 | 1,445 | 812 | 154 | 61 | 225 | 351 | 486 | 1,252 | 78 | 61 |
| Defiance | 1,600 | 2.2 | 42,003 | 1,766 | 726 | 127 | 52 | 127 | 225 | 394 | 1,032 | 65 | 52 |
| Delaware | 15,851 | 4.4 | 75,778 | 85 | 5,770 | 807 | 393 | 1,197 | 2,776 | 1,273 | 8,166 | 471 | 393 |
| Erie | 4,591 | 2.0 | 61,821 | 256 | 1,628 | 306 | 116 | 1,252 | 663 | 849 | 3,303 | 184 | 116 |
| Fairfield | 7,452 | 3.7 | 47,295 | 1,098 | 1,962 | 355 | 138 | 424 | 1,053 | 1,358 | 2,879 | 184 | 138 |
| Fayette | 1,141 | 2.3 | 40,002 | 2,050 | 459 | 81 | 33 | 74 | 162 | 322 | 648 | 41 | 33 |
| Franklin | 68,002 | 3.4 | 51,644 | 711 | 49,279 | 8,105 | 3,240 | 4,709 | 12,355 | 10,380 | 65,332 | 3,430 | 3,240 |
| Fulton | 1,972 | 2.9 | 46,811 | 1,150 | 862 | 154 | 63 | 178 | 296 | 386 | 1,257 | 76 | 63 |
| Gallia | 1,197 | 2.0 | 40,038 | 2,045 | 478 | 113 | 34 | 85 | 173 | 409 | 710 | 45 | 34 |
| Geauga | 6,645 | 2.7 | 70,955 | 121 | 1,753 | 289 | 130 | 735 | 1,497 | 798 | 2,906 | 177 | 130 |
| Greene | 8,702 | 3.4 | 51,509 | 725 | 5,090 | 1,111 | 383 | 414 | 1,703 | 1,489 | 6,997 | 391 | 383 |
| Guernsey | 1,618 | 2.6 | 41,613 | 1,830 | 706 | 138 | 49 | 101 | 245 | 470 | 995 | 63 | 49 |
| Hamilton | 50,464 | 3.5 | 61,732 | 258 | 36,180 | 5,213 | 2,503 | 3,643 | 12,289 | 7,714 | 47,539 | 2,728 | 2,503 |
| Hancock | 3,923 | 3.0 | 51,762 | 701 | 2,599 | 408 | 181 | 475 | 604 | 641 | 3,662 | 214 | 181 |
| Hardin | 1,060 | 2.2 | 33,788 | 2,823 | 347 | 75 | 26 | 81 | 141 | 286 | 529 | 34 | 26 |
| Harrison | 590 | 1.5 | 39,209 | 2,161 | 196 | 43 | 14 | 32 | 78 | 174 | 284 | 20 | 14 |
| Henry | 1,242 | 2.4 | 46,006 | 1,248 | 516 | 98 | 38 | 92 | 185 | 278 | 743 | 45 | 38 |
| Highland | 1,611 | 3.3 | 37,317 | 2,411 | 426 | 97 | 29 | 136 | 212 | 470 | 688 | 46 | 29 |
| Hocking | 1,102 | 2.6 | 38,975 | 2,196 | 250 | 57 | 16 | 68 | 138 | 327 | 391 | 28 | 16 |
| Holmes | 2,061 | 3.1 | 46,876 | 1,146 | 911 | 148 | 68 | 702 | 259 | 245 | 1,829 | 96 | 68 |
| Huron | 2,435 | 2.5 | 41,797 | 1,802 | 1,080 | 192 | 91 | 148 | 348 | 591 | 1,511 | 97 | 91 |
| Jackson | 1,189 | 3.8 | 36,672 | 2,487 | 409 | 80 | 29 | 78 | 154 | 392 | 596 | 40 | 29 |
| Jefferson | 2,668 | 3.4 | 40,844 | 1,942 | 979 | 197 | 72 | 107 | 377 | 848 | 1,356 | 93 | 72 |
| Knox | 2,743 | 2.7 | 44,019 | 1,482 | 1,008 | 182 | 74 | 232 | 446 | 652 | 1,496 | 95 | 74 |
| Lake | 12,077 | 3.2 | 52,475 | 647 | 5,101 | 880 | 368 | 598 | 1,936 | 2,281 | 6,946 | 431 | 368 |
| Lawrence | 2,355 | 2.0 | 39,610 | 2,100 | 542 | 117 | 38 | 97 | 270 | 779 | 794 | 60 | 38 |
| Licking | 8,330 | 3.4 | 47,097 | 1,121 | 3,012 | 518 | 216 | 533 | 1,189 | 1,581 | 4,279 | 264 | 216 |
| Logan | 2,011 | 3.4 | 44,030 | 1,480 | 978 | 164 | 71 | 130 | 259 | 459 | 1,343 | 84 | 71 |
| Lorain | 14,769 | 3.5 | 47,666 | 1,060 | 4,824 | 897 | 347 | 675 | 2,296 | 3,100 | 6,743 | 433 | 347 |
| Lucas | 20,507 | 2.7 | 47,875 | 1,033 | 11,522 | 1,989 | 820 | 1,913 | 3,050 | 4,646 | 16,243 | 934 | 820 |
| Madison | 1,873 | 2.8 | 41,862 | 1,787 | 849 | 163 | 59 | 155 | 261 | 367 | 1,226 | 68 | 59 |
| Mahoning | 10,045 | 1.6 | 43,925 | 1,497 | 4,393 | 802 | 319 | 703 | 1,776 | 2,743 | 6,216 | 396 | 319 |
| Marion | 2,515 | 2.8 | 38,644 | 2,244 | 1,172 | 239 | 85 | 221 | 329 | 702 | 1,717 | 103 | 85 |
| Medina | 10,315 | 3.7 | 57,387 | 390 | 3,056 | 521 | 217 | 674 | 1,562 | 1,528 | 4,468 | 282 | 217 |
| Meigs | 818 | 3.7 | 35,723 | 2,594 | 141 | 35 | 9 | 38 | 100 | 273 | 223 | 19 | 9 |
| Mercer | 2,145 | -1.1 | 52,106 | 674 | 906 | 170 | 64 | 290 | 347 | 342 | 1,431 | 79 | 64 |
| Miami | 5,242 | 3.5 | 48,997 | 924 | 1,996 | 344 | 143 | 272 | 857 | 998 | 2,755 | 174 | 143 |
| Monroe | 501 | 2.7 | 36,672 | 2,487 | 134 | 31 | 9 | 33 | 88 | 165 | 207 | 16 | 9 |
| Montgomery | 25,759 | 3.0 | 48,448 | 976 | 14,149 | 2,361 | 1,018 | 1,881 | 4,764 | 5,590 | 19,408 | 1,153 | 1,018 |
| Morgan | 494 | 2.9 | 34,038 | 2,797 | 120 | 26 | 8 | 26 | 62 | 159 | 180 | 14 | 8 |
| Morrow | 1,421 | 3.5 | 40,221 | 2,024 | 232 | 54 | 15 | 98 | 164 | 330 | 400 | 29 | 15 |
| Muskingum | 3,621 | 2.9 | 41,998 | 1,767 | 1,557 | 283 | 111 | 219 | 487 | 977 | 2,169 | 139 | 111 |
| Noble | 400 | 2.1 | 27,703 | 3,083 | 128 | 33 | 7 | 31 | 65 | 115 | 200 | 12 | 7 |

1. Based on the resident population estimated as of July 1 of the year shown.

| STATE County | Farm (75) | Mining, quarrying, and extractions (76) | Construction (77) | Manu-facturing (78) | Information; professional, scientific, technical services (79) | Retail trade (80) | Finance, insurance, real estate, and leasing (81) | Health care and social assistance (82) | Govern-ment (83) | Number (84) | Rate[1] (85) | Supple-mental Security Income recipients, 2019 (86) | Total (87) | Percent change, 2010-2020 (88) |
|---|---|---|---|---|---|---|---|---|---|---|---|---|---|---|
| **NORTH DAKOTA—Cont'd** | | | | | | | | | | | | | | |
| Ward | 0.3 | 8.7 | 6.2 | 1.0 | 4.5 | 6.5 | 5.8 | 13.5 | 31.1 | 10,250 | 150 | 700 | 33,213 | 24.2 |
| Wells | 24.5 | D | 5.3 | 2.2 | D | 6.2 | 5.7 | D | 10.6 | 1,135 | 299 | 40 | 2,501 | 0.8 |
| Williams | 0.0 | 41.3 | 11.5 | 1.3 | 3.8 | 3.7 | 5.2 | 3.7 | 6.3 | 4,005 | 106 | 210 | 19,705 | 88.3 |
| **OHIO** | 0.2 | 0.4 | 6.0 | 14.2 | 9.3 | 5.6 | 7.7 | 13.1 | 14.6 | 2,386,362 | 204 | 307,783 | 5,248,017 | 2.3 |
| Adams | 0.8 | 0.7 | 10.1 | 19.3 | D | 8.0 | 3.9 | 10.2 | 20.5 | 6,750 | 244 | 1,686 | 12,953 | -0.2 |
| Allen | 0.3 | D | 5.2 | 25.0 | 4.5 | 6.0 | 3.6 | 19.7 | 12.3 | 23,100 | 225 | 2,925 | 45,289 | 0.7 |
| Ashland | 0.7 | D | 7.4 | 21.2 | 9.9 | 6.9 | 2.9 | D | 13.7 | 11,875 | 222 | 776 | 22,412 | 1.2 |
| Ashtabula | -0.1 | 0.1 | 10.2 | 26.9 | 4.6 | 6.0 | 2.8 | 14.7 | 14.6 | 23,690 | 244 | 3,096 | 46,345 | 0.5 |
| Athens | -0.2 | D | 4.8 | 3.0 | 4.4 | 6.7 | 3.6 | 12.9 | 49.3 | 10,025 | 153 | 2,631 | 26,681 | 1.1 |
| Auglaize | 2.1 | D | 5.4 | 45.7 | 3.6 | 5.2 | 3.1 | 8.3 | 9.5 | 9,690 | 212 | 436 | 20,235 | 3.4 |
| Belmont | -0.3 | 16.0 | 8.4 | 3.7 | 4.1 | 9.2 | 4.2 | 13.1 | 17.2 | 16,790 | 251 | 1,898 | 32,195 | -0.8 |
| Brown | D | D | D | 9.9 | D | 8.0 | 3.7 | 11.4 | 27.6 | 10,405 | 239 | 1,158 | 20,746 | 7.5 |
| Butler | -0.1 | D | 8.0 | 20.8 | 4.7 | 7.0 | 8.4 | 9.6 | 11.9 | 69,605 | 181 | 7,514 | 153,837 | 3.7 |
| Carroll | -0.6 | 1.6 | 17.1 | 18.6 | D | 7.2 | 3.3 | D | 14.3 | 6,850 | 254 | 426 | 13,615 | -0.6 |
| Champaign | 1.6 | D | 5.1 | 40.0 | 4.6 | 5.0 | 3.1 | 6.8 | 16.8 | 8,610 | 221 | 657 | 16,888 | 0.8 |
| Clark | 0.2 | 0.4 | 4.0 | 17.2 | 3.5 | 6.4 | 7.8 | 15.0 | 15.3 | 31,555 | 235 | 4,002 | 61,229 | -0.3 |
| Clermont | -0.1 | D | 7.2 | 9.2 | D | 6.8 | 6.8 | 6.9 | 10.0 | 42,220 | 204 | 3,431 | 84,579 | 4.9 |
| Clinton | -0.2 | D | 3.7 | 17.4 | D | 4.2 | 5.5 | 6.0 | 17.9 | 9,335 | 222 | 1,084 | 18,259 | 0.7 |
| Columbiana | 0.4 | 0.7 | 7.8 | 21.7 | 2.9 | 7.8 | 3.0 | 13.1 | 14.1 | 26,245 | 258 | 2,975 | 46,843 | -0.5 |
| Coshocton | 1.2 | 0.9 | 6.7 | 27.7 | D | 5.9 | 3.0 | 11.1 | 15.2 | 8,850 | 242 | 920 | 16,420 | -0.8 |
| Crawford | -0.1 | D | 5.6 | 19.3 | D | 6.9 | 8.4 | D | 13.2 | 11,130 | 269 | 1,206 | 19,975 | |
| Cuyahoga | 0.0 | 0.6 | 3.7 | 9.5 | 13.4 | 4.0 | 9.9 | 15.5 | 13.2 | 255,950 | 207 | 49,322 | 616,725 | -0.8 |
| Darke | 4.0 | D | 8.6 | 25.2 | 4.0 | 4.7 | 4.4 | 10.0 | 11.3 | 12,105 | 237 | 793 | 23,118 | 1.7 |
| Defiance | 1.0 | 0.0 | 3.9 | 27.0 | 3.4 | 10.3 | 6.3 | D | 13.7 | 9,425 | 248 | 696 | 16,840 | 0.7 |
| Delaware | 0.2 | D | 5.6 | 7.1 | 12.0 | 6.5 | 11.9 | 8.4 | 8.4 | 30,485 | 145 | 1,369 | 78,068 | 17.6 |
| Erie | 0.5 | D | 3.2 | 12.8 | 3.3 | 5.5 | 2.8 | 9.9 | 11.6 | 19,295 | 260 | 1,609 | 37,924 | 0.2 |
| Fairfield | -0.1 | 0.1 | 10.3 | 10.1 | 5.0 | 8.3 | 4.3 | 17.8 | 17.6 | 29,990 | 190 | 2,566 | 62,701 | 6.8 |
| Fayette | 2.5 | 0.0 | 5.2 | 18.5 | D | 11.2 | 3.2 | D | 17.7 | 6,700 | 235 | 949 | 12,942 | 2.0 |
| Franklin | 0.0 | 0.1 | 5.3 | 5.2 | 12.6 | 4.4 | 11.1 | 12.0 | 19.3 | 179,925 | 136 | 33,404 | 567,759 | 7.7 |
| Fulton | 1.3 | D | 10.0 | 36.4 | D | 5.3 | 3.4 | D | 12.5 | 9,460 | 224 | 552 | 17,639 | 1.3 |
| Gallia | -0.9 | D | 8.9 | 5.8 | 1.8 | 7.6 | 4.5 | D | 16.3 | 7,285 | 244 | 1,535 | 13,918 | -0.1 |
| Geauga | 0.1 | D | 15.4 | 19.1 | 6.3 | 7.0 | 4.0 | 9.7 | 9.0 | 20,105 | 215 | 719 | 37,519 | 2.6 |
| Greene | -0.1 | 0.2 | 2.9 | 3.2 | 19.6 | 5.6 | 3.1 | 7.4 | 44.5 | 31,165 | 184 | 2,669 | 71,809 | 5.2 |
| Guernsey | -0.3 | 5.7 | 7.8 | 20.7 | 3.6 | 5.7 | 4.9 | 14.7 | 15.5 | 9,735 | 250 | 1,435 | 19,393 | 1.0 |
| Hamilton | 0.0 | 0.1 | 5.0 | 12.5 | 13.1 | 3.9 | 11.1 | 14.1 | 9.9 | 148,030 | 181 | 23,734 | 381,795 | 1.2 |
| Hancock | 0.7 | D | 2.6 | 23.8 | 4.8 | 4.6 | 2.9 | 10.3 | 6.4 | 15,585 | 206 | 1,077 | 34,806 | 4.9 |
| Hardin | 5.3 | D | 3.5 | 23.1 | D | 6.9 | 4.3 | D | 16.0 | 6,535 | 208 | 650 | 13,253 | 1.2 |
| Harrison | -0.7 | D | 9.9 | 17.2 | D | 3.6 | 4.6 | 6.5 | 16.6 | 3,930 | 261 | 531 | 8,078 | -1.1 |
| Henry | 1.5 | D | 10.5 | 32.4 | 2.3 | 5.9 | 4.0 | 9.0 | 16.9 | 6,400 | 238 | 363 | 12,175 | 1.8 |
| Highland | 0.7 | 0.2 | 10.2 | 17.5 | 2.0 | 11.0 | 6.2 | 10.1 | 21.4 | 10,285 | 239 | 1,431 | 19,328 | -0.3 |
| Hocking | -0.9 | D | 12.2 | 13.1 | D | 9.9 | 4.3 | 27.6 | 5.7 | 6,865 | 243 | 1,008 | 13,425 | 0.1 |
| Holmes | 2.0 | 0.5 | 19.4 | 33.2 | 2.0 | 10.9 | 2.8 | D | 11.3 | 5,040 | 114 | 358 | 13,558 | -0.8 |
| Huron | 2.0 | D | 12.3 | 25.1 | D | 5.2 | 4.1 | 10.6 | 17.2 | 13,150 | 226 | 1,301 | 25,314 | 0.5 |
| Jackson | 0.1 | 0.3 | 7.4 | 26.3 | D | 8.8 | 3.4 | D | 16.0 | 7,795 | 241 | 1,470 | 15,035 | 3.1 |
| Jefferson | -0.3 | D | D | 11.3 | D | 7.7 | 3.3 | D | 16.0 | 17,600 | 269 | 2,493 | 32,351 | -1.5 |
| Knox | 0.2 | 0.5 | 12.4 | 27.6 | 2.9 | 5.6 | 3.3 | 11.9 | 11.7 | 13,565 | 218 | 1,104 | 26,105 | 3.9 |
| Lake | 0.5 | D | 5.9 | 27.5 | 6.0 | 7.0 | 4.0 | 9.9 | 11.7 | 53,345 | 232 | 2,978 | 103,759 | 2.5 |
| Lawrence | -0.5 | D | D | 7.8 | 3.6 | 8.3 | 3.1 | 17.0 | 22.7 | 15,155 | 254 | 3,397 | 27,547 | -0.2 |
| Licking | -0.1 | 0.2 | 7.6 | 13.6 | 6.3 | 8.8 | 6.9 | 10.3 | 14.8 | 35,760 | 202 | 3,273 | 71,389 | 3.0 |
| Logan | 1.0 | D | 5.1 | 36.0 | D | 5.2 | 3.2 | 8.4 | 11.1 | 10,505 | 230 | 810 | 23,581 | 1.7 |
| Lorain | 0.6 | 0.0 | 5.9 | 22.2 | 4.8 | 7.1 | 3.4 | 12.1 | 17.7 | 68,505 | 221 | 7,276 | 133,755 | 5.3 |
| Lucas | 0.0 | 0.1 | 6.0 | 15.9 | 7.2 | 6.4 | 8.9 | 16.5 | 14.5 | 87,875 | 205 | 16,928 | 204,322 | 0.8 |
| Madison | 1.5 | 0.0 | 6.8 | 22.6 | 5.3 | 7.0 | 2.1 | D | 20.2 | 8,145 | 182 | 681 | 16,332 | 2.5 |
| Mahoning | 0.1 | 0.4 | 7.1 | 10.3 | 5.8 | 8.0 | 4.5 | 19.6 | 16.0 | 58,995 | 259 | 8,969 | 111,172 | -0.6 |
| Marion | 1.0 | D | 3.5 | 25.7 | 1.5 | 6.0 | 9.9 | 15.5 | 16.5 | 14,815 | 228 | 2,225 | 27,874 | 0.1 |
| Medina | 0.1 | 0.1 | 12.6 | 15.5 | 6.1 | 7.7 | 4.9 | 8.6 | 11.5 | 37,465 | 208 | 1,527 | 73,984 | 6.9 |
| Meigs | 0.4 | D | 10.4 | 3.3 | D | 10.4 | 5.0 | D | 29.2 | 5,705 | 249 | 1,068 | 11,247 | 0.5 |
| Mercer | 6.4 | D | 8.2 | 29.6 | 2.8 | 6.1 | 5.9 | 5.7 | 12.0 | 8,730 | 212 | 387 | 18,349 | 4.0 |
| Miami | 0.2 | 0.2 | 6.7 | 29.0 | D | 6.8 | 3.6 | 9.6 | 12.6 | 23,640 | 221 | 1,706 | 44,430 | 0.4 |
| Monroe | -1.7 | 9.2 | 14.0 | 1.9 | 2.9 | 5.3 | 8.9 | D | 25.6 | 3,995 | 291 | 367 | 7,480 | -1.1 |
| Montgomery | 0.0 | 0.1 | 5.1 | 12.6 | 10.7 | 4.8 | 7.2 | 19.3 | 14.4 | 112,535 | 211 | 15,733 | 255,092 | 0.1 |
| Morgan | -0.2 | D | 8.8 | 17.7 | 2.0 | 6.1 | 2.9 | 11.6 | 22.5 | 3,445 | 238 | 522 | 8,072 | 2.3 |
| Morrow | 1.1 | 0.5 | 13.3 | 17.8 | D | 7.4 | 2.6 | D | 23.2 | 7,535 | 213 | 607 | 14,457 | 2.1 |
| Muskingum | -0.1 | 2.2 | 6.0 | 9.7 | 6.2 | 8.9 | 4.0 | 21.5 | 15.0 | 21,225 | 247 | 3,398 | 37,927 | -0.4 |
| Noble | -1.6 | 3.5 | 5.5 | 6.8 | 7.3 | 5.1 | 4.0 | 6.6 | 35.8 | 2,760 | 191 | 241 | 6,242 | 3.1 |

1. Per 1,000 resident population estimated as of July 1 of the year shown.

# Table B. States and Counties — Housing, Labor Force, and Employment

| STATE County | Housing units, 2015-2019 Occupied units — Total | Percent | Owner-occupied Median value[1] | Median owner cost as a percent of income — With a mortgage | Without a mortgage[2] | Renter-occupied Median rent[3] | Median rent as a percent of income[2] | Sub-standard units[4] (percent) | Civilian labor force, 2020 — Total | Percent change, 2019-2020 | Unemployment — Total | Rate[5] | Civilian employment[6], 2015-2019 — Total | Percent Management, business, science, and arts | Construction, production, and maintenance occupations |
|---|---|---|---|---|---|---|---|---|---|---|---|---|---|---|---|
| | 89 | 90 | 91 | 92 | 93 | 94 | 95 | 96 | 97 | 98 | 99 | 100 | 101 | 102 | 103 |
| NORTH DAKOTA—Cont'd | | | | | | | | | | | | | | | |
| Ward | 28,277 | 59.5 | 213,100 | 18.8 | 10.0 | 981 | 25.3 | 2.7 | 31,847 | -0.3 | 1,899 | 6.0 | 35,164 | 33.8 | 25.0 |
| Wells | 1,943 | 78.7 | 94,800 | 18.5 | 10.0 | 620 | 26.9 | 0.7 | 1,920 | 1.4 | 84 | 4.4 | 1,886 | 35.7 | 26.5 |
| Williams | 14,320 | 52.6 | 247,400 | 17.3 | 10.0 | 968 | 19.7 | 4.3 | 23,001 | -7.3 | 2,205 | 9.6 | 18,813 | 29.0 | 38.8 |
| OHIO | 4,676,358 | 66.1 | 145,700 | 19.2 | 11.4 | 808 | 27.8 | 1.7 | 5,754,286 | -1.5 | 468,802 | 8.1 | 5,595,444 | 37.0 | 24.4 |
| Adams | 10,673 | 70.8 | 105,300 | 22.4 | 12.7 | 576 | 32.6 | 2.6 | 10,981 | 1.2 | 1,013 | 9.2 | 10,396 | 31.7 | 34.8 |
| Allen | 40,650 | 67.1 | 118,100 | 18.6 | 11.6 | 703 | 28.6 | 1.4 | 48,303 | 0.4 | 4,122 | 8.5 | 48,319 | 29.6 | 32.5 |
| Ashland | 20,417 | 73.8 | 128,400 | 18.6 | 11.1 | 725 | 23.7 | 3.0 | 26,492 | -1.6 | 1,709 | 6.5 | 25,570 | 30.3 | 32.6 |
| Ashtabula | 37,832 | 70.5 | 112,700 | 19.7 | 11.4 | 696 | 31.0 | 2.5 | 43,807 | -2.1 | 3,645 | 8.3 | 40,462 | 27.0 | 35.6 |
| Athens | 22,557 | 57.8 | 129,600 | 19.1 | 11.5 | 791 | 34.2 | 2.2 | 27,207 | -2.9 | 1,993 | 7.3 | 28,134 | 36.8 | 20.2 |
| Auglaize | 18,888 | 75.7 | 147,300 | 17.1 | 10.0 | 729 | 23.3 | 2.0 | 24,698 | -1.5 | 1,701 | 6.9 | 23,435 | 30.7 | 37.1 |
| Belmont | 25,919 | 76.4 | 100,500 | 17.2 | 10.0 | 629 | 25.2 | 0.8 | 28,888 | -4.1 | 2,918 | 10.1 | 28,957 | 29.2 | 28.6 |
| Brown | 17,829 | 72.4 | 132,400 | 19.4 | 11.1 | 719 | 28.5 | 1.7 | 19,407 | -1.1 | 1,601 | 8.2 | 19,654 | 28.5 | 33.2 |
| Butler | 139,113 | 68.1 | 172,900 | 19.1 | 10.8 | 883 | 27.9 | 1.6 | 194,077 | -1.2 | 14,048 | 7.2 | 183,889 | 38.3 | 22.5 |
| Carroll | 11,298 | 78.7 | 126,500 | 19.7 | 10.7 | 706 | 24.4 | 1.8 | 12,883 | -0.7 | 1,136 | 8.8 | 12,502 | 27.1 | 35.9 |
| Champaign | 15,159 | 73.2 | 132,800 | 18.9 | 11.2 | 696 | 22.7 | 1.7 | 19,995 | -0.8 | 1,421 | 7.1 | 18,419 | 26.5 | 39.0 |
| Clark | 54,696 | 66.0 | 115,500 | 19.1 | 10.2 | 749 | 26.7 | 1.8 | 62,945 | -1.1 | 5,116 | 8.1 | 60,958 | 28.6 | 31.6 |
| Clermont | 78,009 | 74.2 | 170,800 | 19.3 | 10.7 | 849 | 26.6 | 1.5 | 106,237 | -1.5 | 7,284 | 6.9 | 102,881 | 37.0 | 23.3 |
| Clinton | 16,528 | 64.6 | 128,400 | 19.1 | 10.9 | 755 | 27.2 | 2.3 | 17,778 | 0.0 | 1,479 | 8.3 | 19,738 | 30.1 | 33.3 |
| Columbiana | 41,537 | 73.5 | 109,400 | 18.6 | 10.5 | 663 | 27.0 | 1.8 | 47,380 | -0.6 | 4,452 | 9.4 | 45,391 | 27.6 | 34.0 |
| Coshocton | 14,476 | 74.0 | 99,200 | 18.4 | 11.3 | 615 | 24.6 | 2.8 | 14,138 | -2.2 | 1,258 | 8.9 | 15,877 | 26.8 | 38.2 |
| Crawford | 17,782 | 69.7 | 88,400 | 19.0 | 10.8 | 640 | 25.2 | 1.4 | 18,698 | 1.9 | 1,749 | 9.4 | 18,664 | 27.2 | 36.4 |
| Cuyahoga | 540,965 | 58.2 | 132,800 | 19.9 | 13.0 | 809 | 29.3 | 1.5 | 592,890 | -3.9 | 61,697 | 10.4 | 593,418 | 40.3 | 19.2 |
| Darke | 20,861 | 72.8 | 123,300 | 18.4 | 10.3 | 673 | 22.6 | 1.5 | 25,836 | -0.3 | 1,743 | 6.7 | 24,963 | 26.7 | 39.4 |
| Defiance | 15,297 | 76.3 | 119,900 | 18.2 | 10.8 | 747 | 25.0 | 1.3 | 17,908 | -1.3 | 1,471 | 8.2 | 18,861 | 29.4 | 36.0 |
| Delaware | 69,985 | 81.4 | 308,800 | 19.6 | 11.9 | 1,069 | 25.0 | 0.8 | 111,352 | -1.2 | 5,957 | 5.3 | 103,594 | 54.1 | 11.7 |
| Erie | 31,183 | 69.2 | 139,900 | 18.6 | 11.8 | 756 | 26.7 | 1.3 | 36,560 | -2.6 | 3,826 | 10.5 | 35,976 | 31.3 | 29.0 |
| Fairfield | 56,339 | 73.9 | 185,300 | 19.2 | 11.0 | 867 | 28.9 | 1.9 | 78,604 | -0.4 | 5,198 | 6.6 | 74,141 | 39.4 | 22.6 |
| Fayette | 11,773 | 64.9 | 113,500 | 18.8 | 10.0 | 718 | 26.8 | 1.5 | 14,023 | 0.5 | 1,032 | 7.4 | 12,744 | 28.3 | 35.7 |
| Franklin | 511,447 | 53.4 | 175,100 | 19.7 | 11.9 | 974 | 27.5 | 2.6 | 705,149 | 0.6 | 52,518 | 7.4 | 678,242 | 43.6 | 18.0 |
| Fulton | 16,506 | 79.1 | 138,200 | 18.4 | 10.0 | 729 | 26.2 | 2.2 | 21,991 | -2.5 | 1,740 | 7.9 | 21,240 | 28.2 | 36.0 |
| Gallia | 11,588 | 74.5 | 110,200 | 19.1 | 12.2 | 675 | 27.3 | 2.7 | 12,229 | -0.5 | 976 | 8.0 | 12,210 | 30.0 | 29.0 |
| Geauga | 35,298 | 86.3 | 240,900 | 19.9 | 11.1 | 818 | 26.3 | 2.6 | 46,329 | -6.7 | 3,119 | 6.7 | 47,910 | 40.7 | 22.8 |
| Greene | 65,604 | 66.1 | 173,000 | 18.6 | 10.0 | 910 | 26.1 | 1.2 | 82,640 | -1.6 | 5,303 | 6.4 | 78,781 | 45.6 | 16.1 |
| Guernsey | 16,220 | 71.5 | 114,700 | 19.4 | 10.5 | 669 | 32.7 | 3.0 | 18,296 | -2.2 | 1,588 | 8.7 | 16,732 | 28.1 | 33.0 |
| Hamilton | 341,873 | 57.9 | 155,400 | 19.2 | 12.1 | 810 | 28.7 | 1.8 | 415,866 | -0.7 | 32,306 | 7.8 | 406,743 | 42.9 | 18.3 |
| Hancock | 31,937 | 68.8 | 147,600 | 17.6 | 10.3 | 768 | 25.3 | 0.9 | 41,803 | -0.8 | 2,860 | 6.8 | 38,927 | 33.6 | 30.9 |
| Hardin | 11,678 | 71.0 | 105,300 | 18.8 | 10.0 | 666 | 23.8 | 3.2 | 14,080 | 0.7 | 1,138 | 8.1 | 14,679 | 28.6 | 38.1 |
| Harrison | 6,223 | 78.9 | 92,700 | 17.9 | 11.3 | 607 | 25.4 | 2.3 | 6,754 | -1.5 | 621 | 9.2 | 6,104 | 24.2 | 39.4 |
| Henry | 10,992 | 77.4 | 125,200 | 18.3 | 10.6 | 730 | 23.4 | 0.5 | 13,093 | -1.1 | 1,065 | 8.1 | 13,346 | 29.8 | 36.0 |
| Highland | 16,772 | 68.4 | 117,400 | 19.8 | 12.7 | 681 | 27.7 | 2.4 | 17,504 | 2.0 | 1,516 | 8.7 | 17,642 | 27.8 | 38.2 |
| Hocking | 11,286 | 75.1 | 132,800 | 19.0 | 11.1 | 619 | 25.9 | 3.2 | 12,961 | -0.6 | 959 | 7.4 | 12,504 | 28.9 | 30.6 |
| Holmes | 12,342 | 75.5 | 208,700 | 20.7 | 10.0 | 643 | 22.0 | 4.3 | 20,977 | 0.2 | 814 | 3.9 | 20,286 | 21.9 | 47.2 |
| Huron | 22,935 | 70.9 | 124,600 | 19.2 | 10.9 | 695 | 25.7 | 1.5 | 27,701 | -1.5 | 2,626 | 9.5 | 27,505 | 26.3 | 41.5 |
| Jackson | 12,780 | 69.2 | 97,400 | 21.1 | 12.2 | 704 | 23.7 | 3.4 | 13,157 | 2.6 | 1,149 | 8.7 | 13,553 | 30.4 | 32.9 |
| Jefferson | 27,428 | 69.0 | 93,000 | 17.4 | 11.4 | 675 | 28.4 | 1.1 | 27,496 | -1.7 | 2,775 | 10.1 | 27,769 | 29.1 | 28.8 |
| Knox | 23,248 | 72.0 | 155,500 | 19.9 | 10.8 | 737 | 26.5 | 1.5 | 31,244 | -1.6 | 1,929 | 6.2 | 28,638 | 33.8 | 29.1 |
| Lake | 96,361 | 74.2 | 156,200 | 19.6 | 11.3 | 897 | 26.5 | 1.2 | 120,876 | -5.5 | 10,136 | 8.4 | 118,331 | 37.6 | 23.6 |
| Lawrence | 23,221 | 72.2 | 106,500 | 19.6 | 12.0 | 733 | 29.1 | 1.5 | 23,849 | -0.5 | 2,024 | 8.5 | 24,555 | 29.9 | 27.5 |
| Licking | 64,035 | 72.9 | 170,200 | 19.5 | 11.4 | 852 | 27.5 | 2.0 | 90,688 | -0.5 | 5,935 | 6.5 | 83,927 | 36.5 | 24.7 |
| Logan | 18,677 | 74.2 | 131,900 | 18.0 | 11.0 | 725 | 24.4 | 2.7 | 22,803 | -0.7 | 1,892 | 8.3 | 21,676 | 28.3 | 39.4 |
| Lorain | 120,281 | 72.3 | 150,500 | 19.7 | 12.0 | 774 | 29.9 | 1.2 | 149,061 | -4.5 | 14,728 | 9.9 | 142,923 | 35.1 | 25.1 |
| Lucas | 179,930 | 59.7 | 116,600 | 19.2 | 12.2 | 745 | 28.3 | 1.5 | 209,345 | -0.9 | 21,679 | 10.4 | 200,130 | 34.2 | 25.7 |
| Madison | 15,013 | 71.2 | 166,500 | 18.9 | 11.1 | 858 | 21.8 | 2.4 | 20,834 | -0.6 | 1,236 | 5.9 | 19,293 | 32.2 | 31.8 |
| Mahoning | 98,472 | 68.7 | 105,400 | 19.0 | 11.5 | 682 | 29.1 | 1.5 | 101,288 | -2.4 | 10,298 | 10.2 | 104,974 | 32.8 | 24.6 |
| Marion | 24,737 | 66.4 | 100,800 | 18.5 | 10.9 | 731 | 29.8 | 1.9 | 28,331 | -1.2 | 2,160 | 7.6 | 26,631 | 26.1 | 35.5 |
| Medina | 69,007 | 79.9 | 196,300 | 19.1 | 10.5 | 874 | 26.3 | 1.5 | 93,378 | -5.9 | 7,124 | 7.6 | 93,280 | 41.6 | 21.5 |
| Meigs | 9,045 | 78.6 | 95,000 | 18.4 | 10.6 | 622 | 30.7 | 2.1 | 9,072 | -1.1 | 872 | 9.6 | 8,935 | 26.7 | 32.1 |
| Mercer | 16,234 | 77.0 | 153,800 | 18.1 | 10.0 | 679 | 21.5 | 1.6 | 23,515 | -0.2 | 1,265 | 5.4 | 21,099 | 30.1 | 40.6 |
| Miami | 41,043 | 71.1 | 151,400 | 18.3 | 10.1 | 767 | 25.1 | 1.0 | 54,163 | | 3,783 | 7.0 | 50,939 | 34.5 | 29.2 |
| Monroe | 5,745 | 78.3 | 110,000 | 18.0 | 10.0 | 593 | 33.5 | 2.8 | 5,719 | 6.4 | 606 | 10.6 | 5,138 | 24.2 | 38.3 |
| Montgomery | 224,328 | 61.0 | 119,800 | 19.3 | 12.2 | 793 | 28.1 | 1.6 | 253,421 | 0.1 | 21,730 | 8.6 | 246,370 | 37.2 | 22.4 |
| Morgan | 6,108 | 76.7 | 99,900 | 20.0 | 12.2 | 637 | 29.1 | 3.0 | 6,697 | | 620 | 9.3 | 5,792 | 27.1 | 39.4 |
| Morrow | 12,922 | 81.8 | 152,800 | 19.3 | 11.4 | 767 | 26.5 | 3.0 | 16,890 | -0.2 | 1,204 | 7.1 | 17,420 | 27.8 | 32.7 |
| Muskingum | 33,878 | 69.0 | 120,200 | 19.1 | 11.3 | 703 | 30.2 | 1.9 | 40,389 | 1.2 | 3,132 | 7.8 | 38,331 | 28.4 | 31.4 |
| Noble | 5,067 | 81.9 | 100,500 | 22.6 | 10.0 | 653 | 27.4 | 0.8 | 4,656 | -1.9 | 449 | 9.6 | 4,061 | 25.4 | 38.5 |

1. Specified owner-occupied units.    2. A value of 10.0 represents 10 percent or less; a value of 50.0 represents 50 percent or more.    3. Specified renter-occupied units.    4. Overcrowded or lacking complete plumbing facilities.    5. Percent of civilian labor force.    6. Civilian employed persons 16 years old and over.

| STATE County | Private nonfarm establishments, employment and payroll, 2019 | | | | | | | | | Agriculture, 2017 | | | |
| | | Employment | | | | | | Annual payroll | | Farms | | | Farm producers whose primary occupation is farming (percent) |
| | | | | | | | Professional, scientific, and technical services | | | | Percent with: | | |
| | Number of establish-ments | Total | Health care and social assistance | Manufac-turing | Retail trade | Finance and insurance | | Total (mil dol) | Average per employee (dollars) | Number | Fewer than 50 acres | 1000 acres or more | |
| | 104 | 105 | 106 | 107 | 108 | 109 | 110 | 111 | 112 | 113 | 114 | 115 | 116 |
| NORTH DAKOTA—Cont'd | | | | | | | | | | | | | |
| Ward | 2,043 | 26,390 | 4,992 | 445 | 5,074 | 983 | 1,169 | 1,348 | 51,091 | 718 | 14.5 | 43.6 | 55.8 |
| Wells | 177 | 1,295 | 413 | 63 | 244 | 78 | 17 | 47 | 36,499 | 435 | 5.5 | 47.4 | 62.2 |
| Williams | 1,547 | 22,156 | 1,460 | 337 | 2,240 | 388 | 737 | 1,640 | 74,029 | 569 | 12.3 | 46.2 | 54.1 |
| OHIO | 250,981 | 4,916,956 | 866,115 | 688,950 | 556,445 | 258,816 | 264,129 | 243,907 | 49,605 | 77,805 | 47.4 | 3.5 | 37.3 |
| Adams | 389 | 4,360 | 1,036 | 974 | 884 | 156 | 78 | 163 | 37,417 | 1,194 | 44.0 | 1.3 | 34.2 |
| Allen | 2,345 | 46,156 | 10,028 | 8,749 | 5,564 | 955 | 1,335 | 2,059 | 44,619 | 855 | 43.5 | 4.1 | 36.5 |
| Ashland | 1,038 | 18,909 | 3,039 | 4,242 | 2,164 | 271 | 1,485 | 693 | 36,670 | 1,122 | 44.8 | 1.8 | 44.5 |
| Ashtabula | 1,880 | 24,440 | 5,149 | 6,807 | 3,535 | 508 | 723 | 920 | 37,645 | 1,212 | 51.7 | 1.7 | 36.1 |
| Athens | 1,045 | 13,562 | 3,727 | 459 | 2,649 | 362 | 506 | 449 | 33,123 | 687 | 35.1 | 0.9 | 33.0 |
| Auglaize | 989 | 20,741 | 2,531 | 9,510 | 2,043 | 318 | 546 | 936 | 45,130 | 976 | 39.3 | 3.7 | 38.8 |
| Belmont | 1,375 | 18,953 | 3,287 | 807 | 3,883 | 453 | 517 | 792 | 41,776 | 750 | 30.7 | 1.6 | 38.6 |
| Brown | 545 | 5,621 | 1,161 | 832 | 967 | 172 | 121 | 170 | 30,231 | 1,237 | 42.3 | 3.3 | 38.8 |
| Butler | 7,259 | 141,966 | 16,769 | 21,749 | 17,482 | 9,214 | 5,899 | 6,847 | 48,227 | 997 | 57.1 | 2.5 | 34.0 |
| Carroll | 453 | 5,380 | 948 | 1,229 | 752 | 84 | 138 | 190 | 35,290 | 888 | 44.6 | 1.5 | 40.2 |
| Champaign | 584 | 9,594 | 879 | 4,173 | 909 | 193 | 250 | 417 | 43,445 | 860 | 46.4 | 5.5 | 40.7 |
| Clark | 2,198 | 39,729 | 6,958 | 6,142 | 5,191 | 2,359 | 709 | 1,497 | 37,673 | 742 | 56.7 | 7.0 | 36.2 |
| Clermont | 3,603 | 52,499 | 6,410 | 5,822 | 9,523 | 2,234 | 2,873 | 2,296 | 43,734 | 928 | 66.4 | 2.2 | 35.3 |
| Clinton | 724 | 15,131 | 1,705 | 3,689 | 1,579 | 352 | 409 | 722 | 47,698 | 747 | 47.8 | 8.0 | 39.8 |
| Columbiana | 1,955 | 26,882 | 5,249 | 6,472 | 4,060 | 502 | 436 | 947 | 35,214 | 1,227 | 51.9 | 1.5 | 36.8 |
| Coshocton | 627 | 8,585 | 1,604 | 2,611 | 1,259 | 188 | 192 | 324 | 37,764 | 1,191 | 39.0 | 2.2 | 36.7 |
| Crawford | 772 | 12,435 | 3,040 | 3,326 | 1,308 | 567 | 597 | 462 | 37,148 | 719 | 36.7 | 7.9 | 46.1 |
| Cuyahoga | 31,780 | 669,407 | 144,702 | 66,673 | 60,121 | 44,387 | 43,798 | 38,436 | 57,418 | 111 | 92.8 | NA | 49.2 |
| Darke | 1,161 | 17,073 | 2,140 | 6,028 | 1,934 | 545 | 332 | 727 | 42,590 | 1,658 | 45.2 | 4.4 | 38.9 |
| Defiance | 808 | 13,398 | 1,976 | 3,073 | 2,267 | 651 | 218 | 573 | 42,794 | 907 | 39.1 | 6.7 | 33.2 |
| Delaware | 4,811 | 86,243 | 8,718 | 5,835 | 12,813 | 15,631 | 4,561 | 4,604 | 53,385 | 803 | 63.4 | 3.9 | 35.7 |
| Erie | 1,805 | 32,581 | 5,491 | 6,325 | 4,641 | 660 | 672 | 1,296 | 39,785 | 382 | 49.5 | 4.7 | 41.5 |
| Fairfield | 2,687 | 37,156 | 7,619 | 5,359 | 6,540 | 769 | 1,178 | 1,399 | 37,652 | 1,117 | 57.6 | 3.8 | 37.6 |
| Fayette | 589 | 9,831 | 1,196 | 1,810 | 2,446 | 141 | 66 | 345 | 35,120 | 491 | 46.2 | 12.8 | 45.6 |
| Franklin | 29,192 | 670,124 | 134,570 | 32,159 | 65,517 | 55,503 | 47,489 | 36,275 | 54,131 | 408 | 67.6 | 2.5 | 35.1 |
| Fulton | 935 | 16,107 | 1,916 | 7,685 | 1,695 | 394 | 233 | 718 | 44,556 | 785 | 44.6 | 6.5 | 39.8 |
| Gallia | 571 | 9,549 | 3,103 | 559 | 1,476 | 338 | 89 | 365 | 38,200 | 990 | 36.1 | 0.7 | 33.2 |
| Geauga | 2,753 | 30,039 | 4,977 | 7,467 | 4,172 | 649 | 1,223 | 1,309 | 43,592 | 1,049 | 66.7 | 0.6 | 31.2 |
| Greene | 3,260 | 56,628 | 8,216 | 3,143 | 10,001 | 1,493 | 12,158 | 2,673 | 47,194 | 817 | 59.0 | 6.1 | 38.4 |
| Guernsey | 844 | 13,689 | 2,994 | 3,114 | 1,549 | 231 | 316 | 550 | 40,200 | 1,103 | 35.7 | 0.9 | 37.5 |
| Hamilton | 20,916 | 498,435 | 89,665 | 46,470 | 44,164 | 35,049 | 41,322 | 30,808 | 61,810 | 318 | 74.2 | 0.3 | 35.1 |
| Hancock | 1,663 | 44,225 | 6,194 | 11,296 | 4,203 | 685 | 1,350 | 2,159 | 48,820 | 887 | 40.2 | 7.0 | 34.8 |
| Hardin | 437 | 7,045 | 874 | 1,812 | 863 | 213 | 153 | 237 | 33,711 | 726 | 36.0 | 8.8 | 45.5 |
| Harrison | 266 | 3,093 | 483 | 337 | 322 | 46 | 64 | 135 | 43,495 | 458 | 27.1 | 4.1 | 46.0 |
| Henry | 544 | 8,023 | 1,301 | 2,451 | 861 | 270 | 181 | 331 | 41,203 | 841 | 38.9 | 5.6 | 39.8 |
| Highland | 676 | 8,585 | 1,690 | 1,803 | 1,567 | 585 | 109 | 321 | 37,443 | 1,254 | 37.6 | 6.0 | 38.2 |
| Hocking | 479 | 5,250 | 974 | 887 | 872 | 137 | 77 | 164 | 31,228 | 377 | 44.8 | 0.3 | 32.9 |
| Holmes | 1,305 | 18,572 | 1,603 | 6,615 | 2,314 | 429 | 359 | 724 | 39,001 | 1,673 | 43.5 | 0.7 | 39.1 |
| Huron | 1,085 | 18,201 | 2,697 | 6,106 | 2,027 | 413 | 407 | 816 | 44,824 | 810 | 43.1 | 7.3 | 44.9 |
| Jackson | 619 | 8,457 | 1,316 | 2,891 | 1,431 | 235 | 152 | 296 | 34,964 | 508 | 38.8 | 1.4 | 36.3 |
| Jefferson | 1,232 | 18,659 | 3,846 | 1,616 | 2,834 | 426 | 313 | 753 | 40,345 | 599 | 31.6 | NA | 35.2 |
| Knox | 1,105 | 18,353 | 3,127 | 4,016 | 2,069 | 487 | 361 | 770 | 41,931 | 1,338 | 50.8 | 2.5 | 35.7 |
| Lake | 5,611 | 86,547 | 11,543 | 19,888 | 11,859 | 1,314 | 3,597 | 4,311 | 49,814 | 214 | 70.6 | 0.5 | 41.4 |
| Lawrence | 794 | 10,342 | 2,614 | 870 | 2,004 | 276 | 194 | 354 | 34,235 | 531 | 30.1 | 0.2 | 32.7 |
| Licking | 3,042 | 55,592 | 7,997 | 10,827 | 6,910 | 2,933 | 2,416 | 2,292 | 41,231 | 1,583 | 57.6 | 2.7 | 35.3 |
| Logan | 829 | 16,905 | 1,725 | 4,860 | 1,664 | 295 | 708 | 782 | 46,266 | 1,009 | 51.3 | 5.5 | 37.6 |
| Lorain | 5,516 | 85,744 | 15,436 | 16,412 | 12,965 | 1,882 | 3,299 | 3,810 | 44,438 | 1,001 | 63.1 | 2.2 | 34.4 |
| Lucas | 9,367 | 195,017 | 40,095 | 26,553 | 23,012 | 6,088 | 9,262 | 9,220 | 47,280 | 386 | 65.8 | 3.1 | 43.5 |
| Madison | 679 | 14,507 | 1,166 | 4,308 | 2,424 | 122 | 548 | 659 | 45,403 | 789 | 46.8 | 9.0 | 44.0 |
| Mahoning | 5,360 | 85,266 | 19,702 | 9,233 | 12,003 | 2,176 | 2,527 | 3,346 | 39,247 | 774 | 57.9 | 0.9 | 38.0 |
| Marion | 1,088 | 22,200 | 4,939 | 6,407 | 2,723 | 344 | 276 | 877 | 39,513 | 615 | 44.1 | 8.3 | 41.5 |
| Medina | 3,978 | 54,831 | 7,511 | 9,455 | 8,441 | 2,645 | 2,033 | 2,448 | 44,647 | 1,149 | 71.6 | 1.6 | 36.7 |
| Meigs | 322 | 2,698 | 653 | 27 | 647 | 132 | 64 | 78 | 28,858 | 515 | 30.7 | 1.0 | 37.5 |
| Mercer | 1,028 | 16,907 | 2,120 | 4,556 | 1,888 | 584 | 352 | 681 | 40,271 | 1,231 | 40.4 | 4.4 | 38.9 |
| Miami | 2,101 | 36,404 | 4,411 | 10,957 | 4,731 | 701 | 980 | 1,535 | 42,160 | 1,037 | 57.7 | 3.9 | 34.7 |
| Monroe | 259 | 1,838 | 271 | 116 | 328 | 144 | 44 | 63 | 34,200 | 808 | 27.5 | 0.5 | 35.7 |
| Montgomery | 11,044 | 234,554 | 53,115 | 32,147 | 25,614 | 11,276 | 13,363 | 11,901 | 50,740 | 781 | 63.9 | 2.4 | 38.0 |
| Morgan | 158 | 2,253 | 486 | 815 | 279 | 78 | 29 | 84 | 37,249 | 530 | 29.4 | 1.3 | 39.4 |
| Morrow | 405 | 3,864 | 921 | 892 | 468 | 72 | 207 | 133 | 34,540 | 865 | 53.8 | 5.0 | 38.2 |
| Muskingum | 1,710 | 28,458 | 6,243 | 2,547 | 4,385 | 764 | 481 | 1,223 | 42,974 | 1,263 | 39.6 | 1.6 | 32.0 |
| Noble | 219 | 1,906 | 364 | 156 | 288 | 88 | 47 | 62 | 32,424 | 593 | 32.7 | 1.0 | 36.6 |

| STATE County | Acreage (1,000) [117] | Percent change, 2012-2017 [118] | Average size of farm [119] | Total irrigated (1,000) [120] | Total cropland (1,000) [121] | Average per farm [122] | Average per acre [123] | Value of machinery and equipment, average per farm (dollars) [124] | Total (mil dol) [125] | Average per farm (acres) [126] | Crops [127] | Livestock and poultry products [128] | Organic farms (number) [129] | Farms with internet access (percent) [130] | Total ($1,000) [131] | Percent of farms [132] |
|---|---|---|---|---|---|---|---|---|---|---|---|---|---|---|---|---|
| NORTH DAKOTA—Cont'd | | | | | | | | | | | | | | | | |
| Ward | 1,153 | 7.5 | 1,607 | 0.5 | 967.5 | 2,623,383 | 1,633 | 373,715 | 206.0 | 286,890 | 91.5 | 8.5 | 2 | 88.9 | 20,719 | 76.3 |
| Wells | 786 | 6.4 | 1,806 | 1.0 | 664.8 | 3,089,877 | 1,711 | 628,953 | 208.2 | 478,733 | 88.8 | 11.2 | 4 | 80.9 | 10,874 | 80.2 |
| Williams | 999 | -6.0 | 1,756 | 21.1 | 754.4 | 1,925,239 | 1,096 | 457,175 | 131.8 | 231,601 | 89.1 | 10.9 | NA | 75.0 | 9,204 | 58.2 |
| OHIO | 13,965 | 0.0 | 179 | 50.7 | 10,960.7 | 1,112,700 | 6,199 | 129,614 | 9,341.2 | 120,059 | 58.1 | 41.9 | 872 | 74.6 | 351,125 | 36.7 |
| Adams | 166 | -3.7 | 139 | 0.3 | 89.2 | 539,592 | 3,882 | 76,734 | 40.1 | 33,599 | 62.5 | 37.5 | 6 | 67.5 | 1,814 | 38.1 |
| Allen | 187 | 1.9 | 218 | 0.4 | 172.7 | 1,583,677 | 7,256 | 163,584 | 139.9 | 163,639 | 61.4 | 38.6 | 2 | 77.8 | 5,704 | 67.7 |
| Ashland | 161 | 5.1 | 143 | 0.2 | 120.8 | 870,121 | 6,075 | 100,864 | 113.8 | 101,381 | 38.9 | 61.1 | 57 | 74.8 | 3,821 | 29.9 |
| Ashtabula | 154 | -7.4 | 127 | 0.3 | 101.1 | 554,349 | 4,373 | 92,598 | 57.9 | 47,762 | 66.2 | 33.8 | 12 | 69.1 | 592 | 11.2 |
| Athens | 99 | 9.1 | 144 | 0.1 | 30.4 | 461,747 | 3,213 | 56,167 | 11.4 | 16,640 | 50.4 | 49.6 | 3 | 76.1 | 394 | 13.7 |
| Auglaize | 210 | 0.0 | 215 | 1.2 | 194.6 | 1,751,762 | 8,141 | 189,133 | 206.9 | 211,992 | 47.4 | 52.6 | 3 | 82.8 | 8,649 | 72.7 |
| Belmont | 129 | 14.2 | 172 | 0.1 | 41.4 | 660,470 | 3,829 | 94,858 | 25.4 | 33,809 | 23.2 | 76.8 | 4 | 73.1 | 50 | 3.1 |
| Brown | 208 | 0.7 | 168 | 0.1 | 154.7 | 741,026 | 4,408 | 124,535 | 71.7 | 57,969 | 88.0 | 12.0 | 8 | 73.2 | 4,162 | 38.1 |
| Butler | 124 | -15.2 | 124 | 0.2 | 92.1 | 1,026,003 | 8,255 | 89,736 | 54.9 | 55,068 | 74.3 | 25.7 | 4 | 80.0 | 2,683 | 25.2 |
| Carroll | 111 | 4.2 | 125 | 0.3 | 64.3 | 587,003 | 4,710 | 97,626 | 48.6 | 54,760 | 39.0 | 61.0 | 4 | 65.9 | 877 | 14.0 |
| Champaign | 189 | -0.6 | 220 | 3.8 | 168.7 | 1,575,686 | 7,170 | 159,256 | 119.6 | 139,053 | 82.9 | 17.1 | 4 | 82.0 | 7,634 | 51.2 |
| Clark | 171 | -1.9 | 230 | 1.8 | 151.4 | 1,738,039 | 7,542 | 165,061 | 126.5 | 170,443 | 79.2 | 20.8 | NA | 80.5 | 6,001 | 43.3 |
| Clermont | 97 | -19.6 | 105 | 0.1 | 65.5 | 587,976 | 5,605 | 86,038 | 31.8 | 34,234 | 89.5 | 10.5 | 12 | 79.1 | 2,588 | 12.9 |
| Clinton | 213 | 2.2 | 285 | 0.0 | 190.7 | 1,697,120 | 5,958 | 185,154 | 116.9 | 156,456 | 92.1 | 7.9 | 4 | 79.5 | 8,430 | 53.4 |
| Columbiana | 142 | 11.4 | 116 | 0.3 | 95.9 | 669,684 | 5,769 | 98,981 | 106.7 | 86,932 | 32.4 | 67.6 | 3 | 74.2 | 1,687 | 19.4 |
| Coshocton | 183 | 7.5 | 153 | 1.2 | 97.2 | 809,143 | 5,279 | 97,607 | 99.1 | 83,219 | 33.6 | 66.4 | 29 | 64.7 | 2,541 | 26.4 |
| Crawford | 238 | -0.7 | 331 | 0.1 | 220.9 | 2,135,734 | 6,446 | 242,518 | 234.0 | 325,403 | 49.6 | 50.4 | 1 | 71.9 | 11,326 | 68.3 |
| Cuyahoga | 2 | -13.8 | 20 | 0.1 | 0.6 | 330,393 | 16,314 | 57,328 | 6.2 | 56,072 | 98.6 | 1.4 | 4 | 89.2 | 14 | 3.6 |
| Darke | 344 | 1.1 | 207 | 0.2 | 316.1 | 1,724,791 | 8,319 | 194,900 | 516.2 | 311,335 | 31.3 | 68.7 | 6 | 77.1 | 8,950 | 60.6 |
| Defiance | 228 | 1.4 | 252 | 1.0 | 205.8 | 1,444,799 | 5,736 | 147,092 | 107.3 | 118,279 | 74.8 | 25.2 | 6 | 78.4 | 5,649 | 80.6 |
| Delaware | 133 | -5.7 | 165 | 0.9 | 117.5 | 1,294,690 | 7,824 | 130,793 | 86.9 | 108,172 | 89.2 | 10.8 | 7 | 85.2 | 3,590 | 35.5 |
| Erie | 86 | 3.7 | 226 | D | 77.1 | 1,480,849 | 6,544 | 176,619 | 94.2 | 246,610 | 92.3 | 7.7 | 6 | 82.2 | 3,052 | 42.1 |
| Fairfield | 188 | -8.8 | 169 | 0.1 | 153.6 | 1,299,166 | 7,702 | 121,720 | 99.8 | 89,303 | 77.6 | 22.4 | 5 | 83.0 | 7,261 | 41.6 |
| Fayette | 204 | 3.9 | 416 | 0.1 | 189.9 | 2,971,045 | 7,142 | 259,936 | 127.2 | 259,059 | 83.3 | 16.7 | 3 | 76.4 | 6,900 | 61.9 |
| Franklin | 52 | -15.6 | 128 | 0.4 | 45.0 | 1,031,825 | 8,041 | 106,842 | 52.2 | 127,841 | 85.1 | 14.9 | 9 | 89.5 | 1,272 | 25.7 |
| Fulton | 196 | 0.5 | 250 | 0.9 | 182.2 | 1,746,704 | 6,985 | 191,817 | 173.1 | 220,513 | 71.0 | 29.0 | 2 | 78.3 | 7,158 | 59.6 |
| Gallia | 119 | 2.4 | 120 | 0.2 | 37.6 | 376,628 | 3,143 | 67,632 | 19.0 | 19,167 | 49.0 | 51.0 | NA | 62.3 | 511 | 8.8 |
| Geauga | 70 | 4.6 | 67 | 0.6 | 33.6 | 482,798 | 7,245 | 60,589 | 36.1 | 34,417 | 48.4 | 51.6 | 22 | 61.7 | 370 | 4.3 |
| Greene | 168 | 15.0 | 205 | 0.4 | 149.6 | 1,465,596 | 7,140 | 162,563 | 97.1 | 118,832 | 90.8 | 9.2 | 10 | 82.4 | 7,461 | 41.0 |
| Guernsey | 152 | 5.6 | 138 | 0.1 | 59.1 | 512,427 | 3,722 | 71,042 | 26.8 | 24,287 | 39.6 | 60.4 | 11 | 66.8 | 520 | 10.1 |
| Hamilton | 18 | -16.9 | 57 | 0.2 | 8.4 | 751,176 | 13,293 | 54,805 | 23.0 | 72,443 | 60.8 | 39.2 | 3 | 79.2 | 189 | 3.8 |
| Hancock | 240 | 4.2 | 271 | D | 226.7 | 1,634,840 | 6,042 | 184,379 | 135.8 | 153,091 | 80.1 | 19.9 | 4 | 78.4 | 9,965 | 67.0 |
| Hardin | 262 | 5.6 | 361 | 0.5 | 244.0 | 2,122,858 | 5,888 | 210,499 | 222.9 | 306,978 | 48.1 | 51.9 | 13 | 77.8 | 7,663 | 65.0 |
| Harrison | 99 | 4.1 | 217 | 0.0 | 41.0 | 717,980 | 3,310 | 97,439 | 18.6 | 40,688 | 42.0 | 58.0 | NA | 71.4 | 397 | 7.2 |
| Henry | 235 | -0.4 | 279 | 0.5 | 225.2 | 1,829,400 | 6,550 | 211,986 | 133.4 | 158,595 | 87.7 | 12.3 | 1 | 79.3 | 7,048 | 73.6 |
| Highland | 288 | 8.9 | 230 | 0.2 | 224.7 | 1,153,001 | 5,021 | 136,788 | 122.9 | 98,008 | 77.1 | 22.9 | 20 | 70.7 | 10,846 | 56.1 |
| Hocking | 38 | 0.7 | 102 | 0.1 | 14.1 | 460,594 | 4,527 | 55,840 | 5.1 | 13,504 | 80.6 | 19.4 | NA | 77.5 | 655 | 13.8 |
| Holmes | 174 | -21.3 | 104 | 0.3 | 104.2 | 804,801 | 7,741 | 82,807 | 182.1 | 108,839 | 18.4 | 81.6 | 134 | 38.0 | 1,509 | 9.0 |
| Huron | 241 | 0.9 | 297 | D | 213.9 | 1,806,321 | 6,083 | 226,973 | 200.0 | 246,863 | 67.5 | 32.5 | 6 | 73.1 | 8,556 | 48.4 |
| Jackson | 67 | -5.9 | 133 | 0.1 | 27.3 | 373,579 | 2,814 | 62,710 | 11.0 | 21,734 | 47.1 | 52.9 | 1 | 67.1 | 632 | 23.4 |
| Jefferson | 77 | 12.7 | 129 | 0.0 | 33.0 | 702,325 | 5,464 | 80,469 | 9.2 | 15,351 | 46.0 | 54.0 | NA | 75.6 | 188 | 8.2 |
| Knox | 194 | 4.5 | 145 | 0.1 | 140.0 | 878,830 | 6,047 | 110,224 | 135.1 | 101,004 | 46.0 | 54.0 | 37 | 71.1 | 3,456 | 31.1 |
| Lake | 13 | -23.5 | 61 | 1.6 | 8.3 | 499,558 | 8,162 | 101,442 | 73.6 | 344,028 | 98.7 | 1.3 | NA | 79.9 | 173 | 7.9 |
| Lawrence | 62 | -4.0 | 117 | 0.1 | 16.0 | 383,896 | 3,287 | 57,780 | 4.0 | 7,597 | 59.8 | 40.2 | NA | 68.7 | 238 | 10.5 |
| Licking | 220 | -1.6 | 139 | 0.3 | 160.7 | 921,506 | 6,616 | 120,427 | 185.4 | 117,118 | 45.1 | 54.9 | 13 | 84.5 | 4,217 | 20.2 |
| Logan | 211 | -0.8 | 209 | 0.1 | 183.9 | 1,266,978 | 6,051 | 136,955 | 121.7 | 120,638 | 70.7 | 29.3 | 16 | 76.5 | 6,395 | 45.4 |
| Lorain | 126 | 2.5 | 126 | 0.6 | 105.1 | 939,911 | 7,484 | 118,302 | 133.9 | 133,767 | 87.1 | 12.9 | 11 | 80.8 | 2,362 | 27.9 |
| Lucas | 66 | 4.0 | 170 | 1.2 | 61.9 | 1,405,729 | 8,277 | 157,847 | 50.7 | 131,288 | 93.1 | 6.9 | 7 | 81.9 | 2,319 | 42.7 |
| Madison | 252 | -4.1 | 320 | 0.1 | 235.3 | 2,218,215 | 6,934 | 215,221 | 159.3 | 201,844 | 78.8 | 21.2 | 25 | 84.7 | 7,372 | 56.1 |
| Mahoning | 75 | -0.5 | 96 | 0.6 | 56.4 | 638,290 | 6,626 | 101,413 | 68.6 | 88,630 | 35.8 | 64.2 | 7 | 77.8 | 1,277 | 16.9 |
| Marion | 204 | 7.7 | 331 | 0.2 | 190.3 | 1,935,205 | 5,838 | 231,900 | 135.9 | 221,015 | 63.5 | 36.5 | 7 | 82.9 | 7,767 | 63.6 |
| Medina | 99 | 4.6 | 86 | 0.9 | 78.6 | 677,096 | 7,833 | 80,359 | 51.5 | 44,842 | 74.1 | 25.9 | 19 | 78.7 | 1,563 | 14.7 |
| Meigs | 78 | 3.5 | 152 | 0.1 | 31.4 | 528,124 | 3,467 | 92,831 | 16.6 | 32,254 | 64.5 | 35.5 | 1 | 75.3 | 372 | 10.9 |
| Mercer | 269 | -1.5 | 218 | 0.2 | 248.5 | 2,113,330 | 9,673 | 229,931 | 631.6 | 513,089 | 19.5 | 80.5 | 6 | 82.3 | 8,719 | 67.9 |
| Miami | 173 | -6.0 | 167 | 1.9 | 158.3 | 1,239,097 | 7,421 | 114,326 | 106.7 | 102,889 | 86.2 | 13.8 | 8 | 83.6 | 4,405 | 53.6 |
| Monroe | 108 | -3.1 | 133 | 0.1 | 31.1 | 435,691 | 3,268 | 79,164 | 14.0 | 17,280 | 35.0 | 65.0 | 3 | 66.3 | 69 | 1.2 |
| Montgomery | 113 | -8.9 | 145 | 0.5 | 91.4 | 1,044,818 | 7,214 | 118,999 | 78.7 | 100,784 | 88.0 | 12.0 | 4 | 78.0 | 3,063 | 37.8 |
| Morgan | 99 | 4.2 | 187 | 0.1 | 40.3 | 647,339 | 3,458 | 88,037 | 18.0 | 33,972 | 39.8 | 60.2 | 3 | 73.2 | 244 | 14.5 |
| Morrow | 165 | -1.5 | 191 | 0.3 | 139.1 | 1,146,537 | 6,002 | 132,894 | 84.2 | 97,341 | 73.1 | 26.9 | 12 | 79.7 | 3,800 | 35.6 |
| Muskingum | 189 | 9.1 | 150 | 0.4 | 88.1 | 613,803 | 4,101 | 90,918 | 70.1 | 55,482 | 41.1 | 58.9 | 4 | 70.5 | 2,539 | 16.9 |
| Noble | 80 | -7.0 | 135 | 0.1 | 24.4 | 415,865 | 3,078 | 63,887 | 7.3 | 12,304 | 36.3 | 63.7 | 5 | 62.4 | 27 | 2.2 |

# Table B. States and Counties — Water Use, Wholesale Trade, Retail Trade, and Real Estate

| STATE County | Water use, 2015 | | Wholesale Trade[1], 2017 | | | | Retail Trade[2], 2017 | | | | Real estate and rental and leasing,[2] 2017 | | | |
|---|---|---|---|---|---|---|---|---|---|---|---|---|---|---|
| | Public supply water withdrawn (mil gal/day) | Public supply gallons withdrawn per person per day | Number of establishments | Number of employees | Sales (mil dol) | Average payroll (mil dol) | Number of establishments | Number of employees | Sales (mil dol) | Average payroll (mil dol) | Number of establishments | Number of employees | Sales (mil dol) | Average payroll (mil dol) |
| | 133 | 134 | 135 | 136 | 137 | 138 | 139 | 140 | 141 | 142 | 143 | 144 | 145 | 146 |
| **NORTH DAKOTA—Cont'd** | | | | | | | | | | | | | | |
| Ward | 6.87 | 96.4 | 110 | 1,549 | 2,011.3 | 95.7 | 310 | 5,462 | 1,483.8 | 153.3 | D | D | D | D |
| Wells | 0.36 | 86.4 | D | D | D | D | 32 | 214 | 61.5 | 5.2 | D | D | D | D |
| Williams | 9.17 | 259.8 | 107 | 1,461 | 1,233.9 | 95.3 | 144 | 2,164 | 842.7 | 78.7 | 124 | 679 | 242.6 | 53.9 |
| **OHIO** | 1,306.28 | 112.5 | 11,430 | 193,412 | 172,949.1 | 11,316.0 | 35,500 | 588,060 | 174,299.7 | 14,861.4 | 10,782 | 62,902 | 20,524.3 | 2,986.2 |
| Adams | 2.05 | 73.2 | D | D | D | 4.1 | 88 | 884 | 262.5 | 22.2 | 12 | 34 | 5.9 | 0.7 |
| Allen | 18.17 | 174.0 | 123 | 2,106 | 1,905.2 | 104.2 | 388 | 5,898 | 1,678.2 | 140.1 | 84 | 314 | 64.8 | 9.3 |
| Ashland | 2.99 | 56.2 | 46 | 401 | 301.8 | 21.7 | 150 | 2,126 | 506.2 | 52.5 | 37 | 160 | 18.3 | 4.8 |
| Ashtabula | 7.20 | 73.0 | D | D | D | D | 304 | 3,341 | 1,017.4 | 81.3 | 63 | 210 | 38.0 | 6.8 |
| Athens | 7.51 | 114.0 | D | D | D | D | 175 | 2,678 | 742.3 | 65.3 | 57 | 208 | 37.2 | 5.5 |
| Auglaize | 5.27 | 114.9 | D | D | D | D | 148 | 2,208 | 502.2 | 47.6 | 31 | 174 | 12.5 | 7.3 |
| Belmont | 7.61 | 110.0 | 36 | 361 | 217.6 | 18.5 | 291 | 4,123 | 1,126.0 | 92.9 | 58 | 330 | 69.1 | 11.5 |
| Brown | 3.06 | 69.8 | 19 | 114 | 60.0 | 4.5 | 97 | 1,281 | 334.3 | 23.6 | D | D | D | 0.9 |
| Butler | 50.69 | 134.7 | 453 | 10,303 | 11,235.7 | 612.9 | 1,024 | 20,813 | 8,376.7 | 517.9 | 314 | 1,355 | 370.6 | 54.4 |
| Carroll | 0.84 | 30.2 | 15 | 160 | 73.9 | 7.6 | 65 | 739 | 215.7 | 18.4 | 17 | 108 | 21.0 | 4.1 |
| Champaign | 2.49 | 63.9 | 25 | 201 | 132.8 | 8.3 | 83 | 914 | 309.9 | 22.8 | 25 | 61 | 12.0 | 2.3 |
| Clark | 16.17 | 118.9 | 84 | 1,661 | 1,808.1 | 81.9 | 373 | 5,695 | 1,411.4 | 122.3 | 89 | 380 | 63.1 | 11.0 |
| Clermont | 18.46 | 91.4 | 144 | 1,411 | 919.4 | 92.1 | 505 | 10,682 | 3,014.6 | 256.7 | 187 | 797 | 202.6 | 34.7 |
| Clinton | 0.88 | 21.0 | 23 | 379 | 470.1 | 16.4 | 127 | 1,919 | 480.7 | 42.1 | 28 | 116 | 23.3 | 3.2 |
| Columbiana | 9.18 | 87.6 | D | D | D | 39.3 | 338 | 3,969 | 1,229.4 | 102.9 | 51 | 226 | 36.1 | 7.0 |
| Coshocton | 6.59 | 180.2 | D | D | D | 4.1 | 106 | 1,280 | 341.0 | 29.0 | 16 | 43 | 6.8 | 0.8 |
| Crawford | 2.41 | 57.0 | 28 | 373 | 243.4 | 16.7 | 121 | 1,334 | 360.3 | 33.2 | 23 | 93 | 9.4 | 1.6 |
| Cuyahoga | 222.45 | 177.1 | 1,824 | 33,298 | 24,854.2 | 2,021.8 | 4,125 | 61,527 | 16,920.7 | 1,581.5 | 1,607 | 13,273 | 4,519.0 | 735.5 |
| Darke | 3.00 | 57.6 | 52 | 490 | 412.6 | 25.8 | 164 | 2,075 | 516.6 | 48.3 | 39 | 101 | 16.1 | 2.9 |
| Defiance | 3.40 | 88.7 | 33 | 472 | 374.6 | 24.0 | 141 | 2,535 | 847.1 | 67.0 | 27 | 95 | 14.9 | 2.4 |
| Delaware | 19.66 | 101.9 | 184 | 2,091 | 2,187.9 | 119.6 | 630 | 12,733 | 4,101.9 | 312.2 | 204 | 1,133 | 310.0 | 58.7 |
| Erie | 11.99 | 158.7 | 54 | 1,853 | 866.9 | 99.8 | 307 | 4,780 | 1,244.0 | 107.8 | 70 | 287 | 43.7 | 10.4 |
| Fairfield | 9.20 | 60.8 | D | D | D | 39.5 | 383 | 6,853 | 1,966.6 | 171.0 | 143 | 546 | 106.8 | 20.6 |
| Fayette | 1.99 | 69.4 | 30 | 529 | 4,097.6 | 26.3 | 177 | 2,685 | 611.8 | 52.8 | 13 | 50 | 20.5 | 1.1 |
| Franklin | 155.82 | 124.5 | 1,267 | 28,719 | 32,205.0 | 1,767.3 | 3,597 | 70,803 | 25,665.4 | 2,065.8 | 1,638 | 11,476 | 3,534.9 | 593.4 |
| Fulton | 2.54 | 59.7 | 45 | 424 | 229.7 | 19.5 | 147 | 1,713 | 504.3 | 43.6 | 19 | 85 | 9.4 | 2.7 |
| Gallia | 3.47 | 115.1 | D | D | D | D | 126 | 1,557 | 392.9 | 35.6 | 17 | 39 | 9.8 | 1.2 |
| Geauga | 1.45 | 15.4 | 145 | 1,176 | 585.4 | 68.9 | 311 | 4,382 | 1,504.3 | 127.4 | 80 | 207 | 41.0 | 9.0 |
| Greene | 8.98 | 54.6 | 78 | 757 | 491.3 | 41.8 | 553 | 11,189 | 2,591.4 | 240.8 | 130 | 445 | 126.7 | 14.6 |
| Guernsey | 5.45 | 138.8 | 23 | 461 | 294.6 | 16.3 | 139 | 1,644 | 532.3 | 39.0 | 32 | 166 | 41.2 | 7.2 |
| Hamilton | 114.75 | 142.1 | 1,049 | 20,157 | 16,309.5 | 1,226.0 | 2,683 | 51,158 | 13,273.6 | 1,259.7 | 1,077 | 6,860 | 1,848.1 | 387.1 |
| Hancock | 11.28 | 149.3 | 74 | 1,054 | 1,095.8 | 59.7 | 257 | 4,444 | 1,244.3 | 110.7 | 61 | 492 | 76.9 | 18.0 |
| Hardin | 2.23 | 70.4 | 15 | 101 | 127.3 | 4.6 | 84 | 918 | 220.5 | 19.0 | 12 | 32 | 4.8 | 0.7 |
| Harrison | 0.60 | 38.8 | D | D | D | D | 36 | 311 | 87.1 | 6.6 | 9 | 112 | 6.5 | 2.5 |
| Henry | 2.96 | 106.4 | 25 | 193 | 198.1 | 8.8 | 64 | 857 | 250.8 | 18.5 | 12 | 57 | 18.3 | 2.0 |
| Highland | 2.30 | 53.5 | D | D | D | 4.1 | 132 | 1,720 | 436.2 | 38.3 | 26 | 63 | 11.4 | 1.7 |
| Hocking | 1.67 | 58.6 | D | D | D | D | 72 | 905 | 237.8 | 21.2 | 22 | 100 | 11.5 | 2.5 |
| Holmes | 1.80 | 41.0 | 70 | 816 | 478.0 | 35.1 | 178 | 2,385 | 643.4 | 63.4 | 22 | 133 | 24.3 | 6.2 |
| Huron | 5.54 | 94.8 | D | D | D | D | 162 | 2,070 | 687.0 | 49.7 | 43 | 143 | 23.1 | 4.7 |
| Jackson | 1.47 | 45.1 | 19 | 131 | 102.8 | 5.6 | 127 | 1,485 | 377.9 | 33.8 | 24 | 65 | 12.1 | 1.5 |
| Jefferson | 8.77 | 130.2 | D | D | D | D | 211 | 3,074 | 841.1 | 72.9 | 42 | 179 | 26.9 | 5.5 |
| Knox | 5.47 | 89.6 | 48 | 457 | 218.8 | 19.1 | 168 | 2,178 | 623.0 | 53.1 | 41 | 142 | 25.4 | 4.2 |
| Lake | 22.82 | 99.5 | 282 | 2,622 | 1,278.4 | 146.7 | 751 | 12,390 | 3,755.5 | 319.5 | 206 | 806 | 207.8 | 30.2 |
| Lawrence | 5.43 | 88.9 | D | D | D | 9.3 | 144 | 1,942 | 583.1 | 47.9 | 36 | 211 | 60.5 | 6.4 |
| Licking | 12.57 | 73.7 | 91 | 1,355 | 967.4 | 71.6 | 453 | 6,711 | 2,497.9 | 182.7 | 116 | 399 | 69.7 | 12.2 |
| Logan | 2.80 | 61.7 | 26 | 1,962 | 1,019.1 | 104.1 | 144 | 1,718 | 504.2 | 44.4 | 33 | 156 | 37.0 | 5.6 |
| Lorain | 42.42 | 139.0 | 242 | 2,599 | 2,315.4 | 162.4 | 807 | 13,044 | 3,875.6 | 330.5 | 215 | 787 | 141.7 | 28.2 |
| Lucas | 81.45 | 187.8 | 426 | 8,101 | 5,611.2 | 479.6 | 1,376 | 24,927 | 6,470.5 | 609.5 | 407 | 2,943 | 4,350.7 | 195.8 |
| Madison | 2.05 | 46.5 | D | D | D | 15.9 | 104 | 1,851 | 1,746.0 | 56.0 | 29 | 63 | 15.9 | 2.0 |
| Mahoning | 4.30 | 18.5 | 261 | 3,661 | 1,927.5 | 190.1 | 850 | 12,647 | 3,772.5 | 326.4 | 199 | 863 | 157.4 | 26.2 |
| Marion | 6.14 | 93.9 | 36 | 432 | 375.1 | 22.0 | 164 | 2,833 | 812.0 | 71.7 | 41 | 140 | 27.0 | 3.9 |
| Medina | 3.01 | 17.1 | 218 | 2,554 | 2,460.7 | 147.9 | 485 | 8,601 | 3,119.5 | 226.6 | 155 | 503 | 117.1 | 18.7 |
| Meigs | 1.75 | 75.2 | 5 | D | 15.4 | D | 66 | 570 | 186.3 | 15.5 | 8 | 21 | 2.0 | 0.6 |
| Mercer | 2.88 | 70.3 | D | D | D | D | 170 | 1,896 | 548.3 | 49.6 | 26 | 118 | 19.6 | 3.1 |
| Miami | 10.89 | 104.5 | 83 | 1,266 | 1,088.6 | 66.4 | 301 | 5,305 | 1,417.1 | 127.3 | 75 | 288 | 57.4 | 10.7 |
| Monroe | 1.07 | 74.3 | NA | NA | NA | NA | 38 | 338 | 75.3 | 9.4 | D | D | D | 0.4 |
| Montgomery | 81.14 | 152.4 | 500 | 7,607 | 8,384.6 | 487.9 | 1,584 | 28,432 | 8,441.2 | 662.9 | 539 | 3,374 | 610.2 | 143.9 |
| Morgan | 0.64 | 43.3 | D | D | D | D | 29 | 392 | 59.4 | 5.3 | D | D | D | 0.0 |
| Morrow | 0.55 | 15.7 | D | D | D | D | 50 | 470 | 174.8 | 10.3 | 12 | 60 | 18.0 | 1.6 |
| Muskingum | 9.89 | 114.6 | 57 | 844 | 606.1 | 37.5 | 301 | 4,689 | 1,330.2 | 114.0 | 61 | 290 | 50.4 | 9.1 |
| Noble | 1.04 | 72.6 | D | D | D | D | 33 | 261 | 81.1 | 6.6 | 3 | 3 | 1.7 | 0.1 |

1 Merchant wholesalers, except manufacturers' sales branches and offices.  2. Employer establishments.

| STATE County | Professional, scientific, and technical services, 2017 | | | | Manufacturing, 2017 | | | | Accommodation and food services, 2017 | | | |
|---|---|---|---|---|---|---|---|---|---|---|---|---|
| | Number of establish-ments | Number of employees | Sales (mil dol) | Average payroll (mil dol) | Number of establish-ments | Number of employees | Sales (mil dol) | Average payroll (mil dol) | Number of establis-hments | Number of employees | Sales (mil dol) | Annual payroll (mil dol) |
| | 147 | 148 | 149 | 150 | 151 | 152 | 153 | 154 | 155 | 156 | 157 | 158 |
| NORTH DAKOTA—Cont'd | | | | | | | | | | | | |
| Ward | D | D | D | D | D | 468 | D | 22.4 | 179 | 3,711 | 181.6 | 58.3 |
| Wells | 11 | 21 | 2 | 0.7 | 6 | 37 | 15.9 | 2.3 | 22 | 104 | 5.1 | 1.3 |
| Williams | 123 | 555 | 113 | 51.4 | 36 | 293 | 136.7 | 18.7 | 122 | 2,115 | 219.8 | 48.1 |
| OHIO | 23,745 | 244,541 | 42,662 | 16,630.8 | 13,922 | 652,462 | 306,222.2 | 36,256.8 | 24,346 | 474,616 | 24,560.6 | 7,078.1 |
| Adams | 21 | 86 | 6 | 2.2 | 22 | 1,026 | 183.5 | 47.9 | 33 | 474 | 21.9 | 6.8 |
| Allen | D | D | D | D | 119 | 8,261 | 12,616.0 | 536.1 | 213 | 4,452 | 221.9 | 61.2 |
| Ashland | 66 | 1,089 | 173 | 71.7 | 85 | 3,777 | 1,122.9 | 176.6 | 89 | 1,597 | 76.9 | 22.2 |
| Ashtabula | D | D | D | D | 145 | 6,147 | 2,132.3 | 324.5 | 239 | 2,837 | 153.2 | 40.7 |
| Athens | D | D | D | D | D | 256 | D | 12.9 | 144 | 2,759 | 119.7 | 36.4 |
| Auglaize | 65 | 475 | 75 | 22.2 | 87 | 9,128 | 3,175.5 | 492.7 | 89 | 1,197 | 57.2 | 16.0 |
| Belmont | 87 | 560 | 58 | 22.3 | 40 | 894 | 399.4 | 40.7 | 134 | 2,675 | 131.1 | 39.4 |
| Brown | 25 | 112 | 9 | 3.2 | D | 730 | D | 31.1 | 51 | 955 | 38.7 | 11.2 |
| Butler | D | D | D | D | 383 | 19,334 | 10,184.4 | 1,141.6 | 713 | 15,436 | 811.8 | 229.0 |
| Carroll | 29 | 151 | 17 | 6.5 | 37 | 1,163 | 360.5 | 52.6 | 47 | 617 | 29.0 | 7.8 |
| Champaign | 40 | 255 | 19 | 13.1 | 40 | 3,631 | 1,375.0 | 202.6 | 49 | 616 | 29.4 | 8.1 |
| Clark | D | D | D | D | 152 | 6,187 | 2,405.7 | 274.0 | 223 | 4,568 | 207.9 | 63.6 |
| Clermont | 357 | 2,798 | 702 | 171.5 | 161 | 5,753 | 1,964.6 | 313.1 | 301 | 6,659 | 334.1 | 100.1 |
| Clinton | D | D | D | D | 38 | 2,900 | 1,087.4 | 154.0 | 69 | 1,275 | 63.3 | 17.1 |
| Columbiana | 91 | 395 | 41 | 15.8 | 166 | 5,912 | 1,517.6 | 264.6 | 174 | 2,772 | 123.4 | 33.6 |
| Coshocton | 32 | 147 | 15 | 4.8 | 51 | 2,199 | 1,072.8 | 119.1 | 48 | 727 | 29.7 | 8.4 |
| Crawford | 47 | 650 | 19 | 13.8 | 74 | 2,971 | 1,004.4 | 141.2 | 81 | 1,142 | 50.5 | 13.6 |
| Cuyahoga | D | D | D | D | 1,677 | 65,044 | 23,225.4 | 3,932.7 | 3,037 | 56,378 | 3,332.7 | 955.0 |
| Darke | 62 | 305 | 28 | 9.5 | 76 | 5,161 | 2,236.9 | 262.0 | 91 | 1,012 | 44.5 | 12.2 |
| Defiance | D | D | D | D | 37 | 3,111 | 1,029.9 | 227.2 | 83 | 1,409 | 63.8 | 17.4 |
| Delaware | D | D | D | D | 124 | 5,311 | 3,096.8 | 318.0 | 469 | 11,025 | 582.9 | 174.2 |
| Erie | D | D | D | D | 99 | 5,576 | 2,100.9 | 264.9 | 273 | 5,721 | 344.3 | 92.4 |
| Fairfield | 211 | 993 | 123 | 43.2 | 108 | 3,853 | 1,121.6 | 193.1 | 269 | 5,391 | 252.3 | 75.7 |
| Fayette | 18 | 66 | 7 | 2.1 | 26 | 1,706 | 1,464.6 | 78.4 | 65 | 1,301 | 61.1 | 16.5 |
| Franklin | D | D | D | D | 821 | 31,316 | 13,796.1 | 1,668.7 | 2,950 | 64,725 | 3,728.3 | 1,066.4 |
| Fulton | 45 | 216 | 23 | 8.5 | 95 | 7,330 | 4,118.7 | 388.9 | 77 | 975 | 43.8 | 11.6 |
| Gallia | 27 | 85 | 7 | 1.8 | 19 | 564 | 161.7 | 25.5 | 51 | 1,032 | 50.3 | 14.1 |
| Geauga | D | D | D | D | 195 | 8,025 | 2,875.3 | 406.8 | 170 | 2,710 | 126.7 | 36.4 |
| Greene | 488 | 10,839 | 2,287 | 845.3 | 97 | 3,014 | 934.9 | 182.7 | 346 | 8,340 | 414.2 | 124.2 |
| Guernsey | 37 | 317 | 39 | 19.2 | 55 | 2,774 | 1,576.1 | 143.3 | 82 | 1,448 | 77.6 | 20.3 |
| Hamilton | 2,500 | 38,507 | 7,292 | 2,996.1 | 950 | 44,819 | 23,656.9 | 2,714.2 | 2,012 | 44,753 | 2,445.3 | 732.2 |
| Hancock | D | D | D | D | 97 | 11,347 | 5,114.8 | 569.9 | 188 | 4,540 | 205.9 | 61.2 |
| Hardin | 22 | 100 | 11 | 3.7 | 28 | 1,782 | 588.5 | 83.4 | 44 | 789 | 32.1 | 10.9 |
| Harrison | 16 | 44 | 7 | 2.3 | 14 | 300 | 72.4 | 14.2 | 24 | 199 | 9.0 | 2.4 |
| Henry | 24 | 159 | 17 | 4.8 | 44 | 3,279 | 3,147.4 | 188.4 | 57 | 686 | 27.5 | 8.2 |
| Highland | 34 | 122 | 11 | 3.2 | 28 | 1,786 | 833.0 | 86.0 | 56 | 920 | 44.5 | 12.4 |
| Hocking | 27 | 96 | 9 | 3.2 | 27 | 804 | 300.2 | 36.3 | 67 | 927 | 49.7 | 14.8 |
| Holmes | 39 | 298 | 33 | 12.2 | 305 | 6,695 | 1,751.9 | 276.9 | 72 | 1,577 | 76.9 | 24.2 |
| Huron | 69 | 412 | 47 | 16.0 | 95 | 6,183 | 2,455.7 | 286.4 | 121 | 1,585 | 64.0 | 17.3 |
| Jackson | 33 | 134 | 11 | 3.3 | 26 | 3,222 | 1,846.1 | 116.1 | 55 | 1,004 | 40.4 | 11.4 |
| Jefferson | D | D | 33 | D | 36 | 1,281 | 423.3 | 84.0 | 120 | 1,704 | 86.5 | 23.4 |
| Knox | 57 | 356 | 38 | 13.8 | 69 | 4,208 | 1,592.8 | 233.3 | 103 | 1,712 | 75.8 | 21.2 |
| Lake | D | D | D | D | 571 | 18,802 | 6,531.5 | 1,041.1 | 523 | 9,875 | 497.0 | 135.3 |
| Lawrence | 41 | 202 | 16 | 6.7 | 31 | 713 | 492.6 | 32.9 | 64 | 1,236 | 64.3 | 17.2 |
| Licking | D | D | D | D | 156 | 8,898 | 3,385.2 | 447.3 | 302 | 4,911 | 240.5 | 71.9 |
| Logan | 58 | 747 | 75 | 30.5 | 43 | 4,979 | 6,334.6 | 336.6 | 98 | 2,237 | 101.8 | 29.3 |
| Lorain | D | D | D | D | 355 | 15,443 | 6,058.9 | 939.5 | 522 | 8,874 | 444.4 | 123.5 |
| Lucas | 854 | 10,105 | 1,586 | 644.2 | 438 | 22,153 | 22,683.2 | 1,357.3 | 1,056 | 19,755 | 986.6 | 282.2 |
| Madison | D | D | D | D | 44 | 3,356 | 1,328.5 | 194.3 | 55 | 1,095 | 51.9 | 15.7 |
| Mahoning | D | D | D | D | 308 | 8,491 | 2,140.6 | 425.3 | 536 | 10,319 | 450.6 | 128.3 |
| Marion | 58 | 255 | 25 | 9.8 | 74 | 6,428 | 3,585.0 | 323.1 | 103 | 2,191 | 103.7 | 27.2 |
| Medina | 420 | 2,160 | 280 | 110.2 | 275 | 9,153 | 3,161.9 | 487.5 | 314 | 6,106 | 276.2 | 80.7 |
| Meigs | 13 | 62 | 6 | 1.7 | 6 | 13 | 3.7 | 0.7 | 31 | 427 | 20.2 | 5.9 |
| Mercer | 46 | 324 | 47 | 14.8 | 83 | 4,424 | 1,654.0 | 217.3 | 93 | 1,176 | 46.1 | 13.5 |
| Miami | 150 | 905 | 137 | 50.3 | 209 | 9,672 | 4,005.3 | 531.8 | 182 | 3,688 | 173.2 | 50.3 |
| Monroe | 13 | 46 | 5 | 1.6 | 8 | 57 | 13.7 | 3.0 | D | D | D | 2.5 |
| Montgomery | 1,070 | 12,067 | 1,914 | 801.9 | 672 | 29,693 | 9,405.7 | 1,776.0 | 1,101 | 23,646 | 1,176.5 | 352.2 |
| Morgan | 7 | 51 | 7 | 2.0 | 9 | 747 | 185.4 | 38.2 | 20 | 272 | 9.6 | 2.9 |
| Morrow | 27 | 186 | 19 | 6.4 | 28 | 1,047 | 715.5 | 52.9 | 30 | 353 | 20.2 | 6.1 |
| Muskingum | 104 | 530 | 79 | 24.9 | 75 | 2,800 | 1,012.6 | 143.5 | 193 | 3,395 | 169.6 | 49.4 |
| Noble | 14 | 61 | 11 | 3.0 | 9 | 89 | 49.6 | 3.4 | D | D | D | D |

# Health Care and Social Assistance, Other Services, Nonemployer Businesses, and Residential Construction

| STATE County | Health care and social assistance, 2017 | | | | Other services, 2017 | | | | Nonemployer businesses, 2018 | | Value of residential construction authorized by building permits, 2020 | |
|---|---|---|---|---|---|---|---|---|---|---|---|---|
| | Number of establishments | Number of employees | Receipts (mil dol) | Annual payroll (mil dol) | Number of establishments | Number of employees | Receipts (mil dol) | Annual payroll (mil dol) | Number | Receipts (mil dol) | New construction ($1,000) | Number of housing units |
| | 159 | 160 | 161 | 162 | 163 | 164 | 165 | 166 | 167 | 168 | 169 | 170 |
| NORTH DAKOTA—Cont'd | | | | | | | | | | | | |
| Ward | 163 | 5,176 | 615.0 | 289.3 | 155 | 811 | 99.3 | 26.6 | 4,412 | 210.3 | 37,971 | 160 |
| Wells | 19 | 399 | 25.1 | 11.7 | D | D | 5.2 | D | 349 | 15.7 | 100 | 1 |
| Williams | 78 | 1,417 | 206.6 | 69.9 | 86 | 477 | 94.9 | 20.7 | 2,798 | 161.5 | 22,455 | 84 |
| OHIO | 29,595 | 856,794 | 97,117.4 | 39,468.5 | 18,425 | 126,378 | 15,268.0 | 4,043.9 | 802,331 | 36,988.6 | 6,418,479 | 29,686 |
| Adams | 49 | 951 | 71.9 | 30.4 | D | D | D | 2.4 | 1,790 | 72.2 | 1,175 | 6 |
| Allen | 289 | 11,913 | 1,427.5 | 561.9 | 176 | 1,150 | 101.6 | 29.0 | 5,363 | 237.4 | 34,386 | 175 |
| Ashland | 118 | 2,898 | 262.6 | 110.7 | 93 | 558 | 54.2 | 13.0 | 3,514 | 167.3 | 11,214 | 60 |
| Ashtabula | 196 | 4,910 | 416.9 | 184.3 | 142 | 575 | 50.3 | 12.2 | 6,139 | 279.1 | 26,082 | 149 |
| Athens | 172 | 3,332 | 383.7 | 142.1 | 80 | 371 | 28.9 | 7.2 | 3,260 | 128.0 | 3,343 | 13 |
| Auglaize | 101 | 2,519 | 198.8 | 76.2 | 87 | 531 | 41.1 | 12.7 | 2,757 | 121.6 | 34,259 | 176 |
| Belmont | 194 | 3,895 | 328.0 | 117.9 | 120 | 635 | 51.1 | 15.3 | 3,298 | 139.1 | 465 | 3 |
| Brown | 66 | 1,406 | 73.7 | 31.7 | 39 | 172 | 14.0 | 4.3 | 2,680 | 115.5 | 29,940 | 108 |
| Butler | 789 | 15,809 | 1,832.0 | 670.0 | 545 | 4,419 | 482.2 | 139.7 | 23,568 | 1,121.9 | 246,790 | 1,152 |
| Carroll | 46 | 732 | 49.7 | 21.5 | 37 | 195 | 18.5 | 4.6 | 1,887 | 90.5 | 2,339 | 7 |
| Champaign | 47 | 976 | 86.0 | 33.2 | 53 | 228 | 18.9 | 4.3 | 2,344 | 96.6 | 10,187 | 52 |
| Clark | 290 | 7,102 | 709.7 | 267.6 | 191 | 1,438 | 194.0 | 52.0 | 6,627 | 269.8 | 36,805 | 139 |
| Clermont | 318 | 6,107 | 640.0 | 222.1 | 276 | 1,709 | 176.4 | 52.5 | 13,747 | 625.5 | 184,563 | 958 |
| Clinton | 94 | 1,900 | 196.2 | 75.4 | 54 | 217 | 20.9 | 6.1 | 2,573 | 111.9 | 10,867 | 46 |
| Columbiana | 258 | 5,348 | 455.0 | 174.2 | 177 | 939 | 86.3 | 33.3 | 6,017 | 263.1 | 20,954 | 94 |
| Coshocton | 82 | 1,755 | 151.3 | 53.3 | 58 | 264 | 25.1 | 5.9 | 2,405 | 109.5 | 634 | 3 |
| Crawford | 89 | 2,732 | 318.3 | 112.7 | 64 | 300 | 24.2 | 6.0 | 2,148 | 95.0 | 840 | 4 |
| Cuyahoga | 3,676 | 147,104 | 17,917.0 | 7,895.6 | 2,399 | 18,570 | 2,156.6 | 614.4 | 97,290 | 4,511.1 | 239,544 | 808 |
| Darke | 82 | 2,242 | 185.9 | 80.7 | 98 | 402 | 27.4 | 6.9 | 3,520 | 157.1 | 19,152 | 72 |
| Defiance | 90 | 2,029 | 232.8 | 79.6 | 71 | 389 | 39.5 | 8.5 | 2,117 | 88.3 | 12,354 | 52 |
| Delaware | 531 | 8,103 | 797.8 | 342.6 | 304 | 2,613 | 499.3 | 107.9 | 18,270 | 1,049.6 | 502,163 | 1,902 |
| Erie | 226 | 5,239 | 536.5 | 225.3 | 126 | 599 | 57.1 | 14.8 | 5,031 | 200.2 | 30,823 | 128 |
| Fairfield | 355 | 7,396 | 832.7 | 313.7 | 177 | 1,070 | 117.7 | 32.8 | 11,035 | 487.4 | 204,897 | 744 |
| Fayette | 44 | 1,147 | 95.6 | 42.0 | 40 | 153 | 12.3 | 3.5 | 1,489 | 68.1 | 6,498 | 38 |
| Franklin | 3,778 | 117,393 | 14,875.9 | 5,676.7 | 1,880 | 16,846 | 2,525.6 | 680.7 | 103,069 | 4,903.3 | 1,154,609 | 8,108 |
| Fulton | 108 | 1,905 | 201.5 | 69.6 | 75 | 266 | 29.6 | 7.0 | 2,937 | 138.2 | 15,426 | 53 |
| Gallia | 78 | 2,894 | 325.8 | 117.0 | 33 | 217 | 20.3 | 6.6 | 1,726 | 82.6 | 0 | 0 |
| Geauga | 272 | 4,933 | 495.6 | 202.1 | 201 | 1,186 | 126.7 | 36.1 | 11,169 | 680.4 | 115,651 | 133 |
| Greene | 394 | 7,764 | 788.6 | 345.6 | 220 | 1,344 | 126.2 | 36.4 | 10,544 | 448.7 | 198,920 | 509 |
| Guernsey | 123 | 3,297 | 302.2 | 113.1 | 60 | 251 | 22.2 | 5.9 | 2,223 | 105.2 | 8,935 | 37 |
| Hamilton | 2,381 | 88,940 | 11,839.3 | 4,868.2 | 1,442 | 11,548 | 1,407.7 | 392.4 | 61,203 | 2,934.1 | 378,563 | 1,739 |
| Hancock | 199 | 5,090 | 579.0 | 216.8 | 141 | 972 | 132.2 | 32.2 | 4,438 | 212.5 | 56,920 | 251 |
| Hardin | 55 | 737 | 61.7 | 24.7 | 22 | 99 | 8.6 | 2.4 | 1,486 | 60.9 | 5,460 | 31 |
| Harrison | 24 | 492 | 39.4 | 14.6 | 21 | 73 | 6.0 | 1.5 | 860 | 49.9 | 230 | 1 |
| Henry | 53 | 1,284 | 89.6 | 37.6 | 45 | 227 | 21.4 | 5.2 | 1,638 | 67.0 | 7,748 | 22 |
| Highland | 103 | 1,683 | 147.8 | 57.7 | 41 | 153 | 14.0 | 3.5 | 2,935 | 143.0 | 655 | 4 |
| Hocking | 57 | 1,156 | 95.7 | 37.2 | 38 | 208 | 20.8 | 4.6 | 1,940 | 80.8 | 1,252 | 7 |
| Holmes | 61 | 1,206 | 163.7 | 37.3 | 59 | 297 | 45.6 | 10.5 | 5,715 | 367.4 | 1,958 | 9 |
| Huron | 100 | 2,772 | 293.1 | 118.2 | 100 | 526 | 41.6 | 12.5 | 3,118 | 142.8 | 8,643 | 36 |
| Jackson | 88 | 1,351 | 132.2 | 35.6 | 40 | 159 | 18.8 | 4.0 | 1,538 | 72.1 | 20,291 | 94 |
| Jefferson | 173 | 3,953 | 427.4 | 159.0 | 105 | 665 | 59.1 | 18.4 | 3,037 | 112.2 | 1,316 | 6 |
| Knox | 149 | 3,142 | 299.2 | 116.0 | 75 | 491 | 52.5 | 12.4 | 4,944 | 237.6 | 32,876 | 131 |
| Lake | 607 | 11,627 | 1,078.4 | 462.8 | 454 | 2,501 | 232.6 | 68.1 | 16,558 | 776.2 | 123,065 | 681 |
| Lawrence | 120 | 2,630 | 139.1 | 62.8 | 46 | 194 | 21.4 | 5.6 | 2,682 | 98.7 | 2,202 | 15 |
| Licking | 296 | 7,120 | 742.6 | 291.2 | 211 | 1,190 | 113.1 | 30.4 | 12,173 | 553.9 | 81,536 | 292 |
| Logan | 89 | 1,895 | 193.6 | 74.7 | 59 | 475 | 82.7 | 15.2 | 2,988 | 132.6 | 26,190 | 134 |
| Lorain | 645 | 15,642 | 1,744.5 | 731.8 | 451 | 2,596 | 239.5 | 67.0 | 18,604 | 765.3 | 272,389 | 1,252 |
| Lucas | 1,366 | 39,689 | 4,694.1 | 1,856.8 | 687 | 4,782 | 516.5 | 137.8 | 25,214 | 1,123.5 | 140,836 | 881 |
| Madison | 66 | 1,199 | 98.7 | 40.4 | 37 | 193 | 17.7 | 5.7 | 2,748 | 120.1 | 38,940 | 145 |
| Mahoning | 763 | 22,098 | 2,114.3 | 853.2 | 388 | 2,885 | 247.6 | 72.2 | 16,154 | 669.3 | 36,091 | 151 |
| Marion | 185 | 4,313 | 466.0 | 187.7 | 88 | 603 | 39.7 | 11.9 | 2,947 | 120.7 | 14,930 | 49 |
| Medina | 358 | 7,418 | 676.6 | 283.4 | 298 | 1,561 | 146.3 | 45.9 | 13,927 | 692.3 | 173,953 | 500 |
| Meigs | 45 | 546 | 46.5 | 15.0 | 23 | 84 | 7.5 | 1.8 | 1,128 | 41.2 | 628 | 4 |
| Mercer | 76 | 2,188 | 160.4 | 71.6 | 96 | 504 | 70.3 | 14.1 | 2,838 | 132.0 | 27,762 | 113 |
| Miami | 204 | 4,692 | 387.4 | 158.2 | 179 | 958 | 105.2 | 27.0 | 7,062 | 311.1 | 80,605 | 277 |
| Monroe | 20 | 253 | 13.2 | 6.4 | 22 | 121 | 11.5 | 2.5 | 1,015 | 36.7 | 250 | 3 |
| Montgomery | 1,511 | 52,281 | 6,622.1 | 2,671.1 | 809 | 6,269 | 676.8 | 236.2 | 33,269 | 1,401.4 | 157,473 | 913 |
| Morgan | 15 | 351 | 28.2 | 10.1 | D | D | D | 0.6 | 796 | 27.8 | 11,076 | 45 |
| Morrow | D | D | D | 31.2 | 25 | 67 | 7.3 | 1.9 | 2,624 | 145.2 | 25,671 | 105 |
| Muskingum | 180 | 6,227 | 820.3 | 289.3 | 150 | 968 | 90.8 | 25.3 | 5,016 | 217.3 | 4,935 | 22 |
| Noble | 22 | 398 | 15.2 | 7.0 | 18 | 36 | 4.8 | 0.9 | 718 | 32.4 | 3,860 | 28 |

# Table B. States and Counties — Government Employment and Payroll, and Local Government Finances

| | Government employment and payroll, 2017 | | | | | | | | | Local government finances, 2017 | | | | |
| | | | March payroll (percent of total) | | | | | | | General revenue | | | | |
| | | | | | | | | | | | | Taxes | | |
| | | | | | | | | | | | | | Per capita[1] (dollars) | |
| STATE County | Full-time equivalent employees | March payroll (dollars) | Administration, judicial, and legal | Police and corrections | Fire protection | Highways and transportation | Health and welfare | Natural resources and utilities | Education and libraries | Total (mil dol) | Intergovernmental (mil dol) | Total (mil dol) | Total | Property |
| | 171 | 172 | 173 | 174 | 175 | 176 | 177 | 178 | 179 | 180 | 181 | 182 | 183 | 184 |

| STATE County | 171 | 172 | 173 | 174 | 175 | 176 | 177 | 178 | 179 | 180 | 181 | 182 | 183 | 184 |
|---|---|---|---|---|---|---|---|---|---|---|---|---|---|---|
| **NORTH DAKOTA—Cont'd** | | | | | | | | | | | | | | |
| Ward | 2,226 | 9,838,354 | 5.8 | 8.3 | 3.5 | 5.3 | 6.5 | 6.7 | 61.8 | 387.1 | 207.1 | 119.7 | 1,746 | 1,270 |
| Wells | 163 | 558,834 | 14.4 | 5.1 | 0.1 | 10.2 | 19.1 | 7.9 | 42.3 | 21.1 | 11.7 | 5.7 | 1,438 | 1,313 |
| Williams | 1,323 | 6,301,179 | 5.5 | 12.5 | 4.8 | 8.4 | 5.6 | 12.1 | 45.4 | 373.1 | 180.2 | 153.0 | 4,573 | 3,231 |
| **OHIO** | X | X | X | X | X | X | X | X | X | X | X | X | X | X |
| Adams | 1,341 | 4,584,553 | 7.8 | 3.6 | 0.8 | 3.2 | 23.8 | 4.1 | 55.1 | 100.8 | 61.5 | 29.6 | 1,065 | 927 |
| Allen | 3,994 | 15,615,627 | 8.1 | 10.6 | 5.1 | 4.1 | 8.2 | 6.1 | 56.1 | 411.1 | 190.3 | 147.7 | 1,432 | 973 |
| Ashland | 1,575 | 5,738,974 | 8.7 | 8.6 | 4.0 | 5.4 | 11.4 | 4.7 | 55.6 | 157.9 | 63.1 | 70.8 | 1,319 | 851 |
| Ashtabula | 3,518 | 13,241,030 | 8.3 | 7.8 | 4.0 | 4.6 | 10.7 | 3.3 | 60.4 | 362.9 | 177.7 | 120.3 | 1,231 | 940 |
| Athens | 2,282 | 8,966,284 | 8.2 | 8.7 | 1.9 | 3.4 | 14.7 | 5.9 | 54.0 | 216.6 | 117.5 | 76.6 | 1,152 | 879 |
| Auglaize | 1,740 | 6,878,001 | 7.6 | 9.5 | 3.0 | 4.7 | 7.2 | 8.0 | 59.0 | 184.5 | 74.4 | 76.8 | 1,678 | 851 |
| Belmont | 2,653 | 8,668,632 | 9.8 | 9.2 | 2.2 | 4.7 | 12.0 | 7.1 | 53.2 | 208.2 | 89.1 | 90.3 | 1,328 | 963 |
| Brown | 1,485 | 5,319,265 | 8.7 | 5.5 | 0.6 | 2.8 | 5.3 | 2.5 | 72.2 | 159.0 | 82.0 | 38.4 | 882 | 675 |
| Butler | 12,961 | 55,036,876 | 7.1 | 10.1 | 5.0 | 2.8 | 5.2 | 5.4 | 63.3 | 1,519.8 | 625.6 | 600.2 | 1,577 | 1,137 |
| Carroll | 835 | 2,542,302 | 17.0 | 4.1 | 0.0 | 10.6 | 12.7 | 3.5 | 50.6 | 82.3 | 38.9 | 29.7 | 1,087 | 984 |
| Champaign | 1,612 | 5,783,433 | 8.3 | 8.4 | 2.4 | 3.4 | 9.4 | 1.9 | 64.2 | 152.8 | 75.6 | 56.6 | 1,456 | 918 |
| Clark | 4,730 | 19,251,860 | 8.8 | 10.4 | 4.4 | 2.0 | 9.0 | 3.3 | 60.6 | 483.7 | 238.1 | 185.9 | 1,382 | 876 |
| Clermont | 6,063 | 23,115,330 | 7.4 | 10.6 | 7.1 | 4.2 | 8.7 | 3.2 | 58.5 | 644.1 | 249.7 | 279.8 | 1,370 | 1,139 |
| Clinton | 1,761 | 6,201,390 | 9.9 | 8.5 | 1.9 | 3.5 | 7.7 | 5.3 | 62.3 | 158.2 | 72.3 | 62.3 | 1,483 | 991 |
| Columbiana | 3,346 | 11,818,467 | 7.6 | 6.7 | 1.5 | 5.1 | 9.3 | 4.7 | 64.1 | 317.6 | 170.9 | 107.9 | 1,047 | 695 |
| Coshocton | 1,432 | 5,411,175 | 10.4 | 5.8 | 11.7 | 4.4 | 13.3 | 2.4 | 50.4 | 126.7 | 61.6 | 44.2 | 1,211 | 975 |
| Crawford | 1,408 | 5,187,605 | 9.3 | 9.7 | 4.1 | 2.9 | 4.9 | 6.4 | 60.8 | 162.8 | 76.1 | 57.5 | 1,377 | 861 |
| Cuyahoga | 63,287 | 318,547,930 | 6.3 | 10.5 | 4.8 | 6.5 | 20.5 | 7.1 | 42.1 | 8,930.2 | 2,379.1 | 4,022.8 | 3,224 | 1,789 |
| Darke | 1,666 | 6,251,519 | 6.5 | 9.6 | 2.6 | 3.4 | 8.3 | 3.9 | 63.6 | 183.8 | 78.3 | 75.5 | 1,465 | 848 |
| Defiance | 1,660 | 6,467,187 | 7.0 | 6.1 | 2.2 | 3.5 | 20.3 | 5.1 | 51.5 | 226.9 | 88.5 | 59.5 | 1,559 | 969 |
| Delaware | 5,809 | 25,871,545 | 6.6 | 8.3 | 7.4 | 2.8 | 6.3 | 4.0 | 62.8 | 732.8 | 133.2 | 466.5 | 2,322 | 1,781 |
| Erie | 3,473 | 15,051,571 | 7.3 | 8.1 | 3.3 | 1.6 | 8.6 | 5.7 | 63.5 | 389.5 | 120.0 | 160.3 | 2,144 | 1,520 |
| Fairfield | 4,765 | 19,070,061 | 8.2 | 8.4 | 7.3 | 2.9 | 7.9 | 5.3 | 59.1 | 654.0 | 207.3 | 273.9 | 1,771 | 1,085 |
| Fayette | 1,435 | 5,512,962 | 6.5 | 6.3 | 2.3 | 3.1 | 34.5 | 3.4 | 40.9 | 164.3 | 55.8 | 45.3 | 1,695 | 1,054 |
| Franklin | 46,320 | 235,016,726 | 7.7 | 12.1 | 8.1 | 5.0 | 6.8 | 6.1 | 52.0 | 8,104.3 | 2,525.9 | 3,976.5 | 3,069 | 1,700 |
| Fulton | 2,019 | 7,317,680 | 8.6 | 6.0 | 2.2 | 4.1 | 4.6 | 4.3 | 68.2 | 183.3 | 75.0 | 79.2 | 1,873 | 1,169 |
| Gallia | 1,382 | 4,502,765 | 12.9 | 7.8 | 0.4 | 4.6 | 7.2 | 3.5 | 63.4 | 147.7 | 83.0 | 33.4 | 1,108 | 802 |
| Geauga | 2,969 | 12,687,062 | 8.2 | 10.9 | 1.3 | 5.7 | 8.1 | 5.6 | 58.9 | 360.6 | 92.2 | 201.3 | 2,144 | 1,837 |
| Greene | 5,452 | 22,544,757 | 8.6 | 10.6 | 6.8 | 3.2 | 8.4 | 4.5 | 56.6 | 645.2 | 175.7 | 343.7 | 2,064 | 1,596 |
| Guernsey | 1,593 | 5,180,768 | 9.3 | 7.2 | 2.0 | 5.2 | 8.3 | 5.5 | 55.7 | 136.6 | 65.7 | 49.1 | 1,257 | 807 |
| Hamilton | 30,891 | 144,061,829 | 7.4 | 12.2 | 8.0 | 5.4 | 8.0 | 8.0 | 48.3 | 4,518.2 | 1,287.0 | 2,266.3 | 2,784 | 1,645 |
| Hancock | 2,516 | 9,829,360 | 7.5 | 8.5 | 3.7 | 3.7 | 8.8 | 6.0 | 61.0 | 286.9 | 101.7 | 129.3 | 1,701 | 1,031 |
| Hardin | 1,232 | 4,171,146 | 9.6 | 4.9 | 1.3 | 4.7 | 9.7 | 2.7 | 65.8 | 110.6 | 44.5 | 49.6 | 1,584 | 985 |
| Harrison | 772 | 2,092,052 | 16.3 | 7.3 | 0.0 | 12.4 | 11.8 | 4.5 | 47.0 | 77.3 | 33.5 | 32.4 | 2,128 | 2,016 |
| Henry | 1,326 | 5,105,969 | 7.8 | 4.3 | 0.9 | 3.2 | 10.4 | 6.1 | 66.3 | 126.0 | 55.6 | 56.8 | 2,091 | 1,470 |
| Highland | 2,211 | 7,371,594 | 5.4 | 4.4 | 1.9 | 2.9 | 32.2 | 2.0 | 50.2 | 194.0 | 85.6 | 46.3 | 1,080 | 654 |
| Hocking | 1,068 | 3,400,935 | 10.3 | 8.7 | 1.5 | 4.6 | 7.4 | 2.7 | 62.8 | 124.1 | 42.1 | 33.4 | 1,174 | 914 |
| Holmes | 1,268 | 4,982,748 | 6.4 | 5.7 | 1.2 | 4.4 | 33.4 | 2.1 | 46.1 | 98.5 | 39.1 | 44.8 | 1,021 | 770 |
| Huron | 2,084 | 7,836,280 | 10.1 | 6.5 | 2.7 | 4.7 | 7.5 | 6.8 | 60.6 | 216.9 | 100.2 | 79.8 | 1,366 | 907 |
| Jackson | 984 | 3,483,617 | 5.6 | 5.6 | 2.6 | 2.6 | 0.7 | 7.3 | 73.9 | 108.8 | 65.4 | 26.8 | 826 | 562 |
| Jefferson | 2,663 | 8,769,479 | 7.7 | 11.9 | 2.1 | 4.9 | 12.2 | 5.3 | 54.7 | 256.6 | 117.3 | 91.8 | 1,384 | 902 |
| Knox | 2,409 | 8,636,325 | 6.3 | 6.1 | 4.0 | 4.4 | 10.3 | 3.2 | 63.3 | 200.9 | 85.0 | 84.0 | 1,371 | 1,082 |
| Lake | 9,551 | 42,209,385 | 5.8 | 10.3 | 6.2 | 4.8 | 7.5 | 8.7 | 55.6 | 1,067.3 | 322.2 | 549.2 | 2,386 | 1,635 |
| Lawrence | 2,243 | 7,789,802 | 6.5 | 5.2 | 0.9 | 2.9 | 7.2 | 3.5 | 72.3 | 242.9 | 124.9 | 51.5 | 857 | 633 |
| Licking | 6,034 | 23,746,451 | 7.2 | 9.4 | 6.0 | 3.1 | 5.8 | 3.3 | 63.1 | 636.5 | 229.8 | 309.8 | 1,784 | 1,231 |
| Logan | 1,974 | 7,287,445 | 6.8 | 5.8 | 2.0 | 4.7 | 15.8 | 3.9 | 59.6 | 196.8 | 79.7 | 74.3 | 1,644 | 1,034 |
| Lorain | 11,370 | 47,591,662 | 7.3 | 9.5 | 4.1 | 2.5 | 8.7 | 6.3 | 60.3 | 1,254.7 | 487.5 | 550.9 | 1,792 | 1,213 |
| Lucas | 15,869 | 73,046,493 | 8.8 | 13.3 | 7.6 | 4.6 | 8.7 | 5.2 | 50.5 | 2,114.5 | 798.6 | 913.5 | 2,119 | 1,237 |
| Madison | 1,539 | 6,248,847 | 8.6 | 7.1 | 5.5 | 3.6 | 5.1 | 3.2 | 62.6 | 153.2 | 53.6 | 74.1 | 1,682 | 1,205 |
| Mahoning | 9,303 | 34,495,182 | 6.0 | 12.1 | 3.6 | 3.9 | 8.3 | 6.6 | 58.1 | 884.7 | 358.6 | 384.3 | 1,672 | 1,026 |
| Marion | 2,470 | 9,069,451 | 7.1 | 11.2 | 5.6 | 3.0 | 6.4 | 3.0 | 60.7 | 218.9 | 112.9 | 76.0 | 1,168 | 678 |
| Medina | 5,948 | 25,936,956 | 6.5 | 8.5 | 2.4 | 3.4 | 6.6 | 6.5 | 64.9 | 694.1 | 202.7 | 376.2 | 2,111 | 1,618 |
| Meigs | 951 | 3,024,364 | 8.7 | 6.6 | 0.0 | 4.6 | 10.9 | 7.4 | 60.8 | 69.9 | 47.8 | 16.8 | 730 | 583 |
| Mercer | 2,148 | 8,056,550 | 6.7 | 5.5 | 1.1 | 2.6 | 24.0 | 3.3 | 56.0 | 222.2 | 68.8 | 66.2 | 1,620 | 1,030 |
| Miami | 3,665 | 15,185,653 | 8.1 | 8.5 | 2.8 | 3.3 | 4.7 | 6.6 | 64.3 | 432.2 | 151.7 | 198.5 | 1,886 | 1,022 |
| Monroe | 774 | 2,184,697 | 12.8 | 6.2 | 0.2 | 7.1 | 7.9 | 4.8 | 51.0 | 66.9 | 30.7 | 28.1 | 2,016 | 1,713 |
| Montgomery | 22,646 | 100,833,229 | 8.2 | 10.1 | 4.8 | 6.5 | 8.5 | 7.7 | 52.6 | 2,822.7 | 1,014.1 | 1,220.0 | 2,296 | 1,441 |
| Morgan | 430 | 1,395,660 | 13.6 | 4.6 | 0.0 | 4.1 | 2.2 | 2.8 | 71.2 | 55.5 | 33.4 | 12.2 | 834 | 651 |
| Morrow | 1,325 | 4,591,056 | 9.4 | 5.2 | 0.1 | 4.0 | 28.6 | 0.9 | 50.9 | 117.5 | 52.1 | 43.6 | 1,249 | 865 |
| Muskingum | 3,926 | 13,651,662 | 5.9 | 8.2 | 2.6 | 3.2 | 11.4 | 3.7 | 63.7 | 358.9 | 174.6 | 129.7 | 1,506 | 1,020 |
| Noble | 467 | 1,672,205 | 14.3 | 8.7 | 0.0 | 7.8 | 11.1 | 6.1 | 49.2 | 51.0 | 23.6 | 21.3 | 1,479 | 1,317 |

1. Based on the resident population estimated as of July 1 of the year shown.

# Local Government Finances, Government Employment, and Income Taxes

| STATE County | Local government finances, 2017 (cont.) | | | | | | | | | Government employment, 2019 | | | Individual income tax returns, 2018 | | |
|---|---|---|---|---|---|---|---|---|---|---|---|---|---|---|---|
| | Direct general expenditure | | | | | | | Debt outstanding | | | | | | | |
| | | | Percent of total for: | | | | | | | | | | | Mean | |
| | Total (mil dol) | Per capita[1] (dollars) | Education | Health and hospitals | Police protection | Public welfare | Highways | Total (mil dol) | Per capita[1] (dollars) | Federal civilian | Federal military | State and local | Number of returns | adjusted gross income | Mean income tax |
| | 185 | 186 | 187 | 188 | 189 | 190 | 191 | 192 | 193 | 194 | 195 | 196 | 197 | 198 | 199 |
| NORTH DAKOTA—Cont'd | | | | | | | | | | | | | | | |
| Ward | 414 | 6,038 | 41.2 | 2.6 | 4.1 | 2.1 | 25.9 | 314.9 | 4,596 | 1,233 | 6,033 | 4,462 | 32,820 | 66,509 | 7,181 |
| Wells | 19 | 4,648 | 50.8 | 3.7 | 3.2 | 0.0 | 8.9 | 12.6 | 3,144 | 23 | 23 | 247 | 1,990 | 61,819 | 6,298 |
| Williams | 276 | 8,256 | 37.2 | 1.9 | 4.0 | 1.5 | 14.6 | 677.2 | 20,236 | 105 | 227 | 2,590 | 18,950 | 91,349 | 12,402 |
| OHIO | X | X | X | X | X | X | X | X | X | 79,790 | 35,812 | 698,228 | 5,623,730 | 64,280 | 7,239 |
| Adams | 102 | 3,671 | 61.5 | 4.6 | 4.3 | 8.4 | 2.3 | 53.0 | 1,911 | 67 | 69 | 1,364 | 10,950 | 45,687 | 3,340 |
| Allen | 466 | 4,524 | 51.6 | 3.1 | 8.5 | 4.4 | 3.7 | 1,857.7 | 18,019 | 329 | 247 | 5,697 | 47,650 | 55,083 | 5,300 |
| Ashland | 151 | 2,810 | 52.3 | 7.6 | 7.1 | 5.0 | 6.5 | 61.3 | 1,143 | 107 | 129 | 2,323 | 24,890 | 51,735 | 4,424 |
| Ashtabula | 396 | 4,051 | 49.9 | 8.6 | 4.4 | 5.5 | 5.8 | 154.8 | 1,584 | 200 | 238 | 4,306 | 44,400 | 46,875 | 3,854 |
| Athens | 229 | 3,450 | 53.4 | 2.7 | 3.0 | 13.9 | 3.9 | 123.1 | 1,851 | 234 | 152 | 10,533 | 22,280 | 50,849 | 4,688 |
| Auglaize | 187 | 4,097 | 50.8 | 4.3 | 5.7 | 5.2 | 9.3 | 206.8 | 4,521 | 96 | 114 | 2,108 | 23,650 | 60,409 | 5,715 |
| Belmont | 202 | 2,976 | 51.2 | 1.4 | 2.7 | 7.0 | 8.9 | 68.5 | 1,007 | 157 | 160 | 3,519 | 30,380 | 62,572 | 7,140 |
| Brown | 170 | 3,914 | 63.6 | 2.5 | 3.8 | 2.8 | 4.3 | 82.7 | 1,901 | 104 | 108 | 2,100 | 19,180 | 48,406 | 3,672 |
| Butler | 1,536 | 4,037 | 52.9 | 3.8 | 7.7 | 4.6 | 5.4 | 1,844.7 | 4,847 | 583 | 955 | 20,441 | 177,860 | 65,597 | 6,950 |
| Carroll | 73 | 2,655 | 54.7 | 8.2 | 3.8 | 6.8 | 9.1 | 91.4 | 3,344 | 44 | 67 | 991 | 12,510 | 53,059 | 4,590 |
| Champaign | 174 | 4,488 | 67.7 | 2.1 | 4.1 | 3.4 | 4.6 | 115.6 | 2,976 | 71 | 97 | 1,914 | 18,640 | 52,237 | 4,290 |
| Clark | 529 | 3,930 | 51.2 | 6.1 | 8.1 | 6.3 | 4.2 | 235.9 | 1,753 | 538 | 332 | 6,322 | 62,530 | 50,982 | 4,411 |
| Clermont | 733 | 3,587 | 55.7 | 5.5 | 5.0 | 5.4 | 4.6 | 495.4 | 2,425 | 379 | 518 | 7,381 | 100,260 | 67,725 | 7,300 |
| Clinton | 164 | 3,914 | 51.0 | 2.7 | 5.5 | 5.0 | 5.3 | 62.8 | 1,496 | 137 | 103 | 2,172 | 19,310 | 58,392 | 6,099 |
| Columbiana | 341 | 3,310 | 55.9 | 3.4 | 5.4 | 9.4 | 6.7 | 146.0 | 1,417 | 586 | 248 | 4,391 | 46,430 | 49,105 | 4,101 |
| Coshocton | 125 | 3,433 | 49.4 | 7.1 | 5.2 | 6.4 | 12.5 | 62.9 | 1,722 | 80 | 91 | 1,454 | 16,110 | 45,655 | 3,256 |
| Crawford | 169 | 4,057 | 48.4 | 5.1 | 5.0 | 4.8 | 8.3 | 134.8 | 3,232 | 82 | 103 | 1,730 | 20,360 | 45,680 | 3,338 |
| Cuyahoga | 9,303 | 7,457 | 35.7 | 15.5 | 6.7 | 3.0 | 2.8 | 13,473.2 | 10,799 | 16,745 | 3,513 | 80,373 | 620,660 | 68,840 | 8,866 |
| Darke | 192 | 3,716 | 59.8 | 2.2 | 6.6 | 6.0 | 6.6 | 121.3 | 2,353 | 106 | 128 | 2,090 | 25,070 | 53,084 | 4,715 |
| Defiance | 236 | 6,188 | 46.8 | 23.9 | 3.6 | 1.4 | 4.5 | 127.9 | 3,352 | 91 | 94 | 2,011 | 18,960 | 54,351 | 4,664 |
| Delaware | 805 | 4,006 | 51.5 | 3.9 | 4.3 | 1.0 | 11.8 | 938.3 | 4,671 | 259 | 568 | 8,159 | 99,810 | 112,736 | 16,508 |
| Erie | 408 | 5,460 | 48.3 | 2.1 | 5.0 | 4.4 | 5.0 | 267.8 | 3,582 | 243 | 185 | 5,044 | 39,680 | 56,241 | 5,815 |
| Fairfield | 661 | 4,272 | 58.9 | 5.0 | 4.6 | 3.7 | 4.5 | 543.9 | 3,516 | 249 | 414 | 6,650 | 73,930 | 64,174 | 6,352 |
| Fayette | 159 | 5,559 | 37.9 | 30.9 | 3.8 | 3.8 | 4.0 | 94.3 | 3,291 | 51 | 71 | 1,658 | 13,230 | 46,914 | 3,797 |
| Franklin | 8,615 | 6,650 | 33.2 | 4.8 | 5.9 | 8.1 | 5.2 | 12,166.6 | 9,391 | 13,426 | 3,809 | 127,663 | 635,390 | 66,883 | 7,980 |
| Fulton | 199 | 4,703 | 57.3 | 5.1 | 5.3 | 3.4 | 4.0 | 126.1 | 2,982 | 94 | 106 | 2,477 | 21,290 | 56,986 | 5,105 |
| Gallia | 153 | 5,063 | 61.8 | 2.1 | 4.5 | 5.1 | 6.9 | 85.7 | 2,840 | 66 | 73 | 1,976 | 11,910 | 53,300 | 4,969 |
| Geauga | 373 | 3,973 | 44.1 | 2.8 | 6.1 | 7.4 | 8.9 | 278.9 | 2,971 | 101 | 235 | 3,733 | 47,350 | 104,981 | 15,803 |
| Greene | 598 | 3,590 | 48.0 | 4.8 | 6.4 | 1.2 | 6.4 | 546.1 | 3,279 | 15,124 | 3,362 | 10,320 | 78,180 | 73,145 | 8,373 |
| Guernsey | 130 | 3,320 | 51.1 | 7.5 | 3.8 | 9.2 | 8.1 | 37.6 | 964 | 119 | 97 | 2,183 | 18,060 | 52,756 | 4,695 |
| Hamilton | 4,460 | 5,478 | 39.2 | 4.9 | 7.6 | 5.4 | 4.7 | 6,249.3 | 7,676 | 8,540 | 2,168 | 50,673 | 403,300 | 79,575 | 11,003 |
| Hancock | 297 | 3,907 | 47.3 | 10.5 | 5.6 | 3.0 | 7.0 | 208.6 | 2,746 | 147 | 188 | 3,203 | 37,170 | 67,649 | 7,690 |
| Hardin | 98 | 3,117 | 68.7 | 6.1 | 2.3 | 0.0 | 3.5 | 42.3 | 1,349 | 66 | 75 | 1,371 | 13,290 | 47,900 | 3,676 |
| Harrison | 67 | 4,407 | 44.3 | 2.5 | 3.8 | 10.2 | 3.2 | 45.3 | 2,978 | 48 | 37 | 744 | 6,580 | 54,951 | 5,106 |
| Henry | 164 | 6,034 | 58.6 | 4.7 | 2.2 | 0.3 | 3.9 | 134.8 | 4,958 | 66 | 67 | 1,877 | 13,990 | 55,101 | 4,669 |
| Highland | 195 | 4,542 | 46.9 | 26.4 | 4.8 | 4.4 | 6.4 | 43.4 | 1,013 | 98 | 108 | 2,230 | 18,690 | 44,904 | 3,202 |
| Hocking | 127 | 4,467 | 33.8 | 37.6 | 4.6 | 1.1 | 5.2 | 23.3 | 818 | 48 | 71 | 1,597 | 12,500 | 48,008 | 3,823 |
| Holmes | 89 | 2,018 | 50.3 | 7.4 | 5.5 | 8.0 | 12.8 | 22.2 | 507 | 65 | 109 | 1,559 | 17,900 | 54,670 | 4,522 |
| Huron | 209 | 3,581 | 57.7 | 4.8 | 4.1 | 3.8 | 5.6 | 201.6 | 3,453 | 137 | 146 | 2,456 | 28,920 | 50,695 | 4,296 |
| Jackson | 113 | 3,473 | 55.6 | 0.3 | 8.9 | 10.6 | 5.9 | 60.3 | 1,861 | 72 | 81 | 1,472 | 13,560 | 45,815 | 3,426 |
| Jefferson | 558 | 8,415 | 71.6 | 4.1 | 2.0 | 2.2 | 2.6 | 154.7 | 2,332 | 169 | 159 | 3,561 | 29,850 | 52,643 | 4,967 |
| Knox | 204 | 3,326 | 50.5 | 6.9 | 2.9 | 5.9 | 6.2 | 89.4 | 1,460 | 112 | 148 | 2,763 | 27,500 | 58,249 | 5,365 |
| Lake | 1,064 | 4,622 | 51.6 | 7.2 | 7.9 | 1.9 | 6.1 | 475.5 | 2,066 | 430 | 611 | 10,756 | 124,260 | 63,515 | 6,857 |
| Lawrence | 254 | 4,224 | 68.9 | 5.5 | 2.7 | 3.1 | 3.2 | 49.7 | 827 | 133 | 149 | 2,711 | 25,000 | 49,443 | 4,022 |
| Licking | 662 | 3,814 | 53.8 | 0.8 | 6.1 | 6.4 | 4.6 | 421.2 | 2,426 | 382 | 450 | 8,302 | 84,090 | 62,140 | 6,178 |
| Logan | 195 | 4,317 | 51.4 | 7.7 | 6.2 | 4.4 | 2.3 | 213.4 | 4,721 | 124 | 116 | 2,141 | 22,740 | 54,587 | 4,944 |
| Lorain | 1,328 | 4,318 | 52.3 | 2.8 | 6.4 | 6.5 | 3.3 | 1,410.0 | 4,586 | 1,116 | 765 | 14,732 | 152,410 | 61,931 | 6,545 |
| Lucas | 2,213 | 5,133 | 37.3 | 5.9 | 9.3 | 4.4 | 3.5 | 2,108.5 | 4,892 | 1,872 | 1,161 | 28,678 | 201,890 | 58,117 | 6,295 |
| Madison | 152 | 3,459 | 59.2 | 5.0 | 3.8 | 4.5 | 5.5 | 73.8 | 1,676 | 81 | 101 | 3,091 | 19,240 | 61,652 | 6,090 |
| Mahoning | 940 | 4,090 | 50.7 | 4.3 | 8.4 | 4.1 | 2.8 | 514.9 | 2,241 | 1,170 | 566 | 13,477 | 110,160 | 54,485 | 5,685 |
| Marion | 224 | 3,445 | 59.9 | 1.1 | 4.2 | 0.1 | 3.7 | 357.8 | 5,498 | 111 | 177 | 3,950 | 27,980 | 47,966 | 3,824 |
| Medina | 654 | 3,669 | 51.9 | 4.9 | 5.9 | 2.9 | 6.1 | 367.3 | 2,061 | 381 | 483 | 6,783 | 93,950 | 74,518 | 8,618 |
| Meigs | 70 | 3,020 | 59.7 | 2.5 | 4.4 | 8.8 | 7.3 | 11.7 | 508 | 57 | 57 | 1,022 | 9,260 | 45,932 | 3,353 |
| Mercer | 213 | 5,205 | 47.2 | 32.4 | 1.4 | 0.0 | 2.7 | 134.9 | 3,301 | 97 | 103 | 2,714 | 21,140 | 60,382 | 5,765 |
| Miami | 421 | 4,003 | 52.9 | 4.0 | 6.6 | 3.4 | 6.4 | 204.3 | 1,942 | 197 | 268 | 4,714 | 52,600 | 62,683 | 6,262 |
| Monroe | 73 | 5,211 | 46.1 | 4.2 | 6.4 | 8.4 | 14.8 | 49.3 | 3,536 | 50 | 34 | 841 | 6,210 | 69,075 | 8,607 |
| Montgomery | 2,925 | 5,503 | 50.5 | 0.5 | 6.1 | 6.5 | 2.9 | 3,132.8 | 5,895 | 4,279 | 3,705 | 26,653 | 255,450 | 58,245 | 6,094 |
| Morgan | 55 | 3,786 | 44.6 | 2.3 | 6.1 | 7.7 | 9.5 | 24.6 | 1,681 | 48 | 36 | 615 | 6,110 | 44,644 | 3,234 |
| Morrow | 166 | 4,758 | 34.8 | 39.6 | 2.6 | 5.4 | 4.3 | 83.1 | 2,381 | 48 | 88 | 1,534 | 15,880 | 52,399 | 4,196 |
| Muskingum | 372 | 4,320 | 58.2 | 3.7 | 5.6 | 5.8 | 4.5 | 85.8 | 996 | 232 | 214 | 5,023 | 39,830 | 50,404 | 4,468 |
| Noble | 47 | 3,286 | 52.3 | 3.3 | 4.2 | 6.5 | 11.2 | 4.4 | 307 | 22 | 30 | 977 | 5,420 | 54,689 | 5,098 |

1. Based on the resident population estimated as of July 1 of the year shown.

# Table B. States and Counties — Land Area and Population

| State / county code | CBSA code[1] | County Type code[2] | STATE County | Land area[3] (sq. mi) | Population, 2020 Total persons 2019 | Rank | Per square mile | Race alone or in combination, not Hispanic or Latino (percent) White | Black | American Indian, Alaska Native | Asian and Pacific Islancer | Percent Hispanic or Latino[4] | Age (percent) Under 5 years | 5 to 14 years | 15 to 24 years | 25 to 34 years | 35 to 44 years | 45 to 54 years |
|---|---|---|---|---|---|---|---|---|---|---|---|---|---|---|---|---|---|---|
| | | | | 1 | 2 | 3 | 4 | 5 | 6 | 7 | 8 | 9 | 10 | 11 | 12 | 13 | 14 | 15 |
| | | | **OHIO—Cont'd** | | | | | | | | | | | | | | | |
| 39123 | 45780 | 4 | Ottawa | 254.7 | 40,253 | 1,178 | 158.0 | 93.1 | 1.6 | 0.6 | 0.6 | 5.3 | 4.2 | 10.2 | 10.1 | 9.6 | 10.1 | 12.1 |
| 39125 | | 6 | Paulding | 416.4 | 18,648 | 1,890 | 44.8 | 93.3 | 1.6 | 0.8 | 0.6 | 5.0 | 5.9 | 13.3 | 11.9 | 11.0 | 11.6 | 12.5 |
| 39127 | 18140 | 1 | Perry | 407.9 | 36,215 | 1,271 | 88.8 | 97.8 | 1.0 | 1.1 | 0.6 | 0.9 | 5.9 | 13.5 | 11.5 | 11.9 | 12.0 | 13.1 |
| 39129 | 18140 | 1 | Pickaway | 501.2 | 58,658 | 883 | 117.0 | 93.9 | 4.5 | 0.8 | 0.9 | 1.6 | 5.4 | 12.0 | 12.4 | 13.6 | 13.3 | 13.8 |
| 39131 | | 7 | Pike | 440.3 | 27,695 | 1,507 | 62.9 | 96.6 | 2.0 | 1.5 | 0.6 | 1.3 | 6.3 | 12.9 | 11.9 | 11.7 | 11.3 | 13.0 |
| 39133 | 10420 | 2 | Portage | 487.4 | 162,583 | 409 | 333.6 | 90.8 | 5.9 | 0.7 | 2.7 | 2.1 | 4.4 | 10.4 | 19.0 | 11.7 | 10.8 | 11.9 |
| 39135 | | 6 | Preble | 424.2 | 40,836 | 1,158 | 96.3 | 97.2 | 1.3 | 0.9 | 1.0 | 1.1 | 5.3 | 12.7 | 11.5 | 11.0 | 11.7 | 12.8 |
| 39137 | | 6 | Putnam | 482.5 | 33,654 | 1,335 | 69.7 | 92.5 | 0.8 | 0.3 | 0.4 | 6.7 | 6.5 | 14.8 | 11.6 | 11.0 | 11.6 | 11.5 |
| 39139 | 31900 | 3 | Richland | 495.2 | 120,891 | 525 | 244.1 | 87.5 | 10.8 | 0.7 | 1.3 | 2.1 | 5.7 | 12.3 | 11.8 | 12.9 | 11.8 | 11.8 |
| 39141 | 17060 | 4 | Ross | 689.2 | 76,420 | 728 | 110.9 | 92.1 | 7.1 | 1.2 | 0.9 | 1.4 | 5.6 | 11.9 | 11.4 | 12.7 | 13.0 | 13.6 |
| 39143 | 23380 | 4 | Sandusky | 408.2 | 58,351 | 888 | 142.9 | 85.7 | 4.6 | 0.7 | 0.7 | 10.5 | 5.5 | 12.3 | 11.6 | 11.6 | 12.0 | 12.4 |
| 39145 | 39020 | 4 | Scioto | 610.1 | 74,347 | 743 | 121.9 | 94.9 | 3.6 | 1.3 | 0.7 | 1.5 | 5.6 | 12.1 | 12.4 | 12.7 | 12.0 | 12.6 |
| 39147 | 45660 | 4 | Seneca | 551.0 | 54,938 | 933 | 99.7 | 91.1 | 3.9 | 0.6 | 1.1 | 5.5 | 5.5 | 11.9 | 14.5 | 11.8 | 11.7 | 11.5 |
| 39149 | 43380 | 4 | Shelby | 407.7 | 48,337 | 1,018 | 118.6 | 94.6 | 3.9 | 0.6 | 1.7 | 1.6 | 6.4 | 13.6 | 12.6 | 11.8 | 11.3 | 12.5 |
| 39151 | 15940 | 2 | Stark | 575.3 | 369,772 | 196 | 642.7 | 88.5 | 9.8 | 0.8 | 1.4 | 2.3 | 5.6 | 11.9 | 12.1 | 12.2 | 11.5 | 12.2 |
| 39153 | 10420 | 2 | Summit | 412.8 | 538,866 | 130 | 1,305.4 | 78.2 | 16.6 | 0.8 | 4.8 | 2.4 | 5.5 | 11.6 | 11.8 | 13.5 | 11.9 | 12.4 |
| 39155 | 49660 | 2 | Trumbull | 618.1 | 196,800 | 341 | 318.4 | 88.7 | 9.8 | 0.7 | 0.9 | 2.0 | 5.3 | 11.3 | 11.2 | 11.9 | 10.7 | 12.3 |
| 39157 | 35420 | 4 | Tuscarawas | 567.4 | 91,776 | 645 | 161.7 | 95.0 | 1.6 | 0.6 | 0.7 | 3.4 | 6.3 | 12.6 | 11.7 | 11.7 | 11.7 | 11.8 |
| 39159 | 18140 | 1 | Union | 431.8 | 60,021 | 872 | 139.0 | 89.6 | 3.4 | 0.6 | 6.0 | 2.2 | 6.4 | 13.8 | 11.9 | 13.2 | 15.1 | 14.1 |
| 39161 | 46780 | 6 | Van Wert | 409.2 | 28,159 | 1,492 | 68.8 | 94.4 | 2.0 | 0.5 | 0.5 | 4.0 | 6.3 | 12.8 | 11.6 | 12.0 | 11.9 | 11.6 |
| 39163 | | 8 | Vinton | 412.4 | 12,972 | 2,223 | 31.5 | 97.7 | 1.3 | 1.1 | 0.5 | 1.0 | 5.5 | 12.0 | 11.6 | 11.6 | 11.0 | 13.8 |
| 39165 | 17140 | 1 | Warren | 401.4 | 238,412 | 289 | 594.0 | 86.4 | 4.3 | 0.6 | 7.5 | 3.1 | 5.6 | 13.9 | 12.6 | 11.4 | 13.7 | 14.2 |
| 39167 | 31930 | 4 | Washington | 632.0 | 59,652 | 874 | 94.4 | 96.5 | 2.1 | 1.0 | 1.1 | 1.2 | 5.0 | 11.0 | 12.1 | 11.4 | 11.3 | 12.2 |
| 39169 | 49300 | 4 | Wayne | 554.8 | 115,894 | 540 | 208.5 | 95.1 | 2.4 | 0.6 | 1.4 | 2.1 | 6.5 | 13.5 | 13.5 | 12.0 | 11.1 | 11.3 |
| 39171 | | 6 | Williams | 420.7 | 36,565 | 1,261 | 86.9 | 92.9 | 1.7 | 0.6 | 0.8 | 5.0 | 5.8 | 12.4 | 11.8 | 11.8 | 11.9 | 11.7 |
| 39173 | 45780 | 2 | Wood | 617.2 | 131,113 | 488 | 212.4 | 89.2 | 3.5 | 0.7 | 2.5 | 6.0 | 5.2 | 11.3 | 20.4 | 12.5 | 11.3 | 11.1 |
| 39175 | | 7 | Wyandot | 406.9 | 21,711 | 1,733 | 53.4 | 95.4 | 0.9 | 0.5 | 0.9 | 3.3 | 5.5 | 12.7 | 11.5 | 11.2 | 11.9 | 12.3 |
| 40000 | | 0 | OKLAHOMA | 68,595.9 | 3,980,783 | X | 58.0 | 69.8 | 9.0 | 12.7 | 3.3 | 11.4 | 6.4 | 13.6 | 13.6 | 13.7 | 12.7 | 11.3 |
| 40001 | | 6 | Adair | 573.9 | 21,955 | 1,723 | 38.3 | 48.4 | 1.2 | 52.0 | 1.3 | 7.3 | 7.3 | 14.9 | 12.9 | 11.7 | 11.7 | 12.4 |
| 40003 | | 9 | Alfalfa | 866.6 | 5,718 | 2,764 | 6.6 | 84.9 | 5.5 | 5.0 | 0.8 | 6.2 | 4.8 | 11.5 | 9.0 | 10.6 | 15.9 | 14.6 |
| 40005 | | 9 | Atoka | 975.4 | 13,912 | 2,162 | 14.3 | 76.8 | 5.1 | 21.0 | 1.2 | 4.0 | 5.7 | 13.3 | 10.9 | 13.3 | 12.3 | 11.1 |
| 40007 | | 9 | Beaver | 1,814.8 | 5,207 | 2,808 | 2.9 | 71.5 | 1.7 | 2.9 | 0.6 | 25.9 | 5.5 | 14.8 | 13.5 | 10.6 | 11.2 | 10.9 |
| 40009 | 21120 | 7 | Beckham | 901.7 | 21,468 | 1,748 | 23.8 | 76.7 | 5.0 | 4.3 | 1.3 | 15.4 | 6.9 | 14.1 | 12.3 | 15.1 | 13.7 | 11.0 |
| 40011 | | 6 | Blaine | 928.5 | 9,447 | 2,461 | 10.2 | 74.7 | 5.0 | 11.2 | 1.0 | 13.2 | 6.9 | 14.5 | 11.6 | 10.9 | 10.8 | 11.3 |
| 40013 | 20460 | 6 | Bryan | 904.3 | 48,998 | 1,010 | 54.2 | 76.5 | 2.9 | 20.2 | 1.0 | 6.7 | 6.5 | 13.1 | 13.2 | 14.3 | 11.8 | 11.5 |
| 40015 | | 6 | Caddo | 1,277.8 | 28,684 | 1,470 | 22.4 | 61.3 | 4.1 | 25.4 | 1.0 | 13.9 | 6.6 | 13.8 | 12.2 | 13.5 | 12.5 | 11.4 |
| 40017 | 36420 | 1 | Canadian | 896.6 | 153,192 | 435 | 170.9 | 78.3 | 4.8 | 7.1 | 4.1 | 10.3 | 6.6 | 14.9 | 11.8 | 14.7 | 15.0 | 11.8 |
| 40019 | 11620 | 5 | Carter | 822.2 | 48,353 | 1,016 | 58.8 | 74.0 | 8.5 | 14.2 | 1.8 | 8.3 | 6.9 | 13.8 | 12.3 | 13.1 | 12.2 | 11.4 |
| 40021 | 45140 | 6 | Cherokee | 749.1 | 49,019 | 1,009 | 65.4 | 55.0 | 2.3 | 43.2 | 1.4 | 7.6 | 5.8 | 12.2 | 18.3 | 12.2 | 11.6 | 10.4 |
| 40023 | | 7 | Choctaw | 770.4 | 14,646 | 2,111 | 19.0 | 65.9 | 12.6 | 23.2 | 1.1 | 5.0 | 7.0 | 13.1 | 11.6 | 11.0 | 10.8 | 11.2 |
| 40025 | | 9 | Cimarron | 1,834.8 | 2,145 | 3,032 | 1.2 | 74.1 | 2.2 | 2.2 | 0.9 | 23.7 | 6.2 | 16.1 | 10.9 | 8.5 | 9.7 | 9.8 |
| 40027 | 36420 | 1 | Cleveland | 538.9 | 287,066 | 246 | 532.7 | 75.6 | 6.9 | 8.0 | 5.9 | 9.5 | 5.1 | 12.0 | 18.3 | 14.5 | 13.4 | 11.1 |
| 40029 | | 9 | Coal | 516.8 | 5,587 | 2,781 | 10.8 | 74.7 | 2.3 | 26.6 | 0.8 | 4.8 | 7.0 | 12.4 | 12.5 | 11.0 | 11.0 | 11.3 |
| 40031 | 30020 | 3 | Comanche | 1,069.3 | 121,099 | 523 | 113.3 | 60.8 | 19.2 | 8.1 | 5.1 | 13.8 | 6.8 | 13.0 | 15.8 | 16.5 | 12.9 | 10.2 |
| 40033 | 30020 | 3 | Cotton | 632.6 | 5,676 | 2,775 | 9.0 | 79.9 | 3.3 | 12.1 | 1.1 | 8.9 | 5.4 | 13.6 | 11.5 | 10.7 | 10.9 | 12.3 |
| 40035 | | 6 | Craig | 761.3 | 14,194 | 2,145 | 18.6 | 70.3 | 4.6 | 28.9 | 1.5 | 4.0 | 5.5 | 12.4 | 11.8 | 12.2 | 11.7 | 12.6 |
| 40037 | 46140 | 2 | Creek | 949.9 | 71,485 | 763 | 75.3 | 81.6 | 3.5 | 16.2 | 1.2 | 4.8 | 5.9 | 13.1 | 12.0 | 12.3 | 12.0 | 12.2 |
| 40039 | 48220 | 7 | Custer | 988.8 | 28,648 | 1,472 | 29.0 | 70.5 | 3.7 | 8.9 | 1.8 | 18.7 | 6.8 | 13.8 | 20.8 | 12.3 | 11.7 | 9.2 |
| 40041 | | 6 | Delaware | 737.8 | 43,136 | 1,117 | 58.5 | 70.8 | 0.9 | 30.6 | 1.5 | 4.0 | 4.9 | 11.2 | 10.6 | 10.3 | 10.0 | 11.8 |
| 40043 | | 9 | Dewey | 999.6 | 4,815 | 2,840 | 4.8 | 84.0 | 2.0 | 8.6 | 1.6 | 7.9 | 6.3 | 15.4 | 11.3 | 12.0 | 12.1 | 10.4 |
| 40045 | 49260 | 9 | Ellis | 1,231.5 | 3,830 | 2,908 | 3.1 | 88.0 | 1.4 | 4.0 | 1.0 | 7.8 | 5.7 | 12.1 | 11.7 | 9.8 | 11.5 | 11.6 |
| 40047 | 21420 | 5 | Garfield | 1,058.5 | 60,869 | 865 | 57.5 | 75.7 | 4.4 | 4.4 | 6.0 | 13.6 | 7.1 | 14.4 | 12.9 | 13.6 | 12.6 | 10.3 |
| 40049 | | 6 | Garvin | 802.1 | 27,691 | 1,508 | 34.5 | 78.6 | 3.3 | 12.8 | 0.9 | 10.1 | 6.2 | 14.3 | 12.0 | 11.5 | 12.8 | 11.7 |
| 40051 | 36420 | 1 | Grady | 1,100.5 | 55,906 | 923 | 50.8 | 85.3 | 3.0 | 9.3 | 1.0 | 6.2 | 5.4 | 13.6 | 12.4 | 11.8 | 13.1 | 12.3 |
| 40053 | | 9 | Grant | 1,000.7 | 4,372 | 2,865 | 4.4 | 89.4 | 2.7 | 4.7 | 0.8 | 5.7 | 6.1 | 14.2 | 10.3 | 10.5 | 11.8 | 10.2 |
| 40055 | | 7 | Greer | 639.3 | 5,704 | 2,768 | 8.9 | 75.8 | 9.4 | 4.9 | 0.6 | 12.4 | 4.8 | 11.3 | 12.0 | 15.7 | 14.0 | 12.4 |
| 40057 | | 9 | Harmon | 537.3 | 2,557 | 2,998 | 4.8 | 57.8 | 9.3 | 3.9 | 1.9 | 30.8 | 5.8 | 14.7 | 11.3 | 10.2 | 11.2 | 12.3 |
| 40059 | | 9 | Harper | 1,038.6 | 3,611 | 2,926 | 3.5 | 73.4 | 0.9 | 2.5 | 0.6 | 24.5 | 5.5 | 15.6 | 10.9 | 9.8 | 14.3 | 9.8 |
| 40061 | | 6 | Haskell | 576.7 | 12,652 | 2,241 | 21.9 | 76.0 | 1.5 | 23.9 | 1.2 | 4.7 | 5.4 | 13.7 | 11.7 | 10.9 | 12.3 | 11.4 |
| 40063 | | 7 | Hughes | 804.5 | 13,126 | 2,212 | 16.3 | 68.3 | 6.8 | 25.1 | 0.7 | 6.2 | 6.0 | 12.3 | 12.1 | 14.4 | 12.0 | 11.8 |
| 40065 | 11060 | 7 | Jackson | 802.7 | 24,305 | 1,636 | 30.3 | 64.4 | 8.0 | 3.4 | 2.5 | 25.3 | 7.4 | 13.9 | 13.8 | 15.2 | 11.9 | 10.5 |
| 40067 | | 8 | Jefferson | 759.0 | 5,949 | 2,749 | 7.8 | 80.2 | 2.3 | 9.2 | 0.9 | 12.0 | 5.7 | 15.0 | 10.7 | 11.1 | 10.6 | 11.2 |
| 40069 | | 9 | Johnston | 643.0 | 10,824 | 2,361 | 16.8 | 75.5 | 3.4 | 22.1 | 1.0 | 6.1 | 5.8 | 13.8 | 12.8 | 12.0 | 11.2 | 11.5 |
| 40071 | 38620 | 5 | Kay | 919.6 | 43,274 | 1,111 | 47.1 | 78.0 | 3.2 | 14.1 | 1.5 | 8.6 | 6.3 | 14.2 | 12.7 | 12.1 | 11.7 | 10.4 |
| 40073 | | 6 | Kingfisher | 898.1 | 15,806 | 2,041 | 17.6 | 77.9 | 2.0 | 5.8 | 0.7 | 17.0 | 6.2 | 15.8 | 12.5 | 11.7 | 12.7 | 11.0 |

1. CBSA = Core Based Statistical Area. See Appendix A for explanation. See Appendix B for list of metropolitan areas with component counties.   2. County type code from the Economic Research Service of USDA Rural-Urban Continuum Codes. See Appendix A for definition.   3. Dry land or land partially or temporarily covered by water.   4. May be of any race.

# Table B. States and Counties — Population and Households

| STATE County | 55 to 64 years | 65 to 74 years | 75 years and over | Percent female | Total persons 2000 | Total persons 2010 | Percent change 2000-2010 | Percent change 2010-2020 | Births | Deaths | Net Migration | Number | Persons per household | Family households | Female family householder[1] | One person |
|---|---|---|---|---|---|---|---|---|---|---|---|---|---|---|---|---|
| | 16 | 17 | 18 | 19 | 20 | 21 | 22 | 23 | 24 | 25 | 26 | 27 | 28 | 29 | 30 | 31 |
| **OHIO—Cont'd** | | | | | | | | | | | | | | | | |
| Ottawa | 17.2 | 15.7 | 10.9 | 50.3 | 40,985 | 41,433 | 1.1 | -2.8 | 3,452 | 5,053 | 459 | 18,051 | 2.2 | 66.0 | 9.7 | 28.7 |
| Paulding | 14.3 | 11.8 | 7.9 | 50.3 | 20,293 | 19,610 | -3.4 | -4.9 | 2,252 | 2,070 | -1,150 | 7,705 | 2.43 | 70.5 | 9.1 | 25.4 |
| Perry | 14.6 | 11.0 | 6.6 | 50.0 | 34,078 | 36,036 | 5.7 | 0.5 | 4,348 | 3,670 | -491 | 13,500 | 2.64 | 71.4 | 12.4 | 21.5 |
| Pickaway | 13.1 | 9.7 | 6.8 | 47.7 | 52,727 | 55,680 | 5.6 | 5.3 | 6,210 | 5,752 | 2,551 | 19,710 | 2.7 | 72.3 | 11.9 | 22.7 |
| Pike | 14.2 | 10.7 | 7.9 | 50.5 | 27,695 | 28,568 | 3.2 | -3.1 | 3,551 | 3,600 | -813 | 10,959 | 2.51 | 67.6 | 11.5 | 27.9 |
| Portage | 14.2 | 10.7 | 7.0 | 51.1 | 152,061 | 161,430 | 6.2 | 0.7 | 14,587 | 14,812 | 1,433 | 61,817 | 2.51 | 63.9 | 11.0 | 28.6 |
| Preble | 14.8 | 12.2 | 7.9 | 50.4 | 42,337 | 42,254 | -0.2 | -3.4 | 4,485 | 4,878 | -1,017 | 16,251 | 2.5 | 69.9 | 9.3 | 25.1 |
| Putnam | 14.5 | 10.6 | 7.8 | 50.0 | 34,726 | 34,496 | -0.7 | -2.4 | 4,523 | 3,170 | -2,218 | 13,327 | 2.52 | 73.3 | 7.7 | 22.2 |
| Richland | 13.4 | 11.5 | 8.9 | 49.3 | 128,852 | 124,472 | -3.4 | -2.9 | 14,214 | 14,524 | -3,247 | 48,449 | 2.35 | 63.9 | 11.5 | 30.5 |
| Ross | 14.2 | 10.5 | 7.2 | 47.9 | 73,345 | 78,079 | 6.5 | -2.1 | 8,579 | 8,659 | -1,549 | 28,802 | 2.47 | 68.8 | 13.8 | 25.8 |
| Sandusky | 14.8 | 11.6 | 8.1 | 50.4 | 61,792 | 60,946 | -1.4 | -4.3 | 6,771 | 6,719 | -2,646 | 23,574 | 2.45 | 66.4 | 11.4 | 27.6 |
| Scioto | 13.6 | 10.9 | 8.1 | 50.7 | 79,195 | 79,659 | 0.6 | -6.7 | 8,897 | 10,176 | -4,039 | 29,858 | 2.43 | 65.2 | 13.3 | 29.4 |
| Seneca | 14.0 | 11.2 | 7.9 | 50.0 | 58,683 | 56,743 | -3.3 | -3.2 | 6,008 | 6,142 | -1,667 | 21,648 | 2.41 | 65.3 | 10.8 | 28.2 |
| Shelby | 14.2 | 10.4 | 7.2 | 49.9 | 47,910 | 49,455 | 3.2 | -2.3 | 6,300 | 4,638 | -2,801 | 18,608 | 2.6 | 69.8 | 10.8 | 25.7 |
| Stark | 14.2 | 11.7 | 8.7 | 51.4 | 378,098 | 375,561 | -0.7 | -1.5 | 42,308 | 43,181 | -4,585 | 153,460 | 2.37 | 64.4 | 12.8 | 29.9 |
| Summit | 14.3 | 11.2 | 7.7 | 51.6 | 542,899 | 541,785 | -0.2 | -0.5 | 61,809 | 59,322 | -5,087 | 224,726 | 2.37 | 61.1 | 12.7 | 32.5 |
| Trumbull | 14.8 | 13.0 | 9.5 | 51.3 | 225,116 | 210,331 | -6.6 | -6.4 | 21,459 | 26,330 | -8,604 | 85,621 | 2.3 | 62.6 | 13.5 | 32.0 |
| Tuscarawas | 14.0 | 11.6 | 8.6 | 50.5 | 90,914 | 92,585 | 1.8 | -0.9 | 11,677 | 10,771 | -1,662 | 36,631 | 2.49 | 66.5 | 9.1 | 27.8 |
| Union | 12.3 | 8.1 | 5.1 | 51.7 | 40,909 | 52,334 | 27.9 | 14.7 | 6,649 | 3,778 | 4,813 | 20,212 | 2.65 | 74.9 | 7.3 | 19.4 |
| Van Wert | 14.0 | 11.3 | 8.7 | 50.7 | 29,659 | 28,759 | -3.0 | -2.1 | 3,440 | 3,258 | -774 | 11,544 | 2.42 | 70.7 | 10.1 | 25.8 |
| Vinton | 15.6 | 11.6 | 7.2 | 50.0 | 12,806 | 13,430 | 4.9 | -3.4 | 1,464 | 1,525 | -399 | 5,140 | 2.53 | 67.2 | 12.9 | 28.1 |
| Warren | 13.5 | 9.1 | 6.2 | 49.7 | 158,383 | 212,795 | 34.4 | 12.0 | 24,565 | 17,249 | 18,501 | 82,957 | 2.67 | 74.9 | 9.1 | 20.7 |
| Washington | 14.9 | 12.6 | 9.5 | 50.8 | 63,251 | 61,787 | -2.3 | -3.5 | 6,216 | 7,590 | -712 | 25,197 | 2.33 | 65.1 | 10.1 | 30.2 |
| Wayne | 13.5 | 10.7 | 8.0 | 50.3 | 111,564 | 114,531 | 2.7 | 1.0 | 15,523 | 11,499 | -2,815 | 43,824 | 2.56 | 68.5 | 9.4 | 27.5 |
| Williams | 14.6 | 11.1 | 8.8 | 50.3 | 39,188 | 37,652 | -3.9 | -2.9 | 4,366 | 4,154 | -1,298 | 15,090 | 2.37 | 63.5 | 8.9 | 30.9 |
| Wood | 12.1 | 9.7 | 6.4 | 50.6 | 121,065 | 125,490 | 3.7 | 4.5 | 13,774 | 11,251 | 3,093 | 50,589 | 2.44 | 60.6 | 8.6 | 29.4 |
| Wyandot | 14.5 | 11.5 | 9.0 | 50.5 | 22,908 | 22,616 | -1.3 | -4.0 | 2,530 | 2,543 | -891 | 9,081 | 2.39 | 67.5 | 10.0 | 25.9 |
| **OKLAHOMA** | 12.3 | 9.6 | 6.8 | 50.5 | 3,450,654 | 3,751,582 | 8.7 | 6.1 | 530,768 | 404,645 | 102,731 | 1,480,061 | 2.58 | 65.9 | 12.3 | 28.3 |
| Adair | 12.9 | 9.4 | 6.7 | 50.2 | 21,038 | 22,671 | 7.8 | -3.2 | 3,333 | 2,607 | -1,443 | 7,733 | 2.86 | 72.2 | 16.4 | 23.3 |
| Alfalfa | 13.4 | 10.6 | 9.5 | 40.3 | 6,105 | 5,639 | -7.6 | 1.4 | 577 | 660 | 160 | 1,870 | 2.53 | 70.2 | 6.3 | 27.6 |
| Atoka | 13.0 | 11.4 | 8.9 | 48.0 | 13,879 | 14,165 | 2.1 | -1.8 | 1,650 | 1,632 | -269 | 5,284 | 2.34 | 69.2 | 13.9 | 26.8 |
| Beaver | 13.3 | 11.3 | 9.0 | 48.9 | 5,857 | 5,636 | -3.8 | -7.6 | 603 | 531 | -508 | 1,971 | 2.69 | 73.5 | 8.8 | 22.4 |
| Beckham | 11.6 | 8.8 | 6.3 | 46.8 | 19,799 | 22,119 | 11.7 | -2.9 | 3,451 | 2,503 | -1,634 | 7,578 | 2.66 | 69.5 | 10.8 | 25.7 |
| Blaine | 13.9 | 11.2 | 9.1 | 50.0 | 11,976 | 11,943 | -0.3 | -20.9 | 1,450 | 1,282 | -2,935 | 3,898 | 2.06 | 64.5 | 12.9 | 31.6 |
| Bryan | 11.8 | 10.0 | 7.9 | 51.3 | 36,534 | 42,416 | 16.1 | 15.5 | 6,074 | 5,261 | 5,758 | 17,253 | 2.64 | 66.0 | 12.9 | 27.4 |
| Caddo | 12.6 | 9.8 | 7.5 | 47.1 | 30,150 | 29,605 | -1.8 | -3.1 | 4,140 | 3,731 | -1,320 | 10,320 | 2.7 | 68.2 | 14.1 | 27.3 |
| Canadian | 11.4 | 8.4 | 5.3 | 50.5 | 87,697 | 115,566 | 31.8 | 32.6 | 17,809 | 10,007 | 29,573 | 44,906 | 3.08 | 74.0 | 10.2 | 22.2 |
| Carter | 12.8 | 9.9 | 7.5 | 51.3 | 45,621 | 47,729 | 4.6 | 1.3 | 6,758 | 6,341 | 235 | 18,339 | 2.58 | 65.9 | 12.9 | 29.9 |
| Cherokee | 12.2 | 10.3 | 7.1 | 51.3 | 42,521 | 46,992 | 10.5 | 4.3 | 6,089 | 5,202 | 1,156 | 16,665 | 2.8 | 63.6 | 11.0 | 30.8 |
| Choctaw | 13.9 | 12.3 | 9.1 | 51.9 | 15,342 | 15,199 | -0.9 | -3.6 | 2,061 | 2,244 | -365 | 5,971 | 2.45 | 64.1 | 15.8 | 32.4 |
| Cimarron | 13.0 | 13.7 | 12.1 | 49.1 | 3,148 | 2,475 | -21.4 | -13.3 | 299 | 294 | -340 | 977 | 2.21 | 61.3 | 7.7 | 35.7 |
| Cleveland | 11.2 | 8.6 | 5.8 | 50.2 | 208,016 | 255,991 | 23.1 | 12.1 | 30,164 | 20,133 | 20,944 | 106,172 | 2.52 | 65.3 | 10.5 | 25.8 |
| Coal | 13.5 | 12.1 | 9.2 | 50.6 | 6,031 | 5,927 | -1.7 | -5.7 | 709 | 833 | -220 | 2,227 | 2.48 | 65.5 | 13.3 | 30.7 |
| Comanche | 11.4 | 7.8 | 5.6 | 48.2 | 114,996 | 124,093 | 7.9 | -2.4 | 18,928 | 10,634 | -11,523 | 42,842 | 2.62 | 65.1 | 14.2 | 29.2 |
| Cotton | 15.5 | 11.4 | 8.7 | 50.9 | 6,614 | 6,186 | -6.5 | -8.2 | 668 | 830 | -345 | 2,195 | 2.61 | 73.3 | 14.1 | 23.9 |
| Craig | 13.9 | 10.5 | 9.4 | 49.0 | 14,950 | 15,023 | 0.5 | -5.5 | 1,605 | 2,168 | -265 | 5,422 | 2.44 | 67.9 | 11.3 | 27.9 |
| Creek | 13.8 | 10.7 | 7.9 | 50.6 | 67,367 | 69,985 | 3.9 | 2.1 | 8,653 | 8,819 | 1,707 | 26,427 | 2.67 | 70.9 | 12.6 | 25.0 |
| Custer | 10.7 | 8.1 | 6.6 | 50.1 | 26,142 | 27,542 | 5.4 | 4.0 | 4,470 | 3,019 | -373 | 10,653 | 2.58 | 63.4 | 11.4 | 29.1 |
| Delaware | 15.4 | 15.0 | 10.8 | 50.7 | 37,077 | 41,500 | 11.9 | 3.9 | 4,328 | 5,749 | 3,077 | 16,739 | 2.51 | 67.6 | 11.3 | 27.7 |
| Dewey | 13.7 | 9.7 | 9.0 | 50.9 | 4,743 | 4,810 | 1.4 | 0.1 | 617 | 711 | 103 | 1,689 | 2.86 | 71.5 | 6.2 | 26.6 |
| Ellis | 13.8 | 13.7 | 10.2 | 50.1 | 4,075 | 4,151 | 1.9 | -7.7 | 451 | 570 | -206 | 1,614 | 2.45 | 62.3 | 7.9 | 34.0 |
| Garfield | 12.2 | 9.3 | 7.5 | 49.7 | 57,813 | 60,582 | 4.8 | 0.5 | 9,487 | 7,038 | -2,139 | 23,541 | 2.56 | 65.9 | 10.6 | 29.2 |
| Garvin | 13.0 | 10.3 | 8.1 | 50.5 | 27,210 | 27,571 | 1.3 | 0.4 | 3,685 | 3,797 | 243 | 10,447 | 2.63 | 68.2 | 11.9 | 27.6 |
| Grady | 14.3 | 10.2 | 7.0 | 50.1 | 45,516 | 52,423 | 15.2 | 6.6 | 6,085 | 5,644 | 3,074 | 19,857 | 2.74 | 74.0 | 10.4 | 21.5 |
| Grant | 14.4 | 12.4 | 10.1 | 50.3 | 5,144 | 4,532 | -11.9 | -3.5 | 531 | 619 | -72 | 1,736 | 2.47 | 65.8 | 7.4 | 31.4 |
| Greer | 11.7 | 9.3 | 8.8 | 42.7 | 6,061 | 6,239 | 2.9 | -8.6 | 635 | 773 | -399 | 2,054 | 2.34 | 62.6 | 9.1 | 34.8 |
| Harmon | 13.1 | 12.4 | 9.2 | 51.2 | 3,283 | 2,922 | -11.0 | -12.5 | 347 | 333 | -382 | 1,113 | 2.35 | 68.2 | 9.5 | 25.7 |
| Harper | 14.3 | 11.0 | 8.8 | 49.2 | 3,562 | 3,682 | 3.4 | -1.9 | 493 | 468 | -99 | 1,258 | 2.95 | 64.1 | 3.1 | 34.3 |
| Haskell | 13.3 | 12.0 | 9.2 | 50.6 | 11,792 | 12,775 | 8.3 | | 1,507 | 1,620 | -5 | 4,931 | 2.56 | 70.9 | 11.5 | 24.4 |
| Hughes | 12.4 | 10.5 | 8.6 | 45.9 | 14,154 | 13,999 | -1.1 | -6.2 | 1,548 | 1,952 | -462 | 4,132 | 2.86 | 67.2 | 12.8 | 30.4 |
| Jackson | 11.7 | 8.8 | 6.8 | 50.2 | 28,439 | 26,446 | -7.0 | -8.1 | 4,185 | 2,621 | -3,743 | 9,762 | 2.5 | 66.1 | 11.8 | 29.1 |
| Jefferson | 14.1 | 11.6 | 9.7 | 50.3 | 6,818 | 6,472 | -5.1 | -8.1 | 717 | 972 | -270 | 2,371 | 2.53 | 68.3 | 12.9 | 27.3 |
| Johnston | 13.1 | 11.1 | 8.8 | 50.3 | 10,513 | 10,957 | 4.2 | -1.2 | 1,327 | 1,467 | 9 | 4,267 | 2.53 | 67.3 | 12.1 | 29.8 |
| Kay | 12.7 | 11.0 | 8.9 | 50.3 | 48,080 | 46,562 | -3.2 | -7.1 | 6,119 | 5,984 | -3,418 | 17,610 | 2.47 | 62.6 | 11.7 | 31.2 |
| Kingfisher | 13.1 | 9.4 | 7.5 | 50.1 | 13,926 | 15,036 | 8.0 | 5.1 | 2,010 | 1,541 | 303 | 5,609 | 2.77 | 70.9 | 11.2 | 24.7 |

1. No spouse present.

# Table B. States and Counties — Population, Vital Statistics, Health, and Crime

| STATE County | Persons in group quarters, 2020 | Daytime Population, 2015-2019 Number | Employment/ residence ratio | Births, 2020 Total | Births Rate[1] | Deaths, 2020 Number | Deaths Rate[1] | Persons under 65 with no health insurance, 2019 Number | Percent | Medicare, 2020 Total beneficiaries | Enrolled in Original Medicare | Enrolled in Medicare Advantage | Crimes reported 2019 Violent | Property |
|---|---|---|---|---|---|---|---|---|---|---|---|---|---|---|
| | 32 | 33 | 34 | 35 | 36 | 37 | 38 | 39 | 40 | 41 | 42 | 43 | 44 | 45 |
| **OHIO—Cont'd** | | | | | | | | | | | | | | |
| Ottawa | 495 | 36,139 | 0.76 | 318 | 7.9 | 545 | 13.5 | 2,098 | 7.1 | 11,858 | 7,254 | 4,605 | NA | NA |
| Paulding | 85 | 15,621 | 0.62 | 219 | 11.7 | 224 | 12.0 | 1,189 | 7.9 | 4,272 | 2,703 | 1,569 | 30 | 104 |
| Perry | 307 | 28,619 | 0.49 | 413 | 11.4 | 413 | 11.4 | 2,319 | 7.8 | 7,705 | 4,596 | 3,110 | 13 | 124 |
| Pickaway | 4,460 | 50,233 | 0.69 | 602 | 10.3 | 643 | 11.0 | 3,204 | 7.2 | 11,119 | 5,336 | 5,783 | 22 | 546 |
| Pike | 496 | 28,518 | 1.05 | 345 | 12.5 | 396 | 14.3 | 1,924 | 8.6 | 6,135 | 3,935 | 2,200 | 33 | 464 |
| Portage | 7,139 | 143,532 | 0.76 | 1,348 | 8.3 | 1,571 | 9.7 | 9,786 | 7.6 | 32,668 | 15,193 | 17,476 | NA | 916 |
| Preble | 386 | 34,818 | 0.68 | 426 | 10.4 | 500 | 12.2 | 2,556 | 7.9 | 9,611 | 5,505 | 4,106 | NA | NA |
| Putnam | 303 | 29,424 | 0.74 | 424 | 12.6 | 350 | 10.4 | 1,938 | 7.0 | 6,963 | 5,068 | 1,895 | 15 | 90 |
| Richland | 6,872 | 123,511 | 1.05 | 1,385 | 11.5 | 1,494 | 12.4 | 7,840 | 8.7 | 28,900 | 18,266 | 10,634 | 61 | 757 |
| Ross | 5,243 | 76,619 | 0.99 | 808 | 10.6 | 929 | 12.2 | 4,392 | 7.5 | 16,231 | 9,688 | 6,542 | 114 | 1,041 |
| Sandusky | 902 | 58,930 | 1.00 | 625 | 10.7 | 668 | 11.4 | 3,395 | 7.3 | 13,578 | 9,153 | 4,425 | NA | NA |
| Scioto | 3,148 | 74,420 | 0.94 | 784 | 10.5 | 1,037 | 13.9 | 4,836 | 8.3 | 16,897 | 11,668 | 5,229 | 48 | 633 |
| Seneca | 2,468 | 49,820 | 0.79 | 584 | 10.6 | 611 | 11.1 | 3,089 | 7.2 | 12,375 | 8,994 | 3,381 | NA | NA |
| Shelby | 589 | 53,629 | 1.20 | 610 | 12.6 | 487 | 10.1 | 2,677 | 6.7 | 9,611 | 6,730 | 2,881 | 12 | 131 |
| Stark | 8,792 | 364,450 | 0.95 | 4,040 | 10.9 | 4,443 | 12.0 | 22,294 | 7.7 | 86,515 | 35,067 | 51,449 | 107 | 1,544 |
| Summit | 9,534 | 560,276 | 1.07 | 5,804 | 10.8 | 6,298 | 11.7 | 34,546 | 8.0 | 115,086 | 50,610 | 64,475 | 80 | 944 |
| Trumbull | 3,810 | 189,244 | 0.87 | 2,078 | 10.6 | 2,644 | 13.4 | 13,491 | 8.9 | 51,901 | 24,257 | 27,643 | NA | NA |
| Tuscarawas | 1,251 | 87,397 | 0.89 | 1,125 | 12.3 | 1,118 | 12.2 | 7,482 | 10.3 | 20,924 | 10,393 | 10,531 | 22 | 241 |
| Union | 2,414 | 64,410 | 1.27 | 714 | 11.9 | 416 | 6.9 | 2,867 | 5.8 | 9,093 | 5,128 | 3,964 | 21 | 126 |
| Van Wert | 396 | 26,358 | 0.86 | 340 | 12.1 | 315 | 11.2 | 1,612 | 7.2 | 6,499 | 4,128 | 2,371 | 16 | 106 |
| Vinton | 68 | 10,792 | 0.57 | 125 | 9.6 | 184 | 14.2 | 1,079 | 10.2 | 2,899 | 1,899 | 1,000 | 10 | 78 |
| Warren | 5,248 | 217,162 | 0.89 | 2,360 | 9.9 | 2,019 | 8.5 | 10,432 | 5.3 | 40,350 | 21,999 | 18,351 | NA | NA |
| Washington | 1,587 | 60,585 | 1.01 | 592 | 9.9 | 782 | 13.1 | 3,523 | 7.7 | 15,328 | 10,767 | 4,560 | 34 | 172 |
| Wayne | 3,365 | 115,959 | 1.00 | 1,428 | 12.3 | 1,249 | 10.8 | 11,007 | 12.0 | 23,811 | 12,674 | 11,137 | 44 | 473 |
| Williams | 990 | 38,018 | 1.07 | 419 | 11.5 | 410 | 11.2 | 2,252 | 7.8 | 8,637 | 5,831 | 2,806 | NA | NA |
| Wood | 6,435 | 135,710 | 1.08 | 1,265 | 9.6 | 1,182 | 9.0 | 6,482 | 6.2 | 24,433 | 13,918 | 10,515 | 13 | 301 |
| Wyandot | 251 | 20,340 | 0.85 | 223 | 10.3 | 263 | 12.1 | 1,191 | 6.9 | 5,017 | 3,672 | 1,346 | 3 | 3 |
| **OKLAHOMA** | 109,811 | 3,926,012 | 1.00 | 49,033 | 12.3 | 43,466 | 10.9 | 540,944 | 16.8 | 753,182 | 562,963 | 190,219 | X | X |
| Adair | 92 | 20,105 | 0.73 | 298 | 13.6 | 291 | 13.3 | 4,083 | 22.4 | 4,058 | 3,362 | 696 | 38 | 159 |
| Alfalfa | 1,051 | 5,750 | 0.96 | 49 | 8.6 | 55 | 9.6 | 613 | 17.2 | 1,096 | 1,013 | 83 | 1 | 26 |
| Atoka | 749 | 12,804 | 0.78 | 154 | 11.1 | 139 | 10.0 | 2,114 | 20.8 | 3,170 | 2,853 | 316 | 5 | 68 |
| Beaver | 43 | 4,960 | 0.83 | 49 | 9.4 | 54 | 10.4 | 943 | 22.4 | 971 | 932 | 40 | 0 | 14 |
| Beckham | 1,926 | 23,841 | 1.18 | 295 | 13.7 | 280 | 13.0 | 2,872 | 17.1 | 3,865 | 3,576 | 290 | 9 | 77 |
| Blaine | 107 | 9,615 | 1.02 | 129 | 13.7 | 114 | 12.1 | 1,353 | 18.1 | 2,066 | 1,878 | 189 | 10 | 104 |
| Bryan | 1,137 | 46,360 | 1.00 | 597 | 12.2 | 601 | 12.3 | 7,762 | 20.3 | 9,521 | 8,133 | 1,388 | 30 | 321 |
| Caddo | 1,927 | 26,588 | 0.76 | 368 | 12.8 | 386 | 13.5 | 4,348 | 19.9 | 5,957 | 5,226 | 731 | 12 | 165 |
| Canadian | 2,482 | 113,360 | 0.61 | 1,891 | 12.3 | 1,228 | 8.0 | 15,671 | 12.3 | 23,428 | 16,139 | 7,289 | 19 | 151 |
| Carter | 913 | 51,368 | 1.15 | 654 | 13.5 | 700 | 14.5 | 7,543 | 19.3 | 10,758 | 9,209 | 1,549 | 29 | 208 |
| Cherokee | 2,013 | 45,673 | 0.85 | 581 | 11.9 | 543 | 11.1 | 9,319 | 24.2 | 9,755 | 7,998 | 1,758 | 69 | 461 |
| Choctaw | 172 | 14,531 | 0.95 | 197 | 13.5 | 238 | 16.3 | 2,175 | 19.3 | 3,775 | 3,433 | 342 | 47 | 79 |
| Cimarron | 7 | 2,111 | 0.93 | 28 | 13.1 | 17 | 7.9 | 348 | 22.2 | 576 | 556 | 20 | 0 | 10 |
| Cleveland | 10,547 | 234,682 | 0.68 | 2,800 | 9.8 | 2,346 | 8.2 | 28,188 | 11.9 | 46,832 | 34,642 | 12,190 | 30 | 245 |
| Coal | 61 | 5,190 | 0.81 | 75 | 13.4 | 83 | 14.9 | 898 | 21.0 | 1,120 | 1,015 | 105 | 0 | 4 |
| Comanche | 10,280 | 124,073 | 1.04 | 1,643 | 13.6 | 1,189 | 9.8 | 14,125 | 14.9 | 19,551 | 16,854 | 2,697 | 17 | 215 |
| Cotton | 48 | 5,687 | 0.92 | 51 | 9.0 | 98 | 17.3 | 756 | 16.9 | 1,263 | 1,134 | 130 | 0 | 21 |
| Craig | 1,050 | 14,603 | 1.04 | 141 | 9.9 | 231 | 16.3 | 1,994 | 18.9 | 3,562 | 2,882 | 680 | 13 | 106 |
| Creek | 1,061 | 61,177 | 0.66 | 776 | 10.9 | 985 | 13.8 | 9,140 | 15.8 | 16,083 | 9,791 | 6,292 | 65 | 374 |
| Custer | 1,525 | 29,879 | 1.05 | 394 | 13.8 | 289 | 10.1 | 4,812 | 20.5 | 4,712 | 4,159 | 553 | 12 | 113 |
| Delaware | 356 | 39,471 | 0.81 | 410 | 9.5 | 635 | 14.7 | 7,090 | 22.4 | 11,652 | 8,829 | 2,823 | 46 | 263 |
| Dewey | 93 | 5,068 | 1.07 | 55 | 11.4 | 79 | 16.4 | 686 | 17.3 | 1,135 | 1,030 | 105 | 2 | 40 |
| Ellis | 46 | 3,848 | 0.90 | 47 | 12.3 | 50 | 13.1 | 537 | 18.2 | 946 | 899 | 48 | 7 | 51 |
| Garfield | 1,801 | 62,091 | 1.01 | 818 | 13.4 | 676 | 11.1 | 8,357 | 16.8 | 11,991 | 10,583 | 1,408 | 12 | 157 |
| Garvin | 317 | 27,908 | 1.01 | 327 | 11.8 | 360 | 13.0 | 4,250 | 19.1 | 6,121 | 4,944 | 1,177 | 30 | 269 |
| Grady | 1,075 | 47,360 | 0.68 | 550 | 9.8 | 581 | 10.4 | 6,388 | 14.0 | 10,876 | 8,276 | 2,600 | 24 | 321 |
| Grant | 73 | 4,193 | 0.89 | 51 | 11.7 | 47 | 10.8 | 542 | 16.2 | 1,067 | 978 | 88 | 3 | 31 |
| Greer | 1,136 | 5,335 | 0.73 | 49 | 8.6 | 60 | 10.5 | 506 | 14.1 | 1,247 | 1,180 | 67 | NA | NA |
| Harmon | 100 | 2,501 | 0.82 | 25 | 9.8 | 31 | 12.1 | 487 | 24.1 | 635 | 593 | 41 | 2 | 9 |
| Harper | 41 | 3,437 | 0.78 | 36 | 10.0 | 77 | 21.3 | 668 | 22.9 | 831 | 791 | 40 | NA | NA |
| Haskell | 78 | 11,921 | 0.83 | 119 | 9.4 | 144 | 11.4 | 2,030 | 20.5 | 2,907 | 2,428 | 479 | 7 | 66 |
| Hughes | 1,556 | 12,868 | 0.88 | 151 | 11.5 | 176 | 13.4 | 1,824 | 19.7 | 2,963 | 2,406 | 557 | 8 | 121 |
| Jackson | 687 | 25,381 | 1.03 | 358 | 14.7 | 222 | 9.1 | 3,478 | 17.3 | 4,381 | 4,108 | 273 | 15 | 46 |
| Jefferson | 144 | 5,378 | 0.64 | 59 | 9.9 | 85 | 14.3 | 896 | 19.7 | 1,512 | 1,343 | 169 | 11 | 18 |
| Johnston | 292 | 10,316 | 0.82 | 125 | 11.5 | 185 | 17.1 | 1,530 | 17.9 | 2,506 | 2,140 | 366 | 14 | 32 |
| Kay | 1,275 | 44,926 | 1.03 | 517 | 11.9 | 577 | 13.3 | 5,927 | 17.4 | 10,134 | 8,508 | 1,626 | 24 | 68 |
| Kingfisher | 155 | 16,856 | 1.16 | 186 | 11.8 | 142 | 9.0 | 2,392 | 18.2 | 2,814 | 2,348 | 466 | 13 | 84 |

1. Per 1,000 estimated resident population.

# Table B. States and Counties — Crime, Education, Money Income, and Poverty

| STATE County | County law enforcement employment, 2019 | | Education — School enrollment and attainment, 2015-2019 | | | | Local government expenditures[3] 2017-2018 | | Money income, 2015-2019 — Households | | | | Income and poverty, 2019 | | | |
|---|---|---|---|---|---|---|---|---|---|---|---|---|---|---|---|---|
| | | | Enrollment[1] | | Attainment[2] (percent) | | | | | | Percent | | | Percent below poverty level | | |
| | Officers | Civilians | Total | Percent private | High school graduate or less | Bachelor's degree or more | Total current spending (mil dol) | Current spending per student (dollars) | Per capita income[4] | Median income (dollars) | with income of less than $50,000 | with income of $200,000 or more | Median household income (dollars) | All persons | Children under 18 years | Children 5 to 17 years in families |
| | 46 | 47 | 48 | 49 | 50 | 51 | 52 | 53 | 54 | 55 | 56 | 57 | 58 | 59 | 60 | 61 |
| **OHIO—Cont'd** | | | | | | | | | | | | | | | | |
| Ottawa | NA | NA | 7,848 | 9.9 | 43.5 | 23.1 | 76 | 12,241 | 34,560 | 59,099 | 42.6 | 4.9 | 63,352 | 8.1 | 12.1 | 11.2 |
| Paulding | NA | NA | 4,061 | 12.6 | 57.4 | 15.6 | 41 | 13,331 | 28,300 | 55,330 | 44.6 | 2.4 | 58,100 | 9.8 | 13.2 | 12.0 |
| Perry | NA | NA | 8,147 | 9.1 | 60.1 | 12.2 | 71 | 12,164 | 23,484 | 50,150 | 49.9 | 1.6 | 51,875 | 15.0 | 20.6 | 17.6 |
| Pickaway | 39 | 44 | 12,379 | 11.1 | 55.4 | 18.4 | 108 | 11,606 | 27,760 | 63,633 | 38.9 | 2.9 | 63,931 | 11.5 | 15.2 | 13.4 |
| Pike | NA | NA | 6,264 | 4.9 | 60.0 | 12.6 | 65 | 14,445 | 24,318 | 42,832 | 55.2 | 1.3 | 46,255 | 19.1 | 27.1 | 25.1 |
| Portage | 82 | 71 | 44,976 | 10.6 | 44.2 | 29.0 | 269 | 12,567 | 30,054 | 57,618 | 43.9 | 4.0 | 57,439 | 11.9 | 13.4 | 12.3 |
| Preble | NA | NA | 9,081 | 11.2 | 53.5 | 16.3 | 72 | 11,850 | 28,890 | 58,957 | 42.4 | 2.3 | 62,128 | 8.9 | 13.3 | 12.3 |
| Putnam | NA | NA | 8,297 | 12.6 | 48.5 | 20.9 | 73 | 12,264 | 31,743 | 64,822 | 36.6 | 4.0 | 73,454 | 7.2 | 7.9 | 6.6 |
| Richland | 52 | 82 | 26,115 | 19.7 | 52.4 | 18.3 | 225 | 13,209 | 25,585 | 49,547 | 50.4 | 2.0 | 51,883 | 13.5 | 20.4 | 19.4 |
| Ross | NA | NA | 16,281 | 11.0 | 55.4 | 15.8 | 144 | 13,578 | 24,913 | 51,092 | 49.1 | 2.7 | 54,728 | 15.2 | 20.2 | 20.3 |
| Sandusky | NA | NA | 13,528 | 13.5 | 52.0 | 16.1 | 102 | 12,810 | 27,319 | 54,089 | 45.6 | 2.2 | 58,415 | 9.6 | 14.1 | 12.0 |
| Scioto | NA | NA | 16,967 | 9.0 | 56.3 | 15.7 | 148 | 12,859 | 23,719 | 41,330 | 57.4 | 2.7 | 45,410 | 21.9 | 30.1 | 26.3 |
| Seneca | NA | NA | 14,255 | 25.7 | 53.5 | 16.8 | 78 | 13,402 | 27,100 | 52,500 | 47.7 | 1.9 | 55,664 | 12.1 | 16.8 | 14.1 |
| Shelby | NA | NA | 11,736 | 9.6 | 51.1 | 18.7 | 88 | 10,603 | 29,381 | 63,806 | 39.1 | 2.8 | 64,304 | 9.1 | 11.7 | 10.4 |
| Stark | 128 | 134 | 85,222 | 17.7 | 46.6 | 22.8 | 651 | 11,635 | 29,495 | 53,860 | 46.6 | 3.3 | 55,623 | 13.0 | 18.7 | 16.2 |
| Summit | 326 | 73 | 126,826 | 18.0 | 38.5 | 32.5 | 940 | 12,899 | 33,606 | 57,181 | 44.4 | 5.4 | 57,753 | 14.2 | 21.0 | 19.6 |
| Trumbull | NA | NA | 40,106 | 10.4 | 55.1 | 18.7 | 360 | 13,377 | 26,935 | 47,280 | 52.6 | 2.5 | 47,966 | 15.4 | 24.8 | 23.7 |
| Tuscarawas | NA | NA | 18,898 | 13.1 | 58.7 | 17.4 | 192 | 12,306 | 27,057 | 53,243 | 47.2 | 2.8 | 54,150 | 11.2 | 15.8 | 13.9 |
| Union | 48 | 19 | 14,513 | 15.3 | 38.6 | 35.5 | 83 | 10,586 | 37,384 | 86,715 | 26.7 | 9.3 | 91,597 | 5.4 | 5.7 | 5.1 |
| Van Wert | NA | NA | 6,479 | 12.4 | 53.0 | 16.6 | 62 | 13,044 | 27,115 | 54,254 | 45.4 | 2.0 | 55,361 | 8.5 | 12.0 | 11.0 |
| Vinton | NA | NA | 2,751 | 6.9 | 63.0 | 12.8 | 26 | 12,566 | 21,781 | 45,673 | 52.7 | 1.2 | 46,301 | 18.7 | 28.1 | 26.9 |
| Warren | 102 | 91 | 59,892 | 19.1 | 31.9 | 43.0 | 425 | 11,151 | 41,792 | 87,125 | 25.0 | 12.9 | 91,645 | 4.5 | 5.1 | 4.6 |
| Washington | 49 | 43 | 12,554 | 18.1 | 49.7 | 18.8 | 91 | 11,901 | 28,782 | 50,021 | 50.0 | 2.3 | 52,565 | 11.0 | 15.2 | 14.1 |
| Wayne | NA | NA | 27,493 | 25.2 | 52.5 | 22.2 | 187 | 12,181 | 27,884 | 58,747 | 42.4 | 3.7 | 58,747 | 9.9 | 13.9 | 14.2 |
| Williams | NA | NA | 7,942 | 13.6 | 51.8 | 15.2 | 67 | 12,277 | 27,124 | 53,183 | 46.3 | 2.0 | 54,239 | 9.3 | 13.3 | 11.8 |
| Wood | 56 | 69 | 40,311 | 11.6 | 36.3 | 33.5 | 262 | 13,824 | 32,431 | 62,390 | 40.5 | 5.6 | 64,723 | 11.2 | 9.8 | 8.6 |
| Wyandot | 15 | 16 | 4,909 | 13.0 | 53.8 | 17.2 | 39 | 11,512 | 28,541 | 55,767 | 44.0 | 2.4 | 61,733 | 7.5 | 9.7 | 8.8 |
| **OKLAHOMA** | X | X | 1,002,998 | 11.3 | 43.3 | 25.5 | 5,670 | 8,158 | 28,422 | 52,919 | 47.4 | 4.3 | 54,447 | 15.1 | 19.7 | 18.4 |
| Adair | 6 | 21 | 5,575 | 5.2 | 65.8 | 11.6 | 44 | 10,017 | 17,026 | 34,695 | 63.8 | 0.7 | 39,589 | 23.6 | 33.2 | 33.6 |
| Alfalfa | 7 | 5 | 1,046 | 5.2 | 52.3 | 21.8 | 15 | 15,520 | 27,615 | 61,852 | 38.1 | 3.7 | 57,497 | 15.8 | 19.0 | 16.7 |
| Atoka | 8 | 9 | 2,910 | 5.3 | 57.0 | 15.0 | 22 | 9,764 | 20,443 | 39,316 | 62.2 | 2.2 | 41,872 | 19.2 | 25.6 | 23.9 |
| Beaver | 7 | 6 | 1,285 | 4.5 | 51.1 | 19.9 | 14 | 12,479 | 24,514 | 52,349 | 47.2 | 1.5 | 55,397 | 10.6 | 15.7 | 14.1 |
| Beckham | 14 | 27 | 5,424 | 5.9 | 50.7 | 16.8 | 35 | 8,757 | 24,480 | 50,721 | 49.6 | 2.3 | 50,574 | 20.8 | 28.9 | 27.6 |
| Blaine | 11 | 9 | 1,755 | 6.8 | 53.2 | 16.5 | 23 | 10,245 | 25,851 | 45,792 | 53.2 | 2.1 | 50,574 | 14.9 | 21.8 | 20.4 |
| Bryan | 15 | 3 | 11,995 | 7.1 | 46.4 | 21.9 | 69 | 8,607 | 23,979 | 44,212 | 54.9 | 2.7 | 44,074 | 17.3 | 19.7 | 20.6 |
| Caddo | 15 | 20 | 6,996 | 3.6 | 55.9 | 15.4 | 47 | 9,001 | 23,167 | 46,592 | 52.9 | 2.1 | 45,755 | 20.6 | 29.0 | 27.0 |
| Canadian | 67 | 43 | 37,457 | 10.8 | 38.4 | 27.4 | 203 | 7,141 | 30,755 | 72,056 | 32.8 | 5.1 | 75,111 | 8.1 | 11.2 | 10.6 |
| Carter | 20 | 37 | 12,414 | 8.4 | 48.7 | 20.5 | 78 | 8,652 | 26,094 | 51,419 | 48.7 | 2.9 | 51,895 | 14.3 | 18.3 | 16.8 |
| Cherokee | 23 | 6 | 13,867 | 8.0 | 44.8 | 26.3 | 69 | 9,099 | 22,161 | 42,774 | 56.9 | 2.3 | 45,544 | 21.4 | 27.0 | 24.1 |
| Choctaw | 6 | 9 | 3,217 | 4.3 | 56.2 | 13.4 | 22 | 8,912 | 21,277 | 34,489 | 61.8 | 1.7 | 37,131 | 22.5 | 32.2 | 32.9 |
| Cimarron | 3 | 4 | 465 | 7.7 | 53.0 | 22.8 | 5 | 13,656 | 29,115 | 46,328 | 54.2 | 2.9 | 47,602 | 17.8 | 26.9 | 24.6 |
| Cleveland | 68 | 162 | 84,441 | 9.2 | 32.9 | 33.4 | 343 | 7,404 | 31,571 | 64,016 | 38.5 | 4.8 | 64,706 | 11.7 | 10.9 | 8.7 |
| Coal | 10 | 4 | 1,218 | 4.6 | 59.0 | 17.3 | 13 | 11,675 | 24,055 | 40,938 | 57.6 | 2.5 | 41,273 | 18.1 | 24.5 | 23.4 |
| Comanche | 30 | 11 | 31,128 | 9.6 | 42.7 | 22.2 | 179 | 8,544 | 26,449 | 52,161 | 48.0 | 2.6 | 51,184 | 17.6 | 23.7 | 22.4 |
| Cotton | 5 | 9 | 1,427 | 6.1 | 57.3 | 13.7 | 9 | 8,681 | 23,564 | 50,885 | 48.8 | 1.5 | 46,450 | 18.6 | 25.1 | 24.4 |
| Craig | 16 | 15 | 3,134 | 9.3 | 52.8 | 14.7 | 23 | 8,954 | 21,318 | 43,329 | 58.8 | 1.4 | 42,978 | 18.2 | 22.5 | 21.4 |
| Creek | 34 | 43 | 15,730 | 9.0 | 52.6 | 17.0 | 101 | 7,929 | 26,044 | 51,318 | 48.9 | 2.9 | 54,737 | 13.7 | 18.5 | 17.8 |
| Custer | 19 | 25 | 9,251 | 3.3 | 43.5 | 24.8 | 47 | 8,536 | 26,520 | 49,900 | 50.1 | 3.4 | 51,803 | 15.8 | 16.1 | 15.5 |
| Delaware | 19 | 27 | 8,446 | 11.2 | 50.9 | 18.1 | 60 | 9,294 | 24,070 | 41,696 | 58.6 | 2.8 | 44,602 | 17.5 | 26.9 | 27.0 |
| Dewey | 7 | 9 | 1,203 | 13.2 | 47.7 | 24.0 | 14 | 12,647 | 28,310 | 52,428 | 47.0 | 6.3 | 56,316 | 11.6 | 15.3 | 13.1 |
| Ellis | 8 | 10 | 799 | 4.5 | 48.6 | 20.1 | 11 | 13,714 | 28,195 | 53,245 | 47.1 | 4.6 | 54,013 | 13.0 | 18.9 | 16.8 |
| Garfield | 24 | 8 | 14,698 | 10.1 | 48.3 | 22.9 | 93 | 8,015 | 28,333 | 54,006 | 46.3 | 3.4 | 58,381 | 12.7 | 17.6 | 17.7 |
| Garvin | 20 | 10 | 6,333 | 4.6 | 58.7 | 14.7 | 46 | 8,421 | 24,880 | 47,125 | 52.8 | 3.1 | 46,786 | 15.6 | 20.2 | 18.9 |
| Grady | 22 | 12 | 13,414 | 7.0 | 50.2 | 19.9 | 77 | 8,018 | 30,242 | 60,875 | 41.6 | 5.3 | 63,647 | 12.1 | 16.0 | 15.4 |
| Grant | 6 | 5 | 997 | 6.7 | 46.3 | 24.6 | 13 | 17,093 | 28,815 | 57,727 | 44.2 | 2.0 | 60,303 | 11.2 | 16.3 | 15.2 |
| Greer | 2 | 4 | 1,092 | 2.5 | 52.6 | 12.3 | 9 | 8,505 | 18,907 | 41,488 | 57.2 | 1.4 | 38,641 | 24.1 | 27.4 | 25.3 |
| Harmon | 3 | 1 | 633 | 2.7 | 45.4 | 20.7 | 5 | 8,885 | 26,045 | 48,344 | 52.4 | 3.0 | 37,988 | 23.7 | 33.8 | 33.1 |
| Harper | 4 | 3 | 886 | 2.8 | 51.3 | 19.3 | 7 | 9,359 | 24,130 | 46,154 | 52.9 | 2.1 | 40,265 | 11.1 | 14.1 | 13.1 |
| Haskell | 8 | 12 | 2,946 | 4.5 | 52.7 | 16.2 | 21 | 8,750 | 22,074 | 42,348 | 56.5 | 1.7 | 41,643 | 20.2 | 26.4 | 24.1 |
| Hughes | 13 | 14 | 2,935 | 2.7 | 58.2 | 12.9 | 21 | 9,380 | 20,572 | 39,365 | 59.4 | 1.0 | 37,631 | 21.4 | 25.8 | 25.0 |
| Jackson | 14 | 25 | 6,218 | 6.9 | 41.9 | 23.3 | 36 | 7,914 | 25,555 | 49,703 | 50.3 | 2.3 | 48,430 | 17.3 | 25.0 | 24.5 |
| Jefferson | 3 | 10 | 1,384 | 5.7 | 58.3 | 14.7 | 12 | 10,790 | 24,313 | 39,895 | 60.7 | 3.2 | 39,785 | 20.8 | 30.8 | 28.8 |
| Johnston | 7 | 18 | 2,741 | 1.8 | 48.9 | 20.1 | 17 | 8,970 | 21,197 | 41,332 | 60.1 | 1.7 | 41,805 | 21.0 | 31.1 | 29.9 |
| Kay | 18 | 8 | 10,686 | 11.7 | 45.6 | 18.9 | 68 | 8,486 | 26,189 | 46,809 | 52.6 | 3.2 | 48,522 | 16.1 | 21.4 | 19.8 |
| Kingfisher | 11 | 11 | 3,773 | 8.0 | 50.0 | 22.4 | 39 | 10,558 | 31,724 | 57,777 | 42.5 | 6.8 | 68,839 | 8.9 | 10.7 | 9.6 |

1. All persons 3 years old and over enrolled in nursery school through college.   2. Persons 25 years old and over.   3. Elementary and secondary education expenditures.   4. Based on population estimated by the American Community Survey, 2014–2018.

| STATE County | Personal income, 2019 Total (mil dol) | Percent change 2018-2019 | Per capita[1] Dollars | Per capita[1] Rank | Wages and salaries (mil dol) | Supplements to wages and salaries, employer contributions (mil dol) Pension and insurance | Supplements... Government social insurance | Proprietors' income (mil dol) | Dividends, interest, and rent (mil dol) | Personal transfer receipts (mil dol) | Earnings, 2019 Total (mil dol) | Contributions for government social insurance (mil dol) From employee and self-employed | Contributions... From employer |
|---|---|---|---|---|---|---|---|---|---|---|---|---|---|
| | 62 | 63 | 64 | 65 | 66 | 67 | 68 | 69 | 70 | 71 | 72 | 73 | 74 |
| OHIO—Cont'd | | | | | | | | | | | | | |
| Ottawa | 2,153 | 2.3 | 53,133 | 607 | 668 | 137 | 49 | 118 | 406 | 502 | 972 | 66 | 49 |
| Paulding | 785 | 1.6 | 42,017 | 1,762 | 197 | 44 | 15 | 68 | 120 | 191 | 323 | 19 | 15 |
| Perry | 1,390 | 3.1 | 38,469 | 2,267 | 298 | 63 | 21 | 103 | 141 | 380 | 485 | 35 | 21 |
| Pickaway | 2,421 | 2.7 | 41,422 | 1,858 | 704 | 156 | 44 | 158 | 318 | 519 | 1,062 | 62 | 44 |
| Pike | 1,072 | 2.4 | 38,610 | 2,247 | 530 | 80 | 38 | 137 | 122 | 361 | 785 | 50 | 38 |
| Portage | 7,315 | 3.3 | 45,023 | 1,357 | 2,837 | 573 | 186 | 331 | 1,174 | 1,458 | 3,926 | 228 | 186 |
| Preble | 1,681 | 2.2 | 41,110 | 1,906 | 489 | 97 | 35 | 111 | 224 | 410 | 732 | 50 | 35 |
| Putnam | 1,712 | 2.3 | 50,560 | 794 | 522 | 99 | 38 | 124 | 283 | 290 | 784 | 49 | 38 |
| Richland | 4,908 | 3.0 | 40,507 | 1,990 | 2,243 | 431 | 161 | 273 | 779 | 1,301 | 3,108 | 196 | 161 |
| Ross | 2,956 | 3.2 | 38,563 | 2,252 | 1,457 | 305 | 104 | 147 | 377 | 845 | 2,013 | 120 | 104 |
| Sandusky | 2,469 | 1.4 | 42,188 | 1,739 | 1,175 | 237 | 90 | 108 | 355 | 599 | 1,610 | 101 | 90 |
| Scioto | 3,023 | 2.4 | 40,144 | 2,035 | 1,042 | 233 | 74 | 280 | 385 | 1,024 | 1,629 | 101 | 74 |
| Seneca | 2,241 | 3.3 | 40,607 | 1,973 | 839 | 171 | 61 | 144 | 337 | 589 | 1,215 | 78 | 61 |
| Shelby | 2,271 | 2.5 | 46,739 | 1,155 | 1,521 | 255 | 109 | 207 | 325 | 415 | 2,092 | 122 | 109 |
| Stark | 17,211 | 2.9 | 46,441 | 1,193 | 7,647 | 1,305 | 557 | 991 | 2,879 | 3,893 | 10,499 | 666 | 557 |
| Summit | 28,629 | 3.0 | 52,918 | 621 | 14,970 | 2,411 | 1,053 | 2,029 | 5,140 | 5,288 | 20,463 | 1,220 | 1,053 |
| Trumbull | 8,316 | 1.9 | 42,005 | 1,765 | 2,851 | 530 | 209 | 751 | 1,321 | 2,376 | 4,342 | 295 | 209 |
| Tuscarawas | 4,158 | 3.0 | 45,199 | 1,340 | 1,638 | 318 | 118 | 496 | 618 | 910 | 2,569 | 156 | 118 |
| Union | 3,528 | 4.0 | 59,811 | 311 | 2,178 | 337 | 151 | 221 | 449 | 380 | 2,887 | 160 | 151 |
| Van Wert | 1,221 | 1.7 | 43,166 | 1,590 | 526 | 103 | 37 | 98 | 168 | 272 | 764 | 46 | 37 |
| Vinton | 450 | 2.7 | 34,400 | 2,755 | 95 | 25 | 6 | 23 | 63 | 153 | 149 | 11 | 6 |
| Warren | 14,571 | 5.0 | 62,111 | 247 | 5,986 | 838 | 400 | 722 | 2,183 | 1,723 | 7,946 | 474 | 400 |
| Washington | 2,653 | 1.4 | 44,283 | 1,448 | 1,350 | 242 | 99 | 162 | 426 | 687 | 1,852 | 118 | 99 |
| Wayne | 5,371 | 3.5 | 46,420 | 1,197 | 2,584 | 451 | 180 | 613 | 888 | 984 | 3,827 | 224 | 180 |
| Williams | 1,577 | 3.6 | 42,970 | 1,623 | 802 | 155 | 59 | 128 | 233 | 366 | 1,145 | 69 | 59 |
| Wood | 6,387 | 3.3 | 48,823 | 941 | 3,529 | 634 | 254 | 345 | 1,014 | 1,042 | 4,762 | 269 | 254 |
| Wyandot | 1,011 | 2.0 | 46,428 | 1,195 | 478 | 89 | 34 | 72 | 136 | 202 | 673 | 40 | 34 |
| OKLAHOMA | 187,327 | 3.7 | 47,297 | X | 85,061 | 14,230 | 6,171 | 23,724 | 34,941 | 36,445 | 129,187 | 7,525 | 6,171 |
| Adair | 655 | 2.6 | 29,528 | 3,051 | 169 | 36 | 13 | 63 | 91 | 231 | 282 | 19 | 13 |
| Alfalfa | 224 | 2.6 | 39,316 | 2,145 | 76 | 16 | 5 | 15 | 71 | 45 | 112 | 7 | 5 |
| Atoka | 452 | 3.1 | 32,823 | 2,904 | 124 | 25 | 9 | 39 | 67 | 150 | 198 | 15 | 9 |
| Beaver | 281 | 1.9 | 52,877 | 624 | 83 | 15 | 6 | 81 | 44 | 42 | 185 | 8 | 6 |
| Beckham | 848 | 0.3 | 38,789 | 2,226 | 503 | 73 | 36 | 71 | 154 | 179 | 683 | 42 | 36 |
| Blaine | 398 | 3.7 | 42,256 | 1,729 | 169 | 30 | 12 | 41 | 114 | 91 | 252 | 16 | 12 |
| Bryan | 1,693 | 5.7 | 35,269 | 2,646 | 832 | 153 | 59 | 91 | 280 | 466 | 1,135 | 73 | 59 |
| Caddo | 1,082 | 4.0 | 37,603 | 2,373 | 344 | 75 | 25 | 71 | 171 | 308 | 515 | 34 | 25 |
| Canadian | 6,839 | 5.6 | 46,116 | 1,234 | 1,888 | 291 | 138 | 314 | 969 | 1,053 | 2,631 | 171 | 138 |
| Carter | 2,097 | 1.9 | 43,578 | 1,536 | 1,117 | 190 | 81 | 119 | 428 | 511 | 1,507 | 96 | 81 |
| Cherokee | 1,634 | 4.0 | 33,583 | 2,839 | 610 | 119 | 43 | 85 | 281 | 494 | 856 | 56 | 43 |
| Choctaw | 475 | 4.2 | 32,368 | 2,924 | 158 | 34 | 12 | 22 | 67 | 187 | 226 | 18 | 12 |
| Cimarron | 153 | 6.6 | 71,734 | 112 | 31 | 6 | 2 | 66 | 20 | 22 | 105 | 3 | 2 |
| Cleveland | 12,972 | 3.9 | 45,675 | 1,286 | 3,892 | 702 | 277 | 750 | 2,467 | 2,231 | 5,620 | 355 | 277 |
| Coal | 213 | 7.0 | 38,741 | 2,230 | 62 | 11 | 5 | 27 | 43 | 58 | 105 | 7 | 5 |
| Comanche | 5,153 | 4.6 | 42,676 | 1,661 | 2,582 | 621 | 212 | 177 | 914 | 1,202 | 3,593 | 196 | 212 |
| Cotton | 244 | 5.1 | 43,032 | 1,614 | 71 | 16 | 5 | 16 | 36 | 61 | 108 | 7 | 5 |
| Craig | 550 | 4.0 | 38,861 | 2,213 | 224 | 46 | 16 | 46 | 88 | 182 | 331 | 22 | 16 |
| Creek | 2,977 | 4.1 | 41,624 | 1,828 | 964 | 161 | 72 | 167 | 490 | 743 | 1,364 | 97 | 72 |
| Custer | 1,197 | 3.5 | 41,261 | 1,886 | 647 | 108 | 45 | 85 | 246 | 231 | 886 | 53 | 45 |
| Delaware | 1,424 | 4.0 | 33,113 | 2,881 | 345 | 64 | 25 | 126 | 275 | 473 | 560 | 42 | 25 |
| Dewey | 233 | 5.1 | 47,685 | 1,056 | 107 | 18 | 8 | 30 | 51 | 41 | 163 | 10 | 8 |
| Ellis | 209 | 2.4 | 54,037 | 543 | 57 | 11 | 4 | 51 | 48 | 36 | 123 | 6 | 4 |
| Garfield | 2,741 | 2.8 | 44,899 | 1,373 | 1,269 | 224 | 97 | 198 | 604 | 577 | 1,788 | 109 | 97 |
| Garvin | 1,089 | 3.4 | 39,303 | 2,150 | 517 | 95 | 37 | 84 | 198 | 295 | 732 | 47 | 37 |
| Grady | 2,277 | 3.8 | 40,777 | 1,951 | 609 | 103 | 45 | 139 | 368 | 483 | 896 | 62 | 45 |
| Grant | 202 | 13.2 | 46,669 | 1,163 | 81 | 14 | 6 | 28 | 51 | 41 | 129 | 7 | 6 |
| Greer | 172 | 3.3 | 30,055 | 3,035 | 37 | 9 | 3 | 13 | 31 | 60 | 62 | 5 | 3 |
| Harmon | 112 | 8.3 | 42,281 | 1,721 | 26 | 6 | 2 | 20 | 18 | 30 | 55 | 3 | 2 |
| Harper | 198 | 0.3 | 53,798 | 556 | 38 | 9 | 3 | 67 | 31 | 32 | 117 | 4 | 3 |
| Haskell | 416 | 2.1 | 32,972 | 2,893 | 122 | 24 | 9 | 50 | 58 | 146 | 205 | 14 | 9 |
| Hughes | 566 | 5.8 | 42,601 | 1,671 | 115 | 23 | 9 | 170 | 78 | 149 | 316 | 13 | 9 |
| Jackson | 1,063 | 5.8 | 43,321 | 1,567 | 518 | 126 | 42 | 89 | 195 | 229 | 776 | 42 | 42 |
| Jefferson | 205 | 3.0 | 34,118 | 2,787 | 44 | 9 | 3 | 17 | 34 | 69 | 73 | 6 | 3 |
| Johnston | 387 | -3.3 | 34,931 | 2,696 | 118 | 25 | 8 | 23 | 54 | 128 | 174 | 13 | 8 |
| Kay | 1,805 | 2.6 | 41,469 | 1,852 | 839 | 139 | 60 | 128 | 317 | 479 | 1,167 | 77 | 60 |
| Kingfisher | 857 | 4.3 | 54,342 | 526 | 388 | 61 | 27 | 77 | 226 | 130 | 553 | 32 | 27 |

1. Based on the resident population estimated as of July 1 of the year shown.

| STATE County | Farm | Mining, quarrying, and extractions | Construction | Manufacturing | Information; professional, scientific, technical services | Retail trade | Finance, insurance, real estate, and leasing | Health care and social assistance | Government | Social Security beneficiaries, December 2019 — Number | Rate[1] | Supplemental Security Income recipients, 2019 | Housing units, 2020 — Total | Percent change, 2010-2020 |
|---|---|---|---|---|---|---|---|---|---|---|---|---|---|---|
| | 75 | 76 | 77 | 78 | 79 | 80 | 81 | 82 | 83 | 84 | 85 | 86 | 87 | 88 |
| **OHIO—Cont'd** | | | | | | | | | | | | | | |
| Ottawa | 0.6 | 0.9 | 7.3 | 16.2 | 2.7 | 5.8 | 4.2 | D | 17.5 | 11,920 | 294 | 588 | 28,824 | 3.3 |
| Paulding | 13.1 | D | 4.6 | 22.9 | 3.0 | 4.9 | 3.0 | D | 20.6 | 4,575 | 245 | 309 | 8,824 | 0.9 |
| Perry | -0.1 | 4.9 | 22.5 | 13.2 | 2.3 | 5.0 | 2.7 | D | 19.8 | 8,070 | 223 | 1,274 | 15,426 | 1.4 |
| Pickaway | 0.1 | 0.2 | 12.7 | 18.4 | 2.6 | 5.7 | 3.0 | D | 29.4 | 11,430 | 195 | 1,187 | 21,825 | 2.6 |
| Pike | 0.2 | D | 6.1 | 5.4 | D | 4.5 | 2.5 | D | 12.0 | 6,635 | 239 | 1,425 | 13,116 | 5.5 |
| Portage | -0.2 | 0.8 | 6.6 | 20.8 | 6.1 | 6.4 | 2.7 | 7.4 | 24.1 | 32,005 | 197 | 2,473 | 70,049 | 3.8 |
| Preble | 1.1 | D | 8.0 | 35.5 | D | 6.3 | 2.8 | 6.7 | 15.1 | 9,500 | 233 | 729 | 17,943 | 0.3 |
| Putnam | 2.3 | D | 11.4 | 31.8 | 3.7 | 6.0 | 4.5 | 7.7 | 13.0 | 7,060 | 209 | 335 | 14,034 | 2.2 |
| Richland | 0.4 | D | 7.1 | 22.8 | 3.6 | 8.0 | 3.4 | 14.5 | 17.5 | 29,825 | 246 | 3,566 | 54,041 | |
| Ross | 0.1 | 0.1 | 4.2 | 19.0 | 2.5 | 7.1 | 2.3 | 19.7 | 26.4 | 16,890 | 220 | 2,897 | 32,014 | -0.4 |
| Sandusky | 0.3 | D | 5.5 | 41.0 | 2.4 | 6.2 | 4.0 | D | 12.0 | 14,085 | 241 | 994 | 26,366 | -0.1 |
| Scioto | -0.2 | D | 4.3 | 15.9 | 3.3 | 6.8 | 2.8 | 27.9 | 21.4 | 16,800 | 224 | 5,219 | 34,558 | 1.0 |
| Seneca | 1.2 | 0.9 | 8.4 | 25.0 | D | 6.8 | 3.9 | 10.0 | 14.5 | 12,850 | 233 | 1,225 | 24,233 | 0.5 |
| Shelby | 0.8 | 0.0 | 10.3 | 50.0 | D | 3.6 | 2.9 | 5.3 | 8.0 | 10,045 | 207 | 730 | 20,544 | 1.8 |
| Stark | 0.1 | D | 7.2 | 18.5 | 6.5 | 7.2 | 6.5 | 16.3 | 12.4 | 88,670 | 239 | 9,886 | 167,520 | 1.4 |
| Summit | 0.0 | 0.1 | 6.3 | 11.0 | 9.6 | 7.7 | 6.6 | 14.8 | 11.2 | 113,410 | 210 | 15,260 | 246,560 | 0.6 |
| Trumbull | -0.1 | 0.2 | 8.3 | 15.9 | 3.0 | 9.2 | 7.4 | 13.7 | 15.0 | 53,945 | 272 | 6,298 | 95,566 | -0.6 |
| Tuscarawas | 0.7 | 6.6 | 9.2 | 22.1 | 4.1 | 7.7 | 4.3 | 10.0 | 13.3 | 21,370 | 232 | 1,933 | 40,236 | 0.1 |
| Union | 1.0 | D | 4.6 | 34.1 | D | 4.2 | 2.2 | 3.1 | 10.4 | 8,780 | 149 | 446 | 22,986 | 18.2 |
| Van Wert | 4.6 | D | 5.6 | 24.7 | D | 5.4 | 10.0 | 12.1 | 12.2 | 6,800 | 241 | 422 | 12,704 | 0.7 |
| Vinton | -0.7 | D | 4.7 | 20.0 | D | 3.8 | 5.0 | 12.1 | 27.2 | 2,820 | 216 | 614 | 6,287 | 0.0 |
| Warren | 0.1 | 0.1 | 6.4 | 13.3 | 7.8 | 7.0 | 7.0 | 8.5 | 9.7 | 38,895 | 165 | 1,874 | 91,061 | 12.8 |
| Washington | 0.2 | 3.0 | 9.4 | 18.3 | 4.2 | 6.0 | 5.3 | 20.8 | 10.5 | 15,585 | 260 | 1,886 | 28,209 | -0.6 |
| Wayne | 1.5 | 3.7 | 7.6 | 32.5 | 3.0 | 6.2 | 4.1 | 7.2 | 11.8 | 23,610 | 204 | 1,862 | 47,107 | 2.7 |
| Williams | 1.8 | D | 4.4 | 39.6 | D | 8.8 | 3.3 | D | 11.8 | 8,730 | 237 | 554 | 16,666 | 0.0 |
| Wood | 0.7 | 0.6 | 8.9 | 22.8 | 5.4 | 4.7 | 3.9 | 7.5 | 15.9 | 23,705 | 181 | 1,418 | 54,254 | 1.6 |
| Wyandot | 3.6 | 1.7 | 13.0 | 40.0 | D | 4.5 | 3.0 | D | 14.1 | 5,085 | 233 | 327 | 9,974 | 1.1 |
| **OKLAHOMA** | 1.0 | 7.4 | 6.0 | 9.4 | 7.5 | 5.6 | 5.4 | 10.3 | 19.3 | 802,326 | 203 | 96,804 | 1,759,720 | 5.7 |
| Adair | 9.1 | 0.1 | 6.9 | 19.9 | D | 8.4 | 2.9 | D | 21.1 | 4,990 | 225 | 886 | 9,434 | 3.2 |
| Alfalfa | 10.7 | 6.6 | 6.4 | 1.3 | D | 3.9 | 5.8 | 3.4 | 30.9 | 1,165 | 206 | 72 | 2,713 | -1.7 |
| Atoka | -3.6 | 7.4 | 5.8 | 3.1 | D | 10.1 | 5.8 | D | 31.4 | 3,510 | 255 | 526 | 6,430 | 2.0 |
| Beaver | 37.0 | 9.7 | 17.4 | D | D | 1.6 | D | D | 13.0 | 1,105 | 208 | 42 | 2,689 | 0.7 |
| Beckham | -0.6 | 32.0 | 8.0 | 3.0 | 3.4 | 8.8 | 5.3 | D | 10.9 | 4,290 | 196 | 522 | 10,072 | 4.4 |
| Blaine | 1.2 | 14.2 | 9.5 | 16.7 | 2.5 | 4.0 | D | 4.1 | 15.5 | 2,215 | 234 | 220 | 5,195 | 0.0 |
| Bryan | 0.2 | 0.2 | 3.9 | 9.2 | 3.9 | 7.1 | 3.4 | 9.9 | 42.7 | 10,270 | 214 | 1,547 | 20,763 | 6.0 |
| Caddo | 3.1 | 5.0 | 9.5 | 1.6 | 7.5 | 6.0 | 3.7 | D | 31.4 | 6,600 | 228 | 990 | 13,254 | 0.9 |
| Canadian | 0.3 | 15.4 | 8.7 | 10.9 | 5.7 | 6.5 | 5.9 | 6.1 | 16.8 | 23,575 | 159 | 1,130 | 50,146 | 9.5 |
| Carter | -0.6 | 7.9 | 7.7 | 18.1 | 5.5 | 6.4 | 4.4 | 13.1 | 13.0 | 11,820 | 245 | 1,502 | 21,908 | 3.6 |
| Cherokee | 2.9 | D | 3.9 | 1.5 | 2.5 | 6.6 | 4.1 | 7.3 | 55.8 | 10,435 | 214 | 1,536 | 22,524 | 5.0 |
| Choctaw | -1.7 | 2.4 | 5.4 | 3.5 | 4.4 | 7.7 | 4.2 | D | 30.1 | 4,100 | 280 | 807 | 7,639 | 1.6 |
| Cimarron | 58.4 | D | 1.0 | D | D | 6.0 | D | 0.4 | 11.1 | 630 | 293 | 26 | 1,574 | -0.8 |
| Cleveland | -0.2 | 1.1 | 8.5 | 4.8 | 8.1 | 7.9 | 5.8 | 9.5 | 31.1 | 48,405 | 170 | 3,721 | 118,206 | 12.5 |
| Coal | 9.9 | 14.7 | 18.8 | 4.5 | 1.5 | 4.1 | 3.9 | 11.2 | 19.0 | 1,265 | 230 | 214 | 2,844 | 1.2 |
| Comanche | 0.2 | D | 3.1 | 8.5 | 5.7 | 5.0 | 3.7 | 5.3 | 54.3 | 21,520 | 178 | 3,256 | 51,614 | 1.7 |
| Cotton | 5.3 | D | 9.3 | 1.4 | D | 5.0 | 2.7 | 3.1 | 48.2 | 1,480 | 262 | 152 | 2,985 | -0.9 |
| Craig | 6.2 | D | 3.3 | 3.1 | D | 7.9 | 4.5 | 10.3 | 29.1 | 4,150 | 294 | 721 | 6,741 | -0.1 |
| Creek | -1.4 | 3.4 | 14.1 | 24.5 | D | 5.6 | 3.7 | 9.2 | 15.3 | 17,250 | 242 | 1,668 | 30,952 | 4.0 |
| Custer | 1.1 | 16.3 | 5.9 | 6.0 | D | 6.1 | 6.0 | 6.8 | 18.0 | 5,115 | 177 | 553 | 12,757 | 4.2 |
| Delaware | 12.4 | 0.2 | 8.3 | 4.3 | D | 8.2 | 3.9 | D | 24.9 | 11,725 | 273 | 1,237 | 25,767 | 3.8 |
| Dewey | -0.1 | 31.2 | 5.5 | 4.5 | D | 3.9 | D | D | 14.1 | 1,110 | 227 | 56 | 2,453 | 0.3 |
| Ellis | 30.5 | 5.9 | D | D | D | 7.0 | D | D | 16.3 | 955 | 247 | 52 | 2,265 | -0.9 |
| Garfield | 0.9 | 8.5 | 4.9 | 10.2 | D | 6.7 | 6.1 | 11.7 | 21.0 | 12,680 | 207 | 1,334 | 26,743 | -0.3 |
| Garvin | -0.9 | 11.2 | 9.4 | 15.9 | 3.9 | 8.0 | 4.1 | D | 10.4 | 6,835 | 247 | 794 | 12,869 | 0.3 |
| Grady | 2.2 | 12.5 | 11.6 | 8.9 | 3.8 | 5.7 | 6.3 | 6.9 | 17.6 | 11,345 | 203 | 1,158 | 23,087 | 3.9 |
| Grant | 14.7 | 29.1 | 10.3 | 1.4 | D | 1.8 | D | 4.1 | 14.7 | 1,075 | 248 | 74 | 2,467 | -0.9 |
| Greer | 14.0 | 0.5 | D | D | D | 6.3 | D | 9.0 | 47.7 | 1,370 | 240 | 215 | 2,690 | -1.8 |
| Harmon | 29.8 | 0.6 | D | D | D | 3.7 | D | 4.9 | 27.2 | 695 | 263 | 143 | 1,524 | -1.3 |
| Harper | 54.3 | 5.2 | D | D | D | 2.9 | D | 2.1 | 18.0 | 815 | 220 | 40 | 1,877 | -1.5 |
| Haskell | 14.8 | 6.3 | 4.4 | 3.0 | D | 6.9 | 2.1 | D | 16.5 | 3,525 | 278 | 498 | 6,224 | 3.1 |
| Hughes | 49.3 | 3.8 | 2.2 | 3.0 | 1.7 | 3.7 | 2.1 | D | 15.9 | 3,450 | 261 | 459 | 6,269 | 1.4 |
| Jackson | 2.7 | 0.1 | 2.3 | 7.0 | D | 5.8 | 3.7 | 4.0 | 50.8 | 4,755 | 195 | 648 | 12,129 | 0.4 |
| Jefferson | 8.2 | 5.0 | 4.4 | 1.4 | D | 6.1 | D | D | 30.9 | 1,585 | 265 | 207 | 3,396 | 0.5 |
| Johnston | -0.8 | 11.4 | 5.0 | 17.2 | D | 4.7 | D | 17.6 | 24.3 | 2,785 | 252 | 424 | 5,233 | 2.1 |
| Kay | 0.9 | 3.1 | 7.7 | 12.4 | D | 6.4 | 3.3 | 9.3 | 19.3 | 11,165 | 256 | 1,195 | 21,395 | -1.4 |
| Kingfisher | 5.2 | 17.9 | 10.9 | 7.1 | D | 6.7 | 5.5 | 4.9 | 9.9 | 3,225 | 203 | 193 | 6,612 | 3.2 |

1. Per 1,000 resident population estimated as of July 1 of the year shown.

# Table B. States and Counties — Housing, Labor Force, and Employment

| | Housing units, 2015-2019 | | | | | | | | Civilian labor force, 2020 | | | | Civilian employment[6], 2015-2019 | | |
| | Occupied units | | | | | | | Sub-standard units[4] (percent) | | | Unemployment | | | Percent | |
| | Owner-occupied | | | Median owner cost as a percent of income | | Renter-occupied | | | | | | | | | |
| STATE County | Total | Percent | Median value[1] | With a mortgage | Without a mortgage[2] | Median rent[3] | Median rent as a percent of income[2] | | Total | Percent change, 2019-2020 | Total | Rate[5] | Total | Management, business, science, and arts | Construction, production, and maintenance occupations |
| | 89 | 90 | 91 | 92 | 93 | 94 | 95 | 96 | 97 | 98 | 99 | 100 | 101 | 102 | 103 |
|---|---|---|---|---|---|---|---|---|---|---|---|---|---|---|---|
| **OHIO—Cont'd** | | | | | | | | | | | | | | | |
| Ottawa | 18,051 | 78.8 | 152,000 | 18.6 | 11.3 | 724 | 27.6 | 0.8 | 20,729 | -2.0 | 1,886 | 9.1 | 19,366 | 32.5 | 30.2 |
| Paulding | 7,705 | 79.0 | 97,000 | 17.1 | 11.4 | 704 | 24.2 | 1.7 | 8,784 | 1.2 | 656 | 7.5 | 8,646 | 26.6 | 41.4 |
| Perry | 13,500 | 73.7 | 114,100 | 19.2 | 11.8 | 646 | 27.9 | 2.1 | 15,961 | 0.0 | 1,339 | 8.4 | 15,084 | 24.2 | 38.4 |
| Pickaway | 19,710 | 74.6 | 163,300 | 19.7 | 11.3 | 812 | 27.2 | 1.9 | 26,523 | -0.4 | 1,797 | 6.8 | 24,785 | 33.4 | 28.4 |
| Pike | 10,959 | 64.8 | 117,100 | 17.9 | 11.9 | 681 | 28.9 | 3.0 | 10,915 | 0.5 | 1,025 | 9.4 | 10,496 | 29.8 | 35.7 |
| Portage | 61,817 | 69.5 | 159,200 | 19.6 | 12.0 | 860 | 32.0 | 0.8 | 85,783 | -2.2 | 6,229 | 7.3 | 83,100 | 32.9 | 26.3 |
| Preble | 16,251 | 78.4 | 125,100 | 19.1 | 11.0 | 741 | 25.0 | 1.5 | 21,175 | -0.4 | 1,418 | 6.7 | 19,834 | 30.9 | 34.1 |
| Putnam | 13,327 | 81.6 | 155,800 | 18.0 | 10.0 | 701 | 20.5 | 1.2 | 18,720 | -1.1 | 1,051 | 5.6 | 17,527 | 33.7 | 34.1 |
| Richland | 48,449 | 67.4 | 110,600 | 18.8 | 11.1 | 682 | 26.8 | 1.7 | 52,087 | -1.1 | 4,681 | 9.0 | 52,254 | 29.1 | 30.1 |
| Ross | 28,802 | 70.9 | 123,800 | 18.9 | 11.1 | 741 | 29.3 | 1.9 | 35,399 | -0.2 | 2,839 | 8.0 | 31,060 | 30.7 | 29.3 |
| Sandusky | 23,574 | 72.7 | 115,400 | 18.1 | 11.2 | 701 | 26.3 | 1.2 | 30,192 | -1.8 | 2,455 | 8.1 | 28,340 | 27.4 | 40.2 |
| Scioto | 29,858 | 67.6 | 97,800 | 19.0 | 12.4 | 626 | 29.8 | 2.2 | 29,354 | 0.9 | 2,546 | 8.7 | 28,010 | 34.1 | 27.0 |
| Seneca | 21,648 | 72.3 | 106,500 | 18.2 | 10.9 | 696 | 24.6 | 1.0 | 27,352 | -0.8 | 2,220 | 8.1 | 26,789 | 27.0 | 38.0 |
| Shelby | 18,608 | 71.2 | 145,600 | 18.2 | 10.6 | 733 | 21.0 | 2.4 | 24,162 | -0.6 | 1,867 | 7.7 | 25,031 | 28.0 | 38.8 |
| Stark | 153,460 | 68.2 | 134,300 | 18.4 | 10.8 | 741 | 26.8 | 1.3 | 184,149 | -1.4 | 14,858 | 8.1 | 178,411 | 33.3 | 26.0 |
| Summit | 224,726 | 65.8 | 146,800 | 18.9 | 11.4 | 823 | 28.5 | 1.2 | 270,407 | -1.3 | 22,267 | 8.2 | 267,727 | 38.2 | 20.5 |
| Trumbull | 85,621 | 70.6 | 102,600 | 19.0 | 10.9 | 677 | 30.2 | 1.6 | 85,195 | -2.6 | 8,857 | 10.4 | 86,952 | 28.8 | 31.5 |
| Tuscarawas | 36,631 | 70.7 | 132,100 | 18.5 | 10.8 | 772 | 26.3 | 1.9 | 44,612 | -1.2 | 3,372 | 7.6 | 44,023 | 27.6 | 36.3 |
| Union | 20,212 | 78.5 | 203,800 | 19.3 | 10.9 | 982 | 23.4 | 1.2 | 29,408 | -0.5 | 1,706 | 5.8 | 28,871 | 41.6 | 23.7 |
| Van Wert | 11,544 | 77.1 | 104,900 | 17.6 | 10.0 | 712 | 27.5 | 0.7 | 14,811 | -1.9 | 1,038 | 7.0 | 13,664 | 27.5 | 39.2 |
| Vinton | 5,140 | 76.0 | 91,500 | 19.5 | 11.5 | 602 | 34.8 | 2.9 | 5,655 | 1.0 | 517 | 9.1 | 5,416 | 24.2 | 45.3 |
| Warren | 82,957 | 77.7 | 222,500 | 18.6 | 10.9 | 1,061 | 24.3 | 0.8 | 118,838 | -1.8 | 7,581 | 6.4 | 114,387 | 49.7 | 18.0 |
| Washington | 25,197 | 73.6 | 132,400 | 19.1 | 10.0 | 668 | 28.3 | 1.6 | 26,944 | -2.3 | 2,311 | 8.6 | 27,110 | 31.4 | 28.3 |
| Wayne | 43,824 | 73.0 | 148,900 | 18.8 | 10.2 | 735 | 24.4 | 2.8 | 60,568 | -3.0 | 3,560 | 5.9 | 56,091 | 31.6 | 34.0 |
| Williams | 15,090 | 77.0 | 105,400 | 18.9 | 11.2 | 694 | 25.7 | 1.3 | 18,829 | -1.6 | 1,464 | 7.8 | 17,823 | 28.3 | 37.8 |
| Wood | 50,589 | 65.3 | 163,600 | 18.7 | 11.4 | 800 | 26.1 | 0.6 | 69,444 | -3.3 | 4,911 | 7.1 | 69,780 | 27.4 | 39.2 |
| Wyandot | 9,081 | 74.1 | 122,200 | 17.2 | 10.0 | 631 | 20.9 | 0.9 | 12,964 | -1.1 | 908 | 7.0 | 11,308 | 37.5 | 24.2 |
| **OKLAHOMA** | 1,480,061 | 65.6 | 136,800 | 19.4 | 10.2 | 810 | 27.4 | 3.1 | 1,848,485 | 0.2 | 113,561 | 6.1 | 1,772,123 | 34.8 | 25.6 |
| Adair | 7,733 | 69.3 | 87,600 | 19.0 | 10.0 | 561 | 24.9 | 6.5 | 8,392 | 3.7 | 488 | 5.8 | 8,028 | 24.9 | 41.4 |
| Alfalfa | 1,870 | 80.2 | 85,700 | 14.6 | 10.0 | 640 | 23.8 | 2.9 | 2,766 | | 99 | 3.6 | 2,233 | 30.9 | 33.1 |
| Atoka | 5,284 | 73.4 | 102,900 | 19.1 | 11.8 | 554 | 31.0 | 4.7 | 5,179 | 5.5 | 365 | 7.0 | 4,753 | 30.4 | 30.4 |
| Beaver | 1,971 | 76.5 | 101,200 | 20.2 | 10.5 | 708 | 21.1 | 2.0 | 2,582 | -10.0 | 69 | 2.7 | 2,441 | 31.7 | 34.9 |
| Beckham | 7,578 | 66.2 | 128,400 | 20.3 | 12.3 | 718 | 25.9 | 3.1 | 10,075 | -3.8 | 760 | 7.5 | 8,738 | 28.6 | 32.2 |
| Blaine | 3,898 | 73.4 | 85,100 | 18.3 | 10.0 | 571 | 19.3 | 2.7 | 4,421 | -6.6 | 261 | 5.9 | 3,548 | 32.2 | 30.6 |
| Bryan | 17,253 | 62.2 | 117,400 | 18.8 | 10.8 | 743 | 26.9 | 3.9 | 22,473 | 4.9 | 1,143 | 5.1 | 19,896 | 30.3 | 27.6 |
| Caddo | 10,320 | 71.3 | 84,900 | 17.3 | 10.0 | 582 | 22.0 | 4.0 | 12,049 | 2.1 | 708 | 5.9 | 11,308 | 28.4 | 32.7 |
| Canadian | 44,906 | 75.4 | 165,500 | 19.6 | 10.0 | 941 | 24.6 | 2.6 | 76,026 | 0.2 | 4,544 | 6.0 | 70,773 | 36.9 | 23.6 |
| Carter | 18,339 | 68.3 | 113,000 | 19.0 | 10.1 | 773 | 24.7 | 3.4 | 21,557 | 0.4 | 1,506 | 7.0 | 21,198 | 27.5 | 32.3 |
| Cherokee | 16,665 | 68.4 | 118,000 | 20.6 | 10.0 | 666 | 29.3 | 3.1 | 19,669 | 4.8 | 1,071 | 5.4 | 19,665 | 33.4 | 25.1 |
| Choctaw | 5,971 | 69.3 | 96,800 | 20.8 | 11.0 | 569 | 32.8 | 2.7 | 5,590 | 3.1 | 423 | 7.6 | 5,171 | 30.3 | 31.2 |
| Cimarron | 977 | 72.9 | 56,500 | 21.5 | 10.0 | 537 | 17.1 | 1.0 | 1,420 | 7.1 | 28 | 2.0 | 973 | 36.3 | 31.9 |
| Cleveland | 106,172 | 63.7 | 165,600 | 19.7 | 10.1 | 926 | 27.2 | 2.7 | 142,684 | -0.5 | 7,984 | 5.6 | 141,590 | 41.6 | 18.9 |
| Coal | 2,227 | 74.0 | 83,800 | 18.9 | 10.0 | 625 | 24.5 | 2.8 | 2,313 | -5.7 | 153 | 6.6 | 2,128 | 31.1 | 35.2 |
| Comanche | 42,842 | 53.1 | 124,600 | 19.6 | 10.0 | 835 | 26.7 | 2.9 | 48,267 | 0.3 | 3,239 | 6.7 | 48,028 | 36.8 | 22.8 |
| Cotton | 2,195 | 77.8 | 76,700 | 18.7 | 10.0 | 591 | 22.4 | 1.7 | 2,733 | 2.8 | 192 | 7.0 | 2,330 | 31.2 | 28.3 |
| Craig | 5,422 | 69.5 | 109,000 | 22.1 | 10.1 | 752 | 26.4 | 2.5 | 5,859 | | 315 | 5.4 | 5,702 | 28.2 | 27.3 |
| Creek | 26,427 | 73.6 | 122,700 | 19.4 | 10.5 | 797 | 25.7 | 3.7 | 31,327 | -0.6 | 2,029 | 6.5 | 30,580 | 29.9 | 31.3 |
| Custer | 10,653 | 62.6 | 142,500 | 18.6 | 10.0 | 728 | 27.4 | 4.4 | 15,210 | -2.5 | 822 | 5.4 | 13,872 | 30.9 | 31.0 |
| Delaware | 16,739 | 75.6 | 117,900 | 23.2 | 11.9 | 688 | 27.8 | 4.2 | 18,521 | 2.3 | 981 | 5.3 | 15,956 | 28.4 | 30.4 |
| Dewey | 1,689 | 76.3 | 97,700 | 16.7 | 10.0 | 706 | 23.2 | 2.5 | 2,735 | -5.5 | 131 | 4.8 | 2,127 | 27.6 | 32.0 |
| Ellis | 1,614 | 74.2 | 89,800 | 17.9 | 10.0 | 677 | 24.6 | 0.9 | 2,161 | 1.2 | 97 | 4.5 | 1,785 | 31.0 | 34.7 |
| Garfield | 23,541 | 65.8 | 112,600 | 18.8 | 10.0 | 856 | 24.7 | 3.8 | 27,007 | -0.2 | 1,475 | 5.5 | 28,230 | 31.0 | 33.3 |
| Garvin | 10,447 | 69.3 | 98,500 | 17.2 | 10.0 | 682 | 23.7 | 2.8 | 12,841 | 2.1 | 796 | 6.2 | 11,147 | 28.3 | 35.2 |
| Grady | 19,857 | 77.1 | 134,300 | 18.2 | 10.0 | 717 | 25.5 | 2.4 | 26,633 | 0.5 | 1,622 | 6.1 | 24,682 | 28.9 | 29.2 |
| Grant | 1,736 | 76.6 | 81,000 | 15.6 | 10.0 | 720 | 19.4 | 1.6 | 2,786 | 4.7 | 92 | 3.3 | 1,951 | 33.6 | 34.3 |
| Greer | 2,054 | 69.7 | 76,500 | 18.9 | 12.1 | 539 | 23.9 | 1.9 | 1,904 | 4.4 | 140 | 7.4 | 1,993 | 34.5 | 23.4 |
| Harmon | 1,113 | 73.1 | 66,500 | 14.9 | 10.0 | 647 | 14.2 | 2.2 | 1,199 | 1.8 | 42 | 3.5 | 1,215 | 29.0 | 28.7 |
| Harper | 1,258 | 78.4 | 77,200 | 14.0 | 11.3 | 707 | 26.8 | 3.1 | 1,807 | 4.6 | 60 | 3.3 | 1,546 | 36.2 | 33.2 |
| Haskell | 4,931 | 74.7 | 96,100 | 18.9 | 10.0 | 615 | 34.3 | 4.6 | 4,279 | 1.3 | 334 | 7.8 | 4,656 | 31.0 | 33.6 |
| Hughes | 4,132 | 72.8 | 73,100 | 19.2 | 10.4 | 555 | 27.7 | 2.2 | 5,392 | 1.3 | 403 | 7.5 | 4,164 | 28.9 | 29.4 |
| Jackson | 9,762 | 58.5 | 106,500 | 18.1 | 10.7 | 737 | 25.7 | 2.4 | 10,766 | 1.1 | 439 | 4.1 | 10,561 | 27.4 | 31.4 |
| Jefferson | 2,371 | 72.1 | 66,400 | 19.4 | 10.0 | 521 | 26.1 | 2.0 | 2,578 | 3.5 | 172 | 6.7 | 2,233 | 24.9 | 39.2 |
| Johnston | 4,267 | 73.5 | 80,800 | 20.5 | 10.0 | 624 | 24.9 | 5.2 | 3,753 | 0.0 | 288 | 7.7 | 4,140 | 32.7 | 32.8 |
| Kay | 17,610 | 67.5 | 92,100 | 18.3 | 11.2 | 674 | 24.2 | 3.6 | 18,314 | 0.3 | 1,220 | 6.7 | 18,828 | 27.6 | 32.0 |
| Kingfisher | 5,609 | 76.3 | 144,700 | 18.3 | 10.5 | 788 | 23.7 | 1.7 | 8,808 | -3.1 | 380 | 4.3 | 7,233 | 29.8 | 36.5 |

1. Specified owner-occupied units. lacking complete plumbing facilities.   2. A value of 10.0 represents 10 percent or less; a value of 50.0 represents 50 percent or more.   3. Specified renter-occupied units.   4. Overcrowded or   5. Percent of civilian labor force.   6. Civilian employed persons 16 years old and over.

| STATE County | Number of establish-ments | Total | Health care and social assistance | Manufac-turing | Retail trade | Finance and insurance | Professional, scientific, and technical services | Total (mil dol) | Average per employee (dollars) | Number | Fewer than 50 acres | 1000 acres or more | Farm producers whose primary occupation is farming (percent) |
|---|---|---|---|---|---|---|---|---|---|---|---|---|---|
| | 104 | 105 | 106 | 107 | 108 | 109 | 110 | 111 | 112 | 113 | 114 | 115 | 116 |
| **OHIO—Cont'd** | | | | | | | | | | | | | |
| Ottawa | 1,019 | 10,209 | 1,931 | 1,922 | 1,364 | 275 | 159 | 485 | 47,530 | 551 | 47.2 | 4.2 | 36.3 |
| Paulding | 302 | 3,876 | 496 | 1,634 | 434 | 94 | 66 | 144 | 37,151 | 622 | 33.3 | 10.6 | 44.5 |
| Perry | 425 | 4,965 | 839 | 1,199 | 712 | 121 | 76 | 169 | 34,014 | 762 | 44.4 | 1.4 | 29.4 |
| Pickaway | 804 | 10,783 | 1,735 | 2,234 | 1,514 | 264 | 181 | 417 | 38,656 | 805 | 43.0 | 10.9 | 49.0 |
| Pike | 418 | 7,484 | 1,457 | 754 | 993 | 233 | 1,833 | 336 | 44,889 | 511 | 31.3 | 2.2 | 29.9 |
| Portage | 3,000 | 47,987 | 6,763 | 10,826 | 8,110 | 641 | 1,549 | 2,085 | 43,453 | 1,118 | 68.4 | 0.9 | 29.9 |
| Preble | 660 | 9,552 | 1,097 | 3,871 | 1,108 | 214 | 187 | 401 | 42,013 | 1,055 | 48.6 | 4.8 | 39.9 |
| Putnam | 736 | 10,246 | 1,154 | 3,737 | 1,137 | 307 | 183 | 381 | 37,195 | 1,335 | 33.7 | 4.0 | 34.1 |
| Richland | 2,658 | 40,279 | 6,458 | 8,820 | 6,338 | 959 | 835 | 1,475 | 36,624 | 1,160 | 49.1 | 1.5 | 41.8 |
| Ross | 1,230 | 22,592 | 5,046 | 4,374 | 3,843 | 488 | 430 | 1,013 | 44,826 | 1,121 | 40.3 | 4.6 | 38.9 |
| Sandusky | 1,289 | 22,818 | 3,105 | 8,710 | 2,417 | 471 | 417 | 913 | 40,022 | 768 | 39.6 | 5.5 | 38.4 |
| Scioto | 1,252 | 18,850 | 7,560 | 1,472 | 3,093 | 409 | 617 | 704 | 37,349 | 688 | 42.0 | 0.9 | 36.5 |
| Seneca | 1,033 | 16,947 | 2,366 | 4,394 | 1,952 | 398 | 300 | 605 | 35,728 | 1,156 | 40.1 | 4.5 | 35.3 |
| Shelby | 965 | 25,889 | 2,025 | 13,425 | 1,879 | 316 | 387 | 1,324 | 51,148 | 947 | 39.7 | 3.6 | 36.6 |
| Stark | 8,071 | 147,315 | 29,778 | 25,822 | 20,134 | 6,272 | 4,633 | 6,085 | 41,309 | 1,547 | 67.7 | 1.2 | 32.4 |
| Summit | 13,201 | 248,954 | 45,235 | 27,849 | 30,601 | 10,446 | 13,540 | 12,464 | 50,066 | 392 | 77.6 | 0.8 | 39.6 |
| Trumbull | 3,899 | 59,481 | 9,617 | 9,279 | 11,373 | 1,501 | 1,360 | 2,252 | 37,867 | 1,036 | 52.0 | 1.7 | 33.4 |
| Tuscarawas | 2,119 | 30,714 | 5,236 | 8,003 | 4,271 | 662 | 1,121 | 1,199 | 39,053 | 1,155 | 47.3 | 1.6 | 34.4 |
| Union | 1,068 | 25,554 | 1,886 | 7,125 | 2,415 | 426 | 4,321 | 1,513 | 59,210 | 997 | 50.6 | 5.3 | 36.3 |
| Van Wert | 548 | 10,520 | 1,558 | 3,583 | 1,134 | 880 | 399 | 408 | 38,800 | 772 | 40.5 | 10.1 | 42.9 |
| Vinton | 145 | 1,691 | 372 | 504 | 185 | 100 | 21 | 51 | 30,097 | 227 | 33.5 | 0.9 | 29.3 |
| Warren | 4,419 | 82,565 | 10,156 | 13,270 | 9,409 | 4,531 | 4,843 | 4,593 | 55,631 | 925 | 73.2 | 2.4 | 29.1 |
| Washington | 1,355 | 23,401 | 5,271 | 3,370 | 2,949 | 761 | 689 | 1,085 | 46,383 | 1,106 | 32.5 | 0.8 | 34.6 |
| Wayne | 2,514 | 42,436 | 6,177 | 13,552 | 4,665 | 1,103 | 1,485 | 1,809 | 42,637 | 2,034 | 54.7 | 1.2 | 42.3 |
| Williams | 797 | 15,924 | 1,886 | 7,427 | 1,359 | 246 | 202 | 640 | 40,165 | 881 | 40.3 | 5.6 | 33.8 |
| Wood | 2,718 | 58,059 | 5,873 | 13,580 | 6,524 | 861 | 3,328 | 2,674 | 46,060 | 1,069 | 46.7 | 6.9 | 35.8 |
| Wyandot | 487 | 8,892 | 852 | 4,065 | 774 | 201 | 91 | 390 | 43,828 | 649 | 40.4 | 10.3 | 42.9 |
| **OKLAHOMA** | 93,761 | 1,404,725 | 218,690 | 133,863 | 179,380 | 58,827 | 77,199 | 65,227 | 46,434 | 78,531 | 29.6 | 10.3 | 37.5 |
| Adair | 222 | 3,069 | 496 | 1,018 | 460 | 110 | 67 | 93 | 30,462 | 1,031 | 28.9 | 3.1 | 40.7 |
| Alfalfa | 135 | 913 | 101 | 28 | 182 | 73 | 30 | 34 | 37,504 | 581 | 8.8 | 32.0 | 52.0 |
| Atoka | 258 | 2,081 | 325 | 133 | 523 | 139 | 45 | 68 | 32,671 | 1,057 | 21.6 | 6.4 | 38.1 |
| Beaver | 147 | 1,179 | 89 | NA | 83 | 56 | 22 | 67 | 56,685 | 805 | 5.0 | 31.3 | 37.2 |
| Beckham | 744 | 8,872 | 1,080 | 268 | 1,462 | 291 | 270 | 443 | 49,932 | 896 | 18.8 | 14.7 | 33.0 |
| Blaine | 255 | 2,369 | 303 | 501 | 338 | 153 | 49 | 102 | 42,989 | 731 | 13.5 | 27.4 | 45.4 |
| Bryan | 809 | 10,901 | 2,003 | 1,518 | 1,674 | 357 | 516 | 370 | 33,911 | 1,609 | 27.8 | 4.5 | 39.8 |
| Caddo | 459 | 4,914 | 810 | 72 | 781 | 192 | 404 | 179 | 36,381 | 1,396 | 15.5 | 16.1 | 44.2 |
| Canadian | 2,791 | 30,034 | 2,910 | 3,000 | 4,572 | 876 | 984 | 1,229 | 40,936 | 1,324 | 37.8 | 11.9 | 35.5 |
| Carter | 1,602 | 20,147 | 3,358 | 2,882 | 2,887 | 692 | 456 | 925 | 45,914 | 1,431 | 34.5 | 5.6 | 30.2 |
| Cherokee | 755 | 9,423 | 2,778 | 179 | 1,719 | 325 | 152 | 304 | 32,302 | 1,200 | 33.2 | 2.6 | 36.8 |
| Choctaw | 292 | 3,260 | 1,150 | 124 | 476 | 124 | 109 | 103 | 31,562 | 851 | 20.4 | 9.2 | 42.4 |
| Cimarron | 68 | 328 | 38 | NA | 89 | 28 | 7 | 10 | 29,061 | 447 | 4.7 | 38.3 | 42.0 |
| Cleveland | 5,818 | 72,124 | 13,275 | 3,254 | 12,026 | 2,809 | 3,760 | 2,738 | 37,956 | 1,182 | 59.4 | 1.4 | 31.7 |
| Coal | 85 | 852 | 180 | 90 | 151 | 51 | 11 | 35 | 40,932 | 590 | 16.8 | 11.2 | 45.9 |
| Comanche | 2,123 | 32,384 | 6,172 | 3,345 | 5,422 | 1,393 | 2,114 | 1,232 | 38,047 | 1,055 | 30.6 | 13.2 | 37.2 |
| Cotton | 78 | 1,310 | 118 | 5 | 108 | 34 | 17 | 45 | 34,449 | 448 | 13.2 | 29.0 | 41.0 |
| Craig | 354 | 3,866 | 1,201 | 145 | 569 | 197 | 79 | 141 | 36,482 | 1,179 | 24.8 | 7.0 | 41.6 |
| Creek | 1,374 | 17,142 | 2,350 | 4,432 | 1,789 | 433 | 426 | 768 | 44,799 | 1,893 | 46.5 | 2.0 | 29.9 |
| Custer | 922 | 10,131 | 1,441 | 882 | 1,684 | 368 | 281 | 413 | 40,785 | 773 | 16.4 | 23.4 | 40.5 |
| Delaware | 749 | 7,918 | 1,257 | 643 | 1,428 | 256 | 344 | 267 | 33,704 | 1,377 | 33.7 | 3.3 | 43.3 |
| Dewey | 166 | 1,306 | 141 | 23 | 243 | 68 | 16 | 62 | 47,208 | 728 | 8.2 | 25.8 | 42.6 |
| Ellis | 112 | 961 | 141 | NA | 108 | 46 | 25 | 43 | 45,188 | 677 | 7.2 | 29.1 | 37.5 |
| Garfield | 1,601 | 21,659 | 3,720 | 2,572 | 3,493 | 757 | 606 | 871 | 40,218 | 936 | 22.1 | 20.6 | 44.8 |
| Garvin | 700 | 7,800 | 790 | 1,088 | 1,342 | 269 | 147 | 353 | 45,210 | 1,500 | 27.8 | 6.8 | 36.0 |
| Grady | 1,127 | 11,105 | 1,515 | 1,740 | 1,469 | 391 | 271 | 450 | 40,492 | 1,625 | 33.8 | 9.2 | 36.6 |
| Grant | 125 | 842 | 139 | 7 | 172 | 52 | 12 | 37 | 43,714 | 659 | 10.5 | 26.7 | 47.3 |
| Greer | 75 | 476 | 101 | 55 | 133 | 47 | 20 | 13 | 26,863 | 432 | 8.6 | 20.8 | 35.8 |
| Harmon | 49 | 455 | 125 | NA | 58 | 41 | 4 | 15 | 33,048 | 374 | 2.7 | 31.3 | 48.3 |
| Harper | 87 | 469 | 92 | NA | 117 | 35 | 32 | 16 | 34,629 | 438 | 7.5 | 37.4 | 41.1 |
| Haskell | 217 | 2,538 | 865 | 175 | 389 | 62 | 37 | 85 | 33,666 | 812 | 23.3 | 4.7 | 37.4 |
| Hughes | 203 | 2,033 | 721 | 90 | 328 | 69 | 29 | 58 | 28,753 | 928 | 19.4 | 8.1 | 43.9 |
| Jackson | 519 | 6,663 | 1,316 | 972 | 1,167 | 393 | 299 | 250 | 37,592 | 634 | 20.2 | 24.6 | 35.2 |
| Jefferson | 101 | 970 | 278 | 18 | 148 | 56 | 39 | 35 | 35,741 | 424 | 14.4 | 28.1 | 50.3 |
| Johnston | 170 | 2,131 | 523 | 514 | 215 | 39 | 20 | 85 | 39,896 | 606 | 21.5 | 8.3 | 35.7 |
| Kay | 1,080 | 14,607 | 2,323 | 2,970 | 2,021 | 410 | 454 | 588 | 40,225 | 864 | 28.2 | 18.5 | 46.1 |
| Kingfisher | 493 | 6,448 | 472 | 374 | 776 | 181 | 260 | 341 | 52,856 | 928 | 18.5 | 20.0 | 44.9 |

# Table B. States and Counties — **Agriculture**

| STATE County | Land in farms — Acreage (1,000) [117] | Percent change, 2012-2017 [118] | Acres: Average size of farm [119] | Acres: Total irrigated (1,000) [120] | Acres: Total cropland (1,000) [121] | Value of land and buildings (dollars): Average per farm [122] | Average per acre [123] | Value of machinery and equipment, average per farm (dollars) [124] | Value of products sold: Total (mil dol) [125] | Average per farm (acres) [126] | Percent from: Crops [127] | Percent from: Livestock and poultry products [128] | Organic farms (number) [129] | Farms with internet access (per-cent) [130] | Government payments: Total ($1,000) [131] | Percent of farms [132] |
|---|---|---|---|---|---|---|---|---|---|---|---|---|---|---|---|---|
| **OHIO—Cont'd** | | | | | | | | | | | | | | | | |
| Ottawa | 121 | 7.8 | 221 | 1.2 | 115.1 | 1,192,630 | 5,409 | 173,655 | 59.2 | 107,477 | 92.9 | 7.1 | 2 | 77.3 | 3,064 | 70.8 |
| Paulding | 220 | -0.6 | 353 | 0.4 | 208.2 | 2,210,654 | 6,260 | 220,491 | 173.5 | 278,860 | 46.3 | 53.7 | 4 | 76.2 | 4,880 | 83.6 |
| Perry | 101 | -5.7 | 133 | 0.0 | 58.9 | 607,087 | 4,574 | 75,495 | 33.8 | 44,398 | 68.3 | 31.7 | 4 | 74.0 | 1,427 | 18.8 |
| Pickaway | 297 | 1.1 | 369 | 2.6 | 275.0 | 2,225,570 | 6,033 | 220,849 | 162.7 | 202,050 | 91.4 | 8.6 | 3 | 81.7 | 10,838 | 61.5 |
| Pike | 98 | 0.4 | 191 | 0.6 | 54.3 | 795,490 | 4,156 | 87,790 | 55.1 | 107,771 | 41.2 | 58.8 | 2 | 68.3 | 1,495 | 23.3 |
| Portage | 86 | 3.1 | 77 | 0.3 | 59.5 | 500,149 | 6,511 | 74,045 | 34.5 | 30,849 | 71.0 | 29.0 | 9 | 78.4 | 1,224 | 10.4 |
| Preble | 213 | -4.8 | 202 | 0.3 | 188.3 | 1,294,416 | 6,397 | 133,405 | 146.3 | 138,647 | 67.1 | 32.9 | 5 | 75.3 | 5,681 | 51.3 |
| Putnam | 305 | -0.2 | 228 | 0.8 | 291.2 | 1,528,572 | 6,694 | 186,197 | 214.5 | 160,661 | 65.0 | 35.0 | 7 | 78.4 | 7,728 | 73.5 |
| Richland | 156 | -3.0 | 134 | 0.2 | 113.1 | 1,008,273 | 7,505 | 108,319 | 135.1 | 116,504 | 37.1 | 62.9 | NA | 65.4 | 2,687 | 21.0 |
| | | | | | | | | | | | | | 20 | | | |
| Ross | 248 | 11.8 | 221 | 0.7 | 168.2 | 1,013,522 | 4,583 | 105,244 | 77.7 | 69,354 | 85.8 | 14.2 | 8 | 69.8 | 10,095 | 45.2 |
| Sandusky | 179 | -1.5 | 233 | 1.1 | 166.4 | 1,379,300 | 5,926 | 177,118 | 101.0 | 131,561 | 90.9 | 9.1 | 4 | 80.7 | 5,305 | 69.3 |
| Scioto | 91 | -3.1 | 133 | 0.1 | 45.1 | 416,922 | 3,138 | 89,131 | 17.8 | 25,932 | 78.1 | 21.9 | 1 | 75.3 | 1,081 | 15.0 |
| Seneca | 267 | -8.1 | 231 | 0.3 | 242.8 | 1,359,558 | 5,889 | 176,787 | 140.9 | 121,866 | 82.0 | 18.0 | 3 | 74.4 | 8,588 | 72.1 |
| Shelby | 215 | 4.2 | 227 | D | 197.3 | 1,745,226 | 7,688 | 179,829 | 178.2 | 188,214 | 52.0 | 48.0 | 5 | 83.7 | 7,162 | 68.0 |
| Stark | 133 | -2.1 | 86 | 0.8 | 101.7 | 794,143 | 9,244 | 89,768 | 95.8 | 61,954 | 42.4 | 57.6 | 10 | 76.3 | 3,035 | 13.7 |
| Summit | 19 | 13.3 | 48 | 0.2 | 10.7 | 639,318 | 13,365 | 54,610 | 12.6 | 32,156 | 65.0 | 35.0 | 2 | 86.7 | 275 | 7.1 |
| Trumbull | 124 | 8.6 | 119 | 0.1 | 82.0 | 516,981 | 4,331 | 108,297 | 56.1 | 54,110 | 64.4 | 35.6 | 12 | 71.7 | 1,345 | 20.9 |
| Tuscarawas | 144 | 4.2 | 125 | 0.1 | 80.1 | 691,723 | 5,555 | 100,450 | 125.2 | 108,384 | 17.2 | 82.8 | 26 | 61.1 | 1,454 | 18.5 |
| Union | 218 | -10.1 | 218 | 0.4 | 196.1 | 1,386,021 | 6,350 | 191,902 | 209.3 | 209,954 | 49.4 | 50.6 | 13 | 86.7 | 7,490 | 52.2 |
| Van Wert | 248 | 9.3 | 322 | D | 240.8 | 2,410,706 | 7,494 | 215,085 | 191.3 | 247,791 | 64.6 | 35.4 | 2 | 83.4 | 4,699 | 68.9 |
| Vinton | 31 | -5.8 | 139 | 0.0 | 14.0 | 438,779 | 3,166 | 63,613 | 5.7 | 25,079 | 80.3 | 19.7 | 3 | 66.1 | 261 | 27.8 |
| Warren | 90 | -15.3 | 98 | 0.5 | 71.2 | 763,996 | 7,824 | 86,177 | 47.7 | 51,536 | 93.8 | 6.2 | 4 | 81.8 | 2,580 | 15.7 |
| Washington | 144 | 3.9 | 131 | 0.6 | 57.0 | 416,318 | 3,189 | 70,252 | 42.0 | 38,019 | 55.5 | 44.5 | NA | 71.2 | 684 | 17.4 |
| Wayne | 252 | -7.2 | 124 | 0.9 | 204.0 | 1,107,599 | 8,940 | 116,845 | 327.9 | 161,205 | 24.8 | 75.2 | 98 | 66.3 | 5,140 | 21.1 |
| Williams | 211 | 1.2 | 239 | 3.8 | 189.9 | 1,233,691 | 5,161 | 145,441 | 122.8 | 139,367 | 66.1 | 33.9 | 3 | 73.4 | 7,092 | 64.4 |
| Wood | 269 | 0.3 | 251 | 2.3 | 253.8 | 1,674,194 | 6,659 | 171,875 | 159.3 | 148,985 | 79.4 | 20.6 | 6 | 76.5 | 8,030 | 70.8 |
| Wyandot | 225 | 1.7 | 346 | 0.0 | 205.1 | 1,974,814 | 5,706 | 226,523 | 157.3 | 242,385 | 62.0 | 38.0 | 3 | 74.0 | 8,100 | 75.2 |
| **OKLAHOMA** | 34,156 | -0.6 | 435 | 573.8 | 11,715.7 | 754,099 | 1,734 | 90,442 | 7,465.5 | 95,065 | 20.3 | 79.7 | 54 | 72.9 | 232,018 | 26.3 |
| Adair | 239 | -5.3 | 232 | 0.4 | 47.6 | 554,052 | 2,392 | 76,599 | 163.1 | 158,242 | 2.0 | 98.0 | 1 | 68.1 | 216 | 8.7 |
| Alfalfa | 524 | -3.9 | 902 | 1.4 | 343.2 | 1,535,433 | 1,703 | 241,428 | 103.0 | 177,248 | 40.3 | 59.7 | NA | 74.5 | 4,482 | 75.9 |
| Atoka | 357 | 1.2 | 338 | 1.5 | 66.8 | 619,842 | 1,834 | 65,617 | 39.0 | 36,901 | 8.6 | 91.4 | 1 | 67.3 | 743 | 32.0 |
| Beaver | 1,037 | -7.1 | 1,288 | 21.5 | 296.1 | 1,157,188 | 898 | 104,629 | 155.7 | 193,388 | 12.5 | 87.5 | NA | 66.2 | 7,620 | 67.0 |
| Beckham | 498 | -12.3 | 556 | 10.5 | 175.7 | 740,737 | 1,332 | 103,007 | 52.0 | 58,029 | 50.7 | 49.3 | NA | 70.1 | 4,317 | 41.2 |
| Blaine | 593 | 13.6 | 811 | 4.3 | 320.9 | 1,321,168 | 1,628 | 197,612 | 109.2 | 149,409 | 22.8 | 77.2 | NA | 74.3 | 2,599 | 47.2 |
| Bryan | 433 | -1.9 | 269 | 6.4 | 104.5 | 610,009 | 2,267 | 69,072 | 61.4 | 38,161 | 33.9 | 66.1 | NA | 70.0 | 2,133 | 36.1 |
| Caddo | 756 | 6.8 | 541 | 37.3 | 326.7 | 937,747 | 1,732 | 122,802 | 132.1 | 94,609 | 37.4 | 62.6 | 5 | 69.8 | 12,419 | 46.6 |
| Canadian | 498 | -0.5 | 376 | 5.4 | 255.6 | 884,268 | 2,349 | 129,048 | 135.8 | 102,535 | 20.8 | 79.2 | NA | 75.8 | 5,299 | 38.3 |
| Carter | 396 | -13.2 | 277 | 1.6 | 64.3 | 562,555 | 2,030 | 59,722 | 39.4 | 27,541 | 16.2 | 83.8 | NA | 76.5 | 932 | 7.1 |
| Cherokee | 217 | -8.0 | 181 | 0.6 | 41.4 | 449,283 | 2,483 | 66,353 | 67.6 | 56,327 | 44.5 | 55.5 | 2 | 67.8 | 838 | 24.1 |
| Choctaw | 338 | 2.2 | 397 | 1.6 | 66.2 | 704,003 | 1,774 | 82,113 | 47.0 | 55,197 | 11.3 | 88.7 | 1 | 75.0 | 1,740 | 15.0 |
| Cimarron | 1,097 | -5.2 | 2,455 | 42.0 | 402.7 | 2,026,541 | 825 | 215,397 | 342.4 | 765,922 | 16.5 | 83.5 | 1 | 75.0 | 8,980 | 76.1 |
| Cleveland | 123 | -8.1 | 104 | 2.6 | 31.6 | 504,329 | 4,851 | 52,819 | 16.6 | 14,066 | 44.8 | 55.2 | 3 | 79.4 | 179 | 3.4 |
| Coal | 273 | -0.1 | 463 | 0.5 | 54.4 | 843,799 | 1,821 | 79,114 | 37.9 | 64,205 | 10.6 | 89.4 | NA | 61.9 | 788 | 33.4 |
| Comanche | 467 | 0.9 | 443 | D | 157.7 | 853,568 | 1,928 | 85,581 | 81.0 | 76,787 | 19.0 | 81.0 | 2 | 75.7 | 3,300 | 33.6 |
| Cotton | 405 | 1.2 | 903 | 0.1 | 202.7 | 1,227,562 | 1,359 | 137,210 | 49.8 | 111,170 | 21.9 | 78.1 | NA | 75.2 | 5,750 | 69.9 |
| Craig | 423 | -8.4 | 359 | D | 100.7 | 755,777 | 2,105 | 82,441 | 99.7 | 84,545 | 10.5 | 89.5 | 6 | 74.1 | 1,567 | 18.6 |
| Creek | 327 | -5.7 | 173 | 0.7 | 62.7 | 379,152 | 2,193 | 54,643 | 18.1 | 9,587 | 16.7 | 83.3 | 3 | 70.5 | 404 | 3.4 |
| Custer | 638 | 2.5 | 826 | 7.5 | 253.6 | 1,199,008 | 1,452 | 159,557 | 104.5 | 135,208 | 26.5 | 73.5 | NA | 72.8 | 5,809 | 45.8 |
| Delaware | 292 | 2.9 | 212 | 0.1 | 69.4 | 577,810 | 2,729 | 85,609 | 244.0 | 177,207 | 2.6 | 97.4 | 2 | 73.6 | 453 | 7.3 |
| Dewey | 652 | 4.4 | 896 | 2.8 | 185.4 | 1,228,171 | 1,371 | 125,050 | 41.2 | 56,651 | 18.7 | 81.3 | NA | 76.4 | 2,425 | 56.7 |
| Ellis | 724 | -4.5 | 1,070 | 8.4 | 127.3 | 1,062,853 | 994 | 79,899 | 115.7 | 170,870 | 4.1 | 95.9 | NA | 76.4 | 2,425 | 56.7 |
| Garfield | 675 | 1.3 | 721 | 3.2 | 442.9 | 1,251,739 | 1,736 | 180,572 | 130.4 | 139,318 | 48.1 | 51.9 | NA | 80.0 | 6,660 | 61.3 |
| Garvin | 483 | 4.4 | 322 | 2.0 | 122.1 | 617,731 | 1,917 | 81,615 | 66.9 | 44,607 | 25.3 | 74.7 | 1 | 70.9 | 2,139 | 25.6 |
| Grady | 593 | 1.7 | 365 | 12.7 | 200.9 | 766,106 | 2,098 | 116,307 | 152.7 | 93,954 | 14.6 | 85.4 | NA | 70.3 | 4,232 | 30.3 |
| Grant | 575 | -1.3 | 872 | 1.5 | 414.8 | 1,400,015 | 1,605 | 206,900 | 85.3 | 129,398 | 72.2 | 27.8 | 1 | 75.7 | 6,156 | 78.1 |
| Greer | 328 | -18.3 | 760 | 4.4 | 132.3 | 877,992 | 1,156 | 98,008 | 32.3 | 74,785 | 45.2 | 54.8 | NA | 74.5 | 4,332 | 63.7 |
| Harmon | 342 | 0.3 | 914 | 24.5 | 167.3 | 1,143,452 | 1,252 | 143,175 | 66.2 | 176,952 | 43.7 | 56.3 | NA | 70.1 | 5,079 | 69.0 |
| Harper | 668 | 8.0 | 1,524 | 4.3 | 180.7 | 1,678,361 | 1,101 | 133,044 | 217.1 | 495,710 | 2.8 | 97.2 | NA | 71.9 | 3,424 | 63.7 |
| Haskell | 238 | -7.2 | 293 | 0.5 | 45.9 | 556,633 | 1,902 | 91,479 | 104.2 | 128,365 | 2.5 | 97.5 | 1 | 73.4 | 879 | 15.9 |
| Hughes | 414 | -5.1 | 446 | 2.9 | 63.2 | 710,946 | 1,595 | 76,477 | 306.3 | 330,044 | 1.0 | 99.0 | NA | 67.0 | 1,393 | 19.8 |
| Jackson | 511 | 6.7 | 806 | 50.8 | 335.5 | 1,125,890 | 1,398 | 197,700 | 115.7 | 182,442 | 85.5 | 14.5 | NA | 77.4 | 12,582 | 58.8 |
| Jefferson | 472 | -0.7 | 1,113 | 0.7 | 92.6 | 1,721,210 | 1,546 | 113,136 | 89.7 | 211,564 | 4.2 | 95.8 | NA | 70.3 | 3,447 | 49.8 |
| Johnston | 289 | 1.8 | 477 | 0.7 | 44.0 | 892,334 | 1,872 | 72,388 | 32.9 | 54,211 | 10.6 | 89.4 | NA | 70.8 | 1,206 | 44.9 |
| Kay | 498 | 2.8 | 576 | 1.2 | 346.7 | 947,482 | 1,645 | 131,856 | 88.2 | 102,078 | 68.1 | 31.9 | 1 | 77.2 | 5,685 | 51.7 |
| Kingfisher | 575 | 1.3 | 620 | 5.6 | 353.5 | 1,191,810 | 1,923 | 162,597 | 145.0 | 156,278 | 21.6 | 78.4 | 2 | 74.9 | 5,382 | 59.3 |

# Table B. States and Counties — Water Use, Wholesale Trade, Retail Trade, and Real Estate

| STATE County | Water use 2015: Public supply water withdrawn (mil gal/day) | Water use 2015: Public supply gallons withdrawn per person per day | Wholesale Trade[1] 2017: Number of establishments | Wholesale: Number of employees | Wholesale: Sales (mil dol) | Wholesale: Average payroll (mil dol) | Retail Trade[2] 2017: Number of establishments | Retail: Number of employees | Retail: Sales (mil dol) | Retail: Average payroll (mil dol) | Real estate and rental and leasing[2] 2017: Number of establishments | Real estate: Number of employees | Real estate: Sales (mil dol) | Real estate: Average payroll (mil dol) |
|---|---|---|---|---|---|---|---|---|---|---|---|---|---|---|
| (Item) | 133 | 134 | 135 | 136 | 137 | 138 | 139 | 140 | 141 | 142 | 143 | 144 | 145 | 146 |
| **OHIO—Cont'd** | | | | | | | | | | | | | | |
| Ottawa | 4.58 | 112.0 | 28 | 155 | 82.5 | 7.0 | 131 | 1,411 | 442.7 | 42.2 | 51 | 132 | 23.0 | 4.2 |
| Paulding | 1.31 | 69.0 | 15 | 168 | 119.1 | 7.9 | 51 | 425 | 136.8 | 9.3 | 5 | 11 | 2.1 | 0.3 |
| Perry | 0.96 | 26.7 | 15 | 77 | 52.8 | 3.1 | 68 | 664 | 191.9 | 14.1 | 12 | 27 | 8.1 | 0.8 |
| Pickaway | 3.66 | 64.2 | D | D | D | 30.0 | 128 | 1,583 | 464.7 | 39.2 | 30 | 115 | 16.7 | 3.8 |
| Pike | 2.58 | 91.4 | 12 | 105 | 58.2 | 4.8 | 74 | 998 | 282.4 | 22.7 | 10 | 73 | 12.2 | 1.8 |
| Portage | 43.85 | 270.2 | 141 | 2,891 | 2,552.6 | 208.4 | 433 | 8,475 | 2,382.7 | 201.9 | 119 | 502 | 129.3 | 14.1 |
| Preble | 2.56 | 61.9 | 28 | 189 | 116.4 | 8.0 | 94 | 1,268 | 387.1 | 28.8 | 20 | 160 | 14.9 | 6.9 |
| Putnam | 2.52 | 74.0 | 33 | 311 | 286.4 | 15.0 | 109 | 1,013 | 289.8 | 25.8 | 11 | 26 | 4.0 | 1.0 |
| Richland | 14.45 | 118.7 | 97 | 1,455 | 760.4 | 65.2 | 419 | 6,707 | 1,641.9 | 160.9 | 104 | 452 | 76.5 | 14.7 |
| Ross | 9.48 | 122.8 | 49 | 503 | 428.4 | 26.6 | 244 | 3,993 | 1,128.9 | 95.4 | 45 | 179 | 32.9 | 4.7 |
| Sandusky | 7.89 | 132.2 | 48 | 631 | 652.8 | 29.9 | 201 | 2,421 | 683.9 | 59.3 | 42 | 142 | 19.8 | 4.0 |
| Scioto | 8.60 | 111.9 | 18 | 168 | 109.2 | 8.4 | 259 | 3,208 | 930.9 | 80.2 | 46 | 221 | 31.3 | 6.4 |
| Seneca | 1.85 | 33.3 | 39 | 979 | 687.1 | 40.9 | 153 | 2,032 | 565.7 | 50.9 | 25 | 58 | 10.4 | 1.5 |
| Shelby | 3.49 | 71.4 | 32 | 871 | 458.2 | 38.3 | 140 | 1,914 | 490.6 | 44.4 | 37 | 84 | 16.3 | 2.4 |
| Stark | 30.44 | 81.1 | 325 | 4,607 | 3,874.9 | 241.4 | 1,214 | 19,873 | 5,377.1 | 488.2 | 270 | 1,379 | 287.8 | 59.2 |
| Summit | 11.08 | 20.4 | 802 | 13,008 | 9,120.2 | 835.4 | 1,655 | 29,860 | 8,582.2 | 830.2 | 532 | 3,044 | 667.8 | 125.9 |
| Trumbull | 33.85 | 166.1 | 148 | 2,275 | 1,791.4 | 123.5 | 668 | 11,479 | 2,584.7 | 269.7 | 137 | 1,546 | 281.1 | 61.0 |
| Tuscarawas | 17.67 | 190.2 | 79 | 927 | 526.5 | 42.0 | 329 | 4,441 | 1,194.7 | 106.9 | 65 | 348 | 63.9 | 10.9 |
| Union | 1.17 | 21.6 | 75 | 983 | 2,464.1 | 62.4 | 127 | 2,409 | 793.9 | 67.0 | 48 | 172 | 32.0 | 6.1 |
| Van Wert | 1.87 | 65.5 | 27 | 349 | 208.1 | 16.3 | 90 | 1,140 | 334.3 | 27.5 | 12 | 50 | 11.0 | 1.8 |
| Vinton | 0.18 | 13.8 | NA | NA | NA | NA | 25 | 135 | 30.3 | 2.7 | D | D | D | 0.2 |
| Warren | 15.07 | 67.1 | 182 | 5,568 | 3,800.3 | 409.8 | 516 | 10,913 | 2,934.1 | 258.8 | 214 | 1,068 | 703.4 | 50.4 |
| Washington | 7.63 | 124.9 | 55 | 755 | 398.6 | 34.7 | 230 | 3,047 | 882.0 | 76.5 | 42 | 216 | 40.3 | 6.6 |
| Wayne | 8.06 | 69.4 | D | D | D | D | 376 | 4,951 | 1,367.8 | 125.5 | 68 | 262 | 55.2 | 8.4 |
| Williams | 2.67 | 71.9 | 42 | 587 | 383.0 | 24.0 | 129 | 1,373 | 372.9 | 30.4 | 17 | 69 | 11.3 | 2.0 |
| Wood | 5.21 | 40.2 | 172 | 3,024 | 2,046.3 | 165.7 | 362 | 6,666 | 1,871.8 | 155.3 | 118 | 561 | 138.9 | 25.2 |
| Wyandot | 0.95 | 42.7 | 27 | 333 | 243.6 | 15.2 | 68 | 779 | 272.5 | 21.0 | 12 | 18 | 3.2 | 0.6 |
| **OKLAHOMA** | 611.24 | 156.3 | 3,859 | 50,233 | 42,221.4 | 2,757.0 | 12,963 | 180,451 | 53,382.1 | 4,682.9 | 4,461 | 22,135 | 4,696.2 | 941.8 |
| Adair | 6.91 | 314.0 | 9 | 90 | 47.1 | 2.7 | 46 | 513 | 124.8 | 10.9 | 7 | 16 | 2.2 | 0.4 |
| Alfalfa | 0.70 | 119.3 | D | D | D | 2.2 | 21 | 169 | 50.3 | 4.1 | NA | NA | NA | NA |
| Atoka | 40.42 | 2,930.5 | D | D | D | 0.7 | 51 | 490 | 157.4 | 12.3 | 7 | 20 | 2.9 | 0.7 |
| Beaver | 0.48 | 88.4 | 6 | 41 | 35.3 | 1.0 | 14 | 88 | 24.4 | 1.5 | NA | NA | NA | NA |
| Beckham | 3.25 | 136.7 | 33 | 276 | 165.5 | 12.7 | 118 | 1,511 | 582.0 | 39.8 | 42 | 311 | 77.4 | 16.4 |
| Blaine | 1.07 | 108.8 | 12 | 118 | 88.8 | 6.5 | 41 | 330 | 105.3 | 6.2 | D | D | D | D |
| Bryan | 5.53 | 123.2 | 31 | 670 | 496.6 | 25.9 | 128 | 1,640 | 519.8 | 40.9 | 27 | 79 | 20.9 | 3.1 |
| Caddo | 7.15 | 243.7 | 17 | 214 | 106.3 | 8.0 | 84 | 808 | 251.7 | 20.5 | 10 | 19 | 3.9 | 0.8 |
| Canadian | 4.24 | 31.8 | 96 | 1,228 | 3,082.2 | 80.5 | 293 | 4,278 | 1,469.3 | 116.2 | 164 | 810 | 211.2 | 40.9 |
| Carter | 5.08 | 104.3 | 71 | 958 | 1,092.6 | 38.0 | 258 | 2,964 | 874.6 | 69.1 | 73 | 297 | 55.5 | 12.8 |
| Cherokee | 7.40 | 152.7 | 14 | 771 | 114.3 | 19.9 | 149 | 1,820 | 426.9 | 40.9 | 35 | 154 | 23.9 | 3.5 |
| Choctaw | 2.69 | 179.4 | 7 | 71 | 33.6 | 3.0 | 47 | 450 | 130.8 | 11.7 | 7 | 12 | 3.3 | 0.2 |
| Cimarron | 0.34 | 153.4 | 6 | D | 33.2 | D | 11 | 85 | 28.1 | 1.6 | NA | NA | NA | NA |
| Cleveland | 23.37 | 85.1 | 146 | 1,495 | 820.2 | 82.2 | 758 | 12,342 | 3,409.8 | 305.6 | 391 | 1,575 | 331.7 | 62.1 |
| Coal | 0.46 | 81.4 | D | D | D | D | 17 | 108 | 26.3 | 2.4 | NA | NA | NA | NA |
| Comanche | 20.98 | 168.3 | D | D | D | D | 397 | 5,275 | 1,337.7 | 127.5 | D | D | D | D |
| Cotton | 0.51 | 85.1 | NA | NA | NA | NA | 14 | 96 | 36.5 | 1.9 | 5 | 15 | 2.3 | 0.3 |
| Craig | 0.10 | 6.7 | 19 | 176 | 59.4 | 7.3 | 53 | 583 | 172.8 | 14.8 | 44 | 125 | 22.0 | 3.6 |
| Creek | 5.89 | 83.1 | 63 | 1,488 | 573.2 | 76.3 | 148 | 1,783 | 497.6 | 43.6 | 45 | 245 | 50.4 | 14.1 |
| Custer | 3.65 | 122.7 | 39 | 408 | 267.6 | 22.2 | 143 | 1,595 | 508.9 | 39.6 | 35 | 107 | 20.7 | 3.6 |
| Delaware | 35.69 | 860.9 | 20 | 63 | 24.7 | 2.0 | 139 | 1,447 | 360.5 | 35.5 | NA | NA | NA | NA |
| Dewey | 0.11 | 22.0 | 8 | D | 7.0 | D | 31 | 224 | 105.1 | 5.2 | NA | NA | NA | NA |
| Ellis | 0.65 | 153.6 | 6 | 74 | 26.9 | 3.1 | 24 | 125 | 47.2 | 2.7 | NA | NA | NA | NA |
| Garfield | 2.87 | 45.1 | D | D | D | D | 268 | 3,780 | 944.3 | 98.8 | 93 | 383 | 70.4 | 14.3 |
| Garvin | 1.99 | 71.7 | 21 | 218 | 112.5 | 11.0 | 116 | 1,270 | 510.2 | 38.4 | 21 | 95 | 27.1 | 5.7 |
| Grady | 2.40 | 43.9 | 46 | 539 | 276.4 | 28.8 | 137 | 1,438 | 471.3 | 37.6 | 47 | 194 | 40.3 | 7.6 |
| Grant | 1.26 | 278.6 | 4 | 31 | 35.6 | 1.4 | 23 | 159 | 79.8 | 4.8 | NA | NA | NA | NA |
| Greer | 1.46 | 240.5 | NA | NA | NA | NA | 13 | 130 | 25.4 | 3.1 | NA | NA | NA | NA |
| Harmon | 0.73 | 261.8 | NA | NA | NA | NA | 11 | 59 | 13.4 | 1.1 | D | D | D | D |
| Harper | 0.60 | 159.8 | NA | NA | NA | NA | 19 | 120 | 19.6 | 2.6 | 6 | D | 1.9 | D |
| Haskell | 1.17 | 91.1 | D | D | D | 1.1 | 36 | 477 | 146.4 | 12.1 | 8 | 23 | 2.2 | 0.7 |
| Hughes | 2.57 | 187.1 | 8 | 50 | 13.9 | 1.4 | 37 | 392 | 83.3 | 7.6 | NA | NA | NA | NA |
| Jackson | 0.21 | 8.2 | 20 | 160 | 218.4 | 10.9 | 105 | 1,206 | 390.3 | 32.4 | 24 | 135 | 21.7 | 3.2 |
| Jefferson | 13.79 | 2,197.3 | NA | NA | NA | NA | 21 | 149 | 25.8 | 2.4 | NA | NA | NA | NA |
| Johnston | 1.98 | 180.3 | 7 | 32 | 11.8 | 1.3 | 34 | 235 | 67.1 | 5.4 | 4 | D | 1.0 | D |
| Kay | 18.13 | 399.6 | 48 | 324 | 142.7 | 15.6 | 174 | 2,180 | 653.6 | 54.9 | 41 | 113 | 26.3 | 3.2 |
| Kingfisher | 1.66 | 106.5 | 24 | 366 | 362.0 | 18.8 | 51 | 738 | 344.7 | 23.4 | 8 | 18 | 2.1 | 0.6 |

1 Merchant wholesalers, except manufacturers' sales branches and offices.  2. Employer establishments.

# Table B. States and Counties — Professional Services, Manufacturing, and Accommodation and Food Services

| STATE County | Professional, scientific, and technical services, 2017 | | | | Manufacturing, 2017 | | | | Accommodation and food services, 2017 | | | |
|---|---|---|---|---|---|---|---|---|---|---|---|---|
| | Number of establishments | Number of employees | Sales (mil dol) | Average payroll (mil dol) | Number of establishments | Number of employees | Sales (mil dol) | Average payroll (mil dol) | Number of establishments | Number of employees | Sales (mil dol) | Annual payroll (mil dol) |
| | 147 | 148 | 149 | 150 | 151 | 152 | 153 | 154 | 155 | 156 | 157 | 158 |
| OHIO—Cont'd | | | | | | | | | | | | |
| Ottawa | D | D | D | D | 51 | 2,039 | 633.7 | 123.1 | 175 | 1,734 | 139.3 | 38.3 |
| Paulding | 13 | 67 | 5 | 1.7 | 33 | 1,342 | 314.1 | 59.4 | D | D | D | D |
| Perry | 21 | 90 | 8 | 3.5 | 16 | 722 | 133.9 | 32.6 | 41 | 394 | 18.9 | 4.9 |
| Pickaway | 44 | 183 | 21 | 7.2 | 38 | 2,249 | 1,107.1 | 128.7 | 73 | 1,487 | 69.5 | 20.8 |
| Pike | 24 | 2,119 | 467 | 163.1 | 31 | 743 | 568.1 | 37.3 | 40 | 572 | 27.5 | 7.6 |
| Portage | D | D | D | D | 243 | 9,969 | 3,414.1 | 508.5 | 332 | 5,756 | 268.6 | 78.7 |
| Preble | 40 | 253 | 14 | 5.4 | 51 | 3,397 | 1,229.7 | 187.4 | 64 | 968 | 45.9 | 14.0 |
| Putnam | 37 | 199 | 22 | 7.3 | 53 | 3,673 | 2,598.1 | 179.8 | 58 | 832 | 29.9 | 8.3 |
| Richland | 190 | 844 | 106 | 34.3 | 170 | 8,503 | 3,520.3 | 459.3 | 229 | 4,578 | 214.5 | 63.2 |
| Ross | 59 | 447 | 32 | 14.0 | 30 | 3,683 | 4,320.3 | 258.9 | 123 | 2,929 | 137.3 | 41.4 |
| Sandusky | 81 | 425 | 53 | 16.9 | 100 | 9,365 | 4,040.6 | 424.7 | 125 | 2,076 | 95.2 | 26.1 |
| Scioto | D | D | D | D | 48 | 1,312 | 799.9 | 65.0 | 137 | 2,556 | 117.7 | 35.2 |
| Seneca | 51 | 286 | 29 | 10.7 | 68 | 4,355 | 1,367.0 | 209.0 | 103 | 2,011 | 86.6 | 24.4 |
| Shelby | 49 | 327 | 57 | 19.7 | 123 | 13,631 | 9,069.5 | 765.6 | 81 | 1,394 | 74.8 | 18.3 |
| Stark | D | D | D | D | 473 | 24,308 | 11,181.6 | 1,292.7 | 798 | 15,665 | 745.8 | 214.7 |
| Summit | D | D | D | D | 779 | 28,652 | 9,410.7 | 1,574.2 | 1,243 | 23,496 | 1,190.1 | 335.6 |
| Trumbull | 266 | 1,937 | 194 | 70.8 | 205 | 10,410 | 6,017.7 | 657.3 | 384 | 6,943 | 300.3 | 84.8 |
| Tuscarawas | D | D | D | D | 206 | 7,368 | 2,639.7 | 382.1 | 206 | 3,009 | 135.6 | 39.0 |
| Union | 101 | 3,224 | 1,079 | 300.3 | 53 | 7,285 | 10,397.7 | 505.3 | 79 | 1,624 | 84.7 | 24.3 |
| Van Wert | 28 | 384 | 35 | 13.8 | 40 | 3,186 | 1,387.9 | 144.2 | 45 | 809 | 31.2 | 8.2 |
| Vinton | 6 | 19 | 2 | 0.5 | 17 | 453 | 109.1 | 20.2 | D | D | D | D |
| Warren | 533 | 4,641 | 747 | 257.3 | 215 | 11,739 | 4,514.1 | 669.4 | 378 | 8,669 | 463.2 | 133.3 |
| Washington | 82 | 716 | 62 | 27.2 | 78 | 3,178 | 2,421.7 | 224.4 | 122 | 2,254 | 105.8 | 31.5 |
| Wayne | 142 | 2,044 | 128 | 108.7 | 274 | 12,270 | 4,199.4 | 640.1 | 189 | 3,390 | 157.9 | 47.1 |
| Williams | 32 | 203 | 13 | 8.0 | 109 | 7,349 | 2,215.0 | 344.5 | 71 | 978 | 46.3 | 12.5 |
| Wood | D | D | D | D | 181 | 12,099 | 4,609.8 | 729.8 | 318 | 7,038 | 308.1 | 88.2 |
| Wyandot | 29 | 77 | 11 | 2.5 | 35 | 3,464 | 978.2 | 175.1 | 47 | 426 | 18.4 | 5.0 |
| OKLAHOMA | D | D | 11,118 | D | 3,376 | 123,138 | 61,143.3 | 6,740.0 | 8,397 | 159,826 | 9,250.8 | 2,469.9 |
| Adair | 20 | 67 | 6 | 1.8 | 18 | 873 | 406.2 | 34.6 | 21 | 283 | 13.5 | 3.7 |
| Alfalfa | 9 | 30 | 3 | 1.0 | 4 | 30 | 4.9 | 0.9 | 12 | D | 2.7 | D |
| Atoka | 16 | 50 | 4 | 1.5 | 14 | 118 | 26.9 | 5.1 | 24 | 314 | 14.8 | 3.9 |
| Beaver | 9 | 29 | 3 | 0.9 | NA | NA | NA | NA | D | D | D | 0.3 |
| Beckham | 67 | 243 | 35 | 11.1 | D | 258 | D | 17.7 | 69 | 799 | 42.8 | 10.9 |
| Blaine | 14 | 49 | 6 | 1.4 | 9 | 380 | 220.5 | 28.4 | 21 | 142 | 8.1 | 1.9 |
| Bryan | 57 | 282 | 39 | 11.7 | 36 | 1,240 | 323.4 | 60.9 | 69 | 1,327 | 67.6 | 17.4 |
| Caddo | 35 | 519 | 51 | 25.8 | 14 | 79 | 53.2 | 3.4 | 44 | 402 | 17.8 | 4.9 |
| Canadian | 260 | 828 | 134 | 51.5 | 79 | 3,130 | 1,186.4 | 156.0 | 222 | 4,416 | 231.5 | 60.6 |
| Carter | D | D | D | D | 42 | 2,881 | 3,275.0 | 191.4 | 112 | 2,055 | 106.0 | 28.5 |
| Cherokee | D | D | D | D | 19 | 117 | 18.9 | 4.7 | 84 | 1,188 | 58.8 | 16.2 |
| Choctaw | D | D | D | D | 10 | 79 | 17.7 | 3.1 | 26 | 273 | 12.2 | 3.2 |
| Cimarron | 3 | 10 | 2 | 0.4 | NA | NA | NA | NA | 11 | 69 | 3.6 | 0.8 |
| Cleveland | D | D | D | D | 125 | 3,215 | 1,646.3 | 165.3 | 596 | 12,444 | 650.0 | 181.3 |
| Coal | 3 | 14 | 2 | 0.4 | 5 | 73 | 11.8 | 3.0 | 8 | 56 | 3.0 | 0.7 |
| Comanche | D | D | D | D | D | D | D | D | D | D | D | D |
| Cotton | 10 | 19 | 1 | 0.6 | NA | NA | NA | NA | D | D | D | D |
| Craig | 24 | 69 | 11 | 2.5 | 12 | 156 | 32.3 | 5.6 | 31 | 375 | 19.4 | 4.9 |
| Creek | 106 | 344 | 46 | 15.7 | 115 | 3,563 | 1,161.3 | 201.4 | 91 | 1,290 | 61.0 | 16.2 |
| Custer | D | D | D | D | 29 | 818 | 375.5 | 41.3 | 77 | 1,223 | 67.7 | 17.1 |
| Delaware | 53 | 335 | 41 | 18.3 | 26 | 468 | 86.7 | 23.1 | 79 | 1,784 | 198.5 | 42.7 |
| Dewey | D | D | D | 0.4 | 5 | 62 | 30.7 | 4.8 | D | D | D | 1.7 |
| Ellis | 8 | 27 | 2 | 0.7 | NA | NA | NA | NA | 5 | 88 | 2.6 | 0.9 |
| Garfield | 119 | 581 | 73 | 28.4 | 57 | 2,357 | 1,240.8 | 139.3 | 134 | 2,422 | 128.3 | 32.8 |
| Garvin | 54 | 149 | 17 | 6.0 | 25 | 1,016 | 1,973.1 | 69.4 | 55 | 729 | 39.6 | 9.8 |
| Grady | 98 | 343 | 41 | 14.1 | 62 | 1,475 | 697.9 | 57.7 | 80 | 1,102 | 62.7 | 15.9 |
| Grant | 5 | 16 | 2 | 0.7 | 4 | 9 | 1.6 | 0.3 | 6 | 28 | 1.1 | 0.3 |
| Greer | D | D | D | 2.3 | D | D | D | D | 4 | 48 | 1.4 | 0.4 |
| Harmon | D | D | D | 0.1 | NA | NA | NA | NA | 4 | D | 0.6 | D |
| Harper | 9 | 34 | 3 | 1.3 | NA | NA | NA | NA | 5 | D | 1.1 | D |
| Haskell | 18 | 45 | 4 | 1.0 | D | 81 | D | 4.1 | D | D | D | D |
| Hughes | 9 | 36 | 3 | 1.3 | 7 | 38 | 22.2 | 1.8 | 13 | 96 | 4.4 | 1.1 |
| Jackson | D | D | D | D | 11 | 891 | 388.5 | 42.1 | 55 | 1,123 | 47.0 | 14.5 |
| Jefferson | 7 | 36 | 6 | 1.5 | 4 | 7 | 1.7 | 0.4 | 13 | D | 4.0 | D |
| Johnston | 9 | 22 | 2 | 0.6 | 9 | 541 | 75.0 | 18.0 | D | D | D | D |
| Kay | D | D | D | D | 53 | 2,856 | 7,241.2 | 190.9 | 112 | 1,801 | 114.1 | 26.9 |
| Kingfisher | 29 | 259 | 24 | 10.0 | 14 | 482 | 131.8 | 21.1 | 36 | 343 | 19.5 | 4.9 |

## Table B. States and Counties — Health Care and Social Assistance, Other Services, Nonemployer Businesses, and Residential Construction

| STATE County | Health care and social assistance, 2017 | | | | Other services, 2017 | | | | Nonemployer businesses, 2018 | | Value of residential construction authorized by building permits, 2020 | |
|---|---|---|---|---|---|---|---|---|---|---|---|---|
| | Number of establishments | Number of employees | Receipts (mil dol) | Annual payroll (mil dol) | Number of establishments | Number of employees | Receipts (mil dol) | Annual payroll (mil dol) | Number | Receipts (mil dol) | New construction ($1,000) | Number of housing units |
| | 159 | 160 | 161 | 162 | 163 | 164 | 165 | 166 | 167 | 168 | 169 | 170 |
| **OHIO—Cont'd** | | | | | | | | | | | | |
| Ottawa | 92 | 1,842 | 145.5 | 58.7 | 81 | 305 | 37.6 | 11.1 | 3,020 | 128.9 | 45,722 | 139 |
| Paulding | 28 | 552 | 36.6 | 15.3 | 20 | 74 | 7.1 | 1.6 | 1,077 | 41.3 | 8,256 | 34 |
| Perry | D | D | D | 23.3 | 27 | 127 | 12.9 | 2.2 | 2,033 | 85.4 | 7,650 | 44 |
| Pickaway | 88 | 1,890 | 193.1 | 72.5 | 45 | 138 | 17.0 | 4.3 | 3,447 | 147.0 | 39,677 | 157 |
| Pike | 56 | 1,582 | 146.0 | 52.2 | 18 | 62 | 8.1 | 1.4 | 1,480 | 59.6 | 11,390 | 56 |
| Portage | 265 | 6,348 | 508.0 | 215.2 | 229 | 1,520 | 148.1 | 43.4 | 10,529 | 477.8 | 89,879 | 344 |
| Preble | 62 | 1,413 | 89.0 | 38.6 | 52 | 248 | 27.6 | 7.3 | 2,481 | 111.8 | 8,944 | 48 |
| Putnam | 59 | 1,075 | 63.1 | 25.8 | 52 | 274 | 33.1 | 7.8 | 2,094 | 84.3 | 15,315 | 51 |
| Richland | 375 | 8,617 | 836.7 | 335.5 | 195 | 1,099 | 108.4 | 27.4 | 6,980 | 318.4 | 38,281 | 205 |
| Ross | 163 | 5,004 | 733.3 | 262.7 | 84 | 605 | 53.0 | 16.5 | 3,883 | 164.3 | 3,124 | 15 |
| Sandusky | 188 | 2,922 | 271.0 | 105.7 | 92 | 570 | 50.0 | 15.1 | 3,049 | 117.1 | 11,019 | 49 |
| Scioto | 219 | 7,527 | 765.1 | 287.4 | 75 | 297 | 32.8 | 7.0 | 3,607 | 127.3 | 380 | 4 |
| Seneca | 131 | 2,319 | 185.5 | 68.9 | 100 | 410 | 38.8 | 8.8 | 2,908 | 118.2 | 3,512 | 26 |
| Shelby | 94 | 2,069 | 210.7 | 74.4 | 60 | 412 | 41.4 | 11.5 | 2,805 | 125.0 | 19,616 | 77 |
| Stark | 992 | 28,654 | 2,769.8 | 1,151.2 | 648 | 4,677 | 587.5 | 151.3 | 24,514 | 1,058.3 | 143,588 | 683 |
| Summit | 1,518 | 54,081 | 5,512.0 | 2,485.6 | 1,098 | 7,441 | 1,546.7 | 242.3 | 38,961 | 1,775.6 | 277,189 | 570 |
| Trumbull | 540 | 11,213 | 1,329.6 | 506.7 | 278 | 1,524 | 124.2 | 37.7 | 13,051 | 591.4 | 17,839 | 105 |
| Tuscarawas | 201 | 5,333 | 416.4 | 167.0 | 175 | 1,080 | 120.2 | 30.0 | 6,199 | 285.0 | 18,746 | 69 |
| Union | 83 | 1,814 | 221.9 | 79.1 | 79 | 389 | 50.7 | 13.4 | 3,956 | 188.6 | 169,729 | 854 |
| Van Wert | 68 | 1,722 | 154.7 | 60.1 | 45 | 255 | 22.6 | 4.5 | 1,662 | 69.2 | 9,357 | 39 |
| Vinton | 17 | 286 | 14.7 | 6.0 | 8 | 41 | 4.8 | 1.0 | 653 | 27.0 | 0 | 0 |
| Warren | 484 | 10,712 | 1,091.4 | 418.9 | 299 | 2,270 | 221.0 | 69.7 | 17,563 | 893.7 | 415,436 | 1,640 |
| Washington | 141 | 5,211 | 660.8 | 205.8 | 103 | 465 | 45.9 | 12.2 | 3,784 | 164.6 | 110 | 1 |
| Wayne | 248 | 5,953 | 511.0 | 221.8 | 155 | 846 | 102.7 | 23.0 | 9,355 | 480.3 | 60,387 | 275 |
| Williams | 71 | 1,774 | 161.0 | 78.1 | 63 | 356 | 35.5 | 8.9 | 2,197 | 94.6 | 6,858 | 30 |
| Wood | 290 | 6,193 | 484.1 | 205.6 | 210 | 1,519 | 143.8 | 45.9 | 7,827 | 353.5 | 90,274 | 484 |
| Wyandot | 44 | 890 | 76.2 | 29.8 | 55 | 250 | 26.2 | 6.9 | 1,366 | 58.3 | 4,078 | 17 |
| **OKLAHOMA** | 11,035 | 226,462 | 27,031.0 | 9,952.8 | 5,565 | 31,947 | 4,933.6 | 1,083.6 | 291,452 | 13,907.1 | 2,852,519 | 13,733 |
| Adair | 21 | 500 | 50.6 | 18.1 | D | D | D | 0.3 | 1,383 | 52.5 | 1,470 | 12 |
| Alfalfa | 11 | 133 | 9.2 | 4.1 | D | D | D | 1.2 | 385 | 11.5 | 0 | 0 |
| Atoka | 22 | 374 | 21.3 | 9.0 | D | D | D | 1.2 | 933 | 44.1 | 209 | 1 |
| Beaver | 5 | D | 8.8 | D | D | D | 5.1 | D | 475 | 25.6 | 0 | 0 |
| Beckham | 84 | 1,011 | 103.9 | 38.1 | 35 | 153 | 27.6 | 5.6 | 1,757 | 112.9 | 1,441 | 5 |
| Blaine | 26 | 369 | 27.1 | 12.5 | D | D | D | D | 699 | 30.5 | 0 | 0 |
| Bryan | 133 | 2,161 | 268.6 | 80.4 | 36 | 155 | 18.1 | 4.5 | 3,041 | 149.8 | 12,179 | 88 |
| Caddo | 46 | 730 | 51.2 | 21.7 | 21 | 101 | 21.6 | 5.1 | 1,541 | 71.3 | 418 | 2 |
| Canadian | 265 | 2,858 | 256.2 | 104.3 | 170 | 701 | 79.4 | 21.0 | 11,547 | 530.6 | 95,313 | 435 |
| Carter | 228 | 3,790 | 356.8 | 167.8 | 92 | 821 | 153.7 | 34.8 | 3,413 | 166.4 | 8,138 | 47 |
| Cherokee | 109 | 2,974 | 316.8 | 134.5 | 46 | 227 | 21.6 | 5.4 | 2,869 | 119.5 | 13,160 | 133 |
| Choctaw | 59 | 1,376 | 103.3 | 39.2 | D | D | D | D | 941 | 46.4 | 1,440 | 12 |
| Cimarron | D | D | D | D | D | D | 0.7 | D | 214 | 11.8 | 633 | 1 |
| Cleveland | 795 | 12,647 | 1,344.1 | 539.9 | 320 | 1,696 | 307.1 | 50.6 | 22,282 | 1,034.5 | 238,774 | 1,150 |
| Coal | 11 | 191 | 12.3 | 4.9 | D | D | 4.3 | D | 468 | 28.6 | 100 | 1 |
| Comanche | D | D | D | D | 145 | 742 | 78.1 | 22.1 | 5,058 | 218.6 | 17,373 | 102 |
| Cotton | D | D | D | D | 4 | 10 | 1.4 | 0.4 | 287 | 9.9 | 1,045 | 5 |
| Craig | 67 | 1,005 | 74.5 | 38.5 | 18 | 88 | 8.1 | 2.7 | 900 | 40.1 | 28,132 | 197 |
| Creek | 108 | 2,409 | 185.8 | 71.5 | 78 | 286 | 33.1 | 8.1 | 4,943 | 215.9 | 5,325 | 19 |
| Custer | 109 | 1,492 | 127.3 | 48.3 | 48 | 257 | 32.5 | 9.1 | 2,289 | 116.4 | 6,112 | 45 |
| Delaware | 78 | 1,334 | 125.8 | 45.8 | 52 | 250 | 24.8 | 6.5 | 2,761 | 130.5 | NA | NA |
| Dewey | 7 | 48 | 3.0 | 1.2 | D | D | D | 1.4 | 506 | 24.6 | 0 | 0 |
| Ellis | 10 | D | 11.8 | D | D | D | 1.4 | D | 312 | 12.6 | 7,898 | 23 |
| Garfield | 197 | 3,778 | 386.2 | 141.0 | 108 | 545 | 63.1 | 15.2 | 4,445 | 181.9 | 749 | 8 |
| Garvin | 71 | 1,063 | 73.8 | 31.4 | 29 | 123 | 17.5 | 5.1 | 2,170 | 109.5 | 17,677 | 209 |
| Grady | D | D | D | D | D | D | D | D | 3,869 | 186.4 | 937 | 4 |
| Grant | 7 | 145 | 9.0 | 3.3 | D | D | 5.6 | 0.1 | 350 | 13.7 | 263 | 1 |
| Greer | 9 | 261 | 16.4 | 5.7 | 4 | 5 | 0.7 | NA | 295 | 11.5 | NA | NA |
| Harmon | 6 | 127 | 11.5 | 4.4 | NA | NA | NA | 0.4 | 145 | 7.6 | 0 | 0 |
| Harper | 8 | D | 6.2 | D | D | D | D | D | 295 | 11.1 | 49 | 3 |
| Haskell | 35 | 774 | 53.9 | 22.2 | D | D | D | D | 976 | 41.4 | 49 | 3 |
| Hughes | 37 | 754 | 35.6 | 16.1 | 12 | 22 | 2.6 | 0.6 | 710 | 30.3 | 290 | 2 |
| Jackson | 57 | 1,368 | 128.8 | 53.0 | 28 | 112 | 10.6 | 2.8 | 1,402 | 59.5 | 5,344 | 42 |
| Jefferson | 12 | 259 | 27.2 | 9.7 | NA | NA | NA | NA | 447 | 15.9 | NA | NA |
| Johnston | 21 | 1,818 | 97.7 | 37.3 | 11 | 32 | 29.4 | 1.7 | 653 | 24.0 | 0 | 0 |
| Kay | 132 | 2,119 | 200.0 | 73.0 | 73 | 409 | 40.7 | 10.6 | 2,463 | 91.9 | 3,391 | 13 |
| Kingfisher | 39 | 514 | 38.8 | 14.6 | D | D | D | D | 1,630 | 96.3 | 5,192 | 17 |

## Table B. States and Counties — Government Employment and Payroll, and Local Government Finances

| STATE County | Full-time equivalent employees | March payroll (dollars) | Administration, judicial, and legal | Police and corrections | Fire protection | Highways and transportation | Health and welfare | Natural resources and utilities | Education and libraries | Total (mil dol) | Intergovernmental (mil dol) | Taxes Total (mil dol) | Per capita Total | Per capita Property |
|---|---|---|---|---|---|---|---|---|---|---|---|---|---|---|
| | 171 | 172 | 173 | 174 | 175 | 176 | 177 | 178 | 179 | 180 | 181 | 182 | 183 | 184 |
| **OHIO—Cont'd** | | | | | | | | | | | | | | |
| Ottawa | 1,632 | 6,665,839 | 8.6 | 10.0 | 3.8 | 5.6 | 12.8 | 7.7 | 49.7 | 174.1 | 49.1 | 84.7 | 2,085 | 1,662 |
| Paulding | 914 | 3,212,384 | 6.4 | 3.7 | 0.3 | 3.1 | 24.6 | 3.9 | 55.9 | 85.3 | 30.4 | 26.8 | 1,425 | 920 |
| Perry | 1,338 | 4,367,873 | 8.0 | 4.4 | 0.7 | 5.2 | 12.9 | 2.7 | 65.5 | 120.5 | 75.8 | 32.6 | 907 | 730 |
| Pickaway | 2,259 | 9,272,269 | 4.8 | 5.9 | 1.8 | 2.5 | 26.4 | 2.8 | 55.5 | 274.7 | 87.2 | 83.9 | 1,453 | 944 |
| Pike | 1,321 | 4,492,199 | 8.8 | 4.9 | 0.2 | 4.0 | 16.0 | 1.7 | 63.5 | 119.0 | 79.6 | 24.7 | 880 | 650 |
| Portage | 5,369 | 22,101,287 | 7.5 | 10.2 | 5.0 | 6.1 | 6.3 | 4.8 | 58.8 | 595.5 | 218.7 | 286.7 | 1,764 | 1,194 |
| Preble | 1,642 | 5,078,832 | 10.5 | 7.9 | 0.9 | 5.0 | 8.1 | 4.0 | 61.2 | 153.9 | 64.2 | 60.6 | 1,474 | 840 |
| Putnam | 1,518 | 4,779,575 | 8.8 | 15.5 | 0.1 | 2.2 | 3.9 | 2.3 | 67.0 | 131.9 | 57.3 | 52.6 | 1,554 | 959 |
| Richland | 4,967 | 18,600,904 | 8.1 | 8.4 | 4.6 | 2.9 | 15.2 | 4.0 | 55.0 | 485.5 | 220.4 | 191.9 | 1,593 | 961 |
| Ross | 2,881 | 10,938,361 | 7.0 | 6.5 | 2.7 | 2.5 | 14.0 | 2.9 | 62.8 | 277.9 | 141.6 | 90.4 | 1,169 | 719 |
| Sandusky | 2,131 | 8,612,401 | 7.2 | 8.7 | 1.6 | 3.4 | 7.0 | 6.9 | 64.5 | 229.5 | 99.4 | 94.0 | 1,591 | 929 |
| Scioto | 2,998 | 10,522,705 | 7.0 | 7.2 | 2.6 | 3.2 | 9.1 | 4.6 | 65.4 | 280.4 | 179.4 | 65.8 | 867 | 633 |
| Seneca | 2,133 | 7,696,532 | 8.3 | 10.3 | 4.6 | 3.7 | 7.2 | 5.1 | 60.2 | 191.2 | 86.9 | 69.1 | 1,250 | 749 |
| Shelby | 1,507 | 5,836,231 | 6.0 | 5.4 | 4.0 | 2.5 | 5.8 | 5.2 | 70.3 | 210.0 | 85.8 | 85.1 | 1,747 | 870 |
| Stark | 14,288 | 55,719,833 | 7.5 | 8.6 | 4.2 | 4.4 | 8.2 | 5.1 | 60.9 | 1,374.9 | 607.3 | 560.3 | 1,506 | 1,023 |
| Summit | 19,664 | 87,775,101 | 7.9 | 10.4 | 6.1 | 5.8 | 8.9 | 6.5 | 52.8 | 2,522.8 | 804.7 | 1,251.5 | 2,311 | 1,375 |
| Trumbull | 7,910 | 28,504,175 | 8.0 | 8.3 | 4.3 | 3.0 | 10.8 | 4.6 | 60.0 | 676.1 | 299.6 | 259.1 | 1,294 | 927 |
| Tuscarawas | 3,469 | 12,598,587 | 10.2 | 6.3 | 3.5 | 4.1 | 5.4 | 8.6 | 59.6 | 364.7 | 143.8 | 132.9 | 1,440 | 997 |
| Union | 2,255 | 10,065,693 | 6.6 | 7.3 | 4.4 | 2.7 | 35.0 | 2.7 | 37.0 | 318.8 | 63.0 | 114.5 | 2,016 | 1,294 |
| Van Wert | 1,023 | 3,549,200 | 8.2 | 7.5 | 2.7 | 4.2 | 3.3 | 4.5 | 66.2 | 122.6 | 46.9 | 46.0 | 1,627 | 973 |
| Vinton | 523 | 1,837,655 | 13.5 | 2.8 | 1.6 | 5.8 | 8.9 | 1.5 | 65.3 | 48.6 | 32.7 | 11.2 | 855 | 755 |
| Warren | 7,582 | 29,873,176 | 7.1 | 8.6 | 5.1 | 2.5 | 7.0 | 3.9 | 63.9 | 748.8 | 240.0 | 366.5 | 1,603 | 1,294 |
| Washington | 2,025 | 6,959,332 | 9.8 | 10.2 | 2.7 | 5.0 | 8.7 | 4.2 | 59.0 | 212.5 | 88.7 | 88.8 | 1,468 | 1,008 |
| Wayne | 4,842 | 18,900,576 | 5.4 | 5.5 | 2.4 | 3.9 | 26.8 | 4.0 | 51.7 | 506.9 | 149.4 | 168.7 | 1,452 | 1,023 |
| Williams | 1,637 | 5,528,317 | 7.9 | 6.3 | 0.8 | 4.7 | 14.3 | 7.2 | 57.0 | 157.8 | 64.6 | 60.9 | 1,660 | 884 |
| Wood | 4,973 | 20,084,911 | 8.8 | 10.6 | 3.6 | 3.4 | 8.9 | 7.2 | 55.6 | 547.3 | 157.2 | 287.3 | 2,202 | 1,393 |
| Wyandot | 1,136 | 4,142,231 | 8.0 | 6.9 | 0.5 | 3.1 | 35.5 | 4.6 | 41.1 | 132.0 | 34.3 | 34.5 | 1,563 | 955 |
| **OKLAHOMA** | X | X | X | X | X | X | X | X | X | X | X | X | X | X |
| Adair | 956 | 2,957,411 | 6.2 | 6.2 | 0.1 | 3.9 | 1.4 | 4.7 | 76.2 | 57.4 | 41.8 | 8.4 | 380 | 239 |
| Alfalfa | 317 | 953,473 | 3.7 | 7.2 | 0.0 | 17.5 | 0.0 | 9.0 | 54.7 | 22.4 | 9.8 | 10.2 | 1,726 | 1,334 |
| Atoka | 499 | 1,550,570 | 4.0 | 4.9 | 0.0 | 1.1 | 20.9 | 4.2 | 63.9 | 57.5 | 37.5 | 10.0 | 727 | 387 |
| Beaver | 364 | 1,311,575 | 7.7 | 3.8 | 0.0 | 10.8 | 15.6 | 2.9 | 57.7 | 25.0 | 11.0 | 11.1 | 2,062 | 1,487 |
| Beckham | 856 | 2,534,945 | 7.0 | 11.6 | 4.0 | 6.4 | 4.1 | 6.6 | 58.0 | 85.4 | 26.9 | 39.3 | 1,804 | 925 |
| Blaine | 526 | 1,516,152 | 9.5 | 7.7 | 1.1 | 8.6 | 17.1 | 3.5 | 51.6 | 43.9 | 14.0 | 18.5 | 1,949 | 1,076 |
| Bryan | 1,778 | 6,176,213 | 5.0 | 8.6 | 2.4 | 3.4 | 1.9 | 5.8 | 71.6 | 127.1 | 50.3 | 44.6 | 959 | 483 |
| Caddo | 1,455 | 4,372,847 | 4.8 | 6.8 | 2.4 | 2.9 | 6.0 | 5.0 | 70.6 | 149.5 | 109.6 | 26.7 | 908 | 529 |
| Canadian | 4,034 | 12,941,499 | 5.0 | 12.3 | 4.2 | 2.0 | 0.1 | 4.4 | 71.2 | 345.9 | 132.8 | 166.1 | 1,187 | 748 |
| Carter | 1,780 | 5,804,226 | 5.3 | 9.9 | 3.0 | 4.3 | 0.0 | 6.3 | 69.3 | 157.0 | 59.1 | 71.4 | 1,478 | 692 |
| Cherokee | 2,102 | 7,399,379 | 2.1 | 1.1 | 0.8 | 2.2 | 46.2 | 5.3 | 40.5 | 196.1 | 59.5 | 26.5 | 541 | 247 |
| Choctaw | 712 | 2,090,429 | 5.3 | 7.3 | 2.0 | 4.4 | 25.9 | 4.4 | 49.4 | 39.9 | 23.5 | 11.0 | 743 | 300 |
| Cimarron | 164 | 396,152 | 11.1 | 5.3 | 0.0 | 16.9 | 0.6 | 6.1 | 60.0 | 10.9 | 5.6 | 4.1 | 1,894 | 1,501 |
| Cleveland | 10,169 | 41,206,071 | 3.1 | 7.0 | 4.2 | 1.6 | 35.4 | 2.8 | 44.7 | 1,087.0 | 243.9 | 348.8 | 1,248 | 734 |
| Coal | 316 | 849,977 | 9.2 | 5.7 | 3.2 | 7.1 | 4.5 | 4.5 | 65.8 | 15.9 | 7.7 | 6.2 | 1,093 | 747 |
| Comanche | 5,989 | 21,027,993 | 3.7 | 6.4 | 3.2 | 2.0 | 36.8 | 3.9 | 43.1 | 591.3 | 144.0 | 113.0 | 928 | 422 |
| Cotton | 258 | 690,239 | 9.4 | 5.7 | 1.6 | 7.5 | 1.0 | 9.1 | 62.6 | 15.3 | 9.2 | 3.4 | 586 | 422 |
| Craig | 839 | 2,850,923 | 4.9 | 5.9 | 2.0 | 4.0 | 32.6 | 4.3 | 43.7 | 43.4 | 16.5 | 22.1 | 1,539 | 1,068 |
| Creek | 2,446 | 8,556,457 | 5.1 | 6.5 | 3.5 | 2.4 | 3.2 | 4.3 | 74.5 | 176.9 | 77.8 | 67.9 | 945 | 559 |
| Custer | 1,419 | 4,126,438 | 4.9 | 8.7 | 2.7 | 3.4 | 14.5 | 6.1 | 58.4 | 91.9 | 32.8 | 41.9 | 1,446 | 774 |
| Delaware | 1,352 | 3,655,341 | 6.1 | 7.1 | 1.1 | 3.8 | 0.7 | 4.5 | 74.9 | 89.7 | 46.6 | 34.6 | 812 | 507 |
| Dewey | 350 | 1,140,367 | 10.3 | 6.0 | 0.0 | 12.4 | 19.4 | 3.1 | 47.8 | 32.8 | 12.1 | 14.6 | 2,978 | 2,076 |
| Ellis | 266 | 870,927 | 9.0 | 6.2 | 0.0 | 13.1 | 2.3 | 2.9 | 65.8 | 24.5 | 12.4 | 8.1 | 2,041 | 1,682 |
| Garfield | 2,370 | 8,025,034 | 5.6 | 9.5 | 5.9 | 3.9 | 0.1 | 4.7 | 65.8 | 182.5 | 64.1 | 80.6 | 1,309 | 555 |
| Garvin | 1,447 | 4,063,190 | 5.4 | 6.9 | 1.8 | 4.5 | 28.0 | 6.2 | 46.4 | 103.0 | 34.4 | 29.1 | 1,045 | 582 |
| Grady | 1,930 | 6,362,126 | 4.6 | 5.0 | 2.8 | 3.2 | 30.7 | 2.6 | 49.2 | 130.5 | 49.5 | 58.0 | 1,058 | 604 |
| Grant | 264 | 792,064 | 10.2 | 6.8 | 0.1 | 18.2 | 1.0 | 10.4 | 52.9 | 20.4 | 4.7 | 12.3 | 2,797 | 1,757 |
| Greer | 246 | 883,154 | 7.5 | 6.3 | 2.0 | 0.8 | 5.3 | 5.9 | 71.3 | 17.9 | 10.2 | 3.9 | 669 | 382 |
| Harmon | 205 | 532,277 | 8.8 | 6.7 | 0.0 | 6.3 | 35.2 | 4.7 | 38.2 | 15.4 | 4.4 | 9.0 | 3,311 | 1,894 |
| Harper | 277 | 752,462 | 6.6 | 5.5 | 0.0 | 11.0 | 29.7 | 5.9 | 40.3 | 10.2 | 3.7 | 4.9 | 1,283 | 1,076 |
| Haskell | 442 | 1,327,720 | 5.9 | 10.9 | 0.0 | 6.4 | 1.8 | 1.1 | 70.6 | 44.5 | 27.7 | 11.4 | 899 | 541 |
| Hughes | 693 | 2,010,922 | 4.7 | 4.4 | 1.0 | 4.4 | 27.4 | 3.7 | 50.9 | 32.5 | 14.3 | 13.5 | 1,020 | 647 |
| Jackson | 1,708 | 6,396,652 | 3.3 | 5.1 | 2.1 | 2.0 | 49.1 | 5.3 | 32.4 | 146.1 | 31.4 | 22.9 | 912 | 353 |
| Jefferson | 365 | 1,102,831 | 5.6 | 3.8 | 0.8 | 3.6 | 16.1 | 12.1 | 56.2 | 23.6 | 11.0 | 4.0 | 656 | 404 |
| Johnston | 405 | 1,241,425 | 7.1 | 6.2 | 0.7 | 3.3 | 3.1 | 6.0 | 72.7 | 26.7 | 12.9 | 9.5 | 854 | 522 |
| Kay | 1,922 | 5,806,528 | 6.9 | 9.7 | 7.7 | 4.5 | 1.6 | 12.8 | 54.4 | 140.0 | 48.8 | 62.0 | 1,396 | 708 |
| Kingfisher | 723 | 2,133,869 | 8.2 | 7.0 | 3.4 | 5.3 | 0.1 | 5.0 | 69.3 | 70.3 | 19.8 | 40.1 | 2,555 | 2,071 |

1. Based on the resident population estimated as of July 1 of the year shown.

# Local Government Finances, Government Employment, and Income Taxes

| STATE County | Local government finances, 2017 (cont.) — Direct general expenditure — Total (mil dol) | Per capita¹ (dollars) | Percent of total for: Education | Health and hospitals | Police protection | Public welfare | Highways | Debt outstanding — Total (mil dol) | Per capita¹ (dollars) | Government employment, 2019 — Federal civilian | Federal military | State and local | Individual income tax returns, 2018 — Number of returns | Mean adjusted gross income | Mean income tax |
|---|---|---|---|---|---|---|---|---|---|---|---|---|---|---|---|
| | 185 | 186 | 187 | 188 | 189 | 190 | 191 | 192 | 193 | 194 | 195 | 196 | 197 | 198 | 199 |
| **OHIO—Cont'd** | | | | | | | | | | | | | | | |
| Ottawa | 172 | 4,243 | 41.1 | 2.9 | 8.1 | 16.3 | 5.8 | 108.3 | 2,667 | 280 | 147 | 2,105 | 21,850 | 62,267 | 6,378 |
| Paulding | 85 | 4,513 | 52.1 | 25.8 | 3.3 | 0.0 | 7.5 | 33.0 | 1,752 | 52 | 47 | 1,046 | 9,020 | 48,634 | 3,608 |
| Perry | 118 | 3,292 | 63.6 | 5.7 | 3.4 | 6.3 | 5.5 | 46.2 | 1,284 | 65 | 91 | 1,530 | 15,320 | 49,477 | 3,874 |
| Pickaway | 282 | 4,885 | 38.5 | 25.5 | 3.8 | 2.6 | 5.1 | 133.5 | 2,312 | 97 | 136 | 3,965 | 25,380 | 58,076 | 5,092 |
| Pike | 123 | 4,369 | 58.7 | 3.9 | 5.9 | 5.2 | 9.4 | 34.6 | 1,231 | 57 | 69 | 1,357 | 11,220 | 47,725 | 3,734 |
| Portage | 553 | 3,402 | 51.4 | 6.3 | 6.7 | 4.2 | 4.1 | 275.1 | 1,693 | 351 | 406 | 14,223 | 76,290 | 61,043 | 6,282 |
| Preble | 153 | 3,712 | 52.8 | 5.3 | 5.1 | 6.0 | 7.0 | 72.1 | 1,754 | 77 | 102 | 1,723 | 19,170 | 50,486 | 3,876 |
| Putnam | 125 | 3,689 | 60.0 | 0.0 | 4.3 | 0.8 | 13.4 | 72.8 | 2,149 | 67 | 85 | 1,584 | 17,190 | 62,337 | 5,596 |
| Richland | 494 | 4,099 | 50.7 | 9.7 | 5.4 | 5.5 | 6.3 | 175.1 | 1,454 | 639 | 289 | 7,004 | 57,340 | 49,520 | 4,286 |
| Ross | 287 | 3,707 | 62.2 | 0.7 | 4.1 | 5.2 | 4.7 | 290.5 | 3,758 | 1,728 | 182 | 5,001 | 32,810 | 52,609 | 4,682 |
| Sandusky | 257 | 4,342 | 49.8 | 5.3 | 7.7 | 5.4 | 5.1 | 232.6 | 3,937 | 115 | 146 | 2,794 | 29,670 | 50,522 | 4,122 |
| Scioto | 275 | 3,621 | 58.8 | 5.9 | 3.2 | 4.7 | 4.3 | 340.5 | 4,482 | 174 | 183 | 5,419 | 28,770 | 51,149 | 4,653 |
| Seneca | 222 | 4,018 | 47.9 | 6.8 | 6.7 | 2.1 | 1.9 | 731.9 | 13,242 | 124 | 133 | 2,611 | 26,570 | 48,958 | 3,890 |
| Shelby | 204 | 4,191 | 50.8 | 0.6 | 5.3 | 9.4 | 8.6 | 270.8 | 5,556 | 75 | 121 | 2,354 | 24,230 | 56,164 | 5,121 |
| Stark | 1,394 | 3,746 | 52.3 | 8.6 | 5.8 | 5.4 | 6.1 | 485.2 | 1,304 | 971 | 947 | 18,079 | 183,900 | 57,430 | 5,906 |
| Summit | 2,589 | 4,781 | 42.4 | 4.7 | 5.6 | 4.2 | 4.1 | 2,109.5 | 3,896 | 1,974 | 1,360 | 28,147 | 269,540 | 66,318 | 7,917 |
| Trumbull | 711 | 3,549 | 53.6 | 4.3 | 6.4 | 5.8 | 3.5 | 455.9 | 2,276 | 525 | 524 | 8,914 | 98,380 | 49,117 | 4,355 |
| Tuscarawas | 378 | 4,098 | 48.8 | 5.3 | 4.7 | 3.6 | 4.9 | 171.5 | 1,858 | 267 | 229 | 5,029 | 44,450 | 55,654 | 5,329 |
| Union | 305 | 5,372 | 28.6 | 34.6 | 4.0 | 1.6 | 3.4 | 426.6 | 7,511 | 79 | 142 | 3,780 | 27,910 | 88,684 | 10,998 |
| Van Wert | 113 | 4,006 | 58.1 | 3.5 | 4.9 | 0.2 | 6.1 | 68.4 | 2,419 | 50 | 71 | 1,483 | 14,020 | 51,516 | 4,046 |
| Vinton | 53 | 4,015 | 52.3 | 4.4 | 3.4 | 10.6 | 10.5 | 10.8 | 829 | 17 | 33 | 626 | 5,180 | 43,830 | 2,991 |
| Warren | 775 | 3,388 | 57.8 | 0.2 | 3.7 | 5.3 | 7.5 | 665.8 | 2,911 | 328 | 604 | 9,927 | 114,200 | 89,970 | 11,697 |
| Washington | 211 | 3,481 | 50.2 | 3.7 | 7.6 | 6.8 | 8.5 | 184.2 | 3,045 | 209 | 148 | 2,765 | 28,520 | 55,388 | 5,289 |
| Wayne | 547 | 4,704 | 41.1 | 28.9 | 3.6 | 4.3 | 4.3 | 151.6 | 1,305 | 262 | 284 | 6,756 | 53,600 | 57,293 | 5,146 |
| Williams | 196 | 5,349 | 51.3 | 0.7 | 3.8 | 7.4 | 5.7 | 105.7 | 2,879 | 83 | 90 | 2,118 | 18,340 | 49,976 | 4,074 |
| Wood | 597 | 4,577 | 54.2 | 6.8 | 4.1 | 3.3 | 3.2 | 620.7 | 4,759 | 206 | 332 | 11,117 | 62,310 | 67,447 | 7,372 |
| Wyandot | 137 | 6,189 | 33.3 | 32.7 | 5.2 | 5.5 | 1.8 | 45.5 | 2,061 | 52 | 54 | 1,468 | 11,110 | 53,045 | 4,441 |
| **OKLAHOMA** | X | X | X | X | X | X | X | X | X | 49,767 | 34,740 | 289,272 | 1,639,660 | 62,610 | 6,764 |
| Adair | 59 | 2,646 | 74.0 | 0.6 | 1.0 | 0.0 | 7.4 | 5.3 | 239 | 40 | 80 | 1,084 | 7,680 | 36,692 | 2,045 |
| Alfalfa | 20 | 3,433 | 80.6 | 0.0 | 3.6 | 0.0 | 3.9 | 13.0 | 2,203 | 32 | 17 | 571 | 2,010 | 66,134 | 7,358 |
| Atoka | 63 | 4,555 | 35.7 | 4.8 | 4.5 | 0.0 | 8.6 | 31.3 | 2,269 | 32 | 47 | 1,081 | 5,170 | 43,880 | 3,239 |
| Beaver | 24 | 4,489 | 53.5 | 3.1 | 3.7 | 0.0 | 25.2 | 4.5 | 841 | 31 | 19 | 450 | 2,170 | 48,676 | 4,084 |
| Beckham | 79 | 3,642 | 50.3 | 3.2 | 5.0 | 1.3 | 7.7 | 63.9 | 2,933 | 46 | 73 | 1,243 | 8,400 | 57,643 | 5,825 |
| Blaine | 42 | 4,415 | 47.2 | 15.5 | 5.2 | 0.8 | 10.1 | 13.9 | 1,466 | 46 | 34 | 703 | 3,810 | 79,555 | 10,473 |
| Bryan | 109 | 2,336 | 59.5 | 3.5 | 7.4 | 0.0 | 5.7 | 62.6 | 1,347 | 90 | 171 | 8,558 | 18,240 | 48,950 | 4,200 |
| Caddo | 98 | 3,339 | 54.9 | 1.8 | 5.9 | 0.0 | 10.9 | 27.1 | 922 | 517 | 98 | 2,271 | 10,480 | 48,328 | 3,941 |
| Canadian | 314 | 2,242 | 68.3 | 0.6 | 6.5 | 0.0 | 3.4 | 298.6 | 2,133 | 586 | 567 | 6,079 | 64,860 | 71,103 | 7,160 |
| Carter | 152 | 3,140 | 52.4 | 0.8 | 7.7 | 0.0 | 7.9 | 70.8 | 1,467 | 101 | 173 | 3,318 | 20,970 | 57,616 | 6,040 |
| Cherokee | 188 | 3,850 | 36.3 | 51.6 | 2.2 | 0.0 | 1.8 | 95.9 | 1,962 | 187 | 171 | 7,798 | 17,780 | 46,582 | 3,644 |
| Choctaw | 35 | 2,361 | 64.4 | 4.5 | 3.2 | 0.0 | 8.3 | 20.1 | 1,356 | 43 | 53 | 1,282 | 5,640 | 39,884 | 2,828 |
| Cimarron | 9 | 4,354 | 59.0 | 0.9 | 4.4 | 0.2 | 17.9 | 1.6 | 751 | 14 | 8 | 230 | 1,060 | 38,501 | 3,004 |
| Cleveland | 1,043 | 3,731 | 35.2 | 34.9 | 4.5 | 0.0 | 4.0 | 980.7 | 3,508 | 774 | 1,032 | 23,265 | 123,960 | 67,062 | 7,298 |
| Coal | 14 | 2,527 | 84.0 | 3.8 | 0.2 | 0.0 | 0.9 | 0.8 | 134 | 15 | 20 | 362 | 2,100 | 44,342 | 3,270 |
| Comanche | 594 | 4,880 | 30.0 | 43.1 | 3.7 | 0.0 | 4.8 | 298.4 | 2,450 | 3,955 | 12,364 | 9,881 | 47,030 | 49,061 | 3,973 |
| Cotton | 17 | 2,920 | 52.1 | 0.7 | 3.2 | 0.0 | 21.4 | 9.5 | 1,623 | 25 | 20 | 1,040 | 2,360 | 42,914 | 3,064 |
| Craig | 32 | 2,212 | 79.3 | 0.3 | 5.1 | 0.3 | 1.8 | 11.1 | 771 | 51 | 48 | 1,499 | 5,470 | 43,370 | 3,169 |
| Creek | 167 | 2,324 | 59.6 | 3.5 | 6.0 | 0.0 | 5.8 | 165.9 | 2,308 | 359 | 257 | 3,054 | 29,320 | 58,487 | 5,545 |
| Custer | 91 | 3,143 | 54.6 | 0.0 | 6.2 | 0.3 | 8.1 | 86.3 | 2,977 | 204 | 100 | 2,553 | 11,580 | 59,041 | 6,263 |
| Delaware | 87 | 2,026 | 72.1 | 0.5 | 6.7 | 0.0 | 2.5 | 53.2 | 1,247 | 76 | 155 | 2,650 | 16,040 | 48,139 | 4,148 |
| Dewey | 30 | 6,023 | 49.2 | 1.3 | 4.0 | 0.0 | 20.0 | 11.9 | 2,425 | 29 | 17 | 403 | 2,090 | 59,875 | 6,149 |
| Ellis | 42 | 10,526 | 29.1 | 3.5 | 3.2 | 0.0 | 31.5 | 7.1 | 1,794 | 21 | 14 | 352 | 1,650 | 58,396 | 5,592 |
| Garfield | 171 | 2,772 | 59.3 | 3.2 | 6.6 | 0.1 | 1.8 | 126.3 | 2,051 | 516 | 1,503 | 3,276 | 26,610 | 59,793 | 5,993 |
| Garvin | 107 | 3,825 | 43.2 | 23.8 | 4.4 | 0.0 | 9.0 | 36.6 | 1,315 | 76 | 100 | 1,321 | 10,900 | 53,121 | 4,759 |
| Grady | 132 | 2,410 | 61.2 | 2.8 | 6.0 | 0.0 | 9.7 | 83.3 | 1,520 | 90 | 199 | 2,650 | 21,970 | 66,507 | 6,655 |
| Grant | 20 | 4,656 | 71.7 | 1.2 | 4.7 | 0.0 | 2.5 | 7.0 | 1,601 | 28 | 16 | 317 | 1,870 | 53,310 | 4,919 |
| Greer | 20 | 3,382 | 42.8 | 5.1 | 7.3 | 0.0 | 20.5 | 8.7 | 1,494 | 25 | 17 | 522 | 1,820 | 40,660 | 2,936 |
| Harmon | 10 | 3,744 | 45.7 | 0.6 | 7.4 | 0.0 | 18.7 | 1.2 | 458 | 20 | 9 | 280 | 920 | 34,870 | 2,924 |
| Harper | 12 | 3,094 | 63.4 | 2.3 | 4.5 | 0.0 | 9.5 | 5.7 | 1,518 | 23 | 13 | 384 | 1,510 | 47,529 | 3,854 |
| Haskell | 32 | 2,492 | 64.3 | 0.0 | 4.9 | 0.2 | 3.9 | 17.4 | 1,365 | 54 | 46 | 564 | 4,460 | 39,232 | 2,851 |
| Hughes | 33 | 2,499 | 63.1 | 0.0 | 5.8 | 0.0 | 10.8 | 31.5 | 2,377 | 30 | 43 | 941 | 4,610 | 42,573 | 3,297 |
| Jackson | 141 | 5,622 | 26.1 | 51.0 | 5.2 | 0.0 | 2.3 | 8.0 | 317 | 1,459 | 1,437 | 2,268 | 10,340 | 51,816 | 4,327 |
| Jefferson | 20 | 3,315 | 56.8 | 1.9 | 4.5 | 0.0 | 15.6 | 62.7 | 10,169 | 28 | 21 | 439 | 2,180 | 40,268 | 3,203 |
| Johnston | 26 | 2,354 | 64.6 | 5.6 | 6.0 | 0.0 | 3.6 | 10.4 | 934 | 44 | 39 | 724 | 4,130 | 45,351 | 3,201 |
| Kay | 154 | 3,470 | 44.3 | 2.1 | 8.0 | 0.0 | 9.7 | 96.4 | 2,172 | 109 | 154 | 4,327 | 18,160 | 53,727 | 4,964 |
| Kingfisher | 70 | 4,447 | 58.6 | 0.6 | 5.8 | 0.0 | 16.2 | 26.8 | 1,708 | 44 | 57 | 938 | 6,850 | 89,310 | 12,192 |

1. Based on the resident population estimated as of July 1 of the year shown.

# Table B. States and Counties — Land Area and Population

| State / county code | CBSA code[1] | County Type code[2] | STATE County | Land area[3] (sq. mi) | Total persons 2019 | Rank | Per square mile | White | Black | American Indian, Alaska Native | Asian and Pacific Islander | Percent Hispanic or Latino[4] | Under 5 years | 5 to 14 years | 15 to 24 years | 25 to 34 years | 35 to 44 years | 45 to 54 years |
|---|---|---|---|---|---|---|---|---|---|---|---|---|---|---|---|---|---|---|
| | | | | 1 | 2 | 3 | 4 | 5 | 6 | 7 | 8 | 9 | 10 | 11 | 12 | 13 | 14 | 15 |
| | | | **OKLAHOMA—Cont'd** | | | | | | | | | | | | | | | |
| 40075 | | 6 | Kiowa ............. | 1,015.1 | 8,741 | 2,517 | 8.6 | 77.0 | 5.4 | 9.0 | 0.9 | 12.0 | 6.8 | 13.2 | 11.9 | 11.1 | 10.7 | 10.7 |
| 40077 | | 7 | Latimer ............. | 722.1 | 10,132 | 2,407 | 14.0 | 70.2 | 2.0 | 30.4 | 1.2 | 4.8 | 6.3 | 12.2 | 13.3 | 11.0 | 10.9 | 10.9 |
| 40079 | | 2 | Le Flore ............. | 1,589.4 | 49,935 | 992 | 31.4 | 75.9 | 2.6 | 18.7 | 1.2 | 7.4 | 6.1 | 13.8 | 12.6 | 12.3 | 11.9 | 11.8 |
| 40081 | 36420 | 1 | Lincoln ............. | 952.4 | 35,045 | 1,302 | 36.8 | 87.2 | 2.8 | 11.2 | 1.0 | 3.5 | 5.5 | 13.4 | 11.7 | 11.6 | 11.9 | 12.0 |
| 40083 | 36420 | 1 | Logan ............. | 743.7 | 48,777 | 1,013 | 65.6 | 81.0 | 9.1 | 6.2 | 1.2 | 7.0 | 5.5 | 12.8 | 14.3 | 11.1 | 13.1 | 12.2 |
| 40085 | 11620 | 9 | Love ............. | 513.9 | 10,230 | 2,403 | 19.9 | 73.7 | 3.1 | 9.8 | 0.9 | 17.0 | 6.5 | 13.8 | 12.3 | 12.6 | 11.8 | 11.9 |
| 40087 | 36420 | 1 | McClain ............. | 570.7 | 41,348 | 1,149 | 72.5 | 83.2 | 1.7 | 10.9 | 1.0 | 8.6 | 5.8 | 14.9 | 12.3 | 11.6 | 14.0 | 12.4 |
| 40089 | | 7 | McCurtain ............. | 1,850.8 | 32,772 | 1,362 | 17.7 | 66.6 | 9.6 | 21.6 | 2.9 | 6.4 | 6.8 | 14.5 | 12.2 | 12.3 | 11.3 | 11.6 |
| 40091 | | 6 | McIntosh ............. | 618.3 | 19,635 | 1,844 | 31.8 | 74.5 | 4.4 | 25.1 | 1.1 | 3.0 | 4.9 | 11.2 | 10.4 | 9.9 | 10.5 | 11.2 |
| 40093 | | 9 | Major ............. | 955.0 | 7,579 | 2,621 | 7.9 | 85.9 | 1.4 | 3.7 | 1.3 | 10.3 | 6.5 | 14.9 | 11.6 | 10.1 | 12.0 | 10.4 |
| 40095 | | 6 | Marshall ............. | 371.5 | 17,114 | 1,968 | 46.1 | 69.6 | 2.6 | 14.5 | 0.8 | 18.6 | 6.0 | 13.0 | 11.5 | 10.5 | 10.8 | 11.5 |
| 40097 | | 6 | Mayes ............. | 655.4 | 41,152 | 1,153 | 62.8 | 72.8 | 1.2 | 30.8 | 0.9 | 3.8 | 5.8 | 13.3 | 11.8 | 12.2 | 11.9 | 12.2 |
| 40099 | | 7 | Murray ............. | 416.4 | 13,955 | 2,160 | 33.5 | 76.7 | 2.4 | 19.9 | 1.0 | 7.3 | 5.9 | 13.4 | 11.8 | 11.0 | 12.0 | 11.9 |
| 40101 | 34780 | 4 | Muskogee ............. | 810.6 | 67,610 | 794 | 83.4 | 62.0 | 12.5 | 26.0 | 1.1 | 6.9 | 6.3 | 13.8 | 13.1 | 12.8 | 12.5 | 11.3 |
| 40103 | | 6 | Noble ............. | 731.7 | 11,113 | 2,337 | 15.2 | 84.2 | 3.0 | 12.1 | 1.0 | 4.4 | 5.7 | 13.3 | 11.4 | 10.9 | 11.9 | 12.3 |
| 40105 | | 6 | Nowata ............. | 565.8 | 10,076 | 2,414 | 17.8 | 74.3 | 3.7 | 27.5 | 0.7 | 3.7 | 6.4 | 12.3 | 11.1 | 12.0 | 11.2 | 12.0 |
| 40107 | | 6 | Okfuskee ............. | 618.5 | 11,765 | 2,308 | 19.0 | 66.3 | 8.2 | 26.6 | 1.1 | 4.5 | 5.9 | 12.8 | 11.7 | 12.6 | 12.5 | 12.9 |
| 40109 | 36420 | 1 | Oklahoma ............. | 708.9 | 804,041 | 79 | 1,134.2 | 59.2 | 17.4 | 5.9 | 4.7 | 18.2 | 7.1 | 14.2 | 13.1 | 15.3 | 13.3 | 11.1 |
| 40111 | 46140 | 2 | Okmulgee ............. | 697.4 | 38,234 | 1,219 | 54.8 | 69.3 | 10.2 | 24.0 | 1.1 | 4.5 | 5.6 | 13.5 | 13.4 | 12.3 | 11.4 | 11.2 |
| 40113 | 46140 | 2 | Osage ............. | 2,246.6 | 46,642 | 1,042 | 20.8 | 69.3 | 12.5 | 20.8 | 0.8 | 4.1 | 5.0 | 12.4 | 11.3 | 11.5 | 11.8 | 11.9 |
| 40115 | 33060 | 6 | Ottawa ............. | 470.8 | 30,879 | 1,407 | 65.6 | 71.3 | 1.8 | 26.7 | 2.0 | 6.1 | 6.5 | 14.4 | 13.7 | 12.1 | 10.9 | 11.3 |
| 40117 | 46140 | 2 | Pawnee ............. | 568.4 | 16,381 | 2,013 | 28.8 | 81.6 | 1.8 | 18.5 | 0.9 | 3.6 | 5.8 | 13.2 | 11.4 | 11.4 | 11.7 | 12.6 |
| 40119 | 44660 | 4 | Payne ............. | 684.9 | 81,755 | 699 | 119.4 | 81.8 | 5.0 | 8.5 | 5.4 | 4.9 | 4.9 | 10.6 | 29.9 | 13.2 | 10.5 | 8.4 |
| 40121 | 32540 | 5 | Pittsburg ............. | 1,305.5 | 43,679 | 1,103 | 33.5 | 75.5 | 4.3 | 21.7 | 0.9 | 5.5 | 6.0 | 12.8 | 11.2 | 13.2 | 12.2 | 11.3 |
| 40123 | 10220 | 7 | Pontotoc ............. | 720.4 | 38,397 | 1,213 | 53.3 | 71.0 | 3.9 | 25.8 | 1.5 | 5.9 | 6.6 | 13.7 | 14.6 | 13.9 | 11.9 | 10.6 |
| 40125 | 43060 | 4 | Pottawatomie ............. | 787.8 | 72,998 | 749 | 92.7 | 77.5 | 4.4 | 17.8 | 1.5 | 5.6 | 5.9 | 13.3 | 13.5 | 12.7 | 12.7 | 12.0 |
| 40127 | | 9 | Pushmataha ............. | 1,395.5 | 10,970 | 2,351 | 7.9 | 75.5 | 1.9 | 23.6 | 0.8 | 4.2 | 5.5 | 12.5 | 10.5 | 11.1 | 11.0 | 11.1 |
| 40129 | | 9 | Roger Mills ............. | 1,141.1 | 3,570 | 2,928 | 3.1 | 85.0 | 1.7 | 7.0 | 1.0 | 8.1 | 5.7 | 14.2 | 11.5 | 9.3 | 12.0 | 10.6 |
| 40131 | 46140 | 2 | Rogers ............. | 675.7 | 93,155 | 641 | 137.9 | 78.7 | 1.8 | 20.5 | 2.2 | 5.2 | 5.8 | 13.2 | 12.5 | 12.6 | 12.2 | 12.7 |
| 40133 | | 7 | Seminole ............. | 632.8 | 24,248 | 1,638 | 38.3 | 70.2 | 6.1 | 24.8 | 1.2 | 5.7 | 6.6 | 13.5 | 13.4 | 11.3 | 11.6 | 11.4 |
| 40135 | 22900 | 2 | Sequoyah ............. | 673.6 | 41,538 | 1,146 | 61.7 | 70.6 | 2.6 | 30.6 | 1.3 | 4.7 | 6.5 | 13.0 | 12.1 | 11.2 | 11.5 | 12.5 |
| 40137 | 20340 | 7 | Stephens ............. | 870.2 | 43,100 | 1,120 | 49.5 | 83.5 | 2.8 | 8.9 | 1.1 | 8.3 | 6.0 | 12.6 | 11.6 | 11.6 | 12.0 | 11.1 |
| 40139 | 25100 | 7 | Texas ............. | 2,041.3 | 19,997 | 1,824 | 9.8 | 43.9 | 5.0 | 1.5 | 3.6 | 47.3 | 8.5 | 15.5 | 15.5 | 13.6 | 12.5 | 11.2 |
| 40141 | | 6 | Tillman ............. | 871.1 | 7,229 | 2,640 | 8.3 | 60.5 | 7.9 | 5.5 | 0.9 | 28.5 | 5.5 | 13.5 | 12.2 | 11.0 | 11.5 | 11.7 |
| 40143 | 46140 | 2 | Tulsa ............. | 570.2 | 657,589 | 102 | 1,153.3 | 65.7 | 12.2 | 10.2 | 4.6 | 13.6 | 6.9 | 14.0 | 12.9 | 14.6 | 13.1 | 11.5 |
| 40145 | 46140 | 2 | Wagoner ............. | 561.9 | 82,925 | 687 | 147.6 | 77.1 | 4.9 | 16.1 | 2.6 | 7.0 | 5.8 | 13.7 | 11.6 | 13.1 | 13.3 | 12.3 |
| 40147 | 12780 | 4 | Washington ............. | 415.5 | 52,222 | 967 | 125.7 | 77.8 | 3.8 | 15.9 | 2.9 | 6.7 | 6.2 | 13.6 | 11.8 | 12.4 | 12.1 | 11.0 |
| 40149 | | 7 | Washita ............. | 1,003.1 | 10,830 | 2,360 | 10.8 | 85.0 | 1.9 | 4.9 | 0.7 | 10.3 | 5.9 | 14.4 | 11.4 | 11.8 | 12.1 | 10.9 |
| 40151 | | 7 | Woods ............. | 1,286.5 | 8,687 | 2,521 | 6.8 | 85.3 | 4.6 | 4.5 | 1.7 | 7.2 | 5.6 | 12.2 | 19.2 | 14.0 | 12.1 | 8.7 |
| 40153 | 49260 | 7 | Woodward ............. | 1,242.8 | 19,812 | 1,834 | 15.9 | 80.4 | 2.4 | 4.3 | 1.1 | 14.0 | 6.6 | 13.7 | 12.2 | 13.9 | 13.7 | 11.5 |
| 41000 | | 0 | **OREGON** ............. | 95,988.0 | 4,241,507 | X | 44.2 | 77.9 | 2.9 | 2.4 | 6.9 | 13.7 | 5.2 | 11.6 | 12.0 | 14.3 | 13.6 | 12.1 |
| 41001 | | 7 | Baker ............. | 3,068.0 | 16,284 | 2,017 | 5.3 | 92.0 | 1.4 | 2.6 | 1.7 | 5.0 | 5.0 | 11.6 | 8.9 | 10.7 | 11.0 | 10.4 |
| 41003 | 18700 | 3 | Benton ............. | 675.2 | 93,239 | 640 | 138.1 | 83.1 | 1.9 | 1.6 | 9.3 | 8.0 | 3.9 | 9.0 | 25.5 | 13.1 | 10.4 | 9.7 |
| 41005 | 38900 | 1 | Clackamas ............. | 1,870.7 | 421,596 | 169 | 225.4 | 83.6 | 1.9 | 1.6 | 7.1 | 9.3 | 5.1 | 12.2 | 11.1 | 12.3 | 13.7 | 12.9 |
| 41007 | 11820 | 4 | Clatsop ............. | 828.1 | 40,423 | 1,169 | 48.8 | 87.3 | 1.4 | 2.2 | 3.0 | 9.0 | 4.8 | 10.6 | 10.5 | 11.7 | 12.6 | 10.9 |
| 41009 | 38900 | 1 | Columbia ............. | 658.7 | 52,876 | 957 | 80.3 | 90.8 | 1.3 | 3.0 | 2.7 | 5.9 | 5.0 | 11.7 | 10.6 | 11.9 | 12.7 | 13.0 |
| 41011 | 18300 | 2 | Coos ............. | 1,596.1 | 64,711 | 825 | 40.5 | 88.5 | 1.1 | 5.1 | 2.5 | 7.0 | 4.7 | 10.3 | 9.3 | 11.1 | 10.9 | 10.7 |
| 41013 | 39260 | 6 | Crook ............. | 2,978.9 | 25,105 | 1,603 | 8.4 | 89.8 | 0.8 | 2.6 | 1.2 | 7.8 | 5.3 | 11.2 | 9.3 | 10.7 | 11.2 | 11.3 |
| 41015 | 15060 | 7 | Curry ............. | 1,628.4 | 23,305 | 1,667 | 14.3 | 88.6 | 1.1 | 4.5 | 1.8 | 7.7 | 3.7 | 8.1 | 6.8 | 8.9 | 8.7 | 10.1 |
| 41017 | 13460 | 3 | Deschutes ............. | 3,017.6 | 201,769 | 338 | 66.9 | 89.0 | 1.0 | 1.8 | 2.6 | 8.4 | 4.9 | 11.2 | 10.1 | 13.3 | 13.6 | 12.5 |
| 41019 | 40700 | 4 | Douglas ............. | 5,035.7 | 111,360 | 550 | 22.1 | 90.3 | 1.0 | 3.7 | 2.3 | 6.2 | 4.9 | 10.9 | 9.7 | 11.3 | 10.9 | 10.8 |
| 41021 | | 9 | Gilliam ............. | 1,204.7 | 1,975 | 3,049 | 1.6 | 89.2 | 1.3 | 2.4 | 1.9 | 7.5 | 4.4 | 12.6 | 8.6 | 8.8 | 11.0 | 9.5 |
| 41023 | | 9 | Grant ............. | 4,527.8 | 7,180 | 2,644 | 1.6 | 93.0 | 1.0 | 3.1 | 1.4 | 4.4 | 4.7 | 10.3 | 8.3 | 9.8 | 10.4 | 9.7 |
| 41025 | | 7 | Harney ............. | 10,134.4 | 7,373 | 2,627 | 0.7 | 88.9 | 1.6 | 5.8 | 1.5 | 5.5 | 5.2 | 11.5 | 9.9 | 11.3 | 12.0 | 10.5 |
| 41027 | 26220 | 6 | Hood River ............. | 522.1 | 23,280 | 1,670 | 44.6 | 65.2 | 1.0 | 1.6 | 2.9 | 31.6 | 5.9 | 13.0 | 12.4 | 12.4 | 13.5 | 12.7 |
| 41029 | 32780 | 3 | Jackson ............. | 2,783.3 | 221,844 | 308 | 79.7 | 82.6 | 1.5 | 2.4 | 3.0 | 13.8 | 5.2 | 11.8 | 10.5 | 12.7 | 12.3 | 11.1 |
| 41031 | | 6 | Jefferson ............. | 1,781.7 | 24,856 | 1,607 | 14.0 | 63.2 | 1.2 | 16.5 | 1.7 | 20.4 | 6.2 | 13.3 | 11.2 | 13.0 | 11.3 | 11.4 |
| 41033 | 24420 | 2 | Josephine ............. | 1,638.6 | 88,053 | 664 | 53.7 | 88.9 | 1.1 | 3.0 | 2.2 | 8.0 | 4.8 | 11.1 | 9.7 | 11.1 | 10.8 | 10.8 |
| 41035 | 28900 | 5 | Klamath ............. | 5,950.0 | 68,739 | 784 | 11.6 | 80.2 | 1.5 | 5.8 | 2.2 | 14.1 | 5.8 | 12.3 | 11.3 | 12.9 | 11.0 | 10.9 |
| 41037 | | 7 | Lake ............. | 8,138.6 | 7,949 | 2,595 | 1.0 | 86.6 | 1.2 | 4.3 | 2.0 | 9.5 | 4.7 | 11.3 | 9.0 | 10.5 | 11.5 | 11.9 |
| 41039 | 21660 | 2 | Lane ............. | 4,554.1 | 382,986 | 187 | 84.1 | 85.0 | 2.1 | 2.8 | 5.1 | 9.5 | 4.6 | 10.2 | 15.4 | 13.3 | 12.4 | 11.0 |
| 41041 | 35440 | 5 | Lincoln ............. | 981.0 | 50,583 | 982 | 51.6 | 84.9 | 1.2 | 5.2 | 2.5 | 9.6 | 3.9 | 9.3 | 8.7 | 9.2 | 10.7 | 10.7 |
| 41043 | 10540 | 3 | Linn ............. | 2,289.3 | 131,054 | 489 | 57.2 | 86.8 | 1.2 | 2.9 | 2.5 | 9.8 | 5.9 | 12.4 | 11.4 | 13.8 | 12.3 | 11.5 |
| 41045 | 36620 | 6 | Malheur ............. | 9,887.7 | 30,983 | 1,405 | 3.1 | 61.5 | 1.7 | 1.8 | 2.1 | 34.8 | 6.7 | 14.9 | 13.8 | 13.4 | 12.1 | 11.0 |
| 41047 | 41420 | 2 | Marion ............. | 1,181.1 | 349,204 | 206 | 295.7 | 66.8 | 1.9 | 2.1 | 4.6 | 27.6 | 6.3 | 13.7 | 13.0 | 14.3 | 13.0 | 11.5 |

1. CBSA = Core Based Statistical Area. See Appendix A for explanation. See Appendix B for list of metropolitan areas with component counties. Service of USDA Rural-Urban Continuum Codes. See Appendix A for definition.  3. Dry land or land partially or temporarily covered by water.
2. County type code from the Economic Research
4. May be of any race.

| STATE County | Population, 2020 (cont.) Age (percent) (cont.) 55 to 64 years | 65 to 74 years | 75 years and over | Percent female | Population change, 2000-2020 Total persons 2000 | 2010 | Percent change 2000-2010 | 2010-2020 | Components of change, 2010-2020 Births | Deaths | Net Migration | Households, 2015-2019 Number | Persons per household | Family house-holds | Female family house-holder[1] | One person |
|---|---|---|---|---|---|---|---|---|---|---|---|---|---|---|---|---|
| | 16 | 17 | 18 | 19 | 20 | 21 | 22 | 23 | 24 | 25 | 26 | 27 | 28 | 29 | 30 | 31 |
| **OKLAHOMA—Cont'd** | | | | | | | | | | | | | | | | |
| Kiowa | 15.3 | 11.7 | 8.6 | 50.3 | 10,227 | 9,446 | -7.6 | -7.5 | 1,175 | 1,374 | -506 | 3,571 | 2.43 | 64.2 | 12.2 | 31.3 |
| Latimer | 13.5 | 10.9 | 11.1 | 49.4 | 10,692 | 11,156 | 4.3 | -9.2 | 1,279 | 1,387 | -925 | 4,088 | 2.44 | 70.0 | 13.6 | 26.1 |
| Le Flore | 13.0 | 10.9 | 7.6 | 49.9 | 48,109 | 50,382 | 4.7 | -0.9 | 6,547 | 6,554 | -419 | 18,261 | 2.67 | 69.6 | 13.4 | 25.7 |
| Lincoln | 14.5 | 11.2 | 8.2 | 50.1 | 32,080 | 34,274 | 6.8 | 2.2 | 3,950 | 3,946 | 786 | 12,763 | 2.7 | 69.3 | 8.5 | 27.4 |
| Logan | 13.9 | 10.4 | 6.7 | 50.4 | 33,924 | 41,843 | 23.3 | 16.6 | 5,128 | 3,783 | 5,562 | 15,512 | 2.87 | 73.0 | 9.8 | 22.8 |
| Love | 12.3 | 10.7 | 8.1 | 49.8 | 8,831 | 9,416 | 6.6 | 8.6 | 1,271 | 1,218 | 761 | 3,268 | 3.05 | 69.4 | 9.4 | 26.3 |
| McClain | 13.1 | 9.5 | 6.4 | 50.6 | 27,740 | 34,507 | 24.4 | 19.8 | 4,342 | 3,676 | 6,141 | 14,428 | 2.7 | 76.3 | 9.6 | 20.5 |
| McCurtain | 12.6 | 10.5 | 8.0 | 50.7 | 34,402 | 33,152 | -3.6 | -1.1 | 4,757 | 4,319 | -802 | 12,646 | 2.58 | 69.5 | 14.0 | 27.2 |
| McIntosh | 15.7 | 14.5 | 11.7 | 50.6 | 19,456 | 20,245 | 4.1 | -3.0 | 2,072 | 3,263 | 603 | 8,327 | 2.33 | 66.5 | 11.5 | 29.9 |
| Major | 13.8 | 11.1 | 9.5 | 50.7 | 7,545 | 7,526 | -0.3 | 0.7 | 1,090 | 941 | -93 | 3,123 | 2.43 | 70.8 | 8.5 | 25.2 |
| Marshall | 14.2 | 12.3 | 10.2 | 50.5 | 13,184 | 15,838 | 20.1 | 8.1 | 1,975 | 2,018 | 1,325 | 6,195 | 2.63 | 70.5 | 12.0 | 25.0 |
| Mayes | 13.7 | 11.3 | 7.9 | 50.0 | 38,369 | 41,268 | 7.6 | -0.3 | 4,989 | 5,137 | 59 | 15,983 | 2.53 | 71.4 | 11.7 | 24.1 |
| Murray | 13.9 | 11.2 | 8.9 | 50.0 | 12,623 | 13,488 | 6.9 | 3.5 | 1,611 | 1,951 | 808 | 5,238 | 2.59 | 65.9 | 9.7 | 28.7 |
| Muskogee | 12.8 | 10.1 | 7.3 | 51.3 | 69,451 | 70,991 | 2.2 | -4.8 | 9,190 | 9,485 | -3,074 | 26,234 | 2.47 | 68.0 | 14.7 | 28.7 |
| Noble | 13.7 | 11.5 | 9.2 | 51.0 | 11,411 | 11,566 | 1.4 | -3.9 | 1,319 | 1,398 | -368 | 4,515 | 2.45 | 71.8 | 9.8 | 25.0 |
| Nowata | 15.0 | 10.9 | 9.1 | 50.2 | 10,569 | 10,536 | -0.3 | -4.4 | 1,259 | 1,450 | -259 | 4,099 | 2.47 | 69.1 | 12.0 | 28.8 |
| Okfuskee | 13.1 | 10.2 | 8.2 | 46.1 | 11,814 | 12,191 | 3.2 | -3.5 | 1,600 | 1,778 | -239 | 3,976 | 2.69 | 69.0 | 13.0 | 27.8 |
| Oklahoma | 11.6 | 8.7 | 5.7 | 51.1 | 660,448 | 718,385 | 8.8 | 11.9 | 122,019 | 72,066 | 35,599 | 301,570 | 2.56 | 61.8 | 13.6 | 31.2 |
| Okmulgee | 13.4 | 10.7 | 8.3 | 50.6 | 39,685 | 40,061 | 0.9 | -4.6 | 4,887 | 5,388 | -1,314 | 14,724 | 2.53 | 66.0 | 16.0 | 28.5 |
| Osage | 14.9 | 12.2 | 9.0 | 49.7 | 44,437 | 47,461 | 6.8 | -1.7 | 4,523 | 4,895 | -427 | 18,263 | 2.51 | 69.4 | 11.7 | 27.3 |
| Ottawa | 12.4 | 10.6 | 8.1 | 50.9 | 33,194 | 31,848 | -4.1 | -3.0 | 4,453 | 4,660 | -756 | 11,847 | 2.56 | 67.2 | 12.9 | 28.1 |
| Pawnee | 13.9 | 11.7 | 8.4 | 50.2 | 16,612 | 16,569 | -0.3 | -1.1 | 1,987 | 2,142 | -32 | 6,179 | 2.62 | 72.2 | 11.6 | 24.5 |
| Payne | 9.2 | 7.5 | 5.8 | 49.1 | 68,190 | 77,347 | 13.4 | 5.7 | 9,032 | 5,900 | 1,295 | 30,788 | 2.39 | 54.4 | 8.2 | 31.4 |
| Pittsburg | 13.0 | 11.3 | 9.2 | 49.1 | 43,953 | 45,827 | 4.3 | -4.7 | 5,390 | 6,292 | -1,219 | 17,816 | 2.35 | 66.1 | 11.1 | 29.0 |
| Pontotoc | 11.9 | 9.4 | 7.5 | 51.4 | 35,143 | 37,488 | 6.7 | 2.4 | 5,427 | 4,600 | 105 | 14,278 | 2.58 | 65.4 | 12.1 | 28.1 |
| Pottawatomie | 12.9 | 10.0 | 7.1 | 52.0 | 65,521 | 69,447 | 6.0 | 5.1 | 9,231 | 8,495 | 2,840 | 25,917 | 2.67 | 69.1 | 13.2 | 26.7 |
| Pushmataha | 14.7 | 12.8 | 10.8 | 51.1 | 11,667 | 11,577 | -0.8 | -5.2 | 1,276 | 1,676 | -205 | 4,477 | 2.46 | 66.2 | 10.0 | 30.7 |
| Roger Mills | 14.1 | 12.2 | 10.3 | 50.2 | 3,436 | 3,647 | 6.1 | -2.1 | 477 | 360 | -195 | 1,377 | 2.65 | 64.9 | 12.4 | 29.6 |
| Rogers | 14.0 | 9.9 | 7.2 | 50.2 | 70,641 | 86,910 | 23.0 | 7.2 | 10,300 | 8,814 | 4,838 | 34,555 | 2.59 | 74.1 | 10.0 | 21.7 |
| Seminole | 13.5 | 10.7 | 8.1 | 51.0 | 24,894 | 25,483 | 2.4 | -4.8 | 3,354 | 3,524 | -1,048 | 9,317 | 2.6 | 68.3 | 15.2 | 27.5 |
| Sequoyah | 13.6 | 11.0 | 8.0 | 50.7 | 38,972 | 42,430 | 8.9 | -2.1 | 5,432 | 5,490 | -821 | 15,474 | 2.67 | 71.6 | 15.1 | 25.0 |
| Stephens | 14.1 | 11.9 | 9.0 | 51.5 | 43,182 | 45,051 | 4.3 | -4.3 | 5,446 | 5,807 | -1,574 | 16,720 | 2.58 | 67.4 | 11.1 | 28.3 |
| Texas | 11.1 | 7.4 | 4.7 | 47.0 | 20,107 | 20,640 | 2.7 | -3.1 | 3,765 | 1,673 | -2,775 | 6,768 | 3.0 | 66.8 | 9.2 | 27.8 |
| Tillman | 14.3 | 11.0 | 9.3 | 49.2 | 9,287 | 7,991 | -14.0 | -9.5 | 923 | 973 | -718 | 2,807 | 2.51 | 66.9 | 12.0 | 25.7 |
| Tulsa | 11.8 | 9.0 | 6.2 | 51.2 | 563,299 | 603,451 | 7.1 | 9.0 | 94,763 | 59,943 | 19,551 | 252,661 | 2.53 | 63.1 | 13.0 | 30.4 |
| Wagoner | 12.9 | 10.4 | 6.9 | 50.5 | 57,491 | 73,084 | 27.1 | 13.5 | 9,025 | 6,685 | 7,520 | 29,208 | 2.7 | 74.3 | 11.5 | 21.4 |
| Washington | 12.9 | 10.6 | 9.3 | 51.4 | 48,996 | 50,979 | 4.0 | 2.4 | 6,402 | 6,440 | 1,326 | 20,455 | 2.5 | 67.2 | 10.9 | 28.5 |
| Washita | 14.3 | 10.7 | 8.6 | 50.6 | 11,508 | 11,558 | 0.4 | -6.3 | 1,492 | 1,477 | -745 | 4,224 | 2.59 | 69.6 | 11.1 | 26.5 |
| Woods | 11.1 | 9.2 | 8.7 | 46.0 | 9,089 | 8,878 | -2.3 | -2.2 | 1,073 | 1,012 | -268 | 3,281 | 2.49 | 58.8 | 8.8 | 31.9 |
| Woodward | 12.1 | 8.9 | 7.4 | 47.4 | 18,486 | 20,084 | 8.6 | -1.4 | 2,974 | 2,065 | -1,213 | 7,495 | 2.62 | 71.0 | 10.4 | 25.8 |
| **OREGON** | 12.5 | 11.3 | 7.3 | 50.4 | 3,421,399 | 3,831,083 | 12.0 | 10.7 | 455,763 | 360,602 | 314,917 | 1,611,982 | 2.51 | 63.1 | 10.0 | 27.5 |
| Baker | 15.2 | 15.6 | 11.5 | 49.1 | 16,741 | 16,138 | -3.6 | 0.9 | 1,678 | 2,135 | 623 | 6,921 | 2.23 | 61.1 | 7.3 | 33.0 |
| Benton | 10.9 | 10.7 | 6.7 | 50.0 | 78,153 | 85,581 | 9.5 | 8.9 | 7,378 | 5,893 | 6,147 | 35,718 | 2.39 | 57.5 | 7.1 | 26.8 |
| Clackamas | 13.5 | 11.6 | 7.5 | 50.6 | 338,391 | 375,991 | 11.1 | 12.1 | 41,061 | 34,188 | 38,718 | 157,408 | 2.59 | 69.0 | 8.6 | 23.9 |
| Clatsop | 14.8 | 15.6 | 8.6 | 50.6 | 35,630 | 37,034 | 3.9 | 9.2 | 4,161 | 4,063 | 3,287 | 15,800 | 2.43 | 60.9 | 8.3 | 30.8 |
| Columbia | 14.9 | 12.4 | 7.7 | 50.0 | 43,560 | 49,350 | 13.3 | 7.1 | 5,058 | 4,565 | 3,031 | 19,671 | 2.59 | 67.0 | 9.7 | 26.1 |
| Coos | 15.5 | 16.3 | 11.1 | 50.8 | 62,779 | 63,055 | 0.4 | 2.6 | 6,231 | 9,259 | 4,716 | 27,025 | 2.32 | 60.9 | 9.5 | 32.0 |
| Crook | 15.3 | 15.9 | 9.8 | 50.5 | 19,182 | 20,978 | 9.4 | 19.7 | 2,188 | 2,556 | 4,474 | 9,588 | 2.39 | 67.4 | 9.9 | 26.4 |
| Curry | 18.0 | 21.3 | 14.4 | 50.8 | 21,137 | 22,364 | 5.8 | 4.2 | 1,810 | 3,828 | 2,954 | 10,546 | 2.12 | 59.7 | 7.8 | 33.0 |
| Deschutes | 13.6 | 13.1 | 7.7 | 50.4 | 115,367 | 157,728 | 36.7 | 27.9 | 18,125 | 14,644 | 40,225 | 74,397 | 2.49 | 67.2 | 9.1 | 24.1 |
| Douglas | 14.8 | 15.2 | 11.5 | 50.7 | 100,399 | 107,684 | 7.3 | 3.4 | 11,047 | 14,929 | 7,662 | 45,456 | 2.36 | 66.0 | 11.6 | 26.9 |
| Gilliam | 17.1 | 16.2 | 11.9 | 49.6 | 1,915 | 1,871 | -2.3 | 5.6 | 176 | 202 | 130 | 842 | 2.2 | 58.6 | 12.1 | 35.4 |
| Grant | 14.9 | 17.2 | 14.7 | 49.7 | 7,935 | 7,444 | -6.2 | -3.5 | 632 | 879 | -17 | 3,381 | 2.07 | 57.9 | 6.1 | 33.2 |
| Harney | 14.6 | 14.4 | 10.5 | 48.9 | 7,609 | 7,422 | -2.5 | -0.7 | 790 | 869 | 27 | 3,244 | 2.19 | 68.6 | 12.4 | 21.5 |
| Hood River | 13.0 | 10.3 | 6.8 | 49.8 | 20,411 | 22,346 | 9.5 | 4.2 | 2,795 | 1,839 | -13 | 8,600 | 2.58 | 66.5 | 7.5 | 27.1 |
| Jackson | 13.4 | 13.6 | 9.5 | 51.2 | 181,269 | 203,206 | 12.1 | 9.2 | 23,394 | 23,882 | 19,130 | 88,241 | 2.41 | 63.6 | 10.5 | 27.9 |
| Jefferson | 13.5 | 12.3 | 7.8 | 48.1 | 19,009 | 21,725 | 14.3 | 14.4 | 2,960 | 2,168 | 2,332 | 8,094 | 2.78 | 69.6 | 13.5 | 22.8 |
| Josephine | 14.8 | 15.6 | 11.4 | 51.4 | 75,726 | 82,718 | 9.2 | 6.4 | 8,533 | 12,381 | 9,182 | 36,367 | 2.34 | 65.1 | 10.1 | 27.3 |
| Klamath | 13.6 | 13.7 | 8.4 | 50.1 | 63,775 | 66,383 | 4.1 | 3.5 | 8,128 | 8,029 | 2,284 | 27,886 | 2.36 | 63.4 | 11.5 | 28.3 |
| Lake | 14.9 | 15.7 | 10.5 | 46.7 | 7,422 | 7,885 | 6.2 | 0.8 | 742 | 936 | 262 | 3,515 | 2.06 | 63.0 | 12.2 | 31.5 |
| Lane | 12.7 | 12.5 | 8.0 | 50.9 | 322,959 | 351,705 | 8.9 | 8.9 | 35,750 | 36,152 | 31,823 | 152,312 | 2.4 | 58.9 | 10.3 | 29.6 |
| Lincoln | 16.8 | 19.8 | 10.9 | 51.8 | 44,479 | 46,033 | 3.5 | 9.9 | 4,289 | 6,150 | 6,379 | 21,298 | 2.25 | 59.0 | 9.2 | 32.8 |
| Linn | 13.2 | 11.6 | 7.8 | 50.5 | 103,069 | 116,681 | 13.2 | 12.3 | 15,074 | 12,929 | 12,273 | 47,762 | 2.59 | 67.8 | 10.7 | 24.9 |
| Malheur | 10.8 | 9.8 | 7.4 | 45.8 | 31,615 | 31,316 | -0.9 | -1.1 | 4,304 | 3,059 | -1,590 | 10,062 | 2.65 | 66.6 | 14.5 | 28.7 |
| Marion | 11.8 | 9.8 | 6.7 | 50.1 | 284,834 | 315,342 | 10.7 | 10.7 | 44,565 | 28,377 | 17,839 | 118,038 | 2.79 | 68.1 | 13.3 | 25.2 |

1. No spouse present.

# Table B. States and Counties — Population, Vital Statistics, Health, and Crime

| STATE County | Persons in group quarters, 2020 | Daytime Population, 2015-2019 Number | Employment/residence ratio | Births, 2020 Total | Rate[1] | Deaths, 2020 Number | Rate[1] | Persons under 65 with no health insurance, 2019 Number | Percent | Medicare, 2020 Total beneficiaries | Enrolled in Original Medicare | Enrolled in Medicare Advantage | Crimes reported by county police or sheriff, 2019 Violent | Property |
|---|---|---|---|---|---|---|---|---|---|---|---|---|---|---|
| | 32 | 33 | 34 | 35 | 36 | 37 | 38 | 39 | 40 | 41 | 42 | 43 | 44 | 45 |
| OKLAHOMA—Cont'd | | | | | | | | | | | | | | |
| Kiowa | 171 | 8,163 | 0.80 | 110 | 12.6 | 108 | 12.4 | 1,160 | 17.1 | 2,228 | 1,911 | 318 | 5 | 26 |
| Latimer | 575 | 10,183 | 0.96 | 121 | 11.9 | 121 | 11.9 | 1,421 | 19.1 | 2,694 | 2,305 | 389 | 15 | 85 |
| Le Flore | 1,612 | 46,140 | 0.80 | 596 | 11.9 | 732 | 14.7 | 8,727 | 22.4 | 11,131 | 8,621 | 2,510 | 37 | 281 |
| Lincoln | 386 | 28,839 | 0.57 | 358 | 10.2 | 382 | 10.9 | 4,735 | 16.9 | 7,561 | 5,421 | 2,140 | 48 | 289 |
| Logan | 2,163 | 36,297 | 0.50 | 504 | 10.3 | 437 | 15.0 | 5,823 | 15.0 | 8,946 | 6,307 | 2,639 | 36 | 384 |
| Love | 88 | 12,592 | 1.65 | 121 | 11.8 | 139 | 13.6 | 1,663 | 20.4 | 2,354 | 2,061 | 294 | 18 | 100 |
| McClain | 194 | 32,381 | 0.63 | 419 | 10.1 | 408 | 9.9 | 4,938 | 14.4 | 7,925 | 6,203 | 1,722 | 27 | 280 |
| McCurtain | 454 | 33,087 | 1.01 | 432 | 13.2 | 490 | 15.0 | 5,862 | 22.5 | 7,401 | 6,821 | 580 | 29 | 318 |
| McIntosh | 345 | 18,051 | 0.75 | 177 | 9.0 | 335 | 17.1 | 3,146 | 22.1 | 5,407 | 4,312 | 1,096 | 37 | 221 |
| Major | 77 | 7,410 | 0.91 | 110 | 14.5 | 74 | 9.8 | 1,116 | 18.5 | 1,616 | 1,506 | 110 | 5 | 30 |
| Marshall | 329 | 15,251 | 0.80 | 203 | 11.9 | 185 | 10.8 | 2,963 | 22.8 | 3,966 | 3,391 | 575 | 19 | 209 |
| Mayes | 565 | 39,668 | 0.92 | 458 | 11.1 | 594 | 14.4 | 6,257 | 19.1 | 9,458 | 7,037 | 2,421 | 38 | 327 |
| Murray | 313 | 13,378 | 0.90 | 163 | 11.7 | 217 | 15.5 | 1,944 | 17.6 | 3,203 | 2,846 | 357 | 10 | 10 |
| Muskogee | 3,505 | 72,882 | 1.16 | 795 | 11.8 | 924 | 13.7 | 9,746 | 18.5 | 15,479 | 12,260 | 3,219 | 73 | 295 |
| Noble | 273 | 11,824 | 1.10 | 109 | 9.8 | 130 | 11.7 | 1,314 | 14.8 | 2,594 | 2,324 | 270 | 6 | 76 |
| Nowata | 157 | 8,505 | 0.57 | 131 | 13.0 | 158 | 15.7 | 1,403 | 17.7 | 2,372 | 1,932 | 440 | 12 | 65 |
| Okfuskee | 1,226 | 10,924 | 0.71 | 144 | 12.2 | 171 | 14.5 | 1,710 | 19.4 | 2,636 | 2,171 | 465 | 13 | 145 |
| Oklahoma | 14,800 | 898,582 | 1.30 | 11,194 | 13.9 | 7,683 | 9.6 | 110,449 | 16.5 | 131,001 | 90,080 | 40,921 | 35 | 227 |
| Okmulgee | 1,376 | 35,294 | 0.77 | 415 | 10.9 | 576 | 15.1 | 5,495 | 18.5 | 9,107 | 6,710 | 2,397 | 42 | 216 |
| Osage | 1,497 | 38,961 | 0.56 | 432 | 9.3 | 597 | 12.8 | 5,699 | 15.9 | 10,263 | 7,099 | 3,164 | 56 | 332 |
| Ottawa | 971 | 31,192 | 0.98 | 393 | 12.7 | 531 | 17.2 | 4,801 | 19.7 | 7,216 | 5,528 | 1,688 | 15 | 122 |
| Pawnee | 194 | 13,707 | 0.59 | 185 | 11.3 | 202 | 12.3 | 2,272 | 17.4 | 3,758 | 3,007 | 751 | NA | NA |
| Payne | 7,786 | 83,603 | 1.05 | 813 | 9.9 | 664 | 8.1 | 11,184 | 17.6 | 12,353 | 10,729 | 1,624 | 25 | 221 |
| Pittsburg | 2,417 | 45,246 | 1.06 | 522 | 12.0 | 632 | 14.5 | 5,997 | 18.4 | 10,675 | 8,858 | 1,817 | 55 | 422 |
| Pontotoc | 1,699 | 40,079 | 1.10 | 498 | 13.0 | 541 | 14.1 | 5,708 | 18.7 | 7,967 | 7,024 | 943 | 21 | 143 |
| Pottawatomie | 3,049 | 68,836 | 0.89 | 849 | 11.6 | 883 | 12.1 | 9,736 | 16.9 | 15,317 | 10,804 | 4,513 | 65 | 540 |
| Pushmataha | 110 | 10,264 | 0.78 | 101 | 9.2 | 172 | 15.7 | 1,712 | 20.6 | 2,909 | 2,460 | 449 | 14 | 91 |
| Roger Mills | 11 | 3,588 | 0.95 | 34 | 9.5 | 28 | 7.8 | 567 | 20.5 | 825 | 776 | 48 | 9 | 31 |
| Rogers | 1,217 | 77,695 | 0.69 | 1,046 | 11.2 | 921 | 9.9 | 10,382 | 13.5 | 18,644 | 13,189 | 5,455 | 109 | 472 |
| Seminole | 581 | 24,050 | 0.91 | 305 | 12.6 | 349 | 14.4 | 3,962 | 20.7 | 5,499 | 4,141 | 1,359 | 29 | 257 |
| Sequoyah | 464 | 37,059 | 0.70 | 546 | 13.1 | 599 | 14.4 | 7,025 | 21.3 | 9,865 | 7,374 | 2,491 | 88 | 301 |
| Stephens | 540 | 42,793 | 0.95 | 512 | 11.9 | 589 | 13.7 | 6,170 | 18.2 | 10,389 | 9,141 | 1,248 | 19 | 152 |
| Texas | 574 | 20,352 | 0.96 | 335 | 16.8 | 130 | 6.5 | 4,085 | 24.1 | 2,607 | 2,471 | 136 | 9 | 51 |
| Tillman | 267 | 6,875 | 0.81 | 73 | 10.1 | 81 | 11.2 | 1,098 | 20.1 | 1,645 | 1,518 | 127 | 1 | 27 |
| Tulsa | 9,858 | 712,334 | 1.21 | 8,841 | 13.4 | 6,617 | 10.1 | 90,972 | 16.6 | 114,970 | 71,254 | 43,716 | 228 | 924 |
| Wagoner | 319 | 56,181 | 0.37 | 891 | 10.7 | 845 | 10.2 | 10,613 | 15.7 | 15,702 | 9,770 | 5,932 | 48 | 295 |
| Washington | 798 | 52,632 | 1.03 | 575 | 11.0 | 683 | 13.1 | 7,023 | 17.2 | 12,099 | 10,385 | 1,714 | NA | NA |
| Washita | 206 | 9,164 | 0.56 | 122 | 11.3 | 151 | 13.9 | 1,581 | 18.0 | 2,276 | 2,054 | 222 | 14 | 52 |
| Woods | 997 | 9,406 | 1.08 | 93 | 10.7 | 105 | 12.1 | 955 | 15.0 | 1,712 | 1,626 | 86 | 6 | 23 |
| Woodward | 1,231 | 21,329 | 1.06 | 254 | 12.8 | 288 | 14.5 | 2,885 | 18.3 | 3,578 | 3,369 | 208 | 12 | 68 |
| OREGON | 88,649 | 4,178,438 | 1.03 | 42,070 | 9.9 | 39,735 | 9.4 | 291,036 | 8.6 | 885,725 | 468,190 | 417,535 | X | X |
| Baker | 368 | 16,170 | 1.03 | 160 | 9.8 | 231 | 14.2 | 1,018 | 9.0 | 5,163 | 4,814 | 350 | 10 | 59 |
| Benton | 5,819 | 93,853 | 1.06 | 678 | 7.3 | 650 | 7.0 | 4,861 | 6.7 | 17,553 | 9,045 | 8,507 | NA | NA |
| Clackamas | 2,915 | 372,427 | 0.81 | 3,864 | 9.2 | 3,871 | 9.2 | 25,418 | 7.4 | 87,253 | 32,718 | 54,536 | 446 | 4,651 |
| Clatsop | 859 | 40,200 | 1.06 | 373 | 9.2 | 435 | 10.8 | 2,459 | 8.2 | 10,868 | 9,099 | 1,770 | 28 | 166 |
| Columbia | 465 | 41,248 | 0.54 | 466 | 8.8 | 533 | 10.1 | 3,069 | 7.3 | 12,258 | 5,944 | 6,314 | NA | NA |
| Coos | 1,188 | 63,930 | 1.01 | 570 | 8.8 | 999 | 15.4 | 4,164 | 9.1 | 20,840 | 17,491 | 3,349 | 40 | 506 |
| Crook | 229 | 21,139 | 0.79 | 237 | 9.4 | 266 | 10.6 | 1,817 | 10.1 | 7,304 | 5,250 | 2,054 | 17 | 126 |
| Curry | 304 | 22,541 | 0.99 | 169 | 7.3 | 403 | 17.3 | 1,453 | 10.0 | 9,530 | 8,165 | 1,365 | 42 | 254 |
| Deschutes | 1,378 | 187,992 | 1.02 | 1,866 | 9.2 | 1,674 | 8.3 | 12,432 | 7.9 | 46,658 | 31,826 | 14,832 | 86 | 654 |
| Douglas | 1,836 | 107,881 | 0.97 | 1,048 | 9.4 | 1,592 | 14.3 | 6,911 | 8.7 | 34,711 | 22,035 | 12,676 | 153 | 889 |
| Gilliam | 18 | 1,965 | 1.11 | 14 | 7.1 | 21 | 10.6 | 118 | 8.6 | 564 | 521 | 42 | 9 | 52 |
| Grant | 126 | 7,171 | 0.99 | 59 | 8.2 | 89 | 12.4 | 472 | 9.8 | 2,321 | 1,934 | 386 | NA | NA |
| Harney | 155 | 7,383 | 1.04 | 63 | 8.5 | 83 | 11.3 | 547 | 10.3 | 2,053 | 1,885 | 169 | 1 | 21 |
| Hood River | 791 | 24,906 | 1.15 | 250 | 10.7 | 179 | 7.7 | 2,480 | 12.8 | 4,435 | 3,391 | 1,044 | NA | NA |
| Jackson | 3,484 | 216,604 | 1.00 | 2,159 | 9.7 | 2,524 | 11.4 | 16,303 | 9.8 | 57,969 | 37,447 | 20,522 | 128 | 1,274 |
| Jefferson | 943 | 22,275 | 0.85 | 310 | 12.5 | 274 | 11.0 | 2,246 | 12.2 | 5,606 | 3,914 | 1,692 | 8 | 172 |
| Josephine | 1,615 | 85,375 | 0.97 | 803 | 9.1 | 1,293 | 14.7 | 6,416 | 10.3 | 27,027 | 15,450 | 11,577 | 45 | 394 |
| Klamath | 1,193 | 66,248 | 0.97 | 771 | 11.2 | 894 | 13.0 | 5,494 | 10.7 | 18,116 | 13,058 | 5,058 | NA | NA |
| Lake | 469 | 7,894 | 1.02 | 71 | 8.9 | 107 | 13.5 | 493 | 9.2 | 2,316 | 2,138 | 179 | 9 | 88 |
| Lane | 8,800 | 374,573 | 1.01 | 3,314 | 8.7 | 4,016 | 10.5 | 26,896 | 9.1 | 90,110 | 41,510 | 48,600 | NA | NA |
| Lincoln | 857 | 49,212 | 1.04 | 386 | 7.6 | 690 | 13.6 | 3,761 | 10.9 | 17,123 | 13,174 | 3,949 | 44 | 271 |
| Linn | 1,171 | 119,113 | 0.89 | 1,498 | 11.4 | 1,472 | 11.2 | 8,693 | 8.4 | 30,127 | 13,929 | 16,197 | 56 | 705 |
| Malheur | 3,266 | 33,554 | 1.29 | 383 | 12.4 | 269 | 8.7 | 2,633 | 12.2 | 6,316 | 5,132 | 1,184 | 17 | 192 |
| Marion | 10,478 | 343,127 | 1.02 | 4,119 | 11.8 | 3,161 | 9.1 | 29,108 | 10.4 | 66,040 | 26,920 | 39,120 | NA | 2,409 |

1. Per 1,000 estimated resident population.

# Table B. States and Counties — Crime, Education, Money Income, and Poverty

| STATE County | County law enforcement employment, 2019 — Officers | Civilians | Enrollment¹ Total | Percent private | High school graduate or less | Bachelor's degree or more | Total current spending (mil dol) | Current spending per student (dollars) | Per capita income⁴ | Median income (dollars) | Percent with income of less than $50,000 | Percent with income of $200,000 or more | Median household income (dollars) | All persons | Children under 18 years | Children 5 to 17 years in families |
|---|---|---|---|---|---|---|---|---|---|---|---|---|---|---|---|---|
| | 46 | 47 | 48 | 49 | 50 | 51 | 52 | 53 | 54 | 55 | 56 | 57 | 58 | 59 | 60 | 61 |
| **OKLAHOMA—Cont'd** | | | | | | | | | | | | | | | | |
| Kiowa | 4 | 6 | 1,864 | 5.6 | 52.5 | 17.2 | 15 | 9,346 | 21,482 | 38,019 | 59.1 | 2.6 | 39,788 | 21.9 | 30.3 | 30.6 |
| Latimer | 8 | 8 | 2,313 | 5.5 | 49.0 | 17.2 | 12 | 8,547 | 25,240 | 44,214 | 56.9 | 3.0 | 42,014 | 18.5 | 26.6 | 26.7 |
| Le Flore | 17 | 4 | 11,396 | 4.9 | 57.5 | 14.6 | 80 | 8,314 | 20,902 | 40,677 | 59.2 | 1.8 | 42,378 | 20.5 | 30.1 | 26.7 |
| Lincoln | 19 | 16 | 8,017 | 9.8 | 53.6 | 15.5 | 46 | 8,275 | 25,255 | 50,671 | 49.4 | 2.6 | 51,012 | 13.9 | 20.0 | 18.4 |
| Logan | 38 | 30 | 11,891 | 10.3 | 42.3 | 27.7 | 36 | 7,796 | 31,105 | 65,357 | 40.1 | 8.0 | 66,625 | 12.4 | 14.3 | 13.4 |
| Love | 12 | 10 | 2,505 | 6.0 | 59.2 | 13.5 | 17 | 9,071 | 22,036 | 49,399 | 50.6 | 1.3 | 51,481 | 13.4 | 19.5 | 17.0 |
| McClain | 26 | 14 | 10,354 | 11.0 | 42.7 | 25.0 | 56 | 7,089 | 32,259 | 67,662 | 35.5 | 5.8 | 75,009 | 7.1 | 10.8 | 9.6 |
| McCurtain | 22 | 4 | 7,467 | 3.8 | 60.1 | 13.4 | 60 | 9,071 | 20,671 | 37,061 | 61.3 | 1.3 | 40,803 | 21.9 | 31.5 | 30.9 |
| McIntosh | 12 | 1 | 3,621 | 8.8 | 56.8 | 13.9 | 25 | 8,687 | 23,268 | 39,084 | 60.3 | 2.1 | 40,614 | 19.7 | 29.2 | 28.0 |
| Major | 8 | 11 | 1,712 | 9.2 | 50.0 | 20.0 | 11 | 9,451 | 28,609 | 54,332 | 46.5 | 2.9 | 59,690 | 13.4 | 17.5 | 14.7 |
| Marshall | 7 | 21 | 3,561 | 2.6 | 56.6 | 14.8 | 25 | 7,907 | 22,781 | 45,746 | 55.8 | 1.5 | 45,005 | 18.2 | 23.6 | 23.6 |
| Mayes | 27 | 28 | 9,081 | 8.4 | 52.9 | 14.3 | 68 | 9,563 | 24,472 | 50,345 | 49.8 | 1.8 | 49,293 | 17.7 | 25.6 | 22.9 |
| Murray | 6 | 7 | 3,028 | 5.7 | 53.8 | 17.4 | 18 | 6,964 | 26,943 | 52,478 | 47.9 | 3.2 | 50,965 | 14.9 | 20.4 | 18.8 |
| Muskogee | 27 | 55 | 16,785 | 6.8 | 49.2 | 20.4 | 105 | 7,942 | 23,826 | 43,078 | 55.8 | 2.7 | 43,392 | 22.7 | 32.6 | 31.7 |
| Noble | 9 | 12 | 2,475 | 4.2 | 51.0 | 23.2 | 20 | 9,165 | 28,011 | 56,117 | 42.7 | 3.4 | 56,307 | 12.3 | 16.0 | 15.1 |
| Nowata | 7 | 10 | 2,246 | 8.0 | 55.1 | 11.5 | 15 | 8,769 | 22,930 | 43,145 | 55.9 | 0.9 | 47,456 | 15.0 | 23.7 | 22.8 |
| Okfuskee | 7 | 13 | 2,622 | 8.7 | 60.1 | 11.6 | 21 | 9,748 | 18,437 | 38,411 | 61.4 | 1.9 | 39,265 | 27.4 | 35.9 | 35.9 |
| Oklahoma | 399 | 192 | 203,928 | 14.1 | 38.3 | 32.0 | 1,106 | 7,587 | 31,609 | 54,520 | 45.7 | 5.8 | 56,272 | 14.9 | 19.2 | 18.8 |
| Okmulgee | 15 | 2 | 9,209 | 7.8 | 49.7 | 14.1 | 54 | 8,507 | 23,431 | 42,998 | 57.1 | 1.9 | 41,725 | 18.7 | 26.0 | 25.4 |
| Osage | 37 | 38 | 10,427 | 10.4 | 49.9 | 18.0 | 37 | 9,719 | 25,473 | 49,103 | 50.9 | 2.5 | 50,667 | 14.9 | 22.0 | 19.8 |
| Ottawa | 9 | 17 | 7,644 | 5.7 | 51.0 | 14.5 | 49 | 8,438 | 20,814 | 39,872 | 60.3 | 1.8 | 41,530 | 18.5 | 27.3 | 23.9 |
| Pawnee | 11 | 14 | 3,651 | 7.5 | 55.5 | 16.6 | 22 | 8,358 | 24,303 | 48,009 | 51.7 | 2.1 | 47,646 | 15.2 | 19.6 | 18.2 |
| Payne | 40 | 55 | 33,253 | 5.2 | 35.3 | 37.4 | 95 | 8,553 | 24,057 | 41,603 | 57.3 | 3.7 | 43,025 | 23.0 | 20.2 | 18.3 |
| Pittsburg | 23 | 12 | 9,679 | 8.2 | 51.2 | 16.1 | 71 | 9,211 | 25,524 | 46,784 | 53.2 | 2.8 | 47,215 | 19.7 | 24.5 | 22.8 |
| Pontotoc | 16 | 2 | 10,295 | 6.7 | 43.6 | 28.2 | 61 | 8,484 | 26,187 | 50,392 | 49.7 | 3.1 | 49,510 | 13.9 | 17.8 | 16.8 |
| Pottawatomie | 25 | 5 | 18,797 | 15.8 | 48.5 | 18.3 | 102 | 7,749 | 24,178 | 50,206 | 50.6 | 2.5 | 52,747 | 14.8 | 17.8 | 17.1 |
| Pushmataha | 14 | 2 | 2,426 | 5.6 | 59.3 | 14.3 | 21 | 9,276 | 22,435 | 37,692 | 62.7 | 2.9 | 37,718 | 23.9 | 30.2 | 29.5 |
| Roger Mills | 8 | 5 | 909 | 4.4 | 47.9 | 16.2 | 11 | 14,501 | 30,866 | 51,302 | 49.1 | 4.7 | 59,286 | 13.4 | 19.1 | 18.0 |
| Rogers | 38 | 5 | 22,322 | 14.5 | 40.1 | 24.9 | 104 | 7,728 | 32,148 | 65,434 | 36.8 | 5.2 | 67,054 | 10.4 | 14.7 | 13.6 |
| Seminole | 14 | 19 | 6,223 | 5.4 | 49.7 | 14.3 | 43 | 8,845 | 20,666 | 39,373 | 59.7 | 1.4 | 40,725 | 22.0 | 31.5 | 29.1 |
| Sequoyah | 23 | 23 | 9,184 | 6.1 | 57.4 | 14.4 | 66 | 8,148 | 20,384 | 40,351 | 59.0 | 1.2 | 43,038 | 21.6 | 28.8 | 28.5 |
| Stephens | 22 | 4 | 9,554 | 7.4 | 54.9 | 16.9 | 64 | 8,257 | 25,769 | 47,214 | 52.2 | 3.1 | 47,826 | 14.8 | 18.6 | 17.2 |
| Texas | 13 | 17 | 5,636 | 6.3 | 53.4 | 22.1 | 41 | 8,535 | 22,003 | 52,282 | 47.4 | 1.8 | 56,014 | 14.7 | 19.1 | 18.6 |
| Tillman | 3 | 0 | 1,613 | 7.1 | 55.9 | 17.3 | 13 | 9,655 | 21,974 | 42,280 | 58.6 | 2.4 | 41,251 | 20.4 | 29.1 | 26.5 |
| Tulsa | 221 | 361 | 166,588 | 18.2 | 36.0 | 31.8 | 981 | 8,104 | 32,044 | 55,517 | 45.3 | 6.1 | 57,668 | 14.3 | 19.6 | 17.5 |
| Wagoner | 37 | 40 | 19,270 | 13.2 | 42.3 | 23.2 | 47 | 7,190 | 29,415 | 62,795 | 38.4 | 3.8 | 65,191 | 8.2 | 10.9 | 10.3 |
| Washington | 29 | 30 | 12,054 | 11.9 | 43.6 | 28.5 | 72 | 7,448 | 30,847 | 54,997 | 46.2 | 5.6 | 62,515 | 12.9 | 17.3 | 15.8 |
| Washita | 6 | 22 | 2,407 | 6.8 | 51.4 | 20.8 | 19 | 9,460 | 26,265 | 53,274 | 46.8 | 2.3 | 50,823 | 12.6 | 19.3 | 18.7 |
| Woods | 7 | 1 | 2,328 | 7.7 | 40.1 | 30.3 | 19 | 13,956 | 27,427 | 54,282 | 47.5 | 3.2 | 40,684 | 15.2 | 16.7 | 16.6 |
| Woodward | 12 | 23 | 4,744 | 4.5 | 52.1 | 19.4 | 32 | 8,471 | 28,915 | 60,147 | 41.5 | 4.3 | 73,253 | 12.2 | 15.8 | 15.7 |
| **OREGON** | X | X | 954,416 | 15.4 | 32.0 | 33.7 | 6,897 | 11,878 | 33,763 | 62,818 | 40.1 | 6.5 | 66,955 | 11.5 | 13.6 | 12.8 |
| Baker | 31 | 12 | 3,187 | 21.8 | 40.1 | 23.7 | 27 | 7,504 | 28,700 | 45,998 | 54.6 | 2.8 | 48,530 | 13.6 | 22.0 | 21.1 |
| Benton | 67 | 14 | 33,642 | 7.1 | 17.3 | 54.1 | 102 | 11,154 | 33,817 | 62,077 | 41.1 | 6.9 | 69,148 | 16.5 | 11.4 | 11.1 |
| Clackamas | 206 | 245 | 95,067 | 16.2 | 27.8 | 37.4 | 658 | 11,026 | 41,492 | 80,484 | 29.3 | 10.9 | 80,539 | 7.2 | 7.2 | 6.2 |
| Clatsop | 26 | 49 | 7,694 | 8.9 | 34.6 | 24.4 | 63 | 12,285 | 29,757 | 54,886 | 45.0 | 2.6 | 59,339 | 12.0 | 15.5 | 13.9 |
| Columbia | 44 | 13 | 10,716 | 13.5 | 42.5 | 18.5 | 81 | 10,679 | 31,883 | 62,257 | 39.4 | 4.4 | 72,285 | 8.6 | 11.6 | 10.8 |
| Coos | 60 | 29 | 11,831 | 10.0 | 41.4 | 17.8 | 103 | 10,288 | 27,582 | 45,051 | 53.4 | 2.6 | 49,504 | 15.6 | 20.2 | 20.2 |
| Crook | 13 | 19 | 4,414 | 10.1 | 43.3 | 19.8 | 31 | 10,381 | 27,007 | 49,006 | 50.7 | 2.5 | 58,739 | 10.9 | 17.1 | 16.2 |
| Curry | 16 | 22 | 2,901 | 9.4 | 37.1 | 24.6 | 27 | 11,849 | 30,021 | 48,440 | 51.3 | 2.4 | 51,267 | 13.4 | 19.9 | 19.8 |
| Deschutes | 186 | 67 | 38,066 | 13.6 | 28.1 | 35.4 | 298 | 11,096 | 35,507 | 67,043 | 37.3 | 6.2 | 71,880 | 9.2 | 11.3 | 10.9 |
| Douglas | 64 | 71 | 20,620 | 12.4 | 42.7 | 17.3 | 177 | 12,375 | 26,478 | 47,267 | 52.8 | 2.4 | 48,940 | 11.8 | 17.6 | 16.1 |
| Gilliam | 6 | 1 | 371 | 8.9 | 40.6 | 17.4 | 7 | 24,616 | 26,482 | 47,500 | 53.2 | 1.2 | 61,205 | 10.9 | 16.1 | 15.1 |
| Grant | 18 | 0 | 1,104 | 8.2 | 47.2 | 18.9 | 15 | 17,527 | 27,367 | 44,712 | 54.8 | 2.3 | 49,792 | 15.3 | 20.6 | 19.4 |
| Harney | 17 | 5 | 1,443 | 20.4 | 45.0 | 18.4 | 19 | 14,707 | 26,370 | 40,735 | 57.5 | 2.4 | 39,874 | 14.2 | 20.1 | 19.8 |
| Hood River | 20 | 27 | 5,344 | 10.2 | 40.5 | 32.0 | 51 | 12,544 | 34,926 | 65,679 | 36.9 | 7.5 | 68,835 | 8.4 | 12.1 | 11.7 |
| Jackson | 113 | 35 | 44,972 | 13.8 | 36.0 | 27.8 | 338 | 11,156 | 30,250 | 53,412 | 46.6 | 4.4 | 56,367 | 13.6 | 18.0 | 17.0 |
| Jefferson | 17 | 26 | 5,289 | 5.1 | 43.7 | 19.4 | 50 | 13,532 | 23,987 | 53,277 | 47.1 | 2.3 | 52,543 | 15.0 | 21.5 | 20.6 |
| Josephine | 63 | 22 | 17,107 | 13.7 | 40.6 | 17.5 | 128 | 11,654 | 25,167 | 45,616 | 53.8 | 2.7 | 47,533 | 15.9 | 24.3 | 20.9 |
| Klamath | 27 | 64 | 15,210 | 10.8 | 42.9 | 20.0 | 114 | 11,817 | 25,880 | 46,491 | 52.8 | 2.2 | 53,030 | 16.7 | 22.9 | 21.9 |
| Lake | 20 | 1 | 1,236 | 7.4 | 48.0 | 16.4 | 18 | 15,208 | 24,241 | 37,898 | 60.2 | 1.8 | 47,867 | 17.0 | 22.7 | 21.8 |
| Lane | 61 | 228 | 89,790 | 10.2 | 31.3 | 30.5 | 531 | 11,549 | 29,705 | 52,426 | 47.9 | 3.9 | 57,212 | 13.9 | 13.7 | 13.2 |
| Lincoln | 30 | 61 | 7,709 | 13.3 | 35.8 | 24.9 | 61 | 10,974 | 27,861 | 47,882 | 52.8 | 3.2 | 52,276 | 14.6 | 21.5 | 19.2 |
| Linn | 79 | 100 | 27,801 | 13.8 | 39.5 | 19.3 | 225 | 9,907 | 27,345 | 55,893 | 45.0 | 2.8 | 61,084 | 12.6 | 16.4 | 15.2 |
| Malheur | 40 | 23 | 7,289 | 8.5 | 50.3 | 13.7 | 70 | 13,873 | 18,974 | 43,313 | 56.6 | 2.1 | 42,019 | 21.0 | 24.7 | 23.4 |
| Marion | 94 | 224 | 84,526 | 14.5 | 41.4 | 23.5 | 751 | 11,911 | 27,338 | 59,625 | 42.2 | 3.7 | 64,058 | 12.2 | 14.2 | 13.6 |

1. All persons 3 years old and over enrolled in nursery school through college.   2. Persons 25 years old and over.   3. Elementary and secondary education expenditures.   4. Based on population estimated by the American Community Survey, 2014–2018.

# Table B. States and Counties — **Personal Income**

| STATE County | Personal income, 2019 | | | | | | | | | | Earnings, 2019 | | |
|---|---|---|---|---|---|---|---|---|---|---|---|---|---|
| | Total (mil dol) | Percent change 2018-2019 | Per capita[1] Dollars | Per capita[1] Rank | Wages and salaries (mil dol) | Supplements to wages and salaries, employer contributions (mil dol) Pension and insurance | Supplements to wages and salaries, employer contributions (mil dol) Government social insurance | Proprietors' income (mil dol) | Dividends, interest, and rent (mil dol) | Personal transfer receipts (mil dol) | Total (mil dol) | Contributions for government social insurance (mil dol) From employee and self-employed | Contributions for government social insurance (mil dol) From employer |
| | 62 | 63 | 64 | 65 | 66 | 67 | 68 | 69 | 70 | 71 | 72 | 73 | 74 |
| **OKLAHOMA—Cont'd** | | | | | | | | | | | | | |
| Kiowa | 329 | 8.1 | 37,761 | 2,349 | 74 | 17 | 5 | 46 | 55 | 98 | 142 | 9 | 5 |
| Latimer | 354 | 2.4 | 35,167 | 2,665 | 112 | 26 | 9 | 19 | 57 | 133 | 166 | 12 | 9 |
| Le Flore | 1,656 | 2.9 | 33,222 | 2,874 | 539 | 103 | 41 | 134 | 230 | 542 | 816 | 57 | 41 |
| Lincoln | 1,245 | 2.8 | 35,697 | 2,596 | 312 | 55 | 22 | 52 | 201 | 324 | 442 | 35 | 22 |
| Logan | 2,106 | 4.3 | 43,861 | 1,505 | 312 | 52 | 23 | 115 | 301 | 384 | 502 | 38 | 23 |
| Love | 386 | 2.4 | 37,666 | 2,363 | 234 | 54 | 17 | 14 | 63 | 104 | 318 | 20 | 17 |
| McClain | 1,897 | 4.2 | 46,874 | 1,147 | 421 | 69 | 30 | 137 | 275 | 359 | 658 | 44 | 30 |
| McCurtain | 1,084 | 2.3 | 33,002 | 2,889 | 456 | 84 | 34 | 84 | 136 | 364 | 659 | 44 | 34 |
| McIntosh | 692 | 6.3 | 35,335 | 2,642 | 155 | 28 | 11 | 37 | 112 | 266 | 232 | 21 | 11 |
| Major | 334 | 0.8 | 43,815 | 1,514 | 107 | 19 | 8 | 36 | 71 | 66 | 170 | 10 | 8 |
| Marshall | 573 | 5.1 | 33,857 | 2,818 | 195 | 35 | 15 | 38 | 89 | 184 | 283 | 21 | 15 |
| Mayes | 1,494 | 3.5 | 36,363 | 2,527 | 635 | 107 | 47 | 81 | 221 | 445 | 871 | 59 | 47 |
| Murray | 609 | 3.2 | 43,275 | 1,576 | 233 | 50 | 16 | 44 | 96 | 147 | 342 | 22 | 16 |
| Muskogee | 2,572 | 4.0 | 37,827 | 2,342 | 1,449 | 294 | 110 | 138 | 444 | 833 | 1,991 | 127 | 110 |
| Noble | 490 | 5.0 | 44,037 | 1,479 | 266 | 47 | 19 | 26 | 93 | 111 | 358 | 23 | 19 |
| Nowata | 373 | 2.0 | 37,054 | 2,443 | 76 | 15 | 6 | 19 | 60 | 111 | 115 | 10 | 6 |
| Okfuskee | 374 | 7.6 | 31,212 | 2,992 | 100 | 21 | 7 | 15 | 60 | 134 | 144 | 12 | 7 |
| Oklahoma | 41,505 | 3.4 | 52,048 | 678 | 27,649 | 4,385 | 1,981 | 5,681 | 8,377 | 6,698 | 39,697 | 2,225 | 1,981 |
| Okmulgee | 1,342 | 4.8 | 34,888 | 2,701 | 407 | 79 | 30 | 45 | 183 | 455 | 561 | 44 | 30 |
| Osage | 1,665 | 3.6 | 35,458 | 2,623 | 296 | 59 | 22 | 76 | 263 | 421 | 452 | 38 | 22 |
| Ottawa | 1,192 | 4.3 | 38,291 | 2,294 | 461 | 92 | 33 | 118 | 185 | 378 | 705 | 45 | 33 |
| Pawnee | 598 | 3.3 | 36,538 | 2,503 | 152 | 33 | 12 | 17 | 92 | 183 | 214 | 18 | 12 |
| Payne | 3,130 | 4.0 | 38,276 | 2,298 | 1,582 | 331 | 111 | 192 | 663 | 589 | 2,215 | 130 | 111 |
| Pittsburg | 1,717 | 2.5 | 39,331 | 2,141 | 820 | 174 | 61 | 48 | 323 | 484 | 1,103 | 73 | 61 |
| Pontotoc | 1,712 | 4.8 | 44,712 | 1,394 | 925 | 180 | 66 | 89 | 309 | 414 | 1,259 | 77 | 66 |
| Pottawatomie | 2,762 | 3.7 | 38,053 | 2,322 | 913 | 165 | 66 | 236 | 450 | 740 | 1,380 | 92 | 66 |
| Pushmataha | 364 | 3.5 | 32,830 | 2,903 | 94 | 20 | 7 | 28 | 55 | 144 | 148 | 12 | 7 |
| Roger Mills | 159 | -2.2 | 44,454 | 1,425 | 44 | 10 | 3 | 14 | 50 | 30 | 71 | 5 | 3 |
| Rogers | 4,404 | 4.7 | 47,635 | 1,065 | 1,459 | 238 | 109 | 195 | 678 | 836 | 2,001 | 134 | 109 |
| Seminole | 841 | 3.0 | 34,658 | 2,730 | 310 | 63 | 23 | 47 | 131 | 273 | 442 | 31 | 23 |
| Sequoyah | 1,399 | 3.6 | 33,665 | 2,829 | 342 | 70 | 25 | 68 | 225 | 468 | 506 | 41 | 25 |
| Stephens | 1,886 | 3.7 | 43,723 | 1,524 | 709 | 108 | 52 | 155 | 435 | 472 | 1,023 | 70 | 52 |
| Texas | 1,232 | 5.2 | 61,652 | 261 | 412 | 71 | 31 | 491 | 124 | 122 | 1,005 | 33 | 31 |
| Tillman | 280 | 9.6 | 38,674 | 2,241 | 70 | 16 | 5 | 46 | 47 | 72 | 137 | 8 | 5 |
| Tulsa | 42,155 | 3.3 | 64,699 | 193 | 20,655 | 2,988 | 1,489 | 10,088 | 8,861 | 5,628 | 35,220 | 1,849 | 1,489 |
| Wagoner | 3,321 | 5.5 | 40,851 | 1,939 | 473 | 75 | 35 | 159 | 401 | 686 | 743 | 60 | 35 |
| Washington | 3,442 | 4.8 | 66,805 | 157 | 1,094 | 183 | 77 | 1,204 | 483 | 526 | 2,558 | 126 | 77 |
| Washita | 395 | 1.3 | 36,141 | 2,551 | 96 | 19 | 7 | 29 | 70 | 103 | 152 | 11 | 7 |
| Woods | 391 | 2.0 | 44,427 | 1,428 | 175 | 34 | 12 | 31 | 123 | 72 | 251 | 15 | 12 |
| Woodward | 771 | 1.0 | 38,156 | 2,310 | 468 | 75 | 33 | -24 | 170 | 160 | 553 | 38 | 33 |
| **OREGON** | 224,346 | 4.2 | 53,212 | X | 112,470 | 17,722 | 9,487 | 19,936 | 47,226 | 42,427 | 159,615 | 10,114 | 9,487 |
| Baker | 685 | 3.2 | 42,494 | 1,692 | 229 | 52 | 21 | 34 | 165 | 215 | 336 | 26 | 21 |
| Benton | 4,534 | 4.2 | 48,725 | 947 | 2,176 | 447 | 179 | 323 | 1,220 | 653 | 3,126 | 193 | 179 |
| Clackamas | 25,813 | 4.4 | 61,726 | 259 | 9,880 | 1,378 | 845 | 2,143 | 5,542 | 3,447 | 14,245 | 924 | 845 |
| Clatsop | 1,894 | 3.8 | 47,081 | 1,125 | 832 | 146 | 76 | 238 | 397 | 473 | 1,292 | 86 | 76 |
| Columbia | 2,409 | 4.0 | 46,019 | 1,245 | 518 | 104 | 47 | 110 | 356 | 579 | 779 | 61 | 47 |
| Coos | 2,994 | 4.2 | 46,429 | 1,194 | 1,046 | 224 | 95 | 261 | 603 | 985 | 1,626 | 119 | 95 |
| Crook | 1,039 | 5.0 | 42,584 | 1,674 | 309 | 59 | 26 | 77 | 222 | 320 | 471 | 37 | 26 |
| Curry | 1,043 | 4.2 | 45,474 | 1,301 | 271 | 57 | 25 | 91 | 265 | 372 | 444 | 39 | 25 |
| Deschutes | 11,159 | 4.6 | 56,447 | 424 | 4,367 | 674 | 385 | 1,629 | 2,907 | 2,028 | 7,054 | 459 | 385 |
| Douglas | 4,698 | 3.9 | 42,334 | 1,716 | 1,745 | 342 | 161 | 286 | 900 | 1,595 | 2,534 | 193 | 161 |
| Gilliam | 106 | 8.1 | 55,409 | 477 | 51 | 10 | 5 | 13 | 20 | 23 | 78 | 5 | 5 |
| Grant | 309 | 3.0 | 42,888 | 1,638 | 108 | 31 | 10 | 12 | 73 | 95 | 162 | 12 | 10 |
| Harney | 309 | 4.9 | 41,818 | 1,794 | 104 | 30 | 10 | 34 | 62 | 92 | 178 | 12 | 10 |
| Hood River | 1,317 | 1.2 | 56,325 | 433 | 603 | 96 | 57 | 140 | 336 | 203 | 896 | 54 | 57 |
| Jackson | 10,670 | 4.2 | 48,291 | 993 | 4,275 | 724 | 382 | 1,101 | 2,376 | 2,712 | 6,482 | 447 | 382 |
| Jefferson | 867 | 4.3 | 35,155 | 2,666 | 299 | 70 | 27 | 58 | 148 | 303 | 454 | 32 | 27 |
| Josephine | 3,810 | 4.3 | 43,554 | 1,540 | 1,183 | 208 | 107 | 453 | 729 | 1,290 | 1,951 | 151 | 107 |
| Klamath | 2,840 | 2.9 | 41,613 | 1,830 | 1,017 | 209 | 96 | 237 | 516 | 924 | 1,558 | 111 | 96 |
| Lake | 340 | 3.9 | 43,228 | 1,584 | 111 | 33 | 10 | 35 | 76 | 100 | 189 | 12 | 10 |
| Lane | 18,087 | 3.6 | 47,340 | 1,096 | 7,685 | 1,382 | 678 | 1,717 | 3,908 | 4,224 | 11,463 | 764 | 678 |
| Lincoln | 2,295 | 4.2 | 45,935 | 1,255 | 814 | 153 | 73 | 224 | 528 | 669 | 1,264 | 96 | 73 |
| Linn | 5,817 | 5.3 | 44,830 | 1,380 | 2,293 | 394 | 212 | 460 | 904 | 1,634 | 3,360 | 229 | 212 |
| Malheur | 1,057 | 5.1 | 34,576 | 2,738 | 523 | 121 | 49 | 127 | 187 | 365 | 821 | 50 | 49 |
| Marion | 15,707 | 4.7 | 45,158 | 1,344 | 8,036 | 1,644 | 703 | 1,597 | 2,745 | 3,802 | 11,979 | 729 | 703 |

1. Based on the resident population estimated as of July 1 of the year shown.

Items 62—74

| STATE County | Farm | Mining, quarrying, and extractions | Construction | Manufacturing | Information; professional, scientific, technical services | Retail trade | Finance, insurance, real estate, and leasing | Health care and social assistance | Government | Number | Rate[1] | Supplemental Security Income recipients, 2019 | Total | Percent change, 2010-2020 |
|---|---|---|---|---|---|---|---|---|---|---|---|---|---|---|
| | 75 | 76 | 77 | 78 | 79 | 80 | 81 | 82 | 83 | 84 | 85 | 86 | 87 | 88 |
| **OKLAHOMA—Cont'd** | | | | | | | | | | | | | | |
| Kiowa | 20.1 | 5.3 | 3.2 | 0.7 | D | 5.6 | 5.9 | 8.3 | 23.6 | 2,435 | 279 | 345 | 5,117 | -1.9 |
| Latimer | 2.3 | 8.3 | 8.3 | D | D | 4.4 | D | 4.9 | 27.0 | 2,790 | 278 | 409 | 5,062 | 1.6 |
| Le Flore | 8.4 | 4.7 | 4.0 | 6.0 | D | 5.5 | 3.4 | D | 38.0 | 12,165 | 244 | 1,909 | 22,294 | 3.9 |
| Lincoln | -4.3 | 5.1 | 16.2 | 7.4 | 4.8 | 7.6 | 11.3 | D | 21.1 | 8,525 | 244 | 745 | 15,469 | 1.7 |
| Logan | -1.1 | 5.0 | 15.6 | 5.6 | 5.5 | 8.8 | 5.8 | 10.9 | 15.4 | 9,160 | 190 | 637 | 17,630 | 2.5 |
| Love | -1.3 | 1.4 | 2.5 | 1.8 | D | 3.6 | 1.1 | D | 59.9 | 2,500 | 246 | 227 | 4,605 | 1.5 |
| McClain | 0.1 | 11.1 | 16.4 | 4.7 | 5.7 | 9.8 | 4.9 | 6.0 | 17.4 | 8,345 | 206 | 599 | 16,762 | 19.7 |
| McCurtain | 6.1 | D | 5.9 | 27.2 | D | 7.7 | 3.3 | D | 19.6 | 8,245 | 252 | 1,375 | 15,808 | 1.8 |
| McIntosh | -1.5 | 1.4 | 6.5 | 1.7 | D | 19.6 | 6.1 | 11.1 | 24.8 | 6,440 | 329 | 745 | 13,945 | 4.6 |
| Major | 10.2 | D | 9.6 | 6.4 | D | 5.4 | 4.1 | 6.9 | 11.4 | 1,785 | 234 | 88 | 3,700 | 0.8 |
| Marshall | -1.6 | 1.2 | 4.2 | 38.2 | D | 6.4 | 5.0 | D | 15.8 | 4,275 | 252 | 440 | 10,429 | 4.2 |
| Mayes | 1.7 | D | 14.3 | 23.8 | D | 9.0 | 3.2 | D | 19.9 | 9,965 | 242 | 1,191 | 19,619 | 2.0 |
| Murray | -0.2 | 3.6 | 7.1 | 11.4 | D | 6.0 | 2.8 | D | 37.9 | 3,500 | 249 | 334 | 6,908 | 2.4 |
| Muskogee | 0.0 | 0.2 | 9.3 | 12.9 | 3.3 | 5.8 | 3.7 | 11.4 | 35.0 | 17,075 | 251 | 2,934 | 30,941 | 0.1 |
| Noble | 1.7 | 2.1 | 4.1 | 45.5 | D | 2.8 | D | 15.1 | 24.4 | 2,720 | 244 | 267 | 5,341 | 0.0 |
| Nowata | -0.5 | 1.7 | 7.3 | 13.4 | D | 5.2 | D | D | 39.0 | 2,735 | 270 | 259 | 4,897 | 1.4 |
| Okfuskee | -7.4 | 14.9 | 10.8 | 6.9 | D | 5.0 | 3.2 | D | 19.4 | 2,895 | 242 | 568 | 5,340 | 1.1 |
| Oklahoma | 0.0 | 5.9 | 5.1 | 5.0 | 10.3 | 5.3 | 6.7 | 12.8 | 35.4 | 136,800 | 172 | 19,861 | 346,985 | 8.6 |
| Okmulgee | | 0.7 | 5.9 | 18.0 | 5.1 | 8.4 | 4.4 | 4.0 | 37.8 | 10,235 | 266 | 1,525 | 17,875 | -0.1 |
| Osage | 2.7 | 4.1 | 9.2 | 7.9 | 3.7 | 6.2 | 4.0 | D | 37.7 | 10,680 | 228 | 790 | 22,136 | 4.7 |
| Ottawa | 9.0 | 0.6 | 4.5 | 11.7 | D | 6.2 | 2.8 | D | 30.4 | 8,895 | 285 | 1,451 | 14,103 | 0.3 |
| Pawnee | -4.4 | D | 7.1 | 1.9 | 16.5 | 7.7 | 3.9 | D | 11.9 | 4,385 | 268 | 411 | 7,875 | 1.7 |
| Payne | -0.6 | 3.2 | 7.0 | 5.6 | 5.9 | 6.3 | 3.9 | 5.0 | 45.0 | 12,925 | 158 | 1,399 | 36,974 | 8.8 |
| Pittsburg | -0.4 | 8.2 | 4.9 | 8.5 | D | 6.4 | 3.7 | 6.9 | 40.8 | 11,045 | 253 | 1,471 | 23,157 | 2.4 |
| Pontotoc | -0.5 | 1.9 | 4.5 | 6.8 | 5.8 | 5.1 | 4.3 | 10.8 | 44.2 | 8,655 | 226 | 1,222 | 16,930 | 2.0 |
| Pottawatomie | -0.6 | 1.5 | 4.5 | 12.2 | 6.5 | 6.9 | 4.2 | D | 24.4 | 16,550 | 228 | 1,959 | 30,024 | 3.0 |
| Pushmataha | -2.6 | D | 11.8 | 8.8 | 5.0 | 7.5 | 4.5 | D | 33.9 | 3,455 | 313 | 530 | 6,226 | 1.9 |
| Roger Mills | -3.1 | 11.7 | D | D | D | 4.7 | D | 6.9 | 22.6 | 775 | 216 | 61 | 1,907 | 0.1 |
| Rogers | -0.3 | 0.8 | 13.4 | 20.7 | 5.5 | 6.2 | 3.6 | D | 24.8 | 19,695 | 213 | 1,266 | 39,723 | 13.0 |
| Seminole | -0.4 | 12.9 | 5.1 | 14.4 | 2.1 | 6.2 | 3.6 | D | 31.4 | 6,050 | 249 | 1,015 | 11,739 | 0.8 |
| Sequoyah | 0.7 | 1.0 | 16.2 | 3.3 | D | 9.3 | 3.9 | D | 11.6 | 10,920 | 263 | 1,823 | 19,587 | 5.0 |
| Stephens | -0.8 | 16.5 | 6.0 | 12.2 | D | 5.8 | 8.3 | D | 8.5 | 11,450 | 265 | 1,129 | 20,659 | 0.0 |
| Texas | 44.7 | 1.9 | D | D | 4.3 | 3.4 | 1.9 | 2.0 | 22.3 | 2,810 | 139 | 171 | 8,274 | 0.8 |
| Tillman | 27.3 | D | D | D | 3.1 | 2.6 | D | 2.9 | 22.3 | 1,800 | 248 | 276 | 3,978 | -2.4 |
| Tulsa | 0.0 | 10.8 | 4.8 | 11.8 | 8.5 | 4.8 | 5.4 | 11.1 | 6.7 | 119,470 | 183 | 15,538 | 287,802 | 7.2 |
| Wagoner | 0.8 | D | 21.8 | 17.1 | 3.9 | 7.8 | 3.2 | D | 15.6 | 16,055 | 197 | 1,214 | 33,792 | 13.8 |
| Washington | -0.1 | 24.5 | 1.9 | 25.6 | D | 4.4 | 3.1 | 5.2 | 5.1 | 13,025 | 252 | 1,172 | 23,794 | 1.5 |
| Washita | 3.7 | 11.4 | 9.0 | 3.8 | 4.8 | 4.9 | 5.8 | 5.2 | 26.3 | 2,510 | 229 | 281 | 5,391 | -0.9 |
| Woods | 2.2 | 29.2 | 4.4 | 2.4 | 3.0 | 6.1 | 5.7 | D | 24.6 | 1,670 | 190 | 91 | 4,401 | -1.7 |
| Woodward | -9.5 | 24.5 | 8.0 | 9.5 | 3.6 | 7.4 | 6.2 | 8.3 | 16.3 | 3,930 | 195 | 258 | 9,081 | 2.7 |
| **OREGON** | 0.8 | 0.1 | 7.2 | 11.3 | 10.7 | 6.3 | 7.2 | 12.1 | 16.3 | 891,726 | 212 | 88,912 | 1,830,891 | 9.3 |
| Baker | 4.1 | D | 4.9 | 10.7 | 4.4 | 8.3 | 4.3 | D | 25.2 | 5,300 | 329 | 454 | 9,128 | 3.5 |
| Benton | 0.6 | D | 4.1 | 10.5 | 9.9 | 4.7 | 5.0 | 16.3 | 30.9 | 16,625 | 179 | 1,036 | 39,708 | 9.5 |
| Clackamas | 0.9 | 0.0 | 10.0 | 12.2 | 11.8 | 7.2 | 9.3 | 12.5 | 10.2 | 86,170 | 206 | 5,364 | 172,451 | 9.9 |
| Clatsop | 0.0 | D | 6.7 | 11.1 | 3.3 | 10.2 | 4.6 | 13.0 | 18.3 | 11,135 | 277 | 781 | 22,934 | 6.5 |
| Columbia | 1.0 | 1.2 | 8.8 | 15.4 | 4.6 | 8.2 | 5.1 | 8.8 | 20.8 | 12,915 | 245 | 1,057 | 21,791 | 5.3 |
| Coos | 0.9 | 0.1 | 6.2 | 7.6 | 3.8 | 8.0 | 4.7 | 11.5 | 29.4 | 21,830 | 338 | 2,410 | 31,610 | 3.3 |
| Crook | 0.8 | D | 11.0 | 8.8 | 13.8 | 5.1 | 4.3 | 9.1 | 22.2 | 7,555 | 310 | 487 | 11,597 | 13.7 |
| Curry | 0.8 | D | 8.0 | 10.3 | 4.0 | 9.0 | 5.1 | D | 23.5 | 9,710 | 423 | 652 | 13,157 | 4.3 |
| Deschutes | -0.2 | 0.1 | 14.2 | 5.6 | 11.6 | 7.9 | 8.9 | 16.7 | 11.6 | 46,680 | 236 | 2,322 | 94,859 | 18.4 |
| Douglas | 0.2 | 0.4 | 6.2 | 12.9 | 4.2 | 7.0 | 4.6 | 14.9 | 23.4 | 36,390 | 328 | 3,659 | 51,353 | 5.0 |
| Gilliam | 11.9 | 0.0 | 15.0 | D | D | 3.0 | 1.4 | 2.5 | 18.5 | 565 | 296 | 38 | 1,198 | 3.6 |
| Grant | 0.5 | D | D | D | 4.6 | 5.3 | D | D | 49.1 | 2,380 | 332 | 173 | 4,507 | 3.8 |
| Harney | 9.8 | 0.0 | D | D | 2.8 | 8.0 | 1.8 | 5.1 | 45.0 | 2,120 | 287 | 229 | 3,937 | 2.7 |
| Hood River | 3.6 | 0.0 | 5.8 | 12.5 | 14.8 | 6.4 | 5.7 | 13.4 | 11.7 | 4,490 | 192 | 248 | 10,260 | 10.7 |
| Jackson | 0.1 | 0.1 | 7.1 | 8.7 | 7.1 | 10.0 | 6.6 | 19.4 | 13.9 | 59,210 | 268 | 4,926 | 98,237 | 8.0 |
| Jefferson | 4.5 | D | 3.4 | 16.2 | D | 5.5 | 2.9 | D | 36.0 | 5,890 | 239 | 606 | 10,448 | 6.4 |
| Josephine | 0.0 | D | 7.0 | 10.0 | 4.6 | 12.5 | 7.5 | 18.5 | 13.3 | 28,190 | 322 | 2,896 | 39,900 | 5.0 |
| Klamath | 3.3 | D | 5.2 | 7.1 | 4.4 | 8.7 | 5.9 | 16.5 | 24.8 | 18,725 | 274 | 2,239 | 34,032 | 3.8 |
| Lake | 12.1 | D | 3.7 | 5.7 | 2.8 | 4.4 | 2.2 | D | 47.8 | 2,360 | 297 | 233 | 4,564 | 2.9 |
| Lane | 0.4 | 0.2 | 6.1 | 8.9 | 7.7 | 7.7 | 7.3 | 16.0 | 18.9 | 90,920 | 238 | 9,766 | 165,962 | 6.3 |
| Lincoln | 0.0 | D | 6.2 | 6.7 | 3.8 | 10.0 | 4.9 | 11.3 | 24.8 | 17,300 | 345 | 1,333 | 32,305 | 5.5 |
| Linn | 1.6 | 0.3 | 8.6 | 21.6 | 4.2 | 8.0 | 4.2 | 10.3 | 15.1 | 31,795 | 245 | 3,573 | 52,091 | 6.7 |
| Malheur | 9.0 | D | 2.7 | 7.1 | 3.9 | 9.2 | 3.5 | D | 31.0 | 6,750 | 220 | 1,022 | 11,999 | 2.6 |
| Marion | 2.1 | 0.2 | 8.8 | 5.5 | 5.2 | 6.3 | 7.0 | 14.8 | 29.6 | 68,265 | 196 | 7,710 | 130,107 | 7.6 |

1. Per 1,000 resident population estimated as of July 1 of the year shown.

Items 75—88

# Table B. States and Counties — Housing, Labor Force, and Employment

| STATE County | Housing units, 2015-2019: Occupied units — Total | Percent | Owner-occupied: Median value[1] | Median owner cost as a percent of income — With a mortgage | Without a mortgage[2] | Renter-occupied: Median rent[3] | Median rent as a percent of income[2] | Substandard units[4] (percent) | Civilian labor force, 2020 — Total | Percent change, 2019-2020 | Unemployment — Total | Rate[5] | Civilian employment[6], 2015-2019 — Total | Percent — Management, business, science, and arts | Construction, production, and maintenance occupations |
|---|---|---|---|---|---|---|---|---|---|---|---|---|---|---|---|
| | 89 | 90 | 91 | 92 | 93 | 94 | 95 | 96 | 97 | 98 | 99 | 100 | 101 | 102 | 103 |
| **OKLAHOMA—Cont'd** | | | | | | | | | | | | | | | |
| Kiowa | 3,571 | 72.1 | 62,900 | 19.1 | 10.0 | 545 | 27.8 | 2.5 | 3,544 | 3.3 | 177 | 5.0 | 3,741 | 33.3 | 30.4 |
| Latimer | 4,088 | 69.8 | 87,000 | 17.7 | 10.0 | 622 | 26.4 | 5.0 | 3,172 | 1.7 | 318 | 10.0 | 3,804 | 29.7 | 32.3 |
| Le Flore | 18,261 | 71.0 | 90,300 | 19.6 | 10.0 | 638 | 27.5 | 3.2 | 19,186 | 1.0 | 1,338 | 7.0 | 19,163 | 26.4 | 33.3 |
| Lincoln | 12,763 | 79.7 | 110,800 | 19.9 | 10.0 | 668 | 25.3 | 3.3 | 15,609 | -0.5 | 869 | 5.6 | 14,183 | 29.9 | 30.9 |
| Logan | 15,512 | 82.9 | 169,500 | 18.9 | 10.1 | 780 | 29.3 | 2.2 | 22,207 | -0.5 | 1,139 | 5.1 | 21,043 | 33.4 | 24.6 |
| Love | 3,268 | 76.7 | 111,700 | 19.5 | 10.0 | 715 | 26.1 | 3.4 | 6,605 | 7.0 | 277 | 4.2 | 3,982 | 26.5 | 31.2 |
| McClain | 14,428 | 79.4 | 177,500 | 20.7 | 10.0 | 780 | 27.5 | 2.6 | 19,882 | 0.5 | 1,099 | 5.5 | 18,641 | 34.7 | 25.8 |
| McCurtain | 12,646 | 70.8 | 83,300 | 18.6 | 10.7 | 584 | 24.8 | 6.0 | 15,143 | 5.2 | 1,002 | 6.6 | 12,682 | 23.7 | 39.2 |
| McIntosh | 8,327 | 78.4 | 100,900 | 19.7 | 11.0 | 586 | 25.9 | 5.3 | 7,142 | 0.8 | 601 | 8.4 | 6,777 | 25.6 | 37.0 |
| Major | 3,123 | 79.4 | 101,800 | 16.3 | 10.1 | 732 | 17.5 | 2.8 | 3,726 | -3.5 | 176 | 4.7 | 3,342 | 29.2 | 37.8 |
| Marshall | 6,195 | 72.6 | 89,700 | 19.5 | 10.0 | 688 | 23.0 | 3.3 | 6,850 | -0.6 | 441 | 6.4 | 6,476 | 25.0 | 37.5 |
| Mayes | 15,983 | 72.9 | 112,800 | 20.3 | 10.0 | 745 | 23.3 | 4.3 | 19,460 | 0.4 | 980 | 5.0 | 17,282 | 27.1 | 34.0 |
| Murray | 5,238 | 70.2 | 127,000 | 17.4 | 10.0 | 698 | 22.8 | 5.3 | 6,186 | -2.0 | 382 | 6.2 | 5,835 | 28.8 | 34.9 |
| Muskogee | 26,234 | 65.3 | 104,200 | 19.2 | 10.7 | 681 | 29.9 | 3.5 | 29,421 | -0.5 | 1,778 | 6.0 | 26,517 | 32.0 | 27.5 |
| Noble | 4,515 | 76.8 | 103,500 | 18.1 | 10.9 | 675 | 19.9 | 3.3 | 5,527 | -2.4 | 224 | 4.1 | 5,110 | 37.0 | 30.4 |
| Nowata | 4,099 | 73.7 | 87,700 | 19.1 | 10.8 | 613 | 26.7 | 1.5 | 4,636 | 1.6 | 243 | 5.2 | 4,293 | 24.6 | 36.2 |
| Okfuskee | 3,976 | 71.2 | 81,400 | 19.9 | 10.2 | 549 | 27.2 | 3.9 | 4,457 | -3.1 | 296 | 6.6 | 3,948 | 26.5 | 32.3 |
| Oklahoma | 301,570 | 58.9 | 153,300 | 20.0 | 10.2 | 870 | 28.2 | 2.9 | 385,860 | 0.2 | 24,760 | 6.4 | 375,218 | 37.1 | 22.3 |
| Okmulgee | 14,724 | 71.0 | 85,400 | 19.8 | 11.4 | 655 | 27.2 | 2.8 | 15,612 | -0.7 | 1,128 | 7.2 | 15,215 | 27.4 | 28.2 |
| Osage | 18,263 | 78.3 | 118,600 | 19.7 | 11.4 | 715 | 28.2 | 3.0 | 20,529 | -0.5 | 1,320 | 6.4 | 19,105 | 32.7 | 29.5 |
| Ottawa | 11,847 | 69.0 | 86,300 | 20.1 | 11.3 | 677 | 27.5 | 3.9 | 14,399 | 1.3 | 818 | 5.7 | 12,855 | 27.7 | 29.3 |
| Pawnee | 6,179 | 75.6 | 92,900 | 18.9 | 10.9 | 741 | 24.5 | 3.6 | 7,317 | -0.3 | 475 | 6.5 | 6,595 | 28.5 | 35.2 |
| Payne | 30,788 | 51.7 | 161,700 | 20.4 | 10.4 | 817 | 38.7 | 3.3 | 37,979 | -0.4 | 1,996 | 5.3 | 36,902 | 41.5 | 18.8 |
| Pittsburg | 17,816 | 72.5 | 106,800 | 18.9 | 10.0 | 718 | 27.9 | 2.9 | 16,762 | -2.4 | 1,273 | 7.6 | 17,434 | 30.6 | 28.8 |
| Pontotoc | 14,278 | 65.9 | 129,300 | 18.2 | 10.0 | 707 | 26.1 | 1.9 | 19,341 | 2.1 | 1,000 | 5.2 | 17,464 | 35.3 | 25.2 |
| Pottawatomie | 25,917 | 69.5 | 120,500 | 19.2 | 10.0 | 718 | 26.0 | 2.8 | 32,454 | 0.3 | 1,975 | 6.1 | 30,187 | 30.6 | 26.8 |
| Pushmataha | 4,477 | 75.6 | 80,500 | 18.6 | 10.8 | 551 | 28.4 | 4.7 | 4,474 | 4.7 | 298 | 6.7 | 3,975 | 26.6 | 33.5 |
| Roger Mills | 1,377 | 75.0 | 109,600 | 19.9 | 10.0 | 585 | 18.8 | 1.7 | 1,856 | 0.1 | 90 | 4.8 | 1,574 | 30.0 | 32.0 |
| Rogers | 34,755 | 78.2 | 162,300 | 19.2 | 10.0 | 872 | 26.3 | 2.8 | 44,897 | -0.9 | 2,606 | 5.8 | 44,056 | 34.8 | 27.4 |
| Seminole | 9,317 | 68.0 | 74,400 | 20.5 | 10.3 | 649 | 26.3 | 4.4 | 9,292 | -1.3 | 684 | 7.4 | 9,297 | 25.8 | 31.8 |
| Sequoyah | 15,474 | 71.9 | 97,800 | 19.8 | 11.8 | 681 | 27.5 | 5.2 | 16,555 | 1.4 | 1,124 | 6.8 | 15,919 | 27.0 | 31.0 |
| Stephens | 16,720 | 72.4 | 107,700 | 18.8 | 10.0 | 694 | 25.6 | 1.7 | 18,635 | 0.4 | 1,437 | 7.7 | 18,281 | 27.8 | 32.0 |
| Texas | 6,768 | 68.0 | 115,100 | 22.7 | 10.3 | 771 | 23.1 | 3.9 | 11,361 | 1.4 | 262 | 2.3 | 10,787 | 27.8 | 40.7 |
| Tillman | 2,807 | 75.8 | 56,800 | 18.2 | 11.1 | 616 | 19.7 | 3.7 | 3,054 | 1.9 | 191 | 6.3 | 2,998 | 29.0 | 39.4 |
| Tulsa | 252,661 | 58.8 | 156,400 | 19.3 | 10.7 | 865 | 27.8 | 3.1 | 324,010 | -0.2 | 21,282 | 6.6 | 314,296 | 38.2 | 22.1 |
| Wagoner | 29,208 | 78.6 | 159,000 | 19.3 | 10.3 | 886 | 27.8 | 3.4 | 38,146 | -0.9 | 2,217 | 5.8 | 36,917 | 34.2 | 25.7 |
| Washington | 20,455 | 70.8 | 120,600 | 17.6 | 10.1 | 740 | 26.2 | 2.1 | 22,360 | 0.1 | 1,311 | 5.9 | 22,277 | 37.4 | 23.8 |
| Washita | 4,224 | 70.9 | 84,700 | 17.1 | 10.0 | 688 | 21.3 | 2.8 | 5,154 | -0.1 | 345 | 6.7 | 4,793 | 32.5 | 32.0 |
| Woods | 3,281 | 68.4 | 107,000 | 17.4 | 10.0 | 690 | 24.6 | 1.0 | 4,621 | 0.6 | 159 | 3.4 | 4,385 | 32.5 | 28.8 |
| Woodward | 7,495 | 71.2 | 124,800 | 16.6 | 10.0 | 805 | 24.7 | 1.6 | 8,998 | 0.3 | 603 | 6.7 | 9,372 | 26.7 | 36.5 |
| **OREGON** | 1,611,982 | 62.4 | 312,200 | 22.8 | 11.9 | 1,110 | 30.3 | 3.7 | 2,104,657 | -0.1 | 159,445 | 7.6 | 1,979,043 | 39.4 | 21.6 |
| Baker | 6,921 | 70.7 | 173,100 | 21.5 | 11.2 | 674 | 27.6 | 2.2 | 7,285 | 1.9 | 522 | 7.2 | 6,078 | 34.2 | 29.0 |
| Benton | 35,718 | 57.2 | 331,300 | 21.0 | 11.1 | 1,081 | 36.1 | 1.8 | 46,257 | -3.0 | 2,612 | 5.6 | 44,026 | 50.4 | 13.9 |
| Clackamas | 157,408 | 71.1 | 395,100 | 22.4 | 11.9 | 1,295 | 30.4 | 2.9 | 218,040 | -0.7 | 15,797 | 7.2 | 204,784 | 40.9 | 21.0 |
| Clatsop | 15,800 | 62.2 | 283,900 | 23.0 | 12.6 | 927 | 27.8 | 1.8 | 19,084 | -1.7 | 1,823 | 9.6 | 17,762 | 31.3 | 24.0 |
| Columbia | 19,671 | 75.4 | 264,900 | 21.9 | 11.3 | 921 | 32.7 | 2.4 | 24,061 | -1.1 | 1,906 | 7.9 | 22,601 | 28.8 | 31.8 |
| Coos | 27,025 | 65.9 | 198,100 | 24.4 | 12.8 | 818 | 29.0 | 3.6 | 26,525 | 1.0 | 2,290 | 8.6 | 25,028 | 30.2 | 26.8 |
| Crook | 9,588 | 72.2 | 246,100 | 23.7 | 12.3 | 865 | 30.4 | 4.0 | 10,328 | 4.7 | 911 | 8.8 | 9,177 | 25.6 | 27.6 |
| Curry | 10,546 | 71.4 | 265,400 | 25.5 | 13.2 | 888 | 30.3 | 2.4 | 8,944 | 0.4 | 774 | 8.7 | 7,617 | 31.8 | 19.0 |
| Deschutes | 74,397 | 67.0 | 364,600 | 23.9 | 12.0 | 1,208 | 30.4 | 3.0 | 97,975 | 1.6 | 7,764 | 7.9 | 91,092 | 37.9 | 19.6 |
| Douglas | 45,456 | 68.2 | 199,200 | 23.9 | 12.3 | 824 | 26.8 | 3.2 | 47,415 | 1.5 | 3,658 | 7.7 | 43,291 | 29.1 | 29.0 |
| Gilliam | 842 | 67.3 | 115,900 | 23.8 | 13.1 | 919 | 26.2 | 2.0 | 997 | 5.3 | 61 | 6.1 | 812 | 30.8 | 25.6 |
| Grant | 3,381 | 71.1 | 144,800 | 25.1 | 11.5 | 704 | 27.1 | 3.7 | 3,211 | 0.6 | 267 | 8.3 | 2,972 | 38.9 | 21.9 |
| Harney | 3,244 | 72.2 | 121,300 | 22.7 | 10.3 | 648 | 26.4 | 1.9 | 3,670 | 2.4 | 216 | 5.9 | 2,974 | 39.0 | 19.6 |
| Hood River | 8,600 | 67.2 | 373,600 | 21.2 | 10.0 | 1,133 | 27.1 | 3.8 | 14,154 | -2.2 | 885 | 6.3 | 11,921 | 35.2 | 31.8 |
| Jackson | 88,241 | 63.2 | 280,300 | 24.1 | 12.8 | 993 | 32.5 | 3.9 | 105,147 | 1.1 | 8,210 | 7.8 | 95,367 | 34.4 | 22.8 |
| Jefferson | 8,094 | 68.1 | 224,600 | 21.5 | 10.6 | 839 | 27.7 | 5.5 | 10,330 | 1.5 | 848 | 8.2 | 9,036 | 31.5 | 28.3 |
| Josephine | 36,367 | 66.3 | 265,500 | 26.2 | 11.5 | 908 | 34.0 | 5.3 | 36,124 | 0.7 | 2,847 | 7.9 | 31,985 | 31.2 | 25.2 |
| Klamath | 27,869 | 64.3 | 170,600 | 21.4 | 10.9 | 772 | 29.7 | 3.9 | 29,511 | 0.1 | 2,554 | 8.7 | 26,347 | 32.8 | 26.3 |
| Lake | 3,515 | 61.9 | 146,800 | 21.1 | 10.0 | 709 | 28.4 | 5.9 | 3,647 | 0.6 | 206 | 5.6 | 3,131 | 34.0 | 26.8 |
| Lane | 152,312 | 58.7 | 263,200 | 23.6 | 11.8 | 989 | 32.9 | 3.0 | 178,467 | -1.4 | 14,103 | 7.9 | 173,561 | 35.8 | 21.3 |
| Lincoln | 21,298 | 65.6 | 251,200 | 26.1 | 12.7 | 924 | 30.4 | 2.4 | 20,598 | -1.3 | 2,243 | 10.9 | 19,011 | 27.6 | 22.0 |
| Linn | 47,762 | 64.4 | 221,600 | 22.9 | 11.9 | 964 | 29.5 | 2.9 | 58,819 | 0.0 | 4,568 | 7.8 | 54,874 | 32.4 | 29.9 |
| Malheur | 10,062 | 58.3 | 135,900 | 23.2 | 10.6 | 688 | 28.0 | 4.0 | 12,617 | 2.3 | 654 | 5.2 | 10,965 | 27.6 | 35.8 |
| Marion | 118,038 | 60.2 | 247,100 | 22.6 | 12.0 | 985 | 29.7 | 5.0 | 163,017 | 0.4 | 11,304 | 6.9 | 153,727 | 31.1 | 28.3 |

1. Specified owner-occupied units.   2. A value of 10.0 represents 10 percent or less; a value of 50.0 represents 50 percent or more.   3. Specified renter-occupied units.   4. Overcrowded or lacking complete plumbing facilities.   5. Percent of civilian labor force.   6. Civilian employed persons 16 years old and over.

# Table B. States and Counties — Nonfarm Employment and Agriculture

| STATE County | Private nonfarm establishments, employment and payroll, 2019 | | | | | | | | | Agriculture, 2017 | | | |
| | Number of establishments | Employment | | | | | | Annual payroll | | Farms | | | Farm producers whose primary occupation is farming (percent) |
| | | Total | Health care and social assistance | Manufacturing | Retail trade | Finance and insurance | Professional, scientific, and technical services | Total (mil dol) | Average per employee (dollars) | Number | Percent with: | | |
| | | | | | | | | | | | Fewer than 50 acres | 1000 acres or more | |
| | 104 | 105 | 106 | 107 | 108 | 109 | 110 | 111 | 112 | 113 | 114 | 115 | 116 |

| STATE County | 104 | 105 | 106 | 107 | 108 | 109 | 110 | 111 | 112 | 113 | 114 | 115 | 116 |
|---|---|---|---|---|---|---|---|---|---|---|---|---|---|
| **OKLAHOMA—Cont'd** | | | | | | | | | | | | | |
| Kiowa | 182 | 1,342 | 467 | NA | 269 | 109 | 29 | 43 | 32,090 | 579 | 10.7 | 29.5 | 43.8 |
| Latimer | 153 | 3,264 | 2,145 | NA | 219 | 37 | 79 | 148 | 45,246 | 707 | 30.8 | 5.4 | 38.0 |
| Le Flore | 791 | 6,834 | 1,629 | 230 | 1,467 | 422 | 307 | 202 | 29,549 | 1,672 | 33.4 | 3.8 | 40.4 |
| Lincoln | 562 | 5,898 | 616 | 719 | 797 | 475 | 185 | 233 | 39,516 | 2,231 | 32.9 | 3.5 | 33.0 |
| Logan | 885 | 6,991 | 876 | 593 | 1,234 | 189 | 241 | 258 | 36,836 | 1,262 | 33.4 | 8.0 | 33.6 |
| Love | 157 | 5,313 | 92 | 95 | 115 | 44 | 63 | 188 | 35,367 | 725 | 31.7 | 5.9 | 35.5 |
| McClain | 902 | 8,571 | 1,280 | 388 | 1,472 | 429 | 330 | 306 | 35,692 | 1,296 | 45.2 | 4.9 | 31.4 |
| McCurtain | 610 | 8,457 | 1,292 | 2,629 | 1,242 | 272 | 125 | 305 | 36,043 | 1,479 | 34.8 | 3.0 | 40.0 |
| McIntosh | 347 | 3,151 | 772 | 46 | 901 | 159 | 128 | 110 | 34,835 | 1,013 | 28.8 | 2.4 | 38.9 |
| Major | 256 | 1,954 | 245 | 91 | 281 | 78 | 36 | 85 | 43,664 | 801 | 18.2 | 17.6 | 35.7 |
| Marshall | 303 | 4,383 | 501 | 1,907 | 649 | 167 | 105 | 154 | 35,173 | 588 | 34.4 | 6.6 | 31.2 |
| Mayes | 786 | 11,447 | 1,319 | 3,649 | 1,628 | 259 | 393 | 501 | 43,757 | 1,552 | 41.2 | 2.8 | 35.7 |
| Murray | 277 | 3,692 | 584 | 564 | 608 | 113 | 49 | 141 | 38,245 | 473 | 34.7 | 9.3 | 34.1 |
| Muskogee | 1,370 | 23,210 | 6,513 | 3,329 | 3,206 | 682 | 430 | 965 | 41,595 | 1,586 | 39.3 | 3.2 | 40.8 |
| Noble | 208 | 3,905 | 284 | 1,886 | 279 | 153 | 38 | 211 | 54,004 | 835 | 20.1 | 13.9 | 37.2 |
| Nowata | 149 | 1,445 | 289 | 252 | 185 | 83 | 25 | 50 | 34,572 | 883 | 22.4 | 7.0 | 41.5 |
| Okfuskee | 161 | 1,809 | 882 | 100 | 251 | 63 | 17 | 59 | 32,792 | 934 | 17.8 | 7.1 | 38.3 |
| Oklahoma | 24,047 | 386,898 | 57,129 | 19,291 | 47,262 | 20,337 | 26,349 | 19,204 | 49,636 | 1,103 | 60.7 | 1.8 | 35.8 |
| Okmulgee | 654 | 6,843 | 1,876 | 1,006 | 1,244 | 270 | 225 | 234 | 34,222 | 1,404 | 36.9 | 4.2 | 33.3 |
| Osage | 595 | 5,730 | 707 | 440 | 1,048 | 188 | 185 | 200 | 34,836 | 1,395 | 31.6 | 12.5 | 36.2 |
| Ottawa | 562 | 8,857 | 1,510 | 1,572 | 997 | 235 | 196 | 285 | 32,215 | 947 | 36.4 | 4.1 | 38.5 |
| Pawnee | 255 | 3,221 | 590 | 180 | 391 | 104 | 1,039 | 177 | 55,085 | 818 | 21.5 | 8.7 | 34.8 |
| Payne | 1,843 | 24,015 | 3,679 | 1,678 | 4,102 | 748 | 1,317 | 863 | 35,941 | 1,541 | 39.1 | 5.1 | 31.9 |
| Pittsburg | 919 | 10,973 | 2,320 | 1,138 | 1,967 | 404 | 319 | 436 | 39,719 | 1,623 | 28.3 | 6.2 | 34.8 |
| Pontotoc | 961 | 14,073 | 3,748 | 1,488 | 1,810 | 1,155 | 293 | 566 | 40,238 | 1,438 | 37.0 | 3.9 | 28.9 |
| Pottawatomie | 1,312 | 17,537 | 3,086 | 2,572 | 2,681 | 628 | 795 | 605 | 34,501 | 1,856 | 40.8 | 3.7 | 35.1 |
| Pushmataha | 189 | 1,836 | 659 | 247 | 324 | 115 | 48 | 54 | 29,423 | 695 | 17.1 | 7.1 | 41.7 |
| Roger Mills | 84 | 563 | 94 | NA | 80 | NA | 7 | 19 | 33,259 | 612 | 4.1 | 31.5 | 47.7 |
| Rogers | 1,799 | 30,078 | 2,817 | 7,074 | 2,748 | 912 | 1,638 | 1,553 | 51,621 | 1,776 | 53.9 | 3.3 | 33.8 |
| Seminole | 438 | 5,324 | 1,000 | 1,080 | 795 | 220 | 87 | 205 | 38,412 | 1,143 | 28.2 | 3.3 | 36.8 |
| Sequoyah | 583 | 6,934 | 2,245 | 300 | 1,200 | 294 | 140 | 170 | 24,553 | 1,205 | 40.2 | 3.2 | 36.7 |
| Stephens | 1,083 | 11,928 | 2,431 | 1,123 | 1,839 | 681 | 405 | 466 | 39,072 | 1,226 | 22.5 | 8.3 | 33.7 |
| Texas | 443 | 7,866 | 310 | 2,498 | 781 | 212 | 129 | 391 | 49,693 | 828 | 7.1 | 33.1 | 37.1 |
| Tillman | 125 | 1,024 | 161 | NA | 134 | 82 | 18 | 35 | 34,068 | 456 | 10.3 | 30.9 | 39.0 |
| Tulsa | 19,035 | 343,395 | 52,103 | 38,286 | 38,374 | 13,914 | 21,710 | 17,810 | 51,866 | 1,053 | 63.1 | 1.7 | 33.2 |
| Wagoner | 994 | 8,562 | 1,006 | 2,087 | 1,560 | 223 | 223 | 371 | 43,327 | 1,059 | 53.2 | 3.2 | 35.4 |
| Washington | 1,124 | 18,297 | 2,717 | 972 | 2,258 | 699 | 1,755 | 977 | 53,376 | 899 | 40.0 | 4.7 | 33.2 |
| Washita | 220 | 1,515 | 334 | 118 | 259 | 90 | 159 | 53 | 34,789 | 864 | 13.7 | 26.5 | 42.0 |
| Woods | 270 | 2,548 | 319 | 53 | 430 | 189 | 52 | 101 | 39,498 | 710 | 10.3 | 32.5 | 42.8 |
| Woodward | 763 | 7,465 | 1,128 | 486 | 1,220 | 267 | 173 | 344 | 46,047 | 843 | 17.4 | 25.7 | 35.2 |
| **OREGON** | 119,074 | 1,643,425 | 260,398 | 180,373 | 209,510 | 65,106 | 97,401 | 87,517 | 53,253 | 37,616 | 67.1 | 6.2 | 40.3 |
| Baker | 519 | 4,454 | 721 | 671 | 806 | 110 | 186 | 164 | 36,903 | 705 | 38.7 | 16.6 | 53.9 |
| Benton | 2,182 | 29,149 | 6,432 | 1,742 | 3,826 | 928 | 2,928 | 1,474 | 50,564 | 964 | 72.3 | 2.9 | 36.6 |
| Clackamas | 12,265 | 147,709 | 23,193 | 16,527 | 19,485 | 6,128 | 8,766 | 7,773 | 52,625 | 4,297 | 85.9 | 0.2 | 33.1 |
| Clatsop | 1,532 | 15,577 | 2,401 | 1,686 | 2,970 | 284 | 332 | 594 | 38,146 | 226 | 66.4 | 0.4 | 37.3 |
| Columbia | 966 | 8,854 | 1,318 | 1,441 | 1,589 | 320 | 351 | 333 | 37,608 | 789 | 75.8 | 0.5 | 30.0 |
| Coos | 1,566 | 18,498 | 4,165 | 1,601 | 2,872 | 403 | 454 | 751 | 40,578 | 559 | 45.1 | 4.7 | 48.9 |
| Crook | 554 | 4,357 | 524 | 707 | 640 | 112 | 176 | 181 | 41,570 | 620 | 57.1 | 12.1 | 40.2 |
| Curry | 706 | 5,485 | 1,137 | 700 | 1,059 | 193 | 107 | 203 | 37,010 | 200 | 39.5 | 8.5 | 54.1 |
| Deschutes | 7,669 | 71,896 | 11,101 | 5,774 | 11,632 | 2,271 | 3,834 | 3,199 | 44,498 | 1,484 | 85.4 | 0.8 | 34.7 |
| Douglas | 2,592 | 31,836 | 6,415 | 4,568 | 4,685 | 917 | 860 | 1,425 | 44,765 | 2,009 | 56.6 | 3.7 | 44.2 |
| Gilliam | 69 | 571 | 71 | NA | 52 | 14 | 21 | 30 | 52,180 | 153 | 6.5 | 69.9 | 49.2 |
| Grant | 219 | 1,538 | 408 | 102 | 250 | 55 | 63 | 58 | 37,954 | 383 | 34.5 | 23.8 | 47.0 |
| Harney | 201 | 1,457 | 376 | NA | 319 | 43 | 53 | 54 | 37,268 | 532 | 21.4 | 30.8 | 50.6 |
| Hood River | 1,029 | 10,792 | 1,675 | 1,848 | 1,379 | 155 | 421 | 400 | 37,043 | 578 | 75.4 | 0.5 | 46.4 |
| Jackson | 6,412 | 77,088 | 15,854 | 7,056 | 12,089 | 2,219 | 2,835 | 3,324 | 43,124 | 2,136 | 78.6 | 1.1 | 39.9 |
| Jefferson | 417 | 4,488 | 635 | 1,392 | 618 | 68 | 58 | 171 | 38,153 | 397 | 45.8 | 12.3 | 51.0 |
| Josephine | 2,048 | 24,136 | 5,361 | 2,983 | 4,454 | 781 | 1,004 | 904 | 37,447 | 746 | 85.0 | 0.3 | 43.6 |
| Klamath | 1,567 | 17,211 | 3,587 | 1,559 | 3,113 | 472 | 621 | 704 | 40,915 | 1,005 | 41.9 | 9.0 | 51.6 |
| Lake | 205 | 1,338 | 383 | 185 | 220 | 31 | 62 | 52 | 38,951 | 381 | 25.7 | 29.4 | 61.9 |
| Lane | 10,024 | 127,760 | 24,178 | 14,466 | 20,453 | 5,073 | 5,865 | 5,597 | 43,806 | 2,646 | 78.2 | 1.1 | 35.9 |
| Lincoln | 1,571 | 14,815 | 1,806 | 984 | 3,185 | 263 | 251 | 522 | 35,215 | 384 | 66.4 | 0.3 | 37.2 |
| Linn | 2,688 | 37,607 | 5,535 | 7,977 | 5,275 | 894 | 1,059 | 1,687 | 44,853 | 2,222 | 71.2 | 3.1 | 41.6 |
| Malheur | 710 | 8,387 | 1,400 | 1,045 | 1,926 | 226 | 404 | 300 | 35,781 | 964 | 33.7 | 17.4 | 56.4 |
| Marion | 8,565 | 111,350 | 21,510 | 10,611 | 17,582 | 3,001 | 3,847 | 4,753 | 42,687 | 2,761 | 75.8 | 2.2 | 41.5 |

| STATE County | Acreage (1,000) 117 | Percent change, 2012-2017 118 | Average size of farm 119 | Total irrigated (1,000) 120 | Total cropland (1,000) 121 | Average per farm 122 | Average per acre 123 | Value of machinery and equipment, average per farm (dollars) 124 | Total (mil dol) 125 | Average per farm (acres) 126 | Crops 127 | Livestock and poultry products 128 | Organic farms (number) 129 | Farms with internet access (percent) 130 | Total ($1,000) 131 | Percent of farms 132 |
|---|---|---|---|---|---|---|---|---|---|---|---|---|---|---|---|---|
| OKLAHOMA—Cont'd | | | | | | | | | | | | | | | | |
| Kiowa | 583 | -1.8 | 1,006 | 0.6 | 302.0 | 1,262,052 | 1,254 | 189,899 | 98.2 | 169,539 | 39.3 | 60.7 | 1 | 76.3 | 7,426 | 57.7 |
| Latimer | 214 | -3.1 | 302 | 0.3 | 38.8 | 542,932 | 1,796 | 63,905 | 35.4 | 50,122 | 4.7 | 95.3 | NA | 73.3 | 273 | 4.1 |
| Le Flore | 381 | -3.6 | 228 | 1.8 | 116.2 | 492,429 | 2,162 | 72,515 | 273.6 | 163,654 | 4.2 | 95.8 | NA | 72.5 | 1,040 | 9.4 |
| Lincoln | 482 | 6.0 | 216 | 0.4 | 132.2 | 483,860 | 2,241 | 65,793 | 45.5 | 20,379 | 21.5 | 78.5 | NA | 75.9 | 733 | 9.2 |
| Logan | 393 | 6.9 | 311 | 0.6 | 154.8 | 696,903 | 2,240 | 77,436 | 44.8 | 35,496 | 36.2 | 63.8 | NA | 77.7 | 1,801 | 27.4 |
| Love | 203 | -7.7 | 279 | 1.2 | 42.0 | 685,961 | 2,455 | 65,841 | 22.1 | 30,510 | 17.1 | 82.9 | NA | 68.0 | 763 | 14.3 |
| McClain | 286 | 1.2 | 221 | 0.4 | 85.4 | 582,207 | 2,636 | 67,571 | 42.7 | 30,510 | 24.6 | 75.4 | NA | 77.5 | 1,093 | 12.8 |
| McCurtain | 342 | 8.1 | 231 | 2.6 | 84.9 | 495,237 | 2,141 | 75,986 | 198.1 | 133,921 | 5.7 | 94.3 | NA | 70.1 | 1,705 | 8.4 |
| McIntosh | 226 | -4.4 | 223 | 1.5 | 54.2 | 423,234 | 1,901 | 58,461 | 21.3 | 20,986 | 8.8 | 91.2 | NA | 64.8 | 881 | 14.8 |
| Major | 525 | -2.2 | 655 | 17.9 | 218.6 | 972,803 | 1,484 | 137,306 | 107.3 | 133,939 | 24.0 | 76.0 | 1 | 77.2 | 3,562 | 47.8 |
| Marshall | 174 | -9.1 | 297 | 1.4 | 31.9 | 718,380 | 2,422 | 59,885 | 12.1 | 20,634 | 16.1 | 83.9 | NA | 69.9 | 348 | 11.2 |
| Mayes | 271 | -4.9 | 175 | 0.1 | 90.6 | 485,811 | 2,780 | 62,555 | 79.4 | 51,154 | 9.4 | 90.6 | NA | 73.8 | 722 | 8.1 |
| Murray | 209 | 0.6 | 443 | 0.3 | 28.3 | 946,663 | 2,138 | 84,584 | 16.6 | 35,197 | 9.6 | 90.4 | NA | 72.7 | 997 | 25.4 |
| Muskogee | 312 | -11.0 | 197 | 8.8 | 95.8 | 443,351 | 2,256 | 60,372 | 47.4 | 29,912 | 28.7 | 71.3 | 1 | 72.5 | 1,872 | 25.7 |
| Noble | 449 | 1.4 | 538 | 9.1 | 193.6 | 956,821 | 1,779 | 113,383 | 61.4 | 73,519 | 40.7 | 59.3 | NA | 69.8 | 4,528 | 49.5 |
| Nowata | 347 | 18.6 | 392 | 0.1 | 60.5 | 809,467 | 2,063 | 73,048 | 54.0 | 61,142 | 11.1 | 88.9 | NA | 75.0 | 993 | 16.2 |
| Okfuskee | 347 | 8.6 | 372 | 1.6 | 75.1 | 637,122 | 1,713 | 75,997 | 33.3 | 35,617 | 17.2 | 82.8 | NA | 70.0 | 654 | 13.5 |
| Oklahoma | 133 | -7.5 | 121 | 1.4 | 38.9 | 783,977 | 6,480 | 58,975 | 21.5 | 19,447 | 68.1 | 31.9 | 5 | 78.5 | 261 | 7.0 |
| Okmulgee | 296 | -1.4 | 211 | 0.4 | 72.4 | 466,316 | 2,212 | 58,324 | 31.0 | 22,078 | 20.3 | 79.7 | NA | 72.4 | 337 | 5.0 |
| Osage | 1,101 | -9.5 | 789 | 1.0 | 100.0 | 1,257,014 | 1,592 | 62,815 | 111.9 | 80,201 | 8.5 | 91.5 | NA | 72.7 | 3,951 | 13.7 |
| Ottawa | 206 | 6.6 | 217 | 0.2 | 92.8 | 603,803 | 2,776 | 82,877 | 139.7 | 147,558 | 49.8 | 50.2 | NA | 72.4 | 1,974 | 23.4 |
| Pawnee | 322 | 12.6 | 394 | 0.3 | 66.9 | 701,747 | 1,783 | 75,890 | 34.8 | 42,484 | 16.8 | 83.2 | NA | 68.1 | 1,780 | 25.2 |
| Payne | 341 | -2.6 | 221 | 0.7 | 95.5 | 535,511 | 2,421 | 65,835 | 41.4 | 26,870 | 11.2 | 88.8 | 2 | 75.9 | 1,060 | 13.6 |
| Pittsburg | 519 | | 320 | 0.3 | 92.0 | 559,812 | 1,752 | 67,533 | 42.7 | 26,302 | 10.4 | 89.6 | NA | 68.6 | 573 | 6.8 |
| Pontotoc | 320 | -1.3 | 223 | 0.4 | 69.3 | 482,701 | 2,168 | 54,476 | 26.1 | 18,182 | 16.5 | 83.5 | NA | 71.3 | 431 | 11.9 |
| Pottawatomie | 346 | 3.3 | 186 | 0.7 | 83.7 | 420,405 | 2,254 | 62,144 | 32.6 | 17,566 | 22.2 | 77.8 | NA | 76.8 | 1,272 | 8.9 |
| Pushmataha | 263 | -11.4 | 379 | 0.6 | 34.0 | 584,413 | 1,542 | 55,456 | 16.4 | 23,535 | 5.7 | 94.3 | NA | 69.5 | 377 | 8.6 |
| Roger Mills | 730 | 1.5 | 1,193 | 3.0 | 123.4 | 1,529,868 | 1,282 | 106,486 | 58.5 | 95,533 | 17.8 | 82.2 | NA | 69.8 | 2,259 | 54.7 |
| Rogers | 299 | -0.8 | 169 | 1.2 | 70.5 | 490,477 | 2,910 | 57,314 | 52.1 | 29,346 | 14.2 | 85.8 | NA | 80.0 | 1,091 | 6.6 |
| Seminole | 266 | 9.2 | 232 | 0.4 | 50.7 | 423,540 | 1,823 | 58,027 | 21.6 | 18,891 | 11.8 | 88.2 | NA | 69.9 | 657 | 20.2 |
| Sequoyah | 217 | 0.7 | 180 | 7.1 | 68.0 | 393,733 | 2,191 | 72,790 | 57.7 | 47,909 | 24.9 | 75.1 | 4 | 71.0 | 446 | 4.4 |
| Stephens | 462 | -3.8 | 377 | D | 85.1 | 663,214 | 1,759 | 66,173 | 49.2 | 40,144 | 8.9 | 91.1 | 1 | 67.9 | 2,866 | 18.4 |
| Texas | 1,278 | -0.7 | 1,544 | 168.8 | 680.0 | 1,563,361 | 1,013 | 210,974 | 1,135.7 | 1,371,593 | 11.0 | 89.0 | NA | 70.9 | 16,336 | 72.2 |
| Tillman | 557 | 2.9 | 1,221 | 33.0 | 353.5 | 1,621,033 | 1,327 | 240,298 | 138.0 | 302,697 | 57.3 | 42.7 | NA | 78.9 | 10,974 | 71.5 |
| Tulsa | 113 | 6.6 | 108 | 3.9 | 45.9 | 626,847 | 5,829 | 51,688 | 20.9 | 19,803 | 78.4 | 21.6 | NA | 77.7 | 178 | 5.3 |
| Wagoner | 194 | -2.2 | 184 | 9.5 | 100.6 | 520,155 | 2,833 | 75,163 | 45.9 | 43,381 | 65.8 | 34.2 | NA | 75.8 | 1,580 | 24.6 |
| Washington | 219 | -5.0 | 244 | 0.5 | 50.5 | 574,708 | 2,354 | 62,352 | 35.7 | 39,665 | 18.3 | 81.7 | 2 | 77.4 | 583 | 8.9 |
| Washita | 643 | 1.5 | 744 | 6.2 | 379.1 | 1,105,614 | 1,486 | 162,962 | 119.7 | 138,559 | 39.1 | 60.9 | NA | 78.0 | 9,041 | 58.1 |
| Woods | 830 | 2.7 | 1,169 | 3.0 | 285.0 | 1,549,585 | 1,326 | 155,030 | 79.8 | 112,376 | 23.3 | 76.7 | NA | 72.7 | 4,795 | 53.8 |
| Woodward | 788 | 10.2 | 935 | 5.5 | 173.6 | 1,136,597 | 1,216 | 95,455 | 70.5 | 83,625 | 8.0 | 92.0 | 5 | 74.1 | 1,892 | 50.5 |
| OREGON | 15,962 | -2.1 | 424 | 1,664.9 | 4,726.1 | 1,032,545 | 2,433 | 100,328 | 5,006.8 | 133,103 | 65.6 | 34.4 | 659 | 85.7 | 92,406 | 10.7 |
| Baker | 755 | 6.2 | 1,070 | 108.5 | 130.5 | 1,337,390 | 1,250 | 146,937 | 79.2 | 112,346 | 41.8 | 58.2 | 4 | 82.0 | 3,466 | 28.1 |
| Benton | 128 | 2.9 | 132 | 27.2 | 69.0 | 852,300 | 6,438 | 80,922 | 76.5 | 79,401 | 82.6 | 17.4 | 32 | 86.7 | 1,144 | 6.7 |
| Clackamas | 157 | -3.2 | 37 | 20.5 | 83.7 | 788,181 | 21,514 | 61,085 | 376.3 | 87,575 | 81.9 | 18.1 | 42 | 87.2 | 282 | 1.6 |
| Clatsop | 15 | -8.0 | 67 | 1.9 | 4.4 | 476,979 | 7,153 | 73,613 | 9.7 | 42,743 | 13.5 | 86.5 | 1 | 85.4 | 11 | 3.5 |
| Columbia | 43 | -23.5 | 55 | 2.2 | 12.6 | 475,785 | 8,654 | 47,057 | D | D | D | D | 3 | 84.0 | 135 | 1.9 |
| Coos | 138 | -12.3 | 247 | 10.9 | 16.3 | 753,383 | 3,048 | 74,373 | 45.2 | 80,877 | 18.3 | 81.7 | 19 | 78.9 | 79 | 4.8 |
| Crook | 800 | -2.8 | 1,290 | 67.6 | 49.2 | 1,231,591 | 955 | 98,369 | 44.6 | 71,877 | 27.1 | 72.9 | NA | 85.5 | 869 | 4.2 |
| Curry | 70 | 11.0 | 352 | 3.2 | 6.3 | 1,245,333 | 3,541 | 82,278 | 15.8 | 79,000 | 55.1 | 44.9 | 2 | 73.5 | 307 | 12.5 |
| Deschutes | 135 | 2.7 | 91 | 36.0 | 31.0 | 786,080 | 8,667 | 51,257 | 28.8 | 19,386 | 57.5 | 42.5 | 9 | 93.2 | 90 | 0.8 |
| Douglas | 400 | 4.7 | 199 | 14.7 | 52.3 | 680,086 | 3,414 | 54,494 | 72.5 | 36,092 | 37.8 | 62.2 | 18 | 82.9 | 866 | 4.4 |
| Gilliam | 612 | -15.4 | 3,999 | 7.7 | 249.7 | 2,878,695 | 720 | 320,066 | 26.7 | 174,242 | 69.8 | 30.2 | NA | 81.0 | 7,680 | 83.7 |
| Grant | 629 | -4.2 | 1,642 | 34.5 | 63.0 | 1,699,233 | 1,035 | 103,621 | 24.1 | 63,000 | 13.6 | 86.4 | 6 | 77.3 | 1,295 | 21.1 |
| Harney | 1,557 | 3.4 | 2,927 | 166.5 | 233.5 | 1,996,031 | 682 | 152,672 | 82.3 | 154,692 | 36.0 | 64.0 | 7 | 77.3 | 2,706 | 30.5 |
| Hood River | 28 | 10.2 | 49 | 16.6 | 19.3 | 685,502 | 13,926 | 95,942 | 126.1 | 218,152 | 99.2 | 0.8 | 25 | 89.6 | 365 | 5.9 |
| Jackson | 170 | -20.5 | 80 | 37.5 | 40.7 | 677,191 | 8,494 | 44,300 | 71.0 | 33,262 | 74.7 | 25.3 | 43 | 86.6 | 55 | 0.8 |
| Jefferson | 793 | -3.0 | 1,997 | 44.5 | 77.8 | 1,707,699 | 855 | 196,826 | 67.4 | 169,866 | 81.2 | 18.8 | 6 | 84.9 | 2,162 | 23.2 |
| Josephine | 28 | -1.4 | 37 | 8.0 | 8.4 | 672,056 | 17,992 | 38,200 | 17.5 | 23,456 | 49.2 | 50.8 | 27 | 85.1 | 1 | 1.1 |
| Klamath | 483 | -25.7 | 481 | 165.5 | 147.5 | 1,052,207 | 2,189 | 151,621 | 192.6 | 191,640 | 52.7 | 47.3 | 83 | 84.7 | 2,033 | 13.9 |
| Lake | 756 | 15.0 | 1,983 | 140.3 | 166.7 | 2,143,381 | 1,081 | 255,011 | 93.9 | 246,441 | 47.7 | 52.3 | 21 | 86.4 | 829 | 18.9 |
| Lane | 203 | -7.5 | 77 | 22.3 | 98.0 | 656,860 | 8,556 | 59,094 | 158.4 | 59,873 | 58.0 | 42.0 | 65 | 83.8 | 659 | 2.9 |
| Lincoln | 29 | -4.0 | 76 | 0.4 | 3.6 | 415,466 | 5,498 | 38,118 | D | D | D | D | 4 | 86.5 | 239 | 1.3 |
| Linn | 315 | -4.9 | 142 | 36.9 | 242.6 | 1,005,264 | 7,092 | 100,759 | 243.0 | 109,375 | 74.6 | 25.4 | 39 | 85.8 | 856 | 4.8 |
| Malheur | 1,093 | 1.5 | 1,134 | 174.0 | 210.8 | 1,687,999 | 1,488 | 241,602 | 353.3 | 366,521 | 47.9 | 52.1 | 8 | 86.0 | 4,477 | 33.1 |
| Marion | 289 | 0.9 | 105 | 102.6 | 237.4 | 1,292,998 | 12,367 | 142,613 | 701.6 | 254,104 | 86.0 | 14.0 | 46 | 84.9 | 1,995 | 6.0 |

# Table B. States and Counties — Water Use, Wholesale Trade, Retail Trade, and Real Estate

| STATE County | Water use, 2015 — Public supply water withdrawn (mil gal/day) [133] | Public supply gallons withdrawn per person per day [134] | Wholesale Trade[1], 2017 — Number of establishments [135] | Number of employees [136] | Sales (mil dol) [137] | Average payroll (mil dol) [138] | Retail Trade[2], 2017 — Number of establishments [139] | Number of employees [140] | Sales (mil dol) [141] | Average payroll (mil dol) [142] | Real estate and rental and leasing,[2] 2017 — Number of establishments [143] | Number of employees [144] | Sales (mil dol) [145] | Average payroll (mil dol) [146] |
|---|---|---|---|---|---|---|---|---|---|---|---|---|---|---|
| **OKLAHOMA—Cont'd** | | | | | | | | | | | | | | |
| Kiowa | 6.38 | 697.7 | 8 | 46 | 42.5 | 2.6 | 34 | 266 | 61.4 | 6.5 | 4 | D | 1.1 | D |
| Latimer | 1.40 | 133.5 | D | D | D | 3.1 | 23 | 226 | 49.7 | 4.6 | NA | NA | NA | NA |
| Le Flore | 7.10 | 143.1 | 22 | 175 | 75.2 | 7.1 | 138 | 1,571 | 398.7 | 34.7 | 16 | 23 | 3.6 | 0.7 |
| Lincoln | 1.03 | 29.4 | 24 | 160 | 104.0 | 7.7 | 84 | 829 | 255.7 | 20.8 | 16 | 30 | 7.3 | 1.1 |
| Logan | 2.61 | 56.7 | 19 | 89 | 46.5 | 5.1 | 105 | 1,227 | 454.0 | 34.4 | 46 | 151 | 25.5 | 5.5 |
| Love | 0.83 | 84.1 | 6 | 21 | 7.2 | 0.7 | 18 | 97 | 53.0 | 2.5 | 7 | 13 | 3.7 | 0.3 |
| McClain | 1.38 | 36.3 | 26 | 144 | 45.9 | 5.4 | 126 | 1,500 | 485.0 | 43.2 | 37 | 82 | 12.9 | 2.9 |
| McCurtain | 5.60 | 169.5 | D | D | D | 4.5 | 103 | 1,234 | 332.9 | 28.6 | 20 | 80 | 13.5 | 1.7 |
| McIntosh | 4.59 | 229.6 | 6 | 84 | 18.2 | 1.9 | 70 | 878 | 295.7 | 22.3 | 12 | 30 | 5.6 | 1.5 |
| Major | 5.52 | 710.3 | 12 | 97 | 39.7 | 3.6 | 40 | 272 | 80.3 | 5.8 | 7 | D | 5.7 | D |
| Marshall | 2.29 | 141.1 | 6 | 79 | 21.8 | 4.5 | 55 | 615 | 198.0 | 16.1 | 9 | 25 | 1.9 | 0.5 |
| Mayes | 47.12 | 1,152.4 | 33 | 390 | 403.2 | 23.0 | 124 | 1,623 | 390.6 | 34.9 | 21 | 48 | 9.3 | 1.4 |
| Murray | 7.77 | 557.5 | D | D | D | D | 44 | 613 | 205.4 | 16.3 | D | D | D | D |
| Muskogee | 1.15 | 16.5 | 55 | 734 | 446.5 | 32.5 | 251 | 3,404 | 907.8 | 80.5 | 47 | 174 | 30.0 | 5.1 |
| Noble | 0.28 | 24.2 | 10 | 107 | 53.7 | 4.0 | 31 | 247 | 99.5 | 6.8 | 4 | D | 0.6 | D |
| Nowata | 1.16 | 110.1 | D | D | D | 1.4 | 23 | 176 | 60.4 | 3.4 | 4 | 9 | 1.3 | 0.3 |
| Okfuskee | 0.99 | 81.3 | 4 | 17 | 5.2 | 0.4 | 27 | 240 | 74.2 | 6.2 | NA | NA | NA | NA |
| Oklahoma | 104.31 | 134.3 | 1,141 | 17,335 | 16,731.1 | 995.1 | 2,922 | 45,748 | 14,238.3 | 1,288.0 | 1,321 | 6,844 | 1,779.5 | 341.7 |
| Okmulgee | 5.82 | 148.5 | 18 | 124 | 47.2 | 4.2 | 112 | 1,237 | 354.4 | 29.2 | D | D | D | D |
| Osage | 16.58 | 346.2 | 14 | 82 | 17.8 | 3.7 | 91 | 1,000 | 273.9 | 23.5 | 16 | 36 | 10.2 | 1.7 |
| Ottawa | 2.70 | 84.4 | D | D | D | D | 93 | 1,044 | 254.5 | 23.8 | 21 | 56 | 6.7 | 1.3 |
| Pawnee | 1.02 | 62.1 | 8 | 65 | 10.0 | 1.1 | 36 | 380 | 101.8 | 10.5 | NA | NA | NA | NA |
| Payne | 2.64 | 32.7 | D | D | D | 19.5 | 295 | 4,255 | 1,059.5 | 97.8 | 88 | 373 | 64.2 | 10.8 |
| Pittsburg | 5.90 | 132.3 | D | D | D | D | 163 | 1,978 | 630.1 | 51.0 | 39 | 148 | 31.3 | 5.1 |
| Pontotoc | 4.35 | 113.9 | D | D | D | D | 159 | 1,721 | 465.2 | 40.7 | 40 | 158 | 31.0 | 6.7 |
| Pottawatomie | 7.51 | 104.5 | D | D | D | D | 233 | 2,842 | 752.0 | 66.4 | 49 | 196 | 24.9 | 5.7 |
| Pushmataha | 0.53 | 47.4 | D | D | D | 2.2 | 36 | 348 | 91.0 | 6.9 | 5 | 11 | 8.0 | 0.9 |
| Roger Mills | 0.56 | 147.8 | NA | NA | NA | NA | 18 | 71 | 21.1 | 1.7 | NA | NA | NA | NA |
| Rogers | 72.73 | 801.0 | 74 | 907 | 1,731.5 | 54.4 | 219 | 2,954 | 862.8 | 73.2 | 86 | 248 | 56.1 | 8.1 |
| Seminole | 3.20 | 125.3 | D | D | D | D | 69 | 825 | 224.7 | 20.7 | 17 | 59 | 13.0 | 2.4 |
| Sequoyah | 4.46 | 108.4 | 12 | 69 | 15.6 | 1.4 | 119 | 1,190 | 343.3 | 28.2 | 14 | 32 | 6.3 | 0.7 |
| Stephens | 1.08 | 24.2 | 47 | 398 | 423.6 | 20.9 | 185 | 1,972 | 511.7 | 44.9 | 27 | 86 | 15.4 | 2.7 |
| Texas | 4.75 | 221.0 | D | D | D | D | 78 | 831 | 198.1 | 17.7 | 17 | 37 | 5.4 | 1.0 |
| Tillman | 1.11 | 147.7 | 10 | 117 | 94.9 | 5.2 | 20 | 137 | 25.6 | 2.6 | NA | NA | NA | NA |
| Tulsa | 0.00 | 0.0 | 960 | 14,551 | 10,266.0 | 886.2 | 2,301 | 39,982 | 11,878.0 | 1,063.2 | 1,016 | 7,058 | 1,264.4 | 288.0 |
| Wagoner | 31.31 | 409.0 | 31 | 204 | 81.3 | 12.0 | 113 | 1,521 | 417.6 | 37.2 | 27 | 42 | 7.0 | 1.4 |
| Washington | 1.85 | 35.6 | 26 | 114 | 48.2 | 4.6 | 173 | 2,385 | 681.7 | 59.2 | 51 | 342 | 37.0 | 7.4 |
| Washita | 7.10 | 608.9 | D | D | D | D | 41 | 267 | 69.9 | 6.4 | 3 | D | 1.7 | D |
| Woods | 0.74 | 79.5 | 21 | 300 | 118.3 | 9.7 | 44 | 454 | 119.3 | 11.3 | 11 | 18 | 4.1 | 0.7 |
| Woodward | 6.30 | 292.2 | D | D | D | D | 117 | 1,234 | 382.6 | 32.0 | 35 | 139 | 34.0 | 6.4 |
| **OREGON** | 567.04 | 140.7 | 4,452 | 67,900 | 58,205.2 | 4,306.2 | 14,318 | 211,222 | 61,699.3 | 6,066.6 | 6,771 | 29,773 | 6,773.7 | 1,263.2 |
| Baker | 3.27 | 204.3 | 13 | 74 | 22.0 | 2.7 | 82 | 702 | 182.8 | 18.0 | 15 | 31 | 5.2 | 1.0 |
| Benton | 11.95 | 136.5 | 41 | 278 | 176.7 | 16.6 | 266 | 3,935 | 903.3 | 97.0 | 130 | 493 | 82.4 | 15.3 |
| Clackamas | 134.35 | 334.6 | 563 | 9,140 | 7,174.3 | 608.3 | 1,175 | 19,430 | 5,737.7 | 573.5 | 721 | 3,013 | 767.5 | 139.5 |
| Clatsop | 10.00 | 264.3 | 29 | 183 | 139.5 | 10.4 | 267 | 2,817 | 766.2 | 78.0 | 83 | 445 | 206.8 | 16.2 |
| Columbia | 4.70 | 94.8 | D | D | D | 1.7 | 128 | 1,566 | 395.7 | 42.4 | 45 | 89 | 28.6 | 3.4 |
| Coos | 5.05 | 80.0 | 43 | 405 | 290.7 | 17.3 | 242 | 2,938 | 853.5 | 83.5 | 57 | 211 | 37.3 | 6.1 |
| Crook | 2.11 | 97.5 | 18 | 105 | 76.6 | 6.3 | 68 | 595 | 210.9 | 15.9 | 24 | 46 | 10.1 | 1.5 |
| Curry | 2.85 | 126.8 | D | D | D | D | 111 | 1,080 | 268.6 | 27.9 | 57 | D | 17.2 | D |
| Deschutes | 38.00 | 216.8 | 227 | 1,621 | 1,110.3 | 82.0 | 879 | 11,709 | 3,550.5 | 341.9 | 499 | 1,275 | 298.1 | 49.8 |
| Douglas | 14.2 | 131.9 | D | D | D | D | 390 | 4,761 | 1,268.0 | 121.9 | 122 | 355 | 61.5 | 9.9 |
| Gilliam | 0.55 | 295.9 | D | D | D | 1.5 | 13 | 56 | 18.4 | 1.1 | NA | NA | NA | NA |
| Grant | 0.97 | 135.0 | 7 | D | 11.6 | D | 39 | 272 | 103.2 | 7.2 | D | D | D | 0.1 |
| Harney | 1.72 | 238.9 | 6 | 43 | 23.3 | 1.6 | 30 | 354 | 160.8 | 9.2 | 7 | 19 | 2.4 | 0.5 |
| Hood River | 6.30 | 272.3 | 26 | 721 | 224.4 | 32.2 | 157 | 1,560 | 389.4 | 42.1 | 45 | 137 | 33.1 | 5.7 |
| Jackson | 39.04 | 183.7 | 230 | 2,127 | 1,280.2 | 104.7 | 889 | 12,331 | 3,796.2 | 363.4 | 377 | 1,201 | 244.5 | 39.4 |
| Jefferson | 4.38 | 193.2 | 17 | 170 | 91.7 | 8.7 | 57 | 596 | 178.9 | 15.2 | 22 | 58 | 8.3 | 1.8 |
| Josephine | 7.18 | 84.7 | 56 | 374 | 294.5 | 19.7 | 323 | 4,406 | 1,208.1 | 127.9 | 107 | 431 | 63.9 | 16.7 |
| Klamath | 9.45 | 143.1 | 49 | 447 | 186.9 | 19.3 | 226 | 2,820 | 824.7 | 72.8 | 75 | 249 | 41.4 | 7.6 |
| Lake | 1.56 | 199.3 | D | D | D | 1.4 | 26 | 203 | 108.6 | 6.2 | 7 | 16 | 2.1 | 0.5 |
| Lane | 43.68 | 120.4 | 392 | 5,458 | 2,822.2 | 304.4 | 1,313 | 20,047 | 5,373.5 | 585.6 | 613 | 2,395 | 523.8 | 84.9 |
| Lincoln | 8.65 | 183.9 | D | D | D | D | 294 | 3,167 | 675.8 | 71.9 | 73 | 367 | 62.2 | 10.9 |
| Linn | 9.57 | 79.4 | 116 | 1,634 | 1,248.8 | 86.5 | 368 | 5,189 | 1,424.8 | 140.2 | 115 | 428 | 54.7 | 11.1 |
| Malheur | 6.76 | 222.5 | D | D | D | D | 110 | 1,918 | 598.3 | 54.2 | 26 | 73 | 12.3 | 2.0 |
| Marion | 71.92 | 217.5 | 290 | 3,512 | 2,731.6 | 185.0 | 1,137 | 17,759 | 5,073.6 | 480.0 | 458 | 1,811 | 409.8 | 72.3 |

1  Merchant wholesalers, except manufacturers' sales branches and offices.  2. Employer establishments.

# Table B. States and Counties — Professional Services, Manufacturing, and Accommodation and Food Services

| STATE County | Professional, scientific, and technical services, 2017 | | | | Manufacturing, 2017 | | | | Accommodation and food services, 2017 | | | |
|---|---|---|---|---|---|---|---|---|---|---|---|---|
| | Number of establish-ments | Number of employees | Sales (mil dol) | Average payroll (mil dol) | Number of establish-ments | Number of employees | Sales (mil dol) | Average payroll (mil dol) | Number of establish-ments | Number of employees | Sales (mil dol) | Annual payroll (mil dol) |
| | 147 | 148 | 149 | 150 | 151 | 152 | 153 | 154 | 155 | 156 | 157 | 158 |
| OKLAHOMA—Cont'd | | | | | | | | | | | | |
| Kiowa | 11 | 30 | 4 | 1.1 | NA | NA | NA | NA | D | D | D | D |
| Latimer | 18 | 79 | 28 | 4.8 | D | D | D | D | 7 | 111 | 5.2 | 1.4 |
| Le Flore | D | D | D | D | 25 | 215 | 65.2 | 10.2 | 53 | 740 | 32.1 | 9.0 |
| Lincoln | 41 | 177 | 31 | 8.3 | 30 | 580 | 201.8 | 28.1 | 47 | 566 | 26.3 | 8.0 |
| Logan | 76 | 202 | 29 | 8.4 | 31 | 441 | 123.2 | 25.4 | 57 | 733 | 39.2 | 10.0 |
| Love | 13 | 52 | 6 | 1.7 | 5 | 139 | 32.1 | 8.3 | 29 | 1,280 | 134.7 | 28.2 |
| McClain | 83 | 291 | 45 | 11.8 | 27 | 447 | 82.7 | 27.0 | 72 | 1,137 | 57.7 | 15.0 |
| McCurtain | 31 | 89 | 14 | 4.0 | 32 | 2,627 | 1,906.3 | 129.7 | 62 | 842 | 52.7 | 12.3 |
| McIntosh | 27 | 113 | 15 | 3.6 | 8 | 37 | 5.5 | 1.4 | 39 | 533 | 24.0 | 7.2 |
| Major | 22 | 39 | 5 | 1.2 | 6 | 46 | 10.3 | 2.6 | 10 | 110 | 5.8 | 1.4 |
| Marshall | 22 | 92 | 19 | 3.6 | 19 | 1,309 | 360.6 | 58.0 | 37 | 417 | 22.2 | 5.7 |
| Mayes | 61 | 364 | 39 | 17.7 | 59 | 2,631 | 1,275.4 | 142.5 | 87 | 1,158 | 53.1 | 14.0 |
| Murray | 25 | 64 | 6 | 2.0 | 17 | 439 | 126.6 | 27.1 | 38 | 515 | 32.3 | 9.9 |
| Muskogee | D | D | D | D | 56 | 3,234 | 1,276.8 | 178.5 | 141 | 2,458 | 113.9 | 31.3 |
| Noble | 12 | 44 | 5 | 1.3 | D | D | D | D | 26 | 290 | 15.5 | 4.0 |
| Nowata | 11 | 28 | 5 | 1.0 | 8 | 205 | 59.2 | 12.8 | 11 | 108 | 4.7 | 1.4 |
| Okfuskee | 9 | 15 | 3 | 0.7 | 8 | 110 | 16.1 | 3.8 | 13 | 150 | 7.0 | 1.7 |
| Oklahoma | D | D | D | D | 671 | 19,153 | 6,670.1 | 939.0 | 2,100 | 43,385 | 2,429.7 | 677.8 |
| Okmulgee | 47 | 215 | 23 | 9.8 | 32 | 1,228 | 583.7 | 68.1 | 64 | 1,026 | 42.4 | 13.1 |
| Osage | 44 | 149 | 19 | 9.3 | 30 | 396 | 99.6 | 24.4 | 54 | 801 | 59.6 | 14.4 |
| Ottawa | 40 | 245 | 27 | 9.8 | 37 | 1,332 | 432.5 | 57.1 | 63 | 3,002 | 326.2 | 72.0 |
| Pawnee | 23 | 131 | 44 | 5.7 | 11 | 271 | 50.8 | 13.3 | 21 | 211 | 10.9 | 2.7 |
| Payne | D | D | D | D | 68 | 1,409 | 436.4 | 72.3 | 228 | 4,201 | 194.4 | 52.2 |
| Pittsburg | D | D | D | D | 27 | 1,084 | 340.9 | 54.1 | 96 | 1,445 | 83.1 | 19.7 |
| Pontotoc | D | D | D | D | 28 | 1,125 | 319.1 | 50.1 | 73 | 1,299 | 68.4 | 18.4 |
| Pottawatomie | D | D | D | D | 55 | 2,725 | 806.8 | 131.6 | 140 | 2,831 | 141.2 | 39.3 |
| Pushmataha | 17 | 43 | 4 | 1.4 | 5 | 118 | 37.6 | 5.2 | D | D | D | 1.7 |
| Roger Mills | D | D | D | 0.2 | NA | NA | NA | NA | 6 | D | 0.9 | D |
| Rogers | 137 | 1,223 | 207 | 80.6 | 153 | 6,052 | 2,483.1 | 368.1 | 122 | 3,553 | 356.0 | 80.8 |
| Seminole | 27 | 74 | 9 | 2.4 | 23 | 944 | 237.5 | 48.3 | 42 | 502 | 24.7 | 6.1 |
| Sequoyah | 41 | 130 | 13 | 3.6 | 21 | 264 | 71.7 | 12.1 | 77 | 1,319 | 120.9 | 27.7 |
| Stephens | 91 | 427 | 54 | 17.4 | 49 | 1,098 | 1,369.6 | 71.1 | 83 | 1,118 | 54.5 | 14.8 |
| Texas | D | D | D | D | D | D | D | D | 52 | 661 | 34.3 | 8.2 |
| Tillman | 8 | 21 | 2 | 1.1 | D | D | D | D | 12 | 76 | 3.2 | 0.9 |
| Tulsa | 2,395 | 20,460 | 3,659 | 1,427.5 | 804 | 34,647 | 16,334.4 | 2,019.5 | 1,716 | 36,217 | 2,023.4 | 569.6 |
| Wagoner | 74 | 200 | 23 | 8.8 | 51 | 1,636 | 567.0 | 95.6 | 70 | 937 | 43.7 | 11.7 |
| Washington | D | D | D | D | 32 | 936 | 312.2 | 49.2 | 107 | 1,840 | 91.6 | 25.7 |
| Washita | 22 | 89 | 18 | 6.0 | 7 | 74 | 21.2 | 3.8 | 10 | 60 | 3.8 | 0.8 |
| Woods | 20 | 48 | 5 | 1.3 | 6 | 42 | 23.4 | 2.4 | 26 | 314 | 15.1 | 3.6 |
| Woodward | D | D | D | D | 22 | 505 | 304.9 | 38.3 | 56 | 757 | 47.1 | 11.6 |
| OREGON | 12,527 | 91,027 | 14,580 | 6,829.0 | 5,557 | 172,210 | 62,411.2 | 10,667.7 | 11,708 | 182,613 | 11,803.9 | 3,504.5 |
| Baker | 38 | 178 | 21 | 7.0 | 25 | 501 | 145.0 | 21.5 | 58 | 614 | 35.2 | 9.5 |
| Benton | D | D | D | D | 93 | 1,295 | 421.2 | 65.3 | 224 | 3,900 | 214.6 | 62.0 |
| Clackamas | 1,346 | 8,564 | 1,775 | 753.2 | 568 | 16,224 | 5,073.2 | 969.2 | 879 | 14,375 | 877.9 | 262.6 |
| Clatsop | 91 | 308 | 30 | 10.9 | 51 | 1,713 | 875.1 | 106.7 | 274 | 3,910 | 276.7 | 81.1 |
| Columbia | D | D | D | D | 51 | 1,453 | 447.4 | 78.4 | 100 | 1,151 | 63.5 | 19.2 |
| Coos | D | D | D | D | 58 | 1,435 | 457.3 | 60.4 | 173 | 2,236 | 157.4 | 43.0 |
| Crook | 42 | 130 | 16 | 5.0 | 32 | 784 | 154.8 | 34.6 | 43 | 512 | 36.5 | 10.2 |
| Curry | 33 | 128 | 13 | 4.4 | 20 | 627 | 234.8 | 37.5 | 111 | 995 | 62.0 | 17.4 |
| Deschutes | D | D | D | D | 342 | 4,828 | 1,107.8 | 226.0 | 608 | 10,178 | 707.1 | 224.8 |
| Douglas | D | D | D | D | 134 | 4,164 | 1,372.2 | 192.6 | 268 | 4,299 | 285.2 | 76.5 |
| Gilliam | D | D | D | 1.1 | NA | NA | NA | NA | 12 | D | 2.3 | D |
| Grant | D | D | 5 | D | D | D | D | D | 25 | 151 | 11.3 | 2.9 |
| Harney | 11 | 44 | 4 | 1.7 | D | D | D | D | 30 | 194 | 13.2 | 3.3 |
| Hood River | D | D | D | D | 65 | 1,561 | 455.1 | 75.7 | 107 | 1,664 | 100.7 | 34.1 |
| Jackson | D | D | D | D | 306 | 6,428 | 2,200.2 | 298.8 | 620 | 9,593 | 535.7 | 168.8 |
| Jefferson | 23 | 66 | 6 | 2.3 | 22 | 1,143 | 217.1 | 50.0 | 50 | 602 | 36.1 | 10.1 |
| Josephine | D | D | D | D | 109 | 2,528 | 561.2 | 111.2 | 208 | 2,919 | 169.8 | 49.3 |
| Klamath | D | D | D | D | 47 | 1,951 | 539.4 | 77.1 | 184 | 2,277 | 152.8 | 43.6 |
| Lake | 12 | 64 | 10 | 2.8 | 13 | 189 | 46.8 | 8.6 | 34 | 158 | 9.7 | 2.4 |
| Lane | 985 | 5,099 | 634 | 252.5 | 521 | 13,517 | 5,054.0 | 708.5 | 1,010 | 15,794 | 943.3 | 272.6 |
| Lincoln | D | D | D | D | 49 | 944 | 683.0 | 61.1 | 285 | 3,736 | 262.1 | 73.8 |
| Linn | D | D | D | D | 193 | 7,359 | 2,494.8 | 417.0 | 230 | 3,278 | 175.6 | 53.3 |
| Malheur | D | D | D | D | 37 | 907 | 346.5 | 34.5 | 82 | 1,091 | 63.5 | 16.0 |
| Marion | 671 | 3,652 | 491 | 188.7 | 383 | 10,050 | 3,185.2 | 474.2 | D | D | D | D |

Items 147—158

## Table B. States and Counties — Health Care and Social Assistance, Other Services, Nonemployer Businesses, and Residential Construction

| STATE County | Health care and social assistance, 2017 | | | | Other services, 2017 | | | | Nonemployer businesses, 2018 | | Value of residential construction authorized by building permits, 2020 | |
|---|---|---|---|---|---|---|---|---|---|---|---|---|
| | Number of establishments | Number of employees | Receipts (mil dol) | Annual payroll (mil dol) | Number of establishments | Number of employees | Receipts (mil dol) | Annual payroll (mil dol) | Number | Receipts (mil dol) | New construction ($1,000) | Number of housing units |
| | 159 | 160 | 161 | 162 | 163 | 164 | 165 | 166 | 167 | 168 | 169 | 170 |
| **OKLAHOMA—Cont'd** | | | | | | | | | | | | |
| Kiowa | 21 | 501 | 27.2 | 14.5 | 8 | 46 | 4.4 | 1.1 | 598 | 24.1 | 75 | 1 |
| Latimer | 26 | 2,075 | 213.9 | 90.1 | 10 | 16 | 2.4 | 0.3 | 794 | 38.7 | 0 | 0 |
| Le Flore | 96 | 1,664 | 100.4 | 41.3 | D | D | D | D | 3,084 | 135.8 | 7,024 | 37 |
| Lincoln | D | D | D | D | D | D | D | D | 2,477 | 111.4 | 4,590 | 17 |
| Logan | D | D | D | D | 61 | 195 | 25.1 | 5.3 | 4,022 | 191.7 | 2,055 | 14 |
| Love | 10 | 117 | 5.8 | 2.4 | D | D | D | 1.4 | 655 | 30.6 | 190 | 2 |
| McClain | 83 | 1,297 | 69.8 | 28.6 | 53 | 316 | 47.4 | 14.7 | 3,656 | 185.8 | 116,953 | 615 |
| McCurtain | 74 | 1,328 | 109.3 | 39.1 | 33 | 122 | 9.9 | 3.5 | 2,576 | 114.9 | 4,421 | 40 |
| McIntosh | 49 | 660 | 73.9 | 26.8 | 24 | 69 | 7.0 | 1.7 | 1,307 | 62.8 | 639 | 4 |
| Major | 15 | 261 | 15.6 | 7.7 | 13 | 87 | 7.0 | 3.2 | 699 | 29.7 | 170 | 1 |
| Marshall | 28 | 613 | 57.4 | 21.7 | D | D | D | D | 1,097 | 47.7 | 3,172 | 17 |
| Mayes | 67 | 992 | 105.1 | 38.7 | 44 | 168 | 12.6 | 3.8 | 2,611 | 111.5 | 2,924 | 29 |
| Murray | 33 | 528 | 50.6 | 20.7 | 15 | 62 | 6.1 | 1.6 | 933 | 35.7 | 1,215 | 10 |
| Muskogee | 228 | 6,010 | 778.7 | 317.8 | 84 | 557 | 52.9 | 14.7 | 3,831 | 183.6 | 8,616 | 42 |
| Noble | 19 | 366 | 18.9 | 8.4 | D | D | D | D | 819 | 30.7 | 1,464 | 7 |
| Nowata | 19 | 290 | 15.3 | 7.5 | 6 | 13 | 2.4 | 0.4 | 614 | 26.1 | 321 | 5 |
| Okfuskee | 33 | 1,497 | 106.2 | 49.1 | 5 | 12 | 0.8 | 0.2 | 740 | 31.4 | 270 | 4 |
| Oklahoma | 3,127 | 62,666 | 9,551.0 | 3,214.7 | 1,480 | 9,635 | 1,283.2 | 342.2 | 66,820 | 3,438.8 | 1,189,935 | 4,991 |
| Okmulgee | 106 | 1,674 | 125.0 | 52.9 | D | D | D | D | 2,255 | 93.0 | 10,188 | 77 |
| Osage | 46 | 762 | 52.6 | 21.5 | 21 | 315 | 36.5 | 16.3 | 3,024 | 127.1 | 32,678 | 132 |
| Ottawa | 72 | 1,495 | 122.0 | 49.5 | 33 | 106 | 10.1 | 2.2 | 1,677 | 68.2 | 469 | 2 |
| Pawnee | 29 | 509 | 26.0 | 14.1 | D | D | D | D | 980 | 41.0 | 705 | 3 |
| Payne | 169 | 3,622 | 336.1 | 149.4 | 119 | 870 | 253.7 | 30.4 | 5,442 | 243.7 | 28,161 | 114 |
| Pittsburg | 122 | 2,496 | 221.5 | 98.0 | 50 | 221 | 22.5 | 6.4 | 2,640 | 113.6 | 6,229 | 50 |
| Pontotoc | 128 | 4,198 | 477.0 | 188.2 | 50 | 256 | 29.3 | 7.5 | 2,843 | 135.2 | 13,798 | 101 |
| Pottawatomie | 178 | 3,144 | 259.8 | 113.8 | 77 | 349 | 36.2 | 9.8 | 4,421 | 201.4 | 27,746 | 162 |
| Pushmataha | 28 | 664 | 39.3 | 16.4 | D | D | D | 0.7 | 915 | 41.0 | 0 | 0 |
| Roger Mills | D | D | D | D | D | D | D | 0.0 | 382 | 15.8 | 0 | 0 |
| Rogers | 176 | 2,730 | 307.2 | 113.5 | 90 | 439 | 69.4 | 14.2 | 7,056 | 329.0 | 111,896 | 685 |
| Seminole | 46 | 1,112 | 84.0 | 31.2 | 22 | 67 | 9.0 | 2.0 | 1,333 | 69.3 | 310 | 1 |
| Sequoyah | 76 | 2,521 | 111.2 | 49.6 | D | D | D | D | 2,958 | 127.0 | 11,038 | 112 |
| Stephens | 120 | 2,035 | 217.2 | 76.8 | 66 | 332 | 86.8 | 13.1 | 3,153 | 153.2 | 2,861 | 11 |
| Texas | 39 | 508 | 44.0 | 19.2 | 31 | 107 | 13.2 | 2.9 | 1,173 | 58.3 | 822 | 4 |
| Tillman | 16 | 154 | 13.1 | 3.7 | 7 | 16 | 0.9 | 0.2 | 411 | 13.4 | 0 | 0 |
| Tulsa | 2,112 | 53,768 | 7,176.7 | 2,579.4 | 1,251 | 8,451 | 1,670.9 | 308.1 | 52,636 | 2,619.5 | 660,772 | 3,215 |
| Wagoner | 81 | 971 | 75.1 | 34.2 | 63 | 210 | 24.6 | 5.3 | 5,597 | 249.4 | 120,571 | 615 |
| Washington | 167 | 3,056 | 294.3 | 121.8 | 76 | 463 | 43.9 | 13.3 | 3,264 | 144.7 | 4,692 | 24 |
| Washita | 12 | 314 | 15.5 | 8.5 | D | D | D | D | 846 | 37.1 | 410 | 2 |
| Woods | 17 | 318 | 24.1 | 10.7 | D | D | D | D | 740 | 28.4 | 450 | 3 |
| Woodward | 77 | 1,102 | 109.7 | 40.5 | 44 | 218 | 27.4 | 7.9 | 1,519 | 81.3 | 1,539 | 6 |
| **OREGON** | 13,948 | 263,278 | 33,083.8 | 13,106.2 | 7,414 | 43,543 | 6,288.8 | 1,555.0 | 302,653 | 14,894.4 | 40,147 | 368 |
| Baker | 42 | 573 | 69.1 | 24.8 | 34 | 122 | 12.3 | 3.0 | 1,145 | 42.4 | 6,263 | 27 |
| Benton | 284 | 6,035 | 728.1 | 322.2 | 146 | 994 | 225.5 | 39.6 | 6,036 | 344.7 | 80,154 | 302 |
| Clackamas | 1,325 | 23,578 | 2,861.9 | 1,208.2 | 744 | 3,763 | 384.7 | 118.1 | 32,935 | 1,853.8 | 523,074 | 2,011 |
| Clatsop | 146 | 2,518 | 323.3 | 125.5 | 99 | 386 | 35.6 | 9.8 | 3,024 | 149.6 | 53,849 | 182 |
| Columbia | 120 | 1,301 | 77.6 | 32.2 | D | D | D | D | 2,869 | 114.7 | 49,737 | 209 |
| Coos | 207 | 4,206 | 493.7 | 195.8 | 76 | 406 | 51.4 | 13.4 | 3,791 | 171.5 | 10,574 | 51 |
| Crook | 34 | 540 | 70.8 | 26.5 | 45 | 184 | 19.8 | 5.2 | 1,626 | 77.2 | 66,980 | 235 |
| Curry | 94 | 1,187 | 119.0 | 49.1 | 25 | 117 | 9.3 | 2.6 | 1,876 | 90.3 | 16,512 | 57 |
| Deschutes | 744 | 12,299 | 1,654.6 | 660.7 | 397 | 2,156 | 261.6 | 70.6 | 19,388 | 1,007.9 | 545,756 | 2,338 |
| Douglas | 329 | 6,550 | 942.8 | 406.4 | 143 | 725 | 70.2 | 22.5 | 5,757 | 281.4 | 64,958 | 243 |
| Gilliam | D | D | D | D | D | D | D | 0.3 | 119 | 3.1 | NA | NA |
| Grant | 23 | 377 | 32.7 | 15.9 | D | D | 4.4 | D | 543 | 19.5 | 2,063 | 7 |
| Harney | 19 | 369 | 36.4 | 15.0 | 15 | 49 | 5.4 | 1.5 | 543 | 19.7 | 3,026 | 14 |
| Hood River | 141 | 1,828 | 179.8 | 79.0 | 50 | 251 | 24.6 | 7.5 | 2,233 | 104.9 | 19,739 | 67 |
| Jackson | 758 | 15,189 | 1,902.0 | 753.3 | 328 | 1,967 | 216.7 | 60.8 | 17,655 | 847.1 | 206,466 | 886 |
| Jefferson | 48 | 679 | 70.3 | 30.6 | 27 | 113 | 10.4 | 2.7 | 1,151 | 55.5 | 28,523 | 99 |
| Josephine | 298 | 5,253 | 525.5 | 221.0 | 107 | 482 | 40.7 | 12.6 | 5,993 | 281.0 | 75,240 | 316 |
| Klamath | 187 | 3,439 | 452.3 | 169.2 | 102 | 396 | 40.8 | 11.3 | 3,525 | 153.8 | 32,993 | 152 |
| Lake | 16 | 369 | 31.0 | 18.1 | 11 | 26 | 2.5 | 0.7 | 496 | 20.1 | 1,841 | 8 |
| Lane | 1,260 | 23,686 | 2,982.6 | 1,139.5 | 636 | 3,703 | 552.4 | 118.3 | 25,052 | 1,159.3 | 272,233 | 1,391 |
| Lincoln | 106 | 1,787 | 239.2 | 93.3 | 100 | 367 | 37.6 | 9.1 | 3,607 | 192.5 | 45,217 | 250 |
| Linn | 230 | 5,366 | 575.8 | 238.0 | 163 | 1,004 | 73.8 | 24.8 | 6,357 | 287.2 | 153,187 | 796 |
| Malheur | 92 | 1,645 | 150.4 | 58.4 | D | D | D | D | 1,382 | 60.1 | 7,290 | 25 |
| Marion | 1,065 | 20,875 | 2,427.5 | 1,028.4 | 509 | 2,535 | 267.2 | 78.0 | 18,162 | 883.1 | 351,752 | 1,743 |

# Table B. States and Counties — Government Employment and Payroll, and Local Government Finances

| | Government employment and payroll, 2017 | | March payroll (percent of total) | | | | | | | Local government finances, 2017 — General revenue | | | | |
|---|---|---|---|---|---|---|---|---|---|---|---|---|---|---|
| STATE County | Full-time equivalent employees | March payroll (dollars) | Administration, judicial, and legal | Police and corrections | Fire protection | Highways and transportation | Health and welfare | Natural resources and utilities | Education and libraries | Total (mil dol) | Inter-governmental (mil dol) | Taxes Total (mil dol) | Taxes Per capita[1] Total (dollars) | Taxes Per capita[1] Property (dollars) |
| | 171 | 172 | 173 | 174 | 175 | 176 | 177 | 178 | 179 | 180 | 181 | 182 | 183 | 184 |
| **OKLAHOMA—Cont'd** | | | | | | | | | | | | | | |
| Kiowa | 496 | 1,527,483 | 8.0 | 5.3 | 1.3 | 5.5 | 30.4 | 4.6 | 44.5 | 28.7 | 12.6 | 9.6 | 1,083 | 775 |
| Latimer | 663 | 2,472,658 | 2.7 | 2.0 | 0.0 | 2.8 | 4.2 | 2.6 | 84.1 | 28.8 | 11.6 | 11.4 | 1,106 | 814 |
| Le Flore | 1,741 | 4,832,675 | 6.1 | 7.2 | 0.7 | 4.0 | 4.4 | 6.1 | 70.0 | 134.6 | 72.4 | 37.9 | 757 | 327 |
| Lincoln | 1,029 | 3,037,666 | 8.1 | 7.5 | 1.7 | 5.7 | 2.6 | 5.8 | 66.8 | 116.6 | 34.3 | 65.3 | 1,869 | 1,521 |
| Logan | 861 | 2,505,344 | 6.8 | 10.3 | 5.2 | 4.9 | 1.4 | 5.6 | 64.8 | 73.2 | 29.8 | 21.1 | 450 | 254 |
| Love | 494 | 1,930,140 | 4.5 | 7.3 | 0.0 | 3.1 | 46.1 | 1.7 | 34.4 | 58.6 | 11.5 | 19.4 | 1,921 | 1,401 |
| McClain | 1,394 | 4,237,439 | 6.2 | 7.2 | 3.4 | 3.9 | 10.7 | 4.1 | 61.3 | 100.4 | 41.5 | 45.1 | 1,147 | 633 |
| McCurtain | 1,437 | 4,219,075 | 4.2 | 6.7 | 1.3 | 4.4 | 3.2 | 7.7 | 71.5 | 99.7 | 52.6 | 27.0 | 817 | 462 |
| McIntosh | 597 | 2,089,827 | 6.1 | 4.2 | 0.0 | 1.0 | 0.9 | 5.5 | 82.2 | 52.2 | 29.2 | 18.9 | 960 | 465 |
| Major | 362 | 1,187,800 | 9.8 | 4.9 | 1.1 | 10.6 | 24.9 | 5.3 | 42.2 | 24.2 | 9.8 | 10.2 | 1,322 | 732 |
| Marshall | 528 | 1,612,547 | 14.6 | 3.8 | 1.3 | 1.0 | 4.4 | 3.5 | 71.1 | 35.7 | 17.5 | 13.9 | 852 | 553 |
| Mayes | 1,382 | 4,259,809 | 5.2 | 8.3 | 1.6 | 4.1 | 2.7 | 8.1 | 67.5 | 103.3 | 41.3 | 43.8 | 1,068 | 725 |
| Murray | 509 | 1,611,245 | 7.9 | 8.5 | 3.9 | 2.4 | 36.4 | 13.4 | 26.7 | 50.5 | 14.9 | 15.1 | 1,083 | 532 |
| Muskogee | 3,087 | 9,138,742 | 5.5 | 9.0 | 4.7 | 3.9 | 6.3 | 5.8 | 63.2 | 200.2 | 82.4 | 83.5 | 1,209 | 616 |
| Noble | 594 | 1,743,048 | 5.6 | 6.5 | 3.2 | 7.2 | 21.0 | 7.5 | 48.5 | 61.4 | 13.4 | 33.6 | 2,969 | 2,428 |
| Nowata | 395 | 1,041,445 | 6.6 | 6.6 | 1.5 | 7.1 | 1.8 | 7.1 | 68.9 | 26.8 | 13.8 | 7.4 | 713 | 417 |
| Okfuskee | 426 | 1,153,858 | 7.4 | 5.6 | 0.2 | 6.9 | 0.2 | 3.4 | 75.1 | 29.9 | 16.8 | 9.1 | 755 | 443 |
| Oklahoma | 24,260 | 95,764,486 | 6.0 | 15.3 | 10.1 | 3.3 | 3.6 | 6.4 | 54.6 | 2,696.9 | 767.7 | 1,310.8 | 1,667 | 812 |
| Okmulgee | 1,412 | 4,000,356 | 7.0 | 5.6 | 3.4 | 3.5 | 3.1 | 6.2 | 69.4 | 87.2 | 48.3 | 25.8 | 664 | 310 |
| Osage | 1,083 | 3,317,455 | 7.8 | 7.3 | 0.9 | 13.9 | 14.5 | 4.6 | 49.8 | 69.9 | 32.9 | 24.3 | 514 | 334 |
| Ottawa | 1,388 | 4,235,686 | 4.8 | 5.3 | 2.4 | 2.7 | 4.4 | 5.2 | 73.4 | 102.5 | 54.6 | 25.3 | 807 | 346 |
| Pawnee | 612 | 1,823,700 | 5.8 | 5.7 | 2.0 | 3.4 | 22.5 | 5.7 | 53.3 | 44.2 | 18.7 | 20.0 | 1,220 | 857 |
| Payne | 3,794 | 14,869,696 | 3.9 | 7.1 | 3.8 | 2.1 | 42.3 | 6.8 | 32.7 | 214.4 | 59.7 | 114.1 | 1,393 | 772 |
| Pittsburg | 2,411 | 8,306,313 | 4.0 | 6.4 | 2.4 | 4.6 | 35.7 | 3.7 | 42.2 | 230.4 | 58.4 | 56.4 | 1,277 | 601 |
| Pontotoc | 1,411 | 4,290,560 | 5.4 | 5.8 | 3.2 | 6.4 | 2.1 | 6.5 | 69.2 | 105.6 | 49.3 | 41.8 | 1,088 | 460 |
| Pottawatomie | 2,401 | 7,834,203 | 5.0 | 7.2 | 4.5 | 3.7 | 2.2 | 6.5 | 69.5 | 182.2 | 93.4 | 64.3 | 891 | 369 |
| Pushmataha | 611 | 1,597,765 | 9.8 | 3.8 | 1.5 | 3.8 | 12.4 | 4.7 | 63.7 | 45.5 | 26.1 | 7.8 | 703 | 377 |
| Roger Mills | 235 | 890,831 | 10.5 | 6.0 | 0.0 | 21.3 | 24.0 | 4.2 | 31.9 | 27.1 | 11.1 | 10.0 | 2,718 | 2,132 |
| Rogers | 2,750 | 9,626,821 | 3.3 | 6.5 | 4.9 | 4.1 | 0.7 | 5.9 | 72.0 | 263.1 | 71.3 | 157.9 | 1,727 | 546 |
| Seminole | 1,132 | 3,191,815 | 5.7 | 6.0 | 3.6 | 4.2 | 3.0 | 4.6 | 71.1 | 73.5 | 39.9 | 23.9 | 964 | 436 |
| Sequoyah | 1,602 | 4,807,295 | 4.1 | 7.3 | 0.6 | 3.0 | 11.4 | 4.5 | 67.6 | 120.0 | 58.5 | 29.6 | 708 | 357 |
| Stephens | 1,521 | 4,977,785 | 6.4 | 7.8 | 5.0 | 4.2 | 0.6 | 5.3 | 68.4 | 121.1 | 56.4 | 39.8 | 919 | 565 |
| Texas | 1,006 | 3,006,560 | 5.6 | 8.6 | 3.3 | 6.4 | 15.6 | 1.8 | 57.9 | 70.4 | 31.3 | 27.0 | 1,289 | 668 |
| Tillman | 404 | 1,244,670 | 8.1 | 7.2 | 2.5 | 9.3 | 11.4 | 10.5 | 48.0 | 74.6 | 15.6 | 54.1 | 7,308 | 390 |
| Tulsa | 23,896 | 83,625,515 | 6.5 | 12.2 | 7.6 | 4.0 | 2.5 | 5.5 | 59.3 | 2,498.3 | 677.0 | 1,217.4 | 1,883 | 1,062 |
| Wagoner | 1,202 | 3,956,073 | 7.8 | 7.3 | 2.6 | 4.0 | 1.7 | 8.6 | 66.0 | 86.3 | 37.8 | 32.1 | 406 | 265 |
| Washington | 1,638 | 5,182,295 | 6.3 | 11.0 | 6.9 | 3.1 | 1.1 | 9.0 | 61.2 | 138.9 | 48.2 | 56.6 | 1,089 | 655 |
| Washita | 617 | 2,122,381 | 6.6 | 6.1 | 0.6 | 5.9 | 13.6 | 3.1 | 63.6 | 34.6 | 17.1 | 10.8 | 977 | 698 |
| Woods | 541 | 1,616,617 | 7.3 | 4.5 | 2.5 | 8.1 | 33.4 | 4.1 | 36.6 | 56.4 | 9.5 | 38.5 | 4,241 | 2,833 |
| Woodward | 927 | 3,072,031 | 6.0 | 6.9 | 4.0 | 4.7 | 4.1 | 7.5 | 64.5 | 68.9 | 20.5 | 36.2 | 1,765 | 973 |
| **OREGON** | X | X | X | X | X | X | X | X | X | X | X | X | X | X |
| Baker | 495 | 1,966,935 | 11.7 | 12.7 | 6.2 | 3.2 | 1.3 | 4.8 | 57.0 | 75.8 | 43.4 | 18.9 | 1,178 | 1,092 |
| Benton | 2,152 | 10,442,594 | 7.7 | 11.7 | 5.2 | 3.9 | 12.2 | 5.7 | 51.2 | 268.8 | 91.1 | 134.8 | 1,468 | 1,335 |
| Clackamas | 10,184 | 52,528,397 | 7.4 | 11.5 | 6.1 | 2.6 | 7.3 | 5.4 | 56.3 | 1,724.0 | 585.3 | 764.7 | 1,854 | 1,669 |
| Clatsop | 1,499 | 6,374,922 | 7.5 | 12.9 | 2.5 | 8.4 | 9.2 | 10.0 | 47.1 | 206.1 | 66.7 | 85.4 | 2,184 | 1,754 |
| Columbia | 1,430 | 6,651,910 | 7.8 | 8.8 | 9.0 | 2.6 | 0.5 | 14.3 | 52.6 | 190.6 | 77.1 | 73.7 | 1,425 | 1,335 |
| Coos | 3,259 | 15,847,439 | 2.8 | 6.0 | 1.6 | 1.6 | 48.7 | 3.0 | 34.9 | 441.9 | 145.1 | 77.2 | 1,212 | 1,098 |
| Crook | 661 | 2,838,257 | 9.1 | 14.2 | 5.0 | 3.7 | 4.4 | 8.8 | 51.0 | 93.4 | 35.0 | 30.1 | 1,306 | 1,078 |
| Curry | 924 | 4,269,627 | 6.4 | 8.7 | 0.3 | 3.9 | 46.4 | 4.3 | 27.9 | 116.0 | 32.9 | 27.0 | 1,194 | 1,051 |
| Deschutes | 5,285 | 26,992,573 | 9.1 | 11.0 | 6.1 | 3.2 | 6.3 | 6.9 | 53.4 | 868.5 | 274.6 | 379.2 | 2,031 | 1,737 |
| Douglas | 3,389 | 14,063,815 | 6.1 | 10.3 | 5.0 | 2.7 | 7.1 | 6.0 | 60.9 | 394.2 | 192.0 | 104.8 | 959 | 876 |
| Gilliam | 141 | 574,205 | 24.8 | 0.7 | 1.4 | 0.0 | 9.7 | 10.3 | 42.1 | 31.1 | 5.1 | 14.5 | 7,795 | 5,603 |
| Grant | 454 | 2,197,975 | 5.6 | 6.0 | 0.3 | 5.2 | 43.2 | 3.9 | 33.9 | 73.7 | 26.6 | 8.6 | 1,200 | 1,162 |
| Harney | 494 | 2,025,870 | 5.0 | 6.4 | 0.6 | 3.2 | 47.8 | 1.9 | 34.6 | 66.1 | 26.9 | 8.3 | 1,142 | 1,054 |
| Hood River | 705 | 3,325,908 | 7.9 | 8.0 | 4.5 | 9.5 | 3.3 | 10.1 | 53.3 | 126.6 | 47.8 | 32.7 | 1,396 | 1,167 |
| Jackson | 5,573 | 25,323,949 | 7.9 | 12.8 | 5.8 | 5.8 | 6.8 | 5.7 | 53.0 | 826.7 | 361.6 | 312.0 | 1,440 | 1,189 |
| Jefferson | 823 | 3,287,247 | 6.5 | 8.7 | 2.1 | 1.9 | 3.4 | 11.6 | 63.9 | 113.4 | 69.3 | 26.2 | 1,106 | 970 |
| Josephine | 2,126 | 9,641,417 | 6.2 | 10.1 | 2.6 | 3.2 | 1.6 | 2.9 | 71.5 | 273.1 | 137.7 | 81.5 | 941 | 854 |
| Klamath | 2,208 | 9,029,591 | 4.9 | 7.5 | 4.6 | 5.2 | 5.9 | 6.3 | 62.1 | 255.4 | 122.3 | 78.4 | 1,174 | 1,030 |
| Lake | 557 | 2,329,271 | 4.0 | 4.5 | 1.5 | 5.5 | 53.0 | 2.4 | 27.1 | 66.0 | 23.0 | 13.7 | 1,745 | 1,678 |
| Lane | 11,728 | 57,941,191 | 5.9 | 12.8 | 5.6 | 6.3 | 5.9 | 10.3 | 47.4 | 1,612.9 | 655.1 | 547.6 | 1,459 | 1,212 |
| Lincoln | 1,567 | 7,811,542 | 10.7 | 14.9 | 3.6 | 5.6 | 8.8 | 21.5 | 32.3 | 245.1 | 63.7 | 125.7 | 2,577 | 2,135 |
| Linn | 4,158 | 19,077,155 | 5.7 | 11.1 | 5.7 | 3.7 | 6.5 | 3.4 | 62.8 | 541.6 | 264.8 | 174.1 | 1,392 | 1,262 |
| Malheur | 1,506 | 5,538,601 | 4.1 | 7.9 | 1.3 | 2.0 | 7.6 | 5.8 | 69.9 | 152.8 | 89.1 | 28.7 | 945 | 782 |
| Marion | 11,408 | 59,276,335 | 5.9 | 9.2 | 3.6 | 3.3 | 3.9 | 3.1 | 68.5 | 1,610.8 | 867.8 | 467.2 | 1,372 | 1,218 |

1. Based on the resident population estimated as of July 1 of the year shown.

| STATE County | Direct general expenditure Total (mil dol) | Per capita[1] (dollars) | Percent of total for: Education | Health and hospitals | Police protection | Public welfare | Highways | Debt outstanding Total (mil dol) | Per capita[1] (dollars) | Government employment, 2019 Federal civilian | Federal military | State and local | Individual income tax returns, 2018 Number of returns | Mean adjusted gross income | Mean income tax |
|---|---|---|---|---|---|---|---|---|---|---|---|---|---|---|---|
| | 185 | 186 | 187 | 188 | 189 | 190 | 191 | 192 | 193 | 194 | 195 | 196 | 197 | 198 | 199 |
| **OKLAHOMA—Cont'd** | | | | | | | | | | | | | | | |
| Kiowa | 26 | 2,894 | 59.5 | 0.8 | 5.4 | 0.0 | 2.4 | 18.1 | 2,041 | 40 | 31 | 605 | 3,300 | 40,567 | 2,957 |
| Latimer | 22 | 2,131 | 57.8 | 0.8 | 4.1 | 0.6 | 11.0 | 11.8 | 1,141 | 28 | 35 | 856 | 3,930 | 43,874 | 3,113 |
| Le Flore | 142 | 2,848 | 56.2 | 4.1 | 5.5 | 0.1 | 7.8 | 53.3 | 1,065 | 153 | 176 | 4,945 | 17,620 | 43,644 | 2,989 |
| Lincoln | 81 | 2,306 | 58.0 | 1.3 | 6.2 | 0.0 | 8.8 | 47.3 | 1,354 | 82 | 126 | 1,642 | 13,650 | 49,876 | 3,929 |
| Logan | 85 | 1,806 | 52.9 | 0.6 | 8.5 | 0.0 | 7.5 | 63.1 | 1,348 | 62 | 167 | 1,328 | 19,670 | 67,697 | 6,946 |
| Love | 45 | 4,420 | 38.3 | 44.6 | 0.6 | 0.0 | 4.5 | 14.2 | 1,412 | 21 | 37 | 3,303 | 4,390 | 47,758 | 3,889 |
| McClain | 105 | 2,660 | 58.9 | 2.3 | 7.7 | 0.0 | 9.1 | 42.7 | 1,087 | 72 | 147 | 1,847 | 17,570 | 67,700 | 6,735 |
| McCurtain | 96 | 2,890 | 62.1 | 3.3 | 2.5 | 0.0 | 8.5 | 47.5 | 1,438 | 124 | 118 | 2,258 | 12,450 | 40,934 | 2,971 |
| McIntosh | 45 | 2,305 | 67.1 | 0.7 | 3.5 | 0.0 | 12.2 | 15.7 | 795 | 37 | 70 | 1,076 | 7,370 | 44,069 | 3,670 |
| Major | 24 | 3,112 | 51.1 | 2.6 | 5.8 | 0.0 | 14.5 | 7.5 | 970 | 29 | 28 | 371 | 3,340 | 62,484 | 6,068 |
| Marshall | 36 | 2,167 | 68.9 | 3.2 | 4.1 | 0.0 | 2.5 | 7.8 | 475 | 19 | 60 | 833 | 6,510 | 52,406 | 4,920 |
| Mayes | 109 | 2,651 | 66.1 | 0.4 | 5.7 | 0.0 | 5.1 | 51.2 | 1,248 | 69 | 148 | 2,417 | 15,970 | 49,360 | 3,917 |
| Murray | 49 | 3,506 | 39.5 | 29.7 | 3.8 | 0.0 | 4.5 | 20.2 | 1,448 | 61 | 50 | 2,366 | 5,800 | 52,499 | 4,702 |
| Muskogee | 201 | 2,918 | 53.4 | 1.2 | 6.2 | 0.0 | 5.3 | 119.8 | 1,735 | 3,178 | 235 | 6,732 | 26,270 | 48,530 | 3,744 |
| Noble | 48 | 4,250 | 45.9 | 22.3 | 4.2 | 0.0 | 10.0 | 23.7 | 2,094 | 40 | 40 | 1,280 | 4,680 | 54,069 | 4,606 |
| Nowata | 26 | 2,555 | 60.6 | 0.0 | 5.3 | 0.0 | 12.1 | 11.2 | 1,085 | 32 | 36 | 522 | 4,050 | 48,303 | 3,774 |
| Okfuskee | 31 | 2,601 | 65.6 | 0.9 | 6.4 | 0.0 | 11.9 | 12.0 | 997 | 29 | 39 | 1,024 | 3,980 | 40,643 | 2,540 |
| Oklahoma | 2,505 | 3,186 | 40.9 | 0.4 | 11.7 | 0.0 | 4.7 | 3,658.4 | 4,654 | 27,834 | 8,804 | 57,028 | 341,510 | 70,139 | 8,899 |
| Okmulgee | 81 | 2,086 | 68.3 | 0.1 | 4.6 | 0.0 | 1.1 | 86.8 | 2,236 | 97 | 135 | 3,401 | 14,590 | 45,202 | 3,456 |
| Osage | 70 | 1,482 | 53.1 | 5.7 | 3.3 | 0.0 | 12.1 | 14.8 | 314 | 165 | 173 | 2,754 | 17,760 | 54,453 | 4,962 |
| Ottawa | 113 | 3,607 | 43.6 | 0.4 | 5.3 | 0.0 | 6.5 | 48.5 | 1,547 | 103 | 110 | 5,370 | 12,610 | 40,021 | 2,751 |
| Pawnee | 34 | 2,055 | 60.1 | 3.7 | 7.0 | 0.0 | 9.2 | 10.6 | 644 | 228 | 59 | 876 | 6,260 | 50,404 | 4,074 |
| Payne | 208 | 2,544 | 55.4 | 0.8 | 11.4 | 0.0 | 7.5 | 207.5 | 2,534 | 231 | 281 | 14,643 | 29,620 | 59,421 | 6,121 |
| Pittsburg | 226 | 5,128 | 29.6 | 38.5 | 4.2 | 0.1 | 6.8 | 92.8 | 2,103 | 2,080 | 154 | 3,980 | 17,490 | 49,628 | 4,110 |
| Pontotoc | 108 | 2,825 | 55.6 | 0.6 | 5.9 | 0.1 | 11.1 | 60.9 | 1,588 | 142 | 133 | 7,549 | 16,120 | 55,474 | 5,423 |
| Pottawatomie | 185 | 2,559 | 56.6 | 0.5 | 6.9 | 0.0 | 9.7 | 77.2 | 1,070 | 147 | 254 | 6,301 | 28,880 | 50,696 | 4,245 |
| Pushmataha | 43 | 3,842 | 49.0 | 18.5 | 3.4 | 0.0 | 9.7 | 9.0 | 809 | 24 | 40 | 859 | 4,160 | 42,171 | 2,863 |
| Roger Mills | 32 | 8,774 | 25.7 | 20.0 | 3.6 | 0.0 | 40.9 | 0.5 | 134 | 34 | 13 | 377 | 1,500 | 52,858 | 5,004 |
| Rogers | 204 | 2,229 | 53.2 | 1.1 | 5.8 | 0.0 | 3.6 | 154.3 | 1,688 | 517 | 332 | 6,157 | 41,090 | 69,554 | 7,312 |
| Seminole | 69 | 2,782 | 61.7 | 0.9 | 3.7 | 0.0 | 8.2 | 23.4 | 943 | 173 | 86 | 1,861 | 8,890 | 43,821 | 3,288 |
| Sequoyah | 123 | 2,956 | 55.5 | 14.9 | 4.7 | 0.0 | 6.3 | 66.9 | 1,603 | 124 | 166 | 2,881 | 15,640 | 43,107 | 3,097 |
| Stephens | 119 | 2,744 | 54.4 | 1.7 | 8.2 | 0.0 | 7.0 | 72.6 | 1,675 | 82 | 155 | 2,008 | 17,470 | 60,741 | 6,357 |
| Texas | 74 | 3,537 | 53.3 | 0.5 | 3.6 | 0.1 | 8.2 | 34.8 | 1,665 | 73 | 71 | 1,566 | 9,060 | 48,006 | 3,715 |
| Tillman | 52 | 6,995 | 24.0 | 8.0 | 2.6 | 0.0 | 8.6 | 14.7 | 1,985 | 32 | 25 | 564 | 2,750 | 40,832 | 2,807 |
| Tulsa | 2,368 | 3,663 | 44.4 | 5.6 | 7.3 | 0.9 | 7.1 | 3,702.0 | 5,727 | 3,462 | 2,385 | 31,786 | 285,480 | 74,294 | 9,157 |
| Wagoner | 80 | 1,017 | 60.3 | 4.2 | 7.4 | 0.0 | 4.6 | 66.4 | 841 | 75 | 308 | 1,768 | 33,660 | 62,186 | 5,645 |
| Washington | 139 | 2,676 | 49.9 | 1.4 | 7.3 | 0.0 | 4.7 | 130.5 | 2,509 | 88 | 185 | 2,135 | 22,270 | 66,315 | 7,394 |
| Washita | 41 | 3,713 | 45.1 | 1.1 | 4.8 | 0.0 | 17.8 | 18.2 | 1,648 | 45 | 39 | 660 | 4,410 | 47,636 | 3,542 |
| Woods | 38 | 4,187 | 49.6 | 0.9 | 4.5 | 0.0 | 15.6 | 5.4 | 595 | 29 | 28 | 1,106 | 3,520 | 60,448 | 6,791 |
| Woodward | 64 | 3,097 | 52.4 | 3.2 | 4.1 | 0.0 | 9.3 | 60.0 | 2,923 | 86 | 69 | 1,559 | 8,520 | 54,694 | 5,539 |
| **OREGON** | X | X | X | X | X | X | X | X | X | 28,548 | 11,388 | 256,378 | 1,965,520 | 71,631 | 8,329 |
| Baker | 74 | 4,598 | 50.6 | 6.7 | 5.2 | 1.4 | 6.8 | 15.0 | 933 | 196 | 37 | 940 | 7,110 | 47,428 | 4,044 |
| Benton | 220 | 2,391 | 55.4 | 0.0 | 7.5 | 0.0 | 2.7 | 177.8 | 1,937 | 483 | 253 | 9,933 | 40,530 | 74,921 | 8,800 |
| Clackamas | 1,695 | 4,110 | 47.2 | 3.7 | 7.4 | 1.2 | 3.7 | 3,247.9 | 7,875 | 1,148 | 1,096 | 15,530 | 203,160 | 90,673 | 12,048 |
| Clatsop | 195 | 4,981 | 40.8 | 1.8 | 8.1 | 1.7 | 5.1 | 436.7 | 11,168 | 199 | 544 | 2,494 | 19,170 | 57,431 | 5,517 |
| Columbia | 178 | 3,448 | 49.3 | 2.8 | 4.9 | 0.2 | 3.8 | 231.5 | 3,634 | 309 | 405 | 5,396 | 28,200 | 63,948 | 6,083 |
| Coos | 460 | 7,222 | 32.3 | 39.7 | 3.0 | 0.0 | 2.0 | 267.5 | 5,172 | 73 | 123 | 1,976 | 24,030 | 53,627 | 5,064 |
| Crook | 92 | 4,001 | 41.4 | 1.9 | 8.9 | 0.1 | 6.7 | 78.2 | 3,392 | 282 | 57 | 956 | 11,050 | 57,270 | 5,383 |
| Curry | 122 | 5,377 | 23.6 | 38.7 | 5.5 | 0.2 | 4.4 | 91.1 | 4,028 | 98 | 96 | 1,143 | 11,030 | 53,536 | 5,304 |
| Deschutes | 871 | 4,662 | 41.6 | 3.7 | 6.6 | 0.1 | 6.2 | 1,327.8 | 7,111 | 976 | 469 | 8,633 | 97,360 | 79,913 | 10,042 |
| Douglas | 412 | 3,770 | 51.3 | 7.0 | 5.3 | 0.0 | 5.7 | 200.8 | 1,839 | 1,540 | 307 | 6,044 | 47,040 | 54,299 | 5,081 |
| Gilliam | 45 | 24,036 | 16.4 | 1.9 | 2.0 | 0.6 | 7.6 | 39.9 | 21,427 | 12 | 4 | 227 | 810 | 57,154 | 4,964 |
| Grant | 64 | 8,845 | 26.7 | 37.1 | 3.1 | 0.7 | 13.2 | 14.3 | 1,995 | 279 | 17 | 761 | 3,150 | 46,696 | 4,015 |
| Harney | 64 | 8,818 | 32.4 | 40.2 | 1.6 | 0.5 | 8.1 | 28.3 | 3,895 | 235 | 17 | 789 | 3,120 | 41,112 | 3,200 |
| Hood River | 123 | 5,242 | 44.1 | 2.2 | 3.0 | 0.7 | 6.2 | 214.8 | 9,178 | 123 | 54 | 1,171 | 12,110 | 72,391 | 8,272 |
| Jackson | 767 | 3,538 | 44.2 | 6.2 | 9.3 | 0.0 | 5.4 | 894.6 | 4,127 | 1,825 | 517 | 8,893 | 103,800 | 61,203 | 6,482 |
| Jefferson | 99 | 4,169 | 51.9 | 6.6 | 4.1 | 0.0 | 4.4 | 95.0 | 4,015 | 121 | 56 | 2,196 | 10,200 | 46,780 | 3,763 |
| Josephine | 276 | 3,186 | 64.0 | 3.0 | 8.2 | 0.2 | 5.2 | 96.3 | 1,112 | 260 | 204 | 3,144 | 37,960 | 51,782 | 4,849 |
| Klamath | 267 | 4,000 | 56.9 | 1.5 | 3.9 | 0.0 | 6.1 | 123.8 | 1,853 | 835 | 265 | 4,085 | 29,050 | 49,237 | 4,373 |
| Lake | 71 | 9,034 | 24.0 | 50.0 | 2.4 | 0.2 | 5.3 | 39.3 | 4,985 | 247 | 18 | 895 | 3,190 | 47,131 | 3,918 |
| Lane | 1,561 | 4,159 | 48.0 | 5.4 | 6.7 | 1.0 | 3.5 | 1,918.9 | 5,114 | 1,884 | 985 | 23,860 | 173,440 | 63,169 | 6,847 |
| Lincoln | 232 | 4,759 | 30.4 | 9.6 | 9.1 | 0.1 | 6.7 | 300.5 | 6,163 | 328 | 204 | 3,572 | 23,030 | 56,663 | 5,387 |
| Linn | 579 | 4,630 | 53.2 | 5.0 | 7.4 | 0.3 | 4.8 | 417.7 | 3,341 | 327 | 307 | 6,356 | 57,160 | 56,383 | 4,998 |
| Malheur | 156 | 5,133 | 63.1 | 2.9 | 4.1 | 0.0 | 3.0 | 81.8 | 2,692 | 205 | 65 | 3,049 | 10,910 | 42,512 | 3,007 |
| Marion | 1,580 | 4,639 | 54.6 | 6.6 | 5.3 | 0.1 | 4.6 | 1,384.2 | 4,065 | 1,352 | 814 | 34,563 | 152,570 | 59,755 | 5,739 |

1. Based on the resident population estimated as of July 1 of the year shown.

# Table B. States and Counties — **Land Area and Population**

| State / county code | CBSA code[1] | County Type code[2] | STATE County | Land area[3] (sq. mi) | Total persons 2019 | Rank | Per square mile | White | Black | American Indian, Alaska Native | Asian and Pacific Islander | Percent Hispanic or Latino[4] | Under 5 years | 5 to 14 years | 15 to 24 years | 25 to 34 years | 35 to 44 years | 45 to 54 years |
|---|---|---|---|---|---|---|---|---|---|---|---|---|---|---|---|---|---|---|
| | | | | **1** | **2** | **3** | **4** | **5** | **6** | **7** | **8** | **9** | **10** | **11** | **12** | **13** | **14** | **15** |
| | | | **OREGON—Cont'd** | | | | | | | | | | | | | | | |
| 41049 | 25840 | 6 | Morrow | 2,030.5 | 11,700 | 2,311 | 5.8 | 58.9 | 1.0 | 2.3 | 1.5 | 38.2 | 7.1 | 15.5 | 14.1 | 11.2 | 11.4 | 11.4 |
| 41051 | 38900 | 1 | Multnomah | 431.2 | 815,637 | 78 | 1,891.6 | 72.6 | 7.1 | 1.9 | 10.8 | 12.2 | 4.9 | 10.2 | 10.7 | 18.8 | 16.8 | 13.3 |
| 41053 | 41420 | 2 | Polk | 740.9 | 87,744 | 665 | 118.4 | 79.9 | 1.7 | 3.0 | 4.1 | 14.9 | 5.5 | 12.7 | 16.1 | 12.4 | 12.3 | 10.8 |
| 41055 | | 9 | Sherman | 823.6 | 1,801 | 3,064 | 2.2 | 91.4 | 1.1 | 2.3 | 1.6 | 5.9 | 6.0 | 10.6 | 8.1 | 11.9 | 11.7 | 10.0 |
| 41057 | | 6 | Tillamook | 1,102.4 | 27,442 | 1,514 | 24.9 | 86.3 | 1.0 | 2.5 | 2.4 | 10.7 | 4.5 | 10.7 | 9.0 | 10.8 | 10.9 | 11.1 |
| 41059 | 25840 | 4 | Umatilla | 3,215.4 | 77,752 | 720 | 24.2 | 67.0 | 1.4 | 4.2 | 1.8 | 27.9 | 6.3 | 14.4 | 13.1 | 14.3 | 12.6 | 11.5 |
| 41061 | 29260 | 7 | Union | 2,036.9 | 26,551 | 1,544 | 13.0 | 90.4 | 1.3 | 2.1 | 3.6 | 5.1 | 5.2 | 12.8 | 14.9 | 11.7 | 11.6 | 9.9 |
| 41063 | | 9 | Wallowa | 3,145.9 | 7,181 | 2,643 | 2.3 | 94.2 | 1.0 | 1.9 | 1.3 | 3.9 | 5.0 | 10.9 | 7.9 | 9.4 | 11.4 | 9.7 |
| 41065 | 45520 | 6 | Wasco | 2,381.1 | 26,403 | 1,554 | 11.1 | 75.0 | 1.1 | 3.8 | 2.5 | 20.0 | 5.9 | 12.3 | 10.7 | 13.7 | 12.0 | 10.8 |
| 41067 | 38900 | 1 | Washington | 724.3 | 603,514 | 114 | 833.2 | 67.5 | 3.2 | 1.3 | 14.9 | 17.3 | 5.6 | 12.7 | 11.9 | 15.9 | 15.2 | 13.1 |
| 41069 | | 9 | Wheeler | 1,716.0 | 1,387 | 3,087 | 0.8 | 90.8 | 1.4 | 4.0 | 1.8 | 6.1 | 4.3 | 7.1 | 8.2 | 7.7 | 8.4 | 9.2 |
| 41071 | 38900 | 1 | Yamhill | 715.9 | 107,664 | 566 | 150.4 | 79.3 | 1.5 | 2.5 | 3.3 | 16.3 | 5.4 | 12.3 | 13.6 | 12.8 | 13.0 | 12.0 |
| 42000 | | 0 | **PENNSYLVANIA** | 44,741.7 | 12,783,254 | X | 285.7 | 76.9 | 12.0 | 0.5 | 4.4 | 8.1 | 5.4 | 11.5 | 12.4 | 13.4 | 12.0 | 12.2 |
| 42001 | 23900 | 3 | Adams | 518.7 | 102,742 | 592 | 198.1 | 90.0 | 2.3 | 0.5 | 1.3 | 7.3 | 4.8 | 11.2 | 12.8 | 11.1 | 10.6 | 12.7 |
| 42003 | 38300 | 1 | Allegheny | 730.0 | 1,211,358 | 36 | 1,659.4 | 79.9 | 14.7 | 0.6 | 5.0 | 2.3 | 5.2 | 10.3 | 11.6 | 15.4 | 12.5 | 11.4 |
| 42005 | 38300 | 1 | Armstrong | 653.3 | 64,162 | 836 | 98.2 | 97.9 | 1.5 | 0.4 | 0.5 | 1.9 | 4.6 | 10.0 | 10.7 | 11.1 | 11.1 | 13.1 |
| 42007 | 38300 | 1 | Beaver | 434.7 | 162,575 | 410 | 374.0 | 90.9 | 8.0 | 0.5 | 1.0 | 1.9 | 5.0 | 10.9 | 10.4 | 11.8 | 11.7 | 12.0 |
| 42009 | | 6 | Bedford | 1,012.4 | 47,817 | 1,027 | 47.2 | 97.5 | 1.1 | 0.4 | 0.6 | 1.3 | 5.0 | 10.9 | 10.4 | 10.7 | 11.8 | 13.0 |
| 42011 | 39740 | 2 | Berks | 856.4 | 421,017 | 170 | 491.6 | 70.9 | 5.2 | 0.4 | 1.8 | 23.1 | 5.6 | 13.3 | 12.6 | 11.7 | 10.4 | 12.5 |
| 42013 | 11020 | 1 | Blair | 525.3 | 121,007 | 524 | 230.4 | 95.8 | 2.9 | 0.4 | 1.0 | 1.3 | 5.1 | 11.4 | 11.2 | 12.2 | 11.6 | 12.3 |
| 42015 | 42380 | 6 | Bradford | 1,147.5 | 60,221 | 870 | 52.5 | 96.8 | 1.1 | 0.7 | 1.0 | 1.5 | 5.7 | 12.6 | 10.6 | 11.3 | 10.4 | 11.8 |
| 42017 | 37980 | 1 | Bucks | 604.4 | 627,987 | 107 | 1,039.0 | 84.4 | 5.0 | 0.5 | 6.1 | 5.9 | 4.8 | 11.3 | 11.6 | 11.1 | 12.0 | 13.4 |
| 42019 | 38300 | 1 | Butler | 789.5 | 189,135 | 356 | 239.6 | 95.3 | 1.9 | 0.4 | 2.0 | 1.7 | 5.0 | 11.0 | 11.9 | 11.4 | 12.1 | 13.2 |
| 42021 | 27780 | 3 | Cambria | 687.5 | 128,672 | 498 | 187.2 | 93.9 | 4.8 | 0.4 | 0.9 | 1.8 | 4.8 | 10.7 | 12.4 | 10.6 | 10.5 | 12.2 |
| 42023 | | 7 | Cameron | 396.2 | 4,330 | 2,873 | 10.9 | 96.8 | 1.5 | 0.8 | 0.7 | 1.6 | 4.5 | 9.9 | 8.9 | 8.9 | 9.3 | 11.9 |
| 42025 | 10900 | 2 | Carbon | 381.3 | 64,081 | 837 | 168.1 | 91.5 | 2.4 | 0.6 | 0.9 | 5.7 | 4.6 | 10.7 | 10.3 | 11.1 | 11.4 | 13.5 |
| 42027 | 44300 | 3 | Centre | 1,108.7 | 161,496 | 415 | 145.7 | 86.8 | 4.2 | 0.4 | 7.1 | 3.0 | 3.7 | 8.4 | 25.8 | 13.3 | 10.9 | 10.7 |
| 42029 | 37980 | 1 | Chester | 750.5 | 526,759 | 135 | 701.9 | 80.1 | 6.7 | 0.5 | 7.2 | 7.6 | 5.4 | 12.8 | 13.0 | 11.6 | 12.6 | 13.3 |
| 42031 | | 6 | Clarion | 600.8 | 38,305 | 1,217 | 63.8 | 96.8 | 1.7 | 0.5 | 1.0 | 1.1 | 5.0 | 10.5 | 16.0 | 11.4 | 10.1 | 11.9 |
| 42033 | 20180 | 4 | Clearfield | 1,145.3 | 78,612 | 718 | 68.6 | 93.5 | 3.0 | 0.4 | 0.9 | 3.1 | 4.5 | 10.1 | 10.8 | 12.4 | 11.9 | 13.8 |
| 42035 | 30820 | 1 | Clinton | 888.0 | 37,957 | 1,225 | 42.7 | 95.9 | 1.9 | 0.4 | 1.0 | 1.8 | 5.3 | 11.3 | 17.0 | 11.5 | 10.3 | 11.5 |
| 42037 | 14100 | 3 | Columbia | 483.2 | 64,842 | 820 | 134.2 | 93.7 | 2.2 | 0.5 | 1.4 | 3.3 | 4.2 | 10.1 | 18.0 | 11.2 | 10.3 | 12.0 |
| 42039 | 32740 | 4 | Crawford | 1,012.4 | 83,697 | 683 | 82.7 | 95.8 | 2.6 | 0.5 | 0.9 | 1.5 | 5.4 | 11.3 | 12.3 | 11.2 | 10.8 | 12.3 |
| 42041 | 25420 | 2 | Cumberland | 545.5 | 255,857 | 273 | 469.0 | 85.7 | 5.5 | 0.5 | 5.9 | 4.6 | 5.4 | 11.5 | 12.7 | 13.0 | 12.8 | 12.4 |
| 42043 | 25420 | 2 | Dauphin | 524.9 | 279,874 | 249 | 533.2 | 66.2 | 19.4 | 0.7 | 6.5 | 10.4 | 6.1 | 12.6 | 11.6 | 14.1 | 12.2 | 12.1 |
| 42045 | 37980 | 1 | Delaware | 183.8 | 566,753 | 121 | 3,083.5 | 66.7 | 23.6 | 0.6 | 7.1 | 4.3 | 5.8 | 12.3 | 13.3 | 13.1 | 12.5 | 11.9 |
| 42047 | 41260 | 7 | Elk | 826.9 | 29,607 | 1,446 | 35.8 | 98.0 | 0.9 | 0.5 | 0.7 | 0.9 | 4.8 | 10.9 | 10.3 | 10.1 | 10.2 | 13.7 |
| 42049 | 21500 | 2 | Erie | 798.9 | 268,426 | 257 | 336.0 | 85.8 | 8.9 | 0.5 | 2.5 | 4.6 | 5.4 | 11.8 | 13.3 | 13.0 | 11.5 | 11.8 |
| 42051 | 38300 | 1 | Fayette | 790.8 | 128,126 | 501 | 162.0 | 93.2 | 6.0 | 0.5 | 0.8 | 1.3 | 5.1 | 10.6 | 10.5 | 12.1 | 11.2 | 13.1 |
| 42053 | | 9 | Forest | 427.3 | 6,965 | 2,661 | 16.3 | 73.2 | 19.9 | 0.6 | 0.3 | 6.7 | 0.6 | 1.1 | 12.3 | 17.7 | 12.5 | 13.3 |
| 42055 | 16540 | 3 | Franklin | 772.3 | 155,637 | 431 | 201.5 | 88.7 | 4.7 | 0.6 | 1.5 | 6.4 | 5.7 | 12.4 | 11.4 | 12.1 | 11.6 | 12.7 |
| 42057 | | 8 | Fulton | 437.6 | 14,501 | 2,122 | 33.1 | 96.8 | 2.0 | 0.7 | 0.4 | 1.3 | 5.0 | 11.1 | 10.9 | 10.8 | 11.0 | 13.1 |
| 42059 | | 6 | Greene | 575.9 | 35,621 | 1,288 | 61.9 | 94.3 | 3.9 | 0.8 | 0.6 | 1.6 | 5.0 | 10.9 | 12.4 | 12.5 | 11.7 | 13.1 |
| 42061 | 26500 | 6 | Huntingdon | 874.7 | 44,590 | 1,083 | 51.0 | 91.5 | 6.1 | 0.4 | 1.0 | 2.1 | 4.6 | 10.0 | 12.6 | 12.6 | 11.4 | 12.9 |
| 42063 | 26860 | 4 | Indiana | 827.4 | 83,664 | 684 | 101.1 | 94.8 | 3.1 | 0.4 | 1.4 | 1.4 | 4.7 | 10.0 | 18.9 | 10.6 | 10.2 | 11.1 |
| 42065 | | 7 | Jefferson | 652.4 | 43,108 | 1,119 | 66.1 | 97.9 | 0.9 | 0.6 | 0.5 | 1.0 | 5.5 | 12.1 | 10.8 | 11.3 | 11.2 | 12.1 |
| 42067 | | 6 | Juniata | 391.4 | 24,619 | 1,623 | 62.9 | 94.6 | 1.1 | 0.4 | 0.7 | 4.1 | 5.8 | 12.1 | 11.5 | 11.4 | 11.1 | 12.2 |
| 42069 | 42540 | 2 | Lackawanna | 459.0 | 208,989 | 325 | 455.3 | 84.5 | 4.2 | 0.4 | 3.6 | 8.9 | 5.2 | 11.6 | 12.3 | 12.4 | 11.6 | 12.2 |
| 42071 | 29540 | 2 | Lancaster | 943.9 | 546,192 | 128 | 578.7 | 82.4 | 4.7 | 0.4 | 2.9 | 11.2 | 6.4 | 12.9 | 12.6 | 13.2 | 11.6 | 11.3 |
| 42073 | 35260 | 4 | Lawrence | 357.4 | 85,083 | 678 | 238.1 | 93.5 | 5.8 | 0.5 | 0.8 | 1.7 | 5.2 | 11.0 | 11.5 | 11.1 | 10.8 | 12.3 |
| 42075 | 30140 | 2 | Lebanon | 361.8 | 141,663 | 463 | 391.6 | 81.6 | 2.8 | 0.4 | 2.0 | 14.5 | 5.8 | 12.9 | 12.2 | 11.8 | 11.8 | 11.9 |
| 42077 | 10900 | 2 | Lehigh | 345.3 | 370,802 | 194 | 1,073.9 | 63.0 | 7.0 | 0.5 | 4.2 | 27.1 | 6.0 | 12.6 | 12.9 | 13.2 | 12.7 | 12.2 |
| 42079 | 42540 | 2 | Luzerne | 889.7 | 316,982 | 224 | 356.3 | 79.3 | 5.2 | 0.4 | 1.7 | 14.9 | 5.4 | 11.0 | 12.0 | 12.8 | 11.6 | 12.8 |
| 42081 | 48700 | 3 | Lycoming | 1,228.9 | 113,209 | 546 | 92.1 | 92.1 | 6.3 | 0.6 | 1.1 | 2.2 | 5.3 | 11.6 | 12.3 | 12.8 | 11.5 | 11.8 |
| 42083 | 14620 | 7 | McKean | 979.7 | 40,333 | 1,172 | 41.2 | 94.3 | 2.9 | 0.8 | 1.0 | 2.3 | 4.7 | 11.3 | 12.6 | 12.1 | 11.3 | 12.9 |
| 42085 | 49660 | 2 | Mercer | 672.5 | 108,545 | 563 | 161.4 | 91.6 | 7.0 | 0.6 | 1.1 | 1.6 | 4.9 | 10.4 | 13.1 | 10.7 | 10.7 | 12.3 |
| 42087 | 30380 | 4 | Mifflin | 411.0 | 46,064 | 1,056 | 112.1 | 96.6 | 1.4 | 0.4 | 0.9 | 1.8 | 6.3 | 12.4 | 10.8 | 11.4 | 10.4 | 12.4 |
| 42089 | 20700 | 3 | Monroe | 608.4 | 170,154 | 391 | 279.7 | 65.3 | 15.3 | 0.8 | 3.2 | 17.6 | 4.7 | 10.8 | 13.9 | 11.7 | 10.9 | 13.4 |
| 42091 | 37980 | 1 | Montgomery | 483.0 | 833,869 | 72 | 1,726.4 | 76.4 | 10.7 | 0.5 | 9.0 | 5.6 | 5.4 | 12.1 | 11.6 | 12.5 | 12.9 | 12.8 |
| 42093 | 14100 | 3 | Montour | 130.2 | 18,042 | 1,919 | 138.6 | 91.6 | 2.6 | 0.5 | 3.8 | 2.8 | 5.4 | 11.1 | 10.1 | 13.1 | 11.7 | 11.5 |
| 42095 | 10900 | 2 | Northampton | 369.8 | 305,892 | 228 | 827.2 | 76.3 | 6.9 | 0.5 | 3.6 | 14.6 | 4.8 | 11.0 | 13.4 | 12.2 | 11.7 | 12.5 |
| 42097 | 44980 | 4 | Northumberland | 457.7 | 90,258 | 658 | 197.2 | 92.5 | 3.3 | 0.4 | 0.7 | 4.2 | 5.1 | 10.9 | 10.6 | 12.3 | 11.6 | 12.5 |
| 42099 | 25420 | 2 | Perry | 551.4 | 46,212 | 1,053 | 83.8 | 96.0 | 1.6 | 0.6 | 0.8 | 2.2 | 5.5 | 11.9 | 10.5 | 12.0 | 11.7 | 13.4 |
| 42101 | 37980 | 1 | Philadelphia | 134.4 | 1,578,487 | 23 | 11,744.7 | 35.8 | 41.7 | 0.8 | 8.5 | 15.6 | 6.3 | 11.9 | 12.6 | 19.4 | 13.0 | 10.9 |
| 42103 | 35620 | 1 | Pike | 544.9 | 56,072 | 920 | 102.9 | 80.7 | 6.4 | 0.8 | | 11.9 | 4.1 | 9.5 | 10.9 | 9.9 | 10.4 | 13.4 |

1. CBSA = Core Based Statistical Area. See Appendix A for explanation. See Appendix B for list of metropolitan areas with component counties.
2. County type code from the Economic Research Service of USDA Rural-Urban Continuum Codes. See Appendix A for definition.
3. Dry land or land partially or temporarily covered by water.
4. May be of any race.

# Table B. States and Counties — Population and Households

| STATE County | Population, 2020 (cont.) Age (percent) (cont.) | | | | Population change, 2000-2020 | | | | | | | Households, 2015-2019 | | | | |
|---|---|---|---|---|---|---|---|---|---|---|---|---|---|---|---|---|
| | 55 to 64 years | 65 to 74 years | 75 years and over | Percent female | Total persons 2000 | Total persons 2010 | Percent change 2000-2010 | Percent change 2010-2020 | Births | Deaths | Net Migration | Number | Persons per household | Family households | Female family householder[1] | One person |
| | 16 | 17 | 18 | 19 | 20 | 21 | 22 | 23 | 24 | 25 | 26 | 27 | 28 | 29 | 30 | 31 |
| **OREGON—Cont'd** | | | | | | | | | | | | | | | | |
| Morrow | 12.5 | 10.5 | 6.4 | 49.3 | 10,995 | 11,175 | 1.6 | 4.7 | 1,660 | 823 | -323 | 4,108 | 2.74 | 76.1 | 8.4 | 17.4 |
| Multnomah | 11.1 | 9.0 | 5.2 | 50.5 | 660,486 | 735,152 | 11.3 | 10.9 | 92,054 | 58,676 | 46,535 | 326,229 | 2.41 | 54.0 | 9.8 | 32.5 |
| Polk | 11.4 | 10.8 | 7.9 | 51.1 | 62,380 | 75,403 | 20.9 | 16.4 | 8,935 | 7,043 | 10,466 | 30,305 | 2.68 | 68.4 | 9.0 | 23.8 |
| Sherman | 16.2 | 13.0 | 12.3 | 49.5 | 1,934 | 1,766 | -8.7 | 2.0 | 183 | 173 | 25 | 748 | 2.18 | 59.5 | 9.6 | 33.7 |
| Tillamook | 15.5 | 17.3 | 10.1 | 49.5 | 24,262 | 25,244 | 4.0 | 8.7 | 2,521 | 3,041 | 2,724 | 10,784 | 2.39 | 61.9 | 7.7 | 30.3 |
| Umatilla | 11.5 | 9.6 | 6.8 | 47.7 | 70,548 | 75,887 | 7.6 | 2.5 | 10,503 | 6,702 | -1,915 | 26,908 | 2.69 | 69.0 | 13.7 | 23.9 |
| Union | 12.3 | 12.2 | 9.5 | 50.4 | 24,530 | 25,737 | 4.9 | 3.2 | 3,002 | 2,831 | 654 | 10,708 | 2.39 | 65.7 | 10.7 | 24.6 |
| Wallowa | 15.6 | 18.2 | 12.0 | 51.1 | 7,226 | 7,008 | -3.0 | 2.5 | 616 | 882 | 438 | 3,214 | 2.14 | 64.7 | 5.9 | 29.7 |
| Wasco | 13.3 | 12.7 | 8.6 | 49.8 | 23,791 | 25,211 | 6.0 | 4.7 | 3,090 | 3,205 | 1,313 | 10,305 | 2.42 | 64.1 | 10.0 | 28.2 |
| Washington | 11.4 | 8.6 | 5.7 | 50.4 | 445,342 | 529,865 | 19.0 | 13.9 | 70,825 | 33,819 | 36,713 | 219,053 | 2.66 | 66.9 | 9.3 | 24.7 |
| Wheeler | 17.5 | 19.8 | 17.7 | 50.8 | 1,547 | 1,441 | -6.9 | -3.7 | 106 | 195 | 32 | 640 | 2.18 | 61.9 | 6.7 | 30.5 |
| Yamhill | 12.6 | 10.8 | 7.7 | 50.0 | 84,992 | 99,214 | 16.7 | 8.5 | 11,399 | 9,300 | 6,377 | 36,808 | 2.7 | 69.9 | 11.4 | 23.3 |
| **PENNSYLVANIA** | 14.0 | 11.0 | 8.2 | 51.0 | 12,281,054 | 12,702,891 | 3.4 | 0.6 | 1,431,212 | 1,352,118 | 9,078 | 5,053,106 | 2.45 | 64.0 | 11.8 | 29.7 |
| Adams | 15.3 | 12.6 | 9.0 | 50.7 | 91,292 | 101,419 | 11.1 | 1.3 | 10,251 | 10,432 | 1,578 | 39,345 | 2.5 | 70.3 | 9.1 | 25.2 |
| Allegheny | 13.9 | 11.3 | 8.5 | 51.5 | 1,281,666 | 1,223,288 | -4.6 | | 133,838 | 141,715 | -3,198 | 541,541 | 2.19 | 56.0 | 10.5 | 36.3 |
| Armstrong | 16.4 | 13.3 | 10.0 | 50.2 | 72,392 | 69,108 | -4.5 | -7.2 | 6,485 | 8,759 | -2,634 | 28,137 | 2.31 | 66.9 | 9.3 | 28.6 |
| Beaver | 15.8 | 12.8 | 9.7 | 51.4 | 181,412 | 170,532 | -6.0 | -4.7 | 16,894 | 21,899 | -2,788 | 71,167 | 2.29 | 63.6 | 11.6 | 31.0 |
| Bedford | 15.9 | 12.9 | 11.1 | 50.2 | 49,984 | 49,775 | -0.4 | -3.9 | 4,857 | 5,849 | -929 | 19,882 | 2.41 | 66.3 | 6.8 | 28.4 |
| Berks | 13.8 | 10.3 | 7.7 | 50.7 | 373,638 | 411,569 | 10.2 | 2.3 | 48,972 | 39,880 | 469 | 154,712 | 2.62 | 68.8 | 13.1 | 25.3 |
| Blair | 14.4 | 12.2 | 9.5 | 51.0 | 129,144 | 127,121 | -1.6 | -4.8 | 13,050 | 16,583 | -2,482 | 52,126 | 2.3 | 64.1 | 12.4 | 30.3 |
| Bradford | 15.3 | 12.5 | 9.9 | 50.4 | 62,761 | 62,700 | -0.1 | -4.0 | 7,244 | 7,167 | -2,556 | 25,021 | 2.41 | 67.1 | 10.0 | 27.2 |
| Bucks | 15.9 | 11.5 | 8.4 | 50.9 | 597,635 | 625,426 | 4.7 | 0.4 | 59,206 | 60,501 | 4,188 | 238,830 | 2.58 | 70.7 | 9.2 | 24.4 |
| Butler | 15.4 | 11.7 | 8.4 | 50.4 | 174,083 | 183,851 | 5.6 | 2.9 | 18,482 | 20,019 | 6,989 | 76,502 | 2.37 | 67.1 | 7.4 | 27.2 |
| Cambria | 15.0 | 13.6 | 10.2 | 50.7 | 152,598 | 143,693 | -5.8 | -10.5 | 13,450 | 18,963 | -9,554 | 56,674 | 2.23 | 62.3 | 10.9 | 32.8 |
| Cameron | 17.4 | 17.3 | 12.0 | 49.7 | 5,974 | 5,085 | -14.9 | -14.8 | 440 | 672 | -526 | 2,184 | 2.07 | 54.7 | 7.6 | 39.8 |
| Carbon | 16.1 | 13.1 | 9.1 | 50.2 | 58,802 | 65,244 | 11.0 | -1.8 | 5,937 | 8,317 | 1,271 | 26,043 | 2.42 | 64.7 | 10.0 | 27.2 |
| Centre | 11.7 | 9.0 | 6.5 | 47.5 | 135,758 | 154,008 | 13.4 | 4.9 | 12,487 | 10,114 | 5,109 | 58,201 | 2.46 | 56.5 | 6.2 | 30.3 |
| Chester | 14.1 | 10.1 | 7.2 | 50.7 | 433,501 | 499,141 | 15.1 | 5.5 | 56,172 | 40,259 | 12,111 | 190,980 | 2.65 | 70.8 | 8.7 | 23.2 |
| Clarion | 14.2 | 11.8 | 8.9 | 50.8 | 41,765 | 39,988 | -4.3 | -4.2 | 4,058 | 4,416 | -1,345 | 16,021 | 2.34 | 62.4 | 8.3 | 29.7 |
| Clearfield | 15.1 | 12.0 | 9.4 | 47.3 | 83,382 | 81,609 | -2.1 | -3.7 | 7,535 | 9,714 | -763 | 31,248 | 2.38 | 65.6 | 9.5 | 28.9 |
| Clinton | 13.5 | 11.0 | 8.6 | 50.8 | 37,914 | 39,248 | 3.5 | -3.3 | 4,319 | 4,171 | -1,432 | 14,690 | 2.54 | 63.3 | 8.7 | 28.5 |
| Columbia | 13.9 | 11.7 | 8.7 | 51.7 | 64,151 | 67,296 | 4.9 | -3.6 | 5,945 | 7,201 | -1,152 | 26,372 | 2.32 | 61.5 | 10.8 | 31.1 |
| Crawford | 14.8 | 12.9 | 8.9 | 51.0 | 90,366 | 88,750 | -1.8 | -5.7 | 9,510 | 10,355 | -4,174 | 35,164 | 2.34 | 65.7 | 10.3 | 28.4 |
| Cumberland | 13.2 | 10.9 | 8.4 | 50.4 | 213,674 | 235,403 | 10.2 | 8.7 | 26,295 | 23,830 | 18,150 | 99,804 | 2.37 | 64.1 | 9.0 | 29.7 |
| Dauphin | 13.6 | 10.6 | 7.2 | 51.5 | 251,798 | 268,127 | 6.5 | 4.4 | 35,079 | 25,907 | 2,692 | 112,212 | 2.39 | 62.0 | 13.6 | 31.5 |
| Delaware | 13.7 | 9.9 | 7.4 | 51.9 | 550,864 | 558,776 | 1.4 | 1.4 | 68,104 | 56,471 | -3,469 | 207,257 | 2.62 | 67.1 | 14.5 | 28.4 |
| Elk | 16.9 | 13.0 | 10.1 | 49.7 | 35,112 | 31,946 | -9.0 | -7.3 | 3,025 | 4,005 | -1,354 | 14,020 | 2.13 | 61.8 | 7.4 | 32.5 |
| Erie | 13.9 | 11.3 | 7.9 | 50.6 | 280,843 | 280,586 | -0.1 | -4.3 | 31,336 | 29,490 | -14,014 | 110,318 | 2.36 | 62.1 | 13.0 | 30.5 |
| Fayette | 15.1 | 13.0 | 9.4 | 50.3 | 148,644 | 136,589 | -8.1 | -6.2 | 13,804 | 18,594 | -3,566 | 54,837 | 2.3 | 62.9 | 12.7 | 31.2 |
| Forest | 17.5 | 14.8 | 10.3 | 31.6 | 4,946 | 7,707 | 55.8 | -9.6 | 304 | 870 | -179 | 1,839 | 1.97 | 58.7 | 6.3 | 35.9 |
| Franklin | 13.7 | 11.2 | 9.2 | 50.9 | 129,313 | 149,625 | 15.7 | 4.0 | 18,702 | 15,687 | 3,161 | 60,438 | 2.51 | 69.6 | 9.9 | 25.8 |
| Fulton | 15.0 | 12.5 | 9.8 | 49.7 | 14,261 | 14,844 | 4.1 | -2.3 | 1,400 | 1,576 | -154 | 5,989 | 2.42 | 67.1 | 9.1 | 28.6 |
| Greene | 14.3 | 12.2 | 7.9 | 47.9 | 40,672 | 38,692 | -4.9 | -7.9 | 3,905 | 4,541 | -2,426 | 14,230 | 2.34 | 66.3 | 10.5 | 28.4 |
| Huntingdon | 14.3 | 12.5 | 9.2 | 46.9 | 45,586 | 46,002 | 0.9 | -3.1 | 4,142 | 4,894 | -625 | 16,779 | 2.37 | 63.5 | 9.2 | 26.3 |
| Indiana | 13.9 | 11.9 | 8.7 | 50.0 | 89,605 | 88,881 | -0.8 | -5.9 | 8,407 | 9,395 | -4,228 | 33,246 | 2.38 | 65.9 | 7.6 | 28.5 |
| Jefferson | 15.2 | 12.2 | 9.6 | 50.2 | 45,932 | 45,189 | -1.6 | -4.6 | 5,089 | 5,817 | -1,321 | 18,427 | 2.34 | 65.9 | 9.1 | 28.8 |
| Juniata | 14.7 | 11.6 | 9.5 | 49.9 | 22,821 | 24,637 | 8.0 | -0.1 | 2,874 | 2,602 | -277 | 9,372 | 2.6 | 68.4 | 7.2 | 26.0 |
| Lackawanna | 14.0 | 11.6 | 9.1 | 51.4 | 213,295 | 214,427 | 0.5 | -2.5 | 22,025 | 27,580 | 233 | 87,161 | 2.32 | 61.2 | 12.2 | 32.3 |
| Lancaster | 13.0 | 10.4 | 8.6 | 51.0 | 470,658 | 519,443 | 10.4 | 5.1 | 72,008 | 50,889 | 6,056 | 201,620 | 2.63 | 70.0 | 9.8 | 24.1 |
| Lawrence | 15.1 | 12.8 | 10.1 | 51.5 | 94,643 | 91,156 | -3.7 | -6.7 | 9,058 | 11,829 | -3,271 | 37,055 | 2.28 | 65.7 | 11.7 | 30.2 |
| Lebanon | 13.5 | 11.2 | 8.9 | 50.8 | 120,327 | 133,596 | 11.0 | 6.0 | 16,592 | 15,185 | 6,782 | 53,579 | 2.55 | 68.3 | 11.4 | 26.0 |
| Lehigh | 13.0 | 9.9 | 7.5 | 51.1 | 312,090 | 349,508 | 12.0 | 6.1 | 43,002 | 34,811 | 13,289 | 138,714 | 2.56 | 67.3 | 13.7 | 26.0 |
| Luzerne | 14.1 | 11.4 | 9.0 | 50.5 | 319,250 | 320,917 | 0.5 | -1.2 | 32,908 | 41,492 | 4,920 | 128,660 | 2.37 | 63.0 | 13.5 | 31.4 |
| Lycoming | 14.3 | 11.7 | 8.7 | 51.1 | 120,044 | 116,095 | -3.3 | -2.5 | 12,652 | 13,133 | -2,373 | 45,608 | 2.37 | 65.2 | 10.4 | 29.0 |
| McKean | 14.8 | 11.6 | 8.7 | 48.7 | 45,936 | 43,463 | -5.4 | -7.2 | 4,209 | 5,395 | -1,927 | 17,147 | 2.24 | 63.3 | 10.0 | 29.9 |
| Mercer | 15.1 | 12.6 | 10.2 | 50.6 | 120,293 | 116,663 | -3.0 | -7.0 | 11,329 | 14,595 | -4,804 | 46,340 | 2.25 | 63.3 | 11.2 | 31.0 |
| Mifflin | 14.1 | 11.9 | 10.3 | 51.0 | 46,486 | 46,677 | 0.4 | -1.3 | 5,965 | 5,582 | -959 | 19,043 | 2.39 | 69.5 | 10.6 | 25.3 |
| Monroe | 16.2 | 11.5 | 7.0 | 50.3 | 138,687 | 169,837 | 22.5 | 0.2 | 14,696 | 14,849 | 444 | 57,098 | 2.9 | 70.8 | 11.0 | 23.6 |
| Montgomery | 14.1 | 10.4 | 8.2 | 51.3 | 750,097 | 799,839 | 6.6 | 4.3 | 90,580 | 77,457 | 21,624 | 316,206 | 2.53 | 68.8 | 9.6 | 25.9 |
| Montour | 15.0 | 11.4 | 10.8 | 51.4 | 18,236 | 18,256 | 0.1 | -1.2 | 2,196 | 2,309 | -85 | 7,404 | 2.36 | 63.2 | 9.9 | 31.7 |
| Northampton | 14.2 | 11.2 | 8.8 | 50.8 | 267,066 | 297,704 | 11.5 | 2.8 | 29,266 | 30,492 | 9,706 | 114,185 | 2.55 | 68.3 | 11.1 | 25.3 |
| Northumberland | 14.9 | 12.5 | 9.7 | 49.6 | 94,556 | 94,468 | -0.1 | -4.5 | 9,617 | 12,087 | -1,678 | 39,075 | 2.24 | 64.2 | 10.1 | 30.5 |
| Perry | 15.3 | 12.3 | 7.4 | 49.4 | 43,602 | 45,927 | 5.3 | 0.6 | 5,477 | 4,656 | -513 | 18,231 | 2.49 | 70.7 | 9.1 | 24.5 |
| Philadelphia | 11.5 | 8.4 | 6.0 | 52.7 | 1,517,550 | 1,525,999 | 0.6 | 3.4 | 226,050 | 148,720 | -23,822 | 601,337 | 2.55 | 54.2 | 20.1 | 37.5 |
| Pike | 17.8 | 14.0 | 10.0 | 49.3 | 46,302 | 57,359 | 23.9 | -2.2 | 3,715 | 4,798 | -208 | 22,119 | 2.49 | 69.9 | 8.9 | 25.4 |

1. No spouse present.

# Table B. States and Counties — Population, Vital Statistics, Health, and Crime

| STATE County | Persons in group quarters, 2020 | Daytime Population, 2015-2019 Number | Employment/ residence ratio | Births, 2020 Total | Rate[1] | Deaths, 2020 Number | Rate[1] | Persons under 65 with no health insurance, 2019 Number | Percent | Medicare, 2020 Total beneficiaries | Enrolled in Original Medicare | Enrolled in Medicare Advantage | Crimes reported by county police or sheriff, 2019 Violent | Property |
|---|---|---|---|---|---|---|---|---|---|---|---|---|---|---|
| | 32 | 33 | 34 | 35 | 36 | 37 | 38 | 39 | 40 | 41 | 42 | 43 | 44 | 45 |
| **OREGON—Cont'd** | | | | | | | | | | | | | | |
| Morrow | 23 | 13,180 | 1.39 | 167 | 14.3 | 101 | 8.6 | 1,152 | 12.2 | 2,228 | 2,033 | 194 | 20 | 112 |
| Multnomah | 19,129 | 910,564 | 1.25 | 8,081 | 9.9 | 6,509 | 8.0 | 57,561 | 8.3 | 128,920 | 50,370 | 78,550 | 165 | 1,696 |
| Polk | 1,799 | 69,709 | 0.64 | 846 | 9.6 | 783 | 8.9 | 5,891 | 8.6 | 18,816 | 7,597 | 11,219 | 52 | 295 |
| Sherman | 0 | 1,787 | 1.21 | 23 | 12.8 | 15 | 8.3 | 100 | 7.6 | 538 | 445 | 94 | 0 | 42 |
| Tillamook | 477 | 25,948 | 0.96 | 248 | 9.0 | 318 | 11.6 | 1,921 | 10.0 | 8,478 | 6,468 | 2,011 | NA | NA |
| Umatilla | 4,089 | 74,731 | 0.92 | 956 | 12.3 | 711 | 9.1 | 7,074 | 11.7 | 14,760 | 13,608 | 1,152 | 43 | 410 |
| Union | 878 | 26,305 | 1.00 | 260 | 9.8 | 344 | 13.0 | 1,733 | 8.5 | 6,513 | 5,849 | 664 | 8 | 209 |
| Wallowa | 105 | 7,035 | 1.01 | 64 | 8.9 | 79 | 11.0 | 431 | 8.7 | 2,526 | 2,351 | 175 | NA | NA |
| Wasco | 741 | 25,667 | 0.96 | 282 | 10.7 | 298 | 11.3 | 2,382 | 11.6 | 6,485 | 5,130 | 1,355 | 22 | 132 |
| Washington | 7,407 | 598,260 | 1.03 | 6,426 | 10.6 | 3,878 | 6.4 | 35,674 | 6.9 | 91,679 | 36,500 | 55,179 | 349 | 2,480 |
| Wheeler | 24 | 1,410 | 0.99 | 8 | 5.8 | 17 | 12.3 | 92 | 11.2 | 504 | 414 | 90 | NA | NA |
| Yamhill | 5,250 | 97,061 | 0.84 | 1,078 | 10.0 | 956 | 8.9 | 7,763 | 9.3 | 22,019 | 10,636 | 11,383 | 41 | 473 |
| **PENNSYLVANIA** | 415,344 | 12,730,936 | 0.99 | 133,740 | 10.5 | 143,283 | 11.2 | 702,689 | 7.0 | 2,775,097 | 1,538,816 | 1,236,281 | X | X |
| Adams | 4,038 | 87,403 | 0.70 | 899 | 8.8 | 1,106 | 10.8 | 6,014 | 7.7 | 25,053 | 16,713 | 8,340 | NA | NA |
| Allegheny | 34,557 | 1,307,981 | 1.14 | 12,681 | 10.5 | 14,553 | 12.0 | 48,870 | 5.1 | 267,160 | 99,306 | 167,854 | 3 | 0 |
| Armstrong | 650 | 54,924 | 0.62 | 562 | 8.8 | 947 | 14.8 | 3,236 | 6.5 | 17,425 | 5,781 | 11,645 | NA | NA |
| Beaver | 3,195 | 145,319 | 0.74 | 1,539 | 9.5 | 2,215 | 13.6 | 7,115 | 5.6 | 42,699 | 14,820 | 27,879 | NA | NA |
| Bedford | 551 | 43,765 | 0.79 | 458 | 9.6 | 598 | 12.5 | 3,040 | 8.4 | 13,043 | 6,128 | 6,916 | NA | NA |
| Berks | 11,407 | 394,198 | 0.88 | 4,564 | 10.8 | 4,204 | 10.0 | 25,575 | 7.6 | 86,262 | 53,563 | 32,699 | 5 | 0 |
| Blair | 3,721 | 129,263 | 1.11 | 1,164 | 9.6 | 1,740 | 14.4 | 6,298 | 6.7 | 31,533 | 14,566 | 16,967 | NA | NA |
| Bradford | 664 | 60,603 | 0.99 | 652 | 10.8 | 787 | 13.1 | 3,815 | 8.2 | 15,125 | 10,242 | 4,882 | NA | NA |
| Bucks | 8,250 | 569,021 | 0.82 | 5,615 | 8.9 | 6,632 | 10.6 | 27,303 | 5.4 | 139,626 | 89,981 | 49,645 | 0 | 1 |
| Butler | 5,443 | 190,995 | 1.04 | 1,783 | 9.4 | 2,133 | 11.3 | 7,125 | 4.8 | 43,733 | 16,887 | 26,846 | NA | NA |
| Cambria | 5,924 | 129,224 | 0.93 | 1,171 | 9.1 | 1,837 | 14.3 | 5,487 | 5.8 | 36,760 | 12,543 | 24,217 | NA | NA |
| Cameron | 92 | 4,941 | 1.17 | 33 | 7.6 | 76 | 17.6 | 193 | 6.1 | 1,471 | 890 | 582 | NA | NA |
| Carbon | 700 | 50,904 | 0.55 | 542 | 8.5 | 857 | 13.4 | 3,336 | 6.7 | 16,473 | 11,923 | 4,550 | NA | NA |
| Centre | 19,160 | 170,639 | 1.11 | 1,137 | 7.0 | 1,208 | 7.5 | 9,766 | 8.2 | 25,121 | 12,256 | 12,864 | NA | NA |
| Chester | 13,409 | 512,769 | 0.97 | 5,359 | 10.2 | 4,549 | 8.6 | 25,430 | 5.9 | 97,420 | 71,836 | 25,584 | NA | NA |
| Clarion | 1,774 | 36,635 | 0.88 | 363 | 9.5 | 481 | 12.6 | 2,370 | 8.2 | 9,438 | 5,404 | 4,033 | NA | NA |
| Clearfield | 5,668 | 77,861 | 0.94 | 681 | 8.7 | 1,008 | 12.8 | 4,097 | 7.1 | 20,008 | 11,024 | 8,984 | NA | NA |
| Clinton | 1,753 | 37,043 | 0.89 | 411 | 10.8 | 424 | 11.2 | 2,120 | 7.3 | 8,810 | 4,522 | 4,289 | NA | NA |
| Columbia | 3,509 | 63,005 | 0.91 | 521 | 8.0 | 774 | 11.9 | 3,220 | 6.6 | 15,038 | 8,754 | 6,284 | NA | NA |
| Crawford | 3,685 | 82,908 | 0.93 | 885 | 10.6 | 1,043 | 12.5 | 5,438 | 8.6 | 21,653 | 12,746 | 8,907 | NA | NA |
| Cumberland | 11,803 | 259,263 | 1.08 | 2,592 | 10.1 | 2,510 | 9.8 | 13,403 | 6.9 | 54,408 | 31,251 | 23,157 | NA | NA |
| Dauphin | 7,057 | 325,593 | 1.36 | 3,381 | 12.1 | 2,798 | 10.0 | 17,492 | 7.7 | 57,451 | 27,535 | 29,917 | NA | NA |
| Delaware | 22,261 | 508,523 | 0.79 | 6,418 | 11.3 | 5,884 | 10.4 | 28,497 | 6.3 | 109,261 | 75,944 | 33,317 | NA | NA |
| Elk | 355 | 29,959 | 0.97 | 261 | 8.8 | 386 | 13.0 | 1,338 | 5.8 | 8,244 | 5,879 | 2,365 | 0 | 0 |
| Erie | 12,120 | 278,131 | 1.03 | 2,681 | 10.0 | 3,012 | 11.2 | 14,801 | 7.1 | 60,631 | 29,082 | 31,549 | 5 | 1 |
| Fayette | 4,259 | 117,780 | 0.75 | 1,284 | 10.0 | 1,938 | 15.1 | 6,712 | 6.8 | 34,560 | 14,403 | 20,157 | NA | NA |
| Forest | 2,430 | 8,218 | 1.94 | 15 | 2.2 | 82 | 11.8 | 217 | 7.2 | 1,754 | 897 | 857 | NA | NA |
| Franklin | 2,690 | 145,064 | 0.88 | 1,717 | 11.0 | 1,763 | 11.3 | 10,407 | 8.5 | 36,422 | 25,326 | 11,096 | NA | NA |
| Fulton | 122 | 13,050 | 0.78 | 144 | 9.9 | 162 | 11.2 | 762 | 6.8 | 3,744 | 2,640 | 1,104 | NA | NA |
| Greene | 2,968 | 37,116 | 1.02 | 320 | 9.0 | 461 | 12.9 | 1,682 | 6.4 | 8,647 | 3,572 | 5,075 | NA | NA |
| Huntingdon | 4,778 | 41,573 | 0.79 | 408 | 9.2 | 572 | 12.8 | 2,177 | 7.1 | 10,802 | 5,897 | 4,905 | NA | NA |
| Indiana | 4,926 | 84,029 | 0.97 | 788 | 9.4 | 961 | 11.5 | 5,146 | 8.2 | 19,575 | 7,115 | 12,460 | NA | NA |
| Jefferson | 759 | 42,576 | 0.93 | 459 | 10.6 | 554 | 12.9 | 2,619 | 7.8 | 11,277 | 6,064 | 5,212 | NA | NA |
| Juniata | 292 | 21,466 | 0.72 | 282 | 11.5 | 271 | 11.0 | 2,212 | 11.4 | 5,576 | 2,850 | 2,726 | NA | NA |
| Lackawanna | 7,653 | 211,433 | 1.01 | 1,987 | 9.5 | 2,711 | 13.0 | 11,474 | 7.1 | 52,134 | 35,454 | 16,680 | NA | NA |
| Lancaster | 12,603 | 532,983 | 0.97 | 6,879 | 12.6 | 5,565 | 10.2 | 48,810 | 11.2 | 113,309 | 64,835 | 48,474 | 1 | 0 |
| Lawrence | 2,091 | 78,810 | 0.80 | 840 | 9.9 | 1,158 | 13.6 | 4,250 | 6.6 | 22,779 | 8,581 | 14,198 | NA | NA |
| Lebanon | 3,684 | 129,684 | 0.85 | 1,541 | 10.9 | 1,628 | 11.5 | 9,080 | 8.2 | 32,434 | 18,032 | 14,402 | NA | NA |
| Lehigh | 8,979 | 383,010 | 1.10 | 4,307 | 11.6 | 3,571 | 9.6 | 24,067 | 8.1 | 75,248 | 46,884 | 28,364 | NA | NA |
| Luzerne | 11,699 | 319,403 | 1.01 | 3,269 | 10.3 | 4,204 | 13.3 | 18,116 | 7.4 | 74,635 | 51,262 | 23,373 | NA | NA |
| Lycoming | 5,276 | 115,834 | 1.03 | 1,169 | 10.3 | 1,418 | 12.5 | 5,753 | 6.7 | 27,227 | 16,621 | 10,606 | NA | NA |
| McKean | 2,966 | 39,691 | 0.90 | 354 | 8.8 | 540 | 13.4 | 1,894 | 6.3 | 10,259 | 7,004 | 3,255 | NA | NA |
| Mercer | 6,071 | 112,862 | 1.03 | 1,053 | 9.7 | 1,520 | 14.0 | 5,681 | 7.1 | 29,100 | 13,169 | 15,931 | NA | NA |
| Mifflin | 560 | 43,257 | 0.86 | 603 | 13.1 | 564 | 12.2 | 3,398 | 9.6 | 11,470 | 6,146 | 5,324 | NA | NA |
| Monroe | 4,164 | 156,184 | 0.85 | 1,521 | 8.9 | 1,775 | 10.4 | 11,267 | 8.3 | 34,769 | 24,520 | 10,249 | 0 | 0 |
| Montgomery | 20,808 | 904,673 | 1.19 | 8,467 | 10.2 | 8,416 | 10.1 | 32,798 | 4.9 | 167,387 | 116,708 | 50,679 | 0 | 0 |
| Montour | 845 | 26,425 | 1.95 | 191 | 10.6 | 241 | 13.4 | 868 | 6.2 | 4,395 | 2,091 | 2,303 | NA | NA |
| Northampton | 11,165 | 281,223 | 0.85 | 2,894 | 9.5 | 3,214 | 10.5 | 13,855 | 5.8 | 69,593 | 45,693 | 23,900 | NA | NA |
| Northumberland | 3,566 | 82,731 | 0.78 | 882 | 9.8 | 1,174 | 13.0 | 4,325 | 6.4 | 23,274 | 13,585 | 9,688 | NA | NA |
| Perry | 659 | 33,579 | 0.44 | 493 | 10.7 | 499 | 10.8 | 3,348 | 9.0 | 10,471 | 5,275 | 5,196 | NA | NA |
| Philadelphia | 55,137 | 1,678,869 | 1.15 | 20,478 | 13.0 | 16,685 | 10.6 | 116,921 | 9.0 | 258,677 | 135,925 | 122,752 | NA | NA |
| Pike | 477 | 43,220 | 0.51 | 418 | 7.5 | 569 | 10.1 | 3,128 | 7.4 | 14,564 | 11,506 | 3,058 | NA | NA |

1. Per 1,000 estimated resident population.

| STATE County | Officers | Civilians | Total | Percent private | High school graduate or less | Bachelor's degree or more | Total current spending (mil dol) | Current spending per student (dollars) | Per capita income[4] | Median income (dollars) | with income of less than $50,000 | with income of $200,000 or more | Median household income (dollars) | All persons | Children under 18 years | Children 5 to 17 years in families |
|---|---|---|---|---|---|---|---|---|---|---|---|---|---|---|---|---|
| | 46 | 47 | 48 | 49 | 50 | 51 | 52 | 53 | 54 | 55 | 56 | 57 | 58 | 59 | 60 | 61 |
| **OREGON—Cont'd** | | | | | | | | | | | | | | | | |
| Morrow | 28 | 12 | 2,819 | 3.3 | 57.3 | 9.0 | 32 | 13,235 | 23,682 | 54,269 | 44.6 | 2.4 | 59,781 | 12.7 | 16.6 | 14.1 |
| Multnomah | 131 | 644 | 182,353 | 20.3 | 25.5 | 45.9 | 1,231 | 13,124 | 39,245 | 69,176 | 36.8 | 8.9 | 72,469 | 12.0 | 14.3 | 14.0 |
| Polk | 27 | 9 | 22,387 | 13.2 | 32.6 | 31.0 | 78 | 11,218 | 29,440 | 62,691 | 40.3 | 3.9 | 69,757 | 9.4 | 12.4 | 11.1 |
| Sherman | 6 | 1 | 306 | 8.5 | 42.6 | 17.3 | 4 | 14,283 | 31,615 | 51,071 | 48.4 | 4.4 | 60,035 | 12.5 | 19.0 | 18.4 |
| Tillamook | 45 | 8 | 5,069 | 13.2 | 39.1 | 21.8 | 45 | 13,272 | 27,659 | 49,895 | 50.1 | 2.3 | 52,298 | 12.0 | 17.2 | 15.0 |
| Umatilla | 25 | 62 | 19,224 | 9.1 | 46.1 | 17.2 | 177 | 12,800 | 24,444 | 54,699 | 44.6 | 3.0 | 58,286 | 13.9 | 17.1 | 16.0 |
| Union | 16 | 18 | 6,353 | 13.6 | 39.5 | 24.5 | 44 | 11,203 | 27,646 | 52,171 | 49.1 | 3.4 | 54,264 | 13.6 | 16.5 | 15.4 |
| Wallowa | 5 | 8 | 1,103 | 10.7 | 35.5 | 26.4 | 16 | 18,973 | 30,485 | 51,224 | 48.5 | 1.8 | 52,106 | 12.3 | 18.8 | 17.1 |
| Wasco | 16 | 15 | 5,300 | 13.4 | 42.7 | 19.6 | 48 | 13,382 | 27,445 | 53,105 | 46.4 | 3.0 | 56,993 | 12.2 | 18.2 | 16.2 |
| Washington | 262 | 333 | 145,623 | 19.0 | 25.2 | 44.4 | 1,085 | 12,377 | 39,679 | 82,215 | 28.9 | 10.3 | 85,665 | 8.2 | 9.0 | 8.5 |
| Wheeler | 4 | 1 | 207 | 16.4 | 45.5 | 14.9 | 8 | 8,069 | 21,702 | 40,926 | 63.0 | 0.6 | 42,747 | 17.7 | 31.7 | 30.8 |
| Yamhill | 42 | 43 | 26,343 | 25.2 | 35.7 | 26.9 | 183 | 10,982 | 31,314 | 63,902 | 38.5 | 6.2 | 71,815 | 10.6 | 12.2 | 10.9 |
| **PENNSYLVANIA** | X | X | 2,950,731 | 23.8 | 44.2 | 31.4 | 24,643 | 16,301 | 34,352 | 61,744 | 41.1 | 6.6 | 63,455 | 12.0 | 16.5 | 15.9 |
| Adams | 14 | 4 | 23,301 | 25.9 | 51.1 | 22.3 | 287 | 20,518 | 31,590 | 67,253 | 36.3 | 4.3 | 67,706 | 7.6 | 10.2 | 10.0 |
| Allegheny | 384 | 49 | 274,806 | 24.6 | 32.4 | 41.6 | 2,602 | 17,999 | 38,709 | 61,043 | 41.7 | 7.2 | 64,799 | 10.8 | 14.1 | 13.2 |
| Armstrong | NA | NA | 12,124 | 14.2 | 58.7 | 16.7 | 120 | 17,089 | 27,715 | 51,410 | 48.9 | 2.4 | 56,259 | 11.1 | 15.9 | 14.1 |
| Beaver | 20 | 4 | 33,191 | 16.1 | 43.7 | 24.7 | 475 | 14,929 | 31,305 | 57,807 | 44.2 | 3.1 | 60,323 | 11.7 | 17.8 | 15.1 |
| Bedford | 9 | 2 | 8,778 | 13.6 | 63.4 | 14.7 | 91 | 13,943 | 26,078 | 50,509 | 49.6 | 1.7 | 54,251 | 10.4 | 14.0 | 13.1 |
| Berks | 91 | 12 | 101,836 | 17.6 | 50.0 | 24.9 | 1,061 | 15,648 | 31,571 | 63,728 | 39.0 | 4.9 | 67,418 | 10.2 | 16.2 | 15.9 |
| Blair | 25 | 2 | 26,256 | 13.6 | 53.6 | 21.3 | 248 | 14,248 | 28,119 | 50,204 | 50.8 | 2.7 | 51,004 | 14.9 | 21.9 | 18.9 |
| Bradford | 10 | 2 | 12,395 | 15.8 | 56.9 | 18.7 | 139 | 15,145 | 28,392 | 52,358 | 48.0 | 2.9 | 51,356 | 14.3 | 21.2 | 17.3 |
| Bucks | 65 | 15 | 142,897 | 23.5 | 34.3 | 41.3 | 1,557 | 18,233 | 45,849 | 89,139 | 26.9 | 14.3 | 93,587 | 5.7 | 7.1 | 6.9 |
| Butler | 28 | 3 | 42,192 | 16.8 | 37.8 | 36.0 | 363 | 14,084 | 37,811 | 70,668 | 34.5 | 7.7 | 72,576 | 7.8 | 8.1 | 7.6 |
| Cambria | NA | NA | 28,610 | 22.7 | 53.4 | 21.1 | 238 | 13,677 | 26,008 | 46,659 | 53.0 | 1.7 | 48,798 | 14.9 | 22.5 | 20.8 |
| Cameron | NA | NA | 778 | 9.5 | 57.6 | 13.7 | 9 | 16,774 | 24,528 | 41,165 | 62.4 | 1.0 | 43,704 | 14.0 | 23.1 | 22.5 |
| Carbon | NA | NA | 12,289 | 11.6 | 55.9 | 17.9 | 127 | 15,000 | 29,202 | 57,006 | 44.4 | 2.0 | 59,985 | 9.8 | 14.0 | 13.2 |
| Centre | 25 | 4 | 58,481 | 8.8 | 35.1 | 45.5 | 226 | 17,101 | 30,753 | 60,403 | 42.3 | 6.3 | 61,036 | 15.9 | 8.6 | 7.9 |
| Chester | 72 | 18 | 134,910 | 22.8 | 27.2 | 53.0 | 1,428 | 17,695 | 50,927 | 100,214 | 23.6 | 18.4 | 102,233 | 5.9 | 6.3 | 5.7 |
| Clarion | 8 | 3 | 9,058 | 13.4 | 56.4 | 22.1 | 101 | 18,039 | 25,157 | 46,680 | 52.9 | 1.8 | 49,173 | 14.1 | 16.9 | 15.1 |
| Clearfield | NA | NA | 14,155 | 14.9 | 62.3 | 15.1 | 170 | 16,011 | 24,403 | 49,015 | 50.9 | 2.0 | 46,967 | 13.7 | 18.6 | 17.1 |
| Clinton | NA | NA | 9,096 | 13.1 | 56.2 | 19.5 | 72 | 16,042 | 25,794 | 50,923 | 48.9 | 1.9 | 50,777 | 13.4 | 20.2 | 20.0 |
| Columbia | NA | NA | 17,182 | 9.4 | 52.7 | 22.9 | 99 | 16,345 | 26,787 | 50,550 | 49.4 | 2.8 | 54,377 | 14.7 | 17.3 | 14.7 |
| Crawford | NA | NA | 17,946 | 25.6 | 56.7 | 21.1 | 157 | 14,923 | 26,582 | 50,304 | 49.7 | 2.3 | 49,741 | 12.6 | 19.4 | 19.2 |
| Cumberland | NA | NA | 58,970 | 23.0 | 39.6 | 36.6 | 416 | 15,536 | 37,500 | 71,269 | 34.3 | 6.6 | 75,269 | 7.2 | 9.5 | 8.9 |
| Dauphin | NA | NA | 62,142 | 17.5 | 43.4 | 30.8 | 692 | 14,658 | 33,639 | 60,715 | 40.5 | 5.3 | 60,763 | 11.3 | 18.4 | 18.7 |
| Delaware | NA | NA | 145,871 | 32.1 | 37.1 | 39.0 | 1,303 | 17,278 | 40,026 | 74,477 | 34.2 | 11.4 | 76,783 | 9.9 | 13.8 | 14.0 |
| Elk | 5 | 1 | 5,748 | 19.0 | 55.7 | 19.6 | 49 | 14,449 | 29,772 | 53,440 | 47.0 | 2.0 | 55,462 | 9.8 | 11.9 | 10.8 |
| Erie | 41 | 7 | 66,283 | 24.7 | 47.2 | 27.9 | 533 | 13,939 | 28,306 | 51,529 | 48.6 | 3.5 | 51,923 | 16.6 | 24.2 | 22.2 |
| Fayette | NA | NA | 24,585 | 12.9 | 59.6 | 17.3 | 237 | 14,893 | 27,368 | 47,364 | 52.2 | 2.2 | 48,055 | 17.5 | 26.9 | 25.0 |
| Forest | NA | NA | 776 | 15.7 | 74.5 | 8.1 | 10 | 22,492 | 15,245 | 39,717 | 63.6 | 0.5 | 41,267 | 26.0 | 30.0 | 23.0 |
| Franklin | 22 | 19 | 32,630 | 16.5 | 54.8 | 21.6 | 280 | 12,234 | 31,221 | 63,379 | 39.1 | 3.6 | 62,699 | 8.1 | 12.7 | 12.2 |
| Fulton | NA | NA | 2,767 | 9.8 | 64.4 | 13.6 | 32 | 15,116 | 27,396 | 53,476 | 46.4 | 1.6 | 56,357 | 12.2 | 18.6 | 17.3 |
| Greene | 8 | 7 | 7,489 | 22.1 | 58.5 | 17.6 | 81 | 16,681 | 27,783 | 54,776 | 45.8 | 3.2 | 56,599 | 14.2 | 18.8 | 16.8 |
| Huntingdon | NA | NA | 9,151 | 24.9 | 59.4 | 17.4 | 73 | 13,553 | 25,746 | 51,678 | 48.1 | 1.7 | 54,323 | 13.0 | 17.5 | 15.9 |
| Indiana | 20 | 3 | 22,385 | 12.2 | 53.2 | 23.0 | 178 | 18,903 | 26,509 | 49,320 | 50.5 | 2.5 | 51,299 | 14.0 | 19.3 | 18.2 |
| Jefferson | 7 | 1 | 8,115 | 15.6 | 61.7 | 15.9 | 75 | 16,401 | 25,509 | 47,603 | 52.0 | 2.0 | 50,789 | 16.4 | 28.0 | 25.9 |
| Juniata | 5 | 1 | 4,715 | 26.4 | 65.5 | 14.5 | 35 | 12,500 | 26,239 | 53,879 | 46.1 | 2.1 | 57,837 | 10.4 | 15.4 | 14.8 |
| Lackawanna | NA | NA | 47,976 | 31.0 | 45.8 | 28.3 | 404 | 14,310 | 29,899 | 52,821 | 47.4 | 3.9 | 53,655 | 14.2 | 19.0 | 18.2 |
| Lancaster | 48 | 7 | 121,674 | 27.4 | 50.2 | 27.3 | 1,104 | 16,415 | 32,194 | 66,056 | 36.8 | 5.3 | 67,515 | 10.5 | 14.6 | 14.2 |
| Lawrence | 18 | 5 | 17,862 | 16.2 | 52.6 | 22.1 | 175 | 15,115 | 28,926 | 50,204 | 49.8 | 3.0 | 50,979 | 12.3 | 18.1 | 15.6 |
| Lebanon | NA | NA | 30,394 | 21.2 | 56.9 | 20.7 | 250 | 12,684 | 29,130 | 60,281 | 40.0 | 3.3 | 61,560 | 10.4 | 15.5 | 13.7 |
| Lehigh | NA | NA | 87,808 | 21.8 | 44.0 | 29.9 | 858 | 16,381 | 33,566 | 63,897 | 39.6 | 6.9 | 65,436 | 11.5 | 18.2 | 17.6 |
| Luzerne | NA | NA | 68,730 | 22.6 | 49.2 | 22.9 | 631 | 14,208 | 28,972 | 53,473 | 47.3 | 3.2 | 53,439 | 15.2 | 24.6 | 23.5 |
| Lycoming | 16 | 6 | 24,974 | 17.1 | 48.8 | 23.4 | 246 | 15,577 | 27,950 | 54,241 | 45.8 | 2.8 | 53,762 | 13.6 | 18.4 | 17.3 |
| McKean | NA | NA | 8,547 | 10.4 | 55.8 | 18.9 | 102 | 16,828 | 27,023 | 48,647 | 51.6 | 2.7 | 53,266 | 14.6 | 22.8 | 22.4 |
| Mercer | 14 | 4 | 23,894 | 23.6 | 53.1 | 22.8 | 255 | 17,250 | 27,639 | 50,696 | 49.4 | 2.7 | 54,099 | 13.1 | 18.7 | 17.7 |
| Mifflin | NA | NA | 8,722 | 21.7 | 66.8 | 11.9 | 88 | 17,858 | 25,469 | 50,219 | 49.7 | 2.1 | 50,571 | 13.3 | 19.7 | 18.4 |
| Monroe | 24 | 26 | 39,169 | 12.1 | 45.5 | 24.7 | 475 | 18,681 | 29,662 | 63,934 | 38.5 | 5.4 | 63,012 | 12.2 | 17.4 | 16.9 |
| Montgomery | 100 | 25 | 200,392 | 30.1 | 29.2 | 49.3 | 2,184 | 18,902 | 48,845 | 91,546 | 26.1 | 15.4 | 92,340 | 6.0 | 6.7 | 6.7 |
| Montour | NA | NA | 3,456 | 17.2 | 49.5 | 31.6 | 35 | 14,723 | 34,575 | 58,333 | 44.2 | 5.9 | 65,534 | 9.1 | 14.2 | 12.3 |
| Northampton | NA | NA | 72,474 | 27.2 | 43.6 | 30.3 | 756 | 16,921 | 35,270 | 70,471 | 34.9 | 7.4 | 73,780 | 7.9 | 10.4 | 9.4 |
| Northumberland | 9 | 2 | 15,901 | 15.8 | 61.7 | 16.6 | 201 | 17,776 | 26,389 | 48,671 | 51.1 | 2.0 | 47,967 | 13.3 | 20.3 | 19.2 |
| Perry | NA | NA | 8,835 | 19.0 | 60.0 | 16.7 | 84 | 13,756 | 30,785 | 63,718 | 38.1 | 3.6 | 65,401 | 8.9 | 13.1 | 11.7 |
| Philadelphia | NA | NA | 401,265 | 33.8 | 47.9 | 29.7 | 3,023 | 15,238 | 27,924 | 45,927 | 53.3 | 4.7 | 47,598 | 23.0 | 31.6 | 32.9 |
| Pike | 22 | 4 | 11,318 | 13.2 | 41.4 | 26.8 | 140 | 18,571 | 34,589 | 65,928 | 36.6 | 5.7 | 66,595 | 9.2 | 14.5 | 13.4 |

1. All persons 3 years old and over enrolled in nursery school through college.   2. Persons 25 years old and over.   3. Elementary and secondary education expenditures.   4. Based on population estimated by the American Community Survey, 2014–2018.

# Table B. States and Counties — **Personal Income**

| STATE County | Personal income, 2019 Total (mil dol) | Percent change 2018-2019 | Per capita[1] Dollars | Per capita[1] Rank | Wages and salaries (mil dol) | Supplements to wages and salaries, employer contributions (mil dol) Pension and insurance | Government social insurance | Proprietors' income (mil dol) | Dividends, interest, and rent (mil dol) | Personal transfer receipts (mil dol) | Earnings, 2019 Total (mil dol) | Contributions for government social insurance (mil dol) From employee and self-employed | From employer |
|---|---|---|---|---|---|---|---|---|---|---|---|---|---|
| | 62 | 63 | 64 | 65 | 66 | 67 | 68 | 69 | 70 | 71 | 72 | 73 | 74 |
| **OREGON—Cont'd** | | | | | | | | | | | | | |
| Morrow | 564 | 8.8 | 48,605 | 953 | 353 | 65 | 35 | 117 | 69 | 115 | 570 | 27 | 35 |
| Multnomah | 49,400 | 4.0 | 60,773 | 282 | 34,598 | 5,198 | 2,889 | 4,318 | 10,605 | 6,851 | 47,003 | 2,792 | 2,889 |
| Polk | 3,693 | 5.1 | 42,897 | 1,635 | 882 | 183 | 81 | 196 | 712 | 827 | 1,341 | 101 | 81 |
| Sherman | 101 | 5.8 | 56,940 | 407 | 49 | 12 | 5 | 20 | 19 | 25 | 86 | 5 | 5 |
| Tillamook | 1,240 | 4.6 | 45,879 | 1,264 | 423 | 84 | 39 | 132 | 291 | 356 | 678 | 48 | 39 |
| Umatilla | 3,268 | 3.9 | 41,928 | 1,779 | 1,392 | 285 | 135 | 266 | 514 | 845 | 2,077 | 131 | 135 |
| Union | 1,120 | 2.6 | 41,721 | 1,818 | 458 | 95 | 44 | 61 | 219 | 333 | 657 | 46 | 44 |
| Wallowa | 332 | 4.6 | 46,044 | 1,243 | 106 | 25 | 10 | 21 | 89 | 104 | 161 | 13 | 10 |
| Wasco | 1,163 | 2.0 | 43,596 | 1,535 | 505 | 92 | 46 | 79 | 218 | 321 | 721 | 49 | 46 |
| Washington | 38,528 | 4.2 | 64,043 | 207 | 23,530 | 2,783 | 1,765 | 2,933 | 8,286 | 4,464 | 31,012 | 1,894 | 1,765 |
| Wheeler | 52 | 4.8 | 39,368 | 2,132 | 11 | 3 | 1 | 1 | 14 | 20 | 17 | 2 | 1 |
| Yamhill | 5,087 | 4.3 | 47,494 | 1,077 | 1,688 | 312 | 159 | 393 | 1,005 | 1,064 | 2,551 | 167 | 159 |
| **PENNSYLVANIA** | 742,924 | 3.6 | 58,046 | X | 353,818 | 60,834 | 26,815 | 72,891 | 131,208 | 143,936 | 514,359 | 30,573 | 26,815 |
| Adams | 5,169 | 3.7 | 50,182 | 817 | 1,598 | 313 | 130 | 378 | 906 | 1,046 | 2,420 | 159 | 130 |
| Allegheny | 79,997 | 3.1 | 65,784 | 172 | 48,397 | 7,340 | 3,594 | 9,393 | 14,024 | 13,242 | 68,723 | 3,945 | 3,594 |
| Armstrong | 3,038 | 2.9 | 46,925 | 1,141 | 755 | 166 | 61 | 249 | 437 | 812 | 1,230 | 87 | 61 |
| Beaver | 8,295 | 3.3 | 50,600 | 792 | 3,339 | 614 | 261 | 482 | 1,080 | 2,093 | 4,696 | 312 | 261 |
| Bedford | 2,080 | 3.1 | 43,426 | 1,556 | 631 | 133 | 54 | 286 | 275 | 578 | 1,104 | 73 | 54 |
| Berks | 21,605 | 3.9 | 51,299 | 734 | 9,605 | 1,825 | 763 | 1,639 | 3,429 | 4,461 | 13,832 | 830 | 763 |
| Blair | 5,882 | 3.2 | 48,277 | 994 | 2,743 | 570 | 231 | 504 | 939 | 1,608 | 4,048 | 255 | 231 |
| Bradford | 2,605 | 4.5 | 43,187 | 1,587 | 1,208 | 240 | 94 | 223 | 467 | 680 | 1,765 | 113 | 94 |
| Bucks | 47,663 | 3.9 | 75,864 | 83 | 15,774 | 2,522 | 1,225 | 3,553 | 8,697 | 6,493 | 23,073 | 1,385 | 1,225 |
| Butler | 11,364 | 3.9 | 60,493 | 290 | 5,062 | 930 | 393 | 733 | 1,878 | 1,925 | 7,117 | 432 | 393 |
| Cambria | 5,873 | 2.8 | 45,107 | 1,350 | 2,226 | 487 | 182 | 316 | 918 | 1,873 | 3,211 | 223 | 182 |
| Cameron | 211 | -2.5 | 47,470 | 1,083 | 78 | 19 | 7 | 10 | 40 | 69 | 114 | 8 | 7 |
| Carbon | 3,405 | 3.6 | 53,053 | 611 | 663 | 150 | 55 | 709 | 429 | 756 | 1,577 | 94 | 55 |
| Centre | 7,724 | 3.6 | 47,564 | 1,071 | 4,134 | 1,465 | 306 | 637 | 1,539 | 1,211 | 6,541 | 337 | 306 |
| Chester | 45,966 | 4.3 | 87,557 | 38 | 20,377 | 2,915 | 1,415 | 4,655 | 9,755 | 4,539 | 29,362 | 1,639 | 1,415 |
| Clarion | 1,625 | 2.3 | 42,281 | 1,721 | 526 | 142 | 43 | 168 | 278 | 464 | 881 | 56 | 43 |
| Clearfield | 3,666 | 2.9 | 46,253 | 1,219 | 1,325 | 290 | 108 | 242 | 485 | 971 | 1,964 | 131 | 108 |
| Clinton | 1,571 | 3.0 | 40,669 | 1,967 | 638 | 158 | 50 | 111 | 228 | 419 | 958 | 61 | 50 |
| Columbia | 2,845 | 2.9 | 43,793 | 1,516 | 1,130 | 276 | 91 | 201 | 447 | 698 | 1,698 | 107 | 91 |
| Crawford | 3,575 | 3.2 | 42,237 | 1,733 | 1,309 | 283 | 107 | 394 | 520 | 1,022 | 2,093 | 136 | 107 |
| Cumberland | 14,325 | 4.1 | 56,540 | 419 | 8,017 | 1,387 | 633 | 1,078 | 2,599 | 2,419 | 11,116 | 656 | 633 |
| Dauphin | 14,671 | 3.7 | 52,716 | 636 | 11,519 | 2,284 | 871 | 1,117 | 2,306 | 3,035 | 15,791 | 898 | 871 |
| Delaware | 39,525 | 4.0 | 69,740 | 131 | 15,375 | 2,381 | 1,134 | 3,591 | 7,503 | 6,273 | 22,482 | 1,311 | 1,134 |
| Elk | 1,470 | 0.7 | 49,141 | 912 | 681 | 143 | 57 | 80 | 233 | 370 | 961 | 62 | 57 |
| Erie | 12,510 | 2.3 | 46,379 | 1,201 | 5,723 | 1,190 | 457 | 810 | 2,297 | 3,210 | 8,181 | 510 | 457 |
| Fayette | 5,667 | 2.8 | 43,835 | 1,509 | 1,818 | 387 | 149 | 332 | 781 | 1,763 | 2,686 | 192 | 149 |
| Forest | 175 | 4.5 | 24,113 | 3,104 | 101 | 34 | 8 | 10 | 47 | 77 | 153 | 10 | 8 |
| Franklin | 7,431 | 3.8 | 47,934 | 1,022 | 2,850 | 564 | 236 | 596 | 1,164 | 1,614 | 4,245 | 268 | 236 |
| Fulton | 665 | 2.6 | 45,774 | 1,278 | 294 | 62 | 26 | 78 | 92 | 166 | 460 | 28 | 26 |
| Greene | 1,636 | 2.3 | 45,161 | 1,343 | 847 | 167 | 64 | 112 | 279 | 441 | 1,190 | 75 | 64 |
| Huntingdon | 1,777 | 3.1 | 39,371 | 2,130 | 549 | 140 | 45 | 163 | 266 | 506 | 896 | 60 | 45 |
| Indiana | 3,489 | 3.3 | 41,503 | 1,848 | 1,539 | 380 | 119 | 307 | 567 | 926 | 2,346 | 145 | 119 |
| Jefferson | 1,991 | 3.0 | 45,855 | 1,267 | 681 | 141 | 57 | 216 | 331 | 541 | 1,095 | 72 | 57 |
| Juniata | 1,145 | 3.4 | 46,258 | 1,217 | 262 | 56 | 22 | 225 | 146 | 246 | 565 | 34 | 22 |
| Lackawanna | 10,267 | 2.8 | 48,967 | 925 | 4,527 | 881 | 365 | 659 | 1,742 | 2,610 | 6,433 | 412 | 365 |
| Lancaster | 29,641 | 3.6 | 54,314 | 527 | 12,820 | 2,213 | 1,004 | 4,561 | 4,943 | 5,160 | 20,598 | 1,177 | 1,004 |
| Lawrence | 3,870 | 2.8 | 45,252 | 1,329 | 1,345 | 265 | 110 | 242 | 535 | 1,115 | 1,963 | 137 | 110 |
| Lebanon | 6,937 | 3.2 | 48,924 | 931 | 2,420 | 524 | 197 | 617 | 1,150 | 1,497 | 3,759 | 235 | 197 |
| Lehigh | 20,117 | 4.2 | 54,471 | 521 | 11,839 | 1,919 | 899 | 1,707 | 3,274 | 3,898 | 16,364 | 965 | 899 |
| Luzerne | 14,716 | 2.8 | 46,363 | 1,204 | 7,062 | 1,376 | 576 | 763 | 2,458 | 3,799 | 9,777 | 622 | 576 |
| Lycoming | 5,092 | 2.6 | 44,941 | 1,363 | 2,476 | 530 | 197 | 271 | 841 | 1,284 | 3,474 | 220 | 197 |
| McKean | 1,819 | 0.6 | 44,774 | 1,388 | 651 | 160 | 53 | 161 | 310 | 504 | 1,025 | 66 | 53 |
| Mercer | 4,780 | 1.8 | 43,682 | 1,527 | 2,128 | 430 | 174 | 360 | 781 | 1,416 | 3,092 | 204 | 174 |
| Mifflin | 1,875 | 3.1 | 40,641 | 1,970 | 698 | 146 | 57 | 230 | 236 | 545 | 1,131 | 74 | 57 |
| Monroe | 7,462 | 4.3 | 43,823 | 1,510 | 2,736 | 619 | 220 | 508 | 1,025 | 1,716 | 4,083 | 257 | 220 |
| Montgomery | 68,166 | 4.3 | 82,037 | 56 | 40,160 | 5,505 | 2,859 | 900 | 19,330 | 8,164 | 49,243 | 3,123 | 2,859 |
| Montour | 1,073 | 2.8 | 58,872 | 343 | 1,159 | 183 | 74 | 64 | 153 | 211 | 1,480 | 86 | 74 |
| Northampton | 17,380 | 4.3 | 56,931 | 409 | 6,322 | 1,129 | 507 | 1,427 | 2,852 | 3,380 | 9,385 | 581 | 507 |
| Northumberland | 3,932 | 3.2 | 43,289 | 1,573 | 1,218 | 269 | 101 | 229 | 615 | 1,121 | 1,818 | 128 | 101 |
| Perry | 2,138 | 3.7 | 46,212 | 1,227 | 293 | 79 | 24 | 235 | 287 | 469 | 631 | 44 | 24 |
| Philadelphia | 90,712 | 3.3 | 57,265 | 396 | 52,834 | 8,474 | 3,857 | 20,048 | 11,957 | 21,785 | 85,214 | 4,575 | 3,857 |
| Pike | 2,783 | 4.4 | 49,873 | 838 | 446 | 108 | 37 | 203 | 468 | 665 | 794 | 61 | 37 |

1. Based on the resident population estimated as of July 1 of the year shown.

Items 62—74

# Table B. States and Counties — Earnings, Social Security, and Housing

| STATE County | Earnings, 2019 (cont.) — Percent by selected industries | | | | | | | | | Social Security beneficiaries, December 2019 | | Supplemental Security Income recipients, 2019 | Housing units, 2020 | |
|---|---|---|---|---|---|---|---|---|---|---|---|---|---|---|
| | Farm | Mining, quarrying, and extractions | Construction | Manufacturing | Information; professional, scientific, technical services | Retail trade | Finance, insurance, real estate, and leasing | Health care and social assistance | Government | Number | Rate[1] | | Total | Percent change, 2010-2020 |
| | 75 | 76 | 77 | 78 | 79 | 80 | 81 | 82 | 83 | 84 | 85 | 86 | 87 | 88 |
| **OREGON—Cont'd** | | | | | | | | | | | | | | |
| Morrow | 23.5 | D | 1.6 | 21.6 | D | 1.5 | D | 2.3 | 14.3 | 2,290 | 197 | 243 | 4,835 | 8.8 |
| Multnomah | 0.1 | 0.0 | 6.2 | 6.0 | 16.3 | 4.8 | 8.2 | 11.1 | 17.3 | 124,845 | 154 | 21,283 | 365,733 | 12.6 |
| Polk | 1.1 | D | 8.3 | 9.7 | 4.4 | 5.4 | 4.9 | 13.0 | 28.6 | 19,160 | 223 | 1,523 | 33,540 | 10.7 |
| Sherman | 11.9 | 0.4 | 10.5 | D | D | 5.0 | D | 0.9 | 32.8 | 540 | 302 | 43 | 953 | 3.7 |
| Tillamook | 5.9 | 0.0 | 7.8 | 14.8 | 3.9 | 6.6 | 3.6 | 12.0 | 22.0 | 8,770 | 323 | 539 | 19,437 | 5.9 |
| Umatilla | 4.2 | 0.2 | 5.2 | 8.8 | 4.8 | 6.5 | 3.9 | 11.1 | 26.8 | 15,090 | 194 | 1,688 | 31,279 | 5.3 |
| Union | 1.0 | D | 5.6 | 13.9 | 4.4 | 8.4 | 4.2 | 16.9 | 24.3 | 6,615 | 247 | 632 | 11,979 | 4.3 |
| Wallowa | 1.5 | 0.0 | 9.9 | 3.9 | 4.8 | 7.2 | 4.7 | 11.0 | 30.6 | 2,520 | 352 | 150 | 4,316 | 5.1 |
| Wasco | 2.7 | D | 7.3 | 4.9 | 7.8 | 9.9 | 3.6 | 21.2 | 22.1 | 6,680 | 252 | 709 | 11,793 | 2.7 |
| Washington | 0.3 | 0.1 | 6.3 | 23.0 | 10.4 | 5.3 | 7.1 | 7.8 | 6.9 | 88,955 | 148 | 7,157 | 237,501 | 11.8 |
| Wheeler | 3.4 | 0.0 | 2.8 | 3.3 | D | 6.1 | D | D | 41.1 | 540 | 401 | 32 | 930 | 3.9 |
| Yamhill | 4.1 | 0.3 | 8.0 | 20.3 | 5.6 | 6.5 | 5.7 | 12.3 | 14.2 | 22,450 | 210 | 1,699 | 40,460 | 9.0 |
| **PENNSYLVANIA** | 0.3 | 0.6 | 6.0 | 10.0 | 14.8 | 5.2 | 7.6 | 14.1 | 12.5 | 2,861,155 | 224 | 354,037 | 5,753,305 | 3.3 |
| Adams | 1.4 | D | 8.8 | 21.1 | 3.9 | 6.1 | 3.4 | 13.2 | 15.0 | 25,700 | 250 | 1,077 | 42,965 | 5.2 |
| Allegheny | 0.0 | 0.5 | 5.1 | 6.3 | 15.5 | 4.4 | 10.8 | 14.7 | 9.1 | 270,610 | 223 | 32,478 | 605,895 | 2.8 |
| Armstrong | 0.8 | 12.7 | 6.8 | 10.8 | 4.4 | 6.3 | 3.5 | 14.5 | 14.9 | 18,665 | 288 | 2,059 | 32,851 | 0.8 |
| Beaver | 0.1 | 0.1 | 24.6 | 11.6 | 4.6 | 5.1 | 2.9 | 12.7 | 12.3 | 44,845 | 274 | 4,436 | 79,894 | 2.2 |
| Bedford | 1.5 | D | 11.4 | 13.2 | 2.7 | 8.8 | 3.6 | D | 13.4 | 14,075 | 294 | 1,180 | 24,516 | 2.3 |
| Berks | 1.0 | 0.1 | 6.5 | 19.0 | 6.9 | 5.7 | 5.4 | 15.0 | 12.4 | 89,330 | 213 | 10,740 | 168,408 | 2.2 |
| Blair | 0.6 | 0.3 | 5.3 | 12.7 | 5.3 | 8.5 | 4.0 | 19.5 | 15.4 | 31,195 | 256 | 4,361 | 57,132 | 1.5 |
| Bradford | 1.7 | 6.2 | 4.4 | 15.5 | 3.1 | 5.1 | 4.1 | 22.0 | 12.3 | 16,415 | 272 | 1,758 | 30,785 | 2.7 |
| Bucks | 0.1 | 0.1 | 9.6 | 9.8 | 13.5 | 6.7 | 6.9 | 15.3 | 9.6 | 138,765 | 221 | 7,071 | 252,835 | 2.8 |
| Butler | 0.2 | 1.5 | 7.1 | 15.1 | 9.6 | 6.4 | 4.4 | 12.3 | 13.8 | 45,440 | 241 | 3,061 | 85,630 | 9.6 |
| Cambria | 0.2 | 0.3 | 6.6 | 9.9 | 6.7 | 7.3 | 5.4 | 21.5 | 17.5 | 39,075 | 300 | 4,739 | 66,073 | 0.6 |
| Cameron | 0.4 | D | D | 50.0 | D | D | D | D | 22.2 | 1,595 | 360 | 121 | 4,420 | -0.7 |
| Carbon | 0.1 | 0.1 | 4.2 | 7.5 | 40.9 | 5.1 | 2.6 | 10.9 | 12.1 | 17,645 | 275 | 1,375 | 34,928 | 1.8 |
| Centre | 0.4 | 0.6 | 4.9 | 5.2 | 7.9 | 4.2 | 4.7 | 10.3 | 48.5 | 25,555 | 158 | 1,366 | 67,800 | 7.1 |
| Chester | 1.0 | 0.0 | 5.6 | 6.9 | 21.4 | 7.4 | 17.3 | 7.4 | 7.7 | 94,065 | 179 | 4,260 | 204,830 | 6.4 |
| Clarion | 0.8 | 1.2 | 8.2 | 10.7 | 2.5 | 7.7 | 4.0 | 15.0 | 25.5 | 10,160 | 264 | 1,200 | 20,583 | 3.1 |
| Clearfield | 0.2 | 2.0 | 4.2 | 9.0 | 3.3 | 7.7 | 3.7 | 22.2 | 18.3 | 21,460 | 271 | 2,222 | 39,437 | 2.1 |
| Clinton | 1.0 | 7.1 | 7.3 | 23.6 | 2.2 | 6.6 | 2.4 | D | 22.4 | 9,335 | 244 | 1,008 | 19,315 | 1.2 |
| Columbia | 0.6 | D | 6.0 | 18.2 | 3.9 | 7.1 | 3.1 | 10.7 | 20.7 | 16,030 | 247 | 1,384 | 30,506 | 3.4 |
| Crawford | 0.8 | 1.5 | 6.1 | 24.6 | 4.6 | 6.3 | 3.0 | 16.1 | 14.6 | 22,905 | 271 | 2,748 | 45,037 | 0.8 |
| Cumberland | 0.4 | 0.1 | 4.7 | 6.0 | 12.4 | 5.7 | 9.5 | 12.3 | 15.2 | 54,720 | 216 | 2,785 | 109,154 | 9.2 |
| Dauphin | 0.1 | D | 4.0 | 7.2 | 8.5 | 3.5 | 8.1 | 16.9 | 22.4 | 58,690 | 211 | 8,362 | 126,033 | 4.7 |
| Delaware | 0.0 | D | 6.9 | 12.7 | 9.5 | 4.7 | 11.0 | 12.9 | 9.9 | 109,165 | 193 | 12,701 | 225,204 | 1.0 |
| Elk | 0.0 | 0.2 | 5.1 | 47.0 | 2.8 | 4.7 | 2.6 | 10.7 | 9.2 | 8,845 | 296 | 558 | 17,848 | 1.5 |
| Erie | 0.3 | 0.0 | 4.6 | 18.7 | 5.1 | 6.5 | 8.4 | 18.2 | 16.0 | 64,700 | 240 | 10,342 | 121,964 | 2.4 |
| Fayette | 0.1 | 6.7 | 4.1 | 9.0 | 4.3 | 7.5 | 3.2 | 15.5 | 18.5 | 36,415 | 281 | 7,372 | 64,210 | 2.3 |
| Forest | 0.5 | D | D | D | D | D | D | 12.0 | 62.2 | 1,875 | 267 | 155 | 8,719 | -0.4 |
| Franklin | 1.9 | 0.2 | 5.7 | 16.7 | 5.1 | 6.7 | 3.7 | 15.4 | 16.2 | 37,515 | 241 | 2,476 | 66,453 | 5.1 |
| Fulton | 3.3 | D | 10.2 | 43.2 | 1.2 | 3.1 | D | D | 10.7 | 4,025 | 278 | 332 | 7,308 | 2.6 |
| Greene | 0.2 | 31.7 | 9.0 | 2.0 | 3.2 | 4.9 | 3.4 | D | 17.5 | 9,100 | 252 | 1,613 | 16,840 | 2.3 |
| Huntingdon | 2.6 | 1.1 | 6.2 | 9.9 | 2.9 | 5.8 | 4.5 | D | 27.7 | 11,540 | 257 | 1,073 | 22,836 | 2.2 |
| Indiana | 0.5 | 6.1 | 6.9 | 6.9 | 4.0 | 6.4 | 5.6 | 12.4 | 23.5 | 20,955 | 249 | 2,428 | 39,063 | 2.2 |
| Jefferson | 0.7 | 4.7 | 6.1 | 23.9 | 5.8 | 4.9 | 2.3 | 14.4 | 11.5 | 12,195 | 282 | 1,278 | 22,755 | 1.4 |
| Juniata | 2.6 | D | 7.6 | 31.6 | D | 6.5 | 3.5 | D | 9.5 | 5,855 | 236 | 414 | 11,326 | 3.1 |
| Lackawanna | 0.0 | 0.1 | 5.7 | 10.4 | 6.5 | 7.1 | 8.7 | 19.5 | 13.3 | 54,640 | 260 | 7,027 | 100,988 | 4.3 |
| Lancaster | 1.0 | 0.1 | 11.5 | 14.9 | 7.1 | 7.7 | 5.6 | 13.0 | 8.2 | 115,465 | 212 | 9,261 | 214,683 | 5.8 |
| Lawrence | 0.3 | 0.5 | 12.7 | 14.0 | 4.9 | 6.6 | 5.7 | 14.5 | 13.8 | 24,210 | 283 | 3,329 | 41,417 | 1.1 |
| Lebanon | 0.9 | 0.0 | 6.3 | 17.6 | 6.0 | 7.7 | 3.1 | 11.8 | 18.9 | 33,680 | 238 | 2,667 | 58,564 | 5.3 |
| Lehigh | 0.1 | D | 4.9 | 12.2 | 8.1 | 4.6 | 5.4 | 21.7 | 9.0 | 78,025 | 211 | 9,856 | 147,545 | 3.4 |
| Luzerne | 0.0 | 0.3 | 5.1 | 11.5 | 5.3 | 6.6 | 5.5 | 16.4 | 15.5 | 79,845 | 252 | 10,209 | 150,737 | 1.3 |
| Lycoming | 0.6 | 3.4 | 5.3 | 16.8 | 5.2 | 6.5 | 4.6 | 17.5 | 18.5 | 29,100 | 256 | 3,235 | 53,692 | 2.3 |
| McKean | 0.3 | 4.0 | 6.2 | 21.3 | 2.9 | 5.7 | 2.1 | D | 17.3 | 11,255 | 277 | 1,502 | 21,312 | 0.4 |
| Mercer | 0.3 | 1.2 | 5.9 | 20.8 | 3.6 | 7.6 | 6.0 | 17.7 | 12.9 | 31,115 | 285 | 3,595 | 52,491 | 1.5 |
| Mifflin | 1.4 | 0.0 | 6.8 | 25.5 | 2.1 | 7.5 | 3.0 | 18.4 | 10.6 | 12,370 | 268 | 1,281 | 22,002 | 2.2 |
| Monroe | 0.0 | 0.1 | 5.1 | 13.6 | 3.8 | 7.9 | 3.2 | 12.3 | 24.9 | 37,680 | 222 | 3,081 | 81,980 | 2.0 |
| Montgomery | 0.0 | 0.0 | 7.4 | 11.1 | 24.0 | 5.0 | 7.4 | 12.8 | 7.0 | 164,205 | 198 | 8,834 | 339,888 | 4.3 |
| Montour | 0.4 | D | 1.0 | 2.7 | D | 1.6 | 8.5 | 58.1 | 7.6 | 4,625 | 254 | 385 | 8,297 | 4.2 |
| Northampton | 0.0 | 0.1 | 5.7 | 16.0 | 7.3 | 6.2 | 6.0 | 9.8 | 13.0 | 72,700 | 238 | 6,361 | 124,427 | 3.4 |
| Northumberland | 1.2 | 0.7 | 5.9 | 16.7 | D | 6.0 | 3.0 | 13.4 | 17.0 | 24,895 | 275 | 2,808 | 45,661 | 1.2 |
| Perry | 3.9 | D | 14.6 | 5.5 | 4.5 | 7.8 | 4.6 | 6.6 | 21.0 | 11,010 | 238 | 734 | 21,108 | 3.4 |
| Philadelphia | 0.0 | D | 2.1 | 2.3 | 30.6 | 2.3 | 7.3 | 14.2 | 13.5 | 261,435 | 165 | 104,136 | 695,797 | 3.8 |
| Pike | 0.1 | D | D | D | 7.2 | 7.6 | 4.2 | 8.2 | 26.6 | 15,615 | 280 | 726 | 39,193 | 2.2 |

1. Per 1,000 resident population estimated as of July 1 of the year shown.

# Table B. States and Counties — Housing, Labor Force, and Employment

| STATE County | Housing units, 2015-2019 | | | | | | | | Civilian labor force, 2020 | | | | Civilian employment[6], 2015-2019 | | |
| | Occupied units | | | | | | | Sub-standard units[4] (percent) | | | Unemployment | | | Percent | |
| | Owner-occupied | | | | Renter-occupied | | | | | | | | | | |
| | | | | Median owner cost as a percent of income | | | Median rent as a percent of income[2] | | | | | | | | Management, business, science, and arts | Construction, production, and maintenance occupations |
| | Total | Percent | Median value[1] | With a mort-gage | Without a mort-gage[2] | Median rent[3] | | | Total | Percent change, 2019-2020 | Total | Rate[5] | Total | | |
| | 89 | 90 | 91 | 92 | 93 | 94 | 95 | 96 | 97 | 98 | 99 | 100 | 101 | 102 | 103 |
|---|---|---|---|---|---|---|---|---|---|---|---|---|---|---|---|
| **OREGON—Cont'd** | | | | | | | | | | | | | | | |
| Morrow | 4,108 | 70.0 | 138,100 | 19.8 | 10.0 | 790 | 24.3 | 8.7 | 5,717 | | 295 | 5.2 | 4,923 | 23.5 | 45.9 |
| Multnomah | 326,229 | 54.5 | 386,200 | 22.7 | 12.8 | 1,237 | 30.7 | 4.3 | 457,372 | 0.8 | 39,302 | 8.6 | 440,395 | 46.8 | 16.7 |
| Polk | 30,305 | 66.3 | 265,200 | 22.1 | 10.7 | 994 | 30.3 | 2.2 | 40,205 | -0.1 | 2,639 | 6.6 | 37,400 | 37.4 | 21.8 |
| Sherman | 748 | 64.6 | 159,800 | 20.2 | 11.6 | 742 | 28.1 | 1.9 | 983 | 5.7 | 60 | 6.1 | 715 | 34.0 | 39.0 |
| Tillamook | 10,784 | 69.3 | 254,300 | 23.6 | 12.0 | 910 | 30.9 | 4.4 | 11,710 | -1.8 | 948 | 8.1 | 10,307 | 30.2 | 30.2 |
| Umatilla | 26,908 | 65.7 | 170,900 | 20.5 | 10.5 | 757 | 26.9 | 6.5 | 37,001 | 0.5 | 2,505 | 6.8 | 32,412 | 28.4 | 33.9 |
| Union | 10,708 | 64.5 | 185,400 | 21.4 | 10.9 | 818 | 26.3 | 3.0 | 12,053 | 0.3 | 944 | 7.8 | 11,726 | 30.0 | 26.5 |
| Wallowa | 3,214 | 69.7 | 239,100 | 22.4 | 12.4 | 681 | 27.5 | 4.0 | 3,473 | 1.0 | 246 | 7.1 | 3,079 | 36.5 | 24.5 |
| Wasco | 10,305 | 62.2 | 212,900 | 22.8 | 11.6 | 842 | 24.6 | 3.5 | 13,430 | 0.0 | 953 | 7.1 | 11,819 | 31.9 | 27.4 |
| Washington | 219,053 | 61.6 | 386,600 | 21.9 | 10.4 | 1,359 | 28.8 | 3.9 | 321,579 | -1.3 | 20,837 | 6.5 | 308,937 | 46.3 | 17.7 |
| Wheeler | 640 | 76.1 | 155,300 | 28.8 | 16.9 | 656 | 23.7 | 0.9 | 693 | -2.8 | 30 | 4.3 | 451 | 37.5 | 29.7 |
| Yamhill | 36,808 | 70.0 | 295,900 | 23.4 | 12.4 | 1,052 | 31.6 | 5.2 | 54,221 | -0.8 | 3,664 | 6.8 | 48,740 | 34.2 | 27.5 |
| **PENNSYLVANIA** | 5,053,106 | 68.9 | 180,200 | 20.3 | 12.5 | 938 | 29.0 | 1.8 | 6,387,869 | -1.7 | 579,927 | 9.1 | 6,199,456 | 38.8 | 22.6 |
| Adams | 39,345 | 78.0 | 205,800 | 21.9 | 12.9 | 922 | 28.4 | 1.5 | 54,035 | -2.5 | 3,742 | 6.9 | 50,873 | 30.6 | 32.2 |
| Allegheny | 541,541 | 64.3 | 154,700 | 18.4 | 12.0 | 890 | 27.5 | 1.2 | 635,957 | -2.2 | 57,367 | 9.0 | 629,528 | 46.5 | 15.2 |
| Armstrong | 28,137 | 76.0 | 109,400 | 19.4 | 11.7 | 666 | 26.6 | 1.8 | 31,451 | -2.1 | 3,074 | 9.8 | 29,658 | 28.2 | 34.0 |
| Beaver | 71,167 | 73.1 | 141,100 | 19.0 | 12.9 | 698 | 27.1 | 1.6 | 82,963 | -1.2 | 8,518 | 10.3 | 80,919 | 35.8 | 24.7 |
| Bedford | 19,882 | 79.1 | 136,700 | 20.1 | 11.4 | 691 | 27.4 | 1.8 | 23,383 | -0.8 | 2,133 | 9.1 | 22,141 | 27.0 | 35.9 |
| Berks | 154,712 | 71.5 | 178,700 | 21.5 | 13.7 | 939 | 30.3 | 1.7 | 211,486 | -2.0 | 20,043 | 9.5 | 205,648 | 32.8 | 28.7 |
| Blair | 52,126 | 69.8 | 123,600 | 18.8 | 11.7 | 722 | 29.8 | 1.2 | 59,205 | -0.8 | 5,152 | 8.7 | 56,808 | 32.5 | 26.1 |
| Bradford | 25,021 | 74.0 | 150,600 | 19.3 | 12.0 | 738 | 27.2 | 2.4 | 28,099 | -2.3 | 2,107 | 7.5 | 26,863 | 28.4 | 37.6 |
| Bucks | 238,830 | 77.0 | 330,600 | 22.1 | 13.2 | 1,228 | 30.0 | 1.3 | 338,801 | -1.9 | 28,120 | 8.3 | 331,860 | 44.6 | 18.5 |
| Butler | 76,502 | 75.9 | 205,600 | 19.1 | 10.9 | 869 | 26.5 | 0.9 | 97,152 | -3.0 | 7,638 | 7.9 | 93,689 | 40.7 | 22.2 |
| Cambria | 56,674 | 74.1 | 91,900 | 18.5 | 12.6 | 634 | 28.5 | 1.0 | 56,235 | -2.9 | 5,363 | 9.5 | 56,957 | 32.3 | 26.4 |
| Cameron | 2,184 | 72.0 | 78,500 | 19.0 | 11.8 | 600 | 29.2 | 1.5 | 2,036 | -2.8 | 256 | 12.6 | 2,036 | 22.7 | 36.7 |
| Carbon | 26,043 | 75.8 | 146,400 | 20.7 | 14.0 | 830 | 26.8 | 1.8 | 31,589 | -1.9 | 3,087 | 9.8 | 29,641 | 29.7 | 32.5 |
| Centre | 58,201 | 61.5 | 234,900 | 20.4 | 11.0 | 1,000 | 34.9 | 2.3 | 76,416 | -4.5 | 4,461 | 5.8 | 77,182 | 48.9 | 14.9 |
| Chester | 190,980 | 75.0 | 357,100 | 20.7 | 12.2 | 1,330 | 27.7 | 1.7 | 280,096 | -2.7 | 18,035 | 6.4 | 273,363 | 51.7 | 14.7 |
| Clarion | 16,021 | 69.1 | 116,900 | 18.7 | 11.1 | 649 | 28.6 | 1.6 | 16,871 | -1.4 | 1,389 | 8.2 | 17,377 | 30.5 | 29.5 |
| Clearfield | 31,248 | 76.8 | 96,100 | 19.2 | 12.3 | 666 | 29.3 | 2.0 | 35,582 | -1.1 | 3,275 | 9.2 | 34,016 | 27.3 | 35.1 |
| Clinton | 14,690 | 70.2 | 136,600 | 19.8 | 12.5 | 736 | 26.6 | 1.2 | 17,701 | -3.4 | 1,542 | 8.7 | 17,883 | 29.3 | 31.0 |
| Columbia | 26,372 | 69.2 | 151,700 | 20.1 | 12.4 | 758 | 28.0 | 0.8 | 33,324 | -0.6 | 2,701 | 8.1 | 30,394 | 33.0 | 28.9 |
| Crawford | 35,164 | 72.9 | 113,200 | 19.3 | 11.8 | 664 | 24.9 | 2.6 | 38,254 | | 3,398 | 8.9 | 38,398 | 30.9 | 32.5 |
| Cumberland | 99,804 | 70.3 | 202,900 | 19.9 | 11.4 | 991 | 26.8 | 1.1 | 130,651 | -1.7 | 8,702 | 6.7 | 127,941 | 42.9 | 20.5 |
| Dauphin | 112,212 | 63.2 | 167,900 | 20.1 | 12.0 | 949 | 27.4 | 1.8 | 144,428 | -0.2 | 12,767 | 8.8 | 138,855 | 38.2 | 21.1 |
| Delaware | 207,257 | 68.8 | 244,400 | 21.4 | 13.5 | 1,078 | 29.7 | 1.9 | 294,706 | -1.8 | 26,947 | 9.1 | 279,607 | 44.0 | 16.6 |
| Elk | 14,020 | 78.6 | 103,200 | 16.5 | 10.9 | 591 | 24.9 | 1.4 | 15,208 | -3.0 | 1,871 | 12.3 | 15,374 | 27.3 | 42.2 |
| Erie | 110,318 | 66.1 | 134,100 | 19.4 | 12.2 | 753 | 28.8 | 2.0 | 126,749 | -1.8 | 12,418 | 9.8 | 126,569 | 34.8 | 23.1 |
| Fayette | 54,837 | 74.4 | 101,500 | 18.6 | 12.5 | 675 | 28.6 | 1.7 | 56,337 | | 6,594 | 11.7 | 54,869 | 30.0 | 30.4 |
| Forest | 1,839 | 83.9 | 95,900 | 20.1 | 11.8 | 520 | 33.8 | 3.5 | 1,799 | -0.9 | 191 | 10.6 | 989 | 26.5 | 28.8 |
| Franklin | 60,438 | 71.2 | 180,400 | 20.8 | 11.1 | 883 | 25.3 | 1.7 | 77,184 | -0.7 | 5,876 | 7.6 | 74,289 | 30.6 | 31.5 |
| Fulton | 5,989 | 77.4 | 161,000 | 21.1 | 11.8 | 663 | 23.0 | 1.7 | 7,710 | -2.1 | 890 | 11.5 | 6,721 | 25.8 | 40.6 |
| Greene | 14,230 | 74.8 | 122,900 | 17.2 | 10.6 | 672 | 25.3 | 1.7 | 15,981 | -3.9 | 1,556 | 9.7 | 14,424 | 32.4 | 29.6 |
| Huntingdon | 16,779 | 75.7 | 136,400 | 19.7 | 11.3 | 642 | 24.7 | 1.5 | 19,763 | | 2,063 | 10.4 | 18,615 | 29.6 | 31.6 |
| Indiana | 33,246 | 71.2 | 121,100 | 19.0 | 12.4 | 709 | 31.1 | 2.3 | 37,666 | -3.7 | 3,428 | 9.1 | 37,420 | 31.1 | 29.1 |
| Jefferson | 18,427 | 74.3 | 103,900 | 18.8 | 10.8 | 669 | 27.9 | 2.4 | 20,056 | -2.5 | 1,763 | 8.8 | 19,674 | 25.4 | 38.4 |
| Juniata | 9,372 | 74.5 | 148,900 | 18.7 | 10.0 | 673 | 23.9 | 2.7 | 12,595 | -0.2 | 902 | 7.2 | 11,270 | 25.9 | 40.2 |
| Lackawanna | 87,161 | 64.3 | 152,400 | 20.4 | 13.8 | 792 | 27.7 | 1.6 | 103,832 | -1.4 | 9,962 | 9.6 | 99,414 | 36.0 | 23.7 |
| Lancaster | 201,620 | 68.1 | 209,400 | 21.5 | 11.6 | 1,009 | 29.0 | 2.4 | 281,736 | -1.3 | 21,118 | 7.5 | 272,111 | 34.4 | 28.8 |
| Lawrence | 37,055 | 74.5 | 111,300 | 19.2 | 12.8 | 696 | 31.0 | 1.8 | 40,232 | -1.3 | 4,152 | 10.3 | 39,582 | 31.1 | 28.7 |
| Lebanon | 53,579 | 70.0 | 171,300 | 20.6 | 12.5 | 874 | 27.9 | 3.0 | 71,763 | -0.9 | 5,703 | 7.9 | 68,411 | 30.1 | 29.6 |
| Lehigh | 138,714 | 64.4 | 208,200 | 21.1 | 13.4 | 1,072 | 31.3 | 2.4 | 191,388 | -1.3 | 18,308 | 9.6 | 178,697 | 35.8 | 25.7 |
| Luzerne | 128,660 | 68.0 | 127,000 | 19.4 | 13.0 | 785 | 27.3 | 1.6 | 157,274 | -0.8 | 17,091 | 10.9 | 151,788 | 30.8 | 28.3 |
| Lycoming | 45,608 | 69.2 | 155,800 | 19.8 | 12.2 | 787 | 29.4 | 1.1 | 55,883 | -2.0 | 4,913 | 8.8 | 53,135 | 31.3 | 27.6 |
| McKean | 17,147 | 74.1 | 80,400 | 18.1 | 11.4 | 657 | 26.9 | 0.6 | 17,054 | -1.7 | 1,685 | 9.9 | 18,207 | 30.9 | 31.2 |
| Mercer | 46,340 | 72.5 | 121,100 | 18.9 | 11.5 | 689 | 27.6 | 2.3 | 48,048 | -2.1 | 4,651 | 9.7 | 49,378 | 31.0 | 26.5 |
| Mifflin | 19,043 | 71.3 | 117,900 | 19.6 | 13.2 | 682 | 25.6 | 2.5 | 21,146 | 0.4 | 1,767 | 8.4 | 21,379 | 25.6 | 37.7 |
| Monroe | 57,098 | 77.0 | 168,000 | 24.6 | 14.5 | 1,108 | 31.8 | 2.0 | 81,386 | -2.5 | 9,390 | 11.5 | 80,347 | 31.3 | 24.9 |
| Montgomery | 316,206 | 71.8 | 316,100 | 20.7 | 12.2 | 1,295 | 28.2 | 1.4 | 446,961 | -2.3 | 34,036 | 7.6 | 434,317 | 51.0 | 14.7 |
| Montour | 7,404 | 67.2 | 185,100 | 18.5 | 11.1 | 721 | 26.9 | 2.3 | 9,043 | -1.4 | 554 | 6.1 | 8,731 | 39.3 | 23.0 |
| Northampton | 114,185 | 71.4 | 219,800 | 21.9 | 13.7 | 1,078 | 30.0 | 1.2 | 158,928 | -1.9 | 14,239 | 9.0 | 150,437 | 36.6 | 23.8 |
| Northumberland | 39,075 | 70.8 | 117,400 | 18.8 | 12.4 | 663 | 26.4 | 1.5 | 42,240 | -2.2 | 3,959 | 9.4 | 41,607 | 29.1 | 31.1 |
| Perry | 18,231 | 79.0 | 169,300 | 21.4 | 11.6 | 792 | 24.1 | 1.9 | 24,128 | -1.6 | 1,629 | 6.8 | 22,622 | 29.6 | 34.8 |
| Philadelphia | 601,337 | 53.0 | 163,000 | 22.4 | 13.7 | 1,042 | 32.5 | 3.3 | 723,938 | -0.1 | 89,529 | 12.4 | 702,105 | 39.3 | 17.9 |
| Pike | 22,119 | 83.2 | 185,700 | 22.3 | 13.6 | 1,200 | 36.7 | 1.0 | 24,852 | -2.1 | 2,639 | 10.6 | 25,191 | 33.6 | 23.6 |

1. Specified owner-occupied units. 2. A value of 10.0 represents 10 percent or less; a value of 50.0 represents 50 percent or more. 3. Specified renter-occupied units. 4. Overcrowded or
lacking complete plumbing facilities. 5. Percent of civilian labor force. 6. Civilian employed persons 16 years old and over.

# Table B. States and Counties — Nonfarm Employment and Agriculture

| STATE County | Number of establishments (104) | Employment Total (105) | Health care and social assistance (106) | Manufacturing (107) | Retail trade (108) | Finance and insurance (109) | Professional, scientific, and technical services (110) | Annual payroll Total (mil dol) (111) | Average per employee (dollars) (112) | Farms Number (113) | Percent with: Fewer than 50 acres (114) | 1000 acres or more (115) | Farm producers whose primary occupation is farming (percent) (116) |
|---|---|---|---|---|---|---|---|---|---|---|---|---|---|
| **OREGON—Cont'd** | | | | | | | | | | | | | |
| Morrow | 193 | 3,763 | 378 | 1,582 | 250 | 95 | NA | 172 | 45,756 | 375 | 30.7 | 42.7 | 50.3 |
| Multnomah | 27,789 | 453,756 | 70,779 | 35,789 | 42,607 | 22,483 | 36,272 | 26,236 | 57,819 | 653 | 89.3 | 0.6 | 32.0 |
| Polk | 1,524 | 15,484 | 3,068 | 1,965 | 1,569 | 266 | 418 | 578 | 37,343 | 1,243 | 67.3 | 2.4 | 41.6 |
| Sherman | 54 | 488 | 11 | NA | 91 | NA | NA | 19 | 38,289 | 190 | 7.4 | 62.6 | 47.4 |
| Tillamook | 729 | 7,332 | 1,054 | 1,385 | 1,130 | 119 | 140 | 296 | 40,311 | 293 | 51.2 | 0.7 | 53.4 |
| Umatilla | 1,543 | 23,984 | 3,295 | 3,511 | 3,026 | 508 | 645 | 942 | 39,256 | 1,724 | 59.6 | 15.7 | 38.7 |
| Union | 765 | 7,558 | 1,437 | 1,378 | 1,553 | 162 | 330 | 305 | 40,330 | 820 | 54.6 | 10.2 | 37.6 |
| Wallowa | 365 | 1,780 | 469 | 122 | 260 | 60 | 85 | 69 | 38,629 | 539 | 40.3 | 21.2 | 42.4 |
| Wasco | 709 | 7,906 | 2,161 | 646 | 1,660 | 338 | 249 | 320 | 40,451 | 595 | 33.4 | 20.2 | 43.9 |
| Washington | 15,945 | 290,205 | 32,529 | 41,765 | 32,873 | 13,332 | 21,486 | 21,320 | 73,466 | 1,755 | 81.1 | 1.0 | 35.5 |
| Wheeler | 27 | 140 | NA | 8 | 38 | NA | NA | 4 | 31,164 | 150 | 21.3 | 46.0 | 40.7 |
| Yamhill | 2,615 | 31,947 | 4,794 | 6,541 | 3,934 | 708 | 806 | 1,275 | 39,894 | 2,138 | 78.0 | 1.1 | 45.7 |
| **PENNSYLVANIA** | 303,224 | 5,557,885 | 1,083,226 | 565,502 | 648,298 | 294,330 | 330,945 | 294,727 | 53,029 | 53,157 | 42.1 | 1.4 | 43.4 |
| Adams | 1,957 | 29,542 | 5,055 | 6,528 | 3,553 | 465 | 617 | 1,079 | 36,519 | 1,146 | 50.2 | 2.2 | 43.4 |
| Allegheny | 33,731 | 704,581 | 140,122 | 35,879 | 70,988 | 53,897 | 54,275 | 40,111 | 56,929 | 389 | 60.7 | 0.5 | 40.2 |
| Armstrong | 1,269 | 14,358 | 3,504 | 1,566 | 1,865 | 594 | 317 | 539 | 37,562 | 668 | 28.4 | 4.2 | 43.6 |
| Beaver | 3,350 | 50,861 | 10,006 | 6,936 | 6,773 | 869 | 1,907 | 2,542 | 49,970 | 613 | 46.8 | 0.2 | 40.1 |
| Bedford | 1,066 | 12,662 | 1,752 | 2,233 | 2,185 | 349 | 205 | 458 | 36,169 | 1,159 | 29.2 | 3.1 | 50.3 |
| Berks | 8,354 | 162,134 | 28,467 | 33,344 | 19,822 | 5,472 | 6,198 | 7,941 | 48,977 | 1,809 | 49.0 | 0.9 | 52.9 |
| Blair | 3,205 | 53,539 | 12,987 | 6,922 | 8,710 | 1,072 | 1,763 | 2,215 | 41,375 | 496 | 34.9 | 1.6 | 52.7 |
| Bradford | 1,314 | 18,887 | 4,979 | 3,386 | 2,818 | 508 | 459 | 868 | 45,954 | 1,449 | 24.4 | 1.6 | 46.4 |
| Bucks | 19,456 | 256,809 | 50,543 | 26,230 | 36,775 | 9,084 | 18,617 | 12,739 | 49,605 | 824 | 72.1 | 1.5 | 43.7 |
| Butler | 4,959 | 85,841 | 14,565 | 12,061 | 11,057 | 2,023 | 5,491 | 4,163 | 48,500 | 955 | 40.4 | 1.4 | 42.2 |
| Cambria | 3,049 | 46,181 | 11,893 | 5,056 | 6,375 | 1,889 | 2,109 | 1,725 | 37,353 | 557 | 37.0 | 2.3 | 36.1 |
| Cameron | 114 | 1,671 | 196 | 1,028 | 134 | 18 | 10 | 62 | 37,001 | 37 | 16.2 | NA | 32.8 |
| Carbon | 1,152 | 13,507 | 2,973 | 1,378 | 2,123 | 278 | 310 | 448 | 33,188 | 200 | 53.0 | 1.0 | 30.3 |
| Centre | 3,356 | 46,608 | 9,499 | 4,090 | 7,602 | 1,085 | 3,144 | 1,922 | 41,228 | 1,023 | 41.3 | 1.4 | 47.0 |
| Chester | 14,849 | 260,642 | 39,124 | 15,400 | 27,089 | 29,839 | 24,859 | 19,469 | 74,697 | 1,646 | 60.6 | 1.2 | 52.3 |
| Clarion | 943 | 11,039 | 2,713 | 2,022 | 1,716 | 305 | 270 | 376 | 34,087 | 594 | 25.3 | 1.9 | 38.8 |
| Clearfield | 1,913 | 25,535 | 6,139 | 2,626 | 4,464 | 706 | 551 | 982 | 38,443 | 497 | 35.8 | 0.6 | 35.4 |
| Clinton | 767 | 10,777 | 1,659 | 3,002 | 1,834 | 164 | 158 | 444 | 41,154 | 267 | 35.6 | 1.9 | 50.9 |
| Columbia | 1,394 | 23,240 | 3,176 | 4,793 | 3,517 | 625 | 551 | 949 | 40,841 | 779 | 43.8 | 1.7 | 38.1 |
| Crawford | 1,950 | 27,125 | 5,738 | 7,495 | 3,350 | 566 | 666 | 1,013 | 37,328 | 1,091 | 36.8 | 2.1 | 44.3 |
| Cumberland | 6,139 | 125,876 | 17,402 | 8,955 | 19,634 | 8,391 | 8,830 | 6,052 | 48,081 | 1,260 | 45.1 | 1.5 | 51.4 |
| Dauphin | 6,850 | 151,385 | 35,424 | 8,796 | 15,622 | 8,496 | 8,492 | 8,008 | 52,895 | 692 | 52.6 | 1.4 | 46.9 |
| Delaware | 12,883 | 224,336 | 46,710 | 13,515 | 24,851 | 14,349 | 10,035 | 12,779 | 56,965 | 61 | 77.0 | NA | 43.0 |
| Elk | 885 | 13,952 | 1,605 | 7,016 | 1,404 | 213 | 403 | 580 | 41,564 | 232 | 43.5 | NA | 32.1 |
| Erie | 6,194 | 116,333 | 26,435 | 21,408 | 15,021 | 5,776 | 4,045 | 4,695 | 40,355 | 1,162 | 39.6 | 1.4 | 46.8 |
| Fayette | 2,518 | 33,867 | 7,463 | 2,977 | 5,431 | 490 | 989 | 1,342 | 39,624 | 834 | 38.6 | 1.8 | 35.6 |
| Forest | 96 | 1,029 | 422 | 213 | 77 | 13 | 8 | 33 | 32,402 | 36 | 25.0 | NA | 71.4 |
| Franklin | 3,079 | 51,117 | 8,997 | 7,366 | 7,542 | 1,132 | 1,858 | 2,099 | 41,058 | 1,581 | 37.3 | 1.6 | 55.5 |
| Fulton | 250 | 5,775 | 681 | 3,324 | 404 | 76 | 23 | 329 | 57,038 | 545 | 31.4 | 1.8 | 36.1 |
| Greene | 697 | 12,437 | 1,224 | 331 | 1,989 | 298 | 225 | 589 | 47,332 | 722 | 25.2 | 1.2 | 39.3 |
| Huntingdon | 826 | 9,735 | 2,193 | 1,167 | 1,406 | 420 | 336 | 336 | 34,521 | 714 | 30.5 | 1.4 | 40.4 |
| Indiana | 1,827 | 24,192 | 4,940 | 2,133 | 4,752 | 1,291 | 875 | 1,003 | 41,444 | 951 | 36.9 | 1.6 | 37.9 |
| Jefferson | 1,106 | 15,130 | 2,992 | 4,271 | 1,530 | 271 | 297 | 575 | 38,036 | 468 | 23.9 | 2.1 | 44.4 |
| Juniata | 490 | 6,156 | 448 | 2,618 | 694 | 244 | 59 | 202 | 32,790 | 670 | 43.9 | 1.2 | 49.3 |
| Lackawanna | 5,196 | 91,930 | 20,743 | 10,314 | 11,974 | 4,247 | 2,865 | 3,782 | 41,136 | 263 | 24.3 | 1.5 | 35.1 |
| Lancaster | 13,287 | 242,635 | 38,366 | 36,683 | 32,430 | 8,256 | 12,752 | 11,002 | 45,342 | 5,108 | 50.0 | 0.4 | 56.9 |
| Lawrence | 1,926 | 26,034 | 6,232 | 3,412 | 3,089 | 1,081 | 761 | 1,052 | 40,410 | 587 | 38.0 | 1.7 | 39.3 |
| Lebanon | 2,752 | 48,371 | 8,933 | 9,701 | 6,771 | 1,010 | 1,216 | 1,948 | 40,268 | 1,149 | 54.6 | 0.2 | 50.5 |
| Lehigh | 8,579 | 183,734 | 44,094 | 19,905 | 23,795 | 6,003 | 8,086 | 10,188 | 55,449 | 381 | 55.4 | 4.5 | 52.8 |
| Luzerne | 7,112 | 137,561 | 26,945 | 18,354 | 17,645 | 5,122 | 6,548 | 5,895 | 42,853 | 451 | 39.0 | 1.6 | 40.6 |
| Lycoming | 2,768 | 45,644 | 9,752 | 7,645 | 7,053 | 1,557 | 1,542 | 1,882 | 41,235 | 1,043 | 32.2 | 1.3 | 38.6 |
| McKean | 958 | 12,291 | 2,464 | 2,882 | 1,704 | 256 | 206 | 495 | 40,297 | 259 | 29.3 | 0.8 | 35.6 |
| Mercer | 2,774 | 44,298 | 9,209 | 9,294 | 6,736 | 1,583 | 711 | 1,720 | 38,831 | 1,168 | 38.5 | 1.5 | 41.5 |
| Mifflin | 974 | 14,954 | 3,500 | 3,855 | 2,092 | 439 | 156 | 586 | 39,203 | 711 | 40.4 | 0.6 | 49.7 |
| Monroe | 3,395 | 47,784 | 8,126 | 5,086 | 9,049 | 914 | 1,583 | 1,902 | 39,795 | 233 | 55.4 | 1.7 | 40.4 |
| Montgomery | 26,354 | 528,077 | 88,244 | 44,123 | 55,918 | 40,866 | 45,282 | 35,407 | 67,049 | 565 | 76.3 | 0.7 | 40.1 |
| Montour | 456 | 15,002 | 7,772 | 566 | 648 | 1,362 | 463 | 1,089 | 72,577 | 356 | 43.5 | 0.8 | 42.5 |
| Northampton | 6,353 | 113,005 | 17,380 | 12,082 | 14,284 | 3,619 | 3,806 | 5,211 | 46,113 | 459 | 64.5 | 2.8 | 43.5 |
| Northumberland | 1,649 | 23,402 | 4,444 | 4,146 | 3,387 | 626 | 593 | 933 | 39,872 | 728 | 42.7 | 2.9 | 49.3 |
| Perry | 795 | 5,930 | 643 | 447 | 1,203 | 242 | 176 | 180 | 30,341 | 759 | 38.3 | 1.2 | 45.6 |
| Philadelphia | 29,077 | 668,557 | 172,413 | 21,283 | 51,534 | 35,116 | 52,554 | 41,185 | 61,603 | 43 | 95.3 | NA | 25.9 |
| Pike | 930 | 8,821 | 1,037 | 393 | 1,975 | 150 | 220 | 271 | 30,776 | 53 | 56.6 | 1.9 | 22.3 |

Private nonfarm establishments, employment and payroll, 2019. Agriculture, 2017.

| STATE County | Acreage (1,000) | Percent change, 2012-2017 | Average size of farm | Total irrigated (1,000) | Total cropland (1,000) | Value of land and buildings (dollars) Average per farm | Average per acre | Value of machinery and equipment, average per farm (dollars) | Total (mil dol) | Average per farm (acres) | Crops | Livestock and poultry products | Organic farms (number) | Farms with internet access (percent) | Government payments Total ($1,000) | Percent of farms |
|---|---|---|---|---|---|---|---|---|---|---|---|---|---|---|---|---|
| | 117 | 118 | 119 | 120 | 121 | 122 | 123 | 124 | 125 | 126 | 127 | 128 | 129 | 130 | 131 | 132 |
| OREGON—Cont'd | | | | | | | | | | | | | | | | |
| Morrow | 1,126 | -3.3 | 3,003 | 111.5 | 511.9 | 3,385,469 | 1,127 | 416,278 | 596.5 | 1,590,632 | 32.0 | 68.0 | 6 | 86.4 | 11,659 | 57.6 |
| Multnomah | 25 | -15.2 | 39 | 5.7 | 15.6 | 813,249 | 20,879 | 54,930 | 74.6 | 114,210 | 95.8 | 4.2 | 24 | 91.3 | 84 | 3.2 |
| Polk | 149 | 2.9 | 120 | 20.4 | 107.6 | 852,419 | 7,116 | 93,564 | 134.8 | 108,408 | 77.2 | 22.8 | 26 | 85.9 | 776 | 8.0 |
| Sherman | 525 | 2.2 | 2,762 | 1.1 | 340.9 | 2,349,030 | 850 | 240,036 | 33.8 | 177,879 | 93.7 | 6.3 | 3 | 77.4 | 9,823 | 86.8 |
| Tillamook | 33 | -9.9 | 112 | 3.6 | 12.0 | 876,306 | 7,796 | 168,856 | 125.3 | 427,604 | 1.6 | 98.4 | 5 | 91.1 | 120 | 4.4 |
| Umatilla | 1,352 | 3.4 | 784 | 108.6 | 816.0 | 1,430,953 | 1,824 | 155,916 | 374.7 | 217,314 | 78.5 | 21.5 | 11 | 84.7 | 20,070 | 30.5 |
| Union | 385 | -6.4 | 470 | 44.3 | 121.1 | 851,017 | 1,812 | 123,048 | 57.1 | 69,662 | 73.5 | 26.5 | 2 | 83.2 | 2,731 | 26.6 |
| Wallowa | 520 | 14.9 | 965 | 42.6 | 94.7 | 1,575,439 | 1,632 | 138,191 | 38.9 | 72,217 | 40.3 | 59.7 | 3 | 82.4 | 3,206 | 36.2 |
| Wasco | 1,389 | -2.7 | 2,334 | 21.5 | 237.7 | 2,127,529 | 911 | 115,183 | 93.9 | 157,736 | 85.3 | 14.7 | 9 | 83.5 | 7,274 | 44.2 |
| Washington | 105 | -22.9 | 60 | 18.0 | 75.7 | 1,020,247 | 17,099 | 88,896 | 201.6 | 114,874 | 96.2 | 3.8 | 21 | 89.5 | 1,485 | 9.2 |
| Wheeler | 557 | -14.2 | 3,713 | 8.6 | 25.4 | 3,360,966 | 905 | 77,154 | 11.1 | 74,020 | 33.3 | 66.7 | NA | 80.7 | 921 | 23.3 |
| Yamhill | 169 | -4.5 | 79 | 29.1 | 113.4 | 806,192 | 10,178 | 88,958 | 314.3 | 147,017 | 80.7 | 19.3 | 39 | 88.2 | 1,658 | 8.5 |
| PENNSYLVANIA | 7,279 | -5.5 | 137 | 32.1 | 4,651.2 | 897,125 | 6,552 | 109,024 | 7,758.9 | 145,962 | 35.8 | 64.2 | 1,142 | 69.3 | 74,182 | 20.5 |
| Adams | 166 | -3 | 145 | 2.2 | 128.4 | 998,833 | 6,886 | 132,547 | 207.6 | 181,122 | 54.2 | 45.8 | 17 | 77.7 | 1,503 | 18.4 |
| Allegheny | 29 | -16.8 | 74 | 0.4 | 15.5 | 652,804 | 8,766 | 79,400 | 13.7 | 35,329 | 90.0 | 10.0 | 7 | 81.7 | D | 4.1 |
| Armstrong | 127 | -1.9 | 190 | 0.2 | 73.7 | 690,965 | 3,644 | 111,565 | 39.8 | 59,533 | 57.8 | 42.2 | 6 | 77.2 | 654 | 20.7 |
| Beaver | 54 | -3.5 | 88 | 0.3 | 30.1 | 531,298 | 6,050 | 80,548 | 23.7 | 38,586 | 61.2 | 38.8 | 3 | 72.6 | 177 | 8.0 |
| Bedford | 222 | 5.9 | 192 | 0.4 | 119.5 | 781,610 | 4,076 | 115,752 | 115.3 | 99,459 | 31.8 | 68.2 | 19 | 69.5 | 1,759 | 19.4 |
| Berks | 225 | -3.9 | 124 | 1.5 | 184.5 | 1,392,373 | 11,209 | 138,139 | 554.7 | 306,609 | 43.8 | 56.2 | 72 | 68.4 | 3,038 | 22.1 |
| Blair | 79 | -12.4 | 159 | 0.2 | 55.8 | 1,073,542 | 6,747 | 130,550 | 107.2 | 216,087 | 15.8 | 84.2 | 9 | 61.7 | 1,064 | 25.6 |
| Bradford | 304 | -1.4 | 210 | 0.3 | 183.3 | 759,047 | 3,623 | 106,595 | 132.6 | 91,539 | 28.3 | 71.7 | 40 | 77.0 | 3,408 | 36.1 |
| Bucks | 77 | 20.7 | 94 | 1.0 | 60.0 | 882,088 | 9,408 | 85,710 | 75.8 | 91,938 | 72.5 | 27.5 | 15 | 84.2 | 635 | 10.1 |
| Butler | 134 | -1.7 | 140 | 0.5 | 86.2 | 742,206 | 5,291 | 113,631 | 49.5 | 51,855 | 64.9 | 35.1 | 6 | 76.5 | 840 | 22.6 |
| Cambria | 79 | 3.2 | 142 | 0.1 | 50.5 | 659,661 | 4,631 | 100,086 | 30.1 | 53,982 | 60.1 | 39.9 | 2 | 68.8 | 979 | 30.3 |
| Cameron | 5 | -15.1 | 143 | NA | 1.5 | 405,700 | 2,844 | 77,509 | 0.5 | 14,135 | 69.4 | 30.6 | NA | 59.5 | 61 | 37.8 |
| Carbon | 19 | -7.9 | 97 | 0.0 | 13.1 | 645,238 | 6,619 | 106,544 | 13.0 | 65,145 | 87.7 | 12.3 | 1 | 83.5 | 185 | 25.0 |
| Centre | 150 | -7.5 | 146 | 0.4 | 88.3 | 981,430 | 6,700 | 92,395 | 91.5 | 89,421 | 35.3 | 64.7 | 23 | 68.5 | 2,266 | 24.0 |
| Chester | 151 | -8.5 | 91 | 1.2 | 105.8 | 1,110,075 | 12,140 | 130,513 | 712.5 | 432,848 | 80.1 | 19.9 | 46 | 74.5 | 1,775 | 10.0 |
| Clarion | 100 | -13.5 | 169 | 0.0 | 52.5 | 542,206 | 3,210 | 96,875 | 27.7 | 46,582 | 47.9 | 52.1 | 9 | 74.4 | 835 | 27.6 |
| Clearfield | 61 | -12 | 123 | 0.1 | 31.0 | 442,517 | 3,608 | 66,449 | 28.7 | 57,684 | 26.4 | 73.6 | 2 | 62.4 | 478 | 17.5 |
| Clinton | 40 | -24 | 150 | 0.3 | 26.5 | 980,810 | 6,538 | 125,749 | 45.6 | 170,640 | 28.7 | 71.3 | 18 | 55.8 | 515 | 23.2 |
| Columbia | 107 | -13 | 137 | 0.8 | 78.9 | 826,216 | 6,029 | 92,891 | 67.3 | 86,376 | 56.2 | 43.8 | 9 | 71.2 | 2,290 | 39.0 |
| Crawford | 194 | -14.6 | 178 | 0.3 | 121.0 | 595,435 | 3,341 | 124,042 | 107.3 | 98,323 | 39.8 | 60.2 | 5 | 70.4 | 679 | 15.4 |
| Cumberland | 170 | 9.5 | 135 | 1.4 | 140.8 | 1,025,011 | 7,613 | 130,622 | 219.2 | 173,950 | 28.7 | 71.3 | 22 | 61.1 | 2,132 | 22.1 |
| Dauphin | 81 | -37.2 | 117 | 0.4 | 63.4 | 1,032,232 | 8,791 | 106,338 | 93.1 | 134,499 | 29.2 | 70.8 | 43 | 68.4 | 779 | 21.0 |
| Delaware | 2 | -49.5 | 39 | 0.1 | 0.7 | 562,808 | 14,395 | 69,987 | 9.5 | 155,656 | 97.7 | 2.3 | NA | 98.4 | NA | NA |
| Elk | 23 | -2.2 | 99 | 0.1 | 9.3 | 415,868 | 4,198 | 65,859 | 4.0 | 17,349 | 50.2 | 49.8 | 4 | 71.1 | 50 | 9.9 |
| Erie | 153 | -9 | 132 | 1.3 | 93.5 | 595,515 | 4,511 | 112,026 | 82.0 | 70,602 | 76.4 | 23.6 | 3 | 78.1 | 907 | 15.7 |
| Fayette | 112 | -0.5 | 135 | 0.3 | 59.1 | 557,551 | 4,141 | 75,770 | 28.8 | 34,576 | 56.1 | 43.9 | 1 | 73.3 | 1,113 | 15.8 |
| Forest | 4 | -49.7 | 116 | 0.0 | 1.7 | 567,562 | 4,900 | 71,350 | 2.1 | 57,194 | 14.7 | 85.3 | NA | 94.4 | D | 2.8 |
| Franklin | 270 | 1.9 | 170 | 2.8 | 213.9 | 1,283,405 | 7,528 | 162,878 | 476.5 | 301,372 | 19.2 | 80.8 | 44 | 59.6 | 3,264 | 23.0 |
| Fulton | 100 | -10.5 | 184 | 0.0 | 52.9 | 769,540 | 4,175 | 95,703 | 75.8 | 139,110 | 16.2 | 83.8 | 7 | 72.5 | 686 | 29.2 |
| Greene | 114 | 1.5 | 158 | 0.0 | 37.5 | 575,776 | 3,644 | 85,505 | 16.4 | 22,763 | 51.3 | 48.7 | NA | 78.4 | 84 | 4.4 |
| Huntingdon | 120 | -24.1 | 168 | 0.6 | 67.5 | 810,060 | 4,814 | 115,139 | 92.1 | 129,036 | 22.0 | 78.0 | 11 | 74.4 | 1,706 | 32.1 |
| Indiana | 148 | -3.6 | 156 | 0.4 | 95.2 | 561,399 | 3,600 | 110,847 | 72.0 | 75,693 | 58.1 | 41.9 | 5 | 63.3 | 2,432 | 23.3 |
| Jefferson | 80 | -11.9 | 172 | 0.2 | 46.6 | 569,456 | 3,314 | 82,484 | 22.4 | 47,912 | 50.3 | 49.7 | 3 | 70.7 | 519 | 19.0 |
| Juniata | 86 | -5.9 | 128 | 0.2 | 55.7 | 833,587 | 6,522 | 106,123 | 126.8 | 189,194 | 17.5 | 82.5 | 64 | 63.3 | 904 | 22.8 |
| Lackawanna | 37 | 11.6 | 139 | 0.1 | 20.3 | 736,770 | 5,301 | 108,490 | 16.5 | 62,620 | 60.4 | 39.6 | NA | 77.2 | 119 | 16.7 |
| Lancaster | 394 | -10.4 | 77 | 4.9 | 314.9 | 1,410,238 | 18,285 | 120,108 | 1,507.2 | 295,068 | 15.3 | 84.7 | 246 | 48.9 | 4,834 | 11.8 |
| Lawrence | 82 | 2.1 | 140 | 0.3 | 55.1 | 613,904 | 4,388 | 96,349 | 34.8 | 59,239 | 54.4 | 45.6 | 2 | 74.4 | 655 | 27.3 |
| Lebanon | 108 | -11.4 | 94 | 0.7 | 86.7 | 1,348,192 | 14,400 | 132,813 | 350.8 | 305,312 | 10.6 | 89.4 | 54 | 71.4 | 1,021 | 13.7 |
| Lehigh | 75 | -2.4 | 196 | 0.4 | 63.1 | 1,534,921 | 7,849 | 136,146 | 79.2 | 207,916 | 72.3 | 27.7 | 8 | 79.3 | 1,063 | 18.4 |
| Luzerne | 49 | -19.4 | 109 | 0.3 | 27.6 | 659,692 | 6,063 | 64,776 | 17.8 | 39,452 | 74.7 | 25.3 | NA | 70.3 | 733 | 33.9 |
| Lycoming | 186 | 17.5 | 178 | 0.2 | 79.0 | 913,565 | 5,119 | 69,673 | 63.7 | 61,086 | 48.1 | 51.9 | 8 | 67.0 | 2,683 | 40.6 |
| McKean | 43 | 18.7 | 166 | 0.1 | 17.5 | 473,804 | 2,848 | 55,905 | 5.5 | 21,297 | 51.4 | 48.6 | 1 | 73.4 | 315 | 26.6 |
| Mercer | 156 | -4.1 | 134 | 0.2 | 98.4 | 535,249 | 3,997 | 103,659 | 65.7 | 56,291 | 53.9 | 46.1 | 20 | 73.6 | 1,526 | 26.3 |
| Mifflin | 81 | -10.6 | 114 | 0.1 | 49.9 | 703,683 | 6,179 | 92,455 | 140.0 | 196,896 | 12.0 | 88.0 | 18 | 65.8 | 818 | 17.0 |
| Monroe | 28 | 4.2 | 118 | 0.2 | 13.4 | 698,127 | 5,892 | 85,598 | 9.9 | 42,627 | 64.6 | 35.4 | NA | 79.4 | 190 | 9.4 |
| Montgomery | 31 | 0.4 | 55 | 0.6 | 20.5 | 1,058,678 | 19,360 | 62,127 | 35.4 | 62,609 | 73.9 | 26.1 | 2 | 83.5 | 179 | 4.8 |
| Montour | 39 | -11.2 | 109 | 0.1 | 29.4 | 699,083 | 6,442 | 103,791 | 60.2 | 169,171 | 42.4 | 57.6 | 4 | 64.9 | 643 | 32.9 |
| Northampton | 59 | -10 | 129 | 0.3 | 51.6 | 967,696 | 7,504 | 159,760 | 36.1 | 78,558 | 76.5 | 23.5 | 4 | 80.2 | 1,204 | 16.6 |
| Northumberland | 124 | -4.1 | 171 | 0.9 | 92.7 | 1,001,321 | 5,872 | 143,349 | 154.6 | 212,339 | 38.0 | 62.0 | 18 | 67.7 | 2,710 | 35.0 |
| Perry | 115 | -15.1 | 151 | 0.4 | 77.2 | 1,002,216 | 6,629 | 125,814 | 172.8 | 227,613 | 18.3 | 81.7 | 33 | 68.2 | 1,251 | 22.8 |
| Philadelphia | 0 | -0.4 | 7 | 0.0 | 0.1 | 387,767 | 58,711 | 16,642 | 0.3 | 7,605 | 81.0 | 19.0 | 2 | 90.7 | NA | NA |
| Pike | 25 | -12.6 | 466 | 0.0 | 1.4 | 1,037,171 | 2,226 | 74,827 | 0.9 | 16,830 | 50.0 | 50.0 | NA | 96.2 | NA | NA |

# Table B. States and Counties — Water Use, Wholesale Trade, Retail Trade, and Real Estate

| STATE County | Water use, 2015 — Public supply water withdrawn (mil gal/day) [133] | Water use, 2015 — Public supply gallons withdrawn per person per day [134] | Wholesale Trade[1], 2017 — Number of establishments [135] | Number of employees [136] | Sales (mil dol) [137] | Average payroll (mil dol) [138] | Retail Trade[2], 2017 — Number of establishments [139] | Number of employees [140] | Sales (mil dol) [141] | Average payroll (mil dol) [142] | Real estate and rental and leasing,[2] 2017 — Number of establishments [143] | Number of employees [144] | Sales (mil dol) [145] | Average payroll (mil dol) [146] |
|---|---|---|---|---|---|---|---|---|---|---|---|---|---|---|
| **OREGON—Cont'd** | | | | | | | | | | | | | | |
| Morrow | 5.61 | 501.3 | 16 | 74 | 85.1 | 2.8 | 20 | 235 | 68.8 | 5.2 | 7 | 19 | 2.1 | 0.2 |
| Multnomah | 14.72 | 18.6 | 1,191 | 24,301 | 25,399.7 | 1,632.9 | 3,084 | 43,576 | 11,927.7 | 1,294.8 | 1,683 | 11,186 | 2,347.9 | 530.4 |
| Polk | 4.96 | 62.5 | 39 | 451 | 184.3 | 21.7 | 116 | 1,604 | 411.5 | 41.4 | 78 | 206 | 37.5 | 5.7 |
| Sherman | 0.47 | 279.8 | 6 | 36 | 71.0 | 1.9 | 7 | 119 | 59.9 | 2.8 | NA | NA | NA | NA |
| Tillamook | 4.57 | 178.1 | 13 | 132 | 56.8 | 5.0 | 111 | 1,161 | 274.8 | 27.6 | 42 | 140 | 20.4 | 4.1 |
| Umatilla | 28.17 | 368.1 | 62 | 839 | 607.0 | 40.7 | 225 | 3,055 | 885.4 | 81.1 | 71 | 254 | 44.2 | 9.3 |
| Union | 4.58 | 177.6 | 20 | 221 | 146.1 | 9.6 | 106 | 1,607 | 387.5 | 41.1 | 21 | 77 | 14.0 | 2.3 |
| Wallowa | 1.37 | 199.8 | D | D | D | D | 49 | 276 | 71.1 | 7.9 | 24 | D | 5.8 | D |
| Wasco | 5.34 | 207.2 | 28 | 435 | 329.5 | 13.3 | 105 | 1,571 | 470.3 | 44.1 | 42 | 88 | 9.6 | 2.6 |
| Washington | 50.04 | 87.1 | 745 | 13,135 | 12,095.2 | 980.6 | 1,598 | 33,923 | 11,937.5 | 1,033.1 | 1,006 | 4,218 | 1,267.8 | 199.5 |
| Wheeler | 0.17 | 125.2 | NA | NA | NA | NA | 7 | 43 | 11.2 | 1.0 | NA | NA | NA | NA |
| Yamhill | 8.83 | 86.0 | 65 | 465 | 314.7 | 26.4 | 300 | 3,841 | 1,122.2 | 109.5 | 112 | 273 | 50.0 | 9.5 |
| **PENNSYLVANIA** | 1,391.7 | 108.7 | 12,071 | 204,256 | 199,203.8 | 12,958.8 | 42,514 | 662,560 | 234,836.3 | 17,354.0 | 10,662 | 66,715 | 18,860.6 | 3,472.2 |
| Adams | 12.39 | 121.1 | D | D | D | D | 337 | 3,599 | 979.0 | 87.0 | 48 | 190 | 34.3 | 6.2 |
| Allegheny | 178.17 | 144.8 | 1,414 | 20,546 | 27,374.2 | 1,327.9 | 4,244 | 72,934 | 56,465.7 | 1,980.9 | 1,360 | 10,209 | 2,937.3 | 478.9 |
| Armstrong | 6.65 | 99.2 | 38 | 523 | 212.0 | 25.5 | 200 | 1,975 | 574.0 | 46.5 | 33 | 167 | 25.6 | 6.4 |
| Beaver | 22.14 | 131.1 | 93 | 1,201 | 817.0 | 64.4 | 504 | 7,120 | 1,626.2 | 156.3 | 83 | 400 | 86.0 | 16.1 |
| Bedford | 9.99 | 205.6 | 33 | 292 | 124.3 | 11.2 | 192 | 2,094 | 577.3 | 50.5 | 12 | 41 | 8.2 | 2.0 |
| Berks | 31.85 | 76.7 | 374 | 8,533 | 7,103.8 | 524.2 | 1,186 | 20,446 | 5,772.3 | 527.8 | 367 | 2,985 | 1,173.9 | 217.9 |
| Blair | 13.07 | 104.1 | 124 | 2,112 | 2,076.2 | 101.4 | 530 | 8,700 | 2,721.2 | 227.4 | 111 | 409 | 84.8 | 14.3 |
| Bradford | 3.02 | 49.3 | 47 | 410 | 358.5 | 19.9 | 228 | 2,811 | 857.4 | 69.4 | 35 | 138 | 29.6 | 5.6 |
| Bucks | 98.64 | 157.2 | 1,035 | 15,293 | 10,028.0 | 939.7 | 2,289 | 38,202 | 10,862.1 | 1,031.5 | 668 | 3,256 | 915.6 | 165.2 |
| Butler | 7.66 | 41.0 | 236 | 3,955 | 2,583.2 | 246.5 | 674 | 11,286 | 3,268.6 | 278.5 | 170 | 879 | 237.9 | 33.4 |
| Cambria | 14.93 | 109.4 | 104 | 988 | 573.3 | 44.9 | 534 | 6,781 | 1,773.7 | 160.2 | 83 | 460 | 75.9 | 15.1 |
| Cameron | 0.34 | 71.9 | NA | NA | NA | NA | 14 | 134 | 30.4 | 2.5 | NA | NA | NA | NA |
| Carbon | 23.34 | 364.9 | D | D | D | D | 179 | 2,145 | 602.5 | 54.5 | 37 | 107 | 21.6 | 3.4 |
| Centre | 17.87 | 111.3 | 86 | 703 | 301.3 | 34.9 | 461 | 7,741 | 2,052.5 | 186.4 | 150 | 957 | 233.2 | 37.5 |
| Chester | 42.15 | 81.7 | 715 | 13,706 | 17,031.5 | 1,271.8 | 1,442 | 28,699 | 14,687.9 | 978.2 | 580 | 3,880 | 1,364.2 | 237.1 |
| Clarion | 2.73 | 69.1 | 24 | 337 | 178.2 | 14.9 | 165 | 1,707 | 463.6 | 41.9 | 25 | 114 | 17.3 | 3.3 |
| Clearfield | 5.84 | 72.1 | 53 | 667 | 596.0 | 30.4 | 339 | 4,571 | 1,277.4 | 109.1 | 39 | 322 | 37.2 | 7.4 |
| Clinton | 4.02 | 101.9 | D | D | D | D | 131 | 1,753 | 539.7 | 41.4 | 27 | 148 | 23.8 | 4.9 |
| Columbia | 5.02 | 75.3 | 41 | 465 | 185.8 | 20.0 | 229 | 3,372 | 919.9 | 78.0 | 42 | 151 | 43.0 | 5.8 |
| Crawford | 5.67 | 65.6 | D | D | D | D | 294 | 3,477 | 991.0 | 91.2 | 51 | 258 | 34.7 | 6.9 |
| Cumberland | 13.36 | 54.2 | 208 | 3,126 | 2,501.8 | 156.9 | 863 | 19,806 | 6,437.1 | 549.1 | 249 | 1,384 | 343.1 | 77.5 |
| Dauphin | 31.47 | 115.3 | 279 | 6,109 | 7,259.1 | 371.7 | 918 | 15,228 | 4,185.3 | 368.5 | 254 | 1,522 | 391.2 | 64.9 |
| Delaware | 20.79 | 36.9 | 513 | 6,250 | 4,471.8 | 415.7 | 1,645 | 26,153 | 7,586.1 | 691.4 | 476 | 2,963 | 957.1 | 194.4 |
| Elk | 5.32 | 172.3 | D | D | D | D | 128 | 1,477 | 387.6 | 35.2 | 16 | 62 | 9.8 | 2.4 |
| Erie | 33.34 | 119.9 | 247 | 3,292 | 1,367.7 | 166.6 | 917 | 15,344 | 3,823.5 | 360.9 | 190 | 878 | 169.8 | 30.3 |
| Fayette | 45.89 | 343.4 | 83 | 845 | 531.6 | 36.0 | 449 | 5,798 | 1,691.6 | 136.5 | 72 | 306 | 49.1 | 9.3 |
| Forest | 0.41 | 55.3 | NA | NA | NA | NA | 18 | 89 | 19.6 | 1.8 | NA | NA | NA | NA |
| Franklin | 8.47 | 55.1 | D | D | D | D | 487 | 7,407 | 2,075.0 | 186.5 | 94 | 364 | 61.4 | 11.7 |
| Fulton | 0.37 | 25.3 | D | D | D | D | 37 | 370 | 96.9 | 8.4 | 5 | D | 1.5 | D |
| Greene | 7.19 | 191.6 | 19 | 315 | 240.9 | 14.4 | 125 | 2,007 | 738.6 | 54.4 | 14 | 88 | 31.9 | 4.4 |
| Huntingdon | 2.85 | 62.4 | 25 | 320 | 122.5 | 15.6 | 138 | 1,513 | 372.6 | 34.7 | 17 | 36 | 5.6 | 1.3 |
| Indiana | 4.11 | 47.3 | 56 | 455 | 720.0 | 24.4 | 280 | 5,146 | 1,587.6 | 135.4 | 48 | 141 | 23.5 | 3.8 |
| Jefferson | 2.14 | 48.2 | 35 | 480 | 173.3 | 23.1 | 156 | 1,634 | 492.3 | 38.2 | 24 | 98 | 20.8 | 3.9 |
| Juniata | 0.82 | 33.1 | D | D | D | 6.0 | 71 | 656 | 221.2 | 16.0 | 8 | 20 | 3.2 | 0.4 |
| Lackawanna | 38.7 | 182.6 | 240 | 4,026 | 5,763.9 | 201.7 | 854 | 12,099 | 3,227.7 | 294.3 | 146 | 696 | 155.5 | 24.4 |
| Lancaster | 54.45 | 101.5 | 562 | 13,388 | 11,432.4 | 742.9 | 1,926 | 31,359 | 8,428.8 | 811.5 | 384 | 2,278 | 524.6 | 101.6 |
| Lawrence | 10.07 | 114.3 | 76 | 983 | 4,066.4 | 50.6 | 272 | 3,260 | 995.3 | 85.9 | 48 | 290 | 29.1 | 7.0 |
| Lebanon | 3.55 | 25.9 | 103 | 2,211 | 1,915.5 | 105.0 | 416 | 6,740 | 1,694.5 | 168.3 | 75 | 303 | 54.7 | 10.2 |
| Lehigh | 34.58 | 95.9 | 414 | 10,678 | 10,020.2 | 902.0 | 1,282 | 24,146 | 8,467.5 | 669.5 | 346 | 1,799 | 470.9 | 80.7 |
| Luzerne | 22.82 | 71.7 | 284 | 5,482 | 3,359.9 | 273.2 | 1,214 | 18,064 | 5,166.5 | 452.7 | 234 | 1,061 | 280.7 | 47.3 |
| Lycoming | 8.78 | 75.7 | 114 | 1,788 | 1,083.6 | 75.3 | 451 | 7,031 | 1,896.4 | 167.4 | 109 | 731 | 142.6 | 30.4 |
| McKean | 6.77 | 159.6 | 27 | 349 | 169.0 | 14.5 | 160 | 1,769 | 466.6 | 43.1 | 10 | 57 | 5.8 | 1.5 |
| Mercer | 13.74 | 120.3 | 83 | 1,032 | 631.2 | 48.0 | 483 | 7,364 | 1,565.4 | 161.6 | 81 | 302 | 65.6 | 9.8 |
| Mifflin | 2.69 | 57.8 | 37 | 515 | 149.7 | 20.1 | 164 | 2,059 | 601.1 | 51.8 | 22 | 59 | 12.8 | 1.9 |
| Monroe | 11.35 | 68.2 | 107 | D | 525.9 | D | 605 | 9,469 | 2,448.5 | 225.8 | 144 | 539 | 122.4 | 18.8 |
| Montgomery | 63.31 | 77.3 | 1,236 | 21,055 | 19,824.7 | 1,626.2 | 3,169 | 57,720 | 18,263.6 | 1,657.5 | 1,070 | 8,189 | 2,978.2 | 557.9 |
| Montour | 0.03 | 1.6 | 8 | 98 | 79.4 | 5.5 | 50 | 639 | 177.6 | 14.7 | 7 | 16 | 3.7 | 0.5 |
| Northampton | 9.53 | 31.7 | 242 | 7,442 | 14,195.0 | 391.9 | 854 | 14,171 | 4,142.3 | 397.4 | 195 | 968 | 233.6 | 40.4 |
| Northumberland | 11.05 | 118.5 | 48 | 851 | 677.9 | 42.0 | 288 | 3,314 | 947.0 | 80.6 | 33 | 149 | 16.2 | 4.2 |
| Perry | 0.71 | 15.5 | 16 | 72 | 88.4 | 4.7 | 129 | 1,333 | 338.4 | 33.0 | 14 | 33 | 4.5 | 0.8 |
| Philadelphia | 249.54 | 159.2 | 1,006 | 17,207 | 16,323.4 | 1,169.1 | 4,540 | 53,157 | 14,973.0 | 1,305.4 | 1,293 | 10,704 | 3,027.6 | 617.0 |
| Pike | 3.05 | 54.5 | 18 | D | 5.3 | D | 141 | 1,916 | 509.6 | 47.4 | 41 | 352 | 57.0 | 14.5 |

1  Merchant wholesalers, except manufacturers' sales branches and offices.    2. Employer establishments.

| STATE County | Professional, scientific, and technical services, 2017 | | | | Manufacturing, 2017 | | | | Accommodation and food services, 2017 | | | |
|---|---|---|---|---|---|---|---|---|---|---|---|---|
| | Number of establishments | Number of employees | Sales (mil dol) | Average payroll (mil dol) | Number of establishments | Number of employees | Sales (mil dol) | Average payroll (mil dol) | Number of establishments | Number of employees | Sales (mil dol) | Annual payroll (mil dol) |
| | 147 | 148 | 149 | 150 | 151 | 152 | 153 | 154 | 155 | 156 | 157 | 158 |
| **OREGON—Cont'd** | | | | | | | | | | | | |
| Morrow | NA | NA | NA | NA | 13 | 1,560 | 1,056.5 | 76.5 | 21 | 126 | 8.5 | 2.5 |
| Multnomah | 4,076 | 36,302 | 6,965 | 2,822.4 | 1,197 | 33,161 | 10,957.3 | 1,823.4 | 3,165 | 52,476 | 3,605.4 | 1,094.6 |
| Polk | 117 | 434 | 51 | 16.6 | 73 | 1,448 | 410.5 | 74.1 | D | D | D | D |
| Sherman | NA | NA | NA | NA | 33 | 1,515 | 1,043.3 | 77.2 | 9 | 76 | 5.6 | 1.4 |
| Tillamook | 41 | 148 | 16 | 6.3 | 67 | 4,148 | 1,061.3 | 150.6 | 136 | 1,336 | 86.0 | 25.8 |
| Umatilla | D | D | D | D | | | | | 171 | 3,050 | 234.1 | 63.0 |
| Union | 56 | 242 | 27 | 10.1 | 29 | 1,235 | 339.1 | 59.7 | 69 | 764 | 37.3 | 10.7 |
| Wallowa | 25 | 88 | 9 | 3.3 | D | 84 | D | D | 46 | 184 | 13.1 | 3.5 |
| Wasco | 41 | 240 | 33 | 11.3 | 31 | 472 | 157.9 | 20.7 | 78 | 1,023 | 71.3 | 20.1 |
| Washington | D | D | D | D | 728 | 43,158 | 19,629.8 | 3,934.6 | 1,284 | 22,112 | 1,417.4 | 423.1 |
| Wheeler | NA | NA | NA | NA | NA | NA | NA | NA | 4 | D | 0.8 | D |
| Yamhill | D | D | D | D | 248 | 5,692 | 1,650.2 | 333.4 | 221 | 3,268 | 190.1 | 58.2 |
| **PENNSYLVANIA** | 29,805 | 329,103 | 63,442 | 25,746.2 | 13,537 | 536,967 | 227,812.0 | 30,392.5 | 28,843 | 481,682 | 28,849.4 | 7,931.4 |
| Adams | D | D | D | D | 112 | 6,561 | 2,010.6 | 314.9 | 227 | 3,958 | 220.6 | 62.9 |
| Allegheny | D | D | 10,584 | D | 1,038 | 34,386 | 15,252.8 | 2,094.0 | 3,387 | 62,831 | 3,697.0 | 1,069.7 |
| Armstrong | D | D | D | D | 68 | 1,710 | 348.3 | 77.6 | 111 | 1,156 | 48.4 | 13.0 |
| Beaver | D | D | D | D | 166 | 6,338 | 2,794.3 | 361.8 | 320 | 4,978 | 225.8 | 60.3 |
| Bedford | D | D | D | D | 64 | 1,747 | 575.3 | 86.7 | 96 | 1,739 | 91.7 | 30.0 |
| Berks | D | D | D | D | 483 | 31,959 | 10,989.4 | 1,674.8 | 770 | 13,389 | 671.4 | 186.8 |
| Blair | 204 | 1,753 | 261 | 87.4 | 130 | 6,802 | 2,282.2 | 344.6 | 286 | 4,683 | 230.9 | 63.2 |
| Bradford | D | D | D | D | 59 | 3,038 | 1,420.5 | 154.7 | 127 | 1,261 | 62.7 | 16.5 |
| Bucks | 2,392 | 17,522 | 4,433 | 1,462.6 | 970 | 25,170 | 9,840.0 | 1,422.7 | 1,444 | 23,034 | 1,379.3 | 370.4 |
| Butler | D | D | D | D | 274 | 13,008 | 4,749.4 | 814.9 | 373 | 7,338 | 384.0 | 105.7 |
| Cambria | D | D | D | D | 115 | 5,161 | 4,739.5 | 286.6 | 298 | 4,130 | 190.7 | 49.8 |
| Cameron | D | D | D | D | 22 | 793 | 210.0 | 41.1 | 10 | D | 2.9 | D |
| Carbon | 63 | 417 | 39 | 15.1 | 51 | 1,374 | 424.2 | 65.0 | 113 | 1,630 | 94.1 | 28.1 |
| Centre | D | D | D | D | 166 | 3,786 | 1,041.7 | 206.6 | 327 | 6,750 | 344.7 | 96.9 |
| Chester | 2,339 | 25,903 | 5,835 | 2,473.9 | 515 | 15,657 | 6,202.3 | 994.3 | 1,055 | 17,111 | 1,009.9 | 293.7 |
| Clarion | D | D | 20 | D | 36 | 1,413 | 620.0 | 62.3 | 94 | 1,439 | 63.4 | 15.9 |
| Clearfield | 37 | 165 | 17 | 4.8 | 97 | 2,304 | 1,037.6 | 102.0 | 179 | 2,590 | 120.5 | 32.4 |
| Clinton | D | D | D | D | 47 | 2,921 | 1,679.5 | 155.9 | 83 | 1,192 | 59.9 | 17.6 |
| Columbia | D | D | D | D | 73 | 4,497 | 1,792.7 | 195.8 | 167 | 2,736 | 137.1 | 39.4 |
| Crawford | D | D | D | D | 273 | 7,653 | 1,906.9 | 392.7 | 180 | 2,265 | 111.0 | 29.3 |
| Cumberland | D | D | D | D | 193 | 8,847 | 3,943.8 | 440.2 | 563 | 10,119 | 564.6 | 158.5 |
| Dauphin | D | D | D | D | 186 | 7,991 | 3,690.7 | 430.2 | 694 | 13,316 | 909.0 | 259.6 |
| Delaware | D | D | D | D | 348 | 13,365 | 10,442.9 | 1,046.8 | 1,145 | 18,599 | 1,080.7 | 302.3 |
| Elk | 43 | 386 | 28 | 11.9 | 134 | 6,499 | 1,734.2 | 342.5 | 79 | 818 | 36.2 | 9.4 |
| Erie | 408 | 3,267 | 495 | 170.6 | 476 | 20,344 | 6,147.9 | 1,215.9 | 627 | 11,662 | 540.9 | 151.4 |
| Fayette | D | D | D | D | 99 | 2,789 | 1,191.6 | 126.4 | 261 | 4,598 | 256.7 | 81.2 |
| Forest | 3 | 2 | 0 | 0.1 | D | D | D | D | 21 | 125 | 7.8 | 2.4 |
| Franklin | D | D | D | D | 183 | 7,425 | 2,941.8 | 378.9 | 275 | 4,501 | 217.9 | 62.4 |
| Fulton | D | D | 2 | D | D | D | D | D | D | D | D | D |
| Greene | D | D | D | D | D | 282 | D | 13.1 | 71 | 1,299 | 63.6 | 17.7 |
| Huntingdon | D | D | D | D | 32 | 1,073 | 355.0 | 52.0 | 77 | 953 | 52.5 | 14.2 |
| Indiana | D | D | 120 | D | 92 | 2,196 | 507.2 | 99.8 | 145 | 2,622 | 127.2 | 36.3 |
| Jefferson | 67 | 282 | 27 | 8.8 | 104 | 3,568 | 895.6 | 172.2 | 102 | 980 | 43.9 | 11.0 |
| Juniata | D | D | D | D | 61 | 2,215 | 558.5 | 87.8 | D | D | D | D |
| Lackawanna | D | D | D | D | D | 7,944 | D | 399.6 | 594 | 8,535 | 516.1 | 127.4 |
| Lancaster | 1,026 | 12,314 | 1,840 | 731.7 | 923 | 34,403 | 13,559.4 | 1,845.3 | 1,102 | 20,227 | 1,091.5 | 305.7 |
| Lawrence | D | D | 79 | D | 134 | 3,289 | 1,371.5 | 180.9 | 157 | 2,409 | 101.3 | 29.1 |
| Lebanon | D | D | D | D | 204 | 9,398 | 2,861.1 | 456.8 | 231 | 3,347 | 173.1 | 48.9 |
| Lehigh | 778 | 7,168 | 1,296 | 495.6 | 360 | 18,925 | 11,085.0 | 1,180.9 | 793 | 13,797 | 772.0 | 216.9 |
| Luzerne | 547 | 4,820 | 632 | 257.3 | 297 | 16,989 | 6,565.2 | 871.0 | 776 | 12,441 | 867.1 | 198.0 |
| Lycoming | D | D | D | D | 156 | 7,422 | 3,192.1 | 376.5 | 273 | 4,720 | 253.7 | 76.5 |
| McKean | D | D | D | D | 45 | 2,768 | 1,110.7 | 152.7 | 104 | 1,150 | 45.2 | 12.4 |
| Mercer | D | D | D | D | 177 | 7,455 | 3,156.2 | 384.3 | 270 | 4,393 | 208.6 | 62.9 |
| Mifflin | D | D | D | D | 82 | 3,168 | 1,085.7 | 177.4 | 85 | 1,289 | 54.1 | 15.5 |
| Monroe | D | D | D | D | 117 | 5,173 | 2,776.5 | 428.1 | 402 | 10,069 | 873.9 | 185.4 |
| Montgomery | 3,712 | 49,547 | 9,786 | 4,508.1 | 977 | 40,679 | 20,295.2 | 2,824.0 | 2,029 | 33,907 | 2,240.9 | 612.3 |
| Montour | D | D | D | D | 19 | 486 | 148.5 | 17.4 | 39 | 724 | 34.9 | 11.2 |
| Northampton | D | D | D | D | 327 | 11,922 | 4,534.7 | 690.3 | 698 | 11,807 | 1,163.7 | 247.6 |
| Northumberland | D | D | D | D | 73 | 3,765 | 1,357.2 | 181.8 | 174 | 1,734 | 84.1 | 22.5 |
| Perry | 47 | 190 | 18 | 6.9 | 37 | 549 | 88.6 | 24.5 | 64 | 505 | 24.1 | 6.2 |
| Philadelphia | D | D | D | D | 694 | 21,042 | 14,253.7 | 1,204.6 | 3,953 | 61,893 | 4,480.2 | 1,278.3 |
| Pike | 78 | 208 | 34 | 9.9 | 17 | 277 | 57.1 | 12.8 | 113 | 2,202 | 156.1 | 49.8 |

Items 147—158

# Table B. States and Counties — Health Care and Social Assistance, Other Services, Nonemployer Businesses, and Residential Construction

| STATE County | Health care and social assistance, 2017 | | | | Other services, 2017 | | | | Nonemployer businesses, 2018 | | Value of residential construction authorized by building permits, 2020 | |
|---|---|---|---|---|---|---|---|---|---|---|---|---|
| | Number of establishments | Number of employees | Receipts (mil dol) | Annual payroll (mil dol) | Number of establishments | Number of employees | Receipts (mil dol) | Annual payroll (mil dol) | Number | Receipts (mil dol) | New construction ($1,000) | Number of housing units |
| | 159 | 160 | 161 | 162 | 163 | 164 | 165 | 166 | 167 | 168 | 169 | 170 |
| OREGON—Cont'd | | | | | | | | | | | | |
| Morrow | 18 | 259 | 27.7 | 12.1 | 12 | 32 | 3.6 | 1.3 | 494 | 25.9 | 11,160 | 41 |
| Multnomah | 3,373 | 73,052 | 9,879.7 | 3,775.2 | 2,099 | 14,700 | 2,542.9 | 582.3 | 75,048 | 3,678.8 | 429,291 | 2,709 |
| Polk | 210 | 2,938 | 226.8 | 102.4 | 85 | 354 | 33.4 | 9.9 | 4,582 | 200.8 | 64,209 | 265 |
| Sherman | D | D | D | 0.2 | NA | NA | NA | NA | 126 | 5.8 | NA | NA |
| Tillamook | 54 | 930 | 122.3 | 48.8 | 43 | 148 | 15.7 | 3.7 | 2,078 | 105.3 | 27,088 | 92 |
| Umatilla | 217 | 3,149 | 383.7 | 143.3 | 107 | 518 | 63.4 | 17.2 | 3,348 | 167.9 | 77,026 | 228 |
| Union | 92 | 1,440 | 169.1 | 70.7 | 48 | 162 | 18.3 | 4.6 | 1,627 | 70.3 | 13,894 | 54 |
| Wallowa | 33 | 416 | 44.1 | 16.8 | D | D | 9.1 | D | 875 | 45.3 | 210 | 3 |
| Wasco | 83 | 2,138 | 323.6 | 102.7 | 50 | 224 | 19.0 | 5.9 | 1,383 | 64.9 | 16,272 | 83 |
| Washington | 1,984 | 34,211 | 4,422.1 | 1,712.0 | 915 | 6,476 | 1,136.4 | 286.2 | 40,983 | 1,991.2 | 671,943 | 3,062 |
| Wheeler | D | D | D | D | NA | NA | NA | NA | 147 | 7.8 | NA | NA |
| Yamhill | 299 | 4,982 | 531.0 | 208.3 | 140 | 571 | 53.1 | 15.9 | 6,697 | 310.0 | 98,334 | 351 |
| PENNSYLVANIA | 37,699 | 1,035,971 | 116,621.8 | 47,885.2 | 26,075 | 161,337 | 22,982.2 | 5,136.6 | 865,041 | 42,770.6 | 5,077,342 | 25,706 |
| Adams | 185 | 4,860 | 512.4 | 201.8 | 174 | 1,312 | 124.6 | 33.5 | 6,697 | 302.2 | 48,233 | 226 |
| Allegheny | 4,666 | 131,693 | 15,389.9 | 6,508.2 | 2,946 | 20,665 | 3,032.8 | 704.0 | 85,700 | 4,097.7 | 468,936 | 1,974 |
| Armstrong | 191 | 3,319 | 280.5 | 127.7 | 106 | 563 | 48.4 | 12.4 | 3,529 | 157.4 | 5,736 | 28 |
| Beaver | 496 | 9,892 | 976.8 | 406.8 | 316 | 1,484 | 118.1 | 34.8 | 9,089 | 389.2 | 43,672 | 216 |
| Bedford | 126 | 1,696 | 146.7 | 59.8 | 101 | 408 | 38.2 | 8.9 | 3,245 | 165.5 | 13,625 | 54 |
| Berks | 861 | 26,638 | 2,897.5 | 1,261.5 | 797 | 4,354 | 425.4 | 116.3 | 25,138 | 1,201.5 | 101,312 | 515 |
| Blair | 452 | 12,918 | 1,345.0 | 564.9 | 269 | 1,459 | 122.6 | 37.7 | 6,424 | 304.4 | 19,639 | 96 |
| Bradford | 150 | 5,097 | 681.1 | 286.7 | 124 | 527 | 43.6 | 11.6 | 3,537 | 158.6 | 16,126 | 57 |
| Bucks | 2,083 | 48,951 | 4,681.8 | 2,032.5 | 1,511 | 8,731 | 881.6 | 276.4 | 55,150 | 3,229.8 | 183,112 | 730 |
| Butler | 628 | 14,969 | 1,479.8 | 673.8 | 407 | 2,886 | 336.7 | 102.8 | 12,597 | 615.9 | 305,734 | 1,298 |
| Cambria | 484 | 11,101 | 1,045.2 | 458.7 | 299 | 1,543 | 145.7 | 35.4 | 6,072 | 246.6 | 14,426 | 59 |
| Cameron | D | D | D | 7.0 | D | D | 3.1 | D | 216 | 6.8 | 0 | 0 |
| Carbon | 198 | 3,442 | 263.7 | 117.0 | 90 | 333 | 31.7 | 7.6 | 3,404 | 153.3 | 24,981 | 110 |
| Centre | 403 | 9,501 | 1,078.6 | 456.8 | 253 | 1,481 | 145.1 | 41.4 | 10,321 | 471.0 | 106,787 | 380 |
| Chester | 1,508 | 37,117 | 4,068.3 | 1,671.6 | 1,155 | 8,779 | 2,769.9 | 353.0 | 44,183 | 2,692.7 | 394,616 | 2,242 |
| Clarion | 162 | 2,789 | 194.6 | 91.5 | 78 | 500 | 67.1 | 13.5 | 2,383 | 121.9 | 3,962 | 16 |
| Clearfield | 273 | 5,881 | 627.6 | 247.1 | 177 | 1,225 | 105.9 | 28.9 | 4,308 | 183.8 | 12,170 | 59 |
| Clinton | 73 | 1,290 | 119.1 | 47.1 | 67 | 305 | 37.7 | 7.4 | 1,897 | 87.5 | 4,314 | 26 |
| Columbia | 180 | 3,252 | 312.3 | 128.9 | 113 | 572 | 65.2 | 11.9 | 3,362 | 151.9 | 14,490 | 59 |
| Crawford | 238 | 5,798 | 524.8 | 223.0 | 163 | 865 | 86.9 | 20.3 | 5,396 | 241.4 | 8,095 | 56 |
| Cumberland | 720 | 17,881 | 1,949.5 | 869.5 | 555 | 3,877 | 462.9 | 122.0 | 16,700 | 840.6 | 159,245 | 817 |
| Dauphin | 840 | 34,680 | 4,514.9 | 1,909.6 | 681 | 4,405 | 594.4 | 172.6 | 17,106 | 799.9 | 108,412 | 552 |
| Delaware | 1,666 | 41,170 | 4,379.7 | 1,830.2 | 1,136 | 7,266 | 984.9 | 243.9 | 44,811 | 2,293.2 | 62,584 | 254 |
| Elk | 107 | 1,651 | 150.1 | 60.5 | 77 | 352 | 28.2 | 6.6 | 1,574 | 66.5 | 5,531 | 22 |
| Erie | 878 | 24,543 | 2,475.6 | 1,085.6 | 563 | 3,433 | 321.7 | 81.7 | 13,819 | 618.9 | 41,234 | 201 |
| Fayette | 419 | 7,612 | 655.4 | 276.9 | 232 | 1,157 | 123.7 | 33.3 | 6,213 | 266.4 | 46,832 | 187 |
| Forest | D | D | D | D | 7 | 13 | 0.8 | 0.2 | 257 | 10.2 | 127 | 2 |
| Franklin | 352 | 9,299 | 1,074.0 | 432.7 | 296 | 1,484 | 147.6 | 38.2 | 9,185 | 404.0 | 72,304 | 330 |
| Fulton | D | D | D | 30.5 | 30 | 95 | 7.8 | 2.2 | 904 | 40.1 | 5,439 | 27 |
| Greene | 105 | 1,257 | 109.1 | 44.6 | 72 | 309 | 40.9 | 8.8 | 1,491 | 62.4 | 7,027 | 40 |
| Huntingdon | 110 | 2,406 | 185.9 | 78.5 | 75 | 304 | 32.7 | 7.2 | 2,436 | 104.1 | 18,934 | 102 |
| Indiana | 263 | 5,086 | 422.9 | 184.9 | 163 | 1,017 | 122.0 | 32.1 | 4,851 | 217.6 | 5,262 | 21 |
| Jefferson | 138 | 3,305 | 243.0 | 100.9 | 99 | 405 | 39.3 | 9.1 | 2,977 | 139.1 | 6,706 | 47 |
| Juniata | 33 | 649 | 51.4 | 19.5 | 47 | 120 | 12.9 | 3.6 | 1,813 | 86.3 | 7,170 | 34 |
| Lackawanna | 758 | 20,316 | 2,132.9 | 835.6 | 399 | 2,280 | 228.7 | 68.4 | 12,392 | 633.4 | 40,059 | 172 |
| Lancaster | 1,203 | 38,065 | 4,023.2 | 1,637.5 | 1,118 | 7,149 | 772.6 | 212.7 | 43,855 | 2,313.9 | 239,795 | 1,338 |
| Lawrence | 277 | 7,083 | 549.7 | 240.6 | 170 | 794 | 90.6 | 23.3 | 5,025 | 208.1 | 16,237 | 64 |
| Lebanon | 267 | 8,916 | 993.1 | 417.1 | 252 | 1,209 | 128.9 | 34.2 | 8,503 | 413.8 | 70,882 | 434 |
| Lehigh | 1,135 | 39,656 | 5,451.0 | 2,163.2 | 735 | 4,824 | 515.8 | 152.6 | 23,959 | 1,122.3 | 96,013 | 426 |
| Luzerne | 930 | 25,920 | 2,958.7 | 1,152.4 | 563 | 2,944 | 355.5 | 83.5 | 17,477 | 942.4 | 81,281 | 385 |
| Lycoming | 302 | 8,805 | 1,146.5 | 415.6 | 225 | 1,591 | 143.2 | 36.9 | 6,347 | 292.5 | 12,181 | 52 |
| McKean | 143 | 2,827 | 221.4 | 96.5 | 88 | 371 | 35.7 | 8.3 | 2,100 | 88.9 | 4,834 | 29 |
| Mercer | 437 | 10,109 | 854.1 | 381.3 | 242 | 1,053 | 87.4 | 21.0 | 6,249 | 313.0 | 17,168 | 68 |
| Mifflin | 121 | 3,763 | 359.2 | 147.2 | 68 | 281 | 31.2 | 6.8 | 2,748 | 117.8 | 14,983 | 87 |
| Monroe | 361 | 7,518 | 784.6 | 317.6 | 304 | 1,564 | 137.4 | 43.4 | 10,781 | 510.7 | 55,729 | 212 |
| Montgomery | 3,136 | 86,556 | 9,664.2 | 3,958.8 | 2,015 | 13,020 | 3,667.0 | 445.1 | 74,145 | 4,587.0 | 426,776 | 1,892 |
| Montour | 77 | 7,811 | 1,638.3 | 613.1 | 35 | 230 | 41.4 | 10.5 | 1,008 | 47.4 | 6,251 | 31 |
| Northampton | 768 | 15,757 | 1,650.3 | 658.0 | 560 | 3,095 | 274.1 | 85.7 | 19,509 | 955.0 | 153,009 | 848 |
| Northumberland | 205 | 5,060 | 319.0 | 150.6 | 142 | 769 | 59.8 | 16.9 | 4,531 | 219.0 | 14,540 | 64 |
| Perry | 72 | 925 | 49.9 | 23.7 | 60 | 299 | 31.1 | 6.7 | 2,949 | 140.7 | 23,749 | 91 |
| Philadelphia | 3,992 | 161,976 | 21,515.8 | 8,284.7 | 2,687 | 20,004 | 2,958.1 | 785.8 | 106,028 | 4,106.3 | 823,775 | 5,665 |
| Pike | 94 | 1,027 | 74.6 | 29.9 | 106 | 898 | 82.9 | 24.5 | 4,137 | 194.8 | 34,890 | 119 |

# Table B. States and Counties — Government Employment and Payroll, and Local Government Finances

| STATE County | Government employment and payroll, 2017 | | | | | | | | | Local government finances, 2017 | | | | |
| | | | March payroll (percent of total) | | | | | | | General revenue | | | | |
| | | | | | | | | | | | | Taxes | | |
| | | | | | | | | | | | | | Per capita[1] (dollars) | |
| | Full-time equivalent employees | March payroll (dollars) | Administration, judicial, and legal | Police and corrections | Fire protection | Highways and transportation | Health and welfare | Natural resources and utilities | Education and libraries | Total (mil dol) | Inter-governmental (mil dol) | Total (mil dol) | Total | Property |
| | 171 | 172 | 173 | 174 | 175 | 176 | 177 | 178 | 179 | 180 | 181 | 182 | 183 | 184 |
|---|---|---|---|---|---|---|---|---|---|---|---|---|---|---|
| **OREGON—Cont'd** | | | | | | | | | | | | | | |
| Morrow | 693 | 3,072,624 | 7.5 | 8.3 | 1.5 | 4.3 | 16.1 | 11.3 | 38.1 | 126.2 | 35.2 | 28.7 | 2,566 | 2,416 |
| Multnomah | 30,993 | 177,311,553 | 8.1 | 11.3 | 4.3 | 14.0 | 8.4 | 8.6 | 41.6 | 5,785.4 | 1,728.6 | 2,533.8 | 3,136 | 2,025 |
| Polk | 1,362 | 5,792,452 | 7.8 | 13.6 | 2.0 | 2.5 | 10.4 | 4.2 | 56.0 | 178.2 | 94.9 | 52.9 | 632 | 558 |
| Sherman | 99 | 409,288 | 28.3 | 12.1 | 0.6 | 12.4 | 7.9 | 5.7 | 29.4 | 34.0 | 5.2 | 8.4 | 4,858 | 4,833 |
| Tillamook | 1,173 | 5,111,468 | 9.6 | 7.2 | 1.9 | 6.3 | 5.7 | 22.8 | 43.9 | 147.0 | 47.6 | 60.3 | 2,275 | 2,056 |
| Umatilla | 2,801 | 11,651,569 | 6.1 | 9.5 | 3.7 | 2.7 | 2.8 | 4.8 | 68.9 | 352.9 | 185.8 | 100.5 | 1,303 | 1,140 |
| Union | 838 | 3,141,163 | 8.3 | 11.4 | 2.6 | 4.8 | 3.7 | 6.9 | 59.7 | 102.5 | 54.2 | 25.9 | 983 | 908 |
| Wallowa | 361 | 1,720,830 | 7.2 | 5.3 | 0.6 | 3.7 | 42.8 | 3.3 | 36.4 | 61.9 | 15.7 | 10.2 | 1,454 | 1,277 |
| Wasco | 1,045 | 4,438,442 | 11.4 | 5.4 | 4.3 | 3.6 | 6.3 | 12.4 | 54.1 | 126.3 | 56.6 | 42.1 | 1,601 | 1,428 |
| Washington | 15,411 | 80,053,640 | 7.4 | 10.5 | 7.2 | 2.4 | 2.2 | 6.6 | 61.1 | 2,448.1 | 840.2 | 1,094.6 | 1,851 | 1,608 |
| Wheeler | 95 | 402,914 | 15.6 | 3.7 | 0.0 | 3.9 | 0.0 | 5.9 | 70.3 | 9.9 | 8.5 | 0.8 | 588 | 560 |
| Yamhill | 3,063 | 14,276,675 | 7.6 | 11.3 | 3.2 | 1.5 | 10.5 | 6.7 | 56.3 | 394.2 | 203.7 | 129.8 | 1,233 | 1,079 |
| **PENNSYLVANIA** | X | X | X | X | X | X | X | X | X | X | X | X | X | X |
| Adams | 3,735 | 14,888,774 | 5.7 | 6.9 | 0.6 | 7.9 | 1.4 | 2.5 | 74.3 | 415.6 | 175.3 | 194.5 | 1,897 | 1,435 |
| Allegheny | 42,657 | 215,609,094 | 6.8 | 11.9 | 2.5 | 11.0 | 7.4 | 4.9 | 54.5 | 7,499.1 | 3,018.9 | 3,138.2 | 2,572 | 1,770 |
| Armstrong | 2,027 | 8,445,739 | 6.7 | 5.1 | 0.0 | 4.0 | 6.7 | 2.8 | 73.3 | 282.4 | 142.8 | 99.7 | 1,511 | 1,269 |
| Beaver | 5,157 | 22,428,838 | 6.2 | 9.8 | 0.3 | 4.5 | 7.5 | 5.4 | 65.6 | 764.1 | 361.5 | 270.0 | 1,630 | 1,293 |
| Bedford | 1,256 | 4,514,150 | 6.0 | 6.0 | 0.9 | 2.7 | 2.2 | 4.0 | 77.8 | 156.1 | 86.7 | 51.9 | 1,072 | 796 |
| Berks | 14,681 | 67,582,415 | 5.9 | 10.7 | 1.3 | 2.5 | 4.6 | 3.4 | 70.0 | 2,205.5 | 858.4 | 952.9 | 2,282 | 1,831 |
| Blair | 3,575 | 13,827,996 | 5.4 | 9.6 | 2.2 | 5.1 | 4.9 | 5.8 | 65.7 | 455.6 | 221.6 | 145.0 | 1,178 | 799 |
| Bradford | 2,218 | 8,868,518 | 8.8 | 4.9 | 0.6 | 6.4 | 10.5 | 4.8 | 63.2 | 286.5 | 154.3 | 88.5 | 1,451 | 1,119 |
| Bucks | 19,308 | 104,622,892 | 5.4 | 10.7 | 0.6 | 4.0 | 4.0 | 2.8 | 71.5 | 3,134.9 | 864.8 | 1,687.3 | 2,692 | 2,234 |
| Butler | 4,824 | 22,231,225 | 6.5 | 5.8 | 0.5 | 3.0 | 6.1 | 3.5 | 73.4 | 724.6 | 303.3 | 318.5 | 1,705 | 1,274 |
| Cambria | 4,432 | 17,315,677 | 6.6 | 9.4 | 1.6 | 6.5 | 5.9 | 4.0 | 62.3 | 581.4 | 337.4 | 163.1 | 1,227 | 939 |
| Cameron | 192 | 683,438 | 14.7 | 5.1 | 0.0 | 6.2 | 5.5 | 5.5 | 61.2 | 20.3 | 10.0 | 6.2 | 1,338 | 1,162 |
| Carbon | 1,897 | 7,011,109 | 9.4 | 10.1 | 0.0 | 3.2 | 3.3 | 4.5 | 67.6 | 242.9 | 87.5 | 123.2 | 1,931 | 1,621 |
| Centre | 3,630 | 14,677,666 | 9.0 | 9.2 | 0.0 | 9.4 | 3.5 | 7.0 | 60.1 | 520.6 | 151.4 | 264.0 | 1,626 | 1,174 |
| Chester | 14,012 | 72,047,879 | 6.9 | 8.8 | 0.3 | 2.2 | 3.9 | 2.7 | 73.7 | 2,440.8 | 660.1 | 1,450.7 | 2,796 | 2,265 |
| Clarion | 1,330 | 5,226,887 | 5.9 | 5.3 | 0.0 | 3.1 | 2.8 | 1.3 | 80.1 | 161.1 | 102.7 | 44.5 | 1,142 | 885 |
| Clearfield | 2,512 | 9,811,849 | 5.6 | 4.8 | 0.0 | 3.5 | 2.3 | 4.0 | 78.7 | 297.1 | 164.5 | 95.3 | 1,195 | 955 |
| Clinton | 1,166 | 4,890,713 | 9.0 | 10.2 | 0.3 | 3.1 | 5.2 | 10.5 | 59.7 | 193.6 | 58.2 | 54.5 | 1,404 | 1,074 |
| Columbia | 1,931 | 7,976,300 | 4.8 | 9.7 | 1.0 | 3.1 | 0.2 | 3.2 | 74.2 | 238.6 | 106.1 | 101.5 | 1,547 | 1,159 |
| Crawford | 2,132 | 8,652,359 | 9.5 | 9.5 | 1.2 | 5.1 | 12.5 | 3.8 | 57.6 | 272.7 | 130.0 | 99.1 | 1,155 | 955 |
| Cumberland | 7,465 | 31,998,627 | 6.0 | 8.7 | 0.0 | 2.1 | 6.9 | 4.1 | 70.7 | 1,183.2 | 382.3 | 605.1 | 2,430 | 1,759 |
| Dauphin | 9,952 | 44,975,290 | 6.9 | 13.2 | 1.1 | 5.0 | 2.5 | 4.9 | 64.4 | 1,567.1 | 622.8 | 615.1 | 2,231 | 1,547 |
| Delaware | 17,457 | 84,422,185 | 9.6 | 8.6 | 1.2 | 1.7 | 4.2 | 5.0 | 68.7 | 3,011.6 | 1,066.8 | 1,388.8 | 2,463 | 2,207 |
| Elk | 911 | 3,473,272 | 10.3 | 7.3 | 0.0 | 14.1 | 3.9 | 5.3 | 57.9 | 116.8 | 52.3 | 41.4 | 1,366 | 977 |
| Erie | 8,171 | 34,509,401 | 5.5 | 10.5 | 3.0 | 6.5 | 4.8 | 5.6 | 63.3 | 1,313.5 | 652.3 | 413.7 | 1,512 | 1,209 |
| Fayette | 3,795 | 16,108,272 | 5.2 | 4.7 | 0.4 | 3.3 | 5.8 | 3.8 | 76.4 | 473.9 | 298.4 | 125.8 | 958 | 724 |
| Forest | 170 | 610,591 | 23.1 | 1.9 | 0.0 | 5.2 | 3.7 | 4.3 | 56.8 | 22.7 | 10.9 | 9.2 | 1,255 | 1,055 |
| Franklin | 3,509 | 14,820,555 | 8.8 | 9.9 | 1.0 | 2.8 | 4.9 | 6.8 | 64.5 | 470.5 | 163.0 | 223.0 | 1,445 | 1,114 |
| Fulton | 389 | 1,566,148 | 11.7 | 2.7 | 0.0 | 3.3 | 0.0 | 3.7 | 78.2 | 51.5 | 29.6 | 18.9 | 1,303 | 1,093 |
| Greene | 1,326 | 5,053,380 | 10.3 | 5.9 | 0.0 | 4.0 | 6.7 | 6.2 | 65.0 | 168.8 | 81.8 | 67.5 | 1,831 | 1,474 |
| Huntingdon | 1,038 | 3,741,819 | 8.7 | 6.2 | 0.0 | 3.0 | 4.6 | 2.9 | 72.3 | 128.3 | 67.5 | 41.4 | 912 | 675 |
| Indiana | 2,464 | 10,033,183 | 5.3 | 6.9 | 0.0 | 5.0 | 3.5 | 4.4 | 71.7 | 318.1 | 164.1 | 110.4 | 1,302 | 1,039 |
| Jefferson | 1,243 | 4,369,814 | 8.3 | 6.4 | 0.0 | 6.3 | 2.9 | 4.9 | 70.3 | 138.1 | 78.0 | 42.4 | 968 | 735 |
| Juniata | 548 | 2,134,469 | 12.9 | 4.7 | 1.4 | 1.9 | 0.1 | 2.5 | 74.4 | 59.2 | 26.9 | 25.1 | 1,021 | 809 |
| Lackawanna | 6,002 | 27,420,359 | 6.5 | 10.1 | 3.4 | 4.6 | 7.7 | 4.4 | 61.6 | 919.7 | 346.4 | 409.9 | 1,948 | 1,427 |
| Lancaster | 14,385 | 63,875,933 | 5.5 | 10.2 | 1.0 | 3.2 | 2.9 | 4.1 | 71.4 | 2,211.0 | 804.0 | 1,004.1 | 1,855 | 1,499 |
| Lawrence | 2,411 | 10,328,970 | 7.6 | 9.1 | 1.4 | 5.2 | 1.8 | 2.9 | 71.6 | 329.0 | 177.8 | 114.8 | 1,327 | 1,045 |
| Lebanon | 3,630 | 16,331,706 | 5.8 | 8.8 | 0.8 | 3.1 | 4.1 | 5.6 | 70.8 | 524.1 | 174.2 | 233.6 | 1,674 | 1,348 |
| Lehigh | 12,126 | 60,052,384 | 6.5 | 9.4 | 1.8 | 5.2 | 8.5 | 3.5 | 63.2 | 1,894.9 | 749.7 | 786.2 | 2,151 | 1,697 |
| Luzerne | 9,532 | 40,790,805 | 7.2 | 11.4 | 2.3 | 4.2 | 4.9 | 4.1 | 64.1 | 1,218.2 | 524.3 | 533.9 | 1,680 | 1,258 |
| Lycoming | 3,735 | 16,093,813 | 7.9 | 7.7 | 1.8 | 4.2 | 1.8 | 6.8 | 69.3 | 504.9 | 219.4 | 183.1 | 1,606 | 1,169 |
| McKean | 1,760 | 6,725,097 | 5.7 | 6.4 | 1.5 | 4.2 | 9.4 | 5.8 | 66.5 | 187.5 | 110.4 | 49.0 | 1,185 | 938 |
| Mercer | 3,818 | 15,441,306 | 7.0 | 8.2 | 0.9 | 2.2 | 2.1 | 3.6 | 74.7 | 454.2 | 238.9 | 153.1 | 1,374 | 1,037 |
| Mifflin | 1,414 | 5,760,648 | 6.3 | 7.2 | 1.1 | 2.4 | 2.6 | 4.9 | 74.9 | 169.1 | 85.6 | 56.6 | 1,224 | 946 |
| Monroe | 5,561 | 25,725,163 | 5.3 | 5.2 | 0.0 | 3.2 | 1.6 | 1.5 | 82.6 | 816.0 | 272.4 | 464.3 | 2,766 | 2,480 |
| Montgomery | 23,842 | 129,438,423 | 5.9 | 12.2 | 0.4 | 2.3 | 3.7 | 4.2 | 69.8 | 4,083.5 | 1,074.7 | 2,442.2 | 2,963 | 2,386 |
| Montour | 503 | 1,895,714 | 7.7 | 6.5 | 0.0 | 4.7 | 2.6 | 5.0 | 71.3 | 95.5 | 23.9 | 31.8 | 1,743 | 1,111 |
| Northampton | 11,625 | 54,068,436 | 4.8 | 10.3 | 2.0 | 2.0 | 7.4 | 3.8 | 67.4 | 1,626.5 | 583.7 | 784.6 | 2,591 | 2,100 |
| Northumberland | 2,403 | 10,826,444 | 9.2 | 11.1 | 0.0 | 2.8 | 6.2 | 3.4 | 58.8 | 315.9 | 165.2 | 108.5 | 1,183 | 814 |
| Perry | 1,076 | 4,097,761 | 7.1 | 5.3 | 0.0 | 3.8 | 3.2 | 1.8 | 76.6 | 158.6 | 78.2 | 66.1 | 1,435 | 1,046 |
| Philadelphia | 61,107 | 332,698,900 | 9.0 | 20.2 | 5.6 | 19.0 | 8.1 | 8.3 | 27.6 | 12,639.4 | 6,005.2 | 4,866.1 | 3,079 | 814 |
| Pike | 1,122 | 4,601,891 | 9.2 | 14.0 | 0.0 | 2.4 | 4.5 | 1.7 | 67.8 | 142.5 | 42.4 | 83.5 | 1,507 | 1,406 |

1. Based on the resident population estimated as of July 1 of the year shown.

# Table B. States and Counties — Local Government Finances, Government Employment, and Income Taxes

| STATE County | Local government finances, 2017 (cont.) — Direct general expenditure — Total (mil dol) [185] | Per capita (dollars) [186] | Percent of total for: Education [187] | Health and hospitals [188] | Police protection [189] | Public welfare [190] | Highways [191] | Debt outstanding — Total (mil dol) [192] | Per capita (dollars) [193] | Government employment, 2019 — Federal civilian [194] | Federal military [195] | State and local [196] | Individual income tax returns, 2018 — Number of returns [197] | Mean adjusted gross income [198] | Mean income tax [199] |
|---|---|---|---|---|---|---|---|---|---|---|---|---|---|---|---|
| **OREGON—Cont'd** | | | | | | | | | | | | | | | |
| Morrow | 148 | 13,200 | 21.3 | 8.3 | 3.4 | 0.0 | 4.1 | 1,311.6 | 117,125 | 58 | 34 | 945 | 4,910 | 52,407 | 3,940 |
| Multnomah | 5,468 | 6,768 | 33.0 | 5.1 | 5.1 | 3.3 | 5.1 | 8,745.7 | 10,825 | 12,294 | 2,159 | 64,893 | 406,970 | 78,045 | 10,089 |
| Polk | 170 | 2,035 | 46.8 | 8.8 | 8.4 | 0.9 | 4.3 | 176.9 | 2,113 | 120 | 201 | 5,036 | 37,360 | 63,227 | 6,053 |
| Sherman | 25 | 14,212 | 24.6 | 14.4 | 3.7 | 1.3 | 10.5 | 8.2 | 4,733 | 124 | 4 | 201 | 860 | 60,158 | 5,243 |
| Tillamook | 132 | 4,967 | 42.2 | 6.1 | 4.0 | 0.3 | 5.6 | 136.1 | 5,132 | 121 | 106 | 1,792 | 12,730 | 56,303 | 5,182 |
| Umatilla | 361 | 4,682 | 63.0 | 1.9 | 5.2 | 0.1 | 3.4 | 501.4 | 6,502 | 492 | 175 | 6,684 | 32,530 | 52,319 | 4,388 |
| Union | 123 | 4,677 | 51.2 | 2.6 | 3.9 | 1.7 | 4.8 | 61.1 | 2,315 | 239 | 61 | 1,890 | 11,550 | 55,077 | 4,692 |
| Wallowa | 60 | 8,525 | 26.0 | 35.1 | 3.3 | 0.7 | 6.9 | 25.6 | 3,638 | 89 | 17 | 557 | 3,560 | 50,055 | 4,159 |
| Wasco | 118 | 4,478 | 49.6 | 2.6 | 5.0 | 0.5 | 5.1 | 74.2 | 2,820 | 291 | 62 | 1,580 | 11,820 | 54,030 | 4,785 |
| Washington | 2,579 | 4,361 | 48.9 | 2.0 | 6.6 | 0.0 | 5.8 | 3,906.3 | 6,606 | 928 | 1,411 | 22,155 | 286,070 | 87,752 | 10,894 |
| Wheeler | 13 | 9,335 | 71.3 | 1.2 | 2.4 | 0.0 | 7.0 | 2.0 | 1,488 | 8 | 3 | 110 | 560 | 35,061 | 3,020 |
| Yamhill | 412 | 3,914 | 50.8 | 8.2 | 5.9 | 0.1 | 4.8 | 518.4 | 4,926 | 447 | 241 | 3,929 | 47,540 | 67,782 | 6,981 |
| **PENNSYLVANIA** | X | X | X | X | X | X | X | X | X | 98,276 | 34,958 | 639,029 | 6,259,700 | 72,082 | 8,939 |
| Adams | 491 | 4,786 | 68.4 | 3.2 | 2.6 | 3.9 | 3.4 | 505.8 | 4,931 | 713 | 253 | 3,412 | 52,350 | 63,317 | 6,430 |
| Allegheny | 7,520 | 6,163 | 42.5 | 5.3 | 4.7 | 9.6 | 4.7 | 11,148.7 | 9,137 | 13,098 | 3,513 | 51,531 | 629,010 | 77,895 | 10,356 |
| Armstrong | 276 | 4,188 | 62.5 | 1.3 | 1.6 | 4.9 | 5.1 | 314.8 | 4,770 | 189 | 164 | 2,234 | 31,990 | 53,628 | 4,802 |
| Beaver | 763 | 4,606 | 53.3 | 3.4 | 3.8 | 8.2 | 4.8 | 1,943.2 | 11,730 | 333 | 413 | 7,055 | 84,660 | 58,915 | 5,755 |
| Bedford | 144 | 2,979 | 66.9 | 0.0 | 1.5 | 3.9 | 5.5 | 214.5 | 4,437 | 109 | 121 | 2,059 | 23,060 | 49,683 | 4,046 |
| Berks | 2,253 | 5,396 | 56.5 | 1.4 | 4.5 | 8.7 | 3.5 | 2,740.4 | 6,564 | 939 | 1,050 | 20,924 | 207,350 | 62,386 | 6,596 |
| Blair | 470 | 3,813 | 52.1 | 6.8 | 4.8 | 7.4 | 3.9 | 388.4 | 3,154 | 1,171 | 304 | 7,417 | 59,290 | 56,001 | 5,656 |
| Bradford | 282 | 4,622 | 54.6 | 1.8 | 1.7 | 11.8 | 6.8 | 324.7 | 5,327 | 203 | 153 | 2,758 | 28,370 | 55,985 | 5,348 |
| Bucks | 3,298 | 5,261 | 59.4 | 0.6 | 5.3 | 6.3 | 4.3 | 3,924.0 | 6,260 | 1,164 | 1,607 | 21,751 | 333,760 | 99,834 | 14,774 |
| Butler | 732 | 3,916 | 55.2 | 1.1 | 2.6 | 11.4 | 5.4 | 1,075.4 | 5,756 | 3,063 | 475 | 8,611 | 96,320 | 78,917 | 9,729 |
| Cambria | 626 | 4,709 | 54.7 | 2.4 | 2.1 | 9.5 | 3.2 | 513.7 | 3,864 | 976 | 340 | 6,660 | 62,430 | 52,063 | 4,707 |
| Cameron | 20 | 4,290 | 61.3 | 0.0 | 2.7 | 4.6 | 7.5 | 11.7 | 2,538 | 13 | 11 | 353 | 2,310 | 43,501 | 3,485 |
| Carbon | 288 | 4,511 | 66.7 | 0.0 | 2.2 | 2.7 | 3.7 | 325.4 | 5,098 | 113 | 162 | 2,462 | 31,580 | 53,550 | 4,900 |
| Centre | 539 | 3,323 | 57.5 | 1.9 | 3.3 | 3.2 | 5.3 | 421.2 | 2,595 | 492 | 457 | 45,661 | 60,550 | 72,814 | 8,365 |
| Chester | 2,718 | 5,237 | 59.7 | 5.5 | 4.2 | 2.9 | 3.4 | 2,431.8 | 4,686 | 2,381 | 1,311 | 23,274 | 258,100 | 121,967 | 19,612 |
| Clarion | 164 | 4,196 | 74.5 | 2.0 | 1.4 | 2.5 | 4.0 | 64.8 | 1,660 | 100 | 112 | 3,246 | 16,890 | 50,942 | 4,142 |
| Clearfield | 296 | 3,703 | 64.9 | 0.0 | 3.9 | 2.8 | 5.3 | 350.6 | 4,393 | 242 | 189 | 4,442 | 36,510 | 50,470 | 4,479 |
| Clinton | 151 | 3,878 | 50.6 | 0.2 | 2.4 | 2.9 | 3.8 | 104.0 | 2,678 | 137 | 98 | 2,914 | 17,080 | 50,853 | 4,348 |
| Columbia | 236 | 3,602 | 64.8 | 0.0 | 4.1 | 0.0 | 4.3 | 223.1 | 3,402 | 153 | 157 | 5,047 | 29,640 | 55,154 | 5,209 |
| Crawford | 284 | 3,309 | 50.2 | 1.4 | 3.3 | 10.8 | 6.7 | 217.6 | 2,537 | 254 | 208 | 3,655 | 38,190 | 61,627 | 6,446 |
| Cumberland | 1,164 | 4,676 | 64.6 | 2.4 | 2.8 | 7.1 | 3.3 | 1,003.3 | 4,029 | 4,675 | 1,322 | 12,391 | 128,470 | 74,002 | 8,638 |
| Dauphin | 1,721 | 6,243 | 44.3 | 4.5 | 3.5 | 10.3 | 4.0 | 2,194.0 | 7,958 | 2,698 | 749 | 38,336 | 144,020 | 61,878 | 6,852 |
| Delaware | 3,167 | 5,616 | 49.3 | 1.9 | 5.5 | 11.3 | 2.5 | 4,583.8 | 8,129 | 1,955 | 1,433 | 23,118 | 278,810 | 90,927 | 13,351 |
| Elk | 104 | 3,446 | 50.4 | 0.2 | 3.0 | 1.3 | 5.9 | 170.2 | 5,622 | 86 | 76 | 1,127 | 16,270 | 55,387 | 5,169 |
| Erie | 1,302 | 4,758 | 46.6 | 4.7 | 2.7 | 14.7 | 3.7 | 1,418.4 | 5,185 | 1,652 | 709 | 15,431 | 127,570 | 56,269 | 5,687 |
| Fayette | 480 | 3,657 | 58.0 | 2.9 | 1.4 | 9.7 | 3.7 | 964.6 | 7,346 | 360 | 320 | 5,838 | 60,690 | 51,197 | 4,591 |
| Forest | 22 | 2,982 | 53.1 | 4.2 | 1.3 | 4.6 | 11.1 | 16.3 | 2,222 | 76 | 12 | 967 | 2,130 | 46,631 | 3,815 |
| Franklin | 484 | 3,135 | 59.0 | 3.3 | 2.4 | 2.2 | 4.5 | 740.3 | 4,796 | 2,367 | 394 | 5,744 | 77,830 | 57,429 | 5,271 |
| Fulton | 68 | 4,682 | 56.5 | 0.0 | 2.9 | 4.8 | 3.9 | 195.8 | 13,529 | 27 | 37 | 705 | 7,200 | 52,948 | 4,508 |
| Greene | 175 | 4,756 | 54.5 | 1.5 | 1.0 | 4.0 | 9.4 | 166.6 | 4,521 | 108 | 85 | 2,424 | 15,430 | 66,711 | 7,485 |
| Huntingdon | 131 | 2,889 | 59.6 | 1.2 | 2.7 | 3.7 | 4.7 | 148.1 | 3,260 | 115 | 103 | 2,975 | 19,210 | 51,272 | 4,184 |
| Indiana | 322 | 3,795 | 65.0 | 0.0 | 1.2 | 6.0 | 4.0 | 340.5 | 4,017 | 197 | 207 | 7,198 | 35,670 | 55,491 | 5,462 |
| Jefferson | 136 | 3,103 | 63.5 | 0.0 | 3.6 | 0.0 | 5.9 | 197.1 | 4,497 | 117 | 109 | 1,682 | 20,850 | 52,797 | 4,985 |
| Juniata | 62 | 2,495 | 66.8 | 0.1 | 0.6 | 2.2 | 4.9 | 49.7 | 2,018 | 65 | 63 | 714 | 11,400 | 50,569 | 4,019 |
| Lackawanna | 932 | 4,428 | 47.8 | 0.1 | 4.9 | 5.4 | 3.1 | 1,471.7 | 6,993 | 930 | 528 | 9,482 | 103,150 | 57,023 | 5,964 |
| Lancaster | 2,292 | 4,234 | 55.6 | 5.9 | 4.6 | 5.2 | 3.2 | 3,717.0 | 6,866 | 1,302 | 1,366 | 19,769 | 271,410 | 66,221 | 7,099 |
| Lawrence | 325 | 3,754 | 57.3 | 2.2 | 4.4 | 6.0 | 4.9 | 391.2 | 4,520 | 208 | 236 | 3,238 | 41,320 | 53,456 | 4,833 |
| Lebanon | 521 | 3,730 | 57.8 | 1.5 | 3.4 | 10.7 | 3.0 | 583.4 | 4,180 | 3,190 | 354 | 4,934 | 71,150 | 57,694 | 5,331 |
| Lehigh | 1,933 | 5,288 | 51.6 | 1.2 | 3.6 | 9.3 | 3.8 | 3,232.0 | 8,841 | 871 | 936 | 16,711 | 187,140 | 65,615 | 7,557 |
| Luzerne | 1,252 | 3,940 | 57.2 | 0.3 | 3.5 | 6.0 | 5.1 | 1,110.8 | 3,495 | 3,353 | 799 | 14,610 | 158,210 | 53,834 | 5,371 |
| Lycoming | 511 | 4,479 | 56.7 | 0.0 | 2.8 | 2.6 | 4.3 | 983.6 | 8,627 | 360 | 278 | 8,491 | 54,010 | 54,649 | 5,197 |
| McKean | 191 | 4,610 | 58.9 | 4.7 | 1.5 | 4.7 | 4.1 | 125.1 | 3,025 | 420 | 96 | 1,913 | 18,760 | 50,804 | 4,367 |
| Mercer | 497 | 4,459 | 62.1 | 1.9 | 3.2 | 2.0 | 4.2 | 577.2 | 5,178 | 260 | 264 | 4,905 | 52,160 | 52,304 | 4,813 |
| Mifflin | 171 | 3,686 | 57.9 | 4.8 | 2.1 | 0.0 | 4.3 | 220.3 | 4,760 | 82 | 117 | 1,595 | 21,600 | 46,762 | 3,613 |
| Monroe | 758 | 4,515 | 69.4 | 0.1 | 3.3 | 5.4 | 3.0 | 1,152.8 | 6,866 | 3,415 | 452 | 8,396 | 80,510 | 56,169 | 5,667 |
| Montgomery | 4,165 | 5,053 | 61.2 | 1.9 | 5.0 | 6.1 | 3.3 | 4,473.0 | 5,426 | 2,532 | 2,263 | 33,412 | 425,610 | 118,544 | 18,779 |
| Montour | 90 | 4,922 | 45.3 | 0.0 | 2.0 | 1.2 | 4.2 | 2,335.8 | 127,851 | 36 | 45 | 1,455 | 8,980 | 74,008 | 9,343 |
| Northampton | 1,725 | 5,696 | 54.6 | 2.1 | 4.3 | 11.6 | 3.2 | 2,562.8 | 8,464 | 1,106 | 785 | 13,728 | 155,390 | 69,468 | 7,964 |
| Northumberland | 311 | 3,391 | 53.8 | 3.9 | 4.6 | 7.4 | 4.2 | 227.1 | 2,476 | 158 | 223 | 3,992 | 43,310 | 49,014 | 4,082 |
| Perry | 147 | 3,195 | 63.2 | 0.3 | 0.8 | 10.3 | 4.2 | 101.0 | 2,193 | 86 | 117 | 1,734 | 22,940 | 54,220 | 4,518 |
| Philadelphia | 11,840 | 7,491 | 36.7 | 14.2 | 5.7 | 1.7 | 1.7 | 14,787.8 | 9,356 | 30,341 | 5,109 | 77,173 | 693,220 | 55,395 | 6,326 |
| Pike | 143 | 2,576 | 56.4 | 0.5 | 12.9 | 5.1 | 3.4 | 87.7 | 1,584 | 204 | 142 | 2,372 | 27,720 | 63,752 | 7,014 |

1. Based on the resident population estimated as of July 1 of the year shown.

# Table B. States and Counties — Land Area and Population

| State / county code | CBSA code[1] | County Type code[2] | STATE County | Land area[3] (sq. mi) | Total persons 2019 | Rank | Per square mile | White | Black | American Indian, Alaska Native | Asian and Pacific Islander | Percent Hispanic or Latino[4] | Under 5 years | 5 to 14 years | 15 to 24 years | 25 to 34 years | 35 to 44 years | 45 to 54 years |
|---|---|---|---|---|---|---|---|---|---|---|---|---|---|---|---|---|---|---|
| | | | | 1 | 2 | 3 | 4 | 5 | 6 | 7 | 8 | 9 | 10 | 11 | 12 | 13 | 14 | 15 |
| | | | **PENNSYLVANIA—Cont'd** | | | | | | | | | | | | | | | |
| 42105 | | 9 | Potter | 1,081.2 | 16,453 | 2,009 | 15.2 | 97.1 | 1.0 | 0.7 | 0.7 | 1.5 | 4.7 | 11.5 | 10.5 | 10.1 | 10.3 | 11.7 |
| 42107 | 39060 | 4 | Schuylkill | 778.6 | 140,709 | 465 | 180.7 | 90.5 | 3.6 | 0.4 | 0.9 | 5.8 | 4.8 | 11.2 | 10.7 | 12.1 | 11.8 | 13.5 |
| 42109 | 42780 | 7 | Snyder | 328.8 | 40,317 | 1,174 | 122.6 | 95.7 | 1.5 | 0.4 | 1.0 | 2.4 | 5.4 | 11.4 | 15.5 | 11.2 | 10.6 | 12.3 |
| 42111 | 43740 | 4 | Somerset | 1,075.0 | 72,916 | 751 | 67.8 | 94.8 | 3.2 | 0.4 | 0.7 | 1.6 | 4.9 | 9.8 | 10.3 | 11.4 | 11.6 | 13.2 |
| 42113 | | 8 | Sullivan | 449.9 | 5,913 | 2,752 | 13.1 | 94.1 | 3.3 | 0.9 | 0.6 | 2.0 | 2.8 | 4.7 | 9.1 | 9.8 | 10.4 | 12.0 |
| 42115 | | 6 | Susquehanna | 823.5 | 40,006 | 1,185 | 48.6 | 96.9 | 1.0 | 0.6 | 0.8 | 1.8 | 4.8 | 10.4 | 9.9 | 10.1 | 10.3 | 12.2 |
| 42117 | | 6 | Tioga | 1,133.8 | 40,381 | 1,170 | 35.6 | 97.0 | 1.3 | 0.6 | 0.8 | 1.3 | 5.0 | 11.3 | 11.7 | 11.0 | 10.8 | 11.9 |
| 42119 | 30260 | 4 | Union | 316.0 | 44,294 | 1,092 | 140.2 | 85.6 | 6.7 | 0.6 | 2.2 | 6.1 | 4.6 | 9.7 | 17.0 | 12.8 | 12.9 | 11.9 |
| 42121 | 36340 | 4 | Venango | 674.3 | 50,328 | 985 | 74.6 | 96.9 | 1.9 | 0.6 | 0.8 | 1.1 | 4.7 | 10.6 | 10.1 | 10.4 | 10.8 | 12.3 |
| 42123 | 47620 | 6 | Warren | 884.3 | 38,911 | 1,203 | 44.0 | 97.5 | 0.9 | 0.6 | 0.8 | 1.2 | 5.3 | 10.7 | 10.1 | 10.4 | 10.4 | 12.5 |
| 42125 | 38300 | 1 | Washington | 857.0 | 206,803 | 329 | 241.3 | 93.5 | 4.5 | 0.5 | 1.5 | 1.9 | 5.0 | 11.0 | 11.5 | 11.5 | 11.6 | 12.7 |
| 42127 | | 6 | Wayne | 725.8 | 51,163 | 974 | 70.5 | 90.8 | 3.7 | 0.6 | 1.1 | 5.0 | 3.9 | 9.2 | 9.7 | 11.4 | 11.2 | 12.9 |
| 42129 | 38300 | 1 | Westmoreland | 1,027.7 | 347,098 | 207 | 337.7 | 94.9 | 3.6 | 0.4 | 1.3 | 1.4 | 4.4 | 10.3 | 10.7 | 10.8 | 11.0 | 13.0 |
| 42131 | 42540 | 2 | Wyoming | 397.4 | 26,557 | 1,543 | 66.8 | 96.1 | 1.5 | 0.6 | 0.7 | 2.1 | 4.5 | 11.0 | 11.8 | 10.9 | 10.9 | 12.6 |
| 42133 | 49620 | 2 | York | 904.4 | 450,448 | 157 | 498.1 | 83.9 | 7.1 | 0.5 | 2.0 | 8.5 | 5.5 | 12.4 | 11.9 | 12.5 | 12.2 | 13.0 |
| 44000 | | 0 | **RHODE ISLAND** | 1,033.9 | 1,057,125 | X | 1,022.5 | 72.9 | 7.5 | 1.0 | 4.3 | 16.7 | 5.1 | 10.6 | 13.6 | 14.1 | 12.0 | 12.4 |
| 44001 | 39300 | 1 | Bristol | 24.1 | 48,350 | 1,017 | 2,006.2 | 92.5 | 2.1 | 0.7 | 3.5 | 3.3 | 3.8 | 10.7 | 14.5 | 10.1 | 11.3 | 13.1 |
| 44003 | 39300 | 1 | Kent | 168.6 | 164,646 | 404 | 976.5 | 88.5 | 3.0 | 0.9 | 3.7 | 6.1 | 4.7 | 10.3 | 10.2 | 13.5 | 12.5 | 13.2 |
| 44005 | 39300 | 1 | Newport | 102.4 | 81,836 | 698 | 799.2 | 87.7 | 5.0 | 1.0 | 3.0 | 6.0 | 4.3 | 9.2 | 12.2 | 12.5 | 10.8 | 12.0 |
| 44007 | 39300 | 1 | Providence | 409.5 | 636,547 | 106 | 1,554.4 | 61.6 | 10.4 | 1.1 | 5.1 | 24.4 | 5.7 | 11.2 | 13.8 | 15.5 | 12.5 | 12.3 |
| 44009 | 39300 | 1 | Washington | 329.3 | 125,746 | 509 | 381.9 | 92.2 | 2.1 | 1.3 | 3.5 | 3.7 | 8.8 | 17.4 | 10.0 | 9.7 | 11.8 | |
| 45000 | | 0 | **SOUTH CAROLINA** | 30,063.7 | 5,218,040 | X | 173.6 | 65.2 | 27.4 | 0.9 | 2.4 | 6.1 | 5.6 | 12.2 | 12.6 | 13.2 | 12.1 | 12.2 |
| 45001 | | 6 | Abbeville | 491.2 | 24,404 | 1,633 | 49.7 | 70.6 | 27.7 | 0.7 | 0.6 | 1.8 | 4.8 | 11.1 | 12.8 | 10.8 | 10.4 | 12.5 |
| 45003 | 12260 | 2 | Aiken | 1,070.7 | 172,895 | 383 | 161.5 | 67.4 | 26.0 | 1.1 | 1.7 | 6.0 | 5.6 | 12.0 | 11.5 | 12.7 | 11.6 | 12.0 |
| 45005 | | 6 | Allendale | 408.1 | 8,331 | 2,553 | 20.4 | 22.8 | 73.3 | 0.5 | 0.9 | 3.2 | 4.9 | 10.9 | 12.0 | 13.7 | 11.2 | 12.5 |
| 45007 | 24860 | 2 | Anderson | 713.9 | 204,353 | 333 | 286.2 | 78.5 | 17.2 | 0.7 | 1.4 | 4.1 | 5.2 | 12.9 | 12.0 | 12.5 | 11.8 | 12.8 |
| 45009 | | 7 | Bamberg | 393.4 | 13,906 | 2,164 | 35.3 | 37.0 | 60.1 | 0.7 | 0.9 | 2.3 | 4.6 | 10.3 | 15.6 | 10.9 | 9.4 | 11.4 |
| 45011 | | 6 | Barnwell | 548.4 | 20,805 | 1,779 | 37.9 | 52.5 | 44.2 | 1.1 | 1.2 | 2.8 | 5.9 | 13.5 | 12.0 | 12.2 | 10.6 | 11.9 |
| 45013 | 25940 | 3 | Beaufort | 576.0 | 195,656 | 346 | 339.7 | 70.1 | 17.9 | 0.6 | 2.0 | 11.2 | 4.8 | 10.1 | 11.7 | 10.7 | 9.9 | 10.2 |
| 45015 | 16700 | 2 | Berkeley | 1,103.6 | 235,987 | 293 | 213.8 | 65.2 | 25.5 | 1.2 | 3.8 | 7.2 | 6.3 | 13.5 | 13.0 | 14.5 | 13.5 | 12.2 |
| 45017 | 17900 | 2 | Calhoun | 381.2 | 14,554 | 2,117 | 38.2 | 55.3 | 40.5 | 0.9 | 0.7 | 4.0 | 4.7 | 10.6 | 10.2 | 10.8 | 10.8 | 12.6 |
| 45019 | 16700 | 2 | Charleston | 918.0 | 417,981 | 173 | 455.3 | 67.0 | 26.1 | 0.7 | 2.5 | 5.4 | 5.6 | 10.7 | 11.6 | 16.4 | 13.7 | 11.6 |
| 45021 | 23500 | 4 | Cherokee | 393.0 | 57,316 | 905 | 145.8 | 73.8 | 21.4 | 0.8 | 0.8 | 4.8 | 5.9 | 12.9 | 12.7 | 13.2 | 11.2 | 13.4 |
| 45023 | 16740 | 1 | Chester | 580.7 | 32,232 | 1,373 | 55.5 | 60.2 | 37.6 | 1.1 | 0.9 | 2.2 | 5.9 | 12.8 | 11.3 | 12.5 | 10.7 | 13.1 |
| 45025 | | 6 | Chesterfield | 799.0 | 45,606 | 1,063 | 57.1 | 61.5 | 33.8 | 1.1 | 0.9 | 4.7 | 5.5 | 12.6 | 11.6 | 12.1 | 10.9 | 13.6 |
| 45027 | 44940 | 6 | Clarendon | 607.2 | 33,415 | 1,344 | 55.0 | 49.5 | 46.7 | 0.6 | 0.9 | 3.2 | 4.4 | 10.4 | 12.0 | 11.1 | 10.3 | 11.4 |
| 45029 | 22500 | 6 | Colleton | 1,056.5 | 37,481 | 1,237 | 35.5 | 58.6 | 37.3 | 1.4 | 0.8 | 3.7 | 6.3 | 12.6 | 11.2 | 11.9 | 10.7 | 11.9 |
| 45031 | 22500 | 3 | Darlington | 560.6 | 66,590 | 804 | 118.6 | 55.4 | 42.2 | 0.7 | 0.8 | 2.3 | 5.8 | 12.2 | 12.5 | 12.3 | 10.8 | 12.5 |
| 45033 | | 6 | Dillon | 405.1 | 30,367 | 1,421 | 75.0 | 47.0 | 48.0 | 3.3 | 0.8 | 2.9 | 6.6 | 14.2 | 12.6 | 12.5 | 11.7 | 11.7 |
| 45035 | 16700 | 2 | Dorchester | 568.6 | 165,737 | 401 | 291.5 | 65.6 | 26.8 | 1.2 | 3.3 | 6.0 | 5.8 | 13.9 | 12.0 | 14.0 | 13.8 | 12.8 |
| 45037 | 12260 | 2 | Edgefield | 500.7 | 27,120 | 1,522 | 54.2 | 58.9 | 34.5 | 0.9 | 0.8 | 6.3 | 3.8 | 9.9 | 11.7 | 13.3 | 12.7 | 13.2 |
| 45039 | 17900 | 2 | Fairfield | 686.3 | 22,059 | 1,716 | 32.1 | 40.1 | 57.3 | 0.9 | 1.0 | 2.4 | 4.5 | 10.5 | 10.7 | 11.1 | 9.9 | 12.9 |
| 45041 | 22500 | 3 | Florence | 800.5 | 137,588 | 473 | 171.9 | 52.0 | 43.9 | 0.7 | 1.9 | 2.9 | 5.9 | 13.4 | 12.6 | 12.7 | 12.1 | 12.5 |
| 45043 | 23860 | 4 | Georgetown | 813.6 | 63,353 | 844 | 77.9 | 66.0 | 30.3 | 0.6 | 0.8 | 3.3 | 4.2 | 9.9 | 10.0 | 9.4 | 9.3 | 11.3 |
| 45045 | 24860 | 2 | Greenville | 785.9 | 532,486 | 131 | 677.5 | 69.3 | 18.9 | 0.6 | 3.4 | 9.8 | 6.1 | 12.9 | 12.3 | 14.1 | 12.8 | 12.6 |
| 45047 | 24940 | 4 | Greenwood | 455.6 | 71,074 | 767 | 156.0 | 60.3 | 32.4 | 0.6 | 1.6 | 6.5 | 5.9 | 13.0 | 13.0 | 12.7 | 11.0 | 12.2 |
| 45049 | | 6 | Hampton | 560.0 | 18,053 | 1,917 | 32.2 | 43.7 | 52.7 | 0.7 | 1.0 | 3.2 | 5.5 | 12.7 | 11.6 | 12.1 | 11.3 | 12.3 |
| 45051 | 34820 | 4 | Horry | 1,133.3 | 365,449 | 199 | 322.5 | 79.4 | 13.4 | 1.0 | 2.1 | 6.2 | 4.4 | 9.9 | 9.9 | 11.1 | 10.8 | 12.0 |
| 45053 | 25940 | 3 | Jasper | 655.2 | 31,588 | 1,388 | 48.2 | 46.2 | 39.3 | 0.8 | 1.1 | 14.0 | 5.6 | 10.3 | 11.3 | 13.5 | 10.5 | 11.5 |
| 45055 | 17900 | 2 | Kershaw | 726.6 | 67,472 | 795 | 92.9 | 69.6 | 25.1 | 0.8 | 1.2 | 5.1 | 5.7 | 13.2 | 11.3 | 11.7 | 12.1 | 12.6 |
| 45057 | 16740 | 1 | Lancaster | 549.1 | 100,926 | 601 | 183.8 | 71.2 | 21.6 | 0.6 | 2.1 | 6.1 | 5.5 | 12.5 | 9.6 | 11.9 | 13.0 | 12.7 |
| 45059 | 24860 | 2 | Laurens | 712.9 | 67,883 | 792 | 95.2 | 68.8 | 25.6 | 0.7 | 0.9 | 5.6 | 5.9 | 12.3 | 12.5 | 12.3 | 11.1 | 12.7 |
| 45061 | | 6 | Lee | 410.2 | 16,701 | 1,990 | 40.7 | 33.6 | 63.5 | 0.7 | 0.8 | 2.6 | 4.8 | 10.7 | 12.8 | 14.0 | 11.7 | 11.8 |
| 45063 | 17900 | 2 | Lexington | 699.0 | 303,946 | 230 | 434.8 | 75.3 | 16.5 | 0.9 | 2.9 | 6.4 | 5.6 | 13.1 | 11.5 | 13.4 | 13.1 | 13.1 |
| 45065 | | 8 | McCormick | 358.9 | 9,430 | 2,462 | 26.3 | 54.0 | 44.2 | 0.6 | 0.8 | 1.5 | 2.2 | 6.7 | 6.8 | 10.0 | 8.9 | 11.4 |
| 45067 | | 6 | Marion | 489.4 | 30,158 | 1,427 | 61.6 | 39.0 | 57.2 | 1.0 | 1.0 | 3.2 | 5.6 | 12.8 | 11.4 | 11.8 | 11.4 | 11.9 |
| 45069 | 13500 | 6 | Marlboro | 479.9 | 25,581 | 1,573 | 53.3 | 40.8 | 51.8 | 5.6 | 0.8 | 3.2 | 5.3 | 11.5 | 11.1 | 14.2 | 12.3 | 12.4 |
| 45071 | 35140 | 6 | Newberry | 630.3 | 38,445 | 1,212 | 61.0 | 61.4 | 30.4 | 0.7 | 0.8 | 8.1 | 5.5 | 12.2 | 13.1 | 11.4 | 10.4 | 12.3 |
| 45073 | 42860 | 4 | Oconee | 626.6 | 80,015 | 707 | 127.7 | 85.7 | 8.4 | 0.8 | 1.2 | 5.8 | 4.6 | 11.3 | 10.7 | 11.5 | 10.1 | 12.2 |
| 45075 | 36700 | 4 | Orangeburg | 1,106.4 | 85,343 | 677 | 77.1 | 34.3 | 62.3 | 1.1 | 1.3 | 2.4 | 5.5 | 12.6 | 13.6 | 11.7 | 10.3 | 11.3 |
| 45077 | 24860 | 2 | Pickens | 496.9 | 127,983 | 502 | 257.6 | 86.6 | 7.9 | 0.7 | 2.4 | 4.2 | 4.8 | 10.5 | 21.1 | 12.3 | 10.3 | 11.3 |
| 45079 | 17900 | 2 | Richland | 757.3 | 419,051 | 171 | 553.3 | 42.8 | 49.4 | 0.8 | 3.9 | 5.5 | 5.7 | 11.9 | 18.6 | 15.0 | 12.3 | 11.4 |
| 45081 | 17900 | 2 | Saluda | 452.7 | 20,315 | 1,805 | 44.9 | 60.0 | 24.3 | 0.7 | 0.7 | 15.6 | 5.5 | 12.4 | 10.5 | 11.3 | 11.6 | 12.7 |
| 45083 | 43900 | 2 | Spartanburg | 808.4 | 326,205 | 220 | 403.5 | 69.1 | 21.7 | 0.7 | 3.1 | 7.5 | 6.1 | 13.0 | 12.8 | 14.1 | 11.9 | 12.7 |

1. CBSA = Core Based Statistical Area. See Appendix A for explanation. See Appendix B for list of metropolitan areas with component counties.  2. County type code from the Economic Research Service of USDA Rural-Urban Continuum Codes. See Appendix A for definition.  3. Dry land or land partially or temporarily covered by water.  4. May be of any race.

## Table B. States and Counties — Population and Households

| STATE County | 55 to 64 years (16) | 65 to 74 years (17) | 75 years and over (18) | Percent female (19) | Total persons 2000 (20) | Total persons 2010 (21) | Percent change 2000-2010 (22) | Percent change 2010-2020 (23) | Births (24) | Deaths (25) | Net Migration (26) | Number (27) | Persons per household (28) | Family households (29) | Female family householder[1] (30) | One person (31) |
|---|---|---|---|---|---|---|---|---|---|---|---|---|---|---|---|---|
| **PENNSYLVANIA—Cont'd** | | | | | | | | | | | | | | | | |
| Potter | 16.1 | 14.3 | 10.7 | 50.1 | 18,080 | 17,445 | -3.5 | -5.7 | 1,875 | 2,148 | -716 | 6,480 | 2.55 | 66.1 | 9.2 | 29.2 |
| Schuylkill | 14.7 | 12.1 | 9.1 | 48.7 | 150,336 | 148,281 | -1.4 | -5.1 | 14,070 | 19,867 | -1,678 | 58,749 | 2.3 | 65.3 | 12.0 | 29.7 |
| Snyder | 13.7 | 10.9 | 9.0 | 50.5 | 37,546 | 39,719 | 5.8 | 1.5 | 4,314 | 3,679 | -2 | 14,794 | 2.58 | 71.2 | 8.6 | 24.3 |
| Somerset | 15.4 | 13.3 | 10.2 | 47.7 | 80,023 | 77,736 | -2.9 | -6.2 | 7,082 | 9,955 | -1,890 | 29,644 | 2.33 | 67.3 | 8.6 | 28.2 |
| Sullivan | 21.2 | 17.4 | 12.6 | 48.4 | 6,556 | 6,431 | -1.9 | -8.1 | 501 | 1,081 | 63 | 2,723 | 2.09 | 58.4 | 6.3 | 35.4 |
| Susquehanna | 17.4 | 14.2 | 10.5 | 49.9 | 42,238 | 43,318 | 2.6 | -7.6 | 3,845 | 4,826 | -2,322 | 17,235 | 2.36 | 67.0 | 9.3 | 27.4 |
| Tioga | 15.3 | 12.9 | 10.2 | 50.3 | 41,373 | 41,931 | 1.3 | -3.7 | 4,316 | 4,712 | -1,151 | 16,310 | 2.47 | 66.2 | 9.4 | 27.8 |
| Union | 12.1 | 10.1 | 9.0 | 45.9 | 41,624 | 44,968 | 8.0 | -1.5 | 4,122 | 3,937 | -874 | 14,533 | 2.46 | 65.3 | 7.4 | 31.1 |
| Venango | 16.9 | 14.1 | 10.1 | 50.5 | 57,565 | 54,988 | -4.5 | -8.5 | 5,326 | 6,727 | -3,246 | 22,050 | 2.3 | 65.1 | 9.7 | 29.4 |
| Warren | 16.3 | 13.7 | 10.5 | 50.0 | 43,863 | 41,811 | -4.7 | -6.9 | 4,103 | 5,245 | -1,739 | 17,115 | 2.28 | 64.7 | 8.1 | 30.5 |
| Washington | 15.2 | 12.6 | 8.9 | 50.8 | 202,897 | 207,852 | 2.4 | -0.5 | 20,451 | 26,109 | 4,828 | 84,948 | 2.38 | 65.2 | 9.3 | 29.4 |
| Wayne | 16.6 | 14.8 | 10.3 | 46.9 | 47,722 | 52,844 | 10.7 | -3.2 | 4,175 | 6,313 | 488 | 18,841 | 2.52 | 66.7 | 7.7 | 28.3 |
| Westmoreland | 16.0 | 13.7 | 10.2 | 49.9 | 369,993 | 365,071 | -1.3 | -4.9 | 31,400 | 46,261 | -2,828 | 152,283 | 2.26 | 65.1 | 9.2 | 30.0 |
| Wyoming | 15.6 | 13.4 | 9.3 | 49.9 | 28,080 | 28,291 | 0.8 | -6.1 | 2,794 | 3,220 | -1,299 | 10,790 | 2.46 | 66.6 | 11.0 | 27.2 |
| York | 14.1 | 10.8 | 7.7 | 50.6 | 381,751 | 435,016 | 14.0 | 3.5 | 49,895 | 41,514 | 7,458 | 172,421 | 2.53 | 69.0 | 11.1 | 24.7 |
| **RHODE ISLAND** | 14.1 | 10.4 | 7.8 | 51.3 | 1,048,319 | 1,052,970 | 0.4 | 0.4 | 110,543 | 101,181 | -4,890 | 410,489 | 2.47 | 62.1 | 12.9 | 31.0 |
| Bristol | 15.7 | 11.5 | 9.2 | 51.9 | 50,648 | 49,842 | -1.6 | -3.0 | 3,455 | 5,264 | 345 | 19,217 | 2.34 | 65.2 | 9.4 | 29.5 |
| Kent | 15.6 | 11.6 | 8.3 | 51.6 | 167,090 | 166,109 | -0.6 | -0.9 | 16,075 | 17,926 | 501 | 69,422 | 2.34 | 62.5 | 9.7 | 30.8 |
| Newport | 15.4 | 13.3 | 10.4 | 50.5 | 85,433 | 83,141 | -2.7 | -1.6 | 7,094 | 7,534 | -811 | 34,777 | 2.27 | 62.4 | 10.0 | 30.4 |
| Providence | 13.0 | 9.1 | 6.9 | 51.2 | 621,602 | 626,796 | 0.8 | 1.6 | 74,744 | 58,650 | -6,326 | 237,971 | 2.56 | 60.7 | 15.4 | 32.2 |
| Washington | 16.3 | 13.2 | 9.1 | 51.6 | 123,546 | 127,082 | 2.9 | -1.1 | 9,175 | 11,807 | 1,401 | 49,102 | 2.43 | 66.9 | 9.1 | 26.3 |
| **SOUTH CAROLINA** | 13.4 | 11.4 | 7.3 | 51.6 | 4,012,012 | 4,625,358 | 15.3 | 12.8 | 586,324 | 483,494 | 487,169 | 1,921,862 | 2.54 | 65.6 | 14.1 | 28.9 |
| Abbeville | 14.4 | 13.6 | 9.6 | 51.7 | 26,167 | 25,427 | -2.8 | -4.0 | 2,491 | 3,026 | -475 | 9,660 | 2.46 | 64.7 | 11.7 | 31.6 |
| Aiken | 14.2 | 12.1 | 8.4 | 51.9 | 142,552 | 160,128 | 12.3 | 8.0 | 19,388 | 17,900 | 11,356 | 67,598 | 2.45 | 67.5 | 13.6 | 28.0 |
| Allendale | 13.0 | 13.5 | 8.4 | 48.6 | 11,211 | 10,419 | -7.1 | -20.0 | 910 | 1,191 | -1,823 | 3,365 | 2.35 | 59.8 | 21.4 | 36.7 |
| Anderson | 13.6 | 10.9 | 7.7 | 51.9 | 165,740 | 186,927 | 12.8 | 9.3 | 23,031 | 21,906 | 16,397 | 76,798 | 2.55 | 68.0 | 13.6 | 27.3 |
| Bamberg | 14.4 | 13.6 | 9.9 | 52.4 | 16,658 | 15,962 | -4.2 | -12.9 | 1,472 | 1,933 | -1,611 | 5,334 | 2.49 | 62.9 | 18.6 | 33.7 |
| Barnwell | 14.4 | 11.7 | 7.7 | 52.1 | 23,478 | 22,621 | -3.7 | -8.0 | 2,772 | 2,707 | -1,889 | 8,360 | 2.52 | 62.9 | 19.6 | 32.7 |
| Beaufort | 13.9 | 16.7 | 12.0 | 51.2 | 120,937 | 162,219 | 34.1 | 20.6 | 20,505 | 15,673 | 28,343 | 71,477 | 2.53 | 69.1 | 9.7 | 26.0 |
| Berkeley | 12.3 | 9.3 | 5.4 | 50.4 | 142,651 | 178,375 | 25.0 | 32.3 | 27,552 | 14,456 | 44,187 | 76,881 | 2.75 | 71.7 | 13.8 | 22.9 |
| Calhoun | 15.9 | 14.5 | 9.9 | 51.8 | 15,185 | 15,179 | 0.0 | -4.1 | 1,415 | 1,797 | -247 | 6,179 | 2.34 | 63.3 | 13.6 | 34.1 |
| Charleston | 13.0 | 10.8 | 6.6 | 51.7 | 309,969 | 350,162 | 13.0 | 19.4 | 49,739 | 32,499 | 49,761 | 159,195 | 2.45 | 58.0 | 12.1 | 33.1 |
| Cherokee | 13.2 | 10.5 | 6.9 | 51.2 | 52,537 | 55,484 | 5.6 | 3.3 | 6,885 | 6,562 | 1,550 | 20,699 | 2.71 | 56.0 | 11.7 | 40.8 |
| Chester | 14.4 | 11.6 | 7.8 | 51.9 | 34,068 | 33,159 | -2.7 | -2.8 | 3,929 | 4,226 | -620 | 12,653 | 2.54 | 63.6 | 16.4 | 33.4 |
| Chesterfield | 14.2 | 11.8 | 7.6 | 51.3 | 42,768 | 46,717 | 9.2 | -2.4 | 5,178 | 5,365 | -907 | 17,900 | 2.52 | 69.3 | 15.6 | 26.8 |
| Clarendon | 15.2 | 14.8 | 10.3 | 51.4 | 32,502 | 34,949 | 7.5 | -4.4 | 3,414 | 4,067 | -865 | 13,161 | 2.47 | 67.0 | 18.8 | 29.1 |
| Colleton | 14.6 | 12.7 | 8.2 | 52.1 | 38,264 | 38,885 | 1.6 | -3.6 | 4,666 | 5,359 | -702 | 15,075 | 2.45 | 63.4 | 16.3 | 32.0 |
| Darlington | 13.9 | 12.1 | 8.0 | 52.9 | 67,394 | 68,600 | 1.8 | -3.0 | 8,012 | 8,697 | -1,361 | 26,484 | 2.48 | 64.5 | 19.8 | 31.1 |
| Dillon | 13.2 | 10.6 | 7.0 | 53.0 | 30,722 | 32,059 | 4.4 | -5.3 | 4,324 | 3,884 | -2,142 | 11,029 | 2.75 | 63.6 | 21.2 | 31.1 |
| Dorchester | 12.7 | 9.4 | 5.6 | 51.3 | 96,413 | 136,110 | 41.2 | 21.8 | 18,754 | 11,073 | 21,773 | 55,351 | 2.83 | 69.3 | 14.1 | 26.2 |
| Edgefield | 14.9 | 12.2 | 8.1 | 47.0 | 24,595 | 26,966 | 9.6 | 0.6 | 2,002 | 2,285 | 452 | 9,176 | 2.64 | 70.5 | 16.6 | 26.6 |
| Fairfield | 17.1 | 14.7 | 8.6 | 52.1 | 23,454 | 23,977 | 2.2 | -8.0 | 2,205 | 3,039 | -1,077 | 9,191 | 2.42 | 64.4 | 17.8 | 33.1 |
| Florence | 12.9 | 10.8 | 7.0 | 53.3 | 125,761 | 136,981 | 8.9 | 0.4 | 17,547 | 15,626 | -1,195 | 52,188 | 2.6 | 66.6 | 18.2 | 28.8 |
| Georgetown | 16.0 | 18.1 | 11.7 | 52.8 | 55,797 | 60,324 | 8.1 | 5.0 | 5,766 | 7,792 | 5,085 | 25,498 | 2.41 | 68.0 | 14.2 | 29.2 |
| Greenville | 12.6 | 9.9 | 6.7 | 51.6 | 379,616 | 451,183 | 18.9 | 18.0 | 63,928 | 43,017 | 60,103 | 192,975 | 2.57 | 66.1 | 12.0 | 28.4 |
| Greenwood | 12.9 | 11.0 | 8.3 | 53.3 | 66,271 | 69,683 | 5.1 | 2.0 | 8,745 | 7,785 | 465 | 27,612 | 2.46 | 65.7 | 19.0 | 29.3 |
| Hampton | 13.8 | 12.3 | 8.5 | 51.8 | 21,386 | 21,092 | -1.4 | -14.4 | 2,178 | 2,346 | -2,874 | 6,993 | 2.61 | 59.3 | 17.0 | 38.2 |
| Horry | 15.9 | 17.0 | 9.0 | 52.0 | 196,629 | 269,136 | 36.9 | 35.8 | 31,902 | 34,200 | 97,587 | 131,143 | 2.5 | 64.6 | 11.6 | 28.0 |
| Jasper | 15.5 | 14.3 | 7.4 | 49.7 | 20,678 | 24,791 | 19.9 | 27.4 | 3,559 | 2,493 | 5,686 | 10,269 | 2.74 | 71.1 | 16.5 | 25.2 |
| Kershaw | 14.2 | 11.6 | 7.5 | 51.8 | 52,647 | 61,586 | 17.0 | 9.6 | 7,538 | 6,880 | 5,262 | 24,980 | 2.59 | 68.3 | 14.8 | 27.6 |
| Lancaster | 12.7 | 12.9 | 9.1 | 51.5 | 61,351 | 76,653 | 24.9 | 31.7 | 10,001 | 8,946 | 23,015 | 33,899 | 2.66 | 69.5 | 12.8 | 26.9 |
| Laurens | 14.1 | 11.3 | 7.8 | 51.8 | 69,567 | 66,529 | -4.4 | 2.0 | 7,926 | 8,616 | 2,087 | 25,563 | 2.52 | 68.8 | 15.6 | 27.0 |
| Lee | 14.2 | 12.3 | 7.7 | 49.1 | 20,119 | 19,225 | -4.4 | -13.1 | 1,815 | 2,453 | -1,901 | 6,423 | 2.47 | 61.8 | 18.4 | 36.1 |
| Lexington | 13.5 | 10.2 | 6.6 | 51.4 | 216,014 | 262,454 | 21.5 | 15.8 | 33,172 | 24,823 | 33,102 | 113,104 | 2.54 | 67.9 | 13.1 | 26.4 |
| McCormick | 17.0 | 22.1 | 14.8 | 46.2 | 9,958 | 10,229 | 2.7 | -7.8 | 577 | 1,401 | 34 | 3,957 | 2.11 | 63.5 | 15.5 | 32.1 |
| Marion | 14.0 | 13.1 | 8.1 | 54.0 | 35,466 | 33,055 | -6.8 | -8.8 | 3,848 | 4,450 | -2,296 | 11,600 | 2.69 | 64.9 | 24.1 | 31.4 |
| Marlboro | 13.9 | 11.8 | 7.4 | 48.4 | 28,818 | 28,935 | 0.4 | -11.6 | 3,025 | 3,571 | -2,814 | 9,613 | 2.5 | 63.1 | 21.0 | 30.8 |
| Newberry | 14.2 | 12.2 | 8.5 | 51.0 | 36,108 | 37,510 | 3.9 | 2.5 | 4,448 | 4,478 | 990 | 14,810 | 2.5 | 65.5 | 14.5 | 29.9 |
| Oconee | 15.5 | 14.6 | 9.6 | 50.9 | 66,215 | 74,285 | 12.2 | 7.7 | 7,754 | 9,219 | 7,203 | 31,978 | 2.4 | 67.6 | 10.9 | 28.1 |
| Orangeburg | 14.0 | 12.3 | 8.8 | 53.4 | 91,582 | 92,508 | 1.0 | -7.7 | 10,609 | 11,007 | -6,780 | 33,060 | 2.57 | 60.7 | 19.0 | 36.3 |
| Pickens | 12.5 | 9.8 | 7.4 | 50.3 | 110,757 | 119,384 | 7.8 | 7.2 | 12,393 | 12,179 | 8,389 | 47,934 | 2.45 | 63.2 | 10.4 | 26.5 |
| Richland | 11.4 | 8.5 | 5.2 | 51.9 | 320,677 | 384,425 | 19.9 | 9.0 | 49,165 | 30,690 | 16,336 | 151,853 | 2.51 | 60.3 | 17.3 | 32.5 |
| Saluda | 14.7 | 11.7 | 9.5 | 49.9 | 19,181 | 19,857 | 3.5 | 2.3 | 2,470 | 2,164 | 168 | 7,094 | 2.81 | 72.3 | 15.3 | 25.9 |
| Spartanburg | 12.7 | 10.0 | 6.7 | 51.5 | 253,791 | 284,329 | 12.0 | 14.7 | 37,858 | 31,185 | 35,265 | 116,645 | 2.57 | 67.8 | 14.0 | 27.4 |

1. No spouse present.

# Table B. States and Counties — Population, Vital Statistics, Health, and Crime

| STATE County | Persons in group quarters, 2020 | Daytime Population, 2015-2019 | | Births, 2020 | | Deaths, 2020 | | Persons under 65 with no health insurance, 2019 | | Medicare, 2020 | | | Crimes reported by county police or sheriff, 2019 | |
|---|---|---|---|---|---|---|---|---|---|---|---|---|---|---|
| | | Number | Employment/ residence ratio | Total | Rate[1] | Number | Rate[1] | Number | Percent | Total beneficiaries | Enrolled in Original Medicare | Enrolled in Medicare Advantage | Violent | Property |
| | 32 | 33 | 34 | 35 | 36 | 37 | 38 | 39 | 40 | 41 | 42 | 43 | 44 | 45 |
| PENNSYLVANIA—Cont'd | | | | | | | | | | | | | | |
| Potter | 217 | 16,689 | 0.98 | 158 | 9.6 | 230 | 14.0 | 900 | 7.3 | 4,697 | 2,986 | 1,711 | NA | NA |
| Schuylkill | 6,566 | 132,867 | 0.84 | 1,261 | 9.0 | 1,882 | 13.4 | 6,511 | 6.1 | 35,767 | 23,846 | 11,922 | NA | NA |
| Snyder | 2,684 | 39,834 | 0.97 | 409 | 10.1 | 416 | 10.3 | 2,590 | 8.7 | 8,665 | 4,714 | 3,951 | 0 | 0 |
| Somerset | 4,863 | 69,038 | 0.83 | 705 | 9.7 | 999 | 13.7 | 3,976 | 7.6 | 19,637 | 7,993 | 11,644 | NA | NA |
| Sullivan | 371 | 5,668 | 0.82 | 39 | 6.6 | 110 | 18.6 | 373 | 8.8 | 1,878 | 1,241 | 637 | NA | NA |
| Susquehanna | 282 | 35,609 | 0.7 | 376 | 9.4 | 501 | 12.5 | 2,567 | 8.5 | 10,381 | 7,334 | 3,047 | NA | NA |
| Tioga | 1,314 | 39,679 | 0.93 | 379 | 9.4 | 520 | 12.9 | 2,516 | 8.3 | 10,793 | 6,828 | 3,965 | NA | NA |
| Union | 8,328 | 47,667 | 1.14 | 401 | 9.1 | 463 | 10.5 | 2,383 | 8.4 | 8,478 | 4,999 | 3,479 | NA | NA |
| Venango | 1,302 | 49,660 | 0.9 | 447 | 8.9 | 723 | 14.4 | 2,629 | 6.9 | 14,593 | 6,945 | 7,648 | NA | NA |
| Warren | 759 | 39,213 | 0.97 | 400 | 10.3 | 562 | 14.4 | 1,911 | 6.5 | 10,672 | 6,734 | 3,938 | NA | NA |
| Washington | 5,316 | 204,555 | 0.97 | 2,026 | 9.8 | 2,691 | 13.0 | 9,364 | 5.9 | 50,972 | 19,077 | 31,895 | NA | NA |
| Wayne | 3,748 | 48,535 | 0.86 | 374 | 7.3 | 664 | 13.0 | 2,672 | 7.6 | 14,133 | 10,666 | 3,468 | 0 | 0 |
| Westmoreland | 6,968 | 328,498 | 0.86 | 2,801 | 8.1 | 4,744 | 13.7 | 14,550 | 5.5 | 94,522 | 33,122 | 61,399 | NA | NA |
| Wyoming | 644 | 26,889 | 0.97 | 223 | 8.4 | 340 | 12.8 | 1,217 | 6.0 | 6,860 | 4,458 | 2,402 | NA | NA |
| York | 8,609 | 410,922 | 0.84 | 4,625 | 10.3 | 4,460 | 9.9 | 22,184 | 6.1 | 95,126 | 56,245 | 38,881 | NA | NA |
| RHODE ISLAND | 41,167 | 1,034,477 | 0.96 | 10,447 | 9.9 | 10,470 | 9.9 | 41,789 | 5.0 | 224,629 | 117,402 | 107,228 | X | X |
| Bristol | 2,755 | 40,376 | 0.66 | 310 | 6.4 | 558 | 11.5 | 1,277 | 3.4 | 11,568 | 6,179 | 5,389 | NA | NA |
| Kent | 1,145 | 152,013 | 0.86 | 1,494 | 9.1 | 1,788 | 10.9 | 4,944 | 3.7 | 39,962 | 20,303 | 19,659 | NA | NA |
| Newport | 4,071 | 85,707 | 1.07 | 694 | 8.5 | 874 | 10.7 | 2,537 | 4.2 | 20,202 | 13,069 | 7,132 | NA | NA |
| Providence | 26,365 | 634,168 | 0.99 | 7,081 | 11.1 | 5,904 | 9.3 | 29,775 | 5.8 | 121,265 | 59,172 | 62,093 | NA | NA |
| Washington | 6,831 | 122,213 | 0.94 | 868 | 6.9 | 1,346 | 10.7 | 3,256 | 3.5 | 31,633 | 18,678 | 12,954 | NA | NA |
| SOUTH CAROLINA | 130,657 | 4,983,480 | 0.98 | 56,554 | 10.8 | 55,751 | 10.7 | 540,230 | 13.2 | 1,105,730 | 737,579 | 368,151 | X | X |
| Abbeville | 901 | 21,024 | 0.63 | 232 | 9.5 | 312 | 12.8 | 2,485 | 13.7 | 6,396 | 3,837 | 2,560 | NA | NA |
| Aiken | 2,341 | 163,562 | 0.93 | 1,845 | 10.7 | 2,025 | 11.7 | 17,918 | 13.3 | 39,707 | 27,350 | 12,357 | 549 | 2,639 |
| Allendale | 946 | 8,729 | 0.9 | 64 | 7.7 | 111 | 13.3 | 618 | 10.8 | 2,079 | 1,139 | 940 | 12 | 48 |
| Anderson | 2,956 | 185,192 | 0.85 | 2,205 | 10.8 | 2,414 | 11.8 | 21,364 | 13.1 | 45,523 | 26,655 | 18,869 | 693 | 4,585 |
| Bamberg | 999 | 13,589 | 0.85 | 140 | 10.1 | 205 | 14.7 | 1,281 | 12.7 | 3,487 | 2,054 | 1,433 | 24 | 146 |
| Barnwell | 285 | 19,734 | 0.79 | 247 | 11.9 | 304 | 14.6 | 1,928 | 11.7 | 4,845 | 3,061 | 1,784 | 57 | 292 |
| Beaufort | 5,537 | 189,611 | 1.04 | 1,875 | 9.6 | 1,894 | 9.7 | 19,110 | 14.4 | 53,148 | 40,068 | 13,080 | 577 | 1,879 |
| Berkeley | 3,684 | 180,846 | 0.67 | 2,873 | 12.2 | 1,833 | 7.8 | 23,493 | 12.2 | 38,279 | 26,903 | 11,377 | NA | NA |
| Calhoun | 158 | 12,364 | 0.63 | 124 | 8.5 | 194 | 13.3 | 1,429 | 13.1 | 3,852 | 2,375 | 1,477 | 62 | 294 |
| Charleston | 11,055 | 469,184 | 1.33 | 4,886 | 11.7 | 3,890 | 9.3 | 43,577 | 13.1 | 76,379 | 56,378 | 20,000 | 280 | 1,698 |
| Cherokee | 1,138 | 54,313 | 0.89 | 639 | 11.1 | 723 | 12.6 | 6,303 | 13.7 | 12,645 | 7,125 | 5,520 | NA | NA |
| Chester | 218 | 28,856 | 0.73 | 374 | 11.6 | 486 | 15.1 | 3,169 | 12.4 | 7,882 | 4,856 | 3,027 | 111 | 529 |
| Chesterfield | 872 | 44,728 | 0.94 | 475 | 10.4 | 603 | 13.2 | 5,393 | 14.9 | 9,932 | 6,791 | 3,141 | NA | NA |
| Clarendon | 1,287 | 30,886 | 0.74 | 302 | 9.0 | 441 | 13.2 | 3,471 | 14.6 | 9,031 | 5,791 | 3,240 | 137 | 668 |
| Colleton | 385 | 34,786 | 0.82 | 451 | 12.0 | 614 | 16.4 | 4,497 | 15.4 | 10,006 | 6,385 | 3,621 | 214 | 726 |
| Darlington | 1,441 | 64,115 | 0.89 | 757 | 11.4 | 936 | 14.1 | 6,625 | 12.8 | 15,591 | 11,079 | 4,513 | 386 | 1,853 |
| Dillon | 468 | 29,341 | 0.88 | 404 | 13.3 | 418 | 13.8 | 4,073 | 16.7 | 6,670 | 4,301 | 2,368 | 138 | 576 |
| Dorchester | 2,000 | 128,759 | 0.6 | 1,835 | 11.1 | 1,359 | 8.2 | 16,151 | 11.7 | 28,044 | 19,971 | 8,073 | 292 | 1,871 |
| Edgefield | 2,447 | 23,391 | 0.66 | 191 | 7.0 | 298 | 11.0 | 2,641 | 13.7 | 5,797 | 3,726 | 2,071 | NA | NA |
| Fairfield | 397 | 22,289 | 0.97 | 193 | 8.7 | 326 | 14.8 | 2,044 | 12.1 | 5,682 | 3,429 | 2,253 | 154 | 465 |
| Florence | 2,939 | 147,299 | 1.15 | 1,584 | 11.5 | 1,759 | 12.8 | 13,857 | 12.5 | 29,884 | 22,001 | 7,884 | 516 | 2,325 |
| Georgetown | 553 | 62,453 | 1.02 | 520 | 8.2 | 939 | 14.8 | 6,251 | 14.2 | 20,748 | 14,371 | 6,377 | 187 | 986 |
| Greenville | 10,465 | 537,575 | 1.13 | 6,276 | 11.8 | 4,894 | 9.2 | 57,215 | 13.3 | 100,632 | 61,456 | 39,177 | 1,705 | 7,923 |
| Greenwood | 2,524 | 71,973 | 1.05 | 801 | 11.3 | 883 | 12.4 | 7,827 | 14.3 | 15,949 | 10,473 | 5,476 | NA | NA |
| Hampton | 281 | 18,843 | 0.9 | 188 | 10.4 | 272 | 15.1 | 1,895 | 13.5 | 4,432 | 2,770 | 1,662 | 61 | 269 |
| Horry | 3,733 | 333,174 | 1.01 | 3,135 | 8.6 | 4,357 | 11.9 | 46,935 | 18.0 | 103,809 | 73,489 | 30,320 | 844 | 6,145 |
| Jasper | 1,247 | 26,042 | 0.78 | 372 | 11.8 | 302 | 9.6 | 4,358 | 19.4 | 7,064 | 4,576 | 2,488 | NA | NA |
| Kershaw | 400 | 56,459 | 0.69 | 741 | 11.0 | 841 | 12.5 | 6,595 | 12.3 | 14,891 | 10,422 | 4,469 | 160 | 1,201 |
| Lancaster | 1,814 | 82,862 | 0.75 | 1,016 | 10.1 | 1,123 | 11.1 | 8,978 | 11.9 | 23,627 | 16,748 | 6,879 | 284 | 1,677 |
| Laurens | 2,239 | 62,512 | 0.84 | 778 | 11.5 | 940 | 13.8 | 7,470 | 14.2 | 16,604 | 9,896 | 6,708 | 219 | 978 |
| Lee | 1,657 | 15,033 | 0.6 | 167 | 10.0 | 299 | 17.9 | 1,462 | 12.3 | 4,128 | 2,556 | 1,571 | 64 | 306 |
| Lexington | 2,342 | 275,973 | 0.9 | 3,202 | 10.5 | 2,844 | 9.4 | 30,853 | 12.4 | 56,211 | 40,502 | 15,709 | 554 | 5,086 |
| McCormick | 948 | 9,170 | 0.87 | 48 | 5.1 | 184 | 19.5 | 613 | 12.0 | 3,857 | 2,303 | 1,554 | 8 | 77 |
| Marion | 200 | 28,255 | 0.73 | 317 | 10.5 | 482 | 16.0 | 3,199 | 13.5 | 7,765 | 5,054 | 2,711 | 71 | 462 |
| Marlboro | 2,451 | 24,886 | 0.79 | 271 | 10.6 | 388 | 15.2 | 2,578 | 14.1 | 6,326 | 3,960 | 2,366 | 106 | 342 |
| Newberry | 1,310 | 37,394 | 0.95 | 412 | 10.7 | 497 | 12.9 | 4,354 | 14.9 | 9,238 | 6,187 | 3,051 | 91 | 299 |
| Oconee | 792 | 73,892 | 0.88 | 695 | 8.7 | 1,045 | 13.1 | 8,866 | 14.8 | 22,017 | 14,366 | 7,651 | 199 | 1,553 |
| Orangeburg | 3,032 | 86,997 | 0.98 | 942 | 11.0 | 1,233 | 14.4 | 8,372 | 13.0 | 20,675 | 11,947 | 8,728 | 744 | 2,661 |
| Pickens | 8,297 | 114,395 | 0.82 | 1,212 | 9.5 | 1,369 | 10.7 | 12,827 | 13.2 | 26,692 | 15,503 | 11,188 | 228 | 1,408 |
| Richland | 28,576 | 447,802 | 1.18 | 4,713 | 11.2 | 3,440 | 8.2 | 37,559 | 11.3 | 68,266 | 48,726 | 19,541 | 2,206 | 8,709 |
| Saluda | 298 | 17,640 | 0.68 | 224 | 11.0 | 235 | 11.6 | 2,835 | 17.6 | 4,592 | 3,093 | 1,499 | 43 | 198 |
| Spartanburg | 7,711 | 319,454 | 1.09 | 3,942 | 12.1 | 3,518 | 10.8 | 33,742 | 13.0 | 66,614 | 37,418 | 29,195 | 1,042 | 5,420 |

1. Per 1,000 estimated resident population.

Items 32—45

# Table B. States and Counties — Crime, Education, Money Income, and Poverty

| STATE County | County law enforcement employment, 2019 | | Education — School enrollment and attainment, 2015-2019 | | | | Local government expenditures,[3] 2017-2018 | | Money income, 2015-2019 | | Households Percent | | Income and poverty, 2019 | Percent below poverty level | | |
| | | | Enrollment[1] | | Attainment[2] (percent) | | | | | | | | | | | |
| | Officers | Civilians | Total | Percent private | High school graduate or less | Bachelor's degree or more | Total current spending (mil dol) | Current spending per student (dollars) | Per capita income[4] | Median income (dollars) | with income of less than $50,000 | with income of $200,000 or more | Median household income (dollars) | All persons | Children under 18 years | Children 5 to 17 years in families |
| | 46 | 47 | 48 | 49 | 50 | 51 | 52 | 53 | 54 | 55 | 56 | 57 | 58 | 59 | 60 | 61 |
| **PENNSYLVANIA—Cont'd** | | | | | | | | | | | | | | | | |
| Potter | NA | NA | 3,219 | 13.5 | 59.1 | 15.4 | 36 | 15,683 | 24,797 | 45,419 | 54.5 | 1.4 | 49,054 | 11.7 | 20.9 | 19.7 |
| Schuylkill | 16 | 3 | 27,866 | 14.1 | 57.3 | 16.6 | 250 | 14,575 | 27,431 | 52,280 | 48.1 | 2.3 | 55,951 | 11.7 | 15.5 | 15.0 |
| Snyder | 7 | 0 | 9,370 | 37.6 | 59.0 | 19.8 | 64 | 13,533 | 27,482 | 58,997 | 42.2 | 4.2 | 56,873 | 10.3 | 15.1 | 14.6 |
| Somerset | NA | NA | 12,700 | 15.0 | 61.6 | 15.9 | 132 | 14,736 | 25,781 | 49,089 | 51.1 | 2.1 | 49,082 | 12.5 | 20.2 | 18.1 |
| Sullivan | NA | NA | 859 | 17.2 | 58.1 | 17.2 | 13 | 20,842 | 30,375 | 47,407 | 51.7 | 2.4 | 51,718 | 12.9 | 16.1 | 15.3 |
| Susquehanna | NA | NA | 7,328 | 13.7 | 55.1 | 18.6 | 108 | 17,923 | 31,309 | 54,966 | 45.1 | 4.0 | 60,235 | 11.3 | 18.0 | 16.8 |
| Tioga | 8 | 2 | 7,980 | 13.1 | 54.9 | 19.1 | 78 | 14,346 | 26,301 | 51,324 | 48.8 | 1.9 | 51,324 | 13.4 | 19.1 | 18.5 |
| Union | 7 | 1 | 10,862 | 48.1 | 52.2 | 26.0 | 61 | 15,206 | 27,079 | 59,399 | 42.5 | 4.9 | 61,306 | 12.0 | 11.6 | 11.3 |
| Venango | NA | NA | 9,271 | 13.3 | 57.8 | 18.7 | 88 | 14,761 | 26,969 | 49,945 | 50.1 | 1.9 | 53,619 | 12.5 | 18.2 | 16.6 |
| Warren | NA | NA | 7,268 | 15.3 | 52.6 | 19.5 | 70 | 15,227 | 28,230 | 50,250 | 49.8 | 2.2 | 52,592 | 13.5 | 23.2 | 21.3 |
| Washington | 29 | 3 | 43,624 | 18.2 | 44.2 | 30.0 | 426 | 15,813 | 35,537 | 63,543 | 40.0 | 6.7 | 65,558 | 9.9 | 11.8 | 10.9 |
| Wayne | 12 | 3 | 8,722 | 15.4 | 53.3 | 20.4 | 87 | 19,612 | 27,412 | 56,096 | 44.5 | 2.7 | 56,265 | 12.4 | 17.1 | 16.2 |
| Westmoreland | 56 | 7 | 70,965 | 19.1 | 42.6 | 29.1 | 668 | 14,519 | 34,274 | 60,471 | 41.7 | 5.0 | 59,407 | 10.5 | 13.9 | 12.6 |
| Wyoming | 4 | 2 | 5,463 | 20.4 | 54.2 | 19.6 | 60 | 18,349 | 29,584 | 59,415 | 42.3 | 2.5 | 60,449 | 9.9 | 15.2 | 13.7 |
| York | 105 | 10 | 99,935 | 19.0 | 49.7 | 24.9 | 953 | 14,255 | 32,623 | 66,457 | 35.9 | 4.5 | 69,101 | 9.2 | 12.6 | 12.3 |
| **RHODE ISLAND** | X | X | 257,417 | 26.0 | 39.5 | 34.2 | 2,322 | 16,266 | 36,121 | 67,167 | 38.7 | 7.1 | 70,383 | 11.6 | 16.5 | 15.4 |
| Bristol | NA | NA | 13,492 | 41.3 | 28.5 | 49.0 | 105 | 15,867 | 46,566 | 83,092 | 31.0 | 13.7 | 87,926 | 6.8 | 7.6 | 6.7 |
| Kent | NA | NA | 34,640 | 18.6 | 35.8 | 33.3 | 379 | 17,498 | 39,869 | 73,521 | 33.8 | 6.9 | 74,230 | 8.1 | 10.3 | 9.4 |
| Newport | NA | NA | 17,761 | 31.2 | 28.1 | 48.1 | 164 | 17,342 | 47,594 | 79,454 | 32.7 | 10.6 | 80,862 | 8.5 | 11.2 | 10.3 |
| Providence | NA | NA | 158,342 | 28.2 | 45.2 | 29.0 | 1,403 | 15,554 | 31,522 | 58,974 | 43.6 | 5.3 | 62,165 | 14.0 | 20.3 | 19.2 |
| Washington | NA | NA | 33,182 | 14.8 | 27.9 | 46.1 | 271 | 18,295 | 42,869 | 85,531 | 29.6 | 11.0 | 86,450 | 8.0 | 8.9 | 8.3 |
| **SOUTH CAROLINA** | X | X | 1,192,839 | 14.4 | 41.6 | 28.1 | 8,615 | 11,080 | 29,426 | 53,199 | 47.2 | 4.6 | 56,360 | 13.9 | 19.9 | 19.0 |
| Abbeville | 333 | 323 | 5,342 | 17.5 | 52.4 | 15.6 | 33 | 10,799 | 22,646 | 38,741 | 60.8 | 1.3 | 46,499 | 14.8 | 21.6 | 19.7 |
| Aiken | 132 | 110 | 36,894 | 15.6 | 44.8 | 26.4 | 232 | 9,459 | 28,396 | 51,399 | 48.9 | 3.6 | 56,824 | 12.5 | 19.3 | 18.2 |
| Allendale | 12 | 2 | 1,735 | 7.3 | 66.1 | 9.2 | 20 | 16,595 | 16,985 | 27,185 | 67.3 | 0.2 | 32,147 | 30.2 | 46.9 | 44.7 |
| Anderson | 169 | 144 | 46,723 | 15.2 | 45.9 | 22.2 | 319 | 9,784 | 27,552 | 50,865 | 49.2 | 3.7 | 54,496 | 12.9 | 18.8 | 18.4 |
| Bamberg | 13 | 1 | 3,926 | 22.3 | 52.2 | 18.8 | 27 | 13,122 | 18,479 | 31,422 | 62.9 | 0.5 | 35,364 | 24.2 | 33.9 | 31.9 |
| Barnwell | 34 | 55 | 4,644 | 8.5 | 55.6 | 12.5 | 47 | 12,514 | 20,297 | 35,803 | 59.8 | 1.1 | 36,675 | 24.9 | 38.5 | 39.8 |
| Beaufort | 213 | 96 | 35,835 | 20.0 | 29.5 | 41.2 | 264 | 11,935 | 38,946 | 68,377 | 37.0 | 8.5 | 73,890 | 10.2 | 18.4 | 16.5 |
| Berkeley | 151 | 33 | 51,816 | 15.0 | 40.1 | 25.0 | 340 | 9,674 | 29,662 | 63,309 | 38.1 | 4.0 | 68,690 | 10.7 | 14.8 | 13.3 |
| Calhoun | 25 | 15 | 2,750 | 24.5 | 50.0 | 19.9 | 21 | 11,863 | 26,398 | 46,339 | 54.2 | 2.3 | 47,090 | 18.2 | 28.0 | 25.4 |
| Charleston | 287 | 433 | 90,843 | 16.3 | 31.1 | 43.6 | 573 | 11,548 | 39,914 | 64,022 | 39.4 | 9.7 | 70,980 | 11.7 | 16.1 | 15.6 |
| Cherokee | 60 | 52 | 13,452 | 14.5 | 54.4 | 14.2 | 92 | 10,299 | 21,152 | 36,883 | 63.4 | 1.2 | 46,905 | 16.0 | 22.6 | 20.4 |
| Chester | 49 | 53 | 6,938 | 7.0 | 57.2 | 13.4 | 59 | 11,156 | 22,324 | 42,442 | 56.9 | 1.4 | 45,400 | 17.7 | 25.7 | 25.0 |
| Chesterfield | 52 | 9 | 10,116 | 11.1 | 64.7 | 10.4 | 76 | 10,612 | 21,325 | 41,505 | 56.9 | 1.1 | 42,441 | 19.9 | 28.7 | 26.0 |
| Clarendon | NA | NA | 6,477 | 12.5 | 58.0 | 15.1 | 59 | 11,905 | 22,824 | 40,900 | 58.7 | 2.2 | 39,900 | 25.1 | 41.8 | 40.1 |
| Colleton | NA | NA | 7,960 | 7.6 | 58.9 | 14.6 | 60 | 10,460 | 21,377 | 36,324 | 65.8 | 1.9 | 40,808 | 21.0 | 33.4 | 31.1 |
| Darlington | 59 | 13 | 15,471 | 17.8 | 52.8 | 18.4 | 56 | 10,806 | 23,027 | 38,448 | 60.2 | 2.5 | 44,007 | 19.6 | 31.3 | 29.8 |
| Dillon | NA | NA | 7,658 | 7.6 | 63.7 | 11.1 | 56 | 9,630 | 17,605 | 30,812 | 68.0 | 0.8 | 35,483 | 26.8 | 35.2 | 32.1 |
| Dorchester | 147 | 126 | 41,567 | 12.6 | 37.6 | 28.2 | 272 | 9,521 | 29,853 | 63,080 | 38.8 | 4.7 | 68,051 | 9.3 | 12.6 | 11.7 |
| Edgefield | NA | NA | 5,744 | 17.4 | 53.8 | 16.7 | 39 | 11,088 | 26,228 | 49,127 | 50.9 | 2.7 | 52,591 | 15.7 | 20.7 | 18.9 |
| Fairfield | 51 | 4 | 4,492 | 18.1 | 55.0 | 16.9 | 47 | 17,175 | 23,380 | 38,213 | 60.5 | 1.9 | 42,496 | 19.1 | 29.2 | 27.3 |
| Florence | 215 | 39 | 35,236 | 11.9 | 47.9 | 23.6 | 248 | 10,742 | 26,127 | 47,058 | 52.2 | 3.3 | 49,770 | 17.0 | 23.9 | 21.8 |
| Georgetown | 76 | 70 | 12,415 | 9.0 | 41.6 | 27.3 | 107 | 11,275 | 31,382 | 48,456 | 51.2 | 5.4 | 53,747 | 17.5 | 31.1 | 26.5 |
| Greenville | 451 | 99 | 123,139 | 22.6 | 35.7 | 35.1 | 733 | 9,682 | 32,679 | 60,351 | 41.8 | 6.1 | 64,399 | 10.7 | 14.8 | 14.7 |
| Greenwood | 68 | 76 | 17,336 | 12.7 | 45.3 | 24.6 | 115 | 9,913 | 24,752 | 42,336 | 56.9 | 2.6 | 43,958 | 18.0 | 25.3 | 24.4 |
| Hampton | 30 | 16 | 3,986 | 9.8 | 61.5 | 11.9 | 39 | | 18,424 | 33,429 | 68.7 | 0.7 | 37,560 | 23.2 | 32.2 | 30.0 |
| Horry | 324 | 44 | 66,456 | 8.6 | 42.9 | 23.6 | 493 | 10,994 | 28,202 | 50,704 | 49.2 | 3.2 | 53,648 | 12.7 | 21.0 | 21.1 |
| Jasper | NA | NA | 5,937 | 19.3 | 55.7 | 17.8 | 37 | 13,917 | 24,566 | 45,601 | 55.4 | 2.4 | 50,790 | 17.2 | 28.8 | 29.6 |
| Kershaw | 66 | 11 | 14,796 | 11.1 | 48.7 | 20.3 | 105 | 9,771 | 25,442 | 51,479 | 48.4 | 2.8 | 56,318 | 13.8 | 22.9 | 20.8 |
| Lancaster | 119 | 21 | 19,234 | 13.8 | 44.0 | 26.9 | 128 | 9,675 | 30,742 | 58,849 | 43.5 | 5.4 | 63,842 | 11.8 | 15.3 | 14.9 |
| Laurens | 70 | 38 | 14,935 | 18.4 | 52.9 | 16.2 | 95 | 10,826 | 23,004 | 43,304 | 56.3 | 1.9 | 47,038 | 19.1 | 25.5 | 24.5 |
| Lee | NA | NA | 3,590 | 10.2 | 60.0 | 15.9 | 26 | 13,395 | 19,300 | 32,371 | 67.3 | 2.7 | 37,710 | 25.4 | 34.8 | 31.7 |
| Lexington | 292 | 147 | 69,335 | 11.0 | 38.5 | 30.4 | 461 | 11,285 | 31,671 | 61,173 | 40.5 | 5.0 | 62,059 | 11.9 | 17.3 | 16.5 |
| McCormick | 18 | 21 | 1,290 | 16.4 | 50.5 | 20.5 | 14 | 17,958 | 25,617 | 43,633 | 53.4 | 2.5 | 48,645 | 18.1 | 32.6 | 30.9 |
| Marion | 38 | 34 | 7,097 | 13.5 | 63.1 | 13.5 | 52 | 11,445 | 20,160 | 32,063 | 67.4 | 1.1 | 35,138 | 24.9 | 41.0 | 38.5 |
| Marlboro | NA | NA | 5,319 | 6.1 | 65.0 | 8.9 | 68 | 12,260 | 17,456 | 33,586 | 66.9 | 1.3 | 34,532 | 28.9 | 30.7 | 29.0 |
| Newberry | 47 | 56 | 8,671 | 18.2 | 54.0 | 17.9 | 68 | 11,250 | 24,959 | 44,226 | 55.1 | 2.4 | 50,773 | 14.0 | 22.4 | 20.7 |
| Oconee | 105 | 78 | 16,325 | 11.2 | 44.2 | 25.8 | 120 | 11,400 | 29,844 | 49,134 | 50.7 | 4.2 | 52,240 | 13.1 | 19.9 | 18.8 |
| Orangeburg | 80 | 43 | 23,044 | 18.9 | 45.3 | 19.9 | 168 | 12,950 | 20,716 | 37,955 | 63.2 | 1.5 | 38,736 | 26.3 | 42.8 | 43.9 |
| Pickens | 149 | 30 | 38,071 | 10.3 | 43.7 | 26.0 | 145 | 8,890 | 26,061 | 49,573 | 50.4 | 3.1 | 52,949 | 15.3 | 15.5 | 14.8 |
| Richland | 536 | 109 | 118,541 | 13.3 | 31.0 | 39.1 | 1,377 | 14,341 | 30,175 | 54,767 | 45.7 | 5.2 | 52,905 | 16.2 | 21.1 | 20.8 |
| Saluda | 24 | 38 | 4,115 | 11.1 | 57.0 | 16.8 | 24 | 10,302 | 22,814 | 45,714 | 53.4 | 1.1 | 49,493 | 16.1 | 23.8 | 21.9 |
| Spartanburg | 317 | 29 | 75,520 | 15.6 | 43.6 | 24.5 | 541 | 11,113 | 27,240 | 52,332 | 47.7 | 3.6 | 55,588 | 12.9 | 18.2 | 16.2 |

1. All persons 3 years old and over enrolled in nursery school through college.  2. Persons 25 years old and over.  3. Elementary and secondary education expenditures.  4. Based on population estimated by the American Community Survey, 2014–2018.

# Table B. States and Counties — **Personal Income**

| STATE County | Personal income, 2019 Total (mil dol) | Percent change 2018-2019 | Per capita Dollars | Per capita Rank | Wages and salaries (mil dol) | Supplements — Pension and insurance | Supplements — Government social insurance | Proprietors' income (mil dol) | Dividends, interest, and rent (mil dol) | Personal transfer receipts (mil dol) | Earnings, 2019 Total (mil dol) | Contributions — From employee and self-employed | Contributions — From employer |
|---|---|---|---|---|---|---|---|---|---|---|---|---|---|
| | 62 | 63 | 64 | 65 | 66 | 67 | 68 | 69 | 70 | 71 | 72 | 73 | 74 |
| **PENNSYLVANIA—Cont'd** | | | | | | | | | | | | | |
| Potter | 758 | 3.5 | 45,887 | 1,262 | 250 | 61 | 20 | 151 | 115 | 209 | 483 | 29 | 20 |
| Schuylkill | 6,228 | 3.4 | 44,055 | 1,477 | 2,289 | 483 | 190 | 374 | 986 | 1,719 | 3,336 | 225 | 190 |
| Snyder | 1,748 | 2.8 | 43,286 | 1,574 | 609 | 134 | 52 | 188 | 248 | 485 | 982 | 61 | 52 |
| Somerset | 3,166 | 3.5 | 43,103 | 1,600 | 1,075 | 247 | 88 | 279 | 534 | 865 | 1,689 | 114 | 88 |
| Sullivan | 288 | 2.8 | 47,455 | 1,085 | 61 | 16 | 5 | 29 | 71 | 81 | 111 | 8 | 5 |
| Susquehanna | 1,885 | 3.6 | 46,737 | 1,156 | 435 | 98 | 34 | 200 | 407 | 442 | 767 | 53 | 34 |
| Tioga | 1,704 | 4.5 | 41,990 | 1,769 | 600 | 142 | 47 | 154 | 292 | 467 | 943 | 63 | 47 |
| Union | 1,860 | 2.5 | 41,400 | 1,862 | 880 | 183 | 73 | 219 | 325 | 367 | 1,354 | 80 | 73 |
| Venango | 2,220 | 3.3 | 43,822 | 1,512 | 786 | 190 | 63 | 138 | 353 | 755 | 1,177 | 82 | 63 |
| Warren | 1,701 | 2.6 | 43,401 | 1,559 | 651 | 150 | 53 | 161 | 299 | 486 | 1,014 | 66 | 53 |
| Washington | 12,473 | 3.4 | 60,295 | 292 | 5,661 | 919 | 425 | 1,315 | 2,006 | 2,389 | 8,319 | 505 | 425 |
| Wayne | 2,323 | 3.6 | 45,221 | 1,333 | 684 | 156 | 55 | 235 | 459 | 621 | 1,129 | 77 | 55 |
| Westmoreland | 18,750 | 3.5 | 53,740 | 559 | 6,752 | 1,234 | 543 | 1,062 | 3,024 | 4,332 | 9,590 | 644 | 543 |
| Wyoming | 1,262 | 2.5 | 47,099 | 1,120 | 515 | 98 | 41 | 145 | 208 | 301 | 799 | 50 | 41 |
| York | 23,165 | 3.7 | 51,585 | 715 | 9,659 | 1,758 | 769 | 1,461 | 3,541 | 4,536 | 13,646 | 847 | 769 |
| **RHODE ISLAND** | 59,707 | 3.7 | 56,426 | X | 28,668 | 4,545 | 2,283 | 4,104 | 10,692 | 11,768 | 39,600 | 2,790 | 2,283 |
| Bristol | 3,829 | 2.9 | 78,975 | 68 | 706 | 122 | 57 | 247 | 959 | 477 | 1,132 | 87 | 57 |
| Kent | 10,001 | 4.0 | 60,874 | 278 | 4,309 | 648 | 349 | 597 | 1,586 | 1,879 | 5,903 | 430 | 349 |
| Newport | 5,819 | 2.9 | 70,893 | 122 | 2,700 | 514 | 227 | 498 | 1,580 | 914 | 3,938 | 263 | 227 |
| Providence | 31,558 | 3.8 | 49,392 | 889 | 17,978 | 2,699 | 1,407 | 1,994 | 4,751 | 7,145 | 24,077 | 1,688 | 1,407 |
| Washington | 8,500 | 3.6 | 67,688 | 150 | 2,975 | 563 | 244 | 768 | 1,816 | 1,353 | 4,549 | 321 | 244 |
| **SOUTH CAROLINA** | 233,948 | 4.8 | 45,359 | X | 108,607 | 19,422 | 8,166 | 15,824 | 44,552 | 51,047 | 152,018 | 9,964 | 8,166 |
| Abbeville | 864 | 4.1 | 35,219 | 2,657 | 252 | 57 | 20 | 50 | 118 | 287 | 379 | 31 | 20 |
| Aiken | 7,578 | 4.7 | 44,349 | 1,438 | 3,483 | 541 | 257 | 325 | 1,320 | 1,799 | 4,606 | 321 | 257 |
| Allendale | 292 | 0.5 | 33,582 | 2,840 | 124 | 31 | 9 | 15 | 46 | 108 | 180 | 13 | 9 |
| Anderson | 8,339 | 4.1 | 41,170 | 1,897 | 3,080 | 590 | 232 | 409 | 1,269 | 2,077 | 4,311 | 308 | 232 |
| Bamberg | 495 | 3.3 | 35,203 | 2,659 | 149 | 35 | 11 | 30 | 73 | 176 | 226 | 17 | 11 |
| Barnwell | 743 | 3.7 | 35,606 | 2,604 | 223 | 48 | 17 | 21 | 105 | 238 | 309 | 24 | 17 |
| Beaufort | 11,396 | 5.4 | 59,318 | 327 | 3,831 | 723 | 305 | 585 | 4,380 | 2,143 | 5,444 | 365 | 305 |
| Berkeley | 9,655 | 6.9 | 42,365 | 1,710 | 3,396 | 531 | 248 | 564 | 1,446 | 1,824 | 4,739 | 319 | 248 |
| Calhoun | 599 | 3.6 | 41,136 | 1,902 | 228 | 48 | 18 | 40 | 97 | 169 | 333 | 24 | 18 |
| Charleston | 26,289 | 5.1 | 63,901 | 210 | 15,849 | 2,776 | 1,193 | 2,775 | 7,161 | 3,539 | 22,593 | 1,324 | 1,193 |
| Cherokee | 1,941 | 3.5 | 33,875 | 2,816 | 858 | 157 | 68 | 95 | 262 | 587 | 1,178 | 85 | 68 |
| Chester | 1,136 | 4.0 | 35,227 | 2,655 | 496 | 99 | 36 | 34 | 165 | 366 | 666 | 49 | 36 |
| Chesterfield | 1,472 | 2.8 | 32,249 | 2,928 | 656 | 123 | 52 | 73 | 184 | 475 | 903 | 65 | 52 |
| Clarendon | 1,163 | 3.5 | 34,475 | 2,746 | 257 | 62 | 19 | 57 | 173 | 414 | 396 | 36 | 19 |
| Colleton | 1,414 | 4.4 | 37,521 | 2,385 | 445 | 85 | 33 | 74 | 234 | 474 | 637 | 51 | 33 |
| Darlington | 2,611 | 4.0 | 39,195 | 2,163 | 1,110 | 209 | 83 | 100 | 367 | 757 | 1,502 | 107 | 83 |
| Dillon | 921 | 5.6 | 30,203 | 3,029 | 354 | 66 | 29 | 30 | 109 | 342 | 479 | 37 | 29 |
| Dorchester | 6,663 | 5.7 | 40,926 | 1,928 | 1,653 | 308 | 125 | 322 | 941 | 1,428 | 2,407 | 174 | 125 |
| Edgefield | 1,017 | 4.9 | 37,294 | 2,414 | 249 | 56 | 19 | 47 | 151 | 263 | 371 | 28 | 19 |
| Fairfield | 845 | 3.3 | 37,834 | 2,341 | 316 | 73 | 23 | 24 | 120 | 272 | 437 | 32 | 23 |
| Florence | 6,085 | 4.7 | 44,001 | 1,484 | 3,374 | 637 | 249 | 278 | 1,015 | 1,541 | 4,539 | 295 | 249 |
| Georgetown | 3,093 | 4.4 | 49,341 | 894 | 1,136 | 202 | 83 | 211 | 865 | 856 | 1,632 | 122 | 83 |
| Greenville | 26,073 | 4.2 | 49,801 | 843 | 15,668 | 2,401 | 1,160 | 1,837 | 4,719 | 4,485 | 21,066 | 1,324 | 1,160 |
| Greenwood | 2,704 | 3.4 | 38,180 | 2,307 | 1,359 | 289 | 103 | 108 | 502 | 789 | 1,859 | 126 | 103 |
| Hampton | 660 | 3.7 | 34,322 | 2,768 | 221 | 48 | 16 | 25 | 93 | 221 | 310 | 24 | 16 |
| Horry | 13,819 | 6.2 | 39,028 | 2,189 | 5,496 | 915 | 423 | 1,081 | 2,788 | 4,179 | 7,915 | 596 | 423 |
| Jasper | 947 | 8.2 | 31,488 | 2,979 | 435 | 77 | 33 | 59 | 149 | 301 | 604 | 44 | 33 |
| Kershaw | 2,857 | 4.6 | 42,929 | 1,627 | 774 | 154 | 59 | 161 | 403 | 735 | 1,148 | 85 | 59 |
| Lancaster | 5,135 | 5.7 | 52,396 | 651 | 1,489 | 250 | 106 | 1,200 | 626 | 1,003 | 3,044 | 197 | 106 |
| Laurens | 2,392 | 3.2 | 35,441 | 2,628 | 967 | 188 | 73 | 87 | 327 | 798 | 1,315 | 98 | 73 |
| Lee | 566 | 3.2 | 33,605 | 2,834 | 158 | 34 | 12 | 16 | 72 | 206 | 220 | 19 | 12 |
| Lexington | 14,337 | 4.8 | 47,992 | 1,014 | 6,071 | 1,106 | 457 | 1,071 | 2,223 | 2,615 | 8,705 | 556 | 457 |
| McCormick | 379 | 2.7 | 40,004 | 2,049 | 70 | 18 | 5 | 17 | 96 | 151 | 110 | 12 | 5 |
| Marion | 1,029 | 4.7 | 33,569 | 2,841 | 271 | 61 | 22 | 36 | 132 | 388 | 390 | 33 | 22 |
| Marlboro | 852 | 0.3 | 32,637 | 2,912 | 332 | 69 | 25 | 33 | 107 | 300 | 459 | 35 | 25 |
| Newberry | 1,500 | 2.8 | 39,024 | 2,190 | 630 | 131 | 50 | 35 | 242 | 423 | 846 | 61 | 50 |
| Oconee | 3,580 | 4.2 | 45,004 | 1,359 | 1,290 | 269 | 96 | 163 | 771 | 915 | 1,818 | 134 | 96 |
| Orangeburg | 3,073 | 3.1 | 35,656 | 2,598 | 1,278 | 262 | 99 | 99 | 462 | 1,045 | 1,738 | 127 | 99 |
| Pickens | 5,018 | 4.2 | 39,549 | 2,108 | 1,777 | 373 | 129 | 241 | 949 | 1,192 | 2,520 | 178 | 129 |
| Richland | 19,665 | 4.2 | 47,299 | 1,097 | 12,633 | 2,428 | 943 | 1,637 | 3,473 | 3,795 | 17,641 | 1,042 | 943 |
| Saluda | 703 | 3.0 | 34,353 | 2,762 | 183 | 39 | 15 | 19 | 103 | 209 | 256 | 21 | 15 |
| Spartanburg | 14,125 | 4.4 | 44,169 | 1,461 | 7,729 | 1,338 | 577 | 788 | 2,806 | 2,994 | 10,431 | 678 | 577 |

1. Based on the resident population estimated as of July 1 of the year shown.

# Table B. States and Counties — Earnings, Social Security, and Housing

| STATE County | Farm | Mining, quarrying, and extractions | Construction | Manufacturing | Information; professional, scientific, technical services | Retail trade | Finance, insurance, real estate, and leasing | Health care and social assistance | Government | Number | Rate[1] | Supplemental Security Income recipients, 2019 | Total | Percent change, 2010-2020 |
|---|---|---|---|---|---|---|---|---|---|---|---|---|---|---|
| | 75 | 76 | 77 | 78 | 79 | 80 | 81 | 82 | 83 | 84 | 85 | 86 | 87 | 88 |
| **PENNSYLVANIA—Cont'd** | | | | | | | | | | | | | | |
| Potter | 2.4 | 0.7 | 6.0 | 8.9 | 24.0 | 3.7 | 2.6 | D | 14.1 | 5,055 | 306 | 468 | 13,003 | 0.6 |
| Schuylkill | 0.6 | 1.3 | 4.8 | 24.4 | 3.5 | 6.1 | 3.0 | 13.2 | 17.1 | 38,725 | 274 | 3,707 | 70,058 | 1.1 |
| Snyder | 1.7 | D | 7.9 | 21.5 | 4.0 | 9.5 | 2.8 | D | 16.4 | 9,210 | 228 | 602 | 16,561 | 3.3 |
| Somerset | 1.2 | 5.4 | 6.4 | 11.3 | 4.4 | 6.1 | 4.5 | 12.9 | 20.1 | 20,840 | 284 | 1,956 | 38,632 | 1.4 |
| Sullivan | 2.1 | D | 14.8 | D | D | 6.1 | D | D | 25.2 | 2,000 | 336 | 135 | 6,388 | 1.3 |
| Susquehanna | 1.9 | 9.3 | 11.5 | 3.8 | 8.6 | 6.5 | 3.0 | 7.8 | 17.5 | 11,150 | 277 | 833 | 23,429 | 2.1 |
| Tioga | 1.0 | 7.7 | 5.1 | 12.1 | 5.4 | 7.2 | 4.1 | D | 19.5 | 11,485 | 282 | 1,034 | 21,918 | 2.5 |
| Union | 0.9 | D | 5.9 | 8.1 | 2.7 | 6.1 | 3.5 | D | 20.0 | 8,955 | 201 | 531 | 17,552 | 3.2 |
| Venango | 0.1 | 0.6 | 4.2 | 24.4 | 2.6 | 6.7 | 3.5 | 15.2 | 20.2 | 15,905 | 313 | 1,943 | 27,661 | 0.7 |
| Warren | 0.6 | 3.1 | 3.8 | 18.3 | 3.2 | 7.1 | 7.9 | 14.3 | 14.8 | 11,495 | 293 | 879 | 23,619 | 0.3 |
| Washington | 0.0 | 6.9 | 10.2 | 9.3 | 10.2 | 4.8 | 7.1 | 10.5 | 8.8 | 54,795 | 265 | 4,695 | 96,912 | 4.2 |
| Wayne | 0.5 | 0.4 | 12.9 | 3.5 | 4.8 | 8.6 | 5.8 | 13.1 | 24.4 | 15,005 | 293 | 1,153 | 32,550 | 2.8 |
| Westmoreland | 0.2 | 1.4 | 8.3 | 14.8 | 7.6 | 7.6 | 4.3 | 12.8 | 12.5 | 99,770 | 286 | 7,610 | 171,277 | 1.9 |
| Wyoming | -0.3 | 4.8 | 7.3 | 23.5 | D | 6.0 | 2.5 | D | 9.9 | 7,265 | 271 | 578 | 13,539 | 2.1 |
| York | 0.2 | 0.2 | 9.0 | 18.6 | 6.5 | 5.6 | 4.6 | 14.1 | 13.1 | 99,170 | 221 | 8,353 | 186,871 | 4.6 |
| **RHODE ISLAND** | 0.1 | D | 6.1 | 8.1 | 9.9 | 6.2 | 10.3 | 13.9 | 16.8 | 228,257 | 216 | 32,552 | 471,325 | 1.7 |
| Bristol | 0.0 | 0.0 | 8.2 | 10.5 | 10.2 | 5.2 | 6.4 | 11.4 | 15.8 | 11,450 | 236 | 628 | 21,265 | 2.1 |
| Kent | 0.0 | D | 6.3 | 9.7 | 9.9 | 8.8 | 8.2 | 15.6 | 12.6 | 41,070 | 250 | 3,581 | 74,789 | 1.5 |
| Newport | 0.2 | D | 5.5 | 6.8 | 12.1 | 5.8 | 6.4 | 7.0 | 33.6 | 19,625 | 238 | 1,370 | 42,906 | 2.7 |
| Providence | 0.0 | D | 5.9 | 6.1 | 9.9 | 5.0 | 12.6 | 15.3 | 14.5 | 125,040 | 196 | 25,378 | 267,256 | 0.9 |
| Washington | 0.2 | 0.1 | 6.9 | 16.9 | 7.7 | 9.8 | 4.9 | 10.6 | 19.7 | 31,070 | 247 | 1,595 | 65,109 | 4.7 |
| **SOUTH CAROLINA** | 0.1 | 0.1 | 6.7 | 13.9 | 10.2 | 6.7 | 6.9 | 9.3 | 19.7 | 1,174,399 | 228 | 114,706 | 2,386,385 | 11.6 |
| Abbeville | 0.0 | 0.0 | 7.6 | 36.2 | D | 3.4 | D | D | 23.9 | 6,885 | 280 | 551 | 12,289 | 1.7 |
| Aiken | -0.4 | 0.2 | 12.7 | 15.8 | 7.2 | 6.1 | 4.7 | 8.1 | 11.4 | 42,275 | 247 | 3,841 | 79,427 | 9.9 |
| Allendale | 3.1 | 0.0 | 1.4 | 34.7 | D | 2.7 | D | D | 33.4 | 2,315 | 268 | 576 | 4,513 | 0.6 |
| Anderson | -0.2 | 0.3 | 5.4 | 25.9 | 5.2 | 8.1 | 3.6 | 7.9 | 20.7 | 49,895 | 246 | 4,343 | 90,973 | 7.4 |
| Bamberg | 1.6 | D | D | 18.0 | 6.7 | 5.5 | D | 10.7 | 21.5 | 3,780 | 268 | 677 | 7,720 | 0.2 |
| Barnwell | -0.2 | D | 8.8 | 31.0 | D | 7.4 | 2.2 | D | 25.0 | 5,145 | 246 | 1,085 | 10,622 | 1.3 |
| Beaufort | 0.1 | D | 7.7 | 1.0 | 10.3 | 7.8 | 6.7 | 8.7 | 31.3 | 52,075 | 270 | 1,934 | 104,966 | 12.9 |
| Berkeley | 0.0 | D | 10.3 | 15.1 | 21.8 | 8.1 | 4.7 | 3.6 | 14.5 | 40,520 | 177 | 3,402 | 90,772 | 23.3 |
| Calhoun | 1.5 | 0.0 | 13.4 | 34.1 | D | 1.8 | D | D | 13.2 | 4,115 | 284 | 365 | 7,559 | 3.0 |
| Charleston | 0.0 | 0.0 | 7.6 | 8.2 | 13.5 | 6.0 | 6.4 | 11.1 | 23.1 | 76,635 | 186 | 6,845 | 198,275 | 16.7 |
| Cherokee | 0.3 | D | 7.7 | 36.0 | D | 6.7 | 2.3 | D | 13.3 | 14,225 | 248 | 1,546 | 24,979 | 4.0 |
| Chester | -0.2 | 0.0 | 5.5 | 41.3 | D | 4.0 | D | 3.6 | 17.7 | 8,695 | 270 | 1,152 | 14,844 | 0.9 |
| Chesterfield | 1.1 | 0.4 | 5.2 | 37.1 | D | 4.4 | 2.2 | 8.7 | 13.1 | 11,290 | 247 | 1,531 | 21,985 | 2.4 |
| Clarendon | -0.5 | 0.0 | 4.5 | 5.9 | 5.1 | 13.3 | 3.9 | 8.9 | 34.7 | 9,905 | 293 | 1,441 | 18,066 | 3.5 |
| Colleton | 0.6 | 0.3 | 8.0 | 5.9 | D | 8.0 | 4.1 | 14.5 | 19.3 | 11,015 | 292 | 1,611 | 20,363 | 2.4 |
| Darlington | 0.7 | 0.0 | 7.5 | 26.6 | 1.9 | 5.3 | 2.3 | 9.8 | 13.4 | 17,430 | 261 | 2,775 | 31,135 | 2.9 |
| Dillon | -1.3 | 0.0 | 1.9 | 22.6 | 1.9 | 9.0 | 3.6 | D | 17.2 | 7,415 | 244 | 1,471 | 13,883 | 1.0 |
| Dorchester | -0.1 | 0.1 | 10.0 | 19.5 | 6.9 | 7.7 | 4.2 | 7.2 | 17.5 | 30,030 | 184 | 2,769 | 63,244 | 15.0 |
| Edgefield | 3.3 | D | 7.7 | 17.3 | D | 4.9 | 2.4 | D | 32.3 | 6,170 | 226 | 840 | 11,324 | 7.3 |
| Fairfield | 0.2 | D | 2.8 | 11.1 | D | 3.3 | 1.2 | D | 19.6 | 6,165 | 276 | 782 | 12,067 | 3.2 |
| Florence | 0.3 | 0.0 | 3.3 | 12.2 | 7.2 | 7.3 | 12.0 | 12.4 | 21.9 | 32,535 | 236 | 5,656 | 62,815 | 7.0 |
| Georgetown | 0.3 | D | 6.2 | 13.4 | 6.3 | 6.7 | 10.1 | 12.3 | 21.7 | 21,740 | 345 | 1,778 | 36,506 | 8.2 |
| Greenville | 0.0 | 0.0 | 6.8 | 12.3 | 13.0 | 6.0 | 8.9 | 10.2 | 11.5 | 104,950 | 200 | 9,464 | 222,243 | 13.7 |
| Greenwood | 0.2 | D | 4.3 | 26.1 | 2.9 | 6.7 | 3.7 | 12.4 | 24.5 | 17,515 | 247 | 1,784 | 31,715 | 2.1 |
| Hampton | 1.2 | 0.0 | 5.4 | D | D | 6.4 | D | D | 31.0 | 5,010 | 259 | 968 | 9,236 | 1.0 |
| Horry | -0.1 | 0.1 | 8.4 | 2.7 | 6.8 | 12.0 | 9.9 | 11.3 | 15.7 | 108,015 | 305 | 5,882 | 219,318 | 18.0 |
| Jasper | 0.5 | 0.0 | 19.9 | 3.0 | 4.1 | 14.8 | 3.1 | 13.9 | 14.4 | 7,265 | 240 | 654 | 13,119 | 27.3 |
| Kershaw | 1.2 | 0.4 | 8.2 | 20.3 | 5.1 | 10.9 | 7.0 | 12.0 | 13.8 | 16,285 | 244 | 1,478 | 30,056 | 9.4 |
| Lancaster | 0.1 | D | 2.6 | 8.0 | 47.1 | 4.2 | 3.0 | 6.0 | 9.8 | 24,625 | 251 | 1,772 | 40,779 | 24.8 |
| Laurens | 0.2 | D | 4.5 | 37.7 | D | 4.2 | 2.0 | 7.2 | 17.7 | 18,180 | 270 | 1,940 | 31,958 | 4.1 |
| Lee | 0.0 | 0.0 | 2.3 | 14.9 | D | 7.7 | D | D | 27.0 | 4,655 | 276 | 789 | 7,813 | 0.5 |
| Lexington | 0.2 | 0.1 | 7.3 | 14.3 | 7.2 | 8.1 | 5.6 | 7.0 | 19.5 | 59,955 | 200 | 4,535 | 130,238 | 14.2 |
| McCormick | 1.2 | 0.0 | D | D | D | 3.4 | D | 7.3 | 42.0 | 3,965 | 419 | 302 | 5,806 | 6.5 |
| Marion | 0.9 | D | 3.5 | 10.3 | D | 8.6 | 6.0 | D | 25.8 | 8,690 | 284 | 1,578 | 15,202 | 1.7 |
| Marlboro | 2.7 | D | 1.2 | 39.9 | 1.2 | 7.6 | 2.9 | D | 25.8 | 7,190 | 275 | 1,383 | 12,077 | 0.0 |
| Newberry | -0.3 | D | 7.3 | 41.5 | D | 5.5 | 1.9 | 5.0 | 17.8 | 10,155 | 264 | 951 | 18,542 | 3.4 |
| Oconee | -0.4 | D | 6.0 | 27.6 | 4.7 | 7.5 | 3.9 | 7.3 | 15.3 | 23,390 | 294 | 1,469 | 41,643 | 7.4 |
| Orangeburg | 0.2 | D | 4.1 | 23.6 | 2.8 | 8.1 | 3.3 | 8.2 | 26.4 | 22,810 | 265 | 3,790 | 43,014 | 1.2 |
| Pickens | -0.1 | D | 5.5 | 17.6 | 4.7 | 7.7 | 3.9 | 7.5 | 35.2 | 28,425 | 224 | 2,223 | 56,232 | 9.6 |
| Richland | 0.0 | 0.1 | 4.1 | 5.7 | 11.7 | 5.1 | 12.4 | 11.1 | 28.3 | 73,155 | 175 | 8,604 | 178,810 | 10.6 |
| Saluda | -0.4 | D | 5.0 | 43.1 | D | 3.9 | D | D | 22.7 | 5,005 | 246 | 424 | 9,557 | 3.0 |
| Spartanburg | 0.0 | D | 6.7 | 27.5 | 5.1 | 5.7 | 4.2 | 6.2 | 15.8 | 71,160 | 222 | 7,223 | 136,704 | 11.5 |

1. Per 1,000 resident population estimated as of July 1 of the year shown.

| STATE County | Housing units, 2015-2019 — Occupied units — Owner-occupied — Total | Percent | Median value[1] | Median owner cost as a percent of income — With a mortgage | Without a mortgage[2] | Renter-occupied — Median rent[3] | Median rent as a percent of income[2] | Sub-standard units[4] (percent) | Civilian labor force, 2020 — Total | Percent change, 2019-2020 | Unemployment — Total | Rate[5] | Civilian employment[6], 2015-2019 — Total | Percent — Management, business, science, and arts | Construction, production, and maintenance occupations |
|---|---|---|---|---|---|---|---|---|---|---|---|---|---|---|---|
| | 89 | 90 | 91 | 92 | 93 | 94 | 95 | 96 | 97 | 98 | 99 | 100 | 101 | 102 | 103 |
| **PENNSYLVANIA—Cont'd** | | | | | | | | | | | | | | | |
| Potter | 6,480 | 76.2 | 109,600 | 20.5 | 12.9 | 663 | 27.3 | 1.7 | 7,185 | -0.7 | 669 | 9.3 | 6,654 | 30.1 | 35.1 |
| Schuylkill | 58,749 | 75.7 | 102,100 | 19.0 | 13.5 | 703 | 29.0 | 1.0 | 66,129 | -0.7 | 6,107 | 9.2 | 62,570 | 29.1 | 35.8 |
| Snyder | 14,794 | 72.2 | 166,100 | 20.0 | 10.9 | 727 | 24.1 | 2.2 | 19,389 | -3.6 | 1,483 | 7.6 | 20,143 | 30.0 | 33.3 |
| Somerset | 29,644 | 78.1 | 107,000 | 19.1 | 12.7 | 607 | 27.7 | 1.6 | 32,661 | -2.6 | 3,094 | 9.5 | 32,230 | 30.4 | 31.7 |
| Sullivan | 2,723 | 82.0 | 154,400 | 20.8 | 12.2 | 643 | 24.4 | 1.1 | 2,597 | -2.3 | 221 | 8.5 | 2,718 | 24.9 | 38.6 |
| Susquehanna | 17,235 | 77.1 | 167,700 | 21.3 | 12.0 | 727 | 26.7 | 1.3 | 19,774 | -3.8 | 1,451 | 7.3 | 18,046 | 29.3 | 34.7 |
| Tioga | 16,310 | 72.8 | 142,100 | 20.7 | 12.9 | 728 | 28.4 | 1.3 | 18,831 | -1.1 | 1,703 | 9.0 | 17,977 | 29.4 | 33.3 |
| Union | 14,533 | 72.4 | 189,500 | 19.0 | 12.7 | 769 | 28.7 | 2.3 | 19,314 | -3.1 | 1,305 | 6.8 | 17,908 | 39.1 | 25.0 |
| Venango | 22,050 | 75.1 | 89,400 | 17.5 | 10.2 | 622 | 26.7 | 1.9 | 22,047 | -1.7 | 1,968 | 8.9 | 22,682 | 28.4 | 30.4 |
| Warren | 17,115 | 76.9 | 98,000 | 17.7 | 11.1 | 614 | 24.7 | 1.8 | 18,668 | -1.6 | 1,505 | 8.1 | 18,266 | 30.2 | 33.4 |
| Washington | 84,948 | 75.6 | 167,900 | 18.7 | 10.9 | 772 | 26.1 | 1.3 | 105,089 | -2.1 | 9,871 | 9.4 | 99,980 | 35.9 | 23.9 |
| Wayne | 18,841 | 81.0 | 181,300 | 23.1 | 12.6 | 820 | 28.2 | 1.4 | 22,103 | -1.8 | 2,027 | 9.2 | 21,111 | 31.0 | 29.5 |
| Westmoreland | 152,283 | 77.2 | 153,100 | 18.6 | 11.4 | 721 | 26.9 | 1.1 | 176,723 | -2.4 | 16,077 | 9.1 | 172,201 | 37.1 | 24.5 |
| Wyoming | 10,790 | 77.2 | 167,900 | 21.0 | 12.0 | 799 | 26.7 | 1.4 | 13,254 | -2.3 | 1,110 | 8.4 | 12,835 | 26.2 | 35.1 |
| York | 172,421 | 74.9 | 177,100 | 21.4 | 13.1 | 957 | 29.5 | 1.7 | 232,812 | -1.4 | 18,647 | 8.0 | 224,825 | 34.8 | 27.9 |
| **RHODE ISLAND** | 410,489 | 60.8 | 261,900 | 22.9 | 13.8 | 1,004 | 29.1 | 2.0 | 541,680 | -3.1 | 50,835 | 9.4 | 533,878 | 39.9 | 19.3 |
| Bristol | 19,217 | 70.7 | 358,100 | 22.1 | 14.9 | 1,037 | 29.6 | 0.7 | 24,895 | -4.3 | 1,896 | 7.6 | 25,125 | 50.2 | 12.6 |
| Kent | 69,422 | 70.1 | 236,300 | 22.8 | 14.3 | 1,079 | 28.7 | 1.7 | 88,283 | -3.3 | 7,721 | 8.7 | 88,514 | 42.5 | 18.8 |
| Newport | 34,777 | 63.2 | 387,900 | 23.5 | 13.9 | 1,285 | 28.2 | 1.2 | 42,502 | -3.8 | 3,464 | 8.2 | 41,244 | 45.3 | 15.8 |
| Providence | 237,971 | 54.2 | 233,500 | 23.4 | 13.7 | 967 | 29.3 | 2.6 | 320,264 | -2.4 | 32,616 | 10.2 | 314,000 | 36.6 | 21.2 |
| Washington | 49,102 | 74.0 | 343,000 | 21.8 | 12.9 | 1,133 | 28.9 | 1.1 | 65,736 | -4.9 | 5,139 | 7.8 | 64,995 | 45.0 | 15.8 |
| **SOUTH CAROLINA** | 1,921,862 | 69.4 | 162,300 | 19.9 | 10.0 | 894 | 29.6 | 2.2 | 2,384,590 | 0.7 | 147,183 | 6.2 | 2,275,531 | 34.9 | 25.1 |
| Abbeville | 9,660 | 75.0 | 90,800 | 21.2 | 11.8 | 681 | 32.1 | 1.5 | 10,124 | 1.0 | 670 | 6.6 | 10,104 | 26.9 | 35.9 |
| Aiken | 67,598 | 72.6 | 147,300 | 18.8 | 10.0 | 819 | 30.7 | 1.9 | 75,531 | 0.8 | 3,788 | 5.0 | 71,279 | 34.7 | 26.5 |
| Allendale | 3,365 | 68.2 | 49,200 | 20.7 | 13.2 | 602 | 35.6 | 3.6 | 2,641 | -0.2 | 249 | 9.4 | 2,890 | 26.9 | 31.2 |
| Anderson | 76,798 | 71.7 | 145,800 | 18.2 | 10.0 | 778 | 28.2 | 2.5 | 90,877 | 0.2 | 5,402 | 5.9 | 91,058 | 33.0 | 28.7 |
| Bamberg | 5,334 | 70.6 | 75,400 | 20.8 | 13.5 | 794 | 41.5 | 2.7 | 4,846 | -1.6 | 438 | 9.0 | 5,354 | 29.2 | 33.3 |
| Barnwell | 8,360 | 69.6 | 81,900 | 20.3 | 12.0 | 621 | 30.7 | 4.6 | 7,972 | -2.0 | 587 | 7.4 | 8,004 | 24.4 | 34.5 |
| Beaufort | 71,477 | 73.3 | 298,100 | 24.3 | 10.5 | 1,202 | 31.2 | 1.8 | 77,127 | -0.2 | 4,166 | 5.4 | 76,980 | 37.0 | 19.5 |
| Berkeley | 76,881 | 71.9 | 185,500 | 20.4 | 10.0 | 1,109 | 28.0 | 2.2 | 104,916 | -0.5 | 5,874 | 5.6 | 100,304 | 33.9 | 25.9 |
| Calhoun | 6,179 | 80.3 | 107,200 | 20.6 | 10.2 | 666 | 43.1 | 3.0 | 6,572 | 0.6 | 399 | 6.1 | 6,250 | 36.1 | 29.0 |
| Charleston | 159,195 | 61.6 | 315,600 | 21.9 | 11.3 | 1,190 | 30.6 | 1.5 | 211,939 | 0.5 | 13,387 | 6.3 | 204,353 | 43.1 | 17.0 |
| Cherokee | 20,699 | 68.9 | 101,500 | 20.0 | 10.0 | 712 | 31.3 | 0.8 | 25,302 | 0.7 | 2,039 | 8.1 | 23,119 | 23.7 | 37.6 |
| Chester | 12,653 | 76.1 | 95,600 | 18.8 | 10.0 | 661 | 31.1 | 1.8 | 13,646 | 3.5 | 1,220 | 8.9 | 13,196 | 24.0 | 39.3 |
| Chesterfield | 17,900 | 71.1 | 81,500 | 18.7 | 10.9 | 670 | 24.2 | 4.9 | 21,834 | -0.6 | 1,323 | 6.1 | 19,346 | 23.1 | 44.9 |
| Clarendon | 13,161 | 77.0 | 102,000 | 21.3 | 11.3 | 606 | 29.3 | 3.4 | 12,737 | 0.8 | 863 | 6.8 | 11,923 | 28.4 | 32.5 |
| Colleton | 15,075 | 74.9 | 90,400 | 25.4 | 11.0 | 749 | 27.7 | 1.9 | 16,732 | -0.1 | 1,015 | 6.1 | 15,900 | 24.5 | 36.0 |
| Darlington | 26,484 | 69.6 | 92,600 | 18.9 | 10.9 | 675 | 31.5 | 2.4 | 30,521 | 0.8 | 1,979 | 6.5 | 26,910 | 34.5 | 27.3 |
| Dillon | 11,029 | 59.4 | 79,300 | 19.9 | 13.1 | 590 | 35.2 | 4.3 | 13,238 | 2.5 | 814 | 6.1 | 11,808 | 24.7 | 34.6 |
| Dorchester | 55,351 | 71.7 | 198,800 | 21.9 | 10.7 | 1,099 | 29.5 | 2.0 | 77,693 | -0.2 | 4,519 | 5.8 | 74,909 | 37.4 | 22.9 |
| Edgefield | 9,176 | 74.1 | 129,300 | 19.9 | 10.2 | 637 | 32.0 | 3.1 | 10,799 | 0.8 | 552 | 5.1 | 10,602 | 28.6 | 34.3 |
| Fairfield | 9,191 | 75.3 | 103,000 | 19.5 | 11.5 | 712 | 36.8 | 2.8 | 9,637 | 1.7 | 760 | 7.9 | 9,037 | 28.0 | 30.2 |
| Florence | 52,188 | 65.7 | 131,900 | 18.8 | 10.5 | 772 | 28.1 | 3.1 | 67,150 | 0.4 | 3,744 | 5.6 | 60,708 | 34.9 | 24.8 |
| Georgetown | 25,498 | 78.8 | 188,800 | 24.5 | 12.0 | 901 | 29.5 | 1.4 | 26,175 | 1.9 | 1,965 | 7.5 | 25,080 | 30.9 | 23.6 |
| Greenville | 192,975 | 67.5 | 183,800 | 18.7 | 10.0 | 918 | 27.5 | 2.3 | 255,984 | 0.2 | 14,613 | 5.7 | 247,829 | 39.3 | 23.0 |
| Greenwood | 27,612 | 64.0 | 122,800 | 18.4 | 10.0 | 733 | 30.2 | 2.3 | 31,628 | 0.9 | 2,012 | 6.4 | 30,493 | 30.0 | 30.8 |
| Hampton | 6,993 | 76.0 | 72,300 | 21.0 | 12.5 | 607 | 25.1 | 1.8 | 8,234 | -1.4 | 411 | 5.0 | 7,470 | 26.5 | 31.7 |
| Horry | 131,143 | 71.8 | 181,500 | 22.8 | 10.6 | 959 | 31.2 | 1.6 | 147,697 | 0.2 | 12,741 | 8.6 | 147,486 | 29.0 | 20.5 |
| Jasper | 10,269 | 71.7 | 166,200 | 23.5 | 11.3 | 929 | 33.1 | 4.4 | 12,804 | -0.3 | 657 | 5.1 | 12,309 | 22.3 | 31.4 |
| Kershaw | 24,980 | 82.1 | 131,100 | 20.4 | 10.5 | 775 | 28.9 | 1.3 | 29,490 | 0.5 | 1,566 | 5.3 | 28,257 | 34.4 | 27.6 |
| Lancaster | 33,899 | 79.7 | 199,800 | 19.2 | 11.0 | 776 | 29.5 | 2.0 | 42,533 | 2.1 | 2,923 | 6.9 | 38,913 | 33.7 | 27.3 |
| Laurens | 25,563 | 69.8 | 101,200 | 21.4 | 10.0 | 734 | 29.2 | 3.2 | 30,197 | 0.9 | 2,075 | 6.9 | 28,437 | 27.0 | 34.3 |
| Lee | 6,423 | 74.7 | 78,800 | 20.5 | 12.2 | 656 | 38.6 | 2.8 | 6,812 | 3.9 | 519 | 7.6 | 5,940 | 25.5 | 36.8 |
| Lexington | 113,104 | 74.2 | 157,000 | 18.9 | 10.0 | 917 | 29.4 | 2.4 | 150,439 | 0.2 | 6,557 | 4.4 | 140,808 | 38.2 | 23.7 |
| McCormick | 3,957 | 76.6 | 116,500 | 23.2 | 12.7 | 720 | 26.9 | 1.3 | 3,359 | -0.4 | 224 | 6.7 | 2,803 | 30.1 | 30.2 |
| Marion | 11,600 | 66.0 | 76,700 | 20.0 | 13.2 | 594 | 27.7 | 3.3 | 13,251 | 2.5 | 1,037 | 7.8 | 11,558 | 28.8 | 36.7 |
| Marlboro | 9,613 | 62.9 | 67,500 | 19.7 | 12.8 | 610 | 28.6 | 3.3 | 9,294 | 2.2 | 911 | 9.8 | 9,161 | 25.2 | 38.4 |
| Newberry | 14,810 | 73.3 | 116,300 | 20.1 | 10.9 | 765 | 32.0 | 2.7 | 19,201 | 0.8 | 921 | 4.8 | 16,841 | 25.3 | 35.7 |
| Oconee | 31,978 | 72.7 | 159,800 | 19.0 | 10.6 | 762 | 31.0 | 2.6 | 35,056 | 0.7 | 1,876 | 5.4 | 32,106 | 31.9 | 29.0 |
| Orangeburg | 33,060 | 67.3 | 92,700 | 23.2 | 12.2 | 687 | 31.2 | 2.3 | 34,945 | 1.2 | 3,155 | 9.0 | 35,030 | 27.3 | 32.0 |
| Pickens | 47,934 | 67.1 | 147,700 | 18.9 | 10.0 | 768 | 30.2 | 2.5 | 57,109 | -0.6 | 3,037 | 5.3 | 56,013 | 36.2 | 27.3 |
| Richland | 151,853 | 59.5 | 163,100 | 20.0 | 10.0 | 972 | 31.3 | 2.1 | 199,718 | 1.4 | 11,661 | 5.8 | 195,730 | 40.0 | 17.6 |
| Saluda | 7,094 | 74.1 | 108,600 | 22.8 | 10.0 | 661 | 27.9 | 3.1 | 8,698 | | 370 | 4.3 | 8,392 | 27.3 | 38.1 |
| Spartanburg | 116,645 | 70.6 | 141,400 | 18.8 | 10.0 | 794 | 28.4 | 2.8 | 158,701 | 2.6 | 10,427 | 6.6 | 141,860 | 31.2 | 31.5 |

1. Specified owner-occupied units. lacking complete plumbing facilities.  2. A value of 10.0 represents 10 percent or less; a value of 50.0 represents 50 percent or more.  3. Specified renter-occupied units.  4. Overcrowded or
5. Percent of civilian labor force.  6. Civilian employed persons 16 years old and over.

# Table B. States and Counties — Nonfarm Employment and Agriculture

| | Private nonfarm establishments, employment and payroll, 2019 | | | | | | | | | Agriculture, 2017 | | | |
| STATE County | Employment | | | | | | | Annual payroll | | Farms | | | Farm producers whose primary occupation is farming (percent) |
| | Number of establishments | Total | Health care and social assistance | Manufacturing | Retail trade | Finance and insurance | Professional, scientific, and technical services | Total (mil dol) | Average per employee (dollars) | Number | Fewer than 50 acres | 1000 acres or more | |
| | 104 | 105 | 106 | 107 | 108 | 109 | 110 | 111 | 112 | 113 | 114 | 115 | 116 |
| **PENNSYLVANIA—Cont'd** | | | | | | | | | | | | | |
| Potter | 359 | 4,741 | 890 | 641 | 570 | 80 | 164 | 209 | 44,112 | 447 | 25.3 | 4.0 | 43.5 |
| Schuylkill | 2,708 | 40,831 | 7,760 | 10,102 | 5,304 | 869 | 1,013 | 1,660 | 40,654 | 685 | 44.5 | 1.0 | 44.0 |
| Snyder | 894 | 15,571 | 1,922 | 3,922 | 2,952 | 258 | 236 | 431 | 27,660 | 864 | 49.0 | 0.9 | 49.1 |
| Somerset | 1,651 | 19,890 | 3,387 | 3,168 | 2,579 | 715 | 660 | 755 | 37,945 | 1,152 | 28.6 | 2.1 | 46.0 |
| Sullivan | 157 | 880 | 189 | 69 | 193 | 21 | 27 | 28 | 32,150 | 190 | 21.6 | 2.6 | 42.1 |
| Susquehanna | 811 | 6,781 | 1,062 | 620 | 1,165 | 174 | 259 | 253 | 37,323 | 909 | 29.6 | 0.8 | 45.7 |
| Tioga | 854 | 10,388 | 2,072 | 1,631 | 1,885 | 370 | 252 | 417 | 40,101 | 1,056 | 24.1 | 1.7 | 44.4 |
| Union | 948 | 16,857 | 4,044 | 1,424 | 1,811 | 478 | 459 | 658 | 39,017 | 574 | 38.0 | 0.7 | 53.1 |
| Venango | 1,155 | 15,554 | 3,332 | 4,066 | 2,367 | 348 | 303 | 588 | 37,795 | 409 | 40.6 | 2.0 | 36.2 |
| Warren | 866 | 13,416 | 2,997 | 2,840 | 2,021 | 864 | 294 | 588 | 43,861 | 452 | 33.0 | 1.3 | 46.1 |
| Washington | 5,176 | 86,667 | 14,678 | 8,849 | 8,638 | 1,624 | 3,697 | 4,950 | 57,116 | 1,760 | 37.7 | 0.1 | 41.8 |
| Wayne | 1,300 | 12,550 | 2,503 | 692 | 2,764 | 458 | 332 | 483 | 38,522 | 640 | 27.3 | 0.9 | 42.3 |
| Westmoreland | 8,439 | 127,911 | 20,153 | 18,130 | 18,377 | 2,792 | 6,821 | 5,603 | 43,806 | 1,099 | 42.0 | 2.1 | 38.6 |
| Wyoming | 652 | 10,473 | 601 | 2,068 | 1,205 | 195 | 273 | 545 | 52,056 | 410 | 32.0 | 1.2 | 39.7 |
| York | 8,662 | 168,358 | 25,893 | 30,621 | 21,663 | 4,837 | 6,265 | 7,671 | 45,564 | 2,067 | 62.0 | 2.0 | 39.3 |
| **RHODE ISLAND** | 28,801 | 444,948 | 87,067 | 39,469 | 47,840 | 31,771 | 23,070 | 22,726 | 51,077 | 1,043 | 72.5 | 0.4 | 38.6 |
| Bristol | 1,266 | 14,273 | 3,174 | 1,564 | 1,189 | 251 | 405 | 502 | 35,175 | 40 | 67.5 | NA | 41.7 |
| Kent | 4,696 | 71,197 | 13,094 | 5,481 | 11,328 | 4,603 | 3,007 | 3,573 | 50,184 | 111 | 72.1 | 1.8 | 29.4 |
| Newport | 2,756 | 32,768 | 5,130 | 1,750 | 4,157 | 1,406 | 3,823 | 1,509 | 46,056 | 196 | 71.4 | NA | 44.0 |
| Providence | 16,033 | 278,048 | 58,097 | 22,087 | 24,644 | 23,874 | 12,959 | 14,596 | 52,494 | 377 | 73.2 | NA | 32.5 |
| Washington | 3,738 | 43,260 | 7,359 | 8,587 | 6,521 | 956 | 1,670 | 2,223 | 51,397 | 319 | 73.0 | 0.6 | 47.0 |
| **SOUTH CAROLINA** | 111,926 | 1,949,406 | 249,536 | 250,097 | 251,938 | 75,592 | 99,772 | 83,500 | 42,834 | 24,791 | 49.8 | 3.8 | 36.1 |
| Abbeville | 326 | 4,101 | 455 | 1,792 | 414 | 82 | 32 | 160 | 39,133 | 576 | 39.6 | 1.2 | 30.8 |
| Aiken | 2,844 | 50,104 | 6,452 | 8,649 | 7,066 | 1,057 | 2,067 | 2,453 | 48,967 | 1,249 | 57.1 | 2.1 | 38.9 |
| Allendale | 117 | 1,825 | 383 | 854 | 137 | 39 | 33 | 83 | 45,278 | 165 | 21.8 | 9.1 | 40.7 |
| Anderson | 3,809 | 63,173 | 9,452 | 14,462 | 9,037 | 2,096 | 2,993 | 2,822 | 44,663 | 1,742 | 54.5 | 1.5 | 33.0 |
| Bamberg | 264 | 2,961 | 475 | 955 | 411 | 74 | 68 | 101 | 34,032 | 355 | 29.0 | 6.2 | 30.8 |
| Barnwell | 333 | 4,309 | 381 | 1,578 | 625 | 116 | 101 | 162 | 37,671 | 369 | 42.5 | 4.1 | 34.3 |
| Beaufort | 5,496 | 57,733 | 8,361 | 572 | 10,561 | 1,750 | 2,818 | 2,120 | 36,714 | 161 | 67.7 | 9.3 | 28.4 |
| Berkeley | 3,407 | 55,413 | 3,356 | 6,756 | 8,063 | 1,247 | 5,152 | 3,007 | 54,260 | 339 | 56.0 | 5.3 | 31.9 |
| Calhoun | 231 | 3,404 | 348 | 1,252 | 266 | 33 | 17 | 153 | 45,006 | 480 | 36.9 | 9.6 | 42.6 |
| Charleston | 14,397 | 209,573 | 26,188 | 15,424 | 29,464 | 6,748 | 16,642 | 9,930 | 47,382 | 403 | 68.2 | 1.0 | 39.3 |
| Cherokee | 920 | 18,967 | 1,194 | 6,966 | 2,764 | 261 | 179 | 674 | 35,555 | 415 | 47.0 | 2.2 | 33.0 |
| Chester | 529 | 7,686 | 589 | 3,432 | 894 | 128 | 138 | 335 | 43,594 | 517 | 40.6 | 4.3 | 38.8 |
| Chesterfield | 690 | 13,995 | 1,647 | 5,761 | 1,318 | 218 | 85 | 594 | 42,464 | 636 | 47.8 | 4.4 | 38.5 |
| Clarendon | 483 | 5,636 | 1,311 | 846 | 1,138 | 223 | 88 | 160 | 28,386 | 381 | 36.2 | 8.9 | 50.1 |
| Colleton | 748 | 7,358 | 1,247 | 550 | 1,595 | 246 | 178 | 265 | 35,957 | 459 | 44.4 | 7.0 | 32.9 |
| Darlington | 1,087 | 17,655 | 2,962 | 2,531 | 2,380 | 386 | 240 | 928 | 52,536 | 322 | 45.7 | 12.7 | 48.1 |
| Dillon | 449 | 6,320 | 1,053 | 1,606 | 1,196 | 152 | 134 | 222 | 35,070 | 182 | 22.0 | 14.8 | 54.9 |
| Dorchester | 2,478 | 32,400 | 4,257 | 6,693 | 4,755 | 674 | 1,235 | 1,266 | 39,087 | 358 | 55.3 | 4.2 | 38.7 |
| Edgefield | 311 | 4,945 | 469 | 996 | 423 | 48 | 68 | 192 | 38,871 | 397 | 43.1 | 4.5 | 35.1 |
| Fairfield | 315 | 5,506 | 532 | 827 | 525 | 52 | 1,632 | 345 | 62,655 | 228 | 25.4 | 8.3 | 27.0 |
| Florence | 3,138 | 67,075 | 24,009 | 7,330 | 8,681 | 2,520 | 2,098 | 2,862 | 42,666 | 540 | 42.4 | 6.7 | 38.1 |
| Georgetown | 1,832 | 20,426 | 3,639 | 2,425 | 3,352 | 490 | 820 | 826 | 40,457 | 166 | 36.7 | 8.4 | 32.9 |
| Greenville | 13,859 | 245,614 | 32,009 | 31,315 | 28,322 | 9,930 | 19,044 | 11,266 | 45,867 | 1,036 | 69.5 | 0.3 | 30.4 |
| Greenwood | 1,302 | 22,384 | 4,783 | 5,004 | 3,702 | 477 | 612 | 887 | 39,644 | 466 | 44.2 | 1.1 | 31.4 |
| Hampton | 313 | 3,389 | 658 | 618 | 608 | 120 | 106 | 127 | 37,367 | 242 | 24.4 | 12.0 | 42.4 |
| Horry | 9,369 | 115,376 | 13,207 | 2,981 | 25,170 | 2,925 | 3,783 | 3,817 | 33,082 | 767 | 47.3 | 5.7 | 46.6 |
| Jasper | 702 | 8,447 | 1,006 | 295 | 2,080 | 100 | 258 | 340 | 40,290 | 135 | 55.6 | 11.9 | 45.5 |
| Kershaw | 1,119 | 16,097 | 2,197 | 3,135 | 2,387 | 385 | 397 | 602 | 37,405 | 466 | 50.9 | 3.6 | 39.3 |
| Lancaster | 1,516 | 21,700 | 2,961 | 2,483 | 3,358 | 1,264 | 623 | 1,090 | 50,231 | 534 | 48.5 | 2.1 | 31.3 |
| Laurens | 920 | 18,358 | 1,813 | 7,860 | 1,723 | 277 | 479 | 739 | 40,231 | 840 | 44.9 | 2.1 | 32.6 |
| Lee | 194 | 2,069 | 308 | 276 | 393 | 59 | 44 | 69 | 33,127 | 334 | 42.2 | 9.6 | 38.4 |
| Lexington | 6,764 | 108,960 | 14,149 | 9,888 | 19,241 | 3,252 | 3,393 | 4,366 | 40,068 | 1,137 | 63.5 | 0.9 | 32.3 |
| McCormick | 88 | 1,023 | 257 | 245 | 113 | 11 | NA | 31 | 30,023 | 92 | 34.8 | 5.4 | 51.4 |
| Marion | 480 | 5,760 | 957 | 929 | 1,027 | 314 | 66 | 169 | 29,263 | 197 | 43.7 | 6.1 | 41.2 |
| Marlboro | 316 | 5,251 | 1,067 | 2,150 | 809 | 94 | 39 | 189 | 35,958 | 201 | 38.8 | 10.0 | 38.8 |
| Newberry | 767 | 12,456 | 1,474 | 5,216 | 1,437 | 183 | 163 | 499 | 40,030 | 607 | 40.4 | 1.6 | 35.7 |
| Oconee | 1,604 | 19,924 | 2,254 | 5,419 | 3,270 | 457 | 511 | 905 | 45,422 | 815 | 58.9 | 0.2 | 36.7 |
| Orangeburg | 1,613 | 25,300 | 4,242 | 6,405 | 3,804 | 577 | 390 | 957 | 37,821 | 978 | 38.8 | 8.0 | 46.5 |
| Pickens | 2,079 | 27,097 | 3,373 | 5,152 | 5,302 | 534 | 809 | 914 | 33,723 | 740 | 73.6 | 0.1 | 28.6 |
| Richland | 9,116 | 162,412 | 30,280 | 9,615 | 19,502 | 12,402 | 11,937 | 7,344 | 45,217 | 440 | 62.7 | 1.4 | 42.7 |
| Saluda | 232 | 4,120 | 441 | 2,423 | 330 | 48 | 50 | 137 | 33,356 | 574 | 35.0 | 3.7 | 39.1 |
| Spartanburg | 6,662 | 141,317 | 15,735 | 35,850 | 16,102 | 3,000 | 3,982 | 6,670 | 47,201 | 1,433 | 67.9 | 0.3 | 30.0 |

# Table B. States and Counties — Agriculture

| STATE County | Acreage (1,000) 117 | Percent change, 2012-2017 118 | Average size of farm 119 | Total irrigated (1,000) 120 | Total cropland (1,000) 121 | Value of land and buildings Average per farm (dollars) 122 | Average per acre 123 | Value of machinery and equipment, average per farm (dollars) 124 | Total (mil dol) 125 | Average per farm (acres) 126 | Crops 127 | Livestock and poultry products 128 | Organic farms (number) 129 | Farms with internet access (percent) 130 | Government payments Total ($1,000) 131 | Percent of farms 132 |
|---|---|---|---|---|---|---|---|---|---|---|---|---|---|---|---|---|
| **PENNSYLVANIA—Cont'd** | | | | | | | | | | | | | | | | |
| Potter | 98 | 1.1 | 219 | 0.0 | 45.5 | 750,104 | 3,429 | 110,415 | 39.2 | 87,756 | 27.4 | 72.6 | 7 | 74.3 | 1,139 | 44.7 |
| Schuylkill | 97 | -8.4 | 141 | 0.4 | 69.7 | 939,868 | 6,645 | 133,965 | 143.4 | 209,400 | 46.5 | 53.5 | 15 | 74.2 | 1,655 | 42.2 |
| Snyder | 99 | 8.6 | 115 | 0.7 | 71.4 | 839,894 | 7,332 | 109,041 | 200.4 | 231,888 | 16.1 | 83.9 | 39 | 64.4 | 1,125 | 17.4 |
| Somerset | 219 | 2.1 | 190 | 0.2 | 136.3 | 612,870 | 3,223 | 111,459 | 115.4 | 100,216 | 31.5 | 68.5 | 40 | 62.6 | 1,265 | 17.5 |
| Sullivan | 43 | 15.9 | 229 | 0.0 | 20.9 | 934,952 | 4,091 | 109,248 | 12.2 | 64,121 | 31.4 | 68.6 | NA | 72.1 | 495 | 34.7 |
| Susquehanna | 154 | -7.2 | 170 | 0.2 | 71.7 | 805,158 | 4,740 | 101,463 | 49.8 | 54,758 | 29.5 | 70.5 | 11 | 76.2 | 500 | 24.2 |
| Tioga | 213 | 3.7 | 202 | 0.5 | 123.2 | 765,228 | 3,797 | 98,872 | 92.3 | 87,363 | 27.9 | 72.1 | 20 | 71.8 | 1,925 | 39.8 |
| Union | 66 | -29.5 | 114 | 0.2 | 49.8 | 1,098,979 | 9,599 | 114,872 | 147.4 | 256,820 | 12.7 | 87.3 | 24 | 52.1 | 800 | 19.2 |
| Venango | 53 | -13.3 | 130 | 0.1 | 30.1 | 495,109 | 3,797 | 82,682 | 14.8 | 36,137 | 58.9 | 41.1 | NA | 76.3 | 357 | 14.7 |
| Warren | 68 | -17.3 | 151 | 0.1 | 30.4 | 521,904 | 3,461 | 71,238 | 21.3 | 47,029 | 26.3 | 73.7 | 1 | 68.6 | 157 | 13.9 |
| Washington | 190 | -7.5 | 108 | 1.0 | 90.9 | 704,588 | 6,511 | 85,650 | 37.0 | 21,022 | 64.3 | 35.7 | 6 | 74.2 | 399 | 8.5 |
| Wayne | 101 | -10.9 | 157 | 0.1 | 46.3 | 675,611 | 4,294 | 82,875 | 29.4 | 45,892 | 31.6 | 68.4 | 4 | 78.4 | 211 | 10.2 |
| Westmoreland | 144 | 0.8 | 131 | 0.2 | 93.4 | 775,756 | 5,909 | 108,523 | 66.3 | 60,346 | 50.7 | 49.3 | 13 | 72.2 | 1,265 | 19.1 |
| Wyoming | 61 | -10.8 | 150 | 0.1 | 29.5 | 619,664 | 4,144 | 78,912 | 13.2 | 32,300 | 49.6 | 50.4 | 1 | 73.9 | 407 | 23.7 |
| York | 253 | -3.6 | 122 | 0.7 | 199.2 | 1,002,707 | 8,201 | 105,570 | 260.9 | 126,235 | 51.7 | 48.3 | 25 | 74.5 | 4,781 | 17.8 |
| **RHODE ISLAND** | 57 | -18.3 | 55 | 3.0 | 17.7 | 897,835 | 16,468 | 62,786 | 58.0 | 55,607 | 70.5 | 29.5 | 22 | 84.1 | 1,037 | 7.0 |
| Bristol | 1 | D | 33 | 0.0 | 0.7 | 1,500,196 | 45,085 | 66,112 | 1.0 | 25,125 | 37.2 | 62.8 | NA | 75.0 | D | 2.5 |
| Kent | 10 | D | 87 | 0.1 | 1.1 | 1,227,510 | 14,155 | 49,665 | 3.1 | 27,856 | 74.9 | 25.1 | NA | 81.1 | D | 5.4 |
| Newport | 10 | -16.0 | 50 | 0.4 | 5.3 | 1,209,495 | 24,407 | 80,755 | 19.3 | 98,367 | 62.7 | 37.3 | 1 | 81.1 | 158 | 7.1 |
| Providence | 16 | D | 43 | 0.4 | 4.2 | 617,448 | 14,256 | 46,482 | 12.4 | 32,979 | 72.7 | 27.3 | 7 | 82.5 | 218 | 6.6 |
| Washington | 20 | -27.2 | 62 | 2.0 | 6.3 | 847,467 | 13,608 | 75,161 | 22.2 | 69,561 | 77.0 | 23.0 | 14 | 90.0 | 632 | 8.5 |
| **SOUTH CAROLINA** | 4,745 | -4.6 | 191 | 210.4 | 2,035.3 | 683,873 | 3,573 | 83,077 | 3,008.7 | 121,364 | 36.4 | 63.6 | 75 | 72.6 | 55,192 | 21.4 |
| Abbeville | 89 | -3.8 | 154 | 0.3 | 18.8 | 475,695 | 3,096 | 45,030 | 9.2 | 15,891 | 15.7 | 84.3 | 1 | 78.0 | 643 | 22.9 |
| Aiken | 163 | 5.4 | 130 | 8.5 | 62.9 | 490,520 | 3,767 | 74,216 | 137.4 | 109,990 | 21.2 | 78.8 | NA | 77.3 | 440 | 8.7 |
| Allendale | 87 | -30.4 | 525 | 5.9 | 35.3 | 1,157,218 | 2,206 | 131,114 | 15.3 | 92,739 | 82.0 | 18.0 | NA | 50.9 | 2,259 | 72.1 |
| Anderson | 184 | 15.5 | 105 | 0.6 | 69.9 | 598,456 | 5,675 | 59,960 | 75.2 | 43,144 | 13.4 | 86.6 | NA | 72.1 | 1,896 | 15.6 |
| Bamberg | 103 | 10.9 | 289 | 9.4 | 53.5 | 847,270 | 2,932 | 106,214 | 36.9 | 104,048 | 62.1 | 37.9 | NA | 74.6 | 2,126 | 54.4 |
| Barnwell | 74 | -15.4 | 201 | 4.9 | 33.3 | 561,132 | 2,785 | 85,317 | 35.3 | 95,745 | 43.2 | 56.8 | NA | 69.9 | 1,556 | 30.4 |
| Beaufort | 56 | 32.7 | 348 | 2.0 | 7.1 | 1,258,884 | 3,622 | 58,947 | 20.3 | 126,006 | 94.7 | 5.3 | 3 | 78.3 | 80 | 9.9 |
| Berkeley | 100 | 33.6 | 296 | 0.7 | 18.1 | 790,654 | 2,673 | 80,348 | 5.0 | 14,885 | 85.4 | 14.6 | NA | 66.1 | 403 | 12.7 |
| Calhoun | 149 | 25.6 | 310 | 24.5 | 79.9 | 945,231 | 3,051 | 152,649 | 80.8 | 168,329 | 73.0 | 27.0 | 1 | 80.0 | 3,859 | 40.6 |
| Charleston | 38 | 6.0 | 93 | 0.6 | 9.7 | 873,600 | 9,369 | 63,570 | 22.4 | 55,536 | 83.6 | 16.4 | 1 | 76.7 | D | 2.0 |
| Cherokee | 61 | -5.0 | 148 | D | 20.9 | 497,342 | 3,364 | 69,235 | 33.4 | 80,407 | 26.8 | 73.2 | 1 | 66.7 | 504 | 17.1 |
| Chester | 96 | 0.5 | 186 | 0.2 | 25.0 | 616,832 | 3,323 | 58,274 | 31.1 | 60,077 | 22.4 | 77.6 | NA | 68.1 | 671 | 22.8 |
| Chesterfield | 126 | -4.0 | 198 | 1.6 | 50.7 | 583,564 | 2,952 | 81,283 | 128.1 | 201,390 | 15.2 | 84.8 | NA | 69.5 | 784 | 21.5 |
| Clarendon | 137 | -21.3 | 359 | 8.9 | 102.8 | 816,645 | 2,274 | 128,382 | 108.6 | 284,979 | 47.3 | 52.7 | 2 | 65.9 | 1,608 | 49.6 |
| Colleton | 168 | -10.5 | 366 | 2.4 | 40.7 | 1,109,808 | 3,030 | 86,861 | 20.3 | 44,266 | 76.9 | 23.1 | 6 | 71.9 | 724 | 16.1 |
| Darlington | 144 | -18.5 | 447 | 6.1 | 97.1 | 1,161,402 | 2,596 | 170,002 | 88.2 | 274,019 | 55.6 | 44.4 | 8 | 74.8 | 1,167 | 27.6 |
| Dillon | 91 | -14.4 | 502 | 0.4 | 71.8 | 1,150,361 | 2,291 | 197,314 | 113.1 | 621,159 | 28.5 | 71.5 | NA | 74.2 | 1,230 | 62.6 |
| Dorchester | 74 | -0.9 | 206 | 1.7 | 32.4 | 579,913 | 2,811 | 100,333 | 39.9 | 111,589 | 36.6 | 63.4 | NA | 68.4 | 901 | 23.7 |
| Edgefield | 79 | -3.6 | 198 | 8.9 | 23.2 | 697,738 | 3,527 | 66,385 | 37.7 | 94,967 | 66.6 | 33.4 | NA | 76.3 | 689 | 17.4 |
| Fairfield | 73 | 64.2 | 321 | 0.2 | 10.8 | 918,819 | 2,867 | 54,829 | 16.7 | 73,417 | 10.2 | 89.8 | NA | 64.0 | 98 | 10.1 |
| Florence | 146 | -6.5 | 270 | 1.3 | 103.5 | 855,509 | 3,166 | 102,599 | 45.3 | 83,848 | 97.6 | 2.4 | NA | 67.6 | 1,563 | 36.5 |
| Georgetown | 80 | 21.0 | 484 | 0.2 | 13.4 | 988,048 | 2,043 | 84,996 | 9.3 | 55,861 | 98.2 | 1.8 | 2 | 74.7 | 693 | 42.8 |
| Greenville | 59 | -18.5 | 57 | 1.5 | 17.0 | 526,644 | 9,188 | 40,726 | 13.3 | 12,861 | 75.6 | 24.4 | 9 | 74.1 | 290 | 6.7 |
| Greenwood | 72 | -15.5 | 155 | 0.2 | 15.1 | 529,780 | 3,416 | 53,938 | 9.8 | 20,961 | 17.6 | 82.4 | NA | 79.8 | 367 | 13.7 |
| Hampton | 107 | -23.3 | 441 | 8.5 | 42.3 | 1,074,779 | 2,439 | 129,432 | 30.5 | 125,959 | 98.9 | 1.1 | 1 | 63.6 | 2,346 | 57.4 |
| Horry | 171 | -4.0 | 222 | 2.3 | 117.5 | 1,002,891 | 4,510 | 128,399 | 87.8 | 114,536 | 80.9 | 19.1 | 8 | 71.1 | 2,287 | 37.2 |
| Jasper | 63 | -8.0 | 468 | D | 11.9 | 1,712,283 | 3,661 | 101,313 | D | D | D | D | NA | 73.3 | 123 | 15.6 |
| Kershaw | 79 | -4.6 | 170 | 0.4 | 19.8 | 523,961 | 3,089 | 58,053 | 137.5 | 295,155 | 2.3 | 97.7 | NA | 65.5 | 269 | 8.6 |
| Lancaster | 63 | -3.1 | 118 | 0.5 | 17.2 | 466,819 | 3,952 | 63,920 | 57.1 | 106,919 | 6.1 | 93.9 | NA | 76.4 | 91 | 5.4 |
| Laurens | 122 | -0.3 | 146 | 0.4 | 40.9 | 558,251 | 3,834 | 65,424 | 70.9 | 84,400 | 13.0 | 87.0 | 5 | 77.9 | 957 | 17.9 |
| Lee | 110 | -22.6 | 330 | 15.6 | 75.2 | 797,300 | 2,416 | 156,323 | 95.3 | 285,278 | 38.4 | 61.6 | NA | 65.0 | 2,612 | 44.9 |
| Lexington | 103 | -4.7 | 90 | 13.2 | 47.8 | 499,227 | 5,533 | 77,434 | 222.2 | 195,412 | 32.5 | 67.5 | 9 | 79.6 | 600 | 9.3 |
| McCormick | 41 | 35.5 | 442 | D | 3.9 | 886,483 | 2,004 | 87,978 | D | D | D | D | NA | 72.8 | D | 19.6 |
| Marion | 50 | -37.2 | 256 | 0.5 | 27.9 | 596,308 | 2,331 | 120,416 | 22.4 | 113,761 | 55.1 | 44.9 | 1 | 74.1 | 867 | 46.7 |
| Marlboro | 90 | -20.3 | 449 | 3.5 | 55.8 | 865,333 | 1,927 | 162,975 | 79.8 | 397,060 | 38.1 | 61.9 | 2 | 70.1 | 1,856 | 43.8 |
| Newberry | 95 | -9.3 | 156 | 1.2 | 31.6 | 517,675 | 3,314 | 74,874 | 143.0 | 235,506 | 4.3 | 95.7 | 3 | 68.4 | 850 | 27.0 |
| Oconee | 62 | -7.9 | 77 | 0.4 | 18.9 | 489,668 | 6,385 | 71,668 | 159.4 | 195,610 | 2.9 | 97.1 | 3 | 74.8 | 635 | 16.4 |
| Orangeburg | 294 | 3.8 | 300 | 38.0 | 165.5 | 906,458 | 3,018 | 139,771 | 213.9 | 218,724 | 50.1 | 49.9 | NA | 69.5 | 9,879 | 30.6 |
| Pickens | 39 | -12.5 | 53 | 0.2 | 12.2 | 374,709 | 7,050 | 43,935 | 6.6 | 8,972 | 56.2 | 43.8 | NA | 75.3 | 194 | 5.7 |
| Richland | 52 | -13.9 | 119 | 1.6 | 22.2 | 571,315 | 4,797 | 67,954 | 32.3 | 73,314 | 44.5 | 55.5 | 3 | 75.7 | 960 | 13.4 |
| Saluda | 119 | 10.7 | 208 | 5.4 | 33.3 | 643,937 | 3,093 | 108,040 | 159.5 | 277,916 | 11.8 | 88.2 | 1 | 68.5 | 540 | 18.6 |
| Spartanburg | 96 | -5.9 | 67 | 1.8 | 34.3 | 550,282 | 8,231 | 45,163 | 30.5 | 21,292 | 70.4 | 29.6 | NA | 78.4 | 481 | 7.9 |

| STATE County | Water use, 2015 | | Wholesale Trade[1], 2017 | | | | Retail Trade[2], 2017 | | | | Real estate and rental and leasing,[2] 2017 | | | |
|---|---|---|---|---|---|---|---|---|---|---|---|---|---|---|
| | Public supply water withdrawn (mil gal/day) | Public supply gallons withdrawn per person per day | Number of establishments | Number of employees | Sales (mil dol) | Average payroll (mil dol) | Number of establishments | Number of employees | Sales (mil dol) | Average payroll (mil dol) | Number of establishments | Number of employees | Sales (mil dol) | Average payroll (mil dol) |
| | 133 | 134 | 135 | 136 | 137 | 138 | 139 | 140 | 141 | 142 | 143 | 144 | 145 | 146 |
| **PENNSYLVANIA—Cont'd** | | | | | | | | | | | | | | |
| Potter | 0.91 | 53.2 | D | D | D | 0.6 | 71 | 611 | 164.1 | 14.4 | D | D | D | 0.3 |
| Schuylkill | 23.38 | 161.7 | 92 | 1,418 | 688.7 | 53.7 | 454 | 5,467 | 1,388.2 | 125.7 | 62 | 239 | 49.1 | 10.5 |
| Snyder | 1.60 | 39.6 | 31 | 537 | 227.6 | 26.4 | 186 | 3,035 | 725.1 | 63.8 | 13 | 71 | 13.4 | 2.8 |
| Somerset | 22.61 | 299.4 | 76 | 806 | 296.7 | 35.5 | 248 | 2,628 | 798.3 | 66.3 | 40 | 119 | 24.3 | 4.2 |
| Sullivan | 0.65 | 102.7 | NA | NA | NA | NA | 28 | 169 | 48.3 | 3.7 | D | D | D | D |
| Susquehanna | 1.41 | 33.8 | 28 | 268 | 262.2 | 11.4 | 134 | 1,203 | 440.5 | 31.3 | 16 | 63 | 15.7 | 3.7 |
| Tioga | 2.03 | 48.5 | 23 | 300 | 167.4 | 13.0 | 156 | 1,722 | 502.3 | 42.1 | 26 | 79 | 14.4 | 2.9 |
| Union | 2.87 | 63.8 | 31 | 329 | 102.6 | 12.6 | 137 | 1,848 | 540.6 | 43.0 | 22 | 122 | 19.2 | 3.8 |
| Venango | 4.68 | 88.1 | D | D | D | D | 193 | 2,430 | 623.8 | 54.7 | 28 | 96 | 13.7 | 2.0 |
| Warren | 3.07 | 76.0 | D | D | D | D | 128 | 2,226 | 597.5 | 53.4 | 19 | 33 | 5.4 | 0.9 |
| Washington | 39.71 | 190.7 | 221 | 3,334 | 2,325.8 | 209.3 | 642 | 9,024 | 2,763.2 | 230.3 | 174 | 1,282 | 411.7 | 77.7 |
| Wayne | 2.38 | 46.5 | 29 | 318 | 134.9 | 17.9 | 206 | 2,872 | 3,116.2 | 82.6 | 31 | 112 | 17.0 | 3.7 |
| Westmoreland | 24.18 | 67.6 | 305 | 6,191 | 8,592.3 | 350.3 | 1,206 | 18,468 | 5,515.6 | 488.9 | 271 | 961 | 211.8 | 33.9 |
| Wyoming | 0.66 | 23.7 | 21 | 466 | 138.5 | 22.1 | 111 | 1,280 | 425.5 | 31.4 | 16 | 55 | 13.5 | 3.7 |
| York | 34.82 | 78.6 | 340 | 7,055 | 6,755.3 | 384.1 | 1,212 | 21,792 | 6,119.8 | 538.5 | 291 | 1,983 | 410.2 | 81.0 |
| **RHODE ISLAND** | 97.46 | 92.3 | 1,107 | 16,847 | 13,636.5 | 1,042.1 | 3,769 | 48,753 | 13,843.5 | 1,444.4 | 1,110 | 5,287 | 1,351.1 | 247.7 |
| Bristol | 0.01 | 0.2 | 50 | 584 | 257.4 | 27.3 | 138 | 1,154 | 297.5 | 33.8 | 44 | 163 | 35.5 | 6.5 |
| Kent | 1.04 | 6.3 | 206 | 3,623 | 4,220.6 | 230.4 | 662 | 11,557 | 3,554.6 | 343.4 | 183 | 1,188 | 320.5 | 51.5 |
| Newport | 7.20 | 87.4 | 69 | 605 | 860.7 | 46.3 | 427 | 4,053 | 1,125.2 | 120.7 | 115 | 540 | 130.3 | 25.2 |
| Providence | 80.06 | 126.4 | 661 | 10,895 | 7,632.1 | 660.6 | 2,023 | 24,975 | 6,772.8 | 730.6 | 627 | 3,025 | 745.3 | 148.8 |
| Washington | 9.15 | 72.3 | 121 | 1,140 | 665.7 | 77.4 | 519 | 7,014 | 2,093.4 | 215.9 | 141 | 371 | 119.5 | 15.8 |
| **SOUTH CAROLINA** | 633.39 | 129.4 | 4,360 | 61,421 | 57,337.4 | 3,537.8 | 17,700 | 253,384 | 69,980.1 | 6,205.6 | 5,890 | 26,764 | 6,509.7 | 1,128.5 |
| Abbeville | 1.97 | 79.0 | 4 | D | 17.8 | D | 61 | 534 | 108.0 | 8.5 | NA | NA | NA | NA |
| Aiken | 32.01 | 193.0 | 65 | 629 | 346.3 | 27.3 | 506 | 7,319 | 2,013.6 | 171.6 | 106 | 358 | 67.8 | 13.1 |
| Allendale | 1.19 | 126.2 | 8 | 52 | 66.3 | 2.7 | 28 | 149 | 45.2 | 3.4 | NA | NA | NA | NA |
| Anderson | 19.00 | 97.6 | 170 | 2,971 | 3,006.5 | 177.4 | 701 | 9,535 | 2,445.5 | 227.4 | 159 | 465 | 122.3 | 16.4 |
| Bamberg | 0.92 | 61.8 | D | D | D | D | 56 | 492 | 126.1 | 10.4 | NA | NA | NA | NA |
| Barnwell | 1.54 | 70.9 | D | D | D | D | 68 | 733 | 154.6 | 14.6 | 7 | 14 | 2.7 | 0.3 |
| Beaufort | 12.76 | 71.1 | 125 | 695 | 337.2 | 32.5 | 787 | 10,709 | 2,821.7 | 270.2 | 466 | 1,879 | 444.5 | 85.8 |
| Berkeley | 70.34 | 346.9 | 154 | 2,623 | 4,535.4 | 181.8 | 442 | 7,661 | 2,150.7 | 186.5 | 179 | 690 | 189.7 | 34.2 |
| Calhoun | 1.11 | 75.1 | D | D | D | D | 34 | 269 | 86.8 | 6.5 | D | D | D | 1.0 |
| Charleston | 4.52 | 11.6 | 473 | 5,705 | 4,203.3 | 345.9 | 2,023 | 29,351 | 8,586.0 | 804.8 | 1,038 | 4,541 | 1,173.0 | 222.3 |
| Cherokee | 8.98 | 159.8 | 28 | 456 | 138.3 | 18.1 | 218 | 2,955 | 771.3 | 57.0 | 35 | 132 | 24.5 | 4.6 |
| Chester | 3.18 | 98.6 | 13 | 142 | 205.2 | 9.2 | 99 | 929 | 186.9 | 18.2 | 12 | 23 | 4.8 | 0.8 |
| Chesterfield | 5.83 | 126.7 | 25 | 474 | 240.6 | 21.5 | 133 | 1,357 | 325.3 | 27.8 | 15 | 36 | 35.9 | 1.0 |
| Clarendon | 2.31 | 68.4 | 15 | 91 | 44.3 | 4.2 | 105 | 1,262 | 344.4 | 27.3 | 14 | 21 | 3.4 | 0.5 |
| Colleton | 2.07 | 54.9 | 28 | 183 | 97.9 | 7.9 | 139 | 1,698 | 430.8 | 36.3 | 43 | 107 | 31.1 | 4.2 |
| Darlington | 6.41 | 94.9 | 55 | 495 | 731.5 | 24.2 | 219 | 2,408 | 563.2 | 51.9 | 31 | 91 | 19.9 | 2.7 |
| Dillon | 4.56 | 146.0 | 12 | 219 | 107.8 | 10.8 | 115 | 1,170 | 354.8 | 24.4 | 25 | 47 | 6.1 | 1.3 |
| Dorchester | 33.04 | 216.7 | 75 | 648 | 324.4 | 34.2 | 325 | 4,875 | 1,271.2 | 105.9 | 135 | 419 | 113.0 | 15.7 |
| Edgefield | 4.54 | 171.2 | D | D | D | D | 48 | 438 | 141.4 | 9.6 | 11 | 25 | 3.0 | 0.8 |
| Fairfield | 1.88 | 82.6 | D | D | D | D | 60 | 507 | 218.4 | 13.2 | 12 | 20 | 3.7 | 0.7 |
| Florence | 15.70 | 113.0 | 163 | 2,638 | 2,417.1 | 127.3 | 668 | 9,456 | 2,500.5 | 217.1 | 130 | 546 | 116.9 | 21.9 |
| Georgetown | 8.91 | 145.4 | D | D | D | D | 293 | 3,330 | 833.4 | 83.2 | 94 | 549 | 74.4 | 18.7 |
| Greenville | 45.87 | 93.3 | 776 | 11,331 | 16,046.6 | 730.0 | 1,783 | 27,942 | 7,610.7 | 708.7 | 744 | 3,383 | 937.5 | 144.5 |
| Greenwood | 4.85 | 69.4 | 53 | D | 628.1 | D | 281 | 3,951 | 921.8 | 86.1 | D | D | D | D |
| Hampton | 1.47 | 73.3 | 8 | 180 | 219.3 | 9.3 | 79 | 618 | 168.9 | 13.9 | 6 | D | 2.1 | 0.9 |
| Horry | 50.13 | 162.1 | 258 | 2,517 | 1,071.9 | 111.7 | 1,732 | 24,447 | 6,602.4 | 593.4 | 696 | 4,708 | 769.1 | 162.4 |
| Jasper | 27.30 | 981.2 | 28 | 708 | 444.2 | 39.3 | 105 | 2,233 | 911.7 | 63.0 | 23 | 64 | 20.9 | 3.3 |
| Kershaw | 5.33 | 83.8 | 27 | 124 | 71.4 | 6.4 | 194 | 2,222 | 686.7 | 55.2 | 35 | 96 | 15.7 | 2.9 |
| Lancaster | 20.99 | 244.5 | 54 | 1,265 | 975.7 | 67.1 | 242 | 3,071 | 781.9 | 70.7 | 71 | 148 | 31.0 | 6.4 |
| Laurens | 4.91 | 73.7 | 29 | 205 | 89.6 | 11.9 | 172 | 1,817 | 436.9 | 39.2 | 25 | 59 | 10.1 | 1.6 |
| Lee | 2.54 | 141.9 | 10 | 75 | 111.6 | 4.3 | 41 | 326 | 82.4 | 6.8 | NA | NA | NA | NA |
| Lexington | 52.21 | 185.3 | 307 | 6,339 | 4,583.1 | 350.8 | 1,098 | 19,592 | 5,858.3 | 481.7 | 308 | 1,280 | 350.1 | 53.3 |
| McCormick | 0.96 | 98.9 | NA | NA | NA | NA | 20 | 138 | 27.7 | 2.1 | NA | NA | NA | NA |
| Marion | 3.41 | 107.4 | 19 | 173 | 136.0 | 6.9 | 110 | 1,075 | 235.3 | 23.1 | 15 | 41 | 13.0 | 1.2 |
| Marlboro | 3.90 | 141.8 | D | D | D | 4.5 | 86 | 825 | 177.4 | 16.9 | 11 | 28 | 3.5 | 0.6 |
| Newberry | 6.58 | 173.1 | 28 | 270 | 208.8 | 14.3 | 129 | 1,516 | 431.0 | 34.3 | 23 | 72 | 19.9 | 2.9 |
| Oconee | 10.82 | 142.9 | 40 | 410 | 225.2 | 16.0 | 253 | 3,122 | 879.8 | 74.8 | 94 | 264 | 43.5 | 10.8 |
| Orangeburg | 8.96 | 100.4 | 52 | 771 | 609.2 | 40.1 | 338 | 3,887 | 1,018.7 | 85.7 | 45 | 146 | 24.6 | 4.9 |
| Pickens | 33.01 | 271.3 | 68 | 565 | 300.1 | 27.6 | 346 | 5,359 | 1,476.9 | 133.2 | 75 | 216 | 54.8 | 8.0 |
| Richland | 31.44 | 77.2 | 338 | 5,581 | 4,495.1 | 349.3 | 1,244 | 19,812 | 5,482.7 | 506.7 | 477 | 2,967 | 988.6 | 146.9 |
| Saluda | 0.03 | 1.5 | 9 | 54 | 16.3 | 1.3 | 47 | 360 | 117.4 | 8.2 | D | D | D | 0.7 |
| Spartanburg | 34.72 | 116.8 | 397 | 5,908 | 5,925.6 | 346.8 | 1,032 | 16,413 | 4,906.4 | 419.0 | 265 | 1,110 | 356.2 | 40.1 |

1 Merchant wholesalers, except manufacturers' sales branches and offices.  2. Employer establishments.

# Table B. States and Counties — Professional Services, Manufacturing, and Accommodation and Food Services

| STATE County | Professional, scientific, and technical services, 2017 | | | | Manufacturing, 2017 | | | | Accommodation and food services, 2017 | | | |
|---|---|---|---|---|---|---|---|---|---|---|---|---|
| | Number of establishments | Number of employees | Sales (mil dol) | Average payroll (mil dol) | Number of establishments | Number of employees | Sales (mil dol) | Average payroll (mil dol) | Number of establishments | Number of employees | Sales (mil dol) | Annual payroll (mil dol) |
| | 147 | 148 | 149 | 150 | 151 | 152 | 153 | 154 | 155 | 156 | 157 | 158 |
| PENNSYLVANIA—Cont'd | | | | | | | | | | | | |
| Potter | D | D | 18 | D | 19 | 551 | 122.1 | 22.8 | 43 | 260 | 13.2 | 3.3 |
| Schuylkill | D | D | D | D | 168 | 9,350 | 3,255.8 | 475.8 | 262 | 2,682 | 137.6 | 34.8 |
| Snyder | 35 | 211 | 24 | 11.0 | 71 | 4,096 | 814.6 | 160.8 | 84 | 1,863 | 86.7 | 27.1 |
| Somerset | D | D | D | D | 102 | 2,713 | 903.0 | 132.0 | 170 | 3,462 | 146.1 | 43.0 |
| Sullivan | D | D | 2 | D | D | D | D | D | D | D | D | 0.9 |
| Susquehanna | 57 | 264 | 28 | 10.1 | 60 | 546 | 116.5 | 23.5 | 83 | 764 | 47.6 | 11.9 |
| Tioga | 47 | 220 | 21 | 8.6 | 38 | 1,782 | 404.5 | 85.1 | 95 | 1,288 | 69.5 | 19.6 |
| Union | D | D | D | D | 40 | 1,304 | 389.9 | 73.1 | 96 | 2,048 | 102.5 | 30.8 |
| Venango | D | D | D | D | 83 | 3,787 | 925.5 | 187.4 | 98 | 1,239 | 52.5 | 15.1 |
| Warren | D | D | D | D | 56 | 2,952 | 2,121.0 | 151.8 | 73 | 954 | 42.7 | 12.0 |
| Washington | D | D | D | D | 216 | 8,571 | 3,060.6 | 500.5 | 422 | 7,014 | 354.9 | 95.4 |
| Wayne | 86 | 296 | 36 | 13.3 | 47 | 550 | 112.4 | 25.2 | 167 | 1,722 | 195.9 | 49.8 |
| Westmoreland | D | D | 1,473 | D | 519 | 17,333 | 6,427.6 | 975.3 | 773 | 13,585 | 637.3 | 178.0 |
| Wyoming | D | D | D | D | D | 2,376 | D | 152.8 | 63 | 635 | 36.1 | 9.9 |
| York | 697 | 6,496 | 884 | 397.1 | 517 | 29,074 | 11,501.2 | 1,589.0 | 785 | 14,306 | 706.4 | 195.4 |
| RHODE ISLAND | 3,011 | 23,479 | 3,990 | 1,517.2 | 1,340 | 40,221 | 12,416.3 | 2,390.2 | 3,167 | 50,642 | 3,617.9 | 1,016.4 |
| Bristol | 105 | 393 | 52 | 21.5 | 80 | 1,823 | 399.8 | 92.9 | 130 | 1,618 | 95.3 | 29.4 |
| Kent | 534 | 3,962 | 479 | 183.9 | 180 | 5,515 | 3,168.5 | 394.8 | 471 | 8,824 | 523.1 | 150.3 |
| Newport | 338 | 3,930 | 626 | 253.5 | 67 | 2,258 | 361.3 | 206.5 | 382 | 6,122 | 566.0 | 164.6 |
| Providence | D | D | D | D | 860 | 22,092 | 6,137.3 | 1,165.6 | 1,706 | 28,025 | 1,995.1 | 547.5 |
| Washington | D | D | D | D | 153 | 8,533 | 2,349.4 | 530.4 | 478 | 6,053 | 438.4 | 124.8 |
| SOUTH CAROLINA | 10,818 | 99,953 | 16,958 | 6,465.3 | 3,827 | 225,237 | 138,586.5 | 12,777.4 | 10,847 | 223,081 | 13,385.0 | 3,666.5 |
| Abbeville | 15 | 36 | 3 | 1.0 | 30 | 1,561 | 505.8 | 75.6 | 32 | 345 | 16.3 | 4.9 |
| Aiken | 228 | 1,927 | 295 | 124.2 | 79 | 7,434 | 3,766.6 | 481.1 | 283 | 5,185 | 271.0 | 74.5 |
| Allendale | D | D | 10 | D | 6 | 739 | 617.0 | 45.4 | D | D | D | D |
| Anderson | D | D | D | D | 192 | 12,576 | 6,161.0 | 627.0 | 388 | 7,170 | 360.9 | 98.3 |
| Bamberg | D | D | 7 | D | 20 | 946 | 194.9 | 38.4 | 24 | 202 | 10.1 | 2.5 |
| Barnwell | 24 | 123 | 13 | 5.0 | 16 | 1,172 | 427.6 | 67.8 | 32 | 502 | 22.0 | 5.7 |
| Beaufort | 650 | 2,544 | 406 | 147.6 | 74 | 505 | 112.3 | 23.7 | 568 | 13,544 | 984.4 | 287.2 |
| Berkeley | D | D | D | D | 80 | 5,332 | 5,380.9 | 396.4 | 292 | 5,395 | 289.0 | 77.9 |
| Calhoun | D | D | D | D | 21 | 1,076 | 1,975.4 | 69.2 | 8 | 70 | 2.8 | 0.7 |
| Charleston | 1,902 | 17,127 | 3,169 | 1,245.3 | 313 | 14,857 | 33,467.9 | 989.1 | 1,417 | 34,621 | 2,471.9 | 694.7 |
| Cherokee | D | D | D | D | 63 | 6,418 | 3,630.3 | 309.7 | 90 | 1,764 | 92.3 | 24.6 |
| Chester | 30 | 127 | 15 | 4.6 | 45 | 2,848 | 1,174.5 | 157.3 | 47 | 698 | 33.1 | 9.1 |
| Chesterfield | D | D | 9 | D | 55 | 5,383 | 1,900.0 | 293.9 | 64 | 1,041 | 46.3 | 14.5 |
| Clarendon | 23 | 74 | 11 | 4.6 | 22 | 590 | 403.9 | 26.9 | 57 | 692 | 37.8 | 9.3 |
| Colleton | 48 | 302 | 50 | 13.4 | 27 | 661 | 196.4 | 28.0 | 79 | 1,495 | 94.6 | 24.6 |
| Darlington | 62 | 291 | 28 | 10.6 | 53 | 2,589 | 1,461.6 | 171.3 | 95 | 1,498 | 74.7 | 18.7 |
| Dillon | 23 | 404 | 12 | 5.0 | 13 | 1,538 | 441.7 | 54.2 | 56 | 802 | 39.9 | 10.3 |
| Dorchester | 207 | 1,860 | 205 | 75.7 | 92 | 5,806 | 2,479.6 | 371.8 | 210 | 3,859 | 197.2 | 56.9 |
| Edgefield | D | D | D | D | 21 | 825 | 316.5 | 40.8 | 26 | 199 | 9.5 | 2.5 |
| Fairfield | D | D | D | D | 17 | 1,158 | 511.7 | 62.2 | 21 | 212 | 10.4 | 2.9 |
| Florence | D | D | D | D | 91 | 6,689 | 3,875.4 | 407.0 | 319 | 6,530 | 336.7 | 95.6 |
| Georgetown | 171 | 848 | 152 | 45.9 | 55 | 2,232 | 1,209.0 | 166.5 | 168 | 3,071 | 196.7 | 53.8 |
| Greenville | D | D | 4,664 | D | 534 | 28,456 | 11,987.3 | 1,713.9 | 1,221 | 24,344 | 1,409.4 | 388.2 |
| Greenwood | D | D | D | D | 60 | 4,839 | 2,774.9 | 283.9 | 121 | 2,644 | 130.5 | 35.3 |
| Hampton | 13 | 143 | 31 | 14.6 | 12 | 418 | 142.1 | 23.4 | 33 | 301 | 16.1 | 3.9 |
| Horry | D | D | D | D | 152 | 2,810 | 688.2 | 131.3 | 1,337 | 31,091 | 2,296.6 | 597.3 |
| Jasper | 40 | 268 | 31 | 11.7 | 22 | 298 | 61.3 | 11.7 | 64 | 806 | 53.2 | 13.7 |
| Kershaw | D | D | D | D | 52 | 3,083 | 1,314.0 | 161.4 | 99 | 1,581 | 78.4 | 20.4 |
| Lancaster | 119 | 3,559 | 479 | 166.7 | 47 | 2,266 | 974.6 | 107.8 | 112 | 2,142 | 118.6 | 31.9 |
| Laurens | 51 | 281 | 29 | 14.4 | 75 | 8,565 | 2,305.9 | 396.5 | 80 | 1,354 | 60.6 | 16.3 |
| Lee | D | D | 3 | D | 12 | 216 | 148.1 | 13.5 | D | D | D | D |
| Lexington | D | D | D | D | 198 | 7,670 | 3,428.8 | 435.1 | 568 | 11,565 | 594.2 | 162.7 |
| McCormick | 4 | 16 | 1 | 0.1 | 4 | 267 | 123.8 | 11.6 | D | D | D | D |
| Marion | 25 | 95 | 6 | 2.1 | 13 | 1,155 | 262.7 | 36.0 | 46 | 825 | 37.5 | 9.4 |
| Marlboro | D | D | D | D | 21 | 1,862 | 972.7 | 94.1 | 26 | 333 | 18.1 | 4.9 |
| Newberry | 50 | 168 | 19 | 6.4 | 44 | 4,594 | 1,874.7 | 189.9 | 67 | 935 | 48.0 | 12.7 |
| Oconee | 131 | 507 | 53 | 19.8 | 76 | 4,828 | 2,064.9 | 280.3 | 130 | 1,704 | 92.7 | 24.3 |
| Orangeburg | D | D | D | D | 75 | 5,107 | 2,356.6 | 250.1 | 183 | 3,454 | 175.0 | 43.9 |
| Pickens | D | D | D | D | 94 | 5,247 | 1,929.0 | 249.1 | 231 | 4,993 | 240.2 | 69.8 |
| Richland | 1,191 | 13,054 | 2,559 | 888.4 | 195 | 9,510 | 4,725.3 | 618.8 | 950 | 20,811 | 1,104.5 | 301.6 |
| Saluda | D | D | D | D | 11 | 2,097 | 630.3 | 71.0 | 13 | 100 | 6.4 | 1.6 |
| Spartanburg | D | D | D | D | 407 | 29,294 | 21,377.8 | 1,729.6 | 578 | 11,099 | 592.3 | 163.7 |

Items 147—158

# Table B. States and Counties — Health Care and Social Assistance, Other Services, Nonemployer Businesses, and Residential Construction

| STATE County | Health care and social assistance, 2017 | | | | Other services, 2017 | | | | Nonemployer businesses, 2018 | | Value of residential construction authorized by building permits, 2020 | |
|---|---|---|---|---|---|---|---|---|---|---|---|---|
| | Number of establishments | Number of employees | Receipts (mil dol) | Annual payroll (mil dol) | Number of establishments | Number of employees | Receipts (mil dol) | Annual payroll (mil dol) | Number | Receipts (mil dol) | New construction ($1,000) | Number of housing units |
| | 159 | 160 | 161 | 162 | 163 | 164 | 165 | 166 | 167 | 168 | 169 | 170 |
| **PENNSYLVANIA—Cont'd** | | | | | | | | | | | | |
| Potter | 38 | 933 | 113.8 | 39.9 | D | D | 9.6 | D | 1,201 | 53.1 | 2,967 | 18 |
| Schuylkill | 371 | 8,260 | 826.1 | 317.9 | 243 | 1,079 | 90.4 | 25.0 | 6,646 | 325.3 | 22,790 | 104 |
| Snyder | 94 | 1,849 | 141.4 | 56.7 | 73 | 318 | 32.8 | 7.9 | 2,912 | 148.8 | 8,548 | 41 |
| Somerset | 175 | 3,380 | 295.2 | 125.2 | 153 | 670 | 60.6 | 14.5 | 4,506 | 191.1 | 18,222 | 71 |
| Sullivan | D | D | D | D | D | D | 4.1 | D | 459 | 23.3 | 1,563 | 6 |
| Susquehanna | 61 | 1,141 | 79.7 | 34.3 | 70 | 316 | 37.1 | 7.7 | 3,125 | 162.5 | 9,706 | 44 |
| Tioga | 97 | 1,664 | 167.4 | 60.8 | 66 | 286 | 29.0 | 6.8 | 2,494 | 103.3 | 8,502 | 45 |
| Union | 139 | 3,935 | 416.2 | 189.2 | 73 | 353 | 33.2 | 7.6 | 2,827 | 144.8 | 14,674 | 46 |
| Venango | 178 | 3,157 | 276.6 | 123.5 | 109 | 413 | 41.9 | 8.9 | 2,787 | 114.0 | 5,445 | 24 |
| Warren | 128 | 3,419 | 257.3 | 133.3 | 77 | 334 | 25.7 | 6.3 | 2,083 | 89.7 | 2,826 | 21 |
| Washington | 733 | 14,226 | 1,389.6 | 594.5 | 437 | 2,197 | 252.8 | 66.5 | 13,129 | 670.2 | 146,950 | 587 |
| Wayne | 132 | 2,517 | 243.9 | 92.5 | 107 | 665 | 63.6 | 17.3 | 3,913 | 193.1 | 28,924 | 120 |
| Westmoreland | 1,178 | 20,721 | 1,835.5 | 795.6 | 811 | 4,592 | 574.7 | 127.0 | 21,831 | 993.8 | 122,247 | 534 |
| Wyoming | 63 | 511 | 44.1 | 20.2 | 55 | 207 | 21.9 | 5.6 | 1,772 | 79.9 | 6,145 | 27 |
| York | 988 | 26,954 | 3,178.4 | 1,260.7 | 795 | 5,081 | 513.5 | 146.4 | 26,828 | 1,342.5 | 208,909 | 1,204 |
| **RHODE ISLAND** | 3,177 | 87,546 | 9,574.0 | 4,092.1 | 2,318 | 13,882 | 1,772.9 | 471.0 | 83,145 | 3,887.0 | 298,949 | 1,374 |
| Bristol | 130 | 3,202 | 157.0 | 87.5 | 118 | 541 | 54.7 | 17.8 | 4,609 | 222.7 | 12,511 | 56 |
| Kent | 561 | 12,532 | 1,363.8 | 593.5 | 399 | 2,297 | 278.1 | 74.3 | 12,215 | 609.8 | 44,865 | 241 |
| Newport | 215 | 4,980 | 396.0 | 172.8 | 216 | 1,298 | 142.5 | 43.9 | 8,306 | 419.0 | 48,579 | 123 |
| Providence | 1,897 | 59,341 | 6,916.1 | 2,929.2 | 1,307 | 8,239 | 1,073.6 | 289.6 | 45,883 | 2,010.5 | 99,342 | 620 |
| Washington | 374 | 7,491 | 741.2 | 309.1 | 278 | 1,507 | 224.0 | 45.4 | 12,132 | 625.2 | 93,652 | 334 |
| **SOUTH CAROLINA** | 10,588 | 244,198 | 29,454.9 | 11,050.5 | 7,068 | 46,847 | 5,653.1 | 1,528.9 | 363,971 | 16,174.0 | 9,505,326 | 42,340 |
| Abbeville | 28 | 471 | 43.6 | 18.1 | D | D | D | D | 1,416 | 51.3 | 18,131 | 108 |
| Aiken | 337 | 6,219 | 576.6 | 218.3 | 201 | 1,021 | 114.3 | 27.8 | 11,212 | 423.6 | 264,763 | 1,145 |
| Allendale | D | D | D | D | D | D | D | D | 438 | 10.3 | 457 | 2 |
| Anderson | 361 | 9,714 | 1,123.4 | 413.5 | 233 | 1,189 | 170.1 | 40.4 | 13,025 | 556.2 | 299,492 | 1,569 |
| Bamberg | 37 | 689 | 44.0 | 18.4 | D | D | D | 1.2 | 854 | 19.8 | 2,397 | 15 |
| Barnwell | 24 | 397 | 32.1 | 12.4 | 27 | 127 | 17.1 | 3.7 | 1,389 | 38.9 | 2,901 | 17 |
| Beaufort | 495 | 8,205 | 947.2 | 359.7 | 329 | 2,696 | 327.8 | 93.6 | 17,222 | 937.0 | 849,991 | 2,287 |
| Berkeley | 254 | 3,143 | 262.6 | 117.5 | 214 | 1,159 | 125.3 | 34.2 | 14,968 | 636.1 | 609,647 | 2,967 |
| Calhoun | D | D | D | D | D | D | D | D | 934 | 44.7 | 969 | 8 |
| Charleston | 1,414 | 31,729 | 5,356.1 | 1,759.0 | 888 | 6,113 | 838.6 | 222.7 | 41,296 | 2,294.1 | 874,336 | 4,273 |
| Cherokee | 69 | 1,156 | 89.9 | 38.5 | 58 | 291 | 33.0 | 8.0 | 2,584 | 89.5 | 29,481 | 174 |
| Chester | 52 | 607 | 68.9 | 24.7 | 37 | 145 | 13.7 | 3.8 | 1,719 | 57.1 | 20,373 | 82 |
| Chesterfield | 95 | 1,735 | 127.2 | 54.9 | 37 | 117 | 9.3 | 2.8 | 2,319 | 82.0 | 29,481 | 146 |
| Clarendon | 37 | 1,259 | 71.6 | 33.9 | 24 | 158 | 13.6 | 4.3 | 2,149 | 82.9 | 11,902 | 75 |
| Colleton | 74 | 1,332 | 157.8 | 60.0 | 36 | 149 | 11.7 | 2.8 | 3,140 | 128.6 | 13,582 | 40 |
| Darlington | 109 | 2,967 | 221.9 | 100.7 | 62 | 350 | 33.7 | 8.4 | 3,702 | 124.9 | 20,564 | 134 |
| Dillon | 53 | 1,469 | 132.7 | 72.9 | 32 | 82 | 8.2 | 2.3 | 1,580 | 52.8 | 3,552 | 34 |
| Dorchester | 262 | 4,033 | 379.7 | 144.3 | 182 | 902 | 90.2 | 29.1 | 11,070 | 434.5 | 358,283 | 1,364 |
| Edgefield | 21 | 487 | 44.0 | 15.2 | 20 | 343 | 65.8 | 18.1 | 1,504 | 55.6 | 49,844 | 219 |
| Fairfield | D | D | D | D | D | D | D | D | 1,474 | 41.7 | 13,697 | 72 |
| Florence | 373 | 14,668 | 1,954.1 | 724.6 | 197 | 1,418 | 194.8 | 40.8 | 8,764 | 354.4 | 87,238 | 630 |
| Georgetown | 255 | 3,851 | 497.4 | 190.8 | 122 | 709 | 58.3 | 17.0 | 5,516 | 253.7 | 154,184 | 622 |
| Greenville | 1,223 | 30,021 | 3,252.3 | 1,355.8 | 784 | 6,027 | 846.5 | 210.6 | 41,965 | 1,967.6 | 1,278,146 | 6,122 |
| Greenwood | 131 | 4,709 | 555.7 | 203.3 | D | D | D | D | 3,821 | 143.1 | 29,589 | 136 |
| Hampton | 26 | 713 | 45.8 | 19.6 | D | D | D | D | 1,189 | 38.1 | 4,907 | 20 |
| Horry | 734 | 12,191 | 1,694.7 | 573.6 | 554 | 3,147 | 338.1 | 83.9 | 27,310 | 1,331.7 | 914,879 | 5,419 |
| Jasper | 46 | 1,117 | 120.2 | 38.4 | 48 | 244 | 24.2 | 7.6 | 2,000 | 100.7 | 156,871 | 690 |
| Kershaw | D | D | D | D | 72 | 372 | 31.2 | 12.4 | 4,436 | 183.2 | 82,530 | 494 |
| Lancaster | 149 | 2,528 | 283.3 | 107.2 | 113 | 1,082 | 117.3 | 50.6 | 5,937 | 234.5 | 513,734 | 1,458 |
| Laurens | 81 | 1,787 | 171.3 | 64.9 | 64 | 288 | 23.2 | 8.9 | 3,528 | 139.2 | 24,881 | 186 |
| Lee | 13 | 177 | 13.1 | 5.0 | D | D | 3.8 | D | 809 | 26.1 | 1,476 | 11 |
| Lexington | 570 | 13,906 | 1,453.7 | 549.0 | 508 | 3,350 | 371.3 | 116.0 | 20,798 | 903.0 | 622,040 | 2,396 |
| McCormick | D | D | D | D | D | D | D | D | 521 | 18.7 | 32,255 | 117 |
| Marion | 73 | 861 | 77.7 | 30.4 | 31 | 101 | 11.9 | 2.2 | 1,736 | 53.7 | 8,763 | 81 |
| Marlboro | 40 | 673 | 50.0 | 21.5 | 17 | 74 | 5.1 | 1.3 | 1,023 | 34.5 | 0 | 0 |
| Newberry | 66 | 1,473 | 126.1 | 49.9 | 55 | 229 | 25.2 | 7.2 | 2,016 | 71.8 | 33,611 | 142 |
| Oconee | 126 | 2,255 | 247.8 | 84.2 | 93 | 757 | 86.3 | 19.9 | 5,027 | 199.2 | 117,375 | 424 |
| Orangeburg | 213 | 4,281 | 420.0 | 170.7 | 92 | 410 | 43.1 | 11.1 | 5,181 | 173.4 | 10,759 | 74 |
| Pickens | 188 | 3,662 | 365.8 | 137.6 | 144 | 706 | 107.3 | 18.2 | 8,057 | 356.8 | 228,497 | 857 |
| Richland | 1,006 | 32,281 | 4,425.0 | 1,636.6 | 657 | 5,503 | 599.1 | 182.3 | 29,310 | 1,261.8 | 340,176 | 1,781 |
| Saluda | D | D | D | D | D | D | D | D | 1,042 | 36.4 | 16,799 | 121 |
| Spartanburg | 574 | 16,460 | 1,784.2 | 848.3 | 425 | 3,014 | 399.0 | 96.3 | 21,822 | 1,033.4 | 468,600 | 2,876 |

# Government Employment and Payroll, and Local Government Finances

| STATE County | Full-time equivalent employees | March payroll (dollars) | Administration, judicial, and legal | Police and corrections | Fire protection | Highways and transportation | Health and welfare | Natural resources and utilities | Education and libraries | General revenue Total (mil dol) | Intergovernmental (mil dol) | Taxes Total (mil dol) | Per capita Total (dollars) | Per capita Property (dollars) |
|---|---|---|---|---|---|---|---|---|---|---|---|---|---|---|
| | 171 | 172 | 173 | 174 | 175 | 176 | 177 | 178 | 179 | 180 | 181 | 182 | 183 | 184 |
| **PENNSYLVANIA—Cont'd** | | | | | | | | | | | | | | |
| Potter | 578 | 2,313,263 | 12.7 | 5.7 | 0.0 | 3.9 | 9.4 | 3.3 | 64.6 | 74.4 | 41.7 | 24.7 | 1,464 | 1,231 |
| Schuylkill | 4,341 | 16,802,653 | 7.2 | 7.4 | 0.8 | 5.7 | 5.0 | 5.0 | 68.2 | 524.0 | 249.6 | 194.9 | 1,367 | 993 |
| Snyder | 953 | 3,640,204 | 6.7 | 8.4 | 0.0 | 3.7 | 3.6 | 2.8 | 73.7 | 127.2 | 47.3 | 58.1 | 1,430 | 970 |
| Somerset | 2,152 | 7,566,757 | 8.6 | 5.0 | 0.0 | 4.6 | 2.2 | 3.7 | 70.3 | 244.2 | 127.6 | 89.3 | 1,203 | 967 |
| Sullivan | 179 | 756,018 | 16.5 | 1.8 | 0.0 | 5.6 | 4.2 | 2.6 | 68.6 | 27.6 | 11.8 | 12.9 | 2,097 | 1,876 |
| Susquehanna | 1,356 | 5,322,865 | 6.3 | 4.6 | 0.2 | 3.6 | 1.6 | 1.0 | 80.2 | 176.5 | 96.9 | 59.4 | 1,450 | 1,286 |
| Tioga | 1,397 | 5,193,310 | 8.1 | 6.3 | 0.7 | 3.8 | 13.5 | 3.2 | 62.9 | 173.2 | 87.5 | 60.9 | 1,497 | 1,188 |
| Union | 1,435 | 6,048,458 | 5.9 | 5.0 | 0.0 | 2.1 | 3.2 | 3.7 | 78.7 | 179.8 | 76.5 | 60.2 | 1,349 | 952 |
| Venango | 1,804 | 6,783,791 | 8.2 | 6.3 | 2.2 | 4.4 | 6.9 | 3.2 | 67.7 | 223.0 | 128.8 | 62.2 | 1,201 | 949 |
| Warren | 1,355 | 5,345,706 | 8.1 | 7.5 | 1.0 | 4.8 | 15.5 | 3.2 | 59.0 | 151.0 | 67.9 | 48.8 | 1,230 | 921 |
| Washington | 6,306 | 26,355,798 | 6.3 | 7.8 | 1.4 | 3.4 | 8.5 | 3.6 | 67.7 | 903.9 | 421.9 | 346.7 | 1,674 | 1,252 |
| Wayne | 1,903 | 7,949,235 | 6.3 | 5.2 | 0.0 | 2.9 | 5.8 | 1.5 | 76.3 | 242.5 | 84.1 | 138.7 | 2,710 | 2,585 |
| Westmoreland | 10,133 | 45,166,775 | 5.7 | 8.6 | 0.1 | 3.8 | 7.2 | 7.9 | 65.0 | 1,369.5 | 545.8 | 559.0 | 1,588 | 1,235 |
| Wyoming | 897 | 3,100,744 | 17.5 | 6.4 | 0.0 | 3.5 | 1.1 | 1.4 | 67.4 | 125.5 | 55.3 | 48.8 | 1,782 | 1,505 |
| York | 11,962 | 52,520,313 | 6.6 | 13.8 | 1.8 | 2.1 | 5.9 | 3.5 | 63.6 | 1,957.2 | 680.8 | 911.8 | 2,047 | 1,668 |
| **RHODE ISLAND** | X | X | X | X | X | X | X | X | X | X | X | X | X | X |
| Bristol | 1,410 | 7,796,071 | 3.5 | 7.9 | 2.3 | 3.1 | 1.3 | 5.9 | 75.5 | 235.1 | 48.9 | 162.8 | 3,341 | 3,275 |
| Kent | 4,906 | 26,300,714 | 3.0 | 10.1 | 10.8 | 2.2 | 1.7 | 3.7 | 67.1 | 688.7 | 144.3 | 459.8 | 2,811 | 2,744 |
| Newport | 2,464 | 13,368,879 | 4.9 | 12.1 | 9.0 | 2.5 | 1.7 | 4.9 | 63.5 | 405.5 | 83.5 | 256.3 | 3,090 | 2,914 |
| Providence | 16,666 | 91,532,903 | 3.5 | 10.2 | 10.2 | 2.1 | 3.1 | 3.7 | 66.3 | 2,617.2 | 1,008.4 | 1,292.3 | 2,038 | 1,982 |
| Washington | 3,857 | 20,455,795 | 4.6 | 9.4 | 4.3 | 3.1 | 1.2 | 3.3 | 73.6 | 613.3 | 121.0 | 440.8 | 3,489 | 3,431 |
| **SOUTH CAROLINA** | X | X | X | X | X | X | X | X | X | X | X | X | X | X |
| Abbeville | 907 | 2,554,601 | 7.3 | 9.8 | 1.2 | 0.6 | 8.6 | 11.0 | 60.8 | 65.4 | 30.2 | 24.1 | 980 | 747 |
| Aiken | 4,746 | 16,744,243 | 9.9 | 11.0 | 0.6 | 3.4 | 2.4 | 5.7 | 65.3 | 499.8 | 189.7 | 213.4 | 1,268 | 903 |
| Allendale | 534 | 1,689,533 | 6.7 | 9.2 | 3.8 | 0.9 | 39.6 | 5.0 | 33.2 | 29.9 | 15.2 | 10.9 | 1,210 | 1,122 |
| Anderson | 5,811 | 20,434,065 | 6.5 | 10.1 | 1.6 | 1.7 | 0.7 | 6.4 | 71.6 | 563.8 | 243.3 | 204.1 | 1,030 | 811 |
| Bamberg | 591 | 1,801,219 | 12.7 | 10.4 | 1.0 | 2.5 | 0.0 | 2.5 | 69.2 | 48.0 | 25.1 | 18.0 | 1,252 | 1,089 |
| Barnwell | 987 | 3,123,982 | 5.2 | 10.1 | 0.5 | 1.1 | 7.1 | 3.9 | 69.8 | 91.9 | 50.2 | 21.1 | 988 | 836 |
| Beaufort | 6,493 | 28,262,745 | 6.2 | 8.9 | 5.4 | 1.4 | 29.7 | 4.0 | 40.7 | 699.5 | 148.1 | 417.4 | 2,238 | 1,855 |
| Berkeley | 6,056 | 20,937,267 | 6.4 | 8.5 | 1.8 | 1.7 | 1.6 | 4.8 | 72.8 | 683.5 | 351.4 | 240.3 | 1,120 | 808 |
| Calhoun | 428 | 1,388,007 | 4.8 | 8.4 | 0.9 | 2.1 | 6.0 | 5.1 | 68.3 | 36.7 | 16.2 | 15.1 | 1,026 | 939 |
| Charleston | 13,441 | 53,983,648 | 9.8 | 15.9 | 10.1 | 4.4 | 3.4 | 12.5 | 41.6 | 2,195.3 | 429.8 | 1,273.9 | 3,169 | 1,783 |
| Cherokee | 1,841 | 6,406,214 | 4.1 | 8.1 | 1.6 | 0.4 | 1.0 | 10.3 | 70.3 | 154.1 | 68.6 | 54.9 | 965 | 768 |
| Chester | 1,159 | 3,879,635 | 10.7 | 11.7 | 2.6 | 0.5 | 5.4 | 9.9 | 57.4 | 95.5 | 45.1 | 34.7 | 1,075 | 939 |
| Chesterfield | 1,463 | 4,604,479 | 6.5 | 8.3 | 0.7 | 0.2 | 0.1 | 4.1 | 75.4 | 113.4 | 56.8 | 46.4 | 1,009 | 807 |
| Clarendon | 1,382 | 4,082,937 | 5.3 | 8.7 | 3.1 | 1.1 | 17.1 | 3.6 | 58.0 | 122.1 | 52.9 | 38.7 | 1,137 | 883 |
| Colleton | 1,464 | 4,778,777 | 13.1 | 11.1 | 9.6 | 2.3 | 1.0 | 4.8 | 55.9 | 131.0 | 64.6 | 54.6 | 1,453 | 1,137 |
| Darlington | 2,298 | 7,202,642 | 5.6 | 9.7 | 2.3 | 1.6 | 2.3 | 6.0 | 67.6 | 178.3 | 78.7 | 71.1 | 1,062 | 958 |
| Dillon | 1,233 | 3,617,987 | 6.2 | 11.2 | 3.7 | 2.4 | 2.4 | 5.7 | 70.8 | 94.7 | 52.3 | 25.8 | 847 | 662 |
| Dorchester | 5,513 | 18,312,394 | 5.7 | 8.5 | 4.8 | 1.4 | 0.2 | 4.6 | 72.9 | 422.3 | 198.9 | 177.7 | 1,118 | 885 |
| Edgefield | 1,083 | 3,379,916 | 6.7 | 8.1 | 0.4 | 1.1 | 19.2 | 4.6 | 58.0 | 52.4 | 29.2 | 18.3 | 683 | 617 |
| Fairfield | 1,043 | 3,516,997 | 8.0 | 9.1 | 1.3 | 1.3 | 2.1 | 6.8 | 68.2 | 85.4 | 23.7 | 55.7 | 2,464 | 2,384 |
| Florence | 4,880 | 17,012,925 | 6.3 | 8.7 | 1.8 | 1.9 | 9.5 | 4.5 | 65.9 | 462.4 | 197.3 | 181.3 | 1,309 | 836 |
| Georgetown | 2,324 | 8,144,776 | 6.5 | 8.7 | 8.7 | 1.8 | 2.8 | 8.2 | 58.8 | 234.2 | 71.2 | 124.4 | 2,012 | 1,873 |
| Greenville | 23,509 | 98,678,051 | 3.4 | 5.2 | 3.2 | 1.1 | 43.8 | 5.0 | 37.5 | 3,789.4 | 605.6 | 670.7 | 1,323 | 1,132 |
| Greenwood | 4,398 | 18,097,146 | 3.9 | 4.2 | 1.1 | 1.6 | 51.5 | 5.2 | 30.7 | 581.3 | 105.6 | 73.7 | 1,045 | 922 |
| Hampton | 941 | 3,070,037 | 6.1 | 12.6 | 1.7 | 1.5 | 3.4 | 3.6 | 69.4 | 65.6 | 33.2 | 26.1 | 1,339 | 1,240 |
| Horry | 10,279 | 38,665,530 | 7.7 | 11.1 | 5.5 | 4.3 | 1.7 | 7.6 | 58.2 | 1,246.7 | 310.0 | 653.9 | 1,966 | 1,234 |
| Jasper | 793 | 2,808,151 | 14.0 | 12.5 | 11.3 | 2.2 | 2.3 | 3.1 | 54.6 | 93.1 | 27.2 | 51.3 | 1,797 | 1,441 |
| Kershaw | 2,795 | 11,238,632 | 3.9 | 4.6 | 2.0 | 0.6 | 40.9 | 3.8 | 44.1 | 294.0 | 138.1 | 62.9 | 964 | 827 |
| Lancaster | 2,702 | 9,468,935 | 4.5 | 8.3 | 2.5 | 1.0 | 4.0 | 6.4 | 68.7 | 245.0 | 100.0 | 109.8 | 1,188 | 912 |
| Laurens | 2,181 | 6,252,719 | 5.8 | 10.7 | 2.2 | 1.2 | 6.7 | 7.7 | 60.4 | 182.2 | 98.6 | 57.2 | 856 | 759 |
| Lee | 576 | 1,769,614 | 11.0 | 6.8 | 10.0 | 1.4 | 0.5 | 4.9 | 65.2 | 45.5 | 23.3 | 20.4 | 1,170 | 647 |
| Lexington | 16,001 | 71,004,914 | 2.3 | 3.7 | 1.5 | 0.5 | 42.6 | 2.8 | 45.9 | 2,013.7 | 463.0 | 460.9 | 1,588 | 1,386 |
| McCormick | 491 | 1,209,140 | 13.0 | 17.2 | 1.4 | 2.5 | 13.2 | 13.1 | 34.4 | 23.9 | 8.9 | 12.9 | 1,347 | 1,304 |
| Marion | 1,176 | 3,510,288 | 5.8 | 10.1 | 1.1 | 0.9 | 4.8 | 2.5 | 72.5 | 90.2 | 46.8 | 30.1 | 961 | 724 |
| Marlboro | 1,015 | 2,423,800 | 10.7 | 12.2 | 1.7 | 1.8 | 1.7 | 7.6 | 60.5 | 83.4 | 44.0 | 25.8 | 968 | 789 |
| Newberry | 1,694 | 6,050,329 | 5.4 | 6.6 | 1.4 | 2.0 | 27.6 | 8.9 | 46.5 | 182.3 | 50.8 | 54.0 | 1,407 | 1,230 |
| Oconee | 2,515 | 9,110,006 | 6.5 | 10.1 | 2.6 | 2.3 | 0.2 | 13.5 | 62.5 | 205.4 | 73.7 | 107.7 | 1,391 | 1,298 |
| Orangeburg | 4,783 | 18,322,449 | 3.9 | 5.3 | 0.4 | 1.1 | 41.0 | 4.6 | 40.8 | 521.3 | 120.3 | 110.8 | 1,264 | 1,031 |
| Pickens | 3,139 | 10,514,539 | 6.7 | 9.4 | 1.9 | 3.0 | 3.2 | 8.6 | 63.8 | 303.0 | 125.3 | 125.4 | 1,016 | 790 |
| Richland | 14,141 | 51,288,677 | 7.2 | 10.3 | 4.2 | 2.9 | 3.3 | 8.9 | 61.8 | 1,616.8 | 487.9 | 851.8 | 2,069 | 1,707 |
| Saluda | 546 | 1,726,702 | 8.0 | 12.3 | 0.6 | 2.6 | 6.1 | 4.1 | 65.1 | 46.2 | 22.1 | 19.1 | 940 | 878 |
| Spartanburg | 15,616 | 72,630,964 | 2.9 | 4.1 | 1.5 | 0.4 | 54.2 | 3.4 | 32.7 | 2,111.4 | 443.9 | 383.7 | 1,251 | 1,089 |

1. Based on the resident population estimated as of July 1 of the year shown.

# Table B. States and Counties — Local Government Finances, Government Employment, and Income Taxes

| STATE County | Local government finances, 2017 (cont.) | | | | | | | Debt outstanding | | Government employment, 2019 | | | Individual income tax returns, 2018 | | |
|---|---|---|---|---|---|---|---|---|---|---|---|---|---|---|---|
| | Direct general expenditure | | | | | | | | | | | | | | |
| | | | Percent of total for: | | | | | | | | | | | Mean adjusted gross income | Mean income tax |
| | Total (mil dol) | Per capita[1] (dollars) | Education | Health and hospitals | Police protection | Public welfare | Highways | Total (mil dol) | Per capita[1] (dollars) | Federal civilian | Federal military | State and local | Number of returns | | |
| | 185 | 186 | 187 | 188 | 189 | 190 | 191 | 192 | 193 | 194 | 195 | 196 | 197 | 198 | 199 |
| **PENNSYLVANIA—Cont'd** | | | | | | | | | | | | | | | |
| Potter | 73 | 4,322 | 56.7 | 1.7 | 1.1 | 10.1 | 6.1 | 46.3 | 2,746 | 44 | 42 | 1,011 | 7,590 | 49,948 | 4,543 |
| Schuylkill | 591 | 4,143 | 53.1 | 2.9 | 2.1 | 4.5 | 5.5 | 507.5 | 3,559 | 603 | 346 | 6,931 | 67,160 | 53,114 | 5,093 |
| Snyder | 130 | 3,203 | 62.3 | 3.5 | 5.9 | 0.1 | 5.4 | 180.8 | 4,455 | 85 | 98 | 2,116 | 18,270 | 51,576 | 4,307 |
| Somerset | 255 | 3,432 | 64.7 | 0.0 | 1.6 | 5.4 | 5.1 | 242.5 | 3,266 | 187 | 175 | 4,149 | 34,210 | 51,194 | 4,483 |
| Sullivan | 28 | 4,578 | 49.3 | 2.5 | 2.3 | 2.7 | 12.5 | 8.4 | 1,371 | 19 | 14 | 366 | 2,840 | 57,930 | 5,782 |
| Susquehanna | 175 | 4,285 | 69.8 | 0.8 | 1.2 | 2.2 | 7.2 | 62.4 | 1,525 | 106 | 102 | 1,726 | 19,460 | 61,934 | 6,821 |
| Tioga | 168 | 4,120 | 54.9 | 0.2 | 1.1 | 11.4 | 7.1 | 107.2 | 2,636 | 140 | 102 | 2,588 | 18,490 | 54,000 | 4,829 |
| Union | 205 | 4,595 | 69.6 | 0.7 | 1.9 | 1.2 | 2.9 | 333.5 | 7,471 | 1,379 | 97 | 1,342 | 17,180 | 66,232 | 7,244 |
| Venango | 226 | 4,356 | 57.5 | 2.0 | 1.7 | 7.5 | 5.9 | 145.2 | 2,803 | 122 | 126 | 3,079 | 24,050 | 50,188 | 4,401 |
| Warren | 155 | 3,907 | 54.8 | 0.3 | 2.1 | 12.3 | 5.3 | 178.9 | 4,509 | 165 | 98 | 1,763 | 18,710 | 52,336 | 4,793 |
| Washington | 971 | 4,688 | 56.4 | 1.4 | 2.8 | 11.8 | 5.3 | 1,031.6 | 4,982 | 457 | 539 | 9,090 | 103,670 | 77,249 | 9,763 |
| Wayne | 240 | 4,687 | 67.9 | 2.6 | 1.0 | 4.9 | 3.4 | 231.2 | 4,515 | 497 | 122 | 2,585 | 25,070 | 56,874 | 5,653 |
| Westmoreland | 1,372 | 3,896 | 58.8 | 1.8 | 3.0 | 6.0 | 4.7 | 2,206.0 | 6,267 | 860 | 893 | 13,659 | 181,380 | 66,043 | 7,255 |
| Wyoming | 121 | 4,423 | 54.0 | 0.0 | 3.4 | 4.4 | 9.7 | 50.4 | 1,841 | 57 | 67 | 983 | 13,200 | 59,537 | 6,268 |
| York | 1,968 | 4,417 | 49.6 | 3.5 | 3.5 | 10.0 | 2.9 | 2,310.6 | 5,187 | 4,360 | 1,310 | 15,615 | 227,080 | 66,519 | 7,030 |
| **RHODE ISLAND** | X | X | X | X | X | X | X | X | X | 11,193 | 6,734 | 56,329 | 541,680 | 69,432 | 8,361 |
| Bristol | 226 | 4,634 | 65.8 | 0.4 | 4.3 | 0.1 | 3.0 | 253.3 | 5,197 | 95 | 208 | 1,941 | 24,640 | 110,214 | 17,083 |
| Kent | 688 | 4,209 | 56.9 | 0.1 | 7.0 | 0.3 | 3.3 | 349.5 | 2,137 | 732 | 738 | 7,532 | 90,050 | 71,647 | 8,530 |
| Newport | 417 | 5,027 | 46.2 | 2.6 | 8.3 | 0.4 | 2.0 | 382.4 | 4,610 | 5,170 | 2,457 | 3,597 | 43,400 | 87,606 | 11,741 |
| Providence | 2,535 | 3,998 | 55.2 | 0.2 | 9.0 | 0.1 | 3.0 | 1,660.4 | 2,618 | 4,569 | 2,782 | 31,655 | 317,440 | 59,311 | 6,466 |
| Washington | 602 | 4,769 | 66.1 | 0.7 | 6.6 | 0.4 | 3.5 | 256.7 | 2,032 | 627 | 549 | 11,604 | 66,160 | 87,859 | 11,760 |
| **SOUTH CAROLINA** | X | X | X | X | X | X | X | X | X | 35,084 | 53,366 | 328,637 | 2,275,930 | 62,218 | 6,784 |
| Abbeville | 58 | 2,351 | 55.7 | 3.8 | 8.0 | 0.0 | 2.9 | 19.0 | 773 | 43 | 87 | 1,418 | 9,720 | 46,168 | 3,407 |
| Aiken | 462 | 2,742 | 63.7 | 0.2 | 6.8 | 0.0 | 1.0 | 321.4 | 1,909 | 668 | 622 | 6,848 | 74,750 | 60,039 | 5,952 |
| Allendale | 30 | 3,322 | 62.9 | 1.3 | 4.9 | 0.0 | 1.5 | 44.5 | 4,936 | 19 | 28 | 1,048 | 3,190 | 34,669 | 2,463 |
| Anderson | 570 | 2,874 | 61.7 | 0.4 | 8.2 | 0.3 | 2.3 | 1,356.3 | 6,844 | 385 | 736 | 12,327 | 87,620 | 56,326 | 5,311 |
| Bamberg | 47 | 3,291 | 60.7 | 0.3 | 8.2 | 0.0 | 1.8 | 34.2 | 2,375 | 35 | 48 | 824 | 5,230 | 40,731 | 3,037 |
| Barnwell | 80 | 3,753 | 59.7 | 1.5 | 5.9 | 0.0 | 1.9 | 45.7 | 2,141 | 47 | 76 | 1,353 | 8,490 | 42,728 | 3,194 |
| Beaufort | 644 | 3,454 | 45.5 | 2.2 | 7.8 | 0.6 | 4.3 | 928.7 | 4,979 | 2,218 | 10,668 | 8,120 | 88,060 | 89,642 | 12,113 |
| Berkeley | 727 | 3,389 | 55.2 | 1.0 | 4.0 | 0.3 | 0.3 | 2,699.7 | 12,584 | 838 | 836 | 8,267 | 102,220 | 62,839 | 6,609 |
| Calhoun | 35 | 2,404 | 60.7 | 2.5 | 6.5 | 0.1 | 3.7 | 33.1 | 2,250 | 26 | 53 | 767 | 6,260 | 51,521 | 4,425 |
| Charleston | 1,813 | 4,509 | 37.6 | 3.5 | 10.9 | 0.2 | 2.8 | 2,346.9 | 5,838 | 10,485 | 11,536 | 38,499 | 196,220 | 91,083 | 13,445 |
| Cherokee | 167 | 2,931 | 68.7 | 0.0 | 7.0 | 0.0 | 3.5 | 1,034.7 | 18,181 | 95 | 207 | 2,396 | 23,510 | 44,838 | 3,418 |
| Chester | 101 | 3,121 | 59.7 | 3.4 | 5.1 | 0.0 | 0.3 | 31.3 | 968 | 63 | 118 | 1,725 | 14,000 | 42,700 | 3,055 |
| Chesterfield | 114 | 2,474 | 65.5 | 1.6 | 7.9 | 0.0 | 2.3 | 83.0 | 1,804 | 95 | 165 | 1,985 | 17,990 | 41,694 | 2,923 |
| Clarendon | 118 | 3,481 | 52.2 | 12.7 | 5.2 | 0.0 | 4.2 | 61.4 | 1,806 | 65 | 119 | 2,109 | 13,270 | 40,026 | 2,923 |
| Colleton | 143 | 3,813 | 45.1 | 1.5 | 6.2 | 0.7 | 3.4 | 132.2 | 3,521 | 94 | 145 | 1,892 | 17,110 | 42,787 | 3,588 |
| Darlington | 167 | 2,491 | 68.7 | 1.7 | 7.0 | 0.1 | 0.9 | 76.9 | 1,148 | 156 | 240 | 3,199 | 27,880 | 47,251 | 4,248 |
| Dillon | 94 | 3,094 | 60.1 | 0.4 | 5.4 | 0.3 | 3.9 | 3.4 | 112 | 83 | 111 | 1,257 | 11,990 | 36,639 | 2,367 |
| Dorchester | 420 | 2,639 | 74.5 | 0.2 | 7.1 | 0.0 | 1.3 | 542.6 | 3,413 | 251 | 597 | 6,604 | 73,200 | 58,134 | 5,200 |
| Edgefield | 50 | 1,856 | 76.9 | 0.0 | 9.0 | 0.0 | 2.7 | 11.4 | 426 | 377 | 90 | 1,158 | 10,430 | 54,671 | 4,859 |
| Fairfield | 87 | 3,841 | 56.2 | 5.3 | 7.5 | 0.0 | 0.0 | 40.2 | 1,778 | 40 | 81 | 1,251 | 9,820 | 44,374 | 3,789 |
| Florence | 521 | 3,759 | 51.9 | 4.7 | 7.3 | 0.2 | 3.8 | 509.4 | 3,678 | 559 | 529 | 13,464 | 59,500 | 52,965 | 5,293 |
| Georgetown | 215 | 3,472 | 52.5 | 1.9 | 5.6 | 1.1 | 3.3 | 231.1 | 3,738 | 127 | 268 | 4,907 | 30,870 | 67,329 | 8,164 |
| Greenville | 3,513 | 6,931 | 22.7 | 57.3 | 2.9 | 0.1 | 1.1 | 2,888.8 | 5,700 | 2,165 | 1,956 | 31,213 | 235,410 | 70,838 | 8,187 |
| Greenwood | 510 | 7,237 | 23.7 | 58.5 | 2.8 | 0.1 | 1.5 | 391.1 | 5,547 | 123 | 251 | 6,862 | 29,240 | 50,018 | 4,470 |
| Hampton | 64 | 3,305 | 61.4 | 4.0 | 8.8 | 1.1 | 2.5 | 35.9 | 1,840 | 316 | 66 | 1,110 | 7,710 | 43,372 | 3,703 |
| Horry | 1,318 | 3,962 | 50.3 | 1.7 | 5.9 | 0.0 | 10.1 | 1,246.8 | 3,748 | 811 | 1,296 | 16,797 | 169,470 | 53,486 | 5,549 |
| Jasper | 85 | 2,996 | 43.0 | 1.8 | 7.0 | 0.0 | 7.5 | 74.9 | 2,627 | 56 | 106 | 1,234 | 12,550 | 49,034 | 4,230 |
| Kershaw | 308 | 4,715 | 34.7 | 45.3 | 2.4 | 0.1 | 0.6 | 286.8 | 4,398 | 102 | 244 | 2,439 | 29,150 | 52,936 | 4,590 |
| Lancaster | 302 | 3,266 | 50.6 | 2.6 | 24.2 | 0.0 | 3.3 | 373.2 | 4,038 | 115 | 354 | 4,426 | 41,690 | 63,591 | 6,336 |
| Laurens | 180 | 2,688 | 54.0 | 0.0 | 5.4 | 0.2 | 0.5 | 126.5 | 1,893 | 114 | 243 | 3,783 | 27,490 | 44,111 | 3,174 |
| Lee | 46 | 2,619 | 62.3 | 3.3 | 7.8 | 0.0 | 0.0 | 64.0 | 3,678 | 24 | 56 | 1,006 | 6,810 | 35,745 | 2,357 |
| Lexington | 2,026 | 6,979 | 35.5 | 47.4 | 2.6 | 0.0 | 1.1 | 1,986.1 | 6,841 | 704 | 1,092 | 21,210 | 133,010 | 63,875 | 6,642 |
| McCormick | 21 | 2,216 | 61.7 | 6.6 | 6.9 | 0.0 | 2.4 | 13.1 | 1,371 | 72 | 31 | 722 | 3,890 | 54,915 | 5,212 |
| Marion | 85 | 2,704 | 59.2 | 2.5 | 11.0 | 0.0 | 2.2 | 24.6 | 785 | 68 | 117 | 1,573 | 12,780 | 34,093 | 2,233 |
| Marlboro | 89 | 3,321 | 54.2 | 1.9 | 5.8 | 0.2 | 1.9 | 81.0 | 3,033 | 339 | 86 | 1,316 | 10,360 | 35,450 | 2,188 |
| Newberry | 193 | 5,030 | 36.5 | 30.3 | 3.5 | 0.1 | 1.2 | 126.9 | 3,305 | 113 | 137 | 2,271 | 16,440 | 49,934 | 4,168 |
| Oconee | 220 | 2,842 | 57.7 | 0.3 | 7.9 | 0.0 | 2.4 | 112.0 | 1,447 | 187 | 290 | 3,994 | 33,850 | 63,558 | 6,808 |
| Orangeburg | 470 | 5,355 | 35.6 | 42.3 | 4.4 | 0.0 | 1.1 | 1,175.4 | 13,407 | 232 | 313 | 6,412 | 36,280 | 42,890 | 3,683 |
| Pickens | 275 | 2,223 | 52.5 | 2.5 | 7.4 | 0.0 | 3.9 | 330.1 | 2,672 | 186 | 453 | 11,045 | 51,000 | 57,392 | 5,698 |
| Richland | 1,673 | 4,064 | 43.3 | 1.0 | 5.8 | 0.1 | 4.2 | 3,106.6 | 7,544 | 9,723 | 10,505 | 45,590 | 181,740 | 61,157 | 6,770 |
| Saluda | 48 | 2,354 | 53.4 | 0.4 | 6.5 | 0.4 | 3.6 | 17.3 | 852 | 36 | 74 | 997 | 7,980 | 46,851 | 3,627 |
| Spartanburg | 2,201 | 7,175 | 26.1 | 58.6 | 2.3 | 0.2 | 1.2 | 1,354.9 | 4,417 | 591 | 1,154 | 21,689 | 141,180 | 57,348 | 5,570 |

1. Based on the resident population estimated as of July 1 of the year shown.

## Table B. States and Counties — **Land Area and Population**

| State / county code | CBSA code[1] | County Type code[2] | STATE County | Land area[3] (sq. mi) | Total persons 2019 | Rank | Per square mile | White | Black | American Indian, Alaska Native | Asian and Pacific Islander | Percent Hispanic or Latino[4] | Under 5 years | 5 to 14 years | 15 to 24 years | 25 to 34 years | 35 to 44 years | 45 to 54 years |
|---|---|---|---|---|---|---|---|---|---|---|---|---|---|---|---|---|---|---|
| | | | | 1 | 2 | 3 | 4 | 5 | 6 | 7 | 8 | 9 | 10 | 11 | 12 | 13 | 14 | 15 |
| | | | **SOUTH CAROLINA— Cont'd** | | | | | | | | | | | | | | | |
| 45085 | 44940 | 3 | Sumter | 665.1 | 106,360 | 578 | 159.9 | 46.4 | 48.3 | 0.9 | 2.3 | 4.3 | 6.6 | 13.3 | 13.6 | 14.2 | 11.3 | 11.0 |
| 45087 | 46420 | 2 | Union | 513.6 | 26,991 | 1,525 | 52.6 | 66.0 | 33.0 | 0.6 | 0.7 | 1.8 | 5.5 | 11.9 | 11.2 | 12.0 | 10.6 | 13.1 |
| 45089 | | 6 | Williamsburg | 934.2 | 29,825 | 1,440 | 31.9 | 32.7 | 64.4 | 0.6 | 1.0 | 2.4 | 4.9 | 11.4 | 11.5 | 12.1 | 11.1 | 12.2 |
| 45091 | 16740 | 1 | York | 681.0 | 289,105 | 243 | 424.5 | 70.9 | 20.2 | 1.3 | 3.6 | 6.3 | 5.7 | 14.0 | 12.1 | 12.4 | 14.1 | 13.9 |
| 46000 | | 0 | SOUTH DAKOTA | 75,809.7 | 892,717 | X | 11.8 | 83.2 | 3.0 | 9.5 | 2.2 | 4.4 | 6.8 | 13.7 | 13.3 | 12.9 | 12.1 | 10.5 |
| 46003 | | 9 | Aurora | 708.5 | 2,730 | 2,988 | 3.9 | 88.3 | 1.1 | 3.4 | 0.9 | 7.5 | 6.6 | 13.6 | 11.9 | 10.5 | 12.3 | 10.4 |
| 46005 | 26700 | 7 | Beadle | 1,258.7 | 18,513 | 1,898 | 14.7 | 75.5 | 2.1 | 1.7 | 10.9 | 11.6 | 8.5 | 15.0 | 11.6 | 11.8 | 11.3 | 10.4 |
| 46007 | | 9 | Bennett | 1,184.6 | 3,399 | 2,941 | 2.9 | 38.0 | 1.4 | 58.5 | 1.0 | 6.0 | 8.7 | 18.9 | 14.4 | 12.8 | 10.7 | 9.0 |
| 46009 | | 9 | Bon Homme | 563.5 | 6,848 | 2,671 | 12.2 | 87.4 | 1.8 | 8.6 | 0.6 | 2.9 | 5.5 | 10.6 | 12.1 | 13.4 | 12.9 | 11.3 |
| 46011 | 15100 | 5 | Brookings | 792.2 | 35,603 | 1,290 | 44.9 | 90.7 | 1.9 | 1.6 | 3.6 | 3.8 | 6.1 | 11.4 | 29.0 | 12.6 | 10.5 | 8.2 |
| 46013 | 10100 | 5 | Brown | 1,713.0 | 38,738 | 1,205 | 22.6 | 87.5 | 2.3 | 4.3 | 4.2 | 3.7 | 6.5 | 13.4 | 13.5 | 12.9 | 12.1 | 10.5 |
| 46015 | | 9 | Brule | 817.2 | 5,254 | 2,806 | 6.4 | 86.0 | 1.2 | 12.4 | 1.0 | 3.0 | 6.2 | 15.6 | 12.4 | 10.1 | 10.8 | 11.2 |
| 46017 | | 9 | Buffalo | 471.4 | 1,956 | 3,052 | 4.1 | 18.1 | 1.1 | 78.1 | 0.7 | 4.7 | 9.2 | 21.5 | 17.8 | 12.3 | 11.1 | 9.9 |
| 46019 | | 6 | Butte | 2,250.0 | 10,538 | 2,387 | 4.7 | 92.6 | 0.9 | 3.8 | 1.0 | 4.1 | 6.7 | 14.0 | 11.2 | 11.3 | 11.2 | 10.4 |
| 46021 | | 9 | Campbell | 733.7 | 1,380 | 3,088 | 1.9 | 95.1 | 0.6 | 1.7 | 0.4 | 3.1 | 4.8 | 9.6 | 8.2 | 7.8 | 9.6 | 8.6 |
| 46023 | | 9 | Charles Mix | 1,097.5 | 9,262 | 2,475 | 8.4 | 65.2 | 1.0 | 32.1 | 0.9 | 4.0 | 8.6 | 17.2 | 12.1 | 10.7 | 10.1 | 9.7 |
| 46025 | | 9 | Clark | 957.5 | 3,802 | 2,911 | 4.0 | 94.0 | 1.8 | 1.0 | 0.3 | 3.8 | 10.2 | 14.6 | 8.8 | 12.1 | 9.8 | 8.8 |
| 46027 | 46820 | 9 | Clay | 412.0 | 14,246 | 2,139 | 34.6 | 88.9 | 2.6 | 4.6 | 3.2 | 3.2 | 4.9 | 10.0 | 34.1 | 12.9 | 9.0 | 7.9 |
| 46029 | 47980 | 5 | Codington | 687.6 | 28,186 | 1,490 | 41.0 | 92.8 | 1.1 | 3.0 | 1.2 | 3.4 | 5.9 | 13.5 | 12.0 | 12.4 | 12.3 | 11.4 |
| 46031 | | 9 | Corson | 2,469.7 | 4,031 | 2,893 | 1.6 | 31.8 | 1.1 | 64.3 | 1.1 | 5.1 | 10.7 | 19.6 | 14.8 | 11.3 | 10.9 | 8.8 |
| 46033 | | 9 | Custer | 1,556.9 | 9,017 | 2,495 | 5.8 | 91.0 | 1.1 | 4.7 | 1.0 | 4.0 | 3.5 | 7.4 | 7.8 | 7.3 | 10.3 | 11.3 |
| 46035 | 33580 | 7 | Davison | 435.6 | 19,812 | 1,834 | 45.5 | 91.3 | 1.5 | 3.9 | 1.3 | 4.0 | 6.1 | 13.3 | 13.4 | 12.6 | 11.6 | 10.1 |
| 46037 | | 9 | Day | 1,028.5 | 5,345 | 2,799 | 5.2 | 87.6 | 1.0 | 10.0 | 1.0 | 2.4 | 5.3 | 13.1 | 10.0 | 9.3 | 9.9 | 9.9 |
| 46039 | | 9 | Deuel | 622.7 | 4,346 | 2,869 | 7.0 | 94.5 | 1.2 | 1.4 | 0.5 | 3.7 | 6.2 | 13.5 | 10.4 | 11.4 | 10.9 | 10.5 |
| 46041 | | 9 | Dewey | 2,302.4 | 5,789 | 2,759 | 2.5 | 23.3 | 0.9 | 75.2 | 0.8 | 4.4 | 11.8 | 20.9 | 13.9 | 12.4 | 11.0 | 8.9 |
| 46043 | | 9 | Douglas | 431.8 | 2,906 | 2,970 | 6.7 | 94.6 | 1.0 | 3.4 | 0.4 | 2.1 | 7.9 | 13.6 | 11.4 | 9.0 | 9.8 | 9.1 |
| 46045 | 10100 | 9 | Edmunds | 1,126.0 | 3,817 | 2,909 | 3.4 | 96.0 | 0.8 | 1.5 | 0.6 | 2.1 | 6.0 | 12.6 | 10.6 | 10.1 | 10.7 | 10.7 |
| 46047 | | 6 | Fall River | 1,739.9 | 6,708 | 2,686 | 3.9 | 86.7 | 1.6 | 8.5 | 2.2 | 4.1 | 3.5 | 9.9 | 8.9 | 8.6 | 9.5 | 11.6 |
| 46049 | | 9 | Faulk | 981.7 | 2,306 | 3,015 | 2.3 | 97.8 | 0.5 | 0.7 | 0.6 | 0.9 | 8.2 | 13.5 | 10.5 | 10.2 | 9.8 | 9.1 |
| 46051 | | 7 | Grant | 681.4 | 7,000 | 2,657 | 10.3 | 93.2 | 1.0 | 1.6 | 0.6 | 4.8 | 5.9 | 12.5 | 10.5 | 9.7 | 10.7 | 11.8 |
| 46053 | | 9 | Gregory | 1,015.0 | 4,219 | 2,878 | 4.2 | 89.9 | 1.0 | 9.8 | 0.9 | 1.5 | 6.3 | 13.6 | 9.0 | 9.0 | 10.4 | 10.4 |
| 46055 | | 9 | Haakon | 1,810.5 | 1,861 | 3,060 | 1.0 | 94.3 | 1.2 | 4.2 | 1.6 | 1.9 | 5.6 | 13.4 | 11.1 | 9.2 | 11.1 | 8.9 |
| 46057 | 47980 | 9 | Hamlin | 507.0 | 6,234 | 2,728 | 12.3 | 93.6 | 0.7 | 1.3 | 0.6 | 5.0 | 9.6 | 17.2 | 13.7 | 11.0 | 10.7 | 10.0 |
| 46059 | 33580 | 9 | Hand | 1,436.6 | 3,127 | 2,954 | 2.2 | 97.1 | 0.3 | 1.2 | 0.7 | 1.7 | 5.4 | 12.8 | 10.1 | 9.7 | 10.6 | 9.3 |
| 46061 | | 8 | Hanson | 434.6 | 3,489 | 2,936 | 8.0 | 96.8 | 0.9 | 1.7 | 0.9 | 1.4 | 7.7 | 16.3 | 13.0 | 9.3 | 12.6 | 11.6 |
| 46063 | | 9 | Harding | 2,671.6 | 1,311 | 3,093 | 0.5 | 92.6 | 2.1 | 3.7 | 0.7 | 2.6 | 6.6 | 12.1 | 11.5 | 12.7 | 11.0 | 11.0 |
| 46065 | 38180 | 7 | Hughes | 741.5 | 17,336 | 1,953 | 23.4 | 83.5 | 1.2 | 13.2 | 1.2 | 3.4 | 6.5 | 14.3 | 10.9 | 12.7 | 12.4 | 11.2 |
| 46067 | | 8 | Hutchinson | 813.0 | 7,282 | 2,633 | 9.0 | 95.2 | 1.2 | 1.8 | 0.5 | 2.3 | 8.6 | 13.3 | 10.6 | 9.7 | 10.5 | 10.7 |
| 46069 | | 9 | Hyde | 860.6 | 1,281 | 3,094 | 1.5 | 88.1 | 1.2 | 10.9 | 0.7 | 1.7 | 6.3 | 11.3 | 10.7 | 10.8 | 9.8 | 8.5 |
| 46071 | | 9 | Jackson | 1,863.9 | 3,321 | 2,943 | 1.8 | 44.0 | 1.4 | 55.5 | 0.7 | 3.8 | 11.0 | 18.5 | 14.4 | 11.7 | 9.7 | 9.6 |
| 46073 | 26700 | 9 | Jerauld | 526.1 | 1,985 | 3,048 | 3.8 | 94.2 | 0.5 | 1.5 | 0.5 | 4.6 | 6.2 | 13.9 | 8.4 | 9.2 | 11.6 | 10.0 |
| 46075 | | 9 | Jones | 969.7 | 938 | 3,109 | 1.0 | 91.5 | 1.2 | 7.9 | 0.6 | 2.7 | 6.7 | 12.6 | 8.7 | 9.7 | 8.6 | 12.2 |
| 46077 | | 9 | Kingsbury | 832.2 | 4,987 | 2,827 | 6.0 | 95.1 | 0.7 | 1.5 | 0.8 | 3.0 | 6.9 | 12.8 | 10.3 | 10.6 | 10.8 | 9.5 |
| 46079 | 43940 | 6 | Lake | 562.9 | 12,488 | 2,247 | 22.2 | 94.2 | 1.3 | 1.9 | 1.3 | 2.7 | 5.2 | 11.0 | 14.5 | 9.4 | 10.5 | 9.4 |
| 46081 | | 6 | Lawrence | 800.1 | 26,221 | 1,561 | 32.8 | 92.0 | 1.1 | 3.3 | 1.9 | 3.7 | 4.1 | 9.8 | 14.5 | 11.6 | 11.3 | 10.3 |
| 46083 | 43620 | 3 | Lincoln | 577.3 | 63,019 | 848 | 109.2 | 93.3 | 2.6 | 1.2 | 2.1 | 2.6 | 7.1 | 15.9 | 11.6 | 14.5 | 14.9 | 11.3 |
| 46085 | | 9 | Lyman | 1,642.3 | 3,797 | 2,912 | 2.3 | 58.0 | 1.1 | 40.1 | 1.0 | 2.8 | 8.8 | 15.0 | 13.4 | 11.0 | 11.1 | 9.3 |
| 46087 | 43620 | 6 | McCook | 574.2 | 5,520 | 2,785 | 9.6 | 93.8 | 0.8 | 1.6 | 0.6 | 4.3 | 8.3 | 15.0 | 11.7 | 10.2 | 11.5 | 11.2 |
| 46089 | | 9 | McPherson | 1,136.7 | 2,363 | 3,012 | 2.1 | 97.1 | 1.1 | 1.4 | 0.6 | 1.4 | 6.3 | 14.3 | 9.6 | 8.8 | 9.4 | 9.5 |
| 46091 | | 9 | Marshall | 838.1 | 4,884 | 2,833 | 5.8 | 84.7 | 1.0 | 7.4 | 0.3 | 8.0 | 7.4 | 12.5 | 10.0 | 11.4 | 11.2 | 10.1 |
| 46093 | 39660 | 9 | Meade | 3,470.9 | 28,588 | 1,476 | 8.2 | 89.2 | 2.7 | 4.4 | 2.1 | 4.4 | 5.1 | 13.3 | 14.6 | 14.5 | 12.9 | 10.1 |
| 46095 | | 9 | Mellette | 1,307.3 | 2,089 | 3,042 | 1.6 | 41.6 | 1.6 | 58.4 | 0.8 | 3.7 | 9.7 | 15.9 | 14.0 | 12.4 | 8.9 | 10.1 |
| 46097 | | 8 | Miner | 570.2 | 2,202 | 3,025 | 3.9 | 94.6 | 1.5 | 1.0 | 0.8 | 3.4 | 5.9 | 13.1 | 12.3 | 9.3 | 9.5 | 9.5 |
| 46099 | 43620 | 3 | Minnehaha | 806.8 | 196,659 | 342 | 243.8 | 83.4 | 7.6 | 3.2 | 2.9 | 5.4 | 7.4 | 14.0 | 12.5 | 15.4 | 13.7 | 11.0 |
| 46101 | | 8 | Moody | 519.4 | 6,525 | 2,702 | 12.6 | 79.7 | 1.9 | 14.7 | 2.3 | 4.6 | 7.6 | 14.7 | 11.6 | 9.6 | 11.7 | 10.6 |
| 46102 | | 0 | Oglala Lakota | 2,093.6 | 14,070 | 2,155 | 6.7 | 5.9 | 0.5 | 90.3 | 0.4 | 4.2 | 9.6 | 21.6 | 16.5 | 15.6 | 11.0 | 9.7 |
| 46103 | 39660 | 3 | Pennington | 2,776.8 | 115,926 | 538 | 41.7 | 82.9 | 2.3 | 10.8 | 2.2 | 5.6 | 6.3 | 12.6 | 11.6 | 13.1 | 11.8 | 10.6 |
| 46105 | | 8 | Perkins | 2,870.5 | 2,832 | 2,976 | 1.0 | 94.5 | 1.1 | 3.1 | 0.8 | 1.9 | 6.1 | 12.0 | 9.3 | 7.9 | 11.0 | 10.8 |
| 46107 | | 9 | Potter | 861.1 | 2,163 | 3,029 | 2.5 | 93.0 | 1.1 | 3.6 | 1.7 | 2.6 | 5.4 | 12.5 | 9.6 | 9.8 | 9.3 | 9.2 |
| 46109 | | 9 | Roberts | 1,101.1 | 10,331 | 2,397 | 9.4 | 58.7 | 1.2 | 38.5 | 0.9 | 4.1 | 8.4 | 16.7 | 12.0 | 10.2 | 9.8 | 10.0 |
| 46111 | | 9 | Sanborn | 569.2 | 2,322 | 3,013 | 4.1 | 95.2 | 0.4 | 2.1 | 0.5 | 3.3 | 7.7 | 13.3 | 9.8 | 10.9 | 12.7 | 9.3 |
| 46115 | | 8 | Spink | 1,503.5 | 6,319 | 2,720 | 4.2 | 93.8 | 1.1 | 2.2 | 0.3 | 3.6 | 6.5 | 11.7 | 11.2 | 10.7 | 11.6 | 10.4 |
| 46117 | 38180 | 9 | Stanley | 1,444.4 | 3,121 | 2,955 | 2.2 | 90.1 | 1.4 | 8.9 | 0.5 | 2.4 | 6.8 | 13.0 | 10.5 | 10.2 | 12.4 | 11.1 |
| 46119 | | 9 | Sully | 1,006.7 | 1,391 | 3,085 | 1.4 | 93.5 | 1.2 | 3.7 | 0.2 | 2.9 | 6.4 | 11.7 | 9.7 | 8.8 | 10.6 | 10.4 |
| 46121 | | 9 | Todd | 1,388.6 | 10,313 | 2,399 | 7.4 | 9.3 | 0.8 | 84.2 | 3.5 | 4.1 | 11.7 | 24.2 | 16.3 | 12.7 | 10.7 | 8.5 |

1. CBSA = Core Based Statistical Area. See Appendix A for explanation. See Appendix B for list of metropolitan areas with component counties. 2. County type code from the Economic Research Service of USDA Rural-Urban Continuum Codes. See Appendix A for definition. 3. Dry land or land partially or temporarily covered by water. 4. May be of any race.

Items 1—15

# Population and Households

| STATE County | 55 to 64 years | 65 to 74 years | 75 years and over | Percent female | 2000 | 2010 | 2000-2010 | 2010-2020 | Births | Deaths | Net Migration | Number | Persons per household | Family house-holds | Female family house-holder[1] | One person |
|---|---|---|---|---|---|---|---|---|---|---|---|---|---|---|---|---|
| | 16 | 17 | 18 | 19 | 20 | 21 | 22 | 23 | 24 | 25 | 26 | 27 | 28 | 29 | 30 | 31 |
| **SOUTH CAROLINA— Cont'd** | | | | | | | | | | | | | | | | |
| | 12.7 | 10.1 | 7.2 | 52.0 | 104,646 | 107,485 | 2.7 | -1.0 | 14,906 | 11,266 | -4,741 | 41,776 | 2.5 | 67.3 | 19.1 | 28.5 |
| Sumter | 14.9 | 12.5 | 8.5 | 52.7 | 29,881 | 28,956 | -3.1 | -6.8 | 3,129 | 3,976 | -1,114 | 11,432 | 2.36 | 67.4 | 17.7 | 30.2 |
| Union | 14.2 | 13.7 | 8.9 | 52.8 | 37,217 | 34,405 | -7.6 | -13.3 | 3,348 | 4,341 | -3,618 | 12,686 | 2.35 | 63.6 | 19.2 | 33.9 |
| Williamsburg | 12.7 | 9.3 | 5.7 | 51.8 | 164,614 | 226,033 | 37.3 | 27.9 | 30,073 | 20,945 | 53,625 | 101,211 | 2.59 | 69.6 | 13.1 | 25.3 |
| York | | | | | | | | | | | | | | | | |
| **SOUTH DAKOTA** | 13.1 | 10.5 | 7.1 | 49.5 | 754,844 | 814,198 | 7.9 | 9.6 | 123,483 | 77,930 | 32,742 | 344,397 | 2.43 | 63.4 | 9.3 | 30.0 |
| Aurora | 13.7 | 11.5 | 9.6 | 47.7 | 3,058 | 2,710 | -11.4 | 0.7 | 387 | 287 | -80 | 1,182 | 2.19 | 63.5 | 6.6 | 30.3 |
| Beadle | 13.5 | 9.9 | 8.0 | 49.9 | 17,023 | 17,399 | 2.2 | 6.4 | 3,177 | 1,989 | -84 | 7,557 | 2.35 | 60.7 | 8.1 | 34.7 |
| Bennett | 11.9 | 7.4 | 6.1 | 50.7 | 3,574 | 3,431 | -4.0 | -0.9 | 661 | 347 | -351 | 984 | 3.44 | 74.0 | 22.3 | 23.3 |
| Bon Homme | 13.4 | 10.6 | 10.2 | 41.0 | 7,260 | 7,067 | -2.7 | -3.1 | 682 | 782 | -115 | 2,536 | 2.16 | 58.8 | 5.6 | 34.3 |
| Brookings | 9.4 | 7.7 | 5.1 | 48.4 | 28,220 | 31,966 | 13.3 | 11.4 | 4,297 | 2,010 | 1,355 | 12,896 | 2.4 | 53.3 | 4.8 | 32.9 |
| Brown | 13.0 | 10.0 | 8.1 | 50.8 | 35,460 | 36,532 | 3.0 | 6.0 | 5,109 | 3,779 | 870 | 16,305 | 2.31 | 60.6 | 8.2 | 33.0 |
| Brule | 14.7 | 10.4 | 8.7 | 50.0 | 5,364 | 5,255 | -2.0 | 0.0 | 745 | 552 | -196 | 2,231 | 2.21 | 64.6 | 5.3 | 29.9 |
| Buffalo | 9.2 | 5.5 | 3.4 | 50.1 | 2,032 | 1,913 | -5.9 | 2.2 | 461 | 222 | -195 | 564 | 3.59 | 74.8 | 33.7 | 22.0 |
| Butte | 14.5 | 12.4 | 8.3 | 49.4 | 9,094 | 10,112 | 11.2 | 4.2 | 1,331 | 1,102 | 200 | 4,164 | 2.42 | 61.3 | 10.5 | 31.0 |
| Campbell | 20.8 | 14.8 | 15.9 | 48.0 | 1,782 | 1,466 | -17.7 | -5.9 | 121 | 145 | -64 | 673 | 2.2 | 64.0 | 2.5 | 32.2 |
| Charles Mix | 12.9 | 9.5 | 9.2 | 49.5 | 9,350 | 9,131 | -2.3 | 1.4 | 1,611 | 1,053 | -424 | 3,166 | 2.86 | 65.9 | 11.3 | 30.5 |
| Clark | 13.4 | 12.8 | 9.6 | 48.6 | 4,143 | 3,691 | -10.9 | 3.0 | 598 | 453 | -30 | 1,546 | 2.33 | 60.8 | 6.5 | 32.5 |
| Clay | 8.7 | 7.6 | 4.9 | 50.4 | 13,537 | 13,866 | 2.4 | 2.7 | 1,481 | 1,018 | -93 | 5,233 | 2.23 | 51.9 | 8.0 | 30.2 |
| Codington | 13.8 | 10.4 | 8.3 | 49.8 | 25,897 | 27,225 | 5.1 | 3.5 | 3,685 | 2,516 | -196 | 12,075 | 2.27 | 61.7 | 9.4 | 30.1 |
| Corson | 11.6 | 7.7 | 4.6 | 50.7 | 4,181 | 4,040 | -3.4 | -0.2 | 928 | 497 | -446 | 1,195 | 3.47 | 73.6 | 21.8 | 22.9 |
| Custer | 19.8 | 21.3 | 11.4 | 50.0 | 7,275 | 8,219 | 13.0 | 9.7 | 745 | 928 | 980 | 3,876 | 2.15 | 74.1 | 7.8 | 20.5 |
| Davison | 12.9 | 10.6 | 9.5 | 50.2 | 18,741 | 19,504 | 4.1 | 1.6 | 2,616 | 2,203 | -97 | 8,660 | 2.17 | 60.8 | 6.6 | 33.3 |
| Day | 15.5 | 14.7 | 12.4 | 49.3 | 6,267 | 5,710 | -8.9 | -6.4 | 637 | 780 | -220 | 2,574 | 2.08 | 62.8 | 7.3 | 33.3 |
| Deuel | 14.9 | 12.7 | 9.5 | 48.1 | 4,498 | 4,364 | -3.0 | -0.4 | 519 | 446 | -90 | 1,798 | 2.37 | 69.1 | 3.3 | 27.4 |
| Dewey | 11.1 | 6.0 | 4.0 | 50.7 | 5,972 | 5,301 | -11.2 | 9.2 | 1,462 | 649 | -324 | 1,651 | 3.5 | 72.3 | 21.6 | 23.9 |
| Douglas | 14.9 | 12.3 | 12.0 | 50.6 | 3,458 | 2,997 | -13.3 | -3.0 | 408 | 430 | -66 | 1,220 | 2.31 | 72.5 | 2.0 | 24.0 |
| Edmunds | 16.3 | 12.2 | 10.8 | 49.7 | 4,367 | 4,071 | -6.8 | -6.2 | 447 | 445 | -253 | 1,544 | 2.46 | 69.2 | 5.3 | 26.7 |
| Fall River | 17.3 | 18.8 | 11.8 | 49.7 | 7,453 | 7,094 | -4.8 | -5.4 | 562 | 1,186 | 236 | 2,986 | 2.16 | 61.2 | 13.8 | 34.5 |
| Faulk | 14.2 | 12.3 | 12.1 | 49.7 | 2,640 | 2,364 | -10.5 | -2.5 | 316 | 290 | -82 | 895 | 2.25 | 61.1 | 4.9 | 36.8 |
| Grant | 16.5 | 12.7 | 9.6 | 47.8 | 7,847 | 7,358 | -6.2 | -4.9 | 827 | 868 | -322 | 3,218 | 2.17 | 65.8 | 7.6 | 27.2 |
| Gregory | 14.0 | 13.8 | 12.1 | 48.9 | 4,792 | 4,271 | -10.9 | -1.2 | 515 | 629 | 66 | 1,840 | 2.25 | 60.8 | 8.6 | 35.6 |
| Haakon | 15.5 | 13.4 | 11.8 | 49.9 | 2,196 | 1,937 | -11.8 | -3.9 | 207 | 252 | -30 | 822 | 2.41 | 73.2 | 3.4 | 26.3 |
| Hamlin | 12.0 | 8.7 | 7.2 | 48.4 | 5,540 | 5,899 | 6.5 | 5.7 | 1,221 | 680 | -208 | 2,256 | 2.59 | 74.5 | 5.9 | 17.9 |
| Hand | 15.7 | 12.7 | 13.6 | 50.1 | 3,741 | 3,430 | -8.3 | -8.8 | 366 | 442 | -231 | 1,508 | 2.12 | 64.7 | 7.6 | 30.6 |
| Hanson | 13.4 | 12.1 | 4.2 | 49.4 | 3,139 | 3,332 | 6.1 | 4.7 | 455 | 222 | -76 | 1,045 | 3.25 | 73.0 | 5.5 | 24.8 |
| Harding | 15.0 | 12.8 | 7.8 | 47.0 | 1,353 | 1,255 | -7.2 | 4.5 | 165 | 81 | -34 | 532 | 2.31 | 72.7 | 3.8 | 21.1 |
| Hughes | 13.7 | 10.9 | 7.3 | 50.8 | 16,481 | 17,022 | 3.3 | 1.8 | 2,411 | 1,546 | -548 | 7,460 | 2.2 | 59.2 | 7.1 | 35.8 |
| Hutchinson | 13.8 | 11.1 | 11.8 | 50.8 | 8,075 | 7,346 | -9.0 | -0.9 | 1,068 | 1,131 | 4 | 2,903 | 2.42 | 65.2 | 5.4 | 32.5 |
| Hyde | 16.2 | 13.4 | 12.9 | 49.2 | 1,671 | 1,420 | -15.0 | -9.8 | 160 | 196 | -105 | 597 | 2.14 | 64.8 | 5.2 | 32.3 |
| Jackson | 11.3 | 7.8 | 6.0 | 50.2 | 2,930 | 3,030 | 3.4 | 9.6 | 759 | 325 | -142 | 858 | 3.66 | 73.1 | 21.9 | 24.5 |
| Jerauld | 13.0 | 15.0 | 12.8 | 50.1 | 2,295 | 2,071 | -9.8 | -4.2 | 237 | 280 | -43 | 921 | 2.14 | 62.1 | 2.2 | 31.8 |
| Jones | 15.6 | 14.3 | 11.6 | 49.3 | 1,193 | 1,006 | -15.7 | -6.8 | 116 | 100 | -84 | 408 | 1.94 | 58.3 | 3.4 | 38.7 |
| Kingsbury | 15.5 | 12.8 | 10.7 | 49.3 | 5,815 | 5,147 | -11.5 | -3.1 | 641 | 681 | -127 | 2,279 | 2.13 | 62.3 | 6.1 | 31.8 |
| Lake | 16.1 | 16.4 | 7.5 | 47.7 | 11,276 | 11,200 | -0.7 | 11.5 | 1,325 | 1,143 | 1,086 | 4,700 | 2.55 | 65.3 | 3.1 | 30.1 |
| Lawrence | 15.1 | 14.6 | 8.7 | 49.8 | 21,802 | 24,090 | 10.5 | 8.8 | 2,345 | 2,352 | 2,117 | 11,048 | 2.18 | 56.1 | 6.7 | 34.3 |
| Lincoln | 10.6 | 8.7 | 5.3 | 50.5 | 24,131 | 44,827 | 85.8 | 40.6 | 8,281 | 2,425 | 12,207 | 19,749 | 2.85 | 73.7 | 8.5 | 21.2 |
| Lyman | 13.7 | 9.9 | 7.6 | 47.5 | 3,895 | 3,755 | -3.6 | 1.1 | 707 | 346 | -321 | 1,397 | 2.7 | 69.3 | 13.0 | 23.4 |
| McCook | 13.1 | 10.3 | 8.7 | 49.4 | 5,832 | 5,618 | -3.7 | -1.7 | 812 | 763 | -145 | 2,228 | 2.37 | 65.8 | 3.9 | 25.0 |
| McPherson | 13.8 | 12.9 | 15.4 | 50.5 | 2,904 | 2,459 | -15.3 | -3.9 | 244 | 360 | 21 | 1,030 | 2.09 | 57.9 | 2.4 | 40.2 |
| Marshall | 14.6 | 13.3 | 9.5 | 46.7 | 4,576 | 4,656 | 1.7 | 4.9 | 678 | 510 | 61 | 1,873 | 2.57 | 65.6 | 6.7 | 29.5 |
| Meade | 13.3 | 10.3 | 6.0 | 48.0 | 24,253 | 25,433 | 4.9 | 12.4 | 2,994 | 2,003 | 2,152 | 11,027 | 2.44 | 68.7 | 7.0 | 26.6 |
| Mellette | 13.3 | 9.7 | 6.0 | 49.5 | 2,083 | 2,043 | -1.9 | 2.3 | 363 | 260 | -57 | 661 | 3.04 | 72.9 | 17.4 | 25.1 |
| Miner | 17.4 | 12.1 | 10.9 | 48.7 | 2,884 | 2,389 | -17.2 | -7.8 | 247 | 329 | -105 | 948 | 2.29 | 61.6 | 4.6 | 34.2 |
| Minnehaha | 12.3 | 8.8 | 5.0 | 49.4 | 148,281 | 169,474 | 14.3 | 16.0 | 29,573 | 14,126 | 11,726 | 76,819 | 2.38 | 62.6 | 10.5 | 30.8 |
| Moody | 14.5 | 11.4 | 8.3 | 49.7 | 6,595 | 6,493 | -1.5 | 0.5 | 907 | 559 | -319 | 2,640 | 2.35 | 66.9 | 6.3 | 28.0 |
| Oglala Lakota | 8.7 | 4.9 | 2.5 | 50.9 | 12,466 | 13,586 | 9.0 | 3.6 | 3,227 | 1,487 | -1,259 | 2,729 | 5.14 | 79.8 | 41.3 | 16.8 |
| Pennington | 14.6 | 12.5 | 7.0 | 49.6 | 88,565 | 100,966 | 14.0 | 14.8 | 15,630 | 9,068 | 8,361 | 44,074 | 2.42 | 62.1 | 10.9 | 31.6 |
| Perkins | 16.3 | 14.0 | 12.5 | 49.9 | 3,363 | 2,991 | -11.1 | -5.3 | 361 | 421 | -98 | 1,284 | 2.17 | 63.9 | 3.0 | 31.6 |
| Potter | 15.0 | 14.7 | 14.4 | 51.5 | 2,693 | 2,334 | -13.3 | -7.3 | 237 | 344 | -63 | 982 | 2.27 | 64.1 | 5.1 | 33.2 |
| Roberts | 12.8 | 11.2 | 8.9 | 49.2 | 10,016 | 10,147 | 1.3 | 1.8 | 1,789 | 1,170 | -433 | 3,873 | 2.6 | 66.5 | 15.2 | 29.3 |
| Sanborn | 15.9 | 12.9 | 7.5 | 48.4 | 2,675 | 2,355 | -12.0 | -1.4 | 341 | 272 | -100 | 1,010 | 2.31 | 70.4 | 8.4 | 25.0 |
| Spink | 15.8 | 11.8 | 10.3 | 49.7 | 7,454 | 6,415 | -13.9 | -1.5 | 811 | 757 | -156 | 2,624 | 2.42 | 63.5 | 8.8 | 30.9 |
| Stanley | 13.6 | 12.4 | 9.9 | 49.8 | 2,772 | 2,966 | 7.0 | 5.2 | 386 | 197 | -34 | 1,344 | 2.24 | 72.4 | 7.9 | 22.9 |
| Sully | 16.8 | 14.8 | 10.8 | 48.0 | 1,556 | 1,368 | -12.1 | 1.7 | 171 | 103 | -49 | 569 | 2.28 | 66.3 | 6.9 | 31.1 |
| Todd | 8.4 | 5.1 | 2.4 | 51.5 | 9,050 | 9,610 | 6.2 | 7.3 | 2,689 | 1,026 | -964 | 2,700 | 3.75 | 70.9 | 33.8 | 25.1 |

1. No spouse present.

# Table B. States and Counties — Population, Vital Statistics, Health, and Crime

| STATE County | Persons in group quarters, 2020 | Daytime Population, 2015-2019 | | Births, 2020 | | Deaths, 2020 | | Persons under 65 with no health insurance, 2019 | | Medicare, 2020 | | | Crimes reported by county police or sheriff, 2019 | |
|---|---|---|---|---|---|---|---|---|---|---|---|---|---|---|
| | | Number | Employment/residence ratio | Total | Rate[1] | Number | Rate[1] | Number | Percent | Total beneficiaries | Enrolled in Original Medicare | Enrolled in Medicare Advantage | Violent | Property |
| | 32 | 33 | 34 | 35 | 36 | 37 | 38 | 39 | 40 | 41 | 42 | 43 | 44 | 45 |
| **SOUTH CAROLINA—Cont'd** | | | | | | | | | | | | | | |
| Sumter | 2,544 | 106,799 | 1.00 | 1,366 | 12.8 | 1,200 | 11.3 | 10,652 | 12.5 | 22,329 | 15,440 | 6,889 | 384 | 1,543 |
| Union | 444 | 25,524 | 0.83 | 280 | 10.4 | 411 | 15.2 | 2,642 | 12.5 | 7,397 | 3,956 | 3,441 | 53 | 376 |
| Williamsburg | 919 | 29,896 | 0.87 | 271 | 9.1 | 445 | 14.9 | 3,122 | 13.9 | 8,228 | 4,686 | 3,543 | NA | NA |
| York | 3,426 | 245,879 | 0.85 | 2,969 | 10.3 | 2,465 | 8.5 | 27,603 | 11.6 | 48,781 | 32,408 | 16,372 | 377 | 2,565 |
| **SOUTH DAKOTA** | 34,043 | 874,000 | 1.01 | 11,686 | 13.1 | 7,920 | 8.9 | 85,613 | 12.0 | 180,166 | 138,646 | 41,520 | X | X |
| Aurora | 96 | 2,455 | 0.79 | 38 | 13.9 | 24 | 8.8 | 372 | 17.7 | 599 | 477 | 122 | 5 | 1 |
| Beadle | 620 | 18,217 | 0.99 | 287 | 15.5 | 202 | 10.9 | 2,434 | 16.4 | 3,721 | 2,826 | 896 | 4 | 7 |
| Bennett | 37 | 3,281 | 0.87 | 46 | 13.5 | 30 | 8.8 | 612 | 21.8 | 514 | 487 | 27 | 0 | 6 |
| Bon Homme | 1,534 | 6,304 | 0.77 | 71 | 10.4 | 77 | 11.2 | 543 | 12.9 | 1,546 | 1,214 | 332 | 2 | 13 |
| Brookings | 3,616 | 35,220 | 1.03 | 419 | 11.8 | 217 | 6.1 | 2,662 | 9.7 | 5,062 | 3,963 | 1,099 | 9 | 45 |
| Brown | 1,369 | 40,239 | 1.06 | 477 | 12.3 | 354 | 9.1 | 3,796 | 12.2 | 7,602 | 6,672 | 931 | 9 | 40 |
| Brule | 132 | 5,287 | 1.01 | 65 | 12.4 | 50 | 9.5 | 728 | 17.5 | 1,128 | 900 | 229 | 3 | 1 |
| Buffalo | 3 | 1,975 | 0.91 | 31 | 15.8 | 30 | 15.3 | 318 | 18.4 | 225 | D | D | NA | NA |
| Butte | 111 | 8,945 | 0.73 | 135 | 12.8 | 117 | 11.1 | 1,297 | 15.9 | 2,447 | 1,669 | 778 | 6 | 29 |
| Campbell | 0 | 1,430 | 0.93 | 12 | 8.7 | 9 | 6.5 | 112 | 11.5 | 422 | 337 | 85 | NA | NA |
| Charles Mix | 589 | 9,593 | 1.06 | 147 | 15.9 | 112 | 12.1 | 1,397 | 19.3 | 1,774 | 1,605 | 169 | 11 | 19 |
| Clark | 489 | 3,457 | 0.87 | 67 | 17.6 | 35 | 9.2 | 375 | 13.0 | 840 | 543 | 297 | 1 | 5 |
| Clay | 2,255 | 12,832 | 0.85 | 130 | 9.1 | 88 | 6.2 | 1,120 | 11.0 | 2,094 | 1,678 | 416 | 7 | 50 |
| Codington | 381 | 29,082 | 1.07 | 309 | 11.0 | 265 | 9.4 | 2,321 | 10.2 | 5,837 | 3,553 | 2,284 | 9 | 16 |
| Corson | 1 | 4,160 | 1.01 | 91 | 22.6 | 66 | 16.4 | 695 | 20.2 | 570 | 527 | 43 | 1 | 26 |
| Custer | 252 | 7,687 | 0.74 | 69 | 7.7 | 94 | 10.4 | 757 | 12.7 | 3,035 | 2,373 | 663 | 8 | 135 |
| Davison | 771 | 21,263 | 1.13 | 260 | 13.1 | 205 | 10.3 | 1,567 | 10.2 | 4,355 | 3,389 | 966 | 3 | 14 |
| Day | 139 | 5,234 | 0.90 | 61 | 11.4 | 81 | 15.2 | 610 | 15.5 | 1,546 | 1,117 | 428 | 4 | 33 |
| Deuel | 50 | 3,681 | 0.72 | 56 | 12.9 | 41 | 9.4 | 414 | 12.3 | 959 | 614 | 346 | 1 | 16 |
| Dewey | 19 | 6,091 | 1.13 | 127 | 21.9 | 76 | 13.1 | 1,065 | 20.7 | 805 | 781 | 25 | NA | NA |
| Douglas | 178 | 2,973 | 1.03 | 43 | 14.8 | 35 | 12.0 | 327 | 15.0 | 768 | 632 | 136 | NA | NA |
| Edmunds | 420 | 3,435 | 0.75 | 39 | 10.2 | 34 | 8.9 | 340 | 11.4 | 892 | 802 | 90 | 0 | 0 |
| Fall River | 199 | 7,042 | 1.10 | 46 | 6.9 | 105 | 15.7 | 598 | 13.1 | 2,560 | 2,111 | 449 | NA | NA |
| Faulk | 497 | 2,212 | 0.91 | 31 | 13.4 | 19 | 8.2 | 250 | 14.3 | 561 | 498 | 63 | 0 | 7 |
| Grant | 104 | 7,469 | 1.09 | 76 | 10.9 | 87 | 12.4 | 626 | 11.4 | 1,923 | 1,043 | 879 | NA | NA |
| Gregory | 44 | 4,081 | 0.95 | 49 | 11.6 | 56 | 13.3 | 510 | 16.6 | 1,128 | 955 | 173 | 0 | 7 |
| Haakon | 36 | 2,047 | 1.03 | 17 | 9.1 | 13 | 7.0 | 201 | 14.3 | 518 | 474 | 44 | NA | NA |
| Hamlin | 245 | 5,452 | 0.79 | 131 | 21.0 | 61 | 9.8 | 642 | 12.5 | 1,101 | 739 | 362 | 5 | 29 |
| Hand | 61 | 3,244 | 0.99 | 30 | 9.6 | 27 | 8.6 | 260 | 11.0 | 867 | 781 | 86 | 4 | 1 |
| Hanson | 520 | 2,592 | 0.50 | 44 | 12.6 | 13 | 3.7 | 294 | 10.1 | 1,327 | 1,065 | 262 | 1 | 10 |
| Harding | 29 | 1,322 | 1.02 | 16 | 12.2 | 7 | 5.3 | 198 | 19.1 | 257 | 225 | 32 | NA | NA |
| Hughes | 770 | 18,251 | 1.07 | 224 | 12.9 | 195 | 11.2 | 1,485 | 10.7 | 3,428 | 3,109 | 320 | 4 | 9 |
| Hutchinson | 832 | 7,253 | 0.98 | 112 | 15.4 | 97 | 13.3 | 732 | 13.2 | 1,851 | 1,381 | 470 | 0 | 6 |
| Hyde | 38 | 1,456 | 1.21 | 16 | 12.5 | 15 | 11.7 | 164 | 16.8 | 345 | 323 | 22 | NA | NA |
| Jackson | 42 | 3,246 | 0.96 | 69 | 20.8 | 18 | 5.4 | 681 | 24.4 | 467 | 425 | 41 | NA | NA |
| Jerauld | 176 | 2,381 | 1.39 | 25 | 12.6 | 26 | 13.1 | 179 | 12.4 | 544 | 477 | 68 | 1 | 9 |
| Jones | 0 | 812 | 1.05 | 14 | 14.9 | 6 | 6.4 | 107 | 15.8 | 234 | 207 | 26 | 0 | 2 |
| Kingsbury | 203 | 4,561 | 0.86 | 61 | 12.2 | 52 | 10.4 | 401 | 10.7 | 1,369 | 1,111 | 258 | 0 | 4 |
| Lake | 879 | 11,934 | 0.88 | 122 | 9.8 | 103 | 8.2 | 796 | 8.6 | 2,459 | 1,923 | 536 | 2 | 19 |
| Lawrence | 1,075 | 25,269 | 0.98 | 204 | 7.8 | 230 | 8.8 | 2,283 | 11.9 | 6,600 | 4,788 | 1,812 | 7 | 39 |
| Lincoln | 284 | 47,486 | 0.69 | 854 | 13.6 | 292 | 4.6 | 3,735 | 6.9 | 9,324 | 6,648 | 2,676 | 32 | 274 |
| Lyman | 37 | 3,736 | 0.93 | 68 | 17.9 | 16 | 4.2 | 517 | 16.9 | 662 | 577 | 85 | 2 | 11 |
| McCook | 312 | 4,434 | 0.61 | 90 | 16.3 | 75 | 13.6 | 550 | 12.3 | 1,242 | 936 | 306 | 7 | 19 |
| McPherson | 347 | 2,287 | 0.99 | 25 | 10.6 | 30 | 12.7 | 223 | 13.2 | 705 | 641 | 65 | 3 | 18 |
| Marshall | 373 | 4,724 | 0.93 | 67 | 13.7 | 49 | 10.0 | 627 | 16.2 | 1,102 | 931 | 170 | 6 | 23 |
| Meade | 820 | 21,273 | 0.57 | 253 | 8.8 | 227 | 7.9 | 2,322 | 10.1 | 5,408 | 4,077 | 1,331 | 26 | 151 |
| Mellette | 49 | 1,931 | 0.78 | 31 | 14.8 | 23 | 11.0 | 373 | 22.1 | 335 | 318 | 16 | 0 | 6 |
| Miner | 80 | 1,930 | 0.75 | 24 | 10.9 | 23 | 10.4 | 206 | 11.9 | 600 | 525 | 74 | 0 | 6 |
| Minnehaha | 6,269 | 204,202 | 1.15 | 2,856 | 14.5 | 1,532 | 7.8 | 17,811 | 11.0 | 36,049 | 26,100 | 9,949 | 54 | 319 |
| Moody | 164 | 6,014 | 0.85 | 86 | 13.2 | 54 | 8.3 | 754 | 14.3 | 1,318 | 1,039 | 280 | 3 | 2 |
| Oglala Lakota | 92 | 14,843 | 1.15 | 260 | 18.5 | 172 | 12.2 | 2,132 | 17.1 | 1,327 | 1,312 | 14 | NA | NA |
| Pennington | 2,571 | 118,120 | 1.14 | 1,499 | 12.9 | 1,045 | 9.0 | 11,736 | 12.9 | 26,818 | 20,539 | 6,279 | 122 | 352 |
| Perkins | 65 | 2,924 | 1.02 | 37 | 13.1 | 23 | 8.1 | 342 | 16.5 | 788 | 677 | 110 | 2 | 10 |
| Potter | 63 | 2,315 | 1.00 | 22 | 10.2 | 26 | 12.0 | 175 | 11.5 | 663 | 598 | 65 | 3 | 3 |
| Roberts | 281 | 9,909 | 0.92 | 161 | 15.6 | 115 | 11.1 | 1,553 | 19.0 | 2,137 | 1,623 | 514 | 11 | 7 |
| Sanborn | 174 | 1,901 | 0.63 | 32 | 13.8 | 15 | 6.5 | 217 | 11.8 | 526 | 437 | 89 | 3 | 7 |
| Spink | 404 | 6,298 | 0.94 | 81 | 12.8 | 67 | 10.6 | 570 | 11.8 | 1,558 | 1,315 | 243 | 13 | 84 |
| Stanley | 0 | 2,526 | 0.72 | 35 | 11.2 | 16 | 5.1 | 276 | 11.1 | 699 | 621 | 79 | 0 | 10 |
| Sully | 0 | 1,386 | 1.13 | 16 | 11.5 | 11 | 7.9 | 104 | 10.0 | 339 | 310 | 29 | 4 | 9 |
| Todd | 28 | 10,311 | 1.05 | 230 | 22.3 | 123 | 11.9 | 1,694 | 19.0 | 925 | 909 | 16 | NA | NA |

1. Per 1,000 estimated resident population.

# Table B. States and Counties — Crime, Education, Money Income, and Poverty

| STATE County | Officers [46] | Civilians [47] | Enroll. Total [48] | Percent private [49] | HS grad or less [50] | Bachelor's or more [51] | Total current spending (mil dol) [52] | Current spending per student (dollars) [53] | Per capita income[4] [54] | Median income (dollars) [55] | Pct w/ income <$50,000 [56] | Pct w/ income $200,000+ [57] | Median household income (dollars) [58] | All persons [59] | Children under 18 [60] | Children 5 to 17 in families [61] |
|---|---|---|---|---|---|---|---|---|---|---|---|---|---|---|---|---|
| SOUTH CAROLINA—Cont'd | | | | | | | | | | | | | | | | |
| Sumter | 107 | 94 | 27,926 | 15.8 | 45.3 | 19.3 | 168 | 9,987 | 23,460 | 45,661 | 54.6 | 2.2 | 49,611 | 17.8 | 25.8 | 25.5 |
| Union | 31 | 18 | 5,772 | 6.5 | 58.7 | 11.3 | 39 | 9,647 | 23,470 | 41,186 | 61.1 | 1.4 | 42,851 | 19.9 | 31.5 | 29.7 |
| Williamsburg | 35 | 41 | 7,078 | 11.6 | 61.5 | 12.3 | 45 | 11,351 | 18,257 | 32,485 | 66.2 | 0.9 | 34,409 | 27.8 | 37.7 | 37.1 |
| York | 174 | 35 | 67,292 | 9.0 | 34.3 | 33.3 | 473 | 10,306 | 34,010 | 65,361 | 37.9 | 7.1 | 68,468 | 8.6 | 10.6 | 9.2 |
| SOUTH DAKOTA | X | X | 215,557 | 12.8 | 38.5 | 28.8 | 1,385 | 10,072 | 30,773 | 58,275 | 42.7 | 4.2 | 60,414 | 11.9 | 14.9 | 13.6 |
| Aurora | 3 | 1 | 617 | 10.0 | 46.5 | 21.5 | 6 | 13,611 | 37,086 | 64,083 | 32.6 | 6.0 | 56,360 | 10.2 | 14.4 | 13.2 |
| Beadle | 5 | 21 | 4,224 | 16.0 | 49.6 | 20.5 | 31 | 10,302 | 27,655 | 52,313 | 48.4 | 3.0 | 51,844 | 12.9 | 15.7 | 15.2 |
| Bennett | 3 | 1 | 1,205 | 7.1 | 52.6 | 15.7 | 6 | 12,940 | 17,793 | 47,500 | 53.7 | 3.0 | 36,345 | 31.6 | 44.4 | 41.0 |
| Bon Homme | 3 | 0 | 1,362 | 2.3 | 49.6 | 18.3 | 12 | 11,042 | 24,131 | 54,737 | 46.6 | 2.5 | 52,660 | 14.1 | 16.7 | 16.6 |
| Brookings | 15 | 10 | 13,444 | 5.1 | 30.2 | 40.3 | 47 | 9,664 | 28,330 | 58,136 | 41.9 | 3.1 | 59,036 | 11.9 | 9.7 | 8.4 |
| Brown | 20 | 42 | 9,484 | 19.6 | 38.6 | 29.4 | 51 | 9,178 | 33,122 | 58,216 | 42.6 | 3.9 | 61,447 | 9.4 | 10.2 | 9.2 |
| Brule | NA | NA | 1,153 | 14.3 | 42.5 | 27.1 | 14 | 11,108 | 31,874 | 57,196 | 43.9 | 6.6 | 65,008 | 12.6 | 17.8 | 16.5 |
| Buffalo | 1 | 0 | 685 | 8.9 | 61.5 | 8.5 | NA | NA | 11,995 | 34,808 | 65.8 | 0.7 | 26,671 | 39.8 | 46.3 | 44.3 |
| Butte | 5 | 11 | 2,319 | 6.9 | 51.1 | 19.2 | 16 | 9,585 | 26,961 | 46,508 | 54.2 | 3.3 | 50,105 | 12.9 | 17.5 | 16.3 |
| Campbell | 2 | 0 | 253 | 5.9 | 45.5 | 28.0 | 2 | 14,560 | 32,684 | 54,449 | 45.3 | 3.1 | 56,043 | 10.5 | 10.1 | 10.8 |
| Charles Mix | 7 | 15 | 2,335 | 7.8 | 49.6 | 18.6 | 22 | 11,835 | 22,347 | 50,481 | 49.7 | 2.5 | 51,340 | 21.2 | 29.6 | 26.8 |
| Clark | 3 | 0 | 692 | 6.2 | 51.0 | 19.1 | 7 | 11,106 | 30,991 | 48,980 | 51.3 | 3.3 | 54,158 | 12.1 | 18.5 | 18.8 |
| Clay | 8 | 7 | 6,084 | 5.0 | 25.8 | 49.9 | 12 | 9,524 | 28,192 | 50,724 | 49.5 | 4.9 | 51,039 | 19.0 | 15.0 | 13.0 |
| Codington | 10 | 3 | 6,705 | 11.0 | 45.0 | 21.5 | 42 | 8,978 | 30,043 | 56,376 | 45.4 | 3.3 | 63,442 | 9.7 | 11.5 | 9.7 |
| Corson | 4 | 0 | 1,252 | 1.4 | 49.8 | 17.8 | 15 | 17,812 | 16,449 | 35,759 | 63.3 | 2.5 | 32,848 | 40.3 | 52.8 | 45.8 |
| Custer | 12 | 1 | 1,268 | 4.1 | 41.3 | 23.5 | 10 | 11,384 | 33,649 | 58,522 | 42.5 | 2.8 | 63,665 | 9.6 | 15.5 | 15.1 |
| Davison | 7 | 2 | 4,682 | 21.7 | 39.0 | 26.5 | 31 | 9,518 | 31,560 | 50,570 | 49.2 | 4.0 | 57,367 | 11.7 | 12.9 | 11.3 |
| Day | 4 | 5 | 1,079 | 10.1 | 47.4 | 21.6 | 8 | 10,899 | 29,808 | 46,679 | 52.9 | 2.8 | 51,516 | 13.2 | 17.6 | 14.9 |
| Deuel | 5 | 1 | 865 | 7.5 | 45.9 | 22.2 | 5 | 9,826 | 30,601 | 65,437 | 35.7 | 2.7 | 65,217 | 8.7 | 10.6 | 9.6 |
| Dewey | 3 | 0 | 1,759 | 2.8 | 50.6 | 16.6 | 5 | 14,723 | 17,504 | 47,475 | 53.5 | 2.5 | 47,640 | 27.6 | 32.7 | 29.7 |
| Douglas | 4 | 0 | 555 | 13.3 | 46.7 | 15.5 | 5 | 11,865 | 30,033 | 58,684 | 43.0 | 3.4 | 56,030 | 11.2 | 16.1 | 15.9 |
| Edmunds | 4 | 0 | 862 | 8.7 | 41.4 | 26.3 | 9 | 11,742 | 34,628 | 71,324 | 36.9 | 4.9 | 59,905 | 10.7 | 15.2 | 14.0 |
| Fall River | 7 | 12 | 1,090 | 18.7 | 40.7 | 21.1 | 12 | 10,873 | 27,556 | 50,588 | 49.5 | 0.6 | 49,965 | 13.9 | 17.9 | 15.7 |
| Faulk | 5 | 12 | 395 | 5.1 | 41.2 | 26.9 | 4 | 11,529 | 27,569 | 55,096 | 47.2 | 0.9 | 49,198 | 15.7 | 25.2 | 25.5 |
| Grant | 5 | 4 | 1,410 | 11.4 | 47.5 | 21.9 | 11 | 10,540 | 31,479 | 58,158 | 45.1 | 4.6 | 60,057 | 9.1 | 10.9 | 10.4 |
| Gregory | 4 | 1 | 861 | 3.5 | 46.2 | 20.2 | 9 | 12,497 | 26,546 | 43,438 | 56.4 | 2.7 | 42,868 | 15.5 | 17.6 | 16.5 |
| Haakon | 2 | 0 | 452 | 5.3 | 50.4 | 15.6 | 3 | 11,325 | 24,333 | 42,250 | 56.4 | 3.4 | 50,747 | 11.4 | 14.2 | 13.7 |
| Hamlin | 6 | 0 | 1,528 | 5.5 | 50.7 | 21.0 | 13 | 9,446 | 28,822 | 67,105 | 30.1 | 2.4 | 60,864 | 8.9 | 9.0 | 8.6 |
| Hand | 3 | 1 | 627 | 4.0 | 39.1 | 22.1 | 5 | 11,106 | 33,200 | 54,155 | 45.6 | 3.1 | 56,239 | 10.3 | 11.5 | 10.1 |
| Hanson | 3 | 0 | 936 | 7.3 | 45.1 | 18.5 | 4 | 9,307 | 24,948 | 64,696 | 36.2 | 1.8 | 78,478 | 7.9 | 15.4 | 14.3 |
| Harding | 2 | 1 | 272 | 3.3 | 41.0 | 26.0 | 4 | 17,990 | 31,819 | 59,655 | 32.7 | 2.4 | 55,482 | 11.9 | 18.0 | 18.7 |
| Hughes | 6 | 52 | 3,234 | 8.5 | 35.1 | 34.7 | 23 | 8,652 | 33,677 | 64,783 | 38.4 | 2.7 | 66,952 | 8.9 | 10.6 | 9.6 |
| Hutchinson | 3 | 0 | 1,526 | 6.6 | 45.5 | 25.2 | 16 | 12,324 | 28,872 | 57,089 | 41.4 | 3.8 | 59,336 | 10.5 | 17.2 | 17.3 |
| Hyde | 1 | 0 | 239 | 2.1 | 43.9 | 24.5 | 3 | 12,791 | 34,034 | 57,788 | 41.0 | 2.3 | 63,454 | 11.5 | 17.3 | 16.3 |
| Jackson | 2 | 1 | 893 | 3.0 | 51.1 | 17.8 | 5 | 14,343 | 13,528 | 33,295 | 70.2 | 2.4 | 37,879 | 29.9 | 43.8 | 44.3 |
| Jerauld | 4 | 0 | 401 | 4.5 | 48.5 | 22.5 | 4 | 11,073 | 42,660 | 49,632 | 50.3 | 5.4 | 52,162 | 16.3 | 27.1 | 26.5 |
| Jones | 2 | 0 | 124 | 30.6 | 47.2 | 25.7 | 2 | 12,984 | 27,139 | 46,149 | 56.6 | 1.0 | 52,650 | 12.9 | 23.0 | 22.5 |
| Kingsbury | 5 | 1 | 932 | 6.5 | 48.3 | 21.5 | 12 | 12,441 | 35,773 | 61,699 | 41.4 | 4.3 | 61,992 | 8.5 | 11.7 | 11.7 |
| Lake | 8 | 7 | 3,127 | 2.4 | 40.9 | 33.1 | 20 | 9,310 | 33,728 | 61,846 | 35.9 | 5.7 | 63,946 | 9.5 | 10.0 | 9.5 |
| Lawrence | 17 | 23 | 6,061 | 9.9 | 36.1 | 35.0 | 31 | 9,738 | 31,712 | 52,641 | 47.3 | 2.9 | 53,902 | 12.4 | 14.3 | 12.2 |
| Lincoln | 20 | 4 | 15,155 | 18.8 | 28.1 | 37.6 | 74 | 8,781 | 40,059 | 82,473 | 26.4 | 10.0 | 86,235 | 3.8 | 3.9 | 3.2 |
| Lyman | 4 | 1 | 964 | 7.8 | 47.2 | 20.3 | 5 | 14,378 | 22,850 | 53,862 | 44.6 | 1.7 | 47,813 | 21.1 | 30.3 | 29.2 |
| McCook | 8 | 1 | 1,317 | 5.6 | 46.3 | 24.1 | 15 | 12,047 | 29,020 | 61,507 | 39.2 | 3.2 | 62,535 | 9.2 | 10.8 | 10.5 |
| McPherson | 4 | 0 | 410 | 12.2 | 50.9 | 17.7 | 6 | 15,586 | 28,559 | 44,574 | 52.4 | 3.2 | 43,243 | 19.5 | 25.3 | 21.1 |
| Marshall | 6 | 5 | 1,060 | 5.4 | 38.9 | 24.1 | 7 | 11,137 | 32,472 | 63,723 | 36.5 | 3.5 | 59,103 | 11.8 | 18.7 | 19.6 |
| Meade | 18 | 32 | 6,871 | 12.3 | 34.3 | 27.2 | 27 | 9,055 | 28,809 | 60,578 | 43.3 | 3.3 | 59,981 | 8.7 | 11.2 | 9.8 |
| Mellette | 4 | 0 | 583 | 2.4 | 58.4 | 12.0 | 6 | 13,616 | 13,784 | 30,650 | 69.1 | 0.6 | 36,489 | 33.3 | 41.9 | 44.2 |
| Miner | 3 | 1 | 494 | 6.1 | 48.9 | 18.5 | 5 | 12,332 | 28,381 | 52,910 | 47.8 | 2.6 | 52,230 | 10.5 | 14.5 | 13.6 |
| Minnehaha | 95 | 138 | 45,244 | 19.1 | 33.9 | 33.1 | 305 | 9,186 | 31,716 | 61,772 | 39.8 | 4.0 | 63,162 | 8.7 | 9.4 | 9.2 |
| Moody | 6 | 4 | 1,565 | 7.1 | 43.3 | 21.3 | 9 | 10,011 | 31,300 | 61,538 | 38.0 | 6.3 | 61,029 | 8.7 | 10.9 | 10.3 |
| Oglala Lakota | 1 | 0 | 5,308 | 11.7 | 53.9 | 10.4 | 27 | 18,489 | 10,388 | 31,997 | 65.7 | 2.7 | 34,078 | 40.1 | 44.0 | 40.9 |
| Pennington | 101 | 320 | 26,721 | 14.1 | 34.0 | 31.5 | 170 | 9,624 | 31,874 | 57,039 | 43.1 | 4.5 | 58,919 | 12.0 | 15.6 | 13.3 |
| Perkins | 7 | 1 | 551 | 20.3 | 43.6 | 20.3 | 6 | 14,048 | 33,678 | 57,981 | 42.3 | 5.4 | 49,999 | 14.5 | 21.8 | 20.7 |
| Potter | 2 | 1 | 453 | 9.3 | 43.5 | 22.3 | 5 | 14,592 | 32,379 | 54,583 | 45.9 | 4.1 | 58,012 | 10.4 | 14.5 | 12.7 |
| Roberts | 6 | 29 | 2,351 | 5.1 | 45.9 | 17.3 | 18 | 11,526 | 26,243 | 50,348 | 49.7 | 2.9 | 49,088 | 19.6 | 27.9 | 25.9 |
| Sanborn | 3 | 0 | 454 | 3.5 | 46.8 | 21.6 | 5 | 11,834 | 32,745 | 56,875 | 39.7 | 3.7 | 56,507 | 12.6 | 16.1 | 15.5 |
| Spink | 9 | 5 | 1,533 | 4.2 | 46.1 | 21.8 | 14 | 10,892 | 35,393 | 54,423 | 43.9 | 6.8 | 54,666 | 11.5 | 16.1 | 14.7 |
| Stanley | 5 | 1 | 524 | 9.5 | 35.7 | 32.1 | 5 | 11,514 | 42,838 | 71,382 | 34.0 | 10.8 | 70,472 | 7.7 | 11.9 | 11.1 |
| Sully | 5 | 0 | 303 | 4.6 | 38.1 | 25.0 | 4 | 14,841 | 39,226 | 62,153 | 40.6 | 5.3 | 61,791 | 8.2 | 8.4 | 7.6 |
| Todd | NA | NA | 3,502 | 7.3 | 51.9 | 15.6 | 27 | 13,015 | 10,739 | 24,331 | 74.9 | 1.6 | 34,015 | 43.4 | 49.4 | 44.0 |

1. All persons 3 years old and over enrolled in nursery school through college. 2. Persons 25 years old and over. 3. Elementary and secondary education expenditures. 4. Based on population estimated by the American Community Survey, 2014–2018.

# Table B. States and Counties — **Personal Income**

| STATE County | Personal income, 2019 | | | | | Supplements to wages and salaries, employer contributions (mil dol) | | | | | Earnings, 2019 | Contributions for government social insurance (mil dol) | |
|---|---|---|---|---|---|---|---|---|---|---|---|---|---|
| | Total (mil dol) | Percent change 2018-2019 | Per capita[1] Dollars | Per capita[1] Rank | Wages and salaries (mil dol) | Pension and insurance | Government social insurance | Proprietors' income (mil dol) | Dividends, interest, and rent (mil dol) | Personal transfer receipts (mil dol) | Total (mil dol) | From employee and self-employed | From employer |
| | 62 | 63 | 64 | 65 | 66 | 67 | 68 | 69 | 70 | 71 | 72 | 73 | 74 |
| **SOUTH CAROLINA— Cont'd** | | | | | | | | | | | | | |
| Sumter | 4,271 | 3.9 | 40,019 | 2,046 | 2,095 | 474 | 172 | 196 | 711 | 1,200 | 2,937 | 182 | 172 |
| Union | 938 | 3.8 | 34,341 | 2,765 | 367 | 73 | 29 | 33 | 123 | 344 | 502 | 40 | 29 |
| Williamsburg | 1,063 | 6.0 | 34,993 | 2,688 | 355 | 84 | 28 | 35 | 137 | 394 | 502 | 40 | 28 |
| York | 13,652 | 6.1 | 48,588 | 956 | 5,439 | 884 | 403 | 689 | 1,935 | 2,234 | 7,415 | 484 | 403 |
| **SOUTH DAKOTA** | 47,738 | 3.2 | 53,812 | X | 20,650 | 3,644 | 1,516 | 6,583 | 11,442 | 7,504 | 32,392 | 1,965 | 1,516 |
| Aurora | 112 | 3.8 | 40,709 | 1,962 | 30 | 6 | 2 | 13 | 29 | 22 | 51 | 4 | 2 |
| Beadle | 914 | 3.1 | 49,521 | 876 | 384 | 72 | 29 | 152 | 196 | 155 | 637 | 37 | 29 |
| Bennett | 124 | -0.4 | 36,759 | 2,475 | 33 | 8 | 2 | 22 | 22 | 35 | 65 | 3 | 2 |
| Bon Homme | 281 | 4.9 | 40,713 | 1,960 | 66 | 15 | 5 | 59 | 61 | 57 | 145 | 9 | 5 |
| Brookings | 1,676 | 3.3 | 47,782 | 1,045 | 876 | 188 | 66 | 180 | 410 | 201 | 1,310 | 74 | 66 |
| Brown | 2,152 | 0.2 | 55,397 | 478 | 968 | 177 | 73 | 274 | 564 | 306 | 1,491 | 92 | 73 |
| Brule | 240 | -1.5 | 45,320 | 1,324 | 74 | 14 | 5 | 42 | 65 | 44 | 135 | 8 | 5 |
| Buffalo | 41 | -13.8 | 20,682 | 3,111 | 22 | 6 | 2 | -5 | 9 | 21 | 25 | 2 | 2 |
| Butte | 401 | 2.7 | 38,449 | 2,272 | 106 | 21 | 8 | 49 | 91 | 89 | 184 | 13 | 8 |
| Campbell | 67 | -8.2 | 48,544 | 964 | 20 | 3 | 2 | 8 | 22 | 12 | 33 | 2 | 2 |
| Charles Mix | 386 | 3.1 | 41,567 | 1,835 | 126 | 28 | 9 | 77 | 94 | 87 | 240 | 13 | 9 |
| Clark | 173 | -3.8 | 46,281 | 1,213 | 41 | 8 | 3 | 32 | 53 | 29 | 85 | 5 | 3 |
| Clay | 571 | 4.4 | 40,589 | 1,974 | 252 | 57 | 18 | 65 | 130 | 102 | 392 | 23 | 18 |
| Codington | 1,435 | 3.1 | 51,250 | 737 | 695 | 128 | 52 | 127 | 368 | 235 | 1,002 | 63 | 52 |
| Corson | 111 | -1.1 | 27,114 | 3,093 | 34 | 8 | 2 | 5 | 23 | 37 | 49 | 3 | 2 |
| Custer | 446 | 5.0 | 49,665 | 859 | 95 | 19 | 7 | 36 | 128 | 102 | 157 | 13 | 7 |
| Davison | 1,054 | 3.3 | 53,275 | 593 | 529 | 92 | 39 | 149 | 264 | 174 | 810 | 49 | 39 |
| Day | 250 | -6.3 | 46,087 | 1,238 | 73 | 15 | 6 | 21 | 84 | 59 | 114 | 8 | 6 |
| Deuel | 230 | 3.5 | 52,808 | 633 | 63 | 14 | 5 | 63 | 48 | 37 | 145 | 7 | 5 |
| Dewey | 224 | 3.9 | 38,051 | 2,323 | 95 | 24 | 7 | 21 | 42 | 58 | 147 | 8 | 7 |
| Douglas | 175 | 14.6 | 59,855 | 309 | 43 | 9 | 3 | 59 | 35 | 27 | 114 | 5 | 3 |
| Edmunds | 208 | -6.5 | 54,210 | 535 | 52 | 11 | 4 | 41 | 56 | 32 | 106 | 6 | 4 |
| Fall River | 323 | 3.6 | 48,186 | 1,002 | 116 | 28 | 11 | 31 | 78 | 102 | 185 | 13 | 11 |
| Faulk | 131 | 1.9 | 56,814 | 412 | 26 | 5 | 2 | 44 | 33 | 23 | 77 | 3 | 2 |
| Grant | 405 | 2.5 | 57,365 | 391 | 171 | 31 | 13 | 77 | 90 | 70 | 292 | 17 | 13 |
| Gregory | 188 | 1.0 | 44,981 | 1,361 | 54 | 11 | 4 | 29 | 50 | 43 | 98 | 6 | 4 |
| Haakon | 98 | -9.0 | 51,865 | 691 | 36 | 7 | 3 | 18 | 28 | 17 | 63 | 4 | 3 |
| Hamlin | 272 | 0.4 | 44,145 | 1,465 | 89 | 15 | 7 | 37 | 61 | 43 | 149 | 9 | 7 |
| Hand | 174 | -7.3 | 54,579 | 516 | 55 | 10 | 4 | 30 | 49 | 27 | 99 | 6 | 4 |
| Hanson | 225 | 6.9 | 65,135 | 184 | 28 | 6 | 2 | 41 | 62 | 41 | 77 | 5 | 2 |
| Harding | 65 | 12.3 | 50,170 | 819 | 25 | 5 | 2 | 12 | 19 | 7 | 43 | 2 | 2 |
| Hughes | 953 | 3.1 | 54,352 | 524 | 515 | 97 | 37 | 71 | 226 | 142 | 720 | 45 | 37 |
| Hutchinson | 414 | 10.5 | 56,811 | 413 | 108 | 20 | 8 | 108 | 99 | 68 | 244 | 12 | 8 |
| Hyde | 64 | -1.4 | 48,836 | 939 | 24 | 5 | 2 | 8 | 20 | 12 | 39 | 2 | 2 |
| Jackson | 91 | -0.7 | 27,278 | 3,091 | 27 | 7 | 2 | 8 | 18 | 28 | 45 | 3 | 2 |
| Jerauld | 80 | -3.6 | 39,599 | 2,104 | 59 | 10 | 5 | | 26 | 22 | 73 | 6 | 5 |
| Jones | 53 | 6.5 | 58,796 | 344 | 16 | 3 | 1 | 10 | 17 | 8 | 30 | 2 | 1 |
| Kingsbury | 284 | 3.3 | 57,495 | 388 | 71 | 14 | 5 | 62 | 63 | 48 | 153 | 8 | 5 |
| Lake | 654 | 2.4 | 51,119 | 744 | 227 | 46 | 17 | 105 | 160 | 107 | 395 | 23 | 17 |
| Lawrence | 1,368 | 3.6 | 52,919 | 620 | 491 | 82 | 36 | 118 | 416 | 244 | 727 | 50 | 36 |
| Lincoln | 4,307 | 6.0 | 70,461 | 126 | 1,342 | 196 | 94 | 421 | 1,001 | 324 | 2,053 | 123 | 94 |
| Lyman | 158 | | 41,809 | 1,798 | 49 | 10 | 4 | 26 | 39 | 32 | 88 | 5 | 4 |
| McCook | 298 | 7.7 | 53,322 | 588 | 53 | 10 | 4 | 71 | 53 | 48 | 138 | 7 | 4 |
| McPherson | 104 | -5.9 | 43,635 | 1,532 | 21 | 4 | 2 | 25 | 30 | 23 | 52 | 3 | 2 |
| Marshall | 246 | 3.2 | 49,754 | 851 | 69 | 12 | 5 | 55 | 71 | 37 | 142 | 7 | 5 |
| Meade | 1,257 | 2.4 | 44,365 | 1,436 | 371 | 88 | 29 | 97 | 252 | 233 | 585 | 40 | 29 |
| Mellette | 64 | 0.7 | 31,131 | 2,997 | 11 | 3 | 1 | 8 | 11 | 19 | 23 | 1 | 1 |
| Miner | 112 | 6.1 | 50,603 | 791 | 27 | 5 | 2 | 28 | 25 | 22 | 62 | 3 | 2 |
| Minnehaha | 11,570 | 4.1 | 59,905 | 307 | 6,809 | 1,060 | 488 | 2,053 | 2,288 | 1,529 | 10,410 | 625 | 488 |
| Moody | 341 | 2.6 | 51,866 | 690 | 95 | 21 | 8 | 56 | 77 | 53 | 179 | 9 | 8 |
| Oglala Lakota | 366 | 1.2 | 25,840 | 3,101 | 169 | 42 | 13 | 24 | 40 | 155 | 248 | 14 | 13 |
| Pennington | 6,038 | 4.1 | 53,071 | 609 | 2,835 | 506 | 211 | 514 | 1,633 | 1,135 | 4,067 | 261 | 211 |
| Perkins | 115 | -0.7 | 40,282 | 2,016 | 42 | 10 | 3 | 10 | 30 | 27 | 66 | 5 | 3 |
| Potter | 156 | 2.0 | 72,468 | 106 | 34 | 6 | 3 | 42 | 51 | 26 | 85 | 6 | 3 |
| Roberts | 382 | -3.4 | 36,756 | 2,476 | 137 | 30 | 10 | 28 | 92 | 99 | 205 | 14 | 10 |
| Sanborn | 122 | 6.5 | 52,093 | 675 | 22 | 5 | 2 | 35 | 22 | 18 | 63 | 3 | 2 |
| Spink | 356 | -4.9 | 55,810 | 456 | 101 | 20 | 7 | 58 | 94 | 87 | 186 | 11 | 7 |
| Stanley | 194 | 8.4 | 62,601 | 236 | 49 | 8 | 4 | 24 | 49 | 25 | 85 | 5 | 4 |
| Sully | 102 | -4.1 | 73,244 | 101 | 38 | 5 | 3 | 18 | 30 | 12 | 64 | 3 | 3 |
| Todd | 266 | 2.3 | 26,172 | 3,097 | 119 | 28 | 9 | 10 | 38 | 101 | 167 | 10 | 9 |

1. Based on the resident population estimated as of July 1 of the year shown.

# Table B. States and Counties — Earnings, Social Security, and Housing

| STATE County | Earnings, 2019 (cont.) | | | | | | | | | Social Security beneficiaries, December 2019 | | Supplemental Security Income recipients, 2019 | Housing units, 2020 | |
|---|---|---|---|---|---|---|---|---|---|---|---|---|---|---|
| | Percent by selected industries | | | | | | | | | | | | | |
| | Farm | Mining, quarrying, and extractions | Construction | Manu-facturing | Information; professional, scientific, technical services | Retail trade | Finance, insurance, real estate, and leasing | Health care and social assistance | Govern-ment | Number | Rate[1] | | Total | Percent change, 2010-2020 |
| | 75 | 76 | 77 | 78 | 79 | 80 | 81 | 82 | 83 | 84 | 85 | 86 | 87 | 88 |
| SOUTH CAROLINA—Cont'd | | | | | | | | | | | | | | |
| Sumter | 0.2 | 0.0 | 7.4 | 16.8 | 3.3 | 5.5 | 3.1 | 11.6 | 34.6 | 24,675 | 231 | 4,028 | 48,964 | 6.4 |
| Union | -0.1 | D | 2.8 | 32.5 | 1.7 | 5.5 | 2.4 | 2.7 | 23.4 | 8,325 | 306 | 945 | 14,124 | -0.2 |
| Williamsburg | 0.2 | 0.0 | 8.0 | 19.9 | D | 5.4 | 2.0 | D | 27.5 | 8,920 | 295 | 1,702 | 15,695 | 2.2 |
| York | 0.2 | 0.1 | 6.0 | 12.9 | 11.4 | 6.8 | 8.5 | 8.2 | 11.9 | 51,825 | 184 | 3,847 | 115,213 | 22.3 |
| SOUTH DAKOTA | 4.7 | 0.3 | 7.0 | 9.2 | 6.1 | 6.9 | 12.5 | 15.3 | 15.6 | 182,793 | 206 | 14,438 | 405,869 | 11.7 |
| Aurora | 10.3 | 0.1 | 6.8 | 0.6 | D | 6.9 | 6.9 | 17.7 | 16.0 | 585 | 213 | 19 | 1,390 | 5.0 |
| Beadle | 9.5 | D | 6.0 | 18.3 | 2.8 | 5.9 | 9.1 | D | 14.1 | 3,730 | 203 | 419 | 8,576 | 3.3 |
| Bennett | 27.4 | 0.1 | 8.2 | D | D | 6.2 | D | 2.2 | 35.0 | 530 | 158 | 143 | 1,266 | 0.2 |
| Bon Homme | 22.8 | D | 6.5 | 4.7 | 2.0 | 5.0 | 6.4 | D | 21.4 | 1,570 | 227 | 78 | 3,004 | 2.5 |
| Brookings | 5.9 | D | 6.0 | 27.2 | 4.1 | 5.2 | 6.5 | 3.9 | 23.3 | 4,895 | 138 | 282 | 15,194 | 15.7 |
| Brown | 2.0 | D | 6.7 | 15.5 | 4.5 | 7.7 | 9.7 | 15.7 | 13.1 | 7,665 | 197 | 538 | 18,231 | 9.1 |
| Brule | 18.9 | D | 6.3 | 0.5 | 6.2 | 7.4 | 7.1 | 10.9 | 17.1 | 985 | 189 | 114 | 2,604 | 7.0 |
| Buffalo | -17.4 | 0.0 | D | 0.0 | D | 3.8 | D | 3.7 | 100.2 | 315 | 159 | 89 | 616 | 1.0 |
| Butte | 6.7 | D | 8.6 | 5.2 | D | 12.3 | 5.1 | 6.9 | 17.1 | 2,415 | 232 | 197 | 4,934 | 6.8 |
| Campbell | 16.8 | 0.0 | 4.6 | 9.5 | D | 6.0 | D | 2.2 | 10.1 | 380 | 275 | 11 | 983 | 0.3 |
| Charles Mix | 19.5 | D | 7.2 | 2.3 | 2.6 | 5.2 | 6.2 | D | 29.8 | 1,820 | 196 | 274 | 3,934 | 2.2 |
| Clark | 30.6 | 0.2 | 9.5 | 7.9 | D | 4.5 | 4.3 | D | 13.8 | 815 | 218 | 99 | 1,850 | 8.2 |
| Clay | 8.4 | 0.0 | 4.2 | 3.8 | 1.7 | 5.4 | D | 45.9 | 13.6 | 2,125 | 151 | 168 | 6,320 | 12.1 |
| Codington | 3.3 | D | 6.4 | 21.3 | 4.0 | 8.9 | 8.5 | 13.5 | 62.3 | 6,000 | 214 | 364 | 13,419 | 8.3 |
| Corson | 3.6 | 0.2 | D | 0.0 | D | 3.2 | D | D | 22.4 | 605 | 148 | 260 | 1,542 | 0.4 |
| Custer | 3.2 | D | 8.8 | 0.8 | D | 7.1 | 6.3 | 10.5 | 9.9 | 2,735 | 306 | 77 | 5,484 | 18.4 |
| Davison | 4.8 | D | 8.3 | 16.6 | 5.1 | 8.5 | 6.5 | 15.4 | 18.6 | 4,410 | 222 | 375 | 9,643 | 8.9 |
| Day | 11.1 | 0.1 | 5.6 | 12.0 | 3.0 | 6.6 | 8.4 | D | 7.7 | 1,575 | 291 | 80 | 3,856 | 6.2 |
| Deuel | 33.2 | 0.1 | 8.9 | 12.9 | D | 3.8 | D | D | 63.4 | 1,120 | 258 | 39 | 2,208 | 0.2 |
| Dewey | 10.3 | D | 2.5 | D | D | 3.9 | D | D | | 890 | 152 | 331 | 2,032 | 1.5 |
| Douglas | 36.6 | D | 6.3 | 4.6 | D | 2.9 | 3.5 | 8.7 | 8.7 | 740 | 254 | 45 | 1,488 | 3.5 |
| Edmunds | 19.6 | 0.1 | 6.3 | 5.8 | 2.1 | 4.0 | 14.4 | D | 16.7 | 870 | 226 | 48 | 2,104 | 7.0 |
| Fall River | 8.5 | 0.5 | 5.0 | 0.4 | 3.9 | 5.1 | 2.8 | D | 39.5 | 2,420 | 362 | 161 | 4,240 | 1.2 |
| Faulk | 49.5 | 0.2 | D | 1.5 | 2.6 | 7.0 | D | 9.0 | 510 | 221 | 57 | 1,207 | 6.3 |
| Grant | 11.6 | D | 8.9 | 17.2 | 2.0 | 13.4 | 5.8 | 8.2 | 6.5 | 1,935 | 272 | 80 | 3,643 | 3.3 |
| Gregory | 16.7 | D | 6.4 | 1.6 | 3.1 | 9.7 | 9.0 | D | 13.6 | 1,155 | 276 | 105 | 2,552 | 2.0 |
| Haakon | 12.0 | D | 6.3 | 6.8 | D | 6.0 | 8.3 | D | 11.4 | 450 | 241 | 0 | 1,037 | 2.4 |
| Hamlin | 14.2 | 0.0 | 17.7 | D | 2.2 | 5.9 | D | 2.7 | 14.7 | 1,165 | 187 | 51 | 3,031 | 9.9 |
| Hand | 17.6 | 0.6 | 5.8 | 2.8 | D | 6.5 | D | D | 12.0 | 830 | 262 | 35 | 1,856 | 2.3 |
| Hanson | 28.8 | D | 14.3 | 11.5 | D | 3.4 | D | D | 11.1 | 1,345 | 388 | 42 | 1,228 | 4.2 |
| Harding | 18.6 | 8.5 | 19.8 | 0.0 | D | 6.2 | D | 4.4 | 14.0 | 235 | 180 | 0 | 760 | 4.0 |
| Hughes | 1.2 | 0.0 | 4.8 | 0.4 | 7.1 | 6.8 | 10.4 | 13.1 | 39.1 | 3,525 | 202 | 237 | 8,088 | 6.1 |
| Hutchinson | 36.1 | D | 4.9 | 7.3 | 2.4 | 4.1 | D | D | 8.7 | 1,835 | 252 | 95 | 3,428 | 2.3 |
| Hyde | 20.4 | 0.0 | D | D | D | 6.2 | D | 5.9 | 24.0 | 315 | 243 | 0 | 704 | -0.6 |
| Jackson | 11.7 | 0.1 | D | D | D | 6.9 | D | D | 44.6 | 520 | 156 | 135 | 1,212 | 1.6 |
| Jerauld | -10.1 | D | D | D | D | 2.8 | D | D | 8.2 | 630 | 313 | 25 | 1,082 | 1.1 |
| Jones | 21.2 | 0.0 | D | 0.0 | D | 11.8 | 4.8 | 2.0 | 20.7 | 235 | 258 | 0 | 661 | 12.2 |
| Kingsbury | 28.8 | 0.0 | 7.1 | 9.8 | 1.6 | 3.7 | 10.1 | 6.6 | 10.2 | 1,350 | 275 | 72 | 2,844 | 4.6 |
| Lake | 9.2 | 0.6 | 6.6 | 16.1 | 7.1 | 6.3 | 5.5 | 8.8 | 17.4 | 2,495 | 195 | 133 | 5,887 | 5.9 |
| Lawrence | 0.2 | D | 10.8 | 4.3 | 7.3 | 8.9 | 6.4 | 14.5 | 14.9 | 6,605 | 255 | 315 | 14,587 | 14.4 |
| Lincoln | 2.4 | 0.0 | 10.5 | 6.7 | 8.2 | 6.5 | 20.4 | 15.7 | 5.4 | 8,415 | 137 | 255 | 22,001 | 23.1 |
| Lyman | 21.7 | 0.2 | 3.6 | 0.3 | D | 7.4 | D | D | 36.5 | 710 | 187 | 73 | 1,753 | 2.9 |
| McCook | 38.0 | D | 8.5 | D | 4.1 | 5.0 | 4.8 | D | 10.0 | 1,215 | 219 | 84 | 2,624 | 5.3 |
| McPherson | 36.3 | 0.3 | 4.5 | 2.9 | D | 3.9 | D | 9.2 | 15.3 | 690 | 289 | 37 | 1,444 | 1.8 |
| Marshall | 34.4 | D | 6.3 | 16.5 | D | 3.6 | D | D | 12.2 | 1,050 | 214 | 54 | 2,666 | 5.2 |
| Meade | -1.2 | D | 12.7 | 2.7 | D | 7.4 | 5.3 | 7.1 | 40.5 | 5,990 | 210 | 260 | 12,565 | 14.3 |
| Mellette | 20.4 | 0.3 | D | 0.0 | D | D | D | 1.5 | 44.1 | 360 | 174 | 115 | 864 | 3.2 |
| Miner | 34.2 | 0.0 | 6.9 | 2.0 | D | 4.3 | D | 7.2 | 13.1 | 595 | 268 | 33 | 1,355 | 3.6 |
| Minnehaha | 0.3 | D | 6.5 | 7.3 | 8.2 | 7.3 | 20.5 | 19.8 | 8.6 | 37,370 | 192 | 2,904 | 87,161 | 21.8 |
| Moody | 24.3 | 0.0 | 11.1 | 9.8 | D | 2.1 | D | D | 22.3 | 1,355 | 207 | 46 | 2,938 | 4.0 |
| Oglala Lakota | 6.7 | 0.0 | 1.4 | D | D | D | D | 2.2 | 75.3 | 1,570 | 111 | 1,065 | 3,659 | 1.8 |
| Pennington | 0.1 | 0.2 | 7.5 | 4.0 | 6.6 | 7.5 | 8.5 | 21.1 | 20.2 | 27,440 | 240 | 1,939 | 50,277 | 11.8 |
| Perkins | 4.8 | 0.2 | 8.2 | D | D | 7.2 | D | 7.1 | 18.7 | 790 | 279 | 33 | 1,739 | -0.2 |
| Potter | -6.1 | 0.0 | 5.5 | D | D | 45.3 | D | D | 10.2 | 670 | 312 | 18 | 1,531 | 4.4 |
| Roberts | 5.0 | D | 4.0 | 8.4 | 2.7 | 6.0 | D | D | 42.8 | 2,380 | 229 | 223 | 5,122 | 4.0 |
| Sanborn | 46.1 | 0.0 | D | 4.6 | D | 1.5 | 2.4 | 3.6 | 11.1 | 535 | 228 | 25 | 1,219 | 4.0 |
| Spink | 20.5 | 0.1 | 5.0 | 4.3 | 2.5 | 5.1 | 11.0 | D | 24.8 | 1,535 | 240 | 154 | 3,271 | 4.2 |
| Stanley | 15.6 | D | 23.2 | 0.0 | D | 9.8 | D | D | 13.2 | 725 | 232 | 18 | 1,564 | 12.8 |
| Sully | 18.0 | 0.2 | D | D | D | 5.9 | D | D | 9.2 | 375 | 272 | 0 | 920 | 9.0 |
| Todd | 3.8 | 0.1 | 1.1 | D | D | 3.5 | D | 3.1 | 75.0 | 1,130 | 110 | 591 | 3,190 | 1.6 |

1. Per 1,000 resident population estimated as of July 1 of the year shown.

# Table B. States and Counties — Housing, Labor Force, and Employment

| STATE County | Total (89) | Percent (90) | Median value[1] (91) | With a mortgage (92) | Without a mortgage[2] (93) | Median rent[3] (94) | Median rent as a percent of income[2] (95) | Sub-standard units[4] (percent) (96) | Total (97) | Percent change, 2019-2020 (98) | Unemployment Total (99) | Rate[5] (100) | Total (101) | Management, business, science, and arts (102) | Construction, production, and maintenance occupations (103) |
|---|---|---|---|---|---|---|---|---|---|---|---|---|---|---|---|
| **SOUTH CAROLINA—Cont'd** | | | | | | | | | | | | | | | |
| Sumter | 41,776 | 64.6 | 115,700 | 20.1 | 10.0 | 800 | 29.3 | 2.3 | 43,534 | 0.5 | 2,947 | 6.8 | 43,151 | 28.6 | 31.0 |
| Union | 11,432 | 68.3 | 80,000 | 21.0 | 11.1 | 690 | 26.2 | 2.7 | 12,008 | 4.4 | 1,077 | 9.0 | 11,760 | 23.2 | 41.9 |
| Williamsburg | 12,686 | 73.0 | 68,300 | 24.3 | 13.7 | 625 | 28.1 | 3.7 | 12,004 | 0.1 | 972 | 8.1 | 11,453 | 23.6 | 35.7 |
| York | 101,211 | 72.0 | 201,100 | 18.9 | 10.0 | 980 | 28.8 | 1.9 | 143,890 | 1.9 | 8,747 | 6.1 | 132,601 | 38.4 | 22.7 |
| **SOUTH DAKOTA** | 344,397 | 67.8 | 167,100 | 19.7 | 10.6 | 747 | 25.5 | 2.6 | 463,256 | 0.0 | 21,511 | 4.6 | 443,891 | 36.3 | 24.6 |
| Aurora | 1,182 | 75.3 | 85,800 | 17.3 | 10.0 | 628 | 19.2 | 1.5 | 1,477 | -2.3 | 48 | 3.2 | 1,495 | 39.2 | 29.9 |
| Beadle | 7,557 | 65.7 | 109,700 | 18.8 | 10.2 | 619 | 22.0 | 4.3 | 9,352 | -2.2 | 361 | 3.9 | 9,028 | 28.8 | 39.8 |
| Bennett | 984 | 60.5 | 75,900 | 14.7 | 13.4 | 564 | 20.6 | 11.6 | 1,070 | -0.5 | 54 | 5.0 | 1,184 | 34.4 | 28.6 |
| Bon Homme | 2,536 | 73.5 | 89,600 | 18.7 | 10.0 | 572 | 19.9 | 1.3 | 2,859 | 0.1 | 117 | 4.1 | 2,832 | 40.1 | 29.5 |
| Brookings | 12,896 | 59.2 | 179,000 | 18.6 | 10.3 | 750 | 27.8 | 1.5 | 18,480 | -3.0 | 811 | 4.4 | 19,445 | 38.3 | 24.1 |
| Brown | 16,305 | 64.0 | 166,800 | 19.7 | 10.0 | 695 | 23.8 | 1.4 | 20,582 | -0.6 | 904 | 4.4 | 20,929 | 34.4 | 26.1 |
| Brule | 2,231 | 61.7 | 152,600 | 21.7 | 11.9 | 679 | 17.0 | 0.8 | 2,462 | 0.9 | 109 | 4.4 | 2,914 | 37.4 | 25.4 |
| Buffalo | 564 | 46.3 | 49,000 | 22.5 | 13.8 | 437 | 21.0 | 13.8 | 691 | 3.0 | 55 | 8.0 | 601 | 34.3 | 18.5 |
| Butte | 4,164 | 77.1 | 139,000 | 21.2 | 14.7 | 724 | 30.7 | 2.8 | 4,986 | -1.3 | 268 | 5.4 | 4,919 | 26.3 | 31.1 |
| Campbell | 673 | 86.5 | 68,800 | 14.5 | 10.0 | 534 | 18.4 | 0.6 | 783 | 2.0 | 29 | 3.7 | 753 | 52.6 | 20.7 |
| Charles Mix | 3,166 | 70.9 | 116,700 | 19.8 | 10.7 | 558 | 21.7 | 3.7 | 3,788 | 1.0 | 240 | 6.3 | 3,901 | 33.5 | 23.6 |
| Clark | 1,546 | 79.4 | 102,400 | 20.5 | 12.3 | 554 | 21.9 | 1.7 | 2,181 | 14.6 | 86 | 3.9 | 1,716 | 45.3 | 25.2 |
| Clay | 5,233 | 53.7 | 168,300 | 18.8 | 10.0 | 741 | 35.8 | 0.0 | 7,157 | -3.5 | 323 | 4.5 | 7,840 | 40.1 | 18.4 |
| Codington | 12,075 | 65.3 | 174,500 | 19.9 | 10.0 | 754 | 28.9 | 3.6 | 15,416 | -0.5 | 754 | 4.9 | 15,388 | 31.7 | 30.3 |
| Corson | 1,195 | 53.8 | 58,500 | 20.8 | 10.1 | 443 | 20.1 | 10.0 | 1,343 | -0.4 | 67 | 5.0 | 1,301 | 50.9 | 20.6 |
| Custer | 3,876 | 84.0 | 214,400 | 24.1 | 13.4 | 782 | 26.9 | 2.1 | 4,077 | 0.2 | 243 | 6.0 | 3,997 | 32.1 | 26.5 |
| Davison | 8,660 | 59.7 | 154,700 | 18.4 | 11.8 | 759 | 24.4 | 1.6 | 10,942 | -0.9 | 455 | 4.2 | 10,678 | 32.2 | 29.1 |
| Day | 2,574 | 75.3 | 100,500 | 20.8 | 11.8 | 532 | 24.3 | 0.0 | 2,706 | | 154 | 5.7 | 2,611 | 36.2 | 29.6 |
| Deuel | 1,798 | 79.0 | 133,500 | 19.4 | 10.0 | 629 | 20.8 | 2.9 | 2,336 | 7.5 | 125 | 5.4 | 2,331 | 35.4 | 29.9 |
| Dewey | 1,651 | 61.1 | 64,400 | 18.2 | 10.0 | 579 | 20.0 | 10.4 | 2,255 | 2.0 | 174 | 7.7 | 2,081 | 41.4 | 22.6 |
| Douglas | 1,220 | 79.8 | 94,400 | 17.4 | 10.3 | 537 | 24.3 | 4.1 | 1,558 | -0.7 | 46 | 3.0 | 1,400 | 37.4 | 27.4 |
| Edmunds | 1,544 | 82.0 | 118,400 | 16.9 | 10.0 | 604 | 20.7 | 0.2 | 1,993 | -1.9 | 58 | 2.9 | 1,934 | 40.1 | 26.6 |
| Fall River | 2,986 | 78.1 | 129,400 | 25.3 | 12.4 | 608 | 26.6 | 4.3 | 2,996 | 0.6 | 179 | 6.0 | 2,898 | 34.5 | 27.3 |
| Faulk | 895 | 80.3 | 85,100 | 15.0 | 11.7 | 660 | 27.5 | 0.0 | 1,039 | -1.1 | 36 | 3.5 | 1,080 | 39.4 | 26.7 |
| Grant | 3,218 | 81.2 | 133,700 | 20.4 | 10.3 | 620 | 23.7 | 3.0 | 4,382 | -0.7 | 176 | 4.0 | 3,555 | 33.0 | 29.5 |
| Gregory | 1,840 | 71.6 | 83,400 | 16.4 | 11.6 | 527 | 23.6 | 2.4 | 2,038 | 0.5 | 78 | 3.8 | 1,980 | 36.4 | 25.3 |
| Haakon | 822 | 77.1 | 105,400 | 25.0 | 15.0 | 701 | 24.5 | 0.0 | 1,069 | 2.0 | 27 | 2.5 | 898 | 34.0 | 34.4 |
| Hamlin | 2,256 | 80.5 | 157,100 | 18.7 | 10.0 | 558 | 16.4 | 2.9 | 3,355 | 1.3 | 116 | 3.5 | 2,827 | 35.0 | 35.2 |
| Hand | 1,508 | 68.3 | 128,400 | 19.2 | 10.0 | 515 | 25.1 | 0.3 | 1,768 | 0.0 | 47 | 2.7 | 1,763 | 42.2 | 25.7 |
| Hanson | 1,045 | 87.9 | 130,800 | 20.3 | 10.8 | 782 | 20.8 | 2.0 | 1,750 | -1.4 | 91 | 5.2 | 1,623 | 30.9 | 30.3 |
| Harding | 532 | 67.1 | 101,100 | 16.4 | 10.0 | 596 | 12.5 | 1.7 | 718 | -2.6 | 26 | 3.6 | 693 | 45.0 | 27.7 |
| Hughes | 7,460 | 68.8 | 177,500 | 18.0 | 10.8 | 673 | 22.8 | 3.9 | 9,909 | -0.5 | 335 | 3.4 | 9,334 | 44.0 | 17.3 |
| Hutchinson | 2,903 | 77.2 | 94,100 | 19.1 | 10.0 | 536 | 22.4 | 3.8 | 3,618 | 0.9 | 137 | 3.8 | 3,654 | 34.0 | 27.8 |
| Hyde | 597 | 75.0 | 88,800 | 18.3 | 11.0 | 613 | 17.1 | 1.5 | 680 | -0.9 | 21 | 3.1 | 671 | 47.7 | 27.1 |
| Jackson | 858 | 58.3 | 52,500 | 20.9 | 10.0 | 544 | 19.9 | 10.3 | 1,207 | -5.0 | 52 | 4.3 | 1,004 | 39.5 | 19.2 |
| Jerauld | 921 | 72.6 | 81,900 | 16.7 | 11.1 | 494 | 16.7 | 0.5 | 1,176 | 5.3 | 29 | 2.5 | 950 | 34.0 | 29.4 |
| Jones | 408 | 75.0 | 77,100 | 17.7 | 13.4 | 438 | 22.3 | 2.2 | 508 | -5.6 | 19 | 3.7 | 418 | 48.8 | 27.5 |
| Kingsbury | 2,279 | 79.5 | 109,900 | 17.3 | 10.0 | 500 | 21.5 | 0.8 | 2,661 | -0.2 | 98 | 3.7 | 2,686 | 39.2 | 34.2 |
| Lake | 4,700 | 74.6 | 168,400 | 19.0 | 10.4 | 597 | 24.0 | 1.4 | 6,485 | -2.2 | 246 | 3.8 | 6,684 | 36.5 | 26.5 |
| Lawrence | 11,048 | 64.7 | 207,900 | 22.5 | 11.3 | 761 | 27.0 | 2.0 | 13,567 | 1.5 | 841 | 6.2 | 13,484 | 33.7 | 21.5 |
| Lincoln | 19,749 | 76.7 | 228,800 | 19.6 | 10.0 | 963 | 25.7 | 1.1 | 35,711 | 0.2 | 1,294 | 3.6 | 30,895 | 44.0 | 19.1 |
| Lyman | 1,397 | 67.2 | 102,800 | 21.0 | 11.4 | 568 | 18.5 | 5.3 | 1,648 | -1.7 | 94 | 5.7 | 1,700 | 39.4 | 23.1 |
| McCook | 2,228 | 75.9 | 137,800 | 19.5 | 10.5 | 575 | 19.8 | 1.4 | 3,076 | 0.0 | 113 | 3.7 | 2,859 | 34.6 | 26.0 |
| McPherson | 1,030 | 77.1 | 58,900 | 17.3 | 11.2 | 494 | 26.7 | 0.9 | 1,010 | 1.4 | 52 | 5.1 | 1,075 | 37.1 | 26.2 |
| Marshall | 1,873 | 73.8 | 118,900 | 17.8 | 10.0 | 597 | 18.8 | 2.0 | 2,429 | 3.1 | 116 | 4.8 | 2,478 | 36.4 | 27.6 |
| Meade | 11,027 | 74.4 | 190,500 | 23.2 | 12.1 | 877 | 26.1 | 1.7 | 13,918 | -0.2 | 683 | 4.9 | 14,292 | 36.4 | 25.9 |
| Mellette | 661 | 69.4 | 32,100 | 14.9 | 13.4 | 502 | 24.3 | 12.7 | 708 | -4.5 | 44 | 6.2 | 539 | 34.2 | 22.3 |
| Miner | 948 | 76.7 | 77,000 | 20.1 | 10.0 | 603 | 20.5 | 2.0 | 1,144 | -2.4 | 55 | 4.8 | 1,128 | 40.3 | 34.8 |
| Minnehaha | 76,819 | 63.2 | 186,800 | 19.3 | 10.0 | 813 | 26.5 | 2.2 | 113,168 | 0.9 | 5,135 | 4.5 | 105,355 | 34.4 | 24.0 |
| Moody | 2,640 | 71.4 | 133,700 | 18.7 | 10.0 | 568 | 17.0 | 1.8 | 4,008 | -0.7 | 211 | 5.3 | 3,249 | 36.1 | 27.2 |
| Oglala Lakota | 2,729 | 50.2 | 27,400 | 12.7 | 13.0 | 514 | 21.5 | 40.9 | 3,677 | 4.0 | 373 | 10.1 | 3,576 | 32.2 | 19.8 |
| Pennington | 44,074 | 68.6 | 187,800 | 22.1 | 12.0 | 842 | 28.8 | 1.7 | 57,094 | 0.3 | 3,020 | 5.3 | 54,324 | 34.4 | 20.6 |
| Perkins | 1,284 | 72.6 | 102,000 | 18.9 | 10.0 | 590 | 24.6 | 1.6 | 1,382 | -2.2 | 52 | 3.8 | 1,603 | 41.7 | 25.7 |
| Potter | 982 | 78.8 | 102,300 | 21.7 | 10.8 | 684 | 17.5 | 0.4 | 1,019 | -1.2 | 33 | 3.2 | 1,131 | 36.0 | 29.2 |
| Roberts | 3,873 | 66.5 | 103,600 | 19.9 | 10.8 | 583 | 22.7 | 3.0 | 4,807 | -0.2 | 308 | 6.4 | 4,661 | 36.0 | 28.3 |
| Sanborn | 1,010 | 73.3 | 87,600 | 17.1 | 10.0 | 601 | 20.8 | 0.6 | 1,159 | 2.1 | 47 | 4.1 | 1,320 | 34.3 | 31.4 |
| Spink | 2,624 | 74.2 | 82,800 | 16.9 | 10.0 | 618 | 21.1 | 2.6 | 3,111 | -1.7 | 126 | 4.1 | 3,221 | 36.8 | 26.6 |
| Stanley | 1,344 | 78.9 | 163,300 | 18.1 | 11.0 | 772 | 35.1 | 0.9 | 1,891 | -0.2 | 74 | 3.9 | 1,739 | 41.3 | 25.5 |
| Sully | 569 | 85.9 | 144,100 | 17.8 | 10.0 | 582 | 16.7 | 0.9 | 789 | -1.3 | 22 | 2.8 | 657 | 36.4 | 28.5 |
| Todd | 2,700 | 43.7 | 26,000 | 20.6 | 11.3 | 429 | 20.9 | 14.2 | 3,066 | -2.5 | 197 | 6.4 | 2,433 | 41.0 | 18.4 |

1. Specified owner-occupied units. lacking complete plumbing facilities.   2. A value of 10.0 represents 10 percent or less; a value of 50.0 represents 50 percent or more.   3. Specified renter-occupied units.   4. Overcrowded or

5. Percent of civilian labor force.   6. Civilian employed persons 16 years old and over.

# Table B. States and Counties — Nonfarm Employment and Agriculture

| STATE County | Number of establish-ments | Employment — Total | Health care and social assistance | Manufac-turing | Retail trade | Finance and insurance | Professional, scientific, and technical services | Annual payroll Total (mil dol) | Average per employee (dollars) | Farms Number | Percent with: Fewer than 50 acres | Percent with: 1000 acres or more | Farm producers whose primary occupation is farming (percent) |
|---|---|---|---|---|---|---|---|---|---|---|---|---|---|
| | 104 | 105 | 106 | 107 | 108 | 109 | 110 | 111 | 112 | 113 | 114 | 115 | 116 |
| SOUTH CAROLINA—Cont'd | | | | | | | | | | | | | |
| Sumter | 1,763 | 32,663 | 4,846 | 5,720 | 4,578 | 752 | 881 | 1,263 | 38,668 | 524 | 55.7 | 7.3 | 45.7 |
| Union | 402 | 6,366 | 513 | 1,954 | 838 | 156 | 46 | 242 | 37,937 | 241 | 30.3 | 2.5 | 35.3 |
| Williamsburg | 477 | 6,823 | 950 | 2,223 | 913 | 123 | 100 | 272 | 39,898 | 552 | 26.6 | 8.9 | 34.7 |
| York | 5,391 | 90,388 | 10,603 | 10,684 | 11,848 | 6,981 | 7,321 | 4,212 | 46,604 | 1,000 | 49.4 | 1.5 | 30.2 |
| SOUTH DAKOTA | 27,108 | 358,943 | 68,638 | 46,750 | 52,541 | 25,113 | 12,718 | 15,767 | 43,925 | 29,968 | 19.3 | 32.0 | 52.4 |
| Aurora | 83 | 627 | 204 | 56 | 90 | 26 | 77 | 21 | 33,260 | 392 | 17.1 | 24.2 | 46.2 |
| Beadle | 566 | 6,818 | 1,225 | 1,651 | 970 | 547 | 113 | 261 | 38,215 | 744 | 20.6 | 30.5 | 53.9 |
| Bennett | 58 | 695 | 148 | NA | 154 | 12 | 2 | 22 | 31,528 | 213 | 5.6 | 65.3 | 65.1 |
| Bon Homme | 167 | 1,017 | 311 | 128 | 165 | 62 | 24 | 32 | 31,677 | 583 | 22.0 | 15.4 | 52.0 |
| Brookings | 896 | 14,100 | 1,669 | 4,787 | 1,836 | 708 | 602 | 581 | 41,182 | 886 | 31.0 | 14.3 | 38.9 |
| Brown | 1,254 | 18,313 | 3,165 | 3,665 | 2,734 | 953 | 385 | 769 | 41,976 | 1,034 | 24.3 | 27.4 | 50.5 |
| Brule | 227 | 2,001 | 432 | 15 | 230 | 73 | 53 | 67 | 33,376 | 394 | 17.3 | 37.8 | 54.4 |
| Buffalo | 10 | 162 | 24 | NA | NA | NA | NA | 4 | 24,346 | 68 | 14.7 | 64.7 | 76.6 |
| Butte | 338 | 2,198 | 255 | 242 | 503 | 81 | 95 | 74 | 33,443 | 565 | 17.7 | 30.4 | 48.8 |
| Campbell | 56 | 377 | 16 | 67 | 48 | NA | 8 | 15 | 38,477 | 249 | 7.2 | 48.2 | 57.5 |
| Charles Mix | 278 | 2,283 | 523 | 97 | 437 | 118 | 47 | 69 | 30,187 | 671 | 12.8 | 32.5 | 53.3 |
| Clark | 130 | 586 | 56 | 103 | 117 | 19 | 19 | 22 | 36,770 | 553 | 10.7 | 33.1 | 52.8 |
| Clay | 305 | 3,789 | 793 | 218 | 846 | 88 | 174 | 104 | 27,480 | 472 | 26.3 | 16.7 | 50.5 |
| Codington | 1,110 | 14,252 | 1,817 | 3,849 | 2,695 | 795 | 337 | 572 | 40,138 | 601 | 34.4 | 19.6 | 44.9 |
| Corson | 36 | 170 | 10 | NA | 49 | 18 | NA | 4 | 25,541 | 322 | 5.9 | 67.4 | 73.0 |
| Custer | 292 | 1,398 | 127 | 15 | 297 | 38 | 60 | 53 | 37,786 | 441 | 24.5 | 22.4 | 43.1 |
| Davison | 719 | 10,531 | 1,954 | 1,332 | 1,757 | 381 | 912 | 403 | 38,267 | 463 | 30.7 | 16.6 | 43.4 |
| Day | 201 | 1,408 | 286 | 188 | 278 | 48 | 35 | 49 | 34,649 | 581 | 13.9 | 29.1 | 52.2 |
| Deuel | 132 | 867 | 139 | 215 | 97 | 40 | 22 | 43 | 49,589 | 634 | 24.8 | 18.3 | 45.4 |
| Dewey | 78 | 552 | 169 | NA | 83 | 26 | NA | 22 | 39,949 | 310 | 12.3 | 61.3 | 57.6 |
| Douglas | 103 | 799 | 147 | 107 | 130 | 49 | 12 | 28 | 35,254 | 392 | 21.7 | 22.7 | 53.4 |
| Edmunds | 129 | 819 | 145 | 79 | 207 | 52 | 29 | 32 | 39,327 | 348 | 7.8 | 53.2 | 59.0 |
| Fall River | 209 | 2,511 | 1,549 | 8 | 280 | 36 | 38 | 157 | 62,560 | 314 | 14.0 | 45.5 | 53.8 |
| Faulk | 72 | 429 | 163 | NA | 93 | 26 | NA | 15 | 34,520 | 291 | 7.2 | 56.4 | 72.2 |
| Grant | 285 | 3,366 | 609 | 651 | 500 | 173 | 41 | 145 | 43,204 | 554 | 24.5 | 26.2 | 52.9 |
| Gregory | 187 | 1,056 | 278 | 34 | 281 | 66 | 45 | 32 | 30,774 | 495 | 9.1 | 31.1 | 57.4 |
| Haakon | 82 | 628 | 153 | 84 | 88 | 43 | 13 | 24 | 38,361 | 293 | 8.2 | 67.2 | 62.1 |
| Hamlin | 192 | 1,335 | 50 | 381 | 187 | 59 | 23 | 55 | 40,876 | 476 | 27.7 | 18.1 | 47.4 |
| Hand | 131 | 1,008 | 196 | 62 | 136 | 65 | 150 | 34 | 33,421 | 405 | 12.8 | 50.1 | 70.0 |
| Hanson | 75 | 349 | NA | 143 | 19 | 29 | 11 | 15 | 44,017 | 333 | 18.3 | 26.7 | 47.3 |
| Harding | 47 | 445 | 84 | NA | 53 | NA | NA | 24 | 54,328 | 265 | 6.4 | 79.6 | 70.3 |
| Hughes | 652 | 6,743 | 1,061 | 50 | 1,321 | 503 | 331 | 257 | 38,111 | 315 | 23.2 | 37.8 | 41.3 |
| Hutchinson | 217 | 2,000 | 475 | 305 | 298 | 109 | 37 | 76 | 38,134 | 775 | 20.8 | 19.0 | 49.8 |
| Hyde | 48 | 470 | NA | NA | 65 | 20 | 6 | 19 | 39,815 | 174 | 6.9 | 52.9 | 59.5 |
| Jackson | 53 | 263 | NA | NA | 113 | NA | NA | 8 | 29,141 | 314 | 4.8 | 65.9 | 70.2 |
| Jerauld | 81 | 1,225 | 133 | 725 | 49 | 28 | 22 | 49 | 39,753 | 244 | 13.1 | 27.9 | 46.1 |
| Jones | 51 | 269 | NA | NA | 84 | 13 | NA | 9 | 32,134 | 192 | 4.7 | 60.4 | 60.9 |
| Kingsbury | 156 | 1,175 | 167 | 263 | 142 | 122 | 16 | 48 | 41,273 | 518 | 15.3 | 31.9 | 56.0 |
| Lake | 370 | 3,910 | 647 | 973 | 512 | 230 | 221 | 157 | 40,126 | 463 | 26.8 | 19.4 | 45.8 |
| Lawrence | 1,090 | 10,120 | 1,648 | 413 | 1,498 | 277 | 293 | 349 | 34,496 | 275 | 30.2 | 14.5 | 39.6 |
| Lincoln | 1,731 | 21,375 | 3,723 | 2,178 | 2,591 | 2,094 | 1,169 | 1,069 | 50,002 | 756 | 36.1 | 12.0 | 44.1 |
| Lyman | 74 | 532 | NA | NA | 136 | 32 | NA | 16 | 29,758 | 414 | 5.1 | 43.5 | 51.2 |
| McCook | 187 | 932 | 188 | 23 | 152 | 45 | 39 | 33 | 35,104 | 512 | 23.0 | 26.2 | 47.6 |
| McPherson | 69 | 295 | 46 | 60 | 59 | 40 | NA | 10 | 33,166 | 382 | 5.0 | 45.0 | 59.8 |
| Marshall | 146 | 1,105 | 206 | 328 | 157 | 47 | 29 | 42 | 37,711 | 503 | 12.7 | 27.0 | 49.2 |
| Meade | 707 | 5,453 | 1,450 | 238 | 854 | 151 | 171 | 267 | 48,981 | 835 | 19.9 | 44.3 | 55.2 |
| Mellette | 19 | 161 | NA | NA | 51 | NA | NA | 4 | 22,863 | 219 | 1.4 | 64.8 | 68.6 |
| Miner | 80 | 545 | 119 | 106 | 72 | 26 | NA | 25 | 46,147 | 408 | 15.9 | 26.0 | 39.2 |
| Minnehaha | 5,834 | 120,386 | 25,594 | 13,012 | 15,844 | 10,973 | 4,000 | 5,829 | 48,416 | 1,023 | 42.1 | 11.4 | 39.9 |
| Moody | 156 | 1,544 | 188 | 345 | 106 | 22 | 24 | 60 | 38,542 | 490 | 33.7 | 16.1 | 47.3 |
| Oglala Lakota | 70 | 1,931 | 345 | NA | 184 | 54 | 9 | 68 | 35,232 | 190 | 3.7 | 54.7 | 54.3 |
| Pennington | 3,770 | 48,223 | 9,944 | 2,716 | 8,277 | 2,453 | 1,854 | 2,017 | 41,821 | 656 | 24.2 | 28.4 | 42.6 |
| Perkins | 121 | 723 | 131 | 104 | 112 | 49 | 14 | 25 | 34,246 | 421 | 5.9 | 65.1 | 69.4 |
| Potter | 101 | 529 | 50 | NA | 105 | 43 | 11 | 23 | 43,518 | 221 | 5.4 | 49.3 | 60.4 |
| Roberts | 238 | 2,051 | 358 | 342 | 453 | 77 | 26 | 66 | 32,392 | 782 | 15.0 | 25.2 | 50.3 |
| Sanborn | 51 | 349 | 58 | NA | 42 | 21 | 24 | 12 | 34,338 | 351 | 12.5 | 31.3 | 52.5 |
| Spink | 178 | 1,211 | 235 | 72 | 219 | 96 | 28 | 49 | 40,532 | 556 | 10.6 | 48.4 | 64.5 |
| Stanley | 115 | 881 | 25 | NA | 165 | 29 | 64 | 37 | 42,092 | 172 | 8.7 | 59.3 | 58.1 |
| Sully | 60 | 288 | 7 | NA | 95 | 24 | NA | 13 | 45,427 | 201 | 10.0 | 52.7 | 70.2 |
| Todd | 51 | 955 | 343 | NA | 259 | 16 | NA | 36 | 37,939 | 223 | 3.6 | 58.7 | 68.3 |

| STATE County | Acreage (1,000) | Percent change, 2012-2017 | Average size of farm | Total irrigated (1,000) | Total cropland (1,000) | Average per farm | Average per acre | Value of machinery and equipment, average per farm (dollars) | Total (mil dol) | Average per farm (acres) | Crops | Livestock and poultry products | Organic farms (number) | Farms with internet access (percent) | Total ($1,000) | Percent of farms |
|---|---|---|---|---|---|---|---|---|---|---|---|---|---|---|---|---|
| | 117 | 118 | 119 | 120 | 121 | 122 | 123 | 124 | 125 | 126 | 127 | 128 | 129 | 130 | 131 | 132 |
| SOUTH CAROLINA— Cont'd | | | | | | | | | | | | | | | | |
| Sumter | 168 | -4.7 | 320 | 19.1 | 91.4 | 958,464 | 2,995 | 134,000 | 153.4 | 292,763 | 32.6 | 67.4 | 5 | 79.6 | 1,534 | 30.7 |
| Union | 44 | -7.5 | 182 | 0.0 | 7.3 | 460,174 | 2,534 | 64,867 | 10.3 | 42,859 | 10.4 | 89.6 | NA | 62.7 | 426 | 25.3 |
| Williamsburg | 209 | -7.1 | 378 | 2.6 | 107.7 | 848,443 | 2,245 | 118,529 | 47.2 | 85,562 | 86.7 | 13.3 | NA | 60.1 | 2,191 | 55.3 |
| York | 120 | -3.0 | 120 | 1.2 | 37.9 | 838,682 | 6,979 | 63,006 | 100.5 | 100,504 | 63.2 | 36.8 | NA | 67.7 | 876 | 17.5 |
| SOUTH DAKOTA | 43,244 | 0.0 | 1,443 | 492.5 | 19,813.5 | 2,984,426 | 2,068 | 282,162 | 9,721.5 | 324,397 | 53.1 | 46.9 | 87 | 81.0 | 419,508 | 72.1 |
| Aurora | 355 | -19.8 | 904 | NA | 228.8 | 2,400,283 | 2,654 | 297,809 | 138.2 | 352,541 | 46.7 | 53.3 | NA | 82.1 | 4,879 | 82.9 |
| Beadle | 811 | 2.1 | 1,090 | 20.1 | 548.5 | 2,934,337 | 2,693 | 331,767 | 295.3 | 396,867 | 56.6 | 43.4 | NA | 84.9 | 12,672 | 73.7 |
| Bennett | 637 | 5.0 | 2,990 | 8.5 | 214.3 | 2,846,734 | 952 | 251,150 | 65.9 | 309,545 | 40.6 | 59.4 | NA | 72.3 | 3,641 | 71.8 |
| Bon Homme | 305 | -13.3 | 523 | 8.6 | 228.4 | 2,083,627 | 3,984 | 274,056 | 152.9 | 262,182 | 51.0 | 49.0 | 2 | 79.9 | 6,300 | 86.6 |
| Brookings | 460 | 2.3 | 519 | 20.5 | 345.4 | 2,386,912 | 4,602 | 248,257 | 316.3 | 357,034 | 39.9 | 60.1 | 3 | 84.4 | 6,331 | 59.5 |
| Brown | 1,083 | 0.4 | 1,047 | 4.4 | 884.6 | 3,521,412 | 3,362 | 360,625 | 377.4 | 365,032 | 77.8 | 22.2 | 2 | 78.8 | 11,893 | 62.9 |
| Brule | 518 | 0.7 | 1,314 | 5.4 | 277.9 | 3,386,122 | 2,578 | 312,852 | 153.6 | 389,759 | 32.5 | 67.5 | NA | 78.4 | 5,948 | 77.2 |
| Buffalo | 300 | 1.3 | 4,410 | 9.3 | 85.4 | 6,815,539 | 1,545 | 502,785 | 39.5 | 581,059 | 42.2 | 57.8 | NA | 86.8 | 2,436 | 79.4 |
| Butte | 1,155 | 1.8 | 2,044 | 46.7 | 150.2 | 1,959,536 | 959 | 127,586 | 68.0 | 120,312 | 12.5 | 87.5 | NA | 78.1 | 5,104 | 52.4 |
| Campbell | 434 | 20.4 | 1,742 | 4.6 | 250.8 | 3,190,052 | 1,831 | 339,855 | 96.2 | 386,333 | 57.6 | 42.4 | NA | 73.9 | 3,925 | 87.1 |
| Charles Mix | 686 | -0.9 | 1,022 | 13.4 | 451.0 | 2,850,723 | 2,788 | 350,312 | 262.1 | 390,633 | 45.8 | 54.2 | 1 | 85.5 | 9,733 | 88.5 |
| Clark | 602 | -1.1 | 1,089 | 17.6 | 444.8 | 3,428,717 | 3,150 | 290,526 | 267.6 | 483,870 | 55.6 | 44.4 | NA | 81.7 | 7,415 | 82.3 |
| Clay | 239 | -7.6 | 506 | 28.1 | 221.2 | 2,380,474 | 4,702 | 299,816 | 113.1 | 239,606 | 88.1 | 11.9 | 3 | 79.2 | 4,754 | 81.4 |
| Codington | 383 | 3.8 | 638 | 8.2 | 268.0 | 2,210,441 | 3,467 | 217,944 | 155.9 | 259,364 | 57.1 | 42.9 | 2 | 80.5 | 7,928 | 58.6 |
| Corson | 1,288 | 3.6 | 3,998 | D | 398.6 | 3,675,545 | 919 | 325,133 | 117.0 | 363,205 | 28.6 | 71.4 | 2 | 80.7 | 6,905 | 80.1 |
| Custer | 612 | -1.8 | 1,387 | 3.9 | 49.4 | 2,213,964 | 1,596 | 86,361 | 26.6 | 60,392 | 7.3 | 92.7 | NA | 76.9 | 1,641 | 37.9 |
| Davison | 270 | -1.8 | 584 | 2.7 | 212.4 | 1,983,497 | 3,398 | 224,120 | 107.7 | 232,538 | 61.4 | 38.6 | NA | 84.7 | 4,987 | 60.3 |
| Day | 611 | 7.1 | 1,051 | 1.3 | 456.5 | 2,779,344 | 2,645 | 275,390 | 184.0 | 316,661 | 82.9 | 17.1 | NA | 68.7 | 13,431 | 79.3 |
| Deuel | 335 | -1.9 | 529 | 0.5 | 223.8 | 1,823,780 | 3,448 | 207,041 | 184.9 | 291,655 | 37.9 | 62.1 | 3 | 79.5 | 4,029 | 73.8 |
| Dewey | 1,137 | -3.8 | 3,666 | D | 195.6 | 3,555,413 | 970 | 259,622 | 58.7 | 189,394 | 23.3 | 76.7 | 7 | 73.2 | 5,510 | 83.2 |
| Douglas | 275 | 1.9 | 701 | 1.8 | 214.7 | 2,607,726 | 3,721 | 327,367 | 144.6 | 368,954 | 44.4 | 55.6 | NA | 78.3 | 4,604 | 76.3 |
| Edmunds | 739 | 6.0 | 2,124 | 5.8 | 569.5 | 5,223,055 | 2,460 | 551,208 | 237.7 | 683,029 | 64.6 | 35.4 | 1 | 85.3 | 6,569 | 83.3 |
| Fall River | 898 | -17.5 | 2,860 | 9.3 | 68.7 | 2,048,121 | 716 | 113,245 | 87.1 | 277,347 | 3.7 | 96.3 | 2 | 81.8 | 2,368 | 49.7 |
| Faulk | 627 | 1.9 | 2,155 | D | 408.4 | 5,489,700 | 2,547 | 421,205 | 164.0 | 563,416 | 65.9 | 34.1 | 2 | 86.6 | 12,360 | 82.8 |
| Grant | 426 | -0.7 | 768 | 5.2 | 307.7 | 2,864,773 | 3,729 | 336,697 | 235.4 | 424,986 | 50.3 | 49.7 | 1 | 78.5 | 4,192 | 69.7 |
| Gregory | 562 | -11.5 | 1,136 | 0.3 | 226.1 | 2,134,179 | 1,879 | 208,369 | 93.1 | 188,004 | 45.9 | 54.1 | NA | 82.6 | 4,617 | 80.6 |
| Haakon | 1,158 | 2.1 | 3,951 | D | 323.9 | 3,646,592 | 923 | 257,669 | 77.3 | 263,949 | 33.1 | 66.9 | NA | 87.0 | 7,822 | 84.0 |
| Hamlin | 310 | -0.3 | 651 | 9.0 | 249.6 | 2,557,064 | 3,926 | 304,618 | 179.8 | 377,664 | 54.1 | 45.9 | 1 | 84.5 | 3,100 | 69.3 |
| Hand | 895 | -1.1 | 2,211 | 2.6 | 578.0 | 4,545,489 | 2,056 | 456,429 | 224.5 | 554,210 | 63.5 | 36.5 | 1 | 82.0 | 12,357 | 83.5 |
| Hanson | 271 | -1.1 | 814 | 5.1 | 226.0 | 3,516,095 | 4,322 | 370,625 | 132.8 | 398,778 | 61.8 | 38.2 | NA | 73.9 | 5,966 | 76.3 |
| Harding | 1,468 | 0.0 | 5,539 | 1.1 | 184.6 | 3,683,948 | 665 | 228,551 | 63.4 | 239,125 | 12.0 | 88.0 | NA | 86.4 | 6,699 | 67.2 |
| Hughes | 435 | 1.2 | 1,381 | 11.5 | 263.0 | 2,611,187 | 1,891 | 265,629 | 70.3 | 223,095 | 62.0 | 38.0 | 5 | 84.4 | 6,354 | 63.5 |
| Hutchinson | 456 | -11.2 | 588 | 7.3 | 363.8 | 2,642,567 | 4,495 | 304,494 | 270.0 | 348,405 | 48.6 | 51.4 | 3 | 77.3 | 9,766 | 82.5 |
| Hyde | 506 | -1.7 | 2,906 | D | 219.9 | 4,438,942 | 1,527 | 351,062 | 64.9 | 372,983 | 58.6 | 41.4 | NA | 87.4 | 4,401 | 75.9 |
| Jackson | 1,166 | 0.7 | 3,714 | 0.3 | 185.5 | 3,248,165 | 875 | 164,945 | 52.4 | 167,003 | 14.5 | 85.5 | 1 | 79.3 | 4,724 | 67.2 |
| Jerauld | 342 | 2.6 | 1,400 | 3.0 | 219.8 | 3,348,353 | 2,391 | 378,327 | 112.5 | 461,242 | 45.2 | 54.8 | NA | 76.6 | 3,271 | 80.3 |
| Jones | 618 | 1.0 | 3,221 | D | 233.8 | 3,680,881 | 1,143 | 323,434 | 57.9 | 301,516 | 55.6 | 44.4 | NA | 81.8 | 3,245 | 83.3 |
| Kingsbury | 536 | 2.8 | 1,035 | 2.6 | 391.2 | 3,941,833 | 3,810 | 367,554 | 257.9 | 497,790 | 56.0 | 44.0 | 2 | 82.8 | 3,492 | 70.5 |
| Lake | 273 | 4.1 | 589 | 8.2 | 225.6 | 3,014,550 | 5,121 | 299,337 | 161.9 | 349,624 | 61.4 | 38.6 | 2 | 84.0 | 2,755 | 76.2 |
| Lawrence | 165 | 3.9 | 600 | 4.4 | 33.6 | 1,073,421 | 1,789 | 103,416 | 13.1 | 47,575 | 14.8 | 85.2 | NA | 80.0 | 452 | 18.5 |
| Lincoln | 295 | -19.4 | 390 | 2.1 | 267.9 | 2,692,014 | 6,906 | 189,954 | 204.8 | 270,886 | 59.8 | 40.2 | 2 | 77.5 | 6,210 | 70.0 |
| Lyman | 951 | -7.6 | 2,297 | D | 419.2 | 3,792,074 | 1,651 | 300,717 | 97.0 | 234,401 | 56.0 | 44.0 | 1 | 77.1 | 12,929 | 89.6 |
| McCook | 367 | 1.2 | 717 | 0.0 | 294.3 | 3,131,091 | 4,367 | 271,011 | 195.9 | 382,695 | 62.6 | 37.4 | NA | 82.4 | 5,726 | 84.2 |
| McPherson | 723 | 26.2 | 1,893 | 1.3 | 383.9 | 4,077,571 | 2,154 | 363,067 | 140.2 | 367,047 | 45.2 | 54.8 | 6 | 75.7 | 4,087 | 85.3 |
| Marshall | 525 | -1.3 | 1,045 | 0.7 | 343.4 | 2,939,925 | 2,814 | 321,438 | 270.7 | 538,147 | 39.9 | 60.1 | 2 | 83.7 | 8,489 | 81.3 |
| Meade | 1,999 | -1.7 | 2,394 | 5.3 | 394.8 | 2,277,231 | 951 | 169,000 | 99.2 | 118,764 | 11.4 | 88.6 | 3 | 79.2 | 11,199 | 52.3 |
| Mellette | 753 | 7.7 | 3,436 | 0.7 | 131.1 | 3,314,464 | 965 | 216,199 | 45.4 | 207,411 | 18.4 | 81.6 | 5 | 81.7 | 2,321 | 77.6 |
| Miner | 351 | -1.7 | 861 | D | 239.8 | 3,042,444 | 3,533 | 239,023 | 126.1 | 308,963 | 56.6 | 43.4 | NA | 79.7 | 6,404 | 79.2 |
| Minnehaha | 375 | -8.1 | 366 | 5.2 | 318.0 | 2,362,080 | 6,449 | 232,537 | 253.8 | 248,073 | 58.9 | 41.1 | 2 | 87.4 | 2,507 | 56.0 |
| Moody | 265 | 4.2 | 541 | 3.3 | 223.5 | 3,147,674 | 5,820 | 296,266 | 216.1 | 440,976 | 47.8 | 52.2 | 2 | 83.9 | 3,759 | 59.0 |
| Oglala Lakota | 1,116 | 1.4 | 5,875 | D | 88.9 | 3,960,426 | 674 | 147,320 | 38.7 | 203,926 | 20.0 | 80.0 | NA | 82.6 | 3,643 | 55.8 |
| Pennington | 1,147 | 6.7 | 1,748 | 10.7 | 207.2 | 1,849,403 | 1,058 | 126,171 | 60.5 | 92,155 | 28.0 | 72.0 | NA | 83.1 | 6,415 | 42.7 |
| Perkins | 1,640 | 0.5 | 3,895 | NA | 347.7 | 3,236,529 | 831 | 232,943 | 84.9 | 201,646 | 14.6 | 85.4 | NA | 84.3 | 9,976 | 79.8 |
| Potter | 548 | 1.9 | 2,481 | 1.6 | 386.5 | 5,246,162 | 2,114 | 546,151 | 101.9 | 461,045 | 82.0 | 18.0 | NA | 79.6 | 7,893 | 92.3 |
| Roberts | 595 | -4.5 | 761 | 2.4 | 451.7 | 2,288,827 | 3,007 | 260,731 | 204.4 | 261,366 | 78.1 | 21.9 | 7 | 79.7 | 8,289 | 74.9 |
| Sanborn | 364 | 0.9 | 1,036 | 0.0 | 200.4 | 2,959,610 | 2,857 | 265,179 | 112.1 | 319,262 | 46.4 | 53.6 | NA | 80.1 | 2,914 | 67.2 |
| Spink | 961 | 1.7 | 1,729 | 22.3 | 832.8 | 5,413,741 | 3,132 | 510,061 | 382.5 | 687,917 | 73.7 | 26.3 | 1 | 82.9 | 20,072 | 89.6 |
| Stanley | 812 | 2.7 | 4,722 | NA | 195.7 | 4,349,277 | 921 | 241,233 | 50.6 | 294,256 | 35.7 | 64.3 | NA | 77.9 | 3,863 | 76.7 |
| Sully | 633 | 0.7 | 3,149 | 23.7 | 505.7 | 6,783,106 | 2,154 | 612,046 | 137.2 | 682,781 | 74.2 | 25.8 | NA | 88.6 | 10,351 | 80.1 |
| Todd | 880 | 2.3 | 3,946 | 7.8 | 113.7 | 3,355,405 | 850 | 254,915 | 53.7 | 240,619 | 11.4 | 88.6 | 1 | 82.5 | 686 | 57.0 |

# Table B. States and Counties — Water Use, Wholesale Trade, Retail Trade, and Real Estate

| STATE County | Water use, 2015 | | Wholesale Trade[1], 2017 | | | | Retail Trade[2], 2017 | | | | Real estate and rental and leasing,[2] 2017 | | | |
|---|---|---|---|---|---|---|---|---|---|---|---|---|---|---|
| | Public supply water withdrawn (mil gal/day) | Public supply gallons withdrawn per person per day | Number of establishments | Number of employees | Sales (mil dol) | Average payroll (mil dol) | Number of establishments | Number of employees | Sales (mil dol) | Average payroll (mil dol) | Number of establishments | Number of employees | Sales (mil dol) | Average payroll (mil dol) |
| | 133 | 134 | 135 | 136 | 137 | 138 | 139 | 140 | 141 | 142 | 143 | 144 | 145 | 146 |
| SOUTH CAROLINA—Cont'd | | | | | | | | | | | | | | |
| Sumter | 14.60 | 135.8 | 64 | 634 | 329.7 | 29.9 | 353 | 4,531 | 1,110.3 | 101.7 | 66 | 300 | 43.6 | 8.4 |
| Union | 2.94 | 105.8 | 12 | 97 | 54.4 | 3.5 | 91 | 869 | 188.7 | 17.2 | 15 | 151 | 10.0 | 3.8 |
| Williamsburg | 3.02 | 92.8 | 17 | 190 | 117.4 | 6.5 | 114 | 866 | 216.8 | 17.9 | 13 | 26 | 4.1 | 0.6 |
| York | 20.63 | 82.1 | 256 | 4,176 | 3,063.4 | 250.7 | 682 | 11,255 | 3,169.3 | 270.2 | 302 | 1,499 | 329.4 | 72.5 |
| SOUTH DAKOTA | 71.95 | 81.7 | 1,392 | 16,630 | 17,314.3 | 887.0 | 3,884 | 53,134 | 14,673.7 | 1,372.0 | 1,143 | 4,330 | 829.0 | 155.0 |
| Aurora | 0.19 | 69.5 | 6 | D | 37.6 | D | 9 | 91 | 29.1 | 1.5 | NA | NA | NA | NA |
| Beadle | 2.29 | 124.6 | 33 | 454 | 439.5 | 25.7 | 77 | 1,015 | 260.0 | 24.4 | 31 | 103 | 9.8 | 1.6 |
| Bennett | 0.17 | 49.7 | NA | NA | NA | NA | 15 | 157 | 28.8 | 2.6 | NA | NA | NA | NA |
| Bon Homme | 0.56 | 80.2 | D | D | D | 3.9 | 25 | 151 | 38.8 | 2.9 | 4 | D | 0.7 | D |
| Brookings | 2.77 | 81.7 | 39 | 329 | 369.3 | 17.3 | 108 | 1,753 | 400.7 | 41.2 | 45 | 187 | 22.9 | 5.9 |
| Brown | 3.47 | 89.5 | 80 | 1,070 | 1,680.7 | 54.3 | 184 | 2,916 | 795.4 | 78.1 | 64 | 297 | 51.7 | 11.1 |
| Brule | 0.52 | 98.5 | 13 | 113 | 143.1 | 4.9 | 43 | 276 | 81.5 | 7.2 | NA | NA | NA | NA |
| Buffalo | 0.10 | 47.7 | NA | NA | NA | NA | NA | NA | NA | NA | NA | NA | NA | NA |
| Butte | 0.52 | 50.6 | 9 | 53 | 47.6 | 1.7 | 53 | 510 | 166.7 | 15.4 | D | D | D | D |
| Campbell | 0.08 | 57.3 | 6 | 36 | 43.7 | 1.7 | 7 | D | 6.8 | D | NA | NA | NA | NA |
| Charles Mix | 0.55 | 58.6 | 16 | 135 | 99.8 | 6.3 | 44 | 444 | 78.3 | 7.6 | 5 | D | 0.9 | D |
| Clark | 0.16 | 43.7 | 10 | D | 83.0 | D | 14 | 121 | 29.6 | 2.9 | NA | NA | NA | NA |
| Clay | 1.02 | 73.0 | D | D | D | D | 43 | 692 | 164.1 | 15.2 | 10 | 16 | 4.4 | 0.5 |
| Codington | 3.74 | 133.9 | 63 | 685 | 483.9 | 36.7 | 190 | 2,704 | 678.3 | 66.5 | 63 | 122 | 20.6 | 3.8 |
| Corson | 0.18 | 42.9 | D | D | D | D | 4 | 63 | 13.4 | 1.3 | NA | NA | NA | NA |
| Custer | 0.28 | 33.2 | NA | NA | NA | NA | 33 | 263 | 69.1 | 5.6 | 11 | 28 | 4.1 | 0.8 |
| Davison | 1.92 | 96.7 | 39 | 448 | 572.6 | 19.7 | 131 | 1,910 | 554.2 | 49.9 | D | D | D | D |
| Day | 0.44 | 79.4 | 12 | 143 | 191.1 | 7.0 | 33 | 280 | 82.0 | 6.2 | D | D | D | 0.0 |
| Deuel | 0.17 | 39.2 | 8 | 59 | 69.8 | 2.9 | 18 | 103 | 32.5 | 3.0 | NA | NA | NA | NA |
| Dewey | 0.33 | 58.0 | 4 | 77 | 36.5 | 2.4 | 10 | 80 | 23.2 | 1.2 | 3 | D | 0.8 | D |
| Douglas | 0.18 | 60.5 | D | D | D | 4.9 | 19 | 126 | 43.7 | 2.9 | NA | NA | NA | NA |
| Edmunds | 0.26 | 65.0 | 13 | 220 | 328.2 | 9.6 | 22 | 145 | 42.0 | 3.2 | 4 | 6 | 1.6 | 0.1 |
| Fall River | 0.66 | 96.1 | NA | NA | NA | NA | 35 | 233 | 62.1 | 5.9 | 8 | 12 | 1.7 | 0.3 |
| Faulk | 0.15 | 64.2 | 10 | D | 67.8 | D | 7 | 65 | 16.3 | 1.4 | NA | NA | NA | NA |
| Grant | 0.63 | 88.2 | 12 | 116 | 88.8 | 6.2 | 47 | 534 | 124.3 | 12.4 | 6 | 8 | 0.9 | 0.2 |
| Gregory | 0.26 | 61.9 | 7 | D | 32.4 | D | 34 | 215 | 64.3 | 5.3 | 3 | 8 | 0.5 | 0.2 |
| Haakon | 0.11 | 59.1 | 8 | 117 | 153.7 | 3.8 | 15 | 112 | 28.0 | 2.9 | NA | NA | NA | NA |
| Hamlin | 0.41 | 67.8 | 15 | 144 | 137.9 | 8.2 | 18 | 172 | 56.9 | 4.8 | NA | NA | NA | NA |
| Hand | 0.21 | 62.7 | 11 | 135 | 74.8 | 6.4 | 22 | 140 | 31.6 | 3.0 | NA | NA | NA | NA |
| Hanson | 0.09 | 26.6 | 5 | 29 | 156.1 | 1.5 | 5 | 37 | 5.9 | 0.6 | NA | NA | NA | NA |
| Harding | 0.03 | 23.7 | NA | NA | NA | NA | 8 | 46 | 11.6 | 1.0 | NA | NA | NA | NA |
| Hughes | 2.67 | 152.1 | 29 | 286 | 687.2 | 16.4 | 96 | 1,493 | 378.5 | 35.9 | 29 | 53 | 13.0 | 1.8 |
| Hutchinson | 0.39 | 53.4 | 25 | 336 | 436.4 | 15.5 | 36 | 361 | 83.7 | 7.0 | NA | NA | NA | NA |
| Hyde | 0.11 | 78.7 | 7 | 73 | 89.9 | 4.1 | 6 | 75 | 16.5 | 1.5 | NA | NA | NA | NA |
| Jackson | 0.13 | 39.1 | NA | NA | NA | NA | 14 | 113 | 36.0 | 2.0 | NA | NA | NA | NA |
| Jerauld | 0.54 | 270.4 | 6 | 106 | 48.2 | 4.6 | 9 | 46 | 12.1 | 1.2 | NA | NA | NA | NA |
| Jones | 0.05 | 54.1 | NA | NA | NA | NA | 9 | 74 | 38.2 | 2.2 | NA | NA | NA | NA |
| Kingsbury | 0.22 | 44.1 | 11 | 111 | 168.7 | 6.1 | 22 | 139 | 29.2 | 2.9 | D | D | D | D |
| Lake | 0.81 | 64.2 | 19 | 228 | 290.8 | 13.1 | 46 | 506 | 174.0 | 13.5 | 12 | 25 | 2.6 | 0.5 |
| Lawrence | 2.35 | 94.7 | D | D | D | 2.2 | 143 | 1,485 | 493.1 | 40.7 | 66 | 202 | 25.1 | 5.4 |
| Lincoln | 1.02 | 19.3 | 57 | 577 | 533.3 | 31.9 | 166 | 2,317 | 902.5 | 76.2 | 93 | 506 | 89.8 | 17.1 |
| Lyman | 0.22 | 56.8 | NA | NA | NA | NA | 15 | 185 | 65.2 | 3.6 | NA | NA | NA | NA |
| McCook | 0.71 | 126.8 | D | D | D | D | 22 | 177 | 44.7 | 4.0 | D | D | D | D |
| McPherson | 0.13 | 53.8 | 4 | D | 27.6 | D | 13 | 63 | 12.7 | 1.3 | NA | NA | NA | NA |
| Marshall | 0.22 | 46.1 | 12 | 75 | 104.7 | 3.6 | 24 | 184 | 38.7 | 4.5 | D | D | D | D |
| Meade | 1.42 | 52.6 | D | D | D | D | 91 | 837 | 308.7 | 23.6 | 27 | 69 | 9.2 | 1.8 |
| Mellette | 0.09 | 43.9 | NA | NA | NA | NA | 5 | 45 | 6.8 | 0.7 | NA | NA | NA | NA |
| Miner | 0.11 | 49.2 | 4 | 45 | 7.9 | 1.9 | 14 | 73 | 17.9 | 1.4 | NA | NA | NA | NA |
| Minnehaha | 19.65 | 94.7 | 367 | 5,813 | 4,456.2 | 336.2 | 812 | 15,657 | 4,066.6 | 413.3 | 273 | 1,541 | 313.2 | 70.0 |
| Moody | 0.49 | 76.2 | D | D | D | D | 18 | 146 | 34.4 | 2.4 | D | D | D | D |
| Oglala Lakota | 1.10 | 76.5 | NA | NA | NA | NA | 15 | 186 | 45.7 | 4.0 | NA | NA | NA | NA |
| Pennington | 10.62 | 97.7 | 143 | 1,649 | 1,008.4 | 86.2 | 577 | 9,025 | 2,701.1 | 249.3 | 229 | 681 | 181.3 | 23.1 |
| Perkins | 0.16 | 53.0 | D | D | D | 2.7 | 19 | 145 | 25.8 | 3.2 | NA | NA | NA | NA |
| Potter | 0.20 | 86.2 | 9 | 134 | 324.9 | 7.3 | 18 | 89 | 17.5 | 1.9 | D | D | D | D |
| Roberts | 0.58 | 56.3 | D | D | D | D | 37 | 382 | 81.5 | 7.7 | D | D | D | D |
| Sanborn | 0.13 | 55.2 | 4 | 30 | 38.5 | 1.9 | D | D | D | 0.5 | NA | NA | NA | NA |
| Spink | 0.52 | 79.7 | 19 | 179 | 388.2 | 11.4 | 29 | 213 | 43.5 | 4.7 | 3 | D | 1.0 | D |
| Stanley | 0.32 | 108.3 | D | D | D | D | 20 | 170 | 75.2 | 5.0 | NA | NA | NA | NA |
| Sully | 0.00 | 0.0 | D | D | D | 2.6 | 11 | 95 | 35.1 | 2.5 | NA | NA | NA | NA |
| Todd | 0.42 | 42.2 | NA | NA | NA | NA | 17 | 264 | 46.4 | 3.9 | NA | NA | NA | NA |

1  Merchant wholesalers, except manufacturers' sales branches and offices.      2. Employer establishments.

# Table B. States and Counties — Professional Services, Manufacturing, and Accommodation and Food Services

| STATE County | Professional, scientific, and technical services, 2017 | | | | Manufacturing, 2017 | | | | Accommodation and food services, 2017 | | | |
|---|---|---|---|---|---|---|---|---|---|---|---|---|
| | Number of establishments | Number of employees | Sales (mil dol) | Average payroll (mil dol) | Number of establishments | Number of employees | Sales (mil dol) | Average payroll (mil dol) | Number of establishments | Number of employees | Sales (mil dol) | Annual payroll (mil dol) |
| | 147 | 148 | 149 | 150 | 151 | 152 | 153 | 154 | 155 | 156 | 157 | 158 |
| **SOUTH CAROLINA— Cont'd** | | | | | | | | | | | | |
| Sumter | D | D | D | D | 68 | 5,559 | 1,976.3 | 258.4 | 171 | 3,354 | 164.7 | 44.8 |
| Union | 20 | 43 | 5 | 1.0 | 28 | 1,579 | 757.7 | 80.9 | 35 | 525 | 24.7 | 6.7 |
| Williamsburg | 20 | 107 | 13 | 6.2 | 29 | 2,177 | 1,486.8 | 101.6 | 32 | 307 | 14.5 | 3.2 |
| York | D | D | D | D | 213 | 10,405 | 4,012.9 | 623.7 | 436 | 9,523 | 492.3 | 136.2 |
| **SOUTH DAKOTA** | 1,952 | 12,998 | 1,756 | 651.6 | 1,051 | 42,935 | 16,779.5 | 2,057.4 | 2,495 | 40,704 | 2,315.5 | 659.2 |
| Aurora | D | D | D | 2.8 | NA | NA | NA | NA | 9 | 25 | 1.3 | 0.3 |
| Beadle | 26 | 121 | 12 | 4.6 | 33 | 1,595 | 524.4 | 61.5 | 40 | 514 | 24.2 | 7.3 |
| Bennett | NA | NA | NA | NA | NA | NA | NA | NA | 8 | D | 2.9 | D |
| Bon Homme | 6 | 24 | 3 | 1.0 | 10 | 118 | 30.2 | 5.1 | 8 | D | 1.8 | D |
| Brookings | 76 | 555 | 72 | 26.9 | 35 | 4,950 | 2,634.3 | 267.0 | 90 | 1,808 | 82.0 | 24.2 |
| Brown | 71 | 427 | 53 | 19.0 | D | 3,206 | D | 148.3 | 103 | 1,974 | 100.2 | 29.9 |
| Brule | 17 | 57 | 4 | 1.3 | 7 | 50 | 12.5 | 1.3 | 30 | 275 | 18.9 | 4.3 |
| Buffalo | NA | NA | NA | NA | NA | NA | NA | NA | NA | NA | NA | NA |
| Butte | 27 | 94 | 9 | 2.8 | 17 | 198 | 93.6 | 9.3 | D | D | D | D |
| Campbell | D | D | 1 | D | 3 | 46 | 37.1 | 2.3 | 7 | D | 0.7 | D |
| Charles Mix | 11 | 44 | 5 | 1.2 | 10 | 83 | 17.6 | 3.6 | 21 | 525 | 17.9 | 7.3 |
| Clark | 11 | 26 | 3 | 0.9 | 7 | 94 | 19.7 | 4.6 | 10 | 75 | 2.2 | 0.9 |
| Clay | 19 | 161 | 12 | 4.2 | 11 | 176 | 53.3 | 9.4 | 47 | 876 | 35.8 | 12.3 |
| Codington | 70 | 268 | 40 | 11.8 | 75 | 3,274 | 921.6 | 148.3 | 90 | 1,797 | 90.8 | 27.0 |
| Corson | NA | NA | NA | NA | NA | NA | NA | NA | D | D | D | D |
| Custer | 25 | 54 | 14 | 2.3 | 10 | 17 | 2.5 | 0.6 | 63 | 450 | 45.7 | 11.8 |
| Davison | 41 | 791 | 110 | 40.0 | 41 | 1,986 | 802.6 | 109.3 | 85 | 1,346 | 66.3 | 18.9 |
| Day | 9 | 25 | 3 | 0.9 | 10 | 147 | 36.2 | 7.3 | 20 | 168 | 7.3 | 2.2 |
| Deuel | 7 | 25 | 3 | 1.0 | 5 | 196 | 33.4 | 9.4 | 9 | D | 3.0 | D |
| Dewey | NA | NA | NA | NA | NA | NA | NA | NA | 4 | 11 | 0.7 | 0.1 |
| Douglas | 6 | 12 | 1 | 0.3 | 9 | 93 | 18.9 | 3.9 | 6 | D | 0.9 | D |
| Edmunds | 5 | 26 | 4 | 1.1 | D | 68 | D | 3.6 | 12 | 60 | 2.4 | 0.8 |
| Fall River | 12 | 51 | 5 | 1.3 | 5 | 7 | 1.2 | 0.2 | 40 | 330 | 18.6 | 4.8 |
| Faulk | 3 | 7 | 0 | 0.1 | NA | NA | NA | NA | 12 | 59 | 2.9 | 0.5 |
| Grant | 15 | 42 | 5 | 1.3 | 14 | 628 | 791.0 | 32.0 | 28 | 273 | 12.0 | 2.9 |
| Gregory | 11 | 34 | 6 | 1.1 | 7 | 31 | 4.9 | 1.7 | 20 | 99 | 5.8 | 1.2 |
| Haakon | D | D | 2 | D | D | 85 | D | 3.4 | 7 | D | 2.1 | D |
| Hamlin | 6 | 25 | 1 | 0.7 | D | D | D | 9.5 | 10 | D | 3.5 | D |
| Hand | 12 | 214 | 14 | 8.4 | 7 | 56 | 33.2 | 2.8 | 14 | 85 | 3.1 | 0.8 |
| Hanson | 7 | 11 | 1 | 0.3 | 4 | 100 | 23.6 | 4.7 | 7 | 16 | 1.4 | 0.2 |
| Harding | NA | NA | NA | NA | NA | NA | NA | NA | 4 | D | 1.5 | D |
| Hughes | D | D | D | D | 9 | 47 | 29.2 | 2.7 | 52 | 908 | 45.7 | 13.5 |
| Hutchinson | 11 | 32 | 3 | 1.3 | 14 | 242 | 87.5 | 12.2 | D | D | D | 0.8 |
| Hyde | 3 | 2 | 0 | 0.1 | NA | NA | NA | NA | 3 | 4 | 0.3 | 0.0 |
| Jackson | NA | NA | NA | NA | NA | NA | NA | NA | 13 | 48 | 7.1 | 1.7 |
| Jerauld | 5 | 16 | 2 | 0.6 | D | D | D | D | 9 | 58 | 3.1 | 0.8 |
| Jones | NA | NA | NA | NA | NA | NA | NA | NA | 18 | 76 | 5.7 | 1.6 |
| Kingsbury | 6 | 17 | 2 | 0.8 | 8 | 263 | 68.7 | 12.2 | 6 | 30 | 1.5 | 0.4 |
| Lake | 42 | 240 | 30 | 9.4 | 23 | 769 | 281.8 | 33.9 | 32 | 412 | 16.0 | 4.8 |
| Lawrence | 79 | 283 | 35 | 13.6 | 39 | 428 | 151.6 | 17.8 | 133 | 2,883 | 211.2 | 56.3 |
| Lincoln | D | D | D | D | D | 1,780 | D | 79.5 | 80 | 1,443 | 77.1 | 22.6 |
| Lyman | D | D | D | D | NA | NA | NA | NA | 15 | 220 | 13.5 | 4.1 |
| McCook | 12 | 43 | 6 | 1.5 | 5 | 9 | 1.7 | 0.4 | 19 | 81 | 3.9 | 0.9 |
| McPherson | NA | NA | NA | NA | 8 | 54 | 5.6 | 1.5 | 5 | D | 1.2 | D |
| Marshall | 9 | 24 | 2 | 0.8 | 6 | 381 | 206.7 | 20.2 | 13 | 115 | 4.2 | 1.2 |
| Meade | D | D | D | D | 36 | 184 | 37.9 | 9.5 | 69 | 591 | 50.1 | 11.5 |
| Mellette | NA | NA | NA | NA | NA | NA | NA | NA | 3 | 7 | 0.4 | 0.1 |
| Miner | 4 | 24 | 2 | 0.5 | 7 | 85 | 24.2 | 5.1 | 10 | D | 1.4 | D |
| Minnehaha | D | D | D | D | 185 | 12,175 | 3,956.3 | 608.8 | 485 | 11,738 | 637.8 | 185.6 |
| Moody | 12 | 20 | 2 | 0.5 | 10 | 279 | 70.2 | 12.9 | 12 | 411 | 42.5 | 8.2 |
| Oglala Lakota | NA | NA | NA | NA | NA | NA | NA | NA | 11 | 261 | 19.5 | 6.5 |
| Pennington | D | D | D | D | 126 | 2,325 | 579.9 | 103.4 | 382 | 7,075 | 439.8 | 127.6 |
| Perkins | 8 | 22 | 2 | 0.6 | D | D | D | D | D | D | D | 0.5 |
| Potter | 3 | 10 | 1 | 0.3 | D | D | D | D | 9 | D | 3.6 | D |
| Roberts | D | D | D | D | 12 | 317 | 85.8 | 11.5 | D | D | D | D |
| Sanborn | 7 | 19 | 3 | 0.5 | NA | NA | NA | NA | D | D | 0.9 | D |
| Spink | 7 | 33 | 5 | 1.8 | 7 | 70 | 122.3 | 4.8 | 5 | D | 6.4 | 1.8 |
| Stanley | D | D | D | D | NA | NA | NA | NA | 18 | 126 | D | 1.8 |
| Sully | NA | NA | NA | NA | NA | NA | NA | NA | D | D | D | 3.8 |
| Todd | NA | NA | NA | NA | NA | NA | NA | NA | NA | NA | NA | 2.4 |

Items 147—158

Table B. States and Counties — **Health Care and Social Assistance, Other Services, Nonemployer Businesses, and Residential Construction**

| STATE County | Health care and social assistance, 2017 | | | | Other services, 2017 | | | | Nonemployer businesses, 2018 | | Value of residential construction authorized by building permits, 2020 | |
|---|---|---|---|---|---|---|---|---|---|---|---|---|
| | Number of establishments | Number of employees | Receipts (mil dol) | Annual payroll (mil dol) | Number of establishments | Number of employees | Receipts (mil dol) | Annual payroll (mil dol) | Number | Receipts (mil dol) | New construction ($1,000) | Number of housing units |
| | 159 | 160 | 161 | 162 | 163 | 164 | 165 | 166 | 167 | 168 | 169 | 170 |
| SOUTH CAROLINA— Cont'd | | | | | | | | | | | | |
| Sumter | 189 | 5,237 | 601.9 | 198.2 | 120 | 991 | 95.0 | 27.6 | 6,244 | 234.2 | 42,064 | 333 |
| Union | 27 | 1,012 | 90.4 | 36.0 | 29 | 107 | 10.0 | 2.8 | 1,132 | 38.3 | 6,119 | 31 |
| Williamsburg | 51 | 894 | 80.8 | 25.8 | 28 | 165 | 30.2 | 5.3 | 1,679 | 57.1 | 5,501 | 33 |
| York | 529 | 10,049 | 1,113.4 | 382.8 | 335 | 2,305 | 269.9 | 81.2 | 19,113 | 767.8 | 850,489 | 2,555 |
| SOUTH DAKOTA | 2,420 | 71,821 | 8,714.4 | 3,458.2 | 1,860 | 8,713 | 1,224.4 | 277.6 | 68,405 | 3,339.0 | 1,166,890 | 6,660 |
| Aurora | 8 | 203 | 14.7 | 7.6 | D | D | D | 0.9 | 255 | 9.0 | 700 | 4 |
| Beadle | 62 | 1,300 | 97.8 | 43.8 | 44 | 158 | 18.5 | 4.7 | 1,112 | 52.6 | 2,059 | 11 |
| Bennett | 6 | 142 | 11.8 | 6.3 | NA | NA | NA | NA | 192 | 5.2 | 242 | 6 |
| Bon Homme | 17 | 326 | 23.6 | 10.4 | D | D | D | D | 454 | 19.5 | 3,087 | 18 |
| Brookings | 93 | 1,511 | 132.5 | 49.6 | 65 | 318 | 67.5 | 10.5 | 2,138 | 104.7 | 29,093 | 144 |
| Brown | 111 | 3,135 | 359.2 | 146.1 | D | D | D | D | 2,912 | 193.9 | 14,208 | 79 |
| Brule | 26 | 348 | 34.0 | 14.7 | 19 | 79 | 8.0 | 2.2 | 495 | 26.0 | 2,039 | 10 |
| Buffalo | 4 | 32 | 2.4 | 1.3 | NA | NA | NA | NA | 36 | 0.8 | 0 | 0 |
| Butte | 30 | 298 | 24.2 | 9.2 | D | D | D | D | 919 | 39.8 | 9,333 | 36 |
| Campbell | 5 | D | 1.1 | D | 3 | 5 | 0.6 | 0.1 | 150 | 6.1 | 713 | 3 |
| Charles Mix | 28 | 521 | 40.8 | 16.7 | D | D | D | 1.6 | 704 | 24.1 | 5,752 | 24 |
| Clark | 9 | D | 4.9 | D | D | D | D | 0.8 | 288 | 13.3 | 2,032 | 6 |
| Clay | 32 | 726 | 57.6 | 19.1 | 22 | 85 | 23.4 | 3.4 | 876 | 41.7 | 13,838 | 70 |
| Codington | 79 | 1,932 | 220.3 | 77.6 | 75 | 336 | 36.5 | 9.2 | 2,226 | 97.6 | 15,378 | 90 |
| Corson | 4 | D | 0.7 | D | NA | NA | NA | NA | 145 | 4.3 | 0 | 0 |
| Custer | D | D | D | D | 13 | 43 | 4.5 | 1.2 | 931 | 41.2 | 51,415 | 217 |
| Davison | 69 | 2,093 | 173.8 | 69.9 | D | D | D | 0.7 | 1,517 | 69.8 | 9,804 | 46 |
| Day | 17 | 279 | 25.8 | 8.7 | 16 | 38 | 2.7 | 0.5 | 521 | 19.2 | 5,563 | 28 |
| Deuel | 6 | 173 | 15.4 | 5.7 | 8 | 12 | 1.8 | D | 413 | 17.4 | 2,315 | 10 |
| Dewey | 12 | 151 | 21.0 | 8.2 | D | D | D | D | 274 | 11.2 | 0 | 0 |
| Douglas | 6 | 227 | 14.7 | 7.4 | D | D | 4.2 | D | 289 | 13.5 | 1,613 | 10 |
| Edmunds | 10 | 184 | 10.1 | 4.2 | D | D | D | D | 391 | 25.4 | 3,695 | 14 |
| Fall River | D | D | D | D | 17 | 61 | 6.3 | 1.4 | 565 | 18.8 | 1,220 | 11 |
| Faulk | D | D | D | D | D | D | D | D | 221 | 10.7 | 970 | 4 |
| Grant | 24 | 601 | 52.2 | 23.7 | 17 | 77 | 9.5 | 1.9 | 624 | 21.5 | 4,674 | 26 |
| Gregory | 16 | 246 | 21.2 | 8.0 | 17 | 34 | 3.8 | 0.9 | 505 | 17.6 | 5,162 | 14 |
| Haakon | D | D | D | D | 5 | 11 | 1.9 | 0.5 | 239 | 12.5 | 0 | 0 |
| Hamlin | 10 | 150 | 4.3 | 2.2 | D | D | 4.1 | D | 461 | 19.5 | 9,867 | 45 |
| Hand | 8 | 208 | 16.4 | 6.0 | D | D | 2.2 | D | 342 | 15.1 | 495 | 1 |
| Hanson | 3 | 13 | 0.4 | 0.2 | D | D | D | D | 328 | 15.4 | 1,255 | 5 |
| Harding | 6 | 89 | 2.6 | 1.6 | NA | NA | NA | NA | 168 | 6.9 | 800 | 1 |
| Hughes | 67 | 1,294 | 134.9 | 54.8 | 70 | 272 | 105.9 | 12.2 | 1,533 | 68.6 | 6,793 | 28 |
| Hutchinson | 16 | 536 | 45.2 | 16.9 | D | D | D | D | 583 | 21.4 | 3,905 | 14 |
| Hyde | NA | NA | NA | NA | 3 | 4 | 0.3 | 0.1 | 129 | 5.2 | 700 | 3 |
| Jackson | NA | NA | NA | NA | NA | NA | NA | NA | 195 | 7.0 | 0 | 0 |
| Jerauld | 6 | 118 | 10.1 | 4.5 | D | D | 2.6 | D | 181 | 6.4 | 437 | 4 |
| Jones | NA | NA | NA | NA | D | D | 2.7 | D | 115 | 5.1 | 0 | 4 |
| Kingsbury | 11 | 183 | 13.2 | 5.2 | D | D | D | 2.5 | 517 | 20.3 | 3,673 | 23 |
| Lake | 30 | 625 | 50.5 | 22.5 | 23 | 89 | 8.8 | 6.9 | 927 | 45.2 | 22,081 | 94 |
| Lawrence | 89 | 1,865 | 189.3 | 72.6 | 70 | 212 | 23.1 | D | 2,631 | 125.5 | 76,420 | 250 |
| Lincoln | 173 | 3,092 | 373.2 | 136.5 | D | D | D | D | 5,340 | 298.4 | 77,618 | 583 |
| Lyman | NA | NA | NA | NA | 4 | 6 | 0.6 | 0.1 | 240 | 11.2 | 1,424 | 8 |
| McCook | 19 | 241 | 12.4 | 6.1 | D | D | D | D | 504 | 21.9 | 4,516 | 18 |
| McPherson | 5 | D | 3.5 | D | D | D | D | 0.3 | 187 | 9.5 | 1,072 | 8 |
| Marshall | 12 | 182 | 12.0 | 4.5 | D | D | D | 0.6 | 343 | 17.9 | 3,626 | 20 |
| Meade | D | D | D | D | 50 | 171 | 24.5 | 5.2 | 2,375 | 102.1 | 41,043 | 150 |
| Mellette | NA | NA | NA | NA | D | D | D | NA | 101 | 4.8 | 254 | 2 |
| Miner | 8 | 127 | 6.6 | 3.4 | D | D | D | 0.8 | 209 | 9.7 | 951 | 4 |
| Minnehaha | 480 | 27,862 | 3,669.3 | 1,509.8 | 374 | 2,704 | 394.8 | 93.7 | 14,491 | 780.4 | 456,752 | 2,932 |
| Moody | 13 | 196 | 16.4 | 6.1 | D | D | 3.4 | D | 430 | 23.5 | 2,968 | 13 |
| Oglala Lakota | D | D | D | D | NA | NA | NA | NA | 343 | 8.2 | NA | NA |
| Pennington | 392 | 10,698 | 1,458.5 | 548.1 | 292 | 1,590 | 188.8 | 52.3 | 8,682 | 402.2 | 167,990 | 1,170 |
| Perkins | 18 | 132 | 9.7 | 4.4 | D | D | 4.6 | D | 267 | 9.4 | 0 | 0 |
| Potter | 11 | D | 5.6 | D | D | D | D | 0.9 | 243 | 14.3 | 479 | 1 |
| Roberts | 21 | 418 | 31.4 | 13.9 | 13 | 40 | 4.2 | 1.1 | 644 | 25.3 | 7,861 | 36 |
| Sanborn | D | D | D | D | D | D | 0.7 | D | 192 | 8.1 | 1,165 | 6 |
| Spink | 16 | 299 | 19.7 | 8.6 | 14 | 50 | 5.1 | 1.4 | 508 | 22.5 | 1,888 | 8 |
| Stanley | D | D | D | D | D | D | D | D | 340 | 18.5 | 3,825 | 15 |
| Sully | D | D | D | D | NA | NA | NA | NA | 216 | 10.2 | 5,261 | 10 |
| Todd | 5 | 279 | 44.2 | 18.4 | D | D | D | D | 200 | 4.6 | 475 | 2 |

# Table B. States and Counties — Government Employment and Payroll, and Local Government Finances

| STATE County | Full-time equivalent employees | March payroll (dollars) | Administration, judicial, and legal | Police and corrections | Fire protection | Highways and transportation | Health and welfare | Natural resources and utilities | Education and libraries | Total (mil dol) | Intergovernmental (mil dol) | Taxes Total (mil dol) | Per capita[1] Total | Per capita[1] Property |
|---|---|---|---|---|---|---|---|---|---|---|---|---|---|---|
| | 171 | 172 | 173 | 174 | 175 | 176 | 177 | 178 | 179 | 180 | 181 | 182 | 183 | 184 |
| **SOUTH CAROLINA—Cont'd** | | | | | | | | | | | | | | |
| Sumter | 3,681 | 12,149,418 | 7.0 | 9.5 | 2.9 | 2.1 | 2.6 | 5.3 | 67.4 | 327.7 | 150.8 | 129.3 | 1,215 | 806 |
| Union | 1,044 | 3,525,859 | 8.6 | 11.9 | 0.8 | 1.7 | 2.5 | 8.1 | 64.3 | 81.6 | 37.9 | 27.7 | 1,009 | 872 |
| Williamsburg | 1,109 | 3,202,225 | 11.6 | 5.0 | 6.2 | 3.1 | 6.5 | 5.2 | 61.3 | 102.7 | 59.4 | 30.9 | 990 | 832 |
| York | 7,105 | 29,085,516 | 8.6 | 10.2 | 2.8 | 1.1 | 2.0 | 8.1 | 66.3 | 875.6 | 316.4 | 438.4 | 1,647 | 1,350 |
| **SOUTH DAKOTA** | X | X | X | X | X | X | X | X | X | X | X | X | X | X |
| Aurora | 124 | 323,894 | 13.4 | 4.1 | 0.1 | 10.2 | 0.8 | 5.3 | 59.6 | 12.4 | 4.8 | 6.6 | 2,386 | 2,099 |
| Beadle | 685 | 2,569,177 | 6.1 | 10.0 | 1.8 | 5.8 | 0.8 | 6.3 | 63.6 | 79.5 | 22.5 | 47.5 | 2,560 | 2,120 |
| Bennett | 152 | 425,079 | 7.7 | 7.9 | 0.0 | 5.4 | 2.5 | 2.0 | 72.0 | 10.8 | 6.2 | 3.7 | 1,077 | 826 |
| Bon Homme | 288 | 954,127 | 9.1 | 4.8 | 0.0 | 7.7 | 1.8 | 9.2 | 67.1 | 19.5 | 7.9 | 9.6 | 1,375 | 1,135 |
| Brookings | 1,453 | 5,999,158 | 4.5 | 6.2 | 0.3 | 3.2 | 23.9 | 11.4 | 37.9 | 206.6 | 27.4 | 67.4 | 1,940 | 1,386 |
| Brown | 1,331 | 5,148,388 | 7.0 | 9.9 | 4.2 | 6.0 | 1.4 | 9.3 | 60.3 | 141.8 | 38.5 | 81.3 | 2,069 | 1,460 |
| Brule | 274 | 866,666 | 6.2 | 6.5 | 0.0 | 4.5 | 1.5 | 7.1 | 71.8 | 30.9 | 12.7 | 13.8 | 2,596 | 1,791 |
| Buffalo | 9 | 28,865 | 51.6 | 20.6 | 0.0 | 22.6 | 5.3 | 0.0 | 0.0 | 1.3 | 0.1 | 1.2 | 586 | 488 |
| Butte | 431 | 1,270,170 | 6.7 | 7.1 | 0.0 | 4.6 | 0.5 | 16.3 | 61.7 | 38.5 | 13.2 | 18.1 | 1,794 | 1,424 |
| Campbell | 46 | 155,348 | 21.0 | 5.0 | 0.0 | 20.9 | 0.6 | 3.3 | 44.7 | 6.1 | 1.1 | 3.6 | 2,628 | 2,219 |
| Charles Mix | 487 | 1,599,649 | 6.0 | 3.2 | 0.0 | 3.9 | 2.5 | 6.5 | 74.2 | 38.7 | 16.2 | 17.6 | 1,876 | 1,548 |
| Clark | 185 | 518,044 | 11.0 | 3.7 | 0.0 | 13.2 | 2.8 | 3.6 | 65.0 | 15.6 | 4.0 | 10.3 | 2,812 | 2,457 |
| Clay | 335 | 1,243,105 | 9.8 | 11.8 | 0.7 | 5.4 | 1.5 | 15.1 | 50.2 | 37.6 | 9.4 | 19.6 | 1,396 | 1,067 |
| Codington | 1,155 | 4,716,648 | 5.0 | 7.0 | 4.1 | 3.6 | 0.4 | 14.3 | 63.4 | 117.4 | 28.0 | 59.6 | 2,120 | 1,698 |
| Corson | 252 | 860,565 | 4.5 | 2.8 | 0.0 | 4.1 | 0.0 | 3.0 | 84.5 | 19.3 | 14.6 | 3.9 | 934 | 777 |
| Custer | 245 | 818,492 | 14.0 | 7.1 | 0.0 | 7.3 | 3.5 | 3.3 | 61.6 | 32.4 | 6.1 | 20.0 | 2,287 | 1,865 |
| Davison | 950 | 3,310,784 | 5.3 | 9.6 | 3.8 | 4.4 | 2.6 | 8.0 | 64.9 | 103.0 | 29.5 | 51.8 | 2,608 | 1,903 |
| Day | 191 | 581,367 | 10.7 | 8.3 | 0.0 | 12.4 | 1.1 | 5.9 | 59.6 | 19.5 | 5.1 | 11.4 | 2,074 | 1,758 |
| Deuel | 129 | 484,861 | 17.3 | 8.1 | 0.0 | 16.2 | 0.2 | 3.0 | 50.5 | 14.0 | 2.8 | 8.7 | 2,021 | 1,711 |
| Dewey | 238 | 717,869 | 6.2 | 2.8 | 0.0 | 5.9 | 0.3 | 3.4 | 79.0 | 17.2 | 11.9 | 4.1 | 704 | 519 |
| Douglas | 85 | 312,392 | 30.4 | 6.3 | 0.0 | 14.5 | 0.0 | 5.0 | 36.4 | 12.8 | 2.7 | 8.5 | 2,894 | 2,454 |
| Edmunds | 301 | 888,143 | 7.0 | 3.0 | 0.0 | 8.3 | 35.9 | 0.4 | 41.2 | 29.4 | 4.0 | 13.0 | 3,332 | 2,953 |
| Fall River | 341 | 1,042,827 | 7.0 | 6.1 | 0.0 | 5.9 | 6.1 | 7.1 | 61.5 | 30.8 | 11.3 | 13.2 | 1,976 | 1,587 |
| Faulk | 168 | 625,484 | 9.4 | 4.4 | 0.0 | 6.3 | 60.4 | 0.8 | 18.5 | 19.5 | 1.8 | 7.1 | 3,076 | 2,758 |
| Grant | 271 | 949,862 | 11.8 | 6.8 | 0.0 | 7.1 | 0.6 | 4.5 | 67.0 | 26.1 | 6.4 | 16.3 | 2,272 | 1,783 |
| Gregory | 194 | 691,203 | 10.7 | 4.6 | 0.0 | 10.5 | 0.6 | 1.6 | 71.9 | 16.4 | 5.7 | 9.0 | 2,136 | 1,732 |
| Haakon | 96 | 297,796 | 10.8 | 4.2 | 0.3 | 10.8 | 1.0 | 4.1 | 47.5 | 12.7 | 3.2 | 7.8 | 4,058 | 3,169 |
| Hamlin | 351 | 1,240,059 | 5.0 | 2.0 | 0.0 | 3.3 | 16.4 | 2.3 | 70.7 | 28.2 | 9.0 | 14.7 | 2,457 | 2,210 |
| Hand | 143 | 416,110 | 11.9 | 7.8 | 0.0 | 13.7 | 2.7 | 9.5 | 53.4 | 14.8 | 4.3 | 8.9 | 2,711 | 2,399 |
| Hanson | 155 | 514,848 | 8.0 | 2.2 | 0.0 | 7.8 | 0.1 | 1.5 | 73.3 | 26.7 | 5.0 | 20.3 | 5,962 | 5,803 |
| Harding | 73 | 212,301 | 18.7 | 4.8 | 0.0 | 12.2 | 0.0 | 4.7 | 58.3 | 9.6 | 4.8 | 4.1 | 3,329 | 3,008 |
| Hughes | 634 | 2,522,234 | 9.3 | 14.5 | 0.4 | 4.4 | 0.6 | 11.0 | 58.5 | 65.3 | 17.5 | 27.0 | 1,528 | 1,444 |
| Hutchinson | 288 | 972,279 | 7.9 | 2.6 | 0.2 | 7.7 | 0.1 | 2.1 | 79.4 | 30.2 | 7.6 | 19.8 | 2,694 | 2,325 |
| Hyde | 72 | 248,598 | 6.6 | 2.3 | 0.0 | 7.3 | 0.1 | 1.4 | 78.4 | 8.8 | 2.4 | 5.6 | 4,286 | 3,770 |
| Jackson | 120 | 343,634 | 10.9 | 3.2 | 0.0 | 5.1 | 0.7 | 3.1 | 74.0 | 7.8 | 4.1 | 3.2 | 969 | 805 |
| Jerauld | 89 | 297,125 | 13.6 | 4.5 | 0.0 | 8.9 | 2.2 | 8.1 | 60.2 | 14.8 | 3.7 | 8.0 | 3,971 | 3,190 |
| Jones | 76 | 228,313 | 15.5 | 5.5 | 0.0 | 7.7 | 2.0 | 2.4 | 66.8 | 7.7 | 1.8 | 5.2 | 5,650 | 4,864 |
| Kingsbury | 254 | 773,325 | 10.2 | 2.6 | 0.0 | 7.7 | 0.7 | 4.2 | 74.0 | 25.8 | 5.2 | 17.8 | 3,607 | 3,296 |
| Lake | 467 | 1,748,761 | 6.4 | 7.5 | 2.1 | 5.7 | 2.0 | 14.6 | 59.3 | 51.2 | 13.4 | 30.2 | 2,364 | 1,988 |
| Lawrence | 728 | 2,593,722 | 12.2 | 14.1 | 0.9 | 7.1 | 0.1 | 10.7 | 48.4 | 98.7 | 23.6 | 59.9 | 2,333 | 1,610 |
| Lincoln | 1,125 | 3,735,403 | 9.9 | 4.3 | 0.0 | 3.7 | 0.4 | 1.7 | 78.1 | 122.3 | 40.1 | 67.6 | 1,193 | 1,074 |
| Lyman | 131 | 421,469 | 12.9 | 4.8 | 3.0 | 8.8 | 3.7 | 9.5 | 56.2 | 15.2 | 3.8 | 7.8 | 2,015 | 1,420 |
| McCook | 268 | 829,477 | 8.4 | 4.1 | 0.0 | 7.4 | 2.0 | 1.1 | 74.7 | 23.4 | 6.6 | 13.5 | 2,423 | 2,058 |
| McPherson | 134 | 372,207 | 15.1 | 3.2 | 0.0 | 9.1 | 0.4 | 3.2 | 67.1 | 11.8 | 3.1 | 7.6 | 3,150 | 2,856 |
| Marshall | 184 | 557,061 | 9.5 | 8.0 | 0.0 | 9.5 | 0.7 | 3.7 | 66.4 | 24.9 | 5.3 | 13.0 | 2,639 | 2,333 |
| Meade | 1,069 | 3,798,980 | 6.8 | 7.5 | 0.2 | 3.0 | 1.3 | 3.6 | 75.3 | 79.6 | 23.1 | 37.4 | 1,340 | 1,121 |
| Mellette | 116 | 326,678 | 6.0 | 3.9 | 0.0 | 2.1 | 0.6 | 2.5 | 82.1 | 9.5 | 7.1 | 2.0 | 958 | 893 |
| Miner | 113 | 403,684 | 10.4 | 6.7 | 0.0 | 14.5 | 1.7 | 6.2 | 59.5 | 11.9 | 1.8 | 8.6 | 3,930 | 3,507 |
| Minnehaha | 6,410 | 26,380,232 | 7.6 | 11.0 | 4.1 | 4.0 | 2.9 | 5.8 | 63.2 | 800.1 | 225.5 | 437.3 | 2,307 | 1,442 |
| Moody | 218 | 698,881 | 10.7 | 8.3 | 0.0 | 6.9 | 2.1 | 6.5 | 60.3 | 23.0 | 7.2 | 12.7 | 1,950 | 1,634 |
| Oglala Lakota | 373 | 1,141,302 | 1.0 | 0.4 | 0.0 | 1.0 | 0.1 | 0.1 | 97.3 | 31.9 | 29.3 | 1.0 | 73 | 43 |
| Pennington | 4,271 | 14,327,615 | 5.7 | 14.0 | 5.4 | 6.2 | 3.0 | 9.2 | 53.2 | 445.6 | 120.6 | 244.4 | 2,212 | 1,407 |
| Perkins | 158 | 529,501 | 11.0 | 2.8 | 0.2 | 6.4 | 2.5 | 4.7 | 71.7 | 15.8 | 4.5 | 8.5 | 2,879 | 2,428 |
| Potter | 125 | 368,732 | 10.5 | 4.8 | 0.0 | 10.3 | 0.1 | 2.4 | 67.0 | 14.1 | 2.5 | 10.5 | 4,744 | 4,316 |
| Roberts | 385 | 1,226,313 | 6.7 | 3.9 | 0.9 | 5.6 | 2.1 | 2.5 | 74.3 | 37.9 | 15.9 | 17.7 | 1,723 | 1,448 |
| Sanborn | 123 | 382,548 | 11.8 | 3.3 | 0.0 | 20.4 | 0.8 | 1.3 | 60.8 | 31.0 | 2.5 | 27.6 | 11,344 | 9,662 |
| Spink | 302 | 1,108,758 | 8.9 | 4.1 | 0.0 | 5.7 | 42.1 | 2.2 | 36.2 | 52.4 | 7.0 | 30.0 | 4,606 | 4,021 |
| Stanley | 115 | 415,834 | 11.2 | 5.7 | 0.0 | 10.7 | 0.7 | 9.8 | 60.2 | 14.1 | 3.5 | 8.4 | 2,820 | 2,279 |
| Sully | 83 | 270,364 | 12.6 | 0.0 | 0.0 | 16.1 | 0.0 | 3.8 | 62.1 | 10.0 | 1.4 | 7.9 | 5,629 | 5,167 |
| Todd | 499 | 1,494,157 | 4.2 | 0.8 | 0.0 | 1.9 | 0.0 | 0.7 | 92.4 | 49.5 | 30.8 | 17.7 | 1,719 | 1,656 |

1. Based on the resident population estimated as of July 1 of the year shown.

## Table B. States and Counties — Local Government Finances, Government Employment, and Income Taxes

| STATE County | Local government finances, 2017 (cont.) | | | | | | | | | | Government employment, 2019 | | | Individual income tax returns, 2018 | | |
| | Direct general expenditure | | | | | | | Debt outstanding | | | | | | | | |
| | | | Percent of total for: | | | | | | | | | | | | Mean adjusted gross income | Mean income tax |
| | Total (mil dol) | Per capita[1] (dollars) | Education | Health and hospitals | Police protection | Public welfare | Highways | Total (mil dol) | Per capita[1] (dollars) | Federal civilian | Federal military | State and local | Number of returns | | |
| | 185 | 186 | 187 | 188 | 189 | 190 | 191 | 192 | 193 | 194 | 195 | 196 | 197 | 198 | 199 |
|---|---|---|---|---|---|---|---|---|---|---|---|---|---|---|---|
| **SOUTH CAROLINA— Cont'd** | | | | | | | | | | | | | | | |
| Sumter | 292 | 2,741 | 60.7 | 1.4 | 8.0 | 0.3 | 1.1 | 137.3 | 1,290 | 1,322 | 5,952 | 5,217 | 47,120 | 45,135 | 3,675 |
| Union | 80 | 2,926 | 48.4 | 4.3 | 10.0 | 0.0 | 2.7 | 55.6 | 2,027 | 53 | 99 | 1,749 | 11,100 | 40,326 | 2,680 |
| Williamsburg | 99 | 3,183 | 60.0 | 2.8 | 4.2 | 0.5 | 1.5 | 93.4 | 2,993 | 360 | 108 | 1,736 | 12,590 | 36,035 | 2,394 |
| York | 915 | 3,439 | 63.0 | 0.4 | 4.8 | 0.1 | 2.0 | 1,416.8 | 5,323 | 503 | 1,023 | 12,828 | 125,880 | 73,369 | 8,273 |
| **SOUTH DAKOTA** | X | X | X | X | X | X | X | X | X | 11,316 | 8,216 | 67,551 | 421,090 | 66,733 | 7,337 |
| Aurora | 17 | 6,159 | 53.2 | 0.2 | 2.5 | 0.2 | 21.1 | 13.7 | 4,939 | 15 | 15 | 167 | 1,300 | 46,044 | 3,732 |
| Beadle | 70 | 3,758 | 51.6 | 0.4 | 4.7 | 0.4 | 11.8 | 40.4 | 2,177 | 294 | 100 | 1,068 | 8,570 | 54,697 | 5,047 |
| Bennett | 11 | 3,178 | 64.3 | 0.2 | 5.8 | 0.2 | 7.2 | 1.1 | 323 | 32 | 19 | 353 | 1,110 | 32,230 | 2,487 |
| Bon Homme | 23 | 3,367 | 54.9 | 1.1 | 2.9 | 0.3 | 20.3 | 31.8 | 4,581 | 26 | 30 | 552 | 2,720 | 49,170 | 3,964 |
| Brookings | 217 | 6,257 | 22.2 | 38.3 | 2.7 | 0.2 | 7.0 | 141.6 | 4,075 | 131 | 183 | 5,773 | 14,140 | 65,727 | 6,514 |
| Brown | 139 | 3,543 | 40.9 | 0.3 | 5.5 | 0.5 | 14.4 | 244.9 | 6,233 | 442 | 212 | 2,766 | 18,830 | 67,420 | 7,458 |
| Brule | 31 | 5,812 | 56.8 | 12.3 | 2.6 | 0.2 | 9.5 | 10.8 | 2,031 | 36 | 29 | 388 | 2,480 | 50,561 | 4,944 |
| Buffalo | 1 | 546 | 0.0 | 0.0 | 11.5 | 0.0 | 28.0 | 1.0 | 487 | 148 | 11 | 266 | 610 | 22,069 | 1,075 |
| Butte | 37 | 3,678 | 45.7 | 0.8 | 6.4 | 0.0 | 5.2 | 18.5 | 1,825 | 52 | 58 | 594 | 4,790 | 48,798 | 4,116 |
| Campbell | 9 | 6,629 | 18.1 | 0.4 | 2.8 | 0.0 | 14.1 | 1.6 | 1,158 | 3 | 8 | 79 | 740 | 51,014 | 4,601 |
| Charles Mix | 38 | 4,000 | 67.8 | 1.5 | 3.0 | 0.0 | 13.4 | 16.2 | 1,718 | 189 | 49 | 1,120 | 3,660 | 50,019 | 4,437 |
| Clark | 20 | 5,453 | 44.4 | 0.3 | 1.9 | 0.0 | 19.2 | 5.2 | 1,418 | 23 | 18 | 246 | 1,710 | 54,788 | 5,743 |
| Clay | 33 | 2,321 | 39.0 | 1.6 | 5.1 | 0.2 | 15.1 | 39.0 | 2,781 | 30 | 70 | 3,634 | 5,370 | 57,824 | 5,504 |
| Codington | 117 | 4,171 | 58.5 | 0.2 | 4.3 | 0.2 | 9.1 | 69.3 | 2,464 | 197 | 155 | 2,096 | 13,840 | 62,246 | 6,361 |
| Corson | 21 | 5,029 | 76.4 | 0.2 | 2.4 | 0.0 | 8.6 | 3.4 | 808 | 74 | 23 | 511 | 1,150 | 34,770 | 2,305 |
| Custer | 32 | 3,633 | 33.0 | 17.5 | 5.9 | 0.1 | 20.6 | 27.5 | 3,138 | 143 | 49 | 407 | 4,380 | 63,372 | 6,526 |
| Davison | 90 | 4,514 | 63.3 | 2.4 | 5.8 | 0.2 | 7.4 | 56.3 | 2,832 | 114 | 107 | 1,279 | 9,790 | 60,734 | 6,026 |
| Day | 21 | 3,822 | 38.2 | 0.6 | 3.6 | 0.6 | 22.2 | 11.4 | 2,073 | 61 | 30 | 381 | 2,690 | 48,161 | 4,496 |
| Deuel | 21 | 4,854 | 24.8 | 1.6 | 2.6 | 0.2 | 16.0 | 1.2 | 274 | 24 | 24 | 224 | 2,080 | 50,763 | 4,438 |
| Dewey | 17 | 2,880 | 79.8 | 0.3 | 3.7 | 0.0 | 7.8 | 0.4 | 71 | 402 | 33 | 1,166 | 2,360 | 37,937 | 3,292 |
| Douglas | 13 | 4,528 | 44.4 | 0.8 | 2.6 | 0.2 | 19.1 | 5.7 | 1,951 | 25 | 15 | 199 | 1,350 | 50,917 | 4,101 |
| Edmunds | 33 | 8,343 | 41.0 | 13.0 | 1.5 | 4.2 | 14.1 | 24.2 | 6,173 | 20 | 19 | 342 | 1,750 | 70,375 | 8,188 |
| Fall River | 31 | 4,646 | 42.1 | 5.3 | 4.9 | 0.1 | 8.3 | 15.8 | 2,355 | 502 | 37 | 583 | 3,500 | 47,665 | 4,047 |
| Faulk | 19 | 8,363 | 17.6 | 56.8 | 1.6 | 0.1 | 11.8 | 9.6 | 4,136 | 17 | 10 | 139 | 960 | 56,110 | 5,564 |
| Grant | 28 | 3,872 | 49.7 | 0.4 | 3.9 | 0.4 | 19.7 | 3.3 | 456 | 29 | 39 | 341 | 3,590 | 58,911 | 5,558 |
| Gregory | 17 | 4,122 | 52.7 | 0.5 | 2.9 | 0.2 | 18.9 | 7.8 | 1,866 | 28 | 23 | 233 | 1,960 | 47,622 | 4,361 |
| Haakon | 11 | 5,933 | 41.6 | 0.4 | 2.6 | 0.1 | 18.2 | 20.3 | 10,555 | 21 | 10 | 100 | 990 | 47,194 | 4,152 |
| Hamlin | 31 | 5,120 | 49.4 | 1.2 | 2.3 | 12.3 | 12.3 | 10.1 | 1,684 | 15 | 33 | 471 | 2,700 | 61,418 | 5,776 |
| Hand | 12 | 3,748 | 42.3 | 0.6 | 4.9 | 0.0 | 26.6 | 13.0 | 3,972 | 19 | 18 | 220 | 1,550 | 57,090 | 5,460 |
| Hanson | 24 | 6,996 | 62.9 | 0.2 | 1.5 | 0.0 | 9.8 | 72.5 | 21,262 | 4 | 16 | 192 | 2,010 | 63,664 | 5,911 |
| Harding | 8 | 6,184 | 44.7 | 0.4 | 3.1 | 0.0 | 20.6 | 7.9 | 6,310 | 23 | 7 | 96 | 600 | 57,757 | 5,173 |
| Hughes | 58 | 3,253 | 44.9 | 0.2 | 5.4 | 0.3 | 8.7 | 68.6 | 3,884 | 276 | 95 | 3,662 | 8,780 | 63,083 | 6,203 |
| Hutchinson | 32 | 4,380 | 51.3 | 0.4 | 2.6 | 0.2 | 15.0 | 13.8 | 1,879 | 41 | 36 | 433 | 3,320 | 56,603 | 4,791 |
| Hyde | 10 | 7,894 | 35.4 | 0.1 | 2.6 | 0.2 | 17.2 | 3.0 | 2,333 | 9 | 7 | 183 | 620 | 46,945 | 5,115 |
| Jackson | 9 | 2,598 | 62.5 | 0.6 | 3.9 | 0.1 | 14.4 | 0.1 | 43 | 92 | 19 | 240 | 1,040 | 31,621 | 2,071 |
| Jerauld | 14 | 6,650 | 31.2 | 4.9 | 4.6 | 0.2 | 36.0 | 14.2 | 6,993 | 12 | 10 | 130 | 860 | 45,536 | 3,379 |
| Jones | 5 | 5,650 | 47.6 | 2.2 | 5.4 | 0.0 | 18.7 | 0.5 | 528 | 8 | 5 | 128 | 510 | 41,412 | 3,767 |
| Kingsbury | 24 | 4,902 | 56.8 | 0.5 | 3.4 | 0.1 | 20.8 | 5.0 | 1,020 | 34 | 27 | 286 | 2,600 | 56,380 | 5,436 |
| Lake | 41 | 3,219 | 49.9 | 0.3 | 7.0 | 0.2 | 11.7 | 41.3 | 3,228 | 58 | 67 | 1,241 | 5,390 | 68,579 | 7,166 |
| Lawrence | 95 | 3,710 | 32.3 | 0.3 | 7.1 | 0.2 | 15.6 | 101.5 | 3,955 | 163 | 139 | 1,855 | 12,780 | 67,710 | 7,476 |
| Lincoln | 115 | 2,028 | 64.3 | 0.4 | 3.6 | 0.4 | 11.6 | 284.4 | 5,018 | 58 | 392 | 1,781 | 30,250 | 98,618 | 13,768 |
| Lyman | 14 | 3,600 | 40.2 | 15.7 | 3.5 | 0.1 | 20.9 | 2.1 | 542 | 80 | 21 | 593 | 1,590 | 43,408 | 4,072 |
| McCook | 25 | 4,530 | 52.5 | 1.5 | 3.6 | 0.7 | 14.6 | 14.5 | 2,603 | 25 | 30 | 266 | 2,660 | 61,207 | 5,491 |
| McPherson | 16 | 6,446 | 46.7 | 0.5 | 2.3 | 0.1 | 27.9 | 8.6 | 3,598 | 12 | 11 | 163 | 950 | 50,100 | 3,882 |
| Marshall | 25 | 5,099 | 30.1 | 2.0 | 3.9 | 0.1 | 14.1 | 13.5 | 2,734 | 24 | 26 | 343 | 1,930 | 55,755 | 5,563 |
| Meade | 83 | 2,963 | 52.4 | 0.3 | 5.5 | 0.0 | 12.7 | 85.6 | 3,062 | 1,595 | 159 | 1,385 | 14,220 | 54,767 | 4,645 |
| Mellette | 10 | 5,032 | 61.6 | 0.6 | 3.9 | 0.2 | 11.3 | 0.1 | 34 | 12 | 11 | 217 | 700 | 30,364 | 2,154 |
| Miner | 11 | 4,812 | 44.7 | 2.8 | 3.4 | 0.1 | 20.7 | 4.1 | 1,869 | 18 | 12 | 147 | 1,120 | 51,717 | 4,262 |
| Minnehaha | 759 | 4,003 | 47.4 | 1.5 | 5.5 | 0.6 | 11.8 | 736.9 | 3,888 | 2,653 | 1,071 | 9,558 | 99,950 | 72,700 | 8,174 |
| Moody | 20 | 3,051 | 53.8 | 1.9 | 6.8 | 0.0 | 7.9 | 39.2 | 6,039 | 100 | 36 | 720 | 2,910 | 53,154 | 5,252 |
| Oglala Lakota | 27 | 1,862 | 95.7 | 0.3 | 0.4 | 0.0 | 1.8 | 0.0 | 1 | 615 | 79 | 2,415 | 4,110 | 30,341 | 1,440 |
| Pennington | 382 | 3,452 | 44.9 | 1.6 | 7.9 | 0.5 | 8.4 | 357.8 | 3,238 | 1,397 | 3,975 | 6,652 | 57,390 | 64,972 | 7,062 |
| Perkins | 16 | 5,450 | 41.6 | 1.0 | 5.3 | 0.0 | 21.1 | 1.6 | 535 | 23 | 16 | 226 | 1,300 | 40,175 | 3,449 |
| Potter | 13 | 5,815 | 52.6 | 1.5 | 2.9 | 0.0 | 22.0 | 7.6 | 3,430 | 16 | 12 | 163 | 1,150 | 67,620 | 7,870 |
| Roberts | 38 | 3,660 | 59.7 | 0.1 | 3.6 | 0.4 | 11.9 | 15.8 | 1,541 | 198 | 57 | 1,464 | 4,320 | 51,159 | 4,572 |
| Sanborn | 9 | 3,724 | 56.3 | 0.8 | 3.9 | 0.0 | 4.7 | 1.4 | 591 | 9 | 12 | 154 | 1,080 | 48,944 | 3,996 |
| Spink | 43 | 6,529 | 43.1 | 24.5 | 3.0 | 0.2 | 13.8 | 69.4 | 10,662 | 37 | 34 | 885 | 2,840 | 58,191 | 5,823 |
| Stanley | 18 | 5,954 | 30.3 | 0.7 | 7.3 | 0.1 | 38.6 | 10.1 | 3,396 | 12 | 17 | 182 | 1,590 | 67,234 | 7,306 |
| Sully | 9 | 6,054 | 52.0 | 0.3 | 3.5 | 0.0 | 19.7 | 7.7 | 5,532 | 10 | 8 | 113 | 740 | 60,855 | 8,969 |
| Todd | 34 | 3,294 | 89.9 | 0.1 | 1.0 | 0.0 | 3.6 | 1.3 | 128 | 241 | 57 | 2,165 | 2,770 | 32,655 | 1,777 |

1. Based on the resident population estimated as of July 1 of the year shown.

# Table B. States and Counties — Land Area and Population

| State / county code | CBSA code[1] | County Type code[2] | STATE County | Land area[3] (sq. mi) | Total persons 2019 | Rank | Per square mile | White | Black | American Indian, Alaska Native | Asian and Pacific Islander | Percent Hispanic or Latino[4] | Under 5 years | 5 to 14 years | 15 to 24 years | 25 to 34 years | 35 to 44 years | 45 to 54 years |
|---|---|---|---|---|---|---|---|---|---|---|---|---|---|---|---|---|---|---|
| | | | | 1 | 2 | 3 | 4 | 5 | 6 | 7 | 8 | 9 | 10 | 11 | 12 | 13 | 14 | 15 |
| | | | **SOUTH DAKOTA—Cont'd** | | | | | | | | | | | | | | | |
| 46123 | | 7 | Tripp | 1,612.5 | 5,377 | 2,795 | 3.3 | 82.2 | 1.1 | 16.1 | 0.6 | 2.6 | 7.5 | 12.2 | 11.4 | 10.5 | 9.6 | 10.5 |
| 46125 | 43620 | 3 | Turner | 617.1 | 8,368 | 2,544 | 13.6 | 95.8 | 1.2 | 1.4 | 0.5 | 2.3 | 5.8 | 14.8 | 10.4 | 10.0 | 12.7 | 10.8 |
| 46127 | 43580 | 3 | Union | 460.8 | 16,192 | 2,021 | 35.1 | 92.1 | 1.9 | 1.4 | 2.1 | 4.1 | 6.2 | 14.0 | 11.4 | 11.9 | 12.5 | 12.0 |
| 46129 | | 7 | Walworth | 708.6 | 5,336 | 2,800 | 7.5 | 81.7 | 1.1 | 16.1 | 2.0 | 2.3 | 6.6 | 12.7 | 10.7 | 11.1 | 9.8 | 10.0 |
| 46135 | 49460 | 7 | Yankton | 521.2 | 22,742 | 1,694 | 43.6 | 88.9 | 2.6 | 3.8 | 1.1 | 5.5 | 5.8 | 11.6 | 12.3 | 12.4 | 12.0 | 11.5 |
| 46137 | | 8 | Ziebach | 1,961.2 | 2,656 | 2,989 | 1.4 | 27.9 | 1.3 | 70.3 | 0.5 | 4.0 | 5.2 | 14.9 | 18.6 | 12.0 | 13.2 | 11.3 |
| 47000 | | 0 | TENNESSEE | 41,238.0 | 6,886,834 | X | 167.0 | 75.1 | 17.7 | 0.8 | 2.5 | 5.9 | 5.9 | 12.3 | 12.6 | 14.0 | 12.4 | 12.5 |
| 47001 | 28940 | 2 | Anderson | 337.2 | 77,558 | 724 | 230.0 | 90.9 | 5.2 | 1.0 | 1.8 | 3.4 | 5.3 | 12.0 | 11.1 | 12.4 | 11.6 | 12.7 |
| 47003 | 43180 | 4 | Bedford | 473.6 | 50,179 | 989 | 106.0 | 77.8 | 8.9 | 0.7 | 1.2 | 13.3 | 6.6 | 14.2 | 12.5 | 13.6 | 12.3 | 12.8 |
| 47005 | | 7 | Benton | 394.3 | 16,131 | 2,026 | 40.9 | 93.7 | 3.2 | 1.2 | 0.8 | 2.6 | 5.0 | 11.3 | 9.8 | 10.3 | 10.5 | 13.0 |
| 47007 | | 8 | Bledsoe | 406.6 | 15,223 | 2,081 | 37.4 | 88.7 | 8.1 | 1.1 | 0.5 | 3.0 | 3.6 | 7.8 | 9.6 | 14.0 | 15.0 | 15.0 |
| 47009 | 28940 | 2 | Blount | 558.8 | 134,751 | 482 | 241.1 | 92.0 | 3.6 | 0.9 | 1.3 | 3.8 | 4.9 | 11.2 | 11.2 | 12.2 | 11.4 | 13.3 |
| 47011 | 17420 | 3 | Bradley | 328.8 | 109,071 | 560 | 331.7 | 86.8 | 5.9 | 0.9 | 1.6 | 6.8 | 5.7 | 12.0 | 13.1 | 13.0 | 12.0 | 13.2 |
| 47013 | 28940 | 2 | Campbell | 480.2 | 39,837 | 1,193 | 83.0 | 97.2 | 0.8 | 1.0 | 0.6 | 1.6 | 5.7 | 10.9 | 11.1 | 12.1 | 11.1 | 14.1 |
| 47015 | 34980 | 1 | Cannon | 265.6 | 14,847 | 2,100 | 55.9 | 95.1 | 2.6 | 1.1 | 0.6 | 2.4 | 6.0 | 11.9 | 10.6 | 13.4 | 11.6 | 13.3 |
| 47017 | | 6 | Carroll | 597.7 | 27,779 | 1,504 | 46.5 | 86.1 | 11.3 | 0.9 | 0.7 | 3.1 | 5.5 | 12.6 | 13.0 | 11.4 | 10.4 | 12.5 |
| 47019 | 27740 | 3 | Carter | 341.3 | 56,418 | 916 | 165.3 | 95.6 | 2.2 | 0.8 | 0.6 | 2.1 | 4.3 | 10.0 | 10.5 | 12.5 | 11.0 | 13.7 |
| 47021 | 34980 | 1 | Cheatham | 302.5 | 41,101 | 1,155 | 135.9 | 93.0 | 3.0 | 1.0 | 0.9 | 3.8 | 5.6 | 12.0 | 11.2 | 13.2 | 12.5 | 14.1 |
| 47023 | 27180 | 3 | Chester | 285.7 | 17,432 | 1,948 | 61.0 | 86.7 | 10.5 | 1.0 | 1.0 | 2.9 | 5.3 | 12.3 | 17.0 | 12.0 | 11.0 | 12.1 |
| 47025 | | 6 | Claiborne | 434.6 | 32,023 | 1,382 | 73.7 | 96.1 | 1.6 | 1.0 | 1.1 | 1.5 | 5.0 | 10.4 | 12.4 | 12.9 | 11.1 | 13.3 |
| 47027 | | 9 | Clay | 236.5 | 7,629 | 2,617 | 32.3 | 95.8 | 2.1 | 0.7 | 0.3 | 2.4 | 4.7 | 11.7 | 9.6 | 9.6 | 10.4 | 12.9 |
| 47029 | 35460 | 6 | Cocke | 436.1 | 36,225 | 1,269 | 83.1 | 94.3 | 2.8 | 1.1 | 0.7 | 2.8 | 5.1 | 11.5 | 10.4 | 11.3 | 10.6 | 13.4 |
| 47031 | 46100 | 4 | Coffee | 429.0 | 57,632 | 901 | 134.3 | 89.5 | 5.3 | 0.9 | 1.5 | 5.0 | 6.4 | 13.5 | 11.8 | 12.9 | 11.9 | 12.4 |
| 47033 | 27180 | 3 | Crockett | 265.5 | 14,180 | 2,146 | 53.4 | 74.5 | 14.9 | 0.7 | 0.5 | 11.2 | 5.9 | 13.6 | 11.6 | 12.1 | 11.5 | 12.1 |
| 47035 | 18900 | 4 | Cumberland | 681.3 | 61,603 | 859 | 90.4 | 95.2 | 0.9 | 1.0 | 0.8 | 3.2 | 4.4 | 9.6 | 8.7 | 9.7 | 8.9 | 10.7 |
| 47037 | 34980 | 1 | Davidson | 503.7 | 694,176 | 96 | 1,378.2 | 58.6 | 27.9 | 0.8 | 4.7 | 10.4 | 6.5 | 10.8 | 12.7 | 20.7 | 13.9 | 11.2 |
| 47039 | | 9 | Decatur | 333.9 | 11,601 | 2,314 | 34.7 | 92.8 | 3.3 | 0.9 | 0.9 | 3.5 | 4.8 | 11.7 | 10.3 | 10.3 | 11.0 | 12.5 |
| 47041 | | 6 | DeKalb | 304.4 | 20,837 | 1,777 | 68.5 | 88.9 | 2.0 | 0.9 | 1.2 | 8.4 | 5.6 | 12.0 | 11.1 | 12.3 | 12.0 | 13.1 |
| 47043 | 34980 | 1 | Dickson | 489.9 | 54,376 | 940 | 111.0 | 91.1 | 5.1 | 0.9 | 0.9 | 4.0 | 5.9 | 12.6 | 11.5 | 13.2 | 12.6 | 13.3 |
| 47045 | 20540 | 5 | Dyer | 512.4 | 36,693 | 1,257 | 71.6 | 80.8 | 15.8 | 0.7 | 0.9 | 3.6 | 6.2 | 13.2 | 12.2 | 12.8 | 11.6 | 12.6 |
| 47047 | 32820 | 1 | Fayette | 704.8 | 41,620 | 1,144 | 59.1 | 68.4 | 27.9 | 0.7 | 1.1 | 3.1 | 4.7 | 10.0 | 10.2 | 11.0 | 11.0 | 13.5 |
| 47049 | | 9 | Fentress | 498.6 | 18,787 | 1,885 | 37.7 | 96.9 | 0.8 | 1.0 | 0.7 | 1.8 | 5.4 | 11.5 | 10.7 | 10.5 | 10.8 | 13.0 |
| 47051 | 46100 | 6 | Franklin | 554.5 | 42,485 | 1,131 | 76.6 | 89.9 | 5.9 | 1.2 | 1.3 | 3.7 | 5.1 | 10.8 | 15.1 | 11.0 | 10.9 | 11.9 |
| 47053 | 27180 | 4 | Gibson | 602.7 | 49,159 | 1,005 | 81.6 | 78.4 | 19.2 | 0.7 | 0.6 | 3.0 | 6.1 | 13.9 | 11.6 | 12.0 | 12.6 | 12.2 |
| 47055 | | 6 | Giles | 610.9 | 29,530 | 1,448 | 48.3 | 85.7 | 11.2 | 1.1 | 0.8 | 3.4 | 5.6 | 11.7 | 11.6 | 11.6 | 10.8 | 12.6 |
| 47057 | 34100 | 2 | Grainger | 280.6 | 23,565 | 1,655 | 84.0 | 95.0 | 1.4 | 0.9 | 0.4 | 3.4 | 5.0 | 11.0 | 10.9 | 11.2 | 10.8 | 14.3 |
| 47059 | 24620 | 4 | Greene | 622.2 | 69,571 | 777 | 111.8 | 93.8 | 2.8 | 0.8 | 0.8 | 3.1 | 4.9 | 10.5 | 11.5 | 11.3 | 11.1 | 13.5 |
| 47061 | | 8 | Grundy | 360.4 | 13,485 | 2,191 | 37.4 | 96.6 | 1.1 | 1.6 | 0.7 | 1.5 | 5.5 | 11.6 | 11.2 | 11.8 | 11.6 | 13.0 |
| 47063 | 34100 | 3 | Hamblen | 161.2 | 65,110 | 818 | 403.9 | 82.6 | 4.8 | 0.7 | 1.6 | 12.1 | 6.0 | 13.0 | 11.8 | 12.1 | 11.8 | 13.2 |
| 47065 | 16860 | 1 | Hamilton | 542.2 | 371,662 | 193 | 685.5 | 72.5 | 19.8 | 0.8 | 2.7 | 6.1 | 5.8 | 11.4 | 11.9 | 14.8 | 12.4 | 12.3 |
| 47067 | | 8 | Hancock | 222.3 | 6,493 | 2,706 | 29.2 | 98.0 | 1.0 | 1.3 | 0.4 | 0.7 | 5.6 | 11.6 | 10.1 | 11.3 | 10.8 | 12.5 |
| 47069 | | 6 | Hardeman | 667.8 | 24,836 | 1,610 | 37.2 | 55.2 | 42.4 | 0.7 | 1.1 | 2.0 | 4.7 | 11.0 | 11.8 | 15.1 | 12.9 | 12.2 |
| 47071 | | 6 | Hardin | 577.5 | 25,583 | 1,572 | 44.3 | 93.1 | 4.0 | 1.1 | 0.7 | 2.6 | 4.8 | 11.5 | 10.5 | 11.1 | 10.8 | 12.4 |
| 47073 | 28700 | 2 | Hawkins | 487.1 | 56,775 | 911 | 116.6 | 95.9 | 1.9 | 0.9 | 0.8 | 1.7 | 4.6 | 10.8 | 10.9 | 11.5 | 10.7 | 14.2 |
| 47075 | 15140 | 6 | Haywood | 533.1 | 17,002 | 1,973 | 31.9 | 44.6 | 51.3 | 0.7 | 0.4 | 4.3 | 5.6 | 12.6 | 11.9 | 11.5 | 11.8 | 11.8 |
| 47077 | | 6 | Henderson | 520.0 | 28,076 | 1,495 | 54.0 | 88.9 | 9.3 | 0.7 | 0.5 | 2.7 | 5.7 | 12.9 | 11.5 | 11.8 | 12.6 | 12.7 |
| 47079 | 37540 | 7 | Henry | 561.9 | 32,056 | 1,381 | 57.0 | 88.8 | 8.5 | 0.9 | 0.8 | 2.8 | 4.9 | 11.5 | 10.3 | 10.7 | 11.1 | 12.2 |
| 47081 | | 1 | Hickman | 612.5 | 25,387 | 1,587 | 41.4 | 91.4 | 5.5 | 1.4 | 0.6 | 2.7 | 5.3 | 11.6 | 11.6 | 13.3 | 12.3 | 13.5 |
| 47083 | | 8 | Houston | 200.3 | 8,292 | 2,557 | 41.4 | 93.0 | 4.3 | 1.3 | 0.8 | 2.9 | 5.3 | 12.0 | 11.2 | 11.3 | 11.7 | 12.8 |
| 47085 | | 6 | Humphreys | 530.8 | 18,590 | 1,894 | 35.0 | 93.6 | 3.4 | 1.1 | 1.0 | 2.6 | 5.8 | 11.7 | 11.2 | 12.1 | 11.1 | 12.8 |
| 47087 | 18260 | 8 | Jackson | 308.8 | 11,864 | 2,299 | 38.4 | 96.1 | 1.2 | 1.4 | 0.4 | 2.3 | 4.4 | 10.1 | 9.7 | 11.4 | 10.3 | 13.8 |
| 47089 | 34100 | 3 | Jefferson | 275.1 | 55,307 | 929 | 201.0 | 93.3 | 2.5 | 0.9 | 1.0 | 3.7 | 4.6 | 10.8 | 12.1 | 11.5 | 10.9 | 13.7 |
| 47091 | | 6 | Johnson | 298.4 | 17,849 | 1,932 | 59.8 | 95.0 | 2.6 | 1.0 | 0.5 | 2.2 | 4.1 | 9.4 | 9.8 | 12.1 | 11.3 | 14.4 |
| 47093 | 28940 | 2 | Knox | 508.3 | 475,609 | 149 | 935.7 | 83.8 | 9.9 | 0.8 | 3.1 | 4.7 | 5.6 | 11.6 | 15.2 | 13.9 | 12.6 | 12.1 |
| 47095 | | 9 | Lake | 165.8 | 6,988 | 2,659 | 42.1 | 68.4 | 29.3 | 1.0 | 0.5 | 2.8 | 4.0 | 7.7 | 12.7 | 18.1 | 13.8 | 13.9 |
| 47097 | | 6 | Lauderdale | 471.9 | 25,451 | 1,581 | 53.9 | 61.1 | 35.9 | 1.1 | 0.8 | 2.7 | 5.9 | 12.2 | 12.7 | 13.8 | 13.0 | 12.5 |
| 47099 | 29980 | 6 | Lawrence | 617.1 | 44,432 | 1,086 | 72.0 | 94.7 | 2.5 | 1.1 | 0.7 | 2.5 | 6.5 | 14.0 | 11.7 | 12.5 | 11.3 | 12.6 |
| 47101 | | 6 | Lewis | 282.1 | 12,363 | 2,261 | 43.8 | 94.2 | 2.9 | 1.0 | 0.9 | 2.6 | 5.7 | 12.1 | 11.3 | 11.7 | 11.1 | 12.2 |
| 47103 | | 6 | Lincoln | 570.3 | 34,540 | 1,315 | 60.6 | 88.2 | 7.7 | 1.5 | 0.9 | 3.7 | 5.5 | 12.5 | 11.3 | 11.3 | 11.6 | 12.8 |
| 47105 | 28940 | 2 | Loudon | 229.3 | 54,910 | 935 | 239.5 | 87.8 | 1.8 | 0.8 | 1.1 | 9.6 | 4.7 | 10.9 | 9.7 | 10.0 | 9.6 | 12.0 |
| 47107 | 11940 | 4 | McMinn | 430.1 | 54,208 | 944 | 126.0 | 90.6 | 4.7 | 1.2 | 0.9 | 4.5 | 5.5 | 11.6 | 11.8 | 12.0 | 11.2 | 13.2 |
| 47109 | | 6 | McNairy | 562.8 | 25,696 | 1,570 | 45.7 | 90.9 | 6.9 | 1.0 | 0.6 | 2.4 | 5.3 | 12.0 | 11.9 | 11.2 | 11.2 | 12.9 |
| 47111 | 34980 | 1 | Macon | 307.1 | 24,827 | 1,612 | 80.8 | 92.5 | 1.5 | 1.0 | 1.2 | 5.1 | 6.8 | 13.8 | 11.8 | 13.5 | 11.7 | 13.1 |
| 47113 | 27180 | 3 | Madison | 557.2 | 98,360 | 615 | 176.5 | 57.1 | 38.5 | 0.5 | 1.6 | 4.2 | 6.2 | 12.2 | 14.0 | 12.8 | 11.3 | 11.8 |
| 47115 | 16860 | 2 | Marion | 498.3 | 28,924 | 1,462 | 58.0 | 92.8 | 4.7 | 1.1 | 0.9 | 2.0 | 5.5 | 11.6 | 11.2 | 11.7 | 11.2 | 13.0 |

1. CBSA = Core Based Statistical Area. See Appendix A for explanation. See Appendix B for list of metropolitan areas with component counties. Service of USDA Rural-Urban Continuum Codes. See Appendix A for definition.     3. Dry land or land partially or temporarily covered by water.

2. County type code from the Economic Research     4. May be of any race.

Items 1—15

| STATE County | 55 to 64 years | 65 to 74 years | 75 years and over | Percent female | Total persons 2000 | Total persons 2010 | Percent change 2000-2010 | Percent change 2010-2020 | Births | Deaths | Net Migration | Number | Persons per household | Family households | Female family householder[1] | One person |
|---|---|---|---|---|---|---|---|---|---|---|---|---|---|---|---|---|
| | 16 | 17 | 18 | 19 | 20 | 21 | 22 | 23 | 24 | 25 | 26 | 27 | 28 | 29 | 30 | 31 |
| **SOUTH DAKOTA—Cont'd** | | | | | | | | | | | | | | | | |
| Tripp | 15.2 | 12.1 | 11.1 | 49.6 | 6,430 | 5,649 | -12.1 | -4.8 | 756 | 729 | -297 | 2,356 | 2.25 | 68.1 | 8.8 | 27.6 |
| Turner | 14.0 | 11.9 | 9.6 | 50.1 | 8,849 | 8,349 | -5.7 | 0.2 | 936 | 1,059 | 147 | 3,498 | 2.32 | 67.2 | 5.0 | 27.6 |
| Union | 13.2 | 11.4 | 7.4 | 49.6 | 12,584 | 14,394 | 14.4 | 12.5 | 1,728 | 1,253 | 1,319 | 6,619 | 2.31 | 65.0 | 6.9 | 28.1 |
| Walworth | 13.7 | 13.4 | 12.0 | 50.0 | 5,974 | 5,438 | -9.0 | -1.9 | 740 | 800 | -43 | 2,258 | 2.24 | 60.1 | 7.3 | 31.4 |
| Yankton | 14.1 | 11.4 | 9.0 | 47.6 | 21,652 | 22,438 | 3.6 | 1.4 | 2,729 | 2,363 | -44 | 9,445 | 2.22 | 62.7 | 8.7 | 30.1 |
| Ziebach | 13.6 | 6.9 | 4.4 | 49.7 | 2,519 | 2,801 | 11.2 | -5.2 | 342 | 163 | -323 | 754 | 3.7 | 80.6 | 31.6 | 17.9 |
| **TENNESSEE** | 13.1 | 10.2 | 6.9 | 51.2 | 5,689,283 | 6,346,281 | 11.5 | 8.5 | 825,523 | 678,711 | 392,626 | 2,597,292 | 2.52 | 65.8 | 12.8 | 28.3 |
| Anderson | 14.2 | 11.9 | 8.8 | 51.3 | 71,330 | 75,072 | 5.2 | 3.3 | 8,237 | 9,799 | 4,120 | 30,541 | 2.45 | 67.3 | 13.0 | 27.9 |
| Bedford | 12.5 | 9.3 | 6.2 | 50.6 | 37,586 | 45,058 | 19.9 | 11.4 | 6,536 | 4,917 | 3,503 | 17,029 | 2.8 | 74.6 | 14.0 | 20.1 |
| Benton | 15.6 | 14.4 | 10.1 | 50.8 | 16,537 | 16,491 | -0.3 | -2.2 | 1,658 | 2,580 | 572 | 6,762 | 2.36 | 56.7 | 12.4 | 37.7 |
| Bledsoe | 15.9 | 11.3 | 7.8 | 40.5 | 12,367 | 12,882 | 4.2 | 18.2 | 1,257 | 1,426 | 2,438 | 4,894 | 2.79 | 71.0 | 9.5 | 23.0 |
| Blount | 14.6 | 12.4 | 8.8 | 51.4 | 105,823 | 123,101 | 16.3 | 9.5 | 12,989 | 14,137 | 12,843 | 50,557 | 2.53 | 68.8 | 10.3 | 26.1 |
| Bradley | 13.2 | 10.2 | 7.5 | 51.4 | 87,965 | 98,926 | 12.5 | 10.3 | 12,292 | 10,658 | 8,523 | 40,192 | 2.55 | 68.0 | 10.9 | 26.4 |
| Campbell | 14.0 | 12.1 | 9.0 | 50.9 | 39,854 | 40,722 | 2.2 | -2.2 | 4,431 | 5,878 | 593 | 16,192 | 2.42 | 68.7 | 14.6 | 27.0 |
| Cannon | 14.7 | 10.5 | 7.9 | 50.1 | 12,826 | 13,813 | 7.7 | 7.5 | 1,617 | 1,783 | 1,199 | 5,488 | 2.55 | 71.1 | 11.3 | 24.5 |
| Carroll | 13.8 | 11.9 | 8.9 | 50.8 | 29,475 | 28,486 | -3.4 | -2.5 | 3,177 | 4,279 | 405 | 10,962 | 2.46 | 68.0 | 11.5 | 28.1 |
| Carter | 15.0 | 13.1 | 9.9 | 51.1 | 56,742 | 57,383 | 1.1 | -1.7 | 5,195 | 7,294 | 1,172 | 23,784 | 2.3 | 64.1 | 11.4 | 31.2 |
| Cheatham | 15.4 | 10.4 | 5.6 | 50.3 | 35,912 | 39,101 | 8.9 | 5.1 | 4,628 | 4,076 | 1,465 | 15,089 | 2.64 | 73.0 | 11.4 | 20.7 |
| Chester | 12.6 | 10.2 | 7.5 | 52.1 | 15,540 | 17,145 | 10.3 | 1.7 | 1,901 | 1,873 | 265 | 6,060 | 2.63 | 73.8 | 11.4 | 21.6 |
| Claiborne | 14.3 | 12.4 | 8.3 | 51.1 | 29,862 | 32,209 | 7.9 | -0.6 | 3,246 | 4,429 | 1,014 | 13,281 | 2.29 | 65.4 | 9.4 | 30.4 |
| Clay | 15.3 | 14.4 | 11.4 | 51.2 | 7,976 | 7,856 | -1.5 | -2.9 | 761 | 1,173 | 189 | 3,039 | 2.48 | 65.9 | 7.7 | 28.6 |
| Cocke | 15.6 | 13.5 | 8.6 | 51.4 | 33,565 | 35,642 | 6.2 | 1.6 | 3,852 | 5,225 | 1,984 | 14,060 | 2.5 | 65.4 | 14.7 | 30.0 |
| Coffee | 13.5 | 10.2 | 7.4 | 51.0 | 48,014 | 52,803 | 10.0 | 9.1 | 6,935 | 6,860 | 4,785 | 21,646 | 2.52 | 68.3 | 10.5 | 26.9 |
| Crockett | 14.1 | 10.6 | 8.5 | 52.2 | 14,532 | 14,575 | 0.3 | -2.7 | 1,726 | 1,868 | -243 | 5,491 | 2.58 | 67.4 | 13.8 | 27.2 |
| Cumberland | 15.7 | 18.5 | 13.8 | 51.3 | 46,802 | 56,058 | 19.8 | 9.9 | 5,623 | 8,369 | 8,253 | 25,801 | 2.27 | 68.6 | 10.3 | 26.9 |
| Davidson | 11.3 | 8.0 | 4.9 | 51.9 | 569,891 | 626,577 | 9.9 | 10.8 | 102,198 | 54,407 | 19,424 | 282,366 | 2.36 | 55.4 | 13.5 | 33.7 |
| Decatur | 15.3 | 13.4 | 10.6 | 50.6 | 11,731 | 11,746 | 0.1 | -1.2 | 1,172 | 1,783 | 476 | 4,440 | 2.58 | 68.9 | 10.2 | 28.6 |
| DeKalb | 14.6 | 11.6 | 7.7 | 49.8 | 17,423 | 18,722 | 7.5 | 11.3 | 2,293 | 2,541 | 2,369 | 7,704 | 2.53 | 70.1 | 9.6 | 25.3 |
| Dickson | 14.0 | 10.2 | 6.8 | 50.9 | 43,156 | 49,658 | 15.1 | 9.5 | 6,322 | 5,614 | 4,013 | 19,198 | 2.71 | 67.9 | 10.5 | 26.0 |
| Dyer | 13.3 | 10.7 | 7.5 | 51.6 | 37,279 | 38,331 | 2.8 | -4.3 | 4,815 | 4,704 | -1,740 | 15,120 | 2.44 | 69.9 | 15.7 | 24.3 |
| Fayette | 17.2 | 13.6 | 9.0 | 50.8 | 28,806 | 38,439 | 33.4 | 8.3 | 4,274 | 4,080 | 2,986 | 15,596 | 2.54 | 74.2 | 11.5 | 22.5 |
| Fentress | 15.3 | 13.8 | 8.9 | 50.7 | 16,625 | 17,962 | 8.0 | 4.6 | 1,937 | 2,547 | 1,448 | 7,443 | 2.42 | 66.2 | 10.2 | 29.7 |
| Franklin | 14.4 | 12.1 | 8.8 | 51.2 | 39,270 | 41,064 | 4.6 | 3.5 | 4,084 | 5,111 | 2,459 | 16,326 | 2.43 | 68.6 | 9.5 | 27.5 |
| Gibson | 13.3 | 10.2 | 8.1 | 51.9 | 48,152 | 49,686 | 3.2 | -1.1 | 6,152 | 6,928 | 285 | 19,320 | 2.49 | 67.2 | 15.5 | 29.5 |
| Giles | 15.2 | 12.2 | 8.6 | 51.6 | 29,447 | 29,479 | 0.1 | 0.2 | 3,181 | 3,801 | 673 | 11,904 | 2.39 | 65.6 | 12.3 | 27.5 |
| Grainger | 15.3 | 13.3 | 8.3 | 49.7 | 20,659 | 22,654 | 9.7 | 4.0 | 2,338 | 2,913 | 1,494 | 8,959 | 2.56 | 72.7 | 8.4 | 23.8 |
| Greene | 14.8 | 13.1 | 9.4 | 50.8 | 62,909 | 68,825 | 9.4 | 1.1 | 6,607 | 9,325 | 3,516 | 27,396 | 2.44 | 66.8 | 12.8 | 29.1 |
| Grundy | 14.0 | 12.1 | 9.2 | 50.2 | 14,332 | 13,726 | -4.2 | -1.8 | 1,616 | 2,077 | 226 | 4,820 | 2.73 | 68.9 | 11.0 | 26.5 |
| Hamblen | 13.3 | 10.5 | 8.3 | 51.0 | 58,128 | 62,534 | 7.6 | 4.1 | 7,952 | 7,756 | 2,430 | 24,456 | 2.58 | 67.0 | 11.1 | 27.7 |
| Hamilton | 13.2 | 10.7 | 7.6 | 51.8 | 307,896 | 336,477 | 9.3 | 10.5 | 43,166 | 36,214 | 28,135 | 145,213 | 2.41 | 60.6 | 11.2 | 33.3 |
| Hancock | 15.6 | 14.2 | 8.4 | 50.8 | 6,786 | 6,809 | 0.3 | -4.6 | 711 | 1,028 | 9 | 2,742 | 2.34 | 63.1 | 10.7 | 34.2 |
| Hardeman | 13.4 | 11.1 | 7.8 | 45.0 | 28,105 | 27,245 | -3.1 | -8.8 | 2,620 | 3,024 | -2,025 | 8,891 | 2.41 | 65.4 | 15.8 | 31.2 |
| Hardin | 15.1 | 13.4 | 10.3 | 51.4 | 25,578 | 26,008 | 1.7 | -1.6 | 2,724 | 3,847 | 716 | 10,137 | 2.49 | 71.1 | 12.2 | 26.0 |
| Hawkins | 15.1 | 12.9 | 9.1 | 50.7 | 53,563 | 56,826 | 6.1 | -0.1 | 5,515 | 7,314 | 1,793 | 23,135 | 2.42 | 68.8 | 12.4 | 28.1 |
| Haywood | 14.7 | 12.3 | 7.8 | 53.3 | 19,797 | 18,805 | -5.0 | -9.6 | 2,146 | 2,093 | -1,863 | 7,181 | 2.42 | 65.8 | 20.8 | 31.5 |
| Henderson | 13.6 | 11.2 | 7.9 | 51.0 | 25,522 | 27,786 | 8.9 | 1.0 | 3,271 | 3,385 | 430 | 10,711 | 2.58 | 66.4 | 10.8 | 28.5 |
| Henry | 15.1 | 13.8 | 10.3 | 51.5 | 31,115 | 32,349 | 4.0 | -0.9 | 3,363 | 4,791 | 1,162 | 13,394 | 2.37 | 67.4 | 13.0 | 27.3 |
| Hickman | 14.3 | 10.8 | 7.2 | 47.8 | 22,295 | 24,692 | 10.8 | 2.8 | 2,768 | 2,890 | 814 | 8,636 | 2.67 | 65.0 | 9.8 | 30.7 |
| Houston | 14.3 | 12.4 | 8.9 | 50.9 | 8,088 | 8,428 | 4.2 | -1.6 | 851 | 1,128 | 137 | 2,878 | 2.76 | 60.8 | 7.8 | 36.3 |
| Humphreys | 14.6 | 12.1 | 8.5 | 50.2 | 17,929 | 18,534 | 3.4 | 0.3 | 2,123 | 2,534 | 474 | 6,763 | 2.69 | 64.7 | 7.3 | 31.5 |
| Jackson | 16.8 | 14.2 | 9.2 | 49.7 | 10,984 | 11,636 | 5.9 | 2.0 | 1,009 | 1,596 | 813 | 4,566 | 2.51 | 60.1 | 11.1 | 35.3 |
| Jefferson | 15.5 | 12.3 | 8.6 | 50.8 | 44,294 | 51,673 | 16.7 | 7.0 | 5,221 | 6,288 | 4,707 | 20,154 | 2.57 | 69.5 | 9.0 | 25.1 |
| Johnson | 14.8 | 14.0 | 10.0 | 46.8 | 17,499 | 18,240 | 4.2 | -2.1 | 1,567 | 2,378 | 431 | 6,794 | 2.32 | 68.2 | 11.9 | 27.5 |
| Knox | 12.4 | 9.8 | 6.7 | 51.4 | 382,032 | 432,259 | 13.1 | 10.0 | 53,357 | 44,871 | 34,885 | 187,319 | 2.4 | 61.9 | 11.1 | 30.3 |
| Lake | 12.9 | 9.6 | 7.2 | 37.2 | 7,954 | 7,832 | -1.5 | -10.8 | 676 | 992 | -528 | 2,243 | 2.05 | 66.8 | 21.5 | 27.1 |
| Lauderdale | 13.4 | 9.9 | 6.6 | 48.1 | 27,101 | 27,822 | 2.7 | -8.5 | 3,067 | 3,003 | -2,469 | 9,675 | 2.39 | 69.8 | 19.8 | 26.2 |
| Lawrence | 13.2 | 10.3 | 7.9 | 50.8 | 39,926 | 41,859 | 4.8 | 6.1 | 5,815 | 5,494 | 2,267 | 15,960 | 2.69 | 69.1 | 12.0 | 26.7 |
| Lewis | 14.1 | 12.8 | 9.0 | 51.5 | 11,367 | 12,170 | 7.1 | 1.6 | 1,376 | 1,531 | 345 | 4,715 | 2.5 | 67.3 | 11.8 | 29.8 |
| Lincoln | 15.0 | 11.6 | 8.6 | 51.0 | 31,340 | 33,351 | 6.4 | 3.6 | 3,698 | 4,296 | 1,815 | 13,548 | 2.48 | 68.4 | 13.9 | 27.7 |
| Loudon | 15.4 | 16.2 | 11.4 | 50.7 | 39,086 | 48,569 | 24.3 | 13.1 | 5,341 | 6,519 | 7,484 | 20,669 | 2.51 | 70.7 | 7.9 | 26.3 |
| McMinn | 14.2 | 11.7 | 8.9 | 51.2 | 49,015 | 52,283 | 6.7 | 3.7 | 5,877 | 6,929 | 3,016 | 20,804 | 2.49 | 65.8 | 11.2 | 29.9 |
| McNairy | 14.1 | 12.1 | 9.1 | 51.2 | 24,653 | 26,026 | 5.8 | -1.5 | 2,816 | 3,719 | 533 | 10,022 | 2.54 | 67.1 | 12.1 | 29.8 |
| Macon | 12.9 | 9.9 | 6.6 | 51.0 | 20,386 | 22,225 | 9.0 | 11.7 | 3,321 | 2,782 | 2,065 | 9,170 | 2.56 | 67.8 | 8.4 | 27.2 |
| Madison | 13.6 | 10.8 | 7.3 | 52.7 | 91,837 | 98,302 | 7.0 | 0.1 | 12,708 | 10,201 | -2,409 | 37,944 | 2.47 | 66.2 | 16.3 | 29.4 |
| Marion | 14.9 | 12.7 | 8.2 | 51.0 | 27,776 | 28,226 | 1.6 | 2.5 | 3,197 | 3,781 | 1,296 | 11,477 | 2.46 | 70.7 | 11.3 | 25.9 |

1. No spouse present.

# Table B. States and Counties — Population, Vital Statistics, Health, and Crime

| STATE County | Persons in group quarters, 2020 | Daytime Population, 2015-2019 — Number | Employment/residence ratio | Births, 2020 — Total | Rate[1] | Deaths, 2020 — Number | Rate[1] | Persons under 65 with no health insurance, 2019 — Number | Percent | Medicare, 2020 — Total beneficiaries | Enrolled in Original Medicare | Enrolled in Medicare Advantage | Crimes reported by county police or sheriff, 2019 — Violent | Property |
|---|---|---|---|---|---|---|---|---|---|---|---|---|---|---|
| | 32 | 33 | 34 | 35 | 36 | 37 | 38 | 39 | 40 | 41 | 42 | 43 | 44 | 45 |
| **SOUTH DAKOTA—Cont'd** | | | | | | | | | | | | | | |
| Tripp | 131 | 5,571 | 1.04 | 83 | 15.4 | 74 | 13.8 | 723 | 17.7 | 1,343 | 1,203 | 141 | | |
| Turner | 153 | 6,895 | 0.66 | 90 | 10.8 | 89 | 10.6 | 701 | 10.7 | 1,968 | 1,367 | 601 | 2 | 4 |
| Union | 89 | 17,082 | 1.22 | 175 | 10.8 | 134 | 8.3 | 1,083 | 8.2 | 3,335 | 2,558 | 778 | 18 | 45 |
| Walworth | 152 | 5,282 | 0.94 | 65 | 12.2 | 55 | 10.3 | 583 | 14.4 | 1,499 | 1,361 | 139 | 7 | 46 |
| Yankton | 2,293 | 24,558 | 1.16 | 248 | 10.9 | 232 | 10.2 | 1,852 | 11.1 | 5,060 | 3,862 | 1,198 | 0 | 5 |
| Ziebach | 0 | 2,597 | 0.8 | 31 | 11.7 | 30 | 11.3 | 485 | 20.0 | 163 | 152 | 11 | 7 | 53 |
| **TENNESSEE** | 156,810 | 6,756,511 | 1.02 | 80,474 | 11.7 | 73,874 | 10.7 | 669,850 | 12.1 | 1,384,840 | 796,335 | 588,506 | X | X |
| Anderson | 1,226 | 90,806 | 1.46 | 812 | 10.5 | 1,000 | 12.9 | 7,356 | 12.1 | 18,880 | 10,202 | 8,678 | 144 | 472 |
| Bedford | 524 | 44,734 | 0.84 | 680 | 13.6 | 503 | 10.0 | 6,283 | 15.1 | 9,332 | 5,479 | 3,853 | 75 | 250 |
| Benton | 234 | 14,547 | 0.72 | 170 | 10.5 | 267 | 16.6 | 1,630 | 13.6 | 4,684 | 3,138 | 1,546 | 18 | 114 |
| Bledsoe | 2,746 | 13,055 | 0.66 | 118 | 7.8 | 166 | 10.9 | 1,574 | 16.6 | 3,327 | 1,921 | 1,406 | 11 | 62 |
| Blount | 2,024 | 123,647 | 0.9 | 1,283 | 9.5 | 1,541 | 11.4 | 12,138 | 11.6 | 32,512 | 17,944 | 14,568 | 322 | 1,103 |
| Bradley | 2,532 | 105,056 | 0.99 | 1,224 | 11.2 | 1,199 | 11.0 | 12,174 | 14.0 | 23,032 | 12,260 | 10,772 | 205 | 808 |
| Campbell | 683 | 35,458 | 0.71 | 441 | 11.1 | 628 | 15.8 | 3,962 | 12.9 | 10,465 | 4,481 | 5,984 | 62 | 335 |
| Cannon | 150 | 10,742 | 0.46 | 166 | 11.2 | 158 | 10.6 | 1,473 | 12.5 | 3,305 | 1,998 | 1,308 | 12 | 96 |
| Carroll | 1,212 | 24,510 | 0.69 | 299 | 10.8 | 408 | 14.7 | 2,641 | 12.6 | 7,348 | 5,156 | 2,193 | 44 | 154 |
| Carter | 1,045 | 46,830 | 0.58 | 451 | 8.0 | 835 | 14.8 | 5,950 | 14.0 | 15,454 | 6,543 | 8,912 | 63 | 334 |
| Cheatham | 289 | 29,998 | 0.48 | 466 | 11.3 | 452 | 11.0 | 4,020 | 11.7 | 7,984 | 4,088 | 3,896 | 87 | 280 |
| Chester | 1,140 | 14,876 | 0.67 | 178 | 10.2 | 181 | 10.4 | 1,635 | 12.4 | 3,788 | 2,507 | 1,281 | 25 | 79 |
| Claiborne | 1,263 | 30,247 | 0.87 | 330 | 10.3 | 456 | 14.2 | 2,953 | 12.2 | 8,662 | 3,761 | 4,901 | 58 | 283 |
| Clay | 76 | 6,842 | 0.69 | 65 | 8.5 | 113 | 14.8 | 787 | 14.1 | 2,218 | 1,555 | 663 | 14 | 40 |
| Cocke | 388 | 31,326 | 0.69 | 351 | 9.7 | 558 | 15.4 | 3,544 | 12.9 | 10,438 | 4,763 | 5,675 | NA | NA |
| Coffee | 603 | 57,457 | 1.09 | 703 | 12.2 | 685 | 11.9 | 5,468 | 11.9 | 12,770 | 8,268 | 4,503 | 87 | 403 |
| Crockett | 194 | 12,041 | 0.61 | 165 | 11.6 | 185 | 13.0 | 1,736 | 15.3 | 3,228 | 2,354 | 874 | 29 | 106 |
| Cumberland | 652 | 59,163 | 1.0 | 547 | 8.9 | 926 | 15.0 | 5,862 | 14.3 | 22,785 | 14,381 | 8,405 | 42 | 737 |
| Davidson | 25,409 | 808,216 | 1.32 | 10,034 | 14.5 | 5,962 | 8.6 | 78,123 | 13.4 | 98,157 | 53,302 | 44,855 | NA | NA |
| Decatur | 216 | 10,886 | 0.82 | 106 | 9.1 | 183 | 15.8 | 1,105 | 12.7 | 3,287 | 2,152 | 1,135 | 30 | 100 |
| DeKalb | 274 | 18,944 | 0.89 | 229 | 11.0 | 270 | 13.0 | 2,313 | 14.2 | 4,780 | 2,374 | 2,405 | 17 | 117 |
| Dickson | 683 | 48,543 | 0.82 | 638 | 11.7 | 575 | 10.6 | 5,200 | 11.7 | 10,961 | 6,063 | 4,898 | 121 | 334 |
| Dyer | 517 | 37,527 | 1.01 | 451 | 12.3 | 499 | 13.6 | 3,440 | 11.5 | 8,331 | 5,781 | 2,551 | 18 | 184 |
| Fayette | 442 | 31,199 | 0.48 | 393 | 9.4 | 477 | 11.5 | 3,586 | 11.3 | 10,543 | 7,231 | 3,312 | 55 | 209 |
| Fentress | 261 | 16,734 | 0.79 | 188 | 10.0 | 279 | 14.9 | 1,842 | 13.2 | 5,541 | 3,873 | 1,668 | 27 | 189 |
| Franklin | 2,200 | 39,600 | 0.88 | 405 | 9.5 | 510 | 12.0 | 3,699 | 11.7 | 10,463 | 6,864 | 3,598 | 98 | 238 |
| Gibson | 1,144 | 44,534 | 0.77 | 559 | 11.4 | 703 | 14.3 | 4,596 | 11.7 | 11,715 | 8,054 | 3,661 | 56 | 160 |
| Giles | 660 | 27,679 | 0.87 | 304 | 10.3 | 418 | 14.2 | 2,994 | 13.1 | 7,400 | 5,062 | 2,337 | 51 | 184 |
| Grainger | 151 | 18,407 | 0.5 | 237 | 10.1 | 316 | 13.4 | 2,522 | 13.9 | 6,229 | 2,714 | 3,515 | 43 | 160 |
| Greene | 1,688 | 67,578 | 0.95 | 652 | 9.4 | 1,022 | 14.7 | 6,864 | 13.1 | 19,462 | 9,600 | 9,862 | 168 | 888 |
| Grundy | 195 | 11,699 | 0.67 | 157 | 11.6 | 191 | 14.2 | 1,465 | 14.2 | 3,839 | 2,004 | 1,835 | 77 | 245 |
| Hamblen | 1,002 | 69,184 | 1.19 | 778 | 11.9 | 821 | 12.6 | 8,105 | 15.6 | 14,936 | 7,579 | 7,357 | 120 | 462 |
| Hamilton | 9,945 | 405,448 | 1.26 | 4,290 | 11.5 | 4,042 | 10.9 | 35,273 | 12.0 | 74,659 | 41,303 | 33,357 | NA | NA |
| Hancock | 155 | 5,724 | 0.6 | 70 | 10.8 | 100 | 15.4 | 587 | 11.8 | 1,872 | 853 | 1,020 | 21 | 110 |
| Hardeman | 3,814 | 24,884 | 0.93 | 216 | 8.7 | 306 | 12.3 | 2,033 | 12.3 | 5,975 | 3,722 | 2,253 | 46 | 178 |
| Hardin | 371 | 25,593 | 0.99 | 227 | 8.9 | 397 | 15.5 | 2,370 | 12.2 | 7,474 | 5,312 | 2,162 | 35 | 348 |
| Hawkins | 580 | 48,413 | 0.62 | 527 | 9.3 | 842 | 14.8 | 5,094 | 11.7 | 16,017 | 5,954 | 10,064 | 77 | 587 |
| Haywood | 196 | 15,815 | 0.75 | 186 | 10.9 | 231 | 13.6 | 1,613 | 12.0 | 4,234 | 2,501 | 1,734 | 36 | 133 |
| Henderson | 367 | 25,612 | 0.79 | 299 | 10.6 | 364 | 13.0 | 2,834 | 12.7 | 6,791 | 4,111 | 2,679 | 83 | 295 |
| Henry | 497 | 32,165 | 0.99 | 305 | 9.5 | 525 | 16.4 | 3,121 | 12.9 | 9,148 | 6,482 | 2,666 | 47 | 298 |
| Hickman | 1,477 | 19,960 | 0.5 | 269 | 10.6 | 291 | 11.5 | 2,682 | 14.0 | 5,377 | 2,948 | 2,428 | 84 | 394 |
| Houston | 181 | 6,770 | 0.52 | 80 | 9.6 | 110 | 13.3 | 883 | 13.8 | 2,132 | 1,450 | 682 | 23 | 75 |
| Humphreys | 254 | 17,737 | 0.9 | 207 | 11.1 | 257 | 13.8 | 1,710 | 11.8 | 4,584 | 3,165 | 1,419 | 30 | 93 |
| Jackson | 185 | 9,418 | 0.47 | 89 | 7.5 | 179 | 15.1 | 1,258 | 14.2 | 3,211 | 2,091 | 1,120 | 24 | 83 |
| Jefferson | 1,453 | 47,008 | 0.71 | 485 | 8.8 | 724 | 13.1 | 5,509 | 13.1 | 13,847 | 6,928 | 6,919 | 97 | 428 |
| Johnson | 1,634 | 16,708 | 0.81 | 149 | 8.3 | 257 | 14.4 | 1,476 | 12.4 | 5,247 | 2,545 | 2,701 | 34 | 132 |
| Knox | 12,376 | 486,108 | 1.11 | 5,136 | 10.8 | 4,888 | 10.3 | 41,329 | 10.7 | 90,887 | 49,134 | 41,752 | 577 | 3,413 |
| Lake | 2,225 | 7,500 | 1.06 | 65 | 9.3 | 97 | 13.9 | 444 | 11.9 | 1,365 | 993 | 372 | 4 | 25 |
| Lauderdale | 2,153 | 24,968 | 0.89 | 278 | 10.9 | 295 | 11.6 | 2,374 | 12.5 | 5,373 | 3,405 | 1,969 | 45 | 192 |
| Lawrence | 535 | 39,658 | 0.77 | 555 | 12.5 | 555 | 12.5 | 4,613 | 13.0 | 10,089 | 7,019 | 3,071 | 49 | 267 |
| Lewis | 210 | 10,445 | 0.67 | 139 | 11.2 | 167 | 13.5 | 1,137 | 12.0 | 3,085 | 1,893 | 1,192 | 17 | 71 |
| Lincoln | 287 | 30,006 | 0.74 | 375 | 10.9 | 443 | 12.8 | 3,607 | 13.2 | 8,374 | 4,874 | 3,500 | 62 | 286 |
| Loudon | 481 | 49,758 | 0.88 | 490 | 8.9 | 689 | 12.5 | 5,388 | 13.7 | 17,065 | 9,808 | 7,257 | 72 | 375 |
| McMinn | 1,066 | 51,996 | 0.95 | 582 | 10.7 | 727 | 13.4 | 5,432 | 12.9 | 13,181 | 7,756 | 5,425 | 93 | 549 |
| McNairy | 317 | 23,087 | 0.7 | 256 | 10.0 | 384 | 14.9 | 2,295 | 11.5 | 7,016 | 4,897 | 2,120 | 55 | 276 |
| Macon | 339 | 20,100 | 0.64 | 329 | 13.3 | 278 | 11.2 | 3,011 | 15.0 | 5,177 | 3,148 | 2,030 | 41 | 159 |
| Madison | 4,563 | 113,237 | 1.36 | 1,217 | 12.4 | 1,049 | 10.7 | 8,637 | 11.3 | 21,235 | 14,286 | 6,949 | 103 | 380 |
| Marion | 276 | 24,952 | 0.71 | 307 | 10.6 | 371 | 12.8 | 2,764 | 12.2 | 7,366 | 3,895 | 3,471 | 59 | 220 |

1. Per 1,000 estimated resident population.

| STATE County | County law enforcement employment, 2019 | | Education | | | | | | Money income, 2015-2019 | | | | Income and poverty, 2019 | | | |
|---|---|---|---|---|---|---|---|---|---|---|---|---|---|---|---|---|
| | | | School enrollment and attainment, 2015-2019 | | | | Local government expenditures[3] 2017-2018 | | | Households | | | | Percent below poverty level | | |
| | | | Enrollment[1] | | Attainment[2] (percent) | | | | | | Percent | | | | | |
| | Officers | Civilians | Total | Percent private | High school graduate or less | Bachelor's degree or more | Total current spending (mil dol) | Current spending per student (dollars) | Per capita income[4] | Median income (dollars) | with income of less than $50,000 | with income of $200,000 or more | Median household income (dollars) | All persons | Children under 18 years | Children 5 to 17 years in families |
| | 46 | 47 | 48 | 49 | 50 | 51 | 52 | 53 | 54 | 55 | 56 | 57 | 58 | 59 | 60 | 61 |
| **SOUTH DAKOTA—Cont'd** | | | | | | | | | | | | | | | | |
| Tripp | 2 | 1 | 1,084 | 3.7 | 50.2 | 22.3 | 10 | 10,174 | 27,036 | 57,720 | 44.6 | 2.8 | 50,503 | 18.8 | 27.3 | 26.0 |
| Turner | 8 | 2 | 1,788 | 9.4 | 41.7 | 23.7 | 17 | 10,691 | 30,677 | 59,242 | 43.0 | 3.6 | 63,852 | 10.1 | 10.6 | 10.1 |
| Union | 18 | 10 | 3,610 | 7.0 | 36.8 | 30.6 | 28 | 9,262 | 42,411 | 70,378 | 36.0 | 9.7 | 83,910 | 6.3 | 5.8 | 5.1 |
| Walworth | 4 | 9 | 1,134 | 7.0 | 48.1 | 24.9 | 9 | 10,763 | 32,563 | 55,583 | 45.0 | 3.5 | 51,472 | 15.1 | 19.5 | 18.9 |
| Yankton | 12 | 23 | 4,841 | 18.6 | 44.5 | 25.7 | 28 | 8,840 | 31,772 | 58,342 | 43.6 | 3.5 | 60,868 | 10.1 | 11.0 | 9.7 |
| Ziebach | 3 | 0 | 745 | 3.4 | 57.1 | 16.4 | 13 | 18,073 | 16,555 | 37,400 | 62.6 | 2.0 | 33,508 | 47.7 | 63.4 | 45.9 |
| **TENNESSEE** | X | X | 1,568,653 | 17.6 | 44.6 | 27.3 | 9,555 | 9,537 | 29,859 | 53,320 | 47.1 | 4.8 | 56,047 | 13.8 | 19.4 | 18.0 |
| Anderson | 64 | 110 | 15,922 | 13.1 | 45.9 | 23.3 | 127 | 10,653 | 28,455 | 50,392 | 49.7 | 3.6 | 50,973 | 15.3 | 21.9 | 21.8 |
| Bedford | 52 | 52 | 11,155 | 10.1 | 59.4 | 16.6 | 66 | 7,514 | 24,864 | 50,415 | 49.5 | 3.1 | 50,539 | 13.9 | 19.2 | 18.4 |
| Benton | 21 | 39 | 3,204 | 10.0 | 64.1 | 12.3 | 22 | 10,043 | 22,636 | 37,512 | 63.2 | 2.3 | 42,322 | 17.6 | 27.8 | 25.6 |
| Bledsoe | 14 | 27 | 2,531 | 17.4 | 64.2 | 12.0 | 18 | 9,828 | 21,700 | 44,122 | 55.3 | 3.1 | 42,730 | 27.3 | 40.6 | 39.9 |
| Blount | 173 | 134 | 26,770 | 16.4 | 45.5 | 24.2 | 179 | 9,873 | 30,548 | 56,667 | 44.3 | 3.9 | 59,368 | 10.5 | 14.1 | 12.9 |
| Bradley | 110 | 110 | 26,055 | 23.6 | 46.8 | 22.3 | 140 | 8,760 | 26,655 | 51,331 | 49.0 | 2.9 | 52,534 | 14.5 | 17.4 | 15.0 |
| Campbell | 43 | 44 | 7,810 | 12.4 | 63.4 | 12.4 | 47 | 8,351 | 24,212 | 39,803 | 59.7 | 1.7 | 42,500 | 21.9 | 31.1 | 28.0 |
| Cannon | 12 | 22 | 2,818 | 18.1 | 61.0 | 15.8 | 18 | 8,983 | 26,435 | 55,330 | 45.1 | 2.6 | 52,960 | 13.0 | 19.5 | 18.4 |
| Carroll | 28 | 43 | 6,315 | 21.7 | 58.1 | 18.0 | 42 | 9,195 | 22,394 | 42,637 | 57.4 | 1.6 | 43,258 | 16.8 | 23.6 | 22.6 |
| Carter | 72 | 62 | 10,795 | 16.8 | 55.1 | 17.5 | 75 | 9,587 | 23,205 | 38,092 | 61.6 | 1.4 | 39,964 | 19.3 | 27.6 | 27.1 |
| Cheatham | 48 | 42 | 8,676 | 15.2 | 50.2 | 21.0 | 56 | 8,925 | 27,893 | 61,913 | 38.4 | 2.6 | 62,786 | 9.6 | 12.2 | 11.1 |
| Chester | 16 | 33 | 4,539 | 27.4 | 54.0 | 16.0 | 24 | 8,333 | 22,629 | 51,946 | 48.2 | 0.7 | 50,769 | 14.8 | 19.2 | 18.0 |
| Claiborne | 44 | 41 | 7,104 | 26.9 | 57.9 | 16.1 | 40 | 9,412 | 21,566 | 36,835 | 63.0 | 1.6 | 40,116 | 19.7 | 26.7 | 23.4 |
| Clay | 15 | 9 | 1,417 | 1.6 | 65.9 | 13.9 | 10 | 9,184 | 18,983 | 32,167 | 64.6 | 0.6 | 36,694 | 20.1 | 28.1 | 26.3 |
| Cocke | 40 | 41 | 6,288 | 5.8 | 66.8 | 11.6 | 48 | 9,082 | 21,867 | 36,716 | 63.5 | 1.2 | 40,528 | 22.8 | 41.3 | 33.0 |
| Coffee | 47 | 70 | 12,067 | 8.8 | 53.0 | 20.5 | 92 | 9,715 | 26,557 | 50,351 | 49.7 | 2.8 | 54,931 | 14.4 | 21.0 | 19.6 |
| Crockett | 16 | 23 | 3,317 | 5.7 | 59.3 | 13.8 | 25 | 7,979 | 23,771 | 44,717 | 53.3 | 2.0 | 45,642 | 15.8 | 23.3 | 21.0 |
| Cumberland | 57 | 57 | 9,611 | 9.3 | 51.9 | 18.5 | 62 | 8,379 | 25,832 | 45,958 | 54.3 | 2.0 | 49,142 | 13.3 | 22.9 | 20.9 |
| Davidson | NA | NA | 160,171 | 30.7 | 33.0 | 41.7 | 1,151 | 11,939 | 36,440 | 60,388 | 41.2 | 6.8 | 63,846 | 12.6 | 18.1 | 17.9 |
| Decatur | 13 | 18 | 2,199 | 6.9 | 56.3 | 14.5 | 14 | 8,273 | 23,857 | 42,031 | 59.0 | 1.4 | 45,215 | 16.0 | 23.0 | 21.8 |
| DeKalb | 27 | 24 | 3,937 | 6.9 | 60.7 | 16.7 | 24 | 7,770 | 25,399 | 45,511 | 54.2 | 2.4 | 46,407 | 18.0 | 26.0 | 23.6 |
| Dickson | 63 | 76 | 11,374 | 13.6 | 58.1 | 16.2 | 71 | 8,419 | 27,115 | 53,076 | 46.3 | 3.4 | 59,126 | 10.1 | 13.8 | 13.6 |
| Dyer | 38 | 53 | 8,458 | 9.0 | 55.0 | 16.1 | 63 | 9,430 | 27,710 | 44,185 | 54.7 | 2.8 | 48,770 | 17.3 | 22.6 | 21.6 |
| Fayette | 45 | 48 | 7,413 | 31.3 | 45.9 | 22.2 | 30 | 8,474 | 33,383 | 60,711 | 40.7 | 6.8 | 67,345 | 11.2 | 16.7 | 14.8 |
| Fentress | 21 | 29 | 3,654 | 13.2 | 60.6 | 16.0 | 19 | 6,977 | 20,093 | 36,520 | 66.1 | 1.1 | 38,608 | 20.9 | 28.4 | 25.9 |
| Franklin | 44 | 33 | 9,928 | 26.6 | 52.5 | 20.8 | 50 | 9,264 | 28,317 | 51,585 | 48.2 | 4.3 | 54,319 | 13.9 | 18.2 | 17.2 |
| Gibson | 31 | 50 | 11,168 | 11.2 | 53.8 | 18.4 | 80 | 8,694 | 23,211 | 43,171 | 56.1 | 2.4 | 46,384 | 14.3 | 18.1 | 16.9 |
| Giles | 38 | 33 | 6,001 | 18.1 | 57.3 | 16.5 | 36 | 9,218 | 25,690 | 49,614 | 50.4 | 1.3 | 50,847 | 13.4 | 20.1 | 18.8 |
| Grainger | 26 | 17 | 4,276 | 8.1 | 64.2 | 10.7 | 31 | 8,782 | 23,595 | 44,064 | 55.6 | 1.8 | 44,929 | 17.1 | 25.0 | 22.3 |
| Greene | 74 | 97 | 13,759 | 14.4 | 56.8 | 16.7 | 89 | 9,234 | 25,190 | 42,595 | 56.3 | 2.2 | 44,975 | 15.9 | 24.3 | 23.2 |
| Grundy | 18 | 22 | 2,743 | 8.6 | 66.3 | 13.2 | 19 | 9,263 | 20,592 | 40,516 | 58.9 | 1.7 | 38,447 | 21.4 | 28.5 | 25.2 |
| Hamblen | 43 | 61 | 14,391 | 10.3 | 57.3 | 16.8 | 91 | 8,610 | 23,517 | 43,619 | 55.7 | 1.8 | 46,842 | 14.4 | 20.9 | 19.7 |
| Hamilton | 173 | 227 | 82,382 | 22.8 | 37.4 | 32.2 | 433 | 9,664 | 32,839 | 55,070 | 46.1 | 5.8 | 57,802 | 12.7 | 18.8 | 18.3 |
| Hancock | 17 | 31 | 1,255 | 15.1 | 65.0 | 11.2 | 10 | 10,366 | 23,728 | 30,136 | 66.6 | 3.3 | 31,938 | 26.4 | 35.8 | 33.9 |
| Hardeman | 30 | 40 | 5,342 | 15.9 | 67.1 | 11.2 | 36 | 9,680 | 18,172 | 40,304 | 62.1 | 0.7 | 40,989 | 19.2 | 24.1 | 23.8 |
| Hardin | 25 | 28 | 5,195 | 12.9 | 60.6 | 15.0 | 32 | 8,838 | 24,243 | 40,682 | 57.9 | 2.8 | 44,510 | 17.5 | 26.0 | 25.1 |
| Hawkins | 72 | 46 | 11,008 | 10.3 | 58.4 | 13.7 | 69 | 9,352 | 24,695 | 41,924 | 57.6 | 1.8 | 46,602 | 18.1 | 30.1 | 25.6 |
| Haywood | 19 | 28 | 3,782 | 12.0 | 65.0 | 11.0 | 29 | 9,939 | 21,839 | 37,905 | 60.2 | 1.2 | 38,950 | 20.7 | 29.0 | 26.3 |
| Henderson | 34 | 34 | 5,983 | 9.1 | 59.6 | 14.4 | 43 | 8,944 | 22,695 | 43,305 | 55.0 | 1.5 | 46,826 | 15.8 | 21.9 | 19.7 |
| Henry | 35 | 34 | 6,264 | 10.5 | 59.6 | 16.5 | 43 | 9,125 | 24,124 | 40,502 | 59.2 | 1.6 | 43,159 | 18.3 | 28.0 | 28.1 |
| Hickman | 29 | 21 | 4,898 | 11.9 | 65.8 | 10.9 | 31 | 9,074 | 22,856 | 43,596 | 57.4 | 1.6 | 46,176 | 16.3 | 22.4 | 21.3 |
| Houston | 10 | 17 | 1,835 | 8.4 | 69.9 | 13.0 | 12 | 8,939 | 22,360 | 42,711 | 58.2 | 2.2 | 40,915 | 17.8 | 24.8 | 22.8 |
| Humphreys | 23 | 25 | 3,748 | 11.3 | 63.5 | 14.3 | 26 | 8,761 | 25,428 | 45,667 | 55.4 | 1.9 | 52,121 | 13.1 | 20.7 | 19.7 |
| Jackson | 18 | 30 | 2,108 | 11.1 | 66.4 | 11.0 | 14 | 9,594 | 21,037 | 35,207 | 64.7 | 1.9 | 41,749 | 17.1 | 23.8 | 23.0 |
| Jefferson | 52 | 51 | 11,749 | 20.8 | 53.7 | 16.8 | 63 | 8,611 | 25,159 | 49,139 | 50.8 | 2.6 | 50,332 | 12.5 | 19.3 | 17.5 |
| Johnson | 18 | 31 | 2,944 | 10.8 | 61.8 | 11.6 | 21 | 9,906 | 22,107 | 36,004 | 66.0 | 2.1 | 40,219 | 25.2 | 35.6 | 31.0 |
| Knox | 430 | 585 | 118,028 | 16.4 | 33.6 | 37.6 | 538 | 8,831 | 33,229 | 57,470 | 44.0 | 5.7 | 60,316 | 13.4 | 15.7 | 13.4 |
| Lake | 8 | 13 | 1,229 | 11.0 | 70.2 | 9.4 | 8 | 9,785 | 15,732 | 35,191 | 61.2 | 2.9 | 34,059 | 35.5 | 40.0 | 37.8 |
| Lauderdale | 20 | 39 | 5,815 | 11.0 | 68.3 | 8.3 | 39 | 9,250 | 18,515 | 39,896 | 60.2 | 0.9 | 43,545 | 19.3 | 24.8 | 24.0 |
| Lawrence | 49 | 41 | 10,216 | 18.1 | 60.9 | 14.3 | 60 | 8,562 | 21,720 | 43,614 | 56.6 | 2.0 | 43,448 | 16.7 | 22.6 | 19.9 |
| Lewis | 16 | 18 | 2,636 | 11.8 | 58.2 | 10.4 | 15 | 8,713 | 21,516 | 37,277 | 61.8 | 1.2 | 45,874 | 15.2 | 22.8 | 21.9 |
| Lincoln | 41 | 44 | 7,427 | 10.6 | 57.1 | 18.4 | 47 | 8,809 | 26,965 | 49,485 | 50.4 | 2.9 | 54,558 | 12.8 | 18.6 | 17.2 |
| Loudon | 57 | 49 | 9,730 | 13.9 | 46.2 | 26.4 | 66 | 9,187 | 31,478 | 58,065 | 41.9 | 4.3 | 65,226 | 9.7 | 14.2 | 12.9 |
| McMinn | 41 | 47 | 10,796 | 16.5 | 55.9 | 16.2 | 68 | 8,631 | 23,885 | 43,285 | 56.0 | 1.5 | 49,836 | 14.5 | 20.1 | 19.3 |
| McNairy | 19 | 19 | 5,155 | 12.4 | 62.8 | 12.6 | 37 | 8,562 | 20,934 | 39,161 | 60.2 | 2.0 | 45,469 | 17.0 | 24.3 | 20.7 |
| Macon | 32 | 28 | 5,269 | 13.7 | 65.4 | 10.3 | 33 | 8,196 | 22,527 | 37,430 | 58.4 | 1.3 | 44,334 | 15.5 | 20.6 | 19.6 |
| Madison | 122 | 160 | 23,982 | 29.6 | 46.4 | 25.6 | 119 | 9,206 | 26,722 | 48,161 | 51.7 | 3.6 | 50,401 | 17.6 | 28.9 | 27.9 |
| Marion | 34 | 38 | 5,911 | 9.7 | 57.1 | 12.8 | 38 | 8,771 | 25,467 | 49,432 | 50.5 | 2.3 | 47,675 | 15.3 | 21.8 | 21.1 |

1. All persons 3 years old and over enrolled in nursery school through college.    2. Persons 25 years old and over.    3. Elementary and secondary education expenditures.    4. Based on population estimated by the American Community Survey, 2014–2018.

# Table B. States and Counties — Personal Income

| STATE County | Personal income, 2019 | | | | | | | | | | Earnings, 2019 | | |
| | Total (mil dol) | Percent change 2018-2019 | Per capita[1] Dollars | Per capita[1] Rank | Wages and salaries (mil dol) | Supplements to wages and salaries, employer contributions (mil dol) Pension and insurance | Supplements to wages and salaries, employer contributions (mil dol) Government social insurance | Proprietors' income (mil dol) | Dividends, interest, and rent (mil dol) | Personal transfer receipts (mil dol) | Contributions for government social insurance (mil dol) Total (mil dol) | Contributions for government social insurance (mil dol) From employee and self-employed | Contributions for government social insurance (mil dol) From employer |
|---|---|---|---|---|---|---|---|---|---|---|---|---|---|
| | 62 | 63 | 64 | 65 | 66 | 67 | 68 | 69 | 70 | 71 | 72 | 73 | 74 |
| **SOUTH DAKOTA—Cont'd** | | | | | | | | | | | | | |
| Tripp | 280 | 0.1 | 51,501 | 727 | 86 | 16 | 6 | 53 | 69 | 57 | 161 | 9 | 6 |
| Turner | 456 | 4.6 | 54,445 | 522 | 85 | 18 | 6 | 123 | 76 | 73 | 232 | 12 | 6 |
| Union | 1,865 | 2.6 | 117,055 | 13 | 579 | 85 | 40 | 341 | 706 | 124 | 1,044 | 60 | 40 |
| Walworth | 253 | -0.6 | 46,589 | 1,176 | 81 | 16 | 6 | 24 | 78 | 58 | 127 | 9 | 6 |
| Yankton | 1,164 | 2.7 | 51,011 | 757 | 596 | 107 | 45 | 141 | 262 | 196 | 889 | 55 | 45 |
| Ziebach | 55 | -2.3 | 19,940 | 3,112 | 14 | 3 | 1 | 4 | 13 | 18 | 22 | 1 | 1 |
| **TENNESSEE** | 332,473 | 4.0 | 48,676 | X | 163,370 | 24,551 | 11,164 | 45,906 | 49,331 | 66,256 | 244,991 | 14,791 | 11,164 |
| Anderson | 3,313 | 3.4 | 43,045 | 1,610 | 2,524 | 342 | 170 | 325 | 466 | 852 | 3,361 | 210 | 170 |
| Bedford | 1,922 | 4.4 | 38,667 | 2,243 | 704 | 128 | 49 | 204 | 253 | 457 | 1,084 | 69 | 49 |
| Benton | 557 | 2.4 | 34,482 | 2,744 | 162 | 35 | 12 | 31 | 78 | 217 | 239 | 20 | 12 |
| Bledsoe | 408 | 6.2 | 27,113 | 3,094 | 72 | 20 | 5 | 32 | 46 | 139 | 129 | 11 | 5 |
| Blount | 6,018 | 4.0 | 45,222 | 1,332 | 2,597 | 413 | 177 | 526 | 958 | 1,374 | 3,713 | 244 | 177 |
| Bradley | 4,401 | 4.2 | 40,712 | 1,961 | 1,998 | 330 | 144 | 428 | 614 | 1,098 | 2,901 | 187 | 144 |
| Campbell | 1,381 | 3.0 | 34,655 | 2,731 | 371 | 76 | 27 | 83 | 184 | 515 | 557 | 46 | 27 |
| Cannon | 539 | 5.9 | 36,713 | 2,481 | 93 | 21 | 7 | 35 | 68 | 151 | 155 | 13 | 7 |
| Carroll | 1,042 | 4.6 | 37,529 | 2,383 | 302 | 60 | 22 | 70 | 123 | 365 | 454 | 36 | 22 |
| Carter | 1,956 | 2.5 | 34,691 | 2,728 | 458 | 89 | 32 | 142 | 295 | 659 | 720 | 61 | 32 |
| Cheatham | 1,876 | 4.8 | 46,138 | 1,233 | 429 | 89 | 33 | 225 | 201 | 383 | 776 | 50 | 33 |
| Chester | 602 | 3.6 | 34,800 | 2,715 | 154 | 33 | 11 | 52 | 76 | 174 | 250 | 19 | 11 |
| Claiborne | 1,150 | 2.7 | 35,996 | 2,572 | 369 | 74 | 27 | 82 | 135 | 399 | 553 | 43 | 27 |
| Clay | 236 | 0.0 | 31,056 | 2,999 | 55 | 14 | 4 | 18 | 33 | 97 | 92 | 8 | 4 |
| Cocke | 1,200 | 3.7 | 33,340 | 2,863 | 318 | 62 | 22 | 68 | 140 | 465 | 470 | 41 | 22 |
| Coffee | 2,284 | 5.5 | 40,415 | 1,998 | 1,314 | 213 | 89 | 245 | 301 | 621 | 1,861 | 117 | 89 |
| Crockett | 580 | 5.2 | 40,735 | 1,958 | 178 | 36 | 13 | 87 | 66 | 157 | 314 | 21 | 13 |
| Cumberland | 2,283 | 3.9 | 37,720 | 2,356 | 666 | 117 | 48 | 252 | 400 | 886 | 1,083 | 91 | 48 |
| Davidson | 49,459 | 3.9 | 71,252 | 118 | 33,618 | 4,089 | 2,231 | 13,816 | 7,980 | 5,225 | 53,753 | 2,898 | 2,231 |
| Decatur | 489 | 1.1 | 41,949 | 1,774 | 141 | 29 | 10 | 30 | 66 | 157 | 210 | 17 | 10 |
| DeKalb | 825 | 2.4 | 40,280 | 2,017 | 257 | 49 | 18 | 93 | 101 | 215 | 417 | 28 | 18 |
| Dickson | 2,271 | 4.8 | 42,088 | 1,750 | 792 | 132 | 55 | 274 | 258 | 542 | 1,253 | 81 | 55 |
| Dyer | 1,581 | 4.9 | 42,552 | 1,678 | 668 | 135 | 48 | 182 | 226 | 429 | 1,033 | 65 | 48 |
| Fayette | 2,318 | 4.5 | 56,361 | 430 | 376 | 66 | 26 | 189 | 318 | 432 | 657 | 49 | 26 |
| Fentress | 625 | 4.7 | 33,727 | 2,827 | 187 | 39 | 13 | 73 | 74 | 255 | 311 | 25 | 13 |
| Franklin | 1,704 | 4.3 | 40,370 | 2,005 | 501 | 93 | 36 | 160 | 245 | 468 | 789 | 57 | 36 |
| Gibson | 1,960 | 4.8 | 39,892 | 2,066 | 614 | 121 | 45 | 178 | 237 | 584 | 958 | 67 | 45 |
| Giles | 1,147 | 1.8 | 38,934 | 2,205 | 434 | 79 | 30 | 75 | 163 | 343 | 618 | 44 | 30 |
| Grainger | 816 | 4.1 | 34,983 | 2,689 | 170 | 33 | 13 | 64 | 92 | 261 | 279 | 24 | 13 |
| Greene | 2,575 | 2.7 | 37,287 | 2,417 | 1,090 | 195 | 78 | 139 | 354 | 897 | 1,502 | 110 | 78 |
| Grundy | 452 | 4.9 | 33,635 | 2,831 | 73 | 18 | 5 | 59 | 54 | 180 | 155 | 13 | 5 |
| Hamblen | 2,487 | 3.2 | 38,304 | 2,291 | 1,392 | 250 | 96 | 265 | 331 | 712 | 2,003 | 127 | 96 |
| Hamilton | 19,626 | 3.9 | 53,360 | 587 | 11,521 | 1,788 | 795 | 2,746 | 3,383 | 3,559 | 16,850 | 990 | 795 |
| Hancock | 186 | 5.2 | 28,083 | 3,078 | 32 | 10 | 2 | 2 | 27 | 80 | 46 | 5 | 2 |
| Hardeman | 776 | 4.8 | 30,977 | 3,005 | 285 | 61 | 20 | 44 | 101 | 285 | 411 | 31 | 20 |
| Hardin | 1,035 | 2.2 | 40,331 | 2,010 | 363 | 70 | 25 | 88 | 153 | 344 | 546 | 40 | 25 |
| Hawkins | 1,992 | 2.4 | 35,077 | 2,675 | 592 | 120 | 43 | 102 | 251 | 676 | 856 | 71 | 43 |
| Haywood | 622 | 6.5 | 35,920 | 2,577 | 218 | 49 | 16 | 32 | 87 | 204 | 314 | 23 | 16 |
| Henderson | 1,040 | 4.2 | 36,986 | 2,450 | 342 | 61 | 24 | 99 | 128 | 294 | 526 | 37 | 24 |
| Henry | 1,372 | 4.1 | 42,407 | 1,703 | 479 | 98 | 34 | 213 | 206 | 406 | 823 | 54 | 34 |
| Hickman | 852 | 4.3 | 33,840 | 2,821 | 154 | 35 | 11 | 98 | 91 | 248 | 297 | 23 | 11 |
| Houston | 276 | 4.7 | 33,635 | 2,831 | 56 | 14 | 4 | 23 | 35 | 101 | 97 | 8 | 4 |
| Humphreys | 739 | 4.7 | 39,784 | 2,078 | 314 | 63 | 23 | 58 | 97 | 223 | 458 | 31 | 23 |
| Jackson | 373 | 3.2 | 31,676 | 2,963 | 54 | 14 | 4 | 29 | 45 | 139 | 101 | 10 | 4 |
| Jefferson | 2,043 | 3.7 | 37,494 | 2,389 | 630 | 114 | 45 | 136 | 280 | 614 | 925 | 70 | 45 |
| Johnson | 588 | 3.5 | 33,058 | 2,884 | 193 | 42 | 14 | 28 | 86 | 218 | 277 | 23 | 14 |
| Knox | 24,343 | 3.3 | 51,758 | 702 | 12,853 | 1,918 | 876 | 3,086 | 4,143 | 4,205 | 18,733 | 1,114 | 876 |
| Lake | 181 | 11.0 | 25,772 | 3,102 | 61 | 14 | 4 | 15 | 28 | 77 | 94 | 6 | 4 |
| Lauderdale | 820 | 5.4 | 31,976 | 2,942 | 282 | 61 | 21 | 71 | 105 | 277 | 435 | 30 | 21 |
| Lawrence | 1,588 | 5.2 | 35,983 | 2,573 | 434 | 87 | 31 | 176 | 193 | 480 | 727 | 53 | 31 |
| Lewis | 446 | 6.6 | 36,378 | 2,525 | 109 | 24 | 8 | 59 | 45 | 136 | 199 | 14 | 8 |
| Lincoln | 1,448 | 5.6 | 42,134 | 1,747 | 418 | 82 | 29 | 118 | 207 | 380 | 648 | 47 | 29 |
| Loudon | 2,712 | 4.7 | 50,154 | 822 | 778 | 127 | 56 | 269 | 519 | 664 | 1,230 | 87 | 56 |
| McMinn | 2,023 | 4.4 | 37,601 | 2,374 | 889 | 154 | 63 | 157 | 239 | 613 | 1,264 | 88 | 63 |
| McNairy | 871 | 3.0 | 33,914 | 2,812 | 204 | 50 | 15 | 81 | 102 | 322 | 350 | 29 | 15 |
| Macon | 825 | 3.4 | 33,535 | 2,845 | 194 | 42 | 14 | 80 | 99 | 245 | 330 | 25 | 14 |
| Madison | 4,325 | 3.9 | 44,135 | 1,467 | 2,747 | 489 | 189 | 429 | 646 | 1,089 | 3,854 | 232 | 189 |
| Marion | 1,167 | 4.6 | 40,365 | 2,006 | 317 | 60 | 22 | 94 | 145 | 342 | 494 | 37 | 22 |

1. Based on the resident population estimated as of July 1 of the year shown.

Items 62—74

# Table B. States and Counties — Earnings, Social Security, and Housing

| STATE County | Earnings, 2019 (cont.) Percent by selected industries | | | | | | | | | Social Security beneficiaries, December 2019 | | Supplemental Security Income recipients, 2019 | Housing units, 2020 | |
|---|---|---|---|---|---|---|---|---|---|---|---|---|---|---|
| | Farm | Mining, quarrying, and extractions | Construction | Manufacturing | Information; professional, scientific, technical services | Retail trade | Finance, insurance, real estate, and leasing | Health care and social assistance | Government | Number | Rate[1] | | Total | Percent change, 2010-2020 |
| | 75 | 76 | 77 | 78 | 79 | 80 | 81 | 82 | 83 | 84 | 85 | 86 | 87 | 88 |
| SOUTH DAKOTA—Cont'd | | | | | | | | | | | | | | |
| Tripp | 20.8 | D | 4.4 | 2.0 | 6.4 | 7.7 | D | 14.2 | 13.8 | 1,355 | 250 | 138 | 3,108 | 1.1 |
| Turner | 22.8 | D | 8.1 | 13.8 | D | 3.0 | D | 8.6 | 4.2 | 1,950 | 234 | 77 | 4,101 | 4.1 |
| Union | 2.4 | 0.2 | 3.7 | 12.2 | 7.4 | 3.1 | 14.6 | 21.3 | 4.2 | 3,375 | 211 | 93 | 7,553 | 20.3 |
| Walworth | 4.7 | D | 5.7 | 0.8 | 5.7 | 9.1 | 8.9 | 14.7 | 16.5 | 1,525 | 281 | 106 | 3,084 | 2.7 |
| Yankton | 3.1 | D | 4.8 | 28.2 | D | 5.7 | 6.4 | 16.1 | 13.3 | 5,170 | 226 | 386 | 10,430 | 8.1 |
| Ziebach | 13.7 | 0.0 | D | D | D | D | D | D | 43.9 | 190 | 68 | 91 | 1,005 | 1.8 |
| TENNESSEE | 0.0 | 0.1 | 6.9 | 11.5 | 9.6 | 6.6 | 7.9 | 14.2 | 12.8 | 1,478,145 | 216 | 174,588 | 3,065,735 | 9.0 |
| Anderson | -0.2 | D | 3.8 | 33.4 | 15.1 | 4.5 | 4.2 | 9.1 | 12.4 | 20,000 | 260 | 2,045 | 35,242 | 1.6 |
| Bedford | 0.0 | D | 12.3 | 23.3 | D | 7.5 | 4.3 | D | 13.7 | 10,250 | 206 | 1,057 | 19,831 | 8.0 |
| Benton | -1.3 | D | 5.2 | 17.1 | D | 9.2 | 8.1 | D | 24.5 | 5,220 | 322 | 604 | 9,220 | 2.7 |
| Bledsoe | 0.4 | D | 14.1 | 2.5 | D | 5.0 | 3.2 | 4.2 | 44.9 | 3,490 | 231 | 400 | 5,804 | 1.4 |
| Blount | -0.3 | D | 7.0 | 18.8 | 7.7 | 8.5 | 6.7 | 8.9 | 14.0 | 34,060 | 255 | 2,409 | 59,771 | 8.1 |
| Bradley | 0.0 | D | 7.9 | 22.6 | D | 7.0 | 3.6 | 13.9 | 10.8 | 25,600 | 237 | 2,646 | 45,036 | 8.8 |
| Campbell | -0.4 | 0.7 | 9.5 | 14.8 | D | 8.8 | 5.4 | D | 20.6 | 11,525 | 290 | 2,069 | 21,658 | 8.5 |
| Cannon | -3.4 | 1.2 | 11.1 | 11.1 | 3.6 | 6.5 | 2.2 | D | 23.0 | 3,550 | 243 | 339 | 6,184 | 2.4 |
| Carroll | 0.1 | 0.0 | 4.8 | 9.8 | 2.8 | 6.7 | 10.3 | D | 23.3 | 8,295 | 299 | 1,035 | 13,274 | 0.8 |
| Carter | -0.2 | D | 10.9 | 12.1 | 5.0 | 9.3 | 4.8 | D | 19.9 | 16,495 | 292 | 1,740 | 28,441 | 2.6 |
| Cheatham | -0.8 | D | 15.4 | 28.1 | D | 5.6 | 3.5 | D | 12.9 | 8,705 | 213 | 624 | 17,038 | 8.8 |
| Chester | -2.6 | 0.0 | D | 15.1 | 3.7 | 9.0 | 2.6 | D | 24.9 | 4,240 | 245 | 456 | 7,207 | 3.2 |
| Claiborne | -1.3 | 2.0 | 5.6 | 22.1 | 2.4 | 6.6 | 4.4 | D | 17.3 | 9,765 | 306 | 1,674 | 15,704 | 5.7 |
| Clay | -7.2 | D | 5.8 | 13.7 | D | 5.8 | 1.6 | 7.9 | 32.9 | 2,325 | 306 | 254 | 4,406 | 2.9 |
| Cocke | 0.2 | D | 6.9 | 19.5 | 2.9 | 10.4 | 3.1 | D | 24.1 | 11,525 | 320 | 1,955 | 17,703 | 1.4 |
| Coffee | 0.1 | 0.0 | 6.0 | 20.0 | 20.3 | 7.0 | 3.0 | 10.3 | 13.5 | 13,930 | 246 | 1,433 | 24,788 | 5.8 |
| Crockett | -2.8 | 0.0 | 4.7 | 29.8 | D | 17.7 | 2.2 | D | 15.7 | 3,620 | 255 | 461 | 6,459 | 0.6 |
| Cumberland | 0.3 | 0.7 | 8.5 | 12.1 | 5.4 | 11.0 | 5.0 | 14.0 | 13.2 | 23,715 | 390 | 1,484 | 31,086 | 10.4 |
| Davidson | 0.0 | 0.1 | 6.2 | 2.8 | 15.6 | 5.0 | 8.6 | 20.2 | 7.1 | 100,305 | 145 | 14,293 | 334,555 | 17.8 |
| Decatur | -1.3 | D | 7.6 | 15.5 | D | 6.6 | D | 16.2 | 21.1 | 3,915 | 337 | 466 | 6,941 | 1.1 |
| DeKalb | 2.2 | 0.0 | 4.8 | 34.2 | D | 5.0 | 2.0 | D | 14.9 | 5,235 | 257 | 636 | 9,854 | 4.8 |
| Dickson | -1.1 | D | 14.4 | 19.7 | 4.0 | 8.7 | 4.2 | 14.5 | 13.8 | 11,805 | 219 | 1,159 | 22,722 | 9.2 |
| Dyer | 1.6 | 0.0 | 8.4 | 27.2 | D | 7.6 | 5.1 | D | 18.3 | 9,465 | 255 | 1,470 | 18,124 | 1.8 |
| Fayette | 0.0 | D | 17.1 | 18.2 | 4.9 | 6.1 | 5.0 | 6.5 | 13.1 | 10,890 | 265 | 1,077 | 17,011 | 15.6 |
| Fentress | -1.7 | D | 8.7 | 4.5 | 4.0 | 8.3 | 6.0 | D | 17.4 | 6,105 | 329 | 991 | 9,067 | 1.2 |
| Franklin | 0.7 | 0.9 | 7.2 | 17.2 | D | 8.4 | 3.3 | D | 15.1 | 11,325 | 268 | 946 | 19,877 | 6.3 |
| Gibson | 1.9 | 0.0 | 8.0 | 18.5 | D | 9.8 | 4.5 | D | 18.3 | 13,020 | 265 | 1,613 | 22,814 | 3.7 |
| Giles | -2.0 | D | 5.7 | 33.1 | D | 7.9 | 4.2 | D | 14.0 | 8,160 | 277 | 776 | 14,325 | 3.6 |
| Grainger | -1.4 | D | 13.3 | 25.3 | D | 4.5 | 2.2 | 3.6 | 18.5 | 6,725 | 287 | 989 | 11,269 | 3.4 |
| Greene | -1.2 | D | 4.1 | 26.0 | 3.4 | 7.7 | 3.4 | D | 16.3 | 21,415 | 310 | 2,531 | 32,901 | 2.7 |
| Grundy | 4.6 | D | 9.9 | 8.8 | D | 8.0 | 2.4 | D | 24.4 | 4,310 | 321 | 711 | 6,494 | 1.4 |
| Hamblen | 0.0 | D | 4.6 | 32.6 | 3.2 | 8.5 | 3.2 | 11.4 | 12.8 | 16,590 | 256 | 2,071 | 27,415 | 1.7 |
| Hamilton | 0.0 | 0.1 | 6.6 | 12.0 | 7.2 | 5.0 | 14.6 | 12.3 | 15.7 | 78,010 | 212 | 8,213 | 167,127 | 10.6 |
| Hancock | -7.7 | D | D | D | D | 8.2 | D | 17.3 | 50.5 | 1,660 | 251 | 434 | 3,607 | -0.4 |
| Hardeman | -1.1 | 0.0 | 4.6 | 33.7 | D | 5.0 | D | 8.6 | 23.4 | 6,570 | 262 | 1,251 | 11,067 | 2.0 |
| Hardin | -0.3 | D | 6.1 | 31.5 | 2.8 | 11.1 | 3.9 | D | 19.0 | 8,265 | 323 | 1,053 | 14,156 | 1.6 |
| Hawkins | -1.1 | D | 5.4 | 37.2 | D | 5.3 | 1.9 | D | 18.1 | 17,635 | 310 | 1,836 | 27,647 | 2.9 |
| Haywood | -2.4 | 0.0 | 5.1 | 33.4 | D | 6.4 | 4.6 | 5.8 | 22.3 | 4,680 | 271 | 890 | 8,412 | 1.1 |
| Henderson | -0.5 | D | 7.1 | 16.7 | 2.6 | 7.0 | 12.0 | 8.2 | 16.2 | 6,940 | 247 | 863 | 13,084 | 2.3 |
| Henry | 3.2 | D | 6.2 | 22.8 | 3.7 | 9.1 | 3.0 | 7.5 | 21.9 | 10,005 | 310 | 944 | 17,368 | 1.8 |
| Hickman | -1.7 | D | 16.0 | 14.6 | D | 7.2 | 1.5 | 11.1 | 22.6 | 5,875 | 234 | 667 | 10,699 | 3.7 |
| Houston | -4.9 | 0.0 | D | 12.3 | D | 5.1 | D | 16.5 | 27.7 | 2,130 | 260 | 291 | 4,247 | 1.4 |
| Humphreys | -2.0 | D | 11.5 | 35.6 | 2.0 | 7.7 | 2.5 | D | 16.0 | 5,105 | 275 | 514 | 8,997 | 1.5 |
| Jackson | -3.7 | 0.0 | D | 7.3 | D | 5.9 | D | D | 28.2 | 3,480 | 294 | 387 | 5,942 | 1.7 |
| Jefferson | -0.3 | D | D | 18.0 | D | 7.2 | 2.8 | 8.7 | 15.6 | 15,225 | 279 | 1,397 | 24,729 | 5.1 |
| Johnson | -1.3 | D | 8.1 | 26.4 | D | 4.1 | 3.7 | 16.8 | 19.9 | 5,715 | 322 | 813 | 9,021 | 0.7 |
| Knox | 0.0 | 0.1 | 6.8 | 5.4 | 11.6 | 8.5 | 7.9 | 16.8 | 13.9 | 93,840 | 200 | 10,045 | 211,738 | 8.6 |
| Lake | 8.1 | 0.0 | D | D | D | 5.1 | 1.3 | 9.4 | 40.9 | 1,455 | 208 | 364 | 2,595 | -0.1 |
| Lauderdale | 2.8 | 0.0 | 3.0 | 20.7 | D | 5.4 | 8.0 | D | 26.3 | 6,055 | 236 | 1,237 | 11,355 | 0.9 |
| Lawrence | -0.2 | D | 9.9 | 20.0 | 4.2 | 9.7 | 3.8 | 9.9 | 17.8 | 11,465 | 260 | 1,305 | 18,283 | 0.6 |
| Lewis | -0.2 | 0.0 | 6.1 | 17.1 | 2.9 | 11.7 | 3.1 | D | 20.4 | 3,215 | 262 | 307 | 5,743 | 5.0 |
| Lincoln | -3.5 | D | 7.4 | 31.4 | D | 8.7 | 4.1 | 4.6 | 19.5 | 9,320 | 271 | 874 | 15,828 | 3.9 |
| Loudon | 3.2 | D | 9.5 | 26.6 | 4.9 | 7.2 | 3.3 | 8.1 | 12.4 | 17,110 | 317 | 1,046 | 24,087 | 10.8 |
| McMinn | -0.6 | D | 6.1 | 41.1 | 2.2 | 6.3 | 4.5 | D | 11.4 | 14,360 | 267 | 1,585 | 23,508 | 0.7 |
| McNairy | -0.9 | 0.0 | 7.0 | 18.9 | D | 6.6 | 2.9 | 9.3 | 22.4 | 7,880 | 307 | 1,125 | 12,092 | 1.3 |
| Macon | -0.7 | 0.1 | 8.5 | 13.4 | D | 11.7 | 6.8 | D | 20.0 | 5,690 | 231 | 653 | 10,732 | 8.9 |
| Madison | -0.7 | D | 7.9 | 17.8 | 4.3 | 7.1 | 4.0 | 14.2 | 22.3 | 22,795 | 232 | 3,526 | 43,675 | 4.3 |
| Marion | -0.1 | D | 8.4 | 25.7 | D | 8.5 | 2.5 | 7.0 | 17.6 | 7,995 | 277 | 953 | 13,717 | 5.9 |

1. Per 1,000 resident population estimated as of July 1 of the year shown.

# Table B. States and Counties — Housing, Labor Force, and Employment

| STATE County | Housing units, 2015-2019 | | | | | | | | Civilian labor force, 2020 | | | | Civilian employment[6], 2015-2019 | | |
|---|---|---|---|---|---|---|---|---|---|---|---|---|---|---|---|
| | | Occupied units | | | | | | | | | Unemployment | | | Percent | |
| | | Owner-occupied | | | | Renter-occupied | | | | | | | | | |
| | | | | Median owner cost as a percent of income | | | Median rent as a percent of income[2] | | | | | | | | Construction, production, and maintenance occupations |
| | | | | With a mortgage | Without a mortgage[2] | | | Sub-standard units[4] (percent) | | Percent change, 2019-2020 | | | | Management, business, science, and arts | |
| | Total | Percent | Median value[1] | | | Median rent[3] | | | Total | | Total | Rate[5] | Total | | |
| | 89 | 90 | 91 | 92 | 93 | 94 | 95 | 96 | 97 | 98 | 99 | 100 | 101 | 102 | 103 |
| **SOUTH DAKOTA—Cont'd** | | | | | | | | | | | | | | | |
| Tripp | 2,356 | 76.2 | 101,200 | 18.3 | 11.7 | 579 | 23.9 | 1.4 | 3,013 | -1.6 | 94 | 3.1 | 2,806 | 40.7 | 24.9 |
| Turner | 3,498 | 78.3 | 122,400 | 19.5 | 10.4 | 661 | 26.4 | 1.0 | 4,650 | 0.1 | 183 | 3.9 | 4,261 | 34.2 | 29.2 |
| Union | 6,619 | 69.2 | 179,300 | 16.8 | 10.0 | 854 | 21.5 | 1.4 | 8,522 | | 434 | 5.1 | 7,958 | 38.1 | 28.0 |
| Walworth | 2,258 | 73.9 | 88,200 | 17.8 | 10.0 | 692 | 22.7 | 2.3 | 2,132 | 1.4 | 131 | 6.1 | 2,744 | 44.7 | 19.6 |
| Yankton | 9,445 | 65.2 | 148,300 | 18.9 | 10.9 | 644 | 24.0 | 1.2 | 11,761 | -1.2 | 541 | 4.6 | 11,449 | 33.4 | 27.4 |
| Ziebach | 754 | 57.2 | 57,700 | 14.5 | 12.3 | 576 | 27.9 | 15.5 | 944 | 1.4 | 45 | 4.8 | 958 | 53.2 | 16.2 |
| **TENNESSEE** | 2,597,292 | 66.3 | 167,200 | 19.9 | 10.0 | 869 | 28.9 | 2.2 | 3,289,426 | -1.2 | 245,532 | 7.5 | 3,109,872 | 35.3 | 25.6 |
| Anderson | 30,541 | 67.9 | 146,200 | 19.8 | 10.0 | 750 | 28.3 | 1.9 | 34,360 | | 2,311 | 6.7 | 32,611 | 34.1 | 25.9 |
| Bedford | 17,029 | 68.5 | 151,100 | 19.8 | 10.0 | 767 | 27.3 | 3.4 | 20,956 | -0.6 | 1,663 | 7.9 | 22,010 | 23.3 | 38.8 |
| Benton | 6,762 | 77.1 | 85,600 | 20.7 | 12.2 | 658 | 28.3 | 1.7 | 6,714 | -2.8 | 549 | 8.2 | 5,891 | 25.6 | 36.3 |
| Bledsoe | 4,894 | 73.8 | 134,600 | 19.1 | 11.4 | 656 | 23.1 | 5.2 | 4,264 | -1.4 | 367 | 8.6 | 5,417 | 28.8 | 34.7 |
| Blount | 50,557 | 75.4 | 182,100 | 19.0 | 10.0 | 788 | 27.9 | 1.8 | 63,505 | -0.9 | 4,065 | 6.4 | 60,847 | 31.9 | 26.0 |
| Bradley | 40,192 | 65.5 | 161,000 | 19.4 | 10.0 | 777 | 27.7 | 2.6 | 52,651 | 0.4 | 3,500 | 6.6 | 49,160 | 32.5 | 30.4 |
| Campbell | 16,192 | 68.1 | 104,400 | 19.7 | 12.0 | 614 | 26.3 | 2.0 | 14,717 | | 1,120 | 7.6 | 15,458 | 29.0 | 34.0 |
| Cannon | 5,488 | 74.8 | 164,600 | 19.0 | 10.0 | 635 | 24.9 | 3.8 | 6,506 | -2.2 | 423 | 6.5 | 6,440 | 28.4 | 35.6 |
| Carroll | 10,962 | 73.1 | 88,300 | 17.7 | 10.9 | 622 | 24.8 | 1.9 | 11,851 | -1.4 | 874 | 7.4 | 11,173 | 28.1 | 33.9 |
| Carter | 23,784 | 70.8 | 122,600 | 21.2 | 11.0 | 632 | 27.7 | 2.0 | 23,655 | -1.8 | 1,654 | 7.0 | 23,117 | 32.0 | 27.6 |
| Cheatham | 15,089 | 77.0 | 187,400 | 21.4 | 10.0 | 988 | 26.2 | 2.4 | 21,313 | -2.6 | 1,200 | 5.6 | 20,022 | 33.4 | 29.6 |
| Chester | 6,060 | 77.1 | 121,900 | 17.8 | 10.0 | 682 | 28.7 | 1.4 | 8,390 | -2.4 | 478 | 5.7 | 6,886 | 27.5 | 31.2 |
| Claiborne | 13,281 | 70.9 | 110,000 | 19.9 | 10.0 | 632 | 28.6 | 2.3 | 12,727 | -2.9 | 889 | 7.0 | 12,091 | 30.0 | 35.5 |
| Clay | 3,039 | 78.3 | 98,200 | 24.5 | 12.3 | 479 | 30.3 | 2.1 | 2,721 | -2.3 | 221 | 8.1 | 2,670 | 27.2 | 34.6 |
| Cocke | 14,060 | 67.1 | 112,900 | 22.2 | 10.1 | 646 | 27.0 | 1.8 | 14,713 | -2.1 | 1,508 | 10.2 | 13,907 | 23.1 | 32.3 |
| Coffee | 21,646 | 68.1 | 141,900 | 19.2 | 10.0 | 737 | 27.3 | 3.0 | 25,612 | -0.2 | 1,992 | 7.8 | 24,880 | 26.7 | 35.7 |
| Crockett | 5,491 | 71.6 | 99,600 | 18.6 | 11.5 | 728 | 31.1 | 3.7 | 6,836 | -2.0 | 400 | 5.9 | 6,102 | 32.3 | 33.4 |
| Cumberland | 25,801 | 77.6 | 153,200 | 21.0 | 10.0 | 696 | 28.1 | 2.0 | 22,733 | -3.0 | 1,745 | 7.7 | 21,750 | 32.3 | 34.8 |
| Davidson | 282,366 | 54.3 | 241,700 | 20.9 | 10.0 | 1,102 | 28.9 | 3.0 | 402,058 | -0.2 | 32,271 | 8.0 | 380,627 | 42.6 | 19.2 |
| Decatur | 4,440 | 81.0 | 98,800 | 18.9 | 10.9 | 531 | 29.9 | 1.4 | 4,455 | -2.1 | 383 | 8.6 | 4,409 | 32.7 | 31.9 |
| DeKalb | 7,704 | 68.4 | 148,600 | 19.1 | 10.0 | 644 | 22.4 | 2.9 | 7,651 | -1.8 | 694 | 9.1 | 8,013 | 34.4 | 37.5 |
| Dickson | 19,198 | 75.1 | 164,700 | 20.1 | 10.0 | 782 | 27.9 | 2.4 | 26,354 | -2.9 | 1,555 | 5.9 | 23,532 | 30.9 | 30.5 |
| Dyer | 15,120 | 61.5 | 103,800 | 17.9 | 10.8 | 680 | 27.1 | 2.9 | 15,879 | -3.3 | 1,276 | 8.0 | 16,273 | 29.2 | 33.1 |
| Fayette | 15,596 | 79.9 | 206,700 | 20.2 | 10.0 | 739 | 31.9 | 1.6 | 18,987 | -2.8 | 1,213 | 6.4 | 17,626 | 36.4 | 28.2 |
| Fentress | 7,443 | 75.1 | 110,400 | 20.5 | 10.0 | 526 | 27.0 | 2.0 | 7,503 | -0.1 | 489 | 6.5 | 7,033 | 24.9 | 34.8 |
| Franklin | 16,326 | 74.5 | 139,200 | 20.9 | 10.0 | 678 | 26.1 | 1.8 | 19,782 | -2.2 | 1,569 | 7.9 | 18,283 | 33.9 | 31.6 |
| Gibson | 19,320 | 68.3 | 98,600 | 19.9 | 11.1 | 680 | 29.5 | 1.2 | 21,795 | | 1,521 | 7.0 | 20,127 | 29.2 | 31.1 |
| Giles | 11,904 | 69.6 | 129,700 | 19.2 | 10.0 | 660 | 27.6 | 2.3 | 14,634 | -2.7 | 1,226 | 8.4 | 12,680 | 27.3 | 35.9 |
| Grainger | 8,959 | 79.1 | 119,100 | 19.9 | 10.0 | 641 | 30.1 | 2.4 | 9,480 | -0.7 | 687 | 7.2 | 9,889 | 23.7 | 39.9 |
| Greene | 27,396 | 74.0 | 122,400 | 19.7 | 10.0 | 620 | 27.2 | 1.1 | 28,769 | -3.6 | 2,397 | 8.3 | 28,541 | 29.0 | 32.8 |
| Grundy | 4,820 | 78.0 | 87,200 | 18.7 | 10.8 | 629 | 28.8 | 3.4 | 5,131 | 2.4 | 502 | 9.8 | 5,034 | 27.0 | 42.1 |
| Hamblen | 24,456 | 66.9 | 137,900 | 19.8 | 10.0 | 735 | 29.8 | 3.3 | 28,020 | -0.5 | 1,993 | 7.1 | 26,376 | 28.3 | 34.9 |
| Hamilton | 145,213 | 64.0 | 180,900 | 19.4 | 10.0 | 868 | 29.5 | 1.6 | 182,352 | -1.5 | 13,015 | 7.1 | 173,657 | 38.8 | 21.4 |
| Hancock | 2,742 | 79.6 | 102,600 | 23.0 | 10.9 | 424 | 28.7 | 3.4 | 2,147 | 1.0 | 181 | 8.4 | 2,195 | 28.8 | 32.8 |
| Hardeman | 8,891 | 71.2 | 90,200 | 22.6 | 12.0 | 650 | 27.8 | 1.0 | 9,318 | -1.8 | 766 | 8.2 | 8,055 | 26.1 | 36.8 |
| Hardin | 10,137 | 72.5 | 124,300 | 19.9 | 10.0 | 592 | 29.0 | 2.4 | 10,276 | -0.9 | 728 | 7.1 | 9,750 | 27.0 | 32.4 |
| Hawkins | 23,135 | 73.5 | 124,100 | 18.9 | 10.0 | 638 | 30.2 | 3.4 | 23,468 | -1.8 | 1,805 | 7.7 | 22,268 | 25.5 | 36.2 |
| Haywood | 7,181 | 58.2 | 100,700 | 22.6 | 12.1 | 651 | 28.6 | 1.8 | 7,777 | 0.8 | 693 | 8.9 | 7,328 | 19.2 | 40.3 |
| Henderson | 10,711 | 72.8 | 102,000 | 17.6 | 10.6 | 648 | 31.2 | 2.2 | 12,010 | -1.6 | 959 | 8.0 | 11,272 | 30.6 | 34.8 |
| Henry | 13,394 | 76.2 | 103,300 | 19.8 | 11.0 | 643 | 29.8 | 1.5 | 13,668 | -3.5 | 1,026 | 7.5 | 12,623 | 27.0 | 34.1 |
| Hickman | 8,636 | 80.3 | 116,200 | 21.0 | 11.3 | 708 | 29.8 | 1.8 | 11,103 | -2.8 | 645 | 5.8 | 10,001 | 24.9 | 37.4 |
| Houston | 2,878 | 76.5 | 103,700 | 23.7 | 11.9 | 642 | 26.8 | 0.9 | 3,234 | -1.3 | 264 | 8.2 | 3,095 | 21.7 | 42.2 |
| Humphreys | 6,763 | 77.8 | 113,800 | 19.8 | 10.0 | 670 | 30.7 | 1.4 | 8,670 | -3.2 | 559 | 6.4 | 7,396 | 28.0 | 38.4 |
| Jackson | 4,566 | 77.1 | 126,100 | 23.1 | 13.1 | 548 | 24.8 | 3.2 | 4,740 | 1.5 | 377 | 8.0 | 4,315 | 24.4 | 38.4 |
| Jefferson | 20,154 | 75.2 | 146,300 | 19.7 | 10.5 | 681 | 25.1 | 2.7 | 24,697 | 0.0 | 1,844 | 7.5 | 23,719 | 26.6 | 37.2 |
| Johnson | 6,794 | 75.6 | 133,100 | 21.8 | 10.3 | 545 | 35.5 | 2.5 | 7,682 | -0.6 | 496 | 6.5 | 5,707 | 26.6 | 31.2 |
| Knox | 187,319 | 64.3 | 183,200 | 19.2 | 10.0 | 892 | 29.0 | 1.4 | 242,643 | -1.2 | 14,361 | 5.9 | 232,086 | 25.8 | 34.4 |
| Lake | 2,243 | 54.7 | 81,300 | 19.1 | 10.0 | 519 | 24.6 | 2.3 | 1,691 | -2.2 | 149 | 8.8 | 1,812 | 41.8 | 17.6 |
| Lauderdale | 9,675 | 56.1 | 84,100 | 19.9 | 11.1 | 650 | 26.8 | 3.2 | 9,737 | 0.1 | 891 | 9.2 | 9,172 | 21.5 | 30.7 |
| Lawrence | 15,960 | 75.3 | 112,500 | 21.0 | 11.2 | 661 | 28.8 | 3.2 | 19,092 | -0.5 | 1,489 | 7.8 | 16,807 | 23.9 | 39.1 |
| Lewis | 4,715 | 78.0 | 95,600 | 22.1 | 11.3 | 580 | 30.7 | 1.8 | 5,325 | 2.3 | 401 | 7.5 | 4,920 | 27.6 | 37.2 |
| Lincoln | 13,548 | 75.1 | 135,700 | 19.8 | 10.5 | 685 | 32.5 | 3.2 | 16,026 | -4.6 | 1,371 | 8.6 | 15,088 | 26.8 | 31.0 |
| Loudon | 20,669 | 77.8 | 222,500 | 20.3 | 10.0 | 802 | 27.4 | 3.3 | 23,572 | | 1,500 | 6.4 | 21,396 | 28.5 | 36.2 |
| McMinn | 20,804 | 73.5 | 133,200 | 20.2 | 10.0 | 694 | 31.9 | 2.1 | 23,393 | -0.8 | 1,814 | 7.8 | 21,234 | 28.5 | 31.6 |
| McNairy | 10,022 | 72.7 | 98,100 | 19.9 | 11.5 | 624 | 27.5 | 2.2 | 8,348 | -1.3 | 726 | 8.7 | 9,425 | 27.9 | 35.1 |
| Macon | 9,170 | 74.4 | 115,000 | 19.1 | 11.5 | 673 | 29.2 | 2.1 | 11,156 | -1.9 | 728 | 6.5 | 10,601 | 28.5 | 36.6 |
| Madison | 37,944 | 62.3 | 131,500 | 20.2 | 10.1 | 877 | 30.9 | 1.4 | 49,028 | -0.7 | 3,685 | 7.5 | 43,691 | 29.0 | 39.3 |
| Marion | 11,477 | 74.9 | 125,700 | 19.9 | 10.0 | 696 | 24.3 | 2.1 | 12,344 | -2.0 | 935 | 7.6 | 12,424 | 34.6 | 24.6 |
| | | | | | | | | | | | | | | 27.3 | 37.0 |

1. Specified owner-occupied units lacking complete plumbing facilities.   2. A value of 10.0 represents 10 percent or less; a value of 50.0 represents 50 percent or more.   3. Specified renter-occupied units.   4. Overcrowded or
5. Percent of civilian labor force.   6. Civilian employed persons 16 years old and over.

Items 89—103

## Table B. States and Counties — Nonfarm Employment and Agriculture

| | Private nonfarm establishments, employment and payroll, 2019 | | | | | | | | | Agriculture, 2017 | | | |
| | Employment | | | | | | | Annual payroll | | Farms | | | Farm producers whose primary occupation is farming (percent) |
| STATE County | Number of establish-ments | Total | Health care and social assistance | Manufac-turing | Retail trade | Finance and insurance | Professional, scientific, and technical services | Total (mil dol) | Average per employee (dollars) | Number | Percent with: Fewer than 50 acres | 1000 acres or more | |
| | 104 | 105 | 106 | 107 | 108 | 109 | 110 | 111 | 112 | 113 | 114 | 115 | 116 |
|---|---|---|---|---|---|---|---|---|---|---|---|---|---|
| **SOUTH DAKOTA—Cont'd** | | | | | | | | | | | | | |
| Tripp | 203 | 1,703 | 456 | 34 | 369 | 68 | 67 | 54 | 31,637 | 648 | 13.7 | 42.4 | 58.5 |
| Turner | 243 | 1,419 | 320 | 189 | 205 | 85 | 25 | 55 | 38,956 | 757 | 30.3 | 16.2 | 46.6 |
| Union | 528 | 9,270 | 1,199 | 2,055 | 464 | 1,033 | 268 | 516 | 55,669 | 557 | 29.1 | 18.3 | 50.7 |
| Walworth | 217 | 1,818 | 356 | 59 | 349 | 72 | 56 | 61 | 33,641 | 256 | 11.3 | 39.8 | 52.9 |
| Yankton | 703 | 11,651 | 2,146 | 3,817 | 1,675 | 609 | 213 | 485 | 41,601 | 610 | 23.8 | 17.4 | 51.7 |
| Ziebach | 45 | 309 | 80 | NA | 39 | 27 | NA | 9 | 30,738 | 213 | 1.4 | 74.6 | 70.6 |
| **TENNESSEE** | 139,760 | 2,724,545 | 425,262 | 339,377 | 319,375 | 125,186 | 123,812 | 129,929 | 47,688 | 69,983 | 45.2 | 2.3 | 35.8 |
| Anderson | 1,535 | 41,120 | 4,410 | 12,144 | 3,428 | 1,049 | 8,191 | 2,576 | 62,654 | 538 | 54.3 | 0.2 | 30.5 |
| Bedford | 800 | 14,983 | 1,196 | 4,560 | 1,596 | 571 | 209 | 581 | 38,783 | 1,430 | 41.9 | 2.4 | 40.3 |
| Benton | 288 | 3,287 | 578 | 787 | 628 | 106 | 49 | 99 | 30,061 | 399 | 30.8 | 2.0 | 29.2 |
| Bledsoe | 109 | 820 | 216 | 71 | 112 | 46 | 35 | 28 | 34,478 | 614 | 36.0 | 2.0 | 39.4 |
| Blount | 2,371 | 43,072 | 6,334 | 6,563 | 5,586 | 2,169 | 3,549 | 2,136 | 49,592 | 1,073 | 60.7 | 1.1 | 39.6 |
| Bradley | 1,962 | 41,548 | 5,296 | 8,866 | 6,065 | 1,328 | 849 | 1,712 | 41,198 | 778 | 52.6 | 0.4 | 42.1 |
| Campbell | 564 | 6,982 | 1,678 | 1,177 | 1,477 | 246 | 95 | 215 | 30,776 | 343 | 47.5 | NA | 32.2 |
| Cannon | 211 | 1,767 | 275 | 279 | 275 | 45 | 47 | 59 | 33,456 | 728 | 46.7 | 1.1 | 30.6 |
| Carroll | 419 | 5,467 | 1,493 | 705 | 752 | 263 | 77 | 185 | 33,785 | 662 | 36.3 | 5.6 | 27.8 |
| Carter | 696 | 9,030 | 1,701 | 1,265 | 1,775 | 332 | 189 | 306 | 33,859 | 469 | 58.0 | NA | 33.7 |
| Cheatham | 626 | 7,054 | 786 | 1,834 | 889 | 167 | 212 | 284 | 40,211 | 543 | 48.4 | 1.8 | 34.8 |
| Chester | 238 | 3,109 | 432 | 563 | 368 | 82 | 47 | 89 | 28,522 | 380 | 28.9 | 4.2 | 31.0 |
| Claiborne | 437 | 7,998 | 1,138 | 2,461 | 826 | 265 | 144 | 284 | 35,461 | 966 | 42.1 | 0.6 | 41.5 |
| Clay | 112 | 1,049 | 299 | 216 | 169 | 29 | 12 | 30 | 28,436 | 404 | 32.9 | 2.0 | 42.1 |
| Cocke | 468 | 6,107 | 845 | 1,474 | 1,426 | 158 | 107 | 198 | 32,501 | 645 | 47.6 | 0.5 | 40.7 |
| Coffee | 1,218 | 19,426 | 2,951 | 4,964 | 3,091 | 811 | 1,055 | 848 | 43,648 | 872 | 51.1 | 4.0 | 34.5 |
| Crockett | 225 | 2,071 | 451 | 299 | 311 | 71 | 52 | 69 | 33,240 | 322 | 39.8 | 14.9 | 36.8 |
| Cumberland | 1,094 | 15,831 | 2,854 | 2,626 | 2,506 | 492 | 265 | 545 | 34,452 | 886 | 48.4 | 1.6 | 30.7 |
| Davidson | 20,293 | 464,535 | 78,757 | 20,510 | 36,389 | 26,143 | 27,850 | 27,420 | 59,026 | 414 | 56.8 | 0.2 | 29.7 |
| Decatur | 212 | 2,947 | 1,105 | 537 | 307 | 109 | 47 | 111 | 37,662 | 374 | 24.1 | 1.9 | 33.0 |
| DeKalb | 293 | 4,910 | 540 | 2,514 | 578 | 93 | 69 | 180 | 36,625 | 654 | 43.3 | 1.5 | 39.8 |
| Dickson | 990 | 14,796 | 2,352 | 3,375 | 2,407 | 500 | 240 | 543 | 36,712 | 1,225 | 45.0 | 0.2 | 30.1 |
| Dyer | 786 | 12,975 | 1,888 | 4,077 | 2,027 | 481 | 205 | 490 | 37,730 | 451 | 35.9 | 21.1 | 39.4 |
| Fayette | 630 | 6,990 | 569 | 1,694 | 862 | 210 | 117 | 323 | 46,237 | 892 | 37.6 | 8.1 | 33.4 |
| Fentress | 259 | 4,161 | 816 | 314 | 624 | 175 | 45 | 113 | 27,204 | 620 | 41.9 | 2.3 | 37.6 |
| Franklin | 700 | 12,868 | 1,537 | 3,437 | 1,436 | 257 | 1,271 | 589 | 45,802 | 818 | 51.1 | 1.8 | 34.0 |
| Gibson | 911 | 11,623 | 1,650 | 2,639 | 1,951 | 336 | 198 | 447 | 38,418 | 777 | 49.3 | 10.3 | 37.6 |
| Giles | 530 | 9,139 | 969 | 4,006 | 1,305 | 281 | 143 | 338 | 36,998 | 1,599 | 34.8 | 1.7 | 38.0 |
| Grainger | 226 | 2,556 | 201 | 953 | 454 | 32 | 49 | 90 | 35,235 | 923 | 46.6 | 0.3 | 34.2 |
| Greene | 1,154 | 22,850 | 4,030 | 6,339 | 2,991 | 489 | 272 | 819 | 35,849 | 2,562 | 56.2 | 0.5 | 36.7 |
| Grundy | 159 | 1,284 | 269 | 283 | 260 | 87 | 25 | 38 | 29,685 | 261 | 38.7 | 1.5 | 41.6 |
| Hamblen | 1,291 | 28,567 | 3,752 | 9,883 | 4,080 | 487 | 264 | 1,115 | 39,045 | 559 | 61.2 | 0.7 | 40.1 |
| Hamilton | 9,244 | 195,393 | 32,146 | 24,196 | 23,574 | 13,960 | 8,759 | 8,895 | 45,523 | 547 | 56.7 | 0.5 | 37.4 |
| Hancock | 52 | 403 | 153 | NA | 105 | 12 | 6 | 12 | 30,370 | 408 | 21.6 | 2.9 | 46.0 |
| Hardeman | 337 | 5,687 | 1,169 | 2,042 | 565 | 125 | 32 | 215 | 37,725 | 613 | 32.0 | 6.9 | 30.2 |
| Hardin | 491 | 6,709 | 1,251 | 1,773 | 1,301 | 215 | 77 | 286 | 42,674 | 583 | 32.1 | 5.0 | 34.5 |
| Hawkins | 616 | 10,374 | 1,125 | 4,544 | 1,505 | 188 | 127 | 402 | 38,794 | 1,484 | 47.4 | 0.5 | 37.7 |
| Haywood | 294 | 4,584 | 370 | 2,113 | 584 | 158 | 99 | 189 | 41,255 | 361 | 35.7 | 16.3 | 41.5 |
| Henderson | 498 | 6,484 | 1,063 | 1,420 | 1,042 | 356 | 94 | 219 | 33,792 | 786 | 25.1 | 1.8 | 35.0 |
| Henry | 702 | 8,835 | 1,391 | 2,045 | 1,529 | 336 | 288 | 313 | 35,451 | 710 | 34.5 | 6.2 | 33.6 |
| Hickman | 297 | 2,391 | 628 | 441 | 410 | 52 | 28 | 88 | 36,818 | 706 | 32.2 | 1.8 | 38.8 |
| Houston | 98 | 1,010 | 273 | 178 | 135 | 46 | 35 | 33 | 32,987 | 326 | 29.4 | 1.2 | 38.1 |
| Humphreys | 327 | 4,608 | 617 | 1,653 | 684 | 94 | 70 | 222 | 48,093 | 657 | 35.5 | 2.3 | 31.1 |
| Jackson | 98 | 777 | 138 | 184 | 85 | 31 | 11 | 25 | 32,633 | 538 | 33.3 | 1.3 | 31.1 |
| Jefferson | 725 | 11,633 | 1,372 | 1,890 | 1,841 | 227 | 152 | 436 | 37,518 | 973 | 50.1 | 0.4 | 39.2 |
| Johnson | 235 | 3,348 | 519 | 808 | 370 | 110 | 65 | 141 | 42,162 | 517 | 53.2 | 0.4 | 38.6 |
| Knox | 11,694 | 224,143 | 37,225 | 12,056 | 29,527 | 11,069 | 10,309 | 9,990 | 44,571 | 1,037 | 63.6 | NA | 35.5 |
| Lake | 66 | 624 | 177 | NA | 124 | 17 | NA | 17 | 27,825 | 52 | 15.4 | 48.1 | 68.8 |
| Lauderdale | 300 | 4,740 | 682 | 1,258 | 629 | 169 | 34 | 184 | 38,883 | 404 | 37.4 | 8.7 | 35.5 |
| Lawrence | 738 | 9,137 | 1,463 | 2,113 | 1,554 | 256 | 195 | 305 | 33,378 | 1,394 | 40.0 | 2.3 | 35.8 |
| Lewis | 213 | 2,454 | 504 | 516 | 459 | 71 | 30 | 79 | 31,990 | 272 | 32.7 | 1.1 | 33.7 |
| Lincoln | 574 | 9,230 | 873 | 4,265 | 1,245 | 238 | 133 | 328 | 35,528 | 1,654 | 37.7 | 1.7 | 36.5 |
| Loudon | 945 | 13,672 | 1,554 | 3,164 | 1,964 | 321 | 670 | 593 | 43,396 | 691 | 55.6 | 0.6 | 36.0 |
| McMinn | 912 | 17,218 | 2,152 | 6,761 | 2,326 | 435 | 257 | 699 | 40,568 | 1,054 | 49.1 | 1.4 | 35.4 |
| McNairy | 404 | 4,604 | 780 | 1,133 | 543 | 139 | 67 | 128 | 27,725 | 654 | 34.9 | 3.1 | 33.0 |
| Macon | 308 | 3,659 | 721 | 1,003 | 616 | 299 | 56 | 111 | 30,450 | 912 | 38.3 | 2.1 | 35.5 |
| Madison | 2,532 | 53,679 | 13,192 | 9,326 | 6,992 | 1,275 | 1,175 | 2,149 | 40,031 | 549 | 37.9 | 4.9 | 34.3 |
| Marion | 429 | 6,199 | 668 | 1,780 | 1,242 | 163 | 101 | 214 | 34,546 | 308 | 36.4 | 1.9 | 37.7 |

# Table B. States and Counties — Agriculture

| STATE County | Acreage (1,000) | Percent change, 2012-2017 | Average size of farm | Total irrigated (1,000) | Total cropland (1,000) | Average per farm | Average per acre | Value of machinery and equipment, average per farm (dollars) | Total (mil dol) | Average per farm (acres) | Crops | Livestock and poultry products | Organic farms (number) | Farms with internet access (percent) | Total ($1,000) | Percent of farms |
|---|---|---|---|---|---|---|---|---|---|---|---|---|---|---|---|---|
| | 117 | 118 | 119 | 120 | 121 | 122 | 123 | 124 | 125 | 126 | 127 | 128 | 129 | 130 | 131 | 132 |
| SOUTH DAKOTA—Cont'd | | | | | | | | | | | | | | | | |
| Tripp | 1,037 | 1.7 | 1,600 | 5.1 | 438.1 | 2,591,113 | 1,620 | 271,919 | 202.6 | 312,627 | 24.8 | 75.2 | 2 | 83.5 | 9,317 | 85.0 |
| Turner | 393 | 2.2 | 519 | 26.9 | 350.3 | 2,604,252 | 5,019 | 265,081 | 265.7 | 351,004 | 61.3 | 38.7 | 2 | 78.2 | 9,119 | 72.4 |
| Union | 290 | 0.5 | 521 | 40.7 | 266.7 | 3,059,199 | 5,875 | 315,383 | 187.0 | 335,772 | 71.8 | 28.2 | 1 | 83.3 | 4,214 | 81.7 |
| Walworth | 453 | 1.9 | 1,771 | 2.8 | 273.8 | 3,506,300 | 1,980 | 356,021 | 87.1 | 340,066 | 70.1 | 29.9 | 2 | 87.9 | 6,182 | 76.6 |
| Yankton | 330 | 0.6 | 540 | 38.0 | 252.6 | 2,389,517 | 4,422 | 258,025 | 162.4 | 266,193 | 65.2 | 34.8 | 2 | 80.5 | 7,994 | 76.4 |
| Ziebach | 1,099 | -0.8 | 5,162 | NA | 256.8 | 4,412,411 | 855 | 352,110 | 59.9 | 281,272 | 35.4 | 64.6 | NA | 78.4 | 6,373 | 87.8 |
| TENNESSEE | 10,874 | 0.1 | 155 | 184.9 | 5,286.3 | 608,739 | 3,918 | 80,447 | 3,798.9 | 54,284 | 57.4 | 42.6 | 122 | 72.7 | 115,945 | 26.5 |
| Anderson | 43 | 21.2 | 81 | 0.1 | 14.7 | 441,499 | 5,465 | 59,863 | 4.3 | 8,015 | 31.4 | 68.6 | NA | 74.0 | 205 | 13.4 |
| Bedford | 238 | 2.4 | 166 | 1.2 | 92.6 | 693,089 | 4,167 | 86,121 | 151.6 | 106,013 | 14.3 | 85.7 | NA | 76.9 | 2,460 | 28.3 |
| Benton | 69 | -22.0 | 172 | D | 30.3 | 427,822 | 2,489 | 82,014 | 9.5 | 23,797 | 63.2 | 36.8 | NA | 74.4 | 625 | 30.3 |
| Bledsoe | 94 | -7.9 | 153 | 0.5 | 34.7 | 528,198 | 3,445 | 85,821 | 39.9 | 64,951 | 26.7 | 73.3 | NA | 77.7 | 628 | 32.4 |
| Blount | 95 | -6.1 | 88 | 0.1 | 40.6 | 567,289 | 6,436 | 60,767 | 16.5 | 15,347 | 34.7 | 65.3 | 10 | 77.4 | 488 | 11.6 |
| Bradley | 85 | -2.0 | 109 | 0.1 | 28.3 | 661,318 | 6,065 | 66,853 | 105.9 | 136,067 | 12.6 | 87.4 | 3 | 71.6 | 872 | 16.7 |
| Campbell | 28 | -17.7 | 80 | 0.1 | 10.2 | 332,638 | 4,138 | 55,393 | 2.9 | 8,437 | 43.5 | 56.5 | 1 | 57.1 | 114 | 19.0 |
| Cannon | 89 | -7.4 | 122 | 0.0 | 38.9 | 438,958 | 3,587 | 60,409 | 22.6 | 31,021 | 60.7 | 39.3 | NA | 76.6 | 1,013 | 20.2 |
| Carroll | 170 | -4.7 | 256 | 4.9 | 106.2 | 700,403 | 2,735 | 101,405 | 52.6 | 79,492 | 94.7 | 5.3 | NA | 69.3 | 3,582 | 43.8 |
| Carter | 34 | -14.9 | 73 | 0.1 | 12.0 | 379,965 | 5,199 | 57,459 | 7.7 | 16,377 | 33.4 | 66.6 | 3 | 71.9 | D | 2.3 |
| Cheatham | 67 | 28.4 | 124 | 0.1 | 31.8 | 564,719 | 4,557 | 75,846 | 16.3 | 30,072 | 84.6 | 15.4 | 5 | 76.1 | 313 | 19.2 |
| Chester | 80 | 31.3 | 210 | D | 42.4 | 534,584 | 2,540 | 79,023 | 16.1 | 42,347 | 89.3 | 10.7 | NA | 71.1 | 1,198 | 37.9 |
| Claiborne | 120 | -1.3 | 124 | 0.0 | 35.4 | 388,110 | 3,128 | 55,041 | 17.4 | 17,971 | 17.9 | 82.1 | 1 | 58.7 | 1,282 | 42.3 |
| Clay | 75 | -5.6 | 186 | 0.0 | 24.8 | 620,853 | 3,333 | 63,718 | 60.0 | 148,634 | 9.2 | 90.8 | NA | 69.1 | 438 | 42.1 |
| Cocke | 65 | 7.0 | 101 | D | 23.4 | 448,496 | 4,427 | 69,868 | 36.5 | 56,563 | 32.6 | 67.4 | 1 | 64.2 | 324 | 19.8 |
| Coffee | 139 | -4.3 | 159 | 1.8 | 79.9 | 652,277 | 4,105 | 92,203 | 60.6 | 69,450 | 66.5 | 33.5 | NA | 79.2 | 2,149 | 26.4 |
| Crockett | 149 | 14.3 | 463 | 10.4 | 134.1 | 1,559,863 | 3,365 | 278,689 | 81.6 | 253,388 | 97.0 | 3.0 | NA | 64.3 | 2,486 | 61.2 |
| Cumberland | 129 | 0.1 | 146 | 0.0 | 46.3 | 589,663 | 4,043 | 73,047 | 46.7 | 52,746 | 47.0 | 53.0 | NA | 73.5 | 790 | 19.2 |
| Davidson | 34 | -1.1 | 83 | 0.3 | 13.0 | 747,014 | 8,978 | 62,082 | 11.1 | 26,829 | 78.4 | 21.6 | 4 | 78.5 | 160 | 10.9 |
| Decatur | 74 | -4.4 | 197 | D | 24.8 | 436,208 | 2,209 | 78,589 | 7.4 | 19,826 | 56.5 | 43.5 | NA | 67.6 | 583 | 23.0 |
| DeKalb | 88 | -1.5 | 135 | 0.5 | 29.7 | 456,211 | 3,384 | 63,777 | 26.0 | 39,719 | 78.4 | 21.6 | NA | 72.0 | 633 | 32.9 |
| Dickson | 140 | -5.6 | 114 | 0.2 | 46.9 | 480,574 | 4,201 | 64,723 | 18.6 | 15,208 | 56.6 | 43.4 | 2 | 75.7 | 412 | 15.8 |
| Dyer | 284 | 34.1 | 629 | 25.5 | 257.6 | 2,117,443 | 3,364 | 273,302 | 134.2 | 297,483 | 96.8 | 3.2 | NA | 80.3 | 2,700 | 59.6 |
| Fayette | 306 | 33.8 | 344 | 13.3 | 185.7 | 1,158,789 | 3,373 | 120,662 | 89.9 | 100,784 | 87.0 | 13.0 | NA | 71.0 | 3,742 | 39.3 |
| Fentress | 94 | 3.5 | 152 | 0.0 | 25.7 | 526,303 | 3,472 | 72,186 | 37.6 | 60,584 | 12.9 | 87.1 | NA | 74.4 | 549 | 33.4 |
| Franklin | 112 | -10.4 | 137 | 1.5 | 61.5 | 563,293 | 4,101 | 83,705 | 72.5 | 88,641 | 44.5 | 55.5 | 1 | 72.1 | 1,795 | 30.6 |
| Gibson | 287 | 0.4 | 370 | 9.9 | 248.9 | 1,309,546 | 3,541 | 174,608 | 137.1 | 176,459 | 94.6 | 5.4 | NA | 73.7 | 6,511 | 51.4 |
| Giles | 251 | -7.1 | 157 | 2.1 | 78.8 | 514,156 | 3,275 | 69,966 | 48.9 | 30,551 | 38.3 | 61.7 | NA | 74.2 | 4,230 | 37.3 |
| Grainger | 87 | 2.2 | 94 | 0.3 | 25.8 | 358,497 | 3,798 | 60,980 | 19.5 | 21,085 | 55.5 | 44.5 | 6 | 64.1 | 1,666 | 31.5 |
| Greene | 222 | -1.9 | 86 | D | 97.2 | 388,623 | 4,493 | 64,414 | 61.5 | 24,014 | 30.8 | 69.2 | 1 | 67.1 | 1,112 | 15.8 |
| Grundy | 34 | 1.8 | 129 | 0.1 | 13.3 | 446,403 | 3,459 | 70,611 | 29.6 | 113,586 | 37.2 | 62.8 | NA | 68.6 | 258 | 19.2 |
| Hamblen | 50 | -15.0 | 89 | 0.0 | 24.4 | 492,840 | 5,507 | 67,616 | 15.8 | 28,188 | 24.4 | 75.6 | 1 | 68.0 | 387 | 25.8 |
| Hamilton | 44 | -16.5 | 80 | 0.1 | 14.0 | 786,833 | 9,850 | 59,900 | 20.6 | 37,649 | 12.7 | 87.3 | 3 | 77.7 | 201 | 12.2 |
| Hancock | 73 | 12.9 | 178 | D | 16.5 | 474,077 | 2,662 | 57,539 | 6.7 | 16,341 | 18.3 | 81.7 | NA | 62.5 | 646 | 40.7 |
| Hardeman | 173 | 12.4 | 282 | 2.3 | 83.9 | 645,321 | 2,289 | 75,727 | 31.5 | 51,408 | 90.7 | 9.3 | NA | 63.5 | 2,859 | 39.2 |
| Hardin | 162 | 28.5 | 278 | 2.2 | 76.2 | 620,402 | 2,232 | 101,400 | 29.7 | 50,871 | 81.5 | 18.5 | NA | 64.5 | 2,154 | 43.6 |
| Hawkins | 141 | 6.0 | 95 | 0.0 | 47.8 | 355,038 | 3,727 | 58,427 | 18.8 | 12,648 | 28.5 | 71.5 | NA | 62.9 | 1,252 | 24.7 |
| Haywood | 201 | -8.5 | 556 | 21.6 | 178.6 | 1,845,293 | 3,319 | 211,930 | 97.8 | 270,925 | 99.2 | 0.8 | 2 | 71.2 | 3,391 | 63.7 |
| Henderson | 153 | -5.5 | 195 | 0.0 | 63.7 | 492,153 | 2,522 | 89,292 | 29.2 | 37,104 | 60.2 | 39.8 | NA | 72.8 | 1,940 | 35.4 |
| Henry | 204 | -0.3 | 287 | 4.1 | 127.5 | 954,925 | 3,324 | 139,412 | 94.1 | 132,530 | 62.2 | 37.8 | 1 | 81.5 | 2,783 | 42.4 |
| Hickman | 123 | 2.1 | 175 | 0.0 | 43.9 | 551,572 | 3,154 | 72,096 | 17.1 | 24,183 | 50.6 | 49.4 | 2 | 78.2 | 1,053 | 22.2 |
| Houston | 50 | 0.1 | 154 | 0.0 | 11.7 | 499,292 | 3,233 | 71,786 | D | D | D | D | NA | 81.9 | 76 | 10.4 |
| Humphreys | 115 | -6.5 | 176 | 1.5 | 33.4 | 489,160 | 2,783 | 82,009 | 11.2 | 17,078 | 46.2 | 53.8 | NA | 74.0 | 261 | 6.5 |
| Jackson | 81 | 9.6 | 150 | 0.0 | 18.6 | 466,297 | 3,105 | 58,481 | 4.9 | 9,130 | 36.1 | 63.9 | 8 | 78.4 | 270 | 31.6 |
| Jefferson | 89 | -6.9 | 91 | 0.3 | 37.2 | 470,695 | 5,149 | 61,911 | 24.1 | 24,780 | 21.8 | 78.2 | NA | 67.7 | 1,260 | 25.7 |
| Johnson | 47 | -1.6 | 90 | 0.0 | 16.4 | 358,093 | 3,973 | 62,699 | 7.7 | 14,797 | 29.9 | 70.1 | 2 | 68.9 | 112 | 15.5 |
| Knox | 67 | 3.1 | 65 | 0.2 | 26.7 | 639,694 | 9,845 | 53,723 | 18.7 | 17,987 | 61.6 | 38.4 | 8 | 79.7 | 618 | 15.5 |
| Lake | 88 | 10.7 | 1,698 | 15.0 | 82.0 | 5,959,437 | 3,511 | 793,852 | 45.8 | 880,788 | 99.8 | 0.2 | NA | 84.6 | 2,175 | 76.9 |
| Lauderdale | 156 | -22.5 | 385 | 6.1 | 134.3 | 1,304,233 | 3,385 | 186,207 | 69.6 | 172,260 | 96.0 | 4.0 | NA | 63.4 | 2,608 | 55.4 |
| Lawrence | 230 | -2.3 | 165 | D | 105.7 | 544,725 | 3,296 | 77,410 | 71.1 | 51,024 | 55.6 | 44.4 | NA | 66.9 | 2,770 | 37.2 |
| Lewis | 41 | 32.7 | 150 | 0.0 | 10.4 | 428,585 | 2,859 | 53,189 | 3.0 | 10,996 | 35.2 | 64.8 | 2 | 72.8 | 161 | 11.4 |
| Lincoln | 271 | 2.0 | 164 | 3.4 | 102.6 | 600,890 | 3,663 | 90,043 | 124.9 | 75,524 | 30.9 | 69.1 | NA | 73.1 | 1,905 | 19.4 |
| Loudon | 59 | -15.2 | 85 | 0.1 | 28.4 | 484,550 | 5,690 | 74,092 | D | D | D | D | 1 | 73.7 | 284 | 20.3 |
| McMinn | 129 | 5.6 | 123 | 0.1 | 49.9 | 571,882 | 4,666 | 65,080 | 50.1 | 47,555 | 16.2 | 83.8 | 1 | 69.2 | 1,162 | 24.1 |
| McNairy | 139 | 7.0 | 213 | 0.3 | 65.5 | 468,203 | 2,203 | 70,549 | 22.4 | 34,229 | 86.0 | 14.0 | 2 | 69.7 | 1,212 | 40.7 |
| Macon | 132 | 8.0 | 144 | D | 51.9 | 543,237 | 3,763 | 72,506 | 61.5 | 67,451 | 44.8 | 55.2 | 2 | 74.2 | 1,078 | 32.8 |
| Madison | 151 | -8.9 | 275 | 6.8 | 107.6 | 898,835 | 3,263 | 136,092 | 49.5 | 90,117 | 93.7 | 6.3 | NA | 70.3 | 3,371 | 52.6 |
| Marion | 55 | 8.5 | 179 | 0.0 | 23.1 | 581,773 | 3,254 | 86,733 | 17.1 | 55,406 | 33.9 | 66.1 | NA | 72.7 | 500 | 11.0 |

Items 117—132

# Table B. States and Counties — Water Use, Wholesale Trade, Retail Trade, and Real Estate

| STATE County | Water use, 2015 | | Wholesale Trade[1], 2017 | | | | Retail Trade[2], 2017 | | | | Real estate and rental and leasing,[2] 2017 | | | |
|---|---|---|---|---|---|---|---|---|---|---|---|---|---|---|
| | Public supply water withdrawn (mil gal/day) | Public supply gallons withdrawn per person per day | Number of establishments | Number of employees | Sales (mil dol) | Average payroll (mil dol) | Number of establishments | Number of employees | Sales (mil dol) | Average payroll (mil dol) | Number of establishments | Number of employees | Sales (mil dol) | Average payroll (mil dol) |
| | 133 | 134 | 135 | 136 | 137 | 138 | 139 | 140 | 141 | 142 | 143 | 144 | 145 | 146 |
| SOUTH DAKOTA—Cont'd | | | | | | | | | | | | | | |
| Tripp | 0.49 | 90.2 | 12 | 136 | 82.0 | 8.0 | 44 | 381 | 103.4 | 9.1 | NA | NA | NA | NA |
| Turner | 0.47 | 57.3 | D | D | D | D | 32 | 220 | 44.3 | 4.3 | 26 | 55 | 12.0 | 1.9 |
| Union | 1.59 | 106.6 | D | D | D | D | 46 | 432 | 123.6 | 10.5 | 7 | D | 1.2 | D |
| Walworth | 0.42 | 77.2 | 11 | 83 | 234.9 | 4.1 | 47 | 349 | 100.4 | 7.6 | 28 | 85 | 11.5 | 2.4 |
| Yankton | 1.11 | 48.9 | D | D | D | D | 124 | 1,665 | 419.0 | 40.8 | NA | NA | NA | NA |
| Ziebach | 0.04 | 14.3 | 4 | D | 25.0 | D | 8 | 88 | 20.5 | 1.5 | NA | NA | NA | NA |
| TENNESSEE | 849.69 | 128.7 | 5,864 | 100,044 | 111,030.0 | 5,824.7 | 22,593 | 322,218 | 101,978.3 | 8,574.3 | 6,048 | 36,212 | 10,679.1 | 1,759.7 |
| Anderson | 10.69 | 141.1 | 47 | 528 | 1,101.2 | 23.1 | 236 | 3,336 | 979.5 | 83.5 | 52 | 158 | 36.5 | 6.4 |
| Bedford | 6.03 | 127.8 | D | D | D | D | 144 | 1,498 | 423.7 | 41.3 | 33 | 94 | 19.9 | 2.0 |
| Benton | 1.25 | 77.5 | D | D | D | 2.0 | 57 | 671 | 170.4 | 15.6 | 5 | D | 0.6 | D |
| Bledsoe | 2.19 | 151.0 | NA | NA | NA | NA | 21 | 143 | 35.8 | 2.7 | 3 | 6 | 0.7 | 0.1 |
| Blount | 13.94 | 109.5 | 93 | 1,605 | 1,825.6 | 104.1 | 373 | 5,748 | 1,649.1 | 161.7 | 104 | 461 | 100.4 | 13.5 |
| Bradley | 13.74 | 132.0 | 63 | 694 | 807.3 | 48.6 | 347 | 5,857 | 2,087.4 | 159.9 | D | D | D | D |
| Campbell | 3.33 | 83.8 | 13 | 52 | 34.2 | 2.4 | 132 | 1,543 | 429.4 | 37.1 | 17 | 111 | 13.9 | 2.4 |
| Cannon | 5.28 | 381.5 | D | D | D | 2.5 | 37 | 252 | 68.9 | 5.7 | NA | NA | NA | NA |
| Carroll | 2.48 | 88.9 | 14 | 221 | 81.0 | 8.0 | 88 | 801 | 197.1 | 17.3 | 14 | 46 | 4.2 | 1.4 |
| Carter | 7.07 | 125.2 | D | D | D | D | 140 | 1,780 | 442.3 | 41.0 | 23 | 56 | 10.3 | 1.7 |
| Cheatham | 2.93 | 73.7 | 17 | 107 | 62.2 | 5.2 | 78 | 883 | 241.7 | 21.9 | 25 | 72 | 11.7 | 2.2 |
| Chester | 0.93 | 53.2 | 9 | 46 | 26.1 | 1.9 | 54 | 454 | 130.4 | 10.7 | 8 | 17 | 1.6 | 0.3 |
| Claiborne | 2.98 | 94.0 | 14 | 62 | 19.0 | 1.7 | 89 | 903 | 201.7 | 20.3 | 17 | 37 | 7.1 | 1.1 |
| Clay | 0.86 | 110.7 | D | D | D | 1.8 | 30 | 172 | 42.3 | 3.4 | 5 | D | 1.5 | D |
| Cocke | 4.72 | 134.2 | D | D | D | D | 112 | 1,372 | 396.1 | 32.7 | 17 | 70 | 4.5 | 1.2 |
| Coffee | 5.76 | 106.1 | 36 | 498 | 362.7 | 26.5 | 256 | 2,892 | 844.7 | 73.9 | 41 | 222 | 33.6 | 6.1 |
| Crockett | 1.56 | 106.8 | 21 | 238 | 231.6 | 10.3 | 34 | 219 | 67.2 | 4.7 | 7 | 17 | 6.4 | 0.7 |
| Cumberland | 5.71 | 98.1 | D | D | D | D | 232 | 2,465 | 774.4 | 62.6 | 34 | 118 | 25.0 | 3.3 |
| Davidson | 134.11 | 197.5 | 968 | 19,905 | 18,494.0 | 1,313.8 | 2,603 | 38,670 | 11,329.8 | 1,115.1 | 1,083 | 8,442 | 2,803.8 | 474.5 |
| Decatur | 1.37 | 117.5 | D | D | D | 3.5 | 47 | 349 | 90.7 | 8.0 | NA | NA | NA | NA |
| DeKalb | 2.17 | 113.1 | 4 | 40 | 7.0 | 1.4 | 56 | 564 | 141.4 | 13.5 | 9 | 21 | 1.8 | 0.7 |
| Dickson | 5.24 | 101.8 | 29 | 594 | 1,116.8 | 28.5 | 180 | 2,244 | 722.8 | 57.0 | 25 | 107 | 18.9 | 2.3 |
| Dyer | 2.71 | 71.5 | 42 | 543 | 858.6 | 23.0 | 170 | 1,943 | 573.6 | 47.9 | 23 | 82 | 13.3 | 3.2 |
| Fayette | 1.47 | 37.5 | 34 | 438 | 528.6 | 24.3 | 81 | 926 | 341.0 | 26.3 | 11 | 29 | 7.5 | 1.3 |
| Fentress | 1.65 | 92.1 | D | D | D | 0.5 | 55 | 674 | 163.3 | 16.3 | D | D | D | D |
| Franklin | 4.27 | 103.0 | D | D | D | D | 133 | 1,354 | 399.3 | 35.3 | 17 | 44 | 7.0 | 1.6 |
| Gibson | 3.80 | 76.9 | 36 | 464 | 463.9 | 26.1 | 182 | 1,959 | 502.6 | 46.9 | 25 | 90 | 37.0 | 4.3 |
| Giles | 3.00 | 103.6 | 19 | 249 | 328.7 | 13.8 | 123 | 1,435 | 359.5 | 30.9 | 12 | 41 | 4.3 | 1.2 |
| Grainger | 0.00 | 0.0 | NA | NA | NA | NA | 53 | 482 | 114.1 | 9.6 | 9 | 9 | 0.9 | 0.1 |
| Greene | 8.92 | 130.1 | D | D | D | D | 208 | 2,823 | 744.8 | 67.2 | 36 | 139 | 25.1 | 4.5 |
| Grundy | 1.63 | 121.3 | NA | NA | NA | NA | 40 | 244 | 68.8 | 6.1 | NA | NA | NA | NA |
| Hamblen | 9.18 | 144.8 | 50 | 851 | 785.5 | 39.0 | 273 | 4,148 | 1,175.5 | 108.1 | 51 | 160 | 36.7 | 4.9 |
| Hamilton | 60.03 | 169.5 | 441 | 6,354 | 4,075.0 | 380.6 | 1,367 | 23,791 | 7,171.8 | 652.9 | 401 | 2,072 | 632.3 | 120.4 |
| Hancock | 0.21 | 32.0 | NA | NA | NA | NA | 15 | 105 | 28.2 | 2.4 | 3 | 4 | 1.0 | 0.1 |
| Hardeman | 2.20 | 85.6 | 11 | 73 | 49.0 | 2.9 | 70 | 614 | 138.1 | 13.2 | 4 | 26 | 1.5 | 0.5 |
| Hardin | 2.63 | 102.1 | D | D | D | 4.5 | 111 | 1,368 | 477.4 | 37.2 | 23 | 46 | 8.1 | 1.7 |
| Hawkins | 3.42 | 60.6 | D | D | D | D | 113 | 1,428 | 353.0 | 31.4 | 24 | 108 | 17.2 | 2.8 |
| Haywood | 1.49 | 82.7 | 10 | 109 | 92.7 | 5.4 | 59 | 607 | 212.7 | 15.0 | 10 | 27 | 3.2 | 0.7 |
| Henderson | 0.31 | 11.1 | 17 | 169 | 61.2 | 6.0 | 108 | 1,109 | 301.7 | 25.5 | 10 | 20 | 3.2 | 0.6 |
| Henry | 2.11 | 65.6 | D | D | D | D | 135 | 1,632 | 620.1 | 49.5 | 24 | 99 | 24.7 | 2.5 |
| Hickman | 2.38 | 97.7 | 8 | 89 | 92.1 | 3.5 | 58 | 334 | 122.8 | 9.3 | NA | NA | NA | NA |
| Houston | 1.00 | 122.7 | NA | NA | NA | NA | 25 | 168 | 39.6 | 3.2 | NA | NA | NA | NA |
| Humphreys | 2.22 | 122.4 | 18 | 98 | 53.9 | 5.3 | 66 | 604 | 206.2 | 14.1 | 6 | 8 | 0.9 | 0.3 |
| Jackson | 0.56 | 48.7 | D | D | D | D | 18 | 105 | 28.3 | 2.0 | NA | NA | NA | NA |
| Jefferson | 5.28 | 99.2 | 25 | 200 | 109.9 | 9.1 | 127 | 1,774 | 580.9 | 46.0 | 27 | 50 | 9.3 | 1.4 |
| Johnson | 1.70 | 95.3 | NA | NA | NA | NA | 48 | 417 | 96.3 | 8.3 | 12 | 32 | 3.6 | 0.5 |
| Knox | 64.33 | 142.5 | 639 | 10,093 | 7,554.4 | 551.4 | 1,720 | 29,506 | 8,384.1 | 833.3 | 571 | 3,607 | 870.1 | 162.9 |
| Lake | 1.11 | 146.5 | 5 | 57 | 36.2 | 3.3 | 18 | 132 | 25.0 | 2.7 | NA | NA | NA | NA |
| Lauderdale | 2.53 | 93.9 | 24 | 676 | 1,334.8 | 31.0 | 71 | 653 | 134.5 | 13.5 | 14 | 94 | 39.4 | 3.7 |
| Lawrence | 4.06 | 95.4 | D | D | D | D | 178 | 1,649 | 411.3 | 40.1 | 21 | 52 | 9.5 | 1.4 |
| Lewis | 1.31 | 110.5 | 5 | 10 | 5.5 | 0.4 | 44 | 422 | 212.7 | 11.5 | 6 | 41 | 6.3 | 1.8 |
| Lincoln | 4.33 | 128.3 | 25 | 234 | 388.8 | 15.1 | 127 | 1,401 | 347.7 | 33.4 | 20 | 203 | 10.3 | 3.0 |
| Loudon | 13.14 | 257.0 | 38 | 594 | 669.4 | 26.4 | 143 | 1,954 | 572.6 | 47.8 | 48 | 165 | 30.0 | 5.7 |
| McMinn | 3.81 | 72.4 | 25 | 287 | 247.9 | 11.9 | 178 | 2,170 | 603.4 | 53.7 | 37 | 95 | 15.4 | 2.3 |
| McNairy | 2.54 | 97.4 | 15 | 139 | 64.1 | 5.2 | 82 | 577 | 143.4 | 13.6 | 8 | 27 | 4.5 | 0.8 |
| Macon | 2.09 | 90.2 | D | D | D | 1.5 | 62 | 669 | 174.1 | 17.1 | 10 | 28 | 3.1 | 0.3 |
| Madison | 15.74 | 161.3 | 148 | 2,383 | 1,331.4 | 103.4 | 483 | 7,147 | 2,071.4 | 181.0 | 117 | 578 | 118.2 | 21.7 |
| Marion | 3.42 | 120.1 | 17 | 214 | 100.5 | 13.1 | 98 | 1,244 | 428.5 | 29.7 | 11 | 28 | 6.3 | 1.0 |

1  Merchant wholesalers, except manufacturers' sales branches and offices.    2. Employer establishments.

# Professional Services, Manufacturing, and Accommodation and Food Services

| STATE County | Professional, scientific, and technical services, 2017 | | | | Manufacturing, 2017 | | | | Accommodation and food services, 2017 | | | |
|---|---|---|---|---|---|---|---|---|---|---|---|---|
| | Number of establish-ments | Number of employees | Sales (mil dol) | Average payroll (mil dol) | Number of establish-ments | Number of employees | Sales (mil dol) | Average payroll (mil dol) | Number of establis-hments | Number of employees | Sales (mil dol) | Annual payroll (mil dol) |
| | 147 | 148 | 149 | 150 | 151 | 152 | 153 | 154 | 155 | 156 | 157 | 158 |
| **SOUTH DAKOTA—Cont'd** | | | | | | | | | | | | |
| Tripp | D | D | D | D | D | | | | 23 | 161 | 10.0 | 2.5 |
| Turner | 14 | 39 | 5 | 1.6 | D | 37 | D | 1.7 | 18 | 99 | 3.9 | 0.9 |
| Union | 41 | 318 | 55 | 16.8 | D | 167 | D | 8.6 | 34 | 346 | 21.4 | 4.8 |
| Walworth | 15 | 70 | 10 | 2.9 | D | D | D | D | 25 | 343 | 23.3 | 6.9 |
| Yankton | 44 | 221 | 30 | 9.6 | 32 | 3,048 | 875.0 | 147.9 | 76 | 1,018 | 50.8 | 14.9 |
| Ziebach | NA | NA | NA | NA | NA | NA | NA | NA | 4 | 52 | 3.2 | 0.8 |
| **TENNESSEE** | 11,357 | 110,731 | 17,047 | 7,102.3 | 5,811 | 321,195 | 155,422.4 | 17,021.9 | 13,518 | 287,534 | 17,181.8 | 4,839.4 |
| Anderson | D | D | D | D | 89 | 10,905 | 2,580.0 | 750.3 | 141 | 2,646 | 134.9 | 38.4 |
| Bedford | 47 | 198 | 23 | 6.3 | 52 | 4,501 | 1,488.3 | 199.8 | 66 | 980 | 51.9 | 13.7 |
| Benton | 15 | 44 | 4 | 1.5 | 21 | 688 | 92.9 | 21.3 | 29 | 337 | 17.4 | 4.3 |
| Bledsoe | 7 | 35 | 2 | 0.7 | 8 | 36 | 9.9 | D | 8 | 107 | 5.7 | 1.7 |
| Blount | D | D | D | D | 98 | 7,036 | 3,858.1 | 409.5 | 206 | 5,038 | 303.0 | 89.4 |
| Bradley | 135 | 733 | 87 | 29.8 | 108 | 7,425 | 4,214.8 | 366.4 | 198 | 4,126 | 210.6 | 58.5 |
| Campbell | D | D | D | 4.2 | 34 | 1,257 | 334.8 | 48.4 | 63 | 814 | 46.6 | 11.7 |
| Cannon | 16 | 39 | 4 | 1.2 | 15 | 303 | 43.9 | 11.7 | 14 | 205 | 9.7 | 2.5 |
| Carroll | 25 | 80 | 8 | 2.6 | 21 | 509 | 412.2 | 35.2 | 35 | 505 | 25.1 | 6.1 |
| Carter | 44 | 153 | 15 | 4.2 | 30 | 1,179 | 248.7 | 50.4 | 74 | 1,304 | 57.9 | 17.3 |
| Cheatham | 36 | 198 | 18 | 8.3 | 45 | 1,565 | 759.9 | 76.7 | 51 | 720 | 35.4 | 9.8 |
| Chester | 12 | 56 | 4 | 1.4 | 18 | 534 | 127.3 | 23.6 | D | D | D | D |
| Claiborne | 28 | 158 | 12 | 4.4 | 27 | 2,376 | 505.2 | 84.2 | 36 | 691 | 29.2 | 7.7 |
| Clay | D | D | 0 | D | 10 | 245 | 48.2 | 9.9 | 12 | D | 4.2 | D |
| Cocke | 27 | 83 | 8 | 2.9 | 32 | 1,572 | 592.6 | 76.8 | 68 | 1,056 | 58.4 | 15.0 |
| Coffee | 75 | 812 | 101 | 53.1 | D | 4,408 | D | 234.8 | 122 | 2,252 | 121.3 | 32.2 |
| Crockett | 12 | 46 | 9 | 1.9 | 13 | 200 | 152.6 | 13.4 | D | D | D | D |
| Cumberland | 70 | 225 | 31 | 9.1 | 50 | 2,515 | 768.4 | 116.8 | 99 | 1,707 | 123.8 | 33.8 |
| Davidson | 2,214 | 30,295 | 5,676 | 2,382.9 | 573 | 19,803 | 8,610.8 | 1,015.1 | 2,115 | 52,934 | 4,123.2 | 1,121.8 |
| Decatur | 13 | 45 | 3 | 1.0 | 18 | 517 | 147.1 | 18.9 | 24 | 180 | 8.3 | 2.1 |
| DeKalb | 25 | 81 | 8 | 2.3 | 14 | 2,110 | 680.0 | 92.1 | 30 | 337 | 19.1 | 5.3 |
| Dickson | 59 | 243 | 26 | 9.9 | 43 | 3,579 | 986.5 | 157.0 | 105 | 1,732 | 91.2 | 26.0 |
| Dyer | D | D | D | D | 38 | 3,858 | 1,554.2 | 192.3 | 64 | 1,176 | 59.4 | 15.3 |
| Fayette | 33 | 112 | 13 | 3.8 | 51 | 1,604 | 745.3 | 81.2 | 35 | 605 | 26.6 | 7.0 |
| Fentress | 17 | 41 | 4 | 0.7 | 21 | 255 | 49.1 | 12.2 | 20 | 277 | 11.5 | 3.3 |
| Franklin | 46 | 588 | 62 | 40.8 | 40 | D | 4,939.7 | D | D | D | D | D |
| Gibson | 41 | 220 | 17 | 5.8 | 66 | 2,572 | 940.1 | 151.9 | 75 | 1,070 | 52.9 | 13.8 |
| Giles | 31 | 153 | 15 | 6.5 | 48 | 4,125 | 1,270.6 | 178.9 | 46 | 611 | 33.4 | 8.4 |
| Grainger | D | D | D | D | 27 | 934 | 200.4 | 39.5 | 19 | 152 | 10.2 | 2.1 |
| Greene | D | D | D | D | 88 | 5,822 | 2,134.3 | 284.7 | 115 | 1,991 | 95.3 | 25.8 |
| Grundy | 6 | 22 | 2 | 0.5 | 14 | 213 | 43.2 | 8.3 | 11 | D | 4.0 | D |
| Hamblen | D | D | D | D | 83 | 10,236 | 3,827.1 | 455.0 | 114 | 2,352 | 110.2 | 32.0 |
| Hamilton | D | D | D | D | 409 | 22,895 | 11,828.3 | 1,270.8 | 929 | 19,999 | 1,194.4 | 333.9 |
| Hancock | 4 | 4 | 0 | 0.1 | NA | NA | NA | NA | D | D | D | D |
| Hardeman | 12 | 35 | 3 | 0.7 | 24 | 1,918 | 613.5 | 88.5 | 22 | 243 | 12.8 | 3.1 |
| Hardin | 28 | 77 | 8 | 2.8 | 29 | 1,906 | 950.4 | 122.2 | 49 | 741 | 35.2 | 8.8 |
| Hawkins | 33 | 104 | 11 | 3.8 | 42 | 4,180 | 1,307.2 | 214.9 | 64 | 1,009 | 42.7 | 11.9 |
| Haywood | 16 | 60 | 5 | 2.0 | 15 | 1,877 | 667.0 | 80.5 | 30 | 393 | 21.2 | 5.1 |
| Henderson | 28 | 116 | 11 | 3.9 | 27 | 1,307 | 435.9 | 59.5 | 35 | 758 | 29.8 | 7.9 |
| Henry | 37 | 306 | 25 | 11.8 | 46 | 1,764 | 386.3 | 64.6 | 63 | 947 | 42.1 | 11.4 |
| Hickman | 10 | 40 | 3 | 1.2 | 27 | 525 | 117.9 | 22.3 | D | D | D | 2.7 |
| Houston | 5 | 36 | 4 | 1.0 | 5 | 177 | 29.9 | 8.2 | D | D | D | D |
| Humphreys | 16 | 87 | 7 | 2.5 | 23 | 1,794 | 1,501.5 | 136.5 | 37 | 384 | 21.1 | 6.0 |
| Jackson | 6 | 52 | 1 | 0.7 | 9 | 190 | 40.9 | 5.6 | 14 | 88 | 4.0 | 1.4 |
| Jefferson | 40 | 130 | 15 | 4.3 | 38 | 1,729 | 977.1 | 84.9 | 80 | 1,317 | 64.8 | 18.8 |
| Johnson | 15 | 48 | 8 | 1.6 | 17 | 922 | 213.6 | 49.2 | 28 | 297 | 16.0 | 4.2 |
| Knox | D | D | D | D | 363 | 10,845 | 5,379.3 | 581.5 | 1,052 | 24,754 | 1,377.2 | 416.4 |
| Lake | NA | NA | NA | NA | NA | NA | NA | NA | 10 | 152 | 7.2 | 2.3 |
| Lauderdale | 13 | 34 | 5 | 1.3 | 12 | 1,401 | 296.3 | 50.9 | D | D | D | D |
| Lawrence | 34 | 248 | 27 | 7.9 | 53 | 1,769 | 485.2 | 77.2 | 62 | 1,039 | 50.3 | 12.3 |
| Lewis | 10 | 27 | 3 | 0.7 | 19 | 441 | 81.3 | 17.3 | 21 | D | 11.0 | D |
| Lincoln | 35 | 141 | 15 | 5.2 | 34 | 2,213 | 1,271.0 | 138.0 | D | D | D | D |
| Loudon | 54 | 595 | 60 | 39.4 | 48 | 3,004 | 1,321.9 | 170.5 | 91 | 1,591 | 80.7 | 24.1 |
| McMinn | 48 | 258 | 28 | 6.7 | 72 | 6,401 | 2,698.1 | 334.4 | 91 | 1,740 | 77.7 | 21.2 |
| McNairy | 18 | 71 | 7 | 2.4 | 40 | 1,117 | 385.6 | 54.0 | D | D | D | D |
| Macon | 21 | 61 | 6 | 2.1 | 31 | 945 | 216.7 | 35.8 | 24 | 265 | 14.1 | 3.4 |
| Madison | D | D | D | D | 93 | 9,691 | 4,880.6 | 483.8 | 229 | 5,306 | 276.1 | 77.9 |
| Marion | 27 | 85 | 14 | 4.9 | 28 | 1,947 | 829.7 | 91.8 | 52 | 1,006 | 54.8 | 14.3 |

# Table B. States and Counties — Health Care and Social Assistance, Other Services, Nonemployer Businesses, and Residential Construction

| STATE County | Health care and social assistance, 2017 | | | | Other services, 2017 | | | | Nonemployer businesses, 2018 | | Value of residential construction authorized by building permits, 2020 | |
|---|---|---|---|---|---|---|---|---|---|---|---|---|
| | Number of establishments | Number of employees | Receipts (mil dol) | Annual payroll (mil dol) | Number of establishments | Number of employees | Receipts (mil dol) | Annual payroll (mil dol) | Number | Receipts (mil dol) | New construction ($1,000) | Number of housing units |
| | 159 | 160 | 161 | 162 | 163 | 164 | 165 | 166 | 167 | 168 | 169 | 170 |
| SOUTH DAKOTA—Cont'd | | | | | | | | | | | | |
| Tripp | 25 | 445 | 40.0 | 13.4 | 19 | 74 | 7.1 | 1.5 | 594 | 25.9 | 200 | 1 |
| Turner | 19 | 294 | 17.1 | 8.2 | D | D | 4.6 | D | 772 | 35.4 | 4,412 | 17 |
| Union | 66 | 1,135 | 202.2 | 64.8 | D | D | D | D | 1,451 | 102.1 | 49,597 | 178 |
| Walworth | 15 | 378 | 31.0 | 13.1 | 16 | 77 | 7.8 | 2.1 | 483 | 22.1 | 2,868 | 12 |
| Yankton | 72 | 1,847 | 235.4 | 94.0 | 60 | 238 | 22.4 | 5.2 | 1,661 | 67.8 | 19,318 | 113 |
| Ziebach | 9 | D | 7.3 | D | NA | NA | NA | NA | 87 | 3.7 | NA | NA |
| TENNESSEE | 15,891 | 414,598 | 52,088.8 | 20,219.5 | 8,570 | 61,744 | 8,650.0 | 2,149.1 | 539,987 | 26,244.1 | 9,207,252 | 49,719 |
| Anderson | 210 | 4,597 | 532.3 | 195.9 | 105 | 470 | 42.8 | 13.0 | 4,810 | 211.1 | 59,121 | 436 |
| Bedford | 101 | 1,186 | 116.5 | 42.4 | 45 | 195 | 16.8 | 5.2 | 3,137 | 158.9 | 70,875 | 335 |
| Benton | 35 | 706 | 55.4 | 20.9 | 16 | 83 | 7.6 | 1.7 | 1,031 | 31.8 | 327 | 18 |
| Bledsoe | 14 | 216 | 19.4 | 9.5 | 3 | 16 | 1.2 | 0.3 | 777 | 31.9 | 0 | 0 |
| Blount | 238 | 6,088 | 713.7 | 279.3 | 160 | 964 | 104.9 | 33.1 | 9,267 | 433.3 | 196,760 | 907 |
| Bradley | 241 | 5,172 | 607.7 | 212.3 | 112 | 926 | 88.9 | 24.3 | 7,507 | 411.0 | 97,427 | 500 |
| Campbell | 74 | 2,048 | 200.4 | 73.2 | D | D | D | D | 2,338 | 99.5 | 34,805 | 337 |
| Cannon | 28 | 367 | 34.8 | 12.7 | D | D | D | 1.5 | 1,106 | 45.8 | 195 | 3 |
| Carroll | 65 | 1,534 | 109.1 | 48.0 | D | D | D | D | 1,597 | 63.6 | 1,854 | 8 |
| Carter | 82 | 1,722 | 166.4 | 70.9 | D | D | D | D | 3,224 | 108.7 | 21,589 | 157 |
| Cheatham | 51 | 613 | 55.4 | 22.1 | 35 | 102 | 11.5 | 3.0 | 3,628 | 189.9 | 50,133 | 257 |
| Chester | 29 | 448 | 31.4 | 14.9 | D | D | 4.9 | D | 1,098 | 49.9 | 2,590 | 20 |
| Claiborne | 50 | 999 | 78.2 | 34.9 | 27 | 109 | 15.9 | 3.8 | 1,817 | 78.8 | 11,003 | 75 |
| Clay | 10 | 354 | 27.1 | 11.2 | 4 | 10 | 1.4 | 0.2 | 636 | 24.1 | 3,375 | 22 |
| Cocke | 37 | 1,031 | 112.3 | 43.4 | D | D | D | D | 2,159 | 74.7 | 146 | 3 |
| Coffee | 192 | 3,071 | 337.4 | 119.6 | D | D | D | D | 3,795 | 159.2 | 35,959 | 214 |
| Crockett | 23 | 421 | 29.3 | 11.8 | D | D | D | D | 904 | 34.5 | 7,147 | 38 |
| Cumberland | 134 | 2,793 | 259.6 | 95.2 | 64 | 573 | 45.8 | 16.2 | 4,499 | 198.4 | 94,005 | 443 |
| Davidson | 2,020 | 76,927 | 12,251.2 | 4,700.2 | 1,375 | 13,835 | 1,731.1 | 495.2 | 78,503 | 4,435.8 | 1,741,817 | 13,324 |
| Decatur | 29 | 991 | 71.8 | 41.8 | D | D | 7.8 | D | 821 | 36.5 | 0 | 0 |
| DeKalb | 39 | 591 | 57.0 | 20.9 | D | D | D | D | 1,430 | 64.9 | 18,805 | 121 |
| Dickson | 129 | 2,338 | 325.6 | 104.1 | 55 | 311 | 31.5 | 8.9 | 4,023 | 206.2 | 89,238 | 463 |
| Dyer | 106 | 1,993 | 201.6 | 74.1 | 46 | 186 | 19.6 | 5.5 | 2,454 | 91.8 | 9,361 | 66 |
| Fayette | 47 | 512 | 36.2 | 17.1 | 35 | 184 | 16.4 | 4.1 | 3,674 | 173.0 | 149,311 | 510 |
| Fentress | 33 | 864 | 66.7 | 30.1 | D | D | D | D | 1,496 | 85.9 | 0 | 0 |
| Franklin | 95 | 1,494 | 167.0 | 63.1 | 46 | 222 | 33.5 | 6.5 | 2,708 | 102.7 | 59,045 | 302 |
| Gibson | 100 | 1,576 | 120.2 | 48.3 | 57 | 249 | 24.1 | 6.7 | 2,913 | 123.6 | 35,998 | 165 |
| Giles | 63 | 947 | 99.4 | 36.4 | 30 | 101 | 10.9 | 2.6 | 1,821 | 92.4 | 14,576 | 123 |
| Grainger | 10 | 198 | 13.1 | 5.3 | 17 | 105 | 12.6 | 3.6 | 1,477 | 71.6 | 11,610 | 66 |
| Greene | 139 | 4,207 | 354.5 | 147.5 | 60 | 403 | 33.3 | 10.2 | 4,114 | 169.6 | 40,045 | 210 |
| Grundy | 16 | 332 | 19.3 | 8.8 | 4 | 29 | 1.9 | 0.6 | 1,164 | 46.5 | 1,212 | 4 |
| Hamblen | 173 | 4,263 | 418.4 | 160.1 | 80 | 468 | 36.2 | 11.9 | 3,617 | 189.5 | 44,976 | 270 |
| Hamilton | 1,132 | 30,166 | 3,795.7 | 1,509.9 | 602 | 4,259 | 597.8 | 148.2 | 28,380 | 1,499.5 | 425,615 | 2,364 |
| Hancock | 10 | 156 | 15.4 | 5.1 | NA | NA | NA | NA | 427 | 15.7 | 43 | 1 |
| Hardeman | 42 | 1,310 | 86.8 | 41.1 | 16 | 49 | 5.7 | 1.3 | 1,507 | 56.0 | 5,934 | 32 |
| Hardin | 63 | 1,154 | 115.4 | 42.9 | 24 | 157 | 17.1 | 5.7 | 1,750 | 76.9 | 843 | 8 |
| Hawkins | 58 | 1,239 | 139.2 | 48.2 | 38 | 154 | 17.2 | 4.7 | 2,994 | 103.2 | 17,554 | 149 |
| Haywood | 30 | 430 | 32.8 | 15.2 | D | D | D | D | 1,238 | 48.6 | 3,238 | 18 |
| Henderson | 61 | 1,057 | 79.7 | 32.8 | 35 | 132 | 14.8 | 3.8 | 1,839 | 77.9 | 1,287 | 10 |
| Henry | 78 | 1,382 | 153.5 | 56.1 | 46 | 208 | 16.4 | 4.7 | 2,266 | 100.0 | 1,484 | 18 |
| Hickman | D | D | D | 24.2 | D | D | 7.7 | D | 1,746 | 77.3 | 14,323 | 84 |
| Houston | 12 | 346 | 25.1 | 10.8 | D | D | 1.8 | D | 526 | 20.8 | 0 | 0 |
| Humphreys | 30 | 674 | 51.9 | 18.1 | 17 | 102 | 10.4 | 3.0 | 1,166 | 50.6 | 1,131 | 14 |
| Jackson | 9 | 116 | 6.8 | 2.6 | D | D | D | D | 776 | 29.2 | 0 | 0 |
| Jefferson | 72 | 1,253 | 130.5 | 46.8 | 46 | 212 | 20.3 | 6.5 | 3,249 | 153.0 | 61,615 | 212 |
| Johnson | 22 | 580 | 39.0 | 18.3 | 9 | 28 | 3.5 | 0.8 | 1,050 | 42.0 | 200 | 1 |
| Knox | 1,400 | 38,751 | 4,850.5 | 1,832.6 | 732 | 5,058 | 639.0 | 163.5 | 37,452 | 2,110.5 | 590,113 | 2,978 |
| Lake | 12 | 194 | 15.6 | 7.0 | NA | NA | NA | NA | 310 | 8.1 | 0 | 0 |
| Lauderdale | 33 | 650 | 77.9 | 21.8 | D | D | D | D | 1,238 | 55.9 | 6,298 | 32 |
| Lawrence | 82 | 1,315 | 124.1 | 48.1 | 33 | 107 | 9.9 | 3.2 | 2,994 | 138.9 | 2,120 | 9 |
| Lewis | 29 | 507 | 28.5 | 13.5 | 9 | 69 | 8.2 | 2.3 | 911 | 39.8 | 2,020 | 18 |
| Lincoln | 60 | 855 | 66.1 | 31.4 | 34 | 136 | 14.2 | 3.4 | 2,449 | 105.2 | 325 | 1 |
| Loudon | 110 | 1,560 | 151.2 | 57.6 | 62 | 261 | 35.4 | 8.0 | 3,591 | 177.7 | 129,137 | 499 |
| McMinn | 104 | 2,161 | 195.2 | 74.1 | 49 | 314 | 21.2 | 7.8 | 3,085 | 137.9 | 2,956 | 27 |
| McNairy | 46 | 734 | 56.0 | 22.5 | D | D | 10.8 | D | 1,602 | 76.0 | 377 | 4 |
| Macon | 37 | 631 | 45.2 | 17.1 | D | D | D | D | 1,712 | 84.9 | 38,831 | 177 |
| Madison | 335 | 13,624 | 1,482.7 | 608.6 | D | D | D | D | 6,702 | 304.3 | 58,596 | 252 |
| Marion | 43 | 664 | 57.8 | 22.6 | 29 | 163 | 24.9 | 5.8 | 1,746 | 73.9 | 35,769 | 165 |

# Table B. States and Counties — Government Employment and Payroll, and Local Government Finances

| STATE County | Full-time equivalent employees | March payroll (dollars) | Administration, judicial, and legal | Police and corrections | Fire protection | Highways and transportation | Health and welfare | Natural resources and utilities | Education and libraries | Total (mil dol) | Inter-governmental (mil dol) | Taxes Total (mil dol) | Taxes Per capita[1] Total (dollars) | Taxes Per capita[1] Property (dollars) |
|---|---|---|---|---|---|---|---|---|---|---|---|---|---|---|
| | 171 | 172 | 173 | 174 | 175 | 176 | 177 | 178 | 179 | 180 | 181 | 182 | 183 | 184 |
| **SOUTH DAKOTA—Cont'd** | | | | | | | | | | | | | | |
| Tripp | 337 | 1,063,765 | 7.8 | 14.4 | 0.0 | 5.8 | 5.5 | 9.1 | 54.0 | 24.8 | 6.4 | 12.5 | 2,288 | 1,811 |
| Turner | 309 | 1,219,605 | 10.1 | 1.5 | 0.0 | 6.4 | 0.3 | 4.3 | 76.7 | 38.9 | 8.2 | 27.7 | 3,333 | 3,044 |
| Union | 582 | 2,082,029 | 7.6 | 7.8 | 0.4 | 4.5 | 0.2 | 6.2 | 70.5 | 69.4 | 11.9 | 49.7 | 3,251 | 2,626 |
| Walworth | 225 | 798,106 | 6.6 | 13.6 | 0.2 | 5.6 | 0.2 | 4.0 | 69.4 | 24.5 | 7.6 | 12.9 | 2,342 | 1,840 |
| Yankton | 757 | 2,886,149 | 6.9 | 9.9 | 0.4 | 5.8 | 2.8 | 7.8 | 61.2 | 79.9 | 22.2 | 46.4 | 2,047 | 1,536 |
| Ziebach | 105 | 361,245 | 8.0 | 2.0 | 0.0 | 3.2 | 0.6 | 0.3 | 84.4 | 9.2 | 7.1 | 1.8 | 668 | 621 |
| **TENNESSEE** | X | X | X | X | X | X | X | X | X | X | X | X | X | X |
| Anderson | 3,008 | 11,222,851 | 7.6 | 9.5 | 3.6 | 1.5 | 3.6 | 10.9 | 62.3 | 244.9 | 97.4 | 97.1 | 1,276 | 830 |
| Bedford | 1,718 | 5,739,076 | 5.0 | 10.7 | 3.4 | 4.1 | 4.2 | 10.5 | 61.3 | 125.9 | 68.2 | 35.3 | 733 | 491 |
| Benton | 554 | 2,074,406 | 8.3 | 12.3 | 4.5 | 4.3 | 1.3 | 15.7 | 53.5 | 38.2 | 21.6 | 12.3 | 770 | 447 |
| Bledsoe | 606 | 1,567,248 | 5.9 | 7.9 | 0.0 | 2.9 | 3.7 | 5.0 | 60.7 | 34.8 | 28.2 | 3.9 | 263 | 246 |
| Blount | 5,994 | 22,571,498 | 4.7 | 7.3 | 2.0 | 2.5 | 36.4 | 6.0 | 40.0 | 584.7 | 149.1 | 119.7 | 920 | 825 |
| Bradley | 3,040 | 10,809,028 | 4.4 | 12.1 | 7.3 | 3.4 | 11.0 | 1.7 | 59.0 | 268.8 | 135.3 | 88.1 | 836 | 588 |
| Campbell | 1,680 | 4,210,220 | 6.3 | 9.0 | 2.6 | 4.6 | 9.0 | 20.3 | 48.2 | 108.7 | 54.7 | 31.9 | 802 | 407 |
| Cannon | 569 | 1,512,678 | 4.0 | 8.6 | 0.0 | 2.3 | 4.5 | 3.8 | 72.1 | 27.0 | 19.2 | 5.1 | 363 | 309 |
| Carroll | 1,097 | 3,230,184 | 5.2 | 10.1 | 1.2 | 4.3 | 1.6 | 12.1 | 60.8 | 77.2 | 43.8 | 20.6 | 740 | 489 |
| Carter | 2,193 | 6,104,153 | 5.5 | 9.0 | 1.8 | 4.0 | 1.6 | 8.8 | 67.9 | 121.5 | 65.9 | 35.6 | 630 | 350 |
| Cheatham | 1,143 | 3,615,388 | 6.8 | 9.8 | 1.3 | 4.3 | 3.9 | 5.3 | 66.6 | 94.6 | 45.4 | 34.8 | 861 | 583 |
| Chester | 575 | 1,584,780 | 7.8 | 13.0 | 1.3 | 4.6 | 0.0 | 7.8 | 64.3 | 38.5 | 25.2 | 7.2 | 422 | 317 |
| Claiborne | 1,723 | 5,315,708 | 4.8 | 5.0 | 0.0 | 1.7 | 37.7 | 5.4 | 44.3 | 60.5 | 36.0 | 20.7 | 655 | 434 |
| Clay | 340 | 942,838 | 7.1 | 9.9 | 0.0 | 5.2 | 5.2 | 9.5 | 60.2 | 20.7 | 14.3 | 3.7 | 477 | 333 |
| Cocke | 1,245 | 3,868,897 | 5.8 | 7.6 | 3.8 | 3.2 | 2.5 | 3.8 | 72.5 | 82.6 | 44.4 | 22.8 | 642 | 380 |
| Coffee | 2,388 | 7,175,942 | 5.6 | 8.1 | 3.2 | 2.8 | 2.8 | 13.9 | 59.7 | 154.0 | 69.1 | 57.0 | 1,037 | 765 |
| Crockett | 661 | 1,875,223 | 9.8 | 7.2 | 0.1 | 3.2 | 4.4 | 6.1 | 67.7 | 41.5 | 27.6 | 8.5 | 590 | 402 |
| Cumberland | 1,992 | 5,523,188 | 11.4 | 9.7 | 1.6 | 8.1 | 6.1 | 4.4 | 57.4 | 130.2 | 58.4 | 52.6 | 892 | 391 |
| Davidson | 22,221 | 105,597,426 | 6.1 | 14.0 | 6.3 | 2.7 | 7.4 | 14.7 | 46.7 | 3,001.9 | 764.0 | 1,611.1 | 2,345 | 1,423 |
| Decatur | 559 | 1,619,700 | 7.2 | 9.2 | 0.0 | 3.3 | 23.1 | 7.2 | 47.2 | 37.3 | 15.3 | 8.2 | 697 | 361 |
| DeKalb | 726 | 2,211,919 | 3.9 | 5.3 | 0.0 | 2.5 | 2.4 | 28.2 | 57.0 | 45.3 | 27.8 | 11.1 | 558 | 365 |
| Dickson | 1,840 | 6,148,534 | 7.8 | 8.3 | 3.3 | 3.2 | 3.6 | 19.6 | 53.6 | 144.1 | 61.8 | 56.9 | 1,078 | 627 |
| Dyer | 1,602 | 5,134,617 | 6.4 | 11.9 | 5.4 | 4.3 | 1.2 | 9.7 | 59.3 | 141.9 | 70.0 | 39.8 | 1,067 | 732 |
| Fayette | 1,026 | 2,893,596 | 9.2 | 16.0 | 2.1 | 4.6 | 4.6 | 7.5 | 55.0 | 60.2 | 31.0 | 20.8 | 518 | 306 |
| Fentress | 676 | 1,679,404 | 10.9 | 8.7 | 0.1 | 7.9 | 4.8 | 8.6 | 58.4 | 33.5 | 19.0 | 9.1 | 500 | 364 |
| Franklin | 1,291 | 3,957,206 | 3.3 | 9.3 | 1.7 | 3.8 | 1.0 | 9.2 | 70.1 | 101.9 | 46.1 | 35.4 | 851 | 591 |
| Gibson | 1,931 | 6,018,686 | 4.0 | 12.8 | 2.6 | 6.2 | 1.6 | 8.6 | 63.0 | 146.6 | 75.3 | 38.7 | 785 | 671 |
| Giles | 1,158 | 3,227,522 | 14.1 | 9.8 | 0.7 | 11.4 | 4.5 | 5.8 | 51.9 | 53.1 | 32.5 | 14.9 | 507 | 381 |
| Grainger | 679 | 1,998,640 | 4.1 | 17.0 | 0.0 | 0.6 | 4.4 | 4.2 | 68.8 | 45.1 | 30.6 | 10.7 | 461 | 369 |
| Greene | 2,263 | 7,462,462 | 5.0 | 11.3 | 2.4 | 4.3 | 4.3 | 13.8 | 58.4 | 156.2 | 75.8 | 52.5 | 763 | 460 |
| Grundy | 522 | 1,277,412 | 7.2 | 6.1 | 0.0 | 4.4 | 1.5 | 5.4 | 72.6 | 32.5 | 23.8 | 6.1 | 455 | 352 |
| Hamblen | 2,075 | 7,707,192 | 6.3 | 8.6 | 4.2 | 2.9 | 1.5 | 15.7 | 57.7 | 166.6 | 77.1 | 46.7 | 729 | 596 |
| Hamilton | 16,117 | 72,322,243 | 5.3 | 6.4 | 2.7 | 3.3 | 44.1 | 9.5 | 26.9 | 2,259.7 | 576.1 | 576.9 | 1,598 | 1,217 |
| Hancock | 420 | 1,149,417 | 9.8 | 6.4 | 0.1 | 5.9 | 16.1 | 2.7 | 58.9 | 19.8 | 13.0 | 2.3 | 344 | 263 |
| Hardeman | 1,046 | 3,010,162 | 5.8 | 9.6 | 1.2 | 4.3 | 4.3 | 12.9 | 61.3 | 60.8 | 36.9 | 14.4 | 563 | 430 |
| Hardin | 1,213 | 3,456,931 | 6.1 | 8.9 | 0.9 | 3.5 | 34.4 | 4.3 | 40.4 | 147.9 | 35.4 | 18.1 | 703 | 536 |
| Hawkins | 2,002 | 5,783,824 | 5.4 | 5.8 | 0.4 | 2.7 | 10.8 | 10.0 | 63.8 | 115.8 | 60.5 | 41.5 | 733 | 508 |
| Haywood | 950 | 2,822,431 | 8.8 | 9.9 | 3.2 | 2.7 | 3.2 | 9.7 | 56.0 | 62.4 | 32.9 | 18.5 | 1,050 | 784 |
| Henderson | 1,130 | 3,457,451 | 7.1 | 11.0 | 3.6 | 2.9 | 0.7 | 11.5 | 61.1 | 75.5 | 40.9 | 23.3 | 836 | 379 |
| Henry | 1,075 | 3,623,862 | 6.4 | 10.2 | 2.5 | 1.8 | 0.8 | 14.1 | 62.3 | 158.9 | 46.7 | 23.7 | 732 | 497 |
| Hickman | 1,109 | 2,540,582 | 12.3 | 4.4 | 0.0 | 3.6 | 3.4 | 5.3 | 65.7 | 57.6 | 35.0 | 14.5 | 584 | 411 |
| Houston | 387 | 1,084,439 | 8.3 | 10.7 | 0.5 | 5.6 | 8.2 | 5.8 | 58.1 | 21.9 | 14.7 | 4.3 | 529 | 336 |
| Humphreys | 608 | 1,876,441 | 12.0 | 11.5 | 0.0 | 5.8 | 0.9 | 10.0 | 58.8 | 43.3 | 26.0 | 11.2 | 607 | 496 |
| Jackson | 460 | 1,343,479 | 12.3 | 7.9 | 0.0 | 7.9 | 3.6 | 6.9 | 61.2 | 28.7 | 16.2 | 10.1 | 865 | 762 |
| Jefferson | 1,539 | 4,997,676 | 6.6 | 9.3 | 1.7 | 7.2 | 5.3 | 10.3 | 58.2 | 120.8 | 55.9 | 35.5 | 661 | 397 |
| Johnson | 565 | 1,541,974 | 6.9 | 8.1 | 0.0 | 3.6 | 0.9 | 6.0 | 73.3 | 30.1 | 20.6 | 6.3 | 360 | 253 |
| Knox | 13,331 | 51,598,862 | 5.9 | 12.2 | 3.0 | 3.1 | 3.6 | 16.1 | 54.4 | 1,506.5 | 407.3 | 764.3 | 1,656 | 993 |
| Lake | 320 | 1,046,883 | 15.2 | 14.2 | 0.3 | 4.2 | 2.8 | 9.2 | 54.1 | 17.4 | 11.5 | 3.4 | 451 | 277 |
| Lauderdale | 1,066 | 3,106,646 | 6.6 | 12.0 | 1.9 | 5.2 | 4.2 | 3.4 | 65.5 | 74.9 | 43.5 | 19.1 | 734 | 530 |
| Lawrence | 1,609 | 5,405,378 | 7.1 | 8.6 | 1.8 | 3.9 | 3.4 | 12.6 | 61.7 | 109.1 | 55.1 | 37.1 | 855 | 440 |
| Lewis | 546 | 1,227,740 | 11.9 | 13.0 | 0.6 | 4.8 | 2.6 | 6.3 | 60.7 | 30.7 | 19.5 | 6.3 | 525 | 308 |
| Lincoln | 1,562 | 4,636,209 | 8.8 | 8.2 | 1.9 | 6.1 | 1.5 | 11.4 | 59.9 | 148.7 | 43.6 | 23.3 | 688 | 362 |
| Loudon | 1,512 | 5,432,357 | 5.2 | 9.1 | 2.6 | 2.9 | 1.7 | 19.6 | 57.1 | 126.7 | 51.3 | 48.6 | 930 | 696 |
| McMinn | 2,143 | 8,077,858 | 5.2 | 5.8 | 1.3 | 2.5 | 27.3 | 8.1 | 48.1 | 103.2 | 60.1 | 24.9 | 471 | 378 |
| McNairy | 1,130 | 3,062,012 | 5.4 | 5.7 | 0.6 | 3.7 | 0.3 | 7.5 | 75.0 | 64.4 | 40.7 | 13.9 | 535 | 383 |
| Macon | 738 | 2,075,037 | 7.9 | 10.9 | 1.1 | 0.3 | 7.2 | 5.9 | 65.7 | 54.3 | 34.2 | 11.8 | 495 | 295 |
| Madison | 8,999 | 37,163,340 | 2.0 | 5.6 | 2.3 | 1.4 | 65.9 | 6.0 | 16.5 | 1,048.2 | 100.1 | 131.6 | 1,350 | 821 |
| Marion | 858 | 2,930,021 | 6.6 | 10.6 | 0.6 | 3.3 | 1.6 | 5.4 | 71.5 | 61.4 | 34.8 | 18.2 | 640 | 395 |

1. Based on the resident population estimated as of July 1 of the year shown.

| STATE County | Total (mil dol) [185] | Per capita[1] (dollars) [186] | Education [187] | Health and hospitals [188] | Police protection [189] | Public welfare [190] | Highways [191] | Total (mil dol) [192] | Per capita[1] (dollars) [193] | Federal civilian [194] | Federal military [195] | State and local [196] | Number of returns [197] | Mean adjusted gross income [198] | Mean income tax [199] |
|---|---|---|---|---|---|---|---|---|---|---|---|---|---|---|---|
| **SOUTH DAKOTA—Cont'd** | | | | | | | | | | | | | | | |
| Tripp | 28 | 5,187 | 41.3 | 2.2 | 3.5 | 0.3 | 16.4 | 40.1 | 7,359 | 31 | 30 | 407 | 2,610 | 45,062 | 3,936 |
| Turner | 29 | 3,507 | 47.4 | 0.7 | 4.1 | 0.2 | 22.3 | 12.7 | 1,535 | 25 | 46 | 403 | 3,960 | 55,389 | 4,919 |
| Union | 66 | 4,291 | 53.6 | 0.1 | 4.7 | 0.0 | 12.9 | 42.0 | 2,752 | 52 | 89 | 794 | 7,940 | 150,426 | 25,404 |
| Walworth | 22 | 4,071 | 44.4 | 0.6 | 7.2 | 0.5 | 12.7 | 14.6 | 2,663 | 32 | 30 | 392 | 2,600 | 58,521 | 5,669 |
| Yankton | 74 | 3,262 | 49.0 | 1.7 | 7.7 | 0.3 | 9.3 | 53.9 | 2,375 | 201 | 115 | 1,674 | 10,990 | 60,399 | 5,832 |
| Ziebach | 9 | 3,270 | 68.5 | 0.4 | 2.3 | 0.0 | 18.2 | 1.0 | 368 | 8 | 15 | 145 | 520 | 37,458 | 2,054 |
| **TENNESSEE** | X | X | X | X | X | X | X | X | X | 50,594 | 20,508 | 382,946 | 3,055,310 | 64,740 | 7,595 |
| Anderson | 248 | 3,262 | 54.6 | 3.0 | 6.2 | 0.0 | 3.3 | 216.6 | 2,848 | 855 | 215 | 4,297 | 34,700 | 57,299 | 5,517 |
| Bedford | 132 | 2,729 | 52.0 | 2.8 | 7.2 | 0.1 | 4.7 | 83.6 | 1,735 | 94 | 138 | 2,319 | 21,250 | 50,784 | 4,834 |
| Benton | 38 | 2,389 | 57.2 | 2.3 | 7.8 | 0.4 | 10.5 | 16.8 | 1,053 | 60 | 45 | 923 | 6,650 | 43,249 | 3,182 |
| Bledsoe | 35 | 2,368 | 53.6 | 3.6 | 4.1 | 0.5 | 7.9 | 21.8 | 1,460 | 20 | 35 | 904 | 4,620 | 42,760 | 3,013 |
| Blount | 618 | 4,752 | 29.9 | 42.1 | 5.4 | 0.0 | 2.8 | 1,244.1 | 9,570 | 282 | 391 | 7,662 | 61,690 | 61,805 | 6,442 |
| Bradley | 299 | 2,832 | 48.0 | 2.8 | 6.4 | 4.9 | 4.5 | 188.4 | 1,787 | 215 | 299 | 4,686 | 46,410 | 51,396 | 4,683 |
| Campbell | 131 | 3,279 | 37.1 | 2.9 | 5.0 | 0.1 | 6.9 | 137.4 | 3,454 | 84 | 110 | 1,962 | 15,000 | 45,110 | 3,601 |
| Cannon | 27 | 1,927 | 66.3 | 2.3 | 9.3 | 0.0 | 0.4 | 9.2 | 647 | 32 | 41 | 597 | 6,350 | 47,834 | 3,556 |
| Carroll | 75 | 2,702 | 57.8 | 0.5 | 6.6 | 0.1 | 7.2 | 54.7 | 1,965 | 70 | 75 | 1,751 | 11,350 | 45,354 | 3,414 |
| Carter | 130 | 2,301 | 62.0 | 0.8 | 6.2 | 0.0 | 6.1 | 115.6 | 2,046 | 85 | 156 | 2,480 | 23,080 | 43,895 | 3,410 |
| Cheatham | 95 | 2,342 | 64.4 | 2.9 | 6.1 | 0.3 | 4.8 | 27.6 | 683 | 83 | 114 | 1,574 | 19,660 | 45,821 | 3,233 |
| Chester | 37 | 2,131 | 65.6 | 1.1 | 6.1 | 0.0 | 5.0 | 14.8 | 861 | 42 | 46 | 1,084 | 6,780 | 44,236 | 3,380 |
| Claiborne | 61 | 1,924 | 68.3 | 0.0 | 5.5 | 0.3 | 7.7 | 11.5 | 363 | 67 | 86 | 1,806 | 12,330 | 36,836 | 2,434 |
| Clay | 23 | 3,014 | 44.8 | 7.6 | 6.2 | 0.3 | 16.4 | 13.3 | 1,736 | 49 | 21 | 495 | 2,970 | 38,040 | 2,543 |
| Cocke | 85 | 2,401 | 58.0 | 1.3 | 5.6 | 0.0 | 7.7 | 41.7 | 1,173 | 60 | 100 | 1,847 | 15,160 | 51,355 | 4,584 |
| Coffee | 160 | 2,909 | 58.4 | 2.0 | 5.3 | 0.2 | 3.9 | 253.6 | 4,609 | 540 | 206 | 3,185 | 25,120 | 46,057 | 3,678 |
| Crockett | 40 | 2,786 | 62.9 | 3.9 | 4.8 | 1.0 | 7.9 | 16.0 | 1,110 | 32 | 40 | 840 | 6,060 | 54,625 | 5,364 |
| Cumberland | 133 | 2,249 | 48.6 | 9.7 | 5.2 | 0.0 | 5.3 | 112.0 | 1,898 | 128 | 168 | 2,191 | 27,280 | 83,980 | 12,751 |
| Davidson | 3,228 | 4,698 | 34.6 | 5.0 | 6.8 | 1.5 | 4.3 | 16,254.2 | 23,654 | 8,403 | 2,350 | 37,154 | 348,200 | 45,807 | 3,529 |
| Decatur | 46 | 3,899 | 33.7 | 24.1 | 3.2 | 0.2 | 4.5 | 21.2 | 1,810 | 29 | 32 | 855 | 4,590 | 46,374 | 3,993 |
| DeKalb | 46 | 2,309 | 52.6 | 3.3 | 7.6 | 0.1 | 5.0 | 34.9 | 1,756 | 38 | 57 | 1,002 | 8,100 | 54,926 | 5,002 |
| Dickson | 136 | 2,580 | 55.8 | 2.4 | 7.0 | 0.0 | 4.9 | 97.0 | 1,837 | 97 | 151 | 2,653 | 24,180 | 49,379 | 4,212 |
| Dyer | 131 | 3,519 | 50.6 | 0.3 | 8.6 | 0.6 | 8.0 | 53.4 | 1,431 | 100 | 103 | 2,989 | 15,640 | 72,975 | 8,965 |
| Fayette | 60 | 1,502 | 53.4 | 3.8 | 6.5 | 0.7 | 9.3 | 38.5 | 959 | 66 | 115 | 1,325 | 19,930 | 41,462 | 3,281 |
| Fentress | 37 | 2,043 | 61.4 | 4.9 | 6.5 | 0.0 | 1.0 | 10.3 | 563 | 42 | 51 | 917 | 7,100 | 54,547 | 5,064 |
| Franklin | 93 | 2,234 | 55.5 | 0.6 | 9.0 | 0.2 | 3.4 | 67.9 | 1,632 | 136 | 113 | 1,750 | 18,590 | 47,778 | 3,670 |
| Gibson | 138 | 2,796 | 61.8 | 3.1 | 6.4 | 0.6 | 2.0 | 115.6 | 2,346 | 144 | 135 | 2,824 | 21,070 | 48,867 | 4,033 |
| Giles | 64 | 2,174 | 60.2 | 4.5 | 4.0 | 0.1 | 9.2 | 22.1 | 751 | 71 | 81 | 1,439 | 12,980 | 45,755 | 3,462 |
| Grainger | 44 | 1,896 | 72.1 | 4.3 | 5.4 | 0.0 | 0.8 | 6.7 | 289 | 61 | 65 | 832 | 9,540 | 44,943 | 3,457 |
| Greene | 156 | 2,267 | 59.3 | 3.0 | 5.5 | 0.0 | 4.7 | 61.7 | 897 | 242 | 191 | 3,522 | 29,500 | 41,264 | 2,797 |
| Grundy | 31 | 2,305 | 66.3 | 0.7 | 3.5 | 0.2 | 7.2 | 18.3 | 1,376 | 23 | 37 | 714 | 5,580 | 50,087 | 4,553 |
| Hamblen | 186 | 2,907 | 50.5 | 0.6 | 6.0 | 0.2 | 4.5 | 290.4 | 4,533 | 195 | 181 | 4,000 | 27,050 | 70,045 | 8,347 |
| Hamilton | 2,102 | 5,823 | 20.9 | 45.3 | 5.0 | 0.4 | 2.0 | 1,418.9 | 3,930 | 5,230 | 1,076 | 24,356 | 168,250 | 36,223 | 2,081 |
| Hancock | 22 | 3,311 | 47.1 | 15.9 | 4.1 | 0.0 | 7.1 | 13.8 | 2,089 | 11 | 18 | 492 | 2,290 | 39,390 | 2,659 |
| Hardeman | 96 | 3,784 | 37.4 | 1.9 | 4.4 | 0.1 | 5.4 | 61.1 | 2,398 | 56 | 60 | 1,641 | 9,630 | 49,239 | 4,447 |
| Hardin | 145 | 5,613 | 26.8 | 52.6 | 2.1 | 0.1 | 4.0 | 77.3 | 3,003 | 116 | 71 | 1,728 | 10,610 | 45,967 | 3,534 |
| Hawkins | 118 | 2,079 | 60.6 | 1.7 | 4.3 | 0.2 | 5.6 | 97.2 | 1,716 | 137 | 158 | 2,359 | 23,530 | 39,550 | 2,912 |
| Haywood | 62 | 3,524 | 47.6 | 4.3 | 6.0 | 0.0 | 7.1 | 19.0 | 1,078 | 89 | 48 | 1,039 | 7,870 | 46,282 | 3,649 |
| Henderson | 70 | 2,501 | 66.5 | 0.3 | 7.1 | 0.1 | 5.3 | 68.9 | 2,474 | 56 | 78 | 1,368 | 11,620 | 47,144 | 4,468 |
| Henry | 187 | 5,769 | 27.6 | 55.2 | 2.5 | 0.0 | 3.5 | 68.1 | 2,103 | 115 | 123 | 2,595 | 14,210 | 44,800 | 3,295 |
| Hickman | 56 | 2,254 | 56.7 | 3.4 | 5.4 | 0.0 | 7.4 | 32.2 | 1,296 | 55 | 67 | 1,150 | 9,800 | 46,214 | 3,444 |
| Houston | 22 | 2,659 | 55.2 | 4.7 | 4.3 | 0.4 | 10.1 | 14.1 | 1,727 | 23 | 23 | 555 | 3,310 | 51,577 | 4,298 |
| Humphreys | 48 | 2,600 | 54.2 | 1.3 | 7.6 | 0.0 | 9.6 | 4.6 | 248 | 111 | 52 | 1,059 | 8,090 | 40,881 | 2,707 |
| Jackson | 26 | 2,260 | 58.5 | 2.7 | 3.1 | 0.0 | 8.4 | 23.6 | 2,017 | 26 | 33 | 529 | 4,560 | 51,096 | 4,440 |
| Jefferson | 153 | 2,853 | 42.8 | 3.1 | 5.3 | 7.6 | 3.8 | 126.5 | 2,355 | 129 | 153 | 2,398 | 23,490 | 44,848 | 3,619 |
| Johnson | 33 | 1,868 | 66.3 | 0.9 | 7.2 | 0.1 | 2.3 | 22.0 | 1,247 | 33 | 45 | 936 | 6,280 | 76,805 | 10,225 |
| Knox | 1,322 | 2,864 | 42.5 | 7.2 | 9.7 | 0.1 | 4.3 | 1,738.1 | 3,766 | 3,624 | 1,379 | 30,070 | 213,160 | 38,175 | 2,572 |
| Lake | 18 | 2,391 | 49.3 | 4.0 | 7.7 | 0.6 | 11.4 | 10.2 | 1,376 | 17 | 14 | 654 | 1,970 | 43,287 | 3,537 |
| Lauderdale | 69 | 2,660 | 57.3 | 2.9 | 8.0 | 0.0 | 7.1 | 19.2 | 734 | 52 | 66 | 1,978 | 9,390 | 46,514 | 3,496 |
| Lawrence | 101 | 2,327 | 60.3 | 2.4 | 6.9 | 0.0 | 7.1 | 84.4 | 1,946 | 119 | 123 | 2,026 | 17,300 | 43,179 | 3,294 |
| Lewis | 31 | 2,535 | 52.9 | 0.6 | 6.5 | 0.2 | 7.1 | 9.3 | 774 | 25 | 34 | 729 | 5,110 | 50,092 | 4,152 |
| Lincoln | 127 | 3,750 | 39.0 | 36.8 | 3.6 | 0.0 | 4.2 | 22.4 | 662 | 53 | 96 | 2,036 | 15,390 | 70,378 | 7,662 |
| Loudon | 133 | 2,547 | 53.7 | 0.6 | 6.7 | 0.2 | 4.6 | 616.9 | 11,804 | 159 | 151 | 2,079 | 25,460 | 48,322 | 3,873 |
| McMinn | 119 | 2,240 | 59.1 | 0.7 | 5.1 | 0.1 | 6.5 | 45.1 | 852 | 115 | 149 | 2,208 | 22,620 | 44,007 | 3,138 |
| McNairy | 61 | 2,350 | 61.9 | 0.3 | 7.8 | 0.0 | 6.1 | 20.5 | 788 | 82 | 71 | 1,302 | 10,150 | 42,565 | 2,929 |
| Macon | 57 | 2,400 | 60.4 | 4.0 | 6.3 | 0.1 | 7.5 | 22.0 | 919 | 38 | 68 | 1,118 | 9,760 | 54,699 | 5,533 |
| Madison | 988 | 10,138 | 12.8 | 68.5 | 3.8 | 0.0 | 1.8 | 649.4 | 6,661 | 409 | 266 | 12,152 | 43,720 | 52,092 | 4,591 |
| Marion | 70 | 2,450 | 56.6 | 0.8 | 8.2 | 0.0 | 5.2 | 42.0 | 1,480 | 82 | 81 | 1,332 | 12,320 | | |

1. Based on the resident population estimated as of July 1 of the year shown.

# Table B. States and Counties — Land Area and Population

| State / county code | CBSA code[1] | County Type code[2] | STATE County | Land area[3] (sq. mi) | Total persons 2019 | Rank | Per square mile | White | Black | American Indian, Alaska Native | Asian and Pacific Islander | Percent Hispanic or Latino[4] | Under 5 years | 5 to 14 years | 15 to 24 years | 25 to 34 years | 35 to 44 years | 45 to 54 years |
|---|---|---|---|---|---|---|---|---|---|---|---|---|---|---|---|---|---|---|
| | | | | 1 | 2 | 3 | 4 | 5 | 6 | 7 | 8 | 9 | 10 | 11 | 12 | 13 | 14 | 15 |
| | | | **TENNESSEE—Cont'd** | | | | | | | | | | | | | | | |
| 47117 | 30280 | 6 | Marshall | 375.5 | 35,016 | 1,304 | 93.3 | 86.8 | 7.3 | 0.9 | 1.0 | 5.9 | 6.2 | 13.0 | 11.6 | 13.5 | 12.4 | 13.2 |
| 47119 | 34980 | 1 | Maury | 613.1 | 99,590 | 606 | 162.4 | 80.4 | 13.0 | 0.8 | 1.4 | 6.6 | 6.4 | 12.8 | 11.0 | 14.1 | 13.1 | 12.0 |
| 47121 | | 8 | Meigs | 195.1 | 12,532 | 2,245 | 64.2 | 94.7 | 2.3 | 1.6 | 0.7 | 2.4 | 5.4 | 11.3 | 11.0 | 10.5 | 11.4 | 14.0 |
| 47123 | | 6 | Monroe | 635.8 | 47,177 | 1,033 | 74.2 | 92.3 | 2.6 | 1.4 | 0.8 | 4.6 | 5.3 | 11.6 | 10.8 | 11.6 | 10.7 | 13.0 |
| 47125 | 17300 | 2 | Montgomery | 539.2 | 214,251 | 317 | 397.3 | 65.5 | 23.2 | 1.3 | 4.2 | 10.6 | 8.1 | 14.8 | 14.5 | 18.7 | 13.7 | 10.6 |
| 47127 | 46100 | 9 | Moore | 129.2 | 6,438 | 2,711 | 49.8 | 93.7 | 3.1 | 0.8 | 1.9 | 1.9 | 4.5 | 10.7 | 10.7 | 10.9 | 11.5 | 13.7 |
| 47129 | 28940 | 2 | Morgan | 522.2 | 21,431 | 1,749 | 41.0 | 94.0 | 4.2 | 1.2 | 0.5 | 1.6 | 5.0 | 10.3 | 11.7 | 13.6 | 12.3 | 14.3 |
| 47131 | 46460 | 7 | Obion | 544.8 | 30,131 | 1,430 | 55.3 | 83.6 | 11.6 | 0.7 | 0.6 | 5.2 | 5.9 | 11.9 | 11.5 | 11.5 | 11.6 | 12.7 |
| 47133 | 18260 | 7 | Overton | 433.5 | 22,566 | 1,702 | 52.1 | 96.9 | 1.2 | 1.0 | 0.5 | 1.7 | 5.3 | 11.9 | 11.4 | 12.0 | 11.2 | 13.0 |
| 47135 | | 8 | Perry | 414.8 | 8,099 | 2,577 | 19.5 | 93.1 | 3.5 | 1.7 | 1.2 | 2.7 | 6.5 | 12.1 | 10.8 | 11.9 | 11.1 | 12.2 |
| 47137 | | 9 | Pickett | 163.0 | 5,061 | 2,823 | 31.0 | 96.7 | 0.5 | 0.9 | 0.3 | 2.5 | 4.2 | 9.1 | 10.4 | 8.7 | 9.4 | 13.0 |
| 47139 | 17420 | 3 | Polk | 434.6 | 16,835 | 1,981 | 38.7 | 96.3 | 1.2 | 1.5 | 0.6 | 2.1 | 4.9 | 10.7 | 11.1 | 11.1 | 11.1 | 14.6 |
| 47141 | 18260 | 4 | Putnam | 401.1 | 80,929 | 703 | 201.8 | 89.4 | 3.1 | 0.9 | 1.8 | 6.7 | 5.6 | 11.6 | 17.3 | 13.4 | 11.3 | 11.6 |
| 47143 | 19420 | 6 | Rhea | 315.5 | 33,443 | 1,342 | 106.0 | 91.7 | 2.8 | 1.1 | 0.8 | 5.4 | 5.9 | 12.2 | 13.0 | 11.7 | 11.2 | 12.8 |
| 47145 | 28940 | 2 | Roane | 360.8 | 53,841 | 947 | 149.2 | 93.9 | 3.5 | 1.2 | 1.0 | 2.2 | 4.5 | 10.5 | 10.6 | 10.8 | 10.6 | 13.4 |
| 47147 | 34980 | 1 | Robertson | 476.3 | 72,275 | 754 | 151.7 | 83.7 | 8.4 | 0.8 | 1.0 | 7.6 | 6.2 | 11.7 | 12.9 | 13.1 | 13.1 | 13.3 |
| 47149 | 34980 | 1 | Rutherford | 619.3 | 339,261 | 210 | 547.8 | 70.6 | 17.8 | 0.8 | 4.5 | 9.0 | 6.4 | 13.9 | 16.1 | 14.9 | 14.2 | 12.7 |
| 47151 | | 6 | Scott | 532.3 | 22,090 | 1,714 | 41.5 | 97.9 | 0.5 | 1.0 | 0.5 | 1.1 | 6.1 | 13.1 | 12.3 | 12.2 | 12.7 | 12.8 |
| 47153 | 16860 | 2 | Sequatchie | 265.9 | 15,176 | 2,084 | 57.1 | 94.6 | 1.4 | 1.3 | 0.7 | 3.7 | 5.5 | 11.1 | 10.8 | 11.2 | 11.1 | 13.4 |
| 47155 | 42940 | 2 | Sevier | 592.5 | 99,244 | 608 | 167.5 | 90.3 | 1.7 | 1.0 | 1.6 | 6.8 | 5.3 | 11.4 | 11.0 | 12.2 | 11.1 | 13.2 |
| 47157 | 32820 | 1 | Shelby | 760.6 | 936,017 | 56 | 1,230.6 | 36.2 | 54.8 | 0.6 | 3.4 | 6.7 | 6.9 | 14.0 | 12.9 | 15.1 | 12.5 | 11.9 |
| 47159 | 34980 | 1 | Smith | 314.3 | 20,285 | 1,806 | 64.5 | 93.5 | 3.1 | 1.1 | 0.7 | 3.2 | 6.0 | 12.5 | 11.6 | 12.9 | 12.1 | 12.9 |
| 47161 | 17300 | 8 | Stewart | 459.8 | 13,859 | 2,166 | 30.1 | 92.8 | 2.6 | 1.7 | 1.5 | 3.5 | 5.3 | 11.7 | 11.0 | 11.2 | 11.5 | 13.2 |
| 47163 | 28700 | 2 | Sullivan | 413.4 | 158,755 | 424 | 384.0 | 94.4 | 3.1 | 0.8 | 1.1 | 2.2 | 4.7 | 10.8 | 10.8 | 12.1 | 11.1 | 13.5 |
| 47165 | 34980 | 1 | Sumner | 529.5 | 195,561 | 347 | 369.3 | 84.3 | 9.2 | 0.8 | 2.2 | 5.6 | 5.9 | 13.3 | 11.7 | 12.7 | 13.3 | 13.4 |
| 47167 | 32820 | 1 | Tipton | 455.6 | 61,918 | 856 | 135.9 | 77.6 | 19.3 | 1.1 | 1.4 | 2.9 | 5.9 | 13.6 | 12.6 | 13.1 | 12.7 | 12.9 |
| 47169 | 34980 | 1 | Trousdale | 114.3 | 11,455 | 2,322 | 100.2 | 84.8 | 12.7 | 0.8 | 0.6 | 2.9 | 5.3 | 10.3 | 13.6 | 21.7 | 12.9 | 12.2 |
| 47171 | 27740 | 3 | Unicoi | 186.1 | 17,755 | 1,934 | 95.4 | 93.0 | 1.0 | 0.9 | 0.5 | 5.8 | 4.6 | 9.9 | 10.8 | 10.9 | 11.0 | 13.1 |
| 47173 | 28940 | 2 | Union | 223.6 | 20,187 | 1,813 | 90.3 | 97.0 | 0.9 | 1.1 | 0.3 | 1.9 | 5.4 | 12.0 | 11.0 | 12.2 | 11.4 | 13.4 |
| 47175 | | 9 | Van Buren | 273.4 | 5,947 | 2,750 | 21.8 | 96.5 | 1.5 | 1.0 | 0.4 | 2.0 | 5.6 | 11.0 | 10.0 | 9.9 | 10.7 | 13.9 |
| 47177 | 32660 | 6 | Warren | 432.7 | 41,605 | 1,145 | 96.2 | 86.0 | 4.2 | 0.9 | 1.0 | 9.6 | 6.0 | 13.1 | 11.8 | 12.3 | 12.5 | 12.6 |
| 47179 | 27740 | 3 | Washington | 326.5 | 130,367 | 494 | 399.3 | 89.8 | 5.3 | 0.8 | 2.2 | 3.9 | 4.8 | 10.6 | 15.0 | 12.8 | 11.3 | 12.8 |
| 47181 | | 8 | Wayne | 734.1 | 16,524 | 2,005 | 22.5 | 90.3 | 7.2 | 0.9 | 0.4 | 2.3 | 3.8 | 9.2 | 10.9 | 14.6 | 13.1 | 14.2 |
| 47183 | 32280 | 7 | Weakley | 580.4 | 33,334 | 1,349 | 57.4 | 87.7 | 8.6 | 0.7 | 1.5 | 3.0 | 4.9 | 10.8 | 19.0 | 11.5 | 10.3 | 11.5 |
| 47185 | | 7 | White | 376.7 | 27,707 | 1,505 | 73.6 | 94.4 | 2.6 | 1.0 | 0.7 | 2.8 | 5.7 | 11.9 | 11.1 | 11.9 | 11.3 | 12.8 |
| 47187 | 34980 | 1 | Williamson | 582.9 | 245,348 | 280 | 420.9 | 85.1 | 4.9 | 0.6 | 6.1 | 5.1 | 5.7 | 15.6 | 13.0 | 9.2 | 14.6 | 15.0 |
| 47189 | 34980 | 1 | Wilson | 570.9 | 148,130 | 454 | 259.5 | 85.2 | 8.5 | 0.9 | 2.5 | 4.9 | 6.0 | 13.2 | 11.5 | 12.1 | 13.9 | 13.7 |
| 48000 | | 0 | TEXAS | 261,263.1 | 29,360,759 | X | 112.4 | 42.2 | 12.9 | 0.7 | 6.0 | 39.8 | 6.7 | 14.3 | 14.0 | 14.7 | 13.7 | 12.2 |
| 48001 | 37300 | 7 | Anderson | 1,062.6 | 57,805 | 898 | 54.4 | 59.3 | 22.0 | 0.8 | 1.1 | 18.3 | 5.3 | 10.4 | 11.1 | 16.1 | 16.1 | 14.2 |
| 48003 | 11380 | 6 | Andrews | 1,500.7 | 18,879 | 1,883 | 12.6 | 39.6 | 1.9 | 0.9 | 0.7 | 57.7 | 8.5 | 17.8 | 14.2 | 14.9 | 13.7 | 11.0 |
| 48005 | 31260 | 5 | Angelina | 797.9 | 86,796 | 668 | 108.8 | 60.7 | 15.7 | 0.6 | 1.2 | 22.9 | 6.6 | 14.5 | 13.0 | 12.8 | 11.7 | 12.0 |
| 48007 | 40530 | 2 | Aransas | 252.1 | 23,814 | 1,649 | 94.5 | 67.3 | 1.9 | 1.4 | 2.1 | 28.7 | 4.9 | 10.0 | 9.8 | 9.7 | 9.7 | 11.1 |
| 48009 | 48660 | 3 | Archer | 903.3 | 8,730 | 2,519 | 9.7 | 89.4 | 1.6 | 1.6 | 0.9 | 8.2 | 5.3 | 12.4 | 11.2 | 10.8 | 11.3 | 12.1 |
| 48011 | 11100 | 2 | Armstrong | 909.1 | 1,869 | 3,059 | 2.1 | 87.7 | 1.4 | 1.7 | 0.3 | 10.1 | 5.2 | 13.3 | 10.6 | 9.6 | 13.2 | 10.5 |
| 48013 | 41700 | 1 | Atascosa | 1,219.5 | 51,724 | 969 | 42.4 | 32.9 | 1.2 | 0.7 | 0.7 | 65.3 | 7.0 | 15.3 | 13.7 | 12.7 | 12.8 | 11.7 |
| 48015 | 26420 | 1 | Austin | 646.5 | 29,972 | 1,437 | 46.4 | 61.9 | 9.4 | 0.7 | 1.0 | 28.3 | 5.6 | 13.8 | 11.8 | 11.2 | 11.2 | 11.5 |
| 48017 | | 7 | Bailey | 827.0 | 6,697 | 2,687 | 8.1 | 33.9 | 1.5 | 0.7 | 0.9 | 63.8 | 8.2 | 16.3 | 14.8 | 11.5 | 12.1 | 10.5 |
| 48019 | 41700 | 1 | Bandera | 791.0 | 23,861 | 1,648 | 30.2 | 77.4 | 1.5 | 1.4 | 0.8 | 20.2 | 4.0 | 9.7 | 8.6 | 8.3 | 9.8 | 11.6 |
| 48021 | 12420 | 1 | Bastrop | 888.2 | 91,601 | 646 | 103.1 | 51.4 | 7.2 | 1.0 | 1.2 | 40.8 | 6.5 | 14.4 | 13.0 | 12.4 | 12.2 | 12.1 |
| 48023 | | 8 | Baylor | 867.5 | 3,518 | 2,933 | 4.1 | 82.3 | 3.4 | 1.0 | 0.8 | 14.7 | 5.7 | 13.4 | 10.2 | 11.1 | 11.1 | 10.2 |
| 48025 | 13300 | 6 | Bee | 880.2 | 32,513 | 1,366 | 36.9 | 31.1 | 8.6 | 0.6 | 0.8 | 59.7 | 5.3 | 11.9 | 14.7 | 17.3 | 15.6 | 12.3 |
| 48027 | 28660 | 2 | Bell | 1,053.8 | 369,927 | 195 | 351.0 | 47.1 | 24.7 | 1.1 | 5.3 | 25.9 | 8.0 | 15.3 | 15.4 | 16.3 | 13.4 | 10.4 |
| 48029 | 41700 | 1 | Bexar | 1,240.3 | 2,026,823 | 16 | 1,634.1 | 28.0 | 8.2 | 0.6 | 4.0 | 60.9 | 6.8 | 14.1 | 14.3 | 15.9 | 13.9 | 11.7 |
| 48031 | | 8 | Blanco | 709.3 | 12,269 | 2,266 | 17.3 | 77.1 | 1.8 | 1.0 | 1.1 | 20.3 | 3.8 | 9.9 | 10.4 | 8.9 | 10.3 | 12.4 |
| 48033 | | 8 | Borden | 897.4 | 706 | 3,128 | 0.8 | 80.5 | 1.3 | 1.6 | 0.1 | 17.7 | 4.0 | 11.9 | 11.5 | 9.2 | 10.1 | 15.2 |
| 48035 | | 6 | Bosque | 983.0 | 18,603 | 1,893 | 18.9 | 77.8 | 2.5 | 1.0 | 0.7 | 19.3 | 5.0 | 12.1 | 10.5 | 9.7 | 10.5 | 11.6 |
| 48037 | 45500 | 3 | Bowie | 884.9 | 93,481 | 636 | 105.6 | 64.6 | 26.4 | 1.4 | 1.8 | 8.1 | 6.2 | 13.3 | 12.4 | 13.8 | 12.8 | 11.9 |
| 48039 | 26420 | 1 | Brazoria | 1,363.3 | 380,518 | 189 | 279.1 | 45.8 | 15.7 | 0.8 | 7.7 | 31.8 | 6.5 | 14.9 | 12.7 | 13.8 | 15.0 | 13.1 |
| 48041 | 17780 | 3 | Brazos | 586.1 | 232,555 | 295 | 396.8 | 56.3 | 11.4 | 0.6 | 6.8 | 26.5 | 5.8 | 11.5 | 28.2 | 15.5 | 11.7 | 8.9 |
| 48043 | | 7 | Brewster | 6,183.8 | 9,237 | 2,477 | 1.5 | 51.2 | 2.2 | 1.2 | 1.9 | 45.0 | 5.0 | 10.1 | 10.4 | 11.7 | 13.7 | 11.4 |
| 48045 | | 9 | Briscoe | 900.0 | 1,487 | 3,080 | 1.7 | 70.7 | 4.8 | 1.1 | 0.7 | 25.1 | 5.2 | 12.3 | 8.5 | 10.8 | 11.0 | 10.9 |
| 48047 | | 7 | Brooks | 943.4 | 6,964 | 2,662 | 7.4 | 7.2 | 0.9 | 0.2 | 1.0 | 91.1 | 7.6 | 16.3 | 12.8 | 12.7 | 11.6 | 9.4 |
| 48049 | 15220 | 5 | Brown | 944.5 | 37,633 | 1,233 | 39.8 | 75.9 | 4.3 | 1.1 | 0.9 | 23.1 | 5.0 | 11.6 | 13.0 | 11.8 | 11.7 | 11.9 |
| 48051 | 17780 | 3 | Burleson | 659.1 | 18,514 | 1,897 | 28.1 | 65.7 | 12.8 | 0.8 | 0.7 | 21.4 | 5.7 | 12.1 | 10.8 | 11.4 | 10.6 | 11.9 |
| 48053 | | 6 | Burnet | 994.8 | 49,653 | 997 | 49.9 | 74.3 | 2.1 | 1.1 | 1.2 | 22.5 | 5.2 | 11.8 | 11.2 | 10.8 | 11.1 | 11.1 |

1. CBSA = Core Based Statistical Area. See Appendix A for explanation. See Appendix B for list of metropolitan areas with component counties.   2. County type code from the Economic Research Service of USDA Rural-Urban Continuum Codes. See Appendix A for definition.   3. Dry land or land partially or temporarily covered by water.   4. May be of any race.

Items 1—15

# Table B. States and Counties — **Population and Households**

| STATE County | 55 to 64 years | 65 to 74 years | 75 years and over | Percent female | Total persons 2000 | Total persons 2010 | Percent change 2000-2010 | Percent change 2010-2020 | Births | Deaths | Net Migration | Number | Persons per household | Family households | Female family householder[1] | One person |
|---|---|---|---|---|---|---|---|---|---|---|---|---|---|---|---|---|
| | 16 | 17 | 18 | 19 | 20 | 21 | 22 | 23 | 24 | 25 | 26 | 27 | 28 | 29 | 30 | 31 |
| **TENNESSEE—Cont'd** | | | | | | | | | | | | | | | | |
| Marshall | 13.6 | 10.3 | 6.4 | 51.0 | 26,767 | 30,611 | 14.4 | 14.4 | 3,943 | 3,425 | 3,903 | 12,324 | 2.64 | 70.0 | 11.9 | 25.2 |
| Maury | 13.5 | 10.6 | 6.4 | 51.7 | 69,498 | 80,986 | 16.5 | 23.0 | 11,814 | 9,133 | 15,831 | 34,688 | 2.62 | 71.0 | 13.5 | 24.1 |
| Meigs | 15.0 | 13.2 | 8.2 | 50.4 | 11,086 | 11,766 | 6.1 | 6.5 | 1,251 | 1,644 | 1,162 | 4,938 | 2.43 | 67.5 | 8.1 | 26.8 |
| Monroe | 14.9 | 13.2 | 8.8 | 50.5 | 38,961 | 44,496 | 14.2 | 6.0 | 5,110 | 5,614 | 3,210 | 17,987 | 2.53 | 71.4 | 11.0 | 25.6 |
| Montgomery | 9.7 | 6.2 | 3.6 | 50.2 | 134,768 | 172,361 | 27.9 | 24.3 | 34,732 | 12,755 | 19,502 | 72,617 | 2.71 | 70.5 | 13.6 | 22.9 |
| Moore | 15.2 | 12.4 | 10.3 | 50.2 | 5,740 | 6,345 | 10.5 | 1.5 | 534 | 687 | 249 | 2,592 | 2.44 | 71.8 | 4.6 | 22.2 |
| Morgan | 14.0 | 11.5 | 7.4 | 45.3 | 19,757 | 21,995 | 11.3 | -2.6 | 2,040 | 2,440 | -164 | 7,625 | 2.45 | 74.2 | 12.7 | 30.4 |
| Obion | 13.5 | 12.1 | 9.2 | 51.6 | 32,450 | 31,807 | -2.0 | -5.3 | 3,567 | 4,022 | -1,220 | 12,717 | 2.35 | 66.0 | 12.6 | 29.2 |
| Overton | 14.1 | 12.4 | 8.5 | 50.3 | 20,118 | 22,076 | 9.7 | 2.2 | 2,419 | 3,085 | 1,174 | 9,140 | 2.38 | 68.1 | 10.4 | 29.6 |
| Perry | 13.7 | 12.2 | 9.4 | 49.5 | 7,631 | 7,930 | 3.9 | 2.1 | 1,038 | 1,173 | 309 | 3,073 | 2.54 | 67.3 | 7.1 | 28.5 |
| Pickett | 16.4 | 16.6 | 12.2 | 50.7 | 4,945 | 5,086 | 2.9 | -0.5 | 423 | 733 | 283 | 2,281 | 2.19 | 70.1 | 10.0 | 23.6 |
| Polk | 15.3 | 12.6 | 9.0 | 50.7 | 16,050 | 16,821 | 4.8 | 0.1 | 1,648 | 2,350 | 718 | 7,383 | 2.24 | 71.0 | 11.5 | 29.1 |
| Putnam | 12.0 | 10.0 | 7.2 | 50.4 | 62,315 | 72,340 | 16.1 | 11.9 | 9,122 | 8,121 | 7,619 | 31,424 | 2.36 | 62.6 | 13.1 | 28.0 |
| Rhea | 13.8 | 11.5 | 7.9 | 50.5 | 28,400 | 31,804 | 12.0 | 5.2 | 3,949 | 4,077 | 1,780 | 12,177 | 2.61 | 67.6 | 12.5 | 29.2 |
| Roane | 15.7 | 13.9 | 9.9 | 50.8 | 51,910 | 54,209 | 4.4 | -0.7 | 4,849 | 7,469 | 2,299 | 20,901 | 2.5 | 66.7 | 9.2 | 29.2 |
| Robertson | 14.2 | 9.6 | 5.9 | 50.6 | 54,433 | 66,312 | 21.8 | 9.0 | 9,078 | 6,783 | 3,681 | 25,713 | 2.69 | 77.9 | 12.7 | 18.4 |
| Rutherford | 10.9 | 7.0 | 4.1 | 51.0 | 182,023 | 262,593 | 44.3 | 29.2 | 40,482 | 19,148 | 54,865 | 111,676 | 2.78 | 69.9 | 12.5 | 21.3 |
| Scott | 13.2 | 10.3 | 7.4 | 50.7 | 21,127 | 22,232 | 5.2 | -0.6 | 2,721 | 2,833 | -20 | 8,664 | 2.49 | 69.9 | 15.0 | 24.0 |
| Sequatchie | 15.0 | 12.9 | 8.9 | 50.3 | 11,370 | 14,117 | 24.2 | 7.5 | 1,631 | 1,631 | 1,062 | 5,528 | 2.64 | 75.8 | 9.6 | 19.8 |
| Sevier | 15.0 | 12.6 | 8.1 | 50.7 | 71,170 | 89,712 | 26.1 | 10.6 | 10,740 | 10,470 | 9,223 | 37,210 | 2.58 | 70.3 | 11.2 | 23.2 |
| Shelby | 12.2 | 9.0 | 5.5 | 52.7 | 897,472 | 927,682 | 3.4 | 0.9 | 137,903 | 83,001 | -46,670 | 351,194 | 2.62 | 62.2 | 19.5 | 32.4 |
| Smith | 14.9 | 10.4 | 6.7 | 50.0 | 17,712 | 19,150 | 8.1 | 5.9 | 2,310 | 2,239 | 1,066 | 7,603 | 2.57 | 71.1 | 12.1 | 26.0 |
| Stewart | 15.2 | 12.7 | 8.3 | 50.2 | 12,370 | 13,311 | 7.6 | 4.1 | 1,394 | 1,669 | 827 | 5,178 | 2.57 | 64.8 | 8.7 | 32.5 |
| Sullivan | 14.5 | 12.5 | 10.0 | 51.3 | 153,048 | 156,804 | 2.5 | 1.2 | 15,754 | 21,106 | 7,447 | 66,511 | 2.32 | 64.4 | 11.4 | 31.0 |
| Sumner | 13.3 | 9.9 | 6.6 | 51.3 | 130,449 | 160,639 | 23.1 | 21.7 | 21,109 | 15,524 | 29,214 | 67,089 | 2.71 | 72.3 | 10.9 | 23.8 |
| Tipton | 13.8 | 9.6 | 5.9 | 50.9 | 51,271 | 61,008 | 19.0 | 1.5 | 7,390 | 5,948 | -525 | 21,452 | 2.83 | 76.5 | 15.3 | 19.8 |
| Trousdale | 10.9 | 8.3 | 4.9 | 41.4 | 7,259 | 7,858 | 8.3 | 45.8 | 1,053 | 949 | 3,457 | 3,189 | 2.92 | 65.3 | 9.4 | 30.7 |
| Unicoi | 15.5 | 13.8 | 10.4 | 50.5 | 17,667 | 18,311 | 3.6 | -3.0 | 1,626 | 2,824 | 661 | 7,658 | 2.25 | 64.7 | 11.7 | 29.5 |
| Union | 15.8 | 11.4 | 7.5 | 50.4 | 17,808 | 19,108 | 7.3 | 5.6 | 2,127 | 2,224 | 1,181 | 7,405 | 2.61 | 73.9 | 13.0 | 22.6 |
| Van Buren | 14.9 | 14.8 | 9.1 | 49.5 | 5,508 | 5,556 | 0.9 | 7.0 | 648 | 734 | 473 | 2,141 | 2.64 | 72.2 | 12.3 | 23.4 |
| Warren | 13.4 | 10.8 | 7.5 | 50.4 | 38,276 | 39,824 | 4.0 | 4.5 | 4,934 | 5,034 | 1,913 | 15,727 | 2.55 | 64.6 | 11.0 | 31.3 |
| Washington | 13.3 | 11.2 | 8.1 | 51.2 | 107,198 | 123,063 | 14.8 | 5.9 | 13,325 | 14,366 | 8,359 | 53,859 | 2.28 | 61.3 | 10.4 | 31.4 |
| Wayne | 14.1 | 11.5 | 8.5 | 44.4 | 16,842 | 17,021 | 1.1 | -2.9 | 1,414 | 2,100 | 201 | 5,764 | 2.54 | 69.7 | 13.2 | 27.3 |
| Weakley | 12.7 | 10.8 | 8.4 | 51.3 | 34,895 | 35,015 | 0.3 | -4.8 | 3,498 | 3,914 | -1,277 | 13,640 | 2.27 | 63.6 | 11.1 | 28.8 |
| White | 14.2 | 12.2 | 8.8 | 51.1 | 23,102 | 25,847 | 11.9 | 7.2 | 3,051 | 3,766 | 2,585 | 10,048 | 2.63 | 68.7 | 9.6 | 27.6 |
| Williamson | 12.8 | 8.7 | 5.4 | 50.9 | 126,638 | 183,208 | 44.7 | 33.9 | 22,368 | 12,055 | 51,629 | 77,855 | 2.89 | 80.6 | 7.5 | 15.7 |
| Wilson | 13.4 | 9.9 | 6.3 | 50.8 | 88,809 | 114,063 | 28.4 | 29.9 | 15,222 | 10,826 | 29,447 | 49,664 | 2.72 | 76.2 | 9.6 | 19.8 |
| **TEXAS** | 11.2 | 8.0 | 5.2 | 50.3 | 20,851,820 | 25,146,072 | 20.6 | 16.8 | 3,962,570 | 1,935,156 | 2,173,519 | 9,691,647 | 2.85 | 69.3 | 13.8 | 25.2 |
| Anderson | 11.5 | 9.1 | 6.2 | 38.7 | 55,109 | 58,457 | 6.1 | -1.1 | 6,125 | 6,814 | 63 | 16,677 | 2.66 | 69.6 | 12.7 | 26.7 |
| Andrews | 10.1 | 5.7 | 4.0 | 49.0 | 13,004 | 14,786 | 13.7 | 27.7 | 3,185 | 1,379 | 2,238 | 5,573 | 3.22 | 76.7 | 9.5 | 19.6 |
| Angelina | 12.3 | 9.8 | 7.3 | 51.2 | 80,130 | 86,771 | 8.3 | 0.0 | 12,042 | 9,224 | -2,747 | 31,035 | 2.72 | 73.1 | 16.2 | 23.1 |
| Aransas | 16.5 | 15.8 | 12.4 | 50.5 | 22,497 | 23,165 | 3.0 | 2.8 | 2,520 | 3,557 | 1,626 | 9,548 | 2.52 | 64.2 | 9.5 | 31.5 |
| Archer | 15.9 | 11.9 | 9.4 | 50.7 | 8,854 | 9,061 | 2.3 | -3.7 | 822 | 913 | -242 | 3,452 | 2.5 | 69.6 | 8.0 | 26.9 |
| Armstrong | 14.6 | 13.9 | 9.1 | 51.2 | 2,148 | 1,901 | -11.5 | -1.7 | 200 | 287 | 55 | 694 | 2.76 | 76.2 | 8.9 | 18.4 |
| Atascosa | 11.7 | 8.9 | 6.3 | 50.1 | 38,628 | 44,923 | 16.3 | 15.1 | 6,946 | 4,151 | 3,990 | 15,546 | 3.16 | 74.6 | 16.4 | 20.3 |
| Austin | 14.0 | 12.0 | 8.7 | 50.3 | 23,590 | 28,421 | 20.5 | 5.5 | 3,470 | 2,872 | 968 | 11,301 | 2.61 | 70.1 | 10.7 | 21.4 |
| Bailey | 11.2 | 8.4 | 7.1 | 50.1 | 6,594 | 7,165 | 8.7 | -6.5 | 1,253 | 597 | -1,134 | 2,054 | 3.35 | 69.7 | 14.0 | 26.0 |
| Bandera | 19.0 | 17.7 | 11.3 | 50.5 | 17,645 | 20,473 | 16.0 | 16.5 | 1,731 | 2,258 | 3,912 | 8,399 | 2.59 | 72.8 | 10.4 | 26.1 |
| Bastrop | 13.6 | 10.1 | 5.7 | 49.1 | 57,733 | 74,230 | 28.6 | 23.4 | 10,270 | 6,876 | 13,908 | 25,571 | 3.22 | 61.3 | 12.8 | 22.9 |
| Baylor | 13.8 | 12.3 | 12.1 | 50.9 | 4,093 | 3,726 | -9.0 | -5.6 | 391 | 612 | 12 | 1,530 | 2.29 | 69.3 | 4.6 | 35.2 |
| Bee | 9.9 | 7.6 | 5.6 | 39.2 | 32,359 | 31,858 | -1.5 | 2.1 | 3,686 | 2,591 | -438 | 8,269 | 3.0 | 66.2 | 16.6 | 25.3 |
| Bell | 9.8 | 7.0 | 4.6 | 50.3 | 237,974 | 310,178 | 30.3 | 19.3 | 63,382 | 21,854 | 17,927 | 122,689 | 2.75 | 72.3 | 15.2 | 25.9 |
| Bexar | 10.6 | 7.6 | 5.0 | 50.6 | 1,392,931 | 1,714,781 | 23.1 | 18.2 | 275,579 | 132,360 | 167,751 | 636,245 | 3.02 | 66.2 | 15.8 | 27.9 |
| Blanco | 18.0 | 16.5 | 9.7 | 50.0 | 8,418 | 10,485 | 24.6 | 17.0 | 921 | 1,260 | 2,120 | 4,343 | 2.62 | 67.6 | 9.9 | 23.7 |
| Borden | 12.5 | 12.9 | 12.9 | 47.7 | 729 | 641 | -12.1 | 10.1 | 60 | 51 | 55 | 227 | 2.7 | 74.2 | 8.4 | 24.7 |
| Bosque | 15.0 | 14.2 | 11.5 | 50.9 | 17,204 | 18,214 | 5.9 | 2.1 | 1,851 | 2,573 | 1,110 | 7,211 | 2.49 | 67.6 | 10.1 | 22.7 |
| Bowie | 12.3 | 10.0 | 7.3 | 49.8 | 89,306 | 92,564 | 3.6 | 1.0 | 12,199 | 10,387 | -846 | 34,076 | 2.57 | 62.1 | 16.6 | 29.1 |
| Brazoria | 11.5 | 7.8 | 4.7 | 49.5 | 241,767 | 313,186 | 29.5 | 21.5 | 48,335 | 23,392 | 42,193 | 121,523 | 2.87 | 73.1 | 10.7 | 22.5 |
| Brazos | 8.5 | 5.9 | 4.0 | 49.7 | 152,415 | 194,861 | 27.8 | 19.3 | 27,484 | 10,332 | 20,439 | 79,412 | 2.61 | 56.4 | 12.6 | 28.5 |
| Brewster | 13.5 | 14.2 | 9.9 | 48.9 | 8,866 | 9,232 | 4.1 | 0.1 | 1,053 | 801 | -253 | 4,088 | 2.23 | 62.1 | 8.2 | 37.1 |
| Briscoe | 15.3 | 13.0 | 13.0 | 49.8 | 1,790 | 1,637 | -8.5 | -9.2 | 150 | 187 | -115 | 602 | 2.42 | 67.3 | 6.0 | 30.9 |
| Brooks | 11.0 | 10.0 | 8.6 | 49.0 | 7,976 | 7,227 | -9.4 | -3.6 | 1,201 | 818 | -651 | 2,120 | 3.11 | 73.4 | 18.5 | 30.9 |
| Brown | 13.9 | 11.9 | 9.1 | 50.5 | 37,674 | 38,106 | 1.1 | -1.2 | 4,193 | 4,991 | 353 | 14,409 | 2.5 | 66.4 | 9.3 | 28.4 |
| Burleson | 15.8 | 12.8 | 8.9 | 50.6 | 16,470 | 17,187 | 4.4 | 7.7 | 2,053 | 2,198 | 1,476 | 6,810 | 2.63 | 69.8 | 9.5 | 26.1 |
| Burnet | 15.5 | 13.9 | 9.4 | 51.1 | 34,147 | 42,710 | 25.1 | 16.3 | 4,894 | 4,922 | 6,971 | 16,743 | 2.74 | 73.4 | 9.4 | 23.1 |

1. No spouse present.

# Table B. States and Counties — **Population, Vital Statistics, Health, and Crime**

| STATE County | Persons in group quarters, 2020 | Daytime Population, 2015-2019 | | Births, 2020 | | Deaths, 2020 | | Persons under 65 with no health insurance, 2019 | | Medicare, 2020 | | | Crimes reported by county police or sheriff, 2019 | |
|---|---|---|---|---|---|---|---|---|---|---|---|---|---|---|
| | | Number | Employment/residence ratio | Total | Rate[1] | Number | Rate[1] | Number | Percent | Total beneficiaries | Enrolled in Original Medicare | Enrolled in Medicare Advantage | Violent | Property |
| | 32 | 33 | 34 | 35 | 36 | 37 | 38 | 39 | 40 | 41 | 42 | 43 | 44 | 45 |
| **TENNESSEE—Cont'd** | | | | | | | | | | | | | | |
| Marshall | 384 | 29,798 | 0.78 | 425 | 12.1 | 388 | 11.1 | 3,353 | 11.8 | 6,895 | 3,967 | 2,928 | 28 | 72 |
| Maury | 1,010 | 86,061 | 0.86 | 1,207 | 12.1 | 1,084 | 10.9 | 9,081 | 11.3 | 19,932 | 12,179 | 7,753 | 140 | 431 |
| Meigs | 148 | 9,760 | 0.47 | 130 | 10.4 | 184 | 14.7 | 1,249 | 13.1 | 3,404 | 1,921 | 1,482 | 61 | 284 |
| Monroe | 514 | 44,887 | 0.93 | 507 | 10.7 | 631 | 13.4 | 4,731 | 13.2 | 12,339 | 6,307 | 6,032 | 234 | 873 |
| Montgomery | 3,706 | 174,869 | 0.73 | 3,482 | 16.3 | 1,483 | 6.9 | 17,282 | 9.3 | 27,211 | 19,420 | 7,792 | 104 | 571 |
| Moore | 104 | 5,873 | 0.82 | 57 | 8.9 | 63 | 9.8 | 514 | 10.2 | 1,549 | 1,087 | 463 | 7 | 58 |
| Morgan | 2,265 | 18,596 | 0.59 | 188 | 8.8 | 263 | 12.3 | 1,887 | 12.5 | 5,071 | 2,207 | 2,864 | 40 | 320 |
| Obion | 461 | 30,286 | 0.99 | 347 | 11.5 | 442 | 14.7 | 2,918 | 12.5 | 8,051 | 5,114 | 2,937 | 34 | 177 |
| Overton | 286 | 19,124 | 0.68 | 241 | 10.7 | 337 | 14.9 | 2,296 | 13.3 | 5,842 | 3,981 | 1,861 | 13 | 135 |
| Perry | 129 | 8,152 | 1.07 | 106 | 13.1 | 106 | 13.1 | 901 | 14.5 | 2,155 | 1,292 | 863 | 13 | 59 |
| Pickett | 65 | 4,913 | 0.91 | 41 | 8.1 | 63 | 12.4 | 538 | 14.9 | 1,636 | 1,184 | 452 | 8 | 28 |
| Polk | 277 | 12,992 | 0.45 | 162 | 9.6 | 262 | 15.6 | 1,871 | 14.5 | 4,834 | 2,864 | 1,970 | 25 | 311 |
| Putnam | 2,496 | 83,887 | 1.19 | 909 | 11.2 | 842 | 10.4 | 8,504 | 13.3 | 16,871 | 11,543 | 5,328 | 85 | 518 |
| Rhea | 826 | 33,233 | 1.04 | 376 | 11.2 | 425 | 12.7 | 3,635 | 14.0 | 7,866 | 4,704 | 3,162 | 67 | 191 |
| Roane | 730 | 46,260 | 0.68 | 463 | 8.6 | 803 | 14.9 | 4,305 | 10.6 | 14,722 | 8,029 | 6,694 | 63 | 235 |
| Robertson | 913 | 61,535 | 0.74 | 891 | 12.3 | 725 | 10.0 | 6,963 | 11.6 | 13,675 | 7,121 | 6,554 | 75 | 243 |
| Rutherford | 5,136 | 295,206 | 0.88 | 4,254 | 12.5 | 2,322 | 6.8 | 30,323 | 10.4 | 43,964 | 25,454 | 18,510 | 205 | 746 |
| Scott | 314 | 20,693 | 0.84 | 251 | 11.4 | 292 | 13.2 | 2,143 | 12.0 | 5,258 | 2,505 | 2,753 | 27 | 161 |
| Sequatchie | 188 | 12,152 | 0.56 | 173 | 11.4 | 183 | 12.1 | 1,465 | 12.5 | 4,497 | 2,506 | 1,990 | 31 | 153 |
| Sevier | 1,180 | 99,885 | 1.06 | 1,041 | 10.5 | 1,146 | 11.5 | 13,557 | 17.6 | 23,956 | 12,400 | 11,556 | 153 | 836 |
| Shelby | 17,362 | 1,011,191 | 1.17 | 12,832 | 13.7 | 9,204 | 9.8 | 103,192 | 13.2 | 154,890 | 99,838 | 55,052 | 569 | 2,473 |
| Smith | 236 | 16,914 | 0.69 | 216 | 10.6 | 241 | 11.9 | 1,982 | 12.1 | 4,216 | 2,603 | 1,613 | 43 | 148 |
| Stewart | 153 | 11,759 | 0.68 | 143 | 10.3 | 192 | 13.9 | 1,353 | 12.6 | 3,379 | 2,411 | 968 | 18 | 92 |
| Sullivan | 2,710 | 166,512 | 1.14 | 1,491 | 9.4 | 2,144 | 13.5 | 14,688 | 12.1 | 42,989 | 16,871 | 26,118 | 228 | 1,043 |
| Sumner | 1,435 | 156,585 | 0.70 | 2,156 | 11.0 | 1,857 | 9.5 | 16,587 | 10.4 | 36,283 | 19,185 | 17,098 | 118 | 404 |
| Tipton | 1,024 | 46,621 | 0.46 | 688 | 11.1 | 651 | 10.5 | 5,353 | 10.4 | 11,720 | 8,225 | 3,495 | 123 | 465 |
| Trousdale | 2,612 | 8,210 | 0.56 | 124 | 10.8 | 82 | 7.2 | 933 | 12.9 | 1,865 | 1,049 | 816 | 30 | 142 |
| Unicoi | 453 | 16,683 | 0.84 | 155 | 8.7 | 295 | 16.6 | 1,752 | 13.0 | 5,291 | 2,404 | 2,887 | 21 | 116 |
| Union | 171 | 15,632 | 0.47 | 210 | 10.4 | 235 | 11.6 | 2,188 | 13.7 | 4,696 | 1,846 | 2,850 | 22 | 204 |
| Van Buren | 136 | 4,711 | 0.50 | 68 | 11.4 | 67 | 11.3 | 551 | 12.7 | 1,753 | 1,167 | 586 | 3 | 30 |
| Warren | 542 | 39,793 | 0.95 | 496 | 11.9 | 508 | 12.2 | 5,237 | 15.8 | 9,489 | 6,038 | 3,451 | NA | NA |
| Washington | 4,046 | 135,635 | 1.13 | 1,231 | 9.4 | 1,500 | 11.5 | 12,094 | 11.9 | 30,254 | 14,761 | 15,494 | 159 | 762 |
| Wayne | 2,235 | 15,467 | 0.78 | 123 | 7.4 | 217 | 13.1 | 1,497 | 13.4 | 3,871 | 2,443 | 1,428 | 14 | 93 |
| Weakley | 1,638 | 31,354 | 0.85 | 318 | 9.5 | 400 | 12.0 | 2,938 | 11.5 | 7,395 | 5,190 | 2,205 | 27 | 116 |
| White | 392 | 24,276 | 0.75 | 285 | 10.3 | 369 | 13.3 | 2,717 | 12.8 | 7,097 | 4,429 | 2,668 | 55 | 251 |
| Williamson | 1,181 | 250,171 | 1.22 | 2,310 | 9.4 | 1,501 | 6.1 | 13,215 | 6.3 | 36,396 | 22,768 | 13,628 | 119 | 289 |
| Wilson | 1,592 | 121,700 | 0.78 | 1,669 | 11.3 | 1,312 | 8.9 | 11,543 | 9.5 | 26,892 | 15,399 | 11,493 | 85 | 555 |
| **TEXAS** | 590,686 | 28,259,997 | 1.00 | 378,318 | 12.9 | 220,988 | 7.5 | 5,114,881 | 20.7 | 4,284,123 | 2,454,061 | 1,830,062 | X | X |
| Anderson | 14,250 | 59,414 | 1.08 | 578 | 10.0 | 688 | 11.9 | 7,415 | 20.8 | 10,517 | 6,468 | 4,049 | 30 | 363 |
| Andrews | 77 | 17,583 | 0.94 | 325 | 17.2 | 117 | 6.2 | 3,532 | 20.9 | 2,178 | 1,439 | 739 | NA | NA |
| Angelina | 2,608 | 87,372 | 1.00 | 1,101 | 12.7 | 993 | 11.4 | 15,329 | 22.0 | 17,696 | 11,041 | 6,655 | 97 | 567 |
| Aransas | 437 | 23,298 | 0.88 | 219 | 9.2 | 376 | 15.8 | 4,085 | 24.9 | 7,394 | 3,904 | 3,490 | 60 | 240 |
| Archer | 53 | 6,449 | 0.48 | 73 | 8.4 | 65 | 7.4 | 1,356 | 19.7 | 1,925 | 1,486 | 440 | NA | NA |
| Armstrong | 51 | 1,687 | 0.68 | 16 | 8.6 | 27 | 14.4 | 254 | 18.0 | 448 | 338 | 110 | NA | NA |
| Atascosa | 304 | 46,653 | 0.85 | 701 | 13.6 | 464 | 9.0 | 9,579 | 22.2 | 9,027 | 4,284 | 4,743 | NA | NA |
| Austin | 208 | 27,917 | 0.86 | 332 | 11.1 | 324 | 10.8 | 5,181 | 21.6 | 6,758 | 4,540 | 2,218 | 13 | 87 |
| Bailey | 92 | 6,888 | 0.93 | 117 | 17.5 | 45 | 6.7 | 1,635 | 28.1 | 1,087 | 849 | 239 | 3 | 36 |
| Bandera | 303 | 18,095 | 0.55 | 170 | 7.1 | 246 | 10.3 | 3,371 | 20.5 | 6,870 | 4,537 | 2,332 | 12 | 268 |
| Bastrop | 2,286 | 68,877 | 0.56 | 1,095 | 12.0 | 803 | 8.8 | 16,504 | 22.7 | 15,926 | 9,933 | 5,993 | 212 | 649 |
| Baylor | 64 | 3,626 | 1.03 | 38 | 10.8 | 45 | 12.8 | 432 | 16.3 | 983 | 757 | 226 | NA | NA |
| Bee | 7,753 | 31,871 | 0.92 | 307 | 9.4 | 210 | 6.5 | 3,847 | 18.6 | 4,905 | 2,460 | 2,445 | 21 | 159 |
| Bell | 10,413 | 354,841 | 1.04 | 5,939 | 16.1 | 2,527 | 6.8 | 51,972 | 16.6 | 53,786 | 33,570 | 20,216 | 82 | 631 |
| Bexar | 42,211 | 1,997,311 | 1.05 | 26,476 | 13.1 | 15,038 | 7.4 | 327,450 | 19.1 | 301,500 | 153,912 | 147,588 | 692 | 5,297 |
| Blanco | 53 | 10,189 | 0.75 | 83 | 6.8 | 145 | 11.8 | 1,937 | 21.7 | 3,418 | 2,464 | 953 | NA | NA |
| Borden | 0 | 566 | 0.80 | 4 | 5.7 | 7 | 9.9 | 60 | 12.3 | 132 | 105 | 27 | 1 | 10 |
| Bosque | 298 | 15,154 | 0.59 | 159 | 8.5 | 267 | 14.4 | 3,173 | 23.0 | 5,140 | 3,083 | 2,057 | 27 | 71 |
| Bowie | 5,825 | 100,257 | 1.19 | 1,098 | 11.7 | 1,068 | 11.4 | 12,418 | 17.3 | 19,661 | 14,079 | 5,582 | 102 | 331 |
| Brazoria | 10,301 | 312,136 | 0.71 | 4,721 | 12.4 | 2,816 | 7.4 | 52,277 | 16.4 | 50,792 | 28,373 | 22,419 | 170 | 1,325 |
| Brazos | 14,468 | 230,159 | 1.07 | 2,587 | 11.1 | 1,262 | 5.4 | 33,871 | 17.5 | 24,761 | 17,853 | 6,907 | 29 | 306 |
| Brewster | 44 | 8,918 | 0.93 | 88 | 9.5 | 79 | 8.6 | 1,462 | 21.2 | 2,233 | 1,517 | 716 | 7 | 10 |
| Briscoe | 0 | 1,472 | 1.03 | 20 | 13.4 | 12 | 8.1 | 274 | 24.1 | 399 | 282 | 117 | 0 | 3 |
| Brooks | 54 | 7,310 | 1.07 | 108 | 15.5 | 75 | 10.8 | 1,048 | 18.6 | 1,575 | 834 | 742 | 1 | 0 |
| Brown | 1,579 | 38,790 | 1.06 | 383 | 10.2 | 475 | 12.6 | 5,812 | 20.0 | 9,195 | 6,594 | 2,601 | 31 | 189 |
| Burleson | 160 | 15,339 | 0.64 | 193 | 10.4 | 235 | 12.7 | 2,943 | 20.3 | 4,536 | 2,954 | 1,582 | 20 | 100 |
| Burnet | 1,140 | 44,025 | 0.88 | 477 | 9.6 | 517 | 10.4 | 7,566 | 20.9 | 13,107 | 8,623 | 4,484 | 54 | 219 |

1. Per 1,000 estimated resident population.

| STATE County | County law enforcement employment, 2019 | | Education / School enrollment and attainment, 2015-2019 | | | | Local government expenditures,[3] 2017-2018 | | Money income, 2015-2019 | | | | Income and poverty, 2019 | | | |
|---|---|---|---|---|---|---|---|---|---|---|---|---|---|---|---|---|
| | | | Enrollment[1] | | Attainment[2] (percent) | | | | | Households | | | | Percent below poverty level | | |
| | Officers | Civilians | Total | Percent private | High school graduate or less | Bachelor's degree or more | Total current spending (mil dol) | Current spending per student (dollars) | Per capita income[4] | Median income (dollars) | Percent with income of less than $50,000 | Percent with income of $200,000 or more | Median household income (dollars) | All persons | Children under 18 years | Children 5 to 17 years in families |
| | 46 | 47 | 48 | 49 | 50 | 51 | 52 | 53 | 54 | 55 | 56 | 57 | 58 | 59 | 60 | 61 |
| **TENNESSEE—Cont'd** | | | | | | | | | | | | | | | | |
| Marshall | 27 | 25 | 7,165 | 13.2 | 56.0 | 14.7 | 47 | 8,497 | 25,410 | 53,197 | 47.5 | 1.9 | 58,151 | 13.1 | 20.2 | 19.3 |
| Maury | 82 | 80 | 19,914 | 19.9 | 45.6 | 23.0 | 115 | 8,960 | 28,970 | 57,170 | 43.1 | 3.9 | 66,434 | 8.5 | 13.2 | 11.5 |
| Meigs | 15 | 20 | 2,143 | 11.2 | 61.6 | 9.8 | 16 | 9,114 | 24,525 | 49,167 | 51.0 | 1.6 | 48,334 | 16.1 | 24.7 | 22.8 |
| Monroe | 46 | 37 | 8,631 | 10.0 | 59.1 | 13.4 | 61 | 8,982 | 23,207 | 42,429 | 56.1 | 1.4 | 50,506 | 16.5 | 27.3 | 24.9 |
| Montgomery | 138 | 254 | 56,470 | 11.7 | 35.5 | 28.5 | 307 | 8,823 | 26,923 | 57,541 | 43.4 | 3.1 | 57,606 | 12.0 | 16.4 | 15.4 |
| Moore | 17 | 14 | 1,263 | 12.3 | 54.1 | 22.7 | 10 | 10,739 | 30,658 | 57,708 | 42.6 | 3.5 | 61,804 | 10.0 | 14.5 | 13.6 |
| Morgan | 25 | 25 | 3,877 | 9.3 | 68.2 | 9.4 | 28 | 9,376 | 20,204 | 41,333 | 57.6 | 1.6 | 43,952 | 22.0 | 27.4 | 26.1 |
| Obion | 34 | 35 | 6,441 | 6.7 | 61.3 | 15.3 | 47 | 9,289 | 23,375 | 39,615 | 60.6 | 1.7 | 44,061 | 14.6 | 23.2 | 20.6 |
| Overton | 24 | 33 | 4,730 | 7.8 | 66.0 | 12.8 | 26 | 8,139 | 21,808 | 37,197 | 59.2 | 1.7 | 43,636 | 17.1 | 21.6 | 19.5 |
| Perry | 16 | 13 | 1,681 | 20.1 | 63.6 | 11.5 | 11 | 10,628 | 27,970 | 41,034 | 58.5 | 4.8 | 42,939 | 14.4 | 24.4 | 25.1 |
| Pickett | 12 | 5 | 739 | 4.1 | 63.9 | 9.4 | 6 | 9,325 | 24,907 | 39,554 | 64.8 | 0.6 | 41,860 | 17.8 | 23.4 | 21.5 |
| Polk | 27 | 37 | 2,925 | 15.8 | 57.0 | 11.9 | 21 | 8,932 | 25,405 | 43,306 | 56.1 | 1.7 | 45,922 | 15.3 | 23.2 | 21.0 |
| Putnam | 81 | 85 | 20,421 | 6.9 | 50.0 | 26.4 | 101 | 8,882 | 24,946 | 44,259 | 56.2 | 3.0 | 49,083 | 16.1 | 19.6 | 17.5 |
| Rhea | 55 | 2 | 7,397 | 15.2 | 56.0 | 16.9 | 46 | 8,774 | 22,336 | 42,206 | 56.0 | 2.1 | 47,599 | 15.4 | 22.9 | 20.5 |
| Roane | 42 | 37 | 10,319 | 16.7 | 47.4 | 19.8 | 63 | 9,524 | 30,209 | 53,367 | 47.0 | 3.9 | 50,561 | 14.6 | 21.1 | 19.7 |
| Robertson | 72 | 94 | 15,912 | 17.5 | 50.9 | 19.6 | 98 | 8,567 | 29,524 | 63,307 | 38.8 | 3.9 | 65,790 | 10.5 | 14.7 | 13.0 |
| Rutherford | 239 | 235 | 88,051 | 11.3 | 36.0 | 32.2 | 483 | 8,930 | 30,159 | 67,429 | 35.6 | 4.5 | 69,614 | 10.0 | 12.9 | 11.2 |
| Scott | 30 | 27 | 4,977 | 8.0 | 64.5 | 9.2 | 37 | 8,671 | 20,849 | 38,864 | 60.1 | 1.2 | 38,750 | 22.0 | 30.4 | 32.6 |
| Sequatchie | 22 | 27 | 2,807 | 12.4 | 57.5 | 15.1 | 18 | 8,088 | 23,050 | 49,370 | 50.6 | 1.0 | 46,876 | 17.4 | 30.8 | 28.3 |
| Sevier | 104 | 108 | 18,384 | 11.4 | 52.2 | 17.4 | 146 | 10,091 | 25,283 | 49,610 | 50.3 | 2.4 | 54,476 | 12.9 | 20.4 | 17.1 |
| Shelby | 638 | 1,212 | 241,560 | 19.0 | 39.1 | 31.6 | 1,510 | 10,603 | 30,104 | 51,657 | 48.7 | 5.8 | 52,659 | 17.2 | 25.7 | 24.6 |
| Smith | 28 | 29 | 4,234 | 13.0 | 61.3 | 14.4 | 25 | 8,055 | 28,797 | 48,068 | 52.7 | 4.1 | 54,570 | 15.3 | 22.3 | 20.5 |
| Stewart | 22 | 33 | 2,777 | 4.2 | 57.6 | 15.6 | 18 | 8,750 | 24,113 | 45,809 | 54.3 | 1.3 | 52,177 | 12.9 | 18.9 | 18.3 |
| Sullivan | 148 | 170 | 31,195 | 13.6 | 47.5 | 23.6 | 209 | 9,693 | 28,429 | 46,684 | 52.7 | 3.4 | 51,303 | 15.1 | 23.0 | 21.3 |
| Sumner | 116 | 174 | 43,275 | 17.1 | 41.2 | 28.1 | 261 | 8,794 | 33,851 | 67,204 | 36.0 | 6.1 | 69,391 | 8.6 | 11.7 | 10.8 |
| Tipton | 63 | 45 | 15,112 | 10.8 | 49.8 | 17.3 | 91 | 8,167 | 28,457 | 61,291 | 41.7 | 4.2 | 61,436 | 10.5 | 16.5 | 15.0 |
| Trousdale | 23 | 24 | 2,263 | 18.3 | 53.9 | 13.8 | 11 | 9,118 | 24,795 | 56,321 | 43.5 | 1.3 | 51,982 | 15.8 | 17.9 | 16.6 |
| Unicoi | 26 | 19 | 3,159 | 11.8 | 54.5 | 14.8 | 22 | 9,320 | 23,455 | 41,890 | 59.1 | 1.8 | 44,548 | 15.8 | 22.3 | 20.3 |
| Union | 35 | 25 | 3,861 | 13.3 | 67.7 | 9.9 | 33 | 7,758 | 24,010 | 44,671 | 54.4 | 2.7 | 48,106 | 18.3 | 28.7 | 26.7 |
| Van Buren | 7 | 16 | 1,113 | 9.0 | 71.8 | 7.8 | 8 | 10,981 | 20,634 | 42,724 | 55.0 | 1.4 | 44,451 | 15.5 | 23.5 | 24.6 |
| Warren | 56 | 56 | 8,700 | 7.9 | 63.8 | 14.5 | 56 | 8,627 | 22,802 | 41,125 | 58.8 | 2.6 | 46,144 | 15.9 | 23.7 | 22.7 |
| Washington | 95 | 111 | 31,370 | 12.4 | 40.2 | 31.8 | 149 | 9,056 | 29,656 | 48,334 | 51.3 | 4.1 | 51,879 | 15.8 | 19.4 | 17.5 |
| Wayne | 16 | 34 | 2,973 | 11.5 | 66.4 | 9.7 | 22 | 9,380 | 20,689 | 41,427 | 57.9 | 1.4 | 45,091 | 17.3 | 22.6 | 20.5 |
| Weakley | 22 | 21 | 9,503 | 6.7 | 53.8 | 21.6 | 36 | 8,568 | 22,755 | 39,937 | 59.8 | 1.8 | 42,506 | 17.7 | 23.4 | 20.7 |
| White | 39 | 46 | 5,507 | 10.7 | 62.5 | 12.3 | 34 | 8,454 | 22,040 | 41,998 | 59.0 | 2.0 | 43,583 | 16.0 | 21.5 | 21.0 |
| Williamson | 198 | 125 | 64,475 | 24.8 | 17.9 | 59.8 | 412 | 9,585 | 52,702 | 112,962 | 18.5 | 21.5 | 119,637 | 4.3 | 4.1 | 3.3 |
| Wilson | 134 | 132 | 32,776 | 17.2 | 37.4 | 32.4 | 187 | 8,435 | 34,575 | 75,991 | 31.4 | 7.6 | 80,080 | 7.4 | 10.1 | 9.4 |
| **TEXAS** | X | X | 7,681,758 | 11.2 | 41.3 | 29.9 | 51,634 | 9,560 | 31,277 | 61,874 | 40.8 | 7.4 | 64,044 | 13.6 | 19.2 | 18.0 |
| Anderson | 36 | 34 | 12,774 | 4.6 | 57.5 | 10.6 | 83 | 9,787 | 17,280 | 43,455 | 56.2 | 1.5 | 48,461 | 19.8 | 23.8 | 21.2 |
| Andrews | NA | NA | 4,893 | 3.7 | 62.7 | 12.2 | 38 | 9,382 | 30,673 | 76,158 | 33.6 | 7.6 | 74,918 | 10.2 | 12.7 | 11.6 |
| Angelina | 47 | 63 | 21,625 | 7.2 | 49.6 | 16.9 | 160 | 9,213 | 23,147 | 50,453 | 49.6 | 2.2 | 51,750 | 15.6 | 22.6 | 21.7 |
| Aransas | 23 | 46 | 4,903 | 8.9 | 46.8 | 21.9 | 35 | 13,803 | 30,863 | 45,137 | 54.8 | 6.1 | 53,085 | 18.7 | 32.1 | 30.3 |
| Archer | NA | NA | 1,878 | 6.4 | 45.1 | 23.9 | 18 | 9,601 | 34,617 | 63,835 | 41.3 | 6.0 | 63,731 | 10.3 | 12.8 | 12.2 |
| Armstrong | 3 | 4 | 478 | 4.4 | 35.6 | 22.8 | 4 | 10,427 | 31,749 | 72,500 | 33.4 | 4.6 | 64,424 | 10.6 | 16.7 | 16.2 |
| Atascosa | 44 | 69 | 12,177 | 6.4 | 61.6 | 14.4 | 84 | 9,229 | 23,890 | 55,366 | 44.4 | 3.9 | 57,309 | 14.7 | 19.0 | 18.3 |
| Austin | 42 | 13 | 7,058 | 8.0 | 45.2 | 23.9 | 56 | 9,440 | 30,219 | 66,206 | 39.6 | 4.0 | 68,311 | 11.1 | 17.0 | 15.4 |
| Bailey | 7 | 20 | 1,625 | 1.1 | 58.2 | 17.2 | 15 | 10,543 | 21,945 | 52,273 | 47.6 | 2.4 | 46,174 | 15.9 | 25.8 | 23.9 |
| Bandera | 29 | 36 | 3,737 | 24.0 | 43.0 | 22.8 | 24 | 9,253 | 29,561 | 58,661 | 41.2 | 4.5 | 59,604 | 12.9 | 23.6 | 20.4 |
| Bastrop | 79 | 120 | 20,350 | 12.7 | 48.5 | 20.7 | 163 | 9,393 | 27,773 | 64,597 | 37.9 | 4.7 | 62,627 | 14.0 | 23.8 | 21.4 |
| Baylor | 4 | 5 | 649 | 2.5 | 49.5 | 22.8 | 7 | 11,544 | 22,987 | 40,739 | 59.0 | 0.9 | 41,539 | 15.9 | 24.5 | 22.3 |
| Bee | 24 | 45 | 7,595 | 8.0 | 58.8 | 8.9 | 52 | 9,776 | 18,340 | 44,578 | 55.7 | 3.4 | 44,171 | 24.0 | 27.8 | 27.5 |
| Bell | NA | NA | 100,192 | 12.6 | 35.0 | 25.2 | 666 | 8,943 | 26,677 | 54,884 | 44.6 | 3.5 | 54,831 | 13.8 | 18.4 | 18.4 |
| Bexar | 578 | 1,153 | 539,883 | 13.5 | 41.2 | 28.0 | 3,467 | 9,771 | 27,834 | 57,157 | 43.6 | 5.3 | 58,956 | 15.2 | 22.4 | 21.2 |
| Blanco | 15 | 16 | 2,146 | 8.7 | 40.2 | 26.5 | 18 | 10,767 | 37,998 | 66,390 | 39.3 | 5.8 | 68,404 | 8.6 | 14.7 | 13.5 |
| Borden | 3 | 1 | 100 | 4.0 | 25.4 | 39.5 | 6 | 23,403 | 34,028 | 72,188 | 29.5 | 5.7 | 71,926 | 10.6 | 15.4 | 15.4 |
| Bosque | 19 | 24 | 3,849 | 8.4 | 47.7 | 19.8 | 25 | 10,748 | 26,636 | 52,148 | 47.8 | 2.9 | 55,179 | 13.7 | 19.7 | 17.5 |
| Bowie | 39 | 6 | 22,764 | 6.7 | 46.0 | 20.6 | 168 | 9,396 | 27,095 | 50,164 | 49.8 | 4.3 | 51,207 | 16.8 | 23.3 | 20.8 |
| Brazoria | 139 | 196 | 96,898 | 10.9 | 37.6 | 30.0 | 681 | 9,488 | 34,561 | 81,447 | 29.6 | 9.2 | 83,704 | 8.6 | 11.1 | 10.9 |
| Brazos | 103 | 158 | 92,487 | 6.2 | 32.0 | 41.7 | 289 | 9,384 | 27,632 | 49,181 | 50.5 | 5.8 | 54,471 | 20.0 | 19.4 | 19.7 |
| Brewster | 19 | 17 | 2,067 | 11.9 | 35.7 | 40.0 | 17 | 12,952 | 28,915 | 47,080 | 54.7 | 3.4 | 48,581 | 13.1 | 19.8 | 18.8 |
| Briscoe | 3 | 0 | 285 | 1.1 | 50.4 | 19.8 | 4 | 10,647 | 23,933 | 40,227 | 58.6 | 1.8 | 44,259 | 15.0 | 24.6 | 23.1 |
| Brooks | 17 | 20 | 1,825 | 8.3 | 55.9 | 15.7 | 17 | 10,933 | 14,523 | 28,333 | 70.8 | 0.0 | 31,410 | 29.6 | 44.4 | 43.7 |
| Brown | 32 | 38 | 8,716 | 12.6 | 49.1 | 19.1 | 63 | 9,416 | 27,463 | 48,365 | 51.2 | 3.5 | 56,040 | 13.7 | 21.8 | 20.9 |
| Burleson | 18 | 25 | 3,686 | 8.6 | 55.9 | 17.3 | 28 | 9,865 | 30,086 | 57,731 | 42.1 | 3.4 | 61,745 | 13.8 | 21.8 | 21.3 |
| Burnet | 53 | 94 | 9,354 | 13.3 | 42.4 | 25.4 | 73 | 9,851 | 30,980 | 59,492 | 40.1 | 5.2 | 62,827 | 10.5 | 16.3 | 15.5 |

1. All persons 3 years old and over enrolled in nursery school through college.  2. Persons 25 years old and over.  3. Elementary and secondary education expenditures.  4. Based on population estimated by the American Community Survey, 2014–2018.

# Table B. States and Counties — **Personal Income**

| STATE County | Personal income, 2019 | | | | | | | | | | Earnings, 2019 | | |
|---|---|---|---|---|---|---|---|---|---|---|---|---|---|
| | Total (mil dol) | Percent change 2018-2019 | Per capita[1] Dollars | Per capita[1] Rank | Wages and salaries (mil dol) | Supplements to wages and salaries, employer contributions (mil dol) Pension and insurance | Supplements (mil dol) Government social insurance | Proprietors' income (mil dol) | Dividends, interest, and rent (mil dol) | Personal transfer receipts (mil dol) | Total (mil dol) | Contributions for government social insurance (mil dol) From employee and self-employed | Contributions (mil dol) From employer |
| | 62 | 63 | 64 | 65 | 66 | 67 | 68 | 69 | 70 | 71 | 72 | 73 | 74 |
| **TENNESSEE—Cont'd** | | | | | | | | | | | | | |
| Marshall | 1,347 | 5.7 | 39,173 | 2,167 | 443 | 84 | 31 | 112 | 158 | 328 | 669 | 45 | 31 |
| Maury | 4,265 | 6.1 | 44,246 | 1,452 | 1,916 | 317 | 133 | 408 | 504 | 939 | 2,773 | 174 | 133 |
| Meigs | 428 | 4.9 | 34,472 | 2,747 | 99 | 23 | 8 | 23 | 58 | 141 | 153 | 13 | 8 |
| Monroe | 1,616 | 3.6 | 34,720 | 2,725 | 611 | 122 | 45 | 117 | 216 | 542 | 895 | 66 | 45 |
| Montgomery | 8,992 | 5.0 | 43,028 | 1,615 | 2,492 | 458 | 174 | 692 | 1,341 | 1,893 | 3,816 | 233 | 174 |
| Moore | 270 | 5.2 | 41,565 | 1,838 | 106 | 29 | 8 | 15 | 40 | 69 | 158 | 10 | 8 |
| Morgan | 663 | 3.1 | 30,962 | 3,009 | 119 | 31 | 8 | 34 | 75 | 225 | 193 | 18 | 8 |
| Obion | 1,284 | 3.6 | 42,717 | 1,654 | 448 | 83 | 31 | 126 | 210 | 375 | 687 | 47 | 31 |
| Overton | 780 | 3.6 | 35,085 | 2,674 | 212 | 45 | 15 | 91 | 88 | 257 | 363 | 27 | 15 |
| Perry | 279 | 0.8 | 34,522 | 2,742 | 66 | 16 | 5 | 42 | 44 | 94 | 129 | 9 | 5 |
| Pickett | 200 | 1.3 | 39,569 | 2,107 | 37 | 8 | 3 | 39 | 29 | 69 | 87 | 7 | 3 |
| Polk | 576 | 2.7 | 34,234 | 2,776 | 87 | 20 | 6 | 35 | 68 | 187 | 147 | 14 | 6 |
| Putnam | 3,228 | 3.9 | 40,223 | 2,023 | 1,564 | 307 | 108 | 462 | 506 | 823 | 2,441 | 149 | 108 |
| Rhea | 1,175 | 4.3 | 35,424 | 2,631 | 532 | 115 | 41 | 63 | 140 | 377 | 751 | 51 | 41 |
| Roane | 2,238 | 3.2 | 41,917 | 1,783 | 1,382 | 175 | 92 | 125 | 301 | 671 | 1,775 | 121 | 92 |
| Robertson | 3,149 | 5.2 | 43,856 | 1,506 | 960 | 180 | 70 | 368 | 338 | 654 | 1,577 | 98 | 70 |
| Rutherford | 14,193 | 5.8 | 42,712 | 1,655 | 7,174 | 1,137 | 490 | 1,689 | 1,548 | 2,230 | 10,491 | 606 | 490 |
| Scott | 685 | 2.0 | 31,046 | 3,001 | 207 | 49 | 16 | 50 | 78 | 268 | 322 | 25 | 16 |
| Sequatchie | 588 | 4.8 | 39,154 | 2,169 | 110 | 25 | 8 | 37 | 77 | 190 | 180 | 16 | 8 |
| Sevier | 4,042 | 3.2 | 41,136 | 1,902 | 1,778 | 255 | 129 | 708 | 575 | 989 | 2,870 | 183 | 129 |
| Shelby | 47,556 | 2.8 | 50,744 | 772 | 30,715 | 4,251 | 2,114 | 4,589 | 7,849 | 8,591 | 41,669 | 2,447 | 2,114 |
| Smith | 812 | 4.8 | 40,290 | 2,014 | 242 | 48 | 17 | 75 | 129 | 194 | 382 | 26 | 17 |
| Stewart | 564 | 4.6 | 41,155 | 1,899 | 160 | 37 | 12 | 32 | 77 | 175 | 242 | 18 | 12 |
| Sullivan | 6,906 | 2.1 | 43,610 | 1,533 | 3,549 | 607 | 243 | 648 | 1,073 | 1,855 | 5,048 | 332 | 243 |
| Sumner | 9,605 | 5.4 | 50,213 | 812 | 2,745 | 442 | 190 | 1,290 | 1,164 | 1,647 | 4,667 | 289 | 190 |
| Tipton | 2,496 | 3.9 | 40,514 | 1,989 | 482 | 95 | 33 | 154 | 277 | 582 | 765 | 56 | 33 |
| Trousdale | 359 | 6.8 | 31,835 | 2,955 | 83 | 16 | 6 | 50 | 36 | 93 | 154 | 11 | 6 |
| Unicoi | 667 | 3.3 | 37,278 | 2,418 | 245 | 50 | 19 | 40 | 89 | 243 | 354 | 27 | 19 |
| Union | 656 | 3.9 | 32,853 | 2,902 | 104 | 23 | 7 | 61 | 70 | 210 | 195 | 17 | 7 |
| Van Buren | 186 | 5.5 | 31,628 | 2,967 | 35 | 10 | 2 | 13 | 26 | 73 | 60 | 6 | 2 |
| Warren | 1,463 | 3.7 | 35,453 | 2,625 | 557 | 103 | 39 | 166 | 193 | 458 | 864 | 58 | 39 |
| Washington | 5,769 | 2.8 | 44,593 | 1,408 | 2,825 | 544 | 196 | 584 | 879 | 1,392 | 4,148 | 258 | 196 |
| Wayne | 482 | 3.3 | 28,922 | 3,068 | 148 | 34 | 11 | 40 | 66 | 174 | 233 | 18 | 11 |
| Weakley | 1,282 | 5.1 | 38,465 | 2,268 | 526 | 121 | 37 | 127 | 192 | 360 | 811 | 51 | 37 |
| White | 923 | 4.4 | 33,761 | 2,825 | 280 | 57 | 20 | 93 | 121 | 319 | 450 | 35 | 20 |
| Williamson | 22,662 | 5.4 | 95,053 | 27 | 10,438 | 1,169 | 659 | 5,077 | 3,463 | 1,425 | 17,344 | 936 | 659 |
| Wilson | 7,359 | 6.0 | 50,875 | 764 | 2,297 | 335 | 160 | 797 | 854 | 1,214 | 3,589 | 226 | 160 |
| **TEXAS** | 1,531,346 | 4.6 | 52,829 | X | 786,698 | 113,570 | 51,676 | 200,038 | 269,602 | 224,585 | 1,151,983 | 60,709 | 51,676 |
| Anderson | 2,080 | 4.4 | 36,027 | 2,567 | 1,027 | 183 | 69 | 89 | 294 | 547 | 1,367 | 81 | 69 |
| Andrews | 968 | 5.9 | 51,769 | 699 | 594 | 81 | 37 | 86 | 100 | 119 | 799 | 42 | 37 |
| Angelina | 3,438 | 1.8 | 39,644 | 2,096 | 1,532 | 273 | 103 | 275 | 617 | 972 | 2,183 | 130 | 103 |
| Aransas | 1,213 | 3.1 | 51,614 | 713 | 246 | 40 | 16 | 86 | 306 | 318 | 389 | 30 | 16 |
| Archer | 448 | 3.7 | 52,335 | 655 | 82 | 17 | 6 | 47 | 79 | 87 | 151 | 9 | 6 |
| Armstrong | 101 | 2.8 | 53,422 | 582 | 18 | 4 | 1 | 14 | 18 | 23 | 36 | 2 | 1 |
| Atascosa | 1,926 | 4.3 | 37,644 | 2,367 | 837 | 124 | 54 | 80 | 315 | 457 | 1,095 | 68 | 54 |
| Austin | 1,535 | 3.6 | 51,118 | 745 | 570 | 83 | 39 | 88 | 321 | 298 | 780 | 49 | 39 |
| Bailey | 313 | 2.2 | 44,665 | 1,398 | 94 | 19 | 7 | 78 | 45 | 62 | 198 | 8 | 7 |
| Bandera | 1,038 | 4.3 | 44,925 | 1,367 | 141 | 26 | 10 | 54 | 259 | 256 | 230 | 21 | 10 |
| Bastrop | 3,397 | 7.0 | 38,289 | 2,295 | 837 | 160 | 57 | 327 | 487 | 712 | 1,381 | 86 | 57 |
| Baylor | 164 | 3.6 | 46,615 | 1,171 | 52 | 13 | 4 | 22 | 31 | 50 | 91 | 5 | 4 |
| Bee | 970 | 4.1 | 29,792 | 3,044 | 361 | 81 | 22 | 40 | 185 | 287 | 504 | 29 | 22 |
| Bell | 15,939 | 5.1 | 43,919 | 1,500 | 8,629 | 1,936 | 687 | 992 | 2,560 | 3,490 | 12,244 | 611 | 687 |
| Bexar | 95,830 | 4.0 | 47,830 | 1,041 | 52,332 | 8,507 | 3,576 | 12,273 | 17,072 | 16,303 | 76,688 | 4,056 | 3,576 |
| Blanco | 654 | 5.6 | 54,814 | 506 | 182 | 30 | 13 | 53 | 209 | 120 | 277 | 18 | 13 |
| Borden | 40 | -7.7 | 61,287 | 270 | 12 | 2 | 1 | 17 | 5 | 14 | 32 | 2 | 1 |
| Bosque | 792 | 3.0 | 42,366 | 1,709 | 174 | 37 | 12 | 42 | 152 | 264 | 264 | 14 | 12 |
| Bowie | 3,839 | 2.3 | 41,172 | 1,896 | 1,848 | 371 | 131 | 264 | 724 | 1,016 | 2,614 | 154 | 131 |
| Brazoria | 18,105 | 5.0 | 48,374 | 990 | 7,275 | 1,161 | 482 | 761 | 2,122 | 2,687 | 9,679 | 545 | 482 |
| Brazos | 9,478 | 4.6 | 41,348 | 1,869 | 4,863 | 985 | 300 | 844 | 2,116 | 1,298 | 6,993 | 334 | 300 |
| Brewster | 446 | 4.2 | 48,422 | 985 | 180 | 39 | 12 | 31 | 147 | 89 | 262 | 15 | 12 |
| Briscoe | 69 | 1.6 | 44,413 | 1,430 | 13 | 3 | 1 | 12 | 15 | 17 | 29 | 2 | 1 |
| Brooks | 259 | 4.2 | 36,558 | 2,501 | 127 | 31 | 9 | 5 | 32 | 112 | 172 | 11 | 9 |
| Brown | 1,502 | 1.7 | 39,661 | 2,092 | 642 | 126 | 44 | 83 | 230 | 486 | 895 | 57 | 44 |
| Burleson | 848 | 4.2 | 45,970 | 1,251 | 237 | 38 | 16 | 59 | 174 | 203 | 351 | 24 | 16 |
| Burnet | 2,395 | 4.9 | 49,731 | 855 | 708 | 120 | 47 | 228 | 689 | 527 | 1,104 | 72 | 47 |

1. Based on the resident population estimated as of July 1 of the year shown.

# Table B. States and Counties — Earnings, Social Security, and Housing

Column groups: Items 75–83 = **Earnings, 2019 (cont.) — Percent by selected industries**; Items 84–85 = **Social Security beneficiaries, December 2019**; Item 86 = **Supplemental Security Income recipients, 2019**; Items 87–88 = **Housing units, 2020**.

| STATE County | Farm | Mining, quarrying, and extractions | Construction | Manufacturing | Information; professional, scientific, technical services | Retail trade | Finance, insurance, real estate, and leasing | Health care and social assistance | Government | Number | Rate[1] | Supplemental Security Income recipients, 2019 | Total | Percent change, 2010-2020 |
|---|---|---|---|---|---|---|---|---|---|---|---|---|---|---|
| | 75 | 76 | 77 | 78 | 79 | 80 | 81 | 82 | 83 | 84 | 85 | 86 | 87 | 88 |
| **TENNESSEE—Cont'd** | | | | | | | | | | | | | | |
| Marshall | -1.0 | D | 8.1 | 38.1 | D | 7.7 | 3.3 | D | 16.6 | 7,555 | 219 | 678 | 14,070 | 7.3 |
| Maury | -0.5 | D | 7.2 | 22.6 | 5.1 | 6.4 | 11.0 | 9.4 | 16.8 | 21,505 | 222 | 1,756 | 40,184 | 13.9 |
| Meigs | -1.9 | D | 7.1 | 41.8 | D | 3.1 | D | 8.1 | 20.5 | 3,535 | 286 | 468 | 6,031 | 7.1 |
| Monroe | -0.8 | D | 5.1 | 42.5 | 2.4 | 8.4 | 3.3 | D | 12.7 | 13,660 | 293 | 1,600 | 21,579 | 3.9 |
| Montgomery | 0.0 | 0.7 | 9.5 | 10.7 | 7.1 | 10.4 | 5.4 | 11.3 | 21.8 | 30,745 | 147 | 3,588 | 86,517 | 23.4 |
| Moore | -2.5 | 0.0 | D | 10.8 | D | 1.6 | 1.0 | 4.3 | 31.5 | 1,560 | 242 | 72 | 3,107 | 6.7 |
| Morgan | -1.9 | D | 8.8 | 21.6 | D | 4.1 | 1.9 | 7.7 | 39.0 | 5,685 | 266 | 843 | 9,097 | 2.0 |
| Obion | 3.1 | 0.0 | 5.7 | 16.0 | 6.3 | 13.5 | 3.7 | 9.3 | 15.5 | 8,900 | 294 | 1,100 | 14,585 | -0.5 |
| Overton | -1.1 | 1.0 | 9.7 | 20.4 | D | 8.6 | 5.5 | 12.3 | 18.2 | 6,410 | 288 | 595 | 10,416 | 1.2 |
| Perry | -0.6 | D | 9.3 | D | D | 9.7 | D | 11.3 | 21.3 | 2,305 | 285 | 230 | 4,716 | 2.4 |
| Pickett | -0.7 | 0.5 | 14.5 | D | D | D | 3.8 | 8.8 | 17.8 | 1,710 | 337 | 142 | 3,522 | 1.8 |
| Polk | 0.6 | D | D | 8.5 | D | 8.3 | 3.4 | D | 28.0 | 4,720 | 282 | 500 | 9,456 | 18.3 |
| Putnam | -0.4 | 0.1 | 7.4 | 14.4 | 6.0 | 8.8 | 5.8 | 11.0 | 22.9 | 18,730 | 234 | 2,174 | 35,812 | 12.3 |
| Rhea | -0.3 | 0.3 | 4.3 | 26.0 | D | 5.5 | 2.2 | D | 40.1 | 8,600 | 259 | 1,274 | 15,047 | 4.8 |
| Roane | -0.5 | D | D | 3.9 | 50.5 | 3.8 | 1.4 | 5.9 | 12.8 | 15,995 | 299 | 1,400 | 25,772 | 0.2 |
| Robertson | 1.1 | D | 14.4 | 23.6 | 3.3 | 7.0 | 4.1 | 5.4 | 14.8 | 14,550 | 202 | 1,193 | 28,477 | 9.1 |
| Rutherford | -0.1 | D | 8.5 | 23.0 | 5.7 | 7.3 | 5.8 | 8.2 | 14.1 | 46,935 | 141 | 4,058 | 129,190 | 25.5 |
| Scott | -0.4 | D | 10.6 | 22.0 | 2.0 | 7.8 | 2.4 | D | 24.2 | 5,865 | 265 | 1,293 | 10,076 | 1.6 |
| Sequatchie | -0.6 | D | 6.4 | 14.2 | D | 10.2 | 7.4 | D | 24.2 | 4,525 | 300 | 503 | 6,556 | 2.9 |
| Sevier | 0.0 | D | 8.0 | 4.2 | 3.7 | 12.0 | 7.0 | 5.3 | 12.1 | 25,590 | 260 | 1,889 | 60,316 | 8.0 |
| Shelby | 0.0 | D | 5.0 | 9.2 | 5.4 | 6.1 | 8.6 | 13.7 | 13.2 | 162,965 | 174 | 33,231 | 407,487 | 2.3 |
| Smith | -1.9 | D | D | 23.3 | 2.0 | 7.0 | 4.5 | 7.3 | 17.5 | 4,680 | 233 | 503 | 8,969 | 5.2 |
| Stewart | -0.1 | 0.0 | D | 14.5 | D | D | 1.7 | 3.9 | 42.8 | 3,910 | 285 | 393 | 6,905 | 1.9 |
| Sullivan | -0.1 | -0.1 | 7.0 | 24.5 | 5.7 | 7.9 | 4.3 | 17.2 | 9.7 | 46,850 | 296 | 4,696 | 76,133 | 3.2 |
| Sumner | 0.0 | D | 12.3 | 13.4 | 10.3 | 7.6 | 6.2 | 11.6 | 11.7 | 38,375 | 200 | 2,571 | 76,799 | 16.4 |
| Tipton | 1.1 | D | 12.1 | 13.9 | 3.3 | 7.9 | 3.8 | D | 23.1 | 12,600 | 204 | 1,402 | 24,234 | 4.6 |
| Trousdale | -1.1 | 0.1 | D | 12.9 | 3.9 | 7.6 | 3.5 | D | 19.9 | 2,020 | 179 | 243 | 3,873 | 15.1 |
| Unicoi | -0.1 | D | 4.4 | 39.2 | D | 5.4 | 1.9 | 8.2 | 15.5 | 5,425 | 303 | 574 | 8,915 | 1.0 |
| Union | -2.2 | D | D | 16.8 | D | 6.8 | D | D | 24.1 | 5,230 | 262 | 742 | 9,675 | 8.0 |
| Van Buren | -4.3 | 0.0 | 8.5 | 19.1 | D | D | D | D | 46.4 | 1,870 | 318 | 159 | 2,706 | 1.6 |
| Warren | 3.9 | 0.0 | 6.6 | 26.7 | D | 7.5 | 3.5 | 10.9 | 15.2 | 10,605 | 257 | 1,529 | 18,137 | 1.8 |
| Washington | 0.0 | D | 4.2 | 8.7 | 6.4 | 8.2 | 7.1 | 21.4 | 22.9 | 32,595 | 251 | 3,336 | 61,670 | 7.7 |
| Wayne | -1.8 | D | 4.6 | 8.7 | D | 6.7 | 5.0 | 17.7 | 29.2 | 4,385 | 263 | 415 | 7,387 | 1.4 |
| Weakley | 4.2 | 1.4 | 3.2 | 13.1 | D | 5.9 | 16.6 | D | 27.2 | 8,095 | 243 | 926 | 15,684 | 1.2 |
| White | -2.5 | D | 6.9 | 26.4 | 2.8 | 7.4 | 3.8 | 11.5 | 16.1 | 8,025 | 293 | 943 | 12,260 | 6.4 |
| Williamson | 0.0 | D | 7.6 | 2.1 | 18.1 | 5.7 | 13.2 | 17.3 | 5.1 | 35,020 | 147 | 1,035 | 90,442 | 32.1 |
| Wilson | -0.5 | D | 10.2 | 10.2 | 6.8 | 9.1 | 6.3 | 7.1 | 10.2 | 28,365 | 196 | 1,547 | 58,256 | 27.8 |
| **TEXAS** | 0.3 | 8.4 | 7.5 | 8.5 | 12.0 | 5.4 | 8.7 | 9.0 | 13.6 | 4,338,301 | 150 | 644,093 | 11,487,821 | 15.1 |
| Anderson | -0.2 | 4.0 | 4.6 | 7.6 | 7.6 | 5.5 | 4.6 | 8.8 | 24.8 | 10,900 | 189 | 1,494 | 20,905 | 3.9 |
| Andrews | 0.4 | 34.1 | 17.5 | 2.2 | 3.8 | 3.9 | 2.9 | D | 12.1 | 2,305 | 122 | 258 | 6,526 | 12.2 |
| Angelina | 0.2 | 2.0 | 6.1 | 9.3 | 4.7 | 7.7 | 4.1 | 19.5 | 18.9 | 19,025 | 219 | 2,988 | 37,515 | 5.4 |
| Aransas | -0.3 | 5.4 | 11.2 | 1.7 | 9.2 | 12.1 | 7.1 | 6.4 | 16.5 | 7,480 | 320 | 513 | 16,480 | 7.3 |
| Archer | 8.8 | 10.0 | 8.4 | 3.6 | D | 7.0 | D | D | 23.6 | 2,000 | 230 | 111 | 4,222 | 2.7 |
| Armstrong | 15.7 | 1.9 | 16.2 | D | 9.7 | 3.0 | D | 8.4 | 17.2 | 455 | 246 | 20 | 937 | 3.7 |
| Atascosa | -0.7 | 29.4 | 5.9 | 3.0 | 4.1 | 6.1 | 5.7 | 6.9 | 14.0 | 9,410 | 184 | 1,317 | 18,926 | 7.3 |
| Austin | -1.8 | 0.6 | 12.8 | 14.7 | 10.6 | 11.4 | 8.5 | 4.9 | 12.1 | 6,615 | 220 | 514 | 13,558 | 4.9 |
| Bailey | 35.2 | D | 2.7 | 3.3 | D | 3.8 | D | 4.4 | 12.6 | 1,165 | 170 | 113 | 2,789 | 0.2 |
| Bandera | -3.2 | 1.0 | 19.1 | 1.3 | 7.2 | 6.9 | 4.8 | 10.9 | 19.4 | 6,885 | 297 | 376 | 12,141 | 5.0 |
| Bastrop | -0.7 | 3.7 | 18.1 | 5.8 | D | 9.7 | 3.8 | 7.5 | 21.2 | 16,220 | 183 | 1,806 | 31,348 | 6.9 |
| Baylor | 10.2 | D | 3.3 | 2.4 | D | 4.4 | D | D | 15.2 | 1,025 | 291 | 117 | 2,640 | -0.9 |
| Bee | 0.6 | 6.5 | 2.5 | 1.4 | 4.9 | 8.8 | 5.0 | D | 39.7 | 5,005 | 153 | 939 | 10,771 | 1.2 |
| Bell | 0.0 | 0.3 | 5.0 | 4.0 | 4.7 | 5.5 | 3.7 | 15.3 | 45.8 | 57,255 | 157 | 9,009 | 146,538 | 16.8 |
| Bexar | 0.0 | 8.1 | 6.5 | 4.2 | 9.3 | 5.5 | 12.4 | 10.9 | 20.6 | 310,805 | 155 | 51,096 | 714,668 | 7.8 |
| Blanco | 0.8 | D | 24.2 | 7.8 | D | 4.9 | 3.9 | 3.4 | 12.7 | 3,080 | 257 | 106 | 5,885 | 6.4 |
| Borden | -24.5 | D | D | D | D | D | D | 1.9 | 40.3 | 125 | 191 | 0 | 391 | 1.6 |
| Bosque | 2.9 | D | 13.6 | 13.1 | 4.8 | 5.3 | 4.0 | D | 27.4 | 5,230 | 281 | 402 | 9,832 | 2.2 |
| Bowie | 0.1 | 0.3 | 5.1 | 4.9 | 4.4 | 8.4 | 5.8 | 19.7 | 27.0 | 20,270 | 217 | 3,833 | 40,265 | 4.6 |
| Brazoria | 0.0 | 2.5 | 19.7 | 20.1 | 4.9 | 5.9 | 4.0 | 8.5 | 13.9 | 53,555 | 143 | 5,874 | 145,871 | 23.3 |
| Brazos | 0.0 | 3.7 | 6.9 | 4.6 | 8.6 | 6.5 | 5.1 | 10.4 | 35.2 | 24,205 | 106 | 3,434 | 95,636 | 23.1 |
| Brewster | -0.1 | 1.7 | 9.5 | 0.7 | D | 5.5 | 5.4 | D | 37.2 | 2,210 | 242 | 183 | 5,595 | 3.9 |
| Briscoe | 37.0 | 1.1 | D | D | 1.9 | 3.3 | 9.5 | D | 21.4 | 435 | 282 | 36 | 957 | 0.4 |
| Brooks | 0.2 | 11.4 | 2.5 | 0.1 | D | 5.0 | D | 9.1 | 50.9 | 1,740 | 245 | 448 | 3,241 | 0.0 |
| Brown | -2.1 | 1.1 | 8.3 | 24.5 | D | 7.6 | 4.6 | 13.0 | 19.2 | 9,770 | 259 | 1,207 | 19,500 | 6.6 |
| Burleson | -0.5 | 9.3 | 21.5 | 7.0 | 5.6 | 8.7 | 5.7 | D | 13.1 | 4,475 | 243 | 500 | 9,444 | 6.9 |
| Burnet | -2.1 | 3.8 | 14.6 | 7.8 | 7.5 | 8.2 | 5.0 | 13.7 | 15.4 | 12,875 | 267 | 605 | 24,570 | 17.9 |

1. Per 1,000 resident population estimated as of July 1 of the year shown.

# Table B. States and Counties — Housing, Labor Force, and Employment

| STATE County | Housing units, 2015-2019 — Occupied units — Total [89] | Percent [90] | Owner-occupied Median value[1] [91] | Median owner cost as a percent of income — With a mortgage[2] [92] | Without a mortgage[2] [93] | Renter-occupied Median rent[3] [94] | Median rent as a percent of income[2] [95] | Sub-standard units[4] (percent) [96] | Civilian labor force, 2020 — Total [97] | Percent change, 2019-2020 [98] | Unemployment Total [99] | Rate[5] [100] | Civilian employment[6], 2015-2019 — Total [101] | Percent — Management, business, science, and arts [102] | Construction, production, and maintenance occupations [103] |
|---|---|---|---|---|---|---|---|---|---|---|---|---|---|---|---|
| **TENNESSEE—Cont'd** | | | | | | | | | | | | | | | |
| Marshall | 12,324 | 71.8 | 136,400 | 20.5 | 10.0 | 763 | 24.7 | 2.0 | 15,374 | -2.5 | 1,396 | 9.1 | 14,927 | 24.3 | 38.7 |
| Maury | 34,688 | 69.9 | 184,800 | 19.9 | 10.1 | 895 | 28.3 | 2.0 | 49,915 | -0.7 | 4,061 | 8.1 | 44,397 | 31.9 | 29.6 |
| Meigs | 4,938 | 78.7 | 136,600 | 19.7 | 10.0 | 726 | 30.5 | 2.1 | 5,195 | 0.0 | 449 | 8.6 | 4,517 | 22.0 | 44.0 |
| Monroe | 17,987 | 76.7 | 133,700 | 20.3 | 11.1 | 668 | 25.3 | 3.1 | 20,090 | | 1,611 | 8.0 | 17,730 | 23.1 | 40.0 |
| Montgomery | 72,617 | 59.5 | 165,800 | 20.7 | 10.0 | 971 | 28.1 | 1.9 | 86,234 | -0.4 | 6,568 | 7.6 | 82,421 | 32.5 | 27.1 |
| Moore | 2,592 | 84.8 | 187,100 | 21.8 | 10.0 | 672 | 24.4 | 3.5 | 3,610 | -0.4 | 194 | 5.4 | 2,777 | 28.5 | 41.4 |
| Morgan | 7,625 | 81.6 | 102,000 | 21.0 | 12.1 | 683 | 31.6 | 3.4 | 7,654 | -1.7 | 501 | 6.5 | 7,370 | 26.6 | 30.5 |
| Obion | 12,717 | 64.9 | 91,500 | 18.1 | 11.3 | 621 | 28.6 | 1.8 | 12,180 | -2.4 | 858 | 7.0 | 12,851 | 28.6 | 32.2 |
| Overton | 9,140 | 79.7 | 124,900 | 21.6 | 11.7 | 545 | 27.0 | 1.6 | 9,689 | -2.5 | 615 | 6.3 | 9,336 | 28.2 | 33.8 |
| Perry | 3,073 | 82.4 | 88,100 | 23.2 | 11.1 | 613 | 27.8 | 2.8 | 2,989 | -4.0 | 284 | 9.5 | 2,765 | 27.8 | 41.0 |
| Pickett | 2,281 | 81.7 | 144,400 | 14.8 | 10.8 | 538 | 25.4 | 1.9 | 2,241 | -4.5 | 137 | 6.1 | 1,901 | 33.1 | 30.2 |
| Polk | 7,383 | 73.7 | 112,700 | 19.3 | 10.0 | 688 | 22.2 | 3.2 | 7,455 | 0.1 | 512 | 6.9 | 7,188 | 27.9 | 40.2 |
| Putnam | 31,424 | 61.5 | 159,600 | 20.9 | 10.0 | 715 | 30.5 | 1.1 | 34,970 | -0.5 | 2,309 | 6.6 | 34,503 | 34.9 | 22.9 |
| Rhea | 12,177 | 71.2 | 122,600 | 18.8 | 10.0 | 647 | 29.2 | 2.2 | 13,011 | -2.6 | 1,279 | 9.8 | 13,076 | 25.2 | 35.1 |
| Roane | 20,901 | 77.2 | 150,300 | 19.4 | 10.0 | 672 | 26.6 | 1.9 | 23,021 | -1.6 | 1,513 | 6.6 | 22,195 | 35.1 | 25.5 |
| Robertson | 25,713 | 74.0 | 194,800 | 20.0 | 10.8 | 894 | 27.8 | 2.8 | 37,300 | -2.4 | 2,282 | 6.1 | 33,928 | 33.6 | 30.0 |
| Rutherford | 111,676 | 65.2 | 213,200 | 19.7 | 10.0 | 1,063 | 28.0 | 2.7 | 184,368 | -1.5 | 12,637 | 6.9 | 168,044 | 35.2 | 26.0 |
| Scott | 8,664 | 70.0 | 99,000 | 19.1 | 11.1 | 556 | 26.7 | 1.1 | 8,170 | -2.7 | 700 | 8.6 | 8,163 | 25.9 | 41.3 |
| Sequatchie | 5,528 | 74.7 | 142,500 | 19.3 | 11.1 | 695 | 23.1 | 3.5 | 6,146 | -1.5 | 482 | 7.8 | 6,117 | 27.4 | 32.6 |
| Sevier | 37,210 | 69.4 | 175,500 | 21.0 | 10.0 | 791 | 28.2 | 3.5 | 53,721 | -2.3 | 5,181 | 9.6 | 46,934 | 25.3 | 22.1 |
| Shelby | 351,194 | 55.1 | 150,400 | 20.9 | 11.2 | 942 | 32.0 | 2.5 | 447,769 | 0.0 | 43,262 | 9.7 | 435,649 | 35.6 | 24.2 |
| Smith | 7,603 | 73.9 | 146,900 | 21.2 | 10.0 | 621 | 23.4 | 1.5 | 9,259 | -2.6 | 558 | 6.0 | 9,136 | 28.1 | 32.8 |
| Stewart | 5,178 | 75.0 | 123,100 | 20.1 | 12.2 | 644 | 25.3 | 2.0 | 5,412 | -2.2 | 359 | 6.6 | 5,341 | 26.7 | 31.7 |
| Sullivan | 66,511 | 71.8 | 141,300 | 19.0 | 10.0 | 668 | 27.3 | 1.8 | 69,352 | -1.9 | 4,982 | 7.2 | 66,728 | 34.9 | 24.1 |
| Sumner | 67,089 | 73.6 | 234,200 | 20.4 | 10.0 | 1,026 | 27.9 | 2.4 | 102,069 | -1.9 | 6,645 | 6.5 | 92,522 | 36.6 | 22.9 |
| Tipton | 21,452 | 71.3 | 150,000 | 18.9 | 10.0 | 825 | 27.4 | 2.6 | 27,498 | -2.8 | 1,886 | 6.9 | 27,600 | 30.4 | 30.8 |
| Trousdale | 3,189 | 78.8 | 155,700 | 22.5 | 12.7 | 787 | 23.5 | 0.7 | 5,383 | -2.6 | 328 | 6.1 | 4,605 | 23.8 | 36.3 |
| Unicoi | 7,658 | 71.9 | 129,500 | 20.4 | 10.9 | 630 | 28.3 | 0.6 | 7,083 | -1.2 | 586 | 8.3 | 7,447 | 28.9 | 33.4 |
| Union | 7,405 | 75.4 | 124,700 | 19.9 | 10.0 | 685 | 27.5 | 4.1 | 7,610 | -1.2 | 522 | 6.9 | 7,346 | 22.6 | 40.0 |
| Van Buren | 2,141 | 81.5 | 98,000 | 19.5 | 10.0 | 511 | 21.1 | 6.4 | 2,011 | -2.3 | 175 | 8.7 | 2,134 | 25.2 | 31.6 |
| Warren | 15,727 | 69.5 | 112,500 | 19.7 | 10.0 | 640 | 27.8 | 3.4 | 17,076 | 0.8 | 1,512 | 8.9 | 17,231 | 25.6 | 41.6 |
| Washington | 53,859 | 64.4 | 163,000 | 19.5 | 10.0 | 757 | 28.8 | 1.6 | 59,783 | -1.8 | 3,855 | 6.4 | 60,301 | 38.6 | 21.1 |
| Wayne | 5,764 | 77.4 | 95,200 | 19.5 | 10.8 | 560 | 24.8 | 2.9 | 6,414 | 1.8 | 457 | 7.1 | 5,756 | 26.3 | 32.1 |
| Weakley | 13,640 | 65.3 | 97,800 | 19.5 | 10.0 | 626 | 30.1 | 1.6 | 15,456 | -3.9 | 902 | 5.8 | 14,319 | 33.9 | 28.7 |
| White | 10,048 | 80.0 | 117,000 | 21.7 | 10.6 | 680 | 32.1 | 1.8 | 12,011 | -2.2 | 856 | 7.1 | 10,438 | 24.0 | 37.6 |
| Williamson | 77,855 | 80.6 | 449,000 | 19.0 | 10.0 | 1,500 | 27.1 | 1.3 | 126,997 | -3.3 | 6,123 | 4.8 | 114,816 | 55.0 | 10.5 |
| Wilson | 49,664 | 76.8 | 264,600 | 19.6 | 10.0 | 1,065 | 27.5 | 1.9 | 76,694 | -2.2 | 4,812 | 6.3 | 68,351 | 39.7 | 21.9 |
| **TEXAS** | 9,691,647 | 62.0 | 172,500 | 21.0 | 11.3 | 1,045 | 29.1 | 5.1 | 13,983,319 | -0.4 | 1,067,982 | 7.6 | 13,253,631 | 36.7 | 23.8 |
| Anderson | 16,677 | 69.6 | 101,900 | 23.4 | 13.5 | 770 | 26.6 | 3.3 | 23,325 | -0.6 | 1,312 | 5.6 | 19,213 | 21.1 | 33.2 |
| Andrews | 5,573 | 74.1 | 143,900 | 16.3 | 10.0 | 1,028 | 23.2 | 8.1 | 9,220 | -7.0 | 754 | 8.2 | 8,167 | 22.5 | 39.3 |
| Angelina | 31,035 | 66.2 | 107,600 | 20.6 | 11.3 | 827 | 28.9 | 4.9 | 35,495 | -0.7 | 2,712 | 7.6 | 36,525 | 28.9 | 29.4 |
| Aransas | 9,548 | 80.2 | 156,800 | 22.5 | 14.8 | 1,012 | 30.9 | 6.2 | 9,058 | -2.4 | 767 | 8.5 | 9,814 | 25.2 | 24.0 |
| Archer | 3,452 | 83.7 | 146,200 | 18.3 | 12.3 | 637 | 25.6 | 2.8 | 3,911 | -2.4 | 208 | 5.3 | 4,438 | 34.4 | 27.9 |
| Armstrong | 694 | 88.0 | 139,700 | 18.0 | 10.0 | 963 | 14.7 | 1.2 | 914 | -1.7 | 39 | 4.3 | 888 | 32.9 | 33.8 |
| Atascosa | 15,546 | 75.3 | 97,900 | 18.8 | 10.4 | 878 | 24.0 | 10.2 | 21,822 | 0.4 | 1,843 | 8.4 | 19,876 | 27.3 | 36.3 |
| Austin | 11,301 | 76.2 | 196,300 | 20.5 | 11.7 | 851 | 29.7 | 7.6 | 13,638 | -2.4 | 897 | 6.6 | 13,498 | 31.5 | 27.6 |
| Bailey | 2,054 | 74.6 | 69,700 | 26.6 | 10.0 | 591 | 23.0 | 5.2 | 2,440 | -2.4 | 117 | 4.8 | 3,305 | 19.5 | 46.2 |
| Bandera | 8,399 | 85.0 | 174,200 | 20.5 | 11.0 | 897 | 26.8 | 3.1 | 9,835 | -2.1 | 574 | 5.8 | 9,475 | 31.7 | 24.1 |
| Bastrop | 25,571 | 77.7 | 172,000 | 21.0 | 10.7 | 1,056 | 27.8 | 5.2 | 42,573 | -0.3 | 2,557 | 6.0 | 36,296 | 32.0 | 31.5 |
| Baylor | 1,530 | 72.4 | 62,000 | 21.4 | 13.0 | 523 | 25.5 | 2.4 | 1,767 | 0.9 | 62 | 3.5 | 1,540 | 30.4 | 21.4 |
| Bee | 8,269 | 62.7 | 90,600 | 22.4 | 10.0 | 868 | 30.0 | 5.7 | 9,906 | 0.8 | 974 | 9.8 | 10,038 | 22.5 | 29.1 |
| Bell | 122,689 | 54.3 | 147,700 | 20.9 | 10.8 | 927 | 28.1 | 4.0 | 144,558 | 0.5 | 10,099 | 7.0 | 141,131 | 34.2 | 23.1 |
| Bexar | 636,245 | 58.5 | 161,800 | 21.5 | 11.0 | 1,015 | 29.8 | 4.6 | 942,127 | -0.1 | 71,110 | 7.5 | 907,686 | 35.3 | 20.9 |
| Blanco | 4,343 | 78.6 | 246,400 | 21.3 | 10.0 | 852 | 25.7 | 4.7 | 6,573 | 1.0 | 251 | 3.8 | 5,488 | 31.5 | 24.5 |
| Borden | 227 | 75.8 | 153,300 | 12.3 | 10.0 | 0 | 18.9 | 0.4 | 430 | 6.4 | 14 | 3.3 | 231 | 45.0 | 29.9 |
| Bosque | 7,211 | 77.4 | 121,700 | 20.6 | 10.1 | 795 | 25.7 | 4.5 | 8,330 | 0.2 | 454 | 5.5 | 7,711 | 29.9 | 32.9 |
| Bowie | 34,076 | 63.5 | 119,400 | 18.9 | 11.5 | 761 | 29.3 | 3.9 | 38,755 | -2.2 | 2,864 | 7.4 | 37,008 | 33.0 | 26.7 |
| Brazoria | 121,523 | 72.1 | 197,500 | 19.6 | 10.1 | 1,141 | 27.3 | 4.0 | 177,709 | -1.1 | 15,267 | 8.6 | 169,376 | 42.5 | 25.1 |
| Brazos | 79,412 | 47.1 | 201,800 | 21.9 | 10.1 | 948 | 36.3 | 3.6 | 116,599 | -1.8 | 6,233 | 5.3 | 106,349 | 43.4 | 18.3 |
| Brewster | 4,088 | 57.8 | 151,400 | 16.1 | 10.0 | 703 | 25.1 | 3.3 | 4,083 | -2.6 | 272 | 6.7 | 4,495 | 36.5 | 22.8 |
| Briscoe | 602 | 69.3 | 66,000 | 18.0 | 11.3 | 675 | 22.0 | 2.0 | 567 | -2.1 | 27 | 4.8 | 600 | 26.7 | 33.0 |
| Brooks | 2,120 | 61.3 | 67,800 | 28.0 | 10.0 | 611 | 31.1 | 6.5 | 2,601 | -0.6 | 278 | 10.7 | 2,569 | 24.0 | 23.9 |
| Brown | 14,409 | 72.5 | 102,500 | 19.6 | 11.4 | 765 | 26.4 | 2.3 | 15,245 | -2.7 | 992 | 6.5 | 16,369 | 30.4 | 29.8 |
| Burleson | 6,810 | 78.2 | 121,900 | 21.2 | 10.0 | 733 | 21.5 | 4.6 | 8,180 | -1.2 | 519 | 6.3 | 7,823 | 26.7 | 39.3 |
| Burnet | 16,743 | 77.2 | 192,000 | 23.1 | 12.4 | 809 | 27.8 | 4.2 | 23,686 | 0.4 | 1,119 | 4.7 | 21,041 | 29.6 | 26.2 |

1. Specified owner-occupied units. lacking complete plumbing facilities.   2. A value of 10.0 represents 10 percent or less; a value of 50.0 represents 50 percent or more.   3. Specified renter-occupied units.   4. Overcrowded or
5. Percent of civilian labor force.   6. Civilian employed persons 16 years old and over.

# Table B. States and Counties — Nonfarm Employment and Agriculture

| | Private nonfarm establishments, employment and payroll, 2019 | | | | | | | | | Agriculture, 2017 | | | |
| STATE County | Number of establishments | Employment — Total | Health care and social assistance | Manufacturing | Retail trade | Finance and insurance | Professional, scientific, and technical services | Annual payroll — Total (mil dol) | Average per employee (dollars) | Farms — Number | Percent with: Fewer than 50 acres | Percent with: 1000 acres or more | Farm producers whose primary occupation is farming (percent) |
|---|---|---|---|---|---|---|---|---|---|---|---|---|---|
| | 104 | 105 | 106 | 107 | 108 | 109 | 110 | 111 | 112 | 113 | 114 | 115 | 116 |
| **TENNESSEE—Cont'd** | | | | | | | | | | | | | |
| Marshall | 512 | 8,369 | 713 | 3,708 | 1,143 | 183 | 139 | 314 | 37,469 | 1,096 | 43.2 | 0.7 | 37.1 |
| Maury | 1,837 | 31,853 | 5,956 | 6,599 | 4,389 | 1,830 | 665 | 1,470 | 46,140 | 1,583 | 46.3 | 2.1 | 35.2 |
| Meigs | 106 | 1,770 | 98 | 1,108 | 226 | 25 | NA | 68 | 38,465 | 351 | 38.2 | 2.0 | 40.9 |
| Monroe | 717 | 12,575 | 1,341 | 5,930 | 1,682 | 359 | 145 | 465 | 36,944 | 838 | 50.6 | 2.6 | 38.8 |
| Montgomery | 3,000 | 46,238 | 8,574 | 6,540 | 8,456 | 1,345 | 2,164 | 1,699 | 36,734 | 787 | 45.0 | 2.3 | 36.2 |
| Moore | 76 | 1,142 | 123 | 651 | 112 | 11 | 26 | 69 | 60,429 | 375 | 35.2 | 2.7 | 33.7 |
| Morgan | 154 | 1,298 | 310 | 192 | 233 | 45 | 16 | 45 | 34,810 | 443 | 38.9 | 11.0 | 41.1 |
| Obion | 637 | 8,773 | 1,145 | 2,452 | 1,527 | 395 | 113 | 316 | 36,053 | 553 | 45.5 | 1.0 | 28.9 |
| Overton | 330 | 3,895 | 653 | 833 | 581 | 131 | 113 | 142 | 36,437 | 1,004 | 31.4 | 4.5 | 30.6 |
| Perry | 106 | 1,765 | 351 | 850 | 185 | 49 | 8 | 58 | 32,820 | 287 | 38.3 | 1.0 | 30.9 |
| Pickett | 81 | 668 | 106 | 149 | 99 | 48 | 9 | 21 | 31,928 | 287 | 52.3 | 1.4 | 36.6 |
| Polk | 223 | 1,338 | 216 | 137 | 310 | 55 | 12 | 41 | 30,880 | 287 | 50.8 | 1.4 | 31.0 |
| Putnam | 1,845 | 31,917 | 6,282 | 5,358 | 4,900 | 1,074 | 749 | 1,247 | 39,062 | 1,003 | 50.4 | 0.4 | 34.1 |
| Rhea | 495 | 8,834 | 1,236 | 3,303 | 1,189 | 212 | 89 | 297 | 33,660 | 498 | 54.1 | 0.5 | 32.8 |
| Roane | 745 | 9,169 | 2,144 | 1,243 | 1,598 | 195 | 257 | 326 | 35,553 | 617 | 52.7 | 2.8 | 37.2 |
| Robertson | 1,202 | 20,049 | 1,973 | 6,528 | 2,667 | 437 | 286 | 773 | 38,580 | 1,202 | 56.4 | 1.3 | 35.8 |
| Rutherford | 5,517 | 114,372 | 14,083 | 18,789 | 15,851 | 4,306 | 3,991 | 5,015 | 43,845 | 1,414 | 39.9 | 0.3 | 39.3 |
| Scott | 324 | 3,828 | 718 | 1,206 | 690 | 131 | 30 | 115 | 30,147 | 288 | 48.5 | 1.7 | 34.4 |
| Sequatchie | 180 | 1,959 | 374 | 138 | 439 | 217 | 52 | 56 | 28,837 | 235 | 43.9 | NA | 39.7 |
| Sevier | 2,859 | 43,528 | 2,586 | 2,096 | 8,498 | 895 | 740 | 1,296 | 29,785 | 547 | 58.1 | 4.0 | 36.1 |
| Shelby | 19,536 | 447,904 | 74,825 | 27,051 | 44,666 | 16,004 | 19,697 | 23,525 | 52,521 | 885 | 33.8 | 1.2 | 38.8 |
| Smith | 274 | 4,105 | 440 | 1,354 | 621 | 108 | 65 | 175 | 42,576 | 389 | 28.0 | 3.9 | 33.4 |
| Stewart | 164 | 1,646 | 208 | 592 | 226 | 54 | 37 | 69 | 42,197 | 1,183 | 62.3 | 0.3 | 30.0 |
| Sullivan | 3,362 | 62,926 | 11,951 | 12,164 | 8,685 | 1,985 | 1,506 | 3,025 | 48,080 | 1,428 | 53.0 | 1.5 | 32.9 |
| Sumner | 3,347 | 48,173 | 7,668 | 8,271 | 6,596 | 1,726 | 1,937 | 2,099 | 43,576 | 527 | 47.1 | 8.5 | 43.6 |
| Tipton | 722 | 10,232 | 1,403 | 1,678 | 1,672 | 278 | 187 | 349 | 34,131 | 317 | 40.7 | NA | 32.6 |
| Trousdale | 111 | 1,440 | 217 | 176 | 201 | 50 | 60 | 53 | 36,863 | 100 | 57.0 | NA | 32.9 |
| Unicoi | 235 | 3,633 | 524 | 1,536 | 492 | 74 | 9 | 160 | 44,015 | 505 | 44.4 | 0.2 | 34.2 |
| Union | 212 | 1,742 | 129 | 534 | 367 | 45 | 48 | 59 | 34,032 | 329 | 37.7 | 0.6 | 47.9 |
| Van Buren | 43 | 469 | 112 | 182 | 74 | NA | NA | 13 | 28,239 | 1,133 | 45.3 | 1.9 | 38.9 |
| Warren | 720 | 9,783 | 1,338 | 3,002 | 1,468 | 354 | 157 | 363 | 37,125 | 1,428 | 63.5 | 0.6 | 36.8 |
| Washington | 2,931 | 55,010 | 13,693 | 5,241 | 8,423 | 2,725 | 1,297 | 2,240 | 40,722 | 685 | 25.4 | 2.2 | 34.2 |
| Wayne | 214 | 2,770 | 753 | 514 | 324 | 150 | 13 | 98 | 35,518 | 788 | 44.8 | 6.3 | 40.0 |
| Weakley | 553 | 8,250 | 1,529 | 1,949 | 1,082 | 308 | 101 | 273 | 33,075 | 971 | 46.9 | 0.4 | 38.4 |
| White | 415 | 5,706 | 820 | 2,147 | 826 | 144 | 85 | 194 | 34,056 | 1,224 | 51.0 | 1.3 | 31.2 |
| Williamson | 7,483 | 132,696 | 18,819 | 2,236 | 13,749 | 13,776 | 10,398 | 8,806 | 66,366 | 1,626 | 44.6 | 0.4 | 34.5 |
| Wilson | 2,818 | 43,013 | 4,000 | 4,851 | 6,024 | 1,137 | 2,413 | 1,757 | 40,844 | | | | |
| **TEXAS** | 609,476 | 11,104,054 | 1,586,107 | 822,922 | 1,322,089 | 559,916 | 757,090 | 611,142 | 55,038 | 248,416 | 44.1 | 8.3 | 35.8 |
| Anderson | 871 | 11,749 | 2,067 | 1,333 | 1,934 | 283 | 170 | 459 | 39,026 | 1,754 | 37.6 | 2.1 | 38.3 |
| Andrews | 432 | 6,264 | 513 | 320 | 580 | 117 | 148 | 412 | 65,749 | 156 | 43.6 | 26.9 | 34.1 |
| Angelina | 1,859 | 29,020 | 7,506 | 3,204 | 4,563 | 760 | 836 | 1,071 | 36,905 | 1,028 | 55.7 | 0.4 | 35.7 |
| Aransas | 503 | 4,172 | 363 | 64 | 1,001 | 165 | 149 | 146 | 35,079 | 97 | 68.0 | 4.1 | 34.3 |
| Archer | 218 | 1,168 | 108 | 70 | 109 | 26 | 25 | 46 | 39,074 | 532 | 7.4 | 36.6 | 42.4 |
| Armstrong | 50 | 281 | 69 | NA | 27 | 8 | NA | 10 | 34,014 | 216 | 39.1 | 7.4 | 36.8 |
| Atascosa | 815 | 12,264 | 1,244 | 285 | 2,062 | 258 | 246 | 640 | 52,209 | 1,681 | 44.7 | 2.9 | 32.9 |
| Austin | 673 | 7,433 | 676 | 1,007 | 1,067 | 281 | 250 | 330 | 44,341 | 2,113 | 11.1 | 31.2 | 44.4 |
| Bailey | 144 | 1,375 | 147 | 200 | 229 | 57 | 33 | 46 | 33,288 | 449 | 37.1 | 12.7 | 42.7 |
| Bandera | 388 | 3,162 | 296 | 21 | 408 | 84 | 102 | 103 | 32,685 | 897 | 46.8 | 2.1 | 34.3 |
| Bastrop | 1,468 | 15,490 | 2,059 | 1,143 | 3,563 | 434 | 588 | 523 | 33,737 | 2,120 | 46.8 | 2.1 | 34.3 |
| Baylor | 104 | 840 | 382 | 43 | 131 | 42 | 17 | 26 | 31,000 | 223 | 17.9 | 35.4 | 38.2 |
| Bee | 481 | 5,427 | 1,637 | 145 | 1,151 | 176 | 134 | 183 | 33,634 | 943 | 35.8 | 8.9 | 35.2 |
| Bell | 5,340 | 97,978 | 24,807 | 6,951 | 14,776 | 3,421 | 3,857 | 4,272 | 43,606 | 2,436 | 61.9 | 2.4 | 28.6 |
| Bexar | 36,838 | 778,555 | 127,403 | 35,136 | 93,916 | 63,957 | 48,236 | 36,918 | 47,419 | 2,520 | 62.9 | 1.4 | 36.3 |
| Blanco | 340 | 2,468 | 126 | 320 | 318 | 79 | NA | 118 | 47,905 | 1,032 | 36.7 | 8.3 | 38.0 |
| Borden | 8 | 29 | NA | NA | NA | NA | NA | 1 | 32,690 | 127 | 3.1 | 47.2 | 52.6 |
| Bosque | 298 | 2,340 | 481 | 419 | 450 | 67 | 71 | 91 | 38,850 | 1,305 | 32.0 | 9.2 | 37.5 |
| Bowie | 2,181 | 33,291 | 7,185 | 2,781 | 6,092 | 1,275 | 2,355 | 1,227 | 36,867 | 1,506 | 47.1 | 3.1 | 34.9 |
| Brazoria | 5,911 | 97,006 | 10,840 | 15,444 | 16,117 | 2,040 | 3,209 | 5,262 | 54,239 | 2,851 | 68.6 | 2.5 | 29.7 |
| Brazos | 4,523 | 72,185 | 10,485 | 5,884 | 11,690 | 1,951 | 3,298 | 2,835 | 39,276 | 1,363 | 48.1 | 4.4 | 30.8 |
| Brewster | 273 | 2,555 | 455 | 16 | 493 | 61 | 37 | 73 | 28,551 | 174 | 20.7 | 56.3 | 54.1 |
| Briscoe | 36 | 144 | NA | NA | 45 | 31 | NA | 5 | 33,035 | 249 | 4.4 | 35.7 | 37.0 |
| Brooks | 112 | 1,572 | 408 | NA | 307 | 66 | 11 | 45 | 28,479 | 437 | 25.6 | 7.8 | 31.5 |
| Brown | 846 | 13,552 | 3,107 | 2,850 | 1,754 | 446 | 180 | 481 | 35,490 | 1,838 | 41.3 | 6.1 | 37.9 |
| Burleson | 330 | 3,105 | 277 | 459 | 511 | 109 | 135 | 138 | 44,541 | 1,648 | 33.7 | 3.2 | 39.4 |
| Burnet | 1,297 | 12,805 | 2,195 | 1,109 | 2,424 | 373 | 570 | 536 | 41,864 | 1,623 | 39.4 | 5.9 | 31.8 |

Items 104—116

# Table B. States and Counties — Agriculture

Agriculture, 2017 (cont.)

| STATE County | Land in farms — Acreage (1,000) | Percent change, 2012-2017 | Acres — Average size of farm | Total irrigated (1,000) | Total cropland (1,000) | Value of land and buildings (dollars) — Average per farm | Average per acre | Value of machinery and equipment, average per farm (dollars) | Total (mil dol) | Average per farm (acres) | Percent from: Crops | Livestock and poultry products | Organic farms (number) | Farms with internet access (percent) | Government payments Total ($1,000) | Percent of farms |
|---|---|---|---|---|---|---|---|---|---|---|---|---|---|---|---|---|
| | 117 | 118 | 119 | 120 | 121 | 122 | 123 | 124 | 125 | 126 | 127 | 128 | 129 | 130 | 131 | 132 |
| **TENNESSEE—Cont'd** | | | | | | | | | | | | | | | | |
| Marshall | 153 | -5.8 | 139 | 0.1 | 57.0 | 487,523 | 3,498 | 60,310 | 43.4 | 39,626 | 24.3 | 75.7 | 6 | 75.0 | 518 | 17.2 |
| Maury | 227 | -6.3 | 144 | 0.7 | 80.2 | 579,318 | 4,037 | 62,386 | 45.6 | 28,788 | 48.3 | 51.7 | 2 | 71.4 | 2,408 | 26.0 |
| Meigs | 56 | 5.5 | 159 | 0.1 | 16.3 | 528,608 | 3,327 | 68,874 | 7.7 | 22,009 | 38.2 | 61.8 | NA | 62.4 | 375 | 28.2 |
| Monroe | 108 | -2.2 | 129 | 0.0 | 50.4 | 548,896 | 4,252 | 89,599 | 42.1 | 50,186 | 28.6 | 71.4 | 2 | 67.9 | 1,842 | 23.0 |
| Montgomery | 133 | -9.6 | 169 | 0.8 | 65.8 | 832,256 | 4,917 | 98,150 | 49.8 | 63,321 | 77.1 | 22.9 | 3 | 73.7 | 1,146 | 28.2 |
| Moore | 58 | -0.5 | 156 | D | 17.5 | 563,666 | 3,621 | 97,990 | 20.2 | 53,883 | 5.7 | 94.3 | NA | 78.9 | 371 | 17.6 |
| Morgan | 60 | 7.4 | 135 | 0.0 | 16.9 | 411,449 | 3,058 | 62,700 | 13.1 | 29,596 | 18.1 | 81.9 | NA | 75.2 | 321 | 25.1 |
| Obion | 225 | -11.1 | 406 | 17.1 | 186.7 | 1,462,228 | 3,599 | 219,944 | 137.4 | 248,488 | 72.1 | 27.9 | NA | 73.8 | 4,910 | 54.2 |
| Overton | 135 | 9.5 | 134 | 0.0 | 33.7 | 452,984 | 3,374 | 64,496 | 31.5 | 31,382 | 56.6 | 43.4 | NA | 77.8 | 1,026 | 35.3 |
| Perry | 62 | 29.7 | 215 | 0.0 | 15.7 | 477,584 | 2,216 | 52,838 | 4.6 | 16,056 | 68.5 | 31.5 | NA | 77.8 | 1,026 | 18.8 |
| Pickett | 35 | -17.4 | 120 | D | 11.6 | 421,319 | 3,501 | 68,361 | 14.3 | 49,728 | 7.7 | 92.3 | 1 | 79.8 | 189 | 40.1 |
| Polk | 35 | -0.7 | 123 | 0.1 | 16.6 | 506,140 | 4,113 | 74,965 | 36.5 | 127,087 | 11.8 | 88.2 | 5 | 74.2 | 451 | 26.8 |
| Putnam | 110 | 14.9 | 110 | 0.0 | 35.1 | 509,659 | 4,649 | 61,327 | 15.8 | 15,728 | 34.0 | 66.0 | 1 | 77.7 | 691 | 26.3 |
| Rhea | 46 | -20.2 | 92 | 0.2 | 16.8 | 398,311 | 4,311 | 58,673 | 11.8 | 23,606 | 31.4 | 68.6 | NA | 70.9 | 308 | 23.3 |
| Roane | 47 | 0.9 | 77 | 0.0 | 15.2 | 404,495 | 5,266 | 52,781 | 6.1 | 9,877 | 30.2 | 69.8 | 6 | 79.7 | 259 | 20.4 |
| Robertson | 192 | -8.1 | 160 | 1.3 | 132.1 | 844,471 | 5,285 | 123,828 | 138.7 | 115,384 | 83.2 | 16.8 | 3 | 77.2 | 1,397 | 25.6 |
| Rutherford | 153 | -13.2 | 108 | 0.5 | 60.7 | 796,242 | 7,361 | 58,477 | 27.2 | 19,242 | 57.6 | 42.4 | 2 | 76.2 | 1,051 | 16.8 |
| Scott | 32 | -18.1 | 111 | 0.0 | 10.2 | 297,351 | 2,671 | 48,364 | 1.8 | 6,403 | 45.9 | 54.1 | 2 | 78.8 | 72 | 19.4 |
| Sequatchie | 31 | 2.0 | 133 | 0.0 | 9.5 | 514,760 | 3,864 | 83,769 | 6.9 | 29,174 | 22.2 | 77.8 | NA | 73.2 | 71 | 9.4 |
| Sevier | 50 | -10.6 | 91 | 0.0 | 17.0 | 554,793 | 6,114 | 57,122 | 6.1 | 11,168 | 35.9 | 64.1 | NA | 65.6 | 265 | 18.1 |
| Shelby | 75 | -7.9 | 189 | 2.8 | 49.9 | 989,747 | 5,237 | 98,686 | 27.4 | 68,769 | 94.4 | 5.6 | 3 | 78.2 | 1,668 | 15.8 |
| Smith | 139 | 7.0 | 157 | 0.1 | 44.9 | 540,943 | 3,451 | 71,841 | 26.5 | 29,977 | 56.1 | 43.9 | 1 | 80.0 | 1,169 | 39.2 |
| Stewart | 73 | 20.7 | 188 | 0.0 | 23.0 | 498,744 | 2,650 | 63,795 | 7.8 | 20,149 | 68.1 | 31.9 | 1 | 78.9 | 265 | 8.5 |
| Sullivan | 84 | -1.2 | 71 | 0.1 | 31.8 | 500,710 | 7,065 | 48,527 | 22.0 | 18,596 | 15.2 | 84.8 | NA | 77.2 | 250 | 2.3 |
| Sumner | 161 | -3.8 | 113 | 0.1 | 70.6 | 598,510 | 5,312 | 66,130 | 44.1 | 30,891 | 55.3 | 44.7 | 1 | 75.5 | 1,402 | 24.1 |
| Tipton | 173 | 11.5 | 329 | 7.8 | 147.6 | 1,082,305 | 3,290 | 147,688 | 77.4 | 146,924 | 97.3 | 2.7 | NA | 77.4 | 2,301 | 27.7 |
| Trousdale | 43 | 3.2 | 134 | 0.1 | 10.8 | 461,107 | 3,432 | 59,710 | 6.4 | 20,256 | 18.8 | 81.2 | NA | 70.7 | 447 | 41.6 |
| Unicoi | 6 | 10.3 | 60 | 0.0 | 2.6 | 427,336 | 7,145 | 53,353 | D | D | D | D | 1 | 80.0 | D | 1.0 |
| Union | 44 | -2.3 | 87 | 0.0 | 12.5 | 357,911 | 4,092 | 55,794 | 3.2 | 6,267 | 32.0 | 68.0 | 2 | 61.4 | 370 | 28.1 |
| Van Buren | 53 | 43.8 | 162 | 0.1 | 14.7 | 541,592 | 3,350 | 77,245 | 8.3 | 25,088 | 21.0 | 79.0 | NA | 66.0 | 317 | 30.4 |
| Warren | 154 | -5.9 | 136 | 2.6 | 83.9 | 469,029 | 3,456 | 89,059 | 126.0 | 111,236 | 73.4 | 26.6 | NA | 80.3 | 1,238 | 30.1 |
| Washington | 106 | -5.0 | 74 | 0.7 | 53.9 | 584,921 | 7,871 | 74,250 | 42.5 | 29,779 | 40.2 | 59.8 | 3 | 72.1 | 416 | 7.3 |
| Wayne | 142 | 6.2 | 207 | 0.0 | 34.7 | 489,765 | 2,367 | 71,722 | 37.9 | 55,288 | 12.8 | 87.2 | NA | 60.1 | 1,102 | 43.5 |
| Weakley | 217 | -14.6 | 276 | 6.8 | 171.1 | 996,358 | 3,616 | 140,623 | 139.3 | 176,788 | 57.9 | 42.1 | NA | 71.1 | 3,829 | 46.7 |
| White | 119 | -2.6 | 122 | 0.0 | 41.8 | 466,043 | 3,816 | 73,121 | 30.5 | 31,441 | 23.3 | 76.7 | 2 | 75.8 | 929 | 32.9 |
| Williamson | 142 | 2.2 | 116 | 0.5 | 54.8 | 702,557 | 6,061 | 64,432 | 30.9 | 25,239 | 59.2 | 40.8 | NA | 79.0 | 1,087 | 12.7 |
| Wilson | 188 | 0.0 | 116 | 0.0 | 55.7 | 568,298 | 4,911 | 58,928 | 22.2 | 13,631 | 27.4 | 72.6 | 2 | 74.7 | 1,371 | 25.8 |
| **TEXAS** | 127,036 | -2.4 | 511 | 4,363.3 | 29,360.2 | 980,409 | 1,917 | 83,627 | 24,924 | 100,332 | 27.7 | 72.3 | 466 | 72.6 | 749,231 | 14.4 |
| Anderson | 401 | 6.8 | 228 | 3.1 | 63.8 | 647,610 | 2,836 | 66,640 | 92.9 | 52,989 | 16.7 | 83.3 | NA | 72.6 | 235 | 1.5 |
| Andrews | 887 | 17.9 | 5,684 | 12.8 | 78.3 | 4,473,490 | 787 | 116,341 | 10.6 | 68,045 | 48.3 | 51.7 | NA | 78.8 | 2,325 | 34.0 |
| Angelina | 104 | -11.1 | 101 | 0.5 | 21.6 | 368,428 | 3,644 | 63,521 | 61.4 | 59,736 | 4.2 | 95.8 | NA | 76.1 | 69 | 1.1 |
| Aransas | 54 | 35.3 | 556 | D | 1.6 | 935,727 | 1,684 | 63,828 | 1.9 | 19,969 | 3.8 | 96.2 | NA | 66.0 | NA | NA |
| Archer | 560 | 3.5 | 1,053 | 0.2 | 124.1 | 1,564,565 | 1,486 | 96,119 | 72.4 | 136,164 | 8.0 | 92.0 | NA | 77.8 | 2,826 | 31.8 |
| Armstrong | 443 | 1.9 | 2,050 | 7.3 | 105.7 | 1,903,678 | 929 | 144,216 | 49.3 | 228,315 | 22.4 | 77.6 | NA | 78.2 | 2,848 | 59.7 |
| Atascosa | 746 | 12.1 | 444 | 25.1 | 107.1 | 1,012,562 | 2,283 | 78,838 | 74.3 | 44,192 | 27.6 | 72.4 | 3 | 66.2 | 1,132 | 4.6 |
| Austin | 330 | -10.7 | 156 | 4.0 | 74.1 | 610,855 | 3,906 | 61,808 | 33.1 | 15,687 | 31.7 | 68.3 | 2 | 70.0 | 886 | 4.2 |
| Bailey | 499 | 5.8 | 1,111 | 54.2 | 332.9 | 947,108 | 852 | 229,500 | 357.0 | 795,140 | 15.4 | 84.6 | 6 | 75.1 | 11,628 | 83.7 |
| Bandera | 403 | 0.2 | 450 | 1.2 | 15.4 | 1,138,137 | 2,531 | 44,993 | 6.9 | 7,737 | 11.9 | 88.1 | 1 | 78.9 | 266 | 3.1 |
| Bastrop | 340 | -12.4 | 160 | 3.4 | 49.9 | 659,245 | 4,114 | 53,044 | 44.7 | 21,061 | 30.6 | 69.4 | 9 | 73.3 | 347 | 4.6 |
| Baylor | 555 | 2.1 | 2,490 | 3.3 | 175.1 | 2,568,040 | 1,032 | 154,639 | 53.7 | 241,018 | 12.5 | 87.5 | 1 | 72.2 | 4,551 | 61.9 |
| Bee | 483 | -10.4 | 512 | 5.5 | 77.2 | 1,047,910 | 2,047 | 58,599 | 37.7 | 39,982 | 65.1 | 34.9 | 2 | 63.9 | 1,511 | 12.0 |
| Bell | 487 | 15.6 | 200 | 2.3 | 152.6 | 656,224 | 3,282 | 62,850 | 77.0 | 31,622 | 49.4 | 50.6 | NA | 75.5 | 3,755 | 9.6 |
| Bexar | 332 | -3.2 | 132 | 7.3 | 91.0 | 782,155 | 5,939 | 48,334 | 67.9 | 26,935 | 74.5 | 25.5 | 1 | 68.6 | 950 | 4.6 |
| Blanco | 337 | -7.5 | 326 | 0.6 | 18.4 | 972,845 | 2,982 | 49,514 | 17.0 | 16,461 | 51.6 | 48.4 | NA | 76.1 | 190 | 3.9 |
| Borden | 494 | 6.5 | 3,893 | 2.2 | 90.8 | 3,439,101 | 883 | 194,541 | 28.8 | 226,677 | 59.2 | 40.8 | 1 | 75.6 | 1,211 | 53.5 |
| Bosque | 626 | 9.9 | 480 | 1.4 | 81.2 | 1,191,428 | 2,483 | 64,216 | 45.1 | 34,528 | 15.4 | 84.6 | 5 | 73.8 | 848 | 9.6 |
| Bowie | 294 | -7.1 | 196 | 7.8 | 75.1 | 587,246 | 3,003 | 73,633 | 60.1 | 39,921 | 20.8 | 79.2 | NA | 74.5 | 3,834 | 21.2 |
| Brazoria | 460 | -27.1 | 161 | 20.0 | 131.8 | 755,243 | 4,681 | 63,949 | 79.5 | 27,894 | 53.7 | 46.3 | NA | 74.5 | 7,201 | 5.7 |
| Brazos | 291 | -2.9 | 213 | 12.1 | 50.0 | 1,212,705 | 5,689 | 68,510 | 91.6 | 67,232 | 17.0 | 83.0 | 2 | 76.7 | 817 | 2.8 |
| Brewster | 2,019 | 5.5 | 11,601 | 4.7 | 57.1 | 7,904,621 | 681 | 90,298 | 16.3 | 93,730 | 6.3 | 93.7 | NA | 84.5 | 221 | 6.9 |
| Briscoe | 553 | 5.5 | 2,222 | 22.1 | 151.7 | 1,984,315 | 893 | 151,887 | 36.6 | 147,020 | 73.0 | 27.0 | NA | 64.3 | 5,624 | 84.7 |
| Brooks | 459 | -19.9 | 1,050 | 0.9 | 11.6 | 1,629,772 | 1,552 | 50,945 | 26.2 | 60,048 | 1.0 | 99.0 | NA | 54.9 | 322 | 9.8 |
| Brown | 547 | -8.2 | 297 | 4.1 | 76.6 | 717,128 | 2,411 | 54,661 | 46.0 | 25,011 | 20.1 | 79.9 | 1 | 73.8 | 1,071 | 9.0 |
| Burleson | 333 | -0.6 | 202 | 17.9 | 67.9 | 693,069 | 3,427 | 66,851 | 58.6 | 35,553 | 38.1 | 61.9 | NA | 67.8 | 446 | 2.8 |
| Burnet | 433 | -10.8 | 267 | 3.4 | 25.2 | 790,719 | 2,964 | 53,217 | 14.1 | 8,690 | 24.6 | 75.4 | 2 | 76.2 | 135 | 2.6 |

# Table B. States and Counties — Water Use, Wholesale Trade, Retail Trade, and Real Estate

| STATE County | Water use, 2015 | | Wholesale Trade[1], 2017 | | | | Retail Trade[2], 2017 | | | | Real estate and rental and leasing,[2] 2017 | | | |
|---|---|---|---|---|---|---|---|---|---|---|---|---|---|---|
| | Public supply water withdrawn (mil gal/day) | Public supply gallons withdrawn per person per day | Number of establishments | Number of employees | Sales (mil dol) | Average payroll (mil dol) | Number of establishments | Number of employees | Sales (mil dol) | Average payroll (mil dol) | Number of establishments | Number of employees | Sales (mil dol) | Average payroll (mil dol) |
| | 133 | 134 | 135 | 136 | 137 | 138 | 139 | 140 | 141 | 142 | 143 | 144 | 145 | 146 |
| TENNESSEE—Cont'd | | | | | | | | | | | | | | |
| Marshall | 2.69 | 85.3 | D | D | D | D | 103 | 1,156 | 314.7 | 27.8 | 15 | 36 | 3.8 | 0.7 |
| Maury | 12.33 | 140.5 | 65 | 749 | 377.9 | 38.3 | 324 | 4,093 | 1,291.8 | 111.1 | 77 | 320 | 96.8 | 11.0 |
| Meigs | 0.66 | 55.8 | NA | NA | NA | NA | 29 | 223 | 46.8 | 4.4 | NA | NA | NA | NA |
| Monroe | 5.62 | 122.8 | 22 | 140 | 71.1 | 6.2 | 137 | 1,764 | 480.7 | 43.8 | 24 | 61 | 9.5 | 1.7 |
| Montgomery | 22.4 | 115.8 | 84 | 1,023 | 460.6 | 47.0 | 530 | 8,734 | 2,261.0 | 220.8 | 175 | 815 | 177.0 | 26.7 |
| Moore | 0.54 | 85.4 | NA | NA | NA | NA | 17 | 100 | 19.0 | 1.6 | NA | NA | NA | NA |
| Morgan | 1.35 | 62.8 | D | D | D | 0.6 | 37 | 241 | 55.4 | 4.5 | D | D | D | 2.9 |
| Obion | 4.62 | 150.8 | D | D | D | D | 124 | 1,556 | 463.7 | 40.0 | 6 | 5 | 0.6 | 0.1 |
| Overton | 2.33 | 105.3 | D | D | D | D | 65 | 560 | 178.4 | 14.1 | 3 | 3 | 0.6 | 0.1 |
| Perry | 0.91 | 114.8 | D | D | D | 0.2 | 28 | 186 | 46.0 | 4.1 | NA | NA | NA | NA |
| Pickett | 0.81 | 157.4 | NA | NA | NA | NA | 21 | 94 | 26.9 | 2.2 | NA | NA | NA | NA |
| Polk | 0.87 | 51.9 | 6 | 35 | 22.8 | 1.2 | 49 | 289 | 75.0 | 7.3 | NA | NA | NA | NA |
| Putnam | 12.46 | 167.1 | 72 | 879 | 496.7 | 47.7 | 350 | 4,858 | 1,459.7 | 129.0 | 71 | 243 | 54.5 | 8.2 |
| Rhea | 3.91 | 120.2 | D | D | D | D | 109 | 1,209 | 284.5 | 26.6 | 18 | 43 | 10.8 | 1.5 |
| Roane | 4.86 | 92.1 | 23 | 157 | 96.9 | 6.2 | 139 | 1,447 | 396.2 | 35.8 | 24 | 84 | 15.4 | 2.7 |
| Robertson | 4.83 | 70.4 | 47 | 878 | 869.3 | 32.2 | 174 | 2,384 | 688.7 | 65.3 | 41 | 144 | 31.8 | 7.0 |
| Rutherford | 32.98 | 110.4 | 218 | 4,791 | 14,706.4 | 305.4 | 872 | 15,693 | 4,829.5 | 432.8 | 249 | 1,498 | 468.5 | 70.6 |
| Scott | 2.53 | 115.3 | D | D | D | 0.3 | 74 | 717 | 176.2 | 17.0 | 8 | D | 2.3 | D |
| Sequatchie | 0.81 | 54.7 | D | D | D | D | 36 | 434 | 112.0 | 10.4 | 7 | 16 | 2.7 | 0.4 |
| Sevier | 11.48 | 119.7 | 44 | 512 | 222.2 | 23.7 | 714 | 8,605 | 1,848.6 | 191.9 | 175 | 1,376 | 242.4 | 46.3 |
| Shelby | 142.33 | 151.7 | 1,097 | 27,087 | 32,209.1 | 1,609.8 | 2,992 | 46,337 | 22,284.4 | 1,278.1 | 959 | 8,235 | 2,299.9 | 443.3 |
| Smith | 0.57 | 29.5 | D | D | D | D | 59 | 641 | 178.0 | 16.9 | 5 | 17 | 2.6 | 0.6 |
| Stewart | 0.71 | 53.5 | D | D | D | 0.3 | 26 | 236 | 48.7 | 6.1 | 4 | D | 1.0 | D |
| Sullivan | 23.76 | 151.5 | 167 | 1,798 | 865.9 | 81.8 | 549 | 8,929 | 2,394.1 | 217.8 | 115 | 502 | 105.3 | 14.8 |
| Sumner | 24.90 | 141.5 | 119 | 1,304 | 995.7 | 66.2 | 460 | 6,522 | 1,832.5 | 163.9 | 160 | 1,014 | 293.2 | 61.1 |
| Tipton | 3.35 | 54.1 | 20 | 341 | 235.7 | 14.7 | 150 | 1,672 | 436.5 | 36.6 | 24 | 62 | 18.4 | 2.2 |
| Trousdale | 0.93 | 115.6 | NA | NA | NA | NA | 26 | 222 | 69.2 | 4.6 | NA | NA | NA | NA |
| Unicoi | 1.46 | 81.7 | D | D | D | D | 38 | 469 | 119.0 | 10.9 | 10 | 29 | 2.8 | 0.8 |
| Union | 0.78 | 40.8 | D | D | D | D | 38 | 354 | 90.4 | 7.6 | NA | NA | NA | NA |
| Van Buren | 0.00 | 0.0 | NA | NA | NA | NA | 8 | 41 | 8.8 | 0.7 | NA | NA | NA | NA |
| Warren | 4.77 | 118.0 | 22 | 253 | 85.8 | 11.7 | 157 | 1,546 | 422.2 | 37.3 | 20 | 57 | 12.2 | 1.5 |
| Washington | 19.08 | 151.1 | 102 | 1,770 | 1,240.9 | 102.9 | 518 | 8,606 | 2,324.0 | 207.4 | 131 | 783 | 133.2 | 25.6 |
| Wayne | 1.07 | 63.9 | 5 | D | 42.0 | D | 45 | 303 | 57.7 | 5.8 | 7 | 10 | 1.3 | 0.1 |
| Weakley | 2.26 | 66.5 | 30 | 320 | 266.2 | 17.8 | 102 | 1,099 | 283.9 | 24.0 | 22 | 51 | 7.0 | 1.2 |
| White | 2.81 | 106.0 | 15 | 110 | 51.4 | 4.8 | 82 | 829 | 291.1 | 22.5 | 11 | 25 | 2.6 | 0.7 |
| Williamson | 1.42 | 6.7 | 257 | 3,242 | 7,570.2 | 258.5 | 786 | 13,709 | 4,563.1 | 421.6 | 338 | 1,710 | 1,399.4 | 115.4 |
| Wilson | 15.58 | 120.9 | 107 | 1,608 | 2,274.3 | 95.0 | 437 | 5,976 | 1,724.5 | 150.6 | 118 | 483 | 142.7 | 21.4 |
| TEXAS | 2,885.33 | 105.0 | 28,861 | 435,728 | 779,742.5 | 27,818.6 | 80,874 | 1,304,540 | 417,231.9 | 36,627.8 | 32,290 | 198,712 | 56,443.4 | 10,559.8 |
| Anderson | 9.06 | 157.3 | 30 | 420 | 213.9 | 21.5 | 159 | 1,946 | 598.1 | 48.9 | 40 | 129 | 25.2 | 4.9 |
| Andrews | 2.36 | 130.4 | D | D | D | D | 40 | 752 | 498.6 | 41.4 | 22 | 150 | 43.5 | 12.3 |
| Angelina | 11.30 | 128.0 | 64 | 802 | 396.7 | 40.3 | 312 | 4,953 | 1,396.4 | 126.6 | 81 | 353 | 101.0 | 12.5 |
| Aransas | 0.21 | 8.3 | 7 | 28 | 11.3 | 0.8 | 74 | 1,010 | 315.7 | 31.9 | 30 | 88 | 13.2 | 2.1 |
| Archer | 4.59 | 526.7 | 11 | 44 | 30.5 | 1.7 | 18 | 102 | 22.2 | 2.1 | NA | NA | NA | NA |
| Armstrong | 0.23 | 118.1 | D | D | D | D | NA | NA | NA | NA | NA | NA | NA | NA |
| Atascosa | 5.34 | 110.3 | 49 | 336 | 236.0 | 20.5 | 127 | 1,756 | 590.4 | 49.8 | 41 | 264 | 77.3 | 16.5 |
| Austin | 2.03 | 68.7 | 21 | 319 | 809.7 | 17.7 | 97 | 1,099 | 349.8 | 30.9 | 22 | 57 | 11.5 | 2.6 |
| Bailey | 6.18 | 857.1 | 19 | 170 | 155.4 | 9.2 | 19 | 190 | 46.7 | 4.5 | NA | NA | NA | NA |
| Bandera | 0.63 | 29.6 | 3 | D | 5.3 | D | 60 | 360 | 93.7 | 8.5 | 19 | 47 | 5.8 | 1.2 |
| Bastrop | 10.60 | 131.6 | 35 | 212 | 118.8 | 11.6 | 209 | 3,416 | 1,198.7 | 92.9 | 47 | 153 | 28.1 | 5.3 |
| Baylor | 2.47 | 682.7 | 6 | 29 | 20.1 | 1.4 | 13 | 146 | 33.0 | 3.0 | NA | NA | NA | NA |
| Bee | 1.07 | 32.5 | 11 | 96 | 69.5 | 4.2 | 84 | 1,202 | 389.0 | 33.6 | 25 | 121 | 19.5 | 4.4 |
| Bell | 16.31 | 48.7 | 125 | 2,838 | 4,327.3 | 147.3 | 897 | 14,650 | 4,247.3 | 376.5 | 317 | 1,552 | 264.8 | 63.9 |
| Bexar | 212.33 | 111.9 | 1,457 | 27,152 | 17,631.3 | 1,534.5 | 5,114 | 91,714 | 27,360.9 | 2,485.1 | 2,118 | 13,354 | 3,513.0 | 692.7 |
| Blanco | 0.78 | 70.9 | 9 | 32 | 18.7 | 1.6 | 39 | 272 | 80.3 | 8.3 | 10 | D | 2.0 | D |
| Borden | 0.06 | 92.6 | NA | NA | NA | NA | NA | NA | NA | NA | NA | NA | NA | NA |
| Bosque | 2.33 | 130.2 | 9 | 87 | 13.4 | 2.3 | 58 | 420 | 106.0 | 9.7 | 7 | 21 | 2.8 | 0.5 |
| Bowie | 16.94 | 181.4 | 80 | 916 | 399.6 | 43.2 | 392 | 6,206 | 1,779.3 | 161.5 | 110 | 519 | 95.7 | 19.5 |
| Brazoria | 21.13 | 61.0 | 242 | 2,282 | 1,917.6 | 141.1 | 842 | 15,746 | 4,728.8 | 422.5 | 313 | 1,793 | 482.3 | 99.9 |
| Brazos | 31.46 | 146.3 | 150 | 2,150 | 1,451.0 | 120.2 | 655 | 10,489 | 3,005.1 | 255.9 | 292 | 1,486 | 337.8 | 53.0 |
| Brewster | 1.02 | 111.5 | 8 | 53 | 21.1 | 2.4 | 40 | 411 | 90.0 | 9.4 | 17 | 49 | 4.7 | 0.9 |
| Briscoe | 0.19 | 126.2 | NA | NA | NA | NA | 9 | 33 | 12.3 | 0.8 | NA | NA | NA | NA |
| Brooks | 1.10 | 152.1 | 3 | 7 | 1.3 | 0.1 | 14 | 235 | 71.2 | 6.3 | 5 | 13 | 1.4 | 0.6 |
| Brown | 0.06 | 1.6 | 32 | 447 | 342.5 | 23.6 | 167 | 1,961 | 581.5 | 49.1 | 37 | 115 | 17.8 | 4.1 |
| Burleson | 1.76 | 100.8 | D | D | D | 5.6 | 51 | 640 | 357.1 | 19.1 | D | D | D | D |
| Burnet | 4.09 | 90.0 | 41 | 234 | 110.0 | 12.1 | 166 | 2,234 | 749.3 | 67.1 | 64 | 164 | 38.2 | 7.5 |

1  Merchant wholesalers, except manufacturers' sales branches and offices.   2. Employer establishments.

# Table B. States and Counties — Professional Services, Manufacturing, and Accommodation and Food Services

| STATE County | Professional, scientific, and technical services, 2017 | | | | Manufacturing, 2017 | | | | Accommodation and food services, 2017 | | | |
|---|---|---|---|---|---|---|---|---|---|---|---|---|
| | Number of establishments | Number of employees | Sales (mil dol) | Average payroll (mil dol) | Number of establishments | Number of employees | Sales (mil dol) | Average payroll (mil dol) | Number of establishments | Number of employees | Sales (mil dol) | Annual payroll (mil dol) |
| | 147 | 148 | 149 | 150 | 151 | 152 | 153 | 154 | 155 | 156 | 157 | 158 |
| **TENNESSEE—Cont'd** | | | | | | | | | | | | |
| Marshall | 31 | 128 | 11 | 3.4 | 42 | 3,453 | 1,319.9 | 179.2 | 43 | 595 | 31.2 | 8.3 |
| Maury | D | D | D | D | 90 | 6,995 | 9,151.0 | 539.5 | 169 | 3,506 | 171.5 | 53.9 |
| Meigs | NA | NA | NA | NA | 12 | 878 | 246.1 | 40.1 | D | D | D | D |
| Monroe | 44 | 180 | 20 | 6.5 | 60 | 5,261 | 1,779.6 | 238.8 | 80 | 1,290 | 60.3 | 15.2 |
| Montgomery | D | D | D | D | 75 | 5,710 | 2,153.8 | 300.6 | 365 | 7,741 | 379.4 | 108.9 |
| Moore | 5 | 22 | 2 | 0.5 | D | D | D | D | D | D | D | D |
| Morgan | D | D | 1 | D | 16 | 281 | 152.5 | 10.4 | D | D | D | D |
| Obion | 23 | 113 | 14 | 4.4 | 37 | 2,560 | 1,095.3 | 94.8 | D | D | D | D |
| Overton | 25 | 70 | 11 | 3.6 | 33 | 783 | 192.5 | 33.4 | D | D | D | D |
| Perry | 5 | 13 | 1 | 0.2 | D | 573 | D | 23.1 | 32 | 384 | 20.1 | 5.2 |
| Pickett | 3 | 7 | 1 | 0.2 | 7 | 201 | 23.1 | 10.6 | D | D | D | 0.9 |
| Polk | 6 | 9 | 2 | 0.3 | 9 | 106 | 48.0 | 5.0 | 12 | 158 | 8.9 | 1.8 |
| | | | | | | | | | 19 | 255 | 12.9 | 3.4 |
| Putnam | D | D | D | D | 99 | 5,151 | 1,380.1 | 192.5 | 181 | 4,189 | 210.4 | 61.2 |
| Rhea | 26 | 76 | 7 | 2.3 | 31 | 3,460 | 754.8 | 133.7 | 61 | 923 | 39.7 | 11.5 |
| Roane | D | D | D | D | 23 | 844 | 267.2 | 40.0 | 67 | 1,090 | 54.6 | 15.3 |
| Robertson | 67 | 421 | 49 | 16.7 | 86 | 6,966 | 1,928.5 | 307.6 | 102 | 2,026 | 101.2 | 27.5 |
| Rutherford | D | D | D | D | 195 | 17,522 | 15,353.1 | 1,024.2 | 575 | 13,307 | 711.2 | 208.0 |
| Scott | 12 | 34 | 3 | 1.3 | 32 | 937 | 139.4 | 33.4 | 25 | 438 | 19.3 | 5.0 |
| Sequatchie | 16 | 55 | 6 | 2.7 | 8 | 182 | 60.4 | 4.6 | D | D | D | D |
| Sevier | 163 | 719 | 84 | 24.2 | 99 | 1,257 | 382.3 | 61.1 | 565 | 14,592 | 1,124.8 | 323.3 |
| Shelby | D | D | D | D | 543 | 24,520 | 18,524.9 | 1,584.2 | 1,873 | 40,808 | 2,369.6 | 670.0 |
| Smith | 19 | 69 | 6 | 2.5 | 13 | 1,259 | 590.4 | 59.9 | D | D | D | 4.6 |
| Stewart | 7 | 29 | 2 | 0.7 | 12 | 449 | 205.5 | 20.5 | 16 | 174 | 5.6 | 1.8 |
| Sullivan | 237 | 1,820 | 227 | 84.5 | 149 | 13,015 | 5,220.9 | 867.1 | 349 | 7,690 | 376.0 | 106.9 |
| Sumner | D | D | D | D | 161 | 8,007 | 2,598.2 | 410.7 | 259 | 5,204 | 287.1 | 83.7 |
| Tipton | D | D | D | D | 26 | 1,302 | 545.7 | 62.5 | 54 | 951 | 44.7 | 11.5 |
| Trousdale | 6 | 57 | 4 | 1.4 | 5 | 130 | 30.6 | 8.0 | 9 | 188 | 8.5 | 2.1 |
| Unicoi | 5 | 10 | 1 | 0.3 | 17 | 1,481 | 394.1 | 98.0 | 38 | 449 | 21.0 | 5.8 |
| Union | D | D | D | D | 14 | 493 | 162.3 | 23.1 | D | D | D | D |
| Van Buren | NA | NA | NA | NA | 5 | D | 35.6 | 8.1 | NA | NA | NA | NA |
| Warren | 42 | 142 | 14 | 4.0 | 57 | 3,933 | 1,278.5 | 194.5 | 54 | 859 | 42.2 | 11.5 |
| Washington | D | D | D | D | 116 | 4,593 | 1,289.5 | 207.0 | 308 | 7,336 | 352.8 | 104.9 |
| Wayne | 8 | 18 | 2 | 0.4 | 20 | 506 | 104.7 | 19.5 | 18 | 201 | 9.4 | 2.3 |
| Weakley | 20 | 92 | 8 | 2.6 | 24 | 1,851 | 658.9 | 61.4 | 58 | 876 | 35.7 | 9.6 |
| White | 20 | 71 | 8 | 2.3 | 40 | 2,100 | 747.0 | 88.5 | 37 | 554 | 24.9 | 7.7 |
| Williamson | 978 | 10,699 | 1,916 | 819.0 | 128 | 2,263 | 763.8 | 103.5 | 534 | 12,927 | 808.9 | 226.6 |
| Wilson | 223 | 1,291 | 204 | 62.1 | 127 | 4,104 | 1,812.3 | 217.7 | 250 | 5,102 | 285.8 | 85.1 |
| **TEXAS** | 69,783 | 728,470 | 149,632 | 58,962.0 | 19,844 | 758,722 | 575,217.8 | 48,003.6 | 57,098 | 1,201,419 | 74,369.4 | 20,610.6 |
| Anderson | 64 | 178 | 21 | 7.9 | 23 | 1,415 | 442.6 | 46.8 | 80 | 1,241 | 68.5 | 18.7 |
| Andrews | 21 | 78 | 11 | 3.6 | 8 | 234 | 75.1 | 10.4 | 32 | 399 | 25.7 | 6.2 |
| Angelina | 144 | 814 | 106 | 37.6 | 61 | 3,803 | 1,377.6 | 158.3 | 155 | 3,198 | 181.5 | 50.6 |
| Aransas | 42 | 155 | 21 | 6.5 | 14 | 39 | 9.5 | 1.6 | 102 | 1,278 | 70.1 | 18.3 |
| Archer | 14 | 29 | 5 | 1.4 | 5 | 51 | 10.3 | 1.9 | 6 | 60 | 1.8 | 0.6 |
| Armstrong | NA | NA | NA | NA | NA | NA | NA | NA | 6 | 60 | 1.8 | 0.6 |
| Atascosa | 45 | 194 | 26 | 7.2 | 19 | 237 | 99.1 | 9.4 | 76 | 1,144 | 57.7 | 18.4 |
| Austin | 57 | 227 | 36 | 12.9 | 39 | 994 | 285.1 | 50.4 | 55 | 662 | 40.0 | 10.9 |
| Bailey | 8 | 37 | 6 | 1.6 | 9 | 161 | 63.7 | 6.9 | 15 | 214 | 10.9 | 3.3 |
| Bandera | D | D | D | D | 12 | 23 | 5.0 | 1.0 | 52 | 472 | 31.6 | 9.8 |
| Bastrop | 107 | 514 | 61 | 19.9 | 73 | 1,011 | 332.0 | 55.2 | 140 | 2,797 | 171.6 | 51.7 |
| Baylor | 8 | 20 | 2 | 0.9 | D | 5 | D | D | 9 | 66 | 3.4 | 0.9 |
| Bee | D | D | D | D | 14 | 131 | 35.3 | 6.3 | 54 | 710 | 41.4 | 10.4 |
| Bell | 369 | 2,983 | 387 | 149.8 | 149 | 5,716 | 1,840.9 | 245.5 | 638 | 13,098 | 663.0 | 180.2 |
| Bexar | D | D | D | D | 870 | 31,288 | 18,293.8 | 1,653.8 | 4,026 | 104,088 | 6,517.8 | 1,812.5 |
| Blanco | 32 | 100 | 19 | 5.5 | 27 | 246 | 50.9 | 9.9 | 38 | 374 | 20.4 | 5.7 |
| Borden | 3 | 4 | 0 | 0.0 | NA | NA | NA | NA | NA | NA | NA | NA |
| Bosque | 18 | 69 | 7 | 2.8 | 13 | 427 | 109.9 | 18.4 | 31 | 261 | 16.6 | 4.2 |
| Bowie | D | D | D | D | 65 | 2,209 | 724.1 | 109.7 | 197 | 5,006 | 227.7 | 65.5 |
| Brazoria | D | D | D | 186.3 | 220 | 13,517 | 24,168.0 | 1,409.7 | 622 | 12,865 | 722.7 | 201.0 |
| Brazos | 438 | 3,354 | 468 | 195.0 | 118 | 5,230 | 1,453.9 | 234.2 | 524 | 12,699 | 667.8 | 174.5 |
| Brewster | D | D | D | D | 7 | 16 | 6.1 | 1.3 | 53 | 810 | 52.2 | 13.5 |
| Briscoe | NA | NA | NA | NA | NA | NA | NA | NA | D | D | D | D |
| Brooks | 3 | 9 | 1 | 0.3 | NA | NA | NA | NA | 17 | 268 | 12.7 | 3.7 |
| Brown | D | D | D | D | 31 | 2,702 | 1,355.7 | 149.6 | 91 | 1,575 | 83.2 | 21.3 |
| Burleson | 23 | 144 | 22 | 8.1 | 15 | 336 | 147.5 | 16.1 | 34 | 404 | 20.0 | 5.2 |
| Burnet | D | D | D | D | 70 | 973 | 223.7 | 46.6 | 129 | 1,760 | 123.4 | 38.8 |

# Health Care and Social Assistance, Other Services, Nonemployer Businesses, and Residential Construction

| STATE County | Health care and social assistance, 2017 | | | | Other services, 2017 | | | | Nonemployer businesses, 2018 | | Value of residential construction authorized by building permits, 2020 | |
|---|---|---|---|---|---|---|---|---|---|---|---|---|
| | Number of establishments | Number of employees | Receipts (mil dol) | Annual payroll (mil dol) | Number of establishments | Number of employees | Receipts (mil dol) | Annual payroll (mil dol) | Number | Receipts (mil dol) | New construction ($1,000) | Number of housing units |
| | 159 | 160 | 161 | 162 | 163 | 164 | 165 | 166 | 167 | 168 | 169 | 170 |
| TENNESSEE—Cont'd | | | | | | | | | | | | |
| Marshall | 56 | 707 | 66.0 | 23.3 | 24 | 137 | 20.1 | 4.8 | 2,284 | 108.2 | 35,913 | 189 |
| Maury | 238 | 5,289 | 579.7 | 232.3 | 108 | 681 | 77.7 | 21.0 | 7,224 | 336.1 | 272,230 | 1,891 |
| Meigs | D | D | D | 3.2 | D | D | D | 0.3 | 655 | 28.3 | 14,622 | 72 |
| Monroe | 77 | 1,490 | 125.7 | 50.2 | 42 | 153 | 14.7 | 4.8 | 2,830 | 125.4 | 48,980 | 212 |
| Montgomery | 351 | 8,502 | 836.5 | 357.2 | 209 | 1,060 | 114.0 | 30.1 | 11,842 | 547.5 | 446,246 | 3,634 |
| Moore | 4 | 130 | 7.4 | 4.0 | D | D | D | D | 453 | 22.6 | 6,066 | 25 |
| Morgan | 18 | 294 | 20.0 | 10.5 | 8 | 16 | 2.4 | 0.7 | 1,051 | 41.9 | 0 | 0 |
| Obion | D | D | D | D | D | D | D | D | 1,887 | 75.6 | 4,521 | 20 |
| Overton | 30 | 586 | 64.5 | 24.7 | D | D | D | D | 1,818 | 87.1 | 1,725 | 11 |
| Perry | 12 | 412 | 28.6 | 13.5 | 6 | 140 | 5.5 | 2.4 | 676 | 30.0 | 1,757 | 9 |
| Pickett | 5 | D | 10.4 | D | D | D | D | 0.3 | 455 | 24.8 | NA | NA |
| Polk | 20 | 319 | 23.6 | 10.4 | 11 | 25 | 3.0 | 0.5 | 1,047 | 38.6 | 38,631 | 213 |
| Putnam | 244 | 6,297 | 588.4 | 248.8 | D | D | D | D | 6,333 | 314.0 | 96,013 | 588 |
| Rhea | 67 | 1,146 | 84.8 | 35.6 | D | D | 8.4 | D | 1,771 | 73.3 | 19,023 | 102 |
| Roane | 99 | 2,040 | 150.2 | 63.8 | 42 | 166 | 19.9 | 5.2 | 3,038 | 122.4 | 31,919 | 142 |
| Robertson | 130 | 1,910 | 178.7 | 71.9 | 80 | 303 | 34.7 | 10.3 | 5,727 | 274.5 | 165,144 | 726 |
| Rutherford | 635 | 14,590 | 1,847.1 | 689.5 | 349 | 3,263 | 345.1 | 127.7 | 24,538 | 1,138.7 | 826,405 | 3,520 |
| Scott | 51 | 611 | 50.5 | 21.4 | D | D | D | D | 1,346 | 74.9 | 272 | 3 |
| Sequatchie | D | D | D | D | D | D | D | D | 1,063 | 42.5 | 2,670 | 13 |
| Sevier | 164 | 2,689 | 268.7 | 92.4 | 145 | 894 | 117.7 | 27.8 | 8,232 | 446.7 | 237,095 | 1,155 |
| Shelby | 2,270 | 71,476 | 9,763.1 | 3,674.5 | 1,200 | 10,989 | 2,794.3 | 467.1 | 84,799 | 3,158.8 | 339,871 | 1,878 |
| Smith | D | D | D | 15.4 | D | D | D | D | 1,419 | 66.1 | 18,603 | 101 |
| Stewart | 16 | 151 | 14.9 | 5.3 | 11 | 33 | 3.4 | 0.9 | 759 | 31.9 | 472 | 2 |
| Sullivan | 450 | 11,469 | 1,581.0 | 605.3 | 212 | 1,416 | 157.5 | 51.3 | 10,025 | 445.5 | 82,209 | 482 |
| Sumner | 384 | 7,644 | 985.4 | 396.7 | 237 | 1,477 | 138.2 | 41.7 | 16,607 | 862.9 | 479,773 | 1,980 |
| Tipton | 78 | 1,318 | 160.5 | 50.5 | 38 | 152 | 16.3 | 5.2 | 4,023 | 152.7 | 43,957 | 233 |
| Trousdale | 22 | 273 | 26.7 | 9.3 | 9 | 37 | 5.7 | 1.5 | 636 | 30.6 | 10,785 | 83 |
| Unicoi | 30 | 448 | 30.6 | 14.9 | D | D | D | D | 866 | 38.3 | 2,459 | 14 |
| Union | 14 | 130 | 11.6 | 3.9 | D | D | D | D | 1,325 | 56.3 | 19,386 | 97 |
| Van Buren | D | D | D | D | NA | NA | NA | NA | 427 | 18.1 | 580 | 5 |
| Warren | 100 | 1,430 | 148.5 | 55.2 | 45 | 158 | 13.3 | 3.8 | 3,083 | 128.6 | 21,630 | 115 |
| Washington | 386 | 14,110 | 1,792.7 | 794.6 | 198 | 1,048 | 88.2 | 27.3 | 8,697 | 412.5 | 119,525 | 880 |
| Wayne | 37 | 729 | 66.5 | 28.2 | 13 | 52 | 5.5 | 1.4 | 994 | 40.9 | 95 | 1 |
| Weakley | 72 | 1,480 | 137.8 | 50.6 | 39 | 183 | 16.9 | 4.9 | 1,767 | 76.3 | 8,866 | 61 |
| White | 52 | 989 | 76.3 | 29.6 | D | D | D | D | 2,054 | 86.3 | 16,538 | 159 |
| Williamson | 772 | 16,857 | 2,141.4 | 911.6 | 398 | 3,022 | 332.3 | 105.6 | 29,249 | 2,068.0 | 956,457 | 2,284 |
| Wilson | 304 | 3,911 | 428.3 | 157.9 | 182 | 1,677 | 186.7 | 66.4 | 12,736 | 658.4 | 533,702 | 2,349 |
| TEXAS | 69,952 | 1,581,577 | 186,108.7 | 69,899.0 | 37,506 | 285,777 | 39,008.3 | 10,468.1 | 2,514,301 | 125,732.8 | 42,986,433 | 230,503 |
| Anderson | 132 | 2,153 | 219.5 | 83.7 | 47 | 258 | 22.9 | 8.9 | 3,021 | 139.0 | 11,184 | 176 |
| Andrews | 18 | 498 | 61.2 | 27.5 | D | D | D | D | 1,449 | 91.2 | 3,345 | 16 |
| Angelina | 301 | 7,728 | 630.8 | 263.3 | 119 | 690 | 97.7 | 21.4 | 5,782 | 283.9 | 25,281 | 122 |
| Aransas | 46 | 545 | 55.6 | 15.8 | 47 | 157 | 16.4 | 4.0 | 2,694 | 142.1 | 37,509 | 173 |
| Archer | 15 | 54 | 4.6 | 1.5 | 13 | 37 | 5.0 | 1.2 | 943 | 48.8 | 7,654 | 44 |
| Armstrong | D | D | D | D | D | D | D | D | 200 | 8.9 | 466 | 2 |
| Atascosa | 85 | 1,190 | 111.3 | 44.0 | 51 | 228 | 23.9 | 6.3 | 3,606 | 157.5 | 13,511 | 67 |
| Austin | 49 | 687 | 57.0 | 24.9 | 37 | 133 | 21.4 | 5.5 | 3,212 | 152.0 | 5,720 | 29 |
| Bailey | 11 | 141 | 11.1 | 4.5 | 11 | 38 | 5.3 | 0.9 | 413 | 20.4 | 0 | 0 |
| Bandera | 27 | 363 | 23.9 | 10.9 | 36 | 122 | 12.8 | 3.5 | 2,548 | 115.0 | 237 | 1 |
| Bastrop | 136 | 1,837 | 147.6 | 64.1 | 109 | 428 | 49.8 | 12.7 | 7,100 | 352.5 | 79,916 | 478 |
| Baylor | 12 | 419 | 19.5 | 8.5 | D | D | D | 0.7 | 297 | 11.4 | 0 | 0 |
| Bee | 67 | 970 | 95.3 | 37.2 | 32 | 191 | 15.2 | 4.7 | 1,608 | 63.4 | 5,054 | 31 |
| Bell | 577 | 30,711 | 4,194.0 | 1,928.5 | 449 | 2,772 | 256.9 | 80.4 | 19,882 | 838.9 | 505,431 | 3,108 |
| Bexar | 4,922 | 123,042 | 14,420.4 | 5,248.7 | 2,582 | 19,389 | 1,994.8 | 586.9 | 151,882 | 7,105.5 | 1,782,553 | 10,392 |
| Blanco | 19 | 170 | 8.2 | 4.0 | 18 | 39 | 6.7 | 1.6 | 1,521 | 80.0 | 5,172 | 24 |
| Borden | NA | NA | NA | NA | NA | NA | NA | NA | 67 | 2.8 | NA | NA |
| Bosque | 16 | 523 | 44.7 | 18.1 | 20 | 58 | 7.7 | 1.7 | 1,590 | 74.6 | 941 | 6 |
| Bowie | 302 | 7,272 | 974.8 | 351.8 | 156 | 1,108 | 108.2 | 37.0 | 5,577 | 266.9 | 15,708 | 72 |
| Brazoria | 775 | 10,758 | 1,129.5 | 408.2 | 413 | 3,368 | 509.6 | 138.7 | 28,672 | 1,203.8 | 1,068,150 | 3,897 |
| Brazos | 443 | 10,581 | 1,445.9 | 506.2 | 289 | 2,158 | 628.2 | 77.4 | 14,818 | 682.2 | 299,754 | 1,937 |
| Brewster | 23 | 517 | 31.1 | 12.3 | 18 | 54 | 3.9 | 1.2 | 1,085 | 41.2 | 2,460 | 21 |
| Briscoe | NA | NA | NA | NA | NA | NA | NA | NA | 152 | 4.7 | NA | NA |
| Brooks | 21 | 471 | 16.4 | 7.8 | 7 | 25 | 1.7 | 0.6 | 490 | 11.5 | 110 | 1 |
| Brown | 129 | 3,266 | 224.0 | 89.1 | 67 | 328 | 28.7 | 7.5 | 2,581 | 117.2 | 21,078 | 102 |
| Burleson | 15 | 370 | 32.3 | 16.3 | 26 | 102 | 9.0 | 3.1 | 1,523 | 75.3 | 4,558 | 36 |
| Burnet | 129 | 2,257 | 271.0 | 106.1 | 82 | 367 | 36.2 | 10.3 | 5,404 | 340.8 | 181,040 | 720 |

# Table B. States and Counties — Government Employment and Payroll, and Local Government Finances

| STATE County | Government employment and payroll, 2017 | | March payroll (percent of total) | | | | | | | Local government finances, 2017 | | | | |
| | | | | | | | | | | General revenue | | | | |
| | Full-time equivalent employees | March payroll (dollars) | Administration, judicial, and legal | Police and corrections | Fire protection | Highways and transportation | Health and welfare | Natural resources and utilities | Education and libraries | Total (mil dol) | Inter-governmental (mil dol) | Taxes Total (mil dol) | Taxes Per capita[1] (dollars) Total | Taxes Per capita[1] (dollars) Property |
| | 171 | 172 | 173 | 174 | 175 | 176 | 177 | 178 | 179 | 180 | 181 | 182 | 183 | 184 |
| **TENNESSEE—Cont'd** | | | | | | | | | | | | | | |
| Marshall | 1,289 | 4,249,144 | 7.9 | 8.8 | 3.5 | 2.9 | 5.1 | 16.0 | 54.0 | 82.7 | 40.8 | 29.5 | 892 | 634 |
| Maury | 4,902 | 19,380,592 | 3.4 | 6.6 | 3.4 | 1.8 | 45.9 | 7.6 | 30.0 | 580.8 | 89.4 | 88.0 | 954 | 611 |
| Meigs | 437 | 1,197,065 | 9.2 | 10.5 | 0.1 | 4.0 | 3.1 | 2.8 | 70.1 | 25.0 | 18.5 | 4.4 | 361 | 300 |
| Monroe | 1,560 | 4,538,988 | 7.0 | 12.6 | 1.4 | 3.5 | 5.0 | 6.9 | 61.8 | 112.9 | 56.1 | 37.4 | 812 | 472 |
| Montgomery | 6,642 | 23,037,470 | 4.8 | 10.9 | 3.9 | 3.6 | 3.0 | 6.7 | 65.7 | 611.4 | 250.6 | 238.8 | 1,195 | 744 |
| Moore | 226 | 636,632 | 3.0 | 19.9 | 0.2 | 12.7 | 1.2 | 13.3 | 49.8 | 18.6 | 8.0 | 8.3 | 1,299 | 1,039 |
| Morgan | 709 | 1,992,528 | 7.1 | 9.2 | 0.0 | 3.4 | 3.8 | 6.7 | 69.8 | 70.3 | 54.0 | 11.1 | 514 | 420 |
| Obion | 1,260 | 3,921,376 | 8.2 | 10.4 | 3.4 | 3.1 | 1.1 | 7.8 | 65.2 | 93.0 | 47.2 | 26.4 | 869 | 601 |
| Overton | 779 | 2,099,396 | 6.3 | 9.0 | 0.7 | 4.4 | 6.1 | 6.7 | 65.7 | 57.6 | 30.7 | 13.5 | 613 | 367 |
| Perry | 362 | 956,220 | 11.1 | 4.1 | 0.2 | 4.5 | 0.3 | 9.4 | 64.5 | 23.3 | 13.3 | 5.9 | 744 | 569 |
| Pickett | 225 | 724,507 | 14.3 | 5.4 | 0.0 | 17.3 | 6.1 | 6.2 | 50.3 | 10.7 | 6.9 | 3.2 | 625 | 538 |
| Polk | 548 | 1,472,484 | 11.5 | 17.3 | 0.0 | 7.5 | 1.7 | 10.1 | 51.9 | 30.7 | 17.7 | 11.0 | 658 | 507 |
| Putnam | 5,060 | 17,410,038 | 3.1 | 4.4 | 1.3 | 1.5 | 57.4 | 5.2 | 26.3 | 479.4 | 83.3 | 88.4 | 1,144 | 712 |
| Rhea | 1,366 | 4,219,018 | 5.7 | 6.2 | 1.0 | 2.5 | 17.4 | 7.5 | 55.9 | 101.5 | 42.1 | 30.8 | 943 | 502 |
| Roane | 1,740 | 8,318,872 | 5.8 | 5.8 | 2.1 | 2.4 | 2.6 | 10.6 | 68.6 | 108.8 | 48.6 | 37.5 | 708 | 556 |
| Robertson | 2,113 | 6,058,248 | 11.9 | 12.5 | 2.1 | 5.2 | 0.9 | 7.5 | 57.5 | 188.3 | 82.4 | 77.2 | 1,097 | 633 |
| Rutherford | 9,387 | 36,391,756 | 5.3 | 11.5 | 4.3 | 1.7 | 3.0 | 8.3 | 65.1 | 905.8 | 350.3 | 398.3 | 1,258 | 745 |
| Scott | 1,043 | 2,576,753 | 7.4 | 9.6 | 1.0 | 4.3 | 4.9 | 11.2 | 58.9 | 56.4 | 37.0 | 11.4 | 519 | 358 |
| Sequatchie | 478 | 1,565,295 | 13.9 | 14.6 | 0.0 | 0.5 | 0.7 | 7.9 | 62.1 | 32.5 | 19.9 | 8.3 | 561 | 398 |
| Sevier | 3,716 | 13,950,913 | 4.3 | 9.8 | 3.5 | 3.6 | 2.4 | 9.4 | 61.3 | 432.6 | 97.7 | 262.1 | 2,693 | 1,064 |
| Shelby | 33,448 | 145,003,814 | 8.1 | 15.7 | 7.6 | 3.6 | 11.0 | 15.7 | 37.2 | 4,168.4 | 1,264.8 | 1,740.2 | 1,861 | 1,311 |
| Smith | 773 | 2,111,612 | 6.5 | 11.2 | 0.1 | 1.6 | 4.6 | 9.4 | 65.3 | 49.9 | 27.0 | 13.1 | 666 | 359 |
| Stewart | 489 | 1,406,680 | 8.8 | 10.6 | 0.0 | 6.2 | 6.5 | 3.6 | 61.6 | 33.3 | 21.7 | 8.7 | 649 | 444 |
| Sullivan | 5,275 | 18,910,394 | 7.3 | 9.7 | 6.7 | 4.6 | 3.4 | 7.2 | 59.8 | 431.7 | 144.8 | 186.9 | 1,191 | 826 |
| Sumner | 6,120 | 19,778,438 | 6.6 | 10.5 | 5.4 | 1.6 | 3.3 | 8.7 | 61.7 | 516.7 | 205.1 | 197.9 | 1,077 | 790 |
| Tipton | 2,026 | 6,784,714 | 5.9 | 9.6 | 3.1 | 2.3 | 1.1 | 7.5 | 68.2 | 137.5 | 90.2 | 28.8 | 469 | 377 |
| Trousdale | 292 | 838,005 | 10.1 | 16.9 | 0.0 | 4.6 | 1.8 | 3.2 | 63.3 | 18.8 | 11.6 | 4.6 | 429 | 353 |
| Unicoi | 723 | 2,357,619 | 13.9 | 7.4 | 1.8 | 4.6 | 14.4 | 3.9 | 53.9 | 34.8 | 20.7 | 10.2 | 571 | 451 |
| Union | 595 | 1,586,119 | 9.0 | 9.8 | 0.0 | 3.2 | 0.8 | 3.5 | 72.9 | 50.5 | 36.3 | 5.9 | 307 | 244 |
| Van Buren | 305 | 837,106 | 4.2 | 3.7 | 0.6 | 14.7 | 2.8 | 3.4 | 70.5 | 14.0 | 8.8 | 3.2 | 553 | 365 |
| Warren | 1,333 | 4,289,920 | 6.6 | 10.7 | 2.5 | 2.2 | 5.6 | 12.9 | 58.7 | 93.2 | 53.0 | 27.2 | 669 | 437 |
| Washington | 4,253 | 14,442,914 | 6.2 | 9.6 | 4.0 | 6.0 | 2.3 | 15.3 | 55.7 | 350.1 | 116.8 | 140.1 | 1,098 | 871 |
| Wayne | 633 | 1,747,710 | 7.1 | 10.8 | 0.1 | 5.1 | 2.9 | 5.9 | 66.9 | 46.2 | 24.8 | 9.5 | 569 | 342 |
| Weakley | 1,240 | 3,792,850 | 5.2 | 11.8 | 2.6 | 5.5 | 9.8 | 13.1 | 51.4 | 62.6 | 36.2 | 12.9 | 388 | 348 |
| White | 986 | 2,994,200 | 7.1 | 9.2 | 0.6 | 3.8 | 3.1 | 9.8 | 65.5 | 58.5 | 32.8 | 15.6 | 582 | 308 |
| Williamson | 9,372 | 37,432,304 | 5.1 | 6.5 | 3.6 | 1.6 | 20.2 | 5.3 | 55.5 | 1,052.0 | 212.8 | 494.8 | 2,189 | 1,394 |
| Wilson | 4,218 | 15,396,898 | 6.6 | 11.8 | 4.7 | 2.3 | 1.3 | 10.5 | 62.0 | 378.6 | 139.6 | 166.1 | 1,215 | 815 |
| **TEXAS** | X | X | X | X | X | X | X | X | X | X | X | X | X | X |
| Anderson | 1,959 | 6,274,701 | 8.3 | 11.2 | 2.4 | 3.5 | 0.8 | 3.9 | 69.4 | 144.0 | 59.2 | 70.4 | 1,210 | 1,002 |
| Andrews | 1,209 | 4,076,745 | 6.4 | 8.4 | 0.0 | 2.3 | 22.8 | 6.3 | 53.2 | 158.5 | 19.8 | 91.3 | 5,185 | 4,855 |
| Angelina | 4,460 | 16,971,483 | 4.5 | 7.4 | 2.2 | 1.6 | 9.4 | 3.8 | 70.2 | 342.4 | 146.6 | 116.1 | 1,326 | 1,014 |
| Aransas | 951 | 3,138,633 | 9.3 | 14.6 | 0.2 | 5.7 | 0.9 | 6.4 | 59.1 | 89.7 | 14.3 | 61.3 | 2,414 | 2,082 |
| Archer | 450 | 1,359,344 | 10.1 | 6.9 | 0.5 | 2.6 | 3.0 | 3.0 | 73.1 | 31.6 | 12.2 | 16.1 | 1,835 | 1,639 |
| Armstrong | 90 | 279,208 | 12.2 | 3.6 | 0.0 | 8.5 | 0.0 | 0.2 | 73.5 | 10.1 | 2.5 | 4.3 | 2,303 | 2,058 |
| Atascosa | 2,465 | 7,902,006 | 6.1 | 7.8 | 0.1 | 2.0 | 15.2 | 3.3 | 64.3 | 283.2 | 180.5 | 84.9 | 1,730 | 1,376 |
| Austin | 1,228 | 4,438,530 | 10.8 | 9.9 | 0.0 | 3.1 | 2.6 | 3.5 | 69.6 | 89.4 | 22.5 | 58.0 | 1,952 | 1,731 |
| Bailey | 444 | 1,253,036 | 7.3 | 9.5 | 0.0 | 3.0 | 14.6 | 2.9 | 62.3 | 31.4 | 15.1 | 10.2 | 1,449 | 1,262 |
| Bandera | 600 | 2,038,213 | 10.0 | 12.2 | 0.2 | 4.1 | 3.8 | 3.8 | 64.1 | 51.5 | 6.6 | 37.6 | 1,683 | 1,564 |
| Bastrop | 3,277 | 12,419,081 | 5.2 | 12.7 | 0.1 | 2.5 | 1.0 | 2.6 | 75.1 | 290.1 | 125.5 | 135.0 | 1,596 | 1,390 |
| Baylor | 327 | 1,035,047 | 7.2 | 4.0 | 0.0 | 3.4 | 45.0 | 3.3 | 34.3 | 44.2 | 6.8 | 8.2 | 2,303 | 2,129 |
| Bee | 1,291 | 4,581,237 | 6.0 | 6.2 | 1.5 | 2.2 | 1.2 | 4.8 | 76.7 | 114.6 | 49.1 | 43.1 | 1,324 | 1,125 |
| Bell | 17,603 | 59,057,678 | 4.9 | 8.6 | 4.0 | 1.4 | 3.1 | 3.7 | 72.3 | 1,287.1 | 579.3 | 454.8 | 1,309 | 1,021 |
| Bexar | 83,813 | 366,188,878 | 4.7 | 9.0 | 4.3 | 3.7 | 12.4 | 10.8 | 54.0 | 9,214.1 | 2,872.9 | 4,100.5 | 2,095 | 1,657 |
| Blanco | 428 | 1,638,856 | 11.1 | 8.5 | 0.0 | 2.5 | 0.6 | 2.0 | 74.8 | 37.5 | 4.7 | 28.3 | 2,465 | 2,251 |
| Borden | 71 | 235,914 | 12.5 | 4.8 | 0.0 | 5.7 | 0.0 | 0.6 | 70.6 | 12.4 | 2.2 | 9.8 | 14,673 | 14,673 |
| Bosque | 680 | 2,178,765 | 9.9 | 8.6 | 0.0 | 2.7 | 0.6 | 3.5 | 71.1 | 69.7 | 16.8 | 44.1 | 2,411 | 2,297 |
| Bowie | 4,554 | 15,358,280 | 5.2 | 7.6 | 2.6 | 3.5 | 1.7 | 3.5 | 73.8 | 324.3 | 138.7 | 139.7 | 1,495 | 1,171 |
| Brazoria | 13,471 | 54,531,582 | 5.5 | 10.2 | 1.2 | 2.8 | 3.0 | 4.5 | 71.1 | 1,527.6 | 395.3 | 813.2 | 2,247 | 1,899 |
| Brazos | 8,278 | 28,154,311 | 9.7 | 13.5 | 6.1 | 3.8 | 2.5 | 6.9 | 55.0 | 699.6 | 163.3 | 429.6 | 1,918 | 1,547 |
| Brewster | 444 | 1,448,705 | 10.8 | 16.6 | 0.0 | 3.1 | 0.7 | 4.6 | 62.5 | 33.0 | 15.1 | 14.0 | 1,503 | 1,325 |
| Briscoe | 88 | 281,166 | 26.4 | 3.2 | 0.0 | 3.6 | 0.3 | 12.0 | 52.3 | 9.7 | 1.7 | 6.9 | 4,568 | 4,500 |
| Brooks | 411 | 1,194,673 | 6.9 | 10.7 | 2.7 | 3.9 | 3.1 | 5.4 | 64.3 | 31.2 | 13.3 | 15.5 | 2,173 | 2,055 |
| Brown | 1,813 | 5,801,396 | 6.6 | 10.5 | 2.8 | 2.4 | 9.9 | 8.3 | 56.3 | 136.9 | 57.1 | 60.3 | 1,596 | 1,304 |
| Burleson | 680 | 2,083,075 | 10.9 | 8.9 | 0.0 | 4.8 | 0.2 | 4.6 | 69.4 | 46.3 | 11.6 | 28.2 | 1,560 | 1,439 |
| Burnet | 1,905 | 7,169,911 | 10.6 | 15.6 | 4.7 | 2.8 | 2.7 | 6.9 | 54.7 | 167.1 | 20.3 | 121.4 | 2,604 | 2,269 |

1. Based on the resident population estimated as of July 1 of the year shown.

Items 171—184

# Local Government Finances, Government Employment, and Income Taxes

| STATE County | Direct general expenditure Total (mil dol) | Per capita[1] (dollars) | Percent of total for: Education | Health and hospitals | Police protection | Public welfare | Highways | Debt outstanding Total (mil dol) | Per capita[1] (dollars) | Government employment, 2019 Federal civilian | Federal military | State and local | Individual income tax returns, 2018 Number of returns | Mean adjusted gross income | Mean income tax |
|---|---|---|---|---|---|---|---|---|---|---|---|---|---|---|---|
| | 185 | 186 | 187 | 188 | 189 | 190 | 191 | 192 | 193 | 194 | 195 | 196 | 197 | 198 | 199 |
| **TENNESSEE—Cont'd** | | | | | | | | | | | | | | | |
| Marshall | 80 | 2,407 | 59.7 | 3.7 | 7.8 | 0.2 | 5.7 | 96.8 | 2,931 | 69 | 96 | 1,739 | 15,460 | 50,053 | 4,042 |
| Maury | 600 | 6,501 | 18.6 | 58.0 | 3.6 | 0.0 | 3.2 | 279.8 | 3,033 | 181 | 269 | 6,889 | 45,700 | 58,446 | 5,639 |
| Meigs | 27 | 2,203 | 65.7 | 2.9 | 6.2 | 0.1 | 8.5 | 5.4 | 452 | 30 | 35 | 488 | 5,220 | 49,624 | 3,878 |
| Monroe | 103 | 2,238 | 58.9 | 4.7 | 6.2 | 0.0 | 6.1 | 71.0 | 1,543 | 100 | 130 | 1,883 | 19,110 | 49,893 | 4,160 |
| Montgomery | 562 | 2,811 | 52.4 | 3.0 | 7.4 | 0.0 | 4.0 | 2,867.6 | 14,354 | 1,487 | 623 | 10,269 | 89,260 | 51,604 | 4,311 |
| Moore | 16 | 2,534 | 58.6 | 3.1 | 4.0 | 0.1 | 12.1 | 9.7 | 1,518 | 7 | 18 | 1,063 | 2,830 | 58,240 | 5,005 |
| Morgan | 48 | 2,210 | 60.1 | 4.1 | 4.2 | 0.2 | 6.6 | 36.4 | 1,691 | 38 | 54 | 1,273 | 7,560 | 45,960 | 3,280 |
| Obion | 93 | 3,051 | 53.3 | 0.2 | 6.3 | 4.2 | 7.9 | 28.3 | 932 | 111 | 83 | 1,686 | 13,540 | 46,677 | 3,704 |
| Overton | 61 | 2,789 | 47.6 | 2.5 | 4.0 | 13.0 | 5.2 | 26.9 | 1,224 | 43 | 62 | 1,256 | 9,250 | 43,352 | 3,038 |
| Perry | 23 | 2,836 | 48.7 | 5.6 | 4.6 | 0.0 | 7.8 | 9.3 | 1,175 | 18 | 22 | 476 | 3,230 | 43,659 | 3,437 |
| Pickett | 11 | 2,201 | 60.6 | 6.0 | 9.6 | 0.0 | 1.7 | 2.0 | 396 | 11 | 14 | 300 | 2,210 | 45,034 | 3,951 |
| Polk | 34 | 2,053 | 63.3 | 2.5 | 6.5 | 0.2 | 8.3 | 45.6 | 2,723 | 77 | 47 | 592 | 7,010 | 44,293 | 3,385 |
| Putnam | 479 | 6,199 | 21.3 | 59.6 | 3.3 | 0.0 | 1.6 | 281.3 | 3,639 | 249 | 225 | 8,328 | 33,360 | 55,556 | 5,874 |
| Rhea | 99 | 3,026 | 46.6 | 19.3 | 4.5 | 0.0 | 3.6 | 60.3 | 1,847 | 1,088 | 91 | 1,855 | 13,440 | 47,800 | 3,854 |
| Roane | 101 | 1,908 | 64.4 | 0.1 | 8.6 | 0.0 | 5.1 | 113.6 | 2,142 | 396 | 148 | 3,417 | 23,390 | 58,287 | 5,419 |
| Robertson | 178 | 2,534 | 55.4 | 2.7 | 8.7 | 0.1 | 5.2 | 169.9 | 2,416 | 95 | 200 | 3,732 | 33,940 | 55,202 | 4,797 |
| Rutherford | 1,009 | 3,187 | 52.7 | 1.7 | 10.9 | 1.0 | 3.4 | 1,153.4 | 3,644 | 3,038 | 946 | 17,123 | 151,390 | 59,074 | 5,581 |
| Scott | 55 | 2,510 | 67.9 | 0.6 | 5.0 | 0.2 | 6.0 | 63.2 | 2,878 | 96 | 61 | 1,371 | 8,040 | 40,628 | 2,573 |
| Sequatchie | 31 | 2,100 | 61.4 | 3.4 | 6.6 | 0.0 | 5.6 | 27.5 | 1,861 | 12 | 42 | 824 | 6,270 | 52,505 | 4,779 |
| Sevier | 374 | 3,840 | 42.4 | 1.7 | 6.5 | 0.0 | 6.2 | 656.8 | 6,749 | 393 | 274 | 4,837 | 46,450 | 48,045 | 4,573 |
| Shelby | 3,787 | 4,049 | 33.4 | 13.7 | 12.4 | 1.0 | 2.0 | 4,208.3 | 4,500 | 13,536 | 3,440 | 53,026 | 427,050 | 65,675 | 8,346 |
| Smith | 50 | 2,522 | 52.4 | 4.9 | 7.2 | 0.1 | 6.2 | 36.8 | 1,863 | 123 | 56 | 965 | 8,690 | 51,042 | 4,198 |
| Stewart | 34 | 2,516 | 55.3 | 5.1 | 5.6 | 0.2 | 9.9 | 45.7 | 3,402 | 410 | 38 | 650 | 5,690 | 47,609 | 3,444 |
| Sullivan | 443 | 2,825 | 48.9 | 2.1 | 7.1 | 0.0 | 4.6 | 347.3 | 2,213 | 470 | 439 | 7,193 | 70,940 | 56,449 | 5,774 |
| Sumner | 472 | 2,571 | 55.1 | 3.0 | 8.9 | 0.1 | 3.8 | 428.7 | 2,333 | 496 | 534 | 8,123 | 88,860 | 70,866 | 8,061 |
| Tipton | 148 | 2,408 | 63.0 | 0.7 | 7.5 | 0.0 | 5.8 | 48.3 | 788 | 99 | 171 | 2,743 | 26,670 | 53,049 | 4,408 |
| Trousdale | 24 | 2,177 | 63.0 | 2.1 | 5.5 | 0.2 | 8.4 | 0.0 | 0 | 40 | 24 | 452 | 3,920 | 48,447 | 3,897 |
| Unicoi | 44 | 2,449 | 54.4 | 1.1 | 7.8 | 0.4 | 12.2 | 38.6 | 2,170 | 74 | 49 | 848 | 7,460 | 46,481 | 3,701 |
| Union | 45 | 2,334 | 67.6 | 3.6 | 4.3 | 0.3 | 6.0 | 20.4 | 1,049 | 22 | 56 | 813 | 7,690 | 44,939 | 3,194 |
| Van Buren | 15 | 2,620 | 59.2 | 3.7 | 4.8 | 0.0 | 11.0 | 4.9 | 848 | 13 | 16 | 449 | 2,360 | 40,889 | 2,533 |
| Warren | 97 | 2,389 | 59.5 | 4.1 | 6.4 | 0.1 | 5.0 | 44.4 | 1,089 | 90 | 115 | 2,034 | 16,880 | 43,414 | 3,256 |
| Washington | 348 | 2,727 | 43.7 | 1.1 | 7.0 | 0.1 | 5.3 | 811.3 | 6,358 | 2,889 | 396 | 10,214 | 57,300 | 61,295 | 6,794 |
| Wayne | 47 | 2,799 | 47.5 | 0.4 | 4.7 | 18.6 | 7.0 | 30.1 | 1,812 | 29 | 41 | 1,177 | 5,730 | 43,258 | 3,224 |
| Weakley | 82 | 2,458 | 56.5 | 0.4 | 8.7 | 8.4 | 7.2 | 21.4 | 643 | 129 | 94 | 3,416 | 12,830 | 47,792 | 3,708 |
| White | 56 | 2,093 | 61.9 | 3.0 | 5.3 | 0.3 | 5.3 | 26.5 | 992 | 56 | 76 | 1,314 | 11,360 | 43,010 | 3,078 |
| Williamson | 936 | 4,142 | 50.4 | 20.2 | 4.4 | 0.2 | 3.0 | 1,048.7 | 4,639 | 935 | 669 | 12,133 | 108,820 | 144,804 | 24,384 |
| Wilson | 486 | 3,556 | 59.2 | 0.3 | 5.7 | 0.0 | 5.5 | 1,185.6 | 8,674 | 237 | 403 | 5,527 | 68,220 | 71,864 | 7,823 |
| **TEXAS** | X | X | X | X | X | X | X | X | X | 204,813 | 178,270 | 1,712,188 | 12,637,270 | 73,228 | 9,501 |
| Anderson | 145 | 2,494 | 58.0 | 0.0 | 6.1 | 0.6 | 5.2 | 132.2 | 2,273 | 165 | 84 | 5,247 | 19,630 | 49,663 | 4,283 |
| Andrews | 159 | 9,005 | 34.3 | 37.3 | 2.7 | 0.0 | 7.5 | 170.2 | 9,667 | 15 | 36 | 1,323 | 7,610 | 80,366 | 9,389 |
| Angelina | 318 | 3,632 | 58.1 | 9.0 | 5.2 | 0.5 | 3.7 | 341.8 | 3,903 | 347 | 164 | 6,546 | 36,320 | 52,051 | 4,912 |
| Aransas | 82 | 3,242 | 47.1 | 1.4 | 8.0 | 1.5 | 4.8 | 110.9 | 4,368 | 29 | 45 | 989 | 10,960 | 61,296 | 8,177 |
| Archer | 31 | 3,539 | 72.1 | 1.7 | 2.5 | 0.7 | 2.3 | 126.8 | 14,434 | 20 | 107 | 487 | 3,940 | 68,120 | 6,646 |
| Armstrong | 10 | 5,163 | 39.8 | 34.0 | 3.6 | 0.2 | 4.9 | 5.9 | 3,134 | 4 | 4 | 123 | 860 | 54,490 | 5,198 |
| Atascosa | 271 | 5,529 | 52.4 | 11.5 | 3.1 | 0.5 | 10.6 | 467.8 | 9,530 | 60 | 102 | 2,484 | 21,320 | 54,573 | 5,376 |
| Austin | 103 | 3,449 | 64.8 | 2.5 | 6.4 | 0.1 | 5.2 | 132.3 | 4,452 | 78 | 58 | 1,475 | 14,390 | 73,816 | 9,400 |
| Bailey | 28 | 4,004 | 55.5 | 14.1 | 9.0 | 0.0 | 5.8 | 43.0 | 6,090 | 23 | 13 | 425 | 2,610 | 43,588 | 3,467 |
| Bandera | 47 | 2,120 | 51.7 | 3.5 | 7.4 | 0.7 | 7.1 | 32.1 | 1,439 | 13 | 44 | 747 | 10,180 | 67,073 | 7,725 |
| Bastrop | 264 | 3,122 | 59.8 | 2.4 | 5.4 | 0.3 | 4.3 | 456.7 | 5,399 | 363 | 169 | 3,938 | 38,640 | 55,701 | 5,125 |
| Baylor | 42 | 11,688 | 23.1 | 65.6 | 2.0 | 0.0 | 2.6 | 9.4 | 2,633 | 19 | 7 | 234 | 1,460 | 54,384 | 4,685 |
| Bee | 119 | 3,664 | 74.3 | 1.7 | 3.6 | 0.1 | 3.0 | 115.7 | 3,549 | 34 | 122 | 3,286 | 10,850 | 49,306 | 4,642 |
| Bell | 1,345 | 3,872 | 61.3 | 2.6 | 5.0 | 0.4 | 3.9 | 1,705.4 | 4,909 | 10,564 | 36,831 | 20,398 | 155,360 | 51,431 | 4,504 |
| Bexar | 9,171 | 4,686 | 41.9 | 18.1 | 5.3 | 1.8 | 3.0 | 24,861.0 | 12,704 | 35,875 | 37,925 | 112,175 | 895,170 | 63,108 | 7,323 |
| Blanco | 36 | 3,089 | 58.5 | 0.2 | 8.3 | 0.8 | 4.4 | 30.6 | 2,667 | 54 | 23 | 533 | 5,610 | 77,186 | 9,813 |
| Borden | 11 | 16,416 | 70.4 | 0.0 | 0.0 | 0.0 | 0.0 | 20.2 | 30,221 | 2 | 1 | 81 | 290 | 118,055 | 20,959 |
| Bosque | 79 | 4,336 | 45.5 | 0.0 | 4.9 | 0.1 | 6.1 | 80.2 | 4,386 | 11 | 36 | 1,223 | 7,930 | 55,998 | 5,824 |
| Bowie | 317 | 3,389 | 63.3 | 0.3 | 5.0 | 0.5 | 5.2 | 292.5 | 3,129 | 3,440 | 188 | 6,496 | 39,880 | 57,239 | 5,989 |
| Brazoria | 1,559 | 4,308 | 57.4 | 4.1 | 4.5 | 0.2 | 5.0 | 3,476.1 | 9,606 | 546 | 778 | 19,667 | 157,510 | 75,939 | 8,629 |
| Brazos | 816 | 3,645 | 48.0 | 5.4 | 5.1 | 0.2 | 4.8 | 1,889.5 | 8,438 | 771 | 496 | 38,180 | 84,570 | 66,865 | 7,744 |
| Brewster | 34 | 3,627 | 50.9 | 7.7 | 5.4 | 0.2 | 4.4 | 18.6 | 1,994 | 291 | 18 | 1,099 | 4,360 | 60,983 | 6,066 |
| Briscoe | 7 | 4,631 | 42.8 | 0.0 | 3.2 | 0.1 | 6.6 | 16.8 | 11,109 | 11 | 3 | 115 | 600 | 49,360 | 4,618 |
| Brooks | 28 | 3,962 | 62.2 | 0.0 | 8.2 | 0.2 | 3.9 | 45.2 | 6,332 | 416 | 14 | 566 | 2,820 | 34,235 | 2,008 |
| Brown | 130 | 3,428 | 49.3 | 8.5 | 6.5 | 0.6 | 4.0 | 148.2 | 3,919 | 118 | 71 | 2,771 | 16,030 | 50,128 | 4,355 |
| Burleson | 50 | 2,790 | 55.5 | 5.9 | 3.6 | 0.1 | 8.5 | 42.8 | 2,370 | 51 | 36 | 783 | 8,350 | 59,440 | 6,606 |
| Burnet | 162 | 3,481 | 49.1 | 0.7 | 7.6 | 0.3 | 5.4 | 267.8 | 5,748 | 72 | 92 | 2,650 | 22,960 | 80,353 | 10,837 |

1. Based on the resident population estimated as of July 1 of the year shown.

# Land Area and Population

| State / county code | CBSA code[1] | County Type code[2] | STATE County | Land area[3] (sq. mi) | Total persons 2019 | Rank | Per square mile | White | Black | American Indian, Alaska Native | Asian and Pacific Islander | Percent Hispanic or Latino[4] | Under 5 years | 5 to 14 years | 15 to 24 years | 25 to 34 years | 35 to 44 years | 45 to 54 years |
|---|---|---|---|---|---|---|---|---|---|---|---|---|---|---|---|---|---|---|
| | | | | 1 | 2 | 3 | 4 | 5 | 6 | 7 | 8 | 9 | 10 | 11 | 12 | 13 | 14 | 15 |
| | | | **TEXAS—Cont'd** | | | | | | | | | | | | | | | |
| 48055 | 12420 | 1 | Caldwell | 544.5 | 43,979 | 1,096 | 80.8 | 38.6 | 6.1 | 0.7 | 1.2 | 54.5 | 6.1 | 12.4 | 15.7 | 13.6 | 12.7 | 12.0 |
| 48057 | 38920 | 6 | Calhoun | 506.9 | 21,001 | 1,770 | 41.4 | 42.1 | 2.7 | 0.6 | 5.7 | 49.8 | 6.3 | 13.3 | 12.6 | 12.6 | 11.8 | 11.0 |
| 48059 | 10180 | 3 | Callahan | 899.4 | 14,110 | 2,151 | 15.7 | 86.9 | 2.2 | 1.3 | 1.1 | 10.1 | 5.2 | 12.2 | 10.4 | 11.1 | 12.5 | 11.9 |
| 48061 | 15180 | 2 | Cameron | 891.7 | 424,180 | 167 | 475.7 | 8.7 | 0.5 | 0.2 | 0.8 | 90.1 | 7.6 | 16.7 | 16.0 | 12.7 | 11.8 | 11.4 |
| 48063 | 34420 | 6 | Camp | 195.8 | 13,060 | 2,217 | 66.7 | 56.6 | 16.9 | 0.8 | 1.4 | 26.4 | 7.0 | 14.7 | 12.7 | 12.0 | 10.6 | 10.6 |
| 48065 | 11100 | 2 | Carson | 920.2 | 5,854 | 2,757 | 6.4 | 86.6 | 2.1 | 2.0 | 1.1 | 10.3 | 4.4 | 13.9 | 11.7 | 11.1 | 12.5 | 11.6 |
| 48067 | | 6 | Cass | 936.9 | 29,879 | 1,439 | 31.9 | 77.1 | 17.2 | 1.0 | 0.9 | 5.1 | 5.6 | 12.5 | 11.4 | 10.2 | 11.0 | 11.9 |
| 48069 | | 6 | Castro | 894.4 | 7,396 | 2,626 | 8.3 | 31.3 | 2.2 | 0.7 | 0.8 | 65.5 | 7.2 | 16.5 | 14.9 | 11.5 | 11.4 | 9.7 |
| 48071 | 26420 | 1 | Chambers | 597.1 | 45,590 | 1,065 | 76.4 | 65.5 | 8.2 | 0.7 | 1.7 | 25.0 | 7.2 | 16.1 | 13.3 | 13.7 | 14.7 | 12.6 |
| 48073 | 27380 | 6 | Cherokee | 1,053.0 | 52,875 | 958 | 50.2 | 61.2 | 14.7 | 0.9 | 0.8 | 24.0 | 6.8 | 14.0 | 13.7 | 12.1 | 11.5 | 11.9 |
| 48075 | | 7 | Childress | 696.5 | 7,143 | 2,646 | 10.3 | 56.5 | 11.1 | 1.1 | 1.1 | 31.5 | 4.3 | 10.8 | 17.1 | 20.0 | 13.1 | 9.0 |
| 48077 | 48660 | 3 | Clay | 1,088.8 | 10,550 | 2,385 | 9.7 | 90.3 | 1.5 | 2.0 | 1.2 | 6.9 | 4.4 | 11.1 | 10.7 | 10.3 | 11.3 | 11.7 |
| 48079 | | 9 | Cochran | 775.1 | 2,897 | 2,971 | 3.7 | 34.9 | 3.7 | 1.0 | 0.8 | 60.7 | 6.8 | 14.9 | 13.0 | 13.4 | 12.0 | 11.4 |
| 48081 | | 8 | Coke | 911.6 | 3,323 | 2,942 | 3.6 | 75.1 | 1.4 | 1.8 | 0.4 | 23.0 | 5.3 | 12.0 | 10.8 | 10.6 | 10.5 | 9.8 |
| 48083 | | 6 | Coleman | 1,262.9 | 8,100 | 2,576 | 6.4 | 77.2 | 3.2 | 1.1 | 1.4 | 18.4 | 5.1 | 11.7 | 10.1 | 9.2 | 9.9 | 11.3 |
| 48085 | 19100 | 1 | Collin | 841.3 | 1,072,069 | 43 | 1,274.3 | 56.0 | 11.7 | 0.9 | 18.4 | 15.6 | 6.0 | 14.6 | 13.0 | 12.9 | 15.8 | 14.9 |
| 48087 | | 9 | Collingsworth | 918.4 | 2,877 | 2,972 | 3.1 | 58.6 | 5.7 | 2.1 | 0.5 | 34.7 | 5.8 | 16.4 | 11.6 | 10.8 | 12.7 | 10.6 |
| 48089 | | 6 | Colorado | 960.3 | 21,610 | 1,740 | 22.5 | 56.2 | 12.5 | 0.5 | 0.9 | 31.2 | 6.2 | 13.4 | 11.6 | 11.1 | 9.9 | 11.1 |
| 48091 | 41700 | 1 | Comal | 559.5 | 164,812 | 403 | 294.6 | 67.3 | 3.0 | 0.9 | 1.9 | 28.4 | 5.4 | 12.8 | 11.4 | 11.5 | 12.6 | 13.0 |
| 48093 | | 7 | Comanche | 937.8 | 13,750 | 2,174 | 14.7 | 69.3 | 0.9 | 1.1 | 0.8 | 28.9 | 5.5 | 13.0 | 11.3 | 10.3 | 9.9 | 11.6 |
| 48095 | | 8 | Concho | 983.8 | 2,827 | 2,977 | 2.9 | 61.2 | 2.1 | 1.1 | 2.2 | 34.5 | 5.0 | 9.0 | 9.9 | 9.9 | 10.5 | 12.8 |
| 48097 | 23620 | 1 | Cooke | 874.8 | 41,393 | 1,148 | 47.3 | 75.9 | 4.0 | 1.6 | 1.2 | 18.9 | 6.5 | 12.7 | 12.2 | 12.4 | 11.3 | 11.4 |
| 48099 | 28660 | 2 | Coryell | 1,052.2 | 76,737 | 725 | 72.9 | 59.9 | 18.7 | 1.4 | 4.4 | 19.6 | 6.5 | 12.7 | 15.9 | 19.5 | 14.9 | 10.9 |
| 48101 | | 9 | Cottle | 900.6 | 1,363 | 3,090 | 1.5 | 64.6 | 10.6 | 0.7 | 0.5 | 25.2 | 4.8 | 11.9 | 11.5 | 7.8 | 12.8 | 10.1 |
| 48103 | | 6 | Crane | 785.1 | 4,765 | 2,844 | 6.1 | 29.7 | 3.0 | 1.1 | 0.9 | 66.0 | 6.9 | 16.5 | 14.9 | 12.9 | 13.8 | 11.2 |
| 48105 | | 7 | Crockett | 2,807.3 | 3,513 | 2,934 | 1.3 | 30.8 | 0.9 | 1.0 | 1.0 | 66.8 | 5.7 | 15.3 | 11.6 | 10.6 | 12.4 | 12.1 |
| 48107 | 31180 | 2 | Crosby | 900.2 | 5,567 | 2,783 | 6.2 | 39.6 | 3.2 | 0.7 | 0.5 | 56.9 | 5.9 | 15.0 | 12.8 | 11.3 | 12.3 | 10.6 |
| 48109 | | 9 | Culberson | 3,812.2 | 2,149 | 3,031 | 0.6 | 22.8 | 1.9 | 1.1 | 2.2 | 73.2 | 6.9 | 12.3 | 12.1 | 13.5 | 10.1 | 10.6 |
| 48111 | | 7 | Dallam | 1,503.1 | 7,273 | 2,635 | 4.8 | 49.1 | 2.5 | 1.2 | 1.1 | 47.3 | 10.0 | 17.7 | 13.0 | 13.8 | 11.9 | 11.8 |
| 48113 | 19100 | 1 | Dallas | 873.1 | 2,635,888 | 8 | 3,019.0 | 29.1 | 23.6 | 0.7 | 7.4 | 40.8 | 7.2 | 14.2 | 13.7 | 16.7 | 13.6 | 12.2 |
| 48115 | 29500 | 7 | Dawson | 900.3 | 12,974 | 2,222 | 14.4 | 34.7 | 6.2 | 0.6 | 1.0 | 58.6 | 7.2 | 14.5 | 14.1 | 16.3 | 12.7 | 10.1 |
| 48117 | 25820 | 6 | Deaf Smith | 1,496.8 | 18,277 | 1,910 | 12.2 | 22.8 | 1.3 | 0.6 | 0.6 | 75.2 | 8.6 | 17.4 | 14.9 | 13.5 | 11.8 | 10.6 |
| 48119 | | 8 | Delta | 256.8 | 5,349 | 2,798 | 20.8 | 81.7 | 8.4 | 2.8 | 1.2 | 8.9 | 6.8 | 12.4 | 10.6 | 11.5 | 10.8 | 11.9 |
| 48121 | 19100 | 1 | Denton | 878.5 | 919,324 | 58 | 1,046.5 | 58.8 | 11.9 | 1.0 | 11.1 | 19.6 | 5.9 | 13.6 | 13.5 | 14.7 | 15.4 | 14.3 |
| 48123 | | 6 | DeWitt | 909.0 | 20,174 | 1,814 | 22.2 | 54.8 | 8.7 | 0.5 | 0.5 | 36.3 | 5.9 | 12.9 | 10.8 | 11.6 | 12.8 | 11.9 |
| 48125 | | 8 | Dickens | 901.7 | 2,140 | 3,034 | 2.4 | 62.5 | 4.9 | 1.4 | 1.4 | 30.8 | 4.4 | 8.6 | 12.9 | 14.2 | 11.6 | 10.7 |
| 48127 | | 6 | Dimmit | 1,328.9 | 9,925 | 2,425 | 7.5 | 10.0 | 1.2 | 0.2 | 0.7 | 88.0 | 6.7 | 16.4 | 14.2 | 12.1 | 11.7 | 10.9 |
| 48129 | | 8 | Donley | 926.9 | 3,308 | 2,946 | 3.6 | 80.4 | 7.0 | 1.5 | 1.2 | 11.7 | 4.8 | 12.1 | 15.5 | 8.9 | 10.3 | 9.4 |
| 48131 | 10860 | 7 | Duval | 1,793.5 | 11,058 | 2,342 | 6.2 | 9.0 | 1.1 | 0.4 | 0.4 | 89.2 | 7.2 | 14.4 | 13.9 | 13.1 | 11.5 | 10.6 |
| 48133 | | 6 | Eastland | 926.5 | 18,388 | 1,903 | 19.8 | 79.1 | 2.6 | 1.2 | 1.0 | 17.4 | 5.7 | 11.4 | 13.7 | 10.7 | 10.9 | 10.7 |
| 48135 | 36220 | 3 | Ector | 897.9 | 167,701 | 399 | 186.8 | 30.3 | 5.1 | 0.7 | 1.5 | 63.4 | 8.8 | 17.0 | 14.5 | 16.6 | 13.6 | 10.4 |
| 48137 | | 9 | Edwards | 2,117.9 | 1,923 | 3,056 | 0.9 | 43.4 | 0.7 | 1.1 | 0.5 | 55.1 | 5.9 | 12.6 | 9.4 | 7.9 | 10.0 | 10.6 |
| 48139 | 19100 | 1 | Ellis | 935.7 | 191,760 | 352 | 204.9 | 58.7 | 13.2 | 1.0 | 1.3 | 27.4 | 6.5 | 15.1 | 13.3 | 13.1 | 13.5 | 12.9 |
| 48141 | 21340 | 2 | El Paso | 1,013.2 | 841,286 | 70 | 830.3 | 12.4 | 3.6 | 0.5 | 1.8 | 82.8 | 7.1 | 14.9 | 15.6 | 15.4 | 12.5 | 11.3 |
| 48143 | 44500 | 4 | Erath | 1,083.2 | 43,224 | 1,114 | 39.9 | 74.6 | 2.4 | 1.1 | 1.4 | 21.8 | 5.7 | 11.7 | 23.5 | 12.7 | 10.6 | 9.7 |
| 48145 | 47380 | 2 | Falls | 765.5 | 17,275 | 1,958 | 22.6 | 51.2 | 23.7 | 0.9 | 0.8 | 24.8 | 5.8 | 12.1 | 11.6 | 14.4 | 12.4 | 11.6 |
| 48147 | 14300 | 6 | Fannin | 890.8 | 35,913 | 1,279 | 40.3 | 79.7 | 7.0 | 1.8 | 1.1 | 12.3 | 4.9 | 12.3 | 11.8 | 12.9 | 12.5 | 12.9 |
| 48149 | | 6 | Fayette | 949.9 | 25,547 | 1,576 | 26.9 | 71.4 | 6.3 | 0.6 | 0.6 | 22.0 | 4.9 | 11.5 | 10.4 | 9.8 | 10.5 | 10.4 |
| 48151 | | 8 | Fisher | 899.0 | 3,784 | 2,913 | 4.2 | 66.0 | 4.0 | 1.2 | 0.6 | 29.4 | 4.9 | 13.2 | 9.7 | 11.5 | 10.6 | 11.1 |
| 48153 | | 6 | Floyd | 992.1 | 5,672 | 2,776 | 5.7 | 37.0 | 3.4 | 0.6 | 0.6 | 59.0 | 6.3 | 14.7 | 11.0 | 11.5 | 11.4 | 11.2 |
| 48155 | | 9 | Foard | 704.4 | 1,135 | 3,103 | 1.6 | 75.3 | 5.8 | 0.8 | 1.1 | 18.2 | 4.9 | 9.6 | 11.0 | 9.3 | 9.7 | 11.6 |
| 48157 | 26420 | 1 | Fort Bend | 861.7 | 839,706 | 71 | 974.5 | 32.5 | 21.5 | 0.6 | 22.4 | 25.0 | 6.5 | 15.7 | 13.1 | 11.7 | 15.3 | 14.0 |
| 48159 | | 7 | Franklin | 284.4 | 10,821 | 2,362 | 38.0 | 78.7 | 4.9 | 1.1 | 1.4 | 15.7 | 5.6 | 13.1 | 11.3 | 10.6 | 11.1 | 11.4 |
| 48161 | | 7 | Freestone | 877.7 | 19,874 | 1,832 | 22.6 | 67.4 | 15.5 | 1.0 | 0.9 | 16.5 | 5.8 | 12.7 | 11.0 | 11.1 | 13.2 | 12.4 |
| 48163 | 37770 | 6 | Frio | 1,133.5 | 20,379 | 1,801 | 18.0 | 13.8 | 3.3 | 0.4 | 2.6 | 80.3 | 6.0 | 12.9 | 17.3 | 19.4 | 12.8 | 10.2 |
| 48165 | | 7 | Gaines | 1,502.4 | 21,996 | 1,719 | 14.6 | 54.6 | 1.9 | 0.6 | 0.6 | 43.0 | 10.5 | 19.9 | 15.5 | 13.4 | 12.6 | 9.5 |
| 48167 | 26420 | 1 | Galveston | 379.3 | 345,089 | 208 | 909.8 | 57.8 | 13.3 | 0.8 | 4.1 | 25.7 | 6.0 | 13.7 | 12.6 | 13.1 | 13.5 | 12.6 |
| 48169 | | 6 | Garza | 893.4 | 6,222 | 2,730 | 7.0 | 40.0 | 7.3 | 0.9 | 0.5 | 52.5 | 3.9 | 9.1 | 16.7 | 20.6 | 12.2 | 14.1 |
| 48171 | 23240 | 7 | Gillespie | 1,058.2 | 26,960 | 1,528 | 25.5 | 74.5 | 0.6 | 0.7 | 0.8 | 24.1 | 4.8 | 11.3 | 10.2 | 8.9 | 9.8 | 10.2 |
| 48173 | | 8 | Glasscock | 900.2 | 1,439 | 3,084 | 1.6 | 58.6 | 1.9 | 0.3 | 0.3 | 39.1 | 6.3 | 14.8 | 12.1 | 15.5 | 12.2 | 11.5 |
| 48175 | 47020 | 3 | Goliad | 852.0 | 7,626 | 2,618 | 9.0 | 58.9 | 4.5 | 1.0 | 0.8 | 36.0 | 4.9 | 11.6 | 11.2 | 9.9 | 11.3 | 11.9 |
| 48177 | | 6 | Gonzales | 1,066.7 | 20,948 | 1,772 | 19.6 | 41.4 | 6.3 | 0.6 | 0.7 | 51.8 | 6.9 | 15.0 | 13.4 | 11.9 | 11.1 | 11.2 |
| 48179 | 37420 | 6 | Gray | 926.0 | 21,658 | 1,735 | 23.4 | 63.3 | 5.9 | 1.6 | 0.9 | 30.2 | 6.0 | 14.4 | 12.0 | 13.7 | 13.3 | 12.2 |
| 48181 | 43300 | 3 | Grayson | 932.8 | 138,318 | 472 | 148.3 | 76.8 | 7.0 | 2.3 | 2.1 | 14.4 | 6.1 | 13.5 | 12.5 | 12.1 | 11.8 | 11.9 |
| 48183 | 30980 | 3 | Gregg | 273.4 | 124,229 | 512 | 454.4 | 58.3 | 21.5 | 0.9 | 1.8 | 19.3 | 6.7 | 14.6 | 13.7 | 13.6 | 12.4 | 11.1 |
| 48185 | | 6 | Grimes | 787.5 | 29,614 | 1,445 | 37.6 | 60.1 | 15.2 | 1.0 | 0.7 | 24.5 | 5.6 | 12.5 | 11.9 | 12.9 | 12.2 | 12.4 |

1. CBSA = Core Based Statistical Area. See Appendix A for explanation. See Appendix B for list of metropolitan areas with component counties. Service of USDA Rural-Urban Continuum Codes. See Appendix A for definition. 3. Dry land or land partially or temporarily covered by water. 2. County type code from the Economic Research 4. May be of any race.

Items 1—15

# Table B. States and Counties — Population and Households

| STATE County | 55 to 64 years | 65 to 74 years | 75 years and over | Percent female | Total persons 2000 | Total persons 2010 | Percent change 2000-2010 | Percent change 2010-2020 | Births | Deaths | Net Migration | Number | Persons per household | Family households | Female family householder[1] | One person |
|---|---|---|---|---|---|---|---|---|---|---|---|---|---|---|---|---|
| | 16 | 17 | 18 | 19 | 20 | 21 | 22 | 23 | 24 | 25 | 26 | 27 | 28 | 29 | 30 | 31 |
| **TEXAS—Cont'd** | | | | | | | | | | | | | | | | |
| Caldwell | 12.3 | 9.2 | 6.0 | 49.3 | 32,194 | 38,014 | 18.1 | 15.7 | 5,333 | 3,397 | 4,037 | 13,460 | 2.84 | 71.5 | 14.5 | 23.9 |
| Calhoun | 13.5 | 10.9 | 8.0 | 48.7 | 20,647 | 21,380 | 3.6 | -1.8 | 2,925 | 2,134 | -1,163 | 7,582 | 2.82 | 67.7 | 12.4 | 27.3 |
| Callahan | 14.8 | 12.7 | 9.1 | 50.7 | 12,905 | 13,545 | 5.0 | 4.2 | 1,407 | 1,796 | 963 | 5,367 | 2.57 | 63.5 | 8.3 | 33.9 |
| Cameron | 9.7 | 8.0 | 6.2 | 51.1 | 335,227 | 406,214 | 21.2 | 4.4 | 71,420 | 26,882 | -26,726 | 124,605 | 3.36 | 78.4 | 20.0 | 19.1 |
| Camp | 12.8 | 12.0 | 7.6 | 51.5 | 11,549 | 12,399 | 7.4 | 5.3 | 1,887 | 1,478 | 255 | 4,605 | 2.78 | 72.0 | 13.4 | 23.0 |
| Carson | 14.0 | 12.2 | 8.7 | 50.1 | 6,516 | 6,186 | -5.1 | -5.4 | 560 | 677 | -219 | 2,288 | 2.6 | 70.0 | 6.9 | 27.6 |
| Cass | 14.4 | 12.7 | 10.5 | 51.6 | 30,438 | 30,469 | 0.1 | -1.9 | 3,462 | 4,253 | 214 | 11,934 | 2.49 | 69.2 | 11.8 | 28.4 |
| Castro | 11.4 | 9.8 | 7.6 | 48.6 | 8,285 | 8,063 | -2.7 | -8.3 | 1,201 | 604 | -1,281 | 2,532 | 3.0 | 68.6 | 17.4 | 27.9 |
| Chambers | 10.6 | 7.7 | 4.2 | 49.6 | 26,031 | 35,103 | 34.9 | 29.9 | 5,309 | 2,852 | 8,002 | 14,069 | 2.92 | 80.4 | 10.9 | 17.9 |
| Cherokee | 11.9 | 10.6 | 7.5 | 48.9 | 46,659 | 50,827 | 8.9 | 4.0 | 7,618 | 5,506 | -25 | 18,138 | 2.7 | 73.1 | 14.4 | 23.4 |
| Childress | 10.1 | 8.6 | 7.1 | 39.0 | 7,688 | 7,041 | -8.4 | 1.4 | 730 | 737 | 100 | 2,407 | 2.39 | 64.3 | 10.3 | 32.7 |
| Clay | 16.4 | 14.1 | 10.0 | 50.0 | 11,006 | 10,754 | -2.3 | -1.9 | 873 | 1,254 | 174 | 4,105 | 2.52 | 69.5 | 4.9 | 26.0 |
| Cochran | 12.3 | 8.5 | 7.8 | 49.6 | 3,730 | 3,127 | -16.2 | -7.4 | 445 | 273 | -408 | 998 | 2.78 | 68.8 | 14.1 | 29.9 |
| Coke | 14.5 | 13.9 | 12.4 | 50.7 | 3,864 | 3,316 | -14.2 | 0.2 | 321 | 543 | 229 | 1,644 | 1.95 | 61.3 | 8.9 | 36.6 |
| Coleman | 16.3 | 15.1 | 11.2 | 49.8 | 9,235 | 8,893 | -3.7 | -8.9 | 856 | 1,345 | -296 | 3,423 | 2.42 | 62.1 | 5.3 | 29.8 |
| Collin | 11.4 | 7.1 | 4.4 | 50.7 | 491,675 | 780,894 | 58.8 | 37.3 | 111,337 | 42,343 | 221,172 | 341,163 | 2.84 | 74.1 | 9.6 | 21.4 |
| Collingsworth | 12.8 | 10.8 | 8.6 | 51.3 | 3,206 | 3,057 | -4.6 | -5.9 | 386 | 388 | -184 | 1,035 | 2.81 | 73.2 | 13.2 | 25.4 |
| Colorado | 14.0 | 12.7 | 10.0 | 50.0 | 20,390 | 20,866 | 2.3 | 3.6 | 2,639 | 2,747 | 868 | 7,450 | 2.8 | 67.6 | 11.0 | 29.3 |
| Comal | 14.6 | 11.8 | 7.0 | 50.6 | 78,021 | 108,525 | 39.1 | 51.9 | 14,846 | 11,546 | 52,577 | 51,367 | 2.74 | 74.4 | 8.3 | 21.8 |
| Comanche | 14.3 | 13.2 | 10.9 | 49.9 | 14,026 | 13,960 | -0.5 | -1.5 | 1,562 | 1,817 | 51 | 5,487 | 2.43 | 68.6 | 8.4 | 28.6 |
| Concho | 15.8 | 14.6 | 12.3 | 46.0 | 3,966 | 4,087 | 3.1 | -30.8 | 272 | 328 | -1,232 | 912 | 2.41 | 67.3 | 10.5 | 30.8 |
| Cooke | 14.1 | 11.3 | 8.0 | 50.4 | 36,363 | 38,445 | 5.7 | 7.7 | 5,515 | 4,225 | 1,666 | 15,351 | 2.57 | 70.5 | 10.9 | 24.5 |
| Coryell | 8.8 | 6.4 | 4.4 | 50.0 | 74,978 | 75,475 | 0.7 | 1.7 | 10,016 | 4,649 | -4,274 | 22,322 | 2.71 | 71.6 | 13.7 | 23.9 |
| Cottle | 15.3 | 14.6 | 11.2 | 51.6 | 1,904 | 1,506 | -20.9 | -9.5 | 145 | 235 | -53 | 707 | 2.32 | 59.4 | 18.0 | 40.0 |
| Crane | 11.3 | 7.2 | 5.3 | 49.4 | 3,996 | 4,375 | 9.5 | 8.9 | 700 | 377 | 56 | 1,421 | 3.31 | 79.8 | 11.5 | 18.2 |
| Crockett | 13.2 | 11.0 | 8.1 | 49.9 | 4,099 | 3,719 | -9.3 | -5.5 | 488 | 358 | -337 | 1,354 | 2.52 | 70.0 | 6.4 | 30.0 |
| Crosby | 13.2 | 10.5 | 8.5 | 50.2 | 7,072 | 6,056 | -14.4 | -8.1 | 773 | 736 | -529 | 2,061 | 2.79 | 70.8 | 16.3 | 27.2 |
| Culberson | 13.2 | 11.7 | 9.7 | 50.0 | 2,975 | 2,398 | -19.4 | -10.4 | 294 | 178 | -369 | 580 | 3.76 | 70.9 | 17.2 | 29.1 |
| Dallam | 9.7 | 7.4 | 4.7 | 47.6 | 6,222 | 6,698 | 7.7 | 8.6 | 1,501 | 454 | -472 | 2,364 | 3.07 | 72.4 | 9.9 | 20.9 |
| Dallas | 11.0 | 7.0 | 4.4 | 50.7 | 2,218,899 | 2,368,156 | 6.7 | 11.3 | 400,126 | 162,249 | 31,111 | 928,341 | 2.78 | 64.6 | 15.2 | 28.9 |
| Dawson | 10.4 | 7.8 | 6.8 | 44.6 | 14,985 | 13,833 | -7.7 | -6.2 | 1,919 | 1,401 | -1,422 | 4,322 | 2.55 | 66.8 | 16.4 | 28.1 |
| Deaf Smith | 9.9 | 7.7 | 5.6 | 50.0 | 18,561 | 19,372 | 4.4 | -5.7 | 3,438 | 1,527 | -3,037 | 6,145 | 3.0 | 73.6 | 10.9 | 22.9 |
| Delta | 14.5 | 12.2 | 9.4 | 50.5 | 5,327 | 5,232 | -1.8 | 2.2 | 668 | 752 | 196 | 1,999 | 2.59 | 70.8 | 9.5 | 26.6 |
| Denton | 11.5 | 7.0 | 4.0 | 50.8 | 432,976 | 661,615 | 52.8 | 39.0 | 99,786 | 36,346 | 192,563 | 290,229 | 2.83 | 70.3 | 9.9 | 23.3 |
| DeWitt | 13.7 | 10.9 | 9.4 | 47.6 | 20,013 | 20,097 | 0.4 | 0.4 | 2,540 | 2,618 | 163 | 7,048 | 2.62 | 70.9 | 8.2 | 24.5 |
| Dickens | 13.5 | 12.7 | 11.5 | 44.5 | 2,762 | 2,441 | -11.6 | -12.3 | 196 | 269 | -235 | 845 | 2.47 | 59.4 | 9.1 | 36.0 |
| Dimmit | 10.4 | 9.8 | 7.7 | 51.3 | 10,248 | 9,996 | -2.5 | -0.7 | 1,636 | 912 | -812 | 3,148 | 3.27 | 71.6 | 20.8 | 26.3 |
| Donley | 14.3 | 13.9 | 10.8 | 51.5 | 3,828 | 3,726 | -2.7 | -11.2 | 331 | 502 | -251 | 1,335 | 2.27 | 60.8 | 11.0 | 37.2 |
| Duval | 11.1 | 8.9 | 9.2 | 48.6 | 13,120 | 11,758 | -10.4 | -6.0 | 1,665 | 1,371 | -1,011 | 3,511 | 3.06 | 69.8 | 22.2 | 25.0 |
| Eastland | 14.3 | 12.4 | 10.1 | 50.0 | 18,297 | 18,582 | 1.6 | | 2,055 | 2,622 | 387 | 6,492 | 2.66 | 59.5 | 6.9 | 38.4 |
| Ector | 9.3 | 5.8 | 3.9 | 48.9 | 121,123 | 137,131 | 13.2 | 22.3 | 29,210 | 12,230 | 13,164 | 52,530 | 3.03 | 69.1 | 15.9 | 25.8 |
| Edwards | 14.3 | 17.4 | 12.0 | 47.3 | 2,162 | 2,002 | -7.4 | -3.9 | 243 | 212 | -107 | 789 | 2.42 | 56.4 | 2.7 | 43.6 |
| Ellis | 12.1 | 8.3 | 5.0 | 50.6 | 111,360 | 149,649 | 34.4 | 28.1 | 21,669 | 12,474 | 32,915 | 57,307 | 3.0 | 78.9 | 11.7 | 18.0 |
| El Paso | 10.4 | 7.4 | 5.4 | 50.5 | 679,622 | 800,634 | 17.8 | 5.1 | 132,855 | 53,079 | -39,458 | 268,310 | 3.06 | 72.8 | 18.6 | 23.9 |
| Erath | 11.3 | 8.7 | 6.1 | 51.1 | 33,001 | 37,895 | 14.8 | 14.1 | 4,960 | 3,421 | 3,755 | 13,595 | 2.92 | 63.9 | 6.4 | 27.8 |
| Falls | 13.3 | 10.5 | 8.3 | 52.6 | 18,576 | 17,866 | -3.8 | -3.3 | 2,099 | 1,918 | -781 | 5,199 | 2.99 | 64.2 | 13.4 | 32.8 |
| Fannin | 13.9 | 10.7 | 8.0 | 46.9 | 31,242 | 33,912 | 8.5 | 5.9 | 3,396 | 4,390 | 3,026 | 12,453 | 2.52 | 69.5 | 10.3 | 27.0 |
| Fayette | 15.6 | 14.8 | 12.0 | 50.9 | 21,804 | 24,542 | 12.6 | 4.1 | 2,467 | 3,191 | 1,751 | 9,135 | 2.7 | 71.5 | 7.8 | 24.5 |
| Fisher | 14.3 | 13.4 | 11.3 | 49.9 | 4,344 | 3,978 | -8.4 | -4.9 | 381 | 539 | -38 | 1,601 | 2.39 | 62.0 | 8.4 | 34.3 |
| Floyd | 12.0 | 10.3 | 9.7 | 49.8 | 7,771 | 6,446 | -17.1 | -12.0 | 788 | 657 | -918 | 2,271 | 2.54 | 70.2 | 10.8 | 27.5 |
| Foard | 15.9 | 11.8 | 16.0 | 53.0 | 1,622 | 1,336 | -17.6 | -15.0 | 107 | 196 | -117 | 533 | 2.32 | 64.9 | 7.1 | 31.1 |
| Fort Bend | 11.5 | 8.0 | 4.1 | 51.0 | 354,452 | 584,632 | 64.9 | 43.6 | 92,628 | 31,474 | 192,449 | 237,883 | 3.19 | 82.3 | 11.9 | 14.9 |
| Franklin | 14.8 | 12.7 | 9.4 | 50.3 | 9,458 | 10,598 | 12.1 | 2.1 | 1,171 | 1,226 | 283 | 3,965 | 2.68 | 71.0 | 11.5 | 25.7 |
| Freestone | 13.0 | 11.8 | 8.9 | 47.9 | 17,867 | 19,817 | 10.9 | 0.3 | 2,254 | 2,230 | 48 | 6,758 | 2.67 | 69.2 | 8.6 | 27.8 |
| Frio | 8.9 | 7.0 | 5.4 | 40.4 | 16,252 | 17,217 | 5.9 | 18.4 | 2,584 | 1,412 | 1,973 | 4,542 | 3.44 | 74.2 | 18.3 | 24.3 |
| Gaines | 9.8 | 5.3 | 3.5 | 49.3 | 14,467 | 17,526 | 21.1 | 25.5 | 4,217 | 1,234 | 1,487 | 5,812 | 3.54 | 78.2 | 7.4 | 19.5 |
| Galveston | 13.3 | 9.5 | 5.7 | 50.9 | 250,158 | 291,318 | 16.5 | 18.5 | 41,027 | 27,118 | 39,554 | 121,438 | 2.69 | 70.0 | 13.4 | 24.5 |
| Garza | 11.6 | 6.9 | 5.0 | 34.2 | 4,872 | 6,461 | 32.6 | -3.7 | 618 | 510 | -363 | 1,543 | 2.44 | 67.3 | 8.6 | 30.3 |
| Gillespie | 14.8 | 16.2 | 14.0 | 51.4 | 20,814 | 24,828 | 19.3 | 8.6 | 2,544 | 3,644 | 3,236 | 10,694 | 2.72 | 76.3 | 14.2 | 19.3 |
| Glasscock | 11.8 | 9.6 | 6.2 | 44.6 | 1,406 | 1,226 | -12.8 | 17.4 | 169 | 78 | 122 | 403 | 3.75 | 88.3 | 5.7 | 7.9 |
| Goliad | 14.9 | 14.2 | 10.0 | 50.5 | 6,928 | 7,210 | 4.1 | 5.8 | 721 | 817 | 509 | 2,727 | 2.74 | 70.5 | 8.2 | 26.3 |
| Gonzales | 12.9 | 9.9 | 7.8 | 49.6 | 18,628 | 19,811 | 6.4 | 5.7 | 2,961 | 2,093 | 275 | 7,364 | 2.77 | 73.5 | 13.5 | 25.0 |
| Gray | 11.9 | 8.9 | 7.7 | 46.0 | 22,744 | 22,537 | -0.9 | -3.9 | 3,115 | 2,671 | -1,351 | 7,884 | 2.64 | 65.5 | 11.5 | 28.9 |
| Grayson | 13.7 | 10.9 | 7.5 | 51.3 | 110,595 | 120,869 | 9.3 | 14.4 | 16,034 | 14,524 | 15,990 | 48,454 | 2.65 | 67.3 | 12.3 | 28.5 |
| Gregg | 12.0 | 9.1 | 6.8 | 51.3 | 111,379 | 121,807 | 9.4 | 2.0 | 18,439 | 13,200 | -2,781 | 45,460 | 2.61 | 65.2 | 15.5 | 28.5 |
| Grimes | 13.9 | 11.3 | 7.3 | 45.6 | 23,552 | 26,588 | 12.9 | 11.4 | 3,258 | 2,836 | 2,617 | 9,011 | 2.72 | 71.7 | 13.6 | 24.4 |

1. No spouse present.

# Table B. States and Counties — Population, Vital Statistics, Health, and Crime

| STATE County | Persons in group quarters, 2020 | Daytime Population, 2015-2019 | | Births, 2020 | | Deaths, 2020 | | Persons under 65 with no health insurance, 2019 | | Medicare, 2020 | | | Crimes reported by county police or sheriff, 2019 | |
| | | Number | Employment/ residence ratio | Total | Rate[1] | Number | Rate[1] | Number | Percent | Total beneficiaries | Enrolled in Original Medicare | Enrolled in Medicare Advantage | Violent | Property |
| | 32 | 33 | 34 | 35 | 36 | 37 | 38 | 39 | 40 | 41 | 42 | 43 | 44 | 45 |
|---|---|---|---|---|---|---|---|---|---|---|---|---|---|---|
| **TEXAS—Cont'd** | | | | | | | | | | | | | | |
| Caldwell | 3,361 | 34,266 | 0.57 | 565 | 12.8 | 385 | 8.8 | 9,102 | 25.7 | 7,696 | 4,700 | 2,996 | 51 | 88 |
| Calhoun | 249 | 25,673 | 1.43 | 268 | 12.8 | 237 | 11.3 | 3,480 | 20.3 | 4,248 | 3,132 | 1,116 | 22 | 103 |
| Callahan | 77 | 10,503 | 0.42 | 121 | 8.6 | 198 | 14.0 | 2,016 | 18.4 | 3,355 | 2,274 | 1,081 | 5 | 44 |
| Cameron | 3,427 | 415,956 | 0.96 | 6,223 | 14.7 | 3,006 | 7.1 | 109,621 | 30.8 | 65,439 | 29,523 | 35,917 | 162 | 937 |
| Camp | 67 | 10,783 | 0.6 | 183 | 14.0 | 138 | 10.6 | 2,279 | 21.8 | 2,802 | 1,784 | 1,018 | 9 | 67 |
| Carson | 25 | 5,721 | 0.9 | 43 | 7.3 | 73 | 12.5 | 721 | 15.4 | 1,218 | 852 | 365 | 22 | 25 |
| Cass | 358 | 27,292 | 0.76 | 308 | 10.3 | 489 | 16.4 | 3,849 | 16.8 | 8,078 | 5,244 | 2,835 | 32 | 174 |
| Castro | 45 | 7,346 | 0.9 | 104 | 14.1 | 56 | 7.6 | 1,816 | 29.3 | 1,218 | 897 | 321 | 6 | 47 |
| Chambers | 224 | 37,472 | 0.8 | 604 | 13.2 | 320 | 7.0 | 6,126 | 15.9 | 6,436 | 4,013 | 2,423 | 93 | 544 |
| Cherokee | 2,779 | 48,051 | 0.8 | 699 | 13.2 | 578 | 10.9 | 9,042 | 22.4 | 10,737 | 6,358 | 4,379 | 61 | 362 |
| Childress | 1,729 | 7,283 | 1.01 | 53 | 7.4 | 78 | 10.9 | 1,027 | 23.1 | 1,320 | 891 | 429 | 1 | 7 |
| Clay | 68 | 7,783 | 0.4 | 77 | 7.3 | 147 | 13.9 | 1,498 | 18.8 | 2,658 | 2,004 | 654 | 16 | 103 |
| Cochran | 78 | 2,544 | 0.71 | 39 | 13.5 | 18 | 6.2 | 663 | 28.5 | 559 | 320 | 239 | 3 | 27 |
| Coke | 36 | 2,961 | 0.76 | 34 | 10.2 | 65 | 19.6 | 470 | 19.3 | 943 | 656 | 287 | 2 | 14 |
| Coleman | 16 | 7,305 | 0.71 | 74 | 9.1 | 125 | 15.4 | 1,541 | 25.6 | 2,519 | 1,942 | 576 | NA | NA |
| Collin | 3,940 | 948,447 | 0.95 | 11,413 | 10.6 | 5,752 | 5.4 | 114,952 | 12.5 | 124,503 | 81,936 | 42,567 | 66 | 338 |
| Collingsworth | 54 | 2,840 | 0.9 | 35 | 12.2 | 32 | 11.1 | 751 | 32.3 | 642 | 538 | 104 | NA | NA |
| Colorado | 319 | 20,349 | 0.91 | 259 | 12.0 | 293 | 13.6 | 4,092 | 24.8 | 5,335 | 3,879 | 1,456 | 19 | 117 |
| Comal | 1,155 | 138,684 | 0.95 | 1,626 | 9.9 | 1,318 | 8.0 | 19,100 | 15.0 | 35,261 | 23,017 | 12,244 | 135 | 601 |
| Comanche | 148 | 12,712 | 0.86 | 154 | 11.2 | 205 | 14.9 | 2,462 | 24.1 | 3,589 | 2,618 | 971 | 3 | 58 |
| Concho | 40 | 3,211 | 0.94 | 29 | 10.3 | 38 | 13.4 | 442 | 21.9 | 729 | 482 | 247 | 11 | 8 |
| Cooke | 607 | 37,158 | 0.85 | 541 | 13.1 | 471 | 11.4 | 7,232 | 22.0 | 8,410 | 5,743 | 2,667 | 33 | 100 |
| Coryell | 12,772 | 64,844 | 0.66 | 895 | 11.7 | 508 | 6.6 | 8,941 | 16.0 | 10,736 | 7,179 | 3,557 | 4 | 54 |
| Cottle | 0 | 1,629 | 0.98 | 14 | 10.3 | 11 | 8.1 | 226 | 21.6 | 414 | 283 | 131 | NA | NA |
| Crane | 87 | 4,412 | 0.81 | 56 | 11.8 | 32 | 6.7 | 850 | 20.4 | 676 | 442 | 234 | NA | NA |
| Crockett | 26 | 3,322 | 0.91 | 35 | 10.0 | 21 | 6.0 | 593 | 20.9 | 727 | 527 | 200 | 2 | 18 |
| Crosby | 63 | 4,964 | 0.65 | 73 | 13.1 | 63 | 11.3 | 983 | 21.5 | 1,275 | 737 | 538 | NA | NA |
| Culberson | 12 | 2,345 | 1.13 | 23 | 10.7 | 21 | 9.8 | 405 | 24.3 | 551 | 398 | 153 | 0 | 0 |
| Dallam | 39 | 8,117 | 1.22 | 151 | 20.8 | 51 | 7.0 | 2,143 | 33.8 | 962 | 721 | 240 | 0 | 1 |
| Dallas | 30,844 | 2,935,614 | 1.26 | 37,726 | 14.3 | 18,457 | 7.0 | 576,926 | 25.1 | 324,978 | 185,210 | 139,769 | 43 | 307 |
| Dawson | 1,602 | 13,139 | 1.08 | 196 | 15.1 | 128 | 9.9 | 2,193 | 23.2 | 2,301 | 1,587 | 715 | 2 | 64 |
| Deaf Smith | 359 | 19,259 | 1.06 | 314 | 17.2 | 169 | 9.2 | 4,144 | 26.2 | 2,756 | 1,931 | 825 | 8 | 60 |
| Delta | 54 | 4,083 | 0.44 | 75 | 14.0 | 51 | 9.5 | 778 | 18.6 | 1,379 | 951 | 428 | 1 | 27 |
| Denton | 11,790 | 684,134 | 0.66 | 10,151 | 11.0 | 5,077 | 5.5 | 105,507 | 13.4 | 103,733 | 66,172 | 37,561 | 95 | 440 |
| DeWitt | 1,793 | 20,559 | 1.03 | 226 | 11.2 | 221 | 11.0 | 2,921 | 20.1 | 4,536 | 3,273 | 1,263 | 33 | 120 |
| Dickens | 315 | 2,234 | 1.04 | 16 | 7.5 | 21 | 9.8 | 299 | 22.0 | 541 | 386 | 155 | NA | NA |
| Dimmit | 87 | 13,610 | 1.95 | 131 | 13.2 | 81 | 8.2 | 1,676 | 20.5 | 1,952 | 1,056 | 897 | 11 | 139 |
| Donley | 224 | 3,067 | 0.81 | 28 | 8.5 | 43 | 13.0 | 476 | 21.3 | 891 | 646 | 245 | 4 | 30 |
| Duval | 601 | 10,691 | 0.86 | 149 | 13.5 | 153 | 13.8 | 1,857 | 21.9 | 2,376 | 1,294 | 1,082 | 12 | 104 |
| Eastland | 779 | 17,959 | 0.95 | 189 | 10.3 | 270 | 14.7 | 3,274 | 24.1 | 4,627 | 3,171 | 1,456 | 5 | 22 |
| Ector | 2,505 | 162,493 | 1.03 | 3,059 | 18.2 | 1,324 | 7.9 | 32,601 | 22.0 | 19,620 | 12,792 | 6,829 | 116 | 1,579 |
| Edwards | 5 | 2,007 | 1.11 | 22 | 11.4 | 9 | 4.7 | 350 | 25.8 | 539 | 396 | 143 | 2 | 5 |
| Ellis | 1,489 | 148,768 | 0.7 | 2,262 | 11.8 | 1,546 | 8.1 | 30,187 | 18.9 | 29,393 | 17,970 | 11,422 | 83 | 362 |
| El Paso | 16,662 | 834,752 | 1.0 | 11,572 | 13.8 | 5,844 | 6.9 | 173,169 | 24.3 | 133,285 | 48,379 | 84,906 | 259 | 641 |
| Erath | 3,518 | 40,462 | 0.93 | 478 | 11.1 | 387 | 9.0 | 7,823 | 23.7 | 6,923 | 4,881 | 2,042 | 50 | 93 |
| Falls | 1,936 | 15,330 | 0.67 | 201 | 11.6 | 186 | 10.8 | 2,573 | 21.4 | 3,672 | 2,029 | 1,643 | NA | NA |
| Fannin | 2,996 | 30,403 | 0.7 | 303 | 8.4 | 483 | 13.4 | 5,675 | 21.5 | 7,780 | 5,391 | 2,389 | 42 | 194 |
| Fayette | 420 | 25,475 | 1.03 | 242 | 9.5 | 323 | 12.6 | 4,008 | 21.6 | 7,061 | 5,276 | 1,786 | 11 | 123 |
| Fisher | 22 | 3,420 | 0.74 | 32 | 8.5 | 68 | 18.0 | 555 | 19.5 | 966 | 692 | 274 | NA | NA |
| Floyd | 34 | 5,507 | 0.88 | 62 | 10.9 | 54 | 9.5 | 1,191 | 25.9 | 1,244 | 750 | 493 | 12 | 25 |
| Foard | 37 | 1,222 | 0.91 | 8 | 7.0 | 10 | 8.8 | 191 | 22.9 | 344 | 269 | 76 | NA | NA |
| Fort Bend | 5,055 | 610,267 | 0.57 | 9,482 | 11.3 | 4,331 | 5.2 | 107,300 | 15.0 | 100,565 | 55,853 | 44,711 | 841 | 4,643 |
| Franklin | 56 | 10,238 | 0.89 | 134 | 12.4 | 144 | 13.3 | 1,434 | 17.3 | 2,388 | 1,605 | 783 | NA | NA |
| Freestone | 1,577 | 18,189 | 0.78 | 230 | 11.6 | 223 | 11.2 | 2,962 | 20.8 | 4,406 | 2,883 | 1,523 | 22 | 131 |
| Frio | 3,397 | 21,729 | 1.28 | 255 | 12.5 | 163 | 8.0 | 3,265 | 22.8 | 2,783 | 1,518 | 1,265 | 9 | 60 |
| Gaines | 63 | 20,021 | 0.92 | 449 | 20.4 | 159 | 7.2 | 6,949 | 35.8 | 2,025 | 1,460 | 565 | 11 | 88 |
| Galveston | 4,845 | 295,342 | 0.76 | 3,930 | 11.4 | 3,265 | 9.5 | 49,934 | 17.4 | 57,939 | 34,461 | 23,478 | 123 | 727 |
| Garza | 1,854 | 5,979 | 0.95 | 45 | 7.2 | 38 | 6.1 | 872 | 23.6 | 861 | 539 | 321 | 15 | 30 |
| Gillespie | 246 | 27,474 | 1.09 | 235 | 8.7 | 405 | 15.0 | 4,544 | 24.2 | 8,724 | 6,782 | 1,942 | 12 | 55 |
| Glasscock | 0 | 1,713 | 1.28 | 25 | 17.4 | 14 | 9.7 | 230 | 19.1 | 192 | 158 | 34 | NA | NA |
| Goliad | 93 | 6,600 | 0.69 | 72 | 9.4 | 73 | 9.6 | 982 | 17.0 | 1,765 | 1,202 | 563 | 22 | 79 |
| Gonzales | 313 | 20,231 | 0.94 | 276 | 13.2 | 216 | 10.3 | 4,266 | 25.2 | 4,072 | 2,794 | 1,278 | 25 | 116 |
| Gray | 1,950 | 22,201 | 0.98 | 243 | 11.2 | 251 | 11.6 | 3,922 | 23.9 | 4,091 | 3,108 | 983 | 15 | 59 |
| Grayson | 2,234 | 126,334 | 0.92 | 1,676 | 12.1 | 1,545 | 11.2 | 24,027 | 21.8 | 29,343 | 19,373 | 9,970 | 32 | 401 |
| Gregg | 3,995 | 143,943 | 1.38 | 1,625 | 13.1 | 1,385 | 11.1 | 22,634 | 22.4 | 24,746 | 15,437 | 9,309 | 84 | 373 |
| Grimes | 2,940 | 25,564 | 0.77 | 319 | 10.8 | 258 | 8.7 | 4,771 | 22.6 | 6,029 | 3,767 | 2,261 | NA | NA |

1. Per 1,000 estimated resident population.

# Table B. States and Counties — Crime, Education, Money Income, and Poverty

| STATE County | Officers (46) | Civilians (47) | Enrollment Total (48) | Enrollment Percent private (49) | HS graduate or less (50) | Bachelor's degree or more (51) | Total current spending (mil dol) (52) | Current spending per student (dollars) (53) | Per capita income (54) | Households Median income (dollars) (55) | Households Percent with income less than $50,000 (56) | Households Percent with income of $200,000 or more (57) | Median household income (dollars) (58) | Poverty All persons (59) | Poverty Children under 18 years (60) | Poverty Children 5 to 17 years in families (61) |
|---|---|---|---|---|---|---|---|---|---|---|---|---|---|---|---|---|
| **TEXAS—Cont'd** | | | | | | | | | | | | | | | | |
| Caldwell | 34 | 67 | 10,107 | 7.9 | 59.6 | 14.6 | 66 | 8,725 | 25,086 | 54,152 | 45.2 | 4.0 | 55,301 | 14.4 | 20.5 | 19.9 |
| Calhoun | 29 | 35 | 4,715 | 6.2 | 53.7 | 14.2 | 37 | 9,357 | 27,268 | 58,776 | 43.3 | 2.1 | 55,909 | 14.7 | 22.6 | 21.6 |
| Callahan | 6 | 9 | 2,888 | 7.2 | 47.6 | 21.0 | 25 | 10,081 | 26,051 | 48,651 | 51.0 | 2.2 | 50,812 | 12.1 | 18.7 | 18.6 |
| Cameron | 105 | 338 | 123,483 | 3.9 | 58.5 | 17.3 | 999 | 10,396 | 17,430 | 38,758 | 59.0 | 2.3 | 40,893 | 25.5 | 36.3 | 32.5 |
| Camp | 9 | 11 | 3,092 | 16.9 | 48.1 | 19.4 | 31 | 13,262 | 22,449 | 48,207 | 52.0 | 2.1 | 47,091 | 16.6 | 26.1 | 25.1 |
| Carson | NA | NA | 1,263 | 9.7 | 38.6 | 28.4 | 15 | 11,604 | 34,708 | 74,872 | 34.5 | 6.0 | 69,760 | 8.5 | 11.0 | 9.7 |
| Cass | 21 | 29 | 6,299 | 3.5 | 57.4 | 16.4 | 53 | 9,656 | 23,075 | 44,848 | 55.8 | 1.8 | 44,529 | 17.3 | 23.6 | 21.5 |
| Castro | 7 | 16 | 2,180 | 3.6 | 58.1 | 15.6 | 21 | 12,133 | 21,884 | 50,142 | 49.8 | 1.9 | 51,736 | 15.3 | 22.7 | 20.9 |
| Chambers | NA | NA | 11,340 | 8.4 | 38.4 | 22.7 | 90 | 10,953 | 35,916 | 91,141 | 32.3 | 10.3 | 93,351 | 8.6 | 11.9 | 11.3 |
| Cherokee | 32 | 45 | 12,988 | 10.4 | 51.5 | 17.2 | 102 | 9,130 | 22,391 | 48,186 | 51.8 | 2.1 | 46,576 | 21.9 | 25.9 | 23.0 |
| Childress | NA | NA | 1,030 | 0.0 | 52.9 | 18.3 | 11 | 10,022 | 21,233 | 43,181 | 56.9 | 1.7 | 44,095 | 11.5 | 16.0 | 13.7 |
| Clay | 16 | 10 | 2,276 | 10.0 | 46.4 | 21.4 | 19 | 10,893 | 29,159 | 55,989 | 45.8 | 3.8 | 59,983 | 19.5 | 30.1 | 27.6 |
| Cochran | 9 | 9 | 641 | 4.8 | 63.1 | 8.0 | 10 | 14,246 | 21,463 | 40,962 | 57.2 | 1.2 | 43,742 | 13.1 | 17.9 | 17.3 |
| Coke | 5 | 1 | 692 | 9.8 | 50.4 | 17.8 | 6 | 11,976 | 25,285 | 45,000 | 56.2 | 0.5 | 48,435 | 13.1 | 17.9 | 17.3 |
| Coleman | NA | NA | 1,598 | 10.6 | 50.2 | 15.8 | 15 | 10,998 | 27,752 | 46,743 | 56.9 | 4.4 | 44,834 | 17.7 | 30.4 | 27.9 |
| Collin | 153 | 360 | 275,673 | 12.0 | 21.2 | 52.3 | 1,879 | 8,947 | 44,548 | 96,913 | 23.5 | 15.1 | 96,847 | 6.1 | 6.8 | 6.1 |
| Collingsworth | 9 | 0 | 697 | 10.2 | 52.1 | 20.2 | 6 | 10,711 | 25,681 | 39,120 | 59.5 | 2.7 | 40,314 | 19.8 | 35.8 | 32.1 |
| Colorado | 23 | 30 | 4,878 | 16.2 | 51.3 | 21.0 | 36 | 10,086 | 29,552 | 52,559 | 47.7 | 4.8 | 55,607 | 13.6 | 20.9 | 20.2 |
| Comal | 151 | 158 | 31,969 | 14.8 | 32.4 | 36.8 | 263 | 8,213 | 38,991 | 79,936 | 30.5 | 11.3 | 87,976 | 6.7 | 9.4 | 8.3 |
| Comanche | NA | NA | 2,817 | 8.8 | 50.4 | 20.3 | 23 | 9,978 | 26,728 | 53,516 | 46.8 | 2.1 | 54,122 | 13.5 | 22.1 | 20.4 |
| Concho | NA | NA | 563 | 1.6 | 62.9 | 13.0 | 6 | 11,388 | 21,420 | 47,500 | 50.2 | 3.9 | 43,321 | 15.1 | 27.0 | 27.0 |
| Cooke | 35 | 50 | 9,212 | 12.6 | 45.4 | 20.9 | 63 | 9,461 | 30,704 | 60,202 | 42.9 | 5.2 | 61,186 | 12.7 | 18.4 | 17.7 |
| Coryell | NA | NA | 19,991 | 7.2 | 40.1 | 16.1 | 100 | 8,755 | 22,400 | 52,893 | 47.4 | 1.9 | 53,083 | 11.8 | 15.7 | 14.9 |
| Cottle | 1 | 0 | 444 | 11.3 | 50.7 | 16.8 | 3 | 12,453 | 19,797 | 32,305 | 70.7 | 1.4 | 40,449 | 18.9 | 32.5 | 30.3 |
| Crane | 9 | 7 | 1,417 | 5.5 | 53.3 | 14.7 | 15 | 13,065 | 27,893 | 65,969 | 35.4 | 5.3 | 68,626 | 9.6 | 13.6 | 12.7 |
| Crockett | NA | NA | 780 | 0.0 | 56.5 | 14.0 | 10 | 12,613 | 26,178 | 47,386 | 50.1 | 0.6 | 54,165 | 13.8 | 19.3 | 17.2 |
| Crosby | 6 | 9 | 1,333 | 11.3 | 59.9 | 13.2 | 17 | 14,332 | 19,443 | 40,759 | 60.6 | 2.0 | 38,871 | 22.1 | 33.6 | 31.7 |
| Culberson | NA | NA | 481 | 8.9 | 68.7 | 10.2 | 6 | 13,951 | 15,109 | 37,900 | 59.1 | 0.0 | 44,462 | 19.7 | 30.6 | 29.5 |
| Dallam | 5 | 1 | 1,794 | 14.4 | 57.7 | 15.8 | 19 | 9,467 | 27,491 | 58,811 | 40.5 | 3.1 | 55,793 | 11.4 | 17.4 | 16.9 |
| Dallas | NA | NA | 689,066 | 12.1 | 43.3 | 31.5 | 4,846 | 9,460 | 32,653 | 59,607 | 41.9 | 7.6 | 61,807 | 14.0 | 20.9 | 20.4 |
| Dawson | 9 | 12 | 3,575 | 10.4 | 62.4 | 14.2 | 28 | 11,177 | 20,084 | 43,239 | 57.8 | 3.1 | 48,165 | 20.6 | 28.9 | 29.0 |
| Deaf Smith | 13 | 27 | 5,928 | 6.1 | 62.7 | 12.0 | 40 | 9,176 | 21,616 | 49,701 | 50.4 | 1.6 | 46,178 | 17.0 | 24.8 | 23.3 |
| Delta | 10 | 12 | 1,262 | 11.6 | 44.2 | 25.1 | 8 | 9,791 | 27,667 | 51,038 | 48.8 | 4.0 | 51,348 | 16.0 | 25.5 | 25.0 |
| Denton | NA | NA | 238,788 | 12.0 | 25.3 | 45.1 | 1,399 | 9,213 | 41,153 | 86,913 | 26.3 | 12.8 | 90,910 | 6.5 | 7.1 | 6.1 |
| DeWitt | 19 | 33 | 4,126 | 8.2 | 58.3 | 12.6 | 31 | 10,427 | 28,910 | 55,357 | 46.5 | 6.6 | 52,544 | 17.1 | 24.3 | 21.6 |
| Dickens | 3 | 4 | 388 | 4.6 | 51.4 | 18.5 | 5 | 14,476 | 25,025 | 42,540 | 56.6 | 2.5 | 41,973 | 19.4 | 26.2 | 23.5 |
| Dimmit | 32 | 39 | 3,084 | 7.5 | 69.8 | 13.6 | 25 | 11,493 | 16,925 | 27,161 | 64.0 | 1.7 | 40,315 | 25.3 | 34.8 | 33.3 |
| Donley | 5 | 5 | 874 | 4.1 | 45.4 | 17.9 | 7 | 11,762 | 25,534 | 42,961 | 56.6 | 2.1 | 41,093 | 16.4 | 24.5 | 22.4 |
| Duval | 20 | 18 | 2,966 | 3.8 | 62.1 | 9.0 | 32 | 11,928 | 17,637 | 41,186 | 56.3 | 0.8 | 37,100 | 23.9 | 33.5 | 32.5 |
| Eastland | 12 | 18 | 4,224 | 4.7 | 47.2 | 16.0 | 29 | 10,260 | 24,548 | 37,276 | 61.9 | 3.7 | 46,991 | 16.7 | 24.2 | 20.7 |
| Ector | 91 | 51 | 44,795 | 7.4 | 54.4 | 15.4 | 272 | 8,022 | 28,954 | 62,973 | 39.5 | 4.9 | 65,564 | 12.1 | 17.0 | 16.9 |
| Edwards | NA | NA | 289 | 0.0 | 50.6 | 18.2 | 7 | 11,655 | 24,689 | 40,766 | 52.7 | 4.3 | 40,424 | 20.7 | 34.5 | 32.9 |
| Ellis | 83 | 146 | 47,656 | 11.3 | 41.5 | 24.3 | 367 | 8,927 | 32,571 | 76,871 | 29.5 | 7.6 | 78,341 | 8.7 | 12.8 | 11.2 |
| El Paso | 261 | 730 | 255,026 | 8.0 | 45.2 | 23.3 | 1,730 | 9,933 | 21,683 | 46,871 | 52.6 | 2.6 | 48,629 | 18.8 | 26.2 | 24.5 |
| Erath | 24 | 42 | 14,020 | 9.2 | 38.1 | 31.5 | 55 | 9,186 | 27,639 | 52,742 | 47.7 | 4.1 | 53,044 | 15.2 | 18.9 | 17.5 |
| Falls | NA | NA | 3,316 | 8.5 | 61.5 | 12.4 | 27 | 12,410 | 18,745 | 39,497 | 61.2 | 1.1 | 41,484 | 21.6 | 28.1 | 27.7 |
| Fannin | 20 | 10 | 7,709 | 10.8 | 49.8 | 17.4 | 57 | 10,155 | 27,112 | 54,648 | 45.7 | 4.4 | 56,123 | 13.2 | 16.1 | 14.6 |
| Fayette | NA | NA | 4,953 | 9.9 | 53.0 | 19.7 | 39 | 10,285 | 30,170 | 60,189 | 42.9 | 5.8 | 62,195 | 10.9 | 15.3 | 14.6 |
| Fisher | 4 | 9 | 799 | 7.6 | 48.6 | 18.6 | 7 | 11,609 | 28,750 | 46,146 | 53.0 | 2.6 | 51,053 | 13.3 | 22.2 | 20.6 |
| Floyd | 7 | 3 | 1,337 | 9.2 | 53.5 | 15.9 | 15 | 12,255 | 23,617 | 50,580 | 49.0 | 1.9 | 45,022 | 17.3 | 30.3 | 29.1 |
| Foard | 4 | 1 | 269 | 0.0 | 50.2 | 16.2 | 3 | 13,930 | 27,573 | 43,625 | 56.7 | 3.6 | 38,615 | 17.0 | 26.2 | 24.0 |
| Fort Bend | 551 | 235 | 229,811 | 12.2 | 27.3 | 46.2 | 1,824 | 9,343 | 39,994 | 97,743 | 23.4 | 16.3 | 101,361 | 6.6 | 7.4 | 6.5 |
| Franklin | 27 | 0 | 2,719 | 5.5 | 36.6 | 26.6 | 15 | 9,238 | 28,789 | 53,783 | 46.5 | 5.2 | 50,630 | 14.4 | 22.8 | 19.7 |
| Freestone | 16 | 20 | 4,085 | 7.1 | 50.6 | 15.3 | 39 | 10,525 | 24,621 | 49,471 | 50.3 | 3.8 | 52,511 | 15.5 | 19.5 | 18.0 |
| Frio | 17 | 19 | 4,785 | 10.1 | 68.0 | 7.3 | 36 | 10,862 | 19,256 | 46,729 | 53.7 | 2.7 | 40,360 | 27.7 | 35.9 | 29.9 |
| Gaines | 18 | 21 | 5,430 | 22.7 | 68.0 | 11.4 | 45 | 12,580 | 23,533 | 63,054 | 40.8 | 4.3 | 67,171 | 13.2 | 20.9 | 20.6 |
| Galveston | NA | NA | 88,470 | 11.7 | 35.9 | 31.1 | 781 | 9,396 | 36,819 | 73,330 | 35.3 | 10.5 | 73,214 | 11.3 | 16.1 | 15.0 |
| Garza | 9 | 34 | 1,013 | 8.5 | 67.0 | 13.9 | 12 | 12,232 | 19,997 | 49,627 | 50.6 | 2.5 | 48,975 | 24.2 | 24.4 | 22.8 |
| Gillespie | 34 | 20 | 4,991 | 23.3 | 40.9 | 32.2 | 39 | 10,073 | 34,077 | 59,155 | 42.6 | 6.4 | 62,143 | 8.5 | 12.8 | 12.8 |
| Glasscock | 4 | 1 | 438 | 5.9 | 49.3 | 27.2 | 6 | 19,696 | 31,830 | 78,456 | 37.5 | 8.4 | 93,759 | 7.9 | 10.7 | 10.2 |
| Goliad | NA | NA | 1,715 | 7.5 | 43.5 | 17.3 | 13 | 9,890 | 29,855 | 60,690 | 44.0 | 4.8 | 56,758 | 13.4 | 22.5 | 21.4 |
| Gonzales | 21 | 36 | 4,791 | 6.8 | 67.4 | 13.1 | 43 | 10,179 | 27,407 | 53,577 | 46.3 | 4.7 | 48,425 | 16.2 | 23.7 | 20.6 |
| Gray | 14 | 29 | 5,103 | 9.5 | 52.5 | 13.9 | 37 | 9,135 | 24,686 | 47,952 | 50.9 | 3.1 | 52,543 | 14.5 | 19.0 | 17.8 |
| Grayson | 63 | 119 | 32,121 | 12.0 | 41.4 | 20.4 | 223 | 9,594 | 28,011 | 54,815 | 45.8 | 3.7 | 57,439 | 11.6 | 17.1 | 15.4 |
| Gregg | 105 | 151 | 30,744 | 12.6 | 44.1 | 21.3 | 240 | 9,924 | 26,535 | 50,180 | 49.8 | 3.6 | 53,137 | 16.5 | 24.2 | 21.9 |
| Grimes | NA | NA | 6,508 | 6.9 | 54.2 | 15.7 | 45 | 9,788 | 25,638 | 52,913 | 47.0 | 5.0 | 51,651 | 15.9 | 20.5 | 19.4 |

1. All persons 3 years old and over enrolled in nursery school through college.   2. Persons 25 years old and over.   3. Elementary and secondary education expenditures.   4. Based on population estimated by the American Community Survey, 2014–2018.

# Table B. States and Counties — **Personal Income**

| STATE County | Personal income, 2019 | | | | | | | | | | Earnings, 2019 | | |
| | Total (mil dol) | Percent change 2018-2019 | Per capita[1] Dollars | Per capita Rank | Wages and salaries (mil dol) | Supplements to wages and salaries, employer contributions (mil dol) Pension and insurance | Government social insurance | Proprietors' income (mil dol) | Dividends, interest, and rent (mil dol) | Personal transfer receipts (mil dol) | Total (mil dol) | Contributions for government social insurance (mil dol) From employee and self-employed | From employer |
| | 62 | 63 | 64 | 65 | 66 | 67 | 68 | 69 | 70 | 71 | 72 | 73 | 74 |
|---|---|---|---|---|---|---|---|---|---|---|---|---|---|
| **TEXAS—Cont'd** | | | | | | | | | | | | | |
| Caldwell | 1,512 | 5.9 | 34,617 | 2,735 | 396 | 73 | 27 | 122 | 222 | 389 | 618 | 40 | 27 |
| Calhoun | 984 | 4.7 | 46,208 | 1,229 | 917 | 157 | 62 | 127 | 137 | 223 | 1,263 | 67 | 62 |
| Callahan | 585 | 4.5 | 41,962 | 1,771 | 119 | 21 | 8 | 32 | 89 | 156 | 180 | 14 | 8 |
| Cameron | 12,664 | 3.7 | 29,928 | 3,038 | 5,263 | 1,109 | 367 | 966 | 1,680 | 4,117 | 7,705 | 449 | 367 |
| Camp | 486 | 0.1 | 37,111 | 2,432 | 152 | 30 | 10 | 40 | 72 | 144 | 232 | 15 | 10 |
| Carson | 288 | 7.7 | 48,571 | 962 | 436 | 44 | 30 | 37 | 40 | 52 | 547 | 29 | 30 |
| Cass | 1,128 | 2.4 | 37,566 | 2,380 | 321 | 67 | 22 | 67 | 159 | 410 | 476 | 35 | 22 |
| Castro | 485 | 8.0 | 64,427 | 199 | 110 | 19 | 10 | 203 | 56 | 71 | 341 | 7 | 10 |
| Chambers | 2,482 | 7.4 | 56,610 | 417 | 1,234 | 187 | 81 | 91 | 273 | 318 | 1,592 | 88 | 81 |
| Cherokee | 1,855 | 1.9 | 35,245 | 2,652 | 607 | 130 | 41 | 124 | 301 | 556 | 902 | 56 | 41 |
| Childress | 225 | 2.0 | 30,731 | 3,016 | 108 | 25 | 7 | 8 | 39 | 66 | 148 | 8 | 7 |
| Clay | 464 | 4.7 | 44,295 | 1,447 | 66 | 15 | 4 | 19 | 75 | 119 | 104 | 9 | 4 |
| Cochran | 112 | -0.8 | 39,333 | 2,140 | 34 | 7 | 3 | 10 | 19 | 32 | 54 | 3 | 3 |
| Coke | 141 | 3.4 | 41,669 | 1,821 | 34 | 7 | 2 | 1 | 26 | 46 | 44 | 4 | 2 |
| Coleman | 349 | 3.0 | 42,683 | 1,660 | 74 | 17 | 5 | 22 | 76 | 126 | 119 | 9 | 5 |
| Collin | 70,852 | 6.3 | 68,474 | 143 | 32,006 | 3,836 | 2,059 | 6,562 | 10,807 | 5,320 | 44,462 | 2,351 | 2,059 |
| Collingsworth | 123 | 6.5 | 42,026 | 1,757 | 39 | 8 | 3 | 7 | 23 | 37 | 57 | 4 | 3 |
| Colorado | 1,008 | 4.3 | 46,909 | 1,142 | 337 | 58 | 23 | 73 | 235 | 270 | 492 | 32 | 23 |
| Comal | 9,381 | 7.5 | 60,056 | 301 | 2,919 | 412 | 198 | 735 | 1,706 | 1,531 | 4,263 | 257 | 198 |
| Comanche | 590 | 6.7 | 43,242 | 1,583 | 146 | 30 | 11 | 76 | 97 | 169 | 263 | 16 | 11 |
| Concho | 97 | 8.1 | 35,758 | 2,590 | 37 | 7 | 3 | 7 | 20 | 38 | 53 | 4 | 3 |
| Cooke | 2,181 | 5.2 | 52,875 | 626 | 789 | 134 | 53 | 301 | 413 | 409 | 1,277 | 72 | 53 |
| Coryell | 2,702 | 4.8 | 35,570 | 2,607 | 771 | 157 | 48 | 87 | 494 | 673 | 1,063 | 63 | 48 |
| Cottle | 84 | -1.6 | 60,260 | 296 | 21 | 4 | 1 | 7 | 31 | 20 | 33 | 2 | 1 |
| Crane | 245 | 5.9 | 51,025 | 756 | 70 | 13 | 4 | 24 | 19 | 37 | 112 | 6 | 4 |
| Crockett | 154 | 6.0 | 44,458 | 1,424 | 70 | 15 | 5 | 7 | 36 | 34 | 97 | 5 | 5 |
| Crosby | 216 | 2.6 | 37,580 | 2,378 | 58 | 11 | 5 | 15 | 34 | 78 | 89 | 6 | 5 |
| Culberson | 129 | 7.8 | 59,506 | 324 | 74 | 13 | 5 | 0 | 17 | 29 | 92 | 6 | 5 |
| Dallam | 472 | 7.9 | 64,756 | 191 | 190 | 27 | 15 | 188 | 45 | 57 | 420 | 15 | 15 |
| Dallas | 165,463 | 3.5 | 62,782 | 232 | 133,413 | 15,845 | 8,626 | 31,357 | 36,051 | 17,926 | 189,242 | 9,699 | 8,626 |
| Dawson | 511 | -2.8 | 40,131 | 2,037 | 188 | 38 | 12 | 66 | 90 | 133 | 305 | 15 | 12 |
| Deaf Smith | 971 | 5.2 | 52,368 | 654 | 322 | 55 | 23 | 362 | 103 | 164 | 762 | 24 | 23 |
| Delta | 217 | 4.8 | 40,622 | 1,971 | 31 | 9 | 2 | 26 | 24 | 70 | 69 | 4 | 2 |
| Denton | 52,713 | 6.8 | 59,414 | 325 | 14,469 | 2,069 | 965 | 3,547 | 6,808 | 4,749 | 21,050 | 1,139 | 965 |
| DeWitt | 1,197 | 2.7 | 59,389 | 326 | 403 | 73 | 25 | 57 | 465 | 250 | 558 | 33 | 25 |
| Dickens | 75 | 2.8 | 33,843 | 2,820 | 22 | 5 | 2 | 0 | 17 | 28 | 28 | 2 | 2 |
| Dimmit | 393 | 3.2 | 38,800 | 2,224 | 382 | 57 | 27 | 41 | 78 | 118 | 507 | 28 | 27 |
| Donley | 149 | -3.2 | 45,531 | 1,296 | 32 | 8 | 2 | 19 | 31 | 42 | 62 | 3 | 2 |
| Duval | 435 | 3.0 | 39,029 | 2,188 | 148 | 34 | 10 | 4 | 54 | 171 | 195 | 13 | 10 |
| Eastland | 1,447 | 1.7 | 78,826 | 70 | 439 | 80 | 30 | 440 | 410 | 248 | 989 | 49 | 30 |
| Ector | 8,338 | 6.3 | 50,161 | 821 | 5,552 | 676 | 357 | 842 | 893 | 1,106 | 7,427 | 395 | 357 |
| Edwards | 88 | 11.2 | 45,429 | 1,311 | 25 | 5 | 2 | 3 | 28 | 25 | 35 | 3 | 2 |
| Ellis | 8,496 | 7.0 | 45,968 | 1,252 | 2,573 | 423 | 174 | 505 | 986 | 1,393 | 3,676 | 218 | 174 |
| El Paso | 31,652 | 4.0 | 37,715 | 2,357 | 15,216 | 3,285 | 1,103 | 2,271 | 4,937 | 7,275 | 21,874 | 1,177 | 1,103 |
| Erath | 1,728 | 5.4 | 40,462 | 1,992 | 675 | 131 | 46 | 231 | 333 | 359 | 1,082 | 55 | 46 |
| Falls | 610 | 1.7 | 35,258 | 2,649 | 146 | 38 | 9 | 36 | 100 | 202 | 229 | 15 | 9 |
| Fannin | 1,415 | 5.3 | 39,830 | 2,072 | 386 | 83 | 27 | 87 | 190 | 405 | 582 | 39 | 27 |
| Fayette | 1,383 | 3.5 | 54,552 | 517 | 416 | 73 | 27 | 75 | 388 | 323 | 591 | 40 | 27 |
| Fisher | 171 | -9.0 | 44,630 | 1,403 | 40 | 9 | 3 | 9 | 31 | 50 | 61 | 4 | 3 |
| Floyd | 255 | 6.9 | 44,646 | 1,401 | 65 | 15 | 5 | 57 | 39 | 71 | 141 | 6 | 5 |
| Foard | 52 | -6.2 | 44,895 | 1,374 | 13 | 3 | 1 | 2 | 11 | 19 | 19 | 1 | 1 |
| Fort Bend | 48,420 | 6.2 | 59,653 | 318 | 11,127 | 1,544 | 734 | 4,445 | 6,740 | 4,388 | 17,850 | 964 | 734 |
| Franklin | 443 | 2.1 | 41,307 | 1,881 | 147 | 21 | 11 | 68 | 75 | 119 | 247 | 14 | 11 |
| Freestone | 753 | 2.9 | 38,182 | 2,306 | 228 | 45 | 17 | 36 | 130 | 209 | 326 | 22 | 17 |
| Frio | 614 | 4.1 | 30,223 | 3,028 | 438 | 68 | 31 | 64 | 103 | 167 | 600 | 32 | 31 |
| Gaines | 954 | 11.3 | 44,405 | 1,431 | 392 | 62 | 26 | 271 | 96 | 124 | 751 | 32 | 26 |
| Galveston | 18,561 | 4.6 | 54,250 | 532 | 6,156 | 1,194 | 389 | 950 | 3,118 | 3,012 | 8,688 | 476 | 389 |
| Garza | 203 | 3.8 | 32,601 | 2,914 | 89 | 14 | 6 | 29 | 43 | 49 | 138 | 8 | 6 |
| Gillespie | 1,708 | 3.6 | 63,291 | 224 | 483 | 77 | 34 | 129 | 704 | 335 | 722 | 49 | 34 |
| Glasscock | 119 | -0.2 | 84,623 | 48 | 31 | 5 | 2 | 16 | 53 | 7 | 55 | 3 | 2 |
| Goliad | 349 | 5.1 | 45,589 | 1,294 | 64 | 14 | 4 | 5 | 83 | 84 | 87 | 7 | 4 |
| Gonzales | 933 | -2.1 | 44,789 | 1,386 | 332 | 60 | 23 | 147 | 234 | 210 | 562 | 28 | 23 |
| Gray | 966 | 3.1 | 44,127 | 1,469 | 409 | 73 | 27 | 93 | 221 | 208 | 601 | 33 | 27 |
| Grayson | 5,991 | 5.4 | 43,987 | 1,486 | 2,285 | 377 | 156 | 332 | 925 | 1,466 | 3,151 | 197 | 156 |
| Gregg | 5,839 | 3.3 | 47,109 | 1,118 | 4,010 | 574 | 282 | 437 | 1,079 | 1,365 | 5,304 | 302 | 282 |
| Grimes | 1,066 | 6.1 | 36,909 | 2,455 | 410 | 71 | 28 | 60 | 190 | 277 | 569 | 36 | 28 |

1. Based on the resident population estimated as of July 1 of the year shown.

Items 62—74

| STATE County | Farm | Mining, quarrying, and extractions | Construction | Manufacturing | Information; professional, scientific, technical services | Retail trade | Finance, insurance, real estate, and leasing | Health care and social assistance | Government | Number | Rate[1] | Supplemental Security Income recipients, 2019 | Total | Percent change, 2010-2020 |
|---|---|---|---|---|---|---|---|---|---|---|---|---|---|---|
| | 75 | 76 | 77 | 78 | 79 | 80 | 81 | 82 | 83 | 84 | 85 | 86 | 87 | 88 |
| **TEXAS—Cont'd** | | | | | | | | | | | | | | |
| Caldwell | -1.1 | 2.4 | 14.6 | 5.6 | D | 9.1 | 5.5 | 12.4 | 18.7 | 7,875 | 181 | 1,034 | 16,064 | 16.9 |
| Calhoun | 0.6 | D | 19.7 | 47.4 | 5.1 | 4.7 | 2.6 | 1.7 | 7.1 | 4,440 | 209 | 473 | 12,241 | 7.3 |
| Callahan | -4.4 | 3.3 | 23.5 | 5.1 | 4.5 | 16.2 | 5.5 | 5.4 | 21.0 | 3,455 | 248 | 314 | 6,805 | 3.9 |
| Cameron | 0.6 | 0.1 | 4.0 | 5.5 | 3.9 | 8.4 | 5.1 | 20.1 | 27.0 | 69,310 | 164 | 21,887 | 155,829 | 9.8 |
| Camp | 3.7 | 0.2 | 15.7 | D | D | 6.5 | 4.6 | D | 14.4 | 3,080 | 236 | 466 | 5,877 | 3.9 |
| Carson | 5.1 | D | D | D | D | 1.1 | D | D | 4.5 | 1,220 | 208 | 49 | 2,785 | 0.0 |
| Cass | 0.7 | 2.3 | 7.1 | 23.2 | 4.6 | 5.4 | 3.7 | D | 20.2 | 8,510 | 284 | 1,127 | 14,800 | 2.9 |
| Castro | 64.3 | D | 0.6 | 0.9 | D | 1.9 | D | 0.6 | 10.8 | 1,280 | 171 | 135 | 3,175 | 0.3 |
| Chambers | 0.9 | 3.0 | 21.6 | 29.2 | D | 3.0 | 2.1 | D | 10.6 | 6,890 | 158 | 493 | 17,598 | 32.4 |
| Cherokee | 5.5 | 0.8 | 7.8 | 14.4 | D | 5.8 | 5.8 | D | 26.4 | 11,250 | 214 | 1,349 | 21,488 | 3.1 |
| Childress | 0.4 | D | D | D | 11.5 | 7.6 | 6.0 | 4.3 | 44.6 | 1,350 | 187 | 152 | 2,848 | -1.2 |
| Clay | -3.1 | 5.1 | 7.2 | 4.4 | D | 8.5 | D | D | 28.4 | 2,755 | 263 | 168 | 5,331 | 3.5 |
| Cochran | 19.3 | D | D | D | D | 3.6 | 2.6 | 2.0 | 31.8 | 580 | 205 | 100 | 1,350 | -0.7 |
| Coke | -11.7 | D | 9.2 | 2.2 | D | 8.8 | D | 1.7 | 33.8 | 975 | 294 | 66 | 2,745 | 3.1 |
| Coleman | -5.0 | 6.4 | 10.9 | 5.4 | D | 7.9 | 7.0 | D | 25.5 | 2,515 | 310 | 249 | 5,554 | 0.2 |
| Collin | -0.1 | 2.4 | 5.8 | 8.4 | 20.1 | 6.3 | 15.4 | 9.8 | 8.9 | 117,040 | 113 | 8,924 | 402,884 | 34.2 |
| Collingsworth | 5.3 | D | 3.5 | 2.6 | 5.5 | 4.7 | D | 21.4 | 21.2 | 645 | 221 | 62 | 1,595 | -1.3 |
| Colorado | -1.8 | 4.3 | 14.9 | 18.2 | 3.9 | 6.5 | 5.3 | 10.4 | 13.7 | 5,520 | 257 | 439 | 10,847 | 3.0 |
| Comal | -0.2 | 1.4 | 14.8 | 5.3 | 7.8 | 10.3 | 7.4 | D | 15.8 | 34,795 | 222 | 1,568 | 66,086 | 40.2 |
| Comanche | 14.0 | 0.8 | 4.9 | 3.9 | 4.5 | 10.7 | 4.6 | 12.1 | 25.4 | 3,720 | 273 | 335 | 7,448 | 3.2 |
| Concho | 6.0 | D | D | D | D | 3.6 | 6.4 | D | 25.4 | 750 | 273 | 54 | 1,666 | 1.8 |
| Cooke | -1.5 | 17.8 | 8.6 | 23.9 | 4.1 | 6.0 | 9.0 | D | 14.5 | 8,375 | 204 | 555 | 17,153 | 3.3 |
| Coryell | -1.3 | D | 8.7 | 3.6 | 12.5 | 6.3 | 7.5 | 4.8 | 38.0 | 11,615 | 152 | 1,347 | 27,288 | 8.2 |
| Cottle | 16.3 | 0.7 | D | 0.2 | 1.8 | 5.4 | D | D | 19.0 | 440 | 318 | 49 | 945 | -2.5 |
| Crane | 0.9 | 28.1 | D | D | 2.0 | 4.9 | 8.2 | 3.9 | 21.9 | 720 | 151 | 72 | 1,689 | 3.5 |
| Crockett | 0.1 | 29.0 | D | D | D | 5.4 | 6.5 | 0.4 | 22.4 | 745 | 215 | 52 | 1,891 | 1.3 |
| Crosby | 16.0 | D | D | D | D | 15.7 | 4.4 | 7.1 | 25.0 | 1,340 | 235 | 148 | 2,909 | 0.3 |
| Culberson | -1.7 | 6.1 | D | D | D | 10.2 | D | D | 26.2 | 540 | 252 | 111 | 1,183 | 4.0 |
| Dallam | 30.6 | 0.0 | 8.6 | 11.7 | 3.1 | 4.3 | 6.9 | D | 8.4 | 970 | 134 | 79 | 2,974 | 5.2 |
| Dallas | 0.0 | 6.9 | 5.9 | 6.6 | 17.6 | 4.7 | 13.5 | 8.8 | 8.4 | 325,870 | 124 | 61,425 | 1,055,499 | 11.9 |
| Dawson | 14.2 | 17.1 | 2.8 | 2.0 | 3.1 | 8.4 | 4.9 | 3.9 | 25.2 | 2,360 | 182 | 351 | 5,157 | -1.2 |
| Deaf Smith | 46.2 | 0.0 | 2.6 | 13.2 | 1.5 | 4.9 | 2.4 | 2.0 | 9.9 | 2,850 | 154 | 372 | 7,078 | 0.0 |
| Delta | 26.6 | 0.1 | D | D | D | 1.5 | D | 16.1 | 24.0 | 1,430 | 268 | 194 | 2,508 | 2.0 |
| Denton | 0.1 | 0.4 | 10.3 | 8.2 | 12.6 | 6.8 | 8.0 | 8.7 | 14.0 | 99,175 | 112 | 7,079 | 338,601 | 32.4 |
| DeWitt | -2.4 | 15.8 | 6.0 | 9.8 | 3.9 | 6.2 | 7.7 | 3.7 | 23.5 | 4,660 | 232 | 445 | 9,293 | 1.3 |
| Dickens | -7.8 | D | D | 1.1 | D | 10.1 | 5.2 | 1.2 | 27.0 | 570 | 261 | 53 | 1,286 | 0.4 |
| Dimmit | 0.3 | 33.3 | D | D | D | 3.1 | 3.1 | D | 19.0 | 2,080 | 206 | 526 | 4,477 | 2.9 |
| Donley | 20.1 | 1.8 | D | D | D | 6.1 | 4.8 | 2.8 | 33.4 | 940 | 288 | 73 | 2,158 | 1.0 |
| Duval | -4.4 | 20.1 | D | D | D | 2.8 | 1.2 | 9.9 | 35.2 | 2,630 | 236 | 600 | 5,616 | 1.8 |
| Eastland | -1.6 | 59.4 | 7.4 | 3.6 | 1.4 | 2.6 | 0.5 | D | 10.8 | 4,955 | 271 | 545 | 10,490 | 2.3 |
| Ector | 0.0 | 24.4 | 12.9 | 7.2 | 3.8 | 6.1 | 5.0 | 4.4 | 10.1 | 20,700 | 125 | 3,216 | 61,298 | 15.6 |
| Edwards | -4.6 | D | D | D | D | 7.8 | D | 4.0 | 26.5 | 575 | 300 | 67 | 1,671 | 4.1 |
| Ellis | -0.4 | 0.5 | 13.0 | 22.4 | 5.5 | 6.8 | 4.8 | 7.4 | 14.5 | 29,815 | 161 | 2,923 | 68,083 | 25.2 |
| El Paso | 0.0 | 0.0 | 6.5 | 5.2 | 4.7 | 7.3 | 4.9 | 10.5 | 35.2 | 137,450 | 164 | 29,073 | 304,299 | 12.6 |
| Erath | 7.5 | 1.9 | 10.9 | 10.3 | D | 8.6 | 5.7 | 8.1 | 19.7 | 6,935 | 163 | 525 | 18,994 | 11.8 |
| Falls | 1.5 | D | 7.8 | 3.6 | 5.0 | 8.4 | D | 7.4 | 39.7 | 3,630 | 210 | 669 | 7,802 | 1.0 |
| Fannin | -0.8 | 1.6 | 11.6 | 9.8 | D | 6.7 | 4.8 | D | 32.6 | 8,080 | 227 | 827 | 14,617 | 3.0 |
| Fayette | -4.2 | 8.1 | 8.3 | 8.2 | 5.2 | 9.9 | 10.7 | 9.6 | 18.7 | 6,960 | 275 | 407 | 14,063 | 1.5 |
| Fisher | 2.7 | 1.6 | D | D | D | 4.3 | D | 4.2 | 29.3 | 990 | 260 | 103 | 2,202 | -0.5 |
| Floyd | 37.0 | D | 1.7 | 1.8 | D | 3.1 | D | 2.3 | 20.8 | 1,270 | 221 | 153 | 2,951 | -1.8 |
| Foard | 12.7 | 0.9 | D | D | D | D | D | D | 27.6 | 325 | 281 | 27 | 796 | 0.9 |
| Fort Bend | 0.1 | 6.2 | 11.3 | 14.9 | 11.1 | 8.1 | 5.2 | 9.7 | 11.0 | 94,590 | 116 | 11,509 | 280,125 | 42.4 |
| Franklin | 12.0 | D | 5.9 | D | D | 3.6 | 4.3 | 21.8 | 10.2 | 2,505 | 233 | 169 | 5,862 | 1.6 |
| Freestone | -0.8 | 19.1 | 6.2 | 2.7 | D | 6.7 | 3.8 | 8.6 | 23.9 | 4,390 | 222 | 393 | 9,637 | 4.0 |
| Frio | 7.3 | 22.9 | 5.6 | 1.6 | D | 3.9 | 3.5 | D | 15.9 | 2,965 | 147 | 654 | 6,030 | 3.1 |
| Gaines | 12.8 | 15.2 | 23.6 | 2.7 | 2.3 | 3.7 | 2.5 | 1.4 | 11.2 | 2,130 | 99 | 288 | 6,549 | 3.9 |
| Galveston | 0.0 | 1.3 | 8.3 | 12.5 | 6.4 | 6.6 | 7.7 | 6.3 | 28.6 | 59,395 | 174 | 6,478 | 152,684 | 15.2 |
| Garza | -1.4 | 41.4 | D | D | D | 3.8 | 1.0 | D | 14.7 | 855 | 137 | 98 | 2,216 | -0.9 |
| Gillespie | -2.4 | 0.5 | 15.3 | 8.9 | 8.1 | 8.9 | 9.2 | 16.0 | 10.7 | 8,395 | 312 | 256 | 13,383 | 4.8 |
| Glasscock | -14.6 | 35.8 | D | D | D | 9.9 | 4.3 | 1.3 | 15.2 | 205 | 143 | 0 | 586 | 1.0 |
| Goliad | -15.3 | 12.4 | 16.1 | 2.3 | D | 4.2 | D | D | 28.3 | 1,775 | 233 | 190 | 3,801 | 2.4 |
| Gonzales | 16.6 | 9.4 | 4.9 | 11.4 | D | 5.8 | 4.5 | D | 16.5 | 4,225 | 202 | 478 | 9,125 | 3.8 |
| Gray | 6.2 | 10.8 | 4.7 | 24.8 | D | 6.2 | 3.6 | D | 14.5 | 4,305 | 197 | 373 | 9,935 | -2.2 |
| Grayson | -0.6 | 0.7 | 9.8 | 14.5 | 4.4 | 7.7 | 6.6 | 20.7 | 14.3 | 29,435 | 216 | 2,946 | 58,499 | 8.9 |
| Gregg | -0.1 | 10.7 | 9.6 | 11.6 | 6.4 | 7.1 | 6.2 | 13.8 | 8.9 | 26,235 | 211 | 4,568 | 52,628 | 6.2 |
| Grimes | -1.9 | 2.1 | 10.1 | 22.9 | D | 4.5 | 7.4 | D | 19.4 | 6,235 | 215 | 678 | 11,661 | 6.9 |

1. Per 1,000 resident population estimated as of July 1 of the year shown.

# Table B. States and Counties — Housing, Labor Force, and Employment

| STATE County | Housing units, 2015-2019 Occupied units | | Owner-occupied | | | Renter-occupied | | | Civilian labor force, 2020 | | Unemployment | | Civilian employment[6], 2015-2019 | Percent | |
|---|---|---|---|---|---|---|---|---|---|---|---|---|---|---|---|
| | Total | Percent | Median value[1] | Median owner cost as a percent of income | | Median rent[3] | Median rent as a percent of income[2] | Sub-standard units[4] (percent) | Total | Percent change, 2019-2020 | Total | Rate[5] | Total | Management, business, science, and arts | Construction, production, and maintenance occupations |
| | | | | With a mortgage | Without a mortgage[2] | | | | | | | | | | |
| | 89 | 90 | 91 | 92 | 93 | 94 | 95 | 96 | 97 | 98 | 99 | 100 | 101 | 102 | 103 |
| **TEXAS—Cont'd** | | | | | | | | | | | | | | | |
| Caldwell | 13,460 | 67.4 | 144,800 | 18.7 | 11.5 | 920 | 32.4 | 7.8 | 19,322 | -0.3 | 1,226 | 6.3 | 18,566 | 26.0 | 34.1 |
| Calhoun | 7,582 | 72.8 | 120,200 | 18.3 | 10.0 | 743 | 21.7 | 8.9 | 11,995 | -1.1 | 699 | 5.8 | 9,625 | 23.7 | 41.2 |
| Callahan | 5,367 | 82.1 | 94,900 | 20.3 | 10.7 | 759 | 22.8 | 2.4 | 5,979 | -1.3 | 325 | 5.4 | 5,851 | 33.5 | 31.3 |
| Cameron | 124,605 | 65.9 | 85,800 | 22.2 | 12.2 | 733 | 31.9 | 11.3 | 169,074 | 1.5 | 17,219 | 10.2 | 161,127 | 27.6 | 22.8 |
| Camp | 4,605 | 72.4 | 101,600 | 19.9 | 11.7 | 722 | 32.6 | 4.4 | 5,008 | 0.8 | 369 | 7.4 | 5,306 | 23.2 | 38.5 |
| Carson | 2,288 | 81.0 | 114,300 | 18.1 | 10.0 | 831 | 18.6 | 2.1 | 2,930 | -1.7 | 122 | 4.2 | 2,932 | 41.6 | 26.4 |
| Cass | 11,934 | 79.3 | 87,100 | 21.7 | 10.5 | 601 | 25.4 | 5.6 | 12,269 | -0.9 | 970 | 7.9 | 11,412 | 24.3 | 31.3 |
| Castro | 2,532 | 64.5 | 81,900 | 17.0 | 10.0 | 627 | 21.5 | 6.4 | 3,320 | -2.9 | 120 | 3.6 | 3,466 | 29.3 | 32.7 |
| Chambers | 14,069 | 82.9 | 200,400 | 17.7 | 12.7 | 918 | 24.9 | 4.0 | 19,836 | -0.4 | 1,892 | 9.5 | 18,917 | 38.3 | 32.5 |
| Cherokee | 18,138 | 71.9 | 102,900 | 19.6 | 12.4 | 728 | 25.7 | 7.9 | 21,303 | 1.4 | 1,679 | 7.9 | 20,251 | 28.8 | 33.1 |
| Childress | 2,407 | 59.9 | 85,500 | 21.9 | 10.3 | 745 | 26.3 | 0.5 | 3,054 | 3.2 | 114 | 3.7 | 2,853 | 33.9 | 28.3 |
| Clay | 4,105 | 84.4 | 101,000 | 18.8 | 12.0 | 737 | 26.9 | 3.8 | 4,793 | -2.4 | 265 | 5.5 | 4,419 | 32.7 | 29.7 |
| Cochran | 998 | 78.4 | 38,800 | 15.3 | 10.0 | 536 | 35.4 | 6.3 | 1,114 | -1.2 | 79 | 7.1 | 1,208 | 23.8 | 47.8 |
| Coke | 1,644 | 74.1 | 69,100 | 18.9 | 11.0 | 619 | 26.1 | 1.4 | 1,374 | -5.2 | 76 | 5.5 | 1,448 | 32.3 | 33.8 |
| Coleman | 3,423 | 74.6 | 76,500 | 20.4 | 13.6 | 725 | 22.4 | 1.8 | 2,886 | 0.9 | 213 | 7.4 | 3,542 | 32.2 | 36.1 |
| Collin | 341,163 | 65.0 | 315,300 | 20.7 | 11.3 | 1,389 | 27.1 | 2.6 | 570,623 | -0.2 | 36,006 | 6.3 | 509,180 | 53.6 | 11.7 |
| Collingsworth | 1,035 | 75.3 | 52,200 | 16.3 | 10.0 | 611 | 36.5 | 5.4 | 1,081 | -5.7 | 46 | 4.3 | 1,311 | 30.3 | 33.0 |
| Colorado | 7,450 | 82.4 | 124,400 | 22.7 | 10.7 | 714 | 24.3 | 8.3 | 9,693 | -1.5 | 538 | 5.6 | 9,553 | 28.9 | 33.2 |
| Comal | 51,367 | 74.9 | 282,300 | 20.3 | 10.0 | 1,144 | 28.6 | 2.8 | 74,950 | -1.4 | 4,639 | 6.2 | 66,229 | 41.0 | 20.2 |
| Comanche | 5,487 | 80.0 | 105,000 | 18.0 | 11.7 | 533 | 25.7 | 4.2 | 5,618 | 0.7 | 309 | 5.5 | 5,975 | 33.3 | 33.6 |
| Concho | 912 | 75.2 | 101,500 | 19.8 | 12.0 | 721 | 32.1 | 2.6 | 1,320 | 5.2 | 59 | 4.5 | 905 | 31.5 | 25.9 |
| Cooke | 15,351 | 69.3 | 161,100 | 20.2 | 11.9 | 887 | 27.1 | 4.0 | 19,370 | -2.3 | 1,369 | 7.1 | 19,303 | 29.6 | 31.2 |
| Coryell | 22,322 | 57.9 | 114,300 | 19.8 | 10.4 | 918 | 27.5 | 3.0 | 23,974 | -0.1 | 1,573 | 6.6 | 24,032 | 31.0 | 24.9 |
| Cottle | 707 | 56.2 | 46,600 | 18.7 | 12.5 | 340 | 29.0 | 3.0 | 570 | 3.8 | 28 | 4.9 | 670 | 22.8 | 23.1 |
| Crane | 1,421 | 85.6 | 100,400 | 14.3 | 10.0 | 835 | 28.5 | 5.9 | 1,624 | 3.6 | 221 | 13.6 | 2,144 | 24.1 | 43.6 |
| Crockett | 1,354 | 73.9 | 89,600 | 20.1 | 10.2 | 554 | 31.6 | 6.6 | 1,559 | -4.5 | 138 | 8.9 | 1,738 | 26.2 | 38.2 |
| Crosby | 2,061 | 75.3 | 60,800 | 19.5 | 12.6 | 593 | 28.1 | 7.6 | 2,389 | -2.8 | 124 | 5.2 | 2,489 | 26.4 | 31.3 |
| Culberson | 580 | 66.7 | 62,400 | 22.9 | 12.3 | 581 | 27.5 | 7.6 | 1,150 | 10.5 | 75 | 6.5 | 1,006 | 23.3 | 31.6 |
| Dallam | 2,364 | 64.4 | 89,200 | 14.9 | 10.0 | 770 | 20.1 | 5.0 | 3,750 | -4.8 | 105 | 2.8 | 3,700 | 24.8 | 41.2 |
| Dallas | 928,341 | 50.0 | 174,900 | 22.1 | 12.2 | 1,105 | 28.5 | 7.0 | 1,355,033 | 0.8 | 103,841 | 7.7 | 1,305,009 | 34.9 | 25.8 |
| Dawson | 4,322 | 69.3 | 73,400 | 19.3 | 11.9 | 645 | 28.5 | 6.4 | 4,558 | -1.4 | 398 | 8.7 | 4,489 | 25.5 | 32.8 |
| Deaf Smith | 6,145 | 64.0 | 91,100 | 20.0 | 10.9 | 788 | 21.3 | 5.5 | 8,567 | 1.6 | 310 | 3.6 | 8,450 | 23.0 | 45.6 |
| Delta | 1,999 | 78.2 | 87,700 | 18.6 | 10.0 | 626 | 24.3 | 1.7 | 2,401 | -1.6 | 133 | 5.5 | 2,188 | 38.6 | 28.8 |
| Denton | 290,229 | 65.1 | 277,800 | 20.8 | 11.1 | 1,218 | 28.2 | 2.5 | 503,962 | 0.1 | 32,666 | 6.5 | 453,391 | 46.7 | 15.9 |
| DeWitt | 7,048 | 72.8 | 117,100 | 17.4 | 10.2 | 838 | 28.0 | 4.5 | 9,253 | -5.0 | 558 | 6.0 | 7,710 | 28.6 | 32.5 |
| Dickens | 845 | 76.7 | 51,100 | 20.7 | 12.5 | 509 | 18.2 | 3.8 | 671 | -0.6 | 42 | 6.3 | 769 | 26.9 | 31.6 |
| Dimmit | 3,148 | 66.2 | 72,800 | 18.4 | 13.2 | 724 | 24.8 | 7.9 | 6,487 | -12.8 | 413 | 6.4 | 3,330 | 22.6 | 25.7 |
| Donley | 1,335 | 70.3 | 68,700 | 19.4 | 12.1 | 664 | 26.1 | 2.2 | 1,423 | -1.2 | 63 | 4.4 | 1,496 | 29.7 | 32.0 |
| Duval | 3,511 | 65.5 | 49,400 | 21.7 | 10.0 | 605 | 26.1 | 8.5 | 4,864 | -1.8 | 589 | 12.1 | 4,314 | 19.9 | 35.1 |
| Eastland | 6,492 | 76.6 | 73,900 | 19.1 | 12.8 | 565 | 23.4 | 3.3 | 7,390 | -11.4 | 497 | 6.7 | 7,031 | 35.7 | 27.8 |
| Ector | 52,530 | 64.9 | 140,500 | 19.5 | 10.0 | 1,059 | 25.0 | 5.9 | 82,852 | -6.2 | 9,116 | 11.0 | 74,960 | 23.3 | 37.3 |
| Edwards | 789 | 85.6 | 69,200 | 22.9 | 10.0 | 394 | 22.3 | 5.6 | 1,251 | 19.4 | 44 | 3.5 | 868 | 16.0 | 27.6 |
| Ellis | 57,307 | 74.4 | 191,400 | 20.0 | 10.9 | 1,054 | 28.7 | 4.0 | 93,578 | -0.5 | 5,641 | 6.0 | 85,372 | 34.9 | 27.5 |
| El Paso | 268,310 | 61.1 | 121,500 | 22.6 | 11.4 | 838 | 29.8 | 5.5 | 364,085 | 0.5 | 30,094 | 8.3 | 352,290 | 31.1 | 22.6 |
| Erath | 13,595 | 63.4 | 159,900 | 19.6 | 10.0 | 911 | 29.4 | 3.6 | 20,013 | -4.0 | 1,150 | 5.7 | 20,076 | 31.4 | 26.1 |
| Falls | 5,199 | 72.4 | 75,300 | 21.6 | 12.2 | 509 | 31.3 | 6.0 | 6,531 | 0.8 | 390 | 6.0 | 5,997 | 23.9 | 38.4 |
| Fannin | 12,453 | 73.2 | 123,100 | 19.1 | 11.7 | 788 | 26.0 | 3.1 | 16,772 | -3.3 | 788 | 4.7 | 13,998 | 32.6 | 29.9 |
| Fayette | 9,135 | 81.7 | 182,100 | 17.7 | 11.1 | 809 | 24.9 | 2.6 | 11,793 | -4.9 | 555 | 4.7 | 11,263 | 26.3 | 32.8 |
| Fisher | 1,601 | 73.6 | 66,900 | 18.0 | 10.0 | 605 | 24.0 | 4.1 | 1,605 | -1.6 | 72 | 4.5 | 1,709 | 32.9 | 29.4 |
| Floyd | 2,271 | 74.9 | 64,300 | 18.5 | 10.3 | 724 | 24.5 | 4.7 | 2,516 | -2.0 | 155 | 6.2 | 2,499 | 27.3 | 34.6 |
| Foard | 533 | 83.1 | 46,400 | 21.7 | 10.4 | 491 | 19.9 | 2.6 | 576 | -0.5 | 24 | 4.2 | 570 | 25.1 | 33.0 |
| Fort Bend | 237,883 | 77.2 | 265,900 | 21.5 | 11.3 | 1,431 | 29.2 | 3.3 | 394,948 | -1.3 | 30,538 | 7.7 | 367,035 | 50.5 | 15.0 |
| Franklin | 3,965 | 78.7 | 126,200 | 19.5 | 10.8 | 815 | 26.7 | 4.4 | 4,827 | 3.5 | 257 | 5.3 | 4,798 | 33.9 | 29.1 |
| Freestone | 6,758 | 77.4 | 111,800 | 19.1 | 11.6 | 725 | 25.0 | 3.4 | 6,348 | -0.6 | 528 | 8.3 | 7,036 | 26.8 | 32.1 |
| Frio | 4,542 | 69.0 | 82,200 | 20.3 | 12.2 | 818 | 24.2 | 7.0 | 9,537 | -8.6 | 541 | 5.7 | 6,634 | 24.2 | 36.0 |
| Gaines | 5,812 | 77.3 | 150,900 | 19.9 | 10.2 | 722 | 22.8 | 6.5 | 9,720 | -1.6 | 543 | 5.6 | 8,228 | 28.1 | 41.3 |
| Galveston | 121,438 | 67.5 | 199,200 | 19.9 | 11.0 | 1,078 | 29.1 | 3.1 | 164,608 | -0.8 | 14,390 | 8.7 | 157,014 | 42.4 | 21.1 |
| Garza | 1,543 | 70.5 | 78,800 | 18.9 | 10.0 | 775 | 28.5 | 2.5 | 2,020 | -5.4 | 137 | 6.8 | 1,940 | 32.9 | 23.4 |
| Gillespie | 10,694 | 72.4 | 286,900 | 24.0 | 11.5 | 948 | 30.7 | 4.8 | 13,265 | -2.1 | 635 | 4.8 | 12,087 | 32.8 | 23.5 |
| Glasscock | 403 | 65.8 | 229,200 | 17.2 | 10.0 | 904 | 10.4 | 6.2 | 797 | -0.3 | 29 | 3.6 | 738 | 33.1 | 47.3 |
| Goliad | 2,727 | 80.0 | 137,800 | 16.0 | 10.0 | 629 | 34.2 | 1.6 | 3,161 | -3.3 | 244 | 7.7 | 3,104 | 34.8 | 29.3 |
| Gonzales | 7,364 | 67.3 | 121,300 | 21.4 | 10.0 | 725 | 24.7 | 10.6 | 9,561 | -0.3 | 515 | 5.4 | 8,813 | 22.3 | 42.4 |
| Gray | 7,884 | 73.5 | 77,600 | 17.6 | 10.5 | 742 | 29.4 | 3.2 | 7,783 | -3.3 | 633 | 8.1 | 9,016 | 26.0 | 33.6 |
| Grayson | 48,454 | 67.6 | 137,400 | 20.6 | 12.0 | 874 | 28.2 | 3.4 | 64,214 | -0.1 | 3,761 | 5.9 | 60,989 | 30.5 | 27.7 |
| Gregg | 45,460 | 59.1 | 136,700 | 20.8 | 10.8 | 855 | 28.7 | 3.0 | 56,565 | -2.6 | 4,682 | 8.3 | 55,110 | 29.4 | 29.1 |
| Grimes | 9,011 | 76.5 | 148,500 | 20.9 | 10.0 | 750 | 32.1 | 4.7 | 11,082 | -1.5 | 891 | 8.0 | 10,670 | 27.1 | 35.4 |

1. Specified owner-occupied units. 2. A value of 10.0 represents 10 percent or less; a value of 50.0 represents 50 percent or more. 3. Specified renter-occupied units. 4. Overcrowded or lacking complete plumbing facilities. 5. Percent of civilian labor force. 6. Civilian employed persons 16 years old and over.

Items 89—103

| STATE County | Number of establishments | Employment — Total | Health care and social assistance | Manufacturing | Retail trade | Finance and insurance | Professional, scientific, and technical services | Annual payroll — Total (mil dol) | Average per employee (dollars) | Farms — Number | Percent with: Fewer than 50 acres | 1000 acres or more | Farm producers whose primary occupation is farming (percent) |
|---|---|---|---|---|---|---|---|---|---|---|---|---|---|
| | 104 | 105 | 106 | 107 | 108 | 109 | 110 | 111 | 112 | 113 | 114 | 115 | 116 |
| **TEXAS—Cont'd** | | | | | | | | | | | | | |
| Caldwell | 628 | 6,352 | 1,337 | 666 | 1,165 | 181 | 105 | 232 | 36,447 | 1,517 | 47.9 | 2.8 | 37.2 |
| Calhoun | 438 | 9,970 | 717 | 3,269 | 996 | 170 | 778 | 624 | 62,547 | 290 | 44.1 | 16.6 | 45.7 |
| Callahan | 233 | 1,713 | 141 | 141 | 436 | 60 | 42 | 68 | 39,472 | 961 | 37.4 | 10.0 | 38.2 |
| Cameron | 6,512 | 110,105 | 37,013 | 4,263 | 18,833 | 3,620 | 3,162 | 3,190 | 28,975 | 1,418 | 74.3 | 6.3 | 34.1 |
| Camp | 210 | 2,638 | 408 | 187 | 403 | 106 | 47 | 109 | 41,414 | 484 | 45.5 | 1.4 | 42.1 |
| Carson | 131 | 893 | 40 | NA | 191 | 40 | 18 | 42 | 47,106 | 331 | 19.3 | 34.7 | 45.1 |
| Cass | 530 | 5,917 | 1,179 | 1,214 | 829 | 222 | 128 | 217 | 36,656 | 1,081 | 39.1 | 2.0 | 39.7 |
| Castro | 157 | 1,241 | 169 | 40 | 171 | 41 | 78 | 44 | 35,799 | 411 | 7.3 | 38.9 | 54.1 |
| Chambers | 701 | 14,762 | 667 | 2,430 | 1,382 | 138 | 438 | 1,047 | 70,897 | 562 | 55.5 | 7.8 | 33.6 |
| Cherokee | 793 | 11,542 | 2,377 | 2,600 | 1,338 | 309 | 334 | 429 | 37,145 | 1,587 | 40.3 | 2.0 | 35.4 |
| Childress | 165 | 1,550 | 359 | 13 | 449 | 70 | 38 | 56 | 35,832 | 285 | 7.7 | 20.0 | 41.6 |
| Clay | 140 | 825 | 90 | 78 | 253 | 49 | 10 | 27 | 33,057 | 851 | 20.9 | 18.3 | 39.9 |
| Cochran | 38 | 236 | 75 | NA | 55 | 12 | NA | 8 | 33,068 | 286 | 5.6 | 41.6 | 45.9 |
| Coke | 58 | 242 | NA | NA | 74 | 25 | NA | 7 | 29,161 | 449 | 10.9 | 22.9 | 32.6 |
| Coleman | 198 | 1,459 | 422 | 164 | 254 | 75 | 26 | 48 | 32,716 | 976 | 14.3 | 16.1 | 40.7 |
| Collin | 25,558 | 453,019 | 58,178 | 19,422 | 53,789 | 61,932 | 48,359 | 30,207 | 66,680 | 2,706 | 74.3 | 1.4 | 28.2 |
| Collingsworth | 65 | 486 | 133 | 18 | 83 | 26 | 14 | 18 | 36,720 | 301 | 4.7 | 28.2 | 39.5 |
| Colorado | 544 | 6,273 | 832 | 1,549 | 923 | 170 | 188 | 265 | 42,283 | 1,773 | 37.2 | 5.4 | 32.9 |
| Comal | 3,978 | 55,173 | 6,784 | 3,551 | 8,200 | 1,088 | 1,985 | 2,305 | 41,769 | 1,068 | 54.4 | 5.3 | 32.6 |
| Comanche | 250 | 2,554 | 445 | 134 | 563 | 120 | 76 | 91 | 35,791 | 1,427 | 32.6 | 8.8 | 40.8 |
| Concho | 47 | 382 | 129 | NA | 81 | NA | NA | 12 | 30,953 | 396 | 9.1 | 32.1 | 44.9 |
| Cooke | 943 | 15,303 | 1,227 | 3,482 | 1,832 | 459 | 244 | 647 | 42,268 | 2,284 | 49.7 | 4.1 | 35.4 |
| Coryell | 738 | 9,720 | 1,196 | 364 | 1,998 | 403 | 718 | 295 | 30,371 | 1,479 | 38.2 | 5.7 | 36.4 |
| Cottle | 27 | 124 | NA | NA | 46 | 21 | NA | 3 | 22,476 | 154 | 2.6 | 29.9 | 31.9 |
| Crane | 91 | 1,299 | 252 | NA | 141 | 28 | 17 | 92 | 70,805 | 30 | 30.0 | 53.3 | 34.7 |
| Crockett | 119 | 1,052 | 21 | NA | 193 | 37 | 13 | 52 | 49,358 | 219 | 1.4 | 59.4 | 52.3 |
| Crosby | 92 | 604 | 170 | 25 | 135 | 20 | 13 | 22 | 36,639 | 343 | 5.2 | 36.2 | 44.6 |
| Culberson | 54 | 719 | 82 | NA | 160 | NA | NA | 39 | 54,160 | 66 | 15.2 | 63.6 | 48.4 |
| Dallam | 217 | 1,752 | 46 | 68 | 254 | 77 | 59 | 71 | 40,503 | 340 | 2.4 | 50.9 | 52.0 |
| Dallas | 67,311 | 1,534,430 | 184,046 | 100,039 | 129,023 | 99,658 | 141,211 | 99,305 | 64,718 | 775 | 69.5 | 1.2 | 25.9 |
| Dawson | 275 | 2,843 | 389 | 250 | 514 | 116 | 66 | 118 | 41,470 | 386 | 11.7 | 34.2 | 50.8 |
| Deaf Smith | 374 | 5,471 | 464 | 1,912 | 837 | 173 | 169 | 244 | 44,659 | 562 | 13.5 | 43.4 | 49.0 |
| Delta | 54 | 266 | 90 | NA | 56 | 22 | 7 | 7 | 24,617 | 571 | 45.4 | 4.4 | 36.0 |
| Denton | 16,086 | 234,142 | 30,295 | 17,361 | 31,579 | 12,577 | 11,700 | 11,308 | 48,297 | 3,295 | 80.2 | 2.4 | 29.2 |
| DeWitt | 450 | 5,408 | 953 | 485 | 795 | 268 | 128 | 243 | 44,938 | 1,768 | 27.3 | 4.8 | 41.9 |
| Dickens | 40 | 231 | 22 | NA | 46 | NA | 5 | 9 | 38,515 | 393 | 11.7 | 23.7 | 28.0 |
| Dimmit | 228 | 5,465 | 588 | 55 | 481 | 26 | 102 | 270 | 49,385 | 328 | 21.0 | 25.9 | 32.9 |
| Donley | 71 | 377 | 42 | NA | 90 | 36 | 20 | 9 | 25,156 | 283 | 8.1 | 25.8 | 39.4 |
| Duval | 153 | 2,007 | 523 | 47 | 233 | 45 | 77 | 79 | 39,589 | 1,367 | 11.1 | 11.7 | 29.8 |
| Eastland | 457 | 4,854 | 888 | 609 | 772 | 132 | 114 | 198 | 40,846 | 1,198 | 22.5 | 7.4 | 36.5 |
| Ector | 3,869 | 67,542 | 7,321 | 4,555 | 8,498 | 1,386 | 1,640 | 3,978 | 58,901 | 275 | 68.7 | 14.5 | 28.2 |
| Edwards | 39 | 317 | 19 | NA | 89 | NA | NA | 9 | 28,145 | 380 | 3.7 | 34.5 | 43.8 |
| Ellis | 3,221 | 46,208 | 4,782 | 11,309 | 6,750 | 1,031 | 1,037 | 1,999 | 43,252 | 2,551 | 57.5 | 3.7 | 31.8 |
| El Paso | 14,960 | 244,644 | 47,753 | 16,096 | 36,470 | 7,796 | 9,044 | 8,279 | 33,839 | 656 | 88.7 | 3.4 | 36.6 |
| Erath | 995 | 13,293 | 1,562 | 2,590 | 1,911 | 318 | 466 | 483 | 36,300 | 2,402 | 39.5 | 5.8 | 40.4 |
| Falls | 220 | 1,747 | 297 | 129 | 491 | 51 | 84 | 62 | 35,493 | 1,103 | 36.2 | 8.2 | 39.5 |
| Fannin | 504 | 5,665 | 1,538 | 673 | 849 | 216 | 142 | 223 | 39,371 | 2,255 | 45.8 | 3.3 | 33.0 |
| Fayette | 764 | 7,246 | 1,001 | 1,072 | 1,338 | 301 | 237 | 278 | 38,383 | 3,166 | 36.9 | 2.1 | 36.7 |
| Fisher | 66 | 457 | 100 | 74 | 66 | 49 | 16 | 22 | 48,512 | 486 | 9.7 | 25.7 | 36.1 |
| Floyd | 143 | 894 | 223 | 43 | 100 | 57 | 7 | 33 | 37,347 | 440 | 8.4 | 35.5 | 39.4 |
| Foard | 25 | 171 | 76 | NA | 24 | NA | NA | 5 | 28,585 | 172 | 12.2 | 36.0 | 40.8 |
| Fort Bend | 14,535 | 182,759 | 29,254 | 13,081 | 29,458 | 6,643 | 11,957 | 8,448 | 46,227 | 1,155 | 52.6 | 5.5 | 35.4 |
| Franklin | 173 | 6,718 | 4,403 | 76 | 288 | 90 | 44 | 120 | 17,877 | 493 | 32.7 | 2.8 | 34.5 |
| Freestone | 367 | 3,126 | 406 | 347 | 486 | 117 | 64 | 117 | 37,401 | 1,459 | 36.5 | 6.1 | 40.9 |
| Frio | 319 | 5,847 | 847 | 65 | 727 | 171 | 97 | 264 | 45,069 | 663 | 25.5 | 18.4 | 38.0 |
| Gaines | 482 | 4,991 | 314 | 281 | 525 | 116 | 103 | 272 | 54,552 | 507 | 11.0 | 43.4 | 51.3 |
| Galveston | 6,020 | 88,887 | 15,815 | 5,295 | 14,639 | 4,159 | 4,255 | 3,678 | 41,375 | 633 | 70.8 | 2.4 | 30.0 |
| Garza | 123 | 1,322 | 169 | NA | 172 | 21 | 13 | 58 | 43,509 | 237 | 13.1 | 31.6 | 34.2 |
| Gillespie | 1,078 | 10,207 | 1,657 | 936 | 1,736 | 310 | 328 | 380 | 37,218 | 2,079 | 38.4 | 8.9 | 32.2 |
| Glasscock | 35 | 356 | NA | NA | 23 | NA | NA | 34 | 95,817 | 175 | 6.3 | 54.3 | 57.1 |
| Goliad | 129 | 1,003 | 104 | NA | 148 | 36 | 30 | 41 | 41,251 | 1,255 | 32.3 | 5.4 | 34.5 |
| Gonzales | 412 | 5,013 | 705 | 1,124 | 849 | 189 | 101 | 200 | 39,981 | 1,612 | 26.9 | 8.7 | 42.4 |
| Gray | 535 | 5,900 | 843 | 802 | 957 | 158 | 257 | 276 | 46,738 | 348 | 23.0 | 26.4 | 31.1 |
| Grayson | 2,636 | 41,210 | 8,815 | 7,739 | 6,215 | 1,835 | 1,053 | 1,659 | 40,252 | 2,845 | 61.4 | 2.4 | 32.3 |
| Gregg | 4,022 | 67,774 | 11,004 | 9,371 | 9,206 | 2,044 | 3,246 | 3,149 | 46,459 | 541 | 65.6 | 2.8 | 28.2 |
| Grimes | 436 | 6,121 | 282 | 1,943 | 577 | 138 | 304 | 320 | 52,259 | 1,771 | 46.6 | 3.1 | 37.6 |

| STATE County | Acreage (1,000) [117] | Percent change, 2012-2017 [118] | Average size of farm [119] | Total irrigated (1,000) [120] | Total cropland (1,000) [121] | Value of land and buildings — Average per farm [122] | Average per acre [123] | Value of machinery and equiopment, average per farm (dollars) [124] | Total (mil dol) [125] | Average per farm (acres) [126] | Percent from: Crops [127] | Livestock and poultry products [128] | Organic farms (number) [129] | Farms with internet access (percent) [130] | Government payments — Total ($1,000) [131] | Percent of farms [132] |
|---|---|---|---|---|---|---|---|---|---|---|---|---|---|---|---|---|
| **TEXAS—Cont'd** | | | | | | | | | | | | | | | | |
| Caldwell | 285 | -8.1 | 188 | 0.7 | 67.9 | 718,250 | 3,821 | 57,287 | 53.6 | 35,358 | 27.0 | 73.0 | | 75.2 | 1,191 | 5.7 |
| Calhoun | 190 | 2.9 | 654 | 2.3 | 48.2 | 1,401,219 | 2,144 | 113,299 | 32.1 | 110,845 | 66.3 | 33.7 | 4 | 80.3 | 1,781 | 37.9 |
| Callahan | 478 | -15.2 | 497 | 0.2 | 85.5 | 861,252 | 1,732 | 60,872 | 31.2 | 32,508 | 9.7 | 90.3 | 6 | 75.0 | 711 | 20.8 |
| Cameron | 271 | -12.3 | 191 | 100.9 | 212.5 | 681,492 | 3,560 | 73,661 | 122.6 | 86,428 | 96.2 | 3.8 | NA | 64.3 | 6,669 | 26.0 |
| Camp | 77 | -1.3 | 160 | 0.6 | 20.8 | 525,814 | 3,294 | 76,458 | 114.2 | 235,934 | 1.4 | 98.6 | 10 | 73.6 | 235 | 16.3 |
| Carson | 521 | 7.5 | 1,574 | 64.4 | 322.6 | 2,139,551 | 1,359 | 305,633 | 91.8 | 277,426 | 89.5 | 10.5 | NA | 78.2 | 9,238 | 65.3 |
| Cass | 177 | 5.7 | 164 | 0.2 | 35.1 | 384,391 | 2,345 | 60,321 | 53.4 | 49,439 | 7.7 | 92.3 | NA | 70.3 | 424 | 13.9 |
| Castro | 555 | 1.2 | 1,350 | 121.8 | 394.6 | 1,863,235 | 1,380 | 600,576 | 1,121.6 | 2,728,944 | 8.2 | 91.8 | 2 | 76.4 | 15,189 | 84.9 |
| Chambers | 205 | -19.1 | 365 | 21.0 | 99.0 | 865,025 | 2,367 | 102,607 | 19.3 | 34,256 | 57.5 | 42.5 | NA | 71.4 | 5,885 | 16.0 |
| Cherokee | 276 | -8.6 | 174 | 1.0 | 58.3 | 539,484 | 3,107 | 74,929 | 115.7 | 72,900 | 57.5 | 42.5 | 7 | 72.1 | 49 | 1.1 |
| Childress | 444 | 0.1 | 1,559 | 8.7 | 121.4 | 1,442,095 | 925 | 112,358 | 27.2 | 95,565 | 72.4 | 27.6 | 2 | 65.6 | 2,721 | 72.3 |
| Clay | 648 | 2.4 | 761 | 0.3 | 100.8 | 1,464,315 | 1,924 | 80,343 | 55.7 | 65,394 | 10.7 | 89.3 | NA | 75.6 | 2,054 | 29.7 |
| Cochran | 486 | 8.3 | 1,700 | 113.2 | 373.6 | 1,574,348 | 926 | 320,452 | 87.6 | 306,367 | 94.7 | 5.3 | 2 | 68.2 | 9,199 | 91.3 |
| Coke | 469 | -3.1 | 1,045 | 0.7 | 43.0 | 1,005,986 | 962 | 63,144 | 7.8 | 17,459 | 16.0 | 84.0 | NA | 65.3 | 644 | 15.8 |
| Coleman | 672 | -7.4 | 689 | 0.7 | 146.3 | 1,131,892 | 1,643 | 68,996 | 41.2 | 42,215 | 32.4 | 67.6 | NA | 73.2 | 2,521 | 35.7 |
| Collin | 281 | -10.2 | 104 | 1.0 | 140.3 | 1,032,094 | 9,946 | 57,985 | 66.8 | 24,697 | 44.2 | 55.8 | NA | 82.0 | 2,401 | 4.8 |
| Collingsworth | 407 | -17.7 | 1,353 | 25.8 | 168.0 | 1,370,944 | 1,014 | 183,807 | 39.7 | 131,980 | 78.0 | 22.0 | NA | 70.8 | 8,813 | 74.8 |
| Colorado | 563 | 16.0 | 317 | 34.1 | 132.3 | 954,840 | 3,009 | 79,375 | 71.0 | 40,040 | 54.7 | 45.3 | 1 | 64.6 | 8,199 | 11.1 |
| Comal | 206 | 0.7 | 193 | 0.3 | 23.6 | 660,324 | 3,415 | 45,101 | 9.6 | 8,999 | 9.7 | 90.3 | 21 | 80.0 | 190 | 3.2 |
| Comanche | 487 | -5.8 | 341 | 17.4 | 137.6 | 900,642 | 2,639 | 82,769 | 173.3 | 121,419 | 13.5 | 86.5 | 2 | 73.2 | 2,533 | 16.0 |
| Concho | 561 | 11.8 | 1,417 | 4.3 | 108.5 | 2,089,817 | 1,475 | 123,623 | 28.1 | 71,008 | 47.6 | 52.4 | 4 | 71.5 | 2,775 | 57.3 |
| Cooke | 492 | -2.3 | 216 | 0.9 | 135.6 | 712,860 | 3,307 | 67,174 | 53.8 | 23,568 | 23.8 | 76.2 | NA | 78.2 | 2,342 | 14.8 |
| Coryell | 457 | -1.3 | 309 | 1.4 | 92.1 | 805,175 | 2,606 | 70,764 | 36.3 | 24,527 | 22.5 | 77.5 | 4 | 76.3 | 1,112 | 11.1 |
| Cottle | 579 | 2.5 | 3,759 | 2.3 | 104.4 | 4,174,361 | 1,110 | 133,366 | 27.7 | 180,104 | 35.5 | 64.5 | 2 | 66.9 | 1,829 | 80.5 |
| Crane | 244 | 2.1 | 8,137 | D | 0.2 | 6,902,732 | 848 | 106,427 | D | D | D | D | NA | 93.3 | 9 | 10.0 |
| Crockett | 1,534 | -0.8 | 7,005 | 0.0 | 6.3 | 5,686,480 | 812 | 84,106 | D | D | D | D | 6 | 73.5 | 208 | 9.1 |
| Crosby | 551 | -1.4 | 1,605 | 86.5 | 272.9 | 1,556,556 | 970 | 262,721 | 86.9 | 253,364 | 93.4 | 6.6 | 1 | 69.1 | 5,844 | 79.9 |
| Culberson | 1,504 | -7.0 | 22,791 | 5.7 | 13.5 | 16,763,487 | 736 | 245,880 | 15.9 | 240,864 | 59.5 | 40.5 | NA | 72.7 | 313 | 15.2 |
| Dallam | 895 | 5.1 | 2,633 | 149.9 | 393.6 | 2,986,690 | 1,134 | 451,440 | 634.9 | 1,867,426 | 14.6 | 85.4 | NA | 77.4 | 15,216 | 82.4 |
| Dallas | 64 | -23.6 | 83 | 0.5 | 26.1 | 668,126 | 8,097 | 54,512 | 29.8 | 38,427 | 87.0 | 13.0 | NA | 72.0 | 404 | 3.6 |
| Dawson | 536 | -4.0 | 1,388 | 54.8 | 408.6 | 1,442,556 | 1,040 | 333,933 | 121.3 | 314,236 | 97.6 | 2.4 | 18 | 73.3 | 11,958 | 79.8 |
| Deaf Smith | 967 | 4.8 | 1,722 | 122.4 | 617.3 | 1,942,201 | 1,128 | 349,244 | 1,638.8 | 2,916,004 | 5.7 | 94.3 | 4 | 82.6 | 29,893 | 74.9 |
| Delta | 144 | 9.7 | 252 | D | 80.5 | 495,901 | 1,968 | 78,856 | 36.3 | 63,620 | 69.9 | 30.1 | NA | 71.5 | 1,843 | 30.6 |
| Denton | 359 | -6.3 | 109 | 2.9 | 144.0 | 1,041,728 | 9,550 | 55,272 | 123.2 | 37,393 | 19.7 | 80.3 | 8 | 82.0 | 1,877 | 6.5 |
| DeWitt | 484 | -9.8 | 274 | 3.5 | 50.8 | 840,718 | 3,072 | 74,495 | 38.7 | 21,881 | 17.4 | 82.6 | NA | 68.3 | 186 | 1.4 |
| Dickens | 543 | -5.1 | 1,382 | 8.9 | 166.1 | 1,472,068 | 1,065 | 97,078 | 26.9 | 68,336 | 49.5 | 50.5 | NA | 70.7 | 2,200 | 54.7 |
| Dimmit | 484 | -28.5 | 1,476 | 3.5 | 68.0 | 2,080,509 | 1,410 | 95,610 | 28.5 | 86,756 | 8.3 | 91.7 | NA | 61.0 | 84 | 4.6 |
| Donley | 593 | 1.4 | 2,096 | 17.1 | 57.2 | 1,813,922 | 865 | 112,440 | 94.2 | 332,707 | 14.6 | 85.4 | NA | 70.7 | 2,234 | 52.7 |
| Duval | 836 | -12.9 | 612 | 2.0 | 45.4 | 1,037,873 | 1,697 | 39,468 | 11.0 | 8,045 | 5.9 | 94.1 | NA | 47.4 | 1,741 | 17.5 |
| Eastland | 490 | -2.8 | 409 | 1.9 | 80.1 | 802,949 | 1,964 | 63,040 | 23.5 | 19,632 | 21.1 | 78.9 | NA | 68.4 | 1,837 | 14.6 |
| Ector | 558 | 30.1 | 2,029 | 0.9 | 1.9 | 2,350,073 | 1,158 | 63,273 | 3.4 | 12,298 | 7.6 | 92.4 | NA | 75.6 | 57 | 1.5 |
| Edwards | 1,013 | 4.5 | 2,667 | 0.4 | 12.5 | 3,121,016 | 1,170 | 65,091 | 10.9 | 28,768 | 3.0 | 97.0 | NA | 66.6 | 1,609 | 22.6 |
| Ellis | 473 | -0.1 | 186 | 2.4 | 221.5 | 595,894 | 3,211 | 76,740 | 73.1 | 28,673 | 73.1 | 26.9 | 1 | 75.3 | 5,962 | 17.4 |
| El Paso | 143 | -31.9 | 217 | 31.6 | 39.9 | 873,269 | 4,015 | 87,385 | 46.7 | 71,248 | 86.4 | 13.6 | 4 | 76.7 | 286 | 1.5 |
| Erath | 626 | 3.0 | 260 | 14.3 | 134.3 | 824,386 | 3,166 | 96,022 | 312.3 | 130,007 | 6.1 | 93.9 | 10 | 77.7 | 1,229 | 3.6 |
| Falls | 392 | 2.4 | 355 | 4.0 | 203.8 | 893,002 | 2,513 | 124,859 | 157.9 | 143,191 | 26.8 | 73.2 | NA | 73.2 | 3,201 | 21.8 |
| Fannin | 482 | -6.2 | 214 | 4.9 | 212.4 | 614,054 | 2,873 | 70,196 | 86.3 | 38,267 | 50.8 | 49.2 | NA | 74.9 | 7,452 | 24.1 |
| Fayette | 522 | 6.0 | 165 | 1.5 | 99.8 | 664,110 | 4,032 | 55,919 | 47.4 | 14,966 | 22.5 | 77.5 | NA | 70.3 | 446 | 6.3 |
| Fisher | 478 | -3.4 | 983 | 10.5 | 248.7 | 1,112,312 | 1,131 | 132,231 | 35.7 | 73,537 | 75.5 | 24.5 | NA | 65.8 | 2,874 | 65.0 |
| Floyd | 639 | 9.8 | 1,452 | 120.1 | 437.7 | 1,460,850 | 1,006 | 263,008 | D | D | D | D | NA | 68.2 | 12,373 | 91.4 |
| Foard | 440 | 19.5 | 2,557 | 1.1 | 92.4 | 2,854,518 | 1,116 | 110,669 | 14.9 | 86,860 | 24.8 | 75.2 | NA | 65.7 | 2,360 | 64.5 |
| Fort Bend | 279 | -17.6 | 242 | 9.6 | 140.7 | 750,749 | 3,103 | 102,575 | 85.0 | 73,603 | 82.6 | 17.4 | 1 | 71.7 | 4,623 | 22.3 |
| Franklin | 103 | -9.2 | 208 | 0.9 | 28.4 | 645,217 | 3,100 | 87,557 | 134.1 | 272,000 | 2.1 | 97.9 | NA | 74.8 | 208 | 5.5 |
| Freestone | 414 | -1.7 | 284 | 1.3 | 55.7 | 727,956 | 2,565 | 71,717 | 68.1 | 46,697 | 6.8 | 93.2 | NA | 66.2 | 61 | 1.0 |
| Frio | 678 | -4.9 | 1,023 | 48.6 | 90.9 | 1,889,823 | 1,848 | 103,474 | 124.4 | 187,703 | 56.1 | 43.9 | NA | 69.7 | 1,214 | 11.6 |
| Gaines | 858 | 10.7 | 1,692 | 197.0 | 680.0 | 1,955,071 | 1,155 | 389,035 | 188.8 | 372,373 | 93.9 | 6.1 | 33 | 73.6 | 31,166 | 82.6 |
| Galveston | 73 | -18.3 | 116 | 1.1 | 17.0 | 612,200 | 5,299 | 53,747 | 9.2 | 14,586 | 48.6 | 51.4 | 2 | 69.0 | 722 | 3.8 |
| Garza | 416 | -8.7 | 1,754 | 8.9 | 68.3 | 1,763,789 | 1,005 | 117,857 | 22.1 | 93,338 | 75.6 | 24.4 | NA | 67.1 | 1,005 | 46.0 |
| Gillespie | 680 | 4.3 | 327 | 2.8 | 73.9 | 993,949 | 3,039 | 49,491 | 31.2 | 15,013 | 22.5 | 77.5 | 1 | 75.6 | 1,244 | 11.3 |
| Glasscock | 496 | 14.4 | 2,836 | 39.7 | 180.3 | 2,604,656 | 919 | 515,993 | 50.6 | 289,400 | 93.7 | 6.3 | NA | 85.7 | 1,233 | 60.0 |
| Goliad | 380 | -23.2 | 303 | 0.1 | 36.2 | 804,567 | 2,658 | 53,417 | 17.7 | 14,072 | 25.9 | 74.1 | NA | 70.3 | 991 | 5.7 |
| Gonzales | 614 | 0.7 | 381 | 1.5 | 58.5 | 1,211,682 | 3,180 | 96,128 | 560.8 | 347,909 | 6.9 | 93.1 | 1 | 68.5 | 478 | 4.0 |
| Gray | 483 | -6.2 | 1,389 | 16.9 | 124.8 | 1,588,481 | 1,144 | 111,560 | 154.6 | 444,313 | 15.8 | 84.2 | 10 | 73.9 | 4,042 | 40.8 |
| Grayson | 430 | -0.3 | 151 | 2.3 | 207.0 | 1,023,106 | 6,770 | 68,105 | 66.2 | 23,259 | 60.8 | 39.2 | 1 | 81.0 | 4,729 | 9.9 |
| Gregg | 58 | 20.1 | 107 | 0.3 | 14.5 | 602,572 | 5,649 | 51,338 | 4.1 | 7,584 | 21.7 | 78.3 | 1 | 73.8 | D | 0.4 |
| Grimes | 341 | -18.3 | 192 | 4.0 | 54.3 | 740,438 | 3,847 | 68,597 | 47.5 | 26,826 | 29.6 | 70.4 | NA | 74.5 | 79 | 1.1 |

## Table B. States and Counties — Water Use, Wholesale Trade, Retail Trade, and Real Estate

| STATE County | Water use, 2015 | | Wholesale Trade[1], 2017 | | | | Retail Trade[2], 2017 | | | | Real estate and rental and leasing,[2] 2017 | | | |
|---|---|---|---|---|---|---|---|---|---|---|---|---|---|---|
| | Public supply water withdrawn (mil gal/day) | Public supply gallons withdrawn per person per day | Number of establishments | Number of employees | Sales (mil dol) | Average payroll (mil dol) | Number of establishments | Number of employees | Sales (mil dol) | Average payroll (mil dol) | Number of establishments | Number of employees | Sales (mil dol) | Average payroll (mil dol) |
| | 133 | 134 | 135 | 136 | 137 | 138 | 139 | 140 | 141 | 142 | 143 | 144 | 145 | 146 |
| **TEXAS—Cont'd** | | | | | | | | | | | | | | |
| Caldwell | 3.70 | 91.3 | 22 | 81 | 22.7 | 3.3 | 79 | 1,108 | 355.0 | 31.0 | 26 | 62 | 10.7 | 1.5 |
| Calhoun | 0.25 | 11.4 | 15 | 94 | 85.1 | 4.1 | 65 | 1,098 | 502.4 | 39.0 | 16 | 164 | 56.0 | 8.8 |
| Callahan | 0.25 | 18.4 | 4 | 29 | 9.2 | 0.9 | 37 | 385 | 277.1 | 17.8 | 6 | 25 | 4.2 | 1.0 |
| Cameron | 24.18 | 57.3 | 320 | 2,851 | 2,163.4 | 101.5 | 1,045 | 17,938 | 4,610.8 | 437.5 | 336 | 1,482 | 226.5 | 40.2 |
| Camp | 1.10 | 86.7 | 7 | 91 | 53.8 | 3.7 | 42 | 373 | 101.6 | 9.5 | D | D | D | 0.1 |
| Carson | 9.00 | 1,507.8 | 11 | 132 | 125.7 | 9.5 | 25 | 174 | 53.7 | 3.8 | NA | NA | NA | NA |
| Cass | 1.05 | 34.6 | 14 | 225 | 144.0 | 10.0 | 81 | 828 | 254.0 | 19.6 | 19 | 32 | 6.8 | 1.1 |
| Castro | 1.06 | 138.5 | 12 | 97 | 121.7 | 6.2 | 34 | 180 | 41.8 | 4.1 | 5 | 17 | 2.6 | 0.4 |
| Chambers | 1.59 | 40.9 | 39 | 445 | 240.7 | 25.0 | 104 | 1,089 | 410.6 | 29.5 | 25 | 158 | 97.5 | 9.0 |
| Cherokee | 6.31 | 122.4 | 21 | 191 | 79.8 | 8.2 | 127 | 1,855 | 394.8 | 45.4 | 31 | 52 | 10.3 | 1.5 |
| Childress | 0.00 | 0.0 | D | D | D | D | 30 | 380 | 113.7 | 9.7 | 5 | 11 | 3.0 | 0.4 |
| Clay | 6.30 | 608.1 | 3 | 5 | 2.2 | 0.1 | 24 | 190 | 41.8 | 5.8 | D | D | D | D |
| Cochran | 0.40 | 135.5 | 3 | 39 | 43.8 | 1.5 | 8 | 42 | 9.2 | 0.8 | NA | NA | NA | NA |
| Coke | 0.37 | 114.3 | NA | NA | NA | NA | 11 | 76 | 20.4 | 1.4 | NA | NA | NA | NA |
| Coleman | 1.31 | 157.1 | 5 | 29 | 19.8 | 1.5 | 33 | 259 | 69.1 | 5.2 | 8 | D | 1.4 | D |
| Collin | 3.04 | 3.3 | 870 | 13,979 | 40,258.4 | 1,243.3 | 2,754 | 55,402 | 19,773.5 | 1,669.7 | 1,415 | 8,959 | 2,472.4 | 518.4 |
| Collingsworth | 0.52 | 170.8 | NA | NA | NA | NA | 11 | 81 | 16.8 | 1.7 | NA | NA | NA | NA |
| Colorado | 2.94 | 140.9 | 31 | 324 | 291.0 | 13.6 | 108 | 928 | 303.5 | 25.6 | 14 | 21 | 16.5 | 1.5 |
| Comal | 16.7 | 129.4 | 138 | 2,253 | 2,457.0 | 161.4 | 431 | 7,210 | 2,607.6 | 215.9 | 241 | 879 | 210.7 | 37.2 |
| Comanche | 0.04 | 3.0 | 16 | 260 | 133.5 | 8.4 | 45 | 492 | 168.5 | 13.6 | 5 | 10 | 0.8 | 0.3 |
| Concho | 0.33 | 80.9 | NA | NA | NA | NA | 15 | 75 | 18.7 | 1.4 | NA | NA | NA | NA |
| Cooke | 4.19 | 106.8 | 49 | 615 | 278.2 | 34.2 | 149 | 1,865 | 587.3 | 50.1 | 31 | 86 | 29.2 | 2.9 |
| Coryell | 0.21 | 2.8 | 14 | 110 | 28.1 | 4.0 | 132 | 1,899 | 530.7 | 47.4 | 39 | 138 | 18.7 | 4.9 |
| Cottle | 0.21 | 147.3 | NA | NA | NA | NA | 7 | 52 | 10.9 | 1.1 | NA | NA | NA | NA |
| Crane | 1.41 | 279.3 | D | D | D | 0.8 | 12 | 114 | 34.2 | 3.2 | D | D | D | D |
| Crockett | 0.96 | 258.8 | D | D | D | 1.7 | 18 | 164 | 63.8 | 4.6 | 5 | D | 0.4 | D |
| Crosby | 1.51 | 252.6 | 8 | 128 | 148.4 | 7.7 | 18 | 132 | 46.7 | 4.3 | NA | NA | NA | NA |
| Culberson | 0.81 | 362.3 | NA | NA | NA | NA | 16 | 148 | 92.2 | 4.3 | NA | NA | NA | NA |
| Dallam | 2.05 | 287.9 | 18 | 242 | 190.7 | 13.9 | 32 | 238 | 94.8 | 7.8 | D | D | D | D |
| Dallas | 334.54 | 131.0 | 3,777 | 75,525 | 70,063.1 | 4,970.6 | 7,816 | 129,602 | 43,896.1 | 4,051.7 | 4,096 | 38,463 | 13,117.6 | 2,469.2 |
| Dawson | 0.51 | 37.7 | 19 | 113 | 89.9 | 6.0 | 42 | 632 | 370.7 | 21.4 | D | D | D | D |
| Deaf Smith | 4.25 | 224.3 | 33 | 429 | 920.6 | 22.7 | 53 | 838 | 271.8 | 20.0 | 15 | 60 | 7.9 | 1.3 |
| Delta | 5.37 | 1,029.3 | 3 | 10 | 10.2 | 0.5 | 12 | 34 | 16.6 | 0.8 | NA | NA | NA | NA |
| Denton | 31.52 | 40.4 | 664 | 11,758 | 28,775.7 | 813.2 | 1,826 | 31,567 | 10,212.3 | 877.8 | 843 | 4,745 | 1,250.9 | 248.1 |
| DeWitt | 2.81 | 135.1 | 15 | 253 | 88.1 | 7.7 | 65 | 862 | 243.0 | 22.5 | 15 | 114 | 47.1 | 4.8 |
| Dickens | 0.10 | 45.3 | NA | NA | NA | NA | 8 | 48 | 11.8 | 1.0 | NA | NA | NA | NA |
| Dimmit | 1.69 | 153.9 | 10 | 71 | 43.5 | 3.4 | 29 | 461 | 143.9 | 12.3 | 12 | 88 | 24.5 | 5.4 |
| Donley | 0.38 | 108.6 | NA | NA | NA | NA | 13 | 117 | 23.1 | 2.3 | 5 | 8 | 0.9 | 0.2 |
| Duval | 1.21 | 106.3 | D | D | D | D | 22 | 193 | 66.7 | 4.9 | 4 | 8 | 2.4 | 0.2 |
| Eastland | 0.68 | 37.4 | 14 | 56 | 56.4 | 2.6 | 76 | 830 | 273.9 | 20.8 | 15 | 30 | 4.8 | 0.9 |
| Ector | 0.64 | 4.0 | 301 | 4,598 | 3,087.9 | 298.2 | 460 | 8,193 | 3,004.4 | 270.5 | 188 | 2,007 | 788.5 | 126.2 |
| Edwards | 0.20 | 105.6 | NA | NA | NA | NA | 7 | 86 | 24.7 | 2.6 | 3 | 2 | 0.5 | 0.1 |
| Ellis | 15.24 | 93.1 | 115 | 1,165 | 1,197.9 | 71.2 | 391 | 6,063 | 1,812.4 | 158.4 | 135 | 444 | 115.8 | 14.9 |
| El Paso | 114.33 | 136.8 | D | D | D | D | 2,272 | 38,202 | 9,956.3 | 883.6 | 844 | 4,043 | 958.5 | 152.1 |
| Erath | 2.21 | 53.7 | 48 | 417 | 156.7 | 16.5 | 161 | 1,964 | 557.6 | 50.7 | 41 | 128 | 18.3 | 3.0 |
| Falls | 2.29 | 133.6 | D | D | D | 2.9 | 48 | 475 | 167.5 | 12.7 | 4 | 16 | 1.2 | 0.5 |
| Fannin | 2.60 | 77.2 | 14 | 132 | 240.9 | 6.3 | 79 | 912 | 304.4 | 24.6 | 17 | 55 | 10.1 | 1.2 |
| Fayette | 2.56 | 102.0 | 29 | 408 | 265.2 | 25.9 | 112 | 1,360 | 435.3 | 38.3 | 32 | 91 | 21.1 | 3.4 |
| Fisher | 0.02 | 5.2 | NA | NA | NA | NA | 15 | 59 | 14.8 | 1.6 | NA | NA | NA | NA |
| Floyd | 0.41 | 69.5 | 15 | 133 | 444.7 | 7.3 | 16 | 96 | 20.3 | 1.9 | NA | NA | NA | NA |
| Foard | 0.01 | 8.2 | NA | NA | NA | NA | 5 | 30 | 5.5 | 0.5 | NA | NA | NA | NA |
| Fort Bend | 55.86 | 78.0 | 731 | 8,114 | 11,721.4 | 478.6 | 1,769 | 28,190 | 9,259.7 | 773.8 | 670 | 2,150 | 573.0 | 93.5 |
| Franklin | 2.77 | 260.1 | 3 | D | 3.8 | D | 28 | 294 | 100.2 | 6.9 | 8 | 10 | 3.1 | 0.3 |
| Freestone | 2.10 | 106.6 | 11 | 128 | 77.6 | 8.9 | 56 | 474 | 177.5 | 12.0 | 11 | 32 | 4.6 | 1.2 |
| Frio | 2.72 | 144.7 | 17 | 193 | 227.5 | 12.4 | 51 | 612 | 176.9 | 17.2 | 8 | 22 | 5.6 | 0.9 |
| Gaines | 8.21 | 409.5 | 33 | 347 | 345.3 | 22.1 | 59 | 508 | 138.3 | 13.3 | 15 | 182 | 24.7 | 9.4 |
| Galveston | 0.23 | 0.7 | 218 | 1,952 | 1,445.9 | 107.1 | 944 | 14,163 | 4,498.3 | 384.5 | 338 | 1,920 | 442.9 | 100.9 |
| Garza | 4.70 | 732.7 | D | D | D | D | 19 | 188 | 49.6 | 3.9 | 4 | D | 1.0 | D |
| Gillespie | 2.13 | 82.0 | 32 | 290 | 138.5 | 12.5 | 169 | 1,676 | 402.1 | 44.2 | 47 | 113 | 24.8 | 4.6 |
| Glasscock | 0.00 | 0.0 | 5 | 25 | 17.4 | 1.2 | NA | NA | NA | NA | NA | NA | NA | NA |
| Goliad | 0.34 | 45.1 | D | D | D | 0.3 | 16 | 132 | 45.0 | 3.2 | NA | NA | NA | NA |
| Gonzales | 29.05 | 1,412 | 22 | 242 | 185.0 | 9.9 | 63 | 804 | 251.7 | 21.4 | 11 | 30 | 6.3 | 1.2 |
| Gray | 1.24 | 53.4 | D | D | D | D | 86 | 1,009 | 256.4 | 23.8 | 20 | 61 | 10.8 | 3.2 |
| Grayson | 21.61 | 172.2 | 97 | 1,024 | 736.7 | 48.3 | 428 | 6,146 | 1,912.7 | 170.4 | 107 | 329 | 73.4 | 11.1 |
| Gregg | 7.13 | 57.4 | 247 | 3,182 | 2,073.0 | 181.2 | 636 | 9,260 | 2,928.8 | 263.2 | 191 | 1,020 | 263.3 | 45.9 |
| Grimes | 2.22 | 80.7 | 20 | 188 | 154.2 | 11.5 | 67 | 565 | 194.9 | 15.6 | 19 | 74 | 11.0 | 2.7 |

1  Merchant wholesalers, except manufacturers' sales branches and offices.  2. Employer establishments.

## Table B. States and Counties — Professional Services, Manufacturing, and Accommodation and Food Services

| STATE County | Professional, scientific, and technical services, 2017 | | | | Manufacturing, 2017 | | | | Accommodation and food services, 2017 | | | |
|---|---|---|---|---|---|---|---|---|---|---|---|---|
| | Number of establishments | Number of employees | Sales (mil dol) | Average payroll (mil dol) | Number of establishments | Number of employees | Sales (mil dol) | Average payroll (mil dol) | Number of establishments | Number of employees | Sales (mil dol) | Annual payroll (mil dol) |
| | 147 | 148 | 149 | 150 | 151 | 152 | 153 | 154 | 155 | 156 | 157 | 158 |
| TEXAS—Cont'd | | | | | | | | | | | | |
| Caldwell | 30 | 63 | 8 | 2.7 | 21 | 520 | 178.0 | 25.3 | 66 | 910 | 59.1 | 15.2 |
| Calhoun | 34 | 615 | 77 | 45.1 | 24 | 3,017 | 7,393.0 | 325.9 | 66 | 765 | 45.7 | 10.1 |
| Callahan | 13 | 32 | 6 | 1.4 | 11 | 137 | 48.4 | 5.5 | 21 | 268 | 11.9 | 3.2 |
| Cameron | D | D | D | D | 188 | 3,983 | 1,301.2 | 161.2 | 735 | 15,258 | 753.2 | 204.0 |
| Camp | 11 | 38 | 5 | 1.5 | 7 | 192 | 39.8 | 8.6 | 20 | 190 | 13.5 | 2.7 |
| Carson | D | D | D | D | NA | NA | NA | NA | 11 | 139 | 4.3 | 1.5 |
| Cass | 24 | 100 | 10 | 3.6 | 22 | 1,120 | 731.5 | 118.0 | 53 | 670 | 28.2 | 7.4 |
| Castro | 10 | 85 | 8 | 3.1 | 5 | 41 | 61.1 | 1.9 | 14 | 86 | 6.1 | 1.2 |
| Chambers | 45 | 353 | 49 | 20.7 | 29 | 2,137 | 3,929.4 | 230.4 | 78 | 1,279 | 69.7 | 18.3 |
| Cherokee | 57 | 181 | 28 | 7.1 | 61 | 2,206 | 461.1 | 91.6 | 58 | 918 | 42.6 | 11.3 |
| Childress | 12 | 41 | 7 | 2.0 | 3 |  | 0.4 | 0.1 | 21 | 244 | 14.9 | 3.7 |
| Clay | 8 | 14 | 2 | 0.5 | 6 | 69 | 9.8 | 2.2 | D | D | D | D |
| Cochran | NA | NA | NA | NA | NA | NA | NA | NA | D | D | D | 0.1 |
| Coke | NA | NA | NA | NA | NA | NA | NA | NA | 4 | 31 | 0.8 | 0.3 |
| Coleman | 14 | 30 | 5 | 1.0 | 11 | 70 | 11.8 | 3.5 | 23 | 181 | 7.8 | 2.4 |
| Collin | D | D | D | D | 461 | 18,072 | 8,816.6 | 1,365.4 | 2,191 | 46,044 | 2,869.0 | 831.8 |
| Collingsworth | D | D | 2 | D | 4 | 17 | 1.9 | 0.8 | 8 | 56 | 2.0 | 0.5 |
| Colorado | 33 | 128 | 19 | 7.9 | 32 | 1,642 | 528.4 | 77.8 | 45 | 494 | 30.5 | 8.0 |
| Comal | 400 | 2,125 | 317 | 106.1 | 114 | 3,095 | 964.0 | 154.1 | 356 | 7,419 | 426.3 | 124.2 |
| Comanche | 20 | 73 | 10 | 3.5 | 8 | 113 | 104.9 | 3.5 | D | D | D | D |
| Concho | NA | NA | NA | NA | NA | NA | NA | NA | D | D | D | 0.8 |
| Cooke | 68 | 254 | 31 | 12.0 | 69 | 3,485 | 1,606.2 | 165.5 | 84 | 1,605 | 80.4 | 23.0 |
| Coryell | 52 | 701 | 93 | 34.6 | 22 | 369 | 108.5 | 21.4 | 91 | 1,430 | 69.9 | 17.5 |
| Cottle | NA | NA | NA | NA | NA | NA | NA | NA | NA | NA | NA | NA |
| Crane | 4 | 16 | 1 | 0.5 | NA | NA | NA | NA | 6 | 61 | 2.7 | 0.7 |
| Crockett | 4 | 12 | 1 | 0.4 | NA | NA | NA | NA | 24 | 274 | 13.2 | 3.1 |
| Crosby | D | D | 1 | D | NA | NA | NA | NA | 4 | 36 | 1.8 | 0.5 |
| Culberson | NA | NA | NA | NA | NA | NA | NA | NA | 14 | 177 | 12.1 | 2.8 |
| Dallam | 15 | 65 | 49 | 17.4 | 5 | 55 | 23.9 | 2.7 | D | D | D | 4.5 |
| Dallas | D | D | D | D | 2,234 | 92,173 | 36,535.0 | 5,961.3 | 6,016 | 134,426 | 9,588.6 | 2,677.2 |
| Dawson | 13 | 64 | 10 | 2.7 | 11 | 88 | 20.6 | 4.5 | D | D | D | D |
| Deaf Smith | 26 | 182 | 26 | 9.2 | 24 | 1,591 | 1,169.5 | 87.8 | 36 | 455 | 24.3 | 5.8 |
| Delta | 3 | 5 | 1 | 0.2 | NA | NA | NA | NA | NA | NA | NA | NA |
| Denton | 2,124 | 10,798 | 1,795 | 652.1 | 395 | 13,378 | 7,948.3 | 801.8 | 1,537 | 30,380 | 1,767.6 | 488.8 |
| DeWitt | 40 | 103 | 15 | 4.5 | 15 | 528 | 83.6 | 16.6 | 44 | 402 | 19.4 | 5.4 |
| Dickens | D | D | D | D | NA | NA | NA | NA | 4 | 39 | 1.6 | 0.4 |
| Dimmit | 8 | 45 | 7 | 3.0 | 4 | 40 | 5.7 | 1.9 | 24 | 281 | 18.2 | 4.6 |
| Donley | 7 | 19 | 4 | 0.6 | NA | NA | NA | NA | 9 | 95 | 4.5 | 1.3 |
| Duval | 9 | 27 | 4 | 1.1 | NA | NA | NA | NA | 21 | 153 | 8.8 | 2.7 |
| Eastland | 30 | 268 | 52 | 18.0 | 16 | 558 | 101.5 | 23.5 | 47 | 547 | 29.0 | 8.0 |
| Ector | D | D | D | D | 224 | 4,128 | 1,379.8 | 243.9 | 284 | 6,669 | 462.1 | 116.5 |
| Edwards | NA | NA | NA | NA | NA | NA | NA | NA | D | D | D | 0.4 |
| Ellis | D | D | D | D | 179 | 11,056 | 5,244.7 | 595.1 | 271 | 5,639 | 296.1 | 85.0 |
| El Paso | D | D | D | D | D | D | D | D | 1,585 | 34,685 | 1,673.4 | 465.1 |
| Erath | 82 | 384 | 49 | 16.9 | 35 | 2,428 | 1,134.2 | 110.9 | D | D | D | D |
| Falls | 12 | 52 | 4 | 1.8 | 9 | 130 | 42.2 | 7.0 | 13 | D | 9.3 | D |
| Fannin | 35 | 110 | 11 | 3.9 | 35 | 591 | 247.3 | 26.9 | 39 | 482 | 22.6 | 6.4 |
| Fayette | 61 | 214 | 27 | 10.0 | 43 | 1,007 | 271.2 | 42.3 | 84 | 991 | 55.7 | 15.4 |
| Fisher | 7 | 20 | 2 | 0.6 | D | D | D | D | 3 | 19 | 1.0 | 0.3 |
| Floyd | 3 | 5 | 1 | 0.2 | 6 | 34 | 11.9 | 1.6 | D | D | D | D |
| Foard | NA | NA | NA | NA | NA | NA | NA | NA | D | D | D | D |
| Fort Bend | 2,083 | 15,035 | 4,443 | 1,329.5 | 369 | 13,211 | 4,585.4 | 739.5 | 1,196 | 22,866 | 1,388.5 | 382.2 |
| Franklin | 13 | 36 | 4 | 1.2 | D | D | D | D | D | D | D | D |
| Freestone | 24 | 65 | 13 | 3.9 | 13 | 390 | 142.4 | 20.2 | 38 | 614 | 30.5 | 8.1 |
| Frio | D | D | D | D | 6 | 66 | 17.4 | 3.3 | 45 | 475 | 28.2 | 6.8 |
| Gaines | 19 | 72 | 11 | 4.0 | 14 | 184 | 56.4 | 9.9 | 36 | 406 | 24.1 | 5.9 |
| Galveston | D | D | D | D | 162 | 5,481 | 23,357.8 | 558.1 | 757 | 17,232 | 1,060.9 | 299.2 |
| Garza | 8 | 13 | 1 | 0.4 | NA | NA | NA | NA | 12 | 113 | 6.3 | 1.5 |
| Gillespie | 72 | 302 | 46 | 13.7 | 64 | 715 | 122.9 | 28.6 | 126 | 1,863 | 114.7 | 34.9 |
| Glasscock | NA | NA | NA | NA | NA | NA | NA | NA | NA | NA | NA | NA |
| Goliad | 13 | 25 | 3 | 0.7 | NA | NA | NA | NA | D | D | D | D |
| Gonzales | 34 | 100 | 11 | 4.0 | 18 | 1,081 | 400.5 | 45.5 | 47 | 425 | 27.4 | 6.4 |
| Gray | 36 | 239 | 32 | 13.1 | 19 | 724 | 334.1 | 52.4 | 45 | 681 | 32.4 | 9.6 |
| Grayson | 222 | 1,017 | 159 | 48.2 | 106 | 7,200 | 2,713.7 | 324.9 | 238 | 4,558 | 260.1 | 73.9 |
| Gregg | D | D | D | D | 176 | 8,399 | 3,283.3 | 500.7 | 376 | 7,023 | 364.7 | 107.9 |
| Grimes | 30 | 299 | 42 | 19.0 | 39 | 1,443 | 618.5 | 103.5 | 38 | 394 | 24.1 | 5.8 |

# Table B. States and Counties — Health Care and Social Assistance, Other Services, Nonemployer Businesses, and Residential Construction

| STATE County | Health care and social assistance, 2017 | | | | Other services, 2017 | | | | Nonemployer businesses, 2018 | | Value of residential construction authorized by building permits, 2020 | |
|---|---|---|---|---|---|---|---|---|---|---|---|---|
| | Number of establishments | Number of employees | Receipts (mil dol) | Annual payroll (mil dol) | Number of establishments | Number of employees | Receipts (mil dol) | Annual payroll (mil dol) | Number | Receipts (mil dol) | New construction ($1,000) | Number of housing units |
| | 159 | 160 | 161 | 162 | 163 | 164 | 165 | 166 | 167 | 168 | 169 | 170 |
| TEXAS—Cont'd | | | | | | | | | | | | |
| Caldwell | 78 | 1,343 | 103.4 | 55.8 | 39 | 156 | 18.2 | 4.7 | 3,089 | 149.9 | 83,897 | 434 |
| Calhoun | 34 | 595 | 53.9 | 22.2 | 29 | 178 | 22.5 | 6.1 | 1,556 | 69.3 | 30,204 | 139 |
| Callahan | 10 | 152 | 8.4 | 4.4 | 12 | 33 | 2.6 | 0.9 | 1,257 | 54.9 | 2,697 | 14 |
| Cameron | 1,029 | 37,028 | 2,310.7 | 987.7 | 406 | 2,151 | 199.1 | 50.2 | 32,692 | 1,296.9 | 208,365 | 1,904 |
| Camp | 12 | 356 | 44.7 | 14.5 | D | D | D | D | 831 | 35.1 | 2,331 | 10 |
| Carson | 5 | 21 | 2.0 | 0.6 | 8 | 18 | 7.1 | 0.9 | 421 | 16.0 | 800 | 2 |
| Cass | 46 | 1,083 | 74.7 | 30.3 | 38 | 174 | 32.7 | 6.9 | 1,920 | 80.7 | 2,627 | 18 |
| Castro | D | D | D | D | 14 | 44 | 7.0 | 1.7 | 455 | 22.9 | 47 | 1 |
| Chambers | 41 | 783 | 119.5 | 30.9 | 35 | 194 | 33.7 | 7.5 | 3,016 | 128.7 | 81,355 | 932 |
| Cherokee | 75 | 2,570 | 211.2 | 93.9 | 35 | 183 | 19.7 | 5.6 | 3,357 | 148.2 | 1,812 | 13 |
| Childress | 19 | 384 | 40.2 | 17.4 | 8 | 17 | 1.6 | 0.5 | 382 | 13.7 | 0 | 0 |
| Clay | 10 | 105 | 7.8 | 3.9 | 9 | 37 | 2.9 | 0.6 | 874 | 40.3 | 909 | 7 |
| Cochran | 9 | 103 | 7.0 | 3.4 | 3 | 7 | 0.7 | 0.1 | 188 | 7.5 | 0 | 0 |
| Coke | NA | NA | NA | NA | D | D | D | 0.3 | 288 | 10.3 | 627 | 3 |
| Coleman | 16 | 469 | 38.2 | 16.3 | 15 | 41 | 5.7 | 1.1 | 733 | 39.8 | 0 | 0 |
| Collin | 3,599 | 56,314 | 7,848.0 | 2,683.8 | 1,299 | 9,562 | 1,458.0 | 340.1 | 101,380 | 5,633.6 | 3,611,185 | 14,794 |
| Collingsworth | 5 | 136 | 9.8 | 4.4 | D | D | 1.0 | D | 230 | 8.6 | 0 | 0 |
| Colorado | 47 | 902 | 90.9 | 31.9 | 37 | 107 | 12.1 | 2.9 | 1,948 | 93.6 | 2,825 | 16 |
| Comal | 375 | 6,079 | 685.3 | 260.0 | 258 | 1,628 | 146.5 | 49.4 | 15,676 | 860.5 | 885,480 | 4,197 |
| Comanche | 19 | 566 | 45.5 | 17.3 | D | D | D | D | 1,117 | 57.2 | 75 | 1 |
| Concho | 6 | 137 | 12.0 | 4.9 | NA | NA | NA | NA | 244 | 10.0 | 0 | 0 |
| Cooke | 82 | 1,145 | 104.5 | 45.0 | 54 | 385 | 38.9 | 11.4 | 3,622 | 198.6 | 16,138 | 68 |
| Coryell | 60 | 1,144 | 118.0 | 39.6 | 74 | 365 | 35.8 | 9.4 | 3,221 | 128.0 | 60,218 | 472 |
| Cottle | NA | NA | NA | NA | NA | NA | NA | NA | 137 | 4.3 | 0 | 0 |
| Crane | 9 | 181 | 18.7 | 7.2 | 3 | 11 | 1.4 | 0.4 | 347 | 15.0 | 105 | 1 |
| Crockett | 5 | 26 | 2.7 | 1.3 | D | D | D | D | 320 | 16.7 | NA | NA |
| Crosby | 9 | 172 | 10.1 | 5.0 | 7 | 12 | 0.8 | 0.2 | 347 | 11.6 | 933 | 4 |
| Culberson | D | D | D | D | NA | NA | NA | NA | 208 | 10.5 | 400 | 48 |
| Dallam | 9 | 40 | 2.8 | 0.9 | 18 | 62 | 8.3 | 1.9 | 543 | 38.4 | 1,024 | 5 |
| Dallas | 7,772 | 183,205 | 24,757.4 | 9,273.4 | 3,647 | 35,701 | 6,320.4 | 1,479.8 | 252,821 | 14,027.2 | 2,109,862 | 10,628 |
| Dawson | 13 | 326 | 33.4 | 11.2 | 24 | 96 | 8.5 | 2.2 | 722 | 31.6 | 0 | 0 |
| Deaf Smith | 21 | 572 | 43.0 | 19.6 | 36 | 184 | 35.7 | 7.6 | 1,209 | 68.5 | 390 | 2 |
| Delta | 5 | 527 | 9.7 | 5.0 | 4 | 7 | 0.7 | 0.2 | 394 | 15.2 | 1,651 | 9 |
| Denton | 1,933 | 28,296 | 3,704.7 | 1,385.4 | 942 | 6,019 | 674.3 | 201.2 | 79,667 | 4,054.6 | 2,182,350 | 9,974 |
| DeWitt | 34 | 765 | 68.4 | 32.2 | 39 | 209 | 29.8 | 7.6 | 1,578 | 75.5 | 836 | 4 |
| Dickens | 3 | 10 | 0.3 | 0.1 | D | D | 1.3 | D | 183 | 5.5 | NA | NA |
| Dimmit | 27 | 569 | 48.9 | 19.9 | D | D | D | D | 686 | 27.1 | 0 | 0 |
| Donley | 8 | 21 | 1.1 | 0.5 | D | D | 2.6 | D | 318 | 11.0 | 0 | 0 |
| Duval | 18 | 654 | 22.3 | 11.2 | D | D | D | 1.2 | 799 | 28.4 | NA | NA |
| Eastland | 49 | 841 | 59.5 | 23.3 | 31 | 134 | 13.7 | 3.5 | 1,397 | 62.1 | 80 | 2 |
| Ector | 285 | 8,075 | 996.1 | 374.9 | 272 | 2,188 | 331.9 | 88.9 | 13,754 | 865.2 | 245,144 | 1,188 |
| Edwards | 3 | 23 | 0.7 | 0.4 | 4 | 9 | 0.8 | 0.2 | 210 | 11.1 | NA | NA |
| Ellis | 297 | 4,568 | 482.4 | 172.5 | 201 | 1,134 | 114.3 | 33.2 | 16,101 | 785.2 | 627,193 | 2,731 |
| El Paso | D | D | D | D | D | D | D | D | 64,533 | 2,861.7 | 571,276 | 2,562 |
| Erath | 81 | 1,566 | 164.1 | 59.2 | 76 | 412 | 54.9 | 12.0 | 3,545 | 192.5 | 10,045 | 62 |
| Falls | 28 | 337 | 26.0 | 12.2 | D | D | D | 0.6 | 956 | 42.1 | 2,458 | 13 |
| Fannin | 44 | 1,627 | 157.6 | 65.8 | 29 | 126 | 12.0 | 2.6 | 2,607 | 118.0 | 6,891 | 42 |
| Fayette | 71 | 1,029 | 97.0 | 39.4 | 51 | 213 | 18.3 | 5.7 | 2,669 | 124.3 | 9,308 | 48 |
| Fisher | 3 | 105 | 9.6 | 5.0 | 5 | 21 | 1.1 | 0.5 | 250 | 10.2 | NA | NA |
| Floyd | 15 | 270 | 16.0 | 6.6 | 13 | 42 | 5.2 | 1.4 | 384 | 11.8 | 0 | 0 |
| Foard | D | D | D | D | D | D | D | 0.5 | 108 | 3.9 | NA | NA |
| Fort Bend | 2,045 | 27,849 | 2,872.2 | 1,064.3 | 778 | 5,073 | 596.3 | 161.2 | 80,262 | 4,001.2 | 3,001,054 | 14,093 |
| Franklin | 16 | 4,712 | 72.9 | 36.9 | 18 | 69 | 8.1 | 1.6 | 932 | 45.1 | 1,045 | 5 |
| Freestone | 27 | 695 | 63.6 | 21.8 | 32 | 104 | 13.7 | 3.0 | 1,304 | 54.8 | 836 | 4 |
| Frio | 31 | 576 | 44.4 | 16.2 | D | D | D | D | 1,015 | 40.1 | 840 | 8 |
| Gaines | 18 | 347 | 33.6 | 14.1 | 38 | 228 | 24.5 | 7.7 | 2,336 | 187.4 | 658 | 2 |
| Galveston | 648 | 15,736 | 1,487.7 | 627.9 | 463 | 2,828 | 491.3 | 94.9 | 27,099 | 1,252.4 | 691,262 | 3,278 |
| Garza | 12 | 163 | 14.5 | 5.4 | 4 | 11 | 2.1 | 0.3 | 365 | 19.0 | 439 | 5 |
| Gillespie | 100 | 1,795 | 202.1 | 82.8 | 64 | 308 | 31.3 | 8.3 | 3,918 | 201.9 | 14,530 | 72 |
| Glasscock | NA | NA | NA | NA | NA | NA | NA | NA | 145 | 9.4 | NA | NA |
| Goliad | 11 | 98 | 6.8 | 2.8 | 5 | 10 | 0.6 | 0.1 | 669 | 30.1 | 0 | 0 |
| Gonzales | 30 | 812 | 79.9 | 28.9 | 29 | 102 | 10.9 | 2.9 | 1,424 | 67.0 | 2,733 | 14 |
| Gray | 55 | 822 | 96.5 | 36.0 | 38 | 146 | 19.2 | 4.8 | 1,361 | 63.6 | 45 | 1 |
| Grayson | 360 | 8,454 | 956.3 | 381.4 | 149 | 689 | 81.7 | 21.5 | 10,926 | 582.2 | 195,489 | 1,181 |
| Gregg | 463 | 11,445 | 1,261.8 | 485.6 | 249 | 1,650 | 239.6 | 66.8 | 10,245 | 530.2 | 77,929 | 275 |
| Grimes | 18 | 317 | 29.7 | 13.3 | D | D | D | D | 2,295 | 99.3 | 13,457 | 87 |

# Table B. States and Counties — Government Employment and Payroll, and Local Government Finances

| STATE County | Government employment and payroll, 2017 | | | | | | | | | Local government finances, 2017 | | | | |
| | | | March payroll (percent of total) | | | | | | | General revenue | | | | |
| | | | | | | | | | | | | Taxes | | |
| | | | | | | | | | | | | | Per capita[1] (dollars) | |
| | Full-time equivalent employees | March payroll (dollars) | Administration, judicial, and legal | Police and corrections | Fire protection | Highways and transportation | Health and welfare | Natural resources and utilities | Education and libraries | Total (mil dol) | Intergovernmental (mil dol) | Total (mil dol) | Total | Property |
| | 171 | 172 | 173 | 174 | 175 | 176 | 177 | 178 | 179 | 180 | 181 | 182 | 183 | 184 |
| **TEXAS—Cont'd** | | | | | | | | | | | | | | |
| Caldwell | 1,439 | 5,815,124 | 8.2 | 13.3 | 1.7 | 2.3 | 1.3 | 4.7 | 68.2 | 114.5 | 52.2 | 46.7 | 1,104 | 938 |
| Calhoun | 1,057 | 3,269,430 | 7.7 | 13.0 | 2.6 | 7.9 | 4.9 | 3.6 | 57.7 | 186.4 | 13.0 | 76.1 | 3,507 | 3,125 |
| Callahan | 570 | 1,686,820 | 8.1 | 5.9 | 0.0 | 2.4 | 0.7 | 2.8 | 79.6 | 40.1 | 15.7 | 20.3 | 1,456 | 1,333 |
| Cameron | 22,268 | 74,927,211 | 4.8 | 8.6 | 2.8 | 2.5 | 2.0 | 4.4 | 73.6 | 1,807.5 | 1,026.7 | 493.5 | 1,169 | 939 |
| Camp | 517 | 1,566,426 | 3.6 | 2.2 | 0.0 | 2.9 | 1.2 | 6.2 | 83.3 | 35.5 | 16.4 | 15.8 | 1,227 | 1,027 |
| Carson | 372 | 1,134,116 | 12.8 | 3.3 | 0.0 | 3.5 | 2.3 | 7.4 | 62.2 | 36.6 | 9.8 | 23.2 | 3,860 | 3,750 |
| Cass | 1,340 | 3,760,026 | 6.9 | 12.1 | 1.3 | 2.0 | 0.6 | 2.6 | 74.1 | 100.6 | 40.8 | 48.6 | 1,621 | 1,437 |
| Castro | 568 | 1,682,097 | 4.3 | 7.3 | 0.3 | 2.4 | 22.4 | 2.5 | 60.4 | 37.9 | 16.4 | 13.2 | 1,711 | 1,481 |
| Chambers | 1,835 | 7,504,551 | 8.8 | 8.1 | 0.0 | 2.7 | 6.9 | 6.4 | 65.8 | 240.7 | 57.3 | 143.6 | 3,481 | 3,232 |
| Cherokee | 2,101 | 6,902,634 | 7.2 | 9.7 | 2.0 | 2.6 | 5.9 | 2.4 | 69.3 | 146.9 | 73.6 | 53.8 | 1,033 | 830 |
| Childress | 572 | 2,299,505 | 4.4 | 7.0 | 1.9 | 2.1 | 54.9 | 4.4 | 25.0 | 51.0 | 8.1 | 10.3 | 1,419 | 1,294 |
| Clay | 565 | 1,853,596 | 4.8 | 3.2 | 0.0 | 4.1 | 14.0 | 7.6 | 64.0 | 49.1 | 13.8 | 15.8 | 1,503 | 1,390 |
| Cochran | 330 | 1,037,152 | 10.2 | 5.8 | 0.0 | 3.3 | 15.3 | 4.0 | 60.2 | 27.2 | 10.2 | 13.9 | 4,884 | 4,560 |
| Coke | 234 | 755,845 | 6.0 | 2.2 | 0.0 | 1.6 | 29.7 | 6.4 | 53.7 | 18.7 | 4.9 | 8.0 | 2,423 | 2,257 |
| Coleman | 471 | 1,479,173 | 6.2 | 6.2 | 1.2 | 17.8 | 2.3 | 7.5 | 58.5 | 31.4 | 19.0 | 9.7 | 1,159 | 972 |
| Collin | 36,296 | 160,438,595 | 4.3 | 7.4 | 4.6 | 3.4 | 1.2 | 5.8 | 70.8 | 5,103.0 | 674.0 | 2,869.6 | 2,953 | 2,582 |
| Collingsworth | 182 | 564,742 | 8.5 | 6.3 | 0.0 | 5.4 | 10.1 | 6.2 | 62.3 | 12.8 | 4.8 | 6.1 | 2,043 | 1,925 |
| Colorado | 903 | 2,990,932 | 10.1 | 10.4 | 0.0 | 6.4 | 0.8 | 4.4 | 66.3 | 72.7 | 17.5 | 44.5 | 2,092 | 1,839 |
| Comal | 5,374 | 20,517,352 | 5.6 | 11.0 | 5.4 | 2.6 | 0.9 | 5.1 | 68.0 | 533.0 | 69.6 | 388.3 | 2,759 | 2,299 |
| Comanche | 853 | 2,848,355 | 5.0 | 7.9 | 3.4 | 2.6 | 24.6 | 1.5 | 54.4 | 53.9 | 21.7 | 21.3 | 1,572 | 1,353 |
| Concho | 201 | 741,256 | 10.8 | 4.8 | 0.0 | 3.7 | 37.5 | 4.4 | 37.5 | 22.0 | 5.8 | 6.8 | 2,522 | 2,359 |
| Cooke | 2,234 | 8,897,671 | 3.9 | 7.0 | 2.3 | 1.7 | 20.1 | 3.0 | 60.6 | 211.3 | 65.7 | 84.0 | 2,104 | 1,742 |
| Coryell | 2,532 | 8,957,496 | 5.7 | 8.3 | 1.4 | 1.3 | 18.1 | 4.0 | 59.9 | 246.6 | 104.1 | 68.5 | 916 | 767 |
| Cottle | 82 | 252,144 | 16.8 | 2.7 | 0.0 | 7.4 | 14.6 | 3.6 | 53.8 | 4.2 | 1.3 | 2.7 | 1,940 | 1,940 |
| Crane | 311 | 1,503,929 | 9.0 | 9.2 | 0.0 | 1.5 | 0.7 | 18.2 | 60.8 | 32.9 | 9.7 | 19.8 | 4,236 | 4,049 |
| Crockett | 325 | 978,310 | 9.0 | 4.2 | 1.2 | 6.3 | 20.1 | 6.2 | 51.1 | 38.1 | 6.5 | 27.5 | 7,784 | 7,779 |
| Crosby | 374 | 1,118,291 | 8.0 | 7.8 | 0.0 | 1.9 | 2.9 | 6.2 | 73.1 | 25.2 | 12.7 | 9.7 | 1,666 | 1,370 |
| Culberson | 235 | 800,326 | 13.0 | 7.2 | 0.0 | 3.8 | 33.4 | 8.0 | 34.0 | 38.3 | 17.7 | 18.9 | 8,458 | 8,167 |
| Dallam | 430 | 1,450,727 | 8.7 | 13.7 | 0.6 | 2.7 | 0.1 | 6.2 | 67.0 | 32.1 | 7.6 | 21.3 | 2,930 | 2,522 |
| Dallas | 115,153 | 544,302,614 | 5.1 | 10.5 | 5.6 | 7.3 | 14.9 | 4.7 | 49.8 | 14,882.1 | 3,355.3 | 7,362.7 | 2,810 | 2,155 |
| Dawson | 1,046 | 3,415,451 | 3.6 | 5.2 | 1.3 | 1.8 | 21.9 | 9.8 | 56.2 | 63.1 | 19.1 | 36.3 | 2,851 | 2,544 |
| Deaf Smith | 1,062 | 3,811,938 | 4.8 | 8.2 | 0.3 | 2.8 | 19.9 | 3.6 | 59.3 | 149.5 | 24.8 | 32.9 | 1,752 | 1,501 |
| Delta | 310 | 865,259 | 9.3 | 5.4 | 0.1 | 0.3 | 2.0 | 3.8 | 78.8 | 17.9 | 10.0 | 6.3 | 1,200 | 1,047 |
| Denton | 23,475 | 99,570,425 | 6.2 | 9.2 | 5.0 | 1.6 | 2.3 | 7.1 | 66.8 | 2,592.9 | 452.5 | 1,737.6 | 2,080 | 1,767 |
| DeWitt | 1,593 | 5,458,272 | 4.5 | 4.6 | 0.6 | 3.0 | 26.2 | 3.1 | 56.6 | 168.6 | 31.6 | 106.4 | 5,275 | 5,035 |
| Dickens | 142 | 386,281 | 9.0 | 7.0 | 0.0 | 7.7 | 0.6 | 3.5 | 70.0 | 11.1 | 4.0 | 5.1 | 2,342 | 2,269 |
| Dimmit | 790 | 2,562,510 | 9.7 | 9.1 | 0.0 | 0.8 | 27.9 | 6.2 | 45.5 | 70.5 | 11.1 | 54.4 | 5,295 | 4,809 |
| Donley | 335 | 1,129,574 | 12.0 | 1.2 | 1.0 | 2.4 | 2.3 | 7.5 | 73.1 | 19.2 | 9.0 | 5.5 | 1,652 | 1,457 |
| Duval | 1,043 | 3,227,857 | 6.4 | 10.4 | 0.0 | 3.6 | 30.6 | 4.3 | 43.2 | 109.5 | 38.2 | 24.5 | 2,176 | 2,017 |
| Eastland | 1,371 | 4,215,159 | 5.5 | 7.8 | 0.9 | 1.8 | 12.4 | 2.4 | 67.9 | 94.1 | 31.4 | 33.0 | 1,804 | 1,489 |
| Ector | 8,000 | 32,951,515 | 4.0 | 7.4 | 3.1 | 1.5 | 28.6 | 2.4 | 51.5 | 876.8 | 185.7 | 341.2 | 2,174 | 1,556 |
| Edwards | 178 | 511,085 | 6.8 | 9.8 | 0.0 | 4.7 | 0.4 | 4.7 | 72.9 | 11.8 | 2.1 | 8.8 | 4,564 | 4,448 |
| Ellis | 6,215 | 23,746,559 | 5.9 | 9.9 | 4.5 | 1.8 | 0.8 | 3.9 | 72.0 | 562.2 | 188.5 | 309.1 | 1,783 | 1,570 |
| El Paso | 41,955 | 157,950,113 | 5.1 | 9.1 | 3.4 | 2.4 | 11.3 | 2.7 | 63.9 | 4,047.0 | 1,836.8 | 1,327.6 | 1,585 | 1,238 |
| Erath | 1,701 | 5,485,209 | 7.5 | 10.9 | 3.3 | 2.5 | 15.6 | 3.3 | 55.4 | 114.5 | 43.2 | 56.7 | 1,360 | 1,111 |
| Falls | 646 | 2,047,971 | 6.4 | 10.6 | 1.6 | 4.3 | 2.6 | 4.9 | 69.1 | 50.3 | 24.8 | 19.9 | 1,147 | 989 |
| Fannin | 1,210 | 3,682,491 | 9.3 | 7.3 | 3.9 | 2.9 | 0.4 | 4.0 | 71.2 | 97.7 | 42.6 | 46.0 | 1,331 | 1,134 |
| Fayette | 1,086 | 3,648,038 | 9.7 | 12.7 | 0.0 | 5.7 | 5.1 | 8.9 | 57.4 | 84.0 | 16.6 | 49.6 | 1,975 | 1,698 |
| Fisher | 253 | 842,936 | 10.7 | 1.6 | 0.0 | 3.4 | 46.3 | 1.9 | 35.4 | 18.7 | 4.8 | 5.8 | 1,508 | 1,332 |
| Floyd | 447 | 1,562,702 | 7.3 | 7.8 | 0.0 | 3.6 | 24.3 | 3.7 | 53.2 | 24.9 | 11.9 | 10.0 | 1,712 | 1,537 |
| Foard | 114 | 294,388 | 13.7 | 3.2 | 0.0 | 29.3 | 4.2 | 3.4 | 44.2 | 7.8 | 2.6 | 4.6 | 3,790 | 3,607 |
| Fort Bend | 20,284 | 79,790,202 | 6.3 | 9.9 | 2.1 | 1.6 | 1.9 | 2.3 | 74.9 | 2,136.7 | 450.1 | 1,352.2 | 1,761 | 1,585 |
| Franklin | 404 | 1,326,799 | 7.5 | 9.7 | 0.0 | 3.6 | 0.7 | 10.8 | 67.0 | 31.2 | 6.9 | 21.1 | 1,953 | 1,751 |
| Freestone | 910 | 2,968,110 | 7.1 | 5.4 | 1.2 | 2.4 | 1.8 | 1.2 | 80.8 | 76.4 | 17.0 | 51.1 | 2,600 | 2,509 |
| Frio | 783 | 2,684,048 | 4.5 | 9.6 | 0.0 | 4.4 | 1.1 | 5.5 | 73.9 | 66.4 | 17.3 | 39.9 | 2,007 | 1,858 |
| Gaines | 1,236 | 4,279,287 | 4.8 | 5.9 | 0.0 | 4.4 | 25.4 | 2.9 | 53.9 | 157.4 | 23.3 | 79.0 | 3,842 | 3,737 |
| Galveston | 17,131 | 60,202,014 | 6.2 | 10.4 | 2.3 | 3.2 | 4.1 | 4.7 | 68.3 | 1,767.6 | 546.8 | 976.9 | 2,919 | 2,529 |
| Garza | 299 | 979,400 | 9.1 | 13.9 | 0.0 | 2.6 | 2.3 | 2.9 | 68.4 | 22.3 | 7.8 | 11.1 | 1,714 | 1,539 |
| Gillespie | 1,146 | 5,013,539 | 6.1 | 8.8 | 0.9 | 2.6 | 2.3 | 20.6 | 54.6 | 85.8 | 7.6 | 64.4 | 2,430 | 1,982 |
| Glasscock | 88 | 338,809 | 4.6 | 1.8 | 0.0 | 6.4 | 0.0 | 3.6 | 78.7 | 29.3 | 0.5 | 26.7 | 19,587 | 19,491 |
| Goliad | 343 | 1,232,592 | 8.4 | 9.7 | 4.7 | 3.7 | 1.8 | 1.0 | 69.8 | 21.9 | 6.9 | 12.6 | 1,663 | 1,481 |
| Gonzales | 1,335 | 4,542,193 | 7.0 | 7.9 | 0.6 | 3.2 | 23.9 | 4.1 | 52.6 | 133.9 | 28.6 | 54.2 | 2,613 | 2,290 |
| Gray | 1,019 | 3,361,637 | 11.3 | 10.0 | 3.3 | 3.6 | 0.7 | 4.1 | 65.5 | 73.4 | 22.2 | 36.5 | 1,653 | 1,387 |
| Grayson | 5,654 | 19,812,399 | 6.1 | 10.4 | 4.1 | 2.3 | 5.3 | 5.8 | 64.9 | 487.4 | 166.5 | 228.8 | 1,745 | 1,467 |
| Gregg | 6,692 | 23,257,940 | 4.7 | 11.3 | 5.3 | 1.5 | 8.1 | 4.3 | 62.8 | 552.3 | 182.3 | 277.2 | 2,256 | 1,638 |
| Grimes | 951 | 3,230,359 | 9.7 | 11.2 | 1.0 | 2.5 | 0.3 | 2.2 | 72.5 | 75.8 | 15.6 | 54.0 | 1,932 | 1,788 |

1.  Based on the resident population estimated as of July 1 of the year shown.

# Table B. States and Counties — Local Government Finances, Government Employment, and Income Taxes

| STATE County | Local government finances, 2017 (cont.) | | | | | | | | | Government employment, 2019 | | | Individual income tax returns, 2018 | | |
| | Direct general expenditure | | | | | | | Debt outstanding | | | | | | | |
| | Total (mil dol) | Per capita¹ (dollars) | Education | Health and hospitals | Police protection | Public welfare | Highways | Total (mil dol) | Per capita¹ (dollars) | Federal civilian | Federal military | State and local | Number of returns | Mean adjusted gross income | Mean income tax |
|---|---|---|---|---|---|---|---|---|---|---|---|---|---|---|---|
| | 185 | 186 | 187 | 188 | 189 | 190 | 191 | 192 | 193 | 194 | 195 | 196 | 197 | 198 | 199 |
| **TEXAS—Cont'd** | | | | | | | | | | | | | | | |
| Caldwell | 137 | 3,229 | 61.8 | 2.9 | 4.9 | 0.3 | 3.3 | 120.3 | 2,842 | 64 | 79 | 1,799 | 18,640 | 48,044 | 3,951 |
| Calhoun | 196 | 9,043 | 35.2 | 15.4 | 2.7 | 25.9 | 3.8 | 86.4 | 3,978 | 36 | 95 | 1,351 | 9,330 | 56,022 | 5,432 |
| Callahan | 46 | 3,258 | 73.3 | 0.5 | 5.3 | 0.0 | 2.4 | 39.2 | 2,804 | 44 | 27 | 629 | 6,060 | 51,227 | 4,450 |
| Cameron | 1,674 | 3,966 | 67.1 | 0.8 | 4.6 | 0.4 | 2.9 | 1,644.9 | 3,896 | 3,554 | 928 | 25,958 | 169,320 | 40,894 | 3,149 |
| Camp | 40 | 3,120 | 71.8 | 0.1 | 3.8 | 0.1 | 5.5 | 56.9 | 4,431 | 25 | 25 | 568 | 5,360 | 50,867 | 4,776 |
| Carson | 35 | 5,868 | 53.1 | 0.7 | 3.2 | 0.2 | 4.3 | 35.4 | 5,903 | 13 | 12 | 434 | 2,580 | 64,067 | 5,941 |
| Cass | 90 | 3,011 | 62.3 | 1.9 | 5.5 | 0.5 | 4.5 | 43.2 | 1,443 | 62 | 58 | 1,687 | 12,440 | 49,231 | 4,045 |
| Castro | 31 | 4,016 | 75.8 | 5.4 | 2.5 | 0.1 | 1.6 | 41.7 | 5,421 | 18 | 15 | 687 | 2,850 | 49,865 | 5,167 |
| Chambers | 245 | 5,929 | 50.1 | 8.3 | 4.2 | 5.6 | 5.6 | 444.6 | 10,772 | 61 | 85 | 2,296 | 20,030 | 84,744 | 9,827 |
| Cherokee | 151 | 2,899 | 55.7 | 7.7 | 5.4 | 0.2 | 6.6 | 159.0 | 3,050 | 73 | 97 | 3,996 | 20,850 | 49,348 | 4,349 |
| Childress | 52 | 7,127 | 24.0 | 60.4 | 2.5 | 0.1 | 3.1 | 2.4 | 332 | 20 | 11 | 1,042 | 2,410 | 47,696 | 3,906 |
| Clay | 40 | 3,814 | 51.2 | 15.3 | 6.4 | 0.4 | 9.4 | 24.0 | 2,294 | 23 | 20 | 542 | 4,560 | 56,850 | 4,754 |
| Cochran | 27 | 9,393 | 49.0 | 14.3 | 3.0 | 0.2 | 4.2 | 16.1 | 5,627 | 14 | 5 | 301 | 1,140 | 44,260 | 3,152 |
| Coke | 17 | 5,155 | 35.7 | 0.0 | 3.9 | 33.3 | 5.0 | 9.1 | 2,761 | 11 | 7 | 337 | 1,470 | 52,991 | 4,517 |
| Coleman | 25 | 2,941 | 66.8 | 7.2 | 2.6 | 0.1 | 4.5 | 15.5 | 1,841 | 37 | 16 | 534 | 3,450 | 50,100 | 5,542 |
| Collin | 5,154 | 5,303 | 51.2 | 1.2 | 3.6 | 0.0 | 4.4 | 20,672.9 | 21,271 | 2,018 | 2,116 | 52,371 | 465,170 | 104,645 | 15,175 |
| Collingsworth | 16 | 5,252 | 40.7 | 10.7 | 5.2 | 0.1 | 4.4 | 10.6 | 3,571 | 12 | 6 | 222 | 1,170 | 43,723 | 4,121 |
| Colorado | 69 | 3,245 | 51.7 | 7.0 | 6.8 | 0.2 | 6.9 | 68.4 | 3,214 | 50 | 41 | 1,081 | 9,790 | 58,464 | 6,269 |
| Comal | 595 | 4,230 | 60.2 | 1.0 | 5.7 | 0.1 | 4.9 | 1,068.5 | 7,593 | 232 | 302 | 6,180 | 74,850 | 87,860 | 11,475 |
| Comanche | 47 | 3,439 | 57.5 | 4.6 | 4.6 | 0.4 | 3.2 | 46.8 | 3,460 | 49 | 26 | 786 | 5,720 | 44,543 | 4,001 |
| Concho | 21 | 7,623 | 27.1 | 39.8 | 3.0 | 0.0 | 2.8 | 4.2 | 1,539 | 13 | 5 | 225 | 1,000 | 51,720 | 4,491 |
| Cooke | 233 | 5,838 | 55.9 | 17.6 | 4.8 | 0.1 | 2.8 | 224.6 | 5,625 | 68 | 79 | 2,862 | 18,720 | 66,075 | 7,638 |
| Coryell | 207 | 2,773 | 51.4 | 23.2 | 5.0 | 0.7 | 2.2 | 151.6 | 2,028 | 189 | 409 | 6,039 | 28,260 | 45,088 | 3,065 |
| Cottle | 4 | 2,907 | 63.6 | 0.0 | 3.2 | 0.3 | 8.2 | 1.3 | 950 | 10 | 3 | 108 | 580 | 33,710 | 2,412 |
| Crane | 30 | 6,475 | 66.8 | 0.0 | 6.1 | 0.0 | 3.6 | 6.3 | 1,345 | 6 | 9 | 370 | 2,060 | 67,047 | 6,475 |
| Crockett | 36 | 10,247 | 48.0 | 1.1 | 2.7 | 1.2 | 7.8 | 6.5 | 1,831 | 3 | 7 | 388 | 1,580 | 66,909 | 7,722 |
| Crosby | 25 | 4,337 | 68.5 | 0.0 | 3.9 | 0.3 | 3.5 | 10.9 | 1,866 | 17 | 11 | 433 | 1,060 | 42,128 | 3,129 |
| Culberson | 16 | 7,230 | 74.2 | 11.2 | 0.0 | 0.0 | 0.0 | 39.4 | 17,669 | 85 | 4 | 241 | | 44,504 | 3,564 |
| Dallam | 24 | 3,343 | 78.7 | 0.1 | 2.4 | 0.1 | 2.9 | 20.1 | 2,762 | 20 | 14 | 285 | 3,200 | 53,291 | 6,732 |
| Dallas | 14,439 | 5,511 | 37.9 | 17.3 | 5.7 | 0.4 | 3.1 | 31,348.3 | 11,964 | 25,639 | 5,686 | 152,889 | 1,185,160 | 79,843 | 12,146 |
| Dawson | 61 | 4,818 | 66.0 | 0.2 | 4.9 | 0.8 | 5.0 | 179.9 | 14,120 | 68 | 22 | 1,220 | 4,850 | 61,814 | 6,803 |
| Deaf Smith | 127 | 6,773 | 31.2 | 16.6 | 3.4 | 0.1 | 2.8 | 956.2 | 50,987 | 49 | 35 | 1,231 | 7,800 | 41,696 | 3,659 |
| Delta | 14 | 2,736 | 72.5 | 0.0 | 2.0 | 0.2 | 2.0 | 17.6 | 3,340 | 15 | 10 | 305 | 2,130 | 42,940 | 3,427 |
| Denton | 2,474 | 2,962 | 53.4 | 1.6 | 5.4 | 0.1 | 4.2 | 7,148.9 | 8,558 | 2,042 | 1,789 | 37,155 | 403,990 | 91,415 | 12,368 |
| DeWitt | 195 | 9,641 | 54.7 | 19.2 | 2.1 | 0.2 | 11.4 | 159.5 | 7,906 | 33 | 36 | 2,233 | 8,100 | 95,556 | 17,742 |
| Dickens | 10 | 4,456 | 59.3 | 0.0 | 2.6 | 0.8 | 0.0 | 5.1 | 2,332 | 10 | 4 | 161 | 750 | 48,991 | 4,260 |
| Dimmit | 71 | 6,863 | 76.3 | 0.0 | 3.2 | 0.1 | 2.5 | 57.0 | 5,542 | 262 | 20 | 1,057 | 4,030 | 53,353 | 6,489 |
| Donley | 25 | 7,592 | 63.7 | 0.0 | 2.5 | 2.6 | 2.6 | 111.1 | 33,272 | 12 | 6 | 415 | 1,410 | 37,221 | 3,103 |
| Duval | 89 | 7,906 | 43.0 | 31.7 | 2.6 | 0.4 | 5.4 | 69.1 | 6,131 | 134 | 21 | 955 | 4,430 | 43,380 | 3,480 |
| Eastland | 96 | 5,235 | 61.3 | 15.8 | 2.9 | 0.2 | 2.8 | 81.7 | 4,468 | 52 | 370 | 1,495 | 7,460 | 53,965 | 5,379 |
| Ector | 915 | 5,830 | 35.1 | 40.8 | 3.5 | 0.0 | 2.5 | 447.1 | 2,849 | 195 | 320 | 9,801 | 72,720 | 67,096 | 7,413 |
| Edwards | 10 | 5,316 | 69.6 | 0.0 | 9.0 | 0.0 | 3.0 | 4.3 | 2,230 | 19 | 4 | 137 | 800 | 56,818 | 6,726 |
| Ellis | 601 | 3,465 | 64.9 | 0.7 | 5.2 | 0.1 | 1.9 | 1,415.4 | 8,162 | 278 | 357 | 7,886 | 83,120 | 67,075 | 6,854 |
| El Paso | 3,936 | 4,700 | 49.6 | 17.4 | 4.5 | 0.2 | 1.3 | 6,815.4 | 8,139 | 12,826 | 28,316 | 56,607 | 377,610 | 46,241 | 3,998 |
| Erath | 96 | 2,293 | 57.0 | 1.6 | 8.3 | 0.0 | 7.2 | 69.4 | 1,663 | 78 | 81 | 4,206 | 16,370 | 54,574 | 5,939 |
| Falls | 41 | 2,354 | 69.9 | 5.8 | 3.0 | 0.1 | 2.0 | 34.0 | 1,959 | 405 | 30 | 1,209 | 6,470 | 43,106 | 3,419 |
| Fannin | 99 | 2,853 | 61.0 | 0.0 | 4.5 | 0.5 | 6.2 | 107.7 | 3,117 | 72 | 49 | 1,950 | 14,000 | 54,214 | 5,105 |
| Fayette | 83 | 3,292 | 51.3 | 3.8 | 6.7 | 0.9 | 9.4 | 117.9 | 4,695 | 15 | 7 | 1,550 | 11,850 | 67,432 | 8,340 |
| Fisher | 20 | 5,059 | 33.1 | 48.0 | 0.9 | 0.1 | 1.1 | 12.1 | 3,129 | 31 | 11 | 308 | 1,490 | 55,895 | 5,293 |
| Floyd | 26 | 4,417 | 75.5 | 1.5 | 4.4 | 0.0 | 0.9 | 15.5 | 2,659 | 5 | 2 | 527 | 2,420 | 46,762 | 4,095 |
| Foard | 5 | 4,050 | 66.8 | 2.0 | 3.5 | 0.3 | 7.6 | 0.2 | 154 | 5 | | 115 | 520 | 38,204 | 2,752 |
| Fort Bend | 2,309 | 3,008 | 52.8 | 0.8 | 4.4 | 0.2 | 7.8 | 6,319.0 | 8,231 | 862 | 1,576 | 25,127 | 345,630 | 94,825 | 13,068 |
| Franklin | 29 | 2,660 | 51.3 | 0.4 | 6.3 | 0.5 | 12.8 | 50.8 | 4,702 | 15 | 21 | 440 | 4,390 | 55,816 | 5,520 |
| Freestone | 70 | 3,542 | 64.7 | 3.4 | 6.6 | 0.1 | 6.5 | 68.7 | 3,497 | 41 | 35 | 1,298 | 7,730 | 53,939 | 5,138 |
| Frio | 79 | 3,977 | 62.6 | 6.2 | 4.0 | 0.1 | 4.4 | 125.6 | 6,312 | 170 | 33 | 1,314 | 6,400 | 46,772 | 4,572 |
| Gaines | 158 | 7,701 | 37.5 | 35.0 | 2.0 | 0.1 | 6.8 | 110.5 | 5,377 | 25 | 42 | 1,366 | 8,160 | 54,610 | 5,184 |
| Galveston | 1,689 | 5,046 | 53.8 | 4.8 | 5.7 | 0.3 | 2.8 | 2,873.3 | 8,586 | 1,086 | 875 | 29,003 | 154,650 | 77,249 | 9,749 |
| Garza | 40 | 6,175 | 59.6 | 2.2 | 4.7 | 0.0 | 4.0 | 40.1 | 6,179 | 13 | 18 | 340 | 1,830 | 56,873 | 5,732 |
| Gillespie | 102 | 3,859 | 57.7 | 3.1 | 9.3 | 0.3 | 5.6 | 49.1 | 1,852 | 50 | 52 | 1,180 | 13,780 | 85,794 | 11,817 |
| Glasscock | 33 | 23,817 | 95.0 | 0.0 | 0.2 | 0.0 | 0.2 | 16.2 | 11,867 | 5 | 3 | 136 | 640 | 158,986 | 31,503 |
| Goliad | 21 | 2,800 | 68.8 | 0.0 | 1.6 | 0.2 | 3.1 | 12.8 | 1,694 | 14 | 15 | 399 | 3,220 | 67,745 | 8,486 |
| Gonzales | 141 | 6,789 | 34.3 | 42.7 | 1.6 | 0.2 | 1.1 | 59.1 | 2,849 | 55 | 40 | 1,461 | 9,150 | 59,660 | 7,134 |
| Gray | 72 | 3,255 | 56.7 | 0.5 | 6.5 | 0.2 | 5.5 | 73.7 | 3,333 | 39 | 39 | 1,463 | 8,560 | 54,038 | 5,255 |
| Grayson | 488 | 3,717 | 54.3 | 3.7 | 4.9 | 0.5 | 3.7 | 1,071.3 | 8,168 | 354 | 262 | 6,940 | 59,710 | 60,307 | 6,156 |
| Gregg | 581 | 4,733 | 53.6 | 6.3 | 7.0 | 0.0 | 4.2 | 804.9 | 6,552 | 348 | 235 | 7,261 | 55,700 | 61,459 | 6,840 |
| Grimes | 69 | 2,463 | 60.4 | 0.4 | 5.2 | 0.1 | 8.6 | 93.4 | 3,339 | 57 | 51 | 1,745 | 12,060 | 54,113 | 5,432 |

1. Based on the resident population estimated as of July 1 of the year shown.

# Table B. States and Counties — **Land Area and Population**

| State / county code | CBSA code[1] | County Type code[2] | STATE County | Land area[3] (sq. mi) | Total persons 2019 | Rank | Per square mile | White | Black | American Indian, Alaska Native | Asian and Pacific Islander | Percent Hispanic or Latino[4] | Under 5 years | 5 to 14 years | 15 to 24 years | 25 to 34 years | 35 to 44 years | 45 to 54 years |
|---|---|---|---|---|---|---|---|---|---|---|---|---|---|---|---|---|---|---|
| | | | | 1 | 2 | 3 | 4 | 5 | 6 | 7 | 8 | 9 | 10 | 11 | 12 | 13 | 14 | 15 |
| | | | **TEXAS—Cont'd** | | | | | | | | | | | | | | | |
| 48187 | 41700 | 1 | Guadalupe ............... | 711.3 | 170,608 | 390 | 239.9 | 50.7 | 8.9 | 0.9 | 3.0 | 38.8 | 6.0 | 14.1 | 13.1 | 12.9 | 14.5 | 13.0 |
| 48189 | 38380 | 4 | Hale ............... | 1,004.7 | 32,754 | 1,363 | 32.6 | 33.1 | 5.4 | 0.7 | 0.8 | 61.0 | 6.5 | 15.5 | 15.9 | 14.1 | 12.2 | 11.0 |
| 48191 | | 9 | Hall ............... | 883.5 | 2,939 | 2,967 | 3.3 | 56.8 | 8.8 | 0.8 | 0.4 | 34.3 | 3.9 | 12.7 | 12.5 | 9.4 | 10.2 | 11.5 |
| 48193 | | 6 | Hamilton ............... | 835.9 | 8,557 | 2,528 | 10.2 | 84.1 | 1.2 | 1.0 | 1.0 | 13.6 | 5.3 | 12.3 | 10.7 | 10.8 | 10.7 | 10.1 |
| 48195 | | 7 | Hansford ............... | 919.8 | 5,279 | 2,803 | 5.7 | 49.6 | 1.3 | 0.7 | 0.6 | 48.8 | 7.2 | 16.4 | 14.9 | 12.0 | 10.6 | 11.4 |
| 48197 | | 9 | Hardeman ............... | 695.1 | 4,011 | 2,896 | 5.8 | 68.5 | 5.8 | 1.4 | 1.1 | 25.2 | 4.9 | 12.7 | 12.0 | 10.1 | 12.4 | 12.0 |
| 48199 | 13140 | 2 | Hardin ............... | 890.6 | 58,305 | 892 | 65.5 | 86.9 | 5.9 | 1.0 | 1.1 | 6.5 | 6.1 | 14.1 | 11.7 | 12.6 | 12.7 | 12.0 |
| 48201 | 26420 | 1 | Harris ............... | 1,707.0 | 4,738,253 | 3 | 2,775.8 | 29.5 | 19.7 | 0.5 | 8.0 | 43.8 | 7.2 | 14.7 | 13.5 | 15.9 | 14.4 | 12.3 |
| 48203 | 30980 | 4 | Harrison ............... | 900.1 | 66,386 | 807 | 73.8 | 64.3 | 21.3 | 1.0 | 1.0 | 13.9 | 5.9 | 14.3 | 12.9 | 11.8 | 12.1 | 11.6 |
| 48205 | | 7 | Hartley ............... | 1,462.0 | 5,443 | 2,791 | 3.7 | 64.4 | 7.0 | 0.8 | 0.9 | 27.7 | 5.9 | 11.3 | 11.0 | 14.8 | 14.8 | 15.4 |
| 48207 | | 6 | Haskell ............... | 903.1 | 5,754 | 2,763 | 6.4 | 65.6 | 4.9 | 1.1 | 1.4 | 28.8 | 3.7 | 10.3 | 12.7 | 13.5 | 11.9 | 10.7 |
| 48209 | 12420 | 1 | Hays ............... | 676.9 | 241,365 | 285 | 356.6 | 53.6 | 4.6 | 0.8 | 2.5 | 40.3 | 5.9 | 12.9 | 19.5 | 14.2 | 14.1 | 11.3 |
| 48211 | | 7 | Hemphill ............... | 906.3 | 3,777 | 2,914 | 4.2 | 62.6 | 0.7 | 1.1 | 1.6 | 35.1 | 6.2 | 17.6 | 14.5 | 9.6 | 13.7 | 11.1 |
| 48213 | 11980 | 4 | Henderson ............... | 873.8 | 83,792 | 682 | 95.9 | 78.8 | 6.8 | 1.4 | 0.9 | 13.8 | 5.4 | 12.1 | 11.5 | 11.1 | 10.8 | 11.7 |
| 48215 | 32580 | 2 | Hidalgo ............... | 1,571.0 | 875,200 | 65 | 557.1 | 6.0 | 0.5 | 0.1 | 1.0 | 92.5 | 8.3 | 18.1 | 16.4 | 13.7 | 12.3 | 11.2 |
| 48217 | | 6 | Hill ............... | 958.9 | 37,006 | 1,248 | 38.6 | 71.1 | 6.9 | 1.0 | 1.0 | 21.6 | 5.8 | 13.1 | 12.3 | 11.3 | 11.2 | 11.4 |
| 48219 | 30220 | 6 | Hockley ............... | 908.4 | 22,921 | 1,683 | 25.2 | 45.8 | 3.9 | 0.8 | 0.6 | 49.9 | 6.8 | 14.6 | 15.7 | 13.4 | 12.3 | 10.1 |
| 48221 | 24180 | 1 | Hood ............... | 420.7 | 63,527 | 840 | 151.0 | 84.4 | 1.5 | 1.2 | 1.1 | 13.0 | 5.1 | 11.8 | 10.1 | 10.3 | 10.8 | 11.0 |
| 48223 | 44860 | 6 | Hopkins ............... | 767.4 | 37,170 | 1,245 | 48.4 | 74.1 | 7.8 | 1.1 | 1.0 | 17.6 | 6.3 | 13.7 | 12.2 | 12.4 | 11.3 | 12.0 |
| 48225 | | 7 | Houston ............... | 1,231.0 | 22,835 | 1,689 | 18.5 | 62.4 | 25.2 | 0.8 | 1.0 | 12.1 | 4.9 | 11.1 | 10.1 | 11.5 | 12.8 | 12.8 |
| 48227 | 13700 | 4 | Howard ............... | 900.8 | 36,540 | 1,262 | 40.6 | 48.0 | 6.7 | 1.1 | 1.5 | 44.0 | 6.2 | 11.7 | 13.2 | 16.3 | 13.4 | 14.8 |
| 48229 | 21340 | 2 | Hudspeth ............... | 4,570.5 | 4,906 | 2,831 | 1.1 | 18.1 | 3.9 | 0.9 | 1.3 | 77.1 | 4.4 | 12.0 | 11.9 | 18.8 | 15.9 | 11.1 |
| 48231 | 19100 | 1 | Hunt ............... | 840.4 | 99,807 | 603 | 118.8 | 71.4 | 8.6 | 1.6 | 1.9 | 18.4 | 6.1 | 13.4 | 13.9 | 12.7 | 11.8 | 12.3 |
| 48233 | 14420 | 6 | Hutchinson ............... | 887.4 | 20,677 | 1,783 | 23.3 | 71.1 | 3.2 | 2.3 | 1.0 | 24.3 | 6.3 | 14.1 | 12.7 | 12.0 | 12.9 | 11.2 |
| 48235 | 41660 | 3 | Irion ............... | 1,051.5 | 1,564 | 3,075 | 1.5 | 71.4 | 1.7 | 1.7 | 1.1 | 25.9 | 4.5 | 13.3 | 10.2 | 12.1 | 10.7 | 11.2 |
| 48237 | | 6 | Jack ............... | 911.0 | 9,056 | 2,491 | 9.9 | 77.4 | 4.3 | 0.9 | 0.8 | 17.5 | 4.8 | 13.1 | 13.3 | 14.0 | 12.2 | 12.5 |
| 48239 | | 6 | Jackson ............... | 829.4 | 14,854 | 2,099 | 17.9 | 58.6 | 6.6 | 0.8 | 1.3 | 33.8 | 7.1 | 14.2 | 12.3 | 11.8 | 12.2 | 10.7 |
| 48241 | | 6 | Jasper ............... | 938.7 | 35,375 | 1,293 | 37.7 | 75.7 | 16.8 | 1.0 | 0.9 | 7.1 | 5.8 | 13.3 | 11.5 | 11.3 | 11.2 | 11.9 |
| 48243 | | 9 | Jeff Davis ............... | 2,264.6 | 2,220 | 3,022 | 1.0 | 65.3 | 1.1 | 2.0 | 2.1 | 31.4 | 1.2 | 1.7 | 7.1 | 8.5 | 9.8 | 14.1 |
| 48245 | 13140 | 2 | Jefferson ............... | 876.8 | 250,127 | 277 | 285.3 | 39.9 | 34.1 | 0.7 | 4.3 | 22.4 | 6.9 | 13.6 | 13.0 | 14.5 | 13.2 | 11.4 |
| 48247 | | 6 | Jim Hogg ............... | 1,136.2 | 5,184 | 2,812 | 4.6 | 5.7 | 0.9 | 0.5 | 0.6 | 92.7 | 6.7 | 18.0 | 14.5 | 11.1 | 11.8 | 10.0 |
| 48249 | 10860 | 4 | Jim Wells ............... | 865.2 | 40,452 | 1,168 | 46.8 | 18.2 | 0.9 | 0.5 | 0.6 | 80.2 | 6.9 | 15.8 | 13.3 | 13.0 | 12.4 | 10.9 |
| 48251 | 19100 | 1 | Johnson ............... | 724.8 | 179,575 | 370 | 247.8 | 70.8 | 4.8 | 1.1 | 1.9 | 23.2 | 6.4 | 14.9 | 13.0 | 13.2 | 13.1 | 12.5 |
| 48253 | 10180 | 3 | Jones ............... | 928.6 | 19,875 | 1,831 | 21.4 | 59.1 | 12.2 | 0.8 | 0.9 | 28.0 | 4.2 | 10.2 | 12.8 | 16.9 | 14.9 | 13.0 |
| 48255 | | 6 | Karnes ............... | 747.8 | 15,562 | 2,055 | 20.8 | 34.8 | 7.8 | 0.4 | 0.5 | 57.0 | 5.8 | 12.4 | 14.2 | 17.3 | 14.3 | 10.6 |
| 48257 | 19100 | 1 | Kaufman ............... | 780.8 | 143,198 | 460 | 183.4 | 58.7 | 15.8 | 1.0 | 2.0 | 24.3 | 7.2 | 16.2 | 12.8 | 13.8 | 14.6 | 12.4 |
| 48259 | 41700 | 1 | Kendall ............... | 662.5 | 48,523 | 1,014 | 73.2 | 72.3 | 1.5 | 1.0 | 1.8 | 24.7 | 4.6 | 13.4 | 12.4 | 9.5 | 12.7 | 13.2 |
| 48261 | 28780 | 9 | Kenedy ............... | 1,458.6 | 379 | 3,140 | 0.3 | 23.5 | 2.9 | 3.2 | 0.8 | 71.5 | 2.9 | 10.3 | 14.2 | 9.8 | 12.7 | 13.7 |
| 48263 | | 9 | Kent ............... | 902.5 | 786 | 3,119 | 0.9 | 78.1 | 1.4 | 2.0 | 0.4 | 19.7 | 5.2 | 15.4 | 10.2 | 8.3 | 11.6 | 9.3 |
| 48265 | 28500 | 4 | Kerr ............... | 1,103.3 | 52,869 | 959 | 47.9 | 68.5 | 1.8 | 1.1 | 1.5 | 28.2 | 5.1 | 10.4 | 11.3 | 10.5 | 9.7 | 10.2 |
| 48267 | | 7 | Kimble ............... | 1,251.0 | 4,396 | 2,864 | 3.5 | 72.9 | 0.9 | 0.9 | 1.1 | 24.8 | 5.2 | 9.9 | 9.4 | 9.7 | 10.1 | 10.3 |
| 48269 | | 9 | King ............... | 910.9 | 283 | 3,141 | 0.3 | 77.0 | 2.8 | 1.8 | 0.7 | 20.1 | 6.0 | 16.3 | 9.9 | 9.2 | 13.4 | 12.0 |
| 48271 | | 7 | Kinney ............... | 1,360.5 | 3,670 | 2,923 | 2.7 | 36.5 | 2.2 | 1.3 | 0.5 | 61.0 | 4.7 | 11.2 | 12.0 | 12.2 | 12.3 | 11.4 |
| 48273 | 28780 | 4 | Kleberg ............... | 881.3 | 30,338 | 1,424 | 34.4 | 20.7 | 3.6 | 0.6 | 2.6 | 73.4 | 6.6 | 13.5 | 23.9 | 13.6 | 10.9 | 9.2 |
| 48275 | | 9 | Knox ............... | 850.6 | 3,683 | 2,922 | 4.3 | 58.8 | 5.9 | 1.1 | 0.7 | 35.3 | 6.5 | 15.6 | 11.1 | 11.3 | 11.2 | 10.1 |
| 48277 | 37580 | 5 | Lamar ............... | 907.3 | 49,905 | 993 | 55.0 | 75.7 | 14.5 | 2.7 | 1.4 | 8.8 | 6.4 | 13.4 | 12.0 | 12.6 | 10.9 | 12.2 |
| 48279 | | 6 | Lamb ............... | 1,016.2 | 12,710 | 2,233 | 12.5 | 38.9 | 3.9 | 0.8 | 0.5 | 56.7 | 6.8 | 15.9 | 13.3 | 11.7 | 11.6 | 10.9 |
| 48281 | 28660 | 2 | Lampasas ............... | 712.5 | 21,789 | 1,726 | 30.6 | 73.6 | 4.9 | 1.8 | 2.6 | 19.7 | 5.2 | 12.4 | 11.4 | 10.4 | 11.8 | 12.6 |
| 48283 | | 6 | La Salle ............... | 1,486.7 | 7,500 | 2,623 | 5.0 | 11.5 | 0.9 | 0.6 | 0.4 | 86.8 | 5.4 | 10.6 | 20.0 | 17.7 | 13.4 | 10.2 |
| 48285 | | 6 | Lavaca ............... | 969.7 | 20,216 | 1,810 | 20.8 | 73.6 | 6.4 | 0.5 | 0.8 | 19.7 | 6.0 | 13.4 | 11.4 | 9.9 | 10.5 | 10.7 |
| 48287 | | 6 | Lee ............... | 629.0 | 17,397 | 1,951 | 27.7 | 64.5 | 11.0 | 0.8 | 0.9 | 24.3 | 6.0 | 11.3 | 11.8 | 11.7 | 12.0 | 11.9 |
| 48289 | | 8 | Leon ............... | 1,073.2 | 17,493 | 1,944 | 16.3 | 76.4 | 7.4 | 1.0 | 0.9 | 15.4 | 5.8 | 12.9 | 10.6 | 9.7 | 10.4 | 10.3 |
| 48291 | 26420 | 1 | Liberty ............... | 1,158.4 | 91,547 | 648 | 79.0 | 59.1 | 9.6 | 0.9 | 0.9 | 30.9 | 6.4 | 16.1 | 13.3 | 14.1 | 12.6 | 12.2 |
| 48293 | | 6 | Limestone ............... | 905.4 | 23,340 | 1,665 | 25.8 | 59.1 | 17.7 | 0.8 | 1.1 | 22.7 | 5.8 | 12.4 | 12.1 | 12.8 | 11.8 | 11.0 |
| 48295 | | 9 | Lipscomb ............... | 932.2 | 3,111 | 2,956 | 3.3 | 61.8 | 1.5 | 2.2 | 1.3 | 35.3 | 5.5 | 14.6 | 12.3 | 10.9 | 13.3 | 9.8 |
| 48297 | | 8 | Live Oak ............... | 1,039.7 | 12,324 | 2,265 | 11.9 | 53.1 | 4.6 | 1.1 | 1.0 | 41.1 | 5.9 | 11.0 | 10.6 | 14.1 | 12.5 | 10.5 |
| 48299 | | 7 | Llano ............... | 934.1 | 21,958 | 1,722 | 23.5 | 86.5 | 1.3 | 1.2 | 0.9 | 11.2 | 3.9 | 8.3 | 7.8 | 7.5 | 7.7 | 9.9 |
| 48301 | 37780 | 9 | Loving ............... | 668.8 | 181 | 3,142 | 0.3 | 75.1 | 7.2 | 1.1 | 0.0 | 16.6 | 5.5 | 18.2 | 15.5 | 17.1 | 11.6 | 9.9 |
| 48303 | 31180 | 2 | Lubbock ............... | 895.6 | 314,772 | 225 | 351.5 | 53.4 | 7.7 | 0.8 | 2.9 | 36.6 | 6.3 | 13.3 | 20.5 | 14.5 | 12.3 | 9.8 |
| 48305 | 31180 | 2 | Lynn ............... | 891.9 | 6,025 | 2,744 | 6.8 | 51.9 | 2.6 | 0.7 | 0.4 | 45.5 | 6.6 | 16.1 | 12.0 | 12.4 | 13.3 | 10.2 |
| 48307 | | 7 | McCulloch ............... | 1,065.6 | 7,823 | 2,603 | 7.3 | 63.4 | 2.4 | 1.0 | 0.9 | 33.5 | 5.2 | 12.4 | 11.7 | 11.2 | 10.5 | 11.1 |
| 48309 | 47380 | 2 | McLennan ............... | 1,036.7 | 259,730 | 269 | 250.5 | 56.7 | 14.9 | 0.7 | 2.3 | 27.2 | 6.6 | 13.6 | 18.3 | 13.2 | 11.7 | 10.4 |
| 48311 | | 9 | McMullen ............... | 1,139.8 | 721 | 3,127 | 0.6 | 55.8 | 2.4 | 0.8 | 0.4 | 41.5 | 6.2 | 10.1 | 10.5 | 9.4 | 10.4 | 12.6 |
| 48313 | | 6 | Madison ............... | 466.1 | 14,427 | 2,127 | 31.0 | 55.0 | 19.8 | 0.8 | 1.4 | 24.5 | 5.7 | 12.2 | 15.2 | 17.6 | 13.0 | 10.4 |
| 48315 | | 8 | Marion ............... | 380.9 | 9,960 | 2,421 | 26.1 | 73.1 | 21.2 | 2.1 | 1.3 | 4.6 | 4.3 | 10.7 | 9.2 | 9.8 | 10.1 | 11.7 |
| 48317 | 33260 | 3 | Martin ............... | 915.0 | 5,816 | 2,758 | 6.4 | 49.5 | 2.1 | 0.9 | 0.7 | 47.6 | 7.8 | 18.9 | 13.5 | 13.0 | 13.1 | 11.3 |

1. CBSA = Core Based Statistical Area. See Appendix A for explanation. See Appendix B for list of metropolitan areas with component counties. Service of USDA Rural-Urban Continuum Codes. See Appendix A for definition.    3. Dry land or land partially or temporarily covered by water.    2. County type code from the Economic Research    4. May be of any race.

# Table B. States and Counties — **Population and Households**

| STATE County | Population, 2020 (cont.) Age (percent) (cont.) 55 to 64 years | 65 to 74 years | 75 years and over | Percent female | Population change, 2000-2020 Total persons 2000 | 2010 | Percent change 2000-2010 | 2010-2020 | Components of change, 2010-2020 Births | Deaths | Net Migration | Households, 2015-2019 Number | Persons per household | Family house-holds | Percent Female family house-holder[1] | One person |
|---|---|---|---|---|---|---|---|---|---|---|---|---|---|---|---|---|
| | 16 | 17 | 18 | 19 | 20 | 21 | 22 | 23 | 24 | 25 | 26 | 27 | 28 | 29 | 30 | 31 |
| **TEXAS—Cont'd** | | | | | | | | | | | | | | | | |
| Guadalupe | 11.8 | 8.7 | 5.9 | 50.5 | 89,023 | 131,522 | 47.7 | 29.7 | 18,277 | 10,558 | 31,082 | 54,110 | 2.90 | 76.4 | 12.0 | 19.8 |
| Hale | 10.9 | 7.7 | 6.2 | 47.9 | 36,602 | 36,207 | -1.1 | -9.5 | 4,941 | 3,179 | -5,285 | 10,933 | 2.89 | 72.0 | 14.7 | 25.9 |
| Hall | 14.6 | 12.9 | 12.5 | 50.3 | 3,782 | 3,353 | -11.3 | -12.3 | 288 | 464 | -240 | 1,267 | 2.35 | 63.4 | 13.0 | 33.2 |
| Hamilton | 14.5 | 13.2 | 12.5 | 50.3 | 8,229 | 8,515 | 3.5 | 0.5 | 906 | 1,383 | 523 | 2,989 | 2.70 | 70.1 | 7.3 | 28.3 |
| Hansford | 11.9 | 8.4 | 7.2 | 49.4 | 5,369 | 5,613 | 4.5 | -6.0 | 767 | 511 | -593 | 1,838 | 2.96 | 74.0 | 7.7 | 25.8 |
| Hardeman | 12.8 | 12.5 | 10.5 | 50.2 | 4,724 | 4,139 | -12.4 | -3.1 | 407 | 471 | -68 | 1,596 | 2.45 | 66.0 | 11.4 | 27.6 |
| Hardin | 13.2 | 10.4 | 7.1 | 50.9 | 48,073 | 54,635 | 13.7 | 6.7 | 7,057 | 5,827 | 2,488 | 20,626 | 2.73 | 72.7 | 11.3 | 24.5 |
| Harris | 10.7 | 7.1 | 4.2 | 50.4 | 3,400,578 | 4,093,109 | 20.4 | 15.8 | 704,169 | 261,529 | 202,597 | 1,605,368 | 2.87 | 68.3 | 15.5 | 26.1 |
| Harrison | 13.4 | 10.7 | 7.3 | 51.1 | 62,110 | 65,639 | 5.7 | 1.1 | 8,279 | 6,695 | -814 | 23,292 | 2.80 | 73.3 | 14.5 | 21.9 |
| Hartley | 11.2 | 8.1 | 7.4 | 39.5 | 5,537 | 6,064 | 9.5 | -10.2 | 645 | 407 | -878 | 1,678 | 2.46 | 72.0 | 3.9 | 25.3 |
| Haskell | 14.5 | 12.2 | 10.7 | 47.2 | 6,093 | 5,902 | -3.1 | -2.5 | 460 | 819 | 208 | 2,146 | 2.45 | 66.3 | 13.4 | 31.6 |
| Hays | 10.3 | 7.7 | 4.1 | 50.4 | 97,589 | 156,978 | 60.9 | 53.8 | 24,260 | 10,314 | 69,556 | 73,437 | 2.81 | 65.8 | 9.3 | 22.8 |
| Hemphill | 11.3 | 9.2 | 6.7 | 50.2 | 3,351 | 3,811 | 13.7 | -0.9 | 600 | 352 | -289 | 1,298 | 3.05 | 79.7 | 7.1 | 17.3 |
| Henderson | 14.7 | 13.0 | 9.7 | 51.0 | 73,277 | 78,567 | 7.2 | 6.7 | 9,226 | 10,909 | 6,951 | 30,757 | 2.59 | 69.8 | 12.1 | 24.9 |
| Hidalgo | 8.5 | 6.4 | 5.1 | 50.9 | 569,463 | 774,734 | 36.0 | 13.0 | 157,532 | 42,913 | -14,013 | 238,345 | 3.55 | 81.0 | 21.9 | 16.2 |
| Hill | 13.8 | 11.9 | 9.1 | 50.2 | 32,321 | 35,096 | 8.6 | 5.4 | 4,204 | 4,325 | 2,052 | 12,992 | 2.67 | 70.0 | 11.9 | 25.7 |
| Hockley | 11.9 | 8.7 | 6.5 | 50.4 | 22,716 | 22,898 | 0.8 | 0.1 | 3,240 | 2,199 | -1,015 | 7,997 | 2.79 | 70.5 | 11.1 | 26.4 |
| Hood | 15.1 | 14.8 | 11.1 | 51.2 | 41,100 | 51,161 | 24.5 | 24.2 | 6,041 | 6,927 | 13,199 | 22,152 | 2.60 | 71.7 | 7.8 | 24.7 |
| Hopkins | 13.3 | 10.7 | 8.2 | 50.9 | 31,960 | 35,166 | 10.0 | 5.7 | 4,720 | 4,056 | 1,351 | 13,424 | 2.68 | 72.7 | 11.8 | 22.9 |
| Houston | 13.8 | 12.4 | 10.7 | 46.5 | 23,185 | 23,731 | 2.4 | -3.8 | 2,289 | 3,103 | -74 | 8,252 | 2.40 | 65.7 | 12.5 | 30.6 |
| Howard | 11.3 | 7.7 | 5.4 | 42.6 | 33,627 | 35,006 | 4.1 | 4.4 | 4,681 | 3,915 | 738 | 11,064 | 2.84 | 65.0 | 12.9 | 30.5 |
| Hudspeth | 10.2 | 8.8 | 6.8 | 46.6 | 3,344 | 3,476 | 3.9 | 41.1 | 429 | 196 | 1,180 | 978 | 3.71 | 74.6 | 17.0 | 23.5 |
| Hunt | 13.2 | 9.7 | 6.7 | 50.7 | 76,596 | 86,144 | 12.5 | 15.9 | 11,555 | 9,534 | 11,646 | 33,189 | 2.75 | 68.6 | 11.5 | 26.1 |
| Hutchinson | 13.0 | 10.5 | 7.3 | 49.4 | 23,857 | 22,249 | -6.7 | -7.1 | 2,899 | 2,447 | -2,029 | 6,929 | 3.06 | 69.2 | 11.8 | 28.0 |
| Irion | 16.0 | 11.6 | 10.3 | 48.8 | 1,771 | 1,605 | -9.4 | -2.6 | 153 | 139 | -56 | 682 | 2.38 | 65.2 | 7.9 | 32.7 |
| Jack | 12.6 | 9.8 | 7.6 | 44.2 | 8,763 | 9,041 | 3.2 | 0.2 | 951 | 926 | -6 | 3,168 | 2.44 | 74.0 | 10.9 | 25.3 |
| Jackson | 12.6 | 10.6 | 8.5 | 50.3 | 14,391 | 14,075 | -2.2 | 5.5 | 2,089 | 1,535 | 223 | 4,917 | 2.97 | 72.4 | 11.1 | 23.3 |
| Jasper | 14.3 | 11.9 | 8.8 | 50.5 | 35,604 | 35,710 | 0.3 | -0.9 | 4,380 | 4,346 | -371 | 12,936 | 2.67 | 68.2 | 13.7 | 29.0 |
| Jeff Davis | 20.2 | 22.7 | 14.7 | 49.7 | 2,207 | 2,342 | 6.1 | -5.2 | 153 | 206 | -72 | 1,054 | 2.07 | 57.7 | 0.9 | 39.1 |
| Jefferson | 12.4 | 8.8 | 6.3 | 48.8 | 252,051 | 252,277 | 0.1 | -0.9 | 35,953 | 25,488 | -12,692 | 92,988 | 2.56 | 64.4 | 15.9 | 30.6 |
| Jim Hogg | 10.6 | 9.6 | 7.6 | 48.8 | 5,281 | 5,300 | 0.4 | -2.2 | 806 | 556 | -369 | 1,626 | 3.20 | 72.5 | 20.7 | 25.3 |
| Jim Wells | 11.5 | 9.2 | 6.9 | 50.4 | 39,326 | 40,867 | 3.9 | -1.0 | 6,214 | 4,111 | -2,524 | 12,987 | 3.12 | 71.3 | 16.3 | 23.9 |
| Johnson | 12.3 | 8.9 | 5.8 | 50.1 | 126,811 | 150,947 | 19.0 | 19.0 | 20,993 | 13,873 | 21,592 | 57,310 | 2.87 | 76.0 | 11.9 | 20.1 |
| Jones | 11.9 | 9.2 | 6.9 | 37.6 | 20,785 | 20,192 | -2.9 | -1.6 | 1,692 | 1,939 | -92 | 5,696 | 2.24 | 70.6 | 11.8 | 24.2 |
| Karnes | 10.9 | 8.3 | 6.4 | 42.1 | 15,446 | 14,828 | -4.0 | 5.0 | 1,680 | 1,523 | 558 | 4,282 | 2.84 | 65.7 | 10.6 | 31.2 |
| Kaufman | 11.1 | 7.4 | 4.4 | 50.7 | 71,313 | 103,352 | 44.9 | 38.6 | 16,490 | 9,595 | 32,899 | 38,015 | 3.22 | 78.4 | 13.3 | 18.1 |
| Kendall | 14.5 | 11.4 | 8.2 | 50.9 | 23,743 | 33,385 | 40.6 | 45.3 | 3,704 | 3,763 | 15,077 | 14,253 | 3.04 | 77.4 | 8.9 | 20.0 |
| Kenedy | 13.7 | 13.2 | 9.5 | 47.8 | 414 | 413 | -0.2 | -8.2 | 36 | 28 | -46 | 197 | 2.86 | 77.7 | 31.5 | 22.3 |
| Kent | 15.1 | 11.2 | 13.7 | 50.3 | 859 | 808 | -5.9 | -2.7 | 74 | 126 | 26 | 273 | 2.20 | 59.3 | 6.2 | 36.3 |
| Kerr | 14.5 | 15.2 | 13.2 | 51.8 | 43,653 | 49,643 | 13.7 | 6.5 | 5,294 | 7,469 | 5,401 | 21,403 | 2.34 | 68.0 | 11.1 | 28.5 |
| Kimble | 14.9 | 17.3 | 13.2 | 51.1 | 4,468 | 4,605 | 3.1 | -4.5 | 451 | 584 | -76 | 1,783 | 2.43 | 71.3 | 8.3 | 27.3 |
| King | 14.5 | 9.2 | 9.5 | 48.4 | 356 | 285 | -19.9 | -0.7 | 23 | 13 | -11 | 77 | 3.08 | 61.0 | 15.6 | 28.6 |
| Kinney | 10.6 | 12.7 | 12.9 | 44.7 | 3,379 | 3,598 | 6.5 | 2.0 | 362 | 390 | 94 | 1,337 | 2.55 | 49.5 | 7.6 | 44.0 |
| Kleberg | 8.6 | 7.5 | 6.2 | 49.2 | 31,549 | 32,061 | 1.6 | -5.4 | 4,547 | 2,657 | -3,658 | 10,955 | 2.66 | 63.7 | 17.1 | 28.0 |
| Knox | 13.8 | 10.3 | 10.1 | 50.5 | 4,253 | 3,719 | -12.6 | -1.0 | 487 | 538 | 14 | 1,375 | 2.59 | 70.0 | 11.5 | 28.6 |
| Lamar | 13.1 | 10.9 | 8.6 | 51.8 | 48,499 | 49,791 | 2.7 | 0.0 | 6,699 | 6,414 | -156 | 19,793 | 2.47 | 66.7 | 14.1 | 29.4 |
| Lamb | 12.3 | 9.3 | 8.2 | 50.4 | 14,709 | 13,976 | -5.0 | -9.1 | 1,915 | 1,579 | -1,621 | 4,728 | 2.73 | 69.1 | 13.5 | 28.6 |
| Lampasas | 15.4 | 12.2 | 8.6 | 50.6 | 17,762 | 19,674 | 10.8 | 10.8 | 2,291 | 2,095 | 1,922 | 7,807 | 2.64 | 73.9 | 10.7 | 23.4 |
| La Salle | 8.5 | 7.6 | 6.4 | 40.8 | 5,866 | 6,886 | 17.4 | 8.9 | 966 | 552 | 185 | 2,269 | 2.85 | 71.8 | 12.6 | 25.3 |
| Lavaca | 14.0 | 13.1 | 11.0 | 51.1 | 19,210 | 19,275 | 0.3 | 4.9 | 2,284 | 2,586 | 1,251 | 7,826 | 2.50 | 71.7 | 12.2 | 25.9 |
| Lee | 15.5 | 11.8 | 8.0 | 49.6 | 15,657 | 16,609 | 6.1 | 4.7 | 2,114 | 1,744 | 427 | 6,036 | 2.74 | 74.3 | 11.9 | 23.3 |
| Leon | 15.1 | 14.1 | 11.1 | 50.4 | 15,335 | 16,799 | 9.5 | 4.1 | 2,089 | 2,252 | 865 | 6,443 | 2.66 | 67.7 | 12.3 | 28.4 |
| Liberty | 11.5 | 8.2 | 4.7 | 50.6 | 70,154 | 75,657 | 7.8 | 21.0 | 11,583 | 8,020 | 12,366 | 26,873 | 2.84 | 74.1 | 11.7 | 25.0 |
| Limestone | 13.7 | 11.9 | 8.7 | 48.7 | 22,051 | 23,388 | 6.1 | -0.2 | 2,965 | 2,761 | -238 | 8,377 | 2.68 | 69.9 | 13.7 | 27.2 |
| Lipscomb | 14.6 | 10.4 | 8.6 | 48.6 | 3,057 | 3,302 | 8.0 | -5.8 | 445 | 308 | -330 | 1,192 | 2.82 | 74.6 | 13.7 | 21.0 |
| Live Oak | 13.7 | 11.6 | 10.0 | 45.0 | 12,309 | 11,527 | -6.4 | 6.9 | 1,331 | 1,240 | 698 | 3,752 | 2.85 | 65.2 | 6.6 | 30.0 |
| Llano | 16.9 | 20.9 | 17.0 | 51.5 | 17,044 | 19,303 | 13.3 | 13.8 | 1,645 | 3,231 | 4,205 | 8,678 | 2.40 | 66.2 | 7.1 | 29.5 |
| Loving | 9.9 | 8.8 | 3.3 | 42.0 | 67 | 82 | 22.4 | 120.7 | 15 | 14 | 96 | 30 | 3.27 | 66.7 | 0.0 | 26.7 |
| Lubbock | 10.2 | 7.6 | 5.4 | 50.8 | 242,628 | 278,946 | 15.0 | 12.8 | 41,349 | 24,825 | 19,229 | 113,488 | 2.58 | 61.7 | 13.6 | 28.1 |
| Lynn | 12.6 | 9.1 | 7.8 | 47.9 | 6,550 | 5,915 | -9.7 | 1.9 | 751 | 567 | -78 | 2,154 | 2.68 | 71.3 | 12.2 | 26.5 |
| McCulloch | 13.8 | 13.8 | 10.4 | 50.1 | 8,205 | 8,284 | 1.0 | -5.6 | 922 | 1,132 | -247 | 3,143 | 2.51 | 65.8 | 11.0 | 30.0 |
| McLennan | 11.1 | 8.7 | 6.4 | 51.3 | 213,517 | 234,895 | 10.0 | 10.6 | 35,539 | 21,730 | 11,137 | 90,054 | 2.68 | 66.4 | 15.6 | 26.3 |
| McMullen | 14.4 | 11.4 | 14.8 | 47.6 | 851 | 707 | -16.9 | 2.0 | 88 | 88 | 9 | 270 | 2.87 | 77.8 | 13.3 | 19.3 |
| Madison | 9.8 | 9.1 | 7.1 | 43.0 | 12,940 | 13,669 | 5.6 | 5.5 | 1,661 | 1,378 | 477 | 4,269 | 2.59 | 66.0 | 11.0 | 31.2 |
| Marion | 17.2 | 15.8 | 11.1 | 51.1 | 10,941 | 10,542 | -3.6 | -5.5 | 913 | 1,678 | 190 | 4,715 | 2.10 | 61.8 | 10.1 | 35.5 |
| Martin | 10.8 | 6.7 | 4.9 | 49.6 | 4,746 | 4,799 | 1.1 | 21.2 | 927 | 454 | 531 | 1,694 | 3.30 | 73.5 | 9.8 | 25.3 |

1. No spouse present.

# Table B. States and Counties — Population, Vital Statistics, Health, and Crime

| STATE County | Persons in group quarters, 2020 | Daytime Population, 2015-2019 | | Births, 2020 | | Deaths, 2020 | | Persons under 65 with no health insurance, 2019 | | Medicare, 2020 | | | Crimes reported by county police or sheriff, 2019 | |
|---|---|---|---|---|---|---|---|---|---|---|---|---|---|---|
| | | Number | Employment/residence ratio | Total | Rate[1] | Number | Rate[1] | Number | Percent | Total beneficiaries | Enrolled in Original Medicare | Enrolled in Medicare Advantage | Violent | Property |
| | 32 | 33 | 34 | 35 | 36 | 37 | 38 | 39 | 40 | 41 | 42 | 43 | 44 | 45 |
| TEXAS—Cont'd | | | | | | | | | | | | | | |
| Guadalupe | 1,871 | 129,399 | 0.61 | 1,874 | 11.0 | 1,246 | 7.3 | 22,560 | 15.9 | 27,825 | 18,063 | 9,762 | 69 | 424 |
| Hale | 2,522 | 33,159 | 0.95 | 428 | 13.1 | 318 | 9.7 | 6,257 | 24.1 | 5,516 | 3,501 | 2,016 | 16 | 64 |
| Hall | 42 | 2,879 | 0.86 | 26 | 8.8 | 32 | 10.9 | 668 | 30.2 | 806 | 582 | 224 | 3 | 7 |
| Hamilton | 178 | 8,163 | 0.95 | 80 | 9.3 | 121 | 14.1 | 1,550 | 24.8 | 2,326 | 1,397 | 929 | 4 | 15 |
| Hansford | 61 | 5,301 | 0.91 | 64 | 12.1 | 41 | 7.8 | 1,348 | 29.6 | 928 | 775 | 153 | 1 | 11 |
| Hardeman | 21 | 3,563 | 0.77 | 38 | 9.5 | 44 | 11.0 | 801 | 26.7 | 970 | 724 | 246 | 5 | 13 |
| Hardin | 404 | 45,305 | 0.52 | 671 | 11.5 | 614 | 10.5 | 7,772 | 16.2 | 11,638 | 6,492 | 5,146 | 38 | 285 |
| Harris | 47,049 | 4,963,672 | 1.14 | 66,344 | 14.0 | 31,172 | 6.6 | 998,337 | 24.2 | 562,657 | 278,734 | 283,923 | 7,329 | 42,526 |
| Harrison | 1,054 | 61,742 | 0.83 | 709 | 10.7 | 755 | 11.4 | 10,752 | 20.0 | 13,585 | 8,348 | 5,237 | 52 | 445 |
| Hartley | 1,125 | 6,114 | 1.22 | 71 | 13.0 | 25 | 4.6 | 837 | 22.9 | 664 | 506 | 157 | 0 | 10 |
| Haskell | 531 | 5,655 | 0.97 | 37 | 6.4 | 58 | 10.1 | 1,042 | 26.9 | 1,352 | 981 | 371 | 1 | 8 |
| Hays | 7,477 | 189,608 | 0.78 | 2,642 | 10.9 | 1,277 | 5.3 | 32,880 | 16.7 | 31,817 | 19,990 | 11,827 | 111 | 672 |
| Hemphill | 37 | 4,410 | 1.23 | 43 | 11.4 | 32 | 8.5 | 748 | 23.3 | 626 | 517 | 109 | 2 | 13 |
| Henderson | 1,216 | 71,827 | 0.71 | 903 | 10.8 | 1,222 | 14.6 | 14,596 | 23.2 | 21,109 | 12,513 | 8,597 | 141 | 584 |
| Hidalgo | 7,029 | 838,636 | 0.95 | 14,081 | 16.1 | 5,110 | 5.8 | 248,170 | 33.2 | 107,772 | 43,160 | 64,612 | 518 | 3,241 |
| Hill | 652 | 32,057 | 0.76 | 419 | 11.3 | 434 | 11.7 | 6,420 | 22.6 | 8,834 | 5,276 | 3,558 | 31 | 234 |
| Hockley | 798 | 22,124 | 0.91 | 302 | 13.2 | 241 | 10.5 | 4,151 | 22.1 | 3,953 | 2,346 | 1,608 | 13 | 71 |
| Hood | 673 | 54,128 | 0.83 | 575 | 9.1 | 747 | 11.8 | 8,201 | 17.8 | 17,478 | 10,923 | 6,555 | 55 | 356 |
| Hopkins | 413 | 34,588 | 0.88 | 471 | 12.7 | 432 | 11.6 | 7,026 | 23.5 | 8,016 | 5,758 | 2,257 | 27 | 61 |
| Houston | 2,661 | 21,976 | 0.88 | 191 | 8.4 | 327 | 14.3 | 3,283 | 21.7 | 5,529 | 3,596 | 1,933 | 2 | 102 |
| Howard | 5,820 | 36,987 | 1.05 | 454 | 12.4 | 351 | 9.6 | 5,082 | 19.1 | 5,323 | 3,792 | 1,531 | 53 | 273 |
| Hudspeth | 86 | 4,438 | 1.02 | 39 | 7.9 | 19 | 3.9 | 1,266 | 32.0 | 725 | 352 | 373 | 8 | 14 |
| Hunt | 2,658 | 87,951 | 0.85 | 1,187 | 11.9 | 1,093 | 11.0 | 15,798 | 19.7 | 18,747 | 12,894 | 5,852 | 350 | 570 |
| Hutchinson | 229 | 22,347 | 1.12 | 235 | 11.4 | 229 | 11.1 | 3,472 | 20.2 | 4,117 | 3,151 | 966 | 11 | 71 |
| Irion | 0 | 2,088 | 1.69 | 14 | 9.0 | 10 | 6.4 | 171 | 14.0 | 358 | 270 | 89 | NA | NA |
| Jack | 1,126 | 8,154 | 0.78 | 80 | 8.8 | 76 | 8.4 | 1,550 | 24.5 | 1,635 | 1,188 | 447 | 9 | 36 |
| Jackson | 222 | 14,508 | 0.95 | 215 | 14.5 | 141 | 9.5 | 2,632 | 22.1 | 3,141 | 2,240 | 901 | 6 | 51 |
| Jasper | 931 | 33,092 | 0.82 | 374 | 10.6 | 455 | 12.9 | 4,859 | 17.6 | 8,411 | 5,319 | 3,092 | 72 | 403 |
| Jeff Davis | 87 | 2,488 | 1.33 | 17 | 7.7 | 31 | 14.0 | 373 | 27.1 | 719 | 532 | 187 | NA | NA |
| Jefferson | 15,585 | 280,416 | 1.25 | 3,346 | 13.4 | 2,718 | 10.9 | 42,811 | 21.6 | 43,337 | 22,938 | 20,399 | 59 | 358 |
| Jim Hogg | 17 | 5,156 | 0.96 | 49 | 9.5 | 61 | 11.8 | 774 | 18.3 | 963 | 563 | 401 | 5 | 20 |
| Jim Wells | 358 | 40,692 | 0.98 | 514 | 12.7 | 409 | 10.1 | 7,341 | 21.8 | 7,898 | 3,633 | 4,265 | NA | NA |
| Johnson | 2,588 | 141,975 | 0.67 | 2,158 | 12.0 | 1,579 | 8.8 | 30,040 | 20.2 | 30,912 | 15,225 | 15,687 | 205 | 660 |
| Jones | 5,090 | 19,391 | 0.89 | 154 | 7.7 | 194 | 9.8 | 2,407 | 20.4 | 3,470 | 2,355 | 1,115 | 4 | 60 |
| Karnes | 2,660 | 19,196 | 1.70 | 144 | 9.3 | 160 | 10.3 | 2,028 | 19.5 | 2,706 | 1,817 | 889 | 3 | 99 |
| Kaufman | 1,251 | 102,248 | 0.63 | 1,953 | 13.6 | 1,174 | 8.2 | 20,317 | 17.1 | 21,274 | 12,791 | 8,483 | 126 | 767 |
| Kendall | 510 | 42,799 | 0.95 | 396 | 8.2 | 484 | 10.0 | 6,399 | 16.7 | 10,306 | 7,097 | 3,209 | 19 | 104 |
| Kenedy | 0 | 700 | 1.64 | 2 | 5.3 | 0 | 0.0 | 88 | 26.3 | 48 | 31 | 17 | 0 | 8 |
| Kent | 52 | 738 | 1.38 | 8 | 10.2 | 14 | 17.8 | 77 | 14.1 | 205 | 131 | 73 | NA | NA |
| Kerr | 1,891 | 51,467 | 0.98 | 504 | 9.5 | 783 | 14.8 | 8,656 | 23.8 | 16,530 | 12,512 | 4,017 | 51 | 199 |
| Kimble | 16 | 4,250 | 0.94 | 56 | 12.7 | 59 | 13.4 | 718 | 23.8 | 1,288 | 976 | 313 | 2 | 4 |
| King | 0 | 281 | 1.39 | 2 | 7.1 | 0 | 0.0 | 48 | 21.6 | 36 | 25 | 12 | NA | NA |
| Kinney | 334 | 3,884 | 1.18 | 37 | 10.1 | 41 | 11.2 | 446 | 18.6 | 917 | 590 | 327 | 0 | 1 |
| Kleberg | 1,362 | 30,940 | 1.00 | 406 | 13.4 | 253 | 8.3 | 4,959 | 19.8 | 5,067 | 2,334 | 2,733 | NA | NA |
| Knox | 104 | 3,555 | 0.90 | 47 | 12.8 | 37 | 10.0 | 717 | 25.0 | 853 | 624 | 229 | NA | NA |
| Lamar | 645 | 51,260 | 1.08 | 628 | 12.6 | 723 | 14.5 | 8,631 | 21.8 | 11,546 | 9,151 | 2,394 | 28 | 135 |
| Lamb | 179 | 12,110 | 0.82 | 157 | 12.4 | 153 | 12.0 | 2,810 | 26.9 | 2,636 | 1,699 | 938 | 6 | 61 |
| Lampasas | 214 | 17,757 | 0.65 | 209 | 9.6 | 212 | 9.7 | 3,143 | 18.5 | 5,390 | 3,602 | 1,788 | 12 | 83 |
| La Salle | 1,634 | 9,909 | 1.96 | 92 | 12.3 | 40 | 5.3 | 887 | 18.3 | 1,163 | 638 | 525 | 5 | 44 |
| Lavaca | 420 | 18,069 | 0.78 | 222 | 11.0 | 259 | 12.8 | 2,881 | 18.9 | 5,345 | 4,024 | 1,321 | 10 | 63 |
| Lee | 470 | 16,217 | 0.89 | 228 | 13.1 | 162 | 9.3 | 3,018 | 22.1 | 3,720 | 2,453 | 1,266 | 25 | 86 |
| Leon | 107 | 16,871 | 0.94 | 192 | 11.0 | 240 | 13.7 | 3,408 | 26.3 | 5,036 | 3,229 | 1,807 | 11 | 116 |
| Liberty | 4,845 | 73,914 | 0.68 | 1,314 | 14.4 | 858 | 9.4 | 16,352 | 22.8 | 13,937 | 6,958 | 6,979 | NA | NA |
| Limestone | 1,571 | 22,401 | 0.89 | 244 | 10.5 | 298 | 12.8 | 3,920 | 22.6 | 5,200 | 3,291 | 1,909 | 14 | 76 |
| Lipscomb | 34 | 3,234 | 0.90 | 36 | 11.6 | 43 | 13.8 | 759 | 29.2 | 608 | 486 | 122 | NA | NA |
| Live Oak | 1,156 | 12,883 | 1.17 | 143 | 11.6 | 113 | 9.2 | 1,932 | 22.4 | 2,539 | 1,536 | 1,003 | NA | NA |
| Llano | 127 | 19,753 | 0.83 | 167 | 7.6 | 381 | 17.4 | 3,176 | 23.3 | 7,529 | 5,022 | 2,508 | 29 | 145 |
| Loving | 0 | 807 | 22.48 | 0 | 0.0 | 0 | 0.0 | 17 | 11.4 | 15 | D | D | NA | NA |
| Lubbock | 12,280 | 307,947 | 1.02 | 3,865 | 12.3 | 2,762 | 8.8 | 45,935 | 17.7 | 47,635 | 27,868 | 19,767 | 16 | 331 |
| Lynn | 42 | 4,992 | 0.64 | 67 | 11.1 | 53 | 8.8 | 1,037 | 21.2 | 1,077 | 681 | 396 | 0 | 3 |
| McCulloch | 104 | 7,771 | 0.92 | 86 | 11.0 | 119 | 15.2 | 1,290 | 21.0 | 2,138 | 1,601 | 536 | 3 | 41 |
| McLennan | 8,550 | 258,563 | 1.07 | 3,456 | 13.3 | 2,345 | 9.0 | 38,118 | 18.0 | 45,262 | 26,635 | 18,626 | 69 | 500 |
| McMullen | 0 | 1,683 | 3.95 | 9 | 12.5 | 12 | 16.6 | 76 | 13.6 | 190 | 126 | 64 | NA | NA |
| Madison | 2,582 | 14,093 | 0.98 | 181 | 12.5 | 139 | 9.6 | 2,288 | 24.0 | 2,458 | 1,536 | 922 | 8 | 66 |
| Marion | 144 | 8,707 | 0.65 | 73 | 7.3 | 154 | 15.5 | 1,420 | 20.2 | 3,051 | 1,928 | 1,123 | NA | NA |
| Martin | 20 | 5,853 | 1.08 | 88 | 15.1 | 43 | 7.4 | 1,038 | 20.4 | 753 | 559 | 194 | NA | NA |

1. Per 1,000 estimated resident population.

# Table B. States and Counties — Crime, Education, Money Income, and Poverty

| STATE County | Officers | Civilians | Total | Percent private | High school graduate or less | Bachelor's degree or more | Total current spending (mil dol) | Current spending per student (dollars) | Per capita income[4] | Median income (dollars) | Percent with income of less than $50,000 | Percent with income of $200,000 or more | Median household income (dollars) | All persons | Children under 18 years | Children 5 to 17 years in families |
|---|---|---|---|---|---|---|---|---|---|---|---|---|---|---|---|---|
| | 46 | 47 | 48 | 49 | 50 | 51 | 52 | 53 | 54 | 55 | 56 | 57 | 58 | 59 | 60 | 61 |
| **TEXAS—Cont'd** | | | | | | | | | | | | | | | | |
| Guadalupe | 104 | 119 | 42,170 | 11.3 | 41.8 | 28.2 | 225 | 8,472 | 31,842 | 74,496 | 31.7 | 5.7 | 78,801 | 7.4 | 11.2 | 10.2 |
| Hale | 64 | 0 | 9,394 | 11.1 | 58.1 | 17.1 | 69 | 9,501 | 20,986 | 48,384 | 52.5 | 1.9 | 49,187 | 17.9 | 23.9 | 22.1 |
| Hall | 4 | 6 | 701 | 3.4 | 50.3 | 15.9 | 6 | 11,934 | 21,907 | 34,673 | 62.3 | 2.1 | 35,940 | 25.4 | 40.5 | 35.5 |
| Hamilton | NA | NA | 1,767 | 9.1 | 51.8 | 22.4 | 17 | 9,653 | 26,930 | 50,825 | 49.4 | 4.3 | 52,505 | 14.3 | 21.9 | 20.7 |
| Hansford | 7 | 6 | 1,490 | 9.3 | 58.9 | 15.1 | 16 | 11,313 | 20,410 | 38,000 | 60.2 | 1.0 | 58,703 | 11.7 | 16.8 | 15.5 |
| Hardeman | 8 | 4 | 764 | 3.9 | 53.8 | 13.7 | 10 | 13,377 | 24,078 | 41,859 | 56.3 | 2.4 | 40,057 | 19.2 | 32.9 | 29.1 |
| Hardin | 34 | 36 | 13,194 | 14.7 | 50.2 | 16.8 | 97 | 8,893 | 30,399 | 60,339 | 41.0 | 5.4 | 62,616 | 12.1 | 16.2 | 15.8 |
| Harris | 2,274 | 2,494 | 1,284,203 | 11.8 | 42.0 | 31.5 | 7,951 | 9,544 | 32,765 | 61,705 | 40.9 | 8.7 | 61,638 | 15.0 | 21.3 | 20.2 |
| Harrison | 45 | 63 | 16,468 | 11.9 | 48.7 | 19.1 | 117 | 8,878 | 27,086 | 52,220 | 47.6 | 3.7 | 55,171 | 17.8 | 26.8 | 26.1 |
| Hartley | 6 | 1 | 1,009 | 12.4 | 55.0 | 17.8 | 5 | 11,664 | 20,563 | 58,298 | 43.9 | 1.8 | 72,796 | 8.6 | 8.3 | 8.1 |
| Haskell | 3 | 9 | 1,162 | 7.5 | 64.8 | 11.9 | 12 | 14,584 | 20,394 | 40,313 | 58.6 | 0.9 | 38,613 | 20.7 | 29.0 | 26.7 |
| Hays | 155 | 204 | 67,753 | 7.8 | 33.3 | 37.2 | 339 | 9,022 | 32,118 | 68,717 | 36.5 | 7.6 | 72,890 | 11.2 | 11.3 | 10.4 |
| Hemphill | 8 | 9 | 1,203 | 1.8 | 49.4 | 21.2 | 10 | 11,113 | 31,530 | 70,625 | 36.7 | 5.5 | 71,299 | 9.1 | 11.7 | 10.3 |
| Henderson | 80 | 84 | 16,812 | 5.6 | 50.9 | 17.3 | 95 | 9,436 | 26,121 | 47,355 | 51.9 | 3.3 | 49,896 | 17.5 | 30.3 | 29.7 |
| Hidalgo | 282 | 512 | 275,356 | 5.3 | 57.8 | 18.7 | 2,547 | 10,424 | 17,175 | 40,014 | 58.5 | 2.6 | 41,656 | 26.9 | 37.4 | 36.9 |
| Hill | 34 | 41 | 8,319 | 6.2 | 51.4 | 17.4 | 70 | 10,541 | 26,370 | 53,307 | 47.4 | 4.0 | 53,210 | 13.1 | 18.1 | 17.6 |
| Hockley | 13 | 15 | 6,436 | 5.7 | 50.5 | 15.8 | 52 | 10,603 | 23,604 | 48,248 | 51.8 | 2.9 | 51,149 | 15.5 | 19.7 | 19.4 |
| Hood | 50 | 88 | 10,573 | 14.5 | 40.6 | 26.4 | 73 | 8,774 | 35,606 | 64,041 | 39.9 | 6.4 | 74,534 | 9.2 | 14.8 | 13.2 |
| Hopkins | 28 | 38 | 8,724 | 9.2 | 50.7 | 20.1 | 66 | 9,701 | 26,821 | 52,078 | 46.6 | 4.1 | 50,332 | 14.7 | 20.2 | 19.4 |
| Houston | 21 | 29 | 4,386 | 9.8 | 55.0 | 15.2 | 34 | 10,723 | 19,129 | 37,904 | 63.5 | 1.7 | 43,601 | 20.9 | 28.8 | 25.3 |
| Howard | 22 | 39 | 7,796 | 6.7 | 53.7 | 12.6 | 54 | 9,446 | 24,547 | 54,208 | 46.6 | 4.0 | 61,112 | 17.0 | 18.6 | 17.8 |
| Hudspeth | 14 | 28 | 1,023 | 1.5 | 75.1 | 10.7 | 9 | 15,866 | 14,239 | 31,677 | 75.4 | 1.8 | 39,309 | 28.0 | 41.2 | 38.7 |
| Hunt | 48 | 92 | 23,980 | 8.4 | 49.5 | 19.9 | 145 | 9,593 | 25,926 | 54,959 | 45.5 | 3.3 | 57,513 | 14.8 | 20.3 | 17.7 |
| Hutchinson | 14 | 21 | 5,298 | 7.2 | 48.5 | 14.0 | 43 | 10,069 | 24,434 | 52,524 | 48.4 | 2.6 | 57,936 | 11.4 | 16.3 | 14.9 |
| Irion | NA | NA | 419 | 2.1 | 44.5 | 15.0 | 4 | 15,671 | 28,487 | 47,500 | 50.7 | 1.6 | 68,630 | 7.4 | 9.0 | 8.1 |
| Jack | 15 | 16 | 1,734 | 12.5 | 58.0 | 13.5 | 18 | 10,832 | 24,956 | 52,045 | 47.4 | 4.0 | 55,358 | 15.6 | 19.4 | 17.0 |
| Jackson | 15 | 20 | 3,674 | 9.3 | 48.7 | 16.5 | 36 | 10,064 | 27,051 | 62,806 | 38.9 | 3.9 | 55,800 | 13.8 | 18.4 | 17.6 |
| Jasper | 20 | 4 | 7,543 | 6.8 | 59.7 | 12.9 | 57 | 9,411 | 25,269 | 44,370 | 53.9 | 3.1 | 47,252 | 19.5 | 28.2 | 25.2 |
| Jeff Davis | 4 | 1 | 447 | 13.2 | 41.3 | 30.5 | 4 | 16,205 | 26,531 | 53,088 | 48.5 | 0.4 | 53,362 | 12.7 | 39.0 | 45.8 |
| Jefferson | 104 | 288 | 60,951 | 7.1 | 47.7 | 19.7 | 415 | 9,940 | 27,094 | 51,248 | 48.8 | 4.2 | 55,173 | 16.9 | 24.4 | 24.2 |
| Jim Hogg | NA | NA | 1,306 | 0.3 | 63.1 | 13.4 | 14 | 12,064 | 16,864 | 33,382 | 63.2 | 1.4 | 38,788 | 22.8 | 35.0 | 31.7 |
| Jim Wells | NA | NA | 10,760 | 3.6 | 61.7 | 12.2 | 77 | 9,348 | 21,510 | 41,505 | 55.9 | 2.8 | 44,444 | 21.0 | 29.0 | 28.7 |
| Johnson | 86 | 39 | 34,775 | 13.0 | 50.2 | 18.7 | 320 | 9,027 | 28,579 | 64,359 | 36.8 | 4.2 | 66,538 | 11.0 | 15.9 | 15.7 |
| Jones | 9 | 26 | 3,493 | 9.2 | 60.2 | 11.7 | 30 | 11,104 | 18,081 | 50,344 | 49.7 | 2.4 | 51,094 | 20.4 | 22.0 | 19.9 |
| Karnes | 31 | 31 | 3,048 | 11.8 | 61.2 | 15.3 | 30 | 11,479 | 27,612 | 56,127 | 44.4 | 9.4 | 54,843 | 21.0 | 27.0 | 24.5 |
| Kaufman | 77 | 114 | 33,522 | 12.7 | 46.6 | 20.5 | 251 | 8,667 | 28,634 | 70,107 | 35.0 | 4.9 | 75,009 | 9.8 | 13.8 | 13.6 |
| Kendall | NA | NA | 10,992 | 15.6 | 27.8 | 42.1 | 86 | 8,664 | 45,866 | 84,747 | 25.8 | 15.7 | 94,899 | 7.4 | 10.9 | 8.2 |
| Kenedy | NA | NA | 98 | 1.0 | 93.7 | 1.0 | 2 | 28,577 | 15,211 | 38,021 | 72.6 | 0.0 | 46,132 | 14.2 | 21.4 | 16.9 |
| Kent | 2 | 3 | 145 | 0.0 | 46.8 | 25.4 | 3 | 20,322 | 30,543 | 44,688 | 53.5 | 10.6 | 52,591 | 11.5 | 14.1 | 11.9 |
| Kerr | 48 | 63 | 9,986 | 16.8 | 38.6 | 27.4 | 62 | 8,890 | 32,988 | 55,990 | 45.1 | 4.2 | 57,789 | 12.5 | 20.4 | 19.9 |
| Kimble | NA | NA | 727 | 2.8 | 47.8 | 22.5 | 6 | 10,340 | 30,099 | 43,328 | 57.8 | 5.9 | 44,757 | 17.0 | 32.6 | 31.7 |
| King | 2 | 0 | 74 | 0.0 | 55.1 | 20.9 | 3 | 28,769 | 26,199 | 52,083 | 46.8 | 1.3 | 71,256 | 11.0 | 16.7 | 15.5 |
| Kinney | 7 | 7 | 645 | 11.6 | 58.2 | 12.1 | 7 | 12,219 | 23,934 | 26,738 | 65.1 | 0.8 | 47,076 | 18.9 | 24.7 | 23.4 |
| Kleberg | NA | NA | 10,424 | 6.6 | 48.2 | 24.6 | 55 | 10,494 | 22,646 | 43,730 | 57.3 | 2.2 | 47,169 | 23.3 | 31.0 | 29.1 |
| Knox | NA | NA | 907 | 6.9 | 52.4 | 20.1 | 11 | 13,860 | 23,146 | 48,798 | 50.9 | 1.7 | 40,808 | 19.1 | 30.8 | 28.4 |
| Lamar | NA | NA | 11,025 | 6.5 | 48.8 | 18.7 | 85 | 9,854 | 25,038 | 45,117 | 53.9 | 2.6 | 47,061 | 18.4 | 27.3 | 26.5 |
| Lamb | 10 | 18 | 3,277 | 2.5 | 57.5 | 14.4 | 32 | 10,822 | 22,337 | 45,111 | 53.8 | 2.8 | 42,719 | 18.5 | 26.1 | 23.4 |
| Lampasas | NA | NA | 4,829 | 11.2 | 38.4 | 19.9 | 34 | 9,464 | 29,412 | 60,772 | 41.2 | 3.9 | 58,817 | 12.4 | 19.2 | 18.3 |
| La Salle | 32 | 97 | 1,574 | 9.2 | 75.0 | 7.5 | 18 | 12,946 | 21,110 | 50,151 | 49.8 | 4.2 | 41,683 | 26.6 | 27.4 | 27.3 |
| Lavaca | 15 | 19 | 4,768 | 12.7 | 55.2 | 16.7 | 44 | 10,918 | 32,188 | 54,403 | 45.4 | 6.0 | 52,632 | 11.0 | 14.5 | 13.5 |
| Lee | 14 | 26 | 4,053 | 6.3 | 51.5 | 14.6 | 30 | 9,536 | 27,227 | 54,744 | 44.8 | 3.4 | 59,250 | 10.7 | 16.1 | 15.4 |
| Leon | 26 | 21 | 3,646 | 4.1 | 58.2 | 14.0 | 34 | 10,550 | 30,129 | 43,045 | 55.3 | 4.7 | 53,274 | 14.4 | 22.2 | 20.8 |
| Liberty | 71 | 24 | 19,352 | 8.7 | 60.5 | 9.7 | 160 | 9,412 | 23,461 | 51,494 | 48.2 | 3.9 | 43,644 | 14.4 | 19.0 | 18.7 |
| Limestone | 20 | 56 | 4,729 | 7.6 | 52.1 | 14.8 | 41 | 10,343 | 22,595 | 44,418 | 57.0 | 1.7 | 65,388 | 19.2 | 26.0 | 26.0 |
| Lipscomb | 6 | 7 | 751 | 2.7 | 50.4 | 19.7 | 10 | 12,627 | 28,131 | 59,091 | 40.5 | 4.9 | 54,163 | 12.4 | 17.2 | 14.9 |
| Live Oak | NA | NA | 2,102 | 6.3 | 57.2 | 12.2 | 19 | 10,853 | 25,370 | 53,848 | 46.7 | 4.7 | 55,617 | 16.5 | 23.0 | 22.1 |
| Llano | NA | NA | 2,632 | 14.9 | 40.7 | 25.3 | 20 | 10,992 | 37,050 | 53,411 | 47.3 | 6.3 | 55,617 | 10.8 | 20.7 | 20.2 |
| Loving | NA | NA | 0 | 0.0 | 39.4 | 0.0 | NA | NA | 33,624 | 83,750 | 36.7 | 3.3 | 88,487 | 7.1 | 10.2 | 7.9 |
| Lubbock | 205 | 289 | 101,767 | 9.7 | 39.4 | 30.1 | 476 | 9,472 | 27,922 | 52,429 | 48.0 | 4.4 | 54,733 | 17.9 | 22.5 | 22.0 |
| Lynn | 7 | 14 | 1,535 | 13.2 | 54.1 | 17.3 | 18 | 11,860 | 24,917 | 43,382 | 57.3 | 4.8 | 45,737 | 16.0 | 23.4 | 21.7 |
| McCulloch | NA | NA | 1,529 | 3.9 | 50.8 | 17.3 | 17 | 11,786 | 24,311 | 47,707 | 53.4 | 2.8 | 43,262 | 17.6 | 26.4 | 23.5 |
| McLennan | 140 | 345 | 75,925 | 23.9 | 42.1 | 24.2 | 540 | 9,927 | 25,703 | 49,778 | 50.2 | 4.2 | 51,078 | 18.0 | 22.6 | 21.5 |
| McMullen | 11 | 2 | 210 | 11.0 | 39.9 | 28.8 | 5 | 19,545 | 31,455 | 62,000 | 35.2 | 11.9 | 68,349 | 10.5 | 13.8 | 12.9 |
| Madison | 13 | 18 | 2,812 | 8.6 | 57.7 | 14.7 | 26 | 9,642 | 20,748 | 52,664 | 46.4 | 5.2 | 46,274 | 17.8 | 22.2 | 20.7 |
| Marion | 11 | 11 | 1,654 | 7.1 | 55.9 | 14.7 | 13 | 9,892 | 27,566 | 37,662 | 64.7 | 3.8 | 37,876 | 21.4 | 32.9 | 30.6 |
| Martin | NA | NA | 1,571 | 4.8 | 55.6 | 18.3 | 15 | 11,820 | 30,548 | 66,610 | 38.7 | 7.9 | 68,745 | 12.1 | 17.9 | 16.5 |

1. All persons 3 years old and over enrolled in nursery school through college. 2. Persons 25 years old and over. 3. Elementary and secondary education expenditures. 4. Based on population estimated by the American Community Survey, 2014–2018.

| STATE County | Personal income, 2019 | | | | | | | | | | Earnings, 2019 | | |
|---|---|---|---|---|---|---|---|---|---|---|---|---|---|
| | Total (mil dol) | Percent change 2018-2019 | Per capita[1] Dollars | Rank | Wages and salaries (mil dol) | Supplements to wages and salaries, employer contributions (mil dol) Pension and insurance | Government social insurance | Proprietors' income (mil dol) | Dividends, interest, and rent (mil dol) | Personal transfer receipts (mil dol) | Total (mil dol) | Contributions for government social insurance (mil dol) From employee and self-employed | From employer |
| | 62 | 63 | 64 | 65 | 66 | 67 | 68 | 69 | 70 | 71 | 72 | 73 | 74 |
| **TEXAS—Cont'd** | | | | | | | | | | | | | |
| Guadalupe | 7,641 | 5.5 | 45,797 | 1,272 | 2,043 | 332 | 137 | 464 | 1,182 | 1,464 | 2,976 | 182 | 137 |
| Hale | 1,190 | 5.5 | 35,633 | 2,600 | 485 | 88 | 35 | 157 | 183 | 318 | 766 | 39 | 35 |
| Hall | 98 | -3.6 | 33,095 | 2,882 | 26 | 6 | 2 | 9 | 19 | 39 | 44 | 3 | 2 |
| Hamilton | 513 | 4.2 | 60,584 | 284 | 107 | 22 | 7 | 23 | 242 | 108 | 160 | 11 | 7 |
| Hansford | 353 | 5.7 | 65,330 | 178 | 121 | 23 | 8 | 136 | 53 | 44 | 288 | 9 | 8 |
| Hardeman | 165 | 0.4 | 42,023 | 1,760 | 53 | 11 | 4 | 12 | 33 | 53 | 80 | 5 | 4 |
| Hardin | 2,720 | 3.3 | 47,221 | 1,106 | 633 | 100 | 44 | 79 | 314 | 598 | 857 | 61 | 44 |
| Harris | 282,809 | 3.8 | 60,002 | 303 | 182,213 | 23,153 | 11,542 | 55,765 | 51,570 | 31,948 | 272,673 | 13,761 | 11,542 |
| Harrison | 2,854 | 3.3 | 42,891 | 1,637 | 1,260 | 227 | 85 | 162 | 475 | 673 | 1,734 | 105 | 85 |
| Hartley | 453 | 12.2 | 81,238 | 61 | 113 | 20 | 9 | 263 | 51 | 22 | 405 | 7 | 9 |
| Haskell | 226 | 5.9 | 39,899 | 2,063 | 56 | 12 | 4 | 41 | 36 | 69 | 113 | 6 | 4 |
| Hays | 10,435 | 8.0 | 45,332 | 1,322 | 3,301 | 559 | 219 | 1,001 | 1,716 | 1,436 | 5,080 | 274 | 219 |
| Hemphill | 218 | 2.3 | 57,053 | 404 | 129 | 22 | 8 | 34 | 73 | 24 | 194 | 9 | 8 |
| Henderson | 3,321 | 4.4 | 40,135 | 2,036 | 730 | 142 | 50 | 233 | 530 | 1,008 | 1,155 | 87 | 50 |
| Hidalgo | 23,815 | 3.8 | 27,415 | 3,089 | 9,988 | 2,118 | 674 | 2,614 | 2,754 | 6,883 | 15,394 | 839 | 674 |
| Hill | 1,511 | 4.1 | 41,240 | 1,887 | 427 | 79 | 28 | 136 | 216 | 430 | 670 | 44 | 28 |
| Hockley | 971 | 2.5 | 42,162 | 1,743 | 614 | 89 | 38 | 63 | 130 | 238 | 805 | 45 | 38 |
| Hood | 3,167 | 5.7 | 51,384 | 729 | 818 | 124 | 56 | 225 | 635 | 717 | 1,224 | 86 | 56 |
| Hopkins | 1,541 | 4.5 | 41,562 | 1,839 | 566 | 95 | 38 | 151 | 228 | 404 | 850 | 50 | 38 |
| Houston | 910 | 2.7 | 39,609 | 2,102 | 397 | 67 | 24 | 26 | 158 | 289 | 514 | 34 | 24 |
| Howard | 1,589 | 2.9 | 43,348 | 1,562 | 764 | 153 | 53 | 121 | 296 | 321 | 1,090 | 59 | 53 |
| Hudspeth | 115 | 6.9 | 23,569 | 3,105 | 85 | 19 | 6 | 13 | 18 | 34 | 124 | 7 | 6 |
| Hunt | 3,835 | 5.3 | 38,892 | 2,210 | 1,551 | 294 | 104 | 200 | 439 | 958 | 2,149 | 129 | 104 |
| Hutchinson | 921 | 3.5 | 43,981 | 1,488 | 528 | 109 | 34 | 38 | 134 | 209 | 709 | 40 | 34 |
| Irion | 111 | 2.5 | 72,177 | 108 | 59 | 9 | 4 | 15 | 35 | 14 | 88 | 5 | 4 |
| Jack | 365 | -5.1 | 40,827 | 1,946 | 138 | 26 | 9 | 35 | 78 | 80 | 207 | 12 | 9 |
| Jackson | 688 | 9.2 | 46,596 | 1,175 | 302 | 52 | 20 | 53 | 100 | 159 | 427 | 24 | 20 |
| Jasper | 1,451 | 2.0 | 40,834 | 1,944 | 433 | 80 | 29 | 49 | 194 | 445 | 592 | 42 | 29 |
| Jeff Davis | 98 | 4.3 | 43,080 | 1,605 | 29 | 6 | 2 | 9 | 30 | 22 | 46 | 3 | 2 |
| Jefferson | 11,223 | 1.8 | 44,613 | 1,406 | 7,598 | 1,383 | 516 | 878 | 1,674 | 2,645 | 10,375 | 564 | 516 |
| Jim Hogg | 175 | 4.6 | 33,602 | 2,835 | 75 | 21 | 6 | 2 | 26 | 61 | 103 | 6 | 6 |
| Jim Wells | 1,707 | 3.3 | 42,174 | 1,742 | 870 | 124 | 57 | 76 | 206 | 535 | 1,127 | 68 | 57 |
| Johnson | 7,694 | 6.1 | 43,759 | 1,521 | 2,449 | 385 | 164 | 529 | 882 | 1,512 | 3,528 | 215 | 164 |
| Jones | 655 | 6.4 | 32,639 | 2,911 | 178 | 45 | 10 | 49 | 85 | 196 | 283 | 17 | 10 |
| Karnes | 881 | 5.0 | 56,449 | 423 | 424 | 69 | 27 | 13 | 409 | 162 | 532 | 31 | 27 |
| Kaufman | 5,987 | 9.1 | 43,972 | 1,491 | 1,561 | 270 | 102 | 317 | 590 | 1,008 | 2,250 | 136 | 102 |
| Kendall | 3,884 | 5.5 | 81,882 | 57 | 942 | 136 | 62 | 462 | 1,140 | 451 | 1,602 | 90 | 62 |
| Kenedy | 17 | 0.0 | 42,262 | 1,728 | 43 | 6 | 3 | 2 | 5 | 3 | 54 | 3 | 3 |
| Kent | 42 | 1.1 | 54,630 | 514 | 12 | 3 | 1 | 12 | 7 | 10 | 28 | 1 | 1 |
| Kerr | 2,723 | 3.4 | 51,768 | 700 | 876 | 149 | 61 | 280 | 887 | 632 | 1,365 | 89 | 61 |
| Kimble | 192 | 3.1 | 44,371 | 1,435 | 47 | 11 | 3 | 9 | 53 | 60 | 70 | 5 | 3 |
| King | 21 | -10.6 | 78,849 | 69 | 5 | 1 | 0 | 12 | 2 | 1 | 18 | 0 | 0 |
| Kinney | 118 | 1.4 | 32,219 | 2,931 | 54 | 14 | 4 | 1 | 29 | 37 | 73 | 5 | 4 |
| Kleberg | 1,274 | 3.1 | 41,526 | 1,844 | 550 | 132 | 39 | 80 | 205 | 322 | 802 | 43 | 39 |
| Knox | 145 | -0.9 | 39,587 | 2,106 | 50 | 11 | 3 | 11 | 26 | 50 | 75 | 4 | 3 |
| Lamar | 2,147 | 5.0 | 43,063 | 1,607 | 1,008 | 176 | 70 | 163 | 325 | 621 | 1,415 | 86 | 70 |
| Lamb | 589 | 6.6 | 45,655 | 1,292 | 186 | 35 | 14 | 155 | 59 | 138 | 389 | 15 | 14 |
| Lampasas | 1,085 | 3.1 | 50,656 | 784 | 184 | 35 | 13 | 66 | 223 | 296 | 297 | 23 | 13 |
| La Salle | 300 | 3.3 | 39,913 | 2,058 | 273 | 37 | 17 | 28 | 107 | 59 | 355 | 19 | 17 |
| Lavaca | 1,078 | 3.5 | 53,483 | 575 | 271 | 47 | 18 | 76 | 263 | 263 | 412 | 28 | 18 |
| Lee | 873 | 8.8 | 50,665 | 783 | 454 | 67 | 31 | 88 | 150 | 181 | 640 | 36 | 31 |
| Leon | 697 | 2.3 | 40,056 | 2,042 | 301 | 45 | 21 | 38 | 140 | 226 | 405 | 28 | 21 |
| Liberty | 3,341 | 5.5 | 37,874 | 2,335 | 960 | 168 | 64 | 195 | 338 | 806 | 1,387 | 86 | 64 |
| Limestone | 885 | 4.2 | 37,774 | 2,347 | 392 | 86 | 24 | 37 | 132 | 302 | 539 | 32 | 24 |
| Lipscomb | 252 | 3.3 | 77,810 | 73 | 79 | 13 | 5 | 119 | 39 | 24 | 217 | 9 | 5 |
| Live Oak | 457 | 4.1 | 37,415 | 2,400 | 269 | 51 | 18 | 21 | 131 | 104 | 359 | 21 | 18 |
| Llano | 1,088 | 4.6 | 49,905 | 837 | 212 | 34 | 15 | 109 | 348 | 301 | 370 | 29 | 15 |
| Loving | 9 | 18.6 | 53,734 | 561 | 18 | 2 | 1 | 0 | 3 | 0 | 21 | 1 | 1 |
| Lubbock | 13,762 | 4.0 | 44,311 | 1,443 | 6,857 | 1,180 | 445 | 1,048 | 2,407 | 2,587 | 9,530 | 509 | 445 |
| Lynn | 257 | 9.3 | 43,141 | 1,593 | 83 | 17 | 6 | 17 | 37 | 58 | 124 | 7 | 6 |
| McCulloch | 311 | -0.6 | 38,895 | 2,209 | 117 | 24 | 8 | 20 | 64 | 108 | 169 | 11 | 8 |
| McLennan | 10,819 | 2.8 | 42,159 | 1,744 | 5,780 | 948 | 396 | 819 | 1,726 | 2,357 | 7,943 | 450 | 396 |
| McMullen | 48 | 3.2 | 65,250 | 182 | 36 | 6 | 3 | -8 | 38 | 7 | 37 | 2 | 3 |
| Madison | 466 | 3.5 | 32,648 | 2,910 | 171 | 34 | 11 | 46 | 85 | 139 | 261 | 15 | 11 |
| Marion | 393 | 3.3 | 39,895 | 2,064 | 75 | 16 | 5 | 17 | 67 | 147 | 114 | 10 | 5 |
| Martin | 351 | -4.6 | 60,844 | 280 | 174 | 27 | 12 | 36 | 98 | 47 | 249 | 13 | 12 |

1. Based on the resident population estimated as of July 1 of the year shown.

Items 62—74

# Table B. States and Counties — Earnings, Social Security, and Housing

| STATE County | Earnings, 2019 (cont.) — Percent by selected industries | | | | | | | | | Social Security beneficiaries, December 2019 | | Supplemental Security Income recipients, 2019 | Housing units, 2020 | |
|---|---|---|---|---|---|---|---|---|---|---|---|---|---|---|
| | Farm | Mining, quarrying, and extractions | Construction | Manufacturing | Information; professional, scientific, technical services | Retail trade | Finance, insurance, real estate, and leasing | Health care and social assistance | Government | Number | Rate[1] | | Total | Percent change, 2010-2020 |
| | 75 | 76 | 77 | 78 | 79 | 80 | 81 | 82 | 83 | 84 | 85 | 86 | 87 | 88 |

TEXAS—Cont'd

| STATE County | 75 | 76 | 77 | 78 | 79 | 80 | 81 | 82 | 83 | 84 | 85 | 86 | 87 | 88 |
|---|---|---|---|---|---|---|---|---|---|---|---|---|---|---|
| Guadalupe | -0.3 | 1.7 | 10.2 | 22.9 | 3.3 | 7.2 | 4.5 | 6.7 | 15.2 | 28,535 | 171 | 2,012 | 63,201 | 26.4 |
| Hale | 15.1 | 1.1 | 7.2 | 5.6 | 3.3 | 5.8 | 4.5 | D | 18.2 | 5,830 | 176 | 902 | 13,347 | -1.3 |
| Hall | 17.1 | 1.0 | 1.6 | 3.4 | D | 8.2 | 6.4 | 5.3 | 30.4 | 820 | 274 | 89 | 1,908 | -1.8 |
| Hamilton | -3.6 | 0.1 | 13.6 | 7.6 | 4.8 | 9.5 | 3.7 | 8.6 | 30.0 | 2,420 | 286 | 164 | 4,542 | -0.5 |
| Hansford | 46.0 | 15.2 | 2.2 | 1.1 | D | 2.6 | 3.7 | 0.3 | 12.5 | 930 | 173 | 27 | 2,337 | 0.0 |
| Hardeman | 10.6 | D | D | D | D | 6.7 | 3.5 | | 29.1 | 1,045 | 266 | 136 | 2,389 | -1.2 |
| Hardin | -1.2 | 3.7 | D | D | 5.8 | 10.2 | D | 17.0 | 15.6 | 12,195 | 211 | 1,177 | 25,810 | 14.2 |
| Harris | 0.0 | 11.5 | 6.9 | 9.2 | 13.1 | 3.7 | 7.9 | 7.1 | 9.0 | 561,540 | 119 | 106,860 | 1,852,914 | 15.9 |
| Harrison | -0.4 | 8.4 | 7.2 | 31.8 | 4.7 | 4.9 | 4.9 | D | 10.7 | 14,145 | 213 | 1,866 | 29,134 | 5.1 |
| Hartley | 62.1 | 0.7 | D | D | D | 2.0 | D | D | 10.8 | 600 | 109 | 0 | 2,066 | 6.2 |
| Haskell | 25.8 | 7.0 | 3.2 | 1.4 | D | 12.2 | D | 3.3 | 22.9 | 1,395 | 246 | 167 | 3,481 | 1.1 |
| Hays | -0.2 | 0.6 | 14.0 | 8.7 | 7.6 | 9.0 | 5.4 | 7.9 | 19.4 | 30,795 | 134 | 2,243 | 88,099 | 48.4 |
| Hemphill | 8.2 | 31.8 | 11.4 | 2.3 | 2.3 | 2.2 | 3.2 | D | 15.7 | 610 | 159 | 18 | 1,712 | 5.1 |
| Henderson | -0.4 | 2.5 | 11.0 | 12.8 | 6.5 | 10.4 | 4.7 | 12.0 | 17.9 | 22,290 | 269 | 2,550 | 42,182 | 6.5 |
| Hidalgo | 0.9 | 0.9 | 4.7 | 2.7 | 4.5 | 9.5 | 4.9 | 19.5 | 26.2 | 114,295 | 132 | 40,969 | 291,856 | 17.6 |
| Hill | 0.3 | 1.8 | 19.7 | 11.3 | D | 8.1 | 3.8 | D | 19.0 | 9,075 | 248 | 984 | 16,534 | 2.6 |
| Hockley | 2.5 | 45.3 | 3.9 | 1.5 | D | 4.0 | 5.2 | D | 13.1 | 4,155 | 180 | 487 | 9,443 | 1.7 |
| Hood | -0.6 | 14.3 | 12.4 | 6.0 | D | 8.7 | 8.7 | 12.8 | 11.1 | 17,185 | 278 | 836 | 27,162 | 8.9 |
| Hopkins | 7.7 | D | 6.6 | 14.1 | | 8.0 | 5.8 | 8.2 | 13.7 | 8,440 | 228 | 915 | 15,534 | 3.4 |
| Houston | -1.3 | 6.9 | 5.8 | 13.8 | 7.6 | 5.1 | 12.4 | 6.4 | 19.2 | 5,675 | 248 | 838 | 11,907 | 3.3 |
| Howard | 0.5 | 14.0 | 6.8 | 9.6 | 3.8 | 5.9 | 3.4 | 9.6 | 25.0 | 5,645 | 154 | 776 | 13,225 | 0.8 |
| Hudspeth | 7.7 | 0.4 | D | D | D | 0.8 | D | D | 53.1 | 815 | 167 | 172 | 1,620 | 6.1 |
| Hunt | -1.3 | 0.1 | 7.9 | 32.1 | 4.5 | 6.3 | 2.7 | 8.1 | 22.5 | 19,475 | 198 | 2,408 | 38,780 | 5.6 |
| Hutchinson | -0.3 | 29.6 | 11.3 | 19.8 | 2.7 | 4.6 | 2.3 | D | 12.7 | 4,340 | 208 | 337 | 10,646 | 0.2 |
| Irion | 3.3 | 55.6 | D | D | D | 2.1 | D | D | 8.0 | 375 | 240 | 14 | 866 | 1.2 |
| Jack | -5.1 | 32.4 | 7.8 | 2.0 | D | 2.0 | 10.1 | D | 18.8 | 1,785 | 200 | 105 | 4,194 | 2.5 |
| Jackson | 5.6 | 6.9 | 17.6 | D | 5.9 | 3.3 | 4.1 | D | 15.6 | 3,220 | 218 | 302 | 6,806 | 3.3 |
| Jasper | -1.1 | 1.3 | 5.3 | 22.6 | 4.4 | 8.0 | 4.3 | D | 19.8 | 8,970 | 252 | 1,190 | 17,617 | 4.9 |
| Jeff Davis | 1.4 | D | D | D | D | 4.2 | D | D | 34.6 | 670 | 295 | 33 | 1,664 | 3.2 |
| Jefferson | 0.1 | 0.5 | 14.7 | 23.1 | 7.5 | 5.9 | 3.7 | 10.7 | 12.6 | 46,225 | 184 | 8,453 | 109,757 | 5.1 |
| Jim Hogg | -0.9 | 7.8 | 3.0 | 2.7 | D | 6.1 | D | 11.6 | 54.8 | 1,035 | 201 | 251 | 2,466 | 1.0 |
| Jim Wells | 1.7 | 35.3 | 3.6 | 2.9 | D | 4.9 | 5.4 | D | 11.2 | 8,675 | 214 | 1,677 | 16,451 | 1.8 |
| Johnson | -0.3 | 4.4 | 14.6 | 13.6 | 4.1 | 8.8 | 3.9 | 6.9 | 14.7 | 31,620 | 180 | 3,070 | 64,811 | 14.3 |
| Jones | 2.3 | 7.5 | 7.8 | 3.1 | 1.8 | 5.1 | 3.4 | D | 44.0 | 3,625 | 182 | 355 | 7,471 | 0.7 |
| Karnes | -2.7 | 27.0 | 1.8 | 6.4 | D | 3.7 | 6.1 | 2.4 | 19.4 | 2,820 | 184 | 393 | 6,179 | 9.3 |
| Kaufman | -1.1 | 0.2 | 13.5 | 13.2 | 5.5 | 7.1 | 4.7 | 7.2 | 20.4 | 21,735 | 159 | 2,208 | 44,208 | 15.3 |
| Kendall | -0.8 | 2.3 | 15.6 | 5.2 | 15.7 | 12.3 | 10.7 | 9.0 | 9.6 | 10,420 | 220 | 298 | 17,291 | 23.1 |
| Kenedy | 7.7 | D | 0.5 | D | 0.1 | D | 0.3 | D | 9.0 | 55 | 143 | 0 | 231 | -0.4 |
| Kent | 11.1 | D | D | 2.2 | D | D | D | D | 30.0 | 190 | 251 | 13 | 552 | 0.0 |
| Kerr | -0.7 | 2.1 | 13.0 | 7.1 | 6.2 | 8.1 | 7.1 | D | 16.5 | 16,280 | 311 | 953 | 25,144 | 5.5 |
| Kimble | -4.5 | 2.1 | 12.5 | 6.4 | D | 8.2 | 8.2 | 1.6 | 30.1 | 1,355 | 311 | 112 | 3,431 | 1.8 |
| King | 52.7 | 2.8 | D | 0.9 | D | 3.6 | 2.0 | D | 19.5 | 25 | 94 | 0 | 185 | 0.0 |
| Kinney | -1.5 | 0.2 | D | D | 0.2 | 2.5 | D | 0.8 | 46.6 | 940 | 256 | 104 | 1,998 | 3.0 |
| Kleberg | 1.2 | 4.9 | 3.9 | 4.1 | D | 6.9 | 3.2 | D | 43.8 | 5,265 | 173 | 1,035 | 13,320 | 4.2 |
| Knox | 12.3 | 10.1 | D | D | D | 10.1 | D | D | 32.3 | 910 | 248 | 107 | 2,030 | -0.7 |
| Lamar | 0.1 | 0.0 | 9.3 | 27.6 | 2.8 | 7.5 | 3.7 | 15.1 | 12.4 | 12,310 | 247 | 1,873 | 22,966 | 2.2 |
| Lamb | 38.8 | D | 6.6 | 3.0 | 1.6 | 3.5 | D | 2.7 | 14.4 | 2,670 | 207 | 397 | 6,059 | -1.1 |
| Lampasas | -5.8 | D | 17.4 | 9.5 | 5.2 | 17.2 | 9.1 | D | 20.2 | 5,675 | 265 | 465 | 9,591 | 10.0 |
| La Salle | 0.3 | 51.3 | 6.7 | 0.3 | D | 2.2 | D | 1.6 | 18.3 | 1,225 | 162 | 304 | 3,036 | 10.6 |
| Lavaca | -5.2 | 6.9 | 15.3 | 18.7 | 4.9 | 6.4 | 5.7 | D | 14.0 | 5,340 | 265 | 413 | 10,565 | 2.1 |
| Lee | -0.2 | D | 52.6 | 5.0 | 3.9 | 2.6 | 3.6 | D | 11.6 | 3,810 | 220 | 276 | 7,879 | 5.1 |
| Leon | 2.0 | 7.0 | 20.6 | 26.2 | 4.5 | 4.0 | 5.2 | D | 12.6 | 5,340 | 307 | 437 | 10,006 | 5.2 |
| Liberty | -0.1 | 7.1 | 17.9 | 8.8 | 3.0 | 7.1 | 3.9 | 7.2 | 22.0 | 15,115 | 171 | 2,683 | 33,665 | 17.0 |
| Limestone | -0.5 | 17.8 | 3.8 | 3.5 | 7.4 | 5.2 | 3.1 | 8.0 | 33.0 | 5,400 | 231 | 798 | 10,838 | 2.9 |
| Lipscomb | 10.3 | 6.3 | D | D | D | D | D | 1.8 | 9.2 | 600 | 187 | 30 | 1,501 | -0.7 |
| Live Oak | -3.0 | 22.9 | D | D | 10.5 | 4.3 | 6.3 | 1.5 | 17.5 | 2,210 | 181 | 175 | 6,333 | 4.4 |
| Llano | -2.6 | 0.8 | 12.4 | 4.5 | 6.8 | 5.8 | 8.6 | 6.4 | 14.7 | 7,195 | 330 | 334 | 15,901 | 11.3 |
| Loving | 0.5 | D | D | D | D | 0.1 | D | D | 4.5 | 15 | 91 | 0 | 56 | 12.0 |
| Lubbock | 0.2 | 1.8 | 7.4 | 3.3 | 7.5 | 8.6 | 6.9 | 15.3 | 24.1 | 48,585 | 156 | 6,362 | 132,874 | 15.4 |
| Lynn | 6.9 | 2.1 | 10.0 | 1.8 | D | 1.7 | 9.6 | D | 27.9 | 1,115 | 187 | 129 | 2,661 | -0.6 |
| McCulloch | -3.9 | 16.2 | 7.1 | 5.2 | 3.0 | 8.1 | 4.2 | 6.9 | 23.1 | 2,130 | 266 | 262 | 4,348 | 1.0 |
| McLennan | -0.2 | 0.3 | 8.3 | 16.0 | 6.2 | 6.9 | 9.2 | 10.8 | 15.5 | 47,060 | 183 | 7,169 | 104,314 | 9.7 |
| McMullen | -18.4 | 38.1 | D | D | D | 3.6 | D | 3.8 | 26.1 | 195 | 266 | 13 | 501 | 3.3 |
| Madison | 7.0 | D | 6.6 | 6.3 | D | 12.8 | 3.1 | D | 27.8 | 2,625 | 183 | 278 | 5,397 | 5.9 |
| Marion | 0.1 | D | 2.5 | 23.6 | D | 4.5 | 2.7 | D | 19.8 | 3,180 | 323 | 380 | 6,540 | 5.2 |
| Martin | -2.7 | 12.6 | 33.2 | 4.1 | D | 5.7 | D | D | 16.7 | 760 | 132 | 92 | 1,917 | 3.5 |

1. Per 1,000 resident population estimated as of July 1 of the year shown.

# Table B. States and Counties — Housing, Labor Force, and Employment

Column groups: **Housing units, 2015–2019** → Occupied units → Owner-occupied (items 89–93, incl. "Median owner cost as a percent of income": With a mortgage / Without a mortgage); Renter-occupied (items 94–95); Sub-standard units (item 96). **Civilian labor force, 2020** (items 97–100, incl. Unemployment Total/Rate). **Civilian employment[6], 2015–2019** (items 101–103, Percent).

| STATE County | Total | Percent | Median value[1] | With a mortgage | Without a mortgage[2] | Median rent[3] | Median rent as a percent of income[2] | Sub-standard units[4] (percent) | Total | Percent change 2019-2020 | Unemp. Total | Rate[5] | Total | Management, business, science, and arts | Construction, production, and maintenance occupations |
|---|---|---|---|---|---|---|---|---|---|---|---|---|---|---|---|
| (item) | 89 | 90 | 91 | 92 | 93 | 94 | 95 | 96 | 97 | 98 | 99 | 100 | 101 | 102 | 103 |
| **TEXAS—Cont'd** | | | | | | | | | | | | | | | |
| Guadalupe | 54,110 | 76.7 | 196,000 | 19.8 | 10.2 | 1,140 | 25.5 | 3.3 | 80,610 | -1.4 | 4,933 | 6.1 | 74,834 | 34 | 25.7 |
| Hale | 10,933 | 62.3 | 82,700 | 18.8 | 10.0 | 647 | 23.0 | 5.1 | 11,826 | -2.9 | 799 | 6.8 | 14,355 | 26.7 | 33.3 |
| Hall | 1,267 | 65.3 | 52,800 | 25.3 | 10.0 | 543 | 23.1 | 5.5 | 1,076 | 0.7 | 62 | 5.8 | 1,239 | 24.7 | 32.8 |
| Hamilton | 2,989 | 81.9 | 130,800 | 21.0 | 12.9 | 678 | 26.2 | 4.4 | 3,759 | 2.2 | 170 | 4.5 | 3,409 | 31.2 | 34.3 |
| Hansford | 1,838 | 70.5 | 95,300 | 27.5 | 13.8 | 783 | 29.7 | 4.4 | 2,502 | -9.6 | 87 | 3.5 | 2,374 | 32.3 | 35.1 |
| Hardeman | 1,596 | 67.9 | 43,100 | 16.9 | 12.7 | 539 | 23.9 | 5.0 | 1,710 | -0.6 | 75 | 4.4 | 1,701 | 30.8 | 29.8 |
| Hardin | 20,626 | 81.2 | 124,100 | 18.8 | 10.0 | 808 | 26.9 | 2.4 | 25,019 | -2.2 | 2,251 | 9.0 | 24,076 | 32.8 | 29.2 |
| Harris | 1,605,368 | 54.6 | 175,700 | 21.0 | 10.8 | 1,078 | 30.0 | 6.0 | 2,292,759 | -0.4 | 204,310 | 8.9 | 2,248,663 | 36.1 | 25.5 |
| Harrison | 23,292 | 72.3 | 130,900 | 18.7 | 11.4 | 786 | 28.5 | 4.5 | 29,133 | -1.0 | 2,290 | 7.9 | 27,934 | 34.4 | 30.6 |
| Hartley | 1,678 | 62.7 | 167,200 | 22.6 | 11.3 | 900 | 22.2 | 4.4 | 2,633 | -5.9 | 54 | 2.1 | 2,010 | 35.9 | 28.9 |
| Haskell | 2,146 | 75.6 | 53,900 | 18.0 | 11.5 | 546 | 23.6 | 2.7 | 2,653 | 17.8 | 114 | 4.3 | 2,213 | 24.8 | 34.8 |
| Hays | 73,437 | 62.3 | 238,800 | 22.4 | 11.6 | 1,154 | 32.6 | 4.6 | 121,304 | 0.4 | 7,665 | 6.3 | 109,808 | 37.8 | 20.2 |
| Hemphill | 1,298 | 71.3 | 151,700 | 16.7 | 10.5 | 823 | 19.1 | 5.7 | 1,850 | -7.8 | 82 | 4.4 | 1,842 | 30.4 | 33 |
| Henderson | 30,757 | 75.2 | 108,500 | 21.7 | 12.7 | 785 | 29.3 | 3.4 | 37,123 | 1.3 | 2,377 | 6.4 | 32,219 | 28.7 | 29.9 |
| Hidalgo | 238,345 | 68.0 | 87,100 | 22.9 | 12.1 | 734 | 31.4 | 13.4 | 359,969 | 2.3 | 41,893 | 11.6 | 328,097 | 27.8 | 25.1 |
| Hill | 12,992 | 72.6 | 103,800 | 19.7 | 10.9 | 767 | 27.3 | 5.3 | 16,294 | -0.6 | 1,023 | 6.3 | 15,427 | 24.8 | 35.1 |
| Hockley | 7,997 | 70.5 | 93,600 | 18.7 | 11.9 | 696 | 28.2 | 5.9 | 10,671 | -5.3 | 845 | 7.9 | 10,174 | 28.4 | 33.5 |
| Hood | 22,152 | 77.3 | 193,500 | 19.9 | 10.5 | 961 | 29.9 | 3.5 | 27,578 | -0.6 | 1,833 | 6.6 | 24,812 | 33.9 | 31.7 |
| Hopkins | 13,424 | 69.6 | 114,900 | 19.3 | 10.6 | 811 | 24.2 | 3.9 | 17,484 | -0.3 | 895 | 5.1 | 16,402 | 30.3 | 31.5 |
| Houston | 8,252 | 70.0 | 100,700 | 26.0 | 13.7 | 702 | 40.5 | 3.8 | 10,010 | -2.6 | 529 | 5.3 | 8,227 | 30 | 30.2 |
| Howard | 11,064 | 65.4 | 99,000 | 19.4 | 10.0 | 868 | 27.0 | 4.5 | 13,502 | -3.7 | 1,093 | 8.1 | 14,303 | 28.4 | 31 |
| Hudspeth | 978 | 71.9 | 50,900 | 24.6 | 12.0 | 660 | 18.6 | 12.4 | 1,855 | -0.4 | 149 | 8.0 | 1,162 | 12.7 | 36.8 |
| Hunt | 33,189 | 68.9 | 123,300 | 22.1 | 11.1 | 899 | 30.7 | 5.5 | 43,506 | -0.5 | 2,847 | 6.5 | 41,358 | 33.2 | 26.6 |
| Hutchinson | 6,929 | 80.3 | 78,800 | 17.2 | 10.4 | 758 | 23.9 | 2.1 | 8,539 | -1.5 | 613 | 7.2 | 8,780 | 22.4 | 37.9 |
| Irion | 682 | 77.6 | 118,300 | 31.3 | 10.4 | 1,031 | 23.9 | 7.3 | 752 | -1.3 | 45 | 6.0 | 664 | 29.7 | 41.6 |
| Jack | 3,168 | 81.3 | 86,300 | 19.5 | 12.6 | 647 | 24.1 | 4.5 | 3,393 | -2.2 | 238 | 7.0 | 3,172 | 27.7 | 35.3 |
| Jackson | 4,917 | 69.6 | 114,000 | 17.7 | 10.0 | 826 | 24.8 | 5.5 | 7,329 | -2.7 | 422 | 5.8 | 6,602 | 25.1 | 40 |
| Jasper | 12,936 | 77.6 | 100,900 | 19.2 | 13.1 | 765 | 33.6 | 3.5 | 12,883 | 0.4 | 1,443 | 11.2 | 13,464 | 26.7 | 33.9 |
| Jeff Davis | 1,054 | 81.6 | 111,900 | 18.9 | 10.0 | 672 | 16.2 | 2.2 | 975 | -6.9 | 50 | 5.1 | 763 | 34.5 | 30.1 |
| Jefferson | 92,988 | 61.1 | 112,800 | 19.9 | 11.0 | 871 | 28.6 | 2.8 | 105,592 | -0.3 | 12,570 | 11.9 | 106,092 | 28.8 | 30.6 |
| Jim Hogg | 1,626 | 71.6 | 65,000 | 20.4 | 10.2 | 584 | 40.1 | 11.5 | 1,891 | 0.5 | 180 | 9.5 | 1,940 | 19.8 | 38.5 |
| Jim Wells | 12,987 | 70.6 | 76,200 | 20.0 | 10.2 | 713 | 32.6 | 7.9 | 16,127 | -4.3 | 2,094 | 13.0 | 16,016 | 26.9 | 32.8 |
| Johnson | 57,310 | 73.3 | 155,400 | 19.7 | 10.8 | 1,008 | 27.6 | 4.5 | 82,931 | -0.5 | 5,390 | 6.5 | 77,938 | 28.3 | 38.4 |
| Jones | 5,696 | 74.6 | 82,400 | 17.5 | 11.1 | 727 | 25.0 | 2.5 | 5,738 | -0.8 | 401 | 7.0 | 4,994 | 36.2 | 24.5 |
| Karnes | 4,282 | 74.5 | 102,300 | 14.7 | 11.0 | 807 | 19.8 | 3.4 | 7,007 | -2.1 | 440 | 6.3 | 5,275 | 28.8 | 28.7 |
| Kaufman | 38,015 | 76.6 | 178,300 | 21.9 | 11.8 | 1,044 | 29.9 | 3.7 | 67,263 | -0.1 | 4,377 | 6.5 | 58,675 | 33.3 | 26.4 |
| Kendall | 14,253 | 74.1 | 348,500 | 21.6 | 11.8 | 1,196 | 27.1 | 1.9 | 22,123 | -2.4 | 1,116 | 5.0 | 20,605 | 42.3 | 18.9 |
| Kenedy | 197 | 23.4 | 24,400 | 0.0 | 15.0 | 533 | 25.9 | 7.6 | 184 | -20.3 | 10 | 5.4 | 211 | 3.8 | 52.1 |
| Kent | 273 | 69.6 | 35,000 | 16.3 | 13.1 | 475 | 10.0 | 0.0 | 458 | -3.2 | 18 | 3.9 | 263 | 43.3 | 18.6 |
| Kerr | 21,403 | 70.7 | 180,600 | 21.0 | 11.3 | 874 | 28.1 | 4.0 | 21,636 | -2.7 | 1,271 | 5.9 | 22,320 | 31.1 | 22.2 |
| Kimble | 1,783 | 76.2 | 149,700 | 24.5 | 11.0 | 732 | 28.7 | 3.0 | 1,809 | -2.5 | 92 | 5.1 | 2,051 | 34.7 | 29.1 |
| King | 77 | 33.8 | 36,700 | 0.0 | 10.0 | 658 | 50.0 | 0.0 | 236 | 0.4 | 4 | 1.7 | 113 | 34.5 | 27.4 |
| Kinney | 1,337 | 76.6 | 68,800 | 22.2 | 11.3 | 515 | 27.0 | 1.4 | 1,312 | -4.0 | 79 | 6.0 | 1,202 | 27.4 | 23.4 |
| Kleberg | 10,955 | 52.8 | 91,900 | 18.8 | 11.4 | 859 | 31.0 | 7.5 | 13,402 | -0.6 | 1,125 | 8.4 | 12,854 | 27.4 | 30.3 |
| Knox | 1,375 | 82.9 | 45,000 | 14.8 | 10.0 | 513 | 28.5 | 3.6 | 1,454 | 0.8 | 82 | 5.6 | 1,491 | 28.9 | 43 |
| Lamar | 19,793 | 64.5 | 99,100 | 18.2 | 12.2 | 735 | 28.6 | 1.8 | 23,507 | -1.8 | 1,518 | 6.5 | 21,502 | 28.2 | 31.5 |
| Lamb | 4,728 | 71.5 | 65,500 | 19.2 | 11.4 | 719 | 24.2 | 4.6 | 5,234 | -0.6 | 274 | 5.2 | 5,591 | 34.3 | 30.8 |
| Lampasas | 7,807 | 77.4 | 161,600 | 21.7 | 11.6 | 774 | 28.3 | 3.3 | 9,041 | -0.8 | 485 | 5.4 | 8,951 | 35.2 | 24.2 |
| La Salle | 2,269 | 68.7 | 82,300 | 16.2 | 12.7 | 636 | 18.2 | 5.0 | 3,693 | -15.8 | 251 | 6.8 | 2,684 | 24.7 | 38.7 |
| Lavaca | 7,826 | 74.3 | 162,400 | 17.8 | 10.0 | 752 | 21.5 | 6.1 | 8,406 | -2.5 | 441 | 5.2 | 9,119 | 30 | 35.7 |
| Lee | 6,036 | 80.6 | 138,700 | 21.6 | 12.1 | 873 | 25.8 | 4.8 | 9,158 | -9.6 | 452 | 4.9 | 8,153 | 26.3 | 30 |
| Leon | 6,443 | 77.4 | 111,000 | 18.5 | 10.2 | 773 | 29.8 | 2.6 | 6,121 | -3.1 | 477 | 7.8 | 6,262 | 25.6 | 34.1 |
| Liberty | 26,873 | 75.5 | 111,600 | 20.7 | 11.6 | 850 | 28.4 | 6.2 | 33,582 | 0.3 | 3,571 | 10.6 | 30,771 | 22.8 | 40.7 |
| Limestone | 8,377 | 74.1 | 95,400 | 20.5 | 12.6 | 701 | 29.5 | 5.4 | 8,363 | -2.8 | 596 | 7.1 | 9,101 | 25.8 | 34.7 |
| Lipscomb | 1,192 | 73.8 | 100,900 | 17.6 | 10.0 | 726 | 19.4 | 4.4 | 1,541 | -5.4 | 64 | 4.2 | 1,611 | 27.7 | 37.9 |
| Live Oak | 3,752 | 82.0 | 104,100 | 16.6 | 10.9 | 750 | 24.9 | 4.7 | 5,177 | -6.6 | 386 | 7.5 | 4,504 | 30 | 36.1 |
| Llano | 8,678 | 78.1 | 194,800 | 22.5 | 11.7 | 885 | 28.3 | 3.7 | 8,479 | -2.0 | 486 | 5.7 | 7,682 | 29.3 | 25.8 |
| Loving | 30 | 63.3 | 0 | 0.0 | 10.0 | 0 | 0.0 | 16.7 | 289 | -42.7 | 5 | 1.7 | 33 | 30.3 | 63.6 |
| Lubbock | 113,488 | 55.2 | 140,600 | 19.9 | 11.1 | 940 | 32.5 | 3.9 | 156,625 | -1.0 | 8,996 | 5.7 | 149,699 | 36.9 | 21.2 |
| Lynn | 2,154 | 68.3 | 84,000 | 23.4 | 11.6 | 748 | 30.6 | 3.3 | 2,740 | -1.5 | 149 | 5.4 | 2,398 | 33.2 | 32.5 |
| McCulloch | 3,143 | 72.2 | 93,700 | 20.0 | 12.1 | 727 | 30.3 | 4.3 | 3,260 | -6.2 | 244 | 7.5 | 3,702 | 23.5 | 36.6 |
| McLennan | 90,054 | 59.2 | 142,800 | 20.8 | 12.2 | 871 | 31.8 | 3.6 | 120,347 | 1.0 | 7,410 | 6.2 | 114,524 | 32.5 | 27.2 |
| McMullen | 270 | 75.9 | 110,900 | 17.9 | 10.0 | 826 | 20.9 | 12.6 | 727 | -10.8 | 20 | 2.8 | 310 | 48.7 | 21.6 |
| Madison | 4,269 | 74.2 | 113,100 | 21.8 | 10.0 | 750 | 20.7 | 7.1 | 4,462 | -1.6 | 321 | 7.2 | 4,882 | 27.5 | 33.8 |
| Marion | 4,715 | 77.3 | 96,500 | 19.0 | 10.9 | 738 | 34.3 | 2.3 | 4,271 | 0.9 | 351 | 8.2 | 3,715 | 25.7 | 33.4 |
| Martin | 1,694 | 73.8 | 148,000 | 18.3 | 10.0 | 821 | 20.5 | 6.4 | 2,546 | -9.5 | 155 | 6.1 | 2,483 | 31.7 | 35.9 |

1. Specified owner-occupied units lacking complete plumbing facilities. 2. A value of 10.0 represents 10 percent or less; a value of 50.0 represents 50 percent or more. 3. Specified renter-occupied units. 4. Overcrowded or
5. Percent of civilian labor force. 6. Civilian employed persons 16 years old and over.

| STATE County | Number of establishments | Total | Health care and social assistance | Manufacturing | Retail trade | Finance and insurance | Professional, scientific, and technical services | Total (mil dol) | Average per employee (dollars) | Number | Fewer than 50 acres | 1000 acres or more | Farm producers whose primary occupation is farming (percent) |
|---|---|---|---|---|---|---|---|---|---|---|---|---|---|
| | 104 | 105 | 106 | 107 | 108 | 109 | 110 | 111 | 112 | 113 | 114 | 115 | 116 |
| TEXAS—Cont'd | | | | | | | | | | | | | |
| Guadalupe | 2,222 | 35,064 | 3,671 | 7,731 | 4,896 | 703 | 1,059 | 1,597 | 45,554 | 2,543 | 52.4 | 2.2 | 34.3 |
| Hale | 669 | 9,947 | 1,016 | 644 | 1,255 | 277 | 172 | 355 | 35,736 | 671 | 18.6 | 27.6 | 42.5 |
| Hall | 68 | 403 | 58 | 8 | 106 | 53 | NA | 10 | 25,871 | 327 | 4.3 | 35.8 | 35.6 |
| Hamilton | 222 | 1,982 | 511 | 218 | 408 | 46 | 43 | 73 | 36,820 | 1,163 | 19.4 | 9.7 | 38.9 |
| Hansford | 152 | 1,162 | 279 | 15 | 234 | 89 | 37 | 47 | 40,259 | 181 | 6.1 | 55.2 | 63.8 |
| Hardeman | 90 | 775 | 142 | NA | 139 | 35 | NA | 26 | 34,152 | 275 | 6.2 | 24.0 | 38.9 |
| Hardin | 838 | 8,961 | 884 | 711 | 2,137 | 230 | 330 | 358 | 39,983 | 661 | 64.6 | 1.7 | 29.8 |
| Harris | 104,302 | 2,110,957 | 270,939 | 153,884 | 214,096 | 79,859 | 180,773 | 142,998 | 67,741 | 1,891 | 80.7 | 1.7 | 31.9 |
| Harrison | 1,309 | 19,694 | 1,345 | 5,471 | 1,956 | 803 | 716 | 913 | 46,354 | 1,134 | 51.5 | 2.4 | 32.2 |
| Hartley | 114 | 1,191 | 296 | NA | 194 | 35 | 15 | 47 | 39,799 | 206 | 7.3 | 45.1 | 65.4 |
| Haskell | 136 | 1,022 | 156 | NA | 324 | 43 | 30 | 33 | 32,377 | 488 | 16.0 | 21.9 | 39.3 |
| Hays | 4,597 | 64,405 | 7,553 | 5,139 | 15,958 | 1,127 | 2,951 | 2,348 | 36,461 | 1,128 | 53.9 | 5.9 | 31.0 |
| Hemphill | 134 | 1,369 | 209 | 34 | 148 | NA | 41 | 80 | 58,087 | 230 | 7.8 | 38.3 | 49.5 |
| Henderson | 1,328 | 13,848 | 2,253 | 1,777 | 2,648 | 384 | 866 | 533 | 38,467 | 1,988 | 51.3 | 2.2 | 36.7 |
| Hidalgo | 12,437 | 205,275 | 67,282 | 6,454 | 38,209 | 7,139 | 6,377 | 6,108 | 29,757 | 2,436 | 73.6 | 6.4 | 32.0 |
| Hill | 643 | 7,778 | 906 | 1,765 | 1,369 | 195 | 144 | 284 | 36,456 | 2,003 | 40.8 | 4.6 | 39.3 |
| Hockley | 493 | 7,524 | 944 | 264 | 860 | 193 | 99 | 416 | 55,230 | 661 | 24.2 | 23.3 | 35.1 |
| Hood | 1,414 | 14,295 | 2,399 | 506 | 2,893 | 485 | 481 | 542 | 37,912 | 1,176 | 64.0 | 3.5 | 35.4 |
| Hopkins | 749 | 10,069 | 1,145 | 1,839 | 1,582 | 531 | 170 | 382 | 37,953 | 2,200 | 36.4 | 3.2 | 41.7 |
| Houston | 333 | 3,424 | 517 | 517 | 628 | 142 | 106 | 141 | 41,097 | 1,422 | 27.8 | 5.2 | 43.9 |
| Howard | 764 | 9,797 | 2,372 | 666 | 1,437 | 259 | 183 | 493 | 50,349 | 373 | 26.8 | 22.5 | 34.5 |
| Hudspeth | 34 | 393 | NA | NA | 59 | NA | NA | 18 | 46,949 | 134 | 13.4 | 49.3 | 54.1 |
| Hunt | 1,562 | 25,406 | 3,708 | 8,532 | 3,858 | 325 | 693 | 1,369 | 53,881 | 4,110 | 63.6 | 1.8 | 31.6 |
| Hutchinson | 425 | 5,980 | 549 | 1,537 | 932 | 159 | 267 | 374 | 62,623 | 149 | 26.8 | 37.6 | 45.5 |
| Irion | 55 | 485 | NA | NA | 14 | NA | 9 | 32 | 66,786 | 175 | 18.3 | 40.6 | 42.4 |
| Jack | 218 | 1,752 | 304 | 66 | 178 | 40 | 48 | 81 | 46,404 | 870 | 22.9 | 10.5 | 39.1 |
| Jackson | 340 | 4,928 | 318 | 1,452 | 411 | 148 | 138 | 245 | 49,814 | 788 | 30.7 | 12.6 | 37.4 |
| Jasper | 614 | 8,459 | 2,286 | 1,397 | 1,554 | 260 | 161 | 320 | 37,874 | 896 | 66.2 | 1.5 | 35.8 |
| Jeff Davis | 61 | 386 | 69 | NA | 66 | NA | 15 | 10 | 25,946 | 77 | 28.6 | 49.4 | 43.2 |
| Jefferson | 5,663 | 103,210 | 17,185 | 13,230 | 14,001 | 2,550 | 5,725 | 5,425 | 52,563 | 729 | 62.3 | 7.5 | 44.6 |
| Jim Hogg | 84 | 1,211 | 657 | 44 | 250 | 39 | NA | 28 | 22,782 | 244 | 14.8 | 29.5 | 33.1 |
| Jim Wells | 797 | 14,012 | 5,268 | 352 | 1,716 | 286 | 180 | 532 | 37,954 | 1,224 | 42.2 | 6.5 | 36.4 |
| Johnson | 3,010 | 39,143 | 3,496 | 6,415 | 6,215 | 926 | 1,299 | 1,723 | 44,024 | 3,140 | 67.1 | 2.3 | 33.5 |
| Jones | 258 | 2,112 | 378 | 208 | 267 | 75 | 56 | 97 | 46,086 | 915 | 31.1 | 14.0 | 35.5 |
| Karnes | 336 | 5,485 | 492 | 179 | 559 | 101 | 98 | 361 | 65,786 | 1,213 | 19.5 | 6.8 | 40.6 |
| Kaufman | 2,108 | 27,909 | 3,743 | 3,991 | 4,353 | 715 | 588 | 1,105 | 39,608 | 2,778 | 64.7 | 2.4 | 35.0 |
| Kendall | 1,476 | 15,251 | 2,154 | 1,137 | 2,997 | 692 | 884 | 694 | 45,536 | 1,349 | 48.2 | 6.4 | 31.1 |
| Kenedy | 16 | 140 | NA | NA | NA | NA | NA | 11 | 76,400 | 30 | 3.3 | 56.7 | 67.3 |
| Kent | 12 | 72 | NA | NA | 20 | NA | NA | 6 | 79,986 | 164 | 3.0 | 34.1 | 31.4 |
| Kerr | 1,502 | 17,611 | 4,046 | 998 | 2,908 | 418 | 761 | 698 | 39,651 | 1,128 | 34.2 | 11.5 | 38.3 |
| Kimble | 143 | 889 | 129 | 61 | 267 | 41 | 15 | 30 | 34,200 | 670 | 19.3 | 26.0 | 35.6 |
| King | NA | NA | NA | NA | NA | NA | NA | NA | NA | 43 | NA | 39.5 | 21.5 |
| Kinney | 40 | 535 | 13 | NA | 50 | NA | NA | 26 | 49,450 | 236 | 17.4 | 37.7 | 37.4 |
| Kleberg | 545 | 6,991 | 1,559 | 78 | 1,516 | 298 | 173 | 232 | 33,157 | 459 | 60.3 | 3.7 | 29.8 |
| Knox | 90 | 638 | 180 | NA | 107 | 47 | NA | 30 | 46,351 | 216 | 17.6 | 22.7 | 43.9 |
| Lamar | 1,218 | 18,660 | 3,781 | 5,564 | 2,639 | 492 | 301 | 708 | 37,960 | 1,946 | 38.4 | 3.9 | 33.3 |
| Lamb | 237 | 2,012 | 465 | 198 | 319 | 114 | 36 | 88 | 43,642 | 777 | 13.4 | 22.9 | 42.5 |
| Lampasas | 407 | 4,244 | 478 | 665 | 747 | 99 | 129 | 143 | 33,605 | 1,151 | 42.1 | 8.1 | 41.5 |
| La Salle | 142 | 1,615 | 146 | NA | 279 | NA | 4 | 76 | 46,759 | 383 | 11.5 | 32.4 | 38.3 |
| Lavaca | 480 | 5,594 | 1,147 | 1,414 | 770 | 258 | 186 | 224 | 40,063 | 2,900 | 34.4 | 1.9 | 34.5 |
| Lee | 406 | 6,538 | 276 | 441 | 596 | 230 | 81 | 372 | 56,843 | 1,809 | 41.5 | 2.3 | 33.7 |
| Leon | 352 | 4,323 | 113 | 758 | 613 | 121 | 63 | 250 | 57,775 | 1,951 | 34.8 | 3.3 | 41.9 |
| Liberty | 1,116 | 14,204 | 1,696 | 852 | 2,597 | 400 | 340 | 647 | 45,555 | 1,538 | 59.5 | 3.0 | 33.8 |
| Limestone | 393 | 5,405 | 1,248 | 563 | 882 | 208 | 202 | 240 | 44,398 | 1,284 | 33.9 | 6.5 | 40.0 |
| Lipscomb | 77 | 1,016 | NA | NA | 164 | 69 | 17 | 57 | 56,253 | 299 | 6.7 | 40.5 | 36.9 |
| Live Oak | 291 | 3,687 | 116 | 374 | 473 | 84 | 120 | 220 | 59,644 | 856 | 21.8 | 12.0 | 37.1 |
| Llano | 478 | 3,676 | 577 | 79 | 560 | 174 | 151 | 127 | 34,663 | 835 | 30.5 | 15.1 | 37.8 |
| Loving | 6 | 103 | NA | NA | NA | NA | NA | 11 | 104,553 | 8 | NA | 100.0 | 45.8 |
| Lubbock | 7,400 | 118,487 | 26,126 | 5,283 | 18,478 | 5,325 | 4,617 | 4,668 | 39,393 | 1,033 | 48.2 | 14.4 | 37.3 |
| Lynn | 79 | 645 | 197 | 74 | 48 | 47 | 9 | 30 | 46,733 | 434 | 15.4 | 33.4 | 47.0 |
| McCulloch | 206 | 1,993 | 210 | 95 | 417 | 110 | 63 | 67 | 33,697 | 682 | 17.4 | 19.1 | 41.2 |
| McLennan | 5,359 | 106,901 | 17,337 | 15,331 | 13,145 | 5,224 | 4,343 | 4,295 | 40,179 | 3,366 | 60.2 | 3.2 | 30.5 |
| McMullen | 46 | 425 | 13 | NA | 68 | NA | NA | 49 | 115,471 | 191 | 6.8 | 45.0 | 46.7 |
| Madison | 266 | 2,607 | 253 | 30 | 663 | 89 | 66 | 89 | 34,136 | 977 | 35.7 | 4.4 | 42.3 |
| Marion | 151 | 1,619 | 400 | 349 | 210 | 38 | 25 | 56 | 34,787 | 280 | 39.3 | 5.7 | 27.5 |
| Martin | 117 | 1,320 | 242 | 13 | 214 | 32 | 15 | 79 | 59,848 | 356 | 15.2 | 30.9 | 42.5 |

# Table B. States and Counties — Agriculture

Agriculture, 2017 (cont.)

| STATE County | Land in farms — Acreage (1,000) | Percent change, 2012-2017 | Acres — Average size of farm | Total irrigated (1,000) | Total cropland (1,000) | Value of land and buildings (dollars) — Average per farm | Average per acre | Value of machinery and equipment, average per farm (dollars) | Value of products sold: Total (mil dol) | Average per farm (acres) | Percent from: Crops | Livestock and poultry products | Organic farms (number) | Farms with internet access (percent) | Government payments — Total ($1,000) | Percent of farms |
|---|---|---|---|---|---|---|---|---|---|---|---|---|---|---|---|---|
| | 117 | 118 | 119 | 120 | 121 | 122 | 123 | 124 | 125 | 126 | 127 | 128 | 129 | 130 | 131 | 132 |
| **TEXAS—Cont'd** | | | | | | | | | | | | | | | | |
| Guadalupe | 359 | -6.2 | 141 | 2.2 | 102.5 | 550,283 | 3,893 | 54,175 | 73.6 | 28,937 | 28.1 | 71.9 | 3 | 73.4 | 1,652 | 6.9 |
| Hale | 584 | -8.8 | 870 | 189.3 | 458.9 | 1,131,985 | 1,301 | 249,333 | 411.7 | 613,568 | 33.3 | 66.7 | NA | 70.6 | 14,493 | 73.0 |
| Hall | 494 | -2.9 | 1,511 | 29.7 | 236.5 | 1,389,562 | 919 | 222,564 | 56.4 | 172,502 | 81.0 | 19.0 | NA | 67.9 | 5,295 | 77.1 |
| Hamilton | 484 | 8.5 | 416 | 1.5 | 86.1 | 1,044,354 | 2,510 | 67,913 | 62.0 | 53,333 | 10.3 | 89.7 | 1 | 74.6 | 1,045 | 17.5 |
| Hansford | 587 | 3.6 | 3,243 | 93.7 | 301.1 | 4,353,971 | 1,342 | 539,803 | 737.4 | 4,074,127 | 12.5 | 87.5 | NA | 71.3 | 7,750 | 73.5 |
| Hardeman | 295 | -16.9 | 1,072 | 0.9 | 114.2 | 1,118,742 | 1,044 | 91,771 | 18.0 | 65,436 | 29.9 | 70.1 | NA | 69.8 | 2,537 | 73.8 |
| Hardin | 65 | -5.0 | 98 | 1.1 | 13.1 | 307,897 | 3,127 | 74,026 | 4.7 | 7,101 | 50.4 | 49.6 | 1 | 72.9 | 201 | 1.1 |
| Harris | 219 | -7.5 | 116 | 7.3 | 52.7 | 998,511 | 8,635 | 58,260 | 50.6 | 26,765 | 73.4 | 26.6 | 10 | 75.0 | 1,448 | 2.7 |
| Harrison | 190 | -4.7 | 168 | 1.7 | 38.5 | 440,839 | 2,628 | 66,104 | 15.8 | 13,964 | 26.1 | 73.9 | 2 | 75.4 | 421 | 4.7 |
| Hartley | 835 | -7.6 | 4,052 | 153 | 296.4 | 5,028,050 | 1,241 | 997,785 | 1,221.7 | 5,930,437 | 12.8 | 87.2 | 6 | 74.3 | 6,780 | 59.7 |
| Haskell | 565 | -0.4 | 1,158 | 14.8 | 293 | 1,218,046 | 1,052 | 175,284 | 54.3 | 111,307 | 68.3 | 31.7 | 2 | 69.5 | 6,906 | 73.0 |
| Hays | 263 | 7.4 | 233 | 0.6 | 53 | 2,280,804 | 9,773 | 53,651 | 21.8 | 19,282 | 52.6 | 47.4 | 4 | 79.9 | 324 | 3.6 |
| Hemphill | 528 | -8.3 | 2,296 | 2.3 | 32.1 | 2,558,235 | 1,114 | 122,556 | 138.9 | 603,870 | 1.7 | 98.3 | NA | 62.6 | 1,699 | 46.1 |
| Henderson | 310 | -10.2 | 156 | 1.6 | 86.6 | 497,751 | 3,188 | 64,795 | 40.2 | 20,213 | 29.0 | 71.0 | 1 | 78.6 | 58 | 1.2 |
| Hidalgo | 624 | -21.5 | 256 | 162.5 | 356.9 | 1,106,061 | 4,319 | 92,928 | 311.0 | 127,681 | 94.3 | 5.7 | 26 | 61.8 | 6,631 | 10.5 |
| Hill | 523 | 3.8 | 261 | 1.2 | 256.4 | 660,515 | 2,529 | 89,657 | 114.0 | 56,915 | 56.6 | 43.4 | 1 | 70.0 | 7,927 | 22.8 |
| Hockley | 525 | 8.5 | 794 | 106.4 | 394.1 | 800,066 | 1,008 | 176,801 | 92.0 | 139,215 | 96.6 | 3.4 | NA | 69.9 | 6,956 | 65.2 |
| Hood | 205 | -8.4 | 175 | 2.7 | 39.7 | 598,809 | 3,428 | 53,816 | 18.9 | 16,109 | 40.6 | 59.4 | 1 | 83.3 | 25 | 0.9 |
| Hopkins | 395 | -6.6 | 179 | 2.5 | 127.9 | 523,841 | 2,918 | 78,642 | 253.7 | 115,335 | 4.1 | 95.9 | NA | 73.5 | 1,644 | 14.7 |
| Houston | 395 | -15.7 | 277 | 3.5 | 148.3 | 774,459 | 2,791 | 75,170 | 64.5 | 45,371 | 10.5 | 89.5 | NA | 68.7 | 358 | 3.0 |
| Howard | 521 | 4.5 | 1,397 | 6.9 | 148.3 | 1,291,868 | 925 | 132,948 | 26.9 | 72,027 | 75.4 | 24.6 | NA | 70.2 | 3,272 | 46.6 |
| Hudspeth | 2,276 | 1.1 | 16,985 | 9.7 | 20.5 | 15,341,351 | 903 | 151,528 | 17.4 | 129,881 | 56.4 | 43.6 | NA | 61.9 | 171 | 6.0 |
| Hunt | 483 | 6.2 | 117 | 3.5 | 155 | 363,446 | 3,094 | 54,517 | 55.3 | 13,458 | 46.7 | 53.3 | 5 | 75.8 | 4,192 | 11.5 |
| Hutchinson | 548 | 5.2 | 3,679 | 26.7 | 81.7 | 3,561,262 | 968 | 227,673 | 44.9 | 301,416 | 55.5 | 44.5 | NA | 70.5 | 3,068 | 30.9 |
| Irion | 613 | 23.5 | 3,501 | 0.9 | 4.3 | 3,482,288 | 995 | 81,488 | 9.3 | 53,000 | 3.2 | 96.8 | NA | 81.1 | 318 | 10.9 |
| Jack | 468 | -11.4 | 537 | 0.8 | 34.2 | 1,253,842 | 2,333 | 62,108 | 23.2 | 26,639 | 6.1 | 93.9 | 1 | 71.7 | 414 | 5.5 |
| Jackson | 382 | -13.5 | 485 | 6.4 | 176.9 | 1,411,888 | 2,910 | 146,609 | 85.0 | 107,854 | 79.0 | 21.0 | 1 | 66.6 | 7,988 | 24.0 |
| Jasper | 91 | 3.8 | 102 | 0.3 | 13.4 | 327,242 | 3,207 | 58,063 | 9.1 | 10,200 | 43.8 | 56.2 | 3 | 73.1 | 9 | 0.8 |
| Jeff Davis | 1,378 | 9.8 | 17,896 | 0.1 | 0.6 | 12,302,948 | 687 | 125,618 | D | D | D | D | NA | 67.5 | 192 | 10.4 |
| Jefferson | 359 | 1.4 | 492 | 24.9 | 137.3 | 1,159,300 | 2,355 | 96,054 | 32.3 | 44,331 | 54.7 | 45.3 | 9 | 68.7 | 9,154 | 22.8 |
| Jim Hogg | 692 | 7.3 | 2,836 | 0.3 | 10.4 | 4,926,365 | 1,737 | 76,129 | 10.4 | 42,816 | 2.0 | 98.0 | NA | 59.4 | 529 | 9.0 |
| Jim Wells | 426 | -15.4 | 348 | 2.4 | 158.1 | 720,153 | 2,069 | 70,377 | 121.6 | 99,379 | 30.2 | 69.8 | NA | 65.8 | 3,855 | 11.0 |
| Johnson | 411 | -4.2 | 131 | 3.7 | 136.7 | 517,349 | 3,951 | 57,125 | 57.9 | 18,424 | 29.6 | 70.4 | 1 | 79.3 | 3,814 | 4.1 |
| Jones | 517 | -8.2 | 564 | 4.6 | 313.8 | 644,140 | 1,141 | 91,000 | 41.5 | 45,344 | 72.0 | 28.0 | NA | 74.5 | 4,470 | 37.4 |
| Karnes | 432 | -7.1 | 356 | 0.7 | 75 | 870,685 | 2,446 | 81,068 | 29.4 | 24,267 | 37.4 | 62.6 | NA | 61.7 | 508 | 5.0 |
| Kaufman | 455 | 1.3 | 164 | 1.7 | 133.6 | 493,007 | 3,010 | 59,513 | 57.1 | 20,541 | 27.0 | 73.0 | NA | 72.4 | 1,551 | 2.2 |
| Kendall | 394 | 6.5 | 292 | 0.7 | 32.4 | 904,088 | 3,096 | 41,444 | 12.4 | 9,222 | 9.7 | 90.3 | NA | 77.8 | 1,207 | 6.6 |
| Kenedy | 869 | -5.2 | 28,961 | 0.7 | 1.9 | 21,350,824 | 737 | 105,455 | D | D | D | D | NA | 83.3 | D | 6.7 |
| Kent | 578 | 2.6 | 3,522 | D | 45.2 | 3,017,965 | 857 | 73,840 | 9.9 | 60,159 | 10.4 | 89.6 | NA | 66.5 | 1,025 | 47.6 |
| Kerr | 518 | -11.1 | 459 | 0.8 | 16 | 1,145,687 | 2,497 | 48,881 | 9.3 | 8,268 | 16.9 | 83.1 | 1 | 73.9 | 225 | 3.0 |
| Kimble | 746 | 7.5 | 1,114 | 2.2 | 11.3 | 2,218,386 | 1,992 | 49,335 | 10.9 | 16,197 | 8.2 | 91.8 | NA | 74.6 | 743 | 6.9 |
| King | 417 | -0.1 | 9,700 | D | 10 | 6,723,919 | 693 | 95,161 | 13.8 | 320,140 | 0.9 | 99.1 | NA | 67.4 | 320 | 60.5 |
| Kinney | 587 | 1.8 | 2,487 | 2.3 | 15.8 | 3,086,936 | 1,241 | 69,026 | 5.0 | 21,377 | 10.3 | 89.7 | 3 | 60.6 | 386 | 11.4 |
| Kleberg | 483 | -0.3 | 1,051 | 0 | 65.6 | 1,387,563 | 1,320 | 100,068 | 52.8 | 114,996 | 41.6 | 58.4 | 1 | 67.1 | 1,537 | 15.0 |
| Knox | 489 | 8.5 | 2,263 | 11.2 | 207.4 | 2,520,181 | 1,114 | 211,879 | 60.5 | 280,222 | 22.4 | 77.6 | NA | 76.9 | 3,982 | 70.8 |
| Lamar | 464 | -6.6 | 238 | 5 | 170.7 | 597,545 | 2,507 | 80,089 | 73.4 | 37,738 | 34.0 | 66.0 | NA | 73.0 | 8,480 | 34.2 |
| Lamb | 569 | -7.6 | 733 | 173.9 | 489.3 | 912,685 | 1,246 | 268,352 | 537.3 | 691,529 | 22.1 | 77.9 | 3 | 64.7 | 16,363 | 81.5 |
| Lampasas | 469 | 5.5 | 407 | 0.4 | 40.4 | 1,152,472 | 2,828 | 61,081 | 18.4 | 16,021 | 10.9 | 89.1 | 2 | 72.5 | 381 | 6.7 |
| La Salle | 533 | -16.1 | 1,391 | 2.1 | 19.8 | 2,203,808 | 1,584 | 120,717 | 6.3 | 16,444 | 8.5 | 91.5 | NA | 58.2 | 165 | 3.1 |
| Lavaca | 507 | -7.3 | 175 | 2.6 | 74.2 | 613,235 | 3,511 | 58,515 | 50.5 | 17,430 | 14.8 | 85.2 | 3 | 65.1 | 856 | 7.0 |
| Lee | 329 | 3.3 | 182 | 0.8 | 41.4 | 667,815 | 3,676 | 54,124 | 56.9 | 31,478 | 26.7 | 73.3 | NA | 67.6 | 608 | 6.1 |
| Leon | 488 | -18.0 | 250 | 2 | 71.2 | 759,052 | 3,037 | 71,907 | 169.4 | 86,829 | 5.9 | 94.1 | 1 | 70.3 | 144 | 0.8 |
| Liberty | 252 | -12.0 | 164 | 5.2 | 68.3 | 493,738 | 3,008 | 74,395 | 30.0 | 19,473 | 40.3 | 59.7 | NA | 75.0 | 3,962 | 4.1 |
| Limestone | 493 | 1.2 | 384 | 0.5 | 73 | 706,578 | 1,842 | 71,310 | 66.3 | 51,602 | 15.1 | 84.9 | NA | 68.5 | 678 | 6.2 |
| Lipscomb | 586 | -0.9 | 1,961 | 23.8 | 115.7 | 2,204,190 | 1,124 | 137,377 | 79.3 | 265,204 | 21.5 | 78.5 | NA | 72.2 | 4,833 | 65.2 |
| Live Oak | 455 | -15.9 | 531 | 2.4 | 47.4 | 1,147,237 | 2,159 | 84,738 | 19.5 | 22,723 | 25.8 | 74.2 | NA | 74.2 | 717 | 10.3 |
| Llano | 523 | -0.9 | 627 | 0.5 | 21.6 | 1,642,693 | 2,620 | 54,170 | 15.7 | 18,832 | 9.5 | 90.5 | NA | 76.8 | 355 | 4.9 |
| Loving | 468 | 23.3 | 58,518 | D | D | 11,163,974 | 191 | 146,059 | D | D | NA | D | NA | 62.5 | D | 37.5 |
| Lubbock | 531 | 5.6 | 514 | 166.7 | 451.4 | 874,304 | 1,702 | 186,233 | 219.5 | 212,458 | 58.0 | 42.0 | 2 | 82.0 | 4,685 | 41.7 |
| Lynn | 499 | 5.7 | 1,150 | 83.1 | 456.3 | 1,158,640 | 1,007 | 318,818 | 111.4 | 256,758 | 95.7 | 4.3 | 10 | 70.5 | 6,493 | 73.3 |
| McCulloch | 563 | -8.3 | 826 | 1.9 | 83.7 | 1,627,004 | 1,971 | 73,710 | 22.5 | 32,978 | 30.5 | 69.5 | NA | 77.9 | 2,261 | 31.5 |
| McLennan | 573 | 3.6 | 170 | 2.2 | 262.5 | 614,516 | 3,608 | 67,948 | 179.7 | 53,377 | 33.1 | 66.9 | 4 | 72.5 | 5,860 | 11.7 |
| McMullen | 452 | -12.7 | 2,365 | D | 26.7 | 4,293,308 | 1,816 | 111,599 | 8.3 | 43,586 | 7.5 | 92.5 | NA | 77.0 | 360 | 5.8 |
| Madison | 246 | -15.7 | 251 | 1.3 | 34 | 752,922 | 2,996 | 87,971 | D | D | D | D | 1 | 71.0 | 60 | 1.4 |
| Marion | 50 | 24.5 | 178 | D | 13.6 | 357,305 | 2,003 | 53,364 | 5.9 | 21,021 | 9.2 | 90.8 | NA | 71.1 | 151 | 7.1 |
| Martin | 445 | -2.0 | 1,249 | 12.2 | 298.9 | 1,111,299 | 890 | 191,065 | 54.3 | 152,522 | 96.7 | 3.3 | NA | 79.8 | 5,464 | 68.5 |

Items 117—132

# Table B. States and Counties — Water Use, Wholesale Trade, Retail Trade, and Real Estate

| STATE County | Water use, 2015 Public supply water withdrawn (mil gal/day) | Public supply gallons withdrawn per person per day | Wholesale Trade[1], 2017 Number of establishments | Number of employees | Sales (mil dol) | Average payroll (mil dol) | Retail Trade[2], 2017 Number of establishments | Number of employees | Sales (mil dol) | Average payroll (mil dol) | Real estate and rental and leasing,[2] 2017 Number of establishments | Number of employees | Sales (mil dol) | Average payroll (mil dol) |
|---|---|---|---|---|---|---|---|---|---|---|---|---|---|---|
| | 133 | 134 | 135 | 136 | 137 | 138 | 139 | 140 | 141 | 142 | 143 | 144 | 145 | 146 |
| **TEXAS—Cont'd** | | | | | | | | | | | | | | |
| Guadalupe | 6.85 | 45.3 | 95 | 1,759 | 1,066.4 | 99.9 | 269 | 4,723 | 1,675.0 | 139.9 | 109 | 410 | 95.3 | 17.2 |
| Hale | 1.44 | 41.9 | 50 | 542 | 381.6 | 24.7 | 104 | 1,356 | 453.6 | 33.5 | 26 | 64 | 11.6 | 1.7 |
| Hall | 0.04 | 12.7 | D | D | D | 0.6 | 12 | 91 | 33.4 | 1.8 | NA | NA | NA | NA |
| Hamilton | 0.16 | 19.6 | 7 | 69 | 37.5 | 3.3 | 47 | 381 | 88.4 | 9.4 | NA | NA | NA | NA |
| Hansford | 0.96 | 171.1 | 17 | 110 | 140.2 | 5.3 | 25 | 218 | 72.9 | 6.8 | 5 | D | 0.7 | D |
| Hardeman | 0.07 | 18.2 | D | D | D | 1.0 | 14 | 142 | 60.7 | 2.7 | NA | NA | NA | NA |
| Hardin | 4.03 | 72.1 | D | D | D | D | 153 | 2,248 | 964.9 | 68.1 | D | D | D | D |
| Harris | 287.29 | 63.3 | 6,540 | 107,902 | 386,380.7 | 7,607.6 | 13,165 | 212,528 | 70,018.5 | 6,167.3 | 5,942 | 44,290 | 13,375.1 | 2,422.8 |
| Harrison | 8.53 | 127.8 | D | D | D | 37.7 | 175 | 1,916 | 604.6 | 52.1 | 66 | 308 | 75.0 | 19.4 |
| Hartley | 1.02 | 164.7 | 7 | 75 | 46.1 | 3.9 | 11 | 204 | 53.6 | 3.8 | 6 | 25 | 3.2 | 0.7 |
| Haskell | 0.14 | 24.4 | 3 | 3 | 2.7 | 0.1 | 26 | 346 | 118.6 | 9.4 | 3 | 2 | 0.4 | 0.0 |
| Hays | 8.21 | 42.2 | 146 | 1,444 | 1,153.9 | 99.2 | 658 | 15,180 | 4,967.7 | 372.5 | 261 | 1,246 | 289.8 | 48.8 |
| Hemphill | 0.50 | 117.3 | 14 | 275 | 145.6 | 20.8 | 18 | 149 | 35.6 | 3.6 | 9 | 18 | 6.0 | 1.0 |
| Henderson | 4.55 | 57.2 | 37 | 268 | 127.9 | 10.7 | 238 | 2,738 | 755.3 | 64.8 | 58 | 298 | 39.6 | 8.1 |
| Hidalgo | 64.64 | 76.7 | 871 | 9,035 | 5,467.8 | 353.5 | 2,162 | 37,469 | 9,923.4 | 887.1 | 571 | 2,280 | 490.1 | 68.7 |
| Hill | 2.97 | 85.2 | 23 | 165 | 76.2 | 6.0 | 133 | 1,413 | 430.6 | 33.4 | 27 | 67 | 11.9 | 2.0 |
| Hockley | 0.50 | 21.3 | 32 | 271 | 305.9 | 18.6 | 59 | 828 | 236.0 | 20.1 | 14 | 60 | 10.0 | 3.3 |
| Hood | 4.85 | 87.5 | 43 | 357 | 139.9 | 14.2 | 186 | 2,881 | 990.1 | 79.8 | 70 | 347 | 64.3 | 15.4 |
| Hopkins | 11.05 | 305.1 | 30 | 956 | 1,242.9 | 47.7 | 145 | 1,647 | 576.5 | 42.8 | 28 | 119 | 22.9 | 3.1 |
| Houston | 2.34 | 102.7 | D | D | D | 2.7 | 59 | 707 | 182.1 | 16.5 | 9 | 48 | 7.8 | 1.2 |
| Howard | 0.03 | 0.8 | 31 | 221 | 230.5 | 14.1 | D | D | D | D | 41 | 127 | 36.1 | 4.4 |
| Hudspeth | 0.24 | 71.0 | NA | NA | NA | NA | 8 | 49 | 8.6 | 0.7 | NA | NA | NA | NA |
| Hunt | 4.87 | 54.2 | 60 | 723 | 474.4 | 41.2 | 265 | 3,962 | 1,263.0 | 109.3 | 72 | 264 | 49.9 | 11.3 |
| Hutchinson | 5.30 | 243.9 | 20 | 157 | 93.6 | 7.8 | 68 | 950 | 244.3 | 22.6 | 11 | 61 | 13.3 | 2.1 |
| Irion | 0.08 | 51.5 | NA | NA | NA | NA | 5 | 22 | 5.9 | 0.5 | NA | NA | NA | NA |
| Jack | 0.55 | 62.0 | 8 | 20 | 42.1 | 1.3 | 26 | 168 | 51.6 | 4.2 | 9 | 21 | 3.5 | 0.4 |
| Jackson | 0.90 | 60.7 | 12 | 114 | 103.8 | 5.9 | 50 | 518 | 132.0 | 11.7 | 13 | 23 | 2.4 | 0.4 |
| Jasper | 2.97 | 83.6 | 30 | 215 | 141.1 | 9.6 | 114 | 1,452 | 434.9 | 34.9 | 19 | 67 | 14.1 | 2.7 |
| Jeff Davis | 1.12 | 519.5 | NA | NA | NA | NA | 11 | 68 | 10.4 | 1.1 | NA | NA | NA | NA |
| Jefferson | 23.95 | 94.2 | 275 | 3,601 | 2,854.6 | 203.9 | 950 | 13,584 | 4,419.5 | 388.2 | 294 | 2,203 | 541.5 | 103.5 |
| Jim Hogg | 0.67 | 128.8 | 4 | 8 | 2.8 | 0.3 | 18 | 188 | 66.1 | 4.6 | NA | NA | NA | NA |
| Jim Wells | 3.95 | 95.5 | 37 | 328 | 210.7 | 22.8 | 120 | 1,546 | 473.1 | 40.6 | 31 | 280 | 62.2 | 17.6 |
| Johnson | 10.33 | 64.6 | 139 | 1,306 | 798.2 | 63.2 | 404 | 6,006 | 2,255.8 | 182.8 | 141 | 759 | 159.3 | 38.3 |
| Jones | 8.07 | 404.1 | 15 | 116 | 166.9 | 5.9 | 37 | 239 | 192.7 | 9.1 | 7 | 24 | 2.5 | 0.5 |
| Karnes | 2.75 | 183.6 | 21 | 119 | 110.6 | 5.5 | 41 | 548 | 151.4 | 15.0 | 16 | 102 | 34.0 | 6.1 |
| Kaufman | 0.85 | 7.4 | 77 | 892 | 386.8 | 40.2 | 298 | 4,326 | 1,351.5 | 111.7 | 67 | 217 | 35.4 | 6.7 |
| Kendall | 1.66 | 41.1 | 56 | 498 | 314.6 | 25.5 | 174 | 2,754 | 1,381.1 | 95.7 | 78 | 200 | 40.0 | 7.6 |
| Kenedy | 0.07 | 172.0 | NA | NA | NA | NA | NA | NA | NA | NA | NA | NA | NA | NA |
| Kent | 0.08 | 104.7 | NA | NA | NA | NA | 3 | 15 | 9.2 | 0.4 | NA | NA | NA | NA |
| Kerr | 5.58 | 109.5 | D | D | D | 9.2 | 203 | 2,733 | 887.8 | 81.0 | 83 | 332 | 50.4 | 12.0 |
| Kimble | 0.45 | 102.6 | NA | NA | NA | NA | 34 | 267 | 123.1 | 6.6 | 4 | 14 | 1.8 | 0.5 |
| King | 0.19 | 673.8 | NA | NA | NA | NA | NA | NA | NA | NA | NA | NA | NA | NA |
| Kinney | 0.82 | 231.1 | NA | NA | NA | NA | 8 | 40 | 6.2 | 0.6 | NA | NA | NA | NA |
| Kleberg | 3.31 | 103.9 | D | D | D | D | 98 | 1,409 | 444.2 | 37.8 | D | D | D | D |
| Knox | 0.14 | 36.3 | 10 | 85 | 31.9 | 5.7 | 17 | 114 | 30.5 | 2.7 | NA | NA | NA | NA |
| Lamar | 14.84 | 300.2 | 55 | 419 | 160.5 | 16.5 | 212 | 2,629 | 903.1 | 74.8 | 46 | 173 | 26.0 | 4.5 |
| Lamb | 1.35 | 100.9 | 21 | 173 | 128.1 | 8.3 | 33 | 317 | 91.6 | 7.2 | NA | NA | NA | NA |
| Lampasas | 2.71 | 131.6 | 10 | 43 | 16.5 | 2.3 | 47 | 676 | 254.7 | 20.9 | 16 | 128 | 19.4 | 5.5 |
| La Salle | 1.36 | 178.2 | D | D | D | D | 22 | 303 | 177.3 | 18.1 | 11 | 30 | 5.2 | 0.9 |
| Lavaca | 1.44 | 72.6 | 19 | 181 | 97.5 | 12.3 | 82 | 737 | 177.3 | 18.1 | 11 | 102 | 44.5 | 5.4 |
| Lee | 3.97 | 234.9 | 22 | 152 | 53.4 | 6.8 | 63 | 618 | 158.6 | 16.0 | 16 | 78 | 16.2 | 4.2 |
| Leon | 1.90 | 111.2 | 15 | 125 | 82.4 | 7.5 | 62 | 590 | 182.3 | 12.8 | 40 | 136 | 25.6 | 5.8 |
| Liberty | 6.42 | 80.6 | 43 | 500 | 756.4 | 28.5 | 197 | 2,547 | 842.3 | 72.5 | 12 | 65 | 6.3 | 1.2 |
| Limestone | 1.95 | 83.6 | D | D | D | D | 86 | 941 | 270.2 | 22.6 | NA | NA | NA | NA |
| Lipscomb | 0.81 | 227.0 | NA | NA | NA | NA | 11 | 152 | 62.2 | 5.1 | 9 | 19 | 5.6 | 1.0 |
| Live Oak | 2.76 | 225.7 | 14 | 170 | 97.8 | 9.9 | 41 | 391 | 208.2 | 11.8 | 22 | 62 | 8.4 | 1.7 |
| Llano | 2.87 | 145.0 | 13 | 217 | 108.4 | 10.2 | 67 | 597 | 175.1 | 14.9 | NA | NA | NA | NA |
| Loving | 0.01 | 89.3 | NA | NA | NA | NA | NA | NA | NA | NA | NA | NA | NA | NA |
| Lubbock | 1.32 | 4.4 | 382 | 5,412 | 5,588.6 | 295.7 | 987 | 18,278 | 5,616.9 | 503.5 | 446 | 1,998 | 411.4 | 75.1 |
| Lynn | 0.15 | 26.2 | 7 | 49 | 30.9 | 1.8 | 8 | 72 | 22.5 | 1.6 | NA | NA | NA | NA |
| McCulloch | 2.69 | 322.5 | 4 | 59 | 25.5 | 1.8 | 47 | 465 | 152.9 | 13.0 | 7 | 22 | 1.5 | 0.3 |
| McLennan | 41.41 | 168.6 | 246 | 2,789 | 1,828.0 | 139.6 | 837 | 12,847 | 3,711.8 | 340.0 | 258 | 1,692 | 378.2 | 77.8 |
| McMullen | 0.10 | 122.0 | NA | NA | NA | NA | 5 | 67 | 26.2 | 2.2 | NA | NA | NA | NA |
| Madison | 2.37 | 168.5 | D | D | D | D | 37 | 664 | 299.2 | 21.3 | 12 | 23 | 5.2 | 1.1 |
| Marion | 0.35 | 34.4 | NA | NA | NA | NA | 27 | 228 | 60.7 | 4.6 | 6 | 8 | 1.3 | 0.2 |
| Martin | 1.48 | 262.4 | 8 | 47 | 45.4 | 2.9 | 13 | 168 | 115.4 | 7.4 | D | D | D | 1.0 |

1 Merchant wholesalers, except manufacturers' sales branches and offices.     2. Employer establishments.

| STATE County | Professional, scientific, and technical services, 2017 | | | | Manufacturing, 2017 | | | | Accommodation and food services, 2017 | | | |
|---|---|---|---|---|---|---|---|---|---|---|---|---|
| | Number of establishments | Number of employees | Sales (mil dol) | Average payroll (mil dol) | Number of establishments | Number of employees | Sales (mil dol) | Average payroll (mil dol) | Number of establishments | Number of employees | Sales (mil dol) | Annual payroll (mil dol) |
| | 147 | 148 | 149 | 150 | 151 | 152 | 153 | 154 | 155 | 156 | 157 | 158 |
| **TEXAS—Cont'd** | | | | | | | | | | | | |
| Guadalupe | 151 | 950 | 156 | 49.8 | 115 | 7,103 | 4,583.7 | 416.7 | 204 | 3,949 | 210.3 | 57.3 |
| Hale | D | D | D | D | 18 | 607 | 477.8 | 23.3 | 62 | 1,187 | 55.5 | 15.9 |
| Hall | NA | NA | NA | NA | 4 | 14 | 1.8 | 0.6 | D | D | D | 0.5 |
| Hamilton | 17 | 51 | 5 | 1.9 | 14 | 185 | 74.9 | 8.8 | 24 | 266 | 10.6 | 3.2 |
| Hansford | 9 | 39 | 5 | 1.9 | 4 | 16 | 5.6 | 0.9 | 10 | 70 | 3.7 | 0.9 |
| Hardeman | NA | NA | NA | NA | NA | NA | NA | NA | D | D | D | D |
| Hardin | 46 | 309 | 42 | 20.0 | D | 706 | D | 44.1 | 86 | 1,476 | 67.0 | 19.1 |
| Harris | 14,034 | 189,107 | 44,253 | 17,423.5 | 4,023 | 148,629 | 146,184.2 | 10,003.4 | 9,616 | 205,281 | 13,797.5 | 3,815.8 |
| Harrison | 120 | 672 | 115 | 38.1 | 72 | 3,945 | 2,567.2 | 240.6 | 103 | 1,830 | 93.4 | 27.5 |
| Hartley | D | D | D | D | NA | NA | NA | NA | D | D | D | 0.5 |
| Haskell | 9 | 44 | 3 | 0.9 | NA | NA | NA | NA | 16 | 109 | 5.5 | 1.3 |
| Hays | 450 | 2,200 | 313 | 105.6 | 182 | 4,443 | 1,414.8 | 221.6 | 454 | 10,154 | 541.9 | 156.8 |
| Hemphill | 12 | 51 | 12 | 2.9 | 5 | 21 | 6.6 | 1.0 | D | D | D | D |
| Henderson | 94 | 753 | 86 | 25.2 | 54 | 1,437 | 295.6 | 71.2 | 154 | 2,082 | 104.1 | 29.1 |
| Hidalgo | D | D | 720 | D | 280 | 5,703 | 2,136.8 | 225.8 | 1,112 | 23,976 | 1,254.7 | 329.5 |
| Hill | 32 | 143 | 17 | 6.9 | 42 | 953 | 478.6 | 46.1 | 73 | 1,054 | 60.7 | 15.8 |
| Hockley | 25 | 90 | 9 | 3.2 | 14 | 277 | 230.7 | 16.1 | 42 | 691 | 31.0 | 8.7 |
| Hood | 132 | 426 | 57 | 20.3 | 38 | 379 | 78.9 | 17.1 | 112 | 1,918 | 107.7 | 28.7 |
| Hopkins | 51 | 193 | 25 | 7.7 | 40 | 1,646 | 940.9 | 81.0 | 66 | 1,048 | 52.8 | 16.0 |
| Houston | 29 | 91 | 12 | 3.3 | 13 | 451 | 204.9 | 33.2 | 32 | 391 | 20.0 | 5.1 |
| Howard | D | D | D | D | D | D | D | D | D | D | D | D |
| Hudspeth | NA | NA | NA | NA | NA | NA | NA | NA | 6 | 37 | 1.6 | 0.7 |
| Hunt | 92 | 1,069 | 261 | 57.1 | 70 | 8,659 | 2,974.1 | 705.8 | 156 | 2,708 | 150.4 | 41.1 |
| Hutchinson | 27 | 279 | 40 | 19.7 | 19 | 1,673 | 5,756.7 | 166.5 | 51 | 724 | 32.0 | 8.4 |
| Irion | 6 | 8 | 1 | 0.1 | NA | NA | NA | NA | D | D | D | D |
| Jack | 10 | 46 | 6 | 2.2 | 6 | 33 | 16.2 | 2.4 | D | D | D | D |
| Jackson | 24 | 198 | 13 | 6.4 | D | D | D | D | 24 | 351 | 15.6 | 4.2 |
| Jasper | 36 | 152 | 25 | 5.0 | 23 | 1,186 | 672.3 | 93.5 | 53 | 817 | 40.5 | 11.0 |
| Jeff Davis | D | D | D | 0.2 | NA | NA | NA | NA | 13 | 117 | 6.8 | 1.9 |
| Jefferson | D | D | D | D | 185 | 13,150 | 47,196.4 | 1,352.0 | 512 | 10,395 | 583.1 | 155.2 |
| Jim Hogg | NA | NA | NA | NA | 5 | 40 | 24.9 | 1.8 | 11 | 64 | 3.8 | 0.8 |
| Jim Wells | 52 | 294 | 71 | 13.4 | 18 | 298 | 74.3 | 17.9 | 87 | 1,118 | 61.6 | 16.9 |
| Johnson | 197 | 716 | 103 | 32.8 | 154 | 5,752 | 1,778.9 | 300.7 | 251 | 4,512 | 238.0 | 60.9 |
| Jones | 11 | 34 | 3 | 1.5 | 8 | 126 | 57.4 | 8.2 | 17 | 105 | 5.6 | 1.4 |
| Karnes | 21 | 87 | 9 | 3.6 | 10 | 228 | 75.7 | 9.6 | 44 | 476 | 34.2 | 8.2 |
| Kaufman | 136 | 772 | 96 | 33.3 | 111 | 4,238 | 1,203.0 | 214.4 | 192 | 3,200 | 180.3 | 47.6 |
| Kendall | 178 | 801 | 127 | 50.4 | 42 | 950 | 395.3 | 53.8 | 108 | 1,606 | 99.1 | 31.1 |
| Kenedy | NA | NA | NA | NA | NA | NA | NA | NA | NA | NA | NA | NA |
| Kent | NA | NA | NA | NA | NA | NA | NA | NA | NA | NA | NA | NA |
| Kerr | D | D | D | D | 50 | 795 | 116.2 | 34.4 | 143 | 2,345 | 157.2 | 46.9 |
| Kimble | 7 | 11 | 1 | 0.4 | 7 | 57 | 8.2 | 2.1 | D | D | D | D |
| King | NA | NA | NA | NA | NA | NA | NA | NA | D | D | D | D |
| Kinney | NA | NA | NA | NA | NA | NA | NA | NA | NA | NA | NA | NA |
| Kleberg | 33 | 145 | 19 | 5.1 | 15 | 57 | 25.1 | 3.0 | 6 | D | 1.3 | D |
| Knox | NA | NA | NA | NA | NA | NA | NA | NA | D | D | D | 0.5 |
| Lamar | D | D | D | D | 56 | 4,769 | 2,631.3 | 236.5 | 116 | 1,933 | 95.3 | 26.2 |
| Lamb | 14 | 48 | 8 | 1.6 | 6 | 74 | 15.5 | 3.7 | 18 | 210 | 10.3 | 2.6 |
| Lampasas | 36 | 131 | 16 | 5.8 | 21 | 612 | 167.8 | 25.4 | 38 | 423 | 25.1 | 6.6 |
| La Salle | NA | NA | NA | NA | NA | NA | NA | NA | D | D | D | D |
| Lavaca | 35 | 151 | 23 | 9.5 | 33 | 1,415 | 382.0 | 60.6 | 35 | 385 | 19.1 | 5.7 |
| Lee | 26 | 101 | 10 | 3.4 | 15 | 440 | 92.8 | 19.2 | 38 | 373 | 20.7 | 5.1 |
| Leon | 15 | 55 | 6 | 1.7 | 16 | 650 | 574.3 | 68.3 | 39 | 403 | 20.2 | 4.8 |
| Liberty | D | D | D | D | 34 | 986 | 341.8 | 55.8 | 95 | 1,523 | 92.9 | 23.9 |
| Limestone | 21 | 117 | 21 | 6.3 | 10 | 539 | 88.0 | 19.0 | 38 | 434 | 25.1 | 5.7 |
| Lipscomb | 6 | 21 | 2 | 0.9 | NA | NA | NA | NA | D | D | D | 0.6 |
| Live Oak | 26 | 74 | 12 | 3.1 | D | D | D | D | 41 | 348 | 27.7 | 5.8 |
| Llano | 39 | 147 | 17 | 6.0 | 16 | 63 | 19.1 | 3.7 | 48 | 1,145 | 77.2 | 28.1 |
| Loving | NA | NA | NA | NA | NA | NA | NA | NA | NA | NA | NA | NA |
| Lubbock | 653 | 4,497 | 615 | 231.6 | 225 | 4,583 | 1,503.7 | 232.7 | 706 | 16,429 | 936.1 | 263.6 |
| Lynn | D | D | 0 | D | D | D | D | D | 6 | 17 | 0.7 | 0.1 |
| McCulloch | 17 | 57 | 5 | 1.7 | 10 | 204 | 24.3 | 8.9 | 24 | 276 | 14.7 | 3.7 |
| McLennan | 388 | 2,720 | 521 | 164.1 | 249 | 13,777 | 7,597.6 | 829.8 | 523 | D | 620.5 | D |
| McMullen | NA | NA | NA | NA | NA | NA | NA | NA | 3 | 26 | 2.0 | 0.6 |
| Madison | 16 | 75 | 9 | 2.6 | 7 | 36 | 7.3 | 1.5 | 44 | 609 | 29.0 | 7.4 |
| Marion | 11 | 13 | 2 | 0.7 | 8 | 310 | 203.4 | 18.6 | 24 | 396 | 14.1 | 5.0 |
| Martin | D | D | D | 1.4 | 3 | 11 | 1.3 | 0.3 | D | D | D | D |

# Table B. States and Counties — Health Care and Social Assistance, Other Services, Nonemployer Businesses, and Residential Construction

| STATE County | Health care and social assistance, 2017 | | | | Other services, 2017 | | | | Nonemployer businesses, 2018 | | Value of residential construction authorized by building permits, 2020 | |
|---|---|---|---|---|---|---|---|---|---|---|---|---|
| | Number of establishments | Number of employees | Receipts (mil dol) | Annual payroll (mil dol) | Number of establishments | Number of employees | Receipts (mil dol) | Annual payroll (mil dol) | Number | Receipts (mil dol) | New construction ($1,000) | Number of housing units |
| | 159 | 160 | 161 | 162 | 163 | 164 | 165 | 166 | 167 | 168 | 169 | 170 |
| **TEXAS—Cont'd** | | | | | | | | | | | | |
| Guadalupe | 213 | 3,227 | 353.7 | 137.2 | 161 | 989 | 106.4 | 32.1 | 12,239 | 513.3 | 324,703 | 1,549 |
| Hale | 74 | 1,136 | 115.4 | 39.7 | 53 | 354 | 30.5 | 9.8 | 2,150 | 98.5 | 6,843 | 22 |
| Hall | 6 | D | 3.6 | D | 6 | 20 | 2.4 | 0.5 | 203 | 4.9 | 0 | 0 |
| Hamilton | 21 | 558 | 41.5 | 19.6 | 20 | 50 | 6.5 | 1.4 | 746 | 42.5 | 1,416 | 13 |
| Hansford | D | D | D | D | 11 | 45 | 4.6 | 1.6 | 530 | 27.4 | 615 | 3 |
| Hardeman | 9 | 138 | 13.7 | 6.0 | 11 | 47 | 4.4 | 1.1 | 258 | 9.3 | 0 | 0 |
| Hardin | 57 | 946 | 64.6 | 27.5 | 56 | 335 | 35.6 | 10.4 | 3,947 | 188.7 | 64,716 | 413 |
| Harris | 11,829 | 275,105 | 38,411.2 | 14,179.2 | 6,301 | 65,343 | 9,616.9 | 2,832.9 | 456,877 | 22,852.5 | 5,716,826 | 36,961 |
| Harrison | 116 | 1,799 | 158.7 | 66.8 | 73 | 821 | 82.3 | 20.2 | 4,917 | 237.4 | 19,117 | 82 |
| Hartley | 10 | 317 | 30.9 | 10.5 | D | D | 17.2 | 0.8 | 346 | 20.4 | NA | NA |
| Haskell | 8 | 133 | 11.2 | 5.8 | 17 | 48 | 3.6 | 0.8 | 398 | 18.6 | 0 | 0 |
| Hays | 391 | 7,379 | 854.9 | 332.9 | 267 | 1,531 | 177.7 | 51.0 | 19,461 | 962.4 | 983,300 | 4,925 |
| Hemphill | 7 | 209 | 12.7 | 7.1 | D | D | D | D | 420 | 20.9 | 0 | 0 |
| Henderson | 132 | 2,304 | 359.1 | 135.6 | 77 | 370 | 36.7 | 10.4 | 6,356 | 328.4 | 46,664 | 203 |
| Hidalgo | 2,193 | 64,956 | 4,371.9 | 1,772.6 | 592 | 3,522 | 322.6 | 90.2 | 76,358 | 3,229.1 | 774,933 | 4,894 |
| Hill | 48 | 1,094 | 91.9 | 37.0 | 38 | 140 | 12.1 | 3.5 | 2,577 | 121.6 | 8,106 | 50 |
| Hockley | 44 | 1,043 | 62.1 | 25.5 | D | D | D | D | 1,496 | 70.6 | 2,043 | 9 |
| Hood | 145 | 2,130 | 229.0 | 82.5 | 95 | 679 | 49.2 | 16.3 | 6,070 | 316.8 | 90,918 | 389 |
| Hopkins | 57 | 1,527 | 149.7 | 50.1 | 54 | 210 | 21.9 | 5.8 | 2,975 | 156.5 | 2,699 | 13 |
| Houston | 29 | 730 | 41.9 | 18.7 | 24 | 147 | 16.3 | 4.7 | 1,378 | 56.0 | 0 | 0 |
| Howard | 75 | 2,475 | 329.0 | 135.7 | D | D | D | D | 1,881 | 89.2 | 8,833 | 45 |
| Hudspeth | NA | NA | NA | NA | NA | NA | NA | NA | 270 | 10.1 | NA | NA |
| Hunt | 175 | 3,469 | 358.0 | 149.7 | 101 | 451 | 43.7 | 12.4 | 7,434 | 376.9 | 138,389 | 904 |
| Hutchinson | 42 | 586 | 64.2 | 21.7 | 28 | 197 | 24.2 | 6.8 | 1,011 | 38.4 | 209 | 1 |
| Irion | NA | NA | NA | NA | D | D | 2.3 | D | 197 | 13.1 | 0 | 0 |
| Jack | 12 | 258 | 26.2 | 9.4 | 15 | 36 | 5.1 | 1.1 | 781 | 42.6 | 2,221 | 9 |
| Jackson | 18 | 305 | 26.4 | 12.1 | 25 | 82 | 11.3 | 2.7 | 1,128 | 52.0 | 33,821 | 277 |
| Jasper | 80 | 2,535 | 139.6 | 54.4 | 35 | 163 | 14.3 | 3.7 | 2,501 | 93.2 | NA | NA |
| Jeff Davis | D | D | D | 2.7 | D | D | D | 0.6 | 280 | 9.0 | NA | NA |
| Jefferson | 900 | 17,612 | 1,945.4 | 675.6 | 372 | 3,569 | 367.8 | 164.9 | 16,695 | 764.0 | 189,094 | 1,115 |
| Jim Hogg | 10 | 747 | 15.7 | 10.2 | NA | NA | NA | NA | 414 | 12.1 | NA | NA |
| Jim Wells | 112 | 5,419 | 216.7 | 100.9 | 53 | 243 | 33.3 | 8.9 | 2,950 | 121.7 | 3,018 | 17 |
| Johnson | 268 | 3,488 | 361.2 | 130.7 | 216 | 1,549 | 174.3 | 60.7 | 14,539 | 727.9 | 435,261 | 2,582 |
| Jones | 20 | 548 | 66.5 | 18.7 | 18 | 44 | 3.0 | 1.0 | 1,278 | 57.7 | 300 | 1 |
| Karnes | 18 | 441 | 35.5 | 15.7 | 14 | 43 | 4.9 | 1.2 | 962 | 39.5 | 13,246 | 72 |
| Kaufman | 173 | 3,498 | 292.0 | 130.9 | 142 | 923 | 119.1 | 41.0 | 12,155 | 598.5 | 219,316 | 1,068 |
| Kendall | 158 | 1,912 | 184.5 | 72.5 | 88 | 518 | 68.4 | 18.3 | 5,801 | 358.6 | 73,203 | 314 |
| Kenedy | NA | NA | NA | NA | 5 | 85 | 51.1 | 5.6 | 24 | 0.7 | NA | NA |
| Kent | NA | NA | NA | NA | NA | NA | NA | NA | 61 | 2.5 | NA | NA |
| Kerr | 183 | 3,690 | 478.5 | 194.3 | 104 | 642 | 86.1 | 21.3 | 5,626 | 301.5 | 17,659 | 91 |
| Kimble | 7 | 135 | 9.4 | 4.9 | 9 | 34 | 3.0 | 0.8 | 624 | 22.7 | 0 | 0 |
| King | NA | NA | NA | NA | NA | NA | NA | NA | 38 | 1.4 | 0 | 0 |
| Kinney | 4 | 11 | 1.7 | 0.4 | D | D | 3.5 | D | 216 | 7.4 | 3,609 | 44 |
| Kleberg | 75 | 1,347 | 107.6 | 42.2 | 39 | 161 | 17.0 | 3.2 | 1,726 | 59.7 | 0 | 0 |
| Knox | D | D | D | D | D | D | 2.2 | D | 240 | 9.3 | 0 | 0 |
| Lamar | 196 | 3,655 | 365.0 | 136.5 | 80 | 408 | 44.2 | 11.7 | 4,011 | 177.3 | 5,656 | 34 |
| Lamb | 28 | 551 | 36.0 | 15.6 | 19 | 44 | 5.8 | 1.1 | 795 | 38.7 | 1,599 | 13 |
| Lampasas | 24 | 521 | 49.1 | 20.0 | 23 | 136 | 14.4 | 3.8 | 1,622 | 88.1 | 6,932 | 49 |
| La Salle | 13 | 145 | 13.0 | 4.6 | D | D | D | D | 592 | 20.6 | 2,300 | 11 |
| Lavaca | 39 | 999 | 94.9 | 34.2 | 40 | 139 | 13.5 | 3.6 | 1,864 | 84.9 | 1,882 | 9 |
| Lee | 31 | 288 | 22.1 | 8.5 | 32 | 124 | 12.2 | 3.1 | 1,562 | 71.7 | 3,887 | 23 |
| Leon | 12 | D | 5.8 | D | 24 | 95 | 8.7 | 2.3 | 1,529 | 73.5 | 0 | 0 |
| Liberty | 92 | 2,120 | 222.2 | 62.1 | 68 | 359 | 45.2 | 11.6 | 6,169 | 269.8 | 134,927 | 964 |
| Limestone | 52 | 1,184 | 101.3 | 39.6 | 21 | 74 | 7.7 | 2.0 | 1,320 | 54.0 | 637 | 4 |
| Lipscomb | NA | NA | NA | NA | D | D | 1.0 | D | 291 | 14.5 | 0 | 0 |
| Live Oak | 11 | 149 | 8.9 | 4.1 | 8 | 25 | 1.8 | 0.4 | 1,033 | 51.7 | 577 | 5 |
| Llano | 49 | 624 | 64.7 | 21.2 | 25 | 84 | 9.1 | 2.2 | 2,234 | 127.5 | 76,192 | 300 |
| Loving | NA | NA | NA | NA | NA | NA | NA | NA | 18 | 0.7 | NA | NA |
| Lubbock | 878 | 24,682 | 3,010.3 | 1,033.8 | 469 | 3,871 | 398.8 | 112.4 | 23,724 | 1,236.2 | 602,528 | 3,851 |
| Lynn | 7 | 170 | 17.4 | 6.7 | 7 | 34 | 8.8 | 1.5 | 442 | 18.7 | 466 | 2 |
| McCulloch | 17 | 198 | 22.9 | 6.7 | 17 | 65 | 7.5 | 1.5 | 688 | 28.4 | 0 | 0 |
| McLennan | 583 | 17,869 | 1,804.9 | 761.3 | D | D | D | 79.4 | 17,116 | 818.8 | 230,366 | 1,201 |
| McMullen | NA | NA | NA | NA | NA | NA | NA | NA | 132 | 4.8 | NA | NA |
| Madison | 21 | 269 | 25.6 | 12.3 | 14 | 96 | 9.1 | 3.2 | 1,012 | 45.6 | 1,166 | 5 |
| Marion | 14 | 475 | 29.5 | 12.7 | D | D | D | 0.4 | 705 | 25.7 | 1,335 | 11 |
| Martin | 8 | 289 | 28.3 | 12.5 | 8 | 18 | 4.7 | 0.6 | 474 | 33.6 | 57 | 3 |

# Table B. States and Counties — Government Employment and Payroll, and Local Government Finances

| STATE County | Full-time equivalent employees | March payroll (dollars) | March payroll (percent of total) | | | | | | | General revenue | | | | |
|---|---|---|---|---|---|---|---|---|---|---|---|---|---|---|
| | | | Administration, judicial, and legal | Police and corrections | Fire protection | Highways and transportation | Health and welfare | Natural resources and utilities | Education and libraries | Total (mil dol) | Inter-governmental (mil dol) | Taxes Total (mil dol) | Per capita[1] (dollars) Total | Per capita[1] (dollars) Property |
| | 171 | 172 | 173 | 174 | 175 | 176 | 177 | 178 | 179 | 180 | 181 | 182 | 183 | 184 |

TEXAS—Cont'd

| STATE County | 171 | 172 | 173 | 174 | 175 | 176 | 177 | 178 | 179 | 180 | 181 | 182 | 183 | 184 |
|---|---|---|---|---|---|---|---|---|---|---|---|---|---|---|
| Guadalupe | 5,343 | 21,348,342 | 6.0 | 10.3 | 2.7 | 2.1 | 14.7 | 4.2 | 59.0 | 545.4 | 136.7 | 250.0 | 1,566 | 1,310 |
| Hale | 2,228 | 8,615,470 | 4.1 | 5.0 | 2.1 | 1.2 | 5.1 | 2.5 | 79.6 | 133.0 | 56.9 | 54.9 | 1,617 | 1,293 |
| Hall | 229 | 674,102 | 12.4 | 8.3 | 0.3 | 4.7 | 1.5 | 5.9 | 66.3 | 16.2 | 6.5 | 7.9 | 2,592 | 2,131 |
| Hamilton | 605 | 2,098,033 | 6.4 | 6.2 | 0.0 | 0.2 | 45.5 | 1.6 | 39.9 | 52.2 | 9.6 | 14.8 | 1,759 | 1,525 |
| Hansford | 555 | 1,832,514 | 6.9 | 4.0 | 0.0 | 2.1 | 32.9 | 4.6 | 48.8 | 52.5 | 11.5 | 15.8 | 2,887 | 2,658 |
| Hardeman | 457 | 1,691,737 | 9.7 | 4.0 | 2.7 | 2.3 | 45.6 | 1.6 | 33.9 | 21.5 | 6.1 | 11.7 | 2,951 | 2,531 |
| Hardin | 2,176 | 7,148,427 | 7.0 | 7.5 | 1.1 | 2.3 | 0.8 | 2.7 | 76.7 | 146.0 | 61.5 | 71.7 | 1,256 | 1,094 |
| Harris | 187,579 | 822,114,261 | 4.2 | 10.5 | 3.4 | 4.9 | 8.4 | 2.7 | 64.0 | 24,091.5 | 6,040.7 | 12,838.7 | 2,756 | 2,295 |
| Harrison | 2,588 | 8,365,927 | 5.4 | 9.8 | 3.7 | 2.0 | 0.7 | 4.0 | 72.8 | 218.0 | 58.1 | 131.9 | 1,984 | 1,737 |
| Hartley | 99 | 347,028 | 17.7 | 2.3 | 0.0 | 0.0 | 0.0 | 1.3 | 72.3 | 9.4 | 1.3 | 7.2 | 1,251 | 1,188 |
| Haskell | 379 | 1,056,108 | 8.3 | 4.8 | 0.1 | 3.4 | 28.3 | 5.6 | 49.3 | 30.2 | 7.9 | 12.4 | 2,175 | 1,825 |
| Hays | 8,066 | 29,596,807 | 5.4 | 11.0 | 3.3 | 1.8 | 6.5 | 5.6 | 62.5 | 786.3 | 196.0 | 436.4 | 2,032 | 1,697 |
| Hemphill | 427 | 1,487,981 | 6.6 | 3.6 | 0.0 | 3.9 | 37.8 | 5.7 | 41.6 | 39.9 | 4.2 | 32.0 | 8,152 | 7,590 |
| Henderson | 2,869 | 9,571,842 | 6.4 | 10.8 | 1.7 | 2.0 | 0.3 | 3.6 | 74.0 | 237.5 | 74.3 | 131.5 | 1,624 | 1,437 |
| Hidalgo | 42,017 | 156,936,079 | 3.8 | 6.8 | 1.8 | 1.8 | 4.5 | 4.0 | 76.4 | 3,554.6 | 2,159.8 | 994.9 | 1,162 | 962 |
| Hill | 1,869 | 6,875,341 | 5.4 | 8.2 | 0.9 | 1.4 | 0.7 | 1.9 | 80.8 | 162.0 | 61.0 | 76.9 | 2,154 | 1,793 |
| Hockley | 1,854 | 7,035,365 | 3.9 | 4.4 | 0.7 | 1.5 | 0.1 | 2.0 | 86.9 | 151.1 | 58.3 | 62.1 | 2,701 | 2,445 |
| Hood | 1,876 | 6,914,567 | 8.5 | 10.6 | 0.3 | 2.1 | 1.3 | 5.1 | 70.1 | 158.3 | 22.0 | 112.6 | 1,942 | 1,644 |
| Hopkins | 1,565 | 4,980,595 | 5.3 | 8.6 | 2.9 | 2.8 | 3.1 | 3.6 | 72.6 | 206.5 | 49.2 | 52.0 | 1,427 | 1,145 |
| Houston | 793 | 2,385,024 | 9.6 | 9.3 | 0.5 | 3.0 | 1.7 | 4.2 | 70.8 | 68.3 | 28.7 | 31.6 | 1,370 | 1,129 |
| Howard | 2,197 | 7,472,601 | 5.0 | 7.1 | 4.0 | 2.0 | 12.9 | 7.3 | 61.1 | 153.4 | 51.1 | 67.6 | 1,886 | 1,675 |
| Hudspeth | 287 | 841,839 | 10.1 | 16.1 | 0.0 | 3.7 | 0.1 | 6.2 | 63.0 | 13.1 | 5.2 | 6.1 | 1,323 | 1,222 |
| Hunt | 4,521 | 17,022,379 | 4.3 | 6.6 | 2.2 | 1.1 | 27.4 | 8.2 | 49.5 | 428.0 | 126.2 | 142.0 | 1,510 | 1,250 |
| Hutchinson | 1,316 | 4,530,193 | 8.6 | 8.5 | 4.0 | 2.4 | 1.0 | 5.5 | 67.7 | 115.2 | 36.6 | 54.1 | 2,537 | 2,149 |
| Irion | 116 | 392,701 | 12.5 | 6.2 | 0.0 | 8.0 | 0.3 | 19.0 | 48.3 | 20.8 | 3.1 | 17.4 | 11,499 | 11,409 |
| Jack | 566 | 2,266,420 | 4.5 | 5.2 | 0.2 | 1.6 | 25.0 | 3.7 | 59.0 | 35.9 | 7.6 | 25.1 | 2,847 | 2,652 |
| Jackson | 1,024 | 3,448,643 | 5.4 | 5.9 | 0.0 | 2.4 | 19.3 | 5.6 | 53.2 | 90.6 | 23.0 | 35.6 | 2,403 | 2,135 |
| Jasper | 1,556 | 4,879,242 | 6.6 | 9.1 | 0.1 | 3.0 | 1.9 | 6.7 | 70.3 | 108.7 | 44.7 | 53.6 | 1,507 | 1,386 |
| Jeff Davis | 129 | 453,660 | 9.1 | 4.8 | 0.0 | 0.0 | 1.2 | 0.5 | 83.1 | 11.5 | 4.6 | 4.6 | 2,049 | 1,815 |
| Jefferson | 9,986 | 39,634,101 | 6.3 | 15.7 | 6.3 | 3.8 | 7.2 | 8.7 | 50.7 | 1,131.8 | 279.1 | 608.0 | 2,375 | 1,908 |
| Jim Hogg | 358 | 983,737 | 8.8 | 7.8 | 0.5 | 0.0 | 1.1 | 7.7 | 72.3 | 27.4 | 11.8 | 12.7 | 2,430 | 2,039 |
| Jim Wells | 1,870 | 6,179,342 | 7.9 | 10.4 | 2.6 | 5.9 | 1.2 | 3.4 | 66.7 | 136.1 | 63.7 | 55.5 | 1,357 | 989 |
| Johnson | 6,968 | 23,471,528 | 6.4 | 11.0 | 3.5 | 2.4 | 1.3 | 4.6 | 68.5 | 583.4 | 194.6 | 304.8 | 1,825 | 1,554 |
| Jones | 859 | 2,861,245 | 6.6 | 6.8 | 0.0 | 2.2 | 17.1 | 3.8 | 62.6 | 62.1 | 26.0 | 20.5 | 1,033 | 893 |
| Karnes | 889 | 2,965,577 | 6.2 | 10.8 | 0.0 | 5.2 | 17.6 | 3.7 | 55.6 | 111.6 | 25.4 | 76.7 | 4,931 | 4,682 |
| Kaufman | 5,273 | 21,071,674 | 4.3 | 8.9 | 1.6 | 2.4 | 8.9 | 2.7 | 69.9 | 457.7 | 192.6 | 211.1 | 1,721 | 1,496 |
| Kendall | 1,755 | 6,657,795 | 9.0 | 8.4 | 1.1 | 2.7 | 0.4 | 5.3 | 70.4 | 177.9 | 14.1 | 133.7 | 3,041 | 2,653 |
| Kenedy | 82 | 186,075 | 26.6 | 5.4 | 0.0 | 8.5 | 0.0 | 0.0 | 55.4 | 11.1 | 0.3 | 10.1 | 23,642 | 23,611 |
| Kent | 133 | 417,992 | 10.7 | 6.9 | 0.0 | 12.7 | 28.4 | 4.6 | 35.3 | 9.9 | 1.4 | 7.7 | 10,112 | 9,995 |
| Kerr | 1,814 | 6,748,491 | 8.7 | 14.0 | 6.2 | 2.8 | 0.9 | 5.0 | 60.4 | 136.6 | 26.0 | 93.0 | 1,792 | 1,448 |
| Kimble | 146 | 443,542 | 25.6 | 5.7 | 0.2 | 4.6 | 0.6 | 3.4 | 59.2 | 14.0 | 2.8 | 7.3 | 1,658 | 1,466 |
| King | 65 | 200,785 | 27.6 | 2.1 | 0.0 | 0.0 | 0.0 | 0.5 | 69.9 | 6.6 | 1.4 | 4.8 | 16,734 | 16,734 |
| Kinney | 204 | 626,114 | 11.9 | 9.4 | 0.0 | 1.2 | 7.5 | 3.6 | 65.5 | 19.1 | 8.1 | 7.9 | 2,137 | 1,826 |
| Kleberg | 1,530 | 6,649,322 | 6.9 | 10.2 | 2.4 | 2.5 | 2.7 | 5.1 | 68.3 | 111.7 | 50.8 | 46.3 | 1,506 | 1,211 |
| Knox | 393 | 1,268,571 | 6.4 | 3.3 | 0.0 | 1.4 | 34.3 | 5.1 | 48.8 | 18.7 | 8.9 | 8.2 | 2,220 | 1,968 |
| Lamar | 2,196 | 8,027,047 | 5.9 | 8.7 | 0.2 | 1.5 | 1.7 | 3.9 | 76.5 | 183.8 | 69.7 | 83.7 | 1,689 | 1,328 |
| Lamb | 775 | 2,503,676 | 7.2 | 9.8 | 0.3 | 2.9 | 1.9 | 4.6 | 72.7 | 67.4 | 25.3 | 26.1 | 1,985 | 1,733 |
| Lampasas | 929 | 3,077,108 | 22.7 | 4.7 | 1.9 | 2.0 | 0.4 | 3.6 | 63.1 | 55.6 | 19.2 | 30.4 | 1,456 | 1,273 |
| La Salle | 400 | 1,555,564 | 7.0 | 7.7 | 0.0 | 2.4 | 1.0 | 2.3 | 77.8 | 73.3 | 5.4 | 64.1 | 8,519 | 8,199 |
| Lavaca | 836 | 2,894,533 | 12.3 | 9.2 | 1.6 | 4.0 | 18.7 | 8.3 | 43.1 | 76.1 | 9.7 | 36.9 | 1,845 | 1,648 |
| Lee | 696 | 2,247,761 | 7.9 | 10.6 | 0.0 | 4.8 | 0.1 | 5.5 | 70.7 | 55.3 | 16.8 | 33.0 | 1,929 | 1,677 |
| Leon | 800 | 2,600,523 | 8.0 | 7.3 | 0.0 | 3.3 | 0.9 | 3.2 | 77.0 | 58.2 | 16.6 | 36.0 | 2,088 | 1,872 |
| Liberty | 3,386 | 11,319,873 | 5.7 | 7.6 | 1.0 | 2.2 | 0.5 | 2.6 | 79.2 | 246.8 | 107.9 | 118.3 | 1,415 | 1,254 |
| Limestone | 1,375 | 4,409,125 | 5.8 | 10.8 | 2.5 | 2.9 | 18.5 | 2.6 | 56.1 | 180.5 | 30.8 | 49.6 | 2,119 | 1,984 |
| Lipscomb | 264 | 922,493 | 8.4 | 5.8 | 0.0 | 7.0 | 4.1 | 5.5 | 67.1 | 26.1 | 6.9 | 17.0 | 5,059 | 4,644 |
| Live Oak | 632 | 2,424,236 | 6.4 | 10.8 | 0.0 | 3.5 | 1.3 | 4.2 | 70.7 | 45.0 | 6.1 | 35.6 | 2,935 | 2,694 |
| Llano | 659 | 2,053,537 | 13.3 | 14.0 | 0.0 | 4.1 | 0.4 | 9.0 | 57.3 | 66.4 | 3.8 | 57.7 | 2,726 | 2,636 |
| Loving | 12 | 57,878 | 58.1 | 24.4 | 0.0 | 15.4 | 0.0 | 2.1 | 0.0 | 4.3 | 0.0 | | 331 | 233 |
| Lubbock | 14,801 | 59,464,737 | 4.6 | 10.4 | 4.5 | 1.0 | 32.6 | 5.5 | 40.2 | 1,628.6 | 432.1 | 515.3 | 1,687 | 1,320 |
| Lynn | 483 | 1,591,459 | 5.3 | 5.4 | 0.0 | 1.2 | 29.3 | 2.7 | 55.0 | 36.9 | 14.9 | 10.8 | 1,847 | 1,722 |
| McCulloch | 632 | 2,059,659 | 5.8 | 5.3 | 1.4 | 2.0 | 22.6 | 8.6 | 52.2 | 28.9 | 12.0 | 13.6 | 1,718 | 1,533 |
| McLennan | 11,175 | 42,970,259 | 5.7 | 11.5 | 3.1 | 1.8 | 6.0 | 7.2 | 63.7 | 1,119.6 | 400.8 | 470.3 | 1,869 | 1,493 |
| McMullen | 127 | 458,723 | 17.6 | 15.5 | 1.5 | 8.2 | 0.0 | 3.0 | 52.0 | 24.0 | 1.8 | 21.4 | 28,001 | 27,903 |
| Madison | 551 | 1,796,252 | 9.5 | 6.2 | 0.0 | 2.0 | 0.5 | 2.9 | 75.5 | 41.6 | 13.6 | 22.5 | 1,578 | 1,382 |
| Marion | 348 | 925,317 | 12.5 | 9.6 | 0.0 | 3.5 | 1.3 | 2.3 | 70.4 | 15.4 | 7.4 | 6.8 | 676 | 522 |
| Martin | 454 | 2,092,649 | 7.6 | 3.7 | 0.0 | 7.2 | 40.6 | 2.6 | 37.6 | 74.9 | 4.8 | 53.0 | 9,587 | 9,435 |

1. Based on the resident population estimated as of July 1 of the year shown.

Items 171—184

# Local Government Finances, Government Employment, and Income Taxes

| STATE County | Local government finances, 2017 (cont.) | | | | | | | Debt outstanding | | Government employment, 2019 | | | Individual income tax returns, 2018 | | |
|---|---|---|---|---|---|---|---|---|---|---|---|---|---|---|---|
| | Direct general expenditure | | | | | | | | | | | | | | |
| | | | Percent of total for: | | | | | | | | | | | Mean adjusted gross income | Mean income tax |
| | Total (mil dol) | Per capita[1] (dollars) | Education | Health and hospitals | Police protection | Public welfare | Highways | Total (mil dol) | Per capita[1] (dollars) | Federal civilian | Federal military | State and local | Number of returns | | |
| | 185 | 186 | 187 | 188 | 189 | 190 | 191 | 192 | 193 | 194 | 195 | 196 | 197 | 198 | 199 |
| TEXAS—Cont'd | | | | | | | | | | | | | | | |
| Guadalupe | 627 | 3,929 | 48.1 | 17.7 | 4.3 | 0.6 | 3.4 | 1,494.4 | 9,361 | 222 | 322 | 6,498 | 76,170 | 63,258 | 6,173 |
| Hale | 123 | 3,624 | 55.4 | 7.8 | 8.4 | 0.7 | 4.3 | 84.3 | 2,482 | 76 | 60 | 2,324 | 12,960 | 42,531 | 3,332 |
| Hall | 16 | 5,149 | 52.6 | 1.7 | 5.8 | 0.3 | 6.8 | 5.1 | 1,667 | 17 | 6 | 260 | 1,200 | 34,682 | 2,363 |
| Hamilton | 48 | 5,713 | 31.6 | 47.2 | 2.0 | 0.4 | 2.7 | 21.1 | 2,506 | 29 | 16 | 773 | 3,720 | 59,429 | 7,261 |
| Hansford | 52 | 9,453 | 33.7 | 53.8 | 1.6 | 0.1 | 1.6 | 24.5 | 4,458 | 18 | 10 | 653 | 2,270 | 58,585 | 6,184 |
| Hardeman | 22 | 5,429 | 51.6 | 18.4 | 3.4 | 0.3 | 3.3 | 7.1 | 1,783 | 19 | 8 | 410 | 1,580 | 41,874 | 3,528 |
| Hardin | 140 | 2,451 | 69.5 | 0.5 | 6.5 | 0.6 | 4.0 | 131.0 | 2,293 | 74 | 111 | 2,352 | 24,710 | 67,940 | 6,991 |
| Harris | 23,990 | 5,150 | 47.6 | 12.4 | 5.4 | 0.1 | 3.6 | 57,919.7 | 12,435 | 26,172 | 10,243 | 255,932 | 2,036,830 | 77,767 | 11,110 |
| Harrison | 296 | 4,456 | 69.7 | 0.1 | 3.9 | 0.5 | 3.0 | 352.8 | 5,308 | 124 | 127 | 3,081 | 28,730 | 59,950 | 6,120 |
| Hartley | 7 | 1,298 | 67.9 | 0.0 | 5.8 | 0.0 | 6.7 | 7.9 | 1,376 | 5 | 9 | 713 | 1,750 | 58,726 | 6,229 |
| Haskell | 29 | 5,123 | 39.7 | 29.4 | 4.1 | 0.1 | 8.0 | 9.3 | 1,640 | 25 | 10 | 441 | 2,080 | 45,538 | 4,378 |
| Hays | 746 | 3,476 | 50.7 | 8.0 | 6.0 | 0.0 | 4.9 | 2,097.2 | 9,767 | 276 | 477 | 14,080 | 100,560 | 73,328 | 8,697 |
| Hemphill | 42 | 10,617 | 60.3 | 4.4 | 2.7 | 0.1 | 5.7 | 34.8 | 8,852 | 9 | 7 | 502 | 1,600 | 77,912 | 9,299 |
| Henderson | 260 | 3,211 | 68.3 | 1.2 | 5.0 | 0.0 | 4.0 | 202.0 | 2,495 | 97 | 160 | 3,341 | 34,340 | 52,423 | 5,047 |
| Hidalgo | 3,601 | 4,205 | 68.2 | 4.5 | 3.5 | 0.6 | 2.7 | 3,753.4 | 4,384 | 4,565 | 1,733 | 53,611 | 328,590 | 40,424 | 3,050 |
| Hill | 155 | 4,351 | 58.9 | 0.6 | 6.0 | 1.1 | 6.3 | 131.8 | 3,691 | 106 | 70 | 2,210 | 15,520 | 51,248 | 4,615 |
| Hockley | 166 | 7,216 | 75.2 | 0.5 | 2.5 | 0.3 | 2.9 | 113.4 | 4,936 | 39 | 43 | 1,838 | 9,390 | 57,874 | 5,683 |
| Hood | 158 | 2,718 | 54.1 | 0.5 | 6.5 | 0.1 | 3.4 | 261.3 | 4,507 | 105 | 119 | 2,075 | 27,980 | 75,393 | 8,876 |
| Hopkins | 217 | 5,960 | 36.1 | 42.4 | 3.2 | 0.0 | 3.7 | 179.5 | 4,923 | 72 | 71 | 2,033 | 15,920 | 50,714 | 4,664 |
| Houston | 65 | 2,813 | 50.9 | 2.9 | 8.1 | 0.1 | 6.2 | 69.7 | 3,017 | 71 | 39 | 1,622 | 8,210 | 48,758 | 4,799 |
| Howard | 143 | 3,989 | 56.7 | 15.3 | 5.4 | 0.1 | 4.8 | 399.9 | 11,165 | 987 | 61 | 2,655 | 13,310 | 77,198 | 10,738 |
| Hudspeth | 13 | 2,815 | 71.8 | 0.0 | 1.5 | 0.2 | 1.9 | 3.9 | 852 | 315 | 9 | 316 | 1,310 | 34,620 | 2,360 |
| Hunt | 401 | 4,263 | 39.3 | 33.9 | 2.7 | 0.2 | 3.7 | 444.7 | 4,728 | 275 | 272 | 7,469 | 41,030 | 55,289 | 5,166 |
| Hutchinson | 142 | 6,656 | 42.0 | 4.3 | 4.7 | 0.1 | 2.5 | 196.8 | 9,220 | 67 | 40 | 1,506 | 8,730 | 63,062 | 6,315 |
| Irion | 18 | 12,086 | 64.2 | 0.2 | 4.3 | 2.3 | 14.8 | 0.0 | 0 | 3 | 4 | 127 | 750 | 93,944 | 14,219 |
| Jack | 29 | 3,251 | 72.5 | 0.2 | 3.4 | 0.2 | 2.9 | 66.4 | 7,524 | 18 | 15 | 674 | 3,290 | 54,976 | 5,485 |
| Jackson | 83 | 5,589 | 45.9 | 24.1 | 3.9 | 0.1 | 5.1 | 151.9 | 10,262 | 32 | 28 | 1,126 | 6,610 | 58,226 | 5,458 |
| Jasper | 112 | 3,162 | 59.9 | 1.5 | 5.5 | 1.0 | 6.4 | 112.6 | 3,168 | 82 | 68 | 2,011 | 14,080 | 55,875 | 5,313 |
| Jeff Davis | 10 | 4,188 | 44.6 | 0.0 | 18.6 | 0.5 | 0.5 | 0.0 | 10 | 18 | 4 | 224 | 1,010 | 61,372 | 6,650 |
| Jefferson | 1,119 | 4,372 | 40.3 | 6.6 | 8.2 | 0.6 | 4.7 | 4,712.0 | 18,403 | 1,810 | 669 | 15,535 | 105,140 | 61,122 | 6,641 |
| Jim Hogg | 32 | 6,128 | 54.9 | 0.8 | 5.5 | 1.9 | 10.0 | 14.2 | 2,731 | 251 | 10 | 399 | 1,980 | 40,270 | 2,764 |
| Jim Wells | 135 | 3,299 | 56.3 | 0.8 | 8.5 | 0.1 | 4.7 | 113.6 | 2,777 | 87 | 78 | 2,090 | 16,820 | 49,671 | 4,997 |
| Johnson | 617 | 3,696 | 58.7 | 0.1 | 6.6 | 0.3 | 3.8 | 1,431.6 | 8,572 | 306 | 339 | 7,660 | 78,310 | 60,814 | 5,793 |
| Jones | 60 | 3,012 | 50.0 | 19.2 | 3.1 | 0.1 | 4.6 | 44.2 | 2,229 | 54 | 33 | 1,958 | 6,310 | 48,085 | 3,968 |
| Karnes | 127 | 8,162 | 75.9 | 0.5 | 1.0 | 0.2 | 4.4 | 133.7 | 8,597 | 73 | 25 | 1,623 | 5,670 | 112,605 | 23,605 |
| Kaufman | 544 | 4,438 | 56.4 | 7.3 | 5.9 | 0.0 | 9.0 | 1,055.6 | 8,608 | 207 | 263 | 6,947 | 60,730 | 61,690 | 5,732 |
| Kendall | 202 | 4,586 | 62.2 | 0.9 | 7.7 | 0.2 | 4.8 | 438.0 | 9,961 | 108 | 91 | 2,136 | 20,850 | 127,733 | 21,098 |
| Kenedy | 11 | 25,412 | 89.2 | 0.0 | 0.5 | 0.0 | 0.6 | 3.3 | 7,815 | 1 | 1 | 86 | 150 | 39,753 | 2,567 |
| Kent | 10 | 13,318 | 49.8 | 5.1 | 0.7 | 4.9 | 6.5 | 0.0 | 0 | 6 | 1 | 186 | 340 | 54,144 | 4,874 |
| Kerr | 127 | 2,442 | 51.1 | 0.9 | 8.9 | 0.1 | 3.7 | 118.8 | 2,289 | 527 | 100 | 2,642 | 25,020 | 68,204 | 8,386 |
| Kimble | 12 | 2,732 | 53.0 | 11.4 | 3.2 | 0.3 | 1.1 | 7.6 | 1,735 | 15 | 8 | 373 | 2,080 | 47,885 | 4,531 |
| King | 4 | 14,277 | 76.8 | 0.0 | 1.2 | 0.0 | 4.9 | 3.6 | 12,526 | 3 | 1 | 60 | 110 | 36,545 | 3,691 |
| Kinney | 22 | 5,810 | 42.2 | 3.7 | 9.3 | 0.7 | 4.9 | 5.0 | 1,351 | 124 | 6 | 288 | 1,310 | 56,198 | 5,127 |
| Kleberg | 100 | 3,261 | 61.1 | 1.3 | 8.9 | 1.7 | 3.4 | 121.0 | 3,933 | 833 | 447 | 3,664 | 12,890 | 50,167 | 4,730 |
| Knox | 20 | 5,515 | 69.6 | 3.5 | 4.4 | 0.2 | 2.3 | 16.3 | 4,427 | 25 | 7 | 431 | 1,420 | 37,360 | 3,218 |
| Lamar | 189 | 3,812 | 63.2 | 2.3 | 7.5 | 0.8 | 2.6 | 153.1 | 3,089 | 130 | 96 | 2,852 | 21,250 | 49,921 | 4,600 |
| Lamb | 70 | 5,320 | 50.1 | 20.5 | 5.5 | 0.1 | 5.4 | 23.3 | 1,771 | 40 | 25 | 962 | 5,230 | 43,400 | 3,628 |
| Lampasas | 61 | 2,932 | 57.1 | 0.0 | 7.8 | 0.4 | 4.4 | 60.7 | 2,909 | 45 | 41 | 999 | 9,600 | 61,362 | 6,324 |
| La Salle | 73 | 9,661 | 85.4 | 0.0 | 1.5 | 0.2 | 1.9 | 81.6 | 10,833 | 110 | 11 | 827 | 2,490 | 90,761 | 18,344 |
| Lavaca | 77 | 3,827 | 34.9 | 27.8 | 6.2 | 0.2 | 8.0 | 46.6 | 2,325 | 61 | 39 | 982 | 9,530 | 65,986 | 6,414 |
| Lee | 55 | 3,193 | 54.1 | 0.6 | 6.9 | 0.8 | 7.6 | 77.1 | 4,476 | 27 | 33 | 1,176 | 8,010 | 62,536 | 6,414 |
| Leon | 60 | 3,489 | 64.7 | 0.0 | 3.8 | 0.3 | 6.1 | 77.1 | 4,247 | 44 | 34 | 843 | 7,540 | 53,399 | 5,063 |
| Liberty | 315 | 3,763 | 68.8 | 0.0 | 4.2 | 0.4 | 3.2 | 355.0 | 4,247 | 128 | 162 | 4,958 | 34,240 | 55,209 | 4,967 |
| Limestone | 170 | 7,274 | 27.1 | 9.9 | 4.5 | 45.1 | 2.3 | 73.6 | 3,148 | 46 | 43 | 2,918 | 9,290 | 48,194 | 4,286 |
| Lipscomb | 23 | 6,857 | 53.2 | 0.3 | 5.3 | 0.4 | 13.5 | 8.2 | 2,431 | 26 | 6 | 365 | 1,370 | 59,350 | 6,903 |
| Live Oak | 53 | 4,319 | 75.1 | 0.0 | 3.6 | 0.2 | 2.6 | 84.9 | 6,992 | 242 | 22 | 590 | 4,750 | 65,660 | 7,835 |
| Llano | 71 | 3,370 | 61.8 | 0.0 | 7.7 | 0.1 | 4.8 | 59.3 | 2,802 | 34 | 42 | 798 | 9,250 | 76,717 | 10,415 |
| Loving | 1 | 7,361 | 0.0 | 0.0 | 0.6 | 0.0 | 0.7 | 0.0 | 105 | 0 | 0 | 15 | 50 | 83,560 | 7,780 |
| Lubbock | 1,662 | 5,443 | 32.3 | 35.3 | 6.0 | 0.0 | 2.0 | 2,313.0 | 7,573 | 1,348 | 605 | 29,141 | 131,780 | 62,445 | 7,206 |
| Lynn | 46 | 7,940 | 42.7 | 43.2 | 3.0 | 0.0 | 1.3 | 46.1 | 7,912 | 18 | 12 | 620 | 2,540 | 57,933 | 5,641 |
| McCulloch | 28 | 3,571 | 61.5 | 0.8 | 4.5 | 0.2 | 7.0 | 28.0 | 3,522 | 24 | 15 | 656 | 3,570 | 49,224 | 4,761 |
| McLennan | 977 | 3,883 | 52.2 | 6.3 | 6.9 | 0.9 | 2.1 | 1,253.1 | 4,980 | 2,786 | 602 | 14,844 | 109,480 | 57,722 | 5,982 |
| McMullen | 26 | 33,558 | 97.2 | 0.0 | 0.1 | 0.0 | 0.2 | 32.2 | 42,119 | 3 | 1 | 150 | 440 | 116,570 | 20,884 |
| Madison | 59 | 4,118 | 70.5 | 0.4 | 4.4 | 0.0 | 4.6 | 39.6 | 2,779 | 22 | 23 | 1,143 | 5,290 | 46,300 | 3,917 |
| Marion | 21 | 2,089 | 59.8 | 1.3 | 9.9 | 0.0 | 9.0 | 6.8 | 675 | 37 | 19 | 396 | 3,960 | 47,504 | 3,929 |
| Martin | 88 | 15,828 | 61.4 | 30.4 | 1.0 | 0.0 | 1.0 | 74.1 | 13,393 | 17 | 11 | 583 | 2,270 | 131,188 | 24,568 |

1. Based on the resident population estimated as of July 1 of the year shown.

# Table B. States and Counties — Land Area and Population

| State / county code | CBSA code[1] | County Type code[2] | STATE County | Land area[3] (sq. mi) | Total persons 2019 | Rank | Per square mile | White | Black | American Indian, Alaska Native | Asian and Pacific Islancer | Percent Hispanic or Latino[4] | Under 5 years | 5 to 14 years | 15 to 24 years | 25 to 34 years | 35 to 44 years | 45 to 54 years |
|---|---|---|---|---|---|---|---|---|---|---|---|---|---|---|---|---|---|---|
| | | | | 1 | 2 | 3 | 4 | 5 | 6 | 7 | 8 | 9 | 10 | 11 | 12 | 13 | 14 | 15 |
| | | | TEXAS—Cont'd | | | | | | | | | | | | | | | |
| 48319 | | 9 | Mason | 928.8 | 4,344 | 2,871 | 4.7 | 73.7 | 0.9 | 0.7 | 0.4 | 25.1 | 4.4 | 11.6 | 10.3 | 8.9 | 10.5 | 10.6 |
| 48321 | 13060 | 4 | Matagorda | 1,092.9 | 36,725 | 1,256 | 33.6 | 43.2 | 10.7 | 0.8 | 2.0 | 44.3 | 7.3 | 14.5 | 12.3 | 12.8 | 11.5 | 10.6 |
| 48323 | 20580 | 5 | Maverick | 1,279.5 | 58,378 | 887 | 45.6 | 2.9 | 0.4 | 1.2 | 0.6 | 95.1 | 8.9 | 16.7 | 16.6 | 14.2 | 11.4 | 11.1 |
| 48325 | 41700 | 1 | Medina | 1,325.4 | 52,358 | 964 | 39.5 | 43.2 | 3.0 | 0.8 | 1.0 | 52.9 | 5.8 | 12.7 | 13.9 | 12.0 | 12.0 | 12.7 |
| 48327 | | 9 | Menard | 902.0 | 2,124 | 3,037 | 2.4 | 61.0 | 1.8 | 1.0 | 0.6 | 36.5 | 4.0 | 11.4 | 9.0 | 8.8 | 8.8 | 9.8 |
| 48329 | 33260 | 3 | Midland | 900.4 | 177,863 | 374 | 197.5 | 43.8 | 7.1 | 0.8 | 2.5 | 47.0 | 8.6 | 16.2 | 12.7 | 17.7 | 14.5 | 10.0 |
| 48331 | | 6 | Milam | 1,016.4 | 24,708 | 1,620 | 24.3 | 62.7 | 9.2 | 0.9 | 1.1 | 27.3 | 5.4 | 13.4 | 12.4 | 10.4 | 11.3 | 11.6 |
| 48333 | | 9 | Mills | 748.2 | 4,840 | 2,837 | 6.5 | 79.2 | 1.4 | 0.8 | 0.8 | 19.1 | 4.2 | 12.3 | 11.2 | 8.2 | 9.5 | 11.1 |
| 48335 | | 7 | Mitchell | 911.1 | 8,202 | 2,563 | 9.0 | 48.5 | 9.8 | 1.2 | 1.0 | 40.6 | 5.8 | 12.8 | 14.2 | 16.6 | 13.1 | 11.5 |
| 48337 | | 6 | Montague | 930.9 | 19,962 | 1,826 | 21.4 | 86.3 | 1.3 | 1.7 | 0.7 | 11.5 | 5.4 | 13.0 | 11.0 | 10.9 | 10.9 | 11.9 |
| 48339 | 26420 | 1 | Montgomery | 1,042.2 | 626,351 | 108 | 601.0 | 65.0 | 6.3 | 0.8 | 4.1 | 25.6 | 6.5 | 14.9 | 12.8 | 12.7 | 13.9 | 13.3 |
| 48341 | 20300 | 7 | Moore | 899.7 | 20,654 | 1,786 | 23.0 | 31.4 | 4.5 | 1.0 | 5.8 | 58.3 | 9.1 | 17.3 | 14.4 | 13.8 | 12.6 | 10.9 |
| 48343 | | 6 | Morris | 252.0 | 12,393 | 2,257 | 49.2 | 65.7 | 23.4 | 1.4 | 1.1 | 10.9 | 5.9 | 13.0 | 11.2 | 11.2 | 10.9 | 11.3 |
| 48345 | | 9 | Motley | 989.6 | 1,185 | 3,101 | 1.2 | 77.1 | 2.9 | 1.4 | 0.2 | 19.3 | 5.5 | 10.6 | 10.8 | 10.1 | 7.9 | 9.6 |
| 48347 | 34860 | 4 | Nacogdoches | 946.3 | 64,753 | 823 | 68.4 | 60.6 | 18.3 | 0.9 | 1.6 | 20.3 | 6.2 | 13.2 | 22.2 | 11.7 | 10.4 | 9.7 |
| 48349 | 18620 | 4 | Navarro | 1,009.7 | 50,694 | 979 | 50.2 | 55.5 | 13.2 | 0.9 | 2.7 | 29.3 | 7.0 | 15.1 | 13.2 | 11.8 | 11.2 | 11.7 |
| 48351 | | 2 | Newton | 933.7 | 13,414 | 2,195 | 14.4 | 75.4 | 20.1 | 1.5 | 0.9 | 4.1 | 4.4 | 10.9 | 12.0 | 11.9 | 10.8 | 12.6 |
| 48353 | 45020 | 2 | Nolan | 912.0 | 14,835 | 2,101 | 16.3 | 55.0 | 5.2 | 0.9 | 1.0 | 39.3 | 6.6 | 14.5 | 13.8 | 11.3 | 11.7 | 11.5 |
| 48355 | 18580 | 2 | Nueces | 839.1 | 363,148 | 200 | 432.8 | 29.2 | 3.8 | 0.5 | 2.6 | 64.9 | 6.4 | 13.6 | 14.1 | 14.1 | 13.2 | 11.4 |
| 48357 | | 7 | Ochiltree | 917.7 | 9,598 | 2,446 | 10.5 | 41.9 | 1.0 | 1.0 | 0.7 | 56.4 | 7.7 | 17.9 | 15.0 | 13.2 | 11.9 | 11.1 |
| 48359 | 11100 | 2 | Oldham | 1,500.5 | 2,135 | 3,036 | 1.4 | 77.2 | 3.7 | 1.2 | 2.1 | 17.3 | 3.1 | 11.6 | 14.0 | 12.6 | 14.4 | 13.4 |
| 48361 | 13140 | 2 | Orange | 333.8 | 82,878 | 688 | 248.3 | 80.7 | 9.2 | 1.1 | 1.7 | 8.9 | 6.6 | 14.2 | 12.0 | 13.2 | 12.5 | 12.0 |
| 48363 | 33420 | 6 | Palo Pinto | 952.5 | 29,320 | 1,452 | 30.8 | 75.7 | 3.0 | 1.1 | 1.3 | 20.3 | 6.2 | 13.1 | 11.9 | 12.1 | 10.6 | 11.3 |
| 48365 | | 6 | Panola | 811.4 | 23,187 | 1,674 | 28.6 | 74.5 | 15.9 | 1.1 | 0.9 | 9.1 | 5.7 | 12.9 | 12.6 | 11.5 | 12.1 | 11.2 |
| 48367 | 19100 | 1 | Parker | 903.7 | 148,198 | 453 | 164.0 | 83.6 | 2.0 | 1.4 | 1.2 | 13.5 | 6.1 | 14.2 | 11.8 | 11.8 | 13.1 | 13.0 |
| 48369 | | 7 | Parmer | 880.8 | 9,522 | 2,451 | 10.8 | 32.8 | 1.3 | 0.4 | 0.8 | 65.2 | 7.4 | 15.4 | 14.1 | 12.5 | 11.6 | 11.8 |
| 48371 | | 7 | Pecos | 4,763.8 | 15,718 | 2,048 | 3.3 | 24.7 | 4.3 | 0.7 | 1.3 | 69.8 | 6.4 | 14.4 | 12.5 | 15.3 | 15.0 | 12.2 |
| 48373 | | 6 | Polk | 1,057.0 | 52,995 | 954 | 50.1 | 72.7 | 10.0 | 2.2 | 1.1 | 15.4 | 5.2 | 11.5 | 10.5 | 12.0 | 11.8 | 12.5 |
| 48375 | 11100 | 2 | Potter | 908.4 | 116,004 | 537 | 127.7 | 44.4 | 11.0 | 1.0 | 5.5 | 39.7 | 7.1 | 15.5 | 13.4 | 14.6 | 13.4 | 11.5 |
| 48377 | | 7 | Presidio | 3,855.3 | 6,508 | 2,704 | 1.7 | 13.1 | 1.4 | 0.7 | 2.1 | 83.5 | 7.9 | 15.1 | 12.0 | 10.8 | 10.8 | 9.8 |
| 48379 | | 8 | Rains | 229.5 | 12,552 | 2,242 | 54.7 | 86.2 | 2.9 | 1.9 | 1.3 | 9.6 | 4.8 | 11.7 | 9.6 | 10.0 | 10.8 | 11.6 |
| 48381 | 11100 | 2 | Randall | 912.7 | 139,899 | 468 | 153.3 | 70.6 | 3.9 | 1.0 | 2.3 | 23.7 | 6.0 | 13.8 | 13.9 | 14.4 | 13.7 | 11.0 |
| 48383 | | 6 | Reagan | 1,175.3 | 3,833 | 2,907 | 3.3 | 24.3 | 3.2 | 0.7 | 0.6 | 72.3 | 8.1 | 16.6 | 13.8 | 12.4 | 15.5 | 11.5 |
| 48385 | | 9 | Real | 699.2 | 3,411 | 2,939 | 4.9 | 68.5 | 1.5 | 1.5 | 0.9 | 29.0 | 5.0 | 9.4 | 7.6 | 8.9 | 9.4 | 10.9 |
| 48387 | | 6 | Red River | 1,043.9 | 11,995 | 2,294 | 11.5 | 75.1 | 16.4 | 2.0 | 0.7 | 7.5 | 5.0 | 11.1 | 10.0 | 10.0 | 10.5 | 12.3 |
| 48389 | 37780 | 7 | Reeves | 2,635.4 | 15,949 | 2,038 | 6.1 | 18.6 | 5.1 | 0.5 | 1.6 | 74.8 | 6.0 | 12.8 | 15.5 | 17.7 | 15.8 | 11.8 |
| 48391 | | 6 | Refugio | 770.5 | 6,877 | 2,670 | 8.9 | 42.1 | 5.9 | 0.8 | 0.9 | 51.2 | 5.3 | 13.0 | 11.8 | 10.6 | 11.0 | 11.4 |
| 48393 | 37420 | 9 | Roberts | 924.1 | 813 | 3,115 | 0.9 | 87.9 | 0.5 | 2.1 | 1.0 | 10.2 | 5.5 | 13.9 | 10.2 | 10.5 | 14.8 | 8.9 |
| 48395 | 17780 | 3 | Robertson | 855.2 | 17,155 | 1,967 | 20.1 | 57.9 | 19.6 | 0.8 | 1.2 | 21.8 | 6.1 | 13.4 | 11.0 | 11.8 | 12.4 | 11.1 |
| 48397 | 19100 | 1 | Rockwall | 127.2 | 109,888 | 556 | 863.9 | 69.3 | 8.5 | 0.9 | 4.1 | 19.1 | 6.1 | 15.5 | 13.1 | 11.3 | 15.0 | 14.1 |
| 48399 | | 6 | Runnels | 1,051.1 | 10,401 | 2,395 | 9.9 | 60.9 | 2.7 | 1.0 | 1.0 | 35.3 | 6.0 | 13.1 | 12.5 | 10.6 | 11.9 | 11.6 |
| 48401 | 30980 | 3 | Rusk | 924.2 | 54,324 | 941 | 58.8 | 64.0 | 17.8 | 0.9 | 0.9 | 17.8 | 5.5 | 12.6 | 12.6 | 13.6 | 13.2 | 12.5 |
| 48403 | | 8 | Sabine | 491.7 | 10,507 | 2,388 | 21.4 | 87.1 | 7.7 | 1.5 | 0.8 | 4.8 | 4.8 | 9.6 | 9.3 | 9.0 | 8.7 | 10.3 |
| 48405 | | 9 | San Augustine | 530.7 | 8,248 | 2,561 | 15.5 | 70.2 | 21.9 | 1.0 | 0.6 | 7.8 | 5.1 | 11.4 | 9.3 | 10.6 | 9.2 | 11.0 |
| 48407 | | 8 | San Jacinto | 569.2 | 29,301 | 1,453 | 51.5 | 75.2 | 9.7 | 1.4 | 1.0 | 14.6 | 5.3 | 12.3 | 10.8 | 10.8 | 10.4 | 11.6 |
| 48409 | 18580 | 2 | San Patricio | 693.4 | 67,069 | 799 | 96.7 | 38.3 | 1.9 | 0.6 | 1.5 | 58.6 | 6.9 | 14.7 | 13.2 | 13.4 | 13.2 | 11.4 |
| 48411 | | 7 | San Saba | 1,135.3 | 6,039 | 2,740 | 5.3 | 63.9 | 4.2 | 1.2 | 0.5 | 31.2 | 5.5 | 10.6 | 13.3 | 14.7 | 9.1 | 10.0 |
| 48413 | | 8 | Schleicher | 1,310.6 | 2,761 | 2,983 | 2.1 | 43.5 | 1.7 | 0.5 | 0.5 | 54.7 | 4.2 | 14.7 | 13.4 | 10.7 | 13.2 | 11.1 |
| 48415 | 43660 | 7 | Scurry | 905.4 | 16,662 | 1,996 | 18.4 | 51.9 | 5.5 | 0.8 | 0.8 | 42.1 | 6.7 | 14.1 | 13.6 | 13.7 | 13.6 | 11.4 |
| 48417 | | 8 | Shackelford | 914.3 | 3,300 | 2,947 | 3.6 | 85.6 | 2.0 | 1.0 | 0.8 | 12.2 | 5.1 | 13.4 | 11.3 | 10.8 | 10.2 | 11.5 |
| 48419 | | 7 | Shelby | 795.6 | 24,915 | 1,606 | 31.3 | 62.0 | 17.9 | 0.7 | 1.7 | 19.0 | 7.2 | 14.4 | 12.2 | 12.0 | 11.3 | 11.7 |
| 48421 | | 9 | Sherman | 922.9 | 3,027 | 2,957 | 3.3 | 52.0 | 1.5 | 0.6 | 0.6 | 45.9 | 6.7 | 15.5 | 13.4 | 12.1 | 10.8 | 12.1 |
| 48423 | 46340 | 3 | Smith | 921.5 | 235,806 | 294 | 255.9 | 60.3 | 18.2 | 0.7 | 2.1 | 20.2 | 6.5 | 13.7 | 13.7 | 13.6 | 12.0 | 11.2 |
| 48425 | | 1 | Somervell | 186.4 | 9,139 | 2,483 | 49.0 | 79.3 | 1.8 | 1.7 | 1.2 | 17.8 | 4.9 | 12.4 | 11.6 | 10.9 | 12.2 | 12.9 |
| 48427 | 40100 | 4 | Starr | 1,223.2 | 64,266 | 834 | 52.5 | 3.3 | 0.2 | 0.1 | 0.2 | 96.3 | 9.1 | 18.3 | 16.3 | 13.7 | 11.5 | 11.1 |
| 48429 | | 7 | Stephens | 896.7 | 9,334 | 2,471 | 10.4 | 70.7 | 3.4 | 1.1 | 1.1 | 24.7 | 5.6 | 12.1 | 12.8 | 12.3 | 12.5 | 10.5 |
| 48431 | 41660 | 8 | Sterling | 923.4 | 1,315 | 3,092 | 1.4 | 58.7 | 3.0 | 1.3 | 0.7 | 38.3 | 7.5 | 15.9 | 13.3 | 11.9 | 14.4 | 10.2 |
| 48433 | | 9 | Stonewall | 916.3 | 1,348 | 3,091 | 1.5 | 77.2 | 3.4 | 1.3 | 1.5 | 18.1 | 5.7 | 12.5 | 10.6 | 9.3 | 10.4 | 10.5 |
| 48435 | | 7 | Sutton | 1,453.9 | 3,738 | 2,918 | 2.6 | 33.6 | 0.6 | 0.2 | 0.5 | 65.4 | 6.5 | 12.7 | 12.7 | 11.2 | 10.6 | 13.5 |
| 48437 | | 6 | Swisher | 890.2 | 7,340 | 2,632 | 8.2 | 45.8 | 8.6 | 1.1 | 0.6 | 45.3 | 5.7 | 14.0 | 14.6 | 13.5 | 11.7 | 10.8 |
| 48439 | 19100 | 1 | Tarrant | 865.3 | 2,123,347 | 15 | 2,453.9 | 46.4 | 18.4 | 0.9 | 6.9 | 29.7 | 6.7 | 14.6 | 13.9 | 15.0 | 13.8 | 12.6 |
| 48441 | 10180 | 3 | Taylor | 915.5 | 139,200 | 469 | 152.0 | 64.3 | 8.6 | 0.9 | 3.0 | 25.5 | 7.2 | 13.7 | 17.8 | 14.5 | 11.9 | 9.3 |
| 48443 | | 9 | Terrell | 2,358.0 | 702 | 3,129 | 0.3 | 43.3 | 2.1 | 2.1 | 1.9 | 52.6 | 5.0 | 10.3 | 8.1 | 8.0 | 11.8 | 9.7 |
| 48445 | | 6 | Terry | 888.8 | 12,183 | 2,278 | 13.7 | 38.5 | 4.6 | 0.6 | 0.6 | 56.6 | 7.6 | 15.4 | 14.1 | 13.7 | 12.4 | 10.3 |
| 48447 | | 9 | Throckmorton | 912.6 | 1,487 | 3,080 | 1.6 | 84.3 | 1.3 | 2.0 | 0.8 | 13.2 | 5.4 | 12.7 | 8.5 | 9.4 | 9.6 | 11.0 |
| 48449 | 34420 | 7 | Titus | 406.1 | 32,926 | 1,358 | 81.1 | 44.6 | 9.7 | 0.8 | 1.4 | 44.7 | 7.5 | 16.4 | 14.5 | 12.5 | 10.9 | 12.2 |

1. CBSA = Core Based Statistical Area. See Appendix A for explanation. See Appendix B for list of metropolitan areas with component counties. Service of USDA Rural-Urban Continuum Codes. See Appendix A for definition. 3. Dry land or land partially or temporarily covered by water. 2. County type code from the Economic Research 4. May be of any race.

| STATE County | 55 to 64 years (16) | 65 to 74 years (17) | 75 years and over (18) | Percent female (19) | 2000 (20) | 2010 (21) | 2000-2010 (22) | 2010-2020 (23) | Births (24) | Deaths (25) | Net Migration (26) | Number (27) | Persons per household (28) | Family households (29) | Female family householder[1] (30) | One person (31) |
|---|---|---|---|---|---|---|---|---|---|---|---|---|---|---|---|---|
| TEXAS—Cont'd | | | | | | | | | | | | | | | | |
| Mason | 13.5 | 15.8 | 14.5 | 49.2 | 3,738 | 4,013 | 7.4 | 8.2 | 359 | 474 | 448 | 1,696 | 2.46 | 63.4 | 11.0 | 32.5 |
| Matagorda | 13.5 | 10.4 | 7.2 | 49.9 | 37,957 | 36,702 | -3.3 | 0.1 | 5,455 | 3,882 | -1,547 | 13,848 | 2.63 | 66.9 | 10.2 | 31.2 |
| Maverick | 8.9 | 7.0 | 5.1 | 49.8 | 47,297 | 54,258 | 14.7 | 7.6 | 10,988 | 3,866 | -3,007 | 16,286 | 3.56 | 82.6 | 20.7 | 14.1 |
| Medina | 13.4 | 10.4 | 7.0 | 48.4 | 39,304 | 46,009 | 17.1 | 13.8 | 5,776 | 4,204 | 4,789 | 15,682 | 3.03 | 74.1 | 10.2 | 21.4 |
| Menard | 15.5 | 16.4 | 16.2 | 50.2 | 2,360 | 2,242 | -5.0 | -5.3 | 178 | 350 | 53 | 1,014 | 2.05 | 57.3 | 4.5 | 41.7 |
| Midland | 10.0 | 6.0 | 4.2 | 49.2 | 116,009 | 136,887 | 18.0 | 29.9 | 28,321 | 11,059 | 23,216 | 56,998 | 2.92 | 68.4 | 11.5 | 26.3 |
| Milam | 14.0 | 12.3 | 9.2 | 50.4 | 24,238 | 24,759 | 2.1 | -0.2 | 3,019 | 3,017 | -54 | 9,228 | 2.63 | 67.4 | 12.3 | 28.4 |
| Mills | 14.9 | 15.3 | 13.3 | 49.9 | 5,151 | 4,941 | -4.1 | -2.0 | 415 | 684 | 171 | 1,790 | 2.63 | 64.8 | 10.1 | 33.7 |
| Mitchell | 10.8 | 8.9 | 6.4 | 41.1 | 9,698 | 9,403 | -3.0 | -12.8 | 995 | 969 | -1,255 | 2,382 | 2.72 | 63.6 | 7.0 | 29.2 |
| Montague | 14.1 | 12.8 | 10.1 | 50.8 | 19,117 | 19,722 | 3.2 | 1.2 | 2,254 | 2,779 | 776 | 7,800 | 2.45 | 66.4 | 7.6 | 28.6 |
| Montgomery | 12.3 | 8.6 | 5.2 | 50.5 | 293,768 | 455,740 | 55.1 | 37.4 | 71,546 | 37,044 | 135,016 | 198,649 | 2.87 | 74.7 | 10.3 | 20.8 |
| Moore | 10.0 | 7.1 | 4.8 | 48.0 | 20,121 | 21,904 | 8.9 | -5.7 | 4,223 | 1,460 | -4,058 | 6,733 | 3.15 | 75.3 | 10.0 | 21.6 |
| Morris | 14.2 | 12.5 | 9.8 | 51.8 | 13,048 | 12,934 | -0.9 | -4.2 | 1,546 | 1,753 | -328 | 5,117 | 2.39 | 66.0 | 13.2 | 31.8 |
| Motley | 14.3 | 16.5 | 14.6 | 49.3 | 1,426 | 1,205 | -15.5 | -1.7 | 97 | 180 | 61 | 489 | 2.56 | 64.0 | 14.5 | 33.7 |
| Nacogdoches | 11.1 | 9.3 | 6.3 | 52.1 | 59,203 | 64,524 | 9.0 | 0.4 | 8,761 | 5,903 | -2,654 | 23,757 | 2.52 | 63.8 | 12.6 | 27.3 |
| Navarro | 12.9 | 10.0 | 7.1 | 51.0 | 45,124 | 47,840 | 6.0 | 6.0 | 7,023 | 5,463 | 1,320 | 17,338 | 2.77 | 71.2 | 11.5 | 24.1 |
| Newton | 15.1 | 12.8 | 9.5 | 48.3 | 15,072 | 14,445 | -4.2 | -7.1 | 1,340 | 1,693 | -675 | 5,315 | 2.59 | 73.1 | 15.2 | 24.3 |
| Nolan | 11.7 | 10.5 | 8.4 | 50.0 | 15,802 | 15,214 | -3.7 | -2.5 | 2,058 | 1,823 | -618 | 5,407 | 2.71 | 67.6 | 11.1 | 29.6 |
| Nueces | 11.9 | 9.1 | 6.3 | 50.6 | 313,645 | 340,223 | 8.5 | 6.7 | 48,668 | 29,349 | 3,645 | 129,451 | 2.73 | 67.6 | 15.8 | 26.0 |
| Ochiltree | 11.2 | 7.4 | 4.6 | 50.0 | 9,006 | 10,223 | 13.5 | -6.1 | 1,734 | 830 | -1,543 | 3,483 | 2.89 | 75.1 | 10.5 | 21.4 |
| Oldham | 14.8 | 9.6 | 6.4 | 47.6 | 2,185 | 2,052 | -6.1 | 4.0 | 224 | 153 | 9 | 594 | 2.76 | 72.6 | 4.0 | 22.1 |
| Orange | 13.2 | 9.4 | 6.8 | 50.6 | 84,966 | 81,839 | -3.7 | 1.3 | 11,236 | 9,782 | -354 | 31,694 | 2.63 | 69.9 | 12.1 | 25.5 |
| Palo Pinto | 14.4 | 11.9 | 8.5 | 50.9 | 27,026 | 28,134 | 4.1 | 4.2 | 3,654 | 3,425 | 965 | 10,255 | 2.76 | 67.2 | 11.7 | 26.0 |
| Panola | 13.5 | 12.1 | 8.4 | 50.4 | 22,756 | 23,794 | 4.6 | -2.6 | 2,875 | 2,829 | -648 | 8,637 | 2.65 | 72.2 | 12.9 | 24.6 |
| Parker | 13.8 | 10.0 | 6.1 | 50.1 | 88,495 | 116,959 | 32.2 | 26.7 | 14,778 | 11,115 | 27,524 | 44,263 | 2.99 | 76.4 | 8.3 | 20.4 |
| Parmer | 12.1 | 8.4 | 6.7 | 48.3 | 10,016 | 10,269 | 2.5 | -7.3 | 1,536 | 737 | -1,571 | 3,203 | 3.01 | 72.2 | 9.6 | 24.9 |
| Pecos | 10.8 | 7.7 | 5.6 | 43.4 | 16,809 | 15,507 | -7.7 | 1.4 | 2,164 | 1,203 | -743 | 4,656 | 2.89 | 71.7 | 13.2 | 25.5 |
| Polk | 17.2 | 14.0 | 5.4 | 46.3 | 41,133 | 45,413 | 10.4 | 16.7 | 5,196 | 6,558 | 8,874 | 18,033 | 2.51 | 69.8 | 13.8 | 26.8 |
| Potter | 11.1 | 8.0 | 5.6 | 48.8 | 113,546 | 121,037 | 6.6 | -4.2 | 19,400 | 12,196 | -12,320 | 44,380 | 2.54 | 64.3 | 15.1 | 31.0 |
| Presidio | 9.6 | 11.1 | 13.0 | 49.5 | 7,304 | 7,817 | 7.0 | -16.7 | 1,173 | 493 | -2,019 | 2,543 | 2.74 | 54.1 | 11.0 | 44.5 |
| Rains | 16.4 | 14.9 | 10.3 | 50.1 | 9,139 | 10,902 | 19.3 | 15.1 | 1,093 | 1,451 | 2,001 | 4,330 | 2.70 | 70.7 | 8.8 | 25.9 |
| Randall | 11.7 | 9.4 | 6.3 | 50.9 | 104,312 | 120,761 | 15.8 | 15.8 | 16,873 | 10,694 | 12,949 | 49,389 | 2.66 | 68.4 | 10.7 | 27.1 |
| Reagan | 10.7 | 7.0 | 4.4 | 46.4 | 3,326 | 3,367 | 1.2 | 13.8 | 622 | 256 | 92 | 1,084 | 3.44 | 76.2 | 5.6 | 19.5 |
| Real | 16.4 | 18.4 | 13.9 | 49.6 | 3,047 | 3,309 | 8.6 | 3.1 | 357 | 541 | 282 | 1,126 | 2.92 | 59.9 | 9.9 | 37.1 |
| Red River | 15.0 | 14.3 | 11.8 | 51.4 | 14,314 | 12,862 | -10.1 | -6.7 | 1,295 | 1,856 | -299 | 4,963 | 2.41 | 66.2 | 16.3 | 32.1 |
| Reeves | 8.9 | 6.4 | 5.2 | 38.9 | 13,137 | 13,783 | 4.9 | 15.7 | 1,906 | 1,107 | 1,332 | 3,518 | 3.87 | 72.6 | 15.2 | 24.6 |
| Refugio | 14.2 | 12.0 | 10.8 | 50.7 | 7,828 | 7,383 | -5.7 | -6.9 | 841 | 887 | -462 | 2,547 | 2.74 | 64.8 | 14.6 | 28.9 |
| Roberts | 13.7 | 12.9 | 9.7 | 51.4 | 887 | 925 | 4.3 | -12.1 | 93 | 97 | -109 | 301 | 2.67 | 77.7 | 7.3 | 19.3 |
| Robertson | 13.5 | 12.0 | 8.7 | 51.0 | 16,000 | 16,617 | 3.9 | 3.2 | 2,139 | 1,958 | 362 | 6,444 | 2.61 | 71.0 | 14.7 | 24.9 |
| Rockwall | 12.1 | 7.9 | 4.9 | 50.6 | 43,080 | 78,398 | 82.0 | 40.2 | 10,492 | 5,814 | 26,713 | 33,032 | 2.92 | 80.5 | 8.1 | 16.5 |
| Runnels | 13.2 | 11.3 | 9.8 | 49.2 | 11,495 | 10,502 | -8.6 | -1.0 | 1,131 | 1,396 | 165 | 3,896 | 2.58 | 69.1 | 11.7 | 26.8 |
| Rusk | 12.8 | 10.2 | 7.2 | 46.3 | 47,372 | 53,255 | 12.4 | 2.0 | 6,297 | 5,675 | 445 | 18,108 | 2.70 | 72.2 | 12.7 | 23.3 |
| Sabine | 17.1 | 16.7 | 14.5 | 50.5 | 10,469 | 10,836 | 3.5 | -3.0 | 977 | 1,721 | 413 | 4,311 | 2.41 | 68.2 | 10.6 | 29.7 |
| San Augustine | 16.4 | 14.2 | 12.9 | 50.5 | 8,946 | 8,861 | -1.0 | -6.9 | 917 | 1,401 | -126 | 3,451 | 2.33 | 72.0 | 19.0 | 27.0 |
| San Jacinto | 15.8 | 13.8 | 9.2 | 50.0 | 22,246 | 26,383 | 18.6 | 11.1 | 2,903 | 3,112 | 3,128 | 10,043 | 2.79 | 66.5 | 9.2 | 28.9 |
| San Patricio | 11.7 | 9.1 | 6.3 | 49.6 | 67,138 | 64,792 | -3.5 | 3.5 | 9,850 | 6,350 | -1,242 | 22,898 | 2.89 | 73.6 | 15.5 | 22.4 |
| San Saba | 13.2 | 12.6 | 10.9 | 45.7 | 6,186 | 6,130 | -0.9 | -1.5 | 672 | 725 | -42 | 2,082 | 2.54 | 63.8 | 5.7 | 31.4 |
| Schleicher | 12.5 | 12.2 | 8.1 | 50.2 | 2,935 | 3,461 | 17.9 | -20.2 | 317 | 261 | -762 | 1,075 | 2.77 | 62.2 | 6.6 | 32.3 |
| Scurry | 11.8 | 8.4 | 6.7 | 46.1 | 16,361 | 16,919 | 3.4 | -1.5 | 2,430 | 1,798 | -894 | 5,941 | 2.55 | 67.4 | 7.2 | 29.1 |
| Shackelford | 15.9 | 12.3 | 9.5 | 50.9 | 3,302 | 3,376 | 2.2 | -2.3 | 337 | 328 | -87 | 1,317 | 2.48 | 66.1 | 11.1 | 28.0 |
| Shelby | 12.8 | 10.5 | 7.9 | 50.4 | 25,224 | 25,450 | 0.9 | -2.1 | 3,687 | 3,008 | -1,222 | 9,293 | 2.71 | 72.7 | 15.3 | 26.0 |
| Sherman | 13.4 | 8.1 | 7.8 | 47.3 | 3,186 | 3,034 | -4.8 | -0.2 | 402 | 268 | -143 | 1,019 | 2.97 | 74.0 | 8.6 | 22.7 |
| Smith | 12.0 | 9.9 | 7.5 | 51.7 | 174,706 | 209,729 | 20.0 | 12.4 | 31,078 | 21,241 | 16,290 | 77,678 | 2.86 | 69.4 | 12.1 | 25.6 |
| Somervell | 15.1 | 12.0 | 8.1 | 50.6 | 6,809 | 8,491 | 24.7 | 7.6 | 878 | 935 | 704 | 3,123 | 2.73 | 81.5 | 12.1 | 16.4 |
| Starr | 8.4 | 6.5 | 5.1 | 51.1 | 53,597 | 60,982 | 13.8 | 5.4 | 13,176 | 4,055 | -5,870 | 16,188 | 3.92 | 80.1 | 26.3 | 19.0 |
| Stephens | 13.3 | 11.7 | 9.1 | 47.2 | 9,674 | 9,634 | -0.4 | -3.1 | 1,054 | 1,139 | -224 | 3,247 | 2.75 | 70.3 | 14.7 | 25.5 |
| Sterling | 12.0 | 8.1 | 6.7 | 50.0 | 1,393 | 1,143 | -17.9 | 15.0 | 149 | 163 | 184 | 458 | 2.61 | 77.9 | 8.1 | 19.9 |
| Stonewall | 14.7 | 13.1 | 13.2 | 50.9 | 1,693 | 1,490 | -12.0 | -9.5 | 142 | 219 | -69 | 580 | 2.46 | 64.5 | 7.6 | 35.5 |
| Sutton | 13.4 | 11.5 | 7.9 | 48.8 | 4,077 | 4,128 | 1.3 | -9.4 | 530 | 340 | -591 | 1,412 | 2.69 | 70.5 | 11.4 | 28.3 |
| Swisher | 11.4 | 9.7 | 8.6 | 46.7 | 8,378 | 7,859 | -6.2 | -6.6 | 989 | 815 | -702 | 2,505 | 2.70 | 68.5 | 13.9 | 27.7 |
| Tarrant | 11.6 | 7.4 | 4.6 | 51.1 | 1,446,219 | 1,811,345 | 25.2 | 17.2 | 284,982 | 128,625 | 155,542 | 708,252 | 2.86 | 70.0 | 14.2 | 24.5 |
| Taylor | 10.6 | 8.2 | 6.7 | 51.2 | 126,555 | 131,515 | 3.9 | 5.8 | 21,063 | 13,612 | 285 | 49,868 | 2.61 | 65.7 | 13.5 | 26.4 |
| Terrell | 13.4 | 17.9 | 15.8 | 49.3 | 1,081 | 981 | -9.3 | -28.4 | 99 | 115 | -264 | 418 | 2.14 | 60.5 | 12.2 | 37.3 |
| Terry | 11.2 | 8.6 | 6.7 | 46.9 | 12,761 | 12,651 | -0.9 | -3.7 | 1,948 | 1,296 | -1,124 | 4,061 | 2.84 | 70.8 | 14.2 | 25.7 |
| Throckmorton | 15.6 | 12.6 | 15.2 | 51.8 | 1,850 | 1,641 | -11.3 | -9.4 | 148 | 214 | -91 | 668 | 2.14 | 63.5 | 9.7 | 34.4 |
| Titus | 11.1 | 8.9 | 6.1 | 51.1 | 28,118 | 32,334 | 15.0 | 1.8 | 5,012 | 2,822 | -1,604 | 10,893 | 2.98 | 77.5 | 13.2 | 19.4 |

1. No spouse present.

| STATE County | Persons in group quarters, 2020 | Daytime Population, 2015-2019 | | Births, 2020 | | Deaths, 2020 | | Persons under 65 with no health insurance, 2019 | | Medicare, 2020 | | | Crimes reported by county police or sheriff, 2019 | |
|---|---|---|---|---|---|---|---|---|---|---|---|---|---|---|
| | | Number | Employment/residence ratio | Total | Rate[1] | Number | Rate[1] | Number | Percent | Total beneficiaries | Enrolled in Original Medicare | Enrolled in Medicare Advantage | Violent | Property |
| | 32 | 33 | 34 | 35 | 36 | 37 | 38 | 39 | 40 | 41 | 42 | 43 | 44 | 45 |
| TEXAS—Cont'd | | | | | | | | | | | | | | |
| Mason | 3 | 4,151 | 0.98 | 24 | 5.5 | 47 | 10.8 | 883 | 29.4 | 1,335 | 965 | 371 | 8 | 10 |
| Matagorda | 348 | 35,964 | 0.95 | 517 | 14.1 | 390 | 10.6 | 6,192 | 20.8 | 7,207 | 4,461 | 2,746 | 62 | 332 |
| Maverick | 1,049 | 54,771 | 0.84 | 1,019 | 17.5 | 410 | 7.0 | 14,265 | 28.5 | 10,385 | 5,580 | 4,805 | NA | NA |
| Medina | 2,304 | 42,505 | 0.64 | 576 | 11.0 | 491 | 9.4 | 7,422 | 18.2 | 9,887 | 5,550 | 4,337 | 101 | 351 |
| Menard | 39 | 1,861 | 0.74 | 16 | 7.5 | 24 | 11.3 | 407 | 28.5 | 683 | 487 | 195 | 3 | 1 |
| Midland | 1,553 | 190,612 | 1.27 | 3,025 | 17.0 | 1,222 | 6.9 | 29,022 | 18.4 | 19,329 | 13,771 | 5,558 | NA | NA |
| Milam | 394 | 22,056 | 0.72 | 250 | 10.1 | 303 | 12.3 | 3,704 | 19.3 | 6,078 | 3,464 | 2,614 | 13 | 87 |
| Mills | 145 | 4,867 | 0.99 | 35 | 7.2 | 78 | 16.1 | 908 | 26.9 | 1,410 | 948 | 462 | 14 | 20 |
| Mitchell | 1,439 | 7,882 | 0.77 | 88 | 10.7 | 75 | 9.1 | 1,085 | 19.9 | 1,432 | 935 | 497 | 1 | 25 |
| Montague | 212 | 17,586 | 0.76 | 189 | 9.5 | 293 | 14.7 | 3,321 | 21.7 | 5,264 | 3,975 | 1,289 | 33 | 83 |
| Montgomery | 3,123 | 531,783 | 0.85 | 7,586 | 12.1 | 4,800 | 7.7 | 90,482 | 17.3 | 93,040 | 52,535 | 40,504 | 736 | 4,976 |
| Moore | 146 | 21,844 | 1.05 | 388 | 18.8 | 135 | 6.5 | 4,996 | 27.2 | 2,563 | 1,998 | 565 | 6 | 39 |
| Morris | 138 | 11,269 | 0.78 | 148 | 11.9 | 156 | 12.6 | 1,786 | 18.7 | 3,427 | 2,020 | 1,406 | 10 | 72 |
| Motley | 0 | 1,146 | 0.81 | 14 | 11.8 | 20 | 16.9 | 189 | 22.7 | 361 | 256 | 105 | NA | NA |
| Nacogdoches | 4,586 | 63,820 | 0.94 | 794 | 12.3 | 643 | 9.9 | 10,678 | 21.2 | 11,705 | 7,818 | 3,887 | 58 | 270 |
| Navarro | 681 | 46,827 | 0.89 | 645 | 12.7 | 614 | 12.1 | 9,643 | 23.6 | 10,554 | 6,902 | 3,653 | 58 | 242 |
| Newton | 723 | 10,983 | 0.41 | 106 | 7.9 | 211 | 15.7 | 2,176 | 21.8 | 3,051 | 1,931 | 1,120 | 22 | 98 |
| Nolan | 419 | 15,507 | 1.10 | 195 | 13.1 | 174 | 11.7 | 2,383 | 20.5 | 3,267 | 2,213 | 1,054 | 4 | 31 |
| Nueces | 6,740 | 372,449 | 1.07 | 4,467 | 12.3 | 3,227 | 8.9 | 60,083 | 19.9 | 61,022 | 26,852 | 34,170 | 61 | 160 |
| Ochiltree | 22 | 10,973 | 1.18 | 139 | 14.5 | 73 | 7.6 | 2,360 | 27.3 | 1,304 | 1,085 | 219 | 0 | 22 |
| Oldham | 277 | 2,261 | 1.18 | 20 | 9.4 | 8 | 3.7 | 281 | 18.1 | 442 | 317 | 125 | 0 | 6 |
| Orange | 608 | 73,502 | 0.71 | 1,019 | 12.3 | 982 | 11.8 | 11,627 | 16.8 | 16,272 | 9,055 | 7,216 | NA | NA |
| Palo Pinto | 260 | 26,481 | 0.82 | 344 | 11.7 | 407 | 13.9 | 5,323 | 23.1 | 6,538 | 4,662 | 1,876 | 9 | 68 |
| Panola | 345 | 22,282 | 0.89 | 261 | 11.3 | 280 | 12.1 | 3,606 | 19.7 | 5,256 | 3,338 | 1,918 | 26 | 260 |
| Parker | 1,114 | 114,218 | 0.68 | 1,600 | 10.8 | 1,320 | 8.9 | 20,751 | 17.3 | 27,068 | 17,653 | 9,415 | 100 | 751 |
| Parmer | 64 | 11,357 | 1.38 | 135 | 14.2 | 37 | 3.9 | 2,297 | 28.2 | 1,508 | 1,241 | 267 | NA | NA |
| Pecos | 2,056 | 16,313 | 1.08 | 204 | 13.0 | 102 | 6.5 | 2,871 | 24.8 | 2,304 | 1,597 | 708 | 1 | 98 |
| Polk | 4,227 | 46,096 | 0.85 | 515 | 9.7 | 663 | 12.5 | 8,609 | 23.0 | 18,835 | 11,831 | 7,004 | 76 | 671 |
| Potter | 6,521 | 149,586 | 1.58 | 1,570 | 13.5 | 1,162 | 10.0 | 24,112 | 25.3 | 19,025 | 12,992 | 6,033 | NA | NA |
| Presidio | 0 | 6,992 | 1.01 | 96 | 14.8 | 57 | 8.8 | 1,638 | 32.6 | 1,860 | 1,311 | 549 | NA | NA |
| Rains | 63 | 9,805 | 0.60 | 121 | 9.6 | 143 | 11.4 | 2,071 | 22.1 | 3,202 | 2,101 | 1,100 | 8 | 48 |
| Randall | 2,384 | 102,367 | 0.54 | 1,645 | 11.8 | 1,223 | 8.7 | 15,147 | 13.2 | 23,531 | 16,993 | 6,538 | 61 | 249 |
| Reagan | 22 | 4,814 | 1.64 | 67 | 17.5 | 17 | 4.4 | 676 | 19.9 | 443 | 318 | 126 | 7 | 32 |
| Real | 71 | 3,475 | 1.06 | 39 | 11.4 | 63 | 18.5 | 522 | 22.2 | 1,329 | 948 | 382 | 3 | 59 |
| Red River | 163 | 10,097 | 0.58 | 118 | 9.8 | 206 | 17.2 | 1,886 | 21.3 | 3,510 | 2,501 | 1,010 | NA | NA |
| Reeves | 3,137 | 22,002 | 2.13 | 192 | 12.0 | 115 | 7.2 | 2,435 | 22.2 | 2,113 | 1,570 | 543 | NA | NA |
| Refugio | 121 | 6,715 | 0.84 | 59 | 8.6 | 43 | 6.3 | 986 | 18.8 | 1,737 | 1,159 | 578 | 3 | 52 |
| Roberts | 0 | 760 | 0.87 | 6 | 7.4 | 10 | 12.3 | 112 | 16.7 | 178 | 134 | 44 | 0 | 2 |
| Robertson | 174 | 15,124 | 0.74 | 199 | 11.6 | 186 | 10.8 | 2,693 | 20.0 | 3,861 | 2,561 | 1,300 | 8 | 116 |
| Rockwall | 625 | 83,543 | 0.71 | 1,150 | 10.5 | 689 | 6.3 | 13,643 | 14.9 | 15,481 | 10,249 | 5,233 | 13 | 91 |
| Runnels | 165 | 9,428 | 0.81 | 116 | 11.2 | 121 | 11.6 | 1,617 | 20.2 | 2,562 | 2,056 | 507 | 1 | 11 |
| Rusk | 6,179 | 46,724 | 0.66 | 556 | 10.2 | 623 | 11.5 | 8,119 | 20.7 | 10,270 | 6,189 | 4,081 | 33 | 442 |
| Sabine | 111 | 10,229 | 0.92 | 97 | 9.2 | 153 | 14.6 | 1,450 | 20.1 | 3,377 | 2,481 | 897 | 14 | 73 |
| San Augustine | 182 | 7,669 | 0.78 | 72 | 8.7 | 151 | 18.3 | 1,240 | 21.1 | 2,501 | 1,679 | 821 | 13 | 57 |
| San Jacinto | 127 | 21,858 | 0.37 | 277 | 9.5 | 346 | 11.8 | 5,018 | 22.6 | 6,613 | 3,362 | 3,250 | 44 | 330 |
| San Patricio | 483 | 62,592 | 0.84 | 917 | 13.7 | 598 | 8.9 | 10,456 | 18.7 | 12,963 | 5,482 | 7,481 | 37 | 166 |
| San Saba | 689 | 5,656 | 0.87 | 70 | 11.6 | 82 | 13.6 | 1,210 | 30.7 | 1,529 | 1,059 | 470 | NA | NA |
| Schleicher | 1 | 2,775 | 0.84 | 24 | 8.7 | 19 | 6.9 | 553 | 24.9 | 594 | 463 | 131 | 4 | 13 |
| Scurry | 1,594 | 18,074 | 1.14 | 199 | 11.9 | 188 | 11.3 | 2,820 | 22.5 | 2,951 | 2,031 | 920 | 7 | 44 |
| Shackelford | 11 | 2,988 | 0.79 | 34 | 10.3 | 40 | 12.1 | 578 | 22.5 | 764 | 541 | 223 | NA | NA |
| Shelby | 145 | 24,981 | 0.96 | 340 | 13.6 | 340 | 13.6 | 5,297 | 25.8 | 5,148 | 3,255 | 1,894 | 40 | 189 |
| Sherman | 27 | 2,627 | 0.71 | 42 | 13.9 | 13 | 4.3 | 720 | 28.3 | 388 | 310 | 77 | 0 | 3 |
| Smith | 5,181 | 235,038 | 1.08 | 2,921 | 12.4 | 2,312 | 9.8 | 39,717 | 21.1 | 46,779 | 29,978 | 16,801 | 335 | 1,453 |
| Somervell | 286 | 9,471 | 1.17 | 90 | 9.8 | 124 | 13.6 | 1,303 | 17.8 | 1,993 | 1,261 | 732 | 4 | 30 |
| Starr | 818 | 61,146 | 0.86 | 1,156 | 18.0 | 401 | 6.2 | 17,283 | 31.2 | 9,609 | 4,459 | 5,150 | 79 | 232 |
| Stephens | 668 | 9,154 | 0.94 | 105 | 11.2 | 98 | 10.5 | 1,633 | 24.0 | 2,083 | 1,509 | 574 | 1 | 21 |
| Sterling | 34 | 1,300 | 1.11 | 12 | 9.1 | 30 | 22.8 | 209 | 19.0 | 239 | 180 | 59 | NA | NA |
| Stonewall | 24 | 1,446 | 0.95 | 13 | 9.6 | 15 | 11.1 | 192 | 19.5 | 358 | 273 | 85 | 1 | 0 |
| Sutton | 13 | 3,706 | 0.93 | 49 | 13.1 | 35 | 9.4 | 628 | 20.7 | 781 | 647 | 133 | 0 | 1 |
| Swisher | 639 | 6,800 | 0.76 | 83 | 11.3 | 58 | 7.9 | 1,429 | 26.3 | 1,441 | 1,074 | 367 | 7 | 24 |
| Tarrant | 25,670 | 2,051,504 | 1.00 | 27,220 | 12.8 | 15,107 | 7.1 | 347,063 | 18.9 | 279,704 | 146,186 | 133,518 | 123 | 742 |
| Taylor | 4,866 | 141,294 | 1.07 | 2,029 | 14.6 | 1,485 | 10.7 | 21,168 | 18.5 | 25,111 | 17,179 | 7,932 | 39 | 89 |
| Terrell | 0 | 893 | 0.99 | 4 | 5.7 | 10 | 14.2 | 101 | 19.5 | 246 | 157 | 89 | NA | NA |
| Terry | 1,041 | 12,477 | 0.99 | 178 | 14.6 | 130 | 10.7 | 2,487 | 26.3 | 2,040 | 1,168 | 872 | 1 | 30 |
| Throckmorton | 13 | 1,425 | 0.98 | 14 | 9.4 | 22 | 14.8 | 257 | 24.3 | 401 | 310 | 92 | NA | NA |
| Titus | 426 | 35,941 | 1.23 | 432 | 13.1 | 272 | 8.3 | 6,864 | 25.1 | 5,613 | 3,679 | 1,935 | 50 | 162 |

1. Per 1,000 estimated resident population.

Items 32—45

# Table B. States and Counties — Crime, Education, Money Income, and Poverty

| STATE County | County law enforcement employment, 2019 | | School enrollment and attainment, 2015-2019 | | | | Local government expenditures,[3] 2017-2018 | | Money income, 2015-2019 | | | | Income and poverty, 2019 | | |
| | | | Enrollment[1] | | Attainment[2] (percent) | | | | | Households | | | | Percent below poverty level | |
| | Officers | Civilians | Total | Percent private | High school graduate or less | Bachelor's degree or more | Total current spending (mil dol) | Current spending per student (dollars) | Per capita income[4] | Median income (dollars) | Percent with income of less than $50,000 | Percent with income of $200,000 or more | Median household income (dollars) | All persons | Children under 18 years | Children 5 to 17 years in families |
| | 46 | 47 | 48 | 49 | 50 | 51 | 52 | 53 | 54 | 55 | 56 | 57 | 58 | 59 | 60 | 61 |
|---|---|---|---|---|---|---|---|---|---|---|---|---|---|---|---|---|
| **TEXAS—Cont'd** | | | | | | | | | | | | | | | | |
| Mason | NA | NA | 889 | 1.6 | 39.0 | 27.6 | 8 | 11,111 | 30,388 | 42,276 | 57.1 | 5.3 | 54,914 | 11.7 | 20.4 | 18.8 |
| Matagorda | 39 | 36 | 8,831 | 6.3 | 51.0 | 17.4 | 78 | 10,963 | 25,172 | 48,913 | 51.2 | 3.5 | 49,276 | 17.5 | 24.1 | 23.4 |
| Maverick | NA | NA | 16,450 | 4.0 | 63.1 | 12.7 | 138 | 9,452 | 16,779 | 39,625 | 57.8 | 0.6 | 40,017 | 26.9 | 40.3 | 39.9 |
| Medina | 42 | 41 | 12,126 | 10.1 | 47.8 | 19.8 | 97 | 9,182 | 27,475 | 62,599 | 39.6 | 5.5 | 60,695 | 12.2 | 19.4 | 17.5 |
| Menard | NA | NA | 281 | 6.8 | 58.1 | 18.0 | 4 | 13,848 | 25,742 | 36,395 | 66.0 | 1.8 | 38,425 | 20.7 | 35.4 | 33.1 |
| Midland | NA | NA | 44,336 | 14.2 | 41.2 | 27.4 | 241 | 8,345 | 39,248 | 79,421 | 30.4 | 11.5 | 85,811 | 9.9 | 12.3 | 10.8 |
| Milam | 24 | 44 | 5,641 | 8.4 | 56.7 | 14.1 | 47 | 10,012 | 25,714 | 47,902 | 51.6 | 2.9 | 48,989 | 16.3 | 25.1 | 23.6 |
| Mills | NA | NA | 1,055 | 9.9 | 49.0 | 18.3 | 15 | 14,599 | 27,374 | 49,306 | 50.8 | 5.9 | 46,876 | 13.2 | 24.9 | 22.0 |
| Mitchell | 5 | 13 | 2,070 | 7.5 | 61.6 | 11.3 | 16 | 11,348 | 21,136 | 51,492 | 48.4 | 4.5 | 44,457 | 19.5 | 23.4 | 22.1 |
| Montague | 13 | 19 | 4,097 | 12.8 | 51.9 | 16.9 | 34 | 9,977 | 28,096 | 51,765 | 48.2 | 3.2 | 53,649 | 12.9 | 18.0 | 16.4 |
| Montgomery | 521 | 402 | 150,649 | 14.2 | 35.1 | 34.5 | 951 | 8,674 | 41,211 | 80,902 | 30.2 | 13.7 | 88,833 | 8.9 | 13.2 | 12.6 |
| Moore | 20 | 32 | 5,639 | 1.6 | 58.9 | 14.3 | 46 | 9,481 | 23,001 | 54,871 | 40.8 | 2.9 | 55,099 | 11.9 | 15.1 | 14.2 |
| Morris | 10 | 14 | 2,861 | 7.1 | 52.8 | 14.3 | 19 | 10,126 | 22,971 | 41,359 | 60.2 | 1.7 | 43,304 | 18.1 | 29.0 | 26.8 |
| Motley | 2 | 0 | 285 | 1.4 | 43.8 | 16.8 | 3 | 15,297 | 29,071 | 43,859 | 54.8 | 3.5 | 40,217 | 17.6 | 30.9 | 29.7 |
| Nacogdoches | 39 | 44 | 22,426 | 4.8 | 45.7 | 24.5 | 101 | 9,057 | 23,617 | 44,847 | 54.9 | 3.8 | 46,207 | 20.9 | 26.3 | 24.5 |
| Navarro | 56 | 74 | 11,980 | 7.6 | 50.6 | 15.9 | 97 | 9,698 | 23,786 | 48,529 | 51.3 | 3.6 | 48,649 | 13.4 | 21.1 | 20.6 |
| Newton | NA | NA | 2,864 | 3.2 | 59.5 | 9.7 | 22 | 11,830 | 21,659 | 40,101 | 59.3 | 1.2 | 45,890 | 22.8 | 28.3 | 25.8 |
| Nolan | 16 | 27 | 3,325 | 10.3 | 52.9 | 15.2 | 37 | 11,421 | 25,094 | 45,537 | 55.3 | 2.7 | 46,389 | 17.5 | 23.6 | 23.4 |
| Nueces | 47 | 247 | 91,886 | 5.7 | 46.4 | 21.8 | 604 | 9,758 | 27,740 | 55,919 | 44.9 | 4.6 | 56,079 | 16.5 | 23.1 | 22.0 |
| Ochiltree | NA | NA | 2,774 | 4.6 | 58.6 | 15.6 | 23 | 10,217 | 23,624 | 50,464 | 49.4 | 3.4 | 62,567 | 11.4 | 16.2 | 15.2 |
| Oldham | 7 | 5 | 681 | 4.7 | 46.9 | 22.0 | 15 | 16,427 | 26,408 | 64,545 | 34.2 | 6.9 | 59,404 | 13.1 | 29.5 | 22.4 |
| Orange | 61 | 70 | 20,063 | 10.0 | 49.3 | 15.5 | 153 | 10,295 | 30,114 | 59,399 | 43.1 | 4.4 | 62,976 | 12.5 | 18.7 | 16.8 |
| Palo Pinto | 24 | 33 | 6,280 | 7.8 | 50.0 | 16.6 | 46 | 10,022 | 25,150 | 50,154 | 49.9 | 3.6 | 50,159 | 17.2 | 26.2 | 25.0 |
| Panola | 35 | 26 | 5,526 | 10.4 | 48.6 | 16.1 | 40 | 9,960 | 26,577 | 52,982 | 46.6 | 3.9 | 56,882 | 14.3 | 21.5 | 19.7 |
| Parker | 93 | 38 | 32,906 | 15.5 | 40.2 | 26.4 | 193 | 8,845 | 35,142 | 77,503 | 31.3 | 9.5 | 82,501 | 8.2 | 10.4 | 9.0 |
| Parmer | NA | NA | 2,492 | 10.8 | 60.3 | 17.9 | 26 | 11,074 | 21,120 | 52,397 | 48.0 | 2.6 | 52,763 | 18.7 | 23.1 | 21.9 |
| Pecos | NA | NA | 3,720 | 2.6 | 66.8 | 9.8 | 35 | 10,685 | 23,859 | 55,359 | 44.6 | 2.0 | 50,607 | 16.7 | 26.6 | 26.0 |
| Polk | 47 | 59 | 8,806 | 7.7 | 57.8 | 13.1 | 71 | 9,236 | 25,164 | 49,279 | 50.6 | 3.8 | 50,416 | 20.3 | 26.1 | 23.0 |
| Potter | 96 | 106 | 31,879 | 6.2 | 55.0 | 14.9 | 339 | 12,399 | 22,568 | 42,528 | 56.0 | 3.4 | 41,138 | 19.2 | 29.1 | 29.2 |
| Presidio | NA | NA | 1,613 | 9.4 | 62.5 | 21.2 | 20 | 9,180 | 15,337 | 25,098 | 74.0 | 0.0 | 33,499 | 13.4 | 22.9 | 21.9 |
| Rains | 15 | 20 | 2,082 | 7.3 | 55.2 | 13.7 | 16 | 10,019 | 27,938 | 51,579 | 48.9 | 4.3 | 65,657 | — | — | — |
| Randall | 91 | 118 | 36,367 | 8.3 | 30.5 | 32.4 | 100 | 10,019 | 34,293 | 68,261 | 35.6 | 5.7 | 70,694 | 9.0 | 12.1 | 11.0 |
| Reagan | 15 | 16 | 1,043 | 2.1 | 59.7 | 10.5 | 12 | 13,742 | 25,600 | 74,868 | 32.9 | 3.0 | 67,384 | 8.9 | 12.2 | 11.9 |
| Real | 5 | 6 | 809 | 1.4 | 47.1 | 17.4 | 7 | 12,745 | 18,235 | 35,862 | 67.8 | 1.2 | 40,834 | 17.7 | 32.7 | 30.5 |
| Red River | 13 | 18 | 2,351 | 3.6 | 55.0 | 12.5 | 24 | 11,490 | 22,689 | 39,142 | 57.7 | 1.1 | 41,325 | 19.6 | 27.2 | 23.0 |
| Reeves | NA | NA | 3,154 | 6.7 | 63.0 | 12.3 | 30 | 10,498 | 20,020 | 53,311 | 48.6 | 2.6 | 48,990 | 22.1 | 24.6 | 21.9 |
| Refugio | 17 | 25 | 1,580 | 4.3 | 58.2 | 11.5 | 18 | 13,870 | 24,248 | 50,076 | 49.9 | 1.6 | 46,883 | 18.0 | 27.0 | 24.6 |
| Roberts | 5 | 1 | 208 | 5.3 | 37.7 | 25.0 | 3 | 14,110 | 30,955 | 68,750 | 35.5 | 3.7 | 70,705 | 7.4 | 9.3 | 9.0 |
| Robertson | 15 | 20 | 3,919 | 9.7 | 51.0 | 16.1 | 37 | 10,726 | 26,033 | 52,928 | 48.7 | 3.5 | 47,016 | 20.7 | 31.0 | 24.3 |
| Rockwall | 50 | 74 | 27,497 | 13.1 | 26.0 | 40.7 | 184 | 8,352 | 42,346 | 100,920 | 20.8 | 14.3 | 105,763 | 4.8 | 6.5 | 6.1 |
| Runnels | 6 | 24 | 2,162 | 6.7 | 56.0 | 18.1 | 23 | 11,276 | 24,402 | 44,940 | 55.0 | 2.8 | 48,650 | 17.2 | 23.7 | 21.3 |
| Rusk | 40 | 36 | 11,297 | 8.1 | 51.1 | 15.5 | 76 | 9,347 | 26,658 | 55,234 | 45.9 | 4.8 | 56,036 | 15.4 | 20.4 | 16.5 |
| Sabine | 10 | 10 | 1,867 | 8.9 | 50.9 | 17.6 | 22 | 11,051 | 24,870 | 34,992 | 60.8 | 3.1 | 43,778 | 17.6 | 26.6 | 26.3 |
| San Augustine | 10 | 10 | 1,652 | 3.1 | 57.7 | 14.0 | 14 | 11,509 | 23,683 | 40,353 | 60.4 | 1.8 | 40,094 | 22.5 | 32.1 | 30.3 |
| San Jacinto | 31 | 34 | 5,846 | 3.5 | 63.5 | 11.2 | 36 | 10,224 | 23,312 | 41,614 | 57.1 | 3.2 | 42,633 | 16.8 | 26.2 | 25.1 |
| San Patricio | NA | NA | 16,870 | 4.6 | 52.8 | 15.6 | 148 | 10,299 | 26,054 | 56,556 | 44.9 | 4.2 | 57,302 | 14.4 | 20.6 | 19.7 |
| San Saba | NA | NA | 1,324 | 8.2 | 60.1 | 14.1 | 11 | 11,323 | 24,961 | 45,083 | 53.1 | 1.5 | 46,061 | 20.7 | 34.4 | 30.6 |
| Schleicher | 6 | 6 | 765 | 7.3 | 45.2 | 16.6 | 6 | 11,088 | 26,560 | 53,229 | 49.7 | 1.2 | 53,286 | 15.0 | 23.4 | 19.4 |
| Scurry | 7 | 34 | 4,459 | 9.1 | 55.6 | 16.2 | 34 | 10,665 | 24,552 | 54,326 | 45.8 | 2.6 | 56,299 | 14.7 | 18.3 | 17.6 |
| Shackelford | NA | NA | 784 | 13.3 | 35.4 | 30.5 | 7 | 11,477 | 24,852 | 46,935 | 52.9 | 1.7 | 65,095 | 12.7 | 17.5 | 16.0 |
| Shelby | 17 | 20 | 6,259 | 5.5 | 58.6 | 13.8 | 57 | 10,478 | 21,950 | 42,522 | 57.2 | 3.2 | 39,708 | 21.4 | 32.5 | 31.1 |
| Sherman | 5 | 8 | 718 | 2.8 | 51.7 | 15.0 | 9 | 11,132 | 30,467 | 51,926 | 47.2 | 3.3 | 55,473 | 11.1 | 13.7 | 12.5 |
| Smith | 170 | 212 | 59,232 | 11.9 | 37.7 | 26.4 | 320 | 8,923 | 28,483 | 56,810 | 44.1 | 5.2 | 59,538 | 12.9 | 17.8 | 16.7 |
| Somervell | NA | NA | 1,781 | 6.8 | 48.9 | 26.4 | 23 | 11,353 | 29,449 | 60,632 | 41.7 | 4.9 | 67,897 | 9.4 | 14.2 | 13.5 |
| Starr | 45 | 64 | 19,989 | 3.3 | 73.0 | 9.9 | 186 | 10,886 | 14,179 | 30,387 | 70.8 | 1.7 | 32,516 | 32.5 | 43.2 | 41.4 |
| Stephens | NA | NA | 1,964 | 9.1 | 50.1 | 16.2 | 13 | 8,476 | 25,015 | 46,223 | 56.5 | 4.0 | 46,658 | 17.0 | 26.9 | 25.2 |
| Sterling | 5 | 0 | 325 | 0.0 | 62.6 | 9.5 | 4 | 13,029 | 28,757 | 59,868 | 40.0 | 2.2 | 66,790 | 10.7 | 12.9 | 12.0 |
| Stonewall | 3 | 6 | 397 | 0.0 | 53.1 | 14.8 | 3 | 13,031 | 25,638 | 51,250 | 48.8 | 2.6 | 48,267 | 15.1 | 20.2 | 19.4 |
| Sutton | NA | NA | 800 | 0.0 | 59.1 | 17.1 | 10 | 12,763 | 26,904 | 54,306 | 45.8 | 1.2 | 63,427 | 13.1 | 17.8 | 17.2 |
| Swisher | 5 | 6 | 2,051 | 1.8 | 56.6 | 15.3 | 18 | 11,034 | 19,697 | 39,771 | 62.5 | 1.8 | 44,388 | 20.6 | 29.1 | 27.8 |
| Tarrant | 372 | 1,040 | 568,298 | 13.2 | 37.9 | 32.3 | 3,330 | 9,266 | 33,292 | 67,700 | 36.1 | 8.0 | 70,130 | 10.2 | 13.9 | 12.0 |
| Taylor | 81 | 149 | 38,013 | 26.5 | 41.6 | 25.2 | 217 | 9,133 | 27,162 | 53,143 | 47.4 | 3.8 | 54,465 | 14.0 | 18.9 | 17.7 |
| Terrell | NA | NA | 109 | 15.6 | 40.3 | 18.4 | 3 | 17,076 | 25,687 | 43,208 | 62.2 | 0.0 | 45,677 | 17.5 | 20.7 | 19.4 |
| Terry | 12 | 21 | 2,899 | 9.8 | 63.3 | 11.4 | 25 | 10,542 | 20,627 | 44,627 | 54.8 | 2.2 | 42,953 | 21.6 | 31.0 | 29.3 |
| Throckmorton | 2 | 4 | 212 | 5.7 | 49.1 | 21.7 | 4 | 12,583 | 30,463 | 40,000 | 56.0 | 2.4 | 42,746 | 13.9 | 27.2 | 24.2 |
| Titus | 24 | 35 | 8,712 | 4.8 | 56.7 | 16.3 | 65 | 9,158 | 21,730 | 50,196 | 49.8 | 2.7 | 49,924 | 16.3 | 23.2 | 22.5 |

1. All persons 3 years old and over enrolled in nursery school through college.　2. Persons 25 years old and over.　3. Elementary and secondary education expenditures.　4. Based on population estimated by the American Community Survey, 2014–2018.

# Table B. States and Counties — **Personal Income**

| STATE County | Personal income, 2019 | | | | | | | | | | Earnings, 2019 | | |
| --- | --- | --- | --- | --- | --- | --- | --- | --- | --- | --- | --- | --- | --- |
| | Total (mil dol) | Percent change 2018-2019 | Per capita[1] Dollars | Per capita[1] Rank | Wages and salaries (mil dol) | Supplements to wages and salaries, employer contributions (mil dol) Pension and insurance | Supplements to wages and salaries, employer contributions (mil dol) Government social insurance | Proprietors' income (mil dol) | Dividends, interest, and rent (mil dol) | Personal transfer receipts (mil dol) | Total (mil dol) | Contributions for government social insurance (mil dol) From employee and self-employed | Contributions for government social insurance (mil dol) From employer |
| | 62 | 63 | 64 | 65 | 66 | 67 | 68 | 69 | 70 | 71 | 72 | 73 | 74 |
| TEXAS—Cont'd | | | | | | | | | | | | | |
| Mason | 203 | 2.3 | 47,439 | 1,088 | 43 | 9 | 3 | 28 | 68 | 48 | 83 | 6 | 3 |
| Matagorda | 1,658 | 6.1 | 45,237 | 1,330 | 643 | 131 | 42 | 127 | 258 | 393 | 943 | 52 | 42 |
| Maverick | 1,843 | 3.7 | 31,380 | 2,985 | 672 | 167 | 45 | 146 | 177 | 566 | 1,031 | 60 | 45 |
| Medina | 2,120 | 4.8 | 41,095 | 1,908 | 432 | 88 | 28 | 122 | 351 | 491 | 670 | 44 | 28 |
| Menard | 80 | 4.1 | 37,218 | 2,423 | 15 | 4 | 1 | 4 | 22 | 28 | 24 | 2 | 1 |
| Midland | 23,162 | 3.8 | 130,983 | 10 | 8,933 | 1,058 | 552 | 10,668 | 3,470 | 1,022 | 21,211 | 912 | 552 |
| Milam | 924 | 3.9 | 37,238 | 2,421 | 239 | 44 | 17 | 43 | 158 | 310 | 342 | 26 | 17 |
| Mills | 192 | 2.8 | 39,334 | 2,139 | 51 | 11 | 4 | 7 | 50 | 66 | 73 | 6 | 4 |
| Mitchell | 287 | 0.4 | 33,593 | 2,838 | 92 | 23 | 6 | 9 | 50 | 86 | 129 | 8 | 6 |
| Montague | 837 | 4.2 | 42,230 | 1,734 | 225 | 41 | 15 | 62 | 166 | 254 | 343 | 25 | 15 |
| Montgomery | 38,523 | 5.8 | 63,424 | 220 | 12,090 | 1,605 | 768 | 2,948 | 6,690 | 4,244 | 17,410 | 972 | 768 |
| Moore | 966 | 6.4 | 46,108 | 1,236 | 519 | 103 | 35 | 164 | 107 | 143 | 820 | 36 | 35 |
| Morris | 509 | 1.3 | 41,068 | 1,914 | 212 | 38 | 15 | 58 | 70 | 185 | 323 | 21 | 15 |
| Motley | 40 | 4.5 | 32,988 | 2,891 | 12 | 3 | 1 | 5 | 7 | 14 | 21 | 1 | 1 |
| Nacogdoches | 2,515 | 1.5 | 38,569 | 2,248 | 937 | 185 | 62 | 265 | 436 | 648 | 1,449 | 80 | 62 |
| Navarro | 1,987 | 4.4 | 39,652 | 2,093 | 741 | 141 | 51 | 112 | 318 | 568 | 1,045 | 65 | 51 |
| Newton | 466 | 1.7 | 34,265 | 2,774 | 53 | 14 | 3 | 6 | 51 | 145 | 77 | 8 | 3 |
| Nolan | 678 | 4.4 | 46,066 | 1,241 | 307 | 59 | 21 | 70 | 100 | 182 | 458 | 26 | 21 |
| Nueces | 16,263 | 3.5 | 44,889 | 1,375 | 8,845 | 1,562 | 615 | 1,250 | 2,782 | 3,527 | 12,274 | 674 | 615 |
| Ochiltree | 599 | 3.4 | 60,862 | 279 | 237 | 39 | 16 | 202 | 82 | 65 | 494 | 19 | 16 |
| Oldham | 117 | -0.6 | 55,479 | 474 | 45 | 8 | 3 | 36 | 16 | 17 | 93 | 3 | 3 |
| Orange | 3,808 | 2.1 | 45,663 | 1,288 | 1,335 | 229 | 90 | 144 | 419 | 937 | 1,799 | 115 | 90 |
| Palo Pinto | 1,202 | 4.6 | 41,193 | 1,892 | 435 | 75 | 29 | 71 | 223 | 321 | 610 | 39 | 29 |
| Panola | 1,055 | -0.2 | 45,467 | 1,303 | 445 | 72 | 30 | 74 | 189 | 266 | 621 | 38 | 30 |
| Parker | 7,974 | 6.6 | 55,811 | 455 | 1,736 | 274 | 116 | 508 | 1,242 | 1,165 | 2,634 | 164 | 116 |
| Parmer | 476 | 8.0 | 49,541 | 873 | 252 | 39 | 19 | 146 | 52 | 72 | 457 | 16 | 19 |
| Pecos | 629 | 4.9 | 39,731 | 2,083 | 307 | 58 | 20 | 61 | 92 | 124 | 445 | 23 | 20 |
| Polk | 2,045 | 4.2 | 39,818 | 2,074 | 511 | 95 | 34 | 88 | 472 | 820 | 727 | 66 | 34 |
| Potter | 5,411 | 2.8 | 46,086 | 1,239 | 3,980 | 687 | 281 | 1,009 | 913 | 1,182 | 5,957 | 306 | 281 |
| Presidio | 312 | 3.1 | 46,581 | 1,178 | 102 | 24 | 8 | 47 | 64 | 65 | 182 | 10 | 8 |
| Rains | 436 | 5.6 | 34,819 | 2,713 | 79 | 16 | 5 | 32 | 65 | 144 | 133 | 11 | 5 |
| Randall | 6,823 | 4.7 | 49,544 | 872 | 1,622 | 237 | 102 | 548 | 1,152 | 973 | 2,509 | 151 | 102 |
| Reagan | 200 | 0.1 | 51,945 | 684 | 160 | 25 | 10 | 13 | 36 | 25 | 208 | 11 | 10 |
| Real | 125 | 2.2 | 36,070 | 2,558 | 24 | 5 | 2 | 11 | 33 | 52 | 42 | 4 | 2 |
| Red River | 517 | 3.5 | 43,039 | 1,613 | 104 | 24 | 7 | 22 | 86 | 190 | 157 | 13 | 7 |
| Reeves | 726 | 12.2 | 45,458 | 1,307 | 650 | 88 | 43 | 53 | 111 | 117 | 834 | 44 | 43 |
| Refugio | 323 | 4.4 | 46,464 | 1,189 | 110 | 20 | 7 | 11 | 64 | 93 | 149 | 10 | 7 |
| Roberts | 41 | -5.0 | 48,344 | 991 | 11 | 3 | 1 | 5 | 12 | 6 | 19 | 1 | 1 |
| Robertson | 725 | 2.0 | 42,463 | 1,695 | 217 | 39 | 15 | 43 | 126 | 203 | 314 | 20 | 15 |
| Rockwall | 6,530 | 7.4 | 62,237 | 242 | 1,513 | 231 | 103 | 597 | 976 | 675 | 2,444 | 135 | 103 |
| Runnels | 430 | 6.8 | 41,929 | 1,777 | 120 | 26 | 8 | 32 | 69 | 134 | 185 | 12 | 8 |
| Rusk | 2,051 | 2.6 | 37,697 | 2,359 | 662 | 118 | 44 | 150 | 312 | 525 | 974 | 61 | 44 |
| Sabine | 386 | 5.0 | 36,627 | 2,492 | 98 | 21 | 7 | 26 | 78 | 169 | 152 | 13 | 7 |
| San Augustine | 348 | 7.1 | 42,299 | 1,718 | 102 | 22 | 7 | 19 | 52 | 139 | 150 | 11 | 7 |
| San Jacinto | 1,046 | 4.6 | 36,260 | 2,542 | 92 | 23 | 6 | 45 | 158 | 309 | 166 | 18 | 6 |
| San Patricio | 3,103 | 3.2 | 46,506 | 1,184 | 1,077 | 198 | 76 | 149 | 403 | 722 | 1,500 | 89 | 76 |
| San Saba | 245 | 0.6 | 40,521 | 1,985 | 85 | 24 | 6 | 7 | 53 | 72 | 122 | 8 | 6 |
| Schleicher | 118 | 1.0 | 42,255 | 1,730 | 44 | 9 | 3 | 4 | 20 | 28 | 59 | 4 | 3 |
| Scurry | 717 | | 42,915 | 1,629 | 390 | 65 | 24 | 60 | 135 | 157 | 539 | 30 | 24 |
| Shackelford | 369 | 1.0 | 113,163 | 16 | 84 | 12 | 5 | 194 | 64 | 37 | 295 | 12 | 5 |
| Shelby | 1,056 | -1.7 | 41,767 | 1,808 | 369 | 65 | 25 | 157 | 146 | 293 | 617 | 33 | 25 |
| Sherman | 293 | 13.1 | 97,002 | 25 | 43 | 8 | 3 | 187 | 19 | 18 | 240 | 3 | 3 |
| Smith | 13,102 | 3.2 | 56,292 | 435 | 5,170 | 805 | 354 | 3,654 | 2,070 | 2,275 | 9,983 | 504 | 354 |
| Somervell | 418 | 4.3 | 45,812 | 1,269 | 247 | 54 | 15 | 33 | 62 | 91 | 349 | 19 | 15 |
| Starr | 1,791 | 3.7 | 27,713 | 3,082 | 518 | 157 | 35 | 112 | 157 | 631 | 822 | 49 | 35 |
| Stephens | 412 | 1.8 | 43,971 | 1,492 | 136 | 30 | 9 | 49 | 72 | 111 | 224 | 13 | 9 |
| Sterling | 80 | 0.9 | 61,920 | 251 | 24 | 6 | 2 | 11 | 26 | 12 | 42 | 2 | 2 |
| Stonewall | 79 | -3.5 | 58,541 | 353 | 27 | 5 | 2 | 11 | 19 | 21 | 45 | 2 | 2 |
| Sutton | 233 | -2.0 | 61,646 | 262 | 127 | 18 | 8 | 19 | 46 | 41 | 172 | 9 | 8 |
| Swisher | 383 | 1.4 | 51,779 | 697 | 72 | 18 | 5 | 158 | 45 | 72 | 253 | 6 | 5 |
| Tarrant | 112,047 | 4.4 | 53,292 | 591 | 57,599 | 8,394 | 3,960 | 11,704 | 20,280 | 14,334 | 81,657 | 4,348 | 3,960 |
| Taylor | 6,597 | 5.0 | 47,793 | 1,044 | 3,313 | 591 | 238 | 453 | 1,331 | 1,417 | 4,594 | 253 | 238 |
| Terrell | 38 | 1.8 | 49,591 | 869 | 15 | 4 | 1 | 0 | 13 | 11 | 19 | 1 | 1 |
| Terry | 466 | 9.0 | 37,741 | 2,353 | 177 | 31 | 12 | 78 | 62 | 127 | 298 | 14 | 12 |
| Throckmorton | 62 | 8.3 | 41,454 | 1,853 | 14 | 4 | 1 | 4 | 14 | 23 | 24 | 2 | 1 |
| Titus | 1,214 | 2.7 | 37,070 | 2,439 | 647 | 125 | 43 | 115 | 196 | 310 | 929 | 50 | 43 |

1. Based on the resident population estimated as of July 1 of the year shown.

# Table B. States and Counties — Earnings, Social Security, and Housing

| STATE County | Earnings, 2019 (cont.) — Percent by selected industries | | | | | | | | | Social Security beneficiaries, December 2019 | | Supplemental Security Income recipients, 2019 | Housing units, 2020 | |
|---|---|---|---|---|---|---|---|---|---|---|---|---|---|---|
| | Farm | Mining, quarrying, and extractions | Construction | Manufacturing | Information; professional, scientific, technical services | Retail trade | Finance, insurance, real estate, and leasing | Health care and social assistance | Government | Number | Rate[1] | | Total | Percent change, 2010-2020 |
| | 75 | 76 | 77 | 78 | 79 | 80 | 81 | 82 | 83 | 84 | 85 | 86 | 87 | 88 |

**TEXAS—Cont'd**

| STATE County | 75 | 76 | 77 | 78 | 79 | 80 | 81 | 82 | 83 | 84 | 85 | 86 | 87 | 88 |
|---|---|---|---|---|---|---|---|---|---|---|---|---|---|---|
| Mason | 2.3 | D | 14.4 | 3.1 | D | 3.9 | 15.2 | 3.8 | 19.5 | 1,245 | 290 | 70 | 2,821 | 3.2 |
| Matagorda | 6.9 | 3.3 | 8.4 | 2.1 | D | 4.9 | 3.1 | 6.3 | 16.1 | 7,610 | 208 | 1,066 | 20,031 | 6.5 |
| Maverick | 0.8 | 1.5 | 4.1 | 3.3 | 3.7 | 8.6 | 3.3 | D | 40.7 | 11,165 | 191 | 3,266 | 18,706 | 7.1 |
| Medina | 1.1 | 4.7 | 13.1 | 0.9 | 5.9 | 10.1 | 7.0 | D | 28.9 | 10,115 | 196 | 1,029 | 18,921 | 5.2 |
| Menard | -9.8 | D | D | D | D | 7.8 | 9.1 | D | 42.9 | 655 | 312 | 52 | 1,736 | 1.9 |
| Midland | 0.0 | 65.3 | 3.7 | 2.1 | 3.8 | 2.4 | 3.7 | 2.3 | 3.3 | 19,130 | 108 | 1,893 | 64,519 | 18.7 |
| Milam | 0.9 | 1.6 | 21.3 | 6.6 | 3.1 | 7.1 | 5.4 | D | 21.4 | 6,245 | 253 | 749 | 11,517 | 1.9 |
| Mills | -7 | 0.3 | 8.7 | 6.5 | D | 14.7 | 6.2 | 7.5 | 28.4 | 1,410 | 289 | 106 | 2,878 | 1.1 |
| Mitchell | -1.6 | 17.5 | D | D | 2.9 | 6.0 | 4.3 | D | 44.2 | 1,500 | 184 | 200 | 4,104 | 1.0 |
| Montague | -4.2 | 21.2 | 10.2 | 8.0 | 8.3 | 7.3 | 5.6 | D | 11.4 | 5,465 | 275 | 460 | 10,342 | 2.1 |
| Montgomery | -0.1 | 7.1 | 9.8 | 6.3 | 11.2 | 6.2 | 7.8 | 9.9 | 13.5 | 91,830 | 151 | 7,647 | 237,742 | 33.8 |
| Moore | 15.6 | 2.1 | 2.3 | 44.9 | 9.5 | 4.3 | D | 2.6 | 10.8 | 2,660 | 127 | 243 | 8,160 | 3.5 |
| Morris | 2.7 | D | D | D | D | 9.0 | D | 8.7 | 26.4 | 3,860 | 313 | 538 | 6,055 | 0.5 |
| Motley | 7.6 | 2.9 | D | 11.4 | D | 7.7 | 7.2 | 14.1 | 24.3 | 350 | 292 | 24 | 774 | -0.5 |
| Nacogdoches | 4.0 | 0.6 | 6.8 | 19.6 | 5.1 | 6.4 | 6.1 | D | 18.9 | 12,330 | 190 | 1,999 | 28,656 | 4.6 |
| Navarro | -0.5 | 1.8 | 11.7 | D | 3.6 | 4.1 | D | 5.9 | 41.3 | 11,095 | 222 | 1,620 | 21,730 | 7.4 |
| Newton | -6.5 | 2.8 | D | 14.7 | D | 7.8 | 4.2 | D | 21.8 | 2,800 | 205 | 501 | 7,490 | 4.9 |
| Nolan | 1.3 | 12.3 | 8.3 | 7.9 | 8.0 | 5.5 | 5.8 | 15.2 | 19.0 | 3,380 | 228 | 487 | 7,063 | -1.2 |
| Nueces | 0.2 | 3.1 | 13.2 | 1.1 | 2.1 | 3.3 | D | D | 9.8 | 63,120 | 174 | 11,576 | 151,818 | 7.6 |
| Ochiltree | 24.2 | 32.8 | 6.7 | D | D | 1.3 | 1.1 | D | 20.7 | 1,360 | 139 | 100 | 4,171 | 2.7 |
| Oldham | 37.6 | D | D | D | D | 1.3 | 1.1 | D | 14.3 | 440 | 208 | 26 | 862 | 2.5 |
| Orange | -0.4 | 0.4 | 12.6 | 33.4 | D | 6.6 | 3.5 | 4.1 | 14.3 | 18,160 | 219 | 2,220 | 38,743 | 9.7 |
| Palo Pinto | 0.1 | 9.6 | 4.6 | 23.4 | 2.8 | 8.9 | 5.9 | D | 12.5 | 6,760 | 232 | 687 | 15,596 | 2.4 |
| Panola | 2.5 | 17.1 | 19.8 | 9.5 | D | 10.5 | 6.6 | 8.6 | 14.3 | 5,695 | 247 | 623 | 11,478 | 5.1 |
| Parker | -1.3 | 3.5 | 13.9 | 9.4 | 7.4 | 1.3 | 1.4 | 1.8 | 10.2 | 26,165 | 183 | 1,311 | 52,569 | 12.7 |
| Parmer | 36.4 | 0.0 | D | D | D | 6.6 | 2.4 | 2.7 | 26.1 | 1,500 | 157 | 116 | 3,788 | -0.3 |
| Pecos | 0.9 | 17.2 | 5.9 | 1.2 | 2.7 | 8.2 | 6.9 | 10.2 | 24.3 | 2,380 | 151 | 333 | 5,860 | 4.9 |
| Polk | -1.2 | 1.3 | 8.4 | 12.4 | D | 6.4 | 8.4 | 15.0 | 18.1 | 19,860 | 387 | 1,722 | 26,969 | 18.9 |
| Potter | 0.0 | 3.5 | 5.6 | 12.4 | 6.7 | 5.1 | D | 2.0 | 42.6 | 19,815 | 170 | 2,968 | 50,606 | 7.1 |
| Presidio | 15.0 | 0.1 | 3.9 | 1.3 | D | 9.8 | 4.7 | 4.9 | 20.0 | 1,800 | 272 | 522 | 3,986 | 4.2 |
| Rains | 2.3 | D | 20.6 | 4.2 | 6.9 | 10.9 | 7.3 | 9.4 | 11.2 | 3,270 | 264 | 267 | 5,479 | 4.1 |
| Randall | 4.3 | 0.7 | 11.9 | 4.2 | 6.7 | 3.0 | 7.3 | D | 14.0 | 22,880 | 166 | 1,363 | 55,733 | 8.0 |
| Reagan | -0.2 | 53.0 | 2.8 | 0.9 | 1.4 | 3.0 | D | D | 24.7 | 470 | 122 | 46 | 1,429 | 4.2 |
| Real | -8 | D | 7.9 | 6.7 | D | 14.5 | D | 11.0 | 23.7 | 1,280 | 370 | 111 | 2,697 | 3.9 |
| Red River | 0.7 | 4.0 | 14.4 | 14.1 | 4.5 | 3.2 | 3.7 | 10.2 | 13.0 | 3,730 | 308 | 477 | 6,998 | 2.5 |
| Reeves | 0.7 | 25.5 | 23.5 | 1.9 | D | 5.3 | 2.8 | 0.8 | 25.9 | 2,200 | 138 | 371 | 4,727 | 1.9 |
| Refugio | 2.1 | 21.4 | D | D | D | 5.7 | 8.0 | D | 26.3 | 1,770 | 259 | 200 | 3,755 | 0.8 |
| Roberts | 14.5 | D | D | 1.6 | 3.5 | 3.8 | 2.7 | D | 18.6 | 180 | 215 | 0 | 440 | 0.5 |
| Robertson | 4.2 | 9.0 | 10.4 | 2.0 | 9.2 | 10.3 | 7.4 | 16.2 | 11.4 | 3,895 | 228 | 553 | 9,135 | 7.7 |
| Rockwall | 0.0 | 0.4 | 11.9 | 6.7 | D | 10.8 | 5.3 | 7.0 | 27.1 | 15,135 | 144 | 877 | 38,824 | 38.9 |
| Runnels | 3.0 | 3.4 | 5.4 | 13.9 | 4.6 | 4.7 | 7.1 | D | 15.2 | 2,705 | 264 | 292 | 5,237 | -1.2 |
| Rusk | 1.6 | 12.6 | 9.7 | 11.7 | 6.6 | 6.1 | 3.7 | 14.4 | 20.4 | 10,980 | 202 | 1,177 | 21,869 | 3.3 |
| Sabine | 1.1 | 8.3 | 5.1 | D | D | 4.4 | D | 10.6 | 15.5 | 3,725 | 355 | 286 | 8,511 | 6.5 |
| San Augustine | 7.3 | D | 26.4 | 5.4 | 7.2 | 5.4 | 4.6 | 4.9 | 33.6 | 2,605 | 317 | 404 | 5,511 | 3.2 |
| San Jacinto | -5.7 | 0.7 | 15.0 | 7.2 | 5.8 | 6.2 | 3.3 | 3.3 | 26.1 | 6,670 | 231 | 691 | 15,100 | 14.5 |
| San Patricio | 1.5 | 4.4 | 24.1 | 2.0 | D | 4.1 | D | 2.3 | 23.5 | 13,870 | 208 | 2,108 | 29,355 | 10.7 |
| San Saba | -4.7 | 4.2 | 3.9 | D | 3.0 | 2.2 | D | 6.3 | 22.4 | 1,570 | 261 | 137 | 3,210 | 1.1 |
| Schleicher | -1.8 | 6.0 | 27.9 | 3.4 | 3.0 | 5.6 | 3.9 | D | 19.2 | 620 | 221 | 61 | 1,512 | 1.5 |
| Scurry | 0.2 | 26.6 | 4.4 | D | D | 1.0 | 2.0 | D | 4.6 | 3,005 | 180 | 339 | 7,200 | 3.4 |
| Shackelford | -0.4 | 81.6 | 3.0 | 17.2 | D | 6.0 | 6.6 | 4.5 | 13.1 | 755 | 231 | 56 | 1,782 | 1.5 |
| Shelby | 15.9 | 11.6 | 3.0 | 17.2 | D | 6.0 | 6.6 | 0.1 | 6.5 | 5,545 | 219 | 950 | 12,516 | 5.4 |
| Sherman | 78.5 | 0.2 | D | D | D | 1.5 | 0.9 | 0.1 | 6.5 | 375 | 124 | 15 | 1,358 | 8.5 |
| Smith | 0.1 | 26.0 | 5.1 | 4.0 | 6.6 | 8.8 | 5.5 | 16.4 | 9.6 | 47,630 | 204 | 5,409 | 92,582 | 6.0 |
| Somervell | 1.5 | 5.9 | 6.9 | 2.6 | 4.0 | 2.0 | 1.9 | D | 15.2 | 2,025 | 222 | 115 | 3,912 | 6.4 |
| Starr | 0.8 | 2.5 | 2.2 | 0.3 | 1.5 | 8.2 | 2.1 | D | 49.4 | 10,345 | 161 | 4,610 | 19,957 | 2.2 |
| Stephens | -3.3 | 26.2 | 8.9 | 15.6 | D | 5.7 | 5.0 | 4.5 | 18.3 | 2,225 | 238 | 223 | 4,971 | 0.7 |
| Sterling | 3.2 | 36.2 | 3.9 | 0.8 | D | 3.0 | D | D | 22.2 | 245 | 189 | 17 | 637 | 3.6 |
| Stonewall | 9.5 | 14.7 | 17.2 | 2.2 | D | 6.9 | D | 2.6 | 29.6 | 385 | 288 | 27 | 940 | 1.3 |
| Sutton | 0.0 | 29.6 | 5.1 | 5.2 | D | 3.1 | 3.0 | D | 13.6 | 815 | 216 | 61 | 2,067 | 1.8 |
| Swisher | 61.5 | D | 1.2 | 2.2 | D | 2.2 | D | 8.7 | 16.7 | 1,560 | 211 | 175 | 3,169 | -1.6 |
| Tarrant | 0.0 | 4.7 | 8.7 | 11.1 | 7.9 | 5.8 | 8.7 | 10.1 | 11.4 | 279,400 | 133 | 36,142 | 808,074 | 12.9 |
| Taylor | -0.3 | 3.1 | 6.4 | 3.9 | 7.6 | 7.0 | 7.2 | 16.8 | 23.1 | 26,090 | 189 | 3,664 | 58,226 | 4.4 |
| Terrell | -0.1 | 0.3 | 4.9 | 0.2 | D | 3.6 | D | 4.3 | 58.5 | 205 | 273 | 26 | 696 | -0.4 |
| Terry | 19.4 | 17.4 | 3.5 | 3.4 | D | 4.1 | D | D | 20.3 | 2,095 | 170 | 316 | 4,857 | 0.6 |
| Throckmorton | -2 | 19.4 | D | D | D | 6.4 | D | 4.2 | 37.4 | 395 | 268 | 27 | 1,084 | 0.5 |
| Titus | 3.6 | 0.3 | 4.7 | 34.2 | D | 6.4 | 4.2 | 7.3 | 19.9 | 5,780 | 177 | 771 | 12,618 | 4.7 |

1. Per 1,000 resident population estimated as of July 1 of the year shown.

# Table B. States and Counties — Housing, Labor Force, and Employment

| STATE County | Housing units, 2015-2019 | | | | | | | | Civilian labor force, 2020 | | | | Civilian employment[6], 2015-2019 | | |
| | Occupied units | | | | | | | Substandard units[4] (percent) | | | Unemployment | | | Percent | |
| | Owner-occupied | | | | | Renter-occupied | | | | | | | | | |
| | | | | Median owner cost as a percent of income | | | | | | | | | | Management, business, science, and arts | Construction, production, and maintenance occupations |
| | Total | Percent | Median value[1] | With a mortgage | Without a mortgage[2] | Median rent[3] | Median rent as a percent of income[2] | | Total | Percent change, 2019-2020 | Total | Rate[5] | Total | | |
| | 89 | 90 | 91 | 92 | 93 | 94 | 95 | 96 | 97 | 98 | 99 | 100 | 101 | 102 | 103 |
| TEXAS—Cont'd | | | | | | | | | | | | | | | |
| Mason | 1,696 | 70.6 | 172,800 | 26.4 | 13.5 | 800 | 34.2 | 1.2 | 1,726 | -2.0 | 96 | 5.6 | 2,026 | 28.9 | 30.1 |
| Matagorda | 13,848 | 70.7 | 114,800 | 18.6 | 12.2 | 851 | 25.6 | 4.8 | 16,529 | -1.1 | 1,737 | 10.5 | 15,607 | 26.6 | 36.5 |
| Maverick | 16,286 | 65.7 | 97,900 | 23.2 | 11.9 | 659 | 26.3 | 12.6 | 24,494 | 3.6 | 3,673 | 15.0 | 21,733 | 17.6 | 29.2 |
| Medina | 15,682 | 80.4 | 137,800 | 21.0 | 10.0 | 842 | 24.6 | 3.9 | 21,461 | -1.5 | 1,351 | 6.3 | 21,668 | 32.1 | 31.8 |
| Menard | 1,014 | 69.1 | 59,300 | 27.8 | 13.3 | 602 | 27.9 | 3.5 | 866 | 0.1 | 45 | 5.2 | 1,058 | 23.5 | 46.4 |
| Midland | 56,998 | 66.9 | 213,100 | 19.0 | 10.0 | 1,274 | 27.6 | 6.1 | 98,237 | -7.9 | 7,924 | 8.1 | 85,206 | 35.7 | 27.4 |
| Milam | 9,228 | 71.4 | 107,400 | 20.6 | 10.9 | 733 | 28.7 | 3.9 | 9,750 | 0.0 | 696 | 7.1 | 9,788 | 29.0 | 34.5 |
| Mills | 1,790 | 85.1 | 140,200 | 22.4 | 11.8 | 624 | 31.9 | 1.5 | 1,889 | 0.0 | 78 | 4.1 | 1,939 | 30.1 | 29.2 |
| Mitchell | 2,382 | 79.0 | 69,000 | 14.5 | 10.8 | 670 | 21.5 | 6.6 | 2,381 | 2.0 | 189 | 7.9 | 2,795 | 31.1 | 28.2 |
| Montague | 7,800 | 72.4 | 115,400 | 19.4 | 11.2 | 858 | 29.1 | 3.8 | 9,081 | -2.3 | 602 | 6.6 | 8,132 | 22.8 | 38.5 |
| Montgomery | 198,649 | 71.0 | 235,800 | 20.7 | 10.6 | 1,203 | 27.0 | 3.4 | 284,994 | -1.4 | 21,493 | 7.5 | 272,189 | 41.4 | 21.9 |
| Moore | 6,733 | 65.8 | 121,900 | 20.8 | 10.1 | 816 | 21.4 | 7.6 | 10,578 | -2.2 | 398 | 3.8 | 9,903 | 21.9 | 51.8 |
| Morris | 5,117 | 73.0 | 90,400 | 19.2 | 12.5 | 682 | 29.3 | 3.2 | 4,793 | -0.3 | 545 | 11.4 | 4,992 | 25.4 | 34.1 |
| Motley | 489 | 68.3 | 56,500 | 13.8 | 12.5 | 624 | 14.1 | 4.9 | 437 | -7.4 | 20 | 4.6 | 574 | 44.6 | 32.9 |
| Nacogdoches | 23,757 | 56.6 | 130,100 | 19.8 | 10.0 | 768 | 33.4 | 5.3 | 27,987 | -0.6 | 1,842 | 6.6 | 27,834 | 31.5 | 28.2 |
| Navarro | 17,338 | 68.9 | 97,600 | 21.1 | 12.2 | 817 | 27.3 | 5.2 | 23,245 | -3.3 | 1,442 | 6.2 | 21,085 | 29.0 | 34.1 |
| Newton | 5,315 | 84.6 | 81,200 | 20.4 | 12.9 | 699 | 40.2 | 4.6 | 5,054 | -1.7 | 557 | 11.0 | 5,156 | 25.0 | 37.6 |
| Nolan | 5,407 | 67.2 | 75,500 | 19.6 | 11.4 | 695 | 27.7 | 6.2 | 7,151 | 1.6 | 377 | 5.3 | 6,279 | 26.6 | 33.4 |
| Nueces | 129,451 | 58.0 | 138,700 | 21.3 | 12.0 | 1,017 | 29.0 | 5.1 | 163,920 | -1.8 | 14,688 | 9.0 | 165,051 | 30.5 | 26.3 |
| Ochiltree | 3,483 | 73.4 | 102,700 | 23.7 | 11.2 | 837 | 21.9 | 11.0 | 4,071 | -5.4 | 236 | 5.8 | 4,645 | 22.7 | 42.6 |
| Oldham | 594 | 78.5 | 91,400 | 17.9 | 10.0 | 780 | 18.1 | 2.2 | 890 | -2.2 | 35 | 3.9 | 941 | 33.2 | 28.2 |
| Orange | 31,694 | 75.3 | 113,500 | 17.0 | 10.0 | 825 | 25.2 | 3.5 | 36,027 | -1.1 | 3,786 | 10.5 | 37,674 | 29.9 | 32.7 |
| Palo Pinto | 10,255 | 72.7 | 107,000 | 21.5 | 12.1 | 800 | 35.5 | 4.0 | 13,267 | -2.3 | 928 | 7.0 | 11,763 | 29.2 | 30.8 |
| Panola | 8,637 | 76.7 | 102,200 | 19.2 | 10.0 | 767 | 23.9 | 3.9 | 9,437 | -6.7 | 792 | 8.4 | 9,547 | 30.0 | 32.5 |
| Parker | 44,263 | 78.8 | 214,200 | 19.7 | 10.2 | 1,027 | 26.8 | 3.0 | 68,141 | -0.9 | 4,012 | 5.9 | 62,395 | 37.7 | 25.9 |
| Parmer | 3,203 | 66.9 | 97,600 | 22.9 | 10.0 | 717 | 24.5 | 9.6 | 4,915 | 1.0 | 127 | 2.6 | 4,374 | 28.9 | 39.7 |
| Pecos | 4,656 | 72.2 | 87,300 | 18.5 | 10.0 | 842 | 23.1 | 5.6 | 6,517 | 2.2 | 530 | 8.1 | 6,122 | 25.0 | 34.9 |
| Polk | 18,033 | 76.0 | 110,200 | 20.4 | 10.3 | 763 | 25.5 | 4.4 | 18,357 | 1.8 | 1,602 | 8.7 | 18,869 | 24.0 | 32.6 |
| Potter | 44,380 | 55.2 | 92,500 | 20.7 | 11.2 | 803 | 30.2 | 5.9 | 54,150 | -0.4 | 2,961 | 5.5 | 51,766 | 23.5 | 33.7 |
| Presidio | 2,543 | 57.8 | 75,100 | 24.0 | 14.1 | 429 | 25.2 | 9.6 | 3,231 | 3.8 | 476 | 14.7 | 2,460 | 35.9 | 30.9 |
| Rains | 4,330 | 77.7 | 127,900 | 21.8 | 12.8 | 733 | 24.7 | 4.4 | 6,027 | -0.5 | 283 | 4.7 | 5,002 | 26.3 | 35.4 |
| Randall | 49,389 | 70.8 | 167,800 | 20.2 | 10.8 | 917 | 26.5 | 2.2 | 72,380 | -1.0 | 3,246 | 4.5 | 70,231 | 38.1 | 21.9 |
| Reagan | 1,084 | 67.4 | 98,800 | 21.2 | 10.0 | 842 | 20.6 | 1.4 | 1,815 | -4.1 | 176 | 9.7 | 1,669 | 27.1 | 48.2 |
| Real | 1,126 | 76.5 | 113,200 | 23.0 | 13.7 | 743 | 30.7 | 4.3 | 1,072 | 1.0 | 72 | 6.7 | 1,125 | 29.5 | 20.9 |
| Red River | 4,963 | 76.6 | 71,100 | 20.9 | 10.9 | 598 | 35.8 | 1.1 | 5,144 | 1.3 | 327 | 6.4 | 5,012 | 26.5 | 33.9 |
| Reeves | 3,518 | 72.5 | 64,800 | 17.5 | 10.0 | 871 | 24.5 | 6.8 | 8,858 | -20.6 | 631 | 7.1 | 5,913 | 24.2 | 41.0 |
| Refugio | 2,547 | 72.8 | 85,600 | 20.4 | 10.7 | 644 | 32.0 | 3.4 | 3,083 | -2.0 | 252 | 8.2 | 2,770 | 24.3 | 28.9 |
| Roberts | 301 | 79.7 | 133,000 | 18.3 | 10.0 | 780 | 20.4 | 4.0 | 414 | -0.7 | 19 | 4.6 | 357 | 40.1 | 23.0 |
| Robertson | 6,444 | 76.1 | 108,600 | 17.6 | 10.0 | 709 | 27.6 | 5.8 | 7,345 | -1.6 | 470 | 6.4 | 7,385 | 27.0 | 33.1 |
| Rockwall | 33,032 | 81.8 | 266,200 | 21.1 | 11.5 | 1,429 | 27.6 | 1.7 | 53,209 | -0.5 | 3,200 | 6.0 | 48,291 | 48.5 | 15.0 |
| Runnels | 3,896 | 71.1 | 81,400 | 16.6 | 13.1 | 644 | 24.0 | 2.7 | 4,529 | -1.5 | 223 | 4.9 | 4,438 | 32.1 | 29.2 |
| Rusk | 18,108 | 77.8 | 124,900 | 18.6 | 10.4 | 763 | 27.8 | 3.8 | 22,127 | -3.4 | 1,619 | 7.3 | 21,176 | 29.7 | 35.4 |
| Sabine | 4,311 | 88.2 | 116,300 | 25.7 | 12.4 | 566 | 34.8 | 1.7 | 3,923 | 4.5 | 423 | 10.8 | 3,122 | 24.5 | 26.6 |
| San Augustine | 3,451 | 77.5 | 88,000 | 20.3 | 12.9 | 656 | 32.6 | 4.1 | 3,020 | -10.5 | 290 | 9.6 | 2,799 | 25.4 | 31.9 |
| San Jacinto | 10,043 | 80.6 | 114,100 | 22.1 | 13.2 | 788 | 24.5 | 5.7 | 11,650 | -0.8 | 940 | 8.1 | 10,300 | 23.2 | 38.2 |
| San Patricio | 22,898 | 68.3 | 122,100 | 19.9 | 12.9 | 975 | 29.6 | 7.5 | 29,221 | -1.8 | 2,998 | 10.3 | 28,539 | 30.0 | 31.1 |
| San Saba | 2,082 | 68.4 | 109,400 | 24.5 | 11.1 | 757 | 34.6 | 4.0 | 2,193 | -13.4 | 114 | 5.2 | 2,538 | 27.3 | 36.9 |
| Schleicher | 1,075 | 78.9 | 77,800 | 17.4 | 10.0 | 439 | 24.1 | 4.9 | 1,209 | -7.6 | 84 | 6.9 | 1,327 | 30.7 | 34.4 |
| Scurry | 5,941 | 75.2 | 85,500 | 20.4 | 10.7 | 761 | 25.6 | 5.9 | 6,251 | -6.1 | 512 | 8.2 | 6,880 | 21.0 | 38.9 |
| Shackelford | 1,317 | 79.6 | 91,400 | 22.2 | 14.9 | 758 | 24.0 | 5.6 | 1,782 | -6.5 | 94 | 5.3 | 1,504 | 31.9 | 23.8 |
| Shelby | 9,293 | 73.6 | 78,100 | 19.5 | 10.5 | 564 | 29.4 | 6.1 | 11,274 | 1.1 | 772 | 6.8 | 9,650 | 23.2 | 41.8 |
| Sherman | 1,019 | 72.6 | 99,800 | 24.7 | 10.0 | 790 | 22.6 | 4.2 | 1,307 | -4.5 | 40 | 3.1 | 1,563 | 26.6 | 44.7 |
| Smith | 77,678 | 66.4 | 158,600 | 20.7 | 11.1 | 941 | 28.7 | 3.5 | 108,544 | 0.2 | 7,343 | 6.8 | 101,856 | 34.3 | 24.6 |
| Somervell | 3,123 | 80.3 | 219,000 | 23.1 | 10.0 | 869 | 20.4 | 7.4 | 4,328 | -0.9 | 280 | 6.5 | 3,667 | 38.6 | 31.6 |
| Starr | 16,188 | 75.1 | 71,900 | 25.2 | 12.2 | 588 | 31.5 | 13.2 | 26,319 | 4.7 | 4,559 | 17.3 | 22,056 | 22.6 | 29.1 |
| Stephens | 3,247 | 82.1 | 74,300 | 21.9 | 13.3 | 722 | 30.5 | 1.8 | 4,020 | -1.8 | 245 | 6.1 | 3,725 | 26.1 | 35.0 |
| Sterling | 458 | 85.6 | 59,000 | 15.7 | 10.0 | 850 | 19.6 | 2.4 | 560 | -2.6 | 36 | 6.4 | 631 | 24.9 | 34.1 |
| Stonewall | 580 | 75.5 | 55,600 | 15.5 | 11.6 | 600 | 17.7 | 1.9 | 576 | -1.2 | 29 | 5.0 | 633 | 30.3 | 36.2 |
| Sutton | 1,412 | 69.2 | 91,800 | 21.4 | 12.5 | 635 | 17.3 | 8.0 | 1,229 | -13.3 | 120 | 9.8 | 1,752 | 30.4 | 31.8 |
| Swisher | 2,505 | 71.0 | 80,600 | 20.8 | 12.9 | 752 | 28.0 | 7.1 | 2,533 | -3.1 | 132 | 5.2 | 2,783 | 29.5 | 31.5 |
| Tarrant | 708,252 | 60.5 | 188,500 | 20.8 | 11.7 | 1,095 | 29.0 | 4.6 | 1,082,822 | 0.3 | 79,553 | 7.3 | 1,017,012 | 37.1 | 23.9 |
| Taylor | 49,868 | 58.4 | 124,500 | 19.9 | 10.8 | 925 | 28.8 | 2.6 | 66,148 | -1.0 | 3,608 | 5.5 | 63,441 | 32.7 | 23.6 |
| Terrell | 418 | 93.1 | 76,400 | 19.7 | 13.2 | 0 | 0.0 | 1.0 | 393 | 4.0 | 18 | 4.6 | 357 | 35.0 | 30.0 |
| Terry | 4,061 | 69.6 | 76,500 | 16.8 | 10.0 | 701 | 27.7 | 6.5 | 4,840 | -3.1 | 354 | 7.3 | 4,849 | 23.9 | 37.8 |
| Throckmorton | 668 | 72.9 | 56,300 | 16.0 | 12.0 | 429 | 24.3 | 2.4 | 635 | -2.0 | 31 | 4.9 | 684 | 32.6 | 25.3 |
| Titus | 10,893 | 67.0 | 105,100 | 20.0 | 10.0 | 709 | 24.6 | 9.7 | 13,303 | 2.9 | 891 | 6.7 | 14,466 | 21.4 | 39.0 |

1. Specified owner-occupied units. 2. A value of 10.0 represents 10 percent or less; a value of 50.0 represents 50 percent or more. 3. Specified renter-occupied units. 4. Overcrowded or lacking complete plumbing facilities. 5. Percent of civilian labor force. 6. Civilian employed persons 16 years old and over.

| STATE County | Number of establishments | Total | Health care and social assistance | Manufacturing | Retail trade | Finance and insurance | Professional, scientific, and technical services | Total (mil dol) | Average per employee (dollars) | Number | Fewer than 50 acres | 1000 acres or more | Farm producers whose primary occupation is farming (percent) |
|---|---|---|---|---|---|---|---|---|---|---|---|---|---|
| | 104 | 105 | 106 | 107 | 108 | 109 | 110 | 111 | 112 | 113 | 114 | 115 | 116 |
| **TEXAS—Cont'd** | | | | | | | | | | | | | |
| Mason | 139 | 790 | 106 | 28 | 124 | 42 | 46 | 26 | 32,443 | 680 | 15.4 | 22.5 | 39.3 |
| Matagorda | 732 | 8,156 | 1,183 | 791 | 1,421 | 200 | 243 | 475 | 58,211 | 858 | 36.2 | 14.8 | 40.4 |
| Maverick | 811 | 12,997 | 4,301 | 440 | 2,709 | 421 | 270 | 359 | 27,602 | 339 | 53.1 | 15.0 | 34.8 |
| Medina | 775 | 7,639 | 993 | 403 | 1,639 | 272 | 468 | 269 | 35,167 | 2,281 | 39.9 | 7.9 | 35.1 |
| Menard | 46 | 193 | NA | NA | 48 | 23 | 9 | 5 | 27,394 | 346 | 13.0 | 29.8 | 40.7 |
| Midland | 5,783 | 105,301 | 7,441 | 3,663 | 9,301 | 2,250 | 4,257 | 7,687 | 73,000 | 410 | 59.8 | 11.5 | 24.4 |
| Milam | 407 | 3,620 | 602 | 162 | 623 | 198 | 99 | 139 | 38,296 | 2,053 | 41.7 | 4.1 | 38.2 |
| Mills | 113 | 833 | 93 | 101 | 165 | 58 | 15 | 29 | 35,301 | 896 | 22.1 | 10.8 | 38.2 |
| Mitchell | 131 | 1,217 | 288 | NA | 198 | 38 | 28 | 53 | 43,226 | 362 | 23.5 | 18.2 | 27.7 |
| Montague | 425 | 3,560 | 392 | 311 | 622 | 181 | 155 | 168 | 47,192 | 1,615 | 32.1 | 6.3 | 37.4 |
| Montgomery | 12,375 | 175,555 | 23,761 | 10,269 | 26,801 | 6,333 | 11,772 | 10,317 | 58,769 | 1,614 | 69.1 | 1.8 | 30.4 |
| Moore | 456 | 8,252 | 654 | 3,613 | 1,055 | 144 | 81 | 372 | 45,066 | 250 | 28.4 | 32.4 | 48.1 |
| Morris | 223 | 2,864 | 173 | 1,464 | 355 | 133 | 71 | 155 | 54,048 | 449 | 39.2 | 2.0 | 35.7 |
| Motley | 24 | 152 | NA | NA | 17 | NA | NA | 5 | 30,967 | 237 | 6.3 | 40.9 | 36.2 |
| Nacogdoches | 1,285 | 18,690 | 3,450 | 3,460 | 2,761 | 551 | 478 | 689 | 36,885 | 1,123 | 33.6 | 2.4 | 37.8 |
| Navarro | 952 | 13,820 | 2,452 | 3,196 | 2,261 | 364 | 305 | 480 | 34,716 | 2,471 | 44.7 | 4.0 | 35.9 |
| Newton | 128 | 791 | 137 | 82 | 156 | 16 | 19 | 26 | 33,134 | 430 | 58.6 | 0.2 | 34.0 |
| Nolan | 340 | 5,349 | 574 | 1,023 | 790 | 125 | 90 | 260 | 48,582 | 410 | 18.0 | 20.0 | 35.2 |
| Nueces | 8,084 | 144,897 | 31,196 | 6,472 | 19,485 | 4,442 | 6,427 | 6,363 | 43,913 | 646 | 61.8 | 13.3 | 33.4 |
| Ochiltree | 352 | 3,531 | 249 | 11 | 432 | 136 | 78 | 186 | 52,619 | 282 | 12.1 | 41.8 | 48.6 |
| Oldham | 45 | 553 | 325 | NA | 73 | NA | NA | 22 | 39,107 | 132 | 3.8 | 49.2 | 46.7 |
| Orange | 1,307 | 19,024 | 1,570 | 5,125 | 3,034 | 566 | 716 | 1,003 | 52,709 | 663 | 81.1 | 1.2 | 31.6 |
| Palo Pinto | 602 | 5,973 | 847 | 1,130 | 929 | 169 | 185 | 244 | 40,785 | 1,265 | 49.2 | 8.6 | 36.0 |
| Panola | 473 | 7,153 | 632 | 961 | 779 | 220 | 246 | 311 | 43,411 | 978 | 30.9 | 3.0 | 37.0 |
| Parker | 2,971 | 30,874 | 3,394 | 3,068 | 5,691 | 726 | 1,203 | 1,363 | 44,134 | 4,626 | 73.1 | 1.8 | 29.3 |
| Parmer | 193 | 4,081 | 179 | 2,664 | 180 | 87 | 53 | 184 | 45,055 | 464 | 10.3 | 36.6 | 52.5 |
| Pecos | 349 | 4,242 | 401 | 16 | 845 | 114 | 56 | 193 | 45,400 | 309 | 9.1 | 61.2 | 45.6 |
| Polk | 769 | 8,564 | 1,333 | 1,368 | 1,650 | 360 | 168 | 330 | 38,524 | 742 | 40.2 | 1.6 | 36.1 |
| Potter | 3,483 | 62,170 | 14,044 | 6,735 | 7,787 | 3,951 | 1,920 | 2,798 | 45,012 | 239 | 38.5 | 17.6 | 42.5 |
| Presidio | 130 | 987 | 59 | 11 | 255 | 60 | NA | 25 | 25,052 | 142 | 5.6 | 58.5 | 53.7 |
| Rains | 182 | 1,537 | 108 | 68 | 331 | 45 | 230 | 50 | 32,520 | 787 | 57.6 | 2.2 | 34.1 |
| Randall | 2,644 | 33,847 | 3,490 | 4,946 | 6,086 | 1,228 | 1,053 | 1,526 | 45,083 | 781 | 41.4 | 17.4 | 35.2 |
| Reagan | 127 | 1,757 | 147 | NA | 166 | 13 | 14 | 100 | 56,681 | 112 | 10.7 | 56.3 | 50.3 |
| Real | 90 | 673 | 93 | 33 | 131 | NA | 51 | 16 | 23,321 | 198 | 16.2 | 27.8 | 41.7 |
| Red River | 166 | 1,405 | 229 | 352 | 249 | 73 | 25 | 52 | 37,335 | 1,126 | 26.1 | 8.6 | 41.2 |
| Reeves | 360 | 6,102 | 351 | 30 | 713 | 47 | 108 | 389 | 63,751 | 224 | 19.2 | 27.7 | 43.6 |
| Refugio | 143 | 1,728 | 199 | NA | 305 | 24 | 24 | 77 | 44,458 | 238 | 31.9 | 23.5 | 37.4 |
| Roberts | 19 | 96 | NA | NA | NA | NA | 11 | 5 | 51,552 | 104 | 10.6 | 48.1 | 58.2 |
| Robertson | 266 | 3,128 | 320 | 81 | 388 | 80 | 365 | 138 | 43,985 | 1,471 | 34.2 | 5.2 | 40.6 |
| Rockwall | 2,479 | 28,547 | 5,316 | 1,901 | 5,709 | 753 | 1,739 | 1,154 | 40,410 | 403 | 74.9 | 1.5 | 24.4 |
| Runnels | 220 | 2,023 | 266 | 446 | 397 | 86 | 29 | 80 | 39,681 | 833 | 18.7 | 17.0 | 38.8 |
| Rusk | 794 | 9,879 | 1,530 | 1,296 | 1,169 | 447 | 287 | 433 | 43,792 | 1,441 | 33.2 | 1.8 | 35.1 |
| Sabine | 186 | 1,551 | 406 | 370 | 328 | 53 | 36 | 60 | 38,601 | 200 | 43.5 | 2.5 | 33.5 |
| San Augustine | 134 | 1,282 | 377 | 66 | 250 | 39 | 15 | 49 | 38,222 | 293 | 30.0 | 1.7 | 40.9 |
| San Jacinto | 222 | 1,003 | 125 | 105 | 190 | 38 | 39 | 36 | 35,936 | 786 | 56.6 | 1.0 | 31.6 |
| San Patricio | 1,063 | 15,838 | 1,136 | 3,914 | 2,740 | 324 | 887 | 770 | 48,646 | 656 | 52.4 | 13.1 | 38.8 |
| San Saba | 149 | 823 | 103 | 55 | 185 | 21 | 27 | 23 | 28,448 | 773 | 16.4 | 22.1 | 43.6 |
| Schleicher | 49 | 483 | 55 | NA | 51 | 23 | 10 | 23 | 47,400 | 327 | 12.5 | 44.6 | 42.2 |
| Scurry | 422 | 5,142 | 527 | 99 | 735 | 127 | 136 | 268 | 52,034 | 560 | 21.8 | 17.5 | 33.5 |
| Shackelford | 129 | 1,527 | 45 | 136 | 88 | 50 | 9 | 119 | 77,954 | 223 | 15.7 | 30.9 | 37.1 |
| Shelby | 503 | 6,562 | 469 | 2,194 | 862 | 360 | 377 | 240 | 36,633 | 995 | 30.8 | 1.6 | 39.1 |
| Sherman | 62 | 353 | NA | NA | 98 | 28 | NA | 15 | 42,411 | 245 | 6.9 | 41.2 | 47.5 |
| Smith | 5,940 | 94,185 | 22,062 | 7,234 | 13,030 | 3,382 | 4,671 | 4,044 | 42,940 | 2,928 | 58.8 | 1.1 | 31.7 |
| Somervell | 218 | 3,261 | 527 | 123 | 161 | 49 | 89 | 208 | 63,825 | 352 | 53.7 | 5.1 | 29.5 |
| Starr | 578 | 8,853 | 4,763 | 35 | 1,848 | 317 | 158 | 190 | 21,468 | 1,345 | 31.7 | 8.8 | 34.1 |
| Stephens | 246 | 2,251 | 311 | 418 | 341 | 103 | 42 | 91 | 40,376 | 575 | 13.7 | 20.2 | 34.9 |
| Sterling | 48 | 289 | NA | NA | 44 | 16 | 8 | 16 | 55,502 | 76 | 5.3 | 51.3 | 50.0 |
| Stonewall | 49 | 384 | 139 | NA | 44 | NA | NA | 18 | 46,057 | 314 | 5.1 | 29.6 | 33.1 |
| Sutton | 119 | 1,128 | 90 | 24 | 212 | 10 | 17 | 55 | 48,439 | 261 | 9.2 | 52.1 | 47.1 |
| Swisher | 124 | 951 | 181 | 105 | 163 | 35 | 31 | 31 | 32,597 | 432 | 4.9 | 34.3 | 46.8 |
| Tarrant | 43,417 | 822,823 | 110,567 | 84,390 | 106,165 | 49,515 | 40,104 | 42,386 | 51,513 | 1,173 | 82.5 | 1.8 | 31.1 |
| Taylor | 3,545 | 57,996 | 13,443 | 3,189 | 8,110 | 2,579 | 1,932 | 2,292 | 39,527 | 1,394 | 36.7 | 9.0 | 29.6 |
| Terrell | 15 | 55 | NA | NA | 20 | NA | NA | 2 | 35,364 | 85 | NA | 74.1 | 50.4 |
| Terry | 222 | 2,088 | 385 | 51 | 380 | 101 | 52 | 78 | 37,501 | 558 | 13.4 | 22.4 | 46.2 |
| Throckmorton | 47 | 264 | NA | NA | 43 | 12 | 10 | 9 | 33,485 | 274 | 4.7 | 27.4 | 38.3 |
| Titus | 637 | 14,726 | 2,139 | 6,806 | 1,813 | 416 | 98 | 545 | 37,004 | 812 | 43.2 | 3.0 | 38.4 |

# Table B. States and Counties — Agriculture

| | Agriculture, 2017 (cont.) | | | | | | | | | | | | | | |
| STATE County | Land in farms | | | | | Value of land and buildings (dollars) | | Value of machinery and equipment, average per farm (dollars) | Value of products sold: | | | | Organic farms (number) | Farms with internet access (percent) | Government payments | |
| | Acreage (1,000) | Percent change, 2012-2017 | Average size of farm | Acres Total irrigated (1,000) | Total cropland (1,000) | Average per farm | Average per acre | | Total (mil dol) | Average per farm (acres) | Percent from: Crops | Livestock and poultry products | | | Total ($1,000) | Percent of farms |
| | 117 | 118 | 119 | 120 | 121 | 122 | 123 | 124 | 125 | 126 | 127 | 128 | 129 | 130 | 131 | 132 |
| TEXAS—Cont'd | | | | | | | | | | | | | | | | |
| Mason | 539 | -2.2 | 793 | 3.9 | 21.8 | 1,838,623 | 2,318 | 72,646 | 21.7 | 31,881 | 10.7 | 89.3 | NA | 70.3 | 1,416 | 15.0 |
| Matagorda | 551 | -2.9 | 643 | 25.6 | 176.2 | 1,540,205 | 2,397 | 146,410 | 124.2 | 144,773 | 58.8 | 41.2 | 16 | 70.6 | 9,559 | 26.7 |
| Maverick | 434 | -19.7 | 1,282 | 14.6 | 21.3 | 1,801,266 | 1,405 | 56,810 | 43.0 | 126,737 | 5.8 | 94.2 | NA | 59.3 | 752 | 8.0 |
| Medina | 782 | -6.1 | 343 | 39.4 | 153.5 | 917,319 | 2,674 | 63,914 | 93.9 | 41,170 | 48.8 | 51.2 | 1 | 73.4 | 2,804 | 10.9 |
| Menard | 508 | -5.4 | 1,467 | 1.2 | 10.5 | 2,174,290 | 1,482 | 68,492 | 9.1 | 26,220 | 6.3 | 93.8 | NA | 74.6 | 596 | 17.3 |
| Midland | 345 | -14.7 | 841 | 7.4 | 75.8 | 1,353,752 | 1,609 | 99,126 | 16.3 | 39,851 | 79.6 | 20.4 | NA | 79.5 | 910 | 17.3 |
| Milam | 497 | -5.8 | 242 | 2.7 | 168.0 | 766,554 | 3,163 | 79,326 | 129.5 | 63,087 | 25.7 | 74.3 | 2 | 70.3 | 3,149 | 8.7 |
| Mills | 441 | -6.4 | 492 | 3.1 | 44.6 | 1,235,360 | 2,509 | 65,340 | 30.9 | 34,484 | 7.9 | 92.1 | NA | 76.9 | 580 | 14.2 |
| Mitchell | 583 | 1.7 | 1,611 | 3.0 | 153.1 | 1,604,910 | 997 | 96,432 | 21.7 | 60,061 | 62.5 | 37.5 | NA | 65.2 | 2,422 | 50.8 |
| Montague | 498 | 2.0 | 309 | 1.5 | 82.9 | 842,695 | 2,732 | 67,799 | 33.4 | 20,691 | 16.0 | 84.0 | 1 | 72.6 | 1,075 | 12.0 |
| Montgomery | 145 | -6.8 | 90 | 0.7 | 25.3 | 1,024,558 | 11,414 | 54,308 | 25.8 | 15,963 | 26.5 | 73.5 | 1 | 82.5 | 201 | 1.4 |
| Moore | 478 | -8.8 | 1,911 | 84.8 | 207.6 | 2,167,617 | 1,134 | 373,450 | 478.1 | 1,912,300 | 14.9 | 85.1 | NA | 71.2 | 8,544 | 58.4 |
| Morris | 74 | -19.6 | 165 | 0.2 | 16.1 | 375,751 | 2,283 | 58,729 | 44.2 | 98,354 | 2.5 | 97.5 | 4 | 67.0 | 436 | 16.7 |
| Motley | 580 | -2.7 | 2,446 | 4.4 | 100.3 | 2,072,503 | 847 | 112,229 | 15.2 | 64,249 | 15.9 | 84.1 | NA | 75.9 | 2,016 | 54.4 |
| Nacogdoches | 265 | 0.0 | 236 | 0.3 | 29.5 | 682,387 | 2,895 | 92,115 | 370.7 | 330,135 | 0.9 | 99.1 | 4 | 68.3 | 144 | 1.5 |
| Navarro | 559 | 0.2 | 226 | 1.7 | 178.6 | 531,307 | 2,349 | 68,964 | 73.3 | 29,667 | 45.9 | 54.1 | 4 | 65.9 | 2,195 | 7.2 |
| Newton | 59 | 0.0 | 137 | 0.1 | 5.5 | 303,605 | 2,221 | 54,042 | 1.6 | 3,691 | 30.6 | 69.4 | 4 | 63.3 | 24 | 2.8 |
| Nolan | 466 | 0.3 | 1,137 | 3.5 | 161.9 | 1,253,213 | 1,102 | 107,278 | 36.6 | 89,290 | 70.4 | 29.6 | NA | 72.4 | 1,351 | 44.1 |
| Nueces | 475 | -9.4 | 735 | 1.2 | 332.3 | 2,243,539 | 3,052 | 220,795 | 161.0 | 249,260 | 96.2 | 3.8 | 4 | 63.6 | 8,905 | 27.4 |
| Ochiltree | 569 | 4.4 | 2,016 | 59.0 | 343.8 | 2,484,158 | 1,232 | 353,856 | 349.1 | 1,237,791 | 20.0 | 80.0 | 2 | 79.4 | 7,773 | 72.0 |
| Oldham | 945 | 13.8 | 7,157 | 1.8 | 105.3 | 5,834,781 | 815 | 158,947 | 156.0 | 1,181,909 | 3.5 | 96.5 | NA | 81.8 | 3,798 | 78.8 |
| Orange | 53 | 0.2 | 80 | 0.3 | 4.7 | 368,662 | 4,619 | 55,345 | 5.0 | 7,492 | 30.0 | 70.0 | NA | 78.4 | 142 | 0.8 |
| Palo Pinto | 573 | -3.4 | 453 | 4.4 | 73.0 | 1,104,960 | 2,440 | 58,278 | 43.2 | 34,165 | 23.8 | 76.2 | NA | 72.4 | 249 | 3.3 |
| Panola | 206 | -9.4 | 211 | 0.8 | 39.8 | 562,292 | 2,670 | 78,393 | 100.7 | 102,986 | 4.6 | 95.4 | NA | 70.4 | 96 | 1.6 |
| Parker | 522 | 5.5 | 113 | 1.7 | 95.1 | 484,457 | 4,296 | 48,756 | 65.0 | 14,060 | 18.7 | 81.3 | 2 | 81.3 | 250 | 1.0 |
| Parmer | 549 | -0.9 | 1,183 | 110.1 | 399.2 | 1,330,531 | 1,125 | 346,434 | 893.3 | 1,925,302 | 10.3 | 89.7 | NA | 75.9 | 17,933 | 81.9 |
| Pecos | 2,868 | -2.7 | 9,281 | 12.9 | 50.8 | 6,420,541 | 692 | 136,293 | 46.2 | 149,398 | 52.8 | 47.2 | NA | 73.1 | 1,560 | 23.6 |
| Polk | 125 | -10.1 | 169 | 0.3 | 22.6 | 507,818 | 3,011 | 58,728 | 6.8 | 9,206 | 33.5 | 66.5 | NA | 70.1 | 166 | 0.7 |
| Potter | 424 | -25.5 | 1,773 | 3.6 | 49.6 | 1,802,406 | 1,016 | 76,318 | 24.8 | 103,891 | 10.5 | 89.5 | NA | 67.4 | 1,962 | 20.1 |
| Presidio | 1,841 | 11.2 | 12,967 | 1.8 | 18.0 | 10,152,603 | 783 | 122,932 | D | D | D | D | NA | 77.5 | 644 | 12.0 |
| Rains | 111 | -4.6 | 142 | 0.3 | 27.7 | 492,612 | 3,479 | 61,364 | 22.8 | 28,919 | 37.8 | 62.2 | NA | 69.4 | 213 | 3.9 |
| Randall | 560 | -1.9 | 718 | 16.2 | 233.9 | 1,135,115 | 1,582 | 108,685 | 479.5 | 613,910 | 5.0 | 95.0 | NA | 79.9 | 9,464 | 40.1 |
| Reagan | 736 | 5.4 | 6,574 | 8.1 | 55.6 | 5,510,283 | 838 | 242,344 | 18.2 | 162,527 | 65.6 | 34.4 | NA | 76.8 | 768 | 42.0 |
| Real | 317 | -1.1 | 1,601 | 0.4 | 1.3 | 3,074,966 | 1,921 | 47,687 | 1.3 | 6,379 | 4.9 | 95.1 | NA | 87.4 | 19 | 4.0 |
| Red River | 458 | 2.2 | 407 | 5.1 | 103.1 | 784,848 | 1,927 | 85,573 | 94.0 | 83,515 | 10.9 | 89.1 | 1 | 70.2 | 3,425 | 34.1 |
| Reeves | 1,064 | -13.9 | 4,750 | 8.1 | 54.7 | 2,500,484 | 526 | 80,186 | 10.9 | 48,621 | 47.5 | 52.5 | 1 | 72.8 | 790 | 24.1 |
| Refugio | 489 | 2.9 | 2,053 | 0.5 | 56.3 | 2,565,852 | 1,250 | 170,947 | 35.9 | 150,891 | 69.2 | 30.8 | NA | 73.5 | 1,986 | 27.7 |
| Roberts | 553 | -1.7 | 5,318 | 7.4 | 36.1 | 4,045,126 | 761 | 140,728 | 18.3 | 175,971 | 20.7 | 79.3 | NA | 76.0 | 1,450 | 52.9 |
| Robertson | 475 | 1.5 | 323 | 20.4 | 107.3 | 932,356 | 2,889 | 72,519 | 158.1 | 107,507 | 15.6 | 84.4 | 2 | 70.0 | 2,496 | 4.6 |
| Rockwall | 40 | -11.0 | 100 | 0.1 | 21.0 | 554,781 | 5,536 | 52,992 | 7.8 | 19,429 | 79.0 | 21.0 | 3 | 80.4 | 439 | 3.7 |
| Runnels | 672 | 1.0 | 807 | 5.6 | 256.2 | 1,116,037 | 1,383 | 111,613 | 53.4 | 64,146 | 59.7 | 40.3 | NA | 70.9 | 6,275 | 52.1 |
| Rusk | 243 | -11.5 | 168 | 0.5 | 46.1 | 431,457 | 2,561 | 71,067 | 100.2 | 69,505 | 5.9 | 94.1 | NA | 69.6 | 313 | 1.1 |
| Sabine | 38 | 31.9 | 192 | 0.1 | 5.6 | 432,114 | 2,256 | 85,329 | 17.7 | 88,575 | 2.5 | 97.5 | NA | 73.0 | 36 | 2.5 |
| San Augustine | 62 | -15.2 | 211 | 0.0 | 9.2 | 594,832 | 2,820 | 83,472 | 56.7 | 193,433 | 2.3 | 97.7 | NA | 72.0 | D | 1.0 |
| San Jacinto | 84 | -24.5 | 107 | 1.0 | 16.9 | 440,438 | 4,100 | 75,877 | 7.2 | 9,148 | 34.5 | 65.5 | NA | 70.2 | 130 | 0.5 |
| San Patricio | 371 | -0.9 | 565 | 4.1 | 235.8 | 1,497,600 | 2,650 | 192,771 | 131.3 | 200,216 | 88.4 | 11.6 | NA | 78.4 | 6,745 | 25.8 |
| San Saba | 660 | -1.7 | 854 | 4.8 | 81.3 | 2,091,365 | 2,449 | 87,070 | 35.8 | 46,343 | 18.1 | 81.9 | 2 | 79.8 | 1,777 | 18.2 |
| Schleicher | 811 | -2.7 | 2,480 | 1.4 | 30.6 | 2,929,781 | 1,181 | 90,424 | 17.8 | 54,404 | 19.3 | 80.7 | NA | 74.6 | 1,449 | 22.6 |
| Scurry | 531 | 7.4 | 948 | 5.5 | 201.7 | 991,118 | 1,046 | 124,077 | 45.2 | 80,629 | 54.0 | 46.0 | 5 | 70.4 | 2,730 | 52.1 |
| Shackelford | 537 | 6.3 | 2,407 | 0.3 | 49.4 | 3,105,956 | 1,290 | 75,451 | 16.6 | 74,480 | 4.9 | 95.1 | NA | 76.2 | 610 | 35.4 |
| Shelby | 179 | -9.2 | 180 | 0.4 | 28.6 | 631,226 | 3,507 | 88,864 | 467.6 | 469,907 | 0.6 | 99.4 | NA | 70.7 | 50 | 0.6 |
| Sherman | 590 | 1.2 | 2,409 | 184.7 | 355.2 | 3,446,896 | 1,431 | 645,756 | 838.1 | 3,420,641 | 17.5 | 82.5 | 3 | 75.9 | 11,501 | 78.0 |
| Smith | 272 | -10.1 | 93 | 1.9 | 64.3 | 487,136 | 5,248 | 52,759 | 53.6 | 18,308 | 68.6 | 31.4 | NA | 71.8 | 94 | 1.2 |
| Somervell | 83 | -9.2 | 236 | 0.3 | 15.7 | 738,349 | 3,133 | 58,179 | 4.1 | 11,648 | 28.1 | 71.9 | NA | 75.3 | 40 | 3.4 |
| Starr | 571 | -14.5 | 425 | 3.4 | 74.1 | 779,096 | 1,834 | 49,858 | 47.2 | 35,115 | 19.9 | 80.1 | NA | 42.4 | 1,977 | 8.7 |
| Stephens | 470 | -9.0 | 818 | 0.3 | 37.1 | 1,430,192 | 1,749 | 65,383 | 10.6 | 18,477 | 5.3 | 94.7 | NA | 66.8 | 641 | 21.7 |
| Sterling | 584 | -0.1 | 7,688 | 0.4 | 9.4 | 5,208,932 | 678 | 107,859 | D | D | D | D | NA | 86.8 | 165 | 9.2 |
| Stonewall | 469 | -0.8 | 1,493 | 0.8 | 113.4 | 1,365,112 | 914 | 80,900 | 15.5 | 49,500 | 44.3 | 55.7 | NA | 62.7 | 1,995 | 63.1 |
| Sutton | 901 | -1.1 | 3,452 | 0.3 | 12.4 | 4,384,828 | 1,270 | 67,361 | 10.4 | 39,655 | 1.3 | 98.7 | NA | 74.3 | 331 | 9.2 |
| Swisher | 563 | 3.3 | 1,304 | 71.5 | 399.8 | 1,364,672 | 1,046 | 292,743 | 623.9 | 1,444,259 | 11.0 | 89.0 | NA | 76.4 | 13,019 | 86.1 |
| Tarrant | 191 | 30.9 | 163 | 1.3 | 43.5 | 992,283 | 6,104 | 53,767 | 29.4 | 25,058 | 59.4 | 40.6 | NA | 79.7 | 170 | 2.2 |
| Taylor | 484 | -16.4 | 347 | 1.2 | 159.2 | 713,098 | 2,053 | 64,361 | 31.5 | 22,626 | 35.4 | 64.6 | NA | 73.7 | 2,520 | 23.2 |
| Terrell | 835 | -24.1 | 9,825 | 0.6 | 9.4 | 7,373,840 | 751 | 95,965 | 4.2 | 49,306 | 13.1 | 86.9 | NA | 76.5 | 193 | 10.6 |
| Terry | 496 | 12.1 | 889 | 104.4 | 436.8 | 905,405 | 1,019 | 245,583 | 136.9 | 245,418 | 83.2 | 16.8 | 13 | 64.3 | 16,544 | 80.3 |
| Throckmorton | 507 | -0.2 | 1,850 | 0.4 | 157.2 | 2,604,134 | 1,408 | 111,148 | 27.3 | 99,489 | 20.2 | 79.8 | 3 | 73.0 | 2,844 | 52.9 |
| Titus | 159 | 8.4 | 196 | 2.1 | 33.8 | 493,023 | 2,520 | 74,641 | 149.3 | 183,858 | 2.7 | 97.3 | NA | 69.2 | 501 | 17.0 |

# Table B. States and Counties — Water Use, Wholesale Trade, Retail Trade, and Real Estate

| STATE County | Water use, 2015 | | Wholesale Trade[1], 2017 | | | | Retail Trade[2], 2017 | | | | Real estate and rental and leasing,[2] 2017 | | | |
|---|---|---|---|---|---|---|---|---|---|---|---|---|---|---|
| | Public supply water withdrawn (mil gal/day) | Public supply gallons withdrawn per person per day | Number of establishments | Number of employees | Sales (mil dol) | Average payroll (mil dol) | Number of establishments | Number of employees | Sales (mil dol) | Average payroll (mil dol) | Number of establishments | Number of employees | Sales (mil dol) | Average payroll (mil dol) |
| | 133 | 134 | 135 | 136 | 137 | 138 | 139 | 140 | 141 | 142 | 143 | 144 | 145 | 146 |
| **TEXAS—Cont'd** | | | | | | | | | | | | | | |
| Mason | 0.44 | 109.1 | 7 | 40 | 10.0 | 1.0 | 24 | 149 | 39.6 | 3.3 | 6 | D | 3.2 | D |
| Matagorda | 4.44 | 120.8 | 26 | 108 | 104.6 | 5.2 | 129 | 1,349 | 360.0 | 33.3 | 28 | 148 | 44.6 | 6.6 |
| Maverick | 6.95 | 120.4 | 38 | 188 | 177.2 | 9.6 | 166 | 2,757 | 661.8 | 59.3 | 32 | 101 | 15.0 | 2.7 |
| Medina | 6.65 | 137.3 | 29 | 220 | 290.8 | 12.2 | 108 | 1,611 | 627.1 | 47.6 | 29 | 53 | 7.9 | 1.3 |
| Menard | 0.37 | 171.0 | NA | NA | NA | NA | 9 | 43 | 12.7 | 0.8 | D | D | D | 132.1 |
| Midland | 1.04 | 6.5 | 299 | 4,958 | 6,456.0 | 341.1 | 514 | 8,861 | 3,322.4 | 287.7 | 17 | 44 | 7.4 | 1.3 |
| Milam | 10.89 | 444.3 | 13 | 124 | 62.9 | 6.5 | 58 | 612 | 188.0 | 16.6 | D | D | D | D |
| Mills | 0.28 | 57.1 | 5 | D | 14.2 | D | 18 | 160 | 62.2 | 5.0 | NA | NA | NA | NA |
| Mitchell | 1.28 | 141.2 | 3 | 8 | 1.3 | 0.3 | 28 | 203 | 67.3 | 5.0 | 10 | 35 | 5.7 | 1.0 |
| Montague | 1.07 | 55.5 | 18 | 59 | 34.8 | 2.0 | 72 | 641 | 197.0 | 16.4 | 629 | 2,457 | 628.9 | 122.5 |
| Montgomery | 57.29 | 106.6 | 534 | 5,373 | 11,349.1 | 372.7 | 1,447 | 25,947 | 7,741.9 | 707.3 | 17 | 47 | 10.4 | 1.5 |
| Moore | 5.07 | 227.8 | D | D | D | D | 65 | 970 | 258.1 | 21.6 | 3 | 3 | 0.7 | 0.1 |
| Morris | 0.23 | 18.4 | 8 | 85 | 36.7 | 3.8 | 38 | 291 | 69.1 | 6.5 | NA | NA | NA | NA |
| Motley | 0.15 | 130.7 | NA | NA | NA | NA | 3 | 16 | 4.4 | 0.3 | 57 | 211 | 29.6 | 6.3 |
| Nacogdoches | 11.2 | 170.6 | D | D | D | 18.6 | 221 | 2,822 | 849.8 | 76.6 | 52 | 133 | 25.3 | 4.1 |
| Navarro | 2.33 | 48.2 | 35 | 460 | 386.6 | 22.8 | 161 | 2,203 | 652.7 | 55.0 | NA | NA | NA | NA |
| Newton | 0.97 | 69.4 | D | D | D | D | 29 | 172 | 76.4 | 2.7 | 10 | 51 | 4.0 | 0.8 |
| Nolan | 2.28 | 150.9 | 18 | 109 | 77.3 | 6.0 | 66 | 781 | 257.0 | 19.0 | 474 | 3,047 | 777.8 | 154.1 |
| Nueces | 72.38 | 201.2 | 395 | 5,963 | 3,789.9 | 315.1 | 1,078 | 18,383 | 5,391.2 | 492.6 | 9 | 58 | 22.1 | 4.2 |
| Ochiltree | 1.76 | 163.8 | 28 | 314 | 194.9 | 18.0 | 44 | 448 | 120.3 | 10.5 | NA | NA | NA | NA |
| Oldham | 0.56 | 270.7 | NA | NA | NA | NA | D | D | D | D | 60 | 243 | 39.0 | 8.8 |
| Orange | 8.40 | 99.7 | 35 | 266 | 204.9 | 14.9 | 255 | 3,081 | 966.0 | 80.9 | 33 | 115 | 21.5 | 5.7 |
| Palo Pinto | 1.98 | 71.0 | 26 | 208 | 140.3 | 12.8 | 90 | 986 | 271.7 | 24.2 | 24 | 113 | 14.0 | 4.0 |
| Panola | 1.15 | 48.4 | 19 | 103 | 71.3 | 5.9 | 76 | 846 | 237.8 | 20.9 | 128 | 405 | 108.5 | 18.7 |
| Parker | 7.51 | 59.6 | 111 | 1,039 | 603.4 | 52.1 | 356 | 5,446 | 2,292.8 | 176.7 | 3 | 4 | 0.6 | 0.1 |
| Parmer | 0.86 | 88.2 | 28 | 228 | 915.0 | 10.5 | 24 | 163 | 38.5 | 3.3 | 9 | 44 | 9.2 | 1.6 |
| Pecos | 4.33 | 267.2 | 17 | 128 | 107.7 | 6.4 | 59 | 704 | 287.6 | 19.2 | 28 | 80 | 14.9 | 2.5 |
| Polk | 561.79 | 11,960.1 | 17 | 123 | 78.2 | 6.4 | 132 | 1,764 | 570.5 | 48.4 | 170 | 804 | 172.2 | 35.3 |
| Potter | 5.46 | 44.8 | 162 | 2,712 | 2,356.5 | 145.5 | 489 | 7,967 | 2,026.7 | 191.1 | NA | NA | NA | NA |
| Presidio | 2.54 | 369.4 | NA | NA | NA | NA | 27 | 242 | 51.7 | 4.5 | 9 | D | 2.8 | D |
| Rains | 1.25 | 112.0 | 6 | D | 29.6 | D | 28 | 355 | 136.4 | 10.7 | 158 | 559 | 122.2 | 21.2 |
| Randall | 2.23 | 17.1 | 96 | 1,908 | 2,537.8 | 97.3 | 357 | 6,254 | 1,882.9 | 173.9 | NA | NA | NA | NA |
| Reagan | 0.00 | 0.0 | 13 | 137 | 177.4 | 9.7 | 10 | 117 | 64.0 | 3.9 | 5 | D | 2.1 | D |
| Real | 0.35 | 105.8 | 3 | 7 | 1.5 | 0.2 | 15 | 116 | 22.4 | 2.1 | 4 | 20 | 0.6 | 0.1 |
| Red River | 0.92 | 73.9 | 6 | 34 | 25.5 | 0.8 | 36 | 266 | 61.6 | 5.4 | 10 | 158 | 43.7 | 8.4 |
| Reeves | 3.65 | 247.8 | 15 | 144 | 132.7 | 7.6 | 29 | 448 | 244.6 | 14.3 | 7 | 50 | 11.5 | 3.2 |
| Refugio | 0.65 | 89.2 | 6 | 52 | 32.1 | 2.2 | 18 | 284 | 91.3 | 6.4 | NA | NA | NA | NA |
| Roberts | 0.13 | 141.9 | NA | NA | NA | NA | 52 | 447 | 172.3 | 10.2 | D | D | D | D |
| Robertson | 2.25 | 135.1 | D | D | D | 0.6 | 287 | 5,527 | 1,834.3 | 166.7 | 107 | 343 | 115.5 | 15.2 |
| Rockwall | 0.00 | 0.0 | 67 | 523 | 482.4 | 34.2 | 44 | 428 | 123.9 | 10.4 | NA | NA | NA | NA |
| Runnels | 3.82 | 362.1 | 5 | 19 | 22.0 | 0.7 | 99 | 1,120 | 337.9 | 30.7 | 21 | 95 | 19.0 | 4.2 |
| Rusk | 5.77 | 108.7 | 27 | 396 | 219.1 | 20.1 | 37 | 316 | 61.1 | 5.9 | NA | NA | NA | NA |
| Sabine | 0.76 | 73.3 | NA | NA | NA | NA | 25 | 210 | 60.8 | 5.0 | NA | NA | NA | NA |
| San Augustine | 0.81 | 95.6 | 4 | 14 | 3.4 | 0.4 | 30 | 216 | 54.2 | 4.8 | 6 | 11 | 1.7 | 0.5 |
| San Jacinto | 2.13 | 77.7 | 9 | 50 | 17.5 | 2.7 | 148 | 2,463 | 776.2 | 68.0 | 54 | 170 | 26.9 | 4.8 |
| San Patricio | 1.25 | 18.6 | 34 | 187 | 211.2 | 9.1 | 32 | 158 | 40.8 | 3.5 | NA | NA | NA | NA |
| San Saba | 1.43 | 242.3 | 8 | 46 | 9.0 | 2.1 | 8 | 56 | 10.7 | 1.1 | 18 | 58 | 8.0 | 2.1 |
| Schleicher | 0.38 | 118.3 | NA | NA | NA | NA | 57 | 711 | 263.5 | 19.0 | 8 | 26 | 4.6 | 1.3 |
| Scurry | 0.13 | 7.4 | D | D | D | D | 17 | 117 | 23.0 | 2.6 | 18 | 47 | 7.4 | 1.1 |
| Shackelford | 0.00 | 0.0 | 4 | D | 3.9 | D | 81 | 891 | 233.6 | 22.2 | NA | NA | NA | NA |
| Shelby | 4.78 | 188.2 | 13 | 94 | 79.1 | 6.3 | 10 | 93 | 58.9 | 2.6 | NA | NA | NA | NA |
| Sherman | 0.44 | 143.2 | 9 | D | 94.6 | D | 839 | 13,251 | 4,074.6 | 364.5 | 333 | 1,177 | 262.1 | 51.6 |
| Smith | 42.89 | 192.4 | 226 | 2,841 | 1,180.6 | 145.3 | 28 | 192 | 44.5 | 3.6 | 9 | 22 | 2.0 | 0.4 |
| Somervell | 1.18 | 135.0 | 5 | 26 | 12.4 | 1.2 | 130 | 1,887 | 474.6 | 41.5 | 13 | 31 | 3.9 | 0.8 |
| Starr | 7.81 | 122.4 | 19 | 110 | 127.5 | 2.7 | 40 | 371 | 107.1 | 8.8 | 6 | 21 | 2.6 | 0.4 |
| Stephens | 9.32 | 987.3 | 4 | 32 | 10.6 | 1.8 | 4 | 35 | 22.4 | 0.9 | NA | NA | NA | NA |
| Sterling | 0.19 | 140.5 | NA | NA | NA | NA | 8 | 36 | 9.7 | 0.7 | NA | NA | NA | NA |
| Stonewall | 0.00 | 0.0 | 3 | 47 | 34.5 | 2.7 | 23 | 145 | 59.2 | 4.1 | 4 | D | 3.7 | D |
| Sutton | 0.81 | 207.0 | 4 | 29 | 28.3 | 2.1 | 17 | 176 | 57.0 | 3.6 | NA | NA | NA | NA |
| Swisher | 0.59 | 78.3 | 12 | 53 | 26.1 | 1.9 | 6,078 | 106,920 | 36,282.1 | 2,996.0 | 2,171 | 13,084 | 3,580.7 | 637.9 |
| Tarrant | 26.52 | 13.4 | 1,960 | 37,008 | 32,115.2 | 2,331.4 | 538 | 8,113 | 2,430.4 | 212.7 | 177 | 811 | 205.2 | 32.9 |
| Taylor | 0.46 | 3.4 | 147 | 1,918 | 2,086.0 | 96.8 | 4 | 18 | 4.9 | 0.6 | NA | NA | NA | NA |
| Terrell | 0.13 | 155.3 | NA | NA | NA | NA | 36 | 339 | 104.8 | 9.2 | 4 | 8 | 0.8 | 0.2 |
| Terry | 0.17 | 13.3 | 23 | 344 | 164.8 | 17.7 | 9 | 29 | 6.5 | 0.7 | NA | NA | NA | NA |
| Throckmorton | 0.05 | 31.7 | 4 | 11 | 2.5 | 0.4 | 113 | 1,747 | 488.9 | 44.9 | 25 | 64 | 12.1 | 2.0 |
| Titus | 5.35 | 164.0 | 34 | 275 | 254.7 | 11.3 | | | | | | | | |

1 Merchant wholesalers, except manufacturers' sales branches and offices.   2. Employer establishments.

# Table B. States and Counties — Professional Services, Manufacturing, and Accommodation and Food Services

| STATE County | Professional, scientific, and technical services, 2017 | | | | Manufacturing, 2017 | | | | Accommodation and food services, 2017 | | | |
|---|---|---|---|---|---|---|---|---|---|---|---|---|
| | Number of establishments | Number of employees | Sales (mil dol) | Average payroll (mil dol) | Number of establishments | Number of employees | Sales (mil dol) | Average payroll (mil dol) | Number of establishments | Number of employees | Sales (mil dol) | Annual payroll (mil dol) |
| | 147 | 148 | 149 | 150 | 151 | 152 | 153 | 154 | 155 | 156 | 157 | 158 |
| TEXAS—Cont'd | | | | | | | | | | | | |
| Mason | 13 | 40 | 4 | 1.5 | 6 | 15 | 3.2 | 0.8 | 15 | 124 | 5.5 | 1.7 |
| Matagorda | 40 | 306 | 31 | 12.0 | 32 | 971 | 1,636.9 | 83.9 | 86 | 979 | 59.7 | 16.4 |
| Maverick | D | D | D | D | 14 | 301 | 72.0 | 13.0 | 78 | 1,460 | 75.4 | 20.3 |
| Medina | 69 | 395 | 32 | 13.4 | 23 | 319 | 58.6 | 12.9 | 83 | 1,262 | 57.6 | 16.2 |
| Menard | D | D | 1 | D | NA | NA | NA | NA | 3 | 25 | 1.0 | 0.3 |
| Midland | D | D | D | D | 140 | 2,831 | 1,177.2 | 168.1 | D | D | D | D |
| Milam | D | D | D | D | 14 | 147 | 98.1 | 7.8 | 46 | 466 | 22.0 | 5.6 |
| Mills | 5 | 15 | 2 | 0.5 | 7 | 36 | 7.4 | 2.4 | 8 | 79 | 3.6 | 1.1 |
| Mitchell | 10 | 29 | 5 | 1.4 | NA | NA | NA | NA | 16 | 172 | 7.4 | 2.0 |
| Montague | 47 | 136 | 15 | 5.1 | 17 | 238 | 96.7 | 9.3 | 35 | 452 | 18.4 | 5.6 |
| Montgomery | D | D | D | D | 407 | 10,069 | 3,415.5 | 528.5 | 990 | 22,530 | 1,444.1 | 396.6 |
| Moore | 20 | 94 | 12 | 4.3 | D | D | D | 201.1 | 47 | 532 | 31.7 | 8.0 |
| Morris | 17 | 127 | 17 | 7.6 | 17 | 1,786 | 500.6 | 84.1 | 17 | 204 | 10.3 | 2.5 |
| Motley | NA | NA | NA | NA | NA | NA | NA | NA | D | D | D | 0.1 |
| Nacogdoches | D | D | D | D | 62 | 3,417 | 1,333.4 | 140.6 | 119 | 2,275 | 118.8 | 36.3 |
| Navarro | 66 | 271 | 41 | 10.4 | 51 | 2,731 | 1,021.1 | 120.0 | 68 | 1,251 | 61.2 | 16.6 |
| Newton | 9 | 21 | 2 | 0.7 | D | 44 | D | 1.8 | 5 | 11 | 0.6 | 0.1 |
| Nolan | 31 | 90 | 9 | 3.0 | 9 | 661 | 371.9 | 47.0 | 42 | 577 | 31.3 | 7.8 |
| Nueces | D | D | D | D | 170 | 6,788 | 26,024.0 | 583.0 | 926 | 19,524 | 1,073.9 | 296.4 |
| Ochiltree | 18 | 108 | 14 | 5.6 | 7 | 18 | 4.5 | 0.7 | 18 | 296 | 15.0 | 4.2 |
| Oldham | NA | NA | NA | NA | NA | NA | NA | NA | D | D | D | D |
| Orange | 83 | 548 | 70 | 35.1 | 72 | 4,838 | 5,121.8 | 439.3 | 146 | 2,070 | 112.9 | 29.7 |
| Palo Pinto | 45 | 168 | 20 | 6.6 | 38 | 1,110 | 330.0 | 55.1 | 65 | 816 | 46.0 | 11.7 |
| Panola | 44 | 208 | 30 | 10.0 | 12 | 1,060 | 361.3 | 40.2 | 28 | 407 | 21.6 | 5.9 |
| Parker | D | D | D | D | 137 | 2,671 | 903.3 | 142.8 | 211 | 3,943 | 211.5 | 60.8 |
| Parmer | 12 | 44 | 5 | 1.5 | D | D | D | D | 8 | 85 | 4.4 | 1.0 |
| Pecos | 12 | 80 | 5 | 1.1 | 4 | 13 | 6.1 | 0.5 | 44 | 586 | 45.5 | 9.8 |
| Polk | 71 | 173 | 23 | 7.4 | 16 | 1,168 | 334.6 | 59.9 | 69 | 1,041 | 58.2 | 15.3 |
| Potter | D | D | D | D | 121 | 6,870 | 4,359.1 | 312.1 | 378 | 8,044 | 463.9 | 122.5 |
| Presidio | NA | NA | NA | NA | D | 9 | D | 0.5 | 25 | 320 | 30.2 | 7.4 |
| Rains | 16 | 194 | 20 | 4.3 | 9 | 67 | 10.7 | 3.8 | 16 | 176 | 9.6 | 2.4 |
| Randall | 232 | 993 | 135 | 51.7 | 66 | 4,601 | 1,121.1 | 405.4 | 210 | 4,100 | 221.0 | 61.4 |
| Reagan | 4 | 12 | 2 | 0.3 | NA | NA | NA | NA | 12 | 140 | 10.8 | 2.2 |
| Real | D | D | 3 | D | 5 | 25 | 10.8 | 1.1 | 16 | 165 | 9.7 | 2.0 |
| Red River | 8 | 23 | 2 | 1.0 | 17 | 336 | 74.3 | 13.7 | D | D | D | D |
| Reeves | 20 | 47 | 11 | 2.2 | 5 | 10 | 7.4 | D | 42 | 664 | 84.9 | 14.6 |
| Refugio | 6 | 24 | 2 | 1.1 | 3 | 9 | 1.2 | D | 18 | 246 | 13.1 | 3.2 |
| Roberts | 3 | 9 | 1 | 0.3 | NA | NA | NA | NA | NA | NA | NA | NA |
| Robertson | 17 | 58 | 10 | 1.9 | 6 | 127 | 20.1 | 5.1 | 27 | 486 | 21.0 | 6.4 |
| Rockwall | D | D | D | D | 63 | 1,639 | 559.5 | 91.0 | 211 | 4,954 | 263.3 | 76.8 |
| Runnels | 14 | 29 | 3 | 0.9 | 14 | 486 | 236.1 | 19.6 | 17 | 170 | 6.5 | 1.6 |
| Rusk | 63 | 285 | 39 | 14.4 | 34 | 1,324 | 398.3 | 59.0 | 68 | 879 | 45.5 | 12.2 |
| Sabine | 14 | 32 | 4 | 0.8 | D | 344 | D | 21.4 | 9 | D | 5.0 | D |
| San Augustine | 6 | 9 | 1 | 0.3 | 4 | 51 | 12.7 | 2.2 | D | D | D | 0.7 |
| San Jacinto | 24 | 43 | 5 | 1.9 | 9 | 92 | 23.5 | 5.6 | D | D | D | D |
| San Patricio | D | D | D | D | 38 | 3,261 | 2,677.3 | 214.4 | 142 | 2,233 | 126.2 | 30.6 |
| San Saba | 11 | 32 | 3 | 1.0 | 9 | 41 | 9.9 | 1.9 | D | D | D | D |
| Schleicher | 4 | 11 | 1 | 0.3 | NA | NA | NA | NA | NA | NA | NA | NA |
| Scurry | 19 | 89 | 13 | 2.1 | 10 | 73 | 81.1 | 4.6 | 48 | 622 | 27.4 | 8.2 |
| Shackelford | 3 | 7 | 1 | 0.2 | 7 | 149 | 22.5 | 6.8 | D | D | D | 0.6 |
| Shelby | 33 | 173 | 32 | 7.1 | 15 | 2,261 | 594.5 | 77.8 | 38 | 505 | 42.1 | 9.8 |
| Sherman | NA | NA | NA | NA | NA | NA | NA | NA | NA | NA | NA | NA |
| Smith | 624 | 4,570 | 917 | 301.9 | 184 | 6,614 | 4,305.0 | 340.6 | 490 | 10,739 | 596.4 | 182.9 |
| Somervell | 18 | 109 | 16 | 5.2 | 8 | 101 | 29.3 | 4.4 | 35 | 465 | 44.5 | 9.8 |
| Starr | D | D | D | D | 7 | 19 | 2.4 | 0.7 | D | D | D | D |
| Stephens | 15 | 37 | 4 | 1.3 | 12 | 252 | 64.3 | 11.5 | 19 | 164 | 8.9 | 2.3 |
| Sterling | D | D | 1 | D | NA | NA | NA | NA | 4 | 33 | 1.0 | 0.3 |
| Stonewall | NA | NA | NA | NA | NA | NA | NA | NA | D | D | D | 0.3 |
| Sutton | 13 | 25 | 2 | 0.5 | NA | NA | NA | NA | 13 | 195 | 18.4 | 4.1 |
| Swisher | 6 | 21 | 1 | 0.7 | 9 | 96 | 21.2 | 4.4 | D | D | D | D |
| Tarrant | D | D | D | D | 1,590 | 77,734 | 62,176.0 | 5,393.5 | 4,170 | 99,956 | 6,194.9 | 1,686.5 |
| Taylor | D | D | D | D | 93 | 2,244 | 1,054.7 | 112.3 | 325 | 7,052 | 363.9 | 106.7 |
| Terrell | NA | NA | NA | NA | NA | NA | NA | NA | NA | NA | NA | NA |
| Terry | 11 | 50 | 5 | 1.4 | 5 | 42 | 13.9 | 1.8 | 26 | 307 | 15.9 | 4.1 |
| Throckmorton | NA | NA | NA | NA | NA | NA | NA | NA | 3 | 11 | 0.4 | 0.1 |
| Titus | 22 | 115 | 13 | 4.9 | 37 | 5,757 | 1,703.8 | 222.6 | 68 | 1,104 | 61.3 | 16.2 |

Items 147—158

# Table B. States and Counties — Health Care and Social Assistance, Other Services, Nonemployer Businesses, and Residential Construction

| STATE County | Health care and social assistance, 2017 | | | | Other services, 2017 | | | | Nonemployer businesses, 2018 | | Value of residential construction authorized by building permits, 2020 | |
|---|---|---|---|---|---|---|---|---|---|---|---|---|
| | Number of establishments | Number of employees | Receipts (mil dol) | Annual payroll (mil dol) | Number of establishments | Number of employees | Receipts (mil dol) | Annual payroll (mil dol) | Number | Receipts (mil dol) | New construction ($1,000) | Number of housing units |
| | 159 | 160 | 161 | 162 | 163 | 164 | 165 | 166 | 167 | 168 | 169 | 170 |
| TEXAS—Cont'd | | | | | | | | | | | | |
| Mason | 12 | 109 | 5.6 | 2.6 | D | D | D | 0.5 | 663 | 29.4 | 1,632 | 7 |
| Matagorda | 74 | 1,200 | 146.2 | 46.7 | 65 | 388 | 54.0 | 15.4 | 2,942 | 123.6 | 23,479 | 176 |
| Maverick | 107 | 4,090 | 240.3 | 97.8 | 39 | 167 | 12.3 | 3.7 | 4,855 | 193.2 | 29,039 | 139 |
| Medina | 81 | 1,130 | 77.1 | 36.8 | 59 | 222 | 29.0 | 7.2 | 3,618 | 169.5 | 6,675 | 36 |
| Menard | NA | NA | NA | NA | D | D | 0.9 | D | 290 | 12.7 | NA | NA |
| Midland | 413 | 7,647 | 913.3 | 361.9 | 301 | 2,553 | 455.6 | 105.2 | 19,075 | 1,445.6 | 178,729 | 1,289 |
| Milam | 39 | 1,618 | 174.9 | 80.6 | 31 | 103 | 12.5 | 2.9 | 1,807 | 82.4 | 1,740 | 11 |
| Mills | 12 | 136 | 9.0 | 3.8 | D | D | 7.4 | D | 497 | 19.0 | 0 | 0 |
| Mitchell | 9 | 214 | 21.2 | 9.6 | 7 | 18 | 1.8 | 0.4 | 432 | 17.0 | 256 | 16 |
| Montague | 35 | 531 | 37.5 | 17.5 | 33 | 123 | 12.2 | 3.2 | 1,987 | 100.0 | NA | NA |
| Montgomery | 1,355 | 22,297 | 3,232.5 | 1,157.1 | 733 | 5,074 | 485.8 | 153.8 | 53,981 | 3,025.2 | 1,992,553 | 10,219 |
| Moore | 43 | 657 | 69.7 | 29.3 | 35 | 124 | 17.7 | 3.5 | 1,207 | 70.2 | 2,642 | 12 |
| Morris | 18 | 198 | 10.7 | 5.1 | 12 | 53 | 7.9 | 2.1 | 751 | 30.9 | 615 | 3 |
| Motley | NA | NA | NA | NA | NA | NA | NA | NA | 127 | 4.7 | NA | NA |
| Nacogdoches | 190 | 3,315 | 374.3 | 133.8 | 82 | 364 | 40.2 | 9.0 | 4,218 | 187.4 | 3,941 | 25 |
| Navarro | 134 | 2,644 | 186.6 | 76.2 | 59 | 211 | 22.5 | 5.7 | 3,574 | 165.6 | 11,749 | 70 |
| Newton | 11 | 180 | 9.0 | 4.3 | 3 | 8 | 1.5 | 0.2 | 713 | 24.6 | NA | NA |
| Nolan | 29 | 578 | 54.3 | 23.9 | 17 | 109 | 22.7 | 5.0 | 1,088 | 39.5 | 725 | 4 |
| Nueces | 1,094 | 30,804 | 2,927.6 | 1,159.3 | 540 | 4,227 | 620.3 | 141.6 | 26,708 | 1,208.8 | 359,166 | 1,557 |
| Ochiltree | 19 | 301 | 33.5 | 11.6 | 26 | 114 | 14.3 | 3.9 | 847 | 46.6 | 345 | 0 |
| Oldham | NA | NA | NA | NA | NA | NA | NA | NA | 198 | 9.1 | 0 | 0 |
| Orange | 128 | 1,462 | 118.6 | 46.4 | 81 | 699 | 97.6 | 24.2 | 4,883 | 189.2 | 34,697 | 225 |
| Palo Pinto | 51 | 817 | 109.8 | 36.0 | 39 | 137 | 16.5 | 4.5 | 2,315 | 122.1 | 2,254 | 11 |
| Panola | 35 | 568 | 58.0 | 19.3 | 26 | 127 | 13.7 | 3.9 | 1,725 | 80.0 | 3,031 | 13 |
| Parker | 261 | 3,345 | 375.3 | 135.5 | 154 | 789 | 79.4 | 25.1 | 14,253 | 830.9 | 197,983 | 1,113 |
| Parmer | 8 | 161 | 13.3 | 5.6 | D | D | D | D | 538 | 28.6 | 100 | 4 |
| Pecos | 17 | 390 | 40.2 | 17.5 | 19 | 91 | 10.6 | 2.9 | 1,068 | 50.4 | 2,506 | 47 |
| Polk | 78 | 1,368 | 128.7 | 52.0 | 51 | 256 | 24.1 | 6.4 | 4,316 | 184.1 | 112,910 | 565 |
| Potter | 469 | 13,427 | 1,850.9 | 655.6 | 239 | 1,956 | 252.3 | 64.0 | 8,507 | 491.2 | 164,865 | 845 |
| Presidio | 5 | 58 | 5.4 | 1.8 | NA | NA | NA | NA | 833 | 28.0 | 1,882 | 9 |
| Rains | 13 | D | 6.7 | D | D | D | D | 1.3 | 984 | 49.5 | 5,285 | 30 |
| Randall | D | D | D | D | 197 | 1,272 | 145.4 | 40.2 | 11,199 | 546.5 | 38,831 | 173 |
| Reagan | D | D | D | D | 7 | 24 | 4.3 | 1.1 | 343 | 18.0 | 216 | 1 |
| Real | 6 | D | 6.3 | D | NA | NA | NA | NA | 470 | 23.8 | 0 | 0 |
| Red River | 14 | 163 | 12.5 | 5.0 | D | D | D | 0.6 | 936 | 48.9 | 1,507 | 13 |
| Reeves | 19 | 395 | 32.1 | 13.0 | D | D | D | D | 885 | 53.9 | 14,958 | 73 |
| Refugio | 11 | 248 | 23.4 | 9.1 | 7 | 21 | 2.1 | 0.6 | 484 | 19.5 | 12,557 | 83 |
| Roberts | NA | NA | NA | NA | NA | NA | NA | NA | 99 | 4.0 | NA | NA |
| Robertson | 19 | 388 | 24.2 | 11.3 | 27 | 83 | 11.4 | 2.5 | 1,332 | 59.8 | 21,827 | 130 |
| Rockwall | 312 | 4,762 | 738.1 | 219.2 | 140 | 854 | 85.2 | 27.4 | 10,820 | 660.3 | 773,623 | 2,699 |
| Runnels | D | D | D | D | 9 | 27 | 3.3 | 0.7 | 832 | 36.4 | 800 | 8 |
| Rusk | 79 | 1,229 | 96.9 | 37.5 | 47 | 239 | 22.9 | 7.1 | 3,364 | 149.0 | 815 | 3 |
| Sabine | 23 | 382 | 21.8 | 10.9 | 11 | 30 | 3.6 | 0.9 | 701 | 31.9 | 170 | 1 |
| San Augustine | 20 | 380 | 24.7 | 11.6 | 3 | 5 | 0.3 | 0.1 | 537 | 20.7 | 0 | 0 |
| San Jacinto | 14 | 129 | 7.0 | 3.1 | D | D | 3.4 | D | 2,040 | 83.5 | 100,519 | 457 |
| San Patricio | 104 | 1,608 | 115.8 | 48.1 | 73 | 356 | 42.3 | 11.8 | 4,895 | 193.4 | 53,741 | 366 |
| San Saba | 12 | 126 | 8.7 | 4.1 | D | D | D | D | 618 | 23.5 | 100 | 1 |
| Schleicher | D | D | D | D | NA | NA | NA | NA | 292 | 10.1 | 0 | 0 |
| Scurry | 23 | 581 | 58.9 | 25.0 | 32 | 201 | 35.6 | 7.9 | 1,105 | 50.6 | 800 | 3 |
| Shackelford | 4 | 46 | 4.5 | 2.1 | 3 | 6 | 0.9 | 0.1 | 506 | 27.6 | NA | NA |
| Shelby | 42 | 609 | 43.8 | 18.2 | D | D | D | D | 1,725 | 88.2 | 233 | 1 |
| Sherman | NA | NA | NA | NA | 3 | 6 | 0.6 | 0.1 | 235 | 11.1 | 1,308 | 12 |
| Smith | 685 | 22,449 | 2,985.8 | 1,095.4 | 374 | 2,239 | 239.2 | 67.8 | 19,718 | 1,028.0 | 173,328 | 828 |
| Somervell | 19 | 513 | 54.7 | 19.2 | 16 | 75 | 9.6 | 3.7 | 903 | 40.3 | 4,332 | 21 |
| Starr | 103 | 5,086 | 169.3 | 88.6 | D | D | D | D | 7,161 | 206.1 | 209 | 1 |
| Stephens | 17 | 333 | 25.2 | 10.8 | 18 | 92 | 5.6 | 4.8 | 915 | 50.3 | 129 | 3 |
| Sterling | NA | NA | NA | NA | NA | NA | NA | NA | 162 | 8.1 | NA | NA |
| Stonewall | D | D | D | D | NA | NA | NA | NA | 151 | 8.2 | NA | NA |
| Sutton | D | D | D | D | 6 | 10 | 1.3 | 0.3 | 396 | 14.6 | 0 | 0 |
| Swisher | 10 | 183 | 17.4 | 6.4 | 12 | 37 | 2.9 | 0.7 | 479 | 18.4 | 0 | 0 |
| Tarrant | 5,412 | 114,635 | 15,432.2 | 5,405.9 | 2,743 | 21,725 | 2,796.3 | 684.7 | 186,406 | 9,101.7 | 2,736,916 | 14,224 |
| Taylor | 404 | 12,073 | 1,326.3 | 513.6 | 225 | 1,499 | 194.7 | 43.9 | 10,474 | 506.1 | 104,023 | 648 |
| Terrell | NA | NA | NA | NA | NA | NA | NA | NA | 92 | 2.0 | NA | NA |
| Terry | 18 | 431 | 35.5 | 14.2 | 16 | 68 | 6.7 | 2.0 | 795 | 37.0 | 608 | 5 |
| Throckmorton | D | D | D | D | D | D | 0.6 | D | 221 | 9.3 | NA | NA |
| Titus | 82 | 2,224 | 175.8 | 72.4 | 44 | 203 | 18.4 | 5.4 | 1,981 | 97.1 | 7,855 | 48 |

# Table B. States and Counties — Government Employment and Payroll, and Local Government Finances

| STATE County | Full-time equivalent employees | March payroll (dollars) | Administration, judicial, and legal | Police and corrections | Fire protection | Highways and transportation | Health and welfare | Natural resources and utilities | Education and libraries | Total (mil dol) | Intergovernmental (mil dol) | Taxes Total (mil dol) | Per capita Total | Per capita Property |
|---|---|---|---|---|---|---|---|---|---|---|---|---|---|---|
| | 171 | 172 | 173 | 174 | 175 | 176 | 177 | 178 | 179 | 180 | 181 | 182 | 183 | 184 |
| **TEXAS—Cont'd** | | | | | | | | | | | | | | |
| Mason | 214 | 674,368 | 11.1 | 7.1 | 0.0 | 4.9 | 4.7 | 9.7 | 59.5 | 13.0 | 4.0 | 7.7 | 1,851 | 1,640 |
| Matagorda | 2,161 | 7,462,387 | 5.4 | 8.0 | 0.0 | 2.0 | 22.4 | 4.1 | 56.8 | 319.9 | 47.1 | 103.9 | 2,822 | 2,636 |
| Maverick | 3,062 | 9,994,336 | 3.0 | 7.0 | 2.0 | 3.0 | 4.7 | 3.9 | 75.6 | 210.2 | 122.0 | 59.3 | 1,020 | 827 |
| Medina | 2,350 | 8,037,563 | 4.8 | 6.3 | 0.1 | 2.4 | 12.1 | 4.2 | 68.1 | 220.6 | 82.7 | 105.2 | 2,097 | 1,953 |
| Menard | 146 | 447,532 | 3.2 | 9.3 | 0.0 | 1.5 | 25.2 | 4.1 | 49.4 | 11.0 | 2.7 | 4.6 | 2,191 | 2,067 |
| Midland | 7,596 | 33,664,386 | 4.7 | 7.9 | 3.5 | 2.9 | 29.3 | 2.4 | 48.2 | 982.4 | 134.3 | 478.6 | 2,895 | 2,149 |
| Milam | 1,153 | 4,037,962 | 7.5 | 9.2 | 0.1 | 4.5 | 2.2 | 2.9 | 73.4 | 91.0 | 35.6 | 42.2 | 1,692 | 1,524 |
| Mills | 327 | 1,086,084 | 7.1 | 5.9 | 0.0 | 7.8 | 0.5 | 4.3 | 74.1 | 24.2 | 10.9 | 10.7 | 2,182 | 1,931 |
| Mitchell | 615 | 1,976,389 | 8.2 | 5.8 | 0.3 | 2.7 | 35.8 | 2.6 | 43.7 | 60.1 | 12.5 | 20.5 | 2,499 | 2,308 |
| Montague | 913 | 2,962,949 | 4.9 | 7.1 | 2.3 | 3.2 | 10.9 | 3.4 | 62.2 | 64.9 | 14.7 | 34.4 | 1,771 | 1,614 |
| Montgomery | 18,814 | 73,051,702 | 4.8 | 8.9 | 5.7 | 1.7 | 5.0 | 3.5 | 68.2 | 1,999.4 | 505.7 | 1,257.0 | 2,199 | 1,935 |
| Moore | 1,382 | 5,222,714 | 5.9 | 7.1 | 1.6 | 2.1 | 26.9 | 5.6 | 50.0 | 121.9 | 24.3 | 51.1 | 2,365 | 2,077 |
| Morris | 471 | 1,467,593 | 9.3 | 9.0 | 0.5 | 2.8 | 2.0 | 2.7 | 73.5 | 29.3 | 9.3 | 17.4 | 1,404 | 1,307 |
| Motley | 92 | 322,465 | 19.8 | 1.2 | 0.2 | 1.9 | 10.8 | 0.4 | 65.7 | 7.1 | 1.4 | 5.1 | 4,179 | 3,487 |
| Nacogdoches | 3,205 | 11,993,798 | 4.6 | 6.1 | 2.5 | 1.4 | 25.8 | 2.3 | 56.3 | 287.4 | 82.8 | 97.2 | 1,487 | 1,286 |
| Navarro | 2,758 | 10,157,929 | 4.0 | 8.6 | 3.0 | 2.2 | 2.4 | 3.1 | 75.2 | 243.9 | 94.5 | 98.6 | 2,023 | 1,651 |
| Newton | 539 | 1,594,597 | 9.8 | 5.2 | 0.1 | 2.9 | 0.8 | 5.2 | 73.5 | 53.9 | 18.9 | 21.1 | 1,517 | 1,442 |
| Nolan | 1,238 | 4,211,547 | 4.7 | 7.1 | 1.9 | 1.7 | 31.0 | 3.8 | 49.0 | 120.4 | 25.0 | 46.9 | 3,160 | 2,687 |
| Nueces | 15,544 | 60,072,732 | 6.2 | 10.0 | 4.7 | 5.8 | 3.7 | 6.7 | 61.5 | 1,684.4 | 456.4 | 789.4 | 2,185 | 1,718 |
| Ochiltree | 751 | 2,510,234 | 12.8 | 5.3 | 2.2 | 0.8 | 25.2 | 3.7 | 48.7 | 44.2 | 11.4 | 27.2 | 2,725 | 2,338 |
| Oldham | 256 | 905,109 | 7.0 | 7.6 | 0.0 | 2.0 | 0.0 | 0.0 | 81.3 | 26.4 | 9.5 | 10.4 | 4,924 | 4,456 |
| Orange | 3,505 | 12,384,125 | 7.1 | 12.2 | 2.1 | 3.0 | 0.7 | 9.5 | 63.1 | 283.2 | 90.2 | 136.0 | 1,601 | 1,362 |
| Palo Pinto | 1,510 | 5,351,012 | 5.9 | 7.2 | 1.7 | 4.5 | 27.8 | 3.8 | 47.0 | 155.5 | 41.7 | 70.8 | 2,479 | 2,211 |
| Panola | 1,165 | 4,316,500 | 6.1 | 7.2 | 0.7 | 4.6 | 0.9 | 2.8 | 76.4 | 121.9 | 28.4 | 77.9 | 3,357 | 3,062 |
| Parker | 4,338 | 17,739,986 | 6.5 | 5.7 | 2.6 | 2.4 | 1.7 | 2.9 | 76.3 | 405.5 | 96.8 | 248.8 | 1,863 | 1,652 |
| Parmer | 784 | 2,562,481 | 9.5 | 7.0 | 0.0 | 3.8 | 24.6 | 1.2 | 52.8 | 64.9 | 18.0 | 32.2 | 3,321 | 3,090 |
| Pecos | 1,009 | 3,481,520 | 6.7 | 7.0 | 0.0 | 2.3 | 9.8 | 7.2 | 63.1 | 119.0 | 19.3 | 57.7 | 3,692 | 3,367 |
| Polk | 1,551 | 5,222,803 | 6.9 | 11.5 | 0.1 | 3.0 | 1.2 | 3.1 | 73.0 | 121.2 | 44.0 | 64.4 | 1,314 | 1,198 |
| Potter | 9,282 | 35,583,891 | 4.3 | 11.7 | 5.5 | 1.9 | 4.3 | 4.3 | 66.5 | 823.0 | 311.6 | 351.6 | 2,922 | 2,008 |
| Presidio | 493 | 1,554,231 | 10.5 | 7.9 | 0.0 | 3.5 | 6.2 | 6.6 | 64.6 | 36.4 | 18.9 | 13.9 | 1,955 | 1,652 |
| Rains | 458 | 1,020,204 | 13.1 | 10.6 | 0.0 | 3.0 | 0.6 | 4.7 | 67.0 | 31.8 | 10.9 | 16.6 | 1,415 | 1,158 |
| Randall | 1,792 | 7,254,356 | 9.3 | 21.4 | 1.3 | 0.4 | 0.1 | 1.7 | 64.9 | 146.0 | 35.7 | 95.0 | 709 | 663 |
| Reagan | 446 | 1,424,626 | 9.8 | 7.8 | 0.4 | 3.9 | 24.6 | 4.0 | 48.3 | 61.8 | 8.3 | 42.2 | 11,361 | 10,348 |
| Real | 96 | 314,765 | 17.7 | 13.0 | 0.0 | 5.3 | 1.7 | 5.4 | 56.6 | 11.0 | 10.1 | 8.9 | 2,591 | 2,218 |
| Red River | 547 | 1,541,912 | 6.3 | 6.2 | 0.6 | 1.9 | 1.3 | 2.4 | 81.1 | 47.1 | 20.6 | 17.1 | 1,407 | 1,131 |
| Reeves | 1,529 | 5,932,035 | 4.9 | 36.3 | 1.3 | 2.5 | 24.7 | 3.5 | 26.3 | 191.7 | 73.5 | 71.1 | 4,685 | 4,282 |
| Refugio | 533 | 1,916,846 | 7.3 | 10.0 | 0.0 | 4.6 | 25.7 | 3.6 | 47.5 | 37.2 | 9.1 | 21.6 | 3,006 | 2,742 |
| Roberts | 89 | 335,619 | 25.8 | 7.4 | 1.8 | 14.2 | 0.0 | 2.8 | 45.8 | 14.8 | 3.3 | 10.6 | 11,283 | 10,394 |
| Robertson | 972 | 2,844,712 | 9.1 | 9.1 | 0.0 | 3.6 | 1.1 | 3.5 | 73.3 | 84.4 | 25.6 | 50.1 | 2,920 | 2,683 |
| Rockwall | 3,603 | 13,505,727 | 7.0 | 11.0 | 1.8 | 0.9 | 0.4 | 2.5 | 74.0 | 355.2 | 77.9 | 238.8 | 2,466 | 2,105 |
| Runnels | 642 | 2,056,985 | 7.7 | 6.2 | 0.2 | 3.4 | 26.8 | 2.9 | 52.5 | 50.6 | 19.0 | 16.4 | 1,593 | 1,368 |
| Rusk | 1,793 | 5,977,953 | 7.6 | 10.1 | 1.8 | 3.7 | 1.0 | 2.7 | 70.2 | 144.4 | 52.7 | 71.3 | 1,316 | 1,182 |
| Sabine | 406 | 1,368,020 | 9.7 | 7.0 | 0.0 | 2.4 | 7.2 | 7.2 | 64.4 | 26.3 | 11.2 | 11.3 | 1,088 | 930 |
| San Augustine | 334 | 1,098,697 | 12.1 | 7.4 | 0.0 | 2.4 | 5.8 | 5.0 | 66.6 | 32.5 | 7.7 | 17.2 | 2,075 | 1,365 |
| San Jacinto | 820 | 2,480,988 | 7.9 | 8.3 | 0.1 | 4.3 | 0.0 | 2.6 | 76.6 | 51.6 | 20.3 | 28.8 | 1,022 | 989 |
| San Patricio | 3,809 | 12,018,785 | 5.3 | 9.0 | 1.3 | 2.7 | 8.3 | 6.2 | 66.3 | 306.0 | 116.0 | 146.3 | 2,177 | 1,880 |
| San Saba | 477 | 1,428,900 | 4.4 | 2.8 | 0.0 | 41.1 | 0.7 | 6.3 | 43.2 | 30.0 | 15.0 | 13.1 | 2,192 | 1,809 |
| Schleicher | 254 | 646,199 | 8.2 | 4.4 | 0.0 | 5.4 | 19.0 | 8.3 | 53.6 | 9.9 | 3.5 | 4.4 | 1,487 | 1,418 |
| Scurry | 1,156 | 4,320,011 | 5.7 | 7.0 | 1.3 | 2.2 | 26.9 | 3.8 | 52.3 | 128.2 | 25.8 | 53.0 | 3,117 | 2,804 |
| Shackelford | 189 | 586,844 | 11.7 | 8.2 | 0.0 | 5.5 | 4.4 | 4.8 | 65.2 | 15.2 | 3.6 | 8.7 | 2,643 | 2,284 |
| Shelby | 1,300 | 4,370,836 | 5.7 | 6.5 | 0.5 | 2.6 | 0.7 | 3.1 | 80.4 | 83.1 | 43.9 | 31.1 | 1,232 | 1,007 |
| Sherman | 299 | 1,255,375 | 9.3 | 2.6 | 0.0 | 3.3 | 12.5 | 6.9 | 63.1 | 18.3 | 3.1 | 11.9 | 3,918 | 3,639 |
| Smith | 9,437 | 33,803,613 | 7.5 | 10.5 | 4.0 | 2.3 | 8.3 | 2.9 | 63.7 | 757.5 | 230.3 | 392.1 | 1,726 | 1,338 |
| Somervell | 468 | 1,782,861 | 7.3 | 10.7 | 0.3 | 6.8 | 1.2 | 5.0 | 61.0 | 83.3 | 6.5 | 49.6 | 5,596 | 5,455 |
| Starr | 4,299 | 14,144,787 | 4.3 | 6.2 | 0.6 | 2.1 | 9.0 | 2.3 | 73.7 | 299.5 | 187.5 | 67.0 | 1,044 | 936 |
| Stephens | 561 | 1,797,939 | 7.3 | 8.3 | 1.8 | 2.5 | 27.9 | 3.9 | 46.7 | 31.2 | 10.1 | 17.0 | 1,826 | 1,650 |
| Sterling | 102 | 428,470 | 17.0 | 5.6 | 0.0 | 2.8 | 0.0 | 2.9 | 70.1 | 14.5 | 2.9 | 8.8 | 6,807 | 6,584 |
| Stonewall | 199 | 660,661 | 7.2 | 2.8 | 0.0 | 10.8 | 56.4 | 1.5 | 20.8 | 19.4 | 4.6 | 6.2 | 4,518 | 4,255 |
| Sutton | 378 | 1,571,493 | 11.6 | 4.4 | 6.6 | 4.0 | 20.3 | 2.6 | 46.2 | 24.4 | 5.2 | 10.7 | 2,820 | 2,537 |
| Swisher | 432 | 1,309,586 | 3.0 | 9.1 | 0.0 | 2.8 | 4.8 | 5.4 | 74.7 | 35.5 | 16.3 | 13.8 | 1,856 | 1,513 |
| Tarrant | 79,300 | 343,925,258 | 6.3 | 10.6 | 4.7 | 1.5 | 12.2 | 4.7 | 58.8 | 9,084.7 | 2,433.2 | 4,733.3 | 2,302 | 1,830 |
| Taylor | 5,441 | 20,363,924 | 5.9 | 13.8 | 5.9 | 1.8 | 5.4 | 5.6 | 59.9 | 483.8 | 184.6 | 231.7 | 1,696 | 1,201 |
| Terrell | 101 | 418,512 | 26.1 | 0.0 | 0.0 | 9.1 | 6.2 | 4.9 | 36.1 | 10.3 | 2.1 | 7.3 | 9,012 | 8,631 |
| Terry | 831 | 2,758,686 | 6.4 | 8.9 | 1.3 | 2.9 | 26.0 | 5.4 | 48.2 | 58.0 | 18.9 | 24.2 | 1,947 | 1,776 |
| Throckmorton | 116 | 372,158 | 5.3 | 0.0 | 0.0 | 2.2 | 36.6 | 6.1 | 49.7 | 5.5 | 1.9 | 2.5 | 1,654 | 1,477 |
| Titus | 2,435 | 8,602,385 | 2.7 | 4.7 | 1.4 | 1.1 | 26.8 | 2.3 | 60.0 | 208.6 | 103.6 | 52.9 | 1,621 | 1,340 |

1. Based on the resident population estimated as of July 1 of the year shown.

# Table B. States and Counties — Local Government Finances, Government Employment, and Income Taxes

| STATE County | Direct general expenditure — Total (mil dol) [185] | Per capita (dollars) [186] | Percent of total for: Education [187] | Health and hospitals [188] | Police protection [189] | Public welfare [190] | Highways [191] | Debt outstanding — Total (mil dol) [192] | Per capita (dollars) [193] | Government employment, 2019 — Federal civilian [194] | Federal military [195] | State and local [196] | Individual income tax returns, 2018 — Number of returns [197] | Mean adjusted gross income [198] | Mean income tax [199] |
|---|---|---|---|---|---|---|---|---|---|---|---|---|---|---|---|
| **TEXAS—Cont'd** | | | | | | | | | | | | | | | |
| Mason | 19 | 4,444 | 63.2 | 2.9 | 5.4 | 0.0 | 7.0 | 5.1 | 1,222 | 12 | 8 | 288 | 2,030 | 56,613 | 6,645 |
| Matagorda | 321 | 8,722 | 27.6 | 54.3 | 2.9 | 0.1 | 3.0 | 173.4 | 4,711 | 76 | 71 | 2,449 | 16,060 | 57,534 | 5,593 |
| Maverick | 427 | 7,349 | 31.0 | 1.8 | 2.3 | 0.1 | 2.6 | 224.9 | 3,871 | 894 | 112 | 5,320 | 25,560 | 38,818 | 2,231 |
| Medina | 212 | 4,224 | 65.5 | 11.8 | 3.1 | 0.4 | 2.6 | 220.7 | 4,397 | 62 | 96 | 3,156 | 21,050 | 60,235 | 6,518 |
| Menard | 11 | 5,091 | 39.6 | 0.0 | 5.6 | 24.8 | 3.0 | 2.7 | 1,267 | 7 | 4 | 207 | 910 | 45,954 | 5,457 |
| Midland | 908 | 5,490 | 36.5 | 36.3 | 4.2 | 0.0 | 2.2 | 941.9 | 5,698 | 568 | 343 | 8,455 | 79,140 | 122,910 | 21,443 |
| Milam | 80 | 3,212 | 59.8 | 2.2 | 5.7 | 0.8 | 8.0 | 79.5 | 3,188 | 56 | 48 | 1,273 | 10,450 | 50,704 | 4,672 |
| Mills | 30 | 6,183 | 47.8 | 0.2 | 14.2 | 0.4 | 5.5 | 19.6 | 3,987 | 14 | 9 | 360 | 2,100 | 47,877 | 4,169 |
| Mitchell | 60 | 7,252 | 35.7 | 42.1 | 2.8 | 0.1 | 3.9 | 54.9 | 6,680 | 21 | 13 | 942 | 2,770 | 56,506 | 5,491 |
| Montague | 68 | 3,494 | 53.8 | 13.4 | 5.2 | 0.4 | 5.8 | 40.5 | 2,087 | 53 | 38 | 1,145 | 8,730 | 59,745 | 5,688 |
| Montgomery | 2,275 | 3,980 | 59.1 | 4.8 | 5.3 | 0.1 | 4.6 | 5,114.7 | 8,948 | 1,131 | 1,183 | 28,773 | 261,170 | 99,471 | 15,006 |
| Moore | 119 | 5,492 | 40.2 | 31.6 | 4.3 | 0.1 | 3.1 | 60.4 | 2,798 | 67 | 41 | 1,648 | 9,410 | 44,248 | 3,803 |
| Morris | 89 | 7,169 | 22.5 | 0.0 | 71.9 | 0.2 | 0.3 | 12.5 | 1,011 | 31 | 24 | 643 | 5,270 | 43,722 | 3,495 |
| Motley | 6 | 5,090 | 43.2 | 2.7 | 7.1 | 0.7 | 7.5 | 1.0 | 794 | 8 | 4 | 105 | 480 | 39,831 | 2,894 |
| Nacogdoches | 281 | 4,295 | 37.1 | 32.9 | 4.6 | 0.2 | 3.0 | 211.3 | 3,234 | 139 | 126 | 5,361 | 25,440 | 53,298 | 5,460 |
| Navarro | 278 | 5,710 | 63.2 | 1.3 | 5.8 | 0.3 | 5.4 | 235.2 | 4,825 | 97 | 96 | 3,209 | 21,090 | 48,927 | 4,481 |
| Newton | 64 | 4,592 | 39.1 | 0.0 | 8.4 | 1.7 | 4.2 | 28.9 | 2,075 | 22 | 25 | 594 | 4,900 | 50,106 | 3,973 |
| Nolan | 135 | 9,058 | 36.0 | 36.3 | 3.8 | 0.1 | 2.8 | 100.5 | 6,764 | 40 | 28 | 1,579 | 6,360 | 49,176 | 4,695 |
| Nueces | 1,707 | 4,726 | 47.8 | 8.7 | 5.9 | 0.1 | 3.6 | 3,368.1 | 9,324 | 5,660 | 3,190 | 22,311 | 155,350 | 59,170 | 6,629 |
| Ochiltree | 43 | 4,342 | 57.4 | 0.7 | 8.1 | 0.0 | 8.2 | 28.6 | 2,859 | 22 | 19 | 803 | 4,250 | 65,746 | 7,084 |
| Oldham | 35 | 16,791 | 81.1 | 0.0 | 3.3 | 0.1 | 1.3 | 48.6 | 23,088 | 7 | 4 | 318 | 950 | 56,051 | 5,374 |
| Orange | 284 | 3,344 | 53.6 | 0.5 | 8.0 | 0.2 | 2.9 | 747.9 | 8,802 | 121 | 161 | 4,115 | 35,890 | 59,969 | 5,726 |
| Palo Pinto | 152 | 5,320 | 36.4 | 31.7 | 4.1 | 0.1 | 3.8 | 84.9 | 2,974 | 52 | 56 | 1,780 | 12,420 | 55,397 | 5,852 |
| Panola | 124 | 5,361 | 68.0 | 1.2 | 3.9 | 0.1 | 6.1 | 118.5 | 5,107 | 60 | 44 | 1,218 | 9,850 | 59,756 | 5,889 |
| Parker | 462 | 3,457 | 62.2 | 5.4 | 4.9 | 0.1 | 5.4 | 901.5 | 6,753 | 207 | 277 | 5,594 | 64,570 | 85,037 | 10,970 |
| Parmer | 65 | 6,666 | 43.6 | 24.0 | 4.2 | 0.6 | 5.8 | 14.2 | 1,463 | 66 | 19 | 834 | 4,020 | 42,408 | 3,520 |
| Pecos | 136 | 8,670 | 42.6 | 31.2 | 3.2 | 0.1 | 3.3 | 91.1 | 5,827 | 43 | 27 | 1,760 | 6,260 | 61,700 | 6,808 |
| Polk | 127 | 2,600 | 55.6 | 0.0 | 5.5 | 0.6 | 6.7 | 240.7 | 4,913 | 88 | 92 | 2,842 | 24,830 | 60,918 | 6,180 |
| Potter | 816 | 6,784 | 52.7 | 5.3 | 6.9 | 0.0 | 3.1 | 735.5 | 6,112 | 2,110 | 314 | 12,850 | 49,820 | 54,874 | 6,592 |
| Presidio | 34 | 4,771 | 56.6 | 1.0 | 3.4 | 0.1 | 6.2 | 22.5 | 3,171 | 314 | 13 | 582 | 3,620 | 45,435 | 4,491 |
| Rains | 33 | 2,808 | 54.8 | 3.3 | 4.5 | 1.0 | 8.7 | 31.4 | 2,675 | 20 | 24 | 506 | 4,760 | 53,086 | 4,812 |
| Randall | 150 | 1,118 | 56.6 | 0.1 | 7.6 | 0.1 | 2.0 | 225.7 | 1,684 | 200 | 295 | 4,573 | 62,980 | 71,222 | 8,117 |
| Reagan | 56 | 14,979 | 49.8 | 22.4 | 3.4 | 0.3 | 5.5 | 82.6 | 22,247 | 11 | 7 | 437 | 1,630 | 83,355 | 9,140 |
| Real | 10 | 2,852 | 39.0 | 0.5 | 9.5 | 2.1 | 11.1 | 7.8 | 2,272 | 6 | 7 | 203 | 1,590 | 50,381 | 4,676 |
| Red River | 45 | 3,679 | 53.4 | 10.9 | 4.1 | 0.8 | 6.1 | 17.1 | 1,408 | 39 | 23 | 701 | 5,140 | 45,404 | 4,029 |
| Reeves | 158 | 10,396 | 34.1 | 16.0 | 1.1 | 0.0 | 3.0 | 132.2 | 8,716 | 65 | 25 | 1,531 | 5,620 | 70,504 | 9,617 |
| Refugio | 53 | 7,340 | 69.6 | 8.5 | 2.2 | 0.4 | 2.2 | 58.2 | 8,109 | 27 | 13 | 668 | 3,140 | 51,976 | 4,725 |
| Roberts | 28 | 29,400 | 76.5 | 0.0 | 2.3 | 0.1 | 4.9 | 27.2 | 28,893 | 4 | 2 | 99 | 380 | 68,418 | 7,739 |
| Robertson | 76 | 4,408 | 66.3 | 0.0 | 5.5 | 1.4 | 5.4 | 36.6 | 2,136 | 36 | 33 | 990 | 7,440 | 50,587 | 4,500 |
| Rockwall | 388 | 4,008 | 59.8 | 0.2 | 5.3 | 0.2 | 5.6 | 1,023.9 | 10,574 | 143 | 203 | 3,988 | 46,150 | 99,416 | 13,529 |
| Runnels | 54 | 5,269 | 44.0 | 25.4 | 2.6 | 0.1 | | 15.5 | 1,509 | 42 | 20 | 893 | 4,570 | 47,875 | 3,835 |
| Rusk | 136 | 2,501 | 60.5 | 0.3 | 5.9 | 0.4 | 7.4 | 162.0 | 2,989 | 70 | 94 | 2,499 | 20,820 | 54,395 | 5,076 |
| Sabine | 27 | 2,573 | 60.5 | 9.5 | 3.2 | 0.2 | 2.7 | 39.0 | 3,741 | 46 | 20 | 522 | 4,200 | 55,725 | 5,296 |
| San Augustine | 20 | 2,380 | 64.0 | 5.8 | 9.3 | 0.0 | 0.5 | 18.8 | 2,265 | 19 | 16 | 439 | 3,250 | 43,314 | 3,219 |
| San Jacinto | 43 | 1,523 | 82.8 | 0.0 | 0.0 | 0.0 | 0.3 | 32.2 | 1,139 | 29 | 56 | 960 | 10,690 | 55,635 | 5,295 |
| San Patricio | 336 | 4,994 | 58.8 | 6.7 | 4.8 | 0.3 | 8.0 | 450.9 | 6,709 | 95 | 1,590 | 4,005 | 31,020 | 55,998 | 5,453 |
| San Saba | 24 | 3,946 | 51.6 | 0.6 | 4.8 | 1.0 | 9.3 | 11.3 | 1,880 | 20 | 10 | 486 | 2,420 | 50,183 | 5,176 |
| Schleicher | 9 | 2,999 | 73.5 | 0.3 | 1.4 | 0.0 | 3.5 | 9.7 | 3,250 | 12 | 5 | 249 | 1,120 | 52,596 | 4,800 |
| Scurry | 126 | 7,404 | 45.3 | 30.0 | 2.7 | 0.1 | 4.4 | 126.9 | 7,464 | 35 | 29 | 1,594 | 6,720 | 65,440 | 7,004 |
| Shackelford | 14 | 4,213 | 57.8 | 9.8 | 3.9 | 0.4 | 4.7 | 3.8 | 1,163 | 13 | 6 | 220 | 1,480 | 61,676 | 6,748 |
| Shelby | 87 | 3,435 | 71.7 | 0.0 | 4.2 | 0.2 | 3.7 | 64.7 | 2,565 | 77 | 49 | 1,317 | 10,140 | 51,775 | 4,815 |
| Sherman | 20 | 6,559 | 52.1 | 0.0 | 4.5 | 0.3 | 5.9 | 1.1 | 351 | 8 | 6 | 275 | 1,090 | 59,732 | 5,833 |
| Smith | 736 | 3,240 | 58.0 | 4.9 | 5.9 | 0.1 | 3.6 | 1,714.9 | 7,548 | 651 | 475 | 13,240 | 101,990 | 68,107 | 6,942 |
| Somervell | 70 | 7,895 | 48.5 | 28.0 | 4.1 | 0.6 | 2.7 | 54.8 | 6,191 | 13 | 17 | 882 | 3,980 | 68,042 | 8,213 |
| Starr | 287 | 4,479 | 67.3 | 9.5 | 3.9 | 0.0 | 4.3 | 251.8 | 3,926 | 824 | 124 | 5,171 | 25,810 | 35,257 | 1,819 |
| Stephens | 34 | 3,663 | 47.7 | 2.5 | 5.0 | 0.6 | 4.6 | 36.2 | 3,893 | 15 | 17 | 707 | 3,590 | 50,359 | 4,910 |
| Sterling | 15 | 11,739 | 50.6 | 2.8 | 2.6 | 18.8 | 3.7 | 21.8 | 16,963 | 4 | 2 | 159 | 580 | 89,866 | 12,971 |
| Stonewall | 17 | 12,636 | 16.9 | 63.3 | 1.7 | 0.0 | 7.0 | 10.2 | 7,440 | 12 | 3 | 220 | 590 | 55,486 | 4,303 |
| Sutton | 24 | 6,383 | 59.1 | 17.9 | 3.1 | 0.3 | 4.4 | 5.5 | 1,448 | 6 | 7 | 384 | 1,670 | 70,659 | 8,877 |
| Swisher | 36 | 4,837 | 48.0 | 16.0 | 5.4 | 0.3 | 3.5 | 7.6 | 1,025 | 21 | 13 | 751 | 2,810 | 45,816 | 4,216 |
| Tarrant | 9,003 | 4,378 | 45.2 | 12.1 | 7.0 | 0.1 | 3.7 | 19,751.6 | 9,605 | 15,206 | 5,572 | 97,451 | 951,560 | 76,276 | 9,953 |
| Taylor | 516 | 3,777 | 46.7 | 4.4 | 6.8 | 0.5 | 2.4 | 443.1 | 3,243 | 1,195 | 4,727 | 8,918 | 61,530 | 60,181 | 6,454 |
| Terrell | 10 | 11,811 | 40.4 | 6.4 | 12.8 | 0.0 | 13.8 | 11.9 | 14,661 | 38 | 2 | 111 | 310 | 50,368 | 4,384 |
| Terry | 63 | 5,044 | 49.3 | 30.5 | 1.8 | 0.5 | 2.8 | 31.3 | 2,519 | 25 | 22 | 1,013 | 4,640 | 45,114 | 3,847 |
| Throckmorton | 5 | 3,309 | 76.2 | 0.0 | 1.7 | 0.1 | 2.0 | 4.1 | 2,741 | 6 | 3 | 197 | 640 | 62,217 | 7,805 |
| Titus | 224 | 6,864 | 48.4 | 31.9 | 2.7 | 0.1 | 2.4 | 334.1 | 10,242 | 84 | 63 | 2,968 | 13,630 | 49,032 | 4,135 |

1. Based on the resident population estimated as of July 1 of the year shown.

# Table B. States and Counties — Land Area and Population

| State/ county code | CBSA code[1] | County Type code[2] | STATE County | Land area[3] (sq. mi) | Total persons 2019 | Rank | Per square mile | White | Black | American Indian, Alaska Native | Asian and Pacific Islander | Percent Hispanic or Latino[4] | Under 5 years | 5 to 14 years | 15 to 24 years | 25 to 34 years | 35 to 44 years | 45 to 54 years |
|---|---|---|---|---|---|---|---|---|---|---|---|---|---|---|---|---|---|---|
| | | | | 1 | 2 | 3 | 4 | 5 | 6 | 7 | 8 | 9 | 10 | 11 | 12 | 13 | 14 | 15 |
| | | | **TEXAS—Cont'd** | | | | | | | | | | | | | | | |
| 48451 | 41660 | 3 | Tom Green | 1,522.0 | 120,010 | 526 | 78.9 | 53.1 | 4.3 | 0.7 | 2.0 | 41.3 | 6.4 | 13.5 | 15.3 | 15.1 | 12.3 | 10.0 |
| 48453 | 12420 | 1 | Travis | 994.1 | 1,300,503 | 32 | 1,308.2 | 50.6 | 8.9 | 0.7 | 8.6 | 33.3 | 5.9 | 11.6 | 12.3 | 20.1 | 16.3 | 12.8 |
| 48455 | | 7 | Trinity | 693.8 | 14,883 | 2,097 | 21.5 | 78.9 | 9.4 | 1.1 | 0.7 | 11.3 | 4.9 | 11.4 | 9.6 | 9.7 | 9.9 | 11.5 |
| 48457 | | 6 | Tyler | 924.4 | 21,591 | 1,743 | 23.4 | 79.8 | 11.0 | 1.2 | 0.8 | 8.6 | 4.7 | 11.0 | 11.7 | 13.5 | 11.2 | 11.1 |
| 48459 | 30980 | 3 | Upshur | 583.0 | 42,166 | 1,137 | 72.3 | 81.9 | 8.6 | 1.5 | 1.0 | 9.1 | 5.7 | 13.6 | 11.8 | 11.5 | 11.9 | 11.6 |
| 48461 | | 8 | Upton | 1,241.3 | 3,623 | 2,924 | 2.9 | 40.8 | 3.3 | 1.6 | 0.5 | 55.1 | 7.0 | 16.1 | 12.9 | 12.8 | 13.3 | 11.0 |
| 48463 | 46620 | 6 | Uvalde | 1,551.9 | 26,742 | 1,537 | 17.2 | 25.6 | 0.9 | 0.4 | 1.2 | 72.5 | 7.2 | 15.0 | 14.9 | 13.6 | 11.3 | 10.7 |
| 48465 | 19620 | 5 | Val Verde | 3,144.8 | 49,028 | 1,008 | 15.6 | 15.1 | 1.7 | 0.5 | 1.1 | 82.4 | 7.9 | 15.7 | 15.4 | 15.1 | 11.7 | 10.8 |
| 48467 | | 6 | Van Zandt | 842.6 | 57,533 | 902 | 68.3 | 83.9 | 3.4 | 1.3 | 0.8 | 12.1 | 5.8 | 12.7 | 11.8 | 10.8 | 11.4 | 12.4 |
| 48469 | 47020 | 3 | Victoria | 882.1 | 91,936 | 643 | 104.2 | 44.3 | 6.4 | 0.5 | 1.6 | 48.3 | 6.7 | 14.2 | 13.5 | 13.8 | 12.4 | 10.5 |
| 48471 | 26660 | 4 | Walker | 784.2 | 72,164 | 756 | 92.0 | 57.3 | 23.5 | 0.7 | 1.6 | 18.2 | 4.1 | 8.3 | 21.3 | 14.5 | 13.2 | 13.1 |
| 48473 | 26420 | 1 | Waller | 513.3 | 57,452 | 904 | 111.9 | 43.5 | 23.8 | 0.7 | 1.8 | 31.5 | 6.3 | 13.7 | 23.9 | 11.2 | 11.3 | 10.4 |
| 48475 | | 6 | Ward | 835.6 | 12,097 | 2,289 | 14.5 | 39.2 | 5.0 | 1.0 | 0.7 | 55.6 | 8.3 | 16.2 | 13.4 | 13.8 | 13.4 | 10.5 |
| 48477 | 14780 | 6 | Washington | 604.2 | 35,771 | 1,284 | 59.2 | 64.5 | 17.3 | 0.6 | 1.8 | 17.0 | 5.4 | 12.2 | 14.7 | 10.1 | 10.6 | 10.4 |
| 48479 | 29700 | 2 | Webb | 3,361.5 | 277,681 | 252 | 82.6 | 3.7 | 0.4 | 0.1 | 0.5 | 95.4 | 8.7 | 18.0 | 16.7 | 13.8 | 12.3 | 11.7 |
| 48481 | 20900 | 4 | Wharton | 1,086.1 | 41,685 | 1,143 | 38.4 | 43.9 | 12.9 | 0.4 | 0.6 | 42.9 | 7.1 | 14.1 | 13.4 | 12.2 | 11.9 | 10.7 |
| 48483 | | 9 | Wheeler | 914.5 | 4,946 | 2,829 | 5.4 | 70.4 | 3.2 | 1.1 | 1.0 | 25.7 | 5.6 | 14.7 | 11.0 | 11.4 | 12.0 | 11.2 |
| 48485 | 48660 | 6 | Wichita | 627.6 | 133,205 | 486 | 212.2 | 66.2 | 11.6 | 1.4 | 3.1 | 20.2 | 6.1 | 12.8 | 16.7 | 14.8 | 12.1 | 10.2 |
| 48487 | 46900 | 6 | Wilbarger | 970.9 | 12,552 | 2,242 | 12.9 | 57.5 | 8.7 | 1.6 | 4.1 | 30.3 | 5.9 | 11.6 | 13.2 | 13.2 | 12.6 | 11.0 |
| 48489 | 39700 | 6 | Willacy | 590.6 | 21,161 | 1,764 | 35.8 | 8.8 | 2.1 | 0.2 | 0.8 | 88.4 | 5.9 | 13.2 | 17.0 | 16.7 | 13.1 | 10.3 |
| 48491 | 12420 | 1 | Williamson | 1,115.8 | 617,855 | 110 | 553.7 | 59.3 | 7.8 | 0.8 | 9.7 | 24.9 | 6.2 | 14.3 | 12.2 | 14.0 | 16.5 | 13.5 |
| 48493 | 41700 | 1 | Wilson | 803.7 | 52,023 | 968 | 64.7 | 56.8 | 1.9 | 0.8 | 0.9 | 40.6 | 5.5 | 13.7 | 12.4 | 11.3 | 12.9 | 13.4 |
| 48495 | | 6 | Winkler | 841.3 | 7,887 | 2,598 | 9.4 | 34.1 | 2.6 | 0.8 | 1.0 | 62.5 | 7.7 | 16.6 | 14.6 | 13.0 | 13.0 | 12.1 |
| 48497 | 19100 | 1 | Wise | 904.4 | 71,084 | 766 | 78.6 | 77.1 | 1.8 | 1.5 | 1.0 | 20.2 | 6.0 | 13.9 | 12.1 | 12.2 | 12.6 | 13.1 |
| 48499 | | 6 | Wood | 645.2 | 46,291 | 1,052 | 71.7 | 82.9 | 5.7 | 1.2 | 0.8 | 10.7 | 4.7 | 10.7 | 11.6 | 9.5 | 9.6 | 10.6 |
| 48501 | | 7 | Yoakum | 799.7 | 8,702 | 2,520 | 10.9 | 29.7 | 1.6 | 0.7 | 0.6 | 68.3 | 9.5 | 18.7 | 15.1 | 12.7 | 11.8 | 9.7 |
| 48503 | | 7 | Young | 914.5 | 17,904 | 1,926 | 19.6 | 78.1 | 2.0 | 1.1 | 1.0 | 19.2 | 5.6 | 13.5 | 11.5 | 11.4 | 11.6 | 11.1 |
| 48505 | 49820 | 6 | Zapata | 998.4 | 14,172 | 2,148 | 14.2 | 4.4 | 0.3 | 0.2 | 0.2 | 95.0 | 8.0 | 19.3 | 15.6 | 12.1 | 11.8 | 10.9 |
| 48507 | | 7 | Zavala | 1,297.4 | 11,840 | 2,302 | 9.1 | 5.2 | 0.7 | 0.3 | 0.2 | 93.8 | 7.0 | 16.5 | 16.5 | 13.8 | 11.6 | 10.5 |
| 49000 | | 0 | UTAH | 82,376.9 | 3,249,879 | X | 39.5 | 79.7 | 1.7 | 1.4 | 5.1 | 14.5 | 7.4 | 16.2 | 16.2 | 14.8 | 13.9 | 10.4 |
| 49001 | | 7 | Beaver | 2,582.9 | 6,761 | 2,681 | 2.6 | 85.3 | 0.7 | 1.4 | 1.9 | 12.2 | 7.7 | 18.1 | 15.0 | 10.0 | 12.9 | 9.4 |
| 49003 | 36260 | 2 | Box Elder | 5,745.6 | 57,007 | 907 | 9.9 | 88.2 | 0.8 | 1.2 | 1.8 | 9.8 | 7.7 | 17.9 | 14.3 | 12.6 | 13.8 | 9.9 |
| 49005 | 30860 | 3 | Cache | 1,164.7 | 130,004 | 495 | 111.6 | 85.0 | 1.2 | 1.0 | 3.6 | 10.9 | 7.9 | 16.8 | 23.9 | 13.8 | 11.8 | 8.3 |
| 49007 | 39220 | 7 | Carbon | 1,479.2 | 20,760 | 1,780 | 14.0 | 83.7 | 1.0 | 1.6 | 1.4 | 13.8 | 6.3 | 14.7 | 14.2 | 11.5 | 12.7 | 10.0 |
| 49009 | | 9 | Daggett | 697.0 | 1,026 | 3,107 | 1.5 | 94.2 | 0.9 | 1.0 | 1.5 | 4.5 | 4.7 | 13.1 | 10.2 | 7.1 | 12.9 | 11.0 |
| 49011 | 36260 | 2 | Davis | 299.1 | 359,232 | 204 | 1,201.0 | 85.1 | 1.9 | 0.8 | 4.5 | 10.4 | 7.7 | 18.0 | 14.8 | 13.8 | 15.1 | 10.6 |
| 49013 | | 7 | Duchesne | 3,235.3 | 19,894 | 1,830 | 6.1 | 87.0 | 0.7 | 5.0 | 1.4 | 8.3 | 8.5 | 19.1 | 13.0 | 12.4 | 14.1 | 9.7 |
| 49015 | | 7 | Emery | 4,462.3 | 10,147 | 2,406 | 2.3 | 91.9 | 0.7 | 1.4 | 1.1 | 6.2 | 6.2 | 16.5 | 13.3 | 10.3 | 12.7 | 10.2 |
| 49017 | | 9 | Garfield | 5,175.1 | 5,050 | 2,825 | 1.0 | 89.6 | 1.0 | 2.6 | 2.1 | 6.4 | 5.4 | 12.3 | 12.2 | 10.9 | 12.1 | 9.7 |
| 49019 | | 7 | Grand | 3,672.7 | 9,796 | 2,432 | 2.7 | 82.5 | 1.2 | 4.5 | 2.9 | 10.6 | 5.5 | 12.1 | 10.5 | 13.4 | 13.4 | 11.4 |
| 49021 | 16260 | 4 | Iron | 3,296.3 | 56,814 | 909 | 17.2 | 87.1 | 1.0 | 2.3 | 2.1 | 9.3 | 7.1 | 15.9 | 19.9 | 13.0 | 11.9 | 9.0 |
| 49023 | 39340 | 2 | Juab | 3,391.1 | 12,122 | 2,284 | 3.6 | 92.3 | 0.8 | 1.4 | 1.1 | 5.7 | 8.4 | 19.5 | 15.2 | 12.5 | 13.8 | 9.8 |
| 49025 | | 6 | Kane | 3,989.9 | 7,914 | 2,597 | 2.0 | 91.7 | 1.0 | 2.5 | 1.3 | 5.1 | 5.6 | 13.2 | 11.1 | 10.5 | 12.0 | 10.0 |
| 49027 | 36260 | 2 | Millard | 6,785.5 | 13,327 | 2,199 | 2.0 | 84.4 | 0.7 | 1.7 | 2.1 | 12.7 | 7.7 | 17.7 | 13.8 | 10.3 | 12.2 | 9.4 |
| 49029 | | 9 | Morgan | 609.2 | 12,462 | 2,250 | 20.5 | 95.5 | 0.7 | 0.6 | 1.2 | 3.1 | 6.4 | 21.3 | 15.7 | 8.8 | 14.2 | 11.4 |
| 49031 | | 8 | Piute | 758.4 | 1,473 | 3,082 | 1.9 | 90.4 | 1.1 | 0.7 | 1.1 | 8.1 | 4.3 | 12.4 | 14.1 | 7.8 | 9.7 | 9.2 |
| 49033 | | 1 | Rich | 1,028.8 | 2,452 | 3,006 | 2.4 | 92.5 | 0.7 | 0.8 | 0.4 | 6.5 | 6.9 | 17.0 | 12.0 | 10.4 | 12.3 | 9.5 |
| 49035 | 41620 | 1 | Salt Lake | 742.1 | 1,165,517 | 37 | 1,570.6 | 72.3 | 2.5 | 1.1 | 7.7 | 19.0 | 6.9 | 14.8 | 14.0 | 16.6 | 14.8 | 11.3 |
| 49037 | | 7 | San Juan | 7,819.8 | 15,278 | 2,072 | 2.0 | 46.6 | 0.7 | 47.5 | 1.2 | 6.1 | 6.7 | 16.7 | 15.2 | 12.3 | 11.0 | 10.8 |
| 49039 | | 6 | Sanpete | 1,589.9 | 31,393 | 1,394 | 19.7 | 87.4 | 1.3 | 1.4 | 2.1 | 9.3 | 6.0 | 13.4 | 19.3 | 12.6 | 13.0 | 10.6 |
| 49041 | | 7 | Sevier | 1,910.4 | 21,780 | 1,728 | 11.4 | 92.7 | 0.7 | 1.5 | 0.9 | 5.2 | 7.0 | 16.1 | 13.7 | 11.1 | 13.6 | 9.8 |
| 49043 | 25720 | 4 | Summit | 1,870.6 | 42,499 | 1,129 | 22.7 | 85.7 | 1.2 | 0.7 | 2.8 | 11.0 | 4.9 | 13.6 | 12.9 | 11.5 | 13.0 | 14.6 |
| 49045 | 41620 | 1 | Tooele | 6,941.9 | 74,512 | 742 | 10.7 | 83.5 | 1.3 | 1.3 | 2.4 | 13.4 | 7.5 | 18.6 | 14.4 | 13.7 | 15.4 | 11.5 |
| 49047 | 46860 | 7 | Uintah | 4,482.4 | 35,970 | 1,276 | 8.0 | 83.4 | 0.8 | 7.6 | 1.6 | 8.5 | 7.9 | 18.8 | 13.6 | 13.5 | 14.8 | 9.8 |
| 49049 | 39340 | 2 | Utah | 2,004.1 | 651,059 | 105 | 324.9 | 83.8 | 1.1 | 0.9 | 4.7 | 12.3 | 8.9 | 18.1 | 22.0 | 14.8 | 12.6 | 8.8 |
| 49051 | 25720 | 6 | Wasatch | 1,177.0 | 35,300 | 1,297 | 30.0 | 83.8 | 0.8 | 0.7 | 1.8 | 14.1 | 6.9 | 17.7 | 14.2 | 10.9 | 14.3 | 11.9 |
| 49053 | 41100 | 3 | Washington | 2,427.4 | 184,913 | 361 | 76.2 | 85.4 | 1.1 | 1.5 | 3.0 | 11.0 | 6.2 | 14.3 | 13.3 | 11.4 | 11.9 | 9.5 |
| 49055 | | 9 | Wayne | 2,461.0 | 2,759 | 2,984 | 1.1 | 91.4 | 0.8 | 1.3 | 1.4 | 6.6 | 5.8 | 12.3 | 12.6 | 15.3 | 14.1 | 10.8 |
| 49057 | 36260 | 2 | Weber | 576.3 | 262,658 | 267 | 455.8 | 77.6 | 1.9 | 1.1 | 3.1 | 18.7 | 7.2 | 15.5 | 14.3 | 15.3 | 14.1 | 10.8 |
| 50000 | | 0 | VERMONT | 9,217.9 | 623,347 | X | 67.6 | 94.2 | 1.9 | 1.2 | 2.6 | 2.1 | 4.5 | 10.3 | 13.7 | 12.1 | 11.6 | 12.2 |
| 50001 | | 6 | Addison | 766.1 | 36,851 | 1,254 | 48.1 | 94.2 | 1.9 | 0.9 | 2.6 | 2.5 | 4.1 | 9.0 | 16.5 | 10.4 | 11.0 | 12.1 |
| 50003 | 13540 | 6 | Bennington | 674.8 | 35,338 | 1,294 | 52.4 | 94.8 | 1.9 | 0.8 | 1.7 | 2.3 | 4.4 | 10.5 | 12.8 | 9.9 | 10.2 | 12.1 |
| 50005 | | 7 | Caledonia | 648.8 | 29,705 | 1,442 | 45.8 | 96.0 | 1.2 | 1.3 | 1.2 | 1.8 | 4.6 | 10.4 | 12.7 | 10.7 | 11.1 | 12.7 |

1. CBSA = Core Based Statistical Area. See Appendix A for explanation. See Appendix B for list of metropolitan areas with component counties.   2. County type code from the Economic Research Service of USDA Rural-Urban Continuum Codes. See Appendix A for definition.   3. Dry land or land partially or temporarily covered by water.   4. May be of any race.

Items 1—15

# Table B. States and Counties — Population and Households

| STATE County | Age (percent) (cont.) | | | | Total persons | | Percent change | | Components of change, 2010-2020 | | | Households, 2015-2019 | | Percent | | |
|---|---|---|---|---|---|---|---|---|---|---|---|---|---|---|---|---|
| | 55 to 64 years | 65 to 74 years | 75 years and over | Percent female | 2000 | 2010 | 2000-2010 | 2010-2020 | Births | Deaths | Net Migration | Number | Persons per household | Family households | Female family householder[1] | One person |
| | 16 | 17 | 18 | 19 | 20 | 21 | 22 | 23 | 24 | 25 | 26 | 27 | 28 | 29 | 30 | 31 |
| **TEXAS—Cont'd** | | | | | | | | | | | | | | | | |
| Tom Green | 11.2 | 9.3 | 6.9 | 50.1 | 104,010 | 110,220 | 6.0 | 8.9 | 15,914 | 10,665 | 4,502 | 43,314 | 2.58 | 63.9 | 11.5 | 29.7 |
| Travis | 10.4 | 6.8 | 3.7 | 49.5 | 812,280 | 1,024,314 | 26.1 | 27.0 | 160,931 | 55,798 | 168,651 | 472,361 | 2.54 | 57.0 | 9.6 | 31.2 |
| Trinity | 15.6 | 15.8 | 11.7 | 51.6 | 13,779 | 14,675 | 6.5 | 1.4 | 1,456 | 2,339 | 1,088 | 6,030 | 2.4 | 68.3 | 12.5 | 28.1 |
| Tyler | 13.5 | 12.9 | 10.4 | 46.6 | 20,871 | 21,760 | 4.3 | -0.8 | 2,113 | 2,782 | 520 | 7,100 | 2.66 | 67.2 | 13.4 | 29.8 |
| Upshur | 14.6 | 11.3 | 7.8 | 50.8 | 35,291 | 39,304 | 11.4 | 7.3 | 4,810 | 4,661 | 2,734 | 14,108 | 2.87 | 71.2 | 9.4 | 25.9 |
| Upton | 11.4 | 8.4 | 6.9 | 49.7 | 3,404 | 3,349 | -1.6 | 8.2 | 514 | 331 | 82 | 1,332 | 2.7 | 64.8 | 8.8 | 30.4 |
| Uvalde | 10.4 | 9.4 | 7.6 | 50.8 | 25,926 | 26,403 | 1.8 | 1.3 | 4,087 | 2,581 | -1,145 | 8,841 | 2.99 | 69.7 | 17.4 | 25.8 |
| Val Verde | 9.1 | 7.6 | 6.7 | 48.9 | 44,856 | 48,879 | 9.0 | 0.3 | 8,602 | 3,701 | -4,797 | 15,779 | 2.98 | 74.2 | 14.5 | 23.0 |
| Van Zandt | 14.3 | 11.9 | 8.9 | 51.1 | 48,140 | 52,549 | 9.2 | 9.5 | 6,163 | 6,721 | 5,566 | 20,156 | 2.7 | 72.9 | 11.4 | 22.2 |
| Victoria | 12.0 | 9.7 | 7.2 | 51.0 | 84,088 | 86,795 | 3.2 | 5.9 | 12,970 | 8,237 | 393 | 32,255 | 2.82 | 69.5 | 13.7 | 25.2 |
| Walker | 11.2 | 8.5 | 5.7 | 42.5 | 61,758 | 67,862 | 9.9 | 6.3 | 6,323 | 5,475 | 3,413 | 21,963 | 2.5 | 59.8 | 10.5 | 28.7 |
| Waller | 11.0 | 7.5 | 4.7 | 50.4 | 32,663 | 43,319 | 32.6 | 32.6 | 6,391 | 3,196 | 10,868 | 15,171 | 3.11 | 71.5 | 11.2 | 20.0 |
| Ward | 11.1 | 7.3 | 6.1 | 49.4 | 10,909 | 10,658 | -2.3 | 13.5 | 1,871 | 1,122 | 674 | 3,978 | 2.9 | 70.9 | 9.3 | 23.8 |
| Washington | 13.7 | 12.3 | 10.5 | 51.2 | 30,373 | 33,687 | 10.9 | 6.2 | 4,056 | 3,938 | 1,978 | 12,625 | 2.6 | 68.1 | 10.5 | 28.3 |
| Webb | 8.8 | 5.9 | 4.1 | 50.7 | 193,117 | 250,304 | 29.6 | 10.9 | 53,126 | 13,615 | -12,134 | 74,789 | 3.61 | 80.1 | 21.2 | 17.3 |
| Wharton | 12.6 | 10.1 | 7.7 | 50.8 | 41,188 | 41,280 | 0.2 | 1.0 | 5,799 | 4,353 | -1,021 | 15,199 | 2.7 | 69.4 | 16.0 | 27.7 |
| Wheeler | 13.8 | 11.5 | 8.8 | 49.9 | 5,284 | 5,406 | 2.3 | -8.5 | 703 | 691 | -483 | 2,183 | 2.42 | 69.5 | 12.4 | 26.2 |
| Wichita | 12.0 | 8.7 | 6.6 | 48.3 | 131,664 | 131,659 | 0.0 | 1.2 | 17,489 | 13,928 | -1,980 | 48,356 | 2.41 | 63.0 | 12.9 | 31.0 |
| Wilbarger | 13.5 | 10.6 | 8.4 | 50.7 | 14,676 | 13,535 | -7.8 | -7.3 | 1,690 | 1,607 | -1,079 | 5,180 | 2.34 | 62.8 | 13.4 | 32.9 |
| Willacy | 9.2 | 8.1 | 6.4 | 44.7 | 20,082 | 22,137 | 10.2 | -4.4 | 2,969 | 1,751 | -2,206 | 5,782 | 3.52 | 78.2 | 21.0 | 18.0 |
| Williamson | 10.4 | 7.9 | 5.0 | 50.6 | 249,967 | 422,771 | 69.1 | 46.1 | 65,713 | 26,158 | 154,183 | 180,160 | 3.02 | 72.1 | 9.8 | 22.4 |
| Wilson | 14.0 | 10.3 | 6.5 | 49.8 | 32,408 | 42,907 | 32.4 | 21.2 | 5,249 | 3,974 | 7,795 | 15,733 | 3.09 | 79.5 | 9.0 | 17.2 |
| Winkler | 11.1 | 7.2 | 4.7 | 48.1 | 7,173 | 7,106 | -0.9 | 11.0 | 1,264 | 657 | 155 | 2,619 | 2.97 | 70.7 | 10.3 | 26.8 |
| Wise | 14.3 | 9.5 | 6.1 | 50.0 | 48,793 | 59,082 | 21.1 | 20.3 | 7,957 | 5,862 | 9,916 | 22,369 | 2.92 | 75.3 | 10.0 | 21.3 |
| Wood | 15.4 | 15.8 | 12.1 | 50.4 | 36,752 | 41,973 | 14.2 | 10.3 | 4,160 | 6,454 | 6,587 | 16,510 | 2.59 | 70.5 | 9.7 | 26.7 |
| Yoakum | 10.5 | 6.8 | 5.2 | 48.9 | 7,322 | 7,879 | 7.6 | 10.4 | 1,611 | 613 | -178 | 2,617 | 3.3 | 75.9 | 7.0 | 20.9 |
| Young | 14.2 | 11.9 | 9.1 | 50.2 | 17,943 | 18,550 | 3.4 | -3.5 | 2,197 | 2,634 | -201 | 7,307 | 2.43 | 70.9 | 10.2 | 25.4 |
| Zapata | 8.6 | 7.7 | 5.8 | 50.1 | 12,182 | 14,018 | 15.1 | 1.1 | 2,692 | 951 | -1,594 | 4,503 | 3.17 | 72.9 | 18.6 | 22.7 |
| Zavala | 9.4 | 8.5 | 6.1 | 48.9 | 11,600 | 11,677 | 0.7 | 1.4 | 1,891 | 995 | -732 | 3,571 | 3.33 | 80.7 | 23.7 | 16.0 |
| **UTAH** | 9.4 | 7.1 | 4.7 | 49.6 | 2,233,169 | 2,763,891 | 23.8 | 17.6 | 512,904 | 174,989 | 148,583 | 977,313 | 3.12 | 74.6 | 9.2 | 19.4 |
| Beaver | 11.7 | 9.3 | 5.9 | 49.0 | 6,005 | 6,629 | 10.4 | 2.0 | 1,121 | 583 | -415 | 2,282 | 2.83 | 73.6 | 6.0 | 24.9 |
| Box Elder | 10.6 | 7.7 | 5.6 | 49.2 | 42,745 | 49,983 | 16.9 | 14.1 | 8,758 | 3,836 | 2,113 | 17,569 | 3.05 | 78.4 | 6.5 | 19.4 |
| Cache | 7.4 | 5.9 | 4.2 | 50.1 | 91,391 | 112,656 | 23.3 | 15.4 | 23,488 | 5,499 | -667 | 38,393 | 3.16 | 75.2 | 7.8 | 16.2 |
| Carbon | 12.0 | 11.4 | 7.0 | 50.3 | 20,422 | 21,403 | 4.8 | -3.0 | 2,825 | 2,270 | -1,218 | 7,829 | 2.54 | 68.1 | 10.9 | 27.8 |
| Daggett | 14.5 | 16.5 | 10.0 | 44.1 | 921 | 1,061 | 15.2 | -3.3 | 100 | 91 | -52 | 163 | 3.09 | 77.3 | 0.6 | 22.7 |
| Davis | 9.3 | 6.5 | 4.2 | 49.4 | 238,994 | 306,492 | 28.2 | 17.2 | 57,688 | 17,311 | 12,454 | 104,551 | 3.28 | 81.0 | 9.0 | 15.5 |
| Duchesne | 10.5 | 7.4 | 5.3 | 49.4 | 14,371 | 18,605 | 29.5 | 6.9 | 4,024 | 1,450 | -1,302 | 6,868 | 2.88 | 77.6 | 9.7 | 18.5 |
| Emery | 12.5 | 11.1 | 7.1 | 49.2 | 10,860 | 10,976 | 1.1 | -7.6 | 1,436 | 953 | -1,331 | 3,665 | 2.74 | 74.7 | 5.6 | 23.3 |
| Garfield | 13.3 | 14.9 | 9.3 | 48.2 | 4,735 | 5,172 | 9.2 | -2.4 | 598 | 483 | -242 | 1,760 | 2.69 | 73.1 | 10.6 | 22.4 |
| Grand | 14.0 | 12.5 | 7.3 | 50.2 | 8,485 | 9,212 | 8.6 | 6.3 | 1,191 | 832 | 225 | 4,191 | 2.28 | 64.4 | 13.6 | 30.2 |
| Iron | 9.5 | 8.5 | 5.2 | 50.0 | 33,779 | 46,163 | 36.7 | 23.1 | 8,582 | 3,176 | 5,229 | 17,015 | 2.97 | 72.6 | 8.8 | 20.5 |
| Juab | 8.7 | 7.4 | 4.7 | 48.6 | 8,238 | 10,246 | 24.4 | 18.3 | 1,911 | 761 | 725 | 3,464 | 3.22 | 81.4 | 12.4 | 16.1 |
| Kane | 13.7 | 14.6 | 9.4 | 50.1 | 6,046 | 7,125 | 17.8 | 11.1 | 833 | 771 | 708 | 2,643 | 2.77 | 56.9 | 4.7 | 33.8 |
| Millard | 11.1 | 10.0 | 7.8 | 48.8 | 12,405 | 12,503 | 0.8 | 6.6 | 1,959 | 1,030 | -110 | 4,337 | 2.92 | 78.8 | 5.9 | 19.1 |
| Morgan | 9.8 | 7.4 | 5.0 | 48.4 | 7,129 | 9,469 | 32.8 | 31.6 | 1,583 | 588 | 1,997 | 3,365 | 3.47 | 88.4 | 3.9 | 9.9 |
| Piute | 13.2 | 16.6 | 12.7 | 48.7 | 1,435 | 1,557 | 8.5 | -5.4 | 146 | 162 | -64 | 532 | 3.44 | 71.2 | 3.9 | 27.6 |
| Rich | 12.0 | 11.9 | 8.0 | 47.5 | 1,961 | 2,264 | 15.5 | 8.3 | 337 | 151 | 1 | 635 | 3.75 | 78.3 | 7.4 | 20.8 |
| Salt Lake | 9.9 | 7.1 | 4.4 | 49.8 | 898,387 | 1,029,590 | 14.6 | 13.2 | 177,462 | 66,162 | 25,221 | 374,820 | 2.99 | 69.9 | 10.3 | 22.7 |
| San Juan | 12.4 | 8.7 | 6.2 | 50.1 | 14,413 | 14,757 | 2.4 | 3.5 | 2,266 | 1,097 | -648 | 4,311 | 3.46 | 74.5 | 15.5 | 22.8 |
| Sanpete | 10.2 | 9.0 | 5.9 | 47.3 | 22,763 | 27,822 | 22.2 | 12.8 | 3,952 | 1,947 | 1,566 | 8,643 | 3.1 | 76.7 | 4.8 | 19.9 |
| Sevier | 11.5 | 10.1 | 7.1 | 48.9 | 18,842 | 20,800 | 10.4 | 4.7 | 3,117 | 2,044 | -93 | 7,364 | 2.84 | 75.6 | 9.8 | 20.8 |
| Summit | 15.2 | 9.7 | 4.6 | 48.7 | 29,736 | 36,324 | 22.2 | 17.0 | 4,351 | 1,509 | 3,341 | 14,220 | 2.87 | 75.2 | 7.1 | 19.0 |
| Tooele | 9.3 | 6.0 | 3.4 | 49.2 | 40,735 | 58,218 | 42.9 | 28.0 | 10,089 | 3,653 | 9,885 | 20,478 | 3.27 | 80.1 | 9.1 | 16.8 |
| Uintah | 9.8 | 7.4 | 4.5 | 49.4 | 25,224 | 32,589 | 29.2 | 10.4 | 6,509 | 2,373 | -847 | 10,569 | 3.38 | 72.4 | 10.3 | 23.6 |
| Utah | 6.7 | 4.7 | 3.3 | 49.4 | 368,536 | 516,639 | 40.2 | 26.0 | 121,760 | 24,189 | 36,979 | 165,991 | 3.55 | 81.3 | 7.6 | 11.9 |
| Wasatch | 11.4 | 8.3 | 4.3 | 49.1 | 15,215 | 23,525 | 54.6 | 50.1 | 4,308 | 1,361 | 8,742 | 9,879 | 3.19 | 76.4 | 4.5 | 17.9 |
| Washington | 10.6 | 12.5 | 10.2 | 50.4 | 90,354 | 138,115 | 52.9 | 33.9 | 22,746 | 12,994 | 36,905 | 57,144 | 2.87 | 74.2 | 8.6 | 21.1 |
| Wayne | 14.9 | 14.6 | 8.4 | 49.0 | 2,509 | 2,778 | 10.7 | -0.7 | 305 | 249 | -77 | 1,000 | 2.64 | 67.4 | 3.9 | 30.1 |
| Weber | 10.5 | 7.4 | 4.8 | 49.6 | 196,533 | 231,218 | 17.6 | 13.6 | 39,459 | 17,464 | 9,558 | 83,632 | 2.97 | 73.0 | 10.7 | 22.0 |
| **VERMONT** | 15.0 | 12.5 | 8.2 | 50.6 | 608,827 | 625,727 | 2.8 | -0.4 | 59,672 | 59,104 | -2,579 | 260,029 | 2.3 | 60.1 | 8.6 | 28.7 |
| Addison | 15.2 | 13.4 | 8.3 | 50.1 | 35,974 | 36,809 | 2.3 | 0.1 | 3,154 | 3,233 | 143 | 14,584 | 2.32 | 63.4 | 8.3 | 31.6 |
| Bennington | 16.2 | 13.7 | 10.2 | 51.3 | 36,994 | 37,119 | 0.3 | -4.8 | 3,371 | 4,641 | -492 | 14,581 | 2.35 | 61.8 | 10.7 | 30.0 |
| Caledonia | 15.4 | 14.0 | 8.4 | 50.0 | 29,702 | 31,231 | 5.1 | -4.9 | 2,860 | 3,203 | -1,174 | 12,434 | 2.33 | 61.6 | 8.0 | 30.0 |

1. No spouse present.

| STATE County | Persons in group quarters, 2020 | Daytime Population, 2015-2019 | | Births, 2020 | | Deaths, 2020 | | Persons under 65 with no health insurance, 2019 | | Medicare, 2020 | | | Crimes reported by county police or sheriff, 2019 | |
|---|---|---|---|---|---|---|---|---|---|---|---|---|---|---|
| | | Number | Employment/ residence ratio | Total | Rate[1] | Number | Rate[1] | Number | Percent | Total beneficiaries | Enrolled in Original Medicare | Enrolled in Medicare Advantage | Violent | Property |
| | 32 | 33 | 34 | 35 | 36 | 37 | 38 | 39 | 40 | 41 | 42 | 43 | 44 | 45 |
| **TEXAS—Cont'd** | | | | | | | | | | | | | | |
| Tom Green | 4,984 | 117,603 | 0.99 | 1,465 | 12.2 | 1,097 | 9.1 | 16,924 | 17.6 | 22,364 | 14,341 | 8,023 | 32 | 254 |
| Travis | 22,367 | 1,361,935 | 1.2 | 15,463 | 11.9 | 6,880 | 5.3 | 185,356 | 16.5 | 140,541 | 91,028 | 49,513 | 790 | 3,240 |
| Trinity | 153 | 12,604 | 0.6 | 138 | 9.3 | 221 | 14.8 | 2,283 | 21.5 | 4,254 | 2,496 | 1,758 | NA | NA |
| Tyler | 2,287 | 20,123 | 0.78 | 183 | 8.5 | 277 | 12.8 | 2,983 | 20.1 | 5,107 | 3,248 | 1,859 | 67 | 252 |
| Upshur | 465 | 33,031 | 0.52 | 438 | 10.4 | 467 | 11.1 | 6,255 | 18.6 | 8,930 | 5,481 | 3,449 | NA | NA |
| Upton | 63 | 4,435 | 1.54 | 38 | 10.5 | 22 | 6.1 | 630 | 20.5 | 587 | 400 | 187 | 2 | 31 |
| Uvalde | 529 | 27,331 | 1.04 | 383 | 14.3 | 277 | 10.4 | 5,374 | 24.9 | 5,435 | 3,385 | 2,049 | 16 | 85 |
| Val Verde | 1,892 | 47,767 | 0.94 | 752 | 15.3 | 396 | 8.1 | 9,402 | 23.6 | 8,585 | 5,235 | 3,350 | 29 | 87 |
| Van Zandt | 532 | 47,701 | 0.66 | 650 | 11.3 | 726 | 12.6 | 10,703 | 23.9 | 13,046 | 7,938 | 5,108 | NA | NA |
| Victoria | 1,902 | 94,448 | 1.06 | 1,185 | 12.9 | 918 | 10.0 | 15,208 | 20.2 | 17,809 | 11,014 | 6,795 | 94 | 399 |
| Walker | 16,621 | 74,438 | 1.08 | 578 | 8.0 | 687 | 9.5 | 8,758 | 19.2 | 10,443 | 5,737 | 4,706 | 52 | 186 |
| Waller | 4,611 | 50,305 | 0.93 | 683 | 11.9 | 386 | 6.7 | 10,533 | 23.9 | 7,780 | 4,533 | 3,246 | NA | NA |
| Ward | 124 | 12,528 | 1.18 | 216 | 17.9 | 82 | 6.8 | 2,180 | 21.2 | 1,867 | 1,318 | 550 | 27 | 218 |
| Washington | 2,293 | 37,482 | 1.15 | 363 | 10.1 | 435 | 12.2 | 5,103 | 19.7 | 8,758 | 6,156 | 2,601 | 50 | 140 |
| Webb | 3,647 | 272,213 | 0.99 | 4,638 | 16.7 | 1,482 | 5.3 | 73,316 | 30.2 | 33,804 | 20,866 | 12,938 | NA | NA |
| Wharton | 477 | 39,799 | 0.9 | 598 | 14.3 | 430 | 10.3 | 8,663 | 25.6 | 8,499 | 5,594 | 2,905 | 68 | 230 |
| Wheeler | 45 | 5,555 | 1.09 | 42 | 8.5 | 63 | 12.7 | 1,058 | 26.2 | 1,114 | 892 | 223 | 5 | 11 |
| Wichita | 12,698 | 136,139 | 1.07 | 1,570 | 11.8 | 1,440 | 10.8 | 18,616 | 18.6 | 24,950 | 19,293 | 5,658 | 38 | 126 |
| Wilbarger | 632 | 13,323 | 1.08 | 164 | 13.1 | 153 | 12.2 | 2,192 | 21.8 | 2,811 | 1,937 | 874 | 2 | 13 |
| Willacy | 3,303 | 19,715 | 0.75 | 242 | 11.4 | 202 | 9.5 | 3,443 | 23.4 | 3,630 | 1,612 | 2,017 | 24 | 123 |
| Williamson | 4,187 | 477,247 | 0.74 | 6,979 | 11.3 | 3,288 | 5.3 | 63,856 | 12.4 | 83,220 | 52,941 | 30,279 | 152 | 1,377 |
| Wilson | 441 | 38,808 | 0.52 | 514 | 9.9 | 456 | 8.8 | 7,054 | 16.6 | 9,489 | 5,541 | 3,948 | 20 | 228 |
| Winkler | 97 | 9,095 | 1.38 | 133 | 16.9 | 49 | 6.2 | 1,458 | 21.0 | 1,064 | 765 | 299 | 9 | 80 |
| Wise | 770 | 59,277 | 0.76 | 794 | 11.2 | 693 | 9.7 | 12,717 | 21.6 | 12,283 | 7,557 | 4,726 | 83 | 323 |
| Wood | 1,674 | 41,665 | 0.83 | 392 | 8.5 | 734 | 15.9 | 6,987 | 22.2 | 14,805 | 9,448 | 5,357 | 75 | 225 |
| Yoakum | 54 | 9,142 | 1.14 | 180 | 20.7 | 49 | 5.6 | 2,004 | 26.2 | 1,237 | 861 | 376 | 3 | 18 |
| Young | 267 | 18,272 | 1.03 | 181 | 10.1 | 246 | 13.7 | 3,309 | 23.5 | 4,376 | 3,554 | 823 | 12 | 35 |
| Zapata | 32 | 14,058 | 0.95 | 200 | 14.1 | 115 | 8.1 | 3,521 | 29.3 | 2,007 | 1,275 | 732 | 52 | 183 |
| Zavala | 412 | 11,092 | 0.78 | 163 | 13.8 | 87 | 7.3 | 2,076 | 21.9 | 2,058 | 969 | 1,089 | 12 | 32 |
| **UTAH** | 48,669 | 3,097,665 | 1.0 | 47,611 | 14.7 | 19,080 | 5.9 | 298,395 | 10.7 | 414,730 | 248,617 | 166,113 | X | X |
| Beaver | 23 | 6,545 | 1.01 | 106 | 15.7 | 55 | 8.1 | 747 | 13.2 | 1,263 | 1,159 | 105 | NA | NA |
| Box Elder | 331 | 50,459 | 0.85 | 827 | 14.5 | 409 | 7.2 | 4,863 | 10.1 | 8,637 | 4,999 | 3,638 | 12 | 131 |
| Cache | 3,932 | 123,774 | 0.99 | 2,093 | 16.1 | 601 | 4.6 | 10,396 | 9.4 | 14,643 | 7,294 | 7,350 | 32 | 150 |
| Carbon | 526 | 20,559 | 1.03 | 249 | 12.0 | 195 | 9.4 | 1,611 | 10.0 | 4,515 | 3,735 | 780 | 10 | 93 |
| Daggett | 0 | 617 | 1.01 | 6 | 5.8 | 7 | 6.8 | 64 | 9.3 | 243 | 187 | 57 | NA | NA |
| Davis | 2,990 | 309,184 | 0.78 | 5,223 | 14.5 | 1,859 | 5.2 | 25,911 | 8.1 | 42,003 | 25,339 | 16,664 | 50 | 213 |
| Duchesne | 283 | 20,445 | 1.04 | 336 | 16.9 | 143 | 7.2 | 2,526 | 14.8 | 3,076 | 2,322 | 754 | NA | NA |
| Emery | 43 | 9,638 | 0.88 | 121 | 11.9 | 101 | 10.0 | 806 | 9.9 | 2,101 | 1,766 | 335 | NA | NA |
| Garfield | 181 | 4,940 | 0.97 | 55 | 10.9 | 35 | 6.9 | 632 | 17.4 | 1,199 | 1,054 | 144 | NA | NA |
| Grand | 143 | 10,305 | 1.14 | 97 | 9.9 | 80 | 8.2 | 1,265 | 16.4 | 1,999 | 1,936 | 63 | 16 | 79 |
| Iron | 1,143 | 50,080 | 0.95 | 825 | 14.5 | 347 | 6.1 | 5,683 | 12.3 | 8,754 | 6,273 | 2,481 | 49 | 119 |
| Juab | 122 | 10,173 | 0.77 | 192 | 15.8 | 92 | 7.6 | 1,158 | 11.1 | 1,703 | 1,412 | 291 | 3 | 54 |
| Kane | 255 | 7,965 | 1.15 | 86 | 10.9 | 80 | 10.1 | 586 | 10.2 | 1,999 | 1,893 | 105 | 7 | 24 |
| Millard | 177 | 12,916 | 1.01 | 180 | 13.5 | 104 | 7.8 | 1,581 | 15.0 | 2,449 | 2,161 | 288 | 20 | 213 |
| Morgan | 0 | 9,299 | 0.51 | 133 | 10.7 | 53 | 4.3 | 853 | 7.9 | 1,646 | 1,039 | 607 | 4 | NA |
| Piute | 37 | 1,660 | 0.65 | 20 | 13.6 | 10 | 6.8 | 132 | 13.2 | 422 | 372 | 50 | NA | NA |
| Rich | 1 | 2,255 | 0.82 | 29 | 11.8 | 11 | 4.5 | 213 | 10.6 | 450 | 326 | 125 | 3 | 18 |
| Salt Lake | 14,737 | 1,220,128 | 1.15 | 16,105 | 13.8 | 7,338 | 6.3 | 115,752 | 11.4 | 145,655 | 78,066 | 67,589 | 731 | 8,002 |
| San Juan | 336 | 14,909 | 0.93 | 180 | 11.8 | 135 | 8.8 | 2,197 | 17.5 | 2,257 | 2,204 | 38 | 6 | 38 |
| Sanpete | 2,947 | 27,944 | 0.84 | 374 | 11.9 | 206 | 6.6 | 3,483 | 14.8 | 4,971 | 3,945 | 1,026 | 21 | 94 |
| Sevier | 309 | 21,661 | 1.04 | 281 | 12.9 | 223 | 10.2 | 2,073 | 11.8 | 4,292 | 3,566 | 726 | 8 | 98 |
| Summit | 106 | 45,759 | 1.21 | 392 | 9.2 | 175 | 4.1 | 3,728 | 10.1 | 6,332 | 4,243 | 2,090 | 33 | 399 |
| Tooele | 326 | 55,828 | 0.62 | 1,004 | 13.5 | 402 | 5.4 | 6,823 | 10.5 | 8,207 | 5,414 | 2,793 | 12 | 218 |
| Uintah | 276 | 35,717 | 0.97 | 543 | 15.1 | 260 | 7.2 | 5,076 | 16.2 | 4,759 | 3,477 | 1,283 | 55 | 158 |
| Utah | 14,075 | 588,263 | 0.94 | 11,840 | 18.2 | 2,695 | 4.1 | 53,785 | 9.4 | 58,318 | 31,774 | 26,544 | 87 | 323 |
| Wasatch | 250 | 26,433 | 0.66 | 416 | 11.8 | 160 | 4.5 | 3,419 | 11.5 | 4,571 | 2,700 | 1,871 | 31 | 91 |
| Washington | 2,259 | 166,896 | 1.02 | 2,213 | 12.0 | 1,511 | 8.2 | 20,380 | 15.0 | 40,871 | 27,164 | 13,706 | 28 | 138 |
| Wayne | 9 | 2,713 | 1.02 | 24 | 8.7 | 20 | 7.2 | 327 | 15.7 | 663 | 587 | 77 | NA | NA |
| Weber | 2,852 | 240,600 | 0.91 | 3,661 | 13.9 | 1,773 | 6.8 | 22,325 | 9.9 | 36,734 | 22,215 | 14,519 | 94 | 963 |
| **VERMONT** | 25,691 | 622,480 | 0.99 | 5,331 | 8.6 | 6,129 | 9.8 | 27,121 | 5.7 | 151,251 | 129,392 | 21,859 | X | X |
| Addison | 2,965 | 34,083 | 0.86 | 288 | 7.8 | 349 | 9.5 | 1,650 | 6.2 | 8,902 | 7,153 | 1,749 | NA | NA |
| Bennington | 1,473 | 37,077 | 1.08 | 303 | 8.6 | 474 | 13.4 | 1,598 | 6.1 | 10,078 | 8,571 | 1,507 | 2 | 6 |
| Caledonia | 1,186 | 28,067 | 0.85 | 256 | 8.6 | 321 | 10.8 | 1,320 | 5.9 | 7,780 | 6,804 | 976 | 2 | 1 |

1. Per 1,000 estimated resident population.

# Table B. States and Counties — Crime, Education, Money Income, and Poverty

| STATE County | County law enforcement employment, 2019 | | Education — School enrollment and attainment, 2015-2019 | | | | Local government expenditures,[3] 2017-2018 | | Money income, 2015-2019 | | | | Income and poverty, 2019 | | | |
|---|---|---|---|---|---|---|---|---|---|---|---|---|---|---|---|---|
| | | | Enrollment[1] | | Attainment[2] (percent) | | | | | | Households — Percent | | | Percent below poverty level | | |
| | Officers | Civilians | Total | Percent private | High school graduate or less | Bachelor's degree or more | Total current spending (mil dol) | Current spending per student (dollars) | Per capita income[4] | Median income (dollars) | with income of less than $50,000 | with income of $200,000 or more | Median household income (dollars) | All persons | Children under 18 years | Children 5 to 17 years in families |
| | 46 | 47 | 48 | 49 | 50 | 51 | 52 | 53 | 54 | 55 | 56 | 57 | 58 | 59 | 60 | 61 |
| **TEXAS—Cont'd** | | | | | | | | | | | | | | | | |
| Tom Green | 50 | 124 | 31,182 | 9.7 | 44.0 | 24.5 | 192 | 9,058 | 29,505 | 53,903 | 46.9 | 4.4 | 54,774 | 12.9 | 17.4 | 15.5 |
| Travis | 357 | 1,371 | 312,415 | 14.4 | 27.2 | 50.0 | 1,669 | 10,006 | 43,658 | 75,887 | 32.4 | 12.3 | 80,690 | 10.8 | 13.6 | 12.2 |
| Trinity | 13 | 4 | 2,615 | 8.1 | 52.8 | 12.9 | 24 | 10,506 | 25,026 | 41,357 | 57.1 | 1.8 | 43,788 | 17.9 | 29.8 | 28.9 |
| Tyler | NA | NA | 3,642 | 4.7 | 61.0 | 13.4 | 37 | 9,987 | 20,249 | 44,497 | 53.2 | 1.5 | 48,254 | 17.2 | 23.3 | 22.1 |
| Upshur | NA | NA | 9,667 | 15.9 | 48.6 | 17.1 | 75 | 10,100 | 25,285 | 52,162 | 48.4 | 3.5 | 53,078 | 14.9 | 21.1 | 20.7 |
| Upton | NA | NA | 946 | 1.8 | 65.6 | 9.6 | 13 | 15,038 | 25,447 | 52,171 | 47.5 | 3.2 | 58,908 | 13.8 | 18.8 | 17.7 |
| Uvalde | 32 | 41 | 6,885 | 6.3 | 53.5 | 17.9 | 59 | 10,505 | 20,329 | 41,679 | 59.4 | 2.3 | 44,690 | 19.8 | 29.9 | 29.1 |
| Val Verde | 46 | 24 | 13,166 | 8.5 | 56.2 | 18.4 | 101 | 9,379 | 21,370 | 46,147 | 54.0 | 2.6 | 45,034 | 20.8 | 28.9 | 26.2 |
| Van Zandt | 40 | 42 | 12,058 | 10.2 | 49.6 | 15.8 | 90 | 8,853 | 27,900 | 54,654 | 44.9 | 4.0 | 55,554 | 13.6 | 19.5 | 17.8 |
| Victoria | 110 | 95 | 22,771 | 13.0 | 47.1 | 20.0 | 157 | 10,281 | 29,199 | 56,834 | 44.3 | 5.0 | 57,344 | 14.3 | 22.6 | 20.7 |
| Walker | 35 | 37 | 20,687 | 4.4 | 53.8 | 20.8 | 89 | 9,454 | 18,544 | 43,742 | 55.1 | 2.0 | 47,519 | 21.4 | 22.4 | 21.5 |
| Waller | NA | NA | 17,840 | 5.6 | 50.0 | 21.1 | 115 | 10,311 | 25,401 | 59,642 | 43.6 | 7.5 | 61,822 | 13.2 | 18.2 | 17.3 |
| Ward | 14 | 17 | 2,744 | 3.1 | 57.6 | 12.5 | 22 | 8,768 | 25,931 | 62,986 | 39.2 | 4.0 | 61,396 | 12.4 | 15.8 | 14.9 |
| Washington | 33 | 31 | 9,117 | 13.7 | 44.3 | 26.8 | 52 | 9,439 | 32,625 | 54,971 | 45.3 | 6.5 | 59,039 | 10.6 | 17.5 | 16.6 |
| Webb | 148 | 171 | 89,187 | 6.7 | 57.7 | 18.9 | 679 | 9,986 | 18,466 | 46,475 | 53.1 | 3.4 | 52,576 | 20.9 | 28.7 | 27.5 |
| Wharton | 44 | 34 | 10,302 | 6.8 | 49.3 | 18.0 | 84 | 10,075 | 25,298 | 48,310 | 51.5 | 2.0 | 49,901 | 16.4 | 21.5 | 20.6 |
| Wheeler | 10 | 14 | 1,218 | 6.1 | 49.8 | 16.8 | 15 | 13,095 | 27,093 | 49,315 | 50.5 | 2.9 | 58,544 | 11.6 | 17.5 | 15.9 |
| Wichita | 53 | 134 | 31,955 | 8.1 | 45.1 | 23.1 | 199 | 9,620 | 24,872 | 48,650 | 51.2 | 2.5 | 51,535 | 13.9 | 17.4 | 17.0 |
| Wilbarger | 7 | 11 | 2,939 | 8.8 | 56.8 | 15.2 | 23 | 9,717 | 23,696 | 45,302 | 53.5 | 2.4 | 42,570 | 14.9 | 21.4 | 20.5 |
| Willacy | 16 | 22 | 5,289 | 4.2 | 69.9 | 8.9 | 52 | 12,105 | 14,888 | 35,521 | 63.3 | 1.4 | 35,821 | 30.5 | 38.1 | 36.7 |
| Williamson | 219 | 347 | 147,839 | 14.2 | 27.3 | 41.3 | 1,081 | 8,881 | 37,242 | 76,692 | 32.9 | 7.5 | 92,661 | 5.4 | 6.5 | 6.1 |
| Wilson | 36 | 46 | 11,624 | 11.0 | 48.5 | 21.8 | 81 | 8,945 | 32,312 | 60,366 | 43.8 | 2.9 | 76,905 | 8.3 | 11.7 | 10.7 |
| Winkler | 25 | 0 | 1,894 | 1.5 | 61.1 | 8.2 | 23 | 12,182 | 24,467 | 64,536 | 37.6 | 5.6 | 64,894 | 12.5 | 17.9 | 17.1 |
| Wise | NA | NA | 15,905 | 11.1 | 49.5 | 18.0 | 92 | 9,668 | 29,418 | 53,394 | 47.3 | 3.4 | 70,468 | 8.3 | 11.7 | 11.2 |
| Wood | 32 | 40 | 8,998 | 10.1 | 48.6 | 16.6 | 60 | 9,597 | 29,005 | 70,005 | 32.8 | 5.8 | 56,945 | 14.3 | 22.3 | 21.5 |
| Yoakum | 10 | 12 | 2,308 | 2.9 | 62.7 | 12.7 | 26 | 11,953 | 24,582 | 70,005 | 32.8 | 5.8 | 62,636 | 10.6 | 14.9 | 14.2 |
| Young | 12 | 22 | 4,145 | 7.0 | 50.7 | 19.3 | 32 | 9,752 | 27,423 | 50,635 | 49.8 | 1.9 | 52,643 | 13.6 | 20.4 | 18.8 |
| Zapata | NA | NA | 4,122 | 0.4 | 68.9 | 11.6 | 36 | 10,234 | 20,169 | 33,952 | 65.2 | 3.8 | 36,069 | 30.1 | 43.7 | 41.1 |
| Zavala | NA | NA | 2,871 | 6.2 | 65.8 | 10.9 | 28 | 11,378 | 13,835 | 34,459 | 70.3 | 0.4 | 32,538 | 29.6 | 40.4 | 36.8 |
| **UTAH** | X | X | 984,917 | 14.1 | 30.6 | 34.0 | 5,029 | 7,525 | 29,775 | 71,621 | 32.9 | 6.4 | 75,705 | 8.8 | 9.6 | 8.7 |
| Beaver | 22 | 48 | 1,918 | 6.3 | 43.7 | 21.1 | 15 | 9,669 | 23,382 | 55,221 | 43.7 | 2.5 | 58,309 | 8.9 | 12.4 | 11.6 |
| Box Elder | 23 | 56 | 16,141 | 7.2 | 38.0 | 23.9 | 91 | 7,315 | 24,797 | 62,233 | 38.3 | 2.8 | 69,380 | 7.5 | 9.1 | 8.2 |
| Cache | 45 | 93 | 47,916 | 7.7 | 27.1 | 38.3 | 208 | 7,770 | 23,713 | 59,038 | 42.9 | 3.7 | 63,180 | 13.1 | 11.3 | 10.4 |
| Carbon | 27 | 21 | 5,773 | 6.7 | 35.6 | 17.1 | 36 | 9,107 | 23,862 | 51,158 | 48.4 | 1.7 | 52,110 | 16.4 | 17.6 | 16.3 |
| Daggett | 4 | 1 | 141 | 5.0 | 46.0 | 13.4 | 4 | 20,271 | 29,574 | 75,417 | 28.8 | 0.0 | 63,433 | 6.7 | 7.4 | 6.3 |
| Davis | 181 | 110 | 112,052 | 9.8 | 25.5 | 37.8 | 612 | 7,091 | 31,205 | 83,310 | 24.6 | 6.8 | 87,610 | 5.5 | 6.1 | 5.6 |
| Duchesne | 27 | 37 | 5,906 | 8.5 | 47.9 | 14.7 | 42 | 7,977 | 24,573 | 63,224 | 39.3 | 2.6 | 59,437 | 14.2 | 14.8 | 14.5 |
| Emery | 31 | 9 | 2,928 | 5.0 | 36.4 | 16.3 | 26 | 11,278 | 23,705 | 55,554 | 43.3 | 0.7 | 61,893 | 10.9 | 13.6 | 12.7 |
| Garfield | 14 | 16 | 1,054 | 10.6 | 43.5 | 25.9 | 10 | 10,616 | 25,247 | 54,565 | 48.2 | 2.3 | 54,625 | 9.4 | 13.1 | 12.4 |
| Grand | NA | NA | 1,879 | 23.8 | 35.1 | 28.6 | 17 | 10,618 | 27,431 | 51,557 | 45.6 | 1.5 | 53,535 | 12.2 | 18.6 | 19.6 |
| Iron | 33 | 46 | 17,616 | 8.4 | 30.6 | 29.8 | 76 | 7,149 | 22,909 | 51,807 | 48.7 | 3.3 | 58,307 | 13.1 | 14.6 | 14.0 |
| Juab | 17 | 12 | 3,531 | 7.4 | 47.5 | 16.0 | 24 | 8,613 | 22,830 | 61,463 | 41.1 | 3.2 | 66,056 | 9.0 | 11.7 | 11.5 |
| Kane | 21 | 31 | 1,598 | 16.2 | 29.3 | 29.9 | 15 | 11,076 | 26,070 | 47,044 | 52.8 | 1.4 | 55,887 | 9.4 | 13.3 | 12.4 |
| Millard | 31 | 25 | 3,577 | 8.0 | 42.5 | 20.6 | 32 | 10,583 | 25,225 | 62,242 | 42.0 | 3.0 | 59,069 | 11.8 | 15.9 | 14.6 |
| Morgan | 12 | 2 | 4,539 | 9.8 | 21.4 | 39.4 | 19 | 6,231 | 31,732 | 91,341 | 21.3 | 10.9 | 101,943 | 4.0 | 4.3 | 3.9 |
| Piute | 4 | 0 | 441 | 5.0 | 49.5 | 20.2 | 5 | 16,595 | 19,012 | 42,813 | 53.6 | 1.3 | 42,196 | 15.7 | 30.9 | 27.0 |
| Rich | 4 | 6 | 647 | 3.7 | 36.4 | 23.0 | 8 | 15,494 | 23,464 | 57,902 | 46.0 | 5.7 | 64,583 | 9.3 | 13.7 | 13.0 |
| Salt Lake | 376 | 107 | 328,710 | 13.0 | 31.8 | 35.6 | 1,692 | 7,757 | 33,238 | 74,865 | 31.0 | 7.7 | 79,941 | 9.0 | 10.6 | 9.8 |
| San Juan | 12 | 20 | 4,594 | 5.3 | 47.3 | 18.3 | 40 | 12,846 | 18,987 | 45,394 | 54.5 | 1.9 | 49,438 | 21.9 | 26.4 | 24.6 |
| Sanpete | 27 | 37 | 9,458 | 9.6 | 38.7 | 20.9 | 52 | 8,531 | 20,056 | 53,838 | 46.8 | 2.0 | 54,648 | 13.1 | 13.7 | 13.7 |
| Sevier | 22 | 37 | 6,046 | 6.6 | 42.1 | 18.5 | 38 | 7,998 | 23,514 | 54,799 | 45.1 | 2.2 | 58,983 | 12.1 | 15.4 | 14.4 |
| Summit | 62 | 43 | 10,867 | 11.1 | 21.1 | 55.0 | 96 | 11,384 | 56,575 | 102,958 | 21.2 | 23.5 | 112,482 | 4.7 | 5.6 | 5.0 |
| Tooele | 28 | 86 | 21,009 | 10.1 | 38.0 | 24.2 | 127 | 6,932 | 25,949 | 74,562 | 30.6 | 2.7 | 80,196 | 5.3 | 7.1 | 6.3 |
| Uintah | 27 | 62 | 10,897 | 13.7 | 49.2 | 15.8 | 63 | 7,938 | 25,759 | 65,264 | 39.0 | 4.3 | 62,541 | 11.1 | 12.3 | 11.3 |
| Utah | 163 | 240 | 237,494 | 23.6 | 21.8 | 40.8 | 987 | 6,790 | 26,215 | 74,665 | 31.3 | 6.6 | 79,505 | 9.7 | 8.4 | 7.0 |
| Wasatch | 35 | 33 | 9,900 | 15.3 | 27.2 | 39.9 | 67 | 9,082 | 35,641 | 85,166 | 26.4 | 11.3 | 92,136 | 5.0 | 5.9 | 5.0 |
| Washington | 40 | 127 | 45,575 | 12.6 | 30.5 | 28.1 | 242 | 6,968 | 29,010 | 59,839 | 40.4 | 5.2 | 64,388 | 9.5 | 12.9 | 11.6 |
| Wayne | 5 | 1 | 697 | 2.9 | 36.2 | 23.4 | 6 | 12,491 | 22,038 | 44,245 | 57.0 | 0.3 | 50,555 | 11.8 | 20.2 | 19.1 |
| Weber | NA | NA | 72,013 | 9.0 | 38.9 | 24.5 | 381 | 7,498 | 27,828 | 67,244 | 35.2 | 3.7 | 63,293 | 10.1 | 10.8 | 9.8 |
| **VERMONT** | X | X | 145,649 | 21.1 | 36.1 | 38.0 | 1,679 | 19,463 | 34,577 | 61,973 | 40.3 | 5.5 | 73,574 | 7.9 | 9.3 | 8.4 |
| Addison | 9 | 2 | 9,422 | 39.4 | 35.6 | 39.6 | 94 | 20,828 | 34,741 | 68,825 | 36.1 | 4.9 | 56,948 | 10.0 | 13.4 | 11.3 |
| Bennington | 13 | 3 | 7,969 | 27.3 | 37.9 | 36.2 | 97 | 20,173 | 33,789 | 56,183 | 43.7 | 5.2 | 50,942 | 12.3 | 14.6 | 13.6 |
| Caledonia | 6 | 1 | 6,906 | 21.1 | 44.0 | 29.0 | 67 | 15,855 | 29,769 | 50,563 | 49.5 | 4.0 | | | | |

1. All persons 3 years old and over enrolled in nursery school through college.  2. Persons 25 years old and over.  3. Elementary and secondary education expenditures.  4. Based on population estimated by the American Community Survey, 2014–2018.

| STATE County | Total (mil dol) 62 | Percent change 2018-2019 63 | Per capita[1] Dollars 64 | Per capita Rank 65 | Wages and salaries (mil dol) 66 | Pension and insurance 67 | Government social insurance 68 | Proprietors' income (mil dol) 69 | Dividends, interest, and rent (mil dol) 70 | Personal transfer receipts (mil dol) 71 | Earnings Total (mil dol) 72 | From employee and self-employed 73 | From employer 74 |
|---|---|---|---|---|---|---|---|---|---|---|---|---|---|
| **TEXAS—Cont'd** | | | | | | | | | | | | | |
| Tom Green | 5,826 | 4.1 | 48,876 | 934 | 2,484 | 454 | 174 | 583 | 1,331 | 1,153 | 3,696 | 202 | 174 |
| Travis | 91,300 | 6.0 | 71,666 | 115 | 58,813 | 7,328 | 3,707 | 14,309 | 20,667 | 7,022 | 84,157 | 4,200 | 3,707 |
| Trinity | 528 | 3.2 | 36,062 | 2,560 | 88 | 18 | 6 | 29 | 71 | 197 | 141 | 14 | 6 |
| Tyler | 715 | 3.9 | 32,978 | 2,892 | 158 | 38 | 10 | 18 | 99 | 245 | 224 | 18 | 10 |
| Upshur | 1,568 | 3.6 | 37,563 | 2,381 | 340 | 70 | 23 | 75 | 189 | 434 | 508 | 38 | 23 |
| Upton | 172 | 6.3 | 47,118 | 1,116 | 139 | 24 | 8 | 10 | 32 | 33 | 181 | 10 | 8 |
| Uvalde | 1,099 | 3.0 | 41,116 | 1,905 | 381 | 84 | 26 | 104 | 241 | 314 | 596 | 34 | 26 |
| Val Verde | 1,879 | 5.6 | 38,331 | 2,287 | 871 | 216 | 69 | 97 | 284 | 455 | 1,252 | 70 | 69 |
| Van Zandt | 2,241 | 4.5 | 39,609 | 2,102 | 457 | 85 | 31 | 164 | 295 | 619 | 738 | 54 | 31 |
| Victoria | 4,506 | 4.4 | 48,938 | 929 | 1,980 | 327 | 133 | 345 | 971 | 960 | 2,784 | 160 | 133 |
| Walker | 2,177 | 3.7 | 29,838 | 3,043 | 1,112 | 284 | 63 | 75 | 474 | 547 | 1,534 | 76 | 63 |
| Waller | 2,346 | 6.6 | 42,456 | 1,696 | 991 | 171 | 67 | 177 | 314 | 413 | 1,406 | 74 | 67 |
| Ward | 646 | 11.5 | 53,870 | 551 | 430 | 58 | 27 | 22 | 75 | 97 | 537 | 31 | 27 |
| Washington | 2,000 | 4.2 | 55,735 | 461 | 675 | 124 | 45 | 167 | 560 | 454 | 1,011 | 61 | 45 |
| Webb | 8,982 | 3.7 | 32,466 | 2,920 | 4,180 | 878 | 286 | 997 | 1,278 | 2,088 | 6,341 | 334 | 286 |
| Wharton | 1,879 | 4.4 | 45,221 | 1,333 | 721 | 123 | 51 | 139 | 315 | 456 | 1,033 | 58 | 51 |
| Wheeler | 224 | 0.3 | 44,309 | 1,444 | 103 | 19 | 6 | 26 | 54 | 64 | 154 | 8 | 6 |
| Wichita | 5,881 | 4.0 | 44,479 | 1,420 | 2,736 | 559 | 199 | 445 | 1,291 | 1,376 | 3,939 | 216 | 199 |
| Wilbarger | 591 | 2.2 | 46,314 | 1,211 | 237 | 59 | 15 | 40 | 142 | 160 | 351 | 19 | 15 |
| Willacy | 589 | 0.0 | 27,584 | 3,087 | 159 | 33 | 11 | 60 | 58 | 245 | 263 | 16 | 11 |
| Williamson | 31,385 | 9.1 | 53,145 | 601 | 12,192 | 1,479 | 740 | 2,247 | 4,340 | 3,564 | 16,658 | 942 | 740 |
| Wilson | 2,372 | 4.9 | 46,448 | 1,191 | 410 | 78 | 27 | 114 | 360 | 447 | 628 | 43 | 27 |
| Winkler | 510 | 10.9 | 63,667 | 216 | 262 | 40 | 17 | 51 | 75 | 65 | 370 | 19 | 17 |
| Wise | 3,140 | 6.0 | 44,870 | 1,378 | 1,070 | 189 | 70 | 201 | 416 | 564 | 1,529 | 90 | 70 |
| Wood | 1,813 | 4.5 | 39,803 | 2,075 | 438 | 85 | 30 | 129 | 314 | 654 | 683 | 55 | 30 |
| Yoakum | 391 | 6.7 | 44,932 | 1,365 | 255 | 39 | 16 | 58 | 45 | 62 | 367 | 18 | 16 |
| Young | 914 | 1.5 | 50,732 | 775 | 310 | 61 | 21 | 141 | 196 | 228 | 533 | 31 | 21 |
| Zapata | 410 | 1.5 | 28,936 | 3,065 | 199 | 36 | 13 | 9 | 51 | 126 | 258 | 16 | 13 |
| Zavala | 364 | 5.0 | 30,779 | 3,014 | 141 | 25 | 10 | 9 | 55 | 127 | 184 | 12 | 10 |
| **UTAH** | 156,896 | 5.8 | 48,978 | X | 83,676 | 12,951 | 6,228 | 12,320 | 34,906 | 19,880 | 115,174 | 6,793 | 6,228 |
| Beaver | 274 | 13.4 | 40,889 | 1,932 | 111 | 21 | 12 | 53 | 47 | 59 | 197 | 10 | 12 |
| Box Elder | 2,277 | 8.4 | 40,621 | 1,972 | 1,012 | 163 | 80 | 136 | 388 | 370 | 1,391 | 86 | 80 |
| Cache | 5,364 | 5.2 | 41,811 | 1,795 | 2,449 | 469 | 186 | 690 | 1,114 | 735 | 3,794 | 214 | 186 |
| Carbon | 832 | 4.7 | 40,679 | 1,966 | 408 | 77 | 31 | 42 | 148 | 219 | 558 | 37 | 31 |
| Daggett | 45 | 8.5 | 47,753 | 1,049 | 14 | 4 | 1 | 3 | 12 | 10 | 23 | 1 | 1 |
| Davis | 17,213 | 5.9 | 48,423 | 984 | 7,077 | 1,380 | 557 | 887 | 3,379 | 1,943 | 9,901 | 578 | 557 |
| Duchesne | 755 | 4.7 | 37,869 | 2,337 | 395 | 75 | 30 | 44 | 147 | 146 | 544 | 33 | 30 |
| Emery | 352 | 3.4 | 35,177 | 2,663 | 161 | 35 | 13 | 16 | 65 | 94 | 225 | 15 | 13 |
| Garfield | 202 | 5.2 | 39,900 | 2,062 | 91 | 16 | 8 | 12 | 47 | 44 | 126 | 8 | 8 |
| Grand | 577 | 5.9 | 59,196 | 329 | 237 | 38 | 19 | 106 | 152 | 91 | 400 | 24 | 19 |
| Iron | 1,884 | 7.7 | 34,353 | 2,762 | 780 | 156 | 61 | 127 | 396 | 407 | 1,124 | 70 | 61 |
| Juab | 470 | 5.9 | 39,103 | 2,177 | 150 | 29 | 12 | 44 | 62 | 83 | 234 | 15 | 12 |
| Kane | 327 | 4.8 | 41,502 | 1,849 | 139 | 23 | 10 | 24 | 87 | 73 | 196 | 14 | 10 |
| Millard | 506 | 6.8 | 38,336 | 2,286 | 215 | 45 | 18 | 54 | 95 | 104 | 333 | 19 | 18 |
| Morgan | 679 | 6.2 | 55,967 | 447 | 125 | 20 | 10 | 32 | 141 | 65 | 186 | 13 | 10 |
| Piute | 65 | 2.1 | 44,169 | 1,461 | 9 | 2 | 1 | 21 | 10 | 17 | 33 | 1 | 1 |
| Rich | 101 | 8.7 | 40,845 | 1,941 | 30 | 6 | 3 | 18 | 30 | 17 | 56 | 3 | 3 |
| Salt Lake | 64,342 | 5.5 | 55,446 | 476 | 44,196 | 6,374 | 3,238 | 5,578 | 14,621 | 7,398 | 59,386 | 3,443 | 3,238 |
| San Juan | 430 | 4.2 | 28,074 | 3,079 | 181 | 39 | 14 | 30 | 82 | 118 | 264 | 17 | 14 |
| Sanpete | 946 | 3.8 | 30,592 | 3,020 | 315 | 74 | 25 | 77 | 165 | 227 | 491 | 31 | 25 |
| Sevier | 812 | 4.6 | 37,558 | 2,382 | 374 | 73 | 29 | 67 | 166 | 189 | 543 | 34 | 29 |
| Summit | 6,378 | 4.3 | 151,326 | 5 | 1,484 | 165 | 111 | 507 | 2,932 | 238 | 2,267 | 130 | 111 |
| Tooele | 2,778 | 7.6 | 38,446 | 2,273 | 776 | 142 | 61 | 80 | 360 | 384 | 1,058 | 68 | 61 |
| Uintah | 1,152 | 2.7 | 32,241 | 2,929 | 629 | 109 | 49 | 53 | 231 | 212 | 840 | 52 | 49 |
| Utah | 27,355 | 6.6 | 42,995 | 1,616 | 13,679 | 1,945 | 984 | 2,317 | 5,223 | 3,079 | 18,926 | 1,105 | 984 |
| Wasatch | 2,031 | 6.9 | 59,584 | 322 | 505 | 72 | 38 | 104 | 586 | 177 | 719 | 46 | 38 |
| Washington | 7,260 | 5.6 | 40,886 | 1,933 | 2,916 | 468 | 223 | 613 | 1,955 | 1,559 | 4,220 | 289 | 223 |
| Wayne | 115 | 5.3 | 42,426 | 1,699 | 41 | 8 | 4 | 12 | 32 | 23 | 64 | 4 | 4 |
| Weber | 11,373 | 6.1 | 43,707 | 1,525 | 5,174 | 923 | 402 | 575 | 2,233 | 1,801 | 7,074 | 434 | 402 |
| **VERMONT** | 34,502 | 3.2 | 55,288 | X | 15,952 | 2,625 | 1,262 | 3,002 | 6,961 | 7,030 | 22,840 | 1,523 | 1,262 |
| Addison | 1,948 | 3.1 | 52,966 | 618 | 803 | 126 | 65 | 225 | 418 | 325 | 1,219 | 80 | 65 |
| Bennington | 1,982 | 4.1 | 55,870 | 452 | 832 | 136 | 68 | 170 | 499 | 459 | 1,205 | 84 | 68 |
| Caledonia | 1,293 | 4.9 | 43,113 | 1,598 | 484 | 88 | 39 | 114 | 231 | 350 | 725 | 53 | 39 |

1. Based on the resident population estimated as of July 1 of the year shown.

# Table B. States and Counties — Earnings, Social Security, and Housing

| STATE County | Earnings, 2019 (cont.) Percent by selected industries | | | | | | | | | Social Security beneficiaries, December 2019 | | Supplemental Security Income recipients, 2019 | Housing units, 2020 | |
|---|---|---|---|---|---|---|---|---|---|---|---|---|---|---|
| | Farm | Mining, quarrying, and extractions | Construction | Manufacturing | Information; professional, scientific, technical services | Retail trade | Finance, insurance, real estate, and leasing | Health care and social assistance | Government | Number | Rate[1] | | Total | Percent change, 2010-2020 |
| | 75 | 76 | 77 | 78 | 79 | 80 | 81 | 82 | 83 | 84 | 85 | 86 | 87 | 88 |
| **TEXAS—Cont'd** | | | | | | | | | | | | | | |
| Tom Green | 0.6 | 5.0 | 5.9 | 11.1 | 5.4 | 7.3 | 6.9 | 14.1 | 23.3 | 23,155 | 194 | 2,815 | 49,166 | 5.6 |
| Travis | 0.0 | 8.6 | 5.9 | 6.5 | 24.2 | 4.4 | 9.1 | 7.1 | 13.2 | 135,165 | 106 | 16,278 | 565,804 | 28.2 |
| Trinity | -3.4 | D | 4.6 | 8.9 | D | 4.7 | 4.1 | 8.3 | 25.4 | 4,580 | 312 | 596 | 9,187 | 5.4 |
| Tyler | -0.2 | 2.5 | 5.8 | 4.3 | D | 8.5 | 4.3 | D | 42.2 | 5,380 | 250 | 618 | 11,035 | 4.3 |
| Upshur | -1.5 | 5.8 | 16.7 | 4.7 | 10.5 | 5.2 | 3.9 | D | 20.6 | 9,675 | 232 | 1,115 | 17,209 | 3.6 |
| Upton | -1.6 | 44.6 | D | D | D | D | D | 0.6 | 19.6 | 590 | 162 | 72 | 1,570 | 1.6 |
| Uvalde | 2.3 | 2.9 | 5.0 | 3.1 | D | 8.2 | 5.4 | D | 31.4 | 5,760 | 215 | 1,022 | 11,260 | 4.2 |
| Val Verde | -0.2 | D | 3.1 | 6.9 | 2.2 | 7.9 | 3.6 | 9.5 | 46.8 | 8,810 | 180 | 2,017 | 19,443 | 4.2 |
| Van Zandt | 3.0 | 2.8 | 16.2 | 10.0 | 4.8 | 8.5 | 4.6 | 6.2 | 18.1 | 13,645 | 240 | 1,226 | 23,766 | 4.2 |
| Victoria | 0.3 | 10.8 | 7.2 | 8.5 | 4.1 | 8.2 | 4.6 | 13.4 | 14.6 | 18,540 | 201 | 2,368 | 37,507 | 5.9 |
| Walker | 0.4 | 0.4 | 3.7 | 5.8 | 3.2 | 6.4 | 3.4 | 7.2 | 57.8 | 10,520 | 147 | 1,300 | 27,732 | 15.3 |
| Waller | 1.9 | 1.5 | 9.7 | 25.0 | 4.3 | 4.4 | 2.9 | 2.8 | 21.3 | 7,660 | 138 | 929 | 17,829 | 12.3 |
| Ward | -3.3 | 45.0 | 7.4 | 3.3 | 2.9 | 3.8 | 5.4 | D | 10.5 | 1,920 | 160 | 232 | 4,848 | 3.3 |
| Washington | -2.6 | 1.0 | 6.9 | 23.7 | 5.0 | 7.2 | 10.5 | 8.1 | 18.1 | 8,960 | 252 | 923 | 16,621 | 7.2 |
| Webb | 0.0 | 5.4 | 3.3 | 0.7 | 4.3 | 7.2 | 5.0 | 10.9 | 27.6 | 35,835 | 130 | 11,780 | 87,421 | 18.9 |
| Wharton | 8.9 | 7.3 | 9.3 | 8.1 | 2.6 | 8.9 | 4.8 | 9.6 | 16.0 | 8,815 | 212 | 1,059 | 17,915 | 4.6 |
| Wheeler | 4.4 | 32.3 | 5.4 | 0.7 | D | 6.6 | 3.7 | 1.5 | 22.2 | 1,135 | 226 | 66 | 2,717 | -0.4 |
| Wichita | -0.1 | 5.3 | 4.2 | 10.4 | 4.8 | 6.9 | 5.5 | 15.9 | 29.4 | 26,355 | 199 | 3,998 | 56,193 | 1.1 |
| Wilbarger | 3.2 | 0.7 | 2.9 | 17.1 | 3.3 | 6.9 | 4.3 | 3.9 | 37.0 | 2,910 | 229 | 387 | 6,221 | -1.5 |
| Willacy | 13.3 | D | 2.3 | 1.5 | D | 4.7 | 3.7 | 9.8 | 27.4 | 3,880 | 182 | 1,179 | 7,450 | 5.8 |
| Williamson | 0.0 | 0.5 | 11.1 | 14.2 | 11.6 | 7.4 | 6.4 | 8.3 | 10.1 | 80,125 | 135 | 4,603 | 212,897 | 30.8 |
| Wilson | -0.1 | 9.9 | 14.8 | 6.3 | 6.2 | 9.2 | 4.0 | 6.2 | 23.3 | 9,490 | 186 | 635 | 17,924 | 6.9 |
| Winkler | -1.1 | 37.4 | D | D | D | 3.1 | 4.7 | D | 10.4 | 1,205 | 151 | 192 | 3,102 | 2.5 |
| Wise | -2.6 | 11.4 | 11.6 | 10.3 | D | 7.4 | 4.6 | D | 20.2 | 12,445 | 178 | 768 | 25,477 | 7.2 |
| Wood | 2.5 | 4.0 | 14.5 | 12.1 | 6.2 | 7.5 | 5.3 | 10.2 | 16.3 | 15,015 | 329 | 1,074 | 21,565 | 3.4 |
| Yoakum | 9.6 | 39.3 | 9.0 | 1.6 | 1.8 | 2.9 | 4.1 | D | 13.1 | 1,345 | 155 | 133 | 3,067 | 3.0 |
| Young | -1.4 | 14.8 | 5.2 | 23.1 | 4.3 | 5.0 | 12.1 | 6.6 | 14.3 | 4,545 | 253 | 452 | 8,731 | 1.3 |
| Zapata | -1.3 | 43.8 | 5.0 | 2.1 | 0.4 | 2.7 | D | 4.2 | 27.5 | 2,170 | 153 | 618 | 6,545 | 5.5 |
| Zavala | 0.3 | 6.9 | D | D | D | 2.9 | D | 6.3 | 22.8 | 2,230 | 189 | 729 | 4,452 | 3.9 |
| **UTAH** | 0.4 | 0.8 | 8.3 | 9.3 | 13.9 | 7.4 | 9.6 | 8.4 | 15.7 | 419,037 | 131 | 31,730 | 1,161,934 | 18.6 |
| Beaver | 28.5 | D | 4.5 | 4.2 | 1.2 | 5.6 | D | 1.7 | 22.4 | 1,250 | 186 | 77 | 3,075 | 5.7 |
| Box Elder | 3.8 | 0.1 | 9.9 | 36.3 | D | 4.9 | 3.6 | 6.2 | 12.1 | 9,035 | 161 | 570 | 19,671 | 13.5 |
| Cache | 1.4 | 0.0 | 5.2 | 20.5 | 10.2 | 6.6 | 12.0 | 8.8 | 18.4 | 14,900 | 116 | 869 | 44,688 | 20.7 |
| Carbon | 0.0 | 19.5 | 4.4 | 5.6 | 3.2 | 6.6 | 3.9 | D | 20.0 | 4,980 | 243 | 529 | 9,893 | 3.6 |
| Daggett | 6.4 | 0.0 | D | D | D | D | D | 0.2 | 47.3 | 260 | 270 | 0 | 1,249 | 9.5 |
| Davis | 0.1 | D | 9.5 | 11.8 | 8.9 | 6.5 | 5.6 | 8.0 | 28.6 | 42,060 | 118 | 2,521 | 114,197 | 17.0 |
| Duchesne | 3.6 | 26.4 | 4.2 | 1.8 | D | 6.4 | 2.9 | D | 23.1 | 3,530 | 178 | 324 | 10,623 | 11.9 |
| Emery | 2.6 | 10.3 | 11.3 | 1.3 | D | 6.5 | D | D | 19.6 | 2,320 | 231 | 131 | 4,656 | 3.7 |
| Garfield | 4.7 | D | 3.0 | 2.1 | D | 4.3 | 1.4 | D | 24.6 | 1,235 | 247 | 35 | 3,923 | 5.3 |
| Grand | 0.6 | D | 5.4 | 3.5 | D | 8.7 | 12.3 | 7.3 | 18.0 | 2,105 | 215 | 118 | 5,847 | 21.6 |
| Iron | 4.4 | 0.4 | 8.0 | 10.2 | 4.1 | 8.2 | 7.6 | 9.5 | 25.6 | 9,110 | 165 | 744 | 23,054 | 17.2 |
| Juab | 1.7 | 0.7 | 7.6 | 27.1 | D | 3.3 | 2.6 | D | 18.1 | 1,775 | 148 | 121 | 3,961 | 13.1 |
| Kane | 0.9 | D | 5.0 | 2.7 | D | 6.7 | 2.8 | 3.3 | 21.8 | 2,055 | 261 | 63 | 6,277 | 8.0 |
| Millard | 13.4 | 2.7 | 2.4 | 4.9 | D | 5.4 | D | D | 16.5 | 2,595 | 198 | 131 | 5,182 | 4.9 |
| Morgan | -0.7 | D | 23.8 | 9.2 | D | 7.2 | 7.5 | 4.7 | 15.9 | 1,560 | 129 | 37 | 3,803 | 26.5 |
| Piute | 53.6 | 0.0 | D | D | D | 4.9 | D | D | 17.3 | 455 | 309 | 18 | 1,045 | 16.2 |
| Rich | 25.1 | D | 8.1 | D | D | D | D | D | 20.2 | 470 | 190 | 0 | 3,260 | 15.0 |
| Salt Lake | 0.0 | 0.5 | 7.2 | 7.9 | 15.0 | 7.0 | 12.6 | 7.4 | 14.2 | 146,210 | 126 | 13,420 | 420,940 | 15.6 |
| San Juan | 1.5 | 8.2 | 5.0 | 1.6 | D | 3.3 | D | D | 35.1 | 2,350 | 154 | 567 | 6,112 | 6.5 |
| Sanpete | 7.2 | D | 6.7 | 12.8 | 5.1 | 5.0 | 2.9 | 7.4 | 34.4 | 5,280 | 170 | 350 | 10,910 | 5.1 |
| Sevier | 8.1 | 11.1 | 4.0 | 4.9 | 4.0 | 8.7 | 3.3 | D | 19.5 | 4,580 | 212 | 305 | 9,048 | 7.1 |
| Summit | 0.4 | 0.5 | 8.2 | 3.8 | 13.9 | 7.2 | 15.5 | 5.4 | 8.5 | 5,800 | 138 | 95 | 29,031 | 9.4 |
| Tooele | 0.5 | D | 7.0 | 10.8 | 5.1 | 6.1 | 3.2 | 8.1 | 29.7 | 8,540 | 118 | 709 | 23,973 | 23.2 |
| Uintah | 1.1 | 20.8 | 6.3 | 1.0 | 3.0 | 8.0 | 5.0 | 6.4 | 22.9 | 4,915 | 137 | 399 | 13,881 | 15.9 |
| Utah | 0.2 | 0.1 | 10.9 | 8.6 | 23.3 | 8.9 | 4.3 | 9.0 | 9.8 | 59,750 | 94 | 4,302 | 193,361 | 30.3 |
| Wasatch | 0.1 | D | 25.3 | 3.9 | 8.5 | 7.6 | 2.8 | 7.7 | 16.1 | 4,355 | 127 | 121 | 14,848 | 40.4 |
| Washington | 0.0 | 0.4 | 11.3 | 4.7 | 7.1 | 10.0 | 9.6 | 16.8 | 13.2 | 40,480 | 227 | 1,505 | 77,677 | 34.5 |
| Wayne | 9.3 | D | 16.9 | 0.6 | D | 5.5 | D | 9.4 | 24.8 | 670 | 248 | 20 | 1,726 | 8.5 |
| Weber | 0.2 | 0.0 | 7.8 | 16.2 | 6.2 | 7.1 | 8.2 | 12.8 | 19.9 | 36,410 | 140 | 3,636 | 96,023 | 11.4 |
| **VERMONT** | 1.0 | 0.3 | 7.1 | 10.3 | 10.1 | 7.1 | 5.8 | 14.9 | 18.3 | 153,124 | 245 | 15,009 | 341,366 | 5.8 |
| Addison | 4.2 | 0.3 | 8.7 | 14.2 | 6.3 | 6.5 | 3.7 | D | 10.8 | 8,585 | 232 | 534 | 17,773 | 6.1 |
| Bennington | 0.4 | 0.0 | 6.3 | 12.7 | D | 9.7 | 4.8 | 17.4 | 13.5 | 10,280 | 289 | 1,205 | 21,235 | 1.5 |
| Caledonia | 1.7 | D | 8.8 | 12.1 | D | 8.9 | 3.9 | 16.9 | 18.4 | 8,080 | 271 | 852 | 16,580 | 4.0 |

1. Per 1,000 resident population estimated as of July 1 of the year shown.

# Table B. States and Counties — Housing, Labor Force, and Employment

| STATE County | Housing units, 2015-2019 — Total (89) | Occupied units — Percent (90) | Owner-occupied Median value[1] (91) | Median owner cost as a percent of income — With a mortgage (92) | Without a mortgage[2] (93) | Renter-occupied Median rent[3] (94) | Median rent as a percent of income[2] (95) | Sub-standard units[4] (percent) (96) | Civilian labor force, 2020 — Total (97) | Percent change, 2019-2020 (98) | Unemployment Total (99) | Rate[5] (100) | Civilian employment[6], 2015-2019 — Total (101) | Percent Management, business, science, and arts (102) | Construction, production, and maintenance occupations (103) |
|---|---|---|---|---|---|---|---|---|---|---|---|---|---|---|---|
| **TEXAS—Cont'd** | | | | | | | | | | | | | | | |
| Tom Green | 43,314 | 63.5 | 141,500 | 20.9 | 10.6 | 880 | 28.8 | 3.3 | 53,742 | -1.6 | 3,406 | 6.3 | 54,481 | 31.6 | 25.1 |
| Travis | 472,361 | 52.4 | 324,800 | 21.4 | 11.9 | 1,289 | 28.9 | 4.6 | 741,012 | 0.5 | 46,579 | 6.3 | 688,232 | 49.7 | 14.9 |
| Trinity | 6,030 | 77.4 | 83,700 | 20.3 | 13.0 | 748 | 26.9 | 4.6 | 5,313 | 0.3 | 429 | 8.1 | 5,117 | 30.9 | 31.1 |
| Tyler | 7,100 | 84 | 88,100 | 20.9 | 11.4 | 696 | 28.9 | 5.7 | 7,372 | 3.0 | 727 | 9.9 | 6,600 | 23.7 | 31.8 |
| Upshur | 14,108 | 78.8 | 141,200 | 20.2 | 11.4 | 832 | 26.6 | 4.2 | 17,448 | -3.1 | 1,345 | 7.7 | 17,221 | 28.5 | 30.0 |
| Upton | 1,332 | 73 | 66,900 | 14.9 | 10.0 | 717 | 20.9 | 3.4 | 1,835 | 10.9 | 103 | 5.6 | 1,428 | 20.7 | 42.3 |
| Uvalde | 8,841 | 73.3 | 89,000 | 21.4 | 12.7 | 645 | 24.3 | 4.9 | 11,333 | -1.3 | 728 | 6.4 | 11,188 | 22.7 | 31.4 |
| Val Verde | 15,779 | 63.5 | 103,400 | 20.4 | 12.1 | 731 | 23.3 | 9.2 | 22,589 | -1.3 | 1,794 | 7.9 | 19,297 | 24.2 | 30.7 |
| Van Zandt | 20,156 | 78 | 132,300 | 20.1 | 11.8 | 836 | 27.1 | 4.1 | 26,130 | 0.0 | 1,601 | 6.1 | 22,235 | 30.7 | 29.1 |
| Victoria | 32,255 | 66.2 | 142,800 | 20.4 | 10.2 | 918 | 29.1 | 4.6 | 41,172 | -2.9 | 3,411 | 8.3 | 42,129 | 31.0 | 27.8 |
| Walker | 21,963 | 51.4 | 152,300 | 20.1 | 10.4 | 908 | 36.7 | 3.0 | 24,085 | -0.1 | 1,742 | 7.2 | 26,268 | 30.2 | 21.8 |
| Waller | 15,171 | 68.3 | 197,400 | 21.8 | 11.0 | 937 | 32.4 | 6.1 | 23,804 | -1.5 | 1,846 | 7.8 | 23,441 | 28.5 | 28.1 |
| Ward | 3,978 | 71.5 | 86,800 | 18.3 | 10.0 | 899 | 21 | 6.2 | 6,171 | -16.1 | 582 | 9.4 | 4,946 | 27.6 | 39.6 |
| Washington | 12,625 | 74.1 | 179,100 | 21.6 | 11.5 | 885 | 31.4 | 3.5 | 15,329 | 1.0 | 866 | 5.6 | 15,501 | 34.4 | 29.7 |
| Webb | 74,789 | 62.2 | 125,900 | 24.0 | 12.9 | 844 | 32.1 | 12.4 | 116,195 | -1.9 | 9,819 | 8.5 | 109,409 | 26.8 | 24.5 |
| Wharton | 15,199 | 67 | 128,100 | 20.0 | 12.3 | 775 | 25.8 | 5.4 | 21,376 | -1.3 | 1,448 | 6.8 | 18,127 | 26.4 | 35.1 |
| Wheeler | 2,183 | 66.6 | 84,000 | 19.2 | 10.0 | 705 | 25 | 6.0 | 2,226 | -1.6 | 134 | 6.0 | 2,482 | 24.8 | 28.4 |
| Wichita | 48,356 | 60.3 | 100,000 | 19.6 | 11.4 | 797 | 29.3 | 2.7 | 55,306 | -1.4 | 3,686 | 6.7 | 55,764 | 33.0 | 25.8 |
| Wilbarger | 5,180 | 60.7 | 68,300 | 18.8 | 12.5 | 616 | 21.5 | 5.5 | 4,949 | -0.1 | 290 | 5.9 | 6,153 | 27.7 | 26.4 |
| Willacy | 5,782 | 70.5 | 55,300 | 19.1 | 11.9 | 664 | 29.1 | 11.8 | 6,597 | 0.7 | 790 | 12.0 | 7,639 | 19.6 | 32.2 |
| Williamson | 180,160 | 68.4 | 262,300 | 21.1 | 11.3 | 1,327 | 28.1 | 2.6 | 318,447 | -0.1 | 18,646 | 5.9 | 279,179 | 46.4 | 15.8 |
| Wilson | 15,733 | 84.4 | 199,600 | 20.8 | 10.4 | 921 | 28.2 | 6.0 | 24,473 | -1.8 | 1,396 | 5.7 | 21,799 | 32.7 | 28.2 |
| Winkler | 2,619 | 79.6 | 62,700 | 15.4 | 10.0 | 836 | 18.6 | 4.6 | 4,039 | -0.4 | 376 | 9.3 | 3,346 | 19.8 | 40.8 |
| Wise | 22,369 | 77.9 | 170,300 | 21.6 | 11.5 | 966 | 28.1 | 4.4 | 32,140 | -0.5 | 2,099 | 6.5 | 30,152 | 28.9 | 33.2 |
| Wood | 16,510 | 80.2 | 137,500 | 21.6 | 11.2 | 787 | 33.6 | 3.1 | 17,919 | 1.2 | 1,179 | 6.6 | 16,364 | 29.3 | 32.2 |
| Yoakum | 2,617 | 78 | 116,400 | 17.6 | 10.0 | 820 | 18.9 | 7.6 | 3,528 | -5.8 | 421 | 11.9 | 3,617 | 24.2 | 45.9 |
| Young | 7,307 | 75.6 | 96,800 | 18.9 | 10.9 | 686 | 24.5 | 3.0 | 7,803 | -2.4 | 410 | 5.3 | 8,038 | 29.3 | 33.3 |
| Zapata | 4,503 | 74.8 | 77,800 | 20.7 | 11.9 | 520 | 40.6 | 15.8 | 4,600 | -10.1 | 571 | 12.4 | 5,220 | 19.2 | 38.8 |
| Zavala | 3,571 | 72 | 50,000 | 19.4 | 10.9 | 626 | 34.2 | 11.9 | 3,430 | 2.4 | 484 | 14.1 | 4,285 | 28.0 | 31.5 |
| **UTAH** | 977,313 | 70.2 | 279,100 | 20.9 | 10.0 | 1,037 | 27.7 | 3.8 | 1,632,215 | 1.4 | 76,433 | 4.7 | 1,497,354 | 38.9 | 22.0 |
| Beaver | 2,282 | 74.1 | 177,600 | 17.2 | 10.0 | 717 | 22.5 | 5.1 | 2,831 | -1.7 | 112 | 4.0 | 2,714 | 28.6 | 36.4 |
| Box Elder | 17,569 | 77.8 | 203,600 | 21.1 | 10.0 | 747 | 24.9 | 2.7 | 26,424 | 1.3 | 1,152 | 4.4 | 23,918 | 32.2 | 34.1 |
| Cache | 38,393 | 63 | 236,500 | 20.3 | 10.0 | 803 | 29.1 | 3.4 | 65,407 | 1.0 | 1,910 | 2.9 | 60,820 | 38.6 | 23.7 |
| Carbon | 7,829 | 73.7 | 140,800 | 19.1 | 10.0 | 669 | 31.3 | 2.7 | 8,575 | 0.4 | 445 | 5.2 | 8,679 | 28.6 | 30.0 |
| Daggett | 163 | 87.1 | 170,800 | 15.6 | 10.0 | 375 | 10 | 3.1 | 401 | 2.8 | 18 | 4.5 | 307 | 24.1 | 35.2 |
| Davis | 104,551 | 77 | 290,300 | 20.1 | 10.0 | 1,105 | 25.9 | 2.3 | 175,905 | 1.4 | 7,168 | 4.1 | 164,139 | 42.0 | 20.5 |
| Duchesne | 6,868 | 75.1 | 185,800 | 21.4 | 10.0 | 855 | 24.7 | 5.0 | 7,782 | 0.5 | 603 | 7.7 | 8,129 | 32.8 | 34.8 |
| Emery | 3,665 | 76.9 | 140,200 | 18.1 | 10.0 | 605 | 22.3 | 3.8 | 4,372 | 1.5 | 206 | 4.7 | 4,040 | 32.1 | 31.1 |
| Garfield | 1,760 | 79.5 | 188,900 | 21.1 | 10.0 | 688 | 24.1 | 2.6 | 2,634 | -5.0 | 269 | 10.2 | 2,184 | 29.1 | 20.2 |
| Grand | 4,191 | 65.1 | 247,800 | 21.1 | 10.0 | 872 | 25.6 | 7.3 | 6,161 | -0.8 | 587 | 9.5 | 5,208 | 27.3 | 18.4 |
| Iron | 17,015 | 64 | 213,600 | 20.7 | 10.0 | 813 | 27.8 | 3.0 | 24,994 | 3.3 | 1,115 | 4.5 | 22,635 | 32.7 | 23.5 |
| Juab | 3,464 | 78.6 | 223,000 | 20.5 | 10.0 | 800 | 22.5 | 4.8 | 5,881 | 0.9 | 172 | 2.9 | 5,044 | 30.3 | 34.2 |
| Kane | 2,643 | 77.4 | 204,900 | 22.1 | 10.0 | 925 | 24.3 | 2.5 | 3,835 | 0.1 | 206 | 5.4 | 3,275 | 34.1 | 13.9 |
| Millard | 4,337 | 74.5 | 155,000 | 17.9 | 10.0 | 726 | 20.8 | 3.9 | 5,941 | -1.7 | 192 | 3.2 | 5,449 | 30.0 | 35.2 |
| Morgan | 3,365 | 85.3 | 374,800 | 20.4 | 10.0 | 1,182 | 21.1 | 1.3 | 5,561 | 0.5 | 186 | 3.3 | 4,822 | 46.0 | 21.3 |
| Piute | 532 | 83.6 | 179,500 | 21.6 | 10.0 | 725 | 18.3 | 2.3 | 477 | -0.8 | 30 | 6.3 | 611 | 41.4 | 29.0 |
| Rich | 635 | 71.7 | 202,500 | 17.2 | 10.0 | 702 | 18.9 | 2.0 | 1,162 | -0.7 | 40 | 3.4 | 754 | 23.7 | 31.8 |
| Salt Lake | 374,820 | 67.1 | 305,700 | 20.8 | 10.0 | 1,118 | 27.9 | 3.7 | 642,357 | 1.4 | 32,591 | 5.1 | 590,125 | 40.1 | 21.1 |
| San Juan | 4,311 | 80.6 | 129,700 | 20.4 | 10.8 | 644 | 23.5 | 13.4 | 5,457 | -4.6 | 554 | 10.2 | 5,694 | 32.3 | 30.4 |
| Sanpete | 8,643 | 77.7 | 189,000 | 21.9 | 10.0 | 729 | 26.7 | 6.2 | 12,640 | 1.3 | 472 | 3.7 | 11,855 | 31.4 | 28.8 |
| Sevier | 7,364 | 78.1 | 164,800 | 19.5 | 10.0 | 703 | 25.6 | 2.5 | 9,811 | 0.6 | 416 | 4.2 | 9,059 | 29.3 | 31.5 |
| Summit | 14,220 | 77.4 | 641,900 | 20.9 | 10.0 | 1,429 | 27.2 | 2.3 | 24,314 | -2.5 | 1,760 | 7.2 | 22,839 | 48.3 | 15.6 |
| Tooele | 20,478 | 80.9 | 223,400 | 21.1 | 10.0 | 966 | 28.3 | 2.9 | 35,084 | 0.7 | 1,667 | 4.8 | 31,073 | 32.8 | 30.2 |
| Uintah | 10,569 | 75.9 | 197,000 | 21.3 | 10.0 | 811 | 23.8 | 3.7 | 13,385 | -2.0 | 1,228 | 9.2 | 14,911 | 29.3 | 33.2 |
| Utah | 165,991 | 67.6 | 305,500 | 21.2 | 10.0 | 1,060 | 29 | 5.1 | 314,022 | 1.9 | 11,774 | 3.7 | 280,920 | 42.3 | 17.9 |
| Wasatch | 9,879 | 73 | 419,900 | 21.4 | 10.0 | 1,364 | 30.3 | 3.8 | 16,269 | 1.2 | 1,046 | 6.4 | 15,771 | 38.0 | 18.3 |
| Washington | 57,144 | 70.4 | 283,800 | 23.7 | 10.0 | 1,015 | 28.3 | 5.2 | 79,208 | 3.7 | 4,201 | 5.3 | 68,915 | 33.3 | 21.3 |
| Wayne | 1,000 | 73.9 | 205,700 | 25.4 | 10.0 | 597 | 23.3 | 3.0 | 1,428 | 0.0 | 108 | 7.6 | 1,174 | 29.4 | 32.5 |
| Weber | 83,632 | 73.4 | 220,600 | 20.5 | 10.0 | 891 | 26.7 | 3.5 | 129,907 | 1.6 | 6,210 | 4.8 | 122,290 | 33.0 | 28.6 |
| **VERMONT** | 260,029 | 70.8 | 227,700 | 23.0 | 14.9 | 985 | 30.3 | 2.3 | 330,058 | -4.2 | 18,413 | 5.6 | 329,028 | 42.0 | 20.9 |
| Addison | 14,584 | 76 | 249,700 | 23.2 | 15.0 | 1,032 | 27.6 | 1.4 | 19,936 | -3.5 | 989 | 5.0 | 20,170 | 41.8 | 24.2 |
| Bennington | 14,581 | 74 | 207,600 | 23.4 | 14.9 | 812 | 27.5 | 1.6 | 17,132 | -5.9 | 1,128 | 6.6 | 17,366 | 39.3 | 21.8 |
| Caledonia | 12,434 | 73.1 | 169,500 | 22.7 | 15.3 | 790 | 30.5 | 2.0 | 14,156 | -2.4 | 819 | 5.8 | 14,815 | 36.8 | 25.4 |

1. Specified owner-occupied units lacking complete plumbing facilities.  2. A value of 10.0 represents 10 percent or less; a value of 50.0 represents 50 percent or more.  3. Specified renter-occupied units.  4. Overcrowded or

5. Percent of civilian labor force.  6. Civilian employed persons 16 years old and over.

| STATE County | Number of establishments | Employment: Total | Health care and social assistance | Manufacturing | Retail trade | Finance and insurance | Professional, scientific, and technical services | Annual payroll: Total (mil dol) | Average per employee (dollars) | Farms: Number | Percent with: Fewer than 50 acres | 1000 acres or more | Farm producers whose primary occupation is farming (percent) |
|---|---|---|---|---|---|---|---|---|---|---|---|---|---|
| | 104 | 105 | 106 | 107 | 108 | 109 | 110 | 111 | 112 | 113 | 114 | 115 | 116 |
| **TEXAS—Cont'd** | | | | | | | | | | | | | |
| Tom Green | 2,791 | 40,511 | 7,318 | 3,355 | 6,652 | 1,766 | 1,611 | 1,702 | 42,010 | 1,303 | 56.3 | 10.3 | 29.4 |
| Travis | 37,501 | 638,903 | 74,838 | 28,710 | 64,417 | 30,769 | 86,561 | 42,242 | 66,116 | 1,099 | 57.0 | 4.7 | 39.5 |
| Trinity | 179 | 1,740 | 336 | 182 | 199 | 72 | 34 | 57 | 32,483 | 601 | 43.3 | 2.3 | 39.6 |
| Tyler | 277 | 2,292 | 531 | 141 | 524 | 67 | 70 | 74 | 32,200 | 778 | 61.2 | 2.3 | 35.0 |
| Upshur | 524 | 4,879 | 603 | 590 | 736 | 271 | 388 | 188 | 38,496 | 1,652 | 51.2 | 0.9 | 32.5 |
| Upton | 82 | 1,350 | 208 | NA | 94 | 21 | NA | 87 | 64,252 | 98 | 13.3 | 62.2 | 48.7 |
| Uvalde | 639 | 7,564 | 1,886 | 306 | 1,486 | 239 | 348 | 261 | 34,533 | 592 | 30.9 | 15.4 | 35.4 |
| Val Verde | 817 | 11,602 | 3,180 | 591 | 2,213 | 542 | 247 | 334 | 28,748 | 528 | 46.4 | 25.6 | 38.3 |
| Van Zandt | 925 | 8,982 | 1,108 | 942 | 1,497 | 243 | 291 | 328 | 36,510 | 3,405 | 56.0 | 1.5 | 35.2 |
| Victoria | 2,289 | 33,676 | 6,708 | 2,281 | 5,888 | 814 | 1,195 | 1,545 | 45,869 | 1,286 | 51.5 | 6.3 | 30.1 |
| Walker | 1,033 | 13,143 | 2,446 | 1,066 | 2,939 | 421 | 471 | 420 | 31,993 | 1,441 | 58.3 | 2.8 | 34.2 |
| Waller | 837 | 13,230 | 1,300 | 3,562 | 1,263 | 114 | 530 | 638 | 48,198 | 1,881 | 64.5 | 2.6 | 34.3 |
| Ward | 332 | 4,818 | 170 | 214 | 428 | 79 | 77 | 276 | 57,318 | 102 | 32.4 | 32.4 | 18.9 |
| Washington | 919 | 12,802 | 1,938 | 2,721 | 2,067 | 891 | 411 | 494 | 38,625 | 2,607 | 48.7 | 0.8 | 33.3 |
| Webb | 5,572 | 79,678 | 16,278 | 765 | 13,812 | 2,779 | 2,073 | 2,503 | 31,413 | 656 | 19.1 | 22.6 | 36.3 |
| Wharton | 1,007 | 11,331 | 1,561 | 1,532 | 2,221 | 449 | 293 | 437 | 38,572 | 1,500 | 42.3 | 10.0 | 37.7 |
| Wheeler | 167 | 1,231 | 278 | 8 | 261 | 46 | 56 | 45 | 36,240 | 510 | 9.6 | 24.9 | 35.4 |
| Wichita | 3,023 | 45,482 | 11,746 | 4,908 | 6,779 | 1,770 | 1,107 | 1,716 | 37,726 | 614 | 47.9 | 12.5 | 33.7 |
| Wilbarger | 280 | 4,532 | 1,601 | 816 | 643 | 148 | 68 | 166 | 36,705 | 395 | 18.7 | 23.8 | 39.6 |
| Willacy | 189 | 2,357 | 600 | NA | 391 | 65 | 25 | 78 | 32,892 | 351 | 54.1 | 16.5 | 43.8 |
| Williamson | 11,943 | 171,694 | 24,334 | 8,254 | 28,470 | 8,972 | 21,330 | 9,150 | 53,291 | 2,634 | 57.3 | 4.6 | 31.4 |
| Wilson | 737 | 7,083 | 1,255 | 368 | 1,487 | 179 | 230 | 263 | 37,156 | 2,621 | 45.6 | 2.4 | 36.0 |
| Winkler | 184 | 3,339 | 102 | NA | 311 | 60 | 3 | 284 | 85,025 | 46 | 17.4 | 60.9 | 63.3 |
| Wise | 1,373 | 18,781 | 3,268 | 2,278 | 2,665 | 359 | 393 | 921 | 49,060 | 3,697 | 63.2 | 2.4 | 32.6 |
| Wood | 803 | 8,167 | 1,034 | 889 | 1,494 | 343 | 372 | 318 | 38,987 | 1,587 | 48.0 | 1.6 | 39.8 |
| Yoakum | 185 | 2,270 | 332 | 104 | 293 | 69 | 16 | 132 | 58,268 | 291 | 14.8 | 41.6 | 46.1 |
| Young | 569 | 5,873 | 914 | 1,338 | 792 | 199 | 235 | 247 | 42,123 | 853 | 22.2 | 13.7 | 36.2 |
| Zapata | 137 | 1,419 | 337 | 21 | 289 | 84 | 4 | 46 | 32,471 | 412 | 12.1 | 27.9 | 31.2 |
| Zavala | 95 | 1,329 | 592 | NA | 169 | 44 | 13 | 38 | 28,592 | 281 | 14.2 | 35.2 | 44.6 |
| **UTAH** | 83,924 | 1,373,876 | 149,648 | 128,958 | 156,151 | 72,265 | 95,498 | 67,688 | 49,268 | 18,409 | 62.1 | 6.9 | 32.1 |
| Beaver | 179 | 1,744 | 270 | 162 | 422 | 34 | 18 | 56 | 32,126 | 272 | 41.5 | 9.6 | 45.8 |
| Box Elder | 1,210 | 18,941 | 1,552 | 7,539 | 1,884 | 289 | 280 | 934 | 49,320 | 1,187 | 50.5 | 15.4 | 37.8 |
| Cache | 3,551 | 45,600 | 5,717 | 11,872 | 6,762 | 1,094 | 4,198 | 1,751 | 38,408 | 1,397 | 58.3 | 3.8 | 30.3 |
| Carbon | 468 | 5,762 | 1,052 | 474 | 958 | 241 | 123 | 240 | 41,625 | 309 | 69.9 | 13.3 | 26.4 |
| Daggett | 25 | 108 | NA | NA | 29 | NA | NA | 4 | 34,389 | 52 | 40.4 | 11.5 | 29.2 |
| Davis | 7,722 | 93,376 | 12,190 | 11,851 | 15,401 | 3,258 | 8,503 | 4,099 | 43,893 | 528 | 90.3 | 1.3 | 30.8 |
| Duchesne | 641 | 5,549 | 1,065 | 230 | 803 | 113 | 132 | 370 | 66,672 | 1,063 | 49.6 | 5.3 | 28.9 |
| Emery | 190 | 2,123 | 199 | 16 | 340 | 36 | 70 | 119 | 55,982 | 504 | 51.6 | 6.9 | 34.5 |
| Garfield | 185 | 1,514 | 139 | 49 | 173 | 19 | 23 | 54 | 35,995 | 286 | 51.7 | 7.0 | 36.5 |
| Grand | 508 | 4,649 | 416 | 49 | 860 | 76 | 172 | 159 | 34,259 | 102 | 49.0 | 8.8 | 37.4 |
| Iron | 1,549 | 15,254 | 2,575 | 1,748 | 2,342 | 592 | 595 | 499 | 32,687 | 486 | 37.9 | 16.7 | 42.1 |
| Juab | 236 | 2,788 | 499 | 621 | 303 | 42 | 299 | 99 | 35,398 | 292 | 27.7 | 14.7 | 26.4 |
| Kane | 275 | 3,128 | 265 | 75 | 386 | 76 | 40 | 116 | 37,022 | 182 | 41.2 | 17.0 | 33.3 |
| Millard | 256 | 3,014 | 253 | 566 | 594 | 53 | 161 | 147 | 48,716 | 654 | 28.0 | 18.3 | 39.0 |
| Morgan | 319 | 2,121 | 135 | 229 | 302 | 23 | 144 | 106 | 50,056 | 372 | 63.4 | 7.3 | 26.5 |
| Piute | 25 | 69 | NA | NA | 24 | NA | NA | 1 | 21,029 | 104 | 31.7 | 16.3 | 36.1 |
| Rich | 106 | 599 | NA | NA | 64 | NA | NA | 24 | 39,472 | 160 | 31.9 | 32.5 | 44.5 |
| Salt Lake | 33,028 | 633,710 | 66,939 | 55,136 | 63,604 | 48,065 | 48,809 | 34,809 | 54,928 | 592 | 87.2 | 2.0 | 31.6 |
| San Juan | 249 | 3,025 | 913 | 123 | 418 | 64 | 22 | 112 | 36,975 | 823 | 52.5 | 13.0 | 64.9 |
| Sanpete | 462 | 5,036 | 651 | 1,215 | 893 | 138 | 429 | 167 | 33,238 | 1,003 | 52.9 | 6.3 | 35.4 |
| Sevier | 545 | 6,876 | 947 | 294 | 1,298 | 138 | 270 | 257 | 37,418 | 691 | 67.6 | 2.9 | 30.3 |
| Summit | 2,519 | 28,700 | 1,499 | 623 | 4,123 | 1,324 | 1,127 | 1,185 | 41,273 | 626 | 67.6 | 5.4 | 26.5 |
| Tooele | 936 | 11,700 | 1,535 | 1,418 | 1,856 | 184 | 468 | 485 | 41,447 | 540 | 63.3 | 7.6 | 29.1 |
| Uintah | 1,060 | 9,153 | 801 | 249 | 1,610 | 178 | 386 | 422 | 46,092 | 1,114 | 56.7 | 6.3 | 30.3 |
| Utah | 14,787 | 234,723 | 26,780 | 17,891 | 28,157 | 6,610 | 17,450 | 10,838 | 46,174 | 2,589 | 82.2 | 1.9 | 23.6 |
| Wasatch | 1,045 | 7,415 | 819 | 305 | 1,165 | 164 | 671 | 318 | 42,866 | 475 | 82.3 | 1.9 | 23.2 |
| Washington | 5,622 | 56,729 | 10,550 | 3,036 | 9,487 | 1,360 | 2,633 | 2,137 | 37,664 | 537 | 60.5 | 5.8 | 26.8 |
| Wayne | 98 | 714 | 120 | 6 | 136 | NA | 9 | 23 | 32,066 | 209 | 51.2 | 5.3 | 36.4 |
| Weber | 5,620 | 84,965 | 11,472 | 13,174 | 11,565 | 4,742 | 4,516 | 3,619 | 42,600 | 1,260 | 85.0 | 0.8 | 24.2 |
| **VERMONT** | 20,829 | 261,196 | 49,728 | 30,755 | 37,759 | 9,138 | 12,078 | 11,885 | 45,502 | 6,808 | 41.1 | 2.3 | 42.1 |
| Addison | 1,166 | 14,052 | 2,416 | 1,732 | 1,852 | 304 | 368 | 573 | 40,744 | 720 | 40.0 | 5.1 | 46.6 |
| Bennington | 1,325 | 15,873 | 3,528 | 2,269 | 2,826 | 324 | 368 | 621 | 39,145 | 250 | 56.4 | 2.4 | 35.4 |
| Caledonia | 906 | 8,872 | 1,804 | 1,277 | 1,557 | 322 | 217 | 361 | 40,734 | 585 | 40.9 | 0.5 | 37.8 |

# Table B. States and Counties — **Agriculture**

| STATE County | Acreage (1,000) | Percent change, 2012-2017 | Average size of farm | Total irrigated (1,000) | Total cropland (1,000) | Average per farm | Average per acre | Value of machinery and equipment, average per farm (dollars) | Total (mil dol) | Average per farm (acres) | Crops | Livestock and poultry products | Organic farms (number) | Farms with internet access (percent) | Total ($1,000) | Percent of farms |
|---|---|---|---|---|---|---|---|---|---|---|---|---|---|---|---|---|
| | 117 | 118 | 119 | 120 | 121 | 122 | 123 | 124 | 125 | 126 | 127 | 128 | 129 | 130 | 131 | 132 |
| **TEXAS—Cont'd** | | | | | | | | | | | | | | | | |
| Tom Green | 813 | -15.0 | 624 | 19.6 | 125.0 | 980,735 | 1,572 | 76,726 | 100.0 | 76,769 | 29.9 | 70.1 | 1 | 75.4 | 2,964 | 13.4 |
| Travis | 222 | -12.2 | 202 | 2.0 | 55.1 | 1,263,065 | 6,256 | 62,118 | 28.1 | 25,578 | 71.3 | 28.7 | 8 | 78.3 | 884 | 11.9 |
| Trinity | 99 | -11.1 | 165 | 0.3 | 20.1 | 435,080 | 2,644 | 70,677 | 8.2 | 13,691 | 25.6 | 74.4 | NA | 66.6 | 42 | 1.5 |
| Tyler | 91 | 0.5 | 117 | 0.8 | 18.8 | 399,746 | 3,412 | 61,078 | 14.9 | 19,134 | 64.8 | 35.2 | NA | 77.6 | 96 | 2.7 |
| Upshur | 185 | -8.4 | 112 | 0.4 | 36.9 | 386,205 | 3,443 | 51,583 | 40.7 | 24,645 | 8.5 | 91.5 | 8 | 75.1 | 347 | 7.2 |
| Upton | 725 | 5.7 | 7,399 | 15.8 | 74.9 | 5,980,038 | 808 | 270,042 | 19.1 | 194,520 | 72.8 | 27.2 | NA | 79.6 | 568 | 29.6 |
| Uvalde | 987 | 1.0 | 1,668 | 32.6 | 111.3 | 2,165,457 | 1,299 | 108,251 | 87.1 | 147,130 | 43.3 | 56.7 | NA | 73.3 | 2,202 | 14.7 |
| Val Verde | 1,471 | -1.7 | 2,787 | 0.7 | 7.2 | 2,808,164 | 1,008 | 52,049 | 9.4 | 17,890 | 1.9 | 98.1 | NA | 54.9 | 416 | 2.8 |
| Van Zandt | 380 | 2.5 | 112 | 2.1 | 99.0 | 409,609 | 3,673 | 55,302 | 104.6 | 30,720 | 40.6 | 59.4 | 3 | 73.5 | 295 | 1.7 |
| Victoria | 426 | -2.7 | 331 | 8.1 | 83.2 | 967,562 | 2,920 | 76,948 | 58.4 | 45,392 | 58.6 | 41.4 | NA | 69.8 | 2,630 | 12.1 |
| Walker | 227 | -19.0 | 158 | 0.6 | 27.5 | 580,115 | 3,679 | 58,302 | 33.8 | 23,452 | 56.7 | 43.3 | 2 | 75.8 | 143 | 0.9 |
| Waller | 253 | -19.6 | 135 | 11.6 | 71.4 | 628,544 | 4,670 | 69,415 | 102.4 | 54,423 | 74.5 | 25.5 | 1 | 75.3 | 5,198 | 4.6 |
| Ward | 406 | 3.6 | 3,978 | 3.3 | 6.5 | 4,234,789 | 1,064 | 84,959 | D | D | D | 100.0 | NA | 48.0 | D | 20.6 |
| Washington | 320 | -13.2 | 123 | 2.3 | 68.7 | 587,144 | 4,781 | 57,676 | 35.6 | 13,661 | 15.3 | 84.7 | NA | 64.7 | 417 | 3.2 |
| Webb | 1,845 | -12.1 | 2,812 | 3.3 | 35.0 | 4,817,630 | 1,713 | 69,020 | 28.4 | 43,287 | 1.5 | 98.5 | NA | 47.6 | 525 | 3.2 |
| Wharton | 535 | -19.0 | 357 | 56.7 | 325.8 | 1,035,627 | 2,902 | 149,677 | 208.5 | 139,027 | 86.0 | 14.0 | 17 | 67.4 | 14,777 | 36.7 |
| Wheeler | 529 | 1.9 | 1,038 | 12.3 | 121.7 | 1,129,454 | 1,089 | 95,512 | 70.6 | 138,424 | 12.0 | 88.0 | NA | 71.6 | 3,608 | 46.5 |
| Wichita | 370 | 1.0 | 603 | 10.2 | 153.0 | 889,018 | 1,474 | 88,165 | 33.8 | 54,992 | 43.3 | 56.7 | NA | 79.5 | 3,583 | 23.8 |
| Wilbarger | 620 | 5.7 | 1,571 | 13.7 | 235.1 | 2,009,636 | 1,279 | 187,858 | 51.9 | 131,380 | 68.7 | 31.3 | NA | 80.8 | 5,709 | 64.6 |
| Willacy | 318 | -5.4 | 906 | 19.0 | 197.4 | 1,591,245 | 1,757 | 197,576 | 88.1 | 250,943 | 95.1 | 4.9 | NA | 59.0 | 4,971 | 46.2 |
| Williamson | 559 | 0.1 | 212 | 1.6 | 231.5 | 790,257 | 3,722 | 72,687 | 114.9 | 43,631 | 57.5 | 42.5 | 12 | 78.9 | 6,359 | 18.0 |
| Wilson | 434 | -1.4 | 165 | 11.7 | 87.9 | 525,222 | 3,174 | 62,753 | 68.6 | 26,185 | 18.5 | 81.5 | 2 | 65.2 | 1,923 | 8.9 |
| Winkler | 489 | -8.3 | 10,635 | D | D | 11,922,788 | 1,121 | 108,111 | D | D | D | D | NA | 89.1 | NA | NA |
| Wise | 514 | 5.5 | 139 | 4.5 | 110.6 | 541,645 | 3,896 | 58,259 | 46.3 | 12,515 | 25.0 | 75.0 | 4 | 78.1 | 932 | 5.4 |
| Wood | 210 | -7.6 | 132 | 1.2 | 60.5 | 445,231 | 3,362 | 69,250 | 127.5 | 80,366 | 6.8 | 93.2 | 6 | 76.8 | 77 | 1.2 |
| Yoakum | 518 | 6.0 | 1,779 | 92.8 | 318.0 | 1,670,162 | 939 | 315,408 | 100.2 | 344,333 | 96.0 | 4.0 | NA | 78.4 | 11,936 | 74.6 |
| Young | 575 | 9.8 | 674 | 1.9 | 114.5 | 1,139,664 | 1,691 | 59,190 | 21.7 | 25,433 | 12.0 | 88.0 | 4 | 76.1 | 1,277 | 18.2 |
| Zapata | 438 | -22.2 | 1,063 | 1.5 | 18.9 | 1,500,482 | 1,412 | 68,908 | 6.3 | 15,408 | 6.9 | 93.1 | 6 | 53.9 | 441 | 8.3 |
| Zavala | 729 | 5.2 | 2,595 | 43.8 | 76.0 | 4,517,368 | 1,741 | 205,052 | 66.6 | 237,103 | 35.2 | 64.8 | NA | 61.6 | 1,796 | 14.6 |
| **UTAH** | 10,812 | -1.5 | 587 | 1,097.2 | 1,654.4 | 1,067,323 | 1,817 | 97,789 | 1,838.6 | 99,876 | 30.5 | 69.5 | 1 | 77.7 | 27,868 | 12.0 |
| Beaver | 157 | -17.4 | 577 | 40.2 | 44.4 | 1,284,918 | 2,226 | 217,327 | 258.0 | 948,559 | 7.8 | 92.2 | 95 | 76.8 | 699 | 26.1 |
| Box Elder | 1,221 | 4.3 | 1,028 | 103.8 | 308.3 | 1,577,010 | 1,533 | 147,150 | 134.1 | 112,947 | 42.4 | 57.6 | NA | 83.2 | 8,193 | 34.1 |
| Cache | 276 | 2.9 | 198 | 90.1 | 159.4 | 955,825 | 4,833 | 120,832 | 162.7 | 116,490 | 25.4 | 74.6 | 14 | 81.1 | 3,439 | 25.8 |
| Carbon | 231 | -4.0 | 747 | 9.3 | 15.8 | 924,915 | 1,238 | 68,983 | 6.5 | 20,906 | 35.9 | 64.1 | 8 | 84.8 | 101 | 5.5 |
| Daggett | 18 | D | 340 | 7.2 | 6.6 | 943,662 | 2,777 | 98,797 | 2.4 | 46,212 | 14.1 | 85.9 | 2 | 80.8 | 220 | 7.7 |
| Davis | 52 | -5.9 | 98 | 10.0 | 7.7 | 914,737 | 9,325 | 58,022 | 23.8 | 45,074 | 85.0 | 15.0 | NA | 76.1 | 50 | 2.8 |
| Duchesne | 1,057 | -2.9 | 995 | 96.5 | 77.3 | 970,628 | 976 | 100,563 | 57.9 | 54,461 | 27.9 | 72.1 | NA | 82.0 | 584 | 5.0 |
| Emery | 134 | -14.4 | 265 | 32.8 | 36.9 | 557,310 | 2,101 | 74,404 | 15.4 | 30,464 | 31.8 | 68.2 | 2 | 84.5 | 460 | 14.3 |
| Garfield | 83 | -9.7 | 289 | 21.2 | 16.3 | 903,654 | 3,127 | 91,968 | 21.8 | 76,175 | 16.5 | 83.5 | 5 | 72.4 | D | 3.8 |
| Grand | 231 | D | 2,268 | 11.5 | 14.6 | 1,868,935 | 824 | 139,177 | 7.2 | 70,294 | 65.5 | 34.5 | 5 | 81.4 | D | 3.9 |
| Iron | 513 | -3.7 | 1,055 | 64.4 | 83.4 | 2,063,886 | 1,955 | 179,494 | 133.5 | 274,716 | 40.6 | 59.4 | NA | 75.7 | 530 | 12.1 |
| Juab | 265 | 8.9 | 906 | 23.7 | 56.6 | 1,202,491 | 1,327 | 122,499 | 53.7 | 183,832 | 57.3 | 42.7 | NA | 80.5 | 933 | 34.2 |
| Kane | 129 | 12.8 | 707 | 6.9 | 14.9 | 1,282,578 | 1,814 | 80,345 | 6.3 | 34,440 | 11.8 | 88.2 | NA | 83.0 | 56 | 2.7 |
| Millard | 482 | -16.6 | 736 | 122.7 | 146.0 | 1,504,464 | 2,043 | 236,249 | 180.0 | 275,168 | 47.5 | 52.5 | 6 | 81.5 | 3,031 | 39.8 |
| Morgan | 243 | 6.1 | 652 | 9.0 | 16.6 | 1,434,803 | 2,200 | 70,446 | 17.1 | 46,046 | 14.4 | 85.6 | NA | 80.9 | 392 | 4.3 |
| Piute | 54 | 43.9 | 524 | 15.3 | 14.9 | 1,089,630 | 2,081 | 125,352 | 40.6 | 390,433 | 8.5 | 91.5 | NA | 81.7 | 68 | 7.7 |
| Rich | 375 | -8.4 | 2,343 | 42.4 | 70.6 | 2,046,056 | 873 | 136,133 | 22.1 | 137,963 | 16.1 | 83.9 | NA | 76.3 | 633 | 13.1 |
| Salt Lake | 62 | -20.7 | 105 | 7.4 | 10.9 | 1,013,467 | 9,682 | 52,551 | 19.9 | 33,617 | 85.7 | 14.3 | 10 | 78.7 | 188 | 2.4 |
| San Juan | 1,657 | 3.0 | 2,014 | 7.6 | 130.2 | 738,348 | 367 | 48,406 | 16.8 | 20,383 | 51.3 | 48.7 | 21 | 33.2 | 1,636 | 14.5 |
| Sanpete | 302 | 6.1 | 301 | 76.5 | 71.7 | 848,242 | 2,820 | 109,542 | 171.8 | 171,243 | 12.1 | 87.9 | 2 | 76.7 | 854 | 14.6 |
| Sevier | 109 | -10.9 | 158 | 49.4 | 50.5 | 675,452 | 4,282 | 118,445 | 88.5 | 128,140 | 24.3 | 75.7 | 2 | 78.3 | 432 | 10.1 |
| Summit | 296 | 9.5 | 472 | 21.3 | 26.3 | 1,541,565 | 3,265 | 71,386 | 25.5 | 40,799 | 16.0 | 84.0 | 3 | 79.9 | 833 | 5.0 |
| Tooele | 349 | 0.6 | 646 | 21.9 | 21.7 | 888,625 | 1,375 | 89,898 | 40.8 | 75,469 | 15.9 | 84.1 | 1 | 80.6 | 406 | 2.6 |
| Uintah | 1,825 | D | 1,638 | 65.3 | 70.7 | 1,032,638 | 630 | 88,260 | 42.3 | 37,943 | 38.6 | 61.4 | NA | 75.9 | 1,402 | 6.4 |
| Utah | 304 | -11.4 | 117 | 72.7 | 118.1 | 1,024,882 | 8,734 | 74,291 | 202.6 | 78,246 | 39.4 | 60.6 | 6 | 81.8 | 1,679 | 5.7 |
| Wasatch | 97 | -34.9 | 204 | 11.3 | 8.8 | 1,135,964 | 5,557 | 62,791 | 8.8 | 18,531 | 21.5 | 78.5 | NA | 82.3 | 102 | 2.1 |
| Washington | 155 | 4.8 | 289 | 13.0 | 22.3 | 1,078,856 | 3,737 | 61,328 | 16.5 | 30,648 | 39.3 | 60.7 | 3 | 76.9 | 190 | 3.2 |
| Wayne | 43 | 0.9 | 205 | 16.6 | 14.6 | 928,420 | 4,539 | 89,480 | 12.9 | 61,646 | 15.7 | 84.3 | NA | 90.9 | 87 | 22.0 |
| Weber | 94 | -19.6 | 75 | 27.2 | 28.3 | 697,872 | 9,319 | 59,112 | 49.4 | 39,240 | 50.0 | 50.0 | 5 | 75.0 | 497 | 3.1 |
| **VERMONT** | 1,193 | -4.7 | 175 | 3.0 | 479.7 | 620,691 | 3,541 | 100,672 | 781.0 | 114,713 | 24.0 | 76.0 | 679 | 86.4 | 5,698 | 10.0 |
| Addison | 170 | -18.5 | 236 | 0.4 | 107.8 | 786,959 | 3,335 | 150,113 | 173.4 | 240,889 | 14.7 | 85.3 | 81 | 92.4 | 1,362 | 17.2 |
| Bennington | 33 | -20.1 | 132 | 0.5 | 11.0 | 630,681 | 4,766 | 80,601 | 17.6 | 70,200 | 62.2 | 37.8 | 13 | 88.0 | 63 | 4.8 |
| Caledonia | 87 | 6.3 | 149 | 0.1 | 31.3 | 474,564 | 3,191 | 77,061 | 42.2 | 72,077 | 16.6 | 83.4 | 46 | 87.4 | 263 | 6.5 |

## Table B. States and Counties — Water Use, Wholesale Trade, Retail Trade, and Real Estate

| STATE County | Water use, 2015 — Public supply water withdrawn (mil gal/day) | Public supply gallons withdrawn per person per day | Wholesale Trade[1], 2017 — Number of establishments | Number of employees | Sales (mil dol) | Average payroll (mil dol) | Retail Trade[2], 2017 — Number of establishments | Number of employees | Sales (mil dol) | Average payroll (mil dol) | Real estate and rental and leasing,[2] 2017 — Number of establishments | Number of employees | Sales (mil dol) | Average payroll (mil dol) |
|---|---|---|---|---|---|---|---|---|---|---|---|---|---|---|
| | 133 | 134 | 135 | 136 | 137 | 138 | 139 | 140 | 141 | 142 | 143 | 144 | 145 | 146 |
| **TEXAS—Cont'd** | | | | | | | | | | | | | | |
| Tom Green | 1.16 | 9.8 | D | D | D | D | 421 | 6,626 | 1,964.8 | 191.3 | 151 | 627 | 129.5 | 22.4 |
| Travis | 122.10 | 103.8 | 1,266 | 23,554 | 80,511.6 | 1,765.7 | 3,815 | 62,409 | 19,457.3 | 1,915.0 | 2,409 | 15,454 | 4,406.2 | 934.1 |
| Trinity | 0.96 | 66.7 | 5 | 91 | 52.3 | 5.7 | 28 | 245 | 47.6 | 4.6 | NA | NA | NA | NA |
| Tyler | 2.84 | 133.0 | 9 | 57 | 46.3 | 3.1 | 45 | 519 | 126.4 | 11.4 | 6 | D | 1.5 | D |
| Upshur | 4.29 | 105.7 | 15 | 156 | 50.6 | 5.8 | 75 | 763 | 215.0 | 18.6 | 13 | 26 | 4.3 | 0.6 |
| Upton | 0.00 | 0.0 | 6 | 123 | 28.3 | 8.3 | 8 | 83 | 24.4 | 2.0 | NA | NA | NA | NA |
| Uvalde | 3.63 | 133.2 | 30 | 314 | 227.7 | 15.1 | 104 | 1,442 | 398.5 | 35.3 | 38 | 128 | 21.8 | 3.6 |
| Val Verde | 7.18 | 146.6 | D | D | D | 9.3 | 146 | 2,098 | 645.6 | 50.7 | 38 | 109 | 24.3 | 3.8 |
| Van Zandt | 3.38 | 63.1 | 32 | 417 | 126.9 | 23.9 | 134 | 1,488 | 507.5 | 38.4 | 29 | 62 | 7.0 | 1.7 |
| Victoria | 12.06 | 130.5 | D | D | D | 83.4 | 369 | 5,734 | 1,751.0 | 159.4 | 127 | 1,141 | 527.0 | 61.0 |
| Walker | 2.67 | 37.8 | 28 | 248 | 189.0 | 11.9 | 177 | 2,690 | 915.4 | 67.0 | 64 | 206 | 57.3 | 8.5 |
| Waller | 4.82 | 99.1 | 61 | 1,165 | 887.5 | 61.5 | 96 | 1,132 | 532.9 | 36.4 | 34 | 180 | 36.3 | 5.8 |
| Ward | 5.20 | 443.6 | 11 | 65 | 44.5 | 3.6 | 27 | 336 | 146.1 | 9.1 | 13 | 189 | 45.4 | 14.0 |
| Washington | 0.89 | 25.6 | 37 | 474 | 420.1 | 23.9 | 138 | 2,093 | 583.3 | 55.8 | 46 | 166 | 38.7 | 6.7 |
| Webb | 36.08 | 133.8 | 325 | 2,777 | 2,604.5 | 109.9 | 798 | 13,459 | 3,360.1 | 314.4 | 247 | 890 | 214.1 | 28.8 |
| Wharton | 3.69 | 88.9 | 60 | 608 | 531.0 | 34.2 | 159 | 2,166 | 634.6 | 59.1 | 44 | 177 | 38.1 | 6.9 |
| Wheeler | 1.14 | 201.5 | 9 | 51 | 24.7 | 2.2 | 27 | 228 | 66.5 | 4.6 | 5 | D | 2.0 | D |
| Wichita | 0.81 | 6.2 | 151 | 1,273 | 472.3 | 58.8 | 473 | 6,963 | 1,938.9 | 172.7 | 162 | 783 | 132.2 | 26.3 |
| Wilbarger | 1.85 | 142.0 | D | D | D | D | 43 | 652 | 298.7 | 17.5 | 12 | 73 | 12.7 | 1.7 |
| Willacy | 0.43 | 19.6 | 3 | 52 | 27.5 | 2.7 | 31 | 316 | 88.7 | 7.7 | D | D | D | D |
| Williamson | 35.49 | 69.8 | 343 | 3,263 | 3,249.0 | 345.3 | 1,448 | 26,298 | 9,916.0 | 889.9 | 618 | 2,270 | 684.8 | 112.9 |
| Wilson | 5.69 | 119.7 | 19 | D | 38.6 | D | 92 | 1,454 | 528.1 | 38.0 | 24 | 72 | 12.8 | 3.2 |
| Winkler | 1.45 | 181.1 | D | D | D | D | 21 | 215 | 92.2 | 5.0 | 10 | D | 30.1 | D |
| Wise | 2.23 | 35.4 | 61 | 600 | 517.5 | 37.6 | 170 | 2,346 | 910.7 | 60.1 | 49 | 232 | 43.6 | 8.9 |
| Wood | 4.21 | 97.1 | 24 | 144 | 62.1 | 4.1 | 119 | 1,458 | 503.1 | 42.1 | 40 | 122 | 17.9 | 4.1 |
| Yoakum | 0.97 | 113.5 | 12 | 102 | 55.9 | 8.0 | 30 | 269 | 67.7 | 6.7 | 6 | 60 | 19.7 | 3.1 |
| Young | 2.76 | 151.1 | 26 | 112 | 47.9 | 5.3 | 70 | 803 | 211.9 | 19.4 | 23 | 63 | 50.5 | 3.9 |
| Zapata | 2.20 | 153.1 | NA | NA | NA | NA | 30 | 269 | 69.7 | 5.3 | NA | NA | NA | NA |
| Zavala | 2.27 | 185.5 | 4 | 73 | 18.2 | 2.9 | 17 | 169 | 36.8 | 3.3 | NA | NA | NA | NA |
| **UTAH** | 785.91 | 262.3 | 3,109 | 46,631 | 41,834.1 | 2,880.8 | 9,995 | 153,633 | 50,008.3 | 4,445.9 | 5,577 | 19,457 | 5,296.5 | 876.9 |
| Beaver | 2.25 | 354.1 | D | D | D | D | 34 | 492 | 133.6 | 9.8 | NA | NA | NA | NA |
| Box Elder | 11.7 | 224.6 | D | D | D | D | 159 | 1,763 | 566.5 | 42.4 | 62 | 73 | 11.2 | 1.8 |
| Cache | 25.91 | 214.5 | 127 | 1,034 | 624.8 | 47.0 | 473 | 6,412 | 1,933.8 | 155.7 | 230 | 491 | 88.5 | 14.3 |
| Carbon | 4.35 | 212.4 | 28 | 131 | 63.1 | 6.0 | 70 | 986 | 298.1 | 26.3 | 16 | 49 | 6.1 | 1.3 |
| Daggett | 0.18 | 162.3 | NA | NA | NA | NA | 8 | 35 | 4.8 | 0.8 | NA | NA | NA | NA |
| Davis | 46.73 | 139.1 | 235 | 1,878 | 1,137.5 | 91.7 | 953 | 15,625 | 4,572.2 | 425.6 | 489 | 1,335 | 327.8 | 51.1 |
| Duchesne | 7.03 | 337.0 | 20 | 163 | 83.3 | 6.9 | 77 | 814 | 239.2 | 20.6 | 21 | 147 | 56.4 | 11.4 |
| Emery | 1.73 | 166.8 | D | D | D | D | 43 | 426 | 91.7 | 10.1 | NA | NA | NA | NA |
| Garfield | 1.42 | 283.5 | NA | NA | NA | NA | 21 | 138 | 31.9 | 2.6 | 5 | D | 0.5 | D |
| Grand | 6.05 | 635.8 | 10 | 96 | 25.5 | 3.1 | 77 | 841 | 220.2 | 24.8 | 33 | 129 | 20.9 | 3.7 |
| Iron | 10.64 | 220.0 | 33 | 282 | 223.9 | 13.5 | 195 | 2,320 | 676.2 | 56.1 | 102 | 179 | 40.6 | 6.0 |
| Juab | 4.59 | 433.3 | 7 | 43 | 32.5 | 1.4 | 33 | 374 | 91.7 | 5.6 | NA | NA | NA | NA |
| Kane | 2.33 | 326.7 | D | D | D | D | 38 | 382 | 72.2 | 7.8 | 17 | 33 | 5.7 | 1.1 |
| Millard | 3.91 | 309.2 | 14 | 78 | 250.2 | 3.5 | 56 | 756 | 181.8 | 13.3 | D | D | D | D |
| Morgan | 5.37 | 485.3 | D | D | D | D | 6 | 26 | 4.8 | 0.3 | 12 | 18 | 3.9 | 0.9 |
| Piute | 0.77 | 507.6 | NA | NA | NA | NA | 8 | 83 | 8.9 | 0.9 | NA | NA | NA | NA |
| Rich | 1.22 | 527.9 | 3 | 15 | 3.4 | 0.3 | 3 | 15 | 3.4 | 0.3 | 12 | 40 | 10.0 | 2.3 |
| Salt Lake | 394.81 | 356.5 | 1,629 | 28,690 | 27,264.8 | 1,904.6 | 3,525 | 63,439 | 23,434.7 | 2,041.3 | 2,383 | 10,595 | 3,305.3 | 542.9 |
| San Juan | 1.03 | 65.3 | D | D | D | D | 41 | 458 | 99.0 | 9.8 | 5 | 64 | 3.9 | 1.3 |
| Sanpete | 6.37 | 221.3 | 11 | 76 | 41.0 | 2.9 | 83 | 912 | 207.7 | 19.0 | 16 | 37 | 5.4 | 0.9 |
| Sevier | 5.19 | 247.3 | D | D | D | 5.4 | 89 | 1,339 | 358.8 | 33.6 | 19 | 92 | 34.7 | 4.5 |
| Summit | 8.80 | 222.0 | 49 | 450 | 415.6 | 34.7 | 331 | 3,835 | 1,313.1 | 110.5 | 319 | 1,466 | 310.0 | 60.7 |
| Tooele | 18.35 | 291.5 | 15 | 85 | 42.8 | 5.0 | 102 | 1,752 | 546.5 | 42.9 | 57 | 98 | 19.6 | 3.1 |
| Uintah | 4.93 | 130.0 | 47 | 359 | 247.2 | 19.1 | 135 | 1,709 | 446.3 | 45.9 | 62 | 244 | 65.3 | 15.2 |
| Utah | 127.67 | 222.0 | 428 | 7,417 | 4,043.4 | 449.7 | 1,915 | 26,903 | 7,988.5 | 744.7 | D | D | D | D |
| Wasatch | 4.34 | 148.8 | 20 | 82 | 40.9 | 3.8 | 95 | 999 | 305.1 | 25.7 | 84 | 185 | 102.3 | 8.2 |
| Washington | 49.38 | 317.3 | 164 | 1,343 | 1,444.0 | 64.5 | 628 | 9,079 | 2,651.3 | 259.4 | 388 | 849 | 186.0 | 26.7 |
| Wayne | 1.18 | 438.3 | NA | NA | NA | NA | 16 | 128 | 39.2 | 2.9 | NA | NA | NA | NA |
| Weber | 27.68 | 113.6 | 192 | 3,621 | 4,647.9 | 184.4 | 749 | 11,378 | 3,435.2 | 302.4 | 349 | 924 | 163.6 | 30.4 |
| **VERMONT** | 42.66 | 68.1 | 671 | 9,395 | 6,032.4 | 502.6 | 3,219 | 38,390 | 10,811.3 | 1,102.5 | 784 | 2,938 | 651.8 | 119.3 |
| Addison | 3.45 | 93.2 | 31 | 263 | 188.8 | 15.0 | 174 | 1,923 | 622.5 | 63.9 | 32 | 74 | 12.0 | 2.8 |
| Bennington | 3.30 | 90.9 | D | D | D | D | 251 | 2,765 | 801.2 | 79.6 | 49 | 214 | 35.5 | 7.6 |
| Caledonia | 1.77 | 57.5 | 30 | 321 | 102.5 | 15.0 | 160 | 1,667 | 524.3 | 47.7 | D | D | D | D |

1 Merchant wholesalers, except manufacturers' sales branches and offices.  2. Employer establishments.

# Table B. States and Counties — Professional Services, Manufacturing, and Accommodation and Food Services

| STATE County | Professional, scientific, and technical services, 2017 | | | | Manufacturing, 2017 | | | | Accommodation and food services, 2017 | | | |
|---|---|---|---|---|---|---|---|---|---|---|---|---|
| | Number of establishments | Number of employees | Sales (mil dol) | Average payroll (mil dol) | Number of establishments | Number of employees | Sales (mil dol) | Average payroll (mil dol) | Number of establishments | Number of employees | Sales (mil dol) | Annual payroll (mil dol) |
| | 147 | 148 | 149 | 150 | 151 | 152 | 153 | 154 | 155 | 156 | 157 | 158 |
| **TEXAS—Cont'd** | | | | | | | | | | | | |
| Tom Green | D | D | D | D | 94 | 3,370 | 1,936.4 | 182.0 | 270 | 5,195 | 279.6 | 78.7 |
| Travis | D | D | D | D | 796 | 27,003 | 11,868.1 | 1,971.1 | 3,254 | 79,884 | 5,781.2 | 1,640.7 |
| Trinity | 10 | 35 | 3 | 1.0 | D | 178 | D | 8.8 | 17 | 224 | 16.7 | 3.7 |
| Tyler | 23 | 67 | 8 | 2.4 | 10 | 170 | 43.5 | 6.6 | 27 | 269 | 13.5 | 3.5 |
| Upshur | 52 | 402 | 67 | 18.2 | 29 | 576 | 117.7 | 20.9 | 39 | 557 | 28.0 | 8.0 |
| Upton | NA | NA | NA | NA | NA | NA | NA | NA | D | D | D | 0.3 |
| Uvalde | 33 | 264 | 23 | 9.4 | 17 | 271 | 120.7 | 11.7 | 82 | 1,059 | 66.8 | 16.6 |
| Val Verde | D | D | D | D | 30 | 251 | 92.7 | 11.3 | 91 | 1,844 | 98.5 | 24.7 |
| Van Zandt | 82 | 256 | 28 | 9.5 | 34 | 788 | 179.9 | 45.8 | 84 | 1,254 | 62.9 | 19.2 |
| Victoria | D | D | D | D | D | D | D | D | D | D | D | D |
| Walker | 82 | 458 | 38 | 12.1 | D | 965 | D | 47.8 | 127 | 2,349 | 123.6 | 35.0 |
| Waller | 69 | 307 | 54 | 16.7 | 82 | 2,873 | 1,030.0 | 168.9 | 60 | 1,007 | 59.6 | 16.1 |
| Ward | 12 | 53 | 11 | 3.3 | 6 | 63 | 29.5 | 4.1 | 34 | 418 | 36.6 | 7.4 |
| Washington | 65 | 362 | 51 | 17.1 | 50 | 2,445 | 806.3 | 117.9 | 85 | 1,156 | 71.7 | 20.1 |
| Webb | D | D | D | D | 65 | 588 | 346.3 | 24.8 | 449 | 10,203 | 522.0 | 141.6 |
| Wharton | 55 | 306 | 54 | 13.6 | 42 | 1,386 | 477.6 | 58.9 | 82 | 1,271 | 61.2 | 16.6 |
| Wheeler | 11 | 48 | 6 | 2.0 | 3 | 3 | 0.3 | 0.1 | 25 | 263 | 12.0 | 3.6 |
| Wichita | D | D | D | D | 122 | 4,277 | 1,253.4 | 235.5 | D | D | D | D |
| Wilbarger | 20 | 67 | 6 | 1.7 | D | D | D | D | 31 | 394 | 20.4 | 5.2 |
| Willacy | D | D | D | D | NA | NA | NA | NA | 23 | 296 | 17.9 | 3.5 |
| Williamson | D | D | D | D | 328 | 6,872 | 1,816.3 | 376.5 | 985 | 20,762 | 1,214.2 | 354.9 |
| Wilson | 57 | 244 | 33 | 10.2 | 29 | 319 | 73.4 | 15.3 | 67 | 940 | 43.6 | 11.7 |
| Winkler | 6 | 6 | 2 | 0.4 | NA | NA | NA | NA | D | D | D | D |
| Wise | 91 | 377 | 53 | 16.7 | 70 | 1,617 | 469.7 | 82.7 | 109 | 1,662 | 89.8 | 24.3 |
| Wood | 70 | 592 | 39 | 15.5 | 40 | 935 | 666.8 | 51.3 | 83 | 1,098 | 67.1 | 15.2 |
| Yoakum | 7 | 18 | 2 | 0.5 | 10 | 132 | 41.6 | 7.5 | 23 | 147 | 7.8 | 1.9 |
| Young | 39 | 216 | 29 | 11.6 | 24 | 990 | 334.4 | 55.1 | 42 | 506 | 26.2 | 7.1 |
| Zapata | NA | NA | NA | NA | D | 8 | D | D | D | D | D | D |
| Zavala | 4 | 7 | 1 | 0.1 | NA | NA | NA | NA | 15 | 143 | 6.0 | 1.6 |
| **UTAH** | 10,822 | 91,901 | 15,231 | 5,581.8 | 3,377 | 121,527 | 55,788.0 | 7,171.0 | 5,931 | 118,734 | 7,098.2 | 2,003.5 |
| Beaver | 7 | 13 | 2 | 0.4 | 7 | 105 | 119.0 | 5.1 | 28 | 259 | 15.4 | 4.1 |
| Box Elder | 59 | 226 | 29 | 9.5 | 79 | 7,317 | 3,345.1 | 531.0 | 84 | 1,643 | 62.7 | 20.2 |
| Cache | D | D | D | D | 205 | 11,028 | 5,251.5 | 536.6 | 174 | 3,807 | 169.9 | 49.1 |
| Carbon | 33 | 115 | 17 | 6.1 | 24 | 422 | 133.5 | 22.7 | 34 | 616 | 28.2 | 7.4 |
| Daggett | NA | NA | NA | NA | NA | NA | NA | NA | 10 | 147 | 6.0 | 1.8 |
| Davis | D | D | D | D | 269 | 11,115 | 5,616.5 | 614.6 | 476 | 9,653 | 495.0 | 137.1 |
| Duchesne | 47 | 150 | 20 | 7.1 | 20 | 145 | 35.4 | 8.7 | 36 | 361 | 19.3 | 4.7 |
| Emery | 8 | 52 | 4 | 1.7 | D | 8 | D | 0.7 | 25 | 233 | 15.9 | 4.0 |
| Garfield | D | D | D | D | 7 | 17 | 3.7 | 0.7 | 67 | 707 | 90.1 | 23.7 |
| Grand | 27 | 145 | 51 | 6.3 | 12 | 78 | 30.3 | 4.6 | 94 | 1,811 | 144.3 | 37.7 |
| Iron | 143 | 477 | 55 | 18.8 | 78 | 1,577 | 845.5 | 72.9 | 119 | 2,022 | 100.7 | 27.1 |
| Juab | 20 | 286 | 26 | 13.6 | 14 | 562 | 172.9 | 27.5 | 29 | 293 | 14.5 | 3.6 |
| Kane | 11 | 38 | 3 | 1.1 | D | 103 | D | 3.3 | 58 | 1,022 | 108.3 | 28.7 |
| Millard | 13 | 114 | 18 | 4.2 | 14 | 540 | 643.0 | 31.7 | 27 | 288 | 15.2 | 3.9 |
| Morgan | 39 | 96 | 14 | 4.7 | 15 | 124 | 97.9 | 11.7 | 10 | 151 | 5.3 | 1.9 |
| Piute | NA | NA | NA | NA | NA | NA | NA | NA | 6 | 3 | 0.7 | 0.1 |
| Rich | D | D | 3 | D | NA | NA | NA | NA | 18 | 124 | 17.4 | 3.2 |
| Salt Lake | D | D | D | D | 1,428 | 53,632 | 26,348.7 | 3,415.5 | 2,419 | 50,385 | 3,104.8 | 878.5 |
| San Juan | D | D | D | D | D | 110 | D | 5.6 | 47 | 761 | 61.5 | 18.9 |
| Sanpete | 25 | 609 | 95 | 17.0 | 30 | 763 | 265.4 | 34.6 | 44 | 481 | 15.2 | 4.1 |
| Sevier | 41 | 200 | 28 | 10.5 | 21 | 252 | 64.1 | 11.1 | 57 | 807 | 39.6 | 10.1 |
| Summit | D | D | D | D | 51 | 512 | 159.3 | 30.8 | 210 | 7,276 | 600.8 | 190.3 |
| Tooele | D | D | D | D | 29 | 1,431 | 644.7 | 76.6 | 77 | 1,222 | 62.9 | 16.5 |
| Uintah | 97 | 331 | 48 | 16.6 | 38 | 236 | 41.1 | 10.9 | 70 | 1,050 | 52.5 | 14.8 |
| Utah | 2,142 | 15,278 | 2,474 | 976.1 | 548 | 15,909 | 5,758.0 | 885.1 | 839 | 17,217 | 926.5 | 257.9 |
| Wasatch | 144 | 606 | 72 | 31.0 | 35 | 309 | 60.3 | 13.9 | 58 | 1,253 | 64.2 | 21.2 |
| Washington | D | D | D | D | 163 | 2,634 | 620.3 | 119.0 | 363 | 7,354 | 449.3 | 119.5 |
| Wayne | D | D | D | 0.1 | D | 5 | D | 0.1 | 28 | 181 | 20.2 | 3.9 |
| Weber | D | D | D | D | 265 | 12,589 | 5,483.8 | 695.8 | 424 | 7,607 | 391.7 | 109.6 |
| **VERMONT** | 2,071 | 12,298 | 2,040 | 825.8 | 1,033 | 28,629 | 8,652.4 | 1,610.0 | 1,979 | 32,891 | 2,019.8 | 620.5 |
| Addison | 125 | 564 | 97 | 44.8 | 60 | 1,568 | 829.3 | 98.4 | 91 | 976 | 73.6 | 25.1 |
| Bennington | 118 | 362 | 51 | 18.0 | 67 | 2,093 | 488.3 | 91.1 | 141 | 1,757 | 113.4 | 38.3 |
| Caledonia | D | D | D | D | 52 | 1,172 | 232.4 | 55.7 | 72 | 792 | 42.2 | 13.1 |

# Health Care and Social Assistance, Other Services, Nonemployer Businesses, and Residential Construction

| STATE County | Health care and social assistance, 2017 | | | | Other services, 2017 | | | | Nonemployer businesses, 2018 | | Value of residential construction authorized by building permits, 2020 | |
|---|---|---|---|---|---|---|---|---|---|---|---|---|
| | Number of establishments | Number of employees | Receipts (mil dol) | Annual payroll (mil dol) | Number of establishments | Number of employees | Receipts (mil dol) | Annual payroll (mil dol) | Number | Receipts (mil dol) | New construction ($1,000) | Number of housing units |
| | 159 | 160 | 161 | 162 | 163 | 164 | 165 | 166 | 167 | 168 | 169 | 170 |
| **TEXAS—Cont'd** | | | | | | | | | | | | |
| Tom Green | D | D | D | D | D | D | 170.8 | D | 8,912 | 398.4 | 143,205 | 686 |
| Travis | 3,627 | 76,597 | 9,316.7 | 3,583.8 | 2,505 | 21,189 | 3,149.1 | 897.6 | 132,806 | 7,475.9 | 3,994,543 | 27,110 |
| Trinity | 19 | 254 | 19.0 | 8.3 | 11 | 47 | 2.4 | 0.7 | 1,015 | 46.2 | 329 | 4 |
| Tyler | 32 | 506 | 38.2 | 17.1 | D | D | D | D | 1,264 | 52.8 | 4,600 | 22 |
| Upshur | 40 | 592 | 35.4 | 15.2 | 24 | 78 | 9.6 | 2.8 | 3,080 | 139.9 | 4,003 | 27 |
| Upton | D | D | D | D | NA | NA | NA | NA | 275 | 15.4 | 0 | 0 |
| Uvalde | 75 | 1,754 | 179.5 | 81.2 | D | D | D | D | 2,683 | 134.3 | 2,265 | 19 |
| Val Verde | 90 | 3,129 | 178.0 | 75.4 | 49 | 284 | 19.9 | 5.0 | 3,708 | 127.1 | 22,957 | 109 |
| Van Zandt | 71 | 1,100 | 79.2 | 31.2 | 58 | 323 | 35.7 | 9.4 | 4,686 | 217.2 | 8,095 | 46 |
| Victoria | 295 | 6,854 | 755.1 | 317.4 | 149 | 1,138 | 170.3 | 46.8 | 6,880 | 306.8 | 30,773 | 132 |
| Walker | 103 | 2,378 | 252.7 | 92.8 | 57 | 333 | 29.4 | 8.0 | 4,696 | 184.9 | 99,575 | 523 |
| Waller | 48 | 1,215 | 55.4 | 24.4 | 45 | 395 | 131.8 | 23.0 | 4,483 | 229.0 | 11,096 | 167 |
| Ward | 10 | 188 | 9.0 | 4.4 | 13 | 130 | 20.5 | 6.5 | 792 | 44.6 | 0 | 0 |
| Washington | 85 | 1,862 | 128.2 | 54.4 | 52 | 207 | 22.3 | 6.4 | 3,532 | 168.4 | 25,817 | 596 |
| Webb | 540 | 16,245 | 1,058.6 | 433.7 | 228 | 1,361 | 128.5 | 36.4 | 26,302 | 1,324.9 | 201,512 | 1,447 |
| Wharton | 78 | 1,774 | 109.0 | 46.9 | 72 | 210 | 23.6 | 5.3 | 3,552 | 148.4 | 36,582 | 173 |
| Wheeler | 9 | 273 | 23.1 | 9.8 | D | D | 6.0 | D | 475 | 20.7 | 0 | 0 |
| Wichita | 407 | 11,101 | 1,150.0 | 427.1 | 233 | 1,130 | 154.3 | 34.9 | 7,797 | 390.8 | 42,814 | 187 |
| Wilbarger | 33 | 1,153 | 134.7 | 56.6 | 24 | 86 | 10.0 | 2.0 | 701 | 23.8 | 627 | 3 |
| Willacy | 36 | 525 | 33.4 | 16.1 | 11 | 52 | 5.9 | 1.6 | 1,322 | 44.5 | 8,738 | 52 |
| Williamson | 1,246 | 22,811 | 2,778.7 | 1,129.1 | 804 | 5,282 | 542.7 | 176.1 | 50,155 | 2,267.0 | 2,006,063 | 9,317 |
| Wilson | 63 | 1,184 | 100.9 | 45.1 | 52 | 182 | 18.1 | 4.0 | 3,850 | 203.0 | 36,964 | 141 |
| Winkler | 7 | 20 | 7.4 | 3.0 | D | D | D | D | 753 | 41.5 | 525 | 2 |
| Wise | 119 | 3,181 | 452.0 | 152.3 | 78 | 415 | 44.9 | 12.5 | 5,989 | 352.9 | 20,979 | 95 |
| Wood | 82 | 1,229 | 115.8 | 43.5 | 48 | 206 | 22.2 | 5.7 | 3,799 | 171.8 | 4,454 | 25 |
| Yoakum | 9 | D | 30.6 | D | D | D | D | 1.8 | 579 | 31.0 | 1,118 | 5 |
| Young | 55 | 1,157 | 86.6 | 36.7 | 46 | 147 | 18.0 | 4.1 | 2,084 | 129.5 | 324 | 7 |
| Zapata | 16 | 256 | 8.6 | 4.3 | D | D | D | D | 1,452 | 46.4 | NA | NA |
| Zavala | 12 | 572 | 15.1 | 9.9 | NA | NA | NA | NA | 837 | 29.9 | 0 | 0 |
| **UTAH** | 8,254 | 146,365 | 18,708.2 | 6,703.4 | 4,763 | 29,903 | 3,526.9 | 957.5 | 246,766 | 11,761.1 | 7,287,441 | 31,775 |
| Beaver | D | D | D | D | 9 | 35 | 3.8 | 0.8 | 547 | 21.1 | 11,815 | 42 |
| Box Elder | D | D | D | D | 62 | 260 | 29.6 | 6.9 | 3,538 | 148.4 | 92,714 | 533 |
| Cache | 396 | 5,553 | 641.6 | 212.1 | 211 | 885 | 97.7 | 25.6 | 9,370 | 372.8 | 321,528 | 1,610 |
| Carbon | 73 | 961 | 104.3 | 38.5 | 42 | 279 | 29.7 | 9.5 | 1,124 | 37.5 | 6,817 | 23 |
| Daggett | NA | NA | NA | NA | NA | NA | NA | NA | 82 | 3.1 | 2,258 | 9 |
| Davis | 820 | 11,587 | 1,186.9 | 451.3 | 433 | 2,737 | 259.3 | 73.4 | 25,523 | 1,125.9 | 603,237 | 2,417 |
| Duchesne | 43 | 985 | 115.2 | 39.0 | D | D | D | D | 1,571 | 68.9 | 24,105 | 105 |
| Emery | D | D | D | 6.8 | D | D | 10.6 | D | 655 | 17.9 | 1,629 | 7 |
| Garfield | D | D | D | D | NA | NA | NA | NA | 499 | 19.2 | 12,108 | 58 |
| Grand | D | D | D | 16.9 | D | D | D | D | 1,109 | 45.9 | 22,796 | 107 |
| Iron | 157 | 2,435 | 208.6 | 75.6 | 83 | 276 | 29.2 | 6.9 | 4,078 | 159.5 | 109,378 | 514 |
| Juab | 22 | 510 | 47.6 | 14.6 | D | D | D | D | 812 | 39.7 | 28,712 | 108 |
| Kane | D | D | D | D | D | D | 126.7 | D | 836 | 36.0 | 60,386 | 236 |
| Millard | D | D | D | D | D | D | D | D | 836 | 34.4 | 10,712 | 63 |
| Morgan | D | D | D | D | 11 | 38 | 6.2 | 1.1 | 1,164 | 52.9 | 44,885 | 105 |
| Piute | D | D | D | D | NA | NA | NA | NA | 126 | 7.4 | 4,156 | 26 |
| Rich | NA | NA | NA | NA | D | D | D | D | 277 | 11.3 | 17,193 | 53 |
| Salt Lake | 3,223 | 67,350 | 9,533.4 | 3,510.6 | 2,064 | 14,470 | 1,710.4 | 502.3 | 93,048 | 4,800.9 | 2,014,033 | 10,101 |
| San Juan | D | D | D | D | 14 | 61 | 4.9 | 1.5 | 830 | 25.8 | 7,171 | 46 |
| Sanpete | D | D | D | D | 29 | 136 | 22.5 | 5.0 | 1,922 | 80.2 | 14,797 | 73 |
| Sevier | 66 | 921 | 92.9 | 31.8 | 27 | 125 | 21.7 | 4.9 | 1,516 | 71.4 | 25,115 | 105 |
| Summit | 160 | 1,569 | 206.8 | 70.2 | 119 | 1,173 | 150.5 | 39.0 | 6,616 | 476.1 | 161,870 | 315 |
| Tooele | 107 | 1,371 | 141.2 | 52.7 | 72 | 330 | 30.9 | 9.2 | 3,467 | 121.9 | 213,860 | 932 |
| Uintah | 64 | 709 | 79.1 | 25.2 | 71 | 352 | 47.7 | 11.7 | 1,917 | 80.5 | 16,891 | 59 |
| Utah | 1,456 | 25,794 | 2,900.9 | 1,042.8 | D | D | D | D | 51,419 | 2,281.0 | 2,068,191 | 7,859 |
| Wasatch | 75 | 698 | 83.9 | 27.4 | 43 | 203 | 19.5 | 5.5 | 3,423 | 191.0 | 270,565 | 517 |
| Washington | 583 | 9,409 | 1,212.8 | 395.3 | 265 | 1,401 | 138.9 | 37.6 | 15,006 | 776.7 | 641,709 | 3,314 |
| Wayne | 4 | D | 8.3 | D | D | D | D | D | 377 | 13.8 | 7,760 | 31 |
| Weber | 653 | 11,567 | 1,624.1 | 531.7 | 337 | 2,162 | 223.4 | 62.1 | 15,078 | 640.2 | 471,048 | 2,407 |
| **VERMONT** | 2,102 | 48,285 | 5,451.6 | 2,229.4 | 1,603 | 7,339 | 922.1 | 244.0 | 62,027 | 2,727.2 | 396,203 | 2,077 |
| Addison | 127 | 2,143 | 183.1 | 86.2 | 83 | 356 | 56.6 | 11.8 | 3,927 | 157.3 | 26,560 | 124 |
| Bennington | 149 | 3,447 | 358.3 | 156.7 | 95 | 394 | 36.9 | 9.7 | 3,785 | 171.5 | 12,858 | 40 |
| Caledonia | 107 | 2,144 | 187.1 | 87.8 | 66 | 217 | 25.5 | 5.6 | 3,075 | 128.6 | 10,426 | 68 |

| STATE County | Full-time equivalent employees | March payroll (dollars) | March payroll (percent of total) | | | | | | | Local government finances, 2017 — General revenue | | | | |
|---|---|---|---|---|---|---|---|---|---|---|---|---|---|---|
| | | | Administration, judicial, and legal | Police and corrections | Fire protection | Highways and transportation | Health and welfare | Natural resources and utilities | Education and libraries | Total (mil dol) | Intergovernmental (mil dol) | Taxes Total (mil dol) | Per capita[1] (dollars) Total | Per capita[1] (dollars) Property |
| | 171 | 172 | 173 | 174 | 175 | 176 | 177 | 178 | 179 | 180 | 181 | 182 | 183 | 184 |
| TEXAS—Cont'd | | | | | | | | | | | | | | |
| Tom Green | 4,841 | 15,029,790 | 7.9 | 15.0 | 6.4 | 2.7 | 4.6 | 5.4 | 55.8 | 363.6 | 123.4 | 184.0 | 1,565 | 1,179 |
| Travis | 51,463 | 243,532,757 | 8.3 | 14.1 | 5.4 | 3.9 | 7.4 | 17.4 | 41.9 | 6,766.8 | 1,086.2 | 4,030.3 | 3,283 | 2,711 |
| Trinity | 688 | 2,230,816 | 4.4 | 4.4 | 0.0 | 1.7 | 19.1 | 3.3 | 65.8 | 41.4 | 23.1 | 15.6 | 1,063 | 985 |
| Tyler | 985 | 2,957,565 | 9.8 | 4.0 | 0.0 | 3.3 | 14.9 | 2.6 | 65.4 | 69.8 | 29.4 | 35.3 | 1,643 | 1,550 |
| Upshur | 1,626 | 4,964,608 | 4.4 | 7.9 | 0.6 | 1.9 | 0.6 | 1.7 | 82.4 | 103.6 | 51.1 | 43.2 | 1,051 | 943 |
| Upton | 440 | 1,729,158 | 8.5 | 8.8 | 0.0 | 4.0 | 37.0 | 4.2 | 35.6 | 55.2 | 5.5 | 41.7 | 11,408 | 11,121 |
| Uvalde | 2,248 | 8,364,083 | 3.5 | 5.5 | 0.2 | 2.5 | 23.8 | 2.4 | 61.9 | 362.3 | 67.7 | 59.3 | 2,192 | 1,495 |
| Val Verde | 2,377 | 7,983,288 | 13.0 | 5.2 | 4.4 | 2.6 | 3.2 | 4.1 | 66.1 | 203.8 | 115.8 | 59.0 | 1,203 | 925 |
| Van Zandt | 2,057 | 6,327,002 | 5.5 | 7.5 | 0.4 | 2.2 | 0.5 | 2.4 | 80.7 | 149.2 | 71.7 | 60.3 | 1,092 | 1,003 |
| Victoria | 5,737 | 20,693,284 | 4.2 | 8.5 | 3.5 | 1.7 | 33.2 | 3.0 | 45.1 | 528.8 | 104.2 | 200.0 | 2,173 | 1,705 |
| Walker | 2,264 | 8,416,852 | 12.9 | 13.6 | 0.9 | 2.8 | 0.5 | 20.0 | 47.2 | 188.1 | 64.3 | 76.7 | 1,053 | 811 |
| Waller | 1,917 | 7,261,196 | 7.2 | 9.1 | 0.6 | 2.7 | 0.1 | 3.0 | 76.4 | 171.6 | 59.7 | 99.3 | 1,920 | 1,807 |
| Ward | 596 | 2,124,695 | 13.5 | 14.7 | 0.1 | 4.9 | 3.2 | 7.7 | 54.2 | 76.5 | 11.9 | 45.7 | 4,011 | 3,625 |
| Washington | 2,228 | 8,498,259 | 5.2 | 5.3 | 0.9 | 1.8 | 2.5 | 3.5 | 78.9 | 214.8 | 55.6 | 67.9 | 1,947 | 1,537 |
| Webb | 15,338 | 59,020,428 | 5.3 | 11.4 | 5.2 | 2.6 | 3.5 | 3.6 | 67.9 | 1,369.4 | 592.0 | 500.1 | 1,827 | 1,480 |
| Wharton | 3,030 | 10,927,958 | 4.1 | 6.8 | 1.3 | 3.6 | 21.3 | 1.9 | 60.5 | 199.1 | 70.2 | 81.5 | 1,948 | 1,676 |
| Wheeler | 559 | 1,693,868 | 6.7 | 4.0 | 0.0 | 2.5 | 30.5 | 3.6 | 51.7 | 56.7 | 9.7 | 33.9 | 6,405 | 5,922 |
| Wichita | 5,560 | 21,003,333 | 5.3 | 13.4 | 10.2 | 2.6 | 9.6 | 5.3 | 51.9 | 436.1 | 168.9 | 189.5 | 1,439 | 1,160 |
| Wilbarger | 1,088 | 3,776,963 | 2.2 | 3.4 | 3.3 | 0.6 | 21.1 | 2.7 | 66.3 | 83.6 | 28.7 | 25.5 | 2,011 | 1,766 |
| Willacy | 1,179 | 3,918,839 | 5.7 | 5.1 | 0.0 | 2.2 | 0.7 | 2.6 | 83.4 | 82.4 | 47.9 | 27.1 | 1,261 | 1,095 |
| Williamson | 19,582 | 75,206,562 | 2.5 | 6.6 | 3.5 | 0.8 | 5.5 | 4.0 | 75.0 | 2,121.8 | 335.9 | 1,501.8 | 2,749 | 2,413 |
| Wilson | 2,173 | 8,547,244 | 4.2 | 4.0 | 0.0 | 1.5 | 13.8 | 5.4 | 70.5 | 170.3 | 65.2 | 62.0 | 1,260 | 1,144 |
| Winkler | 531 | 2,078,630 | 7.0 | 8.9 | 0.0 | 1.5 | 15.7 | 5.6 | 60.0 | 76.7 | 12.0 | 49.8 | 6,548 | 6,025 |
| Wise | 2,062 | 7,356,010 | 8.7 | 12.0 | 0.7 | 3.8 | 3.1 | 4.1 | 66.4 | 193.9 | 40.0 | 126.2 | 1,917 | 1,716 |
| Wood | 1,410 | 4,324,130 | 8.8 | 9.5 | 0.5 | 4.2 | 0.6 | 4.4 | 71.8 | 129.6 | 60.7 | 57.0 | 1,289 | 1,147 |
| Yoakum | 807 | 3,005,305 | 7.3 | 8.7 | 0.3 | 3.3 | 26.7 | 3.9 | 49.1 | 49.4 | 14.3 | 30.7 | 3,585 | 3,342 |
| Young | 1,100 | 3,752,022 | 5.6 | 7.4 | 1.3 | 1.9 | 28.9 | 3.1 | 50.8 | 66.8 | 24.9 | 29.1 | 1,622 | 1,357 |
| Zapata | 928 | 3,072,711 | 3.8 | 13.7 | 3.1 | 2.5 | 0.5 | 6.6 | 69.2 | 62.8 | 29.0 | 23.2 | 1,627 | 1,490 |
| Zavala | 671 | 2,155,340 | 3.1 | 7.7 | 2.2 | 2.3 | 1.9 | 2.4 | 77.4 | 45.8 | 20.6 | 20.7 | 1,730 | 1,624 |
| UTAH | X | X | X | X | X | X | X | X | X | X | X | X | X | X |
| Beaver | 450 | 1,782,865 | 9.5 | 17.4 | 0.0 | 2.4 | 23.1 | 7.2 | 38.9 | 312.6 | 14.4 | 15.0 | 2,336 | 1,960 |
| Box Elder | 1,838 | 5,661,061 | 7.3 | 10.2 | 2.2 | 2.7 | 1.8 | 7.2 | 67.0 | 188.5 | 75.6 | 81.3 | 1,507 | 1,219 |
| Cache | 3,436 | 12,872,862 | 6.9 | 7.2 | 3.0 | 4.3 | 9.8 | 8.2 | 59.0 | 396.3 | 171.3 | 130.4 | 1,049 | 691 |
| Carbon | 929 | 3,249,409 | 11.3 | 10.1 | 1.7 | 3.5 | 12.2 | 8.0 | 50.6 | 83.6 | 31.7 | 31.2 | 1,548 | 1,101 |
| Daggett | 89 | 336,348 | 16.7 | 26.7 | 0.6 | 4.5 | 0.0 | 1.9 | 48.6 | 8.9 | 3.7 | 3.7 | 2,960 | 2,603 |
| Davis | 9,122 | 35,252,166 | 7.4 | 9.0 | 3.2 | 1.9 | 4.6 | 7.7 | 64.9 | 1,098.5 | 460.6 | 377.7 | 1,089 | 785 |
| Duchesne | 1,280 | 4,671,224 | 6.5 | 6.4 | 0.4 | 2.5 | 47.0 | 4.3 | 32.0 | 113.7 | 50.8 | 45.2 | 2,276 | 1,684 |
| Emery | 458 | 1,726,217 | 13.3 | 9.6 | 0.1 | 4.9 | 0.4 | 9.8 | 60.5 | 44.5 | 16.8 | 20.3 | 2,027 | 1,742 |
| Garfield | 229 | 883,712 | 14.5 | 15.1 | 0.0 | 9.8 | 2.4 | 4.3 | 53.4 | 34.3 | 14.7 | 12.5 | 2,491 | 1,187 |
| Grand | 539 | 1,976,807 | 13.3 | 11.0 | 0.8 | 5.8 | 13.1 | 11.8 | 38.8 | 89.3 | 16.0 | 52.7 | 5,500 | 3,877 |
| Iron | 1,323 | 5,015,363 | 8.9 | 12.3 | 1.2 | 4.0 | 2.7 | 7.4 | 61.5 | 150.6 | 57.0 | 71.0 | 1,398 | 996 |
| Juab | 418 | 1,619,105 | 10.1 | 9.3 | 0.6 | 4.1 | 0.5 | 11.1 | 61.4 | 53.0 | 25.8 | 14.4 | 1,273 | 1,032 |
| Kane | 432 | 1,643,224 | 10.6 | 10.7 | 1.6 | 2.6 | 26.5 | 7.9 | 36.9 | 59.6 | 18.9 | 24.1 | 3,208 | 1,994 |
| Millard | 527 | 2,164,822 | 10.1 | 13.1 | 0.1 | 5.5 | 2.0 | 5.4 | 61.5 | 68.2 | 24.3 | 31.5 | 2,454 | 2,129 |
| Morgan | 306 | 1,088,010 | 11.5 | 5.4 | 0.8 | 2.0 | 0.0 | 4.7 | 73.3 | 32.5 | 14.1 | 13.4 | 1,132 | 901 |
| Piute | 86 | 283,969 | 16.6 | 4.3 | 1.2 | 4.6 | 0.9 | 0.5 | 71.3 | 8.6 | 6.2 | 1.5 | 1,060 | 864 |
| Rich | 134 | 513,182 | 18.1 | 4.3 | 4.7 | 2.9 | 0.0 | 2.6 | 65.7 | 14.7 | 4.7 | 7.9 | 3,286 | 2,590 |
| Salt Lake | 35,948 | 144,317,443 | 7.7 | 10.4 | 5.6 | 10.9 | 4.6 | 8.5 | 49.2 | 4,296.5 | 1,263.2 | 2,056.0 | 1,809 | 1,168 |
| San Juan | 857 | 3,215,493 | 6.8 | 7.4 | 0.3 | 4.4 | 20.8 | 1.6 | 57.3 | 95.2 | 44.8 | 22.0 | 1,438 | 998 |
| Sanpete | 1,296 | 4,137,485 | 6.3 | 7.7 | 0.4 | 2.0 | 20.0 | 4.2 | 58.2 | 132.6 | 56.0 | 23.2 | 776 | 518 |
| Sevier | 795 | 2,659,519 | 9.4 | 11.6 | 0.2 | 1.4 | 0.9 | 6.3 | 68.4 | 93.9 | 45.3 | 28.7 | 1,348 | 934 |
| Summit | 1,905 | 8,164,223 | 10.9 | 9.1 | 7.5 | 6.3 | 2.9 | 10.6 | 45.7 | 344.8 | 55.7 | 195.2 | 4,722 | 3,687 |
| Tooele | 1,952 | 6,924,862 | 7.5 | 9.0 | 0.7 | 2.3 | 7.2 | 3.8 | 67.4 | 206.7 | 96.0 | 74.1 | 1,098 | 820 |
| Uintah | 1,443 | 5,249,522 | 8.2 | 9.6 | 0.6 | 4.4 | 13.8 | 10.8 | 49.6 | 167.0 | 67.8 | 64.2 | 1,823 | 1,421 |
| Utah | 14,705 | 58,734,102 | 6.8 | 8.9 | 2.7 | 1.6 | 5.8 | 9.0 | 62.8 | 2,025.2 | 807.5 | 729.4 | 1,202 | 783 |
| Wasatch | 1,076 | 4,418,400 | 9.3 | 8.1 | 2.1 | 3.9 | 0.2 | 14.3 | 55.7 | 138.9 | 33.0 | 69.5 | 2,181 | 1,720 |
| Washington | 4,494 | 17,531,308 | 8.8 | 10.7 | 1.8 | 2.4 | 6.9 | 15.0 | 52.2 | 567.9 | 183.6 | 240.2 | 1,448 | 947 |
| Wayne | 122 | 404,320 | 12.0 | 4.6 | 0.0 | 6.1 | 0.0 | 1.9 | 72.1 | 11.8 | 7.6 | 2.6 | 964 | 652 |
| Weber | 7,190 | 28,603,438 | 6.3 | 11.2 | 4.6 | 1.5 | 8.5 | 8.0 | 57.5 | 814.7 | 285.2 | 337.8 | 1,343 | 864 |
| VERMONT | X | X | X | X | X | X | X | X | X | X | X | X | X | X |
| Addison | 991 | 4,172,913 | 6.0 | 3.7 | 0.0 | 6.6 | 0.0 | 4.1 | 79.1 | 150.5 | 96.9 | 37.2 | 1,008 | 977 |
| Bennington | 1,329 | 5,083,941 | 5.5 | 5.8 | 0.0 | 5.7 | 0.9 | 3.1 | 78.2 | 156.1 | 113.3 | 33.0 | 925 | 883 |
| Caledonia | 936 | 3,532,843 | 5.1 | 3.6 | 1.3 | 9.1 | 0.0 | 5.8 | 74.7 | 120.2 | 86.3 | 26.0 | 864 | 860 |

1. Based on the resident population estimated as of July 1 of the year shown.

# Table B. States and Counties — Local Government Finances, Government Employment, and Income Taxes

| STATE County | Direct general expenditure — Total (mil dol) [185] | Per capita (dollars) [186] | Education [187] | Health and hospitals [188] | Police protection [189] | Public welfare [190] | Highways [191] | Debt outstanding — Total (mil dol) [192] | Per capita (dollars) [193] | Gov't employment 2019 — Federal civilian [194] | Federal military [195] | State and local [196] | Individual income tax returns 2018 — Number of returns [197] | Mean adjusted gross income [198] | Mean income tax [199] |
|---|---|---|---|---|---|---|---|---|---|---|---|---|---|---|---|
| **TEXAS—Cont'd** | | | | | | | | | | | | | | | |
| Tom Green | 368 | 3,128 | 49.6 | 1.1 | 6.8 | 0.1 | 2.8 | 469.7 | 3,996 | 1,250 | 3,775 | 7,812 | 53,530 | 64,568 | 7,857 |
| Travis | 7,071 | 5,760 | 38.1 | 9.8 | 6.4 | 0.5 | 11.2 | 17,031.6 | 13,874 | 12,113 | 2,730 | 117,524 | 603,540 | 107,847 | 17,331 |
| Trinity | 41 | 2,767 | 61.5 | 1.6 | 2.4 | 0.7 | 6.3 | 32.8 | 2,236 | 24 | 28 | 623 | 5,870 | 43,851 | 3,562 |
| Tyler | 57 | 2,664 | 62.3 | 0.2 | 6.7 | 0.4 | 7.3 | 30.0 | 1,392 | 44 | 38 | 1,634 | 7,530 | 50,228 | 4,372 |
| Upshur | 107 | 2,604 | 74.0 | 0.0 | 4.1 | 0.2 | 3.9 | 62.3 | 1,517 | 63 | 80 | 1,781 | 16,550 | 53,528 | 4,844 |
| Upton | 88 | 24,032 | 80.1 | 13.8 | 0.2 | 0.0 | 0.3 | 79.7 | 21,793 | 5 | 7 | 541 | 1,440 | 82,047 | 11,013 |
| Uvalde | 345 | 12,772 | 27.9 | 16.6 | 1.8 | 46.5 | 1.8 | 80.4 | 2,971 | 205 | 51 | 2,667 | 11,730 | 47,859 | 4,672 |
| Val Verde | 200 | 4,082 | 53.7 | 0.3 | 5.5 | 0.7 | 4.1 | 119.7 | 2,440 | 2,208 | 1,520 | 2,942 | 21,210 | 46,854 | 3,512 |
| Van Zandt | 159 | 2,878 | 73.3 | 0.0 | 3.4 | 0.2 | 5.1 | 119.6 | 2,169 | 99 | 109 | 2,359 | 23,370 | 54,878 | 5,161 |
| Victoria | 529 | 5,746 | 36.2 | 34.2 | 5.8 | 0.0 | 3.5 | 522.0 | 5,671 | 199 | 185 | 6,101 | 41,460 | 63,658 | 7,309 |
| Walker | 413 | 5,673 | 20.3 | 3.4 | 2.8 | 0.0 | 2.4 | 1,713.4 | 23,530 | 143 | 138 | 13,138 | 24,210 | 52,461 | 5,061 |
| Waller | 214 | 4,131 | 72.2 | 0.0 | 3.9 | 0.2 | 4.0 | 275.7 | 5,330 | 61 | 105 | 4,752 | 21,590 | 69,619 | 8,369 |
| Ward | 77 | 6,730 | 39.9 | 19.4 | 5.8 | 0.5 | 8.7 | 47.5 | 4,175 | 14 | 23 | 838 | 5,290 | 79,277 | 10,418 |
| Washington | 206 | 5,916 | 67.5 | 1.6 | 3.9 | 0.3 | 8.7 | 197.7 | 5,671 | 83 | 65 | 2,982 | 16,360 | 66,808 | 7,749 |
| Webb | 1,395 | 5,099 | 60.8 | 2.6 | 6.2 | 0.5 | 0.9 | 1,627.0 | 5,945 | 3,703 | 534 | 19,346 | 114,170 | 42,963 | 3,498 |
| Wharton | 242 | 5,793 | 51.4 | 26.4 | 4.5 | 0.1 | 4.3 | 109.0 | 2,606 | 78 | 80 | 2,862 | 19,470 | 54,001 | 5,484 |
| Wheeler | 55 | 10,287 | 53.2 | 27.9 | 2.4 | 0.2 | 2.3 | 22.7 | 4,283 | 23 | 10 | 577 | 2,200 | 57,244 | 5,678 |
| Wichita | 497 | 3,774 | 50.5 | 9.2 | 6.5 | 0.9 | 3.2 | 545.0 | 4,139 | 1,878 | 6,469 | 9,314 | 54,200 | 58,489 | 6,117 |
| Wilbarger | 84 | 6,606 | 52.7 | 25.9 | 3.1 | 0.1 | 3.9 | 19.7 | 1,555 | 39 | 24 | 2,476 | 5,570 | 46,988 | 6,680 |
| Willacy | 79 | 3,653 | 70.8 | 1.2 | 2.6 | 0.1 | 1.8 | 180.4 | 8,389 | 32 | 35 | 1,184 | 7,460 | 34,703 | 2,146 |
| Williamson | 2,269 | 4,154 | 57.0 | 3.9 | 4.1 | 0.4 | 5.6 | 5,149.2 | 9,426 | 892 | 1,185 | 24,411 | 270,870 | 82,161 | 9,668 |
| Wilson | 152 | 3,092 | 55.0 | 22.3 | 3.1 | 0.0 | 3.4 | 202.8 | 4,122 | 90 | 99 | 2,363 | 21,930 | 71,395 | 7,995 |
| Winkler | 71 | 9,371 | 49.9 | 17.1 | 4.9 | 0.4 | 2.0 | 73.8 | 9,706 | 9 | 15 | 568 | 3,390 | 73,120 | 8,996 |
| Wise | 186 | 2,817 | 50.9 | 2.4 | 6.3 | 0.5 | 7.0 | 276.0 | 4,192 | 126 | 135 | 4,395 | 30,150 | 67,381 | 6,983 |
| Wood | 100 | 2,253 | 68.4 | 0.3 | 4.7 | 0.0 | 5.8 | 60.6 | 1,370 | 96 | 85 | 1,815 | 19,270 | 53,507 | 4,905 |
| Yoakum | 65 | 7,624 | 80.4 | 1.0 | 2.1 | 0.1 | 2.3 | 113.0 | 13,200 | 14 | 17 | 812 | 3,650 | 67,239 | 6,744 |
| Young | 63 | 3,509 | 54.1 | 8.2 | 6.5 | 0.1 | 5.7 | 62.0 | 3,461 | 42 | 35 | 1,330 | 7,740 | 65,410 | 8,117 |
| Zapata | 58 | 4,044 | 67.2 | 1.5 | 4.3 | 1.2 | 2.5 | 40.6 | 2,848 | 147 | 28 | 907 | 5,180 | 34,290 | 2,026 |
| Zavala | 67 | 5,595 | 77.3 | 0.6 | 2.7 | 0.6 | 2.2 | 58.4 | 4,885 | 9 | 22 | 862 | 4,280 | 39,046 | 2,833 |
| **UTAH** | X | X | X | X | X | X | X | X | X | 37,303 | 16,661 | 211,410 | 1,363,440 | 72,115 | 7,917 |
| Beaver | 303 | 47,188 | 5.2 | 88.9 | 0.2 | 0.1 | 0.1 | 27.1 | 4,213 | 34 | 26 | 724 | 2,860 | 49,741 | 3,211 |
| Box Elder | 168 | 3,105 | 58.6 | 1.8 | 4.7 | 0.7 | 5.0 | 93.0 | 1,722 | 177 | 216 | 2,686 | 23,760 | 58,754 | 4,554 |
| Cache | 412 | 3,313 | 56.5 | 9.0 | 3.9 | 0.3 | 3.6 | 230.4 | 1,854 | 336 | 487 | 11,195 | 51,130 | 61,167 | 5,356 |
| Carbon | 88 | 4,345 | 45.5 | 6.5 | 5.0 | 1.4 | 10.2 | 29.4 | 1,459 | 138 | 77 | 1,964 | 8,120 | 51,795 | 4,024 |
| Daggett | 9 | 8,372 | 46.1 | 1.4 | 22.4 | 0.0 | 5.8 | 5.6 | 5,576 | 54 | 4 | 116 | 410 | 57,978 | 4,598 |
| Davis | 1,072 | 3,094 | 57.1 | 3.7 | 5.6 | 0.0 | 2.6 | 891.0 | 2,570 | 13,821 | 5,469 | 15,425 | 149,600 | 76,186 | 7,758 |
| Duchesne | 141 | 7,113 | 61.7 | 1.4 | 3.1 | 0.1 | 14.5 | 87.2 | 4,388 | 71 | 76 | 2,228 | 7,430 | 57,660 | 4,770 |
| Emery | 38 | 3,829 | 67.8 | 0.2 | 0.0 | 0.0 | 10.1 | 142.0 | 14,196 | 53 | 38 | 778 | 3,910 | 50,353 | 3,395 |
| Garfield | 33 | 6,564 | 33.9 | 7.7 | 11.5 | 1.0 | 16.1 | 3.9 | 784 | 141 | 19 | 377 | 2,200 | 45,777 | 3,352 |
| Grand | 51 | 5,286 | 31.3 | 8.3 | 8.3 | 1.0 | 7.8 | 38.0 | 3,966 | 250 | 37 | 823 | 5,120 | 56,606 | 5,163 |
| Iron | 139 | 2,742 | 58.7 | 0.7 | 7.1 | 0.0 | 5.1 | 43.5 | 858 | 289 | 209 | 4,421 | 21,340 | 50,334 | 3,705 |
| Juab | 45 | 4,006 | 58.6 | 0.9 | 3.6 | 0.3 | 4.6 | 47.3 | 4,186 | 31 | 46 | 834 | 4,630 | 57,810 | 4,417 |
| Kane | 52 | 6,968 | 32.9 | 18.8 | 4.7 | 0.8 | 4.7 | 47.5 | 6,311 | 80 | 29 | 618 | 3,340 | 52,534 | 4,284 |
| Millard | 70 | 5,435 | 47.5 | 2.0 | 7.1 | 0.2 | 9.9 | 23.6 | 1,841 | 81 | 50 | 939 | 5,140 | 52,541 | 3,760 |
| Morgan | 30 | 2,503 | 65.8 | 0.6 | 4.1 | 0.0 | 1.9 | 19.3 | 1,632 | 24 | 47 | 550 | 4,900 | 103,717 | 12,615 |
| Piute | 9 | 6,571 | 58.8 | 1.4 | 4.3 | 0.0 | 10.2 | 5.8 | 4,154 | 3 | 6 | 122 | 550 | 35,904 | 2,071 |
| Rich | 14 | 5,919 | 54.9 | 0.7 | 3.5 | 1.3 | 5.8 | 3.4 | 1,428 | 9 | 10 | 207 | 950 | 58,607 | 4,196 |
| Salt Lake | 4,414 | 3,883 | 36.6 | 1.3 | 5.8 | 3.4 | 7.4 | 9,978.0 | 8,778 | 11,502 | 4,688 | 96,905 | 530,070 | 74,501 | 8,699 |
| San Juan | 99 | 6,455 | 43.2 | 18.6 | 2.4 | 1.3 | 13.1 | 16.3 | 1,068 | 153 | 58 | 1,532 | 4,890 | 45,726 | 2,994 |
| Sanpete | 111 | 3,715 | 49.0 | 22.5 | 3.9 | 0.1 | 3.6 | 49.8 | 1,663 | 84 | 108 | 2,971 | 10,130 | 48,784 | 3,371 |
| Sevier | 92 | 4,298 | 46.0 | 8.2 | 6.6 | 0.0 | 13.5 | 53.7 | 2,521 | 176 | 82 | 1,704 | 8,470 | 55,447 | 4,478 |
| Summit | 336 | 8,132 | 28.6 | 0.5 | 2.1 | 0.0 | 2.5 | 277.9 | 6,723 | 57 | 162 | 2,902 | 23,380 | 169,283 | 32,200 |
| Tooele | 208 | 3,085 | 63.9 | 4.3 | 4.3 | 1.1 | 3.2 | 150.6 | 2,231 | 1,249 | 302 | 2,870 | 29,090 | 60,766 | 4,643 |
| Uintah | 174 | 4,934 | 38.7 | 11.4 | 4.1 | 0.0 | 10.8 | 89.5 | 2,540 | 385 | 137 | 2,867 | 13,160 | 58,622 | 4,876 |
| Utah | 1,833 | 3,021 | 55.9 | 3.8 | 4.3 | 0.3 | 5.4 | 2,304.2 | 3,798 | 1,076 | 2,420 | 30,770 | 245,300 | 72,988 | 7,820 |
| Wasatch | 162 | 5,066 | 62.0 | 1.0 | 4.2 | 0.0 | 6.0 | 140.3 | 4,401 | 52 | 130 | 1,831 | 14,450 | 90,180 | 12,202 |
| Washington | 515 | 3,103 | 51.2 | 5.2 | 8.3 | 0.1 | 2.6 | 527.3 | 3,178 | 595 | 677 | 8,592 | 73,900 | 68,082 | 7,492 |
| Wayne | 11 | 4,086 | 55.1 | 0.7 | 4.0 | 0.0 | 10.1 | 2.1 | 776 | 81 | 10 | 179 | 1,180 | 48,763 | 3,644 |
| Weber | 745 | 2,961 | 47.2 | 2.3 | 6.4 | 0.0 | 6.2 | 837.4 | 3,328 | 6,301 | 1,041 | 14,280 | 114,180 | 60,954 | 5,458 |
| **VERMONT** | X | X | X | X | X | X | X | X | X | 7,158 | 3,953 | 46,951 | 328,520 | 64,648 | 6,979 |
| Addison | 161 | 4,355 | 74.3 | 0.1 | 2.4 | 0.2 | 10.2 | 58.8 | 1,593 | 108 | 214 | 1,819 | 18,840 | 64,417 | 6,428 |
| Bennington | 192 | 5,367 | 76.1 | 0.2 | 3.8 | 0.1 | 7.8 | 45.5 | 1,276 | 186 | 215 | 2,207 | 18,620 | 60,725 | 6,881 |
| Caledonia | 122 | 4,061 | 73.7 | 0.2 | 2.0 | 0.0 | 10.4 | 34.3 | 1,139 | 97 | 182 | 1,955 | 14,830 | 50,804 | 4,430 |

1. Based on the resident population estimated as of July 1 of the year shown.

| State / county code | CBSA code[1] | County Type code[2] | STATE County | Land area[3] (sq. mi) | Total persons 2019 | Rank | Per square mile | White | Black | American Indian, Alaska Native | Asian and Pacific Islander | Percent Hispanic or Latino[4] | Under 5 years | 5 to 14 years | 15 to 24 years | 25 to 34 years | 35 to 44 years | 45 to 54 years |
|---|---|---|---|---|---|---|---|---|---|---|---|---|---|---|---|---|---|---|
| | | | | | Population, 2020 | | | Race alone or in combination, not Hispanic or Latino (percent) | | | | | Age (percent) | | | | | |
| | | | | 1 | 2 | 3 | 4 | 5 | 6 | 7 | 8 | 9 | 10 | 11 | 12 | 13 | 14 | 15 |
| | | | **VERMONT—Cont'd** | | | | | | | | | | | | | | | |
| 50007 | 15540 | 3 | Chittenden | 537.3 | 164,306 | 406 | 305.8 | 90.0 | 3.2 | 0.8 | 5.8 | 2.6 | 4.4 | 9.8 | 18.8 | 14.5 | 12.1 | 11.5 |
| 50009 | | 9 | Essex | 662.5 | 6,123 | 2,738 | 9.2 | 96.8 | 1.0 | 1.6 | 1.0 | 1.4 | 4.1 | 9.7 | 8.4 | 9.2 | 9.4 | 13.3 |
| 50011 | 15540 | 3 | Franklin | 630.7 | 49,685 | 996 | 78.8 | 95.6 | 1.3 | 2.5 | 1.3 | 1.8 | 5.8 | 12.4 | 10.9 | 13.0 | 12.6 | 13.2 |
| 50013 | 15540 | 3 | Grand Isle | 81.8 | 7,169 | 2,645 | 87.6 | 95.1 | 1.4 | 3.2 | 1.2 | 2.0 | 4.4 | 9.6 | 9.7 | 11.1 | 10.8 | 13.0 |
| 50015 | | 8 | Lamoille | 462.3 | 25,341 | 1,593 | 54.8 | 96.0 | 1.6 | 1.2 | 1.0 | 2.0 | 4.8 | 11.7 | 12.1 | 12.9 | 12.8 | 12.8 |
| 50017 | 30100 | 9 | Orange | 687.2 | 28,837 | 1,466 | 42.0 | 97.0 | 1.0 | 1.2 | 0.9 | 1.4 | 4.6 | 10.3 | 10.6 | 10.9 | 11.8 | 12.4 |
| 50019 | | 7 | Orleans | 694.5 | 26,897 | 1,531 | 38.7 | 96.6 | 1.2 | 1.5 | 0.9 | 1.5 | 4.9 | 11.0 | 10.7 | 11.2 | 11.3 | 12.3 |
| 50021 | 40860 | 4 | Rutland | 929.7 | 57,764 | 899 | 62.1 | 96.4 | 1.3 | 0.9 | 1.2 | 1.7 | 4.2 | 9.9 | 12.1 | 10.9 | 10.5 | 12.2 |
| 50023 | 12740 | 4 | Washington | 686.6 | 58,328 | 889 | 85.0 | 95.7 | 1.5 | 1.1 | 1.5 | 2.1 | 4.5 | 10.2 | 13.3 | 11.4 | 12.1 | 12.8 |
| 50025 | | 7 | Windham | 785.5 | 42,015 | 1,140 | 53.5 | 94.5 | 2.2 | 1.2 | 1.7 | 2.5 | 4.3 | 10.1 | 10.5 | 10.8 | 11.1 | 11.6 |
| 50027 | 30100 | 7 | Windsor | 969.8 | 54,988 | 932 | 56.7 | 95.7 | 1.3 | 1.1 | 1.5 | 2.1 | 4.3 | 10.2 | 9.5 | 11.1 | 11.5 | 12.0 |
| 51000 | | 0 | **VIRGINIA** | 39,482.1 | 8,590,563 | X | 217.6 | 63.3 | 20.6 | 0.8 | 8.4 | 10.0 | 5.9 | 12.2 | 13.0 | 13.9 | 13.1 | 12.6 |
| 51001 | | 8 | Accomack | 449.3 | 32,238 | 1,372 | 71.8 | 61.2 | 29.5 | 0.8 | 1.2 | 9.0 | 5.5 | 11.6 | 9.9 | 10.6 | 10.0 | 11.5 |
| 51003 | 16820 | 3 | Albemarle | 720.5 | 110,652 | 552 | 153.6 | 78.9 | 10.6 | 0.6 | 6.8 | 5.8 | 5.2 | 11.1 | 14.9 | 12.9 | 12.1 | 11.2 |
| 51005 | | 6 | Alleghany | 446.6 | 14,701 | 2,108 | 32.9 | 93.0 | 5.8 | 0.6 | 0.7 | 1.7 | 4.5 | 10.3 | 10.7 | 10.2 | 9.7 | 13.2 |
| 51007 | 40060 | 1 | Amelia | 355.4 | 13,014 | 2,219 | 36.6 | 75.4 | 21.2 | 0.9 | 0.8 | 3.5 | 5.1 | 11.6 | 10.5 | 11.4 | 10.9 | 12.7 |
| 51009 | 31340 | 2 | Amherst | 474.0 | 31,667 | 1,387 | 66.8 | 77.0 | 20.4 | 1.5 | 1.1 | 2.5 | 5.2 | 11.0 | 11.3 | 11.5 | 11.0 | 13.1 |
| 51011 | 31340 | 2 | Appomattox | 334.2 | 16,043 | 2,030 | 48.0 | 79.1 | 19.7 | 0.5 | 0.7 | 2.1 | 6.2 | 11.6 | 10.8 | 12.4 | 11.4 | 12.1 |
| 51013 | 47900 | 1 | Arlington | 26.0 | 240,119 | 287 | 9,235.3 | 63.7 | 10.2 | 0.7 | 13.2 | 15.5 | 5.6 | 9.9 | 11.1 | 23.2 | 16.4 | 12.5 |
| 51015 | 44420 | 3 | Augusta | 967.1 | 76,544 | 727 | 79.1 | 91.3 | 5.3 | 0.6 | 1.0 | 3.4 | 4.8 | 10.5 | 10.9 | 11.4 | 11.7 | 12.9 |
| 51017 | | 8 | Bath | 529.2 | 4,119 | 2,886 | 7.8 | 93.1 | 5.0 | 0.5 | 0.7 | 2.4 | 4.3 | 8.4 | 8.6 | 10.5 | 9.3 | 13.4 |
| 51019 | 31340 | 2 | Bedford | 760.1 | 79,811 | 709 | 105.0 | 88.8 | 7.8 | 0.7 | 1.7 | 2.6 | 4.6 | 11.1 | 11.0 | 10.5 | 10.2 | 13.3 |
| 51021 | 14140 | 8 | Bland | 357.7 | 6,239 | 2,726 | 17.4 | 94.3 | 4.6 | 0.5 | 0.7 | 1.0 | 3.7 | 8.2 | 9.2 | 12.3 | 12.9 | 15.2 |
| 51023 | 40220 | 2 | Botetourt | 541.3 | 33,633 | 1,337 | 62.1 | 93.8 | 3.9 | 0.7 | 1.2 | 2.0 | 4.4 | 10.7 | 10.6 | 9.7 | 10.7 | 13.9 |
| 51025 | | 6 | Brunswick | 566.2 | 16,037 | 2,031 | 28.3 | 41.9 | 55.1 | 0.5 | 1.3 | 2.5 | 4.2 | 8.6 | 11.7 | 14.5 | 12.0 | 11.9 |
| 51027 | | 9 | Buchanan | 502.9 | 20,613 | 1,789 | 41.0 | 95.3 | 3.5 | 0.4 | 0.6 | 0.9 | 4.1 | 9.9 | 9.7 | 11.3 | 11.3 | 13.8 |
| 51029 | | 3 | Buckingham | 579.6 | 17,168 | 1,965 | 29.6 | 63.5 | 34.7 | 0.7 | 0.7 | 2.4 | 4.5 | 10.5 | 9.9 | 13.9 | 13.0 | 13.4 |
| 51031 | 31340 | 2 | Campbell | 503.2 | 55,304 | 930 | 109.9 | 80.6 | 16.3 | 0.8 | 1.5 | 3.0 | 5.5 | 11.0 | 11.0 | 13.3 | 11.2 | 12.6 |
| 51033 | | 1 | Caroline | 527.4 | 30,860 | 1,410 | 58.5 | 66.0 | 28.7 | 1.4 | 1.8 | 5.6 | 6.2 | 13.0 | 10.3 | 13.9 | 12.8 | 12.5 |
| 51035 | | 7 | Carroll | 474.7 | 30,074 | 1,433 | 63.4 | 94.6 | 1.4 | 0.5 | 0.4 | 4.1 | 4.3 | 10.0 | 9.8 | 10.2 | 10.2 | 13.7 |
| 51036 | 40060 | 1 | Charles City | 182.9 | 6,821 | 2,675 | 37.3 | 46.4 | 46.1 | 7.4 | 1.4 | 2.1 | 3.6 | 8.1 | 9.0 | 10.6 | 9.6 | 13.1 |
| 51037 | | 8 | Charlotte | 475.3 | 11,820 | 2,305 | 24.9 | 69.3 | 28.3 | 0.8 | 0.6 | 2.7 | 5.6 | 11.7 | 11.0 | 10.6 | 10.0 | 12.1 |
| 51041 | 40060 | 1 | Chesterfield | 423.9 | 358,245 | 205 | 845.9 | 62.1 | 25.4 | 0.8 | 4.7 | 9.8 | 6.0 | 13.3 | 12.8 | 12.2 | 13.5 | 13.3 |
| 51043 | 47900 | 1 | Clarke | 175.9 | 14,622 | 2,113 | 83.1 | 87.5 | 5.5 | 1.0 | 2.0 | 6.3 | 4.2 | 11.0 | 11.1 | 9.7 | 10.6 | 13.9 |
| 51045 | 40220 | 2 | Craig | 328.1 | 5,077 | 2,821 | 15.5 | 97.3 | 0.9 | 0.7 | 0.5 | 1.6 | 4.1 | 9.6 | 10.7 | 10.2 | 9.6 | 14.0 |
| 51047 | 47900 | 1 | Culpeper | 379.2 | 53,569 | 949 | 141.3 | 72.0 | 15.7 | 0.8 | 2.5 | 12.2 | 6.5 | 14.0 | 11.9 | 11.9 | 12.9 | 12.6 |
| 51049 | | 8 | Cumberland | 297.5 | 9,933 | 2,424 | 33.4 | 65.6 | 31.7 | 1.1 | 0.9 | 3.2 | 5.2 | 10.9 | 10.3 | 12.1 | 10.0 | 12.6 |
| 51051 | | 9 | Dickenson | 330.5 | 14,078 | 2,153 | 42.6 | 98.0 | 0.9 | 0.4 | 0.4 | 1.0 | 4.5 | 11.2 | 10.1 | 11.0 | 12.4 | 12.6 |
| 51053 | 40060 | 1 | Dinwiddie | 503.9 | 28,688 | 1,469 | 56.9 | 63.7 | 31.9 | 0.7 | 1.4 | 3.9 | 5.3 | 11.6 | 11.1 | 13.0 | 11.2 | 13.6 |
| 51057 | | 6 | Essex | 257.3 | 10,943 | 2,354 | 42.5 | 56.8 | 39.0 | 1.2 | 1.7 | 4.0 | 4.6 | 9.7 | 10.4 | 11.7 | 10.2 | 12.6 |
| 51059 | 47900 | 1 | Fairfax | 391.0 | 1,150,847 | 39 | 2,943.3 | 52.7 | 11.1 | 0.6 | 22.8 | 16.5 | 6.2 | 13.0 | 12.3 | 13.3 | 14.4 | 13.8 |
| 51061 | 47900 | 1 | Fauquier | 648.0 | 71,361 | 765 | 110.1 | 81.0 | 9.0 | 0.8 | 2.6 | 9.4 | 5.7 | 13.0 | 11.9 | 11.1 | 12.1 | 13.7 |
| 51063 | | 3 | Floyd | 380.9 | 15,777 | 2,046 | 41.4 | 94.7 | 2.5 | 0.7 | 0.9 | 2.7 | 4.8 | 10.6 | 10.0 | 10.0 | 10.9 | 13.4 |
| 51065 | 16820 | 3 | Fluvanna | 287.1 | 27,422 | 1,515 | 95.5 | 79.7 | 16.6 | 0.8 | 1.6 | 3.8 | 5.0 | 10.9 | 10.7 | 12.5 | 12.7 | 12.8 |
| 51067 | 40220 | 2 | Franklin | 690.6 | 56,167 | 917 | 81.3 | 88.5 | 8.9 | 0.6 | 0.8 | 2.9 | 4.6 | 10.7 | 10.9 | 10.1 | 9.6 | 12.9 |
| 51069 | 49020 | 3 | Frederick | 413.1 | 91,119 | 651 | 220.6 | 83.6 | 5.7 | 0.6 | 2.5 | 9.8 | 5.7 | 12.9 | 11.5 | 12.2 | 12.6 | 13.0 |
| 51071 | 13980 | 3 | Giles | 357.2 | 16,663 | 1,995 | 46.6 | 95.6 | 2.4 | 0.7 | 0.8 | 1.9 | 5.0 | 11.1 | 11.0 | 11.5 | 10.8 | 13.8 |
| 51073 | 47260 | 1 | Gloucester | 217.8 | 37,459 | 1,239 | 172.0 | 87.0 | 8.9 | 1.2 | 1.7 | 3.9 | 5.0 | 11.3 | 10.2 | 12.0 | 11.7 | 12.6 |
| 51075 | 40060 | 1 | Goochland | 282.0 | 24,431 | 1,630 | 86.6 | 79.2 | 16.2 | 0.7 | 2.6 | 3.2 | 3.8 | 9.7 | 10.5 | 10.4 | 10.7 | 13.9 |
| 51077 | | 9 | Grayson | 441.8 | 15,493 | 2,059 | 35.1 | 90.3 | 6.2 | 0.7 | 0.3 | 3.6 | 4.0 | 9.4 | 9.7 | 11.0 | 10.4 | 13.6 |
| 51079 | 16820 | 3 | Greene | 155.9 | 20,131 | 1,817 | 129.1 | 83.4 | 8.9 | 0.8 | 2.9 | 6.6 | 5.9 | 13.3 | 11.0 | 12.4 | 12.1 | 12.4 |
| 51081 | | 6 | Greensville | 295.2 | 11,280 | 2,330 | 38.2 | 36.3 | 60.6 | 0.6 | 1.0 | 2.6 | 3.7 | 9.2 | 11.7 | 16.3 | 14.9 | 15.3 |
| 51083 | | 6 | Halifax | 817.7 | 33,633 | 1,337 | 41.1 | 60.8 | 36.6 | 0.7 | 0.9 | 2.3 | 5.3 | 11.2 | 10.8 | 10.6 | 10.0 | 11.9 |
| 51085 | 40060 | 1 | Hanover | 467.6 | 108,262 | 564 | 231.5 | 84.8 | 10.3 | 0.8 | 2.8 | 3.2 | 4.9 | 12.5 | 12.6 | 10.4 | 12.4 | 13.4 |
| 51087 | 40060 | 1 | Henrico | 233.7 | 333,766 | 214 | 1,428.2 | 53.6 | 31.6 | 0.8 | 10.5 | 6.1 | 6.0 | 12.4 | 11.4 | 14.4 | 13.6 | 12.8 |
| 51089 | 32300 | 4 | Henry | 382.4 | 50,309 | 986 | 131.6 | 70.9 | 23.6 | 0.7 | 0.8 | 5.9 | 4.7 | 11.1 | 10.4 | 10.5 | 10.0 | 12.8 |
| 51091 | | 8 | Highland | 415.2 | 2,200 | 3,026 | 5.3 | 96.6 | 1.2 | 0.3 | 0.9 | 1.3 | 4.0 | 6.9 | 6.5 | 9.3 | 7.0 | 9.7 |
| 51093 | 47260 | 1 | Isle of Wight | 315.7 | 37,725 | 1,230 | 119.5 | 72.1 | 23.6 | 1.0 | 2.0 | 3.8 | 5.2 | 12.0 | 10.6 | 11.2 | 12.1 | 13.8 |
| 51095 | 47260 | 1 | James City | 142.3 | 77,612 | 721 | 545.4 | 77.1 | 14.6 | 0.9 | 4.0 | 6.3 | 4.6 | 11.2 | 10.5 | 10.0 | 11.1 | 11.8 |
| 51097 | 40060 | 8 | King and Queen | 315.1 | 6,942 | 2,665 | 22.0 | 68.7 | 27.1 | 2.4 | 1.1 | 3.6 | 4.6 | 9.6 | 9.6 | 10.6 | 10.7 | 12.9 |
| 51099 | | 6 | King George | 179.6 | 27,381 | 1,517 | 152.5 | 76.1 | 17.2 | 1.3 | 2.9 | 6.0 | 5.8 | 14.3 | 12.5 | 12.7 | 13.6 | 13.4 |
| 51101 | 40060 | 1 | King William | 273.9 | 17,641 | 1,937 | 64.4 | 79.2 | 16.4 | 2.2 | 1.9 | 2.9 | 6.1 | 12.7 | 11.1 | 13.4 | 12.5 | 12.9 |
| 51103 | | 9 | Lancaster | 133.1 | 10,618 | 2,381 | 79.7 | 68.1 | 29.3 | 0.7 | 0.9 | 2.3 | 4.0 | 8.5 | 8.3 | 8.4 | 7.5 | 9.1 |
| 51105 | | 8 | Lee | 435.4 | 23,238 | 1,672 | 53.4 | 93.8 | 4.0 | 0.8 | 0.6 | 1.9 | 4.5 | 10.3 | 9.8 | 12.9 | 11.5 | 13.5 |
| 51107 | 47900 | 1 | Loudoun | 515.7 | 422,784 | 168 | 819.8 | 57.2 | 8.9 | 0.6 | 23.3 | 13.9 | 6.7 | 15.9 | 12.4 | 11.3 | 17.0 | 15.6 |

1. CBSA = Core Based Statistical Area. See Appendix A for explanation. See Appendix B for list of metropolitan areas with component counties.    2. County type code from the Economic Research Service of USDA Rural-Urban Continuum Codes. See Appendix A for definition.    3. Dry land or land partially or temporarily covered by water.    4. May be of any race.

Items 1—15

| STATE County | Population, 2020 (cont.) Age (percent) (cont.) | | | | Population change, 2000-2020 Total persons | | Percent change | | Components of change, 2010-2020 | | | Households, 2015-2019 | | | Percent | |
|---|---|---|---|---|---|---|---|---|---|---|---|---|---|---|---|---|
| | 55 to 64 years | 65 to 74 years | 75 years and over | Percent female | 2000 | 2010 | 2000-2010 | 2010-2020 | Births | Deaths | Net Migration | Number | Persons per household | Family households | Female family householder[1] | One person |
| | 16 | 17 | 18 | 19 | 20 | 21 | 22 | 23 | 24 | 25 | 26 | 27 | 28 | 29 | 30 | 31 |
| **VERMONT—Cont'd** | | | | | | | | | | | | | | | | |
| Chittenden | 12.7 | 9.4 | 6.7 | 51.0 | 146,571 | 156,535 | 6.8 | 5.0 | 15,547 | 11,246 | 3,592 | 66,160 | 2.31 | 56.6 | 8.0 | 28.9 |
| Essex | 18.1 | 16.5 | 11.3 | 50.2 | 6,459 | 6,306 | -2.4 | -2.9 | 522 | 673 | -30 | 2,740 | 2.25 | 62.6 | 10.1 | 32.7 |
| Franklin | 14.9 | 10.6 | 6.5 | 50.2 | 45,417 | 47,759 | 5.2 | 4.0 | 5,854 | 4,223 | 334 | 18,563 | 2.62 | 67.2 | 8.9 | 26.8 |
| Grand Isle | 19.1 | 15.2 | 7.0 | 49.2 | 6,901 | 6,970 | 1.0 | 2.9 | 613 | 608 | 200 | 2,956 | 2.38 | 68.2 | 6.7 | 25.0 |
| Lamoille | 14.5 | 10.9 | 7.4 | 50.0 | 23,233 | 24,473 | 5.3 | 3.5 | 2,603 | 2,058 | 336 | 10,546 | 2.34 | 62.6 | 9.0 | 27.4 |
| Orange | 16.7 | 14.4 | 8.3 | 50.1 | 28,226 | 28,938 | 2.5 | -0.3 | 2,704 | 2,736 | -43 | 12,279 | 2.30 | 62.5 | 7.9 | 29.6 |
| Orleans | 15.0 | 14.0 | 9.6 | 50.1 | 26,277 | 27,234 | 3.6 | -1.2 | 2,704 | 3,200 | 191 | 11,779 | 2.21 | 63.8 | 9.5 | 29.8 |
| Rutland | 16.6 | 14.4 | 9.2 | 50.5 | 63,400 | 61,659 | -2.7 | -6.3 | 5,327 | 7,037 | -2,175 | 25,338 | 2.24 | 58.2 | 7.1 | 34.4 |
| Washington | 14.8 | 12.7 | 8.1 | 50.3 | 58,039 | 59,525 | 2.6 | -2.0 | 5,671 | 5,567 | -1,278 | 24,828 | 2.26 | 60.1 | 10.6 | 32.3 |
| Windham | 16.9 | 15.2 | 9.3 | 51.0 | 44,216 | 44,511 | 0.7 | -5.6 | 3,905 | 4,630 | -1,761 | 18,943 | 2.17 | 56.6 | 10.1 | 35.2 |
| Windsor | 16.6 | 14.7 | 10.1 | 51.0 | 57,418 | 56,658 | -1.3 | -2.9 | 4,837 | 6,049 | -422 | 24,298 | 2.24 | 60.1 | 7.8 | 32.4 |
| **VIRGINIA** | 13.0 | 9.6 | 6.7 | 50.8 | 7,078,515 | 8,001,046 | 13.0 | 7.4 | 1,041,285 | 674,614 | 221,973 | 3,151,045 | 2.61 | 66.4 | 11.7 | 27.1 |
| Accomack | 15.7 | 14.8 | 10.4 | 51.1 | 38,305 | 33,162 | -13.4 | -2.8 | 3,964 | 4,484 | -371 | 13,438 | 2.35 | 66.4 | 13.4 | 29.5 |
| Albemarle | 12.8 | 11.3 | 8.5 | 52.2 | 79,236 | 98,984 | 24.9 | 11.8 | 11,530 | 8,447 | 8,634 | 41,496 | 2.42 | 62.6 | 8.9 | 29.1 |
| Alleghany | 15.5 | 14.1 | 11.8 | 51.3 | 17,215 | 16,264 | -5.5 | -9.6 | 1,407 | 2,301 | -665 | 6,600 | 2.24 | 64.1 | 10.6 | 31.7 |
| Amelia | 17.3 | 11.9 | 8.5 | 50.4 | 11,400 | 12,690 | 11.3 | 2.6 | 1,401 | 1,503 | 431 | 4,954 | 2.59 | 69.7 | 8.5 | 25.1 |
| Amherst | 15.4 | 12.4 | 9.5 | 51.6 | 31,894 | 32,354 | 1.4 | -2.1 | 3,363 | 3,609 | -448 | 12,160 | 2.52 | 68.3 | 12.2 | 27.2 |
| Appomattox | 14.4 | 11.9 | 9.2 | 51.3 | 13,705 | 15,022 | 9.6 | 6.8 | 1,870 | 1,782 | 938 | 6,091 | 2.57 | 74.3 | 13.5 | 21.6 |
| Arlington | 9.9 | 6.8 | 4.5 | 49.9 | 189,453 | 207,696 | 9.6 | 15.6 | 31,305 | 9,585 | 10,462 | 107,032 | 2.15 | 46.0 | 5.6 | 39.2 |
| Augusta | 15.3 | 12.5 | 9.5 | 49.4 | 65,615 | 73,759 | 12.4 | 3.8 | 7,010 | 7,335 | 3,154 | 29,539 | 2.40 | 70.9 | 9.1 | 24.8 |
| Bath | 17.2 | 15.9 | 12.5 | 49.5 | 5,048 | 4,727 | -6.4 | -12.9 | 383 | 673 | -317 | 1,801 | 2.28 | 61.1 | 0.0 | 27.3 |
| Bedford | 16.6 | 13.5 | 9.2 | 50.7 | 60,371 | 74,928 | 24.1 | 6.5 | 7,442 | 8,016 | 5,503 | 31,317 | 2.49 | 72.2 | 8.4 | 23.6 |
| Bland | 14.0 | 14.1 | 10.4 | 45.2 | 6,871 | 6,826 | -0.7 | -8.6 | 468 | 856 | -194 | 2,367 | 2.32 | 73.0 | 10.4 | 26.3 |
| Botetourt | 16.4 | 14.0 | 9.5 | 50.4 | 30,496 | 33,149 | 8.7 | 1.5 | 2,675 | 3,478 | 1,318 | 13,264 | 2.49 | 72.7 | 6.9 | 23.9 |
| Brunswick | 15.1 | 12.9 | 9.3 | 46.3 | 18,419 | 17,418 | -5.4 | -7.9 | 1,463 | 2,146 | -684 | 6,037 | 2.34 | 62.2 | 20.8 | 33.1 |
| Buchanan | 16.0 | 13.5 | 10.3 | 49.5 | 26,978 | 24,109 | -10.6 | -14.5 | 1,980 | 2,981 | -2,510 | 8,569 | 2.44 | 67.4 | 13.7 | 28.4 |
| Buckingham | 14.5 | 12.2 | 8.0 | 45.0 | 15,623 | 17,140 | 9.7 | 0.2 | 1,603 | 1,671 | 90 | 5,826 | 2.56 | 67.7 | 14.4 | 27.9 |
| Campbell | 14.6 | 11.8 | 9.1 | 51.4 | 51,078 | 54,815 | 7.3 | 0.9 | 6,120 | 5,716 | 123 | 23,071 | 2.37 | 66.0 | 11.1 | 29.3 |
| Caroline | 14.0 | 10.4 | 6.9 | 50.6 | 22,121 | 28,550 | 29.1 | 8.1 | 3,929 | 2,802 | 1,178 | 10,946 | 2.57 | 71.5 | 15.5 | 22.9 |
| Carroll | 15.7 | 14.9 | 11.1 | 50.3 | 29,245 | 30,072 | 2.8 | 0.0 | 2,582 | 3,683 | 1,133 | 12,357 | 2.40 | 66.0 | 10.4 | 30.2 |
| Charles City | 19.5 | 16.0 | 10.6 | 51.5 | 6,926 | 7,252 | 4.7 | -5.9 | 593 | 864 | -158 | 2,896 | 2.42 | 65.0 | 13.4 | 30.8 |
| Charlotte | 16.5 | 12.6 | 10.0 | 50.4 | 12,472 | 12,597 | 1.0 | -6.2 | 1,354 | 1,656 | -467 | 4,661 | 2.55 | 62.9 | 13.1 | 33.1 |
| Chesterfield | 13.0 | 10.0 | 5.8 | 51.9 | 259,903 | 316,236 | 21.7 | 13.3 | 39,709 | 24,472 | 27,063 | 124,971 | 2.72 | 73.0 | 12.7 | 22.1 |
| Clarke | 17.2 | 13.1 | 9.1 | 50.6 | 12,652 | 14,028 | 10.9 | 4.2 | 1,271 | 1,575 | 903 | 5,612 | 2.54 | 69.4 | 9.2 | 25.7 |
| Craig | 17.3 | 14.3 | 10.3 | 50.2 | 5,091 | 5,175 | 1.6 | -1.9 | 413 | 541 | 31 | 2,297 | 2.22 | 62.6 | 9.8 | 34.8 |
| Culpeper | 13.7 | 9.9 | 6.6 | 50.2 | 34,262 | 46,687 | 36.3 | 14.7 | 6,506 | 4,314 | 4,706 | 17,071 | 2.91 | 74.1 | 10.9 | 20.2 |
| Cumberland | 15.9 | 13.4 | 9.5 | 51.8 | 9,017 | 10,039 | 11.3 | -1.1 | 1,063 | 985 | -185 | 3,975 | 2.46 | 65.2 | 10.8 | 28.7 |
| Dickenson | 14.6 | 13.6 | 10.1 | 48.8 | 16,395 | 15,877 | -3.2 | -11.3 | 1,499 | 2,119 | -1,178 | 5,778 | 2.47 | 66.1 | 11.7 | 31.3 |
| Dinwiddie | 15.8 | 11.0 | 7.5 | 50.9 | 24,533 | 28,011 | 14.2 | 2.4 | 3,075 | 2,708 | 326 | 10,401 | 2.66 | 67.9 | 13.5 | 28.0 |
| Essex | 16.5 | 14.7 | 9.6 | 53.1 | 9,989 | 11,149 | 11.6 | -1.8 | 1,186 | 1,357 | -25 | 4,555 | 2.38 | 65.3 | 13.3 | 30.0 |
| Fairfax | 12.7 | 8.6 | 5.8 | 50.3 | 969,749 | 1,081,681 | 11.5 | 6.4 | 151,795 | 51,808 | -31,392 | 396,501 | 2.87 | 71.5 | 9.1 | 22.5 |
| Fauquier | 15.2 | 10.1 | 7.2 | 50.5 | 55,139 | 65,222 | 18.3 | 9.4 | 7,779 | 5,585 | 3,986 | 24,562 | 2.82 | 73.4 | 9.4 | 21.7 |
| Floyd | 15.9 | 14.3 | 10.0 | 50.0 | 13,874 | 15,293 | 10.2 | 3.2 | 1,530 | 1,632 | 595 | 6,493 | 2.40 | 69.0 | 8.6 | 26.5 |
| Fluvanna | 14.5 | 12.3 | 8.6 | 54.1 | 20,047 | 25,743 | 28.4 | 6.5 | 2,704 | 2,237 | 1,227 | 9,923 | 2.55 | 71.5 | 8.5 | 22.5 |
| Franklin | 16.5 | 14.7 | 10.1 | 50.7 | 47,286 | 56,124 | 18.7 | 0.1 | 5,219 | 6,173 | 1,058 | 22,997 | 2.38 | 70.9 | 11.7 | 24.4 |
| Frederick | 13.8 | 10.6 | 7.7 | 50.2 | 59,209 | 78,266 | 32.2 | 16.4 | 9,812 | 6,945 | 10,003 | 31,503 | 2.70 | 74.9 | 10.9 | 20.5 |
| Giles | 14.5 | 12.7 | 9.6 | 50.8 | 16,657 | 17,288 | 3.8 | -3.6 | 1,752 | 2,327 | -40 | 6,910 | 2.41 | 64.6 | 10.1 | 27.9 |
| Gloucester | 16.8 | 12.1 | 8.4 | 50.8 | 34,780 | 36,859 | 6.0 | 1.6 | 3,690 | 3,961 | 895 | 14,786 | 2.50 | 70.8 | 8.9 | 23.4 |
| Goochland | 17.8 | 14.3 | 8.9 | 50.7 | 16,863 | 21,852 | 29.6 | 11.8 | 1,766 | 1,979 | 2,776 | 8,506 | 2.55 | 72.1 | 5.0 | 24.4 |
| Grayson | 16.1 | 14.7 | 11.1 | 48.5 | 17,917 | 15,557 | -13.2 | -0.4 | 1,381 | 2,126 | 644 | 6,498 | 2.25 | 63.8 | 8.7 | 32.0 |
| Greene | 14.3 | 11.4 | 7.2 | 50.9 | 15,244 | 18,412 | 20.8 | 9.3 | 2,251 | 1,594 | 1,074 | 7,548 | 2.57 | 70.5 | 11.2 | 23.9 |
| Greensville | 12.9 | 10.1 | 6.0 | 37.1 | 11,560 | 12,245 | 5.9 | -7.9 | 954 | 1,399 | -547 | 3,647 | 2.26 | 64.9 | 16.0 | 31.7 |
| Halifax | 14.7 | 14.5 | 11.1 | 52.1 | 37,355 | 36,253 | -3.0 | -7.2 | 3,724 | 4,927 | -1,391 | 14,158 | 2.38 | 65.2 | 14.6 | 31.8 |
| Hanover | 15.1 | 11.1 | 7.6 | 51.0 | 86,320 | 99,833 | 15.7 | 8.4 | 9,497 | 9,329 | 8,340 | 38,987 | 2.66 | 74.7 | 8.2 | 21.2 |
| Henrico | 12.8 | 9.7 | 6.8 | 52.6 | 262,300 | 306,698 | 16.9 | 8.8 | 41,311 | 28,018 | 13,968 | 128,464 | 2.53 | 62.9 | 13.5 | 30.9 |
| Henry | 15.7 | 13.2 | 11.6 | 51.8 | 57,930 | 54,144 | -6.5 | -7.1 | 4,945 | 7,158 | -1,592 | 21,218 | 2.38 | 61.0 | 12.1 | 34.0 |
| Highland | 20.2 | 22.1 | 14.4 | 50.5 | 2,536 | 2,319 | -8.6 | -5.1 | 165 | 292 | 9 | 1,078 | 2.04 | 66.1 | 5.0 | 29.3 |
| Isle of Wight | 16.3 | 12.0 | 8.0 | 51.0 | 29,728 | 35,279 | 18.7 | 6.9 | 3,598 | 3,648 | 2,522 | 14,587 | 2.49 | 70.5 | 11.4 | 25.1 |
| James City | 14.1 | 14.7 | 12.0 | 51.8 | 48,102 | 67,400 | 40.1 | 15.2 | 6,738 | 7,362 | 10,797 | 28,920 | 2.55 | 71.6 | 8.3 | 24.7 |
| King and Queen | 18.1 | 14.2 | 9.8 | 49.0 | 6,630 | 6,940 | 4.7 | 0.0 | 651 | 844 | 199 | 2,707 | 2.60 | 62.5 | 9.1 | 31.2 |
| King George | 13.6 | 8.5 | 5.6 | 49.4 | 16,803 | 23,578 | 40.3 | 16.1 | 3,134 | 1,741 | 2,405 | 9,202 | 2.82 | 75.3 | 8.3 | 18.3 |
| King William | 14.6 | 10.2 | 6.5 | 51.0 | 13,146 | 15,927 | 21.2 | 10.8 | 2,025 | 1,498 | 1,193 | 6,078 | 2.74 | 77.6 | 12.7 | 20.8 |
| Lancaster | 16.9 | 19.5 | 17.7 | 52.6 | 11,567 | 11,390 | -1.5 | -6.8 | 857 | 2,140 | 528 | 5,062 | 2.09 | 63.0 | 9.3 | 34.5 |
| Lee | 14.4 | 13.4 | 9.7 | 47.9 | 23,589 | 25,591 | 8.5 | -9.2 | 2,306 | 2,872 | -1,790 | 9,149 | 2.45 | 63.9 | 13.4 | 32.4 |
| Loudoun | 10.9 | 5.9 | 4.2 | 50.2 | 169,599 | 312,342 | 84.2 | 35.4 | 52,465 | 13,320 | 70,619 | 128,637 | 3.06 | 78.1 | 7.7 | 16.9 |

1. No spouse present.

# Table B. States and Counties — Population, Vital Statistics, Health, and Crime

| STATE County | Persons in group quarters, 2020 | Daytime Population, 2015-2019 Number | Daytime Population Employment/ residence ratio | Births, 2020 Total | Births, 2020 Rate[1] | Deaths, 2020 Number | Deaths, 2020 Rate[1] | Persons under 65 with no health insurance, 2019 Number | Persons under 65... Percent | Medicare, 2020 Total beneficiaries | Medicare Enrolled in Original Medicare | Medicare Enrolled in Medicare Advantage | Crimes reported... 2019 Violent | Crimes... Property |
|---|---|---|---|---|---|---|---|---|---|---|---|---|---|---|
| | 32 | 33 | 34 | 35 | 36 | 37 | 38 | 39 | 40 | 41 | 42 | 43 | 44 | 45 |
| **VERMONT—Cont'd** | | | | | | | | | | | | | | |
| Chittenden | 10,030 | 175,865 | 1.15 | 1,389 | 8.5 | 1,242 | 7.6 | 6,400 | 4.9 | 30,477 | 25,902 | 4,575 | NA | NA |
| Essex | 16 | 4,823 | 0.49 | 43 | 7.0 | 79 | 12.9 | 289 | 6.5 | 1,854 | 1,612 | 242 | 0 | 18 |
| Franklin | 575 | 43,481 | 0.77 | 538 | 10.8 | 430 | 8.7 | 2,199 | 5.4 | 10,370 | 8,897 | 1,473 | 6 | 26 |
| Grand Isle | 0 | 5,078 | 0.45 | 64 | 8.9 | 60 | 8.4 | 301 | 5.3 | 2,071 | 1,726 | 345 | 4 | 27 |
| Lamoille | 738 | 24,361 | 0.93 | 235 | 9.3 | 207 | 8.2 | 1,392 | 6.9 | 5,512 | 4,799 | 713 | 9 | 59 |
| Orange | 661 | 23,339 | 0.63 | 245 | 8.5 | 305 | 10.6 | 1,371 | 6.3 | 7,480 | 6,640 | 840 | 6 | 18 |
| Orleans | 780 | 26,511 | 0.97 | 262 | 9.7 | 319 | 11.9 | 1,337 | 6.7 | 7,646 | 6,585 | 1,061 | 0 | 75 |
| Rutland | 2,186 | 58,133 | 0.97 | 453 | 7.8 | 701 | 12.1 | 2,381 | 5.5 | 16,836 | 14,357 | 2,480 | 2 | 13 |
| Washington | 2,960 | 62,805 | 1.15 | 489 | 8.4 | 579 | 9.9 | 2,310 | 5.2 | 14,215 | 12,656 | 1,559 | 1 | 1 |
| Windham | 1,364 | 44,763 | 1.09 | 335 | 8.0 | 475 | 11.3 | 1,927 | 6.2 | 12,113 | 10,311 | 1,802 | 2 | 23 |
| Windsor | 757 | 54,094 | 0.96 | 431 | 7.8 | 588 | 10.7 | 2,646 | 6.4 | 15,916 | 13,379 | 2,537 | 0 | 15 |
| **VIRGINIA** | 245,393 | 8,368,338 | 0.98 | 98,147 | 11.4 | 75,821 | 8.8 | 640,265 | 9.2 | 1,545,049 | 1,156,714 | 388,335 | X | X |
| Accomack | 429 | 32,924 | 1.02 | 339 | 10.5 | 478 | 14.8 | 3,601 | 15.1 | 9,188 | 7,299 | 1,889 | 100 | 333 |
| Albemarle | 6,680 | 112,326 | 1.09 | 1,094 | 9.9 | 973 | 8.8 | 7,182 | 8.7 | 22,137 | 19,186 | 2,951 | 116 | 1,267 |
| Alleghany | 281 | 14,758 | 0.93 | 118 | 8.0 | 252 | 17.1 | 1,040 | 9.5 | 4,640 | 3,674 | 967 | 14 | 88 |
| Amelia | 128 | 9,350 | 0.43 | 129 | 9.9 | 181 | 13.9 | 1,208 | 11.6 | 3,167 | 2,208 | 959 | 26 | 172 |
| Amherst | 1,401 | 25,436 | 0.56 | 314 | 9.9 | 423 | 13.4 | 2,436 | 10.2 | 7,970 | 5,937 | 2,033 | 74 | 356 |
| Appomattox | 56 | 11,994 | 0.48 | 200 | 12.5 | 194 | 12.1 | 1,485 | 11.9 | 4,054 | 3,072 | 982 | 20 | 85 |
| Arlington | 2,955 | 282,019 | 1.33 | 2,732 | 11.4 | 1,099 | 4.6 | 12,911 | 6.1 | 25,177 | 20,757 | 4,419 | 298 | 3,085 |
| Augusta | 2,915 | 67,724 | 0.79 | 613 | 8.0 | 875 | 11.4 | 5,609 | 10.0 | 18,824 | 15,401 | 3,423 | 90 | 801 |
| Bath | 53 | 4,489 | 1.09 | 43 | 10.4 | 58 | 14.1 | 310 | 10.5 | 1,280 | 1,124 | 156 | 3 | 7 |
| Bedford | 545 | 62,381 | 0.56 | 677 | 8.5 | 892 | 11.2 | 5,649 | 9.2 | 20,398 | 15,992 | 4,406 | 63 | 475 |
| Bland | 676 | 5,912 | 0.79 | 45 | 7.2 | 80 | 12.8 | 327 | 7.9 | 1,825 | 1,296 | 529 | 5 | 37 |
| Botetourt | 287 | 29,123 | 0.73 | 255 | 7.6 | 401 | 11.9 | 1,858 | 7.3 | 8,913 | 6,479 | 2,434 | 29 | 247 |
| Brunswick | 2,190 | 14,129 | 0.59 | 124 | 7.7 | 242 | 15.1 | 1,303 | 12.5 | 4,114 | 3,065 | 1,049 | 9 | 81 |
| Buchanan | 1,028 | 22,340 | 1.09 | 169 | 8.2 | 308 | 14.9 | 1,836 | 12.2 | 7,183 | 3,696 | 3,487 | 47 | 281 |
| Buckingham | 2,218 | 13,902 | 0.52 | 145 | 8.4 | 193 | 11.2 | 1,367 | 11.9 | 3,730 | 2,934 | 796 | 13 | 122 |
| Campbell | 457 | 50,944 | 0.84 | 547 | 9.9 | 619 | 11.2 | 4,462 | 10.3 | 13,334 | 10,092 | 3,242 | NA | NA |
| Caroline | 513 | 22,865 | 0.51 | 338 | 11.0 | 320 | 10.4 | 2,384 | 9.5 | 6,176 | 4,690 | 1,486 | 58 | 223 |
| Carroll | 337 | 25,484 | 0.66 | 240 | 8.0 | 399 | 13.3 | 2,677 | 12.2 | 8,603 | 6,839 | 1,763 | 36 | 166 |
| Charles City | 0 | 5,276 | 0.48 | 52 | 7.6 | 116 | 17.0 | 722 | 14.0 | 1,991 | 1,423 | 568 | 4 | 40 |
| Charlotte | 175 | 10,336 | 0.65 | 131 | 11.1 | 181 | 15.3 | 1,113 | 12.4 | 3,295 | 2,564 | 731 | 20 | 122 |
| Chesterfield | 4,579 | 301,414 | 0.76 | 3,970 | 11.1 | 2,954 | 8.2 | 26,720 | 9.1 | 63,807 | 46,247 | 17,559 | 458 | 5,841 |
| Clarke | 180 | 11,598 | 0.60 | 128 | 8.8 | 164 | 11.2 | 1,080 | 9.5 | 3,439 | 2,905 | 534 | 6 | 64 |
| Craig | 9 | 3,573 | 0.34 | 41 | 8.1 | 75 | 14.8 | 251 | 6.5 | 1,370 | 966 | 404 | 5 | 17 |
| Culpeper | 1,475 | 44,728 | 0.73 | 651 | 12.2 | 484 | 9.0 | 4,744 | 11.1 | 9,851 | 8,104 | 1,747 | NA | NA |
| Cumberland | 37 | 6,868 | 0.38 | 93 | 9.4 | 121 | 12.2 | 797 | 10.5 | 2,410 | 1,753 | 657 | 11 | 41 |
| Dickenson | 512 | 13,929 | 0.81 | 115 | 8.2 | 202 | 14.3 | 1,076 | 10.3 | 4,921 | 2,356 | 2,565 | 17 | 33 |
| Dinwiddie | 864 | 23,500 | 0.63 | 233 | 8.1 | 315 | 11.0 | 2,273 | 10.0 | 6,168 | 4,511 | 1,657 | 86 | 294 |
| Essex | 190 | 9,963 | 0.79 | 108 | 9.9 | 139 | 12.7 | 936 | 11.3 | 3,043 | 2,314 | 729 | 19 | 44 |
| Fairfax | 10,443 | 1,182,446 | 1.06 | 13,523 | 11.8 | 6,215 | 5.4 | 89,030 | 9.1 | 157,621 | 124,773 | 32,849 | 957 | 13,279 |
| Fauquier | 389 | 58,635 | 0.68 | 767 | 10.7 | 650 | 9.1 | 5,392 | 9.1 | 13,199 | 11,307 | 1,891 | 40 | 284 |
| Floyd | 85 | 12,236 | 0.52 | 139 | 8.8 | 175 | 11.1 | 1,443 | 12.1 | 4,266 | 3,224 | 1,041 | 19 | 82 |
| Fluvanna | 1,251 | 19,189 | 0.41 | 255 | 9.3 | 254 | 9.3 | 1,854 | 9.1 | 6,311 | 5,294 | 1,018 | 33 | 160 |
| Franklin | 1,161 | 47,886 | 0.66 | 493 | 8.8 | 719 | 12.8 | 4,259 | 10.3 | 15,395 | 10,475 | 4,920 | 56 | 565 |
| Frederick | 1,176 | 77,742 | 0.80 | 958 | 10.5 | 828 | 9.1 | 6,852 | 9.5 | 17,496 | 14,243 | 3,253 | 80 | 910 |
| Giles | 136 | 14,937 | 0.75 | 151 | 9.1 | 258 | 15.5 | 1,187 | 9.2 | 4,634 | 3,404 | 1,230 | 12 | 127 |
| Gloucester | 293 | 29,360 | 0.56 | 339 | 9.0 | 474 | 12.7 | 2,826 | 9.4 | 9,003 | 7,158 | 1,845 | 32 | 399 |
| Goochland | 983 | 28,019 | 1.47 | 184 | 7.5 | 226 | 9.3 | 1,117 | 6.4 | 6,210 | 4,751 | 1,460 | 13 | 160 |
| Grayson | 1,066 | 13,095 | 0.58 | 126 | 8.1 | 195 | 12.6 | 1,234 | 11.6 | 4,634 | 3,499 | 1,135 | 26 | 169 |
| Greene | 137 | 14,222 | 0.46 | 214 | 10.6 | 173 | 8.6 | 1,742 | 10.8 | 4,128 | 3,468 | 660 | 19 | 195 |
| Greensville | 3,416 | 11,567 | 1.01 | 93 | 8.2 | 123 | 10.9 | 527 | 8.6 | 2,135 | 1,426 | 709 | 10 | 79 |
| Halifax | 757 | 32,868 | 0.88 | 351 | 10.4 | 535 | 15.9 | 2,616 | 10.5 | 9,935 | 7,899 | 2,036 | 45 | 209 |
| Hanover | 1,951 | 98,825 | 0.88 | 952 | 8.8 | 1,032 | 9.5 | 5,789 | 6.7 | 22,402 | 16,402 | 6,000 | 168 | 884 |
| Henrico | 2,750 | 341,153 | 1.08 | 3,897 | 11.7 | 3,067 | 9.2 | 24,306 | 8.8 | 59,234 | 41,105 | 18,129 | 456 | 7,887 |
| Henry | 562 | 48,134 | 0.84 | 428 | 8.5 | 782 | 15.5 | 4,267 | 11.4 | 14,940 | 9,602 | 5,338 | 126 | 1,115 |
| Highland | 0 | 1,948 | 0.70 | 14 | 6.4 | 31 | 14.1 | 205 | 14.6 | 814 | 715 | 98 | 0 | 11 |
| Isle of Wight | 310 | 30,404 | 0.65 | 349 | 9.3 | 428 | 11.3 | 2,443 | 8.2 | 8,740 | 6,604 | 2,135 | 23 | 200 |
| James City | 1,140 | 69,863 | 0.85 | 671 | 8.6 | 770 | 9.9 | 3,887 | 6.9 | 21,732 | 18,025 | 3,707 | 83 | 696 |
| King and Queen | 0 | 5,057 | 0.37 | 60 | 8.6 | 105 | 15.1 | 653 | 12.3 | 1,818 | 1,403 | 415 | 13 | 30 |
| King George | 301 | 27,814 | 1.12 | 306 | 11.2 | 200 | 7.3 | 1,545 | 6.7 | 4,181 | 3,584 | 597 | 29 | 213 |
| King William | 72 | 13,479 | 0.61 | 206 | 11.7 | 164 | 9.3 | 1,219 | 8.5 | 3,619 | 2,841 | 779 | 13 | 85 |
| Lancaster | 183 | 11,185 | 1.11 | 84 | 7.9 | 219 | 20.6 | 782 | 11.8 | 4,415 | 3,667 | 748 | 29 | 50 |
| Lee | 1,457 | 21,364 | 0.66 | 200 | 8.6 | 296 | 12.7 | 1,807 | 10.9 | 6,442 | 3,002 | 3,440 | 27 | 122 |
| Loudoun | 1,693 | 364,024 | 0.85 | 5,075 | 12.0 | 1,749 | 4.1 | 22,946 | 6.1 | 41,134 | 32,457 | 8,677 | 393 | 2,309 |

1. Per 1,000 estimated resident population.

# Table B. States and Counties — Crime, Education, Money Income, and Poverty

| STATE County | County law enforcement employment, 2019 | | School enrollment and attainment, 2015-2019 | | | | Local government expenditures,[3] 2017-2018 | | Money income, 2015-2019 | | | | Income and poverty, 2019 | | | |
| | | | Enrollment[1] | | Attainment[2] (percent) | | | | | | Households | | | Percent below poverty level | | |
| | | | | | | | | | | | | Percent | | | | |
| | Officers | Civilians | Total | Percent private | High school graduate or less | Bachelor's degree or more | Total current spending (mil dol) | Current spending per student (dollars) | Per capita income[4] | Median income (dollars) | with income of less than $50,000 | with income of $200,000 or more | Median household income (dollars) | All persons | Children under 18 years | Children 5 to 17 years in families |
| | 46 | 47 | 48 | 49 | 50 | 51 | 52 | 53 | 54 | 55 | 56 | 57 | 58 | 59 | 60 | 61 |
| **VERMONT—Cont'd** | | | | | | | | | | | | | | | | |
| Chittenden | 15 | 2 | 46,702 | 20.9 | 24.6 | 51.3 | 418 | 18,509 | 39,062 | 73,647 | 33.8 | 8.5 | 76,483 | 10.5 | 8.1 | 7.1 |
| Essex | 4 | 0 | 1,102 | 17.4 | 59.2 | 16.2 | 16 | 22,393 | 25,376 | 44,349 | 56.4 | 1.2 | 45,796 | 14.8 | 19.2 | 17.7 |
| Franklin | 16 | 1 | 9,989 | 7.8 | 47.3 | 24.1 | 157 | 18,868 | 31,195 | 65,485 | 35.9 | 3.4 | 65,056 | 9.6 | 11.1 | 10.0 |
| Grand Isle | 3 | 0 | 1,312 | 12.7 | 34.1 | 39.8 | 14 | 20,182 | 39,770 | 71,587 | 31.7 | 6.4 | 68,364 | 8.1 | 11.1 | 10.2 |
| Lamoille | 13 | 11 | 5,892 | 12.5 | 32.7 | 38.2 | 69 | 18,788 | 36,747 | 64,003 | 40.7 | 5.8 | 60,555 | 9.1 | 11.0 | 10.2 |
| Orange | 6 | 2 | 5,739 | 16.1 | 44.4 | 29.1 | 75 | 18,802 | 31,697 | 60,925 | 40.8 | 3.2 | 59,758 | 9.4 | 11.2 | 10.3 |
| Orleans | 15 | 2 | 5,142 | 14.5 | 51.1 | 21.5 | 78 | 19,641 | 27,023 | 49,168 | 50.7 | 2.5 | 48,826 | 12.7 | 17.7 | 16.7 |
| Rutland | 15 | 5 | 12,448 | 15.0 | 41.8 | 30.8 | 158 | 19,984 | 31,391 | 56,139 | 44.5 | 3.5 | 51,903 | 10.8 | 11.8 | 11.4 |
| Washington | 6 | 4 | 13,809 | 26.4 | 33.3 | 41.8 | 159 | 19,172 | 35,236 | 62,791 | 39.5 | 5.2 | 65,879 | 8.4 | 9.3 | 8.0 |
| Windham | 12 | 7 | 8,681 | 29.7 | 37.1 | 38.1 | 116 | 22,036 | 32,535 | 51,985 | 48.5 | 4.8 | 52,068 | 11.6 | 13.1 | 12.5 |
| Windsor | 14 | 2 | 10,536 | 19.0 | 35.7 | 38.2 | 162 | 22,019 | 36,252 | 60,987 | 40.9 | 5.9 | 61,843 | 9.2 | 9.6 | 8.7 |
| **VIRGINIA** | X | X | 2,126,976 | 17.4 | 34.3 | 38.8 | 15,773 | 12,214 | 39,278 | 74,222 | 34.0 | 10.9 | 76,471 | 9.9 | 13.3 | 12.5 |
| Accomack | 56 | 8 | 6,326 | 12.4 | 55.3 | 19.5 | 58 | 11,063 | 26,018 | 46,073 | 52.7 | 2.7 | 47,335 | 16.4 | 28.5 | 29.6 |
| Albemarle | 146 | 38 | 30,286 | 16.9 | 23.7 | 55.1 | (7) | (7) | 44,799 | 79,880 | 30.8 | 12.8 | 86,332 | 6.7 | 7.3 | 7.0 |
| Alleghany | 51 | 20 | 2,787 | 12.3 | 54.3 | 15.0 | 25 | 11,580 | 27,338 | 47,673 | 52.4 | 2.3 | 53,341 | 11.5 | 18.8 | 17.6 |
| Amelia | 18 | 11 | 2,715 | 25.8 | 56.9 | 17.1 | 18 | 9,883 | 31,688 | 57,946 | 41.5 | 3.8 | 60,096 | 9.3 | 13.8 | 13.4 |
| Amherst | 40 | 6 | 6,917 | 27.7 | 50.3 | 20.9 | 22 | 9,858 | 28,586 | 58,696 | 43.5 | 3.6 | 54,609 | 12.6 | 18.5 | 19.3 |
| Appomattox | 22 | 2 | 3,523 | 20.0 | 48.0 | 19.1 | 44 | 10,929 | 26,712 | 52,888 | 47.5 | 2.4 | 56,218 | 12.0 | 16.9 | 15.5 |
| Arlington | 344 | 101 | 48,465 | 28.4 | 13.5 | 75.3 | 508 | 18,816 | 71,841 | 120,071 | 17.5 | 23.5 | 118,986 | 7.6 | 9.1 | 9.0 |
| Augusta | 74 | 11 | 14,460 | 19.9 | 51.4 | 22.7 | 111 | 10,728 | 30,272 | 62,711 | 39.0 | 3.2 | 63,621 | 7.3 | 10.6 | 9.2 |
| Bath | 15 | 7 | 711 | 21.8 | 52.6 | 15.9 | 11 | 19,092 | 27,527 | 49,738 | 50.6 | 0.0 | 54,385 | 9.9 | 13.6 | 13.0 |
| Bedford | 85 | 8 | 16,788 | 24.0 | 40.2 | 29.2 | 102 | 10,476 | 33,678 | 64,199 | 38.4 | 5.5 | 66,591 | 8.6 | 11.0 | 9.7 |
| Bland | 12 | 6 | 1,026 | 9.2 | 53.9 | 14.2 | 8 | 11,111 | 22,522 | 48,531 | 51.0 | 1.6 | 49,023 | 13.4 | 15.6 | 14.1 |
| Botetourt | 97 | 27 | 7,004 | 16.5 | 39.5 | 28.3 | 52 | 11,249 | 35,893 | 71,110 | 33.4 | 6.4 | 74,178 | 5.7 | 7.9 | 7.3 |
| Brunswick | 34 | 14 | 3,282 | 17.8 | 57.8 | 13.6 | 22 | 13,141 | 21,732 | 44,434 | 55.5 | 1.0 | 43,835 | 20.9 | 26.7 | 25.2 |
| Buchanan | 35 | 13 | 3,942 | 20.5 | 65.0 | 11.8 | 32 | 11,312 | 19,496 | 31,956 | 65.9 | 1.1 | 36,881 | 21.7 | 31.3 | 28.8 |
| Buckingham | 19 | 7 | 3,243 | 13.7 | 62.9 | 12.5 | 23 | 11,138 | 20,605 | 49,025 | 51.2 | 0.9 | 47,202 | 16.8 | 22.2 | 20.9 |
| Campbell | 67 | 9 | 11,752 | 26.0 | 45.0 | 22.5 | 80 | 10,118 | 27,739 | 49,664 | 50.3 | 2.2 | 48,984 | 10.8 | 14.3 | 13.4 |
| Caroline | 50 | 22 | 6,325 | 18.8 | 52.6 | 19.9 | 44 | 10,380 | 31,568 | 65,103 | 36.8 | 5.5 | 64,647 | 9.6 | 14.1 | 13.0 |
| Carroll | 38 | 5 | 5,188 | 9.7 | 55.2 | 15.0 | 43 | 11,311 | 25,449 | 44,835 | 54.9 | 1.6 | 45,698 | 13.9 | 20.3 | 18.0 |
| Charles City | 13 | 7 | 1,107 | 22.5 | 58.6 | 14.7 | 10 | 15,443 | 35,304 | 57,198 | 42.6 | 3.7 | 56,465 | 12.5 | 18.7 | 18.0 |
| Charlotte | 38 | 2 | 2,532 | 6.0 | 57.2 | 10.9 | 24 | 12,644 | 22,018 | 40,573 | 56.7 | 2.1 | 43,001 | 19.3 | 29.5 | 30.3 |
| Chesterfield | 515 | 108 | 88,729 | 15.8 | 31.2 | 40.5 | 602 | 9,886 | 37,658 | 82,599 | 27.7 | 8.4 | 81,641 | 6.6 | 9.1 | 8.3 |
| Clarke | 19 | 11 | 3,034 | 17.3 | 40.5 | 32.2 | 23 | 11,796 | 41,332 | 80,026 | 32.6 | 11.0 | 85,567 | 6.3 | 7.6 | 6.4 |
| Craig | 8 | 5 | 888 | 19.1 | 46.7 | 23.6 | 7 | 11,270 | 30,103 | 55,708 | 47.8 | 1.0 | 53,319 | 11.2 | 19.1 | 18.1 |
| Culpeper | 88 | 14 | 11,824 | 21.2 | 45.1 | 25.7 | 86 | 10,486 | 32,028 | 77,935 | 31.0 | 7.2 | 79,739 | 9.2 | 11.2 | 10.9 |
| Cumberland | 16 | 6 | 1,858 | 17.7 | 54.6 | 14.9 | 15 | 11,103 | 25,138 | 47,469 | 52.5 | 1.4 | 52,005 | 13.8 | 23.4 | 21.8 |
| Dickenson | 26 | 11 | 2,655 | 6.4 | 61.1 | 10.7 | 24 | 11,040 | 24,978 | 29,932 | 68.3 | 1.7 | 37,161 | 24.2 | 27.6 | 25.1 |
| Dinwiddie | 40 | 4 | 6,179 | 14.8 | 52.9 | 18.6 | 46 | 10,584 | 28,922 | 60,346 | 41.2 | 5.2 | 58,474 | 10.7 | 15.1 | 14.0 |
| Essex | 13 | 3 | 2,245 | 21.4 | 55.4 | 17.4 | 18 | 12,473 | 26,847 | 51,954 | 47.8 | 2.2 | 53,538 | 16.2 | 26.9 | 24.1 |
| Fairfax | 1,464 | 264 | 306,125 | 18.5 | 20.3 | 61.6 | (10) 2,816.4 | (10) 14,937 | 56,231 | 124,831 | 16.7 | 25.4 | 127,898 | 6.0 | 7.9 | 7.2 |
| Fauquier | 131 | 40 | 16,931 | 17.2 | 34.1 | 36.4 | 141 | 12,629 | 45,408 | 100,783 | 21.2 | 15.2 | 103,827 | 6.1 | 7.0 | 5.8 |
| Floyd | 27 | 10 | 2,710 | 15.1 | 48.7 | 22.1 | 22 | 10,822 | 26,882 | 51,521 | 48.4 | 1.8 | 52,277 | 11.4 | 16.5 | 15.4 |
| Fluvanna | 34 | 13 | 6,056 | 15.9 | 33.0 | 33.9 | 39 | 10,946 | 39,633 | 76,873 | 30.1 | 6.1 | 75,089 | 7.3 | 9.4 | 8.9 |
| Franklin | 81 | 24 | 11,690 | 21.0 | 46.8 | 21.9 | 81 | 11,191 | 30,487 | 56,254 | 46.1 | 3.8 | 61,878 | 11.5 | 17.8 | 15.8 |
| Frederick | 131 | 10 | 19,975 | 20.1 | 43.2 | 28.3 | (9) | (9) | 35,123 | 78,002 | 28.5 | 5.9 | 83,672 | 6.5 | 9.0 | 8.7 |
| Giles | 26 | 11 | 3,283 | 13.0 | 49.8 | 18.1 | 26 | 10,461 | 27,137 | 54,520 | 44.6 | 1.9 | 53,111 | 11.4 | 16.5 | 15.2 |
| Gloucester | 80 | 16 | 7,870 | 15.1 | 42.8 | 23.3 | 59 | 10,873 | 33,692 | 70,537 | 32.4 | 4.6 | 71,715 | 8.6 | 12.9 | 12.1 |
| Goochland | 39 | 16 | 4,478 | 26.8 | 32.5 | 41.8 | 31 | 11,959 | 51,720 | 93,994 | 25.7 | 16.9 | 100,444 | 6.2 | 7.2 | 6.6 |
| Grayson | 27 | 7 | 2,694 | 11.1 | 56.0 | 13.7 | 20 | 12,515 | 21,663 | 36,544 | 63.7 | 1.0 | 41,312 | 18.8 | 25.5 | 25.5 |
| Greene | 25 | 14 | 4,735 | 17.2 | 42.5 | 29.0 | 34 | 10,808 | 32,508 | 67,398 | 34.6 | 3.0 | 73,345 | 7.6 | 11.1 | 11.2 |
| Greensville | 23 | 13 | 2,386 | 17.1 | 63.5 | 9.2 | (8) | (8) | 19,550 | 50,300 | 49.6 | 1.0 | 47,315 | 24.5 | 26.0 | 25.0 |
| Halifax | 36 | 4 | 6,858 | 12.8 | 54.5 | 15.4 | 56 | 10,874 | 23,909 | 42,669 | 57.4 | 1.3 | 42,619 | 17.1 | 23.6 | 23.3 |
| Hanover | 254 | 19 | 26,706 | 19.7 | 32.4 | 39.8 | 183 | 10,189 | 41,200 | 89,390 | 24.8 | 10.5 | 90,824 | 5.1 | 5.8 | 4.8 |
| Henrico | 656 | 170 | 79,273 | 16.3 | 29.5 | 43.7 | 514 | 9,951 | 39,516 | 70,307 | 35.4 | 9.5 | 68,975 | 8.7 | 12.9 | 11.9 |
| Henry | 119 | 13 | 10,242 | 8.2 | 54.2 | 13.9 | 74 | 9,862 | 22,372 | 37,952 | 62.2 | 1.5 | 41,908 | 14.6 | 22.9 | 20.9 |
| Highland | 8 | 5 | 189 | 18.0 | 50.2 | 22.5 | 4 | 19,531 | 31,485 | 48,587 | 52.4 | 4.2 | 45,917 | 12.4 | 19.7 | 20.6 |
| Isle of Wight | 50 | 9 | 8,516 | 24.9 | 40.0 | 27.0 | 57 | 10,434 | 37,126 | 73,991 | 34.4 | 7.5 | 78,749 | 9.1 | 13.7 | 14.7 |
| James City | 104 | 6 | 17,238 | 11.7 | 22.7 | 50.4 | (11) | (11) | 44,495 | 87,678 | 26.8 | 10.8 | 92,773 | 5.8 | 8.3 | 7.5 |
| King and Queen | 13 | 8 | 1,434 | 17.1 | 49.0 | 23.6 | 10 | 12,554 | 31,174 | 63,982 | 41.6 | 3.3 | 54,185 | 11.7 | 18.0 | 17.3 |
| King George | 38 | 16 | 6,417 | 15.9 | 33.3 | 36.0 | 45 | 9,867 | 38,739 | 94,274 | 23.7 | 10.8 | 85,657 | 5.9 | 7.7 | 7.0 |
| King William | 21 | 14 | 4,207 | 13.3 | 42.2 | 21.0 | 35 | 11,526 | 30,704 | 66,987 | 34.8 | 4.4 | 73,035 | 6.9 | 10.0 | 9.4 |
| Lancaster | 32 | 9 | 1,719 | 6.6 | 40.3 | 31.5 | 16 | 13,967 | 40,062 | 53,711 | 48.6 | 7.9 | 55,072 | 13.3 | 23.9 | 23.4 |
| Lee | 34 | 0 | 4,465 | 5.2 | 57.7 | 11.2 | 39 | 11,992 | 19,720 | 32,888 | 65.6 | 1.8 | 35,878 | 27.1 | 42.8 | 33.7 |
| Loudoun | 558 | 123 | 117,908 | 17.4 | 18.5 | 61.3 | 1,145 | 14,208 | 55,744 | 142,299 | 12.7 | 30.0 | 151,806 | 3.1 | 3.2 | 2.8 |

1. All persons 3 years old and over enrolled in nursery school through college.   2. Persons 25 years old and over.   3. Elementary and secondary education expenditures.   4. Based on population estimated by the American Community Survey, 2014–2018.   7. Charlottesville city is included with Albemarle county.   8. Emporia city is included with Greensville county.   9. Winchester city is included with Frederick county.   10. Fairfax city is included with Fairfax county   11. Williamsburg city is included with James city county.

# Table B. States and Counties — Personal Income

| STATE County | Personal income, 2019 Total (mil dol) | Percent change 2018-2019 | Per capita¹ Dollars | Per capita¹ Rank | Wages and salaries (mil dol) | Supplements — Pension and insurance | Supplements — Government social insurance | Proprietors' income (mil dol) | Dividends, interest, and rent (mil dol) | Personal transfer receipts (mil dol) | Earnings, 2019 Total (mil dol) | Contributions — From employee and self-employed | Contributions — From employer |
|---|---|---|---|---|---|---|---|---|---|---|---|---|---|
| | 62 | 63 | 64 | 65 | 66 | 67 | 68 | 69 | 70 | 71 | 72 | 73 | 74 |
| **VERMONT—Cont'd** | | | | | | | | | | | | | |
| Chittenden | 10,131 | 2.7 | 61,858 | 253 | 6,136 | 935 | 471 | 878 | 2,112 | 1,406 | 8,420 | 528 | 471 |
| Essex | 239 | 6.0 | 38,844 | 2,217 | 41 | 11 | 3 | 25 | 35 | 74 | 81 | 7 | 3 |
| Franklin | 2,366 | 4.2 | 47,889 | 1,031 | 906 | 182 | 74 | 176 | 325 | 459 | 1,338 | 87 | 74 |
| Grand Isle | 433 | 5.0 | 59,863 | 308 | 50 | 10 | 4 | 42 | 89 | 69 | 106 | 9 | 4 |
| Lamoille | 1,389 | 2.6 | 54,761 | 509 | 536 | 84 | 45 | 158 | 350 | 243 | 823 | 55 | 45 |
| Orange | 1,415 | 4.6 | 48,961 | 926 | 347 | 67 | 28 | 141 | 250 | 310 | 583 | 43 | 28 |
| Orleans | 1,205 | 3.8 | 44,579 | 1,411 | 446 | 85 | 38 | 116 | 211 | 365 | 684 | 49 | 38 |
| Rutland | 3,093 | 3.0 | 53,152 | 600 | 1,312 | 225 | 108 | 196 | 483 | 1,025 | 1,841 | 132 | 108 |
| Washington | 3,526 | 2.2 | 60,368 | 291 | 1,884 | 310 | 144 | 287 | 652 | 717 | 2,625 | 171 | 144 |
| Windham | 2,201 | 3.6 | 52,123 | 673 | 1,009 | 162 | 82 | 230 | 478 | 543 | 1,482 | 103 | 82 |
| Windsor | 3,282 | 3.2 | 59,604 | 320 | 1,165 | 206 | 94 | 245 | 828 | 686 | 1,709 | 122 | 94 |
| **VIRGINIA** | 509,201 | 3.3 | 59,509 | X | 262,502 | 40,306 | 18,458 | 28,850 | 107,610 | 71,293 | 350,116 | 21,437 | 18,458 |
| Accomack | 1,387 | 1.8 | 42,923 | 1,628 | 593 | 116 | 43 | 79 | 315 | 391 | 831 | 58 | 43 |
| Albemarle [2] | 12,161 | 2.2 | 77,657 | 74 | 6,131 | 1,112 | 422 | 950 | 4,694 | 1,177 | 8,614 | 504 | 422 |
| Alleghany [3] | 835 | 2.4 | 40,928 | 1,927 | 387 | 66 | 29 | 20 | 146 | 276 | 502 | 38 | 29 |
| Amelia | 582 | 3.1 | 44,297 | 1,445 | 109 | 19 | 8 | 35 | 88 | 141 | 171 | 14 | 8 |
| Amherst | 1,206 | 2.5 | 38,165 | 2,309 | 347 | 65 | 25 | 36 | 201 | 345 | 473 | 39 | 25 |
| Appomattox | 625 | 2.9 | 39,268 | 2,153 | 123 | 24 | 9 | 18 | 92 | 176 | 174 | 16 | 9 |
| Arlington | 23,544 | 3.0 | 99,407 | 22 | 20,118 | 2,912 | 1,377 | 1,297 | 5,448 | 1,122 | 25,703 | 1,437 | 1,377 |
| Augusta [4] | 5,621 | 3.4 | 45,658 | 1,290 | 2,314 | 398 | 166 | 340 | 1,117 | 1,234 | 3,219 | 222 | 166 |
| Bath | 277 | 2.4 | 66,864 | 156 | 103 | 18 | 7 | 8 | 109 | 53 | 136 | 9 | 7 |
| Bedford | 3,759 | 2.3 | 47,590 | 1,070 | 850 | 137 | 61 | 145 | 762 | 816 | 1,194 | 100 | 61 |
| Bland | 239 | 2.8 | 38,126 | 2,318 | 99 | 23 | 7 | 9 | 42 | 76 | 137 | 10 | 7 |
| Botetourt | 1,729 | 2.9 | 51,738 | 703 | 553 | 87 | 39 | 59 | 332 | 352 | 738 | 55 | 39 |
| Brunswick | 548 | 3.8 | 33,784 | 2,824 | 152 | 30 | 11 | 8 | 94 | 196 | 201 | 18 | 11 |
| Buchanan | 778 | 2.6 | 37,048 | 2,445 | 359 | 64 | 25 | 32 | 107 | 318 | 481 | 39 | 25 |
| Buckingham | 531 | 3.6 | 30,966 | 3,007 | 149 | 33 | 11 | 21 | 84 | 162 | 214 | 17 | 11 |
| Campbell [5] | 5,285 | 2.4 | 38,565 | 2,250 | 3,665 | 531 | 261 | 194 | 1,040 | 1,465 | 4,650 | 300 | 261 |
| Caroline | 1,320 | 3.3 | 42,977 | 1,621 | 268 | 57 | 20 | 34 | 205 | 275 | 380 | 30 | 20 |
| Carroll [6] | 1,330 | 2.8 | 36,812 | 2,470 | 436 | 81 | 32 | 57 | 218 | 459 | 606 | 50 | 32 |
| Charles City | 327 | 3.2 | 46,976 | 1,135 | 88 | 14 | 6 | 13 | 67 | 81 | 122 | 10 | 6 |
| Charlotte | 436 | 2.9 | 36,737 | 2,479 | 114 | 24 | 8 | 19 | 83 | 145 | 166 | 15 | 8 |
| Chesterfield | 19,429 | 3.9 | 55,070 | 489 | 7,496 | 1,126 | 529 | 848 | 3,389 | 2,894 | 9,999 | 643 | 529 |
| Clarke | 981 | 3.3 | 67,094 | 153 | 200 | 31 | 14 | 55 | 245 | 131 | 300 | 21 | 14 |
| Craig | 201 | 2.9 | 39,087 | 2,179 | 30 | 7 | 2 | 6 | 37 | 55 | 46 | 5 | 2 |
| Culpeper | 2,509 | 4.6 | 47,689 | 1,055 | 797 | 135 | 57 | 173 | 419 | 436 | 1,162 | 78 | 57 |
| Cumberland | 388 | 4.4 | 39,094 | 2,178 | 50 | 11 | 4 | 22 | 61 | 109 | 86 | 8 | 4 |
| Dickenson | 493 | 4.9 | 34,397 | 2,757 | 156 | 32 | 11 | 10 | 68 | 219 | 209 | 20 | 11 |
| Dinwiddie [7] | 3,194 | 3.0 | 41,337 | 1,872 | 1,448 | 239 | 106 | 59 | 579 | 978 | 1,852 | 129 | 106 |
| Essex | 471 | 3.6 | 43,041 | 1,611 | 143 | 24 | 10 | 15 | 102 | 132 | 193 | 16 | 10 |
| Fairfax [8] | 102,177 | 2.8 | 86,141 | 42 | 67,454 | 7,602 | 4,442 | 5,929 | 24,636 | 7,138 | 85,427 | 5,130 | 4,442 |
| Fauquier | 5,169 | 3.7 | 72,577 | 105 | 1,283 | 197 | 89 | 352 | 1,223 | 554 | 1,920 | 124 | 89 |
| Floyd | 631 | 3.7 | 40,065 | 2,041 | 125 | 23 | 9 | 31 | 116 | 167 | 189 | 17 | 9 |
| Fluvanna | 1,236 | 3.8 | 45,334 | 1,321 | 206 | 41 | 14 | 58 | 239 | 239 | 320 | 27 | 14 |
| Franklin | 2,342 | 3.0 | 41,795 | 1,803 | 603 | 106 | 44 | 117 | 494 | 599 | 869 | 72 | 44 |
| Frederick [9] | 6,142 | 4.2 | 52,317 | 656 | 3,138 | 497 | 224 | 429 | 1,111 | 962 | 4,288 | 268 | 224 |
| Giles | 689 | 3.3 | 41,179 | 1,895 | 211 | 41 | 15 | 24 | 109 | 197 | 291 | 24 | 15 |
| Gloucester | 1,936 | 3.5 | 51,847 | 694 | 389 | 73 | 28 | 65 | 376 | 381 | 555 | 44 | 28 |
| Goochland | 2,439 | 3.7 | 102,690 | 18 | 1,910 | 181 | 113 | 112 | 811 | 240 | 2,317 | 142 | 113 |
| Grayson | 516 | 2.8 | 33,190 | 2,878 | 109 | 26 | 8 | 7 | 98 | 183 | 150 | 17 | 8 |
| Greene | 922 | 3.8 | 46,529 | 1,182 | 157 | 27 | 11 | 70 | 152 | 173 | 266 | 21 | 11 |
| Greensville [10] | 553 | 3.2 | 33,144 | 2,880 | 324 | 63 | 23 | 10 | 81 | 187 | 420 | 28 | 23 |
| Halifax | 1,303 | 4.0 | 38,414 | 2,278 | 519 | 100 | 38 | 49 | 239 | 428 | 705 | 54 | 38 |
| Hanover | 6,663 | 3.6 | 61,832 | 254 | 2,694 | 372 | 190 | 454 | 1,214 | 905 | 3,709 | 237 | 190 |
| Henrico | 22,707 | 3.4 | 68,639 | 141 | 12,261 | 1,546 | 842 | 4,061 | 4,322 | 2,680 | 18,709 | 1,111 | 842 |
| Henry [11] | 2,559 | 2.4 | 40,545 | 1,980 | 959 | 173 | 70 | 113 | 481 | 878 | 1,315 | 103 | 70 |
| Highland | 98 | 1.0 | 44,791 | 1,385 | 18 | 4 | 1 | 6 | 40 | 28 | 30 | 3 | 1 |
| Isle of Wight | 2,139 | 4.1 | 57,645 | 385 | 536 | 80 | 37 | 64 | 343 | 379 | 717 | 53 | 37 |
| James City [12] | 6,066 | 3.4 | 66,306 | 165 | 2,017 | 329 | 144 | 222 | 1,770 | 989 | 2,712 | 186 | 144 |
| King and Queen | 315 | 3.8 | 44,798 | 1,384 | 48 | 9 | 4 | 14 | 51 | 77 | 75 | 7 | 4 |
| King George | 1,443 | 4.2 | 53,777 | 558 | 1,046 | 270 | 83 | 66 | 268 | 194 | 1,465 | 81 | 83 |
| King William | 820 | 4.8 | 47,802 | 1,043 | 204 | 32 | 15 | 28 | 122 | 154 | 278 | 21 | 15 |
| Lancaster | 652 | 2.6 | 61,447 | 266 | 194 | 31 | 14 | 39 | 255 | 164 | 278 | 23 | 14 |
| Lee | 728 | 2.8 | 31,087 | 2,998 | 166 | 44 | 12 | 23 | 113 | 301 | 245 | 24 | 12 |
| Loudoun | 33,461 | 4.8 | 80,914 | 64 | 13,114 | 1,661 | 871 | 1,649 | 4,728 | 1,788 | 17,296 | 1,011 | 871 |

1. Based on the resident population estimated as of July 1 of the year shown. 2. Charlottesville city is included with Albemarle county. 3. Covington city is included with Alleghany county. 4. Staunton and Waynesboro cities are included with Augusta county. 5. Lynchburg city is included with Campbell county. 6. Galax city is included with Carroll county. 7. Petersburg and Colonial Heights cities are included with Dinwiddie county. 8. Fairfax city and Falls Church city are included with Fairfax county. 9. Winchester city is included with Frederick county. 10. Emporia city is included with Greensville county. 11. Martinsville city is included with Henry county. 12. Williamsburg city is included with James City county.

# Table B. States and Counties — Earnings, Social Security, and Housing

Earnings, 2019 (cont.) — Percent by selected industries

| STATE County | Farm | Mining, quarrying, and extractions | Construction | Manufacturing | Information; professional, scientific, technical services | Retail trade | Finance, insurance, real estate, and leasing | Health care and social assistance | Government | Social Security beneficiaries, December 2019 — Number | Rate[1] | Supplemental Security Income recipients, 2019 | Housing units, 2020 — Total | Percent change, 2010-2020 |
|---|---|---|---|---|---|---|---|---|---|---|---|---|---|---|
| | 75 | 76 | 77 | 78 | 79 | 80 | 81 | 82 | 83 | 84 | 85 | 86 | 87 | 88 |
| **VERMONT—Cont'd** | | | | | | | | | | | | | | |
| Chittenden | 0.2 | 0.1 | 5.8 | 10.1 | 15.4 | 6.6 | 6.4 | 15.6 | 17.1 | 30,690 | 187 | 3,118 | 71,966 | 9.5 |
| Essex | 4.4 | 0.1 | 9.6 | 9.2 | D | 2.7 | D | D | 34.7 | 2,090 | 340 | 208 | 5,254 | 4.7 |
| Franklin | 4.3 | 0.1 | 5.6 | 15.3 | D | 7.3 | 2.3 | 12.8 | 29.5 | 10,210 | 207 | 1,251 | 22,951 | 6.3 |
| Grand Isle | 4.4 | D | 14.1 | 2.4 | D | 5.9 | 7.1 | D | 17.8 | 2,035 | 282 | 131 | 5,420 | 7.4 |
| Lamoille | 1.3 | D | 10.9 | 5.2 | 9.1 | 7.6 | 4.5 | 14.5 | 13.9 | 5,470 | 215 | 476 | 14,125 | 8.9 |
| Orange | 3.2 | 0.2 | 13.9 | 6.1 | D | 5.9 | D | 17.3 | 21.4 | 7,350 | 255 | 572 | 15,680 | 5.6 |
| Orleans | 3.6 | D | 9.5 | 11.1 | D | 9.8 | 4.0 | 15.8 | 19.3 | 7,855 | 292 | 890 | 18,147 | 12.3 |
| Rutland | 0.3 | 2.2 | 7.7 | 13.5 | 4.5 | 8.0 | 3.5 | 17.8 | 14.8 | 17,825 | 307 | 1,967 | 34,714 | 2.8 |
| Washington | 0.3 | 0.2 | 5.6 | 5.9 | 7.7 | 6.3 | 12.7 | 12.9 | 24.1 | 14,590 | 250 | 1,361 | 31,149 | 4.0 |
| Windham | 0.6 | D | 7.8 | 12.7 | D | 7.2 | 4.7 | 15.1 | 12.0 | 11,995 | 284 | 1,190 | 31,085 | 4.5 |
| Windsor | 0.1 | 0.2 | 8.8 | 7.0 | 11.2 | 6.9 | 4.0 | 12.1 | 24.0 | 16,070 | 292 | 1,254 | 35,287 | 3.5 |
| **VIRGINIA** | 0.1 | 0.2 | 5.5 | 5.6 | 19.9 | 4.8 | 7.3 | 9.1 | 23.0 | 1,559,858 | 182 | 155,582 | 3,586,972 | 6.6 |
| Accomack | 2.8 | 0.0 | 4.3 | 18.7 | 12.4 | 4.9 | 2.7 | D | 27.0 | 9,820 | 304 | 1,083 | 21,356 | 1.7 |
| Albemarle | (2)0.0 | (2)0.1 | (2)4.1 | (2)2.8 | (2)13.5 | (2)4.5 | (2)8.5 | (2)10.3 | (2)34.9 | 21,115 | 192 | 1,026 | 47,945 | 13.8 |
| Alleghany | (3)-0.1 | (3)D | (3)8.8 | (3)D | (3)D | (3)5.4 | (3)D | (3)10.2 | (3)17.7 | 4,650 | 312 | 298 | 8,013 | -0.8 |
| Amelia | 6.8 | D | 17.2 | 7.7 | 3.2 | 4.4 | 2.4 | 10.0 | 17.8 | 3,425 | 262 | 290 | 5,721 | 6.8 |
| Amherst | -0.2 | 0.0 | 9.0 | 22.9 | 3.5 | 7.8 | 2.5 | D | 23.1 | 8,640 | 273 | 810 | 14,290 | 2.2 |
| Appomattox | -1.6 | 0.1 | 11.7 | 2.5 | 4.3 | 12.9 | 2.7 | D | 26.6 | 4,275 | 269 | 479 | 7,489 | 7.8 |
| Arlington | 0.0 | 0.0 | 1.2 | D | 36.4 | 1.7 | 4.8 | 3.4 | 28.8 | 21,075 | 88 | 2,206 | 119,299 | 13.2 |
| Augusta | (4)0.4 | (4)0.1 | (4)D | (4)18.3 | (4)3.7 | (4)6.3 | (4)4.5 | (4)D | (4)17.2 | 19,385 | 255 | 634 | 33,182 | 6.4 |
| Bath | 1.0 | 0.0 | 5.0 | 3.0 | 13.6 | 1.5 | D | D | 15.6 | 1,345 | 323 | 81 | 3,300 | 1.0 |
| Bedford | -0.6 | 0.0 | 8.8 | 12.6 | 8.6 | 7.1 | 7.3 | 11.2 | 15.5 | 21,380 | 269 | 1,153 | 37,039 | 6.3 |
| Bland | -0.2 | D | 3.2 | 35.1 | D | 3.2 | D | D | 25.3 | 1,985 | 315 | 118 | 3,374 | 3.3 |
| Botetourt | -0.4 | D | 8.1 | 19.7 | 4.2 | 3.8 | 3.9 | D | 13.1 | 8,985 | 268 | 770 | 15,093 | 3.6 |
| Brunswick | 0.2 | D | 6.6 | 9.8 | D | 4.6 | 3.4 | 9.2 | 27.1 | 4,415 | 273 | 611 | 8,237 | 1.0 |
| Buchanan | -0.1 | 31.1 | 5.5 | 4.5 | 4.4 | 4.8 | 2.6 | 11.0 | 17.3 | 8,165 | 388 | 1,518 | 11,540 | -0.4 |
| Buckingham | 0.0 | D | 9.3 | 3.7 | D | 4.8 | 1.9 | D | 36.1 | 3,955 | 231 | 402 | 7,515 | 3.8 |
| Campbell | (5)0.0 | (5)D | (5)D | (5)19.6 | (5)9.8 | (5)6.3 | (5)6.3 | (5)18.2 | (5)9.9 | 14,055 | 254 | 1,064 | 26,140 | 5.6 |
| Caroline | 0.5 | D | 7.4 | 5.2 | D | 5.2 | 2.0 | D | 32.3 | 6,500 | 211 | 497 | 12,499 | 6.5 |
| Carroll | (6)0.1 | (6)D | (6)D | (6)18.2 | (6)9.5 | (6)2.8 | (6)D | (6)D | (6)19.6 | 9,005 | 300 | 555 | 16,680 | 0.6 |
| Charles City | 4.1 | D | 26.3 | 10.0 | D | D | D | 8.4 | 16.7 | 2,085 | 299 | 148 | 3,403 | 5.4 |
| Charlotte | -1.4 | 0.1 | 5.3 | 15.3 | D | 5.9 | 2.1 | D | 27.7 | 3,660 | 309 | 502 | 6,363 | 1.3 |
| Chesterfield | 0.0 | D | 9.5 | 7.6 | 10.0 | 8.0 | 7.4 | 11.1 | 16.4 | 65,240 | 185 | 4,484 | 137,225 | 12.0 |
| Clarke | 1.1 | D | 12.1 | 12.7 | 13.4 | 3.3 | 5.8 | 5.5 | 14.9 | 3,390 | 233 | 145 | 6,551 | 5.2 |
| Craig | -4.0 | 0.2 | D | D | 6.6 | D | D | 12.5 | 29.6 | 1,495 | 292 | 83 | 2,912 | 3.8 |
| Culpeper | 0.3 | 0.6 | 10.5 | 8.7 | 11.4 | 8.1 | 4.6 | D | 20.0 | 10,295 | 195 | 914 | 19,488 | 10.4 |
| Cumberland | 6.3 | 0.1 | 16.6 | 8.0 | D | 7.0 | D | D | 29.0 | 2,450 | 246 | 219 | 4,790 | 3.5 |
| Dickenson | -0.2 | 35.4 | 5.7 | 0.6 | 6.8 | 5.3 | D | 6.0 | 23.1 | 5,430 | 380 | 861 | 7,555 | -0.2 |
| Dinwiddie | (7)-0.3 | (7)D | (7)D | (7)8.4 | (7)2.4 | (7)8.3 | (7)5.3 | (7)19.3 | (7)22.5 | 6,565 | 228 | 431 | 11,927 | 4.4 |
| Essex | -2.8 | 0.0 | 5.8 | 6.8 | 14.8 | 6.6 | 6.6 | D | 16.8 | 3,235 | 296 | 354 | 5,914 | 2.7 |
| Fairfax | (8)0.0 | (8)0.0 | (8)4.3 | (8)0.7 | (8)36.9 | (8)3.7 | (8)8.5 | (8)6.5 | (8)15.9 | 136,325 | 118 | 10,841 | 417,226 | 2.3 |
| Fauquier | 0.5 | D | 14.0 | 3.1 | 16.0 | 6.8 | 7.9 | 8.9 | 20.2 | 12,965 | 182 | 625 | 27,540 | 7.5 |
| Floyd | | D | 11.9 | 12.4 | D | 6.6 | D | 15.5 | 19.9 | 4,490 | 284 | 291 | 8,122 | 4.2 |
| Fluvanna | -0.4 | 0.0 | 17.0 | 3.9 | D | 3.4 | 2.9 | 5.4 | 28.7 | 6,285 | 231 | 221 | 11,267 | 8.3 |
| Franklin | -0.2 | D | 12.2 | 19.5 | 4.2 | 8.0 | 3.6 | 9.5 | 16.2 | 15,560 | 277 | 1,201 | 30,153 | 2.9 |
| Frederick | (9)-0.3 | (9)D | (9)D | (9)13.8 | (9)D | (9)7.7 | (9)8.2 | (9)16.7 | (9)17.9 | 17,985 | 201 | 922 | 36,165 | 15.4 |
| Giles | -0.5 | 0.0 | D | 29.1 | 8.7 | 7.7 | 2.5 | 9.1 | 16.8 | 4,985 | 299 | 551 | 8,372 | 0.6 |
| Gloucester | 0.3 | D | 10.0 | 1.4 | 4.9 | 12.5 | 4.2 | 17.9 | 25.6 | 9,280 | 247 | 656 | 16,923 | 6.7 |
| Goochland | -0.1 | 0.3 | 4.3 | 2.0 | D | 1.3 | D | 1.6 | 4.0 | 6,090 | 254 | 250 | 9,871 | 13.9 |
| Grayson | -10.1 | D | 5.4 | 14.1 | D | 3.2 | 6.7 | D | 36.4 | 5,115 | 328 | 310 | 9,272 | 1.1 |
| Greene | 0.2 | D | 13.2 | 1.5 | 12.1 | 11.0 | 2.2 | D | 20.1 | 4,205 | 211 | 244 | 8,648 | 15.2 |
| Greensville | (10)1.1 | (10)D | (10)1.5 | (10)25.3 | (10)D | (10)5.3 | (10)D | (10)13.9 | (10)27.3 | 2,575 | 228 | 37 | 4,202 | 2.8 |
| Halifax | -0.4 | D | 7.0 | 19.0 | 3.2 | 5.5 | 3.0 | D | 16.8 | 10,690 | 314 | 1,422 | 18,262 | 1.4 |
| Hanover | 0.4 | D | 12.3 | 7.0 | 6.7 | 9.2 | 4.1 | 12.1 | 9.2 | 22,430 | 209 | 984 | 43,176 | 12.6 |
| Henrico | 0.0 | 0.0 | 4.4 | 3.4 | 13.3 | 5.1 | 15.0 | 12.9 | 7.8 | 60,195 | 181 | 5,003 | 140,687 | 6.1 |
| Henry | (11)0.7 | (11)0.2 | (11)D | (11)21.9 | (11)D | (11)10.0 | (11)4.0 | (11)12.6 | (11)17.7 | 16,970 | 335 | 2,456 | 26,147 | -0.5 |
| Highland | 10.3 | 0.0 | 7.7 | 3.0 | D | 2.9 | D | D | 30.8 | 850 | 391 | 28 | 1,898 | 3.5 |
| Isle of Wight | 1.7 | 0.0 | 5.2 | 27.3 | 5.5 | 3.8 | 3.9 | D | 13.9 | 8,950 | 239 | 524 | 16,201 | 10.7 |
| James City | (12)-0.2 | (12)D | (12)5.4 | (12)D | (12)9.9 | (12)6.9 | (12)D | 3.6 | 22.6 | 20,900 | 272 | 879 | 34,419 | 14.5 |
| King and Queen | 1.4 | D | 14.6 | 11.5 | D | 1.9 | D | 1.2 | 67.2 | 1,935 | 276 | 126 | 3,532 | 3.5 |
| King George | 0.0 | D | 1.3 | 2.6 | 17.5 | 2.0 | 1.1 | 4.1 | 16.9 | 4,100 | 152 | 255 | 10,397 | 9.7 |
| King William | -0.2 | D | 8.5 | 25.5 | D | 5.4 | D | D | 11.8 | 3,845 | 224 | 218 | 7,373 | 13.1 |
| Lancaster | 0.1 | 0.0 | 8.4 | 2.1 | 8.5 | 8.7 | 14.1 | 12.3 | 11.8 | 4,450 | 419 | 276 | 7,689 | 3.9 |
| Lee | -2.0 | 0.8 | 5.8 | 3.2 | 2.1 | 9.5 | 3.1 | 5.8 | 44.0 | 7,255 | 308 | 1,530 | 11,783 | 0.3 |
| Loudoun | 0.1 | D | 9.8 | 5.8 | 29.0 | 4.4 | 5.7 | 5.8 | 14.1 | 37,545 | 90 | 2,227 | 142,810 | 30.5 |

1. Per 1,000 resident population estimated as of July 1 of the year shown.   2. Charlottesville city is included with Albemarle county.   3. Covington city is included with Alleghany county.   4. Staunton and Waynesboro cities are included with Augusta county.   5. Lynchburg city is included with Campbell county.   6. Galax city is included with Carroll county.   7. Petersburg and Colonial Heights cities are included with Dinwiddie county.   8. Fairfax city and Falls Church city are included with Fairfax county.   9. Winchester city is included with Frederick county.   10. Emporia city is included with Greensville county.   11. Martinsville city is included with Henry county.   12. Williamsburg city is included with James City county.

# Table B. States and Counties — Housing, Labor Force, and Employment

| STATE County | Housing units, 2015-2019 | | | | | | | | Civilian labor force, 2020 | | | | Civilian employment[6], 2015-2019 | | |
|---|---|---|---|---|---|---|---|---|---|---|---|---|---|---|---|
| | Occupied units | | | | | | | Sub-standard units[4] (percent) | | | Unemployment | | Percent | | |
| | | | Owner-occupied | | | Renter-occupied | | | | | | | | | |
| | | | | Median owner cost as a percent of income | | | Median rent as a percent of income[2] | | | | | | | Management, business, science, and arts | Construction, production, and maintenance occupations |
| | Total | Percent | Median value[1] | With a mortgage | Without a mortgage[2] | Median rent[3] | | | Total | Percent change, 2019-2020 | Total | Rate[5] | Total | | |
| | 89 | 90 | 91 | 92 | 93 | 94 | 95 | 96 | 97 | 98 | 99 | 100 | 101 | 102 | 103 |
| **VERMONT—Cont'd** | | | | | | | | | | | | | | | |
| Chittenden | 66,160 | 62.4 | 297,900 | 22.1 | 14.0 | 1,252 | 32.6 | 1.8 | 93,110 | -4.0 | 4,451 | 4.8 | 92,058 | 49.4 | 15.3 |
| Essex | 2,740 | 81.2 | 136,600 | 23.9 | 14.5 | 731 | 30.5 | 2.6 | 2,596 | -3.3 | 170 | 6.5 | 2,798 | 26.2 | 32.4 |
| Franklin | 18,563 | 75.4 | 216,200 | 22.8 | 15.3 | 999 | 27.2 | 2.4 | 26,329 | -4.5 | 1,383 | 5.3 | 25,562 | 36.6 | 26.7 |
| Grand Isle | 2,956 | 84.0 | 274,800 | 25.6 | 16.4 | 997 | 29.7 | 0.4 | 4,003 | -3.8 | 237 | 5.9 | 3,664 | 40.6 | 23.5 |
| Lamoille | 10,546 | 73.2 | 225,600 | 23.3 | 14.7 | 956 | 29.3 | 2.1 | 13,376 | -4.4 | 1,048 | 7.8 | 13,821 | 37.2 | 22.3 |
| Orange | 12,279 | 80.5 | 192,700 | 23.2 | 14.5 | 898 | 30.4 | 2.9 | 15,395 | -3.8 | 743 | 4.8 | 15,483 | 38.6 | 27.5 |
| Orleans | 11,779 | 78.2 | 162,000 | 24.0 | 16.7 | 796 | 31.9 | 2.5 | 12,702 | -4.7 | 1,008 | 7.9 | 12,629 | 32.7 | 27.5 |
| Rutland | 25,338 | 71.4 | 177,400 | 22.7 | 14.9 | 826 | 30.2 | 5.4 | 29,383 | -3.9 | 2,031 | 6.9 | 29,858 | 34.8 | 23.7 |
| Washington | 24,828 | 70.4 | 226,900 | 22.5 | 15.0 | 916 | 27.5 | 1.4 | 33,025 | -4.5 | 1,596 | 4.8 | 31,131 | 46.5 | 18.0 |
| Windham | 18,943 | 67.3 | 214,200 | 24.7 | 17.2 | 881 | 31.0 | 1.9 | 21,014 | -4.7 | 1,327 | 6.3 | 21,840 | 40.9 | 21.2 |
| Windsor | 24,298 | 73.8 | 224,500 | 23.8 | 14.7 | 935 | 32.0 | 2.3 | 27,903 | -4.0 | 1,483 | 5.3 | 27,833 | 41.1 | 20.3 |
| **VIRGINIA** | 3,151,045 | 66.3 | 273,100 | 21.2 | 10.0 | 1,234 | 29.1 | 2.3 | 4,346,644 | -1.8 | 271,407 | 6.2 | 4,156,018 | 44.4 | 18.8 |
| Accomack | 13,438 | 66.8 | 171,800 | 21.6 | 10.9 | 831 | 28.3 | 3.0 | 16,052 | -3.8 | 1,041 | 6.5 | 13,681 | 28.3 | 37.8 |
| Albemarle | 41,496 | 63.3 | 356,100 | 19.3 | 10.0 | 1,273 | 27.6 | 1.3 | 55,794 | -4.3 | 3,002 | 5.4 | 52,965 | 54.7 | 11.2 |
| Alleghany | 6,600 | 76.2 | 119,700 | 17.9 | 10.3 | 653 | 29.9 | 1.3 | 6,820 | -2.7 | 444 | 6.5 | 5,865 | 27.9 | 35.4 |
| Amelia | 4,954 | 84.4 | 201,400 | 21.4 | 10.0 | 826 | 24.5 | 2.6 | 6,202 | -2.5 | 357 | 5.8 | 6,362 | 28.7 | 33.5 |
| Amherst | 12,160 | 77.1 | 155,600 | 18.3 | 10.0 | 728 | 30.8 | 2.0 | 14,695 | -3.0 | 823 | 5.6 | 14,485 | 29.1 | 29.0 |
| Appomattox | 6,091 | 77.8 | 158,800 | 17.4 | 10.0 | 691 | 31.7 | 1.8 | 7,046 | -3.0 | 410 | 5.8 | 7,230 | 32.5 | 23.7 |
| Arlington | 107,032 | 42.7 | 705,400 | 20.1 | 10.0 | 1,970 | 25.6 | 3.5 | 151,080 | -2.7 | 6,773 | 4.5 | 146,900 | 70.1 | 7.1 |
| Augusta | 29,539 | 79.3 | 214,000 | 21.1 | 10.0 | 890 | 25.3 | 0.9 | 37,046 | -1.9 | 1,702 | 4.6 | 35,140 | 33.5 | 28.3 |
| Bath | 1,801 | 75.3 | 165,300 | 18.9 | 10.0 | 537 | 18.7 | 1.8 | 2,290 | -8.0 | 227 | 9.9 | 1,955 | 21.2 | 27.9 |
| Bedford | 31,317 | 82.1 | 201,900 | 19.3 | 10.0 | 855 | 25.6 | 1.7 | 37,737 | -3.1 | 1,937 | 5.1 | 36,986 | 37.5 | 24.7 |
| Bland | 2,367 | 81.4 | 113,200 | 16.7 | 10.0 | 589 | 34.5 | 0.9 | 2,811 | -1.6 | 158 | 5.6 | 2,305 | 25.1 | 26.8 |
| Botetourt | 13,264 | 86.3 | 227,700 | 20.0 | 10.0 | 908 | 24.5 | 0.9 | 17,140 | -3.1 | 777 | 4.5 | 16,083 | 36.9 | 26.8 |
| Brunswick | 6,037 | 70.7 | 112,200 | 21.0 | 12.2 | 668 | 29.0 | 1.9 | 5,951 | -2.0 | 474 | 8.0 | 6,025 | 21.0 | 36.9 |
| Buchanan | 8,569 | 77.3 | 72,300 | 23.7 | 12.1 | 617 | 24.8 | 1.1 | 6,450 | -5.6 | 614 | 9.5 | 6,405 | 29.2 | 33.6 |
| Buckingham | 5,826 | 75.2 | 130,300 | 21.9 | 10.0 | 779 | 28.0 | 2.3 | 6,317 | -3.7 | 467 | 7.4 | 6,692 | 26.6 | 32.6 |
| Campbell | 23,071 | 74.7 | 157,000 | 19.6 | 10.0 | 757 | 27.5 | 1.2 | 25,507 | -2.7 | 1,506 | 5.9 | 26,429 | 32.8 | 27.1 |
| Caroline | 10,946 | 79.2 | 206,000 | 21.5 | 10.6 | 1,093 | 30.1 | 3.0 | 15,004 | -1.9 | 1,009 | 6.7 | 13,834 | 31.9 | 25.7 |
| Carroll | 12,357 | 78.3 | 112,900 | 20.4 | 10.0 | 565 | 25.3 | 1.4 | 13,153 | -1.6 | 1,075 | 8.2 | 12,997 | 26.6 | 34.7 |
| Charles City | 2,896 | 83.9 | 167,900 | 22.4 | 10.0 | 813 | 35.9 | 2.0 | 3,610 | -2.1 | 235 | 6.5 | 3,359 | 26.5 | 33.0 |
| Charlotte | 4,661 | 71.3 | 111,900 | 19.7 | 11.8 | 620 | 30.9 | 2.2 | 5,084 | -3.9 | 285 | 5.6 | 5,048 | 26.9 | 28.4 |
| Chesterfield | 124,971 | 75.8 | 241,200 | 19.9 | 10.0 | 1,251 | 29.1 | 1.6 | 186,087 | -2.3 | 10,971 | 5.9 | 175,948 | 43.3 | 19.1 |
| Clarke | 5,612 | 75.0 | 347,200 | 21.1 | 10.8 | 1,151 | 35.2 | 0.6 | 7,498 | -2.8 | 341 | 4.5 | 7,193 | 42.1 | 23.4 |
| Craig | 2,297 | 76.4 | 165,800 | 18.8 | 10.0 | 566 | 23.8 | 1.4 | 2,286 | -3.0 | 116 | 5.1 | 2,415 | 29.2 | 29.9 |
| Culpeper | 17,071 | 72.7 | 298,900 | 22.3 | 10.0 | 1,160 | 30.4 | 3.2 | 24,511 | -2.6 | 1,214 | 5.0 | 24,567 | 35.0 | 25.3 |
| Cumberland | 3,975 | 73.9 | 146,000 | 24.4 | 12.5 | 855 | 32.8 | 0.9 | 4,569 | -2.7 | 280 | 6.1 | 4,878 | 25.0 | 33.6 |
| Dickenson | 5,778 | 75.7 | 81,000 | 21.2 | 11.9 | 592 | 36.0 | 1.3 | 4,746 | 0.7 | 378 | 8.0 | 4,352 | 32.6 | 25.7 |
| Dinwiddie | 10,401 | 77.2 | 168,300 | 19.5 | 11.9 | 1,005 | 30.2 | 1.7 | 13,439 | -2.1 | 861 | 6.4 | 13,745 | 26.5 | 32.9 |
| Essex | 4,555 | 67.1 | 202,800 | 23.2 | 11.8 | 885 | 34.3 | 3.3 | 5,488 | -0.9 | 382 | 7.0 | 5,121 | 32.1 | 25.1 |
| Fairfax | 396,501 | 68.0 | 563,100 | 21.2 | 10.0 | 1,881 | 28.3 | 3.6 | 628,397 | -1.7 | 36,515 | 5.8 | 613,908 | 57.5 | 10.7 |
| Fauquier | 24,562 | 78.3 | 395,500 | 22.7 | 10.8 | 1,281 | 25.4 | 1.1 | 36,812 | -2.7 | 1,721 | 4.7 | 36,040 | 42.9 | 20.0 |
| Floyd | 6,493 | 81.1 | 170,700 | 23.3 | 10.0 | 705 | 22.3 | 1.0 | 8,054 | -2.9 | 409 | 5.1 | 7,317 | 36.8 | 36.1 |
| Fluvanna | 9,923 | 85.0 | 234,700 | 20.2 | 10.0 | 1,163 | 25.8 | 1.6 | 13,723 | -4.4 | 722 | 5.3 | 12,651 | 44.4 | 20.9 |
| Franklin | 22,997 | 81.7 | 178,100 | 20.3 | 10.0 | 732 | 27.6 | 1.8 | 25,985 | -2.4 | 1,422 | 5.5 | 24,928 | 35.9 | 29.3 |
| Frederick | 31,503 | 77.6 | 251,200 | 20.1 | 10.0 | 1,125 | 25.7 | 1.7 | 48,636 | -0.6 | 2,260 | 4.6 | 43,083 | 37.7 | 26.0 |
| Giles | 6,910 | 75.1 | 123,200 | 19.2 | 10.0 | 700 | 18.9 | 1.6 | 7,739 | -2.8 | 484 | 6.3 | 7,483 | 27.9 | 34.8 |
| Gloucester | 14,786 | 78.2 | 230,000 | 21.3 | 10.0 | 1,084 | 28.6 | 1.5 | 19,249 | -2.6 | 1,045 | 5.4 | 17,786 | 32.1 | 28.6 |
| Goochland | 8,506 | 84.9 | 375,200 | 18.5 | 10.0 | 1,208 | 23.1 | 1.8 | 11,142 | -3.2 | 542 | 4.9 | 11,064 | 45.4 | 18.4 |
| Grayson | 6,498 | 81.3 | 102,900 | 18.9 | 10.8 | 520 | 25.0 | 1.5 | 8,104 | -0.2 | 460 | 5.7 | 6,444 | 25.4 | 34.4 |
| Greene | 7,548 | 77.6 | 236,400 | 22.4 | 10.1 | 1,165 | 26.3 | 2.2 | 10,198 | -4.3 | 534 | 5.2 | 10,035 | 38.9 | 22.0 |
| Greensville | 3,647 | 73.3 | 117,700 | 19.7 | 12.7 | 854 | 25.0 | 0.8 | 4,239 | -1.6 | 287 | 6.8 | 3,899 | 23.2 | 30.3 |
| Halifax | 14,158 | 74.1 | 116,100 | 19.3 | 10.9 | 671 | 30.0 | 1.3 | 15,176 | -3.0 | 1,067 | 7.0 | 14,561 | 30.1 | 34.2 |
| Hanover | 38,987 | 82.5 | 282,900 | 19.7 | 10.0 | 1,159 | 24.9 | 0.8 | 57,934 | -3.1 | 2,798 | 4.8 | 55,625 | 46.5 | 17.2 |
| Henrico | 128,464 | 62.7 | 242,600 | 20.4 | 10.0 | 1,170 | 29.2 | 1.7 | 180,321 | -1.8 | 11,704 | 6.5 | 173,089 | 45.1 | 15.6 |
| Henry | 21,218 | 71.9 | 93,900 | 21.0 | 11.4 | 604 | 27.7 | 2.0 | 23,810 | -0.4 | 1,816 | 7.6 | 19,967 | 26.6 | 33.5 |
| Highland | 1,078 | 82.9 | 173,900 | 19.2 | 10.0 | 618 | 19.9 | 3.2 | 1,239 | -1.5 | 46 | 3.7 | 874 | 36.8 | 27.7 |
| Isle of Wight | 14,587 | 75.8 | 266,800 | 21.3 | 12.2 | 1,045 | 28.9 | 1.6 | 19,092 | -3.1 | 977 | 5.1 | 17,799 | 39.5 | 26.3 |
| James City | 28,920 | 76.4 | 340,500 | 21.0 | 10.0 | 1,327 | 28.8 | 1.7 | 36,558 | -1.2 | 2,527 | 6.9 | 33,950 | 48.5 | 13.4 |
| King and Queen | 2,707 | 83.5 | 183,200 | 19.1 | 10.0 | 968 | 18.2 | 1.6 | 3,867 | -0.6 | 198 | 5.1 | 3,220 | 29.9 | 28.2 |
| King George | 9,202 | 78.8 | 307,900 | 20.5 | 10.0 | 1,208 | 27.1 | 1.2 | 13,819 | 2.0 | 617 | 4.5 | 13,065 | 44.6 | 18.7 |
| King William | 6,078 | 87.4 | 204,300 | 24.0 | 11.1 | 1,061 | 24.0 | 1.3 | 9,067 | -2.6 | 491 | 5.4 | 8,277 | 36.5 | 26.5 |
| Lancaster | 5,062 | 75.9 | 257,200 | 26.1 | 10.0 | 763 | 26.2 | 0.9 | 5,370 | -1.2 | 389 | 7.2 | 4,267 | 33.0 | 23.1 |
| Lee | 9,149 | 70.8 | 88,100 | 19.4 | 10.0 | 558 | 35.7 | 1.5 | 8,304 | -0.6 | 532 | 6.4 | 7,860 | 28.2 | 31.9 |
| Loudoun | 128,637 | 78.0 | 508,100 | 20.9 | 10.0 | 1,870 | 27.7 | 2.0 | 223,194 | -2.2 | 11,936 | 5.3 | 214,991 | 59.0 | 9.7 |

1. Specified owner-occupied units. lacking complete plumbing facilities.   2. A value of 10.0 represents 10 percent or less; a value of 50.0 represents 50 percent or more.   3. Specified renter-occupied units.   4. Overcrowded or
5. Percent of civilian labor force.   6. Civilian employed persons 16 years old and over.

| STATE County | Number of establish-ments | Total | Health care and social assistance | Manufac-turing | Retail trade | Finance and insurance | Professional, scientific, and technical services | Total (mil dol) | Average per employee (dollars) | Number | Fewer than 50 acres | 1000 acres or more | Farm producers whose primary occupation is farming (percent) |
|---|---|---|---|---|---|---|---|---|---|---|---|---|---|
| | | Employment → | | | | | | Annual payroll → | | Farms → | Percent with: | | |
| | 104 | 105 | 106 | 107 | 108 | 109 | 110 | 111 | 112 | 113 | 114 | 115 | 116 |
| **VERMONT—Cont'd** | | | | | | | | | | | | | |
| Chittenden | 5,662 | 89,488 | 16,391 | 9,365 | 12,403 | 3,013 | 6,664 | 4,586 | 51,252 | 585 | 54.4 | 1.0 | 39.7 |
| Essex | 112 | 547 | 75 | 153 | 62 | 7 | 20 | 18 | 32,282 | 106 | 31.1 | 3.8 | 50.9 |
| Franklin | 1,032 | 13,163 | 2,666 | 2,834 | 2,401 | 288 | 249 | 566 | 42,994 | 729 | 31.3 | 4.0 | 49.1 |
| Grand Isle | 181 | 772 | 61 | 51 | 148 | 21 | 34 | 27 | 34,992 | 119 | 42.0 | 2.5 | 47.4 |
| Lamoille | 960 | 11,138 | 1,723 | 556 | 1,595 | 264 | 272 | 397 | 35,674 | 329 | 39.8 | 1.8 | 35.8 |
| Orange | 703 | 6,366 | 1,848 | 725 | 804 | 137 | 342 | 264 | 41,467 | 569 | 35.5 | 1.4 | 43.6 |
| Orleans | 769 | 8,393 | 1,589 | 1,281 | 1,408 | 193 | 157 | 304 | 36,210 | 558 | 33.2 | 2.9 | 47.8 |
| Rutland | 2,030 | 24,022 | 4,688 | 3,248 | 3,858 | 517 | 609 | 1,022 | 42,546 | 614 | 38.8 | 2.0 | 44.6 |
| Washington | 2,179 | 25,956 | 4,452 | 2,750 | 3,927 | 2,524 | 880 | 1,245 | 47,954 | 553 | 47.2 | 0.9 | 35.6 |
| Windham | 1,655 | 20,695 | 3,534 | 2,437 | 2,220 | 541 | 509 | 849 | 41,006 | 414 | 53.6 | 0.2 | 46.4 |
| Windsor | 1,940 | 20,141 | 4,437 | 2,077 | 2,694 | 489 | 1,078 | 924 | 45,859 | 677 | 38.7 | 2.7 | 33.6 |
| **VIRGINIA** | 203,467 | 3,455,993 | 468,391 | 240,484 | 430,594 | 166,976 | 495,867 | 197,418 | 57,123 | 43,225 | 42.2 | 3.1 | 40.1 |
| Accomack | 714 | 9,075 | 1,145 | 3,082 | 1,337 | 239 | 420 | 295 | 32,546 | 239 | 53.6 | 8.8 | 54.9 |
| Albemarle | 2,838 | 41,641 | 8,526 | 2,605 | 6,924 | 2,322 | 4,153 | 2,104 | 50,531 | 913 | 40.7 | 4.3 | 38.6 |
| Alleghany | 250 | 2,533 | 859 | 261 | 396 | 46 | 36 | 81 | 32,118 | 165 | 35.2 | 3.0 | 40.9 |
| Amelia | 277 | 1,925 | 292 | 167 | 244 | 32 | 67 | 65 | 33,558 | 370 | 36.8 | 6.5 | 44.6 |
| Amherst | 565 | 6,136 | 688 | 1,311 | 1,139 | 115 | 197 | 225 | 36,723 | 369 | 32.5 | 3.8 | 40.3 |
| Appomattox | 281 | 2,376 | 496 | 92 | 690 | 96 | 45 | 66 | 27,600 | 412 | 28.6 | 2.7 | 34.1 |
| Arlington | 6,416 | 148,180 | 10,585 | 206 | 8,318 | 3,962 | 50,399 | 13,125 | 88,577 | 5 | 100.0 | NA | NA |
| Augusta | 1,419 | 21,666 | 3,834 | 5,366 | 2,197 | 307 | 458 | 972 | 44,864 | 1,665 | 45.7 | 2.6 | 44.6 |
| Bath | 122 | 1,872 | 270 | 85 | 78 | NA | 24 | 73 | 38,952 | 110 | 8.2 | 10.0 | 47.7 |
| Bedford | 1,736 | 17,799 | 2,325 | 4,106 | 2,631 | 433 | 920 | 718 | 40,312 | 1,418 | 38.7 | 2.2 | 36.3 |
| Bland | 71 | 1,246 | 135 | 470 | 93 | NA | NA | 69 | 55,583 | 339 | 24.2 | 2.7 | 43.5 |
| Botetourt | 750 | 10,717 | 1,023 | 2,851 | 730 | 237 | 219 | 490 | 45,741 | 551 | 35.2 | 1.8 | 38.6 |
| Brunswick | 245 | 2,247 | 160 | 305 | 271 | 46 | 46 | 84 | 37,390 | 242 | 26.0 | 4.1 | 43.3 |
| Buchanan | 373 | 5,090 | 563 | 230 | 715 | 169 | 198 | 251 | 49,393 | 89 | 52.8 | NA | 50.0 |
| Buckingham | 252 | 1,748 | 334 | 138 | 312 | 23 | 49 | 67 | 38,110 | 408 | 28.7 | 2.5 | 37.9 |
| Campbell | 1,162 | 17,184 | 1,215 | 4,508 | 2,225 | 380 | 739 | 831 | 48,363 | 702 | 30.6 | 2.1 | 37.1 |
| Caroline | 386 | 4,091 | 318 | 294 | 640 | 256 | 140 | 151 | 37,010 | 222 | 47.7 | 8.6 | 40.7 |
| Carroll | 409 | 4,077 | 711 | 878 | 733 | 93 | 124 | 114 | 27,855 | 900 | 38.9 | 0.9 | 35.0 |
| Charles City | 156 | 1,609 | NA | 358 | 120 | 4 | 55 | 74 | 46,071 | 77 | 49.4 | 15.6 | 40.4 |
| Charlotte | 213 | 1,897 | 298 | 500 | 298 | 50 | 31 | 58 | 30,520 | 460 | 20.9 | 4.1 | 40.3 |
| Chesterfield | 7,579 | 118,013 | 13,941 | 8,932 | 22,084 | 4,422 | 10,161 | 5,265 | 44,611 | 427 | 47.8 | 3.7 | 43.5 |
| Clarke | 365 | 2,793 | 404 | 471 | 274 | 83 | 121 | 125 | 44,905 | 179 | 20.1 | 1.1 | 43.4 |
| Craig | 66 | 378 | 50 | NA | 143 | 26 | 13 | 11 | 30,397 | 125 | 49.1 | 4.7 | 39.4 |
| Culpeper | 1,031 | 13,167 | 2,324 | 1,141 | 2,447 | 253 | 575 | 577 | 43,856 | 682 | 36.4 | 3.4 | 41.7 |
| Cumberland | 136 | 771 | 71 | 105 | 164 | 19 | 23 | 24 | 31,169 | 264 | 50.0 | 1.6 | 37.0 |
| Dickenson | 193 | 1,972 | 320 | 21 | 433 | 65 | 335 | 66 | 33,518 | 128 | 33.5 | 6.4 | 40.4 |
| Dinwiddie | 359 | 5,814 | 1,368 | 708 | 548 | 78 | 46 | 260 | 44,754 | 358 | 31.8 | 21.6 | 40.3 |
| Essex | 287 | 3,310 | 579 | 506 | 848 | 116 | 81 | 102 | 30,919 | 88 | 78.6 | NA | 31.8 |
| Fairfax | 31,619 | 656,234 | 66,535 | 5,804 | 51,262 | 31,021 | 218,675 | 55,389 | 84,405 | 117 | 50.3 | 4.0 | 39.3 |
| Fauquier | 1,912 | 19,330 | 2,617 | 996 | 3,043 | 841 | 2,112 | 956 | 49,464 | 1,154 | 37.4 | 0.9 | 36.6 |
| Floyd | 323 | 2,601 | 685 | 464 | 405 | 46 | 71 | 77 | 29,423 | 741 | 37.4 | 0.7 | 42.5 |
| Fluvanna | 396 | 2,787 | 207 | 144 | 341 | 31 | 92 | 104 | 37,339 | 273 | 44.7 | 0.7 | 38.0 |
| Franklin | 1,164 | 13,069 | 1,514 | 2,954 | 2,064 | 293 | 329 | 401 | 30,711 | 1,019 | 35.0 | 1.7 | 34.6 |
| Frederick | 1,552 | 26,243 | 1,825 | 5,601 | 3,789 | 2,029 | 757 | 1,199 | 45,677 | 762 | 55.2 | 2.0 | 32.9 |
| Giles | 286 | 3,464 | 488 | 887 | 649 | 72 | 305 | 145 | 41,957 | 389 | 40.4 | 3.3 | 37.7 |
| Gloucester | 862 | 7,580 | 1,296 | 161 | 2,262 | 232 | 286 | 233 | 30,747 | 166 | 64.5 | 4.8 | 40.4 |
| Goochland | 693 | 18,668 | 470 | 216 | 725 | 6,028 | 337 | 2,392 | 128,147 | 355 | 47.3 | 2.0 | 38.4 |
| Grayson | 158 | 1,403 | 173 | 345 | 160 | 103 | 17 | 43 | 30,502 | 716 | 39.5 | 1.8 | 39.8 |
| Greene | 349 | 2,735 | 212 | 56 | 763 | 48 | 136 | 94 | 34,397 | 214 | 39.3 | 0.5 | 40.7 |
| Greensville | 99 | 2,215 | NA | 883 | 256 | 10 | NA | 80 | 35,906 | 150 | 27.3 | 8.7 | 39.0 |
| Halifax | 680 | 9,883 | 1,717 | 1,938 | 1,403 | 201 | 163 | 367 | 37,110 | 895 | 21.7 | 3.1 | 42.5 |
| Hanover | 3,239 | 49,245 | 6,414 | 3,705 | 6,940 | 893 | 1,644 | 2,091 | 42,469 | 567 | 61.7 | 4.6 | 37.0 |
| Henrico | 9,485 | 187,061 | 25,484 | 5,972 | 24,132 | 26,668 | 16,257 | 10,413 | 55,667 | 99 | 70.7 | 2.0 | 30.6 |
| Henry | 798 | 11,033 | 804 | 3,315 | 1,614 | 468 | 349 | 381 | 34,571 | 212 | 30.2 | 5.7 | 48.6 |
| Highland | 88 | 370 | 71 | 25 | 42 | 27 | 22 | 10 | 28,165 | 275 | 16.7 | 6.2 | 52.8 |
| Isle of Wight | 704 | 9,975 | 634 | 3,589 | 1,025 | 196 | 295 | 508 | 50,968 | 237 | 47.3 | 13.1 | 25.8 |
| James City | 1,736 | 28,207 | 4,033 | 2,503 | 4,136 | 612 | 1,507 | 1,102 | 39,083 | 72 | 59.7 | 2.8 | 43.7 |
| King and Queen | 117 | 634 | 22 | 21 | 75 | NA | 67 | 25 | 39,729 | 151 | 41.7 | 4.3 | 33.3 |
| King George | 488 | 5,469 | 329 | 90 | 890 | 134 | 2,148 | 299 | 54,752 | 141 | 36.2 | 15.6 | 36.8 |
| King William | 358 | 3,501 | 354 | 519 | 485 | 85 | 122 | 170 | 48,639 | 90 | 51.1 | 5.0 | 24.4 |
| Lancaster | 436 | 4,004 | 1,032 | 69 | 731 | 274 | 337 | 153 | 38,312 | 80 | 41.3 | 5.0 | 38.4 |
| Lee | 252 | 2,386 | 535 | 116 | 694 | 161 | 85 | 70 | 29,174 | 830 | 40.5 | 0.6 | 33.3 |
| Loudoun | 11,028 | 168,351 | 14,567 | 5,729 | 19,356 | 3,506 | 31,048 | 10,313 | 61,258 | 1,259 | 69.5 | 1.3 | 33.3 |

# Table B. States and Counties — Agriculture

Items 117—132

| STATE County | Acreage (1,000) [117] | Percent change, 2012-2017 [118] | Average size of farm [119] | Total irrigated (1,000) [120] | Total cropland (1,000) [121] | Value of land and buildings: Average per farm [122] | Average per acre [123] | Value of machinery and equipment, average per farm (dollars) [124] | Total (mil dol) [125] | Average per farm (acres) [126] | Crops [127] | Livestock and poultry products [128] | Organic farms (number) [129] | Farms with internet access (percent) [130] | Government payments Total ($1,000) [131] | Percent of farms [132] |
|---|---|---|---|---|---|---|---|---|---|---|---|---|---|---|---|---|
| **VERMONT—Cont'd** | | | | | | | | | | | | | | | | |
| Chittenden | 64 | -12.7 | 110 | 0.5 | 26.8 | 701,224 | 6,387 | 84,797 | 43.6 | 74,492 | 51.2 | 48.8 | 68 | 90.8 | 545 | 12.1 |
| Essex | 43 | 67.9 | 404 | 0.0 | 8.0 | 789,828 | 1,956 | 137,512 | 12.7 | 119,887 | 39.7 | 60.3 | 10 | 83.0 | 41 | 15.1 |
| Franklin | 190 | 1.8 | 260 | 0.3 | 83.2 | 811,688 | 3,119 | 165,431 | 185.6 | 254,615 | 20.8 | 79.2 | 125 | 84.8 | 965 | 11.7 |
| Grand Isle | 19 | -1.2 | 158 | 0.0 | 12.6 | 679,940 | 4,304 | 122,915 | 17.9 | 150,723 | 16.9 | 83.1 | 11 | 88.2 | 354 | 19.3 |
| Lamoille | 53 | 2.0 | 162 | 0.2 | 14.6 | 563,296 | 3,486 | 72,199 | 27.7 | 84,319 | 36.3 | 63.7 | 48 | 86.6 | 74 | 9.1 |
| Orange | 86 | -18.6 | 150 | 0.2 | 32.6 | 553,914 | 3,681 | 90,079 | 55.1 | 96,749 | 22.8 | 77.2 | 59 | 80.8 | 389 | 8.4 |
| Orleans | 128 | -1.6 | 230 | 0.1 | 58.6 | 623,033 | 2,708 | 119,211 | 92.0 | 164,875 | 13.7 | 86.3 | 52 | 81.9 | 234 | 3.6 |
| Rutland | 99 | -8.9 | 161 | 0.2 | 34.6 | 444,531 | 2,761 | 70,922 | 28.8 | 46,932 | 35.9 | 64.1 | 24 | 81.9 | 657 | 16.9 |
| Washington | 64 | -4.3 | 117 | 0.1 | 21.6 | 499,088 | 4,283 | 86,311 | 30.7 | 55,566 | 29.8 | 70.2 | 78 | 86.8 | 153 | 4.0 |
| Windham | 45 | -11.7 | 108 | 0.4 | 14.4 | 542,832 | 5,011 | 71,071 | 28.6 | 69,048 | 39.8 | 60.2 | 28 | 93.0 | 457 | 8.5 |
| Windsor | 113 | 11.1 | 166 | 0.1 | 22.6 | 643,087 | 3,865 | 74,480 | 25.1 | 37,013 | 34.4 | 65.6 | 36 | 84.8 | 141 | 8.0 |
| **VIRGINIA** | 7,798 | -6.1 | 180 | 63.4 | 3,084.1 | 834,254 | 4,624 | 86,136 | 3,960.5 | 91,625 | 34.4 | 65.6 | 252 | 74.0 | 60,805 | 13.9 |
| Accomack | 77 | -0.8 | 321 | 5.1 | 57.7 | 1,370,847 | 4,268 | 217,830 | 163.3 | 683,134 | 28.7 | 71.3 | 1 | 81.2 | 2,399 | 30.5 |
| Albemarle | 183 | 8.2 | 200 | 0.8 | 51.8 | 1,980,339 | 9,892 | 68,432 | 29.6 | 32,472 | 63.6 | 36.4 | 9 | 81.3 | 477 | 6.8 |
| Alleghany | 31 | -16.5 | 187 | NA | 8.9 | 788,210 | 4,215 | 66,701 | 2.7 | 16,618 | 29.5 | 70.5 | NA | 81.8 | 16 | 5.5 |
| Amelia | 103 | 16.3 | 278 | 0.2 | 34.8 | 858,640 | 3,094 | 108,375 | 86.6 | 233,992 | 10.7 | 89.3 | NA | 70.5 | 545 | 34.1 |
| Amherst | 79 | -20.4 | 214 | 0.2 | 17.3 | 993,967 | 4,654 | 60,527 | 8.1 | 22,068 | 43.2 | 56.8 | NA | 71.5 | 24 | 5.1 |
| Appomattox | 76 | -21.4 | 184 | 0.0 | 20.9 | 568,840 | 3,094 | 72,238 | 10.0 | 24,218 | 33.3 | 66.7 | 3 | 66.5 | 329 | 12.1 |
| Arlington | 0 | -86.1 | 1 | NA | NA | 298,000 | 298,000 | 30,686 | NA | NA | NA | 100.0 | NA | 100.0 | NA | NA |
| Augusta | 291 | 11.8 | 175 | 2.5 | 116.8 | 1,222,854 | 6,999 | 101,481 | 292.5 | 175,704 | 12.8 | 87.2 | 5 | 71.5 | 1,049 | 15.7 |
| Bath | 48 | 15.8 | 435 | D | 15.3 | 1,599,416 | 3,677 | 100,025 | 6.7 | 61,345 | 24.8 | 75.2 | 1 | 66.4 | 91 | 10.0 |
| Bedford | 211 | 2.2 | 149 | 0.1 | 65.9 | 654,822 | 4,399 | 68,521 | 26.5 | 18,667 | 31.9 | 68.1 | 1 | 76.0 | 658 | 10.4 |
| Bland | 70 | -9.2 | 207 | 0.8 | 17.3 | 714,594 | 3,446 | 73,919 | 8.8 | 25,906 | 25.7 | 74.3 | NA | 67.3 | 79 | 11.2 |
| Botetourt | 89 | -0.5 | 161 | 0.0 | 27.0 | 676,797 | 4,198 | 72,657 | 14.1 | 25,506 | 28.8 | 71.2 | 1 | 76.6 | 152 | 9.3 |
| Brunswick | 66 | -26.1 | 274 | 2.1 | 24.4 | 689,842 | 2,514 | 118,499 | 23.1 | 95,438 | 71.7 | 28.3 | 1 | 69.4 | 378 | 50.8 |
| Buchanan | 11 | 15.5 | 124 | D | 1.2 | 300,008 | 2,418 | 33,157 | 0.4 | 4,876 | 54.8 | 45.2 | NA | 51.7 | NA | NA |
| Buckingham | 79 | -5.6 | 194 | 0.1 | 30.5 | 724,596 | 3,731 | 81,081 | 43.4 | 106,480 | 15.7 | 84.3 | 1 | 73.0 | 555 | 12.7 |
| Campbell | 132 | -12.6 | 188 | 0.3 | 42.9 | 578,961 | 3,086 | 78,302 | 25.4 | 36,179 | 25.7 | 74.3 | NA | 76.4 | 204 | 13.1 |
| Caroline | 62 | 9.7 | 279 | 3.4 | 42.4 | 1,029,270 | 3,696 | 134,498 | 22.9 | 103,189 | 95.8 | 4.2 | 4 | 88.3 | 739 | 19.8 |
| Carroll | 119 | -15.3 | 132 | 0.6 | 32.2 | 425,568 | 3,219 | 63,111 | 44.5 | 49,488 | 30.7 | 69.3 | 7 | 65.1 | 113 | 7.9 |
| Charles City | 31 | 0.7 | 408 | 1.4 | 20.6 | 1,375,102 | 3,373 | 187,684 | D | D | D | D | 1 | 96.1 | 1,156 | 22.1 |
| Charlotte | 122 | -18.5 | 264 | 0.5 | 46.1 | 693,206 | 2,621 | 92,833 | 26.0 | 56,470 | 37.8 | 62.2 | 6 | 67.6 | 950 | 27.6 |
| Chesterfield | 18 | -9.8 | 86 | 0.1 | 7.2 | 790,553 | 9,216 | 72,007 | 4.5 | 21,481 | 64.3 | 35.7 | 3 | 81.0 | 122 | 15.7 |
| Clarke | 67 | -0.5 | 156 | 0.1 | 24.7 | 1,099,589 | 7,046 | 72,761 | 16.4 | 38,375 | 41.2 | 58.8 | NA | 85.0 | 75 | 7.5 |
| Craig | 43 | -6.9 | 243 | 0.1 | 12.0 | 725,533 | 2,990 | 80,837 | 4.7 | 26,022 | 18.1 | 81.9 | 5 | 84.4 | 143 | 14.5 |
| Culpeper | 124 | -1.6 | 182 | 0.6 | 60.1 | 1,048,206 | 5,747 | 87,179 | 48.5 | 71,176 | 64.7 | 35.3 | 5 | 81.5 | 568 | 11.0 |
| Cumberland | 53 | -7.9 | 199 | 0.2 | 15.4 | 758,741 | 3,806 | 71,848 | 39.8 | 150,780 | 10.2 | 89.8 | NA | 68.2 | 196 | 15.9 |
| Dickenson | 11 | -25.8 | 87 | NA | 1.8 | 246,038 | 2,820 | 74,934 | 0.6 | 4,375 | 17.7 | 82.3 | NA | 72.7 | 216 | 10.2 |
| Dinwiddie | 93 | 4.0 | 259 | 1.6 | 42.8 | 795,681 | 3,068 | 105,397 | 25.7 | 71,799 | 85.4 | 14.6 | NA | 64.2 | 1,534 | 34.6 |
| Essex | 59 | 3.5 | 667 | D | 42.3 | 2,130,482 | 3,194 | 333,241 | 21.0 | 239,159 | 94.3 | 5.7 | NA | 78.4 | 1,341 | 51.1 |
| Fairfax | 6 | -24.4 | 51 | 0.1 | 0.8 | 852,783 | 16,806 | 45,082 | 1.2 | 10,624 | 81.3 | 18.7 | 2 | 90.6 | 21 | 6.8 |
| Fauquier | 217 | -5.1 | 188 | 0.5 | 83.8 | 1,381,646 | 7,359 | 83,383 | 54.8 | 47,497 | 40.0 | 60.0 | 4 | 85.6 | 403 | 5.6 |
| Floyd | 110 | -23.8 | 149 | 0.2 | 32.3 | 565,896 | 3,802 | 62,225 | 33.7 | 45,511 | 48.3 | 51.7 | 11 | 74.8 | 73 | 3.9 |
| Fluvanna | 44 | -5.7 | 163 | 0.1 | 13.3 | 702,048 | 4,315 | 66,495 | 6.1 | 22,436 | 63.6 | 36.4 | 2 | 81.3 | 170 | 11.7 |
| Franklin | 156 | -5.0 | 153 | 0.4 | 72.8 | 552,387 | 3,602 | 81,393 | 69.2 | 67,909 | 24.6 | 75.4 | 4 | 66.7 | 705 | 8.0 |
| Frederick | 110 | 9.1 | 144 | 0.0 | 45.2 | 1,041,790 | 7,223 | 67,981 | 33.8 | 44,324 | 72.9 | 27.1 | 6 | 77.7 | 78 | 3.7 |
| Giles | 65 | -0.3 | 168 | 0.0 | 13.0 | 516,470 | 3,073 | 66,106 | 9.7 | 24,871 | 13.7 | 86.3 | NA | 74.6 | 58 | 2.3 |
| Gloucester | 26 | 28.1 | 157 | 0.0 | 14.2 | 739,196 | 4,717 | 108,605 | 11.7 | 70,482 | 80.0 | 20.0 | 6 | 81.9 | 580 | 12.0 |
| Goochland | 57 | 13.2 | 160 | 0.1 | 24.0 | 867,877 | 5,430 | 89,076 | 11.7 | 33,070 | 38.6 | 61.4 | NA | 78.3 | 144 | 7.6 |
| Grayson | 119 | -9.5 | 167 | 0.2 | 34.4 | 692,467 | 4,155 | 68,462 | 40.8 | 57,011 | 26.1 | 73.9 | 2 | 73.9 | 157 | 8.7 |
| Greene | 29 | 4.6 | 133 | 0.0 | 10.0 | 870,295 | 6,531 | 60,814 | 7.6 | 35,579 | 23.6 | 76.4 | NA | 78.0 | 211 | 16.4 |
| Greensville | 55 | -6.4 | 364 | D | 33.5 | 986,245 | 2,712 | 147,447 | 19.4 | 129,653 | 98.1 | 1.9 | NA | 72.0 | 1,707 | 58.7 |
| Halifax | 209 | -1.2 | 233 | 1.1 | 57.5 | 619,060 | 2,651 | 70,488 | 31.9 | 35,617 | 51.3 | 48.7 | 15 | 66.9 | 429 | 20.0 |
| Hanover | 89 | -5.4 | 157 | 3.6 | 61.3 | 791,920 | 5,035 | 102,289 | 49.3 | 86,868 | 83.7 | 16.3 | NA | 81.3 | 768 | 9.3 |
| Henrico | 10 | -23.8 | 99 | D | 7.3 | 601,700 | 6,066 | 59,847 | 7.3 | 73,606 | 97.5 | 2.5 | 1 | 75.8 | 36 | 14.1 |
| Henry | 46 | 6.0 | 215 | 0.0 | 11.7 | 635,840 | 2,961 | 55,004 | 14.1 | 66,618 | 16.1 | 83.9 | 6 | 72.6 | 34 | 5.2 |
| Highland | 93 | -0.1 | 338 | D | 15.6 | 1,080,848 | 3,198 | 65,010 | 26.1 | 94,982 | 6.6 | 93.4 | 1 | 68.0 | 279 | 11.3 |
| Isle of Wight | 81 | 6.6 | 340 | 0.9 | 51.5 | 1,264,037 | 3,714 | 173,573 | 64.2 | 270,987 | 61.9 | 38.1 | NA | 89.5 | 2,755 | 36.7 |
| James City | 7 | 19.6 | 92 | 0.0 | 3.6 | 931,231 | 10,113 | 129,707 | 2.0 | 28,458 | 92.6 | 7.4 | NA | 93.1 | 78 | 15.3 |
| King and Queen | 48 | 14.9 | 320 | 1.2 | 33.6 | 1,078,959 | 3,377 | 158,462 | 18.4 | 121,682 | 75.6 | 24.4 | NA | 79.5 | 817 | 37.1 |
| King George | 26 | 8.4 | 187 | D | 13.0 | 954,694 | 5,111 | 99,089 | 11.9 | 84,546 | 95.1 | 4.9 | NA | 73.8 | 134 | 27.7 |
| King William | 47 | -11.4 | 527 | 1.6 | 26.3 | 2,133,522 | 4,046 | 250,259 | 14.2 | 157,411 | 88.1 | 11.9 | NA | 87.8 | 713 | 28.9 |
| Lancaster | 16 | 51.8 | 203 | 0.0 | 11.3 | 860,073 | 4,237 | 123,923 | 5.6 | 69,388 | 91.9 | 8.1 | 3 | 90.0 | 659 | 37.5 |
| Lee | 95 | -19.0 | 114 | 0.0 | 25.2 | 302,584 | 2,646 | 71,083 | 15.3 | 18,393 | 26.0 | 74.0 | 5 | 61.8 | 1,658 | 9.8 |
| Loudoun | 122 | -9.5 | 97 | 0.4 | 60.2 | 950,622 | 9,816 | 59,142 | 44.0 | 34,940 | 70.2 | 29.8 | 11 | 88.6 | 190 | 3.9 |

# Table B. States and Counties — Water Use, Wholesale Trade, Retail Trade, and Real Estate

| STATE County | Water use, 2015 — Public supply water withdrawn (mil gal/day) [133] | Public supply gallons withdrawn per person per day [134] | Wholesale Trade[1], 2017 — Number of establishments [135] | Number of employees [136] | Sales (mil dol) [137] | Average payroll (mil dol) [138] | Retail Trade[2], 2017 — Number of establishments [139] | Number of employees [140] | Sales (mil dol) [141] | Average payroll (mil dol) [142] | Real estate and rental and leasing,[2] 2017 — Number of establishments [143] | Number of employees [144] | Sales (mil dol) [145] | Average payroll (mil dol) [146] |
|---|---|---|---|---|---|---|---|---|---|---|---|---|---|---|
| **VERMONT—Cont'd** | | | | | | | | | | | | | | |
| Chittenden | 13.89 | 86.1 | 242 | 3,834 | 2,709.9 | 232.7 | 813 | 12,376 | 3,361.1 | 357.0 | 253 | 1,083 | 335.9 | 49.9 |
| Essex | 0.25 | 40.6 | NA | NA | NA | NA | 16 | 71 | 13.9 | 1.3 | NA | NA | NA | NA |
| Franklin | 3.00 | 61.5 | 43 | 1,065 | 1,012.6 | 43.7 | 176 | 2,268 | 830.0 | 63.7 | 30 | 73 | 9.4 | 2.0 |
| Grand Isle | 0.50 | 72.9 | D | D | D | 0.4 | 25 | 156 | 42.8 | 3.9 | 7 | 9 | 2.0 | 0.2 |
| Lamoille | 1.29 | 51.1 | D | D | D | 12.8 | 158 | 1,696 | 390.9 | 43.8 | 39 | 155 | 35.2 | 7.0 |
| Orange | 2.16 | 74.7 | 21 | 165 | 61.6 | 6.3 | 82 | 908 | 225.6 | 24.6 | 20 | 25 | 6.6 | 1.2 |
| Orleans | 1.63 | 60.1 | D | D | D | D | 135 | 1,510 | 378.0 | 39.3 | 21 | 89 | 20.3 | 3.8 |
| Rutland | 4.37 | 73.2 | D | D | D | D | 375 | 3,858 | 1,124.5 | 106.4 | 63 | 224 | 34.4 | 8.4 |
| Washington | 1.85 | 31.6 | D | D | D | D | 347 | 4,159 | 1,112.1 | 124.2 | 58 | 228 | 43.4 | 8.8 |
| Windham | 2.10 | 48.4 | 39 | 945 | 568.8 | 46.4 | 232 | 2,296 | 574.1 | 65.3 | 87 | 280 | 45.6 | 9.3 |
| Windsor | 3.10 | 55.6 | 51 | 717 | 418.7 | 41.7 | 275 | 2,737 | 810.2 | 81.6 | 99 | 416 | 61.4 | 15.9 |
| **VIRGINIA** | 695.63 | 83.0 | 5,937 | 86,698 | 98,904.6 | 5,734.1 | 27,134 | 429,072 | 120,162.1 | 11,476.0 | 10,051 | 55,778 | 17,207.8 | 2,949.2 |
| Accomack | 0.88 | 26.7 | 22 | 112 | 76.9 | 4.7 | 161 | 1,351 | 366.9 | 29.4 | 37 | 89 | 13.9 | 2.2 |
| Albemarle | 13.29 | 125.7 | 66 | 480 | 233.9 | 27.1 | 359 | 6,689 | 1,958.6 | 204.8 | 184 | 939 | 186.3 | 37.3 |
| Alleghany | 0.91 | 58.0 | 5 | 31 | 7.0 | 1.1 | 45 | 384 | 101.8 | 9.3 | 5 | D | 3.5 | D |
| Amelia | 0.10 | 7.8 | 8 | 105 | 45.9 | 6.4 | 22 | 203 | 60.7 | 5.5 | 11 | D | 3.4 | D |
| Amherst | 12.03 | 377.0 | 8 | 153 | 154.5 | 5.7 | 88 | 1,141 | 302.8 | 25.0 | 15 | 24 | 2.8 | 0.6 |
| Appomattox | 0.00 | 0.0 | 10 | 63 | 12.1 | 2.2 | 55 | 701 | 168.0 | 16.0 | 14 | 39 | 7.5 | 1.0 |
| Arlington | 0.00 | 0.0 | 74 | 568 | 405.0 | 44.0 | 559 | 9,693 | 2,701.2 | 337.0 | 474 | 4,227 | 1,926.8 | 277.7 |
| Augusta | 7.50 | 100.9 | 52 | 724 | 307.4 | 36.4 | 193 | 2,161 | 663.4 | 61.7 | 61 | 165 | 21.6 | 4.3 |
| Bath | 0.29 | 64.9 | NA | NA | NA | NA | 17 | 75 | 16.5 | 1.5 | 7 | 38 | 4.4 | 1.4 |
| Bedford | 1.54 | 19.8 | 48 | 503 | 293.8 | 23.3 | 221 | 2,680 | 718.8 | 66.9 | 105 | 211 | 52.5 | 7.0 |
| Bland | 0.04 | 6.1 | D | D | D | D | 13 | 64 | 26.0 | 1.5 | NA | NA | NA | NA |
| Botetourt | 5.46 | 163.7 | D | D | D | 27.8 | 82 | 777 | 224.1 | 16.6 | 31 | 139 | 28.8 | 6.3 |
| Brunswick | 24.34 | 1,457.7 | 7 | D | 21.0 | 9.2 | 40 | 292 | 90.1 | 6.6 | 8 | 26 | 1.3 | 0.6 |
| Buchanan | 0.00 | 0.0 | 15 | 177 | 181.3 | D | 74 | 799 | 164.0 | 17.0 | 11 | 26 | 6.4 | 0.8 |
| Buckingham | 0.36 | 21.1 | 6 | D | 8.6 | D | 42 | 289 | 72.9 | 6.7 | 4 | D | 0.8 | D |
| Campbell | 4.46 | 81.0 | 39 | 613 | 339.1 | 34.5 | 189 | 2,072 | 513.4 | 46.4 | 52 | 125 | 22.3 | 3.7 |
| Caroline | 0.83 | 27.7 | D | D | D | D | 56 | 513 | 306.9 | 12.6 | 13 | 42 | 2.8 | 0.5 |
| Carroll | 1.70 | 57.2 | 17 | 251 | 45.0 | 4.9 | 81 | 753 | 237.8 | 15.5 | 15 | 33 | 5.7 | 1.0 |
| Charles City | 0.00 | 0.0 | 11 | 72 | 28.1 | 4.7 | 10 | 67 | 19.3 | 2.4 | 7 | D | 4.9 | D |
| Charlotte | 0.24 | 19.7 | 11 | 75 | 67.2 | 3.0 | 36 | 266 | 59.6 | 5.5 | 6 | 11 | 1.0 | 0.2 |
| Chesterfield | 39.86 | 118.7 | 285 | 4,108 | 2,653.4 | 255.0 | 963 | 21,672 | 6,746.1 | 596.5 | 358 | 1,511 | 2,021.6 | 70.1 |
| Clarke | 0.50 | 34.8 | D | D | D | D | 38 | 304 | 94.5 | 8.4 | D | D | D | D |
| Craig | 0.00 | 0.0 | D | D | D | 0.9 | 9 | 199 | 22.5 | 2.2 | NA | NA | NA | NA |
| Culpeper | 2.33 | 47.1 | 21 | 462 | 276.0 | 20.4 | 156 | 2,310 | 671.2 | 63.5 | 45 | 249 | 36.9 | 9.7 |
| Cumberland | 0.00 | 0.0 | NA | NA | NA | NA | 26 | 171 | 40.4 | 4.2 | NA | NA | NA | NA |
| Dickenson | 4.96 | 328.2 | D | D | D | 1.2 | 48 | 416 | 103.5 | 9.5 | NA | NA | NA | NA |
| Dinwiddie | 0.03 | 1.1 | D | D | D | D | 49 | 527 | 148.1 | 10.7 | 13 | 57 | 13.5 | 2.7 |
| Essex | 0.02 | 1.8 | 7 | 69 | 96.7 | 3.4 | 63 | 841 | 221.2 | 21.2 | 10 | 34 | 5.6 | 0.9 |
| Fairfax | 0.06 | 0.1 | 738 | 13,174 | 32,745.7 | 1,438.2 | 2,681 | 52,389 | 16,030.0 | 1,649.9 | 1,665 | 11,578 | 3,610.8 | 832.9 |
| Fauquier | 3.10 | 45.1 | 33 | 532 | 202.7 | 28.2 | 225 | 3,035 | 1,009.5 | 88.6 | 93 | 225 | 64.6 | 12.3 |
| Floyd | 0.11 | 7.0 | 8 | 27 | 23.0 | 0.7 | 46 | 365 | 74.8 | 6.9 | 5 | 14 | 2.1 | 0.3 |
| Fluvanna | 0.77 | 29.4 | 10 | 131 | 99.4 | 8.1 | 37 | 319 | 85.6 | 7.6 | 21 | 39 | 6.6 | 1.2 |
| Franklin | 1.76 | 31.3 | 39 | 438 | 332.7 | 27.0 | 189 | 1,969 | 515.4 | 50.1 | D | D | D | D |
| Frederick | 5.27 | 63.3 | D | D | D | 78.5 | 209 | 3,724 | 1,297.8 | 107.0 | 63 | D | 60.0 | 0.4 |
| Giles | 1.12 | 67.0 | 8 | 43 | 23.1 | 2.2 | 57 | 699 | 178.4 | 16.6 | 8 | 14 | 2.2 | 2.8 |
| Gloucester | 0.66 | 17.8 | 21 | 81 | 22.7 | 2.6 | 146 | 2,007 | 525.4 | 53.3 | 41 | 96 | 14.1 | 6.6 |
| Goochland | 0.08 | 3.6 | D | D | D | 23.3 | 58 | 542 | 190.0 | 15.5 | 18 | 98 | 32.2 | 0.2 |
| Grayson | 0.11 | 6.9 | D | D | D | 0.2 | 27 | 168 | 40.3 | 3.1 | D | D | D | D |
| Greene | 0.63 | 32.9 | 8 | D | 8.2 | D | 53 | 725 | 193.0 | 17.7 | 13 | D | 5.9 | D |
| Greensville | 0.03 | 2.5 | D | D | D | 1.9 | 23 | 245 | 152.0 | 6.3 | 5 | D | 2.6 | D |
| Halifax | 2.10 | 59.8 | 18 | 174 | 75.6 | 6.4 | 125 | 1,335 | 357.7 | 30.1 | 18 | 56 | 10.0 | 1.5 |
| Hanover | 4.08 | 39.5 | 215 | 3,977 | 2,965.1 | 239.6 | 368 | 7,220 | 2,077.0 | 198.7 | 131 | 549 | 154.7 | 25.5 |
| Henrico | 20.15 | 62.0 | 372 | 7,004 | 10,718.5 | 480.1 | 1,255 | 23,634 | 6,237.0 | 618.4 | 518 | 3,685 | 984.0 | 219.5 |
| Henry | 2.86 | 55.1 | 39 | 803 | 1,236.9 | 34.2 | 155 | 1,476 | 443.8 | 36.5 | 27 | 89 | 12.8 | 2.5 |
| Highland | 0.10 | 45.2 | NA | NA | NA | NA | 16 | 54 | 10.5 | 0.8 | NA | NA | NA | NA |
| Isle of Wight | 1.55 | 42.7 | 19 | 87 | 48.1 | 4.7 | 99 | 866 | 203.8 | 17.6 | 35 | 78 | 19.7 | 2.6 |
| James City | 5.36 | 73.3 | 42 | 366 | 151.3 | 18.6 | 295 | 4,106 | 749.3 | 79.4 | 92 | 626 | 153.3 | 28.0 |
| King and Queen | 0.00 | 0.0 | 8 | D | 35.3 | D | 10 | 49 | 19.3 | 1.4 | D | D | D | 0.1 |
| King George | 0.93 | 36.4 | 12 | 196 | 49.8 | 5.3 | 60 | 879 | 343.2 | 21.2 | 18 | 45 | 12.4 | 1.8 |
| King William | 0.56 | 34.4 | D | D | D | D | 50 | 443 | 141.4 | 11.4 | 12 | 30 | 3.7 | 1.0 |
| Lancaster | 0.12 | 10.9 | 17 | 87 | 68.1 | 3.6 | 80 | 781 | 177.0 | 19.4 | 18 | 33 | 6.5 | 1.1 |
| Lee | 2.34 | 94.6 | 10 | 51 | 16.5 | 1.6 | 70 | 728 | 150.0 | 16.3 | 9 | D | 2.1 | D |
| Loudoun | 6.86 | 18.3 | 251 | 3,080 | 1,889.7 | 230.9 | 1,039 | 19,612 | 6,080.4 | 555.9 | 435 | 1,676 | 601.4 | 107.7 |

1  Merchant wholesalers, except manufacturers' sales branches and offices.  2. Employer establishments.

# Professional Services, Manufacturing, and Accommodation and Food Services

| STATE County | Professional, scientific, and technical services, 2017 | | | | Manufacturing, 2017 | | | | Accommodation and food services, 2017 | | | |
|---|---|---|---|---|---|---|---|---|---|---|---|---|
| | Number of establishments | Number of employees | Sales (mil dol) | Average payroll (mil dol) | Number of establishments | Number of employees | Sales (mil dol) | Average payroll (mil dol) | Number of establishments | Number of employees | Sales (mil dol) | Annual payroll (mil dol) |
| | 147 | 148 | 149 | 150 | 151 | 152 | 153 | 154 | 155 | 156 | 157 | 158 |
| VERMONT—Cont'd | | | | | | | | | | | | |
| Chittenden | 816 | 7,045 | 1,231 | 514.5 | 219 | 9,126 | 2,667.1 | 598.8 | 468 | 9,179 | 611.1 | 182.2 |
| Essex | 4 | 17 | 4 | 1.4 | 8 | 133 | 26.4 | 5.2 | 14 | 71 | 4.9 | 1.4 |
| Franklin | D | D | D | D | 59 | 2,772 | 1,487.9 | 157.2 | 94 | 1,058 | 71.0 | 20.9 |
| Grand Isle | D | D | D | D | 6 | 50 | 5.4 | 1.6 | 22 | 89 | 12.6 | 3.0 |
| Lamoille | 82 | 274 | 35 | 12.4 | 45 | 432 | 225.3 | 23.7 | 124 | 4,155 | 255.6 | 77.8 |
| Orange | D | D | D | D | 50 | 721 | 149.7 | 30.8 | 60 | 613 | 37.9 | 11.8 |
| Orleans | 46 | 162 | 15 | 6.4 | 33 | 1,120 | 196.2 | 46.5 | 56 | 1,720 | 78.1 | 28.8 |
| Rutland | D | D | D | D | 101 | 3,111 | 649.9 | 179.3 | 253 | 2,921 | 169.9 | 49.2 |
| Washington | D | D | D | D | 128 | 2,411 | 658.9 | 123.0 | 171 | 2,571 | 135.6 | 44.5 |
| Windham | D | D | D | D | 90 | 1,968 | 536.5 | 101.9 | 212 | 3,596 | 179.9 | 54.3 |
| Windsor | 179 | 999 | 175 | 66.1 | 115 | 1,952 | 499.1 | 96.8 | 201 | 3,393 | 234.0 | 70.2 |
| VIRGINIA | 31,233 | 457,575 | 98,543 | 40,177.3 | 5,038 | 232,695 | 99,343.9 | 12,987.4 | 18,199 | 358,010 | 22,074.7 | 6,226.9 |
| Accomack | 53 | 503 | 55 | 21.5 | 23 | 3,271 | 470.4 | 99.4 | 100 | 962 | 66.9 | 16.2 |
| Albemarle | 380 | 3,592 | 589 | 263.4 | 81 | 2,278 | 656.0 | 151.9 | 208 | 5,256 | 326.4 | 100.5 |
| Alleghany | 15 | 34 | 4 | 1.1 | 11 | 276 | 71.7 | 11.9 | 25 | 414 | 17.9 | 5.6 |
| Amelia | 17 | 58 | 6 | 2.1 | 13 | 150 | 78.2 | 6.3 | D | D | D | 1.7 |
| Amherst | 38 | 214 | 26 | 11.0 | 39 | 1,270 | 731.3 | 78.8 | 47 | 713 | 33.5 | 8.7 |
| Appomattox | 16 | 48 | 5 | 1.3 | 15 | 106 | 13.9 | 4.0 | 19 | 259 | 12.3 | 3.0 |
| Arlington | 1,887 | 45,652 | 11,880 | 4,816.9 | 40 | 268 | 70.6 | 12.0 | 665 | 18,101 | 1,629.6 | 434.9 |
| Augusta | D | D | D | D | 69 | 5,052 | 2,249.1 | 256.5 | 94 | 1,429 | 76.9 | 20.5 |
| Bath | 13 | 177 | 22 | 11.9 | 4 | D | 2.6 | D | 20 | 1,072 | 66.8 | 29.3 |
| Bedford | 184 | 951 | 134 | 48.7 | 79 | 3,004 | 1,184.5 | 168.6 | 103 | 1,309 | 67.0 | 19.3 |
| Bland | NA | NA | NA | NA | 8 | 378 | 183.3 | 22.0 | D | D | D | 0.5 |
| Botetourt | 60 | 201 | 20 | 7.3 | 28 | D | 871.6 | 137.6 | 51 | 816 | 39.2 | 11.4 |
| Brunswick | 16 | 40 | 4 | 1.2 | 14 | 266 | 90.0 | 14.4 | 10 | D | 6.6 | D |
| Buchanan | 32 | 195 | 13 | 4.9 | D | 215 | D | 15.0 | 22 | 269 | 13.5 | 3.6 |
| Buckingham | 11 | 62 | 10 | 3.9 | 8 | 106 | 24.1 | 4.9 | 9 | 89 | 6.0 | 1.2 |
| Campbell | 95 | 779 | 89 | 57.3 | 56 | 4,349 | 2,135.1 | 317.4 | 77 | 1,241 | 66.5 | 17.7 |
| Caroline | 32 | 134 | 26 | 6.2 | 13 | 317 | 99.8 | 13.5 | 37 | 498 | 27.4 | 8.6 |
| Carroll | 35 | 121 | 14 | 4.9 | 24 | 964 | 245.8 | 32.0 | 38 | 596 | 27.1 | 7.4 |
| Charles City | 7 | 48 | 7 | 3.0 | 21 | 336 | 72.3 | 14.1 | D | D | D | 0.7 |
| Charlotte | 8 | 36 | 5 | 0.8 | 12 | 361 | 104.0 | 14.7 | D | D | D | D |
| Chesterfield | 900 | 10,207 | 1,152 | 492.3 | 169 | 9,481 | 4,049.8 | 658.6 | 599 | 13,435 | 682.9 | 198.3 |
| Clarke | 35 | 110 | 17 | 7.5 | 15 | 578 | 111.2 | 28.1 | 22 | 200 | 12.9 | 4.0 |
| Craig | 6 | 7 | 1 | 0.5 | NA | NA | NA | NA | 4 | 31 | 2.0 | 0.8 |
| Culpeper | D | D | D | D | 41 | 1,274 | 539.5 | 69.6 | 88 | 1,709 | 87.7 | 24.0 |
| Cumberland | 5 | 49 | 6 | 2.1 | 5 | 106 | 24.2 | 4.3 | 7 | 48 | 2.3 | 0.7 |
| Dickenson | D | D | 25 | D | D | 26 | D | 0.9 | 14 | 195 | 8.4 | 2.5 |
| Dinwiddie | D | D | 4 | D | 14 | 618 | 555.0 | 47.4 | 25 | 333 | 20.5 | 4.4 |
| Essex | D | D | D | D | 12 | 590 | 131.3 | 19.1 | 31 | 441 | 20.5 | 6.4 |
| Fairfax | 8,813 | 200,603 | 49,361 | 20,006.0 | 342 | 6,064 | 1,359.8 | 395.4 | 2,421 | 44,519 | 3,519.0 | 945.4 |
| Fauquier | 296 | 2,053 | 396 | 162.2 | 66 | 950 | 177.0 | 47.6 | 136 | 2,614 | 150.2 | 43.0 |
| Floyd | D | D | D | D | 23 | 399 | 84.2 | 15.7 | D | D | D | D |
| Fluvanna | 31 | 60 | 10 | 2.5 | 12 | 112 | 19.7 | 4.7 | 19 | 210 | 8.3 | 2.5 |
| Franklin | 82 | 296 | 38 | 11.9 | D | 3,019 | D | D | 72 | 1,087 | 53.7 | 15.5 |
| Frederick | D | D | D | 53.1 | 86 | 5,314 | 3,605.7 | 289.7 | D | D | D | D |
| Giles | 18 | 432 | 27 | 17.8 | 11 | 817 | 460.8 | 54.1 | D | D | D | D |
| Gloucester | 59 | 285 | 30 | 11.8 | 14 | 174 | 48.4 | 6.3 | 64 | 1,138 | 51.5 | 15.5 |
| Goochland | 80 | 348 | 45 | 18.9 | 18 | 192 | 70.0 | 9.0 | 47 | 573 | 40.0 | 11.1 |
| Grayson | 9 | 11 | 1 | 0.7 | 10 | 299 | 51.1 | 10.8 | 10 | D | 3.9 | D |
| Greene | D | D | D | D | 11 | 40 | 3.5 | 1.1 | 32 | 395 | 23.3 | 5.8 |
| Greensville | NA | NA | NA | NA | 5 | 909 | 311.4 | 37.1 | 12 | 207 | 10.1 | 2.3 |
| Halifax | 43 | 164 | 16 | 6.0 | 38 | 1,845 | 708.8 | 81.9 | 62 | 995 | 42.8 | 11.6 |
| Hanover | 266 | 1,486 | 232 | 76.2 | 134 | 3,574 | 872.5 | 168.7 | 220 | 4,052 | 209.2 | 59.0 |
| Henrico | 1,217 | 14,691 | 2,442 | 998.9 | 168 | 5,514 | 1,974.8 | 304.1 | 777 | 17,256 | 1,007.7 | 279.3 |
| Henry | 27 | 413 | 25 | 10.0 | 62 | 3,315 | 1,054.6 | 153.4 | 62 | 1,017 | 47.2 | 11.3 |
| Highland | D | D | D | 0.8 | 4 | D | 4.1 | D | D | D | D | D |
| Isle of Wight | 55 | 265 | 27 | 9.9 | 16 | 3,601 | 992.9 | 159.5 | 51 | 969 | 49.5 | 14.4 |
| James City | 225 | 1,471 | 223 | 95.4 | 35 | 2,446 | 1,613.4 | 141.6 | 148 | 3,812 | 278.7 | 74.8 |
| King and Queen | 12 | 61 | 11 | 3.4 | 6 | D | 6.4 | D | D | D | D | 0.7 |
| King George | D | D | D | D | 8 | 105 | 32.1 | 4.4 | 40 | 420 | 25.1 | 6.9 |
| King William | 32 | 156 | 11 | 4.3 | 11 | 506 | 299.1 | 30.8 | 22 | 309 | 13.9 | 3.7 |
| Lancaster | 52 | 370 | 42 | 20.3 | 9 | 81 | 24.5 | 3.5 | 32 | 492 | 25.6 | 10.7 |
| Lee | 20 | 71 | 5 | 1.8 | D | 95 | D | 2.5 | D | D | D | D |
| Loudoun | D | D | D | D | 183 | 6,523 | 2,123.1 | 494.8 | 777 | 16,541 | 1,310.5 | 365.0 |

# Table B. States and Counties — Health Care and Social Assistance, Other Services, Nonemployer Businesses, and Residential Construction

| STATE County | Health care and social assistance, 2017 — Number of establishments (159) | Number of employees (160) | Receipts (mil dol) (161) | Annual payroll (mil dol) (162) | Other services, 2017 — Number of establishments (163) | Number of employees (164) | Receipts (mil dol) (165) | Annual payroll (mil dol) (166) | Nonemployer businesses, 2018 — Number (167) | Receipts (mil dol) (168) | Value of residential construction authorized by building permits, 2020 — New construction ($1,000) (169) | Number of housing units (170) |
|---|---|---|---|---|---|---|---|---|---|---|---|---|
| **VERMONT—Cont'd** | | | | | | | | | | | | |
| Chittenden | 599 | 15,593 | 2,123.2 | 714.0 | 439 | 2,422 | 318.2 | 87.4 | 15,230 | 776.1 | 146,612 | 890 |
| Essex | D | D | D | D | 12 | 31 | 4.1 | 0.7 | 492 | 18.2 | 9,847 | 40 |
| Franklin | 104 | 2,566 | 258.1 | 122.2 | 80 | 291 | 28.7 | 7.6 | 3,815 | 161.0 | 39,385 | 196 |
| Grand Isle | 13 | 91 | 3.8 | 2.0 | 14 | 30 | 3.5 | 1.1 | 741 | 32.0 | 4,266 | 17 |
| Lamoille | D | D | D | D | 67 | 297 | 63.7 | 9.5 | 2,819 | 136.4 | 25,258 | 126 |
| Orange | 87 | 1,505 | 146.3 | 67.2 | 45 | 144 | 17.0 | 4.5 | 2,946 | 119.9 | 8,461 | 35 |
| Orleans | 82 | 1,731 | 172.2 | 78.5 | 65 | 210 | 26.3 | 5.7 | 2,614 | 104.6 | 15,224 | 152 |
| Rutland | 196 | 4,810 | 546.0 | 210.7 | 156 | 790 | 68.4 | 19.8 | 4,946 | 187.2 | 9,777 | 46 |
| Washington | 219 | 4,566 | 460.6 | 204.3 | 219 | 1,025 | 155.9 | 45.0 | 6,161 | 254.0 | 37,566 | 153 |
| Windham | 165 | 3,317 | 306.3 | 143.3 | 124 | 469 | 46.7 | 12.7 | 5,257 | 214.5 | 25,435 | 101 |
| Windsor | 161 | 4,563 | 536.5 | 276.1 | 138 | 663 | 70.5 | 22.9 | 6,219 | 265.9 | 24,528 | 89 |
| **VIRGINIA** | 20,522 | 454,501 | 58,792.5 | 22,721.6 | 15,700 | 118,537 | 20,344.4 | 5,188.4 | 632,995 | 28,429.4 | 6,342,240 | 33,813 |
| Accomack | 56 | 1,177 | 71.9 | 33.9 | 59 | 289 | 32.0 | 7.4 | 2,387 | 91.0 | 27,715 | 188 |
| Albemarle | 356 | 8,120 | 1,278.0 | 534.7 | 169 | 1,455 | 255.1 | 56.1 | 9,931 | 492.3 | 235,295 | 982 |
| Alleghany | 37 | 829 | 80.5 | 32.7 | 26 | 116 | 11.9 | 3.6 | 644 | 21.0 | 620 | 7 |
| Amelia | D | D | D | 10.1 | D | D | 7.2 | 3.3 | 870 | 34.3 | 11,814 | 54 |
| Amherst | 47 | 689 | 41.5 | 19.2 | 52 | 138 | 11.9 | 2.6 | 1,599 | 55.4 | 15,705 | 69 |
| Appomattox | 26 | 507 | 24.6 | 11.8 | 23 | 98 | 9.0 | 2.6 | 959 | 35.5 | 14,323 | 75 |
| Arlington | 516 | 11,848 | 1,567.8 | 673.2 | 715 | 14,016 | 3,533.2 | 957.6 | 21,303 | 1,114.2 | 169,186 | 553 |
| Augusta | 117 | 3,940 | 578.9 | 195.6 | 107 | 445 | 54.6 | 15.1 | 4,703 | 242.9 | 64,213 | 322 |
| Bath | 5 | 244 | 23.1 | 11.0 | 8 | 52 | 2.8 | 1.3 | 355 | 11.0 | 2,656 | 7 |
| Bedford | 155 | 2,215 | 195.3 | 83.5 | 132 | 637 | 74.7 | 22.5 | 5,506 | 237.7 | 9,632 | 144 |
| Bland | 7 | 193 | 13.6 | 6.7 | D | D | 9.4 | D | 260 | 7.8 | 1,288 | 8 |
| Botetourt | 55 | 946 | 79.5 | 27.7 | D | D | 21.0 | D | 2,351 | 93.7 | 29,569 | 132 |
| Brunswick | 14 | 175 | 9.3 | 4.2 | D | D | D | 1.4 | 727 | 29.1 | 5,071 | 29 |
| Buchanan | 42 | 630 | 55.7 | 23.0 | 26 | 162 | 14.1 | 5.2 | 802 | 25.7 | 1,264 | 7 |
| Buckingham | D | D | D | 13.8 | 14 | 61 | 11.6 | 3.0 | 788 | 30.5 | 7,724 | 49 |
| Campbell | 89 | 1,092 | 60.7 | 26.6 | 96 | 378 | 39.4 | 10.4 | 3,156 | 102.9 | 53,539 | 271 |
| Caroline | 22 | 333 | 23.2 | 11.0 | 34 | 187 | 16.9 | 5.1 | 1,829 | 63.7 | 35,425 | 188 |
| Carroll | 34 | 690 | 48.1 | 19.9 | D | D | D | D | 1,715 | 62.4 | 13,070 | 63 |
| Charles City | D | D | D | D | D | D | D | D | 410 | 14.1 | 3,584 | 20 |
| Charlotte | 20 | 318 | 20.9 | 9.6 | D | D | 2.8 | D | 670 | 26.9 | 1,622 | 11 |
| Chesterfield | 837 | 14,517 | 1,751.9 | 697.6 | 511 | 3,465 | 318.2 | 103.9 | 24,896 | 1,057.1 | 474,031 | 3,116 |
| Clarke | 26 | 734 | 124.9 | 23.3 | 35 | 162 | 12.2 | 3.9 | 1,426 | 61.7 | 19,809 | 78 |
| Craig | 4 | 57 | 3.7 | 1.7 | D | D | 1.6 | D | 277 | 9.3 | 2,205 | 10 |
| Culpeper | 81 | 2,165 | 231.7 | 89.5 | 95 | 606 | 69.3 | 22.1 | 3,758 | 183.6 | 53,881 | 353 |
| Cumberland | 12 | 95 | 4.5 | 2.1 | 5 | 15 | 1.4 | 0.4 | 537 | 20.2 | 5,028 | 37 |
| Dickenson | 24 | 390 | 22.5 | 11.7 | D | D | D | 1.6 | 415 | 11.3 | 1,685 | 7 |
| Dinwiddie | D | D | D | 8.5 | D | D | D | D | 1,416 | 53.4 | 17,147 | 99 |
| Essex | 38 | 635 | 78.6 | 23.5 | D | D | D | D | 805 | 34.2 | 6,011 | 36 |
| Fairfax | 3,242 | 60,021 | 8,427.1 | 3,213.8 | 2,124 | 16,922 | 3,028.4 | 828.8 | 114,600 | 6,010.3 | 270,826 | 1,605 |
| Fauquier | 115 | 2,586 | 306.5 | 136.0 | 142 | 1,056 | 146.2 | 38.9 | 6,592 | 363.0 | 59,107 | 197 |
| Floyd | 23 | 499 | 67.7 | 15.6 | 28 | 92 | 9.8 | 2.5 | 1,286 | 43.4 | 13,524 | 37 |
| Fluvanna | 25 | 210 | 14.4 | 7.2 | 21 | 133 | 12.4 | 3.8 | 1,825 | 64.4 | 41,961 | 155 |
| Franklin | 76 | 1,394 | 126.7 | 48.3 | 101 | 357 | 36.1 | 9.4 | 3,654 | 152.9 | 51,042 | 147 |
| Frederick | D | D | D | D | D | D | D | 5.8 | 5,797 | 249.7 | 171,046 | 678 |
| Giles | 28 | 486 | 62.7 | 17.8 | 30 | 141 | 21.3 | 8.8 | 885 | 33.1 | 7,036 | 37 |
| Gloucester | 73 | 1,363 | 132.8 | 46.0 | 75 | 322 | 28.6 | 20.3 | 2,443 | 92.5 | 32,322 | 139 |
| Goochland | 30 | 492 | 44.2 | 19.7 | 48 | 346 | 61.8 | D | 2,294 | 143.5 | 72,948 | 301 |
| Grayson | 11 | 160 | 10.6 | 4.4 | D | D | 2.3 | D | 989 | 35.7 | 5,494 | 31 |
| Greene | D | D | D | 7.0 | 30 | 136 | 14.3 | 3.7 | 1,430 | 51.3 | 12,991 | 275 |
| Greensville | NA | NA | NA | NA | D | D | 5.9 | D | 388 | 10.5 | 1,608 | 10 |
| Halifax | 76 | 1,675 | 195.3 | 76.2 | 46 | 172 | 19.6 | 4.4 | 1,812 | 67.9 | 10,167 | 91 |
| Hanover | 279 | 6,179 | 868.6 | 289.2 | 271 | 1,712 | 206.0 | 59.8 | 8,551 | 403.6 | 155,718 | 690 |
| Henrico | 1,160 | 25,688 | 3,422.0 | 1,257.8 | 713 | 5,321 | 795.6 | 199.8 | 25,896 | 1,146.2 | 234,169 | 1,915 |
| Henry | 46 | 811 | 58.3 | 26.8 | 55 | 292 | 24.0 | 6.7 | 2,384 | 77.0 | 3,517 | 18 |
| Highland | 5 | D | 3.4 | D | D | D | 1.8 | D | 249 | 6.7 | 1,194 | 12 |
| Isle of Wight | 60 | 632 | 44.0 | 18.1 | 72 | 261 | 29.4 | 7.9 | 2,375 | 105.4 | 49,569 | 293 |
| James City | 159 | 3,957 | 408.3 | 167.9 | 117 | 901 | 82.3 | 21.7 | 5,800 | 250.1 | 99,547 | 374 |
| King and Queen | D | D | D | 0.4 | D | D | 1.5 | D | 435 | 15.4 | 3,472 | 26 |
| King George | 29 | 347 | 25.4 | 9.8 | D | D | D | D | 1,426 | 51.8 | 34,305 | 165 |
| King William | 22 | 327 | 23.7 | 12.0 | 31 | 103 | 12.4 | 3.1 | 1,109 | 43.8 | 25,726 | 169 |
| Lancaster | 43 | 948 | 99.7 | 39.5 | 41 | 188 | 20.4 | 5.8 | 1,075 | 51.0 | 11,249 | 29 |
| Lee | 30 | 519 | 40.8 | 16.5 | D | D | D | 0.7 | 1,044 | 37.0 | 2,995 | 22 |
| Loudoun | 947 | 14,238 | 1,761.1 | 665.5 | 680 | 4,846 | 657.1 | 202.7 | 37,251 | 1,886.8 | 450,624 | 2,493 |

# Table B. States and Counties — Government Employment and Payroll, and Local Government Finances

| STATE County | Government employment and payroll, 2017 | | | | | | | | | Local government finances, 2017 | | | | |
| | | | March payroll (percent of total) | | | | | | | General revenue | | | | |
| | | | | | | | | | | | | Taxes | Per capita[1] (dollars) | |
| | Full-time equivalent employees | March payroll (dollars) | Administration, judicial, and legal | Police and corrections | Fire protection | Highways and transportation | Health and welfare | Natural resources and utilities | Education and libraries | Total (mil dol) | Inter-governmental (mil dol) | Total (mil dol) | Total | Property |
| | 171 | 172 | 173 | 174 | 175 | 176 | 177 | 178 | 179 | 180 | 181 | 182 | 183 | 184 |
|---|---|---|---|---|---|---|---|---|---|---|---|---|---|---|
| **VERMONT—Cont'd** | | | | | | | | | | | | | | |
| Chittenden | 6,368 | 30,113,712 | 5.6 | 6.4 | 2.5 | 8.2 | 1.4 | 6.7 | 68.5 | 767.4 | 474.2 | 142.1 | 871 | 722 |
| Essex | 311 | 902,691 | 5.8 | 1.0 | 0.5 | 4.7 | 0.0 | 1.8 | 84.2 | 30.7 | 20.9 | 8.6 | 1,383 | 1,382 |
| Franklin | 2,134 | 7,474,467 | 4.7 | 2.6 | 0.3 | 4.8 | 1.1 | 3.4 | 82.8 | 183.5 | 136.4 | 33.7 | 689 | 663 |
| Grand Isle | 192 | 737,545 | 12.9 | 0.0 | 0.0 | 7.0 | 0.0 | 3.3 | 76.1 | 29.2 | 18.1 | 9.9 | 1,418 | 1,413 |
| Lamoille | 476 | 2,115,401 | 10.9 | 7.8 | 0.4 | 11.7 | 2.0 | 11.7 | 54.2 | 115.6 | 74.2 | 29.3 | 1,157 | 1,097 |
| Orange | 901 | 3,237,148 | 8.1 | 1.9 | 0.2 | 8.1 | 0.0 | 1.3 | 80.0 | 132.7 | 90.4 | 33.0 | 1,139 | 1,133 |
| Orleans | 1,237 | 4,506,161 | 4.2 | 2.5 | 0.4 | 5.7 | 0.0 | 1.9 | 84.2 | 118.6 | 81.3 | 28.5 | 1,064 | 1,058 |
| Rutland | 1,781 | 7,144,103 | 6.9 | 5.3 | 2.3 | 9.6 | 0.8 | 6.4 | 68.0 | 257.1 | 166.3 | 66.1 | 1,120 | 1,063 |
| Washington | 2,280 | 8,608,849 | 6.2 | 5.6 | 2.9 | 5.7 | 2.2 | 3.1 | 73.4 | 270.0 | 174.2 | 61.5 | 1,057 | 1,049 |
| Windham | 1,910 | 7,669,353 | 5.7 | 4.4 | 2.0 | 6.7 | 1.0 | 3.5 | 76.1 | 216.3 | 129.4 | 68.2 | 1,591 | 1,552 |
| Windsor | 2,876 | 13,142,052 | 3.7 | 4.0 | 1.8 | 4.3 | 0.8 | 3.6 | 81.2 | 267.5 | 171.3 | 70.1 | 1,269 | 1,254 |
| **VIRGINIA** | X | X | X | X | X | X | X | X | X | X | X | X | X | X |
| Accomack | 1,234 | 3,765,447 | 9.0 | 11.3 | 0.0 | 0.8 | 10.5 | 4.7 | 61.6 | 116.0 | 58.7 | 48.6 | 1,486 | 1,106 |
| Albemarle | 3,581 | 14,557,344 | 5.8 | 11.3 | 3.9 | 0.9 | 4.0 | 4.4 | 64.9 | 445.7 | 142.7 | 220.0 | 2,041 | 1,553 |
| Alleghany | 720 | 2,452,619 | 10.5 | 9.9 | 1.7 | 3.2 | 4.3 | 9.5 | 58.3 | 53.8 | 23.3 | 20.0 | 1,323 | 749 |
| Amelia | 338 | 1,180,965 | 9.5 | 7.6 | 0.0 | 0.0 | 0.0 | 3.6 | 75.4 | 31.7 | 16.9 | 10.0 | 772 | 633 |
| Amherst | 959 | 3,122,016 | 6.4 | 6.6 | 4.1 | 0.0 | 3.1 | 3.0 | 75.2 | 67.8 | 32.9 | 31.7 | 996 | 745 |
| Appomattox | 467 | 1,594,829 | 9.0 | 5.6 | 0.0 | 0.3 | 0.0 | 5.0 | 76.8 | 40.5 | 21.1 | 16.2 | 1,030 | 820 |
| Arlington | 10,302 | 63,767,532 | 6.5 | 9.7 | 4.0 | 22.4 | 7.1 | 5.4 | 41.5 | 3,105.9 | 634.0 | 1,107.8 | 4,721 | 3,678 |
| Augusta | 2,532 | 8,458,386 | 5.0 | 11.5 | 4.4 | 0.7 | 6.0 | 1.6 | 69.9 | 198.6 | 109.3 | 76.3 | 1,014 | 750 |
| Bath | 235 | 902,598 | 7.5 | 6.9 | 0.0 | 0.2 | 0.0 | 2.3 | 80.7 | 17.3 | 3.0 | 14.0 | 3,288 | 2,718 |
| Bedford | 2,651 | 9,414,798 | 4.7 | 17.8 | 1.8 | 0.7 | 14.7 | 4.8 | 53.8 | 247.4 | 130.8 | 92.5 | 1,180 | 945 |
| Bland | 217 | 665,699 | 12.9 | 5.5 | 5.6 | 0.0 | 0.0 | 3.7 | 67.8 | 15.6 | 9.0 | 5.0 | 782 | 725 |
| Botetourt | 1,107 | 3,717,883 | 7.9 | 10.8 | 0.4 | 0.1 | 6.0 | 2.2 | 71.0 | 129.4 | 71.4 | 47.8 | 1,434 | 1,181 |
| Brunswick | 498 | 1,647,277 | 10.1 | 8.4 | 0.0 | 2.7 | 5.5 | 4.3 | 65.8 | 36.8 | 18.0 | 16.7 | 1,008 | 898 |
| Buchanan | 731 | 2,092,784 | 7.5 | 7.1 | 1.2 | 2.2 | 0.5 | 6.3 | 71.9 | 100.3 | 55.4 | 26.7 | 1,237 | 1,095 |
| Buckingham | 622 | 1,948,843 | 5.1 | 30.4 | 0.0 | 0.0 | 0.3 | 4.8 | 57.7 | 103.3 | 29.6 | 16.8 | 987 | 889 |
| Campbell | 1,693 | 5,305,069 | 5.9 | 5.8 | 4.2 | 0.6 | 6.2 | 4.2 | 70.1 | 131.3 | 69.9 | 54.7 | 989 | 761 |
| Caroline | 797 | 2,715,160 | 10.0 | 11.1 | 5.4 | 0.0 | 4.7 | 5.1 | 60.7 | 101.2 | 44.1 | 45.9 | 1,509 | 1,234 |
| Carroll | 1,058 | 3,641,643 | 5.0 | 5.8 | 2.0 | 0.5 | 17.5 | 3.3 | 63.5 | 70.5 | 37.4 | 26.1 | 875 | 715 |
| Charles City | 219 | 739,386 | 13.6 | 7.8 | 0.0 | 0.0 | 2.5 | 3.0 | 64.4 | 17.8 | 5.2 | 12.3 | 1,753 | 1,567 |
| Charlotte | 481 | 1,985,592 | 4.7 | 8.7 | 0.0 | 0.4 | 4.5 | 3.0 | 76.9 | 48.6 | 32.8 | 10.8 | 894 | 741 |
| Chesterfield | 12,037 | 46,482,057 | 7.6 | 9.3 | 5.7 | 0.2 | 7.2 | 4.2 | 61.6 | 1,235.4 | 491.5 | 532.3 | 1,551 | 1,223 |
| Clarke | 465 | 1,666,629 | 7.8 | 9.5 | 2.9 | 0.8 | 4.2 | 5.2 | 69.2 | 48.3 | 18.1 | 24.7 | 1,706 | 1,480 |
| Craig | 170 | 506,390 | 11.9 | 9.2 | 0.0 | 0.0 | 6.5 | 3.8 | 68.6 | 10.4 | 5.4 | 4.4 | 875 | 783 |
| Culpeper | 1,925 | 6,793,854 | 8.7 | 9.8 | 2.9 | 1.5 | 3.9 | 5.3 | 64.9 | 180.0 | 81.0 | 80.6 | 1,572 | 1,252 |
| Cumberland | 317 | 998,010 | 10.0 | 7.9 | 0.0 | 0.0 | 0.7 | 2.2 | 74.5 | 27.7 | 16.1 | 10.3 | 1,046 | 935 |
| Dickenson | 516 | 1,599,967 | 7.7 | 7.8 | 0.0 | 0.1 | 0.8 | 14.0 | 67.4 | 72.1 | 44.4 | 17.2 | 1,167 | 1,042 |
| Dinwiddie | 883 | 3,272,325 | 6.7 | 7.5 | 2.8 | 0.0 | 5.2 | 4.9 | 67.7 | 82.6 | 42.4 | 36.1 | 1,261 | 1,072 |
| Essex | 374 | 1,318,222 | 9.9 | 9.4 | 3.3 | 2.9 | 0.0 | 4.7 | 64.1 | 36.0 | 15.3 | 18.8 | 1,718 | 1,375 |
| Fairfax | 44,022 | 252,473,395 | 7.8 | 8.2 | 5.2 | 0.5 | 6.3 | 5.3 | 59.8 | 6,096.2 | 1,439.5 | 3,826.5 | 3,331 | 2,737 |
| Fauquier | 2,871 | 11,391,634 | 8.6 | 8.4 | 4.3 | 1.0 | 2.2 | 5.0 | 65.2 | 274.9 | 94.7 | 162.2 | 2,332 | 1,984 |
| Floyd | 503 | 1,725,104 | 6.0 | 5.4 | 0.0 | 0.0 | 6.2 | 3.9 | 77.5 | 29.7 | 15.6 | 13.0 | 826 | 680 |
| Fluvanna | 763 | 2,674,700 | 6.2 | 6.9 | 0.0 | 0.0 | 5.7 | 1.9 | 77.1 | 70.0 | 31.6 | 36.8 | 1,390 | 1,248 |
| Franklin | 1,881 | 6,737,473 | 6.2 | 5.9 | 4.9 | 0.7 | 3.3 | 2.5 | 75.5 | 153.2 | 72.9 | 69.7 | 1,237 | 1,015 |
| Frederick | 3,242 | 11,189,825 | 5.7 | 12.3 | 4.0 | 0.6 | 0.3 | 3.5 | 72.5 | 251.4 | 83.3 | 148.3 | 1,716 | 1,329 |
| Giles | 773 | 1,953,500 | 12.3 | 10.4 | 0.0 | 2.2 | 0.2 | 6.8 | 60.1 | 73.0 | 44.1 | 21.5 | 1,284 | 997 |
| Gloucester | 1,186 | 4,098,264 | 9.4 | 10.0 | 0.0 | 0.0 | 3.7 | 3.6 | 73.1 | 105.7 | 43.6 | 56.3 | 1,508 | 1,155 |
| Goochland | 612 | 2,265,158 | 13.8 | 7.7 | 3.4 | 0.0 | 0.9 | 3.6 | 66.9 | 78.3 | 19.2 | 47.9 | 2,108 | 1,749 |
| Grayson | 552 | 1,507,640 | 7.1 | 8.3 | 0.0 | 0.2 | 1.8 | 4.9 | 77.7 | 33.4 | 16.9 | 12.6 | 804 | 670 |
| Greene | 752 | 2,411,800 | 6.5 | 4.7 | 0.0 | 2.0 | 3.1 | 5.6 | 71.9 | 65.5 | 28.7 | 24.6 | 1,258 | 1,043 |
| Greensville | 596 | 1,946,073 | 9.4 | 14.7 | 0.3 | 0.0 | 0.0 | 6.9 | 65.2 | 50.4 | 31.4 | 10.4 | 902 | 721 |
| Halifax | 1,241 | 3,464,577 | 7.1 | 7.7 | 1.2 | 1.2 | 0.0 | 1.9 | 80.5 | 113.1 | 54.8 | 41.4 | 1,197 | 847 |
| Hanover | 4,125 | 16,605,750 | 8.1 | 13.3 | 5.0 | 0.6 | 6.3 | 4.3 | 60.6 | 374.7 | 150.3 | 185.7 | 1,757 | 1,380 |
| Henrico | 12,233 | 52,050,101 | 7.2 | 15.5 | 7.7 | 3.7 | 4.8 | 5.1 | 53.9 | 1,323.7 | 532.3 | 583.6 | 1,782 | 1,227 |
| Henry | 1,715 | 5,150,131 | 7.1 | 8.0 | 2.0 | 0.0 | 8.9 | 1.3 | 68.9 | 151.9 | 90.3 | 36.8 | 716 | 474 |
| Highland | 93 | 325,265 | 14.2 | 14.3 | 0.0 | 0.0 | 4.1 | 3.6 | 59.9 | 9.3 | 4.5 | 3.9 | 1,761 | 1,510 |
| Isle of Wight | 1,032 | 4,780,984 | 7.9 | 7.5 | 1.8 | 1.8 | 1.7 | 3.6 | 69.1 | 183.4 | 73.9 | 68.0 | 1,858 | 1,544 |
| James City | 2,848 | 11,706,927 | 6.0 | 5.5 | 6.6 | 2.1 | 3.2 | 6.9 | 67.5 | 285.2 | 94.0 | 161.7 | 2,148 | 1,702 |
| King and Queen | 352 | 1,181,672 | 7.7 | 46.9 | 0.0 | 0.0 | 2.7 | 0.0 | 42.6 | 22.5 | 11.8 | 7.9 | 1,129 | 1,006 |
| King George | 880 | 2,719,706 | 8.1 | 14.3 | 9.7 | 0.0 | 3.5 | 5.1 | 54.9 | 86.6 | 38.3 | 38.1 | 1,447 | 890 |
| King William | 579 | 2,184,019 | 6.5 | 6.8 | 0.0 | 0.6 | 2.3 | 2.1 | 77.8 | 60.7 | 28.5 | 28.0 | 1,680 | 1,483 |
| Lancaster | 354 | 1,129,167 | 13.5 | 15.9 | 0.0 | 0.6 | 0.3 | 3.3 | 66.1 | 42.6 | 12.9 | 22.4 | 2,084 | 1,669 |
| Lee | 873 | 2,129,145 | 6.4 | 7.6 | 0.0 | 0.1 | 0.2 | 5.3 | 75.9 | 69.3 | 47.7 | 15.4 | 643 | 471 |
| Loudoun | 15,352 | 74,079,202 | 7.2 | 6.8 | 4.4 | 0.6 | 4.5 | 4.0 | 70.5 | 2,109.0 | 505.5 | 1,415.8 | 3,566 | 2,908 |

1. Based on the resident population estimated as of July 1 of the year shown.

| STATE County | Direct general expenditure Total (mil dol) | Per capita[1] (dollars) | Education | Health and hospitals | Police protection | Public welfare | Highways | Debt outstanding Total (mil dol) | Per capita[1] (dollars) | Federal civilian | Federal military | State and local | Number of returns | Mean adjusted gross income | Mean income tax |
|---|---|---|---|---|---|---|---|---|---|---|---|---|---|---|---|
| | 185 | 186 | 187 | 188 | 189 | 190 | 191 | 192 | 193 | 194 | 195 | 196 | 197 | 198 | 199 |
| **VERMONT—Cont'd** | | | | | | | | | | | | | | | |
| Chittenden | 783 | 4,804 | 61.1 | 0.2 | 5.0 | 0.0 | 5.2 | 562.6 | 3,451 | 2,495 | 1,088 | 14,507 | 85,770 | 80,015 | 9,754 |
| Essex | 30 | 4,762 | 74.9 | 0.3 | 1.1 | 0.1 | 13.4 | 4.5 | 735 | 98 | 39 | 298 | 2,930 | 42,290 | 3,067 |
| Franklin | 220 | 4,491 | 77.5 | 0.0 | 2.4 | 0.0 | 9.2 | 89.6 | 1,829 | 1,598 | 309 | 2,962 | 25,080 | 57,040 | 5,133 |
| Grand Isle | 27 | 3,903 | 79.1 | 0.0 | 3.0 | 0.0 | 7.1 | 13.8 | 1,977 | 21 | 46 | 297 | 3,980 | 70,087 | 8,529 |
| Lamoille | 131 | 5,154 | 67.3 | 0.9 | 5.7 | 0.1 | 10.6 | 105.1 | 4,146 | 64 | 156 | 1,642 | 13,840 | 74,133 | 9,196 |
| Orange | 144 | 4,984 | 70.7 | 0.8 | 2.0 | 0.1 | 12.6 | 33.7 | 1,165 | 69 | 179 | 1,978 | 14,950 | 56,039 | 5,130 |
| Orleans | 136 | 5,065 | 82.3 | 0.1 | 1.2 | 0.0 | 6.9 | 33.9 | 1,266 | 230 | 166 | 1,834 | 13,860 | 45,973 | 3,990 |
| Rutland | 277 | 4,696 | 64.0 | 0.1 | 5.2 | 0.0 | 9.6 | 93.6 | 1,586 | 284 | 354 | 3,694 | 31,360 | 54,191 | 5,195 |
| Washington | 292 | 5,012 | 67.2 | 1.0 | 3.4 | 0.1 | 8.9 | 103.6 | 1,778 | 231 | 395 | 7,579 | 31,120 | 63,931 | 6,611 |
| Windham | 235 | 5,472 | 68.8 | 0.3 | 4.1 | 0.1 | 7.4 | 106.9 | 2,495 | 137 | 258 | 2,772 | 23,170 | 53,520 | 5,240 |
| Windsor | 288 | 5,218 | 67.6 | 0.6 | 3.8 | 0.1 | 9.2 | 110.7 | 2,004 | 1,540 | 352 | 3,407 | 30,260 | 66,557 | 7,273 |
| **VIRGINIA** | X | X | X | X | X | X | X | X | X | 201,455 | 138,183 | 553,258 | 3,983,370 | 81,995 | 10,610 |
| Accomack | 110 | 3,353 | 56.2 | 1.1 | 3.8 | 5.0 | 0.9 | 48.6 | 1,487 | 657 | 261 | 2,256 | 15,690 | 46,665 | 3,783 |
| Albemarle | 398 | 3,694 | 55.4 | 1.3 | 4.6 | 5.7 | 0.8 | 449.4 | 4,170 | [2]1,359 | [2]942 | [2]34,022 | 50,520 | 115,648 | 17,894 |
| Alleghany | 49 | 3,235 | 56.9 | 13.4 | 6.8 | 0.5 | 0.8 | 21.2 | 1,402 | [3]63 | [3]64 | [3]1,555 | 6,870 | 51,090 | 4,118 |
| Amelia | 30 | 2,285 | 63.4 | 1.6 | 7.0 | 5.6 | 0.7 | 5.5 | 426 | 26 | 41 | 465 | 6,150 | 57,056 | 4,771 |
| Amherst | 54 | 1,698 | 88.5 | 0.0 | 0.9 | 0.0 | 7.1 | 23.7 | 743 | 49 | 96 | 1,748 | 14,080 | 50,331 | 3,924 |
| Appomattox | 40 | 2,549 | 63.9 | 0.0 | 4.8 | 7.1 | 0.0 | 26.2 | 1,667 | 47 | 50 | 771 | 7,170 | 51,177 | 4,028 |
| Arlington | 2,334 | 9,947 | 24.8 | 2.0 | 2.9 | 3.8 | 4.0 | 9,964.4 | 42,465 | 27,949 | 8,543 | 13,730 | 126,230 | 129,791 | 22,383 |
| Augusta | 224 | 2,977 | 66.6 | 0.5 | 3.6 | 7.1 | 1.6 | 86.8 | 1,154 | [4]291 | [4]379 | [4]8,480 | 35,460 | 58,380 | 5,210 |
| Bath | 12 | 2,695 | 96.0 | 0.0 | 0.0 | 0.0 | 0.0 | 8.9 | 2,087 | 30 | 13 | 345 | 2,200 | 50,128 | 4,174 |
| Bedford | 240 | 3,065 | 50.6 | 1.5 | 5.4 | 7.9 | 0.8 | 106.9 | 1,364 | 129 | 249 | 3,018 | 36,550 | 68,284 | 7,173 |
| Bland | 16 | 2,471 | 54.6 | 1.0 | 7.2 | 8.5 | 1.8 | 17.6 | 2,770 | 14 | 27 | 580 | 2,520 | 49,550 | 3,825 |
| Botetourt | 131 | 3,934 | 45.2 | 3.4 | 4.6 | 2.6 | 0.0 | 46.9 | 1,407 | 58 | 105 | 1,495 | 16,100 | 74,107 | 7,681 |
| Brunswick | 32 | 1,956 | 69.4 | 0.0 | 12.4 | 0.0 | 0.0 | 21.5 | 1,298 | 39 | 49 | 937 | 6,320 | 43,587 | 3,272 |
| Buchanan | 91 | 4,231 | 45.7 | 0.8 | 5.0 | 11.4 | 8.4 | 13.3 | 615 | 64 | 63 | 1,436 | 7,130 | 44,974 | 3,322 |
| Buckingham | 47 | 2,773 | 49.2 | 1.3 | 4.5 | 0.0 | 0.5 | 47.0 | 2,758 | 30 | 48 | 1,278 | 6,380 | 45,712 | 3,558 |
| Campbell | 123 | 2,214 | 67.9 | 2.7 | 5.2 | 7.9 | 0.6 | 43.5 | 786 | [5]342 | [5]440 | [5]6,931 | 25,140 | 52,101 | 4,326 |
| Caroline | 88 | 2,906 | 53.8 | 0.8 | 9.7 | 6.6 | 3.6 | 57.4 | 1,887 | 392 | 128 | 1,280 | 14,430 | 56,445 | 4,943 |
| Carroll | 82 | 2,757 | 53.6 | 9.6 | 7.0 | 7.9 | 0.7 | 55.2 | 1,851 | [6]94 | [6]113 | [6]2,044 | 12,410 | 44,124 | 3,126 |
| Charles City | 17 | 2,473 | 62.7 | 0.6 | 5.1 | 6.4 | 0.0 | 6.1 | 869 | 18 | 22 | 305 | 3,500 | 59,105 | 6,106 |
| Charlotte | 43 | 3,538 | 60.4 | 0.5 | 5.4 | 13.1 | 0.2 | 1.6 | 132 | 38 | 37 | 819 | 4,900 | 43,749 | 3,139 |
| Chesterfield | 1,145 | 3,336 | 54.4 | 8.7 | 6.6 | 2.4 | 2.3 | 260.2 | 758 | 3,351 | 1,159 | 18,499 | 170,480 | 76,509 | 8,725 |
| Clarke | 61 | 4,250 | 43.8 | 1.0 | 5.2 | 4.1 | 1.3 | 134.5 | 9,304 | 29 | 46 | 759 | 7,270 | 89,246 | 11,716 |
| Craig | 36 | 7,023 | 21.4 | 0.4 | 1.8 | 0.5 | 0.2 | 108.9 | 21,510 | 14 | 16 | 245 | 2,240 | 48,900 | 3,581 |
| Culpeper | 176 | 3,434 | 56.9 | 0.9 | 8.4 | 8.9 | 2.4 | 120.2 | 2,347 | 232 | 162 | 3,166 | 23,520 | 67,005 | 6,852 |
| Cumberland | 31 | 3,101 | 65.1 | 0.9 | 5.5 | 7.0 | 0.0 | 48.0 | 4,888 | 12 | 31 | 442 | 4,380 | 44,904 | 3,342 |
| Dickenson | 85 | 5,751 | 40.1 | 8.0 | 23.0 | 8.2 | 0.3 | 18.6 | 1,262 | 35 | 44 | 930 | 4,790 | 42,844 | 2,874 |
| Dinwiddie | 84 | 2,920 | 56.6 | 3.9 | 7.3 | 5.2 | 0.1 | 48.4 | 1,692 | [7]230 | [7]250 | [7]5,820 | 12,900 | 53,545 | 4,424 |
| Essex | 105 | 9,597 | 22.5 | 0.0 | 1.9 | 3.2 | 0.1 | 13.5 | 1,231 | 17 | 34 | 539 | 5,330 | 52,734 | 4,615 |
| Fairfax | 5,706 | 4,967 | 52.6 | 5.8 | 4.9 | 4.4 | 0.9 | 5,631.9 | 4,903 | [8]46,273 | [8]10,772 | [8]62,977 | 555,480 | 127,340 | 20,715 |
| Fauquier | 268 | 3,849 | 52.4 | 0.5 | 7.4 | 4.2 | 3.5 | 104.2 | 1,498 | 632 | 225 | 3,804 | 34,980 | 99,048 | 13,447 |
| Floyd | 31 | 1,986 | 67.7 | 3.5 | 5.7 | 3.5 | 0.0 | 28.7 | 1,824 | 47 | 50 | 629 | 6,780 | 51,569 | 4,185 |
| Fluvanna | 62 | 2,352 | 63.7 | 1.5 | 9.3 | 9.3 | 0.0 | 103.2 | 3,895 | 36 | 83 | 1,371 | 12,700 | 64,118 | 6,033 |
| Franklin | 151 | 2,681 | 64.2 | 0.5 | 3.8 | 7.3 | 0.2 | 33.3 | 591 | 96 | 174 | 2,316 | 24,010 | 58,280 | 5,571 |
| Frederick | 271 | 3,134 | 63.3 | 0.2 | 5.1 | 2.9 | 0.2 | 188.6 | 2,181 | [9]2,355 | [9]385 | [9]6,868 | 42,870 | 69,754 | 7,268 |
| Giles | 58 | 3,469 | 51.2 | 0.3 | 11.4 | 9.2 | 2.5 | 77.9 | 4,650 | 32 | 53 | 889 | 7,420 | 50,322 | 3,926 |
| Gloucester | 108 | 2,899 | 66.9 | 1.1 | 5.1 | 5.0 | 0.1 | 50.3 | 1,349 | 87 | 118 | 2,065 | 18,630 | 62,161 | 5,784 |
| Goochland | 63 | 2,783 | 50.1 | 1.3 | 5.6 | 6.1 | 0.0 | 99.2 | 4,371 | 41 | 72 | 1,444 | 11,850 | 143,663 | 25,058 |
| Grayson | 43 | 2,716 | 49.0 | 0.4 | 6.4 | 4.7 | 0.5 | 2.3 | 150 | 27 | 46 | 1,016 | 6,180 | 40,655 | 2,798 |
| Greene | 67 | 3,415 | 61.6 | 1.0 | 5.5 | 5.5 | 2.0 | 46.9 | 2,396 | 36 | 63 | 831 | 9,490 | 62,407 | 5,850 |
| Greensville | 59 | 5,079 | 51.4 | 0.3 | 4.9 | 5.5 | 0.4 | 44.0 | 3,806 | [10]33 | [10]41 | [10]1,799 | 3,480 | 41,838 | 2,813 |
| Halifax | 120 | 3,459 | 47.9 | 11.3 | 9.0 | 0.0 | 4.5 | 109.8 | 3,174 | 88 | 105 | 2,068 | 14,910 | 46,430 | 3,603 |
| Hanover | 408 | 3,858 | 47.7 | 0.4 | 8.1 | 5.3 | 6.5 | 343.1 | 3,247 | 170 | 346 | 5,138 | 53,580 | 83,619 | 9,738 |
| Henrico | 1,234 | 3,768 | 43.6 | 3.0 | 6.3 | 2.7 | 4.1 | 1,040.2 | 3,176 | 2,251 | 1,042 | 17,688 | 164,610 | 78,207 | 9,939 |
| Henry | 162 | 3,162 | 56.3 | 13.8 | 3.7 | 5.0 | 0.0 | 66.0 | 1,285 | [11]129 | [11]204 | [11]3,910 | 21,490 | 43,495 | 3,330 |
| Highland | 10 | 4,285 | 48.4 | 2.1 | 7.2 | 4.1 | 6.5 | 1.6 | 730 | 6 | 7 | 170 | 1,050 | 45,350 | 3,957 |
| Isle of Wight | 157 | 4,304 | 43.9 | 15.6 | 4.4 | 2.4 | 3.8 | 227.2 | 6,212 | 81 | 117 | 1,406 | 18,280 | 73,663 | 7,732 |
| James City | 252 | 3,353 | 59.7 | 1.5 | 4.9 | 1.8 | 0.0 | 214.4 | 2,848 | [12]251 | [12]565 | [12]8,200 | 38,250 | 95,715 | 12,269 |
| King and Queen | 22 | 3,082 | 48.9 | 0.6 | 7.1 | 0.0 | 0.0 | 3.7 | 531 | 9 | 22 | 291 | 3,190 | 52,170 | 4,335 |
| King George | 78 | 2,956 | 61.2 | 0.9 | 6.5 | 0.0 | 1.4 | 123.6 | 4,693 | 5,268 | 611 | 1,168 | 11,950 | 77,179 | 8,114 |
| King William | 56 | 3,339 | 64.8 | 0.3 | 5.3 | 3.1 | 0.0 | 27.4 | 1,644 | 28 | 54 | 761 | 8,480 | 60,294 | 5,179 |
| Lancaster | 35 | 3,248 | 51.2 | 0.0 | 15.5 | 7.9 | 0.4 | 18.0 | 1,675 | 34 | 33 | 529 | 5,670 | 73,462 | 8,334 |
| Lee | 70 | 2,917 | 64.1 | 4.1 | 5.4 | 10.1 | 0.8 | 6.8 | 287 | 418 | 70 | 1,118 | 7,530 | 39,961 | 2,607 |
| Loudoun | 2,182 | 5,495 | 62.3 | 0.3 | 4.4 | 3.9 | 5.5 | 1,920.0 | 4,836 | 4,598 | 1,320 | 22,835 | 192,060 | 127,193 | 18,981 |

1. Based on the resident population estimated as of July 1 of the year shown.   4. Staunton and Waynesboro cities are included with Augusta county.   7. Petersburg and Colonial Heights cities are included with Dinwiddie county.   10. Emporia city is included with Greensville county.   2. Charlottesville city is included with Albemarle county.   5. Lynchburg city is included with Campbell county.   8. Fairfax city and Falls Church city are included with Fairfax county.   11. Martinsville city is included with Henry county.   3. Covington city is included with Alleghany county.   6. Galax city is included with Carroll county.   9. Winchester city is included with Frederick county.   12. Williamsburg city is included with James City county.

# Table B. States and Counties — Land Area and Population

| State / county code | CBSA code[1] | County Type code[2] | STATE County | Land area[3] (sq. mi) | Total persons 2019 | Rank | Per square mile | White | Black | American Indian, Alaska Native | Asian and Pacific Islander | Percent Hispanic or Latino[4] | Under 5 years | 5 to 14 years | 15 to 24 years | 25 to 34 years | 35 to 44 years | 45 to 54 years |
|---|---|---|---|---|---|---|---|---|---|---|---|---|---|---|---|---|---|---|
| | | | | 1 | 2 | 3 | 4 | 5 | 6 | 7 | 8 | 9 | 10 | 11 | 12 | 13 | 14 | 15 |
| | | | VIRGINIA—Cont'd | | | | | | | | | | | | | | | |
| 51109 | | 8 | Louisa | 495.0 | 38,132 | 1,222 | 77.0 | 80.1 | 16.8 | 1.0 | 1.2 | 3.5 | 5.3 | 11.1 | 9.6 | 12.1 | 11.7 | 12.5 |
| 51111 | 47900 | 9 | Lunenburg | 431.7 | 12,267 | 2,267 | 28.4 | 60.8 | 33.6 | 1.1 | 0.9 | 5.7 | 5.2 | 11.0 | 9.6 | 11.6 | 12.7 | 12.2 |
| 51113 | 47260 | 8 | Madison | 320.6 | 13,312 | 2,200 | 41.5 | 86.7 | 10.7 | 0.9 | 1.1 | 3.2 | 5.1 | 11.4 | 10.2 | 10.9 | 10.8 | 11.7 |
| 51115 | | 1 | Mathews | 85.9 | 8,766 | 2,513 | 102.0 | 87.3 | 9.2 | 1.3 | 1.5 | 3.0 | 3.6 | 8.8 | 9.4 | 8.4 | 8.3 | 12.4 |
| 51117 | | 7 | Mecklenburg | 625.3 | 30,679 | 1,413 | 49.1 | 62.0 | 34.7 | 0.8 | 1.2 | 3.0 | 5.0 | 10.2 | 10.1 | 10.4 | 9.7 | 11.5 |
| 51119 | | 8 | Middlesex | 130.3 | 10,569 | 2,383 | 81.1 | 79.9 | 17.0 | 1.2 | 0.9 | 3.1 | 4.2 | 8.6 | 8.2 | 9.1 | 8.5 | 10.5 |
| 51121 | 13980 | 3 | Montgomery | 386.8 | 98,391 | 614 | 254.4 | 85.9 | 5.0 | 0.6 | 7.5 | 3.5 | 4.0 | 8.7 | 29.3 | 13.5 | 10.3 | 10.3 |
| 51125 | 16820 | 3 | Nelson | 470.7 | 14,755 | 2,104 | 31.3 | 83.1 | 12.0 | 1.0 | 1.1 | 4.7 | 4.6 | 9.4 | 9.6 | 9.3 | 10.2 | 11.5 |
| 51127 | 40060 | 1 | New Kent | 210.0 | 23,648 | 1,653 | 112.6 | 80.0 | 14.7 | 2.1 | 2.1 | 3.9 | 4.9 | 11.2 | 10.3 | 13.3 | 12.6 | 13.7 |
| 51131 | | 8 | Northampton | 211.7 | 11,673 | 2,312 | 55.1 | 57.2 | 33.4 | 0.9 | 1.4 | 8.9 | 5.0 | 10.9 | 9.4 | 9.8 | 9.0 | 10.3 |
| 51133 | | 9 | Northumberland | 191.4 | 12,069 | 2,290 | 63.1 | 70.4 | 25.5 | 0.9 | 0.8 | 4.0 | 3.5 | 8.2 | 7.9 | 7.8 | 7.7 | 9.1 |
| 51135 | | 6 | Nottoway | 314.4 | 15,160 | 2,085 | 48.2 | 55.4 | 39.5 | 0.9 | 1.0 | 4.9 | 4.9 | 10.8 | 11.6 | 14.2 | 12.7 | 12.5 |
| 51137 | | 6 | Orange | 341.1 | 37,695 | 1,231 | 110.5 | 79.4 | 14.6 | 1.0 | 1.9 | 5.8 | 5.6 | 11.7 | 10.8 | 12.0 | 11.5 | 12.6 |
| 51139 | | 6 | Page | 310.0 | 23,933 | 1,645 | 77.2 | 94.7 | 2.8 | 0.7 | 0.9 | 2.3 | 5.5 | 10.8 | 10.4 | 11.3 | 11.2 | 13.1 |
| 51141 | | 8 | Patrick | 483.0 | 17,493 | 1,944 | 36.2 | 90.7 | 5.9 | 0.7 | 0.5 | 3.3 | 4.1 | 9.8 | 9.3 | 8.9 | 9.7 | 13.9 |
| 51143 | 19260 | 4 | Pittsylvania | 969.0 | 59,850 | 873 | 61.8 | 75.1 | 21.9 | 0.6 | 0.8 | 2.9 | 4.2 | 10.4 | 10.8 | 10.3 | 10.5 | 13.4 |
| 51145 | 40060 | 1 | Powhatan | 260.2 | 30,148 | 1,428 | 115.9 | 87.3 | 9.9 | 0.9 | 1.1 | 2.4 | 4.4 | 10.4 | 10.1 | 11.0 | 12.9 | 15.4 |
| 51147 | | 6 | Prince Edward | 349.6 | 23,006 | 1,680 | 65.8 | 62.5 | 33.3 | 0.8 | 1.7 | 3.4 | 4.6 | 8.4 | 28.4 | 11.4 | 8.9 | 9.4 |
| 51149 | 40060 | 1 | Prince George | 265.3 | 38,686 | 1,209 | 145.8 | 56.5 | 33.0 | 1.2 | 3.3 | 9.1 | 5.9 | 13.0 | 11.8 | 15.6 | 15.3 | 11.7 |
| 51153 | 47900 | 1 | Prince William | 335.3 | 475,533 | 150 | 1,418.2 | 44.2 | 22.7 | 0.9 | 11.6 | 24.9 | 7.0 | 15.1 | 13.4 | 13.2 | 15.0 | 14.0 |
| 51155 | 13980 | 3 | Pulaski | 319.8 | 33,935 | 1,329 | 106.1 | 92.1 | 6.1 | 0.7 | 0.9 | 1.9 | 4.7 | 9.8 | 10.2 | 11.9 | 10.4 | 14.2 |
| 51157 | 47900 | 1 | Rappahannock | 266.4 | 7,260 | 2,638 | 27.3 | 90.1 | 5.1 | 1.0 | 1.9 | 4.1 | 4.3 | 9.3 | 9.6 | 9.6 | 9.5 | 11.9 |
| 51159 | | 9 | Richmond | 191.5 | 9,071 | 2,490 | 47.4 | 62.1 | 30.4 | 0.9 | 1.2 | 7.5 | 4.3 | 9.6 | 10.0 | 14.5 | 13.9 | 12.6 |
| 51161 | 40220 | 2 | Roanoke | 250.5 | 94,509 | 631 | 377.3 | 86.4 | 7.2 | 0.6 | 4.2 | 3.5 | 4.6 | 11.4 | 11.2 | 11.7 | 11.9 | 13.0 |
| 51163 | | 6 | Rockbridge | 596.5 | 22,757 | 1,693 | 38.2 | 93.1 | 4.2 | 1.3 | 1.1 | 2.0 | 4.5 | 9.9 | 10.6 | 10.2 | 9.9 | 12.0 |
| 51165 | 25500 | 3 | Rockingham | 849.8 | 82,346 | 691 | 96.9 | 88.7 | 3.0 | 0.6 | 1.4 | 7.8 | 5.8 | 12.3 | 12.7 | 11.8 | 11.6 | 12.1 |
| 51167 | | 7 | Russell | 473.5 | 26,647 | 1,540 | 56.3 | 97.1 | 1.2 | 0.5 | 0.5 | 1.5 | 4.8 | 10.7 | 10.1 | 10.8 | 11.3 | 13.4 |
| 51169 | 28700 | 2 | Scott | 535.8 | 21,629 | 1,737 | 40.4 | 97.2 | 1.2 | 0.7 | 0.4 | 1.4 | 4.0 | 10.4 | 9.9 | 10.6 | 11.2 | 13.1 |
| 51171 | | 6 | Shenandoah | 508.1 | 43,905 | 1,099 | 86.4 | 88.3 | 3.4 | 0.8 | 1.4 | 7.8 | 5.7 | 11.6 | 10.7 | 11.6 | 11.2 | 12.5 |
| 51173 | | 7 | Smyth | 451.4 | 30,090 | 1,432 | 66.7 | 94.6 | 2.9 | 0.6 | 0.7 | 2.2 | 4.5 | 11.0 | 10.9 | 11.6 | 10.7 | 13.8 |
| 51175 | 47260 | 6 | Southampton | 599.2 | 17,636 | 1,938 | 29.4 | 62.1 | 35.3 | 0.9 | 1.1 | 2.3 | 4.8 | 10.5 | 10.0 | 11.0 | 10.6 | 14.9 |
| 51177 | 47900 | 1 | Spotsylvania | 401.4 | 138,449 | 471 | 344.9 | 68.6 | 18.9 | 0.9 | 4.1 | 11.1 | 6.1 | 13.9 | 12.7 | 12.9 | 12.9 | 13.3 |
| 51179 | 47900 | 1 | Stafford | 269.2 | 156,748 | 429 | 582.3 | 61.6 | 21.6 | 1.1 | 5.5 | 14.6 | 6.2 | 14.9 | 14.4 | 13.3 | 14.3 | 13.8 |
| 51181 | | 8 | Surry | 278.9 | 6,385 | 2,715 | 22.9 | 55.2 | 41.8 | 1.3 | 1.0 | 3.0 | 4.7 | 9.1 | 9.3 | 11.4 | 9.3 | 12.0 |
| 51183 | 40060 | 1 | Sussex | 490.2 | 10,925 | 2,355 | 22.3 | 40.4 | 55.9 | 0.8 | 0.8 | 3.6 | 3.9 | 8.8 | 11.4 | 18.4 | 13.0 | 12.7 |
| 51185 | 14140 | 5 | Tazewell | 518.8 | 40,529 | 1,166 | 78.1 | 94.6 | 4.0 | 0.5 | 1.0 | 1.1 | 4.9 | 10.6 | 10.9 | 11.3 | 11.5 | 12.6 |
| 51187 | 47900 | 1 | Warren | 214.6 | 40,475 | 1,167 | 188.6 | 87.8 | 6.1 | 1.0 | 2.0 | 5.7 | 6.1 | 12.0 | 11.8 | 13.1 | 11.7 | 12.9 |
| 51191 | 28700 | 2 | Washington | 561.2 | 53,695 | 948 | 95.7 | 96.1 | 1.9 | 0.5 | 0.8 | 1.6 | 4.3 | 10.1 | 11.3 | 10.5 | 11.0 | 13.2 |
| 51193 | | 6 | Westmoreland | 229.3 | 18,149 | 1,915 | 79.1 | 66.9 | 26.7 | 1.3 | 1.5 | 6.5 | 5.1 | 10.3 | 9.5 | 11.5 | 10.0 | 10.8 |
| 51195 | 13720 | 7 | Wise | 403.4 | 37,206 | 1,244 | 92.2 | 92.2 | 6.4 | 0.5 | 0.7 | 1.3 | 4.8 | 10.9 | 12.5 | 13.6 | 12.1 | 13.2 |
| 51197 | | 6 | Wythe | 462.0 | 28,620 | 1,473 | 61.9 | 95.0 | 3.7 | 0.6 | 0.8 | 1.4 | 4.9 | 10.9 | 10.1 | 11.2 | 11.3 | 14.3 |
| 51199 | 47260 | 1 | York | 104.7 | 69,199 | 782 | 660.9 | 72.4 | 14.8 | 1.0 | 8.2 | 7.3 | 5.4 | 13.7 | 12.5 | 11.7 | 13.8 | 11.9 |
| | | | Independent cities | | | | | | | | | | | | | | | |
| 51510 | 47900 | 1 | Alexandria city | 14.9 | 158,726 | 425 | 10,652.8 | 54.9 | 23.1 | 0.7 | 8.0 | 16.4 | 7.0 | 8.9 | 8.2 | 20.5 | 18.5 | 13.3 |
| 51520 | 28700 | 2 | Bristol city | 12.9 | 17,329 | 1,954 | 1,343.3 | 89.7 | 7.7 | 0.8 | 1.5 | 2.6 | 5.5 | 11.6 | 10.2 | 13.1 | 11.8 | 12.3 |
| 51530 | | 6 | Buena Vista city | 6.4 | 6,402 | 2,714 | 1,000.3 | 89.3 | 6.4 | 1.8 | 1.6 | 3.1 | 5.9 | 11.3 | 18.4 | 12.1 | 10.0 | 11.1 |
| 51540 | 16820 | 3 | Charlottesville city | 10.2 | 46,950 | 1,037 | 4,602.9 | 69.1 | 19.0 | 0.7 | 8.6 | 5.8 | 5.2 | 8.3 | 21.0 | 19.6 | 13.1 | 9.8 |
| 51550 | 47260 | 1 | Chesapeake city | 338.5 | 247,011 | 278 | 729.7 | 59.1 | 31.2 | 1.0 | 5.2 | 7.1 | 6.3 | 13.7 | 12.4 | 14.1 | 14.3 | 12.2 |
| 51570 | 40060 | 1 | Colonial Heights city | 7.5 | 17,205 | 1,963 | 2,294.0 | 71.4 | 18.3 | 1.0 | 4.8 | 7.3 | 6.3 | 12.8 | 11.5 | 13.7 | 11.3 | 11.7 |
| 51580 | | 6 | Covington city | 5.5 | 5,639 | 2,778 | 1,025.3 | 82.2 | 15.6 | 1.1 | 1.4 | 2.5 | 7.3 | 11.5 | 11.5 | 12.3 | 10.6 | 12.5 |
| 51590 | 19260 | 4 | Danville city | 42.8 | 39,869 | 1,191 | 931.5 | 41.8 | 52.7 | 0.6 | 1.8 | 4.7 | 5.8 | 12.5 | 12.4 | 12.5 | 10.3 | 10.9 |
| 51595 | | 6 | Emporia city | 6.9 | 5,257 | 2,805 | 761.9 | 28.2 | 64.8 | 0.7 | 1.3 | 6.6 | 7.1 | 13.7 | 11.9 | 12.3 | 9.9 | 11.5 |
| 51600 | 47900 | 1 | Fairfax city | 6.2 | 23,429 | 1,662 | 3,778.9 | 56.7 | 6.5 | 0.8 | 21.0 | 17.9 | 7.8 | 11.7 | 12.1 | 14.2 | 13.2 | 12.7 |
| 51610 | 47900 | 1 | Falls Church city | 2.0 | 14,631 | 2,112 | 7,315.5 | 74.6 | 4.9 | 0.9 | 12.4 | 10.9 | 5.6 | 14.2 | 12.0 | 12.0 | 14.2 | 14.9 |
| 51620 | 47260 | 6 | Franklin city | 8.3 | 7,833 | 2,601 | 943.7 | 40.4 | 56.6 | 1.0 | 1.6 | 2.8 | 7.2 | 14.1 | 11.4 | 11.8 | 10.1 | 11.0 |
| 51630 | 47900 | 1 | Fredericksburg city | 10.5 | 29,492 | 1,449 | 2,808.8 | 61.7 | 25.5 | 1.0 | 4.7 | 11.5 | 6.4 | 11.5 | 22.2 | 16.2 | 11.1 | 10.5 |
| 51640 | | 7 | Galax city | 8.2 | 6,296 | 2,721 | 767.8 | 76.1 | 7.7 | 0.6 | 1.4 | 16.6 | 6.9 | 13.6 | 12.3 | 11.9 | 10.6 | 13.2 |
| 51650 | 47260 | 1 | Hampton city | 51.5 | 135,464 | 479 | 2,630.4 | 39.7 | 52.6 | 1.4 | 4.0 | 6.4 | 6.2 | 11.5 | 14.8 | 15.8 | 11.9 | 10.3 |
| 51660 | 25500 | 3 | Harrisonburg city | 17.3 | 53,204 | 952 | 3,075.4 | 66.6 | 9.1 | 0.5 | 5.6 | 20.7 | 5.2 | 8.8 | 34.4 | 13.5 | 11.3 | 8.8 |
| 51670 | 40060 | 1 | Hopewell city | 10.4 | 22,375 | 1,706 | 2,151.4 | 45.8 | 44.7 | 1.0 | 2.7 | 9.0 | 7.3 | 14.1 | 11.9 | 14.9 | 11.7 | 11.5 |
| 51678 | | 6 | Lexington city | 2.5 | 7,279 | 2,634 | 2,911.6 | 84.6 | 9.2 | 0.9 | 4.3 | 3.7 | 2.9 | 6.8 | 46.4 | 8.4 | 6.6 | 6.3 |
| 51680 | 31340 | 2 | Lynchburg city | 49.0 | 81,561 | 700 | 1,664.5 | 64.1 | 29.5 | 0.8 | 3.4 | 4.7 | 5.8 | 10.0 | 27.6 | 14.7 | 9.0 | 8.5 |
| 51683 | 47900 | 1 | Manassas city | 9.8 | 40,869 | 1,157 | 4,170.3 | 41.1 | 15.1 | 0.7 | 7.2 | 38.7 | 7.9 | 14.4 | 12.6 | 15.2 | 14.2 | 12.6 |
| 51685 | 47900 | 1 | Manassas Park city | 3.0 | 18,004 | 1,920 | 6,001.3 | 32.4 | 15.3 | 0.8 | 12.3 | 41.6 | 6.7 | 13.6 | 13.1 | 16.3 | 15.1 | 15.1 |

1. CBSA = Core Based Statistical Area. See Appendix A for explanation. See Appendix B for list of metropolitan areas with component counties. Service of USDA Rural-Urban Continuum Codes. See Appendix A for definition.
2. County type code from the Economic Research
3. Dry land or land partially or temporarily covered by water.
4. May be of any race.

# Table B. States and Counties — Population and Households

| STATE County | 55 to 64 years | 65 to 74 years | 75 years and over | Percent female | Total persons 2000 | Total persons 2010 | Percent change 2000-2010 | Percent change 2010-2020 | Births | Deaths | Net Migration | Number | Persons per household | Family households | Female family householder[1] | One person |
|---|---|---|---|---|---|---|---|---|---|---|---|---|---|---|---|---|
| | 16 | 17 | 18 | 19 | 20 | 21 | 22 | 23 | 24 | 25 | 26 | 27 | 28 | 29 | 30 | 31 |
| **VIRGINIA—Cont'd** | | | | | | | | | | | | | | | | |
| Louisa | 16.7 | 13.1 | 7.7 | 50.5 | 25,627 | 33,028 | 28.9 | 15.5 | 3,838 | 3,340 | 4,599 | 13,871 | 2.59 | 69.8 | 8.0 | 23.7 |
| Lunenburg | 14.4 | 14.1 | 9.3 | 47.0 | 13,146 | 12,916 | -1.7 | -5.0 | 1,237 | 1,567 | -326 | 4,293 | 2.63 | 70.6 | 12.2 | 27.3 |
| Madison | 16.8 | 13.6 | 9.5 | 51.8 | 12,520 | 13,306 | 6.3 | 0.0 | 1,310 | 1,483 | 186 | 4,949 | 2.62 | 71.3 | 9.0 | 25.6 |
| Mathews | 17.4 | 17.0 | 14.8 | 51.4 | 9,207 | 8,976 | -2.5 | -2.3 | 603 | 1,239 | 435 | 3,920 | 2.21 | 62.5 | 3.9 | 28.8 |
| Mecklenburg | 16.3 | 15.4 | 11.4 | 51.6 | 32,380 | 32,722 | 1.1 | -6.2 | 3,060 | 4,658 | -459 | 12,075 | 2.46 | 63.8 | 13.8 | 33.3 |
| Middlesex | 17.9 | 17.8 | 15.2 | 50.5 | 9,932 | 10,959 | 10.3 | -3.6 | 853 | 1,602 | 371 | 4,603 | 2.23 | 61.0 | 8.3 | 34.0 |
| Montgomery | 10.0 | 8.2 | 5.6 | 48.5 | 83,629 | 94,419 | 12.9 | 4.2 | 8,735 | 6,439 | 1,685 | 35,685 | 2.49 | 55.0 | 7.8 | 27.5 |
| Nelson | 16.7 | 17.7 | 11.0 | 51.2 | 14,445 | 15,015 | 3.9 | -1.7 | 1,369 | 1,908 | 290 | 6,419 | 2.29 | 70.4 | 12.7 | 24.3 |
| New Kent | 15.6 | 12.4 | 6.0 | 48.9 | 13,462 | 18,432 | 36.9 | 28.3 | 2,025 | 1,650 | 4,798 | 7,961 | 2.65 | 76.9 | 11.2 | 19.6 |
| Northampton | 17.2 | 16.1 | 12.4 | 52.7 | 13,093 | 12,391 | -5.4 | -5.8 | 1,322 | 1,889 | -146 | 5,148 | 2.25 | 60.4 | 14.1 | 35.1 |
| Northumberland | 17.9 | 20.7 | 17.3 | 51.1 | 12,259 | 12,331 | 0.6 | -2.1 | 926 | 1,809 | 631 | 5,584 | 2.18 | 66.5 | 11.7 | 28.3 |
| Nottoway | 14.0 | 11.1 | 8.3 | 45.5 | 15,725 | 15,858 | 0.8 | -4.4 | 1,675 | 2,115 | -257 | 5,446 | 2.45 | 67.4 | 18.9 | 30.4 |
| Orange | 14.9 | 11.8 | 9.0 | 51.3 | 25,881 | 33,541 | 29.6 | 12.4 | 3,925 | 3,939 | 4,165 | 13,679 | 2.59 | 69.3 | 10.9 | 26.1 |
| Page | 15.4 | 12.7 | 9.5 | 50.8 | 23,177 | 24,048 | 3.8 | -0.5 | 2,507 | 3,018 | 419 | 9,286 | 2.54 | 68.7 | 10.0 | 24.2 |
| Patrick | 16.7 | 15.0 | 12.5 | 50.1 | 19,407 | 18,505 | -4.6 | -5.5 | 1,468 | 2,623 | 161 | 7,755 | 2.24 | 64.8 | 9.6 | 32.3 |
| Pittsylvania | 16.4 | 13.8 | 10.3 | 50.7 | 61,745 | 63,471 | 2.8 | -5.7 | 5,269 | 7,394 | -1,476 | 26,267 | 2.29 | 70.3 | 14.9 | 26.0 |
| Powhatan | 16.5 | 12.2 | 7.2 | 48.2 | 22,377 | 28,073 | 25.5 | 7.4 | 2,492 | 2,149 | 1,737 | 10,265 | 2.55 | 78.2 | 6.0 | 17.3 |
| Prince Edward | 11.6 | 9.5 | 7.7 | 49.8 | 19,720 | 23,363 | 18.5 | -1.5 | 2,136 | 2,392 | -183 | 7,185 | 2.45 | 60.6 | 13.3 | 31.4 |
| Prince George | 11.9 | 9.0 | 5.9 | 46.3 | 33,047 | 35,715 | 8.1 | 1.8 | 4,136 | 2,461 | 1,268 | 11,507 | 2.96 | 72.3 | 10.4 | 21.8 |
| Prince William | 11.6 | 6.7 | 4.0 | 50.0 | 280,813 | 402,006 | 43.2 | 18.3 | 66,874 | 18,903 | 25,358 | 142,000 | 3.22 | 77.2 | 11.2 | 18.5 |
| Pulaski | 15.3 | 13.7 | 9.8 | 49.9 | 35,127 | 34,857 | -0.8 | -2.6 | 3,287 | 4,603 | 423 | 14,533 | 2.27 | 65.3 | 10.8 | 29.2 |
| Rappahannock | 17.6 | 16.3 | 11.9 | 50.3 | 6,983 | 7,507 | 7.5 | -3.3 | 637 | 724 | -157 | 2,913 | 2.53 | 65.8 | 8.8 | 28.1 |
| Richmond | 13.3 | 11.1 | 10.8 | 43.5 | 8,809 | 9,250 | 5.0 | -1.9 | 717 | 1,153 | 245 | 3,069 | 2.36 | 69.2 | 8.5 | 27.2 |
| Roanoke | 13.9 | 12.6 | 9.7 | 51.8 | 85,778 | 92,476 | 7.8 | 2.2 | 8,500 | 10,282 | 3,893 | 38,222 | 2.4 | 67.6 | 10.1 | 27.8 |
| Rockbridge | 15.9 | 15.1 | 12.0 | 50.7 | 20,808 | 22,357 | 7.4 | 1.8 | 1,969 | 2,613 | 1,063 | 9,248 | 2.42 | 67.2 | 9.1 | 27.5 |
| Rockingham | 13.8 | 11.2 | 8.8 | 50.8 | 67,725 | 76,324 | 12.7 | 7.9 | 9,060 | 7,434 | 4,474 | 30,994 | 2.53 | 71.2 | 8.9 | 23.8 |
| Russell | 15.9 | 13.4 | 9.7 | 51.1 | 30,308 | 28,917 | -4.6 | -7.9 | 2,647 | 3,641 | -1,273 | 10,812 | 2.46 | 65.2 | 8.4 | 30.8 |
| Scott | 15.1 | 14.0 | 11.6 | 49.6 | 23,403 | 23,166 | | -6.6 | 1,833 | 2,981 | -374 | 8,773 | 2.41 | 62.6 | 8.8 | 33.4 |
| Shenandoah | 14.5 | 12.5 | 9.9 | 50.9 | 35,075 | 41,991 | 19.7 | 4.6 | 4,783 | 4,943 | 2,105 | 17,402 | 2.46 | 65.2 | 10.9 | 29.7 |
| Smyth | 14.2 | 13.0 | 10.2 | 51.2 | 33,081 | 32,211 | -2.6 | -6.6 | 3,019 | 4,491 | -627 | 12,686 | 2.36 | 66.3 | 12.6 | 30.0 |
| Southampton | 16.8 | 12.4 | 9.0 | 48.1 | 17,482 | 18,573 | 6.2 | -5.0 | 1,726 | 1,935 | -736 | 6,562 | 2.50 | 71.9 | 14.0 | 25.2 |
| Spotsylvania | 13.3 | 9.1 | 5.8 | 50.8 | 90,395 | 122,455 | 35.5 | 13.1 | 16,141 | 9,039 | 8,984 | 44,436 | 2.98 | 76.0 | 11.9 | 19.0 |
| Stafford | 12.1 | 6.9 | 4.1 | 49.3 | 92,446 | 128,993 | 39.5 | 21.5 | 17,597 | 7,450 | 17,616 | 47,075 | 3.04 | 80.6 | 10.8 | 15.3 |
| Surry | 19.7 | 14.9 | 9.6 | 50.6 | 6,829 | 7,064 | 3.4 | -9.6 | 634 | 789 | -523 | 2,730 | 2.39 | 70.0 | 12.4 | 25.9 |
| Sussex | 13.4 | 10.9 | 7.7 | 40.5 | 12,504 | 12,071 | -3.5 | -9.5 | 987 | 1,411 | -730 | 3,794 | 2.38 | 66.2 | 15.8 | 28.3 |
| Tazewell | 14.6 | 13.9 | 9.7 | 50.4 | 44,598 | 45,058 | 1.0 | -10.1 | 4,328 | 6,106 | -2,752 | 16,531 | 2.42 | 67.2 | 12.0 | 29.6 |
| Warren | 15.3 | 10.5 | 6.6 | 50.0 | 31,584 | 37,452 | 18.6 | 8.1 | 4,920 | 3,855 | 1,841 | 14,651 | 2.64 | 68.4 | 12.0 | 26.0 |
| Washington | 15.3 | 14.0 | 10.4 | 50.8 | 51,103 | 54,959 | 7.5 | -2.3 | 4,739 | 6,391 | 415 | 22,321 | 2.34 | 66.5 | 10.7 | 30.3 |
| Westmoreland | 16.4 | 15.1 | 11.3 | 51.1 | 16,718 | 17,464 | 4.5 | 3.9 | 1,908 | 2,300 | 1,089 | 7,923 | 2.22 | 60.0 | 11.3 | 32.1 |
| Wise | 13.4 | 12.1 | 7.5 | 47.9 | 40,123 | 41,436 | 3.3 | -10.2 | 4,094 | 5,029 | -3,329 | 14,925 | 2.37 | 66.4 | 14.3 | 28.6 |
| Wythe | 14.9 | 12.9 | 9.5 | 51.2 | 27,599 | 29,246 | 6.0 | -2.1 | 2,855 | 3,889 | 431 | 11,983 | 2.39 | 66.4 | 7.1 | 28.2 |
| York | 13.7 | 9.7 | 7.6 | 50.8 | 56,297 | 65,191 | 15.8 | 6.1 | 6,650 | 4,429 | 1,756 | 25,103 | 2.65 | 75.4 | 9.5 | 20.1 |
| **Independent cities** | | | | | | | | | | | | | | | | |
| Alexandria city | 11.1 | 7.6 | 4.8 | 51.8 | 128,283 | 139,998 | 9.1 | 13.4 | 27,943 | 7,553 | -1,604 | 70,598 | 2.21 | 48.9 | 8.7 | 41.3 |
| Bristol city | 13.3 | 11.9 | 10.3 | 52.6 | 17,367 | 17,738 | 2.1 | -2.3 | 1,981 | 2,498 | 114 | 7,334 | 2.26 | 57.7 | 14.2 | 37.8 |
| Buena Vista city | 11.8 | 10.9 | 8.5 | 53.2 | 6,349 | 6,608 | 4.1 | -3.1 | 856 | 797 | -270 | 2,517 | 2.36 | 69.6 | 18.4 | 25.6 |
| Charlottesville city | 10.2 | 8.2 | 4.6 | 51.7 | 45,049 | 43,425 | -3.6 | 8.1 | 5,312 | 2,906 | 1,125 | 18,617 | 2.37 | 45.5 | 10.7 | 34.6 |
| Chesapeake city | 13.1 | 8.6 | 5.5 | 51.2 | 199,184 | 222,268 | 11.6 | 11.1 | 29,585 | 18,133 | 13,373 | 84,849 | 2.76 | 74.5 | 14.6 | 21.4 |
| Colonial Heights city | 13.1 | 10.2 | 9.5 | 53.8 | 16,897 | 17,409 | 3.0 | -1.2 | 2,079 | 2,455 | 181 | 7,036 | 2.45 | 63.2 | 17.3 | 32.7 |
| Covington city | 14.4 | 11.6 | 8.5 | 52.0 | 6,303 | 5,951 | -5.6 | -5.2 | 717 | 955 | -70 | 2,384 | 2.31 | 61.2 | 9.5 | 34.2 |
| Danville city | 14.2 | 12.0 | 9.3 | 54.0 | 48,411 | 43,071 | -11.0 | -7.4 | 5,182 | 7,100 | -1,257 | 18,293 | 2.16 | 56.9 | 21.0 | 37.1 |
| Emporia city | 13.3 | 10.1 | 10.3 | 53.9 | 5,665 | 5,925 | 4.6 | -11.3 | 788 | 773 | -684 | 2,086 | 2.49 | 63.9 | 27.1 | 34.9 |
| Fairfax city | 13.0 | 8.9 | 6.4 | 51.1 | 21,498 | 22,554 | 4.9 | 3.9 | 3,146 | 2,172 | -117 | 8,577 | 2.67 | 69.0 | 6.8 | 20.5 |
| Falls Church city | 12.5 | 9.0 | 5.6 | 50.7 | 10,377 | 12,267 | 18.2 | 19.3 | 1,522 | 817 | 1,629 | 5,493 | 2.56 | 65.6 | 11.2 | 28.8 |
| Franklin city | 14.4 | 11.2 | 8.8 | 54.4 | 8,346 | 8,578 | 2.8 | -8.7 | 1,166 | 1,379 | -532 | 3,522 | 2.28 | 61.0 | 22.9 | 36.0 |
| Fredericksburg city | 10.2 | 7.0 | 4.8 | 53.3 | 19,279 | 24,178 | 25.4 | 22.0 | 4,189 | 1,984 | 3,017 | 10,762 | 2.42 | 52.3 | 12.9 | 37.1 |
| Galax city | 12.8 | 9.8 | 8.9 | 53.4 | 6,837 | 6,989 | 2.2 | -9.9 | 942 | 1,378 | -254 | 2,663 | 2.31 | 63.7 | 12.4 | 35.1 |
| Hampton city | 13.5 | 9.5 | 6.6 | 51.9 | 146,437 | 137,463 | -6.1 | -1.5 | 18,431 | 12,882 | -7,553 | 54,050 | 2.41 | 59.7 | 16.6 | 32.6 |
| Harrisonburg city | 7.8 | 5.8 | 4.3 | 51.9 | 40,468 | 48,902 | 20.8 | 8.8 | 5,742 | 2,889 | 1,395 | 16,723 | 2.75 | 53.9 | 12.7 | 29.8 |
| Hopewell city | 12.6 | 9.1 | 6.9 | 53.5 | 22,354 | 22,591 | 1.1 | -1.0 | 3,422 | 3,081 | -559 | 9,224 | 2.41 | 62.5 | 23.7 | 30.2 |
| Lexington city | 7.3 | 7.8 | 7.5 | 43.5 | 6,867 | 7,037 | 2.5 | 3.4 | 431 | 744 | 548 | 2,158 | 2.02 | 40.3 | 7.4 | 49.4 |
| Lynchburg city | 12.6 | 9.1 | 6.9 | 53.0 | 65,269 | 75,535 | 15.7 | 8.0 | 10,312 | 8,350 | 4,078 | 28,273 | 2.48 | 57.3 | 14.9 | 32.0 |
| Manassas city | 9.7 | 7.8 | 6.8 | 49.7 | 35,135 | 37,799 | 7.6 | 8.1 | 6,965 | 2,254 | -1,689 | 12,898 | 3.19 | 73.4 | 11.6 | 22.0 |
| Manassas Park city | 11.3 | 5.4 | 3.4 | 48.7 | 10,290 | 14,243 | 38.4 | 26.4 | 2,602 | 366 | 1,484 | 4,641 | 3.66 | 70.6 | 11.7 | 21.8 |

1. No spouse present.

# Table B. States and Counties — Population, Vital Statistics, Health, and Crime

| STATE County | Persons in group quarters, 2020 | Daytime Population, 2015-2019 Number | Daytime Population Employment/ residence ratio | Births, 2020 Total | Births, 2020 Rate[1] | Deaths, 2020 Number | Deaths, 2020 Rate[1] | Persons under 65 with no health insurance, 2019 Number | Percent | Medicare, 2020 Total beneficiaries | Enrolled in Original Medicare | Enrolled in Medicare Advantage | Crimes Violent | Crimes Property |
|---|---|---|---|---|---|---|---|---|---|---|---|---|---|---|
| | 32 | 33 | 34 | 35 | 36 | 37 | 38 | 39 | 40 | 41 | 42 | 43 | 44 | 45 |
| **VIRGINIA—Cont'd** | | | | | | | | | | | | | | |
| Louisa | 211 | 30,776 | 0.68 | 376 | 9.9 | 424 | 11.1 | 3,614 | 12.1 | 8,801 | 6,752 | 2,049 | 39 | 333 |
| Lunenburg | 1,002 | 10,394 | 0.58 | 103 | 8.4 | 177 | 14.4 | 1,135 | 13.6 | 3,178 | 2,519 | 659 | 17 | 40 |
| Madison | 193 | 11,426 | 0.71 | 135 | 10.1 | 149 | 11.2 | 1,138 | 11.2 | 3,379 | 2,768 | 611 | 12 | 67 |
| Mathews | 86 | 7,028 | 0.50 | 60 | 6.8 | 132 | 15.1 | 533 | 8.8 | 2,926 | 2,346 | 580 | 5 | 57 |
| Mecklenburg | 925 | 32,029 | 1.11 | 314 | 10.2 | 497 | 16.2 | 2,628 | 12.0 | 9,422 | 7,211 | 2,212 | 53 | 185 |
| Middlesex | 377 | 9,836 | 0.80 | 78 | 7.4 | 153 | 14.5 | 794 | 11.5 | 3,717 | 3,025 | 692 | 13 | 112 |
| Montgomery | 9,832 | 104,703 | 1.15 | 773 | 7.9 | 725 | 7.4 | 6,689 | 8.8 | 14,448 | 11,158 | 3,290 | 61 | 308 |
| Nelson | 102 | 13,062 | 0.74 | 138 | 9.4 | 185 | 12.5 | 1,355 | 12.7 | 4,792 | 3,936 | 856 | 26 | 166 |
| New Kent | 552 | 15,296 | 0.41 | 221 | 9.3 | 199 | 8.4 | 1,593 | 8.6 | 4,953 | 3,884 | 1,070 | 40 | 198 |
| Northampton | 252 | 11,811 | 0.98 | 116 | 9.9 | 194 | 16.6 | 1,042 | 12.5 | 3,793 | 2,885 | 908 | 12 | 53 |
| Northumberland | 0 | 10,677 | 0.68 | 82 | 6.8 | 216 | 17.9 | 832 | 11.1 | 4,705 | 3,862 | 843 | 20 | 70 |
| Nottoway | 2,244 | 16,828 | 1.24 | 155 | 10.2 | 227 | 15.0 | 1,320 | 12.9 | 3,667 | 2,804 | 863 | 15 | 71 |
| Orange | 601 | 29,405 | 0.61 | 398 | 10.6 | 417 | 11.1 | 3,001 | 10.4 | 8,982 | 7,464 | 1,518 | 23 | 125 |
| Page | 194 | 19,608 | 0.61 | 256 | 10.7 | 340 | 14.2 | 2,080 | 11.3 | 6,322 | 5,220 | 1,102 | 29 | 152 |
| Patrick | 295 | 15,425 | 0.68 | 127 | 7.3 | 240 | 13.7 | 1,475 | 11.7 | 5,550 | 3,998 | 1,552 | 37 | 284 |
| Pittsylvania | 1,066 | 49,058 | 0.55 | 481 | 8.0 | 809 | 13.5 | 5,147 | 11.4 | 16,618 | 11,159 | 5,459 | 79 | 425 |
| Powhatan | 1,119 | 22,867 | 0.55 | 259 | 8.6 | 277 | 9.2 | 2,095 | 9.1 | 6,419 | 4,926 | 1,493 | 25 | 284 |
| Prince Edward | 4,060 | 25,011 | 1.25 | 220 | 9.6 | 263 | 11.4 | 1,475 | 9.8 | 4,738 | 3,640 | 1,098 | 19 | 57 |
| Prince George | 3,990 | 42,671 | 1.27 | 386 | 10.0 | 302 | 7.8 | 2,164 | 7.6 | 6,361 | 5,264 | 1,097 | 73 | 339 |
| Prince William | 3,074 | 374,668 | 0.64 | 6,300 | 13.2 | 2,325 | 4.9 | 47,465 | 11.3 | 52,016 | 40,110 | 11,906 | 787 | 4,723 |
| Pulaski | 1,088 | 34,037 | 0.99 | 296 | 8.7 | 519 | 15.3 | 2,074 | 8.3 | 9,226 | 7,031 | 2,194 | 51 | 728 |
| Rappahannock | 35 | 6,063 | 0.62 | 57 | 7.9 | 86 | 11.8 | 647 | 12.2 | 2,007 | 1,762 | 245 | 4 | 16 |
| Richmond | 1,665 | 8,949 | 1.02 | 72 | 7.9 | 110 | 12.1 | 536 | 9.7 | 1,969 | 1,523 | 447 | NA | NA |
| Roanoke | 2,347 | 85,882 | 0.82 | 773 | 8.2 | 1,064 | 11.3 | 5,400 | 7.5 | 24,209 | 17,555 | 6,655 | 187 | 1,207 |
| Rockbridge | 502 | 19,489 | 0.68 | 147 | 6.5 | 267 | 11.7 | 1,621 | 10.1 | 6,477 | 5,154 | 1,323 | 32 | 238 |
| Rockingham | 1,633 | 76,518 | 0.90 | 889 | 10.8 | 837 | 10.2 | 7,429 | 11.5 | 18,413 | 15,474 | 2,939 | 41 | 369 |
| Russell | 432 | 25,480 | 0.82 | 260 | 9.8 | 366 | 13.7 | 1,982 | 9.8 | 8,025 | 4,449 | 3,576 | 54 | 159 |
| Scott | 794 | 19,105 | 0.66 | 155 | 7.2 | 267 | 12.3 | 1,571 | 10.2 | 6,808 | 2,849 | 3,959 | 21 | 201 |
| Shenandoah | 427 | 38,001 | 0.74 | 478 | 10.9 | 543 | 12.4 | 3,434 | 10.2 | 10,883 | 8,866 | 2,018 | 46 | 237 |
| Smyth | 770 | 30,150 | 0.95 | 256 | 8.5 | 489 | 16.3 | 2,114 | 9.3 | 8,694 | 5,725 | 2,969 | 29 | 186 |
| Southampton | 1,576 | 14,534 | 0.55 | 155 | 8.8 | 219 | 12.4 | 1,105 | 8.8 | 4,043 | 3,213 | 830 | 19 | 203 |
| Spotsylvania | 524 | 106,339 | 0.60 | 1,560 | 11.3 | 1,104 | 8.0 | 10,348 | 8.9 | 22,558 | 18,277 | 4,282 | 168 | 803 |
| Stafford | 4,424 | 118,464 | 0.62 | 1,763 | 11.2 | 920 | 5.9 | 9,461 | 7.1 | 18,820 | 15,801 | 3,019 | 338 | 1,427 |
| Surry | 0 | 5,787 | 0.76 | 60 | 9.4 | 78 | 12.2 | 425 | 8.8 | 1,759 | 1,290 | 469 | 3 | 16 |
| Sussex | 2,601 | 10,994 | 0.91 | 78 | 7.1 | 151 | 13.8 | 658 | 10.1 | 2,532 | 1,790 | 742 | 8 | 74 |
| Tazewell | 1,762 | 41,515 | 0.99 | 370 | 9.1 | 629 | 15.5 | 3,072 | 10.4 | 12,021 | 7,409 | 4,612 | 36 | 288 |
| Warren | 679 | 34,550 | 0.74 | 508 | 12.6 | 438 | 10.8 | 3,338 | 10.1 | 8,084 | 6,749 | 1,335 | 34 | 154 |
| Washington | 1,782 | 53,627 | 0.98 | 461 | 8.6 | 656 | 12.2 | 4,175 | 10.6 | 15,252 | 8,955 | 6,297 | 37 | 603 |
| Westmoreland | 70 | 13,827 | 0.49 | 177 | 9.8 | 254 | 14.0 | 1,414 | 10.7 | 5,209 | 4,317 | 892 | 12 | 68 |
| Wise | 2,909 | 41,169 | 1.22 | 358 | 9.6 | 478 | 12.8 | 3,089 | 11.3 | 10,658 | 5,101 | 5,557 | 64 | 249 |
| Wythe | 260 | 28,923 | 1.01 | 265 | 9.3 | 406 | 14.2 | 2,136 | 9.6 | 7,988 | 5,572 | 2,416 | 26 | 193 |
| York | 646 | 58,843 | 0.73 | 635 | 9.2 | 586 | 8.5 | 3,923 | 7.0 | 12,615 | 10,531 | 2,084 | 78 | 875 |
| **Independent cities** | | | | | | | | | | | | | | |
| Alexandria city | 1,830 | 161,940 | 1.04 | 2,572 | 16.2 | 893 | 5.6 | 12,663 | 9.1 | 18,444 | 14,858 | 3,586 | 288 | 2,517 |
| Bristol city | 224 | 20,032 | 1.45 | 171 | 9.9 | 204 | 11.8 | 1,206 | 9.4 | 4,676 | 2,346 | 2,330 | 52 | 388 |
| Buena Vista city | 489 | 5,889 | 0.79 | 88 | 13.7 | 90 | 14.1 | 375 | 7.9 | 1,499 | 1,144 | 355 | 7 | 18 |
| Charlottesville city | 2,423 | 67,665 | 1.83 | 546 | 11.6 | 347 | 7.4 | 4,204 | 10.7 | 6,364 | 5,362 | 1,002 | 157 | 1,124 |
| Chesapeake city | 4,313 | 222,435 | 0.85 | 2,930 | 11.9 | 2,153 | 8.7 | 16,311 | 7.9 | 40,199 | 29,913 | 10,286 | 1,112 | 5,541 |
| Colonial Heights city | 171 | 19,321 | 1.25 | 236 | 13.7 | 217 | 12.6 | 1,259 | 9.1 | 4,174 | 3,271 | 903 | 71 | 720 |
| Covington city | 87 | 5,582 | 0.99 | 81 | 14.4 | 70 | 12.4 | 345 | 8.0 | 1,579 | 1,236 | 343 | 5 | 116 |
| Danville city | 1,578 | 51,388 | 1.64 | 461 | 11.6 | 710 | 17.8 | 2,929 | 9.8 | 11,156 | 7,220 | 3,936 | 112 | 1,467 |
| Emporia city | 246 | 6,506 | 1.56 | 74 | 14.1 | 87 | 16.5 | 343 | 8.3 | 1,337 | 890 | 447 | 12 | 195 |
| Fairfax city | 521 | 45,375 | 2.76 | 514 | 21.9 | 256 | 10.9 | 1,929 | 9.4 | 3,975 | 3,109 | 866 | 18 | 345 |
| Falls Church city | 42 | 17,138 | 1.39 | 179 | 12.2 | 110 | 7.5 | 486 | 3.8 | 2,134 | 1,726 | 408 | 10 | 230 |
| Franklin city | 129 | 8,493 | 1.10 | 128 | 16.3 | 109 | 13.9 | 466 | 7.4 | 2,197 | 1,696 | 501 | 16 | 292 |
| Fredericksburg city | 2,676 | 41,509 | 1.82 | 383 | 13.0 | 258 | 8.7 | 2,393 | 10.4 | 4,269 | 3,387 | 882 | NA | NA |
| Galax city | 377 | 9,757 | 2.25 | 88 | 14.0 | 93 | 14.8 | 707 | 14.1 | 1,766 | 1,494 | 271 | 29 | 255 |
| Hampton city | 4,915 | 133,195 | 0.97 | 1,665 | 12.3 | 1,429 | 10.5 | 8,475 | 7.9 | 26,039 | 17,900 | 8,139 | 393 | 3,994 |
| Harrisonburg city | 7,690 | 65,769 | 1.50 | 605 | 11.4 | 303 | 5.7 | 5,221 | 13.0 | 5,613 | 4,738 | 874 | 112 | 924 |
| Hopewell city | 237 | 20,315 | 0.77 | 333 | 14.9 | 322 | 14.4 | 1,678 | 9.0 | 4,601 | 3,262 | 1,339 | 75 | 480 |
| Lexington city | 2,753 | 10,563 | 2.47 | 78 | 10.7 | 66 | 9.1 | 218 | 6.0 | 1,363 | 1,153 | 210 | 7 | 27 |
| Lynchburg city | 11,044 | 104,565 | 1.67 | 1,067 | 13.1 | 857 | 10.5 | 5,497 | 9.3 | 14,756 | 11,037 | 3,719 | 316 | 1,802 |
| Manassas city | 46 | 40,983 | 0.99 | 678 | 16.6 | 234 | 5.7 | 5,801 | 16.0 | 4,761 | 3,644 | 1,117 | 103 | 623 |
| Manassas Park city | 6 | 11,857 | 0.44 | 175 | 9.7 | 59 | 3.3 | 2,414 | 15.3 | 1,356 | 948 | 409 | 21 | 154 |

1. Per 1,000 estimated resident population.

# Table B. States and Counties — Crime, Education, Money Income, and Poverty

| STATE County | Officers | Civilians | Enrollment Total | Percent private | High school graduate or less | Bachelor's degree or more | Total current spending (mil dol) | Current spending per student (dollars) | Per capita income | Median income (dollars) | Percent with income of less than $50,000 | Percent with income of $200,000 or more | Median household income (dollars) | All persons | Children under 18 years | Children 5 to 17 years in families |
|---|---|---|---|---|---|---|---|---|---|---|---|---|---|---|---|---|
| | 46 | 47 | 48 | 49 | 50 | 51 | 52 | 53 | 54 | 55 | 56 | 57 | 58 | 59 | 60 | 61 |
| **VIRGINIA—Cont'd** | | | | | | | | | | | | | | | | |
| Louisa | 46 | 17 | 7,147 | 13.3 | 47.5 | 24.1 | 58 | 11,816 | 34,041 | 60,975 | 40.1 | 4.7 | 64,135 | 9.5 | 14.5 | 13.0 |
| Lunenburg | 13 | 6 | 2,154 | 16.5 | 62.6 | 11.3 | 16 | 10,468 | 19,848 | 44,303 | 56.3 | 0.4 | 44,860 | 19.2 | 26.7 | 25.0 |
| Madison | 23 | 6 | 2,709 | 18.0 | 50.2 | 23.4 | 21 | 12,530 | 30,777 | 57,895 | 39.1 | 4.5 | 66,397 | 9.3 | 14.6 | 14.0 |
| Mathews | 11 | 8 | 1,258 | 25.7 | 40.3 | 27.9 | 14 | 12,601 | 35,731 | 64,237 | 39.4 | 8.3 | 69,112 | 8.8 | 15.7 | 15.0 |
| Mecklenburg | 50 | 2 | 5,427 | 10.3 | 49.8 | 20.2 | 46 | 10,663 | 25,056 | 43,207 | 54.8 | 2.4 | 43,128 | 17.3 | 24.3 | 23.0 |
| Middlesex | 19 | 8 | 1,833 | 14.7 | 45.6 | 21.5 | 14 | 11,040 | 31,502 | 57,438 | 44.3 | 3.9 | 58,834 | 13.8 | 23.8 | 22.8 |
| Montgomery | 112 | 7 | 40,559 | 7.5 | 30.2 | 45.5 | 109 | 11,014 | 30,054 | 57,977 | 42.8 | 6.5 | 58,740 | 20.5 | 13.8 | 13.0 |
| Nelson | 23 | 5 | 2,791 | 14.1 | 49.5 | 30.4 | 26 | 14,047 | 35,489 | 64,313 | 40.5 | 5.4 | 55,804 | 11.8 | 18.0 | 17.4 |
| New Kent | 32 | 14 | 4,525 | 15.7 | 41.8 | 26.9 | 32 | 9,857 | 38,080 | 87,904 | 24.3 | 9.0 | 102,619 | 5.0 | 7.3 | 7.1 |
| Northampton | 68 | 17 | 2,111 | 18.4 | 52.5 | 22.6 | 20 | 11,817 | 28,518 | 47,227 | 53.0 | 4.6 | 45,235 | 18.1 | 29.6 | 27.5 |
| Northumberland | 24 | 11 | 2,000 | 9.6 | 41.6 | 28.9 | 18 | 13,492 | 36,261 | 62,632 | 40.5 | 5.9 | 60,385 | 12.7 | 26.1 | 24.0 |
| Nottoway | 14 | 11 | 2,820 | 15.9 | 58.0 | 14.8 | 23 | 11,099 | 24,571 | 45,535 | 56.9 | 1.8 | 45,913 | 21.5 | 27.0 | 24.6 |
| Orange | 40 | 3 | 8,139 | 13.1 | 46.0 | 25.1 | 54 | 10,685 | 32,292 | 71,548 | 32.7 | 4.5 | 71,691 | 9.7 | 14.0 | 13.6 |
| Page | 55 | 16 | 4,690 | 12.6 | 62.9 | 13.7 | 36 | 10,001 | 25,170 | 51,792 | 48.3 | 1.7 | 52,877 | 11.3 | 17.3 | 16.6 |
| Patrick | 54 | 17 | 3,020 | 9.0 | 54.6 | 15.2 | 27 | 10,001 | 24,292 | 43,073 | 55.8 | 2.2 | 43,568 | 15.5 | 22.8 | 21.2 |
| Pittsylvania | 111 | 18 | 12,091 | 11.9 | 52.4 | 14.5 | 89 | 9,931 | 26,032 | 47,690 | 52.9 | 2.3 | 51,682 | 15.1 | 23.8 | 20.6 |
| Powhatan | 41 | 5 | 5,969 | 19.5 | 40.1 | 29.2 | 47 | 10,882 | 37,424 | 89,090 | 23.6 | 10.1 | 94,293 | 5.4 | 6.7 | 6.2 |
| Prince Edward | 33 | 0 | 8,406 | 24.6 | 48.2 | 26.8 | 24 | 11,820 | 21,017 | 47,202 | 52.3 | 2.4 | 44,586 | 20.6 | 24.9 | 25.4 |
| Prince George | 59 | 17 | 9,508 | 13.1 | 43.2 | 23.6 | 66 | 10,439 | 29,420 | 71,912 | 32.1 | 5.0 | 74,518 | 8.3 | 10.2 | 9.7 |
| Prince William | 648 | 139 | 133,799 | 14.5 | 31.5 | 41.1 | 1,041 | 11,497 | 40,932 | 107,132 | 19.4 | 17.0 | 106,208 | 6.1 | 8.4 | 7.7 |
| Pulaski | 51 | 2 | 6,086 | 8.4 | 45.8 | 20.0 | 46 | 11,010 | 28,125 | 53,866 | 44.8 | 2.3 | 54,086 | 14.6 | 20.4 | 20.2 |
| Rappahannock | 16 | 5 | 1,309 | 24.2 | 39.2 | 35.8 | 12 | 14,063 | 41,792 | 74,284 | 31.2 | 11.5 | 79,086 | 8.5 | 13.7 | 13.1 |
| Richmond | 13 | 8 | 1,889 | 13.7 | 54.7 | 17.7 | 15 | 11,614 | 22,853 | 49,517 | 50.3 | 4.4 | 49,758 | 17.8 | 20.2 | 18.9 |
| Roanoke | 142 | 14 | 20,526 | 15.9 | 32.7 | 35.6 | 146 | 10,336 | 36,469 | 68,948 | 34.3 | 6.1 | 71,742 | 6.5 | 7.8 | 7.0 |
| Rockbridge | 37 | 4 | 4,298 | 25.3 | 46.9 | 26.7 | 33 | 11,578 | 31,539 | 54,600 | 44.4 | 4.3 | 58,501 | 11.5 | 16.6 | 14.8 |
| Rockingham | 65 | 17 | 18,717 | 24.3 | 51.2 | 26.3 | (7) | (7) | 31,275 | 61,864 | 40.3 | 4.0 | 67,242 | 8.1 | 11.5 | 11.1 |
| Russell | 36 | 16 | 5,106 | 11.3 | 57.7 | 12.2 | 40 | 10,491 | 21,605 | 39,758 | 61.9 | 1.7 | 40,345 | 20.0 | 22.7 | 20.3 |
| Scott | 30 | 7 | 3,945 | 6.6 | 56.7 | 13.3 | 36 | 10,263 | 22,898 | 39,820 | 61.8 | 1.4 | 41,288 | 14.8 | 20.4 | 19.0 |
| Shenandoah | 62 | 6 | 8,877 | 11.5 | 50.9 | 20.5 | 66 | 10,939 | 28,882 | 57,252 | 43.6 | 2.8 | 59,087 | 10.3 | 15.2 | 14.8 |
| Smyth | 44 | 7 | 5,891 | 9.7 | 54.4 | 14.6 | 47 | 10,497 | 21,605 | 40,932 | 59.6 | 1.0 | 41,964 | 16.5 | 23.0 | 21.1 |
| Southampton | 53 | 14 | 3,489 | 17.7 | 45.0 | 18.4 | 31 | 10,936 | 26,600 | 61,348 | 43.5 | 2.3 | 62,327 | 13.3 | 19.4 | 17.9 |
| Spotsylvania | 200 | 53 | 34,547 | 14.9 | 38.3 | 31.7 | 255 | 10,710 | 37,212 | 88,628 | 24.7 | 10.4 | 90,262 | 6.6 | 9.5 | 9.3 |
| Stafford | 184 | 67 | 42,030 | 12.3 | 29.6 | 39.9 | 292 | 10,093 | 42,126 | 111,108 | 18.1 | 16.9 | 109,090 | 5.4 | 6.9 | 6.2 |
| Surry | 14 | 1 | 1,155 | 20.0 | 51.3 | 20.6 | 16 | 20,191 | 30,260 | 57,962 | 42.2 | 1.8 | 59,069 | 11.9 | 20.2 | 18.6 |
| Sussex | 42 | 3 | 2,125 | 7.8 | 62.2 | 12.7 | 18 | 16,891 | 21,652 | 49,487 | 50.6 | 1.3 | 45,134 | 21.6 | 28.4 | 27.4 |
| Tazewell | 50 | 24 | 7,933 | 14.9 | 54.7 | 14.9 | 57 | 9,834 | 23,738 | 42,099 | 58.3 | 2.2 | 43,619 | 21.7 | 29.3 | 30.0 |
| Warren | 59 | 16 | 8,718 | 19.3 | 49.0 | 21.7 | 57 | 10,593 | 32,086 | 69,116 | 34.4 | 4.6 | 69,878 | 9.7 | 13.6 | 13.7 |
| Washington | 74 | 19 | 10,645 | 22.4 | 48.6 | 22.8 | 75 | 10,388 | 27,524 | 48,495 | 50.7 | 3.0 | 52,387 | 12.3 | 17.2 | 16.0 |
| Westmoreland | 27 | 16 | 3,231 | 8.0 | 49.3 | 19.2 | 29 | 12,996 | 32,996 | 53,853 | 46.0 | 4.9 | 54,885 | 15.9 | 26.6 | 27.6 |
| Wise | 53 | 16 | 8,437 | 11.1 | 53.7 | 15.0 | 54 | 9,222 | 21,500 | 38,888 | 60.2 | 1.5 | 42,372 | 20.4 | 28.7 | 26.0 |
| Wythe | 35 | 9 | 5,690 | 8.6 | 51.7 | 19.0 | 43 | 10,461 | 27,381 | 49,364 | 50.7 | 1.9 | 48,543 | 12.8 | 18.6 | 18.2 |
| York | 74 | 8 | 18,652 | 15.5 | 24.3 | 47.1 | 132 | 10,529 | 41,201 | 92,069 | 23.5 | 11.1 | 91,535 | 5.1 | 5.8 | 5.2 |
| Independent cities | | | | | | | | | | | | | | | | |
| Alexandria city | 322 | 91 | 30,504 | 28.1 | 18.6 | 63.1 | 276 | 17,584 | 62,679 | 100,939 | 21.3 | 19.1 | 102,589 | 8.6 | 14.2 | 15.5 |
| Bristol city | 52 | 21 | 3,556 | 11.1 | 45.0 | 25.4 | 26 | 11,343 | 23,015 | 37,500 | 61.4 | 1.8 | 37,678 | 21.8 | 38.3 | 30.7 |
| Buena Vista city | 16 | 1 | 1,881 | 36.7 | 59.2 | 15.9 | 10 | 10,287 | 21,419 | 32,455 | 63.7 | 1.5 | 44,752 | 17.2 | 26.0 | 25.9 |
| Charlottesville city | 110 | 26 | 16,271 | 12.5 | 25.4 | 54.9 | (8)254.0 | (8)13,762 | 38,328 | 59,471 | 43.9 | 9.9 | 58,717 | 22.1 | 21.6 | 25.3 |
| Chesapeake city | 388 | 189 | 65,001 | 17.6 | 31.9 | 33.2 | 469 | 11,544 | 35,536 | 78,640 | 29.7 | 7.8 | 77,361 | 8.3 | 11.9 | 11.1 |
| Colonial Heights city | 54 | 4 | 3,892 | 10.2 | 39.9 | 25.5 | 36 | 12,836 | 28,364 | 54,550 | 46.5 | 3.6 | 57,688 | 9.4 | 16.2 | 15.0 |
| Covington city | 17 | 9 | 1,147 | 17.7 | 57.1 | 15.6 | 11 | 10,445 | 25,309 | 40,655 | 58.8 | 2.5 | 41,842 | 14.5 | 24.3 | 21.1 |
| Danville city | 120 | 11 | 9,872 | 16.7 | 49.8 | 17.6 | 71 | 12,094 | 22,826 | 37,203 | 64.8 | 2.4 | 36,073 | 23.5 | 37.1 | 36.9 |
| Emporia city | 25 | 11 | 1,462 | 7.3 | 54.4 | 13.8 | (9)27.6 | (9)11,569 | 18,494 | 27,063 | 69.4 | 2.6 | 38,631 | 20.5 | 33.9 | 33.9 |
| Fairfax city | 60 | 19 | 6,105 | 24.9 | 19.4 | 60.8 | (10) | (10) | 50,029 | 116,979 | 20.5 | 22.5 | 106,430 | 7.1 | 6.6 | 7.5 |
| Falls Church city | 31 | 13 | 4,006 | 18.1 | 8.4 | 77.6 | 48 | 18,067 | 72,325 | 127,610 | 15.2 | 30.6 | 137,849 | 3.5 | 3.5 | 2.8 |
| Franklin city | 24 | 10 | 2,032 | 17.1 | 42.8 | 19.1 | 16 | 14,045 | 24,025 | 40,417 | 60.9 | 1.1 | 45,433 | 19.2 | 33.3 | 34.8 |
| Fredericksburg city | 73 | 23 | 8,931 | 10.1 | 33.6 | 40.9 | 49 | 13,417 | 35,731 | 65,641 | 37.4 | 8.8 | 62,121 | 16.9 | 23.8 | 25.8 |
| Galax city | 23 | 16 | 1,460 | 12.7 | 54.7 | 14.8 | 15 | 11,197 | 23,025 | 33,575 | 67.7 | 2.4 | 37,229 | 19.0 | 33.1 | 32.5 |
| Hampton city | 285 | 92 | 35,446 | 22.3 | 37.6 | 26.9 | 224 | 11,253 | 30,135 | 56,287 | 43.9 | 3.3 | 55,816 | 13.8 | 21.0 | 20.4 |
| Harrisonburg city | 109 | 18 | 23,330 | 8.0 | 41.7 | 36.3 | (7)218.2 | (7)12,049 | 23,202 | 46,679 | 54.1 | 3.0 | 48,189 | 24.9 | 19.4 | 18.9 |
| Hopewell city | 66 | 20 | 5,784 | 10.2 | 55.4 | 14.8 | 49 | 11,182 | 21,927 | 39,030 | 59.9 | 2.0 | 38,293 | 19.3 | 30.9 | 29.6 |
| Lexington city | 18 | 3 | 4,227 | 30.9 | 30.4 | 46.7 | 5 | 10,472 | 23,863 | 46,409 | 53.0 | 3.0 | 54,204 | 22.9 | 13.0 | 12.8 |
| Lynchburg city | 165 | 24 | 29,571 | 55.6 | 36.5 | 35.6 | 108 | 12,839 | 25,742 | 46,909 | 52.9 | 2.9 | 50,612 | 16.8 | 21.0 | 19.9 |
| Manassas city | 85 | 33 | 10,864 | 15.3 | 43.1 | 29.6 | 101 | 12,874 | 33,082 | 81,493 | 28.7 | 8.9 | 84,405 | 7.4 | 13.4 | 13.4 |
| Manassas Park city | 30 | 7 | 5,004 | 12.0 | 43.7 | 27.9 | 40 | 10,767 | 34,326 | 88,046 | 22.4 | 11.1 | 83,145 | 6.3 | 11.4 | 9.5 |

1. All persons 3 years old and over enrolled in nursery school through college.   2. Persons 25 years old and over.   3. Elementary and secondary education expenditures.   4. Based on population estimated by the American Community Survey, 2014–2018.   7. Harrisonburg city is included with Rockingham county.   8. Charlottesville city is included with Albemarle county.   9. Emporia city is included with Greensville county.   10. Fairfax city is included with Fairfax county.

| STATE County | Personal income, 2019 | | | | | | | | | | Earnings, 2019 | | |
| | Total (mil dol) | Percent change 2018-2019 | Per capita[1] Dollars | Per capita[1] Rank | Wages and salaries (mil dol) | Supplements to wages and salaries, employer contributions (mil dol) Pension and insurance | Supplements to wages and salaries, employer contributions (mil dol) Government social insurance | Proprietors' income (mil dol) | Dividends, interest, and rent (mil dol) | Personal transfer receipts (mil dol) | Total (mil dol) | Contributions for government social insurance (mil dol) From employee and self-employed | Contributions for government social insurance (mil dol) From employer |
|---|---|---|---|---|---|---|---|---|---|---|---|---|---|
| | 62 | 63 | 64 | 65 | 66 | 67 | 68 | 69 | 70 | 71 | 72 | 73 | 74 |
| VIRGINIA—Cont'd | | | | | | | | | | | | | |
| Louisa | 1,606 | 2.9 | 42,723 | 1,653 | 540 | 99 | 39 | 84 | 271 | 358 | 761 | 54 | 39 |
| Lunenburg | 415 | 3.0 | 33,996 | 2,803 | 107 | 22 | 7 | 6 | 88 | 136 | 143 | 13 | 7 |
| Madison | 644 | 3.7 | 48,577 | 961 | 130 | 23 | 10 | 36 | 149 | 124 | 198 | 16 | 10 |
| Mathews | 511 | 3.3 | 57,844 | 377 | 58 | 11 | 4 | 31 | 146 | 114 | 105 | 10 | 4 |
| Mecklenburg | 1,177 | 2.8 | 38,481 | 2,263 | 459 | 80 | 33 | 38 | 247 | 389 | 609 | 49 | 33 |
| Middlesex | 567 | 3.1 | 53,553 | 568 | 138 | 28 | 10 | 34 | 174 | 146 | 209 | 17 | 10 |
| Montgomery | [2] 4,526 | [2] 2.9 | [2] 38,756 | [2] 2,229 | [2] 2,508 | [2] 555 | [2] 178 | [2] 233 | [2] 1,087 | [2] 750 | [2] 3,474 | [2] 207 | [2] 178 |
| Nelson | 766 | 3.4 | 51,297 | 735 | 162 | 31 | 12 | 49 | 208 | 188 | 254 | 22 | 12 |
| New Kent | 1,737 | 5.9 | 75,221 | 88 | 229 | 39 | 17 | 62 | 230 | 206 | 346 | 26 | 17 |
| Northampton | 598 | 3.9 | 51,101 | 748 | 150 | 29 | 12 | 64 | 139 | 169 | 255 | 17 | 12 |
| Northumberland | 661 | 5.2 | 54,616 | 515 | 140 | 23 | 10 | 31 | 216 | 179 | 204 | 19 | 10 |
| Nottoway | 553 | 3.3 | 36,295 | 2,538 | 248 | 64 | 20 | 11 | 115 | 191 | 343 | 24 | 20 |
| Orange | 1,876 | 4.6 | 50,642 | 785 | 432 | 78 | 31 | 112 | 377 | 370 | 653 | 48 | 31 |
| Page | 947 | 2.5 | 39,633 | 2,098 | 222 | 44 | 18 | 54 | 178 | 259 | 337 | 28 | 18 |
| Patrick | 608 | 1.2 | 34,541 | 2,741 | 159 | 33 | 12 | 29 | 115 | 209 | 232 | 21 | 12 |
| Pittsylvania | [3] 3,955 | [3] 1.9 | [3] 39,392 | [3] 2,127 | [3] 1,596 | [3] 274 | [3] 114 | [3] 121 | [3] 692 | [3] 1,310 | [3] 2,106 | [3] 161 | [3] 114 |
| Powhatan | 1,738 | 4.1 | 58,609 | 349 | 354 | 59 | 25 | 80 | 319 | 250 | 517 | 38 | 25 |
| Prince Edward | 739 | 3.0 | 32,416 | 2,922 | 393 | 77 | 29 | 30 | 142 | 223 | 529 | 36 | 29 |
| Prince George | [4] 2,396 | [4] 3.0 | [4] 39,359 | [4] 2,134 | [4] 1,944 | [4] 516 | [4] 164 | [4] 44 | [4] 468 | [4] 577 | [4] 2,668 | [4] 140 | [4] 164 |
| Prince William | [5] 29,812 | [5] 3.9 | [5] 56,366 | [5] 429 | [5] 10,060 | [5] 1,665 | [5] 734 | [5] 1,651 | [5] 4,501 | [5] 3,088 | [5] 14,109 | [5] 829 | [5] 734 |
| Pulaski | 1,406 | 2.7 | 41,311 | 1,880 | 659 | 117 | 50 | 52 | 236 | 399 | 878 | 62 | 50 |
| Rappahannock | 504 | 2.9 | 68,382 | 144 | 66 | 11 | 5 | 45 | 180 | 72 | 126 | 10 | 5 |
| Richmond | 341 | 4.0 | 37,741 | 2,353 | 129 | 29 | 9 | 12 | 76 | 88 | 179 | 13 | 9 |
| Roanoke | [6] 6,391 | [6] 2.7 | [6] 53,489 | [6] 574 | [6] 3,015 | [6] 494 | [6] 214 | [6] 400 | [6] 1,330 | [6] 1,200 | [6] 4,124 | [6] 272 | [6] 214 |
| Rockbridge | [7] 1,525 | [7] 4.4 | [7] 41,794 | [7] 1,804 | [7] 583 | [7] 109 | [7] 44 | [7] 86 | [7] 405 | [7] 365 | [7] 822 | [7] 59 | [7] 44 |
| Rockingham | [8] 5,576 | [8] 2.6 | [8] 41,312 | [8] 1,878 | [8] 3,083 | [8] 525 | [8] 218 | [8] 570 | [8] 1,098 | [8] 998 | [8] 4,395 | [8] 268 | [8] 218 |
| Russell | 926 | 2.7 | 34,827 | 2,712 | 326 | 57 | 24 | 19 | 153 | 342 | 426 | 37 | 24 |
| Scott | 746 | 1.5 | 34,589 | 2,736 | 196 | 39 | 15 | 23 | 134 | 266 | 273 | 26 | 15 |
| Shenandoah | 2,036 | 3.8 | 46,680 | 1,162 | 619 | 109 | 45 | 118 | 398 | 442 | 891 | 65 | 45 |
| Smyth | 1,147 | 1.9 | 38,090 | 2,320 | 498 | 102 | 36 | 31 | 200 | 377 | 666 | 51 | 36 |
| Southampton | [9] 1,132 | [9] 5.1 | [9] 44,215 | [9] 1,455 | [9] 315 | [9] 67 | [9] 22 | [9] 49 | [9] 223 | [9] 308 | [9] 452 | [9] 33 | [9] 22 |
| Spotsylvania | [10] 8,619 | [10] 4.3 | [10] 52,157 | [10] 668 | [10] 3,017 | [10] 454 | [10] 213 | [10] 414 | [10] 1,492 | [10] 1,330 | [10] 4,097 | [10] 263 | [10] 213 |
| Stafford | 8,614 | 5.0 | 56,346 | 432 | 2,824 | 520 | 201 | 259 | 1,329 | 1,085 | 3,804 | 232 | 201 |
| Surry | 287 | -0.3 | 44,647 | 1,399 | 189 | 46 | 13 | 6 | 45 | 74 | 255 | 16 | 13 |
| Sussex | 361 | 3.8 | 32,383 | 2,923 | 161 | 34 | 12 | 12 | 62 | 114 | 218 | 15 | 12 |
| Tazewell | 1,594 | 1.7 | 39,276 | 2,152 | 590 | 114 | 43 | 73 | 288 | 538 | 822 | 65 | 43 |
| Warren | 1,925 | 3.7 | 47,932 | 1,024 | 589 | 97 | 43 | 81 | 315 | 340 | 810 | 56 | 43 |
| Washington | [11] 2,927 | [11] 1.5 | [11] 41,513 | [11] 1,846 | [11] 1,204 | [11] 224 | [11] 88 | [11] 152 | [11] 680 | [11] 827 | [11] 1,668 | [11] 123 | [11] 88 |
| Westmoreland | 825 | 5.2 | 45,782 | 1,275 | 129 | 27 | 9 | 51 | 202 | 207 | 216 | 18 | 9 |
| Wise | [12] 1,414 | [12] 3.1 | [12] 34,196 | [12] 2,781 | [12] 622 | [12] 131 | [12] 46 | [12] 44 | [12] 222 | [12] 559 | [12] 843 | [12] 65 | [12] 46 |
| Wythe | 1,053 | 3.4 | 36,695 | 2,484 | 455 | 88 | 33 | 44 | 178 | 335 | 620 | 46 | 33 |
| York | [13] 4,938 | [13] 3.8 | [13] 61,308 | [13] 269 | [13] 1,206 | [13] 227 | [13] 91 | [13] 149 | [13] 1,063 | [13] 673 | [13] 1,672 | [13] 110 | [13] 91 |
| Independent cities | | | | | | | | | | | | | |
| Alexandria city | 14,666 | 2.8 | 91,990 | 32 | 7,898 | 1,199 | 575 | 905 | 3,541 | 922 | 10,577 | 611 | 575 |
| Bristol city | [11] | [11] | [11] | [11] | [11] | [11] | [11] | [11] | [11] | [11] | [11] | [11] | [11] |
| Buena Vista city | [7] | [7] | [7] | [7] | [7] | [7] | [7] | [7] | [7] | [7] | [7] | [7] | [7] |
| Charlottesville city | [14] | [14] | [14] | [14] | [14] | [14] | [14] | [14] | [14] | [14] | [14] | [14] | [14] |
| Chesapeake city | 12,701 | 4.1 | 51,874 | 689 | 5,281 | 802 | 376 | 364 | 2,151 | 2,030 | 6,822 | 432 | 376 |
| Colonial Heights city | [20] | [20] | [20] | [20] | [20] | [20] | [20] | [20] | [20] | [20] | [20] | [20] | [20] |
| Covington city | [15] | [15] | [15] | [15] | [15] | [15] | [15] | [15] | [15] | [15] | [15] | [15] | [15] |
| Danville city | [3] | [3] | [3] | [3] | [3] | [3] | [3] | [3] | [3] | [3] | [3] | [3] | [3] |
| Emporia city | [16] | [16] | [16] | [16] | [16] | [16] | [16] | [16] | [16] | [16] | [16] | [16] | [16] |
| Fairfax city | [17] | [17] | [17] | [17] | [17] | [17] | [17] | [17] | [17] | [17] | [17] | [17] | [17] |
| Falls Church city | [17] | [17] | [17] | [17] | [17] | [17] | [17] | [17] | [17] | [17] | [17] | [17] | [17] |
| Franklin city | [9] | [9] | [9] | [9] | [9] | [9] | [9] | [9] | [9] | [9] | [9] | [9] | [9] |
| Fredericksburg city | [10] | [10] | [10] | [10] | [10] | [10] | [10] | [10] | [10] | [10] | [10] | [10] | [10] |
| Galax city | [18] | [18] | [18] | [18] | [18] | [18] | [18] | [18] | [18] | [18] | [18] | [18] | [18] |
| Hampton city | 6,041 | 3.5 | 44,909 | 1,372 | 3,429 | 765 | 271 | 125 | 1,198 | 1,379 | 4,589 | 270 | 271 |
| Harrisonburg city | [8] | [8] | [8] | [8] | [8] | [8] | [8] | [8] | [8] | [8] | [8] | [8] | [8] |
| Hopewell city | [4] | [4] | [4] | [4] | [4] | [4] | [4] | [4] | [4] | [4] | [4] | [4] | [4] |
| Lexington city | [7] | [7] | [7] | [7] | [7] | [7] | [7] | [7] | [7] | [7] | [7] | [7] | [7] |
| Lynchburg city | [19] | [19] | [19] | [19] | [19] | [19] | [19] | [19] | [19] | [19] | [19] | [19] | [19] |
| Manassas city | [5] | [5] | [5] | [5] | [5] | [5] | [5] | [5] | [5] | [5] | [5] | [5] | [5] |
| Manassas Park city | [5] | [5] | [5] | [5] | [5] | [5] | [5] | [5] | [5] | [5] | [5] | [5] | [5] |

1. Based on the resident population estimated as of July 1 of the year shown.  2. Radford city is included with Montgomery county.  3. Danville city is included with Pittsylvania county.  4. Hopewell city is included with Prince George county.  5. Manassas and Manassas Park cities are included with Prince William county.  6. Salem city is included with Roanoke county.  7. Buena Vista and Lexington cities are included with Rockbridge county.  8. Harrisonburg city is included with Rockingham county.  9. Franklin city is included with Southampton county.  10. Fredericksburg city is included with Spotsylvania county.  11. Bristol city is included with Washington county.  12. Norton city is included with Wise county.  13. Poquoson city is included with York county.  14. Charlottesville city is included with Albemarle county.  15. Covington city is included with Alleghany county.  16. Emporia city is included with Greensville county.  17. Fairfax city and Falls Church city are included with Fairfax county.  18. Galax city is included with Carroll county.  19. Lynchburg city is included with Campbell county.  20. Petersburg and Colonial Heights cities are included with Dinwiddie county.

# Table B. States and Counties — Earnings, Social Security, and Housing

Columns 75–83: Earnings, 2019 (cont.) — Percent by selected industries.
Columns 84–85: Social Security beneficiaries, December 2019.
Column 86: Supplemental Security Income recipients, 2019.
Columns 87–88: Housing units, 2020.

| STATE County | Farm [75] | Mining, quarrying, and extractions [76] | Construction [77] | Manufacturing [78] | Information; professional, scientific, technical services [79] | Retail trade [80] | Finance, insurance, real estate, and leasing [81] | Health care and social assistance [82] | Government [83] | SS Number [84] | SS Rate[1] [85] | Suppl. Security Income recipients, 2019 [86] | Housing Total [87] | Housing Pct. change, 2010-2020 [88] |
|---|---|---|---|---|---|---|---|---|---|---|---|---|---|---|
| **VIRGINIA—Cont'd** | | | | | | | | | | | | | | |
| Louisa | 0.0 | D | 11.3 | 12.1 | 5.1 | 5.2 | 2.4 | 3.6 | 13.4 | 8,960 | 239 | 592 | 18,124 | 11.4 |
| Lunenburg | -2.3 | 0.0 | 9.0 | 13.4 | D | 3.8 | D | 8.6 | 31.4 | 3,290 | 269 | 359 | 5,999 | 1.1 |
| Madison | -1.4 | 0.1 | 15.3 | 8.4 | D | 18.4 | 2.6 | D | 18.7 | 3,235 | 245 | 186 | 6,190 | 4.3 |
| Mathews | 1.2 | 0.1 | 11.0 | 4.5 | D | 5.6 | 7.7 | D | 23.2 | 2,980 | 339 | 140 | 5,760 | 1.6 |
| Mecklenburg | -0.5 | D | 4.5 | 8.6 | 7.9 | 9.0 | 4.2 | 18.2 | 18.1 | 10,125 | 330 | 1,128 | 19,146 | 3.0 |
| Middlesex | 0.5 | 0.0 | 11.2 | 4.4 | D | 6.2 | 6.5 | D | 28.8 | 3,840 | 364 | 268 | 7,392 | 3.6 |
| Montgomery | (2)0.1 | (2)0.2 | (2)D | (2)14.6 | (2)D | (2)5.4 | (2)3.1 | (2)D | (2)40.5 | 14,640 | 148 | 1,147 | 41,532 | 7.6 |
| Nelson | 1.3 | D | 8.7 | 10.7 | D | 3.2 | 2.7 | 6.4 | 17.9 | 4,950 | 333 | 400 | 10,277 | 3.5 |
| New Kent | -0.3 | 0.0 | 20.5 | 2.9 | D | 6.6 | 3.2 | 7.7 | 19.4 | 5,085 | 221 | 240 | 9,131 | 25.2 |
| Northampton | 18.8 | 0.0 | 5.5 | 4.5 | D | 5.0 | 4.0 | 9.7 | 23.6 | 3,995 | 339 | 579 | 7,493 | 2.6 |
| Northumberland | 1.2 | 0.1 | 9.3 | 34.2 | D | 4.9 | 3.9 | 1.7 | 15.2 | 4,720 | 391 | 246 | 9,394 | 4.4 |
| Nottoway | 2.2 | D | 3.5 | 5.7 | 2.7 | 5.6 | 2.3 | D | 51.1 | 3,910 | 257 | 535 | 6,880 | 3.4 |
| Orange | 5.1 | D | 8.1 | 16.5 | D | 7.5 | 6.0 | 3.7 | 20.5 | 9,330 | 253 | 617 | 15,612 | 6.6 |
| Page | 4.7 | 0.0 | 7.4 | 10.6 | 7.4 | 7.4 | 4.5 | 9.4 | 23.6 | 6,780 | 283 | 560 | 11,859 | 2.2 |
| Patrick | 1.1 | 0.1 | 6.3 | 21.1 | 7.0 | 8.1 | 2.1 | D | 20.4 | 5,685 | 323 | 492 | 10,264 | 1.8 |
| Pittsylvania | (3)0.3 | (3)D | (3)20.2 | (3)4.3 | (3)D | (3)8.0 | (3)3.8 | (3)16.4 | (3)18.3 | 18,040 | 298 | 1,422 | 31,654 | 1.2 |
| Powhatan | 0.1 | D | 23.4 | 3.5 | 12.0 | 6.6 | 5.7 | 4.5 | 21.6 | 6,495 | 218 | 259 | 11,502 | 14.5 |
| Prince Edward | 0.3 | D | 3.6 | 1.4 | 2.8 | 8.1 | 4.0 | D | 29.2 | 5,075 | 222 | 852 | 9,535 | 4.3 |
| Prince George | (4)-0.1 | (4)0.0 | (4)3.0 | (4)13.5 | (4)D | (4)2.2 | (4)1.3 | (4)D | (4)60.8 | 6,660 | 173 | 327 | 12,666 | 5.2 |
| Prince William | (5)0.0 | (5)0.1 | (5)D | (5)D | (5)D | (5)6.7 | (5)3.8 | (5)D | (5)27.7 | 49,860 | 106 | 4,287 | 152,979 | 11.6 |
| Pulaski | 0.1 | D | 3.2 | 43.3 | D | 6.7 | 1.8 | D | 14.7 | 9,915 | 292 | 858 | 17,312 | 0.5 |
| Rappahannock | -1.8 | 0.2 | 15.4 | 4.1 | 20.6 | 4.1 | 4.4 | 3.5 | 17.2 | 2,060 | 279 | 65 | 4,011 | 2.9 |
| Richmond | 1.7 | 0.0 | 7.6 | 5.7 | D | 4.5 | 3.3 | D | 37.2 | 2,165 | 239 | 254 | 3,999 | 3.9 |
| Roanoke | (6)0.0 | (6)D | (6)D | (6)12.7 | (6)9.3 | (6)6.1 | (6)7.5 | (6)D | (6)16.5 | 23,700 | 251 | 970 | 41,159 | 2.7 |
| Rockbridge | (7)-0.1 | (7)D | (7)D | (7)D | (7)D | (7)7.4 | (7)3.1 | (7)7.7 | (7)21.6 | 6,350 | 279 | 300 | 11,584 | 3.7 |
| Rockingham | (8)2.3 | (8)D | (8)7.6 | (8)17.7 | (8)6.5 | (8)6.1 | (8)4.9 | (8)11.4 | (8)17.6 | 18,815 | 229 | 815 | 36,376 | 8.0 |
| Russell | -0.5 | 5.1 | 7.9 | 6.4 | 11.7 | 7.1 | 3.3 | 15.2 | 20.1 | 8,780 | 328 | 1,298 | 13,517 | 0.2 |
| Scott | -1.1 | 0.1 | 3.0 | 21.1 | 3.8 | 6.8 | 2.3 | D | 26.2 | 7,280 | 337 | 1,136 | 11,946 | 0.3 |
| Shenandoah | 0.6 | D | 6.9 | 21.2 | 10.8 | 6.4 | 3.7 | 9.9 | 15.2 | 11,300 | 259 | 737 | 21,403 | 2.5 |
| Smyth | -0.6 | D | 5.2 | 29.1 | D | 5.1 | 1.6 | D | 25.3 | 9,805 | 325 | 1,157 | 15,324 | -0.7 |
| Southampton | (9)5.6 | (9)D | (9)2.5 | (9)6.9 | (9)D | (9)8.2 | (9)4.4 | (9)D | (9)29.7 | 4,565 | 257 | 364 | 7,696 | 3.0 |
| Spotsylvania | (10)0.1 | (10)D | (10)6.8 | (10)D | (10)12.2 | (10)11.1 | (10)5.7 | (10)19.1 | (10)17.2 | 22,865 | 168 | 1,486 | 50,847 | 12.5 |
| Stafford | -0.1 | D | 6.1 | 0.9 | 13.2 | 5.1 | D | 5.8 | 33.7 | 18,620 | 121 | 1,207 | 52,229 | 18.7 |
| Surry | 0.8 | 0.0 | 3.7 | 2.2 | D | D | D | 0.3 | 12.9 | 1,815 | 283 | 166 | 3,635 | 5.5 |
| Sussex | 2.2 | D | 2.4 | 2.0 | D | 6.8 | 1.3 | D | 37.6 | 2,670 | 241 | 422 | 4,861 | 3.6 |
| Tazewell | 0.2 | 6.3 | 5.2 | 10.2 | 3.6 | 11.1 | 4.1 | D | 20.4 | 13,410 | 329 | 1,804 | 20,669 | -0.7 |
| Warren | -0.2 | 0.0 | 9.0 | 10.3 | 4.7 | 6.8 | 3.8 | 11.3 | 17.3 | 8,135 | 202 | 649 | 16,718 | 4.6 |
| Washington | (11)-0.3 | (11)D | (11)D | (11)17.2 | (11)5.0 | (11)9.7 | (11)5.4 | (11)12.0 | (11)18.0 | 16,515 | 307 | 1,452 | 26,035 | 1.6 |
| Westmoreland | 13.3 | 0.0 | 7.1 | 12.8 | D | 5.8 | 3.3 | D | 24.2 | 5,330 | 296 | 445 | 11,134 | 4.8 |
| Wise | (12)-0.1 | (12)4.0 | (12)2.3 | (12)3.0 | (12)5.8 | (12)9.6 | (12)2.8 | (12)17.5 | (12)30.1 | 12,000 | 321 | 2,260 | 17,900 | -0.1 |
| Wythe | -0.9 | 1.3 | 5.5 | 24.6 | D | 10.2 | 4.0 | 11.2 | 21.7 | 8,610 | 301 | 833 | 14,449 | 2.6 |
| York | (13)0.1 | (13)0.0 | (13)8.2 | (13)1.7 | (13)D | (13)8.3 | (13)4.1 | (13)D | (13)33.5 | 12,505 | 182 | 344 | 28,737 | 7.7 |
| **Independent cities** | | | | | | | | | | | | | | |
| Alexandria city | 0.0 | 0.0 | 2.2 | D | 26.4 | 4.1 | 5.4 | 5.1 | 30.7 | 15,895 | 100 | 1,832 | 76,267 | 5.4 |
| Bristol city | (11) | (11) | (11) | (11) | (11) | (11) | (11) | (11) | (11) | 5,105 | 300 | 817 | 8,683 | -1.2 |
| Buena Vista city | (7) | (7) | (7) | (7) | (7) | (7) | (7) | (7) | (7) | 1,640 | 253 | 264 | 2,852 | -2.4 |
| Charlottesville city | (14) | (14) | (14) | (14) | (14) | (14) | (14) | (14) | (14) | 6,440 | 136 | 995 | 20,913 | 9.0 |
| Chesapeake city | 0.1 | 0.1 | 10.6 | 6.2 | 13.1 | 8.9 | 7.0 | 8.6 | 17.3 | 41,305 | 168 | 3,685 | 93,413 | 12.2 |
| Colonial Heights city | (20) | (20) | (20) | (20) | (20) | (20) | (20) | (20) | (20) | 4,465 | 261 | 520 | 7,684 | -1.9 |
| Covington city | (15) | (15) | (15) | (15) | (15) | (15) | (15) | (15) | (15) | 1,920 | 342 | 436 | 2,968 | -3.1 |
| Danville city | (3) | (3) | (3) | (3) | (3) | (3) | (3) | (3) | (3) | 12,690 | 317 | 2,967 | 21,859 | -2.6 |
| Emporia city | (16) | (16) | (16) | (16) | (16) | (16) | (16) | (16) | (16) | 1,480 | 274 | 578 | 2,596 | 1.1 |
| Fairfax city | (17) | (17) | (17) | (17) | (17) | (17) | (17) | (17) | (17) | 3,565 | 153 | 20 | 9,066 | 4.5 |
| Falls Church city | (17) | (17) | (17) | (17) | (17) | (17) | (17) | (17) | (17) | 1,805 | 124 | 100 | 6,366 | 16.4 |
| Franklin city | (9) | (9) | (9) | (9) | (9) | (9) | (9) | (9) | (9) | 2,275 | 289 | 624 | 3,849 | -1.3 |
| Fredericksburg city | (10) | (10) | (10) | (10) | (10) | (10) | (10) | (10) | (10) | 4,270 | 146 | 600 | 12,344 | 18.3 |
| Galax city | (18) | (18) | (18) | (18) | (18) | (18) | (18) | (18) | (18) | 2,160 | 342 | 581 | 3,168 | -1.9 |
| Hampton city | 0.0 | 0.0 | 3.8 | 3.6 | 10.5 | 4.5 | 2.1 | 10.4 | 48.3 | 27,595 | 204 | 3,440 | 60,150 | 1.0 |
| Harrisonburg city | (8) | (8) | (8) | (8) | (8) | (8) | (8) | (8) | (8) | 5,950 | 111 | 1,060 | 18,391 | 5.5 |
| Hopewell city | (4) | (4) | (4) | (4) | (4) | (4) | (4) | (4) | (4) | 5,060 | 225 | 1,086 | 10,426 | 3.0 |
| Lexington city | (7) | (7) | (7) | (7) | (7) | (7) | (7) | (7) | (7) | 1,700 | 235 | 242 | 2,544 | 0.0 |
| Lynchburg city | (19) | (19) | (19) | (19) | (19) | (19) | (19) | (19) | (19) | 16,155 | 198 | 2,665 | 32,765 | 2.4 |
| Manassas city | (5) | (5) | (5) | (5) | (5) | (5) | (5) | (5) | (5) | 4,820 | 118 | 511 | 14,015 | 6.9 |
| Manassas Park city | (5) | (5) | (5) | (5) | (5) | (5) | (5) | (5) | (5) | 1,420 | 79 | 0 | 4,856 | -0.9 |

1. Based on the resident population estimated as of July 1 of the year shown.   2. Radford city is included with Montgomery county.   3. Danville city is included with Pittsylvania county.   4. Hopewell city is included with Prince George county.   5. Manassas and Manassas Park cities are included with Prince William county.   6. Salem city is included with Roanoke county.   7. Buena Vista and Lexington cities are included with Rockbridge county.   8. Harrisonburg city is included with Rockingham county.   9. Franklin city is included with Southampton county.   10. Fredericksburg city is included with Spotsylvania county.   11. Bristol city is included with Washington county.   12. Norton city is included with Wise county.   13. Poquoson city is included with York county.   14. Charlottesville city is included with Albemarle county.   15. Covington city is included with Alleghany county.   16. Emporia city is included with Greensville county.   17. Fairfax city and Falls Church city are included with Fairfax county.   18. Galax city is included with Carroll county.   19. Lynchburg city is included with Campbell county.   20. Petersburg and Colonial Heights cities are included with Dinwiddie county.

# Table B. States and Counties — Housing, Labor Force, and Employment

| STATE County | Housing units, 2015-2019 | | | | | | | | Civilian labor force, 2020 | | | | Civilian employment[6], 2015-2019 | | |
| | Occupied units | | | | | | | | | | Unemployment | | Percent | | |
| | | Owner-occupied | | | | Renter-occupied | | | | | | | | | |
| | | | | Median owner cost as a percent of income | | | Median rent as a percent of income[2] | Sub-standard units[4] (percent) | | Percent change, 2019-2020 | | | | Management, business, science, and arts | Construction, production, and maintenance occupations |
| | Total | Percent | Median value[1] | With a mortgage | Without a mortgage[2] | Median rent[3] | | | Total | | Total | Rate[5] | Total | | |
| | 89 | 90 | 91 | 92 | 93 | 94 | 95 | 96 | 97 | 98 | 99 | 100 | 101 | 102 | 103 |
|---|---|---|---|---|---|---|---|---|---|---|---|---|---|---|---|
| **VIRGINIA—Cont'd** | | | | | | | | | | | | | | | |
| Louisa | 13,871 | 80 | 223,100 | 21.5 | 10.7 | 937.0 | 29.3 | 2.7 | 19,944 | -1.4 | 1,035 | 5.2 | 16,553 | 34.5 | 24.9 |
| Lunenburg | 4,293 | 71.8 | 119,500 | 23.8 | 10.0 | 683.0 | 31.7 | 1.7 | 5,266 | -1.6 | 278 | 5.3 | 4,632 | 20.0 | 41.0 |
| Madison | 4,949 | 74.5 | 257,900 | 24.5 | 11.7 | 687.0 | 31.1 | 3.7 | 7,306 | -2.4 | 280 | 3.8 | 6,204 | 32.2 | 23.5 |
| Mathews | 3,920 | 85 | 226,300 | 18.8 | 12.8 | 922.0 | 31.2 | 1.8 | 4,052 | -3.4 | 200 | 4.9 | 3,581 | 38.0 | 35.6 |
| Mecklenburg | 12,075 | 71.1 | 132,600 | 21.1 | 10.1 | 695.0 | 27.0 | 2.4 | 12,679 | -0.5 | 861 | 6.8 | 12,145 | 33.0 | 28.4 |
| Middlesex | 4,603 | 78.7 | 261,900 | 25.2 | 11.6 | 844.0 | 25.6 | 2.6 | 5,134 | -3.0 | 262 | 5.1 | 4,394 | 27.3 | 25.6 |
| Montgomery | 35,685 | 55.4 | 227,100 | 18.4 | 10.0 | 992.0 | 30.4 | 1.8 | 48,670 | -4.1 | 2,330 | 4.8 | 45,828 | 48.1 | 16.2 |
| Nelson | 6,419 | 76.4 | 235,000 | 19.7 | 10.0 | 759.0 | 26.4 | 1.7 | 7,304 | -3.8 | 409 | 5.6 | 6,898 | 36.7 | 25.7 |
| New Kent | 7,961 | 86.5 | 281,100 | 19.3 | 10.0 | 1,010 | 31.4 | 2.2 | 12,503 | -3.2 | 573 | 4.6 | 10,933 | 38.6 | 25.7 |
| Northampton | 5,148 | 65.2 | 176,800 | 20.6 | 11.7 | 733.0 | 25.4 | 1.3 | 5,409 | -1.2 | 421 | 7.8 | 4,891 | 31.0 | 24.5 |
| Northumberland | 5,584 | 89.2 | 292,600 | 23.8 | 10.5 | 858.0 | 31.4 | 2.0 | 5,686 | -0.2 | 360 | 6.3 | 4,781 | 33.7 | 22.3 |
| Nottoway | 5,446 | 64.7 | 151,400 | 20.1 | 13.0 | 770.0 | 28.5 | 2.1 | 7,150 | -2.4 | 366 | 5.1 | 5,783 | 25.3 | 31.3 |
| Orange | 13,679 | 78.8 | 244,400 | 22.5 | 10.7 | 987.0 | 29.1 | 1.6 | 17,061 | -2.4 | 959 | 5.6 | 17,007 | 32.1 | 26.4 |
| Page | 9,286 | 71.1 | 169,200 | 23.2 | 10.2 | 790.0 | 28.7 | 2.7 | 12,021 | -0.7 | 874 | 7.3 | 10,882 | 22.3 | 30.3 |
| Patrick | 7,755 | 79 | 120,000 | 18.9 | 10.0 | 573.0 | 28.7 | 2.6 | 7,235 | -2.4 | 496 | 6.9 | 7,438 | 25.0 | 38.6 |
| Pittsylvania | 26,267 | 75.4 | 123,900 | 18.9 | 10.4 | 725.0 | 25.8 | 1.8 | 29,943 | | 1,933 | 6.5 | 27,414 | 26.3 | 34.7 |
| Powhatan | 10,265 | 90.1 | 279,200 | 19.8 | 10.0 | 980.0 | 20.4 | 1.0 | 13,976 | -3.1 | 677 | 4.8 | 13,311 | 45.2 | 21.6 |
| Prince Edward | 7,185 | 64.4 | 159,700 | 21.7 | 10.0 | 765.0 | 28.7 | 0.8 | 10,067 | -4.0 | 652 | 6.5 | 8,687 | 33.6 | 17.2 |
| Prince George | 11,507 | 67.6 | 213,300 | 20.9 | 10.0 | 1,338 | 28.9 | 1.4 | 14,897 | -2.3 | 969 | 6.5 | 15,152 | 35.5 | 24.0 |
| Prince William | 142,000 | 73.3 | 382,400 | 22.2 | 10.0 | 1,713 | 30.6 | 3.5 | 245,929 | -1.2 | 16,028 | 6.5 | 237,220 | 44.0 | 17.7 |
| Pulaski | 14,533 | 71.4 | 149,500 | 19.7 | 10.2 | 711.0 | 22.2 | 0.6 | 16,092 | -1.4 | 1,268 | 7.9 | 15,539 | 32.3 | 30.5 |
| Rappahannock | 2,913 | 74.2 | 377,700 | 22.2 | 10.2 | 1,055 | 30.7 | 2.2 | 3,688 | -2.6 | 167 | 4.5 | 3,602 | 39.8 | 24.5 |
| Richmond | 3,069 | 69.9 | 180,300 | 24.4 | 10.0 | 797.0 | 28.9 | 1.5 | 4,048 | -1.1 | 197 | 4.9 | 3,285 | 34.8 | 31.8 |
| Roanoke | 38,222 | 74.8 | 199,800 | 18.9 | 10.0 | 956.0 | 25.9 | 0.9 | 48,720 | -2.6 | 2,489 | 5.1 | 45,744 | 43.5 | 19.2 |
| Rockbridge | 9,248 | 75.2 | 199,600 | 22.8 | 10.0 | 800.0 | 24.7 | 0.9 | 10,467 | -3.7 | 575 | 5.5 | 9,926 | 35.8 | 27.1 |
| Rockingham | 30,994 | 74.5 | 211,500 | 20.5 | 10.0 | 907.0 | 26.4 | 2.8 | 41,525 | -3.0 | 1,952 | 4.7 | 39,423 | 33.1 | 29.4 |
| Russell | 10,812 | 76.1 | 98,000 | 19.9 | 10.0 | 582.0 | 26.2 | 1.3 | 11,210 | -0.8 | 742 | 6.6 | 9,296 | 30.5 | 30.4 |
| Scott | 8,773 | 77.4 | 94,800 | 19.1 | 10.8 | 563.0 | 27.6 | 2.5 | 8,958 | -2.2 | 474 | 5.3 | 8,342 | 25.9 | 34.2 |
| Shenandoah | 17,402 | 70.9 | 213,200 | 22.3 | 10.0 | 900.0 | 29.1 | 2.3 | 22,129 | -1.3 | 1,166 | 5.3 | 20,580 | 28.4 | 31.1 |
| Smyth | 12,686 | 71.5 | 95,800 | 18.7 | 11.0 | 610.0 | 28.7 | 1.8 | 13,384 | -2.9 | 861 | 6.4 | 12,487 | 26.9 | 34.1 |
| Southampton | 6,562 | 74.8 | 178,500 | 20.9 | 10.0 | 829.0 | 28.9 | 2.0 | 9,063 | -1.3 | 449 | 5.0 | 7,617 | 32.0 | 30.9 |
| Spotsylvania | 44,436 | 78.1 | 284,000 | 21.1 | 10.0 | 1,477 | 29.7 | 2.4 | 67,201 | -2.1 | 4,035 | 6.0 | 66,122 | 40.7 | 19.1 |
| Stafford | 47,075 | 77.4 | 346,100 | 19.7 | 10.0 | 1,546 | 29.0 | 1.3 | 72,341 | -2.4 | 4,099 | 5.7 | 72,033 | 47.6 | 16.2 |
| Surry | 2,730 | 74.3 | 197,800 | 21.1 | 10.7 | 903.0 | 25.3 | 2.0 | 3,603 | -2.4 | 207 | 5.7 | 3,151 | 31.4 | 33.2 |
| Sussex | 3,794 | 69.2 | 125,800 | 20.0 | 12.6 | 807.0 | 26.6 | 1.2 | 3,771 | -0.6 | 321 | 8.5 | 4,151 | 27.0 | 37.9 |
| Tazewell | 16,531 | 74.5 | 99,700 | 19.3 | 10.0 | 656.0 | 23.9 | 1.9 | 15,520 | -1.5 | 1,185 | 7.6 | 15,160 | 30.6 | 29.2 |
| Warren | 14,651 | 76.7 | 244,800 | 20.9 | 10.1 | 1,020 | 28.3 | 1.9 | 20,258 | -2.1 | 1,181 | 5.8 | 19,513 | 32.8 | 26.5 |
| Washington | 22,321 | 75.2 | 151,800 | 20.7 | 10.0 | 716.0 | 26.7 | 1.0 | 25,953 | -2.0 | 1,496 | 5.8 | 23,678 | 34.4 | 27.0 |
| Westmoreland | 7,923 | 73.2 | 199,100 | 21.9 | 12.8 | 948.0 | 35.7 | 2.1 | 9,473 | -1.1 | 558 | 5.9 | 7,677 | 32.4 | 25.4 |
| Wise | 14,925 | 66.6 | 85,500 | 18.9 | 11.2 | 632.0 | 30.8 | 1.6 | 13,076 | -2.6 | 996 | 7.6 | 12,942 | 35.1 | 22.3 |
| Wythe | 11,983 | 78 | 128,400 | 19.0 | 10.0 | 676.0 | 21.8 | 1.5 | 13,827 | 0.1 | 1,067 | 7.7 | 13,340 | 28.9 | 28.9 |
| York | 25,103 | 71.3 | 327,100 | 20.5 | 10.0 | 1,480 | 29.3 | 2.1 | 32,390 | -2.8 | 1,803 | 5.6 | 30,556 | 51.9 | 14.6 |
| **Independent cities** | | | | | | | | | | | | | | | |
| Alexandria city | 70,598 | 43.3 | 572,900 | 21.3 | 11.3 | 1,747 | 27.2 | 4.5 | 100,168 | -1.3 | 6,011 | 6.0 | 96,960 | 59.5 | 11.8 |
| Bristol city | 7,334 | 60.4 | 111,700 | 20.7 | 12.7 | 687.0 | 30.2 | 3.1 | 7,316 | -0.3 | 566 | 7.7 | 6,966 | 35.6 | 21.6 |
| Buena Vista city | 2,517 | 68 | 126,600 | 26.1 | 14.3 | 795.0 | 32.7 | 1.0 | 3,114 | -10.7 | 179 | 5.7 | 2,819 | 21.9 | 32.6 |
| Charlottesville city | 18,617 | 42.8 | 299,600 | 19.4 | 10.0 | 1,142 | 31.6 | 1.8 | 25,424 | -3.4 | 1,639 | 6.4 | 24,865 | 53.5 | 10.7 |
| Chesapeake city | 84,849 | 71.4 | 273,700 | 23.0 | 10.6 | 1,279 | 32.3 | 1.7 | 122,036 | -2.4 | 7,480 | 6.1 | 111,227 | 41.8 | 19.9 |
| Colonial Heights city | 7,036 | 62.9 | 171,700 | 21.6 | 10.0 | 1,038 | 34.6 | 1.9 | 8,480 | -1.6 | 603 | 7.1 | 7,768 | 35.5 | 21.4 |
| Covington city | 2,384 | 74.5 | 72,900 | 17.8 | 11.3 | 619.0 | 26.8 | 2.3 | 2,412 | -0.4 | 231 | 9.6 | 2,554 | 21.8 | 28.7 |
| Danville city | 18,293 | 51.5 | 90,500 | 21.6 | 11.9 | 682.0 | 29.1 | 2.4 | 19,250 | 0.1 | 1,871 | 9.7 | 16,463 | 26.2 | 27.3 |
| Emporia city | 2,086 | 40.1 | 116,800 | 22.4 | 16.0 | 694.0 | 34.9 | 2.3 | 2,344 | 1.2 | 261 | 11.1 | 1,895 | 32.2 | 32.9 |
| Fairfax city | 8,577 | 70 | 560,400 | 22.3 | 10.0 | 1,834 | 29.4 | 2.0 | 13,038 | -1.7 | 734 | 5.6 | 12,705 | 58.1 | 9.1 |
| Falls Church city | 5,493 | 58.3 | 789,300 | 22.1 | 11.3 | 1,867 | 26.1 | 3.8 | 8,284 | -3.5 | 314 | 3.8 | 7,883 | 70.0 | 3.6 |
| Franklin city | 3,522 | 51.6 | 164,100 | 21.1 | 16.1 | 860.0 | 32.8 | 1.1 | 3,640 | 0.9 | 308 | 8.5 | 3,460 | 23.2 | 28.5 |
| Fredericksburg city | 10,762 | 35.5 | 367,200 | 19.4 | 10.0 | 1,258 | 29.5 | 3.7 | 14,020 | -1.4 | 1,003 | 7.2 | 15,717 | 47.5 | 11.9 |
| Galax city | 2,663 | 68.1 | 96,600 | 24.0 | 12.2 | 574.0 | 37.3 | 2.3 | 2,881 | -1.8 | 223 | 7.7 | 2,631 | 33.4 | 30.7 |
| Hampton city | 54,050 | 55.7 | 186,700 | 23.3 | 11.4 | 1,118 | 31.0 | 2.0 | 64,604 | -0.7 | 5,514 | 8.5 | 61,782 | 34.9 | 24.7 |
| Harrisonburg city | 16,723 | 39.6 | 203,600 | 20.2 | 10.0 | 883.0 | 29.6 | 4.1 | 24,259 | -2.4 | 1,547 | 6.4 | 25,772 | 34.3 | 24.1 |
| Hopewell city | 9,224 | 46.7 | 122,900 | 19.6 | 14.5 | 886.0 | 28.4 | 3.3 | 9,811 | 1.3 | 1,053 | 10.7 | 9,515 | 26.5 | 32.0 |
| Lexington city | 2,158 | 52.4 | 235,700 | 26.0 | 10.0 | 788.0 | 29.4 | 0.0 | 2,049 | -2.5 | 143 | 7.0 | 2,520 | 44.3 | 12.5 |
| Lynchburg city | 28,273 | 48.7 | 160,100 | 19.2 | 10.0 | 857.0 | 30.8 | 2.7 | 36,015 | -1.9 | 2,610 | 7.2 | 36,762 | 40.0 | 17.8 |
| Manassas city | 12,898 | 66.1 | 324,600 | 24.0 | 10.0 | 1,528 | 34.2 | 4.7 | 21,684 | -1.2 | 1,406 | 6.5 | 21,799 | 35.3 | 25.6 |
| Manassas Park city | 4,641 | 65.9 | 308,000 | 24.8 | 10.0 | 1,732 | 30.0 | 5.5 | 9,483 | -0.7 | 648 | 6.8 | 9,340 | 36.3 | 29.4 |

1. Specified owner-occupied units.  2. A value of 10.0 represents 10 percent or less; a value of 50.0 represents 50 percent or more.  3. Specified renter-occupied units.  4. Overcrowded or
lacking complete plumbing facilities.  5. Percent of civilian labor force.  6. Civilian employed persons 16 years old and over.

# Table B. States and Counties — Nonfarm Employment and Agriculture

| STATE County | Number of establish-ments [104] | Employment — Total [105] | Health care and social assistance [106] | Manufac-turing [107] | Retail trade [108] | Finance and insurance [109] | Professional, scientific, and technical services [110] | Annual payroll Total (mil dol) [111] | Average per employee (dollars) [112] | Farms Number [113] | Percent with: Fewer than 50 acres [114] | 1000 acres or more [115] | Farm producers whose primary occupation is farming (percent) [116] |
|---|---|---|---|---|---|---|---|---|---|---|---|---|---|
| **VIRGINIA—Cont'd** | | | | | | | | | | | | | |
| Louisa | 621 | 7,582 | 340 | 1,338 | 1,273 | 112 | 168 | 372 | 49,038 | 431 | 38.7 | 1.6 | 38.8 |
| Lunenburg | 159 | 1,747 | 249 | 577 | 289 | 57 | 34 | 56 | 32,239 | 335 | 29.3 | 2.7 | 43.9 |
| Madison | 280 | 2,507 | 228 | 281 | 847 | 23 | 78 | 84 | 33,600 | 533 | 33.4 | 3.6 | 45.6 |
| Mathews | 182 | 1,144 | 213 | 48 | 209 | 14 | 38 | 28 | 24,261 | 43 | 65.1 | 2.3 | 31.8 |
| Mecklenburg | 796 | 9,176 | 1,841 | 1,059 | 1,936 | 244 | 287 | 316 | 34,486 | 512 | 25.6 | 4.9 | 42.7 |
| Middlesex | 312 | 2,190 | 367 | 126 | 421 | 59 | 86 | 71 | 32,326 | 79 | 50.6 | 6.3 | 43.5 |
| Montgomery | 1,987 | 30,444 | 4,436 | 5,529 | 4,944 | 787 | 2,068 | 1,258 | 41,318 | 584 | 41.3 | 2.4 | 39.3 |
| Nelson | 352 | 3,529 | 213 | 654 | 342 | 35 | 200 | 103 | 29,265 | 409 | 34.5 | 1.2 | 41.7 |
| New Kent | 402 | 3,552 | 597 | 186 | 623 | 41 | 133 | 152 | 42,731 | 138 | 64.5 | 4.3 | 39.5 |
| Northampton | 315 | 2,995 | 721 | 301 | 562 | 68 | 76 | 111 | 37,153 | 142 | 47.9 | 10.6 | 43.3 |
| Northumberland | 301 | 1,707 | 147 | 326 | 334 | 52 | 71 | 66 | 38,595 | 134 | 58.2 | 11.9 | 46.3 |
| Nottoway | 312 | 3,355 | 1,012 | 459 | 665 | 99 | 98 | 116 | 34,710 | 311 | 34.1 | 2.9 | 48.2 |
| Orange | 634 | 6,930 | 391 | 1,514 | 1,396 | 139 | 221 | 288 | 41,573 | 417 | 34.1 | 5.3 | 40.6 |
| Page | 422 | 4,315 | 552 | 518 | 810 | 121 | 407 | 140 | 32,330 | 519 | 47.0 | 1.9 | 46.3 |
| Patrick | 283 | 4,445 | 498 | 1,243 | 469 | 82 | 65 | 118 | 26,469 | 483 | 38.5 | 1.9 | 41.2 |
| Pittsylvania | 838 | 8,454 | 1,063 | 1,647 | 1,004 | 151 | 178 | 310 | 36,654 | 1,157 | 30.2 | 2.9 | 41.5 |
| Powhatan | 705 | 5,599 | 232 | 131 | 1,010 | 121 | 400 | 224 | 40,036 | 263 | 52.5 | 1.5 | 47.6 |
| Prince Edward | 518 | 7,367 | 2,009 | 101 | 1,364 | 148 | 208 | 249 | 33,833 | 341 | 27.9 | 2.3 | 37.7 |
| Prince George | 476 | 8,308 | 462 | 1,479 | 1,063 | 143 | 1,038 | 351 | 42,278 | 164 | 37.8 | 7.3 | 43.6 |
| Prince William | 8,536 | 110,456 | 12,102 | 3,913 | 23,386 | 2,150 | 11,856 | 5,295 | 47,941 | 304 | 75.7 | 1.3 | 38.0 |
| Pulaski | 613 | 11,654 | 1,301 | 5,357 | 1,749 | 150 | 140 | 477 | 40,965 | 394 | 40.4 | 3.0 | 31.9 |
| Rappahannock | 202 | 1,098 | 66 | 72 | 130 | 12 | 66 | 42 | 37,870 | 439 | 49.2 | 2.5 | 37.5 |
| Richmond | 168 | 2,019 | 685 | 161 | 251 | 42 | 28 | 63 | 31,049 | 98 | 38.8 | 9.2 | 64.3 |
| Roanoke | 1,987 | 29,390 | 4,350 | 2,858 | 4,049 | 4,095 | 1,368 | 1,236 | 42,069 | 262 | 52.3 | 0.4 | 43.1 |
| Rockbridge | 437 | 4,994 | 417 | 804 | 1,420 | 92 | 109 | 153 | 30,606 | 752 | 36.6 | 2.4 | 38.2 |
| Rockingham | 1,530 | 28,951 | 3,993 | 7,672 | 1,898 | 446 | 593 | 1,313 | 45,346 | 2,026 | 46.2 | 0.6 | 49.3 |
| Russell | 464 | 6,136 | 1,154 | 699 | 958 | 210 | 250 | 760 | 40,716 | 918 | 36.2 | 2.7 | 40.7 |
| Scott | 257 | 3,784 | 694 | 1,000 | 610 | 85 | 129 | 127 | 33,594 | 1,138 | 41.4 | 0.5 | 34.9 |
| Shenandoah | 876 | 12,650 | 1,475 | 3,421 | 1,676 | 249 | 508 | 466 | 36,846 | 965 | 49.4 | 1.6 | 42.5 |
| Smyth | 489 | 9,280 | 1,919 | 3,271 | 1,267 | 181 | 139 | 329 | 35,452 | 663 | 41.0 | 2.9 | 37.3 |
| Southampton | 216 | 2,140 | 169 | 575 | 306 | 16 | 50 | 91 | 42,452 | 257 | 23.0 | 20.2 | 51.4 |
| Spotsylvania | 2,539 | 33,029 | 4,444 | 1,105 | 7,880 | 545 | 2,430 | 1,291 | 39,102 | 338 | 57.4 | 2.7 | 40.7 |
| Stafford | 2,374 | 33,663 | 3,492 | 661 | 4,695 | 4,764 | 4,510 | 1,479 | 43,948 | 243 | 75.5 | 0.8 | 36.6 |
| Surry | 81 | 1,308 | 11 | 116 | 40 | NA | 31 | 128 | 97,714 | 111 | 40.5 | 7.2 | 40.0 |
| Sussex | 188 | 1,881 | 350 | 206 | 283 | 27 | 20 | 65 | 34,703 | 124 | 28.2 | 21.0 | 48.5 |
| Tazewell | 966 | 11,888 | 2,392 | 1,184 | 2,848 | 592 | 360 | 414 | 34,785 | 512 | 37.1 | 8.0 | 35.1 |
| Warren | 833 | 11,341 | 1,550 | 1,183 | 1,803 | 220 | 321 | 447 | 39,415 | 321 | 62.6 | 1.2 | 33.8 |
| Washington | 1,149 | 17,299 | 2,846 | 2,861 | 3,241 | 466 | 404 | 664 | 38,404 | 1,506 | 49.7 | 1.1 | 37.8 |
| Westmoreland | 315 | 2,457 | 208 | 607 | 498 | 78 | 64 | 70 | 28,544 | 183 | 34.4 | 7.7 | 50.5 |
| Wise | 642 | 7,897 | 1,731 | 199 | 1,786 | 208 | 300 | 250 | 31,605 | 147 | 46.3 | 5.4 | 38.1 |
| Wythe | 680 | 9,979 | 1,313 | 2,209 | 1,900 | 245 | 185 | 350 | 35,062 | 819 | 30.9 | 2.0 | 43.4 |
| York | 1,514 | 20,740 | 2,379 | 350 | 3,831 | 479 | 2,764 | 758 | 36,563 | 40 | 85.0 | NA | 23.1 |
| **Independent cities** | | | | | | | | | | | | | |
| Alexandria city | 4,627 | 83,150 | 7,847 | 1,635 | 7,541 | 3,125 | 17,796 | 5,440 | 65,420 | NA | NA | NA | NA |
| Bristol city | 550 | 8,415 | 911 | 1,068 | 1,782 | 319 | 169 | 272 | 32,267 | NA | NA | NA | NA |
| Buena Vista city | 110 | 1,967 | 190 | 721 | 177 | 24 | 25 | 68 | 34,632 | NA | NA | NA | NA |
| Charlottesville city | 2,035 | 35,551 | 9,801 | 752 | 3,364 | 900 | 2,737 | 2,268 | 63,805 | NA | NA | NA | NA |
| Chesapeake city | 5,578 | 94,522 | 9,991 | 4,740 | 15,252 | 4,449 | 9,016 | 4,258 | 45,050 | 248 | 76.2 | 5.2 | 33.8 |
| Colonial Heights city | 652 | 9,805 | 1,716 | 198 | 3,279 | 255 | 300 | 263 | 26,847 | NA | NA | NA | NA |
| Covington city | 205 | 3,690 | 171 | 1,602 | 572 | 104 | 37 | 199 | 53,871 | NA | NA | NA | NA |
| Danville city | 1,254 | 24,443 | 5,479 | 4,856 | 4,323 | 803 | 426 | 845 | 34,573 | NA | NA | NA | NA |
| Emporia city | 231 | 4,083 | 1,153 | 979 | 646 | 61 | 67 | 149 | 36,514 | NA | NA | NA | NA |
| Fairfax city | 1,996 | 26,823 | 4,835 | 126 | 4,474 | 1,350 | 5,417 | 1,467 | 54,696 | NA | NA | NA | NA |
| Falls Church city | 893 | 10,555 | 2,273 | 111 | 1,346 | 191 | 2,096 | 578 | 54,764 | NA | NA | NA | NA |
| Franklin city | 260 | 3,516 | 1,307 | 23 | 1,037 | 174 | 89 | 98 | 27,883 | NA | NA | NA | NA |
| Fredericksburg city | 1,405 | 21,243 | 5,132 | 442 | 4,254 | 526 | 1,393 | 944 | 44,435 | NA | NA | NA | NA |
| Galax city | 308 | 5,789 | 1,557 | 1,270 | 979 | 121 | 110 | 181 | 31,198 | NA | NA | NA | NA |
| Hampton city | 2,349 | 42,945 | 8,071 | 2,099 | 6,619 | 885 | 5,245 | 1,849 | 43,056 | NA | NA | NA | NA |
| Harrisonburg city | 1,641 | 25,946 | 3,125 | 3,017 | 5,281 | 894 | 1,078 | 883 | 34,014 | NA | NA | NA | NA |
| Hopewell city | 411 | 5,999 | 1,109 | 1,433 | 667 | 83 | 224 | 319 | 53,169 | NA | NA | NA | NA |
| Lexington city | 261 | 3,914 | 548 | 16 | 396 | 122 | 127 | 162 | 42,545 | NA | NA | NA | NA |
| Lynchburg city | 2,239 | 57,404 | 10,783 | 6,230 | 6,718 | 2,396 | 3,354 | 2,442 | 42,545 | NA | NA | NA | NA |
| Manassas city | 1,522 | 16,954 | 3,330 | 1,210 | 2,124 | 496 | 1,468 | 891 | 52,555 | NA | NA | NA | NA |
| Manassas Park city | 330 | 3,155 | 153 | 179 | 172 | 20 | 182 | 160 | 50,744 | NA | NA | NA | NA |

# Table B. States and Counties — **Agriculture**

| STATE County | Land in farms Acreage (1,000) | Percent change, 2012-2017 | Acres Average size of farm | Total irrigated (1,000) | Total cropland (1,000) | Value of land and buildings (dollars) Average per farm | Average per acre | Value of machinery and equiopmnet, average per farm (dollars) | Value of products sold: Total (mil dol) | Average per farm (acres) | Percent from: Crops | Livestock and poultry products | Organic farms (number) | Farms with internet access (percent) | Government payments Total ($1,000) | Percent of farms |
|---|---|---|---|---|---|---|---|---|---|---|---|---|---|---|---|---|
| | 117 | 118 | 119 | 120 | 121 | 122 | 123 | 124 | 125 | 126 | 127 | 128 | 129 | 130 | 131 | 132 |
| **VIRGINIA—Cont'd** | | | | | | | | | | | | | | | | |
| Louisa | 68 | -14.6 | 159 | 0.2 | 24.8 | 787,124 | 4,953 | 76,509 | 15 | 34,735 | 48.7 | 51.3 | 5 | 76.3 | 314.0 | 16.5 |
| Lunenburg | 73 | -11.9 | 217 | 1.4 | 25.7 | 579,358 | 2,664 | 81,398 | 17.2 | 51,206 | 77.7 | 22.3 | 7 | 69.9 | 354.0 | 24.5 |
| Madison | 107 | -0.1 | 201 | 0.1 | 45.1 | 1,193,189 | 5,949 | 90,957 | 28.4 | 53,263 | 39.6 | 60.4 | 2 | 74.1 | 1,313 | 17.3 |
| Mathews | 7 | 41.7 | 153 | 0.0 | 4.2 | 655,108 | 4,278 | 63,683 | 3.8 | 87,814 | 55.4 | 44.6 | NA | 81.4 | 31.0 | 18.6 |
| Mecklenburg | 141 | -3 | 276 | 4.2 | 62.9 | 762,052 | 2,764 | 122,271 | 50 | 97,590 | 90.1 | 9.9 | 21 | 64.5 | 334.0 | 30.3 |
| Middlesex | 20 | 1.7 | 247 | 0.3 | 14.7 | 979,876 | 3,967 | 154,776 | 8.8 | 111,620 | 96.7 | 3.3 | 3 | 87.3 | 586.0 | 44.3 |
| Montgomery | 102 | -5.2 | 174 | 0.3 | 28.8 | 842,923 | 4,842 | 78,230 | 24.3 | 41,604 | 30.2 | 69.8 | 2 | 81.8 | 327.0 | 8.2 |
| Nelson | 68 | -15.2 | 166 | 1.0 | 20.2 | 841,397 | 5,073 | 85,319 | 26.7 | 65,328 | 83.0 | 17.0 | 5 | 72.9 | 118.0 | 9.0 |
| New Kent | 18 | -7 | 133 | 0.3 | 11.1 | 711,167 | 5,353 | 85,624 | 5.1 | 37,159 | 93.2 | 6.8 | NA | 83.3 | 558.0 | 18.8 |
| Northampton | 48 | -13.9 | 340 | 5.9 | 37.8 | 1,558,831 | 4,585 | 281,016 | 96 | 675,993 | 70.7 | 29.3 | 3 | 93.0 | 1,180 | 29.6 |
| Northumberland | 43 | 0.5 | 324 | NA | 33.8 | 975,400 | 3,006 | 226,725 | 20.1 | 149,642 | 85.8 | 14.2 | NA | 64.9 | 1,423 | 26.1 |
| Nottoway | 50 | -18.2 | 162 | 0.0 | 20.1 | 514,211 | 3,174 | 76,270 | 50.5 | 162,486 | 7.9 | 92.1 | 1 | 71.4 | 405.0 | 25.7 |
| Orange | 95 | -9.1 | 228 | 0.2 | 39.8 | 1,438,347 | 6,297 | 115,676 | 113.1 | 271,149 | 77.7 | 22.3 | NA | 79.1 | 292.0 | 12.7 |
| Page | 72 | 1.3 | 139 | 0.1 | 30.4 | 904,349 | 6,515 | 90,614 | 150.1 | 289,260 | 4.2 | 95.8 | 2 | 73.0 | 836.0 | 11.8 |
| Patrick | 91 | 15.4 | 189 | 0.2 | 21.6 | 524,237 | 2,775 | 58,469 | 17.3 | 35,762 | 42.5 | 57.5 | NA | 73.3 | 260.0 | 12.6 |
| Pittsylvania | 246 | -14.3 | 213 | 1.5 | 83.4 | 603,542 | 2,835 | 93,701 | 72.7 | 62,816 | 41.1 | 58.9 | 6 | 68.3 | 1,441 | 17.2 |
| Powhatan | 35 | 7.8 | 132 | 0.1 | 11.4 | 696,770 | 5,299 | 47,626 | 11.2 | 42,772 | 22.6 | 77.4 | 1 | 75.3 | 54.0 | 9.9 |
| Prince Edward | 70 | -11.9 | 204 | 0.0 | 20.1 | 580,742 | 2,848 | 65,360 | 23.8 | 69,768 | 11.2 | 88.8 | NA | 71.8 | 198.0 | 36.1 |
| Prince George | 40 | 8.1 | 242 | D | 22.0 | 920,064 | 3,807 | 88,285 | 9.3 | 56,616 | 94.5 | 5.5 | NA | 69.5 | 933.0 | 37.2 |
| Prince William | 23 | -35.8 | 75 | 0.9 | 14.1 | 804,671 | 10,694 | 78,627 | 11.6 | 38,197 | 61.1 | 38.9 | NA | 92.1 | 227.0 | 5.9 |
| Pulaski | 78 | -19.8 | 197 | 0.0 | 23.8 | 692,087 | 3,518 | 84,261 | 33 | 83,716 | 35.9 | 64.1 | NA | 77.4 | 88.0 | 7.6 |
| Rappahannock | 70 | 11.7 | 160 | 0.1 | 25.2 | 1,101,588 | 6,891 | 64,028 | 10.1 | 23,109 | 50.4 | 49.6 | 2 | 78.4 | 90.0 | 8.7 |
| Richmond | 32 | -1.3 | 326 | 0.0 | 26.2 | 1,289,515 | 3,955 | 173,701 | 16.8 | 171,571 | 95.3 | 4.7 | NA | 48.0 | 823.0 | 42.9 |
| Roanoke | 26 | -17.1 | 100 | 0.1 | 6.2 | 572,455 | 5,743 | 55,410 | 2.5 | 9,691 | 67.5 | 32.5 | NA | 66.4 | D | 1.9 |
| Rockbridge | 135 | -19.9 | 179 | 0.1 | 36.5 | 881,466 | 4,918 | 70,368 | 31 | 41,201 | 16.9 | 83.1 | NA | 76.3 | 422.0 | 10.0 |
| Rockingham | 229 | 2.9 | 113 | 5.5 | 121.9 | 997,634 | 8,844 | 118,949 | 795.9 | 392,852 | 6.8 | 93.2 | 37 | 65.5 | 1,852 | 11.5 |
| Russell | 170 | -9.2 | 185 | D | 32.7 | 536,112 | 2,890 | 53,598 | 23.2 | 25,277 | 11.4 | 88.6 | 1 | 61.9 | 269.0 | 8.8 |
| Scott | 125 | -20.8 | 110 | 0.0 | 30.5 | 290,473 | 2,637 | 53,894 | 15.6 | 13,671 | 24.7 | 75.3 | 2 | 69.9 | 79.0 | 7.3 |
| Shenandoah | 131 | -2.1 | 135 | 0.9 | 61.2 | 796,732 | 5,884 | 99,776 | 142.8 | 148,006 | 14.1 | 85.9 | 2 | 76.7 | 593.0 | 7.3 |
| Smyth | 123 | -26.1 | 186 | 0.0 | 25.7 | 544,580 | 2,930 | 75,941 | 37.6 | 56,653 | 8.5 | 91.5 | 1 | 65.6 | 782.0 | 14.6 |
| Southampton | 142 | -7.7 | 552 | 1.6 | 97.6 | 1,704,698 | 3,087 | 237,200 | 75.1 | 292,070 | 82.4 | 17.6 | NA | 76.3 | 6,847 | 72.8 |
| Spotsylvania | 42 | -1.2 | 123 | 0.1 | 16.7 | 695,020 | 5,637 | 70,042 | 9 | 26,660 | 48.5 | 51.5 | NA | 76.6 | 199.0 | 8.6 |
| Stafford | 17 | 13.1 | 71 | 0.0 | 10.1 | 930,372 | 13,104 | 73,999 | 6.1 | 25,206 | 62.6 | 37.4 | 2 | 85.2 | 68.0 | 6.2 |
| Surry | 42 | -6.8 | 379 | 0.9 | 25.4 | 1,280,403 | 3,379 | 175,981 | D | D | D | D | 2 | 84.7 | 1,280 | 39.6 |
| Sussex | 66 | 3.1 | 534 | 0.7 | 44.9 | 1,482,681 | 2,775 | 209,543 | D | D | D | D | 3 | 74.2 | 1,963 | 54.0 |
| Tazewell | 138 | -8.1 | 269 | D | 24.7 | 737,125 | 2,736 | 66,474 | 24.6 | 48,117 | 8.5 | 91.5 | NA | 68.8 | 820.0 | 9.6 |
| Warren | 39 | -19.4 | 121 | 0.0 | 13.9 | 909,921 | 7,548 | 68,431 | 5.9 | 18,355 | 36.1 | 63.9 | 2 | 81.0 | 5.0 | 1.2 |
| Washington | 176 | -8.2 | 117 | 0.1 | 58.3 | 589,890 | 5,038 | 60,654 | 69 | 45,821 | 12.1 | 87.9 | 2 | 68.0 | 596.0 | 11.5 |
| Westmoreland | 53 | -11.4 | 288 | 0.9 | 32.0 | 1,073,155 | 3,732 | 170,589 | D | D | D | D | 2 | 74.3 | 1,175 | 36.1 |
| Wise | 26 | 1.7 | 179 | 0.0 | 5.0 | 554,986 | 3,097 | 66,109 | 2.4 | 16,000 | 33.5 | 66.5 | NA | 66.0 | 15.0 | 2.7 |
| Wythe | 152 | -13 | 185 | 0.0 | 49.1 | 725,643 | 3,921 | 87,090 | 65.5 | 80,018 | 8.4 | 91.6 | NA | 75.5 | 474.0 | 19.4 |
| York | 1 | -67.5 | 23 | 0.0 | 0.1 | 276,375 | 12,095 | 39,402 | 1.5 | 37,800 | 75.1 | 24.9 | NA | 97.5 | D | 2.5 |
| **Independent cities** | | | | | | | | | | | | | | | | |
| Alexandria city | NA | NA | NA | NA | NA | NA | NA | NA | NA | NA | NA | NA | NA | NA | NA | NA |
| Bristol city | NA | NA | NA | NA | NA | NA | NA | NA | NA | NA | NA | NA | NA | NA | NA | NA |
| Buena Vista city | NA | NA | NA | NA | NA | NA | NA | NA | NA | NA | NA | NA | NA | NA | NA | NA |
| Charlottesville city | NA | NA | NA | NA | NA | NA | NA | NA | NA | NA | NA | NA | NA | NA | NA | NA |
| Chesapeake city | 37 | -18.4 | 148 | 0.1 | 32.3 | 867,657 | 5,848 | 102,911 | 31.1 | 125,552 | 94.8 | 5.2 | NA | 82.3 | 646.0 | 16.1 |
| Colonial Heights city | NA | NA | NA | NA | NA | NA | NA | NA | NA | NA | NA | NA | NA | NA | NA | NA |
| Covington city | NA | NA | NA | NA | NA | NA | NA | NA | NA | NA | NA | NA | NA | NA | NA | NA |
| Danville city | NA | NA | NA | NA | NA | NA | NA | NA | NA | NA | NA | NA | NA | NA | NA | NA |
| Emporia city | NA | NA | NA | NA | NA | NA | NA | NA | NA | NA | NA | NA | NA | NA | NA | NA |
| Fairfax city | NA | NA | NA | NA | NA | NA | NA | NA | NA | NA | NA | NA | NA | NA | NA | NA |
| Falls Church city | NA | NA | NA | NA | NA | NA | NA | NA | NA | NA | NA | NA | NA | NA | NA | NA |
| Franklin city | NA | NA | NA | NA | NA | NA | NA | NA | NA | NA | NA | NA | NA | NA | NA | NA |
| Fredericksburg city | NA | NA | NA | NA | NA | NA | NA | NA | NA | NA | NA | NA | NA | NA | NA | NA |
| Galax city | NA | NA | NA | NA | NA | NA | NA | NA | NA | NA | NA | NA | NA | NA | NA | NA |
| Hampton city | NA | NA | NA | NA | NA | NA | NA | NA | NA | NA | NA | NA | NA | NA | NA | NA |
| Harrisonburg city | NA | NA | NA | NA | NA | NA | NA | NA | NA | NA | NA | NA | NA | NA | NA | NA |
| Hopewell city | NA | NA | NA | NA | NA | NA | NA | NA | NA | NA | NA | NA | NA | NA | NA | NA |
| Lexington city | NA | NA | NA | NA | NA | NA | NA | NA | NA | NA | NA | NA | NA | NA | NA | NA |
| Lynchburg city | NA | NA | NA | NA | NA | NA | NA | NA | NA | NA | NA | NA | NA | NA | NA | NA |
| Manassas city | NA | NA | NA | NA | NA | NA | NA | NA | NA | NA | NA | NA | NA | NA | NA | NA |
| Manassas Park city | NA | NA | NA | NA | NA | NA | NA | NA | NA | NA | NA | NA | NA | NA | NA | NA |

## Table B. States and Counties — Water Use, Wholesale Trade, Retail Trade, and Real Estate

| STATE County | Water use, 2015 — Public supply water withdrawn (mil gal/day) | Water use, 2015 — Public supply gallons withdrawn per person per day | Wholesale Trade[1], 2017 — Number of establishments | Number of employees | Sales (mil dol) | Average payroll (mil dol) | Retail Trade[2], 2017 — Number of establishments | Number of employees | Sales (mil dol) | Average payroll (mil dol) | Real estate and rental and leasing,[2] 2017 — Number of establishments | Number of employees | Sales (mil dol) | Average payroll (mil dol) |
|---|---|---|---|---|---|---|---|---|---|---|---|---|---|---|
| | 133 | 134 | 135 | 136 | 137 | 138 | 139 | 140 | 141 | 142 | 143 | 144 | 145 | 146 |
| **VIRGINIA—Cont'd** | | | | | | | | | | | | | | |
| Louisa | 0.53 | 15.3 | 10 | 182 | 91.8 | 9.9 | 83 | 1,259 | 355.4 | 29.6 | 24 | 105 | 30.2 | 4.9 |
| Lunenburg | 0.48 | 39.0 | 8 | 104 | 84.8 | 4.7 | 37 | 283 | 60.5 | 6.1 | NA | NA | NA | NA |
| Madison | 0.00 | 0.0 | 7 | D | 11.6 | D | 39 | 703 | 140.3 | 22.0 | 7 | 11 | 2.1 | 0.2 |
| Mathews | 0.00 | 0.0 | D | D | D | D | 25 | 203 | 45.2 | 4.4 | D | D | D | D |
| Mecklenburg | 1.87 | 60.2 | 27 | 321 | 288.3 | 15.4 | 158 | 1,718 | 498.8 | 39.3 | 31 | 88 | 12.8 | 2.5 |
| Middlesex | 0.07 | 6.6 | 11 | 100 | 27.0 | 3.9 | 54 | 423 | 101.7 | 11.2 | 9 | 21 | 5.1 | 0.5 |
| Montgomery | 6.92 | 70.9 | 33 | 668 | 434.4 | 30.7 | 306 | 5,168 | 1,117.1 | 117.4 | 94 | 510 | 112.3 | 19.4 |
| Nelson | 0.20 | 13.5 | 6 | D | 12.8 | D | 48 | 307 | 201.6 | 8.3 | 10 | D | 2.9 | D |
| New Kent | 19.91 | 976.4 | 6 | D | 9.6 | D | 47 | 481 | 179.7 | 12.5 | 16 | 22 | 3.3 | 0.8 |
| Northampton | 0.40 | 32.9 | 18 | 125 | 76.1 | 6.6 | 61 | 454 | 111.5 | 10.4 | 14 | D | 5.5 | D |
| Northumberland | 0.14 | 11.4 | 10 | D | 28.9 | D | 46 | 357 | 71.6 | 7.6 | 18 | 27 | 5.2 | 1.0 |
| Nottoway | 0.48 | 30.6 | 11 | 89 | 75.6 | 3.8 | 64 | 636 | 135.1 | 15.2 | 9 | 23 | 2.3 | 0.9 |
| Orange | 2.08 | 58.8 | 7 | 151 | 56.1 | 6.6 | 108 | 1,355 | 407.3 | 37.6 | 23 | 40 | 6.2 | 1.0 |
| Page | 1.27 | 53.5 | 5 | 40 | 18.3 | 2.0 | 75 | 764 | 170.4 | 18.2 | 10 | 17 | 3.6 | 0.8 |
| Patrick | 0.29 | 16.1 | 6 | D | 20.2 | D | 48 | 485 | 175.1 | 11.9 | 6 | D | 0.6 | D |
| Pittsylvania | 1.62 | 26.0 | 26 | 742 | 386.1 | 28.3 | 121 | 788 | 260.4 | 17.9 | 23 | 73 | 12.1 | 2.4 |
| Powhatan | 0.11 | 3.9 | D | D | D | 9.4 | 78 | 963 | 332.2 | 27.2 | 27 | 131 | 12.6 | 4.5 |
| Prince Edward | 1.11 | 48.4 | 14 | 144 | 35.5 | 4.6 | 99 | 1,331 | 383.7 | 34.9 | 29 | 125 | 14.2 | 3.2 |
| Prince George | 0.14 | 3.7 | 19 | 605 | 556.0 | 26.1 | 65 | 806 | 287.3 | 23.8 | 20 | 91 | 19.9 | 4.4 |
| Prince William | 79.73 | 176.5 | 203 | 3,151 | 2,727.8 | 203.4 | 1,169 | 22,923 | 6,350.6 | 609.6 | 373 | 1,613 | 521.0 | 83.8 |
| Pulaski | 4.39 | 127.9 | 11 | 144 | 37.1 | 5.4 | 98 | 1,777 | 392.8 | 38.6 | 24 | 90 | 9.6 | 2.3 |
| Rappahannock | 0.03 | 4.1 | D | D | D | D | 25 | 154 | 33.6 | 4.9 | D | D | D | D |
| Richmond | 0.36 | 40.4 | 9 | 91 | 43.1 | 3.1 | 28 | 239 | 59.0 | 6.0 | 6 | D | 1.3 | D |
| Roanoke | 25.13 | 266.2 | 88 | 1,049 | 471.7 | 57.4 | 266 | 4,267 | 1,126.2 | 109.2 | 115 | 528 | 72.0 | 19.2 |
| Rockbridge | 1.74 | 77.8 | 13 | 73 | 28.5 | 2.8 | 72 | 1,563 | 366.3 | 30.9 | 9 | D | 3.6 | D |
| Rockingham | 14.75 | 187.7 | 56 | 1,007 | 1,065.8 | 63.0 | 221 | 1,786 | 455.0 | 42.2 | 41 | 519 | 116.5 | 24.3 |
| Russell | 1.18 | 42.3 | 10 | 34 | 15.5 | 1.1 | 78 | 933 | 276.2 | 22.4 | D | D | D | D |
| Scott | 1.12 | 50.6 | D | D | D | D | 69 | 608 | 168.1 | 11.9 | 5 | 8 | 0.7 | 0.1 |
| Shenandoah | 2.57 | 59.5 | 20 | 465 | 523.8 | 16.7 | 147 | 1,643 | 562.3 | 38.0 | D | D | D | D |
| Smyth | 2.19 | 69.6 | 12 | 172 | 98.0 | 9.5 | 110 | 1,266 | 272.7 | 26.5 | 12 | 38 | 4.7 | 1.1 |
| Southampton | 0.84 | 46.4 | 13 | 212 | 200.3 | 9.9 | 37 | 229 | 61.2 | 5.0 | 9 | D | 2.3 | D |
| Spotsylvania | 11.05 | 84.7 | 76 | 675 | 405.7 | 38.5 | 451 | 7,553 | 2,481.0 | 216.0 | 137 | 690 | 133.3 | 26.2 |
| Stafford | 21.98 | 154.8 | 49 | 1,100 | 1,467.2 | 62.6 | 281 | 4,609 | 1,362.4 | 118.5 | 104 | 390 | 101.7 | 15.0 |
| Surry | 0.15 | 22.4 | NA | NA | NA | NA | 12 | 40 | 12.2 | 0.9 | 4 | 17 | 1.4 | 0.6 |
| Sussex | 0.56 | 47.8 | D | D | D | 5.6 | 41 | 287 | 103.3 | 7.3 | 7 | D | 1.8 | D |
| Tazewell | 3.79 | 88.3 | 41 | 440 | 271.8 | 19.9 | 186 | 2,860 | 721.3 | 65.3 | 42 | 113 | 16.6 | 3.2 |
| Warren | 9.08 | 232.3 | 15 | 148 | 184.3 | 4.7 | 129 | 1,837 | 529.3 | 45.7 | 28 | 80 | 15.5 | 3.5 |
| Washington | 11.77 | 215.6 | 36 | 581 | 353.1 | 25.7 | 191 | 3,118 | 812.7 | 72.7 | 49 | 146 | 43.5 | 5.8 |
| Westmoreland | 0.79 | 44.8 | 11 | 20 | 39.5 | 0.7 | 51 | 446 | 116.0 | 10.1 | 8 | 13 | 2.3 | 0.2 |
| Wise | 4.90 | 123.4 | 28 | 241 | 144.9 | 12.2 | 150 | 1,820 | 454.5 | 39.7 | 21 | 59 | 8.5 | 1.8 |
| Wythe | 4.35 | 149.4 | 18 | 207 | 103.2 | 7.7 | 125 | 1,842 | 699.9 | 44.7 | 25 | 73 | 12.7 | 2.8 |
| York | 19.05 | 280.8 | 33 | 159 | 76.7 | 7.7 | 209 | 3,779 | 960.7 | 95.4 | 58 | 297 | 109.9 | 12.7 |
| **Independent cities** | | | | | | | | | | | | | | |
| Alexandria city | 0.02 | 0.1 | 72 | 872 | 651.9 | 60.4 | 475 | 7,862 | 2,548.9 | 258.9 | 282 | 1,580 | 555.7 | 84.0 |
| Bristol city | 0.00 | 0.0 | 28 | 350 | 794.2 | 14.1 | 139 | 1,845 | 447.5 | 41.0 | 21 | 91 | 13.0 | 2.9 |
| Buena Vista city | 1.06 | 160.2 | NA | NA | NA | NA | 23 | 170 | 37.0 | 3.2 | 3 | 7 | 0.5 | 0.1 |
| Charlottesville city | 0.00 | 0.0 | 44 | 451 | 330.2 | 24.3 | 281 | 3,282 | 714.9 | 77.4 | 84 | 555 | 110.8 | 26.2 |
| Chesapeake city | 7.12 | 30.2 | 241 | 3,552 | 2,790.8 | 205.7 | 785 | 14,925 | 4,182.0 | 387.9 | 307 | 1,421 | 382.9 | 66.8 |
| Colonial Heights city | 0.00 | 0.0 | 9 | 58 | 25.6 | 2.3 | 169 | 3,283 | 738.5 | 69.5 | 35 | 164 | 41.4 | 6.1 |
| Covington city | 2.14 | 378.2 | 5 | 35 | 17.6 | 1.5 | 43 | 598 | 144.5 | 14.4 | 8 | 31 | 7.4 | 1.0 |
| Danville city | 5.36 | 127.4 | 45 | 560 | 255.9 | 29.6 | 302 | 4,558 | 1,144.0 | 104.0 | 53 | 237 | 31.6 | 6.9 |
| Emporia city | 0.69 | 125.5 | D | D | D | 3.3 | 54 | 626 | 125.7 | 13.5 | 14 | 45 | 5.2 | 1.1 |
| Fairfax city | 0.00 | 0.0 | 27 | 250 | 157.5 | 15.7 | 217 | 4,681 | 1,722.0 | 172.0 | 62 | 242 | 86.0 | 13.4 |
| Falls Church city | 0.00 | 0.0 | 16 | 126 | 51.5 | 6.7 | 103 | 1,304 | 457.8 | 50.0 | 38 | 176 | 65.1 | 9.3 |
| Franklin city | 0.87 | 102.5 | 4 | 25 | 14.0 | 1.1 | 58 | 1,039 | 234.8 | 23.8 | D | D | D | D |
| Fredericksburg city | 0.00 | 0.0 | 24 | 254 | 123.6 | 10.8 | 261 | 4,326 | 1,015.0 | 109.9 | 75 | 314 | 76.9 | 14.8 |
| Galax city | 1.74 | 251.7 | D | D | D | 1.4 | 69 | 1,062 | 242.3 | 24.8 | 15 | 53 | 10.6 | 1.8 |
| Hampton city | 0.00 | 0.0 | 60 | 921 | 425.1 | 47.0 | 423 | 6,748 | 1,569.0 | 168.8 | 123 | 784 | 154.5 | 26.7 |
| Harrisonburg city | 0.00 | 0.0 | 58 | 864 | 438.7 | 43.9 | 325 | 5,341 | 1,378.8 | 136.1 | 71 | 373 | 99.1 | 14.0 |
| Hopewell city | 20.55 | 918.3 | D | D | D | 5.8 | 72 | 636 | 162.7 | 15.2 | 19 | 71 | 22.9 | 3.0 |
| Lexington city | 0.00 | 0.0 | NA | NA | NA | NA | 41 | 377 | 94.7 | 9.1 | 14 | 34 | 5.6 | 1.4 |
| Lynchburg city | 0.00 | 0.0 | 74 | 898 | 479.1 | 43.2 | 355 | 6,890 | 2,162.9 | 165.3 | 119 | 458 | 87.4 | 16.6 |
| Manassas city | 0.21 | 5.0 | 47 | 458 | 227.6 | 23.4 | 176 | 2,153 | 961.8 | 77.1 | 45 | 226 | 61.4 | 10.7 |
| Manassas Park city | 0.00 | 0.0 | 21 | 257 | 194.1 | 17.4 | 24 | 143 | 96.2 | 7.8 | 9 | 51 | 19.4 | 2.4 |

1 Merchant wholesalers, except manufacturers' sales branches and offices.  2. Employer establishments.

# Table B. States and Counties — Professional Services, Manufacturing, and Accommodation and Food Services

| STATE County | Professional, scientific, and technical services, 2017 | | | | Manufacturing, 2017 | | | | Accommodation and food services, 2017 | | | |
|---|---|---|---|---|---|---|---|---|---|---|---|---|
| | Number of establishments | Number of employees | Sales (mil dol) | Average payroll (mil dol) | Number of establishments | Number of employees | Sales (mil dol) | Average payroll (mil dol) | Number of establishments | Number of employees | Sales (mil dol) | Annual payroll (mil dol) |
| | 147 | 148 | 149 | 150 | 151 | 152 | 153 | 154 | 155 | 156 | 157 | 158 |
| VIRGINIA—Cont'd | | | | | | | | | | | | |
| Louisa | 46 | 136 | 15 | 4.9 | 32 | 1,022 | 304.0 | 44.6 | 43 | 569 | 31.9 | 8.8 |
| Lunenburg | 7 | 33 | 3 | 1.3 | 7 | 483 | 86.1 | 16.1 | 8 | 90 | 3.0 | 0.8 |
| Madison | 25 | 75 | 14 | 3.5 | 16 | 312 | 117.3 | 11.9 | 21 | 247 | 11.2 | 2.9 |
| Mathews | D | D | D | 1.4 | 5 | 44 | 2.9 | 1.2 | D | D | D | D |
| Mecklenburg | 44 | 341 | 54 | 21.7 | 34 | 1,338 | 276.3 | 55.4 | 70 | 1,109 | 52.7 | 15.0 |
| Middlesex | 28 | 183 | 12 | 6.5 | 12 | 125 | 23.7 | 5.3 | 26 | 317 | 16.1 | 5.6 |
| Montgomery | D | D | D | D | 49 | 5,080 | 1,554.4 | 284.5 | 210 | 4,606 | 231.6 | 66.1 |
| Nelson | 42 | 180 | 19 | 7.4 | 27 | 306 | 82.5 | 15.6 | 22 | 857 | 62.9 | 20.2 |
| New Kent | D | D | D | D | 11 | 172 | 28.7 | 6.2 | D | D | D | D |
| Northampton | 22 | 62 | 7 | 2.6 | 9 | 435 | 72.8 | 17.1 | 39 | 487 | 34.3 | 10.0 |
| Northumberland | 21 | 84 | 10 | 3.2 | 15 | 254 | 119.3 | 17.2 | 19 | 92 | 5.8 | 2.1 |
| Nottoway | 21 | 76 | 7 | 2.4 | 18 | 385 | 120.2 | 14.7 | 30 | 340 | 15.6 | 4.4 |
| Orange | 57 | 236 | 32 | 9.9 | 20 | 1,289 | 232.2 | 67.0 | 62 | 747 | 44.1 | 12.2 |
| Page | 35 | 274 | 33 | 12.7 | 15 | 516 | 193.4 | 18.1 | 55 | 730 | 50.1 | 12.3 |
| Patrick | 22 | 98 | 7 | 2.5 | 29 | 1,212 | 229.5 | 41.7 | D | D | D | D |
| Pittsylvania | 47 | 190 | 36 | 7.9 | 41 | 1,887 | 645.2 | 76.3 | 48 | 544 | 26.7 | 7.5 |
| Powhatan | 64 | 393 | 43 | 17.9 | D | 118 | D | 6.2 | 31 | 531 | 24.7 | 7.9 |
| Prince Edward | D | D | D | D | 14 | 60 | 13.7 | 2.2 | 39 | 1,004 | 46.4 | 15.4 |
| Prince George | D | D | D | D | 25 | 1,160 | 950.3 | 64.9 | 47 | 610 | 39.2 | 11.4 |
| Prince William | D | D | D | D | 111 | 4,144 | 1,576.6 | 350.1 | 743 | 13,817 | 878.2 | 239.0 |
| Pulaski | D | D | D | D | 34 | 4,119 | 2,777.8 | 241.5 | 69 | 1,109 | 56.8 | 15.3 |
| Rappahannock | D | D | D | D | 11 | 87 | 6.5 | 1.5 | 19 | 415 | 20.0 | 7.2 |
| Richmond | 10 | 20 | 3 | 0.9 | 8 | 146 | 41.2 | 6.7 | D | D | D | D |
| Roanoke | D | D | D | D | 65 | 3,098 | 941.8 | 172.3 | 156 | 3,079 | 159.5 | 44.3 |
| Rockbridge | 25 | 115 | 20 | 5.4 | 22 | 978 | 321.1 | 38.0 | 53 | 874 | 57.9 | 15.8 |
| Rockingham | D | D | D | D | 87 | 6,873 | 3,872.0 | 364.0 | 92 | 2,428 | 102.0 | 44.0 |
| Russell | D | D | D | D | 14 | 545 | 136.5 | 20.7 | 30 | 448 | 22.1 | 6.1 |
| Scott | 20 | 98 | 12 | 4.2 | 9 | 818 | 415.1 | 32.6 | 25 | 558 | 21.7 | 5.9 |
| Shenandoah | D | D | D | D | 45 | 3,347 | 893.1 | 141.4 | 87 | 1,444 | 69.1 | 20.1 |
| Smyth | D | D | D | D | 33 | 3,081 | 1,169.6 | 144.0 | 44 | 678 | 31.2 | 9.1 |
| Southampton | 13 | 50 | 5 | 1.7 | 13 | 509 | 151.2 | 25.0 | 8 | 109 | 4.5 | 1.4 |
| Spotsylvania | 247 | 1,973 | 293 | 123.9 | 52 | 1,173 | 417.4 | 72.0 | 215 | 4,454 | 232.7 | 65.7 |
| Stafford | D | D | D | D | 38 | 977 | 252.4 | 42.5 | 202 | 3,841 | 228.4 | 58.3 |
| Surry | D | D | D | 0.5 | 4 | 114 | 36.9 | 4.0 | 4 | 33 | 0.7 | 0.1 |
| Sussex | D | D | D | 0.5 | 9 | 148 | 124.6 | 6.5 | D | D | D | D |
| Tazewell | 65 | 340 | 41 | 15.4 | 50 | 844 | 202.2 | 43.5 | 71 | 1,219 | 62.1 | 17.1 |
| Warren | D | D | D | D | 27 | 891 | 1,035.1 | 53.7 | 87 | 1,302 | 71.5 | 20.9 |
| Washington | 100 | 390 | 56 | 23.7 | 65 | 3,540 | 1,017.6 | 164.1 | 97 | 1,984 | 85.9 | 25.3 |
| Westmoreland | 23 | 79 | 10 | 4.7 | 10 | 420 | 125.4 | 19.0 | 31 | 340 | 15.5 | 5.2 |
| Wise | 61 | 358 | 56 | 13.6 | 15 | 214 | 46.7 | 10.6 | 56 | 1,079 | 48.6 | 13.5 |
| Wythe | 38 | 181 | 19 | 7.5 | 36 | 1,906 | 1,068.6 | 88.8 | 76 | 1,576 | 77.2 | 21.6 |
| York | 170 | 2,115 | 290 | 119.8 | 31 | 301 | 57.5 | 14.8 | 171 | 4,025 | 258.4 | 67.2 |
| Independent cities | | | | | | | | | | | | |
| Alexandria city | 1,200 | 18,080 | 4,201 | 1,784.3 | 58 | 1,480 | 431.3 | 76.7 | 402 | 8,518 | 691.8 | 193.3 |
| Bristol city | 34 | 164 | 19 | 7.6 | 17 | 1,096 | 263.1 | 43.5 | 79 | 1,955 | 99.4 | 28.1 |
| Buena Vista city | 4 | 16 | 2 | 0.6 | 14 | 778 | 223.9 | 35.8 | 12 | 214 | 11.3 | 3.8 |
| Charlottesville city | D | D | D | D | 53 | 667 | 143.1 | 35.9 | 292 | 5,491 | 329.6 | 98.3 |
| Chesapeake city | D | D | D | D | 120 | 4,179 | 1,577.0 | 234.2 | 486 | 10,068 | 530.7 | 148.8 |
| Colonial Heights city | 45 | 435 | 61 | 25.4 | D | 175 | D | 8.0 | 84 | 2,221 | 112.4 | 30.6 |
| Covington city | 13 | 40 | 5 | 1.1 | D | D | D | D | 20 | 252 | 15.3 | 3.4 |
| Danville city | D | D | D | D | 41 | 5,089 | 1,892.3 | 260.1 | 143 | 2,925 | 148.8 | 41.4 |
| Emporia city | D | D | D | D | 6 | 1,269 | 282.9 | 58.4 | 28 | 701 | 31.1 | 9.1 |
| Fairfax city | D | D | D | D | 22 | 169 | 36.2 | 7.0 | 183 | 3,332 | 223.9 | 61.9 |
| Falls Church city | 164 | 1,847 | 340 | 144.5 | 11 | 60 | 14.9 | 4.0 | 114 | 1,213 | 94.3 | 24.2 |
| Franklin city | 20 | 107 | 11 | 4.9 | 6 | D | 6.3 | D | D | D | D | D |
| Fredericksburg city | D | D | D | D | 27 | 332 | 71.2 | 19.4 | 186 | 4,460 | 227.5 | 72.2 |
| Galax city | 23 | 138 | 10 | 4.0 | 15 | 1,297 | 264.2 | 41.4 | 39 | 701 | 29.6 | 8.4 |
| Hampton city | 285 | 4,393 | 778 | 344.4 | 55 | 2,226 | 597.2 | 125.6 | 256 | 5,902 | 304.4 | 89.2 |
| Harrisonburg city | D | D | D | D | 40 | 2,861 | 891.0 | 124.0 | 208 | 5,138 | 259.0 | 81.3 |
| Hopewell city | 41 | 293 | 35 | 11.6 | 18 | 1,624 | 1,737.4 | 147.3 | 50 | 771 | 37.5 | 9.7 |
| Lexington city | D | D | D | D | 4 | D | 1.2 | 0.3 | 50 | 663 | 43.8 | 14.0 |
| Lynchburg city | D | D | D | D | 76 | 6,377 | 2,415.3 | 366.1 | 270 | 6,023 | 292.1 | 83.4 |
| Manassas city | D | D | D | D | 40 | 1,319 | 301.5 | 104.0 | 126 | 1,947 | 117.3 | 32.6 |
| Manassas Park city | 19 | 152 | 26 | 9.1 | 13 | 222 | 31.9 | 7.9 | 14 | 118 | 6.6 | 2.1 |

Items 147—158

# Table B. States and Counties — Health Care and Social Assistance, Other Services, Nonemployer Businesses, and Residential Construction

| STATE County | Health care and social assistance, 2017 | | | | Other services, 2017 | | | | Nonemployer businesses, 2018 | | Value of residential construction authorized by building permits, 2020 | |
|---|---|---|---|---|---|---|---|---|---|---|---|---|
| | Number of establishments | Number of employees | Receipts (mil dol) | Annual payroll (mil dol) | Number of establishments | Number of employees | Receipts (mil dol) | Annual payroll (mil dol) | Number | Receipts (mil dol) | New construction ($1,000) | Number of housing units |
| | 159 | 160 | 161 | 162 | 163 | 164 | 165 | 166 | 167 | 168 | 169 | 170 |
| **VIRGINIA—Cont'd** | | | | | | | | | | | | |
| Louisa | 29 | 342 | 26.1 | 13.4 | 42 | 263 | 26.6 | 7.1 | 2,329 | 99.8 | 90,552 | 435 |
| Lunenburg | 16 | 235 | 18.1 | 7.5 | D | D | D | 0.5 | 616 | 23.0 | 3,625 | 19 |
| Madison | 22 | 324 | 20.1 | 8.4 | 22 | 55 | 8.4 | 1.7 | 1,265 | 49.1 | 10,940 | 42 |
| Mathews | 20 | 176 | 10.3 | 4.5 | D | D | D | D | 833 | 34.4 | 10,714 | 31 |
| Mecklenburg | 76 | 1,771 | 164.0 | 68.8 | 69 | 375 | 28.8 | 8.7 | 1,622 | 71.6 | 27,057 | 128 |
| Middlesex | 23 | 366 | 25.5 | 11.1 | D | D | D | D | 960 | 39.1 | 9,022 | 34 |
| Montgomery | 221 | 4,382 | 642.2 | 198.5 | 147 | 867 | 287.5 | 26.6 | 5,313 | 226.1 | 285,706 | 1,625 |
| Nelson | D | D | D | 8.0 | 21 | 177 | 15.1 | 5.9 | 1,286 | 53.4 | 20,152 | 71 |
| New Kent | D | D | D | 18.9 | 26 | 80 | 10.6 | 2.9 | 1,960 | 81.9 | 117,163 | 321 |
| Northampton | 36 | 755 | 80.6 | 28.9 | 20 | 63 | 4.8 | 1.4 | 1,047 | 38.6 | 17,014 | 58 |
| Northumberland | 11 | D | 6.2 | D | 24 | 63 | 7.0 | 1.9 | 1,103 | 44.3 | 11,442 | 50 |
| Nottoway | 30 | 927 | 63.6 | 29.5 | 30 | 96 | 9.1 | 2.5 | 685 | 20.5 | 7,226 | 58 |
| Orange | 33 | 393 | 25.9 | 11.2 | 62 | 434 | 54.5 | 13.1 | 2,494 | 105.6 | 45,059 | 204 |
| Page | 22 | 559 | 44.8 | 22.1 | 28 | 90 | 10.1 | 2.5 | 1,448 | 54.5 | 11,675 | 80 |
| Patrick | 21 | 559 | 34.1 | 14.9 | D | D | D | 1.4 | 1,032 | 32.3 | 0 | 0 |
| Pittsylvania | 67 | 1,239 | 81.0 | 36.0 | 67 | 289 | 27.4 | 7.4 | 3,169 | 117.2 | 14,055 | 58 |
| Powhatan | 33 | 261 | 24.8 | 11.5 | 63 | 207 | 28.2 | 8.1 | 2,222 | 105.8 | 46,722 | 209 |
| Prince Edward | 80 | 1,693 | 188.3 | 74.5 | 40 | 186 | 28.0 | 5.0 | 1,014 | 34.1 | 9,820 | 64 |
| Prince George | 34 | 469 | 48.8 | 23.6 | 29 | 149 | 13.8 | 4.8 | 1,558 | 54.6 | 10,619 | 64 |
| Prince William | 782 | 11,124 | 1,355.0 | 521.6 | 586 | 4,021 | 705.4 | 130.8 | 41,049 | 1,686.6 | 248,650 | 1,433 |
| Pulaski | 71 | 1,105 | 128.3 | 47.4 | 42 | 216 | 27.4 | 6.9 | 1,553 | 48.8 | 14,142 | 62 |
| Rappahannock | D | D | D | 1.6 | D | D | D | 3.3 | 953 | 47.8 | 9,640 | 22 |
| Richmond | 14 | 867 | 26.4 | 13.4 | 11 | 79 | 10.9 | 3.3 | 573 | 18.8 | 3,213 | 22 |
| Roanoke | 230 | 4,628 | 417.5 | 186.4 | 146 | 804 | 80.5 | 21.1 | 5,919 | 254.5 | 35,642 | 163 |
| Rockbridge | 28 | 409 | 40.1 | 12.9 | 28 | 134 | 14.3 | 3.8 | 1,531 | 62.3 | 1,500 | 52 |
| Rockingham | 129 | 4,097 | 635.3 | 215.0 | 105 | 502 | 53.7 | 17.5 | 5,926 | 277.1 | 157,786 | 536 |
| Russell | 62 | 877 | 68.3 | 31.8 | D | D | D | D | 1,152 | 37.3 | 3,763 | 18 |
| Scott | 34 | 730 | 49.1 | 22.6 | 19 | 139 | 17.5 | 7.1 | 890 | 28.3 | 2,767 | 16 |
| Shenandoah | 82 | 1,471 | 125.6 | 56.6 | 74 | 411 | 42.7 | 11.3 | 2,968 | 111.7 | 27,395 | 135 |
| Smyth | 63 | 1,841 | 162.1 | 72.6 | 36 | 169 | 18.8 | 4.8 | 1,397 | 48.4 | 3,136 | 15 |
| Southampton | 7 | D | 9.7 | D | D | D | D | 2.6 | 948 | 30.3 | 9,206 | 60 |
| Spotsylvania | 279 | 4,247 | 524.0 | 202.0 | 203 | 1,223 | 115.5 | 38.0 | 9,182 | 397.8 | 278,585 | 1,411 |
| Stafford | 193 | 3,429 | 354.0 | 130.8 | 232 | 1,590 | 158.6 | 52.4 | 9,954 | 405.1 | 238,167 | 964 |
| Surry | D | D | D | D | NA | NA | NA | NA | 392 | 10.9 | 3,778 | 25 |
| Sussex | D | D | D | D | 11 | 87 | 5.2 | 1.7 | 461 | 15.7 | 4,792 | 37 |
| Tazewell | 131 | 2,487 | 234.7 | 92.3 | 78 | 511 | 61.6 | 18.6 | 1,964 | 74.3 | 733 | 4 |
| Warren | 61 | 1,698 | 140.6 | 58.0 | 64 | 488 | 94.9 | 13.9 | 2,801 | 117.0 | 44,042 | 191 |
| Washington | 158 | 2,693 | 379.0 | 130.2 | 64 | 346 | 29.8 | 9.2 | 3,428 | 135.0 | 23,633 | 86 |
| Westmoreland | 17 | 214 | 11.8 | 5.7 | 28 | 91 | 8.0 | 2.0 | 1,204 | 41.7 | 20,108 | 116 |
| Wise | 92 | 1,627 | 131.7 | 54.2 | D | D | D | D | 1,510 | 51.1 | 1,290 | 6 |
| Wythe | 74 | 1,318 | 125.9 | 51.6 | 62 | 269 | 32.3 | 8.3 | 1,518 | 59.1 | 6,972 | 39 |
| York | 138 | 2,153 | 285.8 | 112.1 | D | D | D | D | 3,975 | 154.2 | 79,963 | 286 |
| **Independent cities** | | | | | | | | | | | | |
| Alexandria city | 426 | 7,386 | 1,056.3 | 379.4 | 611 | 9,983 | 3,295.4 | 767.9 | 17,430 | 863.9 | 40,575 | 107 |
| Bristol city | 51 | 987 | 70.7 | 30.7 | 41 | 243 | 79.5 | 5.9 | 972 | 39.9 | 230 | 1 |
| Buena Vista city | 11 | 238 | 10.8 | 5.3 | 9 | 21 | 2.4 | 0.6 | 292 | 8.2 | 1,200 | 9 |
| Charlottesville city | 158 | 9,293 | 1,742.6 | 877.9 | 158 | 1,729 | 580.2 | 105.1 | 4,416 | 260.4 | 36,772 | 190 |
| Chesapeake city | 514 | 10,002 | 1,239.6 | 503.6 | 438 | 3,057 | 349.2 | 110.8 | 15,425 | 573.0 | 313,351 | 1,128 |
| Colonial Heights city | 91 | 1,616 | 130.8 | 58.3 | 63 | 517 | 39.7 | 14.9 | 982 | 38.9 | 1,029 | 6 |
| Covington city | 16 | 176 | 5.7 | 3.4 | D | D | D | D | 234 | 8.0 | 566 | 3 |
| Danville city | 180 | 5,096 | 489.0 | 210.3 | 89 | 424 | 59.0 | 11.3 | 2,253 | 81.8 | 1,012 | 7 |
| Emporia city | 41 | 1,170 | 96.8 | 37.8 | 14 | 83 | 7.8 | 2.2 | 253 | 10.6 | 344 | 2 |
| Fairfax city | 265 | 4,300 | 347.8 | 148.4 | 166 | 1,258 | 161.5 | 49.7 | 3,062 | 175.2 | 3,382 | 10 |
| Falls Church city | 123 | 2,335 | 225.3 | 103.0 | 94 | 573 | 113.9 | 25.4 | 1,501 | 100.9 | 7,050 | 14 |
| Franklin city | 55 | 1,136 | 101.1 | 39.0 | 26 | 102 | 11.4 | 3.0 | 485 | 15.6 | NA | NA |
| Fredericksburg city | 225 | 5,189 | 883.6 | 329.9 | 113 | 762 | 68.7 | 23.4 | 2,172 | 110.2 | 25,835 | 104 |
| Galax city | 46 | 1,518 | 187.7 | 59.9 | 22 | 97 | 9.5 | 2.6 | 403 | 14.1 | 877 | 7 |
| Hampton city | 281 | 7,833 | 1,299.2 | 454.5 | 169 | 953 | 111.9 | 28.0 | 7,504 | 201.3 | 13,160 | 211 |
| Harrisonburg city | 185 | 3,080 | 281.4 | 124.0 | 131 | 715 | 84.9 | 25.7 | 3,084 | 144.5 | 11,355 | 60 |
| Hopewell city | 45 | 1,183 | 146.7 | 58.7 | 38 | 231 | 19.6 | 6.9 | 994 | 30.0 | 6,576 | 77 |
| Lexington city | 34 | 554 | 70.8 | 21.8 | 29 | 221 | 67.3 | 9.1 | 471 | 19.6 | 1,730 | 9 |
| Lynchburg city | 289 | 10,123 | 1,393.7 | 516.1 | 171 | 1,013 | 107.3 | 31.5 | 4,607 | 164.1 | 31,510 | 361 |
| Manassas city | 177 | 3,231 | 376.1 | 158.6 | 137 | 902 | 142.6 | 35.8 | 3,745 | 167.9 | 17,106 | 73 |
| Manassas Park city | D | D | D | D | D | D | D | D | 1,594 | 70.3 | 0 | 0 |

# Table B. States and Counties — Government Employment and Payroll, and Local Government Finances

| STATE County | Government employment and payroll, 2017 | | March payroll (percent of total) | | | | | | | Local government finances, 2017 | | | | |
|---|---|---|---|---|---|---|---|---|---|---|---|---|---|---|
| | | | | | | | | | | General revenue | | Taxes | | |
| | Full-time equivalent employees | March payroll (dollars) | Administration, judicial, and legal | Police and corrections | Fire protection | Highways and transportation | Health and welfare | Natural resources and utilities | Education and libraries | Total (mil dol) | Intergovernmental (mil dol) | Total (mil dol) | Per capita[1] (dollars) Total | Per capita[1] (dollars) Property |
| | 171 | 172 | 173 | 174 | 175 | 176 | 177 | 178 | 179 | 180 | 181 | 182 | 183 | 184 |
| **VIRGINIA—Cont'd** | | | | | | | | | | | | | | |
| Louisa | 1,134 | 3,879,432 | 9.3 | 7.3 | 4.0 | 0.1 | 3.0 | 4.1 | 68.6 | 159.5 | 68.5 | 66.6 | 1,854 | 1,635 |
| Lunenburg | 366 | 1,243,228 | 3.7 | 15.7 | 0.0 | 0.0 | 0.0 | 6.5 | 71.9 | 37.3 | 22.5 | 12.1 | 981 | 841 |
| Madison | 397 | 1,613,812 | 7.2 | 7.8 | 0.0 | 0.0 | 4.9 | 0.2 | 77.3 | 39.8 | 18.7 | 18.8 | 1,420 | 1,179 |
| Mathews | 311 | 973,033 | 9.6 | 7.8 | 0.0 | 0.0 | 5.8 | 0.0 | 73.0 | 17.6 | 6.4 | 10.7 | 1,231 | 1,061 |
| Mecklenburg | 1,167 | 3,512,833 | 7.1 | 11.5 | 0.0 | 1.1 | 6.1 | 4.9 | 64.2 | 106.5 | 51.6 | 43.6 | 1,419 | 1,058 |
| Middlesex | 297 | 1,024,349 | 10.4 | 15.5 | 0.0 | 0.0 | 1.4 | 3.0 | 69.6 | 30.5 | 10.1 | 18.9 | 1,779 | 1,531 |
| Montgomery | 2,607 | 9,586,596 | 13.2 | 10.6 | 0.3 | 5.7 | 1.7 | 8.7 | 57.3 | 343.0 | 126.8 | 134.3 | 1,368 | 1,022 |
| Nelson | 470 | 1,633,990 | 8.1 | 5.7 | 0.0 | 0.0 | 1.4 | 2.1 | 77.9 | 51.6 | 18.8 | 28.7 | 1,938 | 1,641 |
| New Kent | 681 | 2,482,291 | 11.0 | 8.9 | 5.7 | 0.3 | 5.0 | 7.1 | 60.6 | 64.1 | 24.8 | 33.2 | 1,536 | 1,311 |
| Northampton | 792 | 3,563,409 | 4.6 | 10.9 | 0.0 | 20.9 | 6.3 | 3.0 | 51.4 | 135.6 | 31.0 | 28.4 | 2,394 | 1,899 |
| Northumberland | 392 | 1,439,407 | 7.2 | 8.0 | 0.0 | 0.0 | 0.9 | 1.0 | 78.7 | 33.4 | 10.8 | 21.1 | 1,722 | 1,553 |
| Nottoway | 595 | 1,856,153 | 11.6 | 7.7 | 0.0 | 3.4 | 3.6 | 6.4 | 65.2 | 48.9 | 24.1 | 19.4 | 1,257 | 782 |
| Orange | 1,300 | 4,302,812 | 6.5 | 18.0 | 5.2 | 1.2 | 3.8 | 3.0 | 61.8 | 121.3 | 59.6 | 52.8 | 1,473 | 1,167 |
| Page | 876 | 2,732,192 | 8.3 | 13.4 | 0.0 | 0.7 | 5.5 | 5.9 | 64.6 | 73.2 | 37.5 | 28.9 | 1,216 | 918 |
| Patrick | 643 | 1,910,080 | 5.5 | 14.9 | 0.0 | 0.0 | 4.1 | 1.7 | 71.7 | 40.6 | 23.6 | 14.8 | 834 | 686 |
| Pittsylvania | 1,637 | 5,762,549 | 5.8 | 9.1 | 0.1 | 0.1 | 0.3 | 2.3 | 79.3 | 140.2 | 88.8 | 45.4 | 740 | 605 |
| Powhatan | 817 | 3,614,068 | 6.4 | 6.5 | 0.6 | 0.0 | 2.0 | 1.1 | 80.1 | 75.6 | 31.7 | 41.6 | 1,450 | 1,364 |
| Prince Edward | 611 | 2,006,507 | 13.8 | 11.9 | 0.2 | 3.3 | 0.5 | 7.5 | 56.0 | 62.9 | 32.1 | 24.7 | 1,087 | 605 |
| Prince George | 1,469 | 5,634,849 | 6.4 | 27.3 | 1.7 | 0.0 | 3.3 | 1.8 | 58.4 | 134.3 | 79.7 | 42.2 | 1,109 | 891 |
| Prince William | 15,874 | 83,539,176 | 5.0 | 9.0 | 5.0 | 3.3 | 4.7 | 4.9 | 67.3 | 2,073.5 | 834.4 | 984.8 | 2,127 | 1,704 |
| Pulaski | 1,199 | 3,741,838 | 8.0 | 11.0 | 3.9 | 2.8 | 0.0 | 11.1 | 61.8 | 130.9 | 63.8 | 41.3 | 1,206 | 848 |
| Rappahannock | 202 | 678,379 | 11.1 | 11.5 | 0.0 | 0.0 | 6.8 | 2.6 | 67.4 | 24.7 | 7.6 | 15.8 | 2,137 | 1,790 |
| Richmond | 411 | 1,469,844 | 8.0 | 27.1 | 0.0 | 0.0 | 5.5 | 2.7 | 55.7 | 29.9 | 15.8 | 11.0 | 1,243 | 981 |
| Roanoke | 3,694 | 14,726,656 | 8.1 | 6.7 | 5.0 | 0.3 | 2.9 | 11.5 | 60.9 | 390.3 | 121.6 | 197.0 | 2,097 | 1,571 |
| Rockbridge | 721 | 2,265,484 | 8.1 | 7.5 | 0.3 | 0.3 | 4.2 | 4.1 | 72.4 | 68.0 | 25.0 | 34.8 | 1,533 | 1,115 |
| Rockingham | 3,116 | 10,412,579 | 7.3 | 8.8 | 3.5 | 0.6 | 5.0 | 5.5 | 68.5 | 248.7 | 107.7 | 96.8 | 1,204 | 1,035 |
| Russell | 990 | 3,567,096 | 5.2 | 8.5 | 0.0 | 0.6 | 0.2 | 3.0 | 76.8 | 81.6 | 41.1 | 25.4 | 938 | 691 |
| Scott | 812 | 2,547,769 | 8.1 | 6.8 | 0.0 | 0.0 | 7.5 | 3.4 | 71.9 | 81.8 | 40.6 | 37.4 | 1,708 | 1,100 |
| Shenandoah | 1,562 | 5,381,072 | 7.3 | 9.5 | 3.8 | 2.2 | 2.9 | 7.7 | 64.0 | 134.4 | 59.7 | 58.6 | 1,354 | 1,073 |
| Smyth | 1,327 | 3,913,619 | 7.5 | 10.1 | 2.2 | 2.1 | 11.0 | 5.8 | 59.4 | 103.7 | 63.2 | 26.8 | 871 | 625 |
| Southampton | 649 | 2,076,317 | 9.7 | 12.9 | 0.0 | 0.2 | 4.5 | 4.0 | 67.4 | 62.8 | 33.6 | 23.4 | 1,314 | 1,173 |
| Spotsylvania | 4,684 | 16,854,269 | 6.7 | 5.6 | 5.8 | 0.0 | 2.5 | 4.8 | 72.7 | 449.7 | 200.9 | 217.2 | 1,636 | 1,257 |
| Stafford | 4,836 | 19,394,927 | 6.6 | 5.2 | 3.1 | 0.2 | 1.7 | 6.4 | 75.2 | 574.8 | 208.6 | 320.7 | 2,185 | 1,663 |
| Surry | 310 | 1,057,019 | 7.8 | 8.0 | 0.0 | 0.0 | 2.6 | 3.0 | 71.5 | 34.9 | 7.3 | 26.1 | 4,016 | 3,829 |
| Sussex | 367 | 1,351,833 | 14.7 | 16.7 | 0.0 | 0.2 | 9.2 | 2.2 | 54.8 | 53.3 | 25.8 | 18.3 | 1,606 | 1,387 |
| Tazewell | 1,448 | 4,226,517 | 10.8 | 9.6 | 1.4 | 2.3 | 1.4 | 6.0 | 64.9 | 139.3 | 71.4 | 40.6 | 983 | 679 |
| Warren | 1,246 | 4,431,292 | 7.5 | 11.4 | 4.3 | 1.0 | 3.6 | 9.1 | 58.9 | 113.5 | 43.0 | 57.4 | 1,456 | 1,213 |
| Washington | 2,379 | 7,519,585 | 4.8 | 22.4 | 0.3 | 1.1 | 2.6 | 7.2 | 59.0 | 182.1 | 94.9 | 57.0 | 1,052 | 729 |
| Westmoreland | 584 | 1,969,880 | 12.5 | 10.4 | 0.0 | 0.5 | 4.6 | 2.6 | 68.4 | 66.0 | 27.4 | 26.1 | 1,474 | 1,192 |
| Wise | 1,326 | 4,670,621 | 9.1 | 10.2 | 0.1 | 1.7 | 0.2 | 9.1 | 67.7 | 137.5 | 82.3 | 43.6 | 1,130 | 827 |
| Wythe | 1,223 | 3,964,070 | 6.6 | 26.7 | 0.7 | 1.5 | 0.0 | 5.5 | 53.0 | 215.7 | 44.3 | 36.5 | 1,268 | 774 |
| York | 2,729 | 10,981,478 | 9.4 | 6.4 | 9.1 | 0.0 | 4.4 | 2.6 | 60.3 | 327.9 | 105.7 | 189.4 | 2,784 | 2,259 |
| **Independent cities** | | | | | | | | | | | | | | |
| Alexandria city | 5,558 | 29,910,334 | 9.3 | 14.2 | 7.3 | 1.9 | 12.9 | 8.3 | 42.1 | 892.3 | 177.2 | 611.7 | 3,842 | 2,970 |
| Bristol city | 848 | 3,102,214 | 5.2 | 13.3 | 5.5 | 2.7 | 4.9 | 8.6 | 39.4 | 115.6 | 40.1 | 32.0 | 1,898 | 1,078 |
| Buena Vista city | 277 | 891,039 | 14.2 | 6.9 | 0.0 | 4.3 | 0.0 | 11.0 | 62.6 | 23.2 | 13.0 | 7.7 | 1,186 | 895 |
| Charlottesville city | 2,178 | 9,337,860 | 10.1 | 7.9 | 5.3 | 9.0 | 7.2 | 7.6 | 46.3 | 258.9 | 90.5 | 116.5 | 2,455 | 1,482 |
| Chesapeake city | 11,046 | 44,150,325 | 4.4 | 9.5 | 4.4 | 1.6 | 24.8 | 3.8 | 48.7 | 1,317.1 | 444.8 | 465.8 | 1,939 | 1,356 |
| Colonial Heights city | 797 | 3,207,146 | 6.7 | 12.4 | 7.5 | 3.8 | 0.5 | 4.2 | 63.0 | 71.9 | 24.0 | 43.5 | 2,496 | 1,284 |
| Covington city | 291 | 1,032,630 | 3.0 | 11.2 | 0.0 | 3.4 | 1.0 | 10.1 | 69.3 | 27.2 | 11.3 | 12.1 | 2,174 | 1,493 |
| Danville city | 2,182 | 7,837,955 | 10.7 | 13.4 | 6.7 | 4.0 | 5.6 | 10.5 | 44.5 | 213.5 | 110.0 | 56.2 | 1,369 | 721 |
| Emporia city | 102 | 426,107 | 28.3 | 33.8 | 6.1 | 5.8 | 0.0 | 25.5 | 0.0 | 24.2 | 6.5 | 12.4 | 2,275 | 1,063 |
| Fairfax city | 420 | 2,753,519 | 18.6 | 21.8 | 23.5 | 18.0 | 0.6 | 13.3 | 0.2 | 146.8 | 17.2 | 113.4 | 4,840 | 3,206 |
| Falls Church city | 792 | 4,186,167 | 12.4 | 7.6 | 0.1 | 2.9 | 1.8 | 4.4 | 65.2 | 104.9 | 13.2 | 71.9 | 5,046 | 3,822 |
| Franklin city | 605 | 2,117,474 | 4.7 | 35.6 | 5.5 | 2.4 | 2.9 | 6.3 | 38.2 | 46.9 | 26.0 | 12.8 | 1,575 | 890 |
| Fredericksburg city | 1,663 | 6,605,878 | 9.0 | 26.4 | 3.6 | 4.1 | 3.2 | 4.7 | 44.3 | 208.1 | 73.7 | 81.0 | 2,842 | 1,463 |
| Galax city | 430 | 1,429,941 | 8.2 | 7.1 | 0.3 | 1.0 | 8.6 | 12.5 | 52.9 | 32.1 | 16.9 | 10.6 | 1,632 | 756 |
| Hampton city | 5,451 | 22,010,831 | 7.1 | 11.1 | 7.1 | 0.9 | 5.8 | 7.4 | 58.0 | 596.1 | 268.0 | 244.4 | 1,814 | 1,233 |
| Harrisonburg city | 1,563 | 6,085,265 | 3.9 | 7.3 | 6.4 | 5.5 | 0.3 | 14.2 | 55.8 | 198.8 | 76.6 | 79.3 | 1,479 | 722 |
| Hopewell city | 1,171 | 4,276,913 | 8.5 | 9.1 | 5.4 | 3.0 | 7.9 | 8.2 | 56.0 | 123.7 | 53.6 | 39.2 | 1,743 | 1,273 |
| Lexington city | 270 | 924,775 | 9.0 | 22.4 | 7.1 | 5.2 | 0.0 | 10.9 | 30.7 | 23.8 | 8.2 | 11.1 | 1,545 | 938 |
| Lynchburg city | 3,052 | 10,139,414 | 10.4 | 8.6 | 5.8 | 4.9 | 5.9 | 6.5 | 56.7 | 375.9 | 182.2 | 136.8 | 1,699 | 1,007 |
| Manassas city | 2,228 | 11,497,197 | 4.1 | 22.2 | 2.9 | 1.8 | 14.3 | 5.6 | 47.6 | 244.2 | 108.0 | 99.7 | 2,423 | 1,844 |
| Manassas Park city | 682 | 3,091,611 | 4.9 | 7.1 | 5.1 | 1.0 | 4.3 | 6.2 | 69.9 | 70.9 | 32.9 | 33.8 | 1,973 | 1,638 |

1. Based on the resident population estimated as of July 1 of the year shown.

## Table B. States and Counties — Local Government Finances, Government Employment, and Income Taxes

| STATE County | Local government finances, 2017 (cont.) | | | | | | | | | Government employment, 2019 | | | Individual income tax returns, 2018 | | |
| | Direct general expenditure | | | | | | | Debt outstanding | | | | | | | |
| | Total (mil dol) | Per capita¹ (dollars) | Percent of total for: | | | | | Total (mil dol) | Per capita¹ (dollars) | Federal civilian | Federal military | State and local | Number of returns | Mean adjusted gross income | Mean income tax |
| | | | Education | Health and hospitals | Police protection | Public welfare | Highways | | | | | | | | |
| | 185 | 186 | 187 | 188 | 189 | 190 | 191 | 192 | 193 | 194 | 195 | 196 | 197 | 198 | 199 |

**VIRGINIA—Cont'd**

| STATE County | 185 | 186 | 187 | 188 | 189 | 190 | 191 | 192 | 193 | 194 | 195 | 196 | 197 | 198 | 199 |
|---|---|---|---|---|---|---|---|---|---|---|---|---|---|---|---|
| Louisa | 145 | 4,026 | 68.3 | 1.0 | 3.8 | 5.4 | 0.3 | 17.6 | 491 | 65 | 119 | 1,634 | 17,180 | 63,795 | 6,233 |
| Lunenburg | 48 | 3,922 | 52.3 | 0.9 | 4.3 | 5.7 | 1.7 | 13.5 | 1,094 | 19 | 36 | 773 | 4,720 | 45,133 | 3,327 |
| Madison | 38 | 2,872 | 55.7 | 5.0 | 7.8 | 15.2 | 0.3 | 22.7 | 1,718 | 62 | 42 | 546 | 6,260 | 62,474 | 6,207 |
| Mathews | 16 | 1,883 | 82.2 | 0.0 | 16.7 | 0.0 | 0.0 | 4.7 | 534 | 16 | 71 | 373 | 4,310 | 67,510 | 7,001 |
| Mecklenburg | 100 | 3,258 | 52.8 | 6.2 | 7.2 | 4.8 | 1.0 | 33.4 | 1,088 | 119 | 94 | 1,749 | 13,410 | 48,591 | 4,199 |
| Middlesex | 28 | 2,627 | 54.7 | 1.9 | 5.6 | 6.0 | 0.0 | 19.5 | 1,838 | 21 | 32 | 1,010 | 5,220 | 63,242 | 6,481 |
| Montgomery | 328 | 3,341 | 34.4 | 15.3 | 7.0 | 2.1 | 4.6 | 198.7 | 2,025 | [2] 316 | [2] 409 | [2] 18,256 | 35,480 | 67,894 | 8,523 |
| Nelson | 47 | 3,183 | 62.2 | 0.5 | 6.2 | 4.4 | 0.7 | 22.7 | 1,532 | 58 | 47 | 706 | 7,400 | 64,606 | 6,692 |
| New Kent | 56 | 2,575 | 56.6 | 1.0 | 4.8 | 4.4 | 0.0 | 48.0 | 2,217 | 49 | 72 | 947 | 11,540 | 102,824 | 16,047 |
| Northampton | 201 | 16,890 | 12.9 | 6.2 | 1.8 | 0.5 | 52.1 | 658.0 | 55,422 | 35 | 65 | 941 | 6,030 | 54,472 | 5,920 |
| Northumberland | 31 | 2,526 | 58.5 | 1.1 | 7.2 | 7.0 | 1.3 | 35.7 | 2,905 | 28 | 38 | 473 | 5,950 | 67,533 | 7,466 |
| Nottoway | 43 | 2,757 | 62.1 | 2.9 | 4.6 | 7.9 | 2.1 | 15.8 | 1,023 | 334 | 51 | 2,467 | 5,980 | 46,458 | 3,578 |
| Orange | 122 | 3,410 | 46.8 | 0.5 | 6.4 | 5.5 | 2.7 | 166.7 | 4,647 | 67 | 116 | 2,237 | 17,550 | 63,183 | 6,054 |
| Page | 88 | 3,716 | 42.2 | 1.8 | 21.5 | 4.0 | 1.6 | 82.6 | 3,473 | 201 | 75 | 1,104 | 10,980 | 47,738 | 3,698 |
| Patrick | 43 | 2,437 | 66.3 | 1.2 | 7.0 | 5.0 | 1.2 | 81.9 | 4,626 | 45 | 55 | 849 | 7,330 | 44,190 | 3,308 |
| Pittsylvania | 143 | 2,331 | 63.3 | 1.8 | 5.2 | 10.2 | 0.0 | 120.1 | 1,959 | [3] 202 | [3] 311 | [3] 6,309 | 26,620 | 49,473 | 4,192 |
| Powhatan | 68 | 2,359 | 67.4 | 0.4 | 7.3 | 3.9 | 0.0 | 187.3 | 6,524 | 56 | 91 | 1,729 | 14,050 | 89,327 | 10,849 |
| Prince Edward | 64 | 2,806 | 50.3 | 0.5 | 8.8 | 7.1 | 5.4 | 18.4 | 811 | 70 | 61 | 2,353 | 7,890 | 48,386 | 3,987 |
| Prince George | 143 | 3,766 | 48.5 | 0.5 | 5.3 | 2.5 | 0.0 | 125.8 | 3,309 | [4] 4,710 | [4] 8,853 | [4] 3,299 | 15,620 | 61,850 | 5,351 |
| Prince William | 2,262 | 4,884 | 60.4 | 2.3 | 4.9 | 1.9 | 2.2 | 1,817.4 | 3,925 | [5] 7,353 | [5] 9,403 | [5] 25,158 | 223,840 | 80,760 | 9,228 |
| Pulaski | 110 | 3,215 | 43.1 | 9.9 | 6.0 | 4.2 | 1.8 | 77.1 | 2,251 | 49 | 105 | 2,276 | 14,610 | 52,531 | 4,661 |
| Rappahannock | 29 | 3,980 | 43.5 | 0.0 | 5.4 | 8.6 | 0.0 | 5.4 | 734 | 19 | 23 | 316 | 3,620 | 84,290 | 11,038 |
| Richmond | 30 | 3,417 | 54.8 | 0.7 | 6.0 | 6.6 | 0.0 | 7.7 | 867 | 30 | 30 | 1,091 | 3,520 | 50,074 | 3,877 |
| Roanoke | 409 | 4,354 | 45.8 | 6.9 | 5.7 | 3.9 | 0.3 | 409.9 | 4,364 | [6] 2,173 | [6] 366 | [6] 6,764 | 46,690 | 71,127 | 7,880 |
| Rockbridge | 65 | 2,883 | 49.7 | 12.0 | 4.2 | 1.8 | 0.0 | 65.6 | 2,892 | [7] 90 | [7] 152 | [7] 2,680 | 10,100 | 60,141 | 5,954 |
| Rockingham | 263 | 3,267 | 56.9 | 4.5 | 3.7 | 9.3 | 1.2 | 186.1 | 2,315 | [8] 357 | [8] 407 | [8] 11,826 | 38,540 | 61,751 | 5,946 |
| Russell | 110 | 4,064 | 41.3 | 0.5 | 15.0 | 6.5 | 3.8 | 79.5 | 2,941 | 64 | 83 | 1,403 | 9,780 | 46,705 | 3,445 |
| Scott | 75 | 3,404 | 48.9 | 6.1 | 27.8 | 0.0 | 0.7 | 10.6 | 484 | 55 | 66 | 1,224 | 8,130 | 45,146 | 3,304 |
| Shenandoah | 128 | 2,957 | 56.5 | 0.4 | 7.3 | 6.3 | 2.2 | 112.7 | 2,605 | 125 | 137 | 2,106 | 21,030 | 57,341 | 5,398 |
| Smyth | 101 | 3,285 | 56.3 | 0.9 | 7.3 | 10.0 | 0.5 | 84.6 | 2,754 | 75 | 93 | 2,948 | 12,340 | 42,644 | 2,990 |
| Southampton | 65 | 3,649 | 54.2 | 0.8 | 3.8 | 3.4 | 0.1 | 65.0 | 3,642 | [9] 73 | [9] 76 | [9] 2,288 | 7,460 | 54,453 | 4,529 |
| Spotsylvania | 443 | 3,337 | 65.3 | 0.1 | 4.8 | 4.5 | 0.0 | 425.1 | 3,203 | [10] 525 | [10] 517 | [10] 9,988 | 64,660 | 71,856 | 7,665 |
| Stafford | 457 | 3,114 | 69.2 | 0.5 | 4.1 | 2.9 | 2.0 | 739.8 | 5,040 | 4,793 | 663 | 6,855 | 69,160 | 83,104 | 9,242 |
| Surry | 32 | 4,867 | 62.5 | 1.3 | 5.0 | 7.6 | 0.0 | 14.2 | 2,186 | 10 | 20 | 547 | 3,120 | 53,357 | 4,551 |
| Sussex | 47 | 4,140 | 57.8 | 1.3 | 6.8 | 8.2 | 0.7 | 23.2 | 2,036 | 52 | 27 | 1,238 | 4,230 | 43,268 | 3,296 |
| Tazewell | 179 | 4,332 | 34.7 | 15.0 | 5.6 | 5.2 | 14.5 | 28.6 | 693 | 72 | 123 | 3,250 | 15,260 | 55,410 | 5,271 |
| Warren | 163 | 4,124 | 50.0 | 0.2 | 6.1 | 3.5 | 6.4 | 199.3 | 5,056 | 219 | 125 | 1,757 | 19,090 | 63,348 | 6,249 |
| Washington | 205 | 3,780 | 38.6 | 11.5 | 4.1 | 3.9 | 1.2 | 145.1 | 2,678 | [11] 232 | [11] 220 | [11] 4,718 | 22,360 | 57,069 | 5,608 |
| Westmoreland | 67 | 3,783 | 61.3 | 0.7 | 6.8 | 5.3 | 1.3 | 18.9 | 1,065 | 55 | 57 | 880 | 8,720 | 56,582 | 5,280 |
| Wise | 133 | 3,458 | 44.1 | 7.6 | 6.8 | 7.8 | 3.3 | 104.6 | 2,711 | [12] 223 | [12] 122 | [12] 4,022 | 13,150 | 44,794 | 3,324 |
| Wythe | 114 | 3,968 | 41.4 | 0.1 | 5.9 | 6.8 | 2.0 | 110.3 | 3,826 | 88 | 90 | 2,342 | 12,440 | 47,116 | 3,656 |
| York | 327 | 4,810 | 68.0 | 0.2 | 2.9 | 2.8 | 0.0 | 134.7 | 1,981 | [13] 1,061 | [13] 2,442 | [13] 3,824 | 32,300 | 82,470 | 9,267 |
| **Independent cities** | | | | | | | | | | | | | | | |
| Alexandria city | 954 | 5,989 | 29.3 | 5.1 | 9.9 | 5.1 | 6.3 | 933.6 | 5,864 | 13,443 | 691 | 8,338 | 85,570 | 115,169 | 18,711 |
| Bristol city | 108 | 6,400 | 24.7 | 0.6 | 5.7 | 6.3 | 7.1 | 152.4 | 9,026 | [11] | [11] | [11] | 7,060 | 43,540 | 3,409 |
| Buena Vista city | 23 | 3,575 | 42.3 | 0.0 | 7.5 | 17.8 | 7.3 | 40.9 | 6,315 | [7] | [7] | [7] | 2,740 | 38,932 | 2,357 |
| Charlottesville city | 279 | 5,868 | 28.6 | 7.9 | 6.5 | 10.5 | 7.0 | 155.8 | 3,283 | [14] | [14] | [14] | 20,560 | 124,278 | 19,057 |
| Chesapeake city | 1,272 | 5,296 | 38.8 | 26.0 | 4.9 | 1.6 | 4.4 | 652.5 | 2,717 | 1,210 | 1,536 | 14,796 | 114,600 | 67,171 | 6,667 |
| Colonial Heights city | 71 | 4,072 | 52.4 | 0.3 | 5.7 | 1.6 | 3.4 | 53.2 | 3,051 | [20] | [20] | [20] | 8,950 | 50,951 | 4,154 |
| Covington city | 25 | 4,473 | 47.2 | 0.4 | 10.7 | 7.1 | 5.6 | 41.2 | 7,390 | [15] | [15] | [15] | 2,760 | 40,834 | 2,842 |
| Danville city | 233 | 5,678 | 39.2 | 0.6 | 5.7 | 5.3 | 9.0 | 142.3 | 3,467 | [3] | [3] | [3] | 17,530 | 43,032 | 3,863 |
| Emporia city | 16 | 2,986 | 26.6 | 1.6 | 23.5 | 2.2 | 8.3 | 0.0 | 0 | [16] | [16] | [16] | 2,410 | 37,120 | 2,628 |
| Fairfax city | 175 | 7,470 | 52.0 | 1.6 | 6.8 | 1.9 | 6.1 | 248.3 | 10,597 | [17] | [17] | [17] | 12,900 | 99,902 | 13,818 |
| Falls Church city | 96 | 6,763 | 54.0 | 0.3 | 7.4 | 2.1 | 6.2 | 0.0 | 0 | [17] | [17] | [17] | 7,110 | 163,342 | 28,793 |
| Franklin city | 45 | 5,542 | 34.8 | 0.4 | 6.7 | 3.6 | 2.7 | 31.1 | 3,816 | [9] | [9] | [9] | 3,690 | 46,531 | 3,491 |
| Fredericksburg city | 219 | 7,677 | 25.1 | 16.5 | 3.8 | 3.1 | 3.4 | 178.8 | 6,274 | [10] | [10] | [10] | 12,890 | 71,804 | 9,055 |
| Galax city | 34 | 5,270 | 48.1 | 0.6 | 8.7 | 5.7 | 7.0 | 8.0 | 1,229 | [18] | [18] | [18] | 2,840 | 41,368 | 3,101 |
| Hampton city | 590 | 4,378 | 38.8 | 1.3 | 5.2 | 5.6 | 3.4 | 309.3 | 2,296 | 7,314 | 7,583 | 7,942 | 64,320 | 49,522 | 4,172 |
| Harrisonburg city | 215 | 4,013 | 54.2 | 0.6 | 4.7 | 1.6 | 4.7 | 6.2 | 115 | [8] | [8] | [8] | 18,200 | 47,389 | 3,907 |
| Hopewell city | 125 | 5,543 | 46.0 | 0.5 | 4.6 | 6.2 | 3.5 | 0.0 | 0 | [4] | [4] | [4] | 10,550 | 39,506 | 2,712 |
| Lexington city | 27 | 3,809 | 24.6 | 0.3 | 6.5 | 1.7 | 4.5 | 48.2 | 6,732 | [7] | [7] | [7] | 2,320 | 78,763 | 9,499 |
| Lynchburg city | 335 | 4,154 | 43.6 | 0.6 | 6.3 | 8.1 | 3.7 | 332.3 | 4,127 | [19] | [19] | [19] | 32,170 | 54,762 | 5,675 |
| Manassas city | 270 | 6,565 | 43.2 | 1.5 | 5.5 | 6.8 | 4.2 | 115.2 | 2,801 | [5] | [5] | [5] | 20,820 | 61,952 | 6,127 |
| Manassas Park city | 74 | 4,314 | 57.8 | 1.4 | 6.0 | 4.1 | 5.6 | 145.4 | 8,481 | [5] | [5] | [5] | 8,480 | 55,990 | 5,012 |

1. Based on the resident population estimated as of July 1 of the year shown.  2. Radford city is included with Montgomery county.  3. Danville city is included with Pittsylvania county.  4. Hopewell city is included with Prince George county.  5. Manassas and Manassas Park cities are included with Prince William county.  6. Salem city is included with Roanoke county.  7. Buena Vista and Lexington cities are included with Rockbridge county.  8. Harrisonburg city is included with Rockingham county.  9. Franklin city is included with Southampton county.  10. Fredericksburg city is included with Spotsylvania county.  11. Bristol city is included with Washington county.  12. Norton city is included with Wise county.  13. Poquoson city is included with York county.  14. Charlottesville city is included with Albemarle county.  15. Covington city is included with Alleghany county.  16. Emporia city is included with Greensville county.  17. Fairfax city and Falls Church city are included with Fairfax county.  18. Galax city is included with Carroll county.  19. Lynchburg city is included with Campbell county.  20. Petersburg and Colonial Heights cities are included with Dinwiddie county.

# Table B. States and Counties — Land Area and Population

| State / county code | CBSA code[1] | County Type code[2] | STATE County | Land area[3] (sq. mi) | Population, 2020 Total persons 2019 | Rank | Per square mile | Race alone or in combination, not Hispanic or Latino (percent) White | Black | American Indian, Alaska Native | Asian and Pacific Islander | Percent Hispanic or Latino[4] | Age (percent) Under 5 years | 5 to 14 years | 15 to 24 years | 25 to 34 years | 35 to 44 years | 45 to 54 years |
|---|---|---|---|---|---|---|---|---|---|---|---|---|---|---|---|---|---|---|
| | | | | 1 | 2 | 3 | 4 | 5 | 6 | 7 | 8 | 9 | 10 | 11 | 12 | 13 | 14 | 15 |
| | | | VIRGINIA—Cont'd | | | | | | | | | | | | | | | |
| 51690 | 32300 | 4 | Martinsville city | 11.0 | 12,355 | 2,264 | 1,123.2 | 45.6 | 47.6 | 0.7 | 1.4 | 7.0 | 6.7 | 14.3 | 11.5 | 12.5 | 10.9 | 11.9 |
| 51700 | 47260 | 1 | Newport News city | 69.0 | 179,062 | 371 | 2,595.1 | 44.7 | 43.6 | 1.2 | 4.9 | 9.8 | 7.2 | 12.6 | 15.0 | 17.1 | 12.5 | 10.2 |
| 51710 | 47260 | 1 | Norfolk city | 53.3 | 242,803 | 283 | 4,555.4 | 46.0 | 42.3 | 1.3 | 5.4 | 8.9 | 6.4 | 10.3 | 20.1 | 19.6 | 11.9 | 9.2 |
| 51720 | 13720 | 7 | Norton city | 7.5 | 3,985 | 2,898 | 531.3 | 87.2 | 8.1 | 1.0 | 2.1 | 4.2 | 6.3 | 12.3 | 11.0 | 14.9 | 12.0 | 12.3 |
| 51730 | 40060 | 1 | Petersburg city | 22.7 | 30,446 | 1,419 | 1,341.2 | 16.7 | 77.7 | 1.0 | 1.8 | 5.4 | 7.2 | 11.8 | 11.5 | 16.9 | 10.5 | 11.0 |
| 51735 | 47260 | 1 | Poquoson city | 15.4 | 12,257 | 2,268 | 795.9 | 92.3 | 2.0 | 0.9 | 3.8 | 3.4 | 4.5 | 13.5 | 11.8 | 10.1 | 13.5 | 12.8 |
| 51740 | 47260 | 1 | Portsmouth city | 33.3 | 95,094 | 628 | 2,855.7 | 39.1 | 55.6 | 1.3 | 2.4 | 4.9 | 7.2 | 12.5 | 12.6 | 16.8 | 12.7 | 10.5 |
| 51750 | 13980 | 3 | Radford city | 9.7 | 18,255 | 1,912 | 1,882.0 | 85.1 | 10.8 | 0.7 | 2.7 | 3.2 | 3.2 | 6.9 | 41.1 | 12.5 | 8.9 | 8.9 |
| 51760 | 40060 | 1 | Richmond city | 59.9 | 232,226 | 296 | 3,876.9 | 44.4 | 46.5 | 0.8 | 3.1 | 7.5 | 5.7 | 8.9 | 13.8 | 22.2 | 12.8 | 10.3 |
| 51770 | 40220 | 2 | Roanoke city | 42.5 | 99,058 | 609 | 2,330.8 | 60.1 | 31.4 | 0.8 | 4.0 | 6.8 | 6.5 | 12.4 | 10.8 | 15.6 | 12.6 | 12.0 |
| 51775 | 40220 | 2 | Salem city | 14.5 | 25,340 | 1,594 | 1,747.6 | 86.0 | 9.3 | 0.7 | 2.3 | 3.9 | 4.4 | 10.6 | 16.8 | 11.3 | 11.3 | 12.2 |
| 51790 | 44420 | 3 | Staunton city | 19.9 | 25,190 | 1,599 | 1,265.8 | 82.9 | 13.8 | 0.9 | 2.1 | 3.8 | 5.8 | 11.0 | 11.0 | 14.1 | 12.8 | 11.6 |
| 51800 | 47260 | 1 | Suffolk city | 399.2 | 93,913 | 634 | 235.3 | 50.7 | 43.5 | 0.8 | 3.1 | 4.7 | 6.4 | 13.4 | 11.7 | 13.7 | 13.1 | 12.8 |
| 51810 | 47260 | 1 | Virginia Beach city | 244.7 | 451,231 | 156 | 1,844.0 | 63.9 | 21.2 | 1.0 | 9.5 | 8.8 | 6.1 | 12.2 | 12.5 | 16.3 | 13.6 | 11.7 |
| 51820 | 44420 | 3 | Waynesboro city | 15.0 | 22,741 | 1,695 | 1,516.1 | 75.9 | 14.5 | 1.0 | 2.3 | 9.6 | 6.6 | 12.7 | 11.0 | 13.9 | 13.3 | 11.3 |
| 51830 | 47260 | 1 | Williamsburg city | 8.9 | 15,259 | 2,075 | 1,714.5 | 71.4 | 16.2 | 1.0 | 7.3 | 7.6 | 3.5 | 6.3 | 35.8 | 11.3 | 7.8 | 7.1 |
| 51840 | 49020 | 3 | Winchester city | 9.2 | 27,700 | 1,506 | 3,010.9 | 68.3 | 12.5 | 0.6 | 3.0 | 18.6 | 6.4 | 12.2 | 14.5 | 14.3 | 11.9 | 11.6 |
| 53000 | | 0 | WASHINGTON | 66,455.1 | 7,693,612 | X | 115.8 | 70.7 | 5.5 | 2.4 | 12.9 | 13.4 | 5.8 | 12.2 | 12.1 | 15.5 | 13.7 | 12.0 |
| 53001 | 36830 | 6 | Adams | 1,925.0 | 20,027 | 1,821 | 10.4 | 32.9 | 0.9 | 0.8 | 1.1 | 65.2 | 9.4 | 20.8 | 15.7 | 12.5 | 10.9 | 9.5 |
| 53003 | 30300 | 3 | Asotin | 636.1 | 22,820 | 1,691 | 35.9 | 91.8 | 1.5 | 2.7 | 2.2 | 4.5 | 5.0 | 11.4 | 10.3 | 10.7 | 10.9 | 9.5 |
| 53005 | 28420 | 2 | Benton | 1,700.1 | 206,426 | 331 | 121.4 | 71.2 | 2.2 | 1.5 | 4.5 | 23.3 | 6.6 | 15.3 | 12.6 | 13.7 | 13.1 | 11.1 |
| 53007 | 48300 | 3 | Chelan | 2,921.2 | 77,574 | 722 | 26.6 | 69.0 | 0.9 | 1.6 | 1.8 | 28.4 | 5.9 | 13.4 | 11.6 | 12.5 | 12.1 | 11.0 |
| 53009 | 38820 | 5 | Clallam | 1,738.7 | 78,067 | 719 | 44.9 | 85.2 | 1.6 | 6.5 | 3.5 | 7.1 | 4.1 | 9.6 | 8.8 | 10.4 | 10.7 | 10.8 |
| 53011 | 38900 | 1 | Clark | 628.5 | 496,865 | 141 | 790.6 | 80.3 | 3.5 | 1.8 | 8.0 | 10.7 | 5.9 | 13.3 | 12.1 | 13.6 | 13.4 | 9.7 |
| 53013 | | 3 | Columbia | 868.6 | 4,048 | 2,889 | 4.7 | 85.8 | 1.6 | 2.4 | 4.2 | 8.5 | 4.2 | 10.3 | 9.3 | 10.1 | 10.5 | 12.7 |
| 53015 | 31020 | 3 | Cowlitz | 1,141.2 | 111,371 | 549 | 97.6 | 86.0 | 1.7 | 3.2 | 3.1 | 9.5 | 6.0 | 12.7 | 11.1 | 12.9 | 11.9 | 10.1 |
| 53017 | 48300 | 3 | Douglas | 1,819.2 | 43,560 | 1,105 | 23.9 | 64.7 | 1.0 | 1.7 | 2.0 | 32.7 | 6.3 | 14.6 | 12.5 | 12.5 | 12.2 | 11.9 |
| 53019 | | 9 | Ferry | 2,203.2 | 7,759 | 2,611 | 3.5 | 77.6 | 1.6 | 18.3 | 2.3 | 5.0 | 4.2 | 9.2 | 10.2 | 8.4 | 9.9 | 11.0 |
| 53021 | 28420 | 2 | Franklin | 1,241.6 | 97,075 | 620 | 78.2 | 41.3 | 2.7 | 1.0 | 3.1 | 53.7 | 8.5 | 18.1 | 15.0 | 14.9 | 14.1 | 11.5 |
| 53023 | | 8 | Garfield | 710.8 | 2,290 | 3,016 | 3.2 | 90.0 | 1.3 | 1.3 | 4.6 | 5.3 | 5.7 | 10.9 | 9.2 | 9.2 | 11.0 | 10.7 |
| 53025 | 34180 | 5 | Grant | 2,679.6 | 99,377 | 607 | 37.1 | 54.3 | 1.5 | 1.7 | 1.7 | 42.7 | 7.6 | 16.7 | 13.9 | 13.9 | 11.8 | 10.5 |
| 53027 | 10140 | 4 | Grays Harbor | 1,901.5 | 75,950 | 733 | 39.9 | 82.0 | 2.0 | 6.1 | 3.0 | 10.7 | 5.1 | 11.6 | 10.3 | 11.8 | 11.6 | 10.7 |
| 53029 | 36020 | 4 | Island | 208.5 | 86,014 | 673 | 412.5 | 82.0 | 4.3 | 2.0 | 7.8 | 8.5 | 5.3 | 9.9 | 10.9 | 14.1 | 10.8 | 11.6 |
| 53031 | | 6 | Jefferson | 1,803.7 | 32,700 | 1,364 | 18.1 | 90.7 | 1.6 | 3.6 | 3.3 | 4.0 | 3.0 | 6.6 | 6.4 | 8.3 | 9.2 | 9.4 |
| 53033 | 42660 | 1 | King | 2,115.2 | 2,274,315 | 12 | 1,075.2 | 61.2 | 8.3 | 1.6 | 24.1 | 10.1 | 5.5 | 11.1 | 11.2 | 18.7 | 15.5 | 9.7 |
| 53035 | 14740 | 2 | Kitsap | 395.1 | 272,787 | 253 | 690.4 | 80.7 | 4.5 | 2.8 | 9.8 | 8.5 | 5.5 | 11.2 | 12.5 | 14.9 | 12.5 | 12.9 |
| 53037 | 21260 | 4 | Kittitas | 2,297.3 | 49,204 | 1,002 | 21.4 | 86.1 | 1.7 | 2.0 | 3.7 | 9.5 | 4.4 | 9.5 | 23.7 | 12.6 | 10.5 | 11.1 |
| 53039 | | 6 | Klickitat | 1,871.6 | 22,697 | 1,697 | 12.1 | 84.5 | 0.8 | 3.5 | 1.8 | 11.8 | 4.8 | 10.8 | 9.3 | 10.7 | 11.7 | 9.8 |
| 53041 | 16500 | 4 | Lewis | 2,402.8 | 82,109 | 695 | 34.2 | 85.5 | 1.5 | 3.0 | 2.3 | 10.9 | 5.9 | 12.0 | 10.6 | 12.3 | 11.9 | 12.1 |
| 53043 | | 8 | Lincoln | 2,310.6 | 11,090 | 2,340 | 4.8 | 93.1 | 1.1 | 3.3 | 1.7 | 3.6 | 5.3 | 12.0 | 10.6 | 8.7 | 10.8 | 11.4 |
| 53045 | 43220 | 4 | Mason | 959.5 | 68,224 | 787 | 71.1 | 82.8 | 2.1 | 5.0 | 3.3 | 10.8 | 5.1 | 10.9 | 10.1 | 12.2 | 11.4 | 10.4 |
| 53047 | | 6 | Okanogan | 5,266.2 | 42,620 | 1,126 | 8.1 | 66.8 | 1.1 | 12.0 | 1.5 | 21.1 | 5.9 | 13.2 | 10.3 | 11.2 | 11.2 | 10.9 |
| 53049 | | 7 | Pacific | 933.6 | 22,984 | 1,681 | 24.6 | 84.4 | 1.3 | 4.2 | 3.1 | 10.4 | 3.7 | 8.6 | 8.2 | 9.1 | 9.9 | 11.1 |
| 53051 | | 2 | Pend Oreille | 1,400.2 | 14,144 | 2,150 | 10.1 | 90.1 | 1.2 | 4.7 | 2.0 | 4.6 | 4.9 | 11.0 | 9.3 | 9.1 | 10.0 | 10.6 |
| 53053 | 42660 | 1 | Pierce | 1,668.0 | 913,890 | 61 | 547.9 | 70.6 | 10.0 | 2.6 | 12.4 | 11.8 | 6.4 | 13.1 | 12.5 | 15.8 | 13.8 | 11.0 |
| 53055 | | 9 | San Juan | 173.9 | 17,492 | 1,946 | 100.6 | 90.0 | 1.0 | 2.0 | 2.5 | 6.7 | 2.8 | 7.3 | 7.5 | 8.4 | 9.9 | 11.9 |
| 53057 | 34580 | 3 | Skagit | 1,730.2 | 130,789 | 490 | 75.6 | 75.9 | 1.4 | 2.7 | 3.5 | 19.1 | 5.7 | 12.1 | 11.0 | 12.8 | 12.3 | 10.9 |
| 53059 | 38900 | 1 | Skamania | 1,658.3 | 12,107 | 2,287 | 7.3 | 89.2 | 1.3 | 3.3 | 2.4 | 7.1 | 4.0 | 10.5 | 8.7 | 10.4 | 12.1 | 10.9 |
| 53061 | 42660 | 1 | Snohomish | 2,086.5 | 830,393 | 73 | 398.0 | 71.0 | 5.0 | 2.2 | 15.8 | 11.0 | 6.2 | 12.5 | 11.0 | 14.9 | 14.8 | 12.9 |
| 53063 | 44060 | 2 | Spokane | 1,764.2 | 528,225 | 134 | 299.4 | 87.2 | 3.3 | 2.7 | 4.7 | 6.4 | 5.8 | 12.3 | 12.7 | 15.2 | 12.8 | 12.9 |
| 53065 | 44060 | 2 | Stevens | 2,477.5 | 46,360 | 1,049 | 18.7 | 89.3 | 1.0 | 7.0 | 1.9 | 3.9 | 4.9 | 12.2 | 10.2 | 9.7 | 10.6 | 11.4 |
| 53067 | 36500 | 2 | Thurston | 722.5 | 294,074 | 237 | 407.0 | 78.4 | 5.0 | 2.7 | 10.0 | 9.8 | 5.7 | 12.0 | 11.4 | 14.6 | 13.9 | 11.8 |
| 53069 | | 8 | Wahkiakum | 262.9 | 4,498 | 2,859 | 17.1 | 90.1 | 1.6 | 3.2 | 2.5 | 6.2 | 3.3 | 9.9 | 9.2 | 8.0 | 8.2 | 10.2 |
| 53071 | 47460 | 3 | Walla Walla | 1,270.0 | 61,292 | 861 | 48.3 | 72.9 | 2.5 | 1.6 | 3.4 | 22.0 | 5.3 | 11.9 | 16.4 | 12.8 | 11.6 | 10.7 |
| 53073 | 13380 | 3 | Whatcom | 2,107.9 | 231,016 | 297 | 109.6 | 81.2 | 2.0 | 3.7 | 7.0 | 10.1 | 4.9 | 10.9 | 16.6 | 13.8 | 12.5 | 11.0 |
| 53075 | 39420 | 4 | Whitman | 2,159.3 | 49,500 | 999 | 22.9 | 81.9 | 3.1 | 1.6 | 10.7 | 6.7 | 3.9 | 8.3 | 36.9 | 14.1 | 9.3 | 7.5 |
| 53077 | 49420 | 3 | Yakima | 4,294.5 | 251,879 | 275 | 58.7 | 43.3 | 1.3 | 4.4 | 2.0 | 50.9 | 7.7 | 16.8 | 14.3 | 13.6 | 11.9 | 10.7 |
| 54000 | | 0 | WEST VIRGINIA | 24,041.1 | 1,784,787 | X | 74.2 | 93.5 | 4.6 | 0.8 | 1.2 | 1.8 | 5.1 | 11.3 | 12.1 | 12.0 | 11.8 | 12.6 |
| 54001 | | 6 | Barbour | 341.1 | 16,444 | 2,010 | 48.2 | 96.8 | 1.8 | 1.3 | 0.5 | 1.1 | 4.9 | 11.0 | 14.1 | 11.3 | 10.8 | 12.8 |
| 54003 | 25180 | 2 | Berkeley | 321.1 | 122,125 | 519 | 380.3 | 85.9 | 9.8 | 0.8 | 1.8 | 5.1 | 6.1 | 12.9 | 11.2 | 14.4 | 13.3 | 13.3 |
| 54005 | 16620 | 3 | Boone | 501.5 | 21,055 | 1,766 | 42.0 | 98.1 | 1.2 | 0.4 | 0.2 | 0.8 | 4.9 | 11.6 | 11.2 | 10.4 | 11.9 | 14.0 |
| 54007 | | 8 | Braxton | 510.7 | 13,702 | 2,177 | 26.8 | 97.6 | 1.2 | 0.9 | 0.4 | 1.0 | 4.3 | 11.2 | 9.7 | 11.7 | 10.8 | 13.1 |
| 54009 | 48260 | 3 | Brooke | 89.2 | 21,674 | 1,734 | 243.0 | 96.9 | 2.3 | 0.6 | 0.7 | 1.0 | 4.4 | 9.8 | 12.1 | 10.7 | 10.4 | 12.6 |
| 54011 | 26580 | 2 | Cabell | 281.0 | 91,589 | 647 | 325.9 | 92.3 | 6.3 | 0.8 | 1.8 | 1.5 | 5.4 | 11.0 | 16.6 | 12.3 | 11.7 | 11.5 |

1. CBSA = Core Based Statistical Area. See Appendix A for explanation. See Appendix B for list of metropolitan areas with component counties. 2. County type code from the Economic Research Service of USDA Rural-Urban Continuum Codes. See Appendix A for definition. 3. Dry land or land partially or temporarily covered by water. 4. May be of any race.

| STATE County | 55 to 64 years (16) | 65 to 74 years (17) | 75 years and over (18) | Percent female (19) | Total persons 2000 (20) | 2010 (21) | Percent change 2000-2010 (22) | Percent change 2010-2020 (23) | Births (24) | Deaths (25) | Net Migration (26) | Households Number (27) | Persons per household (28) | Family households (29) | Female family householder[1] (30) | One person (31) |
|---|---|---|---|---|---|---|---|---|---|---|---|---|---|---|---|---|
| VIRGINIA—Cont'd | | | | | | | | | | | | | | | | |
| Martinsville city | 14.3 | 10.8 | 7.2 | 53.3 | 15,416 | 13,840 | -10.2 | -10.7 | 1,792 | 2,761 | -514 | 5,532 | 2.24 | 59.2 | 21.8 | 35.8 |
| Newport News city | 12.0 | 7.9 | 5.7 | 51.8 | 180,150 | 180,956 | 0.4 | -1.0 | 28,369 | 15,652 | -14,704 | 69,835 | 2.44 | 60.4 | 17.7 | 32.9 |
| Norfolk city | 10.7 | 7.3 | 4.6 | 47.9 | 234,403 | 242,840 | 3.6 | 0.0 | 36,840 | 20,383 | -16,407 | 88,353 | 2.47 | 57.1 | 17.5 | 34.0 |
| Norton city | 12.2 | 10.6 | 8.4 | 52.5 | 3,904 | 3,994 | 2.3 | -0.2 | 505 | 449 | -71 | 1,628 | 2.40 | 59.5 | 15.2 | 35.7 |
| Petersburg city | 13.7 | 10.2 | 7.2 | 54.3 | 33,740 | 32,441 | -3.9 | -6.1 | 5,024 | 4,936 | -2,085 | 13,165 | 2.33 | 50.5 | 23.5 | 42.5 |
| Poquoson city | 14.0 | 10.8 | 8.9 | 49.8 | 11,566 | 12,159 | 5.1 | 0.8 | 974 | 1,150 | 277 | 4,593 | 2.62 | 75.8 | 9.7 | 20.7 |
| Portsmouth city | 12.5 | 9.1 | 6.2 | 52.1 | 100,565 | 95,531 | -5.0 | -0.5 | 15,348 | 10,881 | -4,873 | 36,370 | 2.51 | 58.3 | 19.2 | 34.5 |
| Radford city | 8.9 | 6.0 | 3.8 | 53.0 | 15,859 | 16,395 | 3.4 | 11.3 | 1,218 | 1,135 | 1,743 | 5,573 | 2.64 | 42.3 | 12.2 | 41.4 |
| Richmond city | 12.1 | 8.8 | 5.3 | 52.7 | 197,790 | 204,440 | 3.4 | 13.6 | 30,040 | 19,536 | 16,901 | 90,301 | 2.39 | 44.3 | 15.1 | 43.4 |
| Roanoke city | 12.8 | 10.5 | 6.8 | 52.5 | 94,911 | 96,910 | 2.1 | 2.2 | 14,345 | 12,463 | 285 | 41,740 | 2.33 | 53.9 | 16.4 | 38.2 |
| Salem city | 13.7 | 11.2 | 8.5 | 52.4 | 24,747 | 24,834 | 0.4 | 2.0 | 2,320 | 3,347 | 1,544 | 9,912 | 2.30 | 62.1 | 14.5 | 33.1 |
| Staunton city | 13.1 | 12.1 | 9.2 | 53.7 | 23,853 | 23,745 | -0.5 | 6.1 | 2,969 | 3,743 | 2,208 | 10,597 | 2.18 | 57.0 | 12.1 | 36.9 |
| Suffolk city | 13.7 | 9.1 | 6.2 | 51.6 | 63,677 | 84,606 | 32.9 | 11.0 | 11,566 | 7,817 | 5,585 | 33,774 | 2.65 | 73.2 | 15.2 | 22.5 |
| Virginia Beach city | 12.5 | 8.9 | 6.2 | 51.0 | 425,257 | 437,885 | 3.0 | 3.0 | 60,755 | 31,811 | -15,583 | 170,798 | 2.58 | 68.4 | 12.8 | 24.5 |
| Waynesboro city | 13.0 | 10.1 | 8.1 | 52.3 | 19,520 | 20,994 | 7.6 | 8.3 | 2,955 | 2,745 | 1,542 | 9,193 | 2.39 | 55.1 | 12.9 | 38.1 |
| Williamsburg city | 10.1 | 9.8 | 8.1 | 54.1 | 11,998 | 13,679 | 14.0 | 11.6 | 1,137 | 921 | 1,355 | 4,706 | 2.24 | 46.7 | 9.8 | 40.4 |
| Winchester city | 12.3 | 9.9 | 7.0 | 50.8 | 23,585 | 26,223 | 11.2 | 5.6 | 3,811 | 2,778 | 448 | 10,490 | 2.56 | 57.4 | 13.6 | 34.3 |
| WASHINGTON | 12.4 | 9.9 | 6.4 | 49.9 | 5,894,121 | 6,724,540 | 14.1 | 14.4 | 897,866 | 557,404 | 628,387 | 2,848,396 | 2.55 | 64.7 | 9.8 | 26.7 |
| Adams | 9.5 | 7.0 | 4.8 | 49.0 | 16,428 | 18,731 | 14.0 | 6.9 | 4,031 | 1,230 | -1,517 | 5,973 | 3.24 | 76.5 | 14.3 | 19.1 |
| Asotin | 15.4 | 13.7 | 10.9 | 51.2 | 20,551 | 21,623 | 5.2 | 5.5 | 2,401 | 2,548 | 1,348 | 9,101 | 2.44 | 61.9 | 9.3 | 28.0 |
| Benton | 11.9 | 9.5 | 6.2 | 49.9 | 142,475 | 175,168 | 22.9 | 17.8 | 26,683 | 14,219 | 18,823 | 72,121 | 2.72 | 68.5 | 11.0 | 25.3 |
| Chelan | 13.4 | 12.1 | 8.1 | 49.8 | 66,616 | 72,460 | 8.8 | 7.1 | 9,375 | 7,062 | 2,851 | 28,384 | 2.64 | 65.9 | 9.5 | 28.6 |
| Clallam | 15.5 | 18.4 | 12.8 | 50.8 | 64,525 | 71,396 | 10.6 | 9.3 | 6,677 | 10,155 | 10,148 | 32,958 | 2.25 | 61.6 | 10.3 | 31.4 |
| Clark | 12.6 | 10.2 | 6.2 | 50.6 | 345,238 | 425,360 | 23.2 | 16.8 | 56,541 | 34,842 | 49,783 | 174,661 | 2.69 | 69.8 | 10.0 | 22.8 |
| Columbia | 15.9 | 16.5 | 13.1 | 50.8 | 4,064 | 4,078 | 0.3 | -0.7 | 350 | 551 | 170 | 1,795 | 2.18 | 60.3 | 2.2 | 33.6 |
| Cowlitz | 13.9 | 11.7 | 7.9 | 50.4 | 92,948 | 102,408 | 10.2 | 8.8 | 12,585 | 11,610 | 8,016 | 41,952 | 2.52 | 65.5 | 11.1 | 27.9 |
| Douglas | 12.4 | 10.8 | 7.6 | 49.2 | 32,603 | 38,427 | 17.9 | 13.4 | 5,335 | 3,182 | 3,000 | 15,263 | 2.73 | 73.3 | 11.8 | 20.7 |
| Ferry | 16.8 | 19.3 | 10.5 | 49.2 | 7,260 | 7,554 | 4.0 | 2.7 | 706 | 868 | 373 | 3,060 | 2.41 | 63.6 | 12.9 | 29.6 |
| Franklin | 8.7 | 6.5 | 3.6 | 48.7 | 49,347 | 78,160 | 58.4 | 24.2 | 16,633 | 4,007 | 6,176 | 26,723 | 3.36 | 76.5 | 15.2 | 18.5 |
| Garfield | 15.6 | 16.6 | 11.4 | 51.4 | 2,397 | 2,266 | -5.5 | 1.1 | 232 | 271 | 61 | 984 | 2.24 | 66.5 | 3.5 | 30.2 |
| Grant | 11.0 | 8.6 | 5.8 | 49.0 | 74,698 | 89,124 | 19.3 | 11.5 | 15,508 | 7,078 | 1,864 | 30,818 | 3.07 | 71.7 | 13.7 | 22.3 |
| Grays Harbor | 15.0 | 14.4 | 8.5 | 49.0 | 67,194 | 72,800 | 8.3 | 4.3 | 7,903 | 8,666 | 3,918 | 28,722 | 2.43 | 63.1 | 10.7 | 29.8 |
| Island | 13.7 | 15.7 | 10.2 | 50.0 | 71,558 | 78,508 | 9.7 | 9.6 | 9,310 | 7,357 | 5,539 | 34,768 | 2.31 | 68.0 | 7.9 | 24.9 |
| Jefferson | 17.7 | 24.3 | 14.8 | 51.0 | 25,953 | 29,880 | 15.1 | 9.4 | 1,950 | 3,784 | 4,639 | 14,705 | 2.07 | 60.5 | 7.1 | 29.0 |
| King | 11.5 | 8.2 | 5.5 | 49.6 | 1,737,034 | 1,931,289 | 11.2 | 17.8 | 256,380 | 132,315 | 218,630 | 882,028 | 2.45 | 67.2 | 9.5 | 25.1 |
| Kitsap | 13.4 | 11.6 | 7.2 | 48.9 | 231,969 | 251,143 | 8.3 | 8.6 | 30,811 | 22,396 | 13,139 | 103,913 | 2.48 | 54.1 | 6.7 | 30.6 |
| Kittitas | 12.4 | 10.5 | 6.6 | 49.5 | 33,362 | 40,910 | 22.6 | 20.3 | 4,190 | 3,144 | 7,226 | 18,347 | 2.36 | 63.5 | 7.1 | 28.4 |
| Klickitat | 15.7 | 15.6 | 9.4 | 49.5 | 19,161 | 20,317 | 6.0 | 11.7 | 2,135 | 1,976 | 2,219 | 8,877 | 2.42 | 67.0 | 9.9 | 27.3 |
| Lewis | 14.3 | 13.0 | 8.6 | 49.9 | 68,600 | 75,457 | 10.0 | 8.8 | 9,184 | 8,969 | 6,456 | 30,618 | 2.52 | 65.4 | 9.5 | 27.3 |
| Lincoln | 15.6 | 15.8 | 10.9 | 49.4 | 10,184 | 10,570 | 3.8 | 4.9 | 1,019 | 1,177 | 682 | 4,525 | 2.31 | 70.6 | 7.9 | 24.8 |
| Mason | 15.5 | 14.9 | 9.0 | 48.2 | 49,405 | 60,689 | 22.8 | 12.4 | 6,518 | 6,853 | 7,900 | 24,278 | 2.58 | 65.2 | 9.8 | 27.1 |
| Okanogan | 14.1 | 14.3 | 8.6 | 49.6 | 39,564 | 41,117 | 3.9 | 3.7 | 5,230 | 4,563 | 854 | 17,675 | 2.30 | 67.2 | 12.1 | 27.3 |
| Pacific | 16.8 | 20.7 | 12.4 | 50.7 | 20,984 | 20,919 | -0.3 | 9.9 | 1,933 | 3,276 | 3,400 | 9,279 | 2.31 | 61.6 | 7.9 | 30.9 |
| Pend Oreille | 17.4 | 17.6 | 9.7 | 48.8 | 11,732 | 13,001 | 10.8 | 8.8 | 1,192 | 1,572 | 1,532 | 5,727 | 2.31 | 67.6 | 9.9 | 26.7 |
| Pierce | 12.2 | 8.8 | 5.6 | 50.1 | 700,820 | 795,220 | 13.5 | 14.9 | 116,321 | 66,397 | 68,779 | 323,296 | 2.65 | 67.0 | 11.4 | 25.7 |
| San Juan | 17.6 | 22.1 | 13.4 | 51.4 | 14,077 | 15,768 | 12.0 | 10.9 | 891 | 1,424 | 2,257 | 8,252 | 2.01 | 68.0 | 6.5 | 30.6 |
| Skagit | 13.4 | 13.0 | 8.8 | 50.5 | 102,979 | 116,892 | 13.5 | 11.9 | 14,792 | 12,202 | 11,344 | 48,493 | 2.55 | 67.6 | 9.5 | 25.2 |
| Skamania | 17.5 | 15.2 | 8.6 | 49.5 | 9,872 | 11,070 | 12.1 | 9.4 | 981 | 880 | 935 | 4,816 | 2.43 | 70.6 | 9.9 | 22.3 |
| Snohomish | 13.2 | 9.0 | 5.4 | 49.8 | 606,024 | 713,299 | 17.7 | 16.4 | 98,063 | 53,988 | 72,833 | 293,823 | 2.68 | 69.1 | 9.8 | 22.9 |
| Spokane | 12.7 | 10.3 | 6.6 | 50.4 | 417,939 | 471,220 | 12.7 | 12.1 | 60,875 | 45,860 | 42,237 | 202,811 | 2.41 | 62.9 | 10.8 | 29.1 |
| Stevens | 16.3 | 15.5 | 9.3 | 49.9 | 40,066 | 43,532 | 8.7 | 6.5 | 4,436 | 4,679 | 3,100 | 17,554 | 2.53 | 68.0 | 9.5 | 25.6 |
| Thurston | 12.5 | 11.2 | 7.0 | 51.1 | 207,355 | 252,260 | 21.7 | 16.6 | 31,868 | 22,366 | 32,370 | 109,983 | 2.51 | 66.3 | 10.6 | 25.9 |
| Wahkiakum | 16.9 | 20.8 | 12.9 | 49.6 | 3,824 | 3,979 | 4.1 | 13.0 | 289 | 500 | 732 | 1,897 | 2.22 | 69.8 | 6.2 | 24.6 |
| Walla Walla | 12.2 | 10.9 | 8.3 | 49.2 | 55,180 | 58,781 | 6.5 | 4.3 | 6,711 | 5,870 | 1,692 | 22,646 | 2.44 | 64.8 | 10.7 | 27.9 |
| Whatcom | 12.0 | 11.2 | 7.1 | 50.6 | 166,814 | 201,146 | 20.6 | 14.8 | 22,952 | 16,707 | 23,640 | 86,523 | 2.49 | 60.2 | 8.2 | 26.9 |
| Whitman | 8.7 | 6.5 | 4.7 | 49.2 | 40,740 | 44,778 | 9.9 | 10.5 | 4,297 | 2,727 | 3,063 | 17,999 | 2.35 | 48.5 | 5.5 | 32.6 |
| Yakima | 10.7 | 8.3 | 6.0 | 50.1 | 222,581 | 243,240 | 9.3 | 3.6 | 40,578 | 20,133 | -11,825 | 83,048 | 2.96 | 72.4 | 15.4 | 22.4 |
| WEST VIRGINIA | 14.0 | 12.4 | 8.5 | 50.5 | 1,808,344 | 1,853,008 | 2.5 | -3.7 | 200,151 | 231,570 | -36,295 | 732,585 | 2.42 | 64.7 | 11.1 | 29.7 |
| Barbour | 13.9 | 12.4 | 8.8 | 50.8 | 15,557 | 16,586 | 6.6 | -0.9 | 1,704 | 1,986 | 135 | 6,324 | 2.54 | 64.2 | 10.2 | 28.8 |
| Berkeley | 13.3 | 9.9 | 5.5 | 50.3 | 75,905 | 104,196 | 37.3 | 17.2 | 14,274 | 10,272 | 13,839 | 44,221 | 2.60 | 68.2 | 11.2 | 25.7 |
| Boone | 14.5 | 13.7 | 7.8 | 50.3 | 25,535 | 24,624 | -3.6 | -14.5 | 2,494 | 3,297 | -2,780 | 8,932 | 2.49 | 68.0 | 11.7 | 25.4 |
| Braxton | 14.8 | 14.2 | 10.1 | 49.6 | 14,702 | 14,519 | -1.2 | -5.6 | 1,467 | 1,823 | -448 | 5,624 | 2.46 | 72.5 | 12.6 | 23.2 |
| Brooke | 15.2 | 14.5 | 10.3 | 50.5 | 25,447 | 24,046 | -5.5 | -9.9 | 1,958 | 3,402 | -907 | 9,805 | 2.19 | 64.0 | 10.8 | 30.2 |
| Cabell | 12.0 | 11.3 | 8.4 | 51.3 | 96,784 | 96,246 | -0.6 | -4.8 | 11,327 | 12,508 | -3,409 | 39,064 | 2.31 | 55.0 | 11.4 | 37.5 |

1. No spouse present.

# Table B. States and Counties — Population, Vital Statistics, Health, and Crime

| STATE County | Persons in group quarters, 2020 | Daytime Population, 2015-2019 | | Births, 2020 | | Deaths, 2020 | | Persons under 65 with no health insurance, 2019 | | Medicare, 2020 | | | Crimes reported by county police or sheriff, 2019 | |
|---|---|---|---|---|---|---|---|---|---|---|---|---|---|---|
| | | Number | Employment/residence ratio | Total | Rate[1] | Number | Rate[1] | Number | Percent | Total beneficiaries | Enrolled in Original Medicare | Enrolled in Medicare Advantage | Violent | Property |
| | 32 | 33 | 34 | 35 | 36 | 37 | 38 | 39 | 40 | 41 | 42 | 43 | 44 | 45 |
| VIRGINIA—Cont'd | | | | | | | | | | | | | | |
| Martinsville city | 377 | 16,367 | 1.69 | 190 | 15.4 | 245 | 19.8 | 829 | 8.4 | 3,831 | 2,489 | 1,342 | 66 | 343 |
| Newport News city | 7,958 | 207,350 | 1.32 | 2,674 | 14.9 | 1,721 | 9.6 | 13,869 | 9.5 | 29,296 | 20,007 | 9,289 | 1,056 | 4,484 |
| Norfolk city | 34,025 | 305,083 | 1.47 | 3,358 | 13.8 | 2,266 | 9.3 | 18,670 | 10.4 | 33,922 | 22,972 | 10,949 | 1,325 | 8,405 |
| Norton city | 68 | 3,847 | 0.92 | 43 | 10.8 | 31 | 7.8 | 262 | 8.2 | 1,138 | 613 | 525 | 6 | 134 |
| Petersburg city | 884 | 32,509 | 1.09 | 593 | 19.5 | 526 | 17.3 | 2,671 | 10.6 | 7,227 | 4,640 | 2,587 | 234 | 1,034 |
| Poquoson city | 51 | 8,673 | 0.43 | 97 | 7.9 | 119 | 9.7 | 569 | 5.7 | 2,657 | 2,260 | 398 | 23 | 91 |
| Portsmouth city | 3,013 | 105,020 | 1.22 | 1,418 | 14.9 | 1,088 | 11.4 | 6,725 | 8.8 | 18,665 | 12,667 | 5,999 | 889 | 5,509 |
| Radford city | 3,469 | 17,560 | 0.98 | 115 | 6.3 | 105 | 5.8 | 979 | 7.5 | 2,082 | 1,578 | 504 | 45 | 179 |
| Richmond city | 12,511 | 291,178 | 1.56 | 3,036 | 13.1 | 2,351 | 10.1 | 21,931 | 11.8 | 34,931 | 21,378 | 13,553 | 1,068 | 8,074 |
| Roanoke city | 2,100 | 123,968 | 1.53 | 1,393 | 14.1 | 1,295 | 13.1 | 8,929 | 11.1 | 21,433 | 13,816 | 7,617 | 386 | 4,402 |
| Salem city | 1,996 | 34,149 | 1.72 | 227 | 9 | 348 | 13.7 | 1,359 | 7.3 | 6,133 | 4,574 | 1,559 | 23 | 559 |
| Staunton city | 992 | 24,103 | 0.97 | 348 | 13.8 | 365 | 14.5 | 1,764 | 9.3 | 6,359 | 5,119 | 1,241 | 51 | 473 |
| Suffolk city | 996 | 83,891 | 0.86 | 1,150 | 12.2 | 931 | 9.9 | 6,093 | 7.9 | 17,297 | 12,716 | 4,581 | 270 | 2,266 |
| Virginia Beach city | 9,620 | 420,184 | 0.88 | 5,466 | 12.1 | 3,751 | 8.3 | 30,531 | 8.2 | 75,781 | 59,048 | 16,732 | 581 | 7,906 |
| Waynesboro city | 192 | 21,596 | 0.95 | 306 | 13.5 | 276 | 12.1 | 1,861 | 10.1 | 5,073 | 4,047 | 1,025 | 57 | 475 |
| Williamsburg city | 4,282 | 23,172 | 2.31 | 96 | 6.3 | 126 | 8.3 | 648 | 8.0 | 2,869 | 2,312 | 557 | 22 | 197 |
| Winchester city | 1,079 | 39,849 | 1.86 | 355 | 12.8 | 286 | 10.3 | 2,843 | 12.7 | 5,618 | 4,612 | 1,006 | 80 | 657 |
| WASHINGTON | 148,802 | 7,360,782 | 0.99 | 86,443 | 11.2 | 63,769 | 8.3 | 487,573 | 7.7 | 1,398,290 | 888,216 | 510,074 | X | X |
| Adams | 162 | 20,210 | 1.08 | 383 | 19.1 | 124 | 6.2 | 2,241 | 13.1 | 2,593 | 2,140 | 453 | 18 | 93 |
| Asotin | 174 | 19,889 | 0.73 | 229 | 10 | 291 | 12.8 | 1,253 | 7.4 | 6,428 | 5,443 | 985 | 15 | 126 |
| Benton | 1,421 | 199,419 | 1.02 | 2,590 | 12.5 | 1,623 | 7.9 | 14,210 | 8.3 | 36,807 | 34,143 | 2,665 | 73 | 386 |
| Chelan | 931 | 80,619 | 1.13 | 879 | 11.3 | 760 | 9.8 | 7,203 | 11.8 | 17,951 | 13,271 | 4,680 | 43 | 322 |
| Clallam | 1,815 | 73,996 | 0.95 | 595 | 7.6 | 1,171 | 15.0 | 5,052 | 9.7 | 27,438 | 25,596 | 1,842 | 49 | 461 |
| Clark | 3,437 | 426,741 | 0.79 | 5,621 | 11.3 | 4,207 | 8.5 | 29,670 | 7.3 | 92,294 | 36,689 | 55,606 | 230 | 2,189 |
| Columbia | 75 | 4,228 | 1.14 | 29 | 7.2 | 48 | 11.9 | 186 | 6.6 | 1,316 | 1,236 | 80 | 9 | 72 |
| Cowlitz | 1,208 | 105,237 | 0.96 | 1,256 | 11.3 | 1,275 | 11.4 | 6,458 | 7.4 | 26,672 | 12,182 | 14,490 | 64 | 285 |
| Douglas | 190 | 35,670 | 0.67 | 492 | 11.3 | 363 | 8.3 | 4,053 | 11.6 | 8,257 | 6,254 | 2,003 | 15 | 225 |
| Ferry | 251 | 7,251 | 0.87 | 67 | 8.6 | 84 | 10.8 | 617 | 11.5 | 2,437 | 2,357 | 80 | 3 | 26 |
| Franklin | 2,818 | 86,172 | 0.85 | 1,530 | 15.8 | 486 | 5.0 | 10,977 | 13.4 | 11,009 | 10,028 | 981 | 16 | 126 |
| Garfield | 36 | 2,079 | 0.82 | 19 | 8.3 | 23 | 10.0 | 96 | 5.8 | 729 | 709 | 20 | NA | NA |
| Grant | 1,255 | 96,451 | 1.02 | 1,474 | 14.8 | 754 | 7.6 | 10,953 | 13.4 | 16,505 | 13,261 | 3,244 | 95 | 837 |
| Grays Harbor | 2,717 | 70,554 | 0.92 | 748 | 9.8 | 978 | 12.9 | 5,377 | 9.7 | 20,827 | 19,232 | 1,595 | 47 | 250 |
| Island | 1,998 | 74,284 | 0.77 | 885 | 10.3 | 817 | 9.5 | 4,089 | 6.6 | 24,248 | 17,138 | 7,110 | 56 | 373 |
| Jefferson | 630 | 30,540 | 0.94 | 192 | 5.9 | 437 | 13.4 | 1,664 | 8.6 | 13,641 | 12,865 | 776 | 33 | 222 |
| King | 40,296 | 2,400,770 | 1.17 | 24,583 | 10.8 | 15,094 | 6.6 | 121,875 | 6.3 | 324,580 | 187,877 | 136,703 | 607 | 3,621 |
| Kitsap | 9,538 | 255,711 | 0.92 | 3,001 | 11 | 2,647 | 9.7 | 13,724 | 6.4 | 57,694 | 41,720 | 15,974 | 445 | 3,036 |
| Kittitas | 2,815 | 43,551 | 0.89 | 407 | 8.3 | 363 | 7.4 | 3,225 | 8.7 | 9,166 | 8,303 | 863 | 8 | 142 |
| Klickitat | 198 | 21,660 | 0.99 | 217 | 9.6 | 234 | 10.3 | 1,820 | 10.9 | 6,361 | 6,144 | 216 | 4 | 76 |
| Lewis | 901 | 74,426 | 0.88 | 908 | 11.1 | 964 | 11.7 | 5,884 | 9.5 | 21,445 | 13,070 | 8,374 | 86 | 473 |
| Lincoln | 94 | 9,961 | 0.85 | 102 | 9.2 | 123 | 11.1 | 581 | 7.3 | 3,243 | 3,122 | 120 | 10 | 87 |
| Mason | 2,588 | 56,293 | 0.69 | 638 | 9.4 | 738 | 10.8 | 5,117 | 10.6 | 18,112 | 13,302 | 4,810 | 85 | 956 |
| Okanogan | 640 | 42,788 | 1.06 | 469 | 11 | 530 | 12.4 | 4,665 | 14.6 | 11,151 | 9,600 | 1,551 | 38 | 158 |
| Pacific | 280 | 21,309 | 0.95 | 173 | 7.5 | 364 | 15.8 | 1,604 | 10.8 | 8,483 | 8,046 | 437 | 21 | 137 |
| Pend Oreille | 98 | 12,539 | 0.82 | 119 | 8.4 | 167 | 11.8 | 893 | 9.1 | 4,288 | 4,069 | 219 | 5 | 140 |
| Pierce | 20,763 | 816,107 | 0.85 | 11,333 | 12.4 | 7,842 | 8.6 | 57,627 | 7.6 | 155,500 | 96,843 | 58,657 | 1,152 | 6,822 |
| San Juan | 132 | 16,907 | 1.02 | 93 | 5.3 | 146 | 8.3 | 1,196 | 10.6 | 6,169 | 5,159 | 1,010 | 9 | 105 |
| Skagit | 1,627 | 126,375 | 1.01 | 1,406 | 10.8 | 1,374 | 10.5 | 9,725 | 9.7 | 32,171 | 21,257 | 10,914 | 74 | 757 |
| Skamania | 25 | 9,422 | 0.54 | 86 | 7.1 | 106 | 8.8 | 691 | 7.4 | 2,745 | 2,488 | 256 | 12 | 132 |
| Snohomish | 10,327 | 708,275 | 0.78 | 9,814 | 11.8 | 6,318 | 7.6 | 50,923 | 7.2 | 131,515 | 66,801 | 64,714 | 535 | 4,328 |
| Spokane | 14,257 | 516,063 | 1.05 | 5,863 | 11.1 | 5,147 | 9.7 | 29,350 | 7.0 | 108,154 | 60,411 | 47,743 | 195 | 2,666 |
| Stevens | 266 | 40,768 | 0.76 | 412 | 8.9 | 516 | 11.1 | 3,092 | 9.0 | 13,251 | 10,925 | 2,326 | 35 | 338 |
| Thurston | 4,074 | 267,617 | 0.91 | 3,122 | 10.6 | 2,608 | 8.9 | 16,174 | 6.9 | 62,181 | 37,557 | 24,625 | 246 | 1,546 |
| Wahkiakum | 16 | 3,864 | 0.69 | 30 | 6.7 | 58 | 12.9 | 245 | 8.3 | 1,612 | 1,137 | 475 | 4 | 32 |
| Walla Walla | 4,653 | 64,303 | 1.15 | 614 | 10 | 621 | 10.1 | 4,303 | 9.6 | 13,674 | 11,430 | 2,244 | 20 | 266 |
| Whatcom | 5,861 | 216,806 | 0.96 | 2,122 | 9.2 | 1,931 | 8.4 | 16,675 | 9.1 | 48,015 | 28,643 | 19,371 | NA | NA |
| Whitman | 6,767 | 52,038 | 1.13 | 348 | 7 | 264 | 5.3 | 2,609 | 6.9 | 6,157 | 5,863 | 294 | 7 | 35 |
| Yakima | 3,468 | 249,889 | 1.00 | 3,594 | 14.3 | 2,173 | 8.6 | 31,476 | 15.1 | 43,471 | 31,902 | 11,570 | 153 | 1,466 |
| WEST VIRGINIA | 46,883 | 1,799,554 | 0.98 | 17,571 | 9.8 | 24,297 | 13.6 | 112,089 | 8.1 | 442,600 | 275,087 | 167,513 | X | X |
| Barbour | 853 | 13,784 | 0.55 | 152 | 9.2 | 217 | 13.2 | 1,111 | 9.1 | 4,086 | 2,688 | 1,399 | NA | NA |
| Berkeley | 911 | 98,157 | 0.69 | 1,472 | 12.1 | 1,177 | 9.6 | 8,050 | 8.0 | 22,460 | 16,142 | 6,319 | NA | NA |
| Boone | 133 | 21,638 | 0.88 | 188 | 8.9 | 323 | 15.3 | 1,362 | 8.1 | 5,993 | 2,924 | 3,069 | 55 | 88 |
| Braxton | 383 | 13,773 | 0.92 | 111 | 8.1 | 191 | 13.9 | 978 | 9.6 | 3,625 | 2,097 | 1,528 | 4 | 7 |
| Brooke | 673 | 21,739 | 0.93 | 178 | 8.2 | 326 | 15.0 | 1,114 | 7.0 | 6,170 | 3,626 | 2,543 | NA | NA |
| Cabell | 3,808 | 108,696 | 1.39 | 1,013 | 11.1 | 1,177 | 12.9 | 5,855 | 8.3 | 21,206 | 12,752 | 8,454 | NA | NA |

1. Per 1,000 estimated resident population.

# Table B. States and Counties — Crime, Education, Money Income, and Poverty

| STATE County | County law enforcement employment, 2019 | | Education | | | | | | Money income, 2015-2019 | | | | Income and poverty, 2019 | | | |
|---|---|---|---|---|---|---|---|---|---|---|---|---|---|---|---|---|
| | | | School enrollment and attainment, 2015-2019 | | | | Local government expenditures,[3] 2017-2018 | | | Households | | | | Percent below poverty level | | |
| | | | Enrollment[1] | | Attainment[2] (percent) | | | | | | Percent | | | | | |
| | Officers | Civilians | Total | Percent private | High school graduate or less | Bachelor's degree or more | Total current spending (mil dol) | Current spending per student (dollars) | Per capita income[4] | Median income (dollars) | with income of less than $50,000 | with income of $200,000 or more | Median household income (dollars) | All persons | Children under 18 years | Children 5 to 17 years in families |
| | 46 | 47 | 48 | 49 | 50 | 51 | 52 | 53 | 54 | 55 | 56 | 57 | 58 | 59 | 60 | 61 |
| VIRGINIA—Cont'd | | | | | | | | | | | | | | | | |
| Martinsville city | 43 | 5 | 2,804 | 12.0 | 45.7 | 21.0 | 26 | 12,819 | 22,836 | 34,371 | 64.1 | 3.1 | 37,814 | 19.5 | 31.6 | 30.6 |
| Newport News city | 445 | 158 | 47,851 | 12.9 | 37.5 | 26.3 | 331 | 11,527 | 28,294 | 53,215 | 47.0 | 3.1 | 53,022 | 15.1 | 23.6 | 23.0 |
| Norfolk city | 691 | 92 | 64,492 | 12.6 | 37.6 | 28.8 | 354 | 11,492 | 29,830 | 51,590 | 48.3 | 4.1 | 52,437 | 19.0 | 28.5 | 28.9 |
| Norton city | 14 | 8 | 857 | 3.4 | 40.0 | 22.8 | 7 | 9,221 | 21,741 | 29,000 | 65.3 | 3.0 | 38,062 | 19.3 | 30.7 | 29.1 |
| Petersburg city | 91 | 21 | 6,480 | 10.8 | 52.9 | 19.8 | 51 | 12,173 | 23,611 | 38,679 | 63.4 | 1.5 | 40,240 | 21.6 | 34.6 | 38.3 |
| Poquoson city | 26 | 1 | 3,186 | 8.2 | 31.0 | 42.2 | 23 | 10,948 | 40,258 | 97,118 | 27.8 | 10.3 | 98,217 | 5.3 | 6.3 | 5.4 |
| Portsmouth city | 220 | 54 | 22,729 | 15.1 | 41.8 | 21.9 | 166 | 11,557 | 26,312 | 52,175 | 48.0 | 2.4 | 50,411 | 18.5 | 28.6 | 29.9 |
| Radford city | 35 | 13 | 8,703 | 3.4 | 32.6 | 38.0 | 17 | 10,623 | 21,797 | 36,297 | 60.2 | 2.1 | 41,530 | 30.5 | 18.5 | 17.1 |
| Richmond city | 734 | 117 | 55,683 | 18.3 | 36.4 | 39.6 | 354 | 14,056 | 33,549 | 47,250 | 52.1 | 6.2 | 50,949 | 19.2 | 27.6 | 26.5 |
| Roanoke city | 248 | 43 | 20,997 | 14.4 | 47.9 | 23.4 | 181 | 13,122 | 27,006 | 44,230 | 56.2 | 3.6 | 45,534 | 19.5 | 30.5 | 29.9 |
| Salem city | 65 | 22 | 6,816 | 30.0 | 39.8 | 27.9 | 43 | 10,804 | 31,924 | 57,165 | 41.6 | 4.9 | 54,888 | 9.4 | 12.8 | 12.6 |
| Staunton city | 48 | 16 | 4,639 | 27.3 | 40.1 | 33.4 | 35 | 12,431 | 30,166 | 52,611 | 47.6 | 3.3 | 54,296 | 11.3 | 17.8 | 18.2 |
| Suffolk city | 173 | 60 | 23,410 | 18.0 | 37.3 | 29.5 | 152 | 10,593 | 34,940 | 74,884 | 33.2 | 6.5 | 77,847 | 9.8 | 14.7 | 14.1 |
| Virginia Beach city | 760 | 178 | 114,324 | 19.5 | 27.4 | 36.0 | 801 | 11,606 | 37,776 | 76,610 | 29.8 | 7.2 | 78,491 | 7.3 | 9.8 | 9.9 |
| Waynesboro city | 48 | 11 | 5,042 | 14.7 | 47.3 | 27.1 | 36 | 11,628 | 27,033 | 45,011 | 55.4 | 2.4 | 44,619 | 12.3 | 20.3 | 18.7 |
| Williamsburg city | 41 | 2 | 7,198 | 5.5 | 24.0 | 56.8 | (7)136.8 | (7)11,722 | 29,758 | 57,463 | 40.9 | 10.5 | 56,569 | 18.2 | 22.4 | 20.6 |
| Winchester city | 76 | 10 | 7,271 | 34.0 | 40.5 | 35.1 | (8)222.7 | (8)12,434 | 30,859 | 58,818 | 43.4 | 6.8 | 60,254 | 13.3 | 19.9 | 19.2 |
| WASHINGTON | X | X | 1,750,198 | 14.9 | 30.7 | 36.0 | 14,409 | 12,986 | 38,915 | 73,775 | 33.5 | 9.5 | 78,674 | 9.8 | 12.0 | 11.2 |
| Adams | 15 | 19 | 5,750 | 4.5 | 61.7 | 14.3 | 65 | 12,783 | 20,248 | 48,294 | 50.8 | 2.3 | 53,535 | 16.8 | 22.4 | 21.8 |
| Asotin | 13 | 18 | 4,487 | 8.5 | 36.6 | 21.4 | 42 | 12,646 | 29,695 | 53,715 | 47.0 | 2.9 | 54,776 | 13.8 | 18.5 | 16.8 |
| Benton | 73 | 13 | 51,161 | 12.4 | 34.3 | 30.9 | 452 | 12,023 | 32,882 | 69,023 | 34.8 | 6.7 | 72,847 | 10.6 | 16.0 | 15.6 |
| Chelan | 58 | 16 | 17,481 | 10.7 | 43.7 | 25.8 | 180 | 13,716 | 30,870 | 58,795 | 42.8 | 4.8 | 59,838 | 12.2 | 16.4 | 15.5 |
| Clallam | 37 | 8 | 13,158 | 13.3 | 33.8 | 27.4 | 129 | 12,432 | 30,283 | 52,192 | 47.8 | 2.9 | 57,571 | 11.2 | 16.7 | 15.7 |
| Clark | 141 | 89 | 114,256 | 12.0 | 31.6 | 30.6 | 1,013 | 12,650 | 35,860 | 75,253 | 31.5 | 7.4 | 80,407 | 9.2 | 12.1 | 10.6 |
| Columbia | 8 | 1 | 752 | 10.2 | 30.6 | 25.6 | 7 | 14,714 | 36,551 | 53,423 | 46.1 | 5.9 | 56,338 | 12.3 | 16.8 | 15.8 |
| Cowlitz | 42 | 24 | 24,453 | 12.4 | 41.3 | 17.0 | 213 | 12,217 | 28,876 | 54,506 | 46.0 | 3.6 | 57,316 | 12.6 | 17.5 | 14.8 |
| Douglas | 31 | 6 | 10,444 | 5.6 | 47.5 | 20.5 | 102 | 12,083 | 30,544 | 62,951 | 38.7 | 4.3 | 63,086 | 10.9 | 15.0 | 14.0 |
| Ferry | 7 | 16 | 1,304 | 12.5 | 48.0 | 18.8 | 15 | 16,850 | 25,857 | 41,939 | 58.7 | 2.9 | 47,463 | 18.0 | 25.6 | 26.3 |
| Franklin | 27 | 54 | 27,232 | 8.5 | 50.6 | 17.7 | 266 | 12,956 | 24,380 | 63,584 | 36.8 | 4.5 | 63,575 | 13.5 | 18.0 | 17.1 |
| Garfield | 8 | 7 | 551 | 6.4 | 33.5 | 24.5 | 5 | 14,979 | 26,742 | 55,900 | 45.9 | 1.5 | 54,555 | 12.9 | 17.2 | 18.3 |
| Grant | 56 | 71 | 25,869 | 7.0 | 50.5 | 17.7 | 262 | 12,827 | 24,280 | 55,556 | 43.8 | 3.3 | 57,152 | 13.9 | 18.3 | 17.4 |
| Grays Harbor | 62 | 17 | 14,270 | 7.7 | 44.3 | 16.5 | 144 | 13,631 | 27,210 | 51,240 | 49.0 | 2.7 | 59,346 | 13.1 | 19.5 | 17.9 |
| Island | 39 | 33 | 15,502 | 16.2 | 26.4 | 33.6 | 104 | 12,524 | 37,087 | 68,604 | 34.5 | 5.9 | 72,173 | 7.3 | 9.0 | 8.8 |
| Jefferson | 22 | 20 | 3,949 | 19.0 | 25.6 | 44.3 | 38 | 14,278 | 36,598 | 55,127 | 43.5 | 4.9 | 60,556 | 12.6 | 21.4 | 20.3 |
| King | 214 | 56 | 505,065 | 18.9 | 21.9 | 52.5 | 3,637 | 13,320 | 52,462 | 94,974 | 26.0 | 17.3 | 102,338 | 7.7 | 7.9 | 7.6 |
| Kitsap | 122 | 121 | 57,885 | 12.9 | 27.4 | 33.2 | 474 | 13,016 | 37,493 | 75,411 | 31.3 | 7.6 | 79,268 | 7.5 | 9.3 | 8.2 |
| Kittitas | 33 | 47 | 14,930 | 6.5 | 33.5 | 32.3 | 65 | 12,220 | 29,161 | 56,004 | 45.4 | 3.7 | 59,172 | 14.0 | 11.7 | 10.7 |
| Klickitat | 21 | 19 | 4,035 | 24.8 | 36.8 | 31.5 | 43 | 13,330 | 29,521 | 55,773 | 46.2 | 3.8 | 60,567 | 13.5 | 18.3 | 17.4 |
| Lewis | 42 | 67 | 16,788 | 12.7 | 41.6 | 17.7 | 156 | 12,775 | 27,127 | 53,484 | 46.6 | 2.3 | 58,525 | 12.2 | 14.6 | 13.2 |
| Lincoln | 15 | 13 | 2,115 | 8.2 | 36.5 | 25.1 | 34 | 15,787 | 28,891 | 54,631 | 45.8 | 1.9 | 60,849 | 9.1 | 14.2 | 13.4 |
| Mason | 48 | 44 | 11,824 | 9.5 | 42.5 | 17.6 | 120 | 12,982 | 28,862 | 57,634 | 42.0 | 3.3 | 63,689 | 12.8 | 21.0 | 20.3 |
| Okanogan | 30 | 51 | 8,898 | 9.4 | 46.4 | 20.0 | 115 | 11,272 | 24,876 | 47,240 | 53.1 | 2.0 | 44,777 | 16.4 | 21.9 | 20.9 |
| Pacific | 15 | 28 | 3,619 | 10.2 | 41.6 | 18.9 | 43 | 13,896 | 26,109 | 46,733 | 54.3 | 1.6 | 50,521 | 13.5 | 17.6 | 16.6 |
| Pend Oreille | 15 | 21 | 2,454 | 15.0 | 39.8 | 20.8 | 23 | 13,702 | 28,046 | 50,591 | 49.2 | 2.7 | 51,950 | 16.1 | 26.9 | 24.8 |
| Pierce | 290 | 65 | 209,451 | 14.6 | 35.7 | 27.2 | 1,753 | 12,936 | 34,618 | 72,113 | 33.2 | 6.5 | 78,779 | 9.4 | 11.8 | 11.1 |
| San Juan | 20 | 15 | 2,203 | 23.0 | 20.1 | 49.8 | 26 | 14,201 | 45,777 | 63,622 | 39.4 | 8.5 | 69,113 | 8.1 | 11.9 | 11.4 |
| Skagit | 56 | 80 | 26,905 | 11.4 | 36.2 | 26.2 | 281 | 14,497 | 33,168 | 67,028 | 37.2 | 5.3 | 67,551 | 11.2 | 16.2 | 15.7 |
| Skamania | 33 | 3 | 2,195 | 6.4 | 38.8 | 24.5 | 14 | 12,892 | 32,221 | 65,181 | 38.4 | 3.2 | 69,163 | 10.4 | 15.7 | 13.9 |
| Snohomish | 297 | 73 | 186,567 | 15.4 | 31.0 | 32.8 | 1,718 | 12,822 | 39,527 | 86,691 | 26.1 | 10.0 | 89,119 | 7.0 | 8.1 | 7.5 |
| Spokane | 125 | 26 | 124,821 | 17.4 | 29.4 | 30.8 | 995 | 12,951 | 31,146 | 56,904 | 44.2 | 4.4 | 59,976 | 12.9 | 13.6 | 12.2 |
| Stevens | 28 | 32 | 8,837 | 16.3 | 43.9 | 18.7 | 82 | 14,002 | 26,896 | 51,775 | 48.0 | 3.5 | 55,000 | 14.9 | 20.3 | 20.3 |
| Thurston | 93 | 145 | 64,429 | 14.4 | 27.6 | 35.7 | 557 | 13,042 | 35,169 | 72,003 | 33.6 | 6.0 | 77,890 | 9.1 | 11.8 | 10.0 |
| Wahkiakum | 9 | 11 | 641 | 23.7 | 37.9 | 19.4 | 6 | 12,983 | 31,110 | 53,227 | 45.5 | 5.0 | 61,123 | 10.8 | 20.3 | 18.6 |
| Walla Walla | 28 | 7 | 16,052 | 27.9 | 31.5 | 29.8 | 115 | 13,109 | 29,035 | 57,858 | 44.0 | 4.3 | 60,202 | 12.7 | 14.9 | 14.0 |
| Whatcom | 89 | 102 | 57,933 | 12.8 | 30.5 | 34.3 | 352 | 12,796 | 32,267 | 62,984 | 39.3 | 4.9 | 68,656 | 12.7 | 11.3 | 10.8 |
| Whitman | 17 | 18 | 23,866 | 4.4 | 20.4 | 49.8 | 66 | 13,591 | 23,560 | 42,745 | 56.1 | 3.7 | 45,906 | 26.3 | 13.7 | 12.2 |
| Yakima | 57 | 28 | 68,066 | 8.4 | 54.4 | 16.7 | 699 | 12,786 | 23,459 | 51,637 | 48.2 | 3.2 | 55,674 | 16.7 | 24.5 | 22.4 |
| WEST VIRGINIA | X | X | 387,707 | 10.3 | 53.4 | 20.6 | 3,115 | 11,442 | 26,480 | 46,711 | 52.9 | 3.0 | 48,659 | 16.2 | 20.8 | 19.6 |
| Barbour | 7 | 8 | 3,503 | 14.4 | 62.0 | 15.6 | 24 | 10,136 | 22,899 | 38,459 | 60.8 | 1.4 | 39,710 | 18.4 | 25.1 | 23.4 |
| Berkeley | 57 | 7 | 27,768 | 10.2 | 48.6 | 21.2 | 206 | 10,655 | 29,994 | 62,515 | 38.1 | 3.2 | 63,012 | 11.0 | 14.7 | 14.1 |
| Boone | 19 | 3 | 4,605 | 10.1 | 69.9 | 9.7 | 41 | 10,101 | 21,751 | 40,739 | 59.5 | 1.2 | 43,697 | 18.9 | 24.8 | 22.4 |
| Braxton | 9 | 2 | 2,548 | 8.4 | 65.8 | 14.9 | 22 | 10,663 | 22,269 | 41,466 | 58.0 | 1.4 | 39,651 | 18.3 | 24.7 | 23.8 |
| Brooke | 17 | 0 | 5,130 | 26.0 | 50.3 | 19.0 | 37 | 12,523 | 26,678 | 51,496 | 48.9 | 1.1 | 50,900 | 12.0 | 16.1 | 15.7 |
| Cabell | 43 | 6 | 23,092 | 7.8 | 45.2 | 26.0 | 146 | 11,404 | 25,271 | 40,028 | 59.0 | 3.3 | 42,125 | 18.5 | 22.1 | 19.5 |

1. All persons 3 years old and over enrolled in nursery school through college.   2. Persons 25 years old and over.   3. Elementary and secondary education expenditures.   4. Based on population estimated by the American Community Survey, 2014–2018.   7. Williamsburg city is included with James City county.   8. Winchester city is included with Frederick county.

# Table B. States and Counties — **Personal Income**

| STATE County | Personal income, 2019 Total (mil dol) | Percent change 2018-2019 | Per capita[1] Dollars | Per capita[1] Rank | Wages and salaries (mil dol) | Supplements to wages and salaries, employer contributions (mil dol) Pension and insurance | Government social insurance | Proprietors' income (mil dol) | Dividends, interest, and rent (mil dol) | Personal transfer receipts (mil dol) | Earnings, 2019 Total (mil dol) | Contributions for government social insurance (mil dol) From employee and self-employed | From employer |
|---|---|---|---|---|---|---|---|---|---|---|---|---|---|
| | 62 | 63 | 64 | 65 | 66 | 67 | 68 | 69 | 70 | 71 | 72 | 73 | 74 |
| VIRGINIA—Cont'd | | | | | | | | | | | | | |
| Martinsville city | (2) | (2) | (2) | (2) | (2) | (2) | (2) | (2) | (2) | (2) | (2) | (2) | (2) |
| Newport News city | 7,969 | 3.9 | 44,465 | 1,422 | 6,657 | 1,253 | 510 | 276 | 1,584 | 1,636 | 8,696 | 495 | 510 |
| Norfolk city | 10,408 | 3.0 | 42,875 | 1,639 | 11,787 | 2,682 | 957 | 146 | 2,702 | 2,076 | 15,572 | 814 | 957 |
| Norton city | (3) | (3) | (3) | (3) | (3) | (3) | (3) | (3) | (3) | (3) | (3) | (3) | (3) |
| Petersburg city | (4) | (4) | (4) | (4) | (4) | (4) | (4) | (4) | (4) | (4) | (4) | (4) | (4) |
| Poquoson city | (5) | (5) | (5) | (5) | (5) | (5) | (5) | (5) | (5) | (5) | (5) | (5) | (5) |
| Portsmouth city | 4,022 | 3.2 | 42,605 | 1,669 | 3,149 | 859 | 259 | 106 | 782 | 1,008 | 4,373 | 242 | 259 |
| Radford city | (6) | (6) | (6) | (6) | (6) | (6) | (6) | (6) | (6) | (6) | (6) | (6) | (6) |
| Richmond city | 13,033 | 2.9 | 56,560 | 418 | 11,231 | 1,802 | 765 | 1,014 | 3,704 | 2,046 | 14,812 | 868 | 765 |
| Roanoke city | 4,489 | 2.7 | 45,277 | 1,325 | 3,642 | 556 | 272 | 284 | 1,037 | 1,114 | 4,753 | 295 | 272 |
| Salem city | (7) | (7) | (7) | (7) | (7) | (7) | (7) | (7) | (7) | (7) | (7) | (7) | (7) |
| Staunton city | (8) | (8) | (8) | (8) | (8) | (8) | (8) | (8) | (8) | (8) | (8) | (8) | (8) |
| Suffolk city | 4,950 | 4.7 | 53,740 | 559 | 1,814 | 293 | 131 | 160 | 860 | 906 | 2,398 | 155 | 131 |
| Virginia Beach city | 26,911 | 3.1 | 59,805 | 312 | 10,148 | 1,783 | 753 | 1,428 | 6,345 | 3,805 | 14,111 | 848 | 753 |
| Waynesboro city | (8) | (8) | (8) | (8) | (8) | (8) | (8) | (8) | (8) | (8) | (8) | (8) | (8) |
| Williamsburg city | (9) | (9) | (9) | (9) | (9) | (9) | (9) | (9) | (9) | (9) | (9) | (9) | (9) |
| Winchester city | (10) | (10) | (10) | (10) | (10) | (10) | (10) | (10) | (10) | (10) | (10) | (10) | (10) |
| **WASHINGTON** | 493,128 | 5.4 | 64,766 | X | 254,104 | 36,051 | 20,315 | 38,357 | 111,772 | 68,247 | 348,827 | 20,072 | 20,315 |
| Adams | 868 | 5.8 | 43,429 | 1,555 | 328 | 62 | 35 | 129 | 198 | 195 | 554 | 26 | 35 |
| Asotin | 1,094 | 3.1 | 48,456 | 974 | 289 | 54 | 26 | 82 | 253 | 306 | 451 | 33 | 26 |
| Benton | 10,088 | 5.1 | 49,354 | 892 | 5,522 | 787 | 497 | 850 | 1,728 | 1,836 | 7,656 | 435 | 497 |
| Chelan | 4,343 | 3.7 | 56,253 | 437 | 1,989 | 347 | 193 | 398 | 1,145 | 870 | 2,926 | 170 | 193 |
| Clallam | 3,701 | 4.5 | 47,856 | 1,035 | 1,110 | 245 | 101 | 221 | 1,053 | 1,095 | 1,677 | 129 | 101 |
| Clark | 27,537 | 4.9 | 56,401 | 426 | 9,850 | 1,596 | 856 | 1,701 | 6,011 | 4,322 | 14,003 | 861 | 856 |
| Columbia | 206 | 4.6 | 51,660 | 710 | 73 | 17 | 7 | 26 | 45 | 62 | 122 | 7 | 7 |
| Cowlitz | 5,143 | 5.3 | 46,503 | 1,185 | 2,216 | 364 | 202 | 347 | 870 | 1,328 | 3,129 | 204 | 202 |
| Douglas | 1,812 | 4.5 | 41,725 | 1,817 | 513 | 97 | 52 | 102 | 870 | 1,328 | 764 | 48 | 52 |
| Ferry | 285 | 2.4 | 37,311 | 2,413 | 72 | 21 | 7 | 9 | 387 | 394 | 109 | 9 | 7 |
| Franklin | 3,764 | 5.8 | 39,526 | 2,109 | 1,637 | 298 | 168 | 401 | 550 | 736 | 2,503 | 129 | 168 |
| Garfield | 121 | 10.9 | 54,352 | 524 | 35 | 11 | 3 | 21 | . 24 | 34 | 71 | 4 | 3 |
| Grant | 4,021 | 4.7 | 41,141 | 1,900 | 1,827 | 353 | 186 | 444 | 709 | 935 | 2,810 | 144 | 186 |
| Grays Harbor | 3,147 | 5.1 | 41,928 | 1,779 | 1,096 | 222 | 100 | 187 | 612 | 1,004 | 2,810 | 113 | 100 |
| Island | 4,851 | 5.5 | 56,977 | 406 | 1,255 | 322 | 119 | 323 | 1,384 | 974 | 1,606 | 129 | 119 |
| Jefferson | 1,787 | 4.9 | 55,447 | 475 | 426 | 91 | 38 | 155 | 651 | 477 | 2,019 | 58 | 38 |
| King | 213,957 | 5.3 | 94,974 | 28 | 139,442 | 15,395 | 9,971 | 18,145 | 54,255 | 16,190 | 711 | 58 | 38 |
| Kitsap | 15,983 | 5.4 | 58,874 | 342 | 6,073 | 1,490 | 545 | 844 | 4,061 | 2,598 | 182,953 | 10,227 | 9,971 |
| Kittitas | 2,088 | 4.7 | 43,562 | 1,539 | 735 | 152 | 69 | 173 | 516 | 412 | 8,951 | 522 | 545 |
| Klickitat | 1,109 | 2.5 | 49,458 | 884 | 416 | 81 | 41 | 103 | 283 | 293 | 1,129 | 69 | 69 |
| Lewis | 3,597 | 5.9 | 44,569 | 1,412 | 1,310 | 244 | 124 | 217 | 628 | 1,025 | 642 | 40 | 41 |
| Lincoln | 517 | 6.3 | 47,282 | 1,099 | 124 | 31 | 12 | 66 | 122 | 140 | 1,895 | 129 | 124 |
| Mason | 2,936 | 6.6 | 43,966 | 1,494 | 686 | 154 | 62 | 163 | 627 | 796 | 232 | 15 | 12 |
| Okanogan | 1,829 | 2.2 | 43,295 | 1,571 | 628 | 147 | 63 | 155 | 403 | 544 | 1,064 | 81 | 62 |
| Pacific | 957 | 4.0 | 42,603 | 1,670 | 282 | 62 | 27 | 61 | 235 | 352 | 993 | 63 | 63 |
| Pend Oreille | 610 | 5.1 | 44,461 | 1,423 | 170 | 39 | 14 | 24 | 149 | 199 | 432 | 35 | 27 |
| Pierce | 48,481 | 6.0 | 53,572 | 566 | 20,452 | 3,710 | 1,852 | 3,497 | 8,939 | 8,305 | 247 | 19 | 14 |
| San Juan | 1,430 | 3.1 | 81,351 | 60 | 260 | 44 | 24 | 132 | 788 | 207 | 29,511 | 1,695 | 1,852 |
| Skagit | 7,042 | 5.8 | 54,505 | 518 | 2,940 | 548 | 267 | 576 | 1,696 | 1,447 | 460 | 33 | 24 |
| Skamania | 573 | 4.6 | 47,461 | 1,084 | 93 | 20 | 9 | 27 | 139 | 116 | 4,332 | 262 | 267 |
| | | | | | | | | | | | 150 | 12 | 9 |
| Snohomish | 48,280 | 6.1 | 58,729 | 346 | 19,993 | 3,140 | 1,708 | 3,343 | 8,005 | 6,090 | 28,184 | 1,635 | 1,708 |
| Spokane | 25,322 | 5.1 | 48,436 | 980 | 12,723 | 2,156 | 1,126 | 1,612 | 5,300 | 5,446 | 17,617 | 1,067 | 1,126 |
| Stevens | 1,868 | 4.7 | 40,863 | 1,936 | 485 | 109 | 45 | 103 | 372 | 587 | 742 | 58 | 45 |
| Thurston | 15,348 | 6.3 | 52,828 | 631 | 6,857 | 1,324 | 599 | 907 | 3,337 | 2,893 | 9,686 | 581 | 599 |
| Wahkiakum | 200 | 4.9 | 44,522 | 1,414 | 33 | 8 | 3 | 10 | 51 | 60 | 54 | 5 | 3 |
| Walla Walla | 2,943 | 4.6 | 48,444 | 978 | 1,344 | 261 | 129 | 245 | 710 | 658 | 1,979 | 112 | 129 |
| Whatcom | 11,672 | 4.8 | 50,915 | 763 | 4,937 | 905 | 447 | 1,178 | 2,758 | 2,116 | 7,467 | 444 | 447 |
| Whitman | 2,138 | 3.4 | 42,665 | 1,662 | 1,029 | 288 | 91 | 209 | 512 | 339 | 1,617 | 82 | 91 |
| Yakima | 11,479 | -4.3 | 45,757 | 1,280 | 4,852 | 853 | 501 | 1,172 | 2,199 | 2,762 | 7,377 | 389 | 501 |
| **WEST VIRGINIA** | 75,835 | 2.0 | 42,242 | X | 33,403 | 6,249 | 2,606 | 4,782 | 11,509 | 21,561 | 47,040 | 3,262 | 2,606 |
| Barbour | 552 | 3.0 | 33,599 | 2,836 | 165 | 32 | 12 | 22 | 74 | 187 | 231 | 19 | 12 |
| Berkeley | 5,062 | 4.7 | 42,473 | 1,694 | 1,735 | 361 | 141 | 217 | 597 | 967 | 2,454 | 164 | 141 |
| Boone | 764 | -0.1 | 35,615 | 2,603 | 236 | 46 | 19 | 19 | 79 | 302 | 320 | 28 | 19 |
| Braxton | 458 | 2.0 | 32,816 | 2,905 | 144 | 30 | 12 | 30 | 65 | 160 | 216 | 17 | 12 |
| Brooke | 963 | 2.7 | 43,882 | 1,503 | 364 | 69 | 29 | 45 | 161 | 269 | 506 | 37 | 29 |
| Cabell | 4,062 | 0.9 | 44,177 | 1,460 | 2,558 | 454 | 199 | 266 | 691 | 1,161 | 3,477 | 224 | 199 |

1. Based on the resident population estimated as of July 1 of the year shown.  2. Martinsville city is included with Henry county.  3. Norton city is included with Wise county.  4. Petersburg and Colonial Heights cities are included with Dinwiddie county.  5. Poquoson city is included with York county.  6. Radford city is included with Montgomery county.  7. Salem city is included with Roanoke county.  8. Staunton and Waynesboro cities are included with Augusta county.  9. Williamsburg city is included with James City county.  10. Winchester city is included with Frederick county.

# Table B. States and Counties — Earnings, Social Security, and Housing

| STATE County | Farm [75] | Mining, quarrying, and extractions [76] | Construction [77] | Manufacturing [78] | Information; professional, scientific, technical services [79] | Retail trade [80] | Finance, insurance, real estate, and leasing [81] | Health care and social assistance [82] | Government [83] | Soc. Sec. Number [84] | Rate[1] [85] | Supplemental Security Income recipients, 2019 [86] | Housing Total [87] | Percent change, 2010-2020 [88] |
|---|---|---|---|---|---|---|---|---|---|---|---|---|---|---|
| VIRGINIA—Cont'd | | | | | | | | | | | | | | |
| Martinsville city | (2) | (2) | (2) | (2) | (2) | (2) | (2) | (2) | (2) | 4,125 | 332 | 40 | 7,017 | -2.7 |
| Newport News city | 0.0 | D | 3.2 | 31.4 | 6.8 | 3.9 | 3.2 | 10.6 | 23.7 | 31,335 | 174 | 4,969 | 77,956 | 2.1 |
| Norfolk city | 0.0 | D | 2.4 | 3.4 | 7.4 | 2.5 | 4.3 | 9.6 | 51.5 | 35,675 | 146 | 7,540 | 98,941 | 4.1 |
| Norton city | (3) | (3) | (3) | (3) | (3) | (3) | (3) | (3) | (3) | 1,405 | 353 | 359 | 1,948 | -1.1 |
| Petersburg city | (4) | (4) | (4) | (4) | (4) | (4) | (4) | (4) | (4) | 8,145 | 266 | 2,969 | 16,262 | -0.5 |
| Poquoson city | (5) | (5) | (5) | (5) | (5) | (5) | (5) | (5) | (5) | 2,645 | 216 | 67 | 4,866 | 2.9 |
| Portsmouth city | 0.0 | 0.0 | 4.9 | D | 3.7 | 2.3 | 1.2 | 9.5 | 62.5 | 19,460 | 206 | 3,897 | 40,860 | 0.1 |
| Radford city | (6) | (6) | (6) | (6) | (6) | (6) | (6) | (6) | (6) | 2,360 | 130 | 416 | 6,523 | 1.6 |
| Richmond city | 0.0 | 0.0 | 3.5 | 3.7 | 16.7 | 2.1 | 10.4 | 11.9 | 25.9 | 35,920 | 155 | 9,037 | 102,137 | 3.7 |
| Roanoke city | 0.0 | D | 6.5 | D | 8.1 | 6.0 | 8.8 | 20.7 | 13.1 | 22,045 | 222 | 4,772 | 47,043 | -0.7 |
| Salem city | (7) | (7) | (7) | (7) | (7) | (7) | (7) | (7) | (7) | 6,545 | 260 | 630 | 10,832 | -0.1 |
| Staunton city | (8) | (8) | (8) | (8) | (8) | (8) | (8) | (8) | (8) | 7,050 | 283 | 826 | 11,855 | 1.0 |
| Suffolk city | 1.0 | D | 4.5 | 6.8 | 17.6 | 5.4 | 4.0 | 12.8 | 24.4 | 18,150 | 197 | 2,402 | 38,597 | 16.8 |
| Virginia Beach city | 0.0 | 0.0 | 6.9 | 3.2 | 11.6 | 5.9 | 11.6 | 11.9 | 27.2 | 76,925 | 171 | 5,722 | 187,418 | 5.4 |
| Waynesboro city | (8) | (8) | (8) | (8) | (8) | (8) | (8) | (8) | (8) | 5,395 | 241 | 719 | 10,224 | 5.3 |
| Williamsburg city | (9) | (9) | (9) | (9) | (9) | (9) | (9) | (9) | (9) | 2,580 | 172 | 0 | 5,317 | 8.2 |
| Winchester city | (10) | (10) | (10) | (10) | (10) | (10) | (10) | (10) | (10) | 5,525 | 197 | 708 | 12,096 | 1.8 |
| WASHINGTON | 1.1 | 0.2 | 7.0 | 8.7 | 19.0 | 8.7 | 6.1 | 9.8 | 17.3 | 1,376,287 | 181 | 148,731 | 3,242,804 | 12.4 |
| Adams | 21.6 | 0.0 | 2.3 | 13.8 | 1.4 | 4.4 | 3.9 | D | 21.2 | 2,715 | 136 | 367 | 6,799 | 8.9 |
| Asotin | 0.8 | D | 13.0 | 6.0 | 4.1 | 13.8 | 4.5 | D | 18.0 | 6,770 | 298 | 754 | 10,205 | 3.4 |
| Benton | 4.9 | D | 10.6 | 5.6 | 16.0 | 5.5 | 4.1 | 12.4 | 15.3 | 37,260 | 182 | 4,057 | 80,158 | 16.8 |
| Chelan | 4.2 | D | 8.5 | 4.4 | 4.6 | 8.3 | 5.0 | 18.4 | 20.5 | 18,085 | 235 | 1,369 | 39,262 | 10.7 |
| Clallam | 0.3 | 0.1 | 7.3 | 4.8 | D | 9.2 | 3.6 | 9.1 | 39.7 | 27,275 | 352 | 1,943 | 38,021 | 6.9 |
| Clark | 0.1 | D | 11.2 | 8.2 | 9.7 | 6.7 | 7.1 | 12.9 | 17.8 | 90,805 | 186 | 8,369 | 194,249 | 16.0 |
| Columbia | 14.5 | D | 14.5 | 7.3 | 3.5 | 3.4 | 2.3 | 3.1 | 33.6 | 1,330 | 331 | 145 | 2,181 | 2.1 |
| Cowlitz | 0.6 | D | 10.8 | 21.2 | 3.8 | 6.4 | 4.0 | 12.8 | 16.1 | 28,460 | 258 | 3,879 | 45,621 | 5.0 |
| Douglas | 11.5 | 0.4 | 8.4 | 5.2 | D | 10.5 | 2.9 | 5.9 | 26.7 | 8,390 | 194 | 548 | 17,758 | 11.0 |
| Ferry | 0.4 | D | D | D | D | 4.6 | D | D | 57.8 | 2,500 | 326 | 236 | 4,564 | 3.6 |
| Franklin | 11.3 | D | 8.3 | 9.0 | 2.5 | 7.8 | 2.6 | 7.4 | 22.1 | 11,285 | 118 | 1,805 | 29,929 | 22.5 |
| Garfield | 22.4 | 0.0 | D | 0.6 | D | 3.0 | 2.9 | D | 53.2 | 685 | 302 | 43 | 1,254 | 1.7 |
| Grant | 16.4 | D | 5.0 | 11.7 | 5.1 | 5.7 | 3.5 | 5.8 | 26.3 | 17,015 | 174 | 2,166 | 39,193 | 11.7 |
| Grays Harbor | 1.5 | D | 7.4 | 11.9 | 4.0 | 7.4 | 3.0 | 10.0 | 31.9 | 21,935 | 292 | 3,092 | 37,258 | 6.0 |
| Island | 0.3 | D | 7.1 | 3.0 | 6.0 | 4.9 | 4.3 | 5.2 | 51.2 | 23,735 | 278 | 1,006 | 43,018 | 6.9 |
| Jefferson | 0.8 | D | 10.6 | 7.3 | 9.2 | 7.4 | 4.7 | 6.1 | 31.8 | 13,140 | 406 | 598 | 19,141 | 7.7 |
| King | 0.0 | 0.2 | 5.6 | 6.9 | 29.8 | 10.5 | 6.5 | 7.7 | 10.5 | 300,325 | 133 | 36,354 | 987,044 | 15.9 |
| Kitsap | 0.0 | D | 5.3 | 2.4 | 8.0 | 5.5 | 3.8 | 9.2 | 52.4 | 55,240 | 203 | 4,785 | 115,998 | 8.0 |
| Kittitas | 3.1 | D | 10.3 | 3.2 | D | 7.5 | 4.1 | 6.1 | 35.8 | 9,055 | 189 | 542 | 25,195 | 15.1 |
| Klickitat | 7.3 | D | 5.6 | 27.0 | D | 2.2 | 2.7 | 3.6 | 20.3 | 6,495 | 290 | 645 | 10,875 | 11.1 |
| Lewis | 2.2 | 0.8 | 6.6 | 14.2 | 3.4 | 8.8 | 3.1 | 13.7 | 20.6 | 22,570 | 280 | 2,844 | 35,996 | 5.7 |
| Lincoln | 18.4 | D | 9.4 | D | 5.3 | 4.2 | D | D | 34.0 | 3,310 | 303 | 251 | 6,209 | 7.5 |
| Mason | 2.0 | D | 6.5 | 5.5 | D | 7.7 | 3.7 | 6.4 | 43.3 | 18,515 | 277 | 1,641 | 34,306 | 5.5 |
| Okanogan | 7.9 | D | 5.8 | 1.8 | 3.4 | 9.1 | 2.4 | 9.2 | 37.2 | 11,280 | 266 | 1,261 | 23,622 | 6.2 |
| Pacific | 4.0 | 0.4 | 6.5 | 8.5 | D | 5.5 | 4.2 | D | 35.4 | 8,790 | 390 | 713 | 16,515 | 6.2 |
| Pend Oreille | 0.6 | D | 5.8 | 9.4 | 3.9 | 4.0 | 2.6 | 16.1 | 29.4 | 4,460 | 325 | 502 | 8,422 | 6.1 |
| Pierce | 0.1 | 0.1 | 8.9 | 5.2 | 4.9 | 6.3 | 5.7 | 5.7 | 14.1 | 157,040 | 173 | 19,233 | 360,478 | 10.8 |
| San Juan | 0.2 | D | 17.6 | 3.2 | 10.7 | 9.2 | 7.3 | 5.6 | 23.3 | 5,685 | 328 | 128 | 14,510 | 9.0 |
| Skagit | 3.5 | 0.1 | 12.4 | 13.2 | 5.9 | 8.0 | 6.9 | 7.2 | 23.3 | 32,075 | 248 | 2,350 | 55,427 | 7.7 |
| Skamania | 0.5 | D | 7.4 | 11.8 | D | 3.8 | 2.2 | D | 32.2 | 2,710 | 224 | 184 | 6,064 | 7.7 |
| Snohomish | 0.3 | 0.1 | 10.2 | 27.3 | 8.1 | 7.1 | 6.2 | 8.5 | 14.7 | 127,820 | 155 | 11,960 | 322,557 | 12.5 |
| Spokane | 0.3 | 0.1 | 6.9 | 7.1 | 7.9 | 7.2 | 8.9 | 17.1 | 19.7 | 109,000 | 209 | 14,601 | 226,156 | 12.3 |
| Stevens | 1.0 | D | 7.3 | 10.5 | D | 6.6 | 3.4 | D | 29.2 | 13,635 | 298 | 1,382 | 22,376 | 5.8 |
| Thurston | 0.9 | 0.0 | 7.0 | 2.5 | 8.0 | 6.1 | 4.6 | 12.8 | 37.7 | 63,235 | 218 | 5,395 | 121,284 | 12.1 |
| Wahkiakum | 1.2 | 0.0 | 10.7 | 6.0 | 7.9 | 4.4 | D | D | 33.0 | 1,595 | 361 | 121 | 2,196 | 6.1 |
| Walla Walla | 8.6 | D | 4.2 | 15.7 | D | 4.6 | 4.5 | 14.9 | 24.5 | 13,635 | 223 | 1,429 | 25,283 | 7.8 |
| Whatcom | 2.2 | D | 11.0 | 13.0 | 7.3 | 8.3 | 6.0 | 11.6 | 18.2 | 46,755 | 204 | 4,357 | 101,169 | 11.6 |
| Whitman | 5.7 | D | 2.8 | 17.3 | 3.4 | 4.7 | 2.7 | 5.7 | 45.5 | 6,190 | 123 | 534 | 21,676 | 12.2 |
| Yakima | 13.9 | 0.0 | 4.8 | 7.9 | 3.0 | 6.9 | 3.4 | 14.3 | 19.2 | 45,485 | 181 | 7,142 | 90,835 | 6.3 |
| WEST VIRGINIA | -0.1 | 5.7 | 7.0 | 8.4 | 8.0 | 6.4 | 4.7 | 16.3 | 20.6 | 478,209 | 266 | 70,844 | 896,587 | 1.7 |
| Barbour | -0.6 | D | 4.5 | D | D | 4.5 | 2.4 | D | 17.1 | 4,330 | 263 | 694 | 7,927 | 1.0 |
| Berkeley | 0.1 | D | 5.4 | 9.7 | 10.7 | 5.9 | 4.0 | 12.3 | 29.4 | 23,875 | 200 | 2,504 | 51,532 | 15.1 |
| Boone | 0.0 | D | 2.6 | 0.8 | D | 5.9 | D | D | 27.1 | 6,830 | 320 | 1,249 | 11,201 | 1.2 |
| Braxton | 1.2 | 1.1 | 7.9 | 11.6 | 2.6 | 14.2 | 2.1 | 13.7 | 21.7 | 3,920 | 282 | 565 | 7,427 | 0.2 |
| Brooke | 0.0 | 0.0 | D | 21.2 | 7.1 | 7.1 | D | 10.3 | 27.4 | 6,480 | 296 | 458 | 10,726 | -2.1 |
| Cabell | 0.0 | 1.5 | 5.1 | 10.6 | 6.1 | 7.3 | 4.8 | 27.4 | 16.1 | 22,575 | 246 | 4,365 | 46,289 | 0.3 |

1. Per 1,000 resident population estimated as of July 1 of the year shown.  2. Martinsville city is included with Henry county.  3. Norton city is included with Wise county.  4. Petersburg and Colonial Heights cities are included with Dinwiddie county.  5. Poquoson city is included with York county.  6. Radford city is included with Montgomery county.  7. Salem city is included with Roanoke county.  8. Staunton and Waynesboro cities are included with Augusta county.  9. Williamsburg city is included with James City county.  10. Winchester city is included with Frederick county.

| STATE County | Housing units, 2015-2019 | | | | | | | | Civilian labor force, 2020 | | | | Civilian employment[6], 2015-2019 | | |
|---|---|---|---|---|---|---|---|---|---|---|---|---|---|---|---|
| | Occupied units | | | | | | | Sub-standard units[4] (percent) | Total | Percent change, 2019-2020 | Unemployment | | Total | Percent | |
| | Owner-occupied | | | | | Renter-occupied | | | | | | | | Management, business, science, and arts | Construction, production, and maintenance occupations |
| | Total | Percent | Median value[1] | With a mortgage | Without a mortgage[2] | Median rent[3] | Median rent as a percent of income[2] | | | | Total | Rate[5] | | | |
| | 89 | 90 | 91 | 92 | 93 | 94 | 95 | 96 | 97 | 98 | 99 | 100 | 101 | 102 | 103 |
| VIRGINIA—Cont'd | | | | | | | | | | | | | | | |
| Martinsville city | 5,532 | 55.9 | 87,700 | 21.7 | 12.1 | 656 | 30.3 | 0.0 | 5,670 | 3.1 | 619 | 10.9 | 5,144 | 26.5 | 31.1 |
| Newport News city | 69,835 | 48.9 | 194,000 | 23.3 | 11.0 | 1,057 | 32.0 | 3.4 | 89,715 | -0.3 | 7,786 | 8.7 | 81,407 | 34.6 | 24.8 |
| Norfolk city | 88,353 | 43.4 | 206,700 | 24.3 | 12.7 | 1,059 | 31.4 | 2.4 | 111,825 | -0.2 | 9,751 | 8.7 | 104,945 | 34.2 | 22.7 |
| Norton city | 1,628 | 50.1 | 93,800 | 14.6 | 12.8 | 618 | 34.0 | 3.2 | 1,660 | -1.7 | 124 | 7.5 | 1,607 | 35.7 | 18.7 |
| Petersburg city | 13,165 | 38.8 | 108,100 | 27.1 | 12.7 | 947 | 32.8 | 3.3 | 13,459 | 3.7 | 1,877 | 13.9 | 13,140 | 24.6 | 30.6 |
| Poquoson city | 4,593 | 81.4 | 323,100 | 21.6 | 13.5 | 1,222 | 26.6 | 0.0 | 6,249 | -3.7 | 261 | 4.2 | 5,757 | 49.3 | 20.9 |
| Portsmouth city | 36,370 | 55.0 | 170,900 | 24.1 | 13.9 | 1,048 | 33.2 | 1.6 | 44,701 | 0.3 | 4,273 | 9.6 | 41,396 | 29.8 | 27.4 |
| Radford city | 5,573 | 46.6 | 170,800 | 19.8 | 12.9 | 728 | 35.5 | 2.6 | 8,457 | -3.3 | 535 | 6.3 | 8,288 | 34.9 | 19.7 |
| Richmond city | 90,301 | 42.6 | 230,500 | 22.3 | 13.2 | 1,025 | 32.7 | 2.2 | 119,962 | 0.3 | 10,592 | 8.8 | 116,615 | 42.0 | 16.9 |
| Roanoke city | 41,740 | 51.5 | 135,100 | 21.7 | 11.8 | 814 | 30.4 | 2.0 | 49,174 | -0.3 | 3,786 | 7.7 | 47,605 | 32.1 | 23.7 |
| Salem city | 9,912 | 63.6 | 181,900 | 21.3 | 10.9 | 913 | 26.1 | 1.3 | 12,671 | -2.2 | 733 | 5.8 | 12,277 | 37.1 | 21.6 |
| Staunton city | 10,597 | 57.3 | 169,000 | 19.1 | 10.0 | 861 | 28.1 | 2.2 | 12,247 | -0.6 | 767 | 6.3 | 12,102 | 36.4 | 20.2 |
| Suffolk city | 33,774 | 68.7 | 254,400 | 22.5 | 11.6 | 1,201 | 31.4 | 1.8 | 44,546 | -2.1 | 2,903 | 6.5 | 42,459 | 39.0 | 23.2 |
| Virginia Beach city | 170,798 | 63.7 | 280,800 | 23.4 | 11.0 | 1,367 | 29.5 | 1.5 | 230,322 | -2.1 | 14,331 | 6.2 | 221,998 | 41.5 | 18.2 |
| Waynesboro city | 9,193 | 57.7 | 174,100 | 22.9 | 13.3 | 821 | 28.6 | 1.8 | 10,700 | -0.5 | 698 | 6.5 | 10,362 | 30.2 | 28.1 |
| Williamsburg city | 4,706 | 49.3 | 306,000 | 19.3 | 10.0 | 1,193 | 33.9 | 1.7 | 6,705 | -1.1 | 552 | 8.2 | 6,394 | 44.6 | 9.9 |
| Winchester city | 10,490 | 44.3 | 244,900 | 19.6 | 10.0 | 1,054 | 29.0 | 2.2 | 14,912 | 0.4 | 871 | 5.8 | 14,073 | 39.6 | 21.7 |
| WASHINGTON | 2,848,396 | 63.0 | 339,000 | 22.5 | 11.2 | 1,258 | 29.1 | 3.7 | 3,914,869 | 0.1 | 329,087 | 8.4 | 3,594,279 | 41.8 | 21.7 |
| Adams | 5,973 | 63.1 | 154,500 | 23.1 | 10.1 | 822 | 25.0 | 12.0 | 9,261 | -6.1 | 677 | 7.3 | 7,719 | 23.8 | 49.3 |
| Asotin | 9,101 | 70.7 | 198,100 | 21.2 | 10.3 | 797 | 28.1 | 1.8 | 10,383 | 0.4 | 538 | 5.2 | 9,708 | 31.5 | 26.8 |
| Benton | 72,121 | 68.8 | 235,800 | 19.6 | 10.0 | 974 | 27.8 | 3.3 | 103,143 | 0.3 | 8,461 | 8.2 | 88,737 | 40.2 | 24.3 |
| Chelan | 28,384 | 64.4 | 292,900 | 22.7 | 10.0 | 909 | 24.9 | 4.1 | 44,190 | -1.6 | 3,693 | 8.4 | 35,694 | 31.6 | 29.1 |
| Clallam | 32,958 | 70.4 | 253,800 | 23.6 | 11.2 | 952 | 29.6 | 3.1 | 29,280 | 1.8 | 2,977 | 10.2 | 28,673 | 32.1 | 23.3 |
| Clark | 174,661 | 67.0 | 327,000 | 22.0 | 10.1 | 1,261 | 29.6 | 3.3 | 242,578 | -0.7 | 20,682 | 8.5 | 227,108 | 37.4 | 23.8 |
| Columbia | 1,795 | 67.9 | 179,400 | 22.8 | 12.1 | 763 | 25.7 | 0.4 | 1,724 | -5.0 | 125 | 7.3 | 1,737 | 31.1 | 31.7 |
| Cowlitz | 41,952 | 66.1 | 219,300 | 22.0 | 12.2 | 854 | 30.8 | 3.0 | 48,522 | 1.4 | 4,523 | 9.3 | 44,679 | 27.3 | 32.2 |
| Douglas | 15,263 | 68.7 | 271,500 | 22.1 | 10.0 | 909 | 23.8 | 5.1 | 21,225 | -3.8 | 1,805 | 8.5 | 19,378 | 30.4 | 32.2 |
| Ferry | 3,060 | 72.0 | 169,600 | 22.1 | 10.0 | 672 | 28.3 | 8.8 | 2,469 | -1.1 | 285 | 11.5 | 2,599 | 34.3 | 29.8 |
| Franklin | 26,723 | 68.5 | 202,400 | 20.3 | 10.0 | 913 | 28.1 | 9.5 | 42,497 | | 3,833 | 9.0 | 40,135 | 28.0 | 35.0 |
| Garfield | 984 | 71.5 | 149,400 | 18.4 | 10.0 | 598 | 24.7 | 1.0 | 819 | -9.6 | 69 | 8.4 | 889 | 32.5 | 30.9 |
| Grant | 30,818 | 62.5 | 177,500 | 20.4 | 10.0 | 785 | 23.9 | 6.4 | 47,442 | 0.6 | 4,195 | 8.8 | 40,662 | 28.6 | 41.9 |
| Grays Harbor | 28,722 | 67.0 | 178,800 | 22.3 | 11.6 | 775 | 29.0 | 2.6 | 29,055 | 0.6 | 3,402 | 11.7 | 28,323 | 31.1 | 28.2 |
| Island | 34,768 | 71.5 | 358,800 | 25.6 | 11.3 | 1,218 | 29.7 | 2.0 | 36,303 | 1.2 | 2,998 | 8.3 | 32,421 | 35.3 | 24.7 |
| Jefferson | 14,705 | 74.3 | 338,000 | 25.4 | 11.3 | 932 | 30.0 | 2.2 | 13,081 | 1.9 | 1,231 | 9.4 | 11,607 | 36.3 | 20.7 |
| King | 882,028 | 56.9 | 549,200 | 22.3 | 12.0 | 1,606 | 28.1 | 4.2 | 1,286,608 | -0.2 | 96,440 | 7.5 | 1,206,334 | 53.0 | 14.4 |
| Kitsap | 103,913 | 67.8 | 329,000 | 22.7 | 11.1 | 1,249 | 29.0 | 2.2 | 129,092 | 0.7 | 9,952 | 7.7 | 115,237 | 40.1 | 21.6 |
| Kittitas | 18,347 | 59.1 | 286,900 | 24.0 | 11.6 | 1,004 | 34.6 | 1.8 | 22,085 | -6.2 | 1,998 | 9.0 | 22,372 | 33.1 | 22.0 |
| Klickitat | 8,877 | 68.0 | 261,000 | 22.3 | 10.8 | 847 | 31.6 | 4.9 | 9,810 | -4.4 | 808 | 8.2 | 8,561 | 38.2 | 27.9 |
| Lewis | 30,618 | 70.6 | 209,600 | 22.9 | 11.9 | 876 | 30.3 | 3.5 | 35,103 | 0.8 | 3,284 | 9.4 | 31,386 | 27.8 | 31.2 |
| Lincoln | 4,525 | 78.0 | 162,800 | 21.4 | 11.1 | 788 | 23.4 | 2.3 | 4,681 | -6.9 | 318 | 6.8 | 4,206 | 41.8 | 25.1 |
| Mason | 24,278 | 77.4 | 226,500 | 23.9 | 11.1 | 988 | 30.0 | 3.9 | 25,517 | 1.4 | 2,508 | 9.8 | 24,994 | 29.7 | 28.3 |
| Okanogan | 17,675 | 65.8 | 182,700 | 23.4 | 10.0 | 711 | 27.1 | 5.4 | 19,581 | -4.0 | 1,759 | 9.0 | 17,334 | 31.3 | 30.1 |
| Pacific | 9,279 | 80.0 | 176,800 | 22.3 | 12.2 | 754 | 28.7 | 1.5 | 8,524 | 0.1 | 914 | 10.7 | 7,421 | 29.6 | 31.5 |
| Pend Oreille | 5,727 | 77.9 | 216,700 | 22.9 | 10.0 | 757 | 33.7 | 4.9 | 4,958 | -1.4 | 530 | 10.7 | 4,743 | 34.0 | 30.2 |
| Pierce | 323,296 | 62.1 | 303,200 | 23.3 | 12.1 | 1,250 | 30.3 | 3.1 | 450,196 | 1.9 | 43,374 | 9.6 | 409,724 | 34.4 | 25.8 |
| San Juan | 8,252 | 74.6 | 488,800 | 27.4 | 11.9 | 1,000 | 27.2 | 4.9 | 8,379 | -2.9 | 665 | 7.9 | 7,765 | 38.8 | 22.7 |
| Skagit | 48,493 | 68.6 | 306,000 | 23.6 | 12.1 | 1,102 | 29.5 | 5.2 | 62,498 | | 6,124 | 9.8 | 56,470 | 33.2 | 30.2 |
| Skamania | 4,816 | 74.5 | 296,700 | 22.7 | 11.3 | 850 | 28.2 | 2.9 | 5,515 | -1.6 | 509 | 9.2 | 5,366 | 35.9 | 30.4 |
| Snohomish | 293,823 | 67.1 | 403,000 | 22.9 | 11.1 | 1,438 | 29.7 | 3.6 | 441,156 | 0.6 | 37,293 | 8.5 | 409,889 | 40.5 | 23.1 |
| Spokane | 202,811 | 62.4 | 224,800 | 21.9 | 10.7 | 913 | 30.0 | 2.4 | 257,308 | 1.1 | 22,519 | 8.8 | 232,708 | 37.6 | 20.1 |
| Stevens | 17,554 | 78.1 | 199,200 | 22.4 | 10.2 | 673 | 27.4 | 4.2 | 18,545 | -2.0 | 1,746 | 9.4 | 16,739 | 29.4 | 31.0 |
| Thurston | 109,983 | 65.4 | 287,900 | 22.4 | 10.5 | 1,218 | 31.3 | 2.1 | 144,312 | 1.4 | 11,947 | 8.3 | 128,801 | 42.7 | 19.2 |
| Wahkiakum | 1,897 | 86.7 | 218,800 | 24.2 | 12.2 | 760 | 32.1 | 2.4 | 1,409 | 2.8 | 124 | 8.8 | 1,313 | 29.7 | 33.4 |
| Walla Walla | 22,646 | 64.8 | 224,800 | 20.6 | 11.0 | 926 | 32.8 | 3.2 | 30,165 | 0.1 | 2,069 | 6.9 | 26,646 | 36.2 | 22.8 |
| Whatcom | 86,523 | 61.9 | 345,700 | 23.4 | 11.7 | 1,060 | 32.4 | 3.5 | 114,716 | -0.4 | 10,724 | 9.3 | 107,180 | 35.7 | 23.9 |
| Whitman | 17,999 | 43.6 | 217,600 | 20.4 | 10.2 | 814 | 37.4 | 2.1 | 21,984 | -8.9 | 1,301 | 5.9 | 23,003 | 49.5 | 16.7 |
| Yakima | 83,048 | 62.7 | 175,900 | 21.9 | 10.5 | 825 | 28.1 | 8.7 | 130,759 | -0.8 | 12,688 | 9.7 | 106,018 | 26.1 | 39.7 |
| WEST VIRGINIA | 732,585 | 73.2 | 119,600 | 18.3 | 10.0 | 725 | 28.8 | 1.8 | 792,156 | -0.8 | 66,133 | 8.3 | 740,910 | 33.8 | 24.9 |
| Barbour | 6,324 | 72.0 | 110,300 | 20.3 | 10.9 | 549 | 25.7 | 0.9 | 7,335 | 1.7 | 606 | 8.3 | 6,631 | 27.5 | 30.0 |
| Berkeley | 44,221 | 74.6 | 182,200 | 20.7 | 10.0 | 995 | 27.8 | 2.5 | 58,337 | -1.4 | 3,628 | 6.2 | 55,876 | 34.0 | 27.4 |
| Boone | 8,932 | 76.6 | 71,500 | 19.1 | 10.5 | 646 | 37.4 | 1.6 | 7,469 | 0.9 | 784 | 10.5 | 6,527 | 26.9 | 32.9 |
| Braxton | 5,624 | 80.7 | 86,400 | 18.4 | 10.0 | 593 | 23.7 | 2.6 | 5,220 | -0.6 | 542 | 10.4 | 5,221 | 25.7 | 31.0 |
| Brooke | 9,805 | 76.2 | 93,200 | 16.9 | 10.0 | 607 | 25.1 | 1.7 | 9,901 | -1.1 | 889 | 9.0 | 10,066 | 30.7 | 26.4 |
| Cabell | 39,064 | 61.8 | 124,500 | 18.4 | 10.2 | 748 | 33.9 | 0.9 | 41,946 | 1.0 | 3,342 | 8.0 | 38,148 | 36.9 | 16.8 |

1. Specified owner-occupied units lacking complete plumbing facilities.   2. A value of 10.0 represents 10 percent or less; a value of 50.0 represents 50 percent or more.   3. Specified renter-occupied units.   4. Overcrowded or 5. Percent of civilian labor force.   6. Civilian employed persons 16 years old and over.

# Table B. States and Counties — Nonfarm Employment and Agriculture

| STATE County | Number of establishments (104) | Employment — Total (105) | Health care and social assistance (106) | Manufacturing (107) | Retail trade (108) | Finance and insurance (109) | Professional, scientific, and technical services (110) | Annual payroll Total (mil dol) (111) | Average per employee (dollars) (112) | Farms — Number (113) | Percent with fewer than 50 acres (114) | Percent with 1000 acres or more (115) | Farm producers whose primary occupation is farming (percent) (116) |
|---|---|---|---|---|---|---|---|---|---|---|---|---|---|
| **VIRGINIA—Cont'd** | | | | | | | | | | | | | |
| Martinsville city | 508 | 8,884 | 2,301 | 1,595 | 1,410 | 205 | 154 | 278 | 31,337 | NA | NA | NA | NA |
| Newport News city | 3,726 | 91,021 | 15,511 | 27,487 | 9,866 | 1,671 | 4,498 | 4,855 | 53,344 | NA | NA | NA | NA |
| Norfolk city | 5,386 | 108,816 | 21,124 | 6,975 | 11,595 | 4,496 | 11,295 | 5,548 | 50,982 | NA | NA | NA | NA |
| Norton city | 213 | 4,538 | 1,290 | 710 | 737 | 76 | 116 | 190 | 41,848 | NA | NA | NA | NA |
| Petersburg city | 708 | 12,055 | 3,745 | 1,140 | 2,332 | 140 | 152 | 489 | 40,580 | NA | NA | NA | NA |
| Poquoson city | 222 | 1,454 | 193 | 8 | 272 | 44 | 93 | 39 | 26,534 | NA | NA | NA | NA |
| Portsmouth city | 1,655 | 25,744 | 7,277 | 876 | 3,184 | 383 | 1,343 | 1,126 | 43,727 | NA | NA | NA | NA |
| Radford city | 299 | 4,130 | 587 | 1,170 | 380 | 130 | 151 | 151 | 36,648 | NA | NA | NA | NA |
| Richmond city | 6,323 | 128,474 | 25,476 | 5,223 | 9,742 | 10,637 | 11,798 | 8,025 | 62,466 | NA | NA | NA | NA |
| Roanoke city | 3,132 | 75,238 | 15,524 | 4,207 | 8,857 | 3,453 | 2,822 | 3,440 | 45,727 | NA | NA | NA | NA |
| Salem city | 1,010 | 19,322 | 5,257 | 2,963 | 1,913 | 482 | 594 | 973 | 50,336 | NA | NA | NA | NA |
| Staunton city | 749 | 10,558 | 2,481 | 522 | 1,846 | 310 | 349 | 361 | 34,204 | 270 | 57.4 | 11.1 | 38.6 |
| Suffolk city | 1,684 | 25,065 | 5,069 | 1,711 | 4,119 | 793 | 1,066 | 1,010 | 40,300 | 196 | 67.9 | 2.0 | 35.6 |
| Virginia Beach city | 11,316 | 160,538 | 21,397 | 6,156 | 24,501 | 11,875 | 15,537 | 6,931 | 43,175 | NA | NA | NA | NA |
| Waynesboro city | 611 | 9,103 | 569 | 1,279 | 2,081 | 416 | 367 | 337 | 36,979 | NA | NA | NA | NA |
| Williamsburg city | 524 | 9,416 | 511 | 45 | 2,115 | 160 | 179 | 292 | 31,034 | NA | NA | NA | NA |
| Winchester city | 1,354 | 25,529 | 7,555 | 1,898 | 4,247 | 659 | 815 | 1,064 | 41,663 | NA | NA | NA | NA |
| **WASHINGTON** | 195,105 | 2,898,378 | 444,635 | 269,974 | 340,302 | 109,831 | 217,205 | 191,592 | 66,103 | 35,793 | 66.6 | 7.2 | 39.9 |
| Adams | 382 | 4,546 | 786 | 1,110 | 672 | 91 | 63 | 214 | 47,156 | 586 | 20.8 | 32.8 | 55.2 |
| Asotin | 447 | 4,902 | 1,271 | 434 | 968 | 148 | 227 | 207 | 42,308 | 205 | 42.9 | 35.6 | 37.8 |
| Benton | 4,730 | 71,323 | 11,852 | 4,244 | 10,198 | 1,965 | 7,442 | 4,010 | 56,228 | 1,520 | 81.1 | 4.9 | 34.2 |
| Chelan | 2,613 | 32,096 | 6,796 | 1,499 | 4,488 | 792 | 1,157 | 1,519 | 47,336 | 835 | 72.7 | 0.5 | 47.3 |
| Clallam | 2,081 | 18,421 | 4,414 | 986 | 3,634 | 506 | 741 | 734 | 39,862 | 528 | 83.9 | NA | 39.3 |
| Clark | 11,498 | 149,890 | 24,612 | 14,976 | 18,756 | 6,186 | 8,559 | 7,981 | 53,244 | 1,978 | 87.4 | 0.3 | 29.9 |
| Columbia | 117 | 880 | 171 | 163 | 107 | 12 | 39 | 40 | 45,003 | 257 | 27.6 | 17.9 | 43.2 |
| Cowlitz | 2,226 | 34,670 | 5,935 | 7,427 | 5,059 | 833 | 971 | 1,720 | 49,600 | 403 | 76.9 | 1.5 | 33.3 |
| Douglas | 798 | 8,062 | 854 | 552 | 1,785 | 179 | 195 | 297 | 36,813 | 729 | 41.0 | 25.4 | 49.7 |
| Ferry | 142 | 962 | 149 | 226 | 192 | 20 | 18 | 38 | 39,450 | 252 | 31.3 | 9.1 | 53.3 |
| Franklin | 1,720 | 22,756 | 3,053 | 3,206 | 3,734 | 382 | 585 | 1,066 | 46,849 | 772 | 40.2 | 18.0 | 55.9 |
| Garfield | 41 | 310 | 108 | NA | 58 | 16 | NA | 11 | 36,110 | 226 | 22.1 | 31.9 | 45.9 |
| Grant | 1,875 | 23,351 | 3,055 | 4,327 | 3,639 | 541 | 1,315 | 1,073 | 45,949 | 1,384 | 36.1 | 18.8 | 60.0 |
| Grays Harbor | 1,548 | 15,822 | 2,774 | 2,087 | 2,896 | 492 | 369 | 651 | 41,160 | 469 | 65.0 | 1.9 | 35.6 |
| Island | 1,835 | 13,061 | 2,535 | 781 | 2,326 | 357 | 357 | 523 | 40,022 | 390 | 83.6 | NA | 43.0 |
| Jefferson | 1,080 | 7,093 | 1,419 | 681 | 1,168 | 150 | 304 | 293 | 41,320 | 221 | 63.3 | NA | 39.2 |
| King | 70,619 | 1,273,043 | 158,642 | 87,364 | 108,758 | 51,351 | 125,021 | 109,033 | 85,647 | 1,796 | 93.1 | 0.1 | 35.3 |
| Kitsap | 6,052 | 63,818 | 13,905 | 2,354 | 11,975 | 1,741 | 4,958 | 2,775 | 43,488 | 698 | 96.3 | NA | 29.0 |
| Kittitas | 1,316 | 12,381 | 1,859 | 641 | 1,931 | 232 | 336 | 433 | 34,941 | 1,008 | 71.1 | 3.0 | 37.1 |
| Klickitat | 543 | 3,917 | 587 | 691 | 338 | 66 | 338 | 187 | 47,616 | 750 | 46.3 | 13.3 | 38.9 |
| Lewis | 1,944 | 21,311 | 3,867 | 2,825 | 3,823 | 382 | 639 | 904 | 42,416 | 1,723 | 70.3 | 0.6 | 35.9 |
| Lincoln | 262 | 1,706 | 377 | 44 | 231 | 67 | 107 | 75 | 44,147 | 783 | 15.2 | 43.2 | 53.1 |
| Mason | 1,065 | 9,757 | 1,948 | 857 | 1,898 | 285 | 252 | 403 | 41,327 | 324 | 82.1 | 0.6 | 35.5 |
| Okanogan | 1,121 | 8,090 | 1,706 | 327 | 1,885 | 203 | 207 | 296 | 36,573 | 1,192 | 55.4 | 8.1 | 44.6 |
| Pacific | 589 | 4,360 | 664 | 777 | 578 | 120 | 133 | 161 | 36,820 | 346 | 56.1 | 3.8 | 37.4 |
| Pend Oreille | 230 | 1,776 | 435 | 267 | 264 | 56 | 76 | 81 | 45,601 | 261 | 47.9 | 5.7 | 33.3 |
| Pierce | 18,589 | 270,855 | 54,596 | 17,791 | 37,953 | 9,000 | 9,875 | 13,734 | 50,705 | 1,607 | 87.4 | 0.2 | 31.8 |
| San Juan | 1,034 | 5,232 | 417 | 215 | 826 | 141 | 245 | 218 | 41,603 | 316 | 70.6 | 0.6 | 36.5 |
| Skagit | 3,613 | 44,580 | 7,812 | 7,006 | 7,337 | 1,375 | 1,835 | 2,185 | 49,019 | 1,041 | 75.7 | 1.6 | 40.8 |
| Skamania | 194 | 1,575 | 134 | 329 | 163 | 18 | 65 | 53 | 33,389 | 145 | 80.0 | NA | 24.4 |
| Snohomish | 19,292 | 262,857 | 32,380 | 59,835 | 35,966 | 10,147 | 12,828 | 15,557 | 59,183 | 1,558 | 85.2 | 0.4 | 34.3 |
| Spokane | 13,458 | 197,989 | 43,193 | 15,445 | 26,699 | 10,715 | 8,913 | 9,585 | 48,409 | 2,425 | 63.9 | 5.8 | 33.6 |
| Stevens | 916 | 8,059 | 2,176 | 1,250 | 1,256 | 211 | 204 | 320 | 39,681 | 1,114 | 44.9 | 5.0 | 41.6 |
| Thurston | 6,444 | 75,644 | 15,783 | 2,833 | 13,033 | 2,945 | 4,754 | 3,408 | 45,059 | 1,200 | 77.8 | 0.6 | 35.3 |
| Wahkiakum | 82 | 457 | 55 | 66 | 53 | 8 | 12 | 16 | 36,004 | 145 | 60.7 | 0.7 | 49.2 |
| Walla Walla | 1,416 | 20,805 | 4,627 | 3,537 | 2,467 | 797 | 503 | 942 | 45,289 | 903 | 54.2 | 18.4 | 41.7 |
| Whatcom | 6,838 | 78,090 | 11,024 | 10,583 | 10,593 | 2,595 | 3,370 | 3,693 | 47,296 | 1,712 | 79.6 | 0.7 | 35.4 |
| Whitman | 838 | 10,823 | 2,038 | 2,704 | 1,258 | 203 | 339 | 494 | 45,657 | 1,039 | 27.0 | 35.5 | 50.0 |
| Yakima | 4,852 | 72,140 | 15,698 | 9,534 | 10,702 | 1,556 | 2,283 | 3,149 | 43,649 | 2,952 | 74.1 | 3.8 | 44.3 |
| **WEST VIRGINIA** | 35,795 | 554,433 | 132,802 | 48,901 | 80,482 | 16,633 | 25,177 | 23,907 | 43,119 | 23,622 | 34.7 | 1.5 | 36.7 |
| Barbour | 208 | 2,732 | 894 | 119 | 274 | 67 | 61 | 111 | 40,519 | 594 | 28.1 | 1.3 | 37.7 |
| Berkeley | 1,658 | 27,017 | 6,229 | 3,719 | 3,974 | 543 | 969 | 1,192 | 44,129 | 946 | 67.2 | 0.5 | 28.8 |
| Boone | 255 | 3,829 | 1,020 | 86 | 552 | 69 | 69 | 176 | 46,093 | 35 | 17.1 | NA | 34.5 |
| Braxton | 245 | 2,947 | 871 | 319 | 584 | 57 | 40 | 107 | 36,242 | 381 | 16.8 | 3.1 | 40.8 |
| Brooke | 353 | 6,840 | 2,056 | 1,549 | 959 | 122 | 71 | 280 | 40,953 | 89 | 28.1 | NA | 45.9 |
| Cabell | 2,343 | 46,730 | 14,677 | 4,250 | 6,096 | 998 | 1,854 | 1,972 | 42,208 | 407 | 36.9 | NA | 34.8 |

Items 104—116

# Table B. States and Counties — Agriculture

| STATE County | Land in farms — Acreage (1,000) [117] | Percent change, 2012-2017 [118] | Acres — Average size of farm [119] | Acres — Total irrigated (1,000) [120] | Acres — Total cropland (1,000) [121] | Value of land and buildings — Average per farm [122] | Value of land and buildings — Average per acre [123] | Value of machinery and equipment, average per farm (dollars) [124] | Products sold — Total (mil dol) [125] | Products sold — Average per farm (acres) [126] | Percent from Crops [127] | Percent from Livestock and poultry products [128] | Organic farms (number) [129] | Farms with internet access (percent) [130] | Government payments — Total ($1,000) [131] | Government payments — Percent of farms [132] |
|---|---|---|---|---|---|---|---|---|---|---|---|---|---|---|---|---|
| **VIRGINIA—Cont'd** | | | | | | | | | | | | | | | | |
| Martinsville city | NA | NA | NA | NA | NA | NA | NA | NA | NA | NA | NA | NA | NA | NA | NA | NA |
| Newport News city | NA | NA | NA | NA | NA | NA | NA | NA | NA | NA | NA | NA | NA | NA | NA | NA |
| Norfolk city | NA | NA | NA | NA | NA | NA | NA | NA | NA | NA | NA | NA | NA | NA | NA | NA |
| Norton city | NA | NA | NA | NA | NA | NA | NA | NA | NA | NA | NA | NA | NA | NA | NA | NA |
| Petersburg city | NA | NA | NA | NA | NA | NA | NA | NA | NA | NA | NA | NA | NA | NA | NA | NA |
| Poquoson city | NA | NA | NA | NA | NA | NA | NA | NA | NA | NA | NA | NA | NA | NA | NA | NA |
| Portsmouth city | NA | NA | NA | NA | NA | NA | NA | NA | NA | NA | NA | NA | NA | NA | NA | NA |
| Radford city | NA | NA | NA | NA | NA | NA | NA | NA | NA | NA | NA | NA | NA | NA | NA | NA |
| Richmond city | NA | NA | NA | NA | NA | NA | NA | NA | NA | NA | NA | NA | NA | NA | NA | NA |
| Roanoke city | NA | NA | NA | NA | NA | NA | NA | NA | NA | NA | NA | NA | NA | NA | NA | NA |
| Salem city | NA | NA | NA | NA | NA | NA | NA | NA | NA | NA | NA | NA | NA | NA | NA | NA |
| Staunton city | NA | NA | NA | NA | NA | NA | NA | NA | NA | NA | NA | NA | NA | NA | NA | NA |
| Suffolk city | 79 | 14.1 | 293 | 0.6 | 57.9 | 1,208,063 | 4,127 | 131,386 | 53.7 | 199,041 | 84.2 | 15.8 | NA | NA | NA | NA |
| Virginia Beach city | 23 | -10.8 | 119 | 0.3 | 18.6 | 1,149,351 | 9,648 | 107,202 | 13.7 | 69,796 | 91.1 | 8.9 | NA | 82.6 | 3,185 | 56.3 |
| Waynesboro city | NA | NA | NA | NA | NA | NA | NA | NA | NA | NA | NA | NA | 3 | 82.7 | 382 | 17.3 |
| Williamsburg city | NA | NA | NA | NA | NA | NA | NA | NA | NA | NA | NA | NA | NA | NA | NA | NA |
| Winchester city | NA | NA | NA | NA | NA | NA | NA | NA | NA | NA | NA | NA | NA | NA | NA | NA |
| **WASHINGTON** | 14,680 | -0.5 | 410 | 1,689.4 | 7,488.6 | 1,143,889 | 2,789 | 121,662 | 9,634.5 | 269,172 | 72.5 | 27.5 | 933 | 84.1 | 168,990 | 15.4 |
| Adams | 972 | -6.3 | 1,659 | 127.9 | 745.9 | 2,369,294 | 1,428 | 320,442 | 363.9 | 620,947 | 71.4 | 28.6 | 15 | 80.0 | 16,076 | 67.4 |
| Asotin | 251 | -4.7 | 1,224 | 0.9 | 79.9 | 1,618,697 | 1,323 | 86,684 | 12.9 | 62,961 | 58.5 | 41.5 | NA | 93.7 | 3,920 | 48.3 |
| Benton | 614 | -12.8 | 404 | 204.3 | 473.1 | 1,573,284 | 3,898 | 188,926 | 1,005.3 | 661,374 | 76.5 | 23.5 | 25 | 84.9 | 6,663 | 7.0 |
| Chelan | 60 | -21.2 | 72 | 23.8 | 29.1 | 1,102,315 | 15,400 | 78,413 | 258.4 | 309,501 | 98.9 | 1.1 | 54 | 84.7 | 1,087 | 3.6 |
| Clallam | 17 | -27.3 | 33 | 3.5 | 7.4 | 415,890 | 12,769 | 40,366 | 12 | 22,777 | 42.3 | 57.7 | 20 | 83.5 | 43 | 3.0 |
| Clark | 91 | 21.4 | 46 | 4.9 | 24.3 | 410,366 | 8,946 | 42,816 | 47.7 | 24,116 | 41.7 | 58.3 | 32 | 82.9 | 208 | 1.7 |
| Columbia | 243 | -18.2 | 947 | 3.0 | 149.5 | 1,541,289 | 1,628 | 155,022 | 30.7 | 119,479 | 86.4 | 13.6 | NA | 79.8 | 4,211 | 61.5 |
| Cowlitz | 29 | -26.3 | 71 | 3.0 | 11 | 639,038 | 8,955 | 53,997 | 19 | 47,045 | 52.8 | 47.2 | NA | 80.9 | 27 | 1.5 |
| Douglas | 823 | 1.1 | 1,129 | 16.3 | 544.4 | 1,331,223 | 1,180 | 134,210 | 186 | 255,154 | 96.5 | 3.5 | 23 | 76.5 | 14,978 | 48.7 |
| Ferry | 789 | -0.5 | 3,130 | 2.8 | 19.4 | 1,620,480 | 518 | 56,595 | D | D | D | D | 4 | 71.0 | 135 | 7.1 |
| Franklin | 615 | -1.6 | 797 | 188.1 | 446.8 | 3,662,394 | 4,595 | 365,464 | 631.6 | 818,132 | 74.3 | 25.7 | 31 | 88.0 | 9,867 | 24.7 |
| Garfield | 290 | -6.0 | 1,283 | 1.0 | 182.8 | 2,017,792 | 1,573 | 196,343 | 37.2 | 164,385 | 85.7 | 14.3 | NA | 80.1 | 5,997 | 70.8 |
| Grant | 1,042 | 8.1 | 753 | 448.0 | 800.9 | 2,574,272 | 3,421 | 447,392 | 1,938.9 | 1,400,936 | 76.3 | 23.7 | 89 | 84.8 | 13,885 | 29.1 |
| Grays Harbor | 105 | -11.9 | 224 | 6.3 | 17.1 | 537,286 | 2,395 | 84,525 | 33.6 | 71,635 | 52.3 | 47.7 | 6 | 81.4 | 62 | 2.1 |
| Island | 16 | 3.9 | 41 | 1.9 | 6.9 | 446,211 | 10,979 | 42,783 | 12 | 30,774 | 24.9 | 75.1 | 19 | 85.6 | 85 | 4.9 |
| Jefferson | 14 | -11.6 | 62 | 1.0 | 3.7 | 473,659 | 7,611 | 41,419 | 9.3 | 41,860 | 23.3 | 76.7 | 10 | 89.1 | 30 | 8.6 |
| King | 42 | -10.2 | 23 | 4.1 | 18.7 | 823,790 | 35,248 | 38,650 | 135.5 | 75,425 | 66.9 | 33.1 | 39 | 87.6 | 760 | 2.6 |
| Kitsap | 9 | -6.7 | 13 | 0.5 | 2.3 | 473,099 | 35,164 | 33,113 | 6.6 | 9,463 | 73.2 | 26.8 | 26 | 88.3 | D | 0.6 |
| Kittitas | 173 | -5.8 | 171 | 66.8 | 71.1 | 706,325 | 4,127 | 90,038 | 83 | 82,347 | 76.7 | 23.3 | 7 | 88.1 | 1,257 | 6.7 |
| Klickitat | 574 | 4.1 | 765 | 24.4 | 229.5 | 1,360,462 | 1,778 | 103,084 | 99.2 | 132,212 | 81.4 | 18.6 | 22 | 81.9 | 4,684 | 33.5 |
| Lewis | 123 | -7.5 | 71 | 10.0 | 50.5 | 427,935 | 6,001 | 55,899 | 136.3 | 79,132 | 26.1 | 73.9 | 50 | 83.4 | 442 | 3.6 |
| Lincoln | 1,181 | 5.9 | 1,509 | 29.5 | 817 | 1,843,885 | 1,222 | 233,510 | 130.2 | 166,331 | 91.5 | 8.5 | 4 | 78.2 | 24,298 | 79.3 |
| Mason | 18 | -23.6 | 56 | 1.0 | 3.6 | 479,276 | 8,562 | 52,381 | 48.5 | 149,790 | 5.6 | 94.4 | 4 | 83.0 | 77 | 4.3 |
| Okanogan | 1,232 | 2.2 | 1,033 | 46.0 | 98.7 | 1,247,686 | 1,207 | 91,219 | 338.1 | 283,631 | 88.0 | 12.0 | 84 | 81.9 | 2,307 | 7.1 |
| Pacific | 52 | 0.4 | 151 | 3.7 | 15.5 | 524,322 | 3,464 | 81,240 | 38.9 | 112,361 | 18.7 | 81.3 | 15 | 88.2 | 504 | 5.2 |
| Pend Oreille | 58 | 33.1 | 223 | 1.3 | 19 | 593,012 | 2,665 | 48,783 | 4.7 | 18,130 | 47.4 | 52.6 | NA | 87.4 | 74 | 4.2 |
| Pierce | 46 | -7.5 | 28 | 3.0 | 12.1 | 612,026 | 21,490 | 43,894 | 64.9 | 40,371 | 45.4 | 54.6 | 17 | 87.4 | 71 | 2.2 |
| San Juan | 18 | 17.4 | 58 | 0.3 | 5.8 | 550,693 | 9,457 | 28,800 | 4.1 | 13,035 | 58.6 | 41.4 | 12 | 92.1 | 55 | 2.8 |
| Skagit | 98 | -8.3 | 94 | 23.5 | 65.7 | 950,360 | 10,130 | 129,653 | 287.1 | 275,789 | 66.6 | 33.4 | 61 | 83.0 | 407 | 7.8 |
| Skamania | 6 | -9.3 | 41 | 0.4 | 2.3 | 489,432 | 12,082 | 41,679 | 5.6 | 38,828 | 29.2 | 70.8 | 1 | 90.3 | D | 2.8 |
| Snohomish | 64 | -10.1 | 41 | 8.4 | 33.6 | 790,023 | 19,332 | 60,306 | 157.6 | 101,133 | 48.5 | 51.5 | 31 | 89.0 | 464 | 4.6 |
| Spokane | 549 | 2.1 | 226 | 12.7 | 378.8 | 843,187 | 3,728 | 78,785 | 117 | 48,265 | 83.3 | 16.7 | 4 | 84.6 | 8,095 | 19.6 |
| Stevens | 518 | -1.7 | 465 | 7.2 | 77.1 | 729,604 | 1,569 | 60,364 | 30.2 | 27,105 | 39.8 | 60.2 | 47 | 80.2 | 734 | 7.3 |
| Thurston | 62 | -18.8 | 52 | 6.4 | 22.1 | 616,274 | 11,880 | 58,420 | 176.1 | 146,742 | 32.0 | 68.0 | 34 | 86.8 | 107 | 1.9 |
| Wahkiakum | 14 | 44.8 | 95 | 0.1 | 5.1 | 458,501 | 4,805 | 45,757 | 2.6 | 17,959 | 21.4 | 78.6 | 1 | 91.7 | 52 | 13.1 |
| Walla Walla | 703 | 8.9 | 778 | 101.7 | 565.8 | 1,968,909 | 2,531 | 208,241 | D | D | D | D | 13 | 85.3 | 16,092 | 40.1 |
| Whatcom | 103 | -11.5 | 60 | 36.5 | 75.6 | 1,005,681 | 16,794 | 95,376 | 372.9 | 217,786 | 41.3 | 58.7 | 48 | 83.7 | 1,047 | 12.6 |
| Whitman | 1,288 | 1.0 | 1,240 | 5.1 | 1,032.7 | 2,164,850 | 1,746 | 273,129 | 278.8 | 268,309 | 93.1 | 6.9 | 1 | 87.4 | 24,847 | 70.8 |
| Yakima | 1,781 | 0.1 | 603 | 260.0 | 344 | 1,662,430 | 2,755 | 174,332 | 1,988 | 673,451 | 71.3 | 28.7 | 84 | 79.0 | 5,221 | 6.5 |
| **WEST VIRGINIA** | 3,662 | 1.5 | 155 | 1.7 | 947.7 | 411,482 | 2,654 | 56,120 | 754.3 | 31,931 | 20.3 | 79.7 | 63 | 70.0 | 9,094 | 7.9 |
| Barbour | 95 | 11.7 | 159 | 0.0 | 26.7 | 371,966 | 2,335 | 56,514 | 6 | 10,167 | 33.7 | 66.3 | 3 | 70.9 | 205 | 5.1 |
| Berkeley | 73 | 4.3 | 77 | 0.1 | 37.6 | 413,514 | 5,349 | 47,415 | 25.9 | 27,387 | 72.2 | 27.8 | 2 | 68.8 | 857 | 8.4 |
| Boone | 4 | 61.7 | 103 | 0.0 | 0.5 | 241,209 | 2,339 | 23,954 | 0.1 | 4,114 | 77.1 | 22.9 | NA | 71.4 | 5 | 11.4 |
| Braxton | 90 | 0.8 | 235 | 0.0 | 17.9 | 482,984 | 2,054 | 51,405 | 4.5 | 11,916 | 28.5 | 71.5 | NA | 61.9 | 133 | 11.5 |
| Brooke | 14 | -2.7 | 161 | D | 4.3 | 454,092 | 2,825 | 90,249 | 1.3 | 14,135 | 46.4 | 53.6 | NA | 78.7 | 4 | 4.5 |
| Cabell | 40 | -5.0 | 99 | 0.0 | 7.3 | 352,954 | 3,564 | 42,757 | 2.7 | 6,595 | 64.1 | 35.9 | NA | 61.9 | 9 | 4.2 |

Items 117—132

| STATE County | Water use, 2015 | | Wholesale Trade[1], 2017 | | | | Retail Trade[2], 2017 | | | | Real estate and rental and leasing,[2] 2017 | | | |
|---|---|---|---|---|---|---|---|---|---|---|---|---|---|---|
| | Public supply water withdrawn (mil gal/day) | Public supply gallons withdrawn per person per day | Number of establishments | Number of employees | Sales (mil dol) | Average payroll (mil dol) | Number of establishments | Number of employees | Sales (mil dol) | Average payroll (mil dol) | Number of establishments | Number of employees | Sales (mil dol) | Average payroll (mil dol) |
| | 133 | 134 | 135 | 136 | 137 | 138 | 139 | 140 | 141 | 142 | 143 | 144 | 145 | 146 |
| VIRGINIA—Cont'd | | | | | | | | | | | | | | |
| Martinsville city............ | 2.00 | 146.6 | 18 | 154 | 52.9 | 5.5 | 91 | 1,432 | 301.9 | 32.4 | 22 | 83 | 16.1 | 2.6 |
| Newport News city........... | 22.20 | 121.7 | 112 | 1,887 | 1,539.2 | 129.7 | 648 | 10,091 | 2,596.7 | 242.7 | 280 | 1,419 | 348.7 | 59.2 |
| Norfolk city................ | 0.28 | 1.1 | 187 | 2,714 | 2,296.8 | 146.3 | 830 | 11,729 | 2,696.3 | 293.1 | 317 | 3,133 | 608.6 | 145.4 |
| Norton city................. | 0.53 | 134.6 | D | D | D | D | 45 | 768 | 185.2 | 19.7 | D | D | D | D |
| Petersburg city............. | 0.00 | 0.0 | D | D | D | 15.2 | 141 | 2,454 | 940.7 | 67.1 | 37 | 163 | 30.9 | 4.9 |
| Poquoson city............... | 0.00 | 0.0 | NA | NA | NA | NA | 27 | 307 | 57.4 | 6.4 | 8 | 16 | 3.1 | 0.7 |
| Portsmouth city............. | 60.83 | 632.3 | 43 | 659 | 226.8 | 33.6 | 261 | 3,060 | 627.3 | 69.4 | 77 | 388 | 80.4 | 14.0 |
| Radford city................ | 2.48 | 142.5 | 9 | 36 | 37.1 | 1.7 | 42 | 500 | 95.3 | 10.7 | 24 | 121 | 20.1 | 3.8 |
| Richmond city............... | 70.23 | 318.8 | 253 | 3,815 | 3,581.9 | 212.4 | 790 | 9,088 | 2,260.4 | 241.1 | 329 | 2,197 | 939.5 | 106.3 |
| Roanoke city................ | 4.99 | 50.0 | 155 | 2,211 | 1,381.3 | 116.4 | 533 | 8,966 | 2,603.5 | 215.5 | 174 | 938 | 193.3 | 36.6 |
| Salem city.................. | 3.84 | 151.0 | 68 | 1,552 | 1,120.6 | 100.8 | 151 | 1,995 | 524.5 | 49.7 | 43 | 199 | 60.0 | 9.8 |
| Staunton city............... | 0.00 | 0.0 | 21 | 227 | 80.2 | 8.7 | 133 | 1,834 | 449.7 | 46.3 | 38 | 142 | 17.0 | 5.1 |
| Suffolk city................ | 44.45 | 504.2 | 36 | 1,117 | 1,443.4 | 78.9 | 238 | 3,997 | 1,160.6 | 99.6 | 74 | 279 | 53.5 | 12.5 |
| Virginia Beach city......... | 1.76 | 3.9 | 351 | 4,770 | 5,596.5 | 359.4 | 1,506 | 24,023 | 6,155.3 | 607.2 | 761 | 5,451 | 1,198.7 | 261.4 |
| Waynesboro city............. | 0.78 | 36.3 | 14 | 229 | 106.3 | 13.1 | 122 | 2,161 | 483.8 | 48.5 | 35 | 91 | 15.2 | 3.1 |
| Williamsburg city........... | 0.00 | 0.0 | 8 | 68 | 39.0 | 5.3 | 121 | 1,658 | 310.3 | 36.5 | D | D | D | D |
| Winchester city............. | 0.00 | 0.0 | 36 | 459 | 463.9 | 22.2 | 284 | 4,317 | 1,017.2 | 109.6 | 66 | 306 | 97.6 | 11.5 |
| WASHINGTON................ | 866.53 | 120.8 | 7,755 | 120,320 | 112,352.6 | 7,649.3 | 21,751 | 347,728 | 160,284.8 | 11,412.9 | 11,826 | 53,706 | 15,626.7 | 2,756.8 |
| Adams...................... | 5.58 | 289.8 | D | D | D | D | 53 | 719 | 210.3 | 18.5 | 14 | 29 | 4.2 | 0.6 |
| Asotin..................... | 5.21 | 235.7 | 14 | 106 | 115.4 | 4.8 | 50 | 1,128 | 325.0 | 34.9 | 20 | 59 | 4.3 | 1.4 |
| Benton..................... | 34.48 | 181.2 | 131 | 1,248 | 879.7 | 60.6 | 606 | 10,084 | 2,949.8 | 277.4 | 323 | 1,108 | 264.9 | 39.9 |
| Chelan..................... | 11.29 | 149.3 | 90 | 2,811 | 1,319.3 | 115.5 | 385 | 4,329 | 1,141.6 | 124.8 | 139 | 575 | 105.4 | 22.0 |
| Clallam.................... | 6.16 | 83.8 | D | D | D | D | 279 | 3,519 | 945.9 | 103.5 | 92 | 216 | 40.2 | 7.9 |
| Clark...................... | 46.57 | 101.4 | 449 | 5,314 | 5,654.5 | 300.8 | 1,045 | 18,425 | 5,497.0 | 592.5 | 624 | 2,542 | 579.7 | 114.2 |
| Columbia................... | 1.00 | 253.5 | D | D | D | 0.6 | 17 | 77 | 24.4 | 2.6 | 7 | 8 | 1.3 | 0.2 |
| Cowlitz.................... | 10.31 | 99.6 | 88 | 1,034 | 755.8 | 55.6 | 333 | 4,985 | 1,485.4 | 143.8 | 133 | 498 | 83.7 | 18.8 |
| Douglas.................... | 5.45 | 134.5 | 46 | 1,017 | 433.6 | 45.2 | 95 | 1,685 | 522.9 | 49.8 | 44 | 103 | 24.5 | 3.4 |
| Ferry...................... | 0.58 | 76.5 | NA | NA | NA | NA | 29 | 200 | 45.3 | 5.2 | 7 | 24 | 1.9 | 0.5 |
| Franklin................... | 16.30 | 183.5 | 113 | 1,515 | 1,365.0 | 88.0 | 207 | 3,755 | 1,295.0 | 128.9 | 76 | 337 | 103.4 | 11.7 |
| Garfield................... | 0.57 | 256.9 | D | D | D | 3.3 | 9 | 54 | 10.7 | 1.1 | NA | NA | NA | NA |
| Grant...................... | 23.52 | 252.2 | 124 | 1,828 | 974.6 | 87.7 | 287 | 3,492 | 947.0 | 94.5 | 116 | 286 | 45.0 | 6.5 |
| Grays Harbor............... | 7.63 | 107.3 | 43 | 813 | 314.8 | 36.3 | 281 | 2,990 | 780.6 | 80.7 | 76 | 357 | 38.8 | 9.3 |
| Island..................... | 6.49 | 80.5 | 32 | 176 | 74.1 | 9.4 | 227 | 2,319 | 565.5 | 65.9 | 121 | 300 | 67.1 | 11.1 |
| Jefferson.................. | 2.25 | 73.9 | 19 | 137 | 40.6 | 4.7 | 136 | 1,165 | 259.2 | 32.4 | 57 | 181 | 31.0 | 6.1 |
| King....................... | 197.62 | 93.3 | 3,153 | 57,233 | 59,575.5 | 4,083.3 | 6,581 | 115,960 | 89,586.7 | 4,328.6 | 4,955 | 27,367 | 9,285.8 | 1,710.3 |
| Kitsap..................... | 19.63 | 75.5 | 131 | 978 | 500.7 | 54.1 | 739 | 11,457 | 3,319.1 | 355.0 | 405 | 1,132 | 304.1 | 45.8 |
| Kittitas................... | 6.90 | 159.5 | 40 | 547 | 438.0 | 32.1 | 168 | 1,915 | 616.0 | 51.4 | 71 | 172 | 55.6 | 7.5 |
| Klickitat.................. | 3.09 | 147.0 | 15 | 189 | 111.6 | 12.0 | 49 | 354 | 81.7 | 8.6 | 24 | 71 | 16.7 | 1.5 |
| Lewis...................... | 5.95 | 78.4 | 58 | 413 | 287.1 | 20.4 | 309 | 4,044 | 1,079.4 | 108.4 | 86 | 312 | 64.8 | 10.2 |
| Lincoln.................... | 2.03 | 196.7 | 29 | 307 | 321.1 | 21.6 | 35 | 342 | 73.4 | 7.7 | 8 | D | 3.3 | D |
| Mason...................... | 4.96 | 81.3 | 34 | 440 | 132.7 | 20.4 | 135 | 1,783 | 504.0 | 57.4 | 51 | 122 | 24.6 | 5.4 |
| Okanogan................... | 6.05 | 145.7 | 39 | 338 | 305.5 | 15.8 | 196 | 1,854 | 499.4 | 52.6 | 49 | 105 | 12.9 | 2.3 |
| Pacific.................... | 2.73 | 130.9 | D | D | D | D | 78 | 601 | 139.5 | 17.0 | 18 | 116 | 13.9 | 4.2 |
| Pend Oreille............... | 0.99 | 75.6 | 6 | 10 | 4.9 | 0.5 | 28 | 264 | 61.7 | 6.2 | 7 | 12 | 1.8 | 0.5 |
| Pierce..................... | 110.41 | 130.8 | 706 | 11,357 | 9,287.7 | 650.9 | 2,169 | 38,808 | 13,775.9 | 1,254.5 | 1,170 | 5,276 | 1,357.7 | 230.6 |
| San Juan................... | 0.88 | 54.1 | D | D | D | D | 116 | 752 | 180.2 | 24.9 | 66 | 136 | 30.4 | 4.9 |
| Skagit..................... | 20.31 | 166.7 | 119 | 1,463 | 832.8 | 76.5 | 538 | 7,589 | 2,272.1 | 235.1 | 171 | 476 | 120.7 | 16.6 |
| Skamania................... | 1.06 | 93.5 | D | D | D | 3.3 | 22 | 162 | 35.4 | 3.9 | 10 | 5 | 2.4 | 0.2 |
| Snohomish.................. | 65.52 | 84.8 | 729 | 9,382 | 9,583.3 | 711.3 | 2,192 | 36,696 | 11,584.9 | 1,181.5 | 1,037 | 4,121 | 1,200.0 | 183.7 |
| Spokane.................... | 138.81 | 282.7 | 616 | 9,651 | 10,302.5 | 520.7 | 1,635 | 26,670 | 7,847.0 | 813.2 | 701 | 3,509 | 748.5 | 129.1 |
| Stevens.................... | 6.19 | 141.4 | 19 | 260 | 66.5 | 7.7 | 109 | 1,233 | 321.9 | 34.2 | 23 | 69 | 9.9 | 1.7 |
| Thurston................... | 22.58 | 83.8 | 176 | 1,902 | 1,320.6 | 102.1 | 796 | 13,008 | 3,958.8 | 386.6 | 340 | 1,133 | 302.6 | 43.6 |
| Wahkiakum.................. | 0.32 | 79.2 | NA | NA | NA | NA | 8 | 34 | 8.1 | 0.9 | 4 | D | 0.7 | D |
| Walla Walla................ | 12.72 | 210.8 | D | D | D | 15.1 | 171 | 2,225 | 560.8 | 62.1 | 70 | 234 | 36.5 | 8.4 |
| Whatcom.................... | 18.13 | 85.4 | D | D | D | D | 817 | 11,180 | 3,225.2 | 328.6 | 398 | 1,394 | 421.9 | 56.2 |
| Whitman.................... | 6.26 | 129.9 | 52 | 696 | 906.1 | 43.3 | 91 | 1,456 | 411.3 | 36.5 | 50 | 281 | 36.1 | 6.8 |
| Yakima..................... | 30.00 | 120.6 | 220 | 4,707 | 3,634.3 | 241.2 | 730 | 10,415 | 3,136.8 | 301.2 | 263 | 1,013 | 180.5 | 33.1 |
| WEST VIRGINIA............. | 184.96 | 100.3 | 1,205 | 14,578 | 12,188.8 | 676.7 | 5,963 | 82,985 | 23,057.8 | 2,004.1 | 1,432 | 6,061 | 1,368.8 | 228.9 |
| Barbour.................... | 1.45 | 86.8 | 5 | 16 | 5.7 | 0.3 | 29 | 267 | 77.1 | 6.8 | 8 | 13 | 1.5 | 0.4 |
| Berkeley................... | 5.97 | 53.4 | 33 | 664 | 818.4 | 30.8 | 241 | 4,067 | 1,110.3 | 96.9 | 82 | 293 | 61.3 | 10.3 |
| Boone...................... | 0.19 | 8.1 | 9 | 37 | 31.5 | 1.4 | 60 | 641 | 214.4 | 15.6 | 6 | 7 | 1.0 | 0.2 |
| Braxton.................... | 1.09 | 75.6 | 7 | 42 | 8.4 | 1.5 | 51 | 705 | 200.7 | 16.6 | 8 | 14 | 2.7 | 0.4 |
| Brooke..................... | 5.21 | 223.1 | D | D | D | D | 60 | 1,010 | 284.2 | 25.9 | 5 | D | 4.3 | D |
| Cabell..................... | 12.17 | 125.7 | 97 | 1,416 | 902.5 | 70.6 | 417 | 6,509 | 1,478.4 | 140.4 | 105 | 523 | 149.8 | 23.9 |

1 Merchant wholesalers, except manufacturers' sales branches and offices.   2. Employer establishments.

| STATE County | Professional, scientific, and technical services, 2017 | | | | Manufacturing, 2017 | | | | Accommodation and food services, 2017 | | | |
|---|---|---|---|---|---|---|---|---|---|---|---|---|
| | Number of establish-ments | Number of employees | Sales (mil dol) | Average payroll (mil dol) | Number of establish-ments | Number of employees | Sales (mil dol) | Average payroll (mil dol) | Number of establis-hments | Number of employees | Sales (mil dol) | Annual payroll (mil dol) |
| | 147 | 148 | 149 | 150 | 151 | 152 | 153 | 154 | 155 | 156 | 157 | 158 |
| VIRGINIA—Cont'd | | | | | | | | | | | | |
| Martinsville city | D | D | D | D | 24 | 1,220 | 251.4 | 47.9 | 45 | 710 | 34.4 | 9.3 |
| Newport News city | D | D | D | D | 92 | 25,674 | 6,143.6 | 1,668.9 | 411 | 7,516 | 390.0 | 111.5 |
| Norfolk city | D | D | D | D | 133 | 6,747 | 1,601.9 | 362.2 | 612 | 11,326 | 659.3 | 181.7 |
| Norton city | D | D | 11 | D | D | 758 | D | 29.8 | 26 | 433 | 17.2 | 4.8 |
| Petersburg city | D | D | D | D | 32 | 1,162 | 482.6 | 60.5 | 94 | 930 | 50.5 | 13.8 |
| Poquoson city | 16 | 67 | 6 | 2.5 | 4 | 6 | 0.8 | 0.2 | 21 | 292 | 13.9 | 3.7 |
| Portsmouth city | D | D | D | D | 48 | 1,238 | 295.1 | 63.7 | 163 | 2,414 | 119.9 | 34.2 |
| Radford city | 25 | 139 | 15 | 6.6 | 17 | 1,346 | 242.5 | 76.3 | 41 | 993 | 35.5 | 9.3 |
| Richmond city | 897 | 11,297 | 2,503 | 1,050.3 | 200 | 4,908 | 16,939.3 | 334.1 | 677 | 14,306 | 793.6 | 258.9 |
| Roanoke city | D | D | D | D | 96 | 4,433 | 1,470.3 | 210.0 | 302 | 7,205 | 371.9 | 112.9 |
| Salem city | D | D | D | D | 51 | 3,027 | 971.2 | 173.2 | 97 | 1,885 | 95.9 | 28.0 |
| Staunton city | 64 | 298 | 41 | 15.4 | 18 | 617 | 106.9 | 28.6 | 84 | 1,333 | 66.2 | 21.8 |
| Suffolk city | D | D | D | D | 46 | 1,760 | 1,601.4 | 101.7 | 150 | 3,029 | 163.2 | 45.4 |
| Virginia Beach city | 1,488 | 21,552 | 4,079 | 1,607.1 | 215 | 5,584 | 2,279.7 | 304.1 | 1,292 | 25,108 | 1,494.1 | 419.4 |
| Waynesboro city | 43 | 320 | 35 | 15.6 | 26 | 1,198 | 380.7 | 67.2 | 77 | 1,678 | 87.2 | 24.8 |
| Williamsburg city | 28 | 129 | 16 | 5.7 | 7 | 37 | 8.6 | 1.2 | 155 | 3,667 | 242.6 | 72.5 |
| Winchester city | D | D | D | D | 22 | 1,910 | 1,055.4 | 122.1 | 144 | 3,320 | 160.2 | 48.4 |
| WASHINGTON | D | D | D | D | 7,017 | 263,132 | 140,381.5 | 17,943.8 | 17,828 | 289,371 | 21,068.6 | 6,395.3 |
| Adams | 10 | 51 | 10 | 2.7 | 12 | 1,194 | 377.5 | 55.8 | 35 | 428 | 21.0 | 5.8 |
| Asotin | 36 | 238 | 22 | 7.2 | 26 | 321 | 92.3 | 15.9 | 44 | 638 | 32.7 | 10.8 |
| Benton | D | D | D | D | 157 | 3,952 | 1,782.9 | 234.1 | 421 | 7,444 | 453.9 | 135.7 |
| Chelan | 196 | 994 | 128 | 50.7 | 104 | 1,380 | 308.7 | 59.0 | 325 | 4,189 | 289.6 | 93.4 |
| Clallam | D | D | D | D | 76 | 1,339 | 369.8 | 70.8 | 231 | 2,416 | 151.4 | 48.7 |
| Clark | 1,262 | 8,711 | 1,441 | 570.8 | 440 | 13,645 | 4,805.8 | 844.7 | 853 | 13,605 | 856.5 | 255.1 |
| Columbia | 9 | 24 | 1 | 0.6 | D | D | D | D | 13 | 97 | 5.6 | 1.5 |
| Cowlitz | D | D | D | 34.3 | 123 | 6,610 | 3,270.4 | 431.9 | 224 | 3,190 | 180.4 | 54.7 |
| Douglas | 46 | 194 | 17 | 7.2 | 22 | 499 | 112.5 | 26.7 | 60 | 1,091 | 57.2 | 20.8 |
| Ferry | D | D | D | 0.6 | 6 | 186 | 60.5 | 8.6 | 20 | 71 | 5.0 | 1.3 |
| Franklin | D | D | D | D | 46 | 2,662 | 1,147.4 | 132.3 | 140 | 2,278 | 133.1 | 37.9 |
| Garfield | NA | NA | NA | NA | NA | NA | NA | NA | 3 | 5 | 0.4 | 0.1 |
| Grant | 112 | 443 | 64 | 22.1 | 77 | 4,153 | 1,842.4 | 236.7 | 192 | 2,327 | 140.5 | 39.8 |
| Grays Harbor | D | D | D | D | 79 | 2,777 | 1,193.0 | 149.3 | 228 | 2,244 | 139.5 | 42.0 |
| Island | 183 | 780 | 102 | 39.0 | 59 | 735 | 195.4 | 41.5 | 162 | 1,829 | 120.5 | 37.1 |
| Jefferson | D | D | D | D | 74 | 591 | 233.6 | 34.9 | 100 | 1,064 | 66.8 | 20.5 |
| King | 10,900 | 134,568 | 29,481 | 12,159.5 | 2,180 | 86,121 | 47,209.2 | 6,181.7 | 6,575 | 116,157 | 9,457.2 | 2,930.3 |
| Kitsap | 703 | 4,112 | 618 | 247.4 | 150 | 1,996 | 379.0 | 96.4 | 536 | 9,262 | 640.9 | 191.9 |
| Kittitas | D | D | D | D | 35 | 460 | 94.4 | 22.4 | 164 | 2,519 | 137.9 | 48.7 |
| Klickitat | 49 | 365 | 88 | 18.7 | 37 | 689 | 134.1 | 33.7 | 56 | 345 | 27.4 | 8.0 |
| Lewis | D | D | D | D | 100 | 2,735 | 1,302.6 | 150.6 | 199 | 2,309 | 138.3 | 39.7 |
| Lincoln | D | D | D | D | D | 21 | D | D | D | D | D | 1.0 |
| Mason | D | D | D | D | 53 | 709 | 241.3 | 35.8 | 101 | 1,746 | 161.4 | 47.4 |
| Okanogan | D | D | D | D | 33 | 403 | 91.1 | 16.2 | 124 | 1,257 | 82.0 | 23.9 |
| Pacific | 39 | 126 | 15 | 5.5 | 32 | 603 | 148.2 | 25.3 | 102 | 897 | 75.0 | 17.1 |
| Pend Oreille | D | D | D | D | 11 | 252 | 163.0 | 18.2 | 22 | 163 | 8.2 | 2.5 |
| Pierce | 1,544 | 9,155 | 1,425 | 558.2 | 565 | 18,149 | 6,287.2 | 1,037.9 | 1,655 | 29,002 | 1,945.0 | 584.6 |
| San Juan | D | D | D | D | 39 | 186 | 32.5 | D | 103 | 1,036 | 106.8 | 34.2 |
| Skagit | D | D | D | D | 189 | 6,711 | 11,187.8 | 468.7 | 352 | 5,518 | 411.7 | 125.3 |
| Skamania | 17 | 62 | 10 | 3.9 | 16 | 404 | 81.6 | 17.3 | 24 | 568 | 36.7 | 11.7 |
| Snohomish | 1,841 | 12,783 | 2,220 | 871.1 | 780 | 58,428 | 34,547.2 | 5,059.3 | 1,713 | 26,720 | 2,178.0 | 596.9 |
| Spokane | D | 1,548 | D | D | 518 | 14,449 | 4,161.2 | 792.4 | 1,133 | 19,373 | 1,165.3 | 369.1 |
| Stevens | D | D | D | D | 53 | 1,131 | 364.2 | 58.6 | 83 | 557 | 33.8 | 9.8 |
| Thurston | 638 | 4,426 | 595 | 260.8 | 162 | 2,779 | 1,029.7 | 136.7 | 561 | 9,634 | 574.5 | 175.9 |
| Wahkiakum | D | D | D | 0.3 | D | 42 | D | D | D | D | D | 0.3 |
| Walla Walla | D | D | D | D | D | D | D | D | 127 | 2,030 | 120.1 | 36.8 |
| Whatcom | D | D | D | D | 341 | 11,106 | 11,178.0 | 628.0 | 554 | 8,971 | 609.1 | 191.8 |
| Whitman | 65 | 351 | 38 | 16.2 | D | 2,367 | D | D | 122 | 1,792 | 81.8 | 24.4 |
| Yakima | D | D | D | D | 245 | 9,487 | 3,536.6 | 453.2 | 445 | 6,498 | 418.3 | 118.8 |
| WEST VIRGINIA | 2,761 | 21,085 | 3,023 | 1,132.2 | 1,142 | 48,533 | 24,602.1 | 2,747.7 | 3,614 | 68,102 | 4,069.1 | 1,075.6 |
| Barbour | 12 | 70 | 7 | 2.1 | 11 | 95 | 28.9 | 3.5 | 21 | 298 | 12.8 | 3.5 |
| Berkeley | D | D | D | D | 50 | 2,976 | 1,200.4 | 124.5 | 194 | 3,258 | 164.2 | 47.7 |
| Boone | D | D | D | D | D | D | D | D | 20 | 267 | 12.7 | 3.6 |
| Braxton | 12 | 41 | 3 | 1.2 | 13 | 270 | 245.7 | 15.1 | 48 | 630 | 35.4 | 10.0 |
| Brooke | D | D | 9 | D | 18 | 1,671 | 1,280.6 | 95.2 | D | D | D | D |
| Cabell | D | D | D | D | 87 | 3,946 | 1,546.2 | 228.4 | 284 | 5,400 | 264.4 | 79.3 |

## Table B. States and Counties — Health Care and Social Assistance, Other Services, Nonemployer Businesses, and Residential Construction

| STATE County | Health care and social assistance, 2017 | | | | Other services, 2017 | | | | Nonemployer businesses, 2018 | | Value of residential construction authorized by building permits, 2020 | |
|---|---|---|---|---|---|---|---|---|---|---|---|---|
| | Number of establishments | Number of employees | Receipts (mil dol) | Annual payroll (mil dol) | Number of establishments | Number of employees | Receipts (mil dol) | Annual payroll (mil dol) | Number | Receipts (mil dol) | New construction ($1,000) | Number of housing units |
| | 159 | 160 | 161 | 162 | 163 | 164 | 165 | 166 | 167 | 168 | 169 | 170 |
| **VIRGINIA—Cont'd** | | | | | | | | | | | | |
| Martinsville city | 107 | 2,190 | 232.6 | 92.5 | 37 | 178 | 35.2 | 4.4 | 700 | 30.3 | 652 | 1 |
| Newport News city | 452 | 15,710 | 1,843.3 | 782.2 | 283 | 1,731 | 186.4 | 49.2 | 10,423 | 340.1 | 20,234 | 160 |
| Norfolk city | 582 | 20,755 | 3,181.5 | 1,090.6 | 396 | 3,023 | 459.0 | D | 13,242 | 542.1 | 139,505 | 1,202 |
| Norton city | 47 | 1,032 | 139.9 | 46.5 | D | D | D | D | 191 | 6.1 | 0 | 0 |
| Petersburg city | 126 | 4,361 | 479.6 | 190.8 | 75 | 576 | 67.5 | 15.8 | 1,454 | 48.7 | 7,661 | 63 |
| Poquoson city | 16 | 180 | 13.4 | 5.7 | D | D | D | D | 845 | 36.5 | 5,015 | 25 |
| Portsmouth city | 227 | 7,627 | 1,142.0 | 385.7 | 143 | 1,143 | 140.7 | 42.2 | 5,265 | 137.1 | 31,675 | 280 |
| Radford city | 41 | 561 | 48.5 | 21.3 | 23 | 66 | 14.7 | 1.9 | 698 | 30.4 | 3,029 | 24 |
| Richmond city | 607 | 27,376 | 4,123.1 | 1,644.4 | 520 | 4,950 | 609.4 | 174.3 | 16,969 | 741.1 | 108,842 | 1,023 |
| Roanoke city | 316 | 13,827 | 2,050.3 | 743.0 | 252 | 1,629 | 154.5 | 48.3 | 6,005 | 300.5 | 44,929 | 317 |
| Salem city | 143 | 5,424 | 990.3 | 365.2 | 88 | 474 | 41.1 | 12.5 | 1,509 | 61.6 | 0 | 0 |
| Staunton city | 95 | 2,486 | 190.0 | 89.8 | 86 | 456 | 43.9 | 14.0 | 1,671 | 70.2 | 7,721 | 50 |
| Suffolk city | 240 | 4,979 | 593.0 | 226.0 | 137 | 715 | 65.1 | 19.2 | 5,729 | 196.4 | 215,656 | 1,108 |
| Virginia Beach city | 1,162 | 21,729 | 2,767.4 | 1,052.8 | 943 | 5,762 | 918.6 | 174.6 | 32,959 | 1,544.2 | 147,855 | 938 |
| Waynesboro city | 47 | 630 | 45.6 | 19.7 | 59 | 413 | 49.8 | 14.3 | 1,137 | 49.4 | 15,630 | 129 |
| Williamsburg city | 42 | 666 | 55.8 | 26.0 | 36 | 379 | 91.2 | 15.5 | 894 | 38.5 | 4,597 | 28 |
| Winchester city | 269 | 7,080 | 1,125.3 | 407.1 | 92 | 536 | 45.5 | 14.0 | 2,136 | 142.1 | 2,840 | 18 |
| **WASHINGTON** | 21,264 | 438,835 | 56,442.2 | 22,849.3 | 13,444 | 78,480 | 14,513.7 | 3,004.6 | 491,908 | 25,002.7 | 9,488,095 | 43,881 |
| Adams | 26 | 728 | 75.3 | 36.6 | 30 | 84 | 10.0 | 2.4 | 835 | 52.1 | 18,497 | 99 |
| Asotin | 52 | 1,162 | 138.5 | 53.9 | 26 | 129 | 8.4 | 2.6 | 1,033 | 42.4 | 16,391 | 161 |
| Benton | 604 | 12,310 | 1,500.4 | 631.5 | 300 | 1,624 | 160.2 | 50.9 | 9,479 | 437.8 | 373,673 | 1,345 |
| Chelan | 232 | 6,551 | 946.1 | 417.4 | 180 | 638 | 80.2 | 20.1 | 4,952 | 233.0 | 128,724 | 670 |
| Clallam | 240 | 4,436 | 421.8 | 195.8 | 160 | 682 | 62.1 | 18.9 | 4,825 | 185.6 | 69,404 | 279 |
| Clark | 1,224 | 24,118 | 2,833.8 | 1,179.0 | 797 | 4,499 | 567.4 | 144.7 | 33,464 | 1,716.6 | 1,129,933 | 5,022 |
| Columbia | 10 | 169 | 21.0 | 8.7 | D | D | D | D | 222 | 9.2 | 902 | 10 |
| Cowlitz | 236 | 5,617 | 597.1 | 249.1 | 152 | 816 | 87.2 | 26.6 | 4,618 | 202.5 | 83,979 | 346 |
| Douglas | 68 | 891 | 62.8 | 26.5 | 45 | 183 | 15.2 | 4.9 | 1,885 | 76.5 | 71,483 | 321 |
| Ferry | 13 | 161 | 13.6 | 6.2 | D | D | D | 1.9 | 346 | 11.1 | 3,299 | 26 |
| Franklin | 140 | 2,121 | 205.3 | 96.5 | 106 | 467 | 53.0 | 14.6 | 3,755 | 188.6 | 169,035 | 620 |
| Garfield | D | D | D | 3.7 | NA | NA | NA | NA | 141 | 4.9 | 818 | 4 |
| Grant | 171 | 3,139 | 355.4 | 131.1 | 131 | 456 | 49.4 | 11.6 | 3,991 | 220.6 | 129,485 | 544 |
| Grays Harbor | 186 | 2,881 | 299.6 | 129.3 | 106 | 537 | 49.5 | 11.8 | 3,293 | 138.7 | 71,580 | 342 |
| Island | 174 | 2,432 | 270.0 | 111.8 | 119 | 474 | 42.1 | 13.3 | 6,157 | 266.2 | 105,420 | 445 |
| Jefferson | 103 | 1,666 | 172.4 | 80.9 | 89 | 437 | 46.3 | 14.1 | 3,332 | 126.9 | 35,458 | 157 |
| King | 7,874 | 160,920 | 22,808.7 | 9,090.3 | 4,865 | 31,829 | 9,241.4 | 1,445.8 | 183,205 | 10,416.8 | 2,507,262 | 12,337 |
| Kitsap | 716 | 12,346 | 1,619.1 | 581.2 | 423 | 2,230 | 224.5 | 67.3 | 15,583 | 705.4 | 318,102 | 1,285 |
| Kittitas | 107 | 1,886 | 171.0 | 79.2 | 82 | 361 | 43.7 | 11.6 | 2,922 | 143.8 | 115,525 | 414 |
| Klickitat | 46 | 612 | 85.2 | 31.6 | 40 | 180 | 17.6 | 5.8 | 1,621 | 75.7 | 20,349 | 124 |
| Lewis | 205 | 3,691 | 390.7 | 164.4 | 124 | 569 | 50.7 | 15.3 | 3,733 | 170.9 | 64,695 | 382 |
| Lincoln | 15 | 395 | 39.8 | 17.3 | D | D | D | 2.4 | 688 | 29.7 | 9,921 | 56 |
| Mason | 105 | 1,771 | 202.0 | 86.2 | 75 | 317 | 28.8 | 8.5 | 3,005 | 118.9 | 78,815 | 305 |
| Okanogan | 105 | 1,672 | 197.8 | 75.1 | 74 | 213 | 21.7 | 5.6 | 2,387 | 99.9 | 34,192 | 197 |
| Pacific | 39 | 639 | 70.1 | 33.0 | 37 | 139 | 12.8 | 3.3 | 1,425 | 60.6 | 15,281 | 92 |
| Pend Oreille | 20 | 474 | 41.9 | 20.7 | 13 | 30 | 3.5 | 0.9 | 727 | 27.1 | 12,087 | 80 |
| Pierce | 2,003 | 53,231 | 7,449.2 | 3,027.9 | 1,440 | 9,304 | 1,109.1 | 338.4 | 48,267 | 2,298.1 | 1,111,475 | 4,922 |
| San Juan | 66 | 452 | 45.4 | 18.3 | 55 | 181 | 35.5 | 5.8 | 2,838 | 132.2 | 30,767 | 116 |
| Skagit | 355 | 8,148 | 922.3 | 402.4 | 263 | 1,223 | 139.6 | 41.6 | 8,065 | 414.5 | 116,667 | 561 |
| Skamania | 19 | 191 | 11.2 | 5.6 | D | D | D | D | 752 | 36.7 | 17,990 | 82 |
| Snohomish | 1,961 | 33,323 | 3,705.0 | 1,555.2 | 1,351 | 7,499 | 760.5 | 252.6 | 51,383 | 2,432.9 | 1,244,700 | 5,780 |
| Spokane | 1,625 | 41,292 | 4,918.3 | 2,010.3 | 893 | 5,164 | 571.7 | 167.7 | 32,197 | 1,542.3 | 611,684 | 3,170 |
| Stevens | 91 | 1,951 | 144.3 | 68.2 | 60 | 216 | 23.9 | 6.1 | 2,686 | 105.5 | 50,994 | 192 |
| Thurston | 893 | 15,533 | 1,815.9 | 725.3 | 501 | 3,079 | 430.4 | 129.7 | 16,142 | 714.9 | 236,528 | 1,161 |
| Wahkiakum | D | D | D | 0.8 | D | D | D | 0.1 | 341 | 14.2 | 5,972 | 25 |
| Walla Walla | 155 | 4,730 | 566.2 | 245.5 | D | D | D | D | 3,386 | 156.7 | 42,294 | 154 |
| Whatcom | 699 | 10,929 | 1,334.0 | 469.1 | 431 | 2,552 | 303.1 | 91.8 | 16,490 | 819.5 | 288,718 | 1,382 |
| Whitman | 100 | 2,076 | 201.9 | 90.9 | 64 | 225 | 23.5 | 6.7 | 2,224 | 87.2 | 22,030 | 98 |
| Yakima | 576 | 14,031 | 1,779.2 | 692.5 | 283 | 1,580 | 168.1 | 44.4 | 9,513 | 486.4 | 123,964 | 575 |
| **WEST VIRGINIA** | 4,869 | 132,627 | 15,236.9 | 5,844.2 | 2,524 | 14,894 | 1,896.8 | 465.2 | 88,150 | 3,467.4 | 573,944 | 3,204 |
| Barbour | 34 | 837 | 44.1 | 20.7 | D | D | D | D | 625 | 19.1 | 418 | 2 |
| Berkeley | 211 | 6,186 | 806.2 | 439.3 | 116 | 669 | 71.8 | 20.4 | 5,982 | 244.9 | 273,287 | 1,383 |
| Boone | D | D | D | D | 19 | 94 | 8.6 | 3.5 | 665 | 17.6 | 1,669 | 15 |
| Braxton | 20 | 843 | 63.6 | 20.1 | D | D | D | D | 528 | 18.9 | 0 | 0 |
| Brooke | 58 | 2,102 | 233.1 | 87.1 | 31 | 120 | 8.9 | 2.1 | 986 | 41.7 | 1,887 | 8 |
| Cabell | 369 | 14,250 | 1,889.1 | 688.4 | 142 | 920 | 144.6 | 33.3 | 4,677 | 186.2 | 11,809 | 61 |

# Table B. States and Counties — Government Employment and Payroll, and Local Government Finances

| | Government employment and payroll, 2017 | | March payroll (percent of total) | | | | | | | Local government finances, 2017 General revenue | | Taxes | | |
| STATE County | Full-time equivalent employees | March payroll (dollars) | Administration, judicial, and legal | Police and corrections | Fire protection | Highways and transportation | Health and welfare | Natural resources and utilities | Education and libraries | Total (mil dol) | Intergovernmental (mil dol) | Total (mil dol) | Per capita[1] (dollars) Total | Per capita[1] (dollars) Property |
|---|---|---|---|---|---|---|---|---|---|---|---|---|---|---|
| | 171 | 172 | 173 | 174 | 175 | 176 | 177 | 178 | 179 | 180 | 181 | 182 | 183 | 184 |
| **VIRGINIA—Cont'd** | | | | | | | | | | | | | | |
| Martinsville city | 727 | 2,553,769 | 10.7 | 14.0 | 4.0 | 3.0 | 0.6 | 10.2 | 55.0 | 74.8 | 39.7 | 18.4 | 1,433 | 806 |
| Newport News city | 7,978 | 33,914,885 | 6.2 | 12.4 | 6.1 | 2.4 | 5.3 | 11.0 | 55.6 | 894.8 | 386.1 | 368.0 | 2,051 | 1,455 |
| Norfolk city | 13,561 | 62,833,701 | 5.6 | 10.7 | 3.8 | 9.2 | 9.0 | 6.2 | 53.3 | 1,562.9 | 651.5 | 443.6 | 1,814 | 1,104 |
| Norton city | 209 | 683,863 | 7.5 | 14.9 | 0.4 | 4.8 | 6.2 | 11.5 | 53.1 | 27.7 | 13.2 | 11.0 | 2,798 | 1,335 |
| Petersburg city | 1,297 | 5,006,121 | 11.4 | 10.7 | 11.0 | 3.5 | 5.0 | 6.6 | 51.8 | 148.2 | 78.2 | 55.2 | 1,772 | 1,308 |
| Poquoson city | 420 | 1,601,460 | 7.2 | 7.3 | 8.9 | 3.3 | 0.0 | 3.2 | 66.5 | 44.7 | 17.7 | 21.9 | 1,820 | 1,561 |
| Portsmouth city | 4,141 | 17,014,076 | 6.0 | 17.8 | 6.2 | 1.6 | 7.7 | 4.5 | 54.3 | 428.9 | 213.4 | 166.4 | 1,755 | 1,255 |
| Radford city | 467 | 1,767,278 | 9.3 | 10.5 | 2.7 | 2.6 | 3.6 | 15.7 | 51.6 | 48.8 | 26.1 | 13.2 | 757 | 518 |
| Richmond city | 8,858 | 36,278,451 | 8.5 | 17.3 | 6.1 | 3.1 | 7.8 | 10.0 | 43.8 | 1,324.8 | 522.3 | 525.3 | 2,312 | 1,451 |
| Roanoke city | 3,858 | 15,899,140 | 8.6 | 12.8 | 8.0 | 3.5 | 7.4 | 2.3 | 55.4 | 497.7 | 258.5 | 184.9 | 1,868 | 1,142 |
| Salem city | 1,303 | 4,997,631 | 8.9 | 22.8 | 6.5 | 3.3 | 1.1 | 9.3 | 44.6 | 128.7 | 49.0 | 60.4 | 2,380 | 1,562 |
| Staunton city | 940 | 3,606,993 | 12.0 | 25.7 | 2.9 | 3.9 | 0.0 | 6.1 | 46.7 | 99.8 | 46.4 | 36.9 | 1,518 | 1,018 |
| Suffolk city | 3,408 | 14,357,842 | 10.0 | 8.4 | 10.8 | 3.4 | 3.3 | 7.4 | 55.9 | 389.2 | 167.1 | 168.9 | 1,874 | 1,377 |
| Virginia Beach city | 18,457 | 75,108,285 | 6.0 | 9.8 | 3.9 | 1.4 | 6.5 | 14.4 | 55.5 | 2,190.5 | 730.1 | 948.1 | 2,107 | 1,432 |
| Waynesboro city | 831 | 2,848,288 | 7.2 | 8.1 | 4.7 | 1.9 | 0.6 | 8.0 | 61.8 | 93.3 | 41.4 | 37.9 | 1,703 | 986 |
| Williamsburg city | 444 | 1,647,202 | 10.8 | 36.4 | 10.4 | 2.4 | 5.2 | 9.9 | 20.5 | 62.4 | 13.3 | 34.9 | 2,327 | 972 |
| Winchester city | 1,713 | 6,512,336 | 7.4 | 21.8 | 5.1 | 4.3 | 3.3 | 6.4 | 48.6 | 196.3 | 70.3 | 75.6 | 2,683 | 1,545 |
| **WASHINGTON** | X | X | X | X | X | X | X | X | X | X | X | X | X | X |
| Adams | 1,271 | 5,372,706 | 4.5 | 6.0 | 0.9 | 3.8 | 19.7 | 9.5 | 55.3 | 132.3 | 70.4 | 27.8 | 1,414 | 1,017 |
| Asotin | 773 | 3,356,048 | 6.0 | 7.2 | 6.3 | 4.2 | 3.1 | 7.3 | 64.2 | 79.5 | 42.6 | 23.1 | 1,024 | 732 |
| Benton | 8,892 | 52,858,675 | 4.5 | 5.8 | 2.9 | 3.0 | 17.1 | 28.7 | 36.6 | 1,283.4 | 573.7 | 316.0 | 1,595 | 822 |
| Chelan | 4,127 | 22,588,705 | 4.1 | 5.6 | 2.0 | 5.6 | 11.0 | 31.0 | 38.9 | 503.4 | 186.5 | 164.8 | 2,160 | 1,241 |
| Clallam | 3,927 | 21,708,301 | 5.8 | 6.9 | 3.6 | 4.0 | 40.7 | 9.5 | 26.6 | 493.0 | 133.6 | 100.7 | 1,332 | 805 |
| Clark | 14,969 | 81,711,108 | 4.7 | 6.7 | 4.3 | 5.4 | 2.5 | 7.1 | 66.6 | 2,115.1 | 967.4 | 754.4 | 1,590 | 1,028 |
| Columbia | 333 | 1,418,597 | 8.8 | 2.7 | 3.0 | 9.1 | 43.8 | 1.5 | 25.8 | 40.1 | 13.6 | 9.9 | 2,482 | 2,015 |
| Cowlitz | 3,640 | 18,508,063 | 7.2 | 10.5 | 3.3 | 7.0 | 2.9 | 12.2 | 53.9 | 542.9 | 209.0 | 154.7 | 1,448 | 882 |
| Douglas | 1,547 | 7,927,036 | 4.6 | 5.2 | 1.7 | 5.4 | 0.0 | 22.4 | 59.0 | 197.0 | 116.9 | 52.9 | 1,260 | 869 |
| Ferry | 423 | 1,680,561 | 9.9 | 5.7 | 0.4 | 7.7 | 19.8 | 13.2 | 40.6 | 50.8 | 29.8 | 5.8 | 760 | 553 |
| Franklin | 3,219 | 15,792,848 | 5.0 | 7.3 | 3.8 | 2.8 | 0.6 | 11.7 | 66.8 | 440.5 | 228.7 | 109.5 | 1,192 | 703 |
| Garfield | 202 | 821,892 | 8.3 | 7.3 | 1.4 | 11.0 | 35.7 | 2.0 | 32.7 | 25.5 | 10.4 | 5.9 | 2,673 | 2,318 |
| Grant | 5,811 | 29,561,640 | 3.4 | 5.3 | 1.2 | 2.8 | 20.0 | 23.9 | 41.1 | 712.9 | 266.0 | 151.5 | 1,588 | 1,088 |
| Grays Harbor | 3,828 | 18,659,149 | 5.3 | 7.1 | 4.0 | 7.5 | 26.2 | 10.7 | 37.2 | 500.1 | 155.7 | 113.0 | 1,558 | 920 |
| Island | 2,843 | 15,232,539 | 5.8 | 5.7 | 4.0 | 6.0 | 39.8 | 3.3 | 34.0 | 366.5 | 113.2 | 112.0 | 1,346 | 877 |
| Jefferson | 1,532 | 8,627,335 | 6.5 | 4.4 | 5.5 | 8.0 | 46.7 | 5.6 | 21.9 | 219.4 | 43.0 | 58.8 | 1,884 | 1,200 |
| King | 78,547 | 522,087,826 | 9.9 | 9.4 | 5.1 | 12.4 | 14.4 | 11.7 | 34.4 | 17,468.8 | 4,565.5 | 7,233.2 | 3,282 | 1,678 |
| Kitsap | 7,352 | 40,302,464 | 7.9 | 7.7 | 8.9 | 7.4 | 3.0 | 6.6 | 57.4 | 1,280.4 | 652.7 | 428.5 | 1,609 | 1,052 |
| Kittitas | 1,783 | 9,308,957 | 8.4 | 7.9 | 3.4 | 3.4 | 35.9 | 8.3 | 31.1 | 253.8 | 74.4 | 74.6 | 1,615 | 1,011 |
| Klickitat | 1,142 | 5,843,294 | 5.7 | 5.1 | 0.7 | 5.0 | 33.2 | 12.9 | 35.1 | 155.9 | 49.7 | 31.8 | 1,463 | 1,145 |
| Lewis | 2,942 | 14,215,448 | 7.6 | 8.5 | 4.0 | 5.9 | 8.7 | 11.5 | 50.3 | 357.0 | 155.8 | 97.5 | 1,245 | 775 |
| Lincoln | 846 | 3,995,893 | 6.1 | 5.0 | 0.9 | 7.1 | 36.8 | 1.1 | 39.9 | 97.0 | 41.2 | 17.0 | 1,607 | 1,334 |
| Mason | 2,574 | 13,523,685 | 5.7 | 5.7 | 4.1 | 5.1 | 29.9 | 10.4 | 37.5 | 333.4 | 116.8 | 87.7 | 1,377 | 1,006 |
| Okanogan | 2,321 | 10,618,170 | 5.7 | 6.5 | 0.4 | 6.8 | 25.7 | 9.3 | 45.1 | 282.1 | 136.5 | 52.3 | 1,248 | 802 |
| Pacific | 1,032 | 5,376,385 | 5.9 | 6.0 | 3.2 | 6.5 | 33.2 | 10.5 | 34.3 | 152.7 | 50.1 | 38.0 | 1,749 | 1,214 |
| Pend Oreille | 880 | 4,172,244 | 6.0 | 4.5 | 0.5 | 3.4 | 36.2 | 19.6 | 26.4 | 85.1 | 28.8 | 13.0 | 975 | 716 |
| Pierce | 25,902 | 157,612,022 | 8.6 | 7.9 | 8.1 | 8.5 | 2.8 | 11.9 | 50.6 | 4,276.7 | 1,618.1 | 1,578.9 | 1,795 | 1,196 |
| San Juan | 631 | 3,152,711 | 15.4 | 7.2 | 7.0 | 9.5 | 6.4 | 7.6 | 44.8 | 98.1 | 31.2 | 48.1 | 2,877 | 1,811 |
| Skagit | 6,316 | 34,058,435 | 5.1 | 5.3 | 2.2 | 4.5 | 35.2 | 4.8 | 40.1 | 1,108.2 | 363.5 | 253.5 | 2,014 | 1,281 |
| Skamania | 367 | 1,715,614 | 15.5 | 11.6 | 1.3 | 6.3 | 10.7 | 15.5 | 35.5 | 44.8 | 19.0 | 14.6 | 1,233 | 800 |
| Snohomish | 20,647 | 129,047,276 | 6.0 | 8.2 | 4.5 | 7.6 | 3.3 | 13.4 | 53.7 | 3,639.9 | 1,460.7 | 1,384.1 | 1,726 | 1,093 |
| Spokane | 15,579 | 84,768,081 | 8.5 | 11.6 | 6.9 | 6.5 | 2.6 | 7.0 | 54.6 | 2,353.2 | 1,115.1 | 775.1 | 1,533 | 925 |
| Stevens | 1,363 | 5,882,354 | 7.3 | 7.8 | 2.0 | 4.3 | 7.4 | 4.8 | 64.4 | 155.1 | 89.9 | 35.7 | 801 | 571 |
| Thurston | 8,635 | 45,715,941 | 9.4 | 8.2 | 6.1 | 6.6 | 2.8 | 2.8 | 58.6 | 1,293.7 | 527.4 | 507.6 | 1,811 | 1,121 |
| Wahkiakum | 165 | 754,148 | 15.4 | 9.9 | 0.2 | 7.0 | 13.8 | 6.0 | 32.3 | 85.5 | 75.5 | 4.2 | 995 | 653 |
| Walla Walla | 2,255 | 10,369,781 | 7.3 | 9.6 | 5.1 | 6.8 | 5.0 | 15.5 | 60.2 | 278.3 | 123.5 | 98.1 | 1,622 | 997 |
| Whatcom | 5,979 | 33,006,077 | 9.4 | 9.9 | 8.3 | 8.0 | 3.6 | 3.7 | 52.1 | 923.4 | 347.8 | 387.7 | 1,751 | 1,024 |
| Whitman | 1,782 | 8,963,685 | 6.9 | 5.9 | 4.3 | 5.4 | 36.0 | 5.1 | 34.2 | 283.9 | 93.9 | 65.2 | 1,321 | 911 |
| Yakima | 9,179 | 43,173,358 | 6.3 | 9.8 | 3.4 | 2.4 | 1.2 | 3.8 | 69.1 | 1,163.9 | 708.0 | 278.4 | 1,114 | 636 |
| **WEST VIRGINIA** | X | X | X | X | X | X | X | X | X | X | X | X | X | X |
| Barbour | 511 | 1,634,273 | 7.0 | 3.6 | 0.0 | 1.0 | 5.4 | 8.0 | 70.7 | 36.8 | 20.0 | 8.5 | 513 | 462 |
| Berkeley | 3,761 | 13,282,683 | 4.1 | 4.0 | 1.7 | 1.1 | 0.7 | 4.3 | 82.4 | 278.2 | 125.1 | 101.2 | 879 | 756 |
| Boone | 708 | 2,434,601 | 5.3 | 5.3 | 0.0 | 2.0 | 3.1 | 3.0 | 77.2 | 83.4 | 27.7 | 27.3 | 1,223 | 1,197 |
| Braxton | 476 | 1,356,686 | 8.2 | 3.4 | 5.0 | 1.5 | 1.5 | 4.7 | 75.1 | 28.2 | 15.4 | 8.7 | 609 | 563 |
| Brooke | 676 | 2,157,151 | 6.2 | 8.0 | 0.0 | 2.9 | 2.9 | 6.9 | 72.4 | 61.2 | 23.0 | 26.4 | 1,179 | 1,044 |
| Cabell | 3,013 | 10,461,020 | 4.9 | 7.1 | 3.1 | 4.4 | 7.7 | 3.1 | 67.2 | 294.6 | 99.5 | 124.5 | 1,318 | 1,031 |

1. Based on the resident population estimated as of July 1 of the year shown.

# Table B. States and Counties — Local Government Finances, Government Employment, and Income Taxes

| STATE County | Total (mil dol) [185] | Per capita1 (dollars) [186] | Education [187] | Health and hospitals [188] | Police protection [189] | Public welfare [190] | Highways [191] | Total (mil dol) [192] | Per capita1 (dollars) [193] | Federal civilian [194] | Federal military [195] | State and local [196] | Number of returns [197] | Mean adjusted gross income [198] | Mean income tax [199] |
|---|---|---|---|---|---|---|---|---|---|---|---|---|---|---|---|
| VIRGINIA—Cont'd | | | | | | | | | | | | | | | |
| Martinsville city | 75 | 5,830 | 47.4 | 0.4 | 5.8 | 0.8 | 7.9 | 16.3 | 1,268 | (2) | (2) | (2) | 6,020 | 43,783 | 3,837 |
| Newport News city | 929 | 5,177 | 42.6 | 0.6 | 5.5 | 4.1 | 2.2 | 1,181.2 | 6,582 | 5,652 | 7,093 | 11,390 | 84,950 | 50,140 | 4,438 |
| Norfolk city | 1,539 | 6,293 | 41.6 | 5.5 | 4.8 | 3.0 | 2.4 | 2,033.9 | 8,317 | 20,739 | 39,598 | 19,677 | 102,110 | 52,781 | 5,275 |
| Norton city | 27 | 6,819 | 36.6 | 0.5 | 8.7 | 5.8 | 7.0 | 16.4 | 4,182 | (3) | (3) | (3) | 1,630 | 42,683 | 3,137 |
| Petersburg city | 151 | 4,848 | 38.5 | 0.9 | 8.9 | 11.1 | 4.5 | 35.4 | 1,135 | (4) | (4) | (4) | 15,240 | 35,070 | 2,349 |
| Poquoson city | 42 | 3,520 | 54.9 | 1.0 | 7.7 | 2.5 | 3.7 | 35.5 | 2,946 | (5) | (5) | (5) | 5,920 | 90,317 | 10,627 |
| Portsmouth city | 475 | 5,010 | 35.3 | 2.3 | 6.8 | 4.7 | 1.4 | 641.1 | 6,760 | 14,744 | 6,129 | 5,124 | 44,250 | 44,981 | 3,525 |
| Radford city | 57 | 3,286 | 44.8 | 0.4 | 7.0 | 5.1 | 4.7 | 26.7 | 1,526 | (6) | (6) | (6) | 5,460 | 50,061 | 4,488 |
| Richmond city | 1,274 | 5,607 | 29.8 | 6.1 | 8.1 | 5.6 | 3.8 | 2,281.4 | 10,040 | 6,172 | 1,327 | 38,959 | 102,910 | 75,171 | 10,441 |
| Roanoke city | 464 | 4,686 | 40.0 | 0.4 | 5.2 | 12.7 | 5.3 | 138.3 | 1,397 | 1,510 | 335 | 6,883 | 46,090 | 50,086 | 5,206 |
| Salem city | 142 | 5,611 | 44.7 | 0.6 | 4.8 | 1.4 | 4.4 | 112.5 | 4,434 | (7) | (7) | (7) | 11,750 | 63,281 | 6,595 |
| Staunton city | 99 | 4,057 | 37.1 | 7.3 | 6.0 | 5.9 | 7.2 | 81.3 | 3,342 | (8) | (8) | (8) | 12,160 | 54,699 | 4,950 |
| Suffolk city | 396 | 4,399 | 44.7 | 0.7 | 6.3 | 3.2 | 10.5 | 799.3 | 8,871 | 1,295 | 407 | 5,318 | 42,450 | 68,349 | 7,088 |
| Virginia Beach city | 2,084 | 4,632 | 41.1 | 3.2 | 4.7 | 2.8 | 5.2 | 2,366.3 | 5,260 | 6,475 | 17,514 | 22,389 | 221,390 | 73,233 | 8,724 |
| Waynesboro city | 84 | 3,791 | 47.2 | 0.8 | 5.5 | 5.8 | 8.5 | 64.0 | 2,877 | (8) | (8) | (8) | 10,780 | 50,515 | 4,208 |
| Williamsburg city | 66 | 4,369 | 13.6 | 0.8 | 7.2 | 3.1 | 2.1 | 14.1 | 943 | (9) | (9) | (9) | 5,550 | 74,628 | 9,076 |
| Winchester city | 176 | 6,247 | 43.0 | 1.0 | 4.7 | 4.0 | 2.7 | 298.0 | 10,577 | (10) | (10) | (10) | 13,530 | 62,666 | 7,644 |
| WASHINGTON | X | X | X | X | X | X | X | X | X | 75,860 | 75,113 | 514,147 | 3,622,810 | 89,792 | 12,144 |
| Adams | 133 | 6,773 | 49.9 | 12.7 | 3.2 | 0.0 | 7.7 | 144.2 | 7,338 | 38 | 50 | 1,626 | 8,570 | 48,669 | 3,853 |
| Asotin | 79 | 3,501 | 51.1 | 3.5 | 4.5 | 0.0 | 5.1 | 68.6 | 3,046 | 54 | 56 | 1,150 | 10,020 | 63,239 | 6,641 |
| Benton | 1,260 | 6,358 | 40.3 | 28.7 | 3.6 | 0.2 | 2.6 | 7,605.4 | 38,372 | 754 | 526 | 11,679 | 90,550 | 73,740 | 8,096 |
| Chelan | 450 | 5,892 | 44.2 | 11.0 | 4.4 | 0.3 | 4.8 | 1,051.2 | 13,777 | 615 | 192 | 6,638 | 38,930 | 69,790 | 8,118 |
| Clallam | 487 | 6,442 | 25.2 | 37.8 | 3.2 | 0.1 | 3.7 | 306.2 | 4,048 | 470 | 551 | 7,549 | 36,800 | 62,990 | 6,192 |
| Clark | 1,950 | 4,111 | 55.5 | 1.8 | 4.4 | 0.5 | 5.4 | 2,246.6 | 4,736 | 3,624 | 1,273 | 24,064 | 233,280 | 81,667 | 10,032 |
| Columbia | 37 | 9,221 | 18.7 | 38.0 | 4.0 | 0.1 | 12.2 | 25.6 | 6,401 | 72 | 10 | 478 | 1,710 | 58,404 | 5,037 |
| Cowlitz | 509 | 4,765 | 41.8 | 1.6 | 5.7 | 0.5 | 4.7 | 656.3 | 6,144 | 221 | 275 | 6,308 | 49,250 | 61,456 | 5,994 |
| Douglas | 170 | 4,035 | 51.9 | 15.4 | 5.1 | 0.0 | 10.8 | 330.7 | 7,868 | 249 | 108 | 2,084 | 19,450 | 59,712 | 5,477 |
| Ferry | 41 | 5,434 | 34.4 | 26.2 | 3.0 | 0.0 | 14.7 | 20.9 | 2,764 | 149 | 18 | 727 | 2,800 | 50,325 | 4,305 |
| Franklin | 428 | 4,654 | 58.4 | 1.7 | 4.6 | 0.1 | 2.8 | 416.8 | 4,535 | 482 | 232 | 6,405 | 39,440 | 56,614 | 4,816 |
| Garfield | 23 | 10,573 | 22.2 | 33.0 | 4.3 | 0.1 | 16.6 | 9.2 | 4,156 | 134 | 5 | 332 | 940 | 51,802 | 3,847 |
| Grant | 688 | 7,217 | 45.8 | 21.5 | 3.6 | 0.5 | 4.4 | 1,808.6 | 18,967 | 781 | 242 | 7,891 | 42,320 | 54,455 | 4,742 |
| Grays Harbor | 482 | 6,646 | 29.9 | 32.0 | 4.3 | 0.4 | 4.6 | 432.4 | 5,965 | 187 | 226 | 6,566 | 31,790 | 54,033 | 4,881 |
| Island | 348 | 4,174 | 29.0 | 31.2 | 3.6 | 0.2 | 5.2 | 251.7 | 3,023 | 1,346 | 6,084 | 3,408 | 42,170 | 75,469 | 8,122 |
| Jefferson | 198 | 6,344 | 21.1 | 45.3 | 3.3 | 0.1 | 4.9 | 311.5 | 9,982 | 209 | 90 | 2,423 | 16,970 | 73,693 | 8,517 |
| King | 15,017 | 6,814 | 30.1 | 13.1 | 4.6 | 0.9 | 5.9 | 27,717.5 | 12,577 | 19,549 | 7,270 | 164,721 | 1,130,500 | 128,817 | 21,255 |
| Kitsap | 1,208 | 4,535 | 41.9 | 5.2 | 3.7 | 0.0 | 5.0 | 880.5 | 3,307 | 20,485 | 11,703 | 13,858 | 130,560 | 82,111 | 9,833 |
| Kittitas | 267 | 5,771 | 34.2 | 29.8 | 4.3 | 0.1 | 5.6 | 193.9 | 4,199 | 147 | 122 | 4,369 | 20,320 | 70,192 | 7,939 |
| Klickitat | 137 | 6,285 | 33.1 | 32.8 | 3.6 | 0.0 | 9.7 | 227.8 | 10,472 | 90 | 56 | 1,665 | 10,550 | 67,750 | 7,429 |
| Lewis | 355 | 4,529 | 48.8 | 13.5 | 4.6 | 0.3 | 7.1 | 494.0 | 6,308 | 226 | 200 | 5,242 | 36,090 | 57,125 | 5,077 |
| Lincoln | 94 | 8,892 | 36.5 | 31.2 | 2.8 | 0.1 | 12.1 | 50.4 | 4,765 | 61 | 27 | 1,146 | 4,850 | 63,175 | 5,874 |
| Mason | 336 | 5,277 | 42.2 | 29.3 | 3.8 | 0.1 | 4.4 | 490.5 | 7,696 | 78 | 161 | 5,867 | 28,790 | 61,594 | 5,912 |
| Okanogan | 269 | 6,420 | 43.8 | 26.3 | 3.3 | 0.0 | 6.6 | 180.0 | 4,295 | 413 | 104 | 4,787 | 18,470 | 49,242 | 4,166 |
| Pacific | 148 | 6,804 | 30.4 | 31.5 | 3.9 | 0.0 | 5.4 | 100.1 | 4,606 | 58 | 176 | 1,822 | 10,360 | 54,555 | 5,035 |
| Pend Oreille | 92 | 6,859 | 23.1 | 44.5 | 2.9 | 0.1 | 5.8 | 198.6 | 14,866 | 111 | 34 | 1,492 | 5,770 | 58,618 | 5,909 |
| Pierce | 4,027 | 4,578 | 44.6 | 1.4 | 6.4 | 0.5 | 5.3 | 6,262.6 | 7,119 | 12,106 | 31,701 | 48,575 | 424,740 | 72,956 | 8,079 |
| San Juan | 106 | 6,358 | 30.7 | 6.5 | 3.4 | 0.1 | 11.1 | 92.2 | 5,510 | 61 | 44 | 794 | 9,390 | 104,044 | 15,762 |
| Skagit | 1,116 | 8,865 | 28.0 | 43.7 | 2.5 | 0.1 | 3.4 | 865.1 | 6,873 | 410 | 320 | 11,549 | 61,600 | 70,711 | 7,564 |
| Skamania | 42 | 3,544 | 35.2 | 9.3 | 6.1 | 0.0 | 8.9 | 20.1 | 1,705 | 98 | 30 | 535 | 5,190 | 76,105 | 8,802 |
| Snohomish | 3,439 | 4,287 | 45.9 | 8.3 | 5.0 | 0.6 | 4.1 | 4,264.5 | 5,317 | 2,117 | 6,234 | 38,872 | 393,080 | 84,564 | 10,122 |
| Spokane | 2,330 | 4,607 | 48.0 | 7.3 | 5.2 | 0.5 | 4.6 | 2,153.5 | 4,259 | 4,913 | 4,432 | 32,315 | 242,550 | 66,366 | 7,241 |
| Stevens | 151 | 3,377 | 54.0 | 6.8 | 4.3 | 0.1 | 11.3 | 88.0 | 1,971 | 330 | 114 | 2,876 | 19,120 | 57,136 | 5,128 |
| Thurston | 1,288 | 4,595 | 49.0 | 6.7 | 4.0 | 0.0 | 3.7 | 1,289.3 | 4,600 | 871 | 804 | 38,635 | 140,680 | 71,069 | 7,564 |
| Wahkiakum | 76 | 17,767 | 8.5 | 75.9 | 1.4 | 0.0 | 3.5 | 14.5 | 3,417 | 13 | 11 | 274 | 1,950 | 60,389 | 5,691 |
| Walla Walla | 265 | 4,384 | 46.0 | 3.5 | 5.4 | 0.1 | 9.6 | 233.0 | 3,850 | 1,334 | 146 | 4,364 | 26,460 | 63,775 | 6,547 |
| Whatcom | 918 | 4,146 | 48.9 | 3.2 | 4.9 | 0.0 | 5.6 | 850.4 | 3,841 | 1,482 | 637 | 14,578 | 107,850 | 70,944 | 7,863 |
| Whitman | 243 | 4,913 | 29.1 | 37.8 | 4.0 | 0.2 | 5.9 | 155.2 | 3,142 | 239 | 122 | 9,387 | 17,480 | 60,136 | 5,825 |
| Yakima | 1,134 | 4,536 | 60.9 | 0.9 | 4.8 | 0.2 | 4.6 | 803.9 | 3,217 | 1,283 | 727 | 17,096 | 111,590 | 55,801 | 5,369 |
| WEST VIRGINIA | X | X | X | X | X | X | X | X | X | 24,116 | 8,508 | 120,883 | 764,920 | 55,038 | 5,276 |
| Barbour | 42 | 2,563 | 64.9 | 3.2 | 4.0 | 0.0 | 1.6 | 20.4 | 1,235 | 31 | 74 | 675 | 6,320 | 45,653 | 3,185 |
| Berkeley | 310 | 2,690 | 73.8 | 0.5 | 4.5 | 0.0 | 0.6 | 333.8 | 2,901 | 3,363 | 606 | 5,249 | 56,310 | 55,259 | 4,774 |
| Boone | 86 | 3,838 | 45.2 | 28.3 | 2.5 | 0.4 | 0.9 | 13.3 | 596 | 58 | 102 | 1,419 | 7,800 | 47,028 | 3,459 |
| Braxton | 32 | 2,255 | 63.8 | 1.3 | 6.9 | 0.0 | 0.6 | 76.7 | 5,398 | 59 | 65 | 914 | 5,090 | 46,450 | 3,576 |
| Brooke | 71 | 3,149 | 70.0 | 0.9 | 7.6 | 0.0 | 2.1 | 56.9 | 2,543 | 32 | 102 | 904 | 10,610 | 54,220 | 5,032 |
| Cabell | 303 | 3,211 | 49.0 | 1.1 | 6.4 | 0.4 | 1.5 | 99.2 | 1,050 | 1,020 | 482 | 7,742 | 38,590 | 56,421 | 5,874 |

1. Based on the resident population estimated as of July 1 of the year shown.   2. Martinsville city is included with Henry county.   3. Norton city is included with Wise county.   4. Petersburg and Colonial Heights cities are included with Dinwiddie county.   5. Poquoson city is included with York county.   6. Radford city is included with Montgomery county.   7. Salem city is included with Roanoke county.   8. Staunton and Waynesboro cities are included with Augusta county.   9. Williamsburg city is included with James City county.   10. Winchester city is included with Frederick county.

# Table B. States and Counties — Land Area and Population

| State / county code | CBSA code[1] | County Type code[2] | STATE County | Land area[3] (sq. mi) | Total persons 2019 | Rank | Per square mile | White | Black | American Indian, Alaska Native | Asian and Pacific Islander | Percent Hispanic or Latino[4] | Under 5 years | 5 to 14 years | 15 to 24 years | 25 to 34 years | 35 to 44 years | 45 to 54 years |
|---|---|---|---|---|---|---|---|---|---|---|---|---|---|---|---|---|---|---|
| | | | **WEST VIRGINIA—Cont'd** | | | | | | | | | | | | | | | |
| 54013 | | 8 | Calhoun | 279.3 | 6,945 | 2,664 | 24.9 | 97.9 | 0.7 | 0.8 | 0.3 | 1.2 | 4.3 | 10.4 | 9.4 | 9.5 | 10.8 | 12.6 |
| 54015 | 16620 | 3 | Clay | 341.9 | 8,341 | 2,552 | 24.4 | 98.2 | 0.8 | 0.8 | 0.4 | 0.9 | 5.1 | 12.7 | 10.6 | 9.9 | 11.2 | 12.7 |
| 54017 | 17220 | 9 | Doddridge | 319.7 | 8,368 | 2,544 | 26.2 | 96.2 | 2.9 | 0.9 | 0.7 | 0.6 | 3.7 | 7.6 | 12.2 | 12.6 | 12.9 | 13.4 |
| 54019 | 13220 | 3 | Fayette | 661.6 | 42,062 | 1,139 | 63.6 | 94.0 | 5.2 | 0.8 | 0.5 | 1.3 | 5.0 | 12.0 | 10.9 | 11.0 | 11.8 | 12.7 |
| 54021 | | 7 | Gilmer | 338.5 | 7,811 | 2,604 | 23.1 | 82.8 | 10.9 | 1.2 | 1.1 | 5.5 | 3.8 | 8.6 | 17.3 | 14.2 | 13.8 | 12.1 |
| 54023 | | 7 | Grant | 477.4 | 11,510 | 2,318 | 24.1 | 97.2 | 1.3 | 0.5 | 0.4 | 1.5 | 5.2 | 10.8 | 10.0 | 10.7 | 10.3 | 13.1 |
| 54025 | 49020 | 6 | Greenbrier | 1,019.6 | 34,319 | 1,322 | 33.7 | 94.4 | 3.6 | 0.9 | 1.0 | 1.9 | 4.8 | 11.1 | 10.2 | 11.1 | 11.4 | 12.0 |
| 54027 | | 3 | Hampshire | 640.4 | 23,190 | 1,673 | 36.2 | 96.6 | 1.8 | 0.7 | 0.6 | 1.5 | 4.6 | 9.9 | 10.2 | 10.4 | 11.0 | 13.8 |
| 54029 | 48260 | 6 | Hancock | 82.6 | 28,571 | 1,478 | 345.9 | 95.1 | 3.6 | 0.7 | 0.7 | 1.6 | 4.3 | 10.8 | 10.3 | 10.8 | 11.0 | 13.3 |
| 54031 | | 6 | Hardy | 582.3 | 13,633 | 2,183 | 23.4 | 91.2 | 3.9 | 0.7 | 1.0 | 4.6 | 5.6 | 11.1 | 10.0 | 10.3 | 11.0 | 13.4 |
| 54033 | 17220 | 5 | Harrison | 416.0 | 66,870 | 802 | 160.7 | 95.5 | 2.7 | 0.8 | 0.9 | 1.8 | 5.3 | 12.1 | 10.9 | 12.0 | 12.5 | 12.7 |
| 54035 | 16620 | 6 | Jackson | 464.4 | 28,453 | 1,480 | 61.3 | 97.8 | 1.0 | 0.7 | 0.5 | 1.0 | 5.4 | 11.8 | 10.8 | 11.0 | 11.4 | 13.1 |
| 54037 | 47900 | 1 | Jefferson | 209.3 | 57,486 | 903 | 274.7 | 85.4 | 7.8 | 1.0 | 2.4 | 6.3 | 5.1 | 12.4 | 12.5 | 11.1 | 12.8 | 14.3 |
| 54039 | 16620 | 3 | Kanawha | 901.6 | 176,253 | 377 | 195.5 | 89.9 | 9.2 | 0.8 | 1.6 | 1.2 | 5.1 | 11.2 | 11.1 | 12.4 | 11.9 | 12.2 |
| 54041 | | 7 | Lewis | 386.9 | 15,805 | 2,042 | 40.9 | 97.3 | 1.2 | 0.8 | 0.6 | 1.3 | 5.3 | 12.5 | 9.9 | 12.2 | 11.7 | 13.0 |
| 54043 | 16620 | 2 | Lincoln | 437.1 | 20,043 | 1,820 | 45.9 | 98.5 | 0.8 | 0.8 | 0.4 | 0.7 | 5.7 | 12.5 | 10.7 | 10.5 | 11.7 | 13.6 |
| 54045 | 34350 | 6 | Logan | 453.7 | 31,688 | 1,386 | 69.8 | 96.7 | 2.2 | 0.5 | 0.4 | 1.1 | 5.2 | 11.7 | 10.6 | 11.4 | 12.2 | 13.1 |
| 54047 | | 7 | McDowell | 533.5 | 16,916 | 1,978 | 31.7 | 89.9 | 8.5 | 0.7 | 0.3 | 1.8 | 5.1 | 11.8 | 9.4 | 10.9 | 11.3 | 12.6 |
| 54049 | 21900 | 4 | Marion | 308.8 | 55,962 | 922 | 181.2 | 94.3 | 4.4 | 0.8 | 1.0 | 1.4 | 5.3 | 11.2 | 14.7 | 12.0 | 11.7 | 12.3 |
| 54051 | 48540 | 3 | Marshall | 305.4 | 30,103 | 1,431 | 98.6 | 97.5 | 1.4 | 0.7 | 0.6 | 1.0 | 4.6 | 10.9 | 10.4 | 11.6 | 11.0 | 12.9 |
| 54053 | 38580 | 6 | Mason | 430.8 | 26,335 | 1,556 | 61.1 | 97.8 | 1.5 | 0.7 | 0.5 | 0.7 | 5.2 | 11.8 | 10.5 | 11.2 | 11.8 | 12.2 |
| 54055 | 14140 | 4 | Mercer | 419.0 | 58,258 | 893 | 139.0 | 91.8 | 7.3 | 0.7 | 0.8 | 1.3 | 5.6 | 11.9 | 11.2 | 11.9 | 11.0 | 12.1 |
| 54057 | 19060 | 3 | Mineral | 327.9 | 26,722 | 1,538 | 81.5 | 95.2 | 4.1 | 0.7 | 0.8 | 1.0 | 5.0 | 11.1 | 11.7 | 11.6 | 10.9 | 13.0 |
| 54059 | | 7 | Mingo | 423.1 | 22,951 | 1,682 | 54.2 | 96.9 | 2.4 | 0.6 | 0.4 | 0.8 | 5.8 | 12.5 | 10.4 | 10.6 | 12.1 | 13.3 |
| 54061 | 34060 | 3 | Monongalia | 360.1 | 106,819 | 571 | 296.6 | 90.2 | 5.1 | 0.6 | 4.3 | 2.2 | 4.7 | 9.1 | 22.8 | 16.6 | 12.6 | 10.3 |
| 54063 | | 8 | Monroe | 472.8 | 13,229 | 2,205 | 28.0 | 97.4 | 1.3 | 1.2 | 0.5 | 1.3 | 4.8 | 11.4 | 9.4 | 10.2 | 10.6 | 12.6 |
| 54065 | 25180 | 8 | Morgan | 229.1 | 17,873 | 1,930 | 78.0 | 96.3 | 1.6 | 1.2 | 0.8 | 1.9 | 3.9 | 10.0 | 10.2 | 10.5 | 10.4 | 13.6 |
| 54067 | | 6 | Nicholas | 646.8 | 24,340 | 1,634 | 37.6 | 97.7 | 1.0 | 1.1 | 0.6 | 0.8 | 4.8 | 11.9 | 9.7 | 10.8 | 11.4 | 12.8 |
| 54069 | 48540 | 3 | Ohio | 105.8 | 41,182 | 1,152 | 389.2 | 94.0 | 5.0 | 0.7 | 1.2 | 1.3 | 5.0 | 10.9 | 13.5 | 11.5 | 11.0 | 11.2 |
| 54071 | | 8 | Pendleton | 696.1 | 6,932 | 2,667 | 10.0 | 95.9 | 2.6 | 0.8 | 0.6 | 1.4 | 4.7 | 9.9 | 8.9 | 9.9 | 10.0 | 11.2 |
| 54073 | | 9 | Pleasants | 130.1 | 7,438 | 2,625 | 57.2 | 96.7 | 2.2 | 1.0 | 0.5 | 1.1 | 5.2 | 10.4 | 10.6 | 13.0 | 12.6 | 14.3 |
| 54075 | | 3 | Pocahontas | 940.2 | 8,190 | 2,566 | 8.7 | 96.3 | 1.8 | 1.1 | 0.6 | 1.7 | 4.7 | 9.5 | 8.9 | 10.5 | 11.1 | 11.6 |
| 54077 | 34060 | 3 | Preston | 648.8 | 33,380 | 1,346 | 51.4 | 96.8 | 1.4 | 0.7 | 0.5 | 1.5 | 4.9 | 10.8 | 9.6 | 13.2 | 12.8 | 13.2 |
| 54079 | 26580 | 2 | Putnam | 345.7 | 56,428 | 915 | 163.2 | 96.3 | 1.8 | 0.7 | 1.2 | 1.3 | 4.9 | 12.8 | 11.3 | 11.1 | 13.2 | 13.6 |
| 54081 | 13220 | 3 | Raleigh | 605.4 | 72,920 | 750 | 120.4 | 89.1 | 9.0 | 0.9 | 1.4 | 1.7 | 5.4 | 11.9 | 10.8 | 12.4 | 12.3 | 12.4 |
| 54083 | 21180 | 7 | Randolph | 1,039.7 | 28,387 | 1,484 | 27.3 | 96.5 | 2.2 | 0.7 | 0.7 | 1.0 | 4.9 | 10.2 | 11.5 | 12.3 | 11.3 | 12.6 |
| 54085 | | 8 | Ritchie | 452.0 | 9,499 | 2,453 | 21.0 | 98.1 | 0.8 | 0.7 | 0.5 | 1.0 | 5.1 | 10.7 | 10.3 | 10.3 | 10.9 | 12.9 |
| 54087 | | 6 | Roane | 483.6 | 13,482 | 2,192 | 27.9 | 97.7 | 0.8 | 0.8 | 0.7 | 1.3 | 4.7 | 11.5 | 10.7 | 9.8 | 11.5 | 13.2 |
| 54089 | | 6 | Summers | 360.6 | 12,444 | 2,252 | 34.5 | 93.4 | 5.0 | 1.1 | 0.6 | 1.7 | 4.0 | 8.5 | 9.0 | 10.9 | 12.1 | 12.8 |
| 54091 | 17220 | 6 | Taylor | 172.8 | 16,699 | 1,992 | 96.6 | 97.0 | 1.8 | 0.8 | 0.7 | 1.0 | 4.6 | 11.8 | 9.9 | 12.2 | 12.7 | 13.5 |
| 54093 | | 8 | Tucker | 419.0 | 6,816 | 2,677 | 16.3 | 98.1 | 0.8 | 0.6 | 0.5 | 0.9 | 3.8 | 7.5 | 9.2 | 10.6 | 11.2 | 13.3 |
| 54095 | | 9 | Tyler | 256.3 | 8,533 | 2,531 | 33.3 | 97.8 | 0.8 | 0.9 | 1.0 | 0.8 | 4.6 | 11.4 | 10.0 | 10.1 | 10.2 | 14.3 |
| 54097 | | 7 | Upshur | 354.6 | 24,230 | 1,639 | 68.3 | 96.6 | 1.7 | 0.7 | 0.7 | 1.4 | 5.1 | 11.6 | 14.2 | 11.3 | 11.1 | 11.7 |
| 54099 | 26580 | 2 | Wayne | 506.0 | 39,054 | 1,199 | 77.2 | 98.0 | 1.1 | 0.8 | 0.5 | 0.8 | 4.9 | 11.4 | 11.0 | 11.5 | 11.2 | 13.4 |
| 54101 | | 9 | Webster | 553.5 | 8,058 | 2,584 | 14.6 | 98.5 | 1.0 | 0.8 | 0.3 | 0.8 | 4.7 | 11.6 | 9.2 | 10.3 | 10.3 | 13.5 |
| 54103 | | 6 | Wetzel | 358.1 | 14,904 | 2,096 | 41.6 | 97.9 | 1.0 | 0.6 | 0.5 | 1.1 | 5.5 | 11.1 | 11.0 | 10.7 | 10.4 | 13.1 |
| 54105 | 37620 | 3 | Wirt | 232.5 | 5,705 | 2,767 | 24.5 | 98.1 | 1.1 | 0.8 | 0.5 | 0.9 | 4.9 | 12.3 | 10.1 | 9.7 | 11.9 | 12.6 |
| 54107 | 37620 | 3 | Wood | 366.5 | 82,938 | 686 | 226.3 | 96.7 | 2.2 | 0.7 | 0.9 | 1.3 | 5.4 | 11.8 | 10.8 | 11.5 | 11.7 | 13.0 |
| 54109 | | 6 | Wyoming | 499.5 | 20,123 | 1,818 | 40.3 | 98.1 | 1.2 | 0.8 | 0.3 | 0.8 | 4.6 | 11.7 | 10.8 | 10.8 | 11.6 | 13.2 |
| 55000 | | 0 | **WISCONSIN** | 54,167.4 | 5,832,655 | X | 107.7 | 82.2 | 7.4 | 1.4 | 3.7 | 7.3 | 5.6 | 12.2 | 13.1 | 12.7 | 12.4 | 12.0 |
| 55001 | | 8 | Adams | 645.6 | 20,498 | 1,794 | 31.8 | 90.9 | 3.5 | 1.4 | 1.0 | 4.4 | 3.3 | 7.9 | 8.0 | 9.7 | 9.5 | 11.5 |
| 55003 | | 7 | Ashland | 1,045.0 | 15,415 | 2,066 | 14.8 | 85.1 | 1.0 | 12.7 | 1.1 | 3.2 | 5.6 | 12.2 | 13.1 | 10.5 | 11.4 | 10.8 |
| 55005 | | 6 | Barron | 863.0 | 45,090 | 1,074 | 52.2 | 93.9 | 2.2 | 1.5 | 1.0 | 2.7 | 5.5 | 12.1 | 10.4 | 10.9 | 11.2 | 11.4 |
| 55007 | | 8 | Bayfield | 1,477.9 | 15,242 | 2,076 | 10.3 | 86.9 | 1.0 | 11.4 | 1.0 | 2.5 | 4.4 | 9.7 | 8.8 | 8.1 | 9.7 | 11.1 |
| 55009 | 24580 | 2 | Brown | 530.1 | 264,610 | 263 | 499.2 | 81.7 | 3.8 | 3.3 | 4.0 | 9.4 | 6.1 | 13.3 | 13.1 | 13.4 | 12.9 | 12.1 |
| 55011 | | 8 | Buffalo | 675.8 | 13,033 | 2,218 | 19.3 | 96.1 | 0.9 | 0.7 | 0.7 | 2.6 | 5.0 | 11.3 | 10.7 | 10.6 | 11.2 | 12.1 |
| 55013 | | 8 | Burnett | 821.6 | 15,557 | 2,056 | 18.9 | 92.4 | 1.4 | 5.5 | 0.9 | 2.2 | 4.0 | 9.8 | 8.4 | 8.1 | 9.2 | 12.1 |
| 55015 | 11540 | 3 | Calumet | 318.3 | 50,209 | 988 | 157.7 | 91.7 | 1.4 | 0.9 | 2.9 | 4.6 | 5.1 | 13.2 | 12.0 | 11.2 | 13.4 | 13.9 |
| 55017 | 20740 | 3 | Chippewa | 1,008.4 | 64,737 | 824 | 64.2 | 94.3 | 2.3 | 0.9 | 1.8 | 1.9 | 5.4 | 12.4 | 10.9 | 12.3 | 13.0 | 12.3 |
| 55019 | | 6 | Clark | 1,209.7 | 34,720 | 1,312 | 28.7 | 92.8 | 0.9 | 0.8 | 0.7 | 5.6 | 8.3 | 16.2 | 13.3 | 9.9 | 10.4 | 11.0 |
| 55021 | 31540 | 2 | Columbia | 765.5 | 57,668 | 900 | 75.3 | 93.1 | 2.3 | 0.9 | 1.3 | 3.8 | 5.3 | 11.7 | 11.0 | 12.0 | 12.7 | 12.9 |
| 55023 | | 7 | Crawford | 570.6 | 16,021 | 2,033 | 28.1 | 94.9 | 2.6 | 0.7 | 1.0 | 1.8 | 4.8 | 11.1 | 11.2 | 10.0 | 11.2 | 13.2 |
| 55025 | 31540 | 2 | Dane | 1,196.5 | 552,536 | 125 | 461.8 | 81.2 | 6.8 | 0.7 | 7.4 | 6.7 | 5.3 | 11.2 | 16.7 | 15.7 | 13.7 | 11.2 |
| 55027 | 13180 | 4 | Dodge | 875.7 | 87,336 | 667 | 99.7 | 90.1 | 3.7 | 0.8 | 1.0 | 5.4 | 4.5 | 11.0 | 11.5 | 12.7 | 13.0 | 13.2 |
| 55029 | | 6 | Door | 482.0 | 27,889 | 1,502 | 57.9 | 94.3 | 1.2 | 1.1 | 0.9 | 3.7 | 3.9 | 9.0 | 8.8 | 8.6 | 9.8 | 10.9 |

1. CBSA = Core Based Statistical Area. See Appendix A for explanation. See Appendix B for list of metropolitan areas with component counties. Service of USDA Rural-Urban Continuum Codes. See Appendix A for definition. 2. County type code from the Economic Research 3. Dry land or land partially or temporarily covered by water. 4. May be of any race.

# Table B. States and Counties — **Population and Households**

| STATE County | Age (percent) 55 to 64 years | 65 to 74 years | 75 years and over | Percent female | Total persons 2000 | 2010 | Percent change 2000-2010 | 2010-2020 | Births | Deaths | Net Migration | Number | Persons per household | Family households | Female family householder[1] | One person |
|---|---|---|---|---|---|---|---|---|---|---|---|---|---|---|---|---|
| | 16 | 17 | 18 | 19 | 20 | 21 | 22 | 23 | 24 | 25 | 26 | 27 | 28 | 29 | 30 | 31 |
| **WEST VIRGINIA—Cont'd** | | | | | | | | | | | | | | | | |
| Calhoun | 17.0 | 15.9 | 10.2 | 50.2 | 7,582 | 7,629 | 0.6 | -9.0 | 694 | 959 | -415 | 2,826 | 2.57 | 69.2 | 6.2 | 24.5 |
| Clay | 15.8 | 13.1 | 8.9 | 49.4 | 10,330 | 9,394 | -9.1 | -11.2 | 1,037 | 1,164 | -935 | 3,274 | 2.64 | 69.9 | 13.0 | 22.5 |
| Doddridge | 15.5 | 12.9 | 9.2 | 44.4 | 7,403 | 8,201 | 10.8 | 2.0 | 691 | 889 | 350 | 2,685 | 2.89 | 71.3 | 7.6 | 24.6 |
| Fayette | 14.2 | 13.5 | 8.9 | 49.6 | 47,579 | 46,045 | -3.2 | -8.7 | 5,029 | 6,718 | -2,271 | 17,441 | 2.41 | 67.1 | 11.3 | 30.0 |
| Gilmer | 12.2 | 10.0 | 8.1 | 40.6 | 7,160 | 8,695 | 21.4 | -10.2 | 663 | 801 | -757 | 2,587 | 2.52 | 64.6 | 11.2 | 27.5 |
| Grant | 14.4 | 13.8 | 11.6 | 49.9 | 11,299 | 11,920 | 5.5 | -3.4 | 1,219 | 1,378 | -245 | 4,607 | 2.49 | 62.6 | 7.9 | 32.8 |
| Greenbrier | 15.2 | 14.1 | 10.2 | 50.9 | 34,453 | 35,487 | 3.0 | -3.3 | 3,579 | 4,982 | 252 | 15,188 | 2.27 | 61.7 | 10.9 | 32.8 |
| Hampshire | 16.2 | 14.4 | 9.5 | 49.1 | 20,203 | 23,952 | 18.6 | -3.2 | 2,277 | 2,733 | -295 | 9,288 | 2.45 | 66.1 | 12.0 | 28.4 |
| Hancock | 15.2 | 14.2 | 10.0 | 51.3 | 32,667 | 30,691 | -6.0 | -6.9 | 2,735 | 4,264 | -559 | 12,678 | 2.3 | 59.4 | 12.3 | 35.2 |
| Hardy | 15.7 | 13.5 | 9.5 | 49.4 | 12,669 | 14,027 | 10.7 | -2.8 | 1,575 | 1,595 | -373 | 5,674 | 2.42 | 72.5 | 7.8 | 23.2 |
| Harrison | 14.2 | 11.6 | 8.5 | 50.8 | 68,652 | 69,089 | 0.6 | -3.2 | 7,876 | 8,897 | -1,145 | 27,098 | 2.47 | 65.5 | 10.9 | 28.2 |
| Jackson | 15.1 | 12.1 | 9.3 | 50.2 | 28,000 | 29,215 | 4.3 | -2.6 | 3,254 | 3,671 | -321 | 11,334 | 2.53 | 68.7 | 9.1 | 28.6 |
| Jefferson | 14.4 | 11.0 | 6.4 | 50.4 | 42,190 | 53,490 | 26.8 | 7.5 | 5,896 | 4,904 | 3,023 | 20,891 | 2.64 | 70.7 | 10.1 | 22.9 |
| Kanawha | 14.4 | 13.0 | 8.8 | 51.9 | 200,073 | 193,056 | -3.5 | -8.7 | 20,858 | 25,687 | -11,848 | 79,070 | 2.28 | 61.3 | 13.3 | 33.3 |
| Lewis | 14.4 | 12.4 | 8.6 | 50.0 | 16,919 | 16,369 | -3.3 | -3.4 | 1,922 | 2,332 | -141 | 6,574 | 2.42 | 66.6 | 10.7 | 27.4 |
| Lincoln | 14.7 | 12.4 | 8.2 | 50.5 | 22,108 | 21,711 | -1.8 | -7.7 | 2,544 | 2,891 | -1,308 | 8,208 | 2.53 | 72.1 | 9.5 | 24.7 |
| Logan | 14.2 | 13.8 | 7.8 | 50.7 | 37,710 | 36,750 | -2.5 | -13.8 | 3,925 | 5,553 | -3,455 | 13,816 | 2.36 | 69.7 | 11.3 | 26.1 |
| McDowell | 15.8 | 14.1 | 9.1 | 50.8 | 27,329 | 22,108 | -19.1 | -23.5 | 2,302 | 3,559 | -3,974 | 7,607 | 2.26 | 68.6 | 15.6 | 27.9 |
| Marion | 12.8 | 11.6 | 8.3 | 50.5 | 56,598 | 56,451 | -0.3 | -0.9 | 6,449 | 6,971 | 80 | 22,926 | 2.4 | 65.5 | 10.6 | 28.8 |
| Marshall | 15.2 | 14.6 | 9.2 | 50.3 | 35,519 | 33,128 | -6.7 | -9.1 | 3,140 | 4,203 | -1,937 | 12,308 | 2.51 | 65.2 | 9.6 | 31.1 |
| Mason | 15.1 | 12.8 | 9.4 | 51.7 | 25,957 | 27,358 | 5.4 | -3.7 | 2,807 | 3,568 | -238 | 11,034 | 2.36 | 69.0 | 9.7 | 27.5 |
| Mercer | 13.3 | 13.4 | 9.5 | 51.9 | 62,980 | 62,265 | -1.1 | -6.4 | 7,336 | 9,290 | -2,008 | 25,216 | 2.33 | 64.8 | 14.2 | 30.8 |
| Mineral | 14.1 | 13.2 | 9.4 | 50.4 | 27,078 | 28,225 | 4.2 | -5.3 | 2,812 | 3,413 | -889 | 10,916 | 2.42 | 66.5 | 10.9 | 29.5 |
| Mingo | 14.7 | 13.6 | 7.0 | 50.6 | 28,253 | 26,834 | -5.0 | -14.5 | 3,230 | 3,762 | -3,371 | 10,501 | 2.3 | 66.2 | 13.2 | 31.1 |
| Monongalia | 10.5 | 8.4 | 5.1 | 48.6 | 81,866 | 96,192 | 17.5 | 11.0 | 10,908 | 7,135 | 6,752 | 39,466 | 2.53 | 52.5 | 7.4 | 33.8 |
| Monroe | 15.1 | 14.5 | 11.2 | 50.3 | 14,583 | 13,496 | -7.5 | -2.0 | 1,330 | 1,774 | 185 | 5,718 | 2.33 | 63.7 | 9.6 | 33.3 |
| Morgan | 16.6 | 15.0 | 9.9 | 49.5 | 14,943 | 17,541 | 17.4 | 1.9 | 1,430 | 2,276 | 1,188 | 7,185 | 2.45 | 68.3 | 6.1 | 23.3 |
| Nicholas | 14.8 | 14.0 | 9.6 | 50.7 | 26,562 | 26,225 | -1.3 | -7.2 | 2,734 | 3,482 | -1,131 | 10,069 | 2.48 | 70.2 | 10.9 | 26.4 |
| Ohio | 14.1 | 13.1 | 9.7 | 51.9 | 47,427 | 44,433 | -6.3 | -7.3 | 4,564 | 5,942 | -1,857 | 17,193 | 2.31 | 57.4 | 10.3 | 36.9 |
| Pendleton | 16.3 | 15.5 | 13.6 | 49.6 | 8,196 | 7,695 | -6.1 | -9.9 | 686 | 964 | -491 | 3,174 | 2.16 | 66.2 | 10.6 | 30.6 |
| Pleasants | 14.4 | 11.3 | 8.2 | 45.6 | 7,514 | 7,600 | 1.1 | -2.1 | 714 | 952 | 76 | 2,835 | 2.45 | 72.6 | 9.9 | 24.3 |
| Pocahontas | 16.6 | 15.6 | 11.5 | 48.7 | 9,131 | 8,722 | -4.5 | -6.1 | 834 | 1,110 | -254 | 3,530 | 2.31 | 61.8 | 9.0 | 33.7 |
| Preston | 14.1 | 12.9 | 8.5 | 48.7 | 29,334 | 33,511 | 14.2 | -0.4 | 3,460 | 3,759 | 172 | 12,429 | 2.48 | 69.1 | 9.8 | 24.3 |
| Putnam | 13.8 | 11.4 | 7.9 | 50.6 | 51,589 | 55,488 | 7.6 | 1.7 | 5,822 | 5,945 | 1,098 | 21,613 | 2.61 | 72.9 | 11.5 | 23.6 |
| Raleigh | 13.0 | 13.2 | 8.6 | 49.8 | 79,220 | 78,865 | -0.4 | -7.5 | 8,881 | 10,615 | -4,192 | 31,203 | 2.32 | 67.7 | 13.3 | 27.3 |
| Randolph | 14.3 | 12.9 | 10.0 | 47.8 | 28,262 | 29,402 | 4.0 | -3.5 | 3,022 | 3,760 | -259 | 11,135 | 2.4 | 62.2 | 10.1 | 32.8 |
| Ritchie | 16.8 | 13.3 | 9.8 | 50.1 | 10,343 | 10,449 | 1.0 | -9.1 | 969 | 1,326 | -590 | 4,049 | 2.42 | 69.2 | 10.4 | 27.0 |
| Roane | 15.5 | 13.8 | 9.2 | 50.5 | 15,446 | 14,928 | -3.4 | -9.7 | 1,452 | 2,041 | -848 | 5,562 | 2.5 | 66.0 | 10.7 | 27.6 |
| Summers | 15.7 | 15.9 | 11.0 | 54.7 | 12,999 | 13,926 | 7.1 | -10.6 | 1,101 | 1,964 | -619 | 5,566 | 2.15 | 66.2 | 11.0 | 31.3 |
| Taylor | 14.1 | 12.5 | 8.7 | 49.4 | 16,089 | 16,876 | 4.9 | | 1,686 | 2,058 | 203 | 6,614 | 2.49 | 66.0 | 10.3 | 26.7 |
| Tucker | 16.7 | 15.6 | 12.2 | 49.5 | 7,321 | 7,141 | -2.5 | -4.6 | 653 | 966 | -7 | 3,142 | 2.17 | 65.0 | 8.7 | 29.9 |
| Tyler | 15.6 | 14.0 | 9.9 | 50.1 | 9,592 | 9,227 | -3.8 | -7.5 | 889 | 1,222 | -362 | 3,207 | 2.72 | 63.8 | 11.2 | 34.5 |
| Upshur | 13.5 | 12.5 | 9.0 | 50.6 | 23,404 | 24,260 | 3.7 | -0.1 | 2,698 | 2,808 | 96 | 9,713 | 2.39 | 65.4 | 12.8 | 29.8 |
| Wayne | 14.2 | 12.8 | 9.6 | 51.2 | 42,903 | 42,542 | -0.8 | -8.2 | 4,081 | 5,232 | -2,326 | 15,124 | 2.65 | 65.8 | 9.3 | 31.2 |
| Webster | 15.8 | 15.1 | 9.5 | 50.3 | 9,719 | 9,148 | -5.9 | -11.9 | 928 | 1,269 | -749 | 3,781 | 2.2 | 67.3 | 13.2 | 27.8 |
| Wetzel | 15.0 | 12.7 | 10.5 | 50.6 | 17,693 | 16,566 | -6.4 | -10.0 | 1,731 | 2,338 | -1,054 | 5,762 | 2.65 | 64.0 | 12.3 | 28.5 |
| Wirt | 17.1 | 13.2 | 8.3 | 49.1 | 5,873 | 5,710 | -2.8 | -0.1 | 581 | 634 | 54 | 2,505 | 2.31 | 65.1 | 5.5 | 29.5 |
| Wood | 14.6 | 12.2 | 9.1 | 51.4 | 87,986 | 86,954 | -1.2 | -4.6 | 9,752 | 11,053 | -2,636 | 35,488 | 2.37 | 63.7 | 11.6 | 30.7 |
| Wyoming | 14.5 | 14.7 | 8.0 | 50.5 | 25,708 | 23,804 | -7.4 | -15.5 | 2,171 | 3,473 | -2,394 | 8,805 | 2.41 | 75.1 | 12.7 | 22.2 |
| **WISCONSIN** | 14.1 | 10.6 | 7.3 | 50.2 | 5,363,675 | 5,687,285 | 6.0 | 2.6 | 678,202 | 522,217 | -8,205 | 2,358,156 | 2.39 | 62.9 | 9.6 | 29.5 |
| Adams | 19.3 | 18.6 | 12.2 | 46.2 | 18,643 | 20,867 | 11.9 | -1.8 | 1,336 | 2,642 | 943 | 9,062 | 2.09 | 62.1 | 6.8 | 30.8 |
| Ashland | 15.7 | 12.7 | 8.0 | 50.0 | 16,866 | 16,157 | -4.2 | -4.6 | 1,794 | 1,943 | -596 | 6,565 | 2.29 | 58.8 | 10.6 | 31.9 |
| Barron | 15.4 | 13.2 | 9.8 | 49.7 | 44,963 | 45,876 | 2.0 | -1.7 | 5,193 | 5,264 | -691 | 18,907 | 2.35 | 64.5 | 8.2 | 29.2 |
| Bayfield | 18.5 | 18.7 | 11.0 | 49.5 | 15,013 | 15,008 | 0.0 | 1.6 | 1,282 | 1,671 | 637 | 7,057 | 2.11 | 65.2 | 5.9 | 29.0 |
| Brown | 13.3 | 9.6 | 6.3 | 50.4 | 226,778 | 248,006 | 9.4 | 6.7 | 34,053 | 19,879 | 2,637 | 105,271 | 2.41 | 63.9 | 9.5 | 28.7 |
| Buffalo | 16.0 | 13.3 | 9.9 | 49.2 | 13,804 | 13,588 | -1.6 | -4.1 | 1,399 | 1,345 | -608 | 5,772 | 2.26 | 66.0 | 6.8 | 27.3 |
| Burnett | 18.9 | 18.6 | 11.8 | 49.0 | 15,674 | 15,459 | -1.4 | 0.6 | 1,327 | 1,863 | 644 | 7,305 | 2.07 | 65.1 | 7.8 | 28.1 |
| Calumet | 14.8 | 10.0 | 6.4 | 49.5 | 40,631 | 48,981 | 20.6 | 2.5 | 5,371 | 3,436 | -704 | 19,807 | 2.51 | 74.0 | 7.2 | 21.7 |
| Chippewa | 14.8 | 11.3 | 7.6 | 48.1 | 55,195 | 62,510 | 13.3 | 3.6 | 7,245 | 5,920 | 955 | 25,601 | 2.39 | 67.2 | 8.1 | 26.4 |
| Clark | 13.6 | 9.7 | 7.5 | 49.5 | 33,557 | 34,679 | 3.3 | 0.1 | 5,882 | 3,495 | -2,361 | 12,791 | 2.66 | 68.2 | 6.2 | 26.8 |
| Columbia | 15.4 | 11.5 | 7.4 | 49.0 | 52,468 | 56,856 | 8.4 | 1.4 | 6,155 | 5,569 | 256 | 24,112 | 2.31 | 66.1 | 8.3 | 27.8 |
| Crawford | 15.7 | 14.1 | 10.4 | 48.1 | 17,243 | 16,639 | -3.5 | -3.7 | 1,640 | 1,809 | -439 | 6,654 | 2.32 | 63.1 | 6.5 | 30.7 |
| Dane | 11.4 | 9.0 | 5.6 | 50.3 | 426,526 | 488,091 | 14.4 | 13.2 | 61,742 | 33,061 | 35,779 | 222,929 | 2.34 | 65.6 | 8.0 | 30.6 |
| Dodge | 15.6 | 10.6 | 7.9 | 47.1 | 85,897 | 88,744 | 3.3 | -1.6 | 8,340 | 9,332 | -358 | 34,980 | 2.39 | 65.9 | 8.2 | 27.5 |
| Door | 17.6 | 19.0 | 12.3 | 50.4 | 27,961 | 27,784 | -0.6 | 0.4 | 2,158 | 3,416 | 1,388 | 13,191 | 2.06 | 68.8 | 7.4 | 26.2 |

1. No spouse present.

# Table B. States and Counties — Population, Vital Statistics, Health, and Crime

| STATE County | Persons in group quarters, 2020 | Daytime Population, 2015-2019 Number | Employment/ residence ratio | Births, 2020 Total | Rate[1] | Deaths, 2020 Number | Rate[1] | Persons under 65 with no health insurance, 2019 Number | Percent | Medicare, 2020 Total beneficiaries | Enrolled in Original Medicare | Enrolled in Medicare Advantage | Crimes reported by county police or sheriff, 2019 Violent | Property |
|---|---|---|---|---|---|---|---|---|---|---|---|---|---|---|
| | 32 | 33 | 34 | 35 | 36 | 37 | 38 | 39 | 40 | 41 | 42 | 43 | 44 | 45 |
| **WEST VIRGINIA—Cont'd** | | | | | | | | | | | | | | |
| Calhoun | 22 | 6,461 | 0.60 | 49 | 7.1 | 94 | 13.5 | 474 | 9.0 | 2,000 | 1,328 | 673 | 0 | 10 |
| Clay | 74 | 7,792 | 0.63 | 86 | 10.3 | 115 | 13.8 | 555 | 8.4 | 2,534 | 1,354 | 1,179 | NA | NA |
| Doddridge | 1,096 | 7,296 | 0.58 | 62 | 7.4 | 104 | 12.4 | 475 | 8.6 | 1,691 | 1,020 | 671 | NA | NA |
| Fayette | 1,817 | 39,116 | 0.70 | 396 | 9.4 | 712 | 16.9 | 2,890 | 9.2 | 11,352 | 6,802 | 4,550 | 1 | 32 |
| Gilmer | 1,759 | 7,962 | 0.97 | 60 | 7.7 | 80 | 10.2 | 374 | 7.9 | 1,527 | 923 | 604 | 49 | 347 |
| Grant | 126 | 11,079 | 0.89 | 107 | 9.3 | 140 | 12.2 | 768 | 8.9 | 3,271 | 2,389 | 881 | NA | NA |
| Greenbrier | 611 | 36,851 | 1.12 | 300 | 8.7 | 532 | 15.5 | 2,318 | 8.9 | 9,548 | 6,638 | 2,910 | NA | NA |
| Hampshire | 418 | 19,007 | 0.53 | 217 | 9.4 | 287 | 12.4 | 1,795 | 10.4 | 5,932 | 4,354 | 1,578 | 14 | 63 |
| Hancock | 226 | 27,510 | 0.85 | 245 | 8.6 | 466 | 16.3 | 1,854 | 8.5 | 7,800 | 5,166 | 2,634 | 25 | 118 |
| Hardy | 58 | 14,459 | 1.10 | 162 | 11.9 | 181 | 13.3 | 1,094 | 10.3 | 3,509 | 2,553 | 956 | 7 | 2 |
| Harrison | 884 | 77,206 | 1.31 | 705 | 10.5 | 943 | 14.1 | 4,137 | 7.7 | 15,866 | 10,061 | 5,806 | 33 | 13 |
| Jackson | 180 | 26,579 | 0.79 | 303 | 10.6 | 397 | 14.0 | 1,680 | 7.5 | 7,401 | 4,715 | 2,686 | NA | NA |
| Jefferson | 1,078 | 47,275 | 0.66 | 520 | 9.0 | 576 | 10.0 | 3,464 | 7.4 | 11,330 | 8,476 | 2,854 | 44 | 26 |
| Kanawha | 3,305 | 204,386 | 1.27 | 1,790 | 10.2 | 2,633 | 14.9 | 11,108 | 8.1 | 46,512 | 26,434 | 20,079 | 380 | 1,361 |
| Lewis | 247 | 16,819 | 1.11 | 168 | 10.6 | 243 | 15.4 | 1,007 | 8.2 | 4,223 | 2,464 | 1,759 | 3 | 22 |
| Lincoln | 65 | 16,764 | 0.40 | 214 | 10.7 | 331 | 16.5 | 1,332 | 8.3 | 5,403 | 2,836 | 2,567 | NA | NA |
| Logan | 681 | 33,948 | 1.08 | 318 | 10.0 | 529 | 16.7 | 2,207 | 9.0 | 8,959 | 5,257 | 3,702 | NA | NA |
| McDowell | 139 | 19,215 | 1.15 | 165 | 9.8 | 353 | 20.9 | 1,580 | 11.8 | 5,587 | 3,414 | 2,173 | 131 | 4 |
| Marion | 1,401 | 52,118 | 0.83 | 558 | 10.0 | 716 | 12.8 | 3,426 | 7.8 | 13,134 | 7,854 | 5,280 | 3 | 4 |
| Marshall | 550 | 29,064 | 0.82 | 269 | 8.9 | 417 | 13.9 | 1,603 | 6.9 | 7,833 | 3,843 | 3,989 | NA | NA |
| Mason | 683 | 23,757 | 0.69 | 254 | 9.6 | 387 | 14.7 | 1,395 | 7.0 | 6,775 | 4,082 | 2,692 | 9 | 67 |
| Mercer | 1,114 | 58,628 | 0.94 | 642 | 11.0 | 958 | 16.4 | 3,928 | 8.8 | 16,567 | 10,177 | 6,390 | NA | NA |
| Mineral | 519 | 23,310 | 0.66 | 240 | 9.0 | 373 | 14.0 | 1,389 | 6.8 | 6,926 | 5,355 | 1,572 | 83 | 306 |
| Mingo | 85 | 23,244 | 0.84 | 252 | 11.0 | 400 | 17.4 | 1,797 | 9.7 | 6,713 | 3,795 | 2,918 | 2 | 40 |
| Monongalia | 6,380 | 117,773 | 1.24 | 975 | 9.1 | 788 | 7.4 | 5,943 | 6.9 | 14,803 | 8,646 | 6,157 | 136 | 487 |
| Monroe | 57 | 10,840 | 0.48 | 127 | 9.6 | 175 | 13.2 | 1,021 | 10.4 | 3,696 | 2,524 | 1,172 | 1 | 33 |
| Morgan | 123 | 13,696 | 0.49 | 123 | 6.9 | 240 | 13.4 | 1,336 | 9.8 | 4,757 | 3,516 | 1,242 | 59 | 71 |
| Nicholas | 162 | 23,616 | 0.84 | 219 | 9.0 | 380 | 15.6 | 1,641 | 8.8 | 7,324 | 4,227 | 3,097 | 120 | 13 |
| Ohio | 2,390 | 53,201 | 1.59 | 408 | 9.9 | 623 | 15.1 | 2,032 | 6.7 | 10,578 | 5,808 | 4,769 | NA | NA |
| Pendleton | 117 | 6,181 | 0.72 | 59 | 8.5 | 76 | 11.0 | 485 | 9.8 | 2,190 | 1,518 | 672 | NA | NA |
| Pleasants | 697 | 7,384 | 0.97 | 72 | 9.7 | 80 | 10.8 | 296 | 5.5 | 1,812 | 1,159 | 653 | NA | NA |
| Pocahontas | 319 | 8,370 | 0.98 | 70 | 8.5 | 139 | 17.0 | 502 | 8.7 | 2,481 | 1,701 | 780 | NA | NA |
| Preston | 1,822 | 28,374 | 0.59 | 321 | 9.6 | 387 | 11.6 | 2,095 | 8.5 | 7,841 | 5,033 | 2,808 | 10 | 133 |
| Putnam | 249 | 53,074 | 0.86 | 495 | 8.8 | 641 | 11.4 | 2,698 | 5.9 | 12,748 | 7,217 | 5,531 | NA | NA |
| Raleigh | 3,660 | 78,916 | 1.13 | 766 | 10.5 | 1,112 | 15.2 | 4,349 | 8.0 | 19,981 | 13,031 | 6,951 | NA | NA |
| Randolph | 2,306 | 29,712 | 1.07 | 253 | 8.9 | 409 | 14.4 | 1,811 | 9.0 | 7,123 | 4,775 | 2,348 | 80 | 174 |
| Ritchie | 106 | 10,406 | 1.16 | 90 | 9.5 | 165 | 17.4 | 668 | 9.1 | 2,626 | 1,705 | 921 | NA | NA |
| Roane | 98 | 13,013 | 0.78 | 122 | 9.0 | 203 | 15.1 | 1,059 | 10.1 | 3,991 | 2,408 | 1,583 | NA | NA |
| Summers | 1,025 | 11,715 | 0.73 | 94 | 7.6 | 166 | 13.3 | 699 | 8.4 | 3,604 | 2,475 | 1,129 | 4 | 2 |
| Taylor | 509 | 13,749 | 0.53 | 138 | 8.3 | 223 | 13.4 | 1,033 | 8.1 | 3,959 | 2,616 | 1,344 | 10 | 25 |
| Tucker | 181 | 6,590 | 0.87 | 54 | 7.9 | 136 | 20.0 | 449 | 9.2 | 2,172 | 1,333 | 839 | NA | NA |
| Tyler | 67 | 8,112 | 0.77 | 80 | 9.4 | 117 | 13.7 | 453 | 6.9 | 2,209 | 1,302 | 907 | NA | NA |
| Upshur | 1,225 | 23,327 | 0.88 | 242 | 10.0 | 304 | 12.5 | 1,732 | 9.7 | 6,013 | 3,429 | 2,584 | NA | NA |
| Wayne | 246 | 36,125 | 0.68 | 327 | 8.4 | 582 | 14.9 | 2,723 | 8.9 | 10,501 | 6,233 | 4,268 | 3 | 11 |
| Webster | 58 | 7,933 | 0.84 | 82 | 10.2 | 128 | 15.9 | 489 | 8.1 | 2,519 | 1,500 | 1,019 | NA | NA |
| Wetzel | 124 | 16,507 | 1.20 | 168 | 11.3 | 255 | 17.1 | 896 | 7.8 | 4,094 | 2,233 | 1,861 | 41 | 22 |
| Wirt | 0 | 4,407 | 0.36 | 52 | 9.1 | 58 | 10.2 | 357 | 7.8 | 1,500 | 962 | 538 | 14 | 22 |
| Wood | 1,007 | 89,090 | 1.12 | 870 | 10.5 | 1,159 | 14.0 | 4,781 | 7.3 | 22,041 | 14,815 | 7,225 | 105 | 489 |
| Wyoming | 56 | 19,810 | 0.75 | 168 | 8.3 | 373 | 18.5 | 1,421 | 9.0 | 6,106 | 3,404 | 2,703 | 66 | 70 |
| **WISCONSIN** | 142,304 | 5,749,010 | 0.99 | 63,122 | 10.8 | 54,407 | 9.3 | 320,719 | 6.8 | 1,200,231 | 645,066 | 555,165 | X | X |
| Adams | 1,114 | 17,497 | 0.66 | 133 | 6.5 | 268 | 13.1 | 1,183 | 9.1 | 7,223 | 4,895 | 2,328 | 21 | 255 |
| Ashland | 578 | 16,794 | 1.17 | 158 | 10.2 | 190 | 12.3 | 1,055 | 8.9 | 3,949 | 2,611 | 1,338 | 22 | 41 |
| Barron | 581 | 45,485 | 1.01 | 483 | 10.7 | 534 | 11.8 | 3,083 | 8.9 | 11,957 | 7,144 | 4,813 | 20 | 127 |
| Bayfield | 113 | 12,974 | 0.70 | 130 | 8.5 | 188 | 12.3 | 1,141 | 10.7 | 5,211 | 3,349 | 1,862 | 19 | 80 |
| Brown | 6,397 | 279,693 | 1.13 | 3,101 | 11.7 | 2,062 | 7.8 | 16,988 | 7.8 | 48,443 | 20,029 | 28,414 | 83 | 623 |
| Buffalo | 87 | 10,739 | 0.64 | 122 | 9.4 | 134 | 10.3 | 857 | 8.6 | 3,254 | 2,579 | 675 | 4 | 6 |
| Burnett | 135 | 13,922 | 0.79 | 117 | 7.5 | 182 | 11.7 | 978 | 9.2 | 5,465 | 3,467 | 1,998 | 20 | 237 |
| Calumet | 167 | 38,923 | 0.60 | 477 | 9.5 | 373 | 7.4 | 2,056 | 4.9 | 9,318 | 3,182 | 6,136 | 10 | 74 |
| Chippewa | 2,723 | 60,253 | 0.88 | 689 | 10.6 | 641 | 9.9 | 3,376 | 6.7 | 14,289 | 9,093 | 5,196 | 53 | 199 |
| Clark | 466 | 31,696 | 0.82 | 579 | 16.7 | 338 | 9.7 | 4,530 | 16.0 | 7,085 | 3,695 | 3,390 | 14 | 97 |
| Columbia | 1,362 | 49,356 | 0.74 | 592 | 10.3 | 575 | 10.0 | 2,700 | 5.9 | 12,779 | 8,453 | 4,326 | 34 | 201 |
| Crawford | 759 | 16,609 | 1.05 | 154 | 9.6 | 191 | 11.9 | 818 | 7.1 | 4,472 | 2,841 | 1,630 | 25 | 90 |
| Dane | 13,436 | 577,348 | 1.14 | 5,729 | 10.4 | 3,565 | 6.5 | 22,745 | 4.9 | 89,874 | 62,153 | 27,721 | 98 | 684 |
| Dodge | 6,249 | 78,760 | 0.80 | 772 | 8.8 | 966 | 11.1 | 4,125 | 6.2 | 18,569 | 10,803 | 7,765 | 39 | 149 |
| Door | 339 | 27,566 | 1.01 | 203 | 7.3 | 364 | 13.1 | 1,578 | 8.3 | 9,951 | 6,413 | 3,538 | 10 | 57 |

1. Per 1,000 estimated resident population.

# Table B. States and Counties — Crime, Education, Money Income, and Poverty

| STATE County | County law enforcement employment, 2019 | | Education — School enrollment and attainment, 2015-2019 | | | | Local government expenditures,[3] 2017-2018 | | Money income, 2015-2019 | | | | Income and poverty, 2019 | | | |
|---|---|---|---|---|---|---|---|---|---|---|---|---|---|---|---|---|
| | | | Enrollment[1] | | Attainment[2] (percent) | | | | | Households | | | | Percent below poverty level | | |
| | Officers | Civilians | Total | Percent private | High school graduate or less | Bachelor's degree or more | Total current spending (mil dol) | Current spending per student (dollars) | Per capita income[4] | Median income (dollars) | Percent with income of less than $50,000 | Percent with income of $200,000 or more | Median household income (dollars) | All persons | Children under 18 years | Children 5 to 17 years in families |
| | 46 | 47 | 48 | 49 | 50 | 51 | 52 | 53 | 54 | 55 | 56 | 57 | 58 | 59 | 60 | 61 |
| WEST VIRGINIA—Cont'd | | | | | | | | | | | | | | | | |
| Calhoun | 4 | 2 | 1,320 | 7.6 | 66.2 | 12.2 | 11 | 11,040 | 20,447 | 38,382 | 63.0 | 2.3 | 37,307 | 21.6 | 25.1 | 23.3 |
| Clay | 5 | 0 | 1,802 | 5.0 | 74.6 | 9.2 | 22 | 11,475 | 17,239 | 35,024 | 69.7 | 0.4 | 41,923 | 22.5 | 30.5 | 28.3 |
| Doddridge | 9 | 1 | 1,473 | 9.6 | 61.7 | 17.1 | 20 | 17,878 | 24,053 | 45,545 | 53.6 | 3.1 | 50,642 | 18.8 | 22.4 | 21.0 |
| Fayette | 34 | 7 | 8,593 | 13.8 | 61.1 | 14.8 | 71 | 11,037 | 21,914 | 41,394 | 58.7 | 1.2 | 41,430 | 20.6 | 26.8 | 25.4 |
| Gilmer | 4 | 2 | 1,936 | 12.8 | 59.7 | 17.0 | 11 | 13,491 | 18,067 | 42,636 | 56.7 | 0.3 | 40,373 | 25.5 | 20.8 | 19.7 |
| Grant | 10 | 0 | 2,098 | 7.1 | 64.0 | 13.8 | 17 | 9,903 | 23,351 | 42,216 | 55.0 | 1.6 | 47,400 | 13.2 | 19.6 | 19.0 |
| Greenbrier | 27 | 11 | 6,518 | 10.1 | 51.8 | 19.9 | 58 | 11,641 | 24,581 | 40,200 | 60.9 | 1.5 | 43,564 | 17.2 | 25.6 | 23.2 |
| Hampshire | 18 | 5 | 4,133 | 8.4 | 61.9 | 12.4 | 46 | 14,283 | 24,259 | 47,857 | 52.1 | 0.9 | 54,811 | 14.6 | 22.9 | 22.0 |
| Hancock | 31 | 4 | 5,586 | 8.4 | 51.6 | 17.4 | 44 | 10,657 | 26,435 | 45,763 | 55.3 | 2.2 | 51,017 | 12.5 | 17.5 | 15.6 |
| Hardy | 11 | 1 | 2,606 | 5.4 | 64.4 | 14.5 | 23 | 9,537 | 27,004 | 47,438 | 52.1 | 2.8 | 46,658 | 14.2 | 20.6 | 19.2 |
| Harrison | 53 | 4 | 13,805 | 9.7 | 49.8 | 23.3 | 141 | 13,028 | 28,879 | 53,022 | 47.6 | 3.4 | 53,909 | 11.1 | 16.5 | 15.6 |
| Jackson | 16 | 8 | 5,426 | 8.0 | 55.8 | 17.4 | 52 | 11,070 | 25,341 | 47,837 | 52.0 | 3.4 | 50,808 | 16.7 | 23.0 | 19.4 |
| Jefferson | 30 | 6 | 14,074 | 16.3 | 41.1 | 31.8 | 103 | 11,269 | 36,305 | 80,430 | 30.3 | 9.3 | 75,225 | 9.9 | 12.5 | 10.8 |
| Kanawha | 104 | 29 | 36,532 | 11.9 | 48.5 | 25.5 | 319 | 12,133 | 29,253 | 46,639 | 52.6 | 3.8 | 48,334 | 16.3 | 21.2 | 21.2 |
| Lewis | 13 | 1 | 3,006 | 7.2 | 62.6 | 14.5 | 28 | 10,943 | 23,820 | 39,908 | 57.7 | 3.3 | 46,895 | 19.5 | 28.0 | 25.1 |
| Lincoln | 6 | 1 | 3,953 | 5.7 | 66.1 | 9.0 | 41 | 11,642 | 21,456 | 42,345 | 59.4 | 1.2 | 43,198 | 19.7 | 25.9 | 24.0 |
| Logan | 20 | 2 | 6,237 | 6.4 | 64.1 | 10.0 | 66 | 11,176 | 21,936 | 36,168 | 60.0 | 1.5 | 38,533 | 21.9 | 26.6 | 26.1 |
| McDowell | 13 | 2 | 3,213 | 6.1 | 78.4 | 5.4 | 39 | 12,665 | 15,047 | 27,682 | 77.8 | 1.3 | 29,106 | 33.8 | 44.2 | 41.9 |
| Marion | 29 | 15 | 12,562 | 6.7 | 51.3 | 22.6 | 94 | 11,638 | 26,380 | 50,305 | 49.6 | 2.6 | 51,300 | 14.6 | 19.4 | 16.6 |
| Marshall | 32 | 3 | 5,710 | 9.5 | 52.3 | 18.0 | 65 | 13,867 | 26,472 | 48,557 | 51.2 | 2.8 | 53,145 | 13.7 | 19.2 | 18.0 |
| Mason | 19 | 7 | 5,139 | 6.2 | 59.5 | 14.4 | 47 | 11,236 | 24,844 | 46,078 | 53.2 | 1.4 | 47,753 | 15.5 | 22.1 | 21.5 |
| Mercer | 29 | 12 | 12,519 | 8.5 | 56.0 | 19.4 | 97 | 10,683 | 22,212 | 40,784 | 60.0 | 1.5 | 41,794 | 18.7 | 25.6 | 24.0 |
| Mineral | 16 | 3 | 5,548 | 6.7 | 57.0 | 15.5 | 49 | 11,773 | 25,490 | 49,936 | 50.1 | 2.5 | 49,704 | 12.5 | 19.1 | 19.4 |
| Mingo | 16 | 2 | 5,134 | 5.9 | 67.1 | 9.9 | 47 | 10,864 | 20,060 | 32,764 | 66.8 | 1.3 | 34,698 | 27.3 | 33.5 | 33.0 |
| Monongalia | 41 | 4 | 35,939 | 6.8 | 34.8 | 42.8 | 133 | 11,335 | 32,154 | 52,455 | 48.3 | 7.3 | 55,652 | 19.1 | 14.2 | 13.1 |
| Monroe | 9 | 3 | 2,352 | 12.2 | 64.8 | 15.3 | 20 | 11,207 | 23,923 | 38,540 | 58.9 | 1.7 | 44,655 | 15.7 | 22.9 | 21.4 |
| Morgan | 11 | 1 | 3,380 | 15.8 | 54.5 | 17.0 | 26 | 10,892 | 28,127 | 51,745 | 48.9 | 2.0 | 50,660 | 10.5 | 16.4 | 14.9 |
| Nicholas | 21 | 3 | 4,901 | 9.2 | 61.1 | 15.6 | 44 | 11,742 | 22,335 | 40,086 | 61.8 | 1.8 | 40,023 | 18.8 | 25.1 | 23.2 |
| Ohio | 36 | 2 | 9,259 | 19.5 | 40.7 | 32.0 | 70 | 13,283 | 32,429 | 50,584 | 49.5 | 4.8 | 50,390 | 14.1 | 17.6 | 15.1 |
| Pendleton | 3 | 0 | 1,089 | 9.9 | 69.8 | 14.1 | 13 | 13,653 | 25,059 | 42,312 | 62.8 | 1.5 | 43,451 | 13.5 | 19.3 | 17.6 |
| Pleasants | 6 | 1 | 1,480 | 9.3 | 61.6 | 11.6 | 16 | 13,654 | 29,974 | 56,838 | 47.1 | 2.5 | 54,972 | 13.0 | 15.2 | 14.0 |
| Pocahontas | 6 | 3 | 1,414 | 6.2 | 65.2 | 14.8 | 15 | 9,382 | 24,890 | 41,882 | 59.5 | 1.6 | 50,821 | 16.1 | 24.1 | 23.2 |
| Preston | 21 | 3 | 6,187 | 6.8 | 60.4 | 16.9 | 42 | 9,382 | 24,890 | 51,888 | 48.9 | 2.5 | 50,821 | 14.3 | 18.1 | 17.5 |
| Putnam | 42 | 26 | 12,724 | 13.7 | 44.7 | 26.1 | 107 | 11,124 | 31,587 | 60,097 | 41.8 | 5.3 | 66,991 | 8.9 | 11.4 | 10.5 |
| Raleigh | 53 | 15 | 15,156 | 12.0 | 53.5 | 18.4 | 132 | 10,953 | 25,026 | 43,748 | 54.9 | 1.9 | 45,386 | 17.4 | 22.9 | 22.8 |
| Randolph | 14 | 6 | 5,204 | 14.5 | 63.0 | 17.3 | 41 | 10,053 | 24,842 | 43,320 | 56.9 | 2.9 | 46,285 | 16.8 | 22.4 | 21.8 |
| Ritchie | 8 | 4 | 1,699 | 13.4 | 66.4 | 11.3 | 17 | 11,810 | 25,606 | 43,577 | 57.8 | 1.9 | 46,131 | 17.9 | 25.9 | 25.2 |
| Roane | 7 | 3 | 2,875 | 7.2 | 64.4 | 13.1 | 22 | 9,839 | 21,711 | 37,373 | 60.8 | 2.7 | 41,671 | 19.4 | 26.5 | 23.5 |
| Summers | 7 | 1 | 2,221 | 12.0 | 57.1 | 14.5 | 15 | 9,943 | 21,437 | 38,187 | 63.2 | 1.0 | 37,000 | 23.5 | 31.9 | 31.7 |
| Taylor | 6 | 6 | 3,300 | 17.8 | 57.9 | 17.3 | 26 | 10,501 | 25,214 | 48,578 | 51.4 | 2.2 | 48,294 | 16.7 | 20.3 | 19.4 |
| Tucker | 8 | 2 | 1,024 | 15.9 | 59.0 | 19.4 | 12 | 11,566 | 25,686 | 49,118 | 51.3 | 0.5 | 47,222 | 15.3 | 22.1 | 21.2 |
| Tyler | 11 | 2 | 1,599 | 4.2 | 61.2 | 14.1 | 18 | 13,550 | 24,924 | 43,087 | 57.6 | 1.7 | 48,345 | 15.4 | 20.8 | 19.9 |
| Upshur | 12 | 2 | 5,369 | 24.3 | 63.6 | 16.9 | 39 | 10,197 | 24,070 | 40,322 | 56.9 | 1.2 | 45,074 | 17.5 | 22.9 | 20.5 |
| Wayne | 22 | 3 | 8,160 | 8.3 | 60.8 | 15.1 | 74 | 10,590 | 21,553 | 37,988 | 60.2 | 2.5 | 42,095 | 19.6 | 24.7 | 23.3 |
| Webster | 3 | 3 | 1,538 | 3.9 | 70.3 | 11.4 | 15 | 10,889 | 22,054 | 34,927 | 63.2 | 1.1 | 34,385 | 21.8 | 34.1 | 31.5 |
| Wetzel | 13 | 6 | 2,858 | 2.9 | 64.7 | 12.0 | 40 | 15,692 | 22,546 | 43,107 | 57.7 | 0.9 | 46,824 | 16.3 | 22.9 | 20.5 |
| Wirt | 3 | 0 | 1,105 | 9.1 | 59.1 | 11.2 | 12 | 11,354 | 23,568 | 46,048 | 52.7 | 1.1 | 48,627 | 17.5 | 24.6 | 22.2 |
| Wood | 38 | 23 | 17,394 | 12.3 | 44.5 | 21.9 | 139 | 10,999 | 27,725 | 47,321 | 52.3 | 3.2 | 49,209 | 14.1 | 19.2 | 17.6 |
| Wyoming | 18 | 0 | 3,911 | 1.7 | 68.5 | 9.3 | 45 | 11,190 | 22,703 | 42,332 | 57.5 | 1.4 | 42,761 | 22.9 | 30.4 | 27.0 |
| WISCONSIN | X | X | 1,411,273 | 16.4 | 38.4 | 30.1 | 10,466 | 12,159 | 33,375 | 61,747 | 40.6 | 4.9 | 64,177 | 10.4 | 13.5 | 12.7 |
| Adams | 28 | 30 | 2,947 | 8.3 | 55.1 | 13.9 | 19 | 12,666 | 27,241 | 46,369 | 53.7 | 1.8 | 53,156 | 11.6 | 18.2 | 16.8 |
| Ashland | 24 | 27 | 3,751 | 18.9 | 44.8 | 19.8 | 33 | 12,686 | 23,946 | 42,510 | 55.9 | 2.0 | 47,202 | 17.8 | 23.3 | 21.5 |
| Barron | 29 | 44 | 8,941 | 11.7 | 45.9 | 20.6 | 95 | 12,027 | 29,681 | 52,703 | 47.2 | 2.8 | 54,372 | 8.9 | 11.9 | 10.8 |
| Bayfield | 20 | 23 | 2,570 | 13.5 | 34.1 | 31.1 | 25 | 17,102 | 31,825 | 56,096 | 43.7 | 2.5 | 58,065 | 10.3 | 17.7 | 17.0 |
| Brown | 162 | 306 | 66,526 | 16.6 | 37.8 | 30.3 | 507 | 11,324 | 32,874 | 62,340 | 39.9 | 4.9 | 65,494 | 9.9 | 12.4 | 11.2 |
| Buffalo | 13 | 3 | 2,708 | 9.2 | 49.5 | 19.3 | 24 | 11,694 | 30,503 | 57,829 | 42.6 | 2.4 | 58,407 | 9.5 | 12.4 | 11.9 |
| Burnett | 18 | 18 | 2,689 | 8.4 | 43.2 | 20.8 | 31 | 11,807 | 30,342 | 52,672 | 46.4 | 2.8 | 51,519 | 12.6 | 22.9 | 21.0 |
| Calumet | 28 | 27 | 12,416 | 13.2 | 37.2 | 29.4 | 41 | 10,627 | 35,527 | 75,814 | 31.5 | 4.2 | 80,354 | 5.4 | 6.5 | 5.4 |
| Chippewa | 44 | 30 | 14,116 | 13.2 | 42.6 | 21.0 | 103 | 11,398 | 29,991 | 59,742 | 41.3 | 3.5 | 60,713 | 8.8 | 12.1 | 11.6 |
| Clark | 28 | 32 | 7,613 | 25.8 | 57.5 | 12.6 | 63 | 13,051 | 25,382 | 54,012 | 45.6 | 2.7 | 54,296 | 13.5 | 22.2 | 22.1 |
| Columbia | 43 | 64 | 12,289 | 10.6 | 41.0 | 23.3 | 101 | 11,931 | 34,984 | 68,005 | 35.9 | 4.8 | 69,219 | 6.7 | 8.0 | 7.3 |
| Crawford | 23 | 7 | 3,198 | 10.4 | 48.3 | 18.4 | 29 | 13,272 | 27,615 | 50,595 | 49.5 | 3.5 | 51,919 | 11.6 | 16.1 | 15.3 |
| Dane | 464 | 107 | 150,569 | 10.7 | 22.0 | 51.4 | 981 | 12,600 | 40,614 | 73,893 | 33.2 | 8.4 | 77,828 | 9.4 | 7.2 | 7.0 |
| Dodge | 59 | 98 | 18,490 | 18.8 | 50.1 | 17.0 | 120 | 11,509 | 29,726 | 60,652 | 41.5 | 2.3 | 61,730 | 8.0 | 9.5 | 8.7 |
| Door | 45 | 4 | 4,622 | 13.1 | 36.0 | 33.7 | 48 | 14,069 | 38,744 | 61,560 | 38.9 | 5.1 | 64,217 | 9.3 | 13.8 | 11.3 |

1. All persons 3 years old and over enrolled in nursery school through college.   2. Persons 25 years old and over.   3. Elementary and secondary education expenditures.   4. Based on population estimated by the American Community Survey, 2014–2018.

| STATE County | Personal income, 2019 Total (mil dol) | Percent change 2018-2019 | Per capita¹ Dollars | Per capita¹ Rank | Wages and salaries (mil dol) | Supplements to wages and salaries, employer contributions (mil dol) Pension and insurance | Government social insurance | Proprietors' income (mil dol) | Dividends, interest, and rent (mil dol) | Personal transfer receipts (mil dol) | Earnings, 2019 Total (mil dol) | Contributions for government social insurance (mil dol) From employee and self-employed | From employer |
|---|---|---|---|---|---|---|---|---|---|---|---|---|---|
| | 62 | 63 | 64 | 65 | 66 | 67 | 68 | 69 | 70 | 71 | 72 | 73 | 74 |
| **WEST VIRGINIA—Cont'd** | | | | | | | | | | | | | |
| Calhoun | 222 | 2.0 | 31,229 | 2,991 | 56 | 11 | 4 | 5 | 26 | 93 | 76 | 8 | 4 |
| Clay | 274 | 1.3 | 32,197 | 2,932 | 45 | 10 | 4 | 12 | 28 | 101 | 71 | 8 | 4 |
| Doddridge | 255 | 2.0 | 30,173 | 3,031 | 100 | 21 | 7 | 4 | 55 | 61 | 133 | 10 | 7 |
| Fayette | 1,545 | 3.2 | 36,424 | 2,520 | 442 | 87 | 35 | 61 | 183 | 588 | 625 | 52 | 35 |
| Gilmer | 232 | 2.1 | 29,672 | 3,047 | 83 | 23 | 7 | 3 | 55 | 79 | 117 | 8 | 7 |
| Grant | 426 | 3.0 | 36,862 | 2,461 | 151 | 35 | 12 | 24 | 65 | 151 | 221 | 17 | 12 |
| Greenbrier | 1,346 | 2.1 | 38,842 | 2,218 | 546 | 97 | 45 | 85 | 226 | 463 | 773 | 58 | 45 |
| Hampshire | 836 | 2.9 | 36,093 | 2,557 | 140 | 32 | 11 | 43 | 113 | 238 | 226 | 21 | 11 |
| Hancock | 1,263 | 2.2 | 43,823 | 1,510 | 423 | 89 | 34 | 57 | 172 | 370 | 603 | 45 | 34 |
| Hardy | 474 | 1.2 | 34,417 | 2,753 | 221 | 42 | 19 | -3 | 79 | 134 | 279 | 22 | 19 |
| Harrison | 3,349 | 1.5 | 49,793 | 845 | 2,120 | 413 | 168 | 270 | 545 | 772 | 2,971 | 186 | 168 |
| Jackson | 1,122 | 1.5 | 39,254 | 2,154 | 498 | 79 | 38 | 62 | 134 | 331 | 676 | 50 | 38 |
| Jefferson | 2,945 | 3.7 | 51,527 | 722 | 829 | 158 | 66 | 118 | 453 | 526 | 1,172 | 80 | 66 |
| Kanawha | 8,827 | 1.2 | 49,553 | 870 | 5,284 | 922 | 398 | 781 | 1,523 | 2,442 | 7,384 | 478 | 398 |
| Lewis | 664 | 1.5 | 41,728 | 1,816 | 350 | 59 | 27 | 41 | 111 | 188 | 477 | 33 | 27 |
| Lincoln | 653 | 2.3 | 31,997 | 2,941 | 90 | 20 | 7 | 17 | 75 | 232 | 134 | 16 | 7 |
| Logan | 1,221 | 0.8 | 38,147 | 2,312 | 490 | 89 | 38 | 31 | 132 | 536 | 648 | 50 | 38 |
| McDowell | 569 | 1.9 | 32,294 | 2,926 | 218 | 47 | 17 | 15 | 70 | 285 | 297 | 26 | 17 |
| Marion | 2,440 | 2.6 | 43,522 | 1,543 | 871 | 160 | 68 | 150 | 352 | 656 | 1,250 | 89 | 68 |
| Marshall | 1,300 | -0.6 | 42,595 | 1,672 | 887 | 145 | 63 | 43 | 226 | 347 | 1,138 | 75 | 63 |
| Mason | 924 | 3.3 | 34,833 | 2,709 | 268 | 61 | 21 | 50 | 113 | 331 | 399 | 31 | 21 |
| Mercer | 2,279 | 2.1 | 38,778 | 2,227 | 823 | 154 | 68 | 131 | 317 | 892 | 1,176 | 91 | 68 |
| Mineral | 1,089 | 3.2 | 40,518 | 1,986 | 369 | 77 | 31 | 32 | 141 | 328 | 509 | 38 | 31 |
| Mingo | 786 | 1.5 | 33,553 | 2,844 | 268 | 51 | 23 | 23 | 92 | 360 | 366 | 32 | 23 |
| Monongalia | 4,925 | 2.5 | 46,630 | 1,169 | 3,301 | 605 | 246 | 265 | 929 | 862 | 4,417 | 261 | 246 |
| Monroe | 427 | 4.0 | 32,189 | 2,933 | 87 | 22 | 7 | 27 | 63 | 149 | 143 | 13 | 7 |
| Morgan | 636 | 4.1 | 35,536 | 2,616 | 108 | 23 | 9 | 34 | 96 | 212 | 173 | 17 | 9 |
| Nicholas | 852 | 0.9 | 34,792 | 2,717 | 295 | 57 | 24 | 39 | 129 | 333 | 415 | 34 | 24 |
| Ohio | 2,669 | -0.2 | 64,461 | 197 | 1,357 | 230 | 104 | 553 | 515 | 567 | 2,243 | 140 | 104 |
| Pendleton | 254 | 1.0 | 36,491 | 2,510 | 52 | 12 | 4 | 7 | 58 | 94 | 75 | 8 | 4 |
| Pleasants | 328 | 0.6 | 43,929 | 1,496 | 160 | 36 | 12 | 14 | 43 | 109 | 222 | 15 | 12 |
| Pocahontas | 332 | 2.9 | 40,243 | 2,020 | 114 | 23 | 10 | 22 | 60 | 132 | 169 | 13 | 10 |
| Preston | 1,257 | 3.7 | 37,595 | 2,375 | 340 | 79 | 28 | 64 | 166 | 355 | 510 | 40 | 28 |
| Putnam | 2,701 | 1.9 | 47,849 | 1,038 | 1,242 | 184 | 97 | 157 | 353 | 550 | 1,681 | 113 | 97 |
| Raleigh | 3,217 | 3.5 | 43,846 | 1,507 | 1,527 | 281 | 120 | 235 | 465 | 1,043 | 2,163 | 150 | 120 |
| Randolph | 1,094 | 3.1 | 38,142 | 2,313 | 450 | 90 | 36 | 72 | 161 | 380 | 648 | 46 | 36 |
| Ritchie | 381 | 2.1 | 39,828 | 2,073 | 156 | 29 | 13 | 26 | 71 | 118 | 223 | 16 | 13 |
| Roane | 501 | 1.1 | 36,608 | 2,494 | 127 | 24 | 10 | 23 | 63 | 178 | 184 | 17 | 10 |
| Summers | 439 | 3.1 | 34,933 | 2,695 | 92 | 19 | 9 | 16 | 60 | 187 | 136 | 14 | 9 |
| Taylor | 696 | 1.6 | 41,663 | 1,822 | 161 | 31 | 13 | 24 | 92 | 181 | 228 | 19 | 13 |
| Tucker | 295 | 1.4 | 43,069 | 1,606 | 98 | 21 | 8 | 18 | 48 | 95 | 145 | 11 | 8 |
| Tyler | 342 | 2.1 | 39,863 | 2,067 | 107 | 24 | 8 | 11 | 81 | 100 | 150 | 12 | 8 |
| Upshur | 856 | 5.3 | 35,420 | 2,633 | 333 | 61 | 26 | 49 | 127 | 260 | 470 | 35 | 26 |
| Wayne | 1,372 | 1.5 | 34,811 | 2,714 | 414 | 102 | 35 | 90 | 166 | 410 | 641 | 51 | 35 |
| Webster | 238 | 2.0 | 29,348 | 3,055 | 63 | 14 | 5 | 7 | 37 | 114 | 89 | 9 | 5 |
| Wetzel | 556 | -2.4 | 36,874 | 2,459 | 181 | 36 | 15 | 21 | 106 | 204 | 252 | 20 | 15 |
| Wirt | 190 | 2.4 | 32,680 | 2,909 | 21 | 5 | 2 | 7 | 22 | 60 | 35 | 4 | 2 |
| Wood | 3,738 | 0.8 | 44,754 | 1,391 | 1,660 | 330 | 132 | 326 | 581 | 1,043 | 2,449 | 168 | 132 |
| Wyoming | 639 | -0.9 | 31,315 | 2,988 | 185 | 38 | 15 | 16 | 70 | 280 | 253 | 24 | 15 |
| **WISCONSIN** | 309,909 | 3.4 | 53,207 | X | 155,420 | 29,246 | 11,439 | 21,076 | 59,745 | 53,864 | 217,181 | 13,539 | 11,439 |
| Adams | 826 | 4.6 | 40,841 | 1,943 | 187 | 46 | 16 | 41 | 144 | 276 | 290 | 26 | 16 |
| Ashland | 654 | 4.0 | 42,007 | 1,763 | 357 | 82 | 28 | 51 | 112 | 195 | 518 | 34 | 28 |
| Barron | 2,270 | 1.7 | 50,166 | 820 | 974 | 211 | 77 | 131 | 553 | 517 | 1,393 | 92 | 77 |
| Bayfield | 763 | 3.5 | 50,750 | 771 | 161 | 45 | 13 | 48 | 191 | 195 | 267 | 22 | 13 |
| Brown | 14,309 | 3.0 | 54,090 | 541 | 8,992 | 1,600 | 645 | 1,011 | 2,815 | 2,069 | 12,248 | 739 | 645 |
| Buffalo | 621 | 3.0 | 47,630 | 1,066 | 175 | 44 | 14 | 52 | 109 | 137 | 285 | 19 | 14 |
| Burnett | 694 | 3.7 | 45,023 | 1,357 | 182 | 47 | 14 | 48 | 151 | 221 | 292 | 24 | 14 |
| Calumet | 2,648 | 3.5 | 52,859 | 628 | 654 | 129 | 50 | 140 | 424 | 332 | 973 | 63 | 50 |
| Chippewa | 3,072 | 3.8 | 47,507 | 1,076 | 1,147 | 238 | 88 | 300 | 480 | 616 | 1,774 | 115 | 88 |
| Clark | 1,500 | 5.1 | 43,135 | 1,595 | 496 | 115 | 40 | 262 | 269 | 307 | 913 | 53 | 40 |
| Columbia | 3,074 | 4.9 | 53,423 | 581 | 1,067 | 234 | 84 | 257 | 548 | 510 | 1,642 | 106 | 84 |
| Crawford | 696 | 2.8 | 43,124 | 1,597 | 296 | 66 | 23 | 52 | 124 | 183 | 437 | 30 | 23 |
| Dane | 35,027 | 5.0 | 64,071 | 206 | 21,467 | 4,119 | 1,515 | 2,418 | 7,579 | 3,876 | 29,519 | 1,715 | 1,515 |
| Dodge | 4,049 | 3.7 | 46,099 | 1,237 | 1,763 | 343 | 134 | 282 | 651 | 765 | 2,522 | 163 | 134 |
| Door | 1,778 | 2.6 | 64,249 | 202 | 556 | 117 | 45 | 129 | 567 | 357 | 847 | 60 | 45 |

1. Based on the resident population estimated as of July 1 of the year shown.

# Table B. States and Counties — Earnings, Social Security, and Housing

| STATE County | Earnings, 2019 (cont.) — Percent by selected industries | | | | | | | | | Social Security beneficiaries, December 2019 | | Supplemental Security Income recipients, 2019 | Housing units, 2020 | |
|---|---|---|---|---|---|---|---|---|---|---|---|---|---|---|
| | Farm | Mining, quarrying, and extractions | Construction | Manufacturing | Information; professional, scientific, technical services | Retail trade | Finance, insurance, real estate, and leasing | Health care and social assistance | Government | Number | Rate[1] | | Total | Percent change, 2010-2020 |
| | 75 | 76 | 77 | 78 | 79 | 80 | 81 | 82 | 83 | 84 | 85 | 86 | 87 | 88 |
| **WEST VIRGINIA—Cont'd** | | | | | | | | | | | | | | |
| Calhoun | -0.8 | D | 19.5 | 2.5 | D | D | D | D | 21.6 | 2,255 | 319 | 526 | 4,009 | 1.2 |
| Clay | 0.1 | D | 9.2 | D | 1.8 | 6.5 | D | 17.8 | 36.0 | 2,840 | 334 | 648 | 4,682 | 2.3 |
| Doddridge | -1.6 | D | D | D | D | 3.8 | 2.8 | 3.3 | 24.6 | 1,870 | 221 | 187 | 3,904 | -1.0 |
| Fayette | 0.1 | 8.9 | 3.3 | 6.7 | D | 7.9 | 3.7 | 15.9 | 24.7 | 12,435 | 293 | 2,428 | 21,409 | -1.0 |
| Gilmer | -1.3 | D | 1.8 | 9.3 | D | 4.3 | D | D | 56.2 | 1,580 | 200 | 301 | 3,659 | 6.1 |
| Grant | -3.1 | D | 11.7 | 6.6 | D | 7.2 | 3.3 | 12.6 | 24.0 | 3,550 | 309 | 361 | 6,819 | 7.2 |
| Greenbrier | 0.0 | 3.0 | 5.3 | 7.0 | 4.1 | 9.0 | 4.1 | 19.1 | 18.2 | 10,415 | 301 | 1,327 | 19,368 | 2.0 |
| Hampshire | -1.1 | D | 9.3 | 3.1 | D | 8.1 | 6.0 | 7.7 | 12.6 | 6,445 | 279 | 635 | 14,157 | 3.5 |
| Hancock | -0.1 | D | 4.3 | 33.9 | 6.3 | 4.7 | 6.6 | D | 15.3 | 8,685 | 301 | 922 | 14,273 | -1.9 |
| Hardy | -9.0 | 0.0 | 2.5 | 43.2 | D | 6.9 | 3.7 | 13.9 | 24.9 | 3,845 | 278 | 414 | 8,441 | 4.5 |
| Harrison | -0.1 | 4.9 | 9.2 | 4.6 | 9.9 | 5.8 | 3.7 | D | 11.3 | 16,940 | 252 | 2,634 | 31,757 | 1.0 |
| Jackson | -0.6 | D | 24.0 | 22.2 | 6.0 | 6.3 | 4.6 | D | 13.9 | 8,070 | 282 | 1,152 | 13,381 | 0.6 |
| Jefferson | 0.3 | D | 4.6 | 4.4 | 7.8 | 5.2 | 3.9 | 8.1 | 34.2 | 11,530 | 201 | 850 | 23,728 | 7.7 |
| Kanawha | 0.0 | 3.1 | 5.4 | 4.4 | 12.6 | 5.7 | 8.8 | 19.0 | 18.6 | 50,225 | 282 | 6,697 | 92,175 | -0.5 |
| Lewis | 0.1 | 19.3 | 16.7 | 2.0 | D | 5.6 | 1.3 | D | 15.8 | 4,720 | 297 | 816 | 7,960 | 0.1 |
| Lincoln | | 6.4 | 9.4 | 0.6 | 4.8 | 7.1 | D | 15.8 | 35.4 | 5,940 | 290 | 1,563 | 10,003 | 1.2 |
| Logan | 0.0 | D | 1.7 | 3.8 | 3.0 | 8.7 | 2.2 | 16.5 | 18.0 | 10,395 | 325 | 2,138 | 16,881 | 0.8 |
| McDowell | 0.0 | 33.9 | 1.8 | 0.8 | D | 4.2 | 2.7 | 6.2 | 33.0 | 6,500 | 369 | 2,504 | 11,163 | -1.4 |
| Marion | -0.2 | D | 5.6 | 6.0 | 9.8 | 3.9 | 1.8 | D | 9.7 | 14,635 | 261 | 1,950 | 26,389 | -0.3 |
| Marshall | -0.6 | 23.4 | 26.9 | 11.1 | 2.0 | 4.5 | 2.7 | 14.8 | 19.7 | 7,775 | 254 | 821 | 15,694 | -1.5 |
| Mason | 5.0 | D | D | 12.0 | 2.8 | 10.7 | 3.3 | 17.6 | 23.5 | 7,580 | 285 | 1,227 | 13,092 | 0.6 |
| Mercer | 0.8 | 1.8 | 4.7 | 5.2 | 5.0 | 6.3 | D | 11.7 | 18.0 | 17,730 | 301 | 3,558 | 29,998 | -0.4 |
| Mineral | -0.4 | D | 5.6 | 33.4 | D | 6.3 | 2.4 | D | 16.9 | 7,110 | 264 | 748 | 13,141 | 0.8 |
| Mingo | 0.0 | 32.9 | 3.5 | 1.6 | D | 3.7 | 2.4 | D | 16.9 | 7,750 | 331 | 2,565 | 12,917 | 1.7 |
| Monongalia | 0.0 | 1.3 | 5.0 | 7.6 | 9.0 | 5.1 | 3.1 | 23.9 | 25.0 | 14,960 | 141 | 1,582 | 45,212 | 4.6 |
| Monroe | 1.6 | 0.0 | 7.8 | 22.7 | D | D | 5.5 | 6.7 | 35.5 | 4,010 | 302 | 430 | 7,644 | 0.6 |
| Morgan | 0.7 | D | 7.6 | 6.8 | D | 7.9 | 5.5 | 21.8 | 26.3 | 4,970 | 278 | 329 | 10,064 | 3.2 |
| Nicholas | -0.1 | 9.3 | 4.0 | 12.3 | D | 11.8 | 2.3 | 22.5 | 9.7 | 8,000 | 326 | 1,159 | 13,163 | 0.8 |
| Ohio | -0.1 | D | 3.5 | 3.7 | 24.7 | 5.3 | 7.1 | 19.3 | 9.7 | 11,810 | 284 | 1,289 | 20,768 | -1.9 |
| Pendleton | -8.5 | D | 5.2 | 7.1 | D | 6.3 | D | D | 24.5 | 2,360 | 339 | 163 | 5,238 | 2.1 |
| Pleasants | -0.6 | D | 9.3 | 27.5 | D | 2.9 | 2.8 | D | 14.6 | 1,955 | 264 | 194 | 3,414 | 0.7 |
| Pocahontas | 0.9 | D | 4.3 | 7.2 | 8.3 | 5.9 | 3.3 | D | 38.8 | 2,615 | 316 | 262 | 8,949 | 1.1 |
| Preston | -0.1 | 0.8 | 16.6 | 6.4 | 3.3 | 6.0 | 3.3 | 10.5 | 8.9 | 8,390 | 251 | 1,049 | 15,175 | 0.5 |
| Putnam | 0.3 | 0.2 | 16.5 | 13.5 | 7.1 | 5.5 | 5.2 | 10.5 | 8.9 | 13,870 | 245 | 969 | 24,375 | 4.0 |
| Raleigh | 0.0 | 14.5 | 6.5 | 3.0 | 4.5 | 8.2 | 3.1 | 18.3 | 19.3 | 22,270 | 302 | 3,157 | 36,108 | 0.5 |
| Randolph | 0.5 | 5.1 | 4.9 | 9.7 | 3.5 | 8.6 | 4.2 | D | 18.4 | 7,815 | 273 | 1,147 | 14,284 | 0.7 |
| Ritchie | 0.0 | 31.6 | 5.1 | 21.4 | D | 4.2 | 4.3 | 4.3 | 11.7 | 2,940 | 306 | 457 | 5,961 | 2.1 |
| Roane | -0.7 | D | 10.3 | 5.6 | 3.7 | 9.3 | 6.9 | 20.8 | 17.0 | 4,550 | 333 | 873 | 7,421 | 0.9 |
| Summers | -0.8 | D | 5.8 | D | 3.1 | 7.0 | 4.5 | D | 25.6 | 3,650 | 291 | 686 | 7,719 | 0.5 |
| Taylor | -0.4 | D | 8.1 | D | 2.0 | 7.2 | D | 11.2 | 21.7 | 4,105 | 246 | 621 | 7,502 | -0.4 |
| Tucker | 1.8 | D | D | 12.3 | D | 4.5 | 3.4 | 21.7 | 22.0 | 2,140 | 311 | 204 | 5,360 | 0.3 |
| Tyler | -0.3 | D | D | 40.6 | D | 3.3 | D | 5.0 | 15.6 | 2,585 | 300 | 289 | 5,042 | 0.7 |
| Upshur | -0.4 | 3.3 | 11.4 | 10.9 | 5.3 | 9.5 | 3.8 | D | 41.4 | 6,435 | 265 | 992 | 11,488 | 3.5 |
| Wayne | 0.1 | D | 6.9 | 11.8 | 2.1 | 4.5 | D | D | D | 10,730 | 271 | 1,818 | 19,371 | 0.6 |
| Webster | 0.4 | D | 4.6 | 10.6 | D | 5.7 | D | 11.6 | 35.5 | 2,810 | 344 | 638 | 5,460 | 0.6 |
| Wetzel | -0.2 | D | 9.8 | 3.5 | D | 11.0 | 3.8 | D | 26.3 | 4,340 | 287 | 679 | 8,243 | 1.0 |
| Wirt | -4.2 | 1.7 | 14.4 | D | D | 8.1 | 2.3 | 9.9 | 41.3 | 1,785 | 309 | 320 | 3,338 | 3.4 |
| Wood | -0.2 | D | 5.5 | 10.7 | D | 8.5 | 6.4 | 15.7 | 21.4 | 23,435 | 280 | 3,224 | 40,305 | 0.2 |
| Wyoming | 0.0 | 25.7 | 4.1 | 1.6 | 5.4 | 6.5 | 2.0 | D | 24.1 | 6,810 | 333 | 1,506 | 10,881 | *-0.7 |
| **WISCONSIN** | 1.2 | 0.2 | 6.3 | 17.7 | 8.6 | 5.7 | 7.7 | 12.7 | 14.2 | 1,257,850 | 216 | 116,794 | 2,738,749 | 4.4 |
| Adams | 7.3 | 0.0 | 8.1 | 7.1 | D | 5.6 | D | D | 28.1 | 7,595 | 375 | 395 | 17,965 | 3.1 |
| Ashland | 0.8 | D | 7.1 | 14.4 | 3.3 | 7.5 | 2.9 | D | 22.8 | 4,425 | 285 | 412 | 9,642 | -0.1 |
| Barron | 2.8 | 1.2 | 5.7 | 25.8 | 2.5 | 8.0 | 2.9 | 14.9 | 17.3 | 12,895 | 285 | 928 | 24,547 | 3.9 |
| Bayfield | 0.8 | 0.2 | 17.9 | 4.9 | D | 5.5 | 1.9 | D | 33.8 | 5,225 | 344 | 236 | 13,577 | 4.5 |
| Brown | 0.8 | 0.1 | 5.5 | 16.6 | 7.3 | 4.9 | 10.0 | 14.2 | 11.5 | 51,265 | 194 | 4,693 | 111,976 | 7.3 |
| Buffalo | 6.7 | D | 7.2 | 5.1 | D | 3.0 | 3.9 | 3.8 | 20.5 | 3,485 | 268 | 180 | 6,875 | 3.2 |
| Burnett | 2.7 | D | 6.6 | 22.8 | D | 6.5 | 2.1 | D | 28.2 | 5,845 | 380 | 257 | 15,873 | 3.8 |
| Calumet | 5.7 | D | 5.9 | 30.4 | D | 7.5 | 4.6 | 9.4 | 10.5 | 9,030 | 180 | 224 | 21,253 | 7.9 |
| Chippewa | 2.2 | 0.3 | 11.9 | 24.0 | 3.8 | 10.0 | 2.5 | D | 13.7 | 15,065 | 233 | 1,124 | 29,112 | 7.1 |
| Clark | 12.2 | D | 9.6 | 28.2 | 1.6 | 5.3 | D | D | 14.8 | 7,170 | 206 | 480 | 15,264 | 1.3 |
| Columbia | 1.4 | D | 6.2 | 27.0 | 3.0 | 6.5 | 3.2 | 11.1 | 15.7 | 13,410 | 233 | 796 | 27,179 | 4.0 |
| Crawford | 3.7 | D | 4.3 | 24.2 | D | 12.8 | 3.0 | D | 15.6 | 4,675 | 290 | 303 | 9,038 | 2.7 |
| Dane | 0.4 | 0.1 | 6.0 | 7.3 | 18.3 | 5.2 | 9.1 | 10.6 | 21.0 | 89,845 | 164 | 7,454 | 242,549 | 12.3 |
| Dodge | 2.2 | D | 10.1 | 30.6 | 2.3 | 4.9 | 2.6 | 10.1 | 13.3 | 19,540 | 223 | 834 | 38,241 | 3.4 |
| Door | 2.3 | D | 9.1 | 18.1 | D | 9.5 | 4.2 | 12.1 | 15.0 | 9,970 | 359 | 264 | 25,328 | 5.7 |

1. Per 1,000 resident population estimated as of July 1 of the year shown.

# Table B. States and Counties — Housing, Labor Force, and Employment

| STATE County | Housing units, 2015-2019 | | | | | | | | Civilian labor force, 2020 | | | | Civilian employment[6], 2015-2019 | | |
|---|---|---|---|---|---|---|---|---|---|---|---|---|---|---|---|
| | Occupied units | | | | | | | Sub-standard units[4] (percent) | Total | Percent change, 2019-2020 | Unemployment | | Total | Percent | |
| | Owner-occupied | | | | | Renter-occupied | | | | | | | | | |
| | Total | Percent | Median value[1] | With a mortgage | Without a mortgage[2] | Median rent[3] | Median rent as a percent of income[2] | | | | Total | Rate[5] | | Management, business, science, and arts | Construction, production, and maintenance occupations |
| | 89 | 90 | 91 | 92 | 93 | 94 | 95 | 96 | 97 | 98 | 99 | 100 | 101 | 102 | 103 |
| **WEST VIRGINIA—Cont'd** | | | | | | | | | | | | | | | |
| Calhoun | 2,826 | 78.9 | 83,500 | 18.8 | 10.0 | 611 | 23.2 | 1.1 | 2,535 | -1.2 | 407 | 16.1 | 2,115 | 30.7 | 35.5 |
| Clay | 3,274 | 81.4 | 80,200 | 20.6 | 10.0 | 495 | 30.0 | 2.4 | 3,153 | -0.1 | 394 | 12.5 | 2,559 | 26.3 | 40.1 |
| Doddridge | 2,685 | 87.7 | 117,300 | 17.5 | 10.0 | 455 | 24.9 | 1.3 | 3,847 | -4.1 | 242 | 6.3 | 3,072 | 26.2 | 30.4 |
| Fayette | 17,441 | 77.3 | 91,300 | 19.0 | 10.6 | 639 | 31.2 | 3.0 | 16,264 | -0.1 | 1,622 | 10.0 | 15,411 | 30.2 | 26.6 |
| Gilmer | 2,587 | 68.9 | 93,300 | 14.1 | 10.0 | 584 | 29.2 | 3.2 | 2,342 | -1.6 | 236 | 10.1 | 2,704 | 26.0 | 32.5 |
| Grant | 4,607 | 81.5 | 140,000 | 18.6 | 10.0 | 557 | 24.9 | 1.5 | 5,969 | -0.2 | 389 | 6.5 | 4,939 | 27.1 | 36.8 |
| Greenbrier | 15,188 | 72.5 | 124,000 | 20.3 | 10.0 | 681 | 28.4 | 1.4 | 15,680 | -2.2 | 1,264 | 8.1 | 13,906 | 32.0 | 25.8 |
| Hampshire | 9,288 | 76.3 | 153,100 | 22.2 | 10.0 | 655 | 28.0 | 4.0 | 10,847 | 1.4 | 594 | 5.5 | 9,327 | 26.1 | 38.8 |
| Hancock | 12,678 | 72.7 | 93,000 | 16.9 | 10.0 | 658 | 25.9 | 1.4 | 13,131 | 0.1 | 1,317 | 10.0 | 13,142 | 28.5 | 29.8 |
| Hardy | 5,674 | 73.8 | 137,700 | 18.7 | 10.0 | 755 | 22.2 | 0.9 | 5,865 | 0.8 | 420 | 7.2 | 6,367 | 25.4 | 35.0 |
| Harrison | 27,098 | 73.9 | 117,900 | 16.2 | 10.0 | 767 | 27.8 | 2.1 | 34,090 | -1.6 | 2,709 | 7.9 | 30,358 | 35.6 | 23.3 |
| Jackson | 11,334 | 76.7 | 130,600 | 16.0 | 10.1 | 650 | 24.1 | 2.2 | 12,780 | -4.5 | 1,096 | 8.6 | 11,092 | 33.1 | 31.4 |
| Jefferson | 20,891 | 76.4 | 252,300 | 19.6 | 10.0 | 998 | 28.4 | 1.8 | 30,084 | -1.8 | 1,786 | 5.9 | 27,683 | 42.1 | 20.7 |
| Kanawha | 79,070 | 69.1 | 112,500 | 17.5 | 10.1 | 747 | 28.0 | 1.5 | 82,800 | 0.0 | 7,369 | 8.9 | 78,433 | 38.5 | 16.9 |
| Lewis | 6,574 | 71.5 | 104,200 | 19.7 | 10.0 | 648 | 27.1 | 2.7 | 6,337 | -7.1 | 688 | 10.9 | 6,178 | 25.6 | 31.8 |
| Lincoln | 8,208 | 77.8 | 84,200 | 17.7 | 10.0 | 618 | 35.5 | 2.9 | 7,293 | 0.9 | 729 | 10.0 | 7,145 | 25.4 | 30.7 |
| Logan | 13,816 | 72.6 | 89,100 | 19.4 | 11.3 | 662 | 26.5 | 2.4 | 11,181 | -0.1 | 1,361 | 12.2 | 9,827 | 29.1 | 28.6 |
| McDowell | 7,607 | 78.7 | 35,000 | 22.3 | 11.3 | 575 | 34.5 | 2.4 | 4,478 | 0.2 | 564 | 12.6 | 3,782 | 24.1 | 31.0 |
| Marion | 22,926 | 74.6 | 116,800 | 16.5 | 10.2 | 765 | 28.4 | 2.0 | 25,363 | -1.6 | 2,195 | 8.7 | 24,737 | 33.2 | 25.9 |
| Marshall | 12,308 | 80.1 | 109,300 | 14.9 | 10.4 | 630 | 27.5 | 1.3 | 13,610 | -4.1 | 1,350 | 9.9 | 12,758 | 30.2 | 30.0 |
| Mason | 11,034 | 79.5 | 87,600 | 17.6 | 10.0 | 615 | 23.9 | 1.5 | 10,356 | 0.4 | 877 | 8.5 | 10,057 | 25.1 | 33.1 |
| Mercer | 25,216 | 69.5 | 96,700 | 19.6 | 10.0 | 672 | 31.1 | 1.4 | 20,957 | -0.3 | 1,968 | 9.4 | 22,130 | 31.3 | 23.7 |
| Mineral | 10,916 | 76.0 | 148,200 | 18.1 | 10.0 | 671 | 25.9 | 0.2 | 12,249 | -3.2 | 926 | 7.6 | 11,596 | 28.5 | 31.1 |
| Mingo | 10,501 | 73.5 | 78,100 | 19.2 | 11.6 | 574 | 32.0 | 1.7 | 6,862 | -0.8 | 978 | 14.3 | 7,100 | 27.0 | 33.5 |
| Monongalia | 39,466 | 56.8 | 204,400 | 17.3 | 10.0 | 834 | 31.7 | 2.6 | 54,613 | -0.5 | 3,467 | 6.3 | 51,107 | 47.6 | 14.5 |
| Monroe | 5,718 | 78.7 | 107,900 | 18.8 | 10.0 | 626 | 26.0 | 2.4 | 6,079 | 1.1 | 431 | 7.1 | 5,062 | 26.8 | 37.3 |
| Morgan | 7,185 | 83.3 | 183,200 | 20.8 | 10.0 | 732 | 31.0 | 1.7 | 8,362 | 1.6 | 505 | 6.0 | 7,914 | 24.2 | 30.7 |
| Nicholas | 10,069 | 79.4 | 95,100 | 18.3 | 10.7 | 601 | 31.7 | 2.0 | 9,571 | 0.8 | 984 | 10.3 | 9,550 | 26.2 | 27.8 |
| Ohio | 17,193 | 69.8 | 123,900 | 15.8 | 10.0 | 667 | 30.2 | 1.0 | 20,439 | -3.8 | 1,724 | 8.4 | 19,424 | 35.2 | 20.5 |
| Pendleton | 3,174 | 80.5 | 122,800 | 25.0 | 10.0 | 513 | 18.3 | 1.4 | 3,710 | 0.1 | 181 | 4.9 | 2,942 | 19.9 | 41.2 |
| Pleasants | 2,835 | 82.5 | 106,600 | 16.5 | 10.0 | 572 | 24.0 | 1.2 | 2,720 | -6.8 | 303 | 11.1 | 3,003 | 27.0 | 35.8 |
| Pocahontas | 3,530 | 81.6 | 121,700 | 19.2 | 10.0 | 622 | 25.4 | 1.6 | 3,827 | -2.1 | 354 | 9.3 | 3,355 | 28.5 | 30.3 |
| Preston | 12,429 | 82.9 | 115,300 | 16.3 | 10.0 | 666 | 27.1 | 1.7 | 15,525 | -0.4 | 1,131 | 7.3 | 13,301 | 29.2 | 33.9 |
| Putnam | 21,613 | 81.8 | 167,300 | 18.3 | 10.0 | 831 | 25.9 | 1.5 | 27,080 | 0.3 | 2,005 | 7.4 | 25,385 | 40.9 | 21.5 |
| Raleigh | 31,203 | 73.4 | 108,700 | 18.2 | 10.0 | 693 | 28.9 | 1.5 | 30,524 | -0.2 | 2,724 | 8.9 | 28,557 | 31.9 | 20.5 |
| Randolph | 11,135 | 71.5 | 110,300 | 17.3 | 10.0 | 655 | 27.2 | 1.5 | 12,364 | -0.5 | 1,134 | 9.2 | 11,292 | 28.5 | 28.8 |
| Ritchie | 4,049 | 79.8 | 96,700 | 17.1 | 10.0 | 608 | 22.9 | 2.2 | 4,399 | -0.3 | 368 | 8.4 | 3,541 | 22.7 | 39.0 |
| Roane | 5,562 | 79.3 | 98,900 | 19.5 | 10.0 | 510 | 29.3 | 2.2 | 5,128 | -0.2 | 638 | 12.4 | 4,584 | 34.0 | 32.4 |
| Summers | 5,566 | 76.2 | 92,200 | 18.4 | 10.0 | 685 | 33.4 | 4.3 | 4,376 | 0.4 | 343 | 7.8 | 4,279 | 30.3 | 30.1 |
| Taylor | 6,614 | 78.0 | 107,000 | 17.7 | 10.0 | 652 | 26.3 | 0.8 | 7,816 | -1.5 | 576 | 7.4 | 6,718 | 30.5 | 28.5 |
| Tucker | 3,142 | 78.5 | 121,600 | 18.6 | 10.0 | 552 | 27.0 | 1.5 | 3,292 | -1.9 | 279 | 8.5 | 3,064 | 31.9 | 30.3 |
| Tyler | 3,207 | 82.7 | 96,300 | 17.6 | 10.0 | 664 | 28.5 | 0.7 | 3,173 | 0.1 | 328 | 10.3 | 3,127 | 28.6 | 34.1 |
| Upshur | 9,713 | 73.7 | 119,000 | 19.1 | 10.0 | 699 | 30.1 | 1.7 | 9,650 | -0.2 | 924 | 9.6 | 9,959 | 32.6 | 29.4 |
| Wayne | 15,124 | 73.5 | 99,100 | 18.6 | 11.3 | 662 | 29.6 | 2.0 | 15,838 | 0.2 | 1,328 | 8.4 | 13,322 | 31.1 | 25.4 |
| Webster | 3,781 | 72.6 | 69,500 | 19.6 | 10.0 | 573 | 35.3 | 0.4 | 3,328 | 0.2 | 312 | 9.4 | 2,943 | 24.9 | 38.3 |
| Wetzel | 5,762 | 79.4 | 103,600 | 15.4 | 10.0 | 703 | 28.2 | 1.0 | 6,747 | 1.6 | 664 | 9.8 | 5,518 | 20.6 | 39.3 |
| Wirt | 2,505 | 83.7 | 95,800 | 17.2 | 10.0 | 454 | 28.0 | 1.8 | 2,282 | -1.6 | 245 | 10.7 | 2,228 | 33.8 | 30.3 |
| Wood | 35,488 | 71.5 | 124,200 | 17.9 | 10.0 | 713 | 29.8 | 1.4 | 36,445 | -1.2 | 3,224 | 8.8 | 35,794 | 35.0 | 22.5 |
| Wyoming | 8,805 | 84.2 | 71,500 | 17.2 | 10.0 | 679 | 29.4 | 3.1 | 6,591 | -1.3 | 690 | 10.5 | 6,054 | 32.1 | 33.6 |
| **WISCONSIN** | 2,358,156 | 67.0 | 180,600 | 20.0 | 12.3 | 856 | 27.4 | 2.0 | 3,065,402 | -0.9 | 192,793 | 6.3 | 2,982,359 | 36.6 | 26.2 |
| Adams | 9,062 | 83.3 | 131,100 | 24.8 | 13.8 | 639 | 26.1 | 2.1 | 8,047 | -0.9 | 766 | 9.5 | 7,772 | 22.9 | 34.4 |
| Ashland | 6,565 | 68.3 | 113,100 | 22.1 | 12.8 | 631 | 29.1 | 2.3 | 7,686 | -2.0 | 651 | 8.5 | 7,242 | 28.3 | 27.9 |
| Barron | 18,907 | 74.7 | 153,200 | 21.0 | 12.5 | 705 | 27.1 | 2.7 | 23,873 | -1.7 | 1,460 | 6.1 | 22,117 | 29.2 | 34.8 |
| Bayfield | 7,057 | 83.2 | 170,200 | 20.9 | 12.4 | 653 | 25.9 | 3.7 | 7,399 | -1.8 | 688 | 9.3 | 7,015 | 34.8 | 25.3 |
| Brown | 105,271 | 64.8 | 173,900 | 18.8 | 11.2 | 795 | 25.5 | 2.3 | 140,454 | -0.4 | 8,441 | 6.0 | 139,043 | 35.3 | 26.5 |
| Buffalo | 5,772 | 76.4 | 159,700 | 21.9 | 13.3 | 765 | 25.4 | 1.7 | 6,336 | -1.3 | 471 | 7.4 | 6,841 | 29.9 | 34.1 |
| Burnett | 7,305 | 82.2 | 158,800 | 22.0 | 12.1 | 752 | 24.4 | 2.4 | 7,157 | 0.8 | 648 | 9.1 | 6,708 | 33.0 | 29.0 |
| Calumet | 19,807 | 80.6 | 178,900 | 18.4 | 10.4 | 779 | 25.7 | 1.6 | 27,096 | -1.0 | 1,271 | 4.7 | 27,630 | 36.7 | 29.4 |
| Chippewa | 25,601 | 73.1 | 167,400 | 19.9 | 11.9 | 821 | 27.1 | 1.4 | 32,978 | -0.8 | 2,048 | 6.2 | 31,847 | 31.4 | 32.8 |
| Clark | 12,791 | 78.4 | 127,200 | 20.5 | 12.4 | 623 | 21.6 | 5.3 | 17,401 | -1.4 | 786 | 4.5 | 16,043 | 29.1 | 41.0 |
| Columbia | 24,112 | 74.4 | 192,900 | 20.5 | 12.6 | 801 | 24.6 | 1.6 | 31,281 | -0.9 | 1,843 | 5.9 | 30,469 | 34.3 | 29.7 |
| Crawford | 6,654 | 75.7 | 134,200 | 20.7 | 12.3 | 639 | 25.6 | 3.0 | 7,453 | -1.3 | 545 | 7.3 | 7,412 | 30.3 | 36.6 |
| Dane | 222,929 | 58.2 | 265,600 | 20.5 | 11.8 | 1,083 | 28.4 | 2.0 | 319,433 | -1.1 | 15,279 | 4.8 | 308,106 | 51.5 | 14.3 |
| Dodge | 34,980 | 69.7 | 162,100 | 20.7 | 12.2 | 810 | 24.9 | 1.5 | 47,213 | -0.1 | 2,484 | 5.3 | 45,274 | 27.3 | 37.8 |
| Door | 13,191 | 79.9 | 214,100 | 21.1 | 12.0 | 795 | 26.2 | 2.0 | 14,905 | -2.3 | 1,019 | 6.8 | 13,759 | 33.3 | 28.7 |

1. Specified owner-occupied units. 2. A value of 10.0 represents 10 percent or less; a value of 50.0 represents 50 percent or more. 3. Specified renter-occupied units. 4. Overcrowded or lacking complete plumbing facilities. 5. Percent of civilian labor force. 6. Civilian employed persons 16 years old and over.

# Table B. States and Counties — Nonfarm Employment and Agriculture

| | Private nonfarm establishments, employment and payroll, 2019 | | | | | | | | | Agriculture, 2017 | | | |
| | Employment | | | | | | | Annual payroll | | Farms | | | Farm producers whose primary occupation is farming (percent) |
| STATE County | Number of establish-ments | Total | Health care and social assistance | Manufac-turing | Retail trade | Finance and insurance | Professional, scientific, and technical services | Total (mil dol) | Average per employee (dollars) | Number | Percent with: Fewer than 50 acres | 1000 acres or more | |
| | 104 | 105 | 106 | 107 | 108 | 109 | 110 | 111 | 112 | 113 | 114 | 115 | 116 |
|---|---|---|---|---|---|---|---|---|---|---|---|---|---|
| **WEST VIRGINIA—Cont'd** | | | | | | | | | | | | | |
| Calhoun | 89 | 743 | 246 | 36 | 104 | 39 | 31 | 35 | 46,647 | 296 | 18.9 | 2.4 | 33.6 |
| Clay | 69 | 830 | 306 | NA | 117 | 29 | NA | 26 | 30,876 | 131 | 23.7 | 0.8 | 37.0 |
| Doddridge | 68 | 845 | 107 | NA | 86 | 49 | NA | 56 | 66,689 | 392 | 28.6 | 1.8 | 36.1 |
| Fayette | 719 | 8,231 | 1,940 | 552 | 1,334 | 186 | 175 | 313 | 38,069 | 253 | 49.8 | 0.4 | 34.2 |
| Gilmer | 117 | 956 | 192 | 191 | 150 | 16 | 27 | 32 | 32,976 | 264 | 8.3 | 0.4 | 37.1 |
| Grant | 229 | 2,524 | 797 | 205 | 355 | 100 | 39 | 100 | 39,504 | 522 | 30.5 | 2.9 | 34.9 |
| Greenbrier | 890 | 10,672 | 2,522 | 874 | 1,795 | 246 | 248 | 382 | 35,824 | 891 | 31.9 | 2.8 | 39.6 |
| Hampshire | 319 | 2,591 | 860 | 99 | 472 | 183 | 76 | 79 | 30,586 | 883 | 51.4 | 2.5 | 35.0 |
| Hancock | 564 | 8,533 | 1,242 | 2,534 | 801 | 235 | 234 | 308 | 36,122 | 93 | 43.0 | 1.1 | 35.3 |
| Hardy | 266 | 4,898 | 474 | 2,634 | 600 | 205 | 82 | 170 | 34,676 | 580 | 34.5 | 5.3 | 39.3 |
| Harrison | 1,803 | 31,079 | 6,764 | 1,575 | 4,612 | 1,000 | 2,097 | 1,459 | 46,959 | 810 | 33.2 | 1.1 | 35.2 |
| Jackson | 492 | 6,964 | 1,147 | 1,685 | 1,148 | 160 | 277 | 316 | 45,438 | 982 | 31.4 | 0.6 | 31.8 |
| Jefferson | 889 | 12,762 | 1,248 | 859 | 1,966 | 318 | 358 | 450 | 35,236 | 607 | 63.9 | 1.8 | 39.7 |
| Kanawha | 4,636 | 79,541 | 19,622 | 3,389 | 10,467 | 3,441 | 5,195 | 3,768 | 47,368 | 214 | 34.6 | 0.5 | 36.3 |
| Lewis | 342 | 4,262 | 750 | 148 | 944 | 58 | 67 | 173 | 40,505 | 481 | 21.2 | 0.8 | 41.8 |
| Lincoln | 166 | 1,648 | 808 | NA | 253 | 54 | 40 | 46 | 28,135 | 177 | 17.5 | NA | 42.4 |
| Logan | 592 | 9,029 | 2,041 | 336 | 1,610 | 183 | 186 | 407 | 45,111 | 8 | 50.0 | NA | 15.4 |
| McDowell | 218 | 1,671 | 430 | 8 | 361 | 109 | 31 | 65 | 38,832 | 14 | 100.0 | NA | 77.8 |
| Marion | 1,135 | 14,894 | 2,695 | 822 | 2,172 | 427 | 1,175 | 654 | 43,878 | 599 | 41.6 | NA | 39.5 |
| Marshall | 493 | 8,851 | 1,430 | 513 | 1,169 | 222 | 117 | 441 | 49,852 | 638 | 28.7 | 0.2 | 36.9 |
| Mason | 325 | 4,023 | 1,053 | 503 | 638 | 104 | 79 | 178 | 44,254 | 876 | 32.9 | 1.5 | 35.9 |
| Mercer | 1,188 | 17,233 | 4,514 | 1,228 | 3,089 | 327 | 496 | 648 | 37,582 | 410 | 36.3 | 0.7 | 38.4 |
| Mineral | 437 | 6,187 | 1,508 | 1,472 | 999 | 146 | 127 | 267 | 43,129 | 519 | 35.6 | 2.3 | 35.9 |
| Mingo | 347 | 3,183 | 607 | 105 | 373 | 139 | 145 | 123 | 38,544 | 8 | 62.5 | NA | 40.0 |
| Monongalia | 2,340 | 47,192 | 14,948 | 3,796 | 6,255 | 730 | 2,722 | 2,435 | 51,593 | 542 | 36.2 | 0.2 | 36.7 |
| Monroe | 171 | 1,292 | 191 | 516 | 122 | 32 | 52 | 52 | 40,163 | 929 | 38.2 | 1.6 | 38.4 |
| Morgan | 224 | 2,064 | 574 | 97 | 467 | 85 | 83 | 69 | 33,313 | 207 | 50.7 | NA | 30.1 |
| Nicholas | 556 | 6,567 | 1,898 | 748 | 1,398 | 116 | 129 | 266 | 40,543 | 372 | 43.8 | 1.3 | 36.4 |
| Ohio | 1,343 | 27,226 | 6,600 | 986 | 3,071 | 1,465 | 1,885 | 1,155 | 42,432 | 208 | 28.4 | NA | 39.1 |
| Pendleton | 138 | 1,134 | 356 | 154 | 177 | 61 | 19 | 34 | 30,016 | 584 | 19.3 | 5.5 | 42.7 |
| Pleasants | 122 | 2,228 | 235 | 565 | 95 | 58 | 37 | 113 | 50,917 | 208 | 27.9 | 1.0 | 26.5 |
| Pocahontas | 219 | 3,014 | 401 | 301 | 321 | 37 | 117 | 77 | 25,688 | 500 | 28.4 | 4.4 | 34.2 |
| Preston | 525 | 5,455 | 1,142 | 641 | 885 | 151 | 97 | 209 | 38,355 | 1,142 | 34.1 | 0.6 | 39.6 |
| Putnam | 1,205 | 18,246 | 2,545 | 2,263 | 2,319 | 512 | 738 | 862 | 47,245 | 514 | 32.9 | 0.4 | 30.5 |
| Raleigh | 1,704 | 26,416 | 6,824 | 770 | 4,704 | 518 | 1,486 | 1,039 | 39,326 | 365 | 50.1 | 1.4 | 35.8 |
| Randolph | 650 | 9,018 | 2,750 | 1,078 | 1,276 | 253 | 211 | 298 | 33,080 | 402 | 33.6 | 5.5 | 36.7 |
| Ritchie | 190 | 2,616 | 185 | 868 | 283 | 86 | 51 | 103 | 39,528 | 473 | 23.9 | 3.0 | 38.0 |
| Roane | 223 | 2,191 | 748 | 120 | 463 | 94 | 49 | 81 | 36,957 | 604 | 19.7 | 1.2 | 38.4 |
| Summers | 162 | 1,353 | 387 | 15 | 172 | 54 | 50 | 45 | 33,575 | 357 | 28.0 | 2.5 | 35.6 |
| Taylor | 223 | 2,477 | 708 | 174 | 387 | 50 | 38 | 115 | 46,293 | 413 | 48.4 | 1.2 | 34.7 |
| Tucker | 150 | 1,766 | 290 | 250 | 185 | 39 | 9 | 60 | 34,133 | 159 | 28.3 | 1.3 | 40.1 |
| Tyler | 128 | 1,354 | 330 | 494 | 156 | 57 | 29 | 88 | 65,257 | 305 | 19.0 | 1.0 | 42.7 |
| Upshur | 516 | 6,321 | 1,403 | 810 | 850 | 93 | 216 | 253 | 40,021 | 499 | 37.3 | 1.0 | 36.2 |
| Wayne | 485 | 6,951 | 1,946 | 819 | 1,426 | 93 | 113 | 312 | 44,905 | 237 | 24.1 | 0.4 | 41.8 |
| Webster | 125 | 1,084 | 406 | 249 | 158 | 20 | 2 | 37 | 33,799 | 83 | 31.3 | NA | 57.2 |
| Wetzel | 334 | 4,008 | 752 | 871 | 827 | 116 | 66 | 153 | 38,164 | 261 | 13.0 | 0.4 | 35.5 |
| Wirt | 51 | 273 | 95 | NA | 83 | 16 | 4 | 8 | 29,740 | 256 | 28.5 | 1.2 | 37.7 |
| Wood | 1,943 | 29,749 | 6,801 | 2,408 | 5,803 | 842 | 938 | 1,085 | 36,475 | 881 | 38.4 | 0.2 | 34.5 |
| Wyoming | 268 | 2,294 | 722 | 58 | 535 | 59 | 108 | 73 | 31,731 | 21 | 52.4 | NA | 39.5 |
| **WISCONSIN** | 141,635 | 2,610,712 | 410,987 | 478,852 | 311,087 | 145,180 | 114,850 | 130,985 | 50,172 | 64,793 | 35.3 | 3.6 | 45.6 |
| Adams | 331 | 3,078 | 405 | 435 | 517 | 43 | 52 | 105 | 33,970 | 308 | 31.2 | 7.5 | 50.3 |
| Ashland | 473 | 6,978 | 1,342 | 1,161 | 992 | 182 | 181 | 290 | 41,548 | 263 | 34.6 | 3.0 | 33.9 |
| Barron | 1,322 | 17,571 | 2,937 | 5,276 | 2,793 | 459 | 285 | 685 | 38,983 | 1,200 | 31.6 | 4.7 | 45.9 |
| Bayfield | 424 | 2,191 | 323 | 207 | 379 | 58 | 36 | 73 | 33,147 | 427 | 32.6 | 3.3 | 36.4 |
| Brown | 6,615 | 147,199 | 21,086 | 28,137 | 16,209 | 9,594 | 5,358 | 7,448 | 50,600 | 975 | 52.3 | 3.7 | 45.5 |
| Buffalo | 299 | 3,145 | 298 | 261 | 303 | 130 | 68 | 130 | 41,354 | 966 | 20.5 | 4.9 | 47.8 |
| Burnett | 389 | 3,227 | 650 | 825 | 536 | 83 | 120 | 118 | 36,586 | 369 | 22.2 | 3.8 | 40.2 |
| Calumet | 871 | 12,615 | 998 | 3,794 | 1,758 | 440 | 207 | 525 | 41,591 | 684 | 41.5 | 4.7 | 51.4 |
| Chippewa | 1,595 | 21,856 | 2,996 | 5,937 | 3,548 | 391 | 534 | 945 | 43,255 | 1,409 | 31.3 | 4.7 | 45.3 |
| Clark | 753 | 8,669 | 886 | 3,607 | 908 | 191 | 159 | 348 | 40,187 | 2,095 | 22.5 | 2.2 | 59.5 |
| Columbia | 1,400 | 20,776 | 2,888 | 5,274 | 2,739 | 345 | 502 | 850 | 40,902 | 1,357 | 41.3 | 4.5 | 46.5 |
| Crawford | 370 | 5,775 | 795 | 1,999 | 982 | 136 | 158 | 204 | 35,332 | 1,034 | 24.8 | 2.6 | 39.7 |
| Dane | 14,316 | 299,643 | 52,086 | 26,696 | 32,670 | 24,921 | 27,812 | 17,789 | 59,367 | 2,566 | 44.4 | 3.3 | 43.9 |
| Dodge | 1,725 | 32,302 | 4,667 | 10,230 | 3,317 | 554 | 573 | 1,808 | 55,977 | 1,749 | 35.9 | 3.5 | 48.9 |
| Door | 1,298 | 10,243 | 1,181 | 2,209 | 1,827 | 206 | 237 | 417 | 40,759 | 626 | 41.5 | 2.9 | 43.8 |

# Table B. States and Counties  —  **Agriculture**

| STATE County | Land in farms | | Acres | | | Value of land and buildings (dollars) | | Value of machinery and equiopmnet, average per farm (dollars) | Value of products sold: | | Percent from: | | Organic farms (number) | Farms with internet access (per-cent) | Government payments | |
|---|---|---|---|---|---|---|---|---|---|---|---|---|---|---|---|---|
| | Acreage (1,000) | Percent change, 2012-2017 | Average size of farm | Total irrigated (1,000) | Total cropland (1,000) | Average per farm | Average per acre | | Total (mil dol) | Average per farm (acres) | Crops | Livestock and poultry products | | | Total ($1,000) | Percent of farms |
| | 117 | 118 | 119 | 120 | 121 | 122 | 123 | 124 | 125 | 126 | 127 | 128 | 129 | 130 | 131 | 132 |
| WEST VIRGINIA—Cont'd | | | | | | | | | | | | | | | | |
| Calhoun | 61 | 23.3 | 206 | 0.0 | 12.4 | 375,391 | 1,823 | 62,763 | 2.4 | 8,047 | 39.8 | 60.2 | NA | 67.2 | 82 | 9.1 |
| Clay | 21 | 6.0 | 162 | NA | 3.2 | 279,733 | 1,722 | 39,293 | 0.6 | 4,374 | 45.2 | 54.8 | NA | 74.8 | 20 | 7.6 |
| Doddridge | 67 | 2.9 | 172 | 0.0 | 11.8 | 365,169 | 2,129 | 60,481 | 2.0 | 5,171 | 45.2 | 58.4 | NA | 67.3 | 10 | 1.0 |
| Fayette | 26 | 10.9 | 102 | 0.0 | 6.2 | 295,809 | 2,913 | 45,353 | 1.7 | 6,779 | 37.1 | 62.9 | NA | 81.8 | 48 | 5.5 |
| Gilmer | 65 | -7.6 | 246 | D | 11.1 | 324,619 | 1,317 | 61,671 | 2.8 | 10,538 | 17.3 | 82.7 | NA | 72.7 | 102 | 5.3 |
| Grant | 120 | 6.7 | 230 | 0.0 | 26.2 | 601,970 | 2,622 | 74,196 | 57.1 | 109,316 | 3.0 | 97.0 | NA | 64.8 | 635 | 13.0 |
| Greenbrier | 192 | 1.2 | 216 | 0.0 | 44.1 | 562,728 | 2,606 | 76,127 | 69.3 | 77,797 | 6.0 | 94.0 | 2 | 74.9 | 970 | 18.5 |
| Hampshire | 132 | -7.2 | 149 | 0.1 | 36.9 | 448,587 | 3,004 | 56,173 | 38.7 | 43,847 | 13.1 | 86.9 | NA | 69.4 | 244 | 12.3 |
| Hancock | 8 | -6.1 | 90 | 0.0 | 3.4 | 410,404 | 4,574 | 65,960 | 0.5 | 5,667 | 74.0 | 26.0 | NA | 91.4 | D | 1.1 |
| Hardy | 155 | -0.4 | 267 | 0.0 | 40.5 | 857,054 | 3,213 | 99,840 | 190.6 | 328,583 | 2.8 | 97.2 | 2 | 73.3 | 597 | 14.7 |
| Harrison | 112 | -4.2 | 138 | 0.0 | 28.0 | 346,717 | 2,504 | 51,626 | 7.7 | 9,546 | 33.2 | 66.8 | NA | 78.0 | 142 | 5.7 |
| Jackson | 128 | 22.1 | 130 | 0.0 | 33.4 | 287,864 | 2,211 | 41,845 | 6.9 | 7,059 | 38.5 | 61.5 | NA | 71.4 | 162 | 5.3 |
| Jefferson | 66 | -1.3 | 109 | 0.3 | 44.5 | 639,634 | 5,873 | 57,139 | 28.7 | 47,206 | 62.2 | 37.8 | 7 | 82.9 | 476 | 15.2 |
| Kanawha | 24 | -9.0 | 111 | 0.0 | 4.1 | 429,309 | 3,881 | 38,062 | 1.0 | 4,710 | 39.1 | 60.9 | 1 | 73.8 | 27 | 8.9 |
| Lewis | 100 | 21.2 | 208 | D | 22.4 | 449,021 | 2,160 | 53,052 | 6.3 | 13,035 | 21.3 | 78.7 | NA | 69.6 | 29 | 4.6 |
| Lincoln | 24 | -6.9 | 135 | 0.0 | 4.2 | 248,623 | 1,843 | 54,669 | 0.8 | 4,559 | 55.8 | 44.2 | NA | 73.4 | 23 | 15.8 |
| Logan | 1 | 11.7 | 116 | D | D | 287,031 | 2,474 | 22,010 | 0.0 | 5,125 | 82.9 | 17.1 | NA | 100.0 | D | 12.5 |
| McDowell | 0 | -84.3 | 12 | NA | 0.1 | 72,143 | 6,196 | 38,104 | 0.3 | 21,214 | 98.7 | 1.3 | NA | 100.0 | NA | NA |
| Marion | 52 | -3.4 | 86 | 0.0 | 13.5 | 255,112 | 2,964 | 40,156 | 2.5 | 4,200 | 45.7 | 54.3 | 1 | 65.9 | 22 | 2.8 |
| Marshall | 76 | -11.6 | 119 | 0.0 | 22.6 | 308,256 | 2,587 | 65,533 | 3.8 | 5,966 | 51.1 | 48.9 | 1 | 66.5 | 95 | 3.9 |
| Mason | 125 | -10.1 | 142 | 0.2 | 37.6 | 345,965 | 2,430 | 58,830 | 36.4 | 41,502 | 80.8 | 19.2 | 2 | 65.8 | 829 | 11.0 |
| Mercer | 53 | 2.6 | 130 | 0.0 | 12.4 | 326,001 | 2,515 | 57,803 | 6.9 | 16,763 | 21.5 | 78.5 | NA | 68.8 | 59 | 4.4 |
| Mineral | 99 | 30.0 | 191 | 0.1 | 29.6 | 531,811 | 2,788 | 57,610 | 21.6 | 41,667 | 14.8 | 85.2 | 1 | 66.9 | 134 | 8.1 |
| Mingo | 2 | 16.4 | 295 | D | D | 329,528 | 1,116 | 37,841 | 0.2 | 19,750 | 10.8 | 89.2 | NA | 100.0 | 3 | 50.0 |
| Monongalia | 62 | 7.1 | 115 | 0.0 | 18.6 | 518,881 | 4,530 | 60,065 | 5.0 | 9,273 | 35.1 | 64.9 | 4 | 71.8 | 68 | 5.7 |
| Monroe | 145 | 0.3 | 156 | 0.0 | 31.9 | 415,440 | 2,662 | 51,310 | 22.6 | 24,277 | 15.9 | 84.1 | 15 | 64.2 | 363 | 11.0 |
| Morgan | 17 | -8.4 | 81 | 0.0 | 7.2 | 426,092 | 5,234 | 40,087 | 3.3 | 15,836 | 78.3 | 21.7 | 2 | 69.1 | 148 | 8.7 |
| Nicholas | 46 | -21.6 | 122 | 0.0 | 11.8 | 319,996 | 2,615 | 47,319 | 3.1 | 8,304 | 33.9 | 66.1 | NA | 73.1 | 330 | 11.3 |
| Ohio | 24 | -21.8 | 113 | D | 10.9 | 409,625 | 3,624 | 64,696 | 2.7 | 13,014 | 43.2 | 56.8 | NA | 78.4 | 24 | 7.2 |
| Pendleton | 176 | 3.5 | 302 | 0.0 | 29.6 | 678,517 | 2,250 | 84,763 | 99.9 | 171,074 | 2.7 | 97.3 | 2 | 61.1 | 324 | 12.7 |
| Pleasants | 24 | 11.0 | 115 | 0.0 | 5.3 | 264,777 | 2,308 | 44,817 | 1.1 | 5,524 | 43.4 | 56.6 | NA | 78.8 | D | 1.0 |
| Pocahontas | 131 | 10.9 | 263 | 0.0 | 26.4 | 492,444 | 1,875 | 65,758 | 8.6 | 17,128 | 22.7 | 77.3 | 2 | 67.2 | 372 | 16.4 |
| Preston | 143 | -11.1 | 125 | 0.0 | 48.2 | 343,865 | 2,747 | 58,868 | 16.2 | 14,145 | 38.9 | 61.1 | 2 | 68.1 | 309 | 6.4 |
| Putnam | 52 | -13.9 | 101 | 0.1 | 13.7 | 288,179 | 2,866 | 41,740 | 7.3 | 14,239 | 71.0 | 29.0 | 1 | 63.0 | 126 | 6.0 |
| Raleigh | 44 | 19.4 | 121 | 0.0 | 11.2 | 447,458 | 3,709 | 52,892 | 3.0 | 8,263 | 28.2 | 71.8 | 1 | 70.1 | 12 | 2.5 |
| Randolph | 98 | 3.6 | 243 | 0.0 | 26.7 | 588,534 | 2,425 | 65,505 | 7.9 | 19,704 | 32.0 | 68.0 | 4 | 66.7 | 112 | 8.7 |
| Ritchie | 98 | 9.8 | 206 | 0.0 | 22.6 | 372,079 | 1,805 | 53,870 | 9.6 | 20,334 | 18.4 | 81.6 | 2 | 72.7 | 148 | 3.0 |
| Roane | 113 | 1.9 | 187 | 0.0 | 27.8 | 361,187 | 1,930 | 54,938 | 6.4 | 10,675 | 34.7 | 65.3 | 3 | 72.4 | 118 | 7.0 |
| Summers | 55 | -5.7 | 153 | 0.0 | 12.0 | 347,761 | 2,272 | 46,335 | 3.5 | 9,846 | 38.8 | 61.2 | 2 | 66.9 | 86 | 5.3 |
| Taylor | 46 | -6.9 | 111 | 0.0 | 14.0 | 340,360 | 3,075 | 58,663 | 3.4 | 8,124 | 36.5 | 63.5 | NA | 71.4 | 28 | 2.9 |
| Tucker | 26 | -22.1 | 166 | D | 6.5 | 476,151 | 2,863 | 57,473 | 1.9 | 11,937 | 30.4 | 69.6 | NA | 78.6 | 11 | 3.8 |
| Tyler | 56 | 15.4 | 182 | 0.0 | 12.3 | 348,921 | 1,913 | 51,025 | 2.5 | 8,079 | 40.9 | 59.1 | NA | 67.9 | 82 | 7.9 |
| Upshur | 61 | -10.8 | 122 | 0.1 | 16.3 | 323,626 | 2,645 | 48,952 | 4.8 | 9,615 | 49.6 | 50.4 | 1 | 72.5 | 63 | 3.8 |
| Wayne | 39 | 27.5 | 163 | 0.0 | 5.9 | 326,612 | 2,009 | 46,764 | 1.7 | 7,097 | 44.7 | 55.3 | NA | 71.7 | 56 | 11.0 |
| Webster | 10 | 32.1 | 126 | 0.0 | 2.4 | 290,032 | 2,299 | 40,787 | 0.5 | 5,434 | 55.7 | 44.3 | NA | 65.1 | D | 8.4 |
| Wetzel | 41 | 8.5 | 158 | 0.0 | 8.8 | 305,112 | 1,927 | 45,093 | 1.3 | 4,950 | 54.6 | 45.4 | NA | 60.2 | 11 | 4.2 |
| Wirt | 41 | 9.0 | 162 | 0.0 | 9.3 | 292,717 | 1,808 | 49,820 | 2.3 | 8,949 | 30.2 | 69.8 | NA | 71.5 | 120 | 4.7 |
| Wood | 90 | 2.4 | 102 | 0.0 | 25.0 | 340,345 | 3,333 | 43,873 | 6.3 | 7,120 | 44.4 | 55.6 | NA | 72.3 | 187 | 3.1 |
| Wyoming | 1 | -51.0 | 69 | D | 0.7 | 231,769 | 3,347 | 59,560 | 0.1 | 4,429 | 74.2 | 25.8 | NA | 76.2 | 20 | 19.0 |
| WISCONSIN | 14,319 | -1.7 | 221 | 454.4 | 10,085 | 1,083,640 | 4,904 | 156,689 | 11,427.4 | 176,368 | 35.6 | 64.4 | 1,708 | 76.1 | 126,583 | 42.4 |
| Adams | 117 | -1.0 | 381 | 50.9 | 84.5 | 1,708,212 | 4,489 | 227,089 | D | D | D | D | 6 | 80.8 | 637 | 30.2 |
| Ashland | 52 | 14.4 | 199 | 0.2 | 24.7 | 445,403 | 2,234 | 71,332 | 17.6 | 66,806 | 15.0 | 85.0 | 3 | 74.9 | 273 | 12.5 |
| Barron | 306 | -1.3 | 255 | 15.3 | 208.8 | 870,734 | 3,419 | 170,748 | 266.4 | 222,001 | 28.8 | 71.2 | 6 | 81.7 | 936 | 36.1 |
| Bayfield | 81 | 12.8 | 190 | 0.2 | 42.6 | 415,918 | 2,191 | 75,186 | 15.9 | 37,302 | 51.6 | 48.4 | 9 | 78.5 | 100 | 12.6 |
| Brown | 192 | 6.0 | 197 | 0.5 | 166.6 | 1,672,996 | 8,495 | 205,925 | 292.7 | 300,255 | 13.8 | 86.2 | 17 | 77.7 | 2,847 | 37.4 |
| Buffalo | 293 | -4.0 | 303 | 7.6 | 164.3 | 1,195,520 | 3,940 | 184,155 | 203.1 | 210,203 | 26.7 | 73.3 | 36 | 77.6 | 3,056 | 62.2 |
| Burnett | 89 | 6.7 | 242 | 0.3 | 49.0 | 658,630 | 2,723 | 110,503 | 37.9 | 102,835 | 36.0 | 64.0 | 4 | 71.0 | 360 | 27.6 |
| Calumet | 154 | 8.1 | 225 | D | 131.5 | 1,771,197 | 7,874 | 207,042 | 203.6 | 297,601 | 21.8 | 78.2 | 13 | 81.9 | 924 | 52.5 |
| Chippewa | 356 | -7.4 | 253 | 8.6 | 243.0 | 899,774 | 3,559 | 148,239 | 215.3 | 152,830 | 37.5 | 62.5 | 26 | 74.2 | 2,197 | 42.2 |
| Clark | 451 | -1.6 | 215 | 0.5 | 323.8 | 873,709 | 4,058 | 149,770 | 404.1 | 192,889 | 17.9 | 82.1 | 58 | 54.9 | 1,430 | 27.8 |
| Columbia | 304 | -1.3 | 224 | 5.6 | 243.9 | 1,305,879 | 5,828 | 158,741 | 222.3 | 163,808 | 51.0 | 49.0 | 23 | 78.3 | 1,602 | 36.7 |
| Crawford | 211 | -2.8 | 204 | 0.1 | 99.0 | 647,573 | 3,180 | 82,180 | 73.4 | 70,965 | 43.6 | 56.4 | 50 | 73.6 | 1,616 | 43.3 |
| Dane | 507 | 0.4 | 197 | 4.9 | 410.3 | 1,626,929 | 8,239 | 166,567 | 509.1 | 198,391 | 36.3 | 63.7 | 58 | 81.5 | 10,186 | 50.9 |
| Dodge | 406 | 1.0 | 232 | 0.2 | 340.5 | 1,356,553 | 5,844 | 210,146 | 345.7 | 197,636 | 42.1 | 57.9 | 18 | 78.0 | 5,238 | 49.7 |
| Door | 115 | -13.2 | 183 | 0.6 | 90.1 | 870,229 | 4,757 | 157,612 | 78.7 | 125,786 | 41.5 | 58.5 | 18 | 77.0 | 795 | 40.7 |

| STATE County | Water use, 2015 | | Wholesale Trade[1], 2017 | | | | Retail Trade[2], 2017 | | | | Real estate and rental and leasing,[2] 2017 | | | |
|---|---|---|---|---|---|---|---|---|---|---|---|---|---|---|
| | Public supply water withdrawn (mil gal/day) | Public supply gallons withdrawn per person per day | Number of establishments | Number of employees | Sales (mil dol) | Average payroll (mil dol) | Number of establishments | Number of employees | Sales (mil dol) | Average payroll (mil dol) | Number of establishments | Number of employees | Sales (mil dol) | Average payroll (mil dol) |
| | 133 | 134 | 135 | 136 | 137 | 138 | 139 | 140 | 141 | 142 | 143 | 144 | 145 | 146 |
| WEST VIRGINIA—Cont'd | | | | | | | | | | | | | | |
| Calhoun | 0.32 | 42.8 | NA | NA | NA | NA | 16 | 94 | 22.9 | 2.2 | NA | NA | NA | NA |
| Clay | 0.40 | 44.9 | NA | NA | NA | NA | 15 | 135 | 38.6 | 2.9 | NA | NA | NA | NA |
| Doddridge | 0.22 | 26.9 | NA | NA | NA | NA | 8 | 96 | 25.4 | 1.6 | NA | NA | NA | NA |
| Fayette | 5.37 | 119.3 | 14 | 117 | 135.3 | 6.6 | 128 | 1,393 | 340.4 | 33.3 | 19 | 72 | 6.9 | 1.2 |
| Gilmer | 0.64 | 75.1 | D | D | D | 0.7 | 20 | 149 | 35.1 | 3.4 | 3 | 8 | 2.2 | 0.2 |
| Grant | 0.97 | 82.4 | 3 | 5 | 1.6 | 0.1 | 41 | 390 | 107.0 | 9.7 | 8 | D | 1.2 | D |
| Greenbrier | 3.42 | 96.3 | 19 | 213 | 116.7 | 9.5 | 165 | 1,946 | 575.1 | 51.0 | 41 | 104 | 17.7 | 3.8 |
| Hampshire | 0.53 | 22.7 | D | D | D | 1.5 | 52 | 470 | 122.0 | 10.6 | 10 | D | 2.3 | D |
| Hancock | 1.46 | 49.0 | 10 | 141 | 74.6 | 5.7 | 83 | 800 | 184.6 | 16.4 | 19 | D | 11.9 | D |
| Hardy | 3.01 | 217.3 | 6 | 38 | 5.2 | 1.1 | 42 | 612 | 174.5 | 14.5 | 14 | 31 | 3.8 | 0.9 |
| Harrison | 7.91 | 115.1 | 75 | 1,020 | 437.4 | 46.7 | 303 | 4,621 | 1,340.9 | 111.1 | 57 | D | 105.5 | D |
| Jackson | 1.81 | 61.9 | 16 | 279 | 191.7 | 10.1 | 84 | 1,172 | 368.8 | 29.1 | 17 | 61 | 26.2 | 2.5 |
| Jefferson | 2.70 | 47.8 | D | D | D | D | 139 | 1,952 | 513.1 | 45.1 | 51 | 171 | 31.9 | 5.1 |
| Kanawha | 32.16 | 170.8 | 229 | 3,046 | 1,859.8 | 151.6 | 675 | 10,761 | 3,151.4 | 271.4 | 236 | 1,287 | 341.8 | 54.8 |
| Lewis | 1.08 | 65.7 | 15 | 121 | 88.7 | 5.7 | 63 | 978 | 271.9 | 24.3 | 9 | 31 | 5.7 | 0.8 |
| Lincoln | 0.30 | 14.0 | NA | NA | NA | NA | 36 | 297 | 71.9 | 7.0 | 3 | D | 0.6 | D |
| Logan | 4.29 | 123.6 | 27 | 209 | 105.3 | 8.6 | 116 | 1,662 | 508.5 | 44.5 | 16 | 60 | 6.2 | 1.6 |
| McDowell | 2.90 | 146.2 | 4 | 20 | 6.7 | 1.6 | 49 | 440 | 98.0 | 8.8 | D | D | D | 0.2 |
| Marion | 8.21 | 144.2 | 36 | 381 | 232.3 | 19.0 | 185 | 2,218 | 753.0 | 59.7 | 36 | 117 | 22.1 | 3.6 |
| Marshall | 4.67 | 146.0 | 15 | 142 | 72.2 | 5.5 | 77 | 1,199 | 354.9 | 29.6 | 9 | 35 | 3.7 | 0.7 |
| Mason | 2.33 | 86.2 | D | D | D | D | 60 | 647 | 153.9 | 14.4 | 12 | 38 | 6.1 | 0.9 |
| Mercer | 3.58 | 58.5 | 45 | 550 | 334.4 | 22.1 | 221 | 3,048 | 818.2 | 73.1 | 40 | 144 | 45.8 | 5.3 |
| Mineral | 1.57 | 57.2 | 11 | 97 | 21.2 | 3.3 | 71 | 948 | 263.5 | 23.7 | 13 | 38 | 7.2 | 1.0 |
| Mingo | 4.02 | 158.9 | 10 | 68 | 12.7 | 2.4 | 61 | 395 | 106.3 | 9.3 | 11 | 35 | 4.4 | 0.8 |
| Monongalia | 10.87 | 104.3 | 56 | 428 | 373.5 | 17.6 | 376 | 6,077 | 1,718.1 | 144.0 | 140 | 643 | 129.8 | 22.5 |
| Monroe | 0.49 | 36.3 | 3 | 15 | 1.8 | 0.3 | 25 | 123 | 29.9 | 2.8 | NA | NA | NA | NA |
| Morgan | 0.55 | 31.4 | 3 | 43 | 20.0 | 2.7 | 43 | 431 | 114.1 | 10.7 | 9 | 22 | 1.8 | 0.4 |
| Nicholas | 2.78 | 108.6 | 18 | 189 | 71.5 | 8.4 | 100 | 1,453 | 403.1 | 35.7 | 17 | 57 | 13.7 | 1.9 |
| Ohio | 7.70 | 178.8 | 66 | 1,224 | 3,905.1 | 54.1 | 193 | 3,148 | 877.0 | 77.4 | 56 | 328 | 57.0 | 12.5 |
| Pendleton | 0.31 | 42.9 | 4 | 11 | 1.0 | 0.1 | 26 | 179 | 38.6 | 3.6 | 3 | 2 | 0.8 | 0.1 |
| Pleasants | 0.57 | 74.3 | NA | NA | NA | NA | 15 | 128 | 34.2 | 2.6 | 4 | D | 0.7 | D |
| Pocahontas | 0.38 | 44.2 | NA | NA | NA | NA | 30 | 314 | 75.8 | 6.5 | 7 | 56 | 5.7 | 1.1 |
| Preston | 2.21 | 65.1 | 10 | D | 19.2 | D | 89 | 934 | 253.6 | 21.3 | 21 | 68 | 9.2 | 2.0 |
| Putnam | 2.61 | 45.9 | 68 | 1,087 | 676.7 | 56.9 | 160 | 2,350 | 752.4 | 59.6 | 56 | 215 | 45.2 | 9.4 |
| Raleigh | 9.89 | 127.6 | 89 | 836 | 394.2 | 39.3 | 332 | 4,861 | 1,503.4 | 121.2 | 72 | 344 | 69.2 | 15.8 |
| Randolph | 2.75 | 94.4 | D | D | D | 10.8 | 122 | 1,492 | 379.5 | 33.9 | 24 | 94 | 18.9 | 2.8 |
| Ritchie | 0.48 | 48.1 | D | D | D | D | 37 | 334 | 94.7 | 7.5 | 4 | 4 | 0.9 | 0.5 |
| Roane | 0.85 | 58.9 | 4 | 46 | 21.9 | 1.9 | 42 | 511 | 136.4 | 12.5 | 8 | 19 | 1.6 | 0.3 |
| Summers | 2.49 | 188.1 | 6 | 143 | 54.1 | 7.0 | 28 | 261 | 66.9 | 6.0 | 3 | 12 | 1.0 | 0.2 |
| Taylor | 1.82 | 107.6 | D | D | D | D | 32 | 465 | 122.0 | 11.8 | NA | NA | NA | NA |
| Tucker | 0.61 | 87.6 | NA | NA | NA | NA | 26 | 232 | 64.0 | 5.6 | 10 | 48 | 6.4 | 1.4 |
| Tyler | 0.57 | 63.5 | NA | NA | NA | NA | 20 | 148 | 44.7 | 2.9 | NA | NA | NA | NA |
| Upshur | 2.11 | 85.2 | 11 | 108 | 126.0 | 5.4 | 80 | 861 | 270.1 | 22.4 | 21 | 94 | 10.1 | 2.9 |
| Wayne | 2.91 | 71.0 | 17 | 244 | 119.8 | 11.2 | 101 | 1,464 | 354.9 | 34.3 | 19 | D | 16.1 | D |
| Webster | 0.49 | 56.0 | NA | NA | NA | NA | 25 | 172 | 43.7 | 3.9 | NA | NA | NA | NA |
| Wetzel | 1.78 | 112.5 | 9 | 79 | 27.5 | 2.5 | 72 | 893 | 223.1 | 20.7 | D | D | D | 0.8 |
| Wirt | 0.00 | 0.0 | NA | NA | NA | NA | 10 | 73 | 14.2 | 1.1 | NA | NA | NA | NA |
| Wood | 7.97 | 92.2 | D | D | D | D | 350 | 5,900 | 1,545.3 | 145.2 | D | D | D | D |
| Wyoming | 1.82 | 82.2 | 4 | D | D | 3.8 | 61 | 559 | 133.2 | 12.0 | 10 | 26 | 3.6 | 0.7 |
| WISCONSIN | 479.38 | 83.1 | 5,934 | 104,382 | 81,566.3 | 6,129.1 | 18,908 | 317,668 | 91,763.8 | 8,273.7 | 4,956 | 27,163 | 5,908.1 | 1,105.8 |
| Adams | 0.81 | 40.2 | 6 | 48 | 51.1 | 3.1 | 48 | 413 | 121.1 | 10.7 | 13 | 41 | 2.9 | 1.0 |
| Ashland | 0.99 | 62.5 | 9 | 84 | 46.0 | 3.6 | 87 | 952 | 237.2 | 25.6 | 8 | 25 | 2.3 | 0.5 |
| Barron | 4.48 | 98.3 | 33 | 284 | 92.5 | 10.9 | 224 | 2,946 | 840.6 | 78.1 | 36 | 115 | 19.6 | 3.3 |
| Bayfield | 0.37 | 24.7 | 6 | 54 | 31.9 | 2.8 | 68 | 391 | 85.9 | 9.6 | D | D | D | 0.6 |
| Brown | 19.19 | 74.2 | 331 | 7,479 | 5,440.9 | 414.1 | 854 | 15,612 | 4,385.4 | 407.9 | 219 | 1,448 | 387.9 | 54.7 |
| Buffalo | 0.40 | 30.3 | 9 | 129 | 56.9 | 6.5 | 41 | 359 | 84.1 | 6.6 | 8 | 13 | 2.7 | 0.3 |
| Burnett | 0.36 | 23.7 | D | D | D | 0.2 | 65 | 558 | 111.9 | 11.1 | 17 | D | 6.5 | D |
| Calumet | 10.93 | 219.6 | 40 | 455 | 227.7 | 23.6 | 105 | 1,795 | 454.3 | 40.9 | 18 | 47 | 5.1 | 1.2 |
| Chippewa | 6.96 | 109.6 | 64 | 808 | 538.2 | 34.2 | 208 | 3,744 | 1,322.1 | 104.1 | 34 | 162 | 25.4 | 4.6 |
| Clark | 1.23 | 35.7 | 47 | 469 | 379.0 | 24.6 | 113 | 974 | 290.6 | 21.9 | 4 | D | 0.5 | D |
| Columbia | 3.56 | 62.7 | 44 | 607 | 395.8 | 27.8 | 204 | 2,838 | 885.4 | 74.7 | D | D | D | D |
| Crawford | 1.43 | 87.2 | 9 | 51 | 68.9 | 1.8 | 70 | 1,165 | 246.3 | 26.8 | 8 | 43 | 5.0 | 0.9 |
| Dane | 43.49 | 83.1 | 622 | 12,666 | 9,491.3 | 728.3 | 1,671 | 31,714 | 10,139.2 | 894.8 | 744 | 4,834 | 1,068.6 | 215.6 |
| Dodge | 6.30 | 71.2 | 68 | 1,083 | 544.1 | 47.9 | 233 | 3,143 | 860.6 | 81.0 | 42 | 100 | 34.4 | 3.2 |
| Door | 1.36 | 49.4 | 21 | 109 | 53.5 | 5.8 | 251 | 2,016 | 511.4 | 50.5 | 50 | 221 | 34.2 | 6.3 |

1  Merchant wholesalers, except manufacturers' sales branches and offices.    2. Employer establishments.

| STATE County | Professional, scientific, and technical services, 2017 | | | | Manufacturing, 2017 | | | | Accommodation and food services, 2017 | | | |
|---|---|---|---|---|---|---|---|---|---|---|---|---|
| | Number of establishments | Number of employees | Sales (mil dol) | Average payroll (mil dol) | Number of establishments | Number of employees | Sales (mil dol) | Average payroll (mil dol) | Number of establishments | Number of employees | Sales (mil dol) | Annual payroll (mil dol) |
| | 147 | 148 | 149 | 150 | 151 | 152 | 153 | 154 | 155 | 156 | 157 | 158 |
| **WEST VIRGINIA—Cont'd** | | | | | | | | | | | | |
| Calhoun | 6 | 29 | 4 | 1.7 | 4 | 26 | 3.0 | D | 4 | 37 | 1.3 | 0.4 |
| Clay | NA | NA | NA | NA | NA | NA | NA | NA | 6 | 38 | 1.7 | 0.5 |
| Doddridge | NA | NA | NA | NA | NA | NA | NA | NA | 4 | 76 | 2.5 | 0.9 |
| Fayette | D | D | D | D | 21 | 422 | 221.8 | 26.5 | 77 | 1,150 | 89.0 | 26.6 |
| Gilmer | D | D | D | 0.9 | 4 | 174 | 51.6 | 6.6 | 12 | 105 | 5.5 | 1.8 |
| Grant | D | D | 3 | D | 8 | 294 | 89.3 | 11.0 | 20 | 188 | 10.0 | 2.2 |
| Greenbrier | 65 | 246 | 23 | 7.7 | 29 | 864 | 207.0 | 37.8 | 83 | 2,480 | 190.1 | 62.7 |
| Hampshire | D | D | D | D | 11 | 66 | 11.6 | 2.8 | D | D | D | D |
| Hancock | 49 | 341 | 38 | 14.0 | 18 | 2,640 | 1,248.0 | 144.9 | 67 | 1,631 | 178.8 | 31.9 |
| Hardy | 17 | 74 | 9 | 2.8 | D | 2,909 | D | 98.6 | 31 | 349 | 17.5 | 4.8 |
| Harrison | 143 | 1,697 | 293 | 101.3 | D | D | D | D | 169 | 3,408 | 172.4 | 48.3 |
| Jackson | 31 | 278 | 50 | 25.4 | 16 | 1,486 | 773.5 | 104.0 | 45 | 888 | 43.2 | 12.7 |
| Jefferson | D | D | D | D | 27 | 929 | 270.3 | 40.2 | 124 | 3,616 | 502.0 | 82.7 |
| Kanawha | 535 | 4,727 | 801 | 294.3 | 110 | 3,852 | 2,155.0 | 212.2 | 450 | 9,320 | 577.5 | 153.1 |
| Lewis | 14 | 52 | 8 | 1.9 | 13 | 142 | 15.3 | 5.1 | 34 | 630 | 35.8 | 11.6 |
| Lincoln | 11 | 47 | 6 | 2.6 | NA | NA | NA | NA | D | D | D | D |
| Logan | D | D | D | D | 27 | 470 | 94.4 | 17.4 | 60 | 821 | 43.7 | 12.0 |
| McDowell | D | D | D | D | D | 10 | D | D | D | D | D | 2.6 |
| Marion | D | D | D | D | 41 | 749 | 337.2 | 37.4 | 113 | 1,749 | 83.9 | 23.3 |
| Marshall | D | D | D | D | 19 | 455 | 325.1 | 23.1 | 54 | 718 | 34.6 | 9.6 |
| Mason | 16 | 79 | 12 | 3.9 | 11 | 405 | 792.4 | 27.8 | 26 | 306 | 13.4 | 3.9 |
| Mercer | 77 | 525 | 79 | 19.6 | 47 | 1,025 | 260.9 | 43.2 | 103 | 1,972 | 107.9 | 30.4 |
| Mineral | 20 | 109 | 11 | 4.1 | 14 | 1,542 | 520.8 | 117.4 | 51 | 580 | 22.8 | 6.4 |
| Mingo | D | D | D | D | 11 | 129 | 16.2 | 5.4 | 23 | 209 | 12.7 | 2.9 |
| Monongalia | D | D | D | D | 50 | 4,054 | 3,044.1 | 334.4 | 328 | 6,843 | 312.8 | 92.3 |
| Monroe | 10 | 52 | 4 | 1.2 | D | 589 | D | D | 14 | D | 3.3 | D |
| Morgan | 15 | 73 | 8 | 4.3 | 7 | 160 | 23.3 | 6.1 | 26 | 277 | 15.3 | 3.9 |
| Nicholas | D | D | D | D | 22 | 687 | 252.9 | 34.7 | 54 | 802 | 39.6 | 10.9 |
| Ohio | D | D | D | D | 46 | 906 | 262.5 | 42.8 | 139 | 3,334 | 259.1 | 56.2 |
| Pendleton | D | D | D | 0.4 | 4 | 139 | 69.0 | 6.8 | 12 | 102 | 3.8 | 0.9 |
| Pleasants | 6 | 37 | 6 | 2.6 | 10 | 665 | 266.3 | 42.8 | 10 | 187 | 8.4 | 2.4 |
| Pocahontas | D | D | D | 0.2 | 6 | 246 | 45.4 | 8.9 | 22 | 1,404 | 56.4 | 18.1 |
| Preston | D | D | D | D | 24 | 517 | 155.1 | 22.1 | 37 | 323 | 14.9 | 4.1 |
| Putnam | 93 | 712 | 107 | 38.5 | 39 | 2,262 | 2,213.4 | 143.6 | 101 | 1,830 | 97.7 | 25.6 |
| Raleigh | D | D | D | D | 49 | 738 | 192.5 | 31.0 | 144 | 3,180 | 176.8 | 51.2 |
| Randolph | D | D | D | D | 24 | 1,011 | 160.9 | 34.4 | 61 | 846 | 41.4 | 11.9 |
| Ritchie | 10 | 63 | 6 | 2.5 | 13 | 1,131 | 180.2 | 37.4 | 17 | 211 | 9.4 | 3.0 |
| Roane | 12 | 37 | 4 | 1.3 | 12 | 146 | 33.0 | 5.4 | 10 | 194 | 7.8 | 2.5 |
| Summers | 12 | 46 | 4 | 1.6 | 5 | 10 | 2.3 | 0.5 | 12 | 111 | 8.5 | 2.3 |
| Taylor | D | D | D | D | NA | NA | NA | NA | 19 | 242 | 7.1 | 2.2 |
| Tucker | D | D | D | 0.2 | 12 | 220 | 104.2 | 10.7 | 24 | 429 | 17.2 | 6.0 |
| Tyler | 6 | 21 | 2 | 0.4 | 5 | 886 | 268.0 | 49.3 | 8 | 103 | 2.7 | 0.9 |
| Upshur | 38 | 186 | 20 | 7.1 | 18 | 784 | 298.5 | 41.2 | 50 | 686 | 33.7 | 9.5 |
| Wayne | 20 | 149 | 21 | 6.9 | 19 | D | 483.8 | D | 52 | D | 36.5 | D |
| Webster | D | D | D | 0.1 | 8 | 206 | 44.3 | 8.3 | D | D | D | D |
| Wetzel | D | D | D | D | 11 | 802 | 682.9 | 72.8 | D | D | D | D |
| Wirt | 5 | 8 | 1 | 0.1 | NA | NA | NA | NA | D | D | D | D |
| Wood | D | D | D | D | D | D | D | D | NA | NA | NA | NA |
| Wyoming | 17 | 73 | 7 | 3.0 | 8 | 33 | 7.3 | 1.6 | D | D | D | D |
| **WISCONSIN** | 11,493 | 105,204 | 18,749 | 7,257.3 | 8,837 | 455,538 | 171,896.3 | 25,180.6 | 14,840 | 244,099 | 13,496.3 | 3,688.3 |
| Adams | D | D | 4 | D | 14 | 400 | 168.2 | 19.6 | 59 | 1,707 | 72.0 | 21.1 |
| Ashland | 28 | 166 | 16 | 6.6 | 21 | 997 | 196.0 | 51.3 | 56 | 603 | 32.0 | 8.8 |
| Barron | 62 | 288 | 30 | 12.5 | 87 | 5,333 | 1,829.8 | 227.3 | 153 | 1,767 | 72.2 | 20.2 |
| Bayfield | D | D | D | 0.9 | 20 | 154 | 27.6 | 5.9 | 89 | 490 | 34.2 | 8.9 |
| Brown | D | D | D | D | 433 | 24,744 | 10,956.1 | 1,401.4 | 620 | 12,417 | 600.5 | 175.7 |
| Buffalo | 18 | 68 | 5 | 1.4 | 14 | 186 | 95.4 | 10.4 | 49 | 294 | 14.0 | 3.5 |
| Burnett | 23 | 79 | 8 | 3.0 | 24 | 957 | 316.9 | 44.8 | 68 | 466 | 26.0 | 7.2 |
| Calumet | 60 | 216 | 23 | 10.0 | 62 | 3,440 | 1,517.3 | 149.7 | 88 | 1,446 | 61.5 | 16.9 |
| Chippewa | 91 | 577 | 51 | 22.8 | 125 | 5,443 | 1,996.0 | 259.6 | 175 | 1,937 | 87.2 | 22.5 |
| Clark | 32 | 185 | 16 | 5.2 | 72 | 3,269 | 2,473.5 | 157.6 | 60 | 488 | 16.1 | 4.6 |
| Columbia | D | D | D | D | 94 | 5,287 | 2,615.5 | 292.7 | 181 | 3,579 | 199.7 | 58.8 |
| Crawford | 20 | 132 | 22 | 5.4 | 24 | 2,088 | 911.2 | 87.5 | 49 | 610 | 30.1 | 8.3 |
| Dane | 1,904 | 22,044 | 4,791 | 1,896.8 | 521 | 26,628 | 9,108.6 | 1,556.3 | 1,403 | 27,619 | 1,476.7 | 441.3 |
| Dodge | 84 | 521 | 71 | 26.0 | 134 | 9,631 | 3,431.8 | 485.5 | 156 | 1,815 | 82.1 | 22.1 |
| Door | 66 | 227 | 30 | 10.7 | 66 | 2,576 | 608.9 | 120.9 | 249 | 2,039 | 180.2 | 51.2 |

# Health Care and Social Assistance, Other Services, Nonemployer Businesses, and Residential Construction

| STATE County | Health care and social assistance, 2017 | | | | Other services, 2017 | | | | Nonemployer businesses, 2018 | | Value of residential construction authorized by building permits, 2020 | |
|---|---|---|---|---|---|---|---|---|---|---|---|---|
| | Number of establishments | Number of employees | Receipts (mil dol) | Annual payroll (mil dol) | Number of establishments | Number of employees | Receipts (mil dol) | Annual payroll (mil dol) | Number | Receipts (mil dol) | New construction ($1,000) | Number of housing units |
| | 159 | 160 | 161 | 162 | 163 | 164 | 165 | 166 | 167 | 168 | 169 | 170 |
| WEST VIRGINIA—Cont'd | | | | | | | | | | | | |
| Calhoun | D | D | D | D | D | D | 1.1 | D | 416 | 9.7 | NA | NA |
| Clay | 17 | D | 10.3 | D | 4 | 45 | 6.6 | 1.7 | 365 | 12.3 | 600 | 24 |
| Doddridge | 10 | 113 | 3.4 | 1.9 | D | D | D | D | 250 | 9.3 | 0 | 0 |
| Fayette | 108 | 2,135 | 182.5 | 70.8 | 53 | 217 | 31.6 | 7.1 | 1,648 | 62.3 | 3,209 | 32 |
| Gilmer | 13 | 224 | 13.1 | 6.5 | D | D | 3.8 | D | 349 | 8.0 | 926 | 9 |
| Grant | 24 | 820 | 57.7 | 23.5 | D | D | D | 1.4 | 694 | 28.0 | 5,924 | 29 |
| Greenbrier | 136 | 2,652 | 233.5 | 101.7 | 52 | 168 | 30.7 | 5.0 | 2,141 | 75.9 | 17,541 | 76 |
| Hampshire | D | D | D | D | D | D | D | D | 1,324 | 59.4 | 13,531 | 102 |
| Hancock | 66 | 1,256 | 84.5 | 37.5 | 45 | 232 | 26.9 | 7.0 | 1,244 | 39.9 | 3,570 | 16 |
| Hardy | 32 | 564 | 46.2 | 20.3 | 22 | 66 | 6.0 | 1.5 | 878 | 27.7 | 8,203 | 63 |
| Harrison | 224 | 6,889 | 888.2 | 356.6 | 131 | 674 | 63.6 | 16.8 | 3,825 | 164.7 | 12,162 | 62 |
| Jackson | 60 | 1,124 | 80.5 | 35.8 | 35 | 128 | 20.5 | 3.5 | 1,405 | 53.2 | 200 | 2 |
| Jefferson | D | D | D | D | 79 | 543 | 101.0 | 19.2 | 3,641 | 149.2 | 70,022 | 288 |
| Kanawha | D | D | D | D | 373 | 2,396 | 327.0 | 90.7 | 9,077 | 389.3 | 27,748 | 272 |
| Lewis | 34 | 1,169 | 106.8 | 45.1 | 27 | 88 | 14.4 | 2.9 | 819 | 32.3 | 0 | 0 |
| Lincoln | 25 | 923 | 34.0 | 17.2 | D | D | D | D | 723 | 21.4 | 2,966 | 23 |
| Logan | 96 | 1,982 | 210.3 | 79.4 | 43 | 296 | 32.5 | 9.4 | 1,150 | 46.9 | 100 | 1 |
| McDowell | 31 | 823 | 44.6 | 18.5 | D | D | D | D | 479 | 16.3 | 2,758 | 18 |
| Marion | 155 | 2,934 | 258.6 | 105.9 | 91 | 628 | 59.7 | 18.8 | 2,585 | 96.0 | 2,850 | 29 |
| Marshall | 80 | 1,512 | 98.1 | 46.1 | 42 | 200 | 17.5 | 4.9 | 1,128 | 37.0 | 900 | 3 |
| Mason | 43 | 952 | 110.5 | 40.8 | 27 | 116 | 12.7 | 3.2 | 934 | 31.7 | 209 | 1 |
| Mercer | 228 | 4,746 | 457.5 | 172.7 | 81 | 750 | 103.6 | 23.6 | 3,004 | 119.5 | 817 | 5 |
| Mineral | 71 | 1,464 | 89.4 | 36.2 | 34 | 195 | 15.8 | 4.4 | 1,356 | 44.6 | 11,931 | 75 |
| Mingo | 44 | 610 | 56.9 | 20.1 | D | D | D | D | 859 | 29.6 | 0 | 0 |
| Monongalia | 269 | 16,096 | 2,376.4 | 743.1 | 163 | 1,220 | 244.2 | 39.4 | 5,916 | 267.3 | 6,811 | 23 |
| Monroe | 20 | 277 | 16.6 | 7.6 | D | D | D | 0.5 | 784 | 26.1 | 320 | 1 |
| Morgan | 19 | 522 | 50.3 | 20.5 | 22 | 90 | 6.7 | 1.8 | 1,109 | 42.3 | 9,845 | 75 |
| Nicholas | 66 | 1,491 | 105.9 | 48.9 | 31 | 108 | 11.7 | 3.1 | 1,160 | 49.4 | 441 | 4 |
| Ohio | 230 | 6,200 | 822.1 | 312.2 | 116 | 1,017 | 119.1 | 35.0 | 2,509 | 111.6 | 7,785 | 49 |
| Pendleton | 17 | 333 | 18.8 | 9.0 | D | D | 4.7 | D | 519 | 16.2 | 1,872 | 17 |
| Pleasants | D | D | D | D | D | D | D | 0.4 | 269 | 8.5 | 906 | 4 |
| Pocahontas | 24 | 390 | 26.9 | 14.1 | 17 | 136 | 9.5 | 2.5 | 542 | 21.4 | 449 | 5 |
| Preston | 61 | 1,123 | 83.9 | 35.8 | 31 | 174 | 17.6 | 4.2 | 1,669 | 69.6 | 350 | 2 |
| Putnam | 127 | 2,134 | 247.1 | 102.0 | 59 | 339 | 43.3 | 10.5 | 3,069 | 125.4 | 17,645 | 73 |
| Raleigh | 303 | 7,101 | 764.5 | 307.4 | 112 | 814 | 96.8 | 24.3 | 3,513 | 146.4 | 8,071 | 43 |
| Randolph | 100 | 2,864 | 301.2 | 90.2 | 44 | 236 | 19.2 | 6.3 | 1,453 | 45.0 | 369 | 2 |
| Ritchie | 17 | D | 12.4 | D | D | D | D | 0.9 | 615 | 21.1 | 3,278 | 18 |
| Roane | 28 | 716 | 61.6 | 26.6 | D | D | D | 0.7 | 699 | 27.9 | 0 | 0 |
| Summers | 23 | 408 | 34.6 | 13.8 | 15 | 44 | 3.6 | 1.0 | 471 | 15.9 | 2,755 | 15 |
| Taylor | 32 | 682 | 44.1 | 20.9 | D | D | D | D | 753 | 30.3 | 0 | 0 |
| Tucker | 11 | 305 | 18.9 | 8.9 | 10 | 74 | 11.2 | 2.5 | 446 | 14.1 | 398 | 3 |
| Tyler | 15 | 460 | 24.8 | 12.3 | D | D | 1.8 | D | 376 | 10.8 | 0 | 0 |
| Upshur | 79 | 1,364 | 122.9 | 46.9 | 25 | 113 | 12.5 | 3.1 | 1,246 | 53.2 | 7,148 | 50 |
| Wayne | 60 | 2,115 | 312.7 | 125.8 | 32 | 128 | 11.9 | 3.8 | 1,591 | 50.6 | 6,141 | 78 |
| Webster | 16 | 437 | 31.9 | 16.0 | D | D | D | D | 284 | 11.2 | 0 | 0 |
| Wetzel | 37 | 753 | 59.4 | 23.5 | 34 | 129 | 11.8 | 3.1 | 530 | 16.1 | 5,186 | 59 |
| Wirt | D | D | D | D | NA | NA | NA | NA | 259 | 7.8 | 1,672 | 11 |
| Wood | D | D | D | D | D | D | D | D | 3,914 | 164.9 | 17,114 | 72 |
| Wyoming | 27 | 803 | 44.0 | 20.9 | 11 | 41 | 2.9 | 0.9 | 697 | 21.8 | 150 | 1 |
| WISCONSIN | 15,983 | 417,365 | 48,477.9 | 19,198.4 | 10,351 | 63,120 | 9,240.8 | 2,069.4 | 355,166 | 17,333.5 | 4,733,365 | 21,226 |
| Adams | 29 | 487 | 67.6 | 16.8 | 23 | 119 | 12.3 | 3.4 | 1,139 | 48.0 | 24,346 | 104 |
| Ashland | 63 | 1,368 | 154.6 | 65.1 | 34 | 131 | 10.4 | 3.2 | 1,113 | 53.1 | 1,492 | 13 |
| Barron | 132 | 2,430 | 279.6 | 109.4 | 96 | 331 | 48.1 | 9.6 | 3,120 | 156.7 | 27,278 | 138 |
| Bayfield | 22 | 360 | 17.7 | 9.7 | 25 | 71 | 7.7 | 1.9 | 1,561 | 60.8 | 9,085 | 53 |
| Brown | 724 | 24,177 | 3,453.3 | 1,189.1 | 466 | 3,012 | 317.2 | 88.8 | 14,229 | 752.4 | 255,645 | 1,210 |
| Buffalo | 28 | 315 | 14.1 | 6.1 | 19 | 69 | 6.9 | 1.6 | 1,023 | 46.6 | 4,396 | 33 |
| Burnett | 33 | 635 | 39.8 | 19.6 | D | D | 7.3 | D | 1,198 | 47.7 | 22,797 | 98 |
| Calumet | 90 | 1,168 | 96.3 | 41.6 | 63 | 275 | 28.8 | 7.3 | 2,556 | 117.9 | 77,520 | 488 |
| Chippewa | 161 | 2,754 | 230.9 | 94.2 | 101 | 494 | 60.0 | 16.8 | 4,066 | 226.0 | 73,038 | 309 |
| Clark | 66 | 913 | 77.9 | 32.9 | 54 | 163 | 23.5 | 3.9 | 2,379 | 136.6 | 19,747 | 137 |
| Columbia | 133 | 2,882 | 261.4 | 135.2 | 95 | 441 | 47.6 | 12.7 | 3,805 | 193.6 | 63,848 | 292 |
| Crawford | 37 | 1,113 | 99.0 | 47.9 | 25 | 101 | 10.1 | 2.6 | 1,072 | 51.1 | 7,229 | 31 |
| Dane | 1,364 | 53,120 | 7,128.9 | 2,962.3 | 1,165 | 8,810 | 1,477.2 | 353.2 | 39,043 | 2,011.2 | 948,548 | 4,769 |
| Dodge | 187 | 4,261 | 423.3 | 171.2 | 137 | 464 | 72.9 | 13.7 | 4,595 | 223.4 | 39,775 | 176 |
| Door | 77 | 1,310 | 124.2 | 57.7 | 98 | 444 | 55.9 | 14.2 | 3,005 | 132.9 | 32,050 | 139 |

# Table B. States and Counties — Government Employment and Payroll, and Local Government Finances

Columns are identified by their item numbers (171–184):

- **Government employment and payroll, 2017**
  - 171 — Full-time equivalent employees
  - 172 — March payroll (dollars)
  - *March payroll (percent of total):* 173 Administration, judicial, and legal · 174 Police and corrections · 175 Fire protection · 176 Highways and transportation · 177 Health and welfare · 178 Natural resources and utilities · 179 Education and libraries
- **Local government finances, 2017 — General revenue**
  - 180 — Total (mil dol)
  - 181 — Intergovernmental (mil dol)
  - *Taxes:* 182 Total (mil dol) · Per capita[1] (dollars): 183 Total · 184 Property

| STATE County | 171 | 172 | 173 | 174 | 175 | 176 | 177 | 178 | 179 | 180 | 181 | 182 | 183 | 184 |
|---|---|---|---|---|---|---|---|---|---|---|---|---|---|---|
| **WEST VIRGINIA—Cont'd** | | | | | | | | | | | | | | |
| Calhoun | 311 | 743,775 | 10.7 | 0.2 | 0.0 | 11.3 | 0.0 | 3.5 | 74.3 | 14.7 | 10.2 | 2.9 | 400 | 378 |
| Clay | 436 | 1,168,050 | 3.8 | 1.0 | 0.0 | 0.0 | 6.7 | 2.2 | 85.9 | 21.8 | 15.7 | 3.8 | 440 | 425 |
| Doddridge | 285 | 1,016,940 | 6.9 | 2.7 | 0.0 | 0.9 | 5.3 | 2.1 | 81.4 | 37.3 | 3.7 | 29.6 | 3,471 | 3,463 |
| Fayette | 1,484 | 4,411,432 | 6.4 | 5.3 | 1.4 | 1.4 | 0.6 | 5.7 | 78.6 | 100.1 | 48.7 | 38.6 | 886 | 748 |
| Gilmer | 169 | 640,640 | 10.2 | 2.8 | 0.0 | 0.5 | 1.4 | 0.4 | 84.3 | 15.1 | 8.0 | 5.7 | 711 | 629 |
| Grant | 672 | 2,192,767 | 4.3 | 1.3 | 0.0 | 4.1 | 44.1 | 5.4 | 38.9 | 63.8 | 8.6 | 13.2 | 1,134 | 1,080 |
| Greenbrier | 1,147 | 3,671,555 | 7.2 | 7.3 | 0.3 | 1.0 | 0.8 | 5.8 | 75.3 | 107.4 | 47.4 | 36.1 | 1,027 | 900 |
| Hampshire | 745 | 2,066,072 | 8.4 | 4.0 | 0.0 | 0.6 | 2.2 | 3.5 | 77.9 | 41.4 | 19.9 | 14.9 | 639 | 603 |
| Hancock | 1,047 | 3,245,725 | 5.6 | 9.1 | 2.7 | 3.1 | 1.9 | 7.8 | 64.2 | 91.2 | 29.3 | 36.9 | 1,256 | 908 |
| Hardy | 492 | 1,547,623 | 7.9 | 4.9 | 0.0 | 0.8 | 1.5 | 6.0 | 78.3 | 33.5 | 13.6 | 14.9 | 1,076 | 1,012 |
| Harrison | 2,649 | 9,379,255 | 8.0 | 7.2 | 4.5 | 3.3 | 1.7 | 7.2 | 66.7 | 212.2 | 62.7 | 113.9 | 1,678 | 1,329 |
| Jackson | 930 | 3,264,832 | 5.2 | 4.6 | 0.0 | 0.7 | 6.7 | 3.8 | 77.6 | 73.2 | 29.7 | 30.0 | 1,038 | 943 |
| Jefferson | 1,670 | 6,366,407 | 8.9 | 6.0 | 0.0 | 0.5 | 0.8 | 2.4 | 78.1 | 146.4 | 47.6 | 66.1 | 1,171 | 1,032 |
| Kanawha | 6,558 | 23,843,029 | 7.0 | 8.3 | 4.5 | 4.8 | 5.4 | 3.8 | 63.7 | 656.5 | 207.3 | 291.1 | 1,587 | 1,048 |
| Lewis | 621 | 1,475,391 | 10.1 | 6.0 | 1.1 | 0.9 | 6.6 | 1.9 | 68.2 | 39.0 | 16.5 | 18.1 | 1,116 | 1,001 |
| Lincoln | 669 | 2,192,499 | 5.7 | 1.7 | 0.0 | 2.1 | 0.7 | 3.1 | 85.0 | 41.6 | 28.1 | 11.0 | 529 | 521 |
| Logan | 1,246 | 4,074,664 | 7.7 | 5.5 | 0.8 | 0.5 | 0.9 | 4.6 | 77.2 | 77.2 | 38.2 | 30.6 | 927 | 877 |
| McDowell | 758 | 2,486,004 | 6.1 | 3.8 | 0.0 | 1.1 | 0.5 | 5.2 | 80.0 | 67.7 | 26.4 | 19.2 | 1,036 | 937 |
| Marion | 1,849 | 6,457,746 | 5.1 | 6.8 | 2.4 | 3.7 | 1.6 | 6.1 | 72.3 | 169.4 | 57.2 | 68.9 | 1,222 | 1,030 |
| Marshall | 1,147 | 3,794,118 | 11.3 | 6.4 | 0.4 | 1.5 | 2.6 | 6.0 | 69.3 | 117.0 | 13.2 | 80.3 | 2,570 | 2,417 |
| Mason | 830 | 2,893,931 | 5.4 | 4.9 | 0.0 | 0.4 | 1.6 | 4.7 | 81.2 | 61.8 | 27.2 | 23.0 | 858 | 787 |
| Mercer | 3,184 | 11,269,581 | 2.4 | 2.7 | 1.1 | 1.4 | 41.5 | 3.9 | 46.3 | 268.1 | 64.2 | 47.0 | 784 | 610 |
| Mineral | 904 | 2,717,456 | 6.7 | 3.6 | 0.0 | 1.5 | 2.5 | 3.4 | 80.0 | 63.1 | 31.3 | 21.9 | 805 | 738 |
| Mingo | 855 | 2,826,731 | 10.1 | 4.1 | 0.8 | 0.0 | 3.5 | 5.9 | 75.5 | 69.2 | 37.9 | 23.1 | 959 | 879 |
| Monongalia | 2,619 | 9,242,161 | 6.3 | 9.0 | 3.0 | 4.8 | 2.3 | 8.1 | 63.4 | 255.7 | 67.9 | 132.7 | 1,254 | 942 |
| Monroe | 333 | 1,060,932 | 5.5 | 1.9 | 0.0 | 0.0 | 0.0 | 3.1 | 87.0 | 24.7 | 14.6 | 5.2 | 392 | 366 |
| Morgan | 446 | 1,456,131 | 9.5 | 3.1 | 0.0 | 0.0 | 1.6 | 2.0 | 80.6 | 33.3 | 13.8 | 14.8 | 835 | 786 |
| Nicholas | 1,176 | 4,089,192 | 5.6 | 3.7 | 0.0 | 1.4 | 52.4 | 4.5 | 32.2 | 126.9 | 44.6 | 18.9 | 752 | 612 |
| Ohio | 1,975 | 6,700,199 | 4.8 | 7.7 | 5.5 | 6.0 | 3.2 | 20.1 | 50.6 | 208.6 | 43.4 | 95.6 | 2,276 | 1,337 |
| Pendleton | 214 | 780,616 | 8.9 | 1.3 | 0.0 | 0.8 | 1.8 | 4.4 | 80.2 | 13.8 | 7.6 | 4.7 | 667 | 597 |
| Pleasants | 307 | 1,082,025 | 10.8 | 4.0 | 0.0 | 0.4 | 1.5 | 5.1 | 75.4 | 24.5 | 5.0 | 14.2 | 1,909 | 1,902 |
| Pocahontas | 437 | 1,626,369 | 6.3 | 2.6 | 0.0 | 0.4 | 44.1 | 4.0 | 40.1 | 35.1 | 8.8 | 8.9 | 1,049 | 832 |
| Preston | 938 | 2,591,492 | 8.6 | 4.0 | 0.0 | 0.7 | 0.9 | 6.4 | 77.3 | 61.3 | 30.7 | 20.1 | 594 | 554 |
| Putnam | 1,790 | 6,533,244 | 5.1 | 4.4 | 0.1 | 0.5 | 3.0 | 5.1 | 80.6 | 132.5 | 49.7 | 61.3 | 1,081 | 1,013 |
| Raleigh | 2,410 | 8,096,096 | 6.7 | 6.5 | 2.2 | 1.6 | 1.9 | 6.0 | 74.1 | 217.5 | 92.2 | 88.0 | 1,172 | 911 |
| Randolph | 904 | 2,856,923 | 7.3 | 4.1 | 1.0 | 1.4 | 3.9 | 7.9 | 72.5 | 60.4 | 30.2 | 17.7 | 612 | 508 |
| Ritchie | 308 | 976,199 | 14.1 | 0.6 | 0.0 | 1.4 | 0.0 | 5.7 | 78.1 | 24.2 | 8.2 | 13.3 | 1,355 | 1,286 |
| Roane | 405 | 1,249,394 | 5.9 | 3.8 | 0.0 | 0.8 | 4.0 | 5.6 | 78.5 | 24.7 | 16.5 | 6.2 | 440 | 390 |
| Summers | 334 | 998,777 | 8.0 | 4.9 | 1.6 | 1.3 | 1.0 | 2.7 | 77.0 | 25.2 | 11.7 | 7.8 | 605 | 428 |
| Taylor | 483 | 1,437,988 | 4.1 | 2.8 | 0.9 | 1.3 | 6.0 | 5.6 | 76.5 | 57.7 | 14.8 | 17.3 | 1,022 | 939 |
| Tucker | 274 | 858,636 | 11.1 | 1.8 | 0.0 | 0.6 | 1.5 | 11.1 | 65.3 | 24.8 | 9.3 | 8.2 | 1,177 | 1,031 |
| Tyler | 411 | 1,479,004 | 6.2 | 4.3 | 0.0 | 1.6 | 38.5 | 3.2 | 43.1 | 58.2 | 6.6 | 14.8 | 1,677 | 1,625 |
| Upshur | 761 | 2,564,165 | 9.6 | 4.0 | 1.2 | 1.1 | 1.7 | 6.7 | 73.0 | 56.1 | 26.1 | 19.4 | 789 | 691 |
| Wayne | 1,310 | 4,187,944 | 5.2 | 4.6 | 0.0 | 0.4 | 0.5 | 4.8 | 83.8 | 98.1 | 57.5 | 28.9 | 719 | 663 |
| Webster | 363 | 1,192,118 | 5.2 | 1.8 | 0.0 | 0.6 | 0.9 | 3.3 | 85.3 | 37.2 | 11.3 | 3.7 | 443 | 402 |
| Wetzel | 860 | 2,975,640 | 7.5 | 4.2 | 0.0 | 0.8 | 27.6 | 7.9 | 49.9 | 98.1 | 10.0 | 51.1 | 3,314 | 3,201 |
| Wirt | 202 | 618,211 | 7.1 | 2.1 | 0.0 | 0.0 | 0.0 | 1.6 | 89.2 | 14.5 | 10.6 | 2.6 | 451 | 433 |
| Wood | 2,893 | 9,478,825 | 5.4 | 6.6 | 2.5 | 3.3 | 3.2 | 6.8 | 70.7 | 215.4 | 86.1 | 81.9 | 963 | 780 |
| Wyoming | 797 | 2,514,443 | 12.9 | 0.8 | 0.0 | 0.0 | 0.8 | 3.5 | 81.9 | 52.5 | 28.0 | 18.1 | 852 | 813 |
| **WISCONSIN** | X | X | X | X | X | X | X | X | X | X | X | X | X | X |
| Adams | 572 | 2,264,859 | 11.2 | 15.3 | 0.0 | 9.7 | 14.3 | 7.2 | 41.7 | 78.2 | 30.3 | 35.4 | 1,777 | 1,665 |
| Ashland | 817 | 3,164,080 | 7.9 | 9.7 | 4.4 | 8.9 | 9.9 | 3.9 | 51.9 | 85.9 | 47.8 | 23.9 | 1,540 | 1,415 |
| Barron | 1,857 | 7,487,952 | 5.9 | 8.2 | 1.4 | 5.3 | 7.0 | 3.0 | 67.9 | 203.3 | 94.8 | 81.4 | 1,803 | 1,672 |
| Bayfield | 753 | 2,902,925 | 10.6 | 8.5 | 0.3 | 11.0 | 26.0 | 3.9 | 37.7 | 80.2 | 32.0 | 33.5 | 2,230 | 2,084 |
| Brown | 9,500 | 46,580,783 | 3.8 | 8.7 | 3.4 | 3.2 | 4.7 | 4.9 | 69.5 | 1,229.5 | 575.3 | 401.2 | 1,533 | 1,493 |
| Buffalo | 488 | 1,773,496 | 7.8 | 6.9 | 0.2 | 11.1 | 6.9 | 2.4 | 63.9 | 53.2 | 28.9 | 19.6 | 1,495 | 1,414 |
| Burnett | 598 | 2,153,819 | 11.9 | 9.1 | 0.2 | 9.3 | 8.2 | 2.5 | 57.6 | 62.3 | 28.4 | 28.3 | 1,853 | 1,751 |
| Calumet | 1,222 | 4,658,317 | 8.3 | 8.1 | 0.0 | 6.1 | 12.7 | 4.2 | 59.9 | 121.8 | 50.5 | 51.9 | 1,039 | 1,016 |
| Chippewa | 2,061 | 8,330,286 | 6.0 | 8.1 | 1.7 | 6.8 | 5.9 | 3.9 | 67.0 | 223.5 | 121.6 | 78.8 | 1,236 | 1,108 |
| Clark | 1,408 | 5,217,986 | 8.2 | 6.5 | 0.1 | 5.9 | 27.2 | 3.3 | 47.9 | 149.7 | 75.5 | 40.4 | 1,169 | 1,095 |
| Columbia | 2,633 | 10,398,074 | 6.9 | 9.2 | 0.5 | 5.3 | 6.4 | 4.7 | 66.3 | 274.0 | 118.4 | 117.0 | 2,045 | 1,862 |
| Crawford | 597 | 2,205,288 | 7.7 | 9.7 | 0.2 | 10.1 | 7.7 | 3.3 | 60.3 | 68.7 | 37.2 | 25.1 | 1,547 | 1,415 |
| Dane | 20,439 | 91,082,709 | 5.3 | 11.5 | 2.7 | 5.4 | 5.8 | 5.5 | 61.6 | 2,723.6 | 999.2 | 1,270.9 | 2,365 | 2,183 |
| Dodge | 2,531 | 10,252,861 | 7.6 | 13.7 | 1.7 | 6.8 | 17.8 | 4.1 | 46.3 | 297.1 | 124.2 | 103.0 | 1,175 | 1,072 |
| Door | 1,063 | 4,665,555 | 9.3 | 10.6 | 2.2 | 7.9 | 11.5 | 4.4 | 52.9 | 144.6 | 37.5 | 85.7 | 3,123 | 2,876 |

1. Based on the resident population estimated as of July 1 of the year shown.

# Table B. States and Counties — Local Government Finances, Government Employment, and Income Taxes

| STATE County | Direct general expenditure — Total (mil dol) [185] | Per capita¹ (dollars) [186] | Percent of total for: Education [187] | Health and hospitals [188] | Police protection [189] | Public welfare [190] | Highways [191] | Debt outstanding — Total (mil dol) [192] | Per capita¹ (dollars) [193] | Government employment, 2019 — Federal civilian [194] | Federal military [195] | State and local [196] | Individual income tax returns, 2018 — Number of returns [197] | Mean adjusted gross income [198] | Mean income tax [199] |
|---|---|---|---|---|---|---|---|---|---|---|---|---|---|---|---|
| WEST VIRGINIA—Cont'd | | | | | | | | | | | | | | | |
| Calhoun | 14 | 1,859 | 75.1 | 0.0 | 1.4 | 0.0 | 0.1 | 19.4 | 2,658 | 13 | 34 | 295 | 2,390 | 50,808 | 4,233 |
| Clay | 25 | 2,898 | 79.5 | 7.1 | 3.8 | 0.0 | 0.0 | 4.5 | 516 | 13 | 40 | 456 | 3,020 | 43,138 | 2,819 |
| Doddridge | 29 | 3,429 | 73.0 | 2.5 | 4.7 | 0.0 | 0.1 | 4.6 | 544 | 10 | 35 | 559 | 2,800 | 58,291 | 5,901 |
| Fayette | 102 | 2,341 | 67.6 | 0.6 | 5.9 | 0.0 | 2.3 | 43.8 | 1,005 | 222 | 200 | 2,404 | 16,300 | 44,333 | 3,278 |
| Gilmer | 16 | 2,032 | 76.9 | 1.4 | 1.7 | 0.0 | 0.7 | 8.3 | 1,033 | 290 | 29 | 661 | 2,410 | 49,139 | 5,172 |
| Grant | 63 | 5,424 | 24.8 | 58.4 | 2.4 | 0.0 | 0.1 | 17.7 | 1,522 | 48 | 55 | 873 | 5,110 | 44,501 | 3,248 |
| Greenbrier | 101 | 2,880 | 53.3 | 0.8 | 7.7 | 0.0 | 1.1 | 95.9 | 2,723 | 97 | 163 | 2,312 | 14,900 | 48,259 | 4,322 |
| Hampshire | 47 | 2,015 | 65.5 | 1.1 | 7.5 | 0.5 | 0.6 | 22.4 | 957 | 36 | 109 | 1,219 | 9,870 | 46,545 | 3,725 |
| Hancock | 82 | 2,776 | 51.2 | 2.4 | 8.9 | 0.5 | 5.0 | 86.8 | 2,953 | 63 | 136 | 1,316 | 14,110 | 52,091 | 4,710 |
| Hardy | 35 | 2,539 | 62.0 | 1.3 | 18.8 | 0.0 | 0.3 | 50.1 | 3,615 | 44 | 65 | 762 | 6,590 | 41,750 | 3,040 |
| Harrison | 203 | 2,990 | 59.5 | 0.8 | 5.5 | 0.2 | 3.5 | 162.2 | 2,389 | 4,397 | 337 | 3,670 | 30,710 | 61,450 | 6,476 |
| Jackson | 84 | 2,914 | 62.9 | 3.1 | 6.1 | 0.0 | 0.8 | 55.1 | 1,907 | 75 | 136 | 1,191 | 12,030 | 52,905 | 4,625 |
| Jefferson | 150 | 2,655 | 64.6 | 0.6 | 5.4 | 0.1 | 2.0 | 61.8 | 1,096 | 1,511 | 268 | 2,956 | 26,650 | 70,645 | 7,473 |
| Kanawha | 664 | 3,621 | 47.5 | 0.9 | 7.2 | 0.2 | 1.9 | 284.4 | 1,551 | 2,154 | 853 | 18,982 | 81,860 | 58,255 | 6,312 |
| Lewis | 43 | 2,668 | 65.5 | 2.3 | 3.3 | 0.1 | 0.9 | 11.0 | 680 | 55 | 75 | 1,301 | 7,450 | 52,758 | 4,743 |
| Lincoln | 46 | 2,188 | 83.1 | 0.8 | 0.9 | 0.0 | 0.0 | 6.1 | 292 | 40 | 97 | 789 | 7,260 | 46,016 | 3,322 |
| Logan | 95 | 2,881 | 72.1 | 1.0 | 1.4 | 0.3 | 0.6 | 18.1 | 549 | 90 | 150 | 1,969 | 11,250 | 47,591 | 3,752 |
| McDowell | 68 | 3,657 | 58.1 | 0.5 | 4.3 | 0.0 | 1.3 | 20.6 | 1,115 | 330 | 83 | 1,391 | 5,190 | 39,182 | 2,339 |
| Marion | 172 | 3,052 | 55.1 | 0.7 | 5.8 | 1.0 | 2.0 | 222.9 | 3,955 | 183 | 261 | 4,201 | 25,360 | 54,110 | 4,782 |
| Marshall | 106 | 3,394 | 61.6 | 1.4 | 7.4 | 0.2 | 2.3 | 125.1 | 4,003 | 62 | 143 | 1,817 | 14,360 | 57,383 | 5,753 |
| Mason | 66 | 2,454 | 68.7 | 4.8 | 3.4 | 0.1 | 0.9 | 24.3 | 905 | 95 | 123 | 1,326 | 10,140 | 48,609 | 3,821 |
| Mercer | 276 | 4,613 | 35.6 | 43.3 | 4.4 | 0.0 | 1.0 | 90.8 | 1,517 | 159 | 275 | 4,526 | 23,790 | 46,161 | 3,924 |
| Mineral | 81 | 2,980 | 55.3 | 0.8 | 4.4 | 0.0 | 0.6 | 90.9 | 3,340 | 95 | 126 | 1,443 | 12,470 | 49,589 | 4,018 |
| Mingo | 77 | 3,167 | 58.9 | 1.0 | 2.2 | 0.1 | 0.0 | 30.9 | 1,279 | 72 | 111 | 981 | 7,670 | 43,068 | 2,924 |
| Monongalia | 269 | 2,547 | 52.2 | 1.5 | 6.8 | 1.0 | 2.6 | 385.7 | 3,646 | 1,134 | 489 | 15,559 | 42,350 | 71,904 | 8,877 |
| Monroe | 20 | 1,522 | 94.5 | 0.0 | 0.1 | 0.0 | 0.3 | 9.5 | 711 | 188 | 63 | 537 | 5,270 | 43,911 | 3,012 |
| Morgan | 33 | 1,869 | 72.6 | 1.0 | 7.9 | 0.0 | 0.3 | 13.4 | 756 | 29 | 85 | 678 | 8,050 | 48,610 | 3,899 |
| Nicholas | 125 | 4,977 | 47.5 | 40.8 | 4.3 | 0.0 | 0.8 | 47.4 | 1,888 | 96 | 116 | 1,504 | 9,850 | 46,587 | 3,630 |
| Ohio | 209 | 4,983 | 34.4 | 0.1 | 7.5 | 0.0 | 3.4 | 300.4 | 7,153 | 382 | 187 | 3,241 | 19,890 | 68,592 | 8,388 |
| Pendleton | 15 | 2,191 | 76.7 | 2.3 | 1.1 | 0.0 | 0.0 | 5.3 | 766 | 25 | 33 | 329 | 3,220 | 43,075 | 2,991 |
| Pleasants | 29 | 3,835 | 57.0 | 0.0 | 6.4 | 1.3 | 0.8 | 95.2 | 12,776 | 12 | 32 | 601 | 3,140 | 59,690 | 5,941 |
| Pocahontas | 36 | 4,245 | 39.6 | 39.5 | 2.3 | 0.3 | 0.4 | 5.7 | 671 | 62 | 38 | 798 | 3,790 | 40,716 | 2,933 |
| Preston | 66 | 1,959 | 63.3 | 0.7 | 7.0 | 0.0 | 0.7 | 71.2 | 2,105 | 871 | 151 | 1,639 | 13,830 | 49,824 | 3,999 |
| Putnam | 140 | 2,466 | 72.6 | 0.5 | 5.9 | 0.0 | 0.8 | 140.8 | 2,483 | 228 | 269 | 1,991 | 25,800 | 66,739 | 7,023 |
| Raleigh | 230 | 3,064 | 62.2 | 0.7 | 7.1 | 0.0 | 1.6 | 139.4 | 1,858 | 1,708 | 342 | 3,861 | 29,950 | 52,877 | 4,900 |
| Randolph | 69 | 2,385 | 54.8 | 0.1 | 6.1 | 0.2 | 1.6 | 98.7 | 3,417 | 166 | 126 | 1,803 | 12,040 | 47,655 | 4,030 |
| Ritchie | 26 | 2,677 | 62.9 | 11.1 | 4.0 | 0.0 | 1.1 | 4.5 | 455 | 27 | 45 | 457 | 3,950 | 55,332 | 5,418 |
| Roane | 27 | 1,910 | 80.1 | 0.0 | 2.9 | 0.1 | 2.2 | 13.2 | 944 | 28 | 65 | 526 | 5,230 | 47,089 | 3,738 |
| Summers | 26 | 2,007 | 56.6 | 1.0 | 6.8 | 0.1 | 1.3 | 12.6 | 979 | 39 | 55 | 745 | 4,280 | 42,216 | 3,011 |
| Taylor | 58 | 3,400 | 42.1 | 34.8 | 4.8 | 0.6 | 1.4 | 26.9 | 1,593 | 44 | 77 | 1,025 | 7,170 | 55,656 | 5,063 |
| Tucker | 25 | 3,622 | 54.2 | 0.9 | 6.3 | 0.0 | 3.3 | 18.3 | 2,620 | 55 | 32 | 574 | 3,230 | 45,537 | 3,446 |
| Tyler | 47 | 5,337 | 36.7 | 49.5 | 1.8 | 0.0 | 0.6 | 7.7 | 876 | 15 | 41 | 582 | 3,800 | 61,972 | 7,025 |
| Upshur | 62 | 2,541 | 60.5 | 0.6 | 2.8 | 0.3 | 1.7 | 35.3 | 1,439 | 113 | 110 | 1,180 | 9,970 | 49,342 | 3,930 |
| Wayne | 118 | 2,942 | 76.1 | 1.6 | 4.1 | 0.1 | 0.7 | 66.0 | 1,642 | 1,554 | 187 | 1,649 | 15,160 | 48,166 | 3,727 |
| Webster | 39 | 4,666 | 37.0 | 46.7 | 4.5 | 0.0 | 0.4 | 7.6 | 908 | 11 | 38 | 562 | 2,950 | 41,966 | 3,249 |
| Wetzel | 91 | 5,890 | 47.3 | 27.8 | 3.6 | 0.0 | 1.3 | 27.7 | 1,793 | 34 | 71 | 1,183 | 6,750 | 56,997 | 5,532 |
| Wirt | 15 | 2,637 | 83.8 | 0.0 | 0.0 | 0.0 | 0.0 | 1.5 | 263 | 13 | 28 | 241 | 2,270 | 47,164 | 3,359 |
| Wood | 226 | 2,651 | 60.7 | 0.0 | 6.9 | 0.1 | 4.0 | 256.4 | 3,014 | 2,475 | 396 | 4,269 | 37,980 | 57,309 | 5,952 |
| Wyoming | 57 | 2,694 | 74.3 | 0.3 | 5.3 | 0.1 | 0.5 | 41.0 | 1,927 | 90 | 97 | 1,046 | 6,880 | 45,549 | 3,190 |
| WISCONSIN | X | X | X | X | X | X | X | X | X | 29,389 | 15,890 | 390,441 | 2,874,660 | 68,267 | 7,779 |
| Adams | 92 | 4,642 | 26.0 | 10.7 | 5.7 | 4.2 | 16.6 | 59.7 | 3,001 | 258 | 50 | 780 | 9,740 | 47,726 | 4,045 |
| Ashland | 92 | 5,949 | 41.7 | 3.4 | 6.3 | 7.1 | 15.8 | 119.0 | 7,675 | 147 | 39 | 1,760 | 7,290 | 47,549 | 3,838 |
| Barron | 211 | 4,682 | 49.3 | 1.2 | 4.7 | 5.9 | 15.0 | 116.9 | 2,590 | 136 | 117 | 3,660 | 23,270 | 54,893 | 5,177 |
| Bayfield | 83 | 5,537 | 33.9 | 7.3 | 4.6 | 0.2 | 21.5 | 35.9 | 2,393 | 116 | 60 | 1,495 | 7,910 | 59,828 | 6,194 |
| Brown | 1,363 | 5,205 | 52.3 | 2.9 | 5.5 | 3.1 | 7.9 | 1,188.9 | 4,542 | 1,273 | 699 | 17,276 | 132,000 | 74,804 | 9,385 |
| Buffalo | 56 | 4,297 | 48.2 | 1.4 | 3.5 | 5.4 | 22.7 | 21.0 | 1,600 | 150 | 34 | 707 | 6,640 | 48,717 | 3,970 |
| Burnett | 75 | 4,897 | 40.9 | 3.7 | 4.0 | 3.3 | 20.4 | 35.2 | 2,302 | 26 | 40 | 1,457 | 7,840 | 56,528 | 5,310 |
| Calumet | 122 | 2,439 | 41.6 | 3.9 | 6.1 | 6.2 | 13.3 | 120.8 | 2,416 | 59 | 131 | 1,403 | 25,230 | 72,735 | 7,733 |
| Chippewa | 250 | 3,925 | 48.8 | 4.6 | 4.4 | 3.5 | 16.5 | 172.2 | 2,702 | 183 | 163 | 3,197 | 31,390 | 58,660 | 5,536 |
| Clark | 159 | 4,605 | 46.1 | 4.6 | 4.4 | 13.8 | 10.8 | 103.4 | 2,990 | 102 | 89 | 2,159 | 14,930 | 53,955 | 5,571 |
| Columbia | 324 | 5,656 | 47.4 | 2.0 | 4.3 | 5.3 | 12.0 | 332.2 | 5,804 | 173 | 146 | 3,672 | 30,650 | 63,079 | 6,323 |
| Crawford | 76 | 4,666 | 48.0 | 3.0 | 5.6 | 3.7 | 20.6 | 73.3 | 4,524 | 63 | 40 | 1,042 | 7,770 | 48,229 | 4,068 |
| Dane | 2,756 | 5,128 | 47.7 | 2.1 | 6.1 | 8.6 | 6.0 | 4,031.8 | 7,502 | 5,249 | 1,538 | 77,322 | 276,110 | 86,465 | 11,386 |
| Dodge | 308 | 3,512 | 34.0 | 12.6 | 7.6 | 4.1 | 13.4 | 299.5 | 3,415 | 183 | 213 | 4,649 | 43,060 | 58,147 | 5,375 |
| Door | 151 | 5,482 | 34.7 | 7.3 | 5.8 | 2.1 | 15.4 | 93.4 | 3,403 | 72 | 157 | 1,807 | 16,110 | 69,316 | 7,766 |

1. Based on the resident population estimated as of July 1 of the year shown.

# Table B. States and Counties — Land Area and Population

| State / county code | CBSA code[1] | County Type code[2] | STATE County | Land area[3] (sq. mi) | Total persons 2019 | Rank | Per square mile | White | Black | American Indian, Alaska Native | Asian and Pacific Islancer | Percent Hispanic or Latino[4] | Under 5 years | 5 to 14 years | 15 to 24 years | 25 to 34 years | 35 to 44 years | 45 to 54 years |
|---|---|---|---|---|---|---|---|---|---|---|---|---|---|---|---|---|---|---|
| | | | | 1 | 2 | 3 | 4 | 5 | 6 | 7 | 8 | 9 | 10 | 11 | 12 | 13 | 14 | 15 |
| | | | WISCONSIN—Cont'd | | | | | | | | | | | | | | | |
| 55031 | 20260 | 2 | Douglas | 1,304.3 | 43,702 | 1,102 | 33.5 | 93.8 | 2.3 | 3.3 | 1.7 | 1.8 | 4.9 | 11.0 | 11.9 | 12.5 | 12.4 | 11.9 |
| 55033 | 32860 | 6 | Dunn | 850.2 | 45,452 | 1,069 | 53.5 | 93.5 | 1.3 | 0.8 | 3.6 | 2.1 | 4.9 | 10.9 | 21.9 | 11.4 | 11.3 | 10.8 |
| 55035 | 20740 | 3 | Eau Claire | 637.9 | 105,260 | 581 | 165.0 | 91.0 | 2.0 | 1.0 | 5.2 | 2.8 | 5.4 | 11.2 | 19.2 | 13.7 | 12.0 | 10.3 |
| 55037 | 27020 | 9 | Florence | 488.1 | 4,298 | 2,874 | 8.8 | 96.8 | 1.2 | 1.7 | 0.9 | 1.2 | 3.5 | 8.1 | 6.9 | 8.6 | 10.0 | 12.0 |
| 55039 | 22540 | 3 | Fond du Lac | 719.6 | 102,902 | 591 | 143.0 | 90.0 | 2.8 | 0.8 | 2.2 | 5.6 | 5.2 | 12.0 | 12.3 | 11.8 | 12.4 | 12.2 |
| 55041 | | 9 | Forest | 1,014.2 | 8,960 | 2,501 | 8.8 | 81.6 | 1.9 | 15.9 | 1.0 | 2.8 | 5.6 | 10.5 | 11.2 | 10.5 | 9.4 | 12.0 |
| 55043 | 38420 | 6 | Grant | 1,146.9 | 51,021 | 976 | 44.5 | 95.4 | 1.8 | 0.4 | 1.2 | 2.0 | 5.7 | 11.5 | 20.3 | 11.0 | 10.4 | 9.9 |
| 55045 | 31540 | 2 | Green | 584.0 | 36,603 | 1,258 | 62.7 | 94.9 | 1.3 | 0.5 | 1.0 | 3.4 | 5.1 | 12.4 | 10.9 | 10.8 | 12.2 | 13.0 |
| 55047 | | 6 | Green Lake | 349.5 | 18,908 | 1,880 | 54.1 | 92.5 | 1.3 | 0.9 | 1.0 | 5.4 | 5.7 | 11.9 | 11.3 | 9.6 | 10.7 | 11.5 |
| 55049 | 31540 | 2 | Iowa | 762.7 | 23,640 | 1,654 | 31.0 | 96.0 | 1.5 | 0.7 | 1.2 | 2.0 | 5.5 | 12.5 | 10.7 | 10.5 | 11.9 | 12.6 |
| 55051 | | 9 | Iron | 758.2 | 5,698 | 2,770 | 7.5 | 96.6 | 0.9 | 2.3 | 0.5 | 1.4 | 3.5 | 8.9 | 8.1 | 8.0 | 9.4 | 11.1 |
| 55053 | | 6 | Jackson | 987.9 | 20,630 | 1,787 | 20.9 | 87.6 | 2.7 | 6.9 | 1.0 | 3.6 | 5.6 | 12.1 | 11.0 | 12.4 | 11.8 | 12.4 |
| 55055 | 48020 | 4 | Jefferson | 556.5 | 85,038 | 679 | 152.8 | 90.2 | 1.5 | 0.7 | 1.3 | 7.4 | 4.9 | 11.5 | 13.2 | 11.5 | 12.6 | 13.1 |
| 55057 | | 7 | Juneau | 767.1 | 26,908 | 1,529 | 35.1 | 92.4 | 2.8 | 1.8 | 0.9 | 3.2 | 5.1 | 11.2 | 10.3 | 11.3 | 11.9 | 13.0 |
| 55059 | 16980 | 1 | Kenosha | 271.8 | 169,671 | 394 | 624.2 | 77.0 | 8.3 | 0.9 | 2.5 | 13.9 | 5.5 | 12.5 | 13.6 | 13.1 | 12.5 | 13.1 |
| 55061 | 24580 | 2 | Kewaunee | 342.5 | 20,386 | 1,800 | 59.5 | 95.1 | 0.9 | 1.0 | 0.8 | 3.3 | 5.1 | 11.8 | 11.1 | 10.6 | 11.8 | 12.5 |
| 55063 | 29100 | 3 | La Crosse | 451.8 | 118,502 | 530 | 262.3 | 91.0 | 2.5 | 0.8 | 5.5 | 2.2 | 4.9 | 11.1 | 19.2 | 12.8 | 11.6 | 10.6 |
| 55065 | | 8 | Lafayette | 633.6 | 16,646 | 1,998 | 26.3 | 94.4 | 0.7 | 0.6 | 0.7 | 4.4 | 6.5 | 13.4 | 11.7 | 10.5 | 11.1 | 11.3 |
| 55067 | | 6 | Langlade | 870.7 | 19,119 | 1,872 | 22.0 | 94.7 | 1.9 | 2.0 | 0.8 | 2.4 | 5.1 | 10.7 | 10.2 | 9.4 | 10.4 | 11.3 |
| 55069 | 48140 | 6 | Lincoln | 878.7 | 27,566 | 1,510 | 31.4 | 96.0 | 1.4 | 1.1 | 1.0 | 2.1 | 4.6 | 9.7 | 10.1 | 9.9 | 11.4 | 13.2 |
| 55071 | 31820 | 4 | Manitowoc | 589.3 | 78,757 | 716 | 133.6 | 90.8 | 1.6 | 1.0 | 3.4 | 4.7 | 5.1 | 11.5 | 10.9 | 10.6 | 11.5 | 12.4 |
| 55073 | 48140 | 3 | Marathon | 1,545.2 | 135,593 | 478 | 87.8 | 89.4 | 1.5 | 0.9 | 6.9 | 3.0 | 5.8 | 12.9 | 11.4 | 12.1 | 12.4 | 12.5 |
| 55075 | 31940 | 6 | Marinette | 1,399.5 | 40,262 | 1,176 | 28.8 | 95.8 | 1.0 | 1.2 | 1.0 | 2.2 | 4.7 | 10.7 | 9.7 | 10.1 | 10.9 | 11.5 |
| 55077 | | 8 | Marquette | 455.7 | 15,585 | 2,052 | 34.2 | 94.2 | 1.0 | 1.2 | 1.0 | 3.7 | 4.6 | 10.9 | 9.3 | 8.8 | 10.2 | 11.7 |
| 55078 | 43020 | 8 | Menominee | 357.6 | 4,546 | 2,853 | 12.7 | 13.7 | 1.4 | 78.1 | 2.1 | 7.3 | 9.1 | 19.2 | 14.4 | 11.6 | 8.6 | 10.1 |
| 55079 | 33340 | 1 | Milwaukee | 241.5 | 945,016 | 53 | 3,913.1 | 52.1 | 27.9 | 1.2 | 5.5 | 16.0 | 6.7 | 13.2 | 13.2 | 16.4 | 13.1 | 11.2 |
| 55081 | | 6 | Monroe | 900.9 | 46,582 | 1,043 | 51.7 | 91.0 | 2.4 | 1.6 | 1.6 | 5.1 | 6.4 | 14.4 | 11.2 | 11.3 | 12.9 | 11.8 |
| 55083 | 24580 | 2 | Oconto | 997.5 | 38,383 | 1,214 | 38.5 | 95.7 | 0.8 | 2.1 | 0.7 | 2.0 | 5.1 | 11.3 | 9.8 | 9.9 | 11.5 | 13.0 |
| 55085 | | 7 | Oneida | 1,113.9 | 35,751 | 1,286 | 32.1 | 96.1 | 0.9 | 1.9 | 0.9 | 1.7 | 4.5 | 9.5 | 8.7 | 9.5 | 9.9 | 11.6 |
| 55087 | 11540 | 3 | Outagamie | 637.6 | 188,766 | 358 | 296.1 | 88.4 | 2.2 | 2.1 | 4.4 | 4.6 | 6.1 | 13.0 | 12.3 | 13.3 | 13.2 | 12.4 |
| 55089 | 33340 | 8 | Ozaukee | 233.0 | 90,043 | 659 | 386.5 | 92.0 | 2.3 | 0.6 | 3.3 | 3.3 | 5.2 | 11.8 | 12.6 | 9.6 | 11.9 | 12.7 |
| 55091 | | 8 | Pepin | 232.0 | 7,271 | 2,636 | 31.3 | 96.3 | 0.6 | 0.6 | 0.7 | 2.5 | 5.7 | 12.6 | 9.7 | 10.1 | 11.2 | 11.5 |
| 55093 | 33460 | 1 | Pierce | 574.0 | 42,700 | 1,124 | 74.4 | 94.8 | 1.4 | 1.0 | 1.9 | 2.5 | 4.7 | 11.7 | 19.6 | 10.7 | 11.7 | 11.8 |
| 55095 | | 6 | Polk | 914.3 | 43,794 | 1,101 | 47.9 | 96.0 | 0.9 | 1.5 | 1.0 | 1.9 | 4.7 | 11.6 | 10.5 | 9.9 | 11.5 | 12.7 |
| 55097 | 44620 | 4 | Portage | 800.9 | 71,032 | 768 | 88.7 | 91.8 | 1.5 | 0.7 | 3.8 | 3.6 | 4.8 | 10.7 | 19.0 | 12.0 | 11.2 | 10.9 |
| 55099 | | 9 | Price | 1,254.1 | 13,245 | 2,203 | 10.6 | 94.9 | 1.0 | 1.5 | 2.0 | 2.1 | 4.3 | 9.8 | 9.6 | 7.8 | 9.7 | 12.4 |
| 55101 | 39540 | 3 | Racine | 332.6 | 195,802 | 345 | 588.7 | 72.8 | 12.8 | 0.9 | 1.8 | 14.3 | 6.0 | 12.8 | 12.3 | 12.0 | 12.3 | 12.5 |
| 55103 | | 6 | Richland | 586.2 | 17,258 | 1,960 | 29.4 | 95.2 | 1.3 | 0.8 | 1.1 | 2.6 | 4.9 | 12.2 | 11.3 | 9.6 | 10.5 | 11.8 |
| 55105 | 27500 | 3 | Rock | 718.2 | 163,084 | 407 | 227.1 | 84.0 | 6.4 | 0.7 | 1.9 | 9.4 | 5.9 | 12.7 | 12.4 | 12.3 | 12.5 | 12.6 |
| 55107 | | 6 | Rusk | 913.6 | 14,022 | 2,156 | 15.3 | 95.9 | 1.4 | 1.2 | 0.7 | 2.1 | 5.1 | 11.4 | 10.0 | 8.9 | 10.1 | 11.8 |
| 55109 | 33460 | 1 | St. Croix | 722.2 | 91,838 | 644 | 127.2 | 95.1 | 1.5 | 0.8 | 1.8 | 2.6 | 5.9 | 13.8 | 11.9 | 11.4 | 13.9 | 13.5 |
| 55111 | 12660 | 4 | Sauk | 831.5 | 64,449 | 832 | 77.5 | 91.2 | 1.7 | 1.6 | 1.0 | 5.7 | 5.9 | 12.8 | 11.2 | 11.6 | 12.4 | 12.1 |
| 55113 | | 7 | Sawyer | 1,257.6 | 16,700 | 1,991 | 13.3 | 79.9 | 1.1 | 18.4 | 0.7 | 2.8 | 4.5 | 11.3 | 9.2 | 8.9 | 9.4 | 10.9 |
| 55115 | 43020 | 6 | Shawano | 893.2 | 40,786 | 1,160 | 45.7 | 87.9 | 0.8 | 9.2 | 0.9 | 3.2 | 5.3 | 11.9 | 10.8 | 10.4 | 10.9 | 12.7 |
| 55117 | 43100 | 3 | Sheboygan | 511.5 | 115,240 | 541 | 225.3 | 84.5 | 3.0 | 0.8 | 6.5 | 6.9 | 5.4 | 12.4 | 12.1 | 11.7 | 12.1 | 12.4 |
| 55119 | | 6 | Taylor | 975.1 | 20,318 | 1,803 | 20.8 | 96.1 | 0.8 | 0.7 | 0.8 | 2.4 | 5.6 | 13.2 | 10.9 | 9.8 | 11.5 | 12.0 |
| 55121 | | 6 | Trempealeau | 733.0 | 29,681 | 1,444 | 40.5 | 89.3 | 0.8 | 0.7 | 0.9 | 9.3 | 7.1 | 13.9 | 11.0 | 10.7 | 11.7 | 12.4 |
| 55123 | | 6 | Vernon | 791.6 | 30,861 | 1,409 | 39.0 | 97.1 | 0.9 | 0.6 | 0.8 | 1.6 | 6.7 | 14.5 | 11.5 | 9.3 | 11.0 | 11.2 |
| 55125 | | 9 | Vilas | 857.7 | 22,356 | 1,708 | 26.1 | 86.6 | 0.8 | 10.4 | 0.9 | 2.7 | 4.2 | 9.4 | 8.3 | 8.1 | 8.1 | 10.9 |
| 55127 | 48580 | 4 | Walworth | 555.4 | 103,953 | 585 | 187.2 | 86.4 | 1.5 | 0.6 | 1.4 | 11.3 | 4.7 | 11.4 | 16.7 | 10.4 | 11.0 | 12.0 |
| 55129 | | 7 | Washburn | 797.1 | 15,712 | 2,049 | 19.7 | 95.4 | 0.8 | 2.6 | 0.9 | 2.0 | 4.2 | 10.6 | 9.0 | 8.4 | 9.7 | 11.8 |
| 55131 | 33340 | 1 | Washington | 430.6 | 136,445 | 476 | 316.9 | 93.4 | 2.0 | 0.6 | 2.0 | 3.4 | 5.0 | 12.2 | 11.2 | 10.5 | 12.5 | 13.5 |
| 55133 | 33340 | 1 | Waukesha | 549.7 | 406,172 | 177 | 738.9 | 88.9 | 2.3 | 0.5 | 4.7 | 5.1 | 5.1 | 11.9 | 11.8 | 10.5 | 12.7 | 12.8 |
| 55135 | | 6 | Waupaca | 747.7 | 50,664 | 980 | 67.8 | 95.0 | 0.9 | 1.1 | 0.8 | 3.4 | 5.1 | 11.4 | 10.8 | 10.5 | 11.6 | 12.8 |
| 55137 | | 6 | Waushara | 626.2 | 24,326 | 1,635 | 38.8 | 90.4 | 2.4 | 1.1 | 0.8 | 6.5 | 4.4 | 10.1 | 9.4 | 9.5 | 10.8 | 12.2 |
| 55139 | 36780 | 3 | Winnebago | 434.7 | 171,631 | 384 | 394.8 | 89.1 | 3.4 | 1.0 | 3.8 | 4.5 | 5.4 | 11.3 | 15.0 | 13.4 | 12.3 | 11.9 |
| 55141 | 49220 | 4 | Wood | 793.0 | 72,560 | 753 | 91.5 | 92.9 | 1.5 | 1.1 | 2.3 | 3.4 | 5.6 | 12.2 | 10.8 | 11.2 | 11.3 | 11.9 |
| 56000 | | 0 | WYOMING | 97,088.7 | 582,328 | X | 6.0 | 85.4 | 1.6 | 2.9 | 1.7 | 10.4 | 5.8 | 13.2 | 12.9 | 13.1 | 13.0 | 11.0 |
| 56001 | 29660 | 4 | Albany | 4,274.3 | 38,950 | 1,202 | 9.1 | 84.5 | 1.9 | 1.3 | 4.7 | 9.8 | 4.4 | 8.8 | 29.7 | 15.7 | 10.8 | 8.3 |
| 56003 | | 9 | Big Horn | 3,137.0 | 11,575 | 2,317 | 3.7 | 88.5 | 1.0 | 1.5 | 1.0 | 9.3 | 5.5 | 14.1 | 12.0 | 10.1 | 12.0 | 10.5 |
| 56005 | 23940 | 5 | Campbell | 4,802.1 | 46,676 | 1,041 | 9.7 | 88.9 | 1.0 | 2.0 | 1.3 | 8.7 | 7.0 | 15.4 | 12.6 | 14.4 | 14.7 | 11.1 |
| 56007 | | 7 | Carbon | 7,897.8 | 14,711 | 2,107 | 1.9 | 78.3 | 1.5 | 1.9 | 1.3 | 18.5 | 6.0 | 12.7 | 11.3 | 13.8 | 13.3 | 11.0 |
| 56009 | | 7 | Converse | 4,255.0 | 13,804 | 2,171 | 3.2 | 89.7 | 0.9 | 1.4 | 1.1 | 8.3 | 6.1 | 14.1 | 11.0 | 12.3 | 12.6 | 11.8 |
| 56011 | 23940 | 9 | Crook | 2,854.5 | 7,593 | 2,620 | 2.7 | 95.4 | 1.3 | 1.5 | 0.5 | 2.6 | 6.7 | 14.0 | 9.5 | 10.1 | 10.2 | 10.9 |
| 56013 | 40180 | 7 | Fremont | 9,183.6 | 39,317 | 1,196 | 4.3 | 71.9 | 0.8 | 21.4 | 1.1 | 7.5 | 6.5 | 14.5 | 11.9 | 11.3 | 11.8 | 10.7 |

1. CBSA = Core Based Statistical Area. See Appendix A for explanation. See Appendix B for list of metropolitan areas with component counties.
2. County type code from the Economic Research Service of USDA Rural-Urban Continuum Codes. See Appendix A for definition.
3. Dry land or land partially or temporarily covered by water.
4. May be of any race.

Items 1—15

# Table B. States and Counties — Population and Households

| STATE County | Population, 2020 (cont.) Age (percent) (cont.) 55 to 64 years | 65 to 74 years | 75 years and over | Percent female | Population change, 2000-2020 Total persons 2000 | 2010 | Percent change 2000-2010 | 2010-2020 | Components of change, 2010-2020 Births | Deaths | Net Migration | Households, 2015-2019 Number | Persons per household | Family households | Percent Female family householder[1] | One person |
|---|---|---|---|---|---|---|---|---|---|---|---|---|---|---|---|---|
| | 16 | 17 | 18 | 19 | 20 | 21 | 22 | 23 | 24 | 25 | 26 | 27 | 28 | 29 | 30 | 31 |
| **WISCONSIN—Cont'd** | | | | | | | | | | | | | | | | |
| Douglas | 15.4 | 12.2 | 7.7 | 49.8 | 43,287 | 44,160 | 2.0 | -1.0 | 4,391 | 4,340 | -482 | 19,011 | 2.22 | 60.3 | 11.3 | 31.7 |
| Dunn | 12.2 | 9.7 | 6.8 | 49.8 | 39,858 | 43,865 | 10.1 | 3.6 | 4,547 | 3,403 | 457 | 16,859 | 2.44 | 61.5 | 7.3 | 28.4 |
| Eau Claire | 11.7 | 9.9 | 6.5 | 50.5 | 93,142 | 98,873 | 6.2 | 6.5 | 12,001 | 8,511 | 2,932 | 40,981 | 2.43 | 57.6 | 7.0 | 31.4 |
| Florence | 22.2 | 17.4 | 11.2 | 49.1 | 5,088 | 4,423 | -13.1 | -2.8 | 290 | 581 | 168 | 1,948 | 2.19 | 66.1 | 4.5 | 29.3 |
| Fond du Lac | 14.7 | 11.3 | 8.2 | 50.7 | 97,296 | 101,626 | 4.5 | 1.3 | 11,083 | 9,752 | -19 | 41,542 | 2.39 | 65.6 | 7.3 | 28.4 |
| Forest | 17.0 | 14.3 | 9.6 | 49.1 | 10,024 | 9,304 | -7.2 | -3.7 | 1,094 | 1,211 | -221 | 4,008 | 2.17 | 63.5 | 10.4 | 31.4 |
| Grant | 13.0 | 10.2 | 7.9 | 48.2 | 49,597 | 51,204 | 3.2 | -0.4 | 5,700 | 5,077 | -824 | 19,583 | 2.45 | 61.5 | 6.8 | 28.1 |
| Green | 15.7 | 11.8 | 8.2 | 50.0 | 33,647 | 36,839 | 9.5 | -0.6 | 3,843 | 3,567 | -489 | 15,047 | 2.42 | 65.6 | 7.5 | 28.6 |
| Green Lake | 15.9 | 13.5 | 9.7 | 49.7 | 19,105 | 19,046 | -0.3 | -0.7 | 2,058 | 2,377 | 193 | 7,979 | 2.32 | 64.1 | 5.9 | 30.1 |
| Iowa | 16.2 | 12.4 | 7.6 | 49.6 | 22,780 | 23,691 | 4.0 | -0.2 | 2,699 | 2,112 | -625 | 9,850 | 2.38 | 65.8 | 7.3 | 27.8 |
| Iron | 18.8 | 18.1 | 14.2 | 49.7 | 6,861 | 5,916 | -13.8 | -3.7 | 373 | 873 | 283 | 2,898 | 1.92 | 57.3 | 7.6 | 35.8 |
| Jackson | 14.9 | 11.6 | 8.3 | 46.4 | 19,100 | 20,441 | 7.0 | 0.9 | 2,474 | 2,096 | -172 | 8,199 | 2.36 | 62.0 | 8.9 | 31.4 |
| Jefferson | 14.8 | 11.1 | 7.3 | 50.0 | 74,021 | 83,692 | 13.1 | 1.6 | 8,806 | 6,899 | -514 | 32,965 | 2.46 | 68.6 | 9.2 | 25.0 |
| Juneau | 16.0 | 12.6 | 8.8 | 46.8 | 24,316 | 26,665 | 9.7 | 0.9 | 2,781 | 2,955 | 433 | 10,752 | 2.28 | 62.7 | 8.3 | 29.5 |
| Kenosha | 14.5 | 9.1 | 6.0 | 50.6 | 149,577 | 166,424 | 11.3 | 2.0 | 19,680 | 14,491 | -1,862 | 64,111 | 2.56 | 65.3 | 12.1 | 27.6 |
| Kewaunee | 15.5 | 12.2 | 9.4 | 49.3 | 20,187 | 20,578 | 1.9 | -0.9 | 2,012 | 1,907 | -292 | 8,341 | 2.42 | 68.5 | 6.1 | 27.2 |
| La Crosse | 12.4 | 10.1 | 7.3 | 51.2 | 107,120 | 114,640 | 7.0 | 3.4 | 12,562 | 10,274 | 1,616 | 47,518 | 2.37 | 57.9 | 7.6 | 30.2 |
| Lafayette | 15.4 | 11.5 | 8.5 | 49.5 | 16,137 | 16,825 | 4.3 | -1.1 | 2,185 | 1,411 | -956 | 6,722 | 2.47 | 67.7 | 8.6 | 26.4 |
| Langlade | 17.8 | 14.4 | 10.8 | 49.6 | 20,740 | 19,977 | -3.7 | -4.3 | 2,010 | 2,412 | -453 | 8,612 | 2.2 | 63.0 | 7.3 | 32.1 |
| Lincoln | 18.3 | 13.1 | 9.7 | 49.4 | 29,641 | 28,743 | -3.0 | -4.1 | 2,733 | 3,396 | -499 | 12,625 | 2.14 | 64.3 | 7.2 | 29.2 |
| Manitowoc | 16.2 | 12.7 | 9.2 | 50.0 | 82,887 | 81,444 | -1.7 | -3.3 | 8,197 | 8,846 | -2,028 | 34,563 | 2.26 | 62.3 | 7.4 | 31.7 |
| Marathon | 14.4 | 10.6 | 8.0 | 49.9 | 125,834 | 134,073 | 6.5 | 1.1 | 16,254 | 12,128 | -2,545 | 55,466 | 2.41 | 66.9 | 7.9 | 26.2 |
| Marinette | 17.7 | 14.4 | 10.4 | 49.6 | 43,384 | 41,749 | -3.8 | -3.6 | 3,837 | 5,310 | 21 | 18,473 | 2.14 | 60.9 | 7.9 | 32.4 |
| Marquette | 18.5 | 15.6 | 10.3 | 48.9 | 15,832 | 15,397 | -2.7 | 1.2 | 1,516 | 1,893 | 573 | 6,648 | 2.27 | 62.5 | 7.7 | 30.0 |
| Menominee | 13.5 | 8.6 | 5.0 | 50.4 | 4,562 | 4,232 | -7.2 | 7.4 | 955 | 472 | -175 | 1,408 | 3.2 | 73.8 | 28.9 | 21.1 |
| Milwaukee | 11.8 | 8.6 | 5.8 | 51.6 | 940,164 | 947,712 | 0.8 | -0.3 | 140,467 | 85,158 | -57,860 | 382,175 | 2.43 | 55.9 | 16.2 | 35.7 |
| Monroe | 14.1 | 10.6 | 7.3 | 49.0 | 40,899 | 44,675 | 9.2 | 4.3 | 6,165 | 4,322 | 93 | 17,915 | 2.51 | 64.3 | 10.0 | 29.9 |
| Oconto | 17.8 | 13.4 | 8.1 | 48.6 | 35,634 | 37,657 | 5.7 | 1.9 | 3,751 | 3,706 | 710 | 16,036 | 2.33 | 67.9 | 7.1 | 26.2 |
| Oneida | 18.8 | 16.2 | 11.3 | 49.5 | 36,776 | 36,011 | -2.1 | -0.7 | 3,127 | 4,679 | 1,334 | 15,421 | 2.25 | 62.3 | 5.5 | 31.1 |
| Outagamie | 13.8 | 9.5 | 6.4 | 50.1 | 160,971 | 176,614 | 9.7 | 6.9 | 23,269 | 14,164 | 3,187 | 73,648 | 2.48 | 65.8 | 8.5 | 27.9 |
| Ozaukee | 15.1 | 12.0 | 9.1 | 50.7 | 82,317 | 86,395 | 5.0 | 4.2 | 8,408 | 7,828 | 3,127 | 35,807 | 2.43 | 70.5 | 6.7 | 24.9 |
| Pepin | 15.8 | 13.7 | 9.9 | 49.1 | 7,213 | 7,469 | 3.5 | -2.7 | 818 | 740 | -280 | 3,070 | 2.34 | 65.2 | 6.3 | 29.9 |
| Pierce | 14.0 | 9.8 | 5.9 | 50.1 | 36,804 | 41,029 | 11.5 | 4.1 | 3,950 | 2,800 | 518 | 15,593 | 2.52 | 65.1 | 6.1 | 25.0 |
| Polk | 17.0 | 13.2 | 8.8 | 49.6 | 41,319 | 44,200 | 7.0 | -0.9 | 4,263 | 4,533 | -127 | 18,450 | 2.33 | 66.7 | 7.8 | 27.6 |
| Portage | 13.4 | 10.7 | 7.2 | 49.3 | 67,182 | 70,021 | 4.2 | 1.4 | 6,997 | 5,442 | -511 | 28,675 | 2.35 | 60.6 | 7.6 | 28.1 |
| Price | 18.8 | 16.4 | 11.2 | 49.0 | 15,822 | 14,159 | -10.5 | -6.5 | 1,137 | 1,858 | -186 | 6,688 | 1.98 | 61.0 | 7.2 | 32.0 |
| Racine | 14.6 | 10.4 | 7.1 | 50.4 | 188,831 | 195,434 | 3.5 | 0.2 | 24,302 | 18,401 | -5,481 | 76,974 | 2.47 | 67.5 | 12.5 | 26.7 |
| Richland | 15.2 | 14.0 | 10.5 | 49.6 | 17,924 | 18,028 | 0.6 | -4.3 | 1,791 | 1,880 | -679 | 7,538 | 2.26 | 63.8 | 9.1 | 29.5 |
| Rock | 14.1 | 10.3 | 7.3 | 50.7 | 152,307 | 160,327 | 5.3 | 1.7 | 19,706 | 15,320 | -1,537 | 64,739 | 2.46 | 64.0 | 12.5 | 28.9 |
| Rusk | 16.9 | 15.2 | 10.5 | 49.2 | 15,347 | 14,755 | -3.9 | -5.0 | 1,443 | 1,852 | -330 | 6,376 | 2.18 | 62.7 | 7.1 | 31.4 |
| St. Croix | 14.2 | 9.5 | 5.8 | 49.9 | 63,155 | 84,332 | 33.5 | 8.9 | 10,634 | 6,041 | 2,943 | 34,450 | 2.57 | 71.4 | 7.8 | 23.0 |
| Sauk | 14.4 | 11.5 | 8.1 | 50.1 | 55,225 | 61,955 | 12.2 | 4.0 | 7,786 | 6,333 | 1,101 | 26,222 | 2.41 | 65.4 | 8.9 | 28.4 |
| Sawyer | 18.4 | 16.9 | 10.4 | 48.9 | 16,196 | 16,538 | 2.1 | 1.0 | 1,643 | 2,130 | 664 | 7,796 | 2.06 | 65.8 | 10.5 | 28.0 |
| Shawano | 15.9 | 12.3 | 9.8 | 49.9 | 40,664 | 41,955 | 3.2 | -2.8 | 4,445 | 4,781 | -807 | 17,004 | 2.37 | 65.6 | 7.8 | 28.4 |
| Sheboygan | 14.8 | 11.2 | 7.9 | 49.5 | 112,646 | 115,509 | 2.5 | -0.2 | 12,894 | 11,309 | -1,819 | 47,738 | 2.35 | 64.3 | 8.1 | 30.0 |
| Taylor | 16.7 | 11.3 | 9.0 | 49.1 | 19,680 | 20,689 | 5.1 | -1.8 | 2,283 | 1,890 | -779 | 8,614 | 2.34 | 68.1 | 6.7 | 26.4 |
| Trempealeau | 14.1 | 11.1 | 7.9 | 49.3 | 27,010 | 28,815 | 6.7 | 3.0 | 4,180 | 2,895 | -406 | 12,005 | 2.42 | 64.0 | 7.7 | 29.1 |
| Vernon | 15.3 | 12.1 | 8.4 | 49.7 | 28,056 | 29,774 | 6.1 | 3.7 | 4,289 | 3,057 | -121 | 12,071 | 2.5 | 66.8 | 7.4 | 28.4 |
| Vilas | 19.5 | 18.4 | 13.0 | 49.0 | 21,033 | 21,443 | 1.9 | 4.3 | 1,841 | 3,067 | 2,149 | 10,921 | 1.97 | 63.2 | 8.5 | 31.3 |
| Walworth | 14.8 | 11.5 | 7.6 | 50.0 | 93,759 | 102,222 | 9.0 | 1.7 | 10,261 | 9,515 | 1,041 | 40,874 | 2.45 | 65.8 | 9.2 | 26.4 |
| Washburn | 18.1 | 17.2 | 11.0 | 50.4 | 16,036 | 15,907 | -0.8 | -1.2 | 1,454 | 2,138 | 502 | 7,236 | 2.14 | 64.7 | 5.9 | 30.1 |
| Washington | 15.9 | 11.2 | 8.0 | 50.4 | 117,493 | 131,885 | 12.2 | 3.5 | 13,648 | 11,475 | 2,488 | 54,302 | 2.47 | 69.9 | 7.5 | 25.1 |
| Waukesha | 15.4 | 11.4 | 8.3 | 50.8 | 360,767 | 389,962 | 8.1 | 4.2 | 39,288 | 34,395 | 11,640 | 158,808 | 2.49 | 69.5 | 6.5 | 25.2 |
| Waupaca | 16.4 | 12.1 | 9.4 | 49.4 | 51,731 | 52,409 | 1.3 | -3.3 | 5,294 | 7,392 | 389 | 22,305 | 2.23 | 62.6 | 7.3 | 30.4 |
| Waushara | 17.8 | 15.4 | 10.3 | 47.5 | 23,154 | 24,510 | 5.9 | -0.8 | 2,252 | 2,820 | 398 | 9,934 | 2.31 | 67.7 | 7.5 | 28.2 |
| Winnebago | 13.5 | 9.9 | 7.3 | 49.6 | 156,763 | 167,072 | 6.6 | 2.7 | 19,020 | 15,218 | 875 | 70,594 | 2.3 | 59.2 | 9.2 | 31.5 |
| Wood | 15.5 | 12.0 | 9.5 | 50.6 | 75,555 | 74,750 | -1.1 | -2.9 | 8,398 | 8,139 | -2,443 | 32,332 | 2.23 | 61.4 | 8.3 | 32.1 |
| **WYOMING** | 13.3 | 11.0 | 6.9 | 49.0 | 493,782 | 563,775 | 14.2 | 3.3 | 74,224 | 48,935 | -7,214 | 230,101 | 2.46 | 64.6 | 8.1 | 28.2 |
| Albany | 9.4 | 8.3 | 4.6 | 48.0 | 32,014 | 36,299 | 13.4 | 7.3 | 3,993 | 1,963 | 606 | 15,944 | 2.26 | 47.5 | 4.8 | 31.7 |
| Big Horn | 13.9 | 12.7 | 9.3 | 49.6 | 11,461 | 11,669 | 1.8 | -0.8 | 1,415 | 1,431 | -79 | 4,475 | 2.6 | 67.2 | 7.6 | 29.1 |
| Campbell | 13.2 | 8.1 | 3.5 | 48.9 | 33,698 | 46,133 | 36.9 | 1.2 | 7,210 | 2,653 | -4,092 | 17,574 | 2.66 | 70.5 | 6.4 | 22.8 |
| Carbon | 13.8 | 11.1 | 7.0 | 46.2 | 15,639 | 15,884 | 1.6 | -7.4 | 2,008 | 1,348 | -1,840 | 6,204 | 2.33 | 65.4 | 9.9 | 29.3 |
| Converse | 14.3 | 10.8 | 7.0 | 49.0 | 12,052 | 13,833 | 14.8 | -0.2 | 1,829 | 1,258 | -625 | 5,378 | 2.58 | 70.1 | 4.4 | 24.3 |
| Crook | 16.6 | 13.5 | 8.5 | 50.1 | 5,887 | 7,080 | 20.3 | 7.2 | 1,047 | 649 | 118 | 2,919 | 2.53 | 68.7 | 3.5 | 27.5 |
| Fremont | 13.6 | 12.0 | 7.7 | 49.6 | 35,804 | 40,123 | 12.1 | -2.0 | 5,770 | 4,498 | -2,100 | 14,907 | 2.61 | 64.8 | 9.5 | 28.3 |

1. No spouse present.

# Table B. States and Counties — Population, Vital Statistics, Health, and Crime

| STATE County | Persons in group quarters, 2020 (32) | Daytime Pop. 2015–2019 Number (33) | Daytime Pop. Empl./residence ratio (34) | Births 2020 Total (35) | Births 2020 Rate[1] (36) | Deaths 2020 Number (37) | Deaths 2020 Rate[1] (38) | Under 65 no health ins. 2019 Number (39) | Percent (40) | Medicare 2020 Total beneficiaries (41) | Enrolled in Original Medicare (42) | Enrolled in Medicare Advantage (43) | Crimes Violent (44) | Crimes Property (45) |
|---|---|---|---|---|---|---|---|---|---|---|---|---|---|---|
| **WISCONSIN—Cont'd** | | | | | | | | | | | | | | |
| Douglas | 1,306 | 40,245 | 0.86 | 401 | 9.2 | 438 | 10.0 | 2,360 | 7.0 | 10,353 | 5,119 | 5,235 | 15 | 156 |
| Dunn | 3,282 | 41,485 | 0.86 | 407 | 9.0 | 348 | 7.7 | 2,357 | 6.7 | 9,271 | 6,255 | 3,016 | 47 | 82 |
| Eau Claire | 4,585 | 109,215 | 1.10 | 1,094 | 10.4 | 841 | 8.0 | 5,855 | 7.0 | 20,288 | 12,927 | 7,361 | 29 | 192 |
| Florence | 54 | 3,492 | 0.56 | 25 | 5.8 | 57 | 13.3 | 262 | 8.4 | 1,525 | 997 | 527 | 11 | 16 |
| Fond du Lac | 3,311 | 98,378 | 0.92 | 1,042 | 10.1 | 998 | 9.7 | 4,835 | 6.0 | 22,377 | 9,881 | 12,496 | 26 | 204 |
| Forest | 312 | 8,820 | 0.95 | 97 | 10.8 | 122 | 13.6 | 716 | 10.6 | 2,656 | 1,588 | 1,068 | 8 | 52 |
| Grant | 4,041 | 47,230 | 0.82 | 557 | 10.9 | 526 | 10.3 | 2,990 | 7.8 | 10,856 | 5,839 | 5,017 | 38 | 133 |
| Green | 350 | 33,696 | 0.84 | 343 | 9.4 | 374 | 10.2 | 1,921 | 6.5 | 8,249 | 6,224 | 2,025 | 7 | 56 |
| Green Lake | 177 | 17,154 | 0.82 | 193 | 10.2 | 237 | 12.5 | 1,372 | 9.5 | 5,111 | 2,207 | 2,905 | 4 | 49 |
| Iowa | 176 | 22,804 | 0.93 | 243 | 10.3 | 205 | 8.7 | 1,137 | 6.0 | 5,388 | 3,541 | 1,847 | 8 | 43 |
| Iron | 90 | 5,033 | 0.73 | 37 | 6.5 | 95 | 16.7 | 281 | 7.4 | 2,055 | 1,143 | 913 | 7 | 42 |
| Jackson | 1,299 | 19,937 | 0.94 | 235 | 11.4 | 193 | 9.4 | 1,524 | 9.9 | 4,610 | 2,889 | 1,721 | 11 | 131 |
| Jefferson | 3,159 | 74,586 | 0.78 | 774 | 9.1 | 749 | 8.8 | 4,281 | 6.3 | 17,584 | 11,441 | 6,143 | 35 | 231 |
| Juneau | 1,629 | 24,797 | 0.86 | 254 | 9.4 | 294 | 10.9 | 1,824 | 9.2 | 6,610 | 4,802 | 1,808 | 50 | 141 |
| Kenosha | 4,219 | 153,039 | 0.82 | 1,802 | 10.6 | 1,497 | 8.8 | 10,840 | 7.7 | 30,041 | 18,117 | 11,924 | 50 | 253 |
| Kewaunee | 158 | 17,499 | 0.73 | 190 | 9.3 | 199 | 9.8 | 1,071 | 6.7 | 4,659 | 2,046 | 2,613 | 13 | 42 |
| La Crosse | 5,387 | 127,042 | 1.15 | 1,147 | 9.7 | 1,080 | 9.1 | 4,913 | 5.2 | 23,557 | 13,472 | 10,085 | 15 | 126 |
| Lafayette | 98 | 14,244 | 0.71 | 221 | 13.3 | 156 | 9.4 | 1,536 | 11.6 | 3,509 | 2,150 | 1,359 | NA | NA |
| Langlade | 276 | 18,767 | 0.96 | 200 | 10.5 | 270 | 14.1 | 1,117 | 7.9 | 5,720 | 2,912 | 2,807 | NA | NA |
| Lincoln | 588 | 25,747 | 0.86 | 252 | 9.1 | 323 | 11.7 | 1,246 | 5.9 | 7,572 | 3,918 | 3,654 | 38 | 56 |
| Manitowoc | 983 | 74,291 | 0.88 | 754 | 9.6 | 889 | 11.3 | 4,101 | 6.6 | 20,073 | 9,657 | 10,415 | 32 | 131 |
| Marathon | 1,429 | 137,855 | 1.03 | 1,532 | 11.3 | 1,228 | 9.1 | 7,473 | 6.8 | 27,931 | 13,183 | 14,748 | 41 | 319 |
| Marinette | 589 | 41,490 | 1.06 | 353 | 8.8 | 560 | 13.9 | 2,033 | 6.8 | 12,077 | 6,242 | 5,835 | 10 | 198 |
| Marquette | 137 | 12,888 | 0.65 | 150 | 9.6 | 182 | 11.7 | 999 | 8.7 | 4,595 | 2,810 | 1,785 | 9 | 69 |
| Menominee | 57 | 5,303 | 1.51 | 85 | 18.7 | 73 | 16.1 | 301 | 8.0 | 871 | 578 | 293 | NA | NA |
| Milwaukee | 22,370 | 977,190 | 1.06 | 12,911 | 13.7 | 8,761 | 9.3 | 67,133 | 8.5 | 156,821 | 71,942 | 84,879 | 107 | 92 |
| Monroe | 1,125 | 47,241 | 1.07 | 567 | 12.2 | 449 | 9.6 | 3,344 | 8.9 | 9,444 | 6,484 | 2,961 | 30 | 101 |
| Oconto | 257 | 29,697 | 0.58 | 388 | 10.1 | 405 | 10.6 | 2,051 | 6.9 | 9,566 | 3,676 | 5,890 | 19 | 212 |
| Oneida | 599 | 35,493 | 1.01 | 309 | 8.6 | 479 | 13.4 | 1,821 | 7.1 | 12,133 | 7,276 | 4,858 | 28 | 150 |
| Outagamie | 2,866 | 193,060 | 1.07 | 2,193 | 11.6 | 1,504 | 8.0 | 10,565 | 6.7 | 34,192 | 12,246 | 21,946 | 37 | 265 |
| Ozaukee | 1,929 | 85,098 | 0.92 | 829 | 9.2 | 879 | 9.8 | 2,890 | 4.1 | 20,707 | 11,305 | 9,402 | 13 | 76 |
| Pepin | 61 | 6,576 | 0.81 | 78 | 10.7 | 87 | 12.0 | 419 | 7.6 | 2,076 | 1,632 | 444 | 6 | 30 |
| Pierce | 2,756 | 31,255 | 0.54 | 361 | 8.5 | 302 | 7.1 | 1,831 | 5.4 | 7,577 | 4,269 | 3,308 | 4 | 99 |
| Polk | 385 | 38,864 | 0.78 | 387 | 8.8 | 485 | 11.1 | 2,358 | 6.9 | 11,266 | 6,246 | 5,020 | 35 | 167 |
| Portage | 3,037 | 71,885 | 1.03 | 631 | 8.9 | 551 | 7.8 | 3,638 | 6.5 | 14,456 | 7,751 | 6,705 | 40 | 138 |
| Price | 157 | 13,601 | 1.03 | 108 | 8.2 | 171 | 12.9 | 710 | 7.3 | 4,498 | 2,691 | 1,808 | 4 | 73 |
| Racine | 4,910 | 181,285 | 0.85 | 2,267 | 11.6 | 1,885 | 9.6 | 11,317 | 7.2 | 41,218 | 20,939 | 20,279 | NA | NA |
| Richland | 295 | 15,884 | 0.81 | 162 | 9.4 | 186 | 10.8 | 1,195 | 9.2 | 4,508 | 3,641 | 868 | NA | NA |
| Rock | 2,512 | 151,458 | 0.86 | 1,832 | 11.2 | 1,578 | 9.7 | 9,273 | 7.0 | 33,950 | 20,168 | 13,782 | 56 | 244 |
| Rusk | 161 | 13,815 | 0.95 | 143 | 10.2 | 180 | 12.8 | 955 | 9.1 | 4,132 | 2,452 | 1,679 | 13 | 44 |
| St. Croix | 749 | 77,506 | 0.77 | 995 | 10.8 | 666 | 7.3 | 3,457 | 4.5 | 15,965 | 9,178 | 6,787 | 22 | 278 |
| Sauk | 724 | 66,330 | 1.07 | 725 | 11.2 | 650 | 10.1 | 4,315 | 8.3 | 14,415 | 9,394 | 5,020 | 27 | 304 |
| Sawyer | 314 | 16,528 | 1.02 | 136 | 8.1 | 222 | 13.3 | 1,287 | 10.9 | 5,604 | 3,561 | 2,044 | 38 | 105 |
| Shawano | 661 | 35,804 | 0.74 | 430 | 10.5 | 452 | 11.1 | 2,833 | 9.0 | 10,096 | 3,920 | 6,176 | NA | NA |
| Sheboygan | 2,818 | 116,377 | 1.02 | 1,229 | 10.7 | 1,150 | 10.0 | 5,748 | 6.3 | 25,354 | 12,529 | 12,825 | 27 | 271 |
| Taylor | 186 | 19,599 | 0.93 | 212 | 10.4 | 193 | 9.5 | 1,389 | 8.6 | 4,339 | 2,267 | 2,072 | 21 | 56 |
| Trempealeau | 450 | 29,573 | 1.00 | 404 | 13.6 | 281 | 9.5 | 1,937 | 8.1 | 6,312 | 3,899 | 2,413 | 11 | 82 |
| Vernon | 323 | 27,557 | 0.77 | 409 | 13.3 | 329 | 10.7 | 2,292 | 9.4 | 7,085 | 3,625 | 3,460 | 3 | 77 |
| Vilas | 231 | 21,665 | 0.99 | 179 | 8.0 | 332 | 14.9 | 1,642 | 10.9 | 8,547 | 5,676 | 2,871 | 4 | 101 |
| Walworth | 3,029 | 98,767 | 0.92 | 933 | 9.0 | 1,021 | 9.8 | 6,732 | 8.2 | 21,659 | 14,477 | 7,182 | 10 | 152 |
| Washburn | 201 | 15,330 | 0.95 | 117 | 7.4 | 238 | 15.1 | 869 | 7.7 | 5,532 | 3,319 | 2,212 | 19 | 123 |
| Washington | 977 | 120,007 | 0.80 | 1,280 | 9.4 | 1,259 | 9.2 | 5,082 | 4.6 | 28,799 | 14,667 | 14,132 | 7 | 192 |
| Waukesha | 5,335 | 433,911 | 1.16 | 3,810 | 9.4 | 3,765 | 9.3 | 13,222 | 4.0 | 88,785 | 45,849 | 42,936 | 58 | 305 |
| Waupaca | 1,389 | 47,561 | 0.86 | 527 | 10.4 | 742 | 14.6 | 2,679 | 6.8 | 13,098 | 5,386 | 7,712 | 52 | 298 |
| Waushara | 1,178 | 20,974 | 0.69 | 195 | 8.0 | 283 | 11.6 | 1,542 | 9.0 | 7,046 | 2,942 | 4,104 | 36 | 158 |
| Winnebago | 7,408 | 180,292 | 1.11 | 1,770 | 10.3 | 1,598 | 9.3 | 8,357 | 6.1 | 33,531 | 12,842 | 20,689 | 18 | 155 |
| Wood | 713 | 76,175 | 1.09 | 778 | 10.7 | 819 | 11.3 | 3,278 | 5.7 | 18,146 | 8,173 | 9,973 | 18 | 123 |
| **WYOMING** | 14,365 | 588,073 | 1.02 | 6,367 | 10.9 | 5,464 | 9.4 | 69,044 | 14.7 | 113,787 | 108,193 | 5,594 | X | X |
| Albany | 2,262 | 38,209 | 0.99 | 335 | 8.6 | 213 | 5.5 | 4,179 | 13.1 | 5,405 | 4,984 | 421 | 6 | 28 |
| Big Horn | 183 | 11,673 | 0.96 | 116 | 10.0 | 153 | 13.2 | 1,942 | 21.5 | 2,847 | 2,790 | 57 | 3 | 20 |
| Campbell | 518 | 50,025 | 1.11 | 583 | 12.5 | 367 | 7.9 | 5,531 | 13.5 | 6,114 | 6,011 | 103 | 73 | 118 |
| Carbon | 752 | 15,339 | 1.01 | 168 | 11.4 | 143 | 9.7 | 2,029 | 17.7 | 2,879 | 2,719 | 160 | 9 | 38 |
| Converse | 103 | 13,979 | 1.01 | 161 | 11.7 | 126 | 9.1 | 1,564 | 13.7 | 2,634 | 2,585 | 49 | 11 | 34 |
| Crook | 34 | 7,049 | 0.87 | 104 | 13.7 | 73 | 9.6 | 1,021 | 17.1 | 1,704 | 1,590 | 114 | 4 | 17 |
| Fremont | 1,059 | 39,295 | 0.97 | 493 | 12.5 | 525 | 13.4 | 6,931 | 22.3 | 8,754 | 8,116 | 639 | 13 | 183 |

1. Per 1,000 estimated resident population.

# Table B. States and Counties — Crime, Education, Money Income, and Poverty

Column legend (item numbers):

|  |  |
|---|---|
| 46 | County law enforcement employment, 2019 — Officers |
| 47 | County law enforcement employment, 2019 — Civilians |
| 48 | School enrollment and attainment 2015–2019, Enrollment[1] — Total |
| 49 | Enrollment — Percent private |
| 50 | Attainment[2] (percent) — High school graduate or less |
| 51 | Attainment[2] (percent) — Bachelor's degree or more |
| 52 | Local government expenditures[3] 2017–2018 — Total current spending (mil dol) |
| 53 | Local government expenditures[3] 2017–2018 — Current spending per student (dollars) |
| 54 | Money income 2015–2019 — Per capita income[4] |
| 55 | Money income 2015–2019, Households — Median income (dollars) |
| 56 | Households — Percent with income of less than $50,000 |
| 57 | Households — Percent with income of $200,000 or more |
| 58 | Income and poverty 2019 — Median household income (dollars) |
| 59 | Percent below poverty level — All persons |
| 60 | Percent below poverty level — Children under 18 years |
| 61 | Percent below poverty level — Children 5 to 17 years in families |

| STATE County | 46 | 47 | 48 | 49 | 50 | 51 | 52 | 53 | 54 | 55 | 56 | 57 | 58 | 59 | 60 | 61 |
|---|---|---|---|---|---|---|---|---|---|---|---|---|---|---|---|---|
| **WISCONSIN—Cont'd** | | | | | | | | | | | | | | | | |
| Douglas | 31 | 42 | 9,553 | 10.4 | 37.4 | 24.1 | 76 | 12,072 | 30,153 | 53,986 | 46.3 | 2.4 | 55,651 | 11.5 | 15.8 | 15.2 |
| Dunn | 38 | 30 | 14,408 | 6.0 | 39.6 | 27.9 | 68 | 11,030 | 28,034 | 58,783 | 42.0 | 3.0 | 61,506 | 11.7 | 12.9 | 10.2 |
| Eau Claire | 42 | 66 | 29,973 | 8.4 | 31.0 | 32.3 | 171 | 11,849 | 30,983 | 59,476 | 42.3 | 3.8 | 64,972 | 10.6 | 11.1 | 10.7 |
| Florence | 12 | 9 | 694 | 18.6 | 44.6 | 19.9 | 7 | 17,695 | 32,348 | 52,181 | 47.9 | 2.8 | 58,270 | 10.9 | 16.8 | 15.5 |
| Fond du Lac | 59 | 65 | 23,712 | 22.7 | 44.2 | 23.1 | 150 | 11,377 | 31,653 | 62,391 | 39.2 | 3.2 | 65,905 | 7.3 | 10.1 | 7.1 |
| Forest | 22 | 21 | 1,780 | 8.5 | 50.4 | 14.7 | 22 | 14,383 | 25,700 | 45,536 | 54.4 | 1.4 | 48,964 | 16.3 | 23.2 | 21.9 |
| Grant | 30 | 29 | 15,608 | 9.9 | 42.9 | 23.8 | 99 | 14,055 | 25,892 | 54,800 | 45.3 | 2.1 | 56,418 | 13.4 | 15.4 | 15.1 |
| Green | 41 | 14 | 8,198 | 8.4 | 43.0 | 23.6 | 67 | 12,519 | 33,459 | 64,502 | 38.0 | 4.4 | 64,319 | 6.4 | 8.8 | 7.7 |
| Green Lake | 19 | 25 | 3,735 | 14.8 | 48.9 | 20.2 | 35 | 11,416 | 29,207 | 55,075 | 44.9 | 2.9 | 59,458 | 11.7 | 18.5 | 17.4 |
| Iowa | 19 | 19 | 5,113 | 8.6 | 40.1 | 24.6 | 47 | 13,405 | 34,114 | 64,124 | 37.7 | 3.6 | 67,390 | 9.5 | 11.4 | 10.1 |
| Iron | 11 | 12 | 870 | 7.9 | 39.1 | 20.0 | 10 | 14,393 | 28,857 | 43,798 | 55.8 | 1.3 | 48,813 | 12.7 | 20.8 | 17.2 |
| Jackson | 21 | 26 | 4,201 | 8.9 | 51.7 | 14.6 | 37 | 12,648 | 26,481 | 53,650 | 47.6 | 2.2 | 57,247 | 12.4 | 16.0 | 14.9 |
| Jefferson | 93 | 23 | 20,217 | 16.5 | 41.6 | 25.4 | 161 | 12,292 | 31,992 | 66,291 | 37.8 | 3.7 | 71,098 | 7.1 | 8.4 | 7.7 |
| Juneau | 42 | 56 | 5,063 | 19.0 | 53.0 | 13.7 | 46 | 12,188 | 27,889 | 53,490 | 46.0 | 2.8 | 55,787 | 13.4 | 19.3 | 18.5 |
| Kenosha | 121 | 212 | 43,092 | 14.3 | 40.0 | 26.1 | 379 | 13,250 | 32,298 | 63,733 | 38.6 | 5.3 | 66,460 | 9.8 | 12.3 | 10.9 |
| Kewaunee | 31 | 5 | 4,397 | 18.1 | 46.8 | 19.4 | 37 | 10,381 | 30,921 | 66,192 | 37.5 | 2.5 | 65,879 | 7.4 | 8.6 | 7.8 |
| La Crosse | 40 | 68 | 33,841 | 13.1 | 30.4 | 34.9 | 206 | 12,761 | 32,565 | 57,882 | 42.7 | 4.8 | 59,758 | 11.8 | 9.6 | 8.6 |
| Lafayette | 28 | 2 | 3,709 | 9.0 | 49.7 | 18.8 | 39 | 14,388 | 28,448 | 60,691 | 41.4 | 2.4 | 59,128 | 11.3 | 15.9 | 15.5 |
| Langlade | 18 | 27 | 3,564 | 15.1 | 52.1 | 16.7 | 45 | 16,189 | 27,618 | 49,491 | 50.4 | 1.8 | 49,597 | 13.0 | 19.9 | 17.7 |
| Lincoln | 29 | 32 | 5,206 | 17.2 | 46.1 | 18.4 | 53 | 11,584 | 30,972 | 58,541 | 43.4 | 2.2 | 60,728 | 9.1 | 12.2 | 11.4 |
| Manitowoc | 62 | 37 | 16,440 | 19.6 | 46.5 | 20.2 | 120 | 11,519 | 30,130 | 56,612 | 44.6 | 2.5 | 61,326 | 8.8 | 12.1 | 11.4 |
| Marathon | 70 | 114 | 30,970 | 13.3 | 42.4 | 25.4 | 235 | 12,145 | 33,189 | 62,633 | 40.2 | 4.5 | 66,092 | 8.2 | 11.2 | 9.5 |
| Marinette | 32 | 45 | 7,811 | 9.4 | 50.6 | 15.9 | 69 | 11,494 | 28,282 | 50,330 | 49.6 | 2.0 | 54,404 | 11.5 | 17.0 | 14.8 |
| Marquette | 20 | 1 | 2,703 | 18.0 | 51.8 | 15.1 | 21 | 11,978 | 27,922 | 52,288 | 47.2 | 1.3 | 56,790 | 10.0 | 14.3 | 13.4 |
| Menominee | 9 | 7 | 1,343 | 4.4 | 50.7 | 17.9 | 18 | 19,963 | 20,313 | 40,921 | 56.8 | 3.4 | 40,300 | 25.3 | 44.0 | 44.7 |
| Milwaukee | 330 | 475 | 249,664 | 24.6 | 40.1 | 31.0 | 1,699 | 12,541 | 29,270 | 50,606 | 49.5 | 3.8 | 53,505 | 16.9 | 24.0 | 23.9 |
| Monroe | 27 | 29 | 10,665 | 18.5 | 46.0 | 20.6 | 84 | 11,564 | 29,242 | 59,587 | 41.4 | 3.1 | 60,296 | 10.9 | 16.4 | 16.7 |
| Oconto | 32 | 39 | 7,299 | 7.8 | 51.3 | 16.8 | 49 | 12,077 | 30,927 | 60,908 | 40.5 | 2.8 | 66,401 | 9.3 | 12.7 | 12.5 |
| Oneida | 37 | 49 | 5,997 | 15.2 | 38.8 | 27.2 | 61 | 14,498 | 34,910 | 56,852 | 44.0 | 5.0 | 62,765 | 8.9 | 12.6 | 11.5 |
| Outagamie | 72 | 111 | 45,483 | 14.8 | 37.5 | 29.7 | 395 | 11,128 | 33,221 | 65,572 | 36.9 | 4.3 | 66,994 | 7.1 | 9.3 | 8.1 |
| Ozaukee | 78 | 25 | 22,012 | 29.4 | 23.1 | 49.0 | 145 | 11,447 | 49,030 | 85,215 | 28.3 | 12.4 | 90,446 | 4.5 | 4.4 | 3.8 |
| Pepin | 8 | 11 | 1,387 | 15.1 | 47.0 | 20.6 | 15 | 12,659 | 30,827 | 54,583 | 45.0 | 3.1 | 62,675 | 9.0 | 14.1 | 13.4 |
| Pierce | 36 | 14 | 12,846 | 8.6 | 37.0 | 29.2 | 103 | 13,190 | 33,061 | 72,323 | 32.6 | 4.7 | 75,368 | 9.8 | 7.0 | 6.3 |
| Polk | 29 | 46 | 8,788 | 9.8 | 43.5 | 20.9 | 86 | 12,088 | 30,859 | 59,994 | 41.9 | 2.7 | 63,910 | 8.6 | 11.9 | 11.4 |
| Portage | 51 | 56 | 19,693 | 8.7 | 38.0 | 32.9 | 108 | 11,644 | 31,453 | 58,853 | 41.8 | 3.9 | 58,248 | 11.8 | 10.5 | 9.2 |
| Price | 19 | 6 | 2,334 | 10.3 | 48.6 | 16.5 | 24 | 12,714 | 28,882 | 47,956 | 52.6 | 1.4 | 48,761 | 14.9 | 21.8 | 19.6 |
| Racine | 159 | 102 | 46,845 | 18.1 | 41.2 | 25.2 | 360 | 13,024 | 31,572 | 61,336 | 40.4 | 4.6 | 61,691 | 12.4 | 16.8 | 16.0 |
| Richland | 17 | 15 | 3,642 | 17.3 | 50.7 | 19.0 | 22 | 11,986 | 27,597 | 51,947 | 47.7 | 2.1 | 51,295 | 12.5 | 18.5 | 16.4 |
| Rock | 92 | 107 | 38,765 | 13.6 | 45.4 | 22.1 | 333 | 12,099 | 28,945 | 57,875 | 43.4 | 2.7 | 61,624 | 12.3 | 17.7 | 16.1 |
| Rusk | 28 | 3 | 2,585 | 18.4 | 53.5 | 16.6 | 25 | 13,943 | 25,854 | 47,532 | 52.4 | 1.6 | 49,860 | 12.3 | 21.4 | 19.5 |
| St. Croix | 77 | 7 | 22,464 | 11.9 | 28.4 | 35.4 | 162 | 11,068 | 39,287 | 84,756 | 27.2 | 7.5 | 87,098 | 5.9 | 6.6 | 6.1 |
| Sauk | 47 | 95 | 13,684 | 15.1 | 42.9 | 24.1 | 136 | 11,524 | 30,990 | 59,943 | 40.9 | 3.4 | 63,752 | 10.2 | 15.0 | 14.2 |
| Sawyer | 26 | 21 | 3,214 | 11.1 | 41.7 | 24.2 | 35 | 15,088 | 31,772 | 47,711 | 51.4 | 3.5 | 53,908 | 12.7 | 20.8 | 20.4 |
| Shawano | 38 | 71 | 8,362 | 12.1 | 53.4 | 15.7 | 63 | 13,274 | 29,338 | 56,531 | 44.3 | 2.8 | 58,982 | 9.6 | 14.6 | 11.6 |
| Sheboygan | 71 | 118 | 26,669 | 17.1 | 42.7 | 25.1 | 228 | 12,038 | 31,510 | 60,696 | 40.6 | 3.3 | 62,108 | 8.1 | 11.4 | 10.2 |
| Taylor | 19 | 25 | 4,430 | 12.6 | 55.1 | 15.1 | 39 | 11,576 | 27,506 | 53,020 | 46.5 | 2.5 | 52,419 | 9.7 | 12.5 | 11.9 |
| Trempealeau | 27 | 30 | 6,623 | 10.2 | 48.4 | 19.3 | 71 | 12,021 | 28,902 | 58,548 | 41.9 | 2.4 | 61,687 | 8.9 | 11.3 | 10.5 |
| Vernon | 22 | 29 | 6,589 | 22.2 | 48.3 | 22.4 | 50 | 12,307 | 26,198 | 52,459 | 47.8 | 2.7 | 56,272 | 14.1 | 23.8 | 22.7 |
| Vilas | 39 | 37 | 3,324 | 12.8 | 36.3 | 29.1 | 45 | 17,127 | 33,316 | 47,072 | 53.6 | 4.3 | 55,576 | 11.2 | 19.0 | 18.9 |
| Walworth | 82 | 112 | 27,792 | 10.3 | 40.4 | 28.8 | 193 | 12,215 | 32,302 | 63,776 | 39.2 | 4.9 | 64,682 | 9.5 | 9.3 | 8.4 |
| Washburn | 17 | 18 | 2,986 | 14.6 | 41.4 | 23.0 | 32 | 12,707 | 29,584 | 50,280 | 49.7 | 1.8 | 52,735 | 12.8 | 18.3 | 18.0 |
| Washington | 76 | 92 | 31,056 | 22.5 | 35.4 | 31.5 | 220 | 11,144 | 39,055 | 77,663 | 30.6 | 6.2 | 81,928 | 5.1 | 5.6 | 4.4 |
| Waukesha | 166 | 161 | 94,869 | 21.6 | 26.4 | 44.5 | 710 | 11,478 | 46,073 | 87,277 | 26.6 | 11.5 | 91,070 | 4.7 | 5.2 | 4.7 |
| Waupaca | 94 | 17 | 10,206 | 14.3 | 48.9 | 19.9 | 99 | 11,742 | 32,000 | 58,693 | 42.3 | 3.2 | 61,773 | 8.8 | 12.2 | 11.4 |
| Waushara | 25 | 29 | 4,104 | 9.6 | 55.2 | 16.3 | 30 | 11,704 | 28,580 | 52,810 | 47.0 | 2.4 | 55,476 | 12.4 | 18.8 | 16.7 |
| Winnebago | 140 | 59 | 42,179 | 11.3 | 40.5 | 28.3 | 266 | 11,537 | 32,571 | 58,543 | 42.0 | 3.9 | 59,643 | 9.7 | 11.8 | 11.1 |
| Wood | 41 | 30 | 15,071 | 13.2 | 45.3 | 21.6 | 143 | 11,831 | 31,425 | 54,913 | 44.9 | 3.6 | 57,325 | 10.7 | 15.3 | 14.4 |
| **WYOMING** | X | X | 144,936 | 9.1 | 35.9 | 27.4 | 1,519 | 16,134 | 33,366 | 64,049 | 39.0 | 4.6 | 66,152 | 9.9 | 11.7 | 10.1 |
| Albany | 44 | 3 | 16,549 | 5.5 | 18.7 | 51.8 | 65 | 15,843 | 29,146 | 49,322 | 50.7 | 2.2 | 52,216 | 16.0 | 12.0 | 10.5 |
| Big Horn | 15 | 25 | 2,813 | 9.0 | 39.4 | 19.0 | 45 | 17,483 | 24,693 | 52,804 | 48.0 | 4.8 | 53,018 | 12.1 | 16.5 | 14.1 |
| Campbell | 56 | 95 | 11,577 | 7.2 | 42.7 | 18.9 | 142 | 16,299 | 34,776 | 82,659 | 29.7 | 2.7 | 83,997 | 7.7 | 8.6 | 7.2 |
| Carbon | 17 | 38 | 3,431 | 13.4 | 47.2 | 20.6 | 43 | 18,340 | 29,552 | 60,161 | 42.6 | 4.0 | 61,626 | 11.7 | 13.9 | 11.8 |
| Converse | 18 | 26 | 3,229 | 13.7 | 48.2 | 17.3 | 41 | 17,774 | 33,693 | 69,647 | 36.4 | 4.0 | 72,481 | 9.2 | 12.0 | 10.9 |
| Crook | 14 | 7 | 1,651 | 4.2 | 36.3 | 22.3 | 21 | 18,003 | 29,532 | 65,132 | 39.3 | 4.0 | 64,120 | 8.0 | 10.3 | 10.2 |
| Fremont | 65 | 28 | 9,552 | 7.5 | 37.6 | 24.3 | 138 | 19,326 | 28,324 | 55,896 | 45.1 | 4.4 | 57,953 | 12.9 | 16.9 | 15.1 |

1. All persons 3 years old and over enrolled in nursery school through college.   2. Persons 25 years old and over.   3. Elementary and secondary education expenditures.   4. Based on population estimated by the American Community Survey, 2014–2018.

| STATE County | Total (mil dol) | Percent change 2018-2019 | Per capita[1] Dollars | Per capita[1] Rank | Wages and salaries (mil dol) | Supplements to wages and salaries, employer contributions (mil dol) Pension and insurance | Supplements to wages and salaries, employer contributions (mil dol) Government social insurance | Proprietors' income (mil dol) | Dividends, interest, and rent (mil dol) | Personal transfer receipts (mil dol) | Total (mil dol) | Contributions for government social insurance (mil dol) From employee and self-employed | Contributions for government social insurance (mil dol) From employer |
|---|---|---|---|---|---|---|---|---|---|---|---|---|---|
| | 62 | 63 | 64 | 65 | 66 | 67 | 68 | 69 | 70 | 71 | 72 | 73 | 74 |
| WISCONSIN—Cont'd | | | | | | | | | | | | | |
| Douglas | 1,939 | 4.0 | 44,941 | 1,363 | 808 | 190 | 71 | 97 | 301 | 511 | 1,165 | 80 | 71 |
| Dunn | 1,870 | 3.9 | 41,208 | 1,891 | 842 | 194 | 65 | 103 | 319 | 404 | 1,204 | 77 | 65 |
| Eau Claire | 5,217 | 3.1 | 49,851 | 839 | 3,089 | 585 | 225 | 346 | 1,095 | 919 | 4,245 | 262 | 225 |
| Florence | 229 | 1.7 | 53,310 | 589 | 34 | 10 | 3 | 17 | 50 | 59 | 64 | 6 | 3 |
| Fond du Lac | 5,165 | 3.8 | 49,949 | 833 | 2,485 | 468 | 193 | 384 | 873 | 971 | 3,530 | 223 | 193 |
| Forest | 384 | 4.4 | 42,663 | 1,664 | 137 | 40 | 10 | 18 | 82 | 118 | 204 | 15 | 10 |
| Grant | 2,335 | 4.2 | 45,392 | 1,314 | 805 | 219 | 61 | 334 | 394 | 471 | 1,419 | 85 | 61 |
| Green | 1,988 | 5.2 | 53,782 | 557 | 735 | 156 | 56 | 162 | 389 | 330 | 1,109 | 70 | 56 |
| Green Lake | 875 | 2.9 | 46,260 | 1,216 | 264 | 60 | 21 | 73 | 209 | 194 | 419 | 30 | 21 |
| Iowa | 1,191 | 5.0 | 50,309 | 808 | 523 | 96 | 39 | 85 | 230 | 196 | 743 | 47 | 39 |
| Iron | 286 | 1.9 | 50,249 | 810 | 62 | 15 | 5 | 18 | 70 | 83 | 100 | 9 | 5 |
| Jackson | 956 | 1.4 | 46,296 | 1,212 | 414 | 96 | 33 | 58 | 188 | 205 | 600 | 37 | 33 |
| Jefferson | 3,997 | 2.6 | 47,152 | 1,112 | 1,544 | 327 | 117 | 197 | 639 | 740 | 2,186 | 143 | 117 |
| Juneau | 1,047 | 3.1 | 39,218 | 2,160 | 402 | 98 | 31 | 57 | 171 | 282 | 587 | 40 | 31 |
| Kenosha | 8,240 | 3.5 | 48,596 | 954 | 3,386 | 654 | 254 | 398 | 1,211 | 1,471 | 4,692 | 300 | 254 |
| Kewaunee | 1,012 | 6.0 | 49,539 | 875 | 284 | 63 | 23 | 119 | 167 | 180 | 489 | 28 | 23 |
| La Crosse | 6,115 | 3.1 | 51,813 | 695 | 3,644 | 713 | 278 | 416 | 1,252 | 1,071 | 5,051 | 309 | 278 |
| Lafayette | 756 | 5.7 | 45,339 | 1,320 | 191 | 48 | 15 | 95 | 143 | 141 | 349 | 21 | 15 |
| Langlade | 862 | 3.7 | 44,923 | 1,368 | 300 | 68 | 24 | 73 | 157 | 247 | 465 | 33 | 24 |
| Lincoln | 1,301 | 3.0 | 47,135 | 1,115 | 502 | 110 | 38 | 84 | 195 | 317 | 733 | 51 | 38 |
| Manitowoc | 3,831 | 3.1 | 48,502 | 971 | 1,607 | 336 | 124 | 276 | 727 | 815 | 2,343 | 153 | 124 |
| Marathon | 7,075 | 3.4 | 52,141 | 669 | 3,757 | 710 | 275 | 612 | 1,159 | 1,142 | 5,355 | 328 | 275 |
| Marinette | 1,852 | 2.9 | 45,906 | 1,259 | 859 | 190 | 67 | 110 | 304 | 525 | 1,227 | 84 | 67 |
| Marquette | 662 | 5.4 | 42,513 | 1,689 | 172 | 39 | 13 | 29 | 112 | 177 | 254 | 20 | 13 |
| Menominee | 141 | 1.8 | 30,977 | 3,005 | 81 | 34 | 6 | 3 | 27 | 51 | 124 | 7 | 6 |
| Milwaukee | 46,434 | 2.3 | 49,098 | 915 | 29,566 | 5,155 | 2,142 | 3,335 | 8,534 | 9,956 | 40,197 | 2,450 | 2,142 |
| Monroe | 1,988 | 3.9 | 42,974 | 1,622 | 1,027 | 252 | 83 | 143 | 359 | 429 | 1,504 | 90 | 83 |
| Oconto | 1,787 | 3.6 | 47,104 | 1,119 | 365 | 92 | 29 | 132 | 265 | 374 | 618 | 46 | 29 |
| Oneida | 1,851 | 2.9 | 52,014 | 680 | 751 | 146 | 56 | 144 | 422 | 484 | 1,097 | 79 | 56 |
| Outagamie | 9,890 | 3.7 | 52,640 | 639 | 5,990 | 1,044 | 444 | 636 | 1,679 | 1,417 | 8,114 | 494 | 444 |
| Ozaukee | 7,628 | 2.4 | 85,492 | 43 | 2,309 | 426 | 168 | 361 | 2,147 | 778 | 3,264 | 208 | 168 |
| Pepin | 363 | 3.1 | 49,793 | 845 | 101 | 23 | 8 | 33 | 66 | 85 | 165 | 12 | 8 |
| Pierce | 2,014 | 3.3 | 47,114 | 1,117 | 470 | 138 | 35 | 120 | 346 | 330 | 764 | 50 | 35 |
| Polk | 2,095 | 3.7 | 47,856 | 1,035 | 680 | 161 | 53 | 145 | 348 | 474 | 1,039 | 71 | 53 |
| Portage | 3,346 | 3.8 | 47,278 | 1,100 | 1,812 | 353 | 138 | 219 | 599 | 626 | 2,522 | 157 | 138 |
| Price | 629 | 2.8 | 47,086 | 1,122 | 226 | 54 | 18 | 65 | 125 | 188 | 362 | 26 | 18 |
| Racine | 9,981 | 2.3 | 50,845 | 766 | 4,041 | 800 | 301 | 376 | 2,062 | 1,894 | 5,518 | 364 | 301 |
| Richland | 773 | 3.5 | 44,803 | 1,382 | 247 | 60 | 19 | 64 | 134 | 190 | 389 | 26 | 19 |
| Rock | 7,553 | 4.7 | 46,236 | 1,222 | 3,592 | 673 | 270 | 328 | 1,369 | 1,538 | 4,863 | 318 | 270 |
| Rusk | 632 | 1.2 | 44,556 | 1,413 | 206 | 51 | 17 | 42 | 94 | 177 | 317 | 22 | 17 |
| St. Croix | 5,199 | 3.6 | 57,328 | 394 | 1,666 | 328 | 125 | 288 | 879 | 659 | 2,408 | 152 | 125 |
| Sauk | 3,294 | 4.9 | 51,115 | 747 | 1,682 | 318 | 126 | 404 | 550 | 573 | 2,530 | 156 | 126 |
| Sawyer | 784 | 5.4 | 47,360 | 1,094 | 286 | 67 | 22 | 53 | 182 | 234 | 428 | 32 | 22 |
| Shawano | 1,808 | 4.7 | 44,196 | 1,457 | 526 | 122 | 41 | 166 | 285 | 415 | 855 | 58 | 41 |
| Sheboygan | 6,309 | 2.8 | 54,703 | 513 | 3,349 | 592 | 245 | 590 | 1,204 | 1,010 | 4,776 | 297 | 245 |
| Taylor | 852 | 5.5 | 41,858 | 1,788 | 356 | 78 | 28 | 107 | 123 | 183 | 569 | 35 | 28 |
| Trempealeau | 1,341 | 1.9 | 45,218 | 1,336 | 645 | 135 | 49 | 78 | 200 | 290 | 907 | 58 | 49 |
| Vernon | 1,312 | 4.2 | 42,578 | 1,675 | 378 | 93 | 29 | 146 | 212 | 302 | 645 | 43 | 29 |
| Vilas | 1,215 | 3.2 | 54,734 | 510 | 323 | 74 | 25 | 200 | 333 | 308 | 622 | 47 | 25 |
| Walworth | 5,318 | 3.1 | 51,196 | 740 | 1,944 | 430 | 147 | 304 | 1,129 | 904 | 2,825 | 182 | 147 |
| Washburn | 777 | 3.7 | 49,456 | 885 | 234 | 57 | 18 | 84 | 161 | 235 | 392 | 29 | 18 |
| Washington | 8,159 | 3.2 | 59,979 | 304 | 2,965 | 539 | 218 | 535 | 1,426 | 1,137 | 4,257 | 276 | 218 |
| Waukesha | 29,859 | 2.8 | 73,873 | 97 | 15,815 | 2,433 | 1,135 | 1,671 | 6,500 | 3,458 | 21,054 | 1,306 | 1,135 |
| Waupaca | 2,398 | 3.0 | 47,035 | 1,130 | 830 | 189 | 63 | 153 | 410 | 580 | 1,236 | 85 | 63 |
| Waushara | 1,040 | 5.2 | 42,544 | 1,680 | 244 | 63 | 19 | 78 | 187 | 266 | 404 | 31 | 19 |
| Winnebago | 8,471 | 3.3 | 49,276 | 898 | 5,441 | 961 | 390 | 454 | 1,661 | 1,406 | 7,246 | 450 | 390 |
| Wood | 3,507 | 3.6 | 48,046 | 1,010 | 1,951 | 384 | 144 | 240 | 577 | 786 | 2,719 | 174 | 144 |
| WYOMING | 35,993 | 2.7 | 62,044 | X | 15,081 | 2,835 | 1,377 | 4,163 | 10,508 | 5,043 | 23,456 | 1,368 | 1,377 |
| Albany | 1,660 | 4.0 | 42,698 | 1,656 | 740 | 206 | 66 | 95 | 401 | 245 | 1,107 | 62 | 66 |
| Big Horn | 464 | 2.2 | 39,318 | 2,142 | 194 | 47 | 19 | 25 | 89 | 116 | 285 | 19 | 19 |
| Campbell | 2,565 | 3.4 | 55,348 | 479 | 1,649 | 257 | 147 | 291 | 409 | 305 | 2,345 | 135 | 147 |
| Carbon | 883 | 3.3 | 59,665 | 316 | 392 | 99 | 37 | 56 | 212 | 120 | 585 | 34 | 37 |
| Converse | 881 | 14.0 | 63,736 | 214 | 484 | 82 | 43 | 61 | 199 | 125 | 669 | 40 | 43 |
| Crook | 364 | 3.9 | 47,966 | 1,017 | 127 | 26 | 12 | 28 | 88 | 65 | 192 | 12 | 12 |
| Fremont | 1,848 | 2.6 | 47,071 | 1,126 | 711 | 159 | 65 | 129 | 420 | 424 | 1,064 | 67 | 65 |

1. Based on the resident population estimated as of July 1 of the year shown.

Earnings, 2019 (cont.) — Percent by selected industries | Social Security beneficiaries, December 2019 | Housing units, 2020

| STATE County | Farm (75) | Mining, quarrying, and extractions (76) | Construction (77) | Manufacturing (78) | Information; professional, scientific, technical services (79) | Retail trade (80) | Finance, insurance, real estate, and leasing (81) | Health care and social assistance (82) | Government (83) | Number (84) | Rate[1] (85) | Supplemental Security Income recipients, 2019 (86) | Total (87) | Percent change, 2010-2020 (88) |
|---|---|---|---|---|---|---|---|---|---|---|---|---|---|---|
| **WISCONSIN—Cont'd** | | | | | | | | | | | | | | |
| Douglas | 0.2 | D | 7.0 | 13.1 | 2.8 | 6.2 | 3.2 | 7.9 | 19.8 | 10,425 | 240 | 1,092 | 23,333 | 2.2 |
| Dunn | 2.8 | D | 6.7 | 22.0 | 3.5 | 5.9 | 3.4 | D | 23.1 | 9,810 | 216 | 708 | 18,751 | 4.4 |
| Eau Claire | 0.5 | D | 6.0 | 9.3 | 6.0 | 6.6 | 6.7 | 23.7 | 14.4 | 21,360 | 204 | 1,791 | 45,160 | 7.1 |
| Florence | -0.9 | 0.0 | 5.4 | 15.7 | D | D | 5.4 | 2.4 | 27.6 | 1,650 | 384 | 64 | 4,949 | 3.5 |
| Fond du Lac | 2.9 | 0.6 | 9.3 | 25.0 | 4.3 | 6.0 | 5.4 | 12.8 | 11.5 | 23,370 | 227 | 1,596 | 45,749 | 4.2 |
| Forest | -0.1 | D | 3.2 | 8.5 | D | 3.9 | 1.8 | D | 60.8 | 2,875 | 318 | 235 | 9,316 | 3.9 |
| Grant | 7.8 | D | 5.9 | 13.4 | 3.8 | 5.5 | 3.9 | 9.1 | 25.6 | 11,720 | 228 | 789 | 22,326 | 3.5 |
| Green | 3.9 | D | 5.7 | 25.6 | 6.4 | 11.9 | 3.7 | 12.6 | 12.8 | 8,590 | 233 | 403 | 16,210 | 2.2 |
| Green Lake | 3.1 | 1.5 | 11.5 | 13.4 | D | 7.4 | 5.4 | D | 17.2 | 5,365 | 284 | 301 | 10,805 | 1.8 |
| Iowa | 3.6 | D | 8.9 | 12.3 | 2.7 | 28.8 | 2.0 | D | 12.8 | 5,570 | 236 | 310 | 11,039 | 3.0 |
| Iron | 2.5 | 0.0 | 14.4 | D | D | 6.7 | 4.7 | 14.3 | 24.8 | 2,245 | 395 | 122 | 6,144 | 2.4 |
| Jackson | 5.9 | 3.6 | 14.1 | 8.5 | D | 4.2 | 2.7 | 10.6 | 24.8 | 4,990 | 241 | 354 | 10,024 | 3.1 |
| Jefferson | 1.1 | D | 7.4 | 32.5 | 3.7 | 6.2 | 3.5 | 8.8 | 12.9 | 18,195 | 214 | 965 | 36,349 | 3.4 |
| Juneau | 4.6 | 0.0 | 4.8 | 23.9 | 2.4 | 5.8 | 2.5 | D | 25.0 | 7,235 | 271 | 636 | 15,482 | 5.5 |
| Kenosha | 0.3 | 0.0 | 4.5 | 14.3 | 3.6 | 7.0 | 3.1 | 13.5 | 16.0 | 32,315 | 190 | 3,610 | 71,456 | 3.1 |
| Kewaunee | 23.0 | D | 7.6 | 22.3 | D | 3.8 | D | D | 15.7 | 4,900 | 240 | 235 | 9,510 | 2.2 |
| La Crosse | 0.3 | D | 5.0 | 12.3 | 6.1 | 5.7 | 7.9 | 22.1 | 14.5 | 24,005 | 203 | 2,079 | 51,469 | 6.3 |
| Lafayette | 16.8 | D | 9.1 | 16.5 | D | 3.0 | 5.2 | D | 21.3 | 3,690 | 221 | 185 | 7,352 | 1.7 |
| Langlade | 6.4 | 0.0 | 4.6 | 18.2 | D | 11.9 | 6.2 | D | 15.0 | 6,090 | 317 | 431 | 12,666 | 2.5 |
| Lincoln | 1.4 | D | 7.1 | 23.6 | D | 6.4 | 14.5 | 6.9 | 16.9 | 8,290 | 300 | 437 | 17,390 | 3.6 |
| Manitowoc | 3.8 | 0.4 | 5.7 | 28.1 | 3.7 | 5.4 | 3.5 | 10.5 | 12.9 | 21,415 | 272 | 1,401 | 37,680 | 1.3 |
| Marathon | 2.4 | 0.2 | 5.5 | 24.0 | 5.6 | 5.2 | 6.1 | 15.4 | 10.7 | 29,410 | 217 | 1,925 | 60,366 | 4.5 |
| Marinette | 2.7 | D | 5.0 | 35.8 | 2.3 | 6.1 | 3.1 | D | 12.8 | 13,175 | 327 | 887 | 31,055 | 2.2 |
| Marquette | 3.3 | D | 4.1 | 39.4 | 2.8 | 5.2 | 2.3 | 4.8 | 19.7 | 4,950 | 319 | 309 | 10,078 | 1.9 |
| Menominee | 0.0 | 0.0 | 0.3 | D | D | D | D | 0.2 | 94.2 | 1,025 | 225 | 238 | 2,308 | 2.4 |
| Milwaukee | 0.0 | 0.0 | 3.3 | 11.6 | 11.3 | 4.2 | 10.7 | 15.4 | 12.8 | 166,380 | 176 | 42,716 | 419,808 | 0.4 |
| Monroe | 3.3 | D | 4.7 | 17.9 | D | 4.8 | 2.6 | 7.9 | 30.5 | 9,935 | 213 | 841 | 20,183 | 5.1 |
| Oconto | 7.1 | D | 9.1 | 23.1 | D | 5.7 | 2.5 | 10.9 | 19.4 | 10,205 | 268 | 576 | 24,630 | 4.7 |
| Oneida | 0.6 | 0.0 | 9.0 | 11.6 | 4.8 | 10.6 | 5.1 | 20.1 | 15.2 | 12,735 | 358 | 600 | 31,550 | 4.7 |
| Outagamie | 0.9 | D | 10.3 | 19.9 | 6.0 | 5.9 | 9.5 | 11.3 | 11.0 | 37,195 | 198 | 2,556 | 78,505 | 7.4 |
| Ozaukee | 0.7 | D | 4.5 | 24.3 | 10.5 | 5.5 | 9.1 | 14.1 | 9.1 | 20,505 | 229 | 586 | 38,381 | 5.8 |
| Pepin | 5.2 | 0.0 | 14.3 | 6.0 | D | 8.4 | 4.3 | D | 17.9 | 2,135 | 294 | 100 | 3,718 | 3.9 |
| Pierce | 3.3 | 0.8 | 7.8 | 14.6 | 4.2 | 4.4 | 3.7 | D | 34.2 | 8,120 | 190 | 377 | 17,022 | 5.5 |
| Polk | 2.7 | 0.9 | 7.4 | 24.0 | D | 6.9 | 2.9 | 15.0 | 17.4 | 11,920 | 273 | 621 | 25,152 | 3.7 |
| Portage | 2.6 | D | 4.5 | 13.0 | 5.7 | 6.5 | 19.7 | 9.4 | 15.3 | 14,975 | 211 | 801 | 31,558 | 5.0 |
| Price | 2.5 | 0.4 | 5.0 | 32.7 | 6.6 | 5.3 | 2.6 | 11.2 | 15.7 | 4,820 | 361 | 268 | 11,499 | 3.4 |
| Racine | 0.3 | 0.1 | 5.8 | 31.0 | 4.7 | 6.2 | 4.6 | 12.0 | 13.4 | 43,945 | 224 | 5,790 | 83,326 | 1.4 |
| Richland | 7.2 | D | 6.5 | 25.9 | 2.2 | 7.7 | 2.9 | D | 16.4 | 4,545 | 263 | 338 | 9,073 | 2.3 |
| Rock | 0.7 | D | 7.4 | 17.2 | 5.1 | 5.8 | 4.0 | 15.6 | 13.5 | 36,510 | 224 | 3,835 | 69,961 | 2.3 |
| Rusk | 5.6 | D | 5.4 | 24.4 | D | 6.2 | 2.0 | 11.1 | 18.2 | 4,495 | 319 | 346 | 9,369 | 5.5 |
| St. Croix | 1.6 | D | 8.6 | 21.3 | 5.9 | 7.3 | 5.1 | 12.7 | 13.9 | 16,205 | 179 | 580 | 37,395 | 10.1 |
| Sauk | 1.9 | D | 11.6 | 15.7 | 4.6 | 6.8 | 6.9 | 11.0 | 14.5 | 14,985 | 233 | 900 | 30,859 | 3.9 |
| Sawyer | 1.3 | 0.0 | 8.6 | 11.3 | 3.7 | 9.1 | 4.5 | D | 27.7 | 5,910 | 357 | 373 | 16,801 | 5.2 |
| Shawano | 8.9 | 0.0 | 6.2 | 18.0 | D | 7.2 | 3.0 | D | 20.3 | 10,635 | 261 | 645 | 21,025 | 1.5 |
| Sheboygan | 1.1 | D | 5.0 | 43.4 | 3.0 | 5.4 | 6.9 | 12.0 | 8.6 | 26,885 | 233 | 1,666 | 51,533 | 1.5 |
| Taylor | 7.3 | D | 6.6 | 29.9 | D | 5.2 | 3.7 | 9.6 | 12.6 | 4,705 | 231 | 243 | 10,796 | 2.0 |
| Trempealeau | 0.8 | D | 4.9 | 41.2 | 2.4 | 4.3 | 2.4 | D | 17.2 | 6,825 | 231 | 412 | 13,458 | 6.7 |
| Vernon | 6.2 | D | 8.1 | 10.2 | 4.9 | 7.0 | 4.1 | 15.2 | 18.2 | 7,730 | 250 | 557 | 14,337 | 4.5 |
| Vilas | 0.5 | D | 11.4 | 2.9 | 4.7 | 7.4 | 4.0 | 5.4 | 23.3 | 8,620 | 389 | 321 | 26,539 | 5.6 |
| Walworth | 1.0 | D | 6.9 | 22.5 | 4.2 | 6.5 | 4.0 | 7.8 | 20.5 | 22,665 | 218 | 1,226 | 53,198 | 3.2 |
| Washburn | 2.3 | D | 5.9 | 17.7 | 4.4 | 8.0 | 3.6 | D | 21.1 | 5,595 | 355 | 348 | 13,473 | 3.8 |
| Washington | 0.6 | 0.2 | 6.9 | 28.7 | 4.7 | 6.3 | 7.6 | 11.3 | 9.4 | 29,770 | 218 | 1,001 | 57,910 | 5.9 |
| Waukesha | 0.0 | 0.2 | 8.9 | 19.1 | 11.7 | 6.2 | 9.1 | 10.2 | 6.7 | 89,205 | 221 | 3,111 | 168,726 | 4.9 |
| Waupaca | 3.1 | 0.0 | 5.9 | 35.7 | 2.9 | 6.6 | 2.7 | 8.8 | 18.2 | 14,265 | 280 | 840 | 26,072 | 2.7 |
| Waushara | 7.8 | 0.2 | 7.2 | 19.0 | D | 5.7 | 2.9 | D | 21.2 | 7,445 | 305 | 397 | 15,237 | 2.6 |
| Winnebago | 0.4 | D | 8.1 | 27.7 | 7.5 | 4.4 | 6.2 | 9.2 | 11.4 | 36,000 | 210 | 2,668 | 76,566 | 4.4 |
| Wood | 1.3 | 0.0 | 6.0 | 15.9 | 6.3 | 5.3 | 4.0 | 25.4 | 12.3 | 19,880 | 273 | 1,472 | 35,703 | 4.7 |
| **WYOMING** | 1.0 | 12.2 | 8.3 | 4.2 | 6.2 | 5.1 | 5.0 | 7.1 | 23.7 | 115,406 | 199 | 6,995 | 281,937 | 7.7 |
| Albany | 0.7 | 0.7 | 5.5 | 2.7 | 8.5 | 5.6 | 5.7 | 11.3 | 45.3 | 5,150 | 133 | 330 | 19,676 | 9.7 |
| Big Horn | 4.3 | 13.1 | 9.1 | 7.0 | D | 2.7 | 3.0 | D | 35.7 | 2,770 | 236 | 149 | 5,491 | 2.1 |
| Campbell | 0.2 | 35.1 | 7.9 | 2.3 | 3.4 | 4.6 | 3.0 | 2.9 | 16.3 | 6,310 | 136 | 323 | 20,460 | 7.9 |
| Carbon | 2.2 | 3.8 | 11.5 | D | 2.6 | 5.3 | 2.2 | 3.1 | 23.5 | 2,910 | 196 | 150 | 8,864 | 3.4 |
| Converse | 0.9 | 24.4 | 19.4 | 1.4 | D | 3.7 | 2.4 | 3.4 | 19.7 | 2,730 | 197 | 156 | 6,712 | 4.8 |
| Crook | 2.7 | 15.9 | 10.1 | 9.1 | D | 3.7 | 1.8 | D | 25.3 | 1,760 | 232 | 44 | 3,680 | 2.4 |
| Fremont | 3.0 | 6.4 | 6.7 | 1.4 | 4.5 | 6.6 | 4.5 | 11.2 | 36.6 | 9,385 | 238 | 926 | 18,041 | 1.4 |

1. Per 1,000 resident population estimated as of July 1 of the year shown.

# Table B. States and Counties — Housing, Labor Force, and Employment

| STATE County | Occupied units Total | Percent | Owner-occ. Median value[1] | Median owner cost % income — With a mortgage | Median owner cost % income — Without a mortgage[2] | Renter-occ. Median rent[3] | Median rent as a percent of income[2] | Sub-standard units[4] (percent) | Civ. labor force Total | Percent change, 2019-2020 | Unemployment Total | Rate[5] | Civ. employment Total | Mgmt, business, science, and arts | Construction, production, and maintenance occupations |
|---|---|---|---|---|---|---|---|---|---|---|---|---|---|---|---|
| | 89 | 90 | 91 | 92 | 93 | 94 | 95 | 96 | 97 | 98 | 99 | 100 | 101 | 102 | 103 |
| **WISCONSIN—Cont'd** | | | | | | | | | | | | | | | |
| Douglas | 19,011 | 67.5 | 145,800 | 19.7 | 12.3 | 766 | 26.4 | 1.8 | 22,886 | -0.3 | 2,100 | 9.2 | 21,918 | 31.9 | 26.8 |
| Dunn | 16,859 | 68.2 | 165,200 | 20.3 | 12.8 | 796 | 26.1 | 2.0 | 23,606 | -1.8 | 1,303 | 5.5 | 23,564 | 31.5 | 31.8 |
| Eau Claire | 40,981 | 64.5 | 169,400 | 19.4 | 12.5 | 823 | 28.0 | 2.3 | 58,299 | -0.5 | 3,235 | 5.5 | 56,366 | 36.4 | 23.0 |
| Florence | 1,948 | 86.2 | 129,600 | 19.9 | 12.6 | 480 | 20.2 | 4.1 | 2,191 | -0.5 | 175 | 8.0 | 1,931 | 32.0 | 30.3 |
| Fond du Lac | 41,542 | 70.3 | 157,800 | 19.7 | 12.1 | 747 | 24.7 | 1.8 | 56,114 | -1.6 | 3,286 | 5.9 | 53,973 | 29.7 | 33.6 |
| Forest | 4,008 | 76.2 | 132,300 | 21.8 | 12.4 | 506 | 22.7 | 2.3 | 3,890 | -2.6 | 483 | 12.4 | 3,701 | 27.3 | 29.5 |
| Grant | 19,583 | 70.4 | 147,700 | 19.5 | 11.8 | 709 | 24.8 | 2.3 | 27,311 | -1.3 | 1,436 | 5.3 | 26,132 | 32.3 | 30.8 |
| Green | 15,047 | 75.1 | 171,400 | 20.9 | 12.5 | 754 | 23.7 | 1.7 | 20,555 | -2.0 | 985 | 4.8 | 20,448 | 34.4 | 29.5 |
| Green Lake | 7,979 | 76.1 | 146,600 | 19.7 | 12.6 | 710 | 25.9 | 2.4 | 9,253 | -1.7 | 637 | 6.9 | 8,975 | 28.4 | 37.1 |
| Iowa | 9,850 | 73.8 | 183,200 | 20.6 | 12.6 | 750 | 23.7 | 1.6 | 13,502 | -0.9 | 824 | 6.1 | 12,594 | 34.1 | 30.6 |
| Iron | 2,898 | 77.5 | 119,000 | 21.9 | 12.6 | 538 | 31.6 | 1.9 | 2,547 | 0.7 | 286 | 11.2 | 2,511 | 28.6 | 31.7 |
| Jackson | 8,199 | 73.9 | 135,900 | 20.6 | 12.6 | 674 | 24.8 | 3.5 | 9,877 | -3.2 | 918 | 9.3 | 9,386 | 27.1 | 34.9 |
| Jefferson | 32,965 | 70.2 | 190,400 | 21.0 | 12.6 | 857 | 26.3 | 1.8 | 45,213 | | 2,451 | 5.4 | 46,211 | 33.3 | 30.0 |
| Juneau | 10,752 | 76.2 | 125,800 | 21.7 | 13.5 | 761 | 25.1 | 2.0 | 13,280 | -0.9 | 953 | 7.2 | 12,181 | 25.8 | 34.5 |
| Kenosha | 64,111 | 66.1 | 177,400 | 20.2 | 13.3 | 919 | 28.4 | 1.7 | 88,499 | -1.3 | 6,299 | 7.1 | 85,094 | 32.8 | 27.3 |
| Kewaunee | 8,341 | 79.3 | 164,900 | 19.3 | 11.2 | 693 | 23.1 | 2.1 | 10,661 | -2.7 | 489 | 4.6 | 10,699 | 30.8 | 38.0 |
| La Crosse | 47,518 | 63 | 173,300 | 20.0 | 11.7 | 835 | 27.6 | 1.5 | 64,982 | -1.3 | 3,515 | 5.4 | 62,533 | 37.3 | 21.3 |
| Lafayette | 6,722 | 76.4 | 140,000 | 19.5 | 12.9 | 713 | 24.1 | 2.0 | 9,785 | -2.6 | 420 | 4.3 | 8,565 | 32.7 | 33.9 |
| Langlade | 8,612 | 77.1 | 113,900 | 18.4 | 11.0 | 640 | 28.1 | 2.5 | 9,285 | 0.3 | 601 | 6.5 | 8,779 | 27.3 | 35.2 |
| Lincoln | 12,625 | 78.1 | 139,700 | 18.7 | 11.9 | 662 | 26.5 | 1.6 | 14,677 | -2.1 | 872 | 5.9 | 14,131 | 30.8 | 33.6 |
| Manitowoc | 34,563 | 75.4 | 130,100 | 18.6 | 12.2 | 672 | 23.5 | 1.4 | 40,419 | -0.8 | 2,508 | 6.2 | 40,604 | 29.0 | 35.5 |
| Marathon | 55,466 | 72.2 | 156,300 | 18.8 | 11.2 | 758 | 24.2 | 2.6 | 72,313 | -1.1 | 3,602 | 5.0 | 71,727 | 35.4 | 29.3 |
| Marinette | 18,473 | 74.7 | 123,800 | 19.0 | 12.0 | 679 | 25.5 | 1.6 | 19,504 | 0.9 | 1,523 | 7.8 | 18,689 | 26.8 | 38.1 |
| Marquette | 6,648 | 81.4 | 155,300 | 21.4 | 14.7 | 701 | 24.3 | 1.8 | 7,650 | -0.6 | 514 | 6.7 | 6,967 | 24.7 | 39.6 |
| Menominee | 1,408 | 68 | 101,800 | 19.1 | 10.7 | 539 | 17.6 | 7.0 | 1,649 | 3.9 | 253 | 15.3 | 1,518 | 22.3 | 23.4 |
| Milwaukee | 382,175 | 49.5 | 158,300 | 21.5 | 14.0 | 880 | 30.0 | 3.1 | 463,420 | -0.1 | 37,937 | 8.2 | 459,843 | 36.8 | 22.8 |
| Monroe | 17,915 | 69.7 | 155,000 | 19.9 | 11.8 | 815 | 24.7 | 2.6 | 23,363 | 0.4 | 1,288 | 5.5 | 21,703 | 28.6 | 36.3 |
| Oconto | 16,036 | 82.7 | 157,000 | 19.8 | 10.9 | 683 | 24.0 | 1.7 | 20,241 | -0.8 | 1,279 | 6.3 | 19,070 | 29.9 | 37.4 |
| Oneida | 15,421 | 83.2 | 174,100 | 20.4 | 12.2 | 766 | 27.7 | 0.8 | 17,800 | 0.0 | 1,292 | 7.3 | 16,612 | 33.2 | 27.1 |
| Outagamie | 73,648 | 71.1 | 169,400 | 18.9 | 11.7 | 807 | 24.5 | 1.5 | 102,786 | -0.1 | 5,649 | 5.5 | 100,360 | 35.7 | 28.1 |
| Ozaukee | 35,807 | 75.9 | 282,500 | 20.4 | 11.2 | 927 | 26.0 | 1.1 | 47,963 | -1.8 | 2,597 | 5.4 | 47,213 | 48.6 | 17.4 |
| Pepin | 3,070 | 79.8 | 151,600 | 21.4 | 14.3 | 615 | 22.6 | 1.3 | 4,080 | -1.2 | 241 | 5.9 | 3,710 | 30.7 | 33.3 |
| Pierce | 15,593 | 72.2 | 208,700 | 20.5 | 12.2 | 874 | 24.8 | 1.6 | 24,779 | -1.8 | 1,725 | 7.0 | 23,433 | 35.3 | 28.2 |
| Polk | 18,450 | 78.9 | 169,700 | 21.5 | 10.8 | 756 | 27.0 | 1.7 | 24,408 | -0.1 | 1,750 | 7.2 | 21,516 | 30.3 | 36.2 |
| Portage | 28,675 | 69 | 168,100 | 18.5 | 10.8 | 776 | 27.2 | 1.9 | 37,491 | -2.2 | 2,016 | 5.4 | 37,794 | 35.3 | 26.8 |
| Price | 6,688 | 77.8 | 122,500 | 21.5 | 12.8 | 698 | 27.4 | 3.3 | 6,209 | -3.4 | 424 | 6.8 | 6,438 | 27.4 | 39.6 |
| Racine | 76,974 | 67.6 | 176,000 | 20.1 | 12.9 | 857 | 29.6 | 1.4 | 97,155 | -0.9 | 7,051 | 7.3 | 94,772 | 31.8 | 29.4 |
| Richland | 7,538 | 74.7 | 139,900 | 21.7 | 12.4 | 642 | 25.0 | 3.2 | 9,009 | -1.6 | 458 | 5.1 | 8,295 | 27.3 | 39.0 |
| Rock | 64,739 | 68.4 | 146,200 | 20.1 | 12.6 | 838 | 27.9 | 1.9 | 84,935 | -0.2 | 6,027 | 7.1 | 79,931 | 31.1 | 31.9 |
| Rusk | 6,376 | 79.1 | 113,400 | 22.4 | 13.0 | 661 | 26.2 | 2.5 | 6,581 | -1.7 | 437 | 6.6 | 6,441 | 25.0 | 42.1 |
| St. Croix | 34,182 | 76.8 | 244,500 | 19.7 | 10.5 | 962 | 26.2 | 0.9 | 49,739 | -1.5 | 3,516 | 7.1 | 48,941 | 39.9 | 24.8 |
| Sauk | 26,222 | 68.9 | 182,200 | 20.8 | 12.0 | 801 | 27.6 | 2.4 | 34,355 | -3.2 | 2,585 | 7.5 | 34,130 | 32.5 | 28.1 |
| Sawyer | 7,796 | 74 | 168,400 | 21.9 | 11.5 | 682 | 24.6 | 3.0 | 7,910 | -0.6 | 621 | 7.9 | 7,380 | 29.8 | 27.9 |
| Shawano | 17,004 | 77.3 | 136,700 | 19.2 | 12.6 | 629 | 21.3 | 1.9 | 20,791 | -0.8 | 1,328 | 6.4 | 20,016 | 28.6 | 35.6 |
| Sheboygan | 47,738 | 70 | 159,700 | 18.9 | 11.6 | 727 | 23.4 | 2.3 | 61,510 | -0.7 | 3,514 | 5.7 | 59,602 | 31.3 | 33.9 |
| Taylor | 8,614 | 76.6 | 137,800 | 21.0 | 12.6 | 660 | 25.8 | 3.1 | 10,905 | 0.2 | 533 | 4.9 | 10,262 | 26.2 | 44.8 |
| Trempealeau | 12,005 | 73 | 156,000 | 21.2 | 12.8 | 728 | 25.0 | 2.2 | 15,607 | -1.7 | 1,095 | 7.0 | 15,036 | 29.2 | 37.7 |
| Vernon | 12,071 | 77.8 | 157,200 | 21.4 | 13.4 | 670 | 29.0 | 5.6 | 14,896 | -2.8 | 777 | 5.2 | 13,687 | 31.9 | 31.9 |
| Vilas | 10,921 | 77.7 | 202,900 | 22.5 | 13.0 | 699 | 26.8 | 1.3 | 10,221 | 0.5 | 741 | 7.2 | 9,363 | 29.7 | 22.2 |
| Walworth | 40,874 | 68.6 | 203,400 | 21.4 | 13.0 | 880 | 28.1 | 2.1 | 56,322 | -2.1 | 3,416 | 6.1 | 53,782 | 31.8 | 28.7 |
| Washburn | 7,236 | 80 | 157,400 | 21.5 | 12.7 | 709 | 27.6 | 2.1 | 7,777 | -0.1 | 537 | 6.9 | 7,023 | 30.9 | 29.8 |
| Washington | 54,302 | 77.5 | 236,200 | 20.2 | 11.7 | 908 | 25.2 | 0.7 | 75,700 | -1.7 | 4,255 | 5.6 | 73,819 | 38.7 | 25.0 |
| Waukesha | 158,808 | 76.3 | 282,300 | 19.8 | 11.6 | 1,046 | 26.4 | 1.3 | 220,498 | -1.7 | 12,252 | 5.6 | 214,839 | 47.2 | 17.6 |
| Waupaca | 22,305 | 73.7 | 149,900 | 19.4 | 12.7 | 698 | 25.4 | 1.2 | 25,439 | -1.4 | 1,411 | 5.5 | 26,301 | 27.7 | 35.9 |
| Waushara | 9,934 | 81.6 | 146,100 | 21.7 | 12.7 | 671 | 24.5 | 1.9 | 11,080 | -2.6 | 702 | 6.3 | 10,617 | 25.9 | 38.2 |
| Winnebago | 70,594 | 65.2 | 152,500 | 19.2 | 12.2 | 766 | 25.7 | 1.1 | 91,236 | 0.2 | 4,966 | 5.4 | 87,992 | 33.1 | 26.9 |
| Wood | 32,332 | 72.4 | 131,900 | 19.0 | 10.8 | 718 | 26.0 | 1.2 | 34,538 | -1.2 | 2,305 | 6.7 | 35,963 | 32.5 | 32.4 |
| **WYOMING** | 230,101 | 70.4 | 220,500 | 20.2 | 10.0 | 855 | 26.4 | 2.5 | 296,801 | 0.2 | 17,339 | 5.8 | 288,503 | 34.3 | 28.6 |
| Albany | 15,944 | 51 | 235,000 | 21.0 | 10.0 | 796 | 32.7 | 2.0 | 19,726 | -4.5 | 780 | 4.0 | 21,328 | 44.4 | 18.9 |
| Big Horn | 4,475 | 73.2 | 163,400 | 20.9 | 10.0 | 678 | 22.5 | 2.3 | 5,381 | 2.3 | 269 | 5.0 | 5,081 | 27.9 | 35.1 |
| Campbell | 17,574 | 73.4 | 222,700 | 18.9 | 10.0 | 936 | 23.7 | 3.2 | 23,439 | -0.1 | 1,617 | 6.9 | 24,571 | 28.6 | 37.7 |
| Carbon | 6,204 | 69.6 | 174,900 | 18.7 | 10.0 | 792 | 22.3 | 3.6 | 8,111 | 3.3 | 368 | 4.5 | 7,286 | 28.5 | 36.8 |
| Converse | 5,378 | 73.7 | 206,700 | 19.3 | 10.0 | 739 | 27.6 | 3.0 | 8,163 | -6.3 | 486 | 6.0 | 6,643 | 29.4 | 39.2 |
| Crook | 2,919 | 81.8 | 225,300 | 21.2 | 11.6 | 798 | 26.7 | 4.2 | 3,908 | 3.6 | 151 | 3.9 | 3,455 | 29.6 | 41.9 |
| Fremont | 14,907 | 70.5 | 196,400 | 21.7 | 10.0 | 789 | 24.1 | 4.2 | 19,326 | 2.1 | 1,206 | 6.2 | 18,294 | 34.2 | 24.5 |

1. Specified owner-occupied units.    2. A value of 10.0 represents 10 percent or less; a value of 50.0 represents 50 percent or more.    3. Specified renter-occupied units.    4. Overcrowded or lacking complete plumbing facilities.    5. Percent of civilian labor force.    6. Civilian employed persons 16 years old and over.

| STATE County | Private nonfarm establishments, employment and payroll, 2019 | | | | | | | | | Agriculture, 2017 | | | |
| | Number of establish-ments | Employment | | | | | | Annual payroll | | Farms | | | Farm producers whose primary occupation is farming (percent) |
| | | Total | Health care and social assistance | Manufac-turing | Retail trade | Finance and insurance | Professional, scientific, and technical services | Total (mil dol) | Average per employee (dollars) | Number | Percent with: | | |
| | | | | | | | | | | | Fewer than 50 acres | 1000 acres or more | |
| | 104 | 105 | 106 | 107 | 108 | 109 | 110 | 111 | 112 | 113 | 114 | 115 | 116 |
| **WISCONSIN—Cont'd** | | | | | | | | | | | | | |
| Douglas | 1,028 | 14,296 | 1,845 | 1,889 | 2,104 | 343 | 388 | 599 | 41,907 | 329 | 24.0 | 2.7 | 35.4 |
| Dunn | 910 | 15,400 | 2,380 | 3,635 | 1,928 | 431 | 345 | 671 | 43,542 | 1,288 | 29.1 | 4.9 | 43.1 |
| Eau Claire | 2,646 | 53,542 | 11,808 | 5,422 | 7,111 | 2,888 | 1,493 | 2,625 | 49,018 | 1,069 | 33.6 | 1.0 | 37.1 |
| Florence | 105 | 628 | 49 | 154 | 122 | 24 | 14 | 17 | 27,707 | 101 | 21.8 | NA | 34.9 |
| Fond du Lac | 2,284 | 43,834 | 6,655 | 9,512 | 5,820 | 1,641 | 1,359 | 2,029 | 46,281 | 1,244 | 35.9 | 4.8 | 53.4 |
| Forest | 216 | 1,394 | 157 | 229 | 195 | 62 | 60 | 46 | 33,020 | 140 | 25.0 | 2.9 | 48.9 |
| Grant | 1,191 | 13,556 | 2,157 | 2,333 | 2,414 | 697 | 497 | 527 | 38,870 | 2,482 | 25.7 | 3.2 | 48.6 |
| Green | 981 | 13,217 | 1,936 | 3,839 | 2,355 | 420 | 418 | 564 | 42,660 | 1,428 | 43.8 | 3.4 | 46.3 |
| Green Lake | 446 | 4,680 | 680 | 849 | 890 | 257 | 66 | 199 | 42,597 | 502 | 29.3 | 5.4 | 53.0 |
| Iowa | 551 | 9,007 | 1,090 | 1,378 | 3,603 | 174 | 128 | 446 | 49,512 | 1,576 | 28.3 | 3.4 | 39.9 |
| Iron | 191 | 1,380 | 343 | 166 | 203 | 35 | 42 | 41 | 30,071 | 49 | 28.6 | 2.0 | 31.6 |
| Jackson | 415 | 5,843 | 947 | 610 | 870 | 204 | 79 | 263 | 44,995 | 855 | 22.3 | 5.4 | 44.5 |
| Jefferson | 2,010 | 30,062 | 3,904 | 8,798 | 4,025 | 652 | 643 | 1,291 | 42,930 | 1,098 | 41.5 | 3.6 | 41.5 |
| Juneau | 545 | 6,809 | 1,212 | 2,189 | 1,022 | 162 | 62 | 265 | 38,933 | 715 | 32.3 | 4.3 | 40.4 |
| Kenosha | 3,284 | 62,322 | 9,060 | 8,483 | 13,420 | 822 | 1,369 | 2,646 | 42,456 | 415 | 55.7 | 3.6 | 42.1 |
| Kewaunee | 463 | 4,884 | 431 | 1,906 | 565 | 167 | 159 | 196 | 40,165 | 655 | 33.4 | 4.1 | 51.9 |
| La Crosse | 3,093 | 64,775 | 11,945 | 7,487 | 8,915 | 3,288 | 2,027 | 2,900 | 44,772 | 667 | 29.8 | 2.1 | 46.2 |
| Lafayette | 343 | 3,005 | 259 | 917 | 366 | 132 | 48 | 123 | 40,892 | 1,327 | 39.5 | 5.0 | 52.8 |
| Langlade | 551 | 6,419 | 955 | 1,377 | 1,379 | 305 | 123 | 255 | 39,758 | 432 | 23.8 | 4.4 | 49.7 |
| Lincoln | 696 | 9,022 | 812 | 2,525 | 1,306 | 1,036 | 117 | 357 | 39,605 | 426 | 29.8 | 1.9 | 41.2 |
| Manitowoc | 1,710 | 29,811 | 4,068 | 10,343 | 3,528 | 830 | 569 | 1,287 | 43,166 | 1,171 | 43.2 | 4.1 | 48.4 |
| Marathon | 3,415 | 66,913 | 10,527 | 19,635 | 7,630 | 4,159 | 1,439 | 3,137 | 46,878 | 2,237 | 33.8 | 2.9 | 50.2 |
| Marinette | 1,067 | 15,492 | 2,666 | 5,767 | 2,163 | 407 | 298 | 645 | 41,653 | 515 | 32.6 | 5.2 | 47.5 |
| Marquette | 269 | 3,062 | 296 | 1,396 | 333 | 43 | 67 | 130 | 42,605 | 458 | 33.8 | 5.0 | 40.7 |
| Menominee | 24 | 1,006 | NA | NA | 36 | NA | NA | 31 | 30,810 | 3 | 33.3 | NA | 38.2 |
| Milwaukee | 19,846 | 453,220 | 92,044 | 47,122 | 40,904 | 35,199 | 23,240 | 25,843 | 57,022 | 86 | 83.7 | NA | 45.2 |
| Monroe | 953 | 16,299 | 3,060 | 3,803 | 2,019 | 360 | 342 | 738 | 45,300 | 1,555 | 29.2 | 2.3 | 44.0 |
| Oconto | 789 | 7,055 | 1,371 | 2,310 | 1,014 | 159 | 112 | 258 | 36,540 | 834 | 37.9 | 5.2 | 38.1 |
| Oneida | 1,363 | 13,519 | 2,659 | 1,424 | 3,168 | 324 | 336 | 554 | 40,974 | 131 | 39.7 | 3.8 | 45.8 |
| Outagamie | 5,054 | 105,167 | 12,495 | 19,925 | 13,257 | 5,247 | 3,957 | 5,084 | 48,339 | 1,130 | 44.8 | 3.6 | 49.7 |
| Ozaukee | 2,812 | 40,884 | 6,350 | 9,918 | 4,770 | 1,261 | 2,489 | 1,948 | 47,639 | 316 | 46.2 | 2.8 | 41.3 |
| Pepin | 222 | 1,721 | 274 | 175 | 229 | 60 | 32 | 75 | 43,776 | 448 | 27.2 | 4.0 | 38.9 |
| Pierce | 789 | 7,223 | 876 | 1,612 | 981 | 276 | 310 | 254 | 35,221 | 1,229 | 39.9 | 3.1 | 44.1 |
| Polk | 1,120 | 14,014 | 2,817 | 4,790 | 1,997 | 260 | 378 | 531 | 37,902 | 1,234 | 35.4 | 3.6 | 48.0 |
| Portage | 1,719 | 30,297 | 3,424 | 4,706 | 3,908 | 4,976 | 714 | 1,333 | 43,993 | 982 | 35.0 | 5.5 | 41.1 |
| Price | 390 | 4,318 | 805 | 1,755 | 583 | 116 | 90 | 164 | 37,891 | 410 | 25.1 | 2.0 | 42.6 |
| Racine | 4,021 | 68,432 | 11,349 | 17,032 | 9,215 | 1,982 | 1,955 | 3,396 | 49,624 | 611 | 51.7 | 3.8 | 37.0 |
| Richland | 381 | 4,724 | 1,019 | 1,514 | 810 | 150 | 54 | 172 | 36,454 | 1,103 | 31.1 | 3.0 | 40.6 |
| Rock | 3,360 | 62,793 | 9,191 | 12,530 | 9,514 | 1,416 | 2,397 | 2,905 | 46,261 | 1,587 | 53.7 | 4.7 | 57.6 |
| Rusk | 312 | 3,934 | 765 | 1,305 | 534 | 116 | 30 | 144 | 36,491 | 501 | 14.6 | 5.2 | 40.6 |
| St. Croix | 2,304 | 32,072 | 4,653 | 7,075 | 4,813 | 825 | 1,294 | 1,360 | 42,397 | 1,444 | 41.6 | 3.2 | 43.8 |
| Sauk | 1,832 | 31,475 | 3,950 | 5,797 | 4,179 | 955 | 1,041 | 1,311 | 41,654 | 1,412 | 31.7 | 2.5 | 45.7 |
| Sawyer | 637 | 5,423 | 916 | 728 | 969 | 146 | 153 | 212 | 39,088 | 175 | 22.3 | 4.0 | 50.7 |
| Shawano | 881 | 10,344 | 1,126 | 2,600 | 1,427 | 277 | 179 | 378 | 36,587 | 1,139 | 30.3 | 2.7 | 49.6 |
| Sheboygan | 2,678 | 53,477 | 6,334 | 18,939 | 6,044 | 2,313 | 1,287 | 2,508 | 46,891 | 958 | 43.6 | 3.9 | 47.4 |
| Taylor | 463 | 7,540 | 1,026 | 3,319 | 835 | 253 | 130 | 326 | 43,284 | 893 | 23.7 | 3.1 | 40.0 |
| Trempealeau | 655 | 14,832 | 1,244 | 8,646 | 1,061 | 266 | 242 | 629 | 42,386 | 1,229 | 24.6 | 4.6 | 48.8 |
| Vernon | 613 | 6,624 | 1,749 | 985 | 1,066 | 279 | 183 | 252 | 37,994 | 1,961 | 31.1 | 1.8 | 23.6 |
| Vilas | 968 | 6,082 | 552 | 332 | 1,166 | 197 | 153 | 200 | 32,894 | 67 | 64.2 | 1.5 | 44.1 |
| Walworth | 2,752 | 36,722 | 3,907 | 9,223 | 5,138 | 664 | 1,141 | 1,426 | 38,821 | 941 | 49.5 | 5.2 | 37.2 |
| Washburn | 508 | 4,304 | 840 | 1,013 | 774 | 127 | 197 | 148 | 34,320 | 372 | 36.0 | 4.3 | 53.7 |
| Washington | 3,306 | 53,278 | 6,318 | 15,199 | 6,902 | 2,095 | 1,518 | 2,455 | 46,083 | 578 | 40.8 | 3.5 | 39.0 |
| Waukesha | 12,624 | 245,702 | 29,228 | 41,062 | 26,486 | 15,206 | 14,372 | 13,827 | 56,277 | 574 | 56.8 | 4.2 | 44.5 |
| Waupaca | 1,185 | 16,595 | 2,746 | 6,513 | 2,288 | 413 | 238 | 616 | 37,101 | 1,031 | 33.9 | 3.2 | 39.9 |
| Waushara | 470 | 5,022 | 748 | 1,222 | 795 | 110 | 200 | 166 | 32,984 | 633 | 42.8 | 4.4 | 37.8 |
| Winnebago | 3,632 | 90,618 | 13,057 | 23,079 | 8,558 | 3,896 | 3,940 | 4,809 | 53,074 | 957 | 49.9 | 2.2 | 51.3 |
| Wood | 1,784 | 37,366 | 12,029 | 6,311 | 3,845 | 1,330 | 566 | 1,863 | 49,858 | 1,062 | 32.9 | 3.9 | 43.0 |
| **WYOMING** | 21,578 | 207,016 | 32,643 | 10,511 | 30,007 | 6,943 | 10,556 | 10,423 | 50,347 | 11,938 | 32.7 | 25.0 | |
| Albany | 1,073 | 9,881 | 2,135 | 465 | 1,645 | 511 | 807 | 356 | 35,986 | 451 | 28.2 | 30.2 | 38.6 |
| Big Horn | 303 | 2,416 | 402 | 259 | 326 | 114 | 114 | 108 | 44,786 | 586 | 40.4 | 12.8 | 50.7 |
| Campbell | 1,482 | 19,814 | 1,617 | 601 | 2,427 | 370 | 644 | 1,261 | 63,657 | 643 | 25.5 | 45.4 | 44.6 |
| Carbon | 520 | 4,220 | 556 | 634 | 730 | 133 | 133 | 234 | 55,472 | 345 | 18.8 | 42.6 | 49.1 |
| Converse | 428 | 4,960 | 579 | 99 | 423 | 115 | 107 | 320 | 64,487 | 384 | 23.7 | 40.9 | 57.3 |
| Crook | 249 | 1,529 | 201 | 164 | 191 | 56 | 24 | 77 | 50,362 | 554 | 15.7 | 37.0 | 45.8 |
| Fremont | 1,263 | 9,980 | 1,871 | 290 | 1,877 | 314 | 401 | 402 | 40,293 | 1,152 | 36.3 | 12.1 | 46.1 |

# Table B. States and Counties — Agriculture

| STATE County | Acreage (1,000) | Percent change, 2012-2017 | Average size of farm | Total irrigated (1,000) | Total cropland (1,000) | Value of land and buildings — Average per farm (dollars) | Average per acre | Value of machinery and equipment, average per farm (dollars) | Total (mil dol) | Average per farm (acres) | Crops | Livestock and poultry products | Organic farms (number) | Farms with internet access (percent) | Government payments Total ($1,000) | Percent of farms |
|---|---|---|---|---|---|---|---|---|---|---|---|---|---|---|---|---|
| | 117 | 118 | 119 | 120 | 121 | 122 | 123 | 124 | 125 | 126 | 127 | 128 | 129 | 130 | 131 | 132 |
| **WISCONSIN—Cont'd** | | | | | | | | | | | | | | | | |
| Douglas | 70 | -1.2 | 212 | 0.2 | 26.8 | 415,809 | 1,961 | 58,358 | 8.5 | 25,936 | 40.4 | 59.6 | 2 | 74.8 | 36 | 2.4 |
| Dunn | 348 | -6.4 | 270 | 40.1 | 240.8 | 973,838 | 3,601 | 148,164 | 212.9 | 165,325 | 47.8 | 52.2 | 23 | 82.3 | 1,863 | 41.7 |
| Eau Claire | 172 | -15.4 | 161 | 2.5 | 114.0 | 735,990 | 4,567 | 92,355 | 89.9 | 84,116 | 49.1 | 50.9 | 40 | 73.3 | 2,237 | 45.4 |
| Florence | 19 | 39.0 | 184 | 0.0 | 8.0 | 465,629 | 2,527 | 52,675 | 1.3 | 12,723 | 56.9 | 43.1 | NA | 67.3 | 18 | 10.9 |
| Fond du Lac | 317 | 0.6 | 255 | 1.9 | 270.4 | 1,855,960 | 7,275 | 227,075 | 396.7 | 318,921 | 24.0 | 76.0 | 33 | 77.7 | 3,762 | 56.5 |
| Forest | 38 | 25.9 | 272 | 0.4 | 13.1 | 660,064 | 2,426 | 81,548 | 4.9 | 35,064 | 80.6 | 19.4 | 2 | 75.7 | 24 | 11.4 |
| Grant | 600 | 2.2 | 242 | 1.0 | 380.0 | 1,140,897 | 4,717 | 168,735 | 447.2 | 180,181 | 32.3 | 67.7 | 71 | 73.2 | 6,383 | 53.8 |
| Green | 292 | -3.3 | 205 | 3.0 | 239.5 | 1,111,218 | 5,427 | 158,955 | 221.0 | 154,796 | 41.7 | 58.3 | 18 | 82.8 | 3,648 | 47.4 |
| Green Lake | 127 | -18.0 | 252 | 5.4 | 99.7 | 1,444,204 | 5,720 | 196,500 | 90.6 | 180,512 | 49.9 | 50.1 | 11 | 70.7 | 1,594 | 45.8 |
| Iowa | 360 | 2.7 | 229 | 8.6 | 216.4 | 1,111,469 | 4,864 | 141,447 | 206.7 | 131,149 | 38.1 | 61.9 | 41 | 80.8 | 4,969 | 64.4 |
| Iron | 9 | -9.9 | 188 | 0.6 | 5.4 | 638,606 | 3,401 | D | D | D | D | D | 2 | 93.9 | D | 6.1 |
| Jackson | 248 | 3.5 | 290 | 6.3 | 135.6 | 1,103,760 | 3,800 | 161,242 | 154.2 | 180,294 | 52.2 | 47.8 | 41 | 78.1 | 1,034 | 42.6 |
| Jefferson | 221 | -2.9 | 202 | 11.4 | 183.9 | 1,240,595 | 6,154 | 167,109 | 305.3 | 278,041 | 32.3 | 67.7 | 21 | 76.2 | 1,264 | 45.8 |
| Juneau | 175 | -2.6 | 245 | 6.7 | 114.0 | 908,759 | 3,704 | 149,546 | 116.8 | 163,295 | 55.8 | 44.2 | 10 | 73.1 | 1,244 | 39.2 |
| Kenosha | 78 | 1.5 | 187 | 1.1 | 65.2 | 1,419,789 | 7,575 | 170,756 | 59.9 | 144,222 | 67.4 | 32.6 | 6 | 82.4 | 2,469 | 36.1 |
| Kewaunee | 170 | -3.6 | 260 | 0.7 | 146.2 | 1,462,089 | 5,620 | 241,636 | 314.3 | 479,771 | 16.8 | 83.2 | 7 | 75.7 | 1,879 | 62.1 |
| La Crosse | 144 | -9.1 | 216 | 1.1 | 85.9 | 1,079,187 | 4,987 | 124,594 | 74.6 | 111,867 | 37.6 | 62.4 | 27 | 76.8 | 1,955 | 51.4 |
| Lafayette | 343 | -7.1 | 258 | D | 265.4 | 1,542,860 | 5,977 | 200,412 | 301.5 | 227,199 | 38.0 | 62.0 | 57 | 74.0 | 5,322 | 54.1 |
| Langlade | 116 | 2.2 | 269 | 18.3 | 75.8 | 881,494 | 3,272 | 208,454 | 102.2 | 236,655 | 58.4 | 41.6 | 16 | 76.4 | 542 | 32.2 |
| Lincoln | 78 | 1.9 | 184 | 0.3 | 43.0 | 595,649 | 3,241 | 102,719 | 37.0 | 86,838 | 44.7 | 55.3 | 12 | 74.9 | 206 | 15.0 |
| Manitowoc | 232 | 0.4 | 198 | 1.1 | 188.0 | 1,349,733 | 6,824 | 168,986 | 309.1 | 263,971 | 14.7 | 85.3 | 11 | 73.0 | 1,844 | 51.8 |
| Marathon | 473 | -1.2 | 212 | 7.0 | 322.9 | 1,004,890 | 4,751 | 161,950 | 413.6 | 184,888 | 24.4 | 75.6 | 51 | 73.0 | 1,226 | 29.2 |
| Marinette | 133 | 0.8 | 258 | 1.9 | 88.6 | 1,001,818 | 3,877 | 158,953 | 116.0 | 225,266 | 20.6 | 79.4 | 9 | 79.2 | 1,466 | 36.7 |
| Marquette | 113 | -5.8 | 247 | 10.7 | 82.0 | 1,031,933 | 4,176 | 145,836 | 71.5 | 156,059 | 51.2 | 48.8 | 7 | 72.3 | 432 | 25.5 |
| Menominee | D | D | D | NA | D | 143,544 | 1,736 | D | D | D | D | D | NA | 100.0 | NA | NA |
| Milwaukee | D | D | D | 0.1 | D | 542,871 | 8,928 | 68,761 | 6.8 | 79,047 | 96.2 | 3.8 | 2 | 72.1 | 64 | 18.6 |
| Monroe | 301 | -11.0 | 193 | 4.9 | 165.2 | 767,038 | 3,967 | 138,159 | 202.7 | 130,380 | 34.8 | 65.2 | 134 | 72.0 | 840 | 34.6 |
| Oconto | 190 | 0.3 | 228 | 1.2 | 142.9 | 948,545 | 4,166 | 161,633 | 145.9 | 174,930 | 29.7 | 70.3 | 11 | 73.6 | 1,669 | 46.3 |
| Oneida | 35 | -0.7 | 265 | 1.2 | 11.5 | 1,075,764 | 4,065 | 90,003 | 12.6 | 96,412 | 91.5 | 8.5 | 2 | 79.4 | 102 | 10.7 |
| Outagamie | 237 | -5.5 | 210 | 0.8 | 206.2 | 1,511,380 | 7,207 | 198,453 | 263.7 | 233,388 | 27.3 | 72.7 | 6 | 79.2 | 1,606 | 46.3 |
| Ozaukee | 59 | -8.8 | 188 | 0.4 | 48.3 | 1,229,816 | 6,554 | 158,401 | 75.2 | 238,054 | 25.2 | 74.8 | 7 | 86.1 | 355 | 44.0 |
| Pepin | 107 | 3.2 | 239 | 6.6 | 69.3 | 924,201 | 3,874 | 156,970 | 72.3 | 161,413 | 33.7 | 66.3 | 5 | 79.2 | 979 | 62.9 |
| Pierce | 233 | -5.2 | 190 | 1.3 | 166.3 | 876,620 | 4,620 | 134,135 | 149.4 | 121,535 | 45.9 | 54.1 | 28 | 81.8 | 2,297 | 40.5 |
| Polk | 256 | 0.1 | 208 | 2.1 | 165.0 | 734,098 | 3,537 | 113,889 | 138.3 | 112,064 | 37.6 | 62.4 | 18 | 80.4 | 1,397 | 34.2 |
| Portage | 280 | 0.6 | 286 | 96.2 | 207.1 | 1,170,886 | 4,100 | 221,156 | 280.5 | 285,660 | 73.4 | 26.6 | 20 | 77.7 | 1,383 | 34.7 |
| Price | 89 | -3.4 | 218 | 0.7 | 34.3 | 535,028 | 2,459 | 66,706 | 25.8 | 62,932 | 16.7 | 83.3 | 3 | 72.9 | 105 | 14.9 |
| Racine | 127 | 15.9 | 209 | 1.6 | 109.8 | 1,602,954 | 7,682 | 181,939 | 86.4 | 141,475 | 74.8 | 25.2 | 1 | 82.3 | 2,502 | 44.5 |
| Richland | 221 | -3.1 | 200 | 0.3 | 118.5 | 671,915 | 3,356 | 108,023 | 136.7 | 123,890 | 22.5 | 77.5 | 37 | 67.6 | 2,493 | 50.7 |
| Rock | 354 | -0.1 | 223 | 15.6 | 302.8 | 1,467,760 | 6,589 | 175,571 | 294.5 | 185,539 | 56.0 | 44.0 | 20 | 79.7 | 4,915 | 54.7 |
| Rusk | 136 | 1.8 | 272 | 0.0 | 70.7 | 674,903 | 2,485 | 130,623 | 53.8 | 107,353 | 24.8 | 75.2 | 12 | 70.9 | 689 | 24.2 |
| St. Croix | 279 | 4.3 | 193 | 8.7 | 216.5 | 906,467 | 4,688 | 141,360 | 189.1 | 130,944 | 44.0 | 56.0 | 22 | 79.4 | 1,778 | 49.7 |
| Sauk | 299 | -10.1 | 212 | 13.2 | 197.7 | 890,636 | 4,207 | 150,164 | 188.5 | 133,470 | 37.2 | 62.8 | 27 | 73.7 | 2,048 | 45.5 |
| Sawyer | 46 | 5.6 | 263 | 1.5 | 26.0 | 681,805 | 2,593 | 136,876 | 21.9 | 125,091 | 41.6 | 58.4 | 2 | 72.0 | 152 | 18.3 |
| Shawano | 247 | -5.3 | 217 | 0.3 | 186.2 | 1,012,006 | 4,662 | 173,988 | 250.4 | 219,851 | 20.9 | 79.1 | 20 | 72.7 | 1,571 | 50.2 |
| Sheboygan | 196 | 3.0 | 205 | 0.3 | 164.3 | 1,348,670 | 6,594 | 198,128 | 213.4 | 222,785 | 22.6 | 77.4 | 14 | 84.2 | 1,215 | 39.5 |
| Taylor | 226 | 4.1 | 253 | 0.1 | 133.7 | 742,339 | 2,935 | 119,424 | 113.0 | 126,517 | 24.6 | 75.4 | 20 | 68.6 | 650 | 27.0 |
| Trempealeau | 330 | 2.1 | 268 | 8.6 | 215.6 | 1,062,452 | 3,958 | 181,542 | 291.1 | 236,873 | 24.8 | 75.2 | 37 | 77.1 | 2,725 | 56.1 |
| Vernon | 337 | -2.5 | 172 | 1.5 | 196.6 | 667,687 | 3,884 | 107,179 | 181.5 | 92,577 | 34.1 | 65.9 | 297 | 70.4 | 1,575 | 32.6 |
| Vilas | 6 | -17.9 | 84 | 0.9 | 3.3 | 498,803 | 5,913 | 88,693 | 7.0 | 103,851 | 97.3 | 2.7 | 2 | 94.0 | D | 9.0 |
| Walworth | 192 | 2.5 | 204 | 2.1 | 167.7 | 1,417,724 | 6,933 | 162,989 | 167.4 | 177,865 | 46.1 | 53.9 | 15 | 83.2 | 6,423 | 51.4 |
| Washburn | 74 | -15.6 | 198 | 2.7 | 34.1 | 620,384 | 3,128 | 85,256 | 30.3 | 81,384 | 32.7 | 67.3 | 2 | 75.5 | 283 | 18.3 |
| Washington | 126 | -5.5 | 218 | 0.3 | 108.9 | 1,539,370 | 7,053 | 216,870 | 157.4 | 272,394 | 43.4 | 56.6 | 14 | 79.2 | 565 | 32.5 |
| Waukesha | 97 | 5.7 | 170 | 1.1 | 71.5 | 935,065 | 5,507 | 108,440 | 51.1 | 88,944 | 73.9 | 26.1 | 7 | 79.8 | 1,122 | 27.9 |
| Waupaca | 202 | -6.4 | 196 | 7.7 | 142.6 | 883,093 | 4,516 | 135,842 | 152.1 | 147,516 | 24.8 | 75.2 | 6 | 74.4 | 1,777 | 48.2 |
| Waushara | 135 | -6.8 | 214 | 37.6 | 100.2 | 904,427 | 4,231 | 149,181 | 126.7 | 200,161 | 68.4 | 31.6 | 12 | 75.7 | 1,253 | 25.4 |
| Winnebago | 162 | 4.2 | 169 | 0.5 | 136.2 | 1,180,421 | 6,971 | 140,934 | 122.2 | 127,711 | 40.0 | 60.0 | 7 | 76.6 | 3,326 | 53.1 |
| Wood | 221 | -0.8 | 208 | 6.8 | 128.1 | 688,696 | 3,311 | 157,699 | 141.1 | 132,857 | 40.0 | 60.0 | 35 | 80.4 | 1,105 | 27.0 |
| **WYOMING** | 29,005 | -4.5 | 2,430 | 1,567.6 | 2,587.5 | 1,892,340 | 779 | 126,844 | 1,472.1 | 123,313 | 21.6 | 78.4 | 69 | 80.5 | 30,218 | 17.6 |
| Albany | 1,407 | -28.4 | 3,119 | 109.5 | 106.9 | 2,280,442 | 731 | 96,165 | 50.8 | 112,683 | 12.5 | 87.5 | 4 | 81.2 | 441 | 5.3 |
| Big Horn | 322 | 6.5 | 550 | 107.4 | 117.9 | 874,055 | 1,589 | 161,355 | 74.9 | 127,852 | 59.5 | 40.5 | 2 | 78.2 | 541 | 16.2 |
| Campbell | 2,901 | 0.8 | 4,512 | 5.1 | 184.3 | 2,553,745 | 566 | 134,546 | 69.9 | 108,705 | 3.7 | 96.3 | NA | 79.3 | 3,119 | 26.6 |
| Carbon | 2,812 | 18.4 | 8,150 | 187.4 | 143.9 | 4,486,393 | 550 | 145,397 | 73.2 | 212,293 | 8.4 | 91.6 | 1 | 78.6 | 423 | 6.1 |
| Converse | 2,594 | 6.0 | 6,754 | 65.2 | 106.7 | 3,576,045 | 529 | 159,841 | 56.3 | 146,734 | 12.2 | 87.8 | NA | 75.5 | 657 | 9.6 |
| Crook | 1,466 | -7.6 | 2,646 | 9.4 | 175.5 | 2,590,429 | 979 | 134,976 | 52.9 | 95,552 | 6.8 | 93.2 | NA | 70.6 | 2,959 | 31.6 |
| Fremont | 1,165 | -31.9 | 1,011 | 135.9 | 125.9 | 1,288,331 | 1,274 | 122,650 | 82.4 | 71,551 | 40.6 | 59.4 | 1 | 84.9 | 2,073 | 13.3 |

Items 117—132

# Table B. States and Counties — Water Use, Wholesale Trade, Retail Trade, and Real Estate

| STATE County | Water use, 2015 | | Wholesale Trade[1], 2017 | | | | Retail Trade[2], 2017 | | | | Real estate and rental and leasing,[2] 2017 | | | |
|---|---|---|---|---|---|---|---|---|---|---|---|---|---|---|
| | Public supply water withdrawn (mil gal/day) | Public supply gallons withdrawn per person per day | Number of establishments | Number of employees | Sales (mil dol) | Average payroll (mil dol) | Number of establishments | Number of employees | Sales (mil dol) | Average payroll (mil dol) | Number of establishments | Number of employees | Sales (mil dol) | Average payroll (mil dol) |
| | 133 | 134 | 135 | 136 | 137 | 138 | 139 | 140 | 141 | 142 | 143 | 144 | 145 | 146 |
| **WISCONSIN—Cont'd** | | | | | | | | | | | | | | |
| Douglas | 0.03 | 0.7 | 45 | 694 | 991.0 | 35.4 | 148 | 2,193 | 630.2 | 60.9 | 35 | 132 | 22.1 | 4.1 |
| Dunn | 2.39 | 53.7 | 41 | 639 | 321.4 | 30.5 | 117 | 1,825 | 526.6 | 42.0 | D | D | D | D |
| Eau Claire | 7.40 | 72.5 | 104 | 1,616 | 1,155.8 | 81.1 | 376 | 7,110 | 1,979.2 | 165.4 | 112 | 585 | 86.6 | 18.5 |
| Florence | 0.08 | 17.9 | NA | NA | NA | NA | 15 | 131 | 31.1 | 2.5 | 3 | 6 | 0.4 | 0.1 |
| Fond du Lac | 7.93 | 77.8 | D | D | D | D | 332 | 6,014 | 1,719.9 | 156.8 | 62 | 344 | 52.3 | 13.7 |
| Forest | 0.40 | 44.2 | NA | NA | NA | NA | 29 | 264 | 63.8 | 6.2 | 5 | 20 | 1.3 | 0.4 |
| Grant | 3.39 | 64.9 | 48 | 395 | 363.7 | 16.9 | 186 | 2,437 | 658.6 | 56.4 | 50 | 148 | 21.4 | 5.5 |
| Green | 2.66 | 71.5 | 44 | 434 | 274.8 | 20.1 | 149 | 2,646 | 1,168.7 | 103.5 | 30 | 53 | 10.1 | 1.4 |
| Green Lake | 0.35 | 18.6 | 10 | 119 | 195.9 | 3.7 | 67 | 1,021 | 285.8 | 26.0 | 20 | 44 | 8.6 | 1.5 |
| Iowa | 1.01 | 42.4 | 28 | 509 | 533.9 | 32.3 | 75 | 3,810 | 1,513.9 | 149.1 | D | D | D | D |
| Iron | 0.21 | 36.2 | 12 | 83 | 25.5 | 3.0 | 27 | 199 | 45.2 | 4.7 | 5 | D | 1.5 | D |
| Jackson | 0.96 | 46.7 | 13 | 135 | 50.7 | 3.8 | 62 | 836 | 250.0 | 19.4 | D | D | D | D |
| Jefferson | 5.71 | 67.5 | 74 | 1,710 | 730.2 | 80.5 | 270 | 4,214 | 1,283.0 | 96.4 | 63 | 289 | 48.9 | 6.6 |
| Juneau | 1.70 | 64.8 | 16 | 172 | 158.7 | 6.2 | 88 | 971 | 301.2 | 22.2 | D | D | D | D |
| Kenosha | 14.86 | 88.2 | 114 | 1,875 | 2,397.1 | 148.1 | 509 | 13,201 | 5,498.6 | 362.4 | 118 | 528 | 104.5 | 18.5 |
| Kewaunee | 0.87 | 42.7 | 15 | 149 | 93.7 | 6.1 | 59 | 563 | 139.0 | 12.8 | 6 | 8 | 0.9 | 0.2 |
| La Crosse | 14.23 | 120.4 | 108 | 2,584 | 5,065.8 | 136.8 | 420 | 8,978 | 2,341.8 | 216.5 | 141 | 1,047 | 146.3 | 35.2 |
| Lafayette | 0.97 | 57.6 | 23 | 384 | 382.4 | 15.4 | 41 | 373 | 139.9 | 8.0 | 8 | 10 | 2.2 | 0.3 |
| Langlade | 1.16 | 60.3 | 24 | 202 | 187.5 | 9.0 | 90 | 1,398 | 576.0 | 37.9 | D | D | D | D |
| Lincoln | 1.38 | 49.3 | 16 | 298 | 146.4 | 10.4 | 111 | 1,345 | 341.0 | 28.8 | 13 | 35 | 6.1 | 0.9 |
| Manitowoc | 15.68 | 196.5 | 63 | 1,055 | 593.2 | 53.4 | 252 | 3,890 | 909.3 | 86.6 | 43 | 168 | 15.1 | 2.9 |
| Marathon | 10.98 | 80.8 | 176 | 2,726 | 1,307.9 | 142.3 | 449 | 10,596 | 3,159.3 | 261.3 | 95 | 508 | 93.4 | 18.1 |
| Marinette | 2.57 | 62.9 | 25 | 216 | 175.3 | 8.5 | 180 | 2,226 | 596.6 | 52.9 | 19 | 36 | 6.7 | 1.2 |
| Marquette | 0.16 | 10.6 | 9 | 95 | 84.0 | 2.3 | 40 | 303 | 68.2 | 6.0 | D | D | D | 0.1 |
| Menominee | 0.32 | 70.0 | NA | NA | NA | NA | 5 | 40 | 8.5 | 0.5 | NA | NA | NA | NA |
| Milwaukee | 114.35 | 119.4 | 795 | 18,826 | 11,531.1 | 1,309.5 | 2,629 | 42,827 | 10,997.7 | 1,089.3 | 815 | 5,736 | 1,644.5 | 284.9 |
| Monroe | 3.56 | 78.2 | 38 | 827 | 769.5 | 42.8 | 134 | 2,088 | 599.7 | 52.5 | 29 | 105 | 12.7 | 3.6 |
| Oconto | 1.36 | 36.3 | 17 | 100 | 27.6 | 3.5 | 99 | 979 | 300.2 | 24.7 | 20 | 37 | 4.9 | 1.1 |
| Oneida | 2.03 | 57.1 | 31 | 419 | 226.0 | 22.9 | 228 | 3,064 | 884.2 | 82.4 | 63 | 173 | 28.8 | 4.8 |
| Outagamie | 6.13 | 33.5 | 280 | 4,235 | 8,062.3 | 256.2 | 728 | 13,376 | 3,661.6 | 324.6 | 154 | 750 | 167.0 | 31.8 |
| Ozaukee | 5.11 | 58.2 | 128 | 1,512 | 979.3 | 92.2 | 314 | 5,169 | 1,418.9 | 130.9 | D | D | D | 10.7 |
| Pepin | 0.39 | 53.5 | 13 | 153 | 114.6 | 7.1 | 31 | 234 | 62.7 | 6.8 | 4 | D | 3.3 | D |
| Pierce | 2.00 | 48.9 | D | D | D | D | 95 | 989 | 260.1 | 20.7 | D | D | D | D |
| Polk | 2.02 | 46.5 | 33 | 366 | 99.0 | 14.0 | 156 | 1,923 | 516.1 | 47.6 | 34 | 54 | 9.6 | 1.8 |
| Portage | 8.69 | 123.4 | 71 | 934 | 678.8 | 49.3 | 225 | 4,065 | 1,099.9 | 98.5 | 57 | 274 | 49.1 | 9.9 |
| Price | 1.02 | 74.8 | 12 | D | 14.1 | D | 68 | 564 | 128.7 | 12.2 | 9 | 22 | 2.7 | 0.7 |
| Racine | 19.14 | 98.1 | 179 | 2,684 | 2,227.3 | 150.4 | 559 | 9,038 | 2,409.5 | 234.2 | 117 | 545 | 98.9 | 18.9 |
| Richland | 1.18 | 67.4 | 15 | 80 | 46.8 | 3.4 | 64 | 871 | 228.5 | 20.3 | 12 | 83 | 11.0 | 2.0 |
| Rock | 16.71 | 103.5 | 136 | 2,677 | 1,878.5 | 145.6 | 510 | 10,004 | 2,631.5 | 285.4 | 118 | 452 | 132.7 | 19.6 |
| Rusk | 0.48 | 34.0 | 10 | 58 | 22.7 | 2.2 | 50 | 602 | 152.2 | 14.6 | 4 | 14 | 1.4 | 0.3 |
| St. Croix | 4.43 | 50.6 | 101 | 1,468 | 2,505.1 | 78.5 | 260 | 4,684 | 1,669.0 | 144.1 | 73 | 184 | 35.2 | 6.5 |
| Sauk | 7.23 | 113.6 | 60 | 1,432 | 1,138.7 | 81.9 | 309 | 4,232 | 1,173.5 | 100.7 | 62 | 255 | 35.9 | 7.7 |
| Sawyer | 0.64 | 39.1 | 15 | 96 | 18.3 | 3.4 | 96 | 1,067 | 264.8 | 28.3 | 28 | 76 | 11.1 | 2.4 |
| Shawano | 2.42 | 58.6 | 30 | 1,109 | 582.0 | 71.9 | 115 | 1,435 | 424.8 | 37.7 | 16 | 54 | 5.0 | 1.1 |
| Sheboygan | 15.8 | 136.7 | 95 | 1,484 | 933.0 | 76.7 | 379 | 6,132 | 1,729.4 | 150.8 | 76 | 232 | 44.4 | 7.0 |
| Taylor | 0.60 | 29.3 | 11 | 81 | 31.0 | 3.5 | 69 | 886 | 245.7 | 20.6 | 16 | 55 | 8.6 | 1.3 |
| Trempealeau | 2.94 | 99.5 | 22 | 157 | 116.8 | 8.2 | 100 | 975 | 319.9 | 21.7 | 9 | 16 | 5.5 | 0.9 |
| Vernon | 1.09 | 35.7 | 30 | 567 | 735.9 | 28.7 | 92 | 1,147 | 317.2 | 29.1 | 16 | 31 | 3.7 | 0.8 |
| Vilas | 0.48 | 22.4 | 11 | 139 | 121.6 | 6.6 | 143 | 1,067 | 309.0 | 29.0 | 30 | 66 | 19.0 | 2.5 |
| Walworth | 7.35 | 71.5 | 110 | 1,838 | 1,364.5 | 91.0 | 370 | 4,956 | 1,511.8 | 145.3 | 91 | 427 | 77.7 | 15.1 |
| Washburn | 0.63 | 40.5 | 8 | 54 | 18.6 | 2.5 | 81 | 765 | 189.4 | 19.1 | 21 | 59 | 8.3 | 1.5 |
| Washington | 7.91 | 59.2 | 171 | 3,302 | 2,787.2 | 214.6 | 365 | 7,100 | 2,074.2 | 186.7 | D | D | D | 8.7 |
| Waukesha | 19.63 | 49.5 | 820 | 13,733 | 8,211.5 | 892.9 | 1,296 | 26,194 | 7,073.7 | 705.0 | 466 | 3,992 | 804.0 | 182.2 |
| Waupaca | 4.94 | 95.1 | 32 | 274 | 465.7 | 13.8 | 173 | 2,273 | 585.2 | 51.9 | 39 | 128 | 15.9 | 3.2 |
| Waushara | 1.14 | 47.4 | 18 | 150 | 92.5 | 7.3 | 71 | 766 | 263.8 | 17.2 | 8 | 15 | 3.1 | 0.5 |
| Winnebago | 13.99 | 82.5 | D | D | D | 114.4 | 487 | 8,865 | 2,335.2 | 214.8 | 129 | 752 | 198.2 | 29.0 |
| Wood | 4.73 | 64.4 | 57 | 1,042 | 688.3 | 56.1 | 273 | 4,127 | 1,114.7 | 97.5 | 55 | 195 | 32.5 | 4.7 |
| **WYOMING** | 101.35 | 172.9 | 696 | 5,967 | 6,061.0 | 342.4 | 2,583 | 29,786 | 9,124.4 | 852.8 | 1,199 | 4,777 | 1,216.5 | 220.4 |
| Albany | 6.45 | 169.9 | 20 | 164 | 140.0 | 7.4 | 125 | 1,670 | 458.6 | 38.7 | 59 | 155 | 25.4 | 4.1 |
| Big Horn | 2.20 | 183.0 | D | D | D | 3.1 | 41 | 356 | 76.0 | 8.1 | 8 | D | 0.9 | D |
| Campbell | 5.97 | 121.3 | 82 | 1,023 | 874.7 | 70.6 | 174 | 2,469 | 790.3 | 75.7 | 82 | 366 | 157.0 | 21.7 |
| Carbon | 2.85 | 183.2 | D | D | D | 1.4 | 79 | 799 | 270.3 | 22.1 | 31 | 109 | 17.3 | 4.8 |
| Converse | 2.04 | 143.3 | 9 | 39 | 21.2 | 1.8 | 53 | 411 | 106.7 | 10.3 | 23 | 52 | 7.1 | 1.5 |
| Crook | 6.44 | 865.1 | D | D | D | 1.7 | 25 | 181 | 47.8 | 4.3 | 8 | D | 1.3 | D |
| Fremont | 4.31 | 106.9 | D | D | D | 10.7 | 152 | 1,889 | 618.0 | 54.3 | 68 | 314 | 56.0 | 12.8 |

1 Merchant wholesalers, except manufacturers' sales branches and offices.  2. Employer establishments.

# Table B. States and Counties — Professional Services, Manufacturing, and Accommodation and Food Services

| STATE County | Professional, scientific, and technical services, 2017 | | | | Manufacturing, 2017 | | | | Accommodation and food services, 2017 | | | |
|---|---|---|---|---|---|---|---|---|---|---|---|---|
| | Number of establishments | Number of employees | Sales (mil dol) | Average payroll (mil dol) | Number of establishments | Number of employees | Sales (mil dol) | Average payroll (mil dol) | Number of establishments | Number of employees | Sales (mil dol) | Annual payroll (mil dol) |
| | 147 | 148 | 149 | 150 | 151 | 152 | 153 | 154 | 155 | 156 | 157 | 158 |
| **WISCONSIN—Cont'd** | | | | | | | | | | | | |
| Douglas | 74 | 389 | 33 | 16.2 | 51 | 1,656 | 1,383.9 | 85.5 | 152 | 1,815 | 77.5 | 22.6 |
| Dunn | D | D | D | D | 69 | 2,885 | 1,701.6 | 158.5 | 102 | 1,426 | 58.7 | 16.4 |
| Eau Claire | 191 | 1,477 | 200 | 80.4 | 97 | 5,285 | 1,809.1 | 262.1 | 274 | 5,158 | 228.5 | 68.6 |
| Florence | 5 | 7 | 1 | 0.2 | 11 | 174 | 36.2 | 6.0 | 22 | 142 | 6.6 | 1.4 |
| Fond du Lac | 141 | 1,433 | 157 | 104.1 | 146 | 9,480 | 4,108.5 | 486.2 | 222 | 3,542 | 163.0 | 44.3 |
| Forest | D | D | 6 | D | 14 | 247 | 56.7 | 11.5 | D | D | D | D |
| Grant | D | D | D | D | 67 | 2,680 | 1,132.4 | 125.0 | 130 | 1,378 | 57.5 | 14.5 |
| Green | D | D | D | D | 76 | 3,552 | 1,452.9 | 185.0 | 91 | 1,051 | 45.4 | 13.6 |
| Green Lake | 26 | 82 | 9 | 3.3 | 35 | 956 | 233.8 | 41.1 | 49 | 401 | 18.8 | 5.6 |
| Iowa | 44 | 159 | 22 | 6.4 | 41 | 1,112 | 1,458.0 | 52.8 | 58 | 582 | 28.4 | 7.5 |
| Iron | D | D | 5 | D | 12 | 152 | 46.5 | 6.4 | D | D | D | 2.4 |
| Jackson | 24 | 84 | 8 | 2.3 | 21 | 504 | 193.0 | 25.8 | 57 | 665 | 36.0 | 8.7 |
| Jefferson | 124 | 652 | 94 | 27.1 | 141 | 7,994 | 3,717.1 | 466.7 | 198 | 2,551 | 110.0 | 28.4 |
| Juneau | 14 | 49 | 5 | 2.8 | 48 | 2,130 | 808.0 | 105.2 | 93 | 754 | 39.7 | 10.3 |
| Kenosha | 239 | 1,303 | 168 | 61.6 | 194 | 7,387 | 3,539.6 | 416.0 | 372 | 6,241 | 305.1 | 88.4 |
| Kewaunee | 31 | 154 | 18 | 8.3 | 38 | 1,922 | 747.1 | 93.3 | 48 | 422 | 18.6 | 4.8 |
| La Crosse | D | D | D | D | 154 | 7,288 | 1,990.3 | 367.1 | 347 | 6,995 | 327.9 | 95.7 |
| Lafayette | D | D | 6 | D | 28 | 871 | 433.8 | 43.5 | D | D | D | 2.4 |
| Langlade | 24 | 110 | 10 | 4.9 | 44 | 1,512 | 388.0 | 64.5 | 60 | 591 | 28.2 | 7.6 |
| Lincoln | 36 | 111 | 13 | 4.5 | 45 | 2,358 | 823.7 | 113.4 | 98 | 839 | 38.8 | 10.8 |
| Manitowoc | 94 | 573 | 87 | 27.5 | 170 | 9,122 | 3,229.5 | 454.1 | 185 | 2,395 | 105.7 | 30.4 |
| Marathon | D | D | D | D | 240 | 16,995 | 5,596.1 | 855.2 | 312 | 5,290 | 241.8 | 72.0 |
| Marinette | 53 | 271 | 38 | 15.9 | 92 | 6,103 | 1,730.0 | 327.0 | 156 | 1,372 | 67.5 | 17.3 |
| Marquette | D | D | 8 | D | 21 | 1,977 | 459.7 | 63.6 | D | D | D | 3.9 |
| Menominee | NA | NA | NA | NA | NA | NA | NA | NA | D | D | D | D |
| Milwaukee | 1,987 | 24,542 | 4,673 | 1,923.1 | 966 | 46,242 | 17,269.6 | 2,788.9 | 2,102 | 43,834 | 2,811.7 | 743.7 |
| Monroe | 66 | 379 | 42 | 15.8 | 61 | 3,738 | 1,397.4 | 171.6 | 113 | 1,461 | 68.5 | 17.3 |
| Oconto | 33 | 127 | 13 | 4.5 | 60 | 1,979 | 693.7 | 95.4 | 96 | 657 | 31.8 | 8.5 |
| Oneida | 80 | 347 | 39 | 14.5 | 46 | 1,071 | 433.3 | 53.0 | 197 | 1,677 | 110.6 | 27.7 |
| Outagamie | 370 | 3,865 | 617 | 245.1 | 338 | 19,907 | 7,618.8 | 1,202.0 | 478 | 8,872 | 433.7 | 123.5 |
| Ozaukee | 322 | 2,155 | 389 | 137.4 | 200 | 9,331 | 2,936.9 | 554.7 | 207 | 3,616 | 166.0 | 50.5 |
| Pepin | D | D | 4 | D | 12 | 176 | 68.8 | 7.2 | D | D | D | D |
| Pierce | D | D | 34 | D | 51 | 1,711 | 611.3 | 82.4 | 106 | 1,001 | 47.8 | 13.5 |
| Polk | D | D | D | D | 102 | 3,889 | 1,473.2 | 181.4 | 119 | 1,201 | 54.1 | 14.7 |
| Portage | D | D | D | D | 75 | 4,650 | 1,469.3 | 233.7 | 215 | 3,068 | 138.2 | 41.0 |
| Price | 20 | 80 | 6 | 2.4 | 42 | 2,055 | 563.7 | 95.4 | 37 | 249 | 12.6 | 3.1 |
| Racine | 289 | 1,953 | 289 | 118.4 | 297 | 15,883 | 5,623.1 | 1,006.9 | 412 | 6,540 | 319.4 | 91.5 |
| Richland | 20 | 51 | 4 | 1.5 | 30 | 1,380 | 739.3 | 70.2 | 31 | 314 | 14.5 | 3.8 |
| Rock | 208 | 2,273 | 165 | 154.1 | 229 | 10,607 | 4,961.7 | 598.4 | 366 | 6,304 | 308.1 | 85.0 |
| Rusk | 16 | 38 | 4 | 1.5 | 27 | 1,561 | 480.9 | 65.1 | 31 | 225 | 12.3 | 2.6 |
| St. Croix | 229 | 1,202 | 242 | 72.6 | 173 | 5,925 | 1,561.5 | 326.7 | 208 | 3,763 | 165.5 | 49.3 |
| Sauk | 115 | 1,060 | 135 | 59.2 | 94 | 5,312 | 1,682.1 | 276.5 | 266 | 7,430 | 712.7 | 177.8 |
| Sawyer | 40 | 146 | 14 | 5.2 | 36 | 580 | 357.3 | 31.9 | 120 | 1,100 | 81.9 | 24.4 |
| Shawano | D | D | 17 | D | D | D | D | D | D | D | D | D |
| Sheboygan | D | D | D | D | 206 | 17,576 | 7,887.7 | 1,010.5 | 279 | 5,069 | 258.2 | 78.7 |
| Taylor | 18 | 108 | 10 | 3.9 | 38 | 3,127 | 957.6 | 155.8 | 39 | 473 | 15.2 | 5.9 |
| Trempealeau | 36 | 209 | 24 | 7.0 | 56 | 8,872 | 1,938.1 | 360.0 | 82 | 633 | 31.2 | 8.3 |
| Vernon | 33 | 194 | 18 | 7.2 | 37 | 1,106 | 374.8 | 45.0 | 58 | 647 | 23.8 | 6.9 |
| Vilas | D | D | D | D | 35 | 266 | 59.8 | 11.6 | 214 | 1,874 | 159.3 | 45.3 |
| Walworth | 200 | 1,519 | 175 | 65.5 | 202 | 8,638 | 2,704.9 | 435.5 | 296 | 6,180 | 307.0 | 91.2 |
| Washburn | 28 | 173 | 25 | 8.4 | 30 | 1,153 | 199.1 | 44.6 | 84 | 540 | 30.5 | 7.9 |
| Washington | 235 | 1,647 | 173 | 87.7 | 317 | 14,275 | 4,534.2 | 819.8 | 260 | 4,588 | 217.3 | 62.9 |
| Waukesha | 1,361 | 13,996 | 2,752 | 979.7 | 931 | 42,365 | 14,623.2 | 2,609.9 | 860 | 17,318 | 891.9 | 259.7 |
| Waupaca | 66 | 251 | 30 | 12.1 | 83 | 5,851 | 2,646.1 | 304.0 | 144 | 1,487 | 70.5 | 19.0 |
| Waushara | 17 | 194 | 13 | 6.5 | 26 | 808 | 306.5 | 37.8 | 63 | 493 | 23.6 | 6.0 |
| Winnebago | D | D | D | D | 291 | 21,965 | 9,101.8 | 1,389.8 | 399 | 6,740 | 296.1 | 87.2 |
| Wood | 100 | 626 | 83 | 27.6 | 111 | 6,046 | 2,555.6 | 327.4 | 182 | 2,211 | 106.0 | 27.5 |
| **WYOMING** | 2,387 | 9,511 | 1,527 | 527.7 | 581 | 9,354 | 7,897.7 | 643.3 | 1,839 | 27,248 | 1,946.0 | 556.4 |
| Albany | D | D | D | D | 39 | 339 | 143.8 | 22.1 | 109 | 1,758 | 84.7 | 24.1 |
| Big Horn | D | D | 6 | D | 15 | 235 | 93.0 | 10.3 | 33 | D | 9.2 | D |
| Campbell | 121 | 591 | 76 | 28.2 | 44 | 547 | 212.5 | 34.2 | 112 | 1,852 | 105.0 | 30.3 |
| Carbon | 36 | 89 | 14 | 3.4 | D | 621 | D | D | 80 | 840 | 60.8 | 19.6 |
| Converse | 34 | 104 | 15 | 5.2 | D | 120 | D | 5.6 | D | D | D | D |
| Crook | 15 | 29 | 4 | 1.0 | 12 | 137 | 29.0 | 6.7 | 36 | 173 | 11.6 | 2.8 |
| Fremont | 113 | 418 | 47 | 20.0 | 30 | 214 | 54.4 | 10.9 | 123 | 1,758 | 127.7 | 35.4 |

# Table B. States and Counties — Health Care and Social Assistance, Other Services, Nonemployer Businesses, and Residential Construction

| STATE County | Health care and social assistance, 2017 | | | | Other services, 2017 | | | | Nonemployer businesses, 2018 | | Value of residential construction authorized by building permits, 2020 | |
| --- | --- | --- | --- | --- | --- | --- | --- | --- | --- | --- | --- | --- |
| | Number of establishments | Number of employees | Receipts (mil dol) | Annual payroll (mil dol) | Number of establishments | Number of employees | Receipts (mil dol) | Annual payroll (mil dol) | Number | Receipts (mil dol) | New construction ($1,000) | Number of housing units |
| | 159 | 160 | 161 | 162 | 163 | 164 | 165 | 166 | 167 | 168 | 169 | 170 |
| WISCONSIN—Cont'd | | | | | | | | | | | | |
| Douglas | 114 | 2,033 | 140.1 | 60.0 | 71 | 549 | 47.2 | 14.7 | 2,189 | 93.0 | 18,628 | 96 |
| Dunn | 95 | 2,268 | 199.7 | 73.8 | 58 | 309 | 32.2 | 9.3 | 2,562 | 128.0 | 28,137 | 111 |
| Eau Claire | 341 | 9,985 | 1,153.4 | 422.1 | 202 | 1,356 | 124.4 | 37.2 | 6,326 | 328.4 | 128,135 | 683 |
| Florence | D | D | D | D | 6 | 15 | 1.6 | 0.4 | 314 | 11.9 | 9,669 | 33 |
| Fond du Lac | 295 | 6,619 | 864.7 | 308.7 | 179 | 975 | 99.3 | 26.0 | 5,105 | 256.5 | 47,213 | 148 |
| Forest | D | D | D | 3.5 | D | D | 3.2 | D | 668 | 33.2 | 11,099 | 53 |
| Grant | 110 | 2,233 | 187.8 | 82.9 | 106 | 363 | 48.5 | 10.6 | 3,332 | 158.6 | 22,125 | 91 |
| Green | 78 | 2,159 | 233.5 | 109.5 | 73 | 300 | 29.0 | 8.0 | 2,492 | 116.4 | 31,119 | 119 |
| Green Lake | 48 | 993 | 90.0 | 42.2 | 29 | 101 | 9.0 | 3.2 | 1,425 | 74.7 | 14,153 | 51 |
| Iowa | 56 | 1,155 | 104.7 | 44.7 | 32 | 103 | 12.6 | 3.2 | 1,976 | 82.7 | 22,641 | 88 |
| Iron | 13 | 310 | 14.6 | 8.2 | D | D | D | 0.2 | 507 | 21.9 | 8,321 | 35 |
| Jackson | 51 | 918 | 102.1 | 41.4 | 30 | 96 | 8.7 | 2.0 | 1,152 | 49.7 | 9,217 | 49 |
| Jefferson | 222 | 4,503 | 386.1 | 138.0 | 147 | 1,062 | 199.2 | 44.1 | 5,130 | 266.9 | 60,516 | 347 |
| Juneau | 50 | 1,224 | 110.0 | 53.1 | 45 | 187 | 24.6 | 4.9 | 1,469 | 66.6 | 30,284 | 132 |
| Kenosha | 447 | 9,354 | 1,030.4 | 406.4 | 235 | 1,407 | 111.2 | 35.1 | 8,798 | 394.1 | 69,178 | 226 |
| Kewaunee | 41 | 435 | 26.8 | 14.0 | 29 | 76 | 10.4 | 2.4 | 1,309 | 58.2 | 11,218 | 40 |
| La Crosse | 340 | 12,469 | 1,399.4 | 619.7 | 222 | 1,701 | 189.4 | 54.5 | 6,625 | 295.5 | 74,900 | 330 |
| Lafayette | D | D | D | 11.5 | 27 | 146 | 25.6 | 5.1 | 1,333 | 63.6 | 6,320 | 25 |
| Langlade | 50 | 858 | 158.4 | 44.6 | 48 | 167 | 14.7 | 5.0 | 1,213 | 57.4 | 44,981 | 168 |
| Lincoln | D | D | D | D | 55 | 254 | 24.1 | 6.3 | 1,667 | 77.0 | 33,889 | 134 |
| Manitowoc | 176 | 4,335 | 400.4 | 177.7 | 138 | 528 | 62.0 | 14.3 | 3,913 | 184.0 | 25,283 | 139 |
| Marathon | 390 | 10,447 | 1,418.3 | 529.0 | 215 | 1,253 | 160.7 | 41.3 | 7,841 | 406.6 | 84,311 | 398 |
| Marinette | 123 | 2,686 | 244.4 | 119.5 | 70 | 259 | 24.4 | 5.8 | 2,236 | 103.1 | 19,411 | 119 |
| Marquette | 21 | 468 | 16.3 | 7.8 | D | D | D | D | 1,069 | 54.3 | 10,056 | 42 |
| Menominee | NA | NA | NA | NA | NA | NA | NA | NA | 105 | 3.3 | 2,790 | 11 |
| Milwaukee | 3,195 | 97,178 | 11,669.2 | 4,374.3 | 1,509 | 11,002 | 3,008.8 | 389.1 | 50,963 | 2,193.8 | 145,159 | 749 |
| Monroe | 85 | 3,334 | 371.3 | 197.9 | 72 | 388 | 33.8 | 10.0 | 2,598 | 134.6 | 23,575 | 119 |
| Oconto | 110 | 1,612 | 129.1 | 53.3 | 46 | 142 | 14.9 | 3.7 | 2,286 | 109.0 | 39,810 | 183 |
| Oneida | 163 | 2,956 | 365.2 | 142.7 | 106 | 392 | 41.1 | 11.4 | 3,017 | 128.8 | 64,763 | 227 |
| Outagamie | 501 | 12,746 | 1,437.3 | 629.2 | 360 | 2,853 | 340.3 | 99.2 | 10,683 | 558.4 | 183,246 | 823 |
| Ozaukee | 335 | 7,022 | 772.8 | 315.2 | 196 | 1,191 | 101.6 | 32.0 | 7,021 | 396.5 | 107,983 | 396 |
| Pepin | 18 | 227 | 22.0 | 7.6 | D | D | D | 1.8 | 573 | 30.3 | 5,730 | 23 |
| Pierce | 70 | 837 | 51.2 | 21.2 | 61 | 193 | 24.6 | 5.3 | 2,660 | 115.3 | 66,193 | 367 |
| Polk | 110 | 2,911 | 339.9 | 116.9 | 73 | 244 | 24.1 | 6.7 | 3,306 | 152.9 | 48,399 | 216 |
| Portage | 199 | 4,590 | 449.8 | 178.4 | 124 | 718 | 79.6 | 20.3 | 4,034 | 181.4 | 58,552 | 406 |
| Price | 31 | 746 | 65.0 | 21.4 | 34 | 92 | 9.0 | 2.2 | 1,104 | 43.5 | 10,479 | 60 |
| Racine | 510 | 11,145 | 1,032.8 | 423.6 | 302 | 1,880 | 169.7 | 52.9 | 9,775 | 418.0 | 62,495 | 224 |
| Richland | 49 | 988 | 92.0 | 39.7 | D | D | D | D | 1,205 | 58.4 | 7,944 | 42 |
| Rock | 342 | 8,861 | 1,200.5 | 478.6 | 265 | 1,356 | 121.9 | 34.5 | 8,186 | 368.6 | 121,994 | 671 |
| Rusk | 33 | 748 | 69.5 | 30.5 | D | D | D | D | 1,015 | 45.4 | 16,455 | 72 |
| St. Croix | 238 | 4,882 | 479.1 | 188.8 | 170 | 843 | 89.2 | 22.7 | 6,639 | 331.1 | 168,137 | 742 |
| Sauk | 143 | 4,107 | 425.5 | 187.0 | 140 | 592 | 68.6 | 21.4 | 4,603 | 235.6 | 64,093 | 233 |
| Sawyer | 47 | 911 | 90.0 | 35.0 | 39 | 168 | 18.8 | 4.9 | 1,567 | 69.5 | 23,389 | 100 |
| Shawano | 77 | 1,466 | 107.4 | 56.3 | D | D | D | D | 2,482 | 119.0 | 18,543 | 79 |
| Sheboygan | 315 | 6,645 | 744.0 | 290.0 | 202 | 1,060 | 98.6 | 25.5 | 5,449 | 260.9 | 72,366 | 365 |
| Taylor | 46 | 965 | 108.9 | 38.5 | D | D | D | D | 1,408 | 81.3 | 10,979 | 48 |
| Trempealeau | 56 | 1,614 | 113.9 | 54.3 | 39 | 117 | 22.0 | 4.5 | 2,082 | 98.6 | 25,019 | 121 |
| Vernon | 75 | 1,793 | 147.8 | 67.4 | 41 | 114 | 10.8 | 2.5 | 2,537 | 118.2 | 24,636 | 117 |
| Vilas | 54 | 577 | 50.8 | 19.4 | 58 | 212 | 20.2 | 5.4 | 2,448 | 111.7 | 74,363 | 224 |
| Walworth | 229 | 4,002 | 453.9 | 160.6 | 220 | 979 | 96.1 | 27.1 | 7,105 | 352.9 | 118,645 | 364 |
| Washburn | 52 | 833 | 67.7 | 26.9 | 31 | 89 | 11.3 | 2.5 | 1,458 | 64.0 | 19,456 | 74 |
| Washington | 293 | 6,207 | 625.8 | 242.1 | 248 | 1,472 | 132.1 | 42.4 | 8,576 | 448.6 | 127,023 | 407 |
| Waukesha | 1,448 | 29,204 | 3,423.1 | 1,368.5 | 856 | 6,732 | 794.5 | 248.2 | 28,498 | 1,628.0 | 408,769 | 1,201 |
| Waupaca | 133 | 2,834 | 180.2 | 82.2 | 95 | 420 | 40.4 | 12.0 | 3,083 | 146.5 | 36,708 | 185 |
| Waushara | 37 | 722 | 44.1 | 21.9 | D | D | D | 1.9 | 1,560 | 82.3 | 15,268 | 66 |
| Winnebago | 453 | 12,419 | 1,282.3 | 560.5 | 258 | 2,127 | 240.9 | 72.2 | 8,625 | 412.5 | 106,446 | 554 |
| Wood | 208 | 8,232 | 1,190.2 | 459.1 | 132 | 791 | 96.1 | 22.8 | 3,960 | 194.4 | 26,362 | 110 |
| WYOMING | 2,001 | 33,540 | 3,871.9 | 1,657.4 | 1,368 | 6,245 | 931.2 | 221.4 | 53,042 | 2,778.9 | 703,859 | 2,128 |
| Albany | 125 | 2,095 | 227.4 | 81.9 | 76 | 352 | 94.8 | 10.7 | 2,732 | 110.3 | 32,622 | 211 |
| Big Horn | 28 | 374 | 32.7 | 14.6 | 17 | 46 | 3.8 | 1.0 | 923 | 30.9 | 5,901 | 24 |
| Campbell | 115 | 2,124 | 278.9 | 122.0 | 119 | 627 | 106.6 | 30.6 | 3,386 | 156.5 | 20,742 | 76 |
| Carbon | 47 | 583 | 49.7 | 22.7 | 30 | 121 | 20.2 | 4.1 | 1,165 | 46.3 | 6,851 | 36 |
| Converse | 24 | 530 | 74.0 | 33.4 | 36 | 113 | 12.8 | 3.0 | 1,110 | 51.4 | 1,990 | 10 |
| Crook | 13 | 209 | 15.4 | 8.3 | D | D | 4.2 | D | 765 | 32.1 | 1,516 | 6 |
| Fremont | 149 | 1,827 | 175.5 | 73.0 | 82 | 423 | 41.8 | 14.1 | 3,064 | 114.5 | 5,140 | 19 |

# Table B. States and Counties — Government Employment and Payroll, and Local Government Finances

| STATE County | Government employment and payroll, 2017 | | | | | | | | | Local government finances, 2017 | | | | |
| | | | March payroll (percent of total) | | | | | | | General revenue | | | | |
| | | | | | | | | | | | | Taxes | | |
| | Full-time equivalent employees | March payroll (dollars) | Administration, judicial, and legal | Police and corrections | Fire protection | Highways and transportation | Health and welfare | Natural resources and utilities | Education and libraries | Total (mil dol) | Inter-govern-mental (mil dol) | Total (mil dol) | Per capita[1] (dollars) Total | Property |
| | 171 | 172 | 173 | 174 | 175 | 176 | 177 | 178 | 179 | 180 | 181 | 182 | 183 | 184 |
|---|---|---|---|---|---|---|---|---|---|---|---|---|---|---|
| **WISCONSIN—Cont'd** | | | | | | | | | | | | | | |
| Douglas | 2,220 | 8,531,441 | 5.5 | 9.9 | 2.9 | 4.5 | 4.9 | 4.5 | 65.6 | 255.6 | 133.2 | 80.9 | 1,867 | 1,716 |
| Dunn | 1,480 | 5,930,390 | 7.5 | 8.9 | 2.6 | 7.0 | 16.3 | 2.8 | 53.6 | 176.2 | 80.7 | 64.3 | 1,439 | 1,336 |
| Eau Claire | 3,816 | 17,246,976 | 5.8 | 8.9 | 3.2 | 5.2 | 6.7 | 3.0 | 66.3 | 450.1 | 205.7 | 176.6 | 1,705 | 1,545 |
| Florence | 197 | 719,054 | 9.0 | 12.4 | 0.2 | 9.1 | 11.0 | 7.1 | 41.2 | 22.8 | 10.1 | 9.8 | 2,253 | 2,134 |
| Fond du Lac | 3,803 | 16,957,873 | 4.6 | 8.4 | 2.5 | 4.0 | 10.4 | 2.7 | 65.5 | 454.0 | 217.1 | 165.7 | 1,618 | 1,491 |
| Forest | 455 | 1,696,143 | 7.4 | 11.3 | 0.7 | 8.4 | 8.1 | 1.9 | 61.3 | 43.9 | 22.2 | 18.2 | 2,029 | 1,949 |
| Grant | 2,117 | 7,789,053 | 6.5 | 6.2 | 0.1 | 5.7 | 11.8 | 4.3 | 64.4 | 235.5 | 123.2 | 69.0 | 1,334 | 1,247 |
| Green | 1,445 | 5,678,043 | 5.5 | 9.0 | 0.5 | 5.6 | 15.1 | 4.9 | 58.5 | 156.5 | 67.7 | 58.5 | 1,586 | 1,480 |
| Green Lake | 751 | 2,789,699 | 8.1 | 11.7 | 0.2 | 5.2 | 10.4 | 5.3 | 55.9 | 85.8 | 35.9 | 40.5 | 2,166 | 2,048 |
| Iowa | 889 | 3,494,785 | 8.0 | 7.6 | 0.2 | 8.5 | 11.1 | 2.7 | 61.4 | 100.7 | 46.4 | 38.2 | 1,615 | 1,500 |
| Iron | 258 | 1,032,925 | 11.5 | 12.3 | 1.2 | 13.2 | 9.1 | 7.9 | 44.2 | 36.9 | 15.3 | 13.1 | 2,304 | 2,133 |
| Jackson | 794 | 2,850,058 | 12.2 | 8.9 | 0.3 | 9.2 | 7.9 | 2.6 | 57.6 | 87.8 | 47.0 | 29.7 | 1,448 | 1,346 |
| Jefferson | 2,729 | 11,581,146 | 6.6 | 11.2 | 2.0 | 5.3 | 8.2 | 6.0 | 59.7 | 326.6 | 144.4 | 135.1 | 1,595 | 1,475 |
| Juneau | 979 | 3,580,478 | 9.4 | 10.3 | 0.4 | 8.4 | 9.8 | 3.5 | 56.8 | 112.3 | 58.2 | 40.6 | 1,533 | 1,358 |
| Kenosha | 6,781 | 32,794,450 | 4.4 | 10.9 | 4.3 | 3.2 | 5.4 | 4.3 | 66.1 | 898.3 | 421.5 | 343.2 | 2,039 | 1,893 |
| Kewaunee | 785 | 3,019,472 | 8.3 | 7.9 | 0.5 | 5.4 | 10.7 | 3.9 | 61.9 | 89.5 | 42.3 | 32.1 | 1,573 | 1,542 |
| La Crosse | 4,881 | 21,656,178 | 5.9 | 7.4 | 2.8 | 4.0 | 11.4 | 3.3 | 62.5 | 578.4 | 247.7 | 221.8 | 1,879 | 1,699 |
| Lafayette | 881 | 3,297,072 | 6.1 | 6.2 | 0.0 | 5.3 | 26.9 | 3.1 | 51.4 | 94.1 | 41.2 | 23.7 | 1,419 | 1,353 |
| Langlade | 730 | 2,873,041 | 9.5 | 11.6 | 3.3 | 7.8 | 4.1 | 5.2 | 57.4 | 85.5 | 42.3 | 28.8 | 1,501 | 1,382 |
| Lincoln | 1,141 | 4,801,300 | 7.0 | 10.3 | 1.8 | 6.9 | 16.1 | 3.8 | 51.5 | 125.6 | 50.8 | 43.9 | 1,582 | 1,481 |
| Manitowoc | 2,749 | 11,439,956 | 4.7 | 9.1 | 4.0 | 6.2 | 11.1 | 9.1 | 54.6 | 291.0 | 141.7 | 98.4 | 1,245 | 1,208 |
| Marathon | 5,330 | 23,884,066 | 4.7 | 7.1 | 1.7 | 4.7 | 14.0 | 2.4 | 64.6 | 857.8 | 526.0 | 219.3 | 1,619 | 1,491 |
| Marinette | 1,553 | 5,534,079 | 7.4 | 11.0 | 2.0 | 6.4 | 10.2 | 6.4 | 55.4 | 164.0 | 80.7 | 62.1 | 1,542 | 1,410 |
| Marquette | 499 | 1,877,399 | 10.5 | 12.2 | 3.5 | 8.1 | 11.0 | 1.5 | 51.7 | 53.4 | 21.1 | 27.8 | 1,820 | 1,726 |
| Menominee | 282 | 1,057,076 | 3.5 | 3.6 | 0.6 | 4.6 | 13.0 | 0.7 | 70.4 | 27.9 | 22.3 | 5.0 | 1,093 | 1,074 |
| Milwaukee | 35,769 | 185,214,348 | 5.2 | 15.8 | 5.3 | 5.4 | 7.9 | 5.9 | 53.0 | 5,151.2 | 2,178.9 | 1,698.0 | 1,787 | 1,599 |
| Monroe | 1,704 | 6,405,363 | 6.7 | 9.5 | 0.1 | 5.9 | 11.5 | 3.8 | 61.4 | 183.4 | 94.8 | 60.6 | 1,326 | 1,212 |
| Oconto | 1,233 | 5,197,144 | 6.0 | 8.1 | 1.0 | 6.0 | 17.2 | 4.2 | 56.1 | 126.2 | 59.4 | 49.4 | 1,317 | 1,242 |
| Oneida | 1,404 | 5,979,000 | 6.9 | 9.8 | 2.3 | 5.9 | 4.7 | 3.0 | 67.3 | 182.1 | 66.2 | 95.9 | 2,723 | 2,544 |
| Outagamie | 7,636 | 35,643,735 | 4.2 | 7.1 | 3.3 | 3.0 | 7.0 | 3.3 | 70.2 | 881.1 | 428.0 | 272.2 | 1,465 | 1,401 |
| Ozaukee | 2,505 | 12,111,043 | 5.6 | 14.7 | 1.1 | 5.5 | 10.4 | 4.0 | 57.1 | 328.4 | 101.6 | 169.6 | 1,917 | 1,761 |
| Pepin | 308 | 1,151,936 | 9.5 | 9.0 | 0.6 | 10.7 | 9.1 | 3.1 | 57.1 | 33.5 | 16.0 | 13.1 | 1,813 | 1,726 |
| Pierce | 1,637 | 6,765,716 | 8.0 | 7.9 | 0.3 | 7.1 | 7.0 | 4.0 | 64.8 | 179.0 | 82.6 | 71.7 | 1,707 | 1,618 |
| Polk | 1,820 | 7,076,294 | 5.5 | 7.8 | 0.2 | 4.9 | 12.0 | 4.4 | 64.8 | 201.7 | 93.9 | 81.5 | 1,878 | 1,777 |
| Portage | 2,385 | 10,347,340 | 6.6 | 9.9 | 2.7 | 6.3 | 19.8 | 3.9 | 49.6 | 404.3 | 242.4 | 99.8 | 1,414 | 1,277 |
| Price | 596 | 2,108,423 | 11.0 | 9.6 | 0.1 | 11.8 | 6.8 | 2.4 | 57.3 | 59.8 | 27.9 | 25.3 | 1,890 | 1,799 |
| Racine | 5,884 | 28,010,342 | 3.9 | 12.3 | 5.4 | 3.5 | 5.4 | 4.2 | 63.6 | 810.5 | 394.0 | 295.7 | 1,510 | 1,462 |
| Richland | 714 | 2,572,557 | 8.8 | 8.0 | 0.2 | 7.8 | 21.0 | 9.0 | 44.4 | 69.2 | 35.1 | 18.7 | 1,066 | 974 |
| Rock | 6,384 | 28,667,201 | 5.4 | 9.3 | 3.9 | 4.1 | 10.9 | 3.7 | 61.0 | 783.0 | 426.4 | 256.6 | 1,582 | 1,459 |
| Rusk | 546 | 2,105,341 | 10.6 | 10.4 | 0.3 | 8.4 | 9.5 | 5.2 | 54.3 | 85.4 | 33.0 | 20.1 | 1,423 | 1,325 |
| St. Croix | 3,038 | 12,188,587 | 7.2 | 7.5 | 0.6 | 9.9 | 8.0 | 3.0 | 62.2 | 329.2 | 145.9 | 138.5 | 1,564 | 1,427 |
| Sauk | 2,827 | 10,650,902 | 6.8 | 14.4 | 0.3 | 4.3 | 11.5 | 4.7 | 56.3 | 505.0 | 327.7 | 136.4 | 2,132 | 1,766 |
| Sawyer | 560 | 2,015,604 | 11.0 | 13.9 | 0.2 | 10.5 | 8.7 | 1.9 | 53.6 | 70.4 | 25.3 | 35.6 | 2,174 | 2,010 |
| Shawano | 1,267 | 4,976,919 | 8.4 | 12.2 | 0.6 | 6.2 | 10.2 | 4.6 | 56.5 | 148.3 | 76.0 | 51.7 | 1,265 | 1,177 |
| Sheboygan | 4,307 | 19,613,015 | 4.3 | 9.6 | 2.4 | 5.4 | 8.1 | 3.7 | 65.1 | 501.1 | 247.2 | 183.3 | 1,593 | 1,516 |
| Taylor | 716 | 2,928,731 | 8.1 | 8.2 | 0.0 | 7.0 | 12.9 | 3.8 | 58.4 | 80.3 | 44.9 | 27.4 | 1,350 | 1,273 |
| Trempealeau | 1,638 | 5,929,327 | 7.0 | 6.6 | 0.0 | 5.6 | 17.3 | 4.2 | 55.8 | 169.4 | 81.4 | 44.7 | 1,521 | 1,432 |
| Vernon | 1,104 | 3,794,098 | 9.8 | 10.3 | 0.1 | 6.6 | 7.4 | 4.9 | 58.9 | 123.4 | 64.1 | 37.5 | 1,222 | 1,146 |
| Vilas | 703 | 2,972,578 | 11.5 | 12.3 | 0.3 | 9.9 | 4.9 | 3.2 | 57.1 | 100.7 | 34.9 | 57.8 | 2,667 | 2,469 |
| Walworth | 3,702 | 16,113,003 | 7.1 | 13.5 | 0.5 | 4.2 | 9.2 | 3.8 | 61.0 | 469.7 | 156.1 | 239.3 | 2,327 | 2,157 |
| Washburn | 727 | 2,631,612 | 10.7 | 9.7 | 0.0 | 8.3 | 7.6 | 5.1 | 57.5 | 77.9 | 28.7 | 39.4 | 2,505 | 2,400 |
| Washington | 3,758 | 15,770,528 | 6.0 | 13.1 | 2.1 | 4.2 | 7.9 | 4.3 | 61.3 | 473.3 | 179.5 | 217.3 | 1,610 | 1,465 |
| Waukesha | 11,532 | 58,507,193 | 5.5 | 10.2 | 3.5 | 3.5 | 3.8 | 4.7 | 67.3 | 1,577.5 | 513.5 | 811.5 | 2,024 | 1,946 |
| Waupaca | 1,945 | 7,657,117 | 6.2 | 11.6 | 0.2 | 5.8 | 6.7 | 4.4 | 63.9 | 219.4 | 101.6 | 88.1 | 1,724 | 1,622 |
| Waushara | 796 | 3,102,822 | 11.7 | 10.4 | 0.0 | 6.0 | 16.9 | 4.3 | 49.9 | 80.4 | 33.7 | 37.5 | 1,549 | 1,467 |
| Winnebago | 5,484 | 24,749,186 | 4.5 | 13.5 | 6.2 | 7.2 | 7.0 | 6.5 | 53.4 | 838.6 | 445.2 | 251.9 | 1,478 | 1,433 |
| Wood | 2,868 | 13,617,564 | 3.8 | 7.8 | 3.7 | 4.9 | 11.4 | 2.5 | 63.7 | 360.7 | 172.5 | 126.5 | 1,731 | 1,622 |
| **WYOMING** | X | X | X | X | X | X | X | X | X | X | X | X | X | X |
| Albany | 1,630 | 7,194,969 | 6.4 | 8.2 | 3.9 | 2.5 | 31.7 | 5.6 | 40.5 | 238.6 | 92.0 | 38.5 | 1,000 | 674 |
| Big Horn | 999 | 4,066,930 | 4.4 | 5.9 | 0.4 | 1.7 | 27.5 | 6.0 | 53.2 | 113.1 | 61.7 | 20.4 | 1,724 | 1,517 |
| Campbell | 3,735 | 18,642,479 | 5.3 | 5.7 | 0.9 | 1.5 | 36.3 | 5.3 | 40.8 | 531.7 | 119.8 | 194.4 | 4,189 | 3,769 |
| Carbon | 1,049 | 4,514,716 | 6.6 | 9.2 | 1.4 | 2.7 | 22.3 | 7.3 | 49.0 | 122.7 | 64.3 | 38.8 | 2,543 | 2,288 |
| Converse | 1,129 | 5,492,658 | 4.4 | 7.1 | 0.0 | 2.1 | 41.3 | 4.0 | 39.2 | 150.5 | 23.3 | 62.5 | 4,549 | 4,080 |
| Crook | 484 | 1,956,335 | 5.3 | 6.2 | 0.1 | 2.6 | 21.4 | 3.9 | 60.0 | 57.4 | 33.7 | 12.6 | 1,699 | 1,543 |
| Fremont | 2,263 | 9,038,694 | 4.6 | 8.6 | 0.6 | 1.7 | 0.5 | 5.9 | 76.1 | 323.9 | 235.7 | 58.1 | 1,460 | 1,195 |

1. Based on the resident population estimated as of July 1 of the year shown.

## Table B. States and Counties — Local Government Finances, Government Employment, and Income Taxes

| STATE County | Total (mil dol) [185] | Per capita[1] (dollars) [186] | Education [187] | Health and hospitals [188] | Police protection [189] | Public welfare [190] | Highways [191] | Total (mil dol) [192] | Per capita[1] (dollars) [193] | Federal civilian [194] | Federal military [195] | State and local [196] | Number of returns [197] | Mean adjusted gross income [198] | Mean income tax [199] |
|---|---|---|---|---|---|---|---|---|---|---|---|---|---|---|---|
| WISCONSIN—Cont'd | | | | | | | | | | | | | | | |
| Douglas | 265 | 6,115 | 53.4 | 1.5 | 5.1 | 4.0 | 11.4 | 369.4 | 8,526 | 169 | 109 | 3,226 | 20,890 | 55,174 | 4,857 |
| Dunn | 173 | 3,871 | 42.7 | 1.7 | 5.0 | 12.9 | 13.1 | 156.0 | 3,488 | 90 | 111 | 4,455 | 20,010 | 56,828 | 5,449 |
| Eau Claire | 487 | 4,702 | 54.9 | 4.8 | 7.0 | 3.6 | 9.1 | 501.6 | 4,842 | 356 | 262 | 8,272 | 48,780 | 69,894 | 8,628 |
| Florence | 26 | 5,972 | 31.7 | 6.1 | 5.4 | 1.3 | 16.3 | 41.1 | 9,473 | 14 | 11 | 297 | 2,220 | 52,876 | 4,615 |
| Fond du Lac | 503 | 4,908 | 51.0 | 5.3 | 5.0 | 8.5 | 11.1 | 445.8 | 4,353 | 196 | 263 | 5,353 | 51,240 | 63,920 | 6,668 |
| Forest | 47 | 5,224 | 50.4 | 2.6 | 6.0 | 5.0 | 16.6 | 4.6 | 517 | 91 | 23 | 1,890 | 4,280 | 56,558 | 6,004 |
| Grant | 243 | 4,694 | 55.1 | 3.3 | 3.9 | 6.5 | 9.9 | 185.0 | 3,576 | 146 | 122 | 5,746 | 22,620 | 51,788 | 4,595 |
| Green | 173 | 4,700 | 46.4 | 2.7 | 6.2 | 10.3 | 15.2 | 129.9 | 3,523 | 83 | 95 | 2,051 | 18,990 | 62,751 | 6,492 |
| Green Lake | 86 | 4,616 | 46.1 | 4.5 | 6.4 | 3.8 | 14.6 | 68.9 | 3,678 | 47 | 49 | 1,074 | 9,620 | 54,528 | 5,117 |
| Iowa | 113 | 4,785 | 55.4 | 0.9 | 4.5 | 7.3 | 12.6 | 91.2 | 3,854 | 77 | 61 | 1,330 | 12,030 | 60,228 | 5,775 |
| Iron | 36 | 6,332 | 31.0 | 10.1 | 7.7 | 0.5 | 16.3 | 26.0 | 4,587 | 17 | 15 | 340 | 3,100 | 49,209 | 4,348 |
| Jackson | 97 | 4,755 | 46.0 | 2.4 | 4.2 | 5.8 | 12.0 | 77.3 | 3,773 | 44 | 56 | 2,400 | 9,550 | 49,994 | 4,459 |
| Jefferson | 332 | 3,913 | 49.1 | 5.2 | 6.7 | 2.6 | 9.8 | 339.0 | 4,001 | 176 | 213 | 3,892 | 41,710 | 60,775 | 5,724 |
| Juneau | 116 | 4,389 | 49.7 | 5.2 | 5.2 | 3.8 | 13.0 | 87.3 | 3,301 | 240 | 65 | 1,821 | 12,540 | 49,178 | 3,945 |
| Kenosha | 906 | 5,381 | 53.7 | 3.7 | 6.3 | 6.5 | 5.2 | 892.4 | 5,300 | 285 | 431 | 9,643 | 82,950 | 62,325 | 6,523 |
| Kewaunee | 110 | 5,403 | 51.5 | 3.9 | 3.6 | 5.9 | 14.3 | 102.4 | 5,018 | 69 | 53 | 1,170 | 10,190 | 57,378 | 5,122 |
| La Crosse | 665 | 5,637 | 53.9 | 3.7 | 4.2 | 7.5 | 5.9 | 629.3 | 5,332 | 477 | 299 | 10,047 | 57,160 | 65,768 | 7,090 |
| Lafayette | 102 | 6,126 | 45.0 | 15.1 | 3.5 | 8.7 | 12.0 | 63.6 | 3,813 | 45 | 43 | 1,107 | 8,090 | 50,455 | 4,499 |
| Langlade | 90 | 4,716 | 48.3 | 3.7 | 5.4 | 3.4 | 15.5 | 69.7 | 3,637 | 45 | 49 | 1,002 | 9,560 | 49,519 | 4,313 |
| Lincoln | 145 | 5,212 | 39.7 | 5.1 | 5.8 | 11.1 | 14.0 | 78.0 | 2,811 | 60 | 70 | 1,717 | 14,140 | 53,841 | 4,687 |
| Manitowoc | 292 | 3,694 | 45.9 | 1.2 | 7.2 | 6.3 | 10.1 | 253.8 | 3,208 | 191 | 213 | 3,844 | 40,150 | 58,122 | 5,650 |
| Marathon | 921 | 6,798 | 37.5 | 9.1 | 3.2 | 28.4 | 7.3 | 469.1 | 3,463 | 386 | 352 | 7,575 | 68,290 | 65,895 | 7,313 |
| Marinette | 171 | 4,235 | 44.2 | 4.9 | 6.1 | 3.6 | 16.2 | 99.6 | 2,473 | 169 | 108 | 2,145 | 20,310 | 50,650 | 4,477 |
| Marquette | 57 | 3,736 | 41.9 | 5.4 | 7.2 | 2.6 | 18.9 | 26.3 | 1,724 | 49 | 40 | 709 | 7,850 | 50,916 | 4,879 |
| Menominee | 33 | 7,110 | 60.4 | 0.0 | 4.2 | 19.4 | 7.2 | 2.1 | 459 | 5 | 12 | 2,059 | 1,820 | 36,271 | 2,134 |
| Milwaukee | 5,309 | 5,587 | 40.2 | 8.5 | 9.3 | 1.4 | 5.2 | 6,559.5 | 6,903 | 9,519 | 2,766 | 50,960 | 441,320 | 59,748 | 6,707 |
| Monroe | 192 | 4,209 | 47.2 | 1.4 | 6.5 | 8.8 | 10.9 | 110.1 | 2,407 | 2,656 | 412 | 2,724 | 21,660 | 52,367 | 4,423 |
| Oconto | 147 | 3,907 | 38.5 | 5.4 | 4.4 | 3.7 | 16.1 | 112.8 | 3,007 | 88 | 98 | 1,775 | 18,900 | 58,451 | 5,240 |
| Oneida | 209 | 5,924 | 50.6 | 2.8 | 5.9 | 3.3 | 11.8 | 144.1 | 4,089 | 184 | 92 | 2,060 | 19,720 | 62,578 | 6,685 |
| Outagamie | 998 | 5,369 | 55.5 | 2.5 | 4.8 | 4.7 | 8.5 | 808.4 | 4,351 | 628 | 482 | 10,893 | 95,090 | 69,353 | 7,575 |
| Ozaukee | 390 | 4,407 | 48.2 | 2.2 | 6.8 | 5.6 | 13.3 | 322.1 | 3,642 | 159 | 228 | 3,705 | 46,200 | 123,528 | 19,858 |
| Pepin | 42 | 5,770 | 53.1 | 3.0 | 4.4 | 2.0 | 19.6 | 23.8 | 3,282 | 34 | 19 | 474 | 3,510 | 68,258 | 8,479 |
| Pierce | 203 | 4,828 | 53.8 | 2.6 | 6.0 | 2.8 | 13.2 | 267.2 | 6,360 | 80 | 106 | 4,133 | 19,380 | 70,674 | 7,215 |
| Polk | 216 | 4,978 | 51.7 | 4.0 | 4.4 | 5.6 | 11.8 | 134.9 | 3,109 | 130 | 113 | 2,725 | 22,200 | 58,141 | 5,301 |
| Portage | 467 | 6,620 | 24.4 | 2.9 | 3.6 | 47.2 | 7.7 | 126.7 | 1,796 | 163 | 179 | 5,760 | 34,220 | 62,139 | 6,193 |
| Price | 63 | 4,708 | 42.6 | 2.4 | 5.7 | 6.8 | 19.7 | 48.5 | 3,624 | 83 | 34 | 875 | 6,960 | 48,638 | 4,052 |
| Racine | 855 | 4,365 | 48.3 | 5.0 | 9.0 | 4.9 | 8.3 | 675.9 | 3,452 | 390 | 500 | 8,513 | 97,150 | 63,985 | 6,776 |
| Richland | 76 | 4,346 | 38.7 | 2.7 | 4.4 | 20.3 | 13.6 | 96.5 | 5,511 | 47 | 44 | 1,046 | 7,830 | 48,970 | 4,020 |
| Rock | 800 | 4,931 | 49.1 | 6.3 | 6.0 | 6.9 | 6.7 | 701.7 | 4,326 | 312 | 419 | 8,453 | 80,190 | 57,507 | 5,399 |
| Rusk | 92 | 6,525 | 31.2 | 27.2 | 4.1 | 2.6 | 16.3 | 52.6 | 3,721 | 34 | 37 | 939 | 6,620 | 46,938 | 3,864 |
| St. Croix | 388 | 4,376 | 53.8 | 2.9 | 4.8 | 6.6 | 11.5 | 464.3 | 5,243 | 156 | 234 | 4,656 | 45,160 | 84,144 | 10,162 |
| Sauk | 503 | 7,868 | 25.5 | 3.3 | 3.8 | 43.1 | 5.2 | 252.7 | 3,951 | 168 | 167 | 5,438 | 34,260 | 56,473 | 5,305 |
| Sawyer | 68 | 4,158 | 42.4 | 7.3 | 6.4 | 4.8 | 16.8 | 11.2 | 682 | 78 | 42 | 1,920 | 8,370 | 53,543 | 4,846 |
| Shawano | 191 | 4,670 | 40.9 | 4.3 | 4.6 | 3.5 | 14.6 | 90.8 | 2,224 | 109 | 105 | 2,852 | 19,770 | 51,343 | 4,207 |
| Sheboygan | 521 | 4,522 | 56.5 | 3.9 | 6.8 | 5.2 | 7.5 | 547.7 | 4,759 | 208 | 326 | 5,525 | 59,260 | 62,615 | 6,304 |
| Taylor | 84 | 4,145 | 47.7 | 4.7 | 4.1 | 8.4 | 15.5 | 27.7 | 1,362 | 57 | 52 | 1,075 | 9,280 | 50,848 | 4,385 |
| Trempealeau | 177 | 6,033 | 46.8 | 1.9 | 3.4 | 17.7 | 10.2 | 187.9 | 6,398 | 117 | 76 | 2,298 | 14,710 | 52,704 | 4,490 |
| Vernon | 123 | 4,016 | 45.2 | 2.2 | 3.9 | 12.2 | 14.4 | 89.6 | 2,919 | 117 | 80 | 1,830 | 13,650 | 50,506 | 4,255 |
| Vilas | 98 | 4,526 | 42.7 | 2.8 | 6.2 | 6.0 | 15.2 | 61.9 | 2,856 | 70 | 96 | 2,185 | 11,800 | 58,914 | 6,177 |
| Walworth | 508 | 4,943 | 53.4 | 3.5 | 7.7 | 4.4 | 7.1 | 563.2 | 5,476 | 191 | 263 | 8,422 | 51,000 | 66,376 | 7,535 |
| Washburn | 87 | 5,520 | 44.7 | 1.5 | 3.7 | 4.7 | 17.9 | 57.7 | 3,664 | 76 | 40 | 1,191 | 8,280 | 54,221 | 4,648 |
| Washington | 489 | 3,622 | 52.1 | 3.0 | 7.2 | 5.4 | 8.0 | 554.4 | 4,109 | 258 | 354 | 5,182 | 70,920 | 77,296 | 8,800 |
| Waukesha | 1,668 | 4,160 | 52.7 | 3.0 | 7.0 | 2.0 | 8.1 | 1,782.1 | 4,445 | 777 | 1,042 | 16,751 | 210,110 | 100,220 | 14,015 |
| Waupaca | 228 | 4,452 | 48.2 | 3.9 | 6.1 | 4.5 | 15.0 | 218.7 | 4,278 | 127 | 129 | 3,188 | 25,670 | 55,685 | 5,073 |
| Waushara | 91 | 3,739 | 41.6 | 4.7 | 5.7 | 9.6 | 17.4 | 43.9 | 1,815 | 44 | 61 | 1,274 | 11,510 | 50,048 | 4,253 |
| Winnebago | 873 | 5,121 | 34.9 | 2.9 | 4.7 | 26.1 | 6.4 | 874.7 | 5,134 | 426 | 465 | 11,411 | 84,680 | 64,562 | 6,930 |
| Wood | 394 | 5,396 | 50.8 | 7.8 | 4.8 | 5.6 | 10.7 | 285.6 | 3,910 | 186 | 188 | 4,657 | 37,610 | 56,624 | 5,444 |
| WYOMING | X | X | X | X | X | X | X | X | X | 7,579 | 6,137 | 61,592 | 273,350 | 79,620 | 10,043 |
| Albany | 245 | 6,376 | 38.6 | 29.2 | 3.7 | 0.5 | 3.2 | 79.3 | 2,063 | 189 | 203 | 8,256 | 16,180 | 62,028 | 6,648 |
| Big Horn | 108 | 9,138 | 46.8 | 25.8 | 4.0 | 0.2 | 3.4 | 42.7 | 3,598 | 95 | 61 | 1,385 | 5,020 | 50,006 | 4,830 |
| Campbell | 536 | 11,550 | 33.5 | 32.3 | 3.2 | 1.1 | 2.6 | 272.4 | 5,870 | 85 | 241 | 4,799 | 21,640 | 76,341 | 8,588 |
| Carbon | 134 | 8,775 | 58.4 | 3.9 | 4.7 | 0.2 | 2.3 | 39.4 | 2,584 | 180 | 74 | 1,672 | 6,780 | 79,095 | 9,921 |
| Converse | 178 | 12,972 | 26.7 | 35.1 | 4.1 | 0.2 | 6.8 | 1.5 | 106 | 61 | 72 | 1,536 | 6,440 | 85,232 | 11,550 |
| Crook | 53 | 7,105 | 58.0 | 16.3 | 3.1 | 0.2 | 6.4 | 3.0 | 409 | 82 | 40 | 628 | 3,330 | 73,743 | 8,320 |
| Fremont | 314 | 7,897 | 75.2 | 0.4 | 4.1 | 0.3 | 5.3 | 49.1 | 1,234 | 432 | 201 | 5,048 | 17,260 | 55,438 | 5,682 |

1. Based on the resident population estimated as of July 1 of the year shown.

Items 185—199

# Table B. States and Counties — **Land Area and Population**

| State / county code | CBSA code[1] | County Type code[2] | STATE County | Land area[3] (sq. mi) | Total persons 2019 | Rank | Per square mile | White | Black | American Indian, Alaska Native | Asian and Pacific Islander | Percent Hispanic or Latino[4] | Under 5 years | 5 to 14 years | 15 to 24 years | 25 to 34 years | 35 to 44 years | 45 to 54 years |
|---|---|---|---|---|---|---|---|---|---|---|---|---|---|---|---|---|---|---|
| | | | | 1 | 2 | 3 | 4 | 5 | 6 | 7 | 8 | 9 | 10 | 11 | 12 | 13 | 14 | 15 |
| | | | WYOMING—Cont'd | | | | | | | | | | | | | | | |
| 56015 | | 7 | Goshen ............... | 2,225.6 | 13,235 | 2,204 | 5.9 | 86.6 | 1.2 | 1.4 | 1.2 | 10.9 | 4.9 | 11.2 | 12.6 | 11.4 | 11.5 | 11.2 |
| 56017 | | 7 | Hot Springs ............... | 2,004.4 | 4,425 | 2,863 | 2.2 | 92.5 | 1.0 | 3.0 | 0.9 | 4.5 | 5.1 | 12.6 | 9.5 | 9.4 | 9.8 | 10.5 |
| 56019 | | 7 | Johnson ............... | 4,154.2 | 8,588 | 2,526 | 2.1 | 92.3 | 1.0 | 2.1 | 1.0 | 5.1 | 4.6 | 12.7 | 9.5 | 9.4 | 11.4 | 11.2 |
| 56021 | 16940 | 3 | Laramie ............... | 2,685.9 | 100,595 | 602 | 37.5 | 80.4 | 3.2 | 1.4 | 2.2 | 15.4 | 6.0 | 13.0 | 12.6 | 14.6 | 12.8 | 11.0 |
| 56023 | | 7 | Lincoln ............... | 4,075.3 | 20,253 | 1,807 | 5.0 | 93.2 | 0.7 | 1.5 | 1.0 | 4.9 | 6.2 | 15.1 | 11.2 | 9.7 | 12.9 | 11.3 |
| 56025 | 16220 | 3 | Natrona ............... | 5,340.5 | 80,815 | 705 | 15.1 | 88.0 | 1.9 | 1.7 | 1.4 | 9.0 | 6.2 | 13.7 | 11.7 | 14.2 | 13.8 | 11.0 |
| 56027 | | 9 | Niobrara ............... | 2,626.0 | 2,275 | 3,019 | 0.9 | 93.0 | 1.3 | 2.6 | 1.1 | 4.2 | 5.7 | 9.6 | 9.9 | 13.2 | 12.2 | 11.1 |
| 56029 | | 7 | Park ............... | 6,939.0 | 29,331 | 1,451 | 4.2 | 92.2 | 0.9 | 1.3 | 1.3 | 5.8 | 5.0 | 11.8 | 10.4 | 10.9 | 11.4 | 10.5 |
| 56031 | | 7 | Platte ............... | 2,081.5 | 8,578 | 2,527 | 4.1 | 89.4 | 0.9 | 1.5 | 1.3 | 8.6 | 5.3 | 12.0 | 9.3 | 10.0 | 10.7 | 11.3 |
| 56033 | 43260 | 7 | Sheridan ............... | 2,523.4 | 30,863 | 1,408 | 12.2 | 92.6 | 0.9 | 1.8 | 1.4 | 4.7 | 5.0 | 12.1 | 11.4 | 11.5 | 12.3 | 11.3 |
| 56035 | | 9 | Sublette ............... | 4,886.5 | 9,856 | 2,429 | 2.0 | 90.2 | 1.2 | 1.6 | 1.3 | 7.3 | 4.7 | 12.8 | 10.1 | 10.5 | 13.6 | 12.6 |
| 56037 | 40540 | 5 | Sweetwater ............... | 10,427.0 | 42,673 | 1,125 | 4.1 | 80.8 | 1.7 | 1.5 | 1.6 | 16.2 | 6.1 | 14.9 | 13.2 | 13.4 | 14.7 | 11.2 |
| 56039 | 27220 | 7 | Teton ............... | 3,996.8 | 23,497 | 1,659 | 5.9 | 82.4 | 1.0 | 0.9 | 2.4 | 15.0 | 4.7 | 10.1 | 9.4 | 16.5 | 16.2 | 13.7 |
| 56041 | 21740 | 7 | Uinta ............... | 2,081.7 | 20,215 | 1,811 | 9.7 | 88.2 | 0.9 | 1.4 | 1.2 | 9.9 | 6.4 | 16.3 | 12.5 | 11.6 | 13.4 | 11.0 |
| 56043 | | 7 | Washakie ............... | 2,238.7 | 7,760 | 2,610 | 3.5 | 84.0 | 1.0 | 1.4 | 1.0 | 14.2 | 4.9 | 12.3 | 11.4 | 9.8 | 12.3 | 11.4 |
| 56045 | 23940 | 7 | Weston ............... | 2,398.0 | 6,743 | 2,683 | 2.8 | 92.2 | 1.4 | 2.7 | 1.9 | 4.1 | 4.8 | 11.6 | 9.7 | 10.9 | 13.6 | 10.8 |

1. CBSA = Core Based Statistical Area. See Appendix A for explanation. See Appendix B for list of metropolitan areas with component counties.   2. County type code from the Economic Research Service of USDA Rural-Urban Continuum Codes. See Appendix A for definition.   3. Dry land or land partially or temporarily covered by water.   4. May be of any race.

| STATE County | 55 to 64 years (16) | 65 to 74 years (17) | 75 years and over (18) | Percent female (19) | 2000 (20) | 2010 (21) | 2000-2010 (22) | 2010-2020 (23) | Births (24) | Deaths (25) | Net Migration (26) | Number (27) | Persons per household (28) | Family households (29) | Female family householder[1] (30) | One person (31) |
|---|---|---|---|---|---|---|---|---|---|---|---|---|---|---|---|---|
| | **Population, 2020 (cont.)** — Age (percent) (cont.) | | | | **Population change, 2000-2020** — Total persons | | Percent change | | Components of change, 2010-2020 | | | **Households, 2015-2019** | | Percent | | |
| WYOMING—Cont'd | | | | | | | | | | | | | | | | |
| Goshen | 14.0 | 12.5 | 10.8 | 47.4 | 12,538 | 13,247 | 5.7 | -0.1 | 1,383 | 1,447 | 35 | 5,207 | 2.42 | 68.6 | 9.0 | 27.6 |
| Hot Springs | 15.0 | 16.3 | 11.8 | 50.1 | 4,882 | 4,812 | -1.4 | -8.0 | 496 | 743 | -139 | 2,118 | 2.13 | 64.4 | 8.5 | 32.0 |
| Johnson | 14.7 | 16.0 | 10.7 | 49.5 | 7,075 | 8,569 | 21.1 | 0.2 | 870 | 926 | 77 | 3,845 | 2.18 | 65.0 | 6.6 | 32.2 |
| Laramie | 12.9 | 10.2 | 6.8 | 49.2 | 81,607 | 91,885 | 12.6 | 9.5 | 12,975 | 8,353 | 4,056 | 39,683 | 2.43 | 66.2 | 10.4 | 28.3 |
| Lincoln | 14.4 | 12.6 | 6.5 | 49.0 | 14,573 | 18,106 | 24.2 | 11.9 | 2,484 | 1,351 | 1,008 | 6,908 | 2.77 | 68.0 | 5.2 | 27.4 |
| Natrona | 13.1 | 10.1 | 6.2 | 49.5 | 66,533 | 75,448 | 13.4 | 7.1 | 10,903 | 7,370 | 1,703 | 32,799 | 2.4 | 62.6 | 10.3 | 30.1 |
| Niobrara | 13.9 | 13.0 | 11.5 | 54.5 | 2,407 | 2,484 | 3.2 | -8.4 | 276 | 256 | -234 | 964 | 2.22 | 62.3 | 5.6 | 35.5 |
| Park | 14.9 | 14.8 | 10.3 | 50.3 | 25,786 | 28,207 | 9.4 | 4.0 | 3,105 | 2,874 | 898 | 12,160 | 2.33 | 67.2 | 7.1 | 27.6 |
| Platte | 15.1 | 14.5 | 11.8 | 49.7 | 8,807 | 8,667 | -1.6 | -1.0 | 851 | 1,030 | 95 | 4,030 | 2.1 | 65.4 | 4.8 | 30.5 |
| Sheridan | 14.1 | 13.5 | 8.7 | 49.8 | 26,560 | 29,124 | 9.7 | 6.0 | 3,301 | 3,261 | 1,713 | 13,251 | 2.2 | 61.3 | 7.3 | 30.4 |
| Sublette | 13.9 | 13.4 | 8.3 | 46.2 | 5,920 | 10,244 | 73.0 | -3.8 | 1,169 | 559 | -1,026 | 3,340 | 2.94 | 68.3 | 6.9 | 28.0 |
| Sweetwater | 12.8 | 9.0 | 4.7 | 48.4 | 37,613 | 43,806 | 16.5 | -2.6 | 6,007 | 2,982 | -4,242 | 15,523 | 2.77 | 69.5 | 10.6 | 25.7 |
| Teton | 12.8 | 10.6 | 6.0 | 48.1 | 18,251 | 21,298 | 16.7 | 10.3 | 2,522 | 887 | 535 | 9,019 | 2.47 | 57.5 | 4.3 | 28.7 |
| Uinta | 13.1 | 10.4 | 5.2 | 49.4 | 19,742 | 21,121 | 7.0 | -4.3 | 2,959 | 1,488 | -2,404 | 7,597 | 2.66 | 69.9 | 6.8 | 25.3 |
| Washakie | 14.5 | 12.7 | 10.6 | 49.0 | 8,289 | 8,528 | 2.9 | -9.0 | 914 | 892 | -790 | 3,365 | 2.34 | 65.1 | 6.2 | 29.0 |
| Weston | 16.2 | 13.5 | 9.0 | 47.5 | 6,644 | 7,208 | 8.5 | -6.5 | 737 | 716 | -487 | 2,891 | 2.32 | 68.4 | 5.9 | 27.0 |

1. No spouse present.

# Table B. States and Counties — **Population, Vital Statistics, Health, and Crime**

| STATE County | Persons in group quarters, 2020 | Daytime Population, 2015-2019 | | Births, 2020 | | Deaths, 2020 | | Persons under 65 with no health insurance, 2019 | | Medicare, 2020 | | | Crimes reported by county police or sheriff, 2019 | |
| | | Number | Employment/ residence ratio | Total | Rate[1] | Number | Rate[1] | Number | Percent | Total beneficiaries | Enrolled in Original Medicare | Enrolled in Medicare Advantage | Violent | Property |
| | 32 | 33 | 34 | 35 | 36 | 37 | 38 | 39 | 40 | 41 | 42 | 43 | 44 | 45 |
| WYOMING—Cont'd | | | | | | | | | | | | | | |
| Goshen | 1,239 | 12,916 | 0.93 | 112 | 8.5 | 149 | 11.3 | 1,413 | 15.5 | 3,261 | 3,165 | 96 | 10 | 27 |
| Hot Springs | 86 | 4,537 | 0.97 | 45 | 10.2 | 85 | 19.2 | 476 | 15.2 | 1,437 | 1,381 | 56 | 2 | 8 |
| Johnson | 71 | 8,080 | 0.90 | 71 | 8.3 | 88 | 10.2 | 889 | 14.0 | 2,341 | 2,285 | 56 | 4 | 22 |
| Laramie | 2,076 | 101,129 | 1.06 | 1,212 | 12.0 | 960 | 9.5 | 9,645 | 11.8 | 19,635 | 18,748 | 887 | NA | NA |
| Lincoln | 71 | 18,310 | 0.89 | 231 | 11.4 | 162 | 8.0 | 2,607 | 16.1 | 4,004 | 3,837 | 167 | 16 | 119 |
| Natrona | 1,668 | 80,549 | 1.01 | 931 | 11.5 | 804 | 9.9 | 9,564 | 14.5 | 14,773 | 13,821 | 953 | 48 | 145 |
| Niobrara | 249 | 2,506 | 1.09 | 26 | 11.4 | 14 | 6.2 | 294 | 19.1 | 612 | 598 | 14 | 4 | 10 |
| Park | 547 | 29,451 | 1.02 | 269 | 9.2 | 306 | 10.4 | 3,022 | 14.0 | 8,015 | 7,741 | 274 | 13 | 67 |
| Platte | 102 | 8,733 | 1.04 | 76 | 8.9 | 98 | 11.4 | 916 | 15.0 | 2,473 | 2,373 | 100 | 0 | 44 |
| Sheridan | 1,049 | 29,552 | 0.96 | 293 | 9.5 | 352 | 11.4 | 3,230 | 13.9 | 7,646 | 7,185 | 460 | 3 | 47 |
| Sublette | 550 | 9,795 | 0.98 | 77 | 7.8 | 75 | 7.6 | 1,198 | 15.3 | 1,700 | 1,649 | 51 | 3 | 45 |
| Sweetwater | 807 | 45,308 | 1.08 | 481 | 11.3 | 319 | 7.5 | 5,044 | 13.9 | 6,523 | 6,393 | 131 | 16 | 83 |
| Teton | 269 | 27,697 | 1.31 | 225 | 9.6 | 124 | 5.3 | 3,472 | 17.5 | 3,713 | 3,494 | 219 | NA | NA |
| Uinta | 217 | 19,398 | 0.88 | 232 | 11.5 | 171 | 8.5 | 2,394 | 14.1 | 3,664 | 3,228 | 436 | 2 | 40 |
| Washakie | 140 | 7,942 | 0.98 | 73 | 9.4 | 85 | 11.0 | 943 | 15.9 | 1,986 | 1,960 | 26 | 2 | 22 |
| Weston | 313 | 6,601 | 0.86 | 53 | 7.9 | 72 | 10.7 | 740 | 14.3 | 1,664 | 1,538 | 126 | 0 | 1 |

1. Per 1,000 estimated resident population.

**Crime, Education, Money Income, and Poverty**

| STATE County | County law enforcement employment, 2019 | | Education | | | | | | Money income, 2015-2019 | | | | Income and poverty, 2019 | | | |
|---|---|---|---|---|---|---|---|---|---|---|---|---|---|---|---|---|
| | | | School enrollment and attainment, 2015-2019 | | | | Local government expenditures,[3] 2017-2018 | | | Households | | | | Percent below poverty level | | |
| | | | Enrollment[1] | | Attainment[2] (percent) | | | | | | | Percent | | | | | |
| | | | | | | | | | | | | with income of less than $50,000 | with income of $200,000 or more | | | | |
| | Officers | Civilians | Total | Percent private | High school graduate or less | Bachelor's degree or more | Total current spending (mil dol) | Current spending per student (dollars) | Per capita income[4] | Median income (dollars) | | | Median household income (dollars) | All persons | Children under 18 years | Children 5 to 17 years in families |
| | 46 | 47 | 48 | 49 | 50 | 51 | 52 | 53 | 54 | 55 | 56 | 57 | 58 | 59 | 60 | 61 |
| WYOMING—Cont'd | | | | | | | | | | | | | | | | |
| Goshen | 8 | 18 | 2,909 | 14.6 | 36.0 | 25.9 | 31 | 17,723 | 28,554 | 54,289 | 47.5 | 2.1 | 55,540 | 13.0 | 15.6 | 13.4 |
| Hot Springs | 5 | 8 | 900 | 16.6 | 34.9 | 23.7 | 12 | 17,516 | 30,643 | 51,413 | 49.4 | 2.5 | 52,344 | 10.9 | 15.1 | 13.5 |
| Johnson | 29 | 3 | 1,626 | 16.2 | 30.9 | 31.7 | 21 | 16,144 | 36,084 | 58,132 | 40.8 | 4.8 | 57,224 | 9.0 | 10.4 | 8.9 |
| Laramie | 51 | 123 | 23,284 | 9.6 | 32.8 | 28.5 | 235 | 15,292 | 35,190 | 66,910 | 36.0 | 4.4 | 69,613 | 9.5 | 11.3 | 9.8 |
| Lincoln | 22 | 27 | 4,645 | 7.3 | 39.1 | 23.7 | 54 | 15,291 | 29,747 | 66,964 | 37.3 | 4.1 | 71,269 | 7.0 | 8.8 | 7.8 |
| Natrona | 109 | 40 | 19,722 | 8.9 | 37.9 | 22.8 | 194 | 14,545 | 33,797 | 62,772 | 40.0 | 4.8 | 66,104 | 9.9 | 12.6 | 11.3 |
| Niobrara | 4 | 11 | 551 | 4.2 | 47.6 | 18.0 | 12 | 14,712 | 22,958 | 39,150 | 59.5 | 2.8 | 48,513 | 15.1 | 18.9 | 19.9 |
| Park | 45 | 10 | 6,426 | 7.2 | 30.3 | 34.0 | 63 | 15,722 | 35,147 | 63,582 | 37.6 | 4.5 | 59,494 | 10.3 | 14.0 | 12.2 |
| Platte | 24 | 13 | 1,692 | 11.5 | 47.2 | 20.7 | 26 | 20,833 | 35,312 | 50,903 | 49.7 | 5.1 | 57,425 | 11.5 | 15.2 | 13.3 |
| Sheridan | 20 | 38 | 6,758 | 9.7 | 30.3 | 31.0 | 68 | 14,809 | 35,254 | 60,807 | 40.5 | 4.6 | 64,030 | 8.5 | 10.0 | 8.6 |
| Sublette | 55 | 17 | 2,196 | 12.1 | 38.6 | 22.8 | 28 | 17,496 | 35,874 | 77,403 | 29.6 | 4.9 | 78,055 | 7.2 | 8.2 | 7.4 |
| Sweetwater | 37 | 49 | 11,797 | 10.9 | 40.3 | 22.5 | 117 | 14,402 | 32,603 | 74,843 | 33.3 | 3.7 | 80,639 | 8.3 | 10.1 | 8.2 |
| Teton | NA | NA | 5,154 | 18.9 | 19.7 | 57.0 | 53 | 18,305 | 54,051 | 84,678 | 25.1 | 16.9 | 98,837 | 6.0 | 5.7 | 4.7 |
| Uinta | 35 | 14 | 5,284 | 5.1 | 48.8 | 16.0 | 67 | 15,702 | 28,159 | 63,403 | 38.1 | 2.3 | 70,756 | 8.5 | 9.6 | 7.8 |
| Washakie | 7 | 1 | 1,852 | 6.0 | 40.0 | 23.4 | 26 | 18,600 | 28,101 | 54,158 | 45.3 | 2.7 | 55,122 | 11.1 | 13.9 | 12.3 |
| Weston | 7 | 0 | 1,338 | 3.3 | 43.8 | 20.0 | 19 | 19,003 | 28,531 | 57,031 | 45.2 | 2.6 | 59,410 | 10.5 | 13.8 | 11.8 |

1. All persons 3 years old and over enrolled in nursery school through college.  2. Persons 25 years old and over.  3. Elementary and secondary education expenditures.  4. Based on population estimated by the American Community Survey, 2014–2018.

# Table B. States and Counties — Personal Income

| STATE County | Personal income, 2019 | | | | | | | | | | Earnings, 2019 | | |
| | Total (mil dol) | Percent change 2018-2019 | Per capita[1] Dollars | Per capita[1] Rank | Wages and salaries (mil dol) | Supplements to wages and salaries, employer contributions (mil dol) Pension and insurance | Supplements to wages and salaries, employer contributions (mil dol) Government social insurance | Proprietors' income (mil dol) | Dividends, interest, and rent (mil dol) | Personal transfer receipts (mil dol) | Total (mil dol) | Contributions for government social insurance (mil dol) From employee and self-employed | Contributions for government social insurance (mil dol) From employer |
|---|---|---|---|---|---|---|---|---|---|---|---|---|---|
| | 62 | 63 | 64 | 65 | 66 | 67 | 68 | 69 | 70 | 71 | 72 | 73 | 74 |
| **WYOMING—Cont'd** | | | | | | | | | | | | | |
| Goshen ........................... | 578 | 1.6 | 43,725 | 1,523 | 182 | 43 | 17 | 63 | 122 | 131 | 306 | 20 | 17 |
| Hot Springs ..................... | 252 | 1.4 | 57,101 | 401 | 80 | 19 | 7 | 52 | 48 | 60 | 157 | 10 | 7 |
| Johnson ........................... | 460 | 1.2 | 54,476 | 519 | 143 | 31 | 13 | 38 | 126 | 84 | 225 | 15 | 13 |
| Laramie ........................... | 5,384 | 3.4 | 54,113 | 538 | 2,735 | 575 | 257 | 402 | 1,155 | 934 | 3,969 | 233 | 257 |
| Lincoln ............................ | 879 | 6.0 | 44,331 | 1,441 | 344 | 74 | 30 | 50 | 235 | 157 | 498 | 32 | 30 |
| Natrona ........................... | 5,765 | 2.1 | 72,187 | 107 | 2,212 | 329 | 202 | 1,680 | 1,066 | 697 | 4,424 | 235 | 202 |
| Niobrara........................... | 124 | -0.3 | 52,484 | 645 | 39 | 12 | 4 | 29 | 30 | 25 | 83 | 5 | 4 |
| Park................................. | 1,705 | 1.5 | 58,386 | 359 | 625 | 126 | 58 | 132 | 603 | 304 | 940 | 61 | 58 |
| Platte............................... | 442 | 3.6 | 52,624 | 640 | 199 | 44 | 19 | 34 | 86 | 100 | 296 | 19 | 19 |
| Sheridan........................... | 1,819 | 2.7 | 59,666 | 315 | 660 | 130 | 63 | 142 | 602 | 303 | 995 | 64 | 63 |
| Sublette ....................... | 539 | -0.5 | 54,850 | 503 | 243 | 44 | 21 | 41 | 199 | 65 | 349 | 20 | 21 |
| Sweetwater ..................... | 2,423 | 2.1 | 57,225 | 397 | 1,457 | 247 | 126 | 368 | 317 | 318 | 2,199 | 124 | 126 |
| Teton............................... | 5,393 | 0.7 | 229,825 | 1 | 1,204 | 138 | 109 | 356 | 3,787 | 145 | 1,808 | 102 | 109 |
| Uinta................................ | 848 | 3.3 | 41,912 | 1,784 | 387 | 81 | 36 | 42 | 144 | 166 | 546 | 34 | 36 |
| Washakie ........................ | 396 | 1.9 | 50,735 | 774 | 162 | 35 | 15 | 34 | 104 | 82 | 246 | 15 | 15 |
| Weston............................. | 323 | 3.9 | 46,655 | 1,165 | 112 | 31 | 10 | 16 | 65 | 71 | 168 | 11 | 10 |

1. Based on the resident population estimated as of July 1 of the year shown.

| STATE County | Farm | Mining, quarrying, and extractions | Construction | Manufacturing | Information; professional, scientific, technical services | Retail trade | Finance, insurance, real estate, and leasing | Health care and social assistance | Government | Social Security beneficiaries, December 2019 | | Supplemental Security Income recipients, 2019 | Housing units, 2020 | |
|---|---|---|---|---|---|---|---|---|---|---|---|---|---|---|
| | | | | | | | | | | Number | Rate[1] | | Total | Percent change, 2010-2020 |
| | 75 | 76 | 77 | 78 | 79 | 80 | 81 | 82 | 83 | 84 | 85 | 86 | 87 | 88 |
| **WYOMING—Cont'd** | | | | | | | | | | | | | | |
| Goshen | 8.6 | 0.0 | 4.1 | 3.0 | 2.8 | 4.7 | 4.4 | 15.9 | 31.2 | 3,200 | 242 | 199 | 6,033 | 1.0 |
| Hot Springs | -0.6 | D | D | 2.5 | D | D | 2.1 | 6.6 | 26.5 | 1,505 | 341 | 89 | 2,587 | 0.2 |
| Johnson | 2.6 | 10.9 | 6.2 | 1.6 | 10.8 | 5.5 | 6.9 | D | 32.9 | 2,350 | 275 | 48 | 4,674 | 2.7 |
| Laramie | 0.3 | 2.9 | 7.4 | 3.9 | 8.0 | 5.8 | 6.2 | 7.9 | 37.8 | 19,560 | 196 | 1,480 | 44,797 | 10.7 |
| Lincoln | 1.9 | 16.6 | 12.5 | 1.8 | 5.9 | 5.6 | 3.9 | 3.3 | 28.9 | 4,055 | 203 | 133 | 9,849 | 10.1 |
| Natrona | 0.0 | 8.7 | 6.4 | 3.4 | 4.5 | 4.4 | 4.6 | 10.3 | 10.3 | 15,355 | 193 | 1,307 | 37,517 | 11.0 |
| Niobrara | 8.2 | 6.5 | D | D | D | D | 2.2 | D | 36.9 | 630 | 269 | 31 | 1,370 | 2.4 |
| Park | 1.9 | 5.2 | 9.1 | 4.2 | 8.1 | 6.5 | 5.1 | 10.6 | 28.7 | 8,260 | 283 | 282 | 14,832 | 9.4 |
| Platte | 7.6 | D | 10.0 | 0.9 | 4.2 | 4.8 | 4.5 | D | 22.5 | 2,395 | 283 | 77 | 4,919 | 5.4 |
| Sheridan | 0.6 | 1.2 | 10.6 | 5.2 | 8.7 | 6.4 | 4.6 | 7.9 | 30.0 | 7,440 | 243 | 329 | 15,402 | 10.5 |
| Sublette | 4.8 | 31.3 | 10.0 | 1.1 | 4.1 | 4.4 | 4.5 | D | 22.9 | 1,675 | 170 | 48 | 6,136 | 6.4 |
| Sweetwater | 0.5 | 37 | 6.9 | 8.8 | 2.6 | 4.1 | 2.8 | 3.0 | 15.1 | 7,090 | 165 | 412 | 19,974 | 6.6 |
| Teton | 0.2 | 1.7 | 11.4 | 0.7 | 14.9 | 6.3 | 11.5 | 4.0 | 13.1 | 3,290 | 141 | 38 | 14,338 | 11.9 |
| Uinta | 0.9 | 5.7 | 13.0 | 3.5 | 8.7 | 6.8 | 4.8 | 10.5 | 25.4 | 3,825 | 189 | 298 | 9,148 | 5.0 |
| Washakie | 3.4 | 2.7 | 7.8 | 13.1 | 4.5 | 5.1 | 4.4 | D | 24.3 | 2,025 | 259 | 95 | 3,867 | 0.9 |
| Weston | -1.0 | 11.7 | D | D | 2.4 | 4.3 | 4.1 | D | 33.8 | 1,735 | 252 | 51 | 3,570 | 1.0 |

1. Per 1,000 resident population estimated as of July 1 of the year shown.

# Table B. States and Counties — Housing, Labor Force, and Employment

| STATE County | Housing units, 2015-2019 | | | | | | | | Civilian labor force, 2020 | | | | Civilian employment[6], 2015-2019 | | |
|---|---|---|---|---|---|---|---|---|---|---|---|---|---|---|---|
| | Occupied units | | | | | | | | | | Unemployment | | | Percent | |
| | | Owner-occupied | | | | Renter-occupied | | | | | | | | | |
| | | | | Median owner cost as a percent of income | | | Median rent as a percent of income[2] | Sub-standard units[4] (percent) | | Percent change, 2019-2020 | | | | Management, business, science, and arts | Construction, production, and maintenance occupations |
| | Total | Percent | Median value[1] | With a mort-gage | Without a mort-gage[2] | Median rent[3] | | | Total | | Total | Rate[5] | Total | | |
| | 89 | 90 | 91 | 92 | 93 | 94 | 95 | 96 | 97 | 98 | 99 | 100 | 101 | 102 | 103 |
| **WYOMING—Cont'd** | | | | | | | | | | | | | | | |
| Goshen | 5,207 | 73.7 | 174,000 | 20.3 | 10.4 | 705 | 24.7 | 1.2 | 6,639 | 1.8 | 284 | 4.3 | 6,014 | 35.3 | 27.6 |
| Hot Springs | 2,118 | 75.2 | 147,300 | 20.2 | 10.0 | 730 | 27.4 | 1.1 | 2,249 | 1.8 | 113 | 5.0 | 2,217 | 39.2 | 25.6 |
| Johnson | 3,845 | 69.9 | 252,900 | 20.6 | 11.3 | 942 | 23.8 | 2.1 | 4,192 | -0.1 | 232 | 5.5 | 4,220 | 33.8 | 28.9 |
| Laramie | 39,683 | 69.8 | 227,900 | 21.0 | 10.1 | 950 | 28.4 | 1.5 | 50,826 | 2.1 | 2,595 | 5.1 | 47,158 | 36.8 | 24.8 |
| Lincoln | 6,908 | 82.4 | 246,400 | 19.7 | 10.3 | 852 | 24.5 | 4.8 | 9,373 | 3.0 | 466 | 5.0 | 9,189 | 30.3 | 36.2 |
| Natrona | 32,799 | 68.2 | 208,700 | 20.0 | 10.0 | 860 | 27.6 | 2.2 | 40,710 | 0.8 | 3,181 | 7.8 | 41,101 | 32.7 | 27.5 |
| Niobrara | 964 | 77.5 | 156,300 | 19.7 | 13.9 | 560 | 30.6 | 2.0 | 1,265 | 2.3 | 48 | 3.8 | 924 | 35.2 | 28.0 |
| Park | 12,160 | 73.2 | 268,500 | 21.9 | 10.0 | 882 | 22.3 | 1.3 | 15,380 | -0.5 | 831 | 5.4 | 13,922 | 38.0 | 24.8 |
| Platte | 4,030 | 74.8 | 188,500 | 19.2 | 11.6 | 763 | 25.7 | 2.8 | 4,651 | 1.2 | 233 | 5.0 | 4,464 | 28.7 | 36.2 |
| Sheridan | 13,251 | 68.8 | 272,400 | 22.4 | 10.8 | 836 | 24.8 | 1.8 | 16,210 | 3.3 | 801 | 4.9 | 15,193 | 41.6 | 22.0 |
| Sublette | 3,340 | 82.1 | 275,100 | 20.3 | 10.0 | 895 | 19.6 | 0.7 | 4,059 | -1.2 | 294 | 7.2 | 5,234 | 29.8 | 30.8 |
| Sweetwater | 15,523 | 76.2 | 205,600 | 18.4 | 10.0 | 861 | 23.1 | 2.4 | 20,840 | -2.7 | 1,532 | 7.4 | 21,487 | 30.3 | 35.9 |
| Teton | 9,019 | 60.4 | 866,600 | 21.6 | 10.0 | 1,376 | 25.8 | 5.9 | 15,204 | -4.8 | 912 | 6.0 | 14,430 | 37.2 | 19.4 |
| Uinta | 7,597 | 77.1 | 175,000 | 17.8 | 10.0 | 698 | 23.9 | 3.3 | 9,312 | 0.3 | 582 | 6.3 | 9,294 | 29.6 | 35.9 |
| Washakie | 3,365 | 76.5 | 165,800 | 21.4 | 10.2 | 640 | 25.0 | 2.0 | 4,004 | 2.6 | 211 | 5.3 | 3,777 | 34.6 | 32.9 |
| Weston | 2,891 | 83.0 | 166,300 | 19.0 | 11.3 | 767 | 23.9 | 2.2 | 3,836 | 0.4 | 148 | 3.9 | 3,221 | 34.6 | 36.5 |

1. Specified owner-occupied units.   2. A value of 10.0 represents 10 percent or less; a value of 50.0 represents 50 percent or more.   3. Specified renter-occupied units.   4. Overcrowded or lacking complete plumbing facilities.   5. Percent of civilian labor force.   6. Civilian employed persons 16 years old and over.

Items 89—103

# Table B. States and Counties — Nonfarm Employment and Agriculture

| STATE County | Private nonfarm establishments, employment and payroll, 2019 | | | | | | | | | Agriculture, 2017 | | | |
| | Number of establishments | Employment | | | | | | Annual payroll | | Farms | | | Farm producers whose primary occupation is farming (percent) |
| | | Total | Health care and social assistance | Manufacturing | Retail trade | Finance and insurance | Professional, scientific, and technical services | Total (mil dol) | Average per employee (dollars) | Number | Percent with: | | |
| | | | | | | | | | | | Fewer than 50 acres | 1000 acres or more | |
| | 104 | 105 | 106 | 107 | 108 | 109 | 110 | 111 | 112 | 113 | 114 | 115 | 116 |
| WYOMING—Cont'd | | | | | | | | | | | | | |
| Goshen | 336 | 2,798 | 687 | 306 | 331 | 122 | 166 | 94 | 33,569 | 842 | 19.1 | 24.8 | 48.9 |
| Hot Springs | 182 | 1,546 | 385 | 50 | 188 | 39 | 41 | 65 | 42,258 | 223 | 39.9 | 18.4 | 42.7 |
| Johnson | 509 | 2,561 | 352 | 80 | 384 | 182 | 292 | 98 | 38,319 | 384 | 21.1 | 42.2 | 46.8 |
| Laramie | 3,355 | 34,602 | 6,598 | 1,307 | 6,052 | 1,477 | 2,539 | 1,586 | 45,846 | 999 | 31.7 | 21.9 | 38.8 |
| Lincoln | 761 | 4,891 | 899 | 421 | 752 | 112 | 196 | 245 | 50,062 | 698 | 52.0 | 10.5 | 31.0 |
| Natrona | 2,928 | 32,242 | 5,756 | 1,766 | 4,578 | 917 | 1,623 | 1,699 | 52,709 | 430 | 40.0 | 25.3 | 39.3 |
| Niobrara | 87 | 413 | 57 | NA | 102 | 21 | 10 | 14 | 34,048 | 242 | 5.0 | 69.4 | 67.9 |
| Park | 1,226 | 9,638 | 2,047 | 556 | 1,575 | 407 | 360 | 404 | 41,933 | 1,008 | 45.6 | 9.9 | 40.5 |
| Platte | 276 | 2,227 | 302 | 108 | 432 | 126 | 50 | 102 | 45,701 | 505 | 24.4 | 33.1 | 45.4 |
| Sheridan | 1,435 | 11,136 | 2,903 | 594 | 1,735 | 337 | 840 | 504 | 45,261 | 833 | 45.7 | 17.5 | 35.1 |
| Sublette | 420 | 2,814 | 169 | 55 | 338 | 70 | 122 | 176 | 62,422 | 402 | 35.3 | 28.6 | 34.3 |
| Sweetwater | 1,207 | 15,461 | 1,434 | 1,915 | 2,239 | 285 | 482 | 995 | 64,346 | 219 | 25.6 | 24.7 | 38.3 |
| Teton | 2,221 | 18,721 | 1,428 | 181 | 2,091 | 455 | 1,005 | 885 | 47,293 | 142 | 34.5 | 13.4 | 40.6 |
| Uinta | 566 | 7,513 | 1,310 | 206 | 990 | 152 | 377 | 388 | 51,579 | 403 | 42.9 | 19.9 | 34.8 |
| Washakie | 338 | 2,483 | 594 | 340 | 355 | 98 | 118 | 99 | 39,848 | 246 | 36.2 | 22.8 | 49.6 |
| Weston | 216 | 1,737 | 346 | 113 | 245 | 67 | 25 | 74 | 42,855 | 247 | 17.8 | 45.7 | 41.8 |

| STATE County | Land in farms | | | | | Value of land and buildings (dollars) | | Value of machinery and equiopmnet, average per farm (dollars) | Value of products sold: | | | | Organic farms (number) | Farms with internet access (per-cent) | Government payments | |
| | Acreage (1,000) | Percent change, 2012-2017 | Acres | | | | | | Total (mil dol) | Average per farm (acres) | Percent from: | | | | Total ($1,000) | Percent of farms |
| | | | Average size of farm | Total irrigated (1,000) | Total cropland (1,000) | Average per farm | Average per acre | | | | Crops | Livestock and poultry products | | | | |
| | 117 | 118 | 119 | 120 | 121 | 122 | 123 | 124 | 125 | 126 | 127 | 128 | 129 | 130 | 131 | 132 |
|---|---|---|---|---|---|---|---|---|---|---|---|---|---|---|---|---|
| WYOMING—Cont'd | | | | | | | | | | | | | | | | |
| Goshen | 1,256 | -8.3 | 1,492 | 118.4 | 234.3 | 1,286,023 | 862 | 156,103 | 201.9 | 239,760 | 24.0 | 76.0 | 3 | 78.9 | 4,111 | 40.5 |
| Hot Springs | 528 | 2.1 | 2,368 | 14.8 | 22.7 | 1,469,992 | 621 | 90,195 | 15.2 | 68,278 | 6.6 | 93.4 | 1 | 77.6 | 408 | 13.0 |
| Johnson | 1,974 | -3.0 | 5,142 | 31.7 | 41.0 | 3,020,697 | 588 | 128,544 | 44.1 | 114,966 | 7.0 | 93.0 | NA | 82.6 | 1,088 | 21.6 |
| Laramie | 1,630 | -2.8 | 1,631 | 67.1 | 382.3 | 1,383,473 | 848 | 123,891 | 184.6 | 184,762 | 17.9 | 82.1 | 42 | 77.7 | 4,856 | 29.3 |
| Lincoln | 365 | 6.1 | 523 | 81.3 | 94.0 | 955,622 | 1,828 | 82,429 | 47.9 | 68,566 | 23.3 | 76.7 | NA | 81.2 | 182 | 5.2 |
| Natrona | 1,933 | 14.3 | 4,496 | 44.3 | 55.7 | 2,039,344 | 454 | 109,339 | 43.2 | 100,493 | 12.9 | 87.1 | NA | 76.0 | 318 | 7.9 |
| Niobrara | 1,277 | -6.0 | 5,279 | 10.5 | 54.6 | 3,308,558 | 627 | 156,480 | 49.7 | 205,306 | 8.1 | 91.9 | NA | 71.5 | 1,861 | 37.2 |
| Park | 930 | 14.4 | 923 | 127.5 | 140.5 | 1,378,458 | 1,494 | 124,845 | 85.2 | 84,497 | 53.7 | 46.3 | 5 | 86.9 | 931 | 10.0 |
| Platte | 1,047 | -14.5 | 2,073 | 58.9 | 134.8 | 1,820,452 | 878 | 136,932 | 94.2 | 186,598 | 19.7 | 80.3 | 6 | 82.2 | 2,872 | 27.1 |
| Sheridan | 1,214 | -7.0 | 1,457 | 54.6 | 95.9 | 1,783,962 | 1,224 | 101,776 | 59.7 | 71,637 | 14.5 | 85.5 | 1 | 78.9 | 1,417 | 12.2 |
| Sublette | 546 | -29.7 | 1,359 | 158.7 | 143.5 | 2,243,832 | 1,651 | 128,778 | 47.9 | 119,085 | 12.9 | 87.1 | 2 | 84.3 | 155 | 2.7 |
| Sweetwater | 1,370 | -17.7 | 6,256 | 35.4 | 35.6 | 1,851,640 | 296 | 110,326 | 16.5 | 75,132 | 27.7 | 72.3 | NA | 82.6 | 201 | 5.9 |
| Teton | 68 | 68.4 | 476 | 25.5 | 15.8 | 1,929,058 | 4,049 | 103,147 | 17.6 | 123,817 | 22.6 | 77.4 | NA | 88.7 | 131 | 4.2 |
| Uinta | 657 | 1.0 | 1,630 | 76.2 | 84.8 | 1,747,715 | 1,072 | 102,095 | 26.7 | 66,159 | 13.6 | 86.4 | 1 | 80.9 | 130 | 1.5 |
| Washakie | 316 | -7.4 | 1,285 | 38.5 | 53.4 | 1,502,187 | 1,169 | 211,277 | 43.0 | 174,874 | 37.8 | 62.2 | NA | 91.1 | 508 | 25.2 |
| Weston | 1,227 | -4.9 | 4,968 | 4.2 | 37.5 | 3,321,035 | 669 | 129,296 | 34.3 | 138,761 | 1.8 | 98.2 | NA | 81.0 | 836 | 31.2 |

## Table B. States and Counties — Water Use, Wholesale Trade, Retail Trade, and Real Estate

| STATE County | Water use, 2015 | | Wholesale Trade[1], 2017 | | | | Retail Trade[2], 2017 | | | | Real estate and rental and leasing,[2] 2017 | | | |
|---|---|---|---|---|---|---|---|---|---|---|---|---|---|---|
| | Public supply water withdrawn (mil gal/day) | Public supply gallons withdrawn per person per day | Number of establishments | Number of employees | Sales (mil dol) | Average payroll (mil dol) | Number of establishments | Number of employees | Sales (mil dol) | Average payroll (mil dol) | Number of establishments | Number of employees | Sales (mil dol) | Average payroll (mil dol) |
| | 133 | 134 | 135 | 136 | 137 | 138 | 139 | 140 | 141 | 142 | 143 | 144 | 145 | 146 |
| WYOMING—Cont'd | | | | | | | | | | | | | | |
| Goshen .......................... | 1.85 | 138.2 | 19 | 128 | 178.4 | 5.8 | 42 | 380 | 94.7 | 9.2 | D | D | D | D |
| Hot Springs ...................... | 0.67 | 141.3 | NA | NA | NA | NA | 25 | 152 | 31.7 | 3.3 | 7 | D | 1.8 | D |
| Johnson .......................... | 1.61 | 187.5 | 11 | 23 | 8.8 | 0.8 | 59 | 370 | 87.3 | 9.8 | 17 | 37 | 5.2 | 1.1 |
| Laramie .......................... | 14.00 | 144.2 | 122 | 926 | 604.3 | 49.1 | 372 | 5,390 | 1,707.9 | 151.7 | 183 | 689 | 188.4 | 31.6 |
| Lincoln .......................... | 4.05 | 216.3 | 9 | D | 19.6 | D | 90 | 744 | 213.5 | 19.0 | 28 | 57 | 12.4 | 2.0 |
| Natrona .......................... | 12.26 | 149.2 | 164 | 1,812 | 2,888.1 | 115.8 | 344 | 4,766 | 1,386.0 | 139.6 | 181 | 952 | 285.7 | 54.1 |
| Niobrara ......................... | 1.00 | 393.4 | NA | NA | NA | NA | 11 | 99 | 25.6 | 2.0 | D | D | D | 0.1 |
| Park ............................. | 3.17 | 108.5 | 31 | 230 | 177.2 | 10.4 | 169 | 1,546 | 459.3 | 46.0 | 61 | 170 | 26.8 | 4.7 |
| Platte ........................... | 1.56 | 177.0 | 4 | 22 | 23.5 | 1.1 | 41 | 378 | 113.2 | 9.9 | 10 | 21 | 3.8 | 0.6 |
| Sheridan ......................... | 4.55 | 151.6 | 30 | 158 | 78.9 | 8.0 | 152 | 1,693 | 545.6 | 49.8 | 66 | 170 | 30.9 | 5.6 |
| Sublette ......................... | 2.21 | 223.3 | D | D | D | 1.7 | 43 | 338 | 100.3 | 10.2 | 26 | 110 | 16.0 | 4.2 |
| Sweetwater ....................... | 11.37 | 254.8 | 60 | 519 | 462.5 | 29.1 | 172 | 2,410 | 835.4 | 69.3 | 78 | 638 | 96.6 | 22.1 |
| Teton ............................ | 7.78 | 336.4 | 26 | 168 | 88.0 | 8.1 | 254 | 2,006 | 641.1 | 75.2 | 197 | 717 | 251.0 | 42.5 |
| Uinta ............................ | 3.90 | 187.3 | 14 | 142 | 123.1 | 8.5 | 86 | 1,117 | 347.2 | 26.9 | 30 | 101 | 18.7 | 3.8 |
| Washakie ......................... | 0.09 | 10.8 | 12 | 58 | 25.4 | 2.4 | 43 | 356 | 89.3 | 11.3 | 11 | 43 | 6.4 | 1.5 |
| Weston ........................... | 1.02 | 141.0 | 6 | 18 | 4.9 | 0.7 | 31 | 266 | 78.4 | 6.2 | 4 | D | 0.8 | D |

1 Merchant wholesalers, except manufacturers' sales branches and offices.   2. Employer establishments.

| STATE County | Professional, scientific, and technical services, 2017 | | | | Manufacturing, 2017 | | | | Accommodation and food services, 2017 | | | |
|---|---|---|---|---|---|---|---|---|---|---|---|---|
| | Number of establishments | Number of employees | Sales (mil dol) | Average payroll (mil dol) | Number of establishments | Number of employees | Sales (mil dol) | Average payroll (mil dol) | Number of establishments | Number of employees | Sales (mil dol) | Annual payroll (mil dol) |
| | 147 | 148 | 149 | 150 | 151 | 152 | 153 | 154 | 155 | 156 | 157 | 158 |
| WYOMING—Cont'd | | | | | | | | | | | | |
| Goshen | 23 | 153 | 12 | 5.0 | 7 | 261 | 72.2 | 9.2 | 28 | 257 | 13.5 | 3.1 |
| Hot Springs | 12 | 37 | 5 | 1.6 | D | 51 | D | 2.5 | 21 | 246 | 11.6 | 3.4 |
| Johnson | 74 | 215 | 34 | 11.0 | D | 68 | D | 2.7 | 44 | 347 | 23.1 | 6.5 |
| Laramie | D | D | D | D | 69 | 1,296 | 1,348.3 | 85.5 | 218 | 3,755 | 227.3 | 64.5 |
| Lincoln | 73 | 130 | 17 | 4.7 | 23 | 342 | 336.0 | 28.5 | 64 | 473 | 25.0 | 6.9 |
| Natrona | 266 | 1,440 | 267 | 89.1 | 89 | 1,379 | 1,002.2 | 83.6 | 201 | 3,819 | 211.1 | 65.8 |
| Niobrara | D | D | 0 | D | NA | NA | NA | NA | D | D | D | 1.2 |
| Park | 104 | 373 | 48 | 18.0 | 42 | 497 | 140.5 | 25.2 | 126 | 1,621 | 218.9 | 60.4 |
| Platte | 18 | 46 | 7 | 1.7 | D | 79 | D | 3.7 | 36 | 324 | 20.2 | 4.6 |
| Sheridan | 166 | 817 | 122 | 41.1 | 30 | 484 | 104.7 | 23.3 | 111 | 1,609 | 86.2 | 27.1 |
| Sublette | 42 | 111 | 20 | 6.0 | D | D | D | 1.9 | 43 | 255 | 24.7 | 7.6 |
| Sweetwater | 120 | 449 | 82 | 22.6 | 32 | 1,801 | 1,558.9 | 177.1 | 120 | 1,793 | 98.0 | 28.5 |
| Teton | D | D | D | D | 38 | 140 | 23.8 | 6.4 | 181 | 4,557 | 505.1 | 140.7 |
| Uinta | 48 | 228 | 45 | 18.0 | 17 | 190 | 186.3 | 10.0 | 54 | 646 | 33.4 | 9.5 |
| Washakie | 39 | 135 | 19 | 6.8 | 16 | 399 | 190.4 | 17.2 | 30 | 285 | 10.6 | 3.1 |
| Weston | 9 | 22 | 2 | 0.6 | D | 114 | D | D | 22 | 152 | 7.1 | 1.8 |

# Table B. States and Counties — Health Care and Social Assistance, Other Services, Nonemployer Businesses, and Residential Construction

| STATE County | Health care and social assistance, 2017 | | | | Other services, 2017 | | | | Nonemployer businesses, 2018 | | Value of residential construction authorized by building permits, 2020 | |
|---|---|---|---|---|---|---|---|---|---|---|---|---|
| | Number of establishments | Number of employees | Receipts (mil dol) | Annual payroll (mil dol) | Number of establishments | Number of employees | Receipts (mil dol) | Annual payroll (mil dol) | Number | Receipts (mil dol) | New construction ($1,000) | Number of housing units |
| | 159 | 160 | 161 | 162 | 163 | 164 | 165 | 166 | 167 | 168 | 169 | 170 |
| WYOMING—Cont'd | | | | | | | | | | | | |
| Goshen | 26 | 734 | 69.2 | 29.5 | 26 | 86 | 8.9 | 2.2 | 1,004 | 40.3 | 144 | 1 |
| Hot Springs | 21 | 369 | 36.0 | 14.9 | 13 | 56 | 6.0 | 1.5 | 384 | 14.8 | 277 | 2 |
| Johnson | 29 | 386 | 31.9 | 15.4 | 32 | 151 | 15.0 | 4.8 | 1,476 | 88.4 | 6,236 | 22 |
| Laramie | 361 | 6,859 | 876.7 | 382.6 | 206 | 1,187 | 129.0 | 36.9 | 8,945 | 616.3 | 110,949 | 550 |
| Lincoln | 52 | 776 | 80.4 | 34.2 | 43 | 120 | 20.5 | 4.1 | 2,150 | 92.9 | 52,832 | 184 |
| Natrona | 333 | 5,588 | 751.8 | 310.6 | 198 | 976 | 140.4 | 34.5 | 6,254 | 329.3 | 40,548 | 208 |
| Niobrara | 7 | 55 | 4.5 | 2.3 | D | D | 2.5 | D | 241 | 8.1 | 0 | 0 |
| Park | 126 | 1,978 | 243.7 | 98.3 | 87 | 268 | 27.1 | 7.6 | 3,255 | 148.1 | 51,488 | 194 |
| Platte | 20 | 306 | 28.8 | 11.8 | 12 | 60 | 4.8 | 1.3 | 757 | 26.3 | 7,977 | 36 |
| Sheridan | 141 | 3,364 | 305.8 | 156.4 | 78 | 289 | 44.0 | 9.2 | 3,633 | 210.9 | 51,351 | 204 |
| Sublette | 20 | 162 | 12.5 | 5.6 | 27 | 139 | 27.9 | 7.1 | 1,105 | 48.5 | 12,189 | 41 |
| Sweetwater | 102 | 1,482 | 169.5 | 60.7 | 79 | 412 | 67.7 | 15.6 | 2,215 | 102.5 | 13,089 | 50 |
| Teton | 136 | 1,465 | 209.0 | 86.5 | 123 | 574 | 135.0 | 26.8 | 5,647 | 401.1 | 269,823 | 207 |
| Uinta | 72 | 1,320 | 117.7 | 57.7 | 28 | 82 | 8.6 | 2.5 | 1,537 | 67.1 | 11,116 | 41 |
| Washakie | 32 | 601 | 49.5 | 21.1 | 23 | 74 | 5.8 | 1.4 | 687 | 21.1 | 1,025 | 5 |
| Weston | 22 | 353 | 31.2 | 14.1 | 9 | 33 | 3.6 | 1.1 | 607 | 21.3 | 53 | 1 |

Items 159—170

# Table B. States and Counties — Government Employment and Payroll, and Local Government Finances

| STATE County | Full-time equivalent employees | March payroll (dollars) | Administration, judicial, and legal | Police and corrections | Fire protection | Highways and transportation | Health and welfare | Natural resources and utilities | Education and libraries | Total (mil dol) | Inter-governmental (mil dol) | Taxes Total (mil dol) | Per capita¹ Total | Per capita¹ Property |
|---|---|---|---|---|---|---|---|---|---|---|---|---|---|---|
| | 171 | 172 | 173 | 174 | 175 | 176 | 177 | 178 | 179 | 180 | 181 | 182 | 183 | 184 |
| **WYOMING—Cont'd** | | | | | | | | | | | | | | |
| Goshen | 742 | 2,816,098 | 6.4 | 6.9 | 0.0 | 2.2 | 1.3 | 8.3 | 72.6 | 87.8 | 52.7 | 18.8 | 1,406 | 1,235 |
| Hot Springs | 390 | 1,487,824 | 7.9 | 6.5 | 0.0 | 2.5 | 32.5 | 3.6 | 45.0 | 25.3 | 13.4 | 8.0 | 1,711 | 1,517 |
| Johnson | 633 | 2,884,696 | 5.5 | 9.4 | 0.5 | 3.8 | 35.3 | 3.4 | 40.9 | 73.4 | 17.1 | 30.5 | 3,612 | 3,266 |
| Laramie | 6,340 | 31,230,813 | 2.7 | 5.1 | 1.9 | 1.9 | 39.7 | 2.3 | 45.2 | 848.2 | 316.5 | 110.4 | 1,122 | 833 |
| Lincoln | 1,315 | 5,381,730 | 5.6 | 5.7 | 0.0 | 2.7 | 32.4 | 3.7 | 48.9 | 170.8 | 68.3 | 42.1 | 2,185 | 2,003 |
| Natrona | 3,954 | 16,298,613 | 5.7 | 8.4 | 3.7 | 2.6 | 1.2 | 5.2 | 71.3 | 502.1 | 317.5 | 97.1 | 1,220 | 987 |
| Niobrara | 283 | 1,121,423 | 6.7 | 4.7 | 0.0 | 2.0 | 29.6 | 7.5 | 46.9 | 31.8 | 16.5 | 7.0 | 2,911 | 2,731 |
| Park | 2,146 | 9,400,479 | 4.0 | 4.8 | 0.5 | 2.1 | 30.6 | 6.7 | 49.6 | 270.3 | 91.0 | 49.2 | 1,686 | 1,501 |
| Platte | 486 | 1,802,664 | 6.5 | 9.8 | 0.3 | 2.9 | 1.1 | 9.7 | 67.4 | 51.0 | 29.4 | 11.7 | 1,367 | 1,089 |
| Sheridan | 1,985 | 9,635,803 | 3.5 | 4.2 | 1.0 | 1.5 | 35.0 | 3.6 | 50.8 | 288.8 | 121.4 | 30.8 | 1,022 | 767 |
| Sublette | 613 | 2,777,896 | 8.4 | 13.5 | 1.0 | 5.5 | 10.0 | 6.6 | 50.3 | 117.8 | 22.9 | 74.6 | 7,654 | 7,523 |
| Sweetwater | 3,089 | 14,708,500 | 5.2 | 7.2 | 2.2 | 2.5 | 25.5 | 5.9 | 49.7 | 435.4 | 170.1 | 132.0 | 3,037 | 2,666 |
| Teton | 1,853 | 8,974,820 | 6.3 | 5.4 | 1.4 | 7.3 | 41.5 | 6.5 | 31.0 | 287.9 | 60.3 | 78.9 | 3,373 | 2,475 |
| Uinta | 1,266 | 4,979,667 | 5.3 | 6.5 | 1.0 | 2.4 | 0.8 | 6.1 | 76.6 | 120.6 | 78.5 | 26.1 | 1,278 | 1,122 |
| Washakie | 447 | 1,611,024 | 6.3 | 8.1 | 0.4 | 3.3 | 0.6 | 8.4 | 72.5 | 48.5 | 33.1 | 9.8 | 1,218 | 1,021 |
| Weston | 523 | 1,956,847 | 11.2 | 3.5 | 0.5 | 0.8 | 32.9 | 5.3 | 45.2 | 63.6 | 29.8 | 15.3 | 2,197 | 1,829 |

1. Based on the resident population estimated as of July 1 of the year shown.

# Table B. States and Counties — Local Government Finances, Government Employment, and Income Taxes

| STATE County | Local government finances, 2017 (cont.) — Direct general expenditure Total (mil dol) | Per capita[1] (dollars) | Percent of total for: Education | Health and hospitals | Police protection | Public welfare | Highways | Debt outstanding Total (mil dol) | Per capita[1] (dollars) | Gov. employment, 2019 Federal civilian | Federal military | State and local | Individual income tax returns, 2018 Number of returns | Mean adjusted gross income | Mean income tax |
|---|---|---|---|---|---|---|---|---|---|---|---|---|---|---|---|
| | 185 | 186 | 187 | 188 | 189 | 190 | 191 | 192 | 193 | 194 | 195 | 196 | 197 | 198 | 199 |
| WYOMING—Cont'd | | | | | | | | | | | | | | | |
| Goshen | 87 | 6,536 | 62.3 | 1.0 | 3.8 | 2.9 | 2.3 | 17.6 | 1,320 | 68 | 63 | 1,334 | 5,530 | 52,662 | 4,678 |
| Hot Springs | 24 | 5,152 | 52.7 | 1.1 | 6.2 | 0.0 | 5.1 | 4.5 | 960 | 14 | 23 | 552 | 2,110 | 50,107 | 4,809 |
| Johnson | 77 | 9,114 | 30.7 | 37.5 | 6.0 | 0.8 | 7.4 | 9.3 | 1,107 | 130 | 44 | 858 | 4,220 | 68,282 | 7,712 |
| Laramie | 893 | 9,078 | 39.6 | 37.9 | 2.9 | 0.3 | 2.4 | 143.1 | 1,454 | 2,789 | 3,617 | 11,156 | 48,740 | 67,197 | 7,045 |
| Lincoln | 151 | 7,853 | 38.5 | 29.4 | 3.3 | 0.1 | 5.1 | 36.4 | 1,890 | 113 | 104 | 1,782 | 8,440 | 76,903 | 8,441 |
| Natrona | 518 | 6,511 | 62.5 | 0.4 | 4.9 | 0.3 | 4.4 | 110.2 | 1,385 | 634 | 414 | 4,919 | 37,950 | 75,409 | 9,440 |
| Niobrara | 38 | 15,799 | 38.2 | 20.8 | 2.3 | 0.0 | 4.5 | 193.9 | 81,050 | 10 | 11 | 439 | 1,090 | 52,443 | 4,760 |
| Park | 267 | 9,143 | 38.7 | 34.5 | 3.0 | 0.0 | 3.2 | 79.7 | 2,729 | 790 | 150 | 2,696 | 14,870 | 71,549 | 8,317 |
| Platte | 56 | 6,522 | 50.0 | 3.7 | 3.7 | 0.4 | 5.6 | 6.8 | 793 | 103 | 93 | 846 | 4,200 | 55,992 | 5,635 |
| Sheridan | 312 | 10,363 | 48.8 | 29.5 | 2.7 | 0.0 | 3.6 | 76.7 | 2,545 | 808 | 155 | 2,745 | 15,300 | 76,263 | 9,145 |
| Sublette | 99 | 10,129 | 34.2 | 12.7 | 4.3 | 1.5 | 11.7 | 14.1 | 1,445 | 129 | 49 | 916 | 4,250 | 84,074 | 11,438 |
| Sweetwater | 426 | 9,797 | 48.5 | 19.4 | 4.0 | 0.1 | 3.7 | 82.3 | 1,893 | 222 | 220 | 4,229 | 19,780 | 71,672 | 7,416 |
| Teton | 282 | 12,073 | 22.0 | 35.5 | 3.6 | 0.0 | 3.2 | 36.5 | 1,563 | 400 | 122 | 2,170 | 14,550 | 261,500 | 50,655 |
| Uinta | 122 | 5,980 | 64.4 | 0.7 | 4.7 | 0.7 | 3.1 | 9.3 | 456 | 73 | 105 | 2,128 | 9,000 | 64,480 | 6,114 |
| Washakie | 47 | 5,824 | 66.6 | 0.7 | 5.5 | 0.0 | 3.0 | 14.6 | 1,826 | 119 | 40 | 741 | 3,650 | 62,706 | 5,714 |
| Weston | 65 | 9,287 | 35.3 | 35.0 | 3.3 | 0.1 | 4.0 | 11.5 | 1,646 | 53 | 35 | 757 | 3,090 | 60,381 | 5,865 |

1. Based on the resident population estimated as of July 1 of the year shown.

# Metropolitan Areas

(For explanation of symbols, see page viii)

Page

| 775 | Metropolitan Area Highlights and Rankings |
| 793 | Metropolitan Area Column Headings |
| 797 | Table C |
| 797 | (Abilene, TX)—(Casper, WY) |
| 811 | (Cedar Rapids, IA)—(Eugene, OR) |
| 825 | (Evansville, IN-KY)—(Johnstown, PA) |
| 839 | (Jonesboro, AR)—(Missoula, MT) |
| 853 | (Mobile, AL)—(Rapid City, SD) |
| 867 | (Reading, PA)—(Terre Haute, IN) |
| 881 | (Texarkana, TX-AR)—(Yuma, AZ) |

# Metropolitan Areas

(For explanation of symbols, see page vii.)

| | Page |
|---|---|
| Metropolitan Area Highlights and Rankings | 772 |
| Metropolitan Area Column Headings | 785 |
| Table C | 787 |
| (Abilene, TX)—(Casper, WY) | 787 |
| (Cedar Rapids, IA)—(Eugene, OR) | 811 |
| (Evansville, IN-KY)—(Johnstown, PA) | 825 |
| (Jonesboro, AR)—(Missoula, MT) | 838 |
| (Mobile, AL)—(Rapid City, SD) | 852 |
| (Reading, PA)—(Terre Haute, IN) | 867 |
| (Texarkana, TX-AR)—(Yuma, AZ) | 881 |

# Metropolitan Area Highlights and Rankings

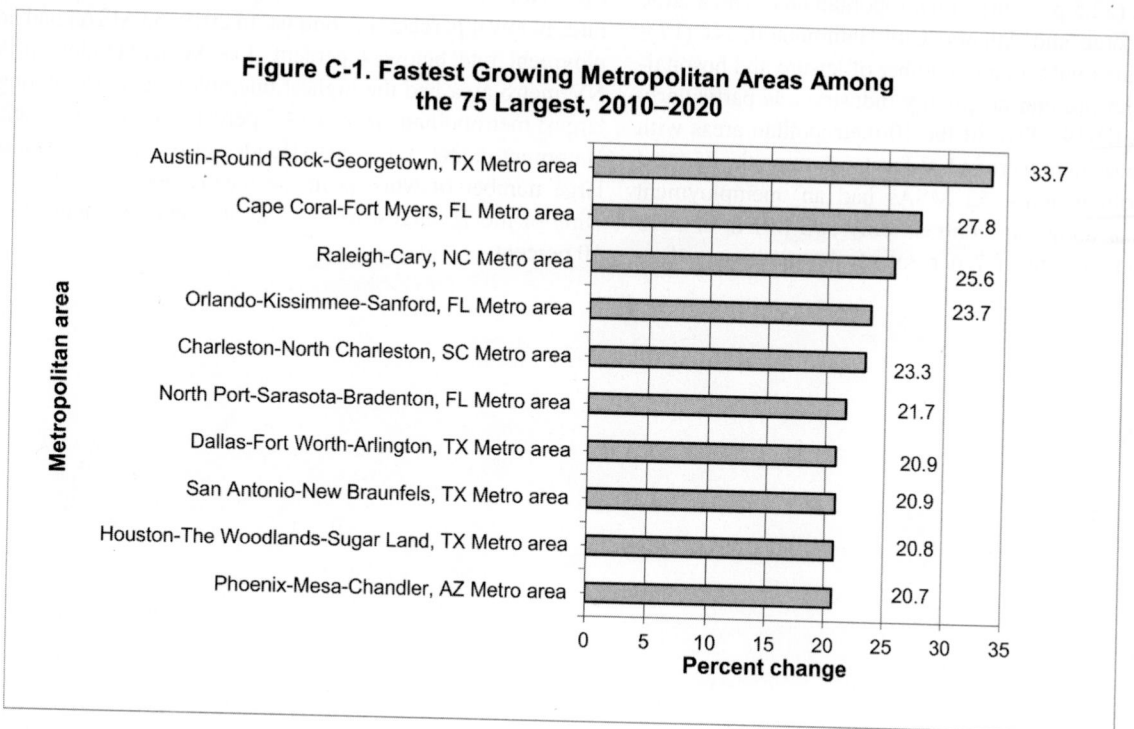

**Figure C-1. Fastest Growing Metropolitan Areas Among the 75 Largest, 2010–2020**

| Metropolitan area | Percent change |
|---|---|
| Austin-Round Rock-Georgetown, TX Metro area | 33.7 |
| Cape Coral-Fort Myers, FL Metro area | 27.8 |
| Raleigh-Cary, NC Metro area | 25.6 |
| Orlando-Kissimmee-Sanford, FL Metro area | 23.7 |
| Charleston-North Charleston, SC Metro area | 23.3 |
| North Port-Sarasota-Bradenton, FL Metro area | 21.7 |
| Dallas-Fort Worth-Arlington, TX Metro area | 20.9 |
| San Antonio-New Braunfels, TX Metro area | 20.9 |
| Houston-The Woodlands-Sugar Land, TX Metro area | 20.8 |
| Phoenix-Mesa-Chandler, AZ Metro area | 20.7 |

In 2020, 86.2 percent of Americans lived in metropolitan areas, but these metropolitan areas made up a mere 28 percent of the nation's land area. After nearly a decade of research and development, the Office of Management and Budget (OMB) first established new rules for defining metropolitan areas and issued a completely new list after the 2000 census. This scheme defines a variety of areas called "Core Based Statistical Areas" (CBSAs). Along with the new definition of metropolitan areas, OMB defined a new type of area—called a micropolitan area—that defines the many American communities with population clusters that are too small to meet the 50,000 minimum that defines a metropolitan area. Appendix C lists these micropolitan areas as of March 2020, but because of size constraints, *County and City Extra* continues to include only metropolitan area data in Table C. In 2013, the Census Bureau released data for new metropolitan and micropolitan areas based on the 2010 census, and these were updated in 2015, 2017, 2018, and again in 2020. This edition of *County and City Extra* uses the 2018 metropolitan area delineations, except in Appendix C which reflects the 2020 addition of one new micropolitan area.

With over 19 million people, the New York metropolitan area was the largest, followed by Los Angeles with a population of over 13 million. Chicago ranked third with 9.4 million people. Another 12 metropolitan areas had more than 4 million residents (Dallas, Houston, Washington, Philadelphia, Miami, Atlanta, Boston, San Francisco, Phoenix, Riverside, Detroit and Seattle), while 40 other metropolitan areas had between 1 million and 4 million people. Nearly 57 percent of the U.S. population lived in these 55 metropolitan areas with one million or more residents.

One hundred ninety-seven metropolitan areas grew by 5 percent or more between 2010 and 2020. The Villages, FL had the highest growth rate, increasing by 48.8 percent to a 2020 population of 139,018 residents. Among the 75 largest metropolitan areas, Austin-Round Rock-Georgetown, TX had the largest increase at 33.7 percent followed by Cape Coral-Fort Myers, FL at 27.8 percent and Raleigh-Cary, NC Metro area at 25.6 percent. Seven of the most populous metropolitan areas lost population since 2010—Pittsburgh, PA; Cleveland-Elyria, OH, New Haven-Milford, CT; Rochester, NY, Hartford-East Hartford-Middletown, CT, Chicago-Naperville-Elgin, IL-IN-WI and Buffalo-Cheektowaga, NY. While many large metropolitan areas in the Midwest lost population or increased only slightly, Columbus, Omaha, Indianapolis, Minneapolis, and Grand Rapids all grew by more than eight percent.

Among metropolitan areas, New York and Los Angeles shared the top spots for density as well as for total population. With 2,860.7 persons per square mile, New York-Newark-Jersey City, NY-NJ-PA was the most densely populated metropolitan area in the country. At the other extreme, nine of the largest metropolitan areas had fewer than 200 persons per square mile. These areas typically had large land areas and were located in the west and south.

In 2020, 36 metropolitan areas had an unemployment rate at 10 percent or higher. In contrast, 131 metropolitan areas had an unemployment rate at 10 percent or higher in 2010 but only two had an unemployment rate of over 10 percent in 2019. Topping the list are El Centro, CA (22.5 percent), a metropolitan area with a large agricultural workforce and Atlantic City-Hammonton, NJ (17.8 percent), a metro area with a large number of leisure and hospitality workers. The leisure and hospitality industry was particularly impacted by COVID-19. Five of the 10 metropolitan areas with the highest unemployment were in California. Among the 75 most populous metropolitan areas, 52 MSAs had an unemployment rate at 7 percent or above. In contrast, only two MSAs in 2019 had an unemployment rate of 7 percent of above. Twenty-four metropolitan areas had unemployment rates below 5 percent in 2020. Many of them were relatively small areas. Logan, UT-ID had the lowest unemployment rate among all metropolitan areas at 3.0 percent followed by Provo-Orem, UT and Ames, IA at 3.7 percent each. None of the 75 largest metropolitan areas had unemployment rates below 4 percent. In contrast, in 2019, 53 MSAs had an unemployment rate below 4 percent. Las Vegas-Henderson-Paradise, NV metro area had the highest unemployment rate among the 75 largest metropolitan areas at 14.7 percent. Similar to Atlantic City-Hammonton, NJ, Las Vegas-Henderson-Paradise, NV also had a large number of workers in the leisure and hospitality industry. Nine of the 75 largest MSAs had an unemployment rate of over 10 percent.

# 75 Largest Metropolitan Areas by 2020 Population
## Selected Rankings

### Population, 2020

| Population rank | Metropolitan area | Population [col 2] |
|---|---|---|
| 1 | New York-Newark-Jersey City, NY-NJ-PA Metro area | 19,124,359 |
| 2 | Los Angeles-Long Beach-Anaheim, CA Metro area | 13,109,903 |
| 3 | Chicago-Naperville-Elgin, IL-IN-WI Metro area | 9,406,638 |
| 4 | Dallas-Fort Worth-Arlington, TX Metro area | 7,694,138 |
| 5 | Houston-The Woodlands-Sugar Land, TX Metro area | 7,154,478 |
| 6 | Washington-Arlington-Alexandria, DC-VA-MD-WV Metro area | 6,324,629 |
| 7 | Miami-Fort Lauderdale-Pompano Beach, FL Metro area | 6,173,008 |
| 8 | Philadelphia-Camden-Wilmington, PA-NJ-DE-MD Metro area | 6,107,906 |
| 9 | Atlanta-Sandy Springs-Alpharetta, GA Metro area | 6,087,762 |
| 10 | Phoenix-Mesa-Chandler, AZ Metro area | 5,059,909 |
| 11 | Boston-Cambridge-Newton, MA-NH Metro area | 4,878,211 |
| 12 | San Francisco-Oakland-Berkeley, CA Metro area | 4,696,902 |
| 13 | Riverside-San Bernardino-Ontario, CA Metro area | 4,678,371 |
| 14 | Detroit-Warren-Dearborn, MI Metro area | 4,304,136 |
| 15 | Seattle-Tacoma-Bellevue, WA Metro area | 4,018,598 |
| 16 | Minneapolis-St. Paul-Bloomington, MN Metro area | 3,657,477 |
| 17 | San Diego-Chula Vista-Carlsbad, CA Metro area | 3,332,427 |
| 18 | Tampa-St. Petersburg-Clearwater, FL Metro area | 3,243,963 |
| 19 | Denver-Aurora-Lakewood, CO Metro area | 2,991,231 |
| 20 | St. Louis, MO-IL Metro area | 2,805,473 |
| 21 | Baltimore-Columbia-Towson, MD Metro area | 2,800,189 |
| 22 | Charlotte-Concord-Gastonia, NC-SC Metro area | 2,684,276 |
| 23 | Orlando-Kissimmee-Sanford, FL Metro area | 2,639,374 |
| 24 | San Antonio-New Braunfels, TX Metro area | 2,590,732 |
| 25 | Portland-Vancouver-Hillsboro, OR-WA Metro area | 2,510,259 |
| 26 | Sacramento-Roseville-Folsom, CA Metro area | 2,374,749 |
| 27 | Las Vegas-Henderson-Paradise, NV Metro area | 2,315,963 |
| 28 | Pittsburgh, PA Metro area | 2,309,246 |
| 29 | Austin-Round Rock-Georgetown, TX Metro area | 2,295,303 |
| 30 | Cincinnati, OH-KY-IN Metro area | 2,232,907 |
| 31 | Kansas City, MO-KS Metro area | 2,173,212 |
| 32 | Columbus, OH Metro area | 2,138,946 |
| 33 | Indianapolis-Carmel-Anderson, IN Metro area | 2,091,019 |
| 34 | Cleveland-Elyria, OH Metro area | 2,043,807 |
| 35 | San Jose-Sunnyvale-Santa Clara, CA Metro area | 1,971,160 |
| 36 | Nashville-Davidson--Murfreesboro--Franklin, TN Metro area | 1,961,232 |
| 37 | Virginia Beach-Norfolk-Newport News, VA-NC Metro area | 1,779,824 |
| 38 | Providence-Warwick, RI-MA Metro area | 1,623,890 |
| 39 | Jacksonville, FL Metro area | 1,587,892 |
| 40 | Milwaukee-Waukesha, WI Metro area | 1,577,676 |
| 41 | Oklahoma City, OK Metro area | 1,425,375 |
| 42 | Raleigh-Cary, NC Metro area | 1,420,376 |
| 43 | Memphis, TN-MS-AR Metro area | 1,348,678 |
| 44 | Richmond, VA Metro area | 1,303,469 |
| 45 | New Orleans-Metairie, LA Metro area | 1,272,258 |
| 46 | Louisville/Jefferson County, KY-IN Metro area | 1,268,993 |
| 47 | Salt Lake City, UT Metro area | 1,240,029 |
| 48 | Hartford-East Hartford-Middletown, CT Metro area | 1,201,483 |
| 49 | Buffalo-Cheektowaga, NY Metro area | 1,125,637 |
| 50 | Birmingham-Hoover, AL Metro area | 1,091,921 |
| 51 | Grand Rapids-Kentwood, MI Metro area | 1,081,372 |
| 52 | Rochester, NY Metro area | 1,067,486 |
| 53 | Tucson, AZ Metro area | 1,061,175 |
| 54 | Tulsa, OK Metro area | 1,006,411 |
| 55 | Fresno, CA Metro area | 1,000,918 |
| 56 | Urban Honolulu, HI Metro area | 963,826 |
| 57 | Omaha-Council Bluffs, NE-IA Metro area | 954,270 |
| 58 | Worcester, MA-CT Metro area | 945,752 |
| 59 | Bridgeport-Stamford-Norwalk, CT Metro area | 942,426 |
| 60 | Greenville-Anderson, SC Metro area | 932,705 |
| 61 | Albuquerque, NM Metro area | 923,630 |
| 62 | Bakersfield, CA Metro area | 901,362 |
| 63 | Albany-Schenectady-Troy, NY Metro area | 878,550 |
| 64 | Knoxville, TN Metro area | 878,124 |
| 65 | McAllen-Edinburg-Mission, TX Metro area | 875,200 |
| 66 | Baton Rouge, LA Metro area | 858,571 |
| 67 | North Port-Sarasota-Bradenton, FL Metro area | 854,684 |
| 68 | New Haven-Milford, CT Metro area | 851,948 |
| 69 | Columbia, SC Metro area | 847,391 |
| 70 | Allentown-Bethlehem-Easton, PA-NJ Metro area | 846,399 |
| 71 | El Paso, TX Metro area | 846,192 |
| 72 | Oxnard-Thousand Oaks-Ventura, CA Metro area | 841,387 |
| 73 | Charleston-North Charleston, SC Metro area | 819,705 |
| 74 | Dayton-Kettering, OH Metro area | 809,248 |
| 75 | Cape Coral-Fort Myers, FL Metro area | 790,767 |

### Total land area, 2020

| Population rank | Land area rank | Metropolitan area | Land area (square miles) [col 1] |
|---|---|---|---|
| 13 | 1 | Riverside-San Bernardino-Ontario, CA Metro area | 27,277.3 |
| 10 | 2 | Phoenix-Mesa-Chandler, AZ Metro area | 14,567.8 |
| 61 | 3 | Albuquerque, NM Metro area | 9,283.4 |
| 53 | 4 | Tucson, AZ Metro area | 9,188.7 |
| 9 | 5 | Atlanta-Sandy Springs-Alpharetta, GA Metro area | 8,685.7 |
| 4 | 6 | Dallas-Fort Worth-Arlington, TX Metro area | 8,675.2 |
| 19 | 7 | Denver-Aurora-Lakewood, CO Metro area | 8,344.7 |
| 5 | 8 | Houston-The Woodlands-Sugar Land, TX Metro area | 8,268.7 |
| 62 | 9 | Bakersfield, CA Metro area | 8,134.6 |
| 27 | 10 | Las Vegas-Henderson-Paradise, NV Metro area | 7,891.7 |
| 20 | 11 | St. Louis, MO-IL Metro area | 7,863.7 |
| 47 | 12 | Salt Lake City, UT Metro area | 7,684.0 |
| 24 | 13 | San Antonio-New Braunfels, TX Metro area | 7,313.2 |
| 31 | 14 | Kansas City, MO-KS Metro area | 7,256.5 |
| 3 | 15 | Chicago-Naperville-Elgin, IL-IN-WI Metro area | 7,194.8 |
| 16 | 16 | Minneapolis-St. Paul-Bloomington, MN Metro area | 7,047.1 |
| 25 | 17 | Portland-Vancouver-Hillsboro, OR-WA Metro area | 6,687.5 |
| 1 | 18 | New York-Newark-Jersey City, NY-NJ-PA Metro area | 6,684.3 |
| 6 | 19 | Washington-Arlington-Alexandria, DC-VA-MD-WV Metro area | 6,567.7 |
| 54 | 20 | Tulsa, OK Metro area | 6,270.2 |
| 55 | 21 | Fresno, CA Metro area | 5,958.4 |
| 15 | 22 | Seattle-Tacoma-Bellevue, WA Metro area | 5,869.7 |
| 36 | 23 | Nashville-Davidson--Murfreesboro--Franklin, TN Metro area | 5,689.4 |
| 22 | 24 | Charlotte-Concord-Gastonia, NC-SC Metro area | 5,597.4 |
| 71 | 25 | El Paso, TX Metro area | 5,583.8 |
| 41 | 26 | Oklahoma City, OK Metro area | 5,511.7 |
| 28 | 27 | Pittsburgh, PA Metro area | 5,282.9 |
| 26 | 28 | Sacramento-Roseville-Folsom, CA Metro area | 5,094.9 |
| 7 | 29 | Miami-Fort Lauderdale-Pompano Beach, FL Metro area | 5,066.9 |
| 2 | 30 | Los Angeles-Long Beach-Anaheim, CA Metro area | 4,852.1 |
| 32 | 31 | Columbus, OH Metro area | 4,796.6 |
| 8 | 32 | Philadelphia-Camden-Wilmington, PA-NJ-DE-MD Metro area | 4,603.2 |
| 43 | 33 | Memphis, TN-MS-AR Metro area | 4,575.2 |
| 30 | 34 | Cincinnati, OH-KY-IN Metro area | 4,546.4 |
| 50 | 35 | Birmingham-Hoover, AL Metro area | 4,488.7 |
| 66 | 36 | Baton Rouge, LA Metro area | 4,370.7 |
| 44 | 37 | Richmond, VA Metro area | 4,364.4 |
| 57 | 38 | Omaha-Council Bluffs, NE-IA Metro area | 4,346.3 |
| 33 | 39 | Indianapolis-Carmel-Anderson, IN Metro area | 4,306.6 |
| 29 | 40 | Austin-Round Rock-Georgetown, TX Metro area | 4,219.5 |
| 17 | 41 | San Diego-Chula Vista-Carlsbad, CA Metro area | 4,210.2 |
| 14 | 42 | Detroit-Warren-Dearborn, MI Metro area | 3,892.4 |
| 69 | 43 | Columbia, SC Metro area | 3,703.1 |
| 37 | 44 | Virginia Beach-Norfolk-Newport News, VA-NC Metro area | 3,530.4 |
| 23 | 45 | Orlando-Kissimmee-Sanford, FL Metro area | 3,490.5 |
| 11 | 46 | Boston-Cambridge-Newton, MA-NH Metro area | 3,486.1 |
| 52 | 47 | Rochester, NY Metro area | 3,266.0 |
| 46 | 48 | Louisville/Jefferson County, KY-IN Metro area | 3,237.0 |
| 64 | 49 | Knoxville, TN Metro area | 3,220.3 |
| 45 | 50 | New Orleans-Metairie, LA Metro area | 3,203.5 |
| 39 | 51 | Jacksonville, FL Metro area | 3,201.8 |
| 63 | 52 | Albany-Schenectady-Troy, NY Metro area | 2,811.7 |
| 60 | 53 | Greenville-Anderson, SC Metro area | 2,709.6 |
| 51 | 54 | Grand Rapids-Kentwood, MI Metro area | 2,689.1 |
| 35 | 55 | San Jose-Sunnyvale-Santa Clara, CA Metro area | 2,679.7 |
| 21 | 56 | Baltimore-Columbia-Towson, MD Metro area | 2,601.5 |
| 73 | 57 | Charleston-North Charleston, SC Metro area | 2,590.2 |
| 18 | 58 | Tampa-St. Petersburg-Clearwater, FL Metro area | 2,515.2 |
| 12 | 59 | San Francisco-Oakland-Berkeley, CA Metro area | 2,470.3 |
| 42 | 60 | Raleigh-Cary, NC Metro area | 2,118.4 |
| 58 | 61 | Worcester, MA-CT Metro area | 2,023.6 |
| 34 | 62 | Cleveland-Elyria, OH Metro area | 1,998.8 |
| 72 | 63 | Oxnard-Thousand Oaks-Ventura, CA Metro area | 1,840.8 |
| 38 | 64 | Providence-Warwick, RI-MA Metro area | 1,587.0 |
| 65 | 65 | McAllen-Edinburg-Mission, TX Metro area | 1,571.0 |
| 49 | 66 | Buffalo-Cheektowaga, NY Metro area | 1,565.1 |
| 48 | 67 | Hartford-East Hartford-Middletown, CT Metro area | 1,514.6 |
| 40 | 68 | Milwaukee-Waukesha, WI Metro area | 1,454.9 |
| 70 | 69 | Allentown-Bethlehem-Easton, PA-NJ Metro area | 1,453.0 |
| 67 | 70 | North Port-Sarasota-Bradenton, FL Metro area | 1,298.8 |
| 74 | 71 | Dayton-Kettering, OH Metro area | 1,281.4 |
| 75 | 72 | Cape Coral-Fort Myers, FL Metro area | 781.0 |
| 59 | 73 | Bridgeport-Stamford-Norwalk, CT Metro area | 625.0 |
| 68 | 74 | New Haven-Milford, CT Metro area | 604.3 |
| 56 | 75 | Urban Honolulu, HI Metro area | 600.6 |

# 75 Largest Metropolitan Areas by 2020 Population
## Selected Rankings

### Population density, 2020

| Population rank | Density rank | Metropolitan area | Density (per square kilometer) [col 4] |
|---|---|---|---|
| 1 | 1 | New York-Newark-Jersey City, NY-NJ-PA Metro area | 2,861 |
| 2 | 2 | Los Angeles-Long Beach-Anaheim, CA Metro area | 2,702 |
| 12 | 3 | San Francisco-Oakland-Berkeley, CA Metro area | 1,901 |
| 56 | 4 | Urban Honolulu, HI Metro area | 1,605 |
| 59 | 5 | Bridgeport-Stamford-Norwalk, CT Metro area | 1,508 |
| 68 | 6 | New Haven-Milford, CT Metro area | 1,409 |
| 11 | 7 | Boston-Cambridge-Newton, MA-NH Metro area | 1,399 |
| 8 | 8 | Philadelphia-Camden-Wilmington, PA-NJ-DE-MD Metro area | 1,327 |
| 3 | 9 | Chicago-Naperville-Elgin, IL-IN-WI Metro area | 1,307 |
| 18 | 10 | Tampa-St. Petersburg-Clearwater, FL Metro area | 1,290 |
| 7 | 11 | Miami-Fort Lauderdale-Pompano Beach, FL Metro area | 1,218 |
| 14 | 12 | Detroit-Warren-Dearborn, MI Metro area | 1,106 |
| 40 | 13 | Milwaukee-Waukesha, WI Metro area | 1,084 |
| 21 | 14 | Baltimore-Columbia-Towson, MD Metro area | 1,076 |
| 38 | 15 | Providence-Warwick, RI-MA Metro area | 1,023 |
| 34 | 16 | Cleveland-Elyria, OH Metro area | 1,022 |
| 75 | 17 | Cape Coral-Fort Myers, FL Metro area | 1,012 |
| 6 | 18 | Washington-Arlington-Alexandria, DC-VA-MD-WV Metro area | 963 |
| 4 | 19 | Dallas-Fort Worth-Arlington, TX Metro area | 887 |
| 5 | 20 | Houston-The Woodlands-Sugar Land, TX Metro area | 865 |
| 48 | 21 | Hartford-East Hartford-Middletown, CT Metro area | 793 |
| 17 | 22 | San Diego-Chula Vista-Carlsbad, CA Metro area | 792 |
| 23 | 23 | Orlando-Kissimmee-Sanford, FL Metro area | 756 |
| 35 | 24 | San Jose-Sunnyvale-Santa Clara, CA Metro area | 736 |
| 49 | 25 | Buffalo-Cheektowaga, NY Metro area | 719 |
| 9 | 26 | Atlanta-Sandy Springs-Alpharetta, GA Metro area | 701 |
| 15 | 27 | Seattle-Tacoma-Bellevue, WA Metro area | 685 |
| 42 | 28 | Raleigh-Cary, NC Metro area | 671 |
| 67 | 29 | North Port-Sarasota-Bradenton, FL Metro area | 658 |
| 74 | 30 | Dayton-Kettering, OH Metro area | 631 |
| 70 | 31 | Allentown-Bethlehem-Easton, PA-NJ Metro area | 583 |
| 65 | 32 | McAllen-Edinburg-Mission, TX Metro area | 557 |
| 29 | 33 | Austin-Round Rock-Georgetown, TX Metro area | 544 |
| 16 | 34 | Minneapolis-St. Paul-Bloomington, MN Metro area | 519 |
| 37 | 35 | Virginia Beach-Norfolk-Newport News, VA-NC Metro area | 504 |
| 39 | 36 | Jacksonville, FL Metro area | 496 |
| 30 | 37 | Cincinnati, OH-KY-IN Metro area | 491 |
| 33 | 38 | Indianapolis-Carmel-Anderson, IN Metro area | 486 |
| 22 | 39 | Charlotte-Concord-Gastonia, NC-SC Metro area | 480 |
| 58 | 40 | Worcester, MA-CT Metro area | 467 |
| 26 | 41 | Sacramento-Roseville-Folsom, CA Metro area | 466 |
| 72 | 42 | Oxnard-Thousand Oaks-Ventura, CA Metro area | 457 |
| 32 | 43 | Columbus, OH Metro area | 446 |
| 28 | 44 | Pittsburgh, PA Metro area | 437 |
| 51 | 45 | Grand Rapids-Kentwood, MI Metro area | 402 |
| 45 | 46 | New Orleans-Metairie, LA Metro area | 397 |
| 46 | 47 | Louisville/Jefferson County, KY-IN Metro area | 392 |
| 25 | 48 | Portland-Vancouver-Hillsboro, OR-WA Metro area | 375 |
| 19 | 49 | Denver-Aurora-Lakewood, CO Metro area | 358 |
| 20 | 50 | St. Louis, MO-IL Metro area | 357 |
| 24 | 51 | San Antonio-New Braunfels, TX Metro area | 354 |
| 10 | 52 | Phoenix-Mesa-Chandler, AZ Metro area | 347 |
| 36 | 53 | Nashville-Davidson--Murfreesboro--Franklin, TN Metro area | 345 |
| 60 | 54 | Greenville-Anderson, SC Metro area | 344 |
| 52 | 55 | Rochester, NY Metro area | 327 |
| 73 | 56 | Charleston-North Charleston, SC Metro area | 317 |
| 63 | 57 | Albany-Schenectady-Troy, NY Metro area | 313 |
| 31 | 58 | Kansas City, MO-KS Metro area | 300 |
| 44 | 59 | Richmond, VA Metro area | 299 |
| 43 | 60 | Memphis, TN-MS-AR Metro area | 295 |
| 27 | 61 | Las Vegas-Henderson-Paradise, NV Metro area | 294 |
| 64 | 62 | Knoxville, TN Metro area | 273 |
| 41 | 63 | Oklahoma City, OK Metro area | 259 |
| 50 | 64 | Birmingham-Hoover, AL Metro area | 243 |
| 69 | 65 | Columbia, SC Metro area | 229 |
| 57 | 66 | Omaha-Council Bluffs, NE-IA Metro area | 220 |
| 66 | 67 | Baton Rouge, LA Metro area | 197 |
| 13 | 68 | Riverside-San Bernardino-Ontario, CA Metro area | 172 |
| 55 | 69 | Fresno, CA Metro area | 168 |
| 47 | 70 | Salt Lake City, UT Metro area | 161 |
| 54 | 71 | Tulsa, OK Metro area | 161 |
| 71 | 72 | El Paso, TX Metro area | 152 |
| 53 | 73 | Tucson, AZ Metro area | 116 |
| 62 | 74 | Bakersfield, CA Metro area | 111 |
| 61 | 75 | Albuquerque, NM Metro area | 100 |

### Percent population change, 2010–2020

| Population rank | Percent change rank | Metropolitan area | Percent change [col 23] |
|---|---|---|---|
| 29 | 1 | Austin-Round Rock-Georgetown, TX Metro area | 33.7 |
| 75 | 2 | Cape Coral-Fort Myers, FL Metro area | 27.8 |
| 42 | 3 | Raleigh-Cary, NC Metro area | 25.6 |
| 23 | 4 | Orlando-Kissimmee-Sanford, FL Metro area | 23.7 |
| 73 | 5 | Charleston-North Charleston, SC Metro area | 23.3 |
| 67 | 6 | North Port-Sarasota-Bradenton, FL Metro area | 21.7 |
| 4 | 7 | Dallas-Fort Worth-Arlington, TX Metro area | 20.9 |
| 24 | 7 | San Antonio-New Braunfels, TX Metro area | 20.9 |
| 5 | 9 | Houston-The Woodlands-Sugar Land, TX Metro area | 20.8 |
| 10 | 10 | Phoenix-Mesa-Chandler, AZ Metro area | 20.7 |
| 22 | 11 | Charlotte-Concord-Gastonia, NC-SC Metro area | 19.6 |
| 36 | 12 | Nashville-Davidson--Murfreesboro--Franklin, TN Metro area | 19.1 |
| 27 | 13 | Las Vegas-Henderson-Paradise, NV Metro area | 18.7 |
| 39 | 14 | Jacksonville, FL Metro area | 18.0 |
| 19 | 15 | Denver-Aurora-Lakewood, CO Metro area | 17.6 |
| 15 | 16 | Seattle-Tacoma-Bellevue, WA Metro area | 16.8 |
| 18 | 17 | Tampa-St. Petersburg-Clearwater, FL Metro area | 16.5 |
| 9 | 18 | Atlanta-Sandy Springs-Alpharetta, GA Metro area | 15.2 |
| 47 | 19 | Salt Lake City, UT Metro area | 14.0 |
| 41 | 20 | Oklahoma City, OK Metro area | 13.8 |
| 60 | 21 | Greenville-Anderson, SC Metro area | 13.2 |
| 65 | 22 | McAllen-Edinburg-Mission, TX Metro area | 13.0 |
| 25 | 23 | Portland-Vancouver-Hillsboro, OR-WA Metro area | 12.8 |
| 32 | 24 | Columbus, OH Metro area | 12.5 |
| 6 | 25 | Washington-Arlington-Alexandria, DC-VA-MD-WV Metro area | 11.9 |
| 7 | 26 | Miami-Fort Lauderdale-Pompano Beach, FL Metro area | 10.9 |
| 13 | 27 | Riverside-San Bernardino-Ontario, CA Metro area | 10.7 |
| 33 | 27 | Indianapolis-Carmel-Anderson, IN Metro area | 10.7 |
| 26 | 29 | Sacramento-Roseville-Folsom, CA Metro area | 10.5 |
| 69 | 30 | Columbia, SC Metro area | 10.4 |
| 57 | 31 | Omaha-Council Bluffs, NE-IA Metro area | 10.3 |
| 44 | 32 | Richmond, VA Metro area | 9.8 |
| 16 | 33 | Minneapolis-St. Paul-Bloomington, MN Metro area | 9.7 |
| 51 | 34 | Grand Rapids-Kentwood, MI Metro area | 8.8 |
| 12 | 35 | San Francisco-Oakland-Berkeley, CA Metro area | 8.3 |
| 53 | 35 | Tucson, AZ Metro area | 8.3 |
| 31 | 37 | Kansas City, MO-KS Metro area | 8.2 |
| 17 | 38 | San Diego-Chula Vista-Carlsbad, CA Metro area | 7.7 |
| 64 | 38 | Knoxville, TN Metro area | 7.7 |
| 55 | 40 | Fresno, CA Metro area | 7.6 |
| 62 | 41 | Bakersfield, CA Metro area | 7.4 |
| 35 | 42 | San Jose-Sunnyvale-Santa Clara, CA Metro area | 7.3 |
| 54 | 42 | Tulsa, OK Metro area | 7.3 |
| 11 | 44 | Boston-Cambridge-Newton, MA-NH Metro area | 7.2 |
| 45 | 45 | New Orleans-Metairie, LA Metro area | 6.9 |
| 46 | 46 | Louisville/Jefferson County, KY-IN Metro area | 5.5 |
| 71 | 47 | El Paso, TX Metro area | 5.2 |
| 30 | 48 | Cincinnati, OH-KY-IN Metro area | 4.5 |
| 61 | 49 | Albuquerque, NM Metro area | 4.1 |
| 66 | 50 | Baton Rouge, LA Metro area | 4.0 |
| 37 | 51 | Virginia Beach-Norfolk-Newport News, VA-NC Metro area | 3.8 |
| 21 | 52 | Baltimore-Columbia-Towson, MD Metro area | 3.3 |
| 58 | 53 | Worcester, MA-CT Metro area | 3.2 |
| 70 | 54 | Allentown-Bethlehem-Easton, PA-NJ Metro area | 3.1 |
| 50 | 55 | Birmingham-Hoover, AL Metro area | 2.9 |
| 59 | 56 | Bridgeport-Stamford-Norwalk, CT Metro area | 2.8 |
| 43 | 57 | Memphis, TN-MS-AR Metro area | 2.5 |
| 8 | 58 | Philadelphia-Camden-Wilmington, PA-NJ-DE-MD Metro area | 2.4 |
| 2 | 59 | Los Angeles-Long Beach-Anaheim, CA Metro area | 2.2 |
| 72 | 59 | Oxnard-Thousand Oaks-Ventura, CA Metro area | 2.2 |
| 38 | 61 | Providence-Warwick, RI-MA Metro area | 1.4 |
| 40 | 61 | Milwaukee-Waukesha, WI Metro area | 1.4 |
| 1 | 63 | New York-Newark-Jersey City, NY-NJ-PA Metro area | 1.2 |
| 74 | 63 | Dayton-Kettering, OH Metro area | 1.2 |
| 56 | 65 | Urban Honolulu, HI Metro area | 1.1 |
| 63 | 66 | Albany-Schenectady-Troy, NY Metro area | 0.9 |
| 20 | 67 | St. Louis, MO-IL Metro area | 0.6 |
| 14 | 68 | Detroit-Warren-Dearborn, MI Metro area | 0.2 |
| 3 | 69 | Chicago-Naperville-Elgin, IL-IN-WI Metro area | -0.6 |
| 48 | 70 | Hartford-East Hartford-Middletown, CT Metro area | -0.9 |
| 49 | 70 | Buffalo-Cheektowaga, NY Metro area | -0.9 |
| 52 | 72 | Rochester, NY Metro area | -1.1 |
| 68 | 73 | New Haven-Milford, CT Metro area | -1.2 |
| 34 | 74 | Cleveland-Elyria, OH Metro area | -1.6 |
| 28 | 75 | Pittsburgh, PA Metro area | -2.0 |

# 75 Largest Metropolitan Areas by 2020 Population
## Selected Rankings

| Population rank | White rank | Metropolitan area — Percent White, not Hispanic or Latino, alone or in combination, 2020 | Percent White [col 5] | Population rank | Black rank | Metropolitan area — Percent Black, not Hispanic or Latino, alone or in combination, 2020 | Percent Black [col 6] |
|---|---|---|---|---|---|---|---|
| 64 | 1 | Knoxville, TN Metro area | 87.7 | 43 | 1 | Memphis, TN-MS-AR Metro area | 48.6 |
| 28 | 2 | Pittsburgh, PA Metro area | 86.6 | 66 | 2 | Baton Rouge, LA Metro area | 36.5 |
| 63 | 3 | Albany-Schenectady-Troy, NY Metro area | 81.0 | 9 | 3 | Atlanta-Sandy Springs-Alpharetta, GA Metro area | 35.9 |
| 30 | 4 | Cincinnati, OH-KY-IN Metro area | 80.6 | 45 | 3 | New Orleans-Metairie, LA Metro area | 35.9 |
| 51 | 5 | Grand Rapids-Kentwood, MI Metro area | 79.8 | 69 | 5 | Columbia, SC Metro area | 35.1 |
| 49 | 6 | Buffalo-Cheektowaga, NY Metro area | 78.2 | 37 | 6 | Virginia Beach-Norfolk-Newport News, VA-NC Metro area | 32.2 |
| 67 | 6 | North Port-Sarasota-Bradenton, FL Metro area | 78.2 | 50 | 7 | Birmingham-Hoover, AL Metro area | 31.2 |
| 74 | 8 | Dayton-Kettering, OH Metro area | 78.1 | 21 | 8 | Baltimore-Columbia-Towson, MD Metro area | 31.1 |
| 58 | 9 | Worcester, MA-CT Metro area | 77.7 | 44 | 9 | Richmond, VA Metro area | 30.8 |
| 52 | 10 | Rochester, NY Metro area | 77.6 | 6 | 10 | Washington-Arlington-Alexandria, DC-VA-MD-WV Metro area | 26.6 |
| 57 | 11 | Omaha-Council Bluffs, NE-IA Metro area | 77.2 | 73 | 11 | Charleston-North Charleston, SC Metro area | 26.1 |
| 16 | 12 | Minneapolis-St. Paul-Bloomington, MN Metro area | 76.7 | 22 | 12 | Charlotte-Concord-Gastonia, NC-SC Metro area | 24.2 |
| 46 | 12 | Louisville/Jefferson County, KY-IN Metro area | 76.7 | 14 | 13 | Detroit-Warren-Dearborn, MI Metro area | 23.3 |
| 38 | 14 | Providence-Warwick, RI-MA Metro area | 76.3 | 39 | 14 | Jacksonville, FL Metro area | 22.5 |
| 25 | 15 | Portland-Vancouver-Hillsboro, OR-WA Metro area | 75.5 | 8 | 15 | Philadelphia-Camden-Wilmington, PA-NJ-DE-MD Metro area | 21.9 |
| 20 | 16 | St. Louis, MO-IL Metro area | 75.2 | 7 | 16 | Miami-Fort Lauderdale-Pompano Beach, FL Metro area | 21.1 |
| 31 | 17 | Kansas City, MO-KS Metro area | 74.3 | 34 | 16 | Cleveland-Elyria, OH Metro area | 21.1 |
| 32 | 18 | Columbus, OH Metro area | 73.7 | 42 | 18 | Raleigh-Cary, NC Metro area | 20.7 |
| 60 | 19 | Greenville-Anderson, SC Metro area | 73.6 | 20 | 19 | St. Louis, MO-IL Metro area | 19.4 |
| 36 | 20 | Nashville-Davidson--Murfreesboro--Franklin, TN Metro area | 73.4 | 5 | 20 | Houston-The Woodlands-Sugar Land, TX Metro area | 18.0 |
| 33 | 21 | Indianapolis-Carmel-Anderson, IN Metro area | 72.9 | 32 | 20 | Columbus, OH Metro area | 18.0 |
| 47 | 21 | Salt Lake City, UT Metro area | 72.9 | 40 | 22 | Milwaukee-Waukesha, WI Metro area | 17.6 |
| 70 | 23 | Allentown-Bethlehem-Easton, PA-NJ Metro area | 72.2 | 74 | 22 | Dayton-Kettering, OH Metro area | 17.6 |
| 11 | 24 | Boston-Cambridge-Newton, MA-NH Metro area | 71.1 | 60 | 24 | Greenville-Anderson, SC Metro area | 17.5 |
| 34 | 25 | Cleveland-Elyria, OH Metro area | 71.0 | 4 | 25 | Dallas-Fort Worth-Arlington, TX Metro area | 17.2 |
| 54 | 26 | Tulsa, OK Metro area | 69.5 | 3 | 26 | Chicago-Naperville-Elgin, IL-IN-WI Metro area | 17.1 |
| 14 | 27 | Detroit-Warren-Dearborn, MI Metro area | 67.7 | 33 | 26 | Indianapolis-Carmel-Anderson, IN Metro area | 17.1 |
| 41 | 27 | Oklahoma City, OK Metro area | 67.7 | 23 | 28 | Orlando-Kissimmee-Sanford, FL Metro area | 16.9 |
| 40 | 29 | Milwaukee-Waukesha, WI Metro area | 67.4 | 1 | 29 | New York-Newark-Jersey City, NY-NJ-PA Metro area | 16.8 |
| 48 | 30 | Hartford-East Hartford-Middletown, CT Metro area | 67.1 | 46 | 30 | Louisville/Jefferson County, KY-IN Metro area | 16.6 |
| 75 | 31 | Cape Coral-Fort Myers, FL Metro area | 66.9 | 36 | 31 | Nashville-Davidson--Murfreesboro--Franklin, TN Metro area | 16.4 |
| 73 | 32 | Charleston-North Charleston, SC Metro area | 66.2 | 68 | 32 | New Haven-Milford, CT Metro area | 14.4 |
| 19 | 33 | Denver-Aurora-Lakewood, CO Metro area | 65.7 | 30 | 33 | Cincinnati, OH-KY-IN Metro area | 13.8 |
| 15 | 34 | Seattle-Tacoma-Bellevue, WA Metro area | 65.4 | 31 | 33 | Kansas City, MO-KS Metro area | 13.8 |
| 39 | 35 | Jacksonville, FL Metro area | 64.1 | 27 | 35 | Las Vegas-Henderson-Paradise, NV Metro area | 13.6 |
| 18 | 36 | Tampa-St. Petersburg-Clearwater, FL Metro area | 62.9 | 49 | 36 | Buffalo-Cheektowaga, NY Metro area | 13.2 |
| 50 | 37 | Birmingham-Hoover, AL Metro area | 62.7 | 18 | 37 | Tampa-St. Petersburg-Clearwater, FL Metro area | 13.1 |
| 68 | 38 | New Haven-Milford, CT Metro area | 62.5 | 52 | 38 | Rochester, NY Metro area | 12.3 |
| 8 | 39 | Philadelphia-Camden-Wilmington, PA-NJ-DE-MD Metro area | 62.4 | 41 | 39 | Oklahoma City, OK Metro area | 12.2 |
| 42 | 39 | Raleigh-Cary, NC Metro area | 62.4 | 48 | 39 | Hartford-East Hartford-Middletown, CT Metro area | 12.2 |
| 59 | 41 | Bridgeport-Stamford-Norwalk, CT Metro area | 61.6 | 59 | 41 | Bridgeport-Stamford-Norwalk, CT Metro area | 12.1 |
| 22 | 42 | Charlotte-Concord-Gastonia, NC-SC Metro area | 61.4 | 16 | 42 | Minneapolis-St. Paul-Bloomington, MN Metro area | 10.7 |
| 44 | 43 | Richmond, VA Metro area | 58.7 | 28 | 43 | Pittsburgh, PA Metro area | 9.8 |
| 37 | 44 | Virginia Beach-Norfolk-Newport News, VA-NC Metro area | 57.4 | 54 | 43 | Tulsa, OK Metro area | 9.8 |
| 21 | 45 | Baltimore-Columbia-Towson, MD Metro area | 57.3 | 63 | 45 | Albany-Schenectady-Troy, NY Metro area | 9.6 |
| 66 | 46 | Baton Rouge, LA Metro area | 57.2 | 57 | 46 | Omaha-Council Bluffs, NE-IA Metro area | 9.0 |
| 69 | 46 | Columbia, SC Metro area | 57.2 | 11 | 47 | Boston-Cambridge-Newton, MA-NH Metro area | 8.9 |
| 10 | 48 | Phoenix-Mesa-Chandler, AZ Metro area | 56.4 | 75 | 48 | Cape Coral-Fort Myers, FL Metro area | 8.8 |
| 26 | 49 | Sacramento-Roseville-Folsom, CA Metro area | 54.6 | 26 | 49 | Sacramento-Roseville-Folsom, CA Metro area | 8.7 |
| 3 | 50 | Chicago-Naperville-Elgin, IL-IN-WI Metro area | 53.5 | 12 | 50 | San Francisco-Oakland-Berkeley, CA Metro area | 8.3 |
| 29 | 51 | Austin-Round Rock-Georgetown, TX Metro area | 53.0 | 51 | 51 | Grand Rapids-Kentwood, MI Metro area | 8.2 |
| 53 | 52 | Tucson, AZ Metro area | 52.7 | 13 | 52 | Riverside-San Bernardino-Ontario, CA Metro area | 8.1 |
| 45 | 53 | New Orleans-Metairie, LA Metro area | 52.3 | 29 | 52 | Austin-Round Rock-Georgetown, TX Metro area | 8.1 |
| 17 | 54 | San Diego-Chula Vista-Carlsbad, CA Metro area | 47.5 | 15 | 54 | Seattle-Tacoma-Bellevue, WA Metro area | 8.0 |
| 9 | 55 | Atlanta-Sandy Springs-Alpharetta, GA Metro area | 47.3 | 24 | 55 | San Antonio-New Braunfels, TX Metro area | 7.4 |
| 6 | 56 | Washington-Arlington-Alexandria, DC-VA-MD-WV Metro area | 47.1 | 2 | 56 | Los Angeles-Long Beach-Anaheim, CA Metro area | 7.1 |
| 23 | 57 | Orlando-Kissimmee-Sanford, FL Metro area | 46.8 | 38 | 56 | Providence-Warwick, RI-MA Metro area | 7.1 |
| 72 | 58 | Oxnard-Thousand Oaks-Ventura, CA Metro area | 46.7 | 67 | 56 | North Port-Sarasota-Bradenton, FL Metro area | 7.1 |
| 1 | 59 | New York-Newark-Jersey City, NY-NJ-PA Metro area | 46.4 | 64 | 59 | Knoxville, TN Metro area | 6.8 |
| 4 | 60 | Dallas-Fort Worth-Arlington, TX Metro area | 46.1 | 10 | 60 | Phoenix-Mesa-Chandler, AZ Metro area | 6.7 |
| 27 | 61 | Las Vegas-Henderson-Paradise, NV Metro area | 44.2 | 19 | 60 | Denver-Aurora-Lakewood, CO Metro area | 6.7 |
| 43 | 62 | Memphis, TN-MS-AR Metro area | 43.7 | 70 | 62 | Allentown-Bethlehem-Easton, PA-NJ Metro area | 6.5 |
| 12 | 63 | San Francisco-Oakland-Berkeley, CA Metro area | 41.7 | 62 | 63 | Bakersfield, CA Metro area | 5.9 |
| 61 | 64 | Albuquerque, NM Metro area | 39.7 | 17 | 64 | San Diego-Chula Vista-Carlsbad, CA Metro area | 5.8 |
| 5 | 65 | Houston-The Woodlands-Sugar Land, TX Metro area | 36.0 | 58 | 65 | Worcester, MA-CT Metro area | 5.6 |
| 24 | 66 | San Antonio-New Braunfels, TX Metro area | 34.3 | 55 | 66 | Fresno, CA Metro area | 5.3 |
| 62 | 67 | Bakersfield, CA Metro area | 33.9 | 53 | 67 | Tucson, AZ Metro area | 4.3 |
| 35 | 68 | San Jose-Sunnyvale-Santa Clara, CA Metro area | 32.6 | 25 | 68 | Portland-Vancouver-Hillsboro, OR-WA Metro area | 4.2 |
| 13 | 69 | Riverside-San Bernardino-Ontario, CA Metro area | 32.0 | 56 | 69 | Urban Honolulu, HI Metro area | 3.7 |
| 56 | 70 | Urban Honolulu, HI Metro area | 31.7 | 71 | 70 | El Paso, TX Metro area | 3.6 |
| 2 | 71 | Los Angeles-Long Beach-Anaheim, CA Metro area | 31.2 | 61 | 71 | Albuquerque, NM Metro area | 3.2 |
| 7 | 72 | Miami-Fort Lauderdale-Pompano Beach, FL Metro area | 30.7 | 35 | 72 | San Jose-Sunnyvale-Santa Clara, CA Metro area | 3.0 |
| 55 | 73 | Fresno, CA Metro area | 29.6 | 47 | 73 | Salt Lake City, UT Metro area | 2.4 |
| 71 | 74 | El Paso, TX Metro area | 12.4 | 72 | 73 | Oxnard-Thousand Oaks-Ventura, CA Metro area | 2.4 |
| 65 | 75 | McAllen-Edinburg-Mission, TX Metro area | 6.0 | 65 | 75 | McAllen-Edinburg-Mission, TX Metro area | 0.5 |

# 75 Largest Metropolitan Areas by 2020 Population
## Selected Rankings

### Percent American Indian, Alaska Native, alone or in combination, 2020

| Population rank | American Indian Alaska native rank | Metropolitan area | Percent American Indian, Alaska Native [col 7] |
|---|---|---|---|
| 54 | 1 | Tulsa, OK Metro area | 13.2 |
| 41 | 2 | Oklahoma City, OK Metro area | 6.9 |
| 61 | 3 | Albuquerque, NM Metro area | 6.1 |
| 53 | 4 | Tucson, AZ Metro area | 3.0 |
| 10 | 5 | Phoenix-Mesa-Chandler, AZ Metro area | 2.4 |
| 15 | 6 | Seattle-Tacoma-Bellevue, WA Metro area | 2.0 |
| 25 | 7 | Portland-Vancouver-Hillsboro, OR-WA Metro area | 1.7 |
| 26 | 8 | Sacramento-Roseville-Folsom, CA Metro area | 1.5 |
| 56 | 8 | Urban Honolulu, HI Metro area | 1.5 |
| 16 | 10 | Minneapolis-St. Paul-Bloomington, MN Metro area | 1.3 |
| 31 | 11 | Kansas City, MO-KS Metro area | 1.2 |
| 62 | 11 | Bakersfield, CA Metro area | 1.2 |
| 19 | 13 | Denver-Aurora-Lakewood, CO Metro area | 1.1 |
| 27 | 13 | Las Vegas-Henderson-Paradise, NV Metro area | 1.1 |
| 37 | 13 | Virginia Beach-Norfolk-Newport News, VA-NC Metro area | 1.1 |
| 47 | 13 | Salt Lake City, UT Metro area | 1.1 |
| 49 | 13 | Buffalo-Cheektowaga, NY Metro area | 1.1 |
| 55 | 13 | Fresno, CA Metro area | 1.0 |
| 57 | 19 | Omaha-Council Bluffs, NE-IA Metro area | 0.9 |
| 4 | 20 | Dallas-Fort Worth-Arlington, TX Metro area | 0.9 |
| 13 | 20 | Riverside-San Bernardino-Ontario, CA Metro area | 0.9 |
| 14 | 20 | Detroit-Warren-Dearborn, MI Metro area | 0.9 |
| 17 | 20 | San Diego-Chula Vista-Carlsbad, CA Metro area | 0.9 |
| 22 | 20 | Charlotte-Concord-Gastonia, NC-SC Metro area | 0.9 |
| 38 | 20 | Providence-Warwick, RI-MA Metro area | 0.9 |
| 40 | 20 | Milwaukee-Waukesha, WI Metro area | 0.9 |
| 42 | 20 | Raleigh-Cary, NC Metro area | 0.9 |
| 44 | 20 | Richmond, VA Metro area | 0.9 |
| 45 | 20 | New Orleans-Metairie, LA Metro area | 0.9 |
| 51 | 20 | Grand Rapids-Kentwood, MI Metro area | 0.9 |
| 64 | 20 | Knoxville, TN Metro area | 0.9 |
| 73 | 20 | Charleston-North Charleston, SC Metro area | 0.9 |
| 74 | 20 | Dayton-Kettering, OH Metro area | 0.9 |
| 6 | 34 | Washington-Arlington-Alexandria, DC-VA-MD-WV Metro area | 0.8 |
| 12 | 34 | San Francisco-Oakland-Berkeley, CA Metro area | 0.8 |
| 21 | 34 | Baltimore-Columbia-Towson, MD Metro area | 0.8 |
| 29 | 34 | Austin-Round Rock-Georgetown, TX Metro area | 0.8 |
| 32 | 34 | Columbus, OH Metro area | 0.8 |
| 36 | 34 | Nashville-Davidson--Murfreesboro--Franklin, TN Metro area | 0.8 |
| 39 | 34 | Jacksonville, FL Metro area | 0.8 |
| 69 | 34 | Columbia, SC Metro area | 0.8 |
| 72 | 34 | Oxnard-Thousand Oaks-Ventura, CA Metro area | 0.8 |
| 9 | 43 | Atlanta-Sandy Springs-Alpharetta, GA Metro area | 0.7 |
| 18 | 43 | Tampa-St. Petersburg-Clearwater, FL Metro area | 0.7 |
| 20 | 43 | St. Louis, MO-IL Metro area | 0.7 |
| 33 | 43 | Indianapolis-Carmel-Anderson, IN Metro area | 0.7 |
| 35 | 43 | San Jose-Sunnyvale-Santa Clara, CA Metro area | 0.7 |
| 46 | 43 | Louisville/Jefferson County, KY-IN Metro area | 0.7 |
| 50 | 43 | Birmingham-Hoover, AL Metro area | 0.7 |
| 66 | 43 | Baton Rouge, LA Metro area | 0.7 |
| 2 | 51 | Los Angeles-Long Beach-Anaheim, CA Metro area | 0.6 |
| 5 | 51 | Houston-The Woodlands-Sugar Land, TX Metro area | 0.6 |
| 8 | 51 | Philadelphia-Camden-Wilmington, PA-NJ-DE-MD Metro area | 0.6 |
| 23 | 51 | Orlando-Kissimmee-Sanford, FL Metro area | 0.6 |
| 24 | 51 | San Antonio-New Braunfels, TX Metro area | 0.6 |
| 30 | 51 | Cincinnati, OH-KY-IN Metro area | 0.6 |
| 34 | 51 | Cleveland-Elyria, OH Metro area | 0.6 |
| 43 | 51 | Memphis, TN-MS-AR Metro area | 0.6 |
| 48 | 51 | Hartford-East Hartford-Middletown, CT Metro area | 0.6 |
| 52 | 51 | Rochester, NY Metro area | 0.6 |
| 58 | 51 | Worcester, MA-CT Metro area | 0.6 |
| 60 | 51 | Greenville-Anderson, SC Metro area | 0.6 |
| 63 | 51 | Albany-Schenectady-Troy, NY Metro area | 0.6 |
| 67 | 51 | North Port-Sarasota-Bradenton, FL Metro area | 0.6 |
| 68 | 51 | New Haven-Milford, CT Metro area | 0.6 |
| 1 | 66 | New York-Newark-Jersey City, NY-NJ-PA Metro area | 0.5 |
| 3 | 66 | Chicago-Naperville-Elgin, IL-IN-WI Metro area | 0.5 |
| 11 | 66 | Boston-Cambridge-Newton, MA-NH Metro area | 0.5 |
| 28 | 66 | Pittsburgh, PA Metro area | 0.5 |
| 70 | 66 | Allentown-Bethlehem-Easton, PA-NJ Metro area | 0.5 |
| 71 | 66 | El Paso, TX Metro area | 0.5 |
| 75 | 66 | Cape Coral-Fort Myers, FL Metro area | 0.4 |
| 59 | 73 | Bridgeport-Stamford-Norwalk, CT Metro area | 0.3 |
| 7 | 74 | Miami-Fort Lauderdale-Pompano Beach, FL Metro area | 0.1 |
| 65 | 75 | McAllen-Edinburg-Mission, TX Metro area | |

### Percent Asian and Pacific Islander, alone or in combination, 2020

| Population rank | Asian and Pacific Islander rank | Metropolitan area | Percent Asian and Pacific Islander [col 8] |
|---|---|---|---|
| 56 | 1 | Urban Honolulu, HI Metro area | 79.0 |
| 35 | 2 | San Jose-Sunnyvale-Santa Clara, CA Metro area | 41.2 |
| 12 | 3 | San Francisco-Oakland-Berkeley, CA Metro area | 31.8 |
| 15 | 4 | Seattle-Tacoma-Bellevue, WA Metro area | 19.7 |
| 2 | 5 | Los Angeles-Long Beach-Anaheim, CA Metro area | 18.7 |
| 26 | 6 | Sacramento-Roseville-Folsom, CA Metro area | 18.1 |
| 17 | 7 | San Diego-Chula Vista-Carlsbad, CA Metro area | 15.3 |
| 27 | 8 | Las Vegas-Henderson-Paradise, NV Metro area | 13.7 |
| 1 | 9 | New York-Newark-Jersey City, NY-NJ-PA Metro area | 13.0 |
| 6 | 10 | Washington-Arlington-Alexandria, DC-VA-MD-WV Metro area | 12.5 |
| 55 | 11 | Fresno, CA Metro area | 12.0 |
| 25 | 12 | Portland-Vancouver-Hillsboro, OR-WA Metro area | 10.1 |
| 11 | 13 | Boston-Cambridge-Newton, MA-NH Metro area | 9.8 |
| 72 | 14 | Oxnard-Thousand Oaks-Ventura, CA Metro area | 9.4 |
| 5 | 15 | Houston-The Woodlands-Sugar Land, TX Metro area | 8.9 |
| 13 | 15 | Riverside-San Bernardino-Ontario, CA Metro area | 8.9 |
| 4 | 17 | Dallas-Fort Worth-Arlington, TX Metro area | 8.5 |
| 16 | 18 | Minneapolis-St. Paul-Bloomington, MN Metro area | 8.2 |
| 3 | 19 | Chicago-Naperville-Elgin, IL-IN-WI Metro area | 8.0 |
| 29 | 20 | Austin-Round Rock-Georgetown, TX Metro area | 7.8 |
| 42 | 21 | Raleigh-Cary, NC Metro area | 7.5 |
| 47 | 22 | Salt Lake City, UT Metro area | 7.4 |
| 9 | 23 | Atlanta-Sandy Springs-Alpharetta, GA Metro area | 7.3 |
| 8 | 24 | Philadelphia-Camden-Wilmington, PA-NJ-DE-MD Metro area | 7.2 |
| 21 | 24 | Baltimore-Columbia-Towson, MD Metro area | 7.2 |
| 59 | 26 | Bridgeport-Stamford-Norwalk, CT Metro area | 6.6 |
| 48 | 27 | Hartford-East Hartford-Middletown, CT Metro area | 6.3 |
| 62 | 28 | Bakersfield, CA Metro area | 6.0 |
| 19 | 29 | Denver-Aurora-Lakewood, CO Metro area | 5.9 |
| 32 | 29 | Columbus, OH Metro area | 5.9 |
| 14 | 31 | Detroit-Warren-Dearborn, MI Metro area | 5.7 |
| 37 | 31 | Virginia Beach-Norfolk-Newport News, VA-NC Metro area | 5.7 |
| 10 | 33 | Phoenix-Mesa-Chandler, AZ Metro area | 5.6 |
| 58 | 33 | Worcester, MA-CT Metro area | 5.6 |
| 63 | 33 | Albany-Schenectady-Troy, NY Metro area | 5.6 |
| 23 | 36 | Orlando-Kissimmee-Sanford, FL Metro area | 5.5 |
| 39 | 37 | Jacksonville, FL Metro area | 5.4 |
| 44 | 38 | Richmond, VA Metro area | 5.2 |
| 22 | 39 | Charlotte-Concord-Gastonia, NC-SC Metro area | 4.9 |
| 40 | 39 | Milwaukee-Waukesha, WI Metro area | 4.9 |
| 68 | 39 | New Haven-Milford, CT Metro area | 4.9 |
| 18 | 42 | Tampa-St. Petersburg-Clearwater, FL Metro area | 4.6 |
| 33 | 43 | Indianapolis-Carmel-Anderson, IN Metro area | 4.5 |
| 41 | 44 | Oklahoma City, OK Metro area | 4.4 |
| 49 | 45 | Buffalo-Cheektowaga, NY Metro area | 4.2 |
| 53 | 45 | Tucson, AZ Metro area | 4.2 |
| 57 | 45 | Omaha-Council Bluffs, NE-IA Metro area | 4.2 |
| 31 | 48 | Kansas City, MO-KS Metro area | 4.1 |
| 38 | 49 | Providence-Warwick, RI-MA Metro area | 3.9 |
| 36 | 50 | Nashville-Davidson--Murfreesboro--Franklin, TN Metro area | 3.8 |
| 70 | 51 | Allentown-Bethlehem-Easton, PA-NJ Metro area | 3.7 |
| 20 | 52 | St. Louis, MO-IL Metro area | 3.6 |
| 30 | 52 | Cincinnati, OH-KY-IN Metro area | 3.6 |
| 52 | 52 | Rochester, NY Metro area | 3.6 |
| 54 | 52 | Tulsa, OK Metro area | 3.6 |
| 24 | 56 | San Antonio-New Braunfels, TX Metro area | 3.5 |
| 45 | 56 | New Orleans-Metairie, LA Metro area | 3.5 |
| 51 | 56 | Grand Rapids-Kentwood, MI Metro area | 3.5 |
| 74 | 59 | Dayton-Kettering, OH Metro area | 3.3 |
| 7 | 60 | Miami-Fort Lauderdale-Pompano Beach, FL Metro area | 3.3 |
| 28 | 60 | Pittsburgh, PA Metro area | 3.2 |
| 61 | 60 | Albuquerque, NM Metro area | 3.2 |
| 34 | 63 | Cleveland-Elyria, OH Metro area | 3.1 |
| 69 | 63 | Columbia, SC Metro area | 3.1 |
| 73 | 63 | Charleston-North Charleston, SC Metro area | 3.1 |
| 46 | 66 | Louisville/Jefferson County, KY-IN Metro area | 2.9 |
| 43 | 67 | Memphis, TN-MS-AR Metro area | 2.8 |
| 60 | 68 | Greenville-Anderson, SC Metro area | 2.6 |
| 66 | 68 | Baton Rouge, LA Metro area | 2.6 |
| 67 | 68 | North Port-Sarasota-Bradenton, FL Metro area | 2.6 |
| 75 | 71 | Cape Coral-Fort Myers, FL Metro area | 2.3 |
| 64 | 72 | Knoxville, TN Metro area | 2.2 |
| 50 | 73 | Birmingham-Hoover, AL Metro area | 2.1 |
| 71 | 74 | El Paso, TX Metro area | 1.8 |
| 65 | 75 | McAllen-Edinburg-Mission, TX Metro area | |

# 75 Largest Metropolitan Areas by 2020 Population
## Selected Rankings

### Percent Hispanic or Latino,[1] 2020

| Population rank | Hispanic or Latino rank | Metropolitan area | Percent Hispanic or Latino [col 9] |
|---|---|---|---|
| 65 | 1 | McAllen-Edinburg-Mission, TX Metro area | 92.5 |
| 71 | 2 | El Paso, TX Metro area | 82.7 |
| 24 | 3 | San Antonio-New Braunfels, TX Metro area | 55.9 |
| 62 | 4 | Bakersfield, CA Metro area | 55.2 |
| 55 | 5 | Fresno, CA Metro area | 54.1 |
| 13 | 6 | Riverside-San Bernardino-Ontario, CA Metro area | 52.6 |
| 61 | 7 | Albuquerque, NM Metro area | 49.8 |
| 7 | 8 | Miami-Fort Lauderdale-Pompano Beach, FL Metro area | 45.9 |
| 2 | 9 | Los Angeles-Long Beach-Anaheim, CA Metro area | 44.9 |
| 72 | 10 | Oxnard-Thousand Oaks-Ventura, CA Metro area | 43.4 |
| 5 | 11 | Houston-The Woodlands-Sugar Land, TX Metro area | 38.1 |
| 53 | 12 | Tucson, AZ Metro area | 38.0 |
| 17 | 13 | San Diego-Chula Vista-Carlsbad, CA Metro area | 34.3 |
| 29 | 14 | Austin-Round Rock-Georgetown, TX Metro area | 32.5 |
| 23 | 15 | Orlando-Kissimmee-Sanford, FL Metro area | 32.2 |
| 27 | 16 | Las Vegas-Henderson-Paradise, NV Metro area | 31.8 |
| 10 | 17 | Phoenix-Mesa-Chandler, AZ Metro area | 31.5 |
| 4 | 18 | Dallas-Fort Worth-Arlington, TX Metro area | 29.3 |
| 35 | 19 | San Jose-Sunnyvale-Santa Clara, CA Metro area | 25.9 |
| 1 | 20 | New York-Newark-Jersey City, NY-NJ-PA Metro area | 25.1 |
| 19 | 21 | Denver-Aurora-Lakewood, CO Metro area | 23.3 |
| 75 | 22 | Cape Coral-Fort Myers, FL Metro area | 23.0 |
| 3 | 23 | Chicago-Naperville-Elgin, IL-IN-WI Metro area | 22.7 |
| 26 | 24 | Sacramento-Roseville-Folsom, CA Metro area | 22.2 |
| 12 | 25 | San Francisco-Oakland-Berkeley, CA Metro area | 21.8 |
| 18 | 26 | Tampa-St. Petersburg-Clearwater, FL Metro area | 20.9 |
| 59 | 26 | Bridgeport-Stamford-Norwalk, CT Metro area | 20.9 |
| 68 | 28 | New Haven-Milford, CT Metro area | 19.6 |
| 70 | 29 | Allentown-Bethlehem-Easton, PA-NJ Metro area | 18.9 |
| 47 | 30 | Salt Lake City, UT Metro area | 18.6 |
| 6 | 31 | Washington-Arlington-Alexandria, DC-VA-MD-WV Metro area | 16.4 |
| 48 | 32 | Hartford-East Hartford-Middletown, CT Metro area | 15.8 |
| 41 | 33 | Oklahoma City, OK Metro area | 14.1 |
| 38 | 34 | Providence-Warwick, RI-MA Metro area | 14.0 |
| 67 | 35 | North Port-Sarasota-Bradenton, FL Metro area | 13.1 |
| 25 | 36 | Portland-Vancouver-Hillsboro, OR-WA Metro area | 12.7 |
| 58 | 37 | Worcester, MA-CT Metro area | 12.5 |
| 11 | 38 | Boston-Cambridge-Newton, MA-NH Metro area | 11.8 |
| 40 | 39 | Milwaukee-Waukesha, WI Metro area | 11.3 |
| 57 | 40 | Omaha-Council Bluffs, NE-IA Metro area | 11.2 |
| 9 | 41 | Atlanta-Sandy Springs-Alpharetta, GA Metro area | 11.0 |
| 42 | 42 | Raleigh-Cary, NC Metro area | 10.9 |
| 22 | 43 | Charlotte-Concord-Gastonia, NC-SC Metro area | 10.8 |
| 15 | 44 | Seattle-Tacoma-Bellevue, WA Metro area | 10.7 |
| 54 | 44 | Tulsa, OK Metro area | 10.7 |
| 56 | 46 | Urban Honolulu, HI Metro area | 10.2 |
| 8 | 47 | Philadelphia-Camden-Wilmington, PA-NJ-DE-MD Metro area | 10.1 |
| 51 | 47 | Grand Rapids-Kentwood, MI Metro area | 10.1 |
| 39 | 49 | Jacksonville, FL Metro area | 9.9 |
| 31 | 50 | Kansas City, MO-KS Metro area | 9.5 |
| 45 | 51 | New Orleans-Metairie, LA Metro area | 9.1 |
| 52 | 52 | Rochester, NY Metro area | 8.0 |
| 36 | 53 | Nashville-Davidson--Murfreesboro--Franklin, TN Metro area | 7.8 |
| 60 | 54 | Greenville-Anderson, SC Metro area | 7.5 |
| 37 | 55 | Virginia Beach-Norfolk-Newport News, VA-NC Metro area | 7.4 |
| 33 | 56 | Indianapolis-Carmel-Anderson, IN Metro area | 7.2 |
| 44 | 57 | Richmond, VA Metro area | 6.9 |
| 21 | 58 | Baltimore-Columbia-Towson, MD Metro area | 6.4 |
| 34 | 59 | Cleveland-Elyria, OH Metro area | 6.3 |
| 16 | 60 | Minneapolis-St. Paul-Bloomington, MN Metro area | 6.1 |
| 73 | 61 | Charleston-North Charleston, SC Metro area | 6.0 |
| 43 | 62 | Memphis, TN-MS-AR Metro area | 5.9 |
| 69 | 62 | Columbia, SC Metro area | 5.9 |
| 63 | 64 | Albany-Schenectady-Troy, NY Metro area | 5.9 |
| 46 | 65 | Louisville/Jefferson County, KY-IN Metro area | 5.6 |
| 49 | 65 | Buffalo-Cheektowaga, NY Metro area | 5.5 |
| 14 | 67 | Detroit-Warren-Dearborn, MI Metro area | 5.5 |
| 50 | 68 | Birmingham-Hoover, AL Metro area | 4.8 |
| 32 | 69 | Columbus, OH Metro area | 4.7 |
| 66 | 70 | Baton Rouge, LA Metro area | 4.6 |
| 64 | 71 | Knoxville, TN Metro area | 4.4 |
| 30 | 72 | Cincinnati, OH-KY-IN Metro area | 4.3 |
| 20 | 73 | St. Louis, MO-IL Metro area | 3.6 |
| 74 | 74 | Dayton-Kettering, OH Metro area | 3.3 |
| 28 | 75 | Pittsburgh, PA Metro area | 2.0 |

### Percent under 15 years old, 2020

| Population rank | Under 18 years old rank | Metropolitan area | Percent Under 18 years old [cols 10 and 11] |
|---|---|---|---|
| 65 | 1 | McAllen-Edinburg-Mission, TX Metro area | 26.4 |
| 62 | 2 | Bakersfield, CA Metro area | 23.9 |
| 55 | 3 | Fresno, CA Metro area | 23.5 |
| 47 | 4 | Salt Lake City, UT Metro area | 22.0 |
| 71 | 4 | El Paso, TX Metro area | 22.0 |
| 5 | 6 | Houston-The Woodlands-Sugar Land, TX Metro area | 21.8 |
| 57 | 7 | Omaha-Council Bluffs, NE-IA Metro area | 21.1 |
| 4 | 8 | Dallas-Fort Worth-Arlington, TX Metro area | 21.0 |
| 13 | 9 | Riverside-San Bernardino-Ontario, CA Metro area | 20.8 |
| 24 | 10 | San Antonio-New Braunfels, TX Metro area | 20.5 |
| 43 | 10 | Memphis, TN-MS-AR Metro area | 20.5 |
| 33 | 12 | Indianapolis-Carmel-Anderson, IN Metro area | 20.2 |
| 54 | 12 | Tulsa, OK Metro area | 20.2 |
| 41 | 14 | Oklahoma City, OK Metro area | 20.1 |
| 31 | 15 | Kansas City, MO-KS Metro area | 19.7 |
| 9 | 16 | Atlanta-Sandy Springs-Alpharetta, GA Metro area | 19.5 |
| 32 | 17 | Columbus, OH Metro area | 19.4 |
| 51 | 17 | Grand Rapids-Kentwood, MI Metro area | 19.4 |
| 66 | 17 | Baton Rouge, LA Metro area | 19.4 |
| 16 | 20 | Minneapolis-St. Paul-Bloomington, MN Metro area | 19.3 |
| 42 | 20 | Raleigh-Cary, NC Metro area | 19.3 |
| 22 | 22 | Charlotte-Concord-Gastonia, NC-SC Metro area | 19.1 |
| 30 | 22 | Cincinnati, OH-KY-IN Metro area | 19.1 |
| 6 | 24 | Washington-Arlington-Alexandria, DC-VA-MD-WV Metro area | 19.0 |
| 10 | 24 | Phoenix-Mesa-Chandler, AZ Metro area | 19.0 |
| 27 | 26 | Las Vegas-Henderson-Paradise, NV Metro area | 18.9 |
| 36 | 26 | Nashville-Davidson--Murfreesboro--Franklin, TN Metro area | 18.9 |
| 40 | 28 | Milwaukee-Waukesha, WI Metro area | 18.8 |
| 50 | 28 | Birmingham-Hoover, AL Metro area | 18.8 |
| 26 | 30 | Sacramento-Roseville-Folsom, CA Metro area | 18.6 |
| 29 | 30 | Austin-Round Rock-Georgetown, TX Metro area | 18.6 |
| 39 | 30 | Jacksonville, FL Metro area | 18.6 |
| 46 | 33 | Louisville/Jefferson County, KY-IN Metro area | 18.4 |
| 3 | 34 | Chicago-Naperville-Elgin, IL-IN-WI Metro area | 18.3 |
| 45 | 34 | New Orleans-Metairie, LA Metro area | 18.3 |
| 60 | 34 | Greenville-Anderson, SC Metro area | 18.3 |
| 72 | 34 | Oxnard-Thousand Oaks-Ventura, CA Metro area | 18.3 |
| 37 | 38 | Virginia Beach-Norfolk-Newport News, VA-NC Metro area | 18.2 |
| 20 | 39 | St. Louis, MO-IL Metro area | 18.1 |
| 21 | 39 | Baltimore-Columbia-Towson, MD Metro area | 18.1 |
| 69 | 41 | Columbia, SC Metro area | 18.0 |
| 73 | 41 | Charleston-North Charleston, SC Metro area | 18.0 |
| 74 | 41 | Dayton-Kettering, OH Metro area | 18.0 |
| 14 | 44 | Detroit-Warren-Dearborn, MI Metro area | 17.9 |
| 19 | 44 | Denver-Aurora-Lakewood, CO Metro area | 17.9 |
| 59 | 44 | Bridgeport-Stamford-Norwalk, CT Metro area | 17.9 |
| 17 | 47 | San Diego-Chula Vista-Carlsbad, CA Metro area | 17.8 |
| 1 | 48 | New York-Newark-Jersey City, NY-NJ-PA Metro area | 17.7 |
| 8 | 48 | Philadelphia-Camden-Wilmington, PA-NJ-DE-MD Metro area | 17.7 |
| 15 | 48 | Seattle-Tacoma-Bellevue, WA Metro area | 17.7 |
| 23 | 48 | Orlando-Kissimmee-Sanford, FL Metro area | 17.7 |
| 35 | 48 | San Jose-Sunnyvale-Santa Clara, CA Metro area | 17.7 |
| 44 | 48 | Richmond, VA Metro area | 17.7 |
| 56 | 48 | Urban Honolulu, HI Metro area | 17.7 |
| 2 | 55 | Los Angeles-Long Beach-Anaheim, CA Metro area | 17.6 |
| 61 | 55 | Albuquerque, NM Metro area | 17.6 |
| 25 | 57 | Portland-Vancouver-Hillsboro, OR-WA Metro area | 17.1 |
| 34 | 58 | Cleveland-Elyria, OH Metro area | 17.0 |
| 70 | 58 | Allentown-Bethlehem-Easton, PA-NJ Metro area | 17.0 |
| 53 | 60 | Tucson, AZ Metro area | 16.8 |
| 52 | 61 | Rochester, NY Metro area | 16.7 |
| 58 | 61 | Worcester, MA-CT Metro area | 16.7 |
| 64 | 61 | Knoxville, TN Metro area | 16.7 |
| 7 | 64 | Miami-Fort Lauderdale-Pompano Beach, FL Metro area | 16.6 |
| 49 | 65 | Buffalo-Cheektowaga, NY Metro area | 16.5 |
| 68 | 66 | New Haven-Milford, CT Metro area | 16.2 |
| 18 | 67 | Tampa-St. Petersburg-Clearwater, FL Metro area | 16.1 |
| 48 | 67 | Hartford-East Hartford-Middletown, CT Metro area | 16.1 |
| 11 | 69 | Boston-Cambridge-Newton, MA-NH Metro area | 16.0 |
| 12 | 69 | San Francisco-Oakland-Berkeley, CA Metro area | 16.0 |
| 38 | 69 | Providence-Warwick, RI-MA Metro area | 16.0 |
| 63 | 72 | Albany-Schenectady-Troy, NY Metro area | 15.9 |
| 28 | 73 | Pittsburgh, PA Metro area | 15.5 |
| 75 | 74 | Cape Coral-Fort Myers, FL Metro area | 14.1 |
| 67 | 75 | North Port-Sarasota-Bradenton, FL Metro area | 12.8 |

# 75 Largest Metropolitan Areas by 2020 Population
## Selected Rankings

| | Percent 65 years old and over, 2020 | | | Percent single-parent family households, 2019 | | | |
|---|---|---|---|---|---|---|---|
| Population rank | 65 years old and over rank | Metropolitan area | Percent 65 years old and over [cols 17 + 18] | Population rank | Single-parent households rank | Metropolitan area | Percent single-parent households [col 30] |
| 67 | 1 | North Port-Sarasota-Bradenton, FL Metro area | 33.6 | 65 | 1 | McAllen-Edinburg-Mission, TX Metro area | 17.8 |
| 75 | 2 | Cape Coral-Fort Myers, FL Metro area | 29.5 | 55 | 2 | Fresno, CA Metro area | 17.3 |
| 28 | 3 | Pittsburgh, PA Metro area | 21.0 | 71 | 3 | El Paso, TX Metro area | 16.2 |
| 53 | 4 | Tucson, AZ Metro area | 20.9 | 62 | 4 | Bakersfield, CA Metro area | 15.0 |
| 18 | 5 | Tampa-St. Petersburg-Clearwater, FL Metro area | 20.4 | 43 | 5 | Memphis, TN-MS-AR Metro area | 14.1 |
| 34 | 6 | Cleveland-Elyria, OH Metro area | 19.4 | 66 | 6 | Baton Rouge, LA Metro area | 13.1 |
| 7 | 7 | Miami-Fort Lauderdale-Pompano Beach, FL Metro area | 19.1 | 24 | 7 | San Antonio-New Braunfels, TX Metro area | 12.7 |
| 49 | 7 | Buffalo-Cheektowaga, NY Metro area | 19.1 | 61 | 7 | Albuquerque, NM Metro area | 12.7 |
| 64 | 7 | Knoxville, TN Metro area | 19.1 | 5 | 9 | Houston-The Woodlands-Sugar Land, TX Metro area | 12.4 |
| 52 | 10 | Rochester, NY Metro area | 19.0 | 13 | 9 | Riverside-San Bernardino-Ontario, CA Metro area | 12.4 |
| 70 | 11 | Allentown-Bethlehem-Easton, PA-NJ Metro area | 18.9 | 37 | 11 | Virginia Beach-Norfolk-Newport News, VA-NC Metro area | 12.1 |
| 56 | 12 | Urban Honolulu, HI Metro area | 18.8 | 69 | 12 | Columbia, SC Metro area | 12.0 |
| 63 | 13 | Albany-Schenectady-Troy, NY Metro area | 18.6 | 27 | 13 | Las Vegas-Henderson-Paradise, NV Metro area | 11.9 |
| 74 | 13 | Dayton-Kettering, OH Metro area | 18.6 | 45 | 14 | New Orleans-Metairie, LA Metro area | 11.6 |
| 68 | 15 | New Haven-Milford, CT Metro area | 18.3 | 7 | 15 | Miami-Fort Lauderdale-Pompano Beach, FL Metro area | 11.5 |
| 48 | 16 | Hartford-East Hartford-Middletown, CT Metro area | 18.2 | 4 | 16 | Dallas-Fort Worth-Arlington, TX Metro area | 11.2 |
| 38 | 17 | Providence-Warwick, RI-MA Metro area | 18.0 | 41 | 16 | Oklahoma City, OK Metro area | 11.2 |
| 61 | 18 | Albuquerque, NM Metro area | 17.9 | 9 | 18 | Atlanta-Sandy Springs-Alpharetta, GA Metro area | 11.1 |
| 20 | 19 | St. Louis, MO-IL Metro area | 17.6 | 32 | 18 | Columbus, OH Metro area | 11.1 |
| 14 | 20 | Detroit-Warren-Dearborn, MI Metro area | 17.4 | 30 | 20 | Cincinnati, OH-KY-IN Metro area | 11.0 |
| 60 | 21 | Greenville-Anderson, SC Metro area | 17.3 | 39 | 20 | Jacksonville, FL Metro area | 11.0 |
| 45 | 22 | New Orleans-Metairie, LA Metro area | 17.0 | 54 | 20 | Tulsa, OK Metro area | 11.0 |
| 8 | 23 | Philadelphia-Camden-Wilmington, PA-NJ-DE-MD Metro area | 16.9 | 46 | 23 | Louisville/Jefferson County, KY-IN Metro area | 10.9 |
| 46 | 24 | Louisville/Jefferson County, KY-IN Metro area | 16.8 | 8 | 24 | Philadelphia-Camden-Wilmington, PA-NJ-DE-MD Metro area | 10.8 |
| 50 | 24 | Birmingham-Hoover, AL Metro area | 16.8 | 22 | 24 | Charlotte-Concord-Gastonia, NC-SC Metro area | 10.8 |
| 59 | 24 | Bridgeport-Stamford-Norwalk, CT Metro area | 16.8 | 23 | 24 | Orlando-Kissimmee-Sanford, FL Metro area | 10.8 |
| 1 | 27 | New York-Newark-Jersey City, NY-NJ-PA Metro area | 16.7 | 68 | 27 | New Haven-Milford, CT Metro area | 10.7 |
| 39 | 27 | Jacksonville, FL Metro area | 16.7 | 40 | 28 | Milwaukee-Waukesha, WI Metro area | 10.6 |
| 58 | 27 | Worcester, MA-CT Metro area | 16.7 | 70 | 28 | Allentown-Bethlehem-Easton, PA-NJ Metro area | 10.6 |
| 72 | 27 | Oxnard-Thousand Oaks-Ventura, CA Metro area | 16.7 | 74 | 30 | Dayton-Kettering, OH Metro area | 10.5 |
| 11 | 31 | Boston-Cambridge-Newton, MA-NH Metro area | 16.6 | 10 | 31 | Phoenix-Mesa-Chandler, AZ Metro area | 10.4 |
| 40 | 31 | Milwaukee-Waukesha, WI Metro area | 16.6 | 38 | 31 | Providence-Warwick, RI-MA Metro area | 10.4 |
| 10 | 33 | Phoenix-Mesa-Chandler, AZ Metro area | 16.5 | 42 | 31 | Raleigh-Cary, NC Metro area | 10.4 |
| 44 | 33 | Richmond, VA Metro area | 16.5 | 49 | 31 | Buffalo-Cheektowaga, NY Metro area | 10.4 |
| 12 | 35 | San Francisco-Oakland-Berkeley, CA Metro area | 16.4 | 2 | 35 | Los Angeles-Long Beach-Anaheim, CA Metro area | 10.3 |
| 21 | 35 | Baltimore-Columbia-Towson, MD Metro area | 16.4 | 21 | 35 | Baltimore-Columbia-Towson, MD Metro area | 10.3 |
| 54 | 37 | Tulsa, OK Metro area | 16.3 | 26 | 35 | Sacramento-Roseville-Folsom, CA Metro area | 10.3 |
| 26 | 38 | Sacramento-Roseville-Folsom, CA Metro area | 16.2 | 31 | 38 | Kansas City, MO-KS Metro area | 10.2 |
| 30 | 39 | Cincinnati, OH-KY-IN Metro area | 16.1 | 34 | 38 | Cleveland-Elyria, OH Metro area | 10.2 |
| 73 | 39 | Charleston-North Charleston, SC Metro area | 16.1 | 63 | 38 | Albany-Schenectady-Troy, NY Metro area | 10.2 |
| 25 | 41 | Portland-Vancouver-Hillsboro, OR-WA Metro area | 15.9 | 20 | 41 | St. Louis, MO-IL Metro area | 10.1 |
| 69 | 42 | Columbia, SC Metro area | 15.8 | 14 | 42 | Detroit-Warren-Dearborn, MI Metro area | 10.0 |
| 31 | 43 | Kansas City, MO-KS Metro area | 15.7 | 53 | 42 | Tucson, AZ Metro area | 10.0 |
| 37 | 44 | Virginia Beach-Norfolk-Newport News, VA-NC Metro area | 15.6 | 36 | 44 | Nashville-Davidson--Murfreesboro--Franklin, TN Metro area | 9.9 |
| 3 | 45 | Chicago-Naperville-Elgin, IL-IN-WI Metro area | 15.5 | 52 | 44 | Rochester, NY Metro area | 9.9 |
| 23 | 45 | Orlando-Kissimmee-Sanford, FL Metro area | 15.5 | 57 | 44 | Omaha-Council Bluffs, NE-IA Metro area | 9.9 |
| 27 | 45 | Las Vegas-Henderson-Paradise, NV Metro area | 15.5 | 33 | 47 | Indianapolis-Carmel-Anderson, IN Metro area | 9.8 |
| 51 | 48 | Grand Rapids-Kentwood, MI Metro area | 15.3 | 50 | 47 | Birmingham-Hoover, AL Metro area | 9.8 |
| 66 | 49 | Baton Rouge, LA Metro area | 15.1 | 58 | 47 | Worcester, MA-CT Metro area | 9.8 |
| 2 | 50 | Los Angeles-Long Beach-Anaheim, CA Metro area | 14.9 | 3 | 50 | Chicago-Naperville-Elgin, IL-IN-WI Metro area | 9.7 |
| 16 | 50 | Minneapolis-St. Paul-Bloomington, MN Metro area | 14.9 | 44 | 51 | Richmond, VA Metro area | 9.6 |
| 17 | 50 | San Diego-Chula Vista-Carlsbad, CA Metro area | 14.9 | 1 | 52 | New York-Newark-Jersey City, NY-NJ-PA Metro area | 9.5 |
| 43 | 53 | Memphis, TN-MS-AR Metro area | 14.8 | 51 | 52 | Grand Rapids-Kentwood, MI Metro area | 9.5 |
| 41 | 54 | Oklahoma City, OK Metro area | 14.6 | 60 | 52 | Greenville-Anderson, SC Metro area | 9.5 |
| 57 | 55 | Omaha-Council Bluffs, NE-IA Metro area | 14.5 | 64 | 55 | Knoxville, TN Metro area | 9.4 |
| 33 | 56 | Indianapolis-Carmel-Anderson, IN Metro area | 14.4 | 72 | 56 | Oxnard-Thousand Oaks-Ventura, CA Metro area | 9.3 |
| 22 | 57 | Charlotte-Concord-Gastonia, NC-SC Metro area | 14.3 | 17 | 57 | San Diego-Chula Vista-Carlsbad, CA Metro area | 9.2 |
| 35 | 58 | San Jose-Sunnyvale-Santa Clara, CA Metro area | 14.2 | 18 | 57 | Tampa-St. Petersburg-Clearwater, FL Metro area | 9.2 |
| 15 | 59 | Seattle-Tacoma-Bellevue, WA Metro area | 14.0 | 59 | 59 | Bridgeport-Stamford-Norwalk, CT Metro area | 9.1 |
| 32 | 59 | Columbus, OH Metro area | 14.0 | 6 | 60 | Washington-Arlington-Alexandria, DC-VA-MD-WV Metro area | 9.0 |
| 36 | 59 | Nashville-Davidson--Murfreesboro--Franklin, TN Metro area | 14.0 | 16 | 61 | Minneapolis-St. Paul-Bloomington, MN Metro area | 8.9 |
| 6 | 62 | Washington-Arlington-Alexandria, DC-VA-MD-WV Metro area | 13.8 | 48 | 62 | Hartford-East Hartford-Middletown, CT Metro area | 8.8 |
| 13 | 62 | Riverside-San Bernardino-Ontario, CA Metro area | 13.8 | 47 | 63 | Salt Lake City, UT Metro area | 8.7 |
| 19 | 64 | Denver-Aurora-Lakewood, CO Metro area | 13.7 | 29 | 64 | Austin-Round Rock-Georgetown, TX Metro area | 8.6 |
| 24 | 65 | San Antonio-New Braunfels, TX Metro area | 13.6 | 25 | 65 | Portland-Vancouver-Hillsboro, OR-WA Metro area | 8.5 |
| 9 | 66 | Atlanta-Sandy Springs-Alpharetta, GA Metro area | 13.1 | 73 | 66 | Charleston-North Charleston, SC Metro area | 8.4 |
| 42 | 67 | Raleigh-Cary, NC Metro area | 12.9 | 19 | 67 | Denver-Aurora-Lakewood, CO Metro area | 8.3 |
| 55 | 67 | Fresno, CA Metro area | 12.9 | 56 | 67 | Urban Honolulu, HI Metro area | 8.3 |
| 71 | 67 | El Paso, TX Metro area | 12.9 | 75 | 67 | Cape Coral-Fort Myers, FL Metro area | 8.3 |
| 5 | 70 | Houston-The Woodlands-Sugar Land, TX Metro area | 11.9 | 11 | 70 | Boston-Cambridge-Newton, MA-NH Metro area | 7.9 |
| 4 | 71 | Dallas-Fort Worth-Arlington, TX Metro area | 11.8 | 15 | 71 | Seattle-Tacoma-Bellevue, WA Metro area | 7.7 |
| 29 | 72 | Austin-Round Rock-Georgetown, TX Metro area | 11.6 | 35 | 71 | San Jose-Sunnyvale-Santa Clara, CA Metro area | 7.7 |
| 62 | 73 | Bakersfield, CA Metro area | 11.5 | 28 | 73 | Pittsburgh, PA Metro area | 7.4 |
| 65 | 73 | McAllen-Edinburg-Mission, TX Metro area | 11.5 | 12 | 74 | San Francisco-Oakland-Berkeley, CA Metro area | 7.1 |
| 47 | 75 | Salt Lake City, UT Metro area | 11.4 | 67 | 75 | North Port-Sarasota-Bradenton, FL Metro area | 5.8 |

# 75 Largest Metropolitan Areas by 2020 Population
## Selected Rankings

### Birth rate, 2020

| Population rank | Birth rate rank | Metropolitan area | Births (per 1,000 population) [col 36] |
|---|---|---|---|
| 65 | 1 | McAllen-Edinburg-Mission, TX Metro area | 16.1 |
| 55 | 2 | Fresno, CA Metro area | 14.2 |
| 62 | 2 | Bakersfield, CA Metro area | 14.2 |
| 47 | 4 | Salt Lake City, UT Metro area | 13.8 |
| 71 | 5 | El Paso, TX Metro area | 13.7 |
| 5 | 6 | Houston-The Woodlands-Sugar Land, TX Metro area | 13.3 |
| 57 | 6 | Omaha-Council Bluffs, NE-IA Metro area | 13.3 |
| 43 | 8 | Memphis, TN-MS-AR Metro area | 13.1 |
| 4 | 9 | Dallas-Fort Worth-Arlington, TX Metro area | 12.7 |
| 32 | 10 | Columbus, OH Metro area | 12.6 |
| 24 | 11 | San Antonio-New Braunfels, TX Metro area | 12.5 |
| 36 | 11 | Nashville-Davidson--Murfreesboro--Franklin, TN Metro area | 12.5 |
| 54 | 11 | Tulsa, OK Metro area | 12.5 |
| 33 | 14 | Indianapolis-Carmel-Anderson, IN Metro area | 12.4 |
| 41 | 14 | Oklahoma City, OK Metro area | 12.4 |
| 66 | 14 | Baton Rouge, LA Metro area | 12.4 |
| 6 | 17 | Washington-Arlington-Alexandria, DC-VA-MD-WV Metro area | 12.2 |
| 13 | 17 | Riverside-San Bernardino-Ontario, CA Metro area | 12.2 |
| 31 | 17 | Kansas City, MO-KS Metro area | 12.2 |
| 37 | 17 | Virginia Beach-Norfolk-Newport News, VA-NC Metro area | 12.2 |
| 56 | 17 | Urban Honolulu, HI Metro area | 12.2 |
| 51 | 22 | Grand Rapids-Kentwood, MI Metro area | 12.1 |
| 16 | 23 | Minneapolis-St. Paul-Bloomington, MN Metro area | 12.0 |
| 30 | 23 | Cincinnati, OH-KY-IN Metro area | 12.0 |
| 40 | 25 | Milwaukee-Waukesha, WI Metro area | 12.0 |
| 1 | 26 | New York-Newark-Jersey City, NY-NJ-PA Metro area | 11.9 |
| 50 | 26 | Birmingham-Hoover, AL Metro area | 11.8 |
| 9 | 28 | Atlanta-Sandy Springs-Alpharetta, GA Metro area | 11.8 |
| 17 | 28 | San Diego-Chula Vista-Carlsbad, CA Metro area | 11.7 |
| 29 | 28 | Austin-Round Rock-Georgetown, TX Metro area | 11.7 |
| 39 | 28 | Jacksonville, FL Metro area | 11.7 |
| 45 | 28 | New Orleans-Metairie, LA Metro area | 11.7 |
| 73 | 28 | Charleston-North Charleston, SC Metro area | 11.7 |
| 22 | 34 | Charlotte-Concord-Gastonia, NC-SC Metro area | 11.6 |
| 27 | 34 | Las Vegas-Henderson-Paradise, NV Metro area | 11.6 |
| 46 | 34 | Louisville/Jefferson County, KY-IN Metro area | 11.6 |
| 74 | 37 | Dayton-Kettering, OH Metro area | 11.5 |
| 15 | 38 | Seattle-Tacoma-Bellevue, WA Metro area | 11.4 |
| 21 | 38 | Baltimore-Columbia-Towson, MD Metro area | 11.4 |
| 44 | 38 | Richmond, VA Metro area | 11.4 |
| 3 | 41 | Chicago-Naperville-Elgin, IL-IN-WI Metro area | 11.3 |
| 10 | 41 | Phoenix-Mesa-Chandler, AZ Metro area | 11.3 |
| 14 | 41 | Detroit-Warren-Dearborn, MI Metro area | 11.3 |
| 20 | 41 | St. Louis, MO-IL Metro area | 11.3 |
| 19 | 45 | Denver-Aurora-Lakewood, CO Metro area | 11.2 |
| 42 | 45 | Raleigh-Cary, NC Metro area | 11.2 |
| 60 | 45 | Greenville-Anderson, SC Metro area | 11.2 |
| 23 | 48 | Orlando-Kissimmee-Sanford, FL Metro area | 11.1 |
| 26 | 48 | Sacramento-Roseville-Folsom, CA Metro area | 11.1 |
| 35 | 48 | San Jose-Sunnyvale-Santa Clara, CA Metro area | 11.1 |
| 8 | 51 | Philadelphia-Camden-Wilmington, PA-NJ-DE-MD Metro area | 11.0 |
| 2 | 52 | Los Angeles-Long Beach-Anaheim, CA Metro area | 10.9 |
| 69 | 52 | Columbia, SC Metro area | 10.9 |
| 7 | 54 | Miami-Fort Lauderdale-Pompano Beach, FL Metro area | 10.8 |
| 34 | 55 | Cleveland-Elyria, OH Metro area | 10.6 |
| 72 | 56 | Oxnard-Thousand Oaks-Ventura, CA Metro area | 10.5 |
| 12 | 57 | San Francisco-Oakland-Berkeley, CA Metro area | 10.4 |
| 49 | 57 | Buffalo-Cheektowaga, NY Metro area | 10.4 |
| 64 | 59 | Knoxville, TN Metro area | 10.3 |
| 70 | 59 | Allentown-Bethlehem-Easton, PA-NJ Metro area | 10.3 |
| 11 | 61 | Boston-Cambridge-Newton, MA-NH Metro area | 10.2 |
| 25 | 61 | Portland-Vancouver-Hillsboro, OR-WA Metro area | 10.2 |
| 52 | 63 | Rochester, NY Metro area | 10.1 |
| 53 | 63 | Tucson, AZ Metro area | 10.1 |
| 59 | 65 | Bridgeport-Stamford-Norwalk, CT Metro area | 10.0 |
| 61 | 65 | Albuquerque, NM Metro area | 10.0 |
| 68 | 67 | New Haven-Milford, CT Metro area | 9.9 |
| 18 | 68 | Tampa-St. Petersburg-Clearwater, FL Metro area | 9.8 |
| 28 | 68 | Pittsburgh, PA Metro area | 9.8 |
| 38 | 68 | Providence-Warwick, RI-MA Metro area | 9.8 |
| 58 | 68 | Worcester, MA-CT Metro area | 9.8 |
| 63 | 72 | Albany-Schenectady-Troy, NY Metro area | 9.7 |
| 48 | 73 | Hartford-East Hartford-Middletown, CT Metro area | 9.6 |
| 75 | 74 | Cape Coral-Fort Myers, FL Metro area | 8.7 |
| 67 | 75 | North Port-Sarasota-Bradenton, FL Metro area | 7.5 |

### Percent under 65 who have no health insurance, 2019

| Population rank | No health insurance rank | Metropolitan area | Percent with no health insurance [col 40] |
|---|---|---|---|
| 65 | 1 | McAllen-Edinburg-Mission, TX Metro area | 33.2 |
| 71 | 2 | El Paso, TX Metro area | 24.4 |
| 5 | 3 | Houston-The Woodlands-Sugar Land, TX Metro area | 21.7 |
| 4 | 4 | Dallas-Fort Worth-Arlington, TX Metro area | 19.5 |
| 24 | 5 | San Antonio-New Braunfels, TX Metro area | 18.6 |
| 7 | 6 | Miami-Fort Lauderdale-Pompano Beach, FL Metro area | 18.5 |
| 75 | 7 | Cape Coral-Fort Myers, FL Metro area | 17.9 |
| 67 | 8 | North Port-Sarasota-Bradenton, FL Metro area | 16.4 |
| 54 | 9 | Tulsa, OK Metro area | 16.2 |
| 29 | 10 | Austin-Round Rock-Georgetown, TX Metro area | 15.8 |
| 18 | 11 | Tampa-St. Petersburg-Clearwater, FL Metro area | 15.2 |
| 9 | 12 | Atlanta-Sandy Springs-Alpharetta, GA Metro area | 15.1 |
| 41 | 13 | Oklahoma City, OK Metro area | 14.9 |
| 23 | 14 | Orlando-Kissimmee-Sanford, FL Metro area | 14.8 |
| 27 | 15 | Las Vegas-Henderson-Paradise, NV Metro area | 13.7 |
| 39 | 16 | Jacksonville, FL Metro area | 13.3 |
| 53 | 16 | Tucson, AZ Metro area | 13.3 |
| 60 | 16 | Greenville-Anderson, SC Metro area | 13.3 |
| 10 | 19 | Phoenix-Mesa-Chandler, AZ Metro area | 13.2 |
| 43 | 20 | Memphis, TN-MS-AR Metro area | 12.8 |
| 22 | 21 | Charlotte-Concord-Gastonia, NC-SC Metro area | 12.7 |
| 73 | 22 | Charleston-North Charleston, SC Metro area | 12.6 |
| 69 | 23 | Columbia, SC Metro area | 12.0 |
| 64 | 24 | Knoxville, TN Metro area | 11.4 |
| 47 | 25 | Salt Lake City, UT Metro area | 11.3 |
| 36 | 26 | Nashville-Davidson--Murfreesboro--Franklin, TN Metro area | 11.1 |
| 42 | 27 | Raleigh-Cary, NC Metro area | 10.9 |
| 61 | 28 | Albuquerque, NM Metro area | 10.8 |
| 45 | 29 | New Orleans-Metairie, LA Metro area | 10.7 |
| 2 | 30 | Los Angeles-Long Beach-Anaheim, CA Metro area | 10.6 |
| 31 | 30 | Kansas City, MO-KS Metro area | 10.6 |
| 50 | 30 | Birmingham-Hoover, AL Metro area | 10.6 |
| 72 | 33 | Oxnard-Thousand Oaks-Ventura, CA Metro area | 10.4 |
| 13 | 34 | Riverside-San Bernardino-Ontario, CA Metro area | 9.9 |
| 33 | 34 | Indianapolis-Carmel-Anderson, IN Metro area | 9.9 |
| 55 | 36 | Fresno, CA Metro area | 9.7 |
| 66 | 36 | Baton Rouge, LA Metro area | 9.7 |
| 3 | 38 | Chicago-Naperville-Elgin, IL-IN-WI Metro area | 9.5 |
| 44 | 39 | Richmond, VA Metro area | 9.3 |
| 59 | 39 | Bridgeport-Stamford-Norwalk, CT Metro area | 9.3 |
| 62 | 39 | Bakersfield, CA Metro area | 9.3 |
| 19 | 42 | Denver-Aurora-Lakewood, CO Metro area | 9.0 |
| 17 | 43 | San Diego-Chula Vista-Carlsbad, CA Metro area | 8.9 |
| 20 | 44 | St. Louis, MO-IL Metro area | 8.6 |
| 37 | 44 | Virginia Beach-Norfolk-Newport News, VA-NC Metro area | 8.6 |
| 57 | 46 | Omaha-Council Bluffs, NE-IA Metro area | 8.5 |
| 74 | 47 | Dayton-Kettering, OH Metro area | 8.2 |
| 6 | 48 | Washington-Arlington-Alexandria, DC-VA-MD-WV Metro area | 8.1 |
| 32 | 49 | Columbus, OH Metro area | 8.0 |
| 1 | 50 | New York-Newark-Jersey City, NY-NJ-PA Metro area | 7.9 |
| 25 | 51 | Portland-Vancouver-Hillsboro, OR-WA Metro area | 7.6 |
| 46 | 52 | Louisville/Jefferson County, KY-IN Metro area | 7.4 |
| 34 | 53 | Cleveland-Elyria, OH Metro area | 7.3 |
| 30 | 54 | Cincinnati, OH-KY-IN Metro area | 7.1 |
| 70 | 54 | Allentown-Bethlehem-Easton, PA-NJ Metro area | 7.1 |
| 51 | 56 | Grand Rapids-Kentwood, MI Metro area | 7.0 |
| 68 | 56 | New Haven-Milford, CT Metro area | 7.0 |
| 8 | 58 | Philadelphia-Camden-Wilmington, PA-NJ-DE-MD Metro area | 6.8 |
| 14 | 58 | Detroit-Warren-Dearborn, MI Metro area | 6.8 |
| 15 | 58 | Seattle-Tacoma-Bellevue, WA Metro area | 6.8 |
| 40 | 58 | Milwaukee-Waukesha, WI Metro area | 6.8 |
| 26 | 62 | Sacramento-Roseville-Folsom, CA Metro area | 6.5 |
| 21 | 63 | Baltimore-Columbia-Towson, MD Metro area | 5.9 |
| 12 | 64 | San Francisco-Oakland-Berkeley, CA Metro area | 5.6 |
| 35 | 64 | San Jose-Sunnyvale-Santa Clara, CA Metro area | 5.6 |
| 48 | 64 | Hartford-East Hartford-Middletown, CT Metro area | 5.6 |
| 16 | 67 | Minneapolis-St. Paul-Bloomington, MN Metro area | 5.4 |
| 28 | 67 | Pittsburgh, PA Metro area | 5.4 |
| 38 | 69 | Providence-Warwick, RI-MA Metro area | 4.6 |
| 52 | 70 | Rochester, NY Metro area | 4.4 |
| 56 | 70 | Urban Honolulu, HI Metro area | 4.4 |
| 49 | 72 | Buffalo-Cheektowaga, NY Metro area | 4.1 |
| 63 | 72 | Albany-Schenectady-Troy, NY Metro area | 4.1 |
| 58 | 74 | Worcester, MA-CT Metro area | 3.9 |
| 11 | 75 | Boston-Cambridge-Newton, MA-NH Metro area | 3.7 |

# 75 Largest Metropolitan Areas by 2020 Population
## Selected Rankings

### Percent college graduates (bachelor's degree or more), 2019

| Population rank | Percent college graduates rank | Metropolitan area | Percent college graduates [col 51] |
|---|---|---|---|
| 35 | 1 | San Jose-Sunnyvale-Santa Clara, CA Metro area | 52.7 |
| 6 | 2 | Washington-Arlington-Alexandria, DC-VA-MD-WV Metro area | 51.4 |
| 12 | 2 | San Francisco-Oakland-Berkeley, CA Metro area | 51.4 |
| 11 | 4 | Boston-Cambridge-Newton, MA-NH Metro area | 49.3 |
| 59 | 5 | Bridgeport-Stamford-Norwalk, CT Metro area | 49.1 |
| 42 | 6 | Raleigh-Cary, NC Metro area | 48.0 |
| 29 | 7 | Austin-Round Rock-Georgetown, TX Metro area | 46.2 |
| 19 | 8 | Denver-Aurora-Lakewood, CO Metro area | 45.8 |
| 15 | 9 | Seattle-Tacoma-Bellevue, WA Metro area | 44.1 |
| 16 | 10 | Minneapolis-St. Paul-Bloomington, MN Metro area | 43.2 |
| 21 | 11 | Baltimore-Columbia-Towson, MD Metro area | 41.9 |
| 1 | 12 | New York-Newark-Jersey City, NY-NJ-PA Metro area | 41.8 |
| 25 | 13 | Portland-Vancouver-Hillsboro, OR-WA Metro area | 40.3 |
| 9 | 14 | Atlanta-Sandy Springs-Alpharetta, GA Metro area | 39.9 |
| 17 | 14 | San Diego-Chula Vista-Carlsbad, CA Metro area | 39.9 |
| 48 | 16 | Hartford-East Hartford-Middletown, CT Metro area | 39.7 |
| 63 | 17 | Albany-Schenectady-Troy, NY Metro area | 39.5 |
| 3 | 18 | Chicago-Naperville-Elgin, IL-IN-WI Metro area | 39.2 |
| 8 | 19 | Philadelphia-Camden-Wilmington, PA-NJ-DE-MD Metro area | 39.0 |
| 36 | 20 | Nashville-Davidson--Murfreesboro--Franklin, TN Metro area | 38.5 |
| 32 | 21 | Columbus, OH Metro area | 37.9 |
| 31 | 22 | Kansas City, MO-KS Metro area | 37.7 |
| 57 | 22 | Omaha-Council Bluffs, NE-IA Metro area | 37.7 |
| 44 | 24 | Richmond, VA Metro area | 37.6 |
| 73 | 25 | Charleston-North Charleston, SC Metro area | 37.5 |
| 52 | 26 | Rochester, NY Metro area | 37.1 |
| 47 | 27 | Salt Lake City, UT Metro area | 36.5 |
| 40 | 28 | Milwaukee-Waukesha, WI Metro area | 36.4 |
| 4 | 29 | Dallas-Fort Worth-Arlington, TX Metro area | 36.3 |
| 22 | 30 | Charlotte-Concord-Gastonia, NC-SC Metro area | 36.2 |
| 28 | 31 | Pittsburgh, PA Metro area | 36.0 |
| 20 | 32 | St. Louis, MO-IL Metro area | 35.8 |
| 56 | 32 | Urban Honolulu, HI Metro area | 35.8 |
| 2 | 34 | Los Angeles-Long Beach-Anaheim, CA Metro area | 35.5 |
| 30 | 35 | Cincinnati, OH-KY-IN Metro area | 35.4 |
| 58 | 36 | Worcester, MA-CT Metro area | 35.3 |
| 33 | 37 | Indianapolis-Carmel-Anderson, IN Metro area | 35.2 |
| 68 | 38 | New Haven-Milford, CT Metro area | 35.1 |
| 72 | 39 | Oxnard-Thousand Oaks-Ventura, CA Metro area | 34.8 |
| 69 | 40 | Columbia, SC Metro area | 34.4 |
| 26 | 41 | Sacramento-Roseville-Folsom, CA Metro area | 34.2 |
| 67 | 42 | North Port-Sarasota-Bradenton, FL Metro area | 33.7 |
| 51 | 43 | Grand Rapids-Kentwood, MI Metro area | 33.4 |
| 5 | 44 | Houston-The Woodlands-Sugar Land, TX Metro area | 33.3 |
| 23 | 44 | Orlando-Kissimmee-Sanford, FL Metro area | 33.3 |
| 38 | 46 | Providence-Warwick, RI-MA Metro area | 33.2 |
| 61 | 46 | Albuquerque, NM Metro area | 33.2 |
| 7 | 48 | Miami-Fort Lauderdale-Pompano Beach, FL Metro area | 33.1 |
| 37 | 49 | Virginia Beach-Norfolk-Newport News, VA-NC Metro area | 32.9 |
| 49 | 50 | Buffalo-Cheektowaga, NY Metro area | 32.8 |
| 53 | 51 | Tucson, AZ Metro area | 32.6 |
| 39 | 52 | Jacksonville, FL Metro area | 32.5 |
| 14 | 53 | Detroit-Warren-Dearborn, MI Metro area | 32.4 |
| 45 | 54 | New Orleans-Metairie, LA Metro area | 32.3 |
| 50 | 54 | Birmingham-Hoover, AL Metro area | 32.3 |
| 10 | 56 | Phoenix-Mesa-Chandler, AZ Metro area | 32.2 |
| 34 | 57 | Cleveland-Elyria, OH Metro area | 31.7 |
| 18 | 58 | Tampa-St. Petersburg-Clearwater, FL Metro area | 31.6 |
| 41 | 59 | Oklahoma City, OK Metro area | 31.5 |
| 60 | 60 | Greenville-Anderson, SC Metro area | 31.3 |
| 46 | 61 | Louisville/Jefferson County, KY-IN Metro area | 30.7 |
| 74 | 62 | Dayton-Kettering, OH Metro area | 30.2 |
| 64 | 63 | Knoxville, TN Metro area | 29.9 |
| 70 | 64 | Allentown-Bethlehem-Easton, PA-NJ Metro area | 29.7 |
| 75 | 65 | Cape Coral-Fort Myers, FL Metro area | 29.0 |
| 54 | 66 | Tulsa, OK Metro area | 28.9 |
| 24 | 67 | San Antonio-New Braunfels, TX Metro area | 28.8 |
| 43 | 67 | Memphis, TN-MS-AR Metro area | 28.8 |
| 66 | 69 | Baton Rouge, LA Metro area | 27.5 |
| 27 | 70 | Las Vegas-Henderson-Paradise, NV Metro area | 25.6 |
| 71 | 71 | El Paso, TX Metro area | 23.3 |
| 13 | 72 | Riverside-San Bernardino-Ontario, CA Metro area | 23.0 |
| 55 | 73 | Fresno, CA Metro area | 22.0 |
| 65 | 74 | McAllen-Edinburg-Mission, TX Metro area | 19.0 |
| 62 | 75 | Bakersfield, CA Metro area | 17.1 |

### Median household income, 2019

| Population rank | Mean income rank | Metropolitan area | Mean income (dollars) [col 55] |
|---|---|---|---|
| 35 | 1 | San Jose-Sunnyvale-Santa Clara, CA Metro area | 130,865 |
| 12 | 2 | San Francisco-Oakland-Berkeley, CA Metro area | 114,696 |
| 6 | 3 | Washington-Arlington-Alexandria, DC-VA-MD-WV Metro area | 105,659 |
| 59 | 4 | Bridgeport-Stamford-Norwalk, CT Metro area | 97,053 |
| 11 | 5 | Boston-Cambridge-Newton, MA-NH Metro area | 94,430 |
| 15 | 6 | Seattle-Tacoma-Bellevue, WA Metro area | 94,027 |
| 72 | 7 | Oxnard-Thousand Oaks-Ventura, CA Metro area | 92,236 |
| 56 | 8 | Urban Honolulu, HI Metro area | 87,470 |
| 19 | 9 | Denver-Aurora-Lakewood, CO Metro area | 85,641 |
| 17 | 10 | San Diego-Chula Vista-Carlsbad, CA Metro area | 83,985 |
| 16 | 11 | Minneapolis-St. Paul-Bloomington, MN Metro area | 83,698 |
| 1 | 12 | New York-Newark-Jersey City, NY-NJ-PA Metro area | 83,160 |
| 21 | 12 | Baltimore-Columbia-Towson, MD Metro area | 83,160 |
| 29 | 14 | Austin-Round Rock-Georgetown, TX Metro area | 80,954 |
| 47 | 15 | Salt Lake City, UT Metro area | 80,196 |
| 42 | 16 | Raleigh-Cary, NC Metro area | 80,096 |
| 25 | 17 | Portland-Vancouver-Hillsboro, OR-WA Metro area | 78,439 |
| 2 | 18 | Los Angeles-Long Beach-Anaheim, CA Metro area | 77,774 |
| 48 | 19 | Hartford-East Hartford-Middletown, CT Metro area | 77,005 |
| 26 | 20 | Sacramento-Roseville-Folsom, CA Metro area | 76,706 |
| 58 | 21 | Worcester, MA-CT Metro area | 76,348 |
| 3 | 22 | Chicago-Naperville-Elgin, IL-IN-WI Metro area | 75,379 |
| 8 | 23 | Philadelphia-Camden-Wilmington, PA-NJ-DE-MD Metro area | 74,533 |
| 63 | 24 | Albany-Schenectady-Troy, NY Metro area | 73,398 |
| 4 | 25 | Dallas-Fort Worth-Arlington, TX Metro area | 72,265 |
| 9 | 26 | Atlanta-Sandy Springs-Alpharetta, GA Metro area | 71,742 |
| 38 | 27 | Providence-Warwick, RI-MA Metro area | 70,967 |
| 13 | 28 | Riverside-San Bernardino-Ontario, CA Metro area | 70,954 |
| 70 | 29 | Allentown-Bethlehem-Easton, PA-NJ Metro area | 70,793 |
| 73 | 30 | Charleston-North Charleston, SC Metro area | 70,505 |
| 57 | 31 | Omaha-Council Bluffs, NE-IA Metro area | 70,373 |
| 36 | 32 | Nashville-Davidson--Murfreesboro--Franklin, TN Metro area | 70,262 |
| 31 | 33 | Kansas City, MO-KS Metro area | 70,215 |
| 68 | 34 | New Haven-Milford, CT Metro area | 69,751 |
| 37 | 35 | Virginia Beach-Norfolk-Newport News, VA-NC Metro area | 69,329 |
| 5 | 36 | Houston-The Woodlands-Sugar Land, TX Metro area | 69,193 |
| 44 | 37 | Richmond, VA Metro area | 68,324 |
| 10 | 38 | Phoenix-Mesa-Chandler, AZ Metro area | 67,896 |
| 32 | 39 | Columbus, OH Metro area | 67,207 |
| 30 | 40 | Cincinnati, OH-KY-IN Metro area | 66,825 |
| 20 | 41 | St. Louis, MO-IL Metro area | 66,417 |
| 22 | 42 | Charlotte-Concord-Gastonia, NC-SC Metro area | 66,399 |
| 39 | 43 | Jacksonville, FL Metro area | 65,880 |
| 40 | 44 | Milwaukee-Waukesha, WI Metro area | 65,845 |
| 51 | 45 | Grand Rapids-Kentwood, MI Metro area | 65,739 |
| 67 | 46 | North Port-Sarasota-Bradenton, FL Metro area | 65,526 |
| 14 | 47 | Detroit-Warren-Dearborn, MI Metro area | 63,474 |
| 28 | 48 | Pittsburgh, PA Metro area | 62,638 |
| 33 | 49 | Indianapolis-Carmel-Anderson, IN Metro area | 62,502 |
| 24 | 50 | San Antonio-New Braunfels, TX Metro area | 62,355 |
| 75 | 51 | Cape Coral-Fort Myers, FL Metro area | 62,240 |
| 27 | 52 | Las Vegas-Henderson-Paradise, NV Metro area | 62,107 |
| 52 | 53 | Rochester, NY Metro area | 62,104 |
| 23 | 54 | Orlando-Kissimmee-Sanford, FL Metro area | 61,876 |
| 46 | 55 | Louisville/Jefferson County, KY-IN Metro area | 61,172 |
| 66 | 56 | Baton Rouge, LA Metro area | 60,746 |
| 41 | 57 | Oklahoma City, OK Metro area | 60,605 |
| 7 | 58 | Miami-Fort Lauderdale-Pompano Beach, FL Metro area | 60,141 |
| 49 | 59 | Buffalo-Cheektowaga, NY Metro area | 60,105 |
| 60 | 60 | Greenville-Anderson, SC Metro area | 58,621 |
| 61 | 61 | Albuquerque, NM Metro area | 58,512 |
| 50 | 62 | Birmingham-Hoover, AL Metro area | 58,366 |
| 74 | 63 | Dayton-Kettering, OH Metro area | 58,169 |
| 18 | 64 | Tampa-St. Petersburg-Clearwater, FL Metro area | 57,906 |
| 54 | 65 | Tulsa, OK Metro area | 57,859 |
| 55 | 66 | Fresno, CA Metro area | 57,518 |
| 34 | 67 | Cleveland-Elyria, OH Metro area | 57,228 |
| 64 | 68 | Knoxville, TN Metro area | 56,623 |
| 53 | 69 | Tucson, AZ Metro area | 56,169 |
| 69 | 70 | Columbia, SC Metro area | 55,725 |
| 45 | 71 | New Orleans-Metairie, LA Metro area | 55,710 |
| 43 | 72 | Memphis, TN-MS-AR Metro area | 54,859 |
| 62 | 73 | Bakersfield, CA Metro area | 53,067 |
| 71 | 74 | El Paso, TX Metro area | 48,823 |
| 65 | 75 | McAllen-Edinburg-Mission, TX Metro area | 41,800 |

# 75 Largest Metropolitan Areas by 2020 Population
## Selected Rankings

### Percent of population below the poverty level, 2019

| Population rank | Poverty rate rank | Metropolitan area | Poverty rate [col 59] |
|---|---|---|---|
| 65 | 1 | McAllen-Edinburg-Mission, TX Metro area | 27.3 |
| 55 | 2 | Fresno, CA Metro area | 20.6 |
| 62 | 3 | Bakersfield, CA Metro area | 19.1 |
| 71 | 4 | El Paso, TX Metro area | 18.7 |
| 45 | 5 | New Orleans-Metairie, LA Metro area | 16.4 |
| 61 | 6 | Albuquerque, NM Metro area | 15.5 |
| 43 | 7 | Memphis, TN-MS-AR Metro area | 15.4 |
| 69 | 8 | Columbia, SC Metro area | 15.1 |
| 66 | 9 | Baton Rouge, LA Metro area | 15.0 |
| 64 | 10 | Knoxville, TN Metro area | 14.1 |
| 50 | 11 | Birmingham-Hoover, AL Metro area | 14.0 |
| 53 | 12 | Tucson, AZ Metro area | 13.8 |
| 54 | 13 | Tulsa, OK Metro area | 13.6 |
| 74 | 13 | Dayton-Kettering, OH Metro area | 13.6 |
| 7 | 15 | Miami-Fort Lauderdale-Pompano Beach, FL Metro area | 13.5 |
| 24 | 15 | San Antonio-New Braunfels, TX Metro area | 13.5 |
| 34 | 15 | Cleveland-Elyria, OH Metro area | 13.5 |
| 41 | 18 | Oklahoma City, OK Metro area | 13.1 |
| 49 | 18 | Buffalo-Cheektowaga, NY Metro area | 13.1 |
| 5 | 20 | Houston-The Woodlands-Sugar Land, TX Metro area | 12.9 |
| 27 | 21 | Las Vegas-Henderson-Paradise, NV Metro area | 12.8 |
| 14 | 22 | Detroit-Warren-Dearborn, MI Metro area | 12.6 |
| 60 | 23 | Greenville-Anderson, SC Metro area | 12.5 |
| 2 | 24 | Los Angeles-Long Beach-Anaheim, CA Metro area | 12.4 |
| 18 | 24 | Tampa-St. Petersburg-Clearwater, FL Metro area | 12.4 |
| 52 | 24 | Rochester, NY Metro area | 12.4 |
| 13 | 27 | Riverside-San Bernardino-Ontario, CA Metro area | 12.2 |
| 68 | 27 | New Haven-Milford, CT Metro area | 12.2 |
| 10 | 29 | Phoenix-Mesa-Chandler, AZ Metro area | 12.1 |
| 23 | 29 | Orlando-Kissimmee-Sanford, FL Metro area | 12.1 |
| 26 | 31 | Sacramento-Roseville-Folsom, CA Metro area | 11.9 |
| 46 | 31 | Louisville/Jefferson County, KY-IN Metro area | 11.9 |
| 8 | 33 | Philadelphia-Camden-Wilmington, PA-NJ-DE-MD Metro area | 11.8 |
| 40 | 33 | Milwaukee-Waukesha, WI Metro area | 11.8 |
| 1 | 35 | New York-Newark-Jersey City, NY-NJ-PA Metro area | 11.6 |
| 32 | 36 | Columbus, OH Metro area | 11.5 |
| 39 | 36 | Jacksonville, FL Metro area | 11.5 |
| 30 | 38 | Cincinnati, OH-KY-IN Metro area | 11.3 |
| 75 | 38 | Cape Coral-Fort Myers, FL Metro area | 11.3 |
| 38 | 40 | Providence-Warwick, RI-MA Metro area | 11.0 |
| 28 | 41 | Pittsburgh, PA Metro area | 10.9 |
| 73 | 42 | Charleston-North Charleston, SC Metro area | 10.7 |
| 3 | 43 | Chicago-Naperville-Elgin, IL-IN-WI Metro area | 10.6 |
| 37 | 43 | Virginia Beach-Norfolk-Newport News, VA-NC Metro area | 10.6 |
| 4 | 45 | Dallas-Fort Worth-Arlington, TX Metro area | 10.5 |
| 9 | 45 | Atlanta-Sandy Springs-Alpharetta, GA Metro area | 10.5 |
| 33 | 45 | Indianapolis-Carmel-Anderson, IN Metro area | 10.5 |
| 51 | 45 | Grand Rapids-Kentwood, MI Metro area | 10.5 |
| 17 | 49 | San Diego-Chula Vista-Carlsbad, CA Metro area | 10.3 |
| 31 | 49 | Kansas City, MO-KS Metro area | 10.3 |
| 63 | 51 | Albany-Schenectady-Troy, NY Metro area | 10.2 |
| 29 | 52 | Austin-Round Rock-Georgetown, TX Metro area | 10.1 |
| 48 | 52 | Hartford-East Hartford-Middletown, CT Metro area | 10.1 |
| 36 | 54 | Nashville-Davidson--Murfreesboro--Franklin, TN Metro area | 10.0 |
| 44 | 54 | Richmond, VA Metro area | 10.0 |
| 20 | 56 | St. Louis, MO-IL Metro area | 9.9 |
| 58 | 57 | Worcester, MA-CT Metro area | 9.8 |
| 22 | 58 | Charlotte-Concord-Gastonia, NC-SC Metro area | 9.7 |
| 25 | 59 | Portland-Vancouver-Hillsboro, OR-WA Metro area | 9.6 |
| 21 | 60 | Baltimore-Columbia-Towson, MD Metro area | 9.4 |
| 70 | 61 | Allentown-Bethlehem-Easton, PA-NJ Metro area | 9.3 |
| 67 | 62 | North Port-Sarasota-Bradenton, FL Metro area | 9.2 |
| 57 | 63 | Omaha-Council Bluffs, NE-IA Metro area | 9.1 |
| 59 | 63 | Bridgeport-Stamford-Norwalk, CT Metro area | 9.1 |
| 42 | 65 | Raleigh-Cary, NC Metro area | 8.9 |
| 11 | 66 | Boston-Cambridge-Newton, MA-NH Metro area | 8.6 |
| 47 | 66 | Salt Lake City, UT Metro area | 8.6 |
| 12 | 68 | San Francisco-Oakland-Berkeley, CA Metro area | 8.2 |
| 16 | 68 | Minneapolis-St. Paul-Bloomington, MN Metro area | 8.2 |
| 56 | 70 | Urban Honolulu, HI Metro area | 8.1 |
| 19 | 71 | Denver-Aurora-Lakewood, CO Metro area | 7.9 |
| 72 | 71 | Oxnard-Thousand Oaks-Ventura, CA Metro area | 7.9 |
| 15 | 73 | Seattle-Tacoma-Bellevue, WA Metro area | 7.8 |
| 6 | 74 | Washington-Arlington-Alexandria, DC-VA-MD-WV Metro area | 7.5 |
| 35 | 75 | San Jose-Sunnyvale-Santa Clara, CA Metro area | 6.3 |

### Percent of children under 18 years old below the poverty level, 2019

| Population rank | Poverty rate rank | Metropolitan area | Poverty rate [col 60] |
|---|---|---|---|
| 65 | 1 | McAllen-Edinburg-Mission, TX Metro area | 24.5 |
| 55 | 2 | Fresno, CA Metro area | 16.8 |
| 71 | 3 | El Paso, TX Metro area | 15.9 |
| 62 | 4 | Bakersfield, CA Metro area | 15.4 |
| 61 | 5 | Albuquerque, NM Metro area | 12.0 |
| 43 | 6 | Memphis, TN-MS-AR Metro area | 11.2 |
| 45 | 7 | New Orleans-Metairie, LA Metro area | 11.0 |
| 69 | 7 | Columbia, SC Metro area | 11.0 |
| 24 | 9 | San Antonio-New Braunfels, TX Metro area | 10.3 |
| 5 | 10 | Houston-The Woodlands-Sugar Land, TX Metro area | 10.2 |
| 66 | 10 | Baton Rouge, LA Metro area | 10.2 |
| 7 | 12 | Miami-Fort Lauderdale-Pompano Beach, FL Metro area | 10.1 |
| 54 | 12 | Tulsa, OK Metro area | 10.1 |
| 64 | 14 | Knoxville, TN Metro area | 10.0 |
| 53 | 15 | Tucson, AZ Metro area | 9.8 |
| 50 | 16 | Birmingham-Hoover, AL Metro area | 9.7 |
| 27 | 17 | Las Vegas-Henderson-Paradise, NV Metro area | 9.6 |
| 13 | 18 | Riverside-San Bernardino-Ontario, CA Metro area | 9.5 |
| 74 | 18 | Dayton-Kettering, OH Metro area | 9.5 |
| 49 | 20 | Buffalo-Cheektowaga, NY Metro area | 9.4 |
| 34 | 21 | Cleveland-Elyria, OH Metro area | 9.1 |
| 2 | 22 | Los Angeles-Long Beach-Anaheim, CA Metro area | 8.9 |
| 14 | 22 | Detroit-Warren-Dearborn, MI Metro area | 8.9 |
| 41 | 24 | Oklahoma City, OK Metro area | 8.6 |
| 60 | 24 | Greenville-Anderson, SC Metro area | 8.6 |
| 68 | 24 | New Haven-Milford, CT Metro area | 8.6 |
| 10 | 27 | Phoenix-Mesa-Chandler, AZ Metro area | 8.5 |
| 23 | 27 | Orlando-Kissimmee-Sanford, FL Metro area | 8.5 |
| 32 | 29 | Columbus, OH Metro area | 8.4 |
| 1 | 30 | New York-Newark-Jersey City, NY-NJ-PA Metro area | 8.3 |
| 46 | 30 | Louisville/Jefferson County, KY-IN Metro area | 8.3 |
| 8 | 32 | Philadelphia-Camden-Wilmington, PA-NJ-DE-MD Metro area | 8.1 |
| 39 | 32 | Jacksonville, FL Metro area | 8.1 |
| 52 | 32 | Rochester, NY Metro area | 8.1 |
| 18 | 35 | Tampa-St. Petersburg-Clearwater, FL Metro area | 8.0 |
| 40 | 36 | Milwaukee-Waukesha, WI Metro area | 7.8 |
| 73 | 36 | Charleston-North Charleston, SC Metro area | 7.8 |
| 4 | 38 | Dallas-Fort Worth-Arlington, TX Metro area | 7.7 |
| 30 | 38 | Cincinnati, OH-KY-IN Metro area | 7.7 |
| 37 | 40 | Virginia Beach-Norfolk-Newport News, VA-NC Metro area | 7.6 |
| 9 | 41 | Atlanta-Sandy Springs-Alpharetta, GA Metro area | 7.5 |
| 75 | 41 | Cape Coral-Fort Myers, FL Metro area | 7.5 |
| 26 | 43 | Sacramento-Roseville-Folsom, CA Metro area | 7.4 |
| 3 | 44 | Chicago-Naperville-Elgin, IL-IN-WI Metro area | 7.3 |
| 38 | 44 | Providence-Warwick, RI-MA Metro area | 7.3 |
| 29 | 46 | Austin-Round Rock-Georgetown, TX Metro area | 7.2 |
| 31 | 46 | Kansas City, MO-KS Metro area | 7.2 |
| 28 | 48 | Pittsburgh, PA Metro area | 7.0 |
| 51 | 48 | Grand Rapids-Kentwood, MI Metro area | 7.0 |
| 59 | 50 | Bridgeport-Stamford-Norwalk, CT Metro area | 6.9 |
| 17 | 51 | San Diego-Chula Vista-Carlsbad, CA Metro area | 6.8 |
| 20 | 51 | St. Louis, MO-IL Metro area | 6.8 |
| 36 | 51 | Nashville-Davidson--Murfreesboro--Franklin, TN Metro area | 6.8 |
| 63 | 54 | Albany-Schenectady-Troy, NY Metro area | 6.7 |
| 22 | 55 | Charlotte-Concord-Gastonia, NC-SC Metro area | 6.6 |
| 33 | 55 | Indianapolis-Carmel-Anderson, IN Metro area | 6.6 |
| 48 | 57 | Hartford-East Hartford-Middletown, CT Metro area | 6.4 |
| 44 | 58 | Richmond, VA Metro area | 6.3 |
| 58 | 58 | Worcester, MA-CT Metro area | 6.3 |
| 70 | 58 | Allentown-Bethlehem-Easton, PA-NJ Metro area | 6.3 |
| 57 | 61 | Omaha-Council Bluffs, NE-IA Metro area | 6.2 |
| 21 | 62 | Baltimore-Columbia-Towson, MD Metro area | 6.0 |
| 25 | 63 | Portland-Vancouver-Hillsboro, OR-WA Metro area | 5.7 |
| 67 | 64 | North Port-Sarasota-Bradenton, FL Metro area | 5.6 |
| 47 | 65 | Salt Lake City, UT Metro area | 5.5 |
| 42 | 66 | Raleigh-Cary, NC Metro area | 5.4 |
| 11 | 67 | Boston-Cambridge-Newton, MA-NH Metro area | 5.2 |
| 19 | 67 | Denver-Aurora-Lakewood, CO Metro area | 5.2 |
| 12 | 69 | San Francisco-Oakland-Berkeley, CA Metro area | 4.9 |
| 56 | 69 | Urban Honolulu, HI Metro area | 4.9 |
| 72 | 69 | Oxnard-Thousand Oaks-Ventura, CA Metro area | 4.9 |
| 6 | 72 | Washington-Arlington-Alexandria, DC-VA-MD-WV Metro area | 4.8 |
| 15 | 73 | Seattle-Tacoma-Bellevue, WA Metro area | 4.7 |
| 16 | 73 | Minneapolis-St. Paul-Bloomington, MN Metro area | 4.7 |
| 35 | 75 | San Jose-Sunnyvale-Santa Clara, CA Metro area | 3.5 |

# 75 Largest Metropolitan Areas by 2020 Population
## Selected Rankings

### Median value of owner-occupied housing units, 2019

| Population rank | Median value rank | Metropolitan area | Median value (dollars) [col 91] |
|---|---|---|---|
| 35 | 1 | San Jose-Sunnyvale-Santa Clara, CA Metro area | 1,116,400 |
| 12 | 2 | San Francisco-Oakland-Berkeley, CA Metro area | 940,900 |
| 56 | 3 | Urban Honolulu, HI Metro area | 739,700 |
| 2 | 4 | Los Angeles-Long Beach-Anaheim, CA Metro area | 666,900 |
| 72 | 5 | Oxnard-Thousand Oaks-Ventura, CA Metro area | 629,600 |
| 17 | 6 | San Diego-Chula Vista-Carlsbad, CA Metro area | 619,300 |
| 15 | 7 | Seattle-Tacoma-Bellevue, WA Metro area | 503,000 |
| 1 | 8 | New York-Newark-Jersey City, NY-NJ-PA Metro area | 482,900 |
| 11 | 9 | Boston-Cambridge-Newton, MA-NH Metro area | 482,700 |
| 6 | 10 | Washington-Arlington-Alexandria, DC-VA-MD-WV Metro area | 446,300 |
| 59 | 11 | Bridgeport-Stamford-Norwalk, CT Metro area | 444,500 |
| 19 | 12 | Denver-Aurora-Lakewood, CO Metro area | 437,800 |
| 26 | 13 | Sacramento-Roseville-Folsom, CA Metro area | 434,400 |
| 25 | 14 | Portland-Vancouver-Hillsboro, OR-WA Metro area | 408,600 |
| 13 | 15 | Riverside-San Bernardino-Ontario, CA Metro area | 378,500 |
| 47 | 16 | Salt Lake City, UT Metro area | 356,400 |
| 29 | 17 | Austin-Round Rock-Georgetown, TX Metro area | 318,400 |
| 7 | 18 | Miami-Fort Lauderdale-Pompano Beach, FL Metro area | 315,400 |
| 21 | 19 | Baltimore-Columbia-Towson, MD Metro area | 313,200 |
| 27 | 20 | Las Vegas-Henderson-Paradise, NV Metro area | 313,100 |
| 38 | 21 | Providence-Warwick, RI-MA Metro area | 296,800 |
| 58 | 22 | Worcester, MA-CT Metro area | 294,300 |
| 55 | 23 | Fresno, CA Metro area | 288,300 |
| 42 | 24 | Raleigh-Cary, NC Metro area | 285,600 |
| 36 | 25 | Nashville-Davidson--Murfreesboro--Franklin, TN Metro area | 285,100 |
| 16 | 26 | Minneapolis-St. Paul-Bloomington, MN Metro area | 284,500 |
| 10 | 27 | Phoenix-Mesa-Chandler, AZ Metro area | 283,500 |
| 67 | 28 | North Port-Sarasota-Bradenton, FL Metro area | 273,500 |
| 73 | 29 | Charleston-North Charleston, SC Metro area | 266,600 |
| 8 | 30 | Philadelphia-Camden-Wilmington, PA-NJ-DE-MD Metro area | 263,700 |
| 37 | 31 | Virginia Beach-Norfolk-Newport News, VA-NC Metro area | 255,900 |
| 23 | 32 | Orlando-Kissimmee-Sanford, FL Metro area | 255,500 |
| 48 | 33 | Hartford-East Hartford-Middletown, CT Metro area | 254,100 |
| 4 | 34 | Dallas-Fort Worth-Arlington, TX Metro area | 253,900 |
| 35 | 35 | Chicago-Naperville-Elgin, IL-IN-WI Metro area | 253,800 |
| 68 | 36 | New Haven-Milford, CT Metro area | 251,500 |
| 44 | 37 | Richmond, VA Metro area | 251,200 |
| 9 | 38 | Atlanta-Sandy Springs-Alpharetta, GA Metro area | 247,200 |
| 75 | 39 | Cape Coral-Fort Myers, FL Metro area | 244,700 |
| 62 | 40 | Bakersfield, CA Metro area | 235,800 |
| 22 | 41 | Charlotte-Concord-Gastonia, NC-SC Metro area | 233,900 |
| 40 | 42 | Milwaukee-Waukesha, WI Metro area | 231,400 |
| 39 | 43 | Jacksonville, FL Metro area | 230,100 |
| 63 | 44 | Albany-Schenectady-Troy, NY Metro area | 227,000 |
| 70 | 45 | Allentown-Bethlehem-Easton, PA-NJ Metro area | 224,900 |
| 18 | 46 | Tampa-St. Petersburg-Clearwater, FL Metro area | 224,300 |
| 5 | 47 | Houston-The Woodlands-Sugar Land, TX Metro area | 219,100 |
| 32 | 48 | Columbus, OH Metro area | 212,600 |
| 45 | 49 | New Orleans-Metairie, LA Metro area | 211,900 |
| 53 | 50 | Tucson, AZ Metro area | 211,600 |
| 31 | 51 | Kansas City, MO-KS Metro area | 209,100 |
| 61 | 52 | Albuquerque, NM Metro area | 206,000 |
| 51 | 53 | Grand Rapids-Kentwood, MI Metro area | 199,400 |
| 24 | 54 | San Antonio-New Braunfels, TX Metro area | 197,600 |
| 66 | 55 | Baton Rouge, LA Metro area | 195,500 |
| 14 | 56 | Detroit-Warren-Dearborn, MI Metro area | 191,800 |
| 57 | 57 | Omaha-Council Bluffs, NE-IA Metro area | 190,800 |
| 64 | 58 | Knoxville, TN Metro area | 186,900 |
| 46 | 59 | Louisville/Jefferson County, KY-IN Metro area | 186,800 |
| 60 | 59 | Greenville-Anderson, SC Metro area | 186,800 |
| 30 | 61 | Cincinnati, OH-KY-IN Metro area | 186,100 |
| 20 | 62 | St. Louis, MO-IL Metro area | 184,600 |
| 33 | 63 | Indianapolis-Carmel-Anderson, IN Metro area | 183,800 |
| 50 | 64 | Birmingham-Hoover, AL Metro area | 177,600 |
| 43 | 65 | Memphis, TN-MS-AR Metro area | 169,400 |
| 41 | 66 | Oklahoma City, OK Metro area | 166,800 |
| 28 | 67 | Pittsburgh, PA Metro area | 165,900 |
| 69 | 68 | Columbia, SC Metro area | 164,700 |
| 34 | 69 | Cleveland-Elyria, OH Metro area | 161,400 |
| 54 | 70 | Tulsa, OK Metro area | 160,900 |
| 49 | 71 | Buffalo-Cheektowaga, NY Metro area | 160,800 |
| 52 | 72 | Rochester, NY Metro area | 150,700 |
| 74 | 73 | Dayton-Kettering, OH Metro area | 141,900 |
| 71 | 74 | El Paso, TX Metro area | 125,100 |
| 65 | 75 | McAllen-Edinburg-Mission, TX Metro area | 93,400 |

### Median rent, 2019

| Population rank | Median gross rent | Metropolitan area | Median rent (dollars) [col 94] |
|---|---|---|---|
| 35 | 1 | San Jose-Sunnyvale-Santa Clara, CA Metro area | 2,374 |
| 12 | 2 | San Francisco-Oakland-Berkeley, CA Metro area | 2,057 |
| 72 | 3 | Oxnard-Thousand Oaks-Ventura, CA Metro area | 1,859 |
| 56 | 4 | Urban Honolulu, HI Metro area | 1,774 |
| 17 | 5 | San Diego-Chula Vista-Carlsbad, CA Metro area | 1,758 |
| 6 | 6 | Washington-Arlington-Alexandria, DC-VA-MD-WV Metro area | 1,708 |
| 2 | 7 | Los Angeles-Long Beach-Anaheim, CA Metro area | 1,655 |
| 15 | 8 | Seattle-Tacoma-Bellevue, WA Metro area | 1,621 |
| 11 | 9 | Boston-Cambridge-Newton, MA-NH Metro area | 1,579 |
| 59 | 10 | Bridgeport-Stamford-Norwalk, CT Metro area | 1,521 |
| 1 | 11 | New York-Newark-Jersey City, NY-NJ-PA Metro area | 1,482 |
| 19 | 12 | Denver-Aurora-Lakewood, CO Metro area | 1,468 |
| 7 | 13 | Miami-Fort Lauderdale-Pompano Beach, FL Metro area | 1,438 |
| 13 | 14 | Riverside-San Bernardino-Ontario, CA Metro area | 1,413 |
| 26 | 15 | Sacramento-Roseville-Folsom, CA Metro area | 1,390 |
| 25 | 16 | Portland-Vancouver-Hillsboro, OR-WA Metro area | 1,356 |
| 29 | 17 | Austin-Round Rock-Georgetown, TX Metro area | 1,327 |
| 21 | 18 | Baltimore-Columbia-Towson, MD Metro area | 1,316 |
| 23 | 19 | Orlando-Kissimmee-Sanford, FL Metro area | 1,303 |
| 67 | 20 | North Port-Sarasota-Bradenton, FL Metro area | 1,284 |
| 9 | 21 | Atlanta-Sandy Springs-Alpharetta, GA Metro area | 1,224 |
| 75 | 22 | Cape Coral-Fort Myers, FL Metro area | 1,220 |
| 73 | 23 | Charleston-North Charleston, SC Metro area | 1,212 |
| 4 | 24 | Dallas-Fort Worth-Arlington, TX Metro area | 1,202 |
| 10 | 25 | Phoenix-Mesa-Chandler, AZ Metro area | 1,188 |
| 27 | 26 | Las Vegas-Henderson-Paradise, NV Metro area | 1,187 |
| 47 | 27 | Salt Lake City, UT Metro area | 1,181 |
| 37 | 28 | Virginia Beach-Norfolk-Newport News, VA-NC Metro area | 1,178 |
| 8 | 29 | Philadelphia-Camden-Wilmington, PA-NJ-DE-MD Metro area | 1,162 |
| 18 | 30 | Tampa-St. Petersburg-Clearwater, FL Metro area | 1,160 |
| 44 | 31 | Richmond, VA Metro area | 1,155 |
| 68 | 32 | New Haven-Milford, CT Metro area | 1,151 |
| 36 | 33 | Nashville-Davidson--Murfreesboro--Franklin, TN Metro area | 1,146 |
| 42 | 33 | Raleigh-Cary, NC Metro area | 1,146 |
| 16 | 35 | Minneapolis-St. Paul-Bloomington, MN Metro area | 1,144 |
| 3 | 36 | Chicago-Naperville-Elgin, IL-IN-WI Metro area | 1,139 |
| 5 | 36 | Houston-The Woodlands-Sugar Land, TX Metro area | 1,139 |
| 39 | 38 | Jacksonville, FL Metro area | 1,121 |
| 48 | 39 | Hartford-East Hartford-Middletown, CT Metro area | 1,113 |
| 70 | 40 | Allentown-Bethlehem-Easton, PA-NJ Metro area | 1,109 |
| 58 | 41 | Worcester, MA-CT Metro area | 1,086 |
| 22 | 42 | Charlotte-Concord-Gastonia, NC-SC Metro area | 1,077 |
| 24 | 43 | San Antonio-New Braunfels, TX Metro area | 1,057 |
| 63 | 44 | Albany-Schenectady-Troy, NY Metro area | 1,043 |
| 55 | 45 | Fresno, CA Metro area | 1,034 |
| 38 | 46 | Providence-Warwick, RI-MA Metro area | 1,003 |
| 31 | 47 | Kansas City, MO-KS Metro area | 989 |
| 45 | 48 | New Orleans-Metairie, LA Metro area | 983 |
| 32 | 49 | Columbus, OH Metro area | 965 |
| 14 | 50 | Detroit-Warren-Dearborn, MI Metro area | 962 |
| 43 | 51 | Memphis, TN-MS-AR Metro area | 949 |
| 57 | 51 | Omaha-Council Bluffs, NE-IA Metro area | 949 |
| 62 | 51 | Bakersfield, CA Metro area | 949 |
| 51 | 54 | Grand Rapids-Kentwood, MI Metro area | 937 |
| 33 | 55 | Indianapolis-Carmel-Anderson, IN Metro area | 932 |
| 69 | 56 | Columbia, SC Metro area | 920 |
| 61 | 57 | Albuquerque, NM Metro area | 916 |
| 40 | 58 | Milwaukee-Waukesha, WI Metro area | 914 |
| 50 | 59 | Birmingham-Hoover, AL Metro area | 913 |
| 66 | 60 | Baton Rouge, LA Metro area | 911 |
| 53 | 61 | Tucson, AZ Metro area | 905 |
| 52 | 62 | Rochester, NY Metro area | 899 |
| 20 | 63 | St. Louis, MO-IL Metro area | 883 |
| 46 | 64 | Louisville/Jefferson County, KY-IN Metro area | 880 |
| 41 | 65 | Oklahoma City, OK Metro area | 876 |
| 64 | 66 | Knoxville, TN Metro area | 872 |
| 60 | 67 | Greenville-Anderson, SC Metro area | 867 |
| 54 | 68 | Tulsa, OK Metro area | 861 |
| 30 | 69 | Cincinnati, OH-KY-IN Metro area | 854 |
| 71 | 70 | El Paso, TX Metro area | 849 |
| 28 | 71 | Pittsburgh, PA Metro area | 847 |
| 49 | 72 | Buffalo-Cheektowaga, NY Metro area | 828 |
| 34 | 73 | Cleveland-Elyria, OH Metro area | 813 |
| 74 | 74 | Dayton-Kettering, OH Metro area | 812 |
| 65 | 75 | McAllen-Edinburg-Mission, TX Metro area | 763 |

# 75 Largest Metropolitan Areas by 2020 Population
## Selected Rankings

### Unemployment rate, 2020

| Population rank | Unemployment rate rank | Metropolitan area | Unemployment rate [col 100] |
|---|---|---|---|
| 27 | 1 | Las Vegas-Henderson-Paradise, NV Metro area | 14.7 |
| 62 | 2 | Bakersfield, CA Metro area | 12.5 |
| 2 | 3 | Los Angeles-Long Beach-Anaheim, CA Metro area | 11.8 |
| 14 | 4 | Detroit-Warren-Dearborn, MI Metro area | 11.7 |
| 65 | 5 | McAllen-Edinburg-Mission, TX Metro area | 11.6 |
| 55 | 6 | Fresno, CA Metro area | 11.3 |
| 1 | 7 | New York-Newark-Jersey City, NY-NJ-PA Metro area | 10.5 |
| 23 | 8 | Orlando-Kissimmee-Sanford, FL Metro area | 10.2 |
| 56 | 8 | Urban Honolulu, HI Metro area | 10.2 |
| 3 | 10 | Chicago-Naperville-Elgin, IL-IN-WI Metro area | 9.9 |
| 13 | 11 | Riverside-San Bernardino-Ontario, CA Metro area | 9.7 |
| 34 | 11 | Cleveland-Elyria, OH Metro area | 9.7 |
| 38 | 11 | Providence-Warwick, RI-MA Metro area | 9.7 |
| 45 | 11 | New Orleans-Metairie, LA Metro area | 9.7 |
| 49 | 11 | Buffalo-Cheektowaga, NY Metro area | 9.7 |
| 8 | 16 | Philadelphia-Camden-Wilmington, PA-NJ-DE-MD Metro area | 9.2 |
| 17 | 16 | San Diego-Chula Vista-Carlsbad, CA Metro area | 9.2 |
| 28 | 16 | Pittsburgh, PA Metro area | 9.2 |
| 70 | 16 | Allentown-Bethlehem-Easton, PA-NJ Metro area | 9.2 |
| 43 | 20 | Memphis, TN-MS-AR Metro area | 8.8 |
| 26 | 21 | Sacramento-Roseville-Folsom, CA Metro area | 8.7 |
| 58 | 21 | Worcester, MA-CT Metro area | 8.7 |
| 5 | 23 | Houston-The Woodlands-Sugar Land, TX Metro area | 8.6 |
| 72 | 23 | Oxnard-Thousand Oaks-Ventura, CA Metro area | 8.6 |
| 11 | 25 | Boston-Cambridge-Newton, MA-NH Metro area | 8.4 |
| 71 | 26 | El Paso, TX Metro area | 8.3 |
| 7 | 27 | Miami-Fort Lauderdale-Pompano Beach, FL Metro area | 8.2 |
| 52 | 27 | Rochester, NY Metro area | 8.2 |
| 61 | 27 | Albuquerque, NM Metro area | 8.2 |
| 12 | 30 | San Francisco-Oakland-Berkeley, CA Metro area | 8.1 |
| 15 | 30 | Seattle-Tacoma-Bellevue, WA Metro area | 8.1 |
| 68 | 32 | New Haven-Milford, CT Metro area | 8.0 |
| 59 | 33 | Bridgeport-Stamford-Norwalk, CT Metro area | 7.9 |
| 74 | 33 | Dayton-Kettering, OH Metro area | 7.9 |
| 25 | 35 | Portland-Vancouver-Hillsboro, OR-WA Metro area | 7.8 |
| 53 | 36 | Tucson, AZ Metro area | 7.7 |
| 48 | 37 | Hartford-East Hartford-Middletown, CT Metro area | 7.6 |
| 19 | 38 | Denver-Aurora-Lakewood, CO Metro area | 7.5 |
| 51 | 38 | Grand Rapids-Kentwood, MI Metro area | 7.5 |
| 66 | 38 | Baton Rouge, LA Metro area | 7.5 |
| 10 | 41 | Phoenix-Mesa-Chandler, AZ Metro area | 7.4 |
| 22 | 42 | Charlotte-Concord-Gastonia, NC-SC Metro area | 7.3 |
| 24 | 42 | San Antonio-New Braunfels, TX Metro area | 7.3 |
| 75 | 42 | Cape Coral-Fort Myers, FL Metro area | 7.3 |
| 18 | 45 | Tampa-St. Petersburg-Clearwater, FL Metro area | 7.2 |
| 63 | 45 | Albany-Schenectady-Troy, NY Metro area | 7.2 |
| 4 | 47 | Dallas-Fort Worth-Arlington, TX Metro area | 7.1 |
| 35 | 47 | San Jose-Sunnyvale-Santa Clara, CA Metro area | 7.1 |
| 37 | 47 | Virginia Beach-Norfolk-Newport News, VA-NC Metro area | 7.1 |
| 40 | 47 | Milwaukee-Waukesha, WI Metro area | 7.1 |
| 30 | 51 | Cincinnati, OH-KY-IN Metro area | 7.0 |
| 32 | 51 | Columbus, OH Metro area | 7.0 |
| 9 | 53 | Atlanta-Sandy Springs-Alpharetta, GA Metro area | 6.9 |
| 36 | 53 | Nashville-Davidson--Murfreesboro--Franklin, TN Metro area | 6.9 |
| 67 | 55 | North Port-Sarasota-Bradenton, FL Metro area | 6.8 |
| 20 | 56 | St. Louis, MO-IL Metro area | 6.7 |
| 44 | 56 | Richmond, VA Metro area | 6.7 |
| 21 | 58 | Baltimore-Columbia-Towson, MD Metro area | 6.6 |
| 33 | 58 | Indianapolis-Carmel-Anderson, IN Metro area | 6.6 |
| 46 | 58 | Louisville/Jefferson County, KY-IN Metro area | 6.6 |
| 6 | 61 | Washington-Arlington-Alexandria, DC-VA-MD-WV Metro area | 6.5 |
| 16 | 62 | Minneapolis-St. Paul-Bloomington, MN Metro area | 6.4 |
| 42 | 62 | Raleigh-Cary, NC Metro area | 6.4 |
| 54 | 62 | Tulsa, OK Metro area | 6.4 |
| 29 | 65 | Austin-Round Rock-Georgetown, TX Metro area | 6.2 |
| 31 | 65 | Kansas City, MO-KS Metro area | 6.2 |
| 39 | 65 | Jacksonville, FL Metro area | 6.2 |
| 64 | 65 | Knoxville, TN Metro area | 6.2 |
| 41 | 69 | Oklahoma City, OK Metro area | 6.1 |
| 73 | 70 | Charleston-North Charleston, SC Metro area | 6.0 |
| 60 | 71 | Greenville-Anderson, SC Metro area | 5.8 |
| 50 | 72 | Birmingham-Hoover, AL Metro area | 5.5 |
| 69 | 73 | Columbia, SC Metro area | 5.3 |
| 47 | 74 | Salt Lake City, UT Metro area | 5.1 |
| 57 | 75 | Omaha-Council Bluffs, NE-IA Metro area | 4.8 |

### Mean income tax, 2018

| Population rank | Mean income tax rank | Metropolitan area | Mean income tax (dollars) [col 199] |
|---|---|---|---|
| 59 | 1 | Bridgeport-Stamford-Norwalk, CT Metro area | 33,039 |
| 35 | 2 | San Jose-Sunnyvale-Santa Clara, CA Metro area | 31,215 |
| 12 | 3 | San Francisco-Oakland-Berkeley, CA Metro area | 25,463 |
| 11 | 4 | Boston-Cambridge-Newton, MA-NH Metro area | 18,113 |
| 1 | 5 | New York-Newark-Jersey City, NY-NJ-PA Metro area | 16,607 |
| 15 | 6 | Seattle-Tacoma-Bellevue, WA Metro area | 16,137 |
| 6 | 7 | Washington-Arlington-Alexandria, DC-VA-MD-WV Metro area | 14,676 |
| 29 | 8 | Austin-Round Rock-Georgetown, TX Metro area | 13,781 |
| 67 | 9 | North Port-Sarasota-Bradenton, FL Metro area | 13,204 |
| 7 | 10 | Miami-Fort Lauderdale-Pompano Beach, FL Metro area | 12,715 |
| 2 | 11 | Los Angeles-Long Beach-Anaheim, CA Metro area | 12,634 |
| 3 | 12 | Chicago-Naperville-Elgin, IL-IN-WI Metro area | 12,380 |
| 19 | 13 | Denver-Aurora-Lakewood, CO Metro area | 12,079 |
| 75 | 14 | Cape Coral-Fort Myers, FL Metro area | 11,892 |
| 8 | 15 | Philadelphia-Camden-Wilmington, PA-NJ-DE-MD Metro area | 11,796 |
| 16 | 16 | Minneapolis-St. Paul-Bloomington, MN Metro area | 11,494 |
| 4 | 17 | Dallas-Fort Worth-Arlington, TX Metro area | 11,451 |
| 5 | 18 | Houston-The Woodlands-Sugar Land, TX Metro area | 11,364 |
| 42 | 19 | Raleigh-Cary, NC Metro area | 11,259 |
| 72 | 20 | Oxnard-Thousand Oaks-Ventura, CA Metro area | 11,252 |
| 17 | 21 | San Diego-Chula Vista-Carlsbad, CA Metro area | 10,997 |
| 48 | 22 | Hartford-East Hartford-Middletown, CT Metro area | 10,884 |
| 36 | 23 | Nashville-Davidson--Murfreesboro--Franklin, TN Metro area | 10,829 |
| 21 | 24 | Baltimore-Columbia-Towson, MD Metro area | 10,610 |
| 63 | 25 | Albany-Schenectady-Troy, NY Metro area | 10,507 |
| 25 | 26 | Portland-Vancouver-Hillsboro, OR-WA Metro area | 10,391 |
| 9 | 27 | Atlanta-Sandy Springs-Alpharetta, GA Metro area | 10,106 |
| 22 | 28 | Charlotte-Concord-Gastonia, NC-SC Metro area | 10,011 |
| 73 | 29 | Charleston-North Charleston, SC Metro area | 9,941 |
| 20 | 30 | St. Louis, MO-IL Metro area | 9,763 |
| 40 | 31 | Milwaukee-Waukesha, WI Metro area | 9,689 |
| 30 | 32 | Cincinnati, OH-KY-IN Metro area | 9,635 |
| 27 | 33 | Las Vegas-Henderson-Paradise, NV Metro area | 9,346 |
| 31 | 34 | Kansas City, MO-KS Metro area | 9,320 |
| 68 | 35 | New Haven-Milford, CT Metro area | 9,283 |
| 14 | 36 | Detroit-Warren-Dearborn, MI Metro area | 9,272 |
| 44 | 37 | Richmond, VA Metro area | 9,258 |
| 39 | 38 | Jacksonville, FL Metro area | 9,240 |
| 50 | 39 | Birmingham-Hoover, AL Metro area | 9,096 |
| 26 | 40 | Sacramento-Roseville-Folsom, CA Metro area | 9,061 |
| 58 | 41 | Worcester, MA-CT Metro area | 9,046 |
| 10 | 42 | Phoenix-Mesa-Chandler, AZ Metro area | 9,037 |
| 28 | 43 | Pittsburgh, PA Metro area | 9,008 |
| 57 | 44 | Omaha-Council Bluffs, NE-IA Metro area | 8,898 |
| 18 | 45 | Tampa-St. Petersburg-Clearwater, FL Metro area | 8,631 |
| 33 | 46 | Indianapolis-Carmel-Anderson, IN Metro area | 8,621 |
| 34 | 47 | Cleveland-Elyria, OH Metro area | 8,579 |
| 47 | 48 | Salt Lake City, UT Metro area | 8,488 |
| 45 | 49 | New Orleans-Metairie, LA Metro area | 8,356 |
| 32 | 50 | Columbus, OH Metro area | 8,355 |
| 64 | 51 | Knoxville, TN Metro area | 8,217 |
| 51 | 52 | Grand Rapids-Kentwood, MI Metro area | 8,118 |
| 66 | 53 | Baton Rouge, LA Metro area | 8,106 |
| 38 | 54 | Providence-Warwick, RI-MA Metro area | 8,090 |
| 41 | 55 | Oklahoma City, OK Metro area | 8,062 |
| 54 | 56 | Tulsa, OK Metro area | 8,014 |
| 56 | 57 | Urban Honolulu, HI Metro area | 7,898 |
| 24 | 58 | San Antonio-New Braunfels, TX Metro area | 7,736 |
| 70 | 59 | Allentown-Bethlehem-Easton, PA-NJ Metro area | 7,589 |
| 23 | 60 | Orlando-Kissimmee-Sanford, FL Metro area | 7,511 |
| 46 | 61 | Louisville/Jefferson County, KY-IN Metro area | 7,442 |
| 43 | 62 | Memphis, TN-MS-AR Metro area | 7,331 |
| 49 | 63 | Buffalo-Cheektowaga, NY Metro area | 7,148 |
| 52 | 64 | Rochester, NY Metro area | 7,123 |
| 60 | 65 | Greenville-Anderson, SC Metro area | 6,900 |
| 37 | 66 | Virginia Beach-Norfolk-Newport News, VA-NC Metro area | 6,815 |
| 53 | 67 | Tucson, AZ Metro area | 6,738 |
| 74 | 68 | Dayton-Kettering, OH Metro area | 6,578 |
| 69 | 69 | Columbia, SC Metro area | 6,363 |
| 61 | 70 | Albuquerque, NM Metro area | 6,144 |
| 55 | 71 | Fresno, CA Metro area | 5,844 |
| 13 | 72 | Riverside-San Bernardino-Ontario, CA Metro area | 5,473 |
| 62 | 73 | Bakersfield, CA Metro area | 5,120 |
| 71 | 74 | El Paso, TX Metro area | 3,993 |
| 65 | 75 | McAllen-Edinburg-Mission, TX Metro area | 3,050 |

# 75 Largest Metropolitan Areas by 2020 Population
## Selected Rankings

### Employment in manufacturing as a percent of total nonfarm employment, 2019

| Population rank | Manu-facturing rank | Metropolitan area | Percent employed in manu-facturing [col 107/col 105] |
|---|---|---|---|
| 51 | 1 | Grand Rapids-Kentwood, MI Metro area | 22.6 |
| 60 | 2 | Greenville-Anderson, SC Metro area | 16.6 |
| 40 | 3 | Milwaukee-Waukesha, WI Metro area | 14.3 |
| 74 | 4 | Dayton-Kettering, OH Metro area | 14.1 |
| 14 | 5 | Detroit-Warren-Dearborn, MI Metro area | 13.5 |
| 46 | 6 | Louisville/Jefferson County, KY-IN Metro area | 13.2 |
| 34 | 7 | Cleveland-Elyria, OH Metro area | 12.9 |
| 54 | 8 | Tulsa, OK Metro area | 12.9 |
| 52 | 9 | Rochester, NY Metro area | 12.1 |
| 48 | 10 | Hartford-East Hartford-Middletown, CT Metro area | 12.0 |
| 58 | 11 | Worcester, MA-CT Metro area | 11.7 |
| 30 | 12 | Cincinnati, OH-KY-IN Metro area | 11.5 |
| 70 | 13 | Allentown-Bethlehem-Easton, PA-NJ Metro area | 10.9 |
| 64 | 14 | Knoxville, TN Metro area | 10.9 |
| 49 | 15 | Buffalo-Cheektowaga, NY Metro area | 10.9 |
| 25 | 16 | Portland-Vancouver-Hillsboro, OR-WA Metro area | 10.8 |
| 16 | 17 | Minneapolis-St. Paul-Bloomington, MN Metro area | 10.6 |
| 38 | 18 | Providence-Warwick, RI-MA Metro area | 9.9 |
| 73 | 19 | Charleston-North Charleston, SC Metro area | 9.7 |
| 55 | 20 | Fresno, CA Metro area | 9.6 |
| 72 | 21 | Oxnard-Thousand Oaks-Ventura, CA Metro area | 9.5 |
| 22 | 22 | Charlotte-Concord-Gastonia, NC-SC Metro area | 9.3 |
| 3 | 23 | Chicago-Naperville-Elgin, IL-IN-WI Metro area | 9.1 |
| 15 | 24 | Seattle-Tacoma-Bellevue, WA Metro area | 9.1 |
| 69 | 25 | Columbia, SC Metro area | 9.0 |
| 37 | 26 | Virginia Beach-Norfolk-Newport News, VA-NC Metro area | 9.0 |
| 20 | 27 | St. Louis, MO-IL Metro area | 9.0 |
| 31 | 28 | Kansas City, MO-KS Metro area | 8.8 |
| 47 | 29 | Salt Lake City, UT Metro area | 8.8 |
| 2 | 30 | Los Angeles-Long Beach-Anaheim, CA Metro area | 8.6 |
| 36 | 31 | Nashville-Davidson--Murfreesboro--Franklin, TN Metro area | 8.5 |
| 33 | 32 | Indianapolis-Carmel-Anderson, IN Metro area | 8.5 |
| 68 | 33 | New Haven-Milford, CT Metro area | 8.3 |
| 50 | 34 | Birmingham-Hoover, AL Metro area | 8.2 |
| 13 | 35 | Riverside-San Bernardino-Ontario, CA Metro area | 8.2 |
| 59 | 36 | Bridgeport-Stamford-Norwalk, CT Metro area | 8.2 |
| 17 | 37 | San Diego-Chula Vista-Carlsbad, CA Metro area | 8.1 |
| 35 | 38 | San Jose-Sunnyvale-Santa Clara, CA Metro area | 7.9 |
| 4 | 39 | Dallas-Fort Worth-Arlington, TX Metro area | 7.9 |
| 28 | 40 | Pittsburgh, PA Metro area | 7.8 |
| 57 | 41 | Omaha-Council Bluffs, NE-IA Metro area | 7.8 |
| 32 | 42 | Columbus, OH Metro area | 7.7 |
| 5 | 43 | Houston-The Woodlands-Sugar Land, TX Metro area | 7.6 |
| 66 | 44 | Baton Rouge, LA Metro area | 7.4 |
| 53 | 45 | Tucson, AZ Metro area | 7.1 |
| 43 | 46 | Memphis, TN-MS-AR Metro area | 7.0 |
| 63 | 47 | Albany-Schenectady-Troy, NY Metro area | 6.9 |
| 62 | 48 | Bakersfield, CA Metro area | 6.6 |
| 71 | 49 | El Paso, TX Metro area | 6.6 |
| 9 | 50 | Atlanta-Sandy Springs-Alpharetta, GA Metro area | 6.5 |
| 8 | 51 | Philadelphia-Camden-Wilmington, PA-NJ-DE-MD Metro area | 6.4 |
| 11 | 52 | Boston-Cambridge-Newton, MA-NH Metro area | 6.3 |
| 10 | 53 | Phoenix-Mesa-Chandler, AZ Metro area | 6.2 |
| 12 | 54 | San Francisco-Oakland-Berkeley, CA Metro area | 6.1 |
| 67 | 55 | North Port-Sarasota-Bradenton, FL Metro area | 5.9 |
| 41 | 56 | Oklahoma City, OK Metro area | 5.6 |
| 44 | 57 | Richmond, VA Metro area | 5.4 |
| 24 | 58 | San Antonio-New Braunfels, TX Metro area | 5.3 |
| 18 | 59 | Tampa-St. Petersburg-Clearwater, FL Metro area | 5.2 |
| 45 | 60 | New Orleans-Metairie, LA Metro area | 5.2 |
| 61 | 61 | Albuquerque, NM Metro area | 5.0 |
| 29 | 62 | Austin-Round Rock-Georgetown, TX Metro area | 4.9 |
| 39 | 63 | Jacksonville, FL Metro area | 4.7 |
| 42 | 64 | Raleigh-Cary, NC Metro area | 4.7 |
| 26 | 65 | Sacramento-Roseville-Folsom, CA Metro area | 4.6 |
| 21 | 66 | Baltimore-Columbia-Towson, MD Metro area | 4.5 |
| 19 | 67 | Denver-Aurora-Lakewood, CO Metro area | 4.3 |
| 1 | 68 | New York-Newark-Jersey City, NY-NJ-PA Metro area | 3.7 |
| 23 | 69 | Orlando-Kissimmee-Sanford, FL Metro area | 3.6 |
| 7 | 70 | Miami-Fort Lauderdale-Pompano Beach, FL Metro area | 3.3 |
| 65 | 71 | McAllen-Edinburg-Mission, TX Metro area | 3.1 |
| 56 | 72 | Urban Honolulu, HI Metro area | 2.6 |
| 27 | 73 | Las Vegas-Henderson-Paradise, NV Metro area | 2.5 |
| 75 | 74 | Cape Coral-Fort Myers, FL Metro area | 2.3 |
| 6 | 75 | Washington-Arlington-Alexandria, DC-VA-MD-WV Metro area | 1.8 |

### Employment in professional, scientific, and technical services as a percent of total nonfarm employment, 2019

| Population rank | Profess-ional services rank | Metropolitan area | Percent employed in services [col 110/col 105] |
|---|---|---|---|
| 6 | 1 | Washington-Arlington-Alexandria, DC-VA-MD-WV Metro area | 20.9 |
| 35 | 2 | San Jose-Sunnyvale-Santa Clara, CA Metro area | 13.1 |
| 29 | 3 | Austin-Round Rock-Georgetown, TX Metro area | 12.4 |
| 12 | 4 | San Francisco-Oakland-Berkeley, CA Metro area | 12.1 |
| 17 | 4 | San Diego-Chula Vista-Carlsbad, CA Metro area | 12.1 |
| 21 | 6 | Baltimore-Columbia-Towson, MD Metro area | 11.5 |
| 42 | 7 | Raleigh-Cary, NC Metro area | 11.2 |
| 11 | 8 | Boston-Cambridge-Newton, MA-NH Metro area | 10.8 |
| 61 | 8 | Albuquerque, NM Metro area | 10.8 |
| 14 | 10 | Detroit-Warren-Dearborn, MI Metro area | 10.5 |
| 9 | 11 | Atlanta-Sandy Springs-Alpharetta, GA Metro area | 9.3 |
| 59 | 12 | Bridgeport-Stamford-Norwalk, CT Metro area | 9.2 |
| 63 | 13 | Albany-Schenectady-Troy, NY Metro area | 9.1 |
| 1 | 14 | New York-Newark-Jersey City, NY-NJ-PA Metro area | 9.0 |
| 19 | 15 | Denver-Aurora-Lakewood, CO Metro area | 8.9 |
| 31 | 15 | Kansas City, MO-KS Metro area | 8.9 |
| 37 | 17 | Virginia Beach-Norfolk-Newport News, VA-NC Metro area | 8.4 |
| 3 | 18 | Chicago-Naperville-Elgin, IL-IN-WI Metro area | 8.3 |
| 18 | 18 | Tampa-St. Petersburg-Clearwater, FL Metro area | 8.3 |
| 15 | 20 | Seattle-Tacoma-Bellevue, WA Metro area | 8.2 |
| 2 | 21 | Los Angeles-Long Beach-Anaheim, CA Metro area | 8.1 |
| 72 | 21 | Oxnard-Thousand Oaks-Ventura, CA Metro area | 8.1 |
| 74 | 21 | Dayton-Kettering, OH Metro area | 8.1 |
| 5 | 24 | Houston-The Woodlands-Sugar Land, TX Metro area | 7.9 |
| 26 | 25 | Sacramento-Roseville-Folsom, CA Metro area | 7.8 |
| 7 | 26 | Miami-Fort Lauderdale-Pompano Beach, FL Metro area | 7.7 |
| 73 | 26 | Charleston-North Charleston, SC Metro area | 7.7 |
| 4 | 28 | Dallas-Fort Worth-Arlington, TX Metro area | 7.6 |
| 8 | 28 | Philadelphia-Camden-Wilmington, PA-NJ-DE-MD Metro area | 7.6 |
| 44 | 28 | Richmond, VA Metro area | 7.6 |
| 47 | 28 | Salt Lake City, UT Metro area | 7.6 |
| 20 | 32 | St. Louis, MO-IL Metro area | 7.1 |
| 33 | 32 | Indianapolis-Carmel-Anderson, IN Metro area | 7.1 |
| 23 | 34 | Orlando-Kissimmee-Sanford, FL Metro area | 7.0 |
| 25 | 34 | Portland-Vancouver-Hillsboro, OR-WA Metro area | 7.0 |
| 66 | 34 | Baton Rouge, LA Metro area | 6.8 |
| 10 | 37 | Phoenix-Mesa-Chandler, AZ Metro area | 6.8 |
| 16 | 37 | Minneapolis-St. Paul-Bloomington, MN Metro area | 6.8 |
| 64 | 37 | Knoxville, TN Metro area | 6.7 |
| 28 | 40 | Pittsburgh, PA Metro area | 6.7 |
| 32 | 40 | Columbus, OH Metro area | 6.7 |
| 60 | 42 | Greenville-Anderson, SC Metro area | 6.6 |
| 52 | 43 | Rochester, NY Metro area | 6.5 |
| 30 | 44 | Cincinnati, OH-KY-IN Metro area | 6.4 |
| 22 | 45 | Charlotte-Concord-Gastonia, NC-SC Metro area | 6.3 |
| 45 | 45 | New Orleans-Metairie, LA Metro area | 6.3 |
| 48 | 45 | Hartford-East Hartford-Middletown, CT Metro area | 6.3 |
| 41 | 48 | Oklahoma City, OK Metro area | 6.2 |
| 54 | 49 | Tulsa, OK Metro area | 6.1 |
| 67 | 49 | North Port-Sarasota-Bradenton, FL Metro area | 6.1 |
| 39 | 51 | Jacksonville, FL Metro area | 6.0 |
| 49 | 51 | Buffalo-Cheektowaga, NY Metro area | 6.0 |
| 58 | 51 | Worcester, MA-CT Metro area | 6.0 |
| 24 | 54 | San Antonio-New Braunfels, TX Metro area | 5.8 |
| 34 | 54 | Cleveland-Elyria, OH Metro area | 5.8 |
| 69 | 54 | Columbia, SC Metro area | 5.8 |
| 75 | 54 | Cape Coral-Fort Myers, FL Metro area | 5.8 |
| 57 | 58 | Omaha-Council Bluffs, NE-IA Metro area | 5.5 |
| 36 | 59 | Nashville-Davidson--Murfreesboro--Franklin, TN Metro area | 5.4 |
| 53 | 60 | Tucson, AZ Metro area | 5.3 |
| 40 | 61 | Milwaukee-Waukesha, WI Metro area | 5.2 |
| 56 | 62 | Urban Honolulu, HI Metro area | 5.1 |
| 50 | 63 | Birmingham-Hoover, AL Metro area | 5.0 |
| 62 | 64 | Bakersfield, CA Metro area | 4.8 |
| 27 | 65 | Las Vegas-Henderson-Paradise, NV Metro area | 4.7 |
| 46 | 66 | Louisville/Jefferson County, KY-IN Metro area | 4.5 |
| 38 | 67 | Providence-Warwick, RI-MA Metro area | 4.3 |
| 51 | 68 | Grand Rapids-Kentwood, MI Metro area | 4.2 |
| 68 | 68 | New Haven-Milford, CT Metro area | 4.2 |
| 43 | 70 | Memphis, TN-MS-AR Metro area | 3.9 |
| 55 | 70 | Fresno, CA Metro area | 3.9 |
| 70 | 70 | Allentown-Bethlehem-Easton, PA-NJ Metro area | 3.9 |
| 71 | 73 | El Paso, TX Metro area | 3.7 |
| 13 | 74 | Riverside-San Bernardino-Ontario, CA Metro area | 3.2 |
| 65 | 75 | McAllen-Edinburg-Mission, TX Metro area | 3.1 |

# 75 Largest Metropolitan Areas by 2020 Population
## Selected Rankings

| | Per capita local government taxes, 2017 | | | Violent crime rate, 2019 (violent crimes known to police) | | | |
|---|---|---|---|---|---|---|---|
| Popu-lation rank | Local taxes rank | Metropolitan area | Local per capita taxes (dollars) [col 183] | Popu-lation rank | Crime rate rank | Metropolitan area | Crime rate (per 100,000 population) [col 45] |
| 1 | 1 | New York-Newark-Jersey City, NY-NJ-PA Metro area | 5,192 | 43 | 1 | Memphis, TN-MS-AR Metro area | 1,120.5 |
| 6 | 2 | Washington-Arlington-Alexandria, DC-VA-MD-WV Metro area | 3,965 | 61 | 2 | Albuquerque, NM Metro area | 1,043.4 |
| 59 | 3 | Bridgeport-Stamford-Norwalk, CT Metro area | 3,926 | 21 | 3 | Baltimore-Columbia-Towson, MD Metro area | 698.7 |
| 35 | 4 | San Jose-Sunnyvale-Santa Clara, CA Metro area | 3,517 | 62 | 4 | Bakersfield, CA Metro area | 621.1 |
| 12 | 5 | San Francisco-Oakland-Berkeley, CA Metro area | 3,419 | 69 | 5 | Columbia, SC Metro area | 583.4 |
| 3 | 6 | Chicago-Naperville-Elgin, IL-IN-WI Metro area | 3,173 | 40 | 6 | Milwaukee-Waukesha, WI Metro area | 577.9 |
| 63 | 7 | Albany-Schenectady-Troy, NY Metro area | 3,142 | 36 | 7 | Nashville-Davidson--Murfreesboro--Franklin, TN Metro area | 570.5 |
| 29 | 8 | Austin-Round Rock-Georgetown, TX Metro area | 2,907 | 66 | 8 | Baton Rouge, LA Metro area | 554.0 |
| 48 | 9 | Hartford-East Hartford-Middletown, CT Metro area | 2,860 | 45 | 9 | New Orleans-Metairie, LA Metro area | 551.5 |
| 52 | 10 | Rochester, NY Metro area | 2,854 | 54 | 10 | Tulsa, OK Metro area | 529.6 |
| 11 | 11 | Boston-Cambridge-Newton, MA-NH Metro area | 2,821 | 27 | 11 | Las Vegas-Henderson-Paradise, NV Metro area | 525.7 |
| 34 | 12 | Cleveland-Elyria, OH Metro area | 2,771 | 14 | 12 | Detroit-Warren-Dearborn, MI Metro area | 516.9 |
| 49 | 13 | Buffalo-Cheektowaga, NY Metro area | 2,736 | 24 | 13 | San Antonio-New Braunfels, TX Metro area | 515.9 |
| 8 | 14 | Philadelphia-Camden-Wilmington, PA-NJ-DE-MD Metro area | 2,726 | 2 | 14 | Los Angeles-Long Beach-Anaheim, CA Metro area | 477.5 |
| 68 | 15 | New Haven-Milford, CT Metro area | 2,680 | 39 | 15 | Jacksonville, FL Metro area | 474.7 |
| 15 | 16 | Seattle-Tacoma-Bellevue, WA Metro area | 2,624 | 12 | 16 | San Francisco-Oakland-Berkeley, CA Metro area | 470.8 |
| 19 | 17 | Denver-Aurora-Lakewood, CO Metro area | 2,616 | 41 | 17 | Oklahoma City, OK Metro area | 461.0 |
| 32 | 18 | Columbus, OH Metro area | 2,597 | 53 | 18 | Tucson, AZ Metro area | 447.4 |
| 5 | 19 | Houston-The Woodlands-Sugar Land, TX Metro area | 2,559 | 46 | 19 | Louisville/Jefferson County, KY-IN Metro area | 436.8 |
| 4 | 20 | Dallas-Fort Worth-Arlington, TX Metro area | 2,492 | 23 | 20 | Orlando-Kissimmee-Sanford, FL Metro area | 429.9 |
| 21 | 21 | Baltimore-Columbia-Towson, MD Metro area | 2,488 | 10 | 21 | Phoenix-Mesa-Chandler, AZ Metro area | 426.5 |
| 70 | 22 | Allentown-Bethlehem-Easton, PA-NJ Metro area | 2,464 | 13 | 22 | Riverside-San Bernardino-Ontario, CA Metro area | 425.9 |
| 2 | 23 | Los Angeles-Long Beach-Anaheim, CA Metro area | 2,356 | 7 | 23 | Miami-Fort Lauderdale-Pompano Beach, FL Metro area | 422.0 |
| 7 | 24 | Miami-Fort Lauderdale-Pompano Beach, FL Metro area | 2,266 | 57 | 24 | Omaha-Council Bluffs, NE-IA Metro area | 413.1 |
| 38 | 25 | Providence-Warwick, RI-MA Metro area | 2,248 | 34 | 25 | Cleveland-Elyria, OH Metro area | 403.3 |
| 74 | 26 | Dayton-Kettering, OH Metro area | 2,194 | 73 | 26 | Charleston-North Charleston, SC Metro area | 398.0 |
| 57 | 27 | Omaha-Council Bluffs, NE-IA Metro area | 2,188 | 47 | 27 | Salt Lake City, UT Metro area | 366.7 |
| 25 | 28 | Portland-Vancouver-Hillsboro, OR-WA Metro area | 2,186 | 49 | 28 | Buffalo-Cheektowaga, NY Metro area | 354.3 |
| 73 | 29 | Charleston-North Charleston, SC Metro area | 2,182 | 26 | 29 | Sacramento-Roseville-Folsom, CA Metro area | 354.1 |
| 45 | 30 | New Orleans-Metairie, LA Metro area | 2,147 | 64 | 30 | Knoxville, TN Metro area | 353.7 |
| 31 | 31 | Kansas City, MO-KS Metro area | 2,097 | 37 | 31 | Virginia Beach-Norfolk-Newport News, VA-NC Metro area | 349.5 |
| 24 | 32 | San Antonio-New Braunfels, TX Metro area | 2,088 | 17 | 32 | San Diego-Chula Vista-Carlsbad, CA Metro area | 341.2 |
| 28 | 33 | Pittsburgh, PA Metro area | 2,086 | 71 | 33 | El Paso, TX Metro area | 333.2 |
| 17 | 34 | San Diego-Chula Vista-Carlsbad, CA Metro area | 2,082 | 67 | 34 | North Port-Sarasota-Bradenton, FL Metro area | 328.4 |
| 66 | 35 | Baton Rouge, LA Metro area | 2,067 | 74 | 34 | Dayton-Kettering, OH Metro area | 328.4 |
| 20 | 36 | St. Louis, MO-IL Metro area | 2,037 | 35 | 36 | San Jose-Sunnyvale-Santa Clara, CA Metro area | 325.0 |
| 67 | 37 | North Port-Sarasota-Bradenton, FL Metro area | 2,003 | 51 | 37 | Grand Rapids-Kentwood, MI Metro area | 321.5 |
| 30 | 38 | Cincinnati, OH-KY-IN Metro area | 1,975 | 58 | 38 | Worcester, MA-CT Metro area | 310.4 |
| 37 | 39 | Virginia Beach-Norfolk-Newport News, VA-NC Metro area | 1,966 | 25 | 39 | Portland-Vancouver-Hillsboro, OR-WA Metro area | 302.6 |
| 58 | 40 | Worcester, MA-CT Metro area | 1,956 | 18 | 40 | Tampa-St. Petersburg-Clearwater, FL Metro area | 292.1 |
| 72 | 41 | Oxnard-Thousand Oaks-Ventura, CA Metro area | 1,866 | 29 | 41 | Austin-Round Rock-Georgetown, TX Metro area | 286.9 |
| 40 | 42 | Milwaukee-Waukesha, WI Metro area | 1,840 | 32 | 42 | Columbus, OH Metro area | 286.5 |
| 26 | 43 | Sacramento-Roseville-Folsom, CA Metro area | 1,802 | 38 | 43 | Providence-Warwick, RI-MA Metro area | 282.6 |
| 50 | 44 | Birmingham-Hoover, AL Metro area | 1,788 | 11 | 44 | Boston-Cambridge-Newton, MA-NH Metro area | 277.7 |
| 69 | 45 | Columbia, SC Metro area | 1,777 | 68 | 45 | New Haven-Milford, CT Metro area | 272.8 |
| 47 | 46 | Salt Lake City, UT Metro area | 1,769 | 16 | 46 | Minneapolis-St. Paul-Bloomington, MN Metro area | 272.5 |
| 44 | 47 | Richmond, VA Metro area | 1,766 | 56 | 47 | Urban Honolulu, HI Metro area | 270.6 |
| 9 | 48 | Atlanta-Sandy Springs-Alpharetta, GA Metro area | 1,753 | 65 | 48 | McAllen-Edinburg-Mission, TX Metro area | 268.9 |
| 75 | 49 | Cape Coral-Fort Myers, FL Metro area | 1,735 | 75 | 49 | Cape Coral-Fort Myers, FL Metro area | 257.8 |
| 23 | 50 | Orlando-Kissimmee-Sanford, FL Metro area | 1,701 | 63 | 50 | Albany-Schenectady-Troy, NY Metro area | 257.5 |
| 36 | 51 | Nashville-Davidson--Murfreesboro--Franklin, TN Metro area | 1,686 | 52 | 51 | Rochester, NY Metro area | 248.7 |
| 16 | 52 | Minneapolis-St. Paul-Bloomington, MN Metro area | 1,670 | 30 | 52 | Cincinnati, OH-KY-IN Metro area | 237.4 |
| 56 | 53 | Urban Honolulu, HI Metro area | 1,662 | 44 | 53 | Richmond, VA Metro area | 227.7 |
| 27 | 54 | Las Vegas-Henderson-Paradise, NV Metro area | 1,657 | 72 | 54 | Oxnard-Thousand Oaks-Ventura, CA Metro area | 216.0 |
| 22 | 55 | Charlotte-Concord-Gastonia, NC-SC Metro area | 1,655 | 48 | 55 | Hartford-East Hartford-Middletown, CT Metro area | 199.0 |
| 43 | 56 | Memphis, TN-MS-AR Metro area | 1,604 | 59 | 56 | Bridgeport-Stamford-Norwalk, CT Metro area | 174.2 |
| 71 | 57 | El Paso, TX Metro area | 1,584 | 42 | 57 | Raleigh-Cary, NC Metro area | 172.6 |
| 42 | 58 | Raleigh-Cary, NC Metro area | 1,579 | 1 | NA | New York-Newark-Jersey City, NY-NJ-PA Metro area | NA |
| 54 | 59 | Tulsa, OK Metro area | 1,559 | 3 | NA | Chicago-Naperville-Elgin, IL-IN-WI Metro area | NA |
| 10 | 60 | Phoenix-Mesa-Chandler, AZ Metro area | 1,550 | 4 | NA | Dallas-Fort Worth-Arlington, TX Metro area | NA |
| 46 | 61 | Louisville/Jefferson County, KY-IN Metro area | 1,533 | 5 | NA | Houston-The Woodlands-Sugar Land, TX Metro area | NA |
| 13 | 62 | Riverside-San Bernardino-Ontario, CA Metro area | 1,510 | 6 | NA | Washington-Arlington-Alexandria, DC-VA-MD-WV Metro area | NA |
| 53 | 63 | Tucson, AZ Metro area | 1,500 | 8 | NA | Philadelphia-Camden-Wilmington, PA-NJ-DE-MD Metro area | NA |
| 18 | 64 | Tampa-St. Petersburg-Clearwater, FL Metro area | 1,475 | 9 | NA | Atlanta-Sandy Springs-Alpharetta, GA Metro area | NA |
| 41 | 65 | Oklahoma City, OK Metro area | 1,459 | 15 | NA | Seattle-Tacoma-Bellevue, WA Metro area | NA |
| 39 | 66 | Jacksonville, FL Metro area | 1,426 | 19 | NA | Denver-Aurora-Lakewood, CO Metro area | NA |
| 14 | 67 | Detroit-Warren-Dearborn, MI Metro area | 1,401 | 20 | NA | St. Louis, MO-IL Metro area | NA |
| 62 | 68 | Bakersfield, CA Metro area | 1,378 | 22 | NA | Charlotte-Concord-Gastonia, NC-SC Metro area | NA |
| 61 | 69 | Albuquerque, NM Metro area | 1,338 | 28 | NA | Pittsburgh, PA Metro area | NA |
| 64 | 70 | Knoxville, TN Metro area | 1,307 | 31 | NA | Kansas City, MO-KS Metro area | NA |
| 55 | 71 | Fresno, CA Metro area | 1,304 | 33 | NA | Indianapolis-Carmel-Anderson, IN Metro area | NA |
| 51 | 72 | Grand Rapids-Kentwood, MI Metro area | 1,235 | 50 | NA | Birmingham-Hoover, AL Metro area | NA |
| 60 | 73 | Greenville-Anderson, SC Metro area | 1,181 | 55 | NA | Fresno, CA Metro area | NA |
| 33 | 74 | Indianapolis-Carmel-Anderson, IN Metro area | 1,165 | 60 | NA | Greenville-Anderson, SC Metro area | NA |
| 65 | 75 | McAllen-Edinburg-Mission, TX Metro area | 1,162 | 70 | NA | Allentown-Bethlehem-Easton, PA-NJ Metro area | NA |

# 75 Largest Metropolitan Areas by 2020 Population
## Selected Rankings

| Nonemployer businesses, 2018 | | | | Value of residential construction authorized by building permits, 2020 | | | |
| --- | --- | --- | --- | --- | --- | --- | --- |
| Population rank | Non-employer businesses rank | Metropolitan area | Non-employer businesses [col 167] | Population rank | New construction rank | Metropolitan area | New construction ($1,000) [col 169] |
| 1 | 1 | New York-Newark-Jersey City, NY-NJ-PA Metro area | 1,974,310 | 4 | 1 | Dallas-Fort Worth-Arlington, TX Metro area | 13,053,058 |
| 2 | 2 | Los Angeles-Long Beach-Anaheim, CA Metro area | 1,430,673 | 5 | 2 | Houston-The Woodlands-Sugar Land, TX Metro area | 12,702,945 |
| 7 | 3 | Miami-Fort Lauderdale-Pompano Beach, FL Metro area | 1,052,550 | 10 | 3 | Phoenix-Mesa-Chandler, AZ Metro area | 10,815,836 |
| 3 | 4 | Chicago-Naperville-Elgin, IL-IN-WI Metro area | 814,273 | 1 | 4 | New York-Newark-Jersey City, NY-NJ-PA Metro area | 7,744,635 |
| 4 | 5 | Dallas-Fort Worth-Arlington, TX Metro area | 701,565 | 9 | 5 | Atlanta-Sandy Springs-Alpharetta, GA Metro area | 7,156,853 |
| 5 | 6 | Houston-The Woodlands-Sugar Land, TX Metro area | 663,771 | 29 | 6 | Austin-Round Rock-Georgetown, TX Metro area | 7,147,718 |
| 9 | 7 | Atlanta-Sandy Springs-Alpharetta, GA Metro area | 629,914 | 2 | 7 | Los Angeles-Long Beach-Anaheim, CA Metro area | 6,225,861 |
| 6 | 8 | Washington-Arlington-Alexandria, DC-VA-MD-WV Metro area | 585,667 | 22 | 8 | Charlotte-Concord-Gastonia, NC-SC Metro area | 5,633,895 |
| 12 | 9 | San Francisco-Oakland-Berkeley, CA Metro area | 463,717 | 23 | 9 | Orlando-Kissimmee-Sanford, FL Metro area | 5,478,188 |
| 8 | 10 | Philadelphia-Camden-Wilmington, PA-NJ-DE-MD Metro area. | 452,633 | 18 | 10 | Tampa-St. Petersburg-Clearwater, FL Metro area | 5,299,528 |
| 11 | 11 | Boston-Cambridge-Newton, MA-NH Metro area | 424,719 | 36 | 11 | Nashville-Davidson--Murfreesboro--Franklin, TN Metro area | 5,183,315 |
| 10 | 12 | Phoenix-Mesa-Chandler, AZ Metro area | 356,955 | 16 | 12 | Minneapolis-St. Paul-Bloomington, MN Metro area | 4,962,093 |
| 14 | 13 | Detroit-Warren-Dearborn, MI Metro area | 351,743 | 15 | 13 | Seattle-Tacoma-Bellevue, WA Metro area | 4,863,438 |
| 13 | 14 | Riverside-San Bernardino-Ontario, CA Metro area | 330,713 | 7 | 14 | Miami-Fort Lauderdale-Pompano Beach, FL Metro area | 4,762,137 |
| 17 | 15 | San Diego-Chula Vista-Carlsbad, CA Metro area | 292,941 | 6 | 15 | Washington-Arlington-Alexandria, DC-VA-MD-WV Metro area | 4,716,373 |
| 18 | 16 | Tampa-St. Petersburg-Clearwater, FL Metro area | 286,933 | 19 | 16 | Denver-Aurora-Lakewood, CO Metro area | 4,431,755 |
| 15 | 17 | Seattle-Tacoma-Bellevue, WA Metro area | 282,855 | 11 | 17 | Boston-Cambridge-Newton, MA-NH Metro area | 3,885,306 |
| 19 | 18 | Denver-Aurora-Lakewood, CO Metro area | 281,622 | 13 | 18 | Riverside-San Bernardino-Ontario, CA Metro area | 3,673,835 |
| 16 | 19 | Minneapolis-St. Paul-Bloomington, MN Metro area | 279,511 | 42 | 19 | Raleigh-Cary, NC Metro area | 3,460,650 |
| 23 | 20 | Orlando-Kissimmee-Sanford, FL Metro area | 270,929 | 39 | 20 | Jacksonville, FL Metro area | 3,262,575 |
| 21 | 21 | Baltimore-Columbia-Towson, MD Metro area | 219,643 | 3 | 21 | Chicago-Naperville-Elgin, IL-IN-WI Metro area | 3,204,665 |
| 29 | 22 | Austin-Round Rock-Georgetown, TX Metro area | 212,611 | 12 | 22 | San Francisco-Oakland-Berkeley, CA Metro area | 3,163,103 |
| 22 | 23 | Charlotte-Concord-Gastonia, NC-SC Metro area | 212,048 | 24 | 23 | San Antonio-New Braunfels, TX Metro area | 3,123,325 |
| 24 | 24 | San Antonio-New Braunfels, TX Metro area | 199,220 | 25 | 24 | Portland-Vancouver-Hillsboro, OR-WA Metro area | 2,920,301 |
| 20 | 25 | St. Louis, MO-IL Metro area | 193,944 | 33 | 25 | Indianapolis-Carmel-Anderson, IN Metro area | 2,827,792 |
| 25 | 26 | Portland-Vancouver-Hillsboro, OR-WA Metro area | 192,748 | 27 | 26 | Las Vegas-Henderson-Paradise, NV Metro area | 2,824,227 |
| 27 | 27 | Las Vegas-Henderson-Paradise, NV Metro area | 189,386 | 26 | 27 | Sacramento-Roseville-Folsom, CA Metro area | 2,728,908 |
| 36 | 28 | Nashville-Davidson--Murfreesboro--Franklin, TN Metro area | 187,108 | 8 | 28 | Philadelphia-Camden-Wilmington, PA-NJ-DE-MD Metro area | 2,466,072 |
| 26 | 29 | Sacramento-Roseville-Folsom, CA Metro area | 179,529 | 67 | 29 | North Port-Sarasota-Bradenton, FL Metro area | 2,341,559 |
| 32 | 30 | Columbus, OH Metro area | 161,295 | 31 | 30 | Kansas City, MO-KS Metro area | 2,238,984 |
| 34 | 31 | Cleveland-Elyria, OH Metro area | 157,548 | 47 | 31 | Salt Lake City, UT Metro area | 2,227,894 |
| 31 | 32 | Kansas City, MO-KS Metro area | 153,741 | 32 | 32 | Columbus, OH Metro area | 2,226,124 |
| 35 | 33 | San Jose-Sunnyvale-Santa Clara, CA Metro area | 152,926 | 17 | 33 | San Diego-Chula Vista-Carlsbad, CA Metro area | 1,997,782 |
| 28 | 34 | Pittsburgh, PA Metro area | 152,088 | 75 | 34 | Cape Coral-Fort Myers, FL Metro area | 1,944,757 |
| 30 | 35 | Cincinnati, OH-KY-IN Metro area | 151,238 | 20 | 35 | St. Louis, MO-IL Metro area | 1,910,035 |
| 33 | 36 | Indianapolis-Carmel-Anderson, IN Metro area | 148,767 | 73 | 36 | Charleston-North Charleston, SC Metro area | 1,842,265 |
| 45 | 37 | New Orleans-Metairie, LA Metro area | 126,876 | 60 | 37 | Greenville-Anderson, SC Metro area | 1,831,017 |
| 39 | 38 | Jacksonville, FL Metro area | 125,333 | 14 | 38 | Detroit-Warren-Dearborn, MI Metro area | 1,793,207 |
| 38 | 39 | Providence-Warwick, RI-MA Metro area | 119,157 | 41 | 39 | Oklahoma City, OK Metro area | 1,665,298 |
| 43 | 40 | Memphis, TN-MS-AR Metro area | 116,228 | 30 | 40 | Cincinnati, OH-KY-IN Metro area | 1,657,987 |
| 41 | 41 | Oklahoma City, OK Metro area | 114,673 | 21 | 41 | Baltimore-Columbia-Towson, MD Metro area | 1,609,153 |
| 42 | 42 | Raleigh-Cary, NC Metro area | 114,345 | 53 | 42 | Tucson, AZ Metro area | 1,394,415 |
| 37 | 43 | Virginia Beach-Norfolk-Newport News, VA-NC Metro area | 112,650 | 37 | 43 | Virginia Beach-Norfolk-Newport News, VA-NC Metro area | 1,322,622 |
| 59 | 44 | Bridgeport-Stamford-Norwalk, CT Metro area | 97,854 | 44 | 44 | Richmond, VA Metro area | 1,302,014 |
| 47 | 45 | Salt Lake City, UT Metro area | 96,515 | 35 | 45 | San Jose-Sunnyvale-Santa Clara, CA Metro area | 1,148,099 |
| 40 | 46 | Milwaukee-Waukesha, WI Metro area | 95,058 | 28 | 46 | Pittsburgh, PA Metro area | 1,140,106 |
| 44 | 47 | Richmond, VA Metro area | 92,477 | 69 | 47 | Columbia, SC Metro area | 1,076,211 |
| 46 | 48 | Louisville/Jefferson County, KY-IN Metro area | 90,753 | 64 | 48 | Knoxville, TN Metro area | 1,061,241 |
| 48 | 49 | Hartford-East Hartford-Middletown, CT Metro area | 84,370 | 50 | 49 | Birmingham-Hoover, AL Metro area | 1,036,613 |
| 50 | 50 | Birmingham-Hoover, AL Metro area | 82,023 | 46 | 50 | Louisville/Jefferson County, KY-IN Metro area | 985,910 |
| 67 | 51 | North Port-Sarasota-Bradenton, FL Metro area | 79,699 | 54 | 51 | Tulsa, OK Metro area | 964,942 |
| 51 | 52 | Grand Rapids-Kentwood, MI Metro area | 76,887 | 45 | 52 | New Orleans-Metairie, LA Metro area | 932,569 |
| 54 | 53 | Tulsa, OK Metro area | 76,491 | 43 | 53 | Memphis, TN-MS-AR Metro area | 932,305 |
| 65 | 54 | McAllen-Edinburg-Mission, TX Metro area | 76,358 | 34 | 54 | Cleveland-Elyria, OH Metro area | 924,603 |
| 75 | 55 | Cape Coral-Fort Myers, FL Metro area | 73,304 | 57 | 55 | Omaha-Council Bluffs, NE-IA Metro area | 905,891 |
| 72 | 56 | Oxnard-Thousand Oaks-Ventura, CA Metro area | 70,479 | 55 | 56 | Fresno, CA Metro area | 862,554 |
| 56 | 57 | Urban Honolulu, HI Metro area | 69,723 | 66 | 57 | Baton Rouge, LA Metro area | 834,122 |
| 53 | 58 | Tucson, AZ Metro area | 69,496 | 51 | 58 | Grand Rapids-Kentwood, MI Metro area | 825,499 |
| 66 | 59 | Baton Rouge, LA Metro area | 68,952 | 40 | 59 | Milwaukee-Waukesha, WI Metro area | 775,816 |
| 52 | 60 | Rochester, NY Metro area | 67,363 | 65 | 60 | McAllen-Edinburg-Mission, TX Metro area | 774,932 |
| 73 | 61 | Charleston-North Charleston, SC Metro area | 67,334 | 62 | 61 | Bakersfield, CA Metro area | 571,907 |
| 60 | 62 | Greenville-Anderson, SC Metro area | 66,575 | 71 | 62 | El Paso, TX Metro area | 571,275 |
| 58 | 63 | Worcester, MA-CT Metro area | 64,841 | 61 | 63 | Albuquerque, NM Metro area | 495,033 |
| 71 | 64 | El Paso, TX Metro area | 64,803 | 59 | 64 | Bridgeport-Stamford-Norwalk, CT Metro area | 492,509 |
| 68 | 65 | New Haven-Milford, CT Metro area | 63,139 | 56 | 65 | Urban Honolulu, HI Metro area | 486,430 |
| 49 | 66 | Buffalo-Cheektowaga, NY Metro area | 63,047 | 38 | 66 | Providence-Warwick, RI-MA Metro area | 439,470 |
| 57 | 67 | Omaha-Council Bluffs, NE-IA Metro area | 62,905 | 74 | 67 | Dayton-Kettering, OH Metro area | 436,999 |
| 64 | 68 | Knoxville, TN Metro area | 62,872 | 63 | 68 | Albany-Schenectady-Troy, NY Metro area | 434,114 |
| 69 | 69 | Columbia, SC Metro area | 57,994 | 52 | 69 | Rochester, NY Metro area | 417,522 |
| 63 | 70 | Albany-Schenectady-Troy, NY Metro area | 56,557 | 58 | 70 | Worcester, MA-CT Metro area | 379,541 |
| 55 | 71 | Fresno, CA Metro area | 55,668 | 49 | 71 | Buffalo-Cheektowaga, NY Metro area | 378,741 |
| 61 | 72 | Albuquerque, NM Metro area | 54,919 | 72 | 72 | Oxnard-Thousand Oaks-Ventura, CA Metro area | 341,012 |
| 70 | 73 | Allentown-Bethlehem-Easton, PA-NJ Metro area | 54,274 | 70 | 73 | Allentown-Bethlehem-Easton, PA-NJ Metro area | 307,274 |
| 74 | 74 | Dayton-Kettering, OH Metro area | 50,875 | 48 | 74 | Hartford-East Hartford-Middletown, CT Metro area | 238,793 |
| 62 | 75 | Bakersfield, CA Metro area | 49,553 | 68 | 75 | New Haven-Milford, CT Metro area | 139,928 |

# 75 Metropolitan Areas with Highest Agricultural Sales
## Selected Rankings

### Value of agricultural products sold, 2017

| Value of sales rank | Metropolitan area | Value of sales (millions of dollars) [col 125] |
|---|---|---|
| 1 | Fresno, CA Metro area | 5,742.8 |
| 2 | Visalia, CA Metro area | 4,474.8 |
| 3 | Salinas, CA Metro area | 4,116.1 |
| 4 | Bakersfield, CA Metro area | 4,076.8 |
| 5 | Merced, CA Metro area | 2,938.4 |
| 6 | Modesto, CA Metro area | 2,526.3 |
| 7 | Stockton, CA Metro area | 2,176.0 |
| 8 | Phoenix-Mesa-Chandler, AZ Metro area | 2,071.1 |
| 9 | Greeley, CO Metro area | 2,047.2 |
| 10 | Yakima, WA Metro area | 1,988.0 |
| 11 | Chicago-Naperville-Elgin, IL-IN-WI Metro area | 1,859.9 |
| 12 | El Centro, CA Metro area | 1,859.7 |
| 13 | Salisbury, MD-DE Metro area | 1,827.9 |
| 14 | Miami-Fort Lauderdale-Pompano Beach, FL Metro area | 1,764.3 |
| 15 | Hanford-Corcoran, CA Metro area | 1,649.3 |
| 16 | Kennewick-Richland, WA Metro area | 1,636.9 |
| 17 | Oxnard-Thousand Oaks-Ventura, CA Metro area | 1,633.3 |
| 18 | Minneapolis-St. Paul-Bloomington, MN Metro area | 1,562.8 |
| 19 | Santa Maria-Santa Barbara, CA Metro area | 1,519.9 |
| 20 | Omaha-Council Bluffs, NE-IA Metro area | 1,509.8 |
| 21 | Lancaster, PA Metro area | 1,507.2 |
| 22 | Madera, CA Metro area | 1,492.6 |
| 23 | Fayetteville-Springdale-Rogers, AR Metro area | 1,382.0 |
| 24 | St. Louis, MO-IL Metro area | 1,375.5 |
| 25 | Philadelphia-Camden-Wilmington, PA-NJ-DE-MD Metro area | 1,358.6 |
| 26 | Grand Rapids-Kentwood, MI Metro area | 1,332.9 |
| 27 | Twin Falls, ID Metro area | 1,319.8 |
| 28 | Riverside-San Bernardino-Ontario, CA Metro area | 1,305.9 |
| 29 | Charlotte-Concord-Gastonia, NC-SC Metro area | 1,305.7 |
| 30 | Madison, WI Metro area | 1,159.1 |
| 31 | Sacramento-Roseville-Folsom, CA Metro area | 1,081.4 |
| 32 | Columbus, OH Metro area | 1,078.5 |
| 33 | Peoria, IL Metro area | 1,038.7 |
| 34 | Des Moines-West Des Moines, IA Metro area | 1,026.7 |
| 35 | Boise City, ID Metro area | 1,021.5 |
| 36 | Portland-Vancouver-Hillsboro, OR-WA Metro area | 1,020.1 |
| 37 | Kansas City, MO-KS Metro area | 985.3 |
| 38 | Rochester, NY Metro area | 956.7 |
| 39 | St. Cloud, MN Metro area | 955.2 |
| 40 | Indianapolis-Carmel-Anderson, IN Metro area | 944.4 |
| 41 | Atlanta-Sandy Springs-Alpharetta, GA Metro area | 931.1 |
| 42 | Rochester, MN Metro area | 930.9 |
| 43 | Sioux Falls, SD Metro area | 920.2 |
| 44 | Santa Rosa-Petaluma, CA Metro area | 919.1 |
| 45 | Sioux City, IA-NE-SD Metro area | 912.3 |
| 46 | Davenport-Moline-Rock Island, IA-IL Metro area | 894.2 |
| 47 | Iowa City, IA Metro area | 891.5 |
| 48 | Cedar Rapids, IA Metro area | 854.5 |
| 49 | Salem, OR Metro area | 836.3 |
| 50 | San Diego-Chula Vista-Carlsbad, CA Metro area | 831.4 |
| 51 | Mankato, MN Metro area | 822.8 |
| 52 | Amarillo, TX Metro area | 801.5 |
| 53 | Harrisonburg, VA Metro area | 795.9 |
| 54 | Grand Island, NE Metro area | 777.9 |
| 55 | Waterloo-Cedar Falls, IA Metro area | 757.9 |
| 56 | Green Bay, WI Metro area | 752.9 |
| 57 | Grand Forks, ND-MN Metro area | 748.1 |
| 58 | Lexington-Fayette, KY Metro area | 723.3 |
| 59 | Fargo, ND-MN Metro area | 717.2 |
| 60 | Lafayette-West Lafayette, IN Metro area | 702.5 |
| 61 | San Luis Obispo-Paso Robles, CA Metro area | 701.6 |
| 62 | Jackson, MS Metro area | 682.5 |
| 63 | Wichita, KS Metro area | 680.7 |
| 64 | Columbia, SC Metro area | 649.0 |
| 65 | Dallas-Fort Worth-Arlington, TX Metro area | 611.7 |
| 66 | Santa Cruz-Watsonville, CA Metro area | 606.5 |
| 67 | Goldsboro, NC Metro area | 592.1 |
| 68 | Yuba City, CA Metro area | 591.3 |
| 69 | Napa, CA Metro area | 573.2 |
| 70 | Reading, PA Metro area | 554.7 |
| 71 | Orlando-Kissimmee-Sanford, FL Metro area | 554.5 |
| 72 | New York-Newark-Jersey City, NY-NJ-PA Metro area | 545.8 |
| 73 | Champaign-Urbana, IL Metro area | 540.8 |
| 74 | Tampa-St. Petersburg-Clearwater, FL Metro area | 534.9 |
| 75 | Athens-Clarke County, GA Metro area | 524.8 |

### Number of farms, 2017

| Value of sales rank | Number of farms rank | Metropolitan area | Number of farms [col 113] |
|---|---|---|---|
| 65 | 1 | Dallas-Fort Worth-Arlington, TX Metro area | 29,254 |
| 37 | 2 | Kansas City, MO-KS Metro area | 12,437 |
| 36 | 3 | Portland-Vancouver-Hillsboro, OR-WA Metro area | 11,755 |
| 18 | 4 | Minneapolis-St. Paul-Bloomington, MN Metro area | 11,525 |
| 24 | 5 | St. Louis, MO-IL Metro area | 11,057 |
| 32 | 6 | Columbus, OH Metro area | 8,506 |
| 41 | 7 | Atlanta-Sandy Springs-Alpharetta, GA Metro area | 7,760 |
| 29 | 8 | Charlotte-Concord-Gastonia, NC-SC Metro area | 7,381 |
| 30 | 9 | Madison, WI Metro area | 6,927 |
| 11 | 10 | Chicago-Naperville-Elgin, IL-IN-WI Metro area | 6,565 |
| 25 | 11 | Philadelphia-Camden-Wilmington, PA-NJ-DE-MD Metro area | 6,506 |
| 40 | 12 | Indianapolis-Carmel-Anderson, IN Metro area | 5,999 |
| 72 | 13 | New York-Newark-Jersey City, NY-NJ-PA Metro area | 5,894 |
| 20 | 14 | Omaha-Council Bluffs, NE-IA Metro area | 5,843 |
| 34 | 15 | Des Moines-West Des Moines, IA Metro area | 5,658 |
| 23 | 16 | Fayetteville-Springdale-Rogers, AR Metro area | 5,444 |
| 21 | 17 | Lancaster, PA Metro area | 5,108 |
| 35 | 17 | Boise City, ID Metro area | 5,108 |
| 50 | 19 | San Diego-Chula Vista-Carlsbad, CA Metro area | 5,082 |
| 1 | 20 | Fresno, CA Metro area | 4,774 |
| 31 | 21 | Sacramento-Roseville-Folsom, CA Metro area | 4,737 |
| 14 | 22 | Miami-Fort Lauderdale-Pompano Beach, FL Metro area | 4,690 |
| 58 | 23 | Lexington-Fayette, KY Metro area | 4,619 |
| 63 | 24 | Wichita, KS Metro area | 4,536 |
| 33 | 25 | Peoria, IL Metro area | 4,468 |
| 74 | 26 | Tampa-St. Petersburg-Clearwater, FL Metro area | 4,325 |
| 38 | 27 | Rochester, NY Metro area | 4,215 |
| 2 | 28 | Visalia, CA Metro area | 4,187 |
| 9 | 29 | Greeley, CO Metro area | 4,062 |
| 26 | 30 | Grand Rapids-Kentwood, MI Metro area | 4,056 |
| 62 | 31 | Jackson, MS Metro area | 4,019 |
| 49 | 32 | Salem, OR Metro area | 4,004 |
| 42 | 33 | Rochester, MN Metro area | 3,960 |
| 39 | 34 | St. Cloud, MN Metro area | 3,767 |
| 28 | 35 | Riverside-San Bernardino-Ontario, CA Metro area | 3,729 |
| 48 | 36 | Cedar Rapids, IA Metro area | 3,632 |
| 6 | 37 | Modesto, CA Metro area | 3,621 |
| 44 | 38 | Santa Rosa-Petaluma, CA Metro area | 3,594 |
| 46 | 39 | Davenport-Moline-Rock Island, IA-IL Metro area | 3,434 |
| 7 | 40 | Stockton, CA Metro area | 3,430 |
| 64 | 41 | Columbia, SC Metro area | 3,325 |
| 71 | 42 | Orlando-Kissimmee-Sanford, FL Metro area | 3,120 |
| 43 | 43 | Sioux Falls, SD Metro area | 3,048 |
| 10 | 44 | Yakima, WA Metro area | 2,952 |
| 55 | 45 | Waterloo-Cedar Falls, IA Metro area | 2,691 |
| 8 | 46 | Phoenix-Mesa-Chandler, AZ Metro area | 2,636 |
| 56 | 47 | Green Bay, WI Metro area | 2,464 |
| 45 | 48 | Sioux City, IA-NE-SD Metro area | 2,428 |
| 47 | 49 | Iowa City, IA Metro area | 2,386 |
| 61 | 50 | San Luis Obispo-Paso Robles, CA Metro area | 2,349 |
| 5 | 51 | Merced, CA Metro area | 2,337 |
| 16 | 52 | Kennewick-Richland, WA Metro area | 2,292 |
| 13 | 53 | Salisbury, MD-DE Metro area | 2,237 |
| 57 | 54 | Grand Forks, ND-MN Metro area | 2,147 |
| 17 | 55 | Oxnard-Thousand Oaks-Ventura, CA Metro area | 2,135 |
| 60 | 56 | Lafayette-West Lafayette, IN Metro area | 2,041 |
| 53 | 57 | Harrisonburg, VA Metro area | 2,026 |
| 68 | 58 | Yuba City, CA Metro area | 1,921 |
| 69 | 59 | Napa, CA Metro area | 1,866 |
| 70 | 60 | Reading, PA Metro area | 1,809 |
| 4 | 61 | Bakersfield, CA Metro area | 1,731 |
| 52 | 62 | Amarillo, TX Metro area | 1,699 |
| 27 | 63 | Twin Falls, ID Metro area | 1,697 |
| 54 | 64 | Grand Island, NE Metro area | 1,682 |
| 51 | 65 | Mankato, MN Metro area | 1,672 |
| 73 | 66 | Champaign-Urbana, IL Metro area | 1,636 |
| 75 | 67 | Athens-Clarke County, GA Metro area | 1,520 |
| 59 | 68 | Fargo, ND-MN Metro area | 1,478 |
| 19 | 69 | Santa Maria-Santa Barbara, CA Metro area | 1,467 |
| 22 | 70 | Madera, CA Metro area | 1,386 |
| 3 | 71 | Salinas, CA Metro area | 1,104 |
| 15 | 72 | Hanford-Corcoran, CA Metro area | 963 |
| 66 | 73 | Santa Cruz-Watsonville, CA Metro area | 625 |
| 67 | 74 | Goldsboro, NC Metro area | 551 |
| 12 | 75 | El Centro, CA Metro area | 396 |

# 75 Metropolitan Areas with Highest Agricultural Sales
## Selected Rankings

| Land in farms, 2017 | | | | Average value of agricultural land and buildings per acre, 2017 | | | |
|---|---|---|---|---|---|---|---|
| Value of sales rank | Land in farms rank | Metropolitan area | Land in farms (1,000 acres) [col 117] | Value of sales rank | Value per Acre rank | Metropolitan area | Value per acre (1,000 acres) [col 123] |
| 65 | 1 | Dallas-Fort Worth-Arlington, TX Metro area | 3,793 | 69 | 1 | Napa, CA Metro area | 44,154 |
| 37 | 2 | Kansas City, MO-KS Metro area | 3,126 | 17 | 2 | Oxnard-Thousand Oaks-Ventura, CA Metro area | 25,498 |
| 52 | 3 | Amarillo, TX Metro area | 2,893 | 50 | 3 | San Diego-Chula Vista-Carlsbad, CA Metro area | 23,209 |
| 24 | 4 | St. Louis, MO-IL Metro area | 2,780 | 44 | 4 | Santa Rosa-Petaluma, CA Metro area | 22,186 |
| 63 | 5 | Wichita, KS Metro area | 2,397 | 66 | 5 | Santa Cruz-Watsonville, CA Metro area | 21,352 |
| 20 | 6 | Omaha-Council Bluffs, NE-IA Metro area | 2,363 | 28 | 6 | Riverside-San Bernardino-Ontario, CA Metro area | 18,504 |
| 4 | 7 | Bakersfield, CA Metro area | 2,295 | 21 | 7 | Lancaster, PA Metro area | 18,285 |
| 11 | 8 | Chicago-Naperville-Elgin, IL-IN-WI Metro area | 2,147 | 6 | 8 | Modesto, CA Metro area | 15,619 |
| 18 | 9 | Minneapolis-St. Paul-Bloomington, MN Metro area | 2,142 | 72 | 9 | New York-Newark-Jersey City, NY-NJ-PA Metro area | 15,483 |
| 9 | 10 | Greeley, CO Metro area | 2,099 | 7 | 10 | Stockton, CA Metro area | 15,020 |
| 57 | 11 | Grand Forks, ND-MN Metro area | 1,822 | 36 | 11 | Portland-Vancouver-Hillsboro, OR-WA Metro area | 14,558 |
| 10 | 12 | Yakima, WA Metro area | 1,781 | 5 | 12 | Merced, CA Metro area | 13,086 |
| 34 | 13 | Des Moines-West Des Moines, IA Metro area | 1,716 | 14 | 13 | Miami-Fort Lauderdale-Pompano Beach, FL Metro area | 12,654 |
| 59 | 14 | Fargo, ND-MN Metro area | 1,703 | 2 | 14 | Visalia, CA Metro area | 11,726 |
| 32 | 15 | Columbus, OH Metro area | 1,666 | 1 | 15 | Fresno, CA Metro area | 11,359 |
| 1 | 16 | Fresno, CA Metro area | 1,647 | 70 | 16 | Reading, PA Metro area | 11,209 |
| 33 | 17 | Peoria, IL Metro area | 1,617 | 12 | 17 | El Centro, CA Metro area | 11,135 |
| 40 | 18 | Indianapolis-Carmel-Anderson, IN Metro area | 1,603 | 22 | 18 | Madera, CA Metro area | 10,958 |
| 8 | 19 | Phoenix-Mesa-Chandler, AZ Metro area | 1,595 | 15 | 19 | Hanford-Corcoran, CA Metro area | 10,815 |
| 30 | 20 | Madison, WI Metro area | 1,463 | 25 | 20 | Philadelphia-Camden-Wilmington, PA-NJ-DE-MD Metro area | 10,688 |
| 43 | 21 | Sioux Falls, SD Metro area | 1,429 | 49 | 21 | Salem, OR Metro area | 10,580 |
| 35 | 22 | Boise City, ID Metro area | 1,351 | 19 | 22 | Santa Maria-Santa Barbara, CA Metro area | 10,381 |
| 3 | 23 | Salinas, CA Metro area | 1,340 | 74 | 23 | Tampa-St. Petersburg-Clearwater, FL Metro area | 9,761 |
| 62 | 24 | Jackson, MS Metro area | 1,258 | 31 | 24 | Sacramento-Roseville-Folsom, CA Metro area | 9,440 |
| 2 | 25 | Visalia, CA Metro area | 1,250 | 73 | 25 | Champaign-Urbana, IL Metro area | 9,300 |
| 16 | 26 | Kennewick-Richland, WA Metro area | 1,229 | 68 | 26 | Yuba City, CA Metro area | 9,107 |
| 45 | 27 | Sioux City, IA-NE-SD Metro area | 1,187 | 53 | 27 | Harrisonburg, VA Metro area | 8,844 |
| 46 | 28 | Davenport-Moline-Rock Island, IA-IL Metro area | 1,146 | 55 | 28 | Waterloo-Cedar Falls, IA Metro area | 8,630 |
| 42 | 29 | Rochester, MN Metro area | 1,140 | 11 | 29 | Chicago-Naperville-Elgin, IL-IN-WI Metro area | 8,547 |
| 48 | 30 | Cedar Rapids, IA Metro area | 1,089 | 60 | 30 | Lafayette-West Lafayette, IN Metro area | 7,808 |
| 29 | 31 | Charlotte-Concord-Gastonia, NC-SC Metro area | 971 | 33 | 31 | Peoria, IL Metro area | 7,799 |
| 5 | 32 | Merced, CA Metro area | 946 | 46 | 32 | Davenport-Moline-Rock Island, IA-IL Metro area | 7,762 |
| 61 | 33 | San Luis Obispo-Paso Robles, CA Metro area | 931 | 61 | 33 | San Luis Obispo-Paso Robles, CA Metro area | 7,547 |
| 31 | 34 | Sacramento-Roseville-Folsom, CA Metro area | 929 | 48 | 34 | Cedar Rapids, IA Metro area | 7,533 |
| 38 | 35 | Rochester, NY Metro area | 900 | 47 | 35 | Iowa City, IA Metro area | 7,457 |
| 60 | 36 | Lafayette-West Lafayette, IN Metro area | 885 | 4 | 36 | Bakersfield, CA Metro area | 7,380 |
| 41 | 37 | Atlanta-Sandy Springs-Alpharetta, GA Metro area | 872 | 3 | 37 | Salinas, CA Metro area | 7,368 |
| 55 | 38 | Waterloo-Cedar Falls, IA Metro area | 862 | 40 | 38 | Indianapolis-Carmel-Anderson, IN Met | 7,349 |
| 71 | 39 | Orlando-Kissimmee-Sanford, FL Metro area | 853 | 13 | 39 | Salisbury, MD-DE Metro area | 7,315 |
| 54 | 40 | Grand Island, NE Metro area | 852 | 58 | 40 | Lexington-Fayette, KY Metro area | 7,245 |
| 39 | 41 | St. Cloud, MN Metro area | 846 | 51 | 41 | Mankato, MN Metro area | 7,121 |
| 73 | 42 | Champaign-Urbana, IL Metro area | 839 | 34 | 42 | Des Moines-West Des Moines, IA Metro area | 6,758 |
| 23 | 43 | Fayetteville-Springdale-Rogers, AR Metro area | 838 | 32 | 43 | Columbus, OH Metro area | 6,557 |
| 26 | 44 | Grand Rapids-Kentwood, MI Metro area | 794 | 30 | 44 | Madison, WI Metro area | 6,346 |
| 7 | 45 | Stockton, CA Metro area | 773 | 20 | 45 | Omaha-Council Bluffs, NE-IA Metro area | 6,243 |
| 58 | 46 | Lexington-Fayette, KY Metro area | 752 | 24 | 46 | St. Louis, MO-IL Metro area | 6,158 |
| 6 | 47 | Modesto, CA Metro area | 723 | 45 | 47 | Sioux City, IA-NE-SD Metro area | 6,132 |
| 19 | 48 | Santa Maria-Santa Barbara, CA Metro area | 715 | 56 | 48 | Green Bay, WI Metro area | 6,120 |
| 25 | 49 | Philadelphia-Camden-Wilmington, PA-NJ-DE-MD Metro area | 656 | 42 | 49 | Rochester, MN Metro area | 6,115 |
| 51 | 50 | Mankato, MN Metro area | 648 | 26 | 50 | Grand Rapids-Kentwood, MI Metro area | 5,874 |
| 22 | 51 | Madera, CA Metro area | 645 | 18 | 51 | Minneapolis-St. Paul-Bloomington, MN Metro area | 5,858 |
| 27 | 52 | Twin Falls, ID Metro area | 640 | 29 | 52 | Charlotte-Concord-Gastonia, NC-SC Metro area | 5,840 |
| 15 | 53 | Hanford-Corcoran, CA Metro area | 616 | 43 | 53 | Sioux Falls, SD Metro area | 5,616 |
| 47 | 54 | Iowa City, IA Metro area | 615 | 75 | 54 | Athens-Clarke County, GA Metro area | 5,541 |
| 36 | 55 | Portland-Vancouver-Hillsboro, OR-WA Metro area | 597 | 71 | 55 | Orlando-Kissimmee-Sanford, FL Metro area | 5,516 |
| 64 | 56 | Columbia, SC Metro area | 575 | 41 | 56 | Atlanta-Sandy Springs-Alpharetta, GA Metro area | 5,337 |
| 14 | 57 | Miami-Fort Lauderdale-Pompano Beach, FL Metro area | 573 | 67 | 57 | Goldsboro, NC Metro area | 4,990 |
| 44 | 58 | Santa Rosa-Petaluma, CA Metro area | 567 | 39 | 58 | St. Cloud, MN Metro area | 4,974 |
| 68 | 59 | Yuba City, CA Metro area | 560 | 27 | 59 | Twin Falls, ID Metro area | 4,868 |
| 56 | 60 | Green Bay, WI Metro area | 552 | 23 | 60 | Fayetteville-Springdale-Rogers, AR Metro area | 4,861 |
| 13 | 61 | Salisbury, MD-DE Metro area | 523 | 65 | 61 | Dallas-Fort Worth-Arlington, TX Metro area | 4,846 |
| 12 | 62 | El Centro, CA Metro area | 522 | 8 | 62 | Phoenix-Mesa-Chandler, AZ Metro area | 4,689 |
| 49 | 63 | Salem, OR Metro area | 438 | 54 | 63 | Grand Island, NE Metro area | 4,556 |
| 74 | 64 | Tampa-St. Petersburg-Clearwater, FL Metro area | 425 | 16 | 64 | Kennewick-Richland, WA Metro area | 4,247 |
| 21 | 65 | Lancaster, PA Metro area | 394 | 38 | 65 | Rochester, NY Metro area | 3,920 |
| 72 | 66 | New York-Newark-Jersey City, NY-NJ-PA Metro area | 349 | 59 | 66 | Fargo, ND-MN Metro area | 3,795 |
| 28 | 67 | Riverside-San Bernardino-Ontario, CA Metro area | 332 | 35 | 67 | Boise City, ID Metro area | 3,783 |
| 17 | 68 | Oxnard-Thousand Oaks-Ventura, CA Metro area | 260 | 64 | 68 | Columbia, SC Metro area | 3,643 |
| 69 | 69 | Napa, CA Metro area | 256 | 37 | 69 | Kansas City, MO-KS Metro area | 3,606 |
| 53 | 70 | Harrisonburg, VA Metro area | 229 | 57 | 70 | Grand Forks, ND-MN Metro area | 3,338 |
| 70 | 71 | Reading, PA Metro area | 225 | 62 | 71 | Jackson, MS Metro area | 2,809 |
| 50 | 72 | San Diego-Chula Vista-Carlsbad, CA Metro area | 222 | 10 | 72 | Yakima, WA Metro area | 2,755 |
| 75 | 73 | Athens-Clarke County, GA Metro area | 185 | 9 | 73 | Greeley, CO Metro area | 2,586 |
| 67 | 74 | Goldsboro, NC Metro area | 165 | 63 | 74 | Wichita, KS Metro area | 2,509 |
| 66 | 75 | Santa Cruz-Watsonville, CA Metro area | 64 | 52 | 75 | Amarillo, TX Metro area | 1,109 |

## Table C. Metropolitan Areas — Land Area and Population

| CBSA/ DIV Code[1] | Area name | Land area[2] (sq mi) | Population, 2020 | | | Population characteristics, 2020 | | | | | | | | | | |
|---|---|---|---|---|---|---|---|---|---|---|---|---|---|---|---|---|
| | | | Total persons 2020 | Rank | Per square mile | Race alone or in combination, not Hispanic or Latino (percent) | | | | Percent Hispanic or Latino[3] | Age (percent) | | | | | |
| | | | | | | White | Black | American Indian, Alaska Native | Asian and Pacific Islander | | Under 5 years | 5 to 14 years | 15 to 24 years | 25 to 34 years | 35 to 44 years | 45 to 54 years |
| | | 1 | 2 | 3 | 4 | 5 | 6 | 7 | 8 | 9 | 10 | 11 | 12 | 13 | 14 | 15 |

1. CBSA = Core Based Statistical Area. DIV = Metropolitan Division. See Appendix A for explanation. See Appendix B for list of metropolitan areas or temporarily covered by water.   2. Dry land or land partially or temporarily covered by water.   3. May be of any race.

## Table C. Metropolitan Areas — Population and Households

| Area name | Population, 2020 (cont.) | | | | Population change and components of change, 2000-2020 | | | | | | | Households, 2019 | | | | |
|---|---|---|---|---|---|---|---|---|---|---|---|---|---|---|---|---|
| | Age (percent) (cont.) | | | | Total persons | | Percent change | | Components of change, 2010-2020 | | | | | | Percent | |
| | 55 to 64 years | 65 to 74 years | 75 years and over | Percent female | 2000 | 2010 | 2000-2010 | 2010-2019 | Births | Deaths | Net migration | Number | Persons per house-hold | Family house-holds | Single parent house-hold | One person |
| | 16 | 17 | 18 | 19 | 20 | 21 | 22 | 23 | 24 | 25 | 26 | 27 | 28 | 29 | 30 | 31 |

## Table C. Metropolitan Areas — Population, Vital Statistics, Health, and Crime

| Area name | Persons in group quarters, 2020 | Daytime population, 2019 | | Births, 2020 | | Deaths, 2020 | | Persons under 65 with no health insurance 2019 | | Medicare, 2020 | | | Serious crimes known to police[2], 2019 | |
|---|---|---|---|---|---|---|---|---|---|---|---|---|---|---|
| | | Number | Employ-ment/ residence ratio | Total | Rate[1] | Number | Rate[1] | Number | Percent | Total Ben-eficiaries | Enrolled in Original Medicare | Enrolled in Medicare Advantage | Violent | |
| | | | | | | | | | | | | | Number | Rate[3] |
| | 32 | 33 | 34 | 35 | 36 | 37 | 38 | 39 | 40 | 41 | 42 | 43 | 44 | 45 |

1. Per 1,000 estimated resident population.   2. Data for serious crimes have not been adjusted for underreporting; this may affect comparability between geographic areas and over time.   3. Per 100,000 population estimated by the FBI.

## Table C. Metropolitan Areas — Crime, Education, Money Income, and Poverty

| Area name | Serious crimes known to police[1], 2019 (cont.) | | Education | | | | | | Income and poverty, 2019 | | | | | | | |
|---|---|---|---|---|---|---|---|---|---|---|---|---|---|---|---|---|
| | Property | | School enrollment and attainment, 2019 | | | | Local government expenditures,[5] 2017-2018 | | | | | Percent of house-holds with income of $200,000 or more | Percent below poverty level | | |
| | | | Enrollment[3] | | Attainment[4] | | | | | | | | | | |
| | Number | Rate[2] | Total | Percent private | High school gradu-ate or less | Bach-elor's degree or more | Total current expendi-tures (mil dol) | Current expen-ditures per student (dollars) | Per capita income[6] (dollars) | Median house-hold income (dollars) | Median family income | Percent of house-holds with income less than $50,000 | | All persons | All fami-lies | Fam-ilies with child-ren under 18 |
| | 46 | 47 | 48 | 49 | 50 | 51 | 52 | 53 | 54 | 55 | 56 | 57 | 58 | 59 | 60 | 61 |

1. Data for serious crimes have not been adjusted for underreporting; this may affect comparability between geographic areas and over time.   2. Per 100,000 population estimated by the FBI.   3. All persons 3 years old and over enrolled in nursery school through college.   4. Persons 25 years old and over.   5. Elementary and secondary education expenditures.   6. Based on population estimated by the American Community Survey, 2017.

## Table C. Metropolitan Areas — **Personal Income and Earnings**

| Area name | Personal income, 2019 | | | | | | | | | | Earnings, 2019 | | |
|---|---|---|---|---|---|---|---|---|---|---|---|---|---|
| | | | Per capita[1] | | | Supplements to wages and salaries, employer contributions (mil dol) | | | | | | Contributions for government social insurance (mil dol) | |
| | Total (mil dol) | Percent change, 2018-2019 | Dollars | Rank | Wages and Salaries (mil dol) | Pension and insurance | Government social insurance | Proprietors' income | Dividends, interest, and rent (mil dol) | Personal transfer receipts (mil dol) | Total (mil dol) | From employee and self-employed | From employer |
| | 62 | 63 | 64 | 65 | 66 | 67 | 68 | 69 | 70 | 71 | 72 | 73 | 74 |

1. Based on the resident population estimated as of July 1 of the year shown.

## Table C. Metropolitan Areas — **Earnings, Social Security, and Housing**

| Area name | Earnings, 2019 (cont.) | | | | | | | | | Social Security beneficiaries, December 2019 | | Supplemental Security Income Recipients, December 2019 | Housing units, 2020 | |
|---|---|---|---|---|---|---|---|---|---|---|---|---|---|---|
| | Percent by selected industries | | | | | | | | | | | | | |
| | Farm | Mining, quarrying, and extracting | Construction | Manufacturing | Information; professional, scientific, and technical serviecs | Retail trade | Finance, insurance, real estate, rental and leasing | Health care and social assistance | Government | Number | Rate[1] | | Total | Percent change, 2010-2020 |
| | 75 | 76 | 77 | 78 | 79 | 80 | 81 | 82 | 83 | 84 | 85 | 86 | 87 | 88 |

1. Per 1,000 resident population estimated as of July 1 of the year shown.

## Table C. Metropolitan Areas — **Housing, Labor Force, and Employment**

| Area name | Occupied Housing units, 2020 | | | | | | | | Civilian labor force, 2020 | | | | Civilian employment, 2019[6] | | |
|---|---|---|---|---|---|---|---|---|---|---|---|---|---|---|---|
| | Occupied units | | | | | | | | | | Unemployment | | Percent | | |
| | | | Owner-occupied | | | | Renter-occupied | | | | | | | | |
| | | | | Median owner cost as a percent of income | | | Median rent as a percent of income[2] | Substandard units[4] (percent) | | Percent change 2019-2020 | | | | Management, business, science, and arts | Construction, production, and maintenance occupations |
| | Total | Percent | Median value[1] | With a mortgage | Without a mortgage[2] | Median rent[3] | | | Total | | Total | Rate[5] | Total | | |
| | 89 | 90 | 91 | 92 | 93 | 94 | 95 | 96 | 97 | 98 | 99 | 100 | 101 | 102 | 103 |

1. Specified owner-occupied units. lacking complete plumbing facilities.　2. A value of 10.0 represents 10 percent or less; a value of 50.0 represents 50 percent or more.　3. Specified renter-occupied units.　4. Overcrowded or 5. Percent of civilian labor force. 6. Civilian employed persons 16 years old and over.

## Table C. Metropolitan Areas — **Nonfarm Employment and Agriculture**

| Area name | Private nonfarm establishments, employment and payroll, 2019 | | | | | | | | | Agriculture, 2017 | | | | Farm producers whose primary occupation is farming (percent) |
|---|---|---|---|---|---|---|---|---|---|---|---|---|---|---|
| | | Employment | | | | | | Annual payroll | | Farms | | | | |
| | | | | | | | | | | | | Percent with: | | |
| | Number of establishments | Total | Health care and social assistance | Manufacturing | Retail trade | Finance and insurance | Professional, scientific, and technical services | Total (mil dol) | Average per employee (dollars) | Number | Fewer than 50 acres | 1000 acres or more | | |
| | 104 | 105 | 106 | 107 | 108 | 109 | 110 | 111 | 112 | 113 | 114 | 115 | | 116 |

## Table C. Metropolitan Areas — Agriculture

| Area name | Land in farms | | | | | Value of land and buildings (dollars) | | Value of machinery and equipmnet, average per farm (dollars) | Value of products sold: | | | | Organic farms (number) | Farms with internet access (percent) | Government payments | |
|---|---|---|---|---|---|---|---|---|---|---|---|---|---|---|---|---|
| | Acreage (1,000) | Percent change, 2012-2017 | Acres | | | Average per farm | Average per acre | | Total (mil dol) | Average per farm (acres) | Percent from: | | | | Total ($1,000) | Percent of farms |
| | | | Average size of farm | Total irrigated (1,000) | Total cropland (1,000) | | | | | | Crops | Livestock and poultry products | | | | | |
| | 117 | 118 | 119 | 120 | 121 | 122 | 123 | 124 | 125 | 126 | 127 | 128 | 129 | 130 | 131 | 132 |

## Table C. Metropolitan Areas — Water Use, Wholesale Trade, Retail Trade, and Real Estate

| Area name | Water use, 2015 | | Wholesale Trade[1], 2017 | | | | Retail Trade[2], 2017 | | | | Real estate and rental and leasing,[2] 2017 | | | |
|---|---|---|---|---|---|---|---|---|---|---|---|---|---|---|
| | Public supply water withdrawn (mil gal/day) | Public supply gallons withdrawn per person per day | Number of establishments | Number of employees | Sales (mil dol) | Average payroll (mil dol) | Number of establishments | Number of employees | Sales (mil dol) | Average payroll (mil dol) | Number of establishments | Number of employees | Sales (mil dol) | Average payroll (mil dol) |
| | 133 | 134 | 135 | 136 | 137 | 138 | 139 | 140 | 141 | 142 | 143 | 144 | 145 | 146 |

1. Merchant wholesalers, except manufacturers' sales branches and offices. 2. Employer establishments.

## Table C. Metropolitan Areas — Professional Services, Manufacturing, and Accommodation and Food Services

| Area name | Professional, scientific, and technical services, 2017 | | | | Manufacturing, 2017 | | | | Accommodation and food services, 2017 | | | |
|---|---|---|---|---|---|---|---|---|---|---|---|---|
| | Number of establishments | Number of employees | Sales (mil dol) | Average payroll (mil dol) | Number of establishments | Number of employees | Receipts (mil dol) | Annual payroll (mil dol) | Number of establishments | Number of employees | Receipts (mil dol) | Annual payroll (mil dol) |
| | 147 | 148 | 149 | 150 | 151 | 152 | 153 | 154 | 155 | 156 | 157 | 158 |

## Table C. Metropolitan Areas — Health Care and Social Assistance, Other Services, Nonemployer Businesses, and Residential Construction

| Area name | Health care and social assistance, 2017 | | | | Other services, 2017 | | | | Nonemployer businesses, 2018 | | Value of residential construction authorized by building permits, 2020 | |
|---|---|---|---|---|---|---|---|---|---|---|---|---|
| | Number of establishments | Number of employees | Receipts (mil dol) | Annual payroll (mil dol) | Number of establishments | Number of employees | Receipts (mil dol) | Annual payroll (mil dol) | Number | Receipts (mil dol) | New construction ($1,000) | Number of housing units |
| | 159 | 160 | 161 | 162 | 163 | 164 | 165 | 166 | 167 | 168 | 169 | 170 |

Table C. Metropolitan Areas — **Government Employment and Payroll, and Local Government Finances**

| Area name | Government employment and payroll, 2017 | | | | | | | | | Local government finances, 2017 | | | | |
|---|---|---|---|---|---|---|---|---|---|---|---|---|---|---|
| | | | March payroll (percent of total) | | | | | | | General revenue | | | | |
| | | | | | | | | | | | | Taxes | | |
| | Full-time equivalent employees | March payroll (dollars) | Adminis-tration, judicial, and legal | Police and corrections | Fire protection | Highways and transpor-tation | Health and welfare | Natural resources and utilities | Education and libraries | Total (mil dol) | Inter-govern-mental (mil dol) | Total (mil dol) | Per capita[1] (dollars) | |
| | | | | | | | | | | | | | Total | Property |
| | 171 | 172 | 173 | 174 | 175 | 176 | 177 | 178 | 179 | 180 | 181 | 182 | 183 | 184 |

1. Based on the resident population estimated as of July 1 of the year shown.

Table C. Metropolitan Areas — **Local Government Finances, Government Employment, and Income Taxes**

| Area name | Local government finances, 2017 (cont.) | | | | | | | Debt outstanding | | Government employment, 2019 | | | Individual income tax returns, 2018 | | |
|---|---|---|---|---|---|---|---|---|---|---|---|---|---|---|---|
| | Direct general expenditure | | | | | | | | | | | | | | |
| | | | Percent of total for: | | | | | | | | | | | | |
| | Total (mil dol) | Per capita[1] (dollars) | Educa-tion | Health and hospitals | Police protection | Public welfare | Highways | Total (mil dol) | Per capita[1] (dollars) | Federal civilian | Federal military | State and local | Number of returns | Mean adjusted gross income | Mean income tax |
| | 185 | 186 | 187 | 188 | 189 | 190 | 191 | 192 | 193 | 194 | 195 | 196 | 197 | 198 | 199 |

1. Based on the resident population estimated as of July 1 of the year shown.

Table C. Metropolitan Areas — **Land Area and Population**

| CBSA/ DIV Code[1] | Area name | Land area[2] (sq mi) | Population, 2020 | | | Population characteristics, 2020 | | | | | | | | | | |
|---|---|---|---|---|---|---|---|---|---|---|---|---|---|---|---|---|
| | | | Total persons 2020 | Rank | Per square mile | Race alone or in combination, not Hispanic or Latino (percent) | | | | | Age (percent) | | | | | |
| | | | | | | White | Black | American Indian, Alaska Native | Asian and Pacific Islander | Percent Hispanic or Latino[3] | Under 5 years | 5 to 14 years | 15 to 24 years | 25 to 34 years | 35 to 44 years | 45 to 54 years |
| | | 1 | 2 | 3 | 4 | 5 | 6 | 7 | 8 | 9 | 10 | 11 | 12 | 13 | 14 | 15 |
| 10180 | Abilene, TX | 2,744 | 173,185 | 246 | 63.1 | 65.6 | 8.5 | 0.9 | 2.6 | 24.5 | 6.7 | 13.2 | 16.6 | 14.5 | 12.3 | 10.0 |
| 10420 | Akron, OH | 900 | 701,449 | 83 | 779.2 | 81.1 | 14.1 | 0.8 | 4.3 | 2.3 | 5.3 | 11.3 | 13.4 | 13.1 | 11.6 | 12.3 |
| 10500 | Albany, GA | 1,591 | 145,206 | 288 | 91.3 | 40.9 | 55.1 | 0.6 | 1.6 | 3.1 | 6.2 | 13.4 | 14.2 | 13.1 | 11.8 | 11.7 |
| 10540 | Albany-Lebanon, OR | 2,289 | 131,054 | 310 | 57.3 | 86.8 | 1.2 | 2.9 | 2.5 | 9.8 | 5.9 | 12.4 | 11.4 | 13.8 | 12.3 | 11.5 |
| 10580 | Albany-Schenectady-Troy, NY | 2,812 | 878,550 | 63 | 312.5 | 81.0 | 9.6 | 0.6 | 5.6 | 5.6 | 5.1 | 10.8 | 14.0 | 13.1 | 12.2 | 12.3 |
| 10740 | Albuquerque, NM | 9,283 | 923,630 | 61 | 99.5 | 39.7 | 3.2 | 6.1 | 3.2 | 49.8 | 5.3 | 12.3 | 12.5 | 14.2 | 13.1 | 11.7 |
| 10780 | Alexandria, LA | 1,964 | 150,821 | 279 | 76.8 | 64.5 | 30.4 | 1.5 | 1.7 | 3.7 | 6.5 | 13.4 | 12.4 | 13.6 | 12.7 | 11.5 |
| 10900 | Allentown-Bethlehem-Easton, PA-NJ | 1,453 | 846,399 | 70 | 582.6 | 72.2 | 6.5 | 0.5 | 3.7 | 18.9 | 5.3 | 11.7 | 12.8 | 12.5 | 12.1 | 12.7 |
| 11020 | Altoona, PA | 525 | 121,007 | 330 | 230.1 | 95.8 | 2.9 | 0.4 | 1.0 | 1.3 | 5.1 | 11.4 | 11.2 | 12.2 | 11.6 | 12.3 |
| 11100 | Amarillo, TX | 5,151 | 265,761 | 185 | 51.6 | 59.7 | 7.0 | 1.0 | 3.7 | 30.3 | 6.4 | 14.5 | 13.6 | 14.4 | 13.5 | 11.2 |
| 11180 | Ames, IA | 1,143 | 124,514 | 319 | 108.8 | 87.1 | 3.2 | 0.5 | 7.2 | 3.7 | 4.5 | 9.7 | 27.9 | 13.1 | 10.7 | 9.1 |
| 11260 | Anchorage, AK | 26,414 | 397,308 | 135 | 15.0 | 69.0 | 6.1 | 11.9 | 12.9 | 8.4 | 6.8 | 13.7 | 13.0 | 16.3 | 13.8 | 11.6 |
| 11460 | Ann Arbor, MI | 706 | 366,473 | 149 | 519.1 | 73.2 | 13.7 | 1.0 | 10.7 | 5.1 | 4.8 | 10.2 | 21.2 | 14.5 | 11.6 | 11.2 |
| 11500 | Anniston-Oxford, AL | 606 | 113,469 | 338 | 187.3 | 73.2 | 22.3 | 1.0 | 1.5 | 4.1 | 5.7 | 12.0 | 13.2 | 12.9 | 12.0 | 12.0 |
| 11540 | Appleton, WI | 956 | 238,975 | 195 | 250.0 | 89.1 | 2.1 | 1.8 | 4.1 | 4.6 | 5.9 | 13.1 | 12.3 | 12.8 | 13.2 | 12.7 |
| 11700 | Asheville, NC | 2,033 | 466,634 | 120 | 229.6 | 86.5 | 5.3 | 1.1 | 1.8 | 7.3 | 4.7 | 10.2 | 10.4 | 12.6 | 12.4 | 12.5 |
| 12020 | Athens-Clarke County, GA | 1,025 | 214,759 | 212 | 209.5 | 67.3 | 20.9 | 0.6 | 4.3 | 8.9 | 5.2 | 11.3 | 21.6 | 14.0 | 12.0 | 11.0 |
| 12060 | Atlanta-Sandy Springs-Alpharetta, GA | 8,686 | 6,087,762 | 9 | 700.9 | 47.3 | 35.9 | 0.7 | 7.3 | 11.0 | 6.0 | 13.5 | 13.3 | 14.4 | 13.7 | 13.8 |
| 12100 | Atlantic City-Hammonton, NJ | 556 | 262,945 | 188 | 473.3 | 57.3 | 15.7 | 0.6 | 8.9 | 19.7 | 5.3 | 11.9 | 12.5 | 12.1 | 11.0 | 12.8 |
| 12220 | Auburn-Opelika, AL | 608 | 166,831 | 261 | 274.6 | 69.0 | 23.4 | 0.7 | 4.8 | 3.8 | 5.6 | 11.8 | 20.9 | 13.9 | 12.2 | 11.6 |
| 12260 | Augusta-Richmond County, GA-SC | 3,481 | 614,312 | 95 | 176.5 | 55.7 | 36.9 | 0.9 | 3.1 | 5.9 | 6.1 | 12.8 | 12.8 | 14.2 | 12.4 | 11.7 |
| 12420 | Austin-Round Rock-Georgetown, TX | 4,220 | 2,295,303 | 29 | 543.7 | 53.0 | 8.1 | 0.8 | 7.8 | 32.5 | 6.0 | 12.6 | 13.1 | 17.4 | 15.9 | 12.8 |
| 12540 | Bakersfield, CA | 8,135 | 901,362 | 62 | 110.8 | 33.9 | 5.9 | 1.2 | 6.0 | 55.2 | 7.4 | 16.5 | 14.5 | 15.6 | 13.0 | 10.9 |
| 12580 | Baltimore-Columbia-Towson, MD | 2,602 | 2,800,189 | 21 | 1,076.4 | 57.3 | 31.1 | 0.8 | 7.2 | 6.4 | 5.9 | 12.2 | 12.2 | 14.3 | 13.1 | 12.4 |
| 12620 | Bangor, ME | 3,397 | 151,655 | 278 | 44.6 | 95.0 | 1.5 | 2.1 | 1.7 | 1.6 | 4.6 | 9.8 | 13.6 | 13.2 | 11.5 | 12.6 |
| 12700 | Barnstable Town, MA | 394 | 213,164 | 213 | 540.8 | 90.9 | 4.2 | 1.2 | 2.2 | 3.5 | 3.6 | 8.1 | 9.7 | 9.1 | 8.8 | 10.9 |
| 12940 | Baton Rouge, LA | 4,371 | 858,571 | 66 | 196.7 | 57.2 | 36.5 | 0.7 | 2.6 | 4.4 | 6.4 | 13.0 | 14.6 | 14.0 | 13.1 | 11.6 |
| 12980 | Battle Creek, MI | 706 | 133,580 | 309 | 189.1 | 80.1 | 13.0 | 1.4 | 3.4 | 5.7 | 5.9 | 12.7 | 12.6 | 12.5 | 11.8 | 12.1 |
| 13020 | Bay City, MI | 442 | 102,387 | 352 | 231.5 | 91.4 | 2.8 | 1.1 | 0.9 | 5.7 | 4.9 | 11.1 | 11.3 | 12.5 | 11.5 | 12.1 |
| 13140 | Beaumont-Port Arthur, TX | 2,101 | 391,310 | 141 | 186.3 | 55.5 | 24.6 | 0.8 | 3.3 | 17.2 | 6.7 | 13.8 | 12.6 | 13.9 | 13.0 | 11.7 |
| 13220 | Beckley, WV | 1,267 | 114,982 | 336 | 90.8 | 90.9 | 7.6 | 0.9 | 1.0 | 1.5 | 5.2 | 12.0 | 10.8 | 11.9 | 12.1 | 12.5 |
| 13380 | Bellingham, WA | 2,108 | 231,016 | 199 | 109.6 | 81.2 | 2.0 | 3.7 | 7.0 | 10.1 | 4.9 | 10.9 | 16.6 | 13.8 | 12.5 | 11.0 |
| 13460 | Bend, OR | 3,018 | 201,769 | 225 | 66.9 | 89.0 | 1.0 | 1.8 | 2.6 | 8.4 | 4.9 | 11.2 | 10.1 | 13.3 | 13.6 | 12.5 |
| 13740 | Billings, MT | 6,478 | 183,799 | 233 | 28.4 | 88.9 | 1.4 | 5.2 | 1.5 | 5.8 | 5.7 | 13.1 | 11.5 | 13.5 | 12.9 | 11.3 |
| 13780 | Binghamton, NY | 1,224 | 237,324 | 196 | 193.8 | 87.0 | 6.1 | 0.7 | 4.6 | 4.2 | 5.0 | 11.0 | 16.2 | 11.3 | 10.7 | 11.0 |
| 13820 | Birmingham-Hoover, AL | 4,489 | 1,091,921 | 50 | 243.3 | 62.7 | 31.2 | 0.7 | 2.1 | 4.7 | 6.0 | 12.8 | 12.2 | 13.8 | 12.9 | 12.5 |
| 13900 | Bismarck, ND | 4,281 | 129,641 | 313 | 30.3 | 90.0 | 2.6 | 4.9 | 1.3 | 3.1 | 6.5 | 13.2 | 12.1 | 13.9 | 13.3 | 10.9 |
| 13980 | Blacksburg-Christiansburg, VA | 1,074 | 167,244 | 260 | 155.8 | 88.0 | 5.6 | 0.6 | 5.0 | 3.0 | 4.1 | 9.0 | 24.9 | 12.9 | 10.2 | 11.3 |
| 14010 | Bloomington, IL | 1,183 | 171,256 | 250 | 144.7 | 81.3 | 9.7 | 0.5 | 5.7 | 5.2 | 5.6 | 11.9 | 21.1 | 12.7 | 12.1 | 10.9 |
| 14020 | Bloomington, IN | 780 | 169,052 | 256 | 216.8 | 86.9 | 4.3 | 0.8 | 7.2 | 3.4 | 4.4 | 9.0 | 26.3 | 13.9 | 10.9 | 9.8 |
| 14100 | Bloomsburg-Berwick, PA | 614 | 82,884 | 374 | 135.2 | 93.2 | 2.3 | 0.5 | 1.9 | 3.2 | 4.4 | 10.3 | 16.3 | 11.6 | 10.6 | 11.9 |
| 14260 | Boise City, ID | 11,767 | 770,353 | 77 | 65.5 | 81.9 | 1.6 | 1.3 | 3.6 | 14.1 | 5.8 | 13.9 | 13.1 | 13.9 | 13.7 | 12.1 |
| 14460 | Boston-Cambridge-Newton, MA-NH | 3,486 | 4,878,211 | 11 | 1,399.3 | 71.1 | 8.9 | 0.5 | 9.8 | 11.8 | 5.2 | 10.8 | 13.4 | 15.3 | 12.8 | 12.7 |
| 14460 | Boston, MA Div 14454 | 1,113 | 2,034,729 | X | 1,828.3 | 65.7 | 14.3 | 0.6 | 9.4 | 12.1 | 5.1 | 10.4 | 13.9 | 16.7 | 12.7 | 12.3 |
| 14460 | Cambridge-Newton-Framingham, MA Div 15754 | 1,310 | 2,400,642 | X | 1,832.0 | 71.6 | 5.6 | 0.4 | 11.4 | 13.1 | 5.3 | 11.1 | 13.0 | 14.6 | 13.1 | 12.9 |
| 14460 | Rockingham County-Strafford County, NH Div 40484 | 1,063 | 442,840 | X | 416.6 | 92.8 | 1.5 | 0.7 | 3.4 | 3.2 | 4.5 | 10.4 | 13.2 | 12.7 | 11.7 | 13.4 |
| 14500 | Boulder, CO | 726 | 327,171 | 156 | 450.4 | 79.8 | 1.6 | 0.9 | 6.4 | 13.8 | 4.2 | 10.6 | 18.0 | 14.1 | 12.5 | 12.6 |
| 14540 | Bowling Green, KY | 1,615 | 180,751 | 237 | 111.9 | 83.5 | 8.6 | 0.7 | 4.6 | 4.8 | 6.2 | 12.6 | 17.2 | 13.3 | 12.1 | 11.7 |
| 14740 | Bremerton-Silverdale-Port Orchard, WA | 395 | 272,787 | 180 | 690.4 | 80.7 | 4.5 | 2.8 | 9.8 | 8.5 | 5.5 | 11.2 | 12.5 | 14.9 | 12.5 | 11.1 |
| 14860 | Bridgeport-Stamford-Norwalk, CT | 625 | 942,426 | 59 | 1,507.9 | 61.6 | 12.1 | 0.4 | 6.6 | 20.9 | 5.4 | 12.5 | 13.2 | 11.5 | 12.5 | 13.9 |
| 15180 | Brownsville-Harlingen, TX | 892 | 424,180 | 128 | 475.8 | 8.7 | 0.5 | 0.2 | 0.8 | 90.1 | 7.6 | 16.7 | 16.0 | 12.7 | 11.8 | 11.4 |
| 15260 | Brunswick, GA | 1,294 | 119,157 | 333 | 92.1 | 69.3 | 24.2 | 0.8 | 1.8 | 5.6 | 5.3 | 11.7 | 11.5 | 11.8 | 11.2 | 12.4 |
| 15380 | Buffalo-Cheektowaga, NY | 1,565 | 1,125,637 | 49 | 719.2 | 78.2 | 13.2 | 1.1 | 4.2 | 5.5 | 5.3 | 11.2 | 12.3 | 14.0 | 11.7 | 12.0 |
| 15500 | Burlington, NC | 424 | 171,346 | 248 | 404.7 | 63.7 | 21.9 | 0.8 | 2.3 | 13.2 | 5.8 | 12.3 | 14.3 | 12.8 | 11.5 | 12.8 |
| 15540 | Burlington-South Burlington, VT | 1,250 | 221,160 | 208 | 176.6 | 91.4 | 2.7 | 1.3 | 4.6 | 2.4 | 4.7 | 10.3 | 16.8 | 14.1 | 12.2 | 11.9 |
| 15680 | California-Lexington Park, MD | 359 | 114,687 | 337 | 319.7 | 76.3 | 16.5 | 0.9 | 4.3 | 5.6 | 6.1 | 13.6 | 13.3 | 14.3 | 12.8 | 12.8 |
| 15940 | Canton-Massillon, OH | 970 | 396,669 | 137 | 409.0 | 89.1 | 9.3 | 0.8 | 1.3 | 2.2 | 5.5 | 11.9 | 12.0 | 12.1 | 11.4 | 12.2 |
| 15980 | Cape Coral-Fort Myers, FL | 781 | 790,767 | 75 | 1,011.9 | 66.9 | 8.8 | 0.5 | 2.3 | 23.0 | 4.5 | 9.6 | 9.8 | 10.9 | 10.3 | 11.2 |
| 16020 | Cape Girardeau, MO-IL | 1,432 | 97,120 | 362 | 67.8 | 87.5 | 9.4 | 0.8 | 2.1 | 2.4 | 5.7 | 11.9 | 16.1 | 12.4 | 11.4 | 11.0 |
| 16060 | Carbondale-Marion, IL | 1,348 | 135,448 | 307 | 100.5 | 84.5 | 10.5 | 0.8 | 2.7 | 3.6 | 5.4 | 11.1 | 16.3 | 12.9 | 11.7 | 11.1 |
| 16180 | Carson City, NV | 145 | 56,034 | 384 | 387.2 | 68.3 | 2.5 | 2.9 | 3.9 | 24.9 | 5.6 | 11.3 | 10.7 | 13.5 | 11.5 | 11.8 |
| 16220 | Casper, WY | 5,341 | 80,815 | 378 | 15.1 | 88.0 | 1.9 | 1.7 | 1.4 | 9.0 | 6.2 | 13.7 | 11.7 | 14.2 | 13.8 | 11.0 |

1. CBSA = Core Based Statistical Area. DIV = Metropolitan Division. See Appendix A for explanation. See Appendix B for list of metropolitan areas or temporarily covered by water. 2. Dry land or land partially or temporarily covered by water. 3. May be of any race.

Table C. Metropolitan Areas — **Population and Households**

| Area name | 55 to 64 years | 65 to 74 years | 75 years and over | Percent female | Total persons 2000 | Total persons 2010 | Percent change 2000-2010 | Percent change 2010-2019 | Births | Deaths | Net migration | Number | Persons per household | Family households | Single parent households | One person |
|---|---|---|---|---|---|---|---|---|---|---|---|---|---|---|---|---|
| | 16 | 17 | 18 | 19 | 20 | 21 | 22 | 23 | 24 | 25 | 26 | 27 | 28 | 29 | 30 | 31 |
| Abilene, TX | 11.1 | 8.7 | 6.9 | 49.6 | 160,245 | 165,252 | 3.1 | 4.8 | 24,162 | 17,347 | 1,156 | 61,750 | 2.56 | 65.4 | 10.5 | 29.1 |
| Akron, OH | 14.2 | 11.1 | 7.6 | 51.4 | 694,960 | 703,215 | 1.2 | -0.3 | 76,396 | 74,134 | -3,654 | 285,647 | 2.40 | 60.8 | 10.4 | 32.0 |
| Albany, GA | 12.6 | 10.2 | 6.7 | 52.8 | 153,759 | 154,033 | 0.2 | -5.7 | 20,510 | 15,136 | -14,306 | 54,128 | 2.61 | 60.7 | 16.5 | 36.5 |
| Albany-Lebanon, OR | 13.2 | 11.6 | 7.8 | 50.5 | 103,069 | 116,681 | 13.2 | 12.3 | 15,074 | 12,929 | 12,273 | 49,407 | 2.60 | 62.7 | 7.7 | 27.0 |
| Albany-Schenectady-Troy, NY | 14.0 | 10.9 | 7.7 | 51.1 | 825,875 | 870,713 | 5.4 | 0.9 | 91,837 | 81,373 | -2,198 | 362,502 | 2.34 | 61.9 | 10.2 | 30.2 |
| Albuquerque, NM | 13.0 | 10.8 | 7.1 | 50.8 | 729,649 | 887,056 | 21.6 | 4.1 | 104,702 | 75,732 | 8,060 | 354,359 | 2.55 | 61.9 | 12.7 | 31.9 |
| Alexandria, LA | 12.9 | 9.8 | 7.1 | 50.5 | 145,035 | 153,910 | 6.1 | -2.0 | 20,794 | 16,998 | -6,812 | 55,591 | 2.60 | 67.2 | 19.2 | 27.9 |
| Allentown-Bethlehem-Easton, PA-NJ | 14.1 | 10.7 | 8.2 | 50.9 | 740,395 | 821,095 | 10.9 | 3.1 | 88,060 | 83,833 | 21,619 | 322,301 | 2.54 | 67.2 | 10.6 | 26.4 |
| Altoona, PA | 14.4 | 12.2 | 9.5 | 51.0 | 129,144 | 127,121 | -1.6 | -4.8 | 13,050 | 16,583 | -2,482 | 52,497 | 2.26 | 66.0 | 12.8 | 27.2 |
| Amarillo, TX | 11.5 | 8.8 | 6.1 | 49.9 | 228,707 | 251,937 | 10.2 | 5.5 | 37,257 | 24,007 | 474.0 | 98,606 | 2.64 | 65.1 | 11.8 | 30.2 |
| Ames, IA | 10.5 | 8.4 | 6.0 | 48.1 | 106,205 | 115,850 | 9.1 | 7.5 | 12,253 | 8,360 | 4,731 | 50,253 | 2.22 | 53.9 | 4.4 | 28.9 |
| Anchorage, AK | 12.2 | 8.3 | 4.2 | 48.8 | 319,605 | 381,233 | 19.3 | 4.2 | 59,705 | 23,339 | -20,661 | 137,934 | 2.81 | 64.2 | 9.4 | 26.6 |
| Ann Arbor, MI | 11.4 | 9.1 | 5.9 | 50.5 | 322,895 | 345,417 | 7.0 | 6.1 | 37,579 | 23,063 | 6,731 | 141,680 | 2.44 | 57.4 | 6.7 | 29.7 |
| Anniston-Oxford, AL | 13.7 | 11.2 | 7.3 | 51.9 | 112,249 | 118,533 | 5.6 | -4.3 | 13,556 | 14,808 | -3,766 | 44,636 | 2.48 | 64.9 | 12.3 | 30.4 |
| Appleton, WI | 14.0 | 9.6 | 6.4 | 50.0 | 201,602 | 225,595 | 11.9 | 5.9 | 28,640 | 17,600 | 2,483 | 94,937 | 2.47 | 64.3 | 7.1 | 28.9 |
| Asheville, NC | 14.0 | 13.5 | 9.7 | 52.0 | 369,171 | 424,864 | 15.1 | 9.8 | 44,832 | 50,760 | 47,368 | 193,451 | 2.34 | 62.2 | 7.5 | 30.2 |
| Athens-Clarke County, GA | 10.7 | 8.6 | 5.6 | 51.9 | 166,079 | 192,558 | 15.9 | 11.5 | 22,561 | 14,493 | 14,008 | 82,655 | 2.42 | 57.2 | 9.7 | 27.8 |
| Atlanta-Sandy Springs-Alpharetta, GA | 12.2 | 8.2 | 4.9 | 51.7 | 4,263,438 | 5,286,689 | 24.0 | 15.2 | 742,606 | 373,246 | 430,885 | 2,189,294 | 2.71 | 66.5 | 11.1 | 27.6 |
| Atlantic City-Hammonton, NJ | 15.1 | 11.2 | 8.0 | 51.6 | 252,552 | 274,532 | 8.7 | -4.2 | 31,062 | 27,578 | -15,198 | 101,975 | 2.53 | 65.0 | 10.4 | 29.8 |
| Auburn-Opelika, AL | 11.1 | 8.0 | 4.9 | 50.9 | 115,092 | 140,300 | 21.9 | 18.9 | 18,526 | 10,814 | 18,627 | 64,448 | 2.46 | 58.7 | 4.2 | 32.4 |
| Augusta-Richmond County, GA-SC | 13.1 | 10.3 | 6.6 | 51.4 | 508,032 | 564,893 | 11.2 | 8.7 | 76,077 | 55,661 | 28,939 | 211,969 | 2.80 | 65.7 | 12.7 | 29.8 |
| Austin-Round Rock-Georgetown, TX | 10.6 | 7.4 | 4.2 | 49.9 | 1,249,763 | 1,716,307 | 37.3 | 33.7 | 266,507 | 102,543 | 410,335 | 818,084 | 2.67 | 62.1 | 8.6 | 27.5 |
| Bakersfield, CA | 10.4 | 7.0 | 4.5 | 48.9 | 661,645 | 839,619 | 26.9 | 7.4 | 140,814 | 60,952 | -17,936 | 272,888 | 3.20 | 73.0 | 15.0 | 22.4 |
| Baltimore-Columbia-Towson, MD | 13.5 | 9.6 | 6.8 | 51.9 | 2,552,994 | 2,710,601 | 6.2 | 3.3 | 342,158 | 251,744 | 997.0 | 1,062,901 | 2.57 | 64.0 | 10.3 | 29.4 |
| Bangor, ME | 15.2 | 11.7 | 7.8 | 50.4 | 144,919 | 153,932 | 6.2 | -1.5 | 14,727 | 16,299 | -615 | 64,096 | 2.27 | 56.9 | 7.8 | 32.1 |
| Barnstable Town, MA | 17.7 | 18.3 | 13.8 | 52.2 | 222,230 | 215,883 | -2.9 | -1.3 | 15,862 | 29,924 | 11,568 | 96,509 | 2.18 | 64.1 | 6.4 | 29.8 |
| Baton Rouge, LA | 12.2 | 9.2 | 5.9 | 51.2 | 729,361 | 825,944 | 13.2 | 4.0 | 114,917 | 72,984 | -9,328 | 305,441 | 2.72 | 63.9 | 13.1 | 29.8 |
| Battle Creek, MI | 13.6 | 10.9 | 7.7 | 51.0 | 137,985 | 136,150 | -1.3 | -1.9 | 16,516 | 15,414 | -3,657 | 53,827 | 2.42 | 61.2 | 12.3 | 31.6 |
| Bay City, MI | 15.2 | 12.5 | 9.1 | 50.9 | 110,157 | 107,773 | -2.2 | -5.0 | 10,569 | 12,501 | -3,402 | 44,887 | 2.27 | 61.4 | 9.4 | 32.6 |
| Beaumont-Port Arthur, TX | 12.7 | 9.2 | 6.5 | 49.5 | 385,090 | 388,751 | 1.0 | 0.7 | 54,246 | 41,097 | -10,558 | 142,135 | 2.64 | 67.7 | 12.3 | 28.2 |
| Beckley, WV | 13.4 | 13.3 | 8.7 | 49.7 | 126,799 | 124,910 | -1.5 | -7.9 | 13,910 | 17,333 | -6,463 | 49,117 | 2.26 | 67.5 | 11.3 | 28.4 |
| Bellingham, WA | 12.0 | 11.2 | 7.1 | 50.6 | 166,814 | 201,146 | 20.6 | 14.8 | 22,952 | 16,707 | 23,640 | 88,794 | 2.53 | 60.8 | 7.2 | 24.1 |
| Bend, OR | 13.6 | 13.1 | 7.7 | 50.4 | 115,367 | 157,728 | 36.7 | 27.9 | 18,125 | 14,644 | 40,225 | 76,528 | 2.57 | 65.7 | 8.0 | 25.4 |
| Billings, MT | 13.5 | 11.1 | 7.6 | 50.6 | 147,099 | 167,165 | 13.6 | 10.0 | 21,528 | 17,014 | 12,126 | 78,108 | 2.28 | 60.8 | 10.1 | 31.1 |
| Binghamton, NY | 14.6 | 11.0 | 9.1 | 50.6 | 252,320 | 251,722 | -0.2 | -5.7 | 25,275 | 26,344 | -13,402 | 99,300 | 2.30 | 59.5 | 10.2 | 32.7 |
| Birmingham-Hoover, AL | 12.9 | 10.1 | 6.7 | 52.1 | 981,525 | 1,061,038 | 8.1 | 2.9 | 138,842 | 111,667 | 4,320 | 428,665 | 2.49 | 65.3 | 9.8 | 29.7 |
| Bismarck, ND | 12.7 | 9.8 | 7.5 | 49.9 | 96,784 | 110,625 | 14.3 | 17.2 | 17,506 | 10,330 | 11,590 | 52,396 | 2.39 | 63.5 | 8.0 | 30.6 |
| Blacksburg-Christiansburg, VA | 11.4 | 9.5 | 6.7 | 49.5 | 151,272 | 162,959 | 7.7 | 2.6 | 14,992 | 14,504 | 3,811 | 62,114 | 2.45 | 57.1 | 7.2 | 30.9 |
| Bloomington, IL | 11.5 | 8.4 | 5.8 | 51.5 | 150,433 | 169,578 | 12.7 | 1.0 | 20,671 | 12,100 | -6,983 | 65,118 | 2.48 | 61.3 | 8.4 | 29.5 |
| Bloomington, IN | 10.9 | 8.9 | 5.9 | 50.2 | 142,349 | 159,530 | 12.1 | 6.0 | 15,484 | 11,985 | 6,053 | 62,229 | 2.47 | 53.9 | 7.1 | 33.4 |
| Bloomsburg-Berwick, PA | 14.1 | 11.6 | 9.1 | 51.7 | 82,387 | 85,552 | 3.8 | -3.1 | 8,141 | 9,510 | -1,237 | 33,738 | 2.32 | 64.8 | 9.7 | 29.6 |
| Boise City, ID | 11.9 | 9.6 | 6.0 | 50.0 | 464,840 | 616,566 | 32.6 | 24.9 | 86,825 | 47,962 | 114,539 | 275,681 | 2.67 | 69.2 | 8.7 | 23.5 |
| Boston-Cambridge-Newton, MA-NH | 13.4 | 9.6 | 7.0 | 51.4 | 4,391,344 | 4,552,598 | 3.7 | 7.2 | 524,112 | 376,593 | 181,265 | 1,855,763 | 2.54 | 62.5 | 7.9 | 28.3 |
| Boston, MA Div 14454 | 12.9 | 9.2 | 6.8 | 51.7 | 1,812,937 | 1,888,056 | 4.1 | 7.8 | 221,853 | 155,557 | 81,628 | 770,703 | 2.54 | 59.8 | 8.2 | 30.4 |
| Cambridge-Newton-Framingham, MA Div 15754 | 13.5 | 9.6 | 7.1 | 51.2 | 2,188,815 | 2,246,189 | 2.6 | 6.9 | 262,877 | 185,631 | 78,868 | 910,264 | 2.55 | 64.2 | 7.8 | 27.1 |
| Rockingham County-Strafford County, NH Div 40484 | 15.9 | 11.1 | 7.1 | 50.6 | 389,592 | 418,353 | 7.4 | 5.9 | 39,382 | 35,405 | 20,769 | 174,796 | 2.45 | 66.0 | 6.9 | 25.3 |
| Boulder, CO | 12.4 | 9.7 | 6.0 | 49.7 | 269,814 | 294,560 | 9.2 | 11.1 | 28,618 | 18,203 | 22,129 | 131,854 | 2.38 | 56.8 | 6.4 | 29.7 |
| Bowling Green, KY | 11.8 | 9.0 | 6.1 | 50.7 | 134,976 | 158,618 | 17.5 | 14.0 | 21,854 | 15,422 | 15,688 | 66,318 | 2.55 | 65.5 | 11.6 | 27.1 |
| Bremerton-Silverdale-Port Orchard, WA | 13.4 | 11.6 | 7.2 | 48.9 | 231,969 | 251,143 | 8.3 | 8.6 | 30,811 | 22,396 | 13,139 | 107,525 | 2.45 | 68.5 | 9.5 | 23.7 |
| Bridgeport-Stamford-Norwalk, CT | 14.3 | 9.3 | 7.5 | 51.3 | 882,567 | 916,910 | 3.9 | 2.8 | 102,311 | 68,987 | -7,760 | 340,557 | 2.72 | 69.2 | 9.1 | 25.0 |
| Brownsville-Harlingen, TX | 9.7 | 8.0 | 6.2 | 51.1 | 335,227 | 406,214 | 21.2 | 4.4 | 71,420 | 26,882 | -26,726 | 129,307 | 3.25 | 78.0 | 18.0 | 19.3 |
| Brunswick, GA | 14.4 | 12.8 | 8.9 | 52.6 | 93,044 | 112,371 | 20.8 | 6.0 | 13,257 | 12,102 | 5,635 | 47,730 | 2.46 | 63.1 | 9.8 | 34.1 |
| Buffalo-Cheektowaga, NY | 14.5 | 11.0 | 8.1 | 51.5 | 1,170,111 | 1,135,608 | -2.9 | -0.9 | 123,684 | 125,060 | -7,949 | 488,951 | 2.24 | 58.6 | 10.4 | 30.4 |
| Burlington, NC | 13.2 | 9.8 | 7.6 | 52.6 | 130,800 | 151,160 | 15.6 | 13.4 | 18,788 | 16,815 | 18,248 | 64,316 | 2.57 | 65.0 | 11.4 | 30.4 |
| Burlington-South Burlington, VT | 13.4 | 9.9 | 6.7 | 50.8 | 198,889 | 211,264 | 6.2 | 4.7 | 22,014 | 16,077 | 4,126 | 89,859 | 2.33 | 60.5 | 8.9 | 27.1 |
| California-Lexington Park, MD | 13.4 | 8.1 | 5.6 | 50.0 | 86,211 | 105,146 | 22.0 | 9.1 | 14,370 | 8,156 | 3,353 | 40,320 | 2.75 | 72.6 | 12.8 | 22.6 |
| Canton-Massillon, OH | 14.3 | 11.8 | 8.7 | 51.3 | 406,934 | 404,408 | -0.6 | -1.9 | 45,144 | 46,447 | -6,106 | 165,718 | 2.34 | 63.1 | 10.8 | 30.2 |
| Cape Coral-Fort Myers, FL | 14.1 | 15.7 | 13.8 | 51.1 | 440,888 | 618,761 | 40.3 | 27.8 | 67,541 | 70,855 | 174,050 | 286,887 | 2.65 | 65.9 | 8.3 | 28.1 |
| Cape Girardeau, MO-IL | 12.9 | 10.5 | 8.0 | 51.4 | 90,312 | 96,279 | 6.6 | 0.9 | 11,527 | 10,319 | -352 | 38,214 | 2.35 | 62.1 | 7.4 | 30.1 |
| Carbondale-Marion, IL | 12.7 | 10.7 | 8.0 | 49.9 | 133,786 | 139,148 | 4.0 | -2.7 | 15,795 | 14,062 | -5,531 | 56,518 | 2.29 | 58.4 | 11.2 | 33.4 |
| Carson City, NV | 14.3 | 12.3 | 8.9 | 49.0 | 52,457 | 55,269 | 5.4 | 1.4 | 6,019 | 7,322 | 2,044 | 22,703 | 2.37 | 63.0 | 13.0 | 32.6 |
| Casper, WY | 13.1 | 10.1 | 6.2 | 49.5 | 66,533 | 75,448 | 13.4 | 7.1 | 10,903 | 7,370 | 1,703 | 33,044 | 2.37 | 67.8 | 9.0 | 25.9 |

| Area name | Persons in group quarters, 2020 | Daytime population, 2019 | | Births, 2020 | | Deaths, 2020 | | Persons under 65 with no health insurance 2019 | | Medicare, 2020 | | | Serious crimes known to police[2], 2019 Violent | |
|---|---|---|---|---|---|---|---|---|---|---|---|---|---|---|
| | | Number | Employment/ residence ratio | Total | Rate[1] | Number | Rate[1] | Number | Percent | Total Beneficiaries | Enrolled in Original Medicare | Enrolled in Medicare Advantage | Number | Rate[3] |
| | 32 | 33 | 34 | 35 | 36 | 37 | 38 | 39 | 40 | 41 | 42 | 43 | 44 | 45 |
| Abilene, TX | 10,033 | 175,759 | 1.05 | 2,304 | 13.3 | 1,877 | 10.8 | 25,591 | 18.7 | 31,936 | 21,808 | 10,128 | 543 | 317.3 |
| Akron, OH | 16,673 | 698,407 | 0.99 | 7,152 | 10.2 | 7,869 | 11.2 | 44,332 | 7.9 | 147,754 | 65,803 | 81,951 | 2,616 | 371.7 |
| Albany, GA | 5,906 | 149,186 | 1.03 | 1,806 | 12.4 | 1,707 | 11.8 | 18,146 | 15.6 | 29,093 | 15,946 | 13,147 | 1,071 | 717.6 |
| Albany-Lebanon, OR | 1,171 | 122,364 | 0.88 | 1,498 | 11.4 | 1,472 | 11.2 | 8,693 | 8.4 | 30,127 | 13,929 | 16,197 | 175 | 136.6 |
| Albany-Schenectady-Troy, NY | 31,567 | 898,098 | 1.04 | 8,493 | 9.7 | 8,845 | 10.1 | 28,782 | 4.1 | 184,309 | 94,190 | 90,117 | 2,266 | 257.5 |
| Albuquerque, NM | 14,802 | 907,658 | 0.98 | 9,234 | 10.0 | 8,629 | 9.3 | 80,326 | 10.8 | 182,827 | 86,143 | 96,686 | 9,580 | 1,043.4 |
| Alexandria, LA | 7,461 | 151,106 | 0.99 | 1,904 | 12.6 | 1,742 | 11.6 | 12,881 | 10.7 | 32,249 | 23,708 | 8,539 | 1,291 | 849.2 |
| Allentown-Bethlehem-Easton, PA-NJ | 22,820 | 811,032 | 0.92 | 8,751 | 10.3 | 8,707 | 10.3 | 47,707 | 7.1 | 184,183 | 121,007 | 63,176 | NA | NA |
| Altoona, PA | 3,721 | 127,179 | 1.09 | 1,164 | 9.6 | 1,740 | 14.4 | 6,298 | 6.7 | 31,533 | 14,566 | 16,967 | 406 | 333.4 |
| Amarillo, TX | 9,258 | 266,780 | 0.98 | 3,294 | 12.4 | 2,493 | 9.4 | 40,515 | 18.6 | 44,664 | 31,492 | 13,171 | 1,584 | 595.4 |
| Ames, IA | 11,236 | 125,347 | 1.03 | 1,088 | 8.7 | 882 | 7.1 | 5,733 | 6.0 | 20,010 | 16,010 | 3,999 | 249 | 199.3 |
| Anchorage, AK | 10,123 | 393,815 | 0.99 | 5,233 | 13.2 | 2,768 | 7.0 | 39,677 | 11.6 | 55,532 | 54,437 | 1,095 | 3,657 | 1,194.6 |
| Ann Arbor, MI | 20,772 | 416,729 | 1.26 | 3,455 | 9.4 | 2,644 | 7.2 | 15,815 | 5.3 | 61,158 | 37,272 | 23,886 | 1,327 | 355.5 |
| Anniston-Oxford, AL | 3,285 | 115,153 | 1.03 | 1,230 | 10.8 | 1,553 | 13.7 | 10,767 | 11.9 | 26,991 | 17,248 | 9,743 | NA | NA |
| Appleton, WI | 3,033 | 233,056 | 0.96 | 2,670 | 11.2 | 1,877 | 7.9 | 12,621 | 6.3 | 43,510 | 15,428 | 28,082 | 392 | 164.2 |
| Asheville, NC | 10,848 | 473,796 | 1.05 | 4,248 | 9.1 | 5,535 | 11.9 | 49,995 | 14.3 | 119,892 | 76,752 | 43,140 | 1,466 | 316.1 |
| Athens-Clarke County, GA | 11,024 | 215,729 | 1.03 | 2,144 | 10.0 | 1,636 | 7.6 | 27,418 | 15.8 | 34,328 | 20,567 | 13,761 | NA | NA |
| Atlanta-Sandy Springs-Alpharetta, GA | 87,507 | 6,051,929 | 1.01 | 71,104 | 11.7 | 44,971 | 7.4 | 780,294 | 15.1 | 874,486 | 478,106 | 396,381 | NA | NA |
| Atlantic City-Hammonton, NJ | 5,995 | 265,636 | 1.02 | 2,761 | 10.5 | 3,055 | 11.6 | 22,420 | 10.7 | 58,468 | 40,777 | 17,692 | 672 | 255.4 |
| Auburn-Opelika, AL | 4,960 | 153,285 | 0.86 | 1,754 | 10.5 | 1,272 | 7.6 | 14,928 | 10.7 | 25,287 | 14,775 | 10,513 | NA | NA |
| Augusta-Richmond County, GA-SC | 18,961 | 615,759 | 1.02 | 7,278 | 11.8 | 6,361 | 10.4 | 66,418 | 13.6 | 119,402 | 73,823 | 45,579 | 2,199 | 361.1 |
| Austin-Round Rock-Georgetown, TX | 39,678 | 2,249,645 | 1.02 | 26,744 | 11.7 | 12,633 | 5.5 | 307,698 | 15.8 | 279,200 | 178,592 | 100,608 | 6,361 | 286.9 |
| Bakersfield, CA | 30,547 | 896,514 | 0.99 | 12,826 | 14.2 | 6,980 | 7.7 | 70,588 | 9.3 | 120,883 | 71,492 | 49,391 | 5,567 | 621.1 |
| Baltimore-Columbia-Towson, MD | 70,446 | 2,786,147 | 0.99 | 32,048 | 11.4 | 27,967 | 10.0 | 134,580 | 5.9 | 502,217 | 432,381 | 69,836 | 19,564 | 698.7 |
| Bangor, ME | 7,264 | 154,090 | 1.03 | 1,350 | 8.9 | 1,728 | 11.4 | 13,631 | 11.7 | 37,389 | 23,162 | 14,228 | 67 | 44.3 |
| Barnstable Town, MA | 4,112 | 208,267 | 0.96 | 1,459 | 6.8 | 3,077 | 14.4 | 5,496 | 3.8 | 78,016 | 64,380 | 13,636 | 691 | 326.8 |
| Baton Rouge, LA | 25,820 | 860,382 | 1.01 | 10,689 | 12.4 | 8,077 | 9.4 | 68,183 | 9.7 | 147,175 | 68,045 | 79,130 | 4,721 | 554.0 |
| Battle Creek, MI | 4,294 | 138,601 | 1.07 | 1,508 | 11.3 | 1,489 | 11.1 | 7,646 | 7.2 | 30,566 | 19,215 | 11,351 | 836 | 624.0 |
| Bay City, MI | 1,438 | 92,683 | 0.78 | 921 | 9.0 | 1,242 | 12.1 | 5,631 | 7.0 | 26,776 | 14,635 | 12,142 | 396 | 383.7 |
| Beaumont-Port Arthur, TX | 16,597 | 393,094 | 1 | 5,036 | 12.9 | 4,314 | 11.0 | 62,210 | 19.7 | 71,247 | 38,485 | 32,761 | 2,133 | 541.2 |
| Beckley, WV | 5,477 | 115,027 | 0.98 | 1,162 | 10.1 | 1,824 | 15.9 | 7,239 | 8.4 | 31,333 | 19,833 | 11,501 | NA | NA |
| Bellingham, WA | 5,861 | 224,772 | 0.96 | 2,122 | 9.2 | 1,931 | 8.4 | 16,675 | 9.1 | 48,015 | 28,643 | 19,371 | 502 | 220.1 |
| Bend, OR | 1,378 | 199,161 | 1.02 | 1,866 | 9.2 | 1,674 | 8.3 | 12,432 | 7.9 | 46,658 | 31,826 | 14,832 | 324 | 165.4 |
| Billings, MT | 3,861 | 181,367 | 1 | 1,966 | 10.7 | 1,791 | 9.7 | 12,340 | 8.4 | 37,755 | 28,108 | 9,647 | NA | NA |
| Binghamton, NY | 11,532 | 238,350 | 1 | 2,250 | 9.5 | 2,779 | 11.7 | 9,141 | 5.0 | 55,657 | 29,153 | 26,505 | 678 | 285.4 |
| Birmingham-Hoover, AL | 23,686 | 1,099,762 | 1.02 | 12,873 | 11.8 | 12,199 | 11.2 | 94,223 | 10.6 | 218,222 | 97,833 | 120,388 | NA | NA |
| Bismarck, ND | 4,645 | D | D | 1,624 | 12.5 | 1,136 | 8.8 | 6,086 | 5.8 | 24,292 | 17,571 | 6,721 | 394 | 305.7 |
| Blacksburg-Christiansburg, VA | 14,525 | 169,922 | 1.05 | 1,335 | 8.0 | 1,607 | 9.6 | 10,929 | 8.6 | 30,390 | 23,171 | 7,218 | 323 | 192.3 |
| Bloomington, IL | 10,506 | 176,344 | 1.06 | 1,752 | 10.2 | 1,318 | 7.7 | 7,970 | 5.8 | 27,167 | 18,010 | 9,157 | 610 | 353.8 |
| Bloomington, IN | 15,203 | 174,736 | 1.07 | 1,471 | 8.7 | 1,310 | 7.7 | 12,824 | 9.8 | 26,879 | 19,155 | 7,724 | NA | NA |
| Bloomsburg-Berwick, PA | 4,354 | 85,340 | 1.05 | 712 | 8.6 | 1,015 | 12.2 | 4,088 | 6.5 | 19,433 | 10,845 | 8,587 | NA | NA |
| Boise City, ID | 15,146 | 748,873 | 1 | 8,301 | 10.8 | 5,495 | 7.1 | 72,805 | 11.6 | 134,794 | 64,913 | 69,881 | 1,880 | 250.9 |
| Boston-Cambridge-Newton, MA-NH | 166,645 | 5,063,751 | 1.07 | 49,616 | 10.2 | 41,839 | 8.6 | 147,365 | 3.7 | 886,227 | 657,125 | 229,104 | 13,556 | 277.7 |
| Boston, MA Div 14454 | 81,215 | 2,241,300 | 1.19 | 21,057 | 10.3 | 17,594 | 8.6 | 57,356 | 3.5 | 355,818 | 264,877 | 90,943 | 8,057 | 396.3 |
| Cambridge-Newton-Framingham, MA Div 15754 | 74,416 | 2,399,791 | 1 | 24,729 | 10.3 | 20,336 | 8.5 | 65,962 | 3.4 | 436,123 | 317,861 | 118,263 | 5,065 | 210.5 |
| Rockingham County-Strafford County, NH Div 40484 | 11,014 | D | D | 3,830 | 8.6 | 3,909 | 8.8 | 24,047 | 6.8 | 94,286 | 74,387 | 19,898 | 434 | 98.3 |
| Boulder, CO | 10,463 | 368,451 | 1.24 | 2,517 | 7.7 | 2,164 | 6.6 | 20,712 | 7.7 | 53,602 | 29,926 | 23,676 | 895 | 272.3 |
| Bowling Green, KY | 7,250 | 174,710 | 0.97 | 2,249 | 12.4 | 1,681 | 9.3 | 13,022 | 9.0 | 31,696 | 20,047 | 11,648 | 275 | 153.5 |
| Bremerton-Silverdale-Port Orchard, WA | 9,538 | 263,117 | 0.94 | 3,001 | 11.0 | 2,647 | 9.7 | 13,724 | 6.4 | 57,694 | 41,720 | 15,974 | 709 | 261.4 |
| Bridgeport-Stamford-Norwalk, CT | 19,496 | 937,379 | 0.99 | 9,461 | 10.0 | 7,423 | 7.9 | 72,442 | 9.3 | 159,488 | 103,589 | 55,899 | 1,619 | 174.2 |
| Brownsville-Harlingen, TX | 3,427 | 414,763 | 0.95 | 6,223 | 14.7 | 3,006 | 7.1 | 109,621 | 30.8 | 65,439 | 29,523 | 35,917 | 1,588 | 375.1 |
| Brunswick, GA | 1,691 | 117,943 | 0.98 | 1,216 | 10.2 | 1,391 | 11.7 | 16,784 | 18.3 | 26,245 | 16,594 | 9,651 | NA | NA |
| Buffalo-Cheektowaga, NY | 30,829 | 1,135,229 | 1.01 | 11,720 | 10.4 | 12,999 | 11.5 | 36,950 | 4.1 | 251,443 | 93,131 | 158,313 | 3,980 | 354.3 |
| Burlington, NC | 6,238 | 157,421 | 0.85 | 1,942 | 11.3 | 1,879 | 11.0 | 19,822 | 14.8 | 33,912 | 13,974 | 19,938 | 686 | 407.6 |
| Burlington-South Burlington, VT | 10,605 | 224,049 | 1.04 | 1,991 | 9.0 | 1,732 | 7.8 | 8,900 | 5.0 | 42,918 | 36,525 | 6,393 | 444 | 200.5 |
| California-Lexington Park, MD | 2,789 | 109,238 | 0.92 | 1,344 | 11.7 | 949 | 8.3 | 4,997 | 5.2 | 17,047 | 16,589 | 458 | 231 | 204.4 |
| Canton-Massillon, OH | 9,197 | 384,862 | 0.93 | 4,309 | 10.9 | 4,783 | 12.1 | 24,057 | 7.7 | 93,244 | 38,227 | 55,019 | 1,345 | 338.6 |
| Cape Coral-Fort Myers, FL | 8,476 | 764,971 | 0.98 | 6,919 | 8.7 | 8,819 | 11.2 | 96,245 | 17.9 | 201,807 | 121,956 | 79,851 | 1,979 | 257.8 |
| Cape Girardeau, MO-IL | 4,122 | 94,002 | 1.01 | 1,063 | 10.9 | 1,036 | 10.7 | 9,114 | 12.0 | 20,739 | 17,189 | 3,551 | 373 | 384.6 |
| Carbondale-Marion, IL | 6,856 | 139,345 | 1.04 | 1,407 | 10.4 | 1,512 | 11.2 | 7,812 | 7.5 | 28,524 | 21,489 | 7,035 | NA | NA |
| Carson City, NV | 2,689 | D | D | 597 | 10.7 | 786 | 14.0 | 6,131 | 14.7 | 14,250 | 11,319 | 2,931 | 193 | 347.8 |
| Casper, WY | 1,668 | 79,692 | 1 | 931 | 11.5 | 804 | 9.9 | 9,564 | 14.5 | 14,773 | 13,821 | 953 | 227 | 285.6 |

1. Per 1,000 estimated resident population.   2. Data for serious crimes have not been adjusted for underreporting; this may affect comparability between geographic areas and over time.   3. Per 100,000 population estimated by the FBI.

Items 32—45

## Table C. Metropolitan Areas — Crime, Education, Money Income, and Poverty

| Area name | Serious crimes known to police[1], 2019 (cont.) Property Number | Rate | Education — School enrollment and attainment, 2019 — Enrollment[3] Total | Percent private | Attainment[4] High school graduate or less | Bachelor's degree or more | Local government expenditures,[5] 2017-2018 Total current expenditures (mil dol) | Current expenditures per student (dollars) | Income and poverty, 2019 Per capita income[6] (dollars) | Median household income (dollars) | Median family income | Percent of households with income less than $50,000 | Percent of households with income of $200,000 or more | Percent below poverty level All persons | All families | Families with children under 18 |
|---|---|---|---|---|---|---|---|---|---|---|---|---|---|---|---|---|
| | 46 | 47 | 48 | 49 | 50 | 51 | 52 | 53 | 54 | 55 | 56 | 57 | 58 | 59 | 60 | 61 |
| Abilene, TX................. | 3,603 | 2,105.5 | 45,764 | 24.6 | 42.2 | 24.1 | 272.6 | 9,398 | 26,603 | 54,808 | 69,218 | 45.2 | 3.2 | 13.7 | 7.4 | 12.8 |
| Akron, OH................... | 14,445 | 2,052.5 | 163,633 | 15.4 | 39.8 | 31.8 | 1,209.2 | 12,823 | 34,122 | 57,158 | 73,770 | 44.5 | 5.8 | 13.9 | 9.5 | 16.9 |
| Albany, GA................. | 5,057 | 3,388.1 | 39,496 | 5.0 | 46.4 | 19.7 | 271.7 | 10,599 | 23,331 | 40,625 | 51,735 | 60.0 | 2.6 | 22.6 | 19.2 | 26.6 |
| Albany-Lebanon, OR.................. | 2,743 | 2,141.2 | 28,856 | 13.7 | 39.0 | 18.8 | 224.9 | 9,907 | 29,151 | 61,488 | 76,343 | 41.2 | 2.7 | 13.3 | 6.7 | 14.4 |
| Albany-Schenectady-Troy, NY.... | 15,367 | 1,746.5 | 206,853 | 19.8 | 32.8 | 39.5 | 2,204.1 | 18,054 | 38,497 | 73,398 | 95,350 | 34.2 | 7.4 | 10.2 | 6.7 | 12.0 |
| Albuquerque, NM.................. | D | D | 222,864 | 13.1 | 35.8 | 33.2 | 1,251.2 | 9,089 | 32,112 | 58,512 | 69,379 | 42.7 | 5.8 | 15.5 | 12.0 | 19.0 |
| Alexandria, LA.............. | 6,900 | 4,538.7 | 37,372 | 17.1 | 53.0 | 17.5 | 274.4 | 10,405 | 26,715 | 50,553 | 57,392 | 49.2 | 2.8 | 17.6 | 14.4 | 22.0 |
| Allentown-Bethlehem-Easton, PA-NJ.. | NA | NA | 196,626 | 21.8 | 44.6 | 29.7 | 2,065.3 | 16,923 | 35,370 | 70,793 | 84,841 | 35.1 | 7.1 | 9.3 | 6.3 | 11.0 |
| Altoona, PA.................. | 1,510 | 1,240.1 | 25,394 | 11.3 | 52.5 | 23.1 | 248.4 | 14,248 | 29,796 | 51,004 | 65,157 | 48.9 | 2.8 | 15.6 | 11.2 | 22.6 |
| Amarillo, TX................ | 8,579 | 3,224.5 | 71,350 | 4.6 | 41.6 | 25.2 | 473.0 | 9,601 | 28,233 | 53,510 | 69,441 | 47.0 | 4.2 | 14.4 | 12.2 | 16.6 |
| Ames, IA.................... | 1,745 | 1,396.6 | 46,537 | 4.8 | 23.9 | 44.6 | 166.2 | 10,779 | 32,402 | 62,181 | 92,145 | 42.3 | 5.2 | 15.6 | 3.7 | 6.7 |
| Anchorage, AK.............. | 13,222 | 4,319.0 | 99,727 | 12.8 | 32.7 | 33.1 | 984.1 | 14,619 | 38,842 | 80,676 | 100,352 | 28.7 | 9.8 | 10.0 | 6.5 | 10.5 |
| Ann Arbor, MI.............. | 6,159 | 1,649.8 | 122,893 | 9.1 | 18.9 | 55.9 | 611.1 | 13,774 | 43,714 | 76,576 | 105,960 | 31.7 | 12.4 | 13.4 | 6.3 | 10.2 |
| Anniston-Oxford, AL........ | NA | NA | 26,109 | 14.0 | 49.9 | 19.3 | 175.3 | 9,981 | 24,579 | 48,156 | 61,918 | 51.3 | 2.2 | 17.3 | 12.6 | 22.9 |
| Appleton, WI............... | 2,599 | 1,088.7 | 58,360 | 14.9 | 37.2 | 30.1 | 436.4 | 11,078 | 33,905 | 68,335 | 86,284 | 34.1 | 4.4 | 7.2 | 5.4 | 8.8 |
| Asheville, NC.............. | 12,815 | 2,763.1 | 89,931 | 17.0 | 33.5 | 35.5 | 534.5 | 9,744 | 32,422 | 57,428 | 72,194 | 43.9 | 4.6 | 11.0 | 8.3 | 14.4 |
| Athens-Clarke County, GA.... | NA | NA | 72,208 | 7.9 | 32.6 | 42.4 | 353.8 | 11,919 | 29,523 | 50,962 | 74,045 | 48.9 | 5.7 | 20.2 | 11.4 | 19.1 |
| Atlanta-Sandy Springs-Alpharetta, GA... | NA | NA | 1,605,816 | 16.5 | 33.4 | 39.9 | 11,027.1 | 10,817 | 37,331 | 71,742 | 86,101 | 33.9 | 9.7 | 10.5 | 7.5 | 11.1 |
| Atlantic City-Hammonton, NJ.. | 6,314 | 2,399.8 | 64,810 | 10.7 | 46.2 | 28.2 | 878.8 | 19,665 | 36,298 | 63,389 | 75,722 | 40.3 | 7.2 | 11.1 | 8.7 | 13.4 |
| Auburn-Opelika, AL......... | NA | NA | 59,113 | 14.2 | 30.1 | 41.4 | 223.5 | 9,839 | 31,454 | 53,712 | 80,982 | 47.6 | 5.8 | 15.8 | 9.0 | 6.2 |
| Augusta-Richmond County, GA-SC... | 15,097 | 2,478.8 | 146,210 | 13.4 | 41.5 | 27.4 | 970.9 | 10,061 | 28,711 | 55,143 | 66,414 | 45.2 | 4.9 | 13.3 | 9.5 | 17.2 |
| Austin-Round Rock-Georgetown, TX... | 53,442 | 2,410.2 | 570,971 | 13.8 | 28.9 | 46.2 | 3,316.9 | 9,453 | 41,957 | 80,954 | 99,227 | 29.5 | 11.7 | 10.1 | 7.2 | 11.0 |
| Bakersfield, CA............ | 29,604 | 3,302.7 | 268,457 | 10.9 | 51.8 | 17.1 | 2,372.5 | 12,490 | 24,524 | 53,067 | 61,124 | 47.0 | 5.1 | 19.1 | 15.4 | 22.3 |
| Baltimore-Columbia-Towson, MD.. | 66,044 | 2,358.5 | 694,054 | 22.3 | 32.9 | 41.9 | 5,863.1 | 14,487 | 43,139 | 83,160 | 104,462 | 29.9 | 13.0 | 9.4 | 6.0 | 9.4 |
| Bangor, ME................. | 2,355 | 1,557.1 | 34,974 | 13.4 | 39.8 | 29.4 | 280.8 | 13,074 | 30,245 | 50,449 | 73,899 | 49.5 | 3.7 | 12.1 | 6.1 | 10.4 |
| Barnstable Town, MA........ | 2,017 | 953.8 | 36,100 | 14.5 | 23.7 | 47.1 | 471.6 | 19,116 | 50,384 | 85,042 | 104,007 | 26.9 | 10.7 | 5.2 | 3.2 | 7.0 |
| Baton Rouge, LA............ | 28,266 | 3,317.0 | 224,516 | 14.9 | 44.8 | 27.5 | 1,596.4 | 12,129 | 31,571 | 60,746 | 79,297 | 41.8 | 6.4 | 15.0 | 10.2 | 15.7 |
| Battle Creek, MI........... | D | D | 30,864 | 12.3 | 47.0 | 20.2 | 267.7 | 13,631 | 28,175 | 49,055 | 66,013 | 50.7 | 3.8 | 13.4 | 8.8 | 16.2 |
| Bay City, MI............... | 1,295 | 1,254.8 | 20,658 | 11.9 | 41.4 | 21.6 | 165.3 | 11,873 | 29,922 | 49,610 | 67,327 | 50.5 | 3.2 | 13.9 | 8.1 | 17.9 |
| Beaumont-Port Arthur, TX.... | 8,644 | 2,193.3 | 91,964 | 11.7 | 46.3 | 20.3 | 665.0 | 9,849 | 29,298 | 58,818 | 72,252 | 42.6 | 5.9 | 15.0 | 11.1 | 18.9 |
| Beckley, WV................ | NA | NA | 23,479 | 13.9 | 55.5 | 17.8 | 203.6 | 10,982 | 25,219 | 44,785 | 58,270 | 55.1 | 2.5 | 17.9 | 13.5 | 23.0 |
| Bellingham, WA............. | 4,792 | 2,100.9 | 54,728 | 11.3 | 29.8 | 35.7 | 352.3 | 12,796 | 33,503 | 69,372 | 87,522 | 34.1 | 5.9 | 12.9 | 7.0 | 7.9 |
| Bend, OR................... | 3,763 | 1,921.3 | 41,984 | 10.6 | 25.8 | 38.7 | 298.0 | 11,096 | 37,159 | 71,643 | 87,855 | 34.9 | 7.6 | 9.7 | 5.9 | 10.9 |
| Billings, MT............... | NA | NA | 42,196 | 18.0 | 38.3 | 31.8 | 284.2 | 10,796 | 37,253 | 60,962 | 76,327 | 40.4 | 6.0 | 11.7 | 8.8 | 17.0 |
| Binghamton, NY............. | 4,773 | 2,009.4 | 60,846 | 10.4 | 41.6 | 26.5 | 679.6 | 19,870 | 29,133 | 53,768 | 68,731 | 47.2 | 3.8 | 16.6 | 11.0 | 22.4 |
| Birmingham-Hoover, AL...... | NA | NA | 258,399 | 14.7 | 37.5 | 32.3 | 1,686.5 | 10,090 | 33,131 | 58,366 | 74,961 | 43.5 | 6.5 | 14.0 | 9.7 | 15.5 |
| Bismarck, ND............... | 2,983 | 2,314.8 | 30,676 | 15.4 | 31.3 | 33.4 | 230.4 | 12,630 | 37,657 | 70,979 | 92,321 | 33.8 | 6.7 | 5.9 | 3.2 | 4.0 |
| Blacksburg-Christiansburg, VA.... | 2,781 | 1,656.0 | 56,499 | 6.1 | 37.8 | 36.2 | 197.9 | 10,903 | 28,336 | 56,092 | 71,637 | 44.4 | 4.6 | 22.9 | 8.0 | 14.0 |
| Bloomington, IL............ | 2,385 | 1,383.2 | 58,467 | 10.3 | 27.7 | 45.8 | 338.0 | 13,267 | 33,809 | 68,784 | 97,607 | 37.3 | 7.5 | 15.7 | 8.2 | 8.6 |
| Bloomington, IN............ | NA | NA | 64,302 | 9.4 | 34.1 | 41 | 179.0 | 10,564 | 27,938 | 52,526 | 79,578 | 46.6 | 5.0 | 20.9 | 8.3 | 14.0 |
| Bloomsburg-Berwick, PA...... | NA | NA | 20,074 | 14.9 | 49.9 | 26.3 | 133.9 | 15,890 | 30,596 | 56,672 | 71,637 | 45.2 | 3.9 | 14.1 | 10.9 | 18.2 |
| Boise City, ID............. | 9,310 | 1,242.6 | 192,968 | 13.5 | 31.2 | 33.3 | 937.0 | 7,510 | 32,955 | 66,466 | 78,279 | 35.4 | 6.8 | 9.9 | 6.4 | 10.0 |
| Boston-Cambridge-Newton, MA-NH... | D | D | 1,206,992 | 30.1 | 29.8 | 49.3 | 11,452.6 | 17,334 | 50,421 | 94,430 | 119,950 | 27.0 | 17.5 | 8.6 | 5.2 | 8.1 |
| Boston, MA Div 14454....... | D | D | 515,956 | 33.7 | 31.0 | 48 | 4,826.7 | 17,948 | 49,404 | 91,959 | 116,458 | 29.1 | 16.7 | 10.4 | 6.6 | 10.1 |
| Cambridge-Newton-Framingham, MA Div 15754.... | 22,543 | 936.8 | 591,476 | 29.1 | 28.3 | 52 | 5,727.3 | 17,144 | 52,088 | 98,595 | 124,782 | 25.5 | 19.2 | 7.5 | 4.5 | 7.3 |
| Rockingham County-Strafford County, NH Div 40484.... | 4,770 | 1,080.8 | 99,560 | 17.2 | 32.3 | 40.7 | 898.7 | 15,578 | 46,024 | 86,619 | 110,155 | 26.2 | 12.3 | 5.8 | 3.1 | 4.7 |
| Boulder, CO................ | 8,038 | 2,445.6 | 97,668 | 11.0 | 15.7 | 64.8 | 693.4 | 10,884 | 51,415 | 88,535 | 112,791 | 28.2 | 15.9 | 10.9 | 5.1 | 8.3 |
| Bowling Green, KY.......... | 4,188 | 2,337.3 | 49,050 | 8.4 | 42.4 | 30 | 268.6 | 9,848 | 27,380 | 51,198 | 63,142 | 48.4 | 3.6 | 16.5 | 10.6 | 18.6 |
| Bremerton-Silverdale-Port Orchard, WA... | 5,392 | 1,988.1 | 59,308 | 13.6 | 26.0 | 34.8 | 474.3 | 13,016 | 39,068 | 79,624 | 92,222 | 26.6 | 8.5 | 6.9 | 5.1 | 8.9 |
| Bridgeport-Stamford-Norwalk, CT | 11,232 | 1,208.4 | 246,040 | 22.7 | 31.7 | 49.1 | 2,923.0 | 20,186 | 58,815 | 97,053 | 122,271 | 27.0 | 22.8 | 9.1 | 6.9 | 10.5 |
| Brownsville-Harlingen, TX.... | 10,705 | 2,528.9 | 122,399 | 3.9 | 57.9 | 17.1 | 999.4 | 10,396 | 18,431 | 41,123 | 48,112 | 57.1 | 2.7 | 25.6 | 22.6 | 30.9 |
| Brunswick, GA.............. | NA | NA | 23,828 | 14.6 | 45.2 | 23.5 | 193.6 | 10,701 | 31,244 | 57,381 | 67,487 | 43.2 | 4.2 | 14.1 | 9.3 | 20.2 |
| Buffalo-Cheektowaga, NY.... | 20,987 | 1,868.4 | 256,841 | 16.3 | 35.8 | 32.8 | 2,731.4 | 17,465 | 34,238 | 60,105 | 78,893 | 42.6 | 5.4 | 13.1 | 9.4 | 16.8 |
| Burlington, NC............. | 4,443 | 2,639.7 | 40,450 | 24.8 | 43.0 | 25.4 | 215.2 | 8,682 | 28,330 | 58,490 | 70,332 | 43.7 | 4.1 | 15.3 | 12.3 | 20.8 |
| Burlington-South Burlington, VT... | 3,786 | 1,709.7 | 58,179 | 12.6 | 30.9 | 44.6 | 588.5 | 18,640 | 38,497 | 74,909 | 98,043 | 31.8 | 8.4 | 11.4 | 5.6 | 10.4 |
| California-Lexington Park, MD.... | 1,880 | 1,663.8 | 28,949 | 17.8 | 43.3 | 30.7 | 238.7 | 13,223 | 41,402 | 87,947 | 102,798 | 26.1 | 11.8 | 7.3 | 7.0 | 9.1 |
| Canton-Massillon, OH....... | 8,832 | 2,223.3 | 84,461 | 16.9 | 49.2 | 21.3 | 688.5 | 11,669 | 29,994 | 55,706 | 69,158 | 45.4 | 3.7 | 12.8 | 9.1 | 15.8 |
| Cape Coral-Fort Myers, FL... | 8,890 | 1,157.9 | 145,714 | 12.9 | 42.0 | 29 | 902.3 | 9,680 | 35,715 | 62,240 | 74,882 | 38.2 | 7.3 | 11.3 | 7.5 | 15.4 |
| Cape Girardeau, MO-IL...... | 2,060 | 2,124.2 | 21,316 | 15.9 | 41.6 | 30.5 | 128.3 | 9,723 | 29,474 | 52,624 | 70,535 | 46.0 | 3.5 | 16.0 | 11.2 | 18.7 |
| Carbondale-Marion, IL...... | NA | NA | 36,126 | 8.9 | 37.8 | 28.2 | 244.6 | 12,756 | 28,551 | 46,882 | 68,740 | 52.7 | 4.3 | 18.1 | 9.3 | 16.4 |
| Carson City, NV............ | 711 | 1,281.3 | 11,889 | 6.7 | 38.5 | 21.3 | 338.4 | 7,507 | 31,733 | 57,270 | 69,901 | 42.8 | 2.7 | 12.5 | 10.2 | 15.6 |
| Casper, WY................. | 1,966 | 2,473.3 | 19,578 | 14.4 | 41.4 | 20.7 | 193.6 | 14,545 | 32,934 | 65,034 | 76,487 | 37.0 | 4.9 | 8.6 | 6.6 | 11.8 |

1. Data for serious crimes have not been adjusted for underreporting; this may affect comparability between geographic areas and over time.  2. Per 100,000 population estimated by the FBI.  3. All persons 3 years old and over enrolled in nursery school through college.  4. Persons 25 years old and over.  5. Elementary and secondary education expenditures.  6. Based on population estimated by the American Community Survey, 2017.

# Table C. Metropolitan Areas — Personal Income and Earnings

| Area name | Personal income, 2019 | | | | | | | | | | Earnings, 2019 | | |
|---|---|---|---|---|---|---|---|---|---|---|---|---|---|
| | Total (mil dol) | Percent change, 2018-2019 | Per capita¹ Dollars | Per capita¹ Rank | Wages and Salaries (mil dol) | Supplements to wages and salaries, employer contributions (mil dol) Pension and insurance | Supplements — Government social insurance | Proprietors' income | Dividends, interest, and rent (mil dol) | Personal transfer receipts (mil dol) | Total (mil dol) | Contributions for government social insurance (mil dol) From employee and self-employed | From employer |
| | 62 | 63 | 64 | 65 | 66 | 67 | 68 | 69 | 70 | 71 | 72 | 73 | 74 |
| Abilene, TX | 7,838 | 5.1 | 45,552 | 233 | 3,610 | 658 | 256 | 533 | 1,505 | 1,770 | 5,057 | 284 | 256 |
| Akron, OH | 35,944 | 3.1 | 51,095 | 128 | 17,806 | 2,984 | 1,239 | 2,360 | 6,315 | 6,746 | 24,389 | 1,448 | 1,239 |
| Albany, GA | 5,858 | 4.2 | 39,922 | 344 | 2,804 | 523 | 195 | 284 | 1,036 | 1,550 | 3,805 | 233 | 195 |
| Albany-Lebanon, OR | 5,817 | 5.3 | 44,830 | 253 | 2,293 | 394 | 212 | 460 | 904 | 1,634 | 3,360 | 229 | 212 |
| Albany-Schenectady-Troy, NY | 53,498 | 3.8 | 60,767 | 41 | 27,437 | 6,138 | 2,117 | 3,097 | 10,726 | 9,305 | 38,789 | 2,115 | 2,117 |
| Albuquerque, NM | 40,380 | 3.9 | 43,986 | 267 | 20,729 | 3,301 | 1,561 | 1,966 | 7,427 | 8,805 | 27,557 | 1,840 | 1,561 |
| Alexandria, LA | 6,757 | 2.8 | 44,446 | 261 | 2,800 | 571 | 183 | 779 | 1,087 | 1,900 | 4,333 | 248 | 183 |
| Allentown-Bethlehem-Easton, PA-NJ | 46,992 | 4.2 | 55,675 | 78 | 20,687 | 3,540 | 1,600 | 4,254 | 7,482 | 9,112 | 30,080 | 1,822 | 1,600 |
| Altoona, PA | 5,882 | 3.2 | 48,277 | 176 | 2,743 | 570 | 231 | 504 | 939 | 1,608 | 4,048 | 255 | 231 |
| Amarillo, TX | 12,740 | 3.9 | 48,065 | 179 | 6,102 | 980 | 417 | 1,643 | 2,139 | 2,247 | 9,143 | 491 | 417 |
| Ames, IA | 5,545 | 2.9 | 44,952 | 245 | 2,894 | 699 | 214 | 379 | 1,254 | 869 | 4,187 | 249 | 214 |
| Anchorage, AK | 25,503 | 3.7 | 64,350 | 26 | 11,782 | 2,836 | 865 | 2,052 | 4,988 | 4,103 | 17,535 | 932 | 865 |
| Ann Arbor, MI | 22,366 | 2.9 | 60,843 | 40 | 13,037 | 2,544 | 914 | 1,394 | 4,963 | 2,796 | 17,888 | 1,043 | 914 |
| Anniston-Oxford, AL | 4,362 | 3.1 | 38,394 | 361 | 1,983 | 417 | 147 | 188 | 775 | 1,284 | 2,734 | 189 | 147 |
| Appleton, WI | 12,538 | 3.6 | 52,686 | 107 | 6,644 | 1,173 | 494 | 776 | 2,104 | 1,749 | 9,087 | 557 | 494 |
| Asheville, NC | 21,946 | 4.4 | 47,432 | 200 | 9,449 | 1,493 | 681 | 1,818 | 5,288 | 4,900 | 13,441 | 919 | 681 |
| Athens-Clarke County, GA | 8,869 | 3.1 | 41,495 | 313 | 4,536 | 934 | 292 | 537 | 2,003 | 1,593 | 6,299 | 354 | 292 |
| Atlanta-Sandy Springs-Alpharetta, GA | 328,450 | 3.9 | 54,557 | 85 | 185,418 | 24,527 | 12,115 | 32,096 | 59,747 | 41,856 | 254,156 | 14,657 | 12,115 |
| Atlantic City-Hammonton, NJ | 13,350 | 4.2 | 50,631 | 134 | 6,852 | 1,204 | 546 | 1,098 | 2,153 | 3,124 | 9,700 | 613 | 546 |
| Auburn-Opelika, AL | 6,546 | 4.2 | 39,781 | 347 | 2,711 | 493 | 193 | 301 | 1,269 | 1,216 | 3,699 | 238 | 193 |
| Augusta-Richmond County, GA-SC | 26,575 | 4.5 | 43,638 | 274 | 13,316 | 2,537 | 952 | 1,451 | 4,657 | 6,033 | 18,255 | 1,103 | 952 |
| Austin-Round Rock-Georgetown, TX | 138,028 | 6.9 | 61,977 | 36 | 75,539 | 9,599 | 4,751 | 18,006 | 27,432 | 13,123 | 107,895 | 5,542 | 4,751 |
| Bakersfield, CA | 37,667 | 5.9 | 41,843 | 305 | 17,845 | 3,894 | 1,299 | 4,226 | 5,992 | 7,605 | 27,263 | 1,460 | 1,299 |
| Baltimore-Columbia-Towson, MD | 179,169 | 3.2 | 63,988 | 28 | 97,201 | 15,675 | 6,997 | 12,489 | 34,103 | 27,860 | 132,362 | 7,808 | 6,997 |
| Bangor, ME | 6,563 | 3.9 | 43,136 | 281 | 3,415 | 651 | 243 | 307 | 981 | 1,718 | 4,616 | 305 | 243 |
| Barnstable Town, MA | 16,493 | 3.4 | 77,435 | 10 | 5,234 | 954 | 386 | 1,594 | 4,890 | 3,139 | 8,169 | 488 | 386 |
| Baton Rouge, LA | 42,264 | 2.7 | 49,439 | 148 | 23,074 | 3,925 | 1,432 | 3,029 | 7,067 | 7,818 | 31,460 | 1,722 | 1,432 |
| Battle Creek, MI | 5,401 | 2.9 | 40,257 | 339 | 3,093 | 538 | 228 | 208 | 867 | 1,537 | 4,068 | 272 | 228 |
| Bay City, MI | 4,502 | 2.5 | 43,657 | 273 | 1,633 | 308 | 124 | 197 | 721 | 1,340 | 2,261 | 166 | 124 |
| Beaumont-Port Arthur, TX | 17,751 | 2.1 | 45,219 | 240 | 9,567 | 1,713 | 650 | 1,100 | 2,407 | 4,179 | 13,030 | 740 | 650 |
| Beckley, WV | 4,761 | 3.4 | 41,128 | 326 | 1,969 | 368 | 155 | 296 | 647 | 1,630 | 2,789 | 202 | 155 |
| Bellingham, WA | 11,672 | 4.8 | 50,915 | 131 | 4,937 | 905 | 447 | 1,178 | 2,758 | 2,116 | 7,467 | 444 | 447 |
| Bend, OR | 11,159 | 4.6 | 56,447 | 67 | 4,367 | 674 | 385 | 1,629 | 2,907 | 2,028 | 7,054 | 459 | 385 |
| Billings, MT | 9,803 | 3.6 | 53,960 | 95 | 4,727 | 721 | 406 | 895 | 2,097 | 1,689 | 6,749 | 443 | 406 |
| Binghamton, NY | 11,095 | 3.9 | 46,484 | 216 | 4,957 | 1,225 | 387 | 678 | 1,825 | 2,706 | 7,246 | 428 | 387 |
| Birmingham-Hoover, AL | 58,200 | 3.4 | 53,374 | 99 | 29,594 | 4,325 | 2,067 | 5,521 | 11,448 | 10,379 | 41,507 | 2,674 | 2,067 |
| Bismarck, ND | 7,512 | 3.2 | 58,254 | 60 | 3,778 | 580 | 298 | 611 | 1,496 | 1,072 | 5,267 | 332 | 298 |
| Blacksburg-Christiansburg, VA | 6,620 | 2.9 | 39,517 | 351 | 3,378 | 713 | 243 | 309 | 1,432 | 1,346 | 4,642 | 292 | 243 |
| Bloomington, IL | 8,508 | 0.2 | 49,602 | 145 | 5,093 | 908 | 312 | 631 | 1,422 | 1,167 | 6,944 | 379 | 312 |
| Bloomington, IN | 7,346 | 4.3 | 43,411 | 276 | 3,538 | 807 | 254 | 510 | 1,586 | 1,292 | 5,108 | 301 | 254 |
| Bloomsburg-Berwick, PA | 3,918 | 2.9 | 47,097 | 206 | 2,289 | 460 | 165 | 265 | 600 | 909 | 3,179 | 193 | 165 |
| Boise City, ID | 35,873 | 5.5 | 47,881 | 185 | 17,685 | 2,673 | 1,426 | 3,554 | 7,483 | 5,791 | 25,338 | 1,620 | 1,426 |
| Boston-Cambridge-Newton, MA-NH | 397,139 | 3.5 | 81,498 | 6 | 235,168 | 31,240 | 15,119 | 36,441 | 84,632 | 45,801 | 317,968 | 16,993 | 15,119 |
| Boston, MA Div 14454 | 170,729 | 3.4 | 84,025 | NA | 113,105 | 15,219 | 7,103 | 17,554 | 37,152 | 20,827 | 152,982 | 7,980 | 7,103 |
| Cambridge-Newton-Framingham, MA Div 15754 | 195,745 | 3.6 | 81,535 | NA | 109,278 | 14,137 | 7,164 | 15,331 | 42,635 | 21,213 | 145,909 | 7,845 | 7,164 |
| Rockingham County-Strafford County, NH Div 40484 | 30,665 | 4.1 | 69,630 | NA | 12,784 | 1,883 | 853 | 3,557 | 4,845 | 3,761 | 19,077 | 1,168 | 853 |
| Boulder, CO | 24,963 | 4.6 | 76,527 | 11 | 14,832 | 1,741 | 950 | 1,718 | 7,445 | 2,247 | 19,241 | 1,078 | 950 |
| Bowling Green, KY | 6,652 | 3.9 | 37,114 | 374 | 3,348 | 610 | 247 | 487 | 950 | 1,626 | 4,692 | 301 | 247 |
| Bremerton-Silverdale-Port Orchard, WA | 15,983 | 5.4 | 58,874 | 52 | 6,073 | 1,490 | 545 | 844 | 4,061 | 2,598 | 8,951 | 522 | 545 |
| Bridgeport-Stamford-Norwalk, CT | 114,518 | 3.3 | 121,397 | 2 | 39,659 | 4,960 | 2,413 | 15,053 | 34,275 | 8,582 | 62,085 | 3,342 | 2,413 |
| Brownsville-Harlingen, TX | 12,664 | 3.7 | 29,928 | 383 | 5,263 | 1,109 | 367 | 966 | 1,680 | 4,117 | 7,705 | 449 | 367 |
| Brunswick, GA | 5,094 | 2.6 | 42,885 | 288 | 2,106 | 376 | 146 | 301 | 1,313 | 1,177 | 2,930 | 191 | 146 |
| Buffalo-Cheektowaga, NY | 59,028 | 4.0 | 52,331 | 111 | 29,107 | 6,244 | 2,268 | 4,068 | 9,632 | 12,892 | 41,688 | 2,361 | 2,268 |
| Burlington, NC | 6,983 | 5.0 | 41,193 | 325 | 2,965 | 443 | 210 | 329 | 1,153 | 1,584 | 3,946 | 271 | 210 |
| Burlington-South Burlington, VT | 12,930 | 3.1 | 58,662 | 55 | 7,093 | 1,127 | 549 | 1,095 | 2,527 | 1,934 | 9,864 | 624 | 549 |
| California-Lexington Park, MD | 6,650 | 3.8 | 58,582 | 57 | 3,628 | 784 | 281 | 267 | 1,172 | 915 | 4,960 | 282 | 281 |
| Canton-Massillon, OH | 18,310 | 2.9 | 46,061 | 223 | 7,926 | 1,359 | 577 | 1,074 | 3,037 | 4,178 | 10,937 | 698 | 577 |
| Cape Coral-Fort Myers, FL | 40,119 | 4.1 | 52,064 | 115 | 14,063 | 2,052 | 957 | 3,508 | 14,398 | 8,341 | 20,580 | 1,469 | 957 |
| Cape Girardeau, MO-IL | 4,313 | 3.8 | 44,571 | 258 | 2,041 | 389 | 145 | 435 | 759 | 967 | 3,009 | 188 | 145 |
| Carbondale-Marion, IL | 5,666 | 2.4 | 41,733 | 311 | 2,692 | 691 | 175 | 312 | 984 | 1,329 | 3,870 | 216 | 175 |
| Carson City, NV | 2,934 | 4.6 | 52,470 | 110 | 1,654 | 398 | 99 | 351 | 623 | 648 | 2,502 | 127 | 99 |
| Casper, WY | 5,765 | 2.1 | 72,187 | 14 | 2,212 | 329 | 202 | 1,680 | 1,066 | 697 | 4,424 | 235 | 202 |

1. Based on the resident population estimated as of July 1 of the year shown.

Items 62—74

Table C. Metropolitan Areas — **Earnings, Social Security, and Housing**

| Area name | Farm | Mining, quarrying, and extracting | Construction | Manufacturing | Information; professional, scientific, and technical serviecs | Retail trade | Finance, insurance, real estate, rental and leasing | Health care and social assistance | Government | Number | Rate[1] | Supplemental Security Income Recipients December 2019 | Total | Percent change, 2010-2020 |
|---|---|---|---|---|---|---|---|---|---|---|---|---|---|---|
| | 75 | 76 | 77 | 78 | 79 | 80 | 81 | 82 | 83 | 84 | 85 | 86 | 87 | 88 |
| Abilene, TX | -0.3 | 3.3 | 7.1 | 3.9 | 7.2 | 7.2 | 6.9 | D | 24.2 | 33,170 | 193 | 4,333 | 72,502 | 4.0 |
| Akron, OH | 0.0 | 0.2 | 6.4 | 12.6 | 9.0 | 7.5 | 6.0 | 13.6 | 13.2 | 145,415 | 207 | 17,733 | 316,609 | 1.3 |
| Albany, GA | 3.2 | D | 4.9 | 8.6 | 5.4 | 7.4 | D | 16.7 | 23.1 | 31,785 | 217 | 6,697 | 65,863 | 2.1 |
| Albany-Lebanon, OR | 1.6 | 0.3 | 8.6 | 21.6 | 4.2 | 8.0 | 4.2 | 10.3 | 15.1 | 31,795 | 245 | 3,573 | 52,091 | 6.7 |
| Albany-Schenectady-Troy, NY | 0.2 | D | 5.8 | 7.3 | 12.7 | 5.3 | 8.5 | 11.7 | 26.5 | 191,110 | 217 | 19,945 | 413,999 | 5.3 |
| Albuquerque, NM | 0.2 | D | 6.9 | 4.6 | 15.9 | 6.2 | D | D | 24.9 | 185,900 | 202 | 23,029 | 397,004 | 6.0 |
| Alexandria, LA | 1.3 | D | 6.6 | 7.7 | 4.5 | 10.3 | D | 18.9 | 23.3 | 33,995 | 223 | 7,436 | 68,648 | 6.3 |
| Allentown-Bethlehem-Easton, PA-NJ | 0.2 | D | D | 13.1 | 9.7 | 5.5 | 5.2 | 16.6 | 11.2 | 191,930 | 227 | 19,152 | 352,624 | 3.1 |
| Altoona, PA | 0.6 | 0.3 | 5.3 | 12.7 | 5.3 | 8.5 | 4.0 | 19.5 | 15.4 | 31,195 | 256 | 4,361 | 57,132 | 1.5 |
| Amarillo, TX | 1.9 | D | D | 13.8 | D | 7.3 | D | 12.6 | 15.4 | 44,810 | 169 | 4,426 | 110,923 | 7.3 |
| Ames, IA | 1.8 | D | 6.5 | 10.5 | 8.1 | 5.1 | 4.3 | 7.8 | 36.5 | 19,755 | 160 | 1,031 | 53,944 | 11.1 |
| Anchorage, AK | 0.1 | 4.2 | 8.2 | 1.0 | 8.9 | 5.9 | 5.1 | 14.9 | 27.8 | 54,775 | 138 | 7,474 | 162,501 | 5.3 |
| Ann Arbor, MI | 0.0 | D | 3.1 | 7.2 | 18.6 | 3.8 | 4.4 | 11.0 | 33.9 | 60,730 | 165 | 5,343 | 153,091 | 3.6 |
| Anniston-Oxford, AL | 0.2 | D | 3.6 | 16.5 | 4.5 | 7.6 | 3.6 | 8.9 | 33.5 | 30,360 | 266 | 4,434 | 53,933 | 1.3 |
| Appleton, WI | 1.4 | D | 9.9 | 21.1 | D | 6.1 | 5.8 | D | 10.9 | 46,225 | 194 | 2,780 | 99,758 | 7.5 |
| Asheville, NC | 0.4 | 0.1 | 7.5 | 12.2 | D | 8.1 | 5.8 | D | 14.1 | 122,395 | 264 | 9,284 | 236,665 | 10.8 |
| Athens-Clarke County, GA | 1.3 | 0.2 | 5.0 | 8.4 | D | 6.5 | 7.9 | 15.3 | 30.4 | 36,450 | 171 | 4,694 | 88,849 | 8.7 |
| Atlanta-Sandy Springs-Alpharetta, GA | 0.0 | D | D | 5.8 | 20.2 | 5.5 | 10.7 | 8.8 | 10.7 | 900,685 | 149 | 116,138 | 2,394,186 | 10.2 |
| Atlantic City-Hammonton, NJ | 0.6 | D | 6.9 | D | 6.4 | 6.7 | 4.5 | 16.0 | 22.6 | 60,700 | 230 | 7,257 | 128,943 | 1.8 |
| Auburn-Opelika, AL | 0.3 | D | 6.6 | 11.0 | 5.3 | 6.8 | 3.7 | 6.2 | 33.1 | 26,955 | 163 | 3,115 | 73,371 | 17.6 |
| Augusta-Richmond County, GA-SC | 0.3 | D | D | 10.3 | D | 5.7 | 5.3 | 10.5 | 27.0 | 129,040 | 211 | 16,957 | 266,383 | 10.2 |
| Austin-Round Rock-Georgetown, TX | 0.0 | 6.9 | 7.3 | 7.8 | D | 5.2 | 8.4 | 7.3 | 13.1 | 270,180 | 121 | 25,964 | 914,212 | 29.4 |
| Bakersfield, CA | 8.3 | 6.2 | 5.8 | 3.8 | 4.8 | 6.1 | 3.1 | 9.1 | 25.0 | 127,085 | 141 | 33,754 | 303,109 | 6.6 |
| Baltimore-Columbia-Towson, MD | 0.1 | 0.0 | 6.2 | 4.6 | 16.6 | 4.5 | 8.5 | 12.0 | 22.0 | 495,410 | 177 | 69,337 | 1,173,310 | 3.6 |
| Bangor, ME | 0.1 | D | 6.0 | 4.5 | 5.3 | 8.5 | 4.8 | 23.9 | 20.0 | 38,390 | 253 | 5,454 | 76,615 | 3.7 |
| Barnstable Town, MA | 0.0 | D | 12.4 | 2.5 | 8.6 | 9.3 | 6.5 | 15.7 | 17.6 | 73,505 | 345 | 3,107 | 165,101 | 3.0 |
| Baton Rouge, LA | 0.1 | 0.5 | 15.2 | D | D | 5.4 | D | 10.4 | 16.7 | 150,650 | 176 | 24,866 | 375,298 | 10.4 |
| Battle Creek, MI | 0.2 | 0.1 | 4.2 | 21.0 | 9.2 | 5.2 | 2.4 | 14.5 | 21.3 | 33,385 | 250 | 4,751 | 60,888 | -0.3 |
| Bay City, MI | -0.1 | 0.1 | 5.4 | 16.8 | 7.1 | 8.1 | 4.1 | 16.6 | 17.2 | 29,605 | 288 | 3,112 | 48,388 | 0.3 |
| Beaumont-Port Arthur, TX | -0.1 | 0.7 | D | D | D | 6.3 | D | 10.2 | 13.0 | 76,580 | 195 | 11,850 | 174,310 | 7.4 |
| Beckley, WV | 0.0 | 13.3 | 5.8 | 3.8 | D | 8.1 | 3.2 | 17.7 | 20.5 | 34,705 | 299 | 5,585 | 57,517 | -0.1 |
| Bellingham, WA | 2.2 | D | 11.0 | 13.0 | 7.3 | 8.3 | 6.0 | 11.6 | 18.2 | 46,755 | 204 | 4,357 | 101,169 | 11.6 |
| Bend, OR | -0.2 | 0.1 | 14.2 | 5.6 | 11.6 | 7.9 | 9.0 | 16.7 | 11.6 | 46,680 | 236 | 2,322 | 94,859 | 18.4 |
| Billings, MT | 1.0 | D | 8.3 | D | 8.6 | D | 7.9 | D | 11.6 | 38,170 | 210 | 2,701 | 86,481 | 15.0 |
| Binghamton, NY | 0.3 | 0.3 | 5.6 | 14.5 | D | 6.4 | 5.1 | 16.3 | 24.7 | 60,410 | 253 | 8,307 | 114,289 | 1.3 |
| Birmingham-Hoover, AL | 0.0 | 1.3 | 7.2 | 7.0 | D | 5.3 | 11.5 | D | 14.0 | 235,095 | 216 | 30,360 | 492,508 | 5.0 |
| Bismarck, ND | 0.5 | D | 7.3 | D | D | 7.0 | D | 17.8 | 19.1 | 24,670 | 191 | 1,182 | 60,414 | 23.9 |
| Blacksburg-Christiansburg, VA | 0.1 | D | 4.1 | 20.9 | D | 5.8 | 2.8 | 8.6 | 34.1 | 31,900 | 190 | 2,972 | 73,739 | 4.5 |
| Bloomington, IL | 0.9 | D | 3.6 | 3.2 | D | 5.9 | 36.0 | 10.0 | 15.8 | 27,920 | 163 | 1,901 | 72,874 | 4.6 |
| Bloomington, IN | 0.0 | 0.4 | 4.6 | 14.3 | D | 5.0 | 4.4 | D | 31.7 | 28,250 | 167 | 2,220 | 73,428 | 6.1 |
| Bloomsburg-Berwick, PA | 0.5 | D | 3.7 | 11.0 | D | 4.5 | 5.6 | 32.8 | 14.6 | 20,655 | 248 | 1,769 | 38,803 | 3.6 |
| Boise City, ID | 1.3 | D | 8.8 | 12.8 | D | 7.8 | D | 12.9 | 13.4 | 137,050 | 183 | 12,028 | 301,444 | 22.5 |
| Boston-Cambridge-Newton, MA-NH | 0.0 | D | 5.7 | 6.9 | 24.1 | 4.1 | 12.5 | 11.5 | 10.0 | 832,875 | 171 | 103,431 | 1,999,809 | 6.2 |
| Boston, MA Div 14454 | 0.0 | D | 5.3 | 3.0 | 20.9 | 4.0 | 19.6 | 13.4 | 10.7 | 333,355 | 164 | 52,919 | 842,668 | 7.2 |
| Cambridge-Newton-Framingham, MA Div 15754 | 0.0 | 0.0 | 5.8 | 10.5 | 28.8 | 3.8 | 5.7 | 9.6 | 9.2 | 405,355 | 169 | 46,481 | 966,165 | 5.2 |
| Rockingham County-Strafford County, NH Div 40484 | 0.0 | D | 7.9 | 10.3 | 14.5 | 8.0 | 8.3 | 10.0 | 10.4 | 94,165 | 214 | 4,031 | 190,976 | 7.0 |
| Boulder, CO | 0.1 | 0.3 | 3.6 | 11.0 | 34.3 | 4.2 | 5.8 | 9.0 | 14.3 | 48,010 | 147 | 2,203 | 138,578 | 9.1 |
| Bowling Green, KY | 0.6 | D | 7.4 | D | D | 6.7 | D | 13.9 | 14.4 | 33,715 | 188 | 5,096 | 78,844 | 14.5 |
| Bremerton-Silverdale-Port Orchard, WA | 0.0 | D | 5.3 | 2.4 | 8.0 | 5.5 | 3.8 | 9.2 | 52.4 | 55,240 | 203 | 4,785 | 115,998 | 8.0 |
| Bridgeport-Stamford-Norwalk, CT | 0.0 | 0.0 | 4.6 | 8.3 | 16.1 | 5.1 | 24.8 | 8.8 | 7.4 | 155,230 | 164 | 12,790 | 377,497 | 4.5 |
| Brownsville-Harlingen, TX | 0.6 | 0.1 | 4.0 | 5.5 | 3.9 | 8.4 | 5.1 | 20.1 | 27.0 | 69,310 | 164 | 21,887 | 155,829 | 9.8 |
| Brunswick, GA | 0.0 | 0.1 | 5.0 | 5.9 | D | 7.0 | D | 14.2 | 24.3 | 27,645 | 233 | 2,803 | 62,333 | 7.4 |
| Buffalo-Cheektowaga, NY | 0.1 | D | 4.5 | 11.5 | 9.0 | 6.0 | 8.6 | 13.2 | 21.0 | 263,045 | 233 | 32,726 | 532,852 | 2.6 |
| Burlington, NC | -0.1 | D | 7.2 | 15.3 | 4.7 | 8.6 | 5.9 | 19.9 | 11.2 | 36,205 | 214 | 3,566 | 74,058 | 11.2 |
| Burlington-South Burlington, VT | 0.8 | D | 5.8 | 10.7 | D | 6.7 | 5.9 | D | 18.8 | 42,935 | 195 | 4,500 | 100,337 | 8.6 |
| California-Lexington Park, MD | 0.1 | D | 4.0 | 1.0 | 23.1 | 3.7 | 2.0 | 6.7 | 47.2 | 17,170 | 151 | 1,699 | 47,425 | 14.9 |
| Canton-Massillon, OH | 0.1 | D | 7.6 | 18.5 | D | 7.2 | 6.3 | D | 12.5 | 95,520 | 240 | 10,312 | 181,135 | 1.2 |
| Cape Coral-Fort Myers, FL | 0.3 | 0.1 | 11.3 | 2.2 | 11.3 | 8.9 | 7.3 | 11.1 | 16.2 | 202,045 | 262 | 12,811 | 417,359 | 12.5 |
| Cape Girardeau, MO-IL | 0.5 | 0.6 | 5.3 | 14.6 | D | 7.8 | 4.4 | 23.4 | 13.5 | 21,935 | 226 | 2,374 | 45,561 | 7.2 |
| Carbondale-Marion, IL | 0.1 | 2.0 | 4.6 | 7.3 | D | 6.5 | 5.0 | 18.1 | 34.4 | 29,425 | 217 | 3,474 | 66,287 | 2.7 |
| Carson City, NV | 0.0 | D | 5.8 | 8.7 | 7.7 | 7.0 | 5.7 | 14.0 | 34.3 | 13,840 | 248 | 1,024 | 24,612 | 4.6 |
| Casper, WY | 0.0 | 8.7 | 6.4 | 3.4 | 4.5 | 4.4 | 4.6 | 10.3 | 10.3 | 15,355 | 193 | 1,307 | 37,517 | 11.0 |

1. Per 1,000 resident population estimated as of July 1 of the year shown.

Table C. Metropolitan Areas — **Housing, Labor Force, and Employment**

| Area name | Occupied Housing units, 2020 | | | | | | | | Civilian labor force, 2020 | | | | Civilian employment, 2019[6] | | |
|---|---|---|---|---|---|---|---|---|---|---|---|---|---|---|---|
| | Occupied units | | | | | | | | | | Unemployment | | | Percent | |
| | | | Owner-occupied | | | Renter-occupied | | | | | | | | | |
| | | | | Median owner cost as a percent of income | | | Median rent as a percent of income[2] | Sub-standard units[4] (percent) | | Percent change 2019-2020 | | | | Manage-ment, business, science, and arts | Construction, production, and mainte-nance occu-pations |
| | Total | Percent | Median value[1] | With a mortgage | Without a mort-gage[2] | Median rent[3] | | | Total | | Total | Rate[5] | Total | | |
| | 89 | 90 | 91 | 92 | 93 | 94 | 95 | 96 | 97 | 98 | 99 | 100 | 101 | 102 | 103 |
| Abilene, TX | 61,750 | 60.6 | 133,500 | 21.2 | 11.5 | 887 | 27.8 | 2.0 | 77,865 | -1.0 | 4,334 | 5.6 | 75,713 | 35.2 | 25.2 |
| Akron, OH | 285,647 | 66.1 | 157,200 | 18.3 | 11.1 | 830 | 29.4 | 0.8 | 356,190 | -1.5 | 28,496 | 8.0 | 353,676 | 36.8 | 23.2 |
| Albany, GA | 54,128 | 53.6 | 125,600 | 22.7 | 11.4 | 744 | 32.7 | 0.0 | 64,816 | -1.0 | 4,662 | 7.2 | 58,845 | 28.9 | 25.7 |
| Albany-Lebanon, OR | 49,407 | 67.4 | 257,200 | 22.8 | 12.1 | 1,034 | 27.6 | 2.6 | 58,819 | 0.0 | 4,568 | 7.8 | 60,491 | 32.1 | 32.0 |
| Albany-Schenectady-Troy, NY | 362,502 | 64.2 | 227,000 | 19.5 | 10.9 | 1,043 | 27.2 | 1.2 | 448,495 | 0.2 | 32,149 | 7.2 | 455,152 | 47.2 | 16.5 |
| Albuquerque, NM | 354,359 | 67.7 | 206,000 | 21.4 | 10.0 | 916 | 29.4 | 2.5 | 432,502 | -1.6 | 35,520 | 8.2 | 432,540 | 42.3 | 17.4 |
| Alexandria, LA | 55,591 | 66.4 | 138,100 | 19.4 | 10.0 | 841 | 27.1 | 0.0 | 63,325 | -0.8 | 3,976 | 6.3 | 64,379 | 31.3 | 24.4 |
| Allentown-Bethlehem-Easton, PA-NJ | 322,301 | 67.3 | 224,900 | 20.3 | 13.3 | 1,109 | 29.8 | 1.5 | 437,575 | -1.4 | 40,357 | 9.2 | 419,565 | 37.0 | 24.8 |
| Altoona, PA | 52,497 | 70.6 | 133,600 | 18.1 | 10.4 | 717 | 31.1 | 0.9 | 59,205 | -0.8 | 5,152 | 8.7 | 58,519 | 35.3 | 28.0 |
| Amarillo, TX | 98,606 | 64.5 | 156,300 | 20.1 | 10.5 | 860 | 28.1 | 3.3 | 131,264 | -0.8 | 6,403 | 4.9 | 126,633 | 31.6 | 27.0 |
| Ames, IA | 50,253 | 61.3 | 185,800 | 18.8 | 10.0 | 901 | 28.9 | 1.0 | 70,926 | -4.1 | 2,624 | 3.7 | 65,063 | 44.7 | 19.5 |
| Anchorage, AK | 137,934 | 64.6 | 300,100 | 22.2 | 10.0 | 1,213 | 28.1 | 5.4 | 196,240 | -0.6 | 14,961 | 7.6 | 188,676 | 39.6 | 22.3 |
| Ann Arbor, MI | 141,680 | 62.5 | 295,300 | 18.9 | 10.9 | 1,139 | 29.8 | 1.2 | 193,659 | -2.7 | 12,365 | 6.4 | 193,750 | 56.7 | 13.5 |
| Anniston-Oxford, AL | 44,636 | 70.5 | 121,600 | 17.9 | 10.0 | 769 | 28.5 | 0.0 | 46,240 | 0.3 | 3,260 | 7.1 | 45,641 | 28.2 | 30.0 |
| Appleton, WI | 94,937 | 73 | 183,900 | 18.4 | 10.8 | 803 | 23.9 | 0.8 | 129,882 | -0.3 | 6,920 | 5.3 | 128,565 | 36.4 | 27.3 |
| Asheville, NC | 193,451 | 67.6 | 248,500 | 19.9 | 10.0 | 975 | 29.3 | 1.7 | 227,184 | -4.3 | 17,805 | 7.8 | 222,916 | 39.0 | 21.5 |
| Athens-Clarke County, GA | 82,655 | 51.6 | 204,300 | 19.6 | 10.0 | 847 | 32.8 | 1.9 | 97,177 | -3.1 | 5,471 | 5.6 | 102,169 | 41.3 | 19.7 |
| Atlanta-Sandy Springs-Alpharetta, GA | 2,189,294 | 64.3 | 247,200 | 19.8 | 10.0 | 1,224 | 29.7 | 2.0 | 3,056,030 | -1.6 | 209,452 | 6.9 | 3,033,891 | 43.9 | 20.7 |
| Atlantic City-Hammonton, NJ | 101,975 | 66.3 | 214,900 | 24.7 | 17.0 | 1,116 | 32.9 | 2.5 | 121,037 | -0.6 | 21,491 | 17.8 | 128,756 | 33.6 | 19.1 |
| Auburn-Opelika, AL | 64,448 | 67.3 | 183,000 | 19.4 | 10.1 | 765 | 29.9 | 0.0 | 75,564 | -1.1 | 3,935 | 5.2 | 81,858 | 46.7 | 18.8 |
| Augusta-Richmond County, GA-SC | 211,969 | 68.8 | 161,800 | 19.5 | 10.0 | 923 | 30.8 | 1.0 | 267,071 | -0.6 | 15,464 | 5.8 | 266,222 | 35.8 | 24.2 |
| Austin-Round Rock-George-town, TX | 818,084 | 57.6 | 318,400 | 20.9 | 11.1 | 1,327 | 28.7 | 4.0 | 1,242,658 | 0.3 | 76,673 | 6.2 | 1,223,139 | 49.0 | 15.6 |
| Bakersfield, CA | 272,888 | 58.8 | 235,800 | 22.7 | 10.9 | 949 | 33.7 | 8.9 | 383,769 | -1.1 | 48,062 | 12.5 | 352,979 | 28.8 | 34.8 |
| Baltimore-Columbia-Towson, MD | 1,062,901 | 66.4 | 313,200 | 20.5 | 10.2 | 1,316 | 29.5 | 1.8 | 1,480,381 | -3.1 | 97,030 | 6.6 | 1,420,003 | 50.0 | 15.1 |
| Bangor, ME | 64,096 | 69.7 | 154,300 | 19.5 | 12.1 | 809 | 28.1 | 2.0 | 74,717 | -2.8 | 4,007 | 5.4 | 75,163 | 38.3 | 21.3 |
| Barnstable Town, MA | 96,509 | 80 | 409,700 | 23.2 | 11.7 | 1,306 | 29.7 | 1.3 | 108,128 | -5.2 | 11,055 | 10.2 | 100,192 | 40.2 | 17.7 |
| Baton Rouge, LA | 305,441 | 69.7 | 195,500 | 18.4 | 10.0 | 911 | 31.0 | 2.2 | 424,649 | -1.7 | 31,785 | 7.5 | 399,942 | 37.4 | 24.0 |
| Battle Creek, MI | 53,827 | 73.2 | 115,700 | 19.2 | 12.2 | 785 | 28.5 | 1.6 | 61,838 | -1.1 | 5,994 | 9.7 | 62,429 | 29.9 | 31.4 |
| Bay City, MI | 44,887 | 74.3 | 108,900 | 19.0 | 11.1 | 697 | 25.7 | 0.0 | 49,573 | -1.2 | 4,564 | 9.2 | 48,131 | 31.0 | 28.4 |
| Beaumont-Port Arthur, TX | 142,135 | 66.7 | 123,600 | 17.9 | 10.0 | 893 | 25.7 | 3.0 | 166,638 | -0.8 | 18,607 | 11.2 | 172,596 | 31.1 | 29.7 |
| Beckley, WV | 49,117 | 72.2 | 99,400 | 18.1 | 11.4 | 718 | 30.5 | 0.0 | 46,788 | -0.1 | 4,346 | 9.3 | 43,018 | 32.7 | 21.4 |
| Bellingham, WA | 88,794 | 64.4 | 398,300 | 23.6 | 10.2 | 1,134 | 31.1 | 3.8 | 114,716 | -0.4 | 10,724 | 9.3 | 109,586 | 38.9 | 24.1 |
| Bend, OR | 76,528 | 70.7 | 413,500 | 24.4 | 11.3 | 1,388 | 27.9 | 3.0 | 97,975 | 1.6 | 7,764 | 7.9 | 96,972 | 40.7 | 19.6 |
| Billings, MT | 78,108 | 67.7 | 241,400 | 19.9 | 11.3 | 863 | 29.0 | 1.9 | 94,102 | 1.2 | 5,131 | 5.5 | 92,939 | 38.3 | 22.3 |
| Binghamton, NY | 99,300 | 66.6 | 119,400 | 19.2 | 11.5 | 778 | 32.2 | 1.0 | 104,039 | -1.6 | 8,831 | 8.5 | 94,490 | 37.8 | 21.3 |
| Birmingham-Hoover, AL | 428,665 | 67.7 | 177,600 | 19.4 | 10.0 | 913 | 29.3 | 1.6 | 523,030 | -0.9 | 28,618 | 5.5 | 509,059 | 40.8 | 23.0 |
| Bismarck, ND | 52,396 | 70.5 | 261,300 | 19.3 | 10.0 | 862 | 27.5 | 0.0 | 68,216 | 1.4 | 3,098 | 4.5 | 68,699 | 40.9 | 20.7 |
| Blacksburg-Christiansburg, VA | 62,114 | 57.8 | 191,800 | 18.9 | 10.0 | 866 | 28.0 | 0.0 | 80,958 | -3.4 | 4,617 | 5.7 | 76,268 | 44.0 | 22.0 |
| Bloomington, IL | 65,118 | 63.8 | 166,800 | 18.1 | 10.0 | 815 | 24.3 | 0.0 | 85,103 | -3.6 | 5,762 | 6.8 | 83,090 | 45.2 | 13.8 |
| Bloomington, IN | 62,229 | 61.6 | 185,200 | 19.1 | 10.4 | 925 | 34.7 | 1.4 | 77,170 | -3.1 | 4,333 | 5.6 | 83,634 | 42.5 | 21.4 |
| Bloomsburg-Berwick, PA | 33,738 | 70.8 | 171,200 | 18.0 | 10.6 | 725 | 28.7 | 0.0 | 42,367 | -0.7 | 3,255 | 7.7 | 39,997 | 36.8 | 27.2 |
| Boise City, ID | 275,681 | 71.8 | 293,700 | 19.9 | 10.0 | 1,049 | 29.0 | 2.6 | 381,182 | 1.5 | 21,224 | 5.6 | 368,751 | 39.4 | 21.9 |
| Boston-Cambridge-Newton, MA-NH | 1,855,763 | 61.5 | 482,700 | 22.2 | 13.5 | 1,579 | 29.4 | 2.3 | 2,659,612 | -3.1 | 223,592 | 8.4 | 2,695,239 | 51.4 | 14.4 |
| Boston, MA Div 14454 | 770,703 | 57.7 | 484,400 | 22.7 | 13.5 | 1,642 | 29.7 | 2.5 | 1,103,352 | -2.8 | 101,057 | 9.2 | 1,124,694 | 50.1 | 13.7 |
| Cambridge-Newton-Framing-ham, MA Div 15754 | 910,264 | 62.3 | 529,200 | 21.7 | 13.2 | 1,580 | 29.3 | 2.2 | 1,297,058 | -3.7 | 104,778 | 8.1 | 1,324,487 | 53.8 | 13.8 |
| Rockingham County-Strafford County, NH Div 40484 | 174,796 | 74.3 | 334,000 | 22.4 | 14.3 | 1,218 | 28.6 | 1.6 | 259,202 | -1.5 | 17,757 | 6.9 | 246,058 | 44.5 | 21.2 |
| Boulder, CO | 131,854 | 61.6 | 592,000 | 20.5 | 10.0 | 1,637 | 32.5 | 2.0 | 192,879 | -1.3 | 11,887 | 6.2 | 181,272 | 55.5 | 12.8 |
| Bowling Green, KY | 66,318 | 62.1 | 170,000 | 18.8 | 10.0 | 762 | 27.5 | 2.6 | 81,548 | -3.8 | 5,318 | 6.5 | 85,884 | 36.6 | 28.8 |
| Bremerton-Silverdale-Port Orchard, WA | 107,525 | 69.9 | 378,800 | 22.6 | 10.1 | 1,433 | 30.3 | 2.2 | 129,092 | 0.7 | 9,952 | 7.7 | 118,480 | 41.8 | 20.6 |
| Bridgeport-Stamford-Norwalk, CT | 340,557 | 66.1 | 444,500 | 23.5 | 14.4 | 1,521 | 29.7 | 2.1 | 466,208 | -3.0 | 37,044 | 7.9 | 479,796 | 47.2 | 15.0 |
| Brownsville-Harlingen, TX | 129,307 | 65 | 89,000 | 22.1 | 11.6 | 787 | 32.1 | 11.5 | 169,074 | 1.5 | 17,219 | 10.2 | 169,450 | 27.4 | 24.4 |
| Brunswick, GA | 47,730 | 70.8 | 166,900 | 18.6 | 11.3 | 778 | 27.7 | 0.0 | 51,159 | -2.0 | 3,322 | 6.5 | 54,273 | 29.6 | 24.0 |
| Buffalo-Cheektowaga, NY | 488,951 | 65.7 | 160,800 | 18.2 | 11.9 | 828 | 29.2 | 1.4 | 537,396 | -0.3 | 51,954 | 9.7 | 560,382 | 40.7 | 20.0 |
| Burlington, NC | 64,316 | 68.5 | 166,500 | 17.8 | 10.0 | 796 | 28.3 | 1.3 | 80,457 | -3.1 | 5,832 | 7.2 | 83,546 | 34.3 | 25.2 |
| Burlington-South Burlington, VT | 89,859 | 66.7 | 291,800 | 21.7 | 12.8 | 1,242 | 35.0 | 2.1 | 123,442 | -4.1 | 6,071 | 4.9 | 121,580 | 49.0 | 17.7 |
| California-Lexington Park, MD | 40,320 | 72.3 | 320,700 | 19.4 | 10.3 | 1,370 | 27.3 | 0.0 | 57,281 | 0.0 | 2,768 | 4.8 | 56,054 | 44.7 | 21.3 |
| Canton-Massillon, OH | 165,718 | 67.7 | 144,000 | 17.9 | 10.0 | 728 | 25.4 | 0.7 | 197,032 | -1.3 | 15,994 | 8.1 | 189,989 | 34.0 | 26.8 |
| Cape Coral-Fort Myers, FL | 286,887 | 75.5 | 244,700 | 22.2 | 11.3 | 1,220 | 32.4 | 2.4 | 345,417 | -1.0 | 25,256 | 7.3 | 321,744 | 33.1 | 21.4 |
| Cape Girardeau, MO-IL | 38,214 | 69.2 | 158,500 | 18.0 | 10.2 | 774 | 28.5 | 0.0 | 47,660 | -0.9 | 2,571 | 5.4 | 47,200 | 36.6 | 22.7 |
| Carbondale-Marion, IL | 56,518 | 63.3 | 115,500 | 17.3 | 11.1 | 686 | 30.3 | 1.3 | 62,882 | -2.3 | 5,210 | 8.3 | 60,098 | 37.1 | 20.0 |
| Carson City, NV | 22,703 | 58.9 | 325,200 | 23.6 | 10.0 | 932 | 29.6 | 4.8 | 26,130 | -1.8 | 2,143 | 8.2 | 25,609 | 24.0 | 29.6 |
| Casper, WY | 33,044 | 72.4 | 225,400 | 19.9 | 10.0 | 777 | 27.9 | 2.1 | 40,710 | 0.8 | 3,181 | 7.8 | 41,245 | 34.3 | 25.1 |

1. Specified owner-occupied units. lacking complete plumbing facilities. 2. A value of 10.0 represents 10 percent or less; a value of 50.0 represents 50 percent or more. 3. Specified renter-occupied units. 4. Overcrowded or 5. Percent of civilian labor force. 6. Civilian employed persons 16 years old and over.

# Table C. Metropolitan Areas — Nonfarm Employment and Agriculture

| Area name | Number of establishments | Employment — Total | Health care and social assistance | Manufacturing | Retail trade | Finance and insurance | Professional, scientific, and technical services | Annual payroll Total (mil dol) | Average per employee (dollars) | Farms Number | Percent with: Fewer than 50 acres | Percent with: 1000 acres or more | Farm producers whose primary occupation is farming (percent) |
|---|---|---|---|---|---|---|---|---|---|---|---|---|---|
| | 104 | 105 | 106 | 107 | 108 | 109 | 110 | 111 | 112 | 113 | 114 | 115 | 116 |
| Abilene, TX | 4,036 | 61,821 | 13,962 | 3,538 | 8,813 | 2,714 | 2,030 | 2,457 | 39,749 | 3,270 | 35.4 | 10.7 | 33.8 |
| Akron, OH | 16,201 | 296,941 | 51,998 | 38,675 | 38,711 | 11,087 | 15,089 | 14,549 | 48,997 | 1,510 | 70.8 | 0.9 | 32.4 |
| Albany, GA | 3,074 | 45,801 | 9,545 | 4,270 | 7,422 | 1,176 | 2,473 | 1,810 | 39,509 | 1,041 | 32.2 | 14.7 | 41.2 |
| Albany-Lebanon, OR | 2,688 | 37,607 | 5,535 | 7,977 | 5,275 | 894 | 1,059 | 1,687 | 44,853 | 2,227 | 71.2 | 3.1 | 41.6 |
| Albany-Schenectady-Troy, NY | 21,574 | 364,490 | 72,193 | 24,983 | 47,342 | 19,515 | 33,051 | 19,096 | 52,390 | 4,331 | 73.8 | 7.3 | 34.3 |
| Albuquerque, NM | 18,828 | 310,722 | 58,715 | 15,629 | 41,384 | 14,284 | 33,440 | 13,831 | 44,513 | 1,063 | 48.1 | 6.6 | 41.9 |
| Alexandria, LA | 3,288 | 47,731 | 14,550 | 3,332 | 7,746 | 1,470 | 1,808 | 1,969 | 41,261 | 1,055 | 40.0 | 1.4 | 46.4 |
| Allentown-Bethlehem-Easton, PA-NJ | 18,472 | 338,802 | 69,521 | 37,036 | 45,548 | 10,395 | 13,208 | 17,176 | 50,695 | 1,958 | 64.5 | 2.2 | 40.4 |
| Altoona, PA | 3,205 | 53,539 | 12,987 | 6,922 | 8,710 | 1,072 | 1,763 | 2,215 | 41,375 | 496 | 34.9 | 1.6 | 52.7 |
| Amarillo, TX | 6,353 | 97,744 | 17,968 | 11,681 | 14,164 | 5,227 | 2,991 | 4,398 | 44,991 | 1,699 | 29.4 | 25.7 | 40.1 |
| Ames, IA | 2,703 | 38,563 | 7,216 | 5,362 | 6,070 | 863 | 1,271 | 1,714 | 44,437 | 1,922 | 46.0 | 10.5 | 37.2 |
| Anchorage, AK | 11,161 | 164,854 | 31,609 | 2,634 | 18,927 | 5,254 | 14,678 | 10,456 | 63,426 | 350 | 64.3 | 0.9 | 44.4 |
| Ann Arbor, MI | 8,227 | 159,215 | 40,021 | 15,905 | 17,742 | 3,397 | 15,347 | 9,480 | 59,543 | 1,245 | 59.7 | 2.7 | 39.8 |
| Anniston-Oxford, AL | 2,245 | 36,752 | 6,297 | 6,889 | 6,222 | 874 | 985 | 1,327 | 36,109 | 643 | 43.4 | 1.4 | 34.5 |
| Appleton, WI | 5,925 | 117,782 | 13,493 | 23,719 | 15,015 | 5,687 | 4,164 | 5,608 | 47,616 | 1,814 | 43.6 | 4.0 | 47.9 |
| Asheville, NC | 13,436 | 177,409 | 36,447 | 22,261 | 28,599 | 3,864 | 6,728 | 7,229 | 40,750 | 2,708 | 62.3 | 0.9 | 38.7 |
| Athens-Clarke County, GA | 4,893 | 67,560 | 12,896 | 8,229 | 10,667 | 1,654 | 2,845 | 2,673 | 39,571 | 1,520 | 47.8 | 1.6 | 38.8 |
| Atlanta-Sandy Springs-Alpharetta, GA | 148,039 | 2,457,928 | 293,257 | 159,049 | 275,106 | 123,466 | 229,279 | 147,609 | 60,054 | 7,760 | 55.7 | 1.0 | 39.4 |
| Atlantic City-Hammonton, NJ | 6,187 | 110,453 | 18,860 | 2,237 | 15,662 | 2,458 | 4,007 | 4,638 | 41,991 | 450 | 75.3 | 0.9 | 52.1 |
| Auburn-Opelika, AL | 2,952 | 48,188 | 6,842 | 7,399 | 7,233 | 955 | 1,781 | 1,700 | 35,281 | 314 | 43.9 | 5.7 | 35.5 |
| Augusta-Richmond County, GA-SC | 10,850 | 190,195 | 36,800 | 22,501 | 26,825 | 3,617 | 7,888 | 8,844 | 46,502 | 2,787 | 48.4 | 4.6 | 37.0 |
| Austin-Round Rock-Georgetown, TX | 56,137 | 896,844 | 110,121 | 43,912 | 113,573 | 41,483 | 111,535 | 54,494 | 60,762 | 8,498 | 52.5 | 3.8 | 34.1 |
| Bakersfield, CA | 13,197 | 199,982 | 30,947 | 13,138 | 30,814 | 5,219 | 9,548 | 9,521 | 47,611 | 1,731 | 40.3 | 17.6 | 51.1 |
| Baltimore-Columbia-Towson, MD | 67,394 | 1,203,546 | 215,499 | 53,640 | 139,307 | 53,295 | 138,533 | 70,707 | 58,749 | 3,704 | 61.7 | 2.6 | 40.3 |
| Bangor, ME | 4,118 | 57,335 | 15,302 | 2,897 | 10,856 | 2,130 | 1,747 | 2,473 | 43,139 | 601 | 41.3 | 3.2 | 44.6 |
| Barnstable Town, MA | 8,679 | 77,116 | 15,761 | 2,069 | 14,998 | 2,017 | 4,757 | 3,814 | 49,458 | 321 | 93.5 | 0.0 | 46.5 |
| Baton Rouge, LA | 18,752 | 352,700 | 48,170 | 26,260 | 42,384 | 14,390 | 24,643 | 18,502 | 52,457 | 2,866 | 53.4 | 8.1 | 40.4 |
| Battle Creek, MI | 2,440 | 51,721 | 9,300 | 13,713 | 5,831 | 1,075 | 2,562 | 2,661 | 51,447 | 958 | 42.6 | 5.8 | 44.2 |
| Bay City, MI | 2,088 | 30,070 | 6,248 | 4,463 | 5,028 | 900 | 1,235 | 1,220 | 40,583 | 726 | 36.8 | 8.8 | 48.8 |
| Beaumont-Port Arthur, TX | 7,808 | 131,195 | 19,639 | 19,066 | 19,172 | 3,346 | 6,771 | 6,786 | 51,725 | 2,053 | 69.1 | 3.6 | 35.5 |
| Beckley, WV | 2,423 | 34,647 | 8,764 | 1,322 | 6,038 | 704 | 1,661 | 1,352 | 39,027 | 618 | 50.0 | 1.0 | 35.1 |
| Bellingham, WA | 6,838 | 78,090 | 11,024 | 10,583 | 10,593 | 2,595 | 3,370 | 3,693 | 47,296 | 1,712 | 79.6 | 0.7 | 35.4 |
| Bend, OR | 7,669 | 71,896 | 11,101 | 5,774 | 11,632 | 2,271 | 3,834 | 3,199 | 44,498 | 1,484 | 85.4 | 0.8 | 34.7 |
| Billings, MT | 6,316 | 75,615 | 14,776 | 4,383 | 10,687 | 4,015 | 3,806 | 3,632 | 48,026 | 2,601 | 35.9 | 20.2 | 41.8 |
| Binghamton, NY | 4,908 | 79,714 | 17,406 | 8,506 | 11,657 | 2,266 | 3,893 | 3,345 | 41,965 | 1,029 | 32.9 | 1.7 | 44.1 |
| Birmingham-Hoover, AL | 24,928 | 456,250 | 75,468 | 37,441 | 54,395 | 31,868 | 22,829 | 24,248 | 53,146 | 3,138 | 45.0 | 1.6 | 36.8 |
| Bismarck, ND | 3,940 | 58,881 | 13,143 | 2,923 | 8,588 | 2,199 | 4,645 | 2,823 | 47,936 | 1,800 | 20.2 | 32.7 | 45.6 |
| Blacksburg-Christiansburg, VA | 3,185 | 49,692 | 6,812 | 12,943 | 7,722 | 1,139 | 2,664 | 2,032 | 40,891 | 1,367 | 40.7 | 2.9 | 46.4 |
| Bloomington, IL | 3,530 | 72,469 | 8,086 | 3,203 | 8,874 | 18,520 | 2,375 | 3,827 | 52,812 | 1,416 | 33.9 | 13.8 | 31.0 |
| Bloomington, IN | 3,466 | 55,744 | 11,429 | 9,118 | 7,109 | 1,588 | 2,048 | 2,303 | 41,314 | 1,135 | 43.7 | 1.4 | 39.5 |
| Bloomsburg-Berwick, PA | 1,850 | 38,242 | 10,948 | 5,359 | 4,165 | 1,987 | 1,014 | 2,038 | 53,291 | 1,139 | 47.5 | 2.4 | 37.5 |
| Boise City, ID | 20,353 | 278,916 | 47,266 | 26,736 | 35,154 | 11,361 | 16,465 | 13,411 | 48,083 | 5,108 | 76.4 | 4.1 | 43.1 |
| Boston-Cambridge-Newton, MA-NH | 132,059 | 2,595,753 | 446,255 | 162,740 | 267,530 | 159,862 | 280,659 | 199,730 | 76,945 | 2,943 | 74.4 | 0.4 | 46.1 |
| Boston, MA Div 14454 | 55,335 | 1,177,232 | 232,917 | 36,630 | 112,563 | 107,415 | 114,429 | 93,184 | 79,155 | 976 | 76.6 | 0.7 | 43.8 |
| Cambridge-Newton-Framingham, MA Div 15754 | 64,158 | 1,230,680 | 188,808 | 104,365 | 121,468 | 41,453 | 154,960 | 96,483 | 78,398 | 1,039 | 75.6 | 0.2 | 43.1 |
| Rockingham County-Strafford County, NH Div 40484 | 12,566 | 187,841 | 24,530 | 21,745 | 33,499 | 10,994 | 11,270 | 10,064 | 53,576 | 928 | 70.9 | 0.2 | 39.1 |
| Boulder, CO | 12,856 | 156,833 | 22,151 | 15,212 | 19,238 | 3,836 | 26,065 | 10,686 | 68,137 | 1,012 | 79.0 | 2.2 | 33.1 |
| Bowling Green, KY | 3,378 | 61,927 | 9,316 | 10,948 | 8,665 | 1,680 | 3,321 | 2,609 | 42,134 | 4,102 | 41.2 | 2.2 | 33.9 |
| Bremerton-Silverdale-Port Orchard, WA | 6,052 | 63,818 | 13,905 | 2,354 | 11,975 | 1,741 | 4,958 | 2,775 | 43,488 | 698 | 96.3 | 0.0 | 29.0 |
| Bridgeport-Stamford-Norwalk, CT | 26,947 | 422,988 | 73,060 | 34,516 | 49,143 | 31,402 | 38,822 | 35,015 | 82,780 | 402 | 81.8 | 2.2 | 49.3 |
| Brownsville-Harlingen, TX | 6,512 | 110,105 | 37,013 | 4,263 | 18,833 | 3,620 | 3,162 | 3,190 | 28,975 | 1,418 | 74.3 | 6.3 | 34.1 |
| Brunswick, GA | 2,969 | 34,807 | 5,476 | 2,258 | 5,804 | 824 | 1,051 | 1,312 | 37,693 | 320 | 55.0 | 0.6 | 32.2 |
| Buffalo-Cheektowaga, NY | 27,352 | 490,650 | 91,677 | 53,293 | 62,702 | 33,072 | 29,632 | 23,024 | 46,926 | 1,630 | 51.2 | 3.7 | 46.3 |
| Burlington, NC | 3,346 | 61,650 | 12,055 | 8,839 | 9,868 | 1,351 | 1,461 | 2,454 | 39,803 | 720 | 45.8 | 0.7 | 40.6 |
| Burlington-South Burlington, VT | 6,875 | 103,423 | 19,118 | 12,250 | 14,952 | 3,322 | 6,947 | 5,179 | 50,080 | 1,433 | 41.6 | 2.7 | 45.0 |
| California-Lexington Park, MD. | 2,014 | 32,995 | 4,627 | 452 | 5,066 | 434 | 4,771 | 1,822 | 55,210 | 615 | 57.2 | 1.5 | 43.7 |
| Canton-Massillon, OH | 8,524 | 152,695 | 30,726 | 27,051 | 20,886 | 6,356 | 6,008 | 6,275 | 41,097 | 2,435 | 59.3 | 1.3 | 35.1 |
| Cape Coral-Fort Myers, FL | 19,540 | 232,778 | 38,748 | 5,412 | 40,336 | 6,356 | 13,582 | 10,024 | 43,062 | 800 | 81.6 | 2.4 | 38.5 |
| Cape Girardeau, MO-IL | 2,608 | 40,149 | 11,230 | 3,698 | 6,326 | 1,054 | 1,308 | 1,590 | 39,611 | 2,035 | 37.8 | 4.8 | 35.6 |
| Carbondale-Marion, IL | 2,997 | 41,997 | 10,681 | 4,290 | 7,310 | 2,198 | 1,378 | 1,625 | 38,693 | 1,993 | 28.8 | 5.1 | 53.3 |
| Carson City, NV | 1,890 | 22,261 | 4,274 | 2,529 | 3,155 | 1,030 | 1,035 | 1,033 | 46,403 | 17 | 70.6 | D | 53.3 |
| Casper, WY | 2,928 | 32,242 | 5,756 | 1,766 | 4,578 | 917 | 1,623 | 1,699 | 52,709 | 430 | 40.0 | 25.3 | 39.3 |

## Table C. Metropolitan Areas — **Agriculture**

Agriculture, 2017 (cont.)

| Area name | Land in farms | | | | | Value of land and buildings (dollars) | | Value of machinery and equipment, average per farm (dollars) | Value of products sold: | | | | Organic farms (number) | Farms with internet access (percent) | Government payments | |
|---|---|---|---|---|---|---|---|---|---|---|---|---|---|---|---|---|
| | Acreage (1,000) | Percent change, 2012-2017 | Acres | | | Average per farm | Average per acre | | Total (mil dol) | Average per farm (acres) | Percent from: | | | | Total ($1,000) | Percent of farms |
| | | | Average size of farm | Total irrigated (1,000) | Total cropland (1,000) | | | | | | Crops | Livestock and poultry products | | | | |
| | 117 | 118 | 119 | 120 | 121 | 122 | 123 | 124 | 125 | 126 | 127 | 128 | 129 | 130 | 131 | 132 |
| Abilene, TX | 1,479 | -13.3 | 452 | 6.0 | 558.5 | 737,343 | 1,631 | 70,790 | 104 | 31,887 | 42.2 | 57.8 | 0 | 74.3 | 7,701 | 26.5 |
| Akron, OH | 105 | 4.8 | 69 | 0.5 | 70.1 | 536,277 | 7,740 | 69,000 | 47 | 31,188 | 70.3 | 29.7 | 7 | 80.5 | 1,499 | 9.5 |
| Albany, GA | 537 | 3.0 | 515 | 118.7 | 302.2 | 1,619,933 | 3,143 | 230,125 | 258 | 247,927 | 85.1 | 14.9 | 0 | 76.9 | 21,748 | 55.8 |
| Albany-Lebanon, OR | 315 | -4.9 | 142 | 36.9 | 242.6 | 1,005,264 | 7,092 | 100,759 | 243 | 109,375 | 74.6 | 25.4 | 39 | 85.8 | 856 | 4.8 |
| Albany-Schenectady-Troy, NY | 331 | -5.2 | 149 | 3.1 | 193.3 | 581,420 | 3,911 | 102,818 | 219 | 98,131 | 47.2 | 52.8 | 69 | 81.2 | 1,939 | 15.3 |
| Albuquerque, NM | 3,084 | -19.6 | 712 | 43.1 | 71.8 | 548,325 | 770 | 41,738 | 114 | 26,245 | 25.3 | 74.7 | 35 | 65.0 | 2,852 | 4.5 |
| Alexandria, LA | 254 | -1.9 | 239 | 26.3 | 148.7 | 847,287 | 3,545 | 122,673 | 155 | 146,070 | 90.5 | 9.5 | 1 | 73.1 | 3,988 | 23.4 |
| Allentown-Bethlehem-Easton, PA-NJ | 227 | -3.6 | 116 | 2.1 | 174.6 | 1,051,135 | 9,064 | 110,824 | 222 | 113,135 | 73.7 | 26.3 | 16 | 80.6 | 3,404 | 15.0 |
| Altoona, PA | 79 | -12.4 | 159 | 0.2 | 55.8 | 1,073,542 | 6,747 | 130,550 | 107 | 216,087 | 15.8 | 84.2 | 9 | 61.7 | 1,064 | 25.6 |
| Amarillo, TX | 2,893 | 0.1 | 1,703 | 93.3 | 817.1 | 1,887,508 | 1,109 | 150,924 | 802 | 471,719 | 15.6 | 84.4 | 0 | 77.8 | 27,310 | 47.7 |
| Ames, IA | 619 | 0.0 | 322 | 0.6 | 572.9 | 2,803,265 | 8,703 | 202,624 | 443 | 230,632 | 73.4 | 26.6 | 16 | 84.8 | 10,626 | 61.2 |
| Anchorage, AK | 34 | -5.5 | 98 | 1.1 | 15.6 | 809,797 | 8,242 | 0 | 38 | 107,246 | 45.6 | 54.4 | 4 | 87.7 | 262 | 20.6 |
| Ann Arbor, MI | 179 | 5.2 | 144 | 4.0 | 150.4 | 1,124,667 | 7,823 | 112,400 | 91 | 73,227 | 76.3 | 23.7 | 29 | 84.3 | 3,438 | 24.6 |
| Anniston-Oxford, AL | 89 | 9.5 | 138 | 1.1 | 28.5 | 587,726 | 4,246 | 74,079 | 87 | 135,003 | 15.0 | 85.0 | 0 | 77.1 | 913 | 28.8 |
| Appleton, WI | 391 | -0.6 | 215 | 0.8 | 337.7 | 1,609,348 | 7,470 | 201,699 | 467 | 257,601 | 24.9 | 75.1 | 19 | 80.2 | 2,530 | 48.6 |
| Asheville, NC | 222 | 4.7 | 82 | 3.7 | 65.0 | 611,878 | 7,451 | 58,730 | 133 | 49,267 | 82.2 | 17.8 | 37 | 74.2 | 1,461 | 15.4 |
| Athens-Clarke County, GA | 185 | -10.2 | 122 | 2.6 | 50.4 | 675,115 | 5,541 | 76,858 | 525 | 345,272 | 7.0 | 93.0 | 6 | 80.7 | 1,945 | 24.5 |
| Atlanta-Sandy Springs-Alpharetta, GA | 872 | 1.9 | 112 | 7.9 | 208.4 | 599,404 | 5,337 | 63,245 | 931 | 119,985 | 10.6 | 89.4 | 31 | 79.1 | 5,463 | 12.9 |
| Atlantic City-Hammonton, NJ | 29 | -1.6 | 64 | 11.6 | 17.8 | 823,031 | 12,764 | 135,393 | 121 | 268,162 | 98.7 | 1.3 | 10 | 85.3 | 198 | 7.8 |
| Auburn-Opelika, AL | 68 | 14.8 | 216 | 0.2 | 9.0 | 895,685 | 4,150 | 95,631 | 0 | 0 | 0.0 | 0.0 | 4 | 79.9 | 432 | 26.8 |
| Augusta-Richmond County, GA-SC | 562 | 15.8 | 202 | 61.4 | 231.2 | 668,247 | 3,312 | 101,967 | 302 | 108,416 | 41.7 | 58.3 | 8 | 76.7 | 8,683 | 17.4 |
| Austin-Round Rock-Georgetown, TX | 1,669 | -4.8 | 196 | 8.2 | 457.4 | 1,003,716 | 5,110 | 61,144 | 263 | 30,957 | 47.8 | 52.2 | 37 | 76.9 | 9,105 | 9.8 |
| Bakersfield, CA | 2,295 | -1.5 | 1,326 | 730.7 | 954.1 | 9,787,244 | 7,380 | 406,230 | 4,077 | 2,355,161 | 84.3 | 15.7 | 30 | 82.7 | 5,101 | 13.8 |
| Baltimore-Columbia-Towson, MD | 520 | 5.8 | 140 | 20.2 | 379.8 | 1,325,778 | 9,451 | 114,805 | 450 | 121,442 | 65.2 | 34.8 | 33 | 83.6 | 11,036 | 26.3 |
| Bangor, ME | 105 | -6.6 | 175 | 1.6 | 41.1 | 432,699 | 2,466 | 93,820 | 51 | 84,717 | 35.6 | 64.4 | 33 | 79.0 | 603 | 9.2 |
| Barnstable Town, MA | 7 | 40.4 | 20 | 1.1 | 1.6 | 624,807 | 30,555 | 60,763 | 23 | 72,019 | 40.4 | 59.6 | 11 | 90.0 | 197 | 4.7 |
| Baton Rouge, LA | 876 | 4.2 | 306 | 6.9 | 433.0 | 985,685 | 3,226 | 136,025 | 275 | 95,941 | 73.2 | 26.8 | 5 | 74.7 | 6,462 | 14.1 |
| Battle Creek, MI | 214 | -4.9 | 223 | 13.3 | 174.6 | 1,185,546 | 5,309 | 138,963 | 114 | 118,861 | 61.1 | 38.9 | 20 | 71.5 | 4,522 | 36.7 |
| Bay City, MI | 210 | 8.3 | 289 | 5.2 | 197.1 | 1,460,329 | 5,052 | 242,521 | 117 | 160,519 | 88.3 | 11.7 | 0 | 69.3 | 4,501 | 67.8 |
| Beaumont-Port Arthur, TX | 477 | 0.3 | 232 | 26.3 | 155.1 | 629,846 | 2,711 | 75,815 | 42 | 20,447 | 51.3 | 48.7 | 10 | 73.2 | 9,497 | 8.7 |
| Beckley, WV | 70 | 16.1 | 113 | 0.0 | 17.3 | 385,375 | 3,416 | 49,804 | 5 | 7,655 | 31.4 | 68.6 | 1 | 74.9 | 60 | 3.7 |
| Bellingham, WA | 103 | -11.5 | 60 | 36.5 | 75.6 | 1,005,680 | 16,794 | 95,376 | 373 | 217,786 | 41.3 | 58.7 | 48 | 83.7 | 1,047 | 12.6 |
| Bend, OR | 135 | 2.7 | 91 | 36.0 | 31.0 | 786,080 | 8,667 | 51,257 | 29 | 19,386 | 57.5 | 42.5 | 9 | 93.2 | 90 | 0.8 |
| Billings, MT | 3,181 | -2.7 | 1,223 | 198.8 | 609.7 | 1,477,760 | 1,208 | 117,787 | 286 | 109,868 | 32.1 | 67.9 | 3 | 81.9 | 11,178 | 29.1 |
| Binghamton, NY | 176 | -6.3 | 171 | 0.8 | 88.3 | 438,814 | 2,571 | 86,513 | 73 | 70,888 | 31.0 | 69.0 | 26 | 78.6 | 1,317 | 20.2 |
| Birmingham-Hoover, AL | 445 | -3.1 | 142 | 4.9 | 129.3 | 524,825 | 3,704 | 76,246 | 341 | 108,757 | 13.2 | 86.8 | 9 | 76.7 | 5,506 | 23.7 |
| Bismarck, ND | 2,364 | -7.9 | 1,313 | 11.4 | 1,134.7 | 2,077,147 | 1,582 | 245,198 | 328 | 181,918 | 46.1 | 53.9 | 3 | 80.5 | 16,907 | 60.8 |
| Blacksburg-Christiansburg, VA | 245 | -9.2 | 179 | 0.4 | 65.5 | 706,552 | 3,949 | 76,518 | 67 | 48,980 | 30.6 | 69.4 | 2 | 78.5 | 473 | 6.4 |
| Bloomington, IL | 620 | -10.4 | 438 | 3.2 | 599.9 | 4,310,555 | 9,844 | 280,498 | 457 | 322,784 | 85.0 | 15.0 | 3 | 81.1 | 9,368 | 73.0 |
| Bloomington, IN | 160 | 7.7 | 140 | 0.3 | 95.8 | 721,922 | 5,149 | 68,376 | 44 | 38,219 | 82.3 | 17.7 | 19 | 76.3 | 2,350 | 27.6 |
| Bloomsburg-Berwick, PA | 145 | -12.5 | 128 | 0.9 | 108.3 | 786,339 | 6,139 | 96,310 | 128 | 112,345 | 49.7 | 50.3 | 13 | 69.3 | 2,933 | 37.1 |
| Boise City, ID | 1,351 | -1.8 | 264 | 420.9 | 450.1 | 1,000,490 | 3,783 | 131,482 | 1,022 | 199,985 | 42.7 | 57.3 | 35 | 86.1 | 6,139 | 9.0 |
| Boston-Cambridge-Newton, MA-NH | 171 | -10.3 | 58 | 15.3 | 61.1 | 687,073 | 11,814 | 62,502 | 202 | 68,484 | 84.2 | 15.8 | 97 | 87.7 | 1,398 | 5.0 |
| Boston, MA Div 14454 | 68 | -7.9 | 69 | 12.4 | 21.3 | 727,928 | 10,497 | 69,057 | 73 | 74,237 | 86.5 | 13.5 | 31 | 88.7 | 380 | 5.4 |
| Cambridge-Newton-Framingham, MA Div 15754 | 48 | -5.1 | 46 | 2.5 | 23.9 | 767,516 | 16,593 | 64,666 | 96 | 92,604 | 87.6 | 12.4 | 32 | 86.6 | 226 | 3.8 |
| Rockingham County-Strafford County, NH Div 40484 | 55 | -16.9 | 60 | 0.4 | 15.8 | 554,039 | 9,278 | 53,184 | 33 | 35,428 | 69.1 | 30.9 | 34 | 87.9 | 792 | 5.9 |
| Boulder, CO | 107 | -19.5 | 106 | 27.2 | 38.1 | 1,329,691 | 12,571 | 67,039 | 44 | 43,378 | 87.4 | 12.6 | 29 | 90.7 | 501 | 5.6 |
| Bowling Green, KY | 659 | 4.7 | 161 | 0.6 | 344.6 | 641,344 | 3,990 | 78,685 | 276 | 67,255 | 44.4 | 55.6 | 8 | 71.8 | 11,581 | 25.2 |
| Bremerton-Silverdale-Port Orchard, WA | 9 | -6.7 | 13 | 0.5 | 2.3 | 473,099 | 35,164 | 33,113 | 7 | 9,463 | 73.2 | 26.8 | 26 | 88.3 | 0 | 0.6 |
| Bridgeport-Stamford-Norwalk, CT | 52 | -3.2 | 130 | 0.2 | 4.6 | 1,402,826 | 10,794 | 57,697 | 42 | 104,649 | 52.7 | 47.3 | 9 | 87.1 | 33 | 3.5 |
| Brownsville-Harlingen, TX | 271 | -12.3 | 191 | 100.9 | 212.5 | 681,492 | 3,560 | 73,661 | 123 | 86,428 | 96.2 | 3.8 | 10 | 64.3 | 6,669 | 26.0 |
| Brunswick, GA | 36 | -18.0 | 113 | 1.0 | 7.0 | 388,278 | 3,436 | 54,650 | 0 | 975 | 26.0 | 74.0 | 2 | 75.6 | 135 | 14.4 |
| Buffalo-Cheektowaga, NY | 283 | -0.8 | 174 | 3.5 | 212.6 | 665,840 | 3,830 | 147,731 | 250 | 153,123 | 55.1 | 44.9 | 47 | 81.9 | 3,500 | 20.1 |
| Burlington, NC | 80 | -4.2 | 111 | 0.7 | 33.8 | 667,068 | 6,000 | 64,708 | 42 | 57,986 | 35.3 | 64.7 | 23 | 76.4 | 147 | 8.6 |
| Burlington-South Burlington, VT | 273 | -2.2 | 190 | 0.8 | 122.6 | 755,652 | 3,970 | 128,983 | 247 | 172,455 | 25.9 | 74.1 | 204 | 87.5 | 1,864 | 12.5 |
| California-Lexington Park, MD | 62 | -7.9 | 100 | 0.7 | 37.0 | 999,805 | 9,949 | 75,424 | 26 | 42,203 | 78.8 | 21.2 | 2 | 58.0 | 970 | 19.5 |
| Canton-Massillon, OH | 244 | 0.6 | 100 | 1.1 | 166.0 | 718,603 | 7,184 | 92,634 | 145 | 59,331 | 41.2 | 58.8 | 14 | 72.5 | 3,912 | 13.8 |
| Cape Coral-Fort Myers, FL | 87 | 0.1 | 109 | 10.5 | 22.2 | 1,330,329 | 12,206 | 53,238 | 104 | 130,449 | 93.8 | 6.2 | 7 | 76.0 | 163 | 1.8 |
| Cape Girardeau, MO-IL | 521 | 1.0 | 261 | 52.5 | 329.5 | 944,908 | 3,618 | 104,757 | 148 | 74,249 | 74.2 | 25.8 | 0 | 67.7 | 6,795 | 44.9 |
| Carbondale-Marion, IL | 431 | 5.8 | 212 | 2.2 | 305.8 | 980,108 | 4,628 | 106,433 | 140 | 68,824 | 82.3 | 17.7 | 3 | 68.5 | 9,107 | 44.5 |
| Carson City, NV | D | D | 57 | 0.4 | 0.6 | D | D | D | D | D | D | D | D | D | D | D |
| Casper, WY | 1,933 | 14.3 | 4,496 | 44.3 | 55.7 | 2,039,344 | 454 | 109,340 | 43 | 100,493 | 12.9 | 87.1 | 0 | 76.0 | 318 | 7.9 |

Items 117—132

# Table C. Metropolitan Areas — Water Use, Wholesale Trade, Retail Trade, and Real Estate

| Area name | Water use, 2015 | | Wholesale Trade[1], 2017 | | | | Retail Trade[2], 2017 | | | | Real estate and rental and leasing,[2] 2017 | | | |
|---|---|---|---|---|---|---|---|---|---|---|---|---|---|---|
| | Public supply water withdrawn (mil gal/day) | Public supply gallons withdrawn per person per day | Number of establishments | Number of employees | Sales (mil dol) | Average payroll (mil dol) | Number of establishments | Number of employees | Sales (mil dol) | Average payroll (mil dol) | Number of establishments | Number of employees | Sales (mil dol) | Average payroll (mil dol) |
| | 133 | 134 | 135 | 136 | 137 | 138 | 139 | 140 | 141 | 142 | 143 | 144 | 145 | 146 |
| Abilene, TX | 8.78 | 51.8 | 166 | 2,063 | 2,262.1 | 103.7 | 612 | 8,737 | 2,900.3 | 239.5 | 190 | 860 | 211.9 | 34.4 |
| Akron, OH | 54.93 | 78.0 | 943 | 15,899 | 11,672.8 | 1,043.8 | 2,088 | 38,335 | 10,964.9 | 1,032.1 | 651 | 3,546 | 797.1 | 140.0 |
| Albany, GA | 15.89 | 105.7 | 166 | 2,545 | 2,519.0 | 139.2 | 596 | 7,620 | 2,031.2 | 176.3 | 142 | 548 | 123.7 | 19.2 |
| Albany-Lebanon, OR | 9.57 | 79.4 | 116 | 1,634 | 1,248.8 | 86.5 | 368 | 5,189 | 1,424.8 | 140.2 | 115 | 428 | 54.7 | 11.1 |
| Albany-Schenectady-Troy, NY | 290.84 | 329.8 | 760 | 11,323 | 11,290.2 | 697.8 | 2,989 | 45,732 | 13,690.3 | 1,272.7 | 911 | 4,971 | 1,339.3 | 216.3 |
| Albuquerque, NM | 105.54 | 116.3 | 784 | 10,308 | 6,322.8 | 526.5 | 2,372 | 40,549 | 11,953.7 | 1,146.9 | 1,130 | 4,568 | 1,096.1 | 180.6 |
| Alexandria, LA | 23.45 | 151.8 | D | D | D | D | 579 | 7,730 | 2,189.4 | 201.3 | D | D | D | D |
| Allentown-Bethlehem-Easton, PA-NJ | 74.71 | 89.8 | 656 | 18,120 | 24,215.1 | 1,293.9 | 2,689 | 46,229 | 14,828.4 | 1,275.2 | 641 | 3,100 | 789.9 | 134.1 |
| Altoona, PA | 13.07 | 104.1 | 124 | 2,112 | 2,076.2 | 101.4 | 530 | 8,700 | 2,721.2 | 227.4 | 111 | 409 | 84.8 | 14.3 |
| Amarillo, TX | 17.48 | 66.7 | 269 | 4,752 | 5,019.9 | 252.3 | 871 | 14,395 | 3,963.3 | 368.8 | 328 | 1,363 | 294.4 | 56.5 |
| Ames, IA | 10.81 | 88.1 | 78 | 1,444 | 1,083.4 | 56.5 | 345 | 6,110 | 1,454.8 | 145.9 | 135 | 575 | 96.0 | 24.3 |
| Anchorage, AK | 47.18 | 118.0 | 380 | 5,363 | 3,898.7 | 325.5 | 1,106 | 19,354 | 6,158.7 | 632.2 | 518 | 2,774 | 730.4 | 132.1 |
| Ann Arbor, MI | 18.36 | 51.2 | 270 | 3,870 | 5,384.1 | 267.6 | 1,100 | 17,534 | 5,121.3 | 473.5 | 373 | 2,686 | 1,101.3 | 140.5 |
| Anniston-Oxford, AL | 25.25 | 218.4 | 90 | 1,614 | 1,347.9 | 69.2 | 466 | 6,380 | 1,652.3 | 151.6 | 76 | 307 | 58.6 | 10.8 |
| Appleton, WI | 17.06 | 73.2 | 320 | 4,690 | 8,290.0 | 279.8 | 833 | 15,171 | 4,115.9 | 365.5 | 172 | 797 | 172.1 | 33.1 |
| Asheville, NC | 39.02 | 87.3 | 444 | 5,241 | 2,793.6 | 247.6 | 1,951 | 27,791 | 7,206.1 | 696.5 | 779 | 2,556 | 488.9 | 95.4 |
| Athens-Clarke County, GA | 12.62 | 62.1 | 123 | 2,257 | 3,090.5 | 114.9 | 706 | 11,026 | 2,696.2 | 258.9 | 290 | 1,871 | 339.4 | 83.7 |
| Atlanta-Sandy Springs-Alpharetta, GA | 480.18 | 84.1 | 7,129 | 116,733 | 144,514.3 | 8,174.5 | 17,594 | 279,506 | 91,665.8 | 7,549.3 | 8,378 | 47,617 | 19,335.1 | 3,187.4 |
| Atlantic City-Hammonton, NJ | 30.20 | 110.1 | 172 | 2,049 | 1,304.6 | 115.8 | 1,078 | 16,535 | 4,631.1 | 417.9 | 233 | 1,358 | 356.2 | 56.3 |
| Auburn-Opelika, AL | 15.83 | 100.8 | 83 | 999 | 465.5 | 48.1 | 478 | 7,212 | 1,901.9 | 169.0 | 157 | 678 | 139.6 | 25.1 |
| Augusta-Richmond County, GA-SC | 95.09 | 161.1 | 352 | 4,208 | 2,389.7 | 207.2 | 1,810 | 27,390 | 7,706.9 | 688.7 | 465 | 1,798 | 462.6 | 75.1 |
| Austin-Round Rock-Georgetown, TX | 180.10 | 90.0 | 1,812 | 28,554 | 85,056.1 | 2,225.0 | 6,209 | 108,411 | 35,894.7 | 3,301.2 | 3,361 | 19,185 | 5,419.7 | 1,102.5 |
| Bakersfield, CA | 168.87 | 191.4 | 561 | 8,010 | 6,451.9 | 443.1 | 1,960 | 31,869 | 9,382.8 | 860.7 | 670 | 3,474 | 759.1 | 147.4 |
| Baltimore-Columbia-Towson, MD | 267.62 | 95.7 | 2,513 | 43,748 | 38,700.5 | 2,908.6 | 8,821 | 142,444 | 41,386.1 | 3,968.6 | 3,244 | 22,667 | 8,672.8 | 1,301.6 |
| Bangor, ME | 4.79 | 31.4 | 156 | 1,816 | 1,056.4 | 93.9 | 697 | 10,573 | 3,261.5 | 276.5 | 184 | 924 | 270.0 | 29.3 |
| Barnstable Town, MA | 31.84 | 148.6 | D | D | D | D | 1,450 | 16,168 | 4,537.5 | 500.4 | 386 | 1,533 | 361.5 | 66.0 |
| Baton Rouge, LA | 108.01 | 126.6 | 834 | 12,108 | 8,984.2 | 721.0 | 2,914 | 45,152 | 12,723.3 | 1,179.3 | 926 | 4,856 | 1,312.1 | 234.8 |
| Battle Creek, MI | 12.53 | 93.3 | 83 | 808 | 1,385.5 | 43.5 | 467 | 6,105 | 1,855.4 | 157.7 | 82 | 387 | 59.1 | 12.5 |
| Bay City, MI | 8.97 | 84.9 | 80 | 1,168 | 691.9 | 50.3 | 385 | 5,258 | 1,465.6 | 139.5 | 60 | 182 | 26.3 | 4.4 |
| Beaumont-Port Arthur, TX | 36.38 | 92.2 | 310 | 3,867 | 3,059.6 | 218.8 | 1,358 | 18,913 | 6,350.4 | 537.2 | 354 | 2,446 | 580.5 | 112.3 |
| Beckley, WV | 15.26 | 124.6 | 103 | 953 | 529.6 | 45.9 | 460 | 6,254 | 1,843.7 | 154.5 | 91 | 416 | 76.1 | 17.0 |
| Bellingham, WA | 18.13 | 85.4 | D | D | D | D | 817 | 11,180 | 3,225.2 | 328.6 | 398 | 1,394 | 421.9 | 56.2 |
| Bend, OR | 38.00 | 216.8 | 227 | 1,621 | 1,110.3 | 82.0 | 879 | 11,709 | 3,550.5 | 341.9 | 499 | 1,275 | 298.1 | 49.8 |
| Billings, MT | 27.23 | 153.9 | 337 | 4,341 | 3,113.5 | 251.1 | 743 | 10,790 | 3,377.2 | 318.1 | 356 | 1,172 | 255.1 | 48.5 |
| Binghamton, NY | 22.18 | 90.2 | 201 | 3,965 | 2,895.4 | 183.2 | 814 | 11,729 | 3,086.3 | 290.5 | 163 | 926 | 191.1 | 33.1 |
| Birmingham-Hoover, AL | 138.56 | 128.3 | 1,377 | 22,432 | 26,930.5 | 1,355.0 | 3,923 | 54,812 | 15,738.7 | 1,451.9 | 1,100 | 7,208 | 2,558.5 | 399.4 |
| Bismarck, ND | 14.70 | 117.5 | 185 | 2,851 | 2,448.9 | 175.6 | 499 | 9,181 | 2,697.4 | 274.6 | 199 | 629 | 164.9 | 24.6 |
| Blacksburg-Christiansburg, VA | 14.91 | 89.8 | 61 | 891 | 531.7 | 39.9 | 503 | 9,562 | 1,783.6 | 183.3 | 150 | 735 | 144.2 | 25.9 |
| Bloomington, IL | 10.50 | 60.6 | 165 | 2,108 | 7,165.7 | 150.4 | 523 | 9,562 | 2,301.8 | 211.7 | 154 | 754 | 151.1 | 26.2 |
| Bloomington, IN | 17.18 | 103.8 | D | D | D | D | 483 | 7,038 | 1,982.4 | 169.8 | 199 | 1,049 | 194.9 | 37.0 |
| Bloomsburg-Berwick, PA | 5.05 | 59.3 | 49 | 563 | 265.2 | 25.5 | 279 | 4,011 | 1,097.5 | 92.7 | 49 | 167 | 46.7 | 6.3 |
| Boise City, ID | 90.70 | 134.0 | 753 | 10,976 | 11,072.9 | 649.3 | 2,100 | 33,223 | 10,352.6 | 995.5 | 1,167 | 3,818 | 860.7 | 152.7 |
| Boston-Cambridge-Newton, MA-NH | 274.73 | 57.5 | 4,846 | 94,229 | 112,588.5 | 8,625.3 | 16,524 | 264,474 | 81,036.0 | 8,038.1 | 5,711 | 43,513 | 15,985.5 | 3,009.6 |
| Boston, MA Div 14454 | 103.97 | 52.4 | 1,923 | 36,261 | 38,799.9 | 3,222.8 | 6,851 | 110,079 | 34,986.3 | 3,436.8 | 2,713 | 23,752 | 8,255.2 | 1,773.6 |
| Cambridge-Newton-Framingham, MA Div 15754 | 147.41 | 62.4 | 2,371 | 49,207 | 65,951.0 | 4,717.3 | 7,686 | 121,358 | 36,306.9 | 3,682.6 | 2,534 | 17,484 | 7,134.8 | 1,125.0 |
| Rockingham County-Strafford County, NH Div 40484 | 23.35 | 54.5 | 552 | 8,761 | 7,837.7 | 685.2 | 1,987 | 33,037 | 9,742.8 | 918.7 | 464 | 2,277 | 595.5 | 111.0 |
| Boulder, CO | 48.30 | 151.2 | 428 | 6,066 | 4,884.6 | 516.0 | 1,187 | 19,230 | 5,635.2 | 602.1 | 775 | 2,894 | 754.0 | 161.8 |
| Bowling Green, KY | 21.30 | 126.5 | 153 | 2,252 | 2,539.9 | 115.3 | 605 | 8,751 | 2,192.4 | 209.2 | 154 | 648 | 124.9 | 20.2 |
| Bremerton-Silverdale-Port Orchard, WA | 19.63 | 75.5 | 131 | 978 | 500.7 | 54.1 | 739 | 11,457 | 3,319.1 | 355.0 | 405 | 1,132 | 304.1 | 45.8 |
| Bridgeport-Stamford-Norwalk, CT | 87.98 | 92.8 | 1,131 | 19,293 | 66,731.7 | 1,916.5 | 3,368 | 51,210 | 16,458.8 | 1,717.6 | 1,165 | 7,605 | 2,604.5 | 536.9 |
| Brownsville-Harlingen, TX | 24.18 | 57.3 | 320 | 2,851 | 2,163.4 | 101.5 | 1,045 | 17,938 | 4,610.8 | 437.5 | 336 | 1,482 | 226.5 | 40.2 |
| Brunswick, GA | 10.59 | 91.3 | 93 | 857 | 784.5 | 33.7 | 551 | 6,183 | 1,660.7 | 144.5 | 165 | 510 | 98.0 | 18.0 |
| Buffalo-Cheektowaga, NY | 220.89 | 194.6 | 1,117 | 19,586 | 24,536.0 | 1,127.8 | 4,016 | 64,323 | 16,340.6 | 1,614.3 | 1,053 | 7,429 | 1,482.3 | 284.0 |
| Burlington, NC | 15.35 | 97.0 | 129 | 2,065 | 814.5 | 115.5 | 642 | 9,115 | 2,313.5 | 206.7 | 134 | 849 | 213.1 | 38.7 |
| Burlington-South Burlington, VT | 17.39 | 80.1 | 285 | 4,899 | 3,722.5 | 276.8 | 1,014 | 14,800 | 4,233.9 | 424.6 | 290 | 1,165 | 347.2 | 52.1 |
| California-Lexington Park, MD | 4.15 | 37.2 | 37 | D | 131.0 | D | 306 | 4,740 | 1,370.3 | 125.0 | 88 | 274 | 95.6 | 11.5 |
| Canton-Massillon, OH | 31.28 | 77.6 | 340 | 4,767 | 3,948.8 | 249.0 | 1,279 | 20,612 | 5,592.8 | 506.6 | 287 | 1,487 | 308.8 | 63.3 |
| Cape Coral-Fort Myers, FL | 64.55 | 92.0 | 638 | 5,832 | 3,164.5 | 278.6 | 2,651 | 39,847 | 12,227.8 | 1,071.9 | 1,562 | 6,205 | 1,535.0 | 384.0 |
| Cape Girardeau, MO-IL | 9.63 | 98.7 | 105 | 1,081 | 662.7 | 60.5 | 458 | 6,380 | 1,661.0 | 157.4 | 115 | 369 | 73.3 | 12.7 |
| Carbondale-Marion, IL | 8.04 | 57.6 | 89 | 924 | 434.2 | 40.3 | 498 | 7,434 | 2,171.0 | 181.5 | 141 | 547 | 98.9 | 14.0 |
| Carson City, NV | 11.5 | 210.9 | 81 | 620 | 335.2 | 29.4 | 220 | 3,141 | 1,170.8 | 102.3 | 118 | 361 | 94.1 | 13.0 |
| Casper, WY | 12.26 | 149.2 | 164 | 1,812 | 2,888.1 | 115.8 | 344 | 4,766 | 1,386.0 | 139.6 | 181 | 952 | 285.7 | 54.1 |

1. Merchant wholesalers, except manufacturers' sales branches and offices. 2. Employer establishments.

# Table C. Metropolitan Areas — Professional Services, Manufacturing, and Accommodation and Food Services

| Area name | Professional, scientific, and technical services, 2017 | | | | Manufacturing, 2017 | | | | Accommodation and food services, 2017 | | | |
|---|---|---|---|---|---|---|---|---|---|---|---|---|
| | Number of establishments | Number of employees | Sales (mil dol) | Average payroll (mil dol) | Number of establishments | Number of employees | Receipts (mil dol) | Annual payroll (mil dol) | Number of establishments | Number of employees | Receipts (mil dol) | Annual payroll (mil dol) |
| | 147 | 148 | 149 | 150 | 151 | 152 | 153 | 154 | 155 | 156 | 157 | 158 |
| Abilene, TX | 24 | 66 | 9.2 | 2.9 | 112 | 2,507 | 1,160.5 | 125.9 | 363 | 7,425 | 381.4 | 111.3 |
| Akron, OH | D | D | D | D | 1,022 | 38,621 | 12,824.8 | 2,082.7 | 1,575 | 29,252 | 1,458.6 | 414.3 |
| Albany, GA | 270 | 2,368 | 295.6 | 114.1 | 91 | 4,240 | 3,307.4 | 228.9 | 246 | 4,753 | 255.2 | 68.2 |
| Albany-Lebanon, OR | D | D | D | D | 193 | 7,359 | 2,494.8 | 417.0 | 230 | 3,278 | 175.6 | 53.3 |
| Albany-Schenectady-Troy, NY | 1,213 | 14,825 | 2,673.3 | 1,101.2 | 588 | 23,895 | 9,803.5 | 1,716.4 | 2,381 | 35,843 | 2,260.6 | 689.5 |
| Albuquerque, NM | 84 | 381 | 40.8 | 14.6 | 663 | 14,320 | 5,185.5 | 792.5 | 1,784 | 41,656 | 2,461.9 | 723.2 |
| Alexandria, LA | 277 | 1,826 | 242.1 | 86.4 | D | D | D | D | 250 | 5,012 | 271.7 | 74.8 |
| Allentown-Bethlehem-Easton, PA-NJ | 841 | 7,585 | 1,335.5 | 510.7 | 845 | 35,697 | 18,064.1 | 2,147.8 | 1,858 | 30,183 | 2,198.8 | 538.7 |
| Altoona, PA | 204 | 1,753 | 261.3 | 87.4 | 130 | 6,802 | 2,282.2 | 344.6 | 286 | 4,683 | 230.9 | 63.2 |
| Amarillo, TX | 232 | 993 | 135.2 | 51.7 | 187 | 11,471 | 5,480.2 | 717.4 | 599 | 12,283 | 689.2 | 185.4 |
| Ames, IA | 38 | 134 | 16.5 | 6.3 | 110 | 5,075 | 3,409.2 | 297.1 | 290 | 5,605 | 267.3 | 76.3 |
| Anchorage, AK | 1,342 | 14,840 | 2,671.2 | 1,080.8 | 228 | 2,021 | 476.5 | 106.2 | 1,034 | 18,108 | 1,440.2 | 438.0 |
| Ann Arbor, MI | 1,257 | 15,568 | 2,912.8 | 1,205.4 | 337 | 14,480 | 4,603.1 | 850.4 | 860 | 18,272 | 1,027.3 | 310.7 |
| Anniston-Oxford, AL | D | D | D | D | 106 | 5,986 | 2,699.2 | 308.1 | 225 | 4,642 | 247.0 | 66.2 |
| Appleton, WI | 430 | 4,081 | 640.5 | 255.1 | 400 | 23,347 | 9,136.1 | 1,351.7 | 566 | 10,318 | 495.2 | 140.4 |
| Asheville, NC | 1,442 | 6,046 | 802.3 | 313.7 | 487 | 20,382 | 7,509.1 | 1,107.5 | 1,261 | 25,293 | 1,558.1 | 459.3 |
| Athens-Clarke County, GA | 525 | 2,544 | 325.2 | 118.9 | 151 | 8,203 | 2,993.2 | 389.1 | 468 | 9,915 | 488.3 | 136.4 |
| Atlanta-Sandy Springs-Alpharetta, GA | 21,301 | 210,839 | 47,674.2 | 17,302.2 | 3,850 | 145,460 | 65,430.9 | 7,923.9 | 12,324 | 255,597 | 16,419.7 | 4,456.4 |
| Atlantic City-Hammonton, NJ | D | D | D | D | 108 | 2,013 | 474.1 | 98.6 | 830 | 35,488 | 3,550.7 | 1,003.8 |
| Auburn-Opelika, AL | 254 | 1,277 | 163.5 | 56.6 | 118 | 7,008 | 2,707.0 | 317.6 | 363 | 7,591 | 371.4 | 100.0 |
| Augusta-Richmond County, GA-SC | 908 | 8,086 | 1,387.9 | 473.4 | 303 | 21,290 | 11,520.5 | 1,255.5 | 1,077 | 21,741 | 1,145.8 | 306.4 |
| Austin-Round Rock-Georgetown, TX | 587 | 2,777 | 381.3 | 128.3 | 1,400 | 39,849 | 15,609.2 | 2,649.7 | 4,899 | 114,507 | 7,768.0 | 2,219.4 |
| Bakersfield, CA | 1,189 | 9,602 | 1,542.5 | 569.4 | 376 | 13,484 | 6,938.0 | 678.5 | 1,467 | 23,929 | 1,507.6 | 410.3 |
| Baltimore-Columbia-Towson, MD | 4,101 | 59,850 | 13,762.4 | 5,751.0 | 1,538 | 53,613 | 23,685.5 | 3,651.5 | 5,935 | 117,407 | 8,180.4 | 2,212.2 |
| Bangor, ME | D | D | D | D | 147 | 2,881 | 686.3 | 142.8 | 332 | 6,001 | 365.7 | 102.4 |
| Barnstable Town, MA | D | D | D | D | 188 | 2,046 | 645.3 | 129.0 | 1,112 | 14,179 | 1,266.5 | 386.3 |
| Baton Rouge, LA | 376 | 3,476 | 465.2 | 228.9 | 577 | D | 55,851.9 | D | 1,624 | 35,832 | 2,060.9 | 562.0 |
| Battle Creek, MI | D | D | D | D | 149 | 14,437 | 6,760.7 | 781.4 | 272 | 6,655 | 567.1 | 126.1 |
| Bay City, MI | 143 | 1,201 | 100.4 | 51.5 | 116 | 3,824 | 1,209.0 | 242.1 | 225 | 3,858 | 176.3 | 50.1 |
| Beaumont-Port Arthur, TX | 129 | 857 | 111.7 | 55.1 | D | 18,694 | D | 1,835.3 | 744 | 13,941 | 763.1 | 204.1 |
| Beckley, WV | D | D | D | D | 70 | 1,160 | 414.4 | 57.5 | 221 | 4,330 | 265.9 | 77.8 |
| Bellingham, WA | D | D | D | D | 341 | 11,106 | 11,178.0 | 628.0 | 554 | 8,971 | 609.1 | 191.8 |
| Bend, OR | D | D | D | D | 342 | 4,828 | 1,107.8 | 226.0 | 608 | 10,178 | 707.1 | 224.8 |
| Billings, MT | 25 | 42 | 5.4 | 239.6 | D | 3,727 | D | 255.6 | 496 | 9,935 | 578.4 | 165.0 |
| Binghamton, NY | D | D | D | D | 200 | 8,335 | 2,922.9 | 499.8 | 626 | 9,694 | 493.1 | 152.0 |
| Birmingham-Hoover, AL | 190 | 1,023 | 130.0 | 357.8 | 868 | 34,725 | 12,516.6 | 1,901.6 | 2,131 | 44,349 | 2,481.8 | 707.6 |
| Bismarck, ND | D | D | D | D | D | D | D | D | 210 | 5,516 | 296.9 | 96.8 |
| Blacksburg-Christiansburg, VA | 43 | 571 | 41.8 | 24.4 | 111 | 11,362 | 5,035.5 | 656.4 | 320 | 6,708 | 323.9 | 90.8 |
| Bloomington, IL | D | D | D | D | 81 | 4,592 | 993.1 | 185.3 | 417 | 8,518 | 413.1 | 123.6 |
| Bloomington, IN | 298 | 1,937 | 250.6 | 98.4 | 125 | 8,487 | 1,619.3 | 422.7 | 417 | 8,381 | 437.1 | 123.0 |
| Bloomsburg-Berwick, PA | D | D | D | D | 92 | 4,983 | 1,941.2 | 213.1 | 206 | 3,460 | 172.0 | 50.6 |
| Boise City, ID | 1,922 | 13,849 | 2,195.9 | 819.9 | D | D | D | 1,810.3 | 1,355 | 26,863 | 1,411.2 | 412.8 |
| Boston-Cambridge-Newton, MA-NH | 16,402 | 256,642 | 67,699.7 | 28,023.6 | 4,313 | 158,541 | 57,287.4 | 11,097.6 | 12,560 | 235,954 | 17,957.0 | 5,356.5 |
| Boston, MA Div 14454 | 7,420 | 116,086 | 31,414.4 | 12,268.2 | 1,333 | 39,241 | 14,957.4 | 2,429.6 | 5,501 | 115,221 | 9,622.9 | 2,833.5 |
| Cambridge-Newton-Framingham, MA Div 15754 | 8,982 | 140,556 | 36,285.2 | 15,755.4 | 2,406 | 98,324 | 35,530.4 | 7,493.2 | 5,857 | 101,292 | 7,093.4 | 2,155.3 |
| Rockingham County-Strafford County, NH Div 40484 | D | D | D | D | 574 | 20,976 | 6,799.6 | 1,174.8 | 1,202 | 19,441 | 1,240.7 | 367.7 |
| Boulder, CO | D | D | D | D | 561 | 14,472 | 5,400.1 | 939.4 | 939 | 19,279 | 1,157.9 | 360.1 |
| Bowling Green, KY | 26 | 122 | 10.1 | 3.2 | D | D | D | D | 316 | 7,217 | 372.6 | 113.3 |
| Bremerton-Silverdale-Port Orchard, WA | 703 | 4,112 | 617.5 | 247.4 | 150 | 1,996 | 379.0 | 96.4 | 536 | 9,262 | 640.9 | 191.9 |
| Bridgeport-Stamford-Norwalk, CT | 3,519 | 47,964 | 10,735.9 | 4,661.4 | 779 | 34,375 | 12,255.6 | 2,634.2 | 2,438 | 35,172 | 2,597.3 | 767.7 |
| Brownsville-Harlingen, TX | D | D | D | D | 188 | 3,983 | 1,301.2 | 161.2 | 735 | 15,258 | 753.2 | 204.0 |
| Brunswick, GA | 23 | 56 | 6.0 | 2.0 | 63 | 2,125 | 1,025.6 | 146.6 | 331 | 8,159 | 622.8 | 173.0 |
| Buffalo-Cheektowaga, NY | D | D | D | D | 1,236 | 52,250 | 21,609.3 | 3,227.4 | 2,925 | 55,202 | 3,208.1 | 929.3 |
| Burlington, NC | D | D | D | D | 193 | 8,100 | 2,841.6 | 380.2 | 326 | 7,220 | 365.7 | 103.2 |
| Burlington-South Burlington, VT | 816 | 7,045 | 1,230.9 | 514.5 | 284 | 11,948 | 4,160.3 | 757.5 | 584 | 10,326 | 694.7 | 206.1 |
| California-Lexington Park, MD | D | D | D | D | 27 | 409 | 117.5 | 24.1 | 190 | 3,797 | 215.6 | 60.3 |
| Canton-Massillon, OH | 29 | 151 | 16.6 | 6.5 | 510 | 25,471 | 11,542.1 | 1,345.2 | 845 | 16,282 | 774.8 | 222.5 |
| Cape Coral-Fort Myers, FL | D | D | 2,197.4 | D | 366 | 5,117 | 1,176.3 | 249.1 | 1,432 | 31,343 | 1,925.7 | 566.6 |
| Cape Girardeau, MO-IL | D | D | D | D | D | 3,984 | D | 212.2 | 187 | 4,214 | 197.1 | 59.8 |
| Carbondale-Marion, IL | 10 | 160 | 4.7 | 1.3 | 98 | 4,156 | 1,430.5 | 182.6 | 308 | 5,619 | 251.8 | 71.7 |
| Carson City, NV | D | D | D | D | 103 | 2,704 | 714.3 | 159.4 | 170 | 3,036 | 193.0 | 54.1 |
| Casper, WY | 266 | 1,440 | 266.6 | 89.1 | 89 | 1,379 | 1,002.2 | 83.6 | 201 | 3,819 | 211.1 | 65.8 |

| Area name | Health care and social assistance, 2017 | | | | Other services, 2017 | | | | Nonemployer businesses, 2018 | | Value of residential construction authorized by building permits, 2020 | |
|---|---|---|---|---|---|---|---|---|---|---|---|---|
| | Number of establishments | Number of employees | Receipts (mil dol) | Annual payroll (mil dol) | Number of establishments | Number of employees | Receipts (mil dol) | Annual payroll (mil dol) | Number | Receipts (mil dol) | New construction ($1,000) | Number of housing units |
| | 159 | 160 | 161 | 162 | 163 | 164 | 165 | 166 | 167 | 168 | 169 | 170 |
| Abilene, TX | 434 | 12,773 | 1,401.2 | 536.7 | 255 | 1,576 | 200 | 46 | 13,009 | 618.7 | 107,019 | 663 |
| Akron, OH | 1,783 | 60,429 | 6,020.0 | 2,700.8 | 1,327 | 8,961 | 1,695 | 286 | 49,490 | 2,253.4 | 367,068 | 914 |
| Albany, GA | 354 | 8,868 | 1,191.3 | 471.6 | 184 | 1,209 | 118 | 36 | 11,015 | 372.1 | 45,419 | 367 |
| Albany-Lebanon, OR | 230 | 5,366 | 575.8 | 238.0 | 163 | 1,004 | 74 | 25 | 6,357 | 287.2 | 153,188 | 796 |
| Albany-Schenectady-Troy, NY | 2,431 | 67,782 | 7,672.9 | 3,101.2 | 1,446 | 9,469 | 1,457 | 370 | 56,557 | 2,693.8 | 434,114 | 2,046 |
| Albuquerque, NM | 2,317 | 58,898 | 6,906.1 | 2,683.5 | 1,295 | 8,297 | 942 | 268 | 54,919 | 2,322.8 | 495,033 | 2,014 |
| Alexandria, LA | 561 | 13,947 | 1,731.7 | 634.7 | 189 | 1,068 | 129 | 34 | 9,300 | 440.1 | 67,604 | 313 |
| Allentown-Bethlehem-Easton, PA-NJ | 2,408 | 63,972 | 7,886.8 | 3,146.6 | 1,616 | 9,336 | 943 | 281 | 54,274 | 2,618.1 | 307,274 | 1,688 |
| Altoona, PA | 452 | 12,918 | 1,345.0 | 564.9 | 269 | 1,459 | 123 | 38 | 6,424 | 304.4 | 19,639 | 96 |
| Amarillo, TX | 474 | 13,448 | 1,852.9 | 656.2 | 444 | 3,246 | 405 | 105 | 20,525 | 1,071.7 | 204,962 | 1,022 |
| Ames, IA | 270 | 7,715 | 816.3 | 337.6 | 206 | 1,310 | 294 | 42 | 7,422 | 309.6 | 71,554 | 270 |
| Anchorage, AK | 1,622 | 31,851 | 5,052.0 | 1,892.6 | 716 | 4,523 | 988 | 171 | 28,632 | 1,490.6 | 284,726 | 1,030 |
| Ann Arbor, MI | 943 | 40,974 | 5,693.3 | 2,382.4 | 553 | 4,465 | 651 | 195 | 30,716 | 1,382.0 | 239,098 | 1,065 |
| Anniston-Oxford, AL | 275 | 6,358 | 622.7 | 249.8 | 166 | 718 | 79 | 20 | 6,811 | 263.3 | 14,985 | 102 |
| Appleton, WI | 591 | 13,914 | 1,533.6 | 670.8 | 423 | 3,128 | 369 | 106 | 13,239 | 676.3 | 260,767 | 1,311 |
| Asheville, NC | 1,452 | 38,066 | 4,682.5 | 1,883.1 | 830 | 4,487 | 499 | 143 | 45,709 | 1,977.1 | 798,933 | 3,432 |
| Athens-Clarke County, GA | 650 | 12,411 | 1,876.0 | 711.5 | 251 | 1,848 | 341 | 68 | 17,012 | 702.7 | 266,739 | 1,218 |
| Atlanta-Sandy Springs-Alpharetta, GA | 14,829 | 284,534 | 40,118.3 | 15,093.9 | 9,366 | 65,573 | 10,639 | 2,358 | 629,914 | 26,677.9 | 7,156,853 | 32,346 |
| Atlantic City-Hammonton, NJ | 801 | 18,708 | 2,202.5 | 951.0 | 540 | 3,437 | 291 | 84 | 17,834 | 887.3 | 155,744 | 667 |
| Auburn-Opelika, AL | 258 | 6,926 | 697.5 | 276.4 | 177 | 1,068 | 92 | 26 | 10,729 | 485.6 | 488,947 | 2,221 |
| Augusta-Richmond County, GA-SC | 1,322 | 37,218 | 5,008.1 | 1,818.1 | 697 | 4,260 | 534 | 140 | 40,947 | 1,509.0 | 710,662 | 3,556 |
| Austin-Round Rock-Georgetown, TX | 5,478 | 109,967 | 13,201.3 | 5,165.8 | 3,724 | 28,586 | 3,938 | 1,142 | 212,611 | 11,207.6 | 7,147,718 | 42,264 |
| Bakersfield, CA | 1,684 | 30,531 | 4,574.8 | 1,580.8 | 865 | 6,057 | 671 | 199 | 49,553 | 2,607.6 | 571,907 | 2,502 |
| Baltimore-Columbia-Towson, MD | 7,848 | 209,638 | 27,814.0 | 10,620.7 | 5,042 | 38,293 | 4,951 | 1,428 | 219,643 | 10,187.4 | 1,609,153 | 8,094 |
| Bangor, ME | 541 | 15,666 | 1,719.6 | 776.9 | 274 | 1,425 | 190 | 45 | 27,281 | 1,451.8 | 274,179 | 337 |
| Barnstable Town, MA | 795 | 17,032 | 2,064.8 | 858.8 | 627 | 3,498 | 396 | 126 | 68,952 | 3,018.2 | 834,122 | 592 |
| Baton Rouge, LA | 1,940 | 49,816 | 5,740.8 | 2,076.4 | 1,306 | 9,608 | 1,466 | 398 | 6,565 | 255.1 | 21,785 | 3,780 |
| Battle Creek, MI | 321 | 9,318 | 1,138.8 | 443.5 | 215 | 1,304 | 520 | 57 | 5,702 | 229.2 | 23,849 | 90 |
| Bay City, MI | 339 | 6,674 | 706.0 | 257.4 | 168 | 900 | 81 | 23 | 25,525 | 1,141.9 | 288,508 | 135 |
| Beaumont-Port Arthur, TX | 1,085 | 20,020 | 2,128.6 | 749.5 | 509 | 4,603 | 501 | 200 | 5,161 | 208.7 | 11,281 | 1,753 |
| Beckley, WV | 411 | 9,236 | 947.0 | 378.2 | 165 | 1,031 | 128 | 31 | 16,490 | 819.5 | 288,718 | 75 |
| Bellingham, WA | 699 | 10,929 | 1,334.0 | 469.1 | 431 | 2,552 | 303 | 92 | 19,388 | 1,007.9 | 545,755 | 1,382 |
| Bend, OR | 744 | 12,299 | 1,654.6 | 660.7 | 397 | 2,156 | 262 | 71 | 14,376 | 726.2 | 184,112 | 2,338 |
| Billings, MT | 601 | 14,926 | 2,089.2 | 816.2 | 397 | 2,244 | 292 | 77 | 12,672 | 525.2 | 58,563 | 1,435 |
| Binghamton, NY | 551 | 17,101 | 1,823.3 | 746.7 | 404 | 1,987 | 187 | 50 | 82,023 | 3,880.9 | 1,036,613 | 446 |
| Birmingham-Hoover, AL | 2,512 | 73,121 | 11,087.9 | 4,016.8 | 1,452 | 11,679 | 2,664 | 470 | 10,406 | 574.9 | 147,301 | 3,845 |
| Bismarck, ND | 371 | 11,838 | 1,366.2 | 536.3 | 336 | 2,028 | 294 | 82 | 8,449 | 338.5 | 309,914 | 715 |
| Blacksburg-Christiansburg, VA | 361 | 6,534 | 881.7 | 285.1 | 242 | 1,290 | 351 | 41 | 9,646 | 404.9 | 49,950 | 1,748 |
| Bloomington, IL | 377 | 8,713 | 1,061.4 | 402.1 | 272 | 2,436 | 263 | 85 | 11,126 | 445.1 | 119,447 | 265 |
| Bloomington, IN | 453 | 11,112 | 1,013.8 | 435.5 | 239 | 1,599 | 436 | 57 | | | | 650 |
| Bloomsburg-Berwick, PA | 257 | 11,063 | 1,950.6 | 742.1 | 148 | 802 | 107 | 22 | 4,370 | 199.3 | 20,741 | 90 |
| Boise City, ID | 2,026 | 44,022 | 4,952.3 | 2,118.5 | 1,101 | 6,319 | 727 | 209 | 58,236 | 2,715.7 | 2,143,718 | 9,762 |
| Boston-Cambridge-Newton, MA-NH | 13,841 | 461,435 | 54,542.1 | 23,902.5 | 10,971 | 77,747 | 10,648 | 2,909 | 424,719 | 24,144.3 | 3,885,306 | 14,248 |
| Boston, MA Div 14454 | 5,650 | 251,894 | 31,442.8 | 13,903.8 | 4,718 | 38,377 | 5,689 | 1,464 | 170,218 | 9,838.1 | 2,408,100 | 6,963 |
| Cambridge-Newton-Framingham, MA Div 15754 | 6,986 | 185,147 | 20,047.2 | 8,781.6 | 5,274 | 33,717 | 4,356 | 1,263 | 218,452 | 12,025.9 | 1,136,063 | 5,802 |
| Rockingham County-Strafford County, NH Div 40484 | 1,205 | 24,394 | 3,052.0 | 1,217.1 | 979 | 5,653 | 603 | 182 | 36,049 | 2,280.3 | 341,144 | 1,483 |
| Boulder, CO | 1,432 | 21,332 | 2,566.9 | 1,012.1 | 823 | 5,173 | 672 | 202 | 40,829 | 2,155.5 | 405,861 | 1,634 |
| Bowling Green, KY | 400 | 9,097 | 1,196.9 | 438.1 | 256 | 1,415 | 130 | 36 | 13,272 | 734.8 | 179,992 | 978 |
| Bremerton-Silverdale-Port Orchard, WA | 716 | 12,346 | 1,619.1 | 581.2 | 423 | 2,230 | 225 | 67 | 15,583 | 705.4 | 318,102 | 1,285 |
| Bridgeport-Stamford-Norwalk, CT | 3,071 | 72,290 | 10,942.1 | 3,753.5 | 2,180 | 14,442 | 2,103 | 531 | 97,854 | 7,149.9 | 492,509 | 1,862 |
| Brownsville-Harlingen, TX | 1,029 | 37,028 | 2,310.7 | 987.7 | 406 | 2,151 | 199 | 50 | 32,692 | 1,296.9 | 208,366 | 1,904 |
| Brunswick, GA | 257 | 5,461 | 670.2 | 260.3 | 172 | 1,012 | 108 | 25 | 9,157 | 411.1 | 165,994 | 542 |
| Buffalo-Cheektowaga, NY | 3,234 | 90,411 | 9,931.9 | 4,158.6 | 2,196 | 13,465 | 1,324 | 388 | 63,047 | 2,788.2 | 378,741 | 1,444 |
| Burlington, NC | 366 | 12,073 | 1,705.6 | 651.3 | 199 | 1,292 | 126 | 36 | 10,616 | 408.9 | 267,600 | 1,784 |
| Burlington-South Burlington, VT | 716 | 18,250 | 2,385.0 | 838.2 | 533 | 2,743 | 350 | 96 | 19,786 | 969.1 | 190,262 | 1,103 |
| California-Lexington Park, MD | 181 | 4,735 | 528.0 | 201.5 | 140 | 816 | 83 | 30 | 6,998 | 297.4 | 90,238 | 555 |
| Canton-Massillon, OH | 1,038 | 29,386 | 2,819.5 | 1,172.8 | 685 | 4,872 | 606 | 156 | 26,401 | 1,148.8 | 145,927 | 690 |
| Cape Coral-Fort Myers, FL | 1,700 | 38,678 | 5,215.6 | 2,158.8 | 1,425 | 7,305 | 834 | 223 | 73,304 | 3,642.3 | 1,944,757 | 10,673 |
| Cape Girardeau, MO-IL | 332 | 10,265 | 1,317.0 | 480.6 | D | D | D | D | 6,484 | 288.2 | 27,912 | 141 |
| Carbondale-Marion, IL | 384 | 11,627 | 1,592.0 | 501.9 | 189 | 1,035 | 112 | 24 | 8,064 | 303.3 | 24,737 | 183 |
| Carson City, NV | 215 | 3,865 | 623.8 | 214.4 | 132 | 618 | 101 | 22 | 4,752 | 392.9 | 57,589 | 263 |
| Casper, WY | 333 | 5,588 | 751.8 | 310.6 | 198 | 976 | 140 | 35 | 6,254 | 329.3 | 40,548 | 208 |

# Table C. Metropolitan Areas — Government Employment and Payroll, and Local Government Finances

| | Government employment and payroll, 2017 | | | | | | | | | Local government finances, 2017 | | | | |
| | | | March payroll (percent of total) | | | | | | | General revenue | | | | |
| Area name | Full-time equivalent employees | March payroll (dollars) | Adminis-tration, judicial, and legal | Police and corrections | Fire protection | Highways and transpor-tation | Health and welfare | Natural resources and utilities | Education and libraries | Total (mil dol) | Inter-govern-mental (mil dol) | Taxes | | |
| | | | | | | | | | | | | Total (mil dol) | Per capita[1] (dollars) | |
| | | | | | | | | | | | | | Total | Property |
| | 171 | 172 | 173 | 174 | 175 | 176 | 177 | 178 | 179 | 180 | 181 | 182 | 183 | 184 |
|---|---|---|---|---|---|---|---|---|---|---|---|---|---|---|
| Abilene, TX | 6,870 | 24,911,989 | 6.2 | 12.5 | 4.8 | 1.9 | 6.4 | 5.2 | 61.5 | 586 | 226 | 273 | 1,599 | 1,176 |
| Akron, OH | 25,033 | 109,876,388 | 7.8 | 10.3 | 5.9 | 5.9 | 8.4 | 6.2 | 54.0 | 3,118 | 1,024 | 1,538 | 2,185 | 1,333 |
| Albany, GA | 6,840 | 22,227,686 | 6.2 | 10.6 | 4.3 | 2.9 | 5.8 | 7.0 | 60.8 | 545 | 223 | 210 | 1,415 | 922 |
| Albany-Lebanon, OR | 4,158 | 19,077,155 | 5.7 | 11.1 | 5.7 | 3.7 | 6.5 | 3.4 | 62.8 | 542 | 265 | 174 | 1,392 | 1,262 |
| Albany-Schenectady-Troy, NY | 36,832 | 183,705,619 | 5.0 | 10.3 | 2.3 | 4.1 | 7.6 | 3.3 | 65.7 | 5,270 | 1,786 | 2,772 | 3,142 | 2,143 |
| Albuquerque, NM | 30,008 | 123,837,946 | 6.5 | 15.7 | 6.3 | 6.0 | 3.4 | 6.1 | 54.0 | 3,376 | 1,620 | 1,221 | 1,338 | 773 |
| Alexandria, LA | 6,161 | 19,052,868 | 8.9 | 13.7 | 5.2 | 3.3 | 1.0 | 6.7 | 59.7 | 477 | 215 | 209 | 1,362 | 526 |
| Allentown-Bethlehem-Easton, PA-NJ | 29,796 | 143,303,661 | 6.1 | 10.1 | 1.5 | 3.8 | 7.0 | 3.4 | 66.0 | 4,355 | 1,587 | 2,065 | 2,464 | 2,061 |
| Altoona, PA | 3,575 | 13,827,996 | 5.4 | 9.6 | 2.2 | 5.1 | 4.9 | 5.8 | 65.7 | 456 | 222 | 145 | 1,178 | 799 |
| Amarillo, TX | 11,792 | 45,156,680 | 5.4 | 12.9 | 4.5 | 1.7 | 3.5 | 3.9 | 66.4 | 1,042 | 369 | 484 | 1,833 | 1,386 |
| Ames, IA | 4,009 | 17,651,859 | 6.0 | 6.9 | 2.2 | 6.9 | 15.5 | 8.7 | 51.4 | 655 | 151 | 197 | 1,591 | 1,355 |
| Anchorage, AK | 12,481 | 69,716,457 | 5.1 | 7.5 | 5.7 | 4.0 | 3.5 | 8.2 | 64.0 | 1,937 | 906 | 770 | 1,923 | 1,635 |
| Ann Arbor, MI | 10,291 | 50,170,923 | 7.5 | 9.9 | 2.6 | 5.9 | 4.7 | 5.2 | 61.5 | 1,624 | 630 | 665 | 1,804 | 1,745 |
| Anniston-Oxford, AL | 5,653 | 18,814,816 | 3.4 | 5.0 | 2.7 | 2.5 | 37.3 | 5.5 | 41.1 | 532 | 161 | 130 | 1,135 | 389 |
| Appleton, WI | 8,858 | 40,302,052 | 4.7 | 7.2 | 2.9 | 3.4 | 7.7 | 3.4 | 69.0 | 1,003 | 479 | 324 | 1,375 | 1,319 |
| Asheville, NC | 16,586 | 62,893,469 | 4.9 | 7.8 | 2.5 | 1.1 | 24.5 | 4.0 | 51.2 | 2,348 | 984 | 651 | 1,430 | 984 |
| Athens-Clarke County, GA | 7,302 | 25,954,698 | 6.9 | 11.1 | 3.3 | 2.5 | 5.5 | 7.0 | 59.7 | 1,093 | 241 | 282 | 1,349 | 938 |
| Atlanta-Sandy Springs-Alpharetta, GA | 205,101 | 883,124,518 | 6.8 | 8.9 | 4.1 | 3.8 | 6.3 | 4.9 | 63.9 | 22,445 | 6,969 | 10,292 | 1,753 | 1,210 |
| Atlantic City-Hammonton, NJ | 13,889 | 75,732,427 | 5.6 | 14.7 | 5.0 | 2.0 | 4.6 | 4.0 | 62.7 | 1,644 | 578 | 930 | 3,503 | 3,445 |
| Auburn-Opelika, AL | 7,104 | 27,907,786 | 2.7 | 6.0 | 2.4 | 1.6 | 44.2 | 5.5 | 36.9 | 793 | 189 | 215 | 1,334 | 592 |
| Augusta-Richmond County, GA-SC | 21,283 | 71,438,497 | 7.4 | 10.3 | 3.9 | 2.7 | 3.7 | 5.7 | 65.0 | 1,927 | 741 | 809 | 1,348 | 909 |
| Austin-Round Rock-George-town, TX | 83,827 | 366,570,331 | 6.8 | 12.3 | 4.6 | 3.0 | 6.6 | 13.0 | 51.9 | 10,080 | 1,796 | 6,150 | 2,907 | 2,443 |
| Bakersfield, CA | 34,360 | 176,954,893 | 4.6 | 10.6 | 4.2 | 2.3 | 11.2 | 4.9 | 61.3 | 5,453 | 3,333 | 1,223 | 1,378 | 1,104 |
| Baltimore-Columbia-Towson, MD | 101,382 | 520,866,759 | 5.2 | 11.8 | 5.3 | 2.5 | 3.8 | 4.9 | 64.6 | 13,262 | 4,437 | 6,963 | 2,488 | 1,365 |
| Bangor, ME | 4,862 | 18,469,574 | 6.5 | 8.5 | 5.4 | 7.2 | 2.8 | 5.2 | 62.9 | 493 | 178 | 234 | 1,544 | 1,531 |
| Barnstable Town, MA | 8,286 | 44,597,245 | 6.6 | 11.6 | 9.9 | 4.4 | 3.2 | 6.7 | 55.0 | 1,120 | 177 | 766 | 3,587 | 3,359 |
| Baton Rouge, LA | 29,467 | 126,557,979 | 20.1 | 10.1 | 3.6 | 4.5 | 6.3 | 4.3 | 50.5 | 3,364 | 1,046 | 1,766 | 2,067 | 882 |
| Battle Creek, MI | 3,902 | 17,403,464 | 10.1 | 10.3 | 3.5 | 2.2 | 6.9 | 4.6 | 59.9 | 615 | 303 | 180 | 1,339 | 1,163 |
| Bay City, MI | 4,219 | 17,395,958 | 6.6 | 5.0 | 2.0 | 4.1 | 15.6 | 3.4 | 61.1 | 526 | 269 | 127 | 1,219 | 1,193 |
| Beaumont-Port Arthur, TX | 15,667 | 59,166,653 | 6.6 | 14.0 | 4.8 | 3.5 | 5.0 | 8.1 | 56.5 | 1,561 | 431 | 816 | 2,049 | 1,675 |
| Beckley, WV | 3,894 | 12,507,528 | 6.6 | 6.1 | 1.9 | 1.5 | 1.4 | 5.9 | 75.7 | 318 | 141 | 127 | 1,067 | 851 |
| Bellingham, WA | 5,979 | 33,006,077 | 9.4 | 9.9 | 8.3 | 8.0 | 3.6 | 5.1 | 52.1 | 923 | 348 | 388 | 1,751 | 1,024 |
| Bend, OR | 5,285 | 26,992,573 | 9.1 | 11.0 | 6.1 | 3.2 | 6.3 | 6.9 | 53.4 | 869 | 275 | 379 | 2,031 | 1,737 |
| Billings, MT | 5,489 | 24,373,446 | 6.3 | 9.4 | 4.2 | 4.3 | 7.6 | 6.2 | 57.5 | 648 | 229 | 239 | 1,330 | 1,243 |
| Binghamton, NY | 13,165 | 51,504,081 | 4.4 | 8.0 | 2.3 | 4.8 | 8.3 | 3.0 | 67.4 | 1,684 | 816 | 675 | 2,792 | 1,928 |
| Birmingham-Hoover, AL | 36,849 | 143,090,750 | 7.4 | 13.9 | 6.9 | 4.1 | 4.4 | 9.6 | 51.3 | 4,111 | 1,413 | 1,941 | 1,788 | 756 |
| Bismarck, ND | 4,212 | 19,062,240 | 5.0 | 10.3 | 2.8 | 4.0 | 5.9 | 8.4 | 60.7 | 562 | 256 | 197 | 1,535 | 1,196 |
| Blacksburg-Christiansburg, VA | 5,046 | 17,049,212 | 11.5 | 10.7 | 1.3 | 4.3 | 1.3 | 9.8 | 58.0 | 596 | 261 | 210 | 1,262 | 931 |
| Bloomington, IL | 6,093 | 27,474,898 | 6.9 | 11.6 | 5.5 | 5.0 | 3.2 | 7.0 | 58.0 | 711 | 161 | 422 | 2,440 | 1,942 |
| Bloomington, IN | 4,258 | 15,173,384 | 10.8 | 13.2 | 5.3 | 5.1 | 2.3 | 6.5 | 54.9 | 442 | 174 | 187 | 1,116 | 1,060 |
| Bloomsburg-Berwick, PA | 2,434 | 9,872,014 | 5.4 | 9.1 | 0.8 | 3.4 | 0.6 | 3.6 | 73.7 | 334 | 130 | 133 | 1,590 | 1,148 |
| Boise City, ID | 21,065 | 76,002,008 | 9.9 | 12.9 | 5.2 | 3.8 | 2.9 | 5.1 | 57.5 | 2,172 | 928 | 767 | 1,080 | 999 |
| Boston-Cambridge-Newton, MA-NH | 162,731 | 880,588,530 | 3.8 | 10.8 | 7.4 | 2.5 | 5.6 | 4.2 | 63.6 | 23,609 | 6,718 | 13,660 | 2,821 | 2,688 |
| Boston, MA Div 14454 | 65,674 | 377,921,263 | 3.5 | 13.1 | 8.4 | 2.3 | 4.8 | 4.6 | 61.5 | 10,121 | 2,946 | 5,898 | 2,925 | 2,741 |
| Cambridge-Newton-Framing-ham, MA Div 15754 | 80,196 | 434,251,985 | 3.8 | 9.0 | 6.9 | 2.7 | 6.6 | 4.0 | 64.8 | 11,517 | 3,328 | 6,461 | 2,705 | 2,599 |
| Rockingham County-Straf-ford County, NH Div 40484 | 16,861 | 68,415,282 | 5.0 | 10.2 | 5.6 | 2.5 | 3.8 | 2.9 | 67.7 | 1,972 | 445 | 1,301 | 2,980 | 2,933 |
| Boulder, CO | 12,053 | 61,752,806 | 8.2 | 10.5 | 5.2 | 2.6 | 6.2 | 8.9 | 54.2 | 1,955 | 451 | 1,118 | 3,466 | 2,375 |
| Bowling Green, KY | 5,736 | 18,566,807 | 2.6 | 8.9 | 3.6 | 2.0 | 5.1 | 8.4 | 65.8 | 421 | 178 | 188 | 1,077 | 536 |
| Bremerton-Silverdale-Port Orchard, WA | 7,352 | 40,302,464 | 7.9 | 7.7 | 8.9 | 7.4 | 3.0 | 6.6 | 57.4 | 1,280 | 653 | 429 | 1,609 | 1,052 |
| Bridgeport-Stamford-Norwalk, CT | 32,456 | 194,415,331 | 3.6 | 10.0 | 5.5 | 3.2 | 3.2 | 2.7 | 70.4 | 5,259 | 1,080 | 3,703 | 3,926 | 3,839 |
| Brownsville-Harlingen, TX | 22,268 | 74,927,211 | 4.8 | 8.6 | 2.8 | 2.5 | 2.0 | 4.4 | 73.6 | 1,808 | 1,027 | 494 | 1,169 | 939 |
| Brunswick, GA | 6,632 | 26,982,644 | 4.2 | 6.4 | 2.7 | 1.6 | 45.3 | 2.7 | 35.8 | 732 | 137 | 199 | 1,690 | 1,104 |
| Buffalo-Cheektowaga, NY | 46,386 | 236,482,046 | 3.8 | 10.1 | 2.7 | 3.2 | 12.6 | 4.5 | 61.7 | 7,259 | 3,000 | 3,091 | 2,736 | 1,656 |
| Burlington, NC | 5,370 | 20,215,032 | 4.4 | 10.8 | 3.3 | 1.7 | 8.5 | 5.8 | 62.4 | 468 | 227 | 184 | 1,128 | 793 |
| Burlington-South Burlington, VT | 8,694 | 38,325,724 | 5.6 | 5.5 | 2.0 | 7.5 | 1.3 | 6.0 | 71.4 | 980 | 629 | 186 | 848 | 731 |
| California-Lexington Park, MD | 3,228 | 15,729,729 | 6.4 | 10.5 | 0.0 | 2.2 | 1.2 | 4.4 | 72.5 | 423 | 140 | 222 | 1,975 | 1,009 |
| Canton-Massillon, OH | 15,123 | 58,262,135 | 7.9 | 8.4 | 4.0 | 4.7 | 8.4 | 5.0 | 60.4 | 1,457 | 646 | 590 | 1,477 | 1,020 |
| Cape Coral-Fort Myers, FL | 33,112 | 154,572,301 | 3.8 | 7.4 | 4.9 | 3.0 | 47.3 | 4.2 | 28.2 | 4,647 | 704 | 1,284 | 1,735 | 1,464 |
| Cape Girardeau, MO-IL | 3,369 | 9,729,545 | 5.8 | 9.1 | 3.8 | 5.8 | 0.8 | 7.7 | 63.8 | 285 | 101 | 142 | 1,470 | 810 |
| Carbondale-Marion, IL | 5,097 | 18,208,421 | 5.7 | 9.3 | 3.3 | 4.1 | 4.6 | 6.0 | 65.8 | 476 | 217 | 186 | 1,354 | 1,134 |
| Carson City, NV | 1,652 | 7,251,836 | 10.3 | 13.8 | 7.8 | 4.2 | 2.8 | 7.3 | 52.0 | 235 | 122 | 63 | 1,159 | 789 |
| Casper, WY | 3,954 | 16,298,613 | 5.7 | 8.4 | 3.7 | 2.6 | 1.2 | 5.2 | 71.3 | 502 | 318 | 97 | 1,220 | 987 |

1. Based on the resident population estimated as of July 1 of the year shown.

## Table C. Metropolitan Areas — Local Government Finances, Government Employment, and Income Taxes

| Area name | Direct general expenditure Total (mil dol) | Per capita[1] (dollars) | Pct Education | Pct Health and hospitals | Pct Police protection | Pct Public welfare | Pct Highways | Debt outstanding Total (mil dol) | Per capita[1] (dollars) | Govt empl. Federal civilian | Federal military | State and local | Income tax Number of returns | Mean adjusted gross income | Mean income tax |
|---|---|---|---|---|---|---|---|---|---|---|---|---|---|---|---|
| | 185 | 186 | 187 | 188 | 189 | 190 | 191 | 192 | 193 | 194 | 195 | 196 | 197 | 198 | 199 |
| Abilene, TX | 621.4 | 3,646 | 49.0 | 5.5 | 6.3 | 0.5 | 2.6 | 526.5 | 3,089 | 1,293 | 4,787 | 11,505 | 73,900 | 58,414 | 6,077 |
| Akron, OH | 3,141.4 | 4,462 | 44.0 | 5.0 | 5.8 | 4.2 | 4.1 | 2,384.6 | 3,387 | 2,325 | 1,766 | 42,370 | 345,830 | 65,154 | 7,557 |
| Albany, GA | 579.7 | 3,915 | 52.9 | 2.5 | 7.3 | 0.0 | 3.6 | 270.3 | 1,825 | 2,771 | 683 | 9,348 | 61,380 | 48,167 | 4,349 |
| Albany-Lebanon, OR | 578.9 | 4,630 | 53.2 | 5.0 | 7.4 | 0.3 | 4.8 | 417.7 | 3,341 | 327 | 307 | 6,356 | 57,160 | 56,383 | 4,998 |
| Albany-Schenectady-Troy, NY | 5,209.6 | 5,905 | 49.8 | 4.4 | 4.2 | 10.6 | 4.2 | 3,588.7 | 4,068 | 6,934 | 3,381 | 92,391 | 443,850 | 80,331 | 10,507 |
| Albuquerque, NM | 3,037.0 | 3,329 | 52.6 | 1.3 | 8.4 | 1.6 | 2.6 | 4,454.7 | 4,883 | 14,553 | 5,881 | 68,363 | 419,860 | 59,133 | 6,144 |
| Alexandria, LA | 527.6 | 3,436 | 52.8 | 0.0 | 10.0 | 0.0 | 2.6 | 473.5 | 3,084 | 2,824 | 567 | 10,763 | 62,380 | 56,173 | 5,538 |
| Allentown-Bethlehem-Easton, PA-NJ | 4,559.2 | 5,441 | 54.8 | 1.6 | 3.8 | 8.8 | 3.5 | 6,298.5 | 7,517 | 2,320 | 2,090 | 38,219 | 428,640 | 67,101 | 7,589 |
| Altoona, PA | 469.6 | 3,813 | 52.1 | 6.8 | 4.8 | 7.4 | 3.9 | 388.4 | 3,154 | 1,171 | 304 | 7,417 | 59,290 | 56,001 | 5,656 |
| Amarillo, TX | 1,046.4 | 3,959 | 54.1 | 4.5 | 6.8 | 0.0 | 2.9 | 1,051.0 | 3,976 | 2,334 | 629 | 18,298 | 117,190 | 63,869 | 7,377 |
| Ames, IA | 636.9 | 5,148 | 31.6 | 40.8 | 3.5 | 0.6 | 4.2 | 657.1 | 5,311 | 1,012 | 439 | 21,376 | 50,270 | 70,595 | 7,491 |
| Anchorage, AK | 1,915.0 | 4,781 | 50.9 | 1.9 | 7.5 | 0.1 | 7.9 | 2,063.7 | 5,152 | 8,556 | 13,192 | 23,753 | 192,410 | 77,861 | 9,702 |
| Ann Arbor, MI | 1,638.2 | 4,442 | 49.2 | 6.2 | 7.2 | 0.6 | 3.7 | 1,748.4 | 4,741 | 4,066 | 620 | 73,942 | 168,360 | 90,653 | 12,815 |
| Anniston-Oxford, AL | 618.0 | 5,387 | 32.3 | 36.7 | 5.3 | 0.0 | 3.4 | 461.6 | 4,024 | 4,174 | 515 | 8,593 | 47,160 | 48,504 | 3,924 |
| Appleton, WI | 1,119.5 | 4,747 | 54.0 | 2.6 | 5.0 | 4.8 | 9.0 | 929.2 | 3,941 | 687 | 613 | 12,296 | 120,320 | 70,062 | 7,609 |
| Asheville, NC | 2,078.3 | 4,569 | 30.6 | 30.2 | 6.2 | 6.6 | 1.5 | 989.0 | 2,174 | 3,809 | 1,058 | 22,292 | 222,350 | 62,050 | 6,710 |
| Athens-Clarke County, GA | 1,124.7 | 5,378 | 31.4 | 40.4 | 3.6 | 0.1 | 2.4 | 844.9 | 4,041 | 1,074 | 611 | 27,364 | 84,600 | 66,744 | 7,616 |
| Atlanta-Sandy Springs-Alpharetta, GA | 22,633.7 | 3,854 | 50.6 | 5.7 | 6.0 | 0.2 | 4.1 | 36,598.7 | 6,232 | 47,394 | 16,763 | 287,197 | 2,709,020 | 77,280 | 10,106 |
| Atlantic City-Hammonton, NJ | 1,640.6 | 6,179 | 58.1 | 1.2 | 6.3 | 1.4 | 4.6 | 847.7 | 3,193 | 2,477 | 841 | 19,912 | 134,970 | 58,260 | 6,138 |
| Auburn-Opelika, AL | 790.2 | 4,895 | 32.9 | 41.4 | 3.9 | 0.0 | 2.8 | 800.7 | 4,960 | 313 | 732 | 17,985 | 64,470 | 63,008 | 6,527 |
| Augusta-Richmond County, GA-SC | 1,915.1 | 3,190 | 56.9 | 2.7 | 5.6 | 0.1 | 4.2 | 2,056.1 | 3,425 | 9,388 | 13,156 | 41,087 | 259,000 | 58,127 | 5,689 |
| Austin-Round Rock-Georgetown, TX | 10,487.2 | 4,957 | 43.9 | 8.1 | 5.9 | 0.5 | 9.3 | 24,855.1 | 11,749 | 13,708 | 4,640 | 161,752 | 1,032,250 | 94,712 | 13,781 |
| Bakersfield, CA | 7,088.4 | 7,988 | 40.8 | 3.7 | 3.6 | 5.7 | 1.6 | 4,158.1 | 4,686 | 11,070 | 3,798 | 56,967 | 343,720 | 55,249 | 5,120 |
| Baltimore-Columbia-Towson, MD | 13,725.1 | 4,904 | 49.9 | 1.9 | 7.9 | 0.3 | 3.0 | 12,275.1 | 4,386 | 79,597 | 26,938 | 167,987 | 1,363,890 | 82,181 | 10,610 |
| Bangor, ME | 508.2 | 3,352 | 50.2 | 1.4 | 5.1 | 0.3 | 5.1 | 383.0 | 2,526 | 1,131 | 457 | 13,257 | 69,070 | 54,375 | 5,139 |
| Barnstable Town, MA | 1,227.3 | 5,746 | 47.1 | 1.0 | 5.2 | 0.4 | 3.7 | 916.3 | 4,290 | 1,612 | 1,211 | 13,462 | 128,690 | 81,733 | 10,493 |
| Baton Rouge, LA | 3,604.4 | 4,219 | 46.3 | 5.9 | 6.3 | 0.2 | 7.0 | 2,994.4 | 3,505 | 2,955 | 3,354 | 68,965 | 366,560 | 68,250 | 8,106 |
| Battle Creek, MI | 634.2 | 4,727 | 46.9 | 2.5 | 5.8 | 5.4 | 12.5 | 651.6 | 4,856 | 2,991 | 266 | 7,905 | 61,210 | 52,794 | 4,693 |
| Bay City, MI | 486.0 | 4,671 | 50.5 | 10.2 | 3.8 | 5.7 | 5.8 | 292.8 | 2,814 | 259 | 231 | 5,586 | 53,160 | 52,172 | 4,875 |
| Beaumont-Port Arthur, TX | 1,543.5 | 3,877 | 45.4 | 5.0 | 8.0 | 0.6 | 4.3 | 5,591.0 | 14,043 | 2,005 | 941 | 22,002 | 165,740 | 61,889 | 6,495 |
| Beckley, WV | 332.1 | 2,799 | 63.9 | 0.6 | 6.7 | 0.0 | 1.8 | 183.3 | 1,544 | 1,930 | 542 | 6,265 | 46,250 | 49,865 | 4,329 |
| Bellingham, WA | 917.9 | 4,146 | 48.9 | 3.2 | 4.9 | 0.0 | 5.6 | 850.4 | 3,841 | 1,482 | 637 | 14,578 | 107,850 | 70,944 | 7,863 |
| Bend, OR | 870.7 | 4,662 | 41.6 | 3.7 | 6.6 | 0.1 | 6.2 | 1,327.8 | 7,111 | 976 | 469 | 8,633 | 97,360 | 79,913 | 10,042 |
| Billings, MT | 721.6 | 4,023 | 46.4 | 7.4 | 5.7 | 0.3 | 7.6 | 594.9 | 3,317 | 1,915 | 820 | 8,438 | 88,360 | 66,253 | 7,282 |
| Binghamton, NY | 1,742.0 | 7,207 | 48.2 | 3.2 | 2.5 | 12.8 | 4.5 | 1,167.1 | 4,829 | 664 | 364 | 19,890 | 111,230 | 57,324 | 5,667 |
| Birmingham-Hoover, AL | 4,024.4 | 3,707 | 45.6 | 2.6 | 7.5 | 0.0 | 4.8 | 8,542.9 | 7,869 | 8,931 | 4,859 | 70,685 | 472,550 | 73,572 | 9,096 |
| Bismarck, ND | 627.3 | 4,896 | 41.3 | 1.3 | 5.0 | 1.9 | 10.8 | 631.2 | 4,927 | 1,234 | 762 | 12,577 | 62,800 | 77,578 | 9,213 |
| Blacksburg-Christiansburg, VA | 553.5 | 3,322 | 39.0 | 11.1 | 7.3 | 3.6 | 3.8 | 380.3 | 2,283 | 397 | 567 | 21,421 | 62,970 | 60,713 | 6,735 |
| Bloomington, IL | 716.1 | 4,145 | 43.6 | 1.2 | 7.8 | 1.4 | 6.3 | 932.9 | 5,400 | 435 | 330 | 15,164 | 76,400 | 75,144 | 8,730 |
| Bloomington, IN | 545.5 | 3,258 | 32.0 | 0.6 | 3.0 | 0.1 | 3.5 | 627.5 | 3,748 | 361 | 479 | 24,413 | 67,700 | 64,215 | 7,269 |
| Bloomsburg-Berwick, PA | 326.1 | 3,889 | 59.4 | 0.0 | 3.5 | 0.3 | 4.3 | 2,558.9 | 30,518 | 189 | 202 | 6,502 | 38,620 | 59,538 | 6,170 |
| Boise City, ID | 1,985.4 | 2,798 | 43.6 | 2.2 | 8.4 | 0.6 | 6.2 | 1,064.8 | 1,500 | 6,598 | 2,391 | 42,203 | 333,940 | 70,481 | 7,966 |
| Boston-Cambridge-Newton, MA-NH | 23,814.8 | 4,919 | 51.4 | 4.1 | 5.6 | 0.5 | 3.3 | 14,786.3 | 3,054 | 35,990 | 14,288 | 270,268 | 2,485,050 | 112,071 | 18,113 |
| Boston, MA Div 14454 | 10,347.4 | 5,130 | 48.7 | 3.0 | 6.7 | 0.2 | 3.2 | 6,658.0 | 3,301 | 19,434 | 5,967 | 127,457 | 1,027,990 | 114,221 | 19,056 |
| Cambridge-Newton-Framingham, MA Div 15754 | 11,669.6 | 4,886 | 52.7 | 5.5 | 4.4 | 0.2 | 3.3 | 7,275.2 | 3,046 | 15,128 | 6,760 | 117,977 | 1,221,610 | 114,529 | 18,464 |
| Rockingham County-Strafford County, NH Div 40484 | 1,797.7 | 4,118 | 58.8 | 0.7 | 6.7 | 4.0 | 4.4 | 853.0 | 1,954 | 1,428 | 1,561 | 24,834 | 235,450 | 89,931 | 12,174 |
| Boulder, CO | 2,018.9 | 6,261 | 43.6 | 1.9 | 5.4 | 2.4 | 6.8 | 2,807.4 | 8,707 | 2,027 | 836 | 34,078 | 160,980 | 110,472 | 16,864 |
| Bowling Green, KY | 374.1 | 2,138 | 58.5 | 1.8 | 5.2 | 0.1 | 4.8 | 781.6 | 4,467 | 648 | 525 | 11,336 | 72,660 | 54,285 | 5,141 |
| Bremerton-Silverdale-Port Orchard, WA | 1,207.7 | 4,535 | 41.9 | 5.2 | 3.7 | 0.0 | 5.0 | 880.5 | 3,307 | 20,485 | 11,703 | 13,858 | 130,560 | 82,111 | 9,833 |
| Bridgeport-Stamford-Norwalk, CT | 4,968.1 | 5,268 | 53.4 | 1.0 | 6.9 | 0.9 | 3.8 | 3,859.8 | 4,093 | 2,854 | 1,885 | 44,011 | 462,580 | 165,244 | 33,039 |
| Brownsville-Harlingen, TX | 1,674.4 | 3,966 | 67.1 | 0.8 | 4.6 | 0.4 | 2.9 | 1,644.9 | 3,896 | 3,554 | 928 | 25,958 | 169,320 | 40,894 | 3,149 |
| Brunswick, GA | 745.2 | 6,338 | 29.1 | 46.1 | 3.3 | 0.0 | 3.8 | 398.2 | 3,386 | 2,061 | 389 | 6,765 | 49,500 | 62,060 | 6,693 |
| Buffalo-Cheektowaga, NY | 7,235.0 | 6,403 | 45.9 | 10.5 | 4.2 | 10.0 | 3.9 | 5,842.0 | 5,170 | 9,497 | 1,984 | 77,432 | 558,930 | 63,918 | 7,148 |
| Burlington, NC | 439.3 | 2,691 | 53.1 | 4.0 | 9.9 | 6.0 | 1.8 | 147.6 | 904 | 255 | 358 | 7,004 | 75,020 | 55,405 | 5,198 |
| Burlington-South Burlington, VT | 1,030.4 | 4,706 | 65.1 | 0.2 | 4.4 | 0.0 | 6.1 | 665.9 | 3,041 | 4,114 | 1,443 | 17,766 | 114,830 | 74,653 | 8,703 |
| California-Lexington Park, MD | 391.5 | 3,481 | 59.9 | 2.0 | 7.1 | 0.0 | 4.1 | 366.3 | 3,256 | 10,334 | 2,504 | 4,928 | 53,080 | 81,417 | 9,381 |
| Canton-Massillon, OH | 1,466.0 | 3,684 | 52.4 | 8.5 | 5.7 | 5.5 | 6.3 | 576.6 | 1,444 | 1,015 | 1,014 | 19,070 | 196,410 | 57,152 | 5,822 |
| Cape Coral-Fort Myers, FL | 4,462.4 | 6,030 | 24.7 | 37.4 | 5.2 | 0.2 | 2.4 | 4,561.8 | 6,165 | 2,432 | 1,400 | 40,901 | 359,450 | 84,592 | 11,892 |
| Cape Girardeau, MO-IL | 294.5 | 3,042 | 49.9 | 0.5 | 6.0 | 0.0 | 6.7 | 169.9 | 1,755 | 441 | 331 | 6,898 | 42,210 | 58,680 | 5,821 |
| Carbondale-Marion, IL | 479.4 | 3,490 | 53.2 | 1.9 | 7.1 | 0.1 | 6.1 | 458.6 | 3,339 | 1,902 | 268 | 16,243 | 56,600 | 55,138 | 5,302 |
| Carson City, NV | 226.4 | 4,152 | 42.3 | 4.1 | 8.4 | 2.9 | 6.4 | 397.7 | 7,292 | 569 | 142 | 9,101 | 28,630 | 59,571 | 6,443 |
| Casper, WY | 518.2 | 6,511 | 62.5 | 0.4 | 4.9 | 0.3 | 4.4 | 110.2 | 1,385 | 634 | 414 | 4,919 | 37,950 | 75,409 | 9,440 |

1. Based on the resident population estimated as of July 1 of the year shown.

# Table C. Metropolitan Areas — Land Area and Population

| CBSA/ DIV Code[1] | Area name | Population, 2020 | | | | Population characteristics, 2020 | | | | | | | | | | |
| | | | | | | Race alone or in combination, not Hispanic or Latino (percent) | | | | | Age (percent) | | | | | |
| | | Land area[2] (sq mi) | Total persons 2020 | Rank | Per square mile | White | Black | American Indian, Alaska Native | Asian and Pacific Islander | Percent Hispanic or Latino[3] | Under 5 years | 5 to 14 years | 15 to 24 years | 25 to 34 years | 35 to 44 years | 45 to 54 years |
| | | 1 | 2 | 3 | 4 | 5 | 6 | 7 | 8 | 9 | 10 | 11 | 12 | 13 | 14 | 15 |
| 16300 | Cedar Rapids, IA | 2,009 | 273,885 | 178 | 136.3 | 88.7 | 6.9 | 0.6 | 3.0 | 3.3 | 6.0 | 12.7 | 12.6 | 13.1 | 13.1 | 12.1 |
| 16540 | Chambersburg-Waynesboro, PA | 772 | 155,637 | 268 | 201.5 | 88.7 | 4.7 | 0.6 | 1.5 | 6.4 | 5.7 | 12.4 | 11.4 | 12.1 | 11.6 | 12.7 |
| 16580 | Champaign-Urbana, IL | 1,435 | 225,547 | 204 | 157.1 | 70.9 | 14.2 | 0.5 | 11.3 | 6.0 | 5.3 | 10.5 | 24.6 | 13.7 | 11.4 | 9.7 |
| 16620 | Charleston, WV | 2,647 | 254,145 | 189 | 96.0 | 92.4 | 6.7 | 0.7 | 1.2 | 1.1 | 5.2 | 11.4 | 11.0 | 11.8 | 11.8 | 12.6 |
| 16700 | Charleston-North Charleston, SC | 2,590 | 819,705 | 73 | 316.5 | 66.2 | 26.1 | 0.9 | 3.1 | 6.0 | 5.9 | 12.1 | 12.1 | 15.4 | 13.7 | 12.0 |
| 16740 | Charlotte-Concord-Gastonia, NC-SC | 5,597 | 2,684,276 | 22 | 479.6 | 61.4 | 24.2 | 0.9 | 4.9 | 10.8 | 6.0 | 13.1 | 12.5 | 14.4 | 13.7 | 13.8 |
| 16820 | Charlottesville, VA | 1,645 | 219,910 | 209 | 133.7 | 77.6 | 13.1 | 0.7 | 5.8 | 5.6 | 5.2 | 10.6 | 15.0 | 14.0 | 12.2 | 11.3 |
| 16860 | Chattanooga, TN-GA | 2,089 | 569,931 | 100 | 272.8 | 79.5 | 14.3 | 0.9 | 2.2 | 5.0 | 5.6 | 11.7 | 11.8 | 13.8 | 12.3 | 12.5 |
| 16940 | Cheyenne, WY | 2,686 | 100,595 | 356 | 37.5 | 80.4 | 3.2 | 1.4 | 2.2 | 15.4 | 6.0 | 13.0 | 12.6 | 14.6 | 12.8 | 11.0 |
| 16980 | Chicago-Naperville-Elgin, IL-IN-WI | 7,195 | 9,406,638 | 3 | 1,307.4 | 53.5 | 17.1 | 0.5 | 8.0 | 22.7 | 5.8 | 12.5 | 12.8 | 14.4 | 13.4 | 12.7 |
| 16980 | Chicago-Naperville-Evanston, IL Div 16984 | 3,130 | 7,071,372 | X | 2,259.2 | 50.1 | 19.2 | 0.5 | 9.0 | 22.9 | 5.8 | 12.3 | 12.4 | 15.1 | 13.6 | 12.6 |
| 16980 | Elgin, IL Div 20994 | 1,471 | 766,139 | X | 520.9 | 62.0 | 6.9 | 0.4 | 4.6 | 27.7 | 6.0 | 14.0 | 14.9 | 12.2 | 13.4 | 13.3 |
| 16980 | Gary, IN Div 23844 | 1,878 | 705,863 | X | 375.8 | 64.2 | 18.1 | 0.6 | 1.9 | 16.8 | 5.7 | 12.9 | 12.8 | 12.3 | 12.8 | 12.4 |
| 16980 | Lake County-Kenosha County, IL-WI Div 29404 | 716 | 863,264 | X | 1,206.2 | 64.5 | 7.8 | 0.5 | 8.2 | 21.0 | 5.5 | 13.3 | 14.4 | 11.7 | 12.6 | 13.3 |
| 17020 | Chico, CA | 1,637 | 212,744 | 214 | 130.0 | 73.5 | 2.8 | 3.0 | 7.0 | 17.9 | 5.4 | 11.4 | 18.0 | 13.1 | 11.5 | 10.2 |
| 17140 | Cincinnati, OH-KY-IN | 4,546 | 2,232,907 | 30 | 491.2 | 80.6 | 13.8 | 0.6 | 3.6 | 3.6 | 6.1 | 13.0 | 13.2 | 13.5 | 12.5 | 12.2 |
| 17300 | Clarksville, TN-KY | 2,158 | 314,364 | 163 | 145.7 | 68.4 | 21.6 | 1.3 | 3.6 | 9.4 | 8.2 | 14.3 | 15.3 | 17.6 | 12.6 | 10.3 |
| 17420 | Cleveland, TN | 763 | 125,906 | 317 | 164.9 | 88.0 | 5.2 | 1.0 | 1.5 | 6.1 | 5.6 | 11.9 | 12.8 | 12.8 | 11.9 | 13.4 |
| 17460 | Cleveland-Elyria, OH | 1,999 | 2,043,807 | 34 | 1,022.2 | 71.0 | 21.1 | 0.6 | 3.1 | 6.3 | 5.4 | 11.6 | 12.0 | 13.3 | 11.8 | 12.2 |
| 17660 | Coeur d'Alene, ID | 1,238 | 170,628 | 251 | 137.8 | 92.3 | 0.8 | 2.2 | 2.0 | 5.1 | 5.7 | 12.8 | 11.2 | 12.9 | 12.4 | 11.7 |
| 17780 | College Station-Bryan, TX | 2,100 | 268,224 | 183 | 127.7 | 57.0 | 12.1 | 0.7 | 6.0 | 25.9 | 5.8 | 11.6 | 25.9 | 15.0 | 11.7 | 9.3 |
| 17820 | Colorado Springs, CO | 2,684 | 753,839 | 79 | 280.9 | 72.4 | 7.6 | 1.5 | 5.2 | 17.7 | 6.3 | 13.1 | 14.4 | 16.0 | 13.1 | 11.2 |
| 17860 | Columbia, MO | 1,714 | 210,094 | 216 | 122.6 | 82.4 | 11.0 | 1.1 | 5.3 | 3.4 | 5.7 | 11.5 | 20.6 | 14.6 | 12.1 | 10.3 |
| 17900 | Columbia, SC | 3,703 | 847,397 | 69 | 228.8 | 57.2 | 35.1 | 0.8 | 3.1 | 5.9 | 5.6 | 12.4 | 14.9 | 13.9 | 12.5 | 12.2 |
| 17980 | Columbus, GA-AL | 2,786 | 322,658 | 161 | 115.8 | 47.5 | 43.0 | 1.0 | 3.3 | 8.0 | 6.6 | 13.3 | 13.3 | 15.8 | 12.6 | 11.3 |
| 18020 | Columbus, IN | 407 | 84,447 | 371 | 207.5 | 81.4 | 3.0 | 0.6 | 8.7 | 7.8 | 6.4 | 13.5 | 11.8 | 14.4 | 12.6 | 12.3 |
| 18140 | Columbus, OH | 4,797 | 2,138,946 | 32 | 445.9 | 73.7 | 18.0 | 0.8 | 5.9 | 4.6 | 6.4 | 13.0 | 12.8 | 15.7 | 13.6 | 12.4 |
| 18580 | Corpus Christi, TX | 1,533 | 430,217 | 126 | 280.9 | 30.6 | 3.5 | 0.5 | 2.4 | 63.9 | 6.4 | 13.8 | 14.0 | 14.0 | 13.2 | 11.4 |
| 18700 | Corvallis, OR | 675 | 93,239 | 366 | 138.1 | 83.1 | 1.9 | 1.6 | 9.3 | 8.0 | 3.9 | 9.0 | 25.5 | 13.1 | 10.4 | 9.7 |
| 18880 | Crestview-Fort Walton Beach-Destin, FL | 1,969 | 289,468 | 170 | 147.1 | 78.8 | 9.9 | 1.4 | 4.6 | 9.0 | 6.2 | 12.2 | 11.5 | 14.7 | 12.7 | 11.1 |
| 19060 | Cumberland, MD-WV | 750 | 96,779 | 363 | 129.0 | 90.2 | 8.1 | 0.5 | 1.3 | 1.8 | 4.7 | 10.0 | 14.2 | 12.6 | 11.4 | 12.3 |
| 19100 | Dallas-Fort Worth-Arlington, TX | 8,675 | 7,694,138 | 4 | 887.0 | 46.1 | 17.2 | 0.9 | 8.5 | 29.3 | 6.6 | 14.4 | 13.5 | 15.0 | 14.2 | 13.0 |
| 19100 | Dallas-Plano-Irving, TX Div 19124 | 5,277 | 5,171,934 | X | 980.1 | 43.5 | 17.8 | 0.8 | 9.8 | 30.0 | 6.6 | 14.3 | 13.5 | 15.2 | 14.4 | 13.2 |
| 19100 | Fort Worth-Arlington-Grapevine, TX Div 23104 | 3,398 | 2,522,204 | X | 742.5 | 51.2 | 16.0 | 0.9 | 6.0 | 28.0 | 6.6 | 14.6 | 13.6 | 14.6 | 13.7 | 12.6 |
| 19140 | Dalton, GA | 635 | 143,869 | 293 | 226.6 | 64.8 | 3.5 | 0.6 | 1.5 | 30.8 | 6.3 | 14.1 | 13.8 | 13.1 | 12.4 | 13.2 |
| 19180 | Danville, IL | 898 | 74,855 | 380 | 83.3 | 79.9 | 15.3 | 0.6 | 1.3 | 5.4 | 6.2 | 13.2 | 11.6 | 11.8 | 11.6 | 11.5 |
| 19300 | Daphne-Fairhope-Foley, AL | 1,590 | 229,287 | 202 | 144.2 | 85.0 | 9.3 | 1.4 | 1.6 | 4.6 | 5.3 | 12.1 | 10.8 | 11.1 | 12.0 | 12.6 |
| 19340 | Davenport-Moline-Rock Island, IA-IL | 2,270 | 377,759 | 144 | 166.4 | 80.2 | 9.6 | 0.7 | 3.0 | 9.2 | 5.9 | 12.9 | 12.1 | 12.3 | 12.6 | 11.8 |
| 19430 | Dayton-Kettering, OH | 1,281 | 809,248 | 74 | 631.4 | 78.1 | 17.6 | 0.9 | 3.3 | 3.2 | 5.9 | 12.1 | 13.1 | 13.5 | 11.8 | 11.6 |
| 19460 | Decatur, AL | 1,270 | 152,740 | 274 | 120.2 | 77.8 | 13.3 | 3.4 | 0.9 | 7.4 | 5.9 | 12.7 | 11.6 | 12.2 | 12.0 | 13.0 |
| 19500 | Decatur, IL | 581 | 103,015 | 348 | 177.4 | 78.1 | 20.3 | 0.6 | 1.7 | 2.5 | 6.1 | 12.3 | 12.3 | 12.0 | 11.4 | 11.2 |
| 19660 | Deltona-Daytona Beach-Ormond Beach, FL | 1,588 | 679,948 | 86 | 428.6 | 72.5 | 11.4 | 0.8 | 2.6 | 14.7 | 4.4 | 9.7 | 10.5 | 11.3 | 10.4 | 11.7 |
| 19740 | Denver-Aurora-Lakewood, CO | 8,345 | 2,991,231 | 19 | 358.4 | 65.7 | 6.7 | 1.1 | 5.9 | 23.3 | 5.7 | 12.2 | 11.8 | 17.2 | 15.0 | 12.7 |
| 19780 | Des Moines-West Des Moines, IA | 3,612 | 707,915 | 82 | 196.0 | 82.2 | 6.7 | 0.6 | 5.2 | 7.5 | 6.7 | 13.9 | 12.4 | 14.5 | 14.1 | 12.1 |
| 19820 | Detroit-Warren-Dearborn, MI | 3,892 | 4,304,136 | 14 | 1,106.0 | 67.7 | 23.3 | 0.9 | 5.7 | 4.8 | 5.8 | 12.1 | 11.9 | 13.8 | 11.9 | 13.1 |
| 19820 | Detroit-Dearborn-Livonia, MI Div 19804 | 612 | 1,740,623 | X | 2,845.1 | 51.3 | 39.6 | 0.9 | 4.4 | 6.3 | 6.5 | 13.0 | 12.3 | 14.7 | 11.7 | 12.5 |
| 19820 | Warren-Troy-Farmington Hills, MI Div 47664 | 3,280 | 2,563,513 | X | 781.6 | 78.9 | 12.3 | 0.9 | 6.5 | 3.8 | 5.3 | 11.4 | 11.7 | 13.3 | 12.1 | 13.5 |
| 20020 | Dothan, AL | 1,716 | 150,214 | 281 | 87.5 | 70.9 | 25.0 | 1.2 | 1.4 | 3.5 | 6.0 | 12.4 | 11.4 | 12.3 | 12.1 | 12.5 |
| 20100 | Dover, DE | 586 | 183,643 | 235 | 313.3 | 62.5 | 28.8 | 1.3 | 3.5 | 7.7 | 6.1 | 12.8 | 13.6 | 13.5 | 11.9 | 11.3 |
| 20220 | Dubuque, IA | 608 | 97,590 | 361 | 160.4 | 91.3 | 4.6 | 0.5 | 2.5 | 2.8 | 6.2 | 12.7 | 13.7 | 12.9 | 11.4 | 10.9 |
| 20260 | Duluth, MN-WI | 10,522 | 288,648 | 171 | 27.4 | 93.0 | 2.4 | 3.7 | 1.6 | 1.8 | 4.9 | 11.0 | 14.3 | 11.7 | 11.9 | 11.1 |
| 20500 | Durham-Chapel Hill, NC | 2,290 | 652,542 | 91 | 284.9 | 57.1 | 27.1 | 1.0 | 5.7 | 11.3 | 5.4 | 11.4 | 14.6 | 14.9 | 12.7 | 12.4 |
| 20700 | East Stroudsburg, PA | 608 | 170,154 | 252 | 279.7 | 65.3 | 15.3 | 0.8 | 3.2 | 17.6 | 4.7 | 10.8 | 13.9 | 11.7 | 10.9 | 13.4 |
| 20740 | Eau Claire, WI | 1,646 | 169,997 | 253 | 103.3 | 92.3 | 2.2 | 1.0 | 3.9 | 2.4 | 5.4 | 11.7 | 16.1 | 13.2 | 12.4 | 11.0 |
| 20940 | El Centro, CA | 4,176 | 180,267 | 238 | 43.2 | 10.1 | 2.6 | 0.9 | 1.6 | 85.4 | 7.6 | 16.2 | 14.5 | 14.9 | 12.4 | 10.7 |
| 21060 | Elizabethtown-Fort Knox, KY | 1,190 | 154,356 | 271 | 129.7 | 82.6 | 11.3 | 1.1 | 3.1 | 5.3 | 6.1 | 13.5 | 12.7 | 13.3 | 13.2 | 12.4 |
| 21140 | Elkhart-Goshen, IN | 463 | 206,161 | 219 | 445.1 | 75.8 | 7.1 | 0.7 | 1.7 | 17.1 | 7.4 | 15.2 | 13.6 | 13.0 | 12.0 | 11.7 |
| 21300 | Elmira, NY | 407 | 82,622 | 376 | 202.8 | 88.8 | 8.2 | 0.8 | 2.0 | 3.5 | 5.4 | 12.0 | 11.7 | 12.5 | 11.9 | 12.1 |
| 21340 | El Paso, TX | 5,584 | 846,192 | 71 | 151.5 | 12.4 | 3.6 | 0.5 | 1.8 | 82.7 | 7.1 | 14.9 | 15.6 | 15.4 | 12.5 | 11.3 |
| 21420 | Enid, OK | 1,059 | 60,869 | 383 | 57.5 | 75.7 | 4.4 | 4.4 | 6.0 | 13.6 | 7.1 | 14.4 | 12.9 | 13.6 | 12.6 | 10.3 |
| 21500 | Erie, PA | 799 | 268,426 | 182 | 335.9 | 85.8 | 8.9 | 0.5 | 2.5 | 4.6 | 5.4 | 11.8 | 13.3 | 13.0 | 11.5 | 11.8 |
| 21660 | Eugene-Springfield, OR | 4,554 | 382,986 | 143 | 84.1 | 85.0 | 2.1 | 2.8 | 5.1 | 9.5 | 4.6 | 10.2 | 15.4 | 13.3 | 12.4 | 11.0 |

1. CBSA = Core Based Statistical Area. DIV = Metropolitan Division. See Appendix A for explanation. See Appendix B for list of metropolitan areas or temporarily covered by water. 2. Dry land or land partially or temporarily covered by water. 3. May be of any race.

Items 1—15

# Table C. Metropolitan Areas — Population and Households

| Area name | 55 to 64 years | 65 to 74 years | 75 years and over | Percent female | Total persons 2000 | 2010 | Percent change 2000-2010 | 2010-2019 | Births | Deaths | Net migration | Number | Persons per household | Family households | Single parent households | One person |
|---|---|---|---|---|---|---|---|---|---|---|---|---|---|---|---|---|
| | 16 | 17 | 18 | 19 | 20 | 21 | 22 | 23 | 24 | 25 | 26 | 27 | 28 | 29 | 30 | 31 |
| Cedar Rapids, IA .................. | 13.1 | 9.8 | 7.4 | 50.5 | 237,230 | 257,947 | 8.7 | 6.2 | 33,328 | 22,304 | 5,108 | 112,373 | 2.37 | 61.5 | 9.0 | 31.2 |
| Chambersburg-Waynesboro, PA ......... | 13.7 | 11.2 | 9.2 | 50.9 | 129,313 | 149,625 | 15.7 | 4.0 | 18,702 | 15,687 | 3,161 | 60,260 | 2.54 | 66.5 | 8.4 | 28.7 |
| Champaign-Urbana, IL ............ | 10.6 | 8.4 | 5.8 | 50.4 | 196,034 | 217,814 | 11.1 | 3.6 | 25,832 | 15,171 | -3,054 | 88,700 | 2.37 | 52.6 | 10.2 | 35.7 |
| Charleston, WV .................. | 14.6 | 12.9 | 8.7 | 51.4 | 286,046 | 278,000 | -2.8 | -8.6 | 30,187 | 36,710 | -17,192 | 109,452 | 2.32 | 63.9 | 10.2 | 31.3 |
| Charleston-North Charleston, SC .............. | 12.8 | 10.0 | 6.1 | 51.2 | 549,033 | 664,647 | 21.1 | 23.3 | 96,045 | 58,028 | 115,721 | 303,976 | 2.58 | 63.9 | 8.4 | 28.0 |
| Charlotte-Concord-Gastonia, NC-SC ................ | 12.3 | 8.7 | 5.6 | 51.5 | 1,742,647 | 2,243,952 | 28.8 | 19.6 | 312,200 | 191,999 | 318,467 | 982,241 | 2.65 | 66.1 | 10.8 | 27.7 |
| Charlottesville, VA ............. | 12.8 | 11.2 | 7.7 | 52.1 | 174,021 | 201,579 | 15.8 | 9.1 | 23,166 | 17,092 | 12,350 | 87,274 | 2.37 | 60.0 | 7.1 | 30.7 |
| Chattanooga, TN-GA ............. | 13.4 | 11.0 | 7.8 | 51.6 | 476,531 | 528,131 | 10.8 | 7.9 | 64,478 | 57,697 | 35,032 | 222,950 | 2.47 | 64.2 | 10.1 | 28.9 |
| Cheyenne, WY .................. | 12.9 | 10.2 | 6.8 | 49.2 | 81,607 | 91,885 | 12.6 | 9.5 | 12,975 | 8,353 | 4,056 | 41,739 | 2.34 | 65.2 | 10.6 | 30.3 |
| Chicago-Naperville-Elgin, IL-IN-WI.............. | 12.8 | 9.1 | 6.4 | 51.0 | 9,098,316 | 9,461,540 | 4.0 | -0.6 | 1,191,992 | 729,094 | -518,000 | 3,539,174 | 2.63 | 63.6 | 9.7 | 30.0 |
| Chicago-Naperville-Evanston, IL Div 16984 | 12.6 | 9.1 | 6.4 | 51.2 | 7,080,780 | 7,148,320 | 1.0 | -1.1 | 913,084 | 552,511 | -437,568 | 2,691,034 | 2.60 | 62.0 | 9.6 | 31.4 |
| Elgin, IL Div 20994.............. | 12.2 | 8.4 | 5.6 | 50.3 | 547,632 | 735,280 | 34.3 | 4.2 | 96,801 | 45,752 | -20,089 | 262,917 | 2.87 | 70.4 | 9.0 | 23.2 |
| Gary, IN Div 23844 .......... | 13.6 | 10.4 | 7.1 | 51.3 | 675,971 | 708,120 | 4.8 | -0.3 | 83,715 | 70,046 | -15,779 | 273,375 | 2.53 | 66.1 | 12.9 | 28.2 |
| Lake County-Kenosha County, IL-WI Div | 13.9 | 9.2 | 6.2 | 50.1 | 793,933 | 869,820 | 9.6 | -0.8 | 98,392 | 60,785 | -44,564 | 311,848 | 2.70 | 70.1 | 9.2 | 25.1 |
| Chico, CA ..................... | 12.1 | 10.9 | 7.5 | 50.5 | 203,171 | 220,005 | 8.3 | -3.3 | 24,618 | 23,905 | -7,835 | 77,651 | 2.75 | 62.1 | 9.6 | 26.7 |
| Cincinnati, OH-KY-IN............ | 13.3 | 9.7 | 6.4 | 50.9 | 2,016,981 | 2,137,718 | 6.0 | 4.5 | 280,088 | 202,033 | 18,303 | 878,645 | 2.47 | 63.8 | 11.0 | 29.1 |
| Clarksville, TN-KY ............. | 10.0 | 7.1 | 4.6 | 49.4 | 232,000 | 273,942 | 18.1 | 14.8 | 52,921 | 22,427 | 9,459 | 115,195 | 2.61 | 67.3 | 12.0 | 27.1 |
| Cleveland, TN ................. | 13.4 | 10.5 | 7.7 | 51.3 | 104,015 | 115,747 | 11.3 | 8.8 | 13,940 | 13,008 | 9,241 | 46,936 | 2.63 | 68.2 | 9.7 | 25.8 |
| Cleveland-Elyria, OH ............ | 14.3 | 11.2 | 8.2 | 51.7 | 2,148,143 | 2,077,270 | -3.3 | -1.6 | 234,158 | 221,913 | -44,912 | 872,626 | 2.30 | 58.0 | 10.2 | 35.6 |
| Coeur d'Alene, ID .............. | 13.5 | 12.1 | 7.7 | 50.5 | 108,685 | 138,466 | 27.4 | 23.2 | 18,081 | 14,096 | 28,134 | 65,283 | 2.51 | 65.8 | 8.9 | 26.8 |
| College Station-Bryan, TX....... | 9.3 | 6.8 | 4.6 | 49.9 | 184,885 | 228,665 | 23.7 | 17.3 | 31,676 | 14,488 | 22,277 | 94,597 | 2.66 | 60.5 | 11.7 | 28.7 |
| Colorado Springs, CO ........... | 12.0 | 8.7 | 5.2 | 49.5 | 537,484 | 645,616 | 20.1 | 16.8 | 96,734 | 45,798 | 57,145 | 275,778 | 2.64 | 68.6 | 8.8 | 23.9 |
| Columbia, MO ................. | 11.1 | 8.5 | 5.6 | 51.4 | 162,336 | 190,396 | 17.3 | 10.3 | 24,689 | 14,046 | 9,057 | 80,127 | 2.46 | 59.1 | 11.7 | 28.7 |
| Columbia, SC ................. | 12.7 | 9.7 | 6.1 | 51.7 | 647,158 | 767,478 | 18.6 | 10.4 | 95,965 | 69,393 | 53,544 | 317,998 | 2.53 | 63.8 | 12.0 | 30.5 |
| Columbus, GA-AL ............... | 12.1 | 9.0 | 6.0 | 50.5 | 293,518 | 308,481 | 5.1 | 4.6 | 46,593 | 31,008 | -2,055 | 119,728 | 2.58 | 65.7 | 13.8 | 30.8 |
| Columbus, IN ................. | 12.1 | 9.7 | 7.2 | 49.7 | 71,435 | 76,782 | 7.5 | 10.0 | 10,804 | 7,615 | 4,497 | 30,383 | 2.72 | 63.0 | 9.0 | 27.6 |
| Columbus, OH ................. | 12.0 | 8.6 | 5.4 | 50.8 | 1,675,013 | 1,902,012 | 13.6 | 12.5 | 278,121 | 161,438 | 120,623 | 817,086 | 2.53 | 63.4 | 11.1 | 28.7 |
| Corpus Christi, TX ............. | 11.9 | 9.1 | 6.3 | 50.4 | 380,783 | 405,015 | 6.4 | 6.2 | 58,518 | 35,699 | 2,403 | 153,066 | 2.74 | 68.4 | 15.2 | 26.0 |
| Corvallis, OR ................. | 10.9 | 10.7 | 6.7 | 50.0 | 78,153 | 85,581 | 9.5 | 8.9 | 7,378 | 5,893 | 6,147 | 37,378 | 2.33 | 56.8 | 5.1 | 25.8 |
| Crestview-Fort Walton Beach-Destin, FL .......... | 14.1 | 10.6 | 7.0 | 49.4 | 211,099 | 235,868 | 11.7 | 22.7 | 35,488 | 24,231 | 41,889 | 102,572 | 2.70 | 65.2 | 9.7 | 28.5 |
| Cumberland, MD-WV ............. | 13.4 | 11.9 | 9.6 | 48.5 | 102,008 | 103,267 | 1.2 | -6.3 | 9,656 | 12,724 | -3,393 | 36,430 | 2.42 | 57.8 | 9.3 | 34.3 |
| Dallas-Fort Worth-Arlington, TX ..... | 11.5 | 7.3 | 4.5 | 50.8 | 5,156,217 | 6,366,541 | 23.5 | 20.9 | 1,000,165 | 437,830 | 763,593 | 2,635,017 | 2.84 | 68.7 | 11.2 | 25.7 |
| Dallas-Plano-Irving, TX Div 19124 ...... | 11.3 | 7.2 | 4.4 | 50.7 | 3,445,899 | 4,228,208 | 22.7 | 22.3 | 671,455 | 278,355 | 549,019 | 1,782,408 | 2.82 | 68.0 | 11.0 | 26.1 |
| Fort Worth-Arlington-Grapevine, TX Div 23104....... | 11.8 | 7.7 | 4.8 | 51.0 | 1,710,318 | 2,138,333 | 25.0 | 18.0 | 328,710 | 159,475 | 214,574 | 852,609 | 2.89 | 70.3 | 11.6 | 24.9 |
| Dalton, GA................... | 12.1 | 8.7 | 6.3 | 50.4 | 120,031 | 142,230 | 18.5 | 1.2 | 19,091 | 12,312 | -5,104 | 50,267 | 2.84 | 71.8 | 11.8 | 24.3 |
| Danville, IL.................. | 13.8 | 11.4 | 8.8 | 50.1 | 83,919 | 81,625 | -2.7 | -8.3 | 10,333 | 9,944 | -7,166 | 31,154 | 2.34 | 58.3 | 15.8 | 37.6 |
| Daphne-Fairhope-Foley, AL .... | 14.6 | 13.0 | 8.6 | 51.6 | 140,415 | 182,263 | 29.8 | 25.8 | 22,965 | 21,613 | 45,301 | 82,325 | 2.68 | 65.5 | 8.8 | 30.8 |
| Davenport-Moline-Rock Island, IA-IL .......... | 13.4 | 11.0 | 8.1 | 50.7 | 376,019 | 379,692 | 1.0 | -0.5 | 47,732 | 38,710 | -10,865 | 155,389 | 2.38 | 62.8 | 10.2 | 30.9 |
| Dayton-Kettering, OH ........... | 13.4 | 10.7 | 7.9 | 51.4 | 805,816 | 799,280 | -0.8 | 1.2 | 98,062 | 87,927 | 209 | 332,258 | 2.34 | 59.5 | 10.5 | 33.1 |
| Decatur, AL................... | 14.1 | 10.8 | 7.7 | 50.9 | 145,867 | 153,827 | 5.5 | -0.7 | 18,286 | 17,468 | -1,832 | 59,315 | 2.53 | 68.7 | 10.7 | 26.2 |
| Decatur, IL................... | 13.7 | 11.7 | 9.2 | 52.1 | 114,706 | 110,777 | -3.4 | -7 | 13,606 | 12,558 | -8,838 | 42,741 | 2.35 | 56.7 | 6.6 | 37.6 |
| Deltona-Daytona Beach-Ormond Beach, FL.......... | 15.4 | 14.9 | 11.5 | 51.4 | 493,175 | 590,283 | 19.7 | 15.2 | 57,686 | 83,613 | 115,169 | 263,915 | 2.49 | 64.1 | 9.3 | 29.0 |
| Denver-Aurora-Lakewood, CO | 11.8 | 8.5 | 5.2 | 49.9 | 2,179,240 | 2,543,597 | 16.7 | 17.6 | 352,718 | 178,135 | 269,661 | 1,152,457 | 2.54 | 62.4 | 8.3 | 28.1 |
| Des Moines-West Des Moines, IA ........ | 11.8 | 8.6 | 5.8 | 50.5 | 518,607 | 606,478 | 16.9 | 16.7 | 94,427 | 50,696 | 57,416 | 280,056 | 2.45 | 65.1 | 10.8 | 27.9 |
| Detroit-Warren-Dearborn, MI... | 14.0 | 10.3 | 7.1 | 51.3 | 4,452,557 | 4,295,968 | -3.5 | 0.2 | 512,175 | 423,334 | -80,851 | 1,720,610 | 2.48 | 62.5 | 10.0 | 31.7 |
| Detroit-Dearborn-Livonia, MI Div 19804 ...... | 13.1 | 9.7 | 6.6 | 51.9 | 2,061,162 | 1,820,216 | -11.7 | -4.4 | 238,334 | 186,785 | -132,905 | 689,270 | 2.51 | 59.1 | 13.1 | 35.1 |
| Warren-Troy-Farmington Hills, MI Div 47664 ........ | 14.7 | 10.7 | 7.4 | 50.9 | 2,391,395 | 2,475,752 | 3.5 | 3.5 | 273,841 | 236,549 | 52,054 | 1,031,340 | 2.47 | 64.7 | 7.9 | 29.3 |
| Dothan, AL................... | 13.6 | 11.4 | 8.3 | 52.0 | 130,861 | 145,641 | 11.3 | 3.1 | 18,277 | 17,180 | 3,579 | 56,795 | 2.60 | 65.6 | 10.6 | 30.5 |
| Dover, DE................... | 12.9 | 10.5 | 7.4 | 51.9 | 126,697 | 162,352 | 28.1 | 13.1 | 22,616 | 16,251 | 14,926 | 68,023 | 2.59 | 66.0 | 12.8 | 27.5 |
| Dubuque, IA.................. | 13.4 | 10.4 | 8.2 | 50.7 | 89,143 | 93,643 | 5.0 | 4.2 | 12,260 | 9,288 | 1,043 | 38,345 | 2.43 | 59.2 | 9.2 | 31.2 |
| Duluth, MN-WI............... | 14.6 | 12.2 | 8.3 | 49.6 | 286,544 | 290,635 | 1.4 | -0.7 | 29,576 | 31,130 | -189 | 124,913 | 2.21 | 58.8 | 9.9 | 31.6 |
| Durham-Chapel Hill, NC ......... | 12.6 | 10.0 | 6.5 | 52.1 | 474,991 | 564,205 | 18.8 | 15.7 | 72,592 | 45,281 | 60,637 | 255,271 | 2.39 | 60.7 | 10.3 | 31.2 |
| East Stroudsburg, PA ........... | 16.2 | 11.5 | 7.0 | 50.3 | 138,687 | 169,837 | 22.5 | 0.2 | 14,696 | 14,849 | 444 | 56,274 | 2.99 | 72.3 | 11.4 | 22.3 |
| Eau Claire, WI ............... | 12.9 | 10.4 | 7.0 | 49.6 | 148,337 | 161,383 | 8.8 | 5.3 | 19,246 | 14,431 | 3,887 | 67,096 | 2.42 | 59.7 | 6.5 | 30.0 |
| El Centro, CA................. | 10.2 | 7.7 | 6.0 | 48.8 | 142,361 | 174,522 | 22.6 | 3.3 | 30,275 | 10,616 | -14,046 | 48,182 | 3.54 | 75.4 | 15.0 | 22.1 |
| Elizabethtown-Fort Knox, KY... | 13.4 | 9.2 | 6.1 | 50.1 | 133,896 | 148,332 | 10.8 | 4.1 | 19,922 | 13,556 | -538 | 58,590 | 2.59 | 67.0 | 11.2 | 27.4 |
| Elkhart-Goshen, IN ............. | 11.7 | 8.9 | 6.6 | 50.6 | 182,791 | 197,569 | 8.1 | 4.3 | 31,576 | 17,207 | -5,702 | 68,553 | 2.96 | 59.6 | 7.6 | 35.4 |
| Elmira, NY ................. | 14.5 | 11.5 | 8.4 | 50.5 | 91,070 | 88,846 | -2.4 | -7 | 9,747 | 9,870 | -6,111 | 33,490 | 2.36 | 62.3 | 10.1 | 31.9 |
| El Paso, TX................... | 10.4 | 7.4 | 5.5 | 50.5 | 682,966 | 804,110 | 17.7 | 5.2 | 133,284 | 53,275 | -38,278 | 270,137 | 3.06 | 70.8 | 16.2 | 25.7 |
| Enid, OK.................... | 12.2 | 9.3 | 7.5 | 49.7 | 57,813 | 60,582 | 4.8 | 0.5 | 9,487 | 7,038 | -2,139 | 23,683 | 2.51 | 68.4 | 11.2 | 24.5 |
| Erie, PA.................... | 13.9 | 11.3 | 7.9 | 50.6 | 280,843 | 280,586 | -0.1 | -4.3 | 31,336 | 29,490 | -14,014 | 110,128 | 2.33 | 59.7 | 10.9 | 32.2 |
| Eugene-Springfield, OR ......... | 12.7 | 12.5 | 8.0 | 50.9 | 322,959 | 351,705 | 8.9 | 8.9 | 35,750 | 36,152 | 31,823 | 155,696 | 2.40 | 59.0 | 10.8 | 29.8 |

# Table C. Metropolitan Areas — **Population, Vital Statistics, Health, and Crime**

| Area name | Persons in group quarters, 2020 | Daytime population, 2019 | | Births, 2020 | | Deaths, 2020 | | Persons under 65 with no health insurance 2019 | | Medicare, 2020 | | | Serious crimes known to police[2], 2019 Violent | |
|---|---|---|---|---|---|---|---|---|---|---|---|---|---|---|
| | | Number | Employment/residence ratio | Total | Rate[1] | Number | Rate[1] | Number | Percent | Total Beneficiaries | Enrolled in Original Medicare | Enrolled in Medicare Advantage | Number | Rate[3] |
| | 32 | 33 | 34 | 35 | 36 | 37 | 38 | 39 | 40 | 41 | 42 | 43 | 44 | 45 |
| Cedar Rapids, IA | 6,640 | 272,726 | 1 | 3,156 | 11.5 | 2,410 | 8.8 | 10,657 | 4.8 | 52,480 | 34,629 | 17,853 | 514 | 188.5 |
| Chambersburg-Waynesboro, PA | 2,690 | 146,232 | 0.88 | 1,717 | 11.0 | 1,763 | 11.3 | 10,407 | 8.5 | 36,422 | 25,326 | 11,096 | NA | NA |
| Champaign-Urbana, IL | 16,456 | 229,428 | 1.03 | 2,380 | 10.6 | 1,717 | 7.6 | 12,559 | 7.0 | 33,918 | 17,320 | 16,598 | 1,099 | 485.4 |
| Charleston, WV | 3,757 | 270,164 | 1.12 | 2,581 | 10.2 | 3,799 | 14.9 | 16,037 | 8.0 | 67,843 | 38,263 | 29,580 | NA | NA |
| Charleston-North Charleston, SC | 16,739 | 805,123 | 1.01 | 9,594 | 11.7 | 7,082 | 8.6 | 83,221 | 12.6 | 142,702 | 103,252 | 39,450 | 3,202 | 398.0 |
| Charlotte-Concord-Gastonia, NC-SC | 37,927 | 2,664,456 | 1.02 | 31,198 | 11.6 | 22,330 | 8.3 | 284,932 | 12.7 | 428,504 | 250,942 | 177,562 | NA | NA |
| Charlottesville, VA | 10,593 | 229,383 | 1.1 | 2,247 | 10.2 | 1,932 | 8.8 | 16,337 | 9.7 | 43,732 | 37,246 | 6,487 | NA | NA |
| Chattanooga, TN-GA | 13,214 | 574,877 | 1.03 | 6,262 | 11.0 | 6,403 | 11.2 | 56,829 | 12.6 | 119,348 | 68,201 | 51,147 | 385 | 175.7 |
| Cheyenne, WY | 2,076 | D | D | 1,212 | 12.0 | 960 | 9.5 | 9,645 | 11.8 | 19,635 | 18,748 | 887 | 2,969 | 525.6 |
| Chicago-Naperville-Elgin, IL-IN-WI | 161,138 | 9,501,189 | 1.01 | 106,292 | 11.3 | 81,256 | 8.6 | 750,044 | 9.5 | 1,557,393 | 1,087,899 | 469,494 | NA | NA |
| Chicago-Naperville-Evanston, IL Div 16984 | 114,403 | 7,288,720 | 1.05 | 80,897 | 11.4 | 61,635 | 8.7 | 577,789 | 9.7 | 1,159,955 | 804,077 | 355,877 | NA | NA |
| Elgin, IL Div 20994 | 12,674 | 685,548 | 0.79 | 8,777 | 11.5 | 5,251 | 6.9 | 59,930 | 9.2 | 113,724 | 81,706 | 32,019 | NA | NA |
| Gary, IN Div 23844 | 11,012 | 657,843 | 0.86 | 7,889 | 11.2 | 7,705 | 10.9 | 54,083 | 9.4 | 139,848 | 93,708 | 46,140 | NA | NA |
| Lake County-Kenosha County, IL-WI Div | 23,049 | D | D | 8,729 | 10.1 | 6,665 | 7.7 | 58,242 | 8.1 | 143,866 | 108,408 | 35,458 | NA | NA |
| Chico, CA | 5,443 | 220,980 | 1.02 | 2,236 | 10.5 | 2,371 | 11.1 | 13,710 | 7.8 | 45,888 | 42,635 | 3,253 | 1,024 | 443.8 |
| Cincinnati, OH-KY-IN | 51,472 | 2,219,955 | 1 | 26,749 | 12.0 | 21,844 | 9.8 | 130,813 | 7.1 | 409,269 | 220,586 | 188,684 | 5,264 | 237.4 |
| Clarksville, TN-KY | 10,137 | 303,123 | 0.94 | 5,320 | 16.9 | 2,523 | 8.0 | 24,317 | 9.3 | 46,085 | 32,746 | 13,340 | 1,201 | 387.5 |
| Cleveland, TN | 2,809 | 119,922 | 0.88 | 1,386 | 11.0 | 1,461 | 11.6 | 14,045 | 14.1 | 27,866 | 15,124 | 12,742 | 698 | 559.8 |
| Cleveland-Elyria, OH | 43,139 | 2,101,024 | 1.05 | 21,667 | 10.6 | 22,897 | 11.2 | 119,385 | 7.3 | 441,362 | 231,118 | 210,245 | 8,271 | 403.3 |
| Coeur d'Alene, ID | 1,467 | 161,707 | 0.95 | 1,785 | 10.5 | 1,675 | 9.8 | 16,442 | 12.4 | 40,392 | 26,300 | 14,092 | 362 | 219.0 |
| College Station-Bryan, TX | 14,802 | 269,895 | 1.01 | 2,979 | 11.1 | 1,683 | 6.3 | 39,507 | 17.9 | 33,158 | 23,368 | 9,789 | 696 | 262.4 |
| Colorado Springs, CO | 19,950 | 739,407 | 0.98 | 9,262 | 12.3 | 5,477 | 7.3 | 55,987 | 9.0 | 118,301 | 71,450 | 46,852 | 3,482 | 465.7 |
| Columbia, MO | 10,737 | 211,384 | 1.03 | 2,369 | 11.3 | 1,533 | 7.3 | 18,931 | 11.2 | 33,912 | 21,633 | 12,278 | 570 | 271.8 |
| Columbia, SC | 32,171 | 841,983 | 1.01 | 9,197 | 10.9 | 7,880 | 9.3 | 81,315 | 12.0 | 153,494 | 108,547 | 44,948 | 4,909 | 583.4 |
| Columbus, GA-AL | 12,352 | 330,472 | 1.06 | 4,232 | 13.1 | 3,499 | 10.8 | 33,486 | 12.8 | 59,382 | 34,984 | 24,398 | NA | NA |
| Columbus, IN | 1,148 | 93,225 | 1.23 | 1,051 | 12.4 | 790 | 9.4 | 6,827 | 9.8 | 15,717 | 11,487 | 4,230 | 122 | 145.8 |
| Columbus, OH | 51,190 | 2,158,514 | 1.03 | 26,863 | 12.6 | 18,200 | 8.5 | 142,669 | 8.0 | 334,043 | 175,174 | 158,868 | 6,100 | 286.5 |
| Corpus Christi, TX | 7,223 | 436,508 | 1.04 | 5,384 | 12.5 | 3,825 | 8.9 | 70,539 | 19.7 | 73,985 | 32,334 | 41,651 | 3,166 | 736.9 |
| Corvallis, OR | 5,819 | 96,837 | 1.08 | 678 | 7.3 | 650 | 7.0 | 4,861 | 6.7 | 17,553 | 9,045 | 8,507 | NA | NA |
| Crestview-Fort Walton Beach-Destin, FL | 7,252 | 295,929 | 1.08 | 3,583 | 12.4 | 2,825 | 9.8 | 34,378 | 15.0 | 58,421 | 43,312 | 15,108 | 885 | 313.2 |
| Cumberland, MD-WV | 8,119 | 97,140 | 1 | 890 | 9.2 | 1,304 | 13.5 | 4,231 | 6.1 | 23,920 | 21,536 | 2,385 | 266 | 274.8 |
| Dallas-Fort Worth-Arlington, TX | 82,739 | 7,623,582 | 1.01 | 97,614 | 12.7 | 52,487 | 6.8 | 1,287,901 | 19.5 | 988,076 | 573,843 | 414,233 | NA | NA |
| Dallas-Plano-Irving, TX Div 19124 | 52,597 | 5,170,692 | 1.03 | 65,842 | 12.7 | 33,788 | 6.5 | 877,330 | 19.7 | 638,109 | 387,222 | 250,887 | NA | NA |
| Fort Worth-Arlington-Grapevine, TX Div 23104 | 30,142 | 2,452,890 | 0.97 | 31,772 | 12.6 | 18,699 | 7.4 | 410,571 | 19.0 | 349,967 | 186,621 | 163,346 | NA | NA |
| Dalton, GA | 1,399 | 147,876 | 1.05 | 1,715 | 11.9 | 1,396 | 9.7 | 26,967 | 22.2 | 25,126 | 18,495 | 6,631 | NA | NA |
| Danville, IL | 2,906 | 76,685 | 1.03 | 880 | 11.8 | 981 | 13.1 | 3,829 | 6.6 | 17,556 | 10,070 | 7,486 | 710 | 935.7 |
| Daphne-Fairhope-Foley, AL | 2,268 | 210,062 | 0.86 | 2,317 | 10.1 | 2,543 | 11.1 | 19,085 | 10.9 | 55,742 | 29,149 | 26,592 | NA | NA |
| Davenport-Moline-Rock Island, IA-IL | 8,870 | 381,439 | 1.01 | 4,350 | 11.5 | 3,900 | 10.3 | 19,970 | 6.6 | 80,113 | 50,219 | 29,895 | 1,392 | 366.4 |
| Dayton-Kettering, OH | 24,730 | 830,970 | 1.06 | 9,272 | 11.5 | 9,277 | 11.5 | 52,791 | 8.2 | 168,623 | 85,529 | 83,095 | 2,647 | 328.4 |
| Decatur, AL | 2,236 | 142,811 | 0.85 | 1,748 | 11.4 | 1,859 | 12.2 | 16,285 | 13.2 | 34,400 | 21,640 | 12,760 | NA | NA |
| Decatur, IL | 4,009 | 110,423 | 1.14 | 1,234 | 12.0 | 1,245 | 12.1 | 4,788 | 6.0 | 24,263 | 19,173 | 5,090 | 434 | 419.3 |
| Deltona-Daytona Beach-Ormond Beach, FL | 15,164 | 629,436 | 0.87 | 5,676 | 8.3 | 9,372 | 13.8 | 79,310 | 16.5 | 191,434 | 91,968 | 99,466 | 2,120 | 319.3 |
| Denver-Aurora-Lakewood, CO | 34,514 | 2,956,347 | 0.99 | 33,645 | 11.2 | 20,940 | 7.0 | 229,794 | 9.0 | 431,530 | 202,564 | 228,966 | NA | NA |
| Des Moines-West Des Moines, IA | 13,621 | 703,595 | 1.01 | 9,081 | 12.8 | 5,653 | 8.0 | 31,630 | 5.3 | 114,377 | 80,321 | 34,056 | NA | NA |
| Detroit-Warren-Dearborn, MI | 48,556 | 4,330,297 | 1.01 | 48,689 | 11.3 | 45,272 | 10.5 | 243,170 | 6.8 | 853,639 | 454,868 | 398,770 | 22,332 | 516.9 |
| Detroit-Dearborn-Livonia, MI Div 19804 | 22,422 | 1,814,113 | 1.08 | 22,477 | 12.9 | 19,638 | 11.3 | 110,013 | 7.6 | 330,614 | 164,264 | 166,350 | 17,147 | 984.3 |
| Warren-Troy-Farmington Hills, MI Div 47664 | 26,134 | 2,516,184 | 0.96 | 26,212 | 10.2 | 25,634 | 10.0 | 133,157 | 6.3 | 523,025 | 290,604 | 232,420 | 5,185 | 201.1 |
| Dothan, AL | 1,846 | 150,190 | 1.01 | 1,789 | 11.9 | 1,861 | 12.4 | 15,970 | 13.4 | 36,194 | 20,879 | 15,315 | NA | NA |
| Dover, DE | 4,615 | 173,894 | 0.92 | 2,151 | 11.7 | 1,947 | 10.6 | 13,058 | 9.0 | 38,492 | 31,219 | 7,273 | 770 | 427.4 |
| Dubuque, IA | 4,139 | 105,019 | 1.15 | 1,196 | 12.3 | 970 | 9.9 | 3,726 | 4.9 | 20,642 | 9,402 | 11,240 | 130 | 134.4 |
| Duluth, MN-WI | 12,325 | 291,329 | 1.01 | 2,710 | 9.4 | 3,182 | 11.0 | 12,444 | 5.6 | 68,672 | 28,832 | 39,841 | 604 | 209.1 |
| Durham-Chapel Hill, NC | 29,702 | 680,507 | 1.11 | 6,940 | 10.6 | 5,323 | 8.2 | 67,123 | 13.0 | 113,987 | 68,058 | 45,930 | 2,830 | 439.1 |
| East Stroudsburg, PA | 4,164 | 161,019 | 0.89 | 1,521 | 8.9 | 1,775 | 10.4 | 11,267 | 8.3 | 34,769 | 24,520 | 10,249 | 400 | 236.4 |
| Eau Claire, WI | 7,308 | 171,512 | 1.03 | 1,783 | 10.5 | 1,482 | 8.7 | 9,231 | 6.9 | 34,577 | 22,020 | 12,557 | 363 | 214.3 |
| El Centro, CA | 8,535 | 176,445 | 0.92 | 2,541 | 14.1 | 1,120 | 6.2 | 14,070 | 9.6 | 32,249 | 24,264 | 7,985 | 615 | 339.4 |
| Elizabethtown-Fort Knox, KY | 3,715 | 151,071 | 0.95 | 1,794 | 11.6 | 1,471 | 9.5 | 8,334 | 6.5 | 29,579 | 20,476 | 9,103 | 153 | 99.8 |
| Elkhart-Goshen, IN | 3,691 | 237,179 | 1.31 | 3,098 | 15.0 | 1,817 | 8.8 | 25,976 | 15.1 | 35,101 | 21,881 | 13,220 | NA | NA |
| Elmira, NY | 3,606 | 84,961 | 1.04 | 846 | 10.2 | 1,014 | 12.3 | 2,636 | 4.1 | 20,120 | 10,557 | 9,563 | 178 | 213.9 |
| El Paso, TX | 16,748 | 840,455 | 0.99 | 11,611 | 13.7 | 5,863 | 6.9 | 174,435 | 24.4 | 134,010 | 48,731 | 85,279 | 2,816 | 333.2 |
| Enid, OK | 1,801 | 61,892 | 1.03 | 818 | 13.4 | 676 | 11.1 | 8,357 | 16.8 | 11,991 | 10,583 | 1,408 | 225 | 370.2 |
| Erie, PA | 12,120 | 274,065 | 1.04 | 2,681 | 10.0 | 3,012 | 11.2 | 14,801 | 7.1 | 60,631 | 29,082 | 31,549 | 809 | 299.0 |
| Eugene-Springfield, OR | 8,800 | 382,100 | 1 | 3,314 | 8.7 | 4,016 | 10.5 | 26,896 | 9.1 | 90,110 | 41,510 | 48,600 | D | D |

1. Per 1,000 estimated resident population.   2. Data for serious crimes have not been adjusted for underreporting; this may affect comparability between geographic areas and over time.   3. Per 100,000 population estimated by the FBI.

# Table C. Metropolitan Areas — Crime, Education, Money Income, and Poverty

| | Serious crimes known to police[1], 2019 (cont.) Property | | School enrollment and attainment, 2019 | | | | Local government expenditures,[5] 2017-2018 | | Income and poverty, 2019 | | | | | Percent below poverty level | | |
|---|---|---|---|---|---|---|---|---|---|---|---|---|---|---|---|---|
| | | | Enrollment[3] | | Attainment[4] | | | | | | | | | | | |
| Area name | Number | Rate | Total | Percent private | High school graduate or less | Bachelor's degree or more | Total current expenditures (mil dol) | Current expenditures per student (dollars) | Per capita income[6] (dollars) | Median household income (dollars) | Median family income | Percent of households with income less than $50,000 | Percent of households with income of $200,000 or more | All persons | All families | Families with children under 18 |
| | 46 | 47 | 48 | 49 | 50 | 51 | 52 | 53 | 54 | 55 | 56 | 57 | 58 | 59 | 60 | 61 |
| Cedar Rapids, IA .................. | 5,566 | 2,040.9 | 65,638 | 15.1 | 35.3 | 30.7 | 546.3 | 12,142 | 33,499 | 64,088 | 80,265 | 38.4 | 4.6 | 11.4 | 8.5 | 13.8 |
| Chambersburg-Waynesboro, PA.. | NA | NA | 31,620 | 17.8 | 56.2 | 22 | 280.0 | 12,234 | 31,022 | 62,484 | 77,162 | 39.7 | 3.3 | 6.8 | 4.6 | 8.1 |
| Champaign-Urbana, IL ................. | 4,507 | 1,990.8 | 82,357 | 9.7 | 30.2 | 44.5 | 415.5 | 14,965 | 30,181 | 53,081 | 82,092 | 45.9 | 5.7 | 20.0 | 10.9 | 15.4 |
| Charleston, WV .................. | NA | NA | 50,435 | 10.0 | 52.3 | 22.6 | 473.5 | 11,735 | 28,107 | 47,500 | 59,256 | 52.4 | 3.9 | 17.1 | 13.9 | 20.9 |
| Charleston-North Charleston, SC . | 21,554 | 2,678.8 | 185,020 | 16.7 | 33.1 | 37.5 | 1,184.7 | 10,456 | 38,050 | 70,505 | 86,720 | 35.1 | 8.3 | 10.7 | 7.8 | 12.6 |
| Charlotte-Concord-Gastonia, NC-SC.............................. | NA | NA | 655,722 | 16.8 | 33.0 | 36.2 | 3,777.4 | 9,097 | 36,374 | 66,399 | 82,452 | 36.9 | 8.9 | 9.7 | 6.6 | 9.9 |
| Charlottesville, VA .................. | 3,104 | 1,416.6 | 62,239 | 14.1 | 29.0 | 49.2 | 353.3 | 13,068 | 42,734 | 75,907 | 99,990 | 33.2 | 10.9 | 11.2 | 4.4 | 6.6 |
| Chattanooga, TN-GA.............. | 16,332 | 2,891.0 | 130,695 | 20.8 | 40.2 | 29.5 | 744.3 | 9,993 | 30,615 | 55,366 | 67,658 | 45.3 | 4.9 | 12.7 | 9.2 | 15.3 |
| Cheyenne, WY ...................... | NA | NA | 23,980 | 11.5 | 33.6 | 30.1 | 235.0 | 15,292 | 39,791 | 70,567 | 86,805 | 33.6 | 6.3 | 9.8 | 6.1 | 11.6 |
| Chicago-Naperville-Elgin, IL-IN-WI.............................. | NA | NA | 2,353,601 | 19.9 | 34.6 | 39.2 | 24,666.8 | 16,267 | 40,144 | 75,379 | 95,110 | 33.8 | 10.9 | 10.6 | 7.3 | 11.3 |
| Chicago-Naperville-Evanston, IL Div 16984 .......................... | NA | NA | 1,746,446 | 21.6 | 33.7 | 41 | 18,338.5 | 16,689 | 40,833 | 75,231 | 95,117 | 34.2 | 11.1 | 11.0 | 7.7 | 11.9 |
| Elgin, IL Div 20994 .................. | NA | NA | 215,010 | 13.6 | 35.8 | 34.6 | 2,157.2 | 15,119 | 36,886 | 84,450 | 100,622 | 28.1 | 10.1 | 8.1 | 5.8 | 9.1 |
| Gary, IN Div 23844 .................. | NA | NA | 167,557 | 14.0 | 46.5 | 23.2 | 1,151.8 | 10,142 | 31,866 | 62,335 | 76,516 | 40.4 | 4.7 | 12.3 | 7.8 | 13.0 |
| Lake County-Kenosha County, IL-WI Div .......................... | NA | NA | 224,588 | 16.6 | 31.4 | 42 | 3,019.3 | 18,718 | 44,085 | 85,092 | 106,332 | 29.0 | 14.6 | 7.5 | 5.1 | 8.1 |
| Chico, CA ........................... | 5,322 | 2,306.4 | 60,729 | 8.9 | 33.3 | 30.5 | 394.4 | 12,417 | 34,250 | 62,563 | 76,497 | 41.9 | 8.7 | 15.9 | 9.3 | 13.9 |
| Cincinnati, OH-KY-IN............. | 43,841 | 1,976.9 | 547,046 | 19.5 | 38.0 | 35.4 | 3,826.1 | 11,863 | 36,878 | 66,825 | 87,591 | 36.8 | 7.9 | 11.3 | 7.7 | 12.8 |
| Clarksville, TN-KY .................. | 7,005 | 2,260.4 | 79,203 | 15.8 | 38.1 | 26.4 | 436.8 | 9,136 | 26,667 | 53,547 | 63,520 | 46.8 | 3.0 | 14.3 | 12.0 | 18.3 |
| Cleveland, TN ...................... | 3,564 | 2,858.4 | 28,377 | 25.0 | 50.5 | 21.2 | 161.4 | 8,782 | 25,199 | 52,178 | 63,779 | 48.2 | 3.2 | 15.2 | 12.6 | 16.3 |
| Cleveland-Elyria, OH .............. | 37,974 | 1,851.4 | 477,937 | 23.6 | 37.6 | 31.7 | 3,994.4 | 14,285 | 34,200 | 57,228 | 76,861 | 43.7 | 5.7 | 13.5 | 9.1 | 16.2 |
| Coeur d'Alene, ID .................. | 2,099 | 1,269.7 | 36,595 | 22.4 | 32.1 | 24.5 | 170.6 | 7,575 | 30,041 | 62,579 | 73,909 | 38.0 | 4.1 | 10.0 | 8.1 | 14.7 |
| College Station-Bryan, TX ....... | 5,228 | 1,971.2 | 98,660 | 7.7 | 36.4 | 38 | 353.0 | 9,544 | 28,798 | 55,670 | 74,655 | 46.2 | 6.2 | 20.5 | 12.2 | 17.6 |
| Colorado Springs, CO ............ | 21,290 | 2,847.7 | 196,413 | 14.5 | 25.6 | 39.2 | 1,181.0 | 9,536 | 35,683 | 72,633 | 86,530 | 32.3 | 7.6 | 8.4 | 5.5 | 8.6 |
| Columbia, MO ...................... | 4,577 | 2,182.2 | 68,993 | 12.1 | 29.5 | 44 | 305.9 | 11,063 | 30,582 | 57,359 | 77,585 | 43.4 | 6.2 | 18.3 | 7.8 | 11.0 |
| Columbia, SC ....................... | 28,376 | 3,372.3 | 213,894 | 11.6 | 36.2 | 34.4 | 2,035.4 | 13,176 | 31,065 | 55,725 | 74,672 | 45.2 | 5.3 | 15.1 | 11.0 | 17.4 |
| Columbus, GA-AL .................. | NA | NA | 86,474 | 11.7 | 39.8 | 23.1 | 512.7 | 10,072 | 27,286 | 49,013 | 60,741 | 51.0 | 4.4 | 16.5 | 14.0 | 19.0 |
| Columbus, IN ....................... | 1,864 | 2,227.2 | 20,065 | 27.0 | 38.4 | 36.5 | 124.6 | 10,057 | 33,731 | 72,555 | 85,504 | 34.9 | 7.3 | 13.9 | 10.4 | 23.6 |
| Columbus, OH ...................... | 50,897 | 2,390.3 | 538,544 | 16.2 | 35.2 | 37.9 | 4,039.0 | 12,023 | 36,285 | 67,207 | 85,957 | 36.1 | 7.7 | 11.5 | 8.4 | 13.1 |
| Corpus Christi, TX ................. | 13,613 | 3,168.7 | 108,529 | 7.0 | 47.9 | 22.3 | 751.7 | 9,860 | 27,286 | 56,991 | 72,789 | 44.4 | 4.3 | 15.6 | 11.4 | 18.9 |
| Corvallis, OR ....................... | NA | NA | 33,159 | 9.8 | 15.6 | 57.9 | 101.5 | 11,154 | 35,237 | 70,835 | 100,075 | 35.7 | 6.2 | 17.9 | 8.4 | 14.7 |
| Crestview-Fort Walton Beach-Destin, FL .......................... | 4,718 | 1,669.9 | 62,086 | 18.3 | 35.6 | 31.1 | 378.7 | 9,201 | 32,299 | 62,627 | 80,729 | 39.0 | 7.0 | 10.9 | 6.6 | 12.4 |
| Cumberland, MD-WV .............. | 1,762 | 1,820.1 | 22,284 | 8.0 | 54.0 | 19.1 | 170.2 | 13,319 | 24,220 | 49,729 | 64,158 | 50.2 | 2.4 | 13.9 | 8.3 | 18.0 |
| Dallas-Fort Worth-Arlington, TX ... | NA | NA | 2,027,416 | 12.5 | 36.4 | 36.3 | 13,005.4 | 9,240 | 36,274 | 72,265 | 85,982 | 33.3 | 9.8 | 10.5 | 7.7 | 11.3 |
| Dallas-Plano-Irving, TX Div 19124 ............................... | NA | NA | 1,363,168 | 11.6 | 35.1 | 38.3 | 9,071.3 | 9,243 | 37,583 | 73,545 | 87,592 | 32.8 | 10.6 | 10.7 | 7.7 | 11.4 |
| Fort Worth-Arlington-Grapevine, TX Div 23104 .................... | NA | NA | 664,248 | 14.2 | 38.9 | 32 | 3,934.1 | 9,234 | 33,605 | 70,573 | 83,095 | 34.4 | 8.2 | 10.1 | 7.5 | 11.1 |
| Dalton, GA .......................... | NA | NA | 36,309 | 6.5 | 59.2 | 15 | 282.5 | 9,944 | 25,208 | 51,967 | 60,111 | 48.9 | 4.0 | 12.7 | 9.0 | 13.1 |
| Danville, IL .......................... | 2,161 | 2,848.0 | 16,544 | 5.8 | 54.8 | 12.1 | 181.9 | 14,416 | 25,592 | 43,111 | 56,233 | 56.9 | 2.5 | 16.5 | 11.9 | 22.4 |
| Daphne-Fairhope-Foley, AL ..... | NA | NA | 47,583 | 18.4 | 35.7 | 32.2 | 295.1 | 9,284 | 32,443 | 56,439 | 76,892 | 45.1 | 6.6 | 10.1 | 7.9 | 12.3 |
| Davenport-Moline-Rock Island, IA-IL ..................... | 8,353 | 2,198.7 | 87,871 | 17.0 | 39.1 | 27.1 | 786.1 | 12,815 | 33,388 | 62,001 | 78,466 | 40.2 | 5.1 | 11.3 | 7.9 | 13.0 |
| Dayton-Kettering, OH .............. | 18,206 | 2,258.9 | 199,680 | 22.2 | 36.0 | 30.2 | 1,476.5 | 13,176 | 32,270 | 58,169 | 75,683 | 43.6 | 4.7 | 13.6 | 9.5 | 16.8 |
| Decatur, AL ......................... | NA | NA | 31,951 | 6.8 | 51.8 | 21 | 244.5 | 10,031 | 28,346 | 53,447 | 63,475 | 46.8 | 3.8 | 15.5 | 12.6 | 17.0 |
| Decatur, IL .......................... | 2,382 | 2,301.3 | 24,180 | 20.6 | 46.5 | 21.5 | 201.9 | 12,533 | 29,772 | 50,839 | 75,694 | 49.2 | 4.4 | 17.4 | 11.9 | 16.4 |
| Deltona-Daytona Beach-Ormond Beach, FL .......................... | 11,988 | 1,805.3 | 129,844 | 17.6 | 42.3 | 24.5 | 650.3 | 8,557 | 30,352 | 54,533 | 66,116 | 46.1 | 4.1 | 12.4 | 9.0 | 16.4 |
| Denver-Aurora-Lakewood, CO ..... | NA | NA | 699,568 | 14.0 | 27.7 | 45.8 | 4,837.7 | 10,216 | 44,806 | 85,641 | 106,000 | 26.9 | 12.4 | 7.9 | 5.2 | 8.1 |
| Des Moines-West Des Moines, IA ......................... | NA | NA | 172,597 | 17.7 | 31.5 | 38 | 1,406.3 | 11,765 | 37,042 | 71,164 | 88,213 | 34.0 | 7.1 | 8.7 | 6.3 | 9.1 |
| Detroit-Warren-Dearborn, MI........ | 72,684 | 1,682.4 | 1,012,880 | 13.3 | 36.3 | 32.4 | 7,649.7 | 11,775 | 35,315 | 63,474 | 81,699 | 39.7 | 7.2 | 12.6 | 8.9 | 15.2 |
| Detroit-Dearborn-Livonia, MI Div 19804 ............................... | 46,878 | 2,691.1 | 425,397 | 11.2 | 43.5 | 25 | 3,218.0 | 11,520 | 28,623 | 50,753 | 65,526 | 49.1 | 4.6 | 19.6 | 15.0 | 24.5 |
| Warren-Troy-Farmington Hills, MI Div 47664 ..................... | 25,806 | 1,000.9 | 587,483 | 14.9 | 31.6 | 37.2 | 4,431.7 | 11,967 | 39,870 | 73,549 | 92,490 | 33.3 | 8.9 | 7.9 | 5.2 | 8.8 |
| Dothan, AL .......................... | NA | NA | 35,074 | 14.5 | 47.8 | 20.1 | 201.1 | 9,129 | 27,121 | 50,010 | 63,047 | 50.0 | 3.3 | 18.6 | 13.8 | 22.8 |
| Dover, DE ........................... | 4,019 | 2,230.7 | 44,386 | 11.9 | 43.2 | 24.5 | 373.3 | 14,053 | 28,004 | 58,001 | 69,439 | 43.1 | 2.1 | 13.0 | 10.5 | 19.1 |
| Dubuque, IA ......................... | 1,103 | 1,140.1 | 24,688 | 26.9 | 40.1 | 32.6 | 169.5 | 11,601 | 33,294 | 62,178 | 78,750 | 39.2 | 5.3 | 11.4 | 5.9 | 9.9 |
| Duluth, MN-WI ...................... | 7,697 | 2,665.2 | 68,354 | 13.1 | 33.2 | 28.4 | 484.2 | 12,235 | 32,589 | 60,316 | 79,332 | 40.7 | 3.9 | 12.4 | 6.9 | 12.0 |
| Durham-Chapel Hill, NC .......... | 16,639 | 2,581.7 | 173,126 | 19.8 | 31.1 | 46.3 | 934.7 | 10,436 | 36,778 | 65,303 | 83,576 | 39.2 | 8.8 | 14.0 | 9.7 | 17.2 |
| East Stroudsburg, PA.............. | 2,551 | 1,507.4 | 37,061 | 12.4 | 44.0 | 26 | 475.0 | 18,681 | 31,155 | 62,413 | 77,039 | 39.1 | 4.8 | 13.0 | 8.9 | 15.8 |
| Eau Claire, WI ...................... | 2,600 | 1,535.2 | 44,418 | 13.8 | 33.8 | 29.9 | 274.1 | 11,676 | 32,991 | 63,460 | 80,963 | 37.7 | 4.5 | 9.3 | 4.5 | 8.0 |
| El Centro, CA........................ | 3,619 | 1,997.3 | 55,760 | 2.5 | 51.6 | 18.4 | 531.9 | 14,102 | 18,800 | 48,472 | 57,467 | 52.4 | 3.0 | 25.1 | 20.0 | 25.9 |
| Elizabethtown-Fort Knox, KY..... | 1,738 | 1,134.1 | 34,828 | 11.1 | 43.7 | 21.4 | 248.0 | 10,045 | 28,963 | 55,246 | 70,863 | 44.1 | 4.0 | 11.1 | 7.1 | 9.3 |
| Elkhart-Goshen, IN ................. | NA | NA | 49,626 | 16.6 | 54.8 | 19.9 | 368.1 | 10,107 | 26,745 | 54,531 | 75,244 | 45.5 | 4.0 | 9.2 | 6.1 | 14.1 |
| Elmira, NY .......................... | 1,445 | 1,736.1 | 16,663 | 14.2 | 40.1 | 25.3 | 204.6 | 17,448 | 29,639 | 60,782 | 70,078 | 41.3 | 2.9 | 13.9 | 9.1 | 14.1 |
| Elmira, NY .......................... | | | | | | | | | | | | 50.9 | 2.4 | 18.7 | 15.9 | 22.4 |
| El Paso, TX ......................... | 11,967 | 1,415.9 | 250,816 | 7.6 | 45.9 | 23.3 | 1,739.1 | 9,953 | 21,915 | 48,823 | 53,413 | 36.2 | 2.7 | 10.1 | 7.7 | 12.6 |
| Enid, OK ............................. | 1,812 | 2,981.6 | 16,169 | 16.0 | 44.8 | 22.3 | 92.6 | 8,015 | 32,346 | 64,929 | 76,017 | 48.0 | 3.5 | 17.4 | 12.5 | 20.6 |
| Erie, PA .............................. | 4,322 | 1,597.2 | 62,444 | 24.3 | 45.7 | 29.5 | 532.5 | 13,939 | 28,502 | 51,818 | 67,505 | 45.0 | 4.5 | 14.0 | 7.3 | 12.3 |
| Eugene-Springfield, OR ........... | 10,440 | 2,737.0 | 89,426 | 9.7 | 30.4 | 32.3 | 530.8 | 11,549 | 32,333 | 57,325 | 75,523 | | | | | |

1. Data for serious crimes have not been adjusted for underreporting; this may affect comparability between geographic areas and over time. 2. Per 100,000 population estimated by the FBI. 3. All persons 3 years old and over enrolled in nursery school through college. 4. Persons 25 years old and over. 5. Elementary and secondary education expenditures. 6. Based on population estimated by the American Community Survey, 2017.

# Table C. Metropolitan Areas — Personal Income and Earnings

| Area name | Personal income, 2019 | | | | | | | | | | Earnings, 2019 | | |
|---|---|---|---|---|---|---|---|---|---|---|---|---|---|
| | Total (mil dol) | Percent change, 2018-2019 | Per capita[1] Dollars | Rank | Wages and Salaries (mil dol) | Supplements to wages and salaries, employer contributions (mil dol) Pension and insurance | Government social insurance | Proprietors' income | Dividends, interest, and rent (mil dol) | Personal transfer receipts (mil dol) | Total (mil dol) | Contributions for government social insurance (mil dol) From employee and self-employed | From employer |
| | 62 | 63 | 64 | 65 | 66 | 67 | 68 | 69 | 70 | 71 | 72 | 73 | 74 |
| Cedar Rapids, IA | 14,510 | 2.1 | 53,143 | 101 | 7,912 | 1,292 | 606 | 899 | 2,801 | 2,440 | 10,709 | 683 | 606 |
| Chambersburg-Waynesboro, PA | 7,431 | 3.8 | 47,934 | 183 | 2,850 | 564 | 236 | 596 | 1,164 | 1,614 | 4,245 | 268 | 236 |
| Champaign-Urbana, IL | 10,537 | 2.7 | 46,616 | 211 | 5,514 | 1,369 | 332 | 691 | 2,167 | 1,493 | 7,906 | 382 | 332 |
| Charleston, WV | 11,639 | 1.2 | 45,276 | 239 | 6,153 | 1,076 | 465 | 891 | 1,839 | 3,408 | 8,586 | 579 | 465 |
| Charleston-North Charleston, SC | 42,608 | 5.6 | 53,119 | 102 | 20,898 | 3,615 | 1,566 | 3,660 | 9,547 | 6,791 | 29,739 | 1,818 | 1,566 |
| Charlotte-Concord-Gastonia, NC-SC | 142,170 | 4.8 | 53,916 | 96 | 81,037 | 10,713 | 5,504 | 14,357 | 22,702 | 20,929 | 111,611 | 6,800 | 5,504 |
| Charlottesville, VA | 15,085 | 2.5 | 69,003 | 17 | 6,656 | 1,211 | 459 | 1,127 | 5,293 | 1,777 | 9,454 | 573 | 459 |
| Chattanooga, TN-GA | 26,777 | 3.7 | 47,376 | 203 | 13,271 | 2,140 | 920 | 3,217 | 4,336 | 5,538 | 19,549 | 1,195 | 920 |
| Cheyenne, WY | 5,384 | 3.4 | 54,113 | 93 | 2,735 | 575 | 257 | 402 | 1,155 | 934 | 3,969 | 233 | 257 |
| Chicago-Naperville-Elgin, IL-IN-WI | 600,617 | 2.7 | 63,500 | 30 | 327,495 | 49,903 | 21,619 | 47,536 | 124,245 | 80,785 | 446,553 | 25,114 | 21,619 |
| Chicago-Naperville-Evanston, IL Div 16984 | 463,215 | 2.6 | 65,033 | NA | 265,475 | 39,393 | 17,453 | 39,089 | 98,306 | 61,889 | 361,410 | 20,201 | 17,453 |
| Elgin, IL Div 20994 | 38,631 | 3.4 | 50,412 | NA | 15,335 | 2,935 | 1,031 | 2,097 | 5,828 | 5,126 | 21,399 | 1,192 | 1,031 |
| Gary, IN Div 23844 | 34,130 | 3.7 | 48,520 | NA | 14,071 | 2,305 | 1,033 | 2,241 | 4,900 | 7,233 | 19,650 | 1,269 | 1,033 |
| Lake County-Kenosha County, IL-WI Div | 64,641 | 2.7 | 74,635 | NA | 32,615 | 5,270 | 2,101 | 4,109 | 15,210 | 6,537 | 44,095 | 2,451 | 2,101 |
| Chico, CA | 10,490 | 2.2 | 47,860 | 187 | 4,052 | 871 | 289 | 1,020 | 2,166 | 2,628 | 6,232 | 383 | 289 |
| Cincinnati, OH-KY-IN | 124,462 | 4.1 | 56,033 | 72 | 66,432 | 9,968 | 4,638 | 9,526 | 23,717 | 19,731 | 90,564 | 5,359 | 4,638 |
| Clarksville, TN-KY | 12,892 | 4.6 | 41,881 | 303 | 5,963 | 1,406 | 493 | 989 | 2,027 | 2,921 | 8,852 | 474 | 493 |
| Cleveland, TN | 4,978 | 4.0 | 39,839 | 346 | 2,085 | 350 | 150 | 463 | 682 | 1,284 | 3,048 | 201 | 150 |
| Cleveland-Elyria, OH | 113,589 | 2.9 | 55,451 | 80 | 62,356 | 9,787 | 4,387 | 7,779 | 22,362 | 21,022 | 84,309 | 4,973 | 4,387 |
| Coeur d'Alene, ID | 7,729 | 5.6 | 46,645 | 210 | 2,920 | 485 | 250 | 578 | 1,729 | 1,588 | 4,233 | 299 | 250 |
| College Station-Bryan, TX | 11,050 | 4.4 | 41,742 | 310 | 5,317 | 1,062 | 332 | 946 | 2,417 | 1,704 | 7,657 | 378 | 332 |
| Colorado Springs, CO | 38,125 | 5.5 | 51,120 | 127 | 18,737 | 3,141 | 1,404 | 2,246 | 7,750 | 6,808 | 25,528 | 1,416 | 1,404 |
| Columbia, MO | 9,884 | 5.0 | 47,482 | 197 | 5,262 | 1,134 | 349 | 504 | 1,923 | 1,644 | 7,249 | 408 | 349 |
| Columbia, SC | 39,007 | 4.4 | 46,523 | 213 | 20,205 | 3,848 | 1,516 | 2,951 | 6,420 | 7,794 | 28,520 | 1,760 | 1,516 |
| Columbus, GA-AL | 13,543 | 3.9 | 42,183 | 299 | 7,178 | 1,494 | 536 | 447 | 3,244 | 3,310 | 9,656 | 557 | 536 |
| Columbus, IN | 4,402 | 3.9 | 52,546 | 108 | 3,167 | 473 | 223 | 274 | 759 | 728 | 4,137 | 251 | 223 |
| Columbus, OH | 111,370 | 3.5 | 52,477 | 109 | 64,534 | 10,616 | 4,292 | 7,665 | 18,843 | 16,894 | 87,106 | 4,731 | 4,292 |
| Corpus Christi, TX | 19,366 | 3.4 | 45,140 | 243 | 9,922 | 1,760 | 692 | 1,399 | 3,185 | 4,249 | 13,773 | 763 | 692 |
| Corvallis, OR | 4,534 | 4.2 | 48,725 | 165 | 2,176 | 447 | 179 | 323 | 1,220 | 653 | 3,126 | 193 | 179 |
| Crestview-Fort Walton Beach-Destin, FL | 15,457 | 5.0 | 54,270 | 91 | 7,153 | 1,389 | 549 | 1,075 | 4,542 | 2,833 | 10,166 | 610 | 549 |
| Cumberland, MD-WV | 4,008 | 3.0 | 41,195 | 324 | 1,720 | 347 | 137 | 171 | 669 | 1,306 | 2,375 | 166 | 137 |
| Dallas-Fort Worth-Arlington, TX | 444,730 | 4.9 | 58,725 | 54 | 249,941 | 32,210 | 16,443 | 56,027 | 79,477 | 49,605 | 354,621 | 18,623 | 16,443 |
| Dallas-Plano-Irving, TX Div 19124 | 313,876 | 5.0 | 61,763 | NA | 187,087 | 22,967 | 12,133 | 43,086 | 56,658 | 32,030 | 265,273 | 13,806 | 12,133 |
| Fort Worth-Arlington-Grapevine, TX Div 23104 | 130,855 | 4.7 | 52,527 | NA | 62,854 | 9,243 | 4,310 | 12,941 | 22,819 | 17,575 | 89,348 | 4,817 | 4,310 |
| Dalton, GA | 5,419 | 2.4 | 37,444 | 373 | 3,096 | 486 | 218 | 525 | 1,047 | 1,183 | 4,325 | 261 | 218 |
| Danville, IL | 3,039 | 1.1 | 40,109 | 342 | 1,321 | 302 | 93 | 180 | 476 | 850 | 1,896 | 117 | 93 |
| Daphne-Fairhope-Foley, AL | 10,600 | 5.3 | 47,485 | 196 | 3,263 | 488 | 237 | 673 | 2,113 | 2,308 | 4,662 | 345 | 237 |
| Davenport-Moline-Rock Island, IA-IL | 19,149 | 2.6 | 50,502 | 135 | 9,952 | 1,767 | 712 | 1,206 | 3,691 | 3,678 | 13,638 | 844 | 712 |
| Dayton-Kettering, OH | 39,703 | 3.2 | 49,161 | 155 | 21,234 | 3,816 | 1,544 | 2,566 | 7,324 | 8,077 | 29,161 | 1,718 | 1,544 |
| Decatur, AL | 6,149 | 3.9 | 40,295 | 338 | 2,750 | 472 | 200 | 301 | 975 | 1,560 | 3,723 | 261 | 200 |
| Decatur, IL | 5,072 | 0.7 | 48,764 | 164 | 2,850 | 509 | 202 | 340 | 883 | 1,158 | 3,900 | 235 | 202 |
| Deltona-Daytona Beach-Ormond Beach, FL | 29,917 | 4.6 | 44,762 | 255 | 9,238 | 1,444 | 643 | 1,428 | 6,928 | 8,104 | 12,752 | 1,017 | 643 |
| Denver-Aurora-Lakewood, CO | 199,504 | 5.2 | 67,236 | 19 | 111,128 | 13,489 | 7,564 | 23,223 | 40,351 | 21,464 | 155,404 | 8,679 | 7,564 |
| Des Moines-West Des Moines, IA | 38,488 | 3.4 | 55,039 | 81 | 22,482 | 3,336 | 1,658 | 2,379 | 6,952 | 5,448 | 29,856 | 1,865 | 1,658 |
| Detroit-Warren-Dearborn, MI | 234,001 | 3.0 | 54,172 | 92 | 127,211 | 18,283 | 9,183 | 17,236 | 40,418 | 44,965 | 171,913 | 10,853 | 9,183 |
| Detroit-Dearborn-Livonia, MI Div 19804 | 77,867 | 3.0 | 44,512 | NA | 49,126 | 7,361 | 3,529 | 4,768 | 12,055 | 21,011 | 64,784 | 4,109 | 3,529 |
| Warren-Troy-Farmington Hills, MI Div 47664 | 156,134 | 3.0 | 60,746 | NA | 78,085 | 10,921 | 5,653 | 12,468 | 28,363 | 23,954 | 107,128 | 6,743 | 5,653 |
| Dothan, AL | 6,444 | 4.1 | 43,143 | 280 | 2,681 | 453 | 193 | 429 | 1,129 | 1,701 | 3,756 | 260 | 193 |
| Dover, DE | 7,791 | 4.8 | 43,097 | 282 | 3,501 | 879 | 268 | 452 | 1,223 | 2,014 | 5,099 | 309 | 268 |
| Dubuque, IA | 5,014 | 3.2 | 51,525 | 121 | 2,882 | 472 | 222 | 339 | 1,144 | 911 | 3,914 | 251 | 222 |
| Duluth, MN-WI | 13,759 | 3.4 | 47,654 | 189 | 6,729 | 1,203 | 536 | 678 | 2,378 | 3,532 | 9,146 | 612 | 536 |
| Durham-Chapel Hill, NC | 35,069 | 4.3 | 54,423 | 88 | 23,829 | 3,642 | 1,621 | 2,247 | 7,820 | 5,214 | 31,338 | 1,879 | 1,621 |
| East Stroudsburg, PA | 7,462 | 4.3 | 43,823 | 271 | 2,736 | 619 | 220 | 508 | 1,025 | 1,716 | 4,083 | 257 | 220 |
| Eau Claire, WI | 8,288 | 3.4 | 48,956 | 157 | 4,236 | 824 | 313 | 646 | 1,575 | 1,535 | 6,019 | 377 | 313 |
| El Centro, CA | 7,330 | 7.4 | 40,447 | 336 | 2,854 | 802 | 212 | 1,228 | 994 | 1,884 | 5,096 | 247 | 212 |
| Elizabethtown-Fort Knox, KY | 6,631 | 4.4 | 43,076 | 283 | 2,888 | 678 | 230 | 412 | 1,067 | 1,573 | 4,208 | 252 | 230 |
| Elkhart-Goshen, IN | 9,693 | 1.7 | 46,975 | 208 | 7,009 | 1,136 | 506 | 1,083 | 1,539 | 1,675 | 9,735 | 573 | 506 |
| Elmira, NY | 3,874 | 3.7 | 46,421 | 217 | 1,760 | 414 | 140 | 192 | 571 | 1,028 | 2,507 | 150 | 140 |
| El Paso, TX | 31,767 | 4.0 | 37,633 | 370 | 15,301 | 3,304 | 1,109 | 2,283 | 4,955 | 7,309 | 21,998 | 1,183 | 1,109 |
| Enid, OK | 2,741 | 2.8 | 44,899 | 249 | 1,269 | 224 | 97 | 198 | 604 | 577 | 1,788 | 109 | 97 |
| Erie, PA | 12,510 | 2.3 | 46,379 | 218 | 5,723 | 1,190 | 457 | 810 | 2,297 | 3,210 | 8,181 | 510 | 457 |
| Eugene-Springfield, OR | 18,087 | 3.6 | 47,340 | 204 | 7,685 | 1,382 | 678 | 1,717 | 3,908 | 4,224 | 11,463 | 764 | 678 |

1. Based on the resident population estimated as of July 1 of the year shown.

# Table C. Metropolitan Areas — Earnings, Social Security, and Housing

| Area name | Earnings, 2019 (cont.) — Percent by selected industries | | | | | | | | | Social Security beneficiaries, December 2019 | | Supplemental Security Income Recipients, December 2019 | Housing units, 2020 | |
|---|---|---|---|---|---|---|---|---|---|---|---|---|---|---|
| | Farm | Mining, quarrying, and extracting | Construction | Manufacturing | Information; professional, scientific, and technical services | Retail trade | Finance, insurance, real estate, rental and leasing | Health care and social assistance | Government | Number | Rate[1] | | Total | Percent change, 2010-2020 |
| | 75 | 76 | 77 | 78 | 79 | 80 | 81 | 82 | 83 | 84 | 85 | 86 | 87 | 88 |
| Cedar Rapids, IA | 1.4 | 0.2 | 7.2 | 20.6 | 8.9 | 5.2 | 9.6 | D | 11.2 | 54,505 | 199 | 4,647 | 119,689 | 6.6 |
| Chambersburg-Waynesboro, PA | 1.9 | 0.2 | 5.7 | 16.7 | 5.0 | 6.7 | 3.7 | 15.4 | 16.2 | 37,515 | 241 | 2,476 | 66,453 | 5.1 |
| Champaign-Urbana, IL | 1.0 | D | 4.9 | 6.0 | 7.4 | 5.4 | 5.4 | D | 38.4 | 32,210 | 143 | 3,326 | 103,230 | 8.8 |
| Charleston, WV | -0.1 | D | 6.8 | D | D | 5.7 | D | D | 18.8 | 73,905 | 287 | 11,309 | 131,442 | 0.0 |
| Charleston-North Charleston, SC | 0.0 | D | 8.2 | 10.2 | 14.3 | 6.5 | 6.0 | 9.6 | 21.3 | 147,185 | 183 | 13,016 | 352,291 | 18.0 |
| Charlotte-Concord-Gastonia, NC-SC | 0.3 | 0.1 | 6.6 | 8.2 | D | 5.4 | D | 6.8 | 11.2 | 449,115 | 170 | 43,083 | 1,107,724 | 16.9 |
| Charlottesville, VA | 0.0 | 0.1 | 4.9 | 3.0 | 13.1 | 4.6 | 8.0 | D | 33.8 | 42,995 | 196 | 2,886 | 99,050 | 11.1 |
| Chattanooga, TN-GA | 0.1 | D | D | 13.2 | D | 5.7 | 13.4 | D | 16.1 | 126,310 | 223 | 12,973 | 253,485 | 8.1 |
| Cheyenne, WY | 0.3 | 2.9 | 7.4 | 3.9 | 8.0 | 5.8 | 6.1 | 7.9 | 37.8 | 19,560 | 196 | 1,480 | 44,797 | 10.7 |
| Chicago-Naperville-Elgin, IL-IN-WI | 0.0 | 0.1 | 4.6 | 9.8 | D | 4.7 | 12.2 | 9.7 | 11.7 | 1,536,920 | 163 | 197,253 | 3,880,877 | 2.2 |
| Chicago-Naperville-Evanston, IL Div 16984 | 0.0 | 0.1 | 4.2 | 7.4 | 17.9 | 4.3 | 13.7 | 9.8 | 11.0 | 1,131,665 | 159 | 162,977 | 2,961,200 | 1.8 |
| Elgin, IL Div 20994 | 0.2 | 0.1 | 8.2 | 16.3 | 7.7 | 5.8 | 5.1 | 9.9 | 19.5 | 112,775 | 147 | 7,119 | 276,498 | 4.9 |
| Gary, IN Div 23844 | 0.5 | 0.1 | 9.2 | 19.1 | 4.7 | 6.3 | 4.2 | 15.2 | 10.5 | 149,900 | 213 | 15,321 | 305,891 | 4.0 |
| Lake County-Kenosha County, IL-WI Div | 0.1 | 0.0 | 4.2 | 21.4 | 9.3 | 6.8 | 6.4 | 7.0 | 13.6 | 142,580 | 164 | 11,836 | 337,288 | 2.3 |
| Chico, CA | 3.9 | 0.1 | 7.5 | 4.8 | 5.9 | 9.1 | 5.5 | 19.1 | 21.7 | 47,060 | 215 | 9,947 | 86,511 | -9.7 |
| Cincinnati, OH-KY-IN | 0.0 | D | D | 13.7 | D | 5.2 | 9.4 | 11.9 | 10.8 | 413,135 | 186 | 47,797 | 955,626 | 3.8 |
| Clarksville, TN-KY | 0.4 | D | 5.1 | 10.2 | D | 6.0 | 3.3 | D | 45.3 | 51,605 | 167 | 6,669 | 131,534 | 15.2 |
| Cleveland, TN | 0.0 | D | D | 21.9 | D | 7.1 | 3.6 | D | 11.6 | 30,320 | 243 | 3,146 | 54,492 | 10.4 |
| Cleveland-Elyria, OH | 0.1 | D | 4.9 | 12.7 | 11.6 | 4.8 | 8.4 | 14.2 | 13.2 | 435,370 | 212 | 61,822 | 965,742 | 1.0 |
| Coeur d'Alene, ID | 0.0 | 0.7 | 10.0 | 8.0 | 7.5 | 9.9 | 7.2 | 13.3 | 19.2 | 40,935 | 247 | 2,602 | 77,143 | 22.1 |
| College Station-Bryan, TX | 0.1 | 4.2 | 7.7 | 4.6 | 8.3 | 6.5 | 5.0 | 9.9 | 33.5 | 32,575 | 123 | 4,487 | 114,215 | 20.2 |
| Colorado Springs, CO | 0.0 | D | 7.7 | D | 15.1 | 5.9 | 7.2 | 10.0 | 30.3 | 118,480 | 159 | 9,354 | 299,519 | 12.8 |
| Columbia, MO | 0.5 | D | 4.5 | 5.2 | 7.1 | 6.3 | 9.1 | 10.8 | 34.5 | 34,910 | 167 | 3,386 | 92,168 | 12.9 |
| Columbia, SC | 0.1 | D | 5.4 | 9.7 | 9.7 | 6.2 | 9.7 | 9.7 | 24.6 | 164,680 | 196 | 16,188 | 368,287 | 11.1 |
| Columbus, GA-AL | 0.1 | D | D | D | D | 4.5 | D | D | 35.8 | 65,995 | 205 | 11,088 | 142,067 | 6.0 |
| Columbus, IN | 0.5 | D | 3.6 | 47.5 | 4.9 | 4.5 | 3.6 | 7.0 | 9.1 | 16,685 | 198 | 1,238 | 34,810 | 5.2 |
| Columbus, OH | 0.1 | 0.1 | 5.8 | 7.5 | 11.7 | 5.1 | 10.1 | 11.3 | 17.9 | 326,985 | 154 | 45,815 | 884,368 | 7.7 |
| Corpus Christi, TX | 0.3 | 3.2 | 14.4 | 7.8 | 7.7 | 5.6 | 5.6 | 13.9 | 19.8 | 76,990 | 179 | 13,684 | 181,173 | 8.1 |
| Corvallis, OR | 0.6 | D | 4.1 | 10.5 | 9.9 | 4.7 | 5.0 | 16.3 | 30.9 | 16,625 | 179 | 1,036 | 39,708 | 9.5 |
| Crestview-Fort Walton Beach-Destin, FL | 0.2 | D | 6.2 | 2.4 | 14.0 | 7.5 | 6.4 | 7.4 | 35.3 | 61,140 | 214 | 4,447 | 156,773 | 14.0 |
| Cumberland, MD-WV | -0.1 | D | 4.8 | 13.6 | D | 7.4 | D | 20.2 | 23.9 | 24,765 | 254 | 3,109 | 45,819 | -1.1 |
| Dallas-Fort Worth-Arlington, TX | 0.0 | 5.2 | 7.1 | 8.4 | D | 5.4 | 11.9 | D | 9.9 | 977,875 | 129 | 127,135 | 2,937,810 | 17.5 |
| Dallas-Plano-Irving, TX Div 19124 | 0.0 | 5.4 | 6.5 | 7.5 | 17.2 | 5.2 | 13.0 | 9.0 | 9.2 | 628,245 | 124 | 85,844 | 1,986,879 | 19.9 |
| Fort Worth-Arlington-Grapevine, TX Div 23104 | -0.1 | 4.8 | 9.1 | 11.2 | 7.6 | 6.1 | 8.4 | 9.8 | 11.8 | 349,630 | 140 | 41,291 | 950,931 | 12.8 |
| Dalton, GA | 0.4 | 0.5 | 2.1 | 34.1 | D | 5.6 | 6.1 | D | 10.6 | 27,930 | 194 | 3,661 | 56,951 | 1.9 |
| Danville, IL | 2.1 | D | 3.4 | 21.5 | 2.0 | 5.6 | 4.4 | 9.3 | 23.9 | 18,595 | 246 | 2,650 | 36,054 | -0.7 |
| Daphne-Fairhope-Foley, AL | 0.4 | 0.2 | 10.1 | 6.7 | 6.2 | 11.6 | 7.6 | 12.7 | 13.1 | 58,505 | 262 | 3,517 | 122,518 | 17.7 |
| Davenport-Moline-Rock Island, IA-IL | 0.7 | 0.1 | 7.3 | 13.7 | 6.5 | 6.3 | 5.0 | D | 15.8 | 83,020 | 219 | 7,542 | 171,356 | 2.5 |
| Dayton-Kettering, OH | 0.0 | 0.1 | 4.7 | 11.9 | D | 5.2 | 5.9 | 15.6 | 21.5 | 167,340 | 207 | 20,108 | 371,331 | 1.1 |
| Decatur, AL | -0.6 | D | 8.9 | 31.6 | 4.9 | 6.1 | 4.5 | D | 13.7 | 38,200 | 251 | 4,574 | 67,986 | 2.4 |
| Decatur, IL | 0.6 | 0.0 | 6.8 | 30.1 | 4.3 | 5.1 | 4.8 | 13.1 | 11.2 | 25,160 | 242 | 3,249 | 50,245 | -0.5 |
| Deltona-Daytona Beach-Ormond Beach, FL | 0.9 | D | 7.8 | 6.7 | 8.2 | 9.7 | 6.3 | 17.3 | 13.5 | 198,745 | 297 | 14,265 | 322,265 | 6.4 |
| Denver-Aurora-Lakewood, CO | 0.0 | 7.0 | 7.3 | D | D | D | D | D | 11.9 | 406,735 | 137 | 35,919 | 1,193,750 | 10.7 |
| Des Moines-West Des Moines, IA | 0.7 | 0.1 | 7.4 | 5.9 | 10.3 | 5.5 | 21.6 | D | 12.8 | 117,050 | 167 | 9,807 | 301,811 | 17.7 |
| Detroit-Warren-Dearborn, MI | 0.0 | 0.1 | 5.2 | 14.4 | 17.1 | 5.4 | 8.6 | 11.7 | 9.5 | 899,305 | 208 | 133,007 | 1,921,668 | 1.9 |
| Detroit-Dearborn-Livonia, MI Div 19804 | 0.0 | 0.1 | 3.9 | 14.4 | 13.2 | 4.3 | 6.3 | 13.0 | 11.7 | 357,785 | 204 | 82,982 | 815,330 | -0.7 |
| Warren-Troy-Farmington Hills, MI Div 47664 | 0.1 | 0.1 | 5.9 | 14.4 | 19.3 | 6.1 | 9.8 | 10.8 | 8.2 | 541,520 | 211 | 50,025 | 1,106,338 | 3.9 |
| Dothan, AL | 2.3 | D | 4.9 | 7.8 | 5.3 | 9.7 | 4.4 | D | 18.5 | 39,490 | 264 | 5,773 | 70,755 | 5.8 |
| Dover, DE | 2.5 | D | 6.0 | D | 4.3 | 7.2 | 4.2 | 12.4 | 38.1 | 39,720 | 219 | 3,602 | 75,742 | 15.9 |
| Dubuque, IA | 2.2 | D | 5.6 | 20.5 | 7.1 | 6.0 | 11.7 | 13.5 | 8.1 | 21,725 | 223 | 1,636 | 41,963 | 7.7 |
| Duluth, MN-WI | 0.0 | 5.5 | 6.4 | 7.6 | 5.6 | 6.8 | 4.9 | 21.1 | 19.0 | 70,935 | 245 | 6,621 | 153,647 | 3.0 |
| Durham-Chapel Hill, NC | 0.1 | D | 3.2 | 13.3 | 16.6 | 3.3 | 6.7 | 12.2 | 21.1 | 113,950 | 177 | 11,000 | 284,926 | 16.0 |
| East Stroudsburg, PA | 0.0 | 0.1 | 5.1 | 13.6 | 3.7 | 7.9 | 3.2 | 12.3 | 24.9 | 37,680 | 222 | 3,081 | 81,980 | 2.0 |
| Eau Claire, WI | 1.0 | D | 7.7 | 13.6 | 5.3 | 7.6 | 5.5 | 19.5 | 14.2 | 36,425 | 215 | 2,915 | 74,272 | 7.1 |
| El Centro, CA | 18.6 | 0.8 | 2.6 | 2.1 | 2.2 | 7.1 | 2.3 | 6.5 | 35.4 | 34,575 | 192 | 10,166 | 58,467 | 4.3 |
| Elizabethtown-Fort Knox, KY | -0.2 | D | 4.0 | 14.6 | D | 6.4 | 4.4 | 6.5 | 40.4 | 32,520 | 211 | 4,344 | 66,734 | 9.1 |
| Elkhart-Goshen, IN | 0.7 | D | 4.1 | 50.2 | 2.3 | 4.2 | 3.0 | 7.5 | 5.5 | 36,720 | 178 | 3,022 | 80,082 | 3.0 |
| Elmira, NY | 0.1 | 0.9 | 5.5 | 16.6 | 4.8 | 6.8 | 4.6 | 16.5 | 24.2 | 21,825 | 262 | 3,019 | 39,271 | 2.3 |
| El Paso, TX | 0.1 | 0.0 | D | D | D | 7.2 | D | D | 35.3 | 138,265 | 164 | 29,245 | 305,919 | 12.5 |
| Enid, OK | 0.9 | 8.5 | 4.9 | 10.2 | D | 6.7 | 6.1 | 11.7 | 21.0 | 12,680 | 207 | 1,334 | 26,743 | -0.3 |
| Erie, PA | 0.3 | 0.0 | 4.6 | 18.7 | 5.1 | 6.5 | 8.4 | 18.2 | 16.0 | 64,700 | 240 | 10,342 | 121,964 | 2.4 |
| Eugene-Springfield, OR | 0.4 | 0.2 | 6.1 | 8.9 | 7.8 | 7.7 | 7.3 | 16.0 | 18.9 | 90,920 | 238 | 9,766 | 165,962 | 6.3 |

1. Per 1,000 resident population estimated as of July 1 of the year shown.

# Table C. Metropolitan Areas — Housing, Labor Force, and Employment

| Area name | Occupied Housing units, 2020 — Total | Owner-occupied Percent | Owner-occupied Median value[1] | Median owner cost as a percent of income — With a mortgage | Median owner cost as a percent of income — Without a mortgage[2] | Renter-occupied Median rent[3] | Renter-occupied Median rent as a percent of income[2] | Substandard units[4] (percent) | Civilian labor force, 2020 — Total | Percent change 2019-2020 | Unemployment Total | Unemployment Rate[5] | Civilian employment, 2019[6] — Total | Percent Management, business, science, and arts | Percent Construction, production, and maintenance occupations |
|---|---|---|---|---|---|---|---|---|---|---|---|---|---|---|---|
| | 89 | 90 | 91 | 92 | 93 | 94 | 95 | 96 | 97 | 98 | 99 | 100 | 101 | 102 | 103 |
| Cedar Rapids, IA | 112,373 | 72.9 | 163,800 | 19.1 | 11.7 | 758 | 27.0 | 2.0 | 142,030 | -4.4 | 8,789 | 6.2 | 143,160 | 37.2 | 27.0 |
| Chambersburg-Waynesboro, PA | 60,260 | 73.4 | 187,500 | 21.2 | 10.4 | 899 | 26.6 | 2.4 | 77,184 | -0.7 | 5,876 | 7.6 | 75,911 | 31.2 | 32.3 |
| Champaign-Urbana, IL | 88,700 | 54.3 | 158,700 | 18.6 | 11.2 | 909 | 29.9 | 0.0 | 117,187 | -1.1 | 7,446 | 6.4 | 109,426 | 48.9 | 16.7 |
| Charleston, WV | 109,452 | 70.9 | 105,200 | 16.5 | 10.0 | 698 | 28.2 | 1.5 | 113,495 | -0.4 | 10,372 | 9.1 | 103,536 | 35.9 | 21.6 |
| Charleston-North Charleston, SC | 303,976 | 67.5 | 266,600 | 20.1 | 10.0 | 1,212 | 28.7 | 2.1 | 394,548 | 0.1 | 23,780 | 6.0 | 398,105 | 41.5 | 20.9 |
| Charlotte-Concord-Gastonia, NC-SC | 982,241 | 65.4 | 233,900 | 18.7 | 10.0 | 1,077 | 28.1 | 2.3 | 1,364,377 | -1.1 | 99,767 | 7.3 | 1,354,242 | 41.1 | 21.9 |
| Charlottesville, VA | 87,274 | 65 | 317,700 | 18.5 | 10.0 | 1,187 | 27.6 | 1.5 | 112,443 | -4.1 | 6,306 | 5.6 | 111,953 | 50.7 | 14.6 |
| Chattanooga, TN-GA | 222,950 | 66 | 185,000 | 19.4 | 10.0 | 852 | 28.2 | 1.3 | 271,914 | -1.8 | 17,666 | 6.5 | 267,878 | 36.7 | 23.6 |
| Cheyenne, WY | 41,739 | 68.2 | 267,200 | 19.7 | 10.0 | 841 | 29.4 | 0.0 | 50,826 | 2.1 | 2,595 | 5.1 | 48,790 | 39.4 | 21.7 |
| Chicago-Naperville-Elgin, IL-IN-WI | 3,539,174 | 64.4 | 253,800 | 21.5 | 13.0 | 1,139 | 28.5 | 2.7 | 4,716,587 | -2.9 | 468,734 | 9.9 | 4,796,817 | 42.1 | 21.0 |
| Chicago-Naperville-Evanston, IL Div 16984 | 2,691,034 | 62 | 266,100 | 22.1 | 13.4 | 1,165 | 28.5 | 2.9 | 3,551,132 | -2.9 | 367,705 | 10.4 | 3,632,930 | 43.4 | 20.1 |
| Elgin, IL Div 20994 | 262,917 | 74.1 | 243,100 | 21.5 | 12.7 | 1,080 | 29.5 | 3.5 | 381,066 | -4.4 | 33,811 | 8.9 | 400,799 | 37.9 | 23.5 |
| Gary, IN Div 23844 | 273,375 | 70.9 | 172,200 | 18.3 | 10.6 | 894 | 28.1 | 1.8 | 331,346 | -1.9 | 31,508 | 9.5 | 326,779 | 32.4 | 29.5 |
| Lake County-Kenosha County, IL-WI Div | 311,848 | 71.5 | 254,600 | 20.2 | 12.3 | 1,145 | 28.3 | 1.9 | 453,043 | -3.1 | 35,710 | 7.9 | 436,309 | 42.3 | 20.5 |
| Chico, CA | 77,651 | 59.1 | 336,600 | 22.1 | 12.3 | 1,087 | 31.9 | 3.1 | 92,604 | -5.4 | 8,509 | 9.2 | 95,672 | 36.2 | 21.9 |
| Cincinnati, OH-KY-IN | 878,645 | 66.9 | 186,100 | 18.0 | 10.1 | 854 | 26.7 | 1.4 | 1,129,383 | -1.5 | 78,831 | 7.0 | 1,126,224 | 42.3 | 21.4 |
| Clarksville, TN-KY | 115,195 | 58.9 | 162,500 | 21.0 | 10.0 | 902 | 27.1 | 2.8 | 122,263 | -0.2 | 9,271 | 7.6 | 118,988 | 32.1 | 26.6 |
| Cleveland, TN | 46,936 | 66.9 | 160,700 | 17.6 | 10.0 | 770 | 25.0 | 0.0 | 60,106 | 0.4 | 4,012 | 6.7 | 58,639 | 30.2 | 33.1 |
| Cleveland-Elyria, OH | 872,626 | 64.2 | 161,400 | 19.2 | 11.9 | 813 | 28.1 | 1.4 | 1,002,534 | -4.5 | 96,804 | 9.7 | 1,000,995 | 40.1 | 20.5 |
| Coeur d'Alene, ID | 65,283 | 69.2 | 306,900 | 21.8 | 10.0 | 966 | 26.7 | 2.4 | 81,286 | 1.8 | 5,632 | 6.9 | 81,399 | 32.8 | 22.5 |
| College Station-Bryan, TX | 94,597 | 51.9 | 213,200 | 22.1 | 10.0 | 952 | 32.6 | 3.8 | 132,124 | -1.7 | 7,222 | 5.5 | 128,953 | 38.4 | 20.1 |
| Colorado Springs, CO | 275,778 | 64.9 | 324,100 | 22.0 | 10.0 | 1,251 | 30.6 | 2.9 | 357,104 | 1.0 | 26,004 | 7.3 | 354,373 | 45.2 | 16.6 |
| Columbia, MO | 80,127 | 60.2 | 192,700 | 19.4 | 10.0 | 910 | 31.5 | 1.4 | 108,422 | -2.0 | 4,671 | 4.3 | 110,467 | 42.5 | 16.7 |
| Columbia, SC | 317,998 | 67.9 | 164,700 | 18.6 | 10.0 | 920 | 30.9 | 2.4 | 404,554 | 0.8 | 21,313 | 5.3 | 391,490 | 41.4 | 20.6 |
| Columbus, GA-AL | 119,728 | 56.3 | 145,900 | 20.8 | 11.1 | 862 | 27.8 | 1.4 | 127,731 | -1.1 | 8,641 | 6.8 | 127,345 | 32.6 | 26.8 |
| Columbus, IN | 30,383 | 72.4 | 174,000 | 16.9 | 10.0 | 935 | 24.3 | 0.0 | 43,802 | -2.8 | 2,827 | 6.5 | 41,882 | 43.3 | 27.2 |
| Columbus, OH | 817,086 | 60.9 | 212,600 | 19.1 | 11.2 | 965 | 26.6 | 2.1 | 1,108,370 | 0.1 | 77,849 | 7.0 | 1,106,820 | 44.2 | 19.7 |
| Corpus Christi, TX | 153,066 | 61.2 | 150,100 | 20.8 | 12.1 | 1,045 | 30.2 | 3.8 | 193,141 | -1.8 | 17,686 | 9.2 | 194,718 | 31.8 | 28.5 |
| Corvallis, OR | 37,378 | 57.3 | 389,000 | 21.7 | 10.0 | 1,158 | 35.1 | 0.0 | 46,257 | -3.0 | 2,612 | 5.6 | 47,115 | 49.3 | 15.4 |
| Crestview-Fort Walton Beach-Destin, FL | 102,572 | 69.1 | 241,600 | 21.8 | 10.0 | 1,160 | 31.9 | 1.6 | 127,967 | -0.6 | 7,168 | 5.6 | 126,838 | 35.9 | 19.6 |
| Cumberland, MD-WV | 36,430 | 70.6 | 131,400 | 18.1 | 10.0 | 628 | 29.8 | 0.0 | 43,084 | -4.1 | 3,324 | 7.7 | 37,440 | 28.4 | 25.1 |
| Dallas-Fort Worth-Arlington, TX | 2,635,017 | 59.5 | 253,900 | 21.6 | 11.6 | 1,202 | 28.7 | 4.5 | 3,953,208 | 0.3 | 279,632 | 7.1 | 3,858,031 | 40.3 | 22.6 |
| Dallas-Plano-Irving, TX Div 19124 | 1,782,408 | 58 | 272,400 | 21.8 | 11.8 | 1,229 | 28.3 | 4.8 | 2,687,174 | 0.3 | 188,578 | 7.0 | 2,624,740 | 41.8 | 21.1 |
| Fort Worth-Arlington-Grapevine, TX Div 23104 | 852,609 | 62.8 | 225,600 | 21.2 | 11.3 | 1,137 | 29.6 | 3.9 | 1,266,034 | 0.2 | 91,054 | 7.2 | 1,233,291 | 37.2 | 25.6 |
| Dalton, GA | 50,267 | 70.9 | 138,600 | 18.0 | 10.0 | 704 | 22.6 | 3.2 | 58,625 | -1.6 | 4,297 | 7.3 | 67,665 | 24.3 | 40.8 |
| Danville, IL | 31,154 | 67.6 | 80,100 | 18.2 | 11.4 | 649 | 24.3 | 0.0 | 32,542 | -2.0 | 2,871 | 8.8 | 30,492 | 27.6 | 32.7 |
| Daphne-Fairhope-Foley, AL | 82,325 | 79.3 | 223,100 | 19.5 | 10.0 | 1,072 | 30.0 | 0.0 | 96,763 | -0.7 | 5,425 | 5.6 | 96,012 | 37.9 | 21.5 |
| Davenport-Moline-Rock Island, IA-IL | 155,389 | 68.4 | 141,400 | 17.2 | 11.2 | 743 | 26.9 | 1.6 | 186,932 | -3.9 | 14,815 | 7.9 | 183,397 | 34.0 | 27.5 |
| Dayton-Kettering, OH | 332,258 | 63.9 | 141,900 | 18.2 | 11.2 | 812 | 26.9 | 1.8 | 390,224 | -0.4 | 30,816 | 7.9 | 381,559 | 38.9 | 22.6 |
| Decatur, AL | 59,315 | 76.1 | 142,900 | 18.2 | 10.0 | 694 | 25.6 | 2.7 | 73,262 | 1.0 | 3,337 | 4.6 | 66,286 | 30.7 | 35.4 |
| Decatur, IL | 42,741 | 68.3 | 105,500 | 18.3 | 10.0 | 668 | 28.5 | 0.0 | 47,177 | -3.3 | 4,665 | 9.9 | 44,586 | 33.0 | 26.9 |
| Deltona-Daytona Beach-Ormond Beach, FL | 263,915 | 72.7 | 217,000 | 23.0 | 11.7 | 1,123 | 32.9 | 1.5 | 296,196 | -1.9 | 22,528 | 7.6 | 292,137 | 32.2 | 21.8 |
| Denver-Aurora-Lakewood, CO | 1,152,457 | 64.3 | 437,800 | 21.3 | 10.0 | 1,468 | 29.5 | 2.8 | 1,669,889 | 0.2 | 125,594 | 7.5 | 1,645,815 | 46.9 | 17.4 |
| Des Moines-West Des Moines, IA | 280,056 | 69.1 | 201,000 | 19.0 | 11.6 | 940 | 26.2 | 2.4 | 374,723 | -3.2 | 20,685 | 5.5 | 372,250 | 44.8 | 20.3 |
| Detroit-Warren-Dearborn, MI | 1,720,610 | 69.6 | 191,800 | 19.3 | 11.7 | 962 | 29.6 | 1.6 | 2,098,859 | -3.2 | 245,473 | 11.7 | 2,083,915 | 40.5 | 22.6 |
| Detroit-Dearborn-Livonia, MI Div 19804 | 689,270 | 62.5 | 134,300 | 19.9 | 12.4 | 901 | 31.7 | 2.0 | 796,974 | -1.6 | 110,086 | 13.8 | 777,705 | 34.2 | 26.8 |
| Warren-Troy-Farmington Hills, MI Div 47664 | 1,031,340 | 74.4 | 227,800 | 19.1 | 11.0 | 1,031 | 27.8 | 1.3 | 1,301,885 | -4.1 | 135,387 | 10.4 | 1,306,210 | 44.2 | 20.1 |
| Dothan, AL | 56,795 | 68.6 | 142,200 | 18.1 | 10.0 | 727 | 27.9 | 1.2 | 64,007 | -0.2 | 3,268 | 5.1 | 62,226 | 31.6 | 26.0 |
| Dover, DE | 68,023 | 66.6 | 225,200 | 22.5 | 11.6 | 1,125 | 34.1 | 2.2 | 79,715 | 0.8 | 6,872 | 8.6 | 80,350 | 30.4 | 25.8 |
| Dubuque, IA | 38,345 | 74.5 | 171,400 | 19.0 | 12.1 | 825 | 27.5 | 0.8 | 54,877 | -3.3 | 3,211 | 5.9 | 52,342 | 39.3 | 25.2 |
| Duluth, MN-WI | 124,913 | 73 | 167,200 | 18.7 | 10.7 | 768 | 29.1 | 1.1 | 146,984 | -0.7 | 11,050 | 7.5 | 145,262 | 35.7 | 26.0 |
| Durham-Chapel Hill, NC | 255,271 | 61.4 | 251,900 | 19.0 | 10.4 | 1,065 | 29.0 | 2.2 | 331,142 | -2.8 | 20,266 | 6.1 | 322,002 | 50.8 | 16.9 |
| East Stroudsburg, PA | 56,274 | 74.9 | 177,600 | 23.0 | 13.4 | 1,199 | 34.1 | 0.0 | 81,386 | -2.5 | 9,390 | 11.5 | 83,893 | 33.5 | 26.5 |
| Eau Claire, WI | 67,096 | 70.1 | 188,700 | 19.4 | 10.8 | 817 | 26.9 | 1.7 | 91,277 | -0.6 | 5,283 | 5.8 | 89,177 | 38.1 | 24.6 |
| El Centro, CA | 48,182 | 55.5 | 229,600 | 23.0 | 11.9 | 810 | 30.8 | 9.8 | 69,602 | -5.5 | 15,653 | 22.5 | 60,600 | 25.9 | 26.6 |
| Elizabethtown-Fort Knox, KY | 58,590 | 64.6 | 155,700 | 19.3 | 10.0 | 832 | 23.7 | 2.4 | 65,168 | -3.4 | 4,642 | 7.1 | 63,726 | 34.8 | 28.0 |
| Elkhart-Goshen, IN | 68,553 | 72.9 | 160,400 | 19.2 | 10.1 | 841 | 29.8 | 1.2 | 110,494 | -1.9 | 8,339 | 7.5 | 99,878 | 26.4 | 35.8 |
| Elmira, NY | 33,490 | 71.9 | 110,900 | 17.7 | 10.1 | 757 | 28.1 | 0.0 | 35,430 | 1.1 | 3,102 | 8.8 | 37,294 | 36.2 | 23.2 |
| El Paso, TX | 270,137 | 60.5 | 125,100 | 22.1 | 11.6 | 849 | 30.3 | 5.8 | 365,940 | 0.5 | 30,243 | 8.3 | 359,390 | 31.9 | 23.5 |
| Enid, OK | 23,683 | 67.3 | 119,400 | 18.7 | 10.0 | 832 | 21.1 | 0.0 | 27,007 | -0.2 | 1,475 | 5.5 | 27,453 | 29.1 | 32.2 |
| Erie, PA | 110,128 | 66.4 | 143,600 | 19.0 | 11.5 | 756 | 29.3 | 2.5 | 126,749 | -1.8 | 12,418 | 9.8 | 125,802 | 36.4 | 23.0 |
| Eugene-Springfield, OR | 155,696 | 58 | 293,300 | 22.6 | 11.7 | 1,034 | 30.7 | 2.7 | 178,467 | -1.4 | 14,103 | 7.9 | 183,052 | 36.0 | 22.7 |

1. Specified owner-occupied units. ing complete plumbing facilities.     2. A value of 10.0 represents 10 percent or less; a value of 50.0 represents 50 percent or more.     3. Specified renter-occupied units.     4. Overcrowded or lack-     5. Percent of civilian labor force. 6. Civilian employed persons 16 years old and over.

# Table C. Metropolitan Areas — Nonfarm Employment and Agriculture

| | Private nonfarm establishments, employment and payroll, 2019 | | | | | | | | | Agriculture, 2017 | | | |
| | | Employment | | | | | | Annual payroll | | Farms | | | Farm producers whose primary occupation is farming (percent) |
| | | | | | | | | | | | Percent with: | | |
| Area name | Number of establishments | Total | Health care and social assistance | Manufacturing | Retail trade | Finance and insurance | Professional, scientific, and technical services | Total (mil dol) | Average per employee (dollars) | Number | Fewer than 50 acres | 1000 acres or more | |
| | 104 | 105 | 106 | 107 | 108 | 109 | 110 | 111 | 112 | 113 | 114 | 115 | 116 |
|---|---|---|---|---|---|---|---|---|---|---|---|---|---|
| Cedar Rapids, IA | 6,750 | 130,186 | 17,980 | 18,934 | 17,195 | 8,954 | 8,882 | 6,767 | 51,983 | 3,632 | 37.7 | 8.6 | 43.7 |
| Chambersburg-Waynesboro, PA | 3,079 | 51,117 | 8,997 | 7,366 | 7,542 | 1,132 | 1,858 | 2,099 | 41,058 | 1,581 | 37.3 | 1.6 | 55.5 |
| Champaign-Urbana, IL | 4,548 | 72,028 | 14,445 | 6,151 | 10,027 | 3,416 | 2,883 | 3,247 | 45,074 | 1,636 | 33.9 | 17.3 | 50.9 |
| Charleston, WV | 5,618 | 92,812 | 22,903 | 5,160 | 12,537 | 3,753 | 5,581 | 4,333 | 46,682 | 1,539 | 29.2 | 0.5 | 34.0 |
| Charleston-North Charleston, SC | 20,282 | 297,386 | 33,801 | 28,873 | 42,282 | 8,669 | 23,029 | 14,203 | 47,760 | 1,100 | 60.3 | 3.4 | 36.9 |
| Charlotte-Concord-Gastonia, NC-SC | 64,262 | 1,118,149 | 142,273 | 104,199 | 127,477 | 90,329 | 70,523 | 63,234 | 56,553 | 7,381 | 51.5 | 2.1 | 36.9 |
| Charlottesville, VA | 5,970 | 86,243 | 18,959 | 4,211 | 11,734 | 3,336 | 7,318 | 4,674 | 54,195 | 1,809 | 39.7 | 2.6 | 40.0 |
| Chattanooga, TN-GA | 11,699 | 230,512 | 35,925 | 32,912 | 30,585 | 15,165 | 9,414 | 10,054 | 43,617 | 2,162 | 47.2 | 1.2 | 37.1 |
| Cheyenne, WY | 3,355 | 34,602 | 6,598 | 1,307 | 6,052 | 1,477 | 2,539 | 1,586 | 45,846 | 999 | 31.7 | 21.9 | 38.8 |
| Chicago-Naperville-Elgin, IL-IN-WI | 248,428 | 4,225,657 | 612,896 | 386,459 | 439,341 | 266,285 | 352,298 | 267,734 | 63,359 | 6,565 | 49.3 | 9.5 | 46.8 |
| Chicago-Naperville-Evanston, IL Div 16984 | 193,335 | 3,363,852 | 489,109 | 266,593 | 323,151 | 236,416 | 299,804 | 218,092 | 64,834 | 2,353 | 55.1 | 8.3 | 46.1 |
| Elgin, IL Div 20994 | 17,266 | 246,304 | 31,769 | 38,725 | 33,592 | 9,042 | 12,086 | 11,584 | 47,033 | 1,697 | 42.5 | 11.9 | 51.4 |
| Gary, IN Div 23844 | 14,711 | 235,499 | 45,463 | 34,370 | 33,240 | 5,518 | 8,521 | 11,254 | 47,786 | 1,798 | 42 | 11.2 | 45.7 |
| Lake County-Kenosha County, IL-WI Div | 23,116 | 380,002 | 46,555 | 46,771 | 49,358 | 15,309 | 31,887 | 26,805 | 70,538 | 717 | 65 | 3.2 | 41.5 |
| Chico, CA | 4,637 | 63,691 | 15,291 | 4,305 | 11,017 | 2,572 | 2,612 | 2,705 | 42,467 | 1,912 | 65.2 | 2.9 | 51.3 |
| Cincinnati, OH-KY-IN | 46,971 | 981,127 | 148,888 | 113,106 | 108,213 | 61,358 | 63,059 | 54,124 | 55,166 | 10,403 | 48.6 | 1.8 | 32.8 |
| Clarksville, TN-KY | 4,702 | 74,553 | 12,718 | 14,605 | 12,161 | 2,102 | 3,238 | 2,795 | 37,492 | 2,718 | 36.1 | 4.8 | 39.0 |
| Cleveland, TN | 2,185 | 42,886 | 5,512 | 9,003 | 6,375 | 1,383 | 861 | 1,753 | 40,876 | 1,065 | 52.5 | 0.7 | 40.7 |
| Cleveland-Elyria, OH | 49,638 | 926,568 | 184,169 | 119,895 | 97,558 | 50,877 | 53,950 | 50,315 | 54,302 | 3,524 | 68.4 | 1.3 | 35.0 |
| Coeur d'Alene, ID | 5,177 | 55,398 | 11,066 | 5,021 | 8,714 | 2,331 | 2,814 | 2,383 | 43,019 | 1,073 | 68.7 | 2.1 | 32.1 |
| College Station-Bryan, TX | 5,119 | 78,418 | 11,082 | 6,424 | 12,589 | 2,140 | 3,798 | 3,111 | 39,673 | 4,482 | 38.3 | 4.2 | 37.2 |
| Colorado Springs, CO | 18,727 | 256,741 | 41,003 | 10,591 | 32,344 | 12,413 | 24,503 | 12,368 | 48,173 | 1,504 | 46.1 | 10.5 | 35.4 |
| Columbia, MO | 5,214 | 84,251 | 20,200 | 5,077 | 12,309 | 7,073 | 4,293 | 3,571 | 42,383 | 2,757 | 31.6 | 5.4 | 33.1 |
| Columbia, SC | 17,777 | 300,499 | 47,947 | 27,140 | 42,251 | 16,172 | 17,426 | 12,947 | 43,086 | 3,325 | 50.3 | 3.6 | 37.0 |
| Columbus, GA-AL | 5,912 | 100,918 | 17,707 | 11,468 | 13,517 | 13,274 | 3,573 | 4,526 | 44,852 | 1,062 | 33.1 | 6.3 | 36.5 |
| Columbus, IN | 1,890 | 45,897 | 5,203 | 12,730 | 4,942 | 941 | 2,822 | 2,464 | 53,689 | 564 | 47 | 8.7 | 36.6 |
| Columbus, OH | 43,592 | 914,038 | 166,425 | 70,825 | 100,185 | 75,978 | 61,054 | 47,625 | 52,104 | 8,506 | 53.3 | 4.6 | 37.5 |
| Corpus Christi, TX | 9,147 | 160,735 | 32,332 | 10,386 | 22,225 | 4,766 | 7,314 | 7,133 | 44,379 | 1,302 | 57.1 | 13.2 | 36.0 |
| Corvallis, OR | 2,182 | 29,149 | 6,432 | 1,742 | 3,826 | 928 | 2,928 | 1,474 | 50,564 | 1,079 | 56.4 | 2.3 | 38.5 |
| Crestview-Fort Walton Beach-Destin, FL | 8,097 | 85,586 | 10,822 | 2,644 | 16,776 | 2,320 | 8,024 | 3,458 | 40,403 | 809 | 38.6 | 1.6 | 35.7 |
| Cumberland, MD-WV | 1,939 | 29,994 | 7,463 | 4,227 | 4,950 | 1,008 | 706 | 1,127 | 37,584 | | | | 31.1 |
| Dallas-Fort Worth-Arlington, TX | 169,096 | 3,261,282 | 410,793 | 258,706 | 355,797 | 228,517 | 248,326 | 192,840 | 59,130 | 29,254 | 68.9 | 2.2 | 31.1 |
| Dallas-Plano-Irving, TX Div 19124 | 118,325 | 2,349,661 | 290,068 | 162,555 | 235,061 | 176,991 | 205,327 | 146,447 | 62,327 | 16,618 | 68.4 | 2.2 | 30.7 |
| Fort Worth-Arlington-Grapevine, TX Div 23104 | 50,771 | 911,621 | 120,725 | 96,151 | 120,736 | 51,526 | 42,999 | 46,393 | 50,891 | 12,636 | 69.6 | 2.1 | 31.5 |
| Dalton, GA | 2,586 | 54,905 | 6,043 | 20,220 | 5,831 | 773 | 1,578 | 2,410 | 43,893 | 664 | 49.2 | 0.9 | 37.3 |
| Danville, IL | 1,342 | 23,295 | 4,766 | 4,608 | 3,424 | 1,091 | 430 | 1,010 | 43,352 | 1,049 | 38.9 | 14.1 | 47.8 |
| Daphne-Fairhope-Foley, AL | 5,644 | 67,084 | 8,814 | 4,564 | 13,553 | 1,700 | 2,185 | 2,373 | 35,369 | 842 | 53.8 | 4.5 | 45.0 |
| Davenport-Moline-Rock Island, IA-IL | 8,823 | 161,357 | 24,326 | 24,033 | 22,322 | 6,125 | 6,386 | 7,779 | 48,210 | 3,434 | 36.1 | 9.5 | 45.3 |
| Dayton-Kettering, OH | 16,405 | 327,586 | 65,742 | 46,247 | 40,346 | 13,470 | 26,501 | 16,109 | 49,173 | 2,635 | 59.9 | 4.1 | 36.8 |
| Decatur, AL | 3,020 | 49,122 | 6,637 | 12,596 | 6,052 | 1,452 | 2,581 | 2,233 | 45,454 | 2,416 | 49 | 1.8 | 36.1 |
| Decatur, IL | 2,311 | 46,084 | 8,137 | 8,286 | 5,088 | 1,591 | 1,061 | 2,438 | 52,913 | 589 | 41.8 | 16.0 | 49.9 |
| Deltona-Daytona Beach-Ormond Beach, FL | 15,493 | 175,111 | 31,519 | 10,180 | 31,382 | 5,892 | 8,681 | 6,575 | 37,546 | 1,691 | 81.3 | 2.5 | 39.9 |
| Denver-Aurora-Lakewood, CO | 87,441 | 1,330,243 | 167,742 | 57,509 | 143,495 | 81,587 | 118,192 | 83,007 | 62,400 | 5,606 | 59.6 | 7.3 | 28.6 |
| Des Moines-West Des Moines, IA | 17,933 | 339,589 | 45,348 | 22,007 | 43,619 | 46,686 | 21,564 | 18,248 | 53,736 | 5,658 | 40.6 | 8.4 | 36.9 |
| Detroit-Warren-Dearborn, MI | 100,010 | 1,817,005 | 296,733 | 245,558 | 202,918 | 90,123 | 190,574 | 104,101 | 57,293 | 3,980 | 58.2 | 2.7 | 44.3 |
| Detroit-Dearborn-Livonia, MI Div 19804 | 32,575 | 661,250 | 114,085 | 90,385 | 69,209 | 34,706 | 45,342 | 38,340 | 57,982 | 248 | 79.8 | 0.0 | 47.5 |
| Warren-Troy-Farmington Hills, MI Div 47664 | 67,435 | 1,155,755 | 182,648 | 155,173 | 133,709 | 55,417 | 145,232 | 65,761 | 56,899 | 3,732 | 56.8 | 2.9 | 44.0 |
| Dothan, AL | 3,392 | 52,664 | 10,403 | 5,818 | 8,468 | 1,422 | 1,349 | 2,147 | 40,765 | 1,973 | 35 | 6.0 | 38.4 |
| Dover, DE | 3,793 | 54,277 | 12,020 | 4,682 | 9,522 | 1,538 | 2,384 | 2,368 | 43,635 | 822 | 53.4 | 6.3 | 48.5 |
| Dubuque, IA | 2,767 | 56,000 | 8,466 | 10,098 | 7,325 | 4,008 | 1,767 | 2,506 | 44,749 | 1,402 | 30.5 | 3.4 | 42.8 |
| Duluth, MN-WI | 7,405 | 113,279 | 28,609 | 8,645 | 15,889 | 4,307 | 3,887 | 5,154 | 45,496 | 1,679 | 26.9 | 1.7 | 35.7 |
| Durham-Chapel Hill, NC | 14,129 | 287,760 | 57,760 | 23,242 | 26,163 | 11,610 | 35,466 | 18,033 | 62,619 | 2,993 | 47 | 1.6 | 41.1 |
| East Stroudsburg, PA | 3,395 | 47,784 | 8,126 | 5,086 | 9,049 | 914 | 1,583 | 1,902 | 39,795 | 233 | 55.4 | 1.7 | 40.4 |
| Eau Claire, WI | 4,241 | 75,398 | 14,804 | 11,359 | 10,659 | 3,279 | 2,027 | 3,570 | 47,348 | 2,478 | 32.3 | 3.1 | 41.7 |
| El Centro, CA | 2,545 | 33,174 | 6,085 | 2,890 | 8,116 | 758 | 908 | 1,240 | 37,379 | 396 | 24.7 | 35.1 | 73.7 |
| Elizabethtown-Fort Knox, KY | 2,652 | 42,861 | 7,851 | 7,901 | 7,109 | 1,795 | 1,528 | 1,717 | 40,063 | 2,804 | 49.8 | 2.8 | 34.4 |
| Elkhart-Goshen, IN | 4,951 | 135,841 | 11,540 | 75,097 | 10,186 | 2,003 | 2,078 | 6,267 | 46,132 | 1,667 | 64.1 | 1.9 | 34.8 |
| Elmira, NY | 1,721 | 31,018 | 7,122 | 5,101 | 4,936 | 883 | 749 | 1,315 | 42,385 | 398 | 29.1 | 1.3 | 37.8 |
| El Paso, TX | 14,994 | 245,037 | 47,753 | 16,096 | 36,529 | 7,796 | 9,044 | 8,297 | 33,860 | 790 | 75.9 | 11.1 | 39.6 |
| Enid, OK | 1,601 | 21,659 | 3,720 | 2,572 | 3,493 | 757 | 606 | 871 | 40,218 | 936 | 22.1 | 20.6 | 44.8 |
| Erie, PA | 6,194 | 116,333 | 26,435 | 21,408 | 15,021 | 5,776 | 4,045 | 4,695 | 40,355 | 1,162 | 39.6 | 1.4 | 46.8 |
| Eugene-Springfield, OR | 10,024 | 127,760 | 24,178 | 14,466 | 20,453 | 5,073 | 5,865 | 5,597 | 43,806 | 2,646 | 78.2 | 1.1 | 35.9 |

# Table C. Metropolitan Areas — Agriculture

| Area name | Land in farms Acreage (1,000) | Percent change, 2012-2017 | Average size of farm | Acres Total irrigated (1,000) | Total cropland (1,000) | Value of land and buildings (dollars) Average per farm | Average per acre | Value of machinery and equipment, average per farm (dollars) | Value of products sold: Total (mil dol) | Average per farm (acres) | Percent from: Crops | Livestock and poultry products | Organic farms (number) | Farms with internet access (percent) | Government payments Total ($1,000) | Percent of farms |
|---|---|---|---|---|---|---|---|---|---|---|---|---|---|---|---|---|
| | 117 | 118 | 119 | 120 | 121 | 122 | 123 | 124 | 125 | 126 | 127 | 128 | 129 | 130 | 131 | 132 |
| Cedar Rapids, IA | 1,089 | 1.2 | 300 | 1.0 | 954.7 | 2,258,419 | 7,533 | 212,260 | 855 | 235,265 | 62.8 | 37.2 | 14 | 82.0 | 23,487 | 71.1 |
| Chambersburg-Waynesboro, PA | 270 | 1.9 | 170 | 2.8 | 213.9 | 1,283,405 | 7,528 | 162,879 | 477 | 301,372 | 19.2 | 80.8 | 44 | 59.6 | 3,264 | 23.0 |
| Champaign-Urbana, IL | 839 | -4.2 | 513 | 15.3 | 817.6 | 4,767,532 | 9,300 | 327,703 | 541 | 330,584 | 97.2 | 2.8 | 7 | 81.7 | 9,705 | 78.6 |
| Charleston, WV | 200 | 12.1 | 130 | 0.0 | 45.4 | 301,266 | 2,315 | 42,169 | 10 | 6,149 | 41.0 | 59.0 | 1 | 72.3 | 237 | 7.3 |
| Charleston-North Charleston, SC | 212 | 14.4 | 192 | 3.1 | 60.1 | 752,456 | 3,909 | 80,705 | 67 | 61,251 | 55.9 | 44.1 | 1 | 70.7 | 1,304 | 12.4 |
| Charlotte-Concord-Gastonia, NC-SC | 971 | -5.1 | 131 | 4.4 | 476.5 | 767,934 | 5,840 | 86,273 | 1,306 | 176,897 | 20.9 | 79.1 | 18 | 73.0 | 8,866 | 13.9 |
| Charlottesville, VA | 324 | 0.1 | 179 | 1.9 | 95.2 | 1,398,609 | 7,820 | 71,057 | 70 | 38,753 | 66.7 | 33.3 | 16 | 79.0 | 976 | 9.2 |
| Chattanooga, TN-GA | 274 | 2.8 | 127 | 0.4 | 85 | 630,797 | 4,977 | 70,938 | 249 | 115,034 | 8.0 | 92.0 | 7 | 76.6 | 2,430 | 18.9 |
| Cheyenne, WY | 1,630 | -2.8 | 1,631 | 67.1 | 382.3 | 1,383,473 | 848 | 123,891 | 185 | 184,762 | 17.9 | 82.1 | 42 | 77.7 | 4,856 | 29.3 |
| Chicago-Naperville-Elgin, IL-IN-WI | 2,147 | -3.8 | 327 | 59.4 | 2,016 | 2,794,684 | 8,547 | 216,405 | 1,860 | 283,303 | 71.1 | 28.9 | 53 | 82.4 | 36,682 | 45.6 |
| Chicago-Naperville-Evanston, IL Div 16984 | 672 | -4.1 | 286 | 10.2 | 635.2 | 2,661,483 | 9,317 | 188,605 | 456 | 193,918 | 87.7 | 12.3 | 14 | 82.7 | 6,760 | 32.2 |
| Elgin, IL Div 20994 | 680 | -2.3 | 401 | 1.8 | 658.1 | 3,859,110 | 9,632 | 267,239 | 667 | 393,094 | 72.2 | 27.8 | 24 | 85.6 | 14,683 | 60.3 |
| Gary, IN Div 23844 | 686 | -5.8 | 382 | 45.8 | 633.6 | 2,544,392 | 6,668 | 234,527 | 638 | 354,611 | 57.6 | 42.4 | 1 | 79.9 | 12,347 | 57.1 |
| Lake County-Kenosha County, IL-WI Div | 108 | 1.6 | 151 | 1.6 | 89.1 | 1,340,176 | 8,866 | 141,879 | 99 | 137,964 | 74.4 | 25.6 | 14 | 80.5 | 2,892 | 25.5 |
| Chico, CA | 348 | -8.6 | 182 | 192.5 | 214.2 | 2,193,552 | 12,042 | 163,066 | 524 | 274,140 | 96.8 | 3.2 | 100 | 80.4 | 6,677 | 6.1 |
| Cincinnati, OH-KY-IN | 1,331 | -0.7 | 128 | 2.1 | 784.7 | 671,715 | 5,249 | 79,548 | 421 | 40,443 | 79.8 | 20.2 | 35 | 75.4 | 19,347 | 19.0 |
| Clarksville, TN-KY | 674 | -3.3 | 248 | 8.2 | 410.5 | 1,109,600 | 4,473 | 120,824 | 322 | 118,376 | 79.0 | 21.0 | 22 | 65.3 | 10,373 | 32.7 |
| Cleveland, TN | 120 | -1.6 | 113 | 0.2 | 44.8 | 619,500 | 5,491 | 69,039 | 142 | 133,647 | 12.4 | 87.6 | 8 | 72.3 | 1,323 | 19.4 |
| Cleveland-Elyria, OH | 310 | 2.0 | 88 | 3.9 | 226.3 | 672,210 | 7,634 | 85,806 | 301 | 85,520 | 83.3 | 16.7 | 56 | 74.6 | 4,482 | 14.6 |
| Coeur d'Alene, ID | 140 | 12.4 | 130 | 13.7 | 62.2 | 719,382 | 5,525 | 49,847 | 22 | 20,057 | 81.0 | 19.0 | 3 | 81.7 | 1,211 | 9.9 |
| College Station-Bryan, TX | 1,099 | -0.3 | 245 | 50.4 | 225.2 | 929,627 | 3,792 | 69,216 | 308 | 68,802 | 20.3 | 79.7 | 4 | 71.2 | 3,759 | 3.4 |
| Colorado Springs, CO | 701 | -2.6 | 466 | 9.8 | 57.1 | 685,392 | 1,470 | 45,681 | 33 | 22,035 | 42.1 | 57.9 | 8 | 81.1 | 1,300 | 6.7 |
| Columbia, MO | 713 | -9.9 | 259 | 8.7 | 429.1 | 1,010,094 | 3,904 | 99,816 | 259 | 93,746 | 58.6 | 41.4 | 15 | 76.5 | 8,096 | 41.8 |
| Columbia, SC | 575 | 10.2 | 173 | 45.3 | 213.9 | 630,372 | 3,643 | 88,055 | 649 | 195,201 | 26.1 | 73.9 | 14 | 74.2 | 6,326 | 15.9 |
| Columbus, GA-AL | 297 | -0.6 | 279 | 6.0 | 73.2 | 775,478 | 2,775 | 77,092 | 58 | 54,683 | 39.3 | 60.7 | 1 | 70.9 | 4,287 | 31.5 |
| Columbus, IN | 160 | -6.5 | 284 | 15.5 | 142.2 | 1,902,512 | 6,688 | 159,738 | 93 | 164,929 | 88.3 | 11.7 | 0 | 79.3 | 3,919 | 55.7 |
| Columbus, OH | 1,666 | -4.6 | 196 | 5.1 | 1,395.2 | 1,284,052 | 6,557 | 141,978 | 1,079 | 126,795 | 69.5 | 30.5 | 89 | 82.9 | 47,922 | 36.9 |
| Corpus Christi, TX | 846 | -5.8 | 649 | 5.3 | 568.2 | 1,867,705 | 2,876 | 206,676 | 292 | 224,550 | 92.7 | 7.3 | 4 | 71.0 | 15,650 | 26.6 |
| Corvallis, OR | 128 | 2.9 | 132 | 27.2 | 69 | 852,301 | 6,438 | 80,922 | 77 | 79,401 | 82.6 | 17.4 | 32 | 86.7 | 1,144 | 6.7 |
| Crestview-Fort Walton Beach-Destin, FL | 136 | -35.2 | 126 | 1.6 | 36.8 | 517,386 | 4,111 | 42,599 | 39 | 36,088 | 30.4 | 69.6 | 0 | 76.5 | 1,999 | 15.7 |
| Cumberland, MD-WV | 134 | 19.4 | 166 | 0.1 | 42.8 | 585,037 | 3,525 | 56,221 | 26 | 31,879 | 24.2 | 75.8 | 2 | 64.9 | 379 | 13.3 |
| Dallas-Fort Worth-Arlington, TX | 3,793 | 0.9 | 130 | 23.3 | 1,227.2 | 628,364 | 4,846 | 57,126 | 612 | 20,911 | 39.0 | 61.0 | 24 | 78.4 | 21,992 | 6.0 |
| Dallas-Plano-Irving, TX Div 19124 | 2,156 | -2.1 | 130 | 12.1 | 841.4 | 683,005 | 5,265 | 59,441 | 413 | 24,863 | 43.7 | 56.3 | 17 | 77.3 | 16,826 | 8.2 |
| Fort Worth-Arlington-Grapevine, TX Div 23104 | 1,637 | 5.2 | 130 | 11.2 | 385.8 | 556,504 | 4,294 | 54,081 | 199 | 15,713 | 29.4 | 70.6 | 7 | 79.7 | 5,166 | 3.2 |
| Dalton, GA | 84 | -2.7 | 126 | 0.9 | 21 | 700,661 | 5,556 | 81,532 | 260 | 390,855 | 2.3 | 97.7 | 0 | 77.9 | 1,035 | 32.8 |
| Danville, IL | 471 | 8.5 | 449 | 0.8 | 443.7 | 3,686,595 | 8,203 | 261,347 | 283 | 269,783 | 95.2 | 4.8 | 6 | 77.8 | 5,125 | 56.7 |
| Daphne-Fairhope-Foley, AL | 175 | -9.1 | 208 | 7.4 | 110.4 | 1,208,663 | 5,822 | 136,496 | 120 | 142,973 | 84.4 | 15.6 | 0 | 78.1 | 6,316 | 27.3 |
| Davenport-Moline-Rock Island, IA-IL | 1,146 | 4.1 | 334 | 21.9 | 1,022.2 | 2,590,577 | 7,762 | 218,339 | 894 | 260,397 | 71.4 | 28.6 | 18 | 79.9 | 29,119 | 71.3 |
| Dayton-Kettering, OH | 454 | 0.0 | 172 | 2.8 | 399.3 | 1,251,741 | 7,266 | 130,667 | 283 | 107,208 | 88.3 | 11.7 | 22 | 81.6 | 14,929 | 45.0 |
| Decatur, AL | 349 | -12.1 | 144 | 6.7 | 166.7 | 581,656 | 4,031 | 89,976 | 313 | 129,400 | 22.3 | 77.7 | 2 | 74.0 | 5,769 | 32.8 |
| Decatur, IL | 277 | -17.6 | 471 | 0.1 | 268 | 4,356,859 | 9,250 | 281,637 | 180 | 305,610 | 97.7 | 2.3 | 5 | 84.6 | 4,815 | 69.4 |
| Deltona-Daytona Beach-Ormond Beach, FL | 194 | 29.3 | 114 | 12.6 | 30.9 | 957,517 | 8,366 | 53,386 | 211 | 124,603 | 93.2 | 6.8 | 9 | 76.1 | 1,194 | 6.0 |
| Denver-Aurora-Lakewood, CO | 2,489 | -0.1 | 444 | 46.4 | 965.9 | 1,006,299 | 2,267 | 66,978 | 223 | 39,705 | 68.1 | 31.9 | 27 | 85.3 | 10,661 | 11.3 |
| Des Moines-West Des Moines, IA | 1,716 | -1.7 | 303 | 2.7 | 1,380.8 | 2,049,378 | 6,758 | 177,961 | 1,027 | 181,462 | 66.5 | 33.5 | 33 | 80.6 | 30,378 | 52.1 |
| Detroit-Warren-Dearborn, MI | 550 | -1.4 | 138 | 9.9 | 448.1 | 767,746 | 5,560 | 115,703 | 341 | 85,581 | 81.1 | 18.9 | 46 | 79.5 | 5,926 | 16.3 |
| Detroit-Dearborn-Livonia, MI Div 19804 | 10 | -36.3 | 40 | 0.6 | 7.8 | 467,879 | 11,561 | 70,633 | 23 | 93,274 | 96.8 | 3.2 | 2 | 88.3 | 56 | 4.8 |
| Warren-Troy-Farmington Hills, MI Div 47664 | 540 | -0.3 | 145 | 9.4 | 440.3 | 787,673 | 5,448 | 118,698 | 318 | 85,069 | 79.9 | 20.1 | 44 | 78.9 | 5,870 | 17.1 |
| Dothan, AL | 506 | -13.8 | 256 | 28.3 | 241.8 | 723,587 | 2,822 | 111,553 | 330 | 167,361 | 33.3 | 66.7 | 1 | 72.8 | 16,021 | 53.2 |
| Dover, DE | 182 | 5.9 | 222 | 57.8 | 155.7 | 1,758,029 | 7,923 | 179,332 | 391 | 476,038 | 27.9 | 72.1 | 4 | 76.6 | 4,531 | 34.3 |
| Dubuque, IA | 313 | 7.6 | 224 | 0.4 | 258.5 | 1,930,243 | 8,633 | 188,850 | 440 | 313,898 | 29.2 | 70.8 | 15 | 82.0 | 5,667 | 73.5 |
| Duluth, MN-WI | 305 | 3.9 | 182 | 0.2 | 136.6 | 373,649 | 2,054 | 63,481 | 36 | 21,450 | 49.2 | 50.8 | 19 | 78.2 | 321 | 3.6 |
| Durham-Chapel Hill, NC | 402 | 4.2 | 134 | 4.6 | 158 | 666,399 | 4,968 | 73,446 | 258 | 86,290 | 31.0 | 69.0 | 73 | 80.1 | 1,177 | 16.5 |
| East Stroudsburg, PA | 28 | 4.2 | 118 | 0.2 | 13.4 | 698,129 | 5,892 | 85,597 | 10 | 42,627 | 64.6 | 35.4 | 0 | 79.4 | 190 | 9.4 |
| Eau Claire, WI | 528 | -10.2 | 213 | 11.0 | 357 | 829,118 | 3,888 | 124,131 | 305 | 123,187 | 40.9 | 59.1 | 66 | 73.8 | 4,434 | 43.6 |
| El Centro, CA | 522 | 1.2 | 1,317 | 456.1 | 504 | 14,670,311 | 11,135 | 884,987 | 1,860 | 4,696,051 | 65.7 | 34.3 | 45 | 90.7 | 3,640 | 27.5 |
| Elizabethtown-Fort Knox, KY | 451 | 3.7 | 161 | 0.0 | 257.6 | 683,751 | 4,255 | 88,940 | 146 | 52,121 | 69.8 | 30.2 | 1 | 78.4 | 4,561 | 24.4 |
| Elkhart-Goshen, IN | 175 | 1.2 | 105 | 25.0 | 146.5 | 1,173,020 | 11,178 | 98,422 | 298 | 178,936 | 23.4 | 76.6 | 86 | 45.5 | 3,082 | 16.3 |
| Elmira, NY | 67 | 15.1 | 168 | 0.1 | 35 | 440,269 | 2,619 | 91,731 | 19 | 47,771 | 50.6 | 49.4 | 5 | 78.6 | 384 | 19.1 |
| El Paso, TX | 2,419 | -1.7 | 3,062 | 41.3 | 60.4 | 3,327,349 | 1,087 | 98,266 | 64 | 81,194 | 78.3 | 21.7 | 4 | 74.2 | 457 | 2.3 |
| Enid, OK | 675 | 1.3 | 721 | 3.2 | 442.9 | 1,251,739 | 1,736 | 180,573 | 130 | 139,318 | 48.1 | 51.9 | 0 | 80.0 | 6,660 | 61.3 |
| Erie, PA | 153 | -9.0 | 132 | 1.3 | 93.5 | 595,515 | 4,511 | 112,026 | 82 | 70,602 | 76.4 | 23.6 | 3 | 78.1 | 907 | 15.7 |
| Eugene-Springfield, OR | 203 | -7.5 | 77 | 22.3 | 98 | 656,860 | 8,556 | 59,094 | 158 | 59,873 | 58.0 | 42.0 | 65 | 83.8 | 659 | 2.9 |

Table C. Metropolitan Areas —

# Water Use, Wholesale Trade, Retail Trade, and Real Estate

| Area name | Water use, 2015 | | Wholesale Trade[1], 2017 | | | | Retail Trade[2], 2017 | | | | Real estate and rental and leasing,[2] 2017 | | | |
|---|---|---|---|---|---|---|---|---|---|---|---|---|---|---|
| | Public supply water withdrawn (mil gal/day) | Public supply gallons withdrawn per person per day | Number of establishments | Number of employees | Sales (mil dol) | Average payroll (mil dol) | Number of establishments | Number of employees | Sales (mil dol) | Average payroll (mil dol) | Number of establishments | Number of employees | Sales (mil dol) | Average payroll (mil dol) |
| | 133 | 134 | 135 | 136 | 137 | 138 | 139 | 140 | 141 | 142 | 143 | 144 | 145 | 146 |
| Cedar Rapids, IA | 44.04 | 165.5 | 364 | 5,819 | 3,988.4 | 358.6 | 834 | 16,751 | 6,868.5 | 414.3 | 251 | 1,123 | 256.0 | 47.3 |
| Chambersburg-Waynesboro, PA | 8.47 | 55.1 | D | D | D | D | 487 | 7,407 | 2,075.0 | 186.5 | 94 | 364 | 61.4 | 11.7 |
| Champaign-Urbana, IL | 25.54 | 113.4 | 189 | 3,221 | 2,719.7 | 150.5 | 655 | 10,285 | 2,601.7 | 243.9 | 231 | 2,083 | 388.5 | 85.4 |
| Charleston, WV | 34.86 | 128.5 | 254 | 3,362 | 2,083.1 | 163.1 | 870 | 13,006 | 3,845.2 | 326.1 | 262 | 1,355 | 369.5 | 57.4 |
| Charleston-North Charleston, SC | 107.90 | 144.9 | 702 | 8,976 | 9,063.0 | 561.8 | 2,790 | 41,887 | 12,008.0 | 1,097.2 | 1,352 | 5,650 | 1,475.6 | 272.3 |
| Charlotte-Concord-Gastonia, NC-SC | 242.84 | 99.0 | 3,323 | 50,727 | 39,242.1 | 3,171.2 | 7,907 | 125,747 | 36,152.5 | 3,279.3 | 3,860 | 17,619 | 5,600.4 | 963.1 |
| Charlottesville, VA | 14.89 | 70.1 | 134 | 1,062 | 684.4 | 59.5 | 778 | 11,322 | 3,153.7 | 315.9 | 312 | 1,533 | 312.5 | 64.6 |
| Chattanooga, TN-GA | 77.58 | 141.6 | 537 | 7,284 | 4,711.6 | 427.1 | 1,895 | 30,420 | 9,150.1 | 815.8 | 476 | 2,299 | 674.0 | 127.1 |
| Cheyenne, WY | 14.00 | 144.2 | 122 | 926 | 604.3 | 49.1 | 372 | 5,390 | 1,707.9 | 151.7 | 183 | 689 | 188.4 | 31.6 |
| Chicago-Naperville-Elgin, IL-IN-WI | 1,142.97 | 119.7 | 11,933 | 220,423 | 254,367.4 | 16,507.1 | 27,260 | 480,326 | 137,141.2 | 12,646.6 | 11,120 | 71,483 | 30,181.7 | 4,429.9 |
| Chicago-Naperville-Evanston, IL Div 16984 | 898.05 | 124.4 | 9,300 | 168,581 | 197,635.3 | 12,332.7 | 20,399 | 358,052 | 99,423.8 | 9,515.5 | 8,928 | 61,271 | 26,228.9 | 3,906.8 |
| Elgin, IL Div 20994 | 77.05 | 101.6 | 878 | 13,769 | 16,835.4 | 864.3 | 1,987 | 35,868 | 9,164.7 | 864.7 | 634 | 3,148 | 1,386.3 | 156.7 |
| Gary, IN Div 23844 | 93.36 | 132.8 | 570 | 6,589 | 6,202.9 | 375.0 | 2,161 | 34,698 | 10,237.1 | 856.1 | 603 | 2,800 | 580.1 | 107.8 |
| Lake County-Kenosha County, IL-WI Div | 74.51 | 85.4 | 1,185 | 31,484 | 33,693.8 | 2,935.1 | 2,713 | 51,708 | 18,315.5 | 1,410.3 | 955 | 4,264 | 1,986.3 | 258.5 |
| Chico, CA | 28.81 | 127.8 | 144 | 2,073 | 975.0 | 102.1 | 693 | 10,652 | 3,137.5 | 307.3 | 262 | 1,136 | 207.2 | 36.2 |
| Cincinnati, OH-KY-IN | 240.45 | 110.3 | 2,253 | 48,234 | 73,070.1 | 3,157.6 | 6,287 | 125,116 | 36,913.2 | 3,042.0 | 2,229 | 12,476 | 3,702.4 | 646.4 |
| Clarksville, TN-KY | 37.39 | 127.1 | 154 | 1,804 | 1,277.6 | 83.9 | 849 | 12,496 | 3,340.8 | 313.1 | 249 | 1,184 | 240.9 | 43.6 |
| Cleveland, TN | 14.61 | 120.9 | 69 | 729 | 830.1 | 49.8 | 396 | 6,146 | 2,162.4 | 167.2 | D | D | D | D |
| Cleveland-Elyria, OH | 292.15 | 141.8 | 2,711 | 42,249 | 31,494.2 | 2,547.8 | 6,479 | 99,944 | 29,175.5 | 2,585.5 | 2,263 | 15,576 | 5,026.6 | 821.7 |
| Coeur d'Alene, ID | 34.69 | 230.7 | 135 | 1,540 | 1,066.9 | 77.1 | 600 | 8,570 | 2,942.5 | 255.1 | 301 | 731 | 181.1 | 28.7 |
| College Station-Bryan, TX | 35.47 | 142.4 | 150 | 2,150 | 1,451.0 | 126.4 | 758 | 11,576 | 3,534.5 | 285.2 | 292 | 1,486 | 337.8 | 53.0 |
| Colorado Springs, CO | 94.52 | 135.4 | 470 | 4,871 | 3,019.3 | 321.6 | 2,166 | 32,833 | 9,962.3 | 949.6 | 1,380 | 4,548 | 1,033.2 | 191.4 |
| Columbia, MO | 20.22 | 99.7 | 157 | 1,838 | 1,014.0 | 97.9 | 684 | 12,618 | 3,737.5 | 315.3 | 286 | 1,266 | 231.8 | 40.3 |
| Columbia, SC | 92.00 | 113.6 | 681 | 12,098 | 9,165.9 | 707.8 | 2,677 | 42,762 | 12,450.4 | 1,071.6 | 832 | 4,363 | 1,358.1 | 205.4 |
| Columbus, GA-AL | 55.31 | 169.7 | 152 | 1,978 | 1,683.1 | 112.5 | 1,062 | 14,281 | 3,646.7 | 327.1 | 242 | 1,314 | 342.8 | 54.5 |
| Columbus, IN | 9.49 | 116.9 | D | D | D | D | 315 | 5,106 | 1,311.6 | 112.8 | 75 | 355 | 91.4 | 12.4 |
| Columbus, OH | 207.31 | 102.5 | 1,632 | 33,225 | 37,877.3 | 2,109.5 | 5,612 | 104,982 | 37,840.9 | 2,939.5 | 2,254 | 14,091 | 4,123.8 | 701.7 |
| Corpus Christi, TX | 73.63 | 172.4 | 429 | 6,150 | 4,001.0 | 324.3 | 1,226 | 20,846 | 6,167.4 | 560.6 | 528 | 3,217 | 804.7 | 158.9 |
| Corvallis, OR | 11.95 | 136.5 | 41 | 278 | 176.7 | 16.6 | 266 | 3,935 | 903.3 | 97.0 | 130 | 493 | 82.4 | 15.3 |
| Crestview-Fort Walton Beach-Destin, FL | 32.98 | 125.8 | 180 | 1,275 | 670.3 | 58.2 | 1,287 | 17,111 | 4,731.8 | 437.0 | 718 | 3,111 | 701.9 | 125.6 |
| Cumberland, MD-WV | 2.24 | 22.4 | 54 | 546 | 162.1 | 22.9 | 341 | 4,804 | 1,281.6 | 114.2 | 74 | 230 | 43.0 | 7.8 |
| Dallas-Fort Worth-Arlington, TX | 436.65 | 62.0 | 7,901 | 144,518 | 175,673.0 | 9,698.2 | 20,645 | 357,167 | 121,884.6 | 10,560.9 | 9,224 | 67,915 | 21,049.7 | 3,987.6 |
| Dallas-Plano-Irving, TX Div 19124 | 390.06 | 82.9 | 5,630 | 104,565 | 141,638.8 | 7,213.9 | 13,637 | 236,449 | 80,143.2 | 7,145.3 | 6,735 | 53,435 | 17,157.7 | 3,283.8 |
| Fort Worth-Arlington-Grapevine, TX Div 23104 | 46.59 | 20.0 | 2,271 | 39,953 | 34,034.3 | 2,484.3 | 7,008 | 120,718 | 41,741.4 | 3,415.6 | 2,489 | 14,480 | 3,892.0 | 703.8 |
| Dalton, GA | 25.85 | 179.8 | 245 | 4,074 | 1,534.2 | 189.1 | 488 | 5,545 | 1,603.4 | 136.2 | 0 | 0 | 0.0 | 15.9 |
| Danville, IL | 8.95 | 112.9 | D | D | D | D | 247 | 3,709 | 934.8 | 84.3 | 46 | 153 | 38.3 | 5.9 |
| Daphne-Fairhope-Foley, AL | 23.67 | 116.2 | 194 | 2,404 | 1,447.4 | 124.6 | 989 | 13,735 | 3,703.4 | 340.2 | 365 | 1,842 | 369.7 | 65.2 |
| Davenport-Moline-Rock Island, IA-IL | 40.06 | 104.4 | 393 | 6,467 | 4,788.4 | 359.0 | 1,249 | 21,925 | 5,758.7 | 561.0 | 352 | 1,412 | 305.6 | 50.2 |
| Dayton-Kettering, OH | 101.01 | 126.1 | 661 | 9,630 | 9,964.5 | 596.1 | 2,438 | 44,926 | 12,449.7 | 1,031.0 | 744 | 4,107 | 794.4 | 169.3 |
| Decatur, AL | 33.33 | 218.3 | 16 | 107 | 62.3 | 98.9 | 546 | 6,313 | 1,977.8 | 162.7 | 100 | 422 | 108.9 | 18.3 |
| Decatur, IL | 20.02 | 186.6 | 95 | 1,303 | 1,363.5 | 74.8 | 395 | 5,594 | 1,515.3 | 143.5 | 88 | 484 | 78.7 | 14.9 |
| Deltona-Daytona Beach-Ormond Beach, FL | 64.62 | 103.7 | 455 | 3,123 | 1,564.8 | 151.3 | 2,242 | 30,781 | 8,827.0 | 804.5 | 922 | 3,178 | 689.9 | 113.4 |
| Denver-Aurora-Lakewood, CO | 453.79 | 161.2 | 3,395 | 53,171 | 62,554.5 | 3,584.1 | 8,815 | 142,779 | 44,500.0 | 4,400.0 | 6,036 | 25,091 | 7,859.5 | 1,425.8 |
| Des Moines-West Des Moines, IA | 64.73 | 98.1 | 838 | 15,078 | 17,616.8 | 927.4 | 2,067 | 40,570 | 11,501.8 | 1,082.5 | 847 | 4,805 | 1,067.6 | 231.0 |
| Detroit-Warren-Dearborn, MI | 634.84 | 147.6 | 4,529 | 75,766 | 76,057.4 | 5,139.7 | 14,853 | 207,041 | 65,578.4 | 5,674.6 | 3,948 | 29,354 | 12,393.7 | 1,382.1 |
| Detroit-Dearborn-Livonia, MI Div 19804 | 466.55 | 265.2 | 1,434 | 26,754 | 29,897.5 | 1,708.7 | 5,927 | 69,229 | 21,293.3 | 1,772.1 | 1,162 | 7,082 | 7,908.7 | 332.8 |
| Warren-Troy-Farmington Hills, MI Div 47664 | 168.29 | 66.2 | 3,095 | 49,012 | 46,159.9 | 3,431.0 | 8,926 | 137,812 | 44,285.1 | 3,902.5 | 2,786 | 22,272 | 4,485.0 | 1,049.3 |
| Dothan, AL | 22.38 | 151.0 | 188 | 3,531 | 11,757.4 | 194.9 | 673 | 8,868 | 2,466.7 | 225.8 | 108 | 510 | 103.8 | 20.5 |
| Dover, DE | 11.81 | 68.1 | D | D | D | D | 600 | 9,606 | 3,059.7 | 263.9 | 161 | 697 | 225.2 | 26.9 |
| Dubuque, IA | 7.71 | 79.4 | 156 | 2,348 | 2,185.6 | 122.0 | 424 | 7,233 | 1,763.9 | 175.4 | 121 | 400 | 94.9 | 14.4 |
| Duluth, MN-WI | 37.60 | 129.6 | 257 | 3,232 | 2,363.2 | 171.3 | 1,147 | 16,861 | 4,359.7 | 424.1 | 279 | 1,144 | 226.9 | 36.5 |
| Durham-Chapel Hill, NC | 67.46 | 110.4 | 420 | 15,465 | 11,563.5 | 1,776.3 | 1,729 | 26,946 | 7,134.2 | 704.5 | 682 | 3,308 | 765.3 | 157.8 |
| East Stroudsburg, PA | 11.35 | 68.2 | 107 | D | 525.9 | D | 605 | 9,469 | 2,448.5 | 225.8 | 144 | 539 | 122.4 | 18.8 |
| Eau Claire, WI | 14.36 | 86.7 | 168 | 2,424 | 1,694.0 | 115.4 | 584 | 10,854 | 3,301.3 | 269.5 | 146 | 747 | 112.0 | 23.0 |
| El Centro, CA | 24.47 | 135.8 | 202 | 1,837 | 1,656.4 | 87.4 | 473 | 8,444 | 1,815.1 | 199.7 | 161 | 736 | 136.0 | 22.6 |
| Elizabethtown-Fort Knox, KY | 15.75 | 106.0 | 77 | 681 | 605.5 | 31.5 | 480 | 7,186 | 2,027.1 | 178.9 | 116 | 556 | 86.8 | 15.3 |
| Elkhart-Goshen, IN | 12.95 | 63.6 | 317 | 6,495 | 4,827.8 | 332.6 | 705 | 9,603 | 2,889.9 | 262.5 | 181 | 767 | 167.3 | 28.7 |
| Elmira, NY | 8.37 | 96.1 | 85 | 1,216 | 608.6 | 56.7 | 316 | 5,032 | 1,378.8 | 126.2 | 89 | 365 | 98.5 | 13.6 |
| El Paso, TX | 114.57 | 136.6 | D | D | D | D | 2,280 | 38,251 | 9,964.9 | 884.3 | 844 | 4,043 | 958.5 | 152.1 |
| Enid, OK | 2.87 | 45.1 | D | D | D | D | 268 | 3,780 | 944.3 | 98.8 | 93 | 383 | 70.4 | 14.3 |
| Erie, PA | 33.34 | 119.9 | 247 | 3,292 | 1,367.7 | 166.6 | 917 | 15,344 | 3,823.5 | 360.9 | 190 | 878 | 169.8 | 30.3 |
| Eugene-Springfield, OR | 43.68 | 120.4 | 392 | 5,458 | 2,822.2 | 304.4 | 1,313 | 20,047 | 5,373.5 | 585.6 | 613 | 2,395 | 523.8 | 84.9 |

1. Merchant wholesalers, except manufacturers' sales branches and offices. 2. Employer establishments.

# Table C. Metropolitan Areas — Professional Services, Manufacturing, and Accommodation and Food Services

| Area name | Professional, scientific, and technical services, 2017 — Number of establishments | Number of employees | Sales (mil dol) | Average payroll (mil dol) | Manufacturing, 2017 — Number of establishments | Number of employees | Receipts (mil dol) | Annual payroll (mil dol) | Accommodation and food services, 2017 — Number of establishments | Number of employees | Receipts (mil dol) | Annual payroll (mil dol) |
|---|---|---|---|---|---|---|---|---|---|---|---|---|
| | 147 | 148 | 149 | 150 | 151 | 152 | 153 | 154 | 155 | 156 | 157 | 158 |
| Cedar Rapids, IA | 577 | 6,694 | 1,060.7 | 475.0 | 255 | 18,227 | 10,947.2 | 1,511.9 | 612 | 10,427 | 511.8 | 157.7 |
| Chambersburg-Waynesboro, PA | D | D | D | D | 183 | 7,425 | 2,941.8 | 378.9 | 275 | 4,501 | 217.9 | 62.4 |
| Champaign-Urbana, IL | 24 | 114 | 10.3 | 4.8 | 135 | 6,526 | 2,615.7 | 322.4 | 607 | 11,535 | 587.6 | 173.5 |
| Charleston, WV | 577 | 5,052 | 856.9 | 322.3 | D | D | D | D | 521 | 10,513 | 635.0 | 169.8 |
| Charleston-North Charleston, SC | 2,109 | 18,987 | 3,374.2 | 1,321.0 | 485 | 25,995 | 41,328.4 | 1,757.4 | 1,919 | 43,875 | 2,958.0 | 829.5 |
| Charlotte-Concord-Gastonia, NC-SC | 1,288 | 9,622 | 1,372.7 | 476.4 | 2,346 | 96,584 | 38,354.0 | 5,091.2 | 5,200 | 111,768 | 6,670.0 | 1,821.5 |
| Charlottesville, VA | 453 | 3,832 | 618.1 | 273.2 | 184 | 3,403 | 904.8 | 209.2 | 573 | 12,209 | 750.5 | 227.2 |
| Chattanooga, TN-GA | 101 | 360 | 45.7 | 16.6 | 544 | 30,926 | 14,901.7 | 1,620.1 | 1,138 | 23,933 | 1,403.9 | 389.3 |
| Cheyenne, WY | D | D | D | D | 69 | 1,296 | 1,348.3 | 85.5 | 218 | 3,755 | 227.3 | 64.5 |
| Chicago-Naperville-Elgin, IL-IN-WI | 20,041 | 249,957 | 61,495.6 | 23,322.9 | 10,004 | 378,300 | 174,495.9 | 22,707.0 | 21,395 | 417,084 | 29,679.8 | 8,554.5 |
| Chicago-Naperville-Evanston, IL Div 16984 | 19,413 | 246,675 | 61,025.3 | 23,166.0 | 7,476 | 271,214 | 115,192.8 | 15,802.4 | 16,671 | 329,866 | 24,695.4 | 7,141.5 |
| Elgin, IL Div 20994 | 375 | 1,929 | 297.5 | 93.7 | 968 | 36,788 | 14,286.3 | 2,065.9 | 1,356 | 25,268 | 1,414.1 | 405.5 |
| Gary, IN Div 23844 | 14 | 50 | 4.5 | 1.6 | 549 | 34,555 | 30,613.4 | 2,671.3 | 1,433 | 27,566 | 1,587.6 | 429.7 |
| Lake County-Kenosha County, IL-WI Div | 239 | 1,303 | 168.3 | 61.6 | 1,011 | 35,743 | 14,403.5 | 2,167.4 | 1,935 | 34,384 | 1,982.8 | 577.8 |
| Chico, CA | D | D | D | D | 180 | 4,287 | 1,340.8 | 207.0 | 452 | 8,241 | 498.3 | 138.1 |
| Cincinnati, OH-KY-IN | 3,783 | 48,350 | 9,135.3 | 3,544.3 | 2,128 | 104,052 | 49,352.9 | 6,130.9 | 4,345 | 95,734 | 5,175.7 | 1,514.1 |
| Clarksville, TN-KY | 18 | 64 | 6.2 | 2.2 | 181 | 12,975 | 4,942.2 | 640.5 | 522 | 10,732 | 524.2 | 150.4 |
| Cleveland, TN | 141 | 742 | 88.3 | 30.0 | 117 | 7,531 | 4,262.8 | 371.3 | 217 | 4,381 | 223.5 | 62.0 |
| Cleveland-Elyria, OH | 420 | 2,160 | 280.2 | 110.2 | 3,073 | 116,467 | 41,853.0 | 6,807.6 | 4,566 | 83,943 | 4,677.0 | 1,331.0 |
| Coeur d'Alene, ID | 499 | 2,576 | 388.7 | 138.5 | 262 | 4,995 | 1,241.2 | 238.2 | 416 | 7,789 | 502.1 | 141.2 |
| College Station-Bryan, TX | 478 | 3,556 | 499.3 | 204.9 | 139 | 5,693 | 1,621.5 | 255.4 | 585 | 13,589 | 708.8 | 186.1 |
| Colorado Springs, CO | 2,754 | 23,029 | 4,532.5 | 1,779.2 | 464 | 11,082 | 3,345.4 | 676.3 | 1,442 | 32,703 | 2,039.0 | 591.1 |
| Columbia, MO | 14 | 58 | 4.9 | 1.8 | 113 | 4,649 | 2,304.2 | 230.8 | 475 | 10,491 | 490.1 | 156.9 |
| Columbia, SC | 1,191 | 13,054 | 2,558.9 | 888.4 | 494 | 24,594 | 12,585.4 | 1,417.7 | 1,659 | 34,339 | 1,796.7 | 489.8 |
| Columbus, GA-AL | 87 | 301 | 33.7 | 10.3 | D | D | D | D | 571 | 12,555 | 668.0 | 194.3 |
| Columbus, IN | D | D | D | D | 140 | 12,239 | 5,531.9 | 637.4 | 176 | 4,414 | 216.3 | 62.5 |
| Columbus, OH | 431 | 4,772 | 1,258.1 | 363.8 | 1,415 | 64,841 | 35,382.7 | 3,577.2 | 4,335 | 91,932 | 5,099.0 | 1,474.7 |
| Corpus Christi, TX | D | D | D | D | 208 | 10,049 | 28,701.3 | 797.4 | 1,068 | 21,757 | 1,200.0 | 327.0 |
| Corvallis, OR | D | D | D | D | 93 | 1,295 | 421.2 | 65.3 | 224 | 3,900 | 214.6 | 62.0 |
| Crestview-Fort Walton Beach-Destin, FL | 265 | 655 | 114.8 | 32.6 | 133 | 1,767 | 545.8 | 89.8 | 757 | 17,576 | 1,221.6 | 346.4 |
| Cumberland, MD-WV | 20 | 109 | 11.0 | 4.1 | 65 | 4,233 | 1,713.3 | 236.5 | 51 | 580 | 22.8 | 61.7 |
| Dallas-Fort Worth-Arlington, TX | 2,640 | 13,732 | 2,308.3 | 792.0 | 5,464 | 236,989 | 128,609.1 | 15,654.5 | 15,315 | 337,424 | 21,849.5 | 6,080.9 |
| Dallas-Plano-Irving, TX Div 19124 | 2,352 | 12,639 | 2,152.6 | 742.6 | 3,513 | 149,215 | 63,281.2 | 9,734.7 | 10,574 | 227,351 | 15,115.3 | 4,248.4 |
| Fort Worth-Arlington-Grapevine, TX Div 23104 | 288 | 1,093 | 155.8 | 49.4 | 1,951 | 87,774 | 65,327.8 | 5,919.8 | 4,741 | 110,073 | 6,734.3 | 1,832.6 |
| Dalton, GA | 16 | 76 | 17.8 | 6.6 | 303 | 20,955 | 7,829.6 | 868.3 | 221 | 4,155 | 241.7 | 63.2 |
| Danville, IL | 76 | 385 | 57.8 | 17.6 | 80 | 4,496 | 1,881.2 | 244.5 | 152 | 2,338 | 101.6 | 28.0 |
| Daphne-Fairhope-Foley, AL | D | D | D | D | 165 | 4,505 | 2,616.0 | 215.9 | 572 | 12,977 | 794.1 | 228.8 |
| Davenport-Moline-Rock Island, IA-IL | 65 | 309 | 24.7 | 9.1 | 354 | 24,961 | 12,055.6 | 1,280.8 | 918 | 17,365 | 972.8 | 262.4 |
| Dayton-Kettering, OH | 1,708 | 23,811 | 4,338.5 | 1,697.5 | 978 | 42,379 | 14,345.9 | 2,490.5 | 1,629 | 35,674 | 1,763.9 | 526.7 |
| Decatur, AL | 224 | 2,560 | 550.0 | 151.9 | 181 | 12,080 | 10,392.2 | 748.4 | D | D | D | D |
| Decatur, IL | 142 | 1,099 | 146.8 | 54.3 | 101 | 6,731 | 6,948.0 | 460.3 | 255 | 4,345 | 206.4 | 62.5 |
| Deltona-Daytona Beach-Ormond Beach, FL | 223 | 591 | 86.7 | 31.9 | 416 | 9,656 | 2,527.1 | 464.2 | 1,367 | 26,056 | 1,544.0 | 427.2 |
| Denver-Aurora-Lakewood, CO | 5,335 | 50,197 | 12,676.7 | 4,755.6 | D | 57,130 | D | D | 6,716 | 146,316 | 10,463.0 | 3,073.9 |
| Des Moines-West Des Moines, IA | 2,021 | 18,985 | 3,583.9 | 1,311.1 | 472 | 21,989 | 9,986.6 | 1,151.6 | 1,502 | 29,033 | 1,755.0 | 505.1 |
| Detroit-Warren-Dearborn, MI | 2,870 | 51,817 | 8,007.6 | 3,606.7 | 5,313 | 232,954 | 125,503.3 | 14,398.8 | 8,755 | 175,640 | 10,886.5 | 2,990.4 |
| Detroit-Dearborn-Livonia, MI Div 19804 | 2,870 | 51,817 | 8,007.6 | 3,606.7 | 1,454 | 88,033 | 67,028.2 | 5,655.0 | 3,354 | 71,006 | 5,250.8 | 1,354.7 |
| Warren-Troy-Farmington Hills, MI Div 47664 | D | D | D | D | 3,859 | 144,921 | 58,475.1 | 8,743.8 | 5,401 | 104,634 | 5,635.7 | 1,635.6 |
| Dothan, AL | 54 | 238 | 28.0 | 9.9 | 139 | 5,672 | 1,709.6 | 229.7 | 256 | 5,206 | 280.2 | 73.1 |
| Dover, DE | D | D | D | D | 70 | 4,635 | 2,139.3 | 210.0 | 285 | 6,916 | 476.3 | 115.0 |
| Dubuque, IA | D | D | D | D | 148 | 9,356 | 4,987.4 | 463.1 | 251 | 4,953 | 270.8 | 78.0 |
| Duluth, MN-WI | 486 | 3,714 | 505.2 | 210.5 | 293 | 7,606 | 3,765.4 | 460.8 | 826 | 13,766 | 710.1 | 216.3 |
| Durham-Chapel Hill, NC | 68 | 273 | 29.0 | 13.9 | 407 | 23,345 | 15,610.5 | 1,477.9 | 1,317 | 26,440 | 1,600.9 | 470.3 |
| East Stroudsburg, PA | D | D | D | D | 117 | 5,173 | 2,776.5 | 428.1 | 402 | 10,069 | 873.9 | 185.4 |
| Eau Claire, WI | 282 | 2,054 | 250.6 | 103.3 | 222 | 10,728 | 3,805.1 | 521.7 | 449 | 7,095 | 315.7 | 91.1 |
| El Centro, CA | 178 | 826 | 87.9 | 34.7 | 65 | 2,679 | 1,291.9 | 121.7 | 281 | 4,212 | 246.3 | 69.9 |
| Elizabethtown-Fort Knox, KY | 193 | 1,361 | 199.3 | 71.6 | D | 7,200 | D | D | 235 | 5,422 | 261.3 | 74.4 |
| Elkhart-Goshen, IN | D | D | D | D | 819 | 70,059 | 24,520.9 | 3,763.1 | 378 | 7,145 | 369.0 | 97.5 |
| Elmira, NY | D | D | D | D | 82 | 5,229 | 1,296.9 | 286.3 | 210 | 3,073 | 157.5 | 50.4 |
| El Paso, TX | D | D | D | D | D | D | D | D | 1,591 | 34,722 | 1,675.0 | 465.8 |
| Enid, OK | 119 | 581 | 73.0 | 28.4 | 57 | 2,357 | 1,240.8 | 139.3 | 134 | 2,422 | 128.3 | 32.8 |
| Erie, PA | 408 | 3,267 | 495.3 | 170.6 | 476 | 20,344 | 6,147.9 | 1,215.9 | 627 | 11,662 | 540.9 | 151.4 |
| Eugene-Springfield, OR | 985 | 5,099 | 634.1 | 252.5 | 521 | 13,517 | 5,054.0 | 708.5 | 1,010 | 15,794 | 943.3 | 272.6 |

# Table C. Metropolitan Areas — Health Care and Social Assistance, Other Services, Nonemployer Businesses, and Residential Construction

Column groups:
- **Health care and social assistance, 2017** — items 159–162
- **Other services, 2017** — items 163–166
- **Nonemployer businesses, 2018** — items 167–168
- **Value of residential construction authorized by building permits, 2020** — items 169–170

| Area name | Number of establish-ments (159) | Number of employees (160) | Receipts (mil dol) (161) | Annual payroll (mil dol) (162) | Number of establish-ments (163) | Number of employees (164) | Receipts (mil dol) (165) | Annual payroll (mil dol) (166) | Number (167) | Receipts (mil dol) (168) | New construction ($1,000) (169) | Number of housing units (170) |
|---|---|---|---|---|---|---|---|---|---|---|---|---|
| Cedar Rapids, IA | 753 | 17,246 | 1,973.8 | 761.6 | 383 | 2,815 | 304 | 98 | 16,748 | 767.0 | 97,013 | 644 |
| Chambersburg-Waynesboro, PA | 352 | 9,299 | 1,074.0 | 432.7 | 296 | 1,484 | 148 | 38 | 9,185 | 404.0 | 72,304 | 330 |
| Champaign-Urbana, IL | D | D | D | D | 321 | 2,236 | 756 | 86 | 13,614 | 534.4 | 263,642 | 1,547 |
| Charleston, WV | 102 | 2,047 | 124.8 | 53.0 | 431 | 2,663 | 363 | 99 | 12,235 | 493.8 | 33,183 | 336 |
| Charleston-North Charleston, SC | 1,930 | 38,905 | 5,998.4 | 2,020.7 | 1,284 | 8,174 | 1,054 | 286 | 67,334 | 3,364.7 | 1,842,265 | 8,604 |
| Charlotte-Concord-Gastonia, NC-SC | 5,824 | 135,459 | 18,268.2 | 6,815.7 | 4,104 | 28,337 | 3,845 | 946 | 212,048 | 9,601.5 | 5,633,895 | 26,548 |
| Charlottesville, VA | 539 | 17,623 | 3,035.0 | 1,434.8 | 399 | 3,630 | 877 | 175 | 18,888 | 921.8 | 347,172 | 1,673 |
| Chattanooga, TN-GA | 1,335 | 34,077 | 4,151.6 | 1,635.1 | 720 | 4,937 | 675 | 169 | 40,771 | 2,090.5 | 560,264 | 3,061 |
| Cheyenne, WY | 361 | 6,859 | 876.7 | 382.6 | 206 | 1,187 | 129 | 37 | 8,945 | 616.3 | 110,949 | 550 |
| Chicago-Naperville-Elgin, IL-IN-WI | 27,004 | 611,322 | 75,246.9 | 29,310.3 | 18,638 | 140,089 | 25,431 | 6,122 | 814,273 | 38,187.5 | 3,204,665 | 14,995 |
| Chicago-Naperville-Evanston, IL Div 16984 | 21,098 | 490,490 | 59,148.1 | 23,565.3 | 14,607 | 114,300 | 22,650 | 5,279 | 656,570 | 30,712.6 | 1,935,971 | 9,737 |
| Elgin, IL Div 20994 | 1,625 | 30,648 | 4,006.1 | 1,491.4 | 1,244 | 7,904 | 869 | 256 | 50,410 | 2,148.9 | 358,374 | 1,983 |
| Gary, IN Div 23844 | 1,757 | 44,036 | 5,703.1 | 2,028.3 | 1,149 | 8,451 | 882 | 281 | 42,420 | 1,764.3 | 608,099 | 2,269 |
| Lake County-Kenosha County, IL-WI Div | 2,524 | 46,148 | 6,389.6 | 2,225.3 | 1,638 | 9,434 | 1,029 | 306 | 64,873 | 3,561.7 | 302,220 | 1,006 |
| Chico, CA | 720 | 15,210 | 1,977.3 | 749.0 | 334 | 3,579 | 221 | 71 | 13,121 | 646.2 | 345,503 | 1,837 |
| Cincinnati, OH-KY-IN | 5,128 | 149,459 | 18,734.5 | 7,481.8 | 3,248 | 25,080 | 2,805 | 810 | 151,238 | 7,117.3 | 1,657,987 | 7,297 |
| Clarksville, TN-KY | 547 | 12,467 | 1,256.9 | 512.9 | 329 | 1,662 | 168 | 46 | 17,295 | 779.3 | 464,051 | 3,805 |
| Cleveland, TN | 261 | 5,491 | 631.3 | 222.7 | 123 | 951 | 92 | 25 | 8,554 | 449.6 | 136,058 | 713 |
| Cleveland-Elyria, OH | 5,558 | 186,724 | 21,912.0 | 9,575.8 | 3,803 | 26,414 | 2,902 | 832 | 157,548 | 7,425.3 | 924,603 | 3,374 |
| Coeur d'Alene, ID | 528 | 10,600 | 1,069.8 | 492.9 | 284 | 1,450 | 139 | 41 | 13,392 | 632.3 | 561,941 | 2,929 |
| College Station-Bryan, TX | 477 | 11,339 | 1,502.4 | 533.8 | 342 | 2,343 | 649 | 83 | 17,673 | 817.3 | 326,140 | 2,103 |
| Colorado Springs, CO | 2,188 | 39,863 | 4,543.4 | 1,867.6 | 1,315 | 10,802 | 2,395 | 458 | 57,251 | 2,426.8 | 2,134,314 | 6,932 |
| Columbia, MO | 731 | 20,566 | 2,665.5 | 924.9 | 383 | 2,274 | 267 | 72 | 14,138 | 672.2 | 237,510 | 934 |
| Columbia, SC | 1,576 | 46,187 | 5,878.7 | 2,185.6 | 1,237 | 9,225 | 1,002 | 311 | 57,994 | 2,470.9 | 1,076,211 | 4,872 |
| Columbus, GA-AL | 708 | 16,170 | 1,854.9 | 708.9 | 58 | 271 | 29 | 72 | 20,596 | 722.6 | 254,029 | 1,828 |
| Columbus, IN | 217 | 5,146 | 608.5 | 229.4 | 112 | 729 | 87 | 22 | 4,312 | 183.9 | 43,019 | 169 |
| Columbus, OH | 5,254 | 146,071 | 17,858.5 | 6,907.9 | 2,823 | 22,841 | 3,382 | 884 | 161,295 | 7,761.4 | 2,226,124 | 12,358 |
| Corpus Christi, TX | 1,198 | 32,412 | 3,043.4 | 1,207.3 | 613 | 4,583 | 663 | 154 | 31,603 | 1,402.2 | 412,907 | 1,923 |
| Corvallis, OR | 284 | 6,035 | 728.1 | 322.2 | 146 | 994 | 226 | 40 | 6,036 | 344.7 | 80,155 | 302 |
| Crestview-Fort Walton Beach-Destin, FL | 725 | 10,454 | 1,392.0 | 494.9 | 483 | 2,021 | 216 | 60 | 26,448 | 1,478.7 | 1,016,718 | 2,755 |
| Cumberland, MD-WV | 306 | 7,278 | 755.2 | 286.8 | 167 | 944 | 82 | 24 | 4,293 | 155.1 | 16,729 | 94 |
| Dallas-Fort Worth-Arlington, TX | 20,321 | 408,761 | 54,801.2 | 19,839.4 | 9,663 | 79,122 | 11,910 | 2,918 | 701,565 | 37,149.7 | 13,053,058 | 60,812 |
| Dallas-Plano-Irving, TX Div 19124 | 14,261 | 284,112 | 38,180.5 | 14,014.9 | 6,472 | 54,644 | 8,815 | 2,135 | 480,378 | 26,136.3 | 9,661,918 | 42,798 |
| Fort Worth-Arlington-Grapevine, TX Div 23104 | 6,060 | 124,649 | 16,620.6 | 5,824.4 | 3,191 | 24,478 | 3,095 | 783 | 221,187 | 11,013.5 | 3,391,139 | 18,014 |
| Dalton, GA | D | D | D | D | 152 | 966 | 128 | 34 | 8,066 | 399.4 | 63,409 | 445 |
| Danville, IL | 146 | 5,046 | 578.9 | 265.4 | 112 | 509 | 53 | 14 | 3,880 | 134.9 | 1,277 | 7 |
| Daphne-Fairhope-Foley, AL | 503 | 8,049 | 848.4 | 327.8 | 323 | 1,544 | 187 | 46 | 20,047 | 1,018.7 | 854,637 | 3,614 |
| Davenport-Moline-Rock Island, IA-IL | 1,039 | 24,097 | 2,457.5 | 994.2 | 629 | 4,007 | 394 | 119 | 21,203 | 910.1 | 126,095 | 582 |
| Dayton-Kettering, OH | 2,109 | 64,737 | 7,798.1 | 3,174.9 | 1,208 | 8,571 | 908 | 300 | 50,875 | 2,161.3 | 436,999 | 1,699 |
| Decatur, AL | 382 | 6,364 | 564.2 | 239.4 | D | D | D | D | 9,684 | 400.3 | 47,130 | 224 |
| Decatur, IL | 287 | 8,451 | 1,014.4 | 362.8 | 172 | 1,030 | 293 | 36 | 5,091 | 180.2 | 8,466 | 31 |
| Deltona-Daytona Beach-Ormond Beach, FL | 1,682 | 32,361 | 3,965.7 | 1,440.2 | 1,211 | 5,345 | 759 | 167 | 54,764 | 2,326.2 | 1,655,917 | 6,443 |
| Denver-Aurora-Lakewood, CO | 8,296 | 171,590 | 22,532.7 | 8,718.4 | 6,017 | 40,996 | 5,795 | 1,567 | 281,622 | 14,414.1 | 4,431,755 | 19,732 |
| Des Moines-West Des Moines, IA | 1,712 | 43,242 | 4,985.7 | 2,144.5 | 1,356 | 9,065 | 1,354 | 359 | 49,347 | 2,400.6 | 1,529,831 | 6,508 |
| Detroit-Warren-Dearborn, MI | 13,074 | 274,923 | 32,971.4 | 12,946.6 | 7,432 | 49,010 | 6,160 | 1,640 | 351,743 | 16,107.9 | 1,793,207 | 7,165 |
| Detroit-Dearborn-Livonia, MI Div 19804 | 4,118 | 111,962 | 14,321.6 | 5,577.6 | 2,725 | 18,581 | 2,525 | 643 | 132,583 | 4,544.5 | 435,269 | 1,996 |
| Warren-Troy-Farmington Hills, MI Div 47664 | 8,956 | 162,961 | 18,649.8 | 7,368.9 | 4,707 | 30,429 | 3,635 | 997 | 219,160 | 11,563.4 | 1,357,937 | 5,169 |
| Dothan, AL | 375 | 10,572 | 1,236.4 | 517.5 | 206 | 974 | 116 | 28 | 10,142 | 459.1 | 117,399 | 488 |
| Dover, DE | 433 | 10,827 | 1,248.9 | 497.3 | 248 | 1,249 | 124 | 37 | 11,282 | 774.2 | 289,710 | 2,136 |
| Dubuque, IA | 283 | 8,469 | 943.9 | 405.4 | 205 | 1,118 | 136 | 31 | 6,400 | 306.5 | 60,977 | 277 |
| Duluth, MN-WI | 1,046 | 30,326 | 3,231.1 | 1,390.3 | 537 | 3,382 | 338 | 93 | 17,008 | 681.6 | 106,517 | 565 |
| Durham-Chapel Hill, NC | 1,606 | 52,532 | 7,487.2 | 2,672.7 | 853 | 8,455 | 2,154 | 447 | 52,263 | 2,166.9 | 1,060,352 | 5,505 |
| East Stroudsburg, PA | 361 | 7,518 | 784.6 | 317.6 | 304 | 1,564 | 137 | 43 | 10,781 | 510.7 | 55,729 | 212 |
| Eau Claire, WI | 502 | 12,739 | 1,384.2 | 516.3 | 303 | 1,850 | 184 | 54 | 10,392 | 554.4 | 201,173 | 992 |
| El Centro, CA | 288 | 5,327 | 652.5 | 237.9 | 143 | 695 | 73 | 20 | 10,098 | 375.2 | 77,094 | 480 |
| Elizabethtown-Fort Knox, KY | 303 | 8,248 | 860.6 | 383.9 | 180 | 1,006 | 99 | 28 | 8,411 | 349.1 | 117,381 | 762 |
| Elkhart-Goshen, IN | 373 | 11,322 | 1,394.3 | 488.3 | 345 | 2,646 | 334 | 99 | 13,005 | 605.0 | 113,306 | 488 |
| Elmira, NY | 210 | 6,480 | 650.8 | 310.8 | 132 | 632 | 68 | 17 | 3,867 | 141.5 | 29,735 | 184 |
| El Paso, TX | D | D | D | D | D | D | D | D | 64,803 | 2,871.8 | 571,275 | 2,562 |
| Enid, OK | 197 | 3,778 | 386.2 | 141.0 | 108 | 545 | 63 | 15 | 4,445 | 181.9 | 7,898 | 23 |
| Erie, PA | 878 | 24,543 | 2,475.6 | 1,085.6 | 563 | 3,433 | 322 | 82 | 13,819 | 618.9 | 41,234 | 201 |
| Eugene-Springfield, OR | 1,260 | 23,686 | 2,982.6 | 1,139.5 | 636 | 3,703 | 552 | 118 | 25,052 | 1,159.3 | 272,233 | 1,391 |

Items 159—170

# Table C. Metropolitan Areas — Government Employment and Payroll, and Local Government Finances

| Area name | Full-time equivalent employees | March payroll (dollars) | March payroll (percent of total) — Administration, judicial, and legal | Police and corrections | Fire protection | Highways and transportation | Health and welfare | Natural resources and utilities | Education and libraries | General revenue Total (mil dol) | Inter-governmental (mil dol) | Taxes Total (mil dol) | Taxes Per capita[1] Total (dollars) | Taxes Per capita[1] Property (dollars) |
|---|---|---|---|---|---|---|---|---|---|---|---|---|---|---|
| | 171 | 172 | 173 | 174 | 175 | 176 | 177 | 178 | 179 | 180 | 181 | 182 | 183 | 184 |
| Cedar Rapids, IA | 10,938 | 48,372,538 | 3.8 | 8.4 | 2.4 | 4.7 | 2.5 | 5.7 | 71.2 | 1,393 | 561 | 544 | 2,011 | 1,680 |
| Chambersburg-Waynesboro, PA | 3,509 | 14,820,555 | 8.8 | 9.9 | 1.0 | 2.8 | 4.9 | 6.8 | 64.5 | 471 | 163 | 223 | 1,445 | 1,114 |
| Champaign-Urbana, IL | 8,074 | 34,122,603 | 6.8 | 10.0 | 4.1 | 6.7 | 6.9 | 5.2 | 58.0 | 869 | 294 | 437 | 1,926 | 1,626 |
| Charleston, WV | 9,301 | 32,903,011 | 6.5 | 7.1 | 3.3 | 3.8 | 5.1 | 3.7 | 68.3 | 877 | 308 | 363 | 1,375 | 987 |
| Charleston-North Charleston, SC | 25,010 | 93,233,309 | 8.2 | 12.8 | 7.2 | 3.2 | 2.4 | 9.2 | 54.8 | 3,301 | 980 | 1,692 | 2,182 | 1,329 |
| Charlotte-Concord-Gastonia, NC-SC | 118,925 | 545,962,852 | 3.5 | 5.8 | 2.3 | 1.7 | 44.9 | 3.9 | 36.5 | 15,716 | 4,005 | 4,221 | 1,655 | 1,196 |
| Charlottesville, VA | 7,744 | 30,615,694 | 7.3 | 9.0 | 3.5 | 3.3 | 4.9 | 5.1 | 61.5 | 892 | 312 | 427 | 1,974 | 1,460 |
| Chattanooga, TN-GA | 22,494 | 94,289,741 | 5.5 | 6.6 | 2.6 | 2.9 | 34.8 | 8.4 | 37.6 | 2,932 | 835 | 721 | 1,298 | 952 |
| Cheyenne, WY | 6,340 | 31,230,813 | 2.7 | 5.1 | 1.9 | 1.9 | 39.7 | 2.3 | 45.2 | 848 | 317 | 110 | 1,122 | 833 |
| Chicago-Naperville-Elgin, IL-IN-WI | 370,392 | 2,008,221,393 | 5.4 | 12.9 | 5.1 | 7.1 | 4.1 | 6.3 | 56.7 | 57,262 | 17,993 | 30,193 | 3,173 | 2,491 |
| Chicago-Naperville-Evanston, IL 16984 | 277,815 | 1,578,142,028 | 5.2 | 13.9 | 5.3 | 8.3 | 4.5 | 6.4 | 53.7 | 45,457 | 14,062 | 23,950 | 3,336 | 2,492 |
| Elgin, IL Div 20994 | 31,708 | 155,550,255 | 5.1 | 9.1 | 4.7 | 2.0 | 1.9 | 5.6 | 70.6 | 4,100 | 1,237 | 2,349 | 3,079 | 2,814 |
| Gary, IN Div 23844 | 24,510 | 89,508,611 | 9.6 | 11.0 | 4.5 | 5.0 | 2.3 | 7.7 | 58.5 | 2,741 | 1,223 | 995 | 1,421 | 1,351 |
| Lake County-Kenosha County, IL-WI Div | 36,359 | 185,020,499 | 4.7 | 8.9 | 4.1 | 2.4 | 3.6 | 5.6 | 69.4 | 4,963 | 1,470 | 2,898 | 3,328 | 3,117 |
| Chico, CA | 8,504 | 42,774,851 | 6.8 | 10.0 | 1.8 | 1.9 | 13.8 | 4.8 | 59.1 | 1,193 | 764 | 283 | 1,237 | 929 |
| Cincinnati, OH-KY-IN | 76,381 | 322,830,314 | 6.7 | 10.9 | 6.5 | 4.9 | 6.3 | 6.2 | 56.5 | 9,541 | 3,098 | 4,351 | 1,975 | 1,295 |
| Clarksville, TN-KY | 9,239 | 32,080,174 | 4.6 | 10.5 | 4.0 | 3.3 | 5.0 | 7.6 | 62.2 | 822 | 361 | 321 | 1,075 | 642 |
| Cleveland, TN | 3,588 | 12,281,512 | 5.2 | 12.8 | 6.4 | 3.9 | 9.9 | 2.7 | 58.2 | 300 | 153 | 99 | 811 | 577 |
| Cleveland-Elyria, OH | 93,125 | 446,972,995 | 6.5 | 10.2 | 4.6 | 5.7 | 16.8 | 7.1 | 47.1 | 12,307 | 3,484 | 5,700 | 2,771 | 1,673 |
| Coeur d'Alene, ID | 7,757 | 33,973,530 | 5.6 | 7.5 | 3.1 | 2.0 | 43.3 | 3.4 | 34.7 | 903 | 236 | 173 | 1,098 | 1,032 |
| College Station-Bryan, TX | 9,930 | 33,082,098 | 9.7 | 12.9 | 5.2 | 3.9 | 2.3 | 6.5 | 57.5 | 830 | 201 | 508 | 1,960 | 1,614 |
| Colorado Springs, CO | 22,053 | 95,494,427 | 5.6 | 12.6 | 5.0 | 2.9 | 4.0 | 16.1 | 51.8 | 2,614 | 1,071 | 1,080 | 1,490 | 765 |
| Columbia, MO | 7,779 | 26,743,399 | 7.7 | 5.9 | 3.2 | 2.8 | 7.3 | 8.8 | 62.8 | 977 | 218 | 330 | 1,602 | 950 |
| Columbia, SC | 34,954 | 140,163,929 | 4.4 | 6.5 | 2.5 | 1.5 | 26.3 | 5.2 | 52.6 | 4,093 | 1,151 | 1,466 | 1,777 | 1,509 |
| Columbus, GA-AL | 13,049 | 43,882,975 | 4.8 | 11.7 | 4.2 | 2.7 | 7.9 | 6.9 | 58.6 | 1,060 | 423 | 410 | 1,298 | 833 |
| Columbus, IN | 4,111 | 16,129,167 | 3.8 | 7.4 | 3.0 | 1.4 | 41.3 | 4.3 | 37.9 | 614 | 105 | 80 | 977 | 940 |
| Columbus, OH | 72,712 | 341,651,456 | 7.6 | 10.8 | 7.3 | 4.4 | 8.5 | 5.3 | 54.1 | 11,236 | 3,470 | 5,409 | 2,597 | 1,539 |
| Corpus Christi, TX | 19,353 | 72,091,517 | 6.0 | 9.9 | 4.1 | 5.3 | 4.5 | 6.6 | 62.3 | 1,991 | 573 | 936 | 2,184 | 1,743 |
| Corvallis, OR | 2,152 | 10,442,594 | 7.7 | 11.7 | 5.2 | 3.9 | 12.2 | 5.7 | 51.2 | 269 | 91 | 135 | 1,468 | 1,335 |
| Crestview-Fort Walton Beach-Destin, FL | 9,679 | 34,202,616 | 6.6 | 12.1 | 7.5 | 3.9 | 1.9 | 6.2 | 57.7 | 994 | 308 | 463 | 1,704 | 1,306 |
| Cumberland, MD-WV | 3,350 | 15,234,499 | 4.1 | 6.8 | 1.6 | 3.4 | 1.5 | 4.9 | 75.5 | 371 | 190 | 115 | 1,165 | 795 |
| Dallas-Fort Worth-Arlington, TX | 287,204 | 1,272,150,755 | 5.5 | 9.9 | 4.9 | 4.3 | 10.6 | 5.0 | 58.0 | 34,649 | 7,832 | 18,284 | 2,492 | 2,012 |
| Dallas-Plano-Irving, TX Div 19124 | 194,536 | 879,657,973 | 5.1 | 9.7 | 5.1 | 5.4 | 10.5 | 5.2 | 57.0 | 24,381 | 5,067 | 12,871 | 2,619 | 2,118 |
| Fort Worth-Arlington-Grapevine, TX Div 23104 | 92,668 | 392,492,782 | 6.4 | 10.4 | 4.5 | 1.7 | 10.9 | 4.6 | 60.4 | 10,268 | 2,765 | 5,413 | 2,234 | 1,798 |
| Dalton, GA | 5,308 | 19,803,932 | 3.7 | 7.4 | 4.6 | 3.0 | 4.5 | 5.2 | 66.0 | 538 | 246 | 176 | 1,224 | 744 |
| Danville, IL | 3,100 | 11,926,933 | 8.2 | 8.8 | 2.8 | 2.9 | 4.1 | 4.3 | 64.0 | 364 | 229 | 93 | 1,193 | 1,097 |
| Daphne-Fairhope-Foley, AL | 7,998 | 27,159,727 | 6.9 | 8.8 | 3.3 | 4.6 | 13.2 | 8.1 | 51.4 | 778 | 222 | 272 | 1,278 | 507 |
| Davenport-Moline-Rock Island, IA-IL | 14,893 | 64,255,844 | 5.5 | 9.4 | 3.6 | 4.8 | 5.7 | 5.5 | 63.6 | 1,734 | 675 | 734 | 1,927 | 1,687 |
| Dayton-Kettering, OH | 31,763 | 138,563,639 | 8.2 | 10.0 | 4.9 | 5.6 | 8.1 | 7.1 | 54.6 | 3,900 | 1,342 | 1,762 | 2,194 | 1,418 |
| Decatur, AL | 5,002 | 17,822,978 | 4.1 | 9.2 | 3.3 | 2.7 | 3.9 | 10.8 | 63.5 | 492 | 208 | 180 | 1,184 | 515 |
| Decatur, IL | 3,903 | 16,584,624 | 6.5 | 13.5 | 5.1 | 3.4 | 3.3 | 7.1 | 58.7 | 460 | 187 | 180 | 1,707 | 1,439 |
| Deltona-Daytona Beach-Ormond Beach, FL | 19,007 | 72,712,023 | 9.0 | 15.4 | 5.7 | 3.5 | 3.4 | 8.9 | 51.0 | 2,859 | 704 | 979 | 1,511 | 1,155 |
| Denver-Aurora-Lakewood, CO | 102,997 | 507,857,612 | 6.9 | 13.2 | 5.3 | 6.4 | 13.1 | 8.2 | 45.3 | 16,650 | 4,071 | 7,564 | 2,616 | 1,550 |
| Des Moines-West Des Moines, IA | 25,500 | 121,156,720 | 5.2 | 7.7 | 3.1 | 4.3 | 8.8 | 5.4 | 64.2 | 3,515 | 1,322 | 1,401 | 2,056 | 1,870 |
| Detroit-Warren-Dearborn, MI | 110,332 | 518,343,461 | 8.1 | 11.9 | 4.1 | 4.4 | 2.1 | 4.7 | 62.2 | 19,769 | 9,174 | 6,054 | 1,401 | 1,222 |
| Detroit-Dearborn-Livonia, MI Div 19804 | 48,307 | 223,215,723 | 8.0 | 14.5 | 4.2 | 6.5 | 1.3 | 6.5 | 55.6 | 9,545 | 4,482 | 2,659 | 1,513 | 1,151 |
| Warren-Troy-Farmington Hills, MI Div 47664 | 62,025 | 295,127,738 | 8.1 | 9.9 | 4.0 | 2.8 | 2.6 | 3.3 | 67.2 | 10,224 | 4,693 | 3,395 | 1,324 | 1,270 |
| Dothan, AL | 7,279 | 37,160,101 | 2.8 | 5.0 | 2.2 | 1.9 | 30.3 | 4.9 | 51.8 | 808 | 190 | 151 | 1,023 | 361 |
| Dover, DE | 4,452 | 19,533,638 | 5.1 | 6.3 | 0.1 | 0.8 | 1.8 | 5.1 | 79.4 | 513 | 313 | 109 | 619 | 544 |
| Dubuque, IA | 3,536 | 14,203,131 | 6.2 | 8.7 | 3.8 | 7.3 | 5.2 | 4.6 | 62.8 | 433 | 186 | 173 | 1,782 | 1,463 |
| Duluth, MN-WI | 13,477 | 58,909,938 | 7.5 | 11.9 | 3.2 | 6.6 | 10.9 | 7.6 | 48.9 | 1,660 | 823 | 448 | 1,550 | 1,366 |
| Durham-Chapel Hill, NC | 21,823 | 86,844,113 | 5.5 | 10.5 | 3.8 | 1.7 | 11.2 | 6.4 | 55.5 | 2,044 | 773 | 981 | 1,568 | 1,216 |
| East Stroudsburg, PA | 5,561 | 25,725,163 | 5.3 | 5.2 | 0.0 | 3.2 | 1.6 | 1.5 | 82.6 | 816 | 272 | 464 | 2,766 | 2,480 |
| Eau Claire, WI | 5,877 | 25,577,262 | 5.8 | 8.6 | 2.7 | 5.8 | 6.4 | 3.3 | 66.5 | 674 | 327 | 255 | 1,526 | 1,379 |
| El Centro, CA | 11,605 | 61,911,631 | 3.6 | 5.2 | 1.6 | 1.1 | 22.1 | 13.1 | 47.5 | 1,462 | 824 | 204 | 1,121 | 778 |
| Elizabethtown-Fort Knox, KY | 6,955 | 26,839,479 | 1.7 | 3.8 | 1.4 | 1.2 | 44.4 | 4.0 | 42.3 | 679 | 182 | 133 | 884 | 529 |
| Elkhart-Goshen, IN | 7,096 | 28,545,949 | 5.0 | 8.8 | 4.0 | 2.1 | 1.6 | 3.1 | 72.4 | 665 | 334 | 220 | 1,077 | 1,000 |
| Elmira, NY | 4,253 | 19,683,127 | 4.4 | 7.8 | 2.9 | 4.7 | 9.4 | 3.1 | 66.7 | 508 | 229 | 195 | 2,303 | 1,450 |
| El Paso, TX | 42,242 | 158,791,952 | 5.1 | 9.2 | 3.4 | 2.4 | 11.2 | 2.7 | 63.8 | 4,060 | 1,842 | 1,334 | 1,584 | 1,238 |
| Enid, OK | 2,370 | 8,025,034 | 5.6 | 9.5 | 5.9 | 3.9 | 0.1 | 4.7 | 67.5 | 183 | 64 | 81 | 1,309 | 555 |
| Erie, PA | 8,171 | 34,509,401 | 5.5 | 10.5 | 3.0 | 6.5 | 4.8 | 5.6 | 63.3 | 1,314 | 652 | 414 | 1,512 | 1,209 |
| Eugene-Springfield, OR | 11,728 | 57,941,191 | 5.9 | 12.8 | 5.6 | 6.3 | 5.9 | 10.3 | 47.4 | 1,613 | 655 | 548 | 1,459 | 1,212 |

1. Based on the resident population estimated as of July 1 of the year shown.

Items 171—184

# Table C. Metropolitan Areas — Local Government Finances, Government Employment, and Income Taxes

| Area name | Local government finances, 2017 (cont.) | | | | | | | | | Government employment, 2019 | | | Individual income tax returns, 2018 | | |
| | Direct general expenditure | | | | | | | Debt outstanding | | | | | | | |
| | | | Percent of total for: | | | | | | | | | | | Mean | |
| | Total (mil dol) | Per capita¹ (dollars) | Education | Health and hospitals | Police protection | Public welfare | Highways | Total (mil dol) | Per capita¹ (dollars) | Federal civilian | Federal military | State and local | Number of returns | adjusted gross income | Mean income tax |
| | 185 | 186 | 187 | 188 | 189 | 190 | 191 | 192 | 193 | 194 | 195 | 196 | 197 | 198 | 199 |
|---|---|---|---|---|---|---|---|---|---|---|---|---|---|---|---|
| Cedar Rapids, IA | 1,437.2 | 5,310 | 51.2 | 2.0 | 5.1 | 0.8 | 4.6 | 1,507.2 | 5,569 | 1,173 | 969 | 15,593 | 129,560 | 71,717 | 7,881 |
| Chambersburg-Waynesboro, PA | 484.0 | 3,135 | 59.0 | 3.3 | 2.4 | 2.2 | 4.5 | 740.3 | 4,796 | 2,367 | 394 | 5,744 | 77,830 | 57,429 | 5,271 |
| Champaign-Urbana, IL | 862.4 | 3,802 | 49.8 | 3.6 | 6.1 | 2.4 | 5.9 | 745.8 | 3,288 | 1,287 | 452 | 37,281 | 92,760 | 68,070 | 7,673 |
| Charleston, WV | 905.0 | 3,425 | 51.4 | 3.9 | 6.3 | 0.2 | 1.6 | 363.5 | 1,376 | 2,340 | 1,228 | 22,837 | 111,970 | 55,696 | 5,644 |
| Charleston-North Charleston, SC | 2,959.4 | 3,816 | 47.2 | 2.4 | 8.6 | 0.2 | 2.0 | 5,589.2 | 7,207 | 11,574 | 12,969 | 53,370 | 371,640 | 76,824 | 9,941 |
| Charlotte-Concord-Gastonia, NC-SC | 14,982.5 | 5,875 | 28.5 | 39.8 | 5.5 | 2.5 | 1.5 | 14,215.1 | 5,574 | 10,870 | 6,507 | 146,306 | 1,196,440 | 76,885 | 10,011 |
| Charlottesville, VA | 852.9 | 3,947 | 48.1 | 3.4 | 5.7 | 7.4 | 2.5 | 777.9 | 3,600 | 1,489 | 1,135 | 36,930 | 100,670 | 102,139 | 14,676 |
| Chattanooga, TN-GA | 2,697.8 | 4,853 | 28.3 | 38.8 | 4.9 | 0.4 | 2.4 | 1,763.9 | 3,173 | 5,546 | 1,601 | 32,669 | 246,990 | 64,013 | 7,051 |
| Cheyenne, WY | 893.1 | 9,078 | 39.6 | 37.9 | 2.9 | 0.3 | 2.4 | 143.1 | 1,454 | 2,789 | 3,617 | 11,156 | 48,740 | 67,197 | 7,045 |
| Chicago-Naperville-Elgin, IL-IN-WI | 53,454.9 | 5,618 | 44.2 | 4.2 | 7.3 | 0.9 | 4.4 | 93,411.1 | 9,818 | 54,014 | 33,696 | 510,921 | 4,639,460 | 86,174 | 12,380 |
| Chicago-Naperville-Evanston, IL Div 16984 | 42,203.6 | 5,878 | 41.4 | 4.9 | 7.6 | 1.0 | 4.3 | 79,270.9 | 11,041 | 44,462 | 14,722 | 382,416 | 3,518,260 | 86,354 | 12,510 |
| Elgin, IL Div 20994 | 3,928.9 | 5,151 | 56.8 | 0.9 | 7.0 | 0.6 | 4.7 | 6,422.2 | 8,420 | 1,935 | 1,526 | 48,955 | 357,290 | 74,744 | 8,661 |
| Gary, IN Div 23844 | 2,375.8 | 3,391 | 47.1 | 0.6 | 5.8 | 0.1 | 3.6 | 3,383.7 | 4,830 | 2,026 | 2,064 | 31,994 | 337,430 | 62,111 | 6,533 |
| Lake County-Kenosha County, IL-WI Div | 4,946.6 | 5,680 | 56.3 | 2.3 | 6.0 | 1.3 | 5.1 | 4,334.1 | 4,977 | 5,591 | 15,384 | 47,556 | 426,480 | 113,306 | 19,048 |
| Chico, CA | 1,303.8 | 5,701 | 50.9 | 7.0 | 4.3 | 9.8 | 3.6 | 710.4 | 3,106 | 572 | 319 | 16,405 | 87,730 | 61,448 | 6,301 |
| Cincinnati, OH-KY-IN | 9,442.0 | 4,287 | 45.6 | 5.6 | 6.5 | 4.2 | 5.0 | 13,555.8 | 6,155 | 14,774 | 5,939 | 116,132 | 1,067,300 | 76,849 | 9,635 |
| Clarksville, TN-KY | 756.6 | 2,534 | 53.3 | 3.2 | 6.8 | 0.0 | 4.4 | 3,381.5 | 11,323 | 5,457 | 28,052 | 14,747 | 130,400 | 49,739 | 4,071 |
| Cleveland, TN | 333.0 | 2,725 | 49.6 | 2.7 | 6.4 | 4.4 | 4.9 | 234.0 | 1,915 | 292 | 346 | 5,278 | 53,420 | 50,464 | 4,512 |
| Cleveland-Elyria, OH | 12,721.0 | 6,184 | 39.8 | 12.5 | 6.7 | 3.4 | 3.5 | 16,005.0 | 7,780 | 18,773 | 5,607 | 116,377 | 1,038,630 | 69,350 | 8,579 |
| Coeur d'Alene, ID | 909.8 | 5,783 | 24.0 | 51.5 | 3.8 | 0.2 | 2.5 | 282.6 | 1,796 | 671 | 514 | 11,075 | 79,490 | 63,810 | 6,669 |
| College Station-Bryan, TX | 942.2 | 3,636 | 49.9 | 5.0 | 5.1 | 0.3 | 5.0 | 1,968.9 | 7,598 | 858 | 565 | 39,953 | 100,360 | 65,041 | 7,409 |
| Colorado Springs, CO | 2,444.4 | 3,373 | 51.0 | 1.4 | 7.5 | 3.0 | 8.3 | 4,656.7 | 6,425 | 12,270 | 38,683 | 43,060 | 346,480 | 68,909 | 7,467 |
| Columbia, MO | 951.9 | 4,627 | 37.3 | 29.9 | 3.5 | 0.4 | 4.5 | 2,633.1 | 12,799 | 2,706 | 688 | 32,454 | 90,850 | 64,650 | 7,127 |
| Columbia, SC | 4,176.9 | 5,063 | 39.4 | 26.9 | 4.1 | 0.1 | 2.3 | 5,470.1 | 6,631 | 10,631 | 12,049 | 72,254 | 367,960 | 60,566 | 6,363 |
| Columbus, GA-AL | 1,129.5 | 3,575 | 50.2 | 6.6 | 6.2 | 3.3 | 3.9 | 954.3 | 3,020 | 6,792 | 21,654 | 18,155 | 132,070 | 53,473 | 5,300 |
| Columbus, IN | 629.6 | 7,658 | 19.8 | 63.2 | 2.4 | 0.1 | 2.0 | 281.4 | 3,422 | 174 | 246 | 6,279 | 39,640 | 71,231 | 7,970 |
| Columbus, OH | 11,893.4 | 5,711 | 37.7 | 6.6 | 5.5 | 6.8 | 5.5 | 14,856.5 | 7,134 | 14,734 | 5,870 | 166,271 | 1,009,450 | 70,379 | 8,355 |
| Corpus Christi, TX | 2,042.7 | 4,768 | 49.6 | 8.4 | 5.7 | 0.1 | 4.3 | 3,819.1 | 8,914 | 5,755 | 4,780 | 26,316 | 186,370 | 58,642 | 6,433 |
| Corvallis, OR | 219.5 | 2,391 | 55.4 | 0.0 | 7.5 | 0.0 | 2.7 | 177.8 | 1,937 | 483 | 253 | 9,933 | 40,530 | 74,921 | 8,800 |
| Crestview-Fort Walton Beach-Destin, FL | 1,007.1 | 3,709 | 44.4 | 3.0 | 8.7 | 0.3 | 5.1 | 558.7 | 2,058 | 9,119 | 17,849 | 11,671 | 138,550 | 80,923 | 11,193 |
| Cumberland, MD-WV | 389.1 | 3,947 | 55.8 | 0.8 | 5.3 | 0.0 | 4.4 | 305.6 | 3,100 | 561 | 337 | 6,996 | 41,000 | 49,916 | 4,185 |
| Dallas-Fort Worth-Arlington, TX | 34,267.9 | 4,671 | 44.8 | 11.4 | 5.7 | 0.2 | 3.7 | 85,470.2 | 11,649 | 46,447 | 17,009 | 383,805 | 3,409,940 | 82,488 | 11,451 |
| Dallas-Plano-Irving, TX Div 19124 | 24,000.8 | 4,884 | 43.9 | 11.6 | 5.1 | 0.3 | 3.6 | 63,109.5 | 12,842 | 30,602 | 10,686 | 268,705 | 2,285,350 | 85,945 | 12,342 |
| Fort Worth-Arlington-Grapevine, TX Div 23104 | 10,267.2 | 4,238 | 46.9 | 10.9 | 6.8 | 0.1 | 3.9 | 22,360.7 | 9,229 | 15,845 | 6,323 | 115,100 | 1,124,590 | 75,464 | 9,642 |
| Dalton, GA | 533.3 | 3,703 | 57.3 | 10.8 | 4.0 | 0.0 | 4.4 | 130.3 | 904 | 255 | 383 | 6,745 | 58,090 | 56,395 | 5,634 |
| Danville, IL | 307.4 | 3,957 | 52.9 | 0.9 | 5.4 | 0.1 | 7.9 | 108.9 | 1,402 | 1,420 | 147 | 4,537 | 32,570 | 48,911 | 4,109 |
| Daphne-Fairhope-Foley, AL | 833.5 | 3,922 | 37.7 | 18.0 | 5.7 | 0.1 | 8.0 | 1,143.7 | 5,382 | 366 | 969 | 9,606 | 100,970 | 68,660 | 7,884 |
| Davenport-Moline-Rock Island, IA-IL | 1,798.5 | 4,719 | 50.9 | 4.7 | 6.4 | 0.5 | 4.7 | 1,622.1 | 4,256 | 5,771 | 1,410 | 20,716 | 182,650 | 64,694 | 6,882 |
| Dayton-Kettering, OH | 3,943.7 | 4,910 | 50.4 | 1.5 | 6.2 | 5.4 | 3.8 | 3,883.2 | 4,834 | 19,600 | 7,335 | 41,687 | 386,230 | 61,866 | 6,578 |
| Decatur, AL | 558.0 | 3,671 | 53.7 | 1.9 | 6.7 | 0.4 | 2.8 | 762.5 | 5,017 | 373 | 634 | 8,005 | 65,810 | 55,827 | 5,119 |
| Decatur, IL | 452.0 | 4,288 | 47.3 | 1.2 | 10.4 | 1.1 | 5.4 | 425.2 | 4,034 | 317 | 212 | 5,570 | 47,630 | 61,080 | 6,419 |
| Deltona-Daytona Beach-Ormond Beach, FL | 2,713.1 | 4,188 | 29.8 | 23.7 | 6.9 | 0.3 | 4.6 | 2,857.4 | 4,410 | 1,537 | 1,265 | 22,094 | 317,970 | 58,867 | 6,568 |
| Denver-Aurora-Lakewood, CO | 15,273.3 | 5,282 | 34.2 | 7.9 | 5.5 | 2.5 | 4.8 | 30,580.8 | 10,575 | 28,420 | 10,227 | 187,724 | 1,467,010 | 88,747 | 12,079 |
| Des Moines-West Des Moines, IA | 3,513.1 | 5,156 | 51.1 | 8.4 | 4.6 | 0.9 | 4.4 | 3,930.5 | 5,768 | 6,378 | 2,623 | 40,895 | 329,390 | 78,374 | 9,452 |
| Detroit-Warren-Dearborn, MI | 20,409.8 | 4,723 | 42.1 | 5.2 | 5.8 | 3.1 | 5.2 | 22,654.5 | 5,242 | 28,013 | 7,749 | 169,482 | 2,122,030 | 72,306 | 9,272 |
| Detroit-Dearborn-Livonia, MI Div 19804 | 9,879.6 | 5,622 | 33.7 | 3.2 | 5.7 | 5.6 | 3.5 | 11,792.4 | 6,711 | 13,945 | 3,207 | 75,892 | 802,460 | 58,925 | 6,786 |
| Warren-Troy-Farmington Hills, MI Div 47664 | 10,530.2 | 4,106 | 50.0 | 7.2 | 5.8 | 0.8 | 6.7 | 10,862.0 | 4,236 | 14,068 | 4,542 | 93,590 | 1,319,570 | 80,444 | 10,784 |
| Dothan, AL | 863.8 | 5,842 | 24.6 | 48.1 | 3.6 | 0.1 | 3.6 | 578.5 | 3,912 | 402 | 666 | 10,516 | 62,420 | 53,902 | 5,608 |
| Dover, DE | 520.7 | 2,950 | 74.0 | 1.3 | 5.4 | 0.0 | 1.4 | 371.1 | 2,103 | 1,679 | 4,429 | 17,451 | 83,340 | 54,201 | 4,807 |
| Dubuque, IA | 417.0 | 4,296 | 46.7 | 1.8 | 5.2 | 0.3 | 5.1 | 473.0 | 4,873 | 257 | 357 | 4,552 | 47,480 | 67,516 | 7,326 |
| Duluth, MN-WI | 1,782.7 | 6,166 | 35.5 | 3.0 | 4.8 | 6.8 | 13.5 | 2,339.7 | 8,093 | 1,647 | 1,061 | 24,012 | 137,440 | 61,218 | 6,029 |
| Durham-Chapel Hill, NC | 1,964.6 | 3,140 | 45.8 | 4.3 | 9.3 | 6.2 | 1.5 | 1,858.3 | 2,970 | 8,303 | 1,537 | 68,573 | 288,370 | 78,883 | 10,128 |
| East Stroudsburg, PA | 758.0 | 4,515 | 69.4 | 0.1 | 3.3 | 5.4 | 3.0 | 1,152.8 | 6,866 | 3,415 | 452 | 8,396 | 80,510 | 56,169 | 5,567 |
| Eau Claire, WI | 737.3 | 4,406 | 52.8 | 4.8 | 6.1 | 3.5 | 11.6 | 673.8 | 4,027 | 539 | 425 | 11,469 | 80,170 | 65,496 | 7,418 |
| El Centro, CA | 1,512.6 | 8,327 | 43.0 | 24.1 | 3.4 | 7.9 | 2.3 | 1,540.3 | 8,480 | 2,157 | 365 | 16,944 | 80,240 | 43,367 | 3,248 |
| Elizabethtown-Fort Knox, KY | 676.0 | 4,495 | 32.9 | 48.2 | 1.4 | 0.0 | 2.3 | 558.9 | 3,716 | 5,231 | 5,131 | 9,775 | 68,580 | 53,001 | 4,392 |
| Elkhart-Goshen, IN | 693.5 | 3,396 | 57.5 | 1.1 | 4.4 | 0.0 | 4.3 | 677.6 | 3,318 | 284 | 603 | 8,391 | 96,080 | 61,165 | 6,485 |
| Elmira, NY | 499.8 | 5,899 | 45.2 | 2.9 | 3.2 | 18.4 | 5.7 | 196.0 | 2,313 | 211 | 127 | 6,301 | 38,820 | 57,619 | 5,681 |
| El Paso, TX | 3,948.7 | 4,690 | 49.7 | 17.3 | 4.5 | 0.2 | 1.3 | 6,819.3 | 8,099 | 13,141 | 28,325 | 56,923 | 378,920 | 46,200 | 3,993 |
| Enid, OK | 170.7 | 2,772 | 59.3 | 3.2 | 6.6 | 0.1 | 1.8 | 126.3 | 2,051 | 516 | 1,503 | 3,276 | 26,610 | 59,793 | 5,993 |
| Erie, PA | 1,301.5 | 4,758 | 46.6 | 4.7 | 2.7 | 14.7 | 3.7 | 1,418.4 | 5,185 | 1,652 | 709 | 15,431 | 127,570 | 56,269 | 5,687 |
| Eugene-Springfield, OR | 1,560.7 | 4,159 | 48.0 | 5.4 | 6.7 | 1.0 | 3.5 | 1,918.9 | 5,114 | 1,884 | 985 | 23,860 | 173,440 | 63,169 | 6,847 |

1. Based on the resident population estimated as of July 1 of the year shown.

# Table C. Metropolitan Areas — **Land Area and Population**

| CBSA/ DIV Code[1] | Area name | Land area[2] (sq mi) | Total persons 2020 | Rank | Per square mile | White | Black | American Indian, Alaska Native | Asian and Pacific Islander | Percent Hispanic or Latino[3] | Under 5 years | 5 to 14 years | 15 to 24 years | 25 to 34 years | 35 to 44 years | 45 to 54 years |
|---|---|---|---|---|---|---|---|---|---|---|---|---|---|---|---|---|
| | | 1 | 2 | 3 | 4 | 5 | 6 | 7 | 8 | 9 | 10 | 11 | 12 | 13 | 14 | 15 |
| 21780 | Evansville, IN-KY | 1,464 | 315,731 | 162 | 215.6 | 88.1 | 8.9 | 0.6 | 2.1 | 2.6 | 5.7 | 12.5 | 12.4 | 13.2 | 12.3 | 11.8 |
| 21820 | Fairbanks, AK | 7,335 | 95,651 | 365 | 13.0 | 75.1 | 6.3 | 11.3 | 6.2 | 8.1 | 7.1 | 13.1 | 15.9 | 18.3 | 13.2 | 9.8 |
| 22020 | Fargo, ND-MN | 2,810 | 248,594 | 193 | 88.5 | 86.5 | 6.7 | 2.1 | 3.6 | 3.4 | 6.7 | 12.9 | 17.2 | 16.1 | 13.5 | 10.1 |
| 22140 | Farmington, NM | 5,517 | 123,312 | 323 | 22.4 | 38.5 | 1.0 | 40.1 | 1.0 | 21.6 | 6.0 | 14.9 | 13.1 | 13.4 | 12.9 | 10.7 |
| 22180 | Fayetteville, NC | 1,638 | 529,252 | 108 | 323.2 | 49.2 | 35.3 | 3.1 | 3.7 | 12.9 | 7.4 | 14.0 | 15.0 | 16.6 | 12.7 | 10.8 |
| 22220 | Fayetteville-Springdale-Rogers, AR | 2,623 | 548,634 | 103 | 209.1 | 73.9 | 3.3 | 2.5 | 6.0 | 16.9 | 6.7 | 14.0 | 15.0 | 14.9 | 13.9 | 11.5 |
| 22380 | Flagstaff, AZ | 18,616 | 142,481 | 294 | 7.7 | 56.2 | 2.0 | 26.6 | 3.2 | 14.6 | 5.1 | 11.4 | 23.3 | 14.2 | 11.2 | 9.8 |
| 22420 | Flint, MI | 637 | 404,794 | 134 | 635.6 | 74.5 | 21.8 | 1.3 | 1.6 | 3.7 | 5.8 | 12.5 | 12.3 | 12.7 | 11.5 | 12.6 |
| 22500 | Florence, SC | 1,361 | 204,097 | 222 | 150.0 | 53.1 | 43.4 | 0.7 | 1.5 | 2.7 | 5.9 | 13.0 | 12.6 | 12.6 | 11.7 | 12.5 |
| 22520 | Florence-Muscle Shoals, AL | 1,261 | 148,779 | 284 | 118.0 | 83.4 | 13.1 | 1.1 | 1.1 | 3.0 | 5.4 | 11.2 | 13.4 | 12.4 | 11.0 | 12.2 |
| 22540 | Fond du Lac, WI | 720 | 102,902 | 349 | 143.0 | 90.0 | 2.8 | 0.8 | 2.2 | 5.6 | 5.2 | 12.0 | 12.3 | 11.8 | 12.4 | 12.2 |
| 22660 | Fort Collins, CO | 2,596 | 360,428 | 150 | 138.9 | 84.0 | 1.7 | 1.1 | 3.5 | 12.1 | 4.6 | 10.9 | 17.3 | 15.0 | 12.7 | 10.8 |
| 22900 | Fort Smith, AR-OK | 2,409 | 250,434 | 192 | 104.0 | 76.7 | 5.3 | 8.0 | 3.6 | 10.8 | 6.3 | 13.3 | 12.5 | 12.8 | 11.9 | 12.3 |
| 23060 | Fort Wayne, IN | 993 | 416,565 | 133 | 419.5 | 77.1 | 12.6 | 0.8 | 5.1 | 7.5 | 6.9 | 14.1 | 13.1 | 14.1 | 12.4 | 11.7 |
| 23420 | Fresno, CA | 5,958 | 1,000,918 | 55 | 168.0 | 29.6 | 5.3 | 1.1 | 12.0 | 54.1 | 7.3 | 16.2 | 14.2 | 15.3 | 13.0 | 10.9 |
| 23460 | Gadsden, AL | 535 | 102,371 | 353 | 191.3 | 78.7 | 16.5 | 1.1 | 1.1 | 4.3 | 5.8 | 11.8 | 11.8 | 12.3 | 11.8 | 13.2 |
| 23540 | Gainesville, FL | 2,344 | 332,317 | 155 | 141.8 | 66.4 | 19.0 | 0.7 | 6.2 | 10.3 | 5.0 | 10.3 | 20.8 | 14.6 | 11.1 | 9.9 |
| 23580 | Gainesville, GA | 393 | 206,591 | 218 | 525.7 | 61.1 | 8.0 | 0.6 | 2.5 | 29.1 | 6.3 | 13.6 | 13.7 | 12.9 | 12.2 | 13.1 |
| 23900 | Gettysburg, PA | 519 | 102,742 | 350 | 198.1 | 90.0 | 2.3 | 0.5 | 1.3 | 7.3 | 4.8 | 11.2 | 12.8 | 11.1 | 10.6 | 12.7 |
| 24020 | Glens Falls, NY | 1,698 | 124,362 | 321 | 73.2 | 93.9 | 2.7 | 0.7 | 1.2 | 2.9 | 4.5 | 10.3 | 10.6 | 12.2 | 11.4 | 13.0 |
| 24140 | Goldsboro, NC | 554 | 123,967 | 322 | 223.8 | 54.1 | 32.6 | 0.9 | 2.1 | 12.8 | 6.5 | 13.3 | 13.5 | 13.6 | 11.4 | 11.6 |
| 24220 | Grand Forks, ND-MN | 3,407 | 100,381 | 357 | 29.5 | 86.2 | 4.5 | 3.3 | 3.2 | 5.5 | 6.7 | 12.2 | 19.1 | 14.7 | 11.3 | 9.2 |
| 24260 | Grand Island, NE | 1,603 | 75,325 | 379 | 47.0 | 70.7 | 3.0 | 0.8 | 1.6 | 25.1 | 7.4 | 14.8 | 12.3 | 12.5 | 12.4 | 11.4 |
| 24300 | Grand Junction, CO | 3,329 | 155,603 | 269 | 46.7 | 82.5 | 1.3 | 1.4 | 1.7 | 15.0 | 5.3 | 11.9 | 12.3 | 13.1 | 12.5 | 10.7 |
| 24340 | Grand Rapids-Kentwood, MI | 2,689 | 1,081,372 | 51 | 402.1 | 79.8 | 8.2 | 0.9 | 3.5 | 10.1 | 6.2 | 13.2 | 14.1 | 14.7 | 12.7 | 11.5 |
| 24420 | Grants Pass, OR | 1,639 | 88,053 | 369 | 53.7 | 88.9 | 1.1 | 3.0 | 2.2 | 8.0 | 4.8 | 11.1 | 9.7 | 11.1 | 10.8 | 10.8 |
| 24500 | Great Falls, MT | 2,698 | 81,346 | 377 | 30.1 | 88.1 | 2.4 | 6.2 | 2.2 | 5.0 | 6.4 | 12.5 | 12.4 | 14.2 | 11.8 | 10.0 |
| 24540 | Greeley, CO | 3,985 | 333,983 | 154 | 83.8 | 66.1 | 1.7 | 1.1 | 2.5 | 30.2 | 7.0 | 14.5 | 13.2 | 15.4 | 14.2 | 11.9 |
| 24580 | Green Bay, WI | 1,870 | 323,379 | 159 | 172.9 | 84.2 | 3.3 | 3.0 | 3.4 | 8.1 | 5.9 | 12.9 | 12.6 | 12.8 | 12.7 | 12.2 |
| 24660 | Greensboro-High Point, NC | 1,994 | 776,363 | 76 | 389.4 | 58.6 | 28.9 | 1.1 | 4.7 | 9.0 | 5.6 | 12.3 | 13.8 | 13.2 | 11.8 | 13.0 |
| 24780 | Greenville, NC | 652 | 182,924 | 236 | 280.4 | 55.4 | 36.7 | 0.8 | 2.7 | 6.6 | 5.6 | 11.8 | 20.8 | 13.9 | 11.7 | 10.9 |
| 24860 | Greenville-Anderson, SC | 2,710 | 932,705 | 60 | 344.2 | 73.6 | 17.5 | 0.6 | 2.6 | 7.5 | 5.8 | 12.5 | 13.5 | 13.4 | 12.1 | 12.5 |
| 25060 | Gulfport-Biloxi, MS | 2,216 | 418,963 | 131 | 189.1 | 69.5 | 23.0 | 1.0 | 3.2 | 5.7 | 5.9 | 12.9 | 12.5 | 13.4 | 12.5 | 12.2 |
| 25180 | Hagerstown-Martinsburg, MD-WV | 1,008 | 291,144 | 169 | 288.8 | 83.2 | 11.6 | 0.7 | 2.2 | 5.4 | 5.7 | 12.3 | 11.6 | 13.5 | 12.6 | 13.3 |
| 25220 | Hammond, LA | 791 | 136,765 | 303 | 172.8 | 64.0 | 31.0 | 0.8 | 1.1 | 4.6 | 7.0 | 13.5 | 13.4 | 14.7 | 12.7 | 11.1 |
| 25260 | Hanford-Corcoran, CA | 1,391 | 152,692 | 275 | 109.8 | 32.9 | 7.1 | 1.4 | 5.3 | 55.6 | 7.4 | 15.3 | 14.9 | 16.9 | 14.1 | 11.0 |
| 25420 | Harrisburg-Carlisle, PA | 1,622 | 581,943 | 98 | 358.8 | 77.1 | 11.9 | 0.6 | 5.8 | 7.2 | 5.7 | 12.1 | 12.0 | 13.5 | 12.4 | 12.3 |
| 25500 | Harrisonburg, VA | 867 | 135,550 | 305 | 156.3 | 80.0 | 5.4 | 0.6 | 3.0 | 12.9 | 5.6 | 10.9 | 21.2 | 12.5 | 11.5 | 10.8 |
| 25540 | Hartford-East Hartford-Middletown, CT | 1,515 | 1,201,483 | 48 | 793.3 | 67.1 | 12.2 | 0.6 | 6.3 | 15.8 | 5.0 | 11.1 | 13.7 | 13.0 | 12.2 | 12.6 |
| 25620 | Hattiesburg, MS | 2,024 | 169,554 | 255 | 83.8 | 65.8 | 30.6 | 0.6 | 1.4 | 2.9 | 6.4 | 13.4 | 14.5 | 14.2 | 12.8 | 11.5 |
| 25860 | Hickory-Lenoir-Morganton, NC | 1,640 | 370,266 | 147 | 225.8 | 81.6 | 8.0 | 0.7 | 3.6 | 7.9 | 5.2 | 11.3 | 11.6 | 12.2 | 11.2 | 13.8 |
| 25940 | Hilton Head Island-Bluffton, SC | 1,231 | 227,244 | 203 | 184.6 | 66.8 | 20.9 | 0.6 | 1.9 | 11.5 | 4.9 | 10.7 | 11.6 | 12.2 | 11.2 | 13.8 |
| 25980 | Hinesville, GA | 917 | 83,175 | 373 | 90.7 | 45.5 | 41.6 | 1.2 | 3.6 | 12.3 | 9.4 | 14.6 | 16.1 | 11.1 | 9.9 | 10.4 |
| 26140 | Homosassa Springs, FL | 582 | 153,010 | 273 | 262.9 | 88.7 | 3.6 | 0.9 | 2.2 | 6.3 | 3.6 | 8.4 | 7.7 | 8.7 | 8.0 | 9.3 |
| 26300 | Hot Springs, AR | 678 | 99,789 | 358 | 147.3 | 83.6 | 9.8 | 1.6 | 1.3 | 6.1 | 5.3 | 10.9 | 10.6 | 11.4 | 11.2 | 10.3 |
| 26380 | Houma-Thibodaux, LA | 2,298 | 207,455 | 217 | 90.3 | 73.1 | 17.6 | 5.4 | 1.4 | 4.9 | 6.4 | 13.6 | 12.1 | 13.7 | 12.8 | 11.9 |
| 26420 | Houston-The Woodlands-Sugar Land, TX | 8,269 | 7,154,478 | 5 | 865.4 | 36.0 | 18.0 | 0.6 | 8.9 | 38.1 | 7.0 | 14.8 | 13.4 | 14.8 | 14.4 | 12.7 |
| 26580 | Huntington-Ashland, WV-KY-OH | 2,500 | 354,085 | 151 | 141.6 | 95.3 | 3.2 | 0.8 | 1.1 | 1.3 | 5.3 | 11.9 | 12.6 | 11.8 | 12.0 | 12.8 |
| 26620 | Huntsville, AL | 1,362 | 481,681 | 115 | 353.8 | 69.0 | 23.3 | 1.5 | 3.4 | 5.4 | 5.7 | 12.2 | 12.6 | 14.1 | 12.7 | 12.8 |
| 26820 | Idaho Falls, ID | 5,196 | 155,361 | 270 | 29.9 | 84.9 | 0.9 | 1.3 | 1.8 | 12.9 | 7.9 | 17.7 | 13.5 | 13.3 | 13.4 | 10.1 |
| 26900 | Indianapolis-Carmel-Anderson, IN | 4,307 | 2,091,019 | 33 | 485.5 | 72.9 | 17.1 | 0.7 | 4.5 | 7.2 | 6.4 | 13.8 | 12.8 | 14.5 | 13.4 | 12.4 |
| 26980 | Iowa City, IA | 1,182 | 175,732 | 244 | 148.7 | 80.9 | 8.0 | 0.6 | 6.8 | 6.1 | 5.7 | 11.3 | 22.2 | 14.8 | 12.1 | 9.9 |
| 27060 | Ithaca, NY | 475 | 101,058 | 355 | 212.9 | 79.9 | 5.4 | 0.9 | 11.8 | 5.4 | 3.7 | 8.1 | 28.6 | 12.6 | 10.7 | 9.8 |
| 27100 | Jackson, MI | 702 | 156,920 | 267 | 223.6 | 87.1 | 9.8 | 1.0 | 1.3 | 3.8 | 5.6 | 11.9 | 12.1 | 13.0 | 11.6 | 12.6 |
| 27140 | Jackson, MS | 5,405 | 589,082 | 97 | 109.0 | 46.1 | 50.5 | 0.4 | 1.5 | 2.5 | 6.0 | 13.2 | 13.3 | 13.7 | 13.1 | 12.1 |
| 27180 | Jackson, TN | 1,711 | 179,131 | 240 | 104.7 | 67.2 | 28.6 | 0.6 | 1.2 | 4.3 | 6.1 | 12.8 | 13.5 | 12.4 | 11.6 | 12.0 |
| 27260 | Jacksonville, FL | 3,202 | 1,587,892 | 39 | 496.0 | 64.1 | 22.5 | 0.8 | 5.4 | 9.9 | 6.1 | 12.5 | 11.7 | 14.4 | 13.0 | 12.4 |
| 27340 | Jacksonville, NC | 762 | 203,943 | 223 | 267.6 | 69.2 | 16.5 | 1.4 | 4.0 | 13.2 | 8.2 | 12.8 | 25.1 | 17.8 | 10.9 | 7.5 |
| 27500 | Janesville-Beloit, WI | 718 | 163,084 | 263 | 227.1 | 84.0 | 6.4 | 0.7 | 1.9 | 9.4 | 5.9 | 12.7 | 12.4 | 12.3 | 12.5 | 12.6 |
| 27620 | Jefferson City, MO | 2,248 | 150,198 | 282 | 66.8 | 88.0 | 8.7 | 1.0 | 1.5 | 2.8 | 5.8 | 12.5 | 12.9 | 13.0 | 12.6 | 12.3 |
| 27740 | Johnson City, TN | 854 | 204,540 | 221 | 239.6 | 91.7 | 4.1 | 0.8 | 1.6 | 3.6 | 4.7 | 10.4 | 13.4 | 12.6 | 11.2 | 13.1 |
| 27780 | Johnstown, PA | 688 | 128,672 | 314 | 186.9 | 93.9 | 4.8 | 0.4 | 0.9 | 1.8 | 4.8 | 10.7 | 12.4 | 10.6 | 10.5 | 12.2 |

1. CBSA = Core Based Statistical Area. DIV = Metropolitan Division. See Appendix A for explanation. See Appendix B for list of metropolitan areas or temporarily covered by water.  3. May be of any race.  2. Dry land or land partially or temporarily covered by water.

# Table C. Metropolitan Areas — Population and Households

| Area name | Population, 2020 (cont.) Age (percent) (cont.) 55 to 64 years | 65 to 74 years | 75 years and over | Percent female | Population change, 2000-2020 Total persons 2000 | Total persons 2010 | Percent change 2000-2010 | Percent change 2010-2019 | Components of change, 2010-2020 Births | Deaths | Net migration | Households, 2019 Number | Persons per household | Percent Family households | Single parent households | One person |
|---|---|---|---|---|---|---|---|---|---|---|---|---|---|---|---|---|
| | 16 | 17 | 18 | 19 | 20 | 21 | 22 | 23 | 24 | 25 | 26 | 27 | 28 | 29 | 30 | 31 |
| Evansville, IN-KY | 13.9 | 10.7 | 7.5 | 51.2 | 296,195 | 311,551 | 5.2 | 1.3 | 37,630 | 33,558 | 329 | 129,971 | 2.35 | 64.3 | 9.6 | 31.0 |
| Fairbanks, AK | 10.7 | 8.1 | 3.7 | 46.1 | 82,840 | 97,585 | 17.8 | -2.0 | 16,769 | 4,930 | -13,939 | 35,556 | 2.61 | 67.5 | 9.9 | 24.9 |
| Fargo, ND-MN | 10.3 | 7.7 | 5.5 | 49.7 | 174,367 | 208,772 | 19.7 | 19.1 | 33,815 | 15,409 | 21,169 | 104,097 | 2.29 | 55.1 | 6.9 | 35.0 |
| Farmington, NM | 12.8 | 9.6 | 6.6 | 50.6 | 113,801 | 130,045 | 14.3 | -5.2 | 17,580 | 10,400 | -13,988 | 42,561 | 2.87 | 70.5 | 17.4 | 24.9 |
| Fayetteville, NC | 10.8 | 7.7 | 5.0 | 50.6 | 427,634 | 481,037 | 12.5 | 10.0 | 85,360 | 39,012 | 478 | 193,936 | 2.61 | 62.3 | 13.0 | 32.6 |
| Fayetteville-Springdale-Rogers, AR | 10.6 | 7.9 | 5.4 | 50.1 | 325,364 | 440,118 | 35.3 | 24.7 | 71,482 | 35,569 | 72,136 | 196,683 | 2.67 | 67.8 | 9.7 | 24.2 |
| Flagstaff, AZ | 11.2 | 8.9 | 4.9 | 50.7 | 116,320 | 134,435 | 15.6 | 6.0 | 16,520 | 8,280 | -213 | 48,993 | 2.61 | 59.8 | 10.0 | 27.0 |
| Flint, MI | 14.2 | 10.8 | 7.6 | 51.8 | 436,141 | 425,790 | -2.4 | -4.9 | 49,570 | 46,214 | -24,602 | 169,247 | 2.37 | 63.2 | 14.2 | 31.9 |
| Florence, SC | 13.2 | 11.2 | 7.4 | 53.2 | 193,155 | 205,581 | 6.4 | -0.7 | 25,559 | 24,323 | -2,556 | 77,848 | 2.58 | 68.2 | 14.7 | 28.0 |
| Florence-Muscle Shoals, AL | 13.7 | 11.8 | 8.9 | 52.1 | 142,950 | 147,139 | 2.9 | 1.1 | 15,815 | 18,705 | 4,649 | 60,837 | 2.39 | 63.4 | 10.4 | 30.6 |
| Fond du Lac, WI | 14.7 | 11.3 | 8.2 | 50.7 | 97,296 | 101,626 | 4.5 | 1.3 | 11,083 | 9,752 | -19 | 42,604 | 2.35 | 61.1 | 6.9 | 32.5 |
| Fort Collins, CO | 12.0 | 10.3 | 6.4 | 50.2 | 251,494 | 299,628 | 19.1 | 20.3 | 34,394 | 22,266 | 48,125 | 98,307 | 2.50 | 60.8 | 6.0 | 25.0 |
| Fort Smith, AR-OK | 13.3 | 10.2 | 7.4 | 50.9 | 225,061 | 248,264 | 10.3 | 0.9 | 32,540 | 27,362 | -2,907 | 146,638 | 2.37 | 64.7 | 12.3 | 29.2 |
| Fort Wayne, IN | 12.2 | 9.4 | 6.2 | 50.9 | 362,556 | 388,626 | 7.2 | 7.2 | 57,513 | 35,665 | 6,374 | 160,462 | 2.54 | 64.6 | 10.9 | 29.0 |
| Fresno, CA | 10.3 | 7.6 | 5.3 | 50.1 | 799,407 | 930,503 | 16.4 | 7.6 | 156,504 | 69,429 | -16,383 | 315,974 | 3.11 | 71.6 | 17.3 | 23.5 |
| Gadsden, AL | 13.6 | 11.8 | 7.8 | 51.6 | 103,459 | 104,412 | 0.9 | -2.0 | 12,174 | 14,514 | 389 | 40,053 | 2.52 | 62.9 | 10.6 | 33.2 |
| Gainesville, FL | 11.4 | 10.0 | 6.8 | 51.5 | 266,842 | 305,084 | 14.3 | 8.9 | 35,115 | 26,979 | 19,128 | 126,736 | 2.46 | 52.9 | 8.0 | 35.1 |
| Gainesville, GA | 12.2 | 9.3 | 6.7 | 50.3 | 139,277 | 179,765 | 29.1 | 14.9 | 26,275 | 14,787 | 15,362 | 65,625 | 3.09 | 74.3 | 9.4 | 21.6 |
| Gettysburg, PA | 15.3 | 12.6 | 9.0 | 50.7 | 91,292 | 101,419 | 11.1 | 1.3 | 10,251 | 10,432 | 1,578 | 38,703 | 2.55 | 68.1 | 5.0 | 26.4 |
| Glens Falls, NY | 15.9 | 13.1 | 9.1 | 50.7 | 124,345 | 128,952 | 3.7 | -3.6 | 11,850 | 13,623 | -2,749 | 53,686 | 2.25 | 61.2 | 7.7 | 29.6 |
| Goldsboro, NC | 12.9 | 10.2 | 7.1 | 51.1 | 113,329 | 122,665 | 8.2 | 1.1 | 17,091 | 12,485 | -3,298 | 48,482 | 2.48 | 63.9 | 11.7 | 30.8 |
| Grand Forks, ND-MN | 11.4 | 8.8 | 6.6 | 48.9 | 97,478 | 98,464 | 1.0 | 1.9 | 14,329 | 8,529 | -3,888 | 44,724 | 2.13 | 53.9 | 9.5 | 34.5 |
| Grand Island, NE | 12.7 | 9.3 | 7.1 | 49.5 | 68,305 | 72,754 | 6.5 | 3.5 | 11,451 | 7,142 | -1,699 | 30,529 | 2.44 | 66.8 | 12.5 | 27.6 |
| Grand Junction, CO | 13.6 | 12.1 | 8.4 | 50.7 | 116,255 | 146,736 | 26.2 | 6.0 | 17,901 | 15,037 | 6,040 | 63,644 | 2.37 | 66.2 | 8.7 | 26.7 |
| Grand Rapids-Kentwood, MI | 12.3 | 9.1 | 6.2 | 50.2 | 935,433 | 993,655 | 6.2 | 8.8 | 138,249 | 78,829 | 28,715 | 396,763 | 2.65 | 67.2 | 9.5 | 25.4 |
| Grants Pass, OR | 14.8 | 15.6 | 11.4 | 51.4 | 75,726 | 82,718 | 9.2 | 6.4 | 8,533 | 12,381 | 9,182 | 35,206 | 2.45 | 67.0 | 11.9 | 27.2 |
| Great Falls, MT | 13.1 | 11.0 | 8.5 | 49.4 | 80,357 | 81,327 | 1.2 | 0.0 | 11,555 | 8,664 | -2,844 | 32,816 | 2.40 | 62.0 | 12.9 | 31.3 |
| Greeley, CO | 11.1 | 8.0 | 4.8 | 49.4 | 180,926 | 252,836 | 39.7 | 32.1 | 41,936 | 17,551 | 56,262 | 111,298 | 2.85 | 73.6 | 9.7 | 20.4 |
| Green Bay, WI | 14.0 | 10.2 | 6.7 | 50.1 | 282,599 | 306,241 | 8.4 | 5.6 | 39,816 | 25,492 | 3,055 | 132,099 | 2.38 | 65.4 | 10.4 | 27.7 |
| Greensboro-High Point, NC | 13.2 | 10.0 | 7.0 | 52.2 | 643,430 | 723,943 | 12.5 | 7.2 | 88,413 | 70,978 | 35,421 | 301,529 | 2.48 | 64.9 | 13.4 | 29.3 |
| Greenville, NC | 11.2 | 8.7 | 5.6 | 53.1 | 133,798 | 168,173 | 25.7 | 8.8 | 21,557 | 12,967 | 6,181 | 70,926 | 2.44 | 58.4 | 12.1 | 27.7 |
| Greenville-Anderson, SC | 12.9 | 10.2 | 7.1 | 51.5 | 725,680 | 824,023 | 13.6 | 13.2 | 107,278 | 85,718 | 86,976 | 349,480 | 2.57 | 66.1 | 9.5 | 28.1 |
| Gulfport-Biloxi, MS | 13.6 | 10.4 | 6.6 | 51.1 | 377,610 | 388,605 | 2.9 | 7.8 | 51,709 | 41,018 | 19,705 | 159,081 | 2.58 | 66.6 | 13.4 | 28.5 |
| Hagerstown-Martinsburg, MD-WV | 13.7 | 10.3 | 7.0 | 49.6 | 222,771 | 269,153 | 20.8 | 8.2 | 33,217 | 28,287 | 17,047 | 110,354 | 2.54 | 66.8 | 12.5 | 27.0 |
| Hammond, LA | 12.3 | 9.6 | 5.6 | 51.6 | 100,588 | 121,083 | 20.4 | 13.0 | 19,674 | 12,629 | 8,635 | 46,526 | 2.82 | 67.5 | 13.6 | 26.6 |
| Hanford-Corcoran, CA | 9.5 | 6.3 | 4.6 | 45.0 | 129,461 | 152,990 | 18.2 | -0.2 | 23,918 | 8,594 | -15,992 | 44,761 | 3.09 | 75.9 | 14.6 | 18.8 |
| Harrisburg-Carlisle, PA | 13.5 | 10.9 | 7.7 | 50.8 | 509,074 | 549,457 | 7.9 | 5.9 | 66,851 | 54,393 | 20,329 | 234,365 | 2.38 | 62.0 | 8.9 | 31.3 |
| Harrisonburg, VA | 11.4 | 9.1 | 7.0 | 51.3 | 108,193 | 125,226 | 15.7 | 8.2 | 14,802 | 10,323 | 5,869 | 47,098 | 2.68 | 65.4 | 9.3 | 22.8 |
| Hartford-East Hartford-Middletown, CT | 14.2 | 10.2 | 8.0 | 51.2 | 1,148,618 | 1,212,434 | 5.6 | -0.9 | 121,906 | 108,941 | -24,107 | 476,867 | 2.43 | 62.6 | 8.8 | 29.8 |
| Hattiesburg, MS | 11.9 | 8.9 | 6.3 | 52.3 | 143,219 | 162,416 | 13.4 | 4.4 | 22,971 | 16,108 | 204 | 59,170 | 2.78 | 66.1 | 11.2 | 29.8 |
| Hickory-Lenoir-Morganton, NC | 14.6 | 11.7 | 8.3 | 50.6 | 341,851 | 365,807 | 7.0 | 1.2 | 38,594 | 41,234 | 7,350 | 146,324 | 2.47 | 67.8 | 11.0 | 27.0 |
| Hilton Head Island-Bluffton, SC | 14.2 | 16.3 | 11.4 | 51.0 | 141,615 | 187,010 | 32.1 | 21.5 | 24,064 | 18,166 | 34,029 | 84,094 | 2.59 | 67.8 | 8.6 | 26.4 |
| Hinesville, GA | 9.6 | 6.6 | 3.5 | 49.5 | 71,914 | 77,929 | 8.4 | 6.7 | 17,285 | 4,554 | -7,971 | 29,973 | 2.63 | 70.7 | 16.1 | 26.3 |
| Homosassa Springs, FL | 16.2 | 19.8 | 17.3 | 51.6 | 118,085 | 141,229 | 19.6 | 8.3 | 10,688 | 25,947 | 27,055 | 64,448 | 2.28 | 65.3 | 7.4 | 27.4 |
| Hot Springs, AR | 14.2 | 14.0 | 10.7 | 51.9 | 88,068 | 96,003 | 9.0 | 3.9 | 11,205 | 14,174 | 6,792 | 41,743 | 2.33 | 62.3 | 9.8 | 32.6 |
| Houma-Thibodaux, LA | 13.5 | 9.4 | 6.5 | 50.9 | 194,477 | 208,180 | 7.0 | -0.3 | 28,924 | 19,918 | -9,668 | 76,932 | 2.66 | 70.2 | 14.5 | 23.9 |
| Houston-The Woodlands-Sugar Land, TX | 11.1 | 7.5 | 4.4 | 50.5 | 4,693,161 | 5,920,485 | 26.2 | 20.8 | 984,458 | 397,497 | 644,013 | 2,436,438 | 2.87 | 70.3 | 12.4 | 24.3 |
| Huntington-Ashland, WV-KY-OH | 13.5 | 11.7 | 8.5 | 51.1 | 367,127 | 370,886 | 1.0 | -4.5 | 41,276 | 46,200 | -11,687 | 136,687 | 2.54 | 66.9 | 10.6 | 28.5 |
| Huntsville, AL | 14.2 | 9.2 | 6.5 | 50.8 | 342,376 | 417,587 | 22.0 | 15.3 | 53,293 | 39,071 | 49,868 | 187,898 | 2.44 | 67.3 | 10.7 | 27.6 |
| Idaho Falls, ID | 10.4 | 8.3 | 5.4 | 49.8 | 104,576 | 133,331 | 27.5 | 16.5 | 24,600 | 10,510 | 7,990 | 51,152 | 2.94 | 69.7 | 9.9 | 25.3 |
| Indianapolis-Carmel-Anderson, IN | 12.4 | 8.7 | 5.7 | 51.1 | 1,658,462 | 1,888,078 | 13.8 | 10.7 | 271,556 | 167,634 | 99,602 | 793,781 | 2.57 | 61.6 | 9.8 | 31.6 |
| Iowa City, IA | 10.2 | 8.3 | 5.4 | 50.4 | 131,676 | 152,585 | 15.9 | 15.2 | 21,264 | 9,707 | 11,594 | 70,169 | 2.36 | 57.7 | 5.2 | 28.3 |
| Ithaca, NY | 10.9 | 9.6 | 6.0 | 50.9 | 96,501 | 101,581 | 5.3 | -0.5 | 8,432 | 6,996 | -2,022 | 40,322 | 2.21 | 48.1 | 6.4 | 34.1 |
| Jackson, MI | 14.4 | 11.1 | 7.7 | 49.2 | 158,422 | 160,233 | 1.1 | -2.1 | 18,145 | 17,154 | -4,218 | 61,403 | 2.43 | 60.0 | 11.0 | 34.1 |
| Jackson, MS | 12.8 | 9.6 | 6.2 | 52.1 | 546,955 | 587,116 | 7.3 | 0.3 | 78,354 | 55,742 | -20,771 | 218,853 | 2.61 | 64.9 | 13.3 | 30.7 |
| Jackson, TN | 13.4 | 10.6 | 7.6 | 52.4 | 170,061 | 179,708 | 5.7 | -0.3 | 22,487 | 20,870 | -2,102 | 70,379 | 2.45 | 66.8 | 13.7 | 29.5 |
| Jacksonville, FL | 13.3 | 10.2 | 6.5 | 51.3 | 1,122,750 | 1,345,594 | 19.8 | 18.0 | 186,250 | 130,026 | 185,756 | 584,381 | 2.61 | 65.6 | 11.0 | 27.6 |
| Jacksonville, NC | 8.0 | 5.9 | 3.9 | 44.1 | 150,355 | 177,801 | 18.3 | 14.7 | 42,537 | 11,014 | -6,585 | 63,604 | 2.81 | 72.6 | 11.4 | 20.1 |
| Janesville-Beloit, WI | 14.1 | 10.3 | 7.3 | 50.7 | 152,307 | 160,327 | 5.3 | 1.7 | 19,706 | 15,320 | -1,537 | 65,316 | 2.46 | 60.5 | 10.0 | 30.2 |
| Jefferson City, MO | 13.4 | 10.4 | 7.1 | 49.3 | 140,052 | 149,822 | 7.0 | 0.3 | 18,059 | 13,623 | -3,999 | 59,727 | 2.34 | 65.4 | 8.5 | 29.8 |
| Johnson City, TN | 14.0 | 11.9 | 8.8 | 51.1 | 181,607 | 198,757 | 9.4 | 2.9 | 20,146 | 24,484 | 10,192 | 89,021 | 2.21 | 61.0 | 9.2 | 32.2 |
| Johnstown, PA | 15.0 | 13.6 | 10.2 | 50.7 | 152,598 | 143,693 | -5.8 | -10.5 | 13,450 | 18,963 | -9,554 | 56,490 | 2.19 | 62.3 | 9.0 | 33.7 |

# Table C. Metropolitan Areas — **Population, Vital Statistics, Health, and Crime**

| Area name | Persons in group quarters, 2020 | Daytime population, 2019 Number | Daytime population, 2019 Employment/residence ratio | Births, 2020 Total | Births, 2020 Rate[1] | Deaths, 2020 Number | Deaths, 2020 Rate[1] | Persons under 65 with no health insurance 2019 Number | Persons under 65 with no health insurance 2019 Percent | Medicare, 2020 Total Beneficiaries | Medicare, 2020 Enrolled in Original Medicare | Medicare, 2020 Enrolled in Medicare Advantage | Serious crimes known to police[2], 2019 Violent Number | Serious crimes known to police[2], 2019 Violent Rate[3] |
|---|---|---|---|---|---|---|---|---|---|---|---|---|---|---|
| | 32 | 33 | 34 | 35 | 36 | 37 | 38 | 39 | 40 | 41 | 42 | 43 | 44 | 45 |
| Evansville, IN-KY | 9,202 | 322,065 | 1.05 | 3,454 | 10.9 | 3,569 | 11.3 | 23,268 | 9.2 | 66,840 | 44,082 | 22,758 | 1,032 | 327.2 |
| Fairbanks, AK | 4,821 | D | D | 1,408 | 14.7 | 540 | 5.6 | 9,282 | 11.3 | 11,962 | 11,838 | 124 | 267 | 794.5 |
| Fargo, ND-MN | 8,290 | 251,065 | 1.03 | 3,285 | 13.2 | 1,698 | 6.8 | 12,299 | 5.9 | 36,213 | 26,101 | 10,112 | 735 | 296.7 |
| Farmington, NM | 1,739 | 123,601 | 0.99 | 1,382 | 11.2 | 1,199 | 9.7 | 15,208 | 14.8 | 21,722 | 17,863 | 3,859 | NA | NA |
| Fayetteville, NC | 20,262 | 512,503 | 0.94 | 8,075 | 15.3 | 4,509 | 8.5 | 59,697 | 13.6 | 83,001 | 53,909 | 29,091 | NA | NA |
| Fayetteville-Springdale-Rogers, AR | 10,109 | 543,134 | 1.03 | 7,290 | 13.3 | 3,916 | 7.1 | 57,369 | 12.5 | 83,968 | 53,955 | 30,012 | NA | NA |
| Flagstaff, AZ | 11,916 | 144,216 | 1.01 | 1,440 | 10.1 | 963 | 6.8 | 17,084 | 15.2 | 21,301 | 17,234 | 4,067 | NA | NA |
| Flint, MI | 5,677 | 389,952 | 0.91 | 4,486 | 11.1 | 4,916 | 12.1 | 21,718 | 6.6 | 90,711 | 40,826 | 49,885 | 664 | 461.1 |
| Florence, SC | 4,380 | 205,967 | 1.01 | 2,341 | 11.5 | 2,695 | 13.2 | 20,482 | 12.6 | 45,475 | 33,080 | 12,397 | 2,340 | 579.7 |
| Florence-Muscle Shoals, AL | 2,885 | 144,345 | 0.94 | 1,547 | 10.4 | 1,941 | 13.0 | 13,292 | 11.5 | 36,586 | 23,623 | 12,963 | 1,688 | 823.4 |
| Fond du Lac, WI | 3,311 | 97,780 | 0.9 | 1,042 | 10.1 | 998 | 9.7 | 4,835 | 6.0 | 22,377 | 9,881 | 12,496 | NA | NA |
| Fort Collins, CO | 9,905 | 356,022 | 1 | 3,241 | 9.0 | 2,488 | 6.9 | 22,447 | 7.7 | 65,205 | 38,640 | 26,565 | 182 | 176.5 |
| Fort Smith, AR-OK | 3,636 | 250,670 | 1.01 | 3,048 | 12.2 | 2,904 | 11.6 | 29,262 | 14.3 | 54,845 | 34,757 | 20,088 | 833 | 234.1 |
| Fort Wayne, IN | 6,561 | 418,273 | 1.02 | 5,642 | 13.5 | 3,754 | 9.0 | 35,922 | 10.4 | 75,275 | 35,433 | 39,841 | 1,462 | 585.7 |
| Fresno, CA | 17,672 | 1,003,673 | 1.01 | 14,251 | 14.2 | 7,941 | 7.9 | 82,348 | 9.7 | 145,804 | 93,653 | 52,151 | 1,259 | 305.0 |
| Gadsden, AL | 2,085 | 95,422 | 0.84 | 1,173 | 11.5 | 1,609 | 15.7 | 10,281 | 12.7 | 26,225 | 13,840 | 12,385 | NA | NA |
| Gainesville, FL | 16,114 | 338,130 | 1.06 | 3,302 | 9.9 | 3,050 | 9.2 | 35,108 | 13.5 | 62,364 | 42,656 | 19,708 | NA | NA |
| Gainesville, GA | 2,962 | 207,799 | 1.04 | 2,546 | 12.3 | 1,773 | 8.6 | 36,780 | 21.6 | 36,023 | 21,578 | 14,444 | 2,282 | 692.1 |
| Gettysburg, PA | 4,038 | 88,500 | 0.72 | 899 | 8.8 | 1,106 | 10.8 | 6,014 | 7.7 | 25,053 | 16,713 | 8,340 | NA | NA |
| Glens Falls, NY | 3,662 | 120,678 | 0.92 | 1,078 | 8.7 | 1,459 | 11.7 | 4,472 | 4.7 | 32,590 | 16,729 | 15,862 | NA | NA |
| Goldsboro, NC | 2,896 | 119,571 | 0.93 | 1,609 | 13.0 | 1,521 | 12.3 | 15,268 | 15.4 | 24,427 | 16,536 | 7,891 | 158 | 127.1 |
| Grand Forks, ND-MN | 4,160 | 101,871 | 1.02 | 1,320 | 13.1 | 868 | 8.6 | 5,037 | 6.1 | 17,595 | 13,369 | 4,226 | 511 | 414.7 |
| Grand Island, NE | 1,084 | 76,018 | 1.01 | 1,102 | 14.6 | 721 | 9.6 | 7,597 | 12.2 | 13,921 | 11,722 | 2,199 | 255 | 250.6 |
| Grand Junction, CO | 4,434 | 152,766 | 0.98 | 1,596 | 10.3 | 1,688 | 10.8 | 13,094 | 10.9 | 35,695 | 22,687 | 13,008 | 285 | 375.7 |
| Grand Rapids-Kentwood, MI | 28,254 | 1,107,864 | 1.06 | 13,099 | 12.1 | 8,639 | 8.0 | 63,094 | 7.0 | 189,036 | 72,557 | 116,480 | 435 | 283.5 |
| Grants Pass, OR | 1,615 | 87,581 | 1 | 803 | 9.1 | 1,293 | 14.7 | 6,416 | 10.3 | 27,027 | 15,450 | 11,577 | 3,472 | 321.5 |
| Great Falls, MT | 2,447 | D | D | 1,028 | 12.6 | 848 | 10.4 | 6,236 | 9.7 | 18,068 | 13,056 | 5,012 | 209 | 238.7 |
| Greeley, CO | 5,724 | 296,486 | 0.83 | 4,428 | 13.3 | 2,042 | 6.1 | 30,273 | 10.9 | 46,091 | 25,768 | 20,323 | D | D |
| Green Bay, WI | 6,812 | 330,364 | 1.04 | 3,679 | 11.4 | 2,666 | 8.2 | 20,110 | 7.6 | 62,668 | 25,751 | 36,917 | 694 | 215.9 |
| Greensboro-High Point, NC | 22,350 | 796,930 | 1.07 | 8,503 | 11.0 | 8,105 | 10.4 | 87,041 | 14.0 | 152,345 | 61,407 | 90,938 | 730 | 225.9 |
| Greenville, NC | 6,612 | 187,176 | 1.08 | 2,062 | 11.3 | 1,548 | 8.5 | 17,828 | 12.0 | 30,816 | 21,944 | 8,872 | 4,098 | 530.2 |
| Greenville-Anderson, SC | 23,957 | 922,741 | 1.01 | 10,471 | 11.2 | 9,617 | 10.3 | 98,876 | 13.3 | 189,451 | 113,510 | 75,942 | 657 | 362.5 |
| Gulfport-Biloxi, MS | 8,306 | 417,866 | 1 | 4,877 | 11.6 | 4,520 | 10.8 | 53,983 | 15.8 | 83,708 | 57,593 | 26,116 | NA | NA |
| Hagerstown-Martinsburg, MD-WV | 9,350 | 267,693 | 0.84 | 3,262 | 11.2 | 3,058 | 10.5 | 17,292 | 7.5 | 58,907 | 47,190 | 11,719 | 924 | 221.3 |
| Hammond, LA | 3,646 | 124,939 | 0.83 | 1,874 | 13.7 | 1,503 | 11.0 | 11,986 | 10.8 | 24,079 | 13,084 | 10,996 | NA | NA |
| Hanford-Corcoran, CA | 15,835 | 155,092 | 1.04 | 2,184 | 14.3 | 986 | 6.5 | 11,212 | 9.4 | 17,921 | 13,576 | 4,345 | 757 | 504.8 |
| Harrisburg-Carlisle, PA | 19,519 | 624,623 | 1.16 | 6,466 | 11.1 | 5,807 | 10.0 | 34,243 | 7.5 | 122,330 | 64,061 | 58,270 | NA | NA |
| Harrisonburg, VA | 9,323 | 143,678 | 1.13 | 1,494 | 11.0 | 1,140 | 8.4 | 12,650 | 12.1 | 24,026 | 20,212 | 3,813 | 172 | 126.7 |
| Hartford-East Hartford-Middletown, CT | 44,606 | 1,234,544 | 1.05 | 11,538 | 9.6 | 11,644 | 9.7 | 54,047 | 5.6 | 239,847 | 116,587 | 123,259 | 2,021 | 199.0 |
| Hattiesburg, MS | 3,557 | 172,905 | 1.06 | 2,156 | 12.7 | 1,624 | 9.6 | 21,465 | 15.3 | 30,360 | 21,594 | 8,766 | NA | NA |
| Hickory-Lenoir-Morganton, NC | 8,043 | 362,059 | 0.96 | 3,723 | 10.1 | 4,483 | 12.1 | 44,242 | 15.2 | 85,044 | 46,231 | 38,813 | NA | NA |
| Hilton Head Island-Bluffton, SC | 6,784 | 223,711 | 1.02 | 2,247 | 9.9 | 2,196 | 9.7 | 23,468 | 15.1 | 60,212 | 44,644 | 15,568 | 882 | 397.5 |
| Hinesville, GA | 2,679 | 82,308 | 1.03 | 1,690 | 20.3 | 506 | 6.1 | 9,527 | 13.5 | 9,642 | 6,218 | 3,424 | NA | NA |
| Homosassa Springs, FL | 2,253 | 141,961 | 0.82 | 1,029 | 6.7 | 2,865 | 18.7 | 15,868 | 17.1 | 60,329 | 33,745 | 26,584 | 373 | 252.5 |
| Hot Springs, AR | 2,164 | 100,375 | 1.02 | 1,043 | 10.5 | 1,445 | 14.5 | 8,848 | 12.0 | 28,940 | 19,585 | 9,356 | 520 | 524.0 |
| Houma-Thibodaux, LA | 3,106 | 208,830 | 1.01 | 2,553 | 12.3 | 2,191 | 10.6 | 21,772 | 12.6 | 40,120 | 24,439 | 15,681 | 779 | 374.5 |
| Houston-The Woodlands-Sugar Land, TX | 80,261 | 7,080,670 | 1 | 94,996 | 13.3 | 48,272 | 6.7 | 1,336,522 | 21.7 | 899,904 | 470,000 | 429,901 | NA | NA |
| Huntington-Ashland, WV-KY-OH | 7,932 | 348,584 | 0.95 | 3,566 | 10.1 | 4,789 | 13.5 | 21,141 | 7.6 | 87,100 | 53,391 | 33,709 | 945 | 265.1 |
| Huntsville, AL | 10,920 | 499,228 | 1.12 | 5,352 | 11.1 | 4,516 | 9.4 | 37,772 | 9.7 | 86,443 | 57,071 | 29,373 | NA | NA |
| Idaho Falls, ID | 1,478 | 158,253 | 1.1 | 2,301 | 14.8 | 1,124 | 7.2 | 14,856 | 11.4 | 24,430 | 17,673 | 6,755 | 310 | 204.5 |
| Indianapolis-Carmel-Anderson, IN | 40,425 | 2,099,293 | 1.02 | 25,906 | 12.4 | 18,139 | 8.7 | 172,296 | 9.9 | 346,038 | 209,551 | 136,486 | NA | NA |
| Iowa City, IA | 8,680 | 181,745 | 1.09 | 2,004 | 11.4 | 1,061 | 6.0 | 8,528 | 6.0 | 26,302 | 20,484 | 5,816 | 485 | 276.7 |
| Ithaca, NY | 12,890 | 115,223 | 1.27 | 721 | 7.1 | 686 | 6.8 | 3,475 | 4.7 | 17,166 | 11,646 | 5,520 | 127 | 124.1 |
| Jackson, MI | 7,634 | 151,757 | 0.9 | 1,701 | 10.8 | 1,744 | 11.1 | 8,320 | 6.9 | 34,986 | 21,058 | 13,929 | 814 | 514.3 |
| Jackson, MS | 23,091 | 602,873 | 1.03 | 7,144 | 12.1 | 6,057 | 10.3 | 67,708 | 14.1 | 111,450 | 78,056 | 33,393 | NA | NA |
| Jackson, TN | 7,041 | 184,174 | 1.07 | 2,119 | 11.8 | 2,118 | 11.8 | 16,604 | 11.8 | 39,966 | 27,201 | 12,765 | 1,093 | 613.6 |
| Jacksonville, FL | 28,353 | 1,564,876 | 1.01 | 18,643 | 11.7 | 15,416 | 9.7 | 170,529 | 13.3 | 300,923 | 185,777 | 115,146 | 7,353 | 474.7 |
| Jacksonville, NC | 25,397 | 196,917 | 0.99 | 3,802 | 18.6 | 1,269 | 6.2 | 16,678 | 10.6 | 26,277 | 21,220 | 5,057 | NA | NA |
| Janesville-Beloit, WI | 2,512 | 155,335 | 0.9 | 1,832 | 11.2 | 1,578 | 9.7 | 9,273 | 7.0 | 33,950 | 20,168 | 13,782 | 380 | 232.7 |
| Jefferson City, MO | 9,149 | 155,867 | 1.06 | 1,680 | 11.2 | 1,516 | 10.1 | 13,309 | 11.5 | 30,353 | 18,952 | 11,402 | 334 | 220.4 |
| Johnson City, TN | 5,544 | 198,747 | 0.95 | 1,837 | 9.0 | 2,630 | 12.9 | 19,796 | 12.5 | 50,999 | 23,708 | 27,293 | 668 | 328.6 |
| Johnstown, PA | 5,924 | 125,646 | 0.92 | 1,171 | 9.1 | 1,837 | 14.3 | 5,487 | 5.8 | 36,760 | 12,543 | 24,217 | NA | NA |

1. Per 1,000 estimated resident population.   2. Data for serious crimes have not been adjusted for underreporting; this may affect comparability between geographic areas and over time.   3. Per 100,000 population estimated by the FBI.

| Area name | Serious crimes known to police[1], 2019 (cont.) — Property | | Education — School enrollment and attainment, 2019 | | | | Local government expenditures,[5] 2017-2018 | | Income and poverty, 2019 | | | | | Percent below poverty level | | |
|---|---|---|---|---|---|---|---|---|---|---|---|---|---|---|---|---|
| | | | Enrollment[3] | | Attainment[4] | | | | | | | | | | | Families with children under 18 |
| | Number | Rate | Total | Percent private | High school graduate or less | Bachelor's degree or more | Total current expenditures (mil dol) | Current expenditures per student (dollars) | Per capita income[6] (dollars) | Median household income (dollars) | Median family income | Percent of households with income less than $50,000 | Percent of households with income of $200,000 or more | All persons | All families | |
| | 46 | 47 | 48 | 49 | 50 | 51 | 52 | 53 | 54 | 55 | 56 | 57 | 58 | 59 | 60 | 61 |
| Evansville, IN-KY | 7,199 | 2,282.5 | 71,309 | 17.0 | 44.4 | 26.7 | 459.3 | 10,278 | 31,953 | 55,584 | 75,315 | 44.7 | 4.4 | 12.9 | 9.2 | 17.0 |
| Fairbanks, AK | 1,491 | 4,437.0 | 26,355 | 15.6 | 25.1 | 33.3 | 255.2 | 16,344 | 39,252 | 72,065 | 87,974 | 29.6 | 12.4 | 5.9 | 3.0 | 3.9 |
| Fargo, ND-MN | 5,794 | 2,338.5 | 70,909 | 9.6 | 26.6 | 40.4 | 439.8 | 12,435 | 36,283 | 62,820 | 90,408 | 39.2 | 6.4 | 13.0 | 7.0 | 12.5 |
| Farmington, NM | NA | NA | 32,358 | 7.9 | 44.8 | 14.4 | 218.6 | 9,218 | 21,626 | 44,321 | 48,713 | 55.6 | 2.2 | 21.2 | 15.4 | 26.3 |
| Fayetteville, NC | NA | NA | 148,035 | 18.5 | 38.0 | 23.8 | 720.2 | 8,793 | 24,457 | 47,597 | 59,712 | 52.4 | 2.2 | 17.9 | 13.4 | 19.2 |
| Fayetteville-Springdale-Rogers, AR | NA | NA | 151,902 | 13.4 | 41.9 | 33 | 891.2 | 9,779 | 31,663 | 61,674 | 75,264 | 41.1 | 7.6 | 12.6 | 8.0 | 11.7 |
| | | | | | | | | | | | | | 5.6 | 15.8 | 7.5 | 9.6 |
| Flagstaff, AZ | 3,527 | 2,449.3 | 51,112 | 4.8 | 26.9 | 39.6 | 181.2 | 9,827 | 29,379 | 58,085 | 76,601 | 42.1 | 3.4 | 16.5 | 12.1 | 18.5 |
| Flint, MI | D | D | 91,687 | 10.9 | 40.3 | 21.5 | 761.9 | 11,787 | 28,503 | 50,389 | 64,374 | 49.6 | 3.4 | 16.9 | 13.2 | 20.5 |
| Florence, SC | 8,321 | 4,059.2 | 49,482 | 14.1 | 48.6 | 23.1 | 359.0 | 10,762 | 25,901 | 48,547 | 58,301 | 51.2 | 1.9 | 15.8 | 10.8 | 20.9 |
| Florence-Muscle Shoals, AL | NA | NA | 34,478 | 8.8 | 51.3 | 20.2 | 213.4 | 10,245 | 26,044 | 45,824 | 61,387 | 53.7 | 3.1 | 6.2 | 2.8 | 5.3 |
| Fond du Lac, WI | 1,131 | 1,096.7 | 23,162 | 23.6 | 43.1 | 23.6 | 150.2 | 11,377 | 33,847 | 65,329 | 87,451 | 36.0 | 9.5 | 11.3 | 4.4 | 6.2 |
| Fort Collins, CO | 6,683 | 1,878.2 | 100,388 | 10.5 | 21.5 | 49 | 467.2 | 9,849 | 40,582 | 75,186 | 100,135 | 31.8 | 2.4 | 19.4 | 13.4 | 21.6 |
| Fort Smith, AR-OK | 7,944 | 3,182.4 | 58,747 | 8.9 | 49.2 | 21 | 410.5 | 9,663 | 24,134 | 45,157 | 56,083 | 54.1 | 4.3 | 10.2 | 6.6 | 9.9 |
| Fort Wayne, IN | 8,726 | 2,113.7 | 101,271 | 23.8 | 38.7 | 28.1 | 612.9 | 10,085 | 30,800 | 57,287 | 71,184 | 43.3 | 5.4 | 20.6 | 16.8 | 24.5 |
| Fresno, CA | NA | NA | 300,979 | 7.1 | 45.3 | 22 | 2,654.7 | 13,178 | 25,260 | 57,518 | 65,568 | 44.4 | 2.3 | 19.7 | 15.3 | 28.8 |
| Gadsden, AL | NA | NA | 22,891 | 9.5 | 48.0 | 17.2 | 140.6 | 9,214 | 25,065 | 41,447 | 53,690 | 57.3 | 5.2 | 18.5 | 8.6 | 12.9 |
| | | | | | | | | | 28,111 | 47,762 | 71,692 | 52.4 | | | | |
| Gainesville, FL | 8,091 | 2,454.0 | 104,278 | 12.6 | 31.7 | 38.9 | 355.3 | 9,071 | 32,667 | 67,467 | 78,870 | 35.7 | 7.9 | 14.5 | 9.4 | 17.2 |
| Gainesville, GA | NA | NA | 49,401 | 10.5 | 47.4 | 25.4 | 353.2 | 9,978 | 32,927 | 67,715 | 81,210 | 36.5 | 5.2 | 7.2 | 5.7 | 10.8 |
| Gettysburg, PA | NA | NA | 23,155 | 24.7 | 53.4 | 21.6 | 287.3 | 20,518 | 33,610 | 61,255 | 77,663 | 41.3 | 5.4 | 10.8 | 6.9 | 12.2 |
| Glens Falls, NY | 906 | 728.7 | 23,438 | 11.6 | 43.7 | 26.7 | 351.4 | 19,995 | 33,601 | 61,255 | 77,663 | 56.3 | 1.8 | 19.4 | 14.5 | 23.9 |
| Goldsboro, NC | 3,376 | 2,739.8 | 30,708 | 10.0 | 46.2 | 18.1 | 173.1 | 8,856 | 24,480 | 44,596 | 53,190 | 43.2 | 4.2 | 13.9 | 8.6 | 15.0 |
| Grand Forks, ND-MN | 2,080 | 2,044.5 | 28,535 | 8.7 | 26.6 | 34.2 | 182.0 | 12,524 | 32,869 | 57,301 | 83,521 | 43.9 | 2.6 | 10.3 | 7.6 | 14.8 |
| Grand Island, NE | 1,456 | 1,919.5 | 17,446 | 9.6 | 46.7 | 21 | 176.0 | 11,914 | 29,504 | 55,907 | 64,924 | 40.9 | 4.9 | 10.7 | 6.6 | 12.1 |
| Grand Junction, CO | 3,893 | 2,537.3 | 35,103 | 10.4 | 36.4 | 29.8 | 226.2 | 9,967 | 32,324 | 60,500 | 75,143 | 36.8 | 5.5 | 10.5 | 7.0 | 11.5 |
| Grand Rapids-Kentwood, MI | D | D | 272,029 | 18.4 | 34.8 | 33.4 | 2,027.0 | 11,936 | 32,539 | 65,739 | 80,009 | 52.1 | 2.6 | 15.7 | 10.4 | 18.6 |
| Grants Pass, OR | 1,643 | 1,876.3 | 15,753 | 11.4 | 44.0 | 16.8 | 127.7 | 11,654 | 25,958 | 47,573 | 54,646 | 48.4 | 2.8 | 15.4 | 10.8 | 22.4 |
| Great Falls, MT | 3,653 | 4,486.7 | 16,901 | 11.8 | 41.0 | 25.6 | 118.3 | 10,149 | 27,116 | 51,227 | 65,397 | 29.7 | 6.8 | 7.9 | 4.7 | 6.7 |
| Greeley, CO | 5,246 | 1,632.3 | 89,413 | 10.6 | 37.4 | 29.9 | 422.7 | 9,645 | 33,863 | 78,615 | 92,214 | 39.1 | 4.4 | 10.0 | 5.6 | 9.6 |
| Green Bay, WI | 3,549 | 1,098.4 | 78,036 | 17.3 | 40.0 | 27.8 | 593.0 | 11,317 | 33,291 | 65,026 | 81,566 | 48.3 | 4.0 | 16.6 | 12.2 | 19.0 |
| Greensboro-High Point, NC | 20,835 | 2,695.9 | 189,506 | 14.0 | 40.8 | 30 | 1,065.3 | 9,267 | 28,712 | 51,770 | 63,279 | 46.2 | 4.4 | 19.6 | 11.8 | 18.0 |
| Greenville, NC | 4,383 | 2,418.3 | 53,986 | 10.4 | 37.3 | 32.1 | 225.2 | 9,038 | 28,878 | 53,401 | 64,962 | 42.9 | 6.2 | 12.5 | 8.6 | 13.8 |
| Greenville-Anderson, SC | NA | NA | 219,675 | 19.5 | 38.7 | 31.3 | 1,291.4 | 9,685 | 31,996 | 58,621 | 72,251 | 49.3 | 3.4 | 16.4 | 13.3 | 21.5 |
| Gulfport-Biloxi, MS | 13,849 | 3,316.5 | 102,296 | 12.5 | 39.8 | 22.7 | 599.7 | 9,121 | 27,495 | 50,642 | 62,518 | 41.0 | 4.5 | 12.4 | 8.9 | 14.2 |
| Hagerstown-Martinsburg, MD-WV | NA | NA | 62,886 | 10.2 | 50.1 | 22.3 | 540.8 | 12,213 | 30,724 | 60,095 | 69,693 | 43.0 | 3.4 | 22.9 | 17.1 | 26.0 |
| Hammond, LA | NA | NA | 36,020 | 19.7 | 51.1 | 24.8 | 199.4 | 10,123 | 24,217 | 47,825 | 62,684 | 50.7 | 3.6 | 15.2 | 13.0 | 17.5 |
| Hanford-Corcoran, CA | 2,466 | 1,644.1 | 45,620 | 16.1 | 51.1 | 17.6 | 366.8 | 13,899 | 22,979 | 58,453 | 61,133 | 43.0 | | 9.2 | 6.2 | 11.4 |
| Harrisburg-Carlisle, PA | NA | NA | 131,408 | 21.9 | 42.7 | 33.2 | 1,191.6 | 14,882 | 36,043 | 67,069 | 84,853 | 36.4 | 6.4 | 10.1 | 6.4 | 13.0 |
| Harrisonburg, VA | 1,583 | 1,165.8 | 42,623 | 14.4 | 47.7 | 29.9 | 218.2 | 12,049 | 29,119 | 60,740 | 72,455 | 42.0 | 5.3 | 14.4 | 6.3 | 10.7 |
| Hartford-East Hartford-Middletown, CT | D | D | 302,375 | 14.5 | 34.8 | 39.7 | 3,525.2 | 19,724 | 41,991 | 77,005 | 101,867 | 33.1 | 10.8 | 10.1 | 6.4 | 25.0 |
| Hattiesburg, MS | NA | NA | 48,095 | 11.0 | 40.9 | 25.5 | 237.8 | 9,087 | 25,247 | 48,359 | 56,884 | 52.2 | 2.9 | 22.5 | 17.4 | 19.4 |
| Hickory-Lenoir-Morganton, NC | NA | NA | 74,422 | 16.5 | 47.4 | 19.6 | 479.6 | 9,021 | 27,348 | 50,631 | 62,621 | 49.4 | 3.2 | 14.9 | 10.7 | 16.2 |
| Hilton Head Island-Bluffton, SC | 3,476 | 1,566.7 | 39,265 | 14.4 | 31.7 | 38 | 301.6 | 12,149 | 41,421 | 71,252 | 85,062 | 35.6 | 10.9 | 10.8 | 6.6 | 25.9 |
| Hinesville, GA | NA | NA | 20,327 | 8.8 | 39.4 | 17.6 | 141.6 | 10,338 | 23,322 | 53,060 | 54,646 | 46.3 | 2.4 | 12.0 | 11.1 | 21.0 |
| Homosassa Springs, FL | 2,055 | 1,391.0 | 20,881 | 14.8 | 43.4 | 22.6 | 142.4 | 9,165 | 29,264 | 50,751 | 61,680 | 49.2 | 3.4 | 15.7 | 9.4 | 29.1 |
| Hot Springs, AR | D | D | 18,480 | 6.7 | 37.6 | 28.6 | 151.2 | 9,841 | 27,105 | 45,265 | 54,414 | 54.6 | 2.6 | 17.4 | 12.1 | |
| Houma-Thibodaux, LA | 6,097 | 2,930.8 | 46,090 | 16.9 | 59.8 | 16.3 | 325.8 | 10,060 | 27,421 | 49,874 | 62,460 | 50.1 | 3.7 | 19.5 | 16.8 | 14.8 |
| Houston-The Woodlands-Sugar Land, TX | NA | NA | 1,915,864 | 11.3 | 38.9 | 33.3 | 12,607.8 | 9,444 | 35,190 | 69,193 | 81,684 | 36.4 | 10.3 | 12.9 | 10.2 | 21.7 |
| Huntington-Ashland, WV-KY-OH | 5,969 | 1,674.8 | 78,412 | 8.9 | 49.2 | 21.4 | 646.3 | 11,403 | 26,452 | 48,329 | 61,176 | 51.9 | 3.7 | 16.9 | 12.6 | 15.1 |
| Huntsville, AL | NA | NA | 114,150 | 19.4 | 30.7 | 40.2 | 658.1 | 9,646 | 35,357 | 67,157 | 83,363 | 39.3 | 7.0 | 12.3 | 9.1 | 8.7 |
| Idaho Falls, ID | 1,651 | 1,089.1 | 44,480 | 22.1 | 33.4 | 30.6 | 214.2 | 6,673 | 29,280 | 62,502 | 82,637 | 38.7 | 5.5 | 8.5 | 5.6 | 9.8 |
| Indianapolis-Carmel-Anderson, IN | NA | NA | 506,855 | 15.1 | 38.0 | 35.2 | 3,578.8 | 10,120 | 35,391 | 62,502 | 82,637 | 39.1 | 7.2 | 10.5 | 6.6 | 6.6 |
| Iowa City, IA | 2,658 | 1,516.5 | 60,642 | 12.5 | 25.3 | 48.8 | 255.2 | 11,269 | 38,843 | 63,761 | 93,839 | 40.2 | 9.1 | 16.2 | 5.2 | 17.8 |
| Ithaca, NY | 1,707 | 1,667.6 | 41,114 | 61.9 | 21.6 | 56.9 | 243.8 | 21,684 | 37,576 | 58,626 | 90,192 | 43.8 | 9.6 | 17.9 | 4.5 | 17.8 |
| Jackson, MI | 3,173 | 2,004.9 | 36,291 | 12.2 | 45.6 | 22.9 | 295.9 | 12,821 | 29,209 | 55,124 | 75,845 | 46.1 | 3.5 | 14.0 | 9.8 | 17.4 |
| Jackson, MS | NA | NA | 153,748 | 20.7 | 37.5 | 31 | 828.1 | 8,896 | 28,265 | 52,426 | 66,588 | 46.5 | 4.7 | 15.5 | 10.9 | 21.5 |
| Jackson, TN | 4,179 | 2,345.9 | 40,640 | 21.7 | 51.1 | 22.5 | 247.0 | 8,813 | 24,648 | 48,700 | 57,844 | 51.2 | 2.1 | 16.0 | 12.1 | 12.7 |
| Jacksonville, FL | 38,810 | 2,505.4 | 359,418 | 18.0 | 36.2 | 32.5 | 1,981.3 | 8,815 | 34,459 | 65,880 | 77,465 | 37.4 | 7.3 | 11.5 | 8.1 | 17.8 |
| Jacksonville, NC | NA | NA | 54,809 | 9.8 | 38.0 | 24.7 | 232.6 | 8,665 | 24,404 | 49,544 | 54,021 | 50.6 | 2.0 | 11.1 | 10.4 | 15.6 |
| Janesville-Beloit, WI | 3,247 | 1,988.5 | 38,284 | 10.9 | 45.4 | 23.2 | 333.0 | 12,099 | 29,985 | 61,243 | 73,079 | 41.4 | 2.6 | 13.5 | 9.2 | 13.7 |
| Jefferson City, MO | 2,826 | 1,864.7 | 33,372 | 22.4 | 41.4 | 28.9 | 196.2 | 8,932 | 29,475 | 60,786 | 75,075 | 38.8 | 3.0 | 10.8 | 8.1 | 21.2 |
| Johnson City, TN | 4,498 | 2,212.6 | 43,836 | 16.3 | 45.4 | 26.5 | 246.3 | 9,236 | 29,265 | 45,917 | 67,748 | 53.5 | 4.2 | 17.9 | 12.5 | 21.2 |
| Johnstown, PA | NA | NA | 26,626 | 23.5 | 52.6 | 20.6 | 237.6 | 13,677 | 27,773 | 49,076 | 66,461 | 50.7 | 2.3 | 15.3 | 10.3 | 21.6 |

1. Data for serious crimes have not been adjusted for underreporting; this may affect comparability between geographic areas and over time.   2. Per 100,000 population estimated by the FBI.   3. All persons 3 years old and over enrolled in nursery school through college.   4. Persons 25 years old and over.   5. Elementary and secondary education expenditures.   6. Based on population estimated by the American Community Survey, 2017.

## Table C. Metropolitan Areas — **Personal Income and Earnings**

| Area name | Personal income, 2019 | | | | | | | | | | Earnings, 2019 | | |
| | Total (mil dol) | Percent change, 2018-2019 | Per capita[1] Dollars | Per capita[1] Rank | Wages and Salaries (mil dol) | Supplements to wages and salaries, employer contributions (mil dol) Pension and insurance | Supplements to wages and salaries, employer contributions (mil dol) Government social insurance | Proprietors' income (mil dol) | Dividends, interest, and rent (mil dol) | Personal transfer receipts (mil dol) | Total (mil dol) | Contributions for government social insurance (mil dol) From employee and self-employed | Contributions for government social insurance (mil dol) From employer |
| | 62 | 63 | 64 | 65 | 66 | 67 | 68 | 69 | 70 | 71 | 72 | 73 | 74 |
|---|---|---|---|---|---|---|---|---|---|---|---|---|---|
| Evansville, IN-KY | 15,406 | 4.0 | 48,896 | 160 | 8,026 | 1,305 | 587 | 940 | 2,685 | 3,292 | 10,857 | 691 | 587 |
| Fairbanks, AK | 5,807 | 2.1 | 59,958 | 46 | 2,759 | 825 | 215 | 218 | 1,124 | 1,003 | 4,017 | 194 | 215 |
| Fargo, ND-MN | 13,521 | 3.8 | 54,932 | 83 | 7,538 | 1,101 | 588 | 1,118 | 2,953 | 1,763 | 10,345 | 628 | 588 |
| Farmington, NM | 4,462 | 0.8 | 35,999 | 377 | 2,373 | 409 | 176 | 134 | 665 | 1,160 | 3,091 | 205 | 176 |
| Fayetteville, NC | 20,123 | 4.5 | 38,204 | 365 | 10,811 | 2,672 | 903 | 731 | 3,825 | 5,522 | 15,116 | 827 | 903 |
| Fayetteville-Springdale-Rogers, AR | 36,251 | 3.2 | 67,771 | 18 | 14,446 | 1,784 | 990 | 1,228 | 16,534 | 3,882 | 18,447 | 1,201 | 990 |
| Flagstaff, AZ | 7,057 | 1.3 | 49,189 | 154 | 2,998 | 617 | 219 | 612 | 1,649 | 1,276 | 4,447 | 267 | 219 |
| Flint, MI | 17,015 | 3.0 | 41,929 | 301 | 6,983 | 1,173 | 523 | 840 | 2,514 | 5,012 | 9,519 | 671 | 523 |
| Florence, SC | 8,696 | 4.5 | 42,439 | 296 | 4,484 | 846 | 333 | 379 | 1,383 | 2,298 | 6,041 | 402 | 333 |
| Florence-Muscle Shoals, AL | 5,786 | 3.7 | 39,102 | 354 | 2,372 | 420 | 176 | 289 | 1,088 | 1,598 | 3,258 | 239 | 176 |
| Fond du Lac, WI | 5,165 | 3.8 | 49,949 | 140 | 2,485 | 468 | 193 | 384 | 873 | 971 | 3,530 | 223 | 193 |
| Fort Collins, CO | 19,945 | 4.9 | 55,884 | 74 | 9,622 | 1,375 | 645 | 1,350 | 4,736 | 2,731 | 12,992 | 736 | 645 |
| Fort Smith, AR-OK | 9,610 | 3.4 | 38,382 | 362 | 4,547 | 693 | 344 | 667 | 1,582 | 2,620 | 6,251 | 437 | 344 |
| Fort Wayne, IN | 19,637 | 3.3 | 47,516 | 195 | 10,758 | 1,691 | 791 | 1,548 | 3,380 | 3,724 | 14,787 | 913 | 791 |
| Fresno, CA | 45,446 | 5.5 | 45,487 | 234 | 19,924 | 4,262 | 1,463 | 4,970 | 7,841 | 10,492 | 30,619 | 1,681 | 1,463 |
| Gadsden, AL | 3,919 | 2.6 | 38,326 | 363 | 1,458 | 248 | 107 | 248 | 575 | 1,269 | 2,061 | 156 | 107 |
| Gainesville, FL | 14,771 | 4.0 | 44,880 | 250 | 7,838 | 1,488 | 543 | 569 | 3,106 | 3,050 | 10,438 | 640 | 543 |
| Gainesville, GA | 9,318 | 4.0 | 45,576 | 232 | 4,877 | 740 | 321 | 724 | 1,722 | 1,644 | 6,662 | 404 | 321 |
| Gettysburg, PA | 5,169 | 3.7 | 50,182 | 138 | 1,598 | 313 | 130 | 378 | 906 | 1,046 | 2,420 | 159 | 130 |
| Glens Falls, NY | 6,110 | 4.4 | 48,823 | 162 | 2,560 | 595 | 204 | 424 | 1,048 | 1,439 | 3,784 | 226 | 204 |
| Goldsboro, NC | 4,912 | 3.7 | 39,894 | 345 | 2,091 | 439 | 161 | 272 | 872 | 1,293 | 2,963 | 185 | 161 |
| Grand Forks, ND-MN | 5,237 | 2.4 | 51,950 | 119 | 2,674 | 485 | 218 | 412 | 1,065 | 904 | 3,790 | 228 | 218 |
| Grand Island, NE | 3,437 | 3.8 | 45,487 | 235 | 1,788 | 337 | 131 | 330 | 659 | 644 | 2,585 | 156 | 131 |
| Grand Junction, CO | 7,205 | 3.7 | 46,719 | 209 | 3,164 | 471 | 234 | 569 | 1,415 | 1,608 | 4,438 | 279 | 234 |
| Grand Rapids-Kentwood, MI | 54,224 | 3.2 | 50,330 | 136 | 30,338 | 4,888 | 2,229 | 3,984 | 10,926 | 8,586 | 41,439 | 2,544 | 2,229 |
| Grants Pass, OR | 3,810 | 4.3 | 43,554 | 275 | 1,183 | 208 | 107 | 453 | 729 | 1,290 | 1,951 | 151 | 107 |
| Great Falls, MT | 4,052 | 3.6 | 49,803 | 141 | 1,849 | 342 | 164 | 243 | 919 | 862 | 2,598 | 170 | 164 |
| Greeley, CO | 16,289 | 7.7 | 50,198 | 137 | 6,614 | 893 | 473 | 1,962 | 2,344 | 2,202 | 9,942 | 543 | 473 |
| Green Bay, WI | 17,108 | 3.2 | 52,981 | 104 | 9,641 | 1,754 | 697 | 1,262 | 3,247 | 2,623 | 13,356 | 812 | 697 |
| Greensboro-High Point, NC | 34,624 | 4.0 | 44,859 | 252 | 18,952 | 2,800 | 1,325 | 2,418 | 6,378 | 7,252 | 25,495 | 1,639 | 1,325 |
| Greenville, NC | 7,831 | 4.2 | 43,325 | 278 | 3,990 | 803 | 279 | 535 | 1,454 | 1,643 | 5,607 | 341 | 279 |
| Greenville-Anderson, SC | 41,822 | 4.2 | 45,436 | 237 | 21,491 | 3,552 | 1,595 | 2,574 | 7,265 | 8,552 | 29,212 | 1,909 | 1,595 |
| Gulfport-Biloxi, MS | 15,784 | 3.4 | 37,792 | 368 | 8,034 | 1,530 | 620 | 821 | 2,944 | 4,068 | 11,005 | 742 | 620 |
| Hagerstown-Martinsburg, MD-WV | 13,046 | 4.2 | 45,281 | 238 | 5,090 | 923 | 391 | 806 | 1,903 | 2,771 | 7,210 | 471 | 391 |
| Hammond, LA | 5,229 | 3.6 | 38,800 | 359 | 1,878 | 420 | 111 | 306 | 672 | 1,455 | 2,714 | 153 | 111 |
| Hanford-Corcoran, CA | 6,031 | 7.8 | 39,433 | 352 | 2,723 | 746 | 206 | 661 | 1,048 | 1,256 | 4,336 | 196 | 206 |
| Harrisburg-Carlisle, PA | 31,135 | 3.9 | 53,871 | 97 | 19,829 | 3,750 | 1,529 | 2,431 | 5,193 | 5,922 | 27,538 | 1,598 | 1,529 |
| Harrisonburg, VA | 5,576 | 2.6 | 41,312 | 319 | 3,083 | 525 | 218 | 570 | 1,098 | 998 | 4,395 | 268 | 218 |
| Hartford-East Hartford-Middletown, CT | 78,476 | 2.5 | 65,132 | 24 | 44,914 | 7,059 | 3,131 | 7,859 | 12,338 | 12,324 | 62,964 | 3,503 | 3,131 |
| Hattiesburg, MS | 6,588 | 2.2 | 39,018 | 356 | 2,990 | 505 | 217 | 530 | 1,138 | 1,601 | 4,242 | 291 | 217 |
| Hickory-Lenoir-Morganton, NC | 15,094 | 3.6 | 40,826 | 331 | 7,108 | 1,194 | 513 | 890 | 2,517 | 3,869 | 9,706 | 661 | 513 |
| Hilton Head Island-Bluffton, SC | 12,343 | 5.6 | 55,551 | 79 | 4,267 | 800 | 338 | 644 | 4,530 | 2,444 | 6,048 | 409 | 338 |
| Hinesville, GA | 2,832 | 3.4 | 34,962 | 379 | 1,897 | 560 | 167 | 41 | 524 | 687 | 2,665 | 122 | 167 |
| Homosassa Springs, FL | 5,845 | 4.1 | 39,055 | 355 | 1,410 | 243 | 98 | 304 | 1,412 | 2,331 | 2,055 | 218 | 98 |
| Hot Springs, AR | 4,082 | 4.2 | 41,069 | 328 | 1,581 | 227 | 120 | 294 | 816 | 1,315 | 2,222 | 173 | 120 |
| Houma-Thibodaux, LA | 9,140 | 2.6 | 43,927 | 269 | 4,775 | 776 | 303 | 710 | 1,628 | 2,086 | 6,564 | 386 | 303 |
| Houston-The Woodlands-Sugar Land, TX | 416,122 | 4.4 | 58,890 | 51 | 222,616 | 29,265 | 14,166 | 65,419 | 71,486 | 48,113 | 331,466 | 17,015 | 14,166 |
| Huntington-Ashland, WV-KY-OH | 14,722 | 1.6 | 41,368 | 318 | 6,565 | 1,218 | 508 | 789 | 2,012 | 4,325 | 9,080 | 626 | 508 |
| Huntsville, AL | 24,587 | 5.7 | 52,110 | 114 | 15,219 | 2,431 | 1,119 | 1,164 | 4,559 | 4,097 | 19,933 | 1,237 | 1,119 |
| Idaho Falls, ID | 7,189 | 5.0 | 47,443 | 198 | 3,437 | 506 | 284 | 1,088 | 1,608 | 1,169 | 5,315 | 328 | 284 |
| Indianapolis-Carmel-Anderson, IN | 116,921 | 4.0 | 56,360 | 68 | 61,842 | 9,138 | 4,421 | 15,898 | 18,629 | 17,384 | 91,300 | 5,338 | 4,421 |
| Iowa City, IA | 9,665 | 2.2 | 55,834 | 75 | 4,896 | 1,297 | 359 | 907 | 2,207 | 1,181 | 7,459 | 414 | 359 |
| Ithaca, NY | 4,751 | 2.8 | 46,496 | 215 | 2,913 | 541 | 236 | 355 | 966 | 795 | 4,045 | 217 | 236 |
| Jackson, MI | 6,402 | 3.1 | 40,387 | 337 | 3,009 | 579 | 224 | 339 | 991 | 1,703 | 4,151 | 279 | 224 |
| Jackson, MS | 26,608 | 2.4 | 44,734 | 256 | 13,350 | 2,118 | 967 | 2,078 | 4,941 | 5,738 | 18,513 | 1,254 | 967 |
| Jackson, TN | 7,466 | 4.2 | 41,794 | 308 | 3,692 | 679 | 257 | 746 | 1,025 | 2,005 | 5,375 | 338 | 257 |
| Jacksonville, FL | 80,192 | 4.0 | 51,421 | 122 | 41,805 | 6,116 | 2,904 | 4,290 | 16,802 | 14,684 | 55,115 | 3,499 | 2,904 |
| Jacksonville, NC | 9,407 | 3.0 | 47,525 | 194 | 4,772 | 1,362 | 436 | 312 | 1,717 | 1,791 | 6,883 | 321 | 436 |
| Janesville-Beloit, WI | 7,553 | 4.7 | 46,236 | 219 | 3,592 | 673 | 270 | 328 | 1,369 | 1,538 | 4,863 | 318 | 270 |
| Jefferson City, MO | 6,676 | 4.4 | 44,141 | 265 | 3,620 | 834 | 245 | 367 | 1,189 | 1,371 | 5,068 | 295 | 245 |
| Johnson City, TN | 8,392 | 2.7 | 41,209 | 323 | 3,528 | 682 | 246 | 766 | 1,263 | 2,294 | 5,222 | 346 | 246 |
| Johnstown, PA | 5,873 | 2.8 | 45,107 | 244 | 2,226 | 487 | 182 | 316 | 918 | 1,873 | 3,211 | 223 | 182 |

1. Based on the resident population estimated as of July 1 of the year shown.

## Table C. Metropolitan Areas — Earnings, Social Security, and Housing

Column groups: **Earnings, 2019 (cont.) — Percent by selected industries** (items 75–83); **Social Security beneficiaries, December 2019** (items 84–85); **Supplemental Security Income Recipients, December 2019** (item 86); **Housing units, 2020** (items 87–88).

| Area name | Farm (75) | Mining, quarrying, and extracting (76) | Construction (77) | Manufacturing (78) | Information; professional, scientific, and technical services (79) | Retail trade (80) | Finance, insurance, real estate, rental and leasing (81) | Health care and social assistance (82) | Government (83) | SS Number (84) | SS Rate[1] (85) | SSI Recipients, Dec 2019 (86) | Housing Total (87) | Housing % change, 2010–2020 (88) |
|---|---|---|---|---|---|---|---|---|---|---|---|---|---|---|
| Evansville, IN-KY | 0.4 | 0.2 | 7.1 | 18.6 | D | 5.9 | 5.0 | D | 9.6 | 71,415 | 227 | 7,132 | 144,135 | 3.9 |
| Fairbanks, AK | 0.1 | 2.8 | 7.8 | 1.0 | 4.4 | 5.6 | 2.9 | 10.6 | 47.7 | 12,055 | 124 | 1,162 | 44,311 | 6.0 |
| Fargo, ND-MN | 0.7 | D | 7.6 | 7.5 | 10.0 | 6.9 | 10.8 | D | 13.6 | 36,535 | 148 | 3,248 | 113,927 | 24.0 |
| Farmington, NM | 0.8 | 16.3 | 9.8 | 2.8 | D | 6.5 | 2.9 | 12.9 | 24.2 | 23,740 | 191 | 3,932 | 51,670 | 4.7 |
| Fayetteville, NC | 0.5 | D | 5.1 | 4.9 | D | 5.1 | 2.7 | 5.8 | 57.5 | 91,570 | 173 | 14,755 | 223,332 | 11.4 |
| Fayetteville-Springdale-Rogers, AR | 0.3 | D | 6.1 | 8.9 | D | 5.4 | 3.5 | 9.3 | 11.3 | 89,120 | 166 | 8,133 | 222,777 | 18.3 |
| Flagstaff, AZ | 0.2 | 0.1 | 4.6 | 8.1 | 5.4 | 7.0 | 4.0 | 15.5 | 29.7 | 21,590 | 150 | 2,390 | 68,354 | 7.9 |
| Flint, MI | 0.1 | 0.0 | 6.3 | 12.1 | 7.2 | 8.4 | 6.1 | 18.3 | 15.4 | 101,555 | 250 | 16,406 | 193,269 | 0.6 |
| Florence, SC | 0.4 | 0.0 | 4.4 | 15.8 | 5.8 | 6.8 | 9.6 | 11.7 | 19.8 | 49,965 | 244 | 8,431 | 93,950 | 5.6 |
| Florence-Muscle Shoals, AL | -0.2 | D | 8.5 | 18.5 | D | 9.2 | 4.6 | 12.8 | 20.0 | 40,050 | 270 | 4,418 | 72,729 | 4.6 |
| Fond du Lac, WI | 2.9 | 0.6 | 9.3 | 25.0 | 4.3 | 6.0 | 5.4 | 11.5 | 11.5 | 23,370 | 227 | 1,596 | 45,749 | 4.2 |
| Fort Collins, CO | 0.3 | 0.5 | 9.1 | 13.9 | 12.4 | 6.4 | 5.8 | 8.6 | 21.1 | 60,740 | 170 | 2,571 | 148,795 | 12.1 |
| Fort Smith, AR-OK | 0.1 | 0.9 | 5.6 | 16.7 | 4.5 | 6.9 | 7.9 | 14.4 | 13.7 | 60,895 | 243 | 8,615 | 112,510 | 4.7 |
| Fort Wayne, IN | 0.3 | D | 7.6 | 19.1 | D | 6.2 | 5.5 | 14.2 | 21.9 | 80,325 | 195 | 8,506 | 177,492 | 6.6 |
| Fresno, CA | 7.3 | 0.2 | 5.7 | 15.4 | 3.8 | 8.5 | 4.9 | D | 15.1 | 143,880 | 144 | 43,771 | 338,360 | 7.2 |
| Gadsden, AL | 0.0 | D | 4.9 | 3.9 | 8.4 | 5.6 | 5.7 | 17.4 | 34.8 | 29,045 | 283 | 4,410 | 47,921 | 1.0 |
| Gainesville, FL | 1.7 | D | 6.6 | 20.0 | 4.5 | 6.1 | 6.7 | 16.7 | 29.7 | 64,740 | 196 | 8,554 | 151,251 | 7.9 |
| Gainesville, GA | -0.1 | D | 8.8 | 21.1 | 3.9 | 6.1 | 3.5 | 13.2 | 15.0 | 38,550 | 189 | 3,130 | 77,669 | 12.8 |
| Gettysburg, PA | 1.4 | 0.8 | 7.0 | 13.7 | 6.6 | 8.3 | 5.0 | 13.7 | 21.9 | 25,700 | 250 | 1,077 | 42,965 | 5.2 |
| Glens Falls, NY | 1.0 | D | 5.5 | 11.2 | 2.9 | 7.0 | 5.0 | 11.1 | 34.2 | 34,300 | 274 | 3,080 | 70,135 | 3.8 |
| Goldsboro, NC | 2.7 | D | 7.3 | 8.2 | 5.0 | 7.1 | 6.0 | D | 26.0 | 26,595 | 215 | 4,047 | 55,010 | 3.9 |
| Grand Forks, ND-MN | 2.2 | 0.5 | 7.3 | 18.6 | D | 7.6 | D | 10.1 | 16.9 | 18,105 | 179 | 1,295 | 48,796 | 11.0 |
| Grand Island, NE | 4.7 | 0.1 | 10.8 | 4.2 | 5.6 | 7.6 | 7.0 | 16.9 | 16.4 | 14,400 | 190 | 1,149 | 32,279 | 6.9 |
| Grand Junction, CO | 0.2 | 6.7 | 6.6 | 4.2 | D | 5.6 | D | 9.1 | 13.3 | 34,995 | 226 | 2,652 | 65,771 | 5.0 |
| Grand Rapids-Kentwood, MI | 0.5 | D | 7.0 | 23.3 | D | 5.6 | 7.5 | 18.5 | 13.3 | 198,160 | 184 | 18,836 | 429,134 | 6.6 |
| Grants Pass, OR | 0.0 | D | 7.0 | 10.0 | 4.5 | 12.5 | D | 17.8 | 26.5 | 28,190 | 322 | 2,896 | 39,900 | 5.0 |
| Great Falls, MT | 1.0 | 0.1 | 7.6 | 3.4 | 5.9 | 7.8 | 7.5 | 6.5 | 10.9 | 18,825 | 231 | 1,822 | 39,317 | 5.5 |
| Greeley, CO | 5.9 | 9.8 | 14.0 | 11.0 | 4.4 | 5.5 | 4.7 | D | 12.0 | 45,050 | 139 | 3,564 | 113,437 | 17.8 |
| Green Bay, WI | 1.9 | 0.1 | 5.8 | 17.1 | D | 6.1 | 9.3 | 11.7 | 11.7 | 66,370 | 206 | 5,504 | 146,116 | 6.5 |
| Greensboro-High Point, NC | 0.2 | D | 6.3 | 15.6 | D | 6.5 | 5.4 | 10.3 | 37.2 | 161,850 | 210 | 18,167 | 340,872 | 5.6 |
| Greenville, NC | 0.6 | D | 6.1 | 10.3 | 4.3 | 6.5 | 5.4 | 15.2 | 15.2 | 32,280 | 178 | 5,767 | 82,201 | 9.6 |
| Greenville-Anderson, SC | 0.0 | 0.1 | 6.4 | 15.9 | D | 6.4 | 7.4 | 9.5 | 29.8 | 201,450 | 219 | 17,970 | 401,406 | 10.8 |
| Gulfport-Biloxi, MS | 0.0 | D | 6.3 | 15.7 | 6.3 | 6.0 | 3.5 | D | 19.8 | 91,495 | 219 | 11,792 | 193,640 | 11.1 |
| Hagerstown-Martinsburg, MD-WV | 0.8 | D | 5.7 | 10.9 | 7.2 | 8.0 | 7.7 | 13.9 | 19.8 | 62,345 | 216 | 6,580 | 123,443 | 7.0 |
| Hammond, LA | 0.1 | 0.8 | 5.8 | 7.4 | 4.6 | 10.5 | 5.3 | 11.1 | 28.7 | 24,925 | 185 | 4,952 | 58,725 | 17.3 |
| Hanford-Corcoran, CA | 15.7 | D | 2.1 | 7.7 | 1.6 | 4.2 | 1.8 | 8.5 | 42.1 | 18,960 | 124 | 4,695 | 46,949 | 7.0 |
| Harrisburg-Carlisle, PA | 0.3 | 0.0 | 4.5 | 6.7 | 10.0 | 4.5 | 8.6 | 14.8 | 19.5 | 124,420 | 215 | 11,881 | 256,295 | 6.4 |
| Harrisonburg, VA | 2.3 | D | 7.6 | 17.7 | 6.6 | 6.1 | 4.9 | 11.4 | 17.6 | 24,765 | 183 | 1,875 | 54,767 | 7.2 |
| Hartford-East Hartford-Middletown, CT | 0.1 | D | 5.0 | D | 11.5 | 4.8 | 18.3 | 11.9 | 14.3 | 240,150 | 199 | 24,487 | 519,331 | 2.4 |
| Hattiesburg, MS | 1.3 | D | 6.2 | 8.9 | D | 8.2 | D | 17.7 | 23.2 | 34,065 | 202 | 5,494 | 72,655 | 3.2 |
| Hickory-Lenoir-Morganton, NC | 0.5 | 0.1 | 4.5 | 27.1 | 4.6 | 7.2 | 3.5 | D | 15.2 | 93,065 | 252 | 7,799 | 167,440 | 2.9 |
| Hilton Head Island-Bluffton, SC | 0.1 | D | 8.9 | 1.2 | 9.6 | 8.5 | 6.3 | 9.2 | 29.6 | 59,340 | 266 | 2,588 | 118,085 | 14.3 |
| Hinesville, GA | 0.1 | D | 1.2 | D | D | D | D | D | 72.5 | 10,865 | 132 | 1,629 | 36,869 | 12.5 |
| Homosassa Springs, FL | 0.1 | 0.3 | 11.2 | 1.7 | 5.9 | 11.3 | 5.5 | 22.2 | 13.1 | 63,015 | 420 | 3,593 | 81,706 | 4.7 |
| Hot Springs, AR | 0.1 | 0.4 | 9.8 | 7.6 | 5.1 | 10.9 | 5.5 | 22.6 | 12.9 | 30,900 | 311 | 3,816 | 51,145 | 1.2 |
| Houma-Thibodaux, LA | 0.4 | 11.5 | 6.7 | 9.6 | 4.9 | 5.8 | 4.9 | 9.9 | 12.6 | 43,640 | 210 | 7,783 | 88,108 | 6.8 |
| Houston-The Woodlands-Sugar Land, TX | 0.0 | 10.4 | 7.8 | 10.0 | D | 4.3 | 7.6 | D | 10.0 | 897,190 | 127 | 142,987 | 2,751,986 | 19.9 |
| Huntington-Ashland, WV-KY-OH | -0.1 | 1.1 | 8.4 | 11.9 | D | 6.9 | D | D | 16.6 | 91,460 | 257 | 15,821 | 168,095 | 0.7 |
| Huntsville, AL | 0.1 | D | 4.3 | 12.0 | 26.0 | 5.4 | 3.4 | 7.0 | 26.9 | 90,000 | 191 | 8,816 | 206,397 | 13.8 |
| Idaho Falls, ID | 2.0 | D | D | D | D | 14.3 | D | 12.4 | 9.7 | 25,870 | 170 | 2,784 | 56,807 | 14.1 |
| Indianapolis-Carmel-Anderson, IN | 0.1 | 0.1 | 6.2 | 10.7 | D | 5.2 | 16.5 | D | 11.1 | 364,845 | 176 | 37,328 | 884,153 | 8.3 |
| Iowa City, IA | 3.1 | D | 5.1 | 6.8 | 4.2 | 5.2 | 4.4 | 7.0 | 44.2 | 25,970 | 149 | 2,169 | 77,232 | 17.9 |
| Ithaca, NY | 0.5 | 0.8 | 2.4 | 6.2 | 8.9 | 4.6 | 3.4 | D | 13.2 | 16,890 | 166 | 1,443 | 45,392 | 8.9 |
| Jackson, MI | 0.0 | 0.2 | 5.1 | 18.3 | 4.6 | 6.0 | 4.9 | 15.6 | 13.9 | 38,195 | 241 | 4,536 | 70,102 | 0.9 |
| Jackson, MS | 0.0 | D | 5.0 | 8.1 | 8.5 | 6.8 | 7.5 | 14.0 | 21.9 | 121,070 | 203 | 21,251 | 255,283 | 5.9 |
| Jackson, TN | 0.5 | D | D | 18.5 | D | 8.3 | 3.9 | 12.5 | 21.3 | 43,675 | 244 | 6,056 | 80,155 | 3.7 |
| Jacksonville, FL | 0.1 | 0.1 | 6.6 | 5.3 | 11.5 | 6.4 | 13.5 | 13.0 | 13.5 | 308,190 | 197 | 34,148 | 682,915 | 14.1 |
| Jacksonville, NC | 0.7 | D | 2.8 | 0.8 | 2.4 | 4.3 | 2.4 | 3.1 | 72.7 | 29,400 | 145 | 3,280 | 82,375 | 20.7 |
| Janesville-Beloit, WI | 0.7 | D | 7.4 | 17.2 | 5.0 | 5.8 | 4.0 | 15.6 | 13.5 | 36,510 | 224 | 3,835 | 69,961 | 2.3 |
| Jefferson City, MO | 0.2 | 0.2 | 7.5 | 9.5 | D | 6.4 | D | 9.2 | 31.1 | 33,500 | 222 | 2,564 | 65,736 | 3.4 |
| Johnson City, TN | 0.0 | D | 5.1 | 11.2 | D | 8.2 | 6.5 | D | 22.0 | 54,515 | 267 | 5,650 | 99,026 | 5.5 |
| Johnstown, PA | 0.2 | 0.3 | 6.6 | 9.9 | 6.7 | 7.3 | 5.4 | 21.5 | 17.5 | 39,075 | 300 | 4,739 | 66,073 | 0.6 |

1. Per 1,000 resident population estimated as of July 1 of the year shown.

# Table C. Metropolitan Areas — Housing, Labor Force, and Employment

| | Occupied Housing units, 2020 | | | | | | | | Civilian labor force, 2020 | | | | Civilian employment, 2019[6] | | |
| | Occupied units | | | | | | | | | | Unemployment | | | Percent | |
| | | Owner-occupied | | | | Renter-occupied | | | | | | | | | |
| Area name | | | | Median owner cost as a percent of income | | | Median rent as a percent of income[2] | Sub-standard units[4] (percent) | | Percent change 2019-2020 | | | | Manage-ment, business, science, and arts | Construction, production, and mainte-nance occu-pations |
| | Total | Percent | Median value[1] | With a mortgage | Without a mort-gage[2] | Median rent[3] | | | Total | | Total | Rate[5] | Total | | |
| | 89 | 90 | 91 | 92 | 93 | 94 | 95 | 96 | 97 | 98 | 99 | 100 | 101 | 102 | 103 |
|---|---|---|---|---|---|---|---|---|---|---|---|---|---|---|---|
| Evansville, IN-KY | 129,971 | 67.4 | 147,300 | 17.9 | 10.0 | 807 | 26.8 | 1.4 | 156,982 | -3.3 | 10,193 | 6.5 | 157,322 | 35.0 | 25.7 |
| Fairbanks, AK | 35,556 | 61.9 | 259,800 | 20.2 | 10.6 | 1,284 | 30.5 | 4.7 | 44,679 | -1.8 | 2,841 | 6.4 | 43,756 | 41.0 | 22.2 |
| Fargo, ND-MN | 104,097 | 52.5 | 236,100 | 18.7 | 10.0 | 815 | 25.7 | 1.3 | 141,249 | 0.9 | 6,153 | 4.4 | 143,120 | 42.4 | 21.6 |
| Farmington, NM | 42,561 | 66.7 | 154,200 | 23.7 | 10.0 | 749 | 26.0 | 6.8 | 50,367 | -3.7 | 5,046 | 10.0 | 46,994 | 26.1 | 30.8 |
| Fayetteville, NC | 193,936 | 55.2 | 144,800 | 21.7 | 11.3 | 940 | 30.7 | 1.7 | 199,019 | -1.9 | 17,668 | 8.9 | 203,390 | 32.0 | 26.4 |
| Fayetteville-Springdale-Rogers, AR | 196,683 | 61.3 | 196,300 | 17.9 | 10.0 | 840 | 26.3 | 4.0 | 273,569 | 0.8 | 12,313 | 4.5 | 258,136 | 39.2 | 24.6 |
| Flagstaff, AZ | 48,993 | 61.5 | 317,700 | 21.9 | 10.0 | 1,231 | 31.4 | 5.5 | 75,280 | -2.4 | 7,295 | 9.7 | 68,508 | 44.8 | 15.2 |
| Flint, MI | 169,247 | 70 | 128,900 | 20.0 | 12.3 | 769 | 29.0 | 1.7 | 179,933 | -0.8 | 20,074 | 11.2 | 173,534 | 32.7 | 28.1 |
| Florence, SC | 77,848 | 66.5 | 128,500 | 17.9 | 10.0 | 777 | 30.1 | 4.3 | 97,671 | 0.5 | 5,723 | 5.9 | 88,464 | 35.1 | 28.5 |
| Florence-Muscle Shoals, AL | 60,837 | 65.9 | 145,500 | 19.2 | 10.0 | 654 | 28.4 | 0.0 | 65,489 | | 3,871 | 5.9 | 63,304 | 29.1 | 31.0 |
| Fond du Lac, WI | 42,604 | 67.6 | 172,700 | 18.7 | 10.9 | 751 | 23.2 | 0.0 | 56,114 | -1.6 | 3,286 | 5.9 | 56,314 | 28.8 | 36.4 |
| Fort Collins, CO | 146,638 | 65.7 | 420,100 | 22.1 | 10.0 | 1,399 | 32.0 | 1.5 | 203,683 | -0.5 | 12,841 | 6.3 | 191,401 | 48.0 | 16.3 |
| Fort Smith, AR-OK | 98,307 | 62.2 | 126,200 | 19.6 | 10.0 | 707 | 27.0 | 3.0 | 106,627 | -0.9 | 6,336 | 5.9 | 104,745 | 29.8 | 32.1 |
| Fort Wayne, IN | 160,462 | 70.8 | 143,800 | 18.0 | 10.0 | 780 | 26.1 | 1.5 | 202,141 | -1.3 | 15,219 | 7.5 | 208,392 | 36.0 | 28.1 |
| Fresno, CA | 315,974 | 53.6 | 288,300 | 23.3 | 10.1 | 1,034 | 33.0 | 10.3 | 445,518 | -1.1 | 50,260 | 11.3 | 410,423 | 31.7 | 28.7 |
| Gadsden, AL | 40,053 | 72.8 | 127,500 | 17.9 | 11.9 | 677 | 28.8 | 0.0 | 40,724 | -4.1 | 3,126 | 7.7 | 43,451 | 29.5 | 30.2 |
| Gainesville, FL | 126,736 | 58.4 | 191,700 | 19.7 | 10.4 | 972 | 35.7 | 2.0 | 158,528 | -2.7 | 8,581 | 5.4 | 151,810 | 44.3 | 14.7 |
| Gainesville, GA | 65,625 | 73.6 | 227,800 | 20.0 | 10.0 | 974 | 26.3 | 5.2 | 101,949 | -0.8 | 4,816 | 4.7 | 94,125 | 31.3 | 33.5 |
| Gettysburg, PA | 38,703 | 78.4 | 206,200 | 21.7 | 13.4 | 911 | 26.6 | 0.0 | 54,035 | -2.5 | 3,742 | 6.9 | 52,337 | 30.1 | 33.4 |
| Glens Falls, NY | 53,686 | 73.1 | 178,200 | 19.3 | 12.0 | 923 | 27.6 | 1.6 | 58,203 | -1.1 | 4,547 | 7.8 | 60,540 | 34.9 | 23.4 |
| Goldsboro, NC | 48,482 | 60.9 | 118,400 | 21.0 | 11.0 | 798 | 27.5 | 2.2 | 50,708 | -3.0 | 3,278 | 6.5 | 48,552 | 28.3 | 31.1 |
| Grand Forks, ND-MN | 44,724 | 55.5 | 202,800 | 19.3 | 10.0 | 809 | 27.0 | 2.9 | 53,133 | -2.0 | 2,617 | 4.9 | 53,864 | 36.2 | 23.0 |
| Grand Island, NE | 30,529 | 65 | 154,300 | 20.9 | 11.5 | 738 | 27.9 | 0.0 | 39,374 | -0.3 | 2,157 | 5.5 | 39,256 | 28.8 | 35.8 |
| Grand Junction, CO | 63,644 | 71.2 | 262,300 | 21.6 | 10.0 | 981 | 30.7 | 2.5 | 75,551 | -0.6 | 5,634 | 7.5 | 76,060 | 36.6 | 22.8 |
| Grand Rapids-Kentwood, MI | 396,763 | 72.3 | 199,400 | 18.3 | 10.0 | 937 | 28.1 | 1.9 | 574,606 | -1.4 | 42,919 | 7.5 | 557,079 | 35.8 | 27.7 |
| Grants Pass, OR | 35,206 | 66.3 | 293,100 | 26.8 | 10.7 | 1,041 | 33.4 | 6.8 | 36,124 | 0.7 | 2,847 | 7.9 | 33,309 | 35.1 | 24.7 |
| Great Falls, MT | 32,816 | 66 | 190,100 | 22.4 | 10.0 | 763 | 29.0 | 0.0 | 37,797 | -0.8 | 2,236 | 5.9 | 36,538 | 32.6 | 20.9 |
| Greeley, CO | 111,298 | 73.6 | 366,800 | 22.0 | 10.5 | 1,211 | 31.7 | 2.8 | 166,666 | -1.5 | 11,624 | 7.0 | 166,014 | 38.0 | 28.7 |
| Green Bay, WI | 132,099 | 65.2 | 190,300 | 17.8 | 10.3 | 788 | 25.5 | 2.0 | 171,356 | -0.6 | 10,209 | 6.0 | 170,992 | 33.4 | 29.1 |
| Greensboro-High Point, NC | 301,529 | 61.8 | 158,600 | 19.2 | 10.0 | 863 | 28.5 | 2.9 | 359,973 | -3.1 | 30,049 | 8.3 | 359,150 | 34.9 | 27.3 |
| Greenville, NC | 70,926 | 50.3 | 155,700 | 18.9 | 10.6 | 829 | 24.7 | 3.1 | 87,450 | -3.1 | 5,946 | 6.8 | 86,219 | 38.0 | 17.2 |
| Greenville-Anderson, SC | 349,480 | 70.9 | 186,800 | 18.3 | 10.0 | 867 | 27.2 | 2.6 | 434,167 | 0.1 | 25,127 | 5.8 | 443,005 | 38.7 | 25.5 |
| Gulfport-Biloxi, MS | 159,081 | 65.5 | 152,700 | 20.8 | 10.0 | 907 | 30.6 | 3.6 | 171,549 | 0.0 | 15,944 | 9.3 | 177,768 | 34.4 | 24.2 |
| Hagerstown-Martinsburg, MD-WV | 110,354 | 69.3 | 201,200 | 20.1 | 10.0 | 924 | 27.6 | 2.1 | 139,840 | -2.0 | 9,037 | 6.5 | 138,651 | 31.4 | 29.3 |
| Hammond, LA | 46,526 | 72.2 | 179,800 | 17.8 | 10.7 | 841 | 34.6 | 0.0 | 56,389 | 2.1 | 5,377 | 9.5 | 58,617 | 37.1 | 22.9 |
| Hanford-Corcoran, CA | 44,761 | 53.3 | 247,600 | 21.4 | 10.0 | 969 | 29.5 | 11.5 | 56,416 | -2.4 | 6,533 | 11.6 | 53,749 | 26.6 | 35.4 |
| Harrisburg-Carlisle, PA | 234,365 | 67.5 | 186,600 | 19.9 | 11.6 | 946 | 26.4 | 1.5 | 299,207 | | 23,098 | 7.7 | 296,655 | 41.0 | 23.9 |
| Harrisonburg, VA | 47,098 | 61.4 | 222,600 | 19.6 | 10.0 | 967 | 25.3 | 2.8 | 65,784 | -2.8 | 3,499 | 5.3 | 67,449 | 35.4 | 28.5 |
| Hartford-East Hartford-Middletown, CT | 476,867 | 65.9 | 254,100 | 21.7 | 13.6 | 1,113 | 28.8 | 1.5 | 652,934 | -2.0 | 49,928 | 7.6 | 626,236 | 46.3 | 16.9 |
| Hattiesburg, MS | 59,170 | 61.6 | 145,300 | 19.1 | 10.0 | 837 | 30.5 | 4.3 | 76,151 | | 5,435 | 7.1 | 71,575 | 34.4 | 25.0 |
| Hickory-Lenoir-Morganton, NC | 146,324 | 74.2 | 145,700 | 18.0 | 10.0 | 681 | 23.3 | 3.6 | 171,188 | -2.3 | 13,229 | 7.7 | 175,378 | 31.9 | 33.9 |
| Hilton Head Island-Bluffton, SC | 84,094 | 76.6 | 304,800 | 22.4 | 10.0 | 1,166 | 28.7 | 1.5 | 89,931 | -0.2 | 4,823 | 5.4 | 94,602 | 35.5 | 22.8 |
| Hinesville, GA | 29,973 | 55.8 | 135,600 | 20.5 | 10.4 | 1,023 | 31.5 | 0.0 | 33,880 | -1.3 | 1,958 | 5.8 | 27,386 | 29.4 | 29.4 |
| Homosassa Springs, FL | 64,448 | 83 | 156,400 | 22.9 | 10.0 | 903 | 34.0 | 0.0 | 46,500 | -1.9 | 3,918 | 8.4 | 44,646 | 32.5 | 27.5 |
| Hot Springs, AR | 41,743 | 69.7 | 139,100 | 21.7 | 10.0 | 763 | 23.5 | 3.4 | 41,347 | -0.3 | 3,295 | 8.0 | 43,181 | 37.1 | 13.7 |
| Houma-Thibodaux, LA | 76,932 | 72 | 162,600 | 18.5 | 10.0 | 807 | 37.1 | 1.4 | 87,699 | -1.5 | 6,250 | 7.1 | 90,860 | 29.2 | 29.0 |
| Houston-The Woodlands-Sugar Land, TX | 2,436,438 | 60.1 | 219,100 | 20.6 | 10.5 | 1,139 | 30.1 | 5.0 | 3,405,878 | -0.6 | 294,204 | 8.6 | 3,420,767 | 39.5 | 24.6 |
| Huntington-Ashland, WV-KY-OH | 136,687 | 71.1 | 128,800 | 19.2 | 10.0 | 757 | 30.9 | 0.7 | 149,088 | -0.2 | 12,261 | 8.2 | 143,070 | 37.1 | 24.0 |
| Huntsville, AL | 187,898 | 68.7 | 192,700 | 17.5 | 10.0 | 858 | 26.6 | 1.5 | 231,871 | 1.2 | 10,568 | 4.6 | 226,964 | 45.3 | 20.1 |
| Idaho Falls, ID | 51,152 | 72.1 | 212,700 | 18.9 | 10.0 | 798 | 29.2 | 3.0 | 73,671 | 1.9 | 2,880 | 3.9 | 71,061 | 33.8 | 23.5 |
| Indianapolis-Carmel-Anderson, IN | 793,781 | 65.2 | 183,800 | 17.9 | 10.0 | 932 | 28.4 | 1.4 | 1,067,065 | | 70,067 | 6.6 | 1,042,945 | 41.0 | 22.4 |
| Iowa City, IA | 70,169 | 59.1 | 241,600 | 19.7 | 10.0 | 932 | 33.7 | 3.0 | 95,578 | -4.3 | 4,501 | 4.7 | 96,872 | 48.9 | 18.3 |
| Ithaca, NY | 40,322 | 50.1 | 205,700 | 19.2 | 11.1 | 1,219 | 36.3 | 0.0 | 47,929 | -2.4 | 2,962 | 6.2 | 50,419 | 56.6 | 10.1 |
| Jackson, MI | 61,403 | 74.2 | 142,900 | 19.4 | 10.6 | 779 | 32.9 | 0.0 | 74,075 | -0.8 | 6,668 | 9.0 | 68,972 | 33.4 | 29.5 |
| Jackson, MS | 218,853 | 66.4 | 153,800 | 19.0 | 10.0 | 863 | 30.3 | 2.0 | 267,755 | -2.4 | 20,612 | 7.7 | 267,888 | 37.9 | 23.4 |
| Jackson, TN | 70,379 | 65 | 121,400 | 19.0 | 10.0 | 772 | 31.5 | 1.4 | 86,049 | | 6,084 | 7.1 | 78,073 | 32.9 | 28.7 |
| Jacksonville, FL | 584,381 | 66 | 230,100 | 20.4 | 10.0 | 1,121 | 29.5 | 1.9 | 777,712 | -0.9 | 48,285 | 6.2 | 743,173 | 39.1 | 19.7 |
| Jacksonville, NC | 63,604 | 54.4 | 165,300 | 21.6 | 10.0 | 900 | 27.9 | 0.0 | 63,899 | -1.7 | 4,623 | 7.2 | 68,223 | 33.1 | 26.3 |
| Janesville-Beloit, WI | 65,316 | 68.6 | 162,800 | 20.3 | 12.0 | 876 | 26.6 | 1.5 | 84,935 | -0.2 | 6,027 | 7.1 | 81,370 | 32.9 | 32.4 |
| Jefferson City, MO | 59,727 | 69.7 | 164,200 | 18.3 | 10.0 | 710 | 24.2 | 1.9 | 74,963 | 0.2 | 3,242 | 4.3 | 69,995 | 39.0 | 23.7 |
| Johnson City, TN | 89,021 | 68.7 | 156,400 | 18.7 | 10.8 | 730 | 27.7 | 1.3 | 90,521 | -1.8 | 6,095 | 6.7 | 95,111 | 37.2 | 22.3 |
| Johnstown, PA | 56,490 | 74.9 | 96,800 | 17.4 | 12.0 | 620 | 29.4 | 0.0 | 56,235 | -2.9 | 5,363 | 9.5 | 57,273 | 32.2 | 27.8 |

1. Specified owner-occupied units. lacking complete plumbing facilities.   2. A value of 10.0 represents 10 percent or less; a value of 50.0 represents 50 percent or more.   5. Percent of civilian labor force. 6. Civilian employed persons 16 years old and over. 3. Specified renter-occupied units.   4. Overcrowded or

# Table C. Metropolitan Areas — Nonfarm Employment and Agriculture

| Area name | Private nonfarm establishments, employment and payroll, 2019 | | | | | | | | | Agriculture, 2017 | | | |
|---|---|---|---|---|---|---|---|---|---|---|---|---|---|
| | Number of establish-ments | Employment | | | | | | Annual payroll | | Farms | | | Farm producers whose primary occupation is farming (percent) |
| | | Total | Health care and social assistance | Manufactur-ing | Retail trade | Finance and insurance | Professional, scientific, and technical services | Total (mil dol) | Average per employee (dollars) | Number | Fewer than 50 acres | 1000 acres or more | |
| | 104 | 105 | 106 | 107 | 108 | 109 | 110 | 111 | 112 | 113 | 114 | 115 | 116 |
| Evansville, IN-KY | 7,733 | 146,005 | 27,271 | 23,796 | 16,631 | 5,450 | 5,225 | 6,808 | 46,630 | 1,564 | 45.7 | 11.1 | 43.6 |
| Fairbanks, AK | 2,515 | 27,029 | 6,347 | 513 | 4,568 | 629 | 1,260 | 1,534 | 56,745 | 274 | 49.6 | 8.0 | 45.0 |
| Fargo, ND-MN | 6,949 | 126,151 | 25,153 | 10,139 | 15,499 | 9,837 | 6,498 | 6,227 | 49,358 | 1,478 | 20.1 | 36.1 | 58.8 |
| Farmington, NM | 2,527 | 36,497 | 7,634 | 1,586 | 5,752 | 852 | 1,213 | 1,622 | 44,447 | 2,965 | 56.5 | 21.0 | 54.9 |
| Fayetteville, NC | 7,918 | 122,806 | 24,059 | 10,375 | 22,656 | 2,421 | 6,665 | 4,457 | 36,293 | 1,168 | 51.5 | 4.5 | 38.3 |
| Fayetteville-Springdale-Rogers, AR | 12,758 | 219,148 | 26,787 | 27,153 | 25,416 | 5,760 | 13,608 | 11,886 | 54,237 | 5,444 | 39.8 | 1.7 | 39.8 |
| Flagstaff, AZ | 3,744 | 55,692 | 9,660 | 4,891 | 8,006 | 925 | 1,800 | 2,292 | 41,163 | 2,142 | 77.8 | 11.6 | 55.7 |
| Flint, MI | 7,614 | 120,486 | 26,327 | 12,292 | 18,844 | 4,005 | 4,566 | 5,371 | 44,580 | 820 | 59.4 | 3.4 | 46.2 |
| Florence, SC | 4,225 | 84,730 | 26,971 | 9,861 | 11,061 | 2,906 | 2,338 | 3,789 | 44,723 | 862 | 43.6 | 8.9 | 42.1 |
| Florence-Muscle Shoals, AL | 3,197 | 50,813 | 8,245 | 10,030 | 7,956 | 1,669 | 1,022 | 1,831 | 36,027 | 1,900 | 45.6 | 4.9 | 36.6 |
| Fond du Lac, WI | 2,284 | 43,834 | 6,655 | 9,512 | 5,820 | 1,641 | 1,359 | 2,029 | 46,281 | 1,244 | 35.9 | 4.8 | 53.4 |
| Fort Collins, CO | 11,236 | 133,269 | 20,253 | 11,755 | 19,749 | 3,647 | 9,802 | 6,446 | 48,369 | 2,043 | 71 | 4.4 | 31.6 |
| Fort Smith, AR-OK | 5,337 | 87,292 | 15,231 | 18,561 | 11,684 | 2,517 | 2,174 | 3,344 | 38,311 | 3,462 | 38.7 | 3.2 | 37.2 |
| Fort Wayne, IN | 10,000 | 197,575 | 38,602 | 33,751 | 24,456 | 9,260 | 6,454 | 9,044 | 45,773 | 2,244 | 55.1 | 5.5 | 32.6 |
| Fresno, CA | 17,351 | 277,632 | 49,427 | 26,580 | 38,020 | 9,477 | 10,933 | 12,531 | 45,135 | 4,774 | 54.1 | 6.6 | 57.1 |
| Gadsden, AL | 1,917 | 30,302 | 7,153 | 5,544 | 4,822 | 1,013 | 634 | 1,069 | 35,286 | 817 | 55.2 | 0.5 | 38.2 |
| Gainesville, FL | 7,305 | 108,278 | 29,883 | 4,641 | 15,820 | 4,283 | 6,580 | 4,705 | 43,456 | 3,234 | 69.4 | 2.9 | 41.3 |
| Gainesville, GA | 4,569 | 83,924 | 14,580 | 22,554 | 9,269 | 2,022 | 2,080 | 3,961 | 47,197 | 551 | 63.5 | 0.4 | 38.1 |
| Gettysburg, PA | 1,957 | 29,542 | 5,055 | 6,528 | 3,553 | 465 | 617 | 1,079 | 36,519 | 1,146 | 50.2 | 2.2 | 43.4 |
| Glens Falls, NY | 3,310 | 42,175 | 8,465 | 6,446 | 7,623 | 1,341 | 1,233 | 1,807 | 42,854 | 995 | 38.9 | 2.6 | 43.4 |
| Goldsboro, NC | 2,148 | 33,298 | 7,049 | 5,053 | 5,845 | 1,114 | 681 | 1,207 | 36,259 | 551 | 39.7 | 8.5 | 55.4 |
| Grand Forks, ND-MN | 2,633 | 44,108 | 10,020 | 4,799 | 7,229 | 1,202 | 1,640 | 1,861 | 42,194 | 2,147 | 14.3 | 25.0 | 53.5 |
| Grand Island, NE | 2,305 | 34,674 | 5,119 | 7,685 | 5,268 | 1,467 | 680 | 1,378 | 39,741 | 1,682 | 27.5 | 15.9 | 52.1 |
| Grand Junction, CO | 4,560 | 54,402 | 11,576 | 2,995 | 8,133 | 1,813 | 2,452 | 2,429 | 44,654 | 2,465 | 80.1 | 2.6 | 32.1 |
| Grand Rapids-Kentwood, MI | 25,153 | 513,320 | 69,422 | 116,168 | 51,472 | 19,091 | 21,719 | 24,839 | 48,390 | 4,056 | 50.4 | 4.1 | 45.3 |
| Grants Pass, OR | 2,048 | 24,136 | 5,361 | 2,983 | 4,454 | 781 | 1,004 | 904 | 37,447 | 746 | 85 | 0.3 | 43.6 |
| Great Falls, MT | 2,447 | 30,486 | 7,342 | 1,249 | 4,806 | 1,460 | 1,125 | 1,253 | 41,107 | 1,027 | 37.7 | 22.6 | 42.1 |
| Greeley, CO | 6,602 | 91,965 | 9,195 | 14,153 | 10,491 | 3,027 | 2,950 | 5,013 | 54,511 | 4,062 | 46.3 | 10.2 | 38.1 |
| Green Bay, WI | 7,867 | 159,138 | 22,888 | 32,353 | 17,788 | 12,447 | 5,629 | 7,902 | 49,656 | 2,464 | 42.4 | 4.3 | 46.7 |
| Greensboro-High Point, NC | 17,768 | 323,523 | 41,100 | 52,137 | 36,390 | 12,447 | 12,733 | 14,673 | 45,355 | 3,066 | 46.7 | 1.1 | 40.7 |
| Greenville, NC | 3,659 | 61,820 | 16,635 | 6,556 | 9,361 | 1,727 | 2,244 | 2,532 | 40,964 | 478 | 35.8 | 12.6 | 55.0 |
| Greenville-Anderson, SC | 20,667 | 354,242 | 46,647 | 58,789 | 44,384 | 12,837 | 23,325 | 15,739 | 44,431 | 4,358 | 59.5 | 1.1 | 31.5 |
| Gulfport-Biloxi, MS | 7,506 | 132,660 | 22,543 | 19,796 | 18,959 | 3,924 | 5,110 | 5,639 | 42,505 | 1,405 | 62.1 | 0.7 | 35.7 |
| Hagerstown-Martinsburg, MD-WV | 5,332 | 89,395 | 16,729 | 10,112 | 14,070 | 5,623 | 2,799 | 3,814 | 42,663 | 2,030 | 58.2 | 0.7 | 36.4 |
| Hammond, LA | 2,492 | 37,041 | 8,959 | 2,528 | 6,785 | 1,694 | 1,062 | 1,377 | 37,163 | 967 | 55 | 1.1 | 37.4 |
| Hanford-Corcoran, CA | 1,682 | 25,541 | 5,008 | 4,337 | 4,333 | 492 | 896 | 1,066 | 41,730 | 963 | 48.9 | 9.2 | 54.1 |
| Harrisburg-Carlisle, PA | 13,784 | 283,191 | 53,469 | 18,198 | 36,459 | 17,129 | 17,498 | 14,240 | 50,283 | 2,711 | 45.1 | 1.4 | 48.6 |
| Harrisonburg, VA | 3,171 | 54,897 | 7,118 | 10,689 | 7,179 | 1,340 | 1,671 | 2,195 | 39,990 | 2,026 | 46.2 | 0.6 | 49.3 |
| Hartford-East Hartford-Middletown, CT | 29,321 | 552,666 | 107,768 | 66,309 | 61,350 | 55,632 | 35,032 | 35,202 | 63,695 | 1,747 | 72.9 | 0.5 | 38.6 |
| Hattiesburg, MS | 3,625 | 58,388 | 12,834 | 5,694 | 10,649 | 1,841 | 3,127 | 2,223 | 38,069 | 1,696 | 44.5 | 1.7 | 38.3 |
| Hickory-Lenoir-Morganton, NC | 7,592 | 138,326 | 22,388 | 42,017 | 16,944 | 2,651 | 3,311 | 5,579 | 40,332 | 2,101 | 54 | 0.8 | 40.1 |
| Hilton Head Island-Bluffton, SC | 6,198 | 66,180 | 9,367 | 867 | 12,641 | 1,850 | 3,076 | 2,460 | 37,170 | 296 | 62.2 | 10.5 | 35.4 |
| Hinesville, GA | 934 | 14,191 | 2,136 | 2,607 | 2,326 | 360 | 409 | 611 | 43,028 | 154 | 53.2 | 0.6 | 26.3 |
| Homosassa Springs, FL | 2,772 | 30,318 | 10,256 | 343 | 5,617 | 676 | 821 | 1,128 | 37,211 | 609 | 75.2 | 2.1 | 43.4 |
| Hot Springs, AR | 2,765 | 35,253 | 8,677 | 2,854 | 5,925 | 1,013 | 1,084 | 1,229 | 34,854 | 357 | 47.6 | 0.3 | 38.5 |
| Houma-Thibodaux, LA | 4,413 | 71,962 | 12,200 | 5,293 | 10,385 | 1,899 | 3,329 | 3,675 | 51,074 | 592 | 54.1 | 7.4 | 33.2 |
| Houston-The Woodlands-Sugar Land, TX | 146,470 | 2,704,793 | 354,948 | 205,824 | 307,420 | 99,967 | 213,524 | 173,363 | 64,095 | 14,238 | 63.5 | 2.9 | 32.2 |
| Huntington-Ashland, WV-KY-OH | 6,915 | 114,029 | 29,852 | 11,496 | 17,209 | 2,842 | 3,892 | 4,815 | 42,230 | 3,194 | 32.7 | 0.4 | 34.9 |
| Huntsville, AL | 9,962 | 191,827 | 28,974 | 21,264 | 22,920 | 4,225 | 44,381 | 10,689 | 55,721 | 2,177 | 54.2 | 4.0 | 35.9 |
| Idaho Falls, ID | 4,296 | 58,775 | 10,337 | 4,292 | 8,296 | 1,507 | 9,231 | 2,821 | 47,995 | 2,048 | 60.5 | 10.7 | 37.5 |
| Indianapolis-Carmel-Anderson, IN | 48,051 | 927,409 | 146,075 | 79,018 | 105,554 | 51,161 | 65,643 | 47,915 | 51,665 | 5,999 | 55.1 | 7.4 | 37.4 |
| Iowa City, IA | 4,007 | 72,896 | 20,158 | 5,149 | 10,923 | 2,991 | 2,957 | 3,210 | 44,039 | 2,386 | 34.7 | 4.9 | 46.3 |
| Ithaca, NY | 2,348 | 50,689 | 6,363 | 2,666 | 4,794 | 1,201 | 2,552 | 2,193 | 43,255 | 523 | 47.6 | 2.9 | 39.6 |
| Jackson, MI | 2,966 | 51,345 | 10,320 | 9,168 | 6,941 | 1,544 | 2,934 | 2,562 | 49,894 | 923 | 52.1 | 3.9 | 41.5 |
| Jackson, MS | 13,522 | 225,471 | 50,973 | 19,348 | 28,885 | 11,311 | 9,835 | 9,999 | 44,347 | 4,019 | 28.4 | 6.9 | 36.7 |
| Jackson, TN | 3,906 | 70,482 | 15,725 | 12,827 | 9,622 | 1,764 | 1,472 | 2,753 | 39,058 | 2,028 | 40.9 | 8.4 | 35.3 |
| Jacksonville, FL | 39,420 | 595,366 | 87,974 | 28,125 | 77,861 | 45,851 | 35,775 | 28,972 | 48,662 | 1,681 | 75.1 | 2.0 | 40.6 |
| Jacksonville, NC | 2,904 | 37,064 | 5,389 | 1,095 | 8,891 | 1,025 | 1,676 | 1,106 | 29,844 | 340 | 50.3 | 3.5 | 55.9 |
| Janesville-Beloit, WI | 3,360 | 62,793 | 9,191 | 12,530 | 9,514 | 1,416 | 2,397 | 2,905 | 46,261 | 1,587 | 53.7 | 4.7 | 40.6 |
| Jefferson City, MO | 3,528 | 55,767 | 9,580 | 7,933 | 7,328 | 2,381 | 1,936 | 2,345 | 42,043 | 5,019 | 26.2 | 2.8 | 33.9 |
| Johnson City, TN | 3,862 | 67,673 | 15,918 | 8,042 | 10,690 | 3,131 | 1,495 | 2,706 | 39,983 | 1,997 | 61.9 | 0.4 | 35.9 |
| Johnstown, PA | 3,049 | 46,181 | 11,893 | 5,056 | 6,375 | 1,889 | 2,109 | 1,725 | 37,353 | 557 | 37 | 2.3 | 36.1 |

Items 104—116

# Table C. Metropolitan Areas — **Agriculture**

| Area name | Acreage (1,000) | Percent change, 2012–2017 | Average size of farm | Total irrigated (1,000) | Total cropland (1,000) | Value of land and buildings, Average per farm | Average per acre | Value of machinery and equipment, average per farm (dollars) | Total (mil dol) | Average per farm (acres) | Crops | Livestock and poultry products | Organic farms (number) | Farms with internet access (percent) | Government payments Total ($1,000) | Percent of farms |
|---|---|---|---|---|---|---|---|---|---|---|---|---|---|---|---|---|
| | 117 | 118 | 119 | 120 | 121 | 122 | 123 | 124 | 125 | 126 | 127 | 128 | 129 | 130 | 131 | 132 |
| Evansville, IN-KY | 544 | -6.4 | 348 | 23.5 | 479.4 | 2,029,390 | 5,839 | 213,324 | 304 | 194,577 | 92.2 | 7.8 | 2 | 77.8 | 16,227 | 57.6 |
| Fairbanks, AK | 102 | 2.4 | 372 | 1.1 | 63.2 | 610,606 | 1,640 | 0 | 10 | 37,927 | 81.6 | 18.4 | 5 | 86.9 | 1,124 | 21.9 |
| Fargo, ND-MN | 1,703 | -0.9 | 1,152 | 16.8 | 1,623.6 | 4,372,599 | 3,795 | 467,605 | 717 | 485,254 | 91.7 | 8.3 | 19 | 84.9 | 14,177 | 72.3 |
| Farmington, NM | 2,551 | -1.1 | 861 | 73.6 | 107.9 | 345,734 | 402 | 39,306 | 74 | 24,998 | 91.5 | 8.5 | 1 | 44.9 | 1,383 | 8.3 |
| Fayetteville, NC | 226 | -13.3 | 193 | 6.2 | 138.1 | 1,001,288 | 5,177 | 127,616 | 377 | 322,909 | 29.1 | 70.9 | 5 | 78.0 | 1,230 | 26.6 |
| Fayetteville-Springdale-Rogers, AR | 838 | -5.4 | 154 | 1.2 | 218.7 | 748,454 | 4,861 | 75,575 | 1,382 | 253,852 | 1.5 | 98.5 | 18 | 76.2 | 1,982 | 6.5 |
| Flagstaff, AZ | 6,139 | 5.6 | 2,866 | 1.3 | 6.9 | 607,822 | 212 | 34,544 | 24 | 11,162 | 4.1 | 95.9 | 2 | 43.9 | 807 | 0.8 |
| Flint, MI | 124 | 0.5 | 151 | 1.5 | 104.3 | 795,230 | 5,262 | 103,909 | 70 | 85,871 | 86.5 | 13.5 | 11 | 84.8 | 2,561 | 25.0 |
| Florence, SC | 290 | -12.9 | 336 | 7.4 | 200.6 | 969,775 | 2,883 | 127,777 | 134 | 154,886 | 69.8 | 30.2 | 8 | 70.3 | 2,730 | 33.2 |
| Florence-Muscle Shoals, AL | 361 | -1.0 | 190 | 0.9 | 194.8 | 612,645 | 3,226 | 91,579 | 133 | 69,754 | 58.9 | 41.1 | 0 | 70.7 | 7,434 | 40.9 |
| Fond du Lac, WI | 317 | 0.6 | 255 | 1.9 | 270.4 | 1,855,960 | 7,275 | 227,075 | 397 | 318,921 | 24.0 | 76.0 | 33 | 77.7 | 3,762 | 56.5 |
| Fort Collins, CO | 482 | 7.1 | 236 | 60.2 | 98.6 | 1,095,047 | 4,637 | 76,905 | 151 | 73,772 | 45.7 | 54.3 | 20 | 85.1 | 633 | 5.5 |
| Fort Smith, AR-OK | 623 | 0.7 | 180 | 11.3 | 205.6 | 478,884 | 2,659 | 72,249 | 411 | 118,784 | 9.4 | 90.6 | 9 | 73.6 | 1,456 | 4.6 |
| Fort Wayne, IN | 458 | 11.4 | 204 | 2.0 | 403.6 | 1,496,008 | 7,332 | 136,852 | 297 | 132,251 | 68.7 | 31.3 | 19 | 69.6 | 8,492 | 48.4 |
| Fresno, CA | 1,647 | -4.3 | 345 | 972.6 | 1,142.7 | 3,917,829 | 11,359 | 234,782 | 5,743 | 1,202,926 | 71.1 | 28.9 | 183 | 75.6 | 8,894 | 7.7 |
| Gadsden, AL | 89 | 3.7 | 109 | 0.6 | 24.8 | 419,479 | 3,836 | 78,982 | 93 | 114,356 | 6.7 | 93.3 | 0 | 73.8 | 804 | 21.5 |
| Gainesville, FL | 448 | 2.0 | 138 | 42.1 | 177.0 | 833,888 | 6,021 | 72,751 | 321 | 99,151 | 48.7 | 51.3 | 25 | 79.7 | 3,694 | 6.6 |
| Gainesville, GA | 41 | -21.7 | 74 | 0.1 | 13.8 | 689,172 | 9,332 | 79,541 | 129 | 233,156 | 2.3 | 97.7 | 1 | 80.9 | 789 | 20.3 |
| Gettysburg, PA | 166 | -3.0 | 145 | 2.2 | 128.4 | 998,832 | 6,886 | 132,547 | 208 | 181,122 | 54.2 | 45.8 | 17 | 77.7 | 1,503 | 18.4 |
| Glens Falls, NY | 195 | -1.8 | 196 | 1.0 | 103.9 | 562,560 | 2,865 | 127,784 | 136 | 136,494 | 18.0 | 82.0 | 19 | 84.0 | 1,418 | 17.4 |
| Goldsboro, NC | 165 | -13.5 | 300 | 4.8 | 130.5 | 1,497,289 | 4,990 | 222,201 | 592 | 1,074,539 | 18.1 | 81.9 | 13 | 79.9 | 3,611 | 46.1 |
| Grand Forks, ND-MN | 1,822 | -4.7 | 848 | 33.1 | 1,682.6 | 2,831,795 | 3,338 | 378,326 | 748 | 348,455 | 93.3 | 6.7 | 19 | 75.5 | 32,024 | 81.8 |
| Grand Island, NE | 852 | -2.9 | 506 | 510.2 | 651.5 | 2,306,951 | 4,556 | 281,613 | 778 | 462,492 | 47.7 | 52.3 | 6 | 81.4 | 29,470 | 66.7 |
| Grand Junction, CO | 343 | -11.5 | 139 | 76.2 | 79.8 | 767,492 | 5,523 | 57,249 | 94 | 38,209 | 48.7 | 51.3 | 12 | 86.2 | 730 | 3.9 |
| Grand Rapids-Kentwood, MI | 794 | -4.3 | 196 | 102.9 | 662.0 | 1,149,541 | 5,874 | 169,142 | 1,333 | 328,622 | 52.3 | 47.7 | 42 | 81.0 | 12,659 | 27.3 |
| Grants Pass, OR | 28 | -1.4 | 37 | 8.0 | 8.4 | 672,056 | 17,992 | 38,200 | 18 | 23,456 | 49.2 | 50.8 | 27 | 85.1 | 1 | 1.1 |
| Great Falls, MT | 1,270 | 1.2 | 1,237 | 35.7 | 420.0 | 1,499,360 | 1,212 | 134,128 | 107 | 104,453 | 47.7 | 52.3 | 9 | 80.2 | 7,780 | 42.7 |
| Greeley, CO | 2,099 | 7.3 | 517 | 323.4 | 923.0 | 1,335,953 | 2,586 | 159,663 | 2,047 | 503,982 | 17.0 | 83.0 | 42 | 82.9 | 19,375 | 26.0 |
| Green Bay, WI | 552 | 0.9 | 224 | 2.4 | 455.7 | 1,371,723 | 6,120 | 200,427 | 753 | 305,556 | 18.1 | 81.9 | 35 | 75.8 | 6,395 | 47.0 |
| Greensboro-High Point, NC | 349 | -3.1 | 114 | 5.4 | 153.8 | 604,582 | 5,315 | 73,041 | 373 | 121,694 | 24.2 | 75.8 | 40 | 74.5 | 1,271 | 11.5 |
| Greenville, NC | 186 | 8.5 | 390 | 2.8 | 149.8 | 1,499,479 | 3,845 | 183,107 | 243 | 507,234 | 41.6 | 58.4 | 4 | 81.8 | 3,291 | 45.6 |
| Greenville-Anderson, SC | 405 | 1.3 | 93 | 2.7 | 140.1 | 535,642 | 5,767 | 53,720 | 166 | 38,095 | 19.9 | 80.1 | 14 | 74.2 | 3,337 | 12.2 |
| Gulfport-Biloxi, MS | 130 | -1.7 | 93 | 0.7 | 30.2 | 421,243 | 4,539 | 60,528 | 28 | 19,864 | 55.3 | 44.7 | 6 | 75.4 | 610 | 7.1 |
| Hagerstown-Martinsburg, MD-WV | 209 | -4.1 | 103 | 0.7 | 127.0 | 709,469 | 6,883 | 78,595 | 183 | 90,104 | 32.4 | 67.6 | 16 | 69.0 | 2,000 | 12.5 |
| Hammond, LA | 98 | -8.1 | 101 | 1.2 | 29.6 | 513,368 | 5,061 | 63,007 | 42 | 43,845 | 47.3 | 52.7 | 5 | 67.6 | 208 | 5.0 |
| Hanford-Corcoran, CA | 616 | -8.6 | 640 | 371.7 | 488.5 | 6,917,469 | 10,815 | 349,654 | 1,649 | 1,712,640 | 50.1 | 49.9 | 25 | 80.4 | 5,849 | 29.8 |
| Harrisburg-Carlisle, PA | 366 | -12.8 | 135 | 2.1 | 281.4 | 1,020,473 | 7,566 | 123,077 | 485 | 178,904 | 25.1 | 74.9 | 98 | 65.0 | 4,162 | 22.0 |
| Harrisonburg, VA | 229 | 2.9 | 113 | 5.5 | 121.9 | 997,634 | 8,844 | 118,949 | 796 | 392,852 | 6.8 | 93.2 | 37 | 65.5 | 1,852 | 11.5 |
| Hartford-East Hartford-Middletown, CT | 100 | -20.6 | 57 | 4.3 | 48.9 | 779,882 | 13,635 | 64,238 | 204 | 117,004 | 87.0 | 13.0 | 22 | 81.2 | 731 | 3.9 |
| Hattiesburg, MS | 255 | -0.6 | 150 | 1.2 | 49.3 | 546,554 | 3,641 | 69,415 | 353 | 208,227 | 4.0 | 96.0 | 0 | 71.0 | 1,954 | 22.0 |
| Hickory-Lenoir-Morganton, NC | 194 | 1.1 | 92 | 3.5 | 88.8 | 529,900 | 5,730 | 83,055 | 383 | 182,363 | 15.9 | 84.1 | 7 | 74.7 | 1,109 | 6.3 |
| Hilton Head Island-Bluffton, SC | 119 | 7.5 | 402 | 2.0 | 19.0 | 1,465,669 | 3,643 | 78,267 | 20 | 68,537 | 94.7 | 5.3 | 3 | 76.0 | 203 | 12.5 |
| Hinesville, GA | 17 | 0.4 | 107 | 0.2 | 4.0 | 441,065 | 4,103 | 55,760 | 8 | 50,331 | 12.1 | 87.9 | 0 | 81.2 | 82 | 10.4 |
| Homosassa Springs, FL | 56 | 37.5 | 92 | 0.9 | 9.6 | 726,031 | 7,929 | 49,552 | 14 | 22,118 | 67.3 | 32.7 | 0 | 81.0 | 197 | 2.6 |
| Hot Springs, AR | 34 | -4.9 | 95 | 0.1 | 8.2 | 496,952 | 5,226 | 46,232 | 10 | 27,972 | 30.5 | 69.5 | 0 | 74.5 | 93 | 5.9 |
| Houma-Thibodaux, LA | 236 | -6.1 | 398 | 0.3 | 94.4 | 1,316,841 | 3,306 | 143,334 | 69 | 116,730 | 62.4 | 37.6 | 0 | 73.6 | 255 | 3.4 |
| Houston-The Woodlands-Sugar Land, TX | 2,218 | -17.2 | 156 | 80.6 | 680.3 | 749,277 | 4,810 | 67,839 | 435 | 30,543 | 62.4 | 37.6 | 24 | 74.4 | 30,126 | 5.9 |
| Huntington-Ashland, WV-KY-OH | 378 | -6.4 | 118 | 0.4 | 88.8 | 306,815 | 2,594 | 50,542 | 30 | 9,260 | 52.9 | 47.1 | 7 | 69.1 | 719 | 6.0 |
| Huntsville, AL | 410 | -10.2 | 188 | 21.1 | 282.6 | 877,456 | 4,662 | 110,397 | 202 | 92,678 | 71.0 | 29.0 | 2 | 76.4 | 6,048 | 30.9 |
| Idaho Falls, ID | 883 | 3.0 | 431 | 399.3 | 567.5 | 1,300,466 | 3,017 | 187,303 | 505 | 246,386 | 63.3 | 36.7 | 14 | 85.2 | 13,659 | 25.9 |
| Indianapolis-Carmel-Anderson, IN | 1,603 | -5.1 | 267 | 10.0 | 1,440.5 | 1,963,457 | 7,349 | 158,827 | 944 | 157,420 | 89.0 | 11.0 | 23 | 82.3 | 39,946 | 42.9 |
| Iowa City, IA | 615 | -4.4 | 258 | 1.2 | 528.6 | 1,921,071 | 7,457 | 191,325 | 892 | 373,645 | 31.5 | 68.5 | 132 | 75.6 | 20,966 | 68.6 |
| Ithaca, NY | 91 | 0.6 | 175 | 0.4 | 62.1 | 661,191 | 3,789 | 138,310 | 65 | 123,715 | 24.6 | 75.4 | 45 | 85.9 | 1,575 | 22.6 |
| Jackson, MI | 160 | -12.4 | 174 | 4.7 | 125.8 | 866,522 | 4,986 | 116,065 | 71 | 76,714 | 61.6 | 38.4 | 6 | 78.2 | 3,526 | 25.9 |
| Jackson, MS | 1,258 | -9.7 | 313 | 122.9 | 472.0 | 879,420 | 2,809 | 93,282 | 683 | 169,823 | 29.7 | 70.3 | 1 | 64.2 | 24,699 | 44.0 |
| Jackson, TN | 668 | 3.8 | 329 | 27.1 | 533.0 | 1,092,897 | 3,319 | 162,797 | 284 | 140,171 | 94.8 | 5.2 | 0 | 70.8 | 13,566 | 50.7 |
| Jacksonville, FL | 152 | -18.0 | 91 | 17.4 | 36.3 | 599,377 | 6,617 | 58,273 | 102 | 60,642 | 69.5 | 30.5 | 6 | 75.4 | 466 | 5.1 |
| Jacksonville, NC | 52 | -9.0 | 154 | 0.4 | 35.6 | 969,753 | 6,284 | 93,974 | 172 | 504,629 | 13.7 | 86.3 | 0 | 75.0 | 862 | 25.6 |
| Janesville-Beloit, WI | 354 | -0.1 | 223 | 15.6 | 302.8 | 1,467,761 | 6,589 | 175,570 | 295 | 185,539 | 56.0 | 44.0 | 20 | 79.7 | 4,915 | 54.7 |
| Jefferson City, MO | 1,029 | 1.8 | 205 | 10.8 | 443.2 | 684,076 | 3,337 | 83,549 | 387 | 77,050 | 31.2 | 68.8 | 17 | 73.6 | 7,408 | 26.5 |
| Johnson City, TN | 146 | -7.0 | 73 | 0.8 | 68.4 | 528,895 | 7,215 | 69,260 | 50 | 25,140 | 39.2 | 60.8 | 7 | 72.5 | 416 | 5.8 |
| Johnstown, PA | 79 | 3.2 | 142 | 0.1 | 50.5 | 659,661 | 4,631 | 100,086 | 30 | 53,982 | 60.1 | 39.9 | 2 | 68.8 | 979 | 30.3 |

Items 117—132

# Table C. Metropolitan Areas — Water Use, Wholesale Trade, Retail Trade, and Real Estate

| Area name | Water use, 2015 | | Wholesale Trade[1], 2017 | | | | Retail Trade[2], 2017 | | | | Real estate and rental and leasing,[2] 2017 | | | |
|---|---|---|---|---|---|---|---|---|---|---|---|---|---|---|
| | Public supply water withdrawn (mil gal/day) | Public supply gallons withdrawn per person per day | Number of establishments | Number of employees | Sales (mil dol) | Average payroll (mil dol) | Number of establishments | Number of employees | Sales (mil dol) | Average payroll (mil dol) | Number of establishments | Number of employees | Sales (mil dol) | Average payroll (mil dol) |
| | 133 | 134 | 135 | 136 | 137 | 138 | 139 | 140 | 141 | 142 | 143 | 144 | 145 | 146 |
| Evansville, IN-KY | 29.78 | 94.3 | 343 | 4,794 | 3,531.5 | 257.4 | 1,096 | 17,298 | 5,002.5 | 438.1 | 337 | 1,695 | 365.6 | 61.1 |
| Fairbanks, AK | 14.12 | 141.7 | 70 | 693 | 470.6 | 38.4 | 294 | 4,749 | 1,484.8 | 159.0 | 158 | 764 | 188.3 | 42.1 |
| Fargo, ND-MN | 19.03 | 81.4 | 417 | 7,514 | 7,293.4 | 448.1 | 846 | 16,144 | 9,241.6 | 484.2 | 375 | 2,004 | 341.5 | 75.8 |
| Farmington, NM | 18.83 | 158.6 | 129 | 1,053 | 502.9 | 57.5 | 413 | 5,940 | 1,744.6 | 163.8 | 100 | 560 | 141.0 | 30.8 |
| Fayetteville, NC | 60.84 | 120.6 | 158 | 2,223 | 999.4 | 82.8 | 1,445 | 21,586 | 5,894.6 | 525.1 | 406 | 2,070 | 461.2 | 73.3 |
| Fayetteville-Springdale-Rogers, AR | 64.93 | 132.3 | 482 | 5,529 | 4,789.3 | 325.8 | 1,574 | 25,929 | 7,401.1 | 691.9 | 622 | 2,374 | 622.0 | 93.6 |
| Flagstaff, AZ | 19.87 | 142.8 | 96 | 832 | 365.3 | 37.9 | 580 | 8,004 | 2,197.1 | 207.2 | 224 | 884 | 201.3 | 30.5 |
| Flint, MI | 4.22 | 10.3 | 269 | 4,989 | 3,403.5 | 270.6 | 1,355 | 19,591 | 8,429.7 | 497.7 | 328 | 2,107 | 338.1 | 73.5 |
| Florence, SC | 22.11 | 107.1 | 218 | 3,133 | 3,148.6 | 151.5 | 887 | 11,864 | 3,063.7 | 269.0 | 161 | 637 | 136.9 | 24.6 |
| Florence-Muscle Shoals, AL | 20.65 | 140.5 | 147 | 1,877 | 1,075.9 | 89.6 | 629 | 8,289 | 2,121.0 | 207.6 | 143 | 528 | 96.2 | 17.4 |
| Fond du Lac, WI | 7.93 | 77.8 | D | D | D | D | 332 | 6,014 | 1,719.9 | 156.8 | 62 | 344 | 52.3 | 13.7 |
| Fort Collins, CO | 30.58 | 91.7 | 336 | 4,724 | 3,723.5 | 349.3 | 1,266 | 19,902 | 5,893.8 | 574.0 | 766 | 2,404 | 663.9 | 109.9 |
| Fort Smith, AR-OK | 36.45 | 146.8 | 264 | 3,769 | 2,238.4 | 189.1 | 904 | 12,227 | 3,297.9 | 300.9 | 225 | 1,404 | 227.6 | 53.9 |
| Fort Wayne, IN | 37.08 | 92.3 | 497 | 8,411 | 8,761.7 | 445.3 | 1,373 | 24,329 | 7,369.0 | 671.8 | 464 | 2,302 | 569.1 | 94.5 |
| Fresno, CA | 157.08 | 161.1 | 827 | 14,164 | 10,890.9 | 780.4 | 2,438 | 38,243 | 11,347.0 | 1,066.6 | 852 | 4,904 | 1,097.9 | 207.2 |
| Gadsden, AL | 16.85 | 163.5 | 70 | 758 | 497.3 | 32.4 | 393 | 4,777 | 1,319.8 | 114.3 | 61 | 232 | 64.4 | 7.8 |
| Gainesville, FL | 25.08 | 79.1 | 230 | 2,191 | 1,827.9 | 122.5 | 1,096 | 15,974 | 4,200.4 | 376.8 | 399 | 2,096 | 357.5 | 75.6 |
| Gainesville, GA | 88.18 | 455.6 | 261 | 4,205 | 7,901.5 | 236.0 | 619 | 9,240 | 2,948.9 | 268.3 | 220 | 549 | 180.7 | 25.3 |
| Gettysburg, PA | 12.39 | 121.1 | D | D | D | D | 337 | 3,599 | 979.0 | 87.0 | 48 | 190 | 34.3 | 6.2 |
| Glens Falls, NY | 14.15 | 111.5 | 79 | 776 | 402.8 | 36.4 | 585 | 7,627 | 2,194.3 | 205.0 | 98 | 304 | 60.7 | 12.5 |
| Goldsboro, NC | 11.27 | 90.8 | 98 | 1,920 | 1,385.3 | 75.9 | 461 | 6,214 | 1,619.3 | 144.9 | 67 | 335 | 53.7 | 10.5 |
| Grand Forks, ND-MN | 17.35 | 169.4 | 134 | 1,686 | 2,146.5 | 93.1 | 422 | 7,560 | 1,922.4 | 189.7 | 99 | 633 | 96.4 | 22.0 |
| Grand Island, NE | 13.3 | 175.3 | 124 | 1,468 | 1,064.1 | 82.7 | 357 | 5,519 | 1,394.8 | 139.7 | 75 | 313 | 58.1 | 10.5 |
| Grand Junction, CO | 17.51 | 117.9 | 200 | 1,731 | 942.8 | 90.2 | 572 | 8,305 | 2,463.2 | 237.0 | 303 | 936 | 208.8 | 39.0 |
| Grand Rapids-Kentwood, MI | 105.29 | 100.9 | 1,347 | 29,171 | 22,290.1 | 1,708.2 | 3,334 | 51,077 | 15,659.9 | 1,404.9 | 968 | 5,609 | 1,452.3 | 245.2 |
| Grants Pass, OR | 7.18 | 84.7 | 56 | 374 | 294.5 | 19.7 | 323 | 4,406 | 1,208.1 | 127.9 | 107 | 431 | 63.9 | 16.7 |
| Great Falls, MT | 12.79 | 155.4 | 116 | 1,152 | 722.0 | 54.0 | 362 | 5,326 | 1,439.0 | 139.0 | 123 | 412 | 82.2 | 13.8 |
| Greeley, CO | 28.30 | 99.2 | 274 | 3,365 | 3,733.5 | 201.6 | 668 | 10,289 | 3,768.8 | 327.5 | 314 | 1,616 | 358.6 | 79.6 |
| Green Bay, WI | 21.42 | 67.7 | 363 | 7,728 | 5,562.2 | 423.7 | 1,012 | 17,154 | 4,824.6 | 445.4 | 245 | 1,493 | 393.8 | 56.0 |
| Greensboro-High Point, NC | 76.29 | 101.4 | 1,156 | 18,749 | 17,580.4 | 1,345.0 | 2,540 | 36,436 | 10,177.9 | 943.8 | 857 | 5,321 | 942.0 | 251.9 |
| Greenville, NC | 13.64 | 77.6 | 131 | 1,514 | 1,021.3 | 73.6 | 625 | 9,154 | 2,556.1 | 229.1 | 184 | 717 | 170.0 | 25.7 |
| Greenville-Anderson, SC | 102.79 | 117.5 | 1,043 | 15,072 | 19,442.8 | 946.9 | 3,002 | 44,653 | 11,970.0 | 1,108.6 | 1,003 | 4,123 | 1,124.8 | 170.5 |
| Gulfport-Biloxi, MS | 41.19 | 101.1 | 230 | 1,786 | 1,017.3 | 87.4 | 1,416 | 19,174 | 5,019.8 | 462.4 | 364 | 1,537 | 339.1 | 52.6 |
| Hagerstown-Martinsburg, MD-WV | 23.79 | 85.3 | 172 | 2,772 | 3,118.1 | 142.6 | 886 | 14,706 | 3,843.0 | 346.6 | 237 | 1,208 | 273.6 | 44.3 |
| Hammond, LA | 15.06 | 117.0 | 89 | 2,288 | 1,747.3 | 111.0 | 454 | 6,671 | 1,984.3 | 168.0 | 113 | 541 | 86.5 | 20.8 |
| Hanford-Corcoran, CA | 29.07 | 192.6 | 61 | 704 | 969.0 | 42.0 | 292 | 4,155 | 1,272.1 | 117.0 | 112 | 436 | 95.6 | 14.5 |
| Harrisburg-Carlisle, PA | 45.54 | 80.6 | 503 | 9,307 | 9,849.3 | 533.3 | 1,910 | 36,367 | 10,960.8 | 950.7 | 517 | 2,939 | 738.8 | 143.3 |
| Harrisonburg, VA | 14.75 | 112.5 | 114 | 1,871 | 1,504.5 | 106.9 | 546 | 7,127 | 1,833.8 | 178.3 | 112 | 892 | 215.6 | 38.3 |
| Hartford-East Hartford-Middletown, CT | 71.46 | 59.0 | 1,144 | 22,088 | 20,949.8 | 1,464.5 | 4,130 | 63,828 | 18,188.9 | 1,786.0 | 1,124 | 6,434 | 1,750.6 | 313.3 |
| Hattiesburg, MS | 21.11 | 125.4 | 88 | 913 | 755.2 | 41.9 | 762 | 10,006 | 2,698.8 | 241.5 | 99 | 474 | 87.4 | 17.6 |
| Hickory-Lenoir-Morganton, NC | 34.48 | 95.1 | 354 | 7,069 | 4,801.9 | 360.6 | 1,242 | 16,890 | 4,808.2 | 420.7 | 257 | 853 | 201.6 | 32.1 |
| Hilton Head Island-Bluffton, SC | 40.06 | 193.1 | 153 | 1,403 | 781.4 | 71.7 | 892 | 12,942 | 3,733.4 | 333.2 | 489 | 1,943 | 465.3 | 89.1 |
| Hinesville, GA | 6.93 | 86.4 | D | D | D | D | 199 | 2,196 | 617.7 | 53.6 | 0 | 0 | 0.0 | 5.6 |
| Homosassa Springs, FL | 14.15 | 100.3 | 74 | 527 | 272.7 | 24.8 | 431 | 5,633 | 1,864.0 | 157.5 | 163 | 1,014 | 94.7 | 21.7 |
| Hot Springs, AR | 16.07 | 165.4 | 75 | 659 | 460.1 | 39.1 | 496 | 6,194 | 1,684.3 | 157.7 | 130 | 490 | 77.5 | 17.1 |
| Houma-Thibodaux, LA | 27.54 | 129.7 | 195 | 1,750 | 824.5 | 87.2 | 770 | 10,780 | 2,902.6 | 273.2 | 222 | 1,711 | 519.7 | 110.9 |
| Houston-The Woodlands-Sugar Land, TX | 436.66 | 65.6 | 8,429 | 128,052 | 415,508.9 | 8,839.8 | 18,661 | 302,441 | 98,382.8 | 8,624.7 | 8,013 | 53,141 | 15,673.0 | 2,862.9 |
| Huntington-Ashland, WV-KY-OH | 42.10 | 114.6 | 254 | 3,709 | 2,328.7 | 185.6 | 1,232 | 17,655 | 4,552.1 | 396.7 | 290 | 1,269 | 340.9 | 53.3 |
| Huntsville, AL | 78.67 | 176.9 | 346 | 5,403 | 5,362.0 | 327.2 | 1,564 | 22,337 | 6,222.2 | 600.9 | 477 | 1,987 | 595.5 | 78.6 |
| Idaho Falls, ID | 39.12 | 279.9 | 216 | 2,305 | 4,476.6 | 112.5 | 586 | 8,543 | 2,476.3 | 221.7 | 151 | 477 | 124.9 | 19.4 |
| Indianapolis-Carmel-Anderson, IN | 202.51 | 101.8 | 2,107 | 41,088 | 37,575.1 | 2,632.4 | 5,868 | 108,446 | 35,674.9 | 2,962.2 | 2,644 | 17,003 | 4,928.0 | 888.9 |
| Iowa City, IA | 11.85 | 71.2 | 144 | 1,982 | 1,756.8 | 108.6 | 557 | 10,148 | 2,465.8 | 260.6 | 158 | 715 | 149.8 | 31.8 |
| Ithaca, NY | 7.87 | 75.0 | D | D | D | D | 346 | 5,033 | 1,194.1 | 121.5 | 129 | 712 | 171.4 | 26.2 |
| Jackson, MI | 10.84 | 68.0 | 128 | 1,327 | 1,023.5 | 78.3 | 497 | 7,174 | 2,031.9 | 187.5 | 109 | 573 | 104.9 | 19.7 |
| Jackson, MS | 99.80 | 167.1 | 642 | 9,867 | 9,420.6 | 565.9 | 2,258 | 30,846 | 8,487.3 | 778.5 | 669 | 3,008 | 636.7 | 125.4 |
| Jackson, TN | 22.03 | 123.0 | 214 | 3,131 | 2,053.1 | 141.7 | 753 | 9,779 | 2,771.5 | 243.2 | 157 | 702 | 163.2 | 27.0 |
| Jacksonville, FL | 147.27 | 101.6 | 1,338 | 24,247 | 22,172.0 | 1,556.0 | 5,167 | 77,260 | 23,699.5 | 2,066.5 | 2,317 | 11,376 | 3,719.8 | 586.3 |
| Jacksonville, NC | 17.87 | 95.9 | 55 | 320 | 144.1 | 13.4 | 547 | 8,416 | 2,334.6 | 206.5 | 200 | 744 | 146.7 | 24.8 |
| Janesville-Beloit, WI | 16.71 | 103.5 | 136 | 2,677 | 1,878.5 | 145.6 | 510 | 10,004 | 2,631.5 | 285.4 | 118 | 452 | 132.7 | 19.6 |
| Jefferson City, MO | 13.80 | 91.3 | 115 | 1,582 | 1,580.8 | 69.2 | 502 | 7,576 | 2,005.3 | 180.7 | 103 | 376 | 83.9 | 13.0 |
| Johnson City, TN | 27.61 | 137.6 | 102 | 1,770 | 1,240.9 | 102.9 | 696 | 10,855 | 2,885.3 | 259.4 | 164 | 868 | 146.2 | 28.1 |
| Johnstown, PA | 14.93 | 109.4 | 104 | 988 | 573.3 | 44.9 | 534 | 6,781 | 1,773.7 | 160.2 | 83 | 460 | 75.9 | 15.1 |

1. Merchant wholesalers, except manufacturers' sales branches and offices. 2. Employer establishments.

# Table C. Metropolitan Areas — Professional Services, Manufacturing, and Accommodation and Food Services

| Area name | Professional, scientific, and technical services, 2017 | | | | Manufacturing, 2017 | | | | Accommodation and food services, 2017 | | | |
|---|---|---|---|---|---|---|---|---|---|---|---|---|
| | Number of establishments | Number of employees | Sales (mil dol) | Average payroll (mil dol) | Number of establishments | Number of employees | Receipts (mil dol) | Annual payroll (mil dol) | Number of establishments | Number of employees | Receipts (mil dol) | Annual payroll (mil dol) |
| | 147 | 148 | 149 | 150 | 151 | 152 | 153 | 154 | 155 | 156 | 157 | 158 |
| Evansville, IN-KY | 215 | 1,157 | 144.2 | 52.8 | 387 | 22,435 | 11,651.4 | 1,241.5 | 654 | 14,431 | 789.7 | 215.4 |
| Fairbanks, AK | 212 | 1,356 | 313.2 | 91.0 | 69 | 454 | 226.6 | 24.4 | 233 | 3,306 | 250.7 | 70.7 |
| Fargo, ND-MN | 616 | 6,448 | 999.8 | 406.5 | 225 | 9,174 | 3,239.3 | 451.7 | 559 | 12,555 | 596.1 | 196.8 |
| Farmington, NM | D | D | D | D | 69 | 776 | 154.0 | 43.1 | 240 | 4,900 | 237.5 | 68.0 |
| Fayetteville, NC | 685 | 7,134 | 868.4 | 351.3 | 182 | 8,895 | 3,570.3 | 464.0 | 902 | 18,570 | 931.2 | 255.0 |
| Fayetteville-Springdale-Rogers, AR | D | D | D | D | 399 | 25,666 | 8,082.6 | 1,070.6 | 1,100 | 22,361 | 1,201.9 | 349.4 |
| Flagstaff, AZ | 349 | 1,471 | 190.8 | 68.8 | 88 | 4,893 | 2,250.3 | 425.7 | 601 | 15,061 | 1,200.0 | 323.8 |
| Flint, MI | D | D | D | D | 279 | 11,516 | 13,080.7 | 738.1 | 730 | 14,666 | 707.3 | 209.7 |
| Florence, SC | 62 | 291 | 28.4 | 10.6 | 144 | 9,278 | 5,337.0 | 578.4 | 414 | 8,028 | 411.4 | 114.3 |
| Florence-Muscle Shoals, AL | 217 | 992 | 106.2 | 38.4 | 176 | 9,012 | 4,634.2 | 458.7 | 291 | 6,313 | 285.6 | 82.3 |
| Fond du Lac, WI | 141 | 1,433 | 156.5 | 104.1 | 146 | 9,480 | 4,108.5 | 486.2 | 222 | 3,542 | 163.0 | 44.3 |
| Fort Collins, CO | 1,642 | 9,285 | 1,637.8 | 668.0 | 444 | 9,931 | 3,909.8 | 646.2 | 949 | 18,194 | 1,087.4 | 338.0 |
| Fort Smith, AR-OK | 146 | 488 | 54.9 | 17.7 | 251 | 19,204 | 5,983.8 | 788.8 | 491 | 9,213 | 511.7 | 143.6 |
| Fort Wayne, IN | 858 | 5,855 | 892.8 | 314.7 | 531 | 31,277 | 21,098.4 | 1,815.6 | 806 | 17,815 | 860.0 | 254.9 |
| Fresno, CA | 1,581 | 10,913 | 1,576.6 | 610.6 | 579 | 22,774 | 9,073.4 | 1,091.3 | 1,647 | 30,013 | 1,783.4 | 501.2 |
| Gadsden, AL | 127 | 656 | 80.8 | 28.2 | 89 | 5,358 | 1,607.9 | 231.1 | 187 | 3,587 | 187.6 | 50.4 |
| Gainesville, FL | 933 | 5,714 | 809.1 | 320.3 | 198 | 4,479 | 1,299.9 | 277.3 | 677 | 14,133 | 739.7 | 216.8 |
| Gainesville, GA | 410 | 1,919 | 331.0 | 107.0 | 223 | 20,766 | 8,353.0 | 932.3 | 312 | 5,862 | 338.5 | 93.2 |
| Gettysburg, PA | D | D | D | D | 112 | 6,561 | 2,010.6 | 314.9 | 227 | 3,958 | 220.6 | 62.9 |
| Glens Falls, NY | 211 | 1,168 | 168.3 | 61.2 | 166 | 6,276 | 2,142.4 | 344.2 | 546 | 5,814 | 463.2 | 128.1 |
| Goldsboro, NC | 130 | 679 | 74.7 | 29.3 | 56 | 4,957 | 1,351.3 | 226.3 | 208 | 4,152 | 204.4 | 56.5 |
| Grand Forks, ND-MN | 171 | 1,563 | 210.6 | 92.2 | 87 | 4,329 | 2,005.0 | 210.3 | 273 | 5,333 | 236.3 | 76.0 |
| Grand Island, NE | 17 | 75 | 10.2 | 2.9 | 83 | 7,034 | 4,911.9 | 343.6 | 165 | 2,662 | 147.2 | 42.9 |
| Grand Junction, CO | 497 | 3,416 | 527.2 | 209.4 | 150 | 2,740 | 552.2 | 124.2 | 335 | 6,462 | 356.5 | 117.5 |
| Grand Rapids-Kentwood, MI | 95 | 326 | 35.5 | 13.2 | 1,818 | 112,568 | 34,055.6 | 6,015.4 | 1,921 | 42,291 | 2,119.3 | 669.4 |
| Grants Pass, OR | D | D | D | D | 109 | 2,528 | 561.2 | 111.2 | 208 | 2,919 | 169.8 | 49.3 |
| Great Falls, MT | D | D | D | D | 74 | 1,318 | 1,377.7 | 71.3 | 239 | 3,989 | 222.4 | 63.5 |
| Greeley, CO | D | D | D | D | 305 | 11,968 | 5,959.4 | 657.2 | 472 | 7,843 | 406.0 | 117.9 |
| Green Bay, WI | 64 | 281 | 31.3 | 12.7 | 531 | 28,645 | 12,397.0 | 1,590.0 | 764 | 13,496 | 651.0 | 189.0 |
| Greensboro-High Point, NC | 139 | 658 | 69.5 | 22.8 | 965 | 53,335 | 23,488.4 | 2,768.4 | 1,584 | 33,522 | 1,736.9 | 495.7 |
| Greenville, NC | D | D | D | D | 93 | 6,726 | 2,690.2 | 375.9 | 369 | 8,445 | 417.1 | 111.3 |
| Greenville-Anderson, SC | 51 | 281 | 4,693.0 | 14.4 | 895 | 54,844 | 22,383.1 | 2,986.5 | 1,920 | 37,861 | 2,071.2 | 572.7 |
| Gulfport-Biloxi, MS | 346 | 3,175 | 461.2 | 190.7 | 206 | 18,526 | 14,515.7 | 1,292.4 | 898 | 29,155 | 2,424.4 | 611.4 |
| Hagerstown-Martinsburg, MD-WV | 15 | 73 | 8.0 | 4.3 | 188 | 8,884 | 5,133.7 | 477.2 | 539 | 9,342 | 493.4 | 141.7 |
| Hammond, LA | 209 | 1,142 | 174.4 | 60.4 | 77 | 2,023 | 625.2 | 87.3 | 252 | 4,913 | 238.7 | 67.2 |
| Hanford-Corcoran, CA | 105 | 965 | 109.9 | 44.9 | 60 | 4,514 | 2,819.7 | 215.2 | 196 | 4,210 | 460.0 | 97.9 |
| Harrisburg-Carlisle, PA | 47 | 190 | 18.2 | 6.9 | 416 | 17,387 | 7,723.1 | 894.9 | 1,321 | 23,940 | 1,497.7 | 424.3 |
| Harrisonburg, VA | D | D | D | D | 127 | 9,734 | 4,763.0 | 488.0 | 300 | 7,566 | 361.0 | 125.3 |
| Hartford-East Hartford-Middletown, CT | D | D | D | D | 1,498 | 64,785 | 23,826.9 | 4,667.5 | 2,840 | 47,390 | 2,999.0 | 888.5 |
| Hattiesburg, MS | 147 | 908 | 115.7 | 41.9 | 106 | 5,628 | 1,721.3 | 243.7 | 4 | 44 | 2.0 | 6.3 |
| Hickory-Lenoir-Morganton, NC | 417 | 2,586 | 843.7 | 133.6 | 675 | 40,756 | 11,367.2 | 1,757.5 | 645 | 12,771 | 639.9 | 179.8 |
| Hilton Head Island-Bluffton, SC | 690 | 2,812 | 437.0 | 159.3 | 96 | 803 | 173.6 | 35.4 | 632 | 14,350 | 1,037.6 | 300.8 |
| Hinesville, GA | D | D | D | D | 19 | 1,904 | 1,276.2 | 123.5 | 114 | 2,078 | 104.3 | 26.4 |
| Homosassa Springs, FL | D | D | D | D | 55 | 338 | 75.1 | 13.6 | 235 | 3,513 | 175.3 | 50.8 |
| Hot Springs, AR | D | D | D | D | 88 | 2,427 | 694.6 | 115.8 | 306 | 5,510 | 291.2 | 85.8 |
| Houma-Thibodaux, LA | 426 | 4,166 | 522.2 | 245.0 | 182 | 6,023 | 1,614.3 | 342.7 | 405 | 7,171 | 388.7 | 110.1 |
| Houston-The Woodlands-Sugar Land, TX | 16,288 | 205,029 | 48,835.2 | 18,989.6 | 5,365 | 197,897 | 207,297.1 | 13,744.8 | 13,469 | 285,245 | 18,675.9 | 5,164.0 |
| Huntington-Ashland, WV-KY-OH | 205 | 1,326 | 165.3 | 60.3 | D | D | D | D | 671 | 12,309 | 643.4 | 172.8 |
| Huntsville, AL | 1,469 | 37,711 | 10,167.3 | 3,380.3 | 331 | 19,339 | 8,186.3 | 1,165.6 | 898 | 19,126 | 992.7 | 281.2 |
| Idaho Falls, ID | 39 | 125 | 15.1 | 3.9 | D | D | D | D | 237 | 5,065 | 257.2 | 77.1 |
| Indianapolis-Carmel-Anderson, IN | 4,017 | 57,436 | 12,082.6 | 4,973.8 | 1,704 | 78,199 | 35,336.4 | 4,903.2 | 4,398 | 99,434 | 5,604.7 | 1,630.1 |
| Iowa City, IA | 344 | 3,072 | 386.2 | 151.8 | 117 | 5,690 | 7,070.3 | 285.6 | D | D | D | D |
| Ithaca, NY | 284 | 2,037 | 318.8 | 123.2 | 88 | 2,873 | 939.4 | 150.9 | 334 | 4,972 | 279.4 | 89.9 |
| Jackson, MI | D | D | D | D | 248 | 9,000 | 3,017.6 | 486.8 | 272 | 4,870 | 243.9 | 69.4 |
| Jackson, MS | 1,481 | 10,532 | 1,870.3 | 648.9 | 342 | 18,208 | 9,831.2 | 953.6 | 1,214 | 25,457 | 1,298.2 | 349.4 |
| Jackson, TN | 65 | 322 | 29.7 | 9.1 | 190 | 12,997 | 6,100.6 | 672.7 | 304 | 6,376 | 328.9 | 91.7 |
| Jacksonville, FL | 3,635 | 38,234 | 6,827.2 | 2,705.8 | 829 | 27,062 | 13,053.4 | 1,581.3 | 3,297 | 70,314 | 4,264.6 | 1,209.7 |
| Jacksonville, NC | D | D | D | D | 47 | 1,040 | 391.8 | 42.3 | 399 | 7,870 | 435.4 | 117.0 |
| Janesville-Beloit, WI | 208 | 2,273 | 165.4 | 154.1 | 229 | 10,607 | 4,961.7 | 598.4 | 366 | 6,304 | 308.1 | 85.0 |
| Jefferson City, MO | 68 | 431 | 32.1 | 15.2 | D | 5,939 | D | 295.0 | 238 | 4,388 | 211.4 | 60.5 |
| Johnson City, TN | 49 | 163 | 16.1 | 4.5 | 163 | 7,253 | 1,932.3 | 355.4 | 420 | 9,089 | 431.6 | 127.9 |
| Johnstown, PA | D | D | D | D | 115 | 5,161 | 4,739.5 | 286.6 | 298 | 4,130 | 190.7 | 49.8 |

## Table C. Metropolitan Areas — Health Care and Social Assistance, Other Services, Nonemployer Businesses, and Residential Construction

| Area name | Health care and social assistance, 2017 | | | | Other services, 2017 | | | | Nonemployer businesses, 2018 | | Value of residential construction authorized by building permits, 2020 | |
|---|---|---|---|---|---|---|---|---|---|---|---|---|
| | Number of establishments | Number of employees | Receipts (mil dol) | Annual payroll (mil dol) | Number of establishments | Number of employees | Receipts (mil dol) | Annual payroll (mil dol) | Number | Receipts (mil dol) | New construction ($1,000) | Number of housing units |
| | 159 | 160 | 161 | 162 | 163 | 164 | 165 | 166 | 167 | 168 | 169 | 170 |
| Evansville, IN-KY | 957 | 26,880 | 3,219.3 | 1,154.0 | 566 | 4,016 | 490 | 127 | 17,603 | 743.2 | 241,971 | 1,121 |
| Fairbanks, AK | 313 | 6,507 | 890.8 | 371.6 | 178 | 896 | 121 | 33 | 5,992 | 236.2 | 4,982 | 19 |
| Fargo, ND-MN | 702 | 25,095 | 3,252.9 | 1,152.7 | 515 | 3,502 | 407 | 102 | 17,507 | 991.6 | 23,144 | 1,976 |
| Farmington, NM | 282 | 7,628 | 874.3 | 352.8 | 202 | 1,305 | 207 | 44 | 5,045 | 201.7 | 347,466 | 98 |
| Fayetteville, NC | 972 | 26,445 | 2,898.1 | 1,156.8 | 556 | 3,131 | 317 | 89 | 29,199 | 1,097.1 |  | 1,960 |
| Fayetteville-Springdale-Rogers, AR | 1,202 | 25,981 | 3,024.6 | 1,265.3 | 628 | 3,936 | 776 | 132 | 39,186 | 1,846.1 | 1,421,304 | 6,365 |
| Flagstaff, AZ | 422 | 8,994 | 1,229.8 | 448.2 | 249 | 1,422 | 143 | 43 | 9,691 | 406.4 | 215,702 | 941 |
| Flint, MI | 1,286 | 27,955 | 3,165.7 | 1,254.4 | 569 | 3,886 | 606 | 117 | 27,859 | 1,060.7 | 112,571 | 485 |
| Florence, SC | 482 | 17,635 | 2,176.1 | 825.3 | 259 | 1,768 | 229 | 49 | 12,466 | 479.3 | 107,802 | 764 |
| Florence-Muscle Shoals, AL | 359 | 8,212 | 912.1 | 336.5 | D | D | D | 33 | 10,244 | 480.7 | 36,608 | 296 |
| Fond du Lac, WI | 295 | 6,619 | 864.7 | 308.7 | 179 | 975 | 99 | 26 | 5,105 | 256.5 | 47,213 | 148 |
| Fort Collins, CO | 1,147 | 19,461 | 2,304.6 | 932.0 | 734 | 4,247 | 640 | 141 | 33,935 | 1,591.3 | 664,597 | 2,565 |
| Fort Smith, AR-OK | 649 | 16,059 | 1,709.5 | 677.0 | 276 | 1,231 | 124 | 34 | 16,793 | 809.6 | 99,949 | 655 |
| Fort Wayne, IN | 1,105 | 37,370 | 4,622.5 | 1,744.1 | 734 | 5,454 | 594 | 179 | 26,533 | 1,154.9 | 440,373 | 1,786 |
| Fresno, CA | 2,398 | 47,713 | 6,660.5 | 2,612.0 | 1,035 | 8,607 | 994 | 274 | 55,668 | 3,044.1 | 862,554 | 3,130 |
| Gadsden, AL | 300 | 7,233 | 761.6 | 303.7 | 114 | 779 | 126 | 24 | 7,520 | 345.0 | 13,933 | 113 |
| Gainesville, FL | 891 | 27,359 | 4,031.2 | 1,509.1 | 478 | 4,073 | 589 | 131 | 23,738 | 957.2 | 294,273 | 2,080 |
| Gainesville, GA | 488 | 13,913 | 1,965.5 | 721.0 | 282 | 1,514 | 178 | 51 | 16,827 | 799.5 | 347,410 | 1,706 |
| Gettysburg, PA | 185 | 4,860 | 512.4 | 201.8 | 174 | 1,312 | 125 | 34 | 6,697 | 302.2 | 48,233 | 226 |
| Glens Falls, NY | 380 | 8,015 | 879.9 | 361.3 | 225 | 1,129 | 134 | 38 | 8,653 | 387.7 | 77,734 | 306 |
| Goldsboro, NC | 259 | 7,604 | 781.9 | 310.8 | 152 | 810 | 71 | 21 | 6,621 | 264.2 | 75,705 | 414 |
| Grand Forks, ND-MN | 256 | 9,848 | 1,051.4 | 476.0 | 179 | 1,028 | 113 | 28 | 6,553 | 295.3 | 76,439 | 488 |
| Grand Island, NE | 209 | 4,829 | 544.8 | 207.4 | 182 | 1,173 | 113 | 30 | 5,080 | 233.9 | 45,963 | 271 |
| Grand Junction, CO | 438 | 11,405 | 1,429.2 | 530.6 | 315 | 1,662 | 188 | 56 | 12,271 | 576.3 | 144,224 | 1,004 |
| Grand Rapids-Kentwood, MI | 2,374 | 71,291 | 8,496.1 | 3,249.0 | 1,812 | 13,259 | 1,397 | 405 | 76,887 | 3,773.5 | 825,499 | 3,436 |
| Grants Pass, OR | 298 | 5,253 | 525.5 | 221.0 | 107 | 482 | 41 | 13 | 5,993 | 281.0 | 75,240 | 316 |
| Great Falls, MT | 285 | 6,971 | 802.3 | 322.9 | 155 | 941 | 105 | 30 | 4,883 | 214.3 | 36,937 | 151 |
| Greeley, CO | 551 | 8,971 | 1,127.3 | 446.2 | 418 | 2,212 | 279 | 74 | 24,243 | 1,206.3 | 965,414 | 4,185 |
| Green Bay, WI | 875 | 26,224 | 3,609.2 | 1,256.5 | 541 | 3,230 | 343 | 95 | 17,824 | 919.6 | 306,672 | 1,433 |
| Greensboro-High Point, NC | 1,681 | 43,095 | 4,953.5 | 1,919.8 | 1,137 | 7,549 | 1,398 | 248 | 57,916 | 2,569.5 | 645,309 | 3,147 |
| Greenville, NC | 499 | 16,338 | 2,158.5 | 773.9 | 220 | 1,287 | 124 | 36 | 10,993 | 433.2 | 162,847 | 719 |
| Greenville-Anderson, SC | 1,853 | 45,184 | 4,912.9 | 1,971.8 | 1,225 | 8,210 | 1,147 | 278 | 66,575 | 3,019.7 | 1,831,017 | 8,734 |
| Gulfport-Biloxi, MS | 868 | 21,560 | 2,686.4 | 1,158.3 | 436 | 2,505 | 282 | 72 | 29,877 | 1,223.6 | 532,653 | 3,021 |
| Hagerstown-Martinsburg, MD-WV | 669 | 16,247 | 1,944.6 | 902.2 | 394 | 2,465 | 227 | 74 | 16,009 | 696.2 | 346,646 | 1,742 |
| Hammond, LA | 321 | 8,978 | 880.7 | 357.3 | 160 | 1,207 | 127 | 37 | 10,518 | 406.7 | 231,959 | 1,577 |
| Hanford-Corcoran, CA | 213 | 4,803 | 631.8 | 221.0 | 116 | 501 | 59 | 14 | 5,119 | 230.5 | 76,662 | 306 |
| Harrisburg-Carlisle, PA | 1,632 | 53,486 | 6,514.3 | 2,802.8 | 1,296 | 8,581 | 1,088 | 301 | 36,755 | 1,781.3 | 291,406 | 1,460 |
| Harrisonburg, VA | 314 | 7,177 | 916.6 | 339.0 | 236 | 1,217 | 139 | 43 | 9,010 | 421.7 | 169,141 | 596 |
| Hartford-East Hartford-Middletown, CT | 3,820 | 111,796 | 11,858.2 | 5,222.4 | 2,522 | 16,629 | 2,127 | 625 | 84,370 | 4,567.4 | 238,793 | 1,399 |
| Hattiesburg, MS | 225 | 8,350 | 1,077.1 | 469.3 | 95 | 553 | 56 | 14 | 12,693 | 572.4 | 26,847 | 144 |
| Hickory-Lenoir-Morganton, NC | 764 | 21,581 | 2,349.2 | 916.4 | 460 | 2,709 | 253 | 72 | 23,375 | 1,037.1 | 280,586 | 1,567 |
| Hilton Head Island-Bluffton, SC | 541 | 9,322 | 1,067.4 | 398.1 | 377 | 2,940 | 352 | 101 | 19,222 | 1,037.8 | 1,006,862 | 2,977 |
| Hinesville, GA | 96 | 2,182 | 262.4 | 105.0 | D | D | D | 20 | 4,004 | 117.6 | 154,361 | 721 |
| Homosassa Springs, FL | 393 | 9,117 | 940.2 | 381.0 | 225 | 964 | 70 | 20 | 10,082 | 402.8 | 184,925 | 927 |
| Hot Springs, AR | 343 | 8,736 | 1,069.9 | 380.5 | 186 | 762 | 67 | 20 | 8,116 | 361.2 | 18,872 | 88 |
| Houma-Thibodaux, LA | 464 | 11,491 | 1,219.8 | 503.8 | 264 | 1,811 | 265 | 84 | 15,574 | 659.7 | 121,769 | 547 |
| Houston-The Woodlands-Sugar Land, TX | 16,882 | 356,550 | 47,587.1 | 17,579.0 | 8,873 | 82,767 | 11,932 | 3,429 | 663,771 | 33,114.7 | 12,702,945 | 70,540 |
| Huntington-Ashland, WV-KY-OH | 982 | 29,947 | 3,545.2 | 1,392.1 | 392 | 2,346 | 295 | 80 | 17,898 | 694.0 | 45,306 | 254 |
| Huntsville, AL | 1,161 | 27,578 | 3,656.9 | 1,322.4 | 578 | 4,408 | 557 | 154 | 31,930 | 1,355.7 | 792,597 | 4,341 |
| Idaho Falls, ID | 578 | 9,005 | 1,077.5 | 377.8 | D | D | D | D | 11,812 | 579.7 | 198,083 | 1,173 |
| Indianapolis-Carmel-Anderson, IN | 5,475 | 146,136 | 18,863.1 | 7,337.2 | 3,285 | 28,490 | 5,226 | 1,094 | 148,767 | 6,701.0 | 2,827,792 | 10,998 |
| Iowa City, IA | 503 | 19,840 | 2,733.7 | 1,016.9 | 273 | 1,530 | 321 | 55 | 12,114 | 576.7 | 196,610 | 1,019 |
| Ithaca, NY | 284 | 6,258 | 643.2 | 273.4 | 164 | 962 | 197 | 31 | 7,936 | 307.9 | 47,603 | 329 |
| Jackson, MI | 361 | 9,998 | 1,123.1 | 492.2 | 211 | 1,210 | 153 | 37 | 9,115 | 364.1 | 33,347 | 159 |
| Jackson, MS | 1,554 | 53,188 | 5,779.9 | 2,503.1 | 776 | 4,996 | 694 | 207 | 52,412 | 2,323.7 | 292,671 | 1,178 |
| Jackson, TN | 487 | 16,069 | 1,663.6 | 683.6 | 57 | 249 | 29 | 7 | 11,617 | 512.4 | 104,332 | 475 |
| Jacksonville, FL | 4,256 | 94,242 | 12,989.2 | 4,792.7 | 2,341 | 17,350 | 3,487 | 660 | 125,333 | 5,299.0 | 3,262,575 | 17,246 |
| Jacksonville, NC | 263 | 5,331 | 642.5 | 218.9 | 211 | 1,071 | 104 | 28 | 10,712 | 387.6 | 215,040 | 1,435 |
| Janesville-Beloit, WI | 342 | 8,861 | 1,200.5 | 478.6 | 265 | 1,356 | 122 | 35 | 8,186 | 368.6 | 121,993 | 671 |
| Jefferson City, MO | 350 | 9,389 | 960.4 | 397.3 | 285 | 1,588 | 252 | 67 | 9,467 | 431.6 | 54,998 | 253 |
| Johnson City, TN | 498 | 16,280 | 1,989.8 | 880.4 | 198 | 1,048 | 88 | 27 | 12,787 | 559.5 | 143,573 | 1,051 |
| Johnstown, PA | 484 | 11,101 | 1,045.2 | 458.7 | 299 | 1,543 | 146 | 35 | 6,072 | 246.6 | 14,426 | 59 |

# Table C. Metropolitan Areas — Government Employment and Payroll, and Local Government Finances

| Area name | Government employment and payroll, 2017 | | | | | | | | | Local government finances, 2017 | | | | |
| | | | March payroll (percent of total) | | | | | | | General revenue | | | | |
| | | | | | | | | | | | | Taxes | | |
| | | | | | | | | | | | | | Per capita[1] (dollars) | |
| | Full-time equivalent employees | March payroll (dollars) | Administration, judicial, and legal | Police and corrections | Fire protection | Highways and transportation | Health and welfare | Natural resources and utilities | Education and libraries | Total (mil dol) | Intergovernmental (mil dol) | Total (mil dol) | Total | Property |
| | 171 | 172 | 173 | 174 | 175 | 176 | 177 | 178 | 179 | 180 | 181 | 182 | 183 | 184 |
|---|---|---|---|---|---|---|---|---|---|---|---|---|---|---|
| Evansville, IN-KY | 9,902 | 36,941,786 | 6.8 | 12.1 | 4.7 | 4.4 | 2.1 | 7.5 | 60.6 | 1,019 | 469 | 335 | 1,064 | 950 |
| Fairbanks, AK | 3,002 | 16,676,902 | 7.1 | 2.7 | 3.0 | 4.7 | 1.3 | 2.3 | 77.7 | 420 | 227 | 155 | 1,556 | 1,370 |
| Fargo, ND-MN | 8,060 | 35,621,185 | 5.8 | 9.4 | 2.6 | 4.8 | 9.4 | 7.1 | 59.9 | 1,228 | 564 | 378 | 1,564 | 1,152 |
| Farmington, NM | 5,621 | 20,270,322 | 4.8 | 12.1 | 2.9 | 2.2 | 1.4 | 10.3 | 64.6 | 519 | 324 | 118 | 930 | 655 |
| Fayetteville, NC | 25,624 | 94,074,518 | 2.7 | 7.1 | 1.6 | 0.9 | 40.2 | 4.2 | 40.6 | 1,322 | 774 | 427 | 826 | 611 |
| Fayetteville-Springdale-Rogers, AR | 16,064 | 54,460,063 | 5.9 | 10.4 | 4.9 | 3.6 | 0.6 | 6.3 | 66.6 | 1,639 | 909 | 472 | 915 | 358 |
| Flagstaff, AZ | 4,429 | 19,039,894 | 12.5 | 13.9 | 10.9 | 3.0 | 3.7 | 6.5 | 46.5 | 547 | 193 | 257 | 1,820 | 982 |
| Flint, MI | 14,500 | 64,991,149 | 5.0 | 5.5 | 1.6 | 4.1 | 26.9 | 2.5 | 53.4 | 2,164 | 1,126 | 370 | 909 | 840 |
| Florence, SC | 7,178 | 24,215,567 | 6.1 | 9.0 | 2.0 | 1.8 | 7.3 | 4.9 | 67.3 | 641 | 276 | 252 | 1,228 | 875 |
| Florence-Muscle Shoals, AL | 5,046 | 17,607,090 | 3.5 | 8.3 | 4.4 | 3.9 | 4.2 | 18.5 | 54.5 | 408 | 183 | 143 | 970 | 437 |
| Fond du Lac, WI | 3,803 | 16,957,873 | 4.6 | 8.4 | 2.5 | 4.0 | 10.4 | 2.7 | 65.5 | 454 | 217 | 166 | 1,618 | 1,491 |
| Fort Collins, CO | 11,808 | 54,787,902 | 10.2 | 11.8 | 1.5 | 3.1 | 9.4 | 17.5 | 42.7 | 1,541 | 373 | 771 | 2,240 | 1,405 |
| Fort Smith, AR-OK | 8,795 | 30,475,057 | 5.0 | 8.0 | 3.5 | 3.2 | 2.9 | 6.9 | 70.1 | 815 | 441 | 224 | 896 | 351 |
| Fort Wayne, IN | 12,071 | 47,809,113 | 6.4 | 13.4 | 4.1 | 4.5 | 2.2 | 5.2 | 63.0 | 1,257 | 594 | 493 | 1,217 | 942 |
| Fresno, CA | 36,934 | 197,609,273 | 3.7 | 9.8 | 2.2 | 3.1 | 10.2 | 4.8 | 65.2 | 7,477 | 5,083 | 1,285 | 1,304 | 863 |
| Gadsden, AL | 3,617 | 13,009,490 | 5.6 | 11.4 | 6.0 | 3.8 | 7.4 | 7.0 | 56.2 | 298 | 139 | 104 | 1,011 | 317 |
| Gainesville, FL | 11,373 | 42,019,036 | 10.7 | 15.7 | 4.5 | 5.0 | 3.9 | 10.6 | 46.7 | 1,136 | 423 | 408 | 1,258 | 1,002 |
| Gainesville, GA | 6,731 | 25,935,061 | 6.3 | 9.9 | 7.4 | 1.6 | 4.5 | 5.5 | 63.2 | 691 | 255 | 302 | 1,518 | 931 |
| Gettysburg, PA | 3,735 | 14,888,774 | 5.7 | 6.9 | 0.6 | 7.9 | 1.4 | 2.5 | 74.3 | 416 | 175 | 195 | 1,897 | 1,435 |
| Glens Falls, NY | 6,598 | 28,307,844 | 6.0 | 7.5 | 1.9 | 6.5 | 5.4 | 3.3 | 67.7 | 832 | 304 | 423 | 3,359 | 2,409 |
| Goldsboro, NC | 4,543 | 14,430,137 | 3.0 | 6.6 | 2.3 | 1.5 | 7.8 | 6.9 | 69.4 | 356 | 206 | 110 | 895 | 614 |
| Grand Forks, ND-MN | 4,216 | 17,580,017 | 6.4 | 10.9 | 3.1 | 4.9 | 9.4 | 9.1 | 53.4 | 545 | 260 | 153 | 1,497 | 1,235 |
| Grand Island, NE | 3,748 | 16,021,253 | 5.3 | 7.0 | 3.1 | 3.3 | 8.8 | 15.7 | 55.7 | 414 | 134 | 167 | 2,210 | 1,731 |
| Grand Junction, CO | 5,190 | 22,403,036 | 4.9 | 11.8 | 4.6 | 3.6 | 5.4 | 9.0 | 57.7 | 578 | 216 | 243 | 1,606 | 965 |
| Grand Rapids-Kentwood, MI | 27,958 | 124,940,548 | 6.9 | 8.4 | 2.2 | 4.4 | 2.8 | 3.5 | 69.2 | 4,227 | 2,172 | 1,314 | 1,235 | 1,096 |
| Grants Pass, OR | 2,126 | 9,641,417 | 6.2 | 10.1 | 2.6 | 3.2 | 1.6 | 2.9 | 71.5 | 273 | 138 | 82 | 941 | 854 |
| Great Falls, MT | 2,747 | 10,563,299 | 7.1 | 12.7 | 4.0 | 6.0 | 4.7 | 5.1 | 59.8 | 271 | 111 | 93 | 1,142 | 1,106 |
| Greeley, CO | 9,283 | 38,633,927 | 9.4 | 12.0 | 4.9 | 4.5 | 6.1 | 8.0 | 51.9 | 1,520 | 370 | 818 | 2,673 | 1,807 |
| Green Bay, WI | 11,518 | 54,797,399 | 4.3 | 8.6 | 3.0 | 3.6 | 6.2 | 4.8 | 67.9 | 1,445 | 677 | 483 | 1,510 | 1,467 |
| Greensboro-High Point, NC | 25,704 | 99,678,689 | 5.1 | 11.0 | 4.2 | 2.2 | 7.9 | 6.3 | 60.2 | 2,537 | 1,164 | 998 | 1,309 | 986 |
| Greenville, NC | 7,258 | 24,893,380 | 7.1 | 10.4 | 2.2 | 2.4 | 12.6 | 10.0 | 49.6 | 931 | 284 | 183 | 1,025 | 720 |
| Greenville-Anderson, SC | 34,640 | 135,879,374 | 4.2 | 6.5 | 2.8 | 1.3 | 32.5 | 5.6 | 45.7 | 4,838 | 1,073 | 1,058 | 1,181 | 986 |
| Gulfport-Biloxi, MS | 20,249 | 79,714,429 | 3.9 | 6.4 | 3.6 | 3.1 | 37.3 | 2.6 | 42.1 | 2,520 | 802 | 549 | 1,331 | 1,205 |
| Hagerstown-Martinsburg, MD-WV | 9,772 | 38,973,336 | 4.3 | 5.3 | 1.7 | 2.3 | 1.8 | 4.8 | 77.2 | 903 | 390 | 373 | 1,318 | 896 |
| Hammond, LA | 6,779 | 25,167,545 | 4.7 | 6.0 | 1.3 | 2.0 | 46.1 | 2.3 | 36.2 | 780 | 229 | 258 | 1,949 | 370 |
| Hanford-Corcoran, CA | 5,524 | 28,583,504 | 5.7 | 10.6 | 2.5 | 2.1 | 9.0 | 5.0 | 62.4 | 741 | 511 | 136 | 910 | 697 |
| Harrisburg-Carlisle, PA | 18,493 | 81,071,678 | 6.6 | 11.0 | 0.6 | 3.8 | 4.3 | 4.5 | 67.5 | 2,909 | 1,083 | 1,286 | 2,253 | 1,599 |
| Harrisonburg, VA | 4,679 | 16,497,844 | 6.1 | 8.2 | 4.6 | 2.4 | 3.2 | 8.7 | 63.8 | 448 | 184 | 176 | 1,314 | 909 |
| Hartford-East Hartford-Middletown, CT | 44,055 | 241,342,339 | 3.5 | 7.5 | 3.4 | 2.8 | 2.8 | 4.2 | 73.5 | 5,986 | 2,043 | 3,452 | 2,860 | 2,833 |
| Hattiesburg, MS | 10,091 | 35,082,133 | 3.9 | 4.9 | 1.9 | 2.5 | 48.3 | 2.3 | 35.5 | 1,021 | 226 | 196 | 1,166 | 1,085 |
| Hickory-Lenoir-Morganton, NC | 14,794 | 55,039,991 | 4.1 | 5.5 | 2.3 | 1.1 | 26.7 | 4.2 | 54.5 | 1,362 | 550 | 392 | 1,069 | 777 |
| Hilton Head Island-Bluffton, SC | 7,286 | 31,070,896 | 6.9 | 9.2 | 5.9 | 1.5 | 27.2 | 3.9 | 41.9 | 793 | 175 | 469 | 2,180 | 1,800 |
| Hinesville, GA | 3,184 | 10,476,566 | 5.6 | 9.1 | 2.0 | 1.1 | 16.5 | 1.8 | 60.7 | 286 | 127 | 77 | 959 | 680 |
| Homosassa Springs, FL | 3,578 | 11,361,255 | 9.3 | 10.8 | 0.3 | 3.5 | 2.1 | 4.9 | 64.4 | 336 | 117 | 151 | 1,038 | 922 |
| Hot Springs, AR | 2,866 | 10,381,906 | 5.0 | 10.8 | 3.8 | 3.3 | 1.5 | 7.3 | 67.5 | 305 | 156 | 87 | 886 | 270 |
| Houma-Thibodaux, LA | 10,191 | 37,994,926 | 5.6 | 9.2 | 0.6 | 2.5 | 34.4 | 4.0 | 42.7 | 1,295 | 401 | 370 | 1,764 | 974 |
| Houston-The Woodlands-Sugar Land, TX | 265,645 | 1,120,213,911 | 4.6 | 10.3 | 3.2 | 4.2 | 7.1 | 3.0 | 65.9 | 32,272 | 8,186 | 17,657 | 2,559 | 2,163 |
| Huntington-Ashland, WV-KY-OH | 12,791 | 42,889,437 | 4.7 | 6.3 | 2.3 | 2.8 | 5.3 | 5.1 | 71.4 | 1,067 | 486 | 374 | 1,034 | 787 |
| Huntsville, AL | 21,033 | 96,058,347 | 3.1 | 5.7 | 2.8 | 2.6 | 46.2 | 5.7 | 30.7 | 2,858 | 1,136 | 566 | 1,241 | 536 |
| Idaho Falls, ID | 5,031 | 16,371,062 | 6.9 | 10.2 | 5.3 | 4.4 | 6.0 | 9.2 | 56.8 | 433 | 231 | 118 | 810 | 773 |
| Indianapolis-Carmel-Anderson, IN | 71,926 | 294,569,457 | 4.6 | 7.4 | 6.2 | 2.3 | 19.6 | 5.7 | 53.1 | 9,945 | 3,485 | 2,362 | 1,165 | 958 |
| Iowa City, IA | 5,447 | 23,291,400 | 6.2 | 7.7 | 1.8 | 5.3 | 10.0 | 5.9 | 60.3 | 714 | 237 | 331 | 1,928 | 1,720 |
| Ithaca, NY | 4,615 | 22,463,397 | 5.1 | 5.9 | 2.8 | 4.9 | 6.6 | 3.9 | 68.1 | 600 | 194 | 318 | 3,102 | 2,262 |
| Jackson, MI | 4,580 | 19,841,345 | 6.9 | 7.2 | 1.5 | 4.1 | 4.5 | 2.4 | 72.5 | 658 | 378 | 160 | 1,006 | 930 |
| Jackson, MS | 22,236 | 73,312,212 | 5.7 | 9.7 | 4.1 | 2.5 | 1.8 | 3.4 | 71.2 | 1,964 | 912 | 695 | 1,159 | 1,103 |
| Jackson, TN | 12,166 | 46,642,029 | 2.8 | 6.8 | 2.2 | 2.2 | 52.9 | 6.4 | 26.2 | 1,275 | 228 | 186 | 1,043 | 697 |
| Jacksonville, FL | 41,501 | 158,263,187 | 5.8 | 13.1 | 6.6 | 2.5 | 1.2 | 7.2 | 58.8 | 5,424 | 1,861 | 2,147 | 1,426 | 1,035 |
| Jacksonville, NC | 5,784 | 20,744,428 | 4.2 | 8.1 | 2.3 | 1.0 | 8.3 | 4.8 | 68.0 | 507 | 241 | 183 | 935 | 624 |
| Janesville-Beloit, WI | 6,384 | 28,667,201 | 5.4 | 9.3 | 3.9 | 4.1 | 10.9 | 3.7 | 61.0 | 783 | 426 | 257 | 1,582 | 1,459 |
| Jefferson City, MO | 4,544 | 14,427,232 | 7.6 | 7.9 | 2.9 | 4.2 | 5.5 | 6.1 | 63.4 | 422 | 136 | 203 | 1,343 | 828 |
| Johnson City, TN | 7,169 | 22,904,686 | 6.8 | 9.2 | 3.2 | 5.3 | 3.4 | 12.4 | 58.8 | 506 | 203 | 186 | 921 | 688 |
| Johnstown, PA | 4,432 | 17,315,677 | 6.6 | 9.4 | 1.6 | 6.5 | 5.9 | 4.0 | 62.3 | 581 | 337 | 163 | 1,227 | 939 |

1. Based on the resident population estimated as of July 1 of the year shown.

Items 171—184

## Table C. Metropolitan Areas — Local Government Finances, Government Employment, and Income Taxes

| Area name | Local government finances, 2017 (cont.) — Direct general expenditure — Total (mil dol) | Per capita[1] (dollars) | Percent of total for: Education | Health and hospitals | Police protection | Public welfare | Highways | Debt outstanding — Total (mil dol) | Per capita[1] (dollars) | Government employment, 2019 — Federal civilian | Federal military | State and local | Individual income tax returns, 2018 — Number of returns | Mean adjusted gross income | Mean income tax |
|---|---|---|---|---|---|---|---|---|---|---|---|---|---|---|---|
| | 185 | 186 | 187 | 188 | 189 | 190 | 191 | 192 | 193 | 194 | 195 | 196 | 197 | 198 | 199 |
| Evansville, IN-KY | 921.8 | 2,928 | 48.4 | 1.2 | 6.4 | 0.2 | 3.1 | 2,100.1 | 6,672 | 1,320 | 948 | 15,519 | 148,380 | 62,031 | 6,574 |
| Fairbanks, AK | 387.5 | 3,888 | 55.4 | 1.1 | 1.9 | 0.0 | 6.2 | 171.3 | 1,719 | 3,134 | 9,131 | 7,220 | 46,180 | 69,879 | 7,702 |
| Fargo, ND-MN | 1,448.2 | 5,996 | 38.9 | 1.6 | 3.8 | 2.3 | 16.5 | 2,933.0 | 12,145 | 2,610 | 1,321 | 17,027 | 117,070 | 71,735 | 8,158 |
| Farmington, NM | 564.6 | 4,449 | 58.2 | 1.5 | 6.0 | 1.7 | 3.7 | 2,061.0 | 16,239 | 1,483 | 314 | 9,703 | 48,350 | 52,504 | 4,832 |
| Fayetteville, NC | 1,318.4 | 2,548 | 60.5 | 2.0 | 8.1 | 6.0 | 1.6 | 759.0 | 1,467 | 16,011 | 47,844 | 30,672 | 216,660 | 48,645 | 3,998 |
| Fayetteville-Springdale-Rogers, AR | 1,568.8 | 3,043 | 57.0 | 0.5 | 5.6 | 0.0 | 7.5 | 2,367.6 | 4,592 | 2,645 | 2,054 | 29,151 | 228,120 | 95,668 | 11,794 |
| Flagstaff, AZ | 510.3 | 3,619 | 36.3 | 2.7 | 7.6 | 0.9 | 3.9 | 201.0 | 1,426 | 2,687 | 291 | 15,858 | 61,230 | 61,614 | 6,615 |
| Flint, MI | 2,110.3 | 5,179 | 40.0 | 28.9 | 4.1 | 1.6 | 3.1 | 912.4 | 2,239 | 1,089 | 640 | 19,174 | 192,980 | 53,034 | 5,197 |
| Florence, SC | 687.4 | 3,346 | 56.0 | 4.0 | 7.2 | 0.2 | 6.6 | 586.2 | 2,853 | 715 | 769 | 16,663 | 87,380 | 51,142 | 4,960 |
| Florence-Muscle Shoals, AL | 410.9 | 2,791 | 54.8 | 3.7 | 5.8 | 0.1 | 11.1 | 411.2 | 2,792 | 881 | 616 | 9,122 | 62,430 | 55,253 | 5,302 |
| Fond du Lac, WI | 502.6 | 4,908 | 51.0 | 5.3 | 5.0 | 8.5 | 10.2 | 445.8 | 4,353 | 196 | 263 | 5,353 | 51,240 | 63,920 | 6,668 |
| Fort Collins, CO | 1,367.1 | 3,973 | 34.9 | 5.0 | 7.1 | 3.4 | 8.1 | 1,497.6 | 4,352 | 2,525 | 864 | 37,343 | 173,750 | 80,261 | 10,074 |
| Fort Smith, AR-OK | 805.4 | 3,218 | 54.6 | 2.7 | 5.0 | 0.1 | 6.4 | 1,523.2 | 6,087 | 1,296 | 1,005 | 12,782 | 99,580 | 51,269 | 4,652 |
| Fort Wayne, IN | 1,159.3 | 2,859 | 50.9 | 1.2 | 8.0 | 0.0 | 2.3 | 1,227.2 | 3,027 | 2,140 | 1,248 | 18,454 | 199,340 | 61,498 | 6,710 |
| Fresno, CA | 7,002.6 | 7,108 | 41.9 | 23.6 | 5.5 | 6.2 | 2.3 | 5,206.5 | 5,285 | 10,192 | 1,659 | 65,280 | 404,380 | 57,417 | 5,844 |
| Gadsden, AL | 302.7 | 2,939 | 49.8 | 2.2 | 10.0 | 0.2 | 4.2 | 235.4 | 2,285 | 308 | 422 | 5,134 | 42,110 | 49,188 | 4,227 |
| Gainesville, FL | 1,164.2 | 3,588 | 39.3 | 3.4 | 7.8 | 1.6 | 4.6 | 2,461.5 | 7,586 | 4,730 | 660 | 39,997 | 140,130 | 62,783 | 7,455 |
| Gainesville, GA | 705.1 | 3,544 | 54.9 | 6.3 | 5.7 | 0.3 | 3.3 | 1,794.7 | 9,021 | 456 | 539 | 10,548 | 90,510 | 64,891 | 7,124 |
| Gettysburg, PA | 490.8 | 4,786 | 68.4 | 3.2 | 2.6 | 3.9 | 3.4 | 505.8 | 4,931 | 713 | 253 | 3,412 | 52,350 | 63,317 | 6,430 |
| Glens Falls, NY | 793.3 | 6,300 | 52.7 | 3.6 | 3.0 | 8.0 | 7.5 | 369.1 | 2,931 | 316 | 194 | 9,536 | 62,690 | 58,160 | 5,874 |
| Goldsboro, NC | 361.9 | 2,941 | 53.6 | 5.6 | 8.1 | 6.3 | 0.9 | 253.1 | 2,057 | 1,213 | 4,882 | 8,118 | 51,600 | 47,597 | 4,002 |
| Grand Forks, ND-MN | 524.8 | 5,139 | 41.9 | 2.4 | 6.0 | 4.5 | 9.3 | 799.5 | 7,829 | 1,112 | 2,173 | 11,654 | 47,340 | 66,451 | 7,090 |
| Grand Island, NE | 448.6 | 5,946 | 47.9 | 7.0 | 4.5 | 0.3 | 5.1 | 351.5 | 4,658 | 706 | 268 | 5,537 | 36,470 | 54,611 | 4,990 |
| Grand Junction, CO | 590.7 | 3,907 | 38.0 | 2.0 | 11.7 | 5.4 | 9.0 | 401.0 | 2,653 | 1,636 | 381 | 8,781 | 71,300 | 61,564 | 6,357 |
| Grand Rapids-Kentwood, MI | 4,654.2 | 4,376 | 53.2 | 5.5 | 3.8 | 0.8 | 8.5 | 5,701.5 | 5,360 | 3,588 | 1,796 | 47,153 | 510,220 | 71,763 | 8,118 |
| Grants Pass, OR | 275.8 | 3,186 | 64.0 | 3.0 | 8.2 | 0.2 | 5.2 | 96.3 | 1,112 | 260 | 204 | 3,144 | 37,960 | 51,782 | 4,849 |
| Great Falls, MT | 276.9 | 3,392 | 49.0 | 2.5 | 10.1 | 1.9 | 4.5 | 235.4 | 2,883 | 1,686 | 3,557 | 3,872 | 39,540 | 56,044 | 5,339 |
| Greeley, CO | 1,346.8 | 4,403 | 41.0 | 1.2 | 6.1 | 2.3 | 8.2 | 1,730.4 | 5,657 | 598 | 775 | 16,960 | 143,900 | 71,647 | 7,795 |
| Green Bay, WI | 1,619.4 | 5,066 | 51.0 | 3.2 | 5.3 | 3.3 | 9.1 | 1,404.1 | 4,392 | 1,430 | 850 | 20,221 | 161,090 | 71,783 | 8,629 |
| Greensboro-High Point, NC | 2,535.0 | 3,326 | 48.8 | 3.1 | 8.3 | 5.4 | 2.7 | 2,337.5 | 3,067 | 4,380 | 1,692 | 38,644 | 344,300 | 60,670 | 6,724 |
| Greenville, NC | 992.6 | 5,557 | 28.8 | 44.4 | 6.0 | 0.0 | 1.4 | 605.9 | 3,392 | 711 | 426 | 26,816 | 72,810 | 58,788 | 6,429 |
| Greenville-Anderson, SC | 4,536.7 | 5,067 | 30.6 | 44.6 | 4.0 | 0.1 | 1.4 | 4,701.6 | 5,251 | 2,850 | 3,388 | 58,368 | 401,520 | 64,134 | 6,900 |
| Gulfport-Biloxi, MS | 2,527.0 | 6,122 | 31.4 | 33.8 | 4.1 | 0.2 | 4.1 | 1,745.8 | 4,230 | 9,135 | 10,215 | 25,721 | 174,380 | 52,190 | 4,871 |
| Hagerstown-Martinsburg, MD-WV | 926.2 | 3,273 | 63.3 | 0.6 | 4.8 | 0.0 | 3.3 | 774.5 | 2,737 | 3,830 | 1,168 | 13,925 | 134,990 | 56,858 | 5,240 |
| Hammond, LA | 688.6 | 5,203 | 29.1 | 46.7 | 3.7 | 0.0 | 2.2 | 306.9 | 2,318 | 383 | 499 | 10,909 | 52,260 | 51,820 | 4,805 |
| Hanford-Corcoran, CA | 800.5 | 5,349 | 48.0 | 4.4 | 4.9 | 9.9 | 2.5 | 315.7 | 2,110 | 1,198 | 5,021 | 14,265 | 55,610 | 49,394 | 3,886 |
| Harrisburg-Carlisle, PA | 3,032.8 | 5,313 | 53.0 | 3.5 | 3.1 | 9.1 | 3.7 | 3,298.3 | 5,779 | 7,459 | 2,188 | 52,461 | 295,430 | 66,555 | 7,447 |
| Harrisonburg, VA | 477.9 | 3,566 | 55.7 | 2.7 | 4.2 | 5.9 | 2.8 | 192.3 | 1,435 | 357 | 407 | 11,826 | 56,740 | 57,144 | 5,292 |
| Hartford-East Hartford-Middletown, CT | 5,753.0 | 4,766 | 58.7 | 0.8 | 4.7 | 0.6 | 3.6 | 4,272.8 | 3,540 | 6,289 | 2,441 | 91,642 | 603,710 | 82,812 | 10,884 |
| Hattiesburg, MS | 979.8 | 5,842 | 25.3 | 51.8 | 3.3 | 0.0 | 4.1 | 538.8 | 3,213 | 838 | 1,352 | 15,349 | 68,110 | 57,434 | 6,208 |
| Hickory-Lenoir-Morganton, NC | 1,318.2 | 3,591 | 42.8 | 23.1 | 5.9 | 6.2 | 1.1 | 640.5 | 1,745 | 698 | 792 | 22,283 | 161,930 | 53,162 | 5,039 |
| Hilton Head Island-Bluffton, SC | 729.6 | 3,393 | 45.2 | 2.2 | 7.7 | 0.6 | 4.7 | 1,003.6 | 4,667 | 2,274 | 10,774 | 9,354 | 100,610 | 84,577 | 11,129 |
| Hinesville, GA | 299.4 | 3,724 | 51.9 | 17.4 | 5.4 | 0.1 | 1.9 | 137.1 | 1,705 | 3,762 | 15,393 | 4,004 | 33,050 | 41,175 | 2,553 |
| Homosassa Springs, FL | 372.5 | 2,562 | 41.6 | 6.4 | 9.6 | 0.6 | 8.5 | 478.1 | 3,288 | 253 | 256 | 4,162 | 68,660 | 53,580 | 5,401 |
| Hot Springs, AR | 292.2 | 2,972 | 52.9 | 0.3 | 5.9 | 0.1 | 3.2 | 550.8 | 5,602 | 479 | 379 | 4,025 | 43,450 | 55,037 | 5,732 |
| Houma-Thibodaux, LA | 1,181.3 | 5,627 | 29.3 | 37.5 | 4.0 | 0.7 | 4.0 | 345.0 | 1,643 | 413 | 879 | 11,623 | 84,680 | 52,638 | 6,027 |
| Houston-The Woodlands-Sugar Land, TX | 32,697.3 | 4,739 | 50.0 | 10.0 | 5.3 | 0.2 | 4.0 | 76,910.3 | 11,146 | 30,125 | 15,065 | 371,983 | 3,046,040 | 81,159 | 11,364 |
| Huntington-Ashland, WV-KY-OH | 1,094.8 | 3,025 | 62.5 | 3.0 | 4.6 | 0.9 | 2.7 | 807.4 | 2,230 | 3,436 | 1,405 | 19,849 | 148,190 | 55,173 | 5,217 |
| Huntsville, AL | 2,934.3 | 6,436 | 25.9 | 49.6 | 3.1 | 0.0 | 2.5 | 3,356.5 | 7,363 | 19,727 | 2,652 | 32,137 | 215,360 | 73,421 | 8,464 |
| Idaho Falls, ID | 391.7 | 2,692 | 49.1 | 5.2 | 7.2 | 0.3 | 4.0 | 275.2 | 1,891 | 837 | 492 | 7,653 | 63,230 | 63,004 | 6,070 |
| Indianapolis-Carmel-Anderson, IN | 9,316.1 | 4,595 | 35.9 | 31.9 | 4.2 | 0.0 | 1.6 | 13,571.8 | 6,694 | 17,410 | 6,672 | 119,947 | 986,790 | 71,292 | 8,621 |
| Iowa City, IA | 745.7 | 4,347 | 46.1 | 7.6 | 5.2 | 0.4 | 8.0 | 872.2 | 5,085 | 2,173 | 641 | 38,670 | 76,520 | 77,789 | 9,715 |
| Ithaca, NY | 619.3 | 6,032 | 49.2 | 4.1 | 2.9 | 6.9 | 6.0 | 621.0 | 6,049 | 261 | 157 | 5,898 | 41,040 | 72,135 | 8,503 |
| Jackson, MI | 668.6 | 4,217 | 50.5 | 11.8 | 4.6 | 3.8 | 6.7 | 543.5 | 3,428 | 325 | 240 | 7,560 | 71,860 | 55,655 | 5,304 |
| Jackson, MS | 2,011.3 | 3,356 | 56.7 | 2.2 | 6.1 | 0.2 | 6.0 | 2,117.6 | 3,533 | 6,242 | 3,362 | 53,442 | 254,400 | 58,969 | 6,352 |
| Jackson, TN | 1,202.8 | 6,745 | 21.7 | 56.8 | 4.2 | 0.1 | 2.2 | 795.7 | 4,462 | 627 | 487 | 16,900 | 77,630 | 51,371 | 4,681 |
| Jacksonville, FL | 5,156.1 | 3,426 | 44.6 | 2.6 | 10.3 | 0.5 | 2.7 | 11,861.3 | 7,881 | 19,022 | 13,435 | 60,193 | 748,890 | 72,864 | 9,240 |
| Jacksonville, NC | 519.7 | 2,664 | 54.1 | 3.9 | 6.9 | 9.7 | 1.1 | 462.0 | 2,368 | 6,521 | 43,835 | 8,077 | 81,540 | 46,790 | 3,436 |
| Janesville-Beloit, WI | 799.9 | 4,931 | 49.1 | 6.3 | 6.0 | 6.9 | 6.7 | 701.7 | 4,326 | 312 | 419 | 8,453 | 80,190 | 57,507 | 5,399 |
| Jefferson City, MO | 386.2 | 2,552 | 52.9 | 4.4 | 7.3 | 0.0 | 6.6 | 281.4 | 1,860 | 833 | 493 | 24,332 | 68,950 | 57,003 | 5,175 |
| Johnson City, TN | 521.6 | 2,583 | 49.2 | 1.0 | 6.9 | 0.1 | 6.1 | 965.5 | 4,782 | 3,048 | 601 | 13,542 | 87,840 | 55,465 | 5,642 |
| Johnstown, PA | 626.1 | 4,709 | 54.7 | 2.4 | 2.1 | 9.5 | 3.2 | 513.7 | 3,864 | 976 | 340 | 6,660 | 62,430 | 52,063 | 4,707 |

1. Based on the resident population estimated as of July 1 of the year shown.

# Table C. Metropolitan Areas — **Land Area and Population**

| CBSA/ DIV Code[1] | Area name | Land area[2] (sq mi) | Total persons 2020 | Rank | Per square mile | White | Black | American Indian, Alaska Native | Asian and Pacific Islander | Percent Hispanic or Latino[3] | Under 5 years | 5 to 14 years | 15 to 24 years | 25 to 34 years | 35 to 44 years | 45 to 54 years |
|---|---|---|---|---|---|---|---|---|---|---|---|---|---|---|---|---|
| | | 1 | 2 | 3 | 4 | 5 | 6 | 7 | 8 | 9 | 10 | 11 | 12 | 13 | 14 | 15 |
| 27860 | Jonesboro, AR | 1,466 | 135,528 | 306 | 92.5 | 77.9 | 16.5 | 0.8 | 1.5 | 5.0 | 6.8 | 13.8 | 14.4 | 14.8 | 12.6 | 11.3 |
| 27900 | Joplin, MO | 1,263 | 180,099 | 239 | 142.6 | 87.0 | 2.8 | 3.4 | 2.4 | 7.6 | 6.4 | 13.7 | 12.9 | 13.2 | 12.3 | 11.7 |
| 27980 | Kahului-Wailuku-Lahaina, HI | 1,162 | 167,902 | 258 | 144.6 | 44.0 | 1.5 | 1.8 | 68.0 | 11.9 | 5.7 | 12.2 | 10.2 | 13.6 | 12.6 | 12.6 |
| 28020 | Kalamazoo-Portage, MI | 562 | 265,988 | 184 | 473.3 | 79.8 | 13.6 | 1.2 | 3.6 | 5.4 | 5.8 | 12.0 | 18.6 | 13.8 | 11.9 | 10.6 |
| 28100 | Kankakee, IL | 677 | 108,594 | 344 | 160.5 | 72.8 | 16.0 | 0.5 | 1.4 | 11.2 | 5.8 | 12.4 | 15.0 | 12.3 | 11.8 | 12.0 |
| 28140 | Kansas City, MO-KS | 7,257 | 2,173,212 | 31 | 299.5 | 74.3 | 13.8 | 1.2 | 4.1 | 9.5 | 6.3 | 13.4 | 12.1 | 14.2 | 13.4 | 12.1 |
| 28420 | Kennewick-Richland, WA | 2,942 | 303,501 | 166 | 103.2 | 61.6 | 2.4 | 1.4 | 4.1 | 33.0 | 7.2 | 16.2 | 13.4 | 14.1 | 13.4 | 10.9 |
| 28660 | Killeen-Temple, TX | 2,819 | 468,453 | 119 | 166.3 | 50.4 | 22.8 | 1.2 | 5.0 | 24.6 | 7.6 | 14.7 | 15.3 | 16.5 | 13.5 | 10.5 |
| 28700 | Kingsport-Bristol, TN-VA | 2,010 | 308,183 | 165 | 153.3 | 94.9 | 2.8 | 0.8 | 1.0 | 2.0 | 4.6 | 10.7 | 10.8 | 11.7 | 11.0 | 13.5 |
| 28740 | Kingston, NY | 1,124 | 177,716 | 242 | 158.1 | 80.4 | 7.4 | 0.8 | 2.9 | 11.0 | 4.4 | 9.7 | 12.0 | 12.6 | 11.9 | 12.9 |
| 28940 | Knoxville, TN | 3,220 | 878,124 | 64 | 272.7 | 87.7 | 6.8 | 0.9 | 2.2 | 4.3 | 5.3 | 11.4 | 13.3 | 13.0 | 11.9 | 12.6 |
| 29020 | Kokomo, IN | 293 | 82,732 | 375 | 282.3 | 86.6 | 9.7 | 0.9 | 1.9 | 3.7 | 6.1 | 12.7 | 11.9 | 12.4 | 11.1 | 11.9 |
| 29100 | La Crosse-Onalaska, WI-MN | 1,004 | 137,134 | 302 | 136.6 | 91.7 | 2.4 | 0.8 | 4.9 | 2.1 | 5.0 | 11.3 | 18.0 | 12.4 | 11.6 | 10.7 |
| 29180 | Lafayette, LA | 3,409 | 489,759 | 114 | 143.7 | 68.4 | 26.2 | 0.7 | 2.2 | 4.1 | 6.6 | 13.6 | 12.5 | 14.4 | 13.1 | 11.6 |
| 29200 | Lafayette-West Lafayette, IN | 1,642 | 233,278 | 198 | 142.1 | 79.5 | 6.0 | 0.6 | 7.7 | 8.2 | 5.6 | 11.5 | 24.4 | 13.9 | 10.9 | 10.0 |
| 29340 | Lake Charles, LA | 2,349 | 210,313 | 215 | 89.5 | 68.6 | 26.2 | 1.1 | 1.9 | 4.3 | 7.1 | 13.8 | 12.4 | 14.5 | 12.7 | 11.2 |
| 29420 | Lake Havasu City-Kingman, AZ | 13,332 | 217,206 | 211 | 16.3 | 78.1 | 1.6 | 2.9 | 2.2 | 17.2 | 4.2 | 9.3 | 8.8 | 10.3 | 9.1 | 10.2 |
| 29460 | Lakeland-Winter Haven, FL | 1,798 | 744,552 | 81 | 414.2 | 57.2 | 15.8 | 0.7 | 2.4 | 25.7 | 5.7 | 12.4 | 12.0 | 13.4 | 12.2 | 11.7 |
| 29540 | Lancaster, PA | 944 | 546,192 | 105 | 578.7 | 82.4 | 4.7 | 0.4 | 2.9 | 11.2 | 6.4 | 12.9 | 12.6 | 13.2 | 11.6 | 11.3 |
| 29620 | Lansing-East Lansing, MI | 2,229 | 548,248 | 104 | 246.0 | 80.6 | 9.7 | 1.2 | 5.1 | 6.6 | 5.4 | 11.3 | 17.3 | 13.5 | 11.8 | 11.4 |
| 29700 | Laredo, TX | 3,362 | 277,681 | 176 | 82.6 | 3.7 | 0.4 | 0.1 | 0.5 | 95.4 | 8.7 | 18.0 | 16.7 | 13.8 | 12.3 | 11.7 |
| 29740 | Las Cruces, NM | 3,808 | 221,262 | 207 | 58.1 | 27.5 | 2.0 | 1.2 | 1.6 | 68.8 | 6.1 | 13.6 | 18.5 | 13.2 | 11.2 | 9.9 |
| 29820 | Las Vegas-Henderson-Paradise, NV | 7,892 | 2,315,963 | 27 | 293.5 | 44.2 | 13.6 | 1.1 | 13.7 | 31.8 | 6.0 | 12.9 | 11.9 | 14.9 | 13.9 | 13.0 |
| 29940 | Lawrence, KS | 456 | 122,530 | 327 | 268.8 | 81.8 | 6.1 | 3.5 | 6.1 | 6.6 | 4.7 | 10.1 | 26.0 | 14.2 | 12.2 | 9.5 |
| 30020 | Lawton, OK | 1,702 | 126,775 | 316 | 74.5 | 61.6 | 18.5 | 8.3 | 4.9 | 13.6 | 6.7 | 13.1 | 15.6 | 16.2 | 12.8 | 10.3 |
| 30140 | Lebanon, PA | 362 | 141,663 | 296 | 391.6 | 81.6 | 2.8 | 0.4 | 2.0 | 14.5 | 5.8 | 12.9 | 12.2 | 11.8 | 11.8 | 11.9 |
| 30300 | Lewiston, ID-WA | 1,484 | 63,575 | 381 | 42.8 | 89.9 | 1.1 | 5.1 | 2.0 | 4.4 | 5.4 | 11.7 | 11.4 | 12.2 | 11.5 | 11.4 |
| 30340 | Lewiston-Auburn, ME | 468 | 108,547 | 345 | 231.9 | 91.9 | 5.8 | 1.1 | 1.6 | 2.0 | 5.7 | 12.1 | 12.1 | 12.5 | 12.0 | 12.6 |
| 30460 | Lexington-Fayette, KY | 1,470 | 520,391 | 109 | 354.0 | 79.0 | 12.9 | 0.6 | 3.7 | 6.3 | 5.9 | 12.2 | 15.5 | 14.3 | 12.9 | 12.0 |
| 30620 | Lima, OH | 403 | 101,980 | 354 | 253.4 | 82.9 | 14.7 | 0.7 | 1.3 | 3.5 | 6.1 | 12.8 | 13.8 | 12.4 | 11.6 | 11.4 |
| 30700 | Lincoln, NE | 1,409 | 337,836 | 152 | 239.8 | 83.6 | 5.4 | 1.2 | 5.3 | 7.4 | 6.0 | 12.7 | 18.5 | 13.7 | 12.6 | 10.5 |
| 30780 | Little Rock-North Little Rock-Conway, AR | 4,084 | 746,564 | 80 | 182.8 | 67.9 | 25.2 | 1.1 | 2.4 | 5.6 | 6.1 | 13.1 | 13.2 | 13.4 | 13.0 | 11.9 |
| 30860 | Logan, UT-ID | 1,828 | 144,219 | 291 | 78.9 | 85.7 | 1.2 | 1.0 | 3.3 | 10.5 | 7.8 | 16.9 | 23.0 | 13.5 | 11.9 | 8.6 |
| 30980 | Longview, TX | 2,681 | 287,105 | 173 | 107.1 | 64.3 | 18.9 | 1.0 | 1.3 | 16.3 | 6.1 | 14.0 | 13.0 | 12.9 | 12.4 | 11.5 |
| 31020 | Longview, WA | 1,141 | 111,371 | 342 | 97.6 | 86.0 | 1.7 | 3.2 | 3.1 | 9.5 | 6.0 | 12.7 | 11.1 | 12.9 | 11.9 | 11.9 |
| 31080 | Los Angeles-Long Beach-Anaheim, CA | 4,852 | 13,109,903 | 2 | 2,702.2 | 31.2 | 7.1 | 0.6 | 18.7 | 44.9 | 5.6 | 12.0 | 12.7 | 15.9 | 13.5 | 13.2 |
| 31080 | Anaheim-Santa Ana-Irvine, CA Div 11244 | 793 | 3,166,857 | X | 3,994.5 | 41.8 | 2.2 | 0.6 | 24.6 | 33.8 | 5.7 | 12.0 | 12.6 | 14.5 | 12.9 | 13.6 |
| 31080 | Los Angeles-Long Beach-Glendale, CA Div 31084 | 4,059 | 9,943,046 | X | 2,449.8 | 27.8 | 8.7 | 0.6 | 16.8 | 48.5 | 5.6 | 11.9 | 12.7 | 16.3 | 13.7 | 13.1 |
| 31140 | Louisville/Jefferson County, KY-IN | 3,237 | 1,268,993 | 46 | 392.1 | 76.7 | 16.6 | 0.7 | 2.9 | 5.5 | 6.0 | 12.4 | 12.0 | 14.0 | 12.9 | 12.4 |
| 31180 | Lubbock, TX | 2,688 | 326,364 | 157 | 121.4 | 53.1 | 7.6 | 0.8 | 2.8 | 37.1 | 6.3 | 13.4 | 20.3 | 14.4 | 12.3 | 9.8 |
| 31340 | Lynchburg, VA | 2,121 | 264,386 | 187 | 124.7 | 77.5 | 18.5 | 0.8 | 2.1 | 3.3 | 5.3 | 10.8 | 16.1 | 12.6 | 10.2 | 11.6 |
| 31420 | Macon-Bibb County, GA | 1,724 | 229,900 | 201 | 133.4 | 49.1 | 46.4 | 0.6 | 2.1 | 3.3 | 5.9 | 13.1 | 13.3 | 13.3 | 11.6 | 11.9 |
| 31460 | Madera, CA | 2,137 | 157,761 | 266 | 73.8 | 34.0 | 3.6 | 1.7 | 2.9 | 59.5 | 7.0 | 15.8 | 13.7 | 13.9 | 12.8 | 11.2 |
| 31540 | Madison, WI | 3,309 | 670,447 | 89 | 202.6 | 83.5 | 5.9 | 0.8 | 6.3 | 6.1 | 5.3 | 11.4 | 15.7 | 15.0 | 13.4 | 11.5 |
| 31700 | Manchester-Nashua, NH | 877 | 418,735 | 132 | 477.7 | 85.0 | 3.4 | 0.7 | 5.3 | 7.6 | 5.2 | 11.1 | 12.0 | 14.3 | 12.4 | 13.4 |
| 31740 | Manhattan, KS | 1,835 | 130,142 | 312 | 70.9 | 77.2 | 9.7 | 1.5 | 5.5 | 10.1 | 7.5 | 12.0 | 25.7 | 17.2 | 11.5 | 7.3 |
| 31860 | Mankato, MN | 1,196 | 102,723 | 351 | 85.9 | 88.6 | 5.3 | 0.7 | 2.9 | 4.4 | 5.4 | 11.7 | 21.9 | 13.0 | 12.1 | 9.6 |
| 31900 | Mansfield, OH | 495 | 120,891 | 331 | 244.1 | 87.5 | 10.8 | 0.7 | 1.3 | 2.1 | 5.7 | 12.3 | 11.8 | 12.9 | 11.8 | 11.8 |
| 32580 | McAllen-Edinburg-Mission, TX | 1,571 | 875,200 | 65 | 557.2 | 6.0 | 0.5 | 0.1 | 1.0 | 92.5 | 8.3 | 18.1 | 16.4 | 13.7 | 12.3 | 11.2 |
| 32780 | Medford, OR | 2,783 | 221,844 | 206 | 79.7 | 82.6 | 1.5 | 2.4 | 3.0 | 13.8 | 5.2 | 11.8 | 10.5 | 12.7 | 12.3 | 11.1 |
| 32820 | Memphis, TN-MS-AR | 4,575 | 1,348,678 | 43 | 294.5 | 43.7 | 48.6 | 0.6 | 2.8 | 5.9 | 6.6 | 13.9 | 12.9 | 14.4 | 12.6 | 12.3 |
| 32900 | Merced, CA | 1,938 | 279,252 | 175 | 144.2 | 27.4 | 3.6 | 0.9 | 8.6 | 61.5 | 7.4 | 16.6 | 16.1 | 14.7 | 12.5 | 10.9 |
| 33100 | Miami-Fort Lauderdale-Pompano Beach, FL | 5,067 | 6,173,008 | 7 | 1,218.3 | 30.7 | 21.1 | 0.3 | 3.2 | 45.9 | 5.5 | 11.7 | 11.2 | 13.2 | 13.0 | 13.5 |
| 33100 | Fort Lauderdale-Pompano Beach-Sunrise, FL Div 22744 | 1,203 | 1,958,105 | X | 1,628.0 | 35.5 | 29.8 | 0.4 | 4.7 | 31.4 | 5.7 | 11.7 | 11.1 | 13.3 | 13.5 | 13.5 |
| 33100 | Miami-Miami Beach-Kendall, FL Div 33124 | 1,900 | 2,707,303 | X | 1,425.0 | 14.0 | 15.7 | 0.2 | 1.9 | 68.9 | 5.7 | 11.1 | 11.5 | 13.9 | 13.6 | 14.2 |
| 33100 | West Palm Beach-Boca Raton-Boynton Beach, FL Div 48424 | 1,964 | 1,507,600 | X | 767.5 | 54.4 | 19.6 | 0.4 | 3.6 | 23.4 | 5.0 | 10.5 | 10.6 | 11.8 | 11.5 | 12.3 |
| 33140 | Michigan City-La Porte, IN | 598 | 109,663 | 343 | 183.3 | 80.5 | 13.0 | 0.8 | 1.1 | 7.1 | 5.8 | 12.0 | 11.6 | 13.4 | 12.1 | 12.4 |
| 33220 | Midland, MI | 518 | 83,441 | 372 | 161.6 | 92.6 | 2.1 | 1.0 | 2.9 | 3.1 | 5.3 | 11.7 | 12.1 | 12.5 | 12.1 | 12.4 |
| 33260 | Midland, TX | 1,815 | 183,679 | 234 | 101.2 | 44.0 | 6.9 | 0.8 | 2.5 | 47.1 | 8.6 | 16.3 | 12.7 | 17.6 | 14.5 | 10.0 |
| 33340 | Milwaukee-Waukesha, WI | 1,455 | 1,577,676 | 40 | 1,084.4 | 67.4 | 17.6 | 0.9 | 4.9 | 11.3 | 6.1 | 12.7 | 12.6 | 14.0 | 12.9 | 11.9 |
| 33460 | Minneapolis-St. Paul-Bloomington, MN | 7,048 | 3,657,477 | 16 | 519.0 | 76.7 | 10.7 | 1.3 | 8.2 | 6.1 | 6.2 | 13.1 | 12.2 | 14.5 | 13.8 | 12.2 |
| 33540 | Missoula, MT | 2,593 | 121,630 | 328 | 46.9 | 91.3 | 1.1 | 3.9 | 3.0 | 3.7 | 4.7 | 10.5 | 16.5 | 15.9 | 13.6 | 10.5 |

1. CBSA = Core Based Statistical Area. DIV = Metropolitan Division. See Appendix A for explanation. See Appendix B for list of metropolitan areas.   2. Dry land or land partially or temporarily covered by water.   3. May be of any race.

# Table C. Metropolitan Areas — **Population and Households**

| Area name | Population, 2020 (cont.) — Age (percent) (cont.) | | | | Population change and components of change, 2000-2020 — Total persons | | Percent change | | Components of change, 2010-2020 | | | Households, 2019 | | Percent | | |
|---|---|---|---|---|---|---|---|---|---|---|---|---|---|---|---|---|
| | 55 to 64 years | 65 to 74 years | 75 years and over | Percent female | 2000 | 2010 | 2000-2010 | 2010-2019 | Births | Deaths | Net migration | Number | Persons per household | Family households | Single parent households | One person |
| | 16 | 17 | 18 | 19 | 20 | 21 | 22 | 23 | 24 | 25 | 26 | 27 | 28 | 29 | 30 | 31 |
| Jonesboro, AR | 11.4 | 8.6 | 6.3 | 51.6 | 107,762 | 121,013 | 12.3 | 12.0 | 18,269 | 13,033 | 9,269 | 51,288 | 2.53 | 66.0 | 13.0 | 29.0 |
| Joplin, MO | 12.6 | 9.8 | 7.3 | 50.9 | 157,322 | 175,509 | 11.6 | 2.6 | 24,351 | 18,777 | -900 | 67,167 | 2.62 | 64.8 | 10.7 | 27.6 |
| Kahului-Wailuku-Lahaina, HI | 14.0 | 11.9 | 7.6 | 50.4 | 128,094 | 154,840 | 20.9 | 8.4 | 19,797 | 12,199 | 5,531 | 54,744 | 3.01 | 67.4 | 9.3 | 24.7 |
| Kalamazoo-Portage, MI | 11.4 | 9.3 | 6.5 | 51.1 | 238,603 | 250,327 | 4.9 | 6.3 | 31,878 | 21,796 | 5,744 | 103,196 | 2.50 | 59.1 | 8.8 | 30.1 |
| Kankakee, IL | 13.1 | 10.0 | 7.5 | 50.6 | 103,833 | 113,450 | 9.3 | -4.3 | 13,470 | 11,623 | -6,722 | 39,583 | 2.64 | 63.0 | 7.9 | 30.5 |
| Kansas City, MO-KS | 12.7 | 9.3 | 6.4 | 50.9 | 1,811,254 | 2,009,353 | 10.9 | 8.2 | 280,570 | 174,930 | 59,380 | 844,310 | 2.52 | 64.5 | 10.2 | 28.8 |
| Kennewick-Richland, WA | 10.9 | 8.5 | 5.4 | 49.5 | 191,822 | 253,328 | 32.1 | 19.8 | 43,316 | 18,226 | 24,999 | 101,840 | 2.90 | 71.4 | 15.7 | 24.6 |
| Killeen-Temple, TX | 9.9 | 7.1 | 4.7 | 50.3 | 330,714 | 405,327 | 22.6 | 15.6 | 75,689 | 28,598 | 15,575 | 157,088 | 2.76 | 67.9 | 14.9 | 27.0 |
| Kingsport-Bristol, TN-VA | 14.7 | 12.9 | 10.0 | 51.1 | 298,484 | 309,493 | 3.7 | -0.4 | 29,822 | 40,290 | 9,395 | 130,928 | 2.30 | 65.1 | 10.2 | 30.8 |
| Kingston, NY | 15.6 | 12.2 | 8.7 | 50.4 | 177,749 | 182,522 | 2.7 | -2.6 | 16,083 | 17,416 | -3,429 | 69,376 | 2.40 | 60.0 | 9.0 | 33.3 |
| Knoxville, TN | 13.5 | 11.2 | 7.9 | 51.1 | 727,600 | 815,035 | 12.0 | 7.7 | 93,371 | 93,337 | 63,241 | 347,927 | 2.44 | 64.0 | 9.4 | 30.0 |
| Kokomo, IN | 13.8 | 11.2 | 8.9 | 51.6 | 84,964 | 82,748 | -2.6 | 0.0 | 10,165 | 10,109 | -14 | 34,151 | 2.38 | 61.2 | 9.9 | 33.6 |
| La Crosse-Onalaska, WI-MN | 12.9 | 10.5 | 7.7 | 51.0 | 126,838 | 133,660 | 5.4 | 2.6 | 14,526 | 12,051 | 1,044 | 56,515 | 2.32 | 56.7 | 7.1 | 31.8 |
| Lafayette, LA | 13.1 | 9.2 | 6.0 | 51.3 | 425,020 | 466,732 | 9.8 | 4.9 | 68,551 | 43,314 | -2,162 | 184,313 | 2.61 | 67.3 | 13.1 | 27.1 |
| Lafayette-West Lafayette, IN | 10.3 | 7.9 | 5.6 | 49.1 | 186,960 | 210,308 | 12.5 | 10.9 | 27,196 | 15,774 | 11,593 | 89,655 | 2.41 | 55.5 | 8.8 | 34.4 |
| Lake Charles, LA | 12.6 | 9.4 | 6.3 | 51.2 | 193,568 | 199,635 | 3.1 | 5.3 | 29,944 | 21,056 | 1,927 | 79,505 | 2.59 | 69.4 | 15.8 | 24.2 |
| Lake Havasu City-Kingman, AZ | 16.3 | 18.1 | 13.9 | 49.6 | 155,032 | 200,181 | 29.1 | 8.5 | 18,687 | 30,248 | 28,568 | 91,095 | 2.29 | 67.1 | 8.8 | 26.5 |
| Lakeland-Winter Haven, FL | 12.2 | 11.3 | 9.2 | 51.0 | 483,924 | 602,068 | 24.4 | 23.7 | 77,998 | 68,161 | 132,537 | 238,629 | 2.98 | 66.5 | 11.3 | 27.0 |
| Lancaster, PA | 13.0 | 10.4 | 8.6 | 51.0 | 470,658 | 519,443 | 10.4 | 5.1 | 72,008 | 50,889 | 6,056 | 204,701 | 2.61 | 69.6 | 8.0 | 23.8 |
| Lansing-East Lansing, MI | 12.8 | 10.1 | 6.5 | 51.0 | 519,415 | 534,687 | 2.9 | 2.5 | 60,285 | 45,675 | -1,086 | 215,854 | 2.44 | 58.9 | 8.7 | 32.9 |
| Laredo, TX | 8.8 | 5.9 | 4.1 | 50.7 | 193,117 | 250,304 | 29.6 | 10.9 | 53,126 | 13,615 | -12,134 | 76,326 | 3.58 | 82.7 | 18.4 | 15.2 |
| Las Cruces, NM | 10.8 | 9.6 | 7.1 | 50.8 | 174,682 | 209,212 | 19.8 | 5.8 | 29,575 | 15,857 | -1,709 | 79,094 | 2.69 | 64.3 | 12.4 | 27.4 |
| Las Vegas-Henderson-Paradise, NV | 11.9 | 9.4 | 6.1 | 50.2 | 1,375,765 | 1,951,278 | 41.8 | 18.7 | 274,679 | 161,068 | 250,044 | 813,607 | 2.76 | 63.4 | 11.9 | 28.8 |
| Lawrence, KS | 10.1 | 8.2 | 5.1 | 50.3 | 99,962 | 110,826 | 10.9 | 10.6 | 12,245 | 7,120 | 6,605 | 48,508 | 2.35 | 55.2 | 7.5 | 26.2 |
| Lawton, OK | 11.6 | 8.0 | 5.7 | 48.4 | 121,610 | 130,279 | 7.1 | -2.7 | 19,596 | 11,464 | -11,868 | 44,127 | 2.65 | 63.9 | 13.6 | 30.4 |
| Lebanon, PA | 13.5 | 11.2 | 8.9 | 50.8 | 120,327 | 133,596 | 11.0 | 6.0 | 16,592 | 15,185 | 6,782 | 53,861 | 2.57 | 67.4 | 11.4 | 25.4 |
| Lewiston, ID-WA | 14.2 | 12.3 | 9.9 | 50.9 | 57,961 | 60,893 | 5.1 | 4.4 | 7,242 | 7,696 | 3,167 | 26,915 | 2.30 | 60.1 | 9.4 | 28.4 |
| Lewiston-Auburn, ME | 14.5 | 10.9 | 7.5 | 51.1 | 103,793 | 107,713 | 3.8 | 0.8 | 13,026 | 11,262 | -890 | 45,924 | 2.29 | 63.1 | 12.4 | 27.1 |
| Lexington-Fayette, KY | 12.1 | 9.2 | 6.0 | 51.2 | 408,326 | 472,095 | 15.6 | 10.2 | 63,960 | 41,005 | 25,409 | 205,914 | 2.43 | 60.5 | 10.2 | 30.8 |
| Lima, OH | 13.3 | 10.8 | 7.7 | 49.5 | 108,473 | 106,295 | -2 | -4.1 | 12,966 | 11,420 | -5,865 | 40,460 | 2.44 | 64.1 | 13.9 | 29.4 |
| Lincoln, NE | 11.0 | 9.1 | 5.9 | 49.8 | 266,787 | 302,157 | 13.3 | 11.8 | 42,896 | 23,411 | 16,288 | 135,008 | 2.38 | 59.7 | 7.9 | 29.3 |
| Little Rock-North Little Rock-Conway, AR | 12.5 | 9.7 | 6.5 | 51.8 | 610,518 | 699,781 | 14.6 | 6.7 | 97,615 | 67,985 | 17,288 | 288,565 | 2.53 | 63.6 | 11.6 | 30.7 |
| Logan, UT-ID | 7.7 | 6.1 | 4.4 | 50.0 | 102,720 | 125,442 | 22.1 | 15.0 | 25,535 | 6,545 | -239 | 45,043 | 3.08 | 77.7 | 7.3 | 14.9 |
| Longview, TX | 12.8 | 10.0 | 7.1 | 50.2 | 256,152 | 280,005 | 9.3 | 2.5 | 37,825 | 30,231 | -416 | 102,964 | 2.67 | 68.4 | 12.2 | 26.2 |
| Longview, WA | 13.9 | 11.7 | 7.9 | 50.4 | 92,948 | 102,408 | 10.2 | 8.8 | 12,585 | 11,610 | 8,016 | 42,790 | 2.56 | 65.5 | 11.8 | 29.4 |
| Los Angeles-Long Beach-Anaheim, CA | 12.3 | 8.5 | 6.4 | 50.7 | 12,365,627 | 12,828,904 | 3.7 | 2.2 | 1,638,228 | 839,477 | -512,836 | 4,372,678 | 2.97 | 67.5 | 10.3 | 24.7 |
| Anaheim-Santa Ana-Irvine, CA Div 11244 | 13.0 | 8.9 | 6.8 | 50.7 | 2,846,289 | 3,010,258 | 5.8 | 5.2 | 382,227 | 200,199 | -23,093 | 1,044,280 | 3.00 | 71.5 | 8.1 | 21.2 |
| Los Angeles-Long Beach-Glendale, CA Div 31084 | 12.1 | 8.3 | 6.2 | 50.7 | 9,519,338 | 9,818,646 | 3.1 | 1.3 | 1,256,001 | 639,278 | -489,743 | 3,328,398 | 2.96 | 66.2 | 11.0 | 25.8 |
| Louisville/Jefferson County, KY-IN | 13.5 | 10.2 | 6.6 | 51.2 | 1,090,024 | 1,202,686 | 10.3 | 5.5 | 156,220 | 123,786 | 34,598 | 495,246 | 2.51 | 63.1 | 10.9 | 30.1 |
| Lubbock, TX | 10.3 | 7.7 | 5.5 | 50.8 | 256,250 | 290,917 | 13.5 | 12.2 | 42,873 | 26,128 | 18,622 | 121,838 | 2.54 | 60.4 | 11.4 | 30.4 |
| Lynchburg, VA | 13.8 | 11.1 | 8.5 | 51.7 | 222,317 | 252,654 | 13.6 | 4.6 | 29,107 | 27,473 | 10,194 | 102,879 | 2.47 | 66.5 | 9.4 | 27.9 |
| Macon-Bibb County, GA | 13.3 | 10.4 | 7.1 | 52.4 | 222,368 | 232,241 | 4.4 | 4.6 | 29,846 | 25,072 | -7,107 | 85,864 | 2.60 | 62.5 | 13.9 | 31.8 |
| Madera, CA | 11.0 | 8.8 | 5.8 | 51.6 | 123,109 | 150,834 | 22.5 | 10.7 | 22,963 | 11,118 | -4,992 | 44,387 | 3.35 | 82.5 | 17.6 | 14.0 |
| Madison, WI | 12.2 | 9.5 | 6.0 | 50.1 | 535,421 | 605,477 | 13.1 | | 74,439 | 44,309 | 34,921 | 279,014 | 2.33 | 56.9 | 6.9 | 31.5 |
| Manchester-Nashua, NH | 15.0 | 9.9 | 6.8 | 50.2 | 380,841 | 400,706 | 5.2 | 4.5 | 43,670 | 33,432 | 7,970 | 162,588 | 2.52 | 64.4 | 8.1 | 26.6 |
| Manhattan, KS | 7.9 | 6.4 | 4.5 | 48.1 | 108,999 | 127,094 | 16.6 | 2.4 | 24,171 | 7,532 | -14,079 | 46,526 | 2.54 | 59.2 | 9.5 | 29.8 |
| Mankato, MN | 10.9 | 8.9 | 6.5 | 49.8 | 85,712 | 96,742 | 12.9 | 6.2 | 11,387 | 7,357 | 1,971 | 39,791 | 2.39 | 59.5 | 10.3 | 29.0 |
| Mansfield, OH | 13.4 | 11.5 | 8.9 | 49.3 | 128,852 | 124,472 | -3.4 | -2.9 | 14,214 | 14,524 | -3,247 | 49,812 | 2.29 | 66.2 | 9.6 | 28.7 |
| McAllen-Edinburg-Mission, TX | 8.5 | 6.4 | 5.1 | 50.9 | 569,463 | 774,734 | 36.0 | 13.0 | 157,532 | 42,913 | -14,013 | 247,544 | 3.47 | 79.3 | 17.8 | 17.6 |
| Medford, OR | 13.4 | 13.6 | 9.5 | 51.2 | 181,269 | 203,206 | 12.1 | 9.2 | 23,394 | 23,882 | 19,130 | 87,800 | 2.48 | 60.7 | 8.2 | 30.2 |
| Memphis, TN-MS-AR | 12.5 | 9.1 | 5.7 | 52.4 | 1,205,204 | 1,316,100 | 9.2 | 2.5 | 188,746 | 120,188 | -36,133 | 505,048 | 2.62 | 63.8 | 14.1 | 31.0 |
| Merced, CA | 10.1 | 6.9 | 4.8 | 49.5 | 210,554 | 255,797 | 21.5 | 9.2 | 42,237 | 17,487 | -1,251 | 80,645 | 3.36 | 78.9 | 18.3 | 16.4 |
| Miami-Fort Lauderdale-Pompano Beach, FL | 13.3 | 10.0 | 9.1 | 51.4 | 5,007,564 | 5,566,285 | 11.2 | 10.9 | 693,271 | 503,171 | 417,395 | 2,187,192 | 2.78 | 65.5 | 11.5 | 27.8 |
| Fort Lauderdale-Pompano Beach-Sunrise, FL Div 22744 | 13.7 | 9.8 | 7.7 | 51.2 | 1,623,018 | 1,748,158 | 7.7 | 12.0 | 223,220 | 154,173 | 142,202 | 705,472 | 2.74 | 63.0 | 11.8 | 30.0 |
| Miami-Miami Beach-Kendall, FL Div 33124 | 12.9 | 9.1 | 7.9 | 51.3 | 2,253,362 | 2,498,003 | 10.9 | 8.4 | 321,239 | 201,879 | 89,523 | 912,805 | 2.93 | 69.4 | 12.8 | 24.6 |
| West Palm Beach-Boca Raton-Boynton Beach, FL Div 48424 | 13.4 | 12.0 | 12.8 | 51.6 | 1,131,184 | 1,320,124 | 16.7 | 14.2 | 148,812 | 147,119 | 185,670 | 568,915 | 2.59 | 62.3 | 9.1 | 30.3 |
| Michigan City-La Porte, IN | 13.9 | 11.3 | 7.5 | 48.4 | 110,106 | 111,466 | 1.2 | -1.6 | 13,352 | 12,298 | -2,797 | 43,100 | 2.36 | 66.4 | 12.9 | 28.4 |
| Midland, MI | 14.3 | 10.9 | 8.7 | 50.5 | 82,874 | 83,623 | 0.9 | -0.2 | 8,810 | 7,532 | -1,455 | 34,350 | 2.38 | 67.3 | 7.7 | 27.3 |
| Midland, TX | 10.0 | 6.1 | 4.2 | 49.2 | 120,755 | 141,686 | 17.3 | 29.6 | 29,248 | 11,513 | 23,747 | 61,159 | 2.96 | 66.4 | 13.2 | 29.2 |
| Milwaukee-Waukesha, WI | 13.3 | 9.8 | 6.8 | 51.2 | 1,500,741 | 1,555,954 | 3.7 | 1.4 | 201,811 | 138,856 | -40,605 | 635,351 | 2.43 | 60.7 | 10.6 | 32.0 |
| Minneapolis-St. Paul-Bloomington, MN | 13.0 | 8.9 | 6.0 | 50.4 | 3,016,562 | 3,333,623 | 10.5 | 9.7 | 462,985 | 238,354 | 101,213 | 1,396,025 | 2.56 | 63.1 | 8.9 | 29.2 |
| Missoula, MT | 11.5 | 10.6 | 6.2 | 50.0 | 95,802 | 109,295 | 14.1 | 11.3 | 12,092 | 8,909 | 9,164 | 50,853 | 2.29 | 59.3 | 7.9 | 29.0 |

| Area name | Persons in group quarters, 2020 | Daytime population, 2019 Number | Employment/ residence ratio | Births, 2020 Total | Rate[1] | Deaths, 2020 Number | Rate[1] | Persons under 65 with no health insurance 2019 Number | Percent | Medicare, 2020 Total Beneficiaries | Enrolled in Original Medicare | Enrolled in Medicare Advantage | Serious crimes known to police[2], 2019 Violent Number | Rate[3] |
|---|---|---|---|---|---|---|---|---|---|---|---|---|---|---|
| | 32 | 33 | 34 | 35 | 36 | 37 | 38 | 39 | 40 | 41 | 42 | 43 | 44 | 45 |
| Jonesboro, AR............... | 4,284 | 135,562 | 1.03 | 1,822 | 13.4 | 1,485 | 11.0 | 11,357 | 10.3 | 24,835 | 17,713 | 7,122 | 743 | 556.0 |
| Joplin, MO..................... | 3,368 | 183,374 | 1.05 | 2,271 | 12.6 | 1,942 | 10.8 | 23,938 | 16.4 | 36,405 | 23,883 | 12,522 | 593 | 331.1 |
| Kahului-Wailuku-Lahaina, HI ... | 2,845 | D | D | 1,794 | 10.7 | 1,555 | 9.3 | 7,904 | 5.9 | 31,643 | 15,326 | 16,317 | 449 | 268.1 |
| Kalamazoo-Portage, MI........ | 8,156 | 275,963 | 1.08 | 3,044 | 11.4 | 2,383 | 9.0 | 14,695 | 6.8 | 49,789 | 23,993 | 25,796 | 1,523 | 572.5 |
| Kankakee, IL.................. | 6,270 | 106,176 | 0.93 | 1,205 | 11.1 | 1,280 | 11.8 | 6,322 | 7.4 | 21,977 | 16,455 | 5,523 | 424 | 388.6 |
| Kansas City, MO-KS........... | 31,200 | 2,176,595 | 1.02 | 26,606 | 12.2 | 19,290 | 8.9 | 191,651 | 10.6 | 377,526 | 232,188 | 145,336 | NA | NA |
| Kennewick-Richland, WA ....... | 4,239 | 298,371 | 0.99 | 4,120 | 13.6 | 2,109 | 6.9 | 25,187 | 10.0 | 47,816 | 44,171 | 3,646 | 660 | 219.5 |
| Killeen-Temple, TX............ | 23,399 | 450,755 | 0.95 | 7,043 | 15.0 | 3,247 | 6.9 | 64,056 | 16.6 | 69,912 | 44,351 | 25,561 | 1,202 | 264.4 |
| Kingsport-Bristol, TN-VA ....... | 6,090 | 309,965 | 1.02 | 2,805 | 9.1 | 4,113 | 13.3 | 26,734 | 11.5 | 85,742 | 36,975 | 48,768 | 920 | 300.8 |
| Kingston, NY.................. | 11,664 | 166,864 | 0.88 | 1,517 | 8.5 | 1,868 | 10.5 | 7,976 | 5.9 | 41,372 | 28,985 | 12,387 | 266 | 150.2 |
| Knoxville, TN................. | 19,956 | 881,902 | 1.03 | 9,023 | 10.3 | 10,047 | 11.4 | 78,553 | 11.4 | 194,298 | 103,651 | 90,647 | 3,064 | 353.7 |
| Kokomo, IN................... | 1,281 | 84,794 | 1.06 | 995 | 12.0 | 1,016 | 12.3 | 6,419 | 9.9 | 19,535 | 12,687 | 6,848 | 458 | 555.3 |
| La Crosse-Onalaska, WI-MN... | 5,644 | 143,351 | 1.09 | 1,329 | 9.7 | 1,252 | 9.1 | 5,643 | 5.2 | 28,206 | 16,265 | 11,941 | 169 | 123.4 |
| Lafayette, LA................. | 8,392 | 494,576 | 1.02 | 6,228 | 12.7 | 5,014 | 10.2 | 41,568 | 10.1 | 88,869 | 64,816 | 24,055 | 2,283 | 466.5 |
| Lafayette-West Lafayette, IN ... | 16,834 | 238,654 | 1.04 | 2,527 | 10.8 | 1,705 | 7.3 | 18,853 | 10.1 | 33,888 | 24,388 | 9,501 | 520 | 223.0 |
| Lake Charles, LA.............. | 3,812 | 225,367 | 1.19 | 3,043 | 14.5 | 2,291 | 10.9 | 16,742 | 9.6 | 37,564 | 28,510 | 9,054 | 1,151 | 547.7 |
| Lake Havasu City-Kingman, AZ | 4,152 | 199,076 | 0.82 | 1,780 | 8.2 | 3,656 | 16.8 | 19,205 | 13.6 | 68,799 | 44,048 | 24,751 | 485 | 230.0 |
| Lakeland-Winter Haven, FL..... | 13,783 | 671,565 | 0.83 | 8,169 | 11.0 | 8,009 | 10.8 | 95,885 | 17.1 | 160,554 | 68,277 | 92,278 | 1,990 | 277.5 |
| Lancaster, PA................. | 12,603 | 538,610 | 0.97 | 6,879 | 12.6 | 5,565 | 10.2 | 48,810 | 11.2 | 113,309 | 64,835 | 48,474 | NA | NA |
| Lansing-East Lansing, MI....... | 20,629 | 550,076 | 1.00 | 5,597 | 10.2 | 4,936 | 9.0 | 30,067 | 6.8 | 104,820 | 52,054 | 52,765 | 2,367 | 429.9 |
| Laredo, TX.................... | 3,647 | 270,898 | 0.95 | 4,638 | 16.7 | 1,482 | 5.3 | 73,316 | 30.2 | 33,804 | 20,866 | 12,938 | 893 | 321.9 |
| Las Cruces, NM................ | 4,412 | 204,960 | 0.85 | 2,598 | 11.7 | 1,777 | 8.0 | 23,882 | 13.5 | 42,625 | 24,542 | 18,084 | 1,215 | 556.9 |
| Las Vegas-Henderson-Para-dise, NV | 22,331 | 2,269,331 | 1.00 | 26,886 | 11.6 | 19,618 | 8.5 | 260,442 | 13.7 | 371,680 | 203,111 | 168,568 | 11,935 | 525.7 |
| Lawrence, KS.................. | 8,814 | 112,850 | 0.86 | 1,101 | 9.0 | 754 | 6.2 | 9,568 | 9.7 | 18,145 | 14,566 | 3,578 | NA | NA |
| Lawton, OK................... | 10,328 | 126,543 | 1.01 | 1,694 | 13.4 | 1,287 | 10.2 | 14,881 | 15.0 | 20,814 | 17,988 | 2,827 | 897 | 716.4 |
| Lebanon, PA.................. | 3,684 | 136,615 | 0.93 | 1,541 | 10.9 | 1,628 | 11.5 | 9,080 | 8.2 | 32,434 | 18,032 | 14,402 | 237 | 166.8 |
| Lewiston, ID-WA............... | 1,236 | 62,486 | 0.98 | 673 | 10.6 | 780 | 12.3 | 5,104 | 10.5 | 16,840 | 13,220 | 3,621 | 88 | 138.9 |
| Lewiston-Auburn, ME.......... | 3,165 | 107,658 | 0.99 | 1,211 | 11.2 | 1,176 | 10.8 | 8,323 | 9.7 | 25,081 | 12,441 | 12,660 | 186 | 172.2 |
| Lexington-Fayette, KY.......... | 16,987 | 541,170 | 1.09 | 6,066 | 11.7 | 4,422 | 8.5 | 35,235 | 8.3 | 87,794 | 49,166 | 38,628 | 1,200 | 230.6 |
| Lima, OH..................... | 5,743 | 107,282 | 1.11 | 1,237 | 12.1 | 1,168 | 11.5 | 5,966 | 7.6 | 22,007 | 13,676 | 8,331 | 322 | 315.6 |
| Lincoln, NE................... | 16,506 | 348,488 | 1.07 | 3,983 | 11.8 | 2,472 | 7.3 | 22,087 | 8.1 | 55,271 | 44,000 | 11,271 | 1,170 | 346.8 |
| Little Rock-North Little Rock-Conway, AR ................ | 14,933 | 757,970 | 1.04 | 8,956 | 12.0 | 7,416 | 9.9 | 57,492 | 9.4 | 146,160 | 107,576 | 38,584 | 5,819 | 782.3 |
| Logan, UT-ID................. | 4,034 | 139,111 | 0.96 | 2,296 | 15.9 | 699 | 4.8 | 11,942 | 9.7 | 17,028 | 9,251 | 7,778 | 183 | 128.4 |
| Longview, TX................. | 11,693 | 287,050 | 1.00 | 3,328 | 11.6 | 3,230 | 11.3 | 47,760 | 21.0 | 57,531 | 35,455 | 22,076 | 924 | 324.1 |
| Longview, WA................. | 1,208 | 107,371 | 0.93 | 1,256 | 11.3 | 1,275 | 11.4 | 6,458 | 7.4 | 26,672 | 12,182 | 14,490 | 213 | 194.6 |
| Los Angeles-Long Beach-Anaheim, CA.............. | 223,881 | 13,458,686 | 1.04 | 142,952 | 10.9 | 96,746 | 7.4 | 1,175,754 | 10.6 | 2,043,382 | 984,867 | 1,058,515 | 63,201 | 477.5 |
| Anaheim-Santa Ana-Irvine, CA Div 11244............... | 43,650 | 3,279,076 | 1.06 | 35,163 | 11.1 | 23,233 | 7.3 | 235,378 | 8.8 | 519,799 | 249,496 | 270,303 | 6,797 | 213.7 |
| Los Angeles-Long Beach-Glendale, CA Div 31084 | 180,231 | 10,179,610 | 1.03 | 107,789 | 10.8 | 73,513 | 7.4 | 940,376 | 11.1 | 1,523,583 | 735,371 | 788,212 | 56,404 | 560.9 |
| Louisville/Jefferson County, KY-IN ...................... | 27,031 | 1,293,206 | 1.04 | 14,663 | 11.6 | 13,054 | 10.3 | 76,564 | 7.4 | 248,011 | 155,005 | 93,006 | 5,545 | 436.8 |
| Lubbock, TX .................. | 12,385 | 322,843 | 1.01 | 4,005 | 12.3 | 2,878 | 8.8 | 47,955 | 17.8 | 49,987 | 29,286 | 20,701 | 2,688 | 838.5 |
| Lynchburg, VA................. | 13,503 | 261,790 | 0.96 | 2,805 | 10.6 | 2,985 | 11.3 | 19,529 | 9.7 | 60,512 | 46,130 | 14,382 | 607 | 230.6 |
| Macon-Bibb County, GA......... | 8,426 | 242,751 | 1.12 | 2,642 | 11.5 | 2,866 | 12.5 | 26,577 | 14.6 | 48,420 | 26,521 | 21,899 | 1,042 | 454.6 |
| Madera, CA................... | 7,131 | 148,843 | 0.85 | 2,042 | 12.9 | 1,178 | 7.5 | 13,807 | 11.0 | 24,911 | 15,639 | 9,272 | 830 | 528.0 |
| Madison, WI.................. | 15,324 | 693,764 | 1.08 | 6,907 | 10.3 | 4,719 | 7.0 | 28,503 | 5.1 | 116,290 | 80,371 | 35,919 | 1,452 | 217.8 |
| Manchester-Nashua, NH........ | 7,678 | 406,838 | 0.96 | 4,196 | 10.0 | 3,708 | 8.9 | 27,581 | 8.0 | 81,114 | 61,914 | 19,200 | 957 | 229.8 |
| Manhattan, KS................ | 10,373 | 134,480 | 1.06 | 2,097 | 16.1 | 794 | 6.1 | 9,940 | 9.4 | 16,440 | 14,871 | 1,569 | 563 | 430.9 |
| Mankato, MN................. | 7,020 | 107,085 | 1.09 | 1,063 | 10.3 | 793 | 7.7 | 4,486 | 5.5 | 18,006 | 11,697 | 6,308 | 182 | 178.3 |
| Mansfield, OH................. | 6,872 | 120,264 | 0.98 | 1,385 | 11.5 | 1,494 | 12.4 | 7,840 | 8.7 | 28,900 | 18,266 | 10,634 | 276 | 229.0 |
| McAllen-Edinburg-Mission, TX | 7,029 | 840,397 | 0.92 | 14,081 | 16.1 | 5,110 | 5.8 | 248,170 | 33.2 | 107,772 | 43,160 | 64,612 | 2,345 | 268.9 |
| Medford, OR.................. | 3,484 | 221,218 | 1.00 | 2,159 | 9.7 | 2,524 | 11.4 | 16,303 | 9.8 | 57,969 | 37,447 | 20,522 | NA | NA |
| Memphis, TN-MS-AR ........... | 23,422 | 1,366,904 | 1.04 | 17,644 | 13.1 | 13,371 | 9.9 | 143,397 | 12.8 | 231,690 | 155,622 | 76,068 | 15,068 | 1,120.5 |
| Merced, CA................... | 6,943 | 253,892 | 0.78 | 3,855 | 13.8 | 1,922 | 6.9 | 26,939 | 11.4 | 37,596 | 31,828 | 5,768 | 1,520 | 553.2 |
| Miami-Fort Lauderdale-Pom-pano Beach, FL............ | 80,368 | 6,189,759 | 1.01 | 66,442 | 10.8 | 57,339 | 9.3 | 910,265 | 18.5 | 1,139,614 | 470,141 | 669,475 | 26,314 | 422.0 |
| Fort Lauderdale-Pompano Beach-Sunrise, FL Div 22744 | 16,287 | 1,876,775 | 0.92 | 21,491 | 11.0 | 17,369 | 8.9 | 282,874 | 17.7 | 331,717 | 137,131 | 194,586 | 7,332 | 373.4 |
| Miami-Miami Beach-Kendall, FL Div 33124 | 42,231 | 2,790,567 | 1.05 | 30,086 | 11.1 | 23,189 | 8.6 | 428,190 | 19.4 | 475,765 | 138,854 | 336,912 | 13,339 | 480.7 |
| West Palm Beach-Boca Raton-Boynton Beach, FL Div 48424 | 21,850 | 1,522,417 | 1.04 | 14,865 | 9.9 | 16,781 | 11.1 | 199,201 | 17.9 | 332,132 | 194,156 | 137,977 | 5,643 | 377.0 |
| Michigan City-La Porte, IN....... | 6,387 | 102,177 | 0.84 | 1,249 | 11.4 | 1,310 | 11.9 | 8,550 | 10.3 | 23,751 | 17,238 | 6,513 | NA | NA |
| Midland, MI.................. | 1,275 | 85,185 | 1.05 | 836 | 10.0 | 803 | 9.6 | 4,200 | 6.3 | 18,636 | 10,009 | 8,626 | 118 | 142.3 |
| Midland, TX.................. | 1,573 | 214,963 | 1.35 | 3,113 | 16.9 | 1,265 | 6.9 | 30,060 | 18.4 | 20,082 | 14,330 | 5,752 | NA | NA |
| Milwaukee-Waukesha, WI ...... | 30,611 | 1,618,187 | 1.05 | 18,830 | 11.9 | 14,664 | 9.3 | 88,327 | 6.8 | 295,112 | 143,763 | 151,349 | 9,111 | 577.9 |
| Minneapolis-St. Paul-Bloom-ington, MN | 63,760 | 3,669,365 | 1.01 | 43,836 | 12.0 | 26,813 | 7.3 | 165,220 | 5.4 | 594,170 | 292,236 | 301,937 | 9,926 | 272.5 |
| Missoula, MT................. | 3,644 | 124,909 | 1.08 | 1,057 | 8.7 | 973 | 8.0 | 8,936 | 9.2 | 23,220 | 18,409 | 4,811 | 418 | 349.3 |

1. Per 1,000 estimated resident population.   2. Data for serious crimes have not been adjusted for underreporting; this may affect comparability between geographic areas and over time.   3. Per 100,000 population estimated by the FBI.

# Table C. Metropolitan Areas — Crime, Education, Money Income, and Poverty

| Area name | Serious crimes known to police[1], 2019 (cont.) Property | | School enrollment and attainment, 2019 Enrollment[3] | | Attainment[4] | | Local government expenditures,[5] 2017-2018 | | Income and poverty, 2019 | | | | | Percent below poverty level | | |
|---|---|---|---|---|---|---|---|---|---|---|---|---|---|---|---|---|
| | Number | Rate | Total | Percent private | High school graduate or less | Bachelor's degree or more | Total current expenditures (mil dol) | Current expenditures per student (dollars) | Per capita income[6] (dollars) | Median household income (dollars) | Median family income | Percent of households with income less than $50,000 | Percent of households with income of $200,000 or more | All persons | All families | Families with children under 18 |
| | 46 | 47 | 48 | 49 | 50 | 51 | 52 | 53 | 54 | 55 | 56 | 57 | 58 | 59 | 60 | 61 |
| Jonesboro, AR | 4,116 | 3,080.3 | 37,305 | 7.8 | 46.2 | 23.5 | 227.5 | 9,636 | 26,754 | 44,998 | 63,500 | 54.1 | 3.8 | 18.9 | 12.6 | 25.1 |
| Joplin, MO | 6,555 | 3,659.9 | 45,075 | 13.8 | 47.6 | 22.8 | 260.2 | 8,551 | 24,850 | 48,909 | 58,460 | 50.9 | 3.0 | 18.0 | 11.8 | 21.2 |
| Kahului-Wailuku-Lahaina, HI | 4,984 | 2,976.0 | 32,855 | 22.7 | 40.1 | 27.7 | NA | NA | 35,210 | 80,754 | 90,983 | 29.1 | 9.1 | 11.9 | 9.2 | 12.1 |
| Kalamazoo-Portage, MI | 8,583 | 3,226.5 | 75,118 | 13.3 | 30.4 | 38.3 | 442.6 | 12,740 | 31,652 | 56,441 | 77,904 | 43.6 | 5.8 | 13.4 | 7.2 | 11.4 |
| Kankakee, IL | 2,096 | 1,921.0 | 25,496 | 17.2 | 47.3 | 19.5 | 229.5 | 12,901 | 28,332 | 60,923 | 76,663 | 42.6 | 2.7 | 12.2 | 7.4 | 16.4 |
| Kansas City, MO-KS | NA | NA | 516,631 | 15.2 | 33.8 | 37.7 | 3,909.5 | 11,016 | 36,358 | 70,215 | 87,555 | 35.3 | 7.3 | 10.3 | 7.2 | 12.3 |
| Kennewick-Richland, WA | 5,851 | 1,945.8 | 79,218 | 9.3 | 39.6 | 27.7 | 717.6 | 12,355 | 31,411 | 68,283 | 78,720 | 34.7 | 7.2 | 12.9 | 11.3 | 18.0 |
| Killeen-Temple, TX | 8,231 | 1,810.4 | 128,213 | 11.2 | 36.5 | 23.8 | 800.3 | 8,940 | 26,927 | 54,370 | 67,032 | 45.1 | 4.3 | 13.7 | 9.9 | 14.6 |
| Kingsport-Bristol, TN-VA | 7,019 | 2,295.0 | 55,890 | 16.6 | 47.3 | 23 | 414.9 | 9,890 | 28,770 | 48,615 | 62,727 | 50.9 | 2.9 | 15.5 | 11.9 | 22.0 |
| Kingston, NY | 1,932 | 1,090.6 | 38,012 | 16.0 | 37.9 | 32.9 | 575.4 | 25,125 | 35,743 | 64,087 | 86,335 | 39.4 | 6.9 | 11.9 | 7.5 | 14.9 |
| Knoxville, TN | 18,423 | 2,126.6 | 195,154 | 17.7 | 41.3 | 29.9 | 1,081.7 | 9,190 | 31,816 | 56,623 | 73,785 | 44.7 | 5.2 | 14.1 | 10.0 | 16.1 |
| Kokomo, IN | 1,651 | 2,001.7 | 18,924 | 6.9 | 44.6 | 18.2 | 135.4 | 9,773 | 27,393 | 53,440 | 65,189 | 45.3 | 2.0 | 11.9 | 9.0 | 17.6 |
| La Crosse-Onalaska, WI-MN | 2,820 | 2,058.4 | 37,866 | 11.9 | 32.4 | 33.7 | 253.5 | 12,401 | 32,957 | 59,608 | 81,708 | 40.5 | 3.7 | 11.6 | 4.1 | 7.5 |
| Lafayette, LA | 14,108 | 2,882.9 | 119,608 | 16.4 | 50.0 | 25.1 | 727.7 | 9,902 | 28,659 | 53,493 | 67,362 | 46.9 | 5.2 | 18.4 | 14.1 | 22.3 |
| Lafayette-West Lafayette, IN | 3,960 | 1,698.6 | 79,859 | 9.5 | 39.8 | 34.9 | 296.6 | 10,070 | 27,904 | 49,908 | 70,216 | 50.1 | 3.9 | 15.6 | 6.5 | 12.8 |
| Lake Charles, LA | 8,189 | 3,896.6 | 50,703 | 20.4 | 49.6 | 22.9 | 433.3 | 12,106 | 29,899 | 51,547 | 67,444 | 48.1 | 5.2 | 20.4 | 15.7 | 24.7 |
| Lake Havasu City-Kingman, AZ | 5,564 | 2,638.8 | 36,087 | 12.9 | 46.8 | 13.5 | 181.4 | 7,779 | 27,575 | 50,179 | 56,751 | 49.8 | 2.5 | 15.7 | 9.6 | 23.9 |
| Lakeland-Winter Haven, FL | 11,797 | 1,644.9 | 165,894 | 19.3 | 47.6 | 20.8 | 962.2 | 9,239 | 25,281 | 51,833 | 60,761 | 47.6 | 3.3 | 14.0 | 9.5 | 17.8 |
| Lancaster, PA | NA | NA | 121,749 | 29.5 | 49.1 | 28.3 | 1,104.4 | 16,415 | 33,658 | 67,376 | 81,132 | 35.5 | 6.5 | 11.1 | 7.4 | 12.5 |
| Lansing-East Lansing, MI | D | D | 160,786 | 10.4 | 32.2 | 34.1 | 973.9 | 12,062 | 31,534 | 59,117 | 78,090 | 42.2 | 4.7 | 13.1 | 7.0 | 12.7 |
| Laredo, TX | 5,037 | 1,815.9 | 87,840 | 7.7 | 53.8 | 20.3 | 678.7 | 9,986 | 20,222 | 56,084 | 59,386 | 45.5 | 4.6 | 20.5 | 17.6 | 24.8 |
| Las Cruces, NM | 5,049 | 2,314.1 | 66,042 | 6.7 | 42.9 | 27.2 | 371.5 | 9,157 | 22,144 | 43,038 | 51,114 | 55.0 | 2.3 | 25.9 | 19.4 | 29.7 |
| Las Vegas-Henderson-Paradise, NV | 58,207 | 2,564.1 | 524,861 | 12.5 | 41.9 | 25.6 | 2,955.5 | 8,943 | 32,511 | 62,107 | 73,527 | 39.7 | 6.5 | 12.8 | 9.6 | 15.8 |
| Lawrence, KS | NA | NA | 41,439 | 9.0 | 25.9 | 47.1 | 161.7 | 10,661 | 33,253 | 64,233 | 89,158 | 40.1 | 7.6 | 15.7 | 6.2 | 10.2 |
| Lawton, OK | 3,338 | 2,666.0 | 31,705 | 9.0 | 43.9 | 20.5 | 187.7 | 8,550 | 26,031 | 51,332 | 60,007 | 48.6 | 2.0 | 17.8 | 13.5 | 20.4 |
| Lebanon, PA | 1,683 | 1,184.4 | 30,499 | 19.8 | 57.5 | 20.9 | 250.0 | 12,684 | 29,629 | 61,204 | 76,965 | 39.0 | 3.5 | 11.5 | 9.7 | 16.6 |
| Lewiston, ID-WA | 1,525 | 2,406.5 | 13,151 | 4.8 | 36.4 | 24.2 | 96.1 | 11,119 | 34,343 | 61,191 | 83,511 | 38.8 | 3.8 | 11.2 | 5.8 | 10.1 |
| Lewiston-Auburn, ME | 1,454 | 1,346.0 | 24,067 | 18.9 | 43.0 | 25.1 | 227.7 | 13,260 | 32,338 | 63,813 | 75,183 | 42.2 | 3.2 | 8.8 | 4.7 | 8.4 |
| Lexington-Fayette, KY | 14,148 | 2,718.5 | 133,392 | 16.3 | 33.0 | 38.3 | 845.7 | 11,671 | 33,582 | 60,492 | 78,496 | 41.4 | 5.5 | 13.3 | 7.6 | 12.2 |
| Lima, OH | 2,480 | 2,430.8 | 25,537 | 22.8 | 51.0 | 19.4 | 177.0 | 12,161 | 27,215 | 60,429 | 69,534 | 43.2 | 2.6 | 12.0 | 9.4 | 16.8 |
| Lincoln, NE | 8,423 | 2,496.9 | 101,488 | 16.6 | 27.7 | 39.7 | 591.1 | 11,784 | 34,273 | 61,539 | 82,049 | 39.2 | 6.5 | 10.6 | 4.9 | 8.2 |
| Little Rock-North Little Rock-Conway, AR | D | D | 186,759 | 15.5 | 39.4 | 30.3 | 1,166.1 | 9,849 | 30,899 | 56,849 | 71,044 | 44.6 | 4.2 | 13.8 | 9.8 | 15.8 |
| Logan, UT-ID | 1,313 | 921.4 | 54,003 | 9.7 | 32.2 | 33.3 | 229.5 | 7,649 | 24,488 | 61,467 | 69,136 | 40.0 | 4.1 | 14.7 | 9.9 | 13.2 |
| Longview, TX | D | D | 70,086 | 13.2 | 44.0 | 19.7 | 508.3 | 9,599 | 28,898 | 55,970 | 66,470 | 44.5 | 5.0 | 17.7 | 13.8 | 22.6 |
| Longview, WA | 2,054 | 1,876.4 | 25,570 | 13.1 | 39.5 | 15.3 | 213.2 | 12,217 | 30,404 | 55,497 | 67,588 | 44.7 | 4.4 | 12.2 | 7.6 | 11.8 |
| Los Angeles-Long Beach-Anaheim, CA | 283,364 | 2,140.7 | 3,397,799 | 14.9 | 38.6 | 35.5 | 24,881.4 | 12,575 | 37,764 | 77,774 | 88,113 | 32.7 | 12.6 | 12.4 | 8.9 | 13.4 |
| Anaheim-Santa Ana-Irvine, CA Div 31244 | 59,186 | 1,860.8 | 836,849 | 14.2 | 31.7 | 41 | 5,836.9 | 12,014 | 43,200 | 95,934 | 107,171 | 25.2 | 17.0 | 9.4 | 6.3 | 9.6 |
| Los Angeles-Long Beach-Glendale, CA Div 31084 | 224,178 | 2,229.3 | 2,560,950 | 15.2 | 40.8 | 33.8 | 19,044.5 | 12,758 | 36,044 | 72,797 | 81,912 | 35.0 | 11.3 | 13.4 | 9.8 | 14.6 |
| Louisville/Jefferson County, KY-IN | 36,356 | 2,864.0 | 291,957 | 22.0 | 38.9 | 30.7 | 2,171.8 | 12,209 | 32,999 | 61,172 | 76,210 | 40.8 | 5.5 | 11.9 | 8.3 | 14.5 |
| Lubbock, TX | 12,941 | 4,036.8 | 100,798 | 8.7 | 39.2 | 31.4 | 510.9 | 9,646 | 28,530 | 54,598 | 72,447 | 45.6 | 4.5 | 18.3 | 13.3 | 19.7 |
| Lynchburg, VA | 3,832 | 1,456.0 | 70,015 | 35.9 | 40.4 | 29.3 | 355.9 | 11,015 | 29,644 | 57,736 | 71,012 | 44.3 | 4.5 | 11.3 | 7.3 | 10.6 |
| Macon-Bibb County, GA | 7,931 | 3,460.4 | 62,163 | 19.0 | 45.6 | 24.1 | 405.1 | 11,060 | 28,797 | 48,435 | 63,038 | 51.0 | 5.0 | 19.2 | 15.3 | 24.3 |
| Madera, CA | 2,693 | 1,713.2 | 47,271 | 12.0 | 52.3 | 14.1 | 398.6 | 12,562 | 24,584 | 64,827 | 67,877 | 40.6 | 4.9 | 17.6 | 13.2 | 20.6 |
| Madison, WI | 11,432 | 1,714.7 | 175,118 | 9.1 | 24.6 | 48.6 | 1,195.1 | 12,565 | 41,254 | 75,545 | 100,556 | 31.9 | 8.2 | 8.9 | 3.4 | 4.5 |
| Manchester-Nashua, NH | 5,435 | 1,304.9 | 91,782 | 21.3 | 34.0 | 37.8 | 825.2 | 14,523 | 41,779 | 83,626 | 102,960 | 28.3 | 10.9 | 7.3 | 4.6 | 7.5 |
| Manhattan, KS | D | D | 45,527 | 6.9 | 26.8 | 37.4 | 217.5 | 11,246 | 27,537 | 59,010 | 71,936 | 43.5 | 3.5 | 18.6 | 9.2 | 17.0 |
| Mankato, MN | 1,938 | 1,898.8 | 31,183 | 15.2 | 29.0 | 37.2 | 173.4 | 12,294 | 32,068 | 63,126 | 85,357 | 39.1 | 4.0 | 16.1 | 8.6 | 15.9 |
| Mansfield, OH | 3,309 | 2,746.0 | 26,064 | 14.9 | 51.5 | 18.8 | 224.5 | 13,209 | 26,583 | 51,783 | 63,585 | 47.8 | 1.6 | 13.1 | 8.9 | 17.7 |
| McAllen-Edinburg-Mission, TX | 19,284 | 2,211.7 | 278,567 | 4.1 | 56.9 | 19 | 2,546.6 | 10,424 | 17,542 | 41,800 | 46,822 | 57.4 | 2.8 | 27.3 | 24.5 | 31.4 |
| Medford, OR | NA | NA | 44,308 | 15.7 | 37.7 | 27.6 | 337.7 | 11,156 | 30,258 | 56,450 | 73,249 | 45.6 | 4.9 | 13.9 | 8.6 | 16.6 |
| Memphis, TN-MS-AR | 54,510 | 4,053.4 | 343,693 | 17.9 | 41.0 | 28.8 | 2,076.1 | 9,804 | 30,690 | 54,859 | 68,543 | 46.0 | 6.1 | 15.4 | 11.2 | 18.1 |
| Merced, CA | 6,898 | 2,510.5 | 84,174 | 4.8 | 55.3 | 14.4 | 755.1 | 12,839 | 24,704 | 61,167 | 65,725 | 41.2 | 5.4 | 16.8 | 13.1 | 17.4 |
| Miami-Fort Lauderdale-Pompano Beach, FL | 167,498 | 2,686.2 | 1,466,400 | 19.2 | 40.9 | 33.1 | 7,923.4 | 9,611 | 33,917 | 60,141 | 70,110 | 42.1 | 8.3 | 13.5 | 10.1 | 14.4 |
| Fort Lauderdale-Pompano Beach-Sunrise, FL Div 22744 | 50,026 | 2,547.9 | 468,358 | 20.5 | 38.9 | 33 | 2,588.7 | 9,464 | 34,357 | 61,502 | 73,848 | 40.8 | 8.0 | 12.2 | 9.3 | 13.2 |
| Miami-Miami Beach-Kendall, FL Div 33124 | 85,544 | 3,082.5 | 672,687 | 18.3 | 45.8 | 30.6 | 3,386.0 | 9,542 | 29,760 | 55,171 | 61,413 | 45.6 | 6.8 | 15.6 | 12.0 | 16.2 |
| West Palm Beach-Boca Raton-Boynton Beach, FL Div 48424 | 31,928 | 2,133.1 | 325,355 | 19.0 | 34.7 | 37.7 | 1,948.7 | 9,942 | 40,888 | 66,623 | 81,212 | 38.2 | 11.0 | 11.4 | 7.8 | 12.5 |
| Michigan City-La Porte, IN | NA | NA | 20,934 | 8.0 | 50.7 | 18 | 177.3 | 10,131 | 28,484 | 56,019 | 71,066 | 43.2 | 3.4 | 13.7 | 9.4 | 20.9 |
| Midland, MI | 660 | 795.7 | 19,607 | 14.8 | 34.6 | 36.3 | 134.1 | 11,212 | 38,087 | 69,872 | 84,768 | 36.6 | 7.3 | 9.3 | 4.8 | 8.2 |
| Midland, TX | NA | NA | 49,556 | 10.8 | 37.5 | 29.3 | 255.4 | 8,491 | 37,108 | 82,650 | 106,748 | 29.9 | 12.8 | 10.4 | 6.4 | 8.6 |
| Milwaukee-Waukesha, WI | 30,333 | 1,923.9 | 387,545 | 23.0 | 34.5 | 36.4 | 2,773.4 | 12,075 | 36,636 | 65,845 | 86,415 | 38.7 | 6.7 | 11.8 | 7.8 | 13.3 |
| Minneapolis-St. Paul-Bloomington, MN | 84,987 | 2,332.8 | 910,340 | 14.9 | 27.2 | 43.2 | 7,542.0 | 13,036 | 42,681 | 83,698 | 105,945 | 28.3 | 11.2 | 8.2 | 4.7 | 7.9 |
| Missoula, MT | 3,355 | 2,803.9 | 30,169 | 8.6 | 26.4 | 44.4 | 155.5 | 11,196 | 34,144 | 57,347 | 72,141 | 42.7 | 5.2 | 11.5 | 4.5 | 6.0 |

1. Data for serious crimes have not been adjusted for underreporting; this may affect comparability between geographic areas and over time. 2. Per 100,000 population estimated by the FBI. 3. All persons 3 years old and over enrolled in nursery school through college. 4. Persons 25 years old and over. 5. Elementary and secondary education expenditures. 6. Based on population estimated by the American Community Survey, 2017.

# Table C. Metropolitan Areas — Personal Income and Earnings

| Area name | Personal income, 2019 | | | | | | | | | | Earnings, 2019 | | |
|---|---|---|---|---|---|---|---|---|---|---|---|---|---|
| | Total (mil dol) | Percent change, 2018-2019 | Per capita[1] Dollars | Per capita[1] Rank | Wages and Salaries (mil dol) | Supplements to wages and salaries, employer contributions (mil dol) Pension and insurance | Supplements to wages and salaries, employer contributions (mil dol) Government social insurance | Proprietors' income | Dividends, interest, and rent (mil dol) | Personal transfer receipts (mil dol) | Total (mil dol) | Contributrions for government social insurance (mil dol) From employee and self-employed | Contributrions for government social insurance (mil dol) From employer |
| | 62 | 63 | 64 | 65 | 66 | 67 | 68 | 69 | 70 | 71 | 72 | 73 | 74 |
| Jonesboro, AR | 5,103 | 5.3 | 38,122 | 367 | 2,614 | 392 | 196 | 433 | 672 | 1,351 | 3,634 | 239 | 196 |
| Joplin, MO | 7,438 | 3.3 | 41,423 | 315 | 3,566 | 680 | 260 | 695 | 1,226 | 1,697 | 5,201 | 326 | 260 |
| Kahului-Wailuku-Lahaina, HI | 8,601 | 4.7 | 51,348 | 123 | 4,081 | 655 | 305 | 972 | 1,866 | 1,381 | 6,013 | 384 | 305 |
| Kalamazoo-Portage, MI | 13,119 | 2.7 | 49,493 | 146 | 7,062 | 1,218 | 513 | 867 | 2,611 | 2,398 | 9,661 | 597 | 513 |
| Kankakee, IL | 4,690 | 3.1 | 42,687 | 293 | 2,197 | 457 | 154 | 194 | 660 | 1,117 | 3,002 | 178 | 154 |
| Kansas City, MO-KS | 118,708 | 3.5 | 55,009 | 82 | 65,606 | 9,806 | 4,703 | 10,888 | 20,481 | 17,870 | 91,002 | 5,459 | 4,703 |
| Kennewick-Richland, WA | 13,851 | 5.3 | 46,231 | 220 | 7,159 | 1,085 | 665 | 1,251 | 2,278 | 2,572 | 10,160 | 564 | 665 |
| Killeen-Temple, TX | 19,726 | 4.9 | 42,855 | 289 | 9,583 | 2,129 | 747 | 1,145 | 3,278 | 4,460 | 13,604 | 697 | 747 |
| Kingsport-Bristol, TN-VA | 12,570 | 2.0 | 40,918 | 329 | 5,542 | 990 | 389 | 926 | 2,138 | 3,623 | 7,846 | 552 | 389 |
| Kingston, NY | 9,412 | 4.3 | 53,006 | 103 | 2,970 | 749 | 235 | 703 | 1,793 | 2,091 | 4,657 | 279 | 235 |
| Knoxville, TN | 41,324 | 3.5 | 47,550 | 193 | 20,728 | 3,106 | 1,414 | 4,509 | 6,715 | 8,714 | 29,757 | 1,858 | 1,414 |
| Kokomo, IN | 3,542 | 2.2 | 42,911 | 287 | 2,103 | 337 | 150 | 151 | 535 | 973 | 2,741 | 183 | 150 |
| La Crosse-Onalaska, WI-MN | 7,109 | 3.2 | 52,035 | 117 | 3,853 | 756 | 295 | 499 | 1,435 | 1,268 | 5,403 | 334 | 295 |
| Lafayette, LA | 21,776 | 2.2 | 44,512 | 259 | 10,176 | 1,609 | 652 | 1,710 | 4,024 | 4,759 | 14,147 | 827 | 652 |
| Lafayette-West Lafayette, IN | 9,358 | 2.2 | 40,164 | 340 | 5,167 | 991 | 370 | 615 | 1,736 | 1,594 | 7,143 | 424 | 370 |
| Lake Charles, LA | 10,445 | 2.4 | 49,642 | 143 | 6,669 | 1,130 | 421 | 720 | 1,533 | 2,090 | 8,940 | 500 | 421 |
| Lake Havasu City-Kingman, AZ | 7,298 | 4.5 | 34,393 | 381 | 2,193 | 381 | 161 | 422 | 1,312 | 2,593 | 3,156 | 287 | 161 |
| Lakeland-Winter Haven, FL | 26,563 | 5.0 | 36,649 | 375 | 11,447 | 1,802 | 789 | 1,357 | 5,185 | 7,332 | 15,395 | 1,102 | 789 |
| Lancaster, PA | 29,641 | 3.6 | 54,314 | 90 | 12,820 | 2,213 | 1,004 | 4,561 | 4,943 | 5,160 | 20,598 | 1,177 | 1,004 |
| Lansing-East Lansing, MI | 23,389 | 3.5 | 42,495 | 295 | 12,140 | 2,246 | 880 | 1,218 | 3,937 | 5,038 | 16,483 | 1,051 | 880 |
| Laredo, TX | 8,982 | 3.7 | 32,466 | 382 | 4,180 | 878 | 286 | 997 | 1,278 | 2,088 | 6,341 | 334 | 286 |
| Las Cruces, NM | 8,238 | 4.8 | 37,756 | 369 | 3,236 | 626 | 250 | 839 | 1,320 | 2,285 | 4,951 | 334 | 250 |
| Las Vegas-Henderson-Paradise, NV | 110,628 | 4.1 | 48,806 | 163 | 55,589 | 8,275 | 4,161 | 7,293 | 25,860 | 18,672 | 75,318 | 4,531 | 4,161 |
| Lawrence, KS | 5,522 | 3.8 | 45,163 | 241 | 2,274 | 437 | 167 | 450 | 1,157 | 782 | 3,327 | 200 | 167 |
| Lawton, OK | 5,397 | 4.6 | 42,692 | 292 | 2,653 | 637 | 217 | 193 | 950 | 1,263 | 3,701 | 203 | 217 |
| Lebanon, PA | 6,937 | 3.2 | 48,924 | 158 | 2,420 | 524 | 197 | 617 | 1,150 | 1,497 | 3,759 | 235 | 197 |
| Lewiston, ID-WA | 3,001 | 3.3 | 47,650 | 190 | 1,269 | 229 | 111 | 242 | 667 | 739 | 1,851 | 127 | 111 |
| Lewiston-Auburn, ME | 4,652 | 4.2 | 42,968 | 284 | 2,426 | 405 | 179 | 253 | 618 | 1,225 | 3,263 | 220 | 179 |
| Lexington-Fayette, KY | 25,649 | 3.7 | 49,606 | 144 | 14,196 | 2,461 | 1,035 | 2,092 | 5,549 | 4,109 | 19,784 | 1,148 | 1,035 |
| Lima, OH | 4,482 | 2.5 | 43,788 | 272 | 2,596 | 473 | 186 | 310 | 673 | 1,091 | 3,565 | 213 | 186 |
| Lincoln, NE | 17,099 | 3.4 | 50,833 | 133 | 9,424 | 1,801 | 699 | 1,008 | 3,554 | 2,583 | 12,933 | 784 | 699 |
| Little Rock-North Little Rock-Conway, AR | 34,565 | 3.6 | 46,560 | 212 | 18,611 | 2,757 | 1,374 | 1,896 | 6,272 | 7,496 | 24,637 | 1,617 | 1,374 |
| Logan, UT-ID | 5,881 | 5.3 | 41,370 | 317 | 2,575 | 498 | 196 | 750 | 1,198 | 834 | 4,019 | 228 | 196 |
| Longview, TX | 12,313 | 3.2 | 42,953 | 285 | 6,271 | 989 | 434 | 825 | 2,055 | 2,995 | 8,519 | 506 | 434 |
| Longview, WA | 5,143 | 5.3 | 46,503 | 214 | 2,216 | 364 | 202 | 347 | 870 | 1,328 | 3,129 | 204 | 202 |
| Los Angeles-Long Beach-Anaheim, CA | 881,215 | 4.1 | 66,684 | 21 | 446,660 | 68,998 | 30,025 | 92,420 | 200,961 | 131,950 | 638,104 | 36,391 | 30,025 |
| Anaheim-Santa Ana-Irvine, CA Div 11244 | 227,733 | 4.0 | 71,711 | NA | 117,087 | 16,944 | 8,147 | 22,868 | 55,784 | 24,906 | 165,046 | 9,536 | 9,536 |
| Los Angeles-Long Beach-Glendale, CA Div 31084 | 653,483 | 4.1 | 65,094 | NA | 329,573 | 52,054 | 21,879 | 69,552 | 145,177 | 107,044 | 473,058 | 26,855 | 26,855 |
| Louisville/Jefferson County, KY-IN | 65,955 | 3.6 | 52,134 | 112 | 36,266 | 5,470 | 2,657 | 4,609 | 11,966 | 12,021 | 49,002 | 3,068 | 2,657 |
| Lubbock, TX | 14,234 | 4.0 | 44,169 | 263 | 6,999 | 1,209 | 455 | 1,080 | 2,478 | 2,724 | 9,743 | 522 | 455 |
| Lynchburg, VA | 10,876 | 2.4 | 41,265 | 321 | 4,985 | 757 | 355 | 393 | 2,095 | 2,802 | 6,491 | 455 | 355 |
| Macon-Bibb County, GA | 9,704 | 3.2 | 42,191 | 298 | 4,924 | 826 | 333 | 570 | 1,817 | 2,472 | 6,654 | 418 | 333 |
| Madera, CA | 6,492 | 5.9 | 41,267 | 320 | 2,465 | 574 | 180 | 1,056 | 1,107 | 1,389 | 4,275 | 216 | 180 |
| Madison, WI | 41,280 | 5.0 | 62,087 | 34 | 23,793 | 4,605 | 1,695 | 2,921 | 8,746 | 4,911 | 33,013 | 1,938 | 1,695 |
| Manchester-Nashua, NH | 26,364 | 4.0 | 63,218 | 32 | 14,443 | 2,051 | 957 | 2,399 | 3,891 | 3,596 | 19,849 | 1,203 | 957 |
| Manhattan, KS | 6,133 | 3.1 | 47,072 | 207 | 3,361 | 827 | 279 | 375 | 1,253 | 842 | 4,843 | 248 | 279 |
| Mankato, MN | 4,915 | 3.8 | 48,219 | 178 | 2,695 | 456 | 206 | 419 | 954 | 857 | 3,776 | 226 | 206 |
| Mansfield, OH | 4,908 | 3.0 | 40,507 | 334 | 2,243 | 431 | 161 | 273 | 779 | 1,301 | 3,108 | 196 | 161 |
| McAllen-Edinburg-Mission, TX | 23,815 | 3.8 | 27,415 | 384 | 9,988 | 2,118 | 674 | 2,614 | 2,754 | 6,883 | 15,394 | 839 | 674 |
| Medford, OR | 10,670 | 4.2 | 48,291 | 175 | 4,275 | 724 | 382 | 1,101 | 2,376 | 2,712 | 6,482 | 447 | 382 |
| Memphis, TN-MS-AR | 64,590 | 3.1 | 47,985 | 182 | 35,784 | 5,015 | 2,491 | 5,804 | 9,756 | 12,212 | 49,094 | 2,987 | 2,491 |
| Merced, CA | 11,406 | 6.6 | 41,077 | 327 | 3,879 | 920 | 295 | 1,440 | 1,810 | 2,785 | 6,533 | 328 | 295 |
| Miami-Fort Lauderdale-Pompano Beach, FL | 375,944 | 3.4 | 60,966 | 39 | 165,311 | 22,421 | 10,994 | 28,440 | 121,137 | 56,821 | 227,166 | 14,423 | 10,994 |
| Fort Lauderdale-Pompano Beach-Sunrise, FL Div 22744 | 102,146 | 3.9 | 52,308 | NA | 51,376 | 6,910 | 3,419 | 4,863 | 25,419 | 16,242 | 66,568 | 4,225 | 4,225 |
| Miami-Miami Beach-Kendall, FL Div 33124 | 149,166 | 3.4 | 54,902 | NA | 74,558 | 10,439 | 4,992 | 16,137 | 38,234 | 25,320 | 106,127 | 6,635 | 6,635 |
| West Palm Beach-Boca Raton-Boynton Beach, FL Div 48424 | 124,633 | 3.2 | 83,268 | NA | 39,377 | 5,072 | 2,582 | 7,440 | 57,484 | 15,259 | 54,471 | 3,564 | 3,564 |
| Michigan City-La Porte, IN | 4,825 | 3.6 | 43,910 | 270 | 1,882 | 328 | 143 | 292 | 786 | 1,139 | 2,646 | 178 | 143 |
| Midland, MI | 4,654 | 2.9 | 55,972 | 73 | 2,495 | 397 | 170 | 342 | 894 | 849 | 3,405 | 220 | 170 |
| Midland, TX | 23,513 | 3.7 | 128,766 | 1 | 9,107 | 1,085 | 564 | 10,704 | 3,567 | 1,070 | 21,460 | 925 | 564 |
| Milwaukee-Waukesha, WI | 92,080 | 2.6 | 58,457 | 58 | 50,655 | 8,553 | 3,662 | 5,901 | 18,607 | 15,329 | 68,772 | 4,241 | 3,662 |
| Minneapolis-St. Paul-Bloomington, MN | 233,890 | 3.0 | 64,255 | 27 | 134,059 | 18,394 | 9,494 | 16,538 | 46,799 | 30,734 | 178,485 | 10,800 | 9,494 |
| Missoula, MT | 6,110 | 3.4 | 51,090 | 129 | 2,936 | 464 | 258 | 531 | 1,710 | 1,034 | 4,190 | 275 | 258 |

1. Based on the resident population estimated as of July 1 of the year shown.

Items 62—74

# Table C. Metropolitan Areas — Earnings, Social Security, and Housing

| Area name | Earnings, 2019 (cont.) — Percent by selected industries | | | | | | | | | Social Security beneficiaries, December 2019 | | Supplemental Security Income Recipients, December 2019 | Housing units, 2020 | |
|---|---|---|---|---|---|---|---|---|---|---|---|---|---|---|
| | Farm | Mining, quarrying, and extracting | Construction | Manufacturing | Information; professional, scientific, and technical serviecs | Retail trade | Finance, insurance, real estate, rental and leasing | Health care and social assistance | Government | Number | Rate[1] | | Total | Percent change, 2010-2020 |
| | 75 | 76 | 77 | 78 | 79 | 80 | 81 | 82 | 83 | 84 | 85 | 86 | 87 | 88 |
| Jonesboro, AR | 2.6 | 0.1 | 5.9 | 13.8 | D | 7.3 | 4.8 | 22.2 | 14.9 | 27,590 | 205 | 5,493 | 58,802 | 14.3 |
| Joplin, MO | 0.5 | D | 4.9 | 24.2 | 4.0 | 7.3 | 3.5 | 15.5 | 10.5 | 39,225 | 218 | 4,561 | 77,771 | 3.7 |
| Kahului-Wailuku-Lahaina, HI | 0.6 | D | 9.0 | 1.2 | 4.8 | 7.8 | 5.9 | 10.2 | 13.4 | 32,060 | 191 | 1,976 | 75,183 | 6.8 |
| Kalamazoo-Portage, MI | 0.8 | D | 6.6 | 23.6 | 6.2 | 6.1 | 8.2 | 15.7 | 12.3 | 52,525 | 198 | 6,369 | 113,565 | 3.2 |
| Kankakee, IL | 1.3 | D | 4.8 | 23.2 | 4.0 | 6.2 | 3.9 | 16.4 | 15.8 | 22,945 | 209 | 2,646 | 45,772 | 1.2 |
| Kansas City, MO-KS | 0.2 | 0.1 | 6.4 | 7.5 | 17.3 | 5.0 | D | 10.9 | 13.6 | 385,245 | 178 | 33,841 | 935,326 | 7.3 |
| Kennewick-Richland, WA | 6.4 | 0.0 | 10.0 | 6.4 | 12.7 | 6.0 | 3.7 | 11.2 | 16.9 | 48,545 | 162 | 5,862 | 110,087 | 18.3 |
| Killeen-Temple, TX | -0.2 | 0.3 | 5.6 | 4.1 | 5.3 | 5.8 | 4.1 | D | 44.6 | 74,545 | 161 | 10,821 | 183,417 | 15.1 |
| Kingsport-Bristol, TN-VA | -0.3 | D | D | 24.2 | D | 8.0 | 4.2 | 14.9 | 13.0 | 93,385 | 303 | 9,937 | 150,444 | 2.4 |
| Kingston, NY | 0.4 | 0.2 | 6.9 | 6.7 | 6.6 | 7.5 | 4.8 | 13.2 | 29.0 | 42,585 | 239 | 3,961 | 85,957 | 2.8 |
| Knoxville, TN | 0.0 | 0.1 | 6.5 | 11.3 | D | 7.7 | D | D | 14.0 | 203,445 | 234 | 20,599 | 397,040 | 7.2 |
| Kokomo, IN | 0.3 | D | 3.6 | 45.3 | 2.6 | 6.0 | 3.8 | 12.6 | 10.0 | 21,645 | 262 | 2,249 | 40,033 | 3.5 |
| La Crosse-Onalaska, WI-MN | 0.7 | D | 5.5 | 12.0 | D | 5.6 | 7.6 | 21.4 | 14.8 | 28,850 | 211 | 2,281 | 60,292 | 5.8 |
| Lafayette, LA | 0.6 | 10.6 | 6.0 | 9.0 | 9.2 | 7.1 | 7.4 | D | 12.5 | 93,260 | 190 | 15,475 | 214,909 | 9.7 |
| Lafayette-West Lafayette, IN | 1.0 | D | 4.9 | D | D | 5.0 | 4.3 | D | 25.4 | 35,855 | 154 | 2,880 | 96,985 | 10.0 |
| Lake Charles, LA | -0.1 | 0.4 | 25.6 | 17.9 | D | 6.1 | D | 10.0 | 11.3 | 41,725 | 198 | 6,126 | 98,928 | 15.5 |
| Lake Havasu City-Kingman, AZ | 0.8 | 0.4 | 7.5 | 6.6 | 4.8 | 12.9 | 5.2 | 18.8 | 17.6 | 72,295 | 338 | 4,646 | 117,650 | 6.1 |
| Lakeland-Winter Haven, FL | 0.9 | 0.6 | 6.4 | 8.9 | 5.4 | 7.7 | 7.8 | 13.4 | 12.8 | 170,180 | 235 | 22,715 | 313,005 | 11.3 |
| Lancaster, PA | 1.0 | 0.1 | 11.5 | 14.9 | 7.0 | 7.7 | 5.6 | 13.0 | 8.2 | 115,465 | 212 | 9,261 | 214,683 | 5.8 |
| Lansing-East Lansing, MI | 0.4 | D | 5.4 | 10.8 | 8.3 | 5.7 | 9.0 | 12.6 | 25.4 | 112,750 | 205 | 12,074 | 235,355 | 2.6 |
| Laredo, TX | 0.0 | 5.4 | 3.3 | 0.7 | 4.3 | 7.2 | 5.1 | 10.9 | 27.6 | 35,835 | 130 | 11,780 | 87,421 | 18.9 |
| Las Cruces, NM | 3.0 | 0.1 | 6.6 | 4.0 | 7.8 | 5.5 | 4.8 | 15.7 | 29.3 | 44,265 | 202 | 8,308 | 91,091 | 11.8 |
| Las Vegas-Henderson-Paradise, NV | 0.0 | 0.1 | 7.6 | 2.6 | 8.9 | 7.1 | 7.6 | 9.3 | 14.2 | 376,685 | 166 | 44,355 | 938,258 | 11.7 |
| Lawrence, KS | 0.7 | 0.0 | 5.2 | 9.6 | 9.8 | 6.2 | 8.1 | 7.7 | 31.7 | 18,235 | 149 | 1,349 | 52,181 | 11.7 |
| Lawton, OK | 0.4 | 0.4 | 3.3 | 8.3 | D | 5.0 | 3.6 | 5.2 | 54.1 | 23,000 | 181 | 3,408 | 54,599 | 1.6 |
| Lebanon, PA | 0.9 | 0.0 | 6.3 | 17.6 | 6.0 | 7.7 | 3.1 | 11.8 | 18.9 | 33,680 | 238 | 2,667 | 58,564 | 5.3 |
| Lewiston, ID-WA | 0.3 | D | 7.6 | 16.3 | 4.3 | 9.0 | 7.7 | D | 18.3 | 17,590 | 278 | 1,737 | 28,300 | 3.6 |
| Lewiston-Auburn, ME | 0.8 | D | 7.9 | 10.5 | 7.0 | 8.0 | 5.8 | 20.5 | 11.4 | 26,475 | 244 | 4,131 | 50,515 | 2.9 |
| Lexington-Fayette, KY | 3.4 | 0.1 | 6.8 | 13.6 | D | 5.8 | 5.0 | 11.4 | 20.9 | 89,965 | 174 | 11,224 | 226,236 | 8.2 |
| Lima, OH | 0.3 | D | 5.2 | 25.0 | 4.5 | 6.0 | 3.6 | 19.7 | 12.3 | 23,100 | 225 | 2,925 | 45,289 | 0.7 |
| Lincoln, NE | 0.5 | D | 6.7 | 8.0 | 10.1 | 5.6 | 8.7 | D | 21.5 | 54,210 | 161 | 5,082 | 144,157 | 12.8 |
| Little Rock-North Little Rock-Conway, AR | 0.1 | 0.3 | 6.1 | 6.2 | D | 6.3 | D | 13.8 | 22.0 | 155,240 | 209 | 24,181 | 332,494 | 8.3 |
| Logan, UT-ID | 2.0 | D | 5.4 | 19.8 | D | 6.7 | 11.6 | D | 18.6 | 17,450 | 123 | 1,037 | 49,721 | 19.7 |
| Longview, TX | 0.0 | 10.2 | 9.5 | 15.3 | 6.1 | 6.3 | 5.9 | D | 10.7 | 61,035 | 213 | 8,726 | 120,840 | 5.0 |
| Longview, WA | 0.6 | D | 10.8 | 21.2 | 3.8 | 6.4 | 4.0 | 12.8 | 16.1 | 28,460 | 258 | 3,879 | 45,621 | 5.0 |
| Los Angeles-Long Beach-Anaheim, CA | 0.0 | 0.1 | 4.9 | 8.0 | 18.1 | 5.1 | 11.2 | 9.7 | 13.5 | 1,862,740 | 141 | 461,842 | 4,717,269 | 5.0 |
| Anaheim-Santa Ana-Irvine, CA Div 11244 | 0.1 | 0.0 | 7.6 | 10.5 | 14.4 | 5.5 | 13.7 | 8.9 | 10.3 | 470,315 | 148 | 72,767 | 1,117,849 | 6.6 |
| Los Angeles-Long Beach-Glendale, CA Div 31084 | 0.0 | 0.1 | 3.9 | 7.1 | 19.5 | 5.0 | 10.3 | 9.9 | 14.6 | 1,392,425 | 139 | 389,075 | 3,599,420 | 4.5 |
| Louisville/Jefferson County, KY-IN | 0.1 | 0.1 | 5.8 | 14.0 | 9.2 | 5.1 | D | 12.9 | 10.8 | 257,945 | 204 | 32,687 | 556,092 | 5.7 |
| Lubbock, TX | 0.4 | D | D | D | 7.5 | 8.6 | 6.9 | D | 24.2 | 51,040 | 158 | 6,639 | 138,444 | 14.7 |
| Lynchburg, VA | -0.2 | D | D | 18.1 | 9.0 | 6.7 | 6.1 | 16.2 | 12.3 | 64,505 | 244 | 6,171 | 117,723 | 4.6 |
| Macon-Bibb County, GA | 0.2 | 0.9 | D | D | D | D | 13.0 | 20.2 | 14.4 | 51,920 | 226 | 9,627 | 103,252 | 1.7 |
| Madera, CA | 16.8 | D | 4.5 | 6.5 | 2.5 | 4.9 | 2.0 | 14.9 | 22.5 | 26,225 | 167 | 4,663 | 51,063 | 3.9 |
| Madison, WI | 0.6 | 0.1 | 6.1 | 9.0 | 16.8 | 6.0 | 8.5 | D | 20.2 | 117,415 | 176 | 8,963 | 296,977 | 10.5 |
| Manchester-Nashua, NH | 0.0 | D | 6.6 | 14.6 | 15.2 | 7.3 | 10.9 | 11.5 | 9.5 | 83,030 | 199 | 6,909 | 174,506 | 5.1 |
| Manhattan, KS | 1.0 | D | 4.1 | 4.8 | D | 4.5 | 5.4 | D | 53.7 | 17,015 | 132 | 1,341 | 43,057 | 11.0 |
| Mankato, MN | 3.6 | D | 6.3 | 15.9 | D | 6.8 | 4.8 | D | 16.7 | 29,825 | 246 | 3,566 | 54,041 | 10.2 |
| Mansfield, OH | 0.4 | D | 7.1 | 22.8 | 3.6 | 8.0 | 3.4 | 14.5 | 17.5 | 114,295 | 132 | 40,969 | 291,866 | 17.6 |
| McAllen-Edinburg-Mission, TX | 0.9 | 0.9 | 4.7 | 2.7 | 4.5 | 9.5 | 4.9 | 19.4 | 13.9 | 59,210 | 268 | 4,926 | 98,237 | 8.0 |
| Medford, OR | 0.1 | 0.1 | 7.1 | 8.7 | 7.1 | 10.0 | 6.5 | 19.4 | 13.3 | 246,575 | 183 | 45,054 | 575,843 | 4.5 |
| Memphis, TN-MS-AR | 0.2 | D | D | 9.5 | D | D | 7.9 | D | 9.2 | 39,405 | 142 | 10,887 | 89,306 | 6.7 |
| Merced, CA | 17.2 | D | 4.2 | 10.1 | 2.3 | 5.9 | 3.1 | D | 24.7 | | | | | |
| Miami-Fort Lauderdale-Pompano Beach, FL | 0.4 | 0.0 | 6.8 | 3.3 | 14.6 | 6.8 | 11.3 | 11.0 | 12.2 | 1,099,425 | 178 | 229,945 | 2,577,430 | 4.6 |
| Fort Lauderdale-Pompano Beach-Sunrise, FL Div 22744 | 0.0 | 0.0 | 7.0 | 3.7 | 14.6 | 7.9 | 9.9 | 10.2 | 13.4 | 333,670 | 171 | 45,801 | 831,992 | 2.7 |
| Miami-Miami Beach-Kendall, FL Div 33124 | 0.4 | 0.1 | 7.0 | 2.9 | 14.8 | 6.3 | 12.2 | 10.8 | 12.4 | 436,570 | 161 | 160,488 | 1,048,012 | 5.9 |
| West Palm Beach-Boca Raton-Boynton Beach, FL Div 48424 | 0.8 | 0.0 | 6.3 | 3.7 | 14.2 | 6.6 | 11.4 | 12.5 | 10.5 | 329,185 | 220 | 23,656 | 697,426 | 4.9 |
| Michigan City-La Porte, IN | 0.9 | 0.1 | 8.4 | 21.8 | 3.4 | 6.6 | 4.4 | 13.5 | 15.0 | 25,535 | 232 | 2,174 | 49,355 | 1.9 |
| Midland, MI | 0.0 | D | 5.9 | 27.8 | 4.0 | 3.8 | 3.0 | 11.0 | 6.6 | 20,100 | 241 | 1,522 | 37,306 | 3.8 |
| Midland, TX | 0.0 | 64.7 | 4.0 | 2.1 | D | 2.4 | D | 2.2 | 3.5 | 19,890 | 109 | 1,985 | 66,436 | 18.2 |
| Milwaukee-Waukesha, WI | 0.1 | D | 5.3 | 15.5 | 10.9 | 5.0 | 9.9 | 13.5 | 10.5 | 305,860 | 194 | 47,414 | 684,825 | 2.2 |
| Minneapolis-St. Paul-Bloomington, MN | 0.1 | 0.1 | 5.8 | 11.1 | D | 4.8 | 11.7 | 10.9 | 11.6 | 587,060 | 161 | 60,272 | 1,489,974 | 8.0 |
| Missoula, MT | 0.0 | 0.1 | 7.7 | 3.3 | 11.1 | 9.1 | 7.4 | 17.9 | 17.3 | 22,940 | 191 | 1,968 | 56,208 | 12.2 |

1. Per 1,000 resident population estimated as of July 1 of the year shown.

# Table C. Metropolitan Areas — Housing, Labor Force, and Employment

| Area name | Occupied Housing units, 2020 — Occupied units — Total (89) | Owner-occupied Percent (90) | Median value[1] (91) | Median owner cost as % of income — With a mortgage (92) | Without a mortgage[2] (93) | Renter-occupied Median rent[3] (94) | Median rent as % of income[2] (95) | Sub-standard units[4] (percent) (96) | Civilian labor force, 2020 — Total (97) | Percent change 2019-2020 (98) | Unemployment Total (99) | Rate[5] (100) | Civilian employment, 2019[6] Total (101) | Management, business, science, and arts (Percent) (102) | Construction, production, and maintenance occupations (103) |
|---|---|---|---|---|---|---|---|---|---|---|---|---|---|---|---|
| Jonesboro, AR | 51,288 | 59.2 | 144,900 | 17.3 | 10.0 | 717 | 31.5 | 2.0 | 65,694 | -0.8 | 3,517 | 5.4 | 59,851 | 32.8 | 26.7 |
| Joplin, MO | 67,167 | 65.4 | 127,800 | 17.7 | 12.4 | 716 | 27.5 | 2.4 | 82,886 | -1.5 | 4,739 | 5.7 | 79,818 | 32.1 | 26.4 |
| Kahului-Wailuku-Lahaina, HI | 54,744 | 60.8 | 697,900 | 28.3 | 10.0 | 1,615 | 31.1 | 9.9 | 82,820 | -5.3 | 14,764 | 17.8 | 85,790 | 29.7 | 20.9 |
| Kalamazoo-Portage, MI | 103,196 | 61.6 | 171,800 | 18.4 | 10.0 | 824 | 28.0 | 1.5 | 131,618 | -1.9 | 9,488 | 7.2 | 135,583 | 39.9 | 20.1 |
| Kankakee, IL | 39,583 | 66.4 | 153,300 | 19.6 | 12.8 | 923 | 31.0 | 1.1 | 53,934 | -3.6 | 4,770 | 8.8 | 53,029 | 27.4 | 33.1 |
| Kansas City, MO-KS | 844,310 | 65.1 | 209,100 | 19.1 | 11.5 | 989 | 27.3 | 2.1 | 1,138,715 | -0.1 | 70,083 | 6.2 | 1,113,921 | 42.1 | 21.5 |
| Kennewick-Richland, WA | 101,840 | 67.6 | 267,500 | 19.8 | 10.0 | 1,027 | 28.7 | 4.6 | 145,640 | -0.1 | 12,294 | 8.4 | 135,225 | 38.3 | 26.7 |
| Killeen-Temple, TX | 157,088 | 55.9 | 159,800 | 20.9 | 11.4 | 906 | 29.1 | 4.0 | 177,573 | 0.4 | 12,157 | 6.8 | 178,977 | 36.9 | 21.7 |
| Kingsport-Bristol, TN-VA | 130,928 | 73 | 149,200 | 18.3 | 10.0 | 661 | 26.8 | 1.2 | 135,047 | -1.8 | 9,323 | 6.9 | 133,392 | 33.9 | 26.1 |
| Kingston, NY | 69,376 | 67.4 | 238,900 | 22.3 | 13.7 | 1,135 | 32.8 | 1.2 | 86,186 | -1.5 | 6,901 | 8.0 | 88,593 | 38.3 | 18.5 |
| Knoxville, TN | 347,927 | 68.9 | 186,900 | 19.1 | 10.0 | 872 | 28.7 | 1.5 | 417,082 | -1.1 | 25,893 | 6.2 | 406,517 | 37.2 | 23.2 |
| Kokomo, IN | 34,151 | 73 | 109,800 | 16.1 | 10.0 | 685 | 25.5 | 0.0 | 36,931 | 1.3 | 4,010 | 10.9 | 37,008 | 25.4 | 42.3 |
| La Crosse-Onalaska, WI-MN | 56,515 | 67.2 | 194,700 | 20.5 | 10.6 | 847 | 27.5 | 0.0 | 75,293 | -1.2 | 3,957 | 5.3 | 72,354 | 36.2 | 24.9 |
| Lafayette, LA | 184,313 | 68.6 | 166,300 | 18.6 | 10.0 | 818 | 28.7 | 3.6 | 210,984 | -1.8 | 15,939 | 7.6 | 219,611 | 36.9 | 24.5 |
| Lafayette-West Lafayette, IN | 89,655 | 58.5 | 150,200 | 17.6 | 10.0 | 882 | 32.5 | 1.0 | 112,075 | -4.0 | 6,956 | 6.2 | 117,854 | 38.0 | 25.3 |
| Lake Charles, LA | 79,505 | 66.7 | 164,700 | 17.4 | 10.0 | 868 | 29.0 | 3.8 | 101,630 | -8.8 | 9,076 | 8.9 | 86,226 | 32.2 | 28.3 |
| Lake Havasu City-Kingman, AZ | 91,095 | 72.2 | 186,200 | 21.1 | 10.0 | 873 | 28.0 | 3.8 | 88,743 | 1.8 | 8,854 | 10.0 | 76,213 | 24.7 | 25.4 |
| Lakeland-Winter Haven, FL | 238,629 | 69.3 | 175,400 | 21.9 | 11.6 | 998 | 30.9 | 2.6 | 316,626 | 3.3 | 28,647 | 9.0 | 312,921 | 28.7 | 24.8 |
| Lancaster, PA | 204,701 | 69.6 | 227,600 | 21.6 | 11.0 | 1,046 | 28.9 | 2.0 | 281,736 | -1.3 | 21,118 | 7.5 | 276,659 | 36.3 | 27.7 |
| Lansing-East Lansing, MI | 215,854 | 67.3 | 159,700 | 18.9 | 11.2 | 852 | 27.7 | 1.5 | 276,020 | -2.7 | 20,909 | 7.6 | 277,742 | 39.5 | 21.3 |
| Laredo, TX | 76,326 | 59.9 | 145,100 | 22.1 | 12.4 | 855 | 28.7 | 9.5 | 116,195 | -1.9 | 9,819 | 8.5 | 117,386 | 27.9 | 25.7 |
| Las Cruces, NM | 79,094 | 64.6 | 157,300 | 20.9 | 10.0 | 752 | 30.0 | 4.7 | 97,153 | -1.5 | 8,073 | 8.3 | 89,732 | 34.8 | 22.2 |
| Las Vegas-Henderson-Paradise, NV | 813,607 | 54.3 | 313,100 | 21.9 | 10.0 | 1,187 | 31.3 | 4.5 | 1,123,582 | -2.3 | 165,513 | 14.7 | 1,088,640 | 30.0 | 20.2 |
| Lawrence, KS | 48,508 | 49.3 | 209,300 | 18.3 | 10.0 | 961 | 25.3 | 0.0 | 64,225 | -2.1 | 3,888 | 6.1 | 70,700 | 43.4 | 20.9 |
| Lawton, OK | 44,127 | 53.3 | 125,500 | 18.7 | 10.4 | 812 | 27.1 | 2.2 | 51,000 | 0.5 | 3,431 | 6.7 | 49,209 | 37.9 | 20.8 |
| Lebanon, PA | 53,861 | 71.2 | 181,300 | 19.2 | 11.6 | 882 | 29.7 | 2.6 | 71,763 | -0.9 | 5,703 | 7.9 | 71,681 | 32.8 | 29.3 |
| Lewiston, ID-WA | 26,915 | 74.8 | 216,600 | 20.6 | 10.0 | 778 | 27.9 | 0.0 | 31,176 | -0.3 | 1,551 | 5.0 | 30,757 | 33.8 | 30.9 |
| Lewiston-Auburn, ME | 45,924 | 65 | 173,900 | 20.0 | 12.5 | 821 | 25.6 | 0.0 | 54,470 | -1.6 | 2,984 | 5.5 | 56,964 | 37.7 | 21.8 |
| Lexington-Fayette, KY | 205,914 | 58.9 | 199,500 | 18.6 | 10.0 | 885 | 28.1 | 2.3 | 269,334 | -1.8 | 15,628 | 5.8 | 260,269 | 42.6 | 20.4 |
| Lima, OH | 40,460 | 69.9 | 128,700 | 17.9 | 11.4 | 735 | 26.3 | 1.9 | 48,303 | 0.4 | 4,122 | 8.5 | 47,165 | 26.7 | 32.2 |
| Lincoln, NE | 135,008 | 60.9 | 197,100 | 19.5 | 10.1 | 856 | 27.2 | 1.1 | 185,858 | -0.6 | 7,862 | 4.2 | 184,358 | 42.0 | 19.7 |
| Little Rock-North Little Rock-Conway, AR | 288,565 | 64 | 160,500 | 18.7 | 10.0 | 829 | 27.8 | 1.8 | 354,751 | -1.2 | 22,861 | 6.4 | 351,295 | 37.6 | 23.8 |
| Logan, UT-ID | 45,043 | 65.2 | 272,800 | 20.2 | 10.0 | 836 | 29.4 | 4.5 | 72,467 | 0.9 | 2,163 | 3.0 | 68,337 | 40.4 | 22.7 |
| Longview, TX | 102,964 | 68 | 150,400 | 18.6 | 10.9 | 851 | 27.3 | 4.0 | 125,273 | -2.4 | 9,936 | 7.9 | 123,030 | 33.0 | 28.3 |
| Longview, WA | 42,790 | 65 | 281,300 | 22.9 | 11.3 | 934 | 29.7 | 1.3 | 48,522 | 1.4 | 4,523 | 9.3 | 44,445 | 27.5 | 32.6 |
| Los Angeles-Long Beach-Anaheim, CA | 4,372,678 | 48.2 | 666,900 | 26.4 | 10.4 | 1,655 | 33.1 | 10.6 | 6,474,800 | -3.9 | 766,374 | 11.8 | 6,659,534 | 40.3 | 20.0 |
| Anaheim-Santa Ana-Irvine, CA Div 11244 | 1,044,280 | 57.1 | 725,100 | 24.8 | 10.0 | 1,929 | 32.1 | 8.7 | 1,553,301 | -3.7 | 136,563 | 8.8 | 1,621,719 | 44.0 | 17.4 |
| Los Angeles-Long Beach-Glendale, CA Div 31084 | 3,328,398 | 45.4 | 644,100 | 27.0 | 10.8 | 1,577 | 33.3 | 11.1 | 4,921,499 | -3.9 | 629,811 | 12.8 | 5,037,815 | 39.0 | 20.8 |
| Louisville/Jefferson County, KY-IN | 495,246 | 67.2 | 186,800 | 19.4 | 10.0 | 880 | 26.9 | 2.1 | 644,042 |  | 42,649 | 6.6 | 628,113 | 38.4 | 25.4 |
| Lubbock, TX | 121,838 | 57.1 | 149,900 | 21.1 | 11.1 | 959 | 30.7 | 4.3 | 161,754 | -2.4 | 9,269 | 5.7 | 161,320 | 37.0 | 22.5 |
| Lynchburg, VA | 102,879 | 71 | 174,000 | 17.7 | 10.0 | 834 | 26.4 | 2.1 | 121,000 | -2.7 | 7,286 | 6.0 | 124,203 | 37.5 | 24.1 |
| Macon-Bibb County, GA | 85,864 | 62.8 | 125,200 | 19.7 | 10.0 | 831 | 32.9 | 1.1 | 101,745 | -1.5 | 6,793 | 6.7 | 100,079 | 32.4 | 26.6 |
| Madera, CA | 44,387 | 63.6 | 294,000 | 21.3 | 11.0 | 1,028 | 30.3 | 11.2 | 61,738 | -0.6 | 6,649 | 10.8 | 60,238 | 26.2 | 39.1 |
| Madison, WI | 279,014 | 60.6 | 276,600 | 19.9 | 11.4 | 1,088 | 27.2 | 1.7 | 384,771 | -1.1 | 18,931 | 4.9 | 377,913 | 51.4 | 16.4 |
| Manchester-Nashua, NH | 162,588 | 67.1 | 293,300 | 21.7 | 13.7 | 1,262 | 30.2 | 1.9 | 240,135 | -1.1 | 16,792 | 7.0 | 231,821 | 43.6 | 20.9 |
| Manhattan, KS | 46,526 | 48.9 | 189,200 | 19.8 | 10.3 | 991 | 29.0 | 0.0 | 57,682 | -2.1 | 3,008 | 5.2 | 59,895 | 40.2 | 18.0 |
| Mankato, MN | 39,791 | 65.2 | 207,000 | 19.7 | 10.0 | 865 | 27.3 | 1.3 | 61,115 | -0.7 | 3,178 | 5.2 | 57,363 | 36.3 | 24.2 |
| Mansfield, OH | 49,812 | 67.3 | 118,600 | 17.3 | 11.4 | 705 | 26.1 | 0.0 | 52,087 | -1.1 | 4,681 | 9.0 | 54,751 | 31.4 | 29.6 |
| McAllen-Edinburg-Mission, TX | 247,544 | 68.2 | 93,400 | 21.9 | 11.6 | 763 | 30.9 | 12.1 | 359,969 | 2.3 | 41,893 | 11.6 | 341,483 | 27.7 | 25.1 |
| Medford, OR | 87,800 | 63 | 318,900 | 22.4 | 12.1 | 1,052 | 31.0 | 4.0 | 105,147 | 1.1 | 8,210 | 7.8 | 95,442 | 34.3 | 21.4 |
| Memphis, TN-MS-AR | 505,048 | 58.8 | 169,400 | 19.7 | 10.0 | 949 | 30.5 | 1.9 | 636,660 | -0.7 | 56,248 | 8.8 | 631,219 | 35.6 | 26.0 |
| Merced, CA | 80,645 | 52.2 | 282,900 | 22.6 | 10.6 | 1,097 | 27.3 | 6.5 | 114,736 | -0.9 | 13,981 | 12.2 | 112,718 | 22.9 | 40.6 |
| Miami-Fort Lauderdale-Pompano Beach, FL | 2,187,192 | 59.2 | 315,400 | 25.5 | 14.1 | 1,438 | 35.9 | 4.7 | 3,029,819 | -3.8 | 248,404 | 8.2 | 3,081,147 | 35.9 | 19.9 |
| Fort Lauderdale-Pompano Beach-Sunrise, FL Div 22744 | 705,472 | 62.2 | 300,400 | 25.2 | 14.5 | 1,448 | 36.0 | 4.1 | 1,020,586 | -2.0 | 90,289 | 8.8 | 1,004,830 | 36.1 | 18.6 |
| Miami-Miami Beach-Kendall, FL Div 33124 | 912,805 | 50.4 | 330,500 | 27.5 | 13.8 | 1,408 | 36.2 | 6.5 | 1,291,854 | -5.8 | 103,681 | 8.0 | 1,366,451 | 34.4 | 21.8 |
| West Palm Beach-Boca Raton-Boynton Beach, FL Div 48424 | 568,915 | 69.5 | 311,300 | 23.6 | 14.0 | 1,498 | 34.9 | 2.7 | 717,379 | -2.4 | 54,434 | 7.6 | 709,866 | 38.5 | 17.9 |
| Michigan City-La Porte, IN | 43,100 | 72.5 | 145,000 | 16.9 | 10.0 | 728 | 31.0 | 1.1 | 47,444 | 0.0 | 4,520 | 9.5 | 50,522 | 27.7 | 28.1 |
| Midland, MI | 34,350 | 81.2 | 152,300 | 17.5 | 11.0 | 812 | 24.5 | 0.0 | 39,706 | -1.2 | 2,964 | 7.5 | 38,153 | 46.0 | 20.7 |
| Midland, TX | 61,159 | 69.6 | 263,500 | 19.5 | 10.0 | 1,417 | 29.4 | 0.0 | 100,783 | -8.0 | 8,079 | 8.0 | 93,204 | 40.6 | 23.3 |
| Milwaukee-Waukesha, WI | 635,351 | 59.9 | 231,400 | 19.6 | 11.3 | 914 | 27.9 | 1.5 | 807,581 | -0.8 | 57,041 | 7.1 | 798,161 | 42.1 | 21.6 |
| Minneapolis-St. Paul-Bloomington, MN | 1,396,025 | 70.2 | 284,500 | 19.3 | 10.0 | 1,144 | 28.0 | 2.2 | 2,010,611 | -0.1 | 128,527 | 6.4 | 1,995,425 | 46.3 | 18.9 |
| Missoula, MT | 50,853 | 62.3 | 308,500 | 23.0 | 13.7 | 909 | 28.9 | 2.8 | 64,123 | -0.2 | 4,017 | 6.3 | 65,342 | 42.9 | 19.6 |

1. Specified owner-occupied units.   2. A value of 10.0 represents 10 percent or less; a value of 50.0 represents 50 percent or more.   3. Specified renter-occupied units.   4. Overcrowded or lacking complete plumbing facilities.   5. Percent of civilian labor force.   6. Civilian employed persons 16 years old and over.

# Table C. Metropolitan Areas — Nonfarm Employment and Agriculture

| | Private nonfarm establishments, employment and payroll, 2019 | | | | | | | | | Agriculture, 2017 | | | |
| | Number of establishments | Employment | | | | | | Annual payroll | | Farms | | | Farm producers whose primary occupation is farming (percent) |
| Area name | | Total | Health care and social assistance | Manufacturing | Retail trade | Finance and insurance | Professional, scientific, and technical services | Total (mil dol) | Average per employee (dollars) | Number | Percent with: Fewer than 50 acres | Percent with: 1000 acres or more | |
| | 104 | 105 | 106 | 107 | 108 | 109 | 110 | 111 | 112 | 113 | 114 | 115 | 116 |
|---|---|---|---|---|---|---|---|---|---|---|---|---|---|
| Jonesboro, AR | 2,891 | 48,989 | 11,995 | 8,175 | 7,431 | 1,286 | 1,623 | 1,969 | 40,194 | 886 | 28.6 | 24.9 | 50.4 |
| Joplin, MO | 3,998 | 73,316 | 13,242 | 13,376 | 10,302 | 1,633 | 1,504 | 2,999 | 40,899 | 2,903 | 39.3 | 2.4 | 38.0 |
| Kahului-Wailuku-Lahaina, HI | 4,838 | 67,706 | 7,605 | 951 | 9,968 | 892 | 1,794 | 2,993 | 44,200 | 1,408 | 89.6 | 2.3 | 43.8 |
| Kalamazoo-Portage, MI | 5,598 | 113,025 | 20,557 | 19,000 | 13,969 | 5,214 | 4,317 | 5,878 | 52,010 | 707 | 61.8 | 5.1 | 43.2 |
| Kankakee, IL | 2,291 | 36,106 | 7,288 | 5,641 | 5,497 | 958 | 674 | 1,532 | 42,441 | 756 | 36.6 | 12.3 | 46.9 |
| Kansas City, MO-KS | 53,038 | 974,340 | 142,723 | 85,416 | 111,311 | 64,625 | 86,678 | 53,421 | 54,828 | 12,437 | 42 | 5.7 | 34.5 |
| Kennewick-Richland, WA | 6,450 | 94,079 | 14,905 | 7,450 | 13,932 | 2,347 | 8,027 | 5,076 | 53,959 | 2,292 | 67.3 | 9.3 | 41.6 |
| Killeen-Temple, TX | 6,485 | 111,942 | 26,481 | 7,980 | 17,521 | 3,923 | 4,704 | 4,710 | 42,077 | 5,066 | 50.5 | 4.7 | 33.8 |
| Kingsport-Bristol, TN-VA | 5,934 | 102,798 | 17,527 | 21,637 | 15,823 | 3,043 | 2,335 | 4,491 | 43,687 | 5,311 | 50.1 | 0.6 | 35.4 |
| Kingston, NY | 4,856 | 46,888 | 9,491 | 3,435 | 8,810 | 2,038 | 1,727 | 1,789 | 38,160 | 421 | 51.8 | 0.7 | 51.0 |
| Knoxville, TN | 18,220 | 341,198 | 53,784 | 37,073 | 44,180 | 15,139 | 23,135 | 15,941 | 46,721 | 5,247 | 54.6 | 0.6 | 35.1 |
| Kokomo, IN | 1,812 | 31,965 | 5,236 | 9,388 | 4,840 | 807 | 796 | 1,423 | 44,522 | 422 | 36.3 | 10.0 | 49.1 |
| La Crosse-Onalaska, WI-MN | 3,499 | 69,053 | 13,150 | 7,932 | 9,439 | 3,391 | 2,126 | 3,049 | 44,155 | 1,558 | 25 | 2.2 | 43.6 |
| Lafayette, LA | 13,177 | 186,414 | 35,161 | 13,744 | 26,441 | 5,776 | 10,501 | 8,946 | 47,991 | 3,511 | 54.1 | 7.7 | 36.4 |
| Lafayette-West Lafayette, IN | 4,383 | 81,180 | 14,134 | 21,094 | 10,010 | 2,175 | 3,228 | 3,647 | 44,919 | 2,041 | 43.6 | 13.4 | 40.5 |
| Lake Charles, LA | 4,847 | 83,407 | 13,857 | 9,199 | 11,104 | 2,140 | 5,496 | 4,084 | 48,970 | 1,225 | 44.1 | 9.4 | 29.0 |
| Lake Havasu City-Kingman, AZ | 3,907 | 45,840 | 8,916 | 3,319 | 10,616 | 1,112 | 1,257 | 1,648 | 35,960 | 317 | 60.6 | 17.4 | 40.7 |
| Lakeland-Winter Haven, FL | 12,161 | 195,939 | 29,740 | 16,967 | 28,222 | 11,374 | 6,505 | 8,217 | 41,936 | 2,080 | 63.2 | 4.8 | 37.6 |
| Lancaster, PA | 13,287 | 242,635 | 38,366 | 36,683 | 32,430 | 8,256 | 12,752 | 11,000 | 45,342 | 5,108 | 50 | 0.4 | 56.9 |
| Lansing-East Lansing, MI | 10,725 | 187,559 | 32,548 | 21,026 | 24,000 | 15,427 | 9,455 | 8,666 | 46,206 | 3,863 | 49.7 | 5.1 | 41.2 |
| Laredo, TX | 5,572 | 79,678 | 16,278 | 765 | 13,812 | 2,779 | 2,073 | 2,503 | 31,413 | 656 | 19.1 | 22.6 | 36.3 |
| Las Cruces, NM | 3,642 | 53,532 | 15,225 | 2,380 | 7,911 | 1,809 | 3,825 | 1,873 | 34,981 | 1,946 | 86.7 | 1.8 | 33.1 |
| Las Vegas-Henderson-Paradise, NV | 47,435 | 915,555 | 97,560 | 23,191 | 109,793 | 30,964 | 43,129 | 40,977 | 44,756 | 179 | 77.1 | 0.6 | 30.5 |
| Lawrence, KS | 2,757 | 40,811 | 6,486 | 4,134 | 6,576 | 1,282 | 3,635 | 1,478 | 36,214 | 998 | 44.7 | 5.6 | 33.5 |
| Lawton, OK | 2,201 | 33,694 | 6,290 | 3,350 | 5,530 | 1,427 | 2,131 | 1,277 | 37,907 | 1,503 | 25.4 | 17.9 | 38.3 |
| Lebanon, PA | 2,752 | 48,371 | 8,933 | 9,701 | 6,771 | 1,010 | 1,216 | 1,948 | 40,268 | 1,149 | 54.6 | 0.2 | 50.5 |
| Lewiston, ID-WA | 1,586 | 21,655 | 4,408 | 3,839 | 3,501 | 1,213 | 730 | 919 | 42,446 | 651 | 45.9 | 27.8 | 41.0 |
| Lewiston-Auburn, ME | 2,804 | 44,696 | 9,674 | 5,487 | 6,048 | 3,185 | 1,608 | 1,934 | 43,264 | 496 | 58.7 | 1.0 | 61.0 |
| Lexington-Fayette, KY | 12,576 | 231,800 | 39,197 | 28,934 | 29,889 | 6,261 | 14,839 | 10,591 | 45,688 | 4,619 | 46.2 | 2.3 | 40.6 |
| Lima, OH | 2,345 | 46,156 | 10,028 | 8,749 | 5,564 | 955 | 1,335 | 2,059 | 44,619 | 855 | 43.5 | 4.1 | 36.5 |
| Lincoln, NE | 9,155 | 151,082 | 27,517 | 13,736 | 19,655 | 14,894 | 9,131 | 6,567 | 43,470 | 2,730 | 49.5 | 9.2 | 36.3 |
| Little Rock-North Little Rock-Conway, AR | 18,401 | 293,932 | 62,146 | 21,575 | 38,404 | 18,255 | 13,466 | 13,788 | 46,908 | 3,353 | 42.8 | 4.8 | 37.6 |
| Logan, UT-ID | 3,885 | 48,016 | 6,156 | 12,238 | 7,303 | 1,165 | 4,255 | 1,832 | 38,153 | 2,184 | 53.4 | 5.2 | 31.6 |
| Longview, TX | 6,649 | 102,226 | 14,482 | 16,728 | 13,067 | 3,565 | 4,637 | 4,682 | 45,801 | 4,768 | 47.5 | 1.7 | 32.7 |
| Longview, WA | 2,226 | 34,670 | 5,935 | 7,427 | 5,059 | 833 | 971 | 1,720 | 49,600 | 403 | 76.9 | 1.5 | 33.3 |
| Los Angeles-Long Beach-Anaheim, CA | 384,960 | 5,437,272 | 765,177 | 466,151 | 563,383 | 244,690 | 437,911 | 342,309 | 62,956 | 1,228 | 88.8 | 1.1 | 45.6 |
| Anaheim-Santa Ana-Irvine, CA Div 11244 | 99,577 | 1,536,500 | 177,947 | 148,163 | 149,310 | 90,182 | 136,382 | 95,656 | 62,256 | 193 | 83.4 | 1.6 | 42.6 |
| Los Angeles-Long Beach-Glendale, CA Div 31084 | 285,383 | 3,900,772 | 587,230 | 317,988 | 414,073 | 154,508 | 301,529 | 246,653 | 63,232 | 1,035 | 89.9 | 1.0 | 46.1 |
| Louisville/Jefferson County, KY-IN | 29,386 | 609,954 | 88,544 | 80,283 | 70,334 | 36,735 | 27,511 | 30,656 | 50,259 | 6,851 | 49.8 | 2.5 | 35.2 |
| Lubbock, TX | 7,571 | 119,736 | 26,493 | 5,382 | 18,661 | 5,392 | 4,639 | 4,720 | 39,418 | 1,810 | 32.2 | 23.1 | 40.8 |
| Lynchburg, VA | 5,983 | 100,899 | 15,507 | 16,247 | 13,403 | 3,420 | 5,255 | 4,282 | 42,436 | 2,901 | 34.5 | 2.4 | 36.7 |
| Macon-Bibb County, GA | 5,146 | 85,910 | 17,395 | 6,476 | 11,424 | 9,991 | 2,734 | 3,701 | 43,078 | 790 | 39.5 | 5.1 | 35.1 |
| Madera, CA | 2,039 | 27,690 | 6,453 | 3,832 | 3,733 | 458 | 484 | 1,308 | 47,242 | 1,386 | 41.6 | 8.7 | 52.7 |
| Madison, WI | 17,248 | 342,643 | 58,000 | 37,187 | 41,367 | 25,860 | 28,860 | 19,649 | 57,344 | 6,927 | 40 | 3.6 | 44.0 |
| Manchester-Nashua, NH | 11,049 | 191,526 | 32,464 | 22,802 | 28,831 | 9,356 | 15,703 | 11,051 | 57,699 | 605 | 63.3 | 0.2 | 42.0 |
| Manhattan, KS | 2,789 | 37,795 | 6,402 | 2,744 | 7,082 | 1,365 | 1,647 | 1,321 | 34,939 | 1,491 | 27.7 | 15.4 | 34.9 |
| Mankato, MN | 2,640 | 49,387 | 10,552 | 8,225 | 7,065 | 1,268 | 1,577 | 2,042 | 41,354 | 1,672 | 27.5 | 9.6 | 50.3 |
| Mansfield, OH | 2,658 | 40,279 | 6,458 | 8,820 | 6,338 | 959 | 835 | 1,475 | 36,624 | 1,160 | 49.1 | 1.5 | 41.8 |
| McAllen-Edinburg-Mission, TX | 12,437 | 205,275 | 67,282 | 6,454 | 38,209 | 7,139 | 6,377 | 6,108 | 29,757 | 2,436 | 73.6 | 6.4 | 32.0 |
| Medford, OR | 6,412 | 77,088 | 15,854 | 7,056 | 12,089 | 2,219 | 2,835 | 3,324 | 43,124 | 2,136 | 78.6 | 1.1 | 39.9 |
| Memphis, TN-MS-AR | 25,774 | 552,098 | 86,972 | 38,900 | 60,951 | 18,302 | 21,372 | 27,249 | 49,356 | 3,796 | 39.7 | 10.3 | 39.4 |
| Merced, CA | 3,177 | 46,799 | 7,339 | 10,037 | 8,249 | 1,150 | 900 | 1,883 | 40,238 | 2,337 | 53.5 | 7.4 | 56.3 |
| Miami-Fort Lauderdale-Pompano Beach, FL | 199,150 | 2,279,212 | 340,689 | 75,324 | 324,275 | 101,011 | 175,556 | 117,880 | 51,720 | 4,690 | 92.6 | 1.0 | 46.7 |
| Fort Lauderdale-Pompano Beach-Sunrise, FL Div 22744 | 62,337 | 733,082 | 103,064 | 24,268 | 107,360 | 31,573 | 58,447 | 37,665 | 51,378 | 640 | 97.2 | 0.2 | 41.5 |
| Miami-Miami Beach-Kendall, FL Div 33124 | 86,855 | 995,962 | 145,933 | 36,882 | 137,727 | 48,733 | 73,675 | 52,226 | 52,438 | 2,752 | 93.2 | 0.3 | 47.1 |
| West Palm Beach-Boca Raton-Boynton Beach, FL Div 48424 | 49,958 | 550,168 | 91,692 | 14,174 | 79,188 | 20,705 | 43,434 | 27,990 | 50,875 | 1,298 | 89.1 | 3.0 | 48.3 |
| Michigan City-La Porte, IN | 2,247 | 35,649 | 5,478 | 7,904 | 5,624 | 779 | 773 | 1,430 | 40,118 | 740 | 42.4 | 9.5 | 45.4 |
| Midland, MI | 1,783 | 33,924 | 6,824 | 4,488 | 3,510 | 1,306 | 3,023 | 2,219 | 65,415 | 530 | 51.3 | 4.2 | 39.2 |
| Midland, TX | 5,900 | 106,621 | 7,683 | 3,676 | 9,515 | 2,282 | 4,272 | 7,766 | 72,838 | 766 | 39 | 20.5 | 32.4 |
| Milwaukee-Waukesha, WI | 38,588 | 793,084 | 133,940 | 113,301 | 79,062 | 53,761 | 41,619 | 44,074 | 55,572 | 1,554 | 50.2 | 3.4 | 46.8 |
| Minneapolis-St. Paul-Bloomington, MN | 97,071 | 1,839,809 | 304,550 | 194,568 | 188,600 | 130,663 | 124,741 | 112,069 | 60,913 | 11,525 | 46.9 | 3.7 | 39.3 |
| Missoula, MT | 4,577 | 51,454 | 9,925 | 2,345 | 8,194 | 2,219 | 3,776 | 2,033 | 39,518 | 576 | 66.1 | 3.6 | 32.9 |

# Table C. Metropolitan Areas — **Agriculture**

Agriculture, 2017 (cont.)

| Area name | Land in farms — Acreage (1,000) | Percent change, 2012-2017 | Acres — Average size of farm | Acres — Total irrigated (1,000) | Acres — Total cropland (1,000) | Value of land and buildings (dollars) — Average per farm | Value of land and buildings (dollars) — Average per acre | Value of machinery and equipment, average per farm (dollars) | Value of products sold — Total (mil dol) | Value of products sold — Average per farm (acres) | Percent from: Crops | Percent from: Live-stock and poultry products | Organic farms (number) | Farms with internet access (percent) | Government payments — Total ($1,000) | Government payments — Percent of farms |
|---|---|---|---|---|---|---|---|---|---|---|---|---|---|---|---|---|
| | 117 | 118 | 119 | 120 | 121 | 122 | 123 | 124 | 125 | 126 | 127 | 128 | 129 | 130 | 131 | 132 |
| Jonesboro, AR.................. | 638 | -11.8 | 720 | 510.0 | 592.3 | 3,038,718 | 4,220 | 385,903 | 381 | 430,029 | 98.8 | 1.2 | 0 | 75.4 | 31,077 | 63.2 |
| Joplin, MO ..................... | 526 | 6.4 | 181 | 5.8 | 232.4 | 621,114 | 3,429 | 84,225 | 343 | 118,235 | 18.1 | 81.9 | 1 | 74.2 | 2,415 | 18.4 |
| Kahului-Wailuku-Lahaina, HI ... | 249 | 8.6 | 177 | 4.8 | 56.6 | 2,027,298 | 11,466 | 51,466 | 82 | 58,385 | 90.3 | 9.7 | 40 | 83.5 | 2,174 | 9.4 |
| Kalamazoo-Portage, MI.......... | 139 | -3.5 | 196 | 43.0 | 111.8 | 1,382,696 | 7,055 | 206,187 | 237 | 335,109 | 74.5 | 25.5 | 2 | 83.7 | 2,664 | 21.6 |
| Kankakee, IL.................... | 313 | -8.7 | 414 | 18.4 | 300.4 | 3,237,466 | 7,822 | 285,392 | 221 | 292,508 | 91.3 | 8.7 | 3 | 75.5 | 2,346 | 45.4 |
| Kansas City, MO-KS............. | 3,126 | 0.1 | 251 | 26.0 | 2,026.9 | 906,374 | 3,606 | 99,443 | 985 | 79,220 | 75.1 | 24.9 | 39 | 76.7 | 29,916 | 34.7 |
| Kennewick-Richland, WA ....... | 1,229 | -7.5 | 536 | 392.4 | 919.9 | 2,276,946 | 4,247 | 248,388 | 1,637 | 714,174 | 75.7 | 24.3 | 56 | 86.0 | 16,530 | 13.0 |
| Killeen-Temple, TX ............. | 1,413 | 6.3 | 279 | 4.1 | 285.1 | 812,458 | 2,913 | 64,759 | 132 | 26,006 | 36.6 | 63.4 | 4 | 75.0 | 5,248 | 9.4 |
| Kingsport-Bristol, TN-VA ........ | 527 | -7.3 | 99 | 0.2 | 168.3 | 440,247 | 4,437 | 55,882 | 125 | 23,599 | 16.7 | 83.3 | 4 | 69.0 | 2,177 | 12.2 |
| Kingston, NY................... | 59 | -17.3 | 140 | 3.7 | 24.1 | 965,682 | 6,899 | 95,713 | 54 | 129,088 | 88.3 | 11.7 | 24 | 86.9 | 164 | 8.1 |
| Knoxville, TN................... | 443 | -2.1 | 84 | 0.7 | 165.1 | 490,013 | 5,804 | 59,431 | 65 | 12,330 | 38.7 | 61.3 | 28 | 74.2 | 2,659 | 18.0 |
| Kokomo, IN..................... | 146 | 1.1 | 345 | 0.0 | 138.0 | 2,681,235 | 7,765 | 219,981 | 97 | 230,078 | 83.1 | 16.9 | 3 | 79.9 | 4,286 | 68.2 |
| La Crosse-Onalaska, WI-MN... | 361 | -6.8 | 232 | 1.3 | 211.6 | 1,059,615 | 4,568 | 128,124 | 191 | 122,458 | 38.2 | 61.8 | 49 | 77.5 | 5,327 | 61.2 |
| Lafayette, LA .................. | 923 | 21.4 | 263 | 165.7 | 606.7 | 793,548 | 3,018 | 142,157 | 337 | 95,939 | 77.2 | 22.8 | 2 | 64.8 | 31,262 | 40.4 |
| Lafayette-West Lafayette, IN ... | 885 | 3.6 | 434 | 20.0 | 824.7 | 3,386,478 | 7,808 | 254,098 | 703 | 344,214 | 73.4 | 26.6 | 11 | 79.4 | 16,217 | 61.6 |
| Lake Charles, LA ............... | 542 | -5.4 | 442 | 18.9 | 128.4 | 1,386,167 | 3,133 | 80,170 | 38 | 30,793 | 49.9 | 50.1 | 0 | 75.0 | 5,190 | 22.4 |
| Lake Havasu City-Kingman, AZ .......................... | 745 | -40.1 | 2,351 | 20.9 | 30.7 | 1,972,874 | 839 | 96,073 | 32 | 101,871 | 71.1 | 28.9 | 0 | 87.4 | 390 | 2.8 |
| Lakeland-Winter Haven, FL..... | 487 | -6.5 | 234 | 88.6 | 131.0 | 1,420,064 | 6,064 | 69,094 | 298 | 143,136 | 81.0 | 19.0 | 5 | 71.5 | 2,242 | 11.5 |
| Lancaster, PA .................. | 394 | -10.4 | 77 | 4.9 | 314.9 | 1,410,238 | 18,285 | 120,108 | 1,507 | 295,068 | 15.3 | 84.7 | 246 | 48.9 | 4,834 | 11.8 |
| Lansing-East Lansing, MI....... | 829 | -7.0 | 214 | 6.6 | 717.4 | 1,054,915 | 4,918 | 156,236 | 519 | 134,316 | 55.0 | 45.0 | 73 | 80.0 | 16,175 | 36.5 |
| Laredo, TX..................... | 1,845 | -12.1 | 2,812 | 3.3 | 35.0 | 4,817,630 | 1,713 | 69,020 | 28 | 43,287 | 1.5 | 98.5 | 0 | 47.6 | 525 | 3.2 |
| Las Cruces, NM................ | 528 | -20.0 | 271 | 73.7 | 93.1 | 674,302 | 2,484 | 91,092 | 370 | 190,284 | 61.8 | 38.2 | 23 | 73.7 | 578 | 4.0 |
| Las Vegas-Henderson-Para-dise, NV ..................... | D | D | D | 3.7 | 4.0 | 1,439,603 | D | 76,922 | 13 | 70,676 | 90.2 | 9.8 | 1 | 74.9 | 16 | 2.2 |
| Lawrence, KS .................. | 230 | 9.3 | 231 | 3.5 | 159.3 | 939,826 | 4,072 | 94,196 | 66 | 65,999 | 76.7 | 23.3 | 11 | 78.3 | 1,316 | 38.8 |
| Lawton, OK..................... | 872 | 1.0 | 580 | 0.1 | 360.4 | 965,045 | 1,664 | 100,970 | 131 | 87,035 | 20.1 | 79.9 | 2 | 75.6 | 9,050 | 44.4 |
| Lebanon, PA ................... | 108 | -11.4 | 94 | 0.7 | 86.7 | 1,348,192 | 14,400 | 132,813 | 351 | 305,312 | 10.6 | 89.4 | 54 | 71.4 | 1,021 | 13.7 |
| Lewiston, ID-WA ............... | 632 | 8.0 | 972 | 2.0 | 313.7 | 1,728,575 | 1,779 | 185,061 | 87 | 133,986 | 78.0 | 22.0 | 1 | 87.1 | 11,239 | 44.7 |
| Lewiston-Auburn, ME .......... | 56 | -6.4 | 112 | 1.0 | 24.6 | 395,349 | 3,526 | 68,417 | 41 | 81,726 | 37.6 | 62.4 | 18 | 85.1 | 481 | 10.5 |
| Lexington-Fayette, KY .......... | 752 | -1.0 | 163 | 1.3 | 305.4 | 1,179,163 | 7,245 | 87,936 | 723 | 156,582 | 15.0 | 85.0 | 15 | 82.1 | 2,574 | 10.6 |
| Lima, OH....................... | 187 | 1.9 | 218 | 0.4 | 172.7 | 1,583,677 | 7,256 | 163,585 | 140 | 163,639 | 61.4 | 38.6 | 2 | 77.8 | 5,704 | 67.7 |
| Lincoln, NE..................... | 786 | -6.8 | 288 | 160.0 | 673.4 | 1,627,411 | 5,650 | 169,707 | 440 | 161,095 | 70.4 | 29.6 | 11 | 83.6 | 20,112 | 59.5 |
| Little Rock-North Little Rock-Conway, AR.................. | 816 | 3.5 | 243 | 258.9 | 451.5 | 885,319 | 3,640 | 104,215 | 302 | 90,088 | 66.2 | 33.8 | 18 | 76.8 | 19,878 | 16.2 |
| Logan, UT-ID................... | 505 | -5.0 | 231 | 155.4 | 291.4 | 854,536 | 3,698 | 123,635 | 246 | 112,417 | 27.4 | 72.6 | 34 | 82.0 | 6,169 | 30.8 |
| Longview, TX................... | 676 | -6.7 | 142 | 2.9 | 136.0 | 437,425 | 3,085 | 60,897 | 161 | 33,727 | 9.0 | 91.0 | 10 | 73.4 | 1,081 | 4.0 |
| Longview, WA.................. | 29 | -26.3 | 71 | 3.0 | 11.0 | 639,037 | 8,955 | 53,998 | 19 | 47,045 | 52.8 | 47.2 | 0 | 80.9 | 27 | 1.5 |
| Los Angeles-Long Beach-Anaheim, CA............... | 90 | -40.7 | 73 | 18.0 | 39.2 | 1,378,476 | 18,765 | 72,823 | 237 | 193,090 | 91.0 | 9.0 | 48 | 78.9 | 529 | 2.3 |
| Anaheim-Santa Ana-Irvine, CA Div 11244 ................ | 32 | -46.4 | 168 | 4.2 | 9.6 | 3,205,503 | 19,094 | 162,435 | 83 | 427,497 | 99.5 | 0.5 | 7 | 79.8 | 6 | 3.1 |
| Los Angeles-Long Beach-Glendale, CA Div 31084 ...... | 58 | -37.0 | 56 | 13.8 | 29.6 | 1,037,785 | 18,580 | 56,113 | 155 | 149,380 | 86.5 | 13.5 | 41 | 78.7 | 523 | 2.1 |
| Louisville/Jefferson County, KY-IN ...................... | 1,002 | 4.2 | 146 | 2.7 | 602.9 | 756,977 | 5,177 | 83,594 | 450 | 65,699 | 56.2 | 43.8 | 13 | 74.8 | 11,063 | 17.6 |
| Lubbock, TX.................... | 1,580 | 3.1 | 873 | 336.3 | 1,180.5 | 1,071,771 | 1,228 | 232,519 | 418 | 230,832 | 75.4 | 24.6 | 13 | 76.8 | 17,022 | 56.5 |
| Lynchburg, VA.................. | 497 | -10.0 | 171 | 0.6 | 147.0 | 667,392 | 3,893 | 70,399 | 70 | 24,126 | 31.1 | 68.9 | 4 | 74.1 | 1,215 | 10.6 |
| Macon-Bibb County, GA......... | 169 | 16.4 | 213 | 6.3 | 41.5 | 677,035 | 3,172 | 83,004 | 130 | 164,133 | 16.2 | 83.8 | 7 | 81.1 | 974 | 17.1 |
| Madera, CA .................... | 645 | -1.3 | 466 | 300.2 | 346.1 | 5,102,429 | 10,958 | 229,882 | 1,493 | 1,076,903 | 77.4 | 22.6 | 74 | 81.0 | 2,192 | 8.4 |
| Madison, WI.................... | 1,463 | -0.2 | 211 | 22.0 | 1,110.1 | 1,340,446 | 6,346 | 157,750 | 1,159 | 167,331 | 40.5 | 59.5 | 140 | 81.0 | 20,405 | 50.5 |
| Manchester-Nashua, NH ....... | 44 | -7.3 | 73 | 0.6 | 11.7 | 568,208 | 7,775 | 67,926 | 19 | 31,030 | 77.2 | 22.8 | 18 | 93.1 | 281 | 7.8 |
| Manhattan, KS ................. | 775 | 0.3 | 520 | 26.2 | 329.9 | 1,344,062 | 2,584 | 132,384 | 184 | 123,655 | 48.4 | 51.6 | 3 | 79.9 | 4,561 | 52.8 |
| Mankato, MN................... | 648 | -0.5 | 387 | 4.3 | 605.0 | 2,757,805 | 7,121 | 290,184 | 823 | 492,108 | 41.4 | 58.6 | 8 | 82.8 | 15,176 | 64.6 |
| Mansfield, OH.................. | 156 | -3.0 | 134 | 0.2 | 113.1 | 1,008,272 | 7,505 | 108,319 | 135 | 116,504 | 37.1 | 62.9 | 20 | 65.4 | 2,687 | 21.0 |
| McAllen-Edinburg-Mission, TX | 624 | -21.5 | 256 | 162.5 | 356.9 | 1,106,061 | 4,319 | 92,928 | 311 | 127,681 | 94.3 | 5.7 | 26 | 61.8 | 6,631 | 10.5 |
| Medford, OR................... | 170 | -20.5 | 80 | 37.5 | 40.7 | 677,191 | 8,494 | 44,300 | 71 | 33,262 | 74.7 | 25.3 | 43 | 86.6 | 55 | 0.8 |
| Memphis, TN-MS-AR............ | 1,547 | 3.7 | 407 | 327.8 | 1,067 | 1,338,096 | 3,284 | 141,950 | 464 | 122,263 | 91.6 | 8.4 | 4 | 70.1 | 32,022 | 36.6 |
| Merced, CA .................... | 946 | -3.3 | 405 | 493.7 | 546.5 | 5,299,308 | 13,086 | 334,860 | 2,938 | 1,257,337 | 43.9 | 56.1 | 68 | 78.5 | 8,726 | 12.5 |
| Miami-Fort Lauderdale-Pom-pano Beach, FL............. | 573 | -6.0 | 122 | 408.2 | 495.8 | 1,546,374 | 12,654 | 84,442 | 1,764 | 376,191 | 98.5 | 1.5 | 45 | 76.4 | 1,865 | 1.4 |
| Fort Lauderdale-Pompano Beach-Sunrise, FL Div 22744 ..................... | 7 | -53.5 | 11 | 0.7 | 1.6 | 348,906 | 33,140 | 31,889 | 25 | 38,883 | 94.2 | 5.8 | 0 | 85.5 | 18 | 0.6 |
| Miami-Miami Beach-Kendall, FL Div 33124............... | 79 | -3.4 | 29 | 36.8 | 55.2 | 1,068,826 | 37,450 | 51,638 | 838 | 304,409 | 98.8 | 1.2 | 35 | 70.6 | 1,733 | 1.8 |
| West Palm Beach-Boca Raton-Boynton Beach, FL Div 48424............... | 488 | -5.1 | 376 | 370.7 | 438.9 | 3,149,297 | 8,379 | 179,904 | 902 | 694,699 | 98.3 | 1.7 | 10 | 84.2 | 114 | 1.1 |
| Michigan City-La Porte, IN....... | 249 | 9.2 | 336 | 68.5 | 230.0 | 2,473,619 | 7,355 | 210,853 | 166 | 224,842 | 82.2 | 17.8 | 7 | 79.6 | 8,644 | 58.4 |
| Midland, MI..................... | 88 | -2.1 | 165 | 0.6 | 69.9 | 916,909 | 5,542 | 121,113 | 47 | 88,483 | 57.1 | 42.9 | 5 | 80.2 | 2,366 | 40.0 |
| Midland, TX..................... | 790 | -8.0 | 1,031 | 19.6 | 374.7 | 1,241,070 | 1,204 | 141,855 | 71 | 92,215 | 92.7 | 7.3 | 0 | 79.6 | 6,374 | 41.1 |
| Milwaukee-Waukesha, WI...... | 283 | -4.2 | 182 | 1.9 | 228.7 | 1,198,064 | 6,581 | 156,733 | 291 | 186,950 | 45.3 | 54.7 | 30 | 80.4 | 2,106 | 32.4 |
| Minneapolis-St. Paul-Bloom-ington, MN................... | 2,142 | -4.9 | 186 | 134.4 | 1,718.7 | 1,088,585 | 5,858 | 144,831 | 1,563 | 135,599 | 63.4 | 36.6 | 166 | 80.4 | 19,621 | 37.7 |
| Missoula, MT................... | 260 | 5.3 | 452 | 15.5 | 21.6 | 1,262,722 | 2,796 | 42,356 | 10 | 17,099 | 57.3 | 42.7 | 5 | 82.8 | 417 | 6.1 |

Items 117—132

| Area name | Water use, 2015 Public supply water withdrawn (mil gal/day) | Public supply gallons withdrawn per person per day | Wholesale Trade[1], 2017 Number of establishments | Number of employees | Sales (mil dol) | Average payroll (mil dol) | Retail Trade[2], 2017 Number of establishments | Number of employees | Sales (mil dol) | Average payroll (mil dol) | Real estate and rental and leasing,[2] 2017 Number of establishments | Number of employees | Sales (mil dol) | Average payroll (mil dol) |
|---|---|---|---|---|---|---|---|---|---|---|---|---|---|---|
| | 133 | 134 | 135 | 136 | 137 | 138 | 139 | 140 | 141 | 142 | 143 | 144 | 145 | 146 |
| Jonesboro, AR...... | 16.30 | 127.0 | 151 | 2,133 | 1,782.1 | 90.7 | 528 | 7,340 | 1,995.1 | 178.5 | 132 | 531 | 122.9 | 19.8 |
| Joplin, MO ........... | 22.87 | 129.1 | 179 | 3,315 | 2,321.0 | 148.2 | 698 | 10,295 | 3,024.9 | 259.7 | 166 | 725 | 120.8 | 20.1 |
| Kahului-Wailuku-Lahaina, HI ... | 42.09 | 255.7 | 146 | 1,354 | 828.7 | 63.7 | D | D | D | D | 343 | 1,984 | 651.4 | 87.5 |
| Kalamazoo-Portage, MI.......... | 24.35 | 93.6 | 243 | 4,193 | 2,158.4 | 235.8 | 876 | 13,882 | 3,706.7 | 350.7 | 224 | 2,412 | 297.0 | 86.6 |
| Kankakee, IL........... | 13.26 | 119.6 | 116 | 2,199 | 1,384.6 | 110.9 | 366 | 5,932 | 1,516.0 | 134.1 | 89 | 331 | 68.4 | 12.2 |
| Kansas City, MO-KS........... | 270.80 | 129.7 | 2,482 | 43,045 | 60,613.7 | 2,832.2 | 6,159 | 111,632 | 33,514.8 | 3,012.7 | 2,663 | 14,229 | 4,703.0 | 728.0 |
| Kennewick-Richland, WA ....... | 50.78 | 181.9 | 244 | 2,763 | 2,244.7 | 148.6 | 813 | 13,839 | 4,244.8 | 406.3 | 399 | 1,445 | 368.3 | 51.6 |
| Killeen-Temple, TX................. | 19.23 | 44.6 | 149 | 2,991 | 4,371.8 | 153.7 | 1,076 | 17,225 | 5,032.7 | 444.8 | 372 | 1,818 | 302.9 | 74.3 |
| Kingsport-Bristol, TN-VA ......... | 40.07 | 130.5 | 231 | 2,729 | 2,013.2 | 121.7 | 1,061 | 15,928 | 4,175.4 | 374.2 | 214 | 855 | 179.7 | 26.4 |
| Kingston, NY................. | 394.23 | 2,188.4 | 153 | 1,565 | 893.5 | 82.5 | 726 | 8,896 | 2,570.5 | 242.2 | 187 | 611 | 107.7 | 21.8 |
| Knoxville, TN ............. | 112.42 | 134.1 | 853 | 13,029 | 11,281.7 | 714.2 | 2,818 | 44,129 | 12,556.7 | 1,211.3 | 816 | 4,586 | 1,066.4 | 193.6 |
| Kokomo, IN................. | 8.47 | 102.6 | D | D | D | D | 341 | 5,141 | 1,382.5 | 117.8 | 78 | 300 | 62.9 | 9.9 |
| La Crosse-Onalaska, WI-MN... | 15.09 | 110.2 | 122 | 2,738 | 5,156.5 | 143.3 | 475 | 9,557 | 2,453.3 | 228.6 | 147 | 1,061 | 148.0 | 35.6 |
| Lafayette, LA .......... | 51.43 | 104.9 | 684 | 9,448 | 5,391.5 | 495.1 | 1,882 | 26,943 | 7,197.4 | 686.4 | 706 | 5,258 | 1,427.2 | 304.0 |
| Lafayette-West Lafayette, IN... | 14.98 | 67.3 | 156 | 1,957 | 1,744.1 | 97.7 | 648 | 10,324 | 2,871.0 | 247.9 | 226 | 1,244 | 213.0 | 45.8 |
| Lake Charles, LA.............. | 29.01 | 141.1 | D | D | D | 115.9 | 808 | 11,384 | 3,659.9 | 291.5 | 0 | 0 | 0.0 | 0.0 |
| Lake Havasu City-Kingman, AZ | 47.86 | 233.8 | 101 | 1,217 | 455.1 | 53.7 | 626 | 9,435 | 3,063.4 | 262.1 | 241 | 736 | 150.2 | 20.1 |
| Lakeland-Winter Haven, FL.... | 67.54 | 103.9 | 528 | 7,950 | 13,505.1 | 402.1 | 1,793 | 27,302 | 8,333.4 | 726.9 | 727 | 3,332 | 734.1 | 127.7 |
| Lancaster, PA........... | 54.45 | 101.5 | 562 | 13,388 | 11,432.4 | 742.2 | 1,926 | 31,359 | 8,428.8 | 811.5 | 384 | 2,278 | 524.6 | 101.6 |
| Lansing-East Lansing, MI...... | 32.77 | 60.6 | 368 | 6,050 | 11,183.6 | 301.2 | 1,626 | 24,711 | 7,077.9 | 619.6 | 459 | 2,988 | 487.5 | 125.8 |
| Laredo, TX................. | 36.08 | 133.8 | 325 | 2,777 | 2,604.5 | 109.9 | 798 | 13,459 | 3,360.1 | 314.4 | 247 | 890 | 214.1 | 28.8 |
| Las Cruces, NM .......... | 33.99 | 158.6 | 112 | 1,069 | 621.0 | 48.3 | 498 | 8,070 | 2,098.3 | 191.8 | 218 | 686 | 197.3 | 19.8 |
| Las Vegas-Henderson-Paradise, NV ...... | 432.47 | 204.5 | 1,743 | 19,687 | 15,088.0 | 1,212.5 | 6,267 | 107,067 | 32,047.7 | 3,038.7 | 3,454 | 24,634 | 6,064.6 | 1,085.2 |
| Lawrence, KS.............. | 12.67 | 107.3 | 78 | 619 | 464.2 | 29.0 | 371 | 6,373 | 1,529.3 | 146.7 | 156 | 711 | 114.2 | 24.2 |
| Lawton, OK................. | 21.49 | 164.5 | D | D | D | D | 411 | 5,371 | 1,374.2 | 129.4 | D | D | D | D |
| Lebanon, PA.............. | 3.55 | 25.9 | 103 | 2,211 | 1,915.5 | 105.0 | 416 | 6,740 | 1,694.5 | 168.3 | 75 | 303 | 54.7 | 10.2 |
| Lewiston, ID-WA......... | 15.28 | 245.8 | 62 | 666 | 602.9 | 30.3 | 255 | 3,685 | 1,042.5 | 106.1 | 65 | 198 | 31.5 | 6.0 |
| Lewiston-Auburn, ME ............ | 7.92 | 73.9 | 104 | 1,274 | 559.5 | 64.1 | 431 | 5,933 | 1,816.8 | 162.8 | 130 | 466 | 86.1 | 17.1 |
| Lexington-Fayette, KY ............ | 55.56 | 111.0 | 456 | 7,339 | 7,628.3 | 408.8 | 1,735 | 31,208 | 9,108.1 | 812.1 | 643 | 2,709 | 675.9 | 102.1 |
| Lima, OH .................. | 18.17 | 174.0 | 123 | 2,106 | 1,905.2 | 104.2 | 388 | 5,898 | 1,678.2 | 140.1 | 84 | 314 | 64.8 | 9.3 |
| Lincoln, NE ............... | 2.91 | 9.0 | 319 | 4,349 | 4,651.0 | 238.0 | 1,058 | 18,931 | 4,928.0 | 489.6 | 426 | 2,015 | 369.6 | 82.2 |
| Little Rock-North Little Rock-Conway, AR................. | 76.47 | 104.5 | 821 | 13,648 | 10,537.2 | 761.5 | 2,719 | 39,385 | 11,877.5 | 1,062.9 | 982 | 4,273 | 967.1 | 180.8 |
| Logan, UT-ID............. | 37.53 | 280.4 | 139 | 1,176 | 918.5 | 54.5 | 518 | 6,987 | 2,070.6 | 167.0 | 243 | 508 | 90.4 | 14.5 |
| Longview, TX............. | 25.72 | 90.4 | 289 | 3,734 | 2,342.8 | 244.9 | 985 | 13,059 | 4,086.3 | 364.6 | 291 | 1,449 | 361.7 | 70.2 |
| Longview, WA............ | 10.31 | 99.6 | 88 | 1,034 | 755.8 | 55.6 | 333 | 4,985 | 1,485.4 | 143.8 | 133 | 498 | 83.7 | 18.8 |
| Los Angeles-Long Beach-Anaheim, CA........... | 1,713.98 | 128.5 | 27,788 | 328,151 | 316,422.3 | 20,186.7 | 39,063 | 590,548 | 202,708.1 | 18,534.1 | 23,461 | 142,715 | 53,009.7 | 8,801.4 |
| Anaheim-Santa Ana-Irvine, CA Div 11244.............. | 457.54 | 144.3 | 6,681 | 86,779 | 107,972.6 | 6,183.7 | 9,707 | 156,535 | 52,345.3 | 4,927.0 | 6,589 | 45,719 | 15,227.2 | 3,118.6 |
| Los Angeles-Long Beach-Glendale, CA Div 31084 | 1,256.44 | 123.5 | 21,107 | 241,372 | 208,449.7 | 14,003.0 | 29,356 | 434,013 | 150,362.8 | 13,607.1 | 16,872 | 96,996 | 37,782.5 | 5,682.8 |
| Louisville/Jefferson County, KY-IN.................. | 169.01 | 135.7 | 1,266 | 19,589 | 14,209.4 | 1,178.3 | 3,996 | 70,621 | 20,612.1 | 1,863.8 | 1,458 | 7,065 | 2,935.8 | 338.2 |
| Lubbock, TX .......... | 2.98 | 9.6 | 397 | 5,589 | 5,767.9 | 305.3 | 1,013 | 18,482 | 5,686.1 | 509.4 | 446 | 1,998 | 411.4 | 75.1 |
| Lynchburg, VA........... | 18.03 | 69.4 | 179 | 2,230 | 1,278.6 | 108.9 | 908 | 13,484 | 3,865.8 | 319.7 | 305 | 857 | 172.4 | 28.8 |
| Macon-Bibb County, GA......... | 29.29 | 127.3 | 178 | 2,913 | 1,943.5 | 154.2 | 917 | 11,911 | 3,062.2 | 293.3 | 224 | 1,089 | 216.2 | 41.5 |
| Madera, CA ............. | 15.63 | 100.8 | 87 | 1,032 | 709.1 | 59.7 | 315 | 3,648 | 1,157.5 | 101.3 | 103 | 390 | 68.6 | 12.7 |
| Madison, WI............ | 50.72 | 79.1 | 738 | 14,216 | 10,695.8 | 808.5 | 2,099 | 41,008 | 13,707.2 | 1,222.1 | 774 | 4,887 | 1,078.7 | 217.0 |
| Manchester-Nashua, NH........ | 38.74 | 95.3 | 531 | 6,897 | 5,718.9 | 498.3 | 1,577 | 28,064 | 9,169.1 | 836.6 | 454 | 2,938 | 616.4 | 142.0 |
| Manhattan, KS............ | 13.74 | 101.3 | 64 | 894 | 348.0 | 38.7 | 414 | 7,135 | 1,575.4 | 170.7 | 204 | 969 | 175.2 | 32.6 |
| Mankato, MN............ | 10.45 | 105.4 | 120 | 1,782 | 2,071.2 | 102.4 | 392 | 7,290 | 1,791.6 | 177.7 | 0 | 0 | 0.0 | 0.0 |
| Mansfield, OH............ | 14.45 | 118.7 | 97 | 1,455 | 760.4 | 65.2 | 419 | 6,707 | 1,641.9 | 160.9 | 104 | 452 | 76.5 | 14.7 |
| McAllen-Edinburg-Mission, TX | 64.64 | 76.7 | 871 | 9,035 | 5,467.8 | 353.5 | 2,162 | 37,469 | 9,923.4 | 887.1 | 571 | 2,280 | 490.1 | 68.7 |
| Medford, OR............ | 39.04 | 183.7 | 230 | 2,127 | 1,280.2 | 104.7 | 889 | 12,331 | 3,796.2 | 363.4 | 377 | 1,201 | 244.5 | 39.4 |
| Memphis, TN-MS-AR ............. | 181.23 | 135.7 | 1,363 | 32,824 | 39,701.4 | 1,895.1 | 4,201 | 62,746 | 27,072.7 | 1,689.4 | 1,185 | 9,022 | 2,521.1 | 472.8 |
| Merced, CA.............. | 50.17 | 186.9 | 105 | 1,759 | 1,948.0 | 87.1 | 544 | 8,453 | 2,366.8 | 214.4 | 156 | 569 | 116.3 | 18.4 |
| Miami-Fort Lauderdale-Pompano Beach, FL............. | 824.30 | 137.1 | 14,028 | 124,740 | 142,034.7 | 6,991.0 | 23,580 | 329,398 | 104,587.7 | 9,458.6 | 13,264 | 59,852 | 18,206.1 | 2,898.0 |
| Fort Lauderdale-Pompano Beach-Sunrise, FL Div 22744 | 233.65 | 123.2 | 3,886 | 38,460 | 43,614.1 | 2,218.3 | 7,220 | 107,986 | 35,031.3 | 3,159.0 | 4,042 | 19,924 | 6,193.7 | 936.6 |
| Miami-Miami Beach-Kendall, FL Div 33124................. | 351.92 | 130.7 | 8,150 | 67,693 | 80,939.0 | 3,590.8 | 10,680 | 141,982 | 45,110.7 | 3,964.1 | 6,073 | 25,412 | 8,315.8 | 1,227.9 |
| West Palm Beach-Boca Raton-Boynton Beach, FL Div 48424................. | 238.73 | 167.8 | 1,992 | 18,587 | 17,481.6 | 1,181.9 | 5,680 | 79,430 | 24,445.7 | 2,335.5 | 3,149 | 14,516 | 3,696.6 | 733.4 |
| Michigan City-La Porte, IN....... | 9.35 | 84.3 | 92 | 1,254 | 756.2 | 53.1 | 428 | 5,969 | 1,507.0 | 131.8 | 84 | 343 | 67.0 | 13.6 |
| Midland, MI............. | 0.14 | 1.7 | 50 | 735 | 2,781.2 | 71.8 | 270 | 3,902 | 1,116.0 | 98.0 | 64 | 338 | 70.3 | 17.3 |
| Midland, TX............. | 2.52 | 15.1 | 307 | 5,005 | 6,501.5 | 344.0 | 527 | 9,029 | 3,437.8 | 295.1 | 0 | 0 | 0.0 | 133.1 |
| Milwaukee-Waukesha, WI...... | 147.00 | 93.3 | 1,914 | 37,373 | 23,509.1 | 2,509.1 | 4,604 | 81,290 | 21,564.4 | 2,111.8 | 1,281 | 9,728 | 2,448.5 | 486.5 |
| Minneapolis-St. Paul-Bloomington, MN................. | 335.84 | 95.7 | 4,227 | 76,420 | 76,682.1 | 5,351.9 | 10,480 | 189,891 | 63,145.8 | 5,304.7 | 5,432 | 30,770 | 9,098.0 | 1,686.5 |
| Missoula, MT................ | 30.41 | 266.3 | 152 | 1,774 | 1,059.9 | 93.6 | 569 | 8,373 | 2,360.2 | 228.3 | 252 | 902 | 178.0 | 31.0 |

1. Merchant wholesalers, except manufacturers' sales branches and offices. 2. Employer establishments.

# Table C. Metropolitan Areas — Professional Services, Manufacturing, and Accommodation and Food Services

| Area name | Professional, scientific, and technical services, 2017 | | | | Manufacturing, 2017 | | | | Accommodation and food services, 2017 | | | |
|---|---|---|---|---|---|---|---|---|---|---|---|---|
| | Number of establishments | Number of employees | Sales (mil dol) | Average payroll (mil dol) | Number of establishments | Number of employees | Receipts (mil dol) | Annual payroll (mil dol) | Number of establishments | Number of employees | Receipts (mil dol) | Annual payroll (mil dol) |
| | 147 | 148 | 149 | 150 | 151 | 152 | 153 | 154 | 155 | 156 | 157 | 158 |
| Jonesboro, AR...................... | 14 | 56 | 4.0 | 1.3 | 111 | 6,989 | 2,800.9 | 328.4 | 280 | 5,514 | 271.5 | 78.6 |
| Joplin, MO .......................... | 254 | 1,479 | 188.4 | 66.2 | 219 | 12,385 | 4,990.2 | 587.9 | 357 | 6,924 | 315.1 | 92.7 |
| Kahului-Wailuku-Lahaina, HI ... | D | D | D | D | 120 | 845 | 220.4 | 40.8 | 546 | 21,971 | 2,929.7 | 812.5 |
| Kalamazoo-Portage, MI.......... | 543 | 4,855 | 723.1 | 282.0 | 294 | 17,105 | 8,166.0 | 1,137.7 | 598 | 13,151 | 622.3 | 198.8 |
| Kankakee, IL........................ | 143 | 767 | 89.4 | 35.3 | 109 | 5,400 | 5,619.1 | 331.0 | 243 | 4,101 | 187.0 | 56.8 |
| Kansas City, MO-KS............. | 3,231 | 32,619 | 6,506.5 | 2,464.8 | 1,689 | D | 44,995.8 | 4,869.3 | 4,091 | 97,190 | 5,812.6 | 1,684.0 |
| Kennewick-Richland, WA ....... | D | D | D | D | 203 | 6,614 | 2,930.2 | 366.4 | 561 | 9,722 | 587.0 | 173.6 |
| Killeen-Temple, TX ............... | 457 | 3,815 | 496.5 | 190.1 | 192 | 6,697 | 2,117.2 | 292.3 | 767 | 14,951 | 758.0 | 204.3 |
| Kingsport-Bristol, TN-VA ....... | 424 | 2,576 | 323.7 | 123.8 | 282 | 22,649 | 8,223.9 | 1,322.3 | 614 | 13,196 | 625.7 | 178.2 |
| Kingston, NY........................ | D | D | D | D | 187 | 3,298 | 806.4 | 177.2 | 591 | 7,299 | 479.1 | 155.9 |
| Knoxville, TN ....................... | 54 | 595 | 60.7 | 43.6 | 685 | 34,665 | 14,056.1 | 2,033.7 | 1,620 | 35,933 | 1,997.1 | 595.5 |
| Kokomo, IN.......................... | 118 | 714 | 93.0 | 29.8 | 62 | 9,251 | 4,005.0 | 629.8 | 195 | 4,095 | 185.4 | 54.0 |
| La Crosse-Onalaska, WI-MN... | 19 | 93 | 9.8 | 3.6 | 182 | 7,666 | 2,074.2 | 384.4 | 381 | 7,261 | 337.9 | 98.1 |
| Lafayette, LA ....................... | 1,735 | 9,966 | 1,758.0 | 616.4 | 529 | 12,515 | 4,092.1 | 664.4 | 1,093 | 21,263 | 1,271.3 | 332.0 |
| Lafayette-West Lafayette, IN ... | 362 | 3,048 | 452.3 | 177.7 | D | D | D | D | 492 | 9,618 | 503.2 | 137.0 |
| Lake Charles, LA.................. | 489 | 4,700 | 721.4 | 323.6 | D | D | D | D | 457 | 14,160 | 1,377.0 | 296.5 |
| Lake Havasu City-Kingman, AZ ................................. | D | D | D | D | 141 | 2,855 | 1,027.6 | 130.0 | 401 | 6,670 | 397.4 | 110.7 |
| Lakeland-Winter Haven, FL.... | 1,077 | 6,200 | 780.3 | 299.4 | 408 | 14,431 | 8,312.3 | 772.9 | 893 | 18,045 | 1,019.0 | 272.7 |
| Lancaster, PA....................... | 1,026 | 12,314 | 1,840.0 | 731.7 | 923 | 34,403 | 13,559.4 | 1,845.3 | 1,102 | 20,227 | 1,091.5 | 305.7 |
| Lansing-East Lansing, MI....... | 1,065 | 8,874 | 1,414.3 | 534.3 | 405 | 20,165 | 15,671.7 | 1,139.6 | 1,092 | 20,714 | 990.9 | 299.6 |
| Laredo, TX.......................... | D | D | D | D | 65 | 588 | 346.3 | 24.8 | 449 | 10,203 | 522.0 | 141.6 |
| Las Cruces, NM.................... | D | D | D | D | 139 | 2,065 | 922.8 | 92.7 | 335 | 7,168 | 361.2 | 106.7 |
| Las Vegas-Henderson-Paradise, NV ........................... | 6,477 | 45,018 | 8,260.8 | 2,991.7 | 992 | 19,985 | 6,795.1 | 974.2 | 4,839 | 268,757 | 30,017.1 | 8,468.5 |
| Lawrence, KS ...................... | 304 | 4,993 | 331.9 | 131.3 | 65 | 3,723 | 1,499.1 | 183.9 | 327 | 7,255 | 326.0 | 96.0 |
| Lawton, OK.......................... | 10 | 19 | 1.3 | 0.6 | D | D | D | D | D | D | D | D |
| Lebanon, PA........................ | D | D | D | D | 204 | 9,398 | 2,861.1 | 456.8 | 231 | 3,347 | 173.1 | 48.9 |
| Lewiston, ID-WA................... | 36 | 238 | 22.0 | 7.2 | 65 | 4,159 | 1,575.5 | 235.4 | 152 | 2,573 | 132.9 | 39.7 |
| Lewiston-Auburn, ME ............ | D | D | D | D | 152 | 5,665 | 1,693.9 | 263.2 | 238 | 3,455 | 207.1 | 61.3 |
| Lexington-Fayette, KY............ | 1,502 | 14,219 | 2,475.7 | 835.7 | 426 | 27,995 | 17,021.1 | 1,652.8 | 1,102 | 25,726 | 1,382.0 | 403.3 |
| Lima, OH ............................ | D | D | D | D | 119 | 8,261 | 12,616.0 | 536.1 | 213 | 4,452 | 221.9 | 61.2 |
| Lincoln, NE ......................... | 919 | 8,635 | 1,488.7 | 500.0 | 251 | 13,137 | 5,260.7 | 739.7 | 808 | 15,463 | 781.1 | 223.3 |
| Little Rock-North Little Rock-Conway, AR ................. | 280 | 1,389 | 204.8 | 69.3 | D | D | D | D | 1,540 | 30,382 | 1,613.6 | 473.9 |
| Logan, UT-ID ....................... | 24 | 50 | 5.4 | 1.7 | 226 | 11,375 | 5,307.9 | 553.1 | 193 | 4,071 | 177.1 | 51.4 |
| Longview, TX ....................... | 235 | 1,359 | 221.8 | 70.8 | 311 | 14,244 | 6,366.5 | 821.2 | 586 | 10,289 | 531.5 | 155.7 |
| Longview, WA....................... | D | D | D | 34.3 | 123 | 6,610 | 3,270.4 | 431.9 | 224 | 3,190 | 180.4 | 54.7 |
| Los Angeles-Long Beach-Anaheim, CA..................... | 51,845 | 442,517 | 97,482.8 | 36,524.9 | 16,366 | 463,171 | 196,848.6 | 28,752.6 | 31,822 | 633,379 | 48,019.3 | 13,582.4 |
| Anaheim-Santa Ana-Irvine, CA Div 11244................... | 15,867 | 136,767 | 27,614.5 | 10,958.5 | 4,572 | 146,707 | 46,800.8 | 9,553.6 | 8,230 | 176,590 | 13,293.4 | 3,749.0 |
| Los Angeles-Long Beach-Glendale, CA Div 31084 ....... | 35,978 | 305,750 | 69,868.3 | 25,566.5 | 11,794 | 316,464 | 150,047.8 | 19,199.0 | 23,592 | 456,789 | 34,725.9 | 9,833.4 |
| Louisville/Jefferson County, KY-IN ........................... | 2,712 | 26,032 | 4,390.3 | 1,616.3 | D | D | D | D | 2,474 | 59,121 | 3,481.6 | 968.1 |
| Lubbock, TX ........................ | 653 | 4,497 | 616.5 | 231.6 | D | D | D | D | 716 | 16,482 | 938.6 | 264.2 |
| Lynchburg, VA...................... | 333 | 1,992 | 253.9 | 118.3 | 265 | 15,106 | 6,480.1 | 934.8 | 516 | 9,545 | 471.5 | 132.0 |
| Macon-Bibb County, GA......... | 60 | 276 | 32.1 | 12.8 | D | 5,831 | D | 320.2 | 492 | 9,174 | 485.4 | 131.8 |
| Madera, CA ......................... | 107 | 401 | 45.2 | 15.7 | 104 | 3,517 | 1,490.1 | 192.4 | 210 | 2,882 | 200.0 | 52.1 |
| Madison, WI......................... | 1,948 | 22,203 | 4,812.8 | 1,903.2 | 732 | 36,579 | 14,635.0 | 2,086.7 | 1,733 | 32,831 | 1,750.1 | 521.3 |
| Manchester-Nashua, NH ........ | 1,334 | 12,550 | 2,564.9 | 1,158.2 | 518 | 21,931 | 7,038.6 | 1,432.2 | 922 | 16,755 | 973.2 | 292.3 |
| Manhattan, KS ..................... | D | D | D | D | 67 | 2,614 | 826.0 | 142.0 | 297 | 6,579 | 275.4 | 81.9 |
| Mankato, MN ....................... | 40 | 452 | 52.9 | 20.8 | 139 | 7,200 | 4,148.9 | 372.2 | 215 | 4,938 | 210.0 | 64.2 |
| Mansfield, OH ...................... | 190 | 844 | 105.7 | 34.3 | 170 | 8,503 | 3,520.3 | 459.3 | 229 | 4,578 | 214.5 | 63.2 |
| McAllen-Edinburg-Mission, TX | D | D | 720.2 | D | 280 | 5,703 | 2,136.8 | 225.8 | 1,112 | 23,976 | 1,254.7 | 329.5 |
| Medford, OR........................ | D | D | D | D | 306 | 6,428 | 2,200.2 | 298.8 | 620 | 9,593 | 535.7 | 168.8 |
| Memphis, TN-MS-AR ............ | 258 | 1,154 | 156.4 | 46.4 | 798 | 34,966 | 22,426.2 | 2,093.3 | 2,485 | 58,133 | 3,590.0 | 991.1 |
| Merced, CA.......................... | 157 | 800 | 87.6 | 30.4 | 116 | 9,822 | 5,969.6 | 459.9 | 340 | 5,365 | 320.4 | 81.5 |
| Miami-Fort Lauderdale-Pompano Beach, FL............... | 32,927 | 169,474 | 33,208.7 | 11,967.0 | 4,555 | 67,769 | 18,515.9 | 3,481.8 | 12,870 | 283,919 | 22,189.4 | 6,011.4 |
| Fort Lauderdale-Pompano Beach-Sunrise, FL Div 22744 ........................... | 10,546 | 54,062 | 9,519.6 | 3,632.7 | 1,432 | 22,969 | 6,516.5 | 1,183.8 | 4,113 | 83,312 | 6,473.1 | 1,691.1 |
| Miami-Miami Beach-Kendall, FL Div 33124................... | 14,033 | 71,986 | 15,104.0 | 5,282.6 | 2,157 | 31,558 | 7,916.7 | 1,515.7 | 5,693 | 131,647 | 11,060.0 | 2,971.1 |
| West Palm Beach-Boca Raton-Boynton Beach, FL Div 48424................... | 8,348 | 43,426 | 8,585.1 | 3,051.8 | 966 | 13,242 | 4,082.6 | 782.3 | 3,064 | 68,960 | 4,656.3 | 1,349.2 |
| Michigan City-La Porte, IN...... | 145 | 706 | 89.6 | 32.7 | 166 | 7,181 | 2,445.0 | 373.8 | 228 | 5,006 | 367.7 | 88.1 |
| Midland, MI ......................... | D | D | D | D | 65 | 4,271 | 2,280.2 | 298.0 | 152 | 3,306 | 170.6 | 51.1 |
| Midland, TX ......................... | D | D | D | 1.4 | 143 | 2,842 | 1,178.5 | 168.4 | D | D | D | D |
| Milwaukee-Waukesha, WI....... | 3,905 | 42,340 | 7,987.3 | 3,127.9 | 2,414 | 112,213 | 39,363.7 | 6,773.3 | 3,429 | 69,356 | 4,086.9 | 1,116.7 |
| Minneapolis-St. Paul-Bloomington, MN.................. | 9,288 | 97,547 | 19,911.1 | 7,932.3 | 4,576 | 191,291 | 70,163.5 | 11,916.8 | 7,039 | 157,648 | 9,496.6 | 2,968.0 |
| Missoula, MT........................ | 553 | 3,053 | 370.1 | 153.3 | 133 | 1,751 | 521.0 | 75.8 | 365 | 7,110 | 394.6 | 115.0 |

## Table C. Metropolitan Areas — Health Care and Social Assistance, Other Services, Nonemployer Businesses, and Residential Construction

| Area name | Health care and social assistance, 2017 | | | | Other services, 2017 | | | | Nonemployer businesses, 2018 | | Value of residential construction authorized by building permits, 2020 | |
|---|---|---|---|---|---|---|---|---|---|---|---|---|
| | Number of establishments | Number of employees | Receipts (mil dol) | Annual payroll (mil dol) | Number of establishments | Number of employees | Receipts (mil dol) | Annual payroll (mil dol) | Number | Receipts (mil dol) | New construction ($1,000) | Number of housing units |
| | 159 | 160 | 161 | 162 | 163 | 164 | 165 | 166 | 167 | 168 | 169 | 170 |
| Jonesboro, AR | 397 | 10,829 | 1,234.0 | 497.1 | 125 | 776 | 81 | 21 | 9,880 | 457.9 | 123,864 | 1,058 |
| Joplin, MO | 464 | 12,928 | 1,320.5 | 532.7 | 271 | 1,488 | 136 | 41 | 10,855 | 469.5 | 117,522 | 973 |
| Kahului-Wailuku-Lahaina, HI | 431 | 7,778 | 932.7 | 403.9 | 409 | 2,392 | 299 | 79 | 17,577 | 901.0 | 204,979 | 526 |
| Kalamazoo-Portage, MI | 685 | 21,274 | 2,818.2 | 1,154.6 | 421 | 3,340 | 410 | 101 | 16,752 | 730.2 | 126,214 | 496 |
| Kankakee, IL | 290 | 7,298 | 720.7 | 336.0 | 175 | 846 | 92 | 27 | 5,872 | 215.7 | 17,610 | 93 |
| Kansas City, MO-KS | 6,027 | 141,853 | 18,567.6 | 7,081.7 | 3,330 | 23,648 | 3,740 | 846 | 153,741 | 7,415.4 | 2,238,984 | 10,728 |
| Kennewick-Richland, WA | 744 | 14,431 | 1,705.7 | 728.0 | 406 | 2,091 | 213 | 66 | 13,234 | 626.4 | 542,708 | 1,965 |
| Killeen-Temple, TX | 661 | 32,376 | 4,361.2 | 1,988.1 | 546 | 3,273 | 307 | 94 | 24,725 | 1,055.0 | 572,580 | 3,629 |
| Kingsport-Bristol, TN-VA | 751 | 17,118 | 2,219.0 | 837.0 | 374 | 2,298 | 302 | 78 | 18,309 | 752.0 | 126,394 | 734 |
| Kingston, NY | 525 | 10,035 | 841.2 | 383.2 | 363 | 1,619 | 149 | 40 | 16,953 | 718.6 | 69,731 | 265 |
| Knoxville, TN | 2,163 | 55,508 | 6,629.8 | 2,516.7 | 1,109 | 6,935 | 844 | 224 | 62,872 | 3,252.7 | 1,061,241 | 5,396 |
| Kokomo, IN | 233 | 5,323 | 589.4 | 222.2 | 126 | 705 | 88 | 17 | 4,225 | 155.5 | 33,740 | 162 |
| La Crosse-Onalaska, WI-MN | 386 | 13,885 | 1,466.5 | 657.2 | 262 | 1,808 | 205 | 57 | 8,009 | 359.6 | 86,028 | 379 |
| Lafayette, LA | 1,716 | 34,628 | 3,764.1 | 1,358.4 | 727 | 4,575 | 551 | 153 | 44,753 | 2,054.5 | 474,052 | 2,550 |
| Lafayette-West Lafayette, IN | 537 | 13,337 | 1,679.3 | 633.4 | 249 | 2,110 | 497 | 74 | 12,134 | 525.6 | 231,323 | 1,397 |
| Lake Charles, LA | 556 | 13,893 | 1,437.4 | 557.3 | D | D | D | D | 15,009 | 684.6 | 190,286 | 1,230 |
| Lake Havasu City-Kingman, AZ | 500 | 8,679 | 1,115.4 | 403.1 | 293 | 1,375 | 134 | 36 | 11,375 | 557.6 | 281,431 | 1,403 |
| Lakeland-Winter Haven, FL | 1,148 | 27,693 | 3,505.9 | 1,282.2 | 717 | 3,805 | 460 | 123 | 49,619 | 1,941.6 | 1,749,382 | 9,492 |
| Lancaster, PA | 1,203 | 38,065 | 4,023.2 | 1,637.5 | 1,118 | 7,149 | 773 | 213 | 43,855 | 2,313.9 | 239,795 | 1,338 |
| Lansing-East Lansing, MI | 1,260 | 31,458 | 3,499.7 | 1,411.6 | 931 | 7,128 | 1,027 | 289 | 36,745 | 1,646.8 | 200,429 | 1,100 |
| Laredo, TX | 540 | 16,245 | 1,058.6 | 433.7 | 228 | 1,361 | 129 | 36 | 26,302 | 1,324.9 | 201,513 | 1,447 |
| Las Cruces, NM | 553 | 14,627 | 1,251.4 | 527.1 | 238 | 1,142 | 93 | 29 | 13,241 | 533.7 | 270,286 | 1,109 |
| Las Vegas-Henderson-Paradise, NV | 5,349 | 94,935 | 13,080.9 | 4,731.3 | 2,784 | 20,781 | 2,256 | 639 | 189,386 | 9,286.9 | 2,824,227 | 14,100 |
| Lawrence, KS | 316 | 6,533 | 649.4 | 260.4 | 197 | 1,251 | 227 | 43 | 8,650 | 366.2 | 109,517 | 414 |
| Lawton, OK | D | D | D | D | 149 | 752 | 79 | 22 | 5,345 | 228.5 | 17,473 | 103 |
| Lebanon, PA | 267 | 8,916 | 993.1 | 417.1 | 252 | 1,209 | 129 | 34 | 8,503 | 413.8 | 70,882 | 434 |
| Lewiston, ID-WA | 191 | 4,450 | 447.9 | 178.4 | 97 | 602 | 48 | 16 | 3,148 | 132.2 | 37,910 | 260 |
| Lewiston-Auburn, ME | 380 | 10,347 | 1,145.9 | 488.6 | 212 | 1,029 | 103 | 31 | 6,622 | 306.0 | 46,088 | 275 |
| Lexington-Fayette, KY | 1,638 | 39,216 | 5,298.1 | 1,941.5 | 847 | 5,999 | 1,068 | 194 | 38,312 | 1,836.8 | 425,440 | 2,330 |
| Lima, OH | 289 | 11,913 | 1,427.5 | 561.9 | 176 | 1,150 | 102 | 29 | 5,363 | 237.4 | 34,385 | 175 |
| Lincoln, NE | 1,094 | 25,780 | 2,959.0 | 1,130.4 | 768 | 4,475 | 797 | 162 | 23,475 | 965.7 | 384,080 | 2,038 |
| Little Rock-North Little Rock-Conway, AR | 2,228 | 58,812 | 7,733.5 | 3,050.6 | 1,161 | 7,742 | 1,171 | 271 | 53,502 | 2,403.0 | 541,324 | 3,081 |
| Logan, UT-ID | 423 | 5,990 | 674.7 | 227.1 | 211 | 885 | 98 | 26 | 10,464 | 424.4 | 338,524 | 1,701 |
| Longview, TX | 698 | 15,065 | 1,552.8 | 605.1 | 393 | 2,788 | 354 | 97 | 21,606 | 1,056.5 | 101,863 | 387 |
| Longview, WA | 236 | 5,617 | 597.1 | 249.1 | 152 | 816 | 87 | 27 | 4,618 | 202.5 | 83,979 | 346 |
| Los Angeles-Long Beach-Anaheim, CA | 45,199 | 745,534 | 110,035.8 | 40,512.0 | 22,793 | 159,460 | 22,673 | 5,462 | 1,430,673 | 80,446.3 | 6,225,861 | 26,930 |
| Anaheim-Santa Ana-Irvine, CA Div 11244 | 12,231 | 176,838 | 25,663.6 | 9,314.5 | 5,568 | 39,646 | 4,742 | 1,311 | 323,593 | 19,312.8 | 1,471,688 | 6,027 |
| Los Angeles-Long Beach-Glendale, CA Div 31084 | 32,968 | 568,696 | 84,372.2 | 31,197.5 | 17,225 | 119,814 | 17,931 | 4,151 | 1,107,080 | 61,133.5 | 4,754,173 | 20,903 |
| Louisville/Jefferson County, KY-IN | 3,497 | 91,220 | 11,561.3 | 4,441.2 | 1,930 | 15,460 | 1,915 | 515 | 90,753 | 4,252.0 | 985,910 | 4,551 |
| Lubbock, TX | 894 | 25,024 | 3,037.8 | 1,045.6 | 483 | 3,917 | 408 | 114 | 24,513 | 1,266.5 | 603,926 | 3,857 |
| Lynchburg, VA | 606 | 14,626 | 1,715.8 | 657.1 | 474 | 2,264 | 242 | 70 | 15,827 | 595.7 | 124,709 | 920 |
| Macon-Bibb County, GA | 657 | 17,340 | 2,156.4 | 769.5 | 317 | 1,886 | 253 | 68 | 18,862 | 666.0 | 96,005 | 522 |
| Madera, CA | 210 | 6,171 | 999.5 | 387.3 | 117 | 576 | 59 | 17 | 7,504 | 404.3 | 214,136 | 852 |
| Madison, WI | 1,631 | 59,316 | 7,728.5 | 3,251.6 | 1,365 | 9,654 | 1,566 | 377 | 47,316 | 2,403.9 | 1,066,156 | 5,268 |
| Manchester-Nashua, NH | 1,167 | 32,354 | 3,938.1 | 1,643.7 | 883 | 6,272 | 746 | 231 | 29,810 | 1,759.0 | 181,798 | 868 |
| Manhattan, KS | 315 | 6,688 | 661.5 | 248.3 | 217 | 1,545 | 331 | 60 | 6,851 | 278.1 | 92,505 | 450 |
| Mankato, MN | 364 | 10,385 | 991.1 | 487.3 | 187 | 1,339 | 262 | 41 | 6,411 | 313.8 | 38,281 | 205 |
| Mansfield, OH | 375 | 8,617 | 836.7 | 335.5 | 195 | 1,099 | 108 | 27 | 6,980 | 318.4 |  |  |
| McAllen-Edinburg-Mission, TX | 2,193 | 64,956 | 4,371.9 | 1,772.6 | 592 | 3,522 | 323 | 90 | 76,358 | 3,229.1 | 774,932 | 4,894 |
| Medford, OR | 758 | 15,189 | 1,902.0 | 753.3 | 328 | 1,967 | 217 | 61 | 17,655 | 847.1 | 206,466 | 886 |
| Memphis, TN-MS-AR | 2,903 | 82,332 | 11,076.9 | 4,107.6 | 1,484 | 12,661 | 2,957 | 514 | 116,228 | 4,496.8 | 932,305 | 4,797 |
| Merced, CA | 444 | 7,595 | 944.4 | 372.0 | 188 | 944 | 96 | 25 | 11,717 | 627.9 | 314,044 | 1,019 |
| Miami-Fort Lauderdale-Pompano Beach, FL | 21,676 | 327,305 | 46,754.6 | 16,274.5 | 13,598 | 74,945 | 9,493 | 2,336 | 1,052,550 | 46,781.2 | 4,762,137 | 21,758 |
| Fort Lauderdale-Pompano Beach-Sunrise, FL Div 22744 | 6,614 | 95,619 | 13,231.2 | 4,736.4 | 4,538 | 22,941 | 2,920 | 750 | 297,518 | 12,828.7 | 776,309 | 4,428 |
| Miami-Miami Beach-Kendall, FL Div 33124 | 9,159 | 144,441 | 21,236.1 | 7,238.0 | 5,335 | 31,051 | 3,953 | 909 | 557,833 | 23,858.8 | 1,772,918 | 9,831 |
| West Palm Beach-Boca Raton-Boynton Beach, FL Div 48424 | 5,903 | 87,245 | 12,287.3 | 4,300.0 | 3,725 | 20,953 | 2,620 | 677 | 197,199 | 10,093.6 | 2,212,910 | 7,499 |
| Michigan City-La Porte, IN | 229 | 5,904 | 777.2 | 299.5 | 198 | 1,031 | 133 | 29 | 5,909 | 224.1 | 34,611 | 120 |
| Midland, MI | 282 | 6,998 | 883.2 | 294.3 | 142 | 980 | 159 | 30 | 5,185 | 198.6 | 26,149 | 123 |
| Midland, TX | 421 | 7,936 | 941.6 | 374.5 | 309 | 2,571 | 460 | 106 | 19,549 | 1,479.2 | 178,786 | 1,292 |
| Milwaukee-Waukesha, WI | 5,271 | 139,611 | 16,490.9 | 6,300.2 | 2,809 | 20,397 | 4,037 | 712 | 95,058 | 4,667.0 | 775,816 | 2,698 |
| Minneapolis-St. Paul-Bloomington, MN | 10,835 | 298,523 | 33,382.1 | 14,058.1 | 7,165 | 56,225 | 7,641 | 1,994 | 279,511 | 13,638.3 | 4,962,093 | 22,341 |
| Missoula, MT | 512 | 10,122 | 1,190.9 | 412.7 | 312 | 2,257 | 362 | 80 | 10,969 | 523.9 | 131,707 | 789 |

## Table C. Metropolitan Areas — Government Employment and Payroll, and Local Government Finances

| Area name | Full-time equivalent employees (171) | March payroll (dollars) (172) | Administration, judicial, and legal (173) | Police and corrections (174) | Fire protection (175) | Highways and transportation (176) | Health and welfare (177) | Natural resources and utilities (178) | Education and libraries (179) | Total (mil dol) (180) | Intergovernmental (mil dol) (181) | Taxes Total (mil dol) (182) | Per capita Total (183) | Per capita Property (184) |
|---|---|---|---|---|---|---|---|---|---|---|---|---|---|---|
| Jonesboro, AR | 4,583 | 15,309,542 | 5.3 | 9.0 | 2.4 | 3.6 | 1.6 | 9.1 | 68.7 | 402 | 246 | 97 | 739 | 361 |
| Joplin, MO | 6,424 | 20,859,051 | 4.2 | 8.0 | 3.3 | 3.0 | 2.5 | 2.9 | 75.1 | 582 | 273 | 223 | 1,249 | 676 |
| Kahului-Wailuku-Lahaina, HI | 2,489 | 15,157,755 | 20.1 | 26.1 | 17.0 | 6.5 | 3.3 | 24.9 | 0.0 | 480 | 61 | 331 | 1,994 | 1,641 |
| Kalamazoo-Portage, MI | 7,505 | 31,550,062 | 7.4 | 11.7 | 2.4 | 2.8 | 2.4 | 2.9 | 69.1 | 1,031 | 526 | 329 | 1,249 | 1,210 |
| Kankakee, IL | 4,432 | 18,124,649 | 7.1 | 11.6 | 4.2 | 3.3 | 1.1 | 3.4 | 67.9 | 414 | 162 | 176 | 1,596 | 1,513 |
| Kansas City, MO-KS | 86,246 | 354,737,276 | 5.4 | 8.9 | 6.1 | 3.5 | 10.2 | 7.1 | 57.5 | 10,706 | 3,047 | 4,461 | 2,097 | 1,211 |
| Kennewick-Richland, WA | 12,111 | 68,651,523 | 4.6 | 6.2 | 3.1 | 2.9 | 13.3 | 24.8 | 43.6 | 1,724 | 802 | 426 | 1,467 | 784 |
| Killeen-Temple, TX | 21,064 | 71,092,282 | 5.8 | 8.4 | 3.6 | 1.4 | 4.9 | 3.7 | 70.4 | 1,589 | 703 | 554 | 1,250 | 990 |
| Kingsport-Bristol, TN-VA | 11,316 | 37,863,786 | 6.4 | 11.8 | 3.9 | 3.1 | 4.8 | 7.5 | 59.4 | 927 | 381 | 355 | 1,158 | 783 |
| Kingston, NY | 7,944 | 40,510,439 | 5.8 | 8.2 | 1.2 | 5.7 | 4.1 | 2.2 | 71.0 | 1,172 | 385 | 705 | 3,946 | 3,148 |
| Knoxville, TN | 28,569 | 106,933,307 | 5.8 | 10.0 | 2.6 | 2.8 | 10.5 | 13.0 | 53.7 | 2,801 | 899 | 1,116 | 1,307 | 849 |
| Kokomo, IN | 3,042 | 10,199,579 | 6.3 | 12.7 | 4.2 | 3.2 | 3.2 | 3.4 | 65.3 | 262 | 141 | 93 | 1,129 | 1,108 |
| La Crosse-Onalaska, WI-MN | 5,706 | 24,820,556 | 6.4 | 7.6 | 2.5 | 4.0 | 10.9 | 3.3 | 62.7 | 674 | 304 | 248 | 1,817 | 1,657 |
| Lafayette, LA | 18,074 | 56,271,873 | 6.4 | 11.6 | 3.3 | 3.2 | 9.8 | 6.9 | 57.9 | 1,681 | 565 | 792 | 1,615 | 696 |
| Lafayette-West Lafayette, IN | 6,189 | 23,040,680 | 7.9 | 11.4 | 4.4 | 4.8 | 1.9 | 5.5 | 62.7 | 705 | 321 | 234 | 1,024 | 956 |
| Lake Charles, LA | 10,222 | 33,914,186 | 5.9 | 12.3 | 3.6 | 5.6 | 9.5 | 4.9 | 55.7 | 1,142 | 290 | 622 | 2,971 | 1,106 |
| Lake Havasu City-Kingman, AZ | 5,008 | 18,841,384 | 14.4 | 15.1 | 13.1 | 5.0 | 2.0 | 7.1 | 41.4 | 586 | 226 | 231 | 1,114 | 793 |
| Lakeland-Winter Haven, FL | 24,085 | 82,688,018 | 8.2 | 12.8 | 5.3 | 2.2 | 2.2 | 10.4 | 55.7 | 2,122 | 833 | 709 | 1,034 | 683 |
| Lancaster, PA | 14,385 | 63,875,933 | 5.5 | 10.2 | 1.0 | 3.2 | 2.9 | 4.1 | 71.4 | 2,211 | 804 | 1,004 | 1,855 | 1,499 |
| Lansing-East Lansing, MI | 17,747 | 80,130,669 | 7.7 | 7.3 | 2.8 | 4.1 | 10.5 | 7.3 | 55.7 | 2,204 | 1,042 | 697 | 1,270 | 1,172 |
| Laredo, TX | 15,338 | 59,020,428 | 5.3 | 11.4 | 5.2 | 2.6 | 3.5 | 3.6 | 67.9 | 1,369 | 592 | 500 | 1,827 | 1,480 |
| Las Cruces, NM | 8,033 | 28,795,832 | 4.8 | 11.2 | 3.2 | 1.7 | 1.5 | 6.9 | 65.2 | 760 | 406 | 261 | 1,206 | 537 |
| Las Vegas-Henderson-Paradise, NV | 61,549 | 328,902,480 | 8.5 | 16.4 | 5.5 | 3.3 | 10.2 | 7.9 | 46.5 | 10,404 | 4,198 | 3,614 | 1,657 | 853 |
| Lawrence, KS | 5,231 | 23,730,724 | 5.5 | 10.3 | 4.9 | 3.3 | 33.9 | 8.7 | 31.5 | 656 | 135 | 204 | 1,697 | 1,155 |
| Lawton, OK | 6,247 | 21,718,232 | 3.9 | 6.4 | 3.1 | 2.2 | 35.7 | 4.1 | 43.7 | 607 | 153 | 116 | 912 | 422 |
| Lebanon, PA | 3,630 | 16,331,706 | 5.8 | 8.8 | 0.8 | 3.1 | 4.1 | 5.6 | 70.8 | 524 | 174 | 234 | 1,674 | 1,348 |
| Lewiston, ID-WA | 2,193 | 9,109,038 | 7.7 | 10.6 | 6.5 | 4.0 | 3.6 | 7.3 | 56.4 | 226 | 105 | 76 | 1,207 | 1,061 |
| Lewiston-Auburn, ME | 4,097 | 15,739,689 | 4.1 | 8.0 | 5.5 | 4.7 | 2.1 | 5.4 | 68.8 | 419 | 172 | 197 | 1,833 | 1,822 |
| Lexington-Fayette, KY | 16,450 | 62,470,924 | 4.6 | 12.2 | 8.2 | 2.5 | 4.1 | 5.6 | 59.8 | 1,660 | 436 | 943 | 1,840 | 886 |
| Lima, OH | 3,994 | 15,615,627 | 8.1 | 10.6 | 5.1 | 4.1 | 8.2 | 6.1 | 56.1 | 411 | 190 | 148 | 1,432 | 973 |
| Lincoln, NE | 12,414 | 55,567,994 | 5.0 | 8.7 | 3.5 | 4.6 | 3.5 | 10.7 | 62.7 | 1,320 | 404 | 652 | 1,971 | 1,508 |
| Little Rock-North Little Rock-Conway, AR | 24,453 | 89,371,093 | 5.8 | 11.4 | 5.6 | 4.3 | 2.7 | 8.7 | 59.7 | 2,436 | 1,217 | 734 | 995 | 471 |
| Logan, UT-ID | 4,086 | 14,948,932 | 6.6 | 6.9 | 2.6 | 4.1 | 13.3 | 7.5 | 57.6 | 432 | 195 | 138 | 1,000 | 674 |
| Longview, TX | 12,699 | 42,566,428 | 5.2 | 10.4 | 3.9 | 2.0 | 4.8 | 3.7 | 68.1 | 1,018 | 344 | 524 | 1,839 | 1,474 |
| Longview, WA | 3,640 | 18,508,063 | 7.2 | 10.5 | 3.3 | 7.0 | 2.9 | 12.2 | 53.9 | 543 | 209 | 155 | 1,448 | 882 |
| Los Angeles-Long Beach-Anaheim, CA | 493,015 | 3,243,212,859 | 8.7 | 12.2 | 4.2 | 5.5 | 14.1 | 8.2 | 44.5 | 93,808 | 40,622 | 31,283 | 2,356 | 1,576 |
| Anaheim-Santa Ana-Irvine, CA Div 11244 | 92,158 | 640,816,778 | 6.4 | 12.9 | 4.5 | 3.4 | 7.1 | 5.6 | 58.9 | 17,779 | 6,804 | 7,027 | 2,214 | 1,642 |
| Los Angeles-Long Beach-Glendale, CA Div 31084 | 400,857 | 2,602,396,081 | 9.3 | 12.0 | 4.1 | 6.0 | 15.8 | 8.9 | 41.0 | 76,030 | 33,818 | 24,256 | 2,401 | 1,555 |
| Louisville/Jefferson County, KY-IN | 41,775 | 174,583,976 | 3.6 | 9.3 | 3.8 | 3.3 | 9.4 | 7.0 | 62.4 | 4,228 | 1,443 | 1,931 | 1,533 | 903 |
| Lubbock, TX | 15,658 | 62,174,487 | 4.7 | 10.2 | 4.3 | 1.0 | 32.0 | 5.4 | 41.2 | 1,691 | 460 | 536 | 1,690 | 1,328 |
| Lynchburg, VA | 8,822 | 29,576,126 | 7.3 | 10.6 | 3.7 | 2.0 | 8.2 | 5.1 | 61.2 | 863 | 437 | 332 | 1,268 | 893 |
| Macon-Bibb County, GA | 8,443 | 29,792,058 | 6.9 | 10.9 | 5.0 | 2.5 | 5.2 | 5.2 | 61.7 | 688 | 317 | 228 | 994 | 718 |
| Madera, CA | 5,262 | 26,791,038 | 8.4 | 9.4 | 1.2 | 1.9 | 12.2 | 3.1 | 62.3 | 765 | 489 | 169 | 1,088 | 800 |
| Madison, WI | 25,406 | 110,653,611 | 5.5 | 11.1 | 2.3 | 5.5 | 6.5 | 5.3 | 61.9 | 3,255 | 1,232 | 1,485 | 2,266 | 2,091 |
| Manchester-Nashua, NH | 13,924 | 65,255,845 | 4.2 | 10.8 | 6.1 | 5.1 | 3.7 | 4.7 | 63.9 | 1,758 | 439 | 1,097 | 2,654 | 2,624 |
| Manhattan, KS | 5,712 | 19,755,413 | 6.6 | 10.0 | 4.2 | 3.6 | 9.4 | 5.3 | 59.6 | 562 | 214 | 222 | 1,688 | 1,111 |
| Mankato, MN | 3,460 | 15,489,530 | 8.4 | 9.5 | 1.1 | 5.0 | 13.1 | 5.2 | 55.1 | 495 | 221 | 138 | 1,367 | 1,189 |
| Mansfield, OH | 4,967 | 18,600,904 | 8.1 | 8.4 | 4.6 | 2.9 | 15.2 | 4.0 | 55.0 | 486 | 220 | 192 | 1,593 | 961 |
| McAllen-Edinburg-Mission, TX | 42,017 | 156,936,079 | 3.8 | 6.8 | 1.8 | 1.8 | 4.5 | 4.0 | 76.4 | 3,555 | 2,160 | 995 | 1,162 | 962 |
| Medford, OR | 5,573 | 25,323,949 | 7.9 | 12.8 | 5.8 | 5.8 | 6.8 | 5.7 | 53.0 | 827 | 362 | 312 | 1,440 | 1,189 |
| Memphis, TN-MS-AR | 47,599 | 190,916,078 | 7.5 | 14.7 | 6.9 | 3.4 | 8.6 | 13.3 | 44.3 | 5,371 | 1,889 | 2,147 | 1,604 | 1,161 |
| Merced, CA | 11,500 | 58,776,936 | 5.1 | 7.4 | 1.2 | 0.8 | 9.4 | 5.9 | 67.6 | 1,627 | 1,064 | 295 | 1,088 | 823 |
| Miami-Fort Lauderdale-Pompano Beach, FL | 222,781 | 1,085,361,286 | 6.6 | 15.5 | 6.8 | 5.4 | 18.7 | 8.2 | 36.0 | 34,185 | 7,177 | 13,866 | 2,266 | 1,783 |
| Fort Lauderdale-Pompano Beach-Sunrise, FL Div 22744 | 78,792 | 389,220,424 | 5.0 | 13.2 | 5.4 | 3.7 | 32.3 | 6.1 | 32.6 | 12,007 | 2,461 | 3,819 | 1,974 | 1,586 |
| Miami-Miami Beach-Kendall, FL Div 33124 | 93,991 | 469,157,165 | 7.3 | 16.4 | 6.7 | 8.4 | 14.5 | 8.1 | 35.3 | 14,969 | 3,265 | 6,119 | 2,255 | 1,664 |
| West Palm Beach-Boca Raton-Boynton Beach, FL Div 48424 | 49,998 | 226,983,697 | 8.2 | 17.4 | 9.4 | 2.1 | 4.2 | 12.1 | 43.4 | 7,209 | 1,451 | 3,928 | 2,672 | 2,261 |
| Michigan City-La Porte, IN | 4,410 | 14,500,631 | 6.4 | 12.0 | 4.0 | 3.7 | 2.8 | 6.2 | 63.3 | 391 | 185 | 126 | 1,146 | 1,100 |
| Midland, MI | 2,197 | 9,909,177 | 10.3 | 8.4 | 3.0 | 5.2 | 3.8 | 5.2 | 62.3 | 320 | 152 | 114 | 1,369 | 1,353 |
| Midland, TX | 8,050 | 35,757,035 | 4.9 | 7.6 | 3.3 | 3.1 | 30.0 | 2.5 | 47.6 | 1,057 | 139 | 532 | 3,112 | 2,385 |
| Milwaukee-Waukesha, WI | 53,564 | 271,603,112 | 5.3 | 14.4 | 4.5 | 4.9 | 7.1 | 5.5 | 56.7 | 7,530 | 2,974 | 2,896 | 1,840 | 1,685 |
| Minneapolis-St. Paul-Bloomington, MN | 124,427 | 637,696,752 | 7.4 | 10.3 | 1.9 | 3.7 | 8.2 | 4.9 | 61.4 | 18,429 | 7,855 | 5,970 | 1,670 | 1,519 |
| Missoula, MT | 3,480 | 14,645,116 | 8.3 | 11.5 | 6.5 | 6.6 | 8.7 | 4.2 | 51.1 | 389 | 154 | 168 | 1,428 | 1,378 |

1. Based on the resident population estimated as of July 1 of the year shown.

| Area name | Local government finances, 2017 (cont.) | | | | | | | | Debt outstanding | | Government employment, 2019 | | | Individual income tax returns, 2018 | | |
|---|---|---|---|---|---|---|---|---|---|---|---|---|---|---|---|---|
| | Direct general expenditure | | | | | | | | | | | | | | | |
| | | | Percent of total for: | | | | | | | | | | | | | |
| | Total (mil dol) | Per capita[1] (dollars) | Educa-tion | Health and hospitals | Police protection | Public welfare | Highways | | Total (mil dol) | Per capita[1] (dollars) | Federal civilian | Federal military | State and local | Number of returns | Mean adjusted gross income | Mean income tax |
| | 185 | 186 | 187 | 188 | 189 | 190 | 191 | | 192 | 193 | 194 | 195 | 196 | 197 | 198 | 199 |
| Jonesboro, AR | 392.9 | 2,993 | 61.3 | 0.4 | 5.4 | 0.1 | 5.3 | | 536.0 | 4,083 | 400 | 508 | 9,417 | 51,650 | 54,861 | 5,330 |
| Joplin, MO | 587.5 | 3,296 | 58.6 | 1.6 | 4.7 | 0.2 | 7.8 | | 423.1 | 2,374 | 456 | 624 | 9,187 | 76,570 | 52,819 | 4,976 |
| Kahului-Wailuku-Lahaina, HI | 467.9 | 2,816 | 0.0 | 0.0 | 11.5 | 13.0 | 10.1 | | 336.6 | 2,026 | 872 | 1,208 | 8,359 | 82,600 | 63,140 | 6,780 |
| Kalamazoo-Portage, MI | 1,030.7 | 3,919 | 55.8 | 9.5 | 3.4 | 0.4 | 5.5 | | 1,217.5 | 4,629 | 697 | 420 | 14,343 | 122,790 | 69,477 | 8,300 |
| Kankakee, IL | 447.3 | 4,046 | 52.8 | 0.5 | 6.7 | 0.1 | 5.4 | | 491.0 | 4,442 | 258 | 214 | 6,058 | 50,160 | 58,111 | 5,506 |
| Kansas City, MO-KS | 10,573.0 | 4,970 | 40.6 | 9.6 | 6.9 | 0.5 | 5.3 | | 18,038.5 | 8,480 | 28,469 | 11,124 | 123,636 | 1,021,150 | 75,537 | 9,320 |
| Kennewick-Richland, WA | 1,687.9 | 5,819 | 44.9 | 21.8 | 3.9 | 0.1 | 2.7 | | 8,022.2 | 27,653 | 1,236 | 758 | 18,084 | 129,990 | 68,544 | 7,101 |
| Killeen-Temple, TX | 1,613.6 | 3,642 | 59.9 | 5.1 | 5.1 | 0.4 | 3.7 | | 752.5 | 2,455 | 894 | 883 | 15,494 | 132,020 | 53,299 | 5,068 |
| Kingsport-Bristol, TN-VA | 948.3 | 3,094 | 45.3 | 4.2 | 7.6 | 1.6 | 3.9 | | 668.9 | 3,745 | 416 | 284 | 12,830 | 87,620 | 66,561 | 7,565 |
| Kingston, NY | 1,209.9 | 6,773 | 55.7 | 1.7 | 3.2 | 9.5 | 6.0 | | 4,123.5 | 4,830 | 5,460 | 2,504 | 51,573 | 388,650 | 68,693 | 8,217 |
| Knoxville, TN | 2,645.2 | 3,099 | 42.6 | 11.8 | 7.8 | 0.1 | 4.1 | | 246.6 | 2,999 | 189 | 243 | 4,898 | 40,740 | 53,268 | 4,688 |
| Kokomo, IN | 263.3 | 3,202 | 53.5 | 0.4 | 6.4 | 0.1 | 6.8 | | 713.5 | 5,219 | 545 | 363 | 11,106 | 66,600 | 65,106 | 6,915 |
| La Crosse-Onalaska, WI-MN | 764.4 | 5,591 | 53.9 | 3.4 | 4.1 | 7.1 | 6.8 | | 1,495.9 | 3,051 | 1,434 | 1,891 | 24,489 | 208,210 | 61,106 | 6,759 |
| Lafayette, LA | 1,734.3 | 3,537 | 46.5 | 9.3 | 9.5 | 0.4 | 5.4 | | | | | | | | | |
| Lafayette-West Lafayette, IN | 619.1 | 2,714 | 49.7 | 0.7 | 4.6 | 0.6 | 7.1 | | 537.2 | 2,355 | 531 | 714 | 26,031 | 94,200 | 62,693 | 6,626 |
| Lake Charles, LA | 1,052.5 | 5,026 | 39.4 | 8.7 | 7.7 | 0.4 | 7.5 | | 1,112.3 | 5,311 | 575 | 867 | 14,155 | 91,690 | 64,933 | 7,250 |
| Lake Havasu City-Kingman, AZ | 532.1 | 2,570 | 33.3 | 4.1 | 9.9 | 0.1 | 6.2 | | 543.6 | 2,626 | 477 | 447 | 7,839 | 86,120 | 47,845 | 4,288 |
| Lakeland-Winter Haven, FL | 2,193.4 | 3,200 | 47.9 | 2.6 | 8.2 | 1.4 | 4.0 | | 2,340.3 | 3,415 | 1,246 | 1,298 | 27,592 | 315,270 | 52,183 | 4,842 |
| Lancaster, PA | 2,292.0 | 4,234 | 55.6 | 5.9 | 4.6 | 5.2 | 3.2 | | 3,717.0 | 6,866 | 1,302 | 1,366 | 19,769 | 271,410 | 66,221 | 7,099 |
| Lansing-East Lansing, MI | 2,344.8 | 4,276 | 48.7 | 8.2 | 5.3 | 4.5 | 5.7 | | 2,585.8 | 4,715 | 2,064 | 1,046 | 51,402 | 252,090 | 60,907 | 6,461 |
| Laredo, TX | 1,395.4 | 5,099 | 60.8 | 2.6 | 6.2 | 0.5 | 0.9 | | 1,627.0 | 5,945 | 3,703 | 534 | 19,346 | 114,170 | 42,963 | 3,498 |
| Las Cruces, NM | 708.9 | 3,279 | 60.6 | 1.6 | 7.6 | 2.3 | 2.1 | | 452.3 | 2,092 | 3,329 | 557 | 16,887 | 94,870 | 46,148 | 4,113 |
| Las Vegas-Henderson-Para-dise, NV | 9,855.5 | 4,517 | 33.6 | 7.7 | 10.6 | 2.6 | 8.8 | | 23,841.2 | 10,928 | 13,733 | 16,332 | 91,703 | 1,049,220 | 71,163 | 9,346 |
| Lawrence, KS | 638.1 | 5,305 | 25.4 | 35.8 | 4.3 | 0.0 | 3.0 | | 783.8 | 6,517 | 430 | 456 | 17,860 | 52,010 | 68,522 | 7,791 |
| Lawton, OK | 611.5 | 4,790 | 30.6 | 41.9 | 3.7 | 0.0 | 5.2 | | 307.9 | 2,412 | 3,980 | 12,384 | 10,921 | 49,390 | 48,767 | 3,930 |
| Lebanon, PA | 520.5 | 3,730 | 57.8 | 1.5 | 3.4 | 10.7 | 3.0 | | 583.4 | 4,180 | 3,190 | 354 | 4,934 | 71,150 | 57,694 | 5,331 |
| Lewiston, ID-WA | 222.1 | 3,537 | 42.1 | 5.2 | 7.0 | 0.1 | 6.7 | | 183.9 | 2,929 | 255 | 181 | 5,209 | 28,710 | 58,969 | 5,519 |
| Lewiston-Auburn, ME | 392.1 | 3,651 | 53.5 | 0.3 | 3.8 | 0.3 | 4.7 | | 282.6 | 2,631 | 277 | 323 | 5,327 | 51,120 | 51,775 | 4,487 |
| Lexington-Fayette, KY | 1,545.1 | 3,015 | 46.0 | 3.0 | 5.5 | 0.7 | 2.0 | | 2,665.5 | 5,202 | 4,567 | 1,579 | 50,044 | 228,630 | 66,579 | 7,743 |
| Lima, OH | 466.4 | 4,524 | 51.6 | 3.1 | 8.5 | 4.4 | 3.7 | | 1,857.7 | 18,019 | 329 | 247 | 5,697 | 47,650 | 55,083 | 5,300 |
| Lincoln, NE | 1,403.0 | 4,239 | 52.6 | 2.7 | 4.2 | 1.2 | 7.4 | | 2,539.9 | 7,674 | 3,491 | 1,130 | 31,788 | 154,880 | 69,088 | 7,582 |
| Little Rock-North Little Rock-Conway, AR | 2,494.1 | 3,382 | 47.7 | 1.8 | 6.7 | 0.0 | 5.9 | | 3,867.0 | 5,243 | 9,684 | 6,329 | 58,567 | 321,760 | 64,379 | 7,178 |
| Logan, UT-ID | 445.9 | 3,238 | 57.0 | 8.4 | 4.2 | 0.3 | 3.9 | | 237.0 | 1,721 | 369 | 530 | 12,261 | 56,740 | 60,183 | 5,167 |
| Longview, TX | 1,120.1 | 3,936 | 60.6 | 3.3 | 5.8 | 0.2 | 4.2 | | 1,382.1 | 4,856 | 605 | 536 | 14,622 | 121,800 | 58,818 | 6,097 |
| Longview, WA | 508.9 | 4,765 | 41.8 | 1.6 | 5.7 | 0.5 | 4.7 | | 656.3 | 6,144 | 221 | 275 | 6,308 | 49,250 | 61,456 | 5,994 |
| Los Angeles-Long Beach-Anaheim, CA | 87,198.5 | 6,567 | 37.0 | 10.4 | 8.6 | 7.8 | 2.8 | | 126,596.1 | 9,534 | 59,226 | 22,303 | 711,699 | 6,229,730 | 84,836 | 12,634 |
| Anaheim-Santa Ana-Irvine, CA Div 11244 | 16,921.3 | 5,331 | 45.3 | 3.6 | 8.0 | 5.5 | 3.9 | | 24,733.3 | 7,792 | 11,219 | 5,079 | 151,444 | 1,527,670 | 94,533 | 14,034 |
| Los Angeles-Long Beach-Glendale, CA Div 31084 | 70,277.2 | 6,956 | 35.1 | 12.0 | 8.8 | 8.4 | 2.5 | | 101,862.8 | 10,082 | 48,007 | 17,224 | 560,255 | 4,702,060 | 81,685 | 12,179 |
| Louisville/Jefferson County, KY-IN | 4,149.0 | 3,294 | 47.0 | 3.3 | 6.2 | 0.5 | 3.3 | | 7,772.8 | 6,171 | 9,538 | 3,993 | 62,718 | 608,040 | 66,082 | 7,442 |
| Lubbock, TX | 1,733.9 | 5,468 | 33.1 | 35.0 | 5.8 | 0.0 | 2.0 | | 2,370.1 | 7,474 | 1,383 | 628 | 30,194 | 136,580 | 62,025 | 7,110 |
| Lynchburg, VA | 791.4 | 3,023 | 53.6 | 1.1 | 5.4 | 7.4 | 1.9 | | 532.6 | 2,034 | 567 | 835 | 12,468 | 115,110 | 57,709 | 5,539 |
| Macon-Bibb County, GA | 750.8 | 3,279 | 56.0 | 7.1 | 5.7 | 0.0 | 2.0 | | 436.3 | 1,905 | 1,105 | 598 | 13,524 | 96,870 | 54,459 | 5,492 |
| Madera, CA | 839.1 | 5,399 | 50.1 | 4.5 | 3.8 | 8.2 | 6.4 | | 657.0 | 4,227 | 302 | 221 | 11,059 | 59,580 | 51,966 | 4,480 |
| Madison, WI | 3,366.3 | 5,137 | 47.9 | 2.1 | 5.9 | 8.3 | 7.2 | | 4,585.2 | 6,998 | 5,582 | 1,840 | 84,375 | 337,780 | 82,075 | 10,452 |
| Manchester-Nashua, NH | 1,549.9 | 3,751 | 54.3 | 0.4 | 6.9 | 4.3 | 5.0 | | 894.8 | 2,165 | 4,205 | 1,415 | 17,479 | 219,230 | 78,654 | 9,841 |
| Manhattan, KS | 517.2 | 3,928 | 43.3 | 10.9 | 10.1 | 0.1 | 6.0 | | 961.1 | 7,300 | 3,798 | 15,704 | 17,588 | 53,620 | 56,310 | 4,935 |
| Mankato, MN | 552.3 | 5,470 | 42.9 | 7.0 | 3.7 | 5.3 | 11.2 | | 508.5 | 5,036 | 297 | 360 | 8,377 | 47,780 | 60,651 | 6,087 |
| Mansfield, OH | 493.6 | 4,099 | 50.7 | 9.7 | 5.4 | 5.5 | 6.3 | | 175.1 | 1,454 | 639 | 289 | 7,004 | 57,340 | 49,520 | 4,286 |
| McAllen-Edinburg-Mission, TX | 3,600.9 | 4,205 | 68.2 | 4.5 | 3.5 | 0.6 | 2.7 | | 3,753.4 | 4,384 | 4,565 | 1,733 | 53,611 | 328,590 | 40,424 | 3,050 |
| Medford, OR | 766.8 | 3,538 | 44.2 | 6.2 | 9.3 | 0.0 | 5.4 | | 894.6 | 4,127 | 1,825 | 517 | 8,893 | 103,800 | 61,203 | 6,482 |
| Memphis, TN-MS-AR | 4,991.9 | 3,729 | 39.3 | 10.7 | 11.3 | 0.8 | 2.9 | | 5,027.1 | 3,756 | 14,229 | 5,389 | 69,867 | 608,640 | 62,330 | 7,331 |
| Merced, CA | 1,616.8 | 5,964 | 57.0 | 4.2 | 4.0 | 8.9 | 3.7 | | 974.0 | 3,593 | 742 | 400 | 17,133 | 107,320 | 48,237 | 3,887 |
| Miami-Fort Lauderdale-Pom-pano Beach, FL | 33,348.0 | 5,451 | 27.5 | 16.7 | 9.7 | 2.3 | 1.6 | | 45,314.7 | 7,407 | 35,216 | 13,965 | 282,100 | 3,063,830 | 80,855 | 12,715 |
| Fort Lauderdale-Pompano Beach-Sunrise, FL Div 22744 | 11,894.2 | 6,148 | 25.1 | 28.0 | 9.9 | 1.2 | 1.3 | | 11,449.4 | 5,919 | 7,150 | 3,920 | 100,925 | 970,530 | 72,413 | 10,195 |
| Miami-Miami Beach-Kendall, FL Div 33124 | 14,469.2 | 5,333 | 28.2 | 13.3 | 9.2 | 3.4 | 1.6 | | 28,281.1 | 10,423 | 20,979 | 7,336 | 122,728 | 1,355,670 | 69,516 | 10,780 |
| West Palm Beach-Boca Raton-Boynton Beach, FL Div 48424 | 6,984.6 | 4,750 | 30.3 | 4.6 | 10.4 | 1.7 | 2.1 | | 5,584.1 | 3,798 | 7,087 | 2,709 | 58,447 | 737,630 | 112,802 | 19,587 |
| Michigan City-La Porte, IN | 385.7 | 3,511 | 46.8 | 0.7 | 5.0 | 0.1 | 3.5 | | 349.2 | 3,179 | 180 | 328 | 6,403 | 51,730 | 55,109 | 5,366 |
| Midland, MI | 324.6 | 3,896 | 55.6 | 2.2 | 4.0 | 2.1 | 7.2 | | 513.6 | 6,165 | 152 | 130 | 3,072 | 40,440 | 77,674 | 9,818 |
| Midland, TX | 995.1 | 5,824 | 38.7 | 35.8 | 3.9 | 0.0 | 2.1 | | 1,016.0 | 5,947 | 585 | 354 | 9,038 | 81,410 | 123,140 | 21,530 |
| Milwaukee-Waukesha, WI | 7,855.3 | 4,989 | 44.0 | 6.7 | 8.6 | 2.0 | 6.4 | | 9,218.1 | 5,855 | 10,713 | 4,390 | 76,598 | 768,550 | 76,266 | 9,689 |
| Minneapolis-St. Paul-Bloom-ington, MN | 20,224.7 | 5,656 | 42.2 | 8.1 | 5.4 | 3.9 | 8.1 | | 25,977.0 | 7,265 | 21,281 | 12,795 | 231,672 | 1,834,370 | 86,774 | 11,494 |
| Missoula, MT | 459.0 | 3,895 | 45.7 | 7.4 | 6.3 | 0.6 | 3.3 | | 347.1 | 2,946 | 1,450 | 538 | 9,151 | 58,930 | 64,394 | 8,318 |

1. Based on the resident population estimated as of July 1 of the year shown.

# Table C. Metropolitan Areas — Land Area and Population

| CBSA/DIV Code[1] | Area name | Land area[2] (sq mi) | Total persons 2020 | Rank | Per square mile | White | Black | American Indian, Alaska Native | Asian and Pacific Islander | Percent Hispanic or Latino[3] | Under 5 years | 5 to 14 years | 15 to 24 years | 25 to 34 years | 35 to 44 years | 45 to 54 years |
|---|---|---|---|---|---|---|---|---|---|---|---|---|---|---|---|---|
| | | 1 | 2 | 3 | 4 | 5 | 6 | 7 | 8 | 9 | 10 | 11 | 12 | 13 | 14 | 15 |
| 33660 | Mobile, AL | 2,310 | 428,692 | 127 | 185.6 | 58.0 | 36.5 | 1.7 | 2.5 | 2.9 | 6.5 | 12.8 | 12.5 | 14.0 | 12.0 | 11.7 |
| 33700 | Modesto, CA | 1,496 | 550,081 | 102 | 367.7 | 41.8 | 3.6 | 1.2 | 8.1 | 48.3 | 7.0 | 15.3 | 13.9 | 14.5 | 12.9 | 11.5 |
| 33740 | Monroe, LA | 2,282 | 198,836 | 226 | 87.1 | 58.6 | 38.2 | 0.6 | 1.2 | 2.5 | 6.5 | 13.5 | 13.1 | 13.5 | 12.2 | 11.5 |
| 33780 | Monroe, MI | 549 | 150,568 | 280 | 274.1 | 92.5 | 3.5 | 0.9 | 1.1 | 3.8 | 5.2 | 11.9 | 11.6 | 12.0 | 11.7 | 13.2 |
| 33860 | Montgomery, AL | 2,714 | 372,583 | 146 | 137.3 | 48.2 | 46.3 | 0.7 | 2.9 | 3.4 | 6.4 | 12.8 | 12.8 | 14.1 | 12.6 | 12.2 |
| 34060 | Morgantown, WV | 1,009 | 140,199 | 297 | 139.0 | 91.8 | 4.2 | 0.6 | 3.4 | 2.0 | 4.7 | 9.5 | 19.6 | 15.8 | 12.6 | 11.0 |
| 34100 | Morristown, TN | 717 | 143,982 | 292 | 200.9 | 88.7 | 3.4 | 0.8 | 1.1 | 7.5 | 5.3 | 11.8 | 11.8 | 11.7 | 11.3 | 13.6 |
| 34580 | Mount Vernon-Anacortes, WA | 1,730 | 130,789 | 311 | 75.6 | 75.9 | 1.4 | 2.7 | 3.5 | 19.1 | 5.7 | 12.1 | 11.0 | 12.8 | 12.3 | 10.9 |
| 34620 | Muncie, IN | 392 | 113,454 | 339 | 289.3 | 88.6 | 8.5 | 0.7 | 2.0 | 2.6 | 4.8 | 9.9 | 22.5 | 11.8 | 10.1 | 11.0 |
| 34740 | Muskegon, MI | 504 | 173,883 | 245 | 345.1 | 78.9 | 15.4 | 1.7 | 1.2 | 6.0 | 6.0 | 12.8 | 11.9 | 13.3 | 12.2 | 11.8 |
| 34820 | Myrtle Beach-Conway-North Myrtle Beach, SC-NC | 1,983 | 514,488 | 110 | 259.6 | 80.8 | 12.5 | 1.0 | 1.8 | 5.8 | 4.2 | 9.4 | 9.3 | 10.5 | 10.3 | 11.6 |
| 34900 | Napa, CA | 748 | 135,965 | 304 | 181.7 | 53.6 | 2.7 | 1.1 | 10.4 | 34.8 | 4.6 | 11.4 | 12.2 | 12.3 | 12.7 | 12.9 |
| 34940 | Naples-Marco Island, FL | 1,997 | 392,973 | 139 | 196.8 | 62.8 | 7.1 | 0.4 | 2.0 | 28.6 | 4.2 | 9.4 | 9.3 | 9.4 | 9.6 | 10.9 |
| 34980 | Nashville-Davidson--Murfreesboro--Franklin, TN | 5,689 | 1,961,232 | 36 | 344.7 | 73.4 | 16.4 | 0.8 | 3.8 | 7.8 | 6.2 | 12.7 | 12.9 | 15.6 | 13.8 | 12.6 |
| 35100 | New Bern, NC | 1,515 | 123,198 | 324 | 81.4 | 68.3 | 22.5 | 1.1 | 3.6 | 7.3 | 5.8 | 11.4 | 13.8 | 12.9 | 10.7 | 10.2 |
| 35300 | New Haven-Milford, CT | 604 | 851,948 | 68 | 1,409.3 | 62.5 | 14.4 | 0.6 | 4.9 | 19.6 | 5.1 | 11.1 | 13.4 | 13.5 | 12.1 | 12.6 |
| 35380 | New Orleans-Metairie, LA | 3,204 | 1,272,258 | 45 | 397.2 | 52.3 | 35.9 | 0.9 | 3.5 | 9.1 | 6.0 | 12.3 | 11.4 | 14.3 | 13.4 | 11.9 |
| 35620 | New York-Newark-Jersey City, NY-NJ-PA | 6,684 | 19,124,359 | 1 | 2,860.7 | 46.4 | 16.8 | 0.5 | 13.0 | 25.1 | 6.0 | 11.7 | 11.9 | 14.8 | 13.1 | 12.8 |
| 35620 | Nassau County-Suffolk County, NY Div 35004 | 1,196 | 2,825,607 | X | 2,361.8 | 63.2 | 10.3 | 0.5 | 8.5 | 19.1 | 5.4 | 11.8 | 12.5 | 12.0 | 11.9 | 13.6 |
| 35620 | Newark, NJ-PA Div 35084 | 2,181 | 2,167,853 | X | 993.8 | 50.3 | 21.6 | 0.4 | 7.4 | 21.9 | 5.8 | 12.4 | 12.3 | 12.4 | 13.1 | 13.9 |
| 35620 | New Brunswick-Lakewood, NJ Div 35154 | 1,708 | 2,384,685 | X | 1,396.5 | 63.8 | 8.0 | 0.4 | 14.1 | 15.2 | 5.8 | 12.3 | 12.2 | 11.8 | 12.1 | 13.0 |
| 35620 | New York-Jersey City-White Plains, NY-NJ Div 35614 | 1,600 | 11,746,214 | X | 7,342.8 | 38.2 | 19.2 | 0.6 | 14.9 | 29.1 | 6.2 | 11.5 | 11.6 | 16.5 | 13.6 | 12.4 |
| 35660 | Niles, MI | 568 | 153,025 | 272 | 269.5 | 76.9 | 15.7 | 1.2 | 2.9 | 5.9 | 5.5 | 12.2 | 11.8 | 11.7 | 11.6 | 12.1 |
| 35840 | North Port-Sarasota-Bradenton, FL | 1,299 | 854,684 | 67 | 657.7 | 78.2 | 7.1 | 0.6 | 2.6 | 13.1 | 3.9 | 8.9 | 8.8 | 9.3 | 9.3 | 11.0 |
| 35980 | Norwich-New London, CT | 665 | 264,999 | 186 | 398.4 | 77.2 | 7.4 | 1.8 | 5.3 | 11.7 | 4.8 | 10.7 | 13.3 | 13.3 | 11.5 | 12.2 |
| 36100 | Ocala, FL | 1,588 | 373,513 | 145 | 235.2 | 70.5 | 13.5 | 0.9 | 2.3 | 14.7 | 4.9 | 10.4 | 9.7 | 11.0 | 10.0 | 10.8 |
| 36140 | Ocean City, NJ | 252 | 91,546 | 367 | 364.0 | 86.3 | 5.0 | 0.5 | 1.5 | 8.3 | 4.4 | 9.9 | 10.0 | 10.5 | 9.4 | 10.9 |
| 36220 | Odessa, TX | 898 | 167,701 | 259 | 186.8 | 30.3 | 5.1 | 0.7 | 1.5 | 63.4 | 8.8 | 17.0 | 14.5 | 16.6 | 13.6 | 10.4 |
| 36260 | Ogden-Clearfield, UT | 7,230 | 691,359 | 85 | 95.6 | 82.7 | 1.8 | 1.0 | 3.6 | 13.3 | 7.5 | 17.1 | 14.6 | 14.2 | 14.6 | 10.6 |
| 36420 | Oklahoma City, OK | 5,512 | 1,425,375 | 41 | 258.6 | 67.7 | 12.2 | 6.9 | 4.4 | 14.1 | 6.4 | 13.7 | 14.0 | 14.6 | 13.5 | 11.3 |
| 36500 | Olympia-Lacey-Tumwater, WA | 723 | 294,074 | 168 | 407.0 | 78.4 | 5.0 | 2.7 | 10.0 | 9.8 | 5.7 | 12.0 | 11.4 | 14.6 | 13.9 | 11.8 |
| 36540 | Omaha-Council Bluffs, NE-IA | 4,346 | 954,270 | 57 | 219.6 | 77.2 | 9.0 | 1.0 | 4.2 | 11.2 | 6.9 | 14.2 | 12.8 | 14.2 | 13.7 | 11.6 |
| 36740 | Orlando-Kissimmee-Sanford, FL | 3,491 | 2,639,374 | 23 | 756.4 | 46.8 | 16.9 | 0.6 | 5.5 | 32.2 | 5.6 | 12.1 | 12.8 | 15.2 | 14.0 | 12.8 |
| 36780 | Oshkosh-Neenah, WI | 435 | 171,631 | 247 | 394.8 | 89.1 | 3.4 | 1.0 | 3.8 | 4.5 | 5.4 | 11.3 | 15.0 | 13.4 | 12.3 | 11.9 |
| 36980 | Owensboro, KY | 899 | 119,795 | 332 | 133.3 | 90.5 | 5.8 | 0.4 | 2.1 | 3.2 | 6.5 | 13.7 | 12.2 | 12.7 | 11.9 | 11.9 |
| 37100 | Oxnard-Thousand Oaks-Ventura, CA | 1,841 | 841,387 | 72 | 456.7 | 46.7 | 2.4 | 0.8 | 9.4 | 43.4 | 5.5 | 12.8 | 13.0 | 13.5 | 12.5 | 12.7 |
| 37340 | Palm Bay-Melbourne-Titusville, FL | 1,015 | 608,459 | 96 | 599.5 | 75.6 | 11.3 | 0.8 | 3.7 | 11.2 | 4.5 | 10.3 | 10.2 | 11.5 | 10.8 | 11.8 |
| 37460 | Panama City, FL | 759 | 171,322 | 249 | 225.8 | 79.0 | 12.1 | 1.4 | 3.8 | 6.9 | 5.5 | 11.5 | 10.7 | 13.3 | 12.4 | 12.6 |
| 37620 | Parkersburg-Vienna, WV | 599 | 88,643 | 368 | 148.0 | 96.8 | 2.1 | 0.8 | 0.9 | 1.3 | 5.4 | 11.8 | 10.8 | 11.4 | 11.7 | 12.9 |
| 37860 | Pensacola-Ferry Pass-Brent, FL | 1,669 | 511,503 | 111 | 306.4 | 73.3 | 17.9 | 1.6 | 4.6 | 6.1 | 5.8 | 11.9 | 13.3 | 14.4 | 12.1 | 11.6 |
| 37900 | Peoria, IL | 3,333 | 396,781 | 136 | 119.0 | 84.4 | 10.5 | 0.6 | 2.8 | 3.7 | 6.0 | 12.8 | 11.9 | 12.4 | 12.5 | 12.0 |
| 37980 | Philadelphia-Camden-Wilmington, PA-NJ-DE-MD | 4,603 | 6,107,906 | 8 | 1,326.9 | 62.4 | 21.9 | 0.6 | 7.2 | 10.1 | 5.6 | 12.1 | 12.4 | 14.4 | 12.7 | 12.4 |
| 37980 | Camden, NJ Div 15804 | 1,343 | 1,246,650 | X | 928.5 | 65.9 | 17.6 | 0.6 | 6.0 | 12.2 | 5.5 | 12.2 | 12.1 | 13.2 | 12.7 | 13.1 |
| 37980 | Montgomery-Bucks-Chester Counties, PA Div 33874 | 1,838 | 1,988,615 | X | 1,082.0 | 79.9 | 7.8 | 0.5 | 7.6 | 6.2 | 5.2 | 12.0 | 12.0 | 11.8 | 12.5 | 13.1 |
| 37980 | Philadelphia, PA Div 37964 | 318 | 2,145,240 | X | 6,743.9 | 43.9 | 36.9 | 0.7 | 8.1 | 12.6 | 6.2 | 12.0 | 12.8 | 17.7 | 12.8 | 11.2 |
| 37980 | Wilmington, DE-MD-NJ Div 48864 | 1,105 | 727,401 | X | 658.6 | 63.0 | 23.3 | 0.7 | 5.6 | 9.8 | 5.6 | 12.0 | 12.5 | 14.0 | 12.5 | 12.7 |
| 38060 | Phoenix-Mesa-Chandler, AZ | 14,568 | 5,059,909 | 10 | 347.4 | 56.4 | 6.7 | 2.4 | 5.6 | 31.5 | 5.9 | 13.1 | 13.1 | 14.6 | 13.0 | 12.2 |
| 38220 | Pine Bluff, AR | 2,030 | 86,278 | 370 | 42.5 | 47.1 | 49.6 | 0.9 | 1.1 | 2.7 | 5.6 | 11.5 | 13.4 | 13.2 | 12.1 | 12.1 |
| 38300 | Pittsburgh, PA | 5,283 | 2,309,246 | 28 | 437.1 | 86.6 | 9.8 | 0.5 | 3.2 | 2.0 | 5.0 | 10.5 | 11.3 | 13.5 | 12.0 | 12.1 |
| 38340 | Pittsfield, MA | 927 | 124,571 | 318 | 134.4 | 89.4 | 4.6 | 0.6 | 2.4 | 5.3 | 4.1 | 9.1 | 12.3 | 10.9 | 10.7 | 11.9 |
| 38540 | Pocatello, ID | 2,516 | 96,438 | 364 | 38.3 | 83.3 | 1.4 | 3.6 | 2.8 | 11.2 | 6.7 | 14.8 | 14.6 | 14.2 | 13.2 | 10.2 |
| 38860 | Portland-South Portland, ME | 2,081 | 543,221 | 106 | 261.0 | 93.3 | 2.8 | 0.9 | 2.7 | 2.1 | 4.7 | 10.2 | 11.1 | 13.2 | 12.2 | 12.7 |
| 38900 | Portland-Vancouver-Hillsboro, OR-WA | 6,688 | 2,510,259 | 25 | 375.4 | 75.5 | 4.2 | 1.7 | 10.1 | 12.7 | 5.3 | 11.8 | 11.4 | 15.5 | 14.9 | 13.0 |
| 38940 | Port St. Lucie, FL | 1,116 | 499,274 | 112 | 447.6 | 63.9 | 16.5 | 0.6 | 2.5 | 18.3 | 4.6 | 10.4 | 10.0 | 10.7 | 10.5 | 11.6 |
| 39100 | Poughkeepsie-Newburgh-Middletown, NY | 1,608 | 678,527 | 88 | 422.0 | 67.2 | 12.0 | 0.7 | 4.0 | 18.4 | 5.8 | 12.5 | 14.4 | 12.1 | 11.9 | 13.1 |
| 39150 | Prescott Valley-Prescott, AZ | 8,123 | 240,226 | 194 | 29.6 | 81.5 | 1.2 | 2.3 | 1.9 | 15.0 | 3.9 | 8.9 | 9.1 | 9.1 | 8.7 | 10.0 |
| 39300 | Providence-Warwick, RI-MA | 1,587 | 1,623,890 | 38 | 1,023.2 | 76.3 | 7.1 | 0.9 | 3.9 | 14.0 | 5.1 | 10.9 | 13.2 | 13.7 | 12.1 | 12.7 |
| 39340 | Provo-Orem, UT | 5,396 | 663,181 | 90 | 122.9 | 83.9 | 1.1 | 0.9 | 4.6 | 12.2 | 8.9 | 18.2 | 21.9 | 14.8 | 12.7 | 8.8 |
| 39380 | Pueblo, CO | 2,387 | 169,823 | 254 | 71.2 | 53.0 | 2.4 | 1.5 | 1.6 | 43.4 | 5.6 | 12.7 | 12.4 | 13.3 | 12.3 | 11.4 |
| 39460 | Punta Gorda, FL | 681 | 194,711 | 228 | 285.9 | 84.7 | 6.2 | 0.7 | 1.9 | 8.0 | 2.8 | 6.7 | 7.0 | 7.8 | 7.5 | 9.8 |
| 39540 | Racine, WI | 333 | 195,802 | 227 | 588.7 | 72.8 | 12.8 | 0.9 | 1.8 | 14.3 | 6.0 | 12.8 | 12.3 | 12.0 | 12.3 | 12.5 |
| 39580 | Raleigh-Cary, NC | 2,118 | 1,420,376 | 42 | 670.7 | 62.4 | 20.7 | 0.9 | 7.5 | 10.9 | 5.9 | 13.4 | 12.9 | 14.4 | 14.5 | 14.1 |
| 39660 | Rapid City, SD | 6,248 | 144,514 | 290 | 23.1 | 84.2 | 2.4 | 9.6 | 2.2 | 5.4 | 6.0 | 12.7 | 12.2 | 13.3 | 12.1 | 10.5 |

1. CBSA = Core Based Statistical Area. DIV = Metropolitan Division. See Appendix A for explanation. See Appendix B for list of metropolitan areas or temporarily covered by water. 2. Dry land or land partially or temporarily covered by water. 3. May be of any race.

Items 1—15

Table C. Metropolitan Areas — **Population and Households**

| Area name | 55 to 64 years | 65 to 74 years | 75 years and over | Percent female | 2000 | 2010 | 2000-2010 | 2010-2019 | Births | Deaths | Net migration | Number | Persons per house-hold | Family house-holds | Single parent house holds | One person |
|---|---|---|---|---|---|---|---|---|---|---|---|---|---|---|---|---|
| | 16 | 17 | 18 | 19 | 20 | 21 | 22 | 23 | 24 | 25 | 26 | 27 | 28 | 29 | 30 | 31 |
| Mobile, AL.................... | 13.2 | 10.3 | 6.9 | 52.4 | 417,940 | 430,719 | 3.1 | -0.5 | 58,742 | 46,827 | -13,735 | 161,217 | 2.61 | 63.1 | 11.4 | 32.5 |
| Modesto, CA.................. | 11.2 | 8.0 | 5.6 | 50.4 | 446,997 | 514,451 | 15.1 | 6.9 | 77,693 | 42,529 | 729 | 174,698 | 3.12 | 74.0 | 13.3 | 20.0 |
| Monroe, LA.................... | 12.7 | 10.0 | 7.0 | 52.0 | 201,074 | 204,492 | 1.7 | -2.8 | 28,863 | 22,586 | -11,847 | 77,194 | 2.47 | 61.9 | 14.7 | 32.3 |
| Monroe, MI.................... | 15.3 | 11.6 | 7.7 | 50.6 | 145,945 | 152,031 | 4.2 | -2.1 | 15,571 | 14,929 | -2,097 | 60,875 | 2.45 | 68.6 | 9.2 | 26.3 |
| Montgomery, AL............. | 12.9 | 9.6 | 6.6 | 52.5 | 346,528 | 374,528 | 8.1 | -0.5 | 49,504 | 36,860 | -14,650 | 145,250 | 2.48 | 62.8 | 12.4 | 31.9 |
| Morgantown, WV............ | 11.4 | 9.5 | 5.9 | 48.7 | 111,200 | 129,703 | 16.6 | 8.1 | 14,368 | 10,894 | 6,924 | 53,129 | 2.45 | 58.5 | 5.3 | 26.8 |
| Morristown, TN.............. | 14.5 | 11.6 | 8.4 | 50.7 | 123,081 | 136,861 | 11.2 | 5.2 | 15,511 | 16,957 | 8,631 | 55,874 | 2.50 | 67.0 | 9.6 | 28.9 |
| Mount Vernon-Anacortes, WA. | 13.4 | 13.0 | 8.8 | 50.5 | 102,979 | 116,892 | 13.5 | 11.9 | 14,792 | 12,202 | 11,344 | 49,043 | 2.60 | 67.4 | 9.6 | 26.5 |
| Muncie, IN.................... | 12.2 | 9.8 | 7.9 | 51.9 | 118,769 | 117,674 | -0.9 | -3.6 | 12,104 | 12,799 | -3,518 | 45,640 | 2.34 | 56.9 | 12.2 | 33.1 |
| Muskegon, MI................ | 14.0 | 10.9 | 7.1 | 50.5 | 170,200 | 172,202 | 1.2 | 1.0 | 21,441 | 17,713 | -2,042 | 66,148 | 2.53 | 60.3 | 11.6 | 31.8 |
| Myrtle Beach-Conway-North Myrtle Beach, SC-NC.. | 16.6 | 18.5 | 9.6 | 52.0 | 269,772 | 376,565 | 39.6 | 36.6 | 42,433 | 48,162 | 142,172 | 198,645 | 2.48 | 65.0 | 7.5 | 27.6 |
| Napa, CA..................... | 13.6 | 11.4 | 8.9 | 50.2 | 124,279 | 136,535 | 9.9 | -0.4 | 14,281 | 12,639 | -2,103 | 48,107 | 2.78 | 65.7 | 7.9 | 28.3 |
| Naples-Marco Island, FL ........ | 13.7 | 16.0 | 17.5 | 50.8 | 251,377 | 321,514 | 27.9 | 22.2 | 32,959 | 33,701 | 72,076 | 140,578 | 2.71 | 69.2 | 5.4 | 25.8 |
| Nashville-Davidson--Murfrees-boro--Franklin, TN.. | 12.2 | 8.6 | 5.4 | 51.2 | 1,358,992 | 1,646,183 | 21.1 | 19.1 | 241,522 | 145,319 | 217,356 | 733,787 | 2.59 | 65.3 | 9.9 | 26.8 |
| New Bern, NC................ | 13.6 | 12.6 | 9.1 | 49.9 | 114,751 | 126,804 | 10.5 | -2.8 | 17,015 | 13,862 | -6,789 | 52,345 | 2.33 | 68.4 | 8.0 | 26.9 |
| New Haven-Milford, CT ...... | 14.0 | 10.3 | 8.0 | 51.8 | 824,008 | 862,476 | 4.7 | -1.2 | 90,367 | 79,982 | -20,996 | 329,006 | 2.51 | 61.5 | 10.7 | 32.4 |
| New Orleans-Metairie, LA ...... | 13.6 | 10.4 | 6.6 | 51.9 | 1,337,726 | 1,189,901 | -11.1 | 6.9 | 160,988 | 115,092 | 35,420 | 485,267 | 2.58 | 60.0 | 11.6 | 33.8 |
| New York-Newark-Jersey City, NY-NJ-PA ........ | 13.1 | 9.4 | 7.3 | 51.6 | 18,323,002 | 18,896,285 | 3.1 | 1.2 | 2,447,587 | 1,439,793 | -778,084 | 7,102,817 | 2.65 | 65.0 | 9.5 | 28.5 |
| Nassau County-Suffolk County, NY Div 35004 ... | 14.7 | 10.2 | 8.0 | 51.0 | 2,753,913 | 2,832,970 | 2.9 | -0.3 | 306,233 | 238,299 | -74,721 | 949,542 | 2.94 | 73.7 | 7.6 | 22.1 |
| Newark, NJ-PA Div 35084 ... | 13.9 | 9.2 | 6.9 | 51.1 | 2,098,843 | 2,146,226 | 2.3 | 1.0 | 249,545 | 167,745 | -60,513 | 797,142 | 2.66 | 69.2 | 10.2 | 26.2 |
| New Brunswick-Lakewood, NJ Div 35154 ......... | 14.2 | 10.4 | 8.1 | 51.1 | 2,173,869 | 2,340,476 | 7.7 | 1.9 | 278,570 | 216,017 | -18,029 | 875,614 | 2.67 | 69.3 | 6.8 | 25.7 |
| New York-Jersey City-White Plains, NY-NJ Div 35614...... | 12.3 | 9.0 | 7.0 | 51.9 | 11,296,377 | 11,576,613 | 2.5 | 1.5 | 1,613,239 | 817,732 | -624,821 | 4,480,519 | 2.58 | 61.5 | 10.4 | 30.8 |
| Niles, MI..................... | 14.4 | 11.9 | 8.8 | 51.0 | 162,453 | 156,810 | -3.5 | -2.4 | 18,137 | 17,778 | -4,121 | 61,809 | 2.43 | 63.9 | 10.9 | 30.4 |
| North Port-Sarasota-Braden-ton, FL....... | 15.2 | 17.3 | 16.3 | 52.1 | 589,959 | 702,306 | 19.0 | 21.7 | 64,826 | 96,705 | 183,136 | 332,582 | 2.49 | 63.1 | 5.8 | 29.8 |
| Norwich-New London, CT ...... | 14.8 | 11.1 | 8.3 | 49.9 | 259,088 | 274,076 | 5.8 | -3.3 | 27,091 | 25,087 | -11,140 | 109,752 | 2.31 | 63.9 | 8.2 | 29.0 |
| Ocala, FL.................... | 13.7 | 15.6 | 13.7 | 52.0 | 258,916 | 331,296 | 28.0 | 12.7 | 35,199 | 50,252 | 57,293 | 145,622 | 2.45 | 66.9 | 9.6 | 29.2 |
| Ocean City, NJ............. | 16.7 | 16.4 | 11.9 | 51.4 | 102,326 | 97,261 | -4.9 | -5.9 | 8,901 | 13,502 | -1,017 | 40,939 | 2.19 | 65.4 | 7.9 | 29.7 |
| Odessa, TX.................. | 9.3 | 5.8 | 3.9 | 48.9 | 121,123 | 137,131 | 13.2 | 22.3 | 29,210 | 12,230 | 13,164 | 53,155 | 3.10 | 69.4 | 13.4 | 23.9 |
| Ogden-Clearfield, UT............. | 9.9 | 6.9 | 4.6 | 49.5 | 485,401 | 597,162 | 23.0 | 15.8 | 107,488 | 39,199 | 26,122 | 218,724 | 3.10 | 76.9 | 9.6 | 18.7 |
| Oklahoma City, OK......... | 11.8 | 8.8 | 5.8 | 50.7 | 1,095,421 | 1,252,989 | 14.4 | 13.8 | 189,497 | 119,255 | 101,679 | 523,919 | 2.63 | 64.2 | 11.2 | 29.0 |
| Olympia-Lacey-Tumwater, WA | 12.5 | 11.2 | 7.0 | 51.1 | 207,355 | 252,260 | 21.7 | 16.6 | 31,868 | 22,366 | 32,370 | 112,909 | 2.54 | 64.9 | 9.3 | 25.8 |
| Omaha-Council Bluffs, NE-IA.. | 12.1 | 8.8 | 5.7 | 50.4 | 767,041 | 865,353 | 12.8 | 10.3 | 134,977 | 69,901 | 24,263 | 367,037 | 2.54 | 65.2 | 9.9 | 27.3 |
| Orlando-Kissimmee-Sanford, FL...... | 12.0 | 9.0 | 6.5 | 51.1 | 1,644,561 | 2,134,401 | 29.8 | 23.7 | 288,764 | 176,710 | 391,639 | 889,822 | 2.89 | 67.3 | 10.8 | 24.7 |
| Oshkosh-Neenah, WI ............ | 13.5 | 9.9 | 7.3 | 49.6 | 156,763 | 167,072 | 6.6 | 2.7 | 19,020 | 15,218 | 875 | 71,238 | 2.31 | 57.7 | 9.1 | 33.7 |
| Owensboro, KY ............. | 13.3 | 10.4 | 7.5 | 51.0 | 109,875 | 114,751 | 4.4 | 4.4 | 15,854 | 12,811 | 2,095 | 48,419 | 2.43 | 67.6 | 12.4 | 28.6 |
| Oxnard-Thousand Oaks-Ventura, CA ....... | 13.3 | 9.6 | 7.1 | 50.5 | 753,197 | 823,446 | 9.3 | 2.2 | 101,757 | 57,839 | -25,563 | 268,524 | 3.10 | 72.1 | 9.3 | 21.7 |
| Palm Bay-Melbourne-Titus-ville, FL....... | 16.4 | 13.4 | 11.1 | 51.1 | 476,230 | 543,376 | 14.1 | 12.0 | 53,029 | 71,908 | 83,982 | 230,043 | 2.59 | 59.6 | 6.3 | 34.3 |
| Panama City, FL............ | 15.1 | 11.1 | 7.8 | 50.4 | 148,217 | 168,843 | 13.9 | 1.5 | 22,772 | 19,089 | -1,269 | 69,195 | 2.49 | 67.0 | 12.3 | 27.5 |
| Parkersburg-Vienna, WV ......... | 14.7 | 12.2 | 9.1 | 51.3 | 93,859 | 92,664 | -1.3 | -4.3 | 10,333 | 11,687 | -2,582 | 36,946 | 2.40 | 61.0 | 9.6 | 32.3 |
| Pensacola-Ferry Pass-Brent, FL....... | 13.7 | 10.3 | 6.8 | 50.1 | 412,153 | 449,003 | 8.9 | 13.9 | 59,040 | 48,958 | 52,272 | 186,805 | 2.56 | 63.3 | 11.1 | 29.6 |
| Peoria, IL..................... | 13.1 | 10.9 | 8.3 | 50.8 | 405,149 | 416,253 | 2.7 | -4.7 | 51,886 | 44,293 | -27,030 | 162,131 | 2.40 | 61.4 | 9.5 | 32.6 |
| Philadelphia-Camden-Wilm-ington, PA-NJ-DE-MD.... | 13.6 | 9.8 | 7.1 | 51.6 | 5,687,147 | 5,965,874 | 4.9 | 2.4 | 725,215 | 564,756 | -15,166 | 2,323,207 | 2.56 | 64.1 | 10.8 | 29.2 |
| Camden, NJ Div 15804......... | 14.1 | 9.9 | 7.1 | 51.3 | 1,186,999 | 1,251,034 | 5.4 | -0.4 | 140,372 | 116,851 | -27,759 | 464,559 | 2.63 | 68.2 | 10.6 | 26.4 |
| Montgomery-Bucks-Chester Counties, PA Div 33874.. | 14.7 | 10.7 | 8.0 | 51.0 | 1,781,233 | 1,924,406 | 8.0 | 3.3 | 205,958 | 178,217 | 37,923 | 755,098 | 2.57 | 69.0 | 7.6 | 25.3 |
| Philadelphia, PA Div 37964 ... | 12.1 | 8.8 | 6.4 | 52.5 | 2,068,414 | 2,084,775 | 0.8 | 2.9 | 294,154 | 205,191 | -27,291 | 829,007 | 2.51 | 57.6 | 14.3 | 34.4 |
| Wilmington, DE-MD-NJ Div 48864 ......... | 13.9 | 9.9 | 6.9 | 51.4 | 650,501 | 705,659 | 8.5 | 3.1 | 84,731 | 64,497 | 1,961 | 274,543 | 2.56 | 63.8 | 9.7 | 29.1 |
| Phoenix-Mesa-Chandler, AZ ... | 11.7 | 9.5 | 7.0 | 50.4 | 3,251,876 | 4,193,137 | 28.9 | 20.7 | 597,279 | 333,347 | 601,059 | 1,765,135 | 2.76 | 66.1 | 10.4 | 26.1 |
| Pine Bluff, AR .............. | 13.5 | 11.0 | 7.6 | 48.9 | 107,341 | 100,262 | -6.6 | -13.9 | 11,244 | 11,080 | -14,280 | 34,290 | 2.27 | 62.1 | 12.6 | 31.8 |
| Pittsburgh, PA.............. | 14.7 | 12.1 | 8.9 | 51.2 | 2,431,087 | 2,356,291 | -3.1 | -2 | 241,354 | 283,356 | -3,197 | 1,028,436 | 2.19 | 57.9 | 7.4 | 34.5 |
| Pittsfield, MA................ | 16.3 | 14.1 | 10.6 | 51.6 | 134,953 | 131,274 | -2.7 | -5.1 | 10,857 | 15,234 | -2,256 | 53,792 | 2.20 | 58.7 | 9.7 | 33.5 |
| Pocatello, ID................ | 11.1 | 9.5 | 5.8 | 50.2 | 83,103 | 90,661 | 9.1 | 6.4 | 14,014 | 7,825 | -424 | 34,540 | 2.67 | 64.3 | 9.9 | 28.5 |
| Portland-South Portland, ME.. | 15.1 | 12.3 | 8.4 | 51.4 | 487,568 | 514,103 | 5.4 | 5.7 | 50,785 | 50,799 | 29,471 | 226,904 | 2.32 | 59.6 | 6.9 | 30.6 |
| Portland-Vancouver-Hillsboro, OR-WA....... | 12.0 | 9.8 | 6.1 | 50.5 | 1,927,881 | 2,226,002 | 15.5 | 12.8 | 277,919 | 176,270 | 182,092 | 964,554 | 2.54 | 62.1 | 8.5 | 27.5 |
| Port St. Lucie, FL............ | 14.7 | 14.4 | 13.1 | 51.0 | 319,426 | 424,108 | 32.8 | 17.7 | 43,722 | 51,605 | 82,838 | 181,914 | 2.65 | 64.6 | 8.6 | 28.7 |
| Poughkeepsie-Newburgh-Middletown, NY............ | 13.9 | 9.5 | 6.9 | 50.1 | 621,517 | 670,267 | 7.8 | 1.2 | 77,589 | 53,312 | -16,116 | 241,950 | 2.69 | 68.2 | 8.6 | 25.6 |
| Prescott Valley-Prescott, AZ... | 16.7 | 20.0 | 13.7 | 51.1 | 167,517 | 211,008 | 26.0 | 13.8 | 18,829 | 29,931 | 40,097 | 102,784 | 2.24 | 61.3 | 6.8 | 31.4 |
| Providence-Warwick, RI-MA... | 14.1 | 10.3 | 7.7 | 51.4 | 1,582,997 | 1,601,209 | 1.2 | 1.4 | 168,454 | 155,647 | 10,668 | 627,702 | 2.50 | 63.0 | 10.4 | 29.7 |
| Provo-Orem, UT............ | 6.7 | 4.8 | 3.3 | 49.4 | 376,774 | 526,885 | 39.8 | 25.9 | 123,671 | 24,950 | 37,704 | 182,219 | 3.48 | 81.4 | 7.5 | 11.3 |
| Pueblo, CO.................. | 13.0 | 11.4 | 8.0 | 50.7 | 141,472 | 159,066 | 12.4 | 6.8 | 19,008 | 17,833 | 9,683 | 66,111 | 2.49 | 64.4 | 14.3 | 30.8 |
| Punta Gorda, FL............ | 17.2 | 21.8 | 19.4 | 51.2 | 141,627 | 159,974 | 13.0 | 21.7 | 10,452 | 26,195 | 50,133 | 76,589 | 2.43 | 62.8 | 4.8 | 30.9 |
| Racine, WI.................. | 14.6 | 10.4 | 7.1 | 50.4 | 188,831 | 195,434 | 3.5 | 0.2 | 24,302 | 18,401 | -5,481 | 78,905 | 2.42 | 64.2 | 11.3 | 28.4 |
| Raleigh-Cary, NC ........... | 11.8 | 8.0 | 4.9 | 51.3 | 797,071 | 1,130,479 | 41.8 | 25.6 | 160,247 | 74,975 | 203,158 | 519,648 | 2.63 | 68.0 | 10.4 | 25.0 |
| Rapid City, SD ............. | 14.4 | 12.0 | 6.8 | 49.3 | 112,818 | 126,399 | 12.0 | 14.3 | 18,624 | 11,071 | 10,513 | 56,456 | 2.43 | 61.4 | 8.3 | 32.5 |

# Table C. Metropolitan Areas — Population, Vital Statistics, Health, and Crime

| Area name | Persons in group quarters, 2020 | Daytime population, 2019 Number | Daytime population, 2019 Employment/residence ratio | Births, 2020 Total | Births, 2020 Rate[1] | Deaths, 2020 Number | Deaths, 2020 Rate[1] | Persons under 65 with no health insurance 2019 Number | Persons under 65 with no health insurance 2019 Percent | Medicare, 2020 Total Beneficiaries | Medicare, 2020 Enrolled in Original Medicare | Medicare, 2020 Enrolled in Medicare Advantage | Serious crimes known to police[2], 2019 Violent Number | Serious crimes known to police[2], 2019 Violent Rate[3] |
|---|---|---|---|---|---|---|---|---|---|---|---|---|---|---|
| | 32 | 33 | 34 | 35 | 36 | 37 | 38 | 39 | 40 | 41 | 42 | 43 | 44 | 45 |
| Mobile, AL | 7,533 | 440,221 | 1.07 | 5,574 | 13.0 | 5,242 | 12.2 | 44,548 | 12.7 | 89,411 | 37,392 | 52,020 | NA | NA |
| Modesto, CA | 6,536 | 520,359 | 0.87 | 7,335 | 13.3 | 4,676 | 8.5 | 39,386 | 8.4 | 88,425 | 45,388 | 43,037 | 2,908 | 529.1 |
| Monroe, LA | 6,790 | 201,723 | 1.02 | 2,510 | 12.6 | 2,406 | 12.1 | 15,264 | 9.5 | 40,607 | 26,567 | 14,041 | 1,631 | 812.5 |
| Monroe, MI | 1,463 | 121,552 | 0.58 | 1,440 | 9.6 | 1,611 | 10.7 | 6,706 | 5.5 | 34,146 | 18,656 | 15,490 | NA | NA |
| Montgomery, AL | 13,478 | 378,277 | 1.03 | 4,682 | 12.6 | 4,038 | 10.8 | 32,096 | 10.7 | 75,396 | 37,211 | 38,186 | NA | NA |
| Morgantown, WV | 8,202 | 142,049 | 1.05 | 1,296 | 9.2 | 1,175 | 8.4 | 8,038 | 7.3 | 22,644 | 13,679 | 8,965 | 288 | 204.4 |
| Morristown, TN | 2,606 | 137,424 | 0.91 | 1,500 | 10.4 | 1,861 | 12.9 | 16,136 | 14.4 | 35,012 | 17,221 | 17,791 | 552 | 387.6 |
| Mount Vernon-Anacortes, WA | 1,627 | 129,242 | 1 | 1,406 | 10.8 | 1,374 | 10.5 | 9,725 | 9.7 | 32,171 | 21,257 | 10,914 | 201 | 155.5 |
| Muncie, IN | 8,601 | 110,714 | 0.93 | 1,064 | 9.4 | 1,266 | 11.2 | 8,911 | 10.3 | 23,875 | 15,762 | 8,113 | NA | NA |
| Muskegon, MI | 4,681 | 166,319 | 0.9 | 2,027 | 11.7 | 1,909 | 11.0 | 9,617 | 6.9 | 39,501 | 16,484 | 23,017 | 786 | 453.4 |
| Myrtle Beach-Conway-North Myrtle Beach, SC-NC | 4,553 | 484,549 | 0.94 | 4,170 | 8.1 | 6,178 | 12.0 | 59,808 | 16.8 | 154,665 | 110,161 | 44,504 | 1,772 | 357.5 |
| Napa, CA | 5,079 | 149,004 | 1.17 | 1,226 | 9.0 | 1,415 | 10.4 | 9,391 | 8.7 | 29,865 | 17,885 | 11,980 | 753 | 543.4 |
| Naples-Marco Island, FL | 4,547 | 391,258 | 1.04 | 3,163 | 8.0 | 4,286 | 10.9 | 54,372 | 21.3 | 104,801 | 74,053 | 30,748 | 896 | 233.7 |
| Nashville-Davidson--Murfreesboro--Franklin, TN | 40,985 | 1,987,951 | 1.05 | 24,460 | 12.5 | 16,549 | 8.4 | 182,454 | 11.1 | 308,807 | 174,357 | 134,452 | 11,078 | 570.5 |
| New Bern, NC | 5,524 | 125,702 | 0.96 | 1,464 | 11.9 | 1,460 | 11.9 | 12,355 | 13.4 | 30,971 | 25,351 | 5,620 | 355 | 284.3 |
| New Haven-Milford, CT | 28,515 | 821,884 | 0.92 | 8,397 | 9.9 | 8,512 | 10.0 | 47,465 | 7.0 | 167,724 | 89,020 | 78,705 | 2,189 | 272.8 |
| New Orleans-Metairie, LA | 19,941 | 1,292,225 | 1.04 | 14,924 | 11.7 | 13,138 | 10.3 | 111,904 | 10.7 | 239,514 | 98,083 | 141,433 | 7,021 | 551.5 |
| New York-Newark-Jersey City, NY-NJ-PA | 384,464 | 19,431,171 | 1.02 | 226,416 | 11.8 | 163,362 | 8.5 | 1,247,823 | 7.9 | 3,342,128 | 2,112,802 | 1,229,328 | NA | NA |
| Nassau County-Suffolk County, NY Div 35004 | 48,845 | D | D | 29,032 | 10.3 | 26,300 | 9.3 | 116,294 | 5.0 | 567,133 | 430,886 | 136,248 | NA | NA |
| Newark, NJ-PA Div 35084 | 43,874 | 2,115,745 | 0.95 | 23,916 | 11.0 | 18,368 | 8.5 | 177,199 | 9.9 | 370,556 | 246,065 | 124,491 | NA | NA |
| New Brunswick-Lakewood, NJ Div 35154 | 42,336 | 2,275,727 | 0.91 | 26,585 | 11.1 | 23,546 | 9.9 | 148,781 | 7.7 | 479,252 | 338,919 | 140,333 | NA | NA |
| New York-Jersey City-White Plains, NY-NJ Div 35614 | 249,409 | 12,403,393 | 1.1 | 146,883 | 12.5 | 95,148 | 8.1 | 805,549 | 8.2 | 1,925,187 | 1,096,932 | 828,256 | NA | NA |
| Niles, MI | 3,512 | 150,823 | 0.96 | 1,644 | 10.7 | 1,818 | 11.9 | 10,145 | 8.5 | 36,249 | 21,594 | 14,654 | 845 | 550.7 |
| North Port-Sarasota-Bradenton, FL | 10,817 | 836,571 | 1 | 6,391 | 7.5 | 11,571 | 13.5 | 91,412 | 16.4 | 254,503 | 160,463 | 94,041 | 2,732 | 328.4 |
| Norwich-New London, CT | 10,858 | 264,576 | 1 | 2,455 | 9.3 | 2,768 | 10.4 | 12,556 | 6.1 | 57,332 | 33,939 | 23,393 | 339 | 193.8 |
| Ocala, FL | 9,218 | 352,224 | 0.9 | 3,532 | 9.5 | 5,608 | 15.0 | 42,094 | 16.9 | 120,124 | 60,208 | 59,916 | 1,556 | 430.9 |
| Ocean City, NJ | 2,711 | 93,486 | 1.03 | 813 | 8.9 | 1,393 | 15.2 | 6,002 | 9.2 | 28,067 | 21,291 | 6,776 | 161 | 175.8 |
| Odessa, TX | 2,505 | 167,454 | 1.02 | 3,059 | 18.2 | 1,324 | 7.9 | 32,601 | 22.0 | 19,620 | 12,792 | 6,829 | 1,451 | 881.8 |
| Ogden-Clearfield, UT | 6,173 | 630,444 | 0.84 | 9,844 | 14.2 | 4,094 | 5.9 | 53,952 | 8.9 | 89,020 | 53,592 | 35,428 | 1,167 | 170.7 |
| Oklahoma City, OK | 31,647 | 1,419,495 | 1.02 | 17,716 | 12.4 | 13,065 | 9.2 | 176,192 | 14.9 | 236,569 | 167,068 | 69,501 | 6,506 | 461.0 |
| Olympia-Lacey-Tumwater, WA | 4,074 | 283,366 | 0.95 | 3,122 | 10.6 | 2,608 | 8.9 | 16,174 | 6.9 | 62,181 | 37,557 | 24,625 | 715 | 246.7 |
| Omaha-Council Bluffs, NE-IA | 16,675 | 947,140 | 0.99 | 12,705 | 13.3 | 7,548 | 7.9 | 68,515 | 8.5 | 157,132 | 109,025 | 48,107 | 3,916 | 413.1 |
| Orlando-Kissimmee-Sanford, FL | 46,515 | 2,683,716 | 1.06 | 29,287 | 11.1 | 21,208 | 8.0 | 321,710 | 14.8 | 444,542 | 215,825 | 228,718 | 11,242 | 429.9 |
| Oshkosh-Neenah, WI | 7,408 | 181,948 | 1.11 | 1,770 | 10.3 | 1,598 | 9.3 | 8,357 | 6.1 | 33,531 | 12,842 | 20,689 | 338 | 197.3 |
| Owensboro, KY | 2,913 | 121,566 | 1.01 | 1,504 | 12.6 | 1,357 | 11.3 | 7,014 | 7.3 | 26,603 | 17,689 | 8,914 | 159 | 133.4 |
| Oxnard-Thousand Oaks-Ventura, CA | 11,350 | 802,170 | 0.89 | 8,820 | 10.5 | 6,741 | 8.0 | 72,745 | 10.4 | 153,894 | 97,896 | 55,998 | 1,830 | 216.0 |
| Palm Bay-Melbourne-Titusville, FL | 6,768 | 593,444 | 0.97 | 5,236 | 8.6 | 8,105 | 13.3 | 62,753 | 13.9 | 159,117 | 90,617 | 68,500 | 2,224 | 371.0 |
| Panama City, FL | 3,645 | 176,511 | 1.02 | 1,894 | 11.1 | 2,065 | 12.1 | 20,640 | 14.8 | 37,013 | 26,004 | 11,009 | 826 | 443.9 |
| Parkersburg-Vienna, WV | 1,007 | 93,833 | 1.11 | 922 | 10.4 | 1,217 | 13.7 | 5,138 | 7.3 | 23,541 | 15,777 | 7,763 | NA | NA |
| Pensacola-Ferry Pass-Brent, FL | 24,181 | 490,093 | 0.94 | 5,779 | 11.3 | 5,780 | 11.3 | 52,551 | 13.3 | 107,851 | 67,988 | 39,864 | 2,037 | 409.7 |
| Peoria, IL | 11,413 | 399,442 | 0.99 | 4,625 | 11.7 | 4,592 | 11.6 | 20,767 | 6.6 | 85,771 | 57,537 | 28,234 | 1,807 | 451.9 |
| Philadelphia-Camden-Wilmington, PA-NJ-DE-MD | 163,470 | 6,063,321 | 0.99 | 67,333 | 11.0 | 61,899 | 10.1 | 340,601 | 6.8 | 1,153,807 | 762,797 | 391,010 | NA | NA |
| Camden, NJ Div 15804 | 23,513 | 1,152,523 | 0.85 | 13,150 | 10.5 | 12,592 | 10.1 | 69,720 | 6.8 | 244,815 | 163,087 | 81,728 | NA | NA |
| Montgomery-Bucks-Chester Counties, PA Div 33874 | 42,467 | D | D | 19,441 | 9.8 | 19,597 | 9.9 | 85,531 | 5.3 | 404,433 | 278,525 | 125,908 | NA | NA |
| Philadelphia, PA Div 37964 | 77,398 | 2,190,456 | 1.04 | 26,896 | 12.5 | 22,569 | 10.5 | 145,418 | 8.3 | 367,938 | 211,869 | 156,069 | NA | NA |
| Wilmington, DE-MD-NJ Div 48864 | 20,092 | 716,294 | 0.98 | 7,846 | 10.8 | 7,141 | 9.8 | 39,932 | 6.8 | 136,621 | 109,316 | 27,305 | NA | NA |
| Phoenix-Mesa-Chandler, AZ | 88,879 | 4,935,605 | 0.99 | 57,251 | 11.3 | 40,024 | 7.9 | 537,313 | 13.2 | 807,482 | 443,026 | 364,457 | 21,110 | 426.5 |
| Pine Bluff, AR | 8,922 | 88,987 | 0.99 | 980 | 11.4 | 1,141 | 13.2 | 5,222 | 8.3 | 19,822 | 13,868 | 5,955 | 788 | 895.4 |
| Pittsburgh, PA | 60,388 | 2,344,369 | 1.02 | 22,676 | 9.8 | 29,221 | 12.7 | 96,972 | 5.4 | 551,071 | 203,396 | 347,675 | NA | NA |
| Pittsfield, MA | 5,794 | 127,229 | 1.04 | 975 | 7.8 | 1,494 | 12.0 | 3,482 | 3.9 | 35,270 | 31,640 | 3,630 | 475 | 380.7 |
| Pocatello, ID | 2,057 | 93,683 | 0.98 | 1,202 | 12.5 | 801 | 8.3 | 9,981 | 12.6 | 17,031 | 11,059 | 5,972 | 287 | 299.4 |
| Portland-South Portland, ME | 13,151 | 535,132 | 0.99 | 4,913 | 9.0 | 5,660 | 10.4 | 34,757 | 8.2 | 128,630 | 71,666 | 56,966 | 619 | 114.6 |
| Portland-Vancouver-Hillsboro, OR-WA | 38,628 | 2,512,097 | 1.01 | 25,622 | 10.2 | 20,060 | 8.0 | 159,846 | 7.6 | 437,168 | 175,345 | 261,824 | 7,565 | 302.6 |
| Port St. Lucie, FL | 7,257 | 454,348 | 0.84 | 4,382 | 8.8 | 6,123 | 12.3 | 61,549 | 17.5 | 130,165 | 76,491 | 53,674 | 1,161 | 238.8 |
| Poughkeepsie-Newburgh-Middletown, NY | 30,562 | 623,315 | 0.83 | 7,621 | 11.2 | 6,137 | 9.0 | 27,899 | 5.1 | 126,177 | 92,579 | 33,598 | 1,286 | 191.2 |
| Prescott Valley-Prescott, AZ | 3,976 | 231,341 | 0.96 | 1,798 | 7.5 | 3,298 | 13.7 | 23,125 | 14.8 | 85,203 | 57,812 | 27,391 | 651 | 277.1 |
| Providence-Warwick, RI-MA | 56,166 | 1,541,882 | 0.9 | 15,889 | 9.8 | 16,215 | 10.0 | 59,370 | 4.6 | 347,225 | 211,954 | 135,269 | 4,580 | 282.6 |
| Provo-Orem, UT | 14,197 | 631,518 | 0.94 | 12,032 | 18.1 | 2,787 | 4.2 | 54,943 | 9.4 | 60,021 | 33,186 | 26,835 | 583 | 90.2 |
| Pueblo, CO | 4,077 | 164,927 | 0.95 | 1,803 | 10.6 | 2,000 | 11.8 | 12,543 | 9.6 | 39,116 | 19,892 | 19,224 | D | D |
| Punta Gorda, FL | 2,885 | 182,203 | 0.89 | 1,002 | 5.1 | 3,117 | 16.0 | 19,097 | 17.4 | 71,649 | 42,202 | 29,447 | 367 | 196.2 |
| Racine, WI | 4,910 | 177,896 | 0.8 | 2,267 | 11.6 | 1,885 | 9.6 | 11,317 | 7.2 | 41,218 | 20,939 | 20,279 | 625 | 318.1 |
| Raleigh-Cary, NC | 24,210 | 1,395,249 | 1.01 | 15,949 | 11.2 | 9,265 | 6.5 | 131,180 | 10.9 | 198,854 | 120,918 | 77,936 | 2,404 | 172.6 |
| Rapid City, SD | 3,391 | 143,236 | 1.01 | 1,752 | 12.1 | 1,272 | 8.8 | 14,058 | 12.3 | 32,226 | 24,616 | 7,610 | 749 | 532.2 |

1. Per 1,000 estimated resident population.    2. Data for serious crimes have not been adjusted for underreporting; this may affect comparability between geographic areas and over time.    3. Per 100,000 population estimated by the FBI.

| Area name | Serious crimes known to police[1], 2019 (cont.) Property | | Education School enrollment and attainment, 2019 Enrollment[3] | | Attainment[4] | | Local government expenditures,[5] 2017-2018 | | Income and poverty, 2019 | | | | | Percent below poverty level | | |
|---|---|---|---|---|---|---|---|---|---|---|---|---|---|---|---|---|
| | Number | Rate | Total | Percent private | High school graduate or less | Bachelor's degree or more | Total current expenditures (mil dol) | Current expenditures per student (dollars) | Per capita income[6] (dollars) | Median household income (dollars) | Median family income | Percent of households with income less than $50,000 | Percent of households with income of $200,000 or more | All persons | All families | Families with children under 18 |
| | 46 | 47 | 48 | 49 | 50 | 51 | 52 | 53 | 54 | 55 | 56 | 57 | 58 | 59 | 60 | 61 |
| Mobile, AL | NA | NA | 95,773 | 18.9 | 48.8 | 22.8 | 604.9 | 9,464 | 25,641 | 49,561 | 61,147 | 50.4 | 3.1 | 17.7 | 11.9 | 20.2 |
| Modesto, CA | 14,274 | 2,597.0 | 150,879 | 7.9 | 50.3 | 17.3 | 1,335.1 | 12,294 | 26,686 | 63,037 | 71,339 | 39.1 | 5.2 | 12.7 | 9.7 | 13.6 |
| Monroe, LA | 8,353 | 4,161.1 | 48,142 | 16.9 | 48.8 | 23.2 | 385.0 | 11,029 | 24,735 | 40,136 | 50,949 | 58.5 | 3.5 | 26.0 | 20.4 | 35.3 |
| Monroe, MI | NA | NA | 33,021 | 12.7 | 41.1 | 24.6 | 264.9 | 11,760 | 33,754 | 62,839 | 77,207 | 38.9 | 4.6 | 10.8 | 8.4 | 16.8 |
| Montgomery, AL | NA | NA | 91,715 | 21.8 | 39.5 | 31.2 | 497.8 | 9,071 | 29,094 | 53,834 | 67,922 | 45.1 | 4.3 | 14.2 | 11.0 | 18.7 |
| Morgantown, WV | 1,862 | 1,321.3 | 40,650 | 5.1 | 38.6 | 37.9 | 175.7 | 10,793 | 31,550 | 56,395 | 84,766 | 44.1 | 7.3 | 20.1 | 9.3 | 14.3 |
| Morristown, TN | 2,919 | 2,049.9 | 30,755 | 11.3 | 53.6 | 19.3 | 184.4 | 8,639 | 26,948 | 47,326 | 63,610 | 51.8 | 3.2 | 11.9 | 6.9 | 11.3 |
| Mount Vernon-Anacortes, WA | 3,178 | 2,459.1 | 27,108 | 12.7 | 35.3 | 27 | 281.3 | 14,497 | 33,941 | 67,175 | 77,430 | 36.1 | 5.1 | 11.9 | 7.1 | 14.7 |
| Muncie, IN | NA | NA | 32,930 | 3.7 | 44.4 | 22.6 | 156.8 | 10,091 | 25,737 | 45,065 | 62,160 | 54.2 | 2.4 | 22.8 | 13.4 | 23.0 |
| Muskegon, MI | D | D | 38,688 | 8.7 | 42.1 | 20.8 | 321.9 | 11,629 | 26,697 | 50,366 | 63,502 | 49.5 | 2.7 | 12.9 | 9.2 | 13.7 |
| Myrtle Beach-Conway-North Myrtle Beach, SC-NC | 14,926 | 3,011.3 | 86,731 | 10.0 | 40.4 | 26.9 | 631.3 | 10,753 | 32,117 | 56,977 | 69,727 | 43.2 | 4.1 | 11.8 | 7.2 | 16.1 |
| Napa, CA | 2,162 | 1,560.2 | 31,381 | 17.4 | 32.9 | 36.2 | 285.1 | 13,975 | 47,729 | 92,769 | 107,419 | 25.1 | 17.9 | 6.8 | 4.6 | 6.6 |
| Naples-Marco Island, FL | 4,308 | 1,123.7 | 67,608 | 20.6 | 38.2 | 35.8 | 515.1 | 10,999 | 47,433 | 76,025 | 88,736 | 33.6 | 13.1 | 9.3 | 6.4 | 8.2 |
| Nashville-Davidson--Murfreesboro--Franklin, TN | 47,565 | 2,449.5 | 462,732 | 21.4 | 34.9 | 38.5 | 2,920.2 | 9,915 | 37,696 | 70,262 | 86,035 | 34.8 | 8.4 | 10.0 | 6.8 | 11.3 |
| New Bern, NC | 2,954 | 2,365.5 | 27,503 | 14.7 | 39.4 | 20.9 | 159.9 | 9,438 | 29,612 | 52,355 | 67,538 | 48.6 | 4.1 | 14.1 | 12.1 | 22.8 |
| New Haven-Milford, CT | 16,938 | 2,110.6 | 212,536 | 24.0 | 41.2 | 35.1 | 2,384.1 | 19,903 | 38,170 | 69,751 | 89,234 | 38.0 | 9.4 | 12.2 | 8.6 | 14.6 |
| New Orleans-Metairie, LA | 40,101 | 3,149.9 | 311,579 | 28.7 | 39.8 | 32.3 | 2,159.7 | 12,603 | 31,889 | 55,710 | 73,579 | 45.2 | 6.4 | 16.4 | 11.0 | 18.0 |
| New York-Newark-Jersey City, NY-NJ-PA | NA | NA | 4,619,686 | 24.1 | 37.2 | 41.8 | 63,059.5 | 22,703 | 46,241 | 83,160 | 101,316 | 31.7 | 16.1 | 11.6 | 8.3 | 12.4 |
| Nassau County-Suffolk County, NY Div 35004 | NA | NA | 685,735 | 19.5 | 32.8 | 42.6 | 11,538.9 | 25,861 | 50,787 | 111,786 | 131,737 | 20.9 | 21.8 | 6.0 | 3.9 | 6.2 |
| Newark, NJ-PA Div 35084 | NA | NA | 537,138 | 15.2 | 36.8 | 41.8 | 7,242.0 | 19,842 | 46,407 | 83,612 | 105,140 | 30.0 | 16.8 | 9.4 | 6.6 | 10.4 |
| New Brunswick-Lakewood, NJ Div 35154 | NA | NA | 598,690 | 24.1 | 33.8 | 43.8 | 6,719.0 | 19,125 | 46,329 | 92,727 | 115,450 | 26.1 | 16.3 | 7.7 | 4.9 | 7.3 |
| New York-Jersey City-White Plains, NY-NJ Div 35614 | NA | NA | 2,798,123 | 27.0 | 39.1 | 41.2 | 37,559.5 | 23,256 | 45,104 | 76,065 | 89,101 | 35.4 | 14.7 | 14.1 | 10.6 | 15.4 |
| Niles, MI | 3,340 | 2,176.9 | 37,360 | 18.4 | 36.9 | 27 | 314.1 | 12,311 | 32,237 | 50,153 | 66,322 | 49.9 | 4.7 | 16.2 | 11.0 | 20.7 |
| North Port-Sarasota-Bradenton, FL | 13,924 | 1,674.0 | 145,141 | 20.2 | 36.7 | 33.7 | 933.7 | 10,165 | 39,643 | 65,526 | 81,361 | 36.9 | 8.1 | 9.2 | 5.6 | 11.6 |
| Norwich-New London, CT | 2,322 | 1,327.8 | 58,017 | 18.3 | 35.1 | 33.3 | 709.9 | 19,842 | 40,584 | 75,633 | 95,350 | 32.4 | 9.7 | 7.2 | 5.2 | 10.2 |
| Ocala, FL | 6,841 | 1,894.4 | 70,228 | 20.7 | 47.6 | 20.6 | 402.1 | 9,325 | 27,033 | 49,576 | 58,615 | 50.3 | 3.0 | 15.0 | 10.6 | 21.0 |
| Ocean City, NJ | 1,909 | 2,084.6 | 15,798 | 12.5 | 41.4 | 32.1 | 275.4 | 22,081 | 43,852 | 69,000 | 90,719 | 36.1 | 10.5 | 8.8 | 6.1 | 11.6 |
| Odessa, TX | 5,205 | 3,163.2 | 45,421 | 7.3 | 53.9 | 15.1 | 272.3 | 8,022 | 28,582 | 67,205 | 83,388 | 37.4 | 5.9 | 10.4 | 8.3 | 13.8 |
| Ogden-Clearfield, UT | D | D | 211,171 | 9.7 | 30.6 | 32 | 1,101.8 | 7,227 | 31,814 | 79,251 | 89,638 | 26.7 | 6.8 | 5.6 | 3.9 | 6.2 |
| Oklahoma City, OK | 43,321 | 3,069.9 | 371,984 | 12.2 | 38.6 | 31.5 | 1,866.0 | 7,521 | 32,577 | 60,605 | 76,371 | 40.8 | 6.0 | 13.1 | 8.6 | 13.5 |
| Olympia-Lacey-Tumwater, WA | 6,005 | 2,071.6 | 65,624 | 15.3 | 29.2 | 34.4 | 557.2 | 13,042 | 37,040 | 78,512 | 93,142 | 31.2 | 7.5 | 9.2 | 6.4 | 11.3 |
| Omaha-Council Bluffs, NE-IA | 25,089 | 2,646.6 | 252,143 | 20.7 | 31.5 | 37.7 | 1,904.8 | 11,954 | 35,223 | 70,373 | 85,509 | 34.5 | 6.6 | 9.1 | 6.2 | 9.4 |
| Orlando-Kissimmee-Sanford, FL | 62,756 | 2,400.0 | 655,424 | 19.4 | 36.6 | 33.3 | 3,461.2 | 8,906 | 31,186 | 61,876 | 72,046 | 39.7 | 6.5 | 12.1 | 8.5 | 13.3 |
| Oshkosh-Neenah, WI | 2,008 | 1,172.2 | 41,572 | 12.5 | 39.9 | 31.1 | 265.7 | 11,537 | 32,702 | 58,347 | 76,425 | 41.9 | 4.5 | 10.0 | 6.1 | 9.8 |
| Owensboro, KY | 2,896 | 2,429.2 | 28,061 | 16.6 | 46.1 | 22.8 | 214.2 | 10,608 | 29,270 | 52,201 | 68,027 | 47.2 | 3.9 | 15.6 | 13.8 | 18.9 |
| Oxnard-Thousand Oaks-Ventura, CA | 12,634 | 1,491.5 | 214,633 | 13.2 | 34.5 | 34.8 | 1,700.1 | 12,341 | 40,293 | 92,236 | 103,818 | 24.9 | 14.0 | 7.9 | 4.9 | 7.6 |
| Palm Bay-Melbourne-Titusville, FL | 12,269 | 2,046.5 | 125,742 | 18.9 | 35.3 | 30.4 | 650.0 | 8,841 | 32,667 | 57,305 | 74,025 | 43.2 | 5.4 | 8.7 | 5.3 | 9.8 |
| Panama City, FL | 5,707 | 3,067.2 | 39,528 | 12.5 | 37.6 | 26.4 | 244.0 | 8,690 | 31,245 | 59,450 | 70,253 | 38.8 | 4.2 | 10.3 | 6.7 | 9.9 |
| Parkersburg-Vienna, WV | NA | NA | 17,797 | 14.7 | 45.4 | 20 | 151.2 | 11,026 | 27,237 | 48,680 | 62,483 | 51.0 | 2.8 | 14.6 | 9.1 | 17.3 |
| Pensacola-Ferry Pass-Brent, FL | 10,827 | 2,177.7 | 115,718 | 18.8 | 39.7 | 25 | 605.3 | 8,863 | 30,394 | 56,507 | 71,502 | 43.9 | 5.1 | 13.4 | 8.8 | 15.0 |
| Peoria, IL | 7,735 | 1,934.5 | 94,563 | 21.6 | 38.0 | 28.8 | 842.9 | 13,512 | 33,904 | 60,372 | 79,812 | 41.6 | 5.6 | 11.3 | 7.4 | 12.1 |
| Philadelphia-Camden-Wilmington, PA-NJ-DE-MD | NA | NA | 1,483,998 | 25.3 | 37.3 | 39 | 15,002.5 | 17,485 | 40,930 | 74,533 | 94,793 | 34.5 | 11.6 | 11.8 | 8.1 | 13.1 |
| Camden, NJ Div 15804 | NA | NA | 300,915 | 17.4 | 37.4 | 35.9 | 3,817.3 | 19,241 | 41,101 | 82,258 | 101,553 | 29.3 | 11.5 | 8.1 | 5.1 | 8.3 |
| Montgomery-Bucks-Chester Counties, PA Div 33874 | NA | NA | 469,739 | 24.2 | 29.2 | 48.7 | 5,168.4 | 18,354 | 50,224 | 95,007 | 118,191 | 24.6 | 16.9 | 5.8 | 3.8 | 6.0 |
| Philadelphia, PA Div 37964 | NA | NA | 538,839 | 33.0 | 43.9 | 33.6 | 4,326.0 | 15,800 | 33,013 | 53,173 | 67,434 | 47.2 | 7.5 | 19.8 | 15.2 | 23.4 |
| Wilmington, DE-MD-NJ Div 48864 | NA | NA | 174,505 | 17.8 | 40.1 | 33.7 | 1,690.7 | 16,218 | 38,683 | 75,248 | 93,510 | 31.8 | 9.2 | 10.7 | 7.0 | 11.5 |
| Phoenix-Mesa-Chandler, AZ | D | D | 1,211,972 | 12.7 | 35.0 | 32.2 | 6,326.0 | 8,101 | 34,074 | 67,896 | 79,893 | 35.5 | 7.9 | 12.1 | 8.5 | 13.6 |
| Pine Bluff, AR | D | D | 19,877 | 20.4 | 55.2 | 16.4 | 151.6 | 10,936 | 21,376 | 41,541 | 53,977 | 58.8 | 3.7 | 23.6 | 20.2 | 36.8 |
| Pittsburgh, PA | NA | NA | 485,241 | 20.8 | 37.2 | 36 | 4,890.1 | 16,411 | 38,400 | 62,638 | 85,283 | 40.3 | 6.9 | 10.9 | 7.0 | 12.9 |
| Pittsfield, MA | 1,730 | 1,386.5 | 26,171 | 24.1 | 37.3 | 33.9 | 278.7 | 17,620 | 36,759 | 58,895 | 78,546 | 40.9 | 6.0 | 10.8 | 7.2 | 11.8 |
| Pocatello, ID | 2,018 | 2,105.4 | 25,638 | 7.6 | 31.9 | 28.3 | 118.6 | 7,174 | 25,010 | 51,612 | 65,904 | 48.4 | 1.8 | 12.9 | 8.2 | 12.5 |
| Portland-South Portland, ME | 6,784 | 1,256.5 | 109,263 | 23.0 | 31.6 | 42.3 | 1,059.0 | 15,070 | 40,148 | 71,913 | 91,427 | 34.0 | 8.0 | 7.7 | 4.7 | 8.0 |
| Portland-Vancouver-Hillsboro, OR-WA | 68,684 | 2,746.9 | 566,826 | 17.9 | 28.1 | 40.3 | 4,264.7 | 12,307 | 40,526 | 78,439 | 95,851 | 30.4 | 9.9 | 9.6 | 5.7 | 9.5 |
| Port St. Lucie, FL | 6,184 | 1,271.9 | 96,380 | 18.4 | 42.1 | 26.3 | 546.4 | 9,134 | 34,456 | 62,608 | 76,011 | 40.3 | 5.4 | 8.9 | 5.6 | 7.2 |
| Poughkeepsie-Newburgh-Middletown, NY | 7,443 | 1,106.7 | 177,535 | 26.7 | 36.7 | 33.2 | 2,272.0 | 22,803 | 39,492 | 85,152 | 104,062 | 30.4 | 12.0 | 10.7 | 7.5 | 12.3 |
| Prescott Valley-Prescott, AZ | 3,143 | 1,338.0 | 47,465 | 24.2 | 34.3 | 28.5 | 197.5 | 8,527 | 32,187 | 53,816 | 68,513 | 46.3 | 4.8 | 12.0 | 8.2 | 11.9 |
| Providence-Warwick, RI-MA | 22,854 | 1,409.9 | 377,770 | 20.8 | 40.7 | 33.2 | 3,528.2 | 15,881 | 37,114 | 70,967 | 89,587 | 36.9 | 7.7 | 11.0 | 7.3 | 12.7 |
| Provo-Orem, UT | 8,264 | 1,278.2 | 254,165 | 22.7 | 22.5 | 40.6 | 1,010.9 | 6,825 | 28,474 | 79,152 | 86,327 | 29.1 | 7.5 | 10.0 | 7.0 | 8.6 |
| Pueblo, CO | 5,859 | 3,490.7 | 40,068 | 4.3 | 37.9 | 23.4 | 243.7 | 9,096 | 26,778 | 51,276 | 61,720 | 49.4 | 2.6 | 17.9 | 12.7 | 21.4 |
| Punta Gorda, FL | 2,114 | 1,130.1 | 25,122 | 9.2 | 45.0 | 22.1 | 152.5 | 9,592 | 33,407 | 54,652 | 68,349 | 46.9 | 4.4 | 12.0 | 7.8 | 20.3 |
| Racine, WI | 2,801 | 1,425.6 | 43,524 | 15.0 | 40.6 | 26.1 | 359.5 | 13,024 | 31,741 | 60,779 | 80,496 | 41.9 | 4.9 | 12.9 | 9.5 | 15.5 |
| Raleigh-Cary, NC | 19,469 | 1,397.7 | 374,795 | 16.3 | 25.2 | 48 | 1,949.4 | 8,839 | 40,209 | 80,096 | 99,031 | 31.5 | 10.5 | 8.9 | 5.4 | 8.0 |
| Rapid City, SD | 3,382 | 2,403.0 | 33,778 | 16.5 | 32.9 | 30.8 | 197.3 | 9,542 | 31,643 | 58,361 | 81,164 | 42.7 | 4.4 | 11.0 | 5.2 | 12.1 |

1. Data for serious crimes have not been adjusted for underreporting; this may affect comparability between geographic areas and over time.   2. Per 100,000 population estimated by the FBI.   3. All persons 3 years old and over enrolled in nursery school through college.   4. Persons 25 years old and over.   5. Elementary and secondary education expenditures.   6. Based on population estimated by the American Community Survey, 2017.

# Table C. Metropolitan Areas — Personal Income and Earnings

| Area name | Personal income, 2019 Total (mil dol) | Percent change, 2018-2019 | Per capita[1] Dollars | Per capita[1] Rank | Wages and Salaries (mil dol) | Supplements to wages and salaries, employer contributions (mil dol) Pension and insurance | Government social insurance | Proprietors' income | Dividends, interest, and rent (mil dol) | Personal transfer receipts (mil dol) | Earnings, 2019 Total (mil dol) | Contributions for government social insurance (mil dol) From employee and self-employed | From employer |
|---|---|---|---|---|---|---|---|---|---|---|---|---|---|
| | 62 | 63 | 64 | 65 | 66 | 67 | 68 | 69 | 70 | 71 | 72 | 73 | 74 |
| Mobile, AL | 17,192 | 3.2 | 40,025 | 343 | 9,452 | 1,508 | 682 | 1,230 | 2,831 | 4,492 | 12,871 | 854 | 682 |
| Modesto, CA | 25,188 | 5.3 | 45,742 | 229 | 10,340 | 2,008 | 768 | 2,378 | 4,181 | 5,317 | 15,495 | 888 | 768 |
| Monroe, LA | 8,376 | 1.5 | 41,825 | 306 | 3,609 | 655 | 233 | 664 | 1,364 | 2,383 | 5,161 | 304 | 233 |
| Monroe, MI | 7,311 | 3.9 | 48,581 | 168 | 2,179 | 400 | 157 | 456 | 977 | 1,516 | 3,192 | 225 | 157 |
| Montgomery, AL | 16,750 | 3.5 | 44,870 | 251 | 8,477 | 1,546 | 624 | 1,035 | 3,199 | 3,867 | 11,681 | 747 | 624 |
| Morgantown, WV | 6,182 | 2.7 | 44,458 | 260 | 3,641 | 683 | 273 | 329 | 1,094 | 1,217 | 4,927 | 301 | 273 |
| Morristown, TN | 5,346 | 3.5 | 37,452 | 372 | 2,192 | 397 | 154 | 465 | 703 | 1,587 | 3,208 | 221 | 154 |
| Mount Vernon-Anacortes, WA | 7,042 | 5.8 | 54,505 | 86 | 2,940 | 548 | 267 | 576 | 1,696 | 1,447 | 4,332 | 262 | 267 |
| Muncie, IN | 4,352 | 3.1 | 38,129 | 366 | 2,133 | 400 | 157 | 197 | 722 | 1,306 | 2,887 | 193 | 157 |
| Muskegon, MI | 6,880 | 3.5 | 39,637 | 349 | 2,916 | 534 | 219 | 311 | 1,005 | 1,964 | 3,981 | 279 | 219 |
| Myrtle Beach-Conway-North Myrtle Beach, SC-NC | 20,122 | 6.3 | 40,494 | 335 | 7,023 | 1,183 | 533 | 1,488 | 4,248 | 6,118 | 10,228 | 801 | 533 |
| Napa, CA | 10,430 | 4.3 | 75,717 | 12 | 4,969 | 893 | 363 | 1,324 | 2,774 | 1,340 | 7,549 | 420 | 363 |
| Naples-Marco Island, FL | 38,252 | 2.9 | 99,382 | 5 | 8,613 | 1,079 | 576 | 2,001 | 21,973 | 4,051 | 12,269 | 856 | 576 |
| Nashville-Davidson--Murfreesboro--Franklin, TN | 117,374 | 4.8 | 60,680 | 42 | 60,980 | 8,017 | 4,063 | 24,185 | 16,642 | 14,944 | 97,245 | 5,433 | 4,063 |
| New Bern, NC | 5,691 | 5.0 | 45,788 | 226 | 2,616 | 627 | 206 | 259 | 1,233 | 1,467 | 3,709 | 230 | 206 |
| New Haven-Milford, CT | 49,360 | 3.3 | 57,748 | 63 | 23,061 | 3,798 | 1,736 | 4,382 | 8,257 | 9,464 | 32,977 | 1,893 | 1,736 |
| New Orleans-Metairie, LA | 69,069 | 2.7 | 54,363 | 89 | 32,821 | 5,342 | 2,095 | 8,421 | 13,625 | 12,637 | 48,680 | 2,732 | 2,095 |
| New York-Newark-Jersey City, NY-NJ-PA | 1,534,294 | 4.0 | 79,844 | 8 | 797,953 | 116,650 | 52,908 | 171,191 | 332,474 | 215,770 | 1,138,702 | 61,977 | 52,908 |
| Nassau County-Suffolk County, NY Div 35004 | 232,705 | 4.0 | 82,126 | NA | 85,920 | 15,714 | 6,489 | 18,209 | 51,658 | 31,231 | 126,332 | 6,848 | 6,848 |
| Newark, NJ-PA Div 35084 | 166,698 | 3.8 | 76,896 | NA | 79,046 | 11,731 | 5,363 | 15,968 | 31,868 | 20,725 | 112,108 | 6,582 | 6,582 |
| New Brunswick-Lakewood, NJ Div 35154 | 171,859 | 4.0 | 72,210 | NA | 75,144 | 11,310 | 5,403 | 18,282 | 30,740 | 23,320 | 110,139 | 6,630 | 6,630 |
| New York-Jersey City-White Plains, NY-NJ Div 35614 | 963,032 | 4.0 | 81,373 | NA | 557,844 | 77,895 | 35,653 | 118,732 | 218,208 | 140,493 | 790,124 | 41,917 | 41,917 |
| Niles, MI | 7,400 | 1.4 | 48,237 | 177 | 3,283 | 629 | 247 | 397 | 1,399 | 1,745 | 4,556 | 299 | 247 |
| North Port-Sarasota-Bradenton, FL | 48,613 | 3.7 | 58,081 | 61 | 15,855 | 2,200 | 1,082 | 3,152 | 18,202 | 10,151 | 22,290 | 1,653 | 1,082 |
| Norwich-New London, CT | 15,837 | 2.7 | 59,717 | 49 | 7,694 | 1,575 | 560 | 1,309 | 2,913 | 2,810 | 11,138 | 610 | 560 |
| Ocala, FL | 13,999 | 4.8 | 38,293 | 364 | 4,743 | 790 | 329 | 598 | 3,166 | 4,851 | 6,460 | 554 | 329 |
| Ocean City, NJ | 5,774 | 3.6 | 62,734 | 33 | 1,892 | 385 | 159 | 669 | 1,369 | 1,353 | 3,106 | 207 | 159 |
| Odessa, TX | 8,338 | 6.3 | 50,161 | 139 | 5,552 | 676 | 357 | 842 | 893 | 1,106 | 7,427 | 395 | 357 |
| Ogden-Clearfield, UT | 31,542 | 6.2 | 46,123 | 222 | 13,389 | 2,486 | 1,048 | 1,630 | 6,142 | 4,179 | 18,552 | 1,111 | 1,048 |
| Oklahoma City, OK | 68,841 | 3.8 | 48,860 | 161 | 35,083 | 5,657 | 2,516 | 7,189 | 12,958 | 11,533 | 50,446 | 2,930 | 2,516 |
| Olympia-Lacey-Tumwater, WA | 15,348 | 6.3 | 52,828 | 106 | 6,857 | 1,324 | 599 | 907 | 3,337 | 2,893 | 9,686 | 581 | 599 |
| Omaha-Council Bluffs, NE-IA | 55,966 | 3.7 | 58,947 | 50 | 28,840 | 4,729 | 2,144 | 6,372 | 10,879 | 7,648 | 42,086 | 2,518 | 2,144 |
| Orlando-Kissimmee-Sanford, FL | 117,774 | 4.6 | 45,156 | 242 | 71,474 | 9,482 | 4,827 | 6,668 | 19,528 | 21,985 | 92,450 | 5,814 | 4,827 |
| Oshkosh-Neenah, WI | 8,471 | 3.3 | 49,276 | 153 | 5,441 | 961 | 390 | 454 | 1,661 | 1,406 | 7,246 | 450 | 390 |
| Owensboro, KY | 5,000 | 2.4 | 41,858 | 304 | 2,462 | 429 | 183 | 256 | 857 | 1,297 | 3,331 | 218 | 183 |
| Oxnard-Thousand Oaks-Ventura, CA | 54,749 | 4.3 | 64,715 | 25 | 21,113 | 3,613 | 1,496 | 4,926 | 12,244 | 7,387 | 31,148 | 1,797 | 1,496 |
| Palm Bay-Melbourne-Titusville, FL | 28,839 | 5.4 | 47,911 | 184 | 12,993 | 1,990 | 895 | 1,244 | 6,191 | 7,013 | 17,122 | 1,200 | 895 |
| Panama City, FL | 7,982 | -0.8 | 45,690 | 230 | 3,823 | 658 | 277 | 432 | 1,679 | 1,880 | 5,190 | 340 | 277 |
| Parkersburg-Vienna, WV | 3,928 | 0.9 | 43,967 | 268 | 1,681 | 336 | 134 | 334 | 603 | 1,102 | 2,484 | 172 | 134 |
| Pensacola-Ferry Pass-Brent, FL | 22,592 | 5.1 | 44,947 | 246 | 9,826 | 1,778 | 714 | 1,085 | 4,462 | 5,203 | 13,404 | 877 | 714 |
| Peoria, IL | 19,666 | 0.7 | 49,097 | 156 | 10,353 | 1,787 | 691 | 970 | 3,566 | 3,817 | 13,801 | 829 | 691 |
| Philadelphia-Camden-Wilmington, PA-NJ-DE-MD | 406,400 | 3.9 | 66,596 | 22 | 200,148 | 31,214 | 14,591 | 40,855 | 75,376 | 67,683 | 286,808 | 16,791 | 14,591 |
| Camden, NJ Div 15804 | 73,055 | 3.9 | 58,752 | NA | 32,026 | 5,471 | 2,431 | 5,037 | 10,792 | 12,895 | 44,966 | 2,793 | 2,793 |
| Montgomery-Bucks-Chester Counties, PA Div 33874 | 161,795 | 4.2 | 81,543 | NA | 76,311 | 10,941 | 5,498 | 9,108 | 37,782 | 19,195 | 101,858 | 6,148 | 6,148 |
| Philadelphia, PA Div 37964 | 130,237 | 3.5 | 60,552 | NA | 68,210 | 10,856 | 4,991 | 23,640 | 19,459 | 28,058 | 107,696 | 5,886 | 5,886 |
| Wilmington, DE-MD-NJ Div 48864 | 41,313 | 3.7 | 57,063 | NA | 23,601 | 3,946 | 1,671 | 3,070 | 7,343 | 7,534 | 32,288 | 1,964 | 1,964 |
| Phoenix-Mesa-Chandler, AZ | 237,837 | 5.4 | 48,065 | 180 | 128,810 | 17,739 | 8,906 | 16,206 | 45,329 | 39,787 | 171,660 | 10,758 | 8,906 |
| Pine Bluff, AR | 3,127 | 2.0 | 35,611 | 378 | 1,471 | 266 | 117 | 169 | 434 | 1,058 | 2,023 | 139 | 117 |
| Pittsburgh, PA | 139,582 | 3.2 | 60,227 | 44 | 71,783 | 11,590 | 5,425 | 13,565 | 23,230 | 26,555 | 102,363 | 6,118 | 5,425 |
| Pittsfield, MA | 7,284 | 2.1 | 58,299 | 59 | 3,119 | 578 | 227 | 561 | 1,561 | 1,785 | 4,484 | 261 | 227 |
| Pocatello, ID | 3,779 | 4.5 | 39,573 | 350 | 1,649 | 323 | 142 | 213 | 659 | 857 | 2,328 | 153 | 142 |
| Portland-South Portland, ME | 32,310 | 4.2 | 60,000 | 45 | 16,094 | 2,614 | 1,174 | 2,541 | 6,459 | 5,393 | 22,422 | 1,460 | 1,174 |
| Portland-Vancouver-Hillsboro, OR-WA | 149,347 | 4.3 | 59,921 | 47 | 80,158 | 11,391 | 6,569 | 11,624 | 31,944 | 20,843 | 109,743 | 6,712 | 6,569 |
| Port St. Lucie, FL | 27,250 | 4.2 | 55,691 | 77 | 7,308 | 1,136 | 503 | 1,183 | 9,348 | 5,627 | 10,130 | 767 | 503 |
| Poughkeepsie-Newburgh-Middletown, NY | 37,859 | 5.0 | 55,745 | 76 | 14,572 | 3,177 | 1,150 | 2,191 | 6,085 | 6,987 | 21,091 | 1,183 | 1,150 |
| Prescott Valley-Prescott, AZ | 9,731 | 3.8 | 41,393 | 316 | 2,900 | 498 | 210 | 675 | 2,664 | 3,081 | 4,282 | 373 | 210 |
| Providence-Warwick, RI-MA | 91,200 | 3.7 | 56,138 | 70 | 41,342 | 6,873 | 3,194 | 6,587 | 14,902 | 18,474 | 57,996 | 3,814 | 3,194 |
| Provo-Orem, UT | 27,825 | 6.6 | 42,923 | 286 | 13,829 | 1,974 | 996 | 2,361 | 5,284 | 3,162 | 19,160 | 1,120 | 996 |
| Pueblo, CO | 6,852 | 4.2 | 40,680 | 332 | 3,066 | 477 | 228 | 337 | 1,230 | 2,023 | 4,108 | 267 | 228 |
| Punta Gorda, FL | 8,084 | 4.6 | 42,793 | 291 | 2,227 | 342 | 156 | 412 | 2,301 | 2,794 | 3,136 | 294 | 156 |
| Racine, WI | 9,981 | 2.3 | 50,845 | 132 | 4,041 | 800 | 301 | 376 | 2,062 | 1,894 | 5,518 | 364 | 301 |
| Raleigh-Cary, NC | 80,458 | 5.1 | 57,851 | 62 | 43,063 | 5,628 | 2,912 | 5,025 | 14,581 | 9,390 | 56,628 | 3,426 | 2,912 |
| Rapid City, SD | 7,295 | 3.8 | 51,335 | 124 | 3,206 | 594 | 240 | 610 | 1,885 | 1,368 | 4,651 | 302 | 240 |

1. Based on the resident population estimated as of July 1 of the year shown.

Items 62—74

# Table C. Metropolitan Areas — Earnings, Social Security, and Housing

| Area name | Earnings, 2019 (cont.) — Percent by selected industries | | | | | | | | | Social Security beneficiaries, December 2019 | | Supplemental Security Income Recipients, December 2019 | Housing units, 2020 | |
| | Farm | Mining, quarrying, and extracting | Construction | Manufacturing | Information; professional, scientific, and technical services | Retail trade | Finance, insurance, real estate, rental and leasing | Health care and social assistance | Government | Number | Rate[1] | | Total | Percent change, 2010-2020 |
| | 75 | 76 | 77 | 78 | 79 | 80 | 81 | 82 | 83 | 84 | 85 | 86 | 87 | 88 |
|---|---|---|---|---|---|---|---|---|---|---|---|---|---|---|
| Mobile, AL | 0.3 | 0.3 | 7.1 | 15.7 | 8.9 | 6.2 | 7.0 | 12.3 | 14.4 | 98,410 | 229 | 15,764 | 194,188 | 4.1 |
| Modesto, CA | 7.0 | 0.0 | 6.5 | 11.5 | 4.3 | 7.1 | 4.3 | 17.1 | 18.2 | 91,100 | 166 | 20,722 | 183,550 | 2.3 |
| Monroe, LA | 1.6 | D | 6.1 | D | D | 8.2 | 5.7 | 16.7 | 16.1 | 43,130 | 215 | 9,719 | 94,773 | 7.4 |
| Monroe, MI | 0.8 | D | 6.5 | 19.4 | 8.2 | 6.0 | 3.5 | 8.3 | 11.9 | 36,480 | 243 | 2,523 | 65,530 | 4.1 |
| Montgomery, AL | 0.4 | 0.4 | 6.0 | 11.5 | D | 6.1 | 5.6 | D | 28.0 | 82,675 | 222 | 13,966 | 169,663 | 5.0 |
| Morgantown, WV | 0.0 | 1.3 | 6.2 | 7.5 | 8.4 | 5.2 | 3.1 | D | 26.4 | 23,350 | 167 | 2,631 | 60,387 | 3.5 |
| Morristown, TN | -0.2 | D | D | 27.8 | D | 7.8 | 2.9 | D | 14.1 | 38,540 | 270 | 4,457 | 63,413 | 3.3 |
| Mount Vernon-Anacortes, WA | 3.5 | 0.1 | 12.4 | 13.2 | 5.9 | 8.0 | 6.8 | 7.2 | 23.3 | 32,075 | 248 | 2,350 | 55,427 | 7.7 |
| Muncie, IN | 0.3 | D | 4.7 | 9.7 | 7.8 | 8.2 | 6.7 | 19.8 | 21.3 | 25,925 | 228 | 2,986 | 52,700 | 0.7 |
| Muskegon, MI | 0.4 | 0.1 | 6.6 | 26.6 | 4.5 | 10.4 | 3.9 | 16.2 | 14.0 | 43,430 | 250 | 6,091 | 74,776 | 1.6 |
| Myrtle Beach-Conway-North Myrtle Beach, SC-NC | 0.0 | D | 9.2 | 3.2 | D | 11.2 | 9.0 | 11.3 | 15.9 | 158,605 | 319 | 8,080 | 317,756 | 20.6 |
| Napa, CA | 3.0 | D | 7.8 | 19.9 | 6.5 | 4.8 | 5.5 | 9.4 | 15.1 | 27,850 | 202 | 2,084 | 55,292 | 1.0 |
| Naples-Marco Island, FL | 0.9 | 0.0 | 11.1 | 3.5 | 12.8 | 8.4 | 10.9 | 12.9 | 8.9 | 100,610 | 261 | 4,145 | 226,087 | 14.6 |
| Nashville-Davidson--Murfreesboro--Franklin, TN | -0.1 | D | D | 7.1 | 13.5 | 5.8 | 8.8 | 16.6 | 8.5 | 321,505 | 166 | 29,974 | 827,421 | 19.5 |
| New Bern, NC | 1.2 | 0.1 | 3.7 | 7.9 | D | 5.6 | D | 8.9 | 49.0 | 32,310 | 261 | 3,087 | 60,952 | 6.2 |
| New Haven-Milford, CT | 0.1 | 0.1 | 6.5 | 8.7 | 9.6 | 6.1 | 6.4 | 15.7 | 14.1 | 168,330 | 197 | 20,933 | 369,956 | 2.2 |
| New Orleans-Metairie, LA | 0.0 | 7.1 | 5.8 | 7.4 | 10.5 | 5.3 | 6.8 | D | 14.1 | 244,555 | 192 | 43,288 | 564,679 | 3.3 |
| New York-Newark-Jersey City, NY-NJ-PA | 0.0 | D | 4.5 | D | 19.3 | 4.6 | 17.6 | 10.6 | 13.1 | 3,256,860 | 169 | 557,251 | 7,837,181 | 4.1 |
| Nassau County-Suffolk County, NY Div 35004 | 0.1 | D | 7.9 | D | 10.6 | 6.8 | 10.2 | 16.4 | 18.2 | 565,295 | 199 | 35,696 | 1,053,440 | 1.5 |
| Newark, NJ-PA Div 35084 | 0.0 | 0.7 | 4.9 | 7.2 | 17.3 | 4.8 | 10.7 | 10.2 | 14.0 | 369,105 | 170 | 45,793 | 874,088 | 2.6 |
| New Brunswick-Lakewood, NJ Div 35154 | 0.0 | D | 6.1 | 11.0 | 17.1 | 6.2 | 7.2 | 11.1 | 12.1 | 478,600 | 201 | 31,770 | 983,255 | 3.0 |
| New York-Jersey City-White Plains, NY-NJ Div 35614 | 0.0 | 0.2 | 3.7 | 1.9 | 21.3 | 3.9 | 21.3 | 9.7 | 12.2 | 1,843,860 | 156 | 443,992 | 4,926,398 | 5.2 |
| Niles, MI | 1.2 | 0.2 | 5.6 | 29.3 | 3.9 | 5.7 | 5.2 | 11.6 | 13.4 | 38,245 | 249 | 4,527 | 77,979 | 1.4 |
| North Port-Sarasota-Bradenton, FL | 0.9 | D | 9.9 | 5.8 | 11.4 | 8.7 | 8.8 | 15.9 | 9.6 | 252,980 | 302 | 10,716 | 457,745 | 14.1 |
| Norwich-New London, CT | 0.7 | 0.1 | 5.1 | 19.4 | 7.8 | 5.6 | 3.1 | 11.1 | 25.5 | 58,025 | 218 | 4,579 | 124,734 | 3.1 |
| Ocala, FL | 0.4 | 0.2 | 8.0 | 9.1 | 6.3 | 10.5 | 5.4 | 18.1 | 15.2 | 122,830 | 336 | 10,591 | 175,527 | 7.0 |
| Ocean City, NJ | 0.2 | D | 11.0 | D | 5.0 | 9.9 | 7.7 | 9.4 | 25.8 | 29,050 | 315 | 1,689 | 99,496 | 1.2 |
| Odessa, TX | 0.0 | 24.4 | 12.9 | 7.2 | 3.9 | 6.1 | 5.0 | 4.4 | 10.1 | 20,700 | 125 | 3,216 | 61,298 | 15.6 |
| Ogden-Clearfield, UT | 0.4 | D | 9.0 | 15.3 | D | 6.6 | 6.5 | 9.7 | 23.9 | 89,065 | 130 | 6,764 | 233,694 | 14.5 |
| Oklahoma City, OK | 0.0 | 6.0 | 6.1 | 5.4 | 9.5 | 5.8 | 6.6 | D | 20.4 | 246,155 | 175 | 27,851 | 588,285 | 9.1 |
| Olympia-Lacey-Tumwater, WA | 0.9 | 0.0 | 7.0 | 2.5 | 8.1 | 6.1 | 4.6 | 12.8 | 37.7 | 63,235 | 218 | 5,395 | 121,284 | 12.1 |
| Omaha-Council Bluffs, NE-IA | 0.7 | 0.1 | 6.0 | 6.1 | D | 5.1 | 11.3 | 11.6 | 13.4 | 157,220 | 166 | 15,395 | 397,729 | 9.8 |
| Orlando-Kissimmee-Sanford, FL | 0.3 | 0.0 | 7.5 | 4.7 | 13.8 | 6.8 | 8.4 | 10.9 | 10.2 | 463,180 | 178 | 60,461 | 1,100,260 | 16.8 |
| Oshkosh-Neenah, WI | 0.4 | D | 8.1 | 27.7 | 7.5 | 4.4 | 6.2 | 9.2 | 11.4 | 36,000 | 210 | 2,668 | 76,566 | 4.4 |
| Owensboro, KY | 2.0 | D | D | 22.0 | D | 6.5 | D | D | 11.3 | 28,840 | 241 | 3,870 | 51,772 | 4.7 |
| Oxnard-Thousand Oaks-Ventura, CA | 4.3 | 0.7 | 6.1 | 8.4 | 10.9 | 7.1 | 8.1 | 9.8 | 16.7 | 145,535 | 172 | 15,803 | 291,563 | 3.5 |
| Palm Bay-Melbourne-Titusville, FL | 0.1 | 0.0 | 6.5 | 18.6 | 11.9 | 6.7 | 4.8 | 13.1 | 14.8 | 166,545 | 277 | 11,920 | 286,637 | 6.2 |
| Panama City, FL | 0.0 | D | 8.7 | 5.0 | 9.7 | 9.3 | 7.0 | D | 23.3 | 40,005 | 230 | 4,077 | 106,426 | 6.8 |
| Parkersburg-Vienna, WV | -0.2 | D | 5.7 | D | D | 8.5 | 6.3 | 15.7 | 21.7 | 25,220 | 282 | 3,544 | 43,643 | 0.5 |
| Pensacola-Ferry Pass-Brent, FL | 0.2 | 0.1 | 6.3 | 4.6 | 8.3 | 6.9 | 9.6 | 15.2 | 24.7 | 114,010 | 226 | 12,732 | 220,546 | 9.5 |
| Peoria, IL | 0.8 | D | 5.9 | 22.2 | D | 5.4 | D | 14.5 | 12.3 | 90,190 | 225 | 8,024 | 183,157 | 1.5 |
| Philadelphia-Camden-Wilmington, PA-NJ-DE-MD | 0.2 | D | D | D | 20.1 | 5.0 | 10.3 | 13.2 | 11.9 | 1,161,730 | 190 | 177,921 | 2,518,213 | 3.5 |
| Camden, NJ Div 15804 | 0.2 | D | 6.3 | 8.5 | 9.5 | 7.8 | 7.8 | 15.1 | 17.7 | 252,515 | 203 | 27,165 | 502,856 | 2.5 |
| Montgomery-Bucks-Chester Counties, PA Div 33874 | 0.3 | 0.1 | 7.4 | 9.6 | 20.9 | 6.1 | 10.1 | 11.8 | 7.8 | 397,035 | 200 | 20,165 | 797,553 | 4.4 |
| Philadelphia, PA Div 37964 | 0.0 | D | 3.1 | 4.4 | 26.3 | 2.8 | 8.0 | 13.9 | 12.8 | 370,600 | 172 | 116,837 | 921,001 | 3.1 |
| Wilmington, DE-MD-NJ Div 48864 | 0.3 | 0.3 | D | D | 12.4 | 4.7 | 22.5 | 12.8 | 14.2 | 141,580 | 195 | 13,754 | 296,803 | 3.8 |
| Phoenix-Mesa-Chandler, AZ | 0.4 | 0.6 | 7.1 | 7.6 | 11.6 | 6.4 | 13.0 | D | 13.8 | 826,095 | 167 | 67,328 | 2,003,726 | 11.4 |
| Pine Bluff, AR | 4.0 | D | 3.1 | 18.4 | | | | | | 21,290 | 242 | 4,867 | 42,508 | 1.4 |
| Pittsburgh, PA | 0.1 | 1.5 | 6.8 | 8.3 | 13.0 | 5.0 | 8.9 | 14.0 | 10.2 | 570,540 | 246 | 61,711 | 1,136,669 | 3.1 |
| Pittsfield, MA | 0.0 | 0.1 | 8.1 | 7.8 | 10.4 | 7.2 | 5.8 | 19.3 | 14.5 | 34,930 | 279 | 3,884 | 69,466 | 1.4 |
| Pocatello, ID | 2.6 | D | 5.2 | 10.5 | 4.5 | 6.9 | 5.8 | 13.9 | 23.0 | 17,285 | 181 | 2,135 | 38,231 | 5.8 |
| Portland-South Portland, ME | 0.1 | D | 6.7 | D | 11.5 | 6.5 | 10.0 | 14.5 | 15.0 | 126,035 | 233 | 9,465 | 280,900 | 6.9 |
| Portland-Vancouver-Hillsboro, OR-WA | 0.3 | 0.1 | 7.4 | 12.3 | D | 5.5 | 7.8 | D | 13.5 | 428,850 | 172 | 45,113 | 1,038,249 | 12.2 |
| Port St. Lucie, FL | 1.2 | D | 8.2 | 5.3 | 8.4 | 8.2 | 6.6 | 17.3 | 14.2 | 133,495 | 273 | 8,763 | 230,881 | 7.3 |
| Poughkeepsie-Newburgh-Middletown, NY | 0.3 | 0.2 | 6.7 | 7.8 | 7.8 | 7.4 | 4.3 | 15.7 | 25.4 | 130,280 | 192 | 12,074 | 268,558 | 5.0 |
| Prescott Valley-Prescott, AZ | 0.2 | 2.7 | 9.4 | 5.9 | 6.7 | 10.3 | 5.0 | 15.5 | 16.1 | 85,265 | 361 | 3,583 | 122,590 | 11.0 |
| Providence-Warwick, RI-MA | 0.0 | D | 6.6 | 9.5 | 9.0 | 6.7 | 8.3 | 14.2 | 16.1 | 352,505 | 217 | 51,137 | 708,781 | 2.1 |
| Provo-Orem, UT | 0.2 | 0.1 | 10.9 | 8.8 | D | 8.9 | 4.3 | D | 9.9 | 61,525 | 95 | 4,423 | 197,322 | 29.9 |
| Pueblo, CO | 0.2 | D | 8.2 | 9.0 | 7.2 | 8.2 | 3.6 | 20.6 | 20.5 | 38,335 | 228 | 6,045 | 71,999 | 3.6 |
| Punta Gorda, FL | 0.2 | 0.3 | 9.2 | 1.7 | 8.4 | 12.3 | 6.8 | 21.8 | 13.4 | 72,580 | 383 | 2,878 | 109,385 | 8.7 |
| Racine, WI | 1.1 | 0.3 | 5.8 | 31.0 | 4.7 | 6.2 | 4.6 | 12.0 | 13.4 | 43,945 | 224 | 5,790 | 83,326 | 1.4 |
| Raleigh-Cary, NC | 0.3 | D | 7.3 | 7.9 | 22.5 | 5.6 | 8.4 | 9.1 | 13.3 | 200,830 | 144 | 16,942 | 575,599 | 23.5 |
| Rapid City, SD | 0.0 | 0.1 | 8.1 | 3.8 | | 7.5 | 8.1 | 19.4 | 22.7 | 33,430 | 234 | 2,199 | 62,842 | 12.3 |

1. Per 1,000 resident population estimated as of July 1 of the year shown.

# Table C. Metropolitan Areas — Housing, Labor Force, and Employment

| Area name | Occupied Housing units, 2020 | | | | | | | | Civilian labor force, 2020 | | | | Civilian employment, 2019[6] | | |
| | Occupied units | | | | | | | | | | Unemployment | | | Percent | |
| | Owner-occupied | | | | | Renter-occupied | | | | | | | | | |
| | | | | Median owner cost as a percent of income | | Median rent[3] | Median rent as a per-cent of income[2] | Sub-standard units[4] (percent) | | | | | | Manage-ment, business, science, and arts | Construction, production, and mainte-nance occu-pations |
| | Total | Percent | Median value[1] | With a mortgage | Without a mort-gage[2] | | | | Total | Percent change 2019-2020 | Total | Rate[5] | Total | | |
| | 89 | 90 | 91 | 92 | 93 | 94 | 95 | 96 | 97 | 98 | 99 | 100 | 101 | 102 | 103 |
|---|---|---|---|---|---|---|---|---|---|---|---|---|---|---|---|
| Mobile, AL | 161,217 | 64.1 | 139,000 | 19.7 | 10.0 | 843 | 32.0 | 1.6 | 197,611 | 1.1 | 15,660 | 7.9 | 182,712 | 34.8 | 27.4 |
| Modesto, CA | 174,698 | 57.4 | 335,400 | 23.9 | 11.3 | 1,231 | 31.7 | 7.2 | 240,641 | -0.7 | 25,756 | 10.7 | 236,924 | 26.7 | 33.8 |
| Monroe, LA | 77,194 | 60.8 | 128,200 | 18.0 | 10.0 | 730 | 33.9 | 1.9 | 88,889 | -1.3 | 6,794 | 7.6 | 82,695 | 32.7 | 23.5 |
| Monroe, MI | 60,875 | 78.1 | 173,200 | 18.0 | 10.7 | 827 | 32.0 | 0.0 | 74,397 | -1.8 | 6,633 | 8.9 | 70,166 | 33.0 | 33.9 |
| Montgomery, AL | 145,250 | 63.9 | 149,500 | 19.1 | 10.0 | 899 | 28.2 | 1.5 | 173,222 | 0.1 | 11,895 | 6.9 | 168,780 | 37.1 | 24.1 |
| Morgantown, WV | 53,129 | 65 | 192,700 | 15.8 | 10.0 | 775 | 31.2 | 0.0 | 70,138 | -0.5 | 4,598 | 6.6 | 67,704 | 49.3 | 16.4 |
| Morristown, TN | 55,874 | 75.4 | 145,700 | 18.7 | 10.0 | 732 | 28.5 | 2.9 | 62,197 | -0.3 | 4,524 | 7.3 | 62,829 | 26.3 | 32.7 |
| Mount Vernon-Anacortes, WA. | 49,043 | 69.9 | 353,300 | 24.1 | 12.9 | 1,084 | 28.2 | 5.2 | 62,498 | | 6,124 | 9.8 | 56,872 | 35.8 | 31.2 |
| Muncie, IN | 45,640 | 61.6 | 101,800 | 17.4 | 10.8 | 709 | 31.2 | 0.0 | 52,466 | -2.0 | 3,833 | 7.3 | 51,668 | 35.5 | 24.5 |
| Muskegon, MI | 66,148 | 74.2 | 131,100 | 18.8 | 12.5 | 727 | 31.1 | 2.7 | 77,540 | 0.3 | 8,859 | 11.4 | 76,075 | 29.3 | 33.1 |
| Myrtle Beach-Conway-North Myrtle Beach, SC-NC.... | 198,645 | 75.8 | 212,100 | 21.8 | 10.1 | 1,016 | 29.8 | 1.2 | 199,943 | -1.2 | 17,346 | 8.7 | 217,445 | 31.8 | 21.0 |
| Napa, CA | 48,107 | 66.1 | 670,000 | 25.5 | 10.0 | 1,835 | 29.9 | 5.9 | 68,867 | -5.9 | 6,014 | 8.7 | 69,450 | 36.7 | 23.3 |
| Naples-Marco Island, FL | 140,578 | 74.2 | 370,800 | 22.9 | 11.4 | 1,397 | 33.4 | 2.5 | 177,497 | -1.7 | 12,210 | 6.9 | 161,368 | 29.6 | 23.3 |
| Nashville-Davidson--Murfrees-boro--Franklin, TN | 733,787 | 65.8 | 285,100 | 19.7 | 10.0 | 1,146 | 28.2 | 2.2 | 1,059,372 | -1.4 | 73,623 | 6.9 | 1,031,164 | 41.5 | 22.0 |
| New Bern, NC | 52,345 | 70.7 | 153,600 | 19.2 | 10.0 | 937 | 30.1 | 0.0 | 50,261 | -2.8 | 3,236 | 6.4 | 51,793 | 30.8 | 28.4 |
| New Haven-Milford, CT | 329,006 | 60.5 | 251,500 | 22.4 | 15.2 | 1,151 | 31.7 | 1.7 | 457,171 | -1.4 | 36,502 | 8.0 | 427,382 | 42.4 | 19.5 |
| New Orleans-Metairie, LA | 485,267 | 63.4 | 211,900 | 20.6 | 10.0 | 983 | 31.2 | 2.1 | 592,743 | -1.8 | 57,345 | 9.7 | 605,541 | 39.1 | 19.6 |
| New York-Newark-Jersey City, NY-NJ-PA | 7,102,817 | 50.8 | 482,900 | 24.3 | 14.3 | 1,482 | 30.0 | 5.8 | 9,412,977 | -2.3 | 987,173 | 10.5 | 9,686,984 | 44.9 | 16.4 |
| Nassau County-Suffolk County, NY Div 35004 ... | 949,542 | 80.7 | 481,900 | 25.1 | 16.2 | 1,793 | 33.4 | 2.4 | 1,463,506 | -1.4 | 123,858 | 8.5 | 1,472,494 | 44.9 | 16.7 |
| Newark, NJ-PA Div 35084 ..... | 797,142 | 60.2 | 391,100 | 23.7 | 14.9 | 1,324 | 29.9 | 4.0 | 1,064,725 | -0.8 | 104,784 | 9.8 | 1,104,568 | 43.4 | 19.1 |
| New Brunswick-Lakewood, NJ Div 35154 | 875,614 | 72 | 368,500 | 23.4 | 14.5 | 1,488 | 30.0 | 2.5 | 1,210,392 | -0.9 | 106,457 | 8.8 | 1,187,939 | 48.2 | 17.4 |
| New York-Jersey City-White Plains, NY-NJ Div 35614 | 4,480,519 | 38.7 | 585,000 | 24.7 | 13.2 | 1,486 | 29.8 | 7.5 | 5,674,354 | -3.0 | 652,074 | 11.5 | 5,921,983 | 44.5 | 15.6 |
| Niles, MI | 61,809 | 73.9 | 156,900 | 19.0 | 10.8 | 775 | 29.4 | 1.6 | 72,765 | -1.7 | 6,289 | 8.6 | 69,777 | 34.3 | 27.9 |
| North Port-Sarasota-Braden-ton, FL | 332,582 | 76.6 | 273,500 | 22.6 | 11.6 | 1,284 | 31.8 | 1.3 | 360,960 | -2.4 | 24,514 | 6.8 | 348,069 | 36.0 | 18.7 |
| Norwich-New London, CT | 109,752 | 65 | 255,400 | 22.1 | 13.3 | 1,140 | 28.2 | 1.2 | 131,992 | -4.0 | 12,679 | 9.6 | 135,915 | 42.0 | 16.5 |
| Ocala, FL | 145,622 | 71.5 | 161,100 | 19.7 | 11.3 | 970 | 29.6 | 3.3 | 138,469 | -0.6 | 9,760 | 7.0 | 136,306 | 31.6 | 23.0 |
| Ocean City, NJ | 40,939 | 74.6 | 320,900 | 23.1 | 13.2 | 1,286 | 36.5 | 0.0 | 45,492 | -1.2 | 6,286 | 13.8 | 42,389 | 37.2 | 19.3 |
| Odessa, TX | 53,155 | 61.4 | 164,100 | 19.1 | 10.0 | 1,141 | 25.0 | 3.1 | 82,852 | -6.2 | 9,116 | 11.0 | 78,322 | 23.6 | 40.9 |
| Ogden-Clearfield, UT | 218,724 | 76.1 | 306,600 | 20.5 | 10.0 | 1,063 | 25.2 | 2.1 | 337,797 | 1.4 | 14,716 | 4.4 | 337,304 | 39.2 | 24.2 |
| Oklahoma City, OK | 523,919 | 63.3 | 166,800 | 19.0 | 10.0 | 876 | 27.6 | 2.4 | 688,901 | 0.0 | 42,017 | 6.1 | 672,690 | 39.6 | 22.2 |
| Olympia-Lacey-Tumwater, WA | 112,909 | 68.7 | 333,100 | 22.3 | 10.6 | 1,255 | 32.7 | 1.6 | 144,312 | 1.4 | 11,947 | 8.3 | 135,360 | 40.8 | 22.1 |
| Omaha-Council Bluffs, NE-IA.. | 367,037 | 66.1 | 190,800 | 19.4 | 12.0 | 949 | 27.4 | 2.0 | 494,098 | -0.8 | 23,677 | 4.8 | 493,485 | 43.0 | 21.2 |
| Orlando-Kissimmee-Sanford, FL | 889,822 | 61.9 | 255,500 | 22.3 | 10.9 | 1,303 | 33.3 | 3.3 | 1,315,136 | -3.4 | 134,463 | 10.2 | 1,311,650 | 37.9 | 18.9 |
| Oshkosh-Neenah, WI | 71,238 | 63.4 | 164,800 | 17.5 | 12.7 | 795 | 25.8 | 2.1 | 91,236 | 0.2 | 4,966 | 5.4 | 89,405 | 32.9 | 25.9 |
| Owensboro, KY | 48,419 | 70.1 | 140,900 | 18.7 | 10.2 | 788 | 29.7 | 2.9 | 53,541 | -3.2 | 3,259 | 6.1 | 54,401 | 32.9 | 32.2 |
| Oxnard-Thousand Oaks-Ventura, CA | 268,524 | 62.8 | 629,600 | 25.1 | 10.0 | 1,859 | 33.2 | 6.3 | 408,867 | -3.0 | 34,964 | 8.6 | 418,904 | 37.2 | 22.3 |
| Palm Bay-Melbourne-Titus-ville, FL | 230,043 | 77.5 | 228,500 | 22.8 | 10.0 | 1,132 | 30.4 | 1.5 | 281,591 | -0.8 | 18,724 | 6.6 | 259,299 | 40.0 | 19.7 |
| Panama City, FL | 69,195 | 70.7 | 203,500 | 22.8 | 10.0 | 1,204 | 28.5 | 3.7 | 82,763 | -2.0 | 5,017 | 6.1 | 83,287 | 35.8 | 23.7 |
| Parkersburg-Vienna, WV | 36,946 | 74.9 | 129,900 | 19.3 | 10.0 | 650 | 28.9 | 0.0 | 38,727 | -1.2 | 3,469 | 9.2 | 37,483 | 31.9 | 26.3 |
| Pensacola-Ferry Pass-Brent, FL | 186,805 | 65.7 | 190,300 | 21.4 | 10.0 | 1,040 | 27.9 | 1.4 | 227,426 | -0.5 | 13,970 | 6.1 | 219,274 | 36.8 | 21.2 |
| Peoria, IL | 162,131 | 72.7 | 134,300 | 18.6 | 11.1 | 732 | 26.7 | 1.0 | 185,746 | -3.0 | 16,591 | 8.9 | 183,787 | 39.7 | 21.9 |
| Philadelphia-Camden-Wilm-ington, PA-NJ-DE-MD.... | 2,323,207 | 66.7 | 263,700 | 20.9 | 12.4 | 1,162 | 30.3 | 1.8 | 3,102,669 | -1.1 | 284,403 | 9.2 | 3,055,821 | 45.6 | 17.4 |
| Camden, NJ Div 15804......... | 464,559 | 72.1 | 229,000 | 22.0 | 15.0 | 1,223 | 30.3 | 1.7 | 637,912 | 0.0 | 58,785 | 9.2 | 629,158 | 43.5 | 19.6 |
| Montgomery-Bucks-Chester Counties, PA Div 33874. | 755,098 | 74.1 | 347,100 | 20.2 | 11.3 | 1,321 | 28.5 | 1.1 | 1,065,858 | -2.3 | 80,191 | 7.5 | 1,053,391 | 51.4 | 15.5 |
| Philadelphia, PA Div 37964 ... | 829,007 | 56.5 | 200,600 | 21.3 | 12.8 | 1,081 | 31.9 | 2.7 | 1,018,644 | -0.6 | 116,476 | 11.4 | 1,006,971 | 41.5 | 16.9 |
| Wilmington, DE-MD-NJ Div 48864 | 274,543 | 67.9 | 262,400 | 19.9 | 10.0 | 1,122 | 29.1 | 1.3 | 380,255 | | 28,951 | 7.6 | 366,301 | 45.3 | 20.2 |
| Phoenix-Mesa-Chandler, AZ.. | 1,765,135 | 64.6 | 283,500 | 20.4 | 10.0 | 1,188 | 28.7 | 4.3 | 2,522,702 | 1.4 | 186,761 | 7.4 | 2,368,094 | 39.0 | 20.5 |
| Pine Bluff, AR | 34,290 | 67.7 | 83,500 | 18.1 | 10.0 | 704 | 30.6 | 0.0 | 34,435 | -1.1 | 2,628 | 7.6 | 33,702 | 30.4 | 28.7 |
| Pittsburgh, PA | 1,028,436 | 68.2 | 165,900 | 17.9 | 10.9 | 847 | 26.0 | 0.8 | 1,185,672 | -2.1 | 109,139 | 9.2 | 1,166,352 | 43.8 | 19.5 |
| Pittsfield, MA | 53,792 | 70 | 229,900 | 23.4 | 13.2 | 864 | 31.5 | 1.0 | 61,645 | -3.6 | 5,629 | 9.1 | 64,358 | 41.7 | 19.2 |
| Pocatello, ID | 34,540 | 68 | 171,400 | 18.9 | 10.0 | 667 | 26.9 | 0.0 | 46,692 | 0.3 | 2,233 | 4.8 | 44,206 | 36.1 | 24.1 |
| Portland-South Portland, ME.. | 226,904 | 71.2 | 284,300 | 20.6 | 13.1 | 1,102 | 28.4 | 1.5 | 291,006 | -2.9 | 15,355 | 5.3 | 294,663 | 44.8 | 18.6 |
| Portland-Vancouver-Hillsboro, OR-WA.. | 964,554 | 61.8 | 408,600 | 22.1 | 11.0 | 1,356 | 29.9 | 3.7 | 1,323,366 | -0.4 | 102,697 | 7.8 | 1,291,570 | 44.0 | 19.6 |
| Port St. Lucie, FL | 181,914 | 76.8 | 242,600 | 22.6 | 11.8 | 1,198 | 30.2 | 1.6 | 217,543 | -1.1 | 15,687 | 7.2 | 218,195 | 35.9 | 20.3 |
| Poughkeepsie-Newburgh-Middletown, NY | 241,950 | 66.6 | 292,400 | 23.0 | 14.2 | 1,291 | 32.6 | 2.9 | 325,326 | -1.7 | 26,196 | 8.1 | 331,102 | 40.5 | 19.3 |
| Prescott Valley-Prescott, AZ.... | 102,784 | 70.7 | 288,500 | 23.0 | 10.7 | 1,005 | 29.2 | 2.1 | 106,985 | 0.4 | 8,007 | 7.5 | 93,799 | 35.1 | 21.6 |
| Providence-Warwick, RI-MA... | 627,702 | 62 | 296,800 | 22.4 | 12.9 | 1,003 | 29.2 | 1.8 | 835,212 | -2.9 | 80,912 | 9.7 | 835,183 | 41.0 | 20.5 |
| Provo-Orem, UT | 182,219 | 67.9 | 360,600 | 21.0 | 10.0 | 1,121 | 29.6 | 4.2 | 319,903 | 1.9 | 11,946 | 3.7 | 315,426 | 43.9 | 18.6 |
| Pueblo, CO | 66,111 | 67.1 | 189,600 | 20.6 | 11.2 | 860 | 32.2 | 2.2 | 77,355 | 2.1 | 6,318 | 8.2 | 69,791 | 33.2 | 20.9 |
| Punta Gorda, FL | 76,589 | 81.9 | 224,600 | 22.5 | 12.7 | 1,006 | 38.6 | 0.8 | 71,182 | -1.1 | 5,214 | 7.3 | 63,569 | 31.4 | 22.2 |
| Racine, WI | 78,905 | 67.1 | 200,800 | 20.1 | 12.2 | 833 | 29.1 | 0.9 | 97,155 | -0.9 | 7,051 | 7.3 | 93,186 | 33.2 | 29.7 |
| Raleigh-Cary, NC | 519,648 | 65.6 | 285,600 | 18.6 | 10.0 | 1,146 | 27.7 | 2.2 | 713,414 | -2.1 | 45,959 | 6.4 | 724,327 | 49.9 | 16.3 |
| Rapid City, SD | 56,456 | 68 | 199,800 | 21.2 | 11.8 | 829 | 29.7 | 0.0 | 71,012 | 0.2 | 3,703 | 5.2 | 71,782 | 36.1 | 21.7 |

1. Specified owner-occupied units. ing complete plumbing facilities.   2. A value of 10.0 represents 10 percent or less; a value of 50.0 represents 50 percent or more.   3. Specified renter-occupied units.   4. Overcrowded or lack-
5. Percent of civilian labor force.   6. Civilian employed persons 16 years old and over.

# Table C. Metropolitan Areas — Nonfarm Employment and Agriculture

| | Private nonfarm establishments, employment and payroll, 2019 | | | | | | | | | Agriculture, 2017 | | | |
| | | Employment | | | | | | Annual payroll | | Farms | | | Farm producers whose primary occupation is farming (percent) |
| | | | | | | | | | | | | Percent with: | |
| Area name | Number of establishments | Total | Health care and social assistance | Manufacturing | Retail trade | Finance and insurance | Professional, scientific, and technical services | Total (mil dol) | Average per employee (dollars) | Number | Fewer than 50 acres | 1000 acres or more | |
| | 104 | 105 | 106 | 107 | 108 | 109 | 110 | 111 | 112 | 113 | 114 | 115 | 116 |
|---|---|---|---|---|---|---|---|---|---|---|---|---|---|
| Mobile, AL | 8,963 | 160,587 | 22,500 | 19,114 | 20,866 | 6,597 | 9,598 | 7,469 | 46,512 | 1,088 | 51.2 | 4.0 | 37.3 |
| Modesto, CA | 9,279 | 145,805 | 26,936 | 19,321 | 23,873 | 3,466 | 6,214 | 7,026 | 48,189 | 3,621 | 65.3 | 3.5 | 51.3 |
| Monroe, LA | 4,968 | 71,371 | 16,213 | 6,482 | 10,512 | 3,847 | 2,794 | 2,779 | 38,933 | 1,323 | 40.1 | 8.1 | 39.9 |
| Monroe, MI | 2,311 | 35,888 | 4,557 | 6,864 | 5,068 | 780 | 1,122 | 1,672 | 46,589 | 1,085 | 57.7 | 4.1 | 40.5 |
| Montgomery, AL | 7,703 | 133,213 | 22,174 | 17,193 | 17,871 | 4,968 | 6,543 | 5,615 | 42,150 | 1,996 | 35.3 | 7.6 | 37.5 |
| Morgantown, WV | 2,865 | 52,647 | 16,090 | 4,437 | 7,140 | 881 | 2,819 | 2,644 | 50,222 | 1,684 | 34.7 | 0.5 | 38.6 |
| Morristown, TN | 2,242 | 42,756 | 5,325 | 12,726 | 6,375 | 746 | 465 | 1,642 | 38,402 | 2,455 | 51.3 | 0.4 | 37.5 |
| Mount Vernon-Anacortes, WA | 3,613 | 44,580 | 7,812 | 7,006 | 7,337 | 1,375 | 1,835 | 2,185 | 49,019 | 1,041 | 75.7 | 1.6 | 40.8 |
| Muncie, IN | 2,398 | 41,543 | 10,693 | 4,033 | 5,738 | 2,482 | 1,297 | 1,599 | 38,501 | 546 | 52.9 | 8.1 | 45.0 |
| Muskegon, MI | 3,089 | 52,799 | 9,992 | 14,465 | 7,749 | 1,062 | 1,542 | 2,322 | 43,982 | 476 | 61.1 | 2.3 | 44.0 |
| Myrtle Beach-Conway-North Myrtle Beach, SC-NC | 11,959 | 143,402 | 17,190 | 4,373 | 31,063 | 3,525 | 4,543 | 4,864 | 33,917 | 998 | 46.1 | 5.6 | 45.9 |
| Napa, CA | 4,328 | 64,170 | 9,905 | 11,708 | 6,733 | 1,428 | 1,995 | 3,679 | 57,328 | 1,866 | 74.9 | 2.8 | 32.2 |
| Naples-Marco Island, FL | 12,563 | 136,239 | 20,517 | 3,739 | 23,084 | 4,006 | 5,480 | 6,260 | 45,949 | 322 | 77.3 | 7.1 | 36.2 |
| Nashville-Davidson--Murfreesboro--Franklin, TN | 45,017 | 887,512 | 136,047 | 75,805 | 90,674 | 50,524 | 48,220 | 48,565 | 54,720 | 13,501 | 47.7 | 1.4 | 34.0 |
| New Bern, NC | 2,527 | 32,129 | 7,056 | 3,762 | 5,503 | 656 | 1,783 | 1,263 | 39,296 | 522 | 38.9 | 9.6 | 47.2 |
| New Haven-Milford, CT | 19,355 | 343,018 | 77,102 | 28,477 | 40,928 | 10,269 | 14,558 | 18,568 | 54,131 | 686 | 81.9 | 0.3 | 38.3 |
| New Orleans-Metairie, LA | 31,276 | 502,236 | 80,167 | 26,242 | 61,344 | 20,059 | 31,419 | 25,358 | 50,490 | 1,376 | 72.6 | 3.6 | 34.6 |
| New York-Newark-Jersey City, NY-NJ-PA | 565,335 | 8,448,209 | 1,654,327 | 309,880 | 881,234 | 577,002 | 758,290 | 637,183 | 75,422 | 5,894 | 78.6 | 0.7 | 38.1 |
| Nassau County-Suffolk County, NY Div 35004 | 98,847 | 1,174,668 | 247,751 | 68,493 | 158,994 | 56,979 | 81,449 | 66,012 | 56,196 | 592 | 74.5 | 0.3 | 56.1 |
| Newark, NJ-PA Div 35084 | 57,397 | 899,990 | 151,758 | 53,011 | 97,055 | 48,714 | 90,602 | 64,098 | 71,221 | 3,114 | 76.2 | 0.6 | 33.0 |
| New Brunswick-Lakewood, NJ Div 35154 | 64,738 | 1,006,436 | 170,201 | 54,928 | 129,897 | 38,951 | 114,739 | 62,648 | 62,248 | 1,767 | 83 | 1.0 | 40.5 |
| New York-Jersey City-White Plains, NY-NJ Div 35614 | 344,353 | 5,367,115 | 1,084,617 | 133,448 | 495,288 | 432,358 | 471,500 | 444,425 | 82,805 | 421 | 83.8 | 0.5 | 37.6 |
| Niles, MI | 3,549 | 55,043 | 8,982 | 9,352 | 6,782 | 1,248 | 2,692 | 2,721 | 49,429 | 872 | 57.8 | 3.3 | 48.8 |
| North Port-Sarasota-Bradenton, FL | 23,843 | 257,330 | 45,592 | 15,194 | 44,875 | 8,144 | 15,653 | 11,245 | 43,699 | 1,045 | 67.7 | 6.7 | 44.6 |
| Norwich-New London, CT | 5,924 | 107,959 | 17,922 | 15,038 | 14,801 | 1,955 | 7,994 | 5,614 | 51,998 | 823 | 63.1 | 0.6 | 43.3 |
| Ocala, FL | 7,360 | 89,254 | 18,251 | 8,221 | 17,513 | 1,989 | 2,998 | 3,326 | 37,266 | 3,985 | 81.2 | 1.2 | 43.8 |
| Ocean City, NJ | 3,807 | 27,578 | 4,322 | 626 | 5,978 | 1,138 | 1,031 | 1,218 | 44,180 | 164 | 73.8 | 0.0 | 48.5 |
| Odessa, TX | 3,869 | 67,542 | 7,321 | 4,555 | 8,498 | 1,386 | 1,640 | 3,978 | 58,901 | 275 | 68.7 | 14.5 | 28.2 |
| Ogden-Clearfield, UT | 14,871 | 199,403 | 25,349 | 32,793 | 29,152 | 8,312 | 13,443 | 8,758 | 43,923 | 3,347 | 71.2 | 6.8 | 30.5 |
| Oklahoma City, OK | 36,132 | 521,621 | 77,601 | 28,985 | 68,832 | 25,506 | 32,120 | 24,417 | 46,810 | 10,023 | 41.5 | 5.9 | 33.9 |
| Olympia-Lacey-Tumwater, WA | 6,444 | 75,644 | 15,783 | 2,833 | 13,033 | 2,945 | 4,754 | 3,408 | 45,059 | 1,200 | 77.8 | 0.6 | 35.3 |
| Omaha-Council Bluffs, NE-IA | 24,152 | 432,260 | 68,805 | 33,583 | 54,300 | 43,224 | 23,612 | 21,698 | 50,196 | 5,843 | 38.9 | 13.3 | 43.7 |
| Orlando-Kissimmee-Sanford, FL | 68,248 | 1,132,174 | 134,685 | 40,588 | 148,343 | 44,527 | 79,230 | 52,716 | 46,562 | 3,120 | 77 | 3.2 | 40.8 |
| Oshkosh-Neenah, WI | 3,632 | 90,618 | 13,057 | 23,079 | 8,558 | 3,896 | 3,940 | 4,809 | 53,074 | 957 | 49.9 | 2.2 | 37.8 |
| Owensboro, KY | 2,577 | 46,302 | 8,685 | 8,317 | 5,611 | 3,131 | 891 | 1,982 | 42,814 | 1,679 | 46.8 | 6.6 | 37.2 |
| Oxnard-Thousand Oaks-Ventura, CA | 21,519 | 265,180 | 40,627 | 25,321 | 39,683 | 11,695 | 21,439 | 15,093 | 56,915 | 2,135 | 78.4 | 2.2 | 43.0 |
| Palm Bay-Melbourne-Titusville, FL | 14,516 | 189,379 | 32,969 | 15,622 | 29,431 | 4,903 | 19,615 | 8,889 | 46,937 | 522 | 81 | 3.1 | 43.5 |
| Panama City, FL | 4,673 | 59,187 | 9,699 | 3,696 | 11,440 | 1,571 | 4,610 | 2,441 | 41,234 | 190 | 69.5 | 4.2 | 23.2 |
| Parkersburg-Vienna, WV | 1,994 | 30,022 | 6,896 | 2,408 | 5,886 | 858 | 942 | 1,093 | 36,413 | 1,137 | 36.1 | 0.4 | 35.3 |
| Pensacola-Ferry Pass-Brent, FL | 9,876 | 141,923 | 25,310 | 5,495 | 21,694 | 11,153 | 9,154 | 5,962 | 42,009 | 1,348 | 67.8 | 2.5 | 37.5 |
| Peoria, IL | 8,863 | 164,282 | 34,800 | 15,996 | 20,090 | 6,149 | 6,499 | 8,764 | 53,349 | 4,468 | 32.5 | 10.6 | 43.5 |
| Philadelphia-Camden-Wilmington, PA-NJ-DE-MD | 149,745 | 2,731,773 | 543,760 | 175,882 | 301,613 | 190,361 | 208,235 | 166,671 | 61,012 | 6,506 | 67.9 | 1.8 | 45.3 |
| Camden, NJ Div 15804 | 27,984 | 474,118 | 91,415 | 35,745 | 65,779 | 23,728 | 31,269 | 24,701 | 52,100 | 1,692 | 73.5 | 1.5 | 46.0 |
| Montgomery-Bucks-Chester Counties, PA Div 33874 | 60,659 | 1,045,528 | 177,911 | 85,753 | 119,782 | 79,789 | 88,758 | 67,615 | 64,671 | 3,035 | 66.6 | 1.2 | 47.7 |
| Philadelphia, PA Div 37964 | 41,960 | 892,893 | 219,123 | 34,798 | 76,385 | 49,465 | 62,589 | 53,965 | 60,438 | 104 | 84.6 | 0.0 | 31.8 |
| Wilmington, DE-MD-NJ Div 48864 | 19,142 | 319,234 | 55,311 | 19,586 | 39,667 | 37,379 | 25,619 | 20,390 | 63,871 | 1,675 | 63.5 | 3.2 | 41.4 |
| Phoenix-Mesa-Chandler, AZ | 101,797 | 1,849,812 | 261,795 | 114,533 | 224,531 | 149,556 | 125,570 | 97,324 | 52,613 | 2,636 | 74.3 | 8.0 | 47.6 |
| Pine Bluff, AR | 1,489 | 21,437 | 4,374 | 4,884 | 3,371 | 739 | 299 | 830 | 38,731 | 1,010 | 33.7 | 13.7 | 46.8 |
| Pittsburgh, PA | 59,442 | 1,104,086 | 210,491 | 86,398 | 123,129 | 62,289 | 73,497 | 59,250 | 53,665 | 6,318 | 40.3 | 1.3 | 40.4 |
| Pittsfield, MA | 3,780 | 52,368 | 11,544 | 5,115 | 7,747 | 1,814 | 2,698 | 2,471 | 47,179 | 475 | 50.9 | 1.3 | 42.6 |
| Pocatello, ID | 2,332 | 27,407 | 5,506 | 2,833 | 4,742 | 2,520 | 936 | 1,000 | 36,499 | 1,052 | 45.5 | 17.5 | 40.4 |
| Portland-South Portland, ME | 18,366 | 246,049 | 48,264 | 23,873 | 35,225 | 16,406 | 13,140 | 12,439 | 50,556 | 1,612 | 61 | 0.7 | 44.3 |
| Portland-Vancouver-Hillsboro, OR-WA | 71,272 | 1,083,936 | 157,359 | 117,368 | 119,407 | 49,175 | 76,305 | 64,970 | 59,939 | 11,755 | 83.4 | 0.5 | 32.5 |
| Port St. Lucie, FL | 11,908 | 128,678 | 27,280 | 6,861 | 22,117 | 2,972 | 5,803 | 5,138 | 39,931 | 1,009 | 69.4 | 8.9 | 44.3 |
| Poughkeepsie-Newburgh-Middletown, NY | 17,351 | 224,206 | 45,707 | 17,388 | 37,387 | 5,495 | 11,893 | 10,499 | 46,829 | 1,241 | 48.3 | 2.7 | 49.1 |
| Prescott Valley-Prescott, AZ | 6,002 | 62,017 | 13,859 | 3,641 | 10,497 | 1,136 | 1,918 | 2,318 | 37,377 | 850 | 73.2 | 8.5 | 46.2 |
| Providence-Warwick, RI-MA | 41,454 | 642,992 | 128,802 | 63,792 | 81,957 | 35,206 | 27,457 | 32,255 | 50,164 | 1,731 | 73.8 | 0.2 | 43.5 |
| Provo-Orem, UT | 15,023 | 237,511 | 27,279 | 18,512 | 28,460 | 6,652 | 17,749 | 10,937 | 46,047 | 2,881 | 76.7 | 3.2 | 23.9 |
| Pueblo, CO | 3,212 | 51,786 | 13,462 | 5,200 | 8,327 | 1,157 | 2,513 | 2,174 | 41,973 | 839 | 41.7 | 14.2 | 37.6 |
| Punta Gorda, FL | 4,079 | 39,400 | 8,836 | 554 | 9,543 | 1,005 | 1,481 | 1,467 | 37,221 | 306 | 66.3 | 6.2 | 37.2 |
| Racine, WI | 4,021 | 68,432 | 11,349 | 17,032 | 9,215 | 1,982 | 1,955 | 3,396 | 49,624 | 611 | 51.7 | 3.8 | 42.6 |
| Raleigh-Cary, NC | 34,900 | 552,277 | 73,379 | 25,701 | 71,079 | 28,669 | 61,752 | 31,847 | 57,666 | 2,292 | 53.5 | 3.3 | 38.3 |
| Rapid City, SD | 4,477 | 53,676 | 11,394 | 2,954 | 9,131 | 2,604 | 2,025 | 2,284 | 42,549 | 1,491 | 21.8 | 37.3 | 49.7 |

Items 104—116

# Table C. Metropolitan Areas — **Agriculture**

| Area name | Land in farms Acreage (1,000) | Percent change, 2012-2017 | Acres Average size of farm | Total irrigated (1,000) | Total cropland (1,000) | Value of land and buildings (dollars) Average per farm | Average per acre | Value of machinery and equipment, average per farm (dollars) | Value of products sold: Total (mil dol) | Average per farm (acres) | Percent from: Crops | Live-stock and poultry products | Organic farms (number) | Farms with internet access (percent) | Government payments Total ($1,000) | Percent of farms |
|---|---|---|---|---|---|---|---|---|---|---|---|---|---|---|---|---|
| | 117 | 118 | 119 | 120 | 121 | 122 | 123 | 124 | 125 | 126 | 127 | 128 | 129 | 130 | 131 | 132 |
| Mobile, AL | 215 | 16.4 | 197 | 3.1 | 60.0 | 636,089 | 3,226 | 79,393 | 122 | 111,994 | 72.0 | 28.0 | 0 | 72.2 | 3,276 | 23.0 |
| Modesto, CA | 723 | -5.9 | 200 | 380.6 | 404.7 | 3,116,617 | 15,619 | 182,695 | 2,526 | 697,690 | 53.0 | 47.0 | 35 | 82.2 | 5,006 | 7.5 |
| Monroe, LA | 425 | -2.2 | 321 | 160.7 | 264.5 | 1,090,082 | 3,393 | 139,435 | 296 | 223,632 | 51.1 | 48.9 | 3 | 74.9 | 12,544 | 29.2 |
| Monroe, MI | 210 | -2.2 | 193 | 4.3 | 194.1 | 1,190,516 | 6,156 | 150,143 | 175 | 160,833 | 96.2 | 3.8 | 0 | 83.5 | 4,555 | 43.7 |
| Montgomery, AL | 643 | 0.5 | 322 | 10.3 | 175.2 | 769,306 | 2,388 | 88,952 | 177 | 88,562 | 32.3 | 67.7 | 3 | 70.8 | 7,305 | 38.3 |
| Morgantown, WV | 205 | -6.2 | 122 | 0.1 | 66.8 | 400,195 | 3,287 | 59,253 | 21 | 12,577 | 38.0 | 62.0 | 6 | 69.3 | 377 | 6.2 |
| Morristown, TN | 226 | -5.6 | 92 | 0.7 | 87.3 | 433,554 | 4,708 | 62,859 | 59 | 24,167 | 33.5 | 66.5 | 7 | 66.4 | 3,313 | 27.9 |
| Mount Vernon-Anacortes, WA. | 98 | -8.3 | 94 | 23.5 | 65.7 | 950,360 | 10,130 | 129,653 | 287 | 275,789 | 66.6 | 33.4 | 61 | 83.0 | 407 | 7.8 |
| Muncie, IN | 168 | -4.3 | 307 | 0.6 | 157.6 | 2,038,397 | 6,633 | 188,658 | 92 | 168,773 | 92.9 | 7.1 | 0 | 83.5 | 4,044 | 56.4 |
| Muskegon, MI | 63 | -14.9 | 133 | 3.6 | 45.8 | 852,880 | 6,425 | 112,540 | 75 | 156,884 | 58.9 | 41.1 | 5 | 82.8 | 173 | 11.1 |
| Myrtle Beach-Conway-North Myrtle Beach, SC-NC | 215 | -3.5 | 216 | 2.8 | 140.3 | 946,291 | 4,388 | 125,160 | 134 | 134,187 | 64.9 | 35.1 | 8 | 73.8 | 2,625 | 32.1 |
| Napa, CA | 256 | 1.0 | 137 | 60.9 | 67.7 | 6,052,361 | 44,154 | 94,303 | 573 | 307,198 | 97.7 | 2.3 | 117 | 84.7 | 924 | 2.1 |
| Naples-Marco Island, FL | 148 | 20.1 | 461 | 37.3 | 82.7 | 2,189,705 | 4,749 | 121,438 | 190 | 588,994 | 96.8 | 3.2 | 2 | 83.5 | 141 | 5.3 |
| Nashville-Davidson--Murfreesboro--Franklin, TN | 1,707 | -2.3 | 126 | 3.9 | 692.4 | 617,818 | 4,886 | 69,490 | 472 | 34,942 | 62.0 | 38.0 | 24 | 75.7 | 13,308 | 23.3 |
| New Bern, NC | 190 | 7.6 | 365 | 4.8 | 148.5 | 1,426,209 | 3,913 | 210,902 | 309 | 591,193 | 28.9 | 71.1 | 4 | 82.2 | 3,579 | 50.2 |
| New Haven-Milford, CT | 27 | -36.3 | 39 | 1.3 | 11.4 | 922,401 | 23,490 | 67,525 | 112 | 162,720 | 95.9 | 4.1 | 26 | 78.4 | 323 | 5.2 |
| New Orleans-Metairie, LA | 263 | 14.5 | 191 | 0.5 | 85.0 | 690,802 | 3,613 | 78,734 | 61 | 44,569 | 74.0 | 26.0 | 2 | 76.3 | 68 | 1.2 |
| New York-Newark-Jersey City, NY-NJ-PA | 349 | -1.7 | 59 | 23.3 | 189.6 | 917,081 | 15,483 | 74,162 | 546 | 92,603 | 86.9 | 13.1 | 114 | 82.4 | 1,643 | 4.2 |
| Nassau County-Suffolk County, NY Div 35004 | 31 | -20.0 | 52 | 12.2 | 23.3 | 646,571 | 12,371 | 146,890 | 226 | 381,044 | 90.4 | 9.6 | 32 | 84.5 | 76 | 2.0 |
| Newark, NJ-PA Div 35084 | 201 | 0.3 | 64 | 3.3 | 99.4 | 855,845 | 13,290 | 61,297 | 136 | 43,757 | 83.1 | 16.9 | 51 | 82.4 | 894 | 4.7 |
| New Brunswick-Lakewood, NJ Div 35154 | 100 | 0.7 | 56 | 7.2 | 59.3 | 1,155,886 | 20,508 | 77,000 | 164 | 92,671 | 85.1 | 14.9 | 21 | 80.4 | 665 | 4.9 |
| New York-Jersey City-White Plains, NY-NJ Div 35614 | 18 | 2.9 | 43 | 0.5 | 7.7 | 748,109 | 17,469 | 55,131 | 20 | 48,024 | 87.9 | 12.1 | 10 | 88.1 | 8 | 1.4 |
| Niles, MI | 145 | -7.6 | 166 | 20.0 | 123.5 | 1,068,560 | 6,445 | 142,166 | 171 | 196,501 | 91.4 | 8.6 | 15 | 76.4 | 2,594 | 20.6 |
| North Port-Sarasota-Bradenton, FL | 264 | | 252 | 61.8 | 82.2 | 2,060,967 | 8,164 | 101,807 | 383 | 366,712 | 88.5 | 11.5 | 4 | 76.4 | 499 | 4.0 |
| Norwich-New London, CT | 60 | -7.7 | 73 | 0.6 | 23.9 | 837,723 | 11,467 | 67,605 | 136 | 164,989 | 55.5 | 44.5 | 22 | 88.0 | 267 | 6.1 |
| Ocala, FL | 331 | 2.9 | 83 | 13.2 | 79.3 | 922,892 | 11,114 | 47,624 | 146 | 36,501 | 41.1 | 58.9 | 5 | 82.3 | 813 | 1.4 |
| Ocean City, NJ | 8 | 10.7 | 50 | 1.4 | 3.8 | 722,280 | 14,561 | 59,085 | 10 | 59,988 | 89.2 | 10.8 | 0 | 81.1 | 0 | 1.2 |
| Odessa, TX | 558 | 30.1 | 2,029 | 0.9 | 1.9 | 2,350,073 | 1,158 | 63,273 | 3 | 12,298 | 7.6 | 92.4 | 0 | 75.6 | 57 | 1.5 |
| Ogden-Clearfield, UT | 1,610 | 2.4 | 481 | 150.0 | 360.9 | 1,125,771 | 2,341 | 91,422 | 224 | 67,057 | 46.5 | 53.5 | 19 | 78.8 | 9,132 | 14.2 |
| Oklahoma City, OK | 2,509 | 1.7 | 250 | 23.5 | 899.4 | 657,494 | 2,627 | 81,754 | 460 | 45,842 | 23.8 | 76.2 | 8 | 76.1 | 13,598 | 18.3 |
| Olympia-Lacey-Tumwater, WA | 62 | -18.8 | 52 | 6.4 | 22.1 | 616,274 | 11,880 | 58,420 | 176 | 146,742 | 32.0 | 68.0 | 34 | 86.8 | 107 | 1.9 |
| Omaha-Council Bluffs, NE-IA | 2,363 | -0.4 | 404 | 245.9 | 2,123.4 | 2,524,750 | 6,243 | 228,075 | 1,510 | 258,390 | 75.9 | 24.1 | 37 | 82.4 | 39,142 | 60.7 |
| Orlando-Kissimmee-Sanford, FL | 853 | 0.0 | 273 | 65.4 | 106.4 | 1,508,463 | 5,516 | 67,498 | 555 | 177,714 | 87.4 | 12.6 | 31 | 79.6 | 2,279 | 5.7 |
| Oshkosh-Neenah, WI | 162 | 4.2 | 169 | 0.5 | 136.2 | 1,180,421 | 6,971 | 140,933 | 122 | 127,711 | 40.0 | 60.0 | 7 | 76.6 | 3,326 | 53.1 |
| Owensboro, KY | 415 | 0.3 | 247 | 10.9 | 320.3 | 1,135,275 | 4,593 | 152,676 | 388 | 231,278 | 47.6 | 52.4 | 0 | 76.4 | 7,178 | 46.9 |
| Oxnard-Thousand Oaks-Ventura, CA | 260 | -7.5 | 122 | 98.1 | 123.4 | 3,106,351 | 25,498 | 127,982 | 1,633 | 765,008 | 99.2 | 0.8 | 128 | 80.6 | 830 | 2.5 |
| Palm Bay-Melbourne-Titusville, FL | 157 | 6.9 | 300 | 10.5 | 21.0 | 1,507,864 | 5,027 | 62,406 | 59 | 112,977 | 76.9 | 23.1 | 1 | 78.7 | 186 | 3.6 |
| Panama City, FL | 74 | 601.5 | 387 | 1.7 | 4.6 | 794,316 | 2,051 | 41,295 | 3 | 15,274 | 69.6 | 30.4 | 0 | 73.2 | 23 | 3.2 |
| Parkersburg-Vienna, WV | 131 | 4.4 | 116 | 0.0 | 34.3 | 329,622 | 2,852 | 45,212 | 9 | 7,532 | 40.6 | 59.4 | 0 | 72.1 | 307 | 3.4 |
| Pensacola-Ferry Pass-Brent, FL | 144 | -16.4 | 107 | 3.2 | 85.9 | 609,257 | 5,707 | 67,962 | 66 | 48,565 | 89.9 | 10.1 | 4 | 78.1 | 6,506 | 17.7 |
| Peoria, IL | 1,617 | -1.6 | 362 | 47.0 | 1,412.8 | 2,823,086 | 7,799 | 212,486 | 1,039 | 232,469 | 84.1 | 15.9 | 53 | 78.0 | 22,120 | 67.1 |
| Philadelphia-Camden-Wilmington, PA-NJ-DE-MD | 656 | 0.4 | 101 | 50.1 | 466.6 | 1,077,298 | 10,688 | 111,066 | 1,359 | 208,822 | 79.0 | 21.0 | 91 | 79.9 | 9,299 | 12.0 |
| Camden, NJ Div 15804 | 155 | 5.9 | 92 | 23.5 | 90.4 | 1,032,028 | 11,270 | 96,392 | 224 | 132,345 | 93.2 | 6.8 | 12 | 81.1 | 2,024 | 8.5 |
| Montgomery-Bucks-Chester Counties, PA Div 33874 | 259 | -0.2 | 85 | 2.8 | 186.2 | 1,038,608 | 12,186 | 105,618 | 824 | 271,367 | 79.2 | 20.8 | 63 | 78.8 | 2,589 | 9.0 |
| Philadelphia, PA Div 37964 | 3 | -46.7 | 26 | 0.1 | 0.9 | 490,433 | 19,110 | 47,933 | 10 | 94,442 | 97.2 | 2.8 | 2 | 95.2 | 0 | 0.0 |
| Wilmington, DE-MD-NJ Div 48864 | 239 | -1.3 | 143 | 23.8 | 189.2 | 1,229,570 | 8,600 | 139,681 | 301 | 179,851 | 67.3 | 32.7 | 14 | 79.6 | 4,686 | 21.8 |
| Phoenix-Mesa-Chandler, AZ | 1,595 | -3.4 | 605 | 412.4 | 551.3 | 2,837,548 | 4,689 | 203,720 | 2,071 | 785,687 | 37.8 | 62.2 | 22 | 85.1 | 9,253 | 12.7 |
| Pine Bluff, AR | 533 | 2.7 | 527 | 348.4 | 421.5 | 1,715,145 | 3,252 | 258,302 | 478 | 472,780 | 49.9 | 50.1 | 0 | 67.1 | 21,923 | 47.0 |
| Pittsburgh, PA | 790 | -3.3 | 125 | 2.8 | 448.9 | 681,802 | 5,450 | 94,414 | 259 | 40,969 | 60.1 | 39.9 | 42 | 74.7 | 4,448 | 14.4 |
| Pittsfield, MA | 59 | -4.9 | 123 | 0.3 | 19.1 | 943,836 | 7,644 | 72,648 | 24 | 49,453 | 42.8 | 57.2 | 16 | 85.5 | 447 | 7.8 |
| Pocatello, ID | 801 | 5.1 | 762 | 187.7 | 562.9 | 1,725,773 | 2,265 | 192,883 | 273 | 259,742 | 85.0 | 15.0 | 4 | 82.9 | 16,223 | 41.1 |
| Portland-South Portland, ME | 129 | -12.6 | 80 | 2.6 | 38.8 | 451,159 | 5,649 | 59,199 | 54 | 33,620 | 72.3 | 27.7 | 106 | 86.8 | 1,496 | 5.6 |
| Portland-Vancouver-Hillsboro, OR-WA | 597 | -7.3 | 51 | 80.8 | 327.7 | 739,269 | 14,558 | 65,709 | 1,020 | 86,784 | 83.2 | 16.8 | 162 | 87.1 | 3,852 | 4.1 |
| Port St. Lucie, FL | 380 | 13.5 | 376 | 77.0 | 118.3 | 2,008,239 | 5,337 | 96,887 | 252 | 249,942 | 90.2 | 9.8 | 5 | 76.3 | 1,434 | 8.4 |
| Poughkeepsie-Newburgh-Middletown, NY | 183 | -8.7 | 148 | 3.2 | 89.3 | 1,196,421 | 8,107 | 98,721 | 132 | 106,223 | 72.1 | 27.9 | 32 | 84.9 | 705 | 9.7 |
| Prescott Valley-Prescott, AZ | 822 | -0.3 | 967 | 7.5 | 8.0 | 1,596,867 | 1,651 | 63,906 | 36 | 42,036 | 40.1 | 59.9 | 4 | 86.4 | 115 | 1.5 |
| Providence-Warwick, RI-MA | 89 | 20.6 | 51 | 4.9 | 30.7 | 877,440 | 17,087 | 65,103 | 93 | 53,738 | 73.7 | 26.3 | 40 | 80.6 | 1,437 | 9.2 |
| Provo-Orem, UT | 568 | -3.0 | 197 | 96.4 | 174.7 | 1,042,883 | 5,286 | 79,177 | 256 | 88,948 | 43.1 | 56.9 | 6 | 81.7 | 2,612 | 8.6 |
| Pueblo, CO | 896 | 0.0 | 1,067 | 18.1 | 72.0 | 1,097,156 | 1,028 | 75,875 | 52 | 62,033 | 41.0 | 59.0 | 6 | 77.5 | 2,117 | 13.9 |
| Punta Gorda, FL | 113 | -48.1 | 368 | 11.6 | 18.5 | 2,644,176 | 7,176 | 66,706 | 44 | 143,402 | 91.8 | 8.2 | 7 | 73.9 | 329 | 5.2 |
| Racine, WI | 127 | 15.9 | 209 | 1.6 | 109.8 | 1,602,954 | 7,682 | 181,939 | 86 | 141,475 | 74.8 | 25.2 | 2 | 82.3 | 2,502 | 44.5 |
| Raleigh-Cary, NC | 368 | -7.0 | 161 | 5.1 | 218.9 | 938,272 | 5,840 | 103,395 | 390 | 170,103 | 61.2 | 38.8 | 1 | 77.6 | 3,070 | 23.5 |
| Rapid City, SD | 3,145 | 1.2 | 2,110 | 16.0 | 602.0 | 2,088,998 | 990 | 150,158 | 160 | 107,057 | 17.7 | 82.3 | 3 | 80.9 | 17,614 | 48.1 |

# Table C. Metropolitan Areas — Water Use, Wholesale Trade, Retail Trade, and Real Estate

| Area name | Water use, 2015 | | Wholesale Trade[1], 2017 | | | | Retail Trade[2], 2017 | | | | Real estate and rental and leasing,[2] 2017 | | | |
|---|---|---|---|---|---|---|---|---|---|---|---|---|---|---|
| | Public supply water withdrawn (mil gal/day) | Public supply gallons withdrawn per person per day | Number of establishments | Number of employees | Sales (mil dol) | Average payroll (mil dol) | Number of establishments | Number of employees | Sales (mil dol) | Average payroll (mil dol) | Number of establishments | Number of employees | Sales (mil dol) | Average payroll (mil dol) |
| | 133 | 134 | 135 | 136 | 137 | 138 | 139 | 140 | 141 | 142 | 143 | 144 | 145 | 146 |
| Mobile, AL................. | 68.91 | 159.4 | 515 | 6,179 | 3,524.8 | 321.7 | 1,525 | 20,598 | 5,678.4 | 533.3 | 438 | 2,424 | 565.7 | 98.4 |
| Modesto, CA................. | 80.48 | 149.5 | 406 | 6,921 | 6,198.1 | 401.0 | 1,462 | 24,398 | 7,486.4 | 680.1 | 481 | 2,047 | 505.1 | 85.9 |
| Monroe, LA................. | 32.05 | 155.9 | 210 | 2,458 | 1,920.9 | 116.4 | 813 | 10,905 | 2,782.3 | 265.7 | 254 | 1,380 | 279.2 | 51.6 |
| Monroe, MI................. | 9.27 | 62.0 | D | D | D | D | 358 | 5,252 | 1,705.6 | 133.1 | 70 | 273 | 49.8 | 8.7 |
| Montgomery, AL................. | 44.78 | 119.8 | 298 | 5,857 | 4,544.7 | 322.1 | 1,319 | 18,404 | 4,965.9 | 460.5 | 366 | 1,825 | 390.6 | 76.0 |
| Morgantown, WV................. | 13.08 | 94.7 | 66 | 428 | 392.7 | 17.6 | 465 | 7,011 | 1,971.6 | 165.3 | 161 | 711 | 139.0 | 24.4 |
| Morristown, TN................. | 14.46 | 103.7 | 75 | 1,051 | 895.4 | 48.1 | 453 | 6,404 | 1,870.6 | 163.7 | 87 | 219 | 46.9 | 6.4 |
| Mount Vernon-Anacortes, WA.. | 20.31 | 166.7 | 119 | 1,463 | 832.8 | 76.5 | 538 | 7,589 | 2,272.1 | 235.1 | 171 | 476 | 120.7 | 16.6 |
| Muncie, IN................. | 9.58 | 82.0 | 89 | 829 | 520.8 | 35.7 | 430 | 5,983 | 1,662.1 | 144.3 | 104 | 430 | 91.5 | 16.7 |
| Muskegon, MI................. | 16.08 | 93.1 | D | D | D | 62.3 | 564 | 7,930 | 2,191.2 | 195.3 | 87 | 599 | 78.3 | 18.8 |
| Myrtle Beach-Conway-North Myrtle Beach, SC-NC..... | 52.93 | 122.5 | 305 | 3,032 | 1,324.8 | 136.7 | 2,153 | 29,668 | 8,016.6 | 726.8 | 865 | 5,501 | 892.1 | 189.6 |
| Napa, CA................. | 15.28 | 107.3 | D | D | D | D | 497 | 6,742 | 2,005.4 | 219.8 | 219 | 1,041 | 256.2 | 50.0 |
| Naples-Marco Island, FL......... | 51.82 | 145.0 | 309 | 3,273 | 3,862.1 | 269.5 | 1,499 | 22,521 | 7,302.9 | 693.1 | 1,189 | 3,758 | 863.4 | 184.6 |
| Nashville-Davidson--Murfreesboro--Franklin, TN.......... | 243.19 | 134.7 | 1,827 | 33,178 | 46,467.0 | 2,147.2 | 6,098 | 91,958 | 27,714.6 | 2,583.6 | 2,131 | 13,835 | 5,272.6 | 766.4 |
| New Bern, NC................. | 12.73 | 100.8 | 76 | 808 | 1,038.0 | 37.1 | 441 | 5,301 | 1,464.2 | 133.0 | 131 | 373 | 58.6 | 11.3 |
| New Haven-Milford, CT.......... | 54.49 | 63.4 | 880 | 16,151 | 12,689.6 | 1,516.0 | 2,870 | 42,459 | 12,606.7 | 1,225.2 | 753 | 4,825 | 1,990.9 | 210.9 |
| New Orleans-Metairie, LA ...... | 261.17 | 206.8 | 1,362 | 18,990 | 22,368.7 | 1,122.3 | 4,483 | 64,251 | 17,973.8 | 1,690.2 | 1,483 | 8,358 | 2,132.0 | 383.6 |
| New York-Newark-Jersey City, NY-NJ-PA.. | 1,497.29 | 76.7 | 31,188 | 425,473 | 546,830.9 | 33,671.9 | 76,902 | 923,874 | 304,381.5 | 28,460.3 | 35,261 | 207,059 | 80,569.3 | 12,715.6 |
| Nassau County-Suffolk County, NY Div 35004 .... | 440.59 | 153.9 | 5,427 | 68,645 | 62,656.8 | 4,478.8 | 12,561 | 162,294 | 54,642.3 | 5,051.1 | 4,455 | 18,871 | 6,496.7 | 1,074.5 |
| Newark, NJ-PA Div 35084 .... | 391.66 | 179.8 | 2,919 | 54,276 | 63,713.9 | 6,386.2 | 7,424 | 100,823 | 32,991.9 | 3,057.2 | 2,436 | 17,521 | 6,798.3 | 1,023.4 |
| New Brunswick-Lakewood, NJ Div 35154 ...... | 165.41 | 69.2 | 3,192 | 62,770 | 78,824.9 | 4,999.1 | 8,287 | 131,760 | 42,601.5 | 3,814.3 | 2,560 | 15,990 | 5,422.4 | 884.7 |
| New York-Jersey City-White Plains, NY-NJ Div 35614 .......... | 499.63 | 41.4 | 19,650 | 239,782 | 341,635.3 | 17,807.7 | 48,630 | 528,997 | 174,145.8 | 16,537.7 | 25,810 | 154,677 | 61,851.9 | 9,733.0 |
| Niles, MI................. | 12.42 | 80.3 | D | D | D | 78.1 | 553 | 7,053 | 1,781.5 | 171.4 | 156 | 656 | 277.4 | 23.3 |
| North Port-Sarasota-Bradenton, FL.. | 63.00 | 81.9 | 760 | 7,355 | 4,958.3 | 398.9 | 3,054 | 43,108 | 12,819.2 | 1,188.8 | 1,663 | 5,459 | 1,257.7 | 233.8 |
| Norwich-New London, CT ...... | 8.11 | 29.8 | 149 | 2,638 | 1,679.6 | 176.0 | 1,015 | 15,140 | 4,030.8 | 422.3 | 228 | 818 | 236.0 | 32.3 |
| Ocala, FL................. | 27.71 | 80.7 | 280 | 3,742 | 2,243.1 | 188.1 | 1,152 | 16,765 | 5,080.9 | 454.7 | 416 | 1,614 | 301.9 | 47.7 |
| Ocean City, NJ................. | 13.61 | 143.7 | D | D | D | D | 639 | 7,307 | 1,935.5 | 182.6 | 202 | 779 | 217.0 | 32.5 |
| Odessa, TX................. | 0.64 | 4.0 | 301 | 4,598 | 3,087.9 | 298.2 | 460 | 8,193 | 3,004.4 | 270.5 | 188 | 2,007 | 788.5 | 126.2 |
| Ogden-Clearfield, UT................. | 91.48 | 142.3 | 427 | 5,499 | 5,785.4 | 276.1 | 1,896 | 28,995 | 8,629.4 | 775.4 | 912 | 2,350 | 506.5 | 84.2 |
| Oklahoma City, OK................. | 139.34 | 102.6 | 1,498 | 20,990 | 21,106.3 | 1,204.7 | 4,425 | 67,362 | 20,783.3 | 1,845.8 | 2,022 | 9,686 | 2,408.4 | 461.7 |
| Olympia-Lacey-Tumwater, WA | 22.58 | 83.8 | 176 | 1,902 | 1,320.6 | 102.1 | 796 | 13,008 | 3,958.8 | 386.6 | 340 | 1,133 | 302.6 | 43.6 |
| Omaha-Council Bluffs, NE-IA.. | 179.36 | 196.0 | 1,048 | 15,694 | 19,179.8 | 1,041.2 | 2,648 | 53,252 | 15,986.5 | 1,444.1 | 1,296 | 7,543 | 1,621.1 | 348.3 |
| Orlando-Kissimmee-Sanford, FL.. | 370.98 | 155.4 | 2,688 | 35,041 | 40,665.2 | 2,278.9 | 8,756 | 145,107 | 45,062.7 | 3,814.5 | 4,807 | 34,984 | 11,083.6 | 1,796.1 |
| Oshkosh-Neenah, WI.............. | 13.99 | 82.5 | D | D | D | 114.4 | 487 | 8,865 | 2,335.2 | 214.8 | 129 | 752 | 198.2 | 29.0 |
| Owensboro, KY................. | 14.58 | 124.1 | 97 | 1,267 | 1,008.3 | 65.5 | 448 | 5,876 | 1,618.5 | 141.3 | 97 | 549 | 90.2 | 15.7 |
| Oxnard-Thousand Oaks-Ventura, CA.. | 130.02 | 152.9 | 1,028 | 15,216 | 24,894.7 | 1,221.6 | 2,569 | 40,494 | 13,115.6 | 1,238.4 | 1,204 | 4,751 | 1,300.4 | 235.6 |
| Palm Bay-Melbourne-Titusville, FL.. | 31.45 | 55.4 | 454 | 3,798 | 2,494.8 | 207.9 | 2,026 | 29,154 | 7,884.4 | 747.1 | 860 | 2,623 | 512.5 | 92.3 |
| Panama City, FL.. | 47.62 | 262.2 | 155 | 1,728 | 746.4 | 81.2 | 826 | 11,802 | 3,219.3 | 305.2 | 313 | 1,138 | 272.3 | 37.7 |
| Parkersburg-Vienna, WV......... | 7.97 | 86.3 | D | D | D | D | 360 | 5,973 | 1,559.5 | 146.3 | D | D | D | D |
| Pensacola-Ferry Pass-Brent, FL.. | 52.48 | 109.8 | 337 | 3,078 | 1,974.5 | 156.9 | 1,454 | 21,088 | 6,042.6 | 549.9 | 576 | 2,120 | 504.8 | 74.0 |
| Peoria, IL................. | 50.77 | 122.7 | 329 | 5,263 | 3,644.3 | 321.5 | 1,316 | 21,262 | 5,542.0 | 538.7 | 314 | 1,452 | 259.6 | 54.2 |
| Philadelphia-Camden-Wilmington, PA-NJ-DE-MD.... | 657.43 | 108.3 | 5,833 | 101,779 | 100,537.2 | 7,216.1 | 19,413 | 312,616 | 98,646.8 | 8,549.0 | 6,066 | 41,640 | 16,396.5 | 2,469.1 |
| Camden, NJ Div 15804......... | 114.55 | 91.4 | 1,292 | 28,268 | 31,164.3 | 1,793.7 | 3,950 | 67,733 | 19,847.0 | 1,797.2 | 1,028 | 8,034 | 2,422.4 | 465.7 |
| Montgomery-Bucks-Chester Counties, PA Div 33874.... | 204.10 | 104.0 | 2,986 | 50,054 | 46,884.2 | 3,837.7 | 6,900 | 124,621 | 43,813.6 | 3,667.3 | 2,318 | 15,325 | 5,257.9 | 960.2 |
| Philadelphia, PA Div 37964 ... | 270.33 | 126.8 | 1,519 | 23,457 | 20,795.2 | 1,584.8 | 6,185 | 79,310 | 22,559.1 | 1,996.8 | 1,769 | 13,667 | 3,984.8 | 811.4 |
| Wilmington, DE-MD-NJ Div 48864.. | 68.45 | 94.6 | 36 | D | 1,693.4 | D | 2,378 | 40,952 | 12,427.2 | 1,087.7 | 951 | 4,614 | 4,731.3 | 231.8 |
| Phoenix-Mesa-Chandler, AZ ... | 839.78 | 183.6 | 4,065 | 65,234 | 62,385.0 | 4,121.9 | 11,378 | 224,360 | 76,911.2 | 6,833.3 | 7,158 | 37,383 | 10,784.1 | 1,874.9 |
| Pine Bluff, AR................. | 12.89 | 137.6 | 57 | 465 | 314.3 | 21.8 | 326 | 3,399 | 859.8 | 82.5 | 56 | 202 | 34.5 | 6.5 |
| Pittsburgh, PA................. | 324.40 | 137.9 | 2,390 | 36,595 | 42,436.1 | 2,259.8 | 7,919 | 126,605 | 71,904.9 | 3,317.9 | 2,163 | 14,204 | 3,959.2 | 655.7 |
| Pittsfield, MA................. | 14.33 | 112.1 | 98 | 1,276 | 491.8 | 66.1 | 653 | 8,151 | 2,049.7 | 221.0 | 116 | 496 | 92.7 | 18.7 |
| Pocatello, ID................. | 19.06 | 208.6 | 96 | 1,079 | 817.7 | 46.9 | 338 | 4,776 | 1,322.1 | 117.7 | 108 | 270 | 57.0 | 7.6 |
| Portland-South Portland, ME.... | 42.09 | 80.0 | 590 | 6,995 | 6,307.0 | 404.8 | 2,451 | 34,119 | 10,265.6 | 962.7 | 942 | 3,733 | 799.0 | 171.2 |
| Portland-Vancouver-Hillsboro, OR-WA.. | 260.27 | 108.9 | 3,013 | 52,355 | 50,638.4 | 3,554.0 | 7,352 | 120,923 | 36,653.3 | 3,649.8 | 4,201 | 21,326 | 5,043.9 | 996.7 |
| Port St. Lucie, FL................. | 44.98 | 98.9 | 381 | 2,848 | 1,432.5 | 140.8 | 1,514 | 22,235 | 6,884.6 | 607.2 | 662 | 3,122 | 670.2 | 112.0 |
| Poughkeepsie-Newburgh-Middletown, NY.. | 52.01 | 77.2 | 661 | 9,064 | 9,623.4 | 483.9 | 2,661 | 38,850 | 10,933.3 | 1,013.9 | 712 | 3,015 | 771.5 | 125.4 |
| Prescott Valley-Prescott, AZ... | 21.14 | 95.1 | 141 | 1,447 | 952.8 | 72.9 | 815 | 10,820 | 2,952.7 | 294.1 | 414 | 1,075 | 245.0 | 42.3 |
| Providence-Warwick, RI-MA.... | 124.06 | 76.9 | 1,617 | 29,125 | 26,968.7 | 1,832.2 | 5,870 | 82,444 | 23,459.9 | 2,387.6 | 1,563 | 7,413 | 1,790.9 | 323.6 |
| Provo-Orem, UT................. | 132.26 | 225.8 | 435 | 7,460 | 4,075.9 | 451.1 | 1,948 | 27,277 | 8,080.2 | 750.3 | D | D | D | D |
| Pueblo, CO................. | 41.29 | 252.4 | 94 | 1,037 | 684.5 | 57.8 | 506 | 8,124 | 2,144.4 | 214.1 | 148 | 658 | 116.3 | 19.4 |
| Punta Gorda, FL................. | 7.51 | 43.4 | 102 | 1,037 | 578.8 | 53.3 | 599 | 9,574 | 2,757.2 | 254.8 | 304 | 963 | 208.8 | 29.7 |
| Racine, WI................. | 19.14 | 98.1 | 179 | 2,684 | 2,227.3 | 150.4 | 559 | 9,038 | 2,409.5 | 234.2 | 117 | 545 | 98.9 | 18.9 |
| Raleigh-Cary, NC................. | 73.38 | 57.6 | 1,318 | 24,121 | 31,532.3 | 1,884.1 | 4,194 | 68,770 | 20,699.1 | 1,828.2 | 1,888 | 9,555 | 2,707.5 | 528.6 |
| Rapid City, SD................. | 12.04 | 88.7 | 143 | 1,649 | 1,008.4 | 86.2 | 668 | 9,862 | 3,009.8 | 273.0 | 256 | 750 | 190.6 | 24.9 |

1. Merchant wholesalers, except manufacturers' sales branches and offices. 2. Employer establishments.

| Area name | Professional, scientific, and technical services, 2017 | | | | Manufacturing, 2017 | | | | Accommodation and food services, 2017 | | | |
|---|---|---|---|---|---|---|---|---|---|---|---|---|
| | Number of establishments | Number of employees | Sales (mil dol) | Average payroll (mil dol) | Number of establishments | Number of employees | Receipts (mil dol) | Annual payroll (mil dol) | Number of establishments | Number of employees | Receipts (mil dol) | Annual payroll (mil dol) |
| | 147 | 148 | 149 | 150 | 151 | 152 | 153 | 154 | 155 | 156 | 157 | 158 |
| Mobile, AL | 10 | 29 | 7.4 | 2.2 | 337 | 19,403 | 16,284.0 | 1,343.2 | 705 | 15,002 | 791.9 | 218.0 |
| Modesto, CA | 682 | 5,805 | 761.7 | 314.5 | 406 | 19,982 | 12,685.7 | 1,202.1 | 926 | 16,562 | 971.6 | 273.6 |
| Monroe, LA | 463 | 2,813 | 425.1 | 144.6 | D | D | D | 323.3 | 30 | D | 24.0 | D |
| Monroe, MI | D | D | D | D | 127 | 6,664 | 2,660.6 | 337.7 | 275 | 4,478 | 216.9 | 60.9 |
| Montgomery, AL | 155 | 717 | 121.8 | 32.7 | 260 | 17,272 | 14,003.8 | 939.7 | 620 | 14,446 | 1,316.5 | 245.5 |
| Morgantown, WV | D | D | D | D | 74 | 4,571 | 3,199.2 | 356.5 | 365 | 7,166 | 327.8 | 96.5 |
| Morristown, TN | 40 | 130 | 14.9 | 4.3 | 148 | 12,899 | 5,004.6 | 579.3 | 213 | 3,821 | 185.2 | 52.9 |
| Mount Vernon-Anacortes, WA. | D | D | D | D | 189 | 6,711 | 11,187.8 | 468.7 | 352 | 5,518 | 411.7 | 125.3 |
| Muncie, IN | D | D | D | D | 118 | 4,234 | 1,723.0 | 219.5 | 220 | 5,143 | 225.4 | 66.9 |
| Muskegon, MI | D | D | D | D | 267 | 13,190 | 4,003.7 | 712.1 | 351 | 6,148 | 288.0 | 86.0 |
| Myrtle Beach-Conway-North Myrtle Beach, SC-NC | D | D | D | D | 223 | 4,042 | 1,082.1 | 206.8 | 1,632 | 35,235 | 2,556.2 | 666.4 |
| Napa, CA | 430 | 2,044 | 363.0 | 130.9 | 492 | 13,127 | 5,873.9 | 890.0 | 408 | 11,705 | 1,068.5 | 339.4 |
| Naples-Marco Island, FL | 1,602 | 5,768 | 1,103.6 | 381.5 | 209 | 3,397 | 797.0 | 173.7 | 881 | 24,719 | 1,957.8 | 527.6 |
| Nashville-Davidson--Murfreesboro--Franklin, TN | 3,639 | 43,373 | 7,909.5 | 3,306.1 | 1,512 | 73,441 | 42,845.7 | 3,967.3 | 4,207 | 98,116 | 6,648.0 | 1,854.8 |
| New Bern, NC | 23 | 64 | 7.7 | 2.3 | 84 | 3,526 | 1,448.4 | 184.2 | 239 | 4,258 | 235.6 | 62.9 |
| New Haven-Milford, CT | 1,814 | 14,470 | 2,620.3 | 1,218.0 | 1,048 | 29,004 | 10,764.2 | 1,792.6 | 2,047 | 29,719 | 1,854.2 | 534.1 |
| New Orleans-Metairie, LA | 147 | 1,723 | 239.9 | 108.1 | D | 25,628 | D | 2,117.0 | 3,572 | 80,244 | 6,016.3 | 1,643.3 |
| New York-Newark-Jersey City, NY-NJ-PA | 63,412 | 717,551 | 192,112.8 | 70,035.3 | 14,477 | 310,527 | 124,174.4 | 18,682.7 | 52,169 | 767,183 | 66,206.2 | 19,133.1 |
| Nassau County-Suffolk County, NY Div 35004 | 12,754 | 76,397 | 14,410.7 | 5,301.1 | 2,842 | 65,194 | 21,718.2 | 3,828.9 | 7,718 | 109,408 | 8,003.4 | 2,333.7 |
| Newark, NJ-PA Div 35084 | 6,203 | 86,984 | 17,858.9 | 8,652.1 | 1,984 | 50,572 | 27,074.7 | 3,202.3 | 4,985 | 71,941 | 5,302.9 | 1,448.3 |
| New Brunswick-Lakewood, NJ Div 35154 | 8,253 | 104,529 | 21,457.8 | 9,515.9 | 1,633 | 56,281 | 23,217.2 | 3,837.1 | 5,712 | 78,013 | 5,645.8 | 1,489.5 |
| New York-Jersey City-White Plains, NY-NJ Div 35614 | 36,202 | 449,641 | 138,385.4 | 46,566.2 | 8,018 | 138,480 | 52,164.2 | 7,814.3 | 33,754 | 507,821 | 47,254.0 | 13,861.6 |
| Niles, MI | D | D | D | D | 281 | 8,819 | 2,277.9 | 455.5 | 379 | 6,404 | 331.2 | 101.2 |
| North Port-Sarasota-Bradenton, FL | D | D | D | D | 628 | 14,185 | 3,777.6 | 756.7 | 1,636 | 32,440 | 1,917.9 | 578.5 |
| Norwich-New London, CT | 527 | 8,313 | 973.2 | 972.2 | 175 | 14,024 | 6,228.6 | 1,105.9 | 750 | 25,852 | 2,816.6 | 722.2 |
| Ocala, FL | D | D | D | D | 188 | 7,109 | 3,330.8 | 337.8 | 516 | 9,478 | 531.8 | 146.7 |
| Ocean City, NJ | D | D | D | D | 76 | 592 | 110.1 | 26.9 | 931 | 6,065 | 757.0 | 206.6 |
| Odessa, TX | D | D | D | D | 224 | 4,128 | 1,379.8 | 243.9 | 284 | 6,669 | 462.1 | 116.5 |
| Ogden-Clearfield, UT | 98 | 322 | 42.5 | 14.2 | 628 | 31,145 | 14,543.3 | 1,853.1 | 994 | 19,054 | 954.7 | 268.8 |
| Oklahoma City, OK | 558 | 1,841 | 279.6 | 94.1 | 1,025 | 28,441 | 10,608.4 | 1,398.6 | 3,174 | 63,783 | 3,497.1 | 968.8 |
| Olympia-Lacey-Tumwater, WA | 638 | 4,426 | 594.6 | 260.8 | 162 | 2,779 | 1,029.7 | 136.7 | 561 | 9,634 | 574.5 | 175.9 |
| Omaha-Council Bluffs, NE-IA | 2,035 | 20,039 | 3,345.9 | 1,322.3 | 634 | 31,207 | 18,159.4 | 1,586.5 | 2,008 | 40,812 | 2,529.1 | 680.7 |
| Orlando-Kissimmee-Sanford, FL | 8,881 | 67,714 | 10,861.4 | 4,613.2 | 1,469 | 36,717 | 11,560.5 | 2,144.5 | 5,334 | 172,151 | 14,317.5 | 3,685.7 |
| Oshkosh-Neenah, WI | D | D | D | D | 291 | 21,965 | 9,101.8 | 1,389.8 | 399 | 6,740 | 296.1 | 87.2 |
| Owensboro, KY | 16 | 73 | 6.6 | 2.7 | 121 | 7,538 | 5,441.0 | 470.0 | 13 | 72 | 3.4 | 0.9 |
| Oxnard-Thousand Oaks-Ventura, CA | 2,816 | 23,034 | 3,556.9 | 2,558.7 | 872 | 25,626 | 9,751.9 | 1,585.0 | 1,718 | 32,483 | 2,147.4 | 629.8 |
| Palm Bay-Melbourne-Titusville, FL | 1,920 | 19,218 | 3,795.1 | 1,440.5 | 447 | 19,579 | 5,939.2 | 1,622.5 | 1,202 | 23,254 | 1,343.2 | 384.3 |
| Panama City, FL | D | D | D | D | 106 | 3,728 | 1,488.1 | 208.0 | 515 | 11,608 | 757.4 | 215.8 |
| Parkersburg-Vienna, WV | 5 | 8 | 0.5 | 0.1 | D | D | D | D | D | D | D | D |
| Pensacola-Ferry Pass-Brent, FL | 324 | 1,687 | 258.9 | 101.6 | 211 | 4,320 | 3,414.7 | 297.0 | 807 | 18,949 | 1,044.2 | 288.9 |
| Peoria, IL | 293 | 1,974 | 223.2 | 97.3 | 322 | D | 6,068.2 | 863.6 | 951 | 16,034 | 833.1 | 240.0 |
| Philadelphia-Camden-Wilmington, PA-NJ-DE-MD | 11,681 | 124,620 | 25,861.8 | 10,648.2 | 4,897 | 170,097 | 92,492.1 | 11,186.0 | 13,637 | 223,664 | 14,234.5 | 3,978.5 |
| Camden, NJ Div 15804 | 3,152 | 31,071 | 5,723.6 | 2,172.7 | 951 | 35,242 | 17,811.1 | 2,309.0 | 2,523 | 41,094 | 2,377.8 | 654.3 |
| Montgomery-Bucks-Chester Counties, PA Div 33874 | 8,443 | 92,972 | 20,054.3 | 8,444.6 | 2,462 | 81,506 | 36,337.5 | 5,241.0 | 4,528 | 74,052 | 4,630.1 | 1,276.4 |
| Philadelphia, PA Div 37964 | D | D | D | D | 1,042 | 34,407 | 24,696.6 | 2,251.4 | 5,098 | 80,492 | 5,560.9 | 1,580.6 |
| Wilmington, DE-MD-NJ Div 48864 | 86 | 577 | 83.9 | 30.9 | 442 | 18,942 | 13,646.9 | 1,384.6 | 1,488 | 28,026 | 1,665.8 | 467.2 |
| Phoenix-Mesa-Chandler, AZ | D | D | D | D | 3,065 | 103,689 | 41,122.9 | 6,269.0 | 8,373 | 202,441 | 13,558.4 | 3,860.7 |
| Pine Bluff, AR | D | D | D | D | 65 | 5,207 | 1,679.3 | 252.2 | D | D | D | 23.9 |
| Pittsburgh, PA | D | D | 12,056.5 | D | 2,380 | 84,135 | 33,824.7 | 4,950.3 | 5,647 | 101,500 | 5,604.1 | 1,603.2 |
| Pittsfield, MA | D | D | D | D | 142 | 5,115 | 1,554.7 | 324.2 | 516 | 7,311 | 487.9 | 156.4 |
| Pocatello, ID | 10 | 28 | 3.4 | 1.1 | D | D | D | D | 215 | 3,523 | 179.6 | 48.0 |
| Portland-South Portland, ME | 424 | 1,756 | 260.5 | 96.4 | D | D | D | D | 1,989 | 27,304 | 2,026.3 | 612.1 |
| Portland-Vancouver-Hillsboro, OR-WA | 6,701 | 53,639 | 10,191.4 | 4,150.3 | 3,248 | 113,737 | 42,645.4 | 8,000.8 | 6,526 | 107,555 | 7,047.3 | 2,124.4 |
| Port St. Lucie, FL | 1,312 | 6,156 | 829.0 | 287.6 | 345 | 5,658 | 1,566.2 | 284.0 | 828 | 15,683 | 896.4 | 254.2 |
| Poughkeepsie-Newburgh-Middletown, NY | D | D | D | D | 490 | 13,673 | 4,978.4 | 862.7 | 1,783 | 21,524 | 1,370.2 | 397.5 |
| Prescott Valley-Prescott, AZ | 565 | 1,909 | 235.6 | 87.9 | 190 | 3,261 | 716.3 | 160.0 | 575 | 9,695 | 563.9 | 187.5 |
| Providence-Warwick, RI-MA | 2,016 | 14,277 | 2,015.2 | 797.7 | 1,957 | 67,080 | 21,326.0 | 4,033.2 | 4,433 | 72,327 | 4,816.3 | 1,386.6 |
| Provo-Orem, UT | 2,162 | 15,564 | 2,499.7 | 989.7 | 562 | 16,471 | 5,930.8 | 912.7 | 868 | 17,510 | 941.0 | 261.5 |
| Pueblo, CO | D | D | D | D | 94 | 4,380 | 2,103.4 | 256.3 | 336 | 5,761 | 289.7 | 84.7 |
| Punta Gorda, FL | D | D | D | D | 78 | 512 | 101.9 | 21.2 | 294 | 6,067 | 336.5 | 96.2 |
| Racine, WI | 289 | 1,953 | 288.6 | 118.4 | 297 | 15,883 | 5,623.1 | 1,006.9 | 412 | 6,540 | 319.4 | 91.5 |
| Raleigh-Cary, NC | 5,268 | 60,308 | 12,556.0 | 4,997.6 | 767 | 22,427 | 16,858.7 | 1,362.5 | 2,774 | 60,533 | 3,560.9 | 995.6 |
| Rapid City, SD | D | D | D | D | 162 | 2,509 | 617.9 | 112.9 | 451 | 7,666 | 490.0 | 139.1 |

Table C. Metropolitan Areas —

# Health Care and Social Assistance, Other Services, Nonemployer Businesses, and Residential Construction

| Area name | Health care and social assistance, 2017 | | | | Other services, 2017 | | | | Nonemployer businesses, 2018 | | Value of residential construction authorized by building permits, 2020 | |
|---|---|---|---|---|---|---|---|---|---|---|---|---|
| | Number of establishments | Number of employees | Receipts (mil dol) | Annual payroll (mil dol) | Number of establishments | Number of employees | Receipts (mil dol) | Annual payroll (mil dol) | Number | Receipts (mil dol) | New construction ($1,000) | Number of housing units |
| | 159 | 160 | 161 | 162 | 163 | 164 | 165 | 166 | 167 | 168 | 169 | 170 |
| Mobile, AL | 809 | 21,605 | 2,670.4 | 996.8 | 547 | 4,061 | 453 | 121 | 31,470 | 1,212.1 | 196,094 | 937 |
| Modesto, CA | 1,150 | 26,140 | 4,257.6 | 1,674.9 | 691 | 4,587 | 592 | 172 | 30,181 | 1,594.8 | 130,340 | 552 |
| | | | | | | | | | 16,047 | 645.4 | 96,066 | 466 |
| Monroe, LA | 774 | 17,038 | 1,679.2 | 618.0 | 33 | 116 | 9 | 2 | 8,448 | 378.0 | 64,956 | 367 |
| Monroe, MI | 297 | 4,744 | 435.4 | 188.8 | 158 | 701 | 78 | 21 | 25,138 | 1,075.5 | 223,305 | 979 |
| Montgomery, AL | 788 | 20,474 | 2,300.3 | 956.2 | 498 | 2,991 | 441 | 117 | 7,585 | 336.9 | 7,161 | 25 |
| Morgantown, WV | 330 | 17,219 | 2,460.3 | 778.9 | 194 | 1,394 | 262 | 44 | 8,343 | 414.1 | 118,200 | 548 |
| Morristown, TN | 255 | 5,714 | 561.9 | 212.2 | 143 | 785 | 69 | 22 | 8,065 | 414.5 | 116,666 | 561 |
| Mount Vernon-Anacortes, WA. | 355 | 8,148 | 922.3 | 402.4 | 263 | 1,223 | 140 | 42 | 5,665 | 214.4 | 35,868 | 171 |
| Muncie, IN | 379 | 10,377 | 1,061.6 | 418.1 | 169 | 1,102 | 134 | 32 | 9,534 | 391.5 | 55,179 | 268 |
| Muskegon, MI | 370 | 10,922 | 1,416.0 | 541.0 | 248 | 1,332 | 133 | 36 | | | | |
| Myrtle Beach-Conway-North Myrtle Beach, SC-NC.... | 954 | 16,093 | 2,095.2 | 726.6 | 699 | 3,801 | 407 | 103 | 38,763 | 1,840.7 | 1,724,276 | 9,139 |
| Napa, CA | 433 | 10,558 | 1,653.4 | 671.2 | 264 | 1,645 | 204 | 54 | 12,334 | 792.1 | 159,548 | 770 |
| Naples-Marco Island, FL | 1,143 | 20,535 | 2,704.8 | 980.5 | 1,011 | 6,309 | 663 | 192 | 43,851 | 2,768.9 | 1,471,129 | 4,473 |
| Nashville-Davidson--Murfreesboro--Franklin, TN | 4,750 | 131,350 | 18,899.5 | 7,340.7 | 2,828 | 24,708 | 2,895 | 883 | 187,108 | 10,397.9 | 5,183,315 | 27,158 |
| New Bern, NC | 303 | 8,358 | 865.1 | 360.7 | 147 | 603 | 62 | 17 | 7,846 | 309.7 | 131,660 | 735 |
| New Haven-Milford, CT | 2,585 | 76,203 | 9,164.6 | 3,610.4 | 1,772 | 10,522 | 1,230 | 348 | 63,139 | 3,324.7 | 139,928 | 1,373 |
| New Orleans-Metairie, LA | 3,493 | 78,207 | 10,146.2 | 3,813.7 | 1,951 | 12,853 | 1,715 | 449 | 126,876 | 5,805.9 | 932,569 | 4,348 |
| New York-Newark-Jersey City, NY-NJ-PA | 61,986 | 1,607,087 | 197,257.4 | 80,838.4 | 50,412 | 312,707 | 54,285 | 12,459 | 1,974,310 | 115,369.7 | 7,744,635 | 54,835 |
| Nassau County-Suffolk County, NY Div 35004 ... | 10,868 | 228,324 | 28,660.4 | 12,195.2 | 8,566 | 45,888 | 5,173 | 1,420 | 287,069 | 18,620.4 | 932,327 | 1,928 |
| Newark, NJ-PA Div 35084 .... | 6,670 | 152,580 | 19,122.8 | 7,672.3 | 4,938 | 31,183 | 3,670 | 1,043 | 195,368 | 11,614.5 | 848,452 | 7,643 |
| New Brunswick-Lakewood, NJ Div 35154 | 7,726 | 165,453 | 20,023.8 | 7,994.6 | 4,957 | 30,883 | 4,198 | 1,033 | 195,332 | 12,272.1 | 1,463,768 | 10,933 |
| New York-Jersey City-White Plains, NY-NJ Div 35614 | 36,722 | 1,060,730 | 129,450.4 | 52,976.3 | 31,951 | 204,753 | 41,243 | 8,964 | 1,296,541 | 72,862.8 | 4,500,089 | 34,331 |
| Niles, MI | 387 | 9,499 | 987.6 | 414.4 | 258 | 1,325 | 159 | 40 | 10,074 | 422.5 | 87,516 | 260 |
| North Port-Sarasota-Bradenton, FL | 2,552 | 44,802 | 5,709.9 | 2,026.0 | 1,671 | 8,063 | 1,018 | 243 | 79,699 | 4,144.6 | 2,341,559 | 9,489 |
| Norwich-New London, CT | 798 | 17,766 | 1,817.6 | 795.5 | 483 | 2,577 | 313 | 71 | 17,321 | 839.7 | 89,949 | 456 |
| Ocala, FL | 909 | 15,902 | 2,033.6 | 716.2 | 516 | 2,545 | 263 | 72 | 27,086 | 1,189.9 | 573,152 | 3,827 |
| Ocean City, NJ | 244 | 4,578 | 467.7 | 198.8 | 308 | 1,268 | 124 | 36 | 8,493 | 536.4 | 243,196 | 675 |
| Odessa, TX | 285 | 8,075 | 996.1 | 374.9 | 272 | 2,188 | 332 | 89 | 13,754 | 865.2 | 245,144 | 1,188 |
| Ogden-Clearfield, UT | 1,473 | 23,154 | 2,811.0 | 983.0 | 843 | 5,197 | 519 | 143 | 45,303 | 1,967.4 | 1,211,883 | 5,462 |
| Oklahoma City, OK | 4,270 | 79,468 | 11,221.1 | 3,887.6 | 2,084 | 12,543 | 1,742 | 434 | 114,673 | 5,679.2 | 1,665,298 | 7,431 |
| Olympia-Lacey-Tumwater, WA | 893 | 15,533 | 1,815.9 | 725.3 | 501 | 3,079 | 430 | 130 | 16,142 | 714.9 | 236,528 | 1,161 |
| Omaha-Council Bluffs, NE-IA .. | 2,712 | 68,611 | 8,457.0 | 3,319.1 | 1,686 | 10,564 | 2,045 | 350 | 62,905 | 2,989.2 | 905,891 | 5,473 |
| Orlando-Kissimmee-Sanford, FL | 6,613 | 127,510 | 18,014.7 | 6,236.4 | 4,184 | 28,846 | 4,750 | 932 | 270,929 | 10,989.1 | 5,478,188 | 24,499 |
| Oshkosh-Neenah, WI | 453 | 12,419 | 1,282.3 | 560.5 | 258 | 2,127 | 241 | 72 | 8,625 | 412.5 | 106,446 | 554 |
| Owensboro, KY | 362 | 9,931 | 1,100.6 | 481.7 | 146 | 1,072 | 88 | 31 | 7,013 | 344.3 | 33,446 | 348 |
| Oxnard-Thousand Oaks-Ventura, CA | 2,785 | 39,199 | 5,017.7 | 1,883.6 | 1,319 | 6,955 | 795 | 205 | 70,479 | 3,935.7 | 341,012 | 1,501 |
| Palm Bay-Melbourne-Titusville, FL | 1,600 | 32,031 | 3,775.1 | 1,369.8 | 1,023 | 4,793 | 467 | 135 | 46,350 | 1,883.5 | 1,284,437 | 4,739 |
| Panama City, FL | 549 | 10,448 | 1,288.9 | 495.7 | 319 | 1,922 | 176 | 47 | 14,652 | 735.6 | 473,592 | 2,846 |
| Parkersburg-Vienna, WV | D | D | D | D | D | D | D | D | 4,173 | 172.7 | 18,786 | 83 |
| Pensacola-Ferry Pass-Brent, FL | 1,119 | 25,833 | 3,481.5 | 1,315.0 | 618 | 3,828 | 376 | 103 | 36,006 | 1,540.4 | 884,554 | 4,391 |
| Peoria, IL | 933 | 33,858 | 4,018.6 | 1,667.9 | 610 | 5,909 | 630 | 256 | 21,264 | 839.6 | 62,937 | 272 |
| Philadelphia-Camden-Wilmington, PA-NJ-DE-MD.... | 18,016 | 518,895 | 61,725.7 | 25,143.8 | 12,061 | 80,754 | 13,637 | 2,806 | 452,633 | 24,330.1 | 2,466,072 | 16,201 |
| Camden, NJ Div 15804 | 3,617 | 88,640 | 10,461.0 | 4,333.1 | 2,272 | 14,344 | 1,335 | 416 | 80,153 | 4,291.8 | 331,145 | 3,353 |
| Montgomery-Bucks-Chester Counties, PA Div 33874. | 6,727 | 172,624 | 18,414.2 | 7,662.8 | 4,681 | 30,530 | 7,319 | 1,075 | 173,478 | 10,509.5 | 1,004,504 | 4,864 |
| Philadelphia, PA Div 37964 ... | 5,658 | 203,146 | 25,895.5 | 10,114.9 | 3,823 | 27,270 | 3,943 | 1,030 | 150,839 | 6,399.5 | 886,359 | 5,919 |
| Wilmington, DE-MD-NJ Div 48864 | 2,014 | 54,485 | 6,955.0 | 3,033.0 | 1,285 | 8,610 | 1,041 | 286 | 48,163 | 3,129.3 | 244,066 | 2,065 |
| Phoenix-Mesa-Chandler, AZ | 12,900 | 258,648 | 33,804.6 | 12,851.6 | 6,178 | 49,883 | 6,731 | 1,787 | 356,955 | 17,743.8 | 10,815,836 | 48,219 |
| Pine Bluff, AR | 226 | 4,363 | 428.2 | 179.5 | 87 | 435 | 47 | 12 | 4,505 | 175.6 | 7,369 | 73 |
| Pittsburgh, PA | 8,311 | 202,432 | 22,007.5 | 9,383.5 | 5,255 | 33,544 | 4,487 | 1,081 | 152,088 | 7,190.6 | 1,140,106 | 4,824 |
| Pittsfield, MA | 436 | 11,773 | 1,277.1 | 569.5 | 267 | 1,591 | 167 | 44 | 10,395 | 472.8 | 61,892 | 211 |
| Pocatello, ID | 401 | 5,751 | 634.9 | 219.9 | 125 | 547 | 75 | 17 | 5,767 | 247.9 | 78,524 | 772 |
| Portland-South Portland, ME... | 2,082 | 45,765 | 5,157.5 | 2,069.0 | 1,256 | 6,794 | 863 | 232 | 51,647 | 2,628.2 | 691,489 | 2,855 |
| Portland-Vancouver-Hillsboro, OR-WA | 8,344 | 161,433 | 20,617.4 | 8,120.5 | 4,695 | 30,009 | 4,685 | 1,147 | 192,748 | 9,701.7 | 2,920,301 | 13,446 |
| Port St. Lucie, FL | 1,373 | 24,465 | 3,110.2 | 1,134.0 | 851 | 3,897 | 405 | 117 | 45,931 | 2,072.8 | 1,189,293 | 5,238 |
| Poughkeepsie-Newburgh-Middletown, NY | 1,875 | 44,489 | 5,409.9 | 2,160.1 | 1,435 | 7,729 | 957 | 246 | 49,609 | 2,421.2 | 356,619 | 1,696 |
| Prescott Valley-Prescott, AZ... | 790 | 13,844 | 1,568.4 | 579.8 | 390 | 1,672 | 179 | 50 | 19,895 | 868.6 | 522,372 | 1,891 |
| Providence-Warwick, RI-MA.... | 4,611 | 129,505 | 13,976.1 | 6,005.0 | 3,430 | 19,761 | 2,334 | 641 | 119,157 | 5,607.6 | 439,470 | 2,068 |
| Provo-Orem, UT | 1,478 | 26,304 | 2,948.5 | 1,057.3 | D | D | D | D | 52,231 | 2,320.7 | 2,096,903 | 7,967 |
| Pueblo, CO | 407 | 13,427 | 1,403.4 | 602.5 | 235 | 1,348 | 125 | 36 | 8,929 | 376.3 | 93,369 | 579 |
| Punta Gorda, FL | 514 | 8,529 | 1,259.5 | 400.0 | 335 | 1,418 | 147 | 40 | 14,383 | 681.5 | 763,036 | 2,933 |
| Racine, WI | 510 | 11,145 | 1,032.8 | 423.6 | 302 | 1,880 | 170 | 53 | 9,775 | 418.0 | 62,495 | 224 |
| Raleigh-Cary, NC | 3,447 | 70,395 | 8,388.3 | 3,377.3 | 2,273 | 14,746 | 1,837 | 547 | 114,345 | 5,326.7 | 3,460,650 | 16,958 |
| Rapid City, SD | 392 | 10,698 | 1,458.5 | 548.1 | 342 | 1,761 | 213 | 58 | 11,057 | 504.4 | 209,033 | 1,320 |

Items 159—170

## Table C. Metropolitan Areas — Government Employment and Payroll, and Local Government Finances

| Area name | Government employment and payroll, 2017 | | March payroll (percent of total) | | | | | | | Local government finances, 2017 — General revenue | | | | |
|---|---|---|---|---|---|---|---|---|---|---|---|---|---|---|
| | Full-time equivalent employees | March payroll (dollars) | Administration, judicial, and legal | Police and corrections | Fire protection | Highways and transportation | Health and welfare | Natural resources and utilities | Education and libraries | Total (mil dol) | Inter-governmental (mil dol) | Taxes Total (mil dol) | Taxes Per capita¹ Total (dollars) | Taxes Per capita¹ Property (dollars) |
| | 171 | 172 | 173 | 174 | 175 | 176 | 177 | 178 | 179 | 180 | 181 | 182 | 183 | 184 |
| Mobile, AL | 16,158 | 54,780,612 | 6.0 | 12.7 | 4.5 | 4.1 | 11.9 | 6.8 | 51.3 | 1,552 | 658 | 627 | 1,457 | 578 |
| Modesto, CA | 21,814 | 117,420,075 | 5.4 | 8.3 | 2.5 | 1.5 | 10.8 | 8.8 | 60.3 | 3,253 | 1,917 | 609 | 1,117 | 790 |
| Monroe, LA | 8,813 | 27,842,880 | 7.1 | 11.6 | 5.5 | 2.1 | 7.9 | 4.6 | 60.3 | 824 | 331 | 365 | 1,792 | 669 |
| Monroe, MI | 4,140 | 17,254,732 | 6.2 | 6.0 | 1.8 | 3.3 | 2.1 | 3.4 | 75.6 | 493 | 233 | 176 | 1,181 | 1,139 |
| Montgomery, AL | 11,956 | 42,267,586 | 4.9 | 13.6 | 6.0 | 3.2 | 2.5 | 9.8 | 54.3 | 1,679 | 485 | 412 | 1,102 | 331 |
| Morgantown, WV | 3,557 | 11,833,653 | 6.8 | 7.9 | 2.3 | 3.9 | 2.0 | 7.7 | 66.4 | 317 | 99 | 153 | 1,094 | 848 |
| Morristown, TN | 4,293 | 14,703,508 | 6.1 | 10.0 | 2.8 | 4.1 | 3.2 | 12.3 | 59.4 | 333 | 164 | 93 | 659 | 483 |
| Mount Vernon-Anacortes, WA | 6,316 | 34,058,435 | 5.1 | 5.3 | 2.2 | 4.5 | 35.2 | 4.8 | 40.1 | 1,108 | 364 | 254 | 2,014 | 1,281 |
| Muncie, IN | 3,844 | 13,695,351 | 5.6 | 8.5 | 3.6 | 3.7 | 3.0 | 4.5 | 69.0 | 411 | 186 | 89 | 769 | 723 |
| Muskegon, MI | 5,624 | 23,739,095 | 7.6 | 7.6 | 2.2 | 3.3 | 9.2 | 3.3 | 65.9 | 735 | 410 | 175 | 1,008 | 917 |
| Myrtle Beach-Conway-North Myrtle Beach, SC-NC | 14,122 | 53,770,704 | 7.4 | 10.8 | 4.4 | 3.5 | 5.6 | 7.3 | 56.9 | 1,714 | 449 | 859 | 1,854 | 1,217 |
| Napa, CA | 5,188 | 34,539,364 | 11.5 | 13.0 | 3.1 | 2.6 | 10.1 | 4.7 | 51.6 | 1,004 | 287 | 481 | 3,439 | 2,612 |
| Naples-Marco Island, FL | 10,447 | 46,883,097 | 6.7 | 17.1 | 7.6 | 3.2 | 5.8 | 8.0 | 48.7 | 1,527 | 243 | 884 | 2,372 | 2,131 |
| Nashville-Davidson--Murfreesboro--Franklin, TN | 63,688 | 256,336,916 | 5.9 | 11.3 | 4.8 | 2.3 | 10.5 | 10.7 | 52.6 | 7,013 | 2,043 | 3,160 | 1,686 | 1,048 |
| New Bern, NC | 6,552 | 26,838,499 | 3.4 | 4.7 | 1.5 | 0.9 | 49.2 | 4.6 | 33.8 | 712 | 185 | 108 | 869 | 630 |
| New Haven-Milford, CT | 29,926 | 153,376,297 | 3.4 | 10.0 | 6.1 | 3.3 | 3.0 | 4.9 | 68.4 | 4,281 | 1,606 | 2,299 | 2,680 | 2,641 |
| New Orleans-Metairie, LA | 40,736 | 175,113,122 | 8.3 | 17.9 | 3.2 | 3.0 | 17.7 | 7.5 | 39.7 | 6,352 | 1,655 | 2,728 | 2,147 | 1,070 |
| New York-Newark-Jersey City, NY-NJ-PA | 869,812 | 5,463,315,906 | 3.9 | 14.9 | 3.7 | 9.3 | 11.5 | 4.4 | 49.4 | 183,332 | 54,175 | 100,331 | 5,192 | 3,360 |
| Nassau County-Suffolk County, NY 35004 | 127,355 | 841,354,618 | 3.7 | 12.1 | 0.9 | 2.4 | 5.3 | 3.6 | 70.3 | 23,561 | 6,076 | 14,947 | 5,262 | 4,224 |
| Newark, NJ-PA Div 35084 | 83,369 | 489,592,894 | 5.1 | 14.9 | 4.0 | 2.5 | 4.6 | 4.1 | 62.9 | 12,547 | 3,123 | 7,760 | 3,582 | 3,487 |
| New Brunswick-Lakewood, NJ Div 35154 | 86,747 | 512,134,932 | 4.6 | 13.0 | 1.5 | 2.5 | 3.0 | 4.6 | 68.4 | 12,037 | 2,614 | 7,940 | 3,343 | 3,281 |
| New York-Jersey City-White Plains, NY-NJ Div 35614 | 572,341 | 3,620,233,462 | 3.7 | 15.8 | 4.6 | 12.8 | 15.0 | 4.7 | 40.0 | 135,186 | 42,362 | 69,684 | 5,836 | 3,147 |
| Niles, MI | 4,916 | 20,205,038 | 9.5 | 10.0 | 1.7 | 2.9 | 2.8 | 4.4 | 66.9 | 643 | 301 | 221 | 1,433 | 1,406 |
| North Port-Sarasota-Bradenton, FL | 29,030 | 122,562,501 | 7.4 | 11.3 | 5.8 | 3.5 | 24.8 | 5.8 | 38.2 | 4,037 | 655 | 1,613 | 2,003 | 1,617 |
| Norwich-New London, CT | 9,464 | 49,779,946 | 3.9 | 7.1 | 3.6 | 3.2 | 2.1 | 5.2 | 70.6 | 1,248 | 412 | 703 | 2,629 | 2,606 |
| Ocala, FL | 10,693 | 35,122,480 | 4.8 | 11.0 | 9.8 | 2.1 | 0.8 | 6.7 | 62.3 | 963 | 384 | 312 | 882 | 738 |
| Ocean City, NJ | 6,404 | 30,726,844 | 9.5 | 11.4 | 3.5 | 4.4 | 5.6 | 8.4 | 56.0 | 972 | 144 | 573 | 6,152 | 5,996 |
| Odessa, TX | 8,000 | 32,951,515 | 4.0 | 7.4 | 3.1 | 1.5 | 28.6 | 2.4 | 51.5 | 877 | 186 | 341 | 2,174 | 1,556 |
| Ogden-Clearfield, UT | 18,456 | 70,604,675 | 7.0 | 9.9 | 3.7 | 1.8 | 5.9 | 7.7 | 62.2 | 2,134 | 836 | 810 | 1,220 | 852 |
| Oklahoma City, OK | 43,677 | 166,054,631 | 5.2 | 12.2 | 7.5 | 2.9 | 12.4 | 5.1 | 53.7 | 4,551 | 1,299 | 2,015 | 1,459 | 776 |
| Olympia-Lacey-Tumwater, WA | 8,635 | 45,715,941 | 9.4 | 8.2 | 6.1 | 6.6 | 2.8 | 6.0 | 58.6 | 1,294 | 527 | 508 | 1,811 | 1,121 |
| Omaha-Council Bluffs, NE-IA | 36,806 | 166,779,572 | 4.4 | 9.9 | 4.0 | 3.5 | 4.6 | 15.6 | 56.8 | 4,361 | 1,458 | 2,039 | 2,188 | 1,659 |
| Orlando-Kissimmee-Sanford, FL | 83,077 | 317,326,498 | 6.6 | 13.4 | 7.4 | 4.2 | 1.8 | 8.5 | 53.8 | 10,685 | 3,164 | 4,282 | 1,701 | 1,189 |
| Oshkosh-Neenah, WI | 5,484 | 24,749,186 | 4.5 | 13.5 | 6.2 | 7.2 | 7.0 | 6.5 | 53.4 | 839 | 445 | 252 | 1,478 | 1,433 |
| Owensboro, KY | 4,973 | 16,279,975 | 4.4 | 6.6 | 3.5 | 3.6 | 4.4 | 12.0 | 60.9 | 391 | 150 | 140 | 1,183 | 693 |
| Oxnard-Thousand Oaks-Ventura, CA | 30,149 | 195,583,899 | 9.7 | 10.1 | 4.8 | 2.6 | 14.9 | 6.1 | 50.1 | 5,323 | 2,073 | 1,584 | 1,866 | 1,524 |
| Palm Bay-Melbourne-Titusville, FL | 19,876 | 71,754,734 | 7.2 | 12.8 | 6.8 | 4.6 | 7.5 | 8.1 | 51.1 | 2,102 | 686 | 749 | 1,275 | 944 |
| Panama City, FL | 8,252 | 28,582,062 | 5.1 | 9.8 | 2.7 | 3.8 | 24.7 | 5.2 | 47.3 | 727 | 246 | 300 | 1,625 | 1,070 |
| Parkersburg-Vienna, WV | 3,095 | 10,097,036 | 5.5 | 6.3 | 2.4 | 3.1 | 3.0 | 6.5 | 71.8 | 230 | 97 | 85 | 930 | 758 |
| Pensacola-Ferry Pass-Brent, FL | 14,655 | 50,092,302 | 6.8 | 14.7 | 2.9 | 2.4 | 2.0 | 8.3 | 60.5 | 1,634 | 624 | 525 | 1,078 | 723 |
| Peoria, IL | 15,588 | 63,363,344 | 5.6 | 10.9 | 4.0 | 4.2 | 3.5 | 5.9 | 64.5 | 1,680 | 590 | 776 | 1,908 | 1,667 |
| Philadelphia-Camden-Wilmington, PA-NJ-DE-MD | 206,721 | 1,109,212,623 | 6.6 | 13.5 | 2.5 | 7.8 | 4.9 | 5.0 | 57.7 | 35,204 | 12,958 | 16,569 | 2,726 | 1,901 |
| Camden, NJ Div 15804 | 50,457 | 284,416,906 | 4.4 | 11.2 | 1.9 | 3.3 | 3.2 | 3.4 | 69.7 | 7,088 | 2,088 | 3,609 | 2,904 | 2,867 |
| Montgomery-Bucks-Chester Counties, PA Div 33874 | 57,162 | 306,109,194 | 5.9 | 10.9 | 0.5 | 2.9 | 3.9 | 3.4 | 71.3 | 9,659 | 2,600 | 5,580 | 2,833 | 2,306 |
| Philadelphia, PA Div 37964 | 78,564 | 417,121,085 | 9.1 | 17.8 | 4.7 | 15.5 | 7.3 | 7.7 | 35.9 | 15,651 | 7,072 | 6,255 | 2,917 | 1,180 |
| Wilmington, DE-MD-NJ Div 48864 | 20,538 | 101,565,438 | 5.1 | 10.3 | 1.1 | 3.2 | 2.7 | 3.7 | 72.2 | 2,806 | 1,199 | 1,125 | 1,559 | 1,270 |
| Phoenix-Mesa-Chandler, AZ | 133,626 | 644,963,932 | 9.1 | 15.0 | 1.1 | 3.1 | 6.2 | 13.9 | 45.1 | 17,177 | 6,050 | 7,376 | 1,550 | 954 |
| Pine Bluff, AR | 3,046 | 9,287,297 | 6.2 | 10.9 | 6.5 | 3.1 | 6.2 | 1.8 | 72.2 | 270 | 165 | 73 | 802 | 374 |
| Pittsburgh, PA | 74,899 | 356,345,741 | 6.5 | 10.2 | 3.1 | 3.8 | 1.7 | 5.0 | 60.1 | 12,017 | 5,093 | 4,858 | 2,086 | 1,496 |
| Pittsfield, MA | 4,678 | 20,287,892 | 4.8 | 7.3 | 1.7 | 8.1 | 7.3 | 4.1 | 72.2 | 590 | 247 | 294 | 2,326 | 2,238 |
| Pocatello, ID | 3,290 | 11,919,347 | 7.3 | 13.8 | 3.1 | 4.6 | 2.3 | 7.4 | 51.8 | 298 | 131 | 104 | 1,111 | 1,075 |
| Portland-South Portland, ME | 19,499 | 81,364,976 | 5.7 | 8.9 | 4.9 | 5.5 | 6.4 | 6.4 | 63.1 | 2,115 | 478 | 1,314 | 2,470 | 2,433 |
| Portland-Vancouver-Hillsboro, OR-WA | 76,417 | 414,248,897 | 7.2 | 10.2 | 5.1 | 8.0 | 5.8 | 7.6 | 52.8 | 12,702 | 4,421 | 5,366 | 2,186 | 1,611 |
| Port St. Lucie, FL | 14,593 | 56,559,821 | 6.7 | 13.4 | 9.8 | 2.0 | 1.7 | 7.6 | 54.0 | 1,824 | 468 | 836 | 1,767 | 1,539 |
| Poughkeepsie-Newburgh-Middletown, NY | 28,050 | 155,754,491 | 4.9 | 9.6 | 2.0 | 3.7 | 5.6 | 2.1 | 71.0 | 4,560 | 1,556 | 2,551 | 3,789 | 2,887 |
| Prescott Valley-Prescott, AZ | 6,183 | 24,573,588 | 10.4 | 15.0 | 8.4 | 6.0 | 4.2 | 5.0 | 41.7 | 672 | 233 | 333 | 1,460 | 889 |
| Providence-Warwick, RI-MA | 47,406 | 249,888,612 | 3.5 | 10.2 | 8.2 | 2.3 | 2.8 | 3.8 | 67.9 | 6,853 | 2,374 | 3,635 | 2,248 | 2,177 |
| Provo-Orem, UT | 15,123 | 60,353,207 | 6.9 | 8.9 | 2.6 | 1.6 | 5.6 | 9.1 | 62.8 | 2,078 | 833 | 744 | 1,203 | 788 |
| Pueblo, CO | 5,844 | 27,151,176 | 5.3 | 12.1 | 10.8 | 2.9 | 7.3 | 6.0 | 53.1 | 632 | 279 | 262 | 1,578 | 1,002 |
| Punta Gorda, FL | 4,751 | 17,974,279 | 14.2 | 16.9 | 10.5 | 6.1 | 1.4 | 9.3 | 39.5 | 619 | 101 | 292 | 1,607 | 1,248 |
| Racine, WI | 5,884 | 28,010,342 | 3.9 | 12.3 | 5.4 | 3.5 | 5.4 | 4.2 | 63.6 | 811 | 394 | 296 | 1,510 | 1,462 |
| Raleigh-Cary, NC | 43,575 | 186,320,481 | 3.6 | 8.5 | 3.4 | 3.4 | 9.5 | 7.2 | 60.4 | 4,973 | 1,777 | 2,107 | 1,579 | 1,135 |
| Rapid City, SD | 5,340 | 18,126,595 | 5.9 | 12.7 | 4.3 | 5.5 | 2.6 | 8.0 | 57.8 | 525 | 144 | 282 | 2,036 | 1,350 |

1. Based on the resident population estimated as of July 1 of the year shown.

| Area name | Direct general expenditure Total (mil dol) 185 | Per capita[1] (dollars) 186 | Education 187 | Health and hospitals 188 | Police protection 189 | Public welfare 190 | Highways 191 | Debt outstanding Total (mil dol) 192 | Per capita[1] (dollars) 193 | Federal civilian 194 | Federal military 195 | State and local 196 | Number of returns 197 | Mean adjusted gross income 198 | Mean income tax 199 |
|---|---|---|---|---|---|---|---|---|---|---|---|---|---|---|---|
| Mobile, AL | 1,587.7 | 3,688 | 40.6 | 11.1 | 7.2 | 0.2 | 5.1 | 1,496.5 | 3,476 | 2,717 | 2,703 | 23,412 | 178,660 | 54,335 | 5,494 |
| Modesto, CA | 3,266.2 | 5,996 | 49.6 | 7.6 | 5.3 | 9.3 | 2.5 | 6,781.2 | 12,449 | 831 | 807 | 30,271 | 231,670 | 58,213 | 5,722 |
| Monroe, LA | 816.3 | 4,003 | 54.2 | 6.0 | 6.6 | 0.2 | 3.2 | 766.3 | 3,759 | 585 | 736 | 11,930 | 83,660 | 53,610 | 5,363 |
| Monroe, MI | 543.6 | 3,638 | 53.6 | 6.9 | 2.8 | 0.2 | 7.7 | 514.5 | 3,443 | 246 | 237 | 5,471 | 75,290 | 62,405 | 6,208 |
| Montgomery, AL | 1,652.3 | 4,416 | 32.0 | 39.6 | 5.1 | 0.2 | 2.9 | 1,290.9 | 3,450 | 6,547 | 4,370 | 33,240 | 161,300 | 57,015 | 5,619 |
| Morgantown, WV | 335.7 | 2,405 | 54.4 | 1.4 | 6.9 | 0.8 | 2.2 | 456.9 | 3,272 | 2,005 | 640 | 17,198 | 56,180 | 66,469 | 7,676 |
| Morristown, TN | 383.3 | 2,721 | 49.9 | 2.0 | 5.7 | 3.1 | 3.8 | 423.6 | 3,007 | 385 | 399 | 7,230 | 60,080 | 49,794 | 4,336 |
| Mount Vernon-Anacortes, WA | 1,115.7 | 8,865 | 28.0 | 43.7 | 2.5 | 0.1 | 3.4 | 865.1 | 6,873 | 410 | 320 | 11,549 | 61,600 | 70,711 | 7,564 |
| Muncie, IN | 644.3 | 5,590 | 24.6 | 0.6 | 2.4 | 0.0 | 1.8 | 330.8 | 2,870 | 307 | 320 | 10,444 | 47,870 | 49,385 | 4,443 |
| Muskegon, MI | 779.6 | 4,489 | 51.7 | 8.7 | 3.7 | 2.8 | 4.5 | 744.5 | 4,287 | 356 | 281 | 7,333 | 79,760 | 51,293 | 4,634 |
| Myrtle Beach-Conway-North Myrtle Beach, SC-NC | 1,752.4 | 3,781 | 46.2 | 4.4 | 7.0 | 1.2 | 8.1 | 1,680.4 | 3,625 | 1,319 | 1,683 | 21,773 | 235,680 | 57,439 | 6,053 |
| Napa, CA | 964.8 | 6,897 | 37.4 | 6.8 | 8.1 | 4.4 | 5.0 | 1,190.6 | 8,512 | 227 | 195 | 10,728 | 68,540 | 99,868 | 14,637 |
| Naples-Marco Island, FL | 1,422.5 | 3,817 | 38.1 | 3.4 | 12.7 | 0.5 | 5.2 | 1,944.8 | 5,219 | 756 | 661 | 13,159 | 190,930 | 158,523 | 29,256 |
| Nashville-Davidson--Murfreesboro--Franklin, TN | 7,298.3 | 3,893 | 42.6 | 10.3 | 6.9 | 0.9 | 4.0 | 20,712.8 | 11,050 | 13,798 | 5,825 | 98,040 | 917,690 | 80,109 | 10,829 |
| New Bern, NC | 682.6 | 5,474 | 27.0 | 53.7 | 2.7 | 4.2 | 0.2 | 167.1 | 1,340 | 6,122 | 7,253 | 8,589 | 56,360 | 57,057 | 5,525 |
| New Haven-Milford, CT | 4,329.0 | 5,047 | 54.3 | 0.8 | 5.3 | 0.3 | 3.3 | 3,393.1 | 3,956 | 5,522 | 1,917 | 43,653 | 416,540 | 74,374 | 9,283 |
| New Orleans-Metairie, LA | 6,044.5 | 4,758 | 33.8 | 12.6 | 7.1 | 0.8 | 4.3 | 8,601.4 | 6,771 | 13,256 | 7,770 | 66,876 | 558,450 | 66,289 | 8,356 |
| New York-Newark-Jersey City, NY-NJ-PA | 178,961.0 | 9,262 | 38.8 | 7.8 | 5.8 | 9.4 | 2.8 | 230,434.8 | 11,926 | 104,831 | 35,246 | 1,174,052 | 9,792,530 | 99,960 | 16,607 |
| Nassau County-Suffolk County, NY Div 35004 | 24,029.7 | 8,459 | 52.6 | 4.5 | 7.2 | 5.2 | 3.1 | 17,428.0 | 6,135 | 15,931 | 5,121 | 172,723 | 1,518,340 | 106,069 | 17,218 |
| Newark, NJ-PA Div 35084 | 12,542.8 | 5,790 | 52.2 | 2.2 | 6.5 | 2.0 | 2.4 | 7,966.5 | 3,677 | 16,685 | 4,438 | 138,874 | 1,084,730 | 102,582 | 16,429 |
| New Brunswick-Lakewood, NJ Div 35154 | 12,540.2 | 5,279 | 57.2 | 1.2 | 6.0 | 1.9 | 3.1 | 8,207.6 | 3,455 | 9,552 | 5,248 | 130,223 | 1,202,610 | 94,114 | 13,719 |
| New York-Jersey City-White Plains, NY-NJ Div 35614 | 129,848.3 | 10,875 | 33.2 | 9.5 | 5.4 | 11.6 | 2.7 | 196,832.8 | 16,485 | 62,663 | 20,439 | 732,232 | 5,986,850 | 99,110 | 17,065 |
| Niles, MI | 712.3 | 4,622 | 50.9 | 8.7 | 3.9 | 1.3 | 5.1 | 534.7 | 3,469 | 342 | 263 | 8,872 | 73,510 | 61,416 | 6,929 |
| North Port-Sarasota-Bradenton, FL | 3,730.1 | 4,633 | 31.1 | 21.9 | 6.7 | 0.2 | 4.2 | 3,396.5 | 4,218 | 2,077 | 1,527 | 26,336 | 405,830 | 89,292 | 13,204 |
| Norwich-New London, CT | 1,142.3 | 4,271 | 59.0 | 0.8 | 6.3 | 0.5 | 6.0 | 611.7 | 2,288 | 2,842 | 6,810 | 27,127 | 135,780 | 73,015 | 8,582 |
| Ocala, FL | 965.7 | 2,733 | 49.5 | 3.7 | 7.4 | 0.9 | 5.5 | 629.5 | 1,782 | 752 | 623 | 14,753 | 166,940 | 51,068 | 5,269 |
| Ocean City, NJ | 1,325.6 | 14,234 | 23.2 | 1.0 | 3.2 | 2.4 | 6.9 | 959.9 | 10,307 | 440 | 1,081 | 8,632 | 49,920 | 68,351 | 8,023 |
| Odessa, TX | 915.0 | 5,830 | 35.1 | 40.8 | 3.5 | 0.0 | 2.5 | 447.1 | 2,849 | 195 | 320 | 9,801 | 72,720 | 67,096 | 7,413 |
| Ogden-Clearfield, UT | 2,014.5 | 3,034 | 53.7 | 3.0 | 5.8 | 0.1 | 4.1 | 1,840.6 | 2,772 | 20,323 | 6,773 | 32,941 | 292,440 | 69,284 | 6,681 |
| Oklahoma City, OK | 4,263.3 | 3,086 | 43.2 | 9.0 | 9.1 | 0.0 | 4.8 | 5,174.1 | 3,745 | 29,500 | 11,042 | 93,839 | 603,190 | 68,869 | 8,062 |
| Olympia-Lacey-Tumwater, WA | 1,287.9 | 4,595 | 49.0 | 6.7 | 4.0 | 0.0 | 3.7 | 1,289.3 | 4,600 | 871 | 804 | 38,635 | 140,680 | 71,069 | 7,564 |
| Omaha-Council Bluffs, NE-IA | 4,264.8 | 4,576 | 55.8 | 2.7 | 5.6 | 0.6 | 5.3 | 8,245.8 | 8,848 | 9,636 | 9,398 | 54,287 | 448,880 | 75,329 | 8,898 |
| Orlando-Kissimmee-Sanford, FL | 10,770.3 | 4,278 | 40.8 | 1.8 | 7.5 | 0.5 | 4.0 | 14,561.9 | 5,784 | 14,644 | 4,716 | 110,599 | 1,257,020 | 62,377 | 7,511 |
| Oshkosh-Neenah, WI | 872.6 | 5,121 | 34.9 | 2.9 | 4.7 | 26.1 | 6.4 | 874.7 | 5,134 | 426 | 465 | 11,411 | 84,680 | 64,562 | 6,930 |
| Owensboro, KY | 375.0 | 3,165 | 48.1 | 6.2 | 3.0 | 0.2 | 4.2 | 2,274.1 | 19,198 | 314 | 374 | 6,102 | 53,570 | 55,616 | 5,067 |
| Oxnard-Thousand Oaks-Ventura, CA | 5,425.7 | 6,390 | 40.7 | 12.5 | 7.4 | 4.4 | 2.7 | 4,701.3 | 5,537 | 7,500 | 4,953 | 38,698 | 405,950 | 84,283 | 11,252 |
| Palm Bay-Melbourne-Titusville, FL | 2,232.0 | 3,797 | 35.9 | 9.2 | 7.0 | 0.2 | 4.6 | 1,844.7 | 3,138 | 6,576 | 2,901 | 22,506 | 292,710 | 65,100 | 7,575 |
| Panama City, FL | 718.8 | 3,891 | 43.3 | 2.6 | 8.1 | 0.0 | 4.5 | 596.2 | 3,227 | 3,787 | 2,626 | 8,801 | 82,060 | 59,521 | 6,572 |
| Parkersburg-Vienna, WV | 240.7 | 2,650 | 62.2 | 0.0 | 6.4 | 0.1 | 3.7 | 257.9 | 2,839 | 2,488 | 424 | 4,510 | 40,250 | 56,736 | 5,806 |
| Pensacola-Ferry Pass-Brent, FL | 1,706.2 | 3,501 | 42.6 | 1.7 | 7.4 | 0.0 | 3.7 | 4,496.6 | 9,228 | 6,749 | 12,205 | 21,383 | 229,690 | 61,175 | 6,694 |
| Peoria, IL | 1,660.1 | 4,080 | 50.8 | 1.5 | 6.6 | 1.0 | 6.9 | 1,282.1 | 3,151 | 2,209 | 869 | 21,438 | 191,090 | 68,559 | 8,017 |
| Philadelphia-Camden-Wilmington, PA-NJ-DE-MD | 35,039.2 | 5,764 | 51.2 | 6.0 | 5.5 | 3.6 | 2.8 | 38,405.9 | 6,318 | 51,591 | 23,088 | 295,717 | 2,959,450 | 85,162 | 11,796 |
| Camden, NJ Div 15804 | 7,006.6 | 5,638 | 59.6 | 1.3 | 5.2 | 2.3 | 2.9 | 5,817.5 | 4,681 | 7,563 | 7,952 | 73,648 | 620,370 | 76,140 | 9,251 |
| Montgomery-Bucks-Chester Counties, PA Div 33874 | 10,180.1 | 5,168 | 60.2 | 2.5 | 4.9 | 5.3 | 3.7 | 10,828.7 | 5,497 | 6,077 | 5,181 | 78,437 | 1,017,470 | 113,275 | 17,677 |
| Philadelphia, PA Div 37964 | 15,006.8 | 6,998 | 39.4 | 11.6 | 5.6 | 3.7 | 1.9 | 19,371.5 | 9,033 | 32,296 | 6,542 | 100,291 | 972,030 | 65,587 | 8,341 |
| Wilmington, DE-MD-NJ Div 48864 | 2,845.7 | 3,945 | 60.5 | 0.9 | 7.5 | 0.2 | 4.2 | 2,388.2 | 3,311 | 5,655 | 3,413 | 43,341 | 349,580 | 73,777 | 8,804 |
| Phoenix-Mesa-Chandler, AZ | 16,004.3 | 3,363 | 40.1 | 5.0 | 9.4 | 1.6 | 3.8 | 28,954.9 | 6,085 | 22,894 | 15,578 | 220,561 | 2,105,930 | 72,955 | 9,037 |
| Pine Bluff, AR | 255.1 | 2,808 | 56.7 | 0.1 | 8.7 | 0.2 | 4.8 | 216.6 | 2,384 | 1,413 | 334 | 8,082 | 34,510 | 45,260 | 3,579 |
| Pittsburgh, PA | 12,113.1 | 5,201 | 47.9 | 4.1 | 4.0 | 9.3 | 4.7 | 18,684.2 | 8,022 | 18,360 | 6,317 | 98,018 | 1,187,720 | 72,741 | 9,008 |
| Pittsfield, MA | 681.9 | 5,397 | 58.0 | 0.3 | 3.3 | 0.4 | 6.4 | 435.0 | 3,443 | 391 | 288 | 7,994 | 64,630 | 67,844 | 7,792 |
| Pocatello, ID | 272.4 | 2,927 | 41.5 | 4.8 | 8.0 | 0.5 | 7.2 | 71.6 | 769 | 597 | 294 | 8,694 | 38,960 | 52,933 | 4,350 |
| Portland-South Portland, ME | 2,149.5 | 4,041 | 50.8 | 0.8 | 5.0 | 1.9 | 5.4 | 1,864.2 | 3,505 | 9,766 | 4,065 | 28,899 | 287,590 | 73,275 | 8,586 |
| Portland-Vancouver-Hillsboro, OR-WA | 12,324.2 | 5,020 | 42.7 | 3.8 | 5.6 | 1.7 | 5.1 | 18,952.7 | 7,721 | 18,612 | 6,333 | 133,082 | 1,206,240 | 82,481 | 10,391 |
| Port St. Lucie, FL | 1,815.4 | 3,839 | 38.4 | 2.7 | 8.7 | 1.4 | 7.6 | 2,383.6 | 5,041 | 1,065 | 920 | 18,877 | 231,550 | 78,976 | 10,772 |
| Poughkeepsie-Newburgh-Middletown, NY | 4,629.3 | 6,876 | 54.6 | 4.3 | 3.6 | 8.8 | 3.9 | 2,458.4 | 3,651 | 5,782 | 7,252 | 40,492 | 330,090 | 75,542 | 9,151 |
| Prescott Valley-Prescott, AZ | 710.2 | 3,114 | 35.4 | 3.3 | 9.4 | 0.1 | 10.8 | 559.6 | 2,454 | 1,554 | 507 | 9,440 | 110,410 | 61,135 | 6,549 |
| Providence-Warwick, RI-MA | 6,766.0 | 4,185 | 57.5 | 0.6 | 7.2 | 0.4 | 3.1 | 4,172.2 | 2,581 | 12,370 | 8,066 | 84,521 | 823,390 | 68,784 | 8,090 |
| Provo-Orem, UT | 1,877.9 | 3,039 | 55.9 | 3.8 | 4.2 | 0.3 | 5.3 | 2,351.5 | 3,805 | 1,107 | 2,466 | 31,604 | 249,930 | 72,707 | 7,757 |
| Pueblo, CO | 625.0 | 3,759 | 39.5 | 1.4 | 7.3 | 4.1 | 4.1 | 578.0 | 3,476 | 1,083 | 415 | 11,641 | 72,210 | 52,552 | 4,740 |
| Punta Gorda, FL | 621.4 | 3,423 | 26.1 | 3.8 | 10.9 | 1.3 | 15.4 | 578.0 | 3,184 | 350 | 325 | 5,770 | 88,530 | 62,218 | 6,737 |
| Racine, WI | 854.8 | 4,365 | 48.3 | 5.0 | 9.0 | 4.9 | 8.3 | 675.9 | 3,452 | 390 | 500 | 8,513 | 97,150 | 63,985 | 6,776 |
| Raleigh-Cary, NC | 4,736.3 | 3,550 | 50.0 | 8.4 | 5.7 | 3.5 | 2.4 | 8,779.2 | 6,580 | 6,251 | 3,706 | 93,410 | 631,770 | 86,491 | 11,259 |
| Rapid City, SD | 464.3 | 3,353 | 46.2 | 1.4 | 7.5 | 0.4 | 9.2 | 443.4 | 3,202 | 2,992 | 4,134 | 8,037 | 71,610 | 62,945 | 6,582 |

1. Based on the resident population estimated as of July 1 of the year shown.

# Table C. Metropolitan Areas — Land Area and Population

| CBSA/ DIV Code[1] | Area name | Land area[2] (sq mi) | Total persons 2020 | Rank | Per square mile | White | Black | American Indian, Alaska Native | Asian and Pacific Islander | Percent Hispanic or Latino[3] | Under 5 years | 5 to 14 years | 15 to 24 years | 25 to 34 years | 35 to 44 years | 45 to 54 years |
|---|---|---|---|---|---|---|---|---|---|---|---|---|---|---|---|---|
| | | 1 | 2 | 3 | 4 | 5 | 6 | 7 | 8 | 9 | 10 | 11 | 12 | 13 | 14 | 15 |
| 39740 | Reading, PA ...................... | 856 | 421,017 | 130 | 491.6 | 70.9 | 5.2 | 0.4 | 1.8 | 23.1 | 5.6 | 12.5 | 13.3 | 12.6 | 11.7 | 12.5 |
| 39820 | Redding, CA ...................... | 3,776 | 179,027 | 241 | 47.4 | 82.4 | 2.1 | 4.1 | 4.8 | 10.9 | 5.7 | 12.2 | 11.1 | 12.9 | 11.9 | 10.9 |
| 39900 | Reno, NV .......................... | 6,580 | 481,289 | 116 | 73.1 | 64.7 | 3.3 | 2.0 | 8.1 | 25.2 | 5.6 | 11.9 | 12.2 | 15.2 | 12.7 | 11.9 |
| 40060 | Richmond, VA.................... | 4,364 | 1,303,469 | 44 | 298.7 | 58.7 | 30.8 | 0.9 | 5.2 | 6.9 | 5.8 | 11.9 | 12.2 | 14.7 | 13.0 | 12.5 |
| 40140 | Riverside-San Bernardino-Ontario, CA ....... | 27,277 | 4,678,371 | 13 | 171.5 | 32.0 | 8.1 | 0.9 | 8.9 | 52.6 | 6.4 | 14.4 | 14.0 | 14.7 | 13.1 | 12.4 |
| 40220 | Roanoke, VA ..................... | 1,868 | 313,784 | 164 | 168.0 | 79.4 | 14.9 | 0.7 | 3.0 | 4.3 | 5.2 | 11.4 | 11.4 | 12.4 | 11.5 | 12.7 |
| 40340 | Rochester, MN ................... | 2,477 | 223,062 | 205 | 90.0 | 84.8 | 6.1 | 0.7 | 5.8 | 4.8 | 6.4 | 13.6 | 11.6 | 13.4 | 13.3 | 11.2 |
| 40380 | Rochester, NY.................... | 3,266 | 1,067,486 | 52 | 326.9 | 77.6 | 12.3 | 0.6 | 3.6 | 8.0 | 5.3 | 11.4 | 13.2 | 13.4 | 11.6 | 12.4 |
| 40420 | Rockford, IL ...................... | 794 | 334,072 | 153 | 420.8 | 69.9 | 13.4 | 0.7 | 3.2 | 15.3 | 6.1 | 13.1 | 12.5 | 12.4 | 11.7 | 12.3 |
| 40580 | Rocky Mount, NC ............... | 1,046 | 145,688 | 287 | 139.3 | 45.0 | 47.7 | 1.1 | 1.1 | 6.7 | 5.6 | 12.3 | 12.4 | 12.2 | 10.9 | 12.3 |
| 40660 | Rome, GA.......................... | 510 | 98,604 | 360 | 193.4 | 71.9 | 15.6 | 0.7 | 1.9 | 11.8 | 5.9 | 12.9 | 14.1 | 13.0 | 12.0 | 12.3 |
| 40900 | Sacramento-Roseville-Folsom, CA .......................... | 5,095 | 2,374,749 | 26 | 466.1 | 54.6 | 8.7 | 1.5 | 18.1 | 22.2 | 5.8 | 12.8 | 12.9 | 14.3 | 13.4 | 12.1 |
| 40980 | Saginaw, MI ...................... | 801 | 189,868 | 229 | 237.1 | 70.7 | 19.9 | 0.8 | 1.7 | 9.0 | 5.8 | 11.7 | 13.0 | 12.8 | 10.9 | 11.8 |
| 41060 | St. Cloud, MN.................... | 1,751 | 202,996 | 224 | 115.9 | 86.1 | 8.7 | 0.7 | 2.8 | 3.6 | 6.6 | 13.2 | 17.0 | 12.6 | 12.0 | 10.8 |
| 41100 | St. George, UT................... | 2,427 | 184,913 | 232 | 76.2 | 85.4 | 1.1 | 1.5 | 3.0 | 11.0 | 6.2 | 14.3 | 13.3 | 11.4 | 11.9 | 9.5 |
| 41140 | St. Joseph, MO-KS............. | 1,656 | 122,556 | 326 | 74.0 | 87.3 | 6.3 | 1.1 | 2.0 | 5.6 | 5.9 | 12.3 | 12.5 | 13.6 | 12.6 | 11.9 |
| 41180 | St. Louis, MO-IL................. | 7,864 | 2,805,473 | 20 | 356.8 | 75.2 | 19.4 | 0.7 | 3.6 | 3.3 | 5.8 | 12.3 | 11.8 | 13.6 | 12.8 | 12.0 |
| 41420 | Salem, OR ........................ | 1,922 | 436,948 | 124 | 227.4 | 69.4 | 1.9 | 2.3 | 4.5 | 25.1 | 6.1 | 13.5 | 13.6 | 13.9 | 12.8 | 11.4 |
| 41500 | Salinas, CA........................ | 3,282 | 430,906 | 125 | 131.3 | 31.0 | 3.1 | 0.8 | 7.9 | 59.7 | 6.8 | 14.8 | 14.0 | 14.1 | 13.1 | 11.6 |
| 41540 | Salisbury, MD-DE .............. | 2,099 | 423,481 | 129 | 201.8 | 73.0 | 18.6 | 0.9 | 2.3 | 7.5 | 5.1 | 10.6 | 12.0 | 10.7 | 10.0 | 10.8 |
| 41620 | Salt Lake City, UT.............. | 7,684 | 1,240,029 | 47 | 161.4 | 72.9 | 2.4 | 1.1 | 7.4 | 18.6 | 7.0 | 15.0 | 14.0 | 16.5 | 14.8 | 11.3 |
| 41660 | San Angelo, TX ................. | 3,497 | 122,889 | 325 | 35.1 | 53.4 | 4.3 | 0.8 | 2.0 | 41.1 | 6.4 | 13.5 | 15.3 | 15.0 | 12.3 | 10.0 |
| 41700 | San Antonio-New Braunfels, TX ................................... | 7,313 | 2,590,732 | 24 | 354.3 | 34.3 | 7.4 | 0.6 | 3.5 | 55.9 | 6.5 | 14.0 | 13.9 | 15.0 | 13.7 | 12.0 |
| 41740 | San Diego-Chula Vista-Carlsbad, CA ........................... | 4,210 | 3,332,427 | 17 | 791.5 | 47.5 | 5.8 | 0.9 | 15.3 | 34.3 | 5.9 | 11.9 | 13.4 | 16.4 | 13.7 | 12.0 |
| 41860 | San Francisco-Oakland-Berkeley, CA................. | 2,470 | 4,696,902 | 12 | 1,901.3 | 41.7 | 8.3 | 0.8 | 31.8 | 21.8 | 5.2 | 10.8 | 10.8 | 16.1 | 14.8 | 13.3 |
| 41860 | Oakland-Berkeley-Livermore, CA Div 36084 | 1,454 | 2,814,656 | X | 1,935.3 | 38.5 | 10.9 | 0.9 | 30.6 | 23.9 | 5.5 | 11.8 | 11.5 | 14.9 | 14.8 | 13.3 |
| 41860 | San Francisco-San Mateo-Redwood City, CA Div 41884 | 496 | 1,624,914 | X | 3,279.3 | 42.1 | 4.7 | 0.7 | 37.4 | 19.2 | 4.9 | 8.9 | 9.6 | 19.2 | 15.3 | 13.1 |
| 41860 | San Rafael, CA Div 42034...... | 520 | 257,332 | X | 494.5 | 73.9 | 3.4 | 0.8 | 9.1 | 16.4 | 4.4 | 11.2 | 10.6 | 8.7 | 11.6 | 15.1 |
| 41940 | San Jose-Sunnyvale-Santa Clara, CA ........................... | 2,680 | 1,971,160 | 35 | 735.6 | 32.6 | 3.0 | 0.7 | 41.2 | 25.9 | 5.7 | 12.0 | 12.0 | 16.3 | 14.4 | 13.3 |
| 42020 | San Luis Obispo-Paso Robles, CA .............................. | 3,301 | 282,249 | 174 | 85.5 | 70.5 | 2.4 | 1.2 | 5.4 | 23.2 | 4.4 | 9.9 | 17.9 | 11.3 | 11.5 | 10.4 |
| 42100 | Santa Cruz-Watsonville, CA.... | 445 | 269,925 | 181 | 606.4 | 59.6 | 1.7 | 1.2 | 6.9 | 34.0 | 4.7 | 10.7 | 18.0 | 12.1 | 11.7 | 11.8 |
| 42140 | Santa Fe, NM .................... | 1,910 | 151,946 | 277 | 79.5 | 44.5 | 1.1 | 3.2 | 1.9 | 50.6 | 4.0 | 9.9 | 10.3 | 11.1 | 11.5 | 11.8 |
| 42200 | Santa Maria-Santa Barbara, CA ................................... | 2,734 | 444,766 | 123 | 162.6 | 45.6 | 2.4 | 1.0 | 7.2 | 46.4 | 6.1 | 12.4 | 19.0 | 13.5 | 11.5 | 10.4 |
| 42220 | Santa Rosa-Petaluma, CA ...... | 1,576 | 489,819 | 113 | 310.9 | 65.1 | 2.5 | 1.6 | 6.5 | 27.6 | 4.7 | 10.9 | 11.4 | 12.6 | 12.8 | 12.3 |
| 42340 | Savannah, GA .................... | 1,350 | 395,983 | 138 | 293.4 | 56.7 | 34.8 | 0.8 | 3.5 | 6.6 | 6.2 | 12.7 | 13.6 | 15.4 | 13.3 | 11.6 |
| 42540 | Scranton--Wilkes-Barre, PA .... | 1,746 | 552,528 | 101 | 316.5 | 82.1 | 4.7 | 0.4 | 2.3 | 12.0 | 5.2 | 11.2 | 12.1 | 12.5 | 11.6 | 12.6 |
| 42660 | Seattle-Tacoma-Bellevue, WA | 5,870 | 4,018,598 | 15 | 684.6 | 65.4 | 8.0 | 2.0 | 19.7 | 10.7 | 5.9 | 11.8 | 11.4 | 17.2 | 15.0 | 12.7 |
| 42660 | Seattle-Bellevue-Kent, WA Div 42644....................... | 4,202 | 3,104,708 | X | 738.9 | 63.8 | 7.4 | 1.7 | 21.9 | 10.3 | 5.7 | 11.5 | 11.1 | 17.7 | 15.3 | 12.9 |
| 42660 | Tacoma-Lakewood, WA Div 45704 | 1,668 | 913,890 | X | 547.9 | 70.6 | 10.0 | 2.6 | 12.4 | 11.8 | 6.4 | 13.1 | 12.5 | 15.8 | 13.8 | 11.9 |
| 42680 | Sebastian-Vero Beach, FL ....... | 503 | 162,518 | 264 | 323.2 | 76.1 | 9.7 | 0.6 | 2.1 | 12.9 | 4.0 | 8.4 | 8.8 | 9.4 | 8.7 | 10.5 |
| 42700 | Sebring-Avon Park, FL ......... | 1,018 | 106,639 | 347 | 104.8 | 67.0 | 10.4 | 0.8 | 1.9 | 21.3 | 4.2 | 9.6 | 8.7 | 9.6 | 8.8 | 9.4 |
| 43100 | Sheboygan, WI .................. | 512 | 115,240 | 334 | 225.3 | 84.5 | 3.0 | 0.8 | 6.5 | 6.9 | 5.4 | 12.4 | 12.1 | 11.7 | 12.1 | 12.4 |
| 43300 | Sherman-Denison, TX........... | 933 | 138,318 | 300 | 148.3 | 76.8 | 7.0 | 2.3 | 2.1 | 14.4 | 6.1 | 13.5 | 12.5 | 12.1 | 11.8 | 11.9 |
| 43340 | Shreveport-Bossier City, LA .... | 2,595 | 392,404 | 140 | 151.2 | 53.4 | 41.1 | 1.0 | 2.1 | 4.3 | 6.4 | 13.5 | 12.3 | 13.7 | 12.8 | 11.3 |
| 43420 | Sierra Vista-Douglas, AZ........ | 6,210 | 127,450 | 315 | 20.5 | 57.1 | 4.9 | 1.6 | 3.7 | 35.5 | 5.5 | 12.0 | 11.9 | 12.5 | 11.3 | 10.1 |
| 43580 | Sioux City, IA-NE-SD.............. | 2,074 | 144,996 | 289 | 69.9 | 71.3 | 5.7 | 2.2 | 3.8 | 19.3 | 7.1 | 14.8 | 13.7 | 12.8 | 12.3 | 11.3 |
| 43620 | Sioux Falls, SD................... | 2,575 | 273,566 | 179 | 106.2 | 86.3 | 6.1 | 2.7 | 2.6 | 4.6 | 7.3 | 14.5 | 12.2 | 14.9 | 13.9 | 11.0 |
| 43780 | South Bend-Mishawaka, IN-MI | 948 | 323,068 | 160 | 340.8 | 76.6 | 13.9 | 1.1 | 3.1 | 8.5 | 6.1 | 12.8 | 14.3 | 13.0 | 11.8 | 11.6 |
| 43900 | Spartanburg, SC.................. | 808 | 326,205 | 158 | 403.9 | 69.1 | 21.7 | 0.7 | 3.1 | 7.5 | 6.1 | 13.0 | 12.8 | 14.1 | 11.9 | 12.7 |
| 44060 | Spokane-Spokane Valley, WA | 4,241 | 574,585 | 99 | 135.5 | 87.4 | 3.1 | 3.0 | 4.5 | 6.2 | 5.7 | 12.3 | 12.5 | 14.8 | 12.7 | 11.5 |
| 44100 | Springfield, IL .................... | 1,183 | 205,950 | 220 | 174.1 | 82.7 | 14.1 | 0.6 | 2.7 | 2.5 | 5.7 | 12.3 | 11.8 | 12.3 | 12.6 | 12.1 |
| 44140 | Springfield, MA .................. | 1,843 | 695,654 | 84 | 377.4 | 70.7 | 7.1 | 0.7 | 3.9 | 19.7 | 4.7 | 10.8 | 16.2 | 12.6 | 11.4 | 11.8 |
| 44180 | Springfield, MO .................. | 3,007 | 475,220 | 117 | 158.0 | 91.6 | 3.5 | 1.5 | 2.4 | 3.6 | 6.1 | 12.5 | 15.2 | 13.4 | 12.2 | 11.3 |
| 44220 | Springfield, OH .................. | 397 | 133,593 | 308 | 336.6 | 86.4 | 10.8 | 0.9 | 1.2 | 3.7 | 5.9 | 12.5 | 12.6 | 11.9 | 11.0 | 12.1 |
| 44300 | State College, PA ............... | 1,109 | 161,496 | 265 | 145.7 | 86.8 | 4.2 | 0.4 | 7.1 | 3.0 | 3.7 | 8.4 | 25.8 | 13.3 | 10.9 | 10.7 |
| 44420 | Staunton, VA ..................... | 1,002 | 124,475 | 320 | 124.2 | 86.8 | 8.7 | 0.7 | 1.5 | 4.6 | 5.3 | 10.9 | 10.9 | 12.7 | 12.2 | 12.4 |
| 44700 | Stockton, CA...................... | 1,392 | 767,967 | 78 | 551.6 | 32.4 | 8.3 | 1.2 | 19.8 | 42.3 | 6.8 | 15.3 | 14.0 | 14.1 | 13.4 | 11.9 |
| 44940 | Sumter, SC ....................... | 1,272 | 139,775 | 298 | 109.9 | 47.1 | 47.9 | 0.8 | 2.0 | 4.1 | 6.1 | 12.6 | 13.3 | 13.5 | 11.0 | 11.1 |
| 45060 | Syracuse, NY..................... | 2,385 | 646,038 | 92 | 270.9 | 83.4 | 9.7 | 1.2 | 3.7 | 4.6 | 5.5 | 11.6 | 13.9 | 12.9 | 11.5 | 12.0 |
| 45220 | Tallahassee, FL.................. | 2,389 | 389,599 | 142 | 163.2 | 56.8 | 33.9 | 0.8 | 3.7 | 7.0 | 5.1 | 10.6 | 21.0 | 13.8 | 11.6 | 10.9 |
| 45300 | Tampa-St. Petersburg-Clearwater, FL ........................ | 2,515 | 3,243,963 | 18 | 1,289.7 | 62.9 | 13.1 | 0.7 | 4.6 | 20.9 | 5.1 | 11.0 | 10.9 | 13.4 | 12.6 | 12.8 |
| 45460 | Terre Haute, IN.................. | 1,910 | 185,632 | 231 | 97.2 | 90.8 | 6.2 | 0.8 | 1.8 | 2.4 | 5.6 | 11.5 | 15.4 | 13.0 | 11.8 | 11.9 |

1. CBSA = Core Based Statistical Area. DIV = Metropolitan Division. See Appendix A for explanation. See Appendix B for list of metropolitan areas or temporarily covered by water. 3. May be of any race. 2. Dry land or land partially or temporarily covered by water.

Items 1—15

# Table C. Metropolitan Areas — Population and Households

Population, 2020 (cont.) — Age (percent) (cont.) [cols 16–19]; Population change and components of change, 2000–2020 [cols 20–26]; Households, 2019 [cols 27–31]

| Area name | 55 to 64 years | 65 to 74 years | 75 years and over | Percent female | Total persons 2000 | Total persons 2010 | Percent change 2000–2010 | Percent change 2010–2019 | Births | Deaths | Net migration | Number | Persons per household | Family households | Single parent households | One person |
|---|---|---|---|---|---|---|---|---|---|---|---|---|---|---|---|---|
| | 16 | 17 | 18 | 19 | 20 | 21 | 22 | 23 | 24 | 25 | 26 | 27 | 28 | 29 | 30 | 31 |
| Reading, PA | 13.8 | 10.3 | 7.7 | 50.7 | 373,638 | 411,569 | 10.2 | 2.3 | 48,972 | 39,880 | 469 | 154,696 | 2.63 | 68.1 | 12.8 | 25.9 |
| Redding, CA | 14.1 | 12.5 | 8.8 | 50.9 | 163,256 | 177,222 | 8.6 | 1.0 | 20,960 | 22,660 | 3,697 | 72,714 | 2.43 | 62.3 | 9.8 | 29.0 |
| Reno, NV | 13.0 | 10.9 | 6.5 | 49.5 | 342,885 | 425,442 | 24.1 | 13.1 | 55,222 | 39,487 | 40,008 | 192,402 | 2.44 | 62.5 | 9.4 | 26.9 |
| Richmond, VA | 13.4 | 10.0 | 6.5 | 51.8 | 1,040,192 | 1,186,611 | 14.1 | 9.8 | 150,233 | 108,894 | 75,649 | 491,891 | 2.56 | 62.8 | 9.6 | 30.6 |
| Riverside-San Bernardino-Ontario, CA | 11.5 | 8.1 | 5.7 | 50.1 | 3,254,821 | 4,224,957 | 29.8 | 10.7 | 616,100 | 306,646 | 145,185 | 1,379,706 | 3.31 | 74.7 | 12.4 | 20.0 |
| Roanoke, VA | 14.3 | 12.4 | 8.7 | 51.7 | 288,309 | 308,668 | 7.1 | 1.7 | 33,472 | 36,284 | 8,129 | 126,469 | 2.41 | 63.3 | 10.5 | 31.6 |
| Rochester, MN | 13.1 | 9.5 | 7.8 | 50.9 | 184,740 | 206,890 | 12.0 | 7.8 | 29,427 | 16,380 | 3,218 | 92,282 | 2.37 | 64.7 | 8.1 | 28.3 |
| Rochester, NY | 14.3 | 11.0 | 8.0 | 51.4 | 1,062,452 | 1,079,702 | 1.6 | -1.1 | 116,740 | 100,998 | -27,869 | 437,296 | 2.35 | 59.3 | 9.9 | 32.9 |
| Rockford, IL | 13.5 | 10.4 | 7.7 | 51.0 | 320,204 | 349,431 | 9.1 | -4.4 | 42,563 | 33,589 | -24,544 | 133,538 | 2.48 | 65.9 | 12.9 | 27.6 |
| Rocky Mount, NC | 14.3 | 12.0 | 7.9 | 52.8 | 143,026 | 152,369 | 6.5 | -4.4 | 17,430 | 17,393 | -6,745 | 58,531 | 2.44 | 71.5 | 16.3 | 25.3 |
| Rome, GA | 12.5 | 10.0 | 7.3 | 51.5 | 90,565 | 96,318 | 6.4 | 2.4 | 12,044 | 11,091 | 1,390 | 35,443 | 2.66 | 66.5 | 8.6 | 29.1 |
| Sacramento-Roseville-Folsom, CA | 12.5 | 9.6 | 6.6 | 51.1 | 1,796,857 | 2,149,151 | 19.6 | 10.5 | 278,110 | 176,733 | 125,360 | 854,279 | 2.72 | 66.1 | 10.3 | 25.3 |
| Saginaw, MI | 13.9 | 11.5 | 8.6 | 49.7 | 210,039 | 200,169 | -4.7 | -5.1 | 22,928 | 21,691 | -11,627 | 79,050 | 2.33 | 63.2 | 11.3 | 30.7 |
| St. Cloud, MN | 12.2 | 8.8 | 6.8 | 49.7 | 167,392 | 189,097 | 13.0 | 7.4 | 26,142 | 14,015 | 1,808 | 76,473 | 2.54 | 63.5 | 9.4 | 27.6 |
| St. George, UT | 10.6 | 12.5 | 10.2 | 50.4 | 90,354 | 138,115 | 52.9 | 33.9 | 22,746 | 12,994 | 36,905 | 63,144 | 2.78 | 75.2 | 7.8 | 18.8 |
| St. Joseph, MO-KS | 13.6 | 10.2 | 7.8 | 49.1 | 122,336 | 127,323 | 4.1 | -3.7 | 15,603 | 13,561 | -6,783 | 46,958 | 2.49 | 60.0 | 11.0 | 34.5 |
| St. Louis, MO-IL | 14.0 | 10.3 | 7.3 | 51.5 | 2,675,343 | 2,787,786 | 4.2 | 0.6 | 342,967 | 272,411 | -52,185 | 1,145,583 | 2.40 | 67.6 | 10.1 | 31.4 |
| Salem, OR | 11.7 | 10.0 | 6.9 | 50.3 | 347,214 | 390,745 | 12.5 | 11.8 | 53,500 | 35,420 | 28,305 | 152,285 | 2.77 | 67.6 | 10.0 | 26.2 |
| Salinas, CA | 11.0 | 8.5 | 6.0 | 49.2 | 401,762 | 415,058 | 3.3 | 3.8 | 64,707 | 25,951 | -22,797 | 128,227 | 3.25 | 72.4 | 12.6 | 21.4 |
| Salisbury, MD-DE | 15.2 | 15.6 | 10.1 | 51.6 | 312,572 | 373,751 | 19.6 | 13.3 | 42,859 | 44,668 | 51,314 | 166,834 | 2.41 | 65.6 | 9.7 | 28.0 |
| Salt Lake City, UT | 9.9 | 7.0 | 4.4 | 49.7 | 939,122 | 1,087,808 | 15.8 | 14.0 | 187,551 | 69,815 | 35,106 | 405,984 | 3.00 | 69.6 | 8.7 | 22.6 |
| San Angelo, TX | 11.3 | 9.3 | 6.9 | 50.0 | 107,174 | 112,968 | 5.4 | 8.8 | 16,216 | 10,967 | 4,630 | 44,657 | 2.57 | 64.8 | 12.1 | 30.1 |
| San Antonio-New Braunfels, TX | 11.2 | 8.2 | 5.4 | 50.5 | 1,711,703 | 2,142,525 | 25.2 | 20.9 | 332,108 | 172,814 | 286,973 | 830,598 | 3.03 | 67.7 | 12.7 | 26.4 |
| San Diego-Chula Vista-Carlsbad, CA | 11.8 | 8.7 | 6.2 | 49.6 | 2,813,833 | 3,095,352 | 10.0 | 7.7 | 436,256 | 217,725 | 21,049 | 1,132,434 | 2.87 | 67.4 | 9.2 | 23.8 |
| San Francisco-Oakland-Berkeley, CA | 12.6 | 9.4 | 7.0 | 50.5 | 4,123,740 | 4,335,608 | 5.1 | 8.3 | 524,805 | 308,729 | 145,933 | 1,721,576 | 2.70 | 62.7 | 7.1 | 27.4 |
| Oakland-Berkeley-Livermore, CA Div 36084 | 12.6 | 9.1 | 6.4 | 50.9 | 2,392,557 | 2,559,471 | 7.0 | 10.0 | 320,360 | 179,859 | 115,181 | 985,424 | 2.83 | 67.6 | 8.3 | 24.7 |
| San Francisco-San Mateo-Redwood City, CA Div 41884 | 12.2 | 9.4 | 7.4 | 49.7 | 1,483,894 | 1,523,712 | 2.7 | 6.6 | 181,212 | 108,981 | 29,038 | 630,854 | 2.57 | 55.3 | 5.3 | 30.8 |
| San Rafael, CA Div 42034 | 15.0 | 13.3 | 10.3 | 51.1 | 247,289 | 252,425 | 2.1 | 1.9 | 23,233 | 19,889 | 1,714 | 105,298 | 2.40 | 61.3 | 6.7 | 31.2 |
| San Jose-Sunnyvale-Santa Clara, CA | 12.0 | 7.9 | 6.3 | 49.3 | 1,735,819 | 1,836,945 | 5.8 | 7.3 | 242,199 | 106,077 | -683 | 662,842 | 2.95 | 70.6 | 7.7 | 21.0 |
| San Luis Obispo-Paso Robles, CA | 13.1 | 12.7 | 8.7 | 49.5 | 246,681 | 269,599 | 9.3 | 4.7 | 26,424 | 24,329 | 10,603 | 106,512 | 2.51 | 63.2 | 5.7 | 26.7 |
| Santa Cruz-Watsonville, CA | 13.0 | 11.4 | 6.6 | 50.6 | 255,602 | 262,345 | 2.6 | 2.9 | 29,038 | 18,157 | -3,207 | 97,710 | 2.66 | 62.6 | 8.2 | 28.0 |
| Santa Fe, NM | 15.1 | 16.4 | 9.9 | 51.7 | 129,292 | 144,237 | 11.6 | 5.3 | 13,243 | 11,800 | 6,407 | 62,182 | 2.38 | 59.2 | 10.4 | 33.0 |
| Santa Maria-Santa Barbara, CA | 11.1 | 8.9 | 7.1 | 50.0 | 399,347 | 423,947 | 6.2 | 4.9 | 57,494 | 31,893 | -4,565 | 146,466 | 2.90 | 62.9 | 11.6 | 27.9 |
| Santa Rosa-Petaluma, CA | 14.1 | 12.8 | 8.4 | 51.2 | 458,614 | 483,866 | 5.5 | 1.2 | 50,154 | 42,154 | -1,784 | 190,689 | 2.55 | 64.1 | 9.5 | 26.9 |
| Savannah, GA | 12.0 | 9.2 | 6.0 | 51.6 | 293,000 | 347,608 | 18.6 | 13.9 | 52,104 | 31,443 | 27,585 | 142,579 | 2.67 | 64.3 | 12.7 | 29.0 |
| Scranton--Wilkes-Barre, PA | 14.1 | 11.6 | 9.0 | 50.8 | 560,625 | 563,635 | 0.5 | -2 | 57,727 | 72,292 | 3,854 | 230,210 | 2.32 | 62.2 | 11.8 | 31.6 |
| Seattle-Tacoma-Bellevue, WA | 12.0 | 8.5 | 5.5 | 49.8 | 3,043,878 | 3,439,808 | 13.0 | 16.8 | 470,764 | 252,700 | 360,242 | 1,541,245 | 2.54 | 62.6 | 7.7 | 27.4 |
| Seattle-Bellevue-Kent, WA Div 42644 | 12.0 | 8.4 | 5.5 | 49.7 | 2,343,058 | 2,644,588 | 12.9 | 17.4 | 354,443 | 186,303 | 291,463 | 1,207,588 | 2.50 | 61.6 | 7.0 | 27.8 |
| Tacoma-Lakewood, WA Div 45704 | 12.2 | 8.8 | 5.6 | 50.1 | 700,820 | 795,220 | 13.5 | 14.9 | 116,321 | 66,397 | 68,779 | 333,657 | 2.66 | 66.1 | 10.4 | 25.9 |
| Sebastian-Vero Beach, FL | 15.6 | 18.0 | 16.5 | 52.0 | 112,947 | 138,028 | 22.2 | 17.7 | 12,969 | 20,372 | 31,790 | 63,917 | 2.47 | 65.5 | 7.1 | 29.1 |
| Sebring-Avon Park, FL | 13.2 | 17.1 | 19.4 | 51.3 | 87,366 | 98,786 | 13.1 | 7.9 | 9,120 | 15,875 | 14,684 | 43,351 | 2.41 | 62.4 | 5.6 | 29.7 |
| Sheboygan, WI | 14.8 | 11.2 | 7.9 | 49.5 | 112,646 | 115,509 | 2.5 | -0.2 | 12,894 | 11,309 | -1,819 | 48,691 | 2.31 | 61.2 | 6.6 | 32.3 |
| Sherman-Denison, TX | 13.7 | 10.9 | 7.5 | 51.3 | 110,595 | 120,869 | 9.3 | 14.4 | 16,034 | 14,524 | 15,990 | 49,024 | 2.73 | 64.9 | 9.5 | 30.3 |
| Shreveport-Bossier City, LA | 12.7 | 10.1 | 7.2 | 52.0 | 375,965 | 398,614 | 6.0 | -1.6 | 57,200 | 41,944 | -21,580 | 153,824 | 2.52 | 63.9 | 14.4 | 30.7 |
| Sierra Vista-Douglas, AZ | 13.0 | 13.5 | 10.3 | 49.2 | 117,755 | 131,359 | 11.6 | -3 | 15,790 | 13,272 | -6,564 | 51,982 | 2.31 | 62.5 | 9.5 | 30.8 |
| Sioux City, IA-NE-SD | 12.1 | 9.5 | 6.4 | 49.9 | 143,053 | 143,576 | 0.4 | 1.0 | 21,568 | 13,365 | -6,935 | 56,092 | 2.52 | 65.7 | 13.6 | 26.3 |
| Sioux Falls, SD | 12.0 | 8.9 | 5.3 | 49.7 | 187,093 | 228,268 | 22.0 | 19.8 | 39,602 | 18,373 | 23,935 | 107,499 | 2.43 | 64.6 | 8.8 | 28.3 |
| South Bend-Mishawaka, IN-MI | 12.8 | 10.4 | 7.1 | 51.0 | 316,663 | 319,201 | 0.8 | 1.2 | 41,027 | 32,268 | -4,785 | 126,331 | 2.45 | 62.7 | 11.4 | 31.5 |
| Spartanburg, SC | 12.7 | 10.0 | 6.7 | 51.5 | 253,791 | 284,329 | 12.0 | 14.7 | 37,858 | 31,185 | 35,265 | 121,256 | 2.57 | 68.8 | 11.9 | 27.3 |
| Spokane-Spokane Valley, WA | 13.0 | 10.7 | 6.8 | 50.3 | 458,005 | 514,752 | 12.4 | 11.6 | 65,311 | 50,539 | 45,337 | 229,880 | 2.40 | 61.8 | 9.0 | 28.8 |
| Springfield, IL | 14.2 | 11.3 | 7.8 | 52.0 | 201,437 | 210,168 | 4.3 | -2 | 24,672 | 21,541 | -7,275 | 88,394 | 2.29 | 59.1 | 11.2 | 34.3 |
| Springfield, MA | 13.8 | 11.1 | 7.6 | 52.1 | 680,014 | 693,059 | 1.9 | 0.4 | 69,448 | 65,819 | -1,024 | 270,899 | 2.43 | 60.8 | 12.6 | 31.1 |
| Springfield, MO | 12.2 | 9.7 | 7.4 | 51.1 | 368,374 | 436,756 | 18.6 | 8.8 | 58,176 | 44,152 | 24,520 | 198,019 | 2.29 | 60.5 | 9.3 | 29.8 |
| Springfield, OH | 13.7 | 11.7 | 8.5 | 51.4 | 144,742 | 138,339 | -4.4 | -3.4 | 16,273 | 17,744 | -3,203 | 54,406 | 2.41 | 65.9 | 13.6 | 29.3 |
| State College, PA | 11.7 | 9.0 | 6.5 | 47.5 | 135,758 | 154,008 | 13.4 | 4.9 | 12,487 | 10,114 | 5,109 | 58,963 | 2.44 | 56.3 | 4.4 | 31.6 |
| Staunton, VA | 14.4 | 12.0 | 9.2 | 50.8 | 108,988 | 118,498 | 8.7 | 5.0 | 12,934 | 13,823 | 6,904 | 51,142 | 2.28 | 62.3 | 8.7 | 32.0 |
| Stockton, CA | 11.2 | 7.9 | 5.4 | 50.1 | 563,598 | 685,298 | 21.6 | 12.1 | 103,227 | 55,216 | 34,946 | 230,241 | 3.22 | 74.6 | 14.3 | 20.2 |
| Sumter, SC | 13.3 | 11.2 | 7.9 | 51.9 | 137,148 | 142,434 | 3.9 | -1.9 | 18,320 | 15,333 | -5,606 | 56,459 | 2.42 | 67.3 | 15.2 | 28.9 |
| Syracuse, NY | 14.4 | 10.6 | 7.6 | 51.3 | 650,154 | 662,634 | 1.9 | -2.5 | 73,874 | 61,293 | -29,262 | 255,844 | 2.42 | 60.1 | 10.9 | 32.7 |
| Tallahassee, FL | 11.5 | 9.5 | 6.0 | 52.0 | 320,304 | 368,762 | 15.1 | 5.7 | 41,172 | 28,175 | 7,616 | 152,648 | 2.38 | 55.9 | 12.3 | 31.7 |
| Tampa-St. Petersburg-Clearwater, FL | 13.8 | 11.4 | 9.0 | 51.5 | 2,395,997 | 2,783,491 | 16.2 | 16.5 | 325,766 | 317,478 | 450,537 | 1,228,481 | 2.56 | 60.8 | 9.2 | 31.4 |
| Terre Haute, IN | 12.8 | 10.5 | 7.6 | 49.5 | 188,184 | 189,778 | 0.8 | -2.2 | 22,038 | 21,382 | -4,761 | 72,899 | 2.37 | 64.1 | 11.6 | 27.9 |

# Table C. Metropolitan Areas — Population, Vital Statistics, Health, and Crime

| Area name | Persons in group quarters, 2020 | Daytime population, 2019 — Number | Daytime population, 2019 — Employment/residence ratio | Births, 2020 — Total | Births, 2020 — Rate[1] | Deaths, 2020 — Number | Deaths, 2020 — Rate[1] | Persons under 65 with no health insurance 2019 — Number | Persons under 65 with no health insurance 2019 — Percent | Medicare, 2020 — Total Beneficiaries | Medicare, 2020 — Enrolled in Original Medicare | Medicare, 2020 — Enrolled in Medicare Advantage | Serious crimes known to police[2], 2019 Violent — Number | Serious crimes known to police[2], 2019 Violent — Rate[3] |
|---|---|---|---|---|---|---|---|---|---|---|---|---|---|---|
| | 32 | 33 | 34 | 35 | 36 | 37 | 38 | 39 | 40 | 41 | 42 | 43 | 44 | 45 |
| Reading, PA | 11,407 | 395,668 | 0.88 | 4,564 | 10.8 | 4,204 | 10.0 | 25,575 | 7.6 | 86,262 | 53,563 | 32,699 | NA | NA |
| Redding, CA | 2,924 | 179,504 | 0.99 | 1,918 | 10.7 | 2,381 | 13.3 | 11,387 | 8.1 | 47,823 | 44,986 | 2,838 | NA | NA |
| Reno, NV | 5,295 | 480,834 | 1.02 | 5,410 | 11.2 | 4,476 | 9.3 | 48,639 | 12.4 | 92,693 | 61,192 | 31,501 | 2,231 | 469.0 |
| Richmond, VA | 33,392 | 1,304,572 | 1.02 | 14,825 | 11.4 | 12,505 | 9.6 | 98,056 | 9.3 | 239,614 | 169,306 | 70,310 | 2,932 | 227.7 |
| Riverside-San Bernardino-Ontario, CA | 72,691 | 4,401,539 | 0.88 | 57,191 | 12.2 | 36,229 | 7.7 | 388,520 | 9.9 | 694,999 | 279,106 | 415,893 | 19,730 | 425.9 |
| Roanoke, VA | 7,900 | 325,647 | 1.09 | 3,182 | 10.1 | 3,902 | 12.4 | 22,056 | 9.1 | 77,453 | 53,865 | 23,589 | 748 | 238.9 |
| Rochester, MN | 3,190 | 230,201 | 1.07 | 2,773 | 12.4 | 1,754 | 7.9 | 10,379 | 5.7 | 41,474 | 27,320 | 14,155 | 337 | 152.5 |
| Rochester, NY | 39,779 | 1,073,183 | 1.01 | 10,737 | 10.1 | 10,757 | 10.1 | 36,842 | 4.4 | 237,459 | 78,030 | 159,430 | 2,647 | 248.7 |
| Rockford, IL | 5,052 | 333,657 | 0.98 | 3,941 | 11.8 | 3,386 | 10.1 | 22,270 | 8.2 | 69,167 | 41,501 | 27,666 | 2,124 | 634.5 |
| Rocky Mount, NC | 3,115 | 141,732 | 0.93 | 1,633 | 11.2 | 1,904 | 13.1 | 14,134 | 12.4 | 34,438 | 23,381 | 11,056 | NA | NA |
| Rome, GA | 3,725 | 99,276 | 1.02 | 1,088 | 11.0 | 1,252 | 12.7 | 14,101 | 18.1 | 20,813 | 12,918 | 7,895 | NA | NA |
| Sacramento-Roseville-Folsom, CA | 38,818 | 2,351,164 | 0.99 | 26,438 | 11.1 | 20,367 | 8.6 | 126,618 | 6.5 | 430,051 | 216,882 | 213,168 | 8,322 | 354.1 |
| Saginaw, MI | 6,736 | 203,325 | 1.16 | 2,132 | 11.2 | 2,276 | 12.0 | 8,608 | 5.8 | 45,811 | 21,892 | 23,920 | 1,180 | 623.6 |
| St. Cloud, MN | 7,892 | 207,035 | 1.05 | 2,559 | 12.6 | 1,508 | 7.4 | 10,752 | 6.5 | 35,730 | 19,923 | 15,807 | 371 | 184.8 |
| St. George, UT | 2,259 | 179,741 | 1.03 | 2,213 | 12.0 | 1,511 | 8.2 | 20,380 | 15.0 | 40,871 | 27,164 | 13,706 | 299 | 169.9 |
| St. Joseph, MO-KS | 6,541 | 124,601 | 0.99 | 1,425 | 11.6 | 1,304 | 10.6 | 12,364 | 12.9 | 25,253 | 20,310 | 4,943 | 511 | 404.8 |
| St. Louis, MO-IL | 54,284 | 2,818,470 | 1.01 | 31,780 | 11.3 | 29,617 | 10.6 | 197,062 | 8.6 | 555,017 | 302,105 | 252,914 | NA | NA |
| Salem, OR | 12,277 | 420,859 | 0.93 | 4,965 | 11.4 | 3,944 | 9.0 | 34,999 | 10.0 | 84,856 | 34,517 | 50,339 | 1,176 | 270.1 |
| Salinas, CA | 17,092 | 428,768 | 0.97 | 5,799 | 13.5 | 2,892 | 6.7 | 42,273 | 11.9 | 66,756 | 59,648 | 7,108 | 1,433 | 329.9 |
| Salisbury, MD-DE | 13,763 | 407,898 | 0.96 | 4,153 | 9.8 | 5,368 | 12.7 | 28,033 | 9.3 | 114,993 | 99,986 | 15,009 | 1,370 | 331.5 |
| Salt Lake City, UT | 15,063 | 1,306,731 | 1.11 | 17,109 | 13.8 | 7,740 | 6.2 | 122,575 | 11.3 | 153,862 | 83,480 | 70,382 | 4,536 | 366.7 |
| San Angelo, TX | 5,018 | 122,396 | 1.02 | 1,491 | 12.1 | 1,137 | 9.3 | 17,304 | 17.6 | 22,961 | 14,791 | 8,171 | 398 | 328.3 |
| San Antonio-New Braunfels, TX | 49,099 | 2,528,719 | 0.98 | 32,333 | 12.5 | 19,743 | 7.6 | 402,935 | 18.6 | 410,165 | 222,001 | 188,163 | 13,166 | 515.9 |
| San Diego-Chula Vista-Carlsbad, CA | 110,390 | 3,370,234 | 1.02 | 39,080 | 11.7 | 25,040 | 7.5 | 247,271 | 8.9 | 544,476 | 274,208 | 270,268 | 11,417 | 341.2 |
| San Francisco-Oakland-Berkeley, CA | 90,437 | 4,827,553 | 1.04 | 48,795 | 10.4 | 35,813 | 7.6 | 220,183 | 5.6 | 787,282 | 427,318 | 359,964 | 22,317 | 470.8 |
| Oakland-Berkeley-Livermore, CA Div 36084 | 46,994 | 2,638,576 | 0.86 | 29,957 | 10.6 | 21,176 | 7.5 | 134,869 | 5.7 | 450,152 | 238,357 | 211,794 | 13,872 | 490.9 |
| San Francisco-San Mateo-Redwood City, CA Div 41884 | 35,027 | 1,934,473 | 1.31 | 16,773 | 10.3 | 12,454 | 7.7 | 74,908 | 5.4 | 277,490 | 154,161 | 123,329 | 7,929 | 479.0 |
| San Rafael, CA Div 42034 | 8,416 | 254,504 | 0.97 | 2,065 | 8.0 | 2,183 | 8.5 | 10,406 | 5.3 | 59,640 | 34,800 | 24,841 | 516 | 199.7 |
| San Jose-Sunnyvale-Santa Clara, CA | 30,805 | 2,134,690 | 1.14 | 21,878 | 11.1 | 12,493 | 6.3 | 96,132 | 5.6 | 282,579 | 156,129 | 126,450 | 6,508 | 325.0 |
| San Luis Obispo-Paso Robles, CA | 16,154 | 286,222 | 1.02 | 2,423 | 8.6 | 2,641 | 9.4 | 14,763 | 7.0 | 65,649 | 53,715 | 11,934 | 630 | 222.3 |
| Santa Cruz-Watsonville, CA | 13,687 | 255,405 | 0.87 | 2,402 | 8.9 | 2,096 | 7.8 | 17,026 | 7.9 | 52,198 | 43,837 | 8,362 | 1,106 | 404.6 |
| Santa Fe, NM | 2,634 | 151,912 | 1.02 | 1,163 | 7.7 | 1,336 | 8.8 | 15,703 | 14.2 | 41,546 | 26,525 | 15,021 | NA | NA |
| Santa Maria-Santa Barbara, CA | 19,942 | 456,435 | 1.05 | 5,390 | 12.1 | 3,550 | 8.0 | 43,058 | 12.0 | 78,006 | 64,839 | 13,167 | NA | NA |
| Santa Rosa-Petaluma, CA | 10,527 | 478,940 | 0.94 | 4,353 | 8.9 | 4,751 | 9.7 | 30,207 | 7.8 | 108,116 | 56,246 | 51,870 | 2,016 | 405.1 |
| Savannah, GA | 15,463 | 402,732 | 1.05 | 4,899 | 12.4 | 3,644 | 9.2 | 49,440 | 15.4 | 66,959 | 37,624 | 29,336 | NA | NA |
| Scranton--Wilkes-Barre, PA | 19,996 | 553,388 | 1 | 5,479 | 9.9 | 7,255 | 13.1 | 30,807 | 7.2 | 133,629 | 91,174 | 42,455 | 2,220 | 401.0 |
| Seattle-Tacoma-Bellevue, WA | 71,386 | 4,037,391 | 1.03 | 45,730 | 11.4 | 29,254 | 7.3 | 230,425 | 6.8 | 611,595 | 351,521 | 260,074 | NA | NA |
| Seattle-Bellevue-Kent, WA Div 42644 | 50,623 | 3,201,052 | 1.08 | 34,397 | 11.1 | 21,412 | 6.9 | 172,798 | 6.5 | 456,095 | 254,678 | 201,417 | NA | NA |
| Tacoma-Lakewood, WA Div 45704 | 20,763 | 836,339 | 0.84 | 11,333 | 12.4 | 7,842 | 8.6 | 57,627 | 7.6 | 155,500 | 96,843 | 58,657 | NA | NA |
| Sebastian-Vero Beach, FL | 1,326 | 163,731 | 1.06 | 1,277 | 7.9 | 2,416 | 14.9 | 19,417 | 18.6 | 54,766 | 36,923 | 17,843 | 352 | 221.6 |
| Sebring-Avon Park, FL | 1,734 | 104,863 | 0.96 | 835 | 7.8 | 1,807 | 16.9 | 13,536 | 20.4 | 34,125 | 19,311 | 14,814 | 280 | 265.3 |
| Sheboygan, WI | 2,818 | 118,801 | 1.06 | 1,229 | 10.7 | 1,150 | 10.0 | 5,748 | 6.3 | 25,354 | 12,529 | 12,825 | 207 | 179.5 |
| Sherman-Denison, TX | 2,234 | 132,910 | 0.95 | 1,676 | 12.1 | 1,545 | 11.2 | 24,027 | 21.8 | 29,343 | 19,373 | 9,970 | 342 | 253.6 |
| Shreveport-Bossier City, LA | 9,021 | 396,726 | 1.01 | 4,869 | 12.4 | 4,448 | 11.3 | 31,250 | 9.8 | 78,179 | 53,038 | 25,140 | 2,225 | 563.2 |
| Sierra Vista-Douglas, AZ | 5,393 | D | D | 1,335 | 10.5 | 1,442 | 11.3 | 10,599 | 11.5 | 33,438 | 22,351 | 11,088 | 275 | 217.9 |
| Sioux City, IA-NE-SD | 2,929 | 145,766 | 1.02 | 2,006 | 13.8 | 1,291 | 8.9 | 10,631 | 8.9 | 26,621 | 18,703 | 7,919 | 452 | 315.6 |
| Sioux Falls, SD | 7,018 | 273,051 | 1.03 | 3,890 | 14.2 | 1,988 | 7.3 | 22,797 | 10.0 | 48,583 | 35,051 | 13,532 | 1,043 | 388.1 |
| South Bend-Mishawaka, IN-MI | 12,105 | 316,774 | 0.95 | 3,971 | 12.3 | 3,222 | 10.0 | 25,282 | 9.8 | 63,330 | 37,971 | 25,359 | NA | NA |
| Spartanburg, SC | 7,711 | 330,469 | 1.07 | 3,942 | 12.1 | 3,518 | 10.8 | 33,742 | 13.0 | 66,614 | 37,418 | 29,195 | 1,599 | 503.0 |
| Spokane-Spokane Valley, WA | 14,523 | 574,354 | 1.02 | 6,275 | 10.9 | 5,663 | 9.9 | 32,442 | 7.1 | 121,405 | 71,336 | 50,069 | 2,148 | 381.0 |
| Springfield, IL | 4,171 | 220,203 | 1.14 | 2,270 | 11.0 | 2,463 | 12.0 | 10,719 | 6.5 | 44,474 | 26,066 | 18,409 | 1,205 | 584.0 |
| Springfield, MA | 38,964 | 684,844 | 0.96 | 6,236 | 9.0 | 6,832 | 9.8 | 21,121 | 3.9 | 156,832 | 106,956 | 49,876 | 3,294 | 471.7 |
| Springfield, MO | 14,308 | 474,944 | 1.04 | 5,683 | 12.0 | 4,877 | 10.3 | 50,950 | 13.5 | 96,132 | 47,271 | 48,861 | 2,976 | 632.7 |
| Springfield, OH | 3,293 | 123,831 | 0.83 | 1,530 | 11.5 | 1,785 | 13.4 | 8,896 | 8.5 | 31,048 | 14,935 | 16,114 | 330 | 246.5 |
| State College, PA | 19,160 | 172,445 | 1.13 | 1,137 | 7.0 | 1,208 | 7.5 | 9,766 | 8.2 | 25,121 | 12,256 | 12,864 | 200 | 122.2 |
| Staunton, VA | 4,099 | 117,996 | 0.92 | 1,267 | 10.2 | 1,516 | 12.2 | 9,234 | 9.9 | 30,256 | 24,567 | 5,689 | 210 | 170.9 |
| Stockton, CA | 15,882 | 726,546 | 0.89 | 9,957 | 13.0 | 6,339 | 8.3 | 53,036 | 8.2 | 114,638 | 64,537 | 50,101 | 5,952 | 788.5 |
| Sumter, SC | 3,831 | 139,789 | 0.99 | 1,668 | 11.9 | 1,641 | 11.7 | 14,123 | 12.9 | 31,360 | 21,231 | 10,129 | 1,006 | 718.6 |
| Syracuse, NY | 25,514 | 656,077 | 1.02 | 6,977 | 10.8 | 6,406 | 9.9 | 24,966 | 4.9 | 136,568 | 68,458 | 68,110 | 1,807 | 280.0 |
| Tallahassee, FL | 22,799 | 393,462 | 1.04 | 3,846 | 9.9 | 3,191 | 8.2 | 38,627 | 12.6 | 67,231 | 31,182 | 36,049 | 2,028 | 527.5 |
| Tampa-St. Petersburg-Clearwater, FL | 51,059 | 3,189,821 | 1 | 31,898 | 9.8 | 36,719 | 11.3 | 382,967 | 15.2 | 687,983 | 311,273 | 376,711 | 9,254 | 292.1 |
| Terre Haute, IN | 14,008 | 184,931 | 0.99 | 2,009 | 10.8 | 2,191 | 11.8 | 14,045 | 10.0 | 40,187 | 29,246 | 10,943 | NA | NA |

1. Per 1,000 estimated resident population.  2. Data for serious crimes have not been adjusted for underreporting; this may affect comparability between geographic areas and over time.  3. Per 100,000 population estimated by the FBI.

## Table C. Metropolitan Areas — Crime, Education, Money Income, and Poverty

| Area name | Serious crimes known to police[1], 2019 (cont.) Property Number | Rate | School enrollment and attainment, 2019 Enrollment[3] Total | Percent private | Attainment[4] High school graduate or less | Bachelor's degree or more | Local government expenditures,[5] 2017-2018 Total current expenditures (mil dol) | Current expenditures per student (dollars) | Income and poverty, 2019 Per capita income[6] (dollars) | Median household income (dollars) | Median family income | Percent of households with income less than $50,000 | Percent of households with income of $200,000 or more | Percent below poverty level All persons | All families | Families with children under 18 |
|---|---|---|---|---|---|---|---|---|---|---|---|---|---|---|---|---|
| | 46 | 47 | 48 | 49 | 50 | 51 | 52 | 53 | 54 | 55 | 56 | 57 | 58 | 59 | 60 | 61 |
| Reading, PA | NA | NA | 103,842 | 17.2 | 50.2 | 25.4 | 1,061.4 | 15,648 | 32,252 | 67,708 | 81,575 | 37.2 | 5.2 | 10.1 | 7.0 | 13.9 |
| Redding, CA | NA | NA | 40,899 | 15.1 | 31.6 | 21.8 | 347.7 | 12,909 | 32,864 | 63,091 | 80,765 | 41.4 | 5.4 | 12.6 | 7.5 | 11.5 |
| Reno, NV | 8,883 | 1,867.2 | 113,668 | 11.4 | 33.7 | 31.6 | 633.5 | 9,366 | 39,619 | 72,132 | 86,834 | 33.7 | 7.8 | 10.4 | 5.2 | 8.3 |
| Richmond, VA | 27,824 | 2,161.2 | 310,994 | 15.9 | 35.4 | 37.6 | 2,101.8 | 10,765 | 37,590 | 68,324 | 90,921 | 36.4 | 9.0 | 10.0 | 6.3 | 10.2 |
| Riverside-San Bernardino-Ontario, CA | 105,402 | 2,275.1 | 1,294,542 | 10.9 | 44.0 | 23 | 10,375.9 | 12,470 | 28,763 | 70,954 | 78,627 | 35.1 | 7.4 | 12.2 | 9.5 | 13.6 |
| Roanoke, VA | 7,487 | 2,391.4 | 65,432 | 16.1 | 41.2 | 28.9 | 510.5 | 11,491 | 32,576 | 60,471 | 77,537 | 41.2 | 5.8 | 11.8 | 8.8 | 17.9 |
| Rochester, MN | 2,813 | 1,272.8 | 51,616 | 14.8 | 28.9 | 41.2 | 411.5 | 11,562 | 40,989 | 75,926 | 98,058 | 30.7 | 9.0 | 6.9 | 3.6 | 6.4 |
| Rochester, NY | 17,819 | 1,674.5 | 250,596 | 23.6 | 34.3 | 37.1 | 3,085.9 | 19,750 | 35,147 | 62,104 | 82,078 | 40.2 | 6.0 | 12.4 | 8.1 | 14.5 |
| Rockford, IL | 7,233 | 2,160.9 | 81,635 | 18.4 | 42.9 | 24.5 | 828.2 | 15,248 | 30,798 | 61,078 | 72,411 | 41.6 | 4.7 | 16.1 | 12.4 | 23.1 |
| Rocky Mount, NC | NA | NA | 32,239 | 12.9 | 50.4 | 17.4 | 223.6 | 9,477 | 26,179 | 46,466 | 55,365 | 52.7 | 1.8 | 17.2 | 13.7 | 22.0 |
| Rome, GA | NA | NA | 23,098 | 18.2 | 51.7 | 22.9 | 181.7 | 11,170 | 25,581 | 50,880 | 63,562 | 49.5 | 3.2 | 16.6 | 10.5 | 19.1 |
| Sacramento-Roseville-Folsom, CA | 51,565 | 2,193.9 | 621,345 | 10.5 | 31.8 | 34.2 | 4,379.1 | 11,808 | 37,974 | 76,706 | 92,148 | 32.1 | 10.6 | 11.9 | 7.4 | 12.0 |
| Saginaw, MI | 2,731 | 1,443.3 | 44,961 | 12.7 | 43.3 | 21.8 | 322.5 | 12,043 | 29,451 | 48,303 | 61,301 | 52.0 | 3.3 | 19.0 | 13.1 | 23.3 |
| St. Cloud, MN | 4,102 | 2,043.1 | 55,814 | 17.0 | 37.9 | 26.6 | 378.5 | 12,005 | 31,236 | 66,076 | 79,572 | 37.5 | 3.8 | 11.3 | 5.9 | 10.8 |
| St. George, UT | 2,536 | 1,441.1 | 46,155 | 10.2 | 28.2 | 27.3 | 242.4 | 6,968 | 32,087 | 63,595 | 75,507 | 36.8 | 6.4 | 9.8 | 6.9 | 11.4 |
| St. Joseph, MO-KS | 4,889 | 3,872.5 | 26,659 | 8.5 | 52.5 | 20 | 186.1 | 10,343 | 25,947 | 50,425 | 65,695 | 49.7 | 3.0 | 14.3 | 11.8 | 17.8 |
| St. Louis, MO-IL | NA | NA | 669,832 | 24.9 | 33.5 | 35.8 | 5,131.2 | 12,646 | 37,365 | 66,417 | 87,044 | 37.6 | 7.5 | 9.9 | 6.8 | 11.4 |
| Salem, OR | 12,057 | 2,769.0 | 106,824 | 15.1 | 36.9 | 26.6 | 829.4 | 11,842 | 29,065 | 65,689 | 80,126 | 37.4 | 3.9 | 10.9 | 7.2 | 12.3 |
| Salinas, CA | 7,735 | 1,780.8 | 123,121 | 5.1 | 48.4 | 25.7 | 1,037.9 | 13,332 | 31,647 | 77,514 | 81,037 | 31.7 | 10.6 | 12.7 | 9.8 | 15.4 |
| Salisbury, MD-DE | 8,572 | 2,074.2 | 94,802 | 11.7 | 43.3 | 29.3 | 823.7 | 14,649 | 34,571 | 61,283 | 80,088 | 41.3 | 5.5 | 13.9 | 7.9 | 18.1 |
| Salt Lake City, UT | 41,240 | 3,333.6 | 351,860 | 13.0 | 31.5 | 36.5 | 1,818.0 | 7,694 | 34,445 | 80,196 | 91,939 | 27.8 | 8.2 | 8.6 | 5.5 | 7.7 |
| San Angelo, TX | 3,491 | 2,879.3 | 32,755 | 7.6 | 42.5 | 23.8 | 200.5 | 9,198 | 29,184 | 55,097 | 78,547 | 43.9 | 3.9 | 11.9 | 7.4 | 12.6 |
| San Antonio-New Braunfels, TX | 83,725 | 3,280.7 | 680,614 | 14.2 | 40.5 | 28.8 | 4,327.5 | 9,517 | 29,802 | 62,355 | 75,605 | 39.5 | 6.0 | 13.5 | 10.3 | 15.7 |
| San Diego-Chula Vista-Carlsbad, CA | 55,242 | 1,650.9 | 851,945 | 14.7 | 30.1 | 39.9 | 5,870.0 | 11,551 | 40,389 | 83,985 | 96,164 | 29.5 | 13.3 | 10.3 | 6.8 | 9.9 |
| San Francisco-Oakland-Berkeley, CA | 167,013 | 3,523.6 | 1,099,253 | 19.4 | 26.1 | 51.4 | 7,600.3 | 12,748 | 60,223 | 114,696 | 135,968 | 22.5 | 26.1 | 8.2 | 4.9 | 7.2 |
| Oakland-Berkeley-Livermore, CA Div 36084 | 95,286 | 3,371.7 | 701,532 | 15.5 | 28.1 | 47.5 | 4,803.2 | 11,818 | 51,847 | 107,809 | 128,467 | 22.9 | 22.3 | 8.5 | 5.4 | 8.0 |
| San Francisco-San Mateo-Redwood City, CA Div 41884 | 66,284 | 4,003.9 | 339,657 | 27.1 | 24.1 | 56.2 | 2,282.7 | 14,628 | 72,231 | 130,357 | 148,972 | 21.6 | 31.8 | 7.9 | 4.0 | 5.6 |
| San Rafael, CA Div 42034 | 5,443 | 2,107.0 | 58,064 | 21.9 | 18.6 | 59.7 | 514.5 | 15,249 | 75,175 | 110,843 | 151,376 | 23.9 | 28.6 | 6.7 | 3.8 | 6.1 |
| San Jose-Sunnyvale-Santa Clara, CA | 48,078 | 2,400.6 | 514,570 | 21.7 | 25.4 | 52.7 | 3,678.9 | 13,007 | 61,400 | 130,865 | 149,856 | 18.7 | 31.1 | 6.3 | 3.5 | 4.6 |
| San Luis Obispo-Paso Robles, CA | 5,119 | 1,806.1 | 75,133 | 9.5 | 28.7 | 37.2 | 431.8 | 12,432 | 38,308 | 77,265 | 98,277 | 34.1 | 9.2 | 11.9 | 5.9 | 9.2 |
| Santa Cruz-Watsonville, CA | 7,005 | 2,562.6 | 80,223 | 10.3 | 27.9 | 43.8 | 522.0 | 12,762 | 46,714 | 89,269 | 107,298 | 29.8 | 17.3 | 10.1 | 5.2 | 6.6 |
| Santa Fe, NM | NA | NA | 27,133 | 15.8 | 31.1 | 40.1 | 186.7 | 9,679 | 38,172 | 61,298 | 72,542 | 41.6 | 8.3 | 13.0 | 9.5 | 14.7 |
| Santa Maria-Santa Barbara, CA | NA | NA | 135,149 | 9.5 | 36.9 | 34.5 | 878.6 | 12,735 | 38,805 | 75,653 | 90,020 | 32.8 | 13.6 | 12.0 | 6.1 | 9.6 |
| Santa Rosa-Petaluma, CA | 6,760 | 1,358.4 | 113,552 | 14.3 | 29.3 | 37.4 | 930.0 | 13,201 | 46,006 | 87,828 | 101,449 | 25.5 | 13.3 | 6.8 | 3.1 | 4.8 |
| Savannah, GA | NA | NA | 102,950 | 20.5 | 35.8 | 32.4 | 645.0 | 10,911 | 31,541 | 60,371 | 75,041 | 41.0 | 6.1 | 12.4 | 8.4 | 12.7 |
| Scranton--Wilkes-Barre, PA | 6,974 | 1,259.6 | 122,037 | 27.1 | 47.9 | 24.4 | 1,095.0 | 14,426 | 29,882 | 54,304 | 66,665 | 46.4 | 3.8 | 14.9 | 11.4 | 20.6 |
| Seattle-Tacoma-Bellevue, WA | NA | NA | 917,875 | 16.7 | 26.4 | 44.1 | 7,107.7 | 13,101 | 49,184 | 94,027 | 112,094 | 25.1 | 16.3 | 7.8 | 4.7 | 7.3 |
| Seattle-Bellevue-Kent, WA Div 42644 | NA | NA | 702,549 | 17.7 | 23.7 | 48.6 | 5,354.5 | 13,156 | 52,840 | 99,750 | 121,089 | 24.0 | 18.7 | 7.4 | 4.4 | 6.5 |
| Tacoma-Lakewood, WA Div 45704 | NA | NA | 215,326 | 13.6 | 35.8 | 28.2 | 1,753.3 | 12,936 | 36,760 | 79,243 | 91,617 | 29.1 | 7.7 | 9.1 | 5.9 | 9.9 |
| Sebastian-Vero Beach, FL | 2,095 | 1,318.9 | 27,633 | 15.7 | 38.1 | 31.3 | 171.2 | 9,624 | 37,521 | 59,782 | 71,865 | 43.0 | 6.3 | 12.3 | 7.6 | 14.4 |
| Sebring-Avon Park, FL | 2,341 | 2,218.0 | 18,103 | 28.1 | 50.7 | 16.7 | 115.6 | 9,313 | 28,896 | 48,698 | 57,168 | 50.9 | 2.4 | 14.4 | 10.2 | 19.7 |
| Sheboygan, WI | 1,272 | 1,103.2 | 25,203 | 18.4 | 39.9 | 27.9 | 228.0 | 12,038 | 32,928 | 60,706 | 76,098 | 38.9 | 3.8 | 8.0 | 3.7 | 8.8 |
| Sherman-Denison, TX | 2,194 | 1,626.9 | 32,533 | 10.3 | 41.4 | 21.7 | 223.3 | 9,330 | 28,576 | 57,476 | 70,323 | 43.2 | 4.9 | 11.2 | 8.0 | 12.1 |
| Shreveport-Bossier City, LA | 13,978 | 3,538.4 | 99,795 | 10.9 | 46.8 | 22.8 | 781.0 | 11,650 | 27,807 | 47,447 | 58,959 | 52.4 | 4.5 | 22.2 | 17.5 | 27.6 |
| Sierra Vista-Douglas, AZ | 2,729 | 2,161.9 | 29,796 | 8.3 | 38.2 | 25 | 164.2 | 9,076 | 28,373 | 48,484 | 57,991 | 51.7 | 2.8 | 16.3 | 10.2 | 20.0 |
| Sioux City, IA-NE-SD | 4,013 | 2,801.7 | 38,501 | 12.3 | 46.9 | 22.6 | 329.1 | 12,115 | 28,415 | 60,132 | 77,450 | 42.3 | 3.9 | 11.5 | 10.6 | 16.2 |
| Sioux Falls, SD | 6,522 | 2,427.1 | 67,301 | 19.5 | 33.2 | 34.6 | 410.9 | 9,243 | 34,878 | 65,566 | 81,629 | 35.6 | 5.6 | 7.7 | 4.8 | 6.3 |
| South Bend-Mishawaka, IN-MI | NA | NA | 86,895 | 31.0 | 42.3 | 28.4 | 473.5 | 10,312 | 29,148 | 53,471 | 69,737 | 46.0 | 3.8 | 15.9 | 11.9 | 19.6 |
| Spartanburg, SC | 8,729 | 2,745.7 | 75,305 | 17.6 | 41.9 | 26.3 | 540.9 | 11,113 | 27,969 | 55,339 | 67,229 | 43.9 | 4.0 | 12.8 | 9.5 | 15.0 |
| Spokane-Spokane Valley, WA | 21,682 | 3,846.2 | 136,542 | 17.5 | 30.2 | 29.8 | 1,076.9 | 13,025 | 32,168 | 59,646 | 75,380 | 41.8 | 5.1 | 13.2 | 7.2 | 10.3 |
| Springfield, IL | 6,047 | 2,930.5 | 48,881 | 12.8 | 35.1 | 33.7 | 410.7 | 12,973 | 35,802 | 62,205 | 85,809 | 39.9 | 5.2 | 12.2 | 8.3 | 14.6 |
| Springfield, MA | 12,211 | 1,748.7 | 185,395 | 17.4 | 40.5 | 34.2 | 1,643.9 | 16,494 | 33,702 | 62,346 | 82,165 | 40.8 | 6.3 | 12.6 | 7.7 | 11.9 |
| Springfield, MO | 17,261 | 3,669.8 | 114,012 | 16.7 | 40.7 | 27.5 | 631.0 | 9,330 | 27,630 | 47,034 | 62,228 | 51.9 | 2.8 | 15.4 | 9.3 | 13.6 |
| Springfield, OH | 3,554 | 2,654.5 | 30,727 | 17.4 | 51.4 | 16.2 | 256.9 | 12,400 | 27,212 | 50,128 | 62,358 | 49.9 | 2.3 | 14.6 | 11.5 | 16.6 |
| State College, PA | 1,466 | 895.6 | 57,844 | 12.6 | 35.4 | 43.8 | 225.6 | 17,101 | 31,470 | 60,706 | 87,538 | 40.4 | 6.8 | 16.3 | 5.1 | 5.5 |
| Staunton, VA | 1,756 | 1,428.8 | 22,947 | 25.5 | 49.2 | 24.2 | 181.7 | 11,191 | 30,079 | 57,844 | 75,232 | 42.2 | 1.6 | 7.3 | 5.4 | 8.9 |
| Stockton, CA | 21,409 | 2,836.2 | 222,889 | 10.8 | 49.5 | 20 | 1,718.6 | 11,932 | 28,919 | 68,997 | 76,424 | 35.2 | 7.8 | 13.7 | 10.4 | 15.5 |
| Sumter, SC | 4,411 | 3,150.7 | 36,221 | 14.4 | 46.6 | 20.1 | 226.1 | 10,422 | 24,594 | 49,263 | 56,098 | 50.5 | 2.2 | 19.4 | 16.4 | 31.5 |
| Syracuse, NY | 10,822 | 1,676.8 | 166,148 | 23.1 | 37.9 | 32.5 | 1,909.8 | 19,102 | 33,615 | 61,213 | 81,983 | 41.2 | 6.3 | 13.9 | 10.1 | 17.9 |
| Tallahassee, FL | 11,266 | 2,930.5 | 119,491 | 13.0 | 32.6 | 38.7 | 413.0 | 8,516 | 29,728 | 52,729 | 73,224 | 47.5 | 5.2 | 20.5 | 12.0 | 19.9 |
| Tampa-St. Petersburg-Clearwater, FL | 53,394 | 1,685.6 | 690,293 | 19.3 | 38.8 | 31.6 | 3,743.2 | 9,019 | 33,116 | 57,906 | 73,874 | 42.8 | 6.5 | 12.4 | 8.0 | 13.7 |
| Terre Haute, IN | NA | NA | 43,704 | 11.0 | 46.9 | 20.4 | 278.9 | 10,497 | 26,046 | 50,129 | 62,043 | 49.9 | 3.0 | 18.0 | 12.5 | 19.8 |

1. Data for serious crimes have not been adjusted for underreporting; this may affect comparability between geographic areas and over time. 2. Per 100,000 population estimated by the FBI. 3. All persons 3 years old and over enrolled in nursery school through college. 4. Persons 25 years old and over. 5. Elementary and secondary education expenditures. 6. Based on population estimated by the American Community Survey, 2017.

# Table C. Metropolitan Areas — **Personal Income and Earnings**

| Area name | Personal income, 2019 | | | | | | | | | | Earnings, 2019 | | |
|---|---|---|---|---|---|---|---|---|---|---|---|---|---|
| | Total (mil dol) | Percent change, 2018-2019 | Per capita[1] Dollars | Rank | Wages and Salaries (mil dol) | Supplements to wages and salaries, employer contributions (mil dol) Pension and insurance | Government social insurance | Proprietors' income | Dividends, interest, and rent (mil dol) | Personal transfer receipts (mil dol) | Total (mil dol) | Contributions for government social insurance (mil dol) From employee and self-employed | From employer |
| | 62 | 63 | 64 | 65 | 66 | 67 | 68 | 69 | 70 | 71 | 72 | 73 | 74 |
| Reading, PA .................... | 21,605 | 3.9 | 51,299 | 125 | 9,605 | 1,825 | 763 | 1,639 | 3,429 | 4,461 | 13,832 | 830 | 763 |
| Redding, CA .................... | 8,723 | 4.2 | 48,438 | 170 | 3,365 | 709 | 243 | 701 | 1,710 | 2,565 | 5,017 | 335 | 243 |
| Reno, NV .................... | 30,121 | 3.9 | 63,328 | 31 | 13,475 | 2,227 | 994 | 2,359 | 9,080 | 4,046 | 19,055 | 1,129 | 994 |
| Richmond, VA .................... | 75,742 | 3.5 | 58,628 | 56 | 40,178 | 5,987 | 2,793 | 6,835 | 15,426 | 11,342 | 55,793 | 3,401 | 2,793 |
| Riverside-San Bernardino-Ontario, CA .................... | 196,453 | 5.5 | 42,242 | 297 | 81,700 | 16,199 | 5,893 | 14,479 | 31,129 | 39,051 | 118,272 | 7,124 | 5,893 |
| Roanoke, VA .................... | 15,152 | 2.8 | 48,374 | 172 | 7,842 | 1,249 | 571 | 867 | 3,229 | 3,321 | 10,529 | 699 | 571 |
| Rochester, MN .................... | 12,439 | 3.0 | 56,050 | 71 | 7,575 | 1,035 | 548 | 787 | 2,101 | 1,930 | 9,945 | 611 | 548 |
| Rochester, NY .................... | 56,937 | 3.9 | 53,230 | 100 | 27,495 | 5,644 | 2,163 | 4,377 | 9,794 | 12,050 | 39,679 | 2,236 | 2,163 |
| Rockford, IL .................... | 15,052 | 1.4 | 44,783 | 254 | 7,864 | 1,442 | 548 | 638 | 2,276 | 3,177 | 10,492 | 631 | 548 |
| Rocky Mount, NC .................... | 6,041 | 3.6 | 41,444 | 314 | 2,580 | 475 | 184 | 296 | 1,106 | 1,689 | 3,536 | 240 | 184 |
| Rome, GA .................... | 3,872 | 3.0 | 39,314 | 353 | 1,891 | 319 | 132 | 313 | 661 | 1,042 | 2,655 | 168 | 132 |
| Sacramento-Roseville-Folsom, CA .................... | 139,089 | 5.1 | 58,843 | 53 | 68,221 | 14,260 | 4,457 | 11,178 | 26,049 | 24,049 | 98,116 | 5,387 | 4,457 |
| Saginaw, MI .................... | 7,723 | 2.4 | 40,533 | 333 | 4,167 | 733 | 312 | 408 | 1,163 | 2,362 | 5,620 | 380 | 312 |
| St. Cloud, MN .................... | 9,580 | 3.5 | 47,436 | 199 | 5,328 | 882 | 410 | 785 | 1,643 | 1,838 | 7,405 | 450 | 410 |
| St. George, UT .................... | 7,260 | 5.6 | 40,886 | 330 | 2,916 | 468 | 223 | 613 | 1,955 | 1,559 | 4,220 | 289 | 223 |
| St. Joseph, MO-KS .................... | 5,026 | 3.8 | 40,135 | 341 | 2,731 | 509 | 199 | 321 | 746 | 1,234 | 3,760 | 233 | 199 |
| St. Louis, MO-IL .................... | 159,567 | 2.9 | 56,923 | 65 | 83,474 | 13,472 | 5,794 | 9,583 | 33,902 | 26,530 | 112,323 | 6,845 | 5,794 |
| Salem, OR .................... | 19,399 | 4.7 | 44,709 | 257 | 8,917 | 1,827 | 783 | 1,793 | 3,457 | 4,629 | 13,320 | 831 | 783 |
| Salinas, CA .................... | 25,973 | 5.7 | 59,838 | 48 | 10,889 | 2,244 | 847 | 4,205 | 5,875 | 3,671 | 18,184 | 891 | 847 |
| Salisbury, MD-DE .................... | 20,567 | 3.2 | 49,473 | 147 | 7,385 | 1,377 | 556 | 2,190 | 4,188 | 5,442 | 11,508 | 754 | 556 |
| Salt Lake City, UT .................... | 67,120 | 5.6 | 54,450 | 87 | 44,972 | 6,516 | 3,299 | 5,658 | 14,980 | 7,782 | 60,444 | 3,511 | 3,299 |
| San Angelo, TX .................... | 6,017 | 4.0 | 49,308 | 150 | 2,567 | 469 | 180 | 609 | 1,392 | 1,179 | 3,825 | 209 | 180 |
| San Antonio-New Braunfels, TX .................... | 124,192 | 4.4 | 48,684 | 166 | 60,054 | 9,703 | 4,092 | 14,303 | 22,385 | 21,401 | 88,153 | 4,761 | 4,092 |
| San Diego-Chula Vista-Carlsbad, CA .................... | 212,749 | 4.4 | 63,729 | 29 | 110,232 | 19,530 | 7,805 | 16,487 | 48,053 | 29,821 | 154,055 | 8,692 | 7,805 |
| San Francisco-Oakland-Berkeley, CA .................... | 496,467 | 4.8 | 104,921 | 4 | 271,330 | 32,646 | 15,316 | 48,258 | 124,062 | 41,882 | 367,550 | 20,539 | 15,316 |
| Oakland-Berkeley-Livermore, CA Div 36084 .................... | 234,087 | 5.2 | 82,867 | NA | 96,900 | 14,798 | 6,412 | 19,804 | 48,607 | 25,501 | 137,914 | 7,787 | 7,787 |
| San Francisco-San Mateo-Redwood City, CA Div 41884 .................... | 225,695 | 4.5 | 136,941 | NA | 165,098 | 16,460 | 8,305 | 23,651 | 62,768 | 13,936 | 213,514 | 11,850 | 11,850 |
| San Rafael, CA Div 42034 ...... | 36,685 | 3.8 | 141,735 | NA | 9,332 | 1,388 | 599 | 4,804 | 12,688 | 2,445 | 16,123 | 901 | 901 |
| San Jose-Sunnyvale-Santa Clara, CA .................... | 227,095 | 4.9 | 114,080 | 3 | 164,451 | 14,633 | 8,024 | 17,109 | 47,712 | 14,910 | 204,217 | 11,616 | 8,024 |
| San Luis Obispo-Paso Robles, CA .................... | 17,271 | 4.9 | 61,004 | 38 | 6,648 | 1,393 | 451 | 2,094 | 4,478 | 2,602 | 10,586 | 622 | 451 |
| Santa Cruz-Watsonville, CA .... | 19,560 | 4.6 | 71,592 | 16 | 6,400 | 1,183 | 435 | 2,322 | 4,633 | 2,433 | 10,340 | 577 | 435 |
| Santa Fe, NM .................... | 9,063 | 3.1 | 60,276 | 43 | 3,152 | 496 | 228 | 594 | 3,066 | 1,570 | 4,470 | 318 | 228 |
| Santa Maria-Santa Barbara, CA .................... | 29,503 | 4.5 | 66,076 | 23 | 12,614 | 2,344 | 899 | 3,927 | 8,940 | 3,671 | 19,784 | 1,073 | 899 |
| Santa Rosa-Petaluma, CA .... | 32,972 | 4.3 | 66,700 | 20 | 13,247 | 2,208 | 945 | 3,794 | 8,504 | 4,866 | 20,195 | 1,204 | 945 |
| Savannah, GA .................... | 18,830 | 2.8 | 47,869 | 186 | 9,678 | 1,619 | 685 | 1,159 | 3,684 | 3,384 | 13,141 | 768 | 685 |
| Scranton--Wilkes-Barre, PA .... | 26,246 | 2.8 | 47,385 | 202 | 12,105 | 2,356 | 981 | 1,567 | 4,408 | 6,709 | 17,009 | 1,084 | 981 |
| Seattle-Tacoma-Bellevue, WA .................... | 310,718 | 5.5 | 78,073 | 9 | 179,887 | 22,245 | 13,530 | 24,985 | 71,199 | 30,584 | 240,648 | 13,557 | 13,530 |
| Seattle-Bellevue-Kent, WA Div 42644 .................... | 262,237 | 5.5 | 85,284 | NA | 159,435 | 18,535 | 11,678 | 21,488 | 62,260 | 22,279 | 211,137 | 11,862 | 11,862 |
| Tacoma-Lakewood, WA Div 45704 .................... | 48,481 | 6.0 | 53,572 | NA | 20,452 | 3,710 | 1,852 | 3,497 | 8,939 | 8,305 | 29,511 | 1,695 | 1,695 |
| Sebastian-Vero Beach, FL ..... | 12,925 | 3.6 | 80,818 | 7 | 2,679 | 384 | 184 | 661 | 6,121 | 2,151 | 3,908 | 305 | 184 |
| Sebring-Avon Park, FL .......... | 3,663 | 4.0 | 34,480 | 380 | 1,138 | 204 | 82 | 177 | 765 | 1,418 | 1,601 | 140 | 82 |
| Sheboygan, WI .................... | 6,309 | 2.8 | 54,703 | 84 | 3,349 | 592 | 245 | 590 | 1,204 | 1,010 | 4,776 | 297 | 245 |
| Sherman-Denison, TX .......... | 5,991 | 5.4 | 43,987 | 266 | 2,285 | 377 | 156 | 332 | 925 | 1,466 | 3,151 | 197 | 156 |
| Shreveport-Bossier City, LA .... | 19,068 | 2.1 | 48,310 | 174 | 8,561 | 1,599 | 572 | 1,500 | 4,026 | 4,418 | 12,232 | 694 | 572 |
| Sierra Vista-Douglas, AZ ...... | 5,259 | 4.2 | 41,766 | 309 | 2,005 | 477 | 159 | 242 | 1,051 | 1,749 | 2,883 | 198 | 159 |
| Sioux City, IA-NE-SD .............. | 7,644 | 3.8 | 52,829 | 105 | 3,666 | 626 | 275 | 837 | 1,577 | 1,289 | 5,404 | 334 | 275 |
| Sioux Falls, SD .................... | 16,631 | 4.7 | 62,003 | 35 | 8,290 | 1,284 | 593 | 2,667 | 3,418 | 1,974 | 12,833 | 767 | 593 |
| South Bend-Mishawaka, IN-MI | 15,545 | 1.8 | 48,035 | 181 | 6,911 | 1,111 | 511 | 1,492 | 2,735 | 3,096 | 10,025 | 630 | 511 |
| Spartanburg, SC .................... | 14,125 | 4.4 | 44,169 | 264 | 7,729 | 1,338 | 577 | 788 | 2,806 | 2,994 | 10,431 | 678 | 577 |
| Spokane-Spokane Valley, WA | 27,191 | 5.1 | 47,827 | 188 | 13,208 | 2,265 | 1,171 | 1,715 | 5,672 | 6,033 | 18,359 | 1,125 | 1,171 |
| Springfield, IL .................... | 10,199 | 2.3 | 49,301 | 151 | 5,504 | 1,119 | 362 | 565 | 1,974 | 1,962 | 7,550 | 421 | 362 |
| Springfield, MA .................... | 37,643 | 2.8 | 53,978 | 94 | 16,138 | 3,370 | 1,120 | 2,543 | 5,971 | 9,250 | 23,172 | 1,253 | 1,120 |
| Springfield, MO .................... | 20,037 | 4.2 | 42,605 | 294 | 10,000 | 1,810 | 712 | 1,715 | 3,449 | 4,292 | 14,238 | 899 | 712 |
| Springfield, OH .................... | 5,606 | 2.7 | 41,811 | 307 | 2,224 | 402 | 158 | 265 | 849 | 1,538 | 3,049 | 200 | 158 |
| State College, PA .................... | 7,724 | 3.6 | 47,564 | 192 | 4,134 | 1,465 | 306 | 637 | 1,539 | 1,211 | 6,541 | 337 | 306 |
| Staunton, VA .................... | 5,621 | 3.4 | 45,658 | 231 | 2,314 | 398 | 166 | 340 | 1,117 | 1,234 | 3,219 | 222 | 166 |
| Stockton, CA .................... | 35,927 | 6.8 | 47,139 | 205 | 13,780 | 2,759 | 1,018 | 3,239 | 5,482 | 7,890 | 20,796 | 1,189 | 1,018 |
| Sumter, SC .................... | 5,434 | 3.9 | 38,687 | 360 | 2,352 | 536 | 191 | 253 | 885 | 1,614 | 3,332 | 217 | 191 |
| Syracuse, NY .................... | 33,760 | 3.9 | 52,051 | 116 | 16,578 | 3,718 | 1,294 | 2,787 | 5,302 | 7,091 | 24,377 | 1,347 | 1,294 |
| Tallahassee, FL .................... | 17,128 | 3.6 | 44,232 | 262 | 8,905 | 1,683 | 609 | 904 | 3,435 | 3,230 | 12,101 | 739 | 609 |
| Tampa-St. Petersburg-Clearwater, FL .................... | 156,253 | 4.6 | 48,908 | 159 | 79,280 | 11,326 | 5,348 | 7,519 | 31,133 | 33,004 | 103,473 | 6,828 | 5,348 |
| Terre Haute, IN .................... | 7,241 | 3.1 | 38,851 | 358 | 3,183 | 612 | 234 | 362 | 1,218 | 2,080 | 4,391 | 297 | 234 |

1. Based on the resident population estimated as of July 1 of the year shown.

# Table C. Metropolitan Areas — Earnings, Social Security, and Housing

| Area name | Earnings, 2019 (cont.) | | | | | | | | | Social Security beneficiaries, December 2019 | | Supplemental Security Income Recipients, December 2019 | Housing units, 2020 | |
|---|---|---|---|---|---|---|---|---|---|---|---|---|---|---|
| | Percent by selected industries | | | | | | | | | | | | | |
| | Farm | Mining, quarrying, and extracting | Construction | Manufacturing | Information; professional, scientific, and technical services | Retail trade | Finance, insurance, real estate, rental and leasing | Health care and social assistance | Government | Number | Rate[1] | | Total | Percent change, 2010-2020 |
| | 75 | 76 | 77 | 78 | 79 | 80 | 81 | 82 | 83 | 84 | 85 | 86 | 87 | 88 |
| Reading, PA | 1.0 | 0.1 | 6.5 | 19.0 | 7.0 | 5.7 | 5.3 | 15.0 | 12.4 | 89,330 | 213 | 10,740 | 168,408 | 2.2 |
| Redding, CA | 1.1 | D | 7.3 | 3.7 | 6.4 | 9.2 | 5.6 | 19.1 | 22.2 | 49,295 | 274 | 9,270 | 78,280 | 1.3 |
| Reno, NV | 0.0 | D | 10.4 | 11.0 | D | D | D | 13.5 | 13.5 | 90,780 | 190 | 6,869 | 212,752 | 13.9 |
| Richmond, VA | 0.1 | 0.1 | 5.8 | 5.2 | 11.7 | D | D | 11.4 | 17.5 | 246,310 | 190 | 26,794 | 537,584 | 7.7 |
| Riverside-San Bernardino-Ontario, CA | 0.7 | 0.3 | 8.8 | 6.6 | 5.1 | 7.7 | 4.2 | 11.9 | 22.9 | 705,475 | 152 | 133,870 | 1,588,149 | 5.9 |
| Roanoke, VA | -0.1 | D | D | 10.9 | D | 6.0 | D | 16.7 | 14.8 | 78,330 | 250 | 8,426 | 147,192 | 1.5 |
| Rochester, MN | 1.2 | 0.1 | 5.6 | 10.7 | 3.4 | 4.7 | D | 48.0 | 9.3 | 42,070 | 190 | 2,785 | 97,962 | 11.1 |
| Rochester, NY | 0.5 | 1.0 | 6.0 | 12.8 | D | 5.5 | 5.8 | 14.0 | 16.9 | 247,060 | 231 | 32,807 | 485,888 | 3.6 |
| Rockford, IL | 0.0 | D | 5.5 | 25.8 | 4.5 | 5.7 | 6.0 | 16.7 | 12.7 | 72,795 | 217 | 8,113 | 145,877 | 0.0 |
| Rocky Mount, NC | 2.3 | 0.1 | 6.8 | 21.0 | D | 7.8 | D | D | 19.2 | 36,440 | 250 | 6,310 | 68,549 | 2.1 |
| Rome, GA | 0.3 | D | 2.8 | 17.1 | D | 6.0 | 4.7 | 26.2 | 13.6 | 22,795 | 232 | 3,244 | 40,964 | 1.0 |
| Sacramento-Roseville-Folsom, CA | 0.5 | D | 7.8 | 3.4 | 10.6 | 5.4 | 7.7 | 12.4 | 30.5 | 412,895 | 175 | 78,661 | 923,993 | 6.0 |
| Saginaw, MI | 0.1 | 0.3 | 4.8 | 18.9 | 6.4 | 7.6 | 6.3 | 19.2 | 13.7 | 50,150 | 263 | 8,483 | 88,162 | 1.5 |
| St. Cloud, MN | 2.5 | 0.2 | 10.2 | 14.0 | D | 7.5 | 6.2 | 15.3 | 13.8 | 36,840 | 182 | 3,332 | 83,774 | 7.2 |
| St. George, UT | 0.0 | 0.4 | 11.3 | 4.7 | 7.1 | 10.0 | 9.6 | 16.8 | 13.2 | 40,480 | 227 | 1,505 | 77,677 | 34.5 |
| St. Joseph, MO-KS | 2.1 | D | 7.3 | 24.3 | D | 6.2 | 5.5 | D | 14.2 | 26,925 | 218 | 2,755 | 54,234 | 1.1 |
| St. Louis, MO-IL | 0.1 | 0.2 | D | D | 12.6 | 5.1 | 9.8 | 13.1 | 12.0 | 573,935 | 205 | 56,768 | 1,273,084 | 3.9 |
| Salem, OR | 2.0 | D | 8.8 | 5.9 | 5.2 | 6.2 | 6.8 | 14.6 | 29.5 | 87,425 | 202 | 9,233 | 163,647 | 8.2 |
| Salinas, CA | 13.6 | 0.3 | 4.2 | 2.2 | 4.9 | 5.2 | 4.8 | 7.6 | 23.2 | 65,270 | 151 | 8,074 | 143,139 | 2.9 |
| Salisbury, MD-DE | 3.3 | 0.0 | 8.8 | 7.9 | D | 8.5 | 8.0 | 15.0 | 16.1 | 117,305 | 281 | 7,908 | 258,423 | 11.8 |
| Salt Lake City, UT | 0.0 | D | 7.2 | 8.0 | 14.8 | 7.0 | 12.4 | 7.4 | 14.5 | 154,750 | 126 | 14,129 | 444,913 | 16.0 |
| San Angelo, TX | 0.6 | 6.5 | D | D | D | 7.1 | D | D | 23.0 | 23,775 | 194 | 2,846 | 50,669 | 5.5 |
| San Antonio-New Braunfels, TX | 0.0 | 7.7 | 7.3 | 4.8 | 9.0 | 6.0 | 11.7 | D | 19.7 | 420,455 | 165 | 58,331 | 929,158 | 10.9 |
| San Diego-Chula Vista-Carlsbad, CA | 0.4 | 0.0 | 5.5 | 8.7 | 18.9 | 4.8 | 7.5 | 8.6 | 23.2 | 517,095 | 155 | 81,167 | 1,232,928 | 5.8 |
| San Francisco-Oakland-Berkeley, CA | 0.1 | D | 5.1 | 6.0 | 30.5 | 4.0 | 11.5 | 7.6 | 11.2 | 706,155 | 149 | 125,807 | 1,834,655 | 5.3 |
| Oakland-Berkeley-Livermore, CA Div 36084 | 0.1 | 0.2 | 7.5 | 9.3 | 19.1 | 5.0 | 7.6 | 11.3 | 14.3 | 410,415 | 145 | 73,063 | 1,035,618 | 5.4 |
| San Francisco-San Mateo-Redwood City, CA Div 41884 | 0.0 | D | 3.3 | 3.8 | 38.5 | 3.2 | 14.1 | 4.9 | 9.2 | 242,475 | 148 | 49,622 | 686,372 | 5.9 |
| San Rafael, CA Div 42034 | 0.3 | D | 7.6 | 6.3 | 19.8 | 6.3 | 11.1 | 10.4 | 11.4 | 53,265 | 206 | 3,122 | 112,665 | 1.3 |
| San Jose-Sunnyvale-Santa Clara, CA | 0.1 | D | 3.6 | 21.6 | D | 3.0 | 5.2 | 6.1 | 5.9 | 247,665 | 125 | 44,082 | 697,389 | 7.3 |
| San Luis Obispo-Paso Robles, CA | 2.5 | 0.1 | 10.4 | 5.8 | 10.9 | 8.2 | 5.6 | 10.4 | 19.9 | 62,275 | 220 | 4,277 | 124,014 | 5.7 |
| Santa Cruz-Watsonville, CA | 4.5 | 0.0 | 7.0 | 5.9 | 11.9 | 7.2 | 6.6 | 12.7 | 18.4 | 49,005 | 180 | 5,440 | 106,249 | 1.7 |
| Santa Fe, NM | 0.0 | 0.2 | D | D | 12.6 | 8.5 | 6.4 | 13.6 | 26.6 | 40,045 | 265 | 2,516 | 75,053 | 5.3 |
| Santa Maria-Santa Barbara, CA | 5.1 | 1.1 | 4.9 | 6.9 | 14.4 | 5.3 | 7.3 | 10.5 | 19.3 | 75,035 | 168 | 8,350 | 161,754 | 5.8 |
| Santa Rosa-Petaluma, CA | 1.4 | 0.1 | 10.8 | 12.4 | 10.3 | 6.6 | 6.8 | 13.4 | 13.3 | 101,715 | 207 | 8,533 | 206,444 | 0.9 |
| Savannah, GA | 0.1 | D | 4.9 | 16.1 | D | 6.6 | 6.1 | D | 17.8 | 70,770 | 180 | 8,493 | 169,055 | 11.9 |
| Scranton--Wilkes-Barre, PA | 0.0 | 0.4 | 5.4 | 11.6 | D | 6.7 | 6.6 | D | 14.4 | 141,750 | 256 | 17,814 | 265,264 | 2.5 |
| Seattle-Tacoma-Bellevue, WA | 0.1 | 0.2 | 6.5 | 9.1 | 24.2 | 9.6 | 6.4 | 8.8 | 13.3 | 585,185 | 147 | 67,547 | 1,670,079 | 14.1 |
| Seattle-Bellevue-Kent, WA Div 42644 | 0.1 | 0.2 | 6.2 | 9.6 | 26.9 | 10.1 | 6.5 | 7.8 | 11.1 | 428,145 | 139 | 48,314 | 1,309,601 | 15.1 |
| Tacoma-Lakewood, WA Div 45704 | 0.1 | 0.1 | 8.9 | 5.2 | 4.9 | 6.3 | 5.8 | 16.1 | 29.4 | 157,040 | 173 | 19,233 | 360,478 | 10.8 |
| Sebastian-Vero Beach, FL | 2.1 | 0.0 | 7.3 | 4.5 | 12.9 | 8.7 | 9.4 | 16.7 | 9.6 | 55,280 | 346 | 2,640 | 83,611 | 9.5 |
| Sebring-Avon Park, FL | 6.6 | D | 5.3 | 2.3 | 4.0 | 10.6 | 4.2 | 22.9 | 16.2 | 35,675 | 337 | 2,952 | 56,320 | 1.7 |
| Sheboygan, WI | 1.1 | D | 5.0 | 43.4 | 3.0 | 5.4 | 6.8 | 12.0 | 8.6 | 26,885 | 233 | 1,666 | 51,533 | 1.5 |
| Sherman-Denison, TX | -0.6 | 0.7 | 9.8 | 14.5 | 4.3 | 7.7 | 6.6 | 20.7 | 14.3 | 29,435 | 216 | 2,946 | 58,499 | 8.9 |
| Shreveport-Bossier City, LA | -0.1 | 6.7 | 4.6 | 6.7 | D | 7.2 | 5.1 | 16.7 | 22.7 | 80,910 | 205 | 18,058 | 185,972 | 7.1 |
| Sierra Vista-Douglas, AZ | 0.8 | 0.5 | 4.6 | 1.2 | 10.5 | 6.5 | 2.2 | 9.4 | 48.3 | 34,710 | 273 | 3,158 | 61,982 | 5.0 |
| Sioux City, IA-NE-SD | 2.3 | D | 7.5 | D | 4.5 | 5.9 | 7.0 | D | 11.7 | 28,040 | 194 | 2,381 | 61,019 | 5.1 |
| Sioux Falls, SD | 1.4 | D | 7.2 | D | 8.1 | 7.0 | D | 18.8 | 8.1 | 48,950 | 182 | 3,320 | 115,887 | 20.9 |
| South Bend-Mishawaka, IN-MI | 0.1 | 0.0 | 5.5 | 15.0 | 9.8 | 6.3 | 5.5 | D | 9.9 | 66,820 | 207 | 6,619 | 144,251 | 2.5 |
| Spartanburg, SC | 0.0 | D | 6.7 | 27.5 | 5.1 | 5.7 | 4.2 | 6.2 | 15.8 | 71,160 | 222 | 7,223 | 136,704 | 11.5 |
| Spokane-Spokane Valley, WA | 0.3 | D | 6.9 | 7.2 | D | 7.2 | 8.7 | D | 20.1 | 122,635 | 216 | 15,983 | 248,532 | 11.7 |
| Springfield, IL | 0.6 | D | 4.3 | 2.9 | 10.6 | 5.7 | 8.1 | D | 24.9 | 46,755 | 226 | 4,684 | 98,219 | 2.8 |
| Springfield, MA | 0.0 | 0.1 | 6.0 | 9.1 | 5.8 | 6.2 | 7.5 | 18.3 | 21.2 | 153,855 | 221 | 35,267 | 293,114 | 1.6 |
| Springfield, MO | 0.0 | 0.1 | 6.0 | 9.4 | 9.8 | 7.4 | 5.7 | D | 13.3 | 100,120 | 212 | 9,882 | 208,572 | 8.4 |
| Springfield, OH | 0.2 | 0.4 | 4.0 | 17.2 | 3.5 | 6.4 | 7.8 | 15.0 | 15.3 | 31,555 | 235 | 4,002 | 61,229 | -0.3 |
| State College, PA | 0.4 | 0.6 | 4.9 | 5.2 | 7.9 | 4.2 | 4.6 | 10.3 | 48.5 | 25,555 | 158 | 1,366 | 67,800 | 7.1 |
| Staunton, VA | 0.4 | 0.1 | D | 18.3 | 3.7 | 6.3 | 4.6 | D | 17.2 | 31,830 | 258 | 2,179 | 55,261 | 5.0 |
| Stockton, CA | 5.1 | 0.1 | 6.6 | 7.1 | 3.4 | 6.5 | 4.7 | 11.4 | 20.7 | 115,645 | 152 | 27,610 | 250,918 | 7.3 |
| Sumter, SC | 0.1 | 0.0 | 7.1 | 15.5 | 3.5 | 6.4 | 3.1 | 11.3 | 34.6 | 34,580 | 246 | 5,469 | 67,030 | 5.6 |
| Syracuse, NY | 0.4 | D | 4.9 | 10.0 | D | 5.8 | 7.0 | D | 21.5 | 143,490 | 221 | 18,796 | 297,497 | 3.4 |
| Tallahassee, FL | 0.9 | D | 5.1 | 2.1 | D | 5.7 | 5.9 | D | 34.2 | 70,450 | 182 | 10,200 | 177,455 | 8.8 |
| Tampa-St. Petersburg-Clearwater, FL | 0.3 | 0.0 | 6.1 | 5.6 | D | 7.2 | 12.0 | 13.2 | 12.0 | 711,155 | 222 | 80,013 | 1,471,870 | 8.8 |
| Terre Haute, IN | 0.4 | 3.1 | 6.8 | 17.5 | D | 7.1 | 4.5 | 15.5 | 18.9 | 43,635 | 234 | 4,829 | 84,099 | 2.3 |

1. Per 1,000 resident population estimated as of July 1 of the year shown.

# Table C. Metropolitan Areas — Housing, Labor Force, and Employment

| Area name | Occupied Housing units, 2020 | | | | | | | | Civilian labor force, 2020 | | | | Civilian employment, 2019[6] | | |
|---|---|---|---|---|---|---|---|---|---|---|---|---|---|---|---|
| | Occupied units | | | | | | | | | | | | | | |
| | Owner-occupied | | | | | Renter-occupied | | | | | Unemployment | | | Percent | |
| | | | | Median owner cost as a percent of income | | | | | | | | | | | |
| | Total | Percent | Median value[1] | With a mortgage | Without a mortgage[2] | Median rent[3] | Median rent as a percent of income[2] | Substandard units[4] (percent) | Total | Percent change 2019-2020 | Total | Rate[5] | Total | Management, business, science, and arts | Construction, production, and maintenance occupations |
| | 89 | 90 | 91 | 92 | 93 | 94 | 95 | 96 | 97 | 98 | 99 | 100 | 101 | 102 | 103 |
| Reading, PA | 154,696 | 69.6 | 188,100 | 20.0 | 12.9 | 936 | 27.3 | 1.2 | 211,486 | -2.0 | 20,043 | 9.5 | 210,794 | 32.6 | 28.2 |
| Redding, CA | 72,714 | 63.2 | 279,200 | 22.3 | 13.0 | 1,116 | 27.2 | 2.7 | 72,827 | 0.0 | 6,343 | 8.7 | 77,273 | 39.5 | 19.1 |
| Reno, NV | 192,402 | 58.5 | 383,000 | 21.4 | 10.0 | 1,195 | 28.8 | 3.9 | 256,328 | -2.0 | 19,905 | 7.8 | 243,079 | 33.8 | 25.3 |
| Richmond, VA | 491,891 | 66.3 | 251,200 | 19.8 | 10.0 | 1,155 | 31.4 | 1.2 | 668,528 | -1.7 | 44,822 | 6.7 | 663,567 | 42.2 | 19.7 |
| Riverside-San Bernardino-Ontario, CA | 1,379,706 | 63.9 | 378,500 | 24.9 | 11.2 | 1,413 | 33.1 | 7.4 | 2,073,929 | 0.2 | 201,293 | 9.7 | 2,028,548 | 31.5 | 27.2 |
| Roanoke, VA | 126,469 | 69.3 | 191,800 | 18.3 | 10.0 | 835 | 29.4 | 1.4 | 155,976 | -1.9 | 9,323 | 6.0 | 150,659 | 37.6 | 23.4 |
| Rochester, MN | 92,282 | 74.3 | 226,700 | 19.1 | 10.0 | 949 | 27.3 | 1.4 | 127,165 | 1.9 | 6,828 | 5.4 | 121,343 | 48.2 | 19.6 |
| Rochester, NY | 437,296 | 66.3 | 150,700 | 18.8 | 11.5 | 899 | 30.0 | 1.5 | 515,144 | -0.9 | 42,189 | 8.2 | 536,602 | 43.9 | 18.4 |
| Rockford, IL | 133,538 | 68.1 | 132,900 | 18.6 | 10.8 | 836 | 27.0 | 1.6 | 161,339 | -2.9 | 18,366 | 11.4 | 159,971 | 32.7 | 31.7 |
| Rocky Mount, NC | 58,531 | 62.8 | 113,900 | 19.9 | 11.5 | 783 | 29.9 | 3.1 | 62,831 | -2.7 | 5,730 | 9.1 | 62,259 | 30.0 | 32.8 |
| Rome, GA | 35,443 | 60.9 | 138,900 | 18.5 | 10.0 | 780 | 30.2 | 0.0 | 43,234 | -2.1 | 2,712 | 6.3 | 43,328 | 28.7 | 34.8 |
| Sacramento-Roseville-Folsom, CA | 854,279 | 60.4 | 434,400 | 23.3 | 10.0 | 1,390 | 32.3 | 4.4 | 1,087,852 | | 94,536 | 8.7 | 1,125,380 | 41.0 | 18.4 |
| Saginaw, MI | 79,050 | 72.1 | 108,300 | 18.5 | 12.1 | 761 | 31.7 | 1.3 | 84,807 | -2.3 | 8,410 | 9.9 | 84,299 | 33.6 | 23.4 |
| St. Cloud, MN | 76,473 | 67.8 | 204,800 | 19.1 | 10.0 | 849 | 24.7 | 2.7 | 112,925 | 0.3 | 6,678 | 5.9 | 108,232 | 35.0 | 27.2 |
| St. George, UT | 63,144 | 72.1 | 323,800 | 23.4 | 10.0 | 1,002 | 27.0 | 4.6 | 79,208 | 3.7 | 4,201 | 5.3 | 78,146 | 32.9 | 21.7 |
| St. Joseph, MO-KS | 46,958 | 63.3 | 122,400 | 17.8 | 11.0 | 762 | 28.6 | 3.0 | 62,520 | -0.8 | 2,948 | 4.7 | 55,421 | 31.7 | 34.5 |
| St. Louis, MO-IL | 1,145,583 | 69.2 | 184,600 | 18.8 | 10.6 | 883 | 27.3 | 1.3 | 1,455,197 | -1.8 | 97,428 | 6.7 | 1,407,091 | 42.9 | 20.6 |
| Salem, OR | 152,285 | 63.9 | 295,000 | 22.7 | 11.4 | 1,083 | 29.3 | 3.5 | 203,222 | 0.3 | 13,943 | 6.9 | 195,319 | 34.4 | 25.4 |
| Salinas, CA | 128,227 | 52.8 | 602,900 | 26.9 | 11.8 | 1,658 | 32.8 | 13.4 | 213,471 | -3.5 | 23,082 | 10.8 | 191,694 | 31.3 | 31.2 |
| Salisbury, MD-DE | 166,834 | 73.8 | 244,400 | 21.2 | 10.3 | 1,027 | 32.3 | 1.9 | 189,196 | -2.7 | 15,194 | 8.0 | 184,225 | 35.4 | 21.9 |
| Salt Lake City, UT | 405,984 | 68.2 | 356,400 | 20.4 | 10.0 | 1,181 | 28.0 | 3.3 | 677,441 | 1.3 | 34,258 | 5.1 | 651,786 | 41.4 | 22.7 |
| San Angelo, TX | 44,657 | 66.8 | 149,700 | 19.9 | 10.0 | 952 | 29.3 | 4.6 | 55,054 | -1.6 | 3,487 | 6.3 | 57,008 | 32.7 | 25.7 |
| San Antonio-New Braunfels, TX | 830,598 | 62.7 | 197,600 | 21.1 | 10.7 | 1,057 | 29.0 | 4.1 | 1,197,401 | -0.4 | 86,962 | 7.3 | 1,217,550 | 35.6 | 22.1 |
| San Diego-Chula Vista-Carlsbad, CA | 1,132,434 | 53.8 | 619,300 | 25.8 | 10.3 | 1,758 | 32.7 | 7.0 | 1,538,361 | -2.6 | 141,814 | 9.2 | 1,624,854 | 43.9 | 17.0 |
| San Francisco-Oakland-Berkeley, CA | 1,721,576 | 54 | 940,900 | 23.7 | 10.0 | 2,057 | 27.3 | 6.9 | 2,475,133 | -3.9 | 201,351 | 8.1 | 2,554,202 | 52.6 | 13.7 |
| Oakland-Berkeley-Livermore, CA Div 36084 | 985,424 | 58.1 | 798,500 | 23.6 | 10.0 | 1,970 | 29.6 | 6.6 | 1,355,063 | -3.3 | 119,416 | 8.8 | 1,462,723 | 49.8 | 16.0 |
| San Francisco-San Mateo-Redwood City, CA Div 41884 | 630,854 | 46.4 | 1,226,000 | 23.5 | 10.0 | 2,199 | 23.7 | 7.5 | 989,932 | -4.4 | 73,273 | 7.4 | 960,619 | 56.8 | 10.4 |
| San Rafael, CA Div 42034 | 105,298 | 62 | 1,078,800 | 26.0 | 12.5 | 2,096 | 32.1 | 5.7 | 130,138 | -6.1 | 8,662 | 6.7 | 130,860 | 53.5 | 11.9 |
| San Jose-Sunnyvale-Santa Clara, CA | 662,842 | 55.2 | 1,116,400 | 22.9 | 10.0 | 2,374 | 27.7 | 7.7 | 1,052,221 | -2.6 | 74,459 | 7.1 | 1,050,543 | 55.7 | 14.4 |
| San Luis Obispo-Paso Robles, CA | 106,512 | 62.4 | 637,000 | 26.4 | 11.9 | 1,654 | 33.1 | 3.5 | 132,690 | -4.9 | 10,195 | 7.7 | 134,530 | 39.4 | 18.3 |
| Santa Cruz-Watsonville, CA | 97,710 | 59.9 | 839,500 | 24.6 | 11.6 | 1,785 | 32.2 | 6.6 | 133,646 | -5.3 | 12,714 | 9.5 | 140,781 | 43.5 | 21.2 |
| Santa Fe, NM | 62,182 | 71.4 | 314,700 | 23.9 | 10.0 | 1,030 | 27.4 | 4.1 | 71,346 | -3.8 | 5,891 | 8.3 | 71,465 | 43.6 | 16.4 |
| Santa Maria-Santa Barbara, CA | 146,466 | 52.2 | 593,800 | 24.4 | 11.3 | 1,660 | 33.0 | 9.9 | 217,510 | -2.1 | 17,335 | 8.0 | 215,213 | 37.9 | 24.7 |
| Santa Rosa-Petaluma, CA | 190,689 | 62.2 | 664,600 | 25.8 | 10.4 | 1,757 | 31.5 | 5.6 | 245,091 | -4.7 | 19,300 | 7.9 | 258,846 | 41.4 | 19.3 |
| Savannah, GA | 142,579 | 60.5 | 201,000 | 20.7 | 11.2 | 1,097 | 29.5 | 1.6 | 190,749 | | 13,272 | 7.0 | 185,066 | 36.0 | 23.5 |
| Scranton--Wilkes-Barre, PA | 230,210 | 66.2 | 142,000 | 20.0 | 12.6 | 798 | 27.1 | 1.0 | 274,360 | -1.1 | 28,163 | 10.3 | 265,248 | 33.5 | 26.7 |
| Seattle-Tacoma-Bellevue, WA | 1,541,245 | 59.7 | 503,000 | 22.5 | 11.2 | 1,621 | 28.9 | 3.7 | 2,177,960 | 0.4 | 177,107 | 8.1 | 2,121,773 | 48.3 | 18.1 |
| Seattle-Bellevue-Kent, WA Div 42644 | 1,207,588 | 58.9 | 582,900 | 22.2 | 11.3 | 1,690 | 28.6 | 3.8 | 1,727,764 | 0.0 | 133,733 | 7.7 | 1,688,717 | 51.6 | 16.0 |
| Tacoma-Lakewood, WA Div 45704 | 333,657 | 62.8 | 362,100 | 23.3 | 10.9 | 1,362 | 30.5 | 3.2 | 450,196 | 1.9 | 43,374 | 9.6 | 433,056 | 35.6 | 26.2 |
| Sebastian-Vero Beach, FL | 63,917 | 77.3 | 235,300 | 22.3 | 10.0 | 1,004 | 32.3 | 0.0 | 64,233 | -2.1 | 4,751 | 7.4 | 64,578 | 32.7 | 19.9 |
| Sebring-Avon Park, FL | 43,351 | 80 | 144,900 | 20.5 | 10.0 | 807 | 29.7 | 0.0 | 34,641 | -3.3 | 2,674 | 7.7 | 36,416 | 28.7 | 24.3 |
| Sheboygan, WI | 48,691 | 71.5 | 169,100 | 19.2 | 11.8 | 750 | 22.2 | 1.5 | 61,510 | -0.7 | 3,514 | 5.7 | 60,390 | 33.2 | 33.2 |
| Sherman-Denison, TX | 49,024 | 69.2 | 167,900 | 22.4 | 11.8 | 906 | 28.9 | 3.2 | 64,214 | -0.1 | 3,761 | 5.9 | 62,938 | 28.8 | 29.3 |
| Shreveport-Bossier City, LA | 153,824 | 65 | 160,500 | 20.3 | 10.0 | 852 | 34.8 | 1.5 | 169,988 | -2.0 | 13,614 | 8.0 | 160,781 | 33.5 | 21.6 |
| Sierra Vista-Douglas, AZ | 51,982 | 67.8 | 154,000 | 20.3 | 10.0 | 735 | 24.4 | 3.0 | 51,995 | 3.4 | 3,627 | 7.0 | 44,200 | 35.7 | 20.3 |
| Sioux City, IA-NE-SD | 56,092 | 66.4 | 138,400 | 18.6 | 10.0 | 830 | 25.5 | 2.6 | 76,532 | -3.0 | 3,899 | 5.1 | 74,422 | 31.3 | 32.0 |
| Sioux Falls, SD | 107,499 | 65.3 | 222,300 | 19.4 | 10.5 | 851 | 25.3 | 2.0 | 156,605 | 0.7 | 6,725 | 4.3 | 148,269 | 39.8 | 23.4 |
| South Bend-Mishawaka, IN-MI | 126,331 | 70.7 | 140,700 | 18.0 | 10.0 | 854 | 31.8 | 1.8 | 157,855 | -3.1 | 13,331 | 8.4 | 152,256 | 36.7 | 25.0 |
| Spartanburg, SC | 121,256 | 72.6 | 158,800 | 18.7 | 10.0 | 841 | 29.6 | 2.4 | 158,701 | 2.6 | 10,427 | 6.6 | 146,410 | 33.6 | 31.4 |
| Spokane-Spokane Valley, WA | 229,880 | 62.7 | 258,500 | 21.6 | 10.0 | 934 | 28.2 | 2.2 | 275,853 | 0.9 | 24,265 | 8.8 | 256,460 | 39.8 | 20.0 |
| Springfield, IL | 88,394 | 67.5 | 147,000 | 17.9 | 10.0 | 809 | 27.9 | 1.3 | 102,970 | -3.2 | 8,344 | 8.1 | 96,479 | 46.3 | 16.6 |
| Springfield, MA | 270,899 | 63.1 | 241,600 | 22.2 | 12.7 | 952 | 31.7 | 1.5 | 345,421 | -3.7 | 32,007 | 9.3 | 346,129 | 41.0 | 20.8 |
| Springfield, MO | 198,019 | 58.9 | 161,700 | 19.3 | 10.0 | 761 | 29.6 | 2.2 | 234,694 | -0.1 | 12,305 | 5.2 | 213,766 | 34.9 | 22.9 |
| Springfield, OH | 54,406 | 67 | 118,300 | 18.4 | 10.1 | 739 | 28.9 | 1.6 | 62,945 | -1.1 | 5,116 | 8.1 | 60,900 | 28.0 | 30.9 |
| State College, PA | 58,963 | 60.6 | 261,400 | 19.7 | 10.0 | 995 | 33.4 | 2.0 | 76,416 | -4.5 | 4,461 | 5.8 | 79,550 | 49.1 | 15.4 |
| Staunton, VA | 51,142 | 71.9 | 199,900 | 19.3 | 10.0 | 836 | 25.3 | 0.0 | 59,993 | -1.4 | 3,167 | 5.3 | 60,916 | 33.2 | 28.3 |
| Stockton, CA | 230,241 | 58.4 | 385,600 | 24.3 | 10.9 | 1,260 | 31.5 | 8.6 | 331,828 | 1.6 | 37,374 | 11.3 | 326,476 | 29.3 | 30.9 |
| Sumter, SC | 56,459 | 66 | 126,600 | 19.8 | 10.2 | 786 | 27.4 | 0.0 | 56,271 | 0.5 | 3,810 | 6.8 | 55,660 | 29.8 | 30.5 |
| Syracuse, NY | 255,844 | 68.2 | 142,700 | 18.6 | 11.1 | 848 | 29.5 | 1.9 | 305,129 | 0.1 | 25,635 | 8.4 | 315,857 | 39.2 | 19.2 |
| Tallahassee, FL | 152,648 | 57.9 | 195,400 | 19.7 | 10.0 | 982 | 33.7 | 2.5 | 189,279 | -2.7 | 10,955 | 5.8 | 188,074 | 43.3 | 15.5 |
| Tampa-St. Petersburg-Clearwater, FL | 1,228,481 | 66.7 | 224,300 | 21.7 | 11.5 | 1,160 | 31.4 | 2.0 | 1,548,743 | -0.4 | 110,763 | 7.2 | 1,505,886 | 40.0 | 18.8 |
| Terre Haute, IN | 72,899 | 68.7 | 101,000 | 17.3 | 10.0 | 686 | 32.8 | 2.8 | 78,601 | -2.6 | 5,800 | 7.4 | 79,803 | 27.8 | 33.5 |

1. Specified owner-occupied units.   2. A value of 10.0 represents 10 percent or less; a value of 50.0 represents 50 percent or more.   3. Specified renter-occupied units.   4. Overcrowded or lacking complete plumbing facilities.   5. Percent of civilian labor force.   6. Civilian employed persons 16 years old and over.

# Table C. Metropolitan Areas — Nonfarm Employment and Agriculture

| Area name | Private nonfarm establishments, employment and payroll, 2019 — Employment | | | | | | | Annual payroll | | Agriculture, 2017 — Farms | | | Farm producers whose primary occupation is farming (percent) |
|---|---|---|---|---|---|---|---|---|---|---|---|---|---|
| | Number of establishments | Total | Health care and social assistance | Manufacturing | Retail trade | Finance and insurance | Professional, scientific, and technical services | Total (mil dol) | Average per employee (dollars) | Number | Fewer than 50 acres | 1000 acres or more | |
| | 104 | 105 | 106 | 107 | 108 | 109 | 110 | 111 | 112 | 113 | 114 | 115 | 116 |
| Reading, PA | 8,354 | 162,134 | 28,467 | 33,344 | 19,822 | 5,472 | 6,198 | 7,941 | 48,977 | 1,809 | 49 | 0.9 | 52.9 |
| Redding, CA | 4,337 | 50,946 | 10,801 | 2,188 | 9,423 | 2,061 | 2,248 | 2,252 | 44,209 | 1,337 | 71.4 | 4.6 | 31.5 |
| Reno, NV | 13,052 | 205,149 | 27,879 | 17,752 | 26,045 | 5,500 | 11,665 | 10,088 | 49,172 | 355 | 69.9 | 7.3 | 43.5 |
| Richmond, VA | 32,128 | 562,143 | 82,032 | 30,594 | 74,352 | 49,657 | 42,821 | 30,645 | 54,515 | 2,966 | 48.8 | 5.8 | 42.0 |
| Riverside-San Bernardino-Ontario, CA | 77,626 | 1,237,028 | 188,155 | 101,069 | 188,382 | 27,161 | 40,104 | 54,995 | 44,457 | 3,729 | 86.4 | 1.9 | 41.2 |
| Roanoke, VA | 8,109 | 148,114 | 27,718 | 15,833 | 17,756 | 8,586 | 5,345 | 6,552 | 44,239 | 2,011 | 36 | 1.5 | 39.3 |
| Rochester, MN | 5,136 | 115,028 | 23,160 | 8,204 | 13,294 | 2,185 | 36,284 | 6,476 | 56,295 | 3,960 | 34.6 | 6.6 | 45.2 |
| Rochester, NY | 24,610 | 457,157 | 88,984 | 55,449 | 57,854 | 14,918 | 29,700 | 21,754 | 47,585 | 4,215 | 41.9 | 4.9 | 53.2 |
| Rockford, IL | 7,134 | 134,792 | 21,318 | 34,645 | 15,415 | 3,515 | 4,716 | 6,314 | 46,842 | 1,193 | 52.1 | 6.1 | 42.6 |
| Rocky Mount, NC | 2,804 | 49,310 | 8,113 | 10,277 | 6,597 | 1,700 | 959 | 1,918 | 38,905 | 674 | 39.5 | 10.8 | 47.6 |
| Rome, GA | 2,012 | 37,509 | 8,352 | 6,710 | 4,249 | 720 | 758 | 1,498 | 39,945 | 547 | 51.7 | 2.0 | 35.2 |
| Sacramento-Roseville-Folsom, CA | 50,935 | 786,301 | 129,824 | 36,081 | 99,969 | 44,619 | 61,599 | 45,026 | 57,263 | 4,737 | 73 | 3.9 | 43.9 |
| Saginaw, MI | 4,180 | 79,213 | 17,570 | 11,867 | 12,348 | 2,761 | 2,422 | 3,396 | 42,867 | 1,250 | 44.2 | 5.8 | 42.1 |
| St. Cloud, MN | 5,338 | 101,218 | 20,541 | 15,942 | 13,284 | 4,296 | 2,644 | 4,747 | 46,900 | 3,767 | 26.1 | 2.7 | 48.0 |
| St. George, UT | 5,622 | 56,729 | 10,550 | 3,036 | 9,487 | 1,360 | 2,633 | 2,137 | 37,664 | 537 | 60.5 | 5.8 | 26.8 |
| St. Joseph, MO-KS | 2,752 | 49,610 | 8,914 | 12,709 | 6,728 | 1,876 | 1,388 | 2,271 | 45,785 | 2,641 | 32.3 | 7.6 | 37.9 |
| St. Louis, MO-IL | 71,770 | 1,253,072 | 192,186 | 112,592 | 139,523 | 69,626 | 88,407 | 67,253 | 53,670 | 11,057 | 40.9 | 6.3 | 37.1 |
| Salem, OR | 10,089 | 126,834 | 24,578 | 12,576 | 19,151 | 3,267 | 4,265 | 5,331 | 42,034 | 4,004 | 73.2 | 2.3 | 41.5 |
| Salinas, CA | 8,830 | 119,472 | 17,563 | 8,917 | 17,687 | 2,760 | 8,232 | 5,928 | 49,622 | 1,104 | 45.6 | 20.2 | 44.3 |
| Salisbury, MD-DE | 10,792 | 134,180 | 25,034 | 15,434 | 24,243 | 3,660 | 4,391 | 5,356 | 39,920 | 2,237 | 51.7 | 6.6 | 54.3 |
| Salt Lake City, UT | 33,964 | 645,410 | 68,474 | 56,554 | 65,460 | 48,249 | 49,277 | 35,294 | 54,684 | 1,132 | 75.8 | 4.7 | 30.5 |
| San Angelo, TX | 2,894 | 41,285 | 7,318 | 3,355 | 6,710 | 1,782 | 1,628 | 1,750 | 42,395 | 1,554 | 49.5 | 15.7 | 31.8 |
| San Antonio-New Braunfels, TX | 47,229 | 914,191 | 143,800 | 48,632 | 115,605 | 67,233 | 53,210 | 42,790 | 46,807 | 14,960 | 48.4 | 4.8 | 35.4 |
| San Diego-Chula Vista-Carlsbad, CA | 87,334 | 1,313,497 | 187,372 | 105,850 | 148,910 | 58,696 | 159,564 | 81,122 | 61,761 | 5,082 | 90.8 | 0.8 | 36.2 |
| San Francisco-Oakland-Berkeley, CA | 131,744 | 2,282,577 | 288,464 | 139,208 | 219,341 | 130,263 | 276,352 | 228,096 | 99,929 | 1,499 | 63.4 | 8.5 | 42.6 |
| Oakland-Berkeley-Livermore, CA Div 36084 | 65,325 | 1,056,776 | 161,680 | 102,311 | 115,132 | 43,259 | 101,815 | 78,908 | 74,668 | 905 | 69.2 | 7.1 | 41.1 |
| San Francisco-San Mateo-Redwood City, CA Div 41884 | 56,391 | 1,123,115 | 109,916 | 32,780 | 89,093 | 83,117 | 165,929 | 141,662 | 126,133 | 251 | 67.7 | 6.0 | 47.7 |
| San Rafael, CA Div 42034 | 10,028 | 102,686 | 16,868 | 4,117 | 15,116 | 3,887 | 8,608 | 7,526 | 73,293 | 343 | 44.9 | 14.0 | 42.6 |
| San Jose-Sunnyvale-Santa Clara, CA | 50,043 | 1,115,736 | 119,565 | 88,594 | 83,382 | 26,600 | 146,087 | 149,074 | 133,611 | 1,500 | 70.7 | 7.5 | 45.2 |
| San Luis Obispo-Paso Robles, CA | 8,538 | 96,358 | 15,676 | 7,219 | 14,544 | 2,061 | 5,368 | 4,785 | 49,659 | 2,349 | 56.4 | 7.8 | 40.8 |
| Santa Cruz-Watsonville, CA | 7,067 | 82,282 | 14,124 | 5,446 | 12,691 | 2,467 | 4,545 | 4,412 | 53,615 | 625 | 80.5 | 2.1 | 43.2 |
| Santa Fe, NM | 4,806 | 47,235 | 8,651 | 747 | 8,992 | 1,525 | 2,488 | 2,050 | 43,396 | 639 | 67.9 | 9.7 | 35.7 |
| Santa Maria-Santa Barbara, CA | 11,859 | 156,903 | 23,462 | 13,364 | 18,491 | 4,245 | 11,703 | 8,972 | 57,182 | 1,467 | 60.7 | 9.5 | 44.0 |
| Santa Rosa-Petaluma, CA | 14,297 | 178,775 | 27,440 | 22,213 | 24,905 | 6,139 | 8,611 | 10,109 | 56,548 | 3,594 | 73.7 | 3.5 | 40.6 |
| Savannah, GA | 9,567 | 162,286 | 20,797 | 18,990 | 21,899 | 2,708 | 5,725 | 7,171 | 44,186 | 416 | 59.1 | 6.0 | 36.9 |
| Scranton--Wilkes-Barre, PA | 12,960 | 239,964 | 48,289 | 30,736 | 30,824 | 9,564 | 9,686 | 10,222 | 42,597 | 1,124 | 33 | 1.4 | 38.9 |
| Seattle-Tacoma-Bellevue, WA | 108,500 | 1,806,755 | 245,618 | 164,990 | 182,677 | 70,498 | 147,724 | 138,323 | 76,559 | 4,961 | 88.8 | 0.2 | 33.9 |
| Seattle-Bellevue-Kent, WA Div 42644 | 89,911 | 1,535,900 | 191,022 | 147,199 | 144,724 | 61,498 | 137,849 | 124,590 | 81,118 | 3,354 | 89.4 | 0.3 | 34.9 |
| Tacoma-Lakewood, WA Div 45704 | 18,589 | 270,855 | 54,596 | 17,791 | 37,953 | 9,000 | 9,875 | 13,734 | 50,705 | 1,607 | 87.4 | 0.2 | 31.8 |
| Sebastian-Vero Beach, FL | 4,409 | 47,529 | 9,545 | 2,105 | 8,957 | 1,512 | 2,357 | 1,928 | 40,569 | 450 | 76.2 | 5.1 | 42.6 |
| Sebring-Avon Park, FL | 1,959 | 20,941 | 5,925 | 795 | 4,657 | 453 | 727 | 706 | 33,728 | 989 | 66.4 | 6.1 | 45.7 |
| Sheboygan, WI | 2,678 | 53,477 | 6,334 | 18,939 | 6,044 | 2,313 | 1,287 | 2,508 | 46,891 | 958 | 43.6 | 3.9 | 49.6 |
| Sherman-Denison, TX | 2,636 | 41,210 | 8,815 | 7,739 | 6,215 | 1,835 | 1,053 | 1,659 | 40,252 | 2,845 | 61.4 | 2.4 | 32.3 |
| Shreveport-Bossier City, LA | 9,119 | 143,113 | 35,477 | 7,491 | 20,169 | 4,100 | 6,046 | 5,992 | 41,867 | 1,824 | 49.9 | 7.0 | 39.5 |
| Sierra Vista-Douglas, AZ | 2,136 | 26,186 | 5,066 | 308 | 5,045 | 487 | 3,025 | 989 | 37,757 | 1,083 | 39.6 | 17.7 | 47.4 |
| Sioux City, IA-NE-SD | 3,693 | 70,475 | 11,636 | 15,238 | 8,912 | 2,988 | 1,466 | 3,064 | 43,471 | 2,428 | 29.7 | 15.0 | 45.4 |
| Sioux Falls, SD | 7,995 | 144,112 | 29,825 | 15,402 | 18,792 | 13,197 | 5,233 | 6,985 | 48,472 | 3,048 | 34.5 | 15.3 | 44.0 |
| South Bend-Mishawaka, IN-MI | 6,592 | 131,254 | 20,715 | 17,991 | 17,212 | 4,051 | 5,308 | 5,715 | 43,544 | 1,376 | 53.3 | 5.5 | 42.2 |
| Spartanburg, SC | 6,662 | 141,317 | 15,735 | 35,850 | 16,102 | 3,000 | 3,982 | 6,670 | 47,201 | 1,433 | 67.9 | 0.3 | 30.0 |
| Spokane-Spokane Valley, WA | 14,374 | 206,048 | 45,369 | 16,695 | 27,955 | 10,926 | 9,117 | 9,904 | 48,068 | 3,539 | 57.9 | 5.6 | 36.0 |
| Springfield, IL | 5,079 | 83,625 | 22,200 | 2,937 | 12,057 | 5,504 | 4,654 | 3,673 | 43,917 | 1,469 | 42.7 | 12.9 | 45.0 |
| Springfield, MA | 14,617 | 246,565 | 56,523 | 25,938 | 31,896 | 10,323 | 9,406 | 11,255 | 45,647 | 2,045 | 55.4 | 0.3 | 40.3 |
| Springfield, MO | 11,934 | 193,678 | 37,053 | 17,032 | 24,922 | 8,469 | 9,007 | 8,065 | 41,640 | 7,601 | 40.2 | 1.6 | 38.9 |
| Springfield, OH | 2,198 | 39,729 | 6,958 | 6,142 | 5,191 | 2,359 | 709 | 1,497 | 37,673 | 742 | 56.7 | 7.0 | 36.2 |
| State College, PA | 3,356 | 46,608 | 9,499 | 4,090 | 7,602 | 1,085 | 3,144 | 1,922 | 41,228 | 1,023 | 41.3 | 1.4 | 47.0 |
| Staunton, VA | 2,779 | 41,327 | 6,884 | 7,167 | 6,124 | 1,033 | 1,174 | 1,670 | 40,404 | 1,665 | 45.7 | 2.6 | 44.6 |
| Stockton, CA | 11,989 | 195,234 | 29,698 | 17,872 | 26,907 | 6,270 | 5,066 | 9,144 | 46,836 | 3,430 | 62.1 | 5.4 | 55.0 |
| Sumter, SC | 2,246 | 38,299 | 6,157 | 6,566 | 5,716 | 975 | 969 | 1,423 | 37,155 | 905 | 47.5 | 8.0 | 47.4 |
| Syracuse, NY | 15,026 | 261,828 | 51,265 | 22,682 | 35,024 | 11,204 | 15,964 | 12,337 | 47,118 | 1,926 | 37 | 3.6 | 49.1 |
| Tallahassee, FL | 9,084 | 117,467 | 20,830 | 3,091 | 18,804 | 5,870 | 10,167 | 5,083 | 43,273 | 1,648 | 62.1 | 2.7 | 34.3 |
| Tampa-St. Petersburg-Clearwater, FL | 81,384 | 1,169,494 | 197,388 | 61,294 | 158,023 | 86,852 | 96,958 | 58,521 | 50,039 | 4,325 | 80.3 | 1.8 | 39.7 |
| Terre Haute, IN | 3,733 | 59,009 | 11,577 | 10,327 | 9,005 | 1,454 | 1,380 | 2,341 | 39,666 | 2,392 | 42.6 | 9.2 | 41.5 |

# Table C. Metropolitan Areas — **Agriculture**

Agriculture, 2017 (cont.)

| Area name | Acreage (1,000) [117] | Percent change, 2012–2017 [118] | Average size of farm [119] | Total irrigated (1,000) [120] | Total cropland (1,000) [121] | Value of land and buildings: Average per farm [122] | Value of land and buildings: Average per acre [123] | Value of machinery and equipment, average per farm (dollars) [124] | Value of products sold: Total (mil dol) [125] | Average per farm (acres) [126] | Percent from: Crops [127] | Percent from: Livestock and poultry products [128] | Organic farms (number) [129] | Farms with internet access (percent) [130] | Government payments: Total ($1,000) [131] | Percent of farms [132] |
|---|---|---|---|---|---|---|---|---|---|---|---|---|---|---|---|---|
| Reading, PA | 225 | -3.9 | 124 | 1.5 | 184.5 | 1,392,374 | 11,209 | 138,139 | 555 | 306,609 | 43.8 | 56.2 | 72 | 68.4 | 3,038 | 22.1 |
| Redding, CA | 410 | 8.9 | 307 | 48.8 | 38.9 | 895,229 | 2,919 | 51,324 | 62 | 46,546 | 35.0 | 65.0 | 16 | 81.5 | 547 | 2.0 |
| Reno, NV | 501 | 13.2 | 1,412 | 43.6 | 22.2 | 1,449,321 | 1,026 | 61,699 | 20 | 56,166 | 45.8 | 54.2 | 14 | 86.5 | 174 | 5.1 |
| Richmond, VA | 655 | 3.3 | 221 | 10.8 | 347.3 | 897,575 | 4,063 | 105,935 | 243 | 82,022 | 53.2 | 46.8 | 6 | 76.4 | 9,343 | 22.1 |
| Riverside-San Bernardino-Ontario, CA | 332 | -21.2 | 89 | 148.4 | 209.0 | 1,647,577 | 18,504 | 95,567 | 1,306 | 350,203 | 61.4 | 38.6 | 253 | 79.1 | 2,268 | 2.4 |
| Roanoke, VA | 315 | -5.2 | 156 | 0.6 | 117.9 | 604,500 | 3,864 | 75,565 | 91 | 44,978 | 26.1 | 73.9 | 10 | 71.0 | 1,000 | 8.2 |
| Rochester, MN | 1,140 | -1.5 | 288 | 7.5 | 934.5 | 1,760,727 | 6,115 | 212,289 | 931 | 235,069 | 51.7 | 48.3 | 76 | 80.0 | 30,766 | 65.6 |
| Rochester, NY | 900 | -3.0 | 214 | 7.0 | 701.5 | 836,920 | 3,920 | 180,390 | 957 | 226,983 | 54.5 | 45.5 | 261 | 70.7 | 16,376 | 22.4 |
| Rockford, IL | 292 | -8.0 | 245 | 2.5 | 267.0 | 1,811,732 | 7,399 | 155,547 | 186 | 155,583 | 77.0 | 23.0 | 15 | 82.6 | 10,252 | 48.6 |
| Rocky Mount, NC | 278 | 4.2 | 413 | 4.7 | 203.6 | 1,562,533 | 3,783 | 258,892 | 368 | 545,798 | 59.4 | 40.6 | 17 | 68.0 | 4,904 | 44.7 |
| Rome, GA | 75 | 6.7 | 137 | 0.7 | 21.7 | 665,898 | 4,866 | 88,750 | 53 | 97,700 | 10.2 | 89.8 | 0 | 77.3 | 442 | 13.5 |
| Sacramento-Roseville-Folsom, CA | 929 | 0.2 | 196 | 360.9 | 468.5 | 1,852,189 | 9,440 | 103,724 | 1,081 | 228,294 | 84.7 | 15.3 | 178 | 86.4 | 5,225 | 7.3 |
| Saginaw, MI | 327 | 5.6 | 262 | 1.7 | 300.5 | 1,518,681 | 5,805 | 162,378 | 170 | 136,206 | 87.9 | 12.1 | 12 | 79.4 | 6,853 | 63.1 |
| St. Cloud, MN | 846 | -10.6 | 224 | 68.5 | 671.7 | 1,116,661 | 4,974 | 184,892 | 955 | 253,558 | 25.4 | 74.6 | 62 | 77.4 | 7,628 | 53.6 |
| St. George, UT | 155 | 4.8 | 289 | 13.0 | 22.3 | 1,078,857 | 3,737 | 61,328 | 17 | 30,648 | 39.3 | 60.7 | 3 | 76.9 | 190 | 3.2 |
| St. Joseph, MO-KS | 768 | -5.1 | 291 | 4.2 | 583.3 | 1,061,151 | 3,648 | 121,612 | 288 | 109,001 | 84.3 | 15.7 | 1 | 75.9 | 14,004 | 54.8 |
| St. Louis, MO-IL | 2,780 | -4.6 | 251 | 12.2 | 2,129.8 | 1,547,992 | 6,158 | 144,778 | 1,376 | 124,397 | 74.7 | 25.3 | 24 | 76.0 | 44,123 | 46.8 |
| Salem, OR | 438 | 1.5 | 109 | 123.0 | 345.0 | 1,156,225 | 10,580 | 127,386 | 836 | 208,874 | 84.6 | 15.4 | 72 | 85.2 | 2,771 | 6.6 |
| Salinas, CA | 1,340 | 5.7 | 1,214 | 294.6 | 366.7 | 8,944,364 | 7,368 | 805,557 | 4,116 | 3,728,396 | 99.1 | 0.9 | 191 | 75.5 | 1,269 | 7.4 |
| Salisbury, MD-DE | 523 | 0.4 | 234 | 117.5 | 418.0 | 1,709,210 | 7,315 | 200,515 | 1,828 | 817,121 | 17.0 | 83.0 | 13 | 77.0 | 17,204 | 43.7 |
| Salt Lake City, UT | 411 | -3.4 | 363 | 29.3 | 32.6 | 953,913 | 2,628 | 70,367 | 61 | 53,581 | 38.8 | 61.2 | 11 | 79.6 | 594 | 2.5 |
| San Angelo, TX | 2,010 | -1.4 | 1,293 | 20.9 | 138.8 | 1,469,225 | 1,136 | 78,784 | 109 | 70,338 | 27.6 | 72.4 | 1 | 76.6 | 3,447 | 12.9 |
| San Antonio-New Braunfels, TX | 3,657 | 0.4 | 244 | 87.8 | 613.5 | 767,866 | 3,141 | 56,604 | 407 | 27,225 | 37.6 | 62.4 | 13 | 71.5 | 10,124 | 6.7 |
| San Diego-Chula Vista-Carlsbad, CA | 222 | 0.3 | 44 | 42.7 | 64.1 | 1,014,281 | 23,209 | 43,529 | 831 | 163,601 | 93.5 | 6.5 | 419 | 84.2 | 558 | 1.1 |
| San Francisco-Oakland-Berkeley, CA | 525 | 0.1 | 350 | 38.1 | 79.4 | 2,439,344 | 6,965 | 81,559 | 304 | 202,908 | 62.2 | 37.8 | 109 | 85.7 | 1,416 | 6.9 |
| Oakland-Berkeley-Livermore, CA Div 36084 | 339 | 10.9 | 374 | 30.1 | 58.8 | 2,629,149 | 7,022 | 76,063 | 129 | 142,977 | 77.4 | 22.6 | 22 | 84.5 | 704 | 6.2 |
| San Francisco-San Mateo-Redwood City, CA Div 41884 | 46 | -4.4 | 184 | 3.0 | 6.9 | 1,817,211 | 9,902 | 73,187 | 79 | 316,430 | 98.2 | 1.8 | 20 | 88.0 | 0 | 0.8 |
| San Rafael, CA Div 42034 | 140 | -18.0 | 408 | 5.0 | 13.7 | 2,393,810 | 5,862 | 102,187 | 95 | 277,962 | 11.6 | 88.4 | 67 | 86.9 | 712 | 13.1 |
| San Jose-Sunnyvale-Santa Clara, CA | 808 | -3.1 | 539 | 37.3 | 69.9 | 2,800,069 | 5,197 | 93,026 | 473 | 315,395 | 89.6 | 10.4 | 109 | 84.2 | 1,093 | 6.1 |
| San Luis Obispo-Paso Robles, CA | 931 | -30.4 | 396 | 75.8 | 246.4 | 2,991,940 | 7,547 | 87,783 | 702 | 298,676 | 95.2 | 4.8 | 132 | 84.4 | 4,285 | 9.4 |
| Santa Cruz-Watsonville, CA | 64 | -36.1 | 102 | 20.1 | 23.5 | 2,183,024 | 21,352 | 116,842 | 607 | 970,464 | 98.7 | 1.3 | 130 | 85.1 | 37 | 1.9 |
| Santa Fe, NM | D | D | D | 15.6 | 23.7 | 932,061 | D | 65,568 | 25 | 39,798 | 54.5 | 45.5 | 14 | 77.0 | 367 | 5.5 |
| Santa Maria-Santa Barbara, CA | 715 | 2.0 | 487 | 119.9 | 146.3 | 5,060,151 | 10,381 | 183,019 | 1,520 | 1,036,090 | 98.0 | 2.0 | 163 | 87.7 | 854 | 5.2 |
| Santa Rosa-Petaluma, CA | 567 | -3.8 | 158 | 86.4 | 129.9 | 3,501,852 | 22,186 | 77,693 | 919 | 255,719 | 66.9 | 33.1 | 332 | 87.7 | 2,410 | 3.8 |
| Savannah, GA | 81 | 36.2 | 195 | 3.5 | 33.7 | 642,800 | 3,300 | 100,149 | 32 | 75,671 | 87.9 | 12.1 | 5 | 81.3 | 402 | 15.1 |
| Scranton--Wilkes-Barre, PA | 147 | -9.5 | 131 | 0.5 | 77.4 | 663,210 | 5,073 | 80,161 | 48 | 42,264 | 62.7 | 37.3 | 1 | 73.2 | 1,259 | 26.2 |
| Seattle-Tacoma-Bellevue, WA | 151 | -9.4 | 31 | 15.5 | 64.3 | 744,589 | 24,396 | 47,150 | 358 | 72,144 | 54.9 | 45.1 | 87 | 88.0 | 1,295 | 3.1 |
| Seattle-Bellevue-Kent, WA Div 42644 | 106 | -10.1 | 31 | 12.5 | 52.3 | 808,104 | 25,655 | 48,710 | 293 | 87,367 | 57.0 | 43.0 | 70 | 88.3 | 1,224 | 3.5 |
| Tacoma-Lakewood, WA Div 45704 | 46 | -7.5 | 28 | 3.0 | 12.1 | 612,026 | 21,490 | 43,894 | 65 | 40,371 | 45.4 | 54.6 | 17 | 87.4 | 71 | 2.2 |
| Sebastian-Vero Beach, FL | 183 | 12.4 | 406 | 52.9 | 55.5 | 2,330,718 | 5,745 | 90,644 | 107 | 236,611 | 89.3 | 10.7 | 9 | 68.2 | 426 | 7.1 |
| Sebring-Avon Park, FL | 376 | -23.3 | 380 | 73.3 | 87.6 | 1,484,269 | 3,906 | 92,897 | 197 | 198,865 | 72.5 | 27.5 | 2 | 71.1 | 2,471 | 11.0 |
| Sheboygan, WI | 196 | 3.0 | 205 | 0.3 | 164.3 | 1,348,670 | 6,594 | 198,127 | 213 | 222,785 | 22.6 | 77.4 | 14 | 84.2 | 1,215 | 39.5 |
| Sherman-Denison, TX | 430 | -0.3 | 151 | 2.3 | 207.0 | 1,023,105 | 6,770 | 68,105 | 66 | 23,259 | 60.8 | 39.2 | 1 | 81.0 | 4,729 | 9.9 |
| Shreveport-Bossier City, LA | 443 | 14.8 | 243 | 24.8 | 147.1 | 730,792 | 3,011 | 95,830 | 96 | 52,671 | 52.4 | 47.6 | 0 | 74.0 | 2,725 | 8.9 |
| Sierra Vista-Douglas, AZ | 973 | 6.2 | 899 | 86.0 | 152.9 | 1,802,553 | 2,005 | 99,038 | 145 | 133,648 | 56.9 | 43.1 | 4 | 78.9 | 3,119 | 10.6 |
| Sioux City, IA-NE-SD | 1,187 | -0.4 | 489 | 102.6 | 1,035.2 | 2,996,651 | 6,132 | 265,159 | 912 | 375,747 | 56.4 | 43.6 | 4 | 81.5 | 29,267 | 74.0 |
| Sioux Falls, SD | 1,429 | -6.0 | 469 | 34.3 | 1,230.4 | 2,633,238 | 5,616 | 236,521 | 920 | 301,909 | 60.6 | 39.4 | 6 | 81.8 | 23,562 | 68.3 |
| South Bend-Mishawaka, IN-MI | 348 | 2.3 | 253 | 99.4 | 297.5 | 1,548,314 | 6,113 | 178,869 | 257 | 187,018 | 71.1 | 28.9 | 13 | 77.3 | 8,438 | 43.5 |
| Spartanburg, SC | 96 | -5.9 | 67 | 1.8 | 34.3 | 550,283 | 8,231 | 45,163 | 31 | 21,292 | 70.4 | 29.6 | 0 | 78.4 | 481 | 7.9 |
| Spokane-Spokane Valley, WA | 1,066 | 0.2 | 301 | 19.9 | 455.9 | 807,433 | 2,679 | 72,986 | 147 | 41,604 | 74.4 | 25.6 | 51 | 83.2 | 8,829 | 15.7 |
| Springfield, IL | 699 | 4.1 | 476 | 4.6 | 635.8 | 4,123,080 | 8,661 | 259,955 | 442 | 301,075 | 92.6 | 7.4 | 3 | 79.2 | 10,299 | 65.3 |
| Springfield, MA | 175 | -4.1 | 86 | 3.6 | 56.8 | 648,777 | 7,586 | 66,977 | 141 | 68,855 | 76.2 | 23.8 | 67 | 83.5 | 1,311 | 8.1 |
| Springfield, MO | 1,209 | -0.6 | 159 | 3.3 | 360.6 | 517,580 | 3,255 | 58,141 | 273 | 35,876 | 13.2 | 86.8 | 61 | 72.1 | 2,387 | 6.2 |
| Springfield, OH | 171 | -1.9 | 230 | 1.8 | 151.4 | 1,738,039 | 7,542 | 165,062 | 127 | 170,443 | 79.2 | 20.8 | 0 | 80.5 | 6,001 | 43.3 |
| State College, PA | 150 | -7.5 | 146 | 0.4 | 88.3 | 981,430 | 6,700 | 92,395 | 92 | 89,421 | 35.3 | 64.7 | 23 | 68.5 | 2,266 | 24.0 |
| Staunton, VA | 291 | 11.8 | 175 | 2.5 | 116.8 | 1,222,854 | 6,999 | 101,481 | 293 | 175,704 | 12.8 | 87.2 | 5 | 71.5 | 1,049 | 15.7 |
| Stockton, CA | 773 | -1.8 | 225 | 487.1 | 524.4 | 3,384,002 | 15,020 | 189,084 | 2,176 | 634,404 | 74.8 | 25.2 | 35 | 81.7 | 5,788 | 6.9 |
| Sumter, SC | 305 | -13.0 | 336 | 28.0 | 194.2 | 898,759 | 2,671 | 131,634 | 262 | 289,486 | 38.7 | 61.3 | 7 | 73.8 | 3,142 | 38.7 |
| Syracuse, NY | 419 | -3.1 | 217 | 2.5 | 258.9 | 676,931 | 3,113 | 142,481 | 333 | 173,037 | 26.6 | 73.4 | 95 | 81.9 | 3,359 | 20.8 |
| Tallahassee, FL | 350 | 19.5 | 212 | 9.3 | 64.5 | 811,049 | 3,823 | 54,877 | 135 | 81,627 | 64.7 | 35.3 | 12 | 76.2 | 1,600 | 8.5 |
| Tampa-St. Petersburg-Clearwater, FL | 425 | -5.6 | 98 | 31.5 | 111.1 | 958,142 | 9,761 | 57,092 | 535 | 123,682 | 82.5 | 17.5 | 30 | 76.9 | 1,773 | 5.7 |
| Terre Haute, IN | 746 | 0.0 | 312 | 13.4 | 623.9 | 1,683,833 | 5,400 | 168,563 | 390 | 163,056 | 88.7 | 11.3 | 28 | 72.2 | 14,469 | 57.9 |

Items 117—132

# Table C. Metropolitan Areas — Water Use, Wholesale Trade, Retail Trade, and Real Estate

| Area name | Water use, 2015 | | Wholesale Trade[1], 2017 | | | | Retail Trade[2], 2017 | | | | Real estate and rental and leasing,[2] 2017 | | | |
|---|---|---|---|---|---|---|---|---|---|---|---|---|---|---|
| | Public supply water withdrawn (mil gal/day) | Public supply gallons withdrawn per person per day | Number of establishments | Number of employees | Sales (mil dol) | Average payroll (mil dol) | Number of establishments | Number of employees | Sales (mil dol) | Average payroll (mil dol) | Number of establishments | Number of employees | Sales (mil dol) | Average payroll (mil dol) |
| | 133 | 134 | 135 | 136 | 137 | 138 | 139 | 140 | 141 | 142 | 143 | 144 | 145 | 146 |
| Reading, PA | 31.85 | 76.7 | 374 | 8,533 | 7,103.8 | 524.2 | 1,186 | 20,446 | 5,772.3 | 527.8 | 367 | 2,985 | 1,173.9 | 217.9 |
| Redding, CA | 33.47 | 186.4 | 157 | 1,664 | 1,056.6 | 85.1 | 634 | 9,391 | 2,847.7 | 274.5 | 220 | 887 | 188.2 | 32.5 |
| Reno, NV | 58.21 | 129.1 | D | D | D | D | 1,495 | 25,312 | 8,518.6 | 796.6 | D | D | D | D |
| Richmond, VA | 176.36 | 141.3 | 1,186 | 19,744 | 20,619.6 | 1,287.0 | 4,188 | 72,355 | 20,643.9 | 1,924.2 | 1,558 | 8,769 | 4,290.4 | 455.4 |
| Riverside-San Bernardino-Ontario, CA | 778.43 | 173.4 | 4,508 | 66,395 | 67,695.8 | 3,619.0 | 10,203 | 191,153 | 64,394.3 | 5,603.5 | 4,227 | 19,656 | 5,332.5 | 877.0 |
| Roanoke, VA | 41.18 | 130.9 | 350 | 5,250 | 3,306.2 | 330.3 | 1,230 | 18,173 | 5,016.1 | 443.3 | 363 | 1,804 | 354.1 | 71.8 |
| Rochester, MN | 16.91 | 79.1 | 170 | 2,124 | 1,646.2 | 116.8 | 790 | 13,325 | 3,392.5 | 345.0 | 213 | 845 | 167.5 | 29.3 |
| Rochester, NY | 114.43 | 105.8 | 1,010 | 15,106 | 10,659.6 | 877.0 | 3,482 | 57,893 | 14,650.1 | 1,460.6 | 1,115 | 7,242 | 1,497.7 | 310.2 |
| Rockford, IL | 32.79 | 96.3 | 337 | 4,561 | 2,920.9 | 244.4 | 1,026 | 15,411 | 4,342.0 | 384.2 | 256 | 1,376 | 293.4 | 59.1 |
| Rocky Mount, NC | 12.42 | 83.9 | 130 | 2,741 | 3,808.9 | 126.7 | 510 | 6,346 | 1,639.2 | 146.0 | 112 | 520 | 92.3 | 18.1 |
| Rome, GA | 11.80 | 122.3 | 67 | 1,010 | 655.3 | 50.1 | 389 | 4,269 | 1,214.8 | 106.6 | 83 | 260 | 56.2 | 9.3 |
| Sacramento-Roseville-Folsom, CA | 340.07 | 149.5 | 1,769 | 28,073 | 39,577.8 | 1,721.5 | 5,845 | 100,416 | 31,776.5 | 3,056.7 | 3,121 | 15,546 | 4,049.3 | 761.1 |
| Saginaw, MI | 0.42 | 2.2 | 160 | 2,070 | 1,307.0 | 111.1 | 826 | 12,804 | 3,233.8 | 309.9 | 123 | 652 | 139.2 | 20.5 |
| St. Cloud, MN | 16.66 | 85.7 | 237 | 5,250 | 3,376.0 | 280.7 | 795 | 13,568 | 3,812.2 | 351.6 | 225 | 1,052 | 210.1 | 38.8 |
| St. George, UT | 49.38 | 317.3 | 164 | 1,343 | 1,444.0 | 64.5 | 628 | 9,079 | 2,651.3 | 259.4 | 388 | 849 | 186.0 | 26.7 |
| St. Joseph, MO-KS | 18.41 | 145.1 | 102 | 1,239 | 1,683.3 | 77.9 | 399 | 6,762 | 1,734.7 | 161.0 | 117 | 418 | 67.6 | 11.7 |
| St. Louis, MO-IL | 459.39 | 163.4 | 2,955 | 48,659 | 58,992.5 | 3,282.5 | 8,758 | 142,553 | 51,349.8 | 3,910.1 | 3,142 | 18,794 | 4,935.3 | 894.6 |
| Salem, OR | 76.88 | 187.5 | 329 | 3,963 | 2,915.8 | 206.7 | 1,253 | 19,363 | 5,485.1 | 521.4 | 536 | 2,017 | 447.3 | 77.9 |
| Salinas, CA | 38.23 | 88.1 | 349 | 5,528 | 8,040.7 | 353.6 | 1,292 | 17,597 | 5,629.5 | 529.7 | 474 | 1,987 | 587.0 | 90.9 |
| Salisbury, MD-DE | 29.56 | 74.8 | 102 | 1,051 | 1,110.8 | 59.1 | 1,886 | 24,042 | 6,750.0 | 610.2 | 617 | 2,531 | 536.8 | 93.1 |
| Salt Lake City, UT | 413.16 | 353.0 | 1,644 | 28,775 | 27,307.6 | 1,909.6 | 3,627 | 65,191 | 23,981.2 | 2,084.2 | 2,440 | 10,693 | 3,324.9 | 546.0 |
| San Angelo, TX | 1.43 | 11.8 | D | D | D | D | 430 | 6,683 | 1,993.1 | 192.7 | 151 | 627 | 129.5 | 22.4 |
| San Antonio-New Braunfels, TX | 255.85 | 107.3 | 1,846 | 32,218 | 22,040.0 | 1,854.0 | 6,375 | 111,582 | 34,864.0 | 3,080.4 | 2,659 | 15,279 | 3,962.9 | 777.0 |
| San Diego-Chula Vista-Carlsbad, CA | 408.82 | 123.9 | 3,905 | 46,390 | 38,374.0 | 3,219.4 | 9,455 | 152,542 | 46,665.9 | 4,557.3 | 6,381 | 29,423 | 11,062.2 | 1,670.4 |
| San Francisco-Oakland-Berkeley, CA | 472.83 | 101.5 | 5,322 | 83,209 | 96,647.3 | 6,810.7 | 13,232 | 221,632 | 86,926.9 | 8,134.4 | 7,706 | 48,614 | 21,031.2 | 3,357.6 |
| Oakland-Berkeley-Livermore, CA Div 36084 | 308.71 | 111.7 | 3,073 | 51,058 | 53,727.0 | 3,814.2 | 6,807 | 118,589 | 41,875.1 | 4,015.7 | 3,667 | 18,584 | 6,389.5 | 1,079.1 |
| San Francisco-San Mateo-Redwood City, CA Div 41884 | 135.93 | 83.4 | 1,932 | 29,819 | 41,089.6 | 2,833.3 | 5,414 | 87,494 | 39,426.6 | 3,473.5 | 3,397 | 27,061 | 12,767.2 | 2,048.6 |
| San Rafael, CA Div 42034 | 28.19 | 107.9 | 317 | 2,332 | 1,830.8 | 163.2 | 1,011 | 15,549 | 5,625.2 | 645.1 | 642 | 2,969 | 1,874.6 | 229.9 |
| San Jose-Sunnyvale-Santa Clara, CA | 199.83 | 101.1 | 2,287 | 74,875 | 137,070.7 | 13,096.3 | 4,784 | 87,843 | 47,932.7 | 3,400.1 | 2,831 | 15,891 | 6,399.3 | 1,028.2 |
| San Luis Obispo-Paso Robles, CA | 29.10 | 103.4 | 276 | 2,519 | 1,282.1 | 131.0 | 1,130 | 14,708 | 4,372.0 | 442.0 | 494 | 2,024 | 420.2 | 80.3 |
| Santa Cruz-Watsonville, CA | 18.45 | 67.3 | D | D | D | D | 893 | 12,885 | 4,650.4 | 364.9 | 370 | 1,539 | 351.3 | 69.8 |
| Santa Fe, NM | 10.57 | 71.1 | 99 | 856 | 923.5 | 44.1 | 797 | 9,053 | 2,588.1 | 269.2 | 278 | 815 | 200.7 | 41.4 |
| Santa Maria-Santa Barbara, CA | 49.94 | 112.3 | 405 | 5,968 | 5,075.3 | 490.4 | 1,479 | 18,889 | 5,267.8 | 547.6 | 747 | 3,072 | 712.4 | 130.7 |
| Santa Rosa-Petaluma, CA | 40.49 | 80.6 | 560 | 8,903 | 6,419.4 | 684.6 | 1,815 | 25,679 | 7,760.5 | 843.1 | 712 | 3,334 | 817.5 | 152.7 |
| Savannah, GA | 66.47 | 175.3 | 369 | 5,323 | 6,526.0 | 316.1 | 1,528 | 21,949 | 5,824.1 | 534.5 | 535 | 2,567 | 632.0 | 98.9 |
| Scranton--Wilkes-Barre, PA | 62.18 | 111.4 | 545 | 9,974 | 9,262.4 | 497.0 | 2,179 | 31,443 | 8,819.8 | 778.3 | 396 | 1,812 | 449.7 | 75.3 |
| Seattle-Tacoma-Bellevue, WA | 373.55 | 100.1 | 4,588 | 77,972 | 78,446.5 | 5,445.9 | 10,942 | 191,464 | 114,947.5 | 6,764.6 | 7,162 | 36,764 | 11,843.6 | 2,124.6 |
| Seattle-Bellevue-Kent, WA Div 42644 | 263.14 | 91.1 | 3,882 | 66,615 | 69,158.8 | 4,794.6 | 8,773 | 152,656 | 101,171.6 | 5,510.2 | 5,992 | 31,488 | 10,485.9 | 1,894.0 |
| Tacoma-Lakewood, WA Div 45704 | 110.41 | 130.8 | 706 | 11,357 | 9,287.7 | 650.9 | 2,169 | 38,808 | 13,775.9 | 1,254.5 | 1,170 | 5,276 | 1,357.7 | 230.6 |
| Sebastian-Vero Beach, FL | 16.94 | 114.5 | 133 | 856 | 5,125.0 | 96.6 | 684 | 9,009 | 2,267.7 | 222.7 | 276 | 1,569 | 257.8 | 52.5 |
| Sebring-Avon Park, FL | 7.49 | 75.3 | 53 | 434 | 200.2 | 20.7 | 329 | 4,589 | 1,261.2 | 109.9 | 98 | 302 | 60.5 | 8.3 |
| Sheboygan, WI | 15.80 | 136.7 | 95 | 1,484 | 933.0 | 76.7 | 379 | 6,132 | 1,729.4 | 150.8 | 76 | 232 | 44.4 | 7.0 |
| Sherman-Denison, TX | 21.61 | 172.2 | 97 | 1,024 | 736.7 | 48.3 | 428 | 6,146 | 1,912.7 | 170.4 | 107 | 329 | 73.4 | 11.1 |
| Shreveport-Bossier City, LA | 59.24 | 146.7 | 441 | 6,716 | 8,809.0 | 363.9 | 1,399 | 21,366 | 6,200.1 | 553.4 | 486 | 2,480 | 534.3 | 98.2 |
| Sierra Vista-Douglas, AZ | 16.15 | 127.7 | D | D | D | 11.4 | 386 | 5,160 | 1,324.4 | 126.6 | 107 | 391 | 57.8 | 10.8 |
| Sioux City, IA-NE-SD | 18.33 | 127.1 | 163 | 2,368 | 2,167.6 | 132.3 | 532 | 8,901 | 2,359.5 | 210.3 | 121 | 552 | 92.9 | 17.5 |
| Sioux Falls, SD | 21.85 | 79.7 | 424 | 6,390 | 4,989.5 | 368.1 | 1,032 | 18,371 | 5,058.4 | 497.7 | 366 | 2,047 | 402.9 | 87.1 |
| South Bend-Mishawaka, IN-MI | 22.91 | 71.6 | 323 | 4,677 | 4,091.3 | 267.9 | 1,035 | 17,248 | 5,671.1 | 462.3 | 265 | 1,657 | 295.6 | 62.2 |
| Spartanburg, SC | 34.72 | 116.8 | 397 | 5,908 | 5,925.6 | 346.8 | 1,032 | 16,413 | 4,906.4 | 419.0 | 265 | 1,110 | 356.2 | 40.1 |
| Spokane-Spokane Valley, WA | 145.00 | 271.2 | 635 | 9,911 | 10,369.1 | 528.4 | 1,744 | 27,903 | 8,169.0 | 847.4 | 724 | 3,578 | 758.4 | 130.8 |
| Springfield, IL | 24.02 | 113.8 | 185 | 2,517 | 3,179.5 | 133.1 | 750 | 12,390 | 3,089.5 | 295.1 | 246 | 895 | 228.1 | 29.8 |
| Springfield, MA | 67.20 | 95.6 | 450 | 8,122 | 9,053.6 | 474.3 | 2,297 | 32,820 | 8,634.4 | 900.4 | 556 | 3,232 | 685.8 | 123.4 |
| Springfield, MO | 42.64 | 93.4 | 522 | 9,951 | 7,437.6 | 473.4 | 1,635 | 24,757 | 7,136.1 | 656.8 | 615 | 3,400 | 564.1 | 112.4 |
| Springfield, OH | 16.17 | 118.9 | 84 | 1,661 | 1,808.1 | 81.9 | 373 | 5,695 | 1,411.4 | 122.3 | 89 | 380 | 63.1 | 11.0 |
| State College, PA | 17.87 | 111.3 | 86 | 703 | 301.3 | 34.9 | 461 | 7,741 | 2,052.5 | 186.4 | 150 | 957 | 233.2 | 37.5 |
| Staunton, VA | 8.28 | 68.9 | 87 | 1,180 | 493.8 | 58.2 | 448 | 6,156 | 1,597.0 | 156.5 | 134 | 398 | 53.9 | 12.6 |
| Stockton, CA | 93.39 | 128.6 | 528 | 11,280 | 14,265.8 | 728.0 | 1,635 | 27,256 | 8,856.3 | 789.3 | 566 | 2,832 | 634.9 | 119.8 |
| Sumter, SC | 16.91 | 119.7 | 79 | 725 | 374.0 | 34.1 | 458 | 5,793 | 1,454.8 | 129.0 | 80 | 321 | 47.0 | 9.0 |
| Syracuse, NY | 94.13 | 142.5 | 635 | 12,341 | 15,751.9 | 706.2 | 2,213 | 35,026 | 9,617.4 | 913.6 | 684 | 3,952 | 802.4 | 174.2 |
| Tallahassee, FL | 35.54 | 94.0 | 247 | 3,191 | 2,417.8 | 199.1 | 1,257 | 18,840 | 4,773.6 | 450.6 | 486 | 2,922 | 465.0 | 99.4 |
| Tampa-St. Petersburg-Clearwater, FL | 299.64 | 100.7 | 3,259 | 42,190 | 33,298.6 | 2,400.0 | 10,377 | 158,678 | 50,609.6 | 4,447.7 | 5,057 | 23,347 | 6,130.6 | 996.3 |
| Terre Haute, IN | 16.07 | 85.5 | 143 | 1,551 | 833.9 | 65.0 | 633 | 9,040 | 2,433.8 | 209.4 | 108 | 611 | 107.1 | 19.0 |

1. Merchant wholesalers, except manufacturers' sales branches and offices. 2. Employer establishments.

# Table C. Metropolitan Areas — Professional Services, Manufacturing, and Accommodation and Food Services

| Area name | Professional, scientific, and technical services, 2017 | | | | Manufacturing, 2017 | | | | Accommodation and food services, 2017 | | | |
|---|---|---|---|---|---|---|---|---|---|---|---|---|
| | Number of establishments | Number of employees | Sales (mil dol) | Average payroll (mil dol) | Number of establishments | Number of employees | Receipts (mil dol) | Annual payroll (mil dol) | Number of establishments | Number of employees | Receipts (mil dol) | Annual payroll (mil dol) |
| | 147 | 148 | 149 | 150 | 151 | 152 | 153 | 154 | 155 | 156 | 157 | 158 |
| Reading, PA | D | D | D | D | 483 | 31,959 | 10,989.4 | 1,674.8 | 770 | 13,389 | 671.4 | 186.8 |
| Redding, CA | 357 | 2,043 | 272.7 | 116.7 | 130 | 2,158 | 656.3 | 106.5 | 373 | 6,351 | 428.5 | 118.5 |
| Reno, NV | 13 | 21 | 3.8 | 1.4 | 491 | 16,653 | 6,232.1 | 1,068.1 | 1,159 | 31,563 | 2,421.4 | 684.4 |
| Richmond, VA | 3,578 | 39,473 | 6,551.6 | 2,705.0 | D | D | D | D | 2,673 | 55,327 | 3,032.1 | 891.2 |
| Riverside-San Bernardino-Ontario, CA | 6,359 | 39,530 | 5,689.0 | 2,054.1 | 3,342 | 93,958 | 36,097.3 | 4,948.4 | 7,586 | 158,600 | 11,100.0 | 3,111.0 |
| Roanoke, VA | 148 | 504 | 58.3 | 19.7 | D | D | D | D | 682 | 14,103 | 722.2 | 212.9 |
| Rochester, MN | D | D | D | 36.0 | 185 | 7,594 | 3,865.1 | 395.1 | 510 | 9,456 | 557.3 | 170.2 |
| Rochester, NY | 2,168 | 27,039 | 4,463.9 | 1,719.8 | 1,296 | 52,874 | 18,209.3 | 3,035.9 | 2,534 | 39,333 | 2,118.9 | 661.6 |
| Rockford, IL | 577 | 5,044 | 817.5 | 286.8 | 643 | 31,314 | 13,918.0 | 2,065.5 | 643 | 12,291 | 633.7 | 183.0 |
| Rocky Mount, NC | 174 | 1,015 | 137.8 | 42.3 | 121 | 9,686 | 4,924.3 | 440.0 | 245 | 4,832 | 240.1 | 63.7 |
| Rome, GA | 178 | 808 | 123.2 | 31.6 | 98 | 6,153 | 3,872.4 | 358.9 | 202 | 4,241 | 227.3 | 64.8 |
| Sacramento-Roseville-Folsom, CA | 5,225 | 52,037 | 9,468.0 | 4,711.8 | 1,362 | 32,814 | 12,106.9 | 1,928.3 | 4,804 | 95,400 | 6,944.8 | 1,865.9 |
| Saginaw, MI | 298 | 2,414 | 375.2 | 142.5 | 194 | 12,288 | 4,461.3 | 799.8 | 400 | 9,735 | 493.3 | 137.5 |
| St. Cloud, MN | D | D | D | D | 313 | 15,313 | 5,138.8 | 777.8 | 435 | 8,084 | 369.1 | 109.8 |
| St. George, UT | D | D | D | D | 163 | 2,634 | 620.3 | 119.0 | 363 | 7,354 | 449.3 | 119.5 |
| St. Joseph, MO-KS | D | D | D | D | 112 | 12,344 | 8,051.1 | 606.3 | 235 | 4,481 | 214.7 | 62.8 |
| St. Louis, MO-IL | 6,245 | 73,385 | 13,619.6 | 5,137.3 | 2,451 | 107,501 | 53,307.7 | 6,867.3 | 6,011 | 129,726 | 7,542.4 | 2,140.9 |
| Salem, OR | 788 | 4,086 | 541.9 | 205.3 | 456 | 11,498 | 3,595.6 | 548.3 | D | D | D | D |
| Salinas, CA | D | D | D | D | 280 | 9,070 | 4,346.2 | 429.1 | 1,095 | 22,103 | 1,894.5 | 524.6 |
| Salisbury, MD-DE | 169 | 820 | 99.9 | 39.6 | 258 | 14,136 | 5,151.3 | 621.5 | 1,338 | 20,757 | 1,633.1 | 472.6 |
| Salt Lake City, UT | D | D | D | D | 1,457 | 55,063 | 26,993.4 | 3,492.0 | 2,496 | 51,607 | 3,167.7 | 895.0 |
| San Angelo, TX | 6 | 8 | 1.9 | 0.1 | 94 | 3,370 | 1,936.4 | 182.0 | 274 | 5,228 | 280.6 | 79.0 |
| San Antonio-New Braunfels, TX | 900 | 4,709 | 691.0 | 237.1 | 1,224 | 43,334 | 24,472.8 | 2,317.0 | 4,972 | 120,880 | 7,444.0 | 2,081.3 |
| San Diego-Chula Vista-Carlsbad, CA | 14,150 | 143,999 | 37,220.4 | 14,266.1 | 2,972 | 100,773 | 31,443.8 | 7,335.4 | 7,810 | 184,771 | 14,854.7 | 4,139.0 |
| San Francisco-Oakland-Berkeley, CA | 21,738 | 259,352 | 79,228.5 | 30,551.3 | 3,898 | 139,885 | 89,194.6 | 12,056.5 | 13,754 | 244,786 | 21,854.4 | 6,260.6 |
| Oakland-Berkeley-Livermore, CA Div 36084 | 9,586 | 92,928 | 22,371.4 | 9,198.5 | 2,351 | 96,246 | 56,672.8 | 8,022.8 | 6,265 | 100,517 | 7,490.7 | 2,115.1 |
| San Francisco-San Mateo-Redwood City, CA Div 41884 | 10,356 | 158,424 | 55,077.6 | 20,680.8 | 1,315 | 40,677 | 31,740.9 | 3,866.3 | 6,726 | 130,584 | 13,338.0 | 3,824.2 |
| San Rafael, CA Div 42034 | 1,796 | 8,000 | 1,779.5 | 671.9 | 232 | 2,962 | 780.9 | 167.4 | 763 | 13,685 | 1,025.8 | 321.3 |
| San Jose-Sunnyvale-Santa Clara, CA | 8,834 | 144,293 | 42,491.3 | 18,201.6 | 2,258 | 95,476 | 37,428.3 | 8,281.7 | 5,203 | 92,242 | 7,351.8 | 2,169.9 |
| San Luis Obispo-Paso Robles, CA | 929 | 4,853 | 797.0 | 298.6 | 453 | 6,433 | 2,341.3 | 343.5 | 985 | 17,024 | 1,148.8 | 331.1 |
| Santa Cruz-Watsonville, CA | 878 | 4,262 | 648.4 | 315.9 | 314 | 5,049 | 1,707.6 | 301.1 | 711 | 12,032 | 831.9 | 251.5 |
| Santa Fe, NM | 654 | 2,391 | 386.8 | 146.3 | 120 | 725 | 165.9 | 32.3 | 429 | 9,749 | 743.9 | 226.0 |
| Santa Maria-Santa Barbara, CA | 1,389 | 10,005 | 2,285.7 | 777.4 | 495 | 12,939 | 4,208.2 | 916.0 | 1,215 | 25,145 | 2,149.9 | 607.0 |
| Santa Rosa-Petaluma, CA | 1,585 | 8,289 | 1,357.1 | 544.3 | 891 | 21,445 | 7,370.0 | 1,344.2 | 1,306 | 24,035 | 2,085.6 | 534.8 |
| Savannah, GA | 74 | 262 | 40.1 | 12.4 | 218 | 18,300 | 11,185.5 | 1,530.7 | 1,158 | 24,242 | 1,597.6 | 410.2 |
| Scranton--Wilkes-Barre, PA | 547 | 4,820 | 632.3 | 257.3 | D | 27,309 | D | 1,423.4 | 1,433 | 21,611 | 1,419.3 | 335.3 |
| Seattle-Tacoma-Bellevue, WA | 14,285 | 156,506 | 33,125.0 | 13,588.8 | 3,525 | 162,698 | 88,043.6 | 12,278.9 | 9,943 | 171,879 | 13,580.2 | 4,111.8 |
| Seattle-Bellevue-Kent, WA Div 42644 | 12,741 | 147,351 | 31,700.2 | 13,030.7 | 2,960 | 144,549 | 81,756.4 | 11,241.0 | 8,288 | 142,877 | 11,635.2 | 3,527.2 |
| Tacoma-Lakewood, WA Div 45704 | 1,544 | 9,155 | 1,424.7 | 558.2 | 565 | 18,149 | 6,287.2 | 1,037.9 | 1,655 | 29,002 | 1,945.0 | 584.6 |
| Sebastian-Vero Beach, FL | 518 | 2,076 | 286.9 | 116.0 | 87 | 1,538 | 362.3 | 82.6 | 282 | 5,660 | 334.0 | 97.0 |
| Sebring-Avon Park, FL | D | D | D | D | 46 | 707 | 284.3 | 33.2 | 143 | 2,450 | 125.6 | 35.8 |
| Sheboygan, WI | D | D | D | D | 206 | 17,576 | 7,887.7 | 1,010.5 | 279 | 5,069 | 258.2 | 78.7 |
| Sherman-Denison, TX | 222 | 1,017 | 159.1 | 48.2 | 106 | 7,200 | 2,713.7 | 324.9 | 238 | 4,558 | 260.1 | 73.9 |
| Shreveport-Bossier City, LA | 28 | 127 | 14.1 | 4.7 | D | D | D | D | 802 | 21,392 | 1,571.1 | 370.9 |
| Sierra Vista-Douglas, AZ | 223 | 3,433 | 501.8 | 227.8 | 41 | 268 | 118.0 | 13.6 | 261 | 3,522 | 183.7 | 55.3 |
| Sioux City, IA-NE-SD | 41 | 318 | 54.7 | 16.8 | D | D | D | D | 343 | 6,697 | 437.4 | 106.4 |
| Sioux Falls, SD | 26 | 82 | 11.0 | 3.1 | D | 14,131 | D | 697.3 | 602 | 13,361 | 722.7 | 210.0 |
| South Bend-Mishawaka, IN-MI | 553 | 4,962 | 1,586.6 | 303.2 | 404 | 17,830 | 7,055.2 | 969.1 | 632 | 12,192 | 625.7 | 176.2 |
| Spartanburg, SC | D | D | D | D | 407 | 29,294 | 21,377.8 | 1,729.6 | 578 | 11,099 | 592.3 | 163.7 |
| Spokane-Spokane Valley, WA | D | D | 1,547.7 | D | 571 | 15,580 | 4,525.4 | 851.0 | 1,216 | 19,930 | 1,199.1 | 378.9 |
| Springfield, IL | 512 | 5,360 | 764.8 | 324.2 | 97 | 2,742 | 854.6 | 140.3 | 572 | 10,167 | 497.4 | 150.1 |
| Springfield, MA | 1,142 | 8,091 | 1,209.6 | 481.1 | 788 | 25,166 | 8,695.7 | 1,440.3 | 1,458 | 22,607 | 1,248.6 | 387.4 |
| Springfield, MO | 272 | 2,356 | 393.7 | 178.0 | 462 | 14,911 | 5,705.2 | 747.2 | 1,011 | 19,698 | 986.7 | 286.0 |
| Springfield, OH | D | D | D | D | 152 | 6,187 | 2,405.7 | 274.0 | 223 | 4,568 | 207.9 | 63.6 |
| State College, PA | D | D | D | D | 166 | 3,786 | 1,041.7 | 206.6 | 327 | 6,750 | 344.7 | 96.9 |
| Staunton, VA | 107 | 618 | 75.6 | 31.0 | 113 | 6,867 | 2,736.7 | 352.3 | 255 | 4,440 | 230.3 | 67.1 |
| Stockton, CA | 808 | 4,858 | 616.0 | 259.4 | 533 | 19,833 | 9,574.9 | 1,001.7 | 1,185 | 19,118 | 1,206.4 | 332.1 |
| Sumter, SC | 23 | 74 | 11.0 | 4.6 | 90 | 6,149 | 2,380.2 | 285.3 | 228 | 4,046 | 202.6 | 54.1 |
| Syracuse, NY | 1,191 | 13,590 | 3,097.4 | 946.2 | 559 | 23,735 | 10,529.8 | 1,430.8 | 1,637 | 26,188 | 1,429.1 | 440.8 |
| Tallahassee, FL | 21 | 135 | 12.9 | 5.0 | D | D | D | D | 837 | 17,137 | 918.2 | 251.2 |
| Tampa-St. Petersburg-Clearwater, FL | 6,139 | 63,153 | 11,602.0 | 4,758.1 | 2,145 | 52,836 | 21,574.8 | 2,884.3 | 5,726 | 121,393 | 8,280.6 | 2,213.3 |
| Terre Haute, IN | 64 | 292 | 37.3 | 15.1 | 188 | 10,704 | 5,228.3 | 564.5 | 387 | 7,113 | 344.1 | 100.3 |

Items 147—158

# Table C. Metropolitan Areas — Health Care and Social Assistance, Other Services, Nonemployer Businesses, and Residential Construction

| Area name | Health care and social assistance, 2017 | | | | Other services, 2017 | | | | Nonemployer businesses, 2018 | | Value of residential construction authorized by building permits, 2020 | |
|---|---|---|---|---|---|---|---|---|---|---|---|---|
| | Number of establishments | Number of employees | Receipts (mil dol) | Annual payroll (mil dol) | Number of establishments | Number of employees | Receipts (mil dol) | Annual payroll (mil dol) | Number | Receipts (mil dol) | New construction ($1,000) | Number of housing units |
| | 159 | 160 | 161 | 162 | 163 | 164 | 165 | 166 | 167 | 168 | 169 | 170 |
| Reading, PA | 861 | 26,638 | 2,897.5 | 1,261.5 | 797 | 4,354 | 425 | 116 | 25,138 | 1,201.5 | 101,312 | 515 |
| Redding, CA | 670 | 11,319 | 1,553.8 | 575.1 | 300 | 1,659 | 212 | 51 | 11,945 | 587.4 | 109,818 | 393 |
| Reno, NV | D | D | D | D | D | D | D | D | 34,299 | 2,104.9 | 948,610 | 4,504 |
| Richmond, VA | 3,264 | 82,469 | 11,064.2 | 4,301.5 | 2,399 | 17,744 | 2,197 | 616 | 92,477 | 4,028.1 | 1,302,014 | 8,190 |
| Riverside-San Bernardino-Ontario, CA | 9,041 | 178,897 | 25,064.1 | 9,349.5 | 5,187 | 33,554 | 3,492 | 1,003 | 330,713 | 15,142.5 | 3,673,835 | 15,232 |
| Roanoke, VA | 824 | 26,276 | 3,667.9 | 1,372.3 | 587 | 3,264 | 335 | 91 | 19,715 | 872.4 | 163,387 | 769 |
| Rochester, MN | 479 | 21,541 | 2,776.3 | 1,017.1 | 397 | 2,595 | 258 | 74 | 14,859 | 709.4 | 203,658 | 906 |
| Rochester, NY | 2,649 | 88,306 | 8,488.1 | 3,665.2 | 1,857 | 10,406 | 1,215 | 322 | 67,363 | 3,050.7 | 417,522 | 2,242 |
| Rockford, IL | 719 | 21,226 | 2,723.0 | 1,047.1 | 570 | 3,435 | 382 | 10.0 | 20,492 | 770.3 | 33,793 | 192 |
| Rocky Mount, NC | 325 | 8,492 | 747.3 | 307.6 | 142 | 706 | 183 | 2.0 | 8,690 | 334.0 | 51,309 | 339 |
| Rome, GA | 322 | 8,395 | 1,147.7 | 397.0 | 115 | 808 | 8.0 | 24 | 6,962 | 270.4 | 46,376 | 282 |
| Sacramento-Roseville-Folsom, CA | 5,763 | 126,716 | 21,303.3 | 8,492.8 | 3,658 | 25,778 | 3,335 | 993 | 179,529 | 9,221.7 | 2,728,908 | 11,164 |
| Saginaw, MI | 597 | 17,800 | 2,040.7 | 811.3 | 325 | 1,989 | 183 | 5.0 | 10,640 | 436.5 | 44,779 | 245 |
| St. Cloud, MN | 571 | 20,317 | 2,345.5 | 988.0 | 443 | 2,584 | 287 | 79 | 13,753 | 706.8 | 153,834 | 615 |
| St. George, UT | 583 | 9,409 | 1,212.8 | 395.3 | 265 | 1,401 | 139 | 38 | 15,006 | 776.7 | 641,709 | 3,314 |
| St. Joseph, MO-KS | 330 | 9,556 | 1,032.8 | 420.7 | 38 | 194 | 207 | 5 | 6,541 | 268.2 | 34,549 | 256 |
| St. Louis, MO-IL | 9,408 | 198,887 | 22,481.5 | 8,907.1 | 4,985 | 37,262 | 4,419 | 1,31 | 193,944 | 8,750.3 | 1,910,035 | 7,863 |
| Salem, OR | 1,275 | 23,813 | 2,654.2 | 1,130.8 | 594 | 2,889 | 301 | 88 | 22,744 | 1,083.8 | 415,960 | 2,008 |
| Salinas, CA | 999 | 17,169 | 2,803.9 | 1,072.6 | 607 | 4,058 | 573 | 129 | 25,963 | 1,502.6 | 137,709 | 753 |
| Salisbury, MD-DE | 1,099 | 24,107 | 2,852.2 | 1,143.7 | 706 | 4,073 | 435 | 125 | 31,000 | 1,652.4 | 740,434 | 5,108 |
| Salt Lake City, UT | 3,330 | 68,721 | 9,674.6 | 3,563.4 | 2,136 | 14,800 | 1,741 | 512 | 96,515 | 4,922.8 | 2,227,894 | 11,033 |
| San Angelo, TX | D | D | D | D | D | D | 173 | D | 9,271 | 419.6 | 143,205 | 686 |
| San Antonio-New Braunfels, TX | 5,924 | 138,127 | 15,957.1 | 5,855.3 | 3,287 | 23,278 | 2,40 | 708 | 199,220 | 9,483.0 | 3,123,325 | 16,697 |
| San Diego-Chula Vista-Carlsbad, CA | 9,592 | 175,430 | 26,431.3 | 10,143.4 | 5,823 | 42,595 | 4,889 | 1,38 | 292,941 | 15,032.5 | 1,997,782 | 9,472 |
| San Francisco-Oakland-Berkeley, CA | 15,015 | 283,893 | 47,686.4 | 19,336.5 | 9,486 | 76,980 | 14,304 | 3,343 | 463,717 | 28,178.7 | 3,163,103 | 10,156 |
| Oakland-Berkeley-Livermore, CA Div 36084 | 8,045 | 158,335 | 26,234.4 | 10,497.3 | 4,815 | 34,868 | 5,126 | 1,431 | 251,224 | 13,749.1 | 1,852,508 | 6,923 |
| San Francisco-San Mateo-Redwood City, CA Div 41884 | 5,812 | 108,759 | 19,101.3 | 7,774.4 | 3,968 | 37,222 | 8,331 | 1,708 | 174,286 | 11,444.6 | 1,237,008 | 3,078 |
| San Rafael, CA Div 42034 | 1,158 | 16,799 | 2,350.7 | 1,064.8 | 703 | 4,890 | 847 | 204 | 38,207 | 2,985.1 | 73,588 | 155 |
| San Jose-Sunnyvale-Santa Clara, CA | 5,993 | 116,645 | 22,363.4 | 9,137.7 | 3,380 | 22,818 | 5,506 | 983 | 152,926 | 9,322.3 | 1,148,099 | 5,851 |
| San Luis Obispo-Paso Robles, CA | 1,055 | 15,066 | 1,912.6 | 788.2 | 450 | 2,986 | 266 | 79 | 26,143 | 1,417.6 | 227,992 | 911 |
| Santa Cruz-Watsonville, CA | 940 | 15,312 | 2,175.2 | 831.2 | 473 | 3,232 | 414 | 128 | 25,378 | 1,337.0 | 39,102 | 244 |
| Santa Fe, NM | 510 | 9,791 | 1,131.9 | 426.1 | 351 | 2,038 | 295 | 82 | 16,612 | 772.0 | 117,591 | 577 |
| Santa Maria-Santa Barbara, CA | 1,428 | 23,163 | 3,662.0 | 1,207.5 | 785 | 5,846 | 1,759 | 211 | 35,272 | 2,084.9 | 205,525 | 1,027 |
| Santa Rosa-Petaluma, CA | 1,537 | 26,382 | 3,737.9 | 1,632.2 | 959 | 5,810 | 685 | 207 | 46,153 | 2,559.9 | 403,557 | 1,395 |
| Savannah, GA | 893 | 22,833 | 2,694.6 | 1,037.6 | 571 | 4,578 | 536 | 163 | 30,203 | 1,340.2 | 562,919 | 3,026 |
| Scranton--Wilkes-Barre, PA | 1,751 | 46,747 | 5,135.6 | 2,008.2 | 1,017 | 5,431 | 606 | 158 | 31,641 | 1,655.6 | 127,485 | 584 |
| Seattle-Tacoma-Bellevue, WA | 11,838 | 247,474 | 33,962.9 | 13,673.4 | 7,656 | 48,632 | 11,111 | 2,037 | 282,855 | 15,147.7 | 4,863,438 | 23,039 |
| Seattle-Bellevue-Kent, WA Div 42644 | 9,835 | 194,243 | 26,513.7 | 10,645.5 | 6,216 | 39,328 | 10,002 | 1,698 | 234,588 | 12,849.7 | 3,751,962 | 18,117 |
| Tacoma-Lakewood, WA Div 45704 | 2,003 | 53,231 | 7,449.2 | 3,027.9 | 1,440 | 9,304 | 1,109 | 338 | 48,267 | 2,298.1 | 1,111,475 | 4,922 |
| Sebastian-Vero Beach, FL | 495 | 9,718 | 1,155.2 | 454.1 | 313 | 1,726 | 167 | 51 | 14,992 | 815.1 | 462,031 | 1,162 |
| Sebring-Avon Park, FL | 309 | 6,442 | 729.3 | 284.6 | 125 | 403 | 4.0 | 1.0 | 6,295 | 266.0 | 85,682 | 384 |
| Sheboygan, WI | 315 | 6,645 | 744.0 | 290.0 | 202 | 1,060 | 99 | 26 | 5,449 | 260.9 | 72,367 | 365 |
| Sherman-Denison, TX | 360 | 8,454 | 956.3 | 381.4 | 149 | 689 | 82 | 22 | 10,926 | 582.2 | 195,489 | 1,181 |
| Shreveport-Bossier City, LA | 1,118 | 33,275 | 4,640.4 | 1,691.0 | 130 | 1,055 | 107 | 31 | 31,865 | 1,392.0 | 204,295 | 1,000 |
| Sierra Vista-Douglas, AZ | 283 | 4,921 | 454.5 | 188.2 | 160 | 661 | 66 | 17 | 6,885 | 221.8 | 56,265 | 337 |
| Sioux City, IA-NE-SD | 441 | 11,199 | 1,300.2 | 494.3 | 209 | 1,323 | 17.0 | 44 | 8,594 | 443.6 | 130,354 | 779 |
| Sioux Falls, SD | 691 | 31,489 | 4,072.0 | 1,660.6 | 374 | 2,704 | 399 | 94 | 21,107 | 1,136.1 | 543,298 | 3,550 |
| South Bend-Mishawaka, IN-MI | 728 | 20,510 | 2,480.6 | 865.6 | 487 | 3,316 | 384 | 109 | 19,052 | 796.4 | 133,060 | 555 |
| Spartanburg, SC | 574 | 16,460 | 1,784.2 | 848.3 | 425 | 3,014 | 399 | 96 | 21,822 | 1,033.4 | 468,600 | 2,876 |
| Spokane-Spokane Valley, WA | 1,716 | 43,243 | 5,062.6 | 2,078.5 | 953 | 5,380 | 596 | 174 | 34,883 | 1,647.8 | 662,679 | 3,362 |
| Springfield, IL | 478 | 22,613 | 3,039.2 | 1,054.0 | 483 | 3,317 | 413 | 133 | 12,923 | 500.3 | 73,817 | 335 |
| Springfield, MA | 1,832 | 57,305 | 6,096.0 | 2,644.7 | 1,232 | 8,090 | 939 | 254 | 46,503 | 2,083.9 | 180,462 | 760 |
| Springfield, MO | 1,195 | 34,893 | 4,220.1 | 1,578.9 | 838 | 5,502 | 567 | 168 | 36,109 | 1,775.9 | 463,317 | 2,213 |
| Springfield, OH | 290 | 7,102 | 709.7 | 267.6 | 191 | 1,438 | 194 | 52 | 6,627 | 269.8 | 36,806 | 139 |
| State College, PA | 403 | 9,501 | 1,078.6 | 456.8 | 253 | 1,481 | 145 | 41 | 10,321 | 471.0 | 106,787 | 380 |
| Staunton, VA | 259 | 7,056 | 814.5 | 305.1 | 252 | 1,314 | 148 | 43 | 7,511 | 362.5 | 87,564 | 501 |
| Stockton, CA | 1,513 | 32,650 | 4,236.8 | 1,600.6 | 890 | 5,434 | 565 | 164 | 42,899 | 2,395.1 | 1,001,931 | 3,914 |
| Sumter, SC | 226 | 6,496 | 673.6 | 232.1 | 144 | 1,149 | 109 | 32 | 8,393 | 317.1 | 53,966 | 408 |
| Syracuse, NY | 1,678 | 49,957 | 5,879.3 | 2,441.7 | 1,219 | 7,073 | 789 | 218 | 39,662 | 1,814.6 | 161,769 | 777 |
| Tallahassee, FL | 839 | 21,217 | 2,577.5 | 1,050.1 | 701 | 4,744 | 755 | 202 | 29,009 | 1,130.7 | 383,803 | 2,445 |
| Tampa-St. Petersburg-Clearwater, FL | 9,483 | 185,514 | 27,593.2 | 9,516.2 | 5,435 | 32,574 | 3,897 | 97.0 | 286,933 | 12,879.9 | 5,299,528 | 20,348 |
| Terre Haute, IN | 491 | 11,401 | 1,467.2 | 512.8 | 284 | 1,487 | 172 | 41 | 9,113 | 351.2 | 22,817 | 140 |

# Table C. Metropolitan Areas — Government Employment and Payroll, and Local Government Finances

| Area name | Full-time equivalent employees (171) | March payroll (dollars) (172) | Administration, judicial, and legal (173) | Police and corrections (174) | Fire protection (175) | Highways and transportation (176) | Health and welfare (177) | Natural resources and utilities (178) | Education and libraries (179) | Total (mil dol) (180) | Intergovernmental (mil dol) (181) | Taxes Total (mil dol) (182) | Per capita¹ Total (dollars) (183) | Per capita¹ Property (dollars) (184) |
|---|---|---|---|---|---|---|---|---|---|---|---|---|---|---|
| Reading, PA | 14,681 | 67,582,415 | 5.9 | 10.7 | 1.3 | 2.5 | 4.6 | 3.4 | 70.0 | 2,206 | 858 | 953 | 2,282 | 1,831 |
| Redding, CA | 7,243 | 36,294,479 | 6.8 | 9.0 | 2.6 | 3.2 | 14.4 | 8.3 | 53.1 | 1,173 | 746 | 233 | 1,300 | 1,022 |
| Reno, NV | 13,528 | 64,676,757 | 10.2 | 13.6 | 5.9 | 4.3 | 4.4 | 8.6 | 50.3 | 2,110 | 945 | 720 | 1,563 | 1,008 |
| Richmond, VA | 46,835 | 187,813,105 | 7.9 | 13.7 | 5.9 | 2.0 | 5.9 | 5.6 | 56.1 | 5,224 | 2,138 | 2,242 | 1,766 | 1,282 |
| Riverside-San Bernardino-Ontario, CA | 152,935 | 948,566,442 | 7.2 | 10.6 | 2.8 | 2.2 | 13.3 | 5.0 | 57.0 | 31,726 | 18,093 | 6,893 | 1,510 | 1,066 |
| Roanoke, VA | 12,013 | 46,585,173 | 8.1 | 10.7 | 5.7 | 1.8 | 4.6 | 6.0 | 60.3 | 1,310 | 579 | 564 | 1,802 | 1,280 |
| Rochester, MN | 7,887 | 37,423,777 | 9.7 | 10.7 | 2.3 | 4.1 | 10.5 | 8.5 | 53.2 | 1,143 | 504 | 353 | 1,619 | 1,414 |
| Rochester, NY | 51,885 | 245,268,182 | 4.4 | 9.1 | 2.7 | 3.2 | 5.9 | 3.4 | 69.6 | 6,835 | 2,882 | 3,059 | 2,854 | 1,914 |
| Rockford, IL | 11,726 | 52,483,712 | 5.2 | 12.4 | 5.9 | 3.5 | 4.0 | 6.7 | 61.2 | 1,493 | 647 | 638 | 1,887 | 1,647 |
| Rocky Mount, NC | 7,611 | 29,393,778 | 3.6 | 7.0 | 2.6 | 1.7 | 33.5 | 6.4 | 43.6 | 486 | 261 | 152 | 1,033 | 780 |
| Rome, GA | 3,978 | 13,144,682 | 5.5 | 10.4 | 4.8 | 3.7 | 2.2 | 6.5 | 65.2 | 378 | 155 | 150 | 1,537 | 984 |
| Sacramento-Roseville-Folsom, CA | 79,610 | 495,084,789 | 6.2 | 12.4 | 4.7 | 3.6 | 8.0 | 11.9 | 49.8 | 14,636 | 6,526 | 4,181 | 1,802 | 1,283 |
| Saginaw, MI | 5,450 | 22,845,829 | 8.3 | 7.8 | 1.9 | 2.7 | 12.0 | 4.9 | 60.6 | 802 | 465 | 162 | 845 | 736 |
| St. Cloud, MN | 6,869 | 30,909,140 | 8.9 | 11.4 | 1.7 | 5.1 | 8.6 | 3.7 | 57.9 | 872 | 459 | 261 | 1,314 | 1,173 |
| St. George, UT | 4,494 | 17,531,308 | 8.8 | 10.7 | 1.8 | 2.4 | 6.9 | 15.0 | 52.2 | 568 | 184 | 240 | 1,448 | 947 |
| St. Joseph, MO-KS | 4,416 | 15,972,895 | 5.2 | 7.9 | 3.8 | 3.4 | 2.7 | 5.1 | 71.0 | 429 | 161 | 195 | 1,541 | 892 |
| St. Louis, MO-IL | 99,482 | 431,149,817 | 5.7 | 11.4 | 5.6 | 5.8 | 2.8 | 5.5 | 61.4 | 11,390 | 3,645 | 5,715 | 2,037 | 1,313 |
| Salem, OR | 12,770 | 65,068,787 | 6.1 | 9.6 | 3.4 | 3.3 | 4.5 | 3.2 | 67.4 | 1,789 | 963 | 520 | 1,226 | 1,088 |
| Salinas, CA | 19,396 | 124,781,670 | 5.0 | 7.7 | 2.4 | 3.5 | 27.9 | 4.3 | 45.9 | 3,512 | 1,417 | 905 | 2,083 | 1,363 |
| Salisbury, MD-DE | 13,385 | 59,486,106 | 5.7 | 9.7 | 0.8 | 1.9 | 2.5 | 6.0 | 71.4 | 1,674 | 705 | 667 | 1,650 | 1,143 |
| Salt Lake City, UT | 37,900 | 151,242,305 | 7.7 | 10.3 | 5.4 | 10.5 | 4.7 | 8.3 | 50.0 | 4,503 | 1,359 | 2,130 | 1,769 | 1,148 |
| San Angelo, TX | 5,059 | 15,850,961 | 8.3 | 14.6 | 6.1 | 2.9 | 4.4 | 5.7 | 56.0 | 399 | 129 | 210 | 1,746 | 1,365 |
| San Antonio-New Braunfels, TX | 103,873 | 441,237,393 | 4.9 | 9.0 | 4.0 | 3.5 | 11.8 | 9.8 | 55.9 | 11,196 | 3,428 | 5,162 | 2,088 | 1,678 |
| San Diego-Chula Vista-Carlsbad, CA | 106,292 | 629,893,092 | 7.5 | 10.0 | 3.7 | 3.2 | 10.6 | 6.2 | 54.8 | 19,788 | 7,387 | 6,914 | 2,082 | 1,537 |
| San Francisco-Oakland-Berkeley, CA | 174,248 | 1,251,034,774 | 6.7 | 12.0 | 5.3 | 11.3 | 17.0 | 8.3 | 37.1 | 40,740 | 12,173 | 16,112 | 3,419 | 2,252 |
| Oakland-Berkeley-Livermore, CA Div 36084 | 96,003 | 686,338,466 | 5.2 | 11.7 | 4.6 | 9.3 | 16.9 | 8.1 | 42.1 | 22,204 | 7,558 | 7,786 | 2,775 | 1,845 |
| San Francisco-San Mateo-Redwood City, CA Div 41884 | 69,282 | 500,163,023 | 8.6 | 12.7 | 6.1 | 15.0 | 18.1 | 8.2 | 29.0 | 16,691 | 4,186 | 7,334 | 4,453 | 2,880 |
| San Rafael, CA Div 42034 | 8,963 | 64,533,285 | 8.9 | 10.4 | 7.2 | 3.3 | 9.5 | 10.9 | 46.6 | 1,844 | 429 | 992 | 3,820 | 2,667 |
| San Jose-Sunnyvale-Santa Clara, CA | 66,718 | 513,779,027 | 7.3 | 9.0 | 2.9 | 4.9 | 24.0 | 5.0 | 45.1 | 17,165 | 4,145 | 7,009 | 3,517 | 2,571 |
| San Luis Obispo-Paso Robles, CA | 8,813 | 50,653,491 | 7.9 | 12.7 | 2.9 | 2.9 | 9.1 | 5.7 | 55.0 | 1,483 | 507 | 703 | 2,490 | 1,941 |
| Santa Cruz-Watsonville, CA | 9,852 | 61,097,077 | 8.1 | 9.9 | 4.0 | 7.1 | 13.0 | 8.3 | 47.4 | 1,680 | 751 | 584 | 2,124 | 1,525 |
| Santa Fe, NM | 5,045 | 19,287,175 | 8.7 | 13.7 | 6.2 | 5.6 | 3.3 | 12.0 | 47.2 | 639 | 266 | 265 | 1,769 | 1,075 |
| Santa Maria-Santa Barbara, CA | 21,940 | 144,599,650 | 4.5 | 8.3 | 4.1 | 2.5 | 11.8 | 5.0 | 62.2 | 3,757 | 1,907 | 1,050 | 2,358 | 1,793 |
| Santa Rosa-Petaluma, CA | 17,488 | 111,265,857 | 7.9 | 11.7 | 4.0 | 2.5 | 14.7 | 6.4 | 51.1 | 3,030 | 1,129 | 1,172 | 2,332 | 1,749 |
| Savannah, GA | 14,092 | 53,803,266 | 8.5 | 13.4 | 4.5 | 4.6 | 5.5 | 6.7 | 53.1 | 2,228 | 467 | 799 | 2,067 | 1,318 |
| Scranton--Wilkes-Barre, PA | 16,431 | 71,311,908 | 7.4 | 10.7 | 2.7 | 4.3 | 5.8 | 4.1 | 63.3 | 2,263 | 926 | 993 | 1,786 | 1,334 |
| Seattle-Tacoma-Bellevue, WA | 125,096 | 808,747,124 | 9.0 | 8.9 | 5.6 | 10.9 | 10.4 | 12.0 | 40.6 | 25,386 | 7,644 | 10,196 | 2,624 | 1,448 |
| Seattle-Bellevue-Kent, WA Div 42644 | 99,194 | 651,135,102 | 9.2 | 9.1 | 5.0 | 11.5 | 12.2 | 12.1 | 38.2 | 21,109 | 6,026 | 8,617 | 2,867 | 1,522 |
| Tacoma-Lakewood, WA Div 45704 | 25,902 | 157,612,022 | 8.6 | 7.9 | 8.1 | 8.5 | 2.8 | 11.9 | 50.6 | 4,277 | 1,618 | 1,579 | 1,795 | 1,196 |
| Sebastian-Vero Beach, FL | 4,114 | 16,229,719 | 5.2 | 17.3 | 4.8 | 4.9 | 3.8 | 12.9 | 48.2 | 515 | 109 | 298 | 1,931 | 1,523 |
| Sebring-Avon Park, FL | 3,221 | 10,134,596 | 7.5 | 15.0 | 2.0 | 3.4 | 1.9 | 4.9 | 63.1 | 303 | 127 | 102 | 980 | 722 |
| Sheboygan, WI | 4,307 | 19,613,015 | 4.3 | 9.6 | 2.4 | 5.4 | 8.1 | 3.7 | 65.1 | 501 | 247 | 183 | 1,593 | 1,516 |
| Sherman-Denison, TX | 5,654 | 19,812,399 | 6.1 | 10.4 | 4.1 | 2.3 | 5.3 | 5.8 | 64.9 | 487 | 167 | 229 | 1,745 | 1,467 |
| Shreveport-Bossier City, LA | 16,212 | 56,514,633 | 7.0 | 16.1 | 6.7 | 2.8 | 2.1 | 4.4 | 58.7 | 1,693 | 520 | 893 | 2,230 | 1,159 |
| Sierra Vista-Douglas, AZ | 4,703 | 17,332,719 | 11.3 | 11.0 | 5.9 | 3.1 | 2.2 | 4.0 | 60.1 | 436 | 209 | 154 | 1,235 | 895 |
| Sioux City, IA-NE-SD | 6,257 | 27,360,833 | 5.0 | 8.7 | 3.0 | 4.6 | 2.0 | 5.0 | 70.4 | 722 | 322 | 285 | 1,988 | 1,584 |
| Sioux Falls, SD | 8,112 | 32,164,717 | 8.0 | 9.6 | 3.4 | 4.2 | 2.5 | 5.1 | 65.8 | 985 | 280 | 546 | 2,100 | 1,426 |
| South Bend-Mishawaka, IN-MI | 9,840 | 36,684,987 | 6.6 | 12.2 | 7.1 | 4.6 | 2.3 | 6.3 | 59.5 | 1,239 | 545 | 449 | 1,395 | 1,137 |
| Spartanburg, SC | 15,616 | 72,630,964 | 2.9 | 4.1 | 1.5 | 0.4 | 54.2 | 3.4 | 32.7 | 2,111 | 444 | 384 | 1,251 | 1,089 |
| Spokane-Spokane Valley, WA | 16,942 | 90,650,435 | 8.4 | 11.3 | 6.6 | 6.3 | 2.9 | 6.8 | 55.3 | 2,508 | 1,205 | 811 | 1,474 | 896 |
| Springfield, IL | 8,706 | 40,071,271 | 4.7 | 9.8 | 4.4 | 5.2 | 3.0 | 14.5 | 58.2 | 887 | 321 | 410 | 1,958 | 1,668 |
| Springfield, MA | 28,210 | 133,872,927 | 3.7 | 9.3 | 6.2 | 2.8 | 2.7 | 6.2 | 66.3 | 2,963 | 1,322 | 1,284 | 1,836 | 1,777 |
| Springfield, MO | 16,317 | 60,529,281 | 4.8 | 7.6 | 3.3 | 4.0 | 9.6 | 9.6 | 55.0 | 1,527 | 534 | 628 | 1,359 | 770 |
| Springfield, OH | 4,730 | 19,251,860 | 8.8 | 10.4 | 4.4 | 2.0 | 9.0 | 3.3 | 60.6 | 484 | 238 | 186 | 1,382 | 876 |
| State College, PA | 3,630 | 14,677,666 | 9.0 | 9.2 | 0.0 | 9.4 | 3.5 | 7.0 | 60.1 | 521 | 151 | 264 | 1,626 | 1,174 |
| Staunton, VA | 4,303 | 14,913,667 | 7.1 | 14.3 | 4.1 | 1.7 | 3.5 | 3.9 | 62.8 | 392 | 197 | 151 | 1,241 | 847 |
| Stockton, CA | 26,849 | 155,624,972 | 5.8 | 10.0 | 3.2 | 3.1 | 16.7 | 4.3 | 55.8 | 4,671 | 2,332 | 1,193 | 1,605 | 1,058 |
| Sumter, SC | 5,063 | 16,232,355 | 6.6 | 9.3 | 3.0 | 1.8 | 6.2 | 4.9 | 65.0 | 450 | 204 | 168 | 1,196 | 824 |
| Syracuse, NY | 30,422 | 143,144,961 | 3.3 | 8.1 | 2.3 | 4.6 | 5.2 | 4.3 | 71.0 | 4,082 | 1,835 | 1,742 | 2,676 | 1,808 |
| Tallahassee, FL | 13,633 | 51,557,334 | 10.9 | 11.8 | 4.0 | 4.6 | 3.3 | 14.7 | 46.9 | 1,366 | 505 | 466 | 1,216 | 834 |
| Tampa-St. Petersburg-Clearwater, FL | 98,460 | 389,854,339 | 7.4 | 14.8 | 6.3 | 4.3 | 2.9 | 9.0 | 52.3 | 11,723 | 4,167 | 4,583 | 1,475 | 1,105 |
| Terre Haute, IN | 5,972 | 21,153,016 | 7.9 | 9.9 | 4.6 | 4.2 | 6.6 | 3.7 | 61.9 | 617 | 277 | 195 | 1,044 | 917 |

1. Based on the resident metpopulation estimated as of July 1 of the year shown.

Items 171—184

# Table C. Metropolitan Areas — Local Government Finances, Government Employment, and Income Taxes

| Area name | Direct general expenditure — Total (mil dol) | Per capita[1] (dollars) | Education | Health and hospitals | Police protection | Public welfare | Highways | Debt outstanding — Total (mil dol) | Per capita[1] (dollars) | Government employment 2019 — Federal civilian | Federal military | State and local | Number of returns | Mean adjusted gross income | Mean income tax |
|---|---|---|---|---|---|---|---|---|---|---|---|---|---|---|---|
| | 185 | 186 | 187 | 188 | 189 | 190 | 191 | 192 | 193 | 194 | 195 | 196 | 197 | 198 | 199 |
| Reading, PA | 2,252.8 | 5,396 | 56.5 | 1.4 | 4.5 | 8.7 | 3.5 | 2,740.4 | 6,564 | 939 | 1,050 | 20,924 | 207,350 | 62,386 | 6,596 |
| Redding, CA | 1,007.3 | 5,615 | 46.0 | 8.4 | 6.0 | 11.3 | 2.6 | 573.9 | 3,199 | 1,295 | 308 | 11,850 | 77,700 | 59,273 | 5,752 |
| Reno, NV | 1,990.9 | 4,322 | 34.2 | 1.4 | 7.6 | 4.8 | 8.5 | 3,144.9 | 6,828 | 3,770 | 1,264 | 24,479 | 239,750 | 88,821 | 12,567 |
| Richmond, VA | 4,992.3 | 3,932 | 44.4 | 4.5 | 7.0 | 4.1 | 3.3 | 4,588.2 | 3,614 | 17,163 | 13,378 | 96,583 | 617,830 | 75,684 | 9,258 |
| Riverside-San Bernardino-Ontario, CA | 32,108.9 | 7,032 | 38.0 | 19.7 | 6.0 | 6.4 | 3.5 | 29,154.1 | 6,385 | 21,390 | 22,475 | 243,682 | 1,957,200 | 57,513 | 5,473 |
| Roanoke, VA | 1,332.7 | 4,258 | 45.0 | 2.7 | 5.0 | 6.9 | 2.4 | 849.8 | 2,715 | 3,851 | 996 | 17,703 | 146,880 | 61,785 | 6,473 |
| Rochester, MN | 1,256.6 | 5,764 | 38.9 | 2.0 | 4.9 | 6.1 | 10.8 | 3,594.2 | 16,488 | 929 | 765 | 11,756 | 111,440 | 75,813 | 8,932 |
| Rochester, NY | 7,078.3 | 6,603 | 51.5 | 4.0 | 3.9 | 9.3 | 4.0 | 4,323.1 | 4,033 | 5,026 | 1,733 | 71,033 | 524,180 | 64,975 | 7,123 |
| Rockford, IL | 1,552.6 | 4,591 | 52.1 | 0.9 | 9.2 | 2.4 | 6.4 | 1,090.1 | 3,223 | 904 | 697 | 15,626 | 161,810 | 58,547 | 5,972 |
| Rocky Mount, NC | 483.9 | 3,297 | 54.4 | 4.4 | 7.7 | 6.8 | 1.2 | 319.3 | 2,175 | 380 | 314 | 10,355 | 64,620 | 48,120 | 4,563 |
| Rome, GA | 407.1 | 4,176 | 48.2 | 8.6 | 5.5 | 0.1 | 2.8 | 185.7 | 1,904 | 215 | 255 | 5,467 | 40,330 | 56,323 | 5,283 |
| Sacramento-Roseville-Folsom, CA | 13,996.0 | 6,034 | 40.1 | 5.8 | 5.9 | 6.5 | 3.7 | 26,562.1 | 11,451 | 14,378 | 4,279 | 239,371 | 1,080,800 | 75,740 | 9,061 |
| Saginaw, MI | 842.5 | 4,389 | 39.0 | 18.6 | 4.0 | 0.6 | 8.2 | 484.2 | 2,522 | 1,517 | 309 | 9,264 | 90,190 | 52,225 | 5,102 |
| St. Cloud, MN | 920.3 | 4,638 | 49.4 | 0.8 | 5.2 | 5.6 | 9.7 | 1,592.9 | 8,027 | 2,469 | 681 | 11,163 | 95,370 | 62,411 | 6,432 |
| St. George, UT | 514.8 | 3,103 | 51.2 | 5.2 | 8.3 | 0.1 | 2.6 | 527.3 | 3,178 | 595 | 677 | 8,592 | 73,900 | 68,082 | 7,492 |
| St. Joseph, MO-KS | 417.0 | 3,301 | 53.9 | 1.8 | 5.7 | 0.1 | 6.3 | 902.3 | 7,142 | 620 | 408 | 8,694 | 54,410 | 53,565 | 4,888 |
| St. Louis, MO-IL | 11,216.3 | 3,998 | 48.1 | 2.8 | 7.6 | 0.9 | 4.6 | 15,335.4 | 5,466 | 28,544 | 13,519 | 137,135 | 1,358,380 | 76,697 | 9,763 |
| Salem, OR | 1,750.1 | 4,125 | 53.9 | 6.8 | 5.6 | 0.1 | 4.6 | 1,561.0 | 3,680 | 1,472 | 1,015 | 39,599 | 189,930 | 60,438 | 5,801 |
| Salinas, CA | 3,623.3 | 8,338 | 38.0 | 22.1 | 4.4 | 5.0 | 2.3 | 2,222.3 | 5,114 | 5,239 | 5,309 | 29,711 | 197,090 | 71,269 | 8,424 |
| Salisbury, MD-DE | 1,709.1 | 4,229 | 56.0 | 1.6 | 5.5 | 0.6 | 3.0 | 1,053.4 | 2,606 | 1,198 | 1,925 | 22,449 | 200,270 | 62,565 | 6,816 |
| Salt Lake City, UT | 4,622.2 | 3,838 | 37.8 | 1.4 | 5.8 | 3.3 | 7.2 | 10,128.6 | 8,411 | 12,751 | 4,990 | 99,775 | 559,160 | 73,787 | 8,488 |
| San Angelo, TX | 401.1 | 3,333 | 50.3 | 1.2 | 6.5 | 0.9 | 3.4 | 491.6 | 4,084 | 1,257 | 3,781 | 8,098 | 54,860 | 65,237 | 7,998 |
| San Antonio-New Braunfels, TX | 11,278.0 | 4,562 | 44.4 | 16.6 | 5.2 | 1.5 | 3.4 | 28,785.3 | 11,644 | 36,662 | 38,981 | 135,739 | 1,141,520 | 65,904 | 7,736 |
| San Diego-Chula Vista-Carlsbad, CA | 20,668.2 | 6,223 | 39.3 | 9.4 | 5.6 | 4.4 | 3.1 | 41,671.1 | 12,547 | 48,266 | 98,986 | 206,249 | 1,598,530 | 83,078 | 10,997 |
| San Francisco-Oakland-Berkeley, CA | 40,686.9 | 8,634 | 25.3 | 18.9 | 5.7 | 6.8 | 4.4 | 68,469.0 | 14,529 | 31,089 | 9,201 | 293,035 | 2,360,920 | 143,197 | 25,463 |
| Oakland-Berkeley-Livermore, CA Div 36084 | 21,848.4 | 7,787 | 28.9 | 17.0 | 5.9 | 4.9 | 6.6 | 35,862.7 | 12,782 | 13,619 | 5,690 | 151,979 | 1,355,310 | 112,684 | 17,332 |
| San Francisco-San Mateo-Redwood City, CA Div 41884 | 17,017.5 | 10,333 | 19.8 | 22.7 | 5.4 | 9.6 | 1.6 | 29,692.6 | 18,029 | 16,791 | 2,998 | 126,191 | 873,580 | 181,500 | 35,827 |
| San Rafael, CA Div 42034 | 1,820.9 | 7,010 | 33.9 | 7.4 | 6.6 | 3.6 | 3.4 | 2,913.7 | 11,217 | 679 | 513 | 14,865 | 132,030 | 202,985 | 40,355 |
| San Jose-Sunnyvale-Santa Clara, CA | 15,971.7 | 8,015 | 31.1 | 24.9 | 4.9 | 5.6 | 1.9 | 20,746.3 | 10,411 | 10,104 | 3,399 | 90,174 | 963,330 | 168,730 | 31,215 |
| San Luis Obispo-Paso Robles, CA | 1,449.8 | 5,132 | 41.1 | 6.4 | 6.0 | 7.8 | 4.0 | 1,526.6 | 5,404 | 566 | 527 | 22,523 | 132,030 | 78,930 | 9,761 |
| Santa Cruz-Watsonville, CA | 1,650.8 | 6,006 | 38.1 | 7.8 | 5.8 | 7.8 | 3.0 | 1,276.1 | 4,643 | 530 | 381 | 19,884 | 130,370 | 94,197 | 13,737 |
| Santa Fe, NM | 553.0 | 3,699 | 43.8 | 0.0 | 7.4 | 1.4 | 2.9 | 844.6 | 5,650 | 953 | 384 | 14,577 | 78,170 | 75,175 | 9,788 |
| Santa Maria-Santa Barbara, CA | 3,448.8 | 7,745 | 33.0 | 26.8 | 4.7 | 4.2 | 1.9 | 1,732.6 | 3,891 | 3,521 | 3,107 | 32,924 | 202,420 | 83,343 | 11,736 |
| Santa Rosa-Petaluma, CA | 3,042.2 | 6,054 | 40.6 | 9.0 | 6.5 | 6.9 | 3.8 | 3,352.6 | 6,672 | 1,363 | 1,487 | 26,682 | 245,380 | 85,708 | 11,379 |
| Savannah, GA | 2,217.1 | 5,738 | 32.4 | 28.2 | 6.2 | 0.7 | 5.0 | 844.0 | 2,184 | 2,954 | 5,505 | 21,557 | 173,950 | 63,947 | 6,838 |
| Scranton--Wilkes-Barre, PA | 2,305.3 | 4,149 | 53.2 | 0.2 | 4.1 | 5.7 | 4.5 | 2,632.9 | 4,738 | 4,340 | 1,394 | 25,075 | 274,560 | 55,306 | 5,637 |
| Seattle-Tacoma-Bellevue, WA | 22,482.0 | 5,786 | 35.1 | 10.2 | 5.0 | 0.7 | 5.5 | 38,244.6 | 9,843 | 33,772 | 45,205 | 252,168 | 1,948,320 | 107,711 | 16,137 |
| Seattle-Bellevue-Kent, WA Div 42644 | 18,455.2 | 6,140 | 33.1 | 12.2 | 4.7 | 0.8 | 5.6 | 31,982.0 | 10,640 | 21,666 | 13,504 | 203,593 | 1,523,580 | 117,400 | 18,383 |
| Tacoma-Lakewood, WA Div 45704 | 4,026.7 | 4,578 | 44.6 | 1.4 | 6.4 | 0.5 | 5.3 | 6,262.6 | 7,119 | 12,106 | 31,701 | 48,575 | 424,740 | 72,956 | 8,079 |
| Sebastian-Vero Beach, FL | 533.9 | 3,462 | 35.9 | 4.1 | 7.3 | 0.9 | 7.3 | 253.7 | 1,645 | 390 | 276 | 4,711 | 78,500 | 106,913 | 17,196 |
| Sebring-Avon Park, FL | 317.5 | 3,058 | 44.6 | 3.2 | 10.1 | 0.6 | 5.4 | 109.0 | 1,050 | 285 | 183 | 3,969 | 43,010 | 45,263 | 3,849 |
| Sheboygan, WI | 520.5 | 4,522 | 56.5 | 3.9 | 6.8 | 5.2 | 7.5 | 547.7 | 4,759 | 208 | 326 | 5,525 | 59,260 | 62,615 | 6,304 |
| Sherman-Denison, TX | 487.5 | 3,717 | 54.3 | 3.7 | 4.9 | 0.5 | 3.7 | 1,071.3 | 8,168 | 354 | 262 | 6,940 | 59,710 | 60,307 | 6,156 |
| Shreveport-Bossier City, LA | 1,664.9 | 4,161 | 51.3 | 2.4 | 7.2 | 0.2 | 2.6 | 1,377.2 | 3,442 | 4,830 | 6,690 | 23,252 | 171,930 | 58,931 | 6,526 |
| Sierra Vista-Douglas, AZ | 399.0 | 3,196 | 48.4 | 1.3 | 10.9 | 2.7 | 7.5 | 162.4 | 1,301 | 4,899 | 4,315 | 6,006 | 53,160 | 50,605 | 4,304 |
| Sioux City, IA-NE-SD | 741.3 | 5,178 | 55.9 | 1.4 | 5.3 | 0.9 | 6.0 | 552.4 | 3,858 | 801 | 543 | 8,376 | 68,290 | 66,589 | 7,459 |
| Sioux Falls, SD | 928.0 | 3,568 | 49.6 | 1.3 | 5.1 | 0.5 | 12.2 | 1,048.6 | 4,032 | 2,761 | 1,539 | 12,008 | 136,820 | 77,706 | 9,264 |
| South Bend-Mishawaka, IN-MI | 1,068.9 | 3,324 | 44.6 | 2.4 | 5.8 | 0.9 | 4.9 | 859.3 | 2,672 | 952 | 937 | 14,855 | 151,140 | 60,503 | 6,533 |
| Spartanburg, SC | 2,201.0 | 7,175 | 26.1 | 58.6 | 2.3 | 0.2 | 1.2 | 1,354.9 | 4,417 | 591 | 1,154 | 21,689 | 141,180 | 57,348 | 5,570 |
| Spokane-Spokane Valley, WA | 2,480.3 | 4,507 | 48.4 | 7.2 | 5.2 | 0.5 | 5.0 | 2,241.5 | 4,073 | 5,243 | 4,546 | 35,191 | 261,670 | 65,691 | 7,087 |
| Springfield, IL | 929.9 | 4,442 | 45.6 | 2.4 | 8.8 | 0.2 | 8.3 | 3,149.4 | 15,042 | 1,774 | 466 | 19,163 | 104,250 | 71,456 | 8,326 |
| Springfield, MA | 3,054.5 | 4,367 | 57.5 | 0.6 | 4.6 | 0.3 | 3.7 | 1,705.7 | 2,439 | 5,409 | 1,780 | 53,496 | 331,420 | 63,158 | 6,785 |
| Springfield, MO | 1,481.7 | 3,207 | 52.3 | 9.5 | 7.1 | 0.4 | 5.9 | 1,749.8 | 3,787 | 2,524 | 1,543 | 27,028 | 207,890 | 57,211 | 6,051 |
| Springfield, OH | 528.8 | 3,930 | 51.2 | 6.1 | 8.1 | 6.3 | 4.2 | 235.9 | 1,753 | 538 | 332 | 6,322 | 62,530 | 50,982 | 4,411 |
| State College, PA | 539.4 | 3,323 | 57.5 | 1.9 | 3.3 | 3.2 | 5.3 | 421.2 | 2,595 | 492 | 457 | 45,661 | 60,550 | 72,814 | 8,365 |
| Staunton, VA | 406.8 | 3,341 | 55.4 | 2.2 | 4.6 | 6.6 | 4.4 | 232.1 | 1,906 | 291 | 379 | 8,480 | 58,400 | 56,162 | 4,971 |
| Stockton, CA | 4,399.2 | 5,918 | 46.3 | 10.7 | 6.0 | 7.0 | 4.7 | 5,064.4 | 6,813 | 3,212 | 1,168 | 41,374 | 324,270 | 62,118 | 6,116 |
| Sumter, SC | 410.2 | 2,921 | 58.3 | 4.6 | 7.2 | 0.2 | 2.0 | 198.7 | 1,415 | 1,387 | 6,071 | 7,326 | 60,390 | 44,012 | 3,510 |
| Syracuse, NY | 4,287.4 | 6,584 | 50.0 | 3.0 | 3.1 | 9.0 | 5.2 | 2,729.0 | 4,191 | 5,046 | 1,143 | 50,084 | 308,960 | 65,461 | 7,307 |
| Tallahassee, FL | 1,356.5 | 3,540 | 41.1 | 2.8 | 8.8 | 0.4 | 7.2 | 4,923.4 | 12,849 | 2,145 | 691 | 57,403 | 169,750 | 63,148 | 7,303 |
| Tampa-St. Petersburg-Clearwater, FL | 11,968.6 | 3,852 | 37.8 | 3.6 | 8.6 | 0.9 | 3.1 | 11,598.0 | 3,733 | 24,655 | 12,380 | 124,284 | 1,502,530 | 68,370 | 8,631 |
| Terre Haute, IN | 646.1 | 3,458 | 42.3 | 5.7 | 4.8 | 0.0 | 4.3 | 541.1 | 2,896 | 1,288 | 534 | 12,516 | 80,760 | 50,483 | 4,445 |

1. Based on the resident population estimated as of July 1 of the year shown.

# Table C. Metropolitan Areas — Land Area and Population

| CBSA/ DIV Code[1] | Area name | Land area[2] (sq mi) | Population, 2018 — Total persons | Rank | Per square mile | White | Black | American Indian, Alaska Native | Asian and Pacific Islander | Percent Hispanic or Latino[3] | Under 5 years | 5 to 14 years | 15 to 24 years | 25 to 34 years | 35 to 44 years | 45 to 54 years |
|---|---|---|---|---|---|---|---|---|---|---|---|---|---|---|---|---|
| | | 1 | 2 | 3 | 4 | 5 | 6 | 7 | 8 | 9 | 10 | 11 | 12 | 13 | 14 | 15 |
| 45500 | Texarkana, TX-AR | 2,040 | 148,838 | 283 | 72.9 | 66.9 | 26.0 | 1.5 | 1.4 | 6.4 | 6.3 | 13.2 | 12.1 | 13.4 | 12.6 | 12.1 |
| 45540 | The Villages, FL | 557 | 139,018 | 299 | 249.5 | 86.0 | 7.0 | 0.7 | 1.3 | 6.0 | 1.8 | 4.0 | 4.0 | 5.7 | 5.8 | 6.3 |
| 45780 | Toledo, OH | 1,617 | 641,549 | 94 | 396.8 | 77.0 | 15.5 | 0.8 | 2.1 | 7.3 | 5.9 | 12.4 | 13.9 | 13.5 | 11.5 | 11.8 |
| 45820 | Topeka, KS | 3,233 | 230,878 | 200 | 71.4 | 80.2 | 7.9 | 2.5 | 1.9 | 10.9 | 5.9 | 13.1 | 12.3 | 11.8 | 12.0 | 11.4 |
| 45940 | Trenton-Princeton, NJ | 224 | 367,239 | 148 | 1,636.5 | 48.8 | 20.5 | 0.5 | 13.3 | 18.9 | 5.6 | 11.9 | 14.8 | 12.5 | 12.6 | 13.4 |
| 46060 | Tucson, AZ | 9,189 | 1,061,175 | 53 | 115.5 | 52.7 | 4.3 | 3.0 | 4.2 | 38.0 | 5.3 | 11.5 | 15.0 | 13.1 | 11.5 | 10.6 |
| 46140 | Tulsa, OK | 6,270 | 1,006,411 | 54 | 160.5 | 69.5 | 9.8 | 13.2 | 3.6 | 10.7 | 6.5 | 13.7 | 12.6 | 13.8 | 12.8 | 11.8 |
| 46220 | Tuscaloosa, AL | 3,493 | 253,211 | 190 | 72.5 | 58.3 | 36.7 | 0.6 | 1.8 | 3.9 | 5.8 | 11.7 | 18.1 | 14.3 | 11.8 | 11.3 |
| 46300 | Twin Falls, ID | 2,519 | 112,989 | 341 | 44.9 | 75.3 | 1.0 | 1.3 | 2.2 | 21.7 | 7.0 | 16.0 | 13.0 | 13.8 | 13.0 | 10.4 |
| 46340 | Tyler, TX | 922 | 235,806 | 197 | 255.9 | 60.3 | 18.2 | 0.7 | 2.1 | 20.2 | 6.5 | 13.7 | 13.7 | 13.6 | 12.0 | 11.2 |
| 46520 | Urban Honolulu, HI | 601 | 963,826 | 56 | 1,604.8 | 31.7 | 3.7 | 1.5 | 79.0 | 10.2 | 6.1 | 11.6 | 12.1 | 14.9 | 12.9 | 11.6 |
| 46540 | Utica-Rome, NY | 2,624 | 288,291 | 172 | 109.9 | 85.5 | 6.3 | 0.6 | 4.0 | 5.6 | 5.4 | 11.9 | 12.7 | 12.3 | 11.3 | 12.1 |
| 46660 | Valdosta, GA | 1,607 | 148,364 | 285 | 92.3 | 55.8 | 36.0 | 0.8 | 2.6 | 6.9 | 6.6 | 13.6 | 18.2 | 14.5 | 11.7 | 10.5 |
| 46700 | Vallejo, CA | 822 | 446,935 | 122 | 543.8 | 40.8 | 15.9 | 1.4 | 20.5 | 27.8 | 5.9 | 12.3 | 12.0 | 14.6 | 13.2 | 12.0 |
| 47020 | Victoria, TX | 1,734 | 99,562 | 359 | 57.4 | 45.4 | 6.3 | 0.6 | 1.5 | 47.4 | 6.6 | 14.0 | 13.3 | 13.5 | 12.3 | 10.6 |
| 47220 | Vineland-Bridgeton, NJ | 483 | 147,008 | 286 | 304.1 | 46.4 | 20.2 | 1.4 | 1.8 | 32.5 | 6.3 | 13.9 | 12.1 | 13.9 | 12.9 | 12.5 |
| 47260 | Virginia Beach-Norfolk-Newport News, VA-NC | 3,530 | 1,779,824 | 37 | 504.1 | 57.4 | 32.2 | 1.1 | 5.7 | 7.4 | 6.1 | 12.1 | 13.8 | 15.4 | 12.8 | 11.2 |
| 47300 | Visalia, CA | 4,824 | 468,680 | 118 | 97.1 | 28.4 | 1.7 | 1.2 | 4.3 | 66.0 | 7.6 | 17.5 | 15.1 | 14.3 | 12.9 | 10.8 |
| 47380 | Waco, TX | 1,802 | 277,005 | 177 | 153.7 | 56.4 | 15.5 | 0.7 | 2.2 | 27.0 | 6.6 | 13.5 | 17.8 | 13.3 | 11.7 | 10.5 |
| 47460 | Walla Walla, WA | 1,270 | 61,292 | 382 | 48.3 | 72.9 | 2.5 | 1.6 | 3.4 | 22.0 | 5.3 | 11.9 | 16.4 | 12.8 | 11.6 | 10.7 |
| 47580 | Warner Robins, GA | 526 | 188,060 | 230 | 357.3 | 55.2 | 35.9 | 0.9 | 4.0 | 7.0 | 6.4 | 14.3 | 13.6 | 14.1 | 13.3 | 11.7 |
| 47900 | Washington-Arlington-Alexandria, DC-VA-MD-WV | 6,568 | 6,324,629 | 6 | 963.0 | 47.1 | 26.6 | 0.8 | 12.5 | 16.4 | 6.3 | 12.7 | 12.3 | 14.7 | 14.5 | 13.3 |
| 47900 | Frederick-Gaithersburg-Rockville, MD Div 23224 | 1,154 | 1,316,977 | X | 1,141.6 | 50.6 | 18.5 | 0.7 | 15.1 | 18.2 | 6.0 | 13.0 | 11.8 | 12.5 | 13.8 | 13.5 |
| 47900 | Washington-Arlington-Alexandria, DC-VA-MD-WV Div 47894 | 5,414 | 5,007,652 | X | 924.9 | 46.1 | 28.7 | 0.8 | 11.8 | 15.9 | 6.3 | 12.6 | 12.4 | 15.3 | 14.7 | 13.3 |
| 48060 | Waterloo-Cedar Falls, IA | 1,503 | 168,314 | 257 | 112.0 | 85.5 | 9.0 | 0.6 | 3.2 | 4.0 | 6.0 | 12.3 | 16.8 | 12.5 | 11.8 | 10.4 |
| 48140 | Watertown-Fort Drum, NY | 1,269 | 108,095 | 346 | 85.2 | 83.9 | 7.2 | 1.1 | 2.9 | 7.8 | 7.5 | 13.0 | 15.3 | 16.6 | 11.9 | 9.9 |
| 48260 | Wausau-Weston, WI | 2,424 | 163,159 | 262 | 67.3 | 90.5 | 1.5 | 0.9 | 5.9 | 2.8 | 5.6 | 12.3 | 11.1 | 11.7 | 12.3 | 12.6 |
| 48300 | Weirton-Steubenville, WV-OH | 580 | 115,184 | 335 | 198.6 | 93.7 | 5.3 | 0.7 | 0.8 | 1.6 | 4.8 | 10.6 | 11.9 | 11.2 | 10.5 | 12.5 |
| 48540 | Wenatchee, WA | 4,740 | 121,134 | 329 | 25.6 | 67.4 | 0.9 | 1.6 | 1.9 | 30.0 | 6.0 | 13.9 | 11.9 | 12.5 | 12.2 | 10.9 |
| 48620 | Wheeling, WV-OH | 943 | 137,217 | 301 | 145.4 | 94.8 | 4.2 | 0.6 | 0.9 | 1.2 | 4.8 | 10.8 | 11.4 | 12.1 | 11.3 | 12.2 |
| 48660 | Wichita, KS | 4,148 | 643,768 | 93 | 155.2 | 73.9 | 9.1 | 2.0 | 4.7 | 13.9 | 6.5 | 14.4 | 13.4 | 13.7 | 12.6 | 11.0 |
| 48700 | Wichita Falls, TX | 2,620 | 152,485 | 276 | 58.2 | 69.2 | 10.3 | 1.4 | 2.8 | 18.6 | 6.0 | 12.7 | 15.9 | 14.3 | 12.0 | 10.4 |
| 48900 | Williamsport, PA | 1,229 | 113,209 | 340 | 92.1 | 92.1 | 6.3 | 0.6 | 1.1 | 2.2 | 5.3 | 11.6 | 12.3 | 12.8 | 11.5 | 11.8 |
| 49020 | Wilmington, NC | 1,064 | 301,284 | 167 | 283.3 | 78.8 | 14.0 | 1.0 | 2.0 | 6.4 | 4.8 | 10.7 | 14.4 | 12.6 | 12.6 | 12.5 |
| 49180 | Winchester, VA-WV | 1,063 | 142,009 | 295 | 133.6 | 82.7 | 6.4 | 0.6 | 2.3 | 10.2 | 5.6 | 12.3 | 11.9 | 12.3 | 12.2 | 12.8 |
| 49340 | Winston-Salem, NC | 2,009 | 679,731 | 87 | 338.4 | 68.7 | 19.0 | 0.8 | 2.4 | 10.9 | 5.5 | 12.4 | 12.6 | 12.6 | 11.6 | 13.4 |
| 49420 | Worcester, MA-CT | 2,024 | 945,752 | 58 | 467.4 | 77.7 | 5.6 | 0.6 | 5.6 | 12.5 | 5.2 | 11.5 | 13.2 | 13.1 | 12.3 | 13.4 |
| 49620 | Yakima, WA | 4,295 | 251,879 | 191 | 58.7 | 43.3 | 1.3 | 4.4 | 2.0 | 50.9 | 7.7 | 16.8 | 14.3 | 13.6 | 11.9 | 10.7 |
| 49660 | York-Hanover, PA | 904 | 450,448 | 121 | 498.1 | 83.9 | 7.1 | 0.5 | 2.0 | 8.5 | 5.5 | 12.4 | 11.9 | 12.5 | 12.2 | 13.0 |
| 49700 | Youngstown-Warren-Boardman, OH-PA | 1,702 | 531,420 | 107 | 312.2 | 84.3 | 12.1 | 0.7 | 1.1 | 4.0 | 5.3 | 11.0 | 11.8 | 11.8 | 10.9 | 12.1 |
| 49740 | Yuba City, CA | 1,235 | 176,545 | 243 | 143.0 | 51.8 | 4.2 | 2.6 | 15.0 | 31.0 | 7.1 | 14.9 | 13.1 | 14.8 | 12.8 | 10.8 |
| 49740 | Yuma, AZ | 5,514 | 217,824 | 210 | 39.5 | 31.0 | 2.3 | 1.4 | 1.9 | 64.7 | 7.0 | 13.7 | 14.9 | 14.0 | 11.0 | 9.7 |

1. CBSA = Core Based Statistical Area. DIV = Metropolitan Division. See Appendix A for explanation. See Appendix B for list of metropolitan areas or temporarily covered by water.   2. Dry land or land partially or temporarily covered by water.   3. May be of any race.

Table C. Metropolitan Areas — **Population and Households**

| Area name | Age (percent) (cont.) 55 to 64 years | 65 to 74 years | 75 years and over | Percent female | Total persons 2000 | 2010 | Percent change 2000-2010 | 2010-2019 | Components of change, 2010-2020 Births | Deaths | Net migration | Households, 2019 Number | Persons per household | Family households | Female family householder[1] | One person |
|---|---|---|---|---|---|---|---|---|---|---|---|---|---|---|---|---|
| | 16 | 17 | 18 | 19 | 20 | 21 | 22 | 23 | 24 | 25 | 26 | 27 | 28 | 29 | 30 | 31 |
| Texarkana, TX-AR............... | 12.6 | 10.2 | 7.5 | 50.3 | 143,377 | 149,194 | 4.1 | -0.2 | 19,696 | 16,992 | -2,990 | 55,028 | 2.57 | 68.9 | 12.0 | 27.0 |
| The Villages, FL................... | 13.3 | 32.1 | 27.0 | 50.4 | 53,345 | 93,420 | 75.1 | 48.8 | 4,768 | 17,198 | 57,449 | 59,458 | 2.06 | 65.1 | 1.8 | 31.1 |
| Toledo, OH......................... | 13.5 | 10.6 | 7.1 | 51.3 | 659,188 | 651,434 | -1.2 | -1.5 | 79,105 | 66,802 | -22,293 | 268,089 | 2.33 | 59.9 | 11.6 | 31.9 |
| Topeka, KS......................... | 14.0 | 11.4 | 8.1 | 51.1 | 224,551 | 233,860 | 4.1 | -1.3 | 29,495 | 24,756 | -7,653 | 95,552 | 2.37 | 64.0 | 10.1 | 30.4 |
| Trenton-Princeton, NJ ........... | 13.2 | 9.1 | 6.9 | 51.0 | 350,761 | 367,485 | 4.8 | -0.1 | 42,239 | 30,039 | -12,588 | 130,851 | 2.66 | 66.3 | 8.8 | 27.8 |
| Tucson, AZ......................... | 12.2 | 11.8 | 9.1 | 50.8 | 843,746 | 980,266 | 16.2 | 8.3 | 117,433 | 96,509 | 60,898 | 410,404 | 2.48 | 60.3 | 10.0 | 32.1 |
| Tulsa, OK.......................... | 12.5 | 9.6 | 6.7 | 50.9 | 859,532 | 937,521 | 9.1 | 7.3 | 134,138 | 96,686 | 31,843 | 387,176 | 2.54 | 65.0 | 11.0 | 29.0 |
| Tuscaloosa, AL .................... | 11.7 | 9.3 | 5.9 | 51.9 | 212,983 | 239,225 | 12.3 | 5.8 | 30,423 | 23,456 | 6,936 | 90,861 | 2.64 | 64.4 | 10.6 | 29.1 |
| Twin Falls, ID...................... | 11.2 | 9.1 | 6.6 | 50.2 | 82,626 | 99,595 | 20.5 | 13.4 | 16,277 | 9,341 | 6,487 | 39,654 | 2.75 | 66.0 | 8.1 | 26.6 |
| Tyler, TX........................... | 12.0 | 9.9 | 7.5 | 51.7 | 174,706 | 209,729 | 20.0 | 12.4 | 31,078 | 21,241 | 16,290 | 78,600 | 2.88 | 68.5 | 10.2 | 25.4 |
| Urban Honolulu, HI ............... | 11.9 | 10.0 | 8.8 | 49.7 | 876,156 | 953,203 | 8.8 | 1.1 | 133,124 | 79,117 | -42,716 | 316,456 | 2.97 | 68.3 | 8.3 | 25.2 |
| Utica-Rome, NY ................... | 14.3 | 11.3 | 8.7 | 50.2 | 299,896 | 299,320 | -0.2 | -3.7 | 32,683 | 33,117 | -10,577 | 115,904 | 2.39 | 60.7 | 10.4 | 32.2 |
| Valdosta, GA ...................... | 11.0 | 8.4 | 5.5 | 51.6 | 119,560 | 139,670 | 16.8 | 6.2 | 20,669 | 12,305 | 127 | 54,425 | 2.59 | 62.9 | 16.0 | 30.0 |
| Vallejo, CA ........................ | 13.3 | 10.3 | 6.5 | 50.2 | 394,542 | 413,343 | 4.8 | 8.1 | 52,637 | 33,374 | 14,593 | 150,393 | 2.90 | 70.1 | 11.4 | 23.0 |
| Victoria, TX........................ | 12.2 | 10.0 | 7.4 | 50.9 | 91,016 | 94,005 | 3.3 | 5.9 | 13,691 | 9,054 | 902 | 32,627 | 2.99 | 68.4 | 11.2 | 27.1 |
| Vineland-Bridgeton, NJ........... | 12.4 | 9.3 | 6.7 | 49.3 | 146,438 | 156,610 | 6.9 | -6.1 | 20,374 | 15,718 | -14,342 | 51,360 | 2.68 | 66.8 | 16.0 | 28.3 |
| Virginia Beach-Norfolk-Newport News, VA-NC .. | 12.9 | 9.2 | 6.4 | 50.8 | 1,612,770 | 1,713,955 | 6.3 | 3.8 | 231,756 | 147,831 | -17,949 | 675,311 | 2.51 | 65.4 | 12.1 | 27.7 |
| Visalia, CA......................... | 9.9 | 7.0 | 4.8 | 50.1 | 368,021 | 442,176 | 20.1 | 6.0 | 75,989 | 30,822 | -18,658 | 144,109 | 3.20 | 79.2 | 17.2 | 16.6 |
| Waco, TX........................... | 11.2 | 8.8 | 6.5 | 51.3 | 232,093 | 252,761 | 8.9 | 9.6 | 37,638 | 23,648 | 10,356 | 98,944 | 2.63 | 66.1 | 12.5 | 27.6 |
| Walla Walla, WA................... | 12.2 | 10.9 | 8.3 | 49.2 | 55,180 | 58,781 | 6.5 | 4.3 | 6,711 | 5,870 | 1,692 | 23,167 | 2.40 | 64.4 | 7.6 | 25.9 |
| Warner Robins, GA ............... | 12.7 | 8.5 | 5.4 | 51.8 | 134,433 | 167,627 | 24.7 | 12.2 | 23,900 | 14,460 | 11,021 | 68,831 | 2.64 | 66.7 | 13.2 | 28.5 |
| Washington-Arlington-Alexandria, DC-VA-MD-WV ...................... | 12.3 | 8.2 | 5.6 | 51.1 | 4,849,948 | 5,649,698 | 16.5 | 11.9 | 820,653 | 356,952 | 209,149 | 2,251,002 | 2.74 | 65.0 | 9.0 | 27.8 |
| Frederick-Gaithersburg-Rockville, MD Div 23224 | 13.2 | 9.3 | 7.0 | 51.4 | 1,068,618 | 1,204,694 | 12.7 | 9.3 | 160,810 | 79,399 | 31,133 | 462,665 | 2.80 | 69.5 | 8.7 | 24.8 |
| Washington-Arlington-Alexandria, DC-VA-MD-WV Div 47894 ........ | 12.1 | 8.0 | 5.3 | 51.0 | 3,781,330 | 4,445,004 | 17.6 | 12.7 | 659,843 | 277,553 | 178,016 | 1,788,337 | 2.73 | 63.9 | 9.1 | 28.6 |
| Waterloo-Cedar Falls, IA ........ | 12.0 | 10.4 | 7.8 | 50.8 | 163,706 | 167,819 | 2.5 | 0.3 | 21,342 | 16,121 | -4,621 | 67,137 | 2.42 | 60.2 | 9.3 | 30.1 |
| Watertown-Fort Drum, NY ...... | 11.2 | 8.6 | 6.1 | 47.5 | 111,738 | 116,234 | 4.0 | -7.0 | 20,811 | 9,304 | -19,763 | 41,214 | 2.51 | 67.1 | 12.7 | 26.9 |
| Wausau-Weston, WI .............. | 15.1 | 11.1 | 8.2 | 49.8 | 155,475 | 162,816 | 4.7 | 0.2 | 18,987 | 15,524 | -3,044 | 67,342 | 2.39 | 65.7 | 7.3 | 27.8 |
| Weirton-Steubenville, WV-OH . | 15.2 | 13.7 | 9.6 | 51.1 | 132,008 | 124,455 | -5.7 | -7.4 | 11,466 | 17,438 | -3,187 | 50,065 | 2.24 | 62.7 | 10.2 | 31.4 |
| Wenatchee, WA.................... | 13.1 | 11.6 | 7.9 | 49.6 | 99,219 | 110,887 | 11.8 | 9.2 | 14,710 | 10,244 | 5,851 | 44,478 | 2.67 | 69.7 | 11.8 | 24.6 |
| Wheeling, WV-OH ................ | 14.9 | 13.3 | 9.2 | 50.2 | 153,172 | 147,963 | -3.4 | -7.3 | 14,630 | 19,354 | -5,914 | 55,796 | 2.37 | 59.0 | 7.7 | 36.3 |
| Wichita, KS ........................ | 12.6 | 9.4 | 6.5 | 50.5 | 571,166 | 623,061 | 9.1 | 3.3 | 89,475 | 58,245 | -10,344 | 246,836 | 2.55 | 64.4 | 10.3 | 30.2 |
| Wichita Falls, TX.................. | 12.6 | 9.2 | 7.0 | 48.5 | 151,524 | 151,474 | 0.0 | 0.7 | 19,184 | 16,095 | -2,048 | 56,673 | 2.44 | 62.1 | 7.5 | 32.5 |
| Williamsport, PA .................. | 14.3 | 11.7 | 8.7 | 51.1 | 120,044 | 116,095 | -3.3 | -2.5 | 12,652 | 13,133 | -2,373 | 44,842 | 2.39 | 64.1 | 12.0 | 30.6 |
| Wilmington, NC ................... | 13.4 | 11.4 | 7.5 | 52.0 | 201,389 | 254,878 | 26.6 | 18.2 | 29,169 | 24,939 | 41,770 | 121,734 | 2.37 | 59.2 | 9.2 | 31.3 |
| Winchester, VA-WV .............. | 13.9 | 11.1 | 7.9 | 50.2 | 102,997 | 128,441 | 24.7 | 10.6 | 15,900 | 12,456 | 10,156 | 50,062 | 2.72 | 72.2 | 8.9 | 22.8 |
| Winston-Salem, NC ............... | 13.9 | 10.7 | 7.6 | 52.0 | 569,207 | 640,489 | 12.5 | 6.1 | 75,790 | 66,442 | 30,224 | 270,273 | 2.45 | 64.8 | 11.6 | 29.8 |
| Worcester, MA-CT................. | 14.5 | 9.9 | 6.8 | 50.7 | 860,054 | 916,764 | 6.6 | 3.2 | 99,345 | 83,112 | 13,090 | 363,177 | 2.51 | 66.2 | 9.8 | 26.1 |
| Yakima, WA........................ | 10.7 | 8.3 | 6.0 | 50.1 | 222,581 | 243,240 | 9.3 | 3.6 | 40,578 | 20,133 | -11,825 | 83,992 | 2.94 | 72.4 | 16.1 | 22.3 |
| York-Hanover, PA................. | 14.1 | 10.8 | 7.7 | 50.6 | 381,751 | 435,016 | 14.0 | 3.5 | 49,895 | 41,514 | 7,458 | 175,441 | 2.51 | 67.4 | 9.3 | 25.7 |
| Youngstown-Warren-Boardman, OH-PA ........ | 14.7 | 12.8 | 9.6 | 51.0 | 602,964 | 565,782 | -6.2 | -6.1 | 57,343 | 72,228 | -19,278 | 230,643 | 2.25 | 60.9 | 12.0 | 34.5 |
| Yuba City, CA...................... | 11.6 | 8.8 | 6.1 | 49.9 | 139,149 | 166,893 | 19.9 | 5.8 | 25,448 | 14,413 | -1,390 | 59,866 | 2.89 | 70.6 | 13.8 | 22.2 |
| Yuma, AZ........................... | 9.7 | 9.7 | 10.4 | 48.4 | 160,026 | 195,750 | 22.3 | 11.3 | 31,510 | 15,302 | 5,621 | 74,042 | 2.81 | 72.4 | 15.9 | 22.5 |

# Table C. Metropolitan Areas — Population, Vital Statistics, Health, and Crime

| Area name | Persons in group quarters, 2020 | Daytime population, 2019 — Number | Daytime population, 2019 — Employment/residence ratio | Births, 2020 — Total | Births, 2020 — Rate[1] | Deaths, 2020 — Number | Deaths, 2020 — Rate[1] | Persons under 65 with no health insurance 2019 — Number | Persons under 65 with no health insurance 2019 — Percent | Medicare, 2020 — Total Beneficiaries | Medicare, 2020 — Enrolled in Original Medicare | Medicare, 2020 — Enrolled in Medicare Advantage | Serious crimes known to police[2], 2019 Violent — Number | Serious crimes known to police[2], 2019 Violent — Rate[3] |
|---|---|---|---|---|---|---|---|---|---|---|---|---|---|---|
| | 32 | 33 | 34 | 35 | 36 | 37 | 38 | 39 | 40 | 41 | 42 | 43 | 44 | 45 |
| Texarkana, TX-AR | 7,287 | 150,166 | 1.01 | 1,777 | 11.9 | 1,745 | 11.7 | 16,567 | 14.3 | 31,697 | 23,016 | 8,682 | 676 | 451.9 |
| The Villages, FL | 8,264 | 142,124 | 1.41 | 469 | 3.4 | 2,269 | 16.3 | 5,957 | 12.5 | 74,807 | 40,166 | 34,641 | 249 | 187.3 |
| Toledo, OH | 16,784 | 661,204 | 1.06 | 7,287 | 11.4 | 6,994 | 10.9 | 39,230 | 7.6 | 132,163 | 70,820 | 61,344 | 3,026 | 471.7 |
| Topeka, KS | 5,041 | 231,537 | 1 | 2,673 | 11.6 | 2,532 | 11.0 | 16,747 | 9.1 | 51,988 | 42,232 | 9,757 | 1,143 | 492.8 |
| Trenton-Princeton, NJ | 19,977 | 420,930 | 1.31 | 3,920 | 10.7 | 3,219 | 8.8 | 28,211 | 9.6 | 65,853 | 40,457 | 25,396 | 1,277 | 346.6 |
| Tucson, AZ | 24,584 | 1,047,933 | 1 | 10,667 | 10.1 | 11,132 | 10.5 | 107,748 | 13.3 | 236,346 | 118,826 | 117,520 | 4,685 | 447.4 |
| Tulsa, OK | 15,522 | 1,001,089 | 1.01 | 12,586 | 12.5 | 10,743 | 10.7 | 134,573 | 16.2 | 188,527 | 120,820 | 67,707 | 5,286 | 529.6 |
| Tuscaloosa, AL | 12,806 | 256,382 | 1.04 | 2,830 | 11.2 | 2,640 | 10.4 | 21,043 | 10.4 | 48,253 | 27,668 | 20,586 | NA | NA |
| Twin Falls, ID | 1,114 | 109,130 | 0.99 | 1,511 | 13.4 | 1,020 | 9.0 | 14,528 | 15.7 | 20,538 | 12,723 | 7,815 | 321 | 286.8 |
| Tyler, TX | 5,181 | 241,540 | 1.08 | 2,921 | 12.4 | 2,312 | 9.8 | 39,717 | 21.1 | 46,779 | 29,978 | 16,801 | 773 | 334.1 |
| Urban Honolulu, HI | 34,383 | 975,713 | 1 | 11,793 | 12.2 | 9,226 | 9.6 | 33,662 | 4.4 | 187,165 | 94,183 | 92,982 | 2,638 | 270.6 |
| Utica-Rome, NY | 14,208 | 287,067 | 0.98 | 3,008 | 10.4 | 3,352 | 11.6 | 11,536 | 5.2 | 67,781 | 37,745 | 30,036 | 784 | 271.4 |
| Valdosta, GA | 6,378 | 146,053 | 1 | 1,957 | 13.2 | 1,408 | 9.5 | 21,010 | 17.5 | 24,931 | 16,462 | 8,469 | NA | NA |
| Vallejo, CA | 11,825 | 390,913 | 0.74 | 5,023 | 11.2 | 3,953 | 8.8 | 21,935 | 6.0 | 81,509 | 43,610 | 37,899 | 2,114 | 472.9 |
| Victoria, TX | 1,995 | 100,726 | 1.04 | 1,257 | 12.6 | 991 | 10.0 | 16,190 | 20.0 | 19,574 | 12,216 | 7,358 | 463 | 464.4 |
| Vineland-Bridgeton, NJ | 8,760 | 148,420 | 0.98 | 1,850 | 12.6 | 1,630 | 11.1 | 14,106 | 12.1 | 28,789 | 18,381 | 10,408 | 631 | 421.8 |
| Virginia Beach-Norfolk-Newport News, VA-NC | 73,545 | 1,764,787 | 1 | 21,657 | 12.2 | 16,712 | 9.4 | 122,102 | 8.6 | 318,072 | 237,465 | 80,607 | 6,157 | 349.5 |
| Visalia, CA | 4,977 | 455,325 | 0.94 | 6,787 | 14.5 | 3,404 | 7.3 | 38,062 | 9.5 | 64,059 | 48,976 | 15,083 | 1,692 | 364.0 |
| Waco, TX | 10,486 | 278,285 | 1.06 | 3,657 | 13.2 | 2,531 | 9.1 | 40,691 | 18.2 | 48,934 | 28,664 | 20,269 | 1,166 | 427.8 |
| Walla Walla, WA | 4,653 | 62,759 | 1.08 | 614 | 10.0 | 621 | 10.1 | 4,303 | 9.6 | 13,674 | 11,430 | 2,244 | 145 | 237.9 |
| Warner Robins, GA | 2,986 | 180,724 | 0.95 | 2,328 | 12.4 | 1,717 | 9.1 | 21,654 | 13.8 | 31,700 | 21,793 | 9,906 | 765 | 414.7 |
| Washington-Arlington-Alexandria, DC-VA-MD-WV | 106,391 | 6,416,286 | 1.04 | 76,870 | 12.2 | 42,699 | 6.8 | 431,946 | 8.1 | 887,680 | 729,521 | 158,160 | NA | NA |
| Frederick-Gaithersburg-Rockville, MD Div 23224 | 13,274 | 1,245,583 | 0.91 | 14,958 | 11.4 | 8,919 | 6.8 | 82,692 | 7.6 | 214,257 | 185,843 | 28,413 | 2,167 | 164.9 |
| Washington-Arlington-Alexandria, DC-VA-MD-WV Div 47894 | 93,117 | 5,170,703 | 1.07 | 61,912 | 12.4 | 33,780 | 6.7 | 349,254 | 8.2 | 673,423 | 543,678 | 129,747 | NA | NA |
| Waterloo-Cedar Falls, IA | 5,982 | 174,660 | 1.07 | 1,985 | 11.8 | 1,661 | 9.9 | 7,043 | 5.3 | 35,181 | 22,517 | 12,664 | NA | NA |
| Watertown-Fort Drum, NY | 6,015 | 111,679 | 1.04 | 1,750 | 16.2 | 988 | 9.1 | 4,191 | 4.8 | 20,410 | 12,884 | 7,525 | 243 | 219.8 |
| Wausau-Weston, WI | 2,017 | 164,839 | 1.02 | 1,784 | 10.9 | 1,551 | 9.5 | 8,719 | 6.6 | 35,503 | 17,101 | 18,402 | 360 | 220.9 |
| Weirton-Steubenville, WV-OH | 3,252 | 108,289 | 0.84 | 1,082 | 9.4 | 1,782 | 15.5 | 6,801 | 7.8 | 30,692 | 18,840 | 11,851 | 179 | 154.5 |
| Wenatchee, WA | 1,121 | 118,436 | 0.96 | 1,371 | 11.3 | 1,123 | 9.3 | 11,256 | 11.8 | 26,208 | 19,525 | 6,683 | 128 | 106.1 |
| Wheeling, WV-OH | 6,492 | 144,242 | 1.09 | 1,316 | 9.6 | 1,947 | 14.2 | 7,329 | 7.1 | 34,509 | 17,949 | 16,559 | NA | NA |
| Wichita, KS | 11,262 | 643,193 | 1.01 | 7,993 | 12.4 | 6,199 | 9.6 | 62,582 | 11.8 | 116,533 | 84,213 | 32,321 | 4,985 | 781.2 |
| Wichita Falls, TX | 12,819 | 150,452 | 0.99 | 1,720 | 11.3 | 1,652 | 10.8 | 21,470 | 18.7 | 29,533 | 22,783 | 6,752 | 479 | 318.7 |
| Williamsport, PA | 5,276 | 114,293 | 1.02 | 1,169 | 10.3 | 1,418 | 12.5 | 5,753 | 6.7 | 27,227 | 16,621 | 10,606 | NA | NA |
| Wilmington, NC | 9,152 | 315,366 | 1.12 | 2,756 | 9.1 | 2,926 | 9.7 | 29,208 | 12.4 | 60,819 | 43,689 | 17,130 | 1,138 | 380.0 |
| Winchester, VA-WV | 2,673 | 137,161 | 0.97 | 1,530 | 10.8 | 1,401 | 9.9 | 11,490 | 10.3 | 29,046 | 23,209 | 5,837 | 226 | 160.8 |
| Winston-Salem, NC | 13,448 | 649,764 | 0.91 | 7,284 | 10.7 | 7,274 | 10.7 | 76,632 | 14.1 | 142,463 | 59,053 | 83,408 | NA | NA |
| Worcester, MA-CT | 32,476 | 877,221 | 0.85 | 9,312 | 9.8 | 8,942 | 9.5 | 29,695 | 3.9 | 184,005 | 114,612 | 69,393 | 2,702 | 310.4 |
| Yakima, WA | 3,468 | 249,449 | 0.99 | 3,594 | 14.3 | 2,173 | 8.6 | 31,476 | 15.1 | 43,471 | 31,902 | 11,570 | 703 | 279.6 |
| York-Hanover, PA | 8,609 | 412,706 | 0.84 | 4,625 | 10.3 | 4,460 | 9.9 | 22,184 | 6.1 | 95,126 | 56,245 | 38,881 | NA | NA |
| Youngstown-Warren-Boardman, OH-PA | 18,855 | 523,043 | 0.94 | 5,481 | 10.3 | 7,280 | 13.7 | 32,483 | 8.1 | 138,421 | 62,535 | 75,885 | NA | NA |
| Yuba City, CA | 1,929 | 167,343 | 0.88 | 2,382 | 13.5 | 1,503 | 8.5 | 12,637 | 8.6 | 30,846 | 28,019 | 2,827 | 644 | 369.3 |
| Yuma, AZ | 9,391 | D | D | 3,008 | 13.8 | 1,880 | 8.6 | 28,673 | 17.7 | 37,910 | 26,323 | 11,587 | 595 | 277.8 |

1. Per 1,000 estimated resident population.  2. Data for serious crimes have not been adjusted for underreporting; this may affect comparability between geographic areas and over time.  3. Per 100,000 population estimated by the FBI.

# Table C. Metropolitan Areas — Crime, Education, Money Income, and Poverty

| Area name | Serious crimes known to police[1], 2019 (cont.) Property | | Education School enrollment and attainment, 2019 Enrollment[3] | | Attainment[4] | | Local government expenditures,[5] 2017-2018 | | Income and poverty, 2019 | | | | Percent below poverty level | | |
|---|---|---|---|---|---|---|---|---|---|---|---|---|---|---|---|---|
| | Number | Rate | Total | Percent private | High school graduate or less | Bachelor's degree or more | Total current expenditures (mil dol) | Current expenditures per student (dollars) | Per capita income[6] (dollars) | Median household income (dollars) | Median family income | Percent of households with income less than $50,000 | Percent of households with income of $200,000 or more | All persons | All families | Families with children under 18 |
| | 46 | 47 | 48 | 49 | 50 | 51 | 52 | 53 | 54 | 55 | 56 | 57 | 58 | 59 | 60 | 61 |
| Texarkana, TX-AR | 3,968 | 2,652.5 | 34,556 | 6.9 | 48.8 | 17.5 | 254.5 | 9,587 | 26,165 | 51,544 | 62,838 | 48.7 | 3.7 | 14.4 | 10.8 | 20.0 |
| The Villages, FL | 1,139 | 856.9 | 8,955 | 22.1 | 37.4 | 33.2 | 84.2 | 9,741 | 39,187 | 60,287 | 74,976 | 39.3 | 5.1 | 8.1 | 3.6 | 7.1 |
| Toledo, OH | 15,736 | 2,452.9 | 158,718 | 14.8 | 39.8 | 28.2 | 1,311.0 | 12,651 | 31,276 | 53,563 | 70,852 | 46.8 | 4.6 | 15.3 | 10.3 | 19.8 |
| Topeka, KS | D | D | 51,987 | 10.4 | 38.7 | 29.1 | 458.8 | 12,122 | 31,691 | 59,567 | 74,411 | 41.7 | 4.4 | 9.6 | 7.0 | 14.1 |
| Trenton-Princeton, NJ | 5,571 | 1,512.3 | 104,431 | 23.6 | 34.6 | 44.5 | 1,175.2 | 18,518 | 42,338 | 79,492 | 107,209 | 32.2 | 14.9 | 12.8 | 7.7 | 11.1 |
| Tucson, AZ | 30,170 | 2,881.4 | 259,244 | 11.6 | 33.8 | 32.6 | 1,195.9 | 8,536 | 31,004 | 56,169 | 72,920 | 45.4 | 5.0 | 13.8 | 9.8 | 15.6 |
| Tulsa, OK | 32,217 | 3,227.8 | 245,248 | 16.5 | 39.2 | 28.9 | 1,346.3 | 8,081 | 32,380 | 57,859 | 73,097 | 43.3 | 6.0 | 13.6 | 10.1 | 16.1 |
| Tuscaloosa, AL | NA | NA | 68,986 | 8.5 | 44.9 | 28.2 | 355.7 | 10,045 | 26,650 | 49,721 | 65,682 | 50.2 | 4.4 | 17.5 | 11.6 | 17.2 |
| Twin Falls, ID | 1,519 | 1,357.1 | 27,806 | 17.3 | 43.9 | 20 | 152.5 | 7,290 | 26,923 | 56,667 | 67,015 | 42.0 | 4.1 | 11.7 | 7.7 | 11.2 |
| Tyler, TX | 5,034 | 2,175.8 | 58,451 | 14.4 | 36.5 | 27.4 | 320.4 | 8,923 | 29,053 | 59,584 | 73,944 | 41.8 | 4.7 | 12.4 | 8.2 | 12.5 |
| Urban Honolulu, HI | 29,263 | 3,001.6 | 225,660 | 26.6 | 33.2 | 35.8 | 2,756.3 | 15,242 | 38,461 | 87,470 | 101,922 | 26.3 | 12.2 | 8.1 | 4.9 | 8.2 |
| Utica-Rome, NY | 4,544 | 1,573.1 | 66,634 | 18.1 | 42.1 | 27 | 789.3 | 17,817 | 30,953 | 57,778 | 74,145 | 43.2 | 3.9 | 14.2 | 10.4 | 17.4 |
| Valdosta, GA | NA | NA | 39,128 | 8.1 | 43.3 | 23.4 | 238.9 | 9,853 | 25,479 | 43,787 | 57,334 | 54.3 | 3.5 | 20.9 | 15.4 | 24.4 |
| Vallejo, CA | 13,161 | 2,943.8 | 109,346 | 13.0 | 34.4 | 28.9 | 744.9 | 11,321 | 36,968 | 86,652 | 97,754 | 26.8 | 10.1 | 8.9 | 6.3 | 9.6 |
| Victoria, TX | 2,452 | 2,459.3 | 24,432 | 14.3 | 48.4 | 18.2 | 170.4 | 10,249 | 27,903 | 56,627 | 67,174 | 44.7 | 4.8 | 14.1 | 12.3 | 19.9 |
| Vineland-Bridgeton, NJ | 4,047 | 2,705.4 | 38,144 | 10.7 | 59.1 | 16.1 | 523.9 | 17,840 | 29,711 | 54,587 | 65,022 | 41.6 | 4.3 | 13.2 | 9.8 | 13.4 |
| Virginia Beach-Norfolk-Newport News, VA-NC | 42,452 | 2,410.0 | 447,634 | 17.7 | 33.0 | 32.9 | 3,040.7 | 11,363 | 35,032 | 69,329 | 83,854 | 35.5 | 6.4 | 10.6 | 7.6 | 13.3 |
| Visalia, CA | 10,625 | 2,285.7 | 141,310 | 7.7 | 51.5 | 13.6 | 1,360.1 | 13,071 | 23,096 | 57,692 | 60,190 | 43.5 | 3.4 | 18.8 | 15.4 | 21.6 |
| Waco, TX | 6,976 | 2,559.2 | 83,372 | 21.5 | 41.0 | 24.7 | 566.3 | 10,021 | 25,448 | 50,368 | 66,426 | 49.7 | 5.0 | 18.8 | 12.8 | 21.4 |
| Walla Walla, WA | 1,479 | 2,426.1 | 15,651 | 25.5 | 30.8 | 32.2 | 114.9 | 13,109 | 31,896 | 61,285 | 75,191 | 42.0 | 6.0 | 11.3 | 7.0 | 10.6 |
| Warner Robins, GA | 5,732 | 3,107.4 | 55,491 | 8.6 | 39.7 | 28.2 | 340.2 | 10,266 | 28,953 | 61,936 | 72,732 | 40.6 | 4.2 | 12.1 | 9.6 | 11.4 |
| Washington-Arlington-Alexandria, DC-VA-MD-WV | NA | NA | 1,614,781 | 20.4 | 27.0 | 51.4 | 14,733.2 | 15,072 | 51,437 | 105,659 | 126,378 | 21.5 | 20.3 | 7.5 | 4.8 | 7.4 |
| Frederick-Gaithersburg-Rockville, MD Div 23224 | 17,671 | 1,344.4 | 333,491 | 21.4 | 25.7 | 54.6 | 3,132.3 | 15,378 | 52,911 | 108,282 | 129,446 | 21.0 | 21.8 | 7.1 | 4.6 | 6.9 |
| Washington-Arlington-Alexandria, DC-VA-MD-WV Div 47894 | NA | NA | 1,281,290 | 20.1 | 27.3 | 50.5 | 11,600.8 | 14,991 | 51,048 | 104,821 | 125,633 | 21.6 | 20.0 | 7.7 | 4.9 | 7.5 |
| Waterloo-Cedar Falls, IA | NA | NA | 44,850 | 14.3 | 39.1 | 30 | 338.6 | 12,435 | 33,454 | 63,131 | 81,054 | 38.9 | 5.1 | 11.6 | 7.0 | 12.7 |
| Watertown-Fort Drum, NY | 1,615 | 1,460.7 | 25,067 | 10.0 | 45.4 | 22.4 | 310.3 | 17,235 | 26,701 | 53,917 | 65,378 | 45.3 | 2.8 | 15.5 | 10.0 | 17.6 |
| Wausau-Weston, WI | 1,499 | 919.9 | 35,000 | 15.3 | 43.2 | 23.5 | 287.9 | 12,038 | 32,924 | 65,094 | 78,197 | 39.0 | 4.3 | 8.0 | 4.8 | 8.9 |
| Weirton-Steubenville, WV-OH | 1,566 | 1,351.9 | 22,661 | 17.7 | 53.0 | 17.5 | 181.4 | 11,494 | 27,264 | 49,510 | 64,359 | 50.5 | 1.9 | 14.7 | 8.5 | 14.1 |
| Wenatchee, WA | 1,380 | 1,143.8 | 29,296 | 8.3 | 45.5 | 24.7 | 281.5 | 13,079 | 32,765 | 60,532 | 70,913 | 39.7 | 5.0 | 13.0 | 8.1 | 15.2 |
| Wheeling, WV-OH | NA | NA | 27,102 | 13.9 | 50.8 | 21.3 | 233.8 | 12,446 | 30,331 | 50,440 | 67,799 | 49.4 | 4.4 | 11.4 | 6.5 | 11.4 |
| Wichita, KS | D | D | 169,271 | 17.2 | 35.9 | 31.1 | 1,234.2 | 10,905 | 30,730 | 59,779 | 77,054 | 41.4 | 4.6 | 11.6 | 7.4 | 12.4 |
| Wichita Falls, TX | 3,856 | 2,565.2 | 36,892 | 8.4 | 46.3 | 22.4 | 236.0 | 9,709 | 27,206 | 52,233 | 70,660 | 47.0 | 3.5 | 13.1 | 7.8 | 13.1 |
| Williamsport, PA | NA | NA | 24,290 | 12.0 | 49.8 | 24.3 | 245.6 | 15,577 | 29,134 | 53,881 | 70,079 | 45.5 | 3.5 | 14.0 | 10.6 | 17.1 |
| Wilmington, NC | 6,534 | 2,181.6 | 73,325 | 13.8 | 29.1 | 39.3 | 360.3 | 9,624 | 34,372 | 57,667 | 80,086 | 44.1 | 7.1 | 12.5 | 6.8 | 12.7 |
| Winchester, VA-WV | 1,803 | 1,283.1 | 30,687 | 25.3 | 46.2 | 26.7 | 268.2 | 12,714 | 33,494 | 76,583 | 86,719 | 28.8 | 5.6 | 8.6 | 6.2 | 9.5 |
| Winston-Salem, NC | NA | NA | 156,184 | 17.8 | 42.2 | 27.1 | 934.4 | 9,281 | 29,146 | 52,322 | 67,316 | 47.9 | 4.4 | 16.0 | 12.4 | 20.7 |
| Worcester, MA-CT | 9,273 | 1,065.2 | 234,106 | 20.6 | 37.5 | 35.3 | 2,235.1 | 15,614 | 38,315 | 76,348 | 97,088 | 33.8 | 10.0 | 9.8 | 6.3 | 9.9 |
| Yakima, WA | 6,536 | 2,599.1 | 65,095 | 8.3 | 53.2 | 17.2 | 698.8 | 12,786 | 24,781 | 56,233 | 65,021 | 45.0 | 3.7 | 16.9 | 13.7 | 21.6 |
| York-Hanover, PA | NA | NA | 100,802 | 19.3 | 48.0 | 26.4 | 952.6 | 14,255 | 33,979 | 69,172 | 83,957 | 35.0 | 5.4 | 9.4 | 5.5 | 9.8 |
| Youngstown-Warren-Boardman, OH-PA | NA | NA | 111,570 | 14.6 | 50.1 | 22.8 | 1,047.8 | 14,478 | 28,064 | 48,558 | 61,285 | 51.2 | 3.0 | 16.2 | 12.5 | 22.0 |
| Yuba City, CA | 4,291 | 2,460.9 | 48,297 | 8.0 | 44.0 | 19.6 | 423.5 | 11,055 | 28,757 | 61,307 | 68,244 | 40.2 | 6.3 | 13.4 | 10.4 | 14.1 |
| Yuma, AZ | 3,707 | 1,730.6 | 54,610 | 4.1 | 51.7 | 14.2 | 275.1 | 7,471 | 21,766 | 46,419 | 51,480 | 52.8 | 1.8 | 21.4 | 18.3 | 27.1 |

1. Data for serious crimes have not been adjusted for underreporting; this may affect comparability between geographic areas and over time.   2. Per 100,000 population estimated by the FBI.   3. All persons 3 years old and over enrolled in nursery school through college.   4. Persons 25 years old and over.   5. Elementary and secondary education expenditures.   6. Based on population estimated by the American Community Survey, 2017.

# Table C. Metropolitan Areas — Personal Income and Earnings

| Area name | Personal income, 2019 — Total (mil dol) | Percent change, 2018-2019 | Per capita¹ Dollars | Per capita¹ Rank | Wages and Salaries (mil dol) | Supplements to wages and salaries, employer contributions — Pension and insurance | Supplements — Government social insurance | Proprietors' income | Dividends, interest, and rent (mil dol) | Personal transfer receipts (mil dol) | Earnings, 2019 — Total (mil dol) | Contributions for government social insurance — From employee and self-employed | Contributions — From employer |
|---|---|---|---|---|---|---|---|---|---|---|---|---|---|
| | 62 | 63 | 64 | 65 | 66 | 67 | 68 | 69 | 70 | 71 | 72 | 73 | 74 |
| Texarkana, TX-AR | 5,788 | 2.7 | 38,910 | 357 | 2,645 | 488 | 191 | 381 | 1,004 | 1,603 | 3,705 | 233 | 191 |
| The Villages, FL | 6,407 | 5.1 | 48,387 | 171 | 1,519 | 268 | 109 | 226 | 2,092 | 2,636 | 2,122 | 250 | 109 |
| Toledo, OH | 31,019 | 2.8 | 48,330 | 173 | 16,581 | 2,914 | 1,185 | 2,555 | 4,766 | 6,575 | 23,234 | 1,344 | 1,185 |
| Topeka, KS | 11,000 | 2.4 | 47,418 | 201 | 5,553 | 899 | 418 | 909 | 1,913 | 2,387 | 7,780 | 501 | 418 |
| Trenton-Princeton, NJ | 26,378 | 3.7 | 71,790 | 15 | 18,574 | 2,835 | 1,280 | 2,014 | 5,479 | 3,626 | 24,704 | 1,443 | 1,280 |
| Tucson, AZ | 47,605 | 4.3 | 45,456 | 236 | 20,428 | 3,658 | 1,467 | 3,231 | 10,814 | 11,294 | 28,784 | 1,924 | 1,467 |
| Tulsa, OK | 56,462 | 3.6 | 56,540 | 66 | 24,407 | 3,632 | 1,768 | 10,746 | 10,968 | 8,952 | 40,554 | 2,238 | 1,768 |
| Tuscaloosa, AL | 10,016 | 3.7 | 39,738 | 348 | 5,261 | 939 | 374 | 487 | 1,824 | 2,424 | 7,060 | 460 | 374 |
| Twin Falls, ID | 4,663 | 6.1 | 41,899 | 302 | 1,956 | 333 | 169 | 865 | 833 | 911 | 3,323 | 189 | 169 |
| Tyler, TX | 13,102 | 3.2 | 56,292 | 69 | 5,170 | 805 | 354 | 3,654 | 2,070 | 2,275 | 9,983 | 504 | 354 |
| Urban Honolulu, HI | 59,618 | 2.6 | 61,174 | 37 | 30,330 | 6,413 | 2,412 | 4,615 | 12,400 | 8,483 | 43,770 | 2,575 | 2,412 |
| Utica-Rome, NY | 13,383 | 4.5 | 46,151 | 221 | 5,753 | 1,505 | 455 | 810 | 2,141 | 3,479 | 8,523 | 503 | 455 |
| Valdosta, GA | 5,542 | 4.1 | 37,623 | 371 | 2,651 | 569 | 190 | 300 | 1,042 | 1,308 | 3,710 | 212 | 190 |
| Vallejo, CA | 23,951 | 5.9 | 53,505 | 98 | 10,041 | 1,950 | 699 | 1,246 | 4,066 | 4,358 | 13,936 | 827 | 699 |
| Victoria, TX | 4,856 | 4.4 | 48,681 | 167 | 2,044 | 341 | 137 | 349 | 1,054 | 1,044 | 2,871 | 167 | 137 |
| Vineland-Bridgeton, NJ | 6,166 | 3.4 | 41,237 | 322 | 3,090 | 632 | 246 | 498 | 854 | 1,744 | 4,466 | 279 | 246 |
| Virginia Beach-Norfolk-Newport News, VA-NC | 92,002 | 3.6 | 52,011 | 118 | 47,200 | 9,298 | 3,613 | 3,305 | 19,893 | 16,144 | 63,416 | 3,741 | 3,613 |
| Visalia, CA | 19,974 | 6.9 | 42,845 | 290 | 7,454 | 1,740 | 558 | 2,798 | 3,097 | 4,718 | 12,549 | 630 | 558 |
| Waco, TX | 11,429 | 2.7 | 41,723 | 312 | 5,925 | 986 | 405 | 856 | 1,826 | 2,559 | 8,172 | 465 | 405 |
| Walla Walla, WA | 2,943 | 4.6 | 48,444 | 169 | 1,344 | 261 | 129 | 245 | 710 | 658 | 1,979 | 112 | 129 |
| Warner Robins, GA | 8,039 | 4.9 | 43,358 | 277 | 3,926 | 981 | 296 | 304 | 1,478 | 1,701 | 5,507 | 314 | 296 |
| Washington-Arlington-Alexandria, DC-VA-MD-WV | 467,176 | 3.4 | 74,385 | 13 | 279,316 | 43,554 | 19,528 | 36,830 | 94,002 | 44,093 | 379,228 | 21,442 | 19,528 |
| Frederick-Gaithersburg-Rockville, MD Div 23224 | 111,358 | 3.3 | 84,991 | NA | 47,096 | 7,173 | 3,333 | 14,545 | 24,690 | 9,376 | 72,146 | 4,026 | 4,026 |
| Washington-Arlington-Alexandria, DC-VA-MD-WV Div 47894 | 355,819 | 3.4 | 71,590 | NA | 232,220 | 36,382 | 16,195 | 22,286 | 69,312 | 34,718 | 307,082 | 17,416 | 17,416 |
| Waterloo-Cedar Falls, IA | 8,026 | 2.2 | 47,623 | 191 | 4,353 | 769 | 338 | 480 | 1,603 | 1,628 | 5,940 | 382 | 338 |
| Watertown-Fort Drum, NY | 5,422 | 3.6 | 49,365 | 149 | 2,776 | 830 | 249 | 235 | 954 | 1,152 | 4,089 | 201 | 249 |
| Wausau-Weston, WI | 8,376 | 3.4 | 51,295 | 126 | 4,259 | 819 | 313 | 696 | 1,353 | 1,459 | 6,087 | 379 | 313 |
| Weirton-Steubenville, WV-OH | 4,893 | 2.9 | 42,158 | 300 | 1,766 | 355 | 135 | 209 | 710 | 1,488 | 2,464 | 175 | 135 |
| Wenatchee, WA | 6,155 | 3.9 | 51,023 | 130 | 2,502 | 444 | 245 | 499 | 1,532 | 1,264 | 3,690 | 218 | 245 |
| Wheeling, WV-OH | 6,850 | -0.5 | 49,301 | 152 | 3,281 | 559 | 239 | 714 | 1,229 | 1,661 | 4,794 | 311 | 239 |
| Wichita, KS | 33,374 | 3.9 | 52,129 | 113 | 15,661 | 2,547 | 1,203 | 3,673 | 7,719 | 5,326 | 23,084 | 1,399 | 1,203 |
| Wichita Falls, TX | 6,793 | 4.0 | 44,910 | 248 | 2,884 | 591 | 209 | 511 | 1,444 | 1,581 | 4,195 | 234 | 209 |
| Williamsport, PA | 5,092 | 2.6 | 44,941 | 247 | 2,476 | 530 | 197 | 271 | 841 | 1,284 | 3,474 | 220 | 197 |
| Wilmington, NC | 13,679 | 4.2 | 45,975 | 224 | 6,715 | 1,053 | 466 | 864 | 3,367 | 2,833 | 9,097 | 590 | 466 |
| Winchester, VA-WV | 6,978 | 4.0 | 49,643 | 142 | 3,279 | 529 | 235 | 471 | 1,224 | 1,199 | 4,514 | 290 | 235 |
| Winston-Salem, NC | 31,032 | 3.9 | 45,904 | 225 | 14,500 | 2,092 | 1,022 | 2,028 | 5,593 | 6,622 | 19,643 | 1,305 | 1,022 |
| Worcester, MA-CT | 54,204 | 3.8 | 57,214 | 64 | 23,029 | 4,302 | 1,620 | 3,909 | 7,861 | 9,454 | 32,859 | 1,767 | 1,620 |
| Yakima, WA | 11,479 | 4.3 | 45,757 | 228 | 4,852 | 853 | 501 | 1,172 | 2,199 | 2,762 | 7,377 | 389 | 501 |
| York-Hanover, PA | 23,165 | 3.7 | 51,585 | 120 | 9,659 | 1,758 | 769 | 1,461 | 3,541 | 4,536 | 13,646 | 847 | 769 |
| Youngstown-Warren-Boardman, OH-PA | 23,141 | 1.8 | 43,167 | 279 | 9,373 | 1,761 | 702 | 1,814 | 3,879 | 6,535 | 13,650 | 894 | 702 |
| Yuba City, CA | 8,041 | 5.9 | 45,782 | 227 | 2,821 | 684 | 211 | 759 | 1,307 | 2,000 | 4,476 | 253 | 211 |
| Yuma, AZ | 7,818 | 5.5 | 36,570 | 376 | 3,196 | 664 | 275 | 1,180 | 1,173 | 1,847 | 5,316 | 297 | 275 |

1. Based on the resident population estimated as of July 1 of the year shown.

Table C. Metropolitan Areas — **Earnings, Social Security, and Housing**

| Area name | Farm | Mining, quarrying, and extracting | Construction | Manufacturing | Information; professional, scientific, and technical services | Retail trade | Finance, insurance, real estate, rental and leasing | Health care and social assistance | Government | Social Security beneficiaries, Dec. 2019 — Number | Rate[1] | Supplemental Security Income Recipients, December 2017 | Housing units, 2020 — Total | Percent change, 2010-2020 |
|---|---|---|---|---|---|---|---|---|---|---|---|---|---|---|
| | 75 | 76 | 77 | 78 | 79 | 80 | 81 | 82 | 83 | 84 | 85 | 86 | 87 | 88 |
| Texarkana, TX-AR | 0.7 | D | 5.5 | 12.0 | 3.9 | 7.4 | 5.0 | 15.8 | 23.5 | 33,470 | 224 | 6,287 | 66,421 | 3.4 |
| The Villages, FL | 0.8 | 0.5 | 10.5 | 5.8 | 6.3 | 8.6 | 7.8 | 16.2 | 18.9 | 74,860 | 560 | 1,680 | 77,152 | 45.5 |
| Toledo, OH | 0.3 | D | 6.9 | 18.5 | D | 5.9 | 7.3 | D | 14.8 | 132,960 | 207 | 19,486 | 305,039 | 1.2 |
| Topeka, KS | 1.3 | 0.4 | 6.1 | 7.2 | D | 4.9 | D | 14.9 | 22.2 | 53,085 | 229 | 5,594 | 105,757 | 1.9 |
| Trenton-Princeton, NJ | 0.0 | 0.9 | 2.9 | 7.4 | 19.3 | 3.4 | 12.6 | 8.7 | 16.7 | 66,675 | 181 | 9,557 | 145,739 | 1.8 |
| Tucson, AZ | 0.1 | 0.9 | 5.3 | 11.3 | 9.8 | 6.8 | 6.2 | 14.0 | 23.5 | 235,840 | 225 | 21,131 | 470,626 | 6.7 |
| Tulsa, OK | -0.1 | 9.6 | 5.9 | 12.7 | D | 5.0 | D | D | 8.8 | 197,770 | 198 | 22,412 | 440,155 | 7.4 |
| Tuscaloosa, AL | 0.4 | 2.7 | 5.7 | 20.6 | D | 5.8 | D | 13.7 | 26.8 | 53,345 | 211 | 8,633 | 117,047 | 9.4 |
| Twin Falls, ID | 14.1 | D | 4.8 | 14.8 | D | 7.5 | D | 16.4 | 10.7 | 21,585 | 193 | 2,232 | 43,692 | 11.5 |
| Tyler, TX | 0.1 | 26.0 | 5.1 | 4.0 | 6.6 | 8.8 | 5.5 | 10.5 | 9.6 | 47,630 | 204 | 5,409 | 92,582 | 6.0 |
| Urban Honolulu, HI | 0.2 | 0.1 | 7.5 | 1.8 | 7.7 | 5.3 | 6.7 | 16.3 | 31.9 | 458,588 | 236 | 37,588 | 356,788 | 5.9 |
| Utica-Rome, NY | 0.6 | 0.5 | 4.3 | 9.5 | 6.0 | 6.3 | 7.2 | D | 30.6 | 72,065 | 249 | 9,459 | 139,779 | 1.6 |
| Valdosta, GA | 1.3 | D | D | D | 6.3 | 6.8 | 4.8 | 16.2 | 33.9 | 26,975 | 183 | 5,391 | 64,588 | 12.5 |
| Vallejo, CA | 1.0 | 0.4 | 12.2 | 12.8 | 4.2 | 6.0 | 3.5 | D | 23.7 | 81,465 | 182 | 11,490 | 161,002 | 5.4 |
| Victoria, TX | -0.1 | 10.8 | 7.5 | 8.4 | D | 8.1 | D | 16.4 | 15.0 | 20,315 | 204 | 2,558 | 41,308 | 5.6 |
| Vineland-Bridgeton, NJ | 2.0 | 0.3 | 6.8 | 14.6 | D | 7.4 | 2.8 | D | 25.3 | 31,065 | 208 | 5,888 | 56,454 | 1.1 |
| Virginia Beach-Norfolk-Newport News, VA-NC | 0.1 | D | D | 7.9 | 9.3 | 4.9 | D | 10.4 | 35.3 | 327,775 | 185 | 36,006 | 747,954 | 6.3 |
| Visalia, CA | 16.0 | D | 4.7 | 7.7 | 3.5 | 6.0 | 3.3 | 7.1 | 23.4 | 67,340 | 145 | 18,834 | 152,855 | 7.9 |
| Waco, TX | -0.1 | D | 8.3 | 15.6 | 6.1 | 7.0 | D | 10.7 | 16.1 | 50,690 | 185 | 7,838 | 112,116 | 9.0 |
| Walla Walla, WA | 8.6 | D | 4.2 | 15.7 | D | 4.6 | 4.5 | 14.9 | 24.5 | 13,635 | 223 | 1,429 | 25,283 | 7.8 |
| Warner Robins, GA | 0.6 | D | D | D | 8.5 | 5.4 | D | 5.8 | 51.5 | 33,275 | 180 | 4,849 | 77,994 | 12.4 |
| Washington-Arlington-Alexandria, DC-VA-MD-WV | 0.0 | D | D | 1.6 | D | 3.5 | D | D | 26.7 | 806,174 | 128 | 88,663 | 2,422,053 | 8.1 |
| Frederick-Gaithersburg-Rockville, MD Div 23224 | 0.1 | D | D | 3.5 | 20.9 | 4.2 | D | D | 20.7 | 187,460 | 143 | 16,467 | 496,215 | 6.6 |
| Washington-Arlington-Alexandria, DC-VA-MD-WV Div 47894 | 0.0 | 0.0 | D | 1.2 | 27.1 | 3.3 | 5.6 | 6.4 | 28.1 | 618,714 | 124 | 72,196 | 1,925,838 | 8.5 |
| Waterloo-Cedar Falls, IA | 2.2 | 0.1 | 5.5 | 23.5 | 4.7 | 6.4 | D | 12.8 | 15.5 | 36,520 | 217 | 3,585 | 75,052 | 5.2 |
| Watertown-Fort Drum, NY | 1.5 | 0.4 | 4.1 | 3.4 | 2.6 | 6.0 | 2.3 | 11.0 | 56.9 | 22,640 | 206 | 2,713 | 60,327 | 4.1 |
| Wausau-Weston, WI | 2.3 | 0.2 | 5.7 | 24.0 | 5.0 | 5.3 | 10.6 | 14.4 | 11.4 | 37,700 | 231 | 2,362 | 77,756 | 4.3 |
| Weirton-Steubenville, WV-OH | -0.2 | D | D | 18.9 | D | 6.9 | D | 17.2 | 14.0 | 32,765 | 282 | 3,873 | 57,350 | -1.7 |
| Wenatchee, WA | 5.7 | D | 8.5 | 4.6 | D | 8.7 | 4.5 | 15.8 | 21.8 | 26,475 | 220 | 1,917 | 57,020 | 10.8 |
| Wheeling, WV-OH | -0.3 | D | 10.5 | 5.5 | 13.3 | 6.1 | 4.9 | D | 11.9 | 36,375 | 261 | 4,008 | 68,657 | -1.3 |
| Wichita, KS | 0.8 | 2.9 | 6.3 | 21.5 | D | 5.7 | 8.1 | D | 12.8 | 122,475 | 191 | 12,601 | 277,163 | 5.4 |
| Wichita Falls, TX | 0.2 | 5.4 | 4.5 | 10.0 | D | 6.9 | 5.5 | 15.3 | 29.2 | 31,110 | 205 | 4,277 | 65,746 | 1.4 |
| Williamsport, PA | 0.6 | 3.4 | 5.3 | 16.8 | 5.2 | 6.5 | 4.6 | 17.5 | 18.5 | 29,100 | 256 | 3,235 | 53,692 | 2.3 |
| Wilmington, NC | 0.7 | D | 9.3 | 5.8 | 14.5 | 8.2 | 7.3 | 10.3 | 20.0 | 62,480 | 210 | 5,087 | 148,178 | 15.6 |
| Winchester, VA-WV | -0.3 | D | D | 13.3 | 5.0 | 7.7 | 8.1 | D | 18.5 | 29,995 | 213 | 2,265 | 62,418 | 9.7 |
| Winston-Salem, NC | 0.2 | 0.1 | 5.9 | 12.1 | 7.1 | 7.1 | 9.8 | D | 10.4 | 150,620 | 223 | 14,350 | 306,806 | 6.9 |
| Worcester, MA-CT | 0.0 | D | D | 12.3 | 10.0 | 6.1 | 6.5 | 15.9 | 16.3 | 181,205 | 192 | 24,525 | 389,571 | 3.7 |
| Yakima, WA | 13.9 | 0.0 | 4.8 | 7.9 | 3.0 | 6.9 | 3.4 | 14.3 | 19.2 | 45,485 | 181 | 7,142 | 90,835 | 6.3 |
| York-Hanover, PA | 0.2 | 0.2 | 9.0 | 18.6 | 6.5 | 5.6 | 4.6 | 14.1 | 13.1 | 99,170 | 221 | 8,353 | 186,871 | 4.6 |
| Youngstown-Warren-Boardman, OH-PA | 0.1 | 0.5 | 7.2 | 14.5 | 4.4 | 8.3 | 5.7 | 17.3 | 15.0 | 144,055 | 269 | 18,862 | 259,229 | -0.2 |
| Yuba City, CA | 6.3 | 0.4 | 7.0 | 4.0 | D | 6.7 | 3.7 | D | 33.0 | 31,905 | 182 | 7,665 | 63,897 | 3.9 |
| Yuma, AZ | 13.4 | 0.0 | 3.8 | 4.2 | 5.2 | 8.9 | 3.4 | 10.6 | 28.1 | 40,845 | 191 | 4,811 | 95,924 | 9.2 |

1. Per 1,000 resident population estimated as of July 1 of the year shown.

# Table C. Metropolitan Areas — Housing, Labor Force, and Employment

| Area name | Occupied Housing units, 2020 — Occupied units — Owner-occupied — Total | Percent | Median value[1] | Median owner cost as a percent of income — With a mortgage | Without a mortgage[2] | Renter-occupied — Median rent[3] | Median rent as a percent of income[2] | Substandard units[4] (percent) | Civilian labor force, 2020 — Total | Percent change 2019-2020 | Unemployment — Total | Rate[5] | Civilian employment, 2019[6] — Total | Percent — Management, business, science, and arts | Construction, production, and maintenance occupations |
|---|---|---|---|---|---|---|---|---|---|---|---|---|---|---|---|
| | 89 | 90 | 91 | 92 | 93 | 94 | 95 | 96 | 97 | 98 | 99 | 100 | 101 | 102 | 103 |
| Texarkana, TX-AR | 55,028 | 64.7 | 117,200 | 17.6 | 10.0 | 771 | 26.7 | 3.2 | 63,519 | -2.1 | 4,543 | 7.2 | 60,465 | 29.3 | 31.4 |
| The Villages, FL | 59,458 | 88.5 | 282,000 | 22.8 | 10.0 | 1,185 | 46.8 | 0.0 | 32,225 | -1.3 | 2,534 | 7.9 | 0 | 0.0 | 0.0 |
| Toledo, OH | 268,089 | 64 | 139,200 | 18.5 | 11.5 | 760 | 27.7 | 1.0 | 321,509 | -1.6 | 30,216 | 9.4 | 311,414 | 34.6 | 26.7 |
| Topeka, KS | 95,552 | 69.3 | 138,600 | 18.7 | 11.7 | 804 | 25.2 | 2.6 | 120,449 | 0.7 | 6,889 | 5.7 | 115,868 | 37.0 | 25.4 |
| Trenton-Princeton, NJ | 130,851 | 62.6 | 291,600 | 22.6 | 15.2 | 1,234 | 29.8 | 1.8 | 205,405 | 0.5 | 15,255 | 7.4 | 176,443 | 48.6 | 16.4 |
| Tucson, AZ | 410,404 | 63 | 211,600 | 20.4 | 10.0 | 905 | 29.9 | 3.4 | 495,991 | -0.3 | 38,308 | 7.7 | 464,801 | 38.1 | 18.1 |
| Tulsa, OK | 387,176 | 64.8 | 160,900 | 19.0 | 10.8 | 861 | 27.8 | 2.9 | 481,838 | -0.4 | 31,057 | 6.4 | 477,716 | 37.5 | 24.0 |
| Tuscaloosa, AL | 90,861 | 66 | 165,400 | 18.9 | 10.0 | 858 | 34.2 | 1.2 | 118,972 | -2.3 | 8,148 | 6.8 | 112,219 | 34.5 | 30.8 |
| Twin Falls, ID | 39,654 | 69.2 | 194,300 | 22.0 | 10.5 | 761 | 24.2 | 2.0 | 53,448 | 0.1 | 2,654 | 5.0 | 50,729 | 30.0 | 31.5 |
| Tyler, TX | 78,600 | 66.8 | 167,800 | 21.4 | 10.8 | 994 | 27.9 | 3.5 | 108,544 | 0.2 | 7,343 | 6.8 | 107,677 | 33.7 | 25.9 |
| Urban Honolulu, HI | 316,456 | 57.4 | 739,700 | 26.3 | 10.0 | 1,774 | 32.6 | 8.5 | 440,915 | -2.8 | 44,779 | 10.2 | 459,379 | 38.3 | 17.0 |
| Utica-Rome, NY | 115,904 | 69.4 | 131,800 | 18.7 | 11.3 | 728 | 27.3 | 1.7 | 128,125 | -0.7 | 10,425 | 8.1 | 130,676 | 37.3 | 20.8 |
| Valdosta, GA | 54,425 | 56.3 | 131,200 | 20.1 | 10.0 | 817 | 26.7 | 0.0 | 64,063 | -0.3 | 3,737 | 5.8 | 64,423 | 25.8 | 24.1 |
| Vallejo, CA | 150,393 | 62.2 | 460,500 | 24.3 | 10.0 | 1,723 | 30.6 | 4.8 | 202,812 | -2.5 | 19,205 | 9.5 | 215,481 | 34.2 | 25.3 |
| Victoria, TX | 32,627 | 68.9 | 148,500 | 19.7 | 12.5 | 966 | 27.7 | 0.0 | 44,333 | -3.0 | 3,655 | 8.2 | 44,351 | 29.3 | 28.3 |
| Vineland-Bridgeton, NJ | 51,360 | 65.8 | 164,900 | 24.2 | 15.3 | 1,143 | 36.9 | 4.1 | 66,277 | 1.2 | 7,129 | 10.8 | 60,564 | 27.4 | 28.9 |
| Virginia Beach-Norfolk-Newport News, VA-NC | 675,311 | 61.1 | 255,900 | 22.2 | 10.0 | 1,178 | 29.2 | 1.6 | 868,241 | -1.6 | 61,576 | 7.1 | 819,433 | 40.4 | 20.4 |
| Visalia, CA | 144,109 | 59.2 | 244,700 | 23.7 | 10.2 | 1,014 | 29.4 | 9.7 | 199,151 | -1.5 | 26,299 | 13.2 | 189,088 | 27.0 | 36.4 |
| Waco, TX | 98,944 | 62.3 | 161,300 | 21.0 | 11.3 | 877 | 35.5 | 2.8 | 126,878 | 1.0 | 7,800 | 6.1 | 119,328 | 33.2 | 28.6 |
| Walla Walla, WA | 23,167 | 67.1 | 273,900 | 20.8 | 11.6 | 858 | 32.6 | 3.5 | 30,165 | 0.1 | 2,069 | 6.9 | 27,848 | 34.9 | 23.0 |
| Warner Robins, GA | 68,831 | 63.1 | 154,800 | 19.5 | 10.0 | 910 | 27.1 | 0.0 | 82,232 | -1.1 | 4,379 | 5.3 | 84,873 | 36.6 | 23.8 |
| Washington-Arlington-Alexandria, DC-VA-MD-WV | 2,251,002 | 63.5 | 446,300 | 21.3 | 10.0 | 1,708 | 28.3 | 3.1 | 3,414,893 | -2.1 | 221,444 | 6.5 | 3,383,232 | 54.0 | 13.7 |
| Frederick-Gaithersburg-Rockville, MD Div 23224 | 462,665 | 67 | 457,300 | 20.8 | 10.0 | 1,740 | 29.9 | 2.7 | 681,265 | -3.1 | 42,633 | 6.3 | 705,826 | 55.3 | 13.2 |
| Washington-Arlington-Alexandria, DC-VA-MD-WV Div 47894 | 1,788,337 | 62.6 | 442,900 | 21.4 | 10.0 | 1,700 | 27.9 | 3.1 | 2,733,628 | -1.8 | 178,811 | 6.5 | 2,677,406 | 53.7 | 13.8 |
| Waterloo-Cedar Falls, IA | 67,137 | 69.8 | 158,500 | 17.9 | 10.0 | 826 | 26.1 | 1.6 | 87,091 | -4.3 | 4,952 | 5.7 | 88,222 | 34.1 | 28.9 |
| Watertown-Fort Drum, NY | 41,214 | 56.6 | 150,600 | 20.0 | 11.5 | 1,025 | 27.9 | 2.4 | 43,295 | | 3,753 | 8.7 | 38,093 | 33.1 | 21.5 |
| Wausau-Weston, WI | 67,342 | 73.3 | 163,000 | 18.2 | 10.6 | 750 | 24.3 | 1.7 | 86,990 | -1.2 | 4,474 | 5.1 | 86,322 | 33.8 | 33.0 |
| Weirton-Steubenville, WV-OH | 50,065 | 70.9 | 95,400 | 16.4 | 10.0 | 654 | 26.3 | 0.0 | 50,528 | -1.1 | 4,981 | 9.9 | 49,090 | 29.5 | 29.4 |
| Wenatchee, WA | 44,478 | 64.9 | 327,500 | 21.7 | 10.0 | 959 | 22.7 | 4.7 | 65,415 | -2.3 | 5,498 | 8.4 | 56,414 | 28.9 | 31.4 |
| Wheeling, WV-OH | 55,796 | 74.1 | 115,500 | 16.1 | 10.0 | 685 | 27.5 | 0.0 | 62,937 | -4.0 | 5,992 | 9.5 | 61,009 | 31.1 | 27.0 |
| Wichita, KS | 246,836 | 64.5 | 152,600 | 19.0 | 11.0 | 838 | 27.7 | 2.0 | 317,697 | 1.2 | 26,369 | 8.3 | 309,608 | 36.8 | 26.5 |
| Wichita Falls, TX | 56,673 | 63.7 | 117,000 | 19.1 | 11.7 | 776 | 27.1 | 2.7 | 64,010 | -1.5 | 4,159 | 6.5 | 67,860 | 35.0 | 24.2 |
| Williamsport, PA | 44,842 | 69.6 | 161,800 | 19.5 | 11.9 | 755 | 30.5 | 0.0 | 55,883 | -2.0 | 4,913 | 8.8 | 50,871 | 31.8 | 27.5 |
| Wilmington, NC | 121,734 | 62.5 | 257,100 | 21.6 | 11.2 | 1,093 | 31.8 | 1.8 | 148,840 | -2.5 | 10,605 | 7.1 | 146,790 | 42.4 | 18.8 |
| Winchester, VA-WV | 50,062 | 72.2 | 247,300 | 20.1 | 10.0 | 1,120 | 23.5 | 0.0 | 74,395 | -0.1 | 3,725 | 5.0 | 67,877 | 36.6 | 25.7 |
| Winston-Salem, NC | 270,273 | 67.7 | 164,800 | 18.9 | 10.0 | 792 | 29.4 | 1.8 | 321,440 | -3.2 | 23,259 | 7.2 | 310,031 | 36.7 | 26.6 |
| Worcester, MA-CT | 363,177 | 64.9 | 294,300 | 21.2 | 13.6 | 1,086 | 28.5 | 2.0 | 494,223 | -3.1 | 42,773 | 8.7 | 489,005 | 41.9 | 20.6 |
| Yakima, WA | 83,992 | 63.4 | 213,800 | 20.4 | 10.0 | 839 | 29.1 | 9.1 | 130,759 | -0.8 | 12,688 | 9.7 | 109,942 | 27.4 | 39.5 |
| York-Hanover, PA | 175,441 | 75 | 191,000 | 20.4 | 12.6 | 950 | 29.3 | 1.2 | 232,812 | -1.4 | 18,647 | 8.0 | 227,753 | 36.7 | 27.0 |
| Youngstown-Warren-Boardman, OH-PA | 230,643 | 70.6 | 108,900 | 18.2 | 10.3 | 675 | 29.0 | 1.3 | 234,531 | -2.4 | 23,806 | 10.2 | 239,888 | 32.4 | 27.6 |
| Yuba City, CA | 59,866 | 61 | 300,400 | 22.6 | 10.0 | 1,087 | 32.1 | 6.9 | 75,509 | 0.4 | 8,158 | 10.8 | 69,654 | 28.8 | 28.7 |
| Yuma, AZ | 74,042 | 66.3 | 139,100 | 21.7 | 10.0 | 829 | 28.4 | 8.2 | 98,120 | -2.1 | 16,810 | 17.1 | 74,056 | 26.6 | 32.7 |

1. Specified owner-occupied units. lacking complete plumbing facilities.   2. A value of 10.0 represents 10 percent or less; a value of 50.0 represents 50 percent or more.   5. Percent of civilian labor force.   6. Civilian employed persons 16 years old and over.   3. Specified renter-occupied units.   4. Overcrowded or

Items 89—103

# Table C. Metropolitan Areas — Nonfarm Employment and Agriculture

| Area name | Private nonfarm establishments, employment and payroll, 2019 | | | | | | | | | Agriculture, 2017 | | | |
|---|---|---|---|---|---|---|---|---|---|---|---|---|---|
| | | Employment | | | | | | Annual payroll | | Farms | | | Farm producers whose primary occupation is farming (percent) |
| | Number of establishments | Total | Health care and social assistance | Manufacturing | Retail trade | Finance and insurance | Professional, scientific, and technical services | Total (mil dol) | Average per employee (dollars) | Number | Fewer than 50 acres | 1000 acres or more | |
| | 104 | 105 | 106 | 107 | 108 | 109 | 110 | 111 | 112 | 113 | 114 | 115 | 116 |
| Texarkana, TX-AR | 3,059 | 48,123 | 8,486 | 6,432 | 7,750 | 1,556 | 2,610 | 1,878 | 39,024 | 2,433 | 40.1 | 4.3 | 36.1 |
| The Villages, FL | 1,665 | 26,554 | 5,118 | 1,040 | 4,320 | 672 | 666 | 1,014 | 38,168 | 1,307 | 75.7 | 1.9 | 37.0 |
| Toledo, OH | 14,039 | 279,392 | 49,815 | 49,740 | 32,595 | 7,618 | 12,982 | 13,097 | 46,879 | 2,791 | 48.8 | 5.7 | 38.1 |
| Topeka, KS | 5,012 | 84,459 | 18,885 | 8,295 | 10,831 | 6,014 | 3,659 | 3,892 | 46,083 | 4,511 | 29 | 9.4 | 34.9 |
| Trenton-Princeton, NJ | 9,694 | 199,859 | 30,193 | 6,738 | 24,701 | 14,247 | 17,635 | 14,129 | 70,692 | 323 | 72.1 | 0.6 | 35.1 |
| Tucson, AZ | 20,400 | 331,057 | 63,841 | 23,565 | 46,672 | 13,284 | 25,446 | 14,743 | 44,534 | 661 | 79.7 | 7.9 | 42.9 |
| Tulsa, OK | 24,706 | 414,971 | 61,449 | 53,505 | 47,154 | 16,044 | | 21,113 | 50,879 | 9,398 | 44.7 | 4.8 | 33.5 |
| Tuscaloosa, AL | 4,623 | 86,614 | 13,948 | 18,853 | 10,630 | 2,039 | 2,348 | 3,902 | 45,045 | 1,652 | 33.4 | 8.2 | 37.5 |
| Twin Falls, ID | 3,273 | 37,760 | 6,820 | 5,181 | 6,174 | 1,052 | 1,274 | 1,395 | 36,934 | 1,697 | 51.1 | 6.4 | 51.2 |
| Tyler, TX | 5,940 | 94,185 | 22,062 | 7,234 | 13,030 | 3,382 | 4,671 | 4,044 | 42,940 | 2,928 | 58.8 | 1.1 | 31.7 |
| Urban Honolulu, HI | 21,420 | 368,904 | 54,323 | 9,411 | 47,291 | 18,468 | 18,773 | 17,932 | 48,610 | 927 | 91.3 | 1.7 | 55.4 |
| Utica-Rome, NY | 5,982 | 101,677 | 23,711 | 12,305 | 13,694 | 6,588 | 4,144 | 4,188 | 41,185 | 1,563 | 30.4 | 2.0 | 46.2 |
| Valdosta, GA | 3,099 | 45,266 | 8,653 | 3,801 | 7,143 | 1,323 | 4,048 | 1,627 | 35,947 | 909 | 42.2 | 9.2 | 37.5 |
| Vallejo, CA | 7,250 | 116,890 | 22,790 | 9,703 | 18,956 | 3,548 | 1,225 | 6,267 | 53,611 | 849 | 65.3 | 8.1 | 44.6 |
| Victoria, TX | 2,418 | 34,679 | 6,812 | 2,281 | 6,036 | 850 | 1,021 | 1,586 | 45,735 | 2,541 | 42 | 5.9 | 32.3 |
| Vineland-Bridgeton, NJ | 2,835 | 47,003 | 10,299 | 7,946 | 6,772 | 1,030 | | 1,930 | 41,066 | 560 | 63.6 | 2.0 | 48.0 |
| Virginia Beach-Norfolk-Newport News, VA-NC | 38,812 | 639,355 | 99,604 | 57,563 | 91,573 | 26,594 | 53,491 | 28,962 | 45,299 | 1,840 | 54.3 | 10.2 | 41.6 |
| Visalia, CA | 6,531 | 99,961 | 17,063 | 13,594 | 16,231 | 3,046 | 2,275 | 4,149 | 41,505 | 4,187 | 57.6 | 5.4 | 52.3 |
| Waco, TX | 5,579 | 108,648 | 17,634 | 15,460 | 13,636 | 5,275 | 4,427 | 4,357 | 40,104 | 4,469 | 54.3 | 4.4 | 32.7 |
| Walla Walla, WA | 1,416 | 20,805 | 4,627 | 3,537 | 2,467 | 797 | 503 | 942 | 45,289 | 903 | 54.2 | 18.4 | 41.7 |
| Warner Robins, GA | 3,055 | 48,513 | 7,544 | 7,418 | 8,585 | 1,242 | 4,379 | 1,775 | 36,596 | 505 | 53.5 | 3.2 | 30.7 |
| Washington-Arlington-Alexandria, DC-VA-MD-WV | 156,193 | 2,766,974 | 351,524 | 49,031 | 274,902 | 105,264 | 578,151 | 194,832 | 70,413 | 9,392 | 58.4 | 2.5 | 38.7 |
| Frederick-Gaithersburg-Rockville, MD Div 23224 | 33,612 | 536,801 | 90,771 | 13,379 | 58,826 | 25,227 | 91,432 | 34,895 | 65,006 | 1,931 | 58.2 | 2.5 | 40.3 |
| Washington-Arlington-Alexandria, DC-VA-MD-WV Div 47894 | 122,581 | 2,230,173 | 260,753 | 35,652 | 216,076 | 80,037 | 486,719 | 159,937 | 71,715 | 7,461 | 58.5 | 2.5 | 38.2 |
| Waterloo-Cedar Falls, IA | 4,055 | 75,813 | 13,709 | 14,122 | 10,946 | 3,213 | 863 | 3,410 | 44,979 | 2,691 | 38.2 | 8.2 | 42.6 |
| Watertown-Fort Drum, NY | 2,420 | 30,055 | 6,871 | 2,388 | 6,388 | 748 | 1,556 | 1,183 | 39,345 | 792 | 26.5 | 5.6 | 51.5 |
| Wausau-Weston, WI | 4,111 | 75,935 | 11,339 | 22,160 | 8,936 | 5,195 | 618 | 3,494 | 46,014 | 2,663 | 33.2 | 2.7 | 48.7 |
| Weirton-Steubenville, WV-OH | 2,149 | 34,032 | 7,144 | 5,699 | 4,594 | 783 | 1,352 | 1,341 | 39,408 | 781 | 32.5 | 0.1 | 36.3 |
| Wenatchee, WA | 3,411 | 40,158 | 7,650 | 2,051 | 6,273 | 971 | 2,519 | 1,816 | 45,224 | 1,564 | 57.9 | 12.1 | 48.4 |
| Wheeling, WV-OH | 3,211 | 55,030 | 11,317 | 2,306 | 8,123 | 2,140 | | 2,388 | 43,399 | 1,596 | 29.6 | 0.8 | 38.0 |
| Wichita, KS | 14,720 | 265,327 | 39,750 | 52,664 | 32,437 | 8,063 | 13,030 | 12,746 | 48,039 | 4,536 | 35.5 | 15.0 | 36.7 |
| Wichita Falls, TX | 3,381 | 47,475 | 11,944 | 5,056 | 7,141 | 1,845 | 1,142 | 1,789 | 37,678 | 1,997 | 28.8 | 18.1 | 36.5 |
| Williamsport, PA | 2,768 | 45,644 | 9,752 | 7,645 | 7,053 | 1,557 | 1,542 | 1,882 | 41,235 | 1,043 | 32.2 | 1.3 | 38.6 |
| Wilmington, NC | 8,983 | 112,188 | 19,382 | 6,696 | 18,158 | 3,808 | 6,214 | 4,948 | 44,107 | 395 | 58.2 | 3.3 | 47.5 |
| Winchester, VA-WV | 3,225 | 54,363 | 10,240 | 7,598 | 8,508 | 2,871 | 1,648 | 2,342 | 43,073 | 1,645 | 53.2 | 2.2 | 34.8 |
| Winston-Salem, NC | 13,570 | 248,716 | 50,579 | 32,474 | 31,421 | 11,225 | 8,328 | 11,796 | 47,428 | 3,825 | 52.9 | 1.2 | 40.0 |
| Worcester, MA-CT | 20,681 | 347,013 | 78,087 | 40,677 | 43,412 | 16,114 | 20,722 | 17,678 | 50,944 | 2,214 | 66.4 | 0.3 | 38.5 |
| Yakima, WA | 4,852 | 72,140 | 15,698 | 9,534 | 10,702 | 1,556 | 2,283 | 3,149 | 43,649 | 2,952 | 74.1 | 3.8 | 44.3 |
| York-Hanover, PA | 8,662 | 168,358 | 25,893 | 30,621 | 21,663 | 4,837 | 6,265 | 7,671 | 45,564 | 2,067 | 62 | 2.0 | 39.3 |
| Youngstown-Warren-Boardman, OH-PA | 12,033 | 189,045 | 38,528 | 27,806 | 30,112 | 5,260 | 4,598 | 7,319 | 38,715 | 2,978 | 48.3 | 1.4 | 37.7 |
| Yuba City, CA | 2,677 | 33,026 | 6,575 | 2,182 | 5,899 | 1,034 | 1,147 | 1,525 | 46,162 | 1,921 | 53.5 | 5.2 | 53.1 |
| Yuma, AZ | 3,113 | 45,290 | 8,297 | 3,257 | 8,225 | 1,261 | 1,237 | 1,678 | 37,055 | 456 | 60.7 | 13.2 | 50.4 |

Table C. Metropolitan Areas — **Agriculture**

| Area name | Land in farms Acreage (1,000) 117 | Percent change, 2012-2017 118 | Acres Average size of farm 119 | Total irrigated (1,000) 120 | Total cropland (1,000) 121 | Value of land and buildings (dollars) Average per farm 122 | Average per acre 123 | Value of machinery and equipment, average per farm (dollars) 124 | Value of products sold: Total (mil dol) 125 | Average per farm (acres) 126 | Percent from: Crops 127 | Live-stock and poultry products 128 | Organic farms (number) 129 | Farms with internet access (percent) 130 | Government payments Total ($1,000) 131 | Percent of farms 132 |
|---|---|---|---|---|---|---|---|---|---|---|---|---|---|---|---|---|
| Texarkana, TX-AR | 580 | -4.6 | 239 | 24.8 | 191.5 | 655,739 | 2,749 | 84,649 | 181 | 74,199 | 24.9 | 75.1 | 0 | 73.4 | 6,849 | 17.8 |
| The Villages, FL | 177 | -3.4 | 135 | 2.0 | 21.3 | 816,314 | 6,025 | 50,217 | 55 | 41,666 | 36.7 | 63.3 | 8 | 74.7 | 1,099 | 16.7 |
| Toledo, OH | 652 | 2.1 | 234 | 5.6 | 613.0 | 1,562,388 | 6,687 | 175,895 | 442 | 158,461 | 79.5 | 20.5 | 17 | 77.9 | 20,571 | 63.8 |
| Topeka, KS | 1,610 | 0.3 | 357 | 29.6 | 815.2 | 852,691 | 2,389 | 103,730 | 352 | 77,915 | 61.6 | 38.4 | 7 | 77.1 | 10,303 | 43.2 |
| Trenton-Princeton, NJ | 25 | 27.8 | 78 | 1.0 | 15.8 | 1,414,873 | 18,114 | 83,437 | 25 | 77,344 | 80.1 | 19.9 | 7 | 77.1 | 10,303 | 43.2 |
| Tucson, AZ | 2,618 | NA | 3,960 | 30.0 | 40.7 | 2,088,809 | 527 | 79,142 | 76 | 114,174 | 84.3 | 15.7 | 12 | 73.4 | 149 | 8.0 |
| Tulsa, OK | 2,653 | -3.7 | 282 | 16.9 | 518.9 | 615,238 | 2,179 | 60,741 | 315 | 33,485 | 25.0 | 75.0 | 3 | 72.0 | 1,234 | 5.4 |
| | | | | | | | | | | | | | 3 | 74.1 | 9,321 | 10.3 |
| Tuscaloosa, AL | 511 | 8.8 | 309 | 3.9 | 91.8 | 750,702 | 2,427 | 86,005 | 217 | 131,242 | 9.3 | 90.7 | 4 | 70.8 | 5,338 | 41.8 |
| Twin Falls, ID | 640 | -4.7 | 377 | 376.3 | 391.9 | 1,837,189 | 4,868 | 280,963 | 1,320 | 777,738 | 22.0 | 78.0 | 42 | 87.0 | 6,506 | 30.8 |
| Tyler, TX | 272 | -10.1 | 93 | 1.9 | 64.3 | 487,136 | 5,248 | 52,759 | 54 | 18,308 | 68.6 | 31.4 | 0 | 71.8 | 94 | 1.2 |
| Urban Honolulu, HI | 72 | 3.8 | 77 | 11.7 | 23.1 | 1,920,260 | 24,794 | 92,037 | 151 | 163,305 | 90.4 | 9.6 | 25 | 71.1 | 350 | 10.8 |
| Utica-Rome, NY | 311 | -10.1 | 199 | 0.9 | 188.7 | 500,447 | 2,519 | 115,581 | 158 | 101,364 | 27.4 | 72.6 | 43 | 73.4 | 1,533 | 19.1 |
| Valdosta, GA | 310 | 15.6 | 341 | 39.8 | 150.6 | 1,283,208 | 3,764 | 154,355 | 195 | 214,726 | 73.2 | 26.8 | 3 | 73.3 | 5,651 | 41.6 |
| Vallejo, CA | 343 | -15.8 | 404 | 110.4 | 152.1 | 3,691,320 | 9,148 | 178,302 | 297 | 349,337 | 83.8 | 16.2 | 63 | 85.2 | 2,554 | 13.9 |
| Victoria, TX | 806 | -13.6 | 317 | 8.2 | 119.4 | 887,059 | 2,796 | 65,326 | 76 | 29,923 | 51.0 | 49.0 | 0 | 70.1 | 3,621 | 8.9 |
| Vineland-Bridgeton, NJ | 66 | 2.7 | 118 | 20.0 | 49.6 | 1,159,638 | 9,801 | 150,220 | 213 | 379,730 | 97.5 | 2.5 | 8 | 77.5 | 665 | 9.6 |
| Virginia Beach-Norfolk-Newport News, VA-NC | 564 | 2.3 | 307 | 7.3 | 415.8 | 1,278,266 | 4,170 | 163,276 | 388 | 210,824 | 73.4 | 26.6 | 9 | 82.6 | 19,010 | 38.2 |
| Visalia, CA | 1,250 | 0.9 | 299 | 568.2 | 721.4 | 3,501,053 | 11,726 | 218,462 | 4,475 | 1,068,739 | 49.7 | 50.3 | 104 | 78.2 | 13,824 | 10.4 |
| Waco, TX | 965 | 3.1 | 216 | 6.2 | 466.3 | 683,249 | 3,164 | 81,994 | 338 | 75,544 | 30.2 | 69.8 | 4 | 72.7 | 9,061 | 14.2 |
| Walla Walla, WA | 703 | 8.9 | 778 | 101.7 | 565.8 | 1,968,909 | 2,531 | 208,241 | 0 | 0 | 0.0 | 0.0 | 13 | 85.3 | 16,092 | 40.1 |
| Warner Robins, GA | 97 | 17.5 | 193 | 24.0 | 52.1 | 975,172 | 5,064 | 133,780 | 84 | 165,396 | 80.9 | 19.1 | 0 | 82.2 | 699 | 17.4 |
| Washington-Arlington-Alexandria, DC-VA-MD-WV | 1,254 | -3.5 | 134 | 7.2 | 645.0 | 1,047,921 | 7,849 | 82,594 | 477 | 50,780 | 58.4 | 41.6 | 78 | 82.9 | 9,865 | 11.6 |
| Frederick-Gaithersburg-Rockville, MD Div 23224 | 254 | 3.7 | 132 | 2.3 | 189.4 | 1,208,668 | 9,185 | 116,267 | 174 | 90,193 | 58.2 | 41.8 | 35 | 83.1 | 5,159 | 23.0 |
| Washington-Arlington-Alexandria, DC-VA-MD-WV Div 47894 | 1,000 | -5.2 | 134 | 4.8 | 455.6 | 1,006,318 | 7,509 | 73,879 | 303 | 40,580 | 58.5 | 41.5 | 43 | 82.9 | 4,706 | 8.6 |
| Waterloo-Cedar Falls, IA | 862 | -2.7 | 320 | 0.9 | 812.7 | 2,765,372 | 8,630 | 252,938 | 758 | 281,635 | 64.3 | 35.7 | 15 | 82.2 | 22,822 | 76.4 |
| Watertown-Fort Drum, NY | 247 | -14.9 | 312 | 0.4 | 168.8 | 836,501 | 2,677 | 170,553 | 165 | 208,404 | 21.6 | 78.4 | 31 | 74.7 | 1,986 | 22.3 |
| Wausau-Weston, WI | 551 | -0.8 | 207 | 7.3 | 365.9 | 939,424 | 4,537 | 152,474 | 451 | 169,203 | 26.1 | 73.9 | 63 | 73.3 | 1,432 | 27.0 |
| Weirton-Steubenville, WV-OH | 100 | 8.4 | 128 | 0.1 | 40.7 | 639,277 | 5,011 | 79,855 | 11 | 14,059 | 47.4 | 52.6 | 0 | 77.8 | 192 | 6.9 |
| Wenatchee, WA | 883 | -0.8 | 564 | 40.1 | 573.5 | 1,209,012 | 2,143 | 104,421 | 444 | 284,169 | 97.9 | 2.1 | 77 | 80.9 | 16,065 | 24.6 |
| Wheeling, WV-OH | 229 | -0.2 | 143 | 0.1 | 75.0 | 486,981 | 3,396 | 79,205 | 32 | 19,969 | 28.2 | 71.8 | 5 | 71.1 | 169 | 3.9 |
| Wichita, KS | 2,397 | 3.6 | 528 | 109.2 | 1,660.7 | 1,325,671 | 2,509 | 155,663 | 681 | 150,073 | 57.4 | 42.6 | 16 | 79.5 | 26,736 | 50.6 |
| Wichita Falls, TX | 1,578 | 2.4 | 790 | 10.7 | 378.0 | 1,314,140 | 1,663 | 86,950 | 162 | 81,049 | 16.3 | 83.7 | 0 | 77.4 | 8,463 | 28.4 |
| Williamsport, PA | 186 | 17.5 | 178 | 0.2 | 79.0 | 913,566 | 5,119 | 69,673 | 64 | 61,086 | 48.1 | 51.9 | 8 | 67.0 | 2,683 | 40.6 |
| Wilmington, NC | 65 | 11.4 | 165 | 1.9 | 41.4 | 948,987 | 5,735 | 100,043 | 201 | 509,147 | 19.1 | 80.9 | 9 | 79.2 | 583 | 17.7 |
| Winchester, VA-WV | 242 | -0.4 | 147 | 0.2 | 82.1 | 723,372 | 4,922 | 61,643 | 73 | 44,068 | 40.9 | 59.1 | 6 | 73.3 | 322 | 8.3 |
| Winston-Salem, NC | 385 | 1.3 | 101 | 2.1 | 181.5 | 551,408 | 5,485 | 63,812 | 267 | 69,798 | 29.0 | 71.0 | 15 | 73.6 | 1,509 | 10.4 |
| Worcester, MA-CT | 147 | -8.0 | 67 | 1.9 | 56.8 | 751,336 | 11,293 | 57,343 | 110 | 49,813 | 57.0 | 43.0 | 32 | 81.4 | 1,147 | 7.3 |
| Yakima, WA | 1,781 | 0.1 | 603 | 260.0 | 344.0 | 1,662,431 | 2,755 | 174,332 | 1,988 | 673,451 | 71.3 | 28.7 | 84 | 79.0 | 5,221 | 6.5 |
| York-Hanover, PA | 253 | -3.6 | 122 | 0.7 | 199.2 | 1,002,707 | 8,201 | 105,569 | 261 | 126,235 | 51.7 | 48.3 | 25 | 74.5 | 4,781 | 17.8 |
| Youngstown-Warren-Boardman, OH-PA | 355 | 0.7 | 119 | 1.0 | 236.8 | 555,675 | 4,667 | 104,689 | 190 | 63,938 | 50.5 | 49.5 | 39 | 74.0 | 4,148 | 22.0 |
| Yuba City, CA | 560 | -0.4 | 292 | 271.2 | 336.7 | 2,657,292 | 9,107 | 207,693 | 591 | 307,805 | 96.0 | 4.0 | 63 | 80.4 | 10,515 | 10.3 |
| Yuma, AZ | 247 | 15.2 | 542 | 181.4 | 234.3 | 5,006,515 | 9,235 | 462,680 | 0 | 0 | 0.0 | 0.0 | 34 | 84.4 | 3,789 | 14.9 |

| Area name | Water use, 2015 | | Wholesale Trade[1], 2017 | | | | Retail Trade[2], 2017 | | | | Real estate and rental and leasing,[2] 2017 | | | |
|---|---|---|---|---|---|---|---|---|---|---|---|---|---|---|
| | Public supply water withdrawn (mil gal/day) | Public supply gallons withdrawn per person per day | Number of establishments | Number of employees | Sales (mil dol) | Average payroll (mil dol) | Number of establishments | Number of employees | Sales (mil dol) | Average payroll (mil dol) | Number of establishments | Number of employees | Sales (mil dol) | Average payroll (mil dol) |
| | 133 | 134 | 135 | 136 | 137 | 138 | 139 | 140 | 141 | 142 | 143 | 144 | 145 | 146 |
| Texarkana, TX-AR.................. | 19.83 | 132.4 | 121 | 1,323 | 651.1 | 61.1 | 563 | 7,969 | 2,287.6 | 204.8 | 110 | 519 | 95.7 | 19.5 |
| The Villages, FL..................... | 24.13 | 203.0 | 49 | 450 | 267.0 | 22.4 | 241 | 3,726 | 1,089.7 | 88.2 | 137 | 652 | 90.5 | 18.3 |
| Toledo, OH.......................... | 93.78 | 145.0 | 671 | 11,704 | 7,969.7 | 671.8 | 2,016 | 34,717 | 9,289.3 | 850.5 | 595 | 3,721 | 4522.0 | 227.9 |
| Topeka, KS.......................... | 18.66 | 79.8 | 180 | 2,231 | 2,067.4 | 127.1 | 757 | 10,726 | 2,656.9 | 249.9 | 236 | 999 | 181.3 | 33.4 |
| Trenton-Princeton, NJ ............ | 37.62 | 101.3 | 352 | D | 10,156.9 | D | 1,249 | 25,101 | 7,573.9 | 702.8 | 355 | 2,142 | 721.0 | 109.9 |
| Tucson, AZ.......................... | 175.40 | 173.7 | 666 | 6,524 | 3,723.2 | 321.7 | 2,721 | 46,749 | 12,230.9 | 1,271.8 | 1,347 | 6,396 | 1271.3 | 243.8 |
| Tulsa, OK............................ | 133.35 | 135.9 | 1,168 | 17,421 | 12,726.9 | 1,037.9 | 3,020 | 48,857 | 14,386.1 | 1,280.6 | 1,189 | 7,509 | 1359.6 | 302.8 |
| Tuscaloosa, AL...................... | 38.92 | 156.7 | 152 | 1,532 | 926.0 | 83.9 | 836 | 10,974 | 2,963.9 | 267.2 | 215 | 1,870 | 251.1 | 62.4 |
| Twin Falls, ID....................... | 19.41 | 184.5 | 163 | 1,628 | 1,333.3 | 81.5 | 443 | 6,070 | 1,830.9 | 165.2 | 157 | 370 | 75.2 | 11.0 |
| Tyler, TX............................. | 42.89 | 192.4 | 226 | 2,841 | 1,180.6 | 145.3 | 839 | 13,251 | 4,074.6 | 364.5 | 333 | 1,177 | 262.1 | 51.6 |
| Urban Honolulu, HI ................ | 168.78 | 169.0 | 1,036 | 13,217 | 9,376.4 | 689.4 | 2,856 | 48,609 | 14,266.1 | 1,437.5 | 1,288 | 8,864 | 3150.9 | 483.1 |
| Utica-Rome, NY..................... | 61.66 | 208.6 | 211 | 3,044 | 1,488.1 | 142.4 | 958 | 13,710 | 3,721.6 | 348.9 | 197 | 749 | 147.0 | 25.5 |
| Valdosta, GA ........................ | 14.47 | 101.3 | 120 | 1,216 | 1,012.0 | 56.4 | 586 | 7,496 | 2,230.2 | 177.8 | 143 | 701 | 124.2 | 23.0 |
| Vallejo, CA........................... | 52.18 | 119.7 | 253 | 4,043 | 3,119.4 | 261.9 | 1,061 | 20,104 | 6,244.8 | 601.4 | 393 | 1,603 | 503.6 | 68.8 |
| Victoria, TX......................... | 12.40 | 124.1 | D | D | D | 83.7 | 385 | 5,866 | 1,796.0 | 162.6 | 127 | 1,141 | 527.0 | 61.0 |
| Vineland-Bridgeton, NJ........... | 15.72 | 100.9 | 153 | 4,405 | 2,833.3 | 200.5 | 492 | 7,027 | 2,011.0 | 182.1 | 102 | 486 | 108.7 | 18.9 |
| Virginia Beach-Norfolk-<br>    Newport News, VA-NC .. | 169.82 | 96.4 | 1,190 | 16,618 | 14,933.4 | 1,050.5 | 5,874 | 90,028 | 22,507.3 | 2,224.1 | 2,225 | 14,425 | 3195.9 | 650.5 |
| Visalia, CA........................... | 70.33 | 152.9 | 338 | 4,898 | 8,105.4 | 270.5 | 1,098 | 16,622 | 4,599.0 | 429.3 | 330 | 1,500 | 256.8 | 49.1 |
| Waco, TX............................ | 43.70 | 166.3 | 246 | 2,789 | 1,828.0 | 142.6 | 885 | 13,322 | 3,879.3 | 352.7 | 262 | 1,708 | 379.5 | 78.3 |
| Walla Walla, WA.................... | 12.72 | 210.8 | D | D | D | 15.1 | 171 | 2,225 | 560.8 | 62.1 | 70 | 234 | 36.5 | 8.4 |
| Warner Robins, GA ................ | 26.06 | 147.4 | 43 | 320 | 175.8 | 13.7 | 587 | 8,555 | 2,432.6 | 209.0 | 148 | 514 | 127.2 | 18.2 |
| Washington-Arlington-<br>    Alexandria, DC-VA-<br>    MD-WV ....... | 568.53 | 93.0 | 3,488 | 51,603 | 63,173.9 | 4,094.5 | 16,174 | 279,083 | 82,640.2 | 8,372.9 | 7,968 | 57,264 | 21101.4 | 3809.5 |
| Frederick-Gaithersburg-<br>    Rockville, MD Div 23224 | 368.73 | 286.9 | 879 | 11,155 | 9,027.7 | 930.3 | 3,299 | 59,294 | 18,565.8 | 1,866.4 | 1,743 | 13,221 | 6231.6 | 927.0 |
| Washington-Arlington-<br>    Alexandria, DC-VA-<br>    MD-WV Div 47894 ........ | 199.80 | 41.4 | 2,609 | 40,448 | 54,146.2 | 3,164.3 | 12,875 | 219,789 | 64,074.5 | 6,506.5 | 6,225 | 44,043 | 14869.8 | 2882.5 |
| Waterloo-Cedar Falls, IA ........ | 18.33 | 107.4 | 187 | 3,034 | 2,533.2 | 161.0 | 629 | 10,706 | 2,597.8 | 264.8 | 152 | 786 | 143.2 | 26.8 |
| Watertown-Fort Drum, NY ...... | 9.29 | 79.0 | 56 | 678 | 285.3 | 33.0 | 484 | 6,485 | 1,911.8 | 171.6 | 128 | 624 | 102.3 | 24.3 |
| Wausau-Weston, WI................ | 12.36 | 75.4 | 192 | 3,024 | 1,454.3 | 152.7 | 560 | 11,941 | 3,500.3 | 290.2 | 108 | 543 | 99.5 | 19.0 |
| Weirton-Steubenville, WV-OH . | 15.44 | 128.1 | 10 | 141 | 74.6 | 5.7 | 354 | 4,884 | 1,310.0 | 115.2 | 66 | 179 | 43.2 | 5.5 |
| Wenatchee, WA..................... | 16.74 | 144.1 | 136 | 3,828 | 1,752.9 | 160.7 | 480 | 6,014 | 1,664.5 | 174.6 | 183 | 678 | 129.8 | 25.5 |
| Wheeling, WV-OH ................. | 19.98 | 138.6 | 117 | 1,727 | 4,194.9 | 78.2 | 561 | 8,470 | 2,357.8 | 199.9 | 123 | 693 | 129.8 | 24.8 |
| Wichita, KS.......................... | 73.06 | 114.7 | 678 | 9,590 | 8,643.7 | 573.4 | 2,064 | 34,237 | 9,245.6 | 876.1 | 729 | 3,463 | 802.0 | 130.5 |
| Wichita Falls, TX.................... | 11.70 | 77.6 | 165 | 1,322 | 505.0 | 60.7 | 515 | 7,255 | 2,002.9 | 180.6 | 162 | 783 | 132.2 | 26.3 |
| Williamsport, PA.................... | 8.78 | 75.7 | 114 | 1,788 | 1,083.6 | 75.3 | 451 | 7,031 | 1,896.4 | 167.4 | 109 | 731 | 142.6 | 30.4 |
| Wilmington, NC..................... | 9.02 | 32.4 | 339 | 3,331 | 1,801.0 | 183.1 | 1,239 | 17,899 | 5,344.2 | 489.6 | 563 | 2,153 | 531.0 | 93.2 |
| Winchester, VA-WV................. | 5.80 | 43.3 | 36 | 459 | 463.9 | 102.2 | 545 | 8,511 | 2,437.0 | 227.2 | 139 | 306 | 159.9 | 11.5 |
| Winston-Salem, NC ................ | 68.60 | 104.0 | 516 | 7,702 | 6,868.0 | 413.0 | 2,173 | 31,179 | 8,631.3 | 809.9 | 560 | 2,296 | 813.8 | 104.9 |
| Worcester, MA-CT.................. | 259.35 | 277.2 | 801 | 13,268 | 8,527.3 | 796.7 | 2,861 | 44,753 | 14,689.5 | 1,262.5 | 735 | 3,479 | 841.4 | 171.0 |
| Yakima, WA......................... | 30.00 | 120.6 | 220 | 4,707 | 3,634.3 | 241.2 | 730 | 10,415 | 3,136.8 | 301.2 | 263 | 1,013 | 180.5 | 33.1 |
| York-Hanover, PA................... | 34.82 | 78.6 | 340 | 7,055 | 6,755.3 | 384.1 | 1,212 | 21,792 | 6,119.8 | 538.5 | 291 | 1,983 | 410.2 | 81.0 |
| Youngstown-Warren-<br>    Boardman, OH-PA........ | 51.89 | 94.4 | 492 | 6,968 | 4,350.1 | 361.6 | 2,001 | 31,490 | 7,922.6 | 757.7 | 417 | 2,711 | 504.2 | 97.0 |
| Yuba City, CA....................... | 22.56 | 132.0 | 93 | 2,073 | 1,368.5 | 118.3 | 408 | 5,904 | 1,795.0 | 167.5 | 124 | 460 | 92.1 | 15.1 |
| Yuma, AZ............................ | 36.29 | 177.7 | 128 | 1,856 | 1,438.4 | 102.1 | 472 | 8,292 | 2,219.1 | 210.0 | 177 | 761 | 148.3 | 23.3 |

1. Merchant wholesalers, except manufacturers' sales branches and offices. 2. Employer establishments.

Table C. Metropolitan Areas

# Professional Services, Manufacturing, and Accommodation and Food Services

| Area name | Professional, scientific, and technical services, 2017 | | | | Manufacturing, 2017 | | | | Accommodation and food services, 2017 | | | |
|---|---|---|---|---|---|---|---|---|---|---|---|---|
| | Number of establishments | Number of employees | Sales (mil dol) | Average payroll (mil dol) | Number of establishments | Number of employees | Receipts (mil dol) | Annual payroll (mil dol) | Number of establishments | Number of employees | Receipts (mil dol) | Annual payroll (mil dol) |
| | 147 | 148 | 149 | 150 | 151 | 152 | 153 | 154 | 155 | 156 | 157 | 158 |
| Texarkana, TX-AR | 48 | 222 | 23.5 | 7.9 | 105 | 5,572 | 2,438.6 | 323.0 | 197 | 5,006 | 227.7 | 65.5 |
| The Villages, FL | 160 | 753 | 103.6 | 40.1 | 39 | 943 | 474.7 | 49.1 | 143 | 3,614 | 177.1 | 56.6 |
| Toledo, OH | 899 | 10,321 | 1,609.1 | 652.7 | 765 | 43,621 | 32,045.4 | 2,599.0 | 1,626 | 29,502 | 1,477.7 | 420.4 |
| Topeka, KS | 34 | 131 | 11.5 | 3.9 | 119 | D | 3,436.3 | D | 43 | 1,117 | 151.3 | 28.1 |
| Trenton-Princeton, NJ | 1,614 | 24,223 | 5,681.5 | 2,299.6 | 231 | 6,743 | 2,217.8 | 423.3 | 867 | 14,034 | 912.8 | 260.6 |
| Tucson, AZ | 2,554 | 17,645 | 2,636.6 | 1,069.3 | 617 | 23,392 | 9,958.9 | 1,965.6 | 1,852 | 44,608 | 2,877.5 | 808.1 |
| Tulsa, OK | 2,826 | 22,722 | 4,022.0 | 1,557.3 | 1,196 | 47,793 | 21,280.0 | 2,790.4 | 2,138 | 44,035 | 2,597.0 | 708.5 |
| Tuscaloosa, AL | 25 | 124 | 10.6 | 3.2 | 175 | 16,030 | 21,097.0 | 977.2 | 21 | 194 | 9.8 | 2.3 |
| Twin Falls, ID | 259 | 1,171 | 153.5 | 51.3 | 116 | 5,158 | 2,623.2 | 235.0 | 236 | 4,079 | 203.9 | 55.7 |
| Tyler, TX | 624 | 4,570 | 917.0 | 301.9 | 184 | 6,614 | 4,305.0 | 340.6 | 490 | 10,739 | 596.4 | 182.9 |
| Urban Honolulu, HI | 2,473 | 18,527 | 3,231.9 | 1,238.8 | 501 | 9,240 | 5,325.4 | 439.0 | 2,569 | 67,611 | 6,805.4 | 1,767.8 |
| Utica-Rome, NY | 477 | 4,081 | 676.7 | 231.5 | 289 | 11,866 | 3,714.7 | 625.3 | 773 | 13,624 | 1,092.3 | 282.4 |
| Valdosta, GA | 18 | 66 | 4.8 | 1.5 | 245 | 9,085 | 7,191.2 | 610.6 | 290 | 6,087 | 315.3 | 83.5 |
| Vallejo, CA | 590 | 3,726 | 627.2 | 233.5 | D | D | D | D | 775 | 13,222 | 848.6 | 237.8 |
| Victoria, TX | 13 | 25 | 2.5 | 0.7 | D | D | D | D | 243 | 3,541 | 184.5 | 49.1 |
| Vineland-Bridgeton, NJ | D | D | D | D | 158 | 8,509 | 2,612.5 | 408.6 | D | D | D | D |
| Virginia Beach-Norfolk-Newport News, VA-NC | 2,420 | 30,613 | 5,486.4 | 2,214.5 | 857 | D | 17,037.4 | D | 4,078 | 79,461 | 4,617.9 | 1,281.0 |
| Visalia, CA | D | D | D | D | 231 | 13,534 | 8,119.7 | 719.1 | 640 | 9,898 | 641.4 | 173.1 |
| Waco, TX | 400 | 2,772 | 525.7 | 165.9 | 258 | 13,907 | 7,639.8 | 836.8 | 536 | D | 629.7 | D |
| Walla Walla, WA | D | D | D | D | D | D | D | D | 127 | 2,030 | 120.1 | 36.8 |
| Warner Robins, GA | 28 | 134 | 15.5 | 5.8 | D | D | D | D | 329 | 7,252 | 347.1 | 103.8 |
| Washington-Arlington-Alexandria, DC-VA-MD-WV | 26,766 | 486,249 | 122,883.5 | 48,710.3 | 2,127 | 50,350 | 16,724.2 | 3,298.7 | 13,545 | 290,254 | 23,267.5 | 6,515.6 |
| Frederick-Gaithersburg-Rockville, MD Div 23224 | 6,618 | 88,743 | 16,077.2 | 7,187.8 | 523 | 14,072 | 5,274.6 | 988.7 | 2,425 | 44,590 | 3,130.8 | 891.9 |
| Washington-Arlington-Alexandria, DC-VA-MD-WV Div 47894 | 20,148 | 397,506 | 106,806.4 | 41,522.4 | 1,604 | 36,278 | 11,449.6 | 2,310.0 | 11,120 | 245,664 | 20,136.8 | 5,623.7 |
| Waterloo-Cedar Falls, IA | 41 | 258 | 28.0 | 8.1 | 191 | 15,029 | 6,653.4 | 750.2 | 390 | 7,436 | 394.8 | 105.4 |
| Watertown-Fort Drum, NY | D | D | D | D | 68 | 2,334 | 906.1 | 113.5 | 333 | 4,376 | 249.2 | 81.6 |
| Wausau-Weston, WI | 36 | 111 | 12.8 | 4.5 | 285 | 19,353 | 6,419.8 | 968.6 | 410 | 6,129 | 280.7 | 82.8 |
| Weirton-Steubenville, WV-OH | 49 | 341 | 80.4 | 14.0 | 72 | 5,592 | 2,952.0 | 324.1 | 235 | 3,965 | 300.7 | 65.4 |
| Wenatchee, WA | 242 | 1,188 | 145.0 | 57.8 | 126 | 1,879 | 421.2 | 85.7 | 385 | 5,280 | 346.8 | 114.2 |
| Wheeling, WV-OH | 87 | 560 | 58.2 | 22.3 | 105 | 2,255 | 987.0 | 106.5 | 327 | 6,727 | 424.8 | 105.1 |
| Wichita, KS | 1,264 | 11,911 | 1,954.9 | 728.2 | 632 | 46,542 | 22,259.5 | 3,216.9 | 1,393 | 27,549 | 1,306.2 | 385.8 |
| Wichita Falls, TX | 22 | 43 | 7.5 | 1.9 | 133 | 4,397 | 1,273.5 | 239.6 | 6 | 60 | 1.8 | 0.6 |
| Williamsport, PA | D | D | D | D | 156 | 7,422 | 3,192.1 | 376.5 | 273 | 4,720 | 253.7 | 76.5 |
| Wilmington, NC | D | D | D | D | 225 | 6,515 | 2,315.1 | 485.4 | 881 | 16,694 | 934.8 | 266.5 |
| Winchester, VA-WV | D | D | D | 53.1 | 119 | 7,290 | 4,672.7 | 414.7 | 144 | 3,320 | 160.2 | 48.4 |
| Winston-Salem, NC | 1,222 | 7,732 | 1,178.7 | 437.2 | 660 | 30,149 | 30,921.3 | 1,536.0 | 1,225 | 24,340 | 1,266.0 | 355.7 |
| Worcester, MA-CT | 1,833 | 18,742 | 4,100.8 | 1,706.7 | 1,087 | 40,300 | 14,068.7 | 2,587.4 | 2,047 | 29,986 | 1,835.8 | 533.3 |
| Yakima, WA | D | D | D | D | 245 | 9,487 | 3,536.6 | 453.2 | 445 | 6,498 | 418.3 | 118.8 |
| York-Hanover, PA | 697 | 6,496 | 884.4 | 397.1 | 517 | 29,074 | 11,501.2 | 1,589.0 | 785 | 14,306 | 706.4 | 195.4 |
| Youngstown-Warren-Boardman, OH-PA | 266 | 1,937 | 193.7 | 70.8 | 690 | 26,356 | 11,314.5 | 1,467.0 | 1,190 | 21,655 | 959.5 | 275.9 |
| Yuba City, CA | 210 | 1,113 | 166.5 | 62.4 | 92 | 1,971 | 686.0 | 102.3 | 233 | 3,409 | 225.9 | 59.7 |
| Yuma, AZ | D | D | D | D | 75 | 3,095 | 967.4 | 132.4 | 345 | 6,823 | 371.6 | 101.7 |

| Area name | Health care and social assistance, 2017 | | | | Other services, 2017 | | | | Nonemployer businesses, 2018 | | Value of residential construction authorized by building permits, 2020 | |
|---|---|---|---|---|---|---|---|---|---|---|---|---|
| | Number of establishments | Number of employees | Receipts (mil dol) | Annual payroll (mil dol) | Number of establishments | Number of employees | Receipts (mil dol) | Annual payroll (mil dol) | Number | Receipts (mil dol) | New construction ($1,000) | Number of housing units |
| | 159 | 160 | 161 | 162 | 163 | 164 | 165 | 166 | 167 | 168 | 169 | 170 |
| Texarkana, TX-AR.................. | 401 | 8,720 | 1,075.3 | 390.8 | 156 | 1,108 | 108 | 37 | 8,746 | 407.9 | 21,474 | 96 |
| The Villages, FL...................... | 202 | 4,682 | 635.1 | 217.7 | 82 | 436 | 35 | 11 | 38,998 | 1,744.1 | 292,258 | 3,698 |
| Toledo, OH............................. | 1,856 | 49,629 | 5,525.2 | 2,190.7 | 1,053 | 6,872 | 728 | 202 | 13,227 | 583.7 | 103,082 | 1,557 |
| Topeka, KS............................. | 584 | 18,286 | 2,078.8 | 874.9 | 389 | 2,534 | 342 | 98 | 27,402 | 1,492.9 | 115,256 | 544 |
| Trenton-Princeton, NJ ............ | 1,192 | 30,342 | 3,364.8 | 1,424.9 | 854 | 6,011 | 1,423 | 277 | 69,496 | 2,808.0 | 1,394,415 | 832 |
| Tucson, AZ.............................. | 2,862 | 61,632 | 7,137.9 | 2,861.4 | 1,485 | 10,873 | 1,229 | 305 | 76,491 | 3,674.9 | 964,942 | 4,958 |
| Tulsa, OK................................ | 2,658 | 62,823 | 7,948.4 | 2,887.0 | 1,503 | 9,701 | 1,834 | 352 | 15,319 | 709.1 | 251,813 | 4,924 |
| Tuscaloosa, AL........................ | 503 | 14,657 | 1,562.5 | 655.4 | 241 | 1,463 | 168 | 48 | 7,223 | 348.3 | 170,270 | 1,269 |
| Twin Falls, ID.......................... | 432 | 6,980 | 723.5 | 263.4 | 194 | 1,106 | 110 | 32 | 19,718 | 1,028.0 | 173,329 | 1,053 |
| Tyler, TX................................. | 685 | 22,449 | 2,985.8 | 1,095.4 | 374 | 2,239 | 239 | 68 | 69,723 | 3,464.2 | 486,430 | 828 |
| Urban Honolulu, HI.................. | 2,578 | 54,227 | 7,522.5 | 2,994.2 | 2,064 | 15,687 | 1,966 | 501 | 15,663 | 662.7 | 85,858 | 1,621 |
| Utica-Rome, NY ...................... | 719 | 23,182 | 2,122.9 | 974.0 | 523 | 5,660 | 350 | 119 | 9,337 | 447.6 | 238,471 | 414 |
| Valdosta, GA .......................... | 479 | 8,429 | 941.7 | 347.0 | 147 | 670 | 70 | 17 | 26,699 | 1,152.7 | 381,135 | 1,366 |
| Vallejo, CA............................. | 931 | 22,704 | 3,667.2 | 1,608.3 | 576 | 3,920 | 631 | 160 | 7,549 | 336.9 | 30,773 | 1,733 |
| Victoria, TX............................ | 306 | 6,952 | 761.9 | 320.2 | 154 | 1,148 | 171 | 47 | 6,139 | 287.3 | 46,103 | 132 |
| Vineland-Bridgeton, NJ........... | 405 | 11,257 | 1,121.9 | 531.6 | 240 | 1,246 | 121 | 32 | | | | 179 |
| Virginia Beach-Norfolk-Newport News, VA-NC .. | 4,069 | 99,251 | 13,154.4 | 4,927.8 | 2,882 | 18,665 | 2,504 | 596 | 112,650 | 4,442.7 | 1,322,622 | 6,951 |
| Visalia, CA............................. | 884 | 16,822 | 1,970.5 | 745.0 | 405 | 2,313 | 267 | 75 | 21,098 | 1,068.9 | 348,741 | 1,575 |
| Waco, TX................................ | 611 | 18,206 | 1,830.9 | 773.5 | D | D | D | 80 | 18,072 | 860.9 | 232,825 | 1,214 |
| Walla Walla, WA...................... | 155 | 4,730 | 566.2 | 245.5 | D | D | D | D | 3,386 | 156.7 | 42,295 | 154 |
| Warner Robins, GA ................. | 382 | 7,960 | 796.9 | 286.7 | 191 | 1,103 | 104 | 29 | 12,392 | 413.1 | 251,460 | 1,307 |
| Washington-Arlington-Alexandria, DC-VA-MD-WV ........................... | 16,579 | 335,348 | 44,708.1 | 17,766.5 | 13,529 | 167,501 | 42,934 | 10,506 | 585,667 | 27,749.9 | 4,716,373 | 25,166 |
| Frederick-Gaithersburg-Rockville, MD Div 23224 | 4,406 | 84,304 | 10,678.2 | 4,591.1 | 2,395 | 23,025 | 4,920 | 1,377 | 138,908 | 7,199.8 | 934,738 | 4,063 |
| Washington-Arlington-Alexandria, DC-VA-MD-WV Div 47894 ........ | 12,173 | 251,044 | 34,029.9 | 13,175.4 | 11,134 | 144,476 | 38,014 | 9,129 | 446,759 | 20,550.1 | 3,781,636 | 21,103 |
| Waterloo-Cedar Falls, IA ........ | 462 | 13,506 | 1,354.2 | 565.2 | 281 | 1,762 | 181 | 49 | 10,111 | 487.2 | 81,960 | 351 |
| Watertown-Fort Drum, NY ...... | 273 | 6,535 | 660.4 | 304.3 | 183 | 866 | 102 | 23 | 5,293 | 213.4 | 16,547 | 129 |
| Wausau-Weston, WI................ | 390 | 10,447 | 1,418.3 | 529.0 | 270 | 1,507 | 185 | 48 | 9,508 | 483.6 | 118,200 | 532 |
| Weirton-Steubenville, WV-OH . | 297 | 7,311 | 745.1 | 283.7 | 181 | 1,017 | 95 | 28 | 5,267 | 193.9 | 6,773 | 30 |
| Wenatchee, WA....................... | 300 | 7,442 | 1,008.9 | 444.0 | 225 | 821 | 95 | 25 | 6,837 | 309.6 | 200,207 | 991 |
| Wheeling, WV-OH ................... | 504 | 11,607 | 1,248.2 | 476.2 | 278 | 1,852 | 188 | 55 | 6,935 | 287.7 | 9,150 | 55 |
| Wichita, KS............................. | 1,812 | 43,963 | 4,786.3 | 1,922.0 | 960 | 5,960 | 808 | 188 | 41,266 | 1,890.5 | 458,587 | 2,024 |
| Wichita Falls, TX..................... | 432 | 11,260 | 1,162.4 | 432.5 | 255 | 1,204 | 162 | 37 | 9,614 | 479.9 | 51,376 | 238 |
| Williamsport, PA .................... | 302 | 8,805 | 1,146.5 | 415.6 | 225 | 1,591 | 143 | 37 | 6,347 | 292.5 | 12,181 | 52 |
| Wilmington, NC....................... | 936 | 20,349 | 2,432.9 | 951.6 | 586 | 3,449 | 347 | 103 | 25,972 | 1,279.7 | 832,624 | 3,857 |
| Winchester, VA-WV................. | 269 | 7,080 | 1,125.3 | 407.1 | 92 | 536 | 46 | 14 | 9,257 | 451.2 | 187,417 | 798 |
| Winston-Salem, NC................. | 1,180 | 44,180 | 5,013.7 | 1,776.7 | 946 | 4,851 | 769 | 152 | 47,760 | 2,024.4 | 892,457 | 4,216 |
| Worcester, MA-CT................... | 2,409 | 75,856 | 9,076.9 | 3,576.6 | 1,628 | 9,090 | 983 | 281 | 64,841 | 3,143.4 | 379,541 | 1,830 |
| Yakima, WA............................ | 576 | 14,031 | 1,779.2 | 692.5 | 283 | 1,580 | 168 | 44 | 9,513 | 486.4 | 123,963 | 575 |
| York-Hanover, PA.................... | 988 | 26,954 | 3,178.4 | 1,260.7 | 795 | 5,081 | 514 | 146 | 26,828 | 1,342.5 | 208,909 | 1,204 |
| Youngstown-Warren-Boardman, OH-PA........ | 1,740 | 43,420 | 4,298.0 | 1,741.2 | 908 | 5,462 | 459 | 131 | 35,454 | 1,573.7 | 71,098 | 324 |
| Yuba City, CA......................... | 323 | 6,566 | 938.1 | 354.1 | 180 | 808 | 88 | 25 | 9,993 | 611.6 | 148,835 | 586 |
| Yuma, AZ................................ | 411 | 8,367 | 1,074.7 | 365.5 | 214 | 1,124 | 103 | 31 | 9,896 | 391.3 | 208,931 | 1,290 |

# Table C. Metropolitan Areas — Government Employment and Payroll, and Local Government Finances

| Area name | Full-time equivalent employees | March payroll (dollars) | Administration, judicial, and legal | Police and corrections | Fire protection | Highways and transportation | Health and welfare | Natural resources and utilities | Education and libraries | Total (mil dol) | Inter-governmental (mil dol) | Taxes Total (mil dol) | Per capita[1] Total (dollars) | Per capita[1] Property (dollars) |
|---|---|---|---|---|---|---|---|---|---|---|---|---|---|---|
| | | | 173 | 174 | 175 | 176 | 177 | 178 | 179 | 180 | 181 | 182 | 183 | 184 |
| | 171 | 172 | | | | | | | | | | | | |
| Texarkana, TX-AR | 6,643 | 21,508,073 | 5.4 | 9.3 | 3.1 | 3.7 | 4.9 | 3.0 | 68.5 | 527 | 242 | 197 | 1,314 | 971 |
| The Villages, FL | 1,805 | 6,376,591 | 5.2 | 22.2 | 6.8 | 3.6 | 0.3 | 3.5 | 57.8 | 325 | 41 | 148 | 1,180 | 970 |
| Toledo, OH | 24,493 | 107,114,923 | 8.8 | 12.1 | 6.3 | 4.4 | 8.7 | 5.6 | 52.6 | 3,019 | 1,080 | 1,365 | 2,118 | 1,291 |
| Topeka, KS | 11,945 | 42,626,815 | 5.2 | 10.2 | 3.8 | 3.0 | 2.4 | 5.2 | 69.1 | 998 | 428 | 389 | 1,670 | 1,193 |
| Trenton-Princeton, NJ | 14,557 | 89,476,151 | 6.3 | 11.2 | 3.5 | 2.2 | 5.1 | 4.7 | 64.1 | 2,187 | 670 | 1,244 | 3,378 | 3,332 |
| Tucson, AZ | 31,752 | 127,346,777 | 11.2 | 14.7 | 7.2 | 3.8 | 2.7 | 7.1 | 50.9 | 3,447 | 1,196 | 1,540 | 1,500 | 1,074 |
| Tulsa, OK | 33,401 | 114,906,377 | 6.2 | 10.6 | 6.4 | 4.2 | 3.1 | 5.6 | 61.7 | 3,226 | 964 | 1,546 | 1,559 | 847 |
| Tuscaloosa, AL | 11,664 | 43,638,673 | 3.8 | 7.8 | 3.4 | 3.7 | 41.3 | 4.9 | 33.7 | 1,238 | 312 | 235 | 938 | 418 |
| Twin Falls, ID | 4,105 | 14,151,701 | 9.2 | 10.2 | 2.8 | 3.4 | 2.9 | 5.0 | 65.3 | 407 | 195 | 124 | 1,140 | 1,068 |
| Tyler, TX | 9,437 | 33,803,613 | 7.5 | 10.5 | 4.0 | 2.3 | 8.3 | 2.9 | 63.7 | 758 | 230 | 392 | 1,726 | 1,338 |
| Urban Honolulu, HI | 10,620 | 59,188,205 | 13.8 | 33.1 | 16.0 | 4.1 | 4.2 | 22.8 | 0.0 | 2,960 | 436 | 1,640 | 1,662 | 1,115 |
| Utica-Rome, NY | 13,452 | 58,809,562 | 5.8 | 7.7 | 2.6 | 5.3 | 5.3 | 4.4 | 67.7 | 1,787 | 886 | 698 | 2,389 | 1,565 |
| Valdosta, GA | 7,071 | 26,126,999 | 3.8 | 6.9 | 2.0 | 1.5 | 39.9 | 2.8 | 42.7 | 762 | 191 | 181 | 1,245 | 813 |
| Vallejo, CA | 14,181 | 85,279,873 | 8.8 | 16.3 | 3.9 | 2.8 | 11.4 | 6.1 | 48.9 | 2,297 | 1,058 | 777 | 1,751 | 1,193 |
| Victoria, TX | 6,080 | 21,925,876 | 4.5 | 8.6 | 3.6 | 1.8 | 31.4 | 2.9 | 46.5 | 551 | 111 | 213 | 2,134 | 1,688 |
| Vineland-Bridgeton, NJ | 7,181 | 38,072,311 | 3.2 | 10.2 | 0.7 | 1.1 | 5.7 | 5.9 | 71.6 | 898 | 508 | 274 | 1,812 | 1,776 |
| Virginia Beach-Norfolk-Newport News, VA-NC | 75,966 | 315,130,045 | 6.3 | 10.6 | 5.0 | 3.1 | 9.0 | 8.1 | 55.3 | 8,675 | 3,342 | 3,462 | 1,966 | 1,368 |
| Visalia, CA | 23,861 | 111,821,006 | 5.1 | 7.4 | 1.9 | 1.1 | 31.0 | 4.0 | 48.3 | 3,799 | 1,983 | 538 | 1,163 | 700 |
| Waco, TX | 11,821 | 45,018,230 | 5.7 | 11.5 | 3.0 | 1.9 | 5.9 | 7.1 | 63.9 | 1,170 | 426 | 490 | 1,822 | 1,460 |
| Walla Walla, WA | 2,255 | 10,369,781 | 7.3 | 9.6 | 5.1 | 6.8 | 5.0 | 3.7 | 60.2 | 278 | 124 | 98 | 1,622 | 997 |
| Warner Robins, GA | 6,934 | 24,839,122 | 5.3 | 9.1 | 3.4 | 1.7 | 1.6 | 4.1 | 71.8 | 607 | 234 | 248 | 1,376 | 805 |
| Washington-Arlington-Alexandria, DC-VA-MD-WV | 258,342 | 1,509,069,206 | 6.9 | 10.2 | 4.2 | 8.5 | 7.5 | 6.0 | 52.1 | 43,753 | 12,391 | 24,631 | 3,965 | 2,276 |
| Frederick-Gaithersburg-Rockville, MD Div 23224 | 48,208 | 318,667,676 | 3.8 | 7.8 | 4.6 | 3.1 | 6.6 | 3.5 | 66.2 | 8,253 | 1,640 | 5,031 | 3,877 | 1,996 |
| Washington-Arlington-Alexandria, DC-VA-MD-WV Div 47894 | 210,134 | 1,190,401,530 | 7.7 | 10.9 | 4.0 | 9.9 | 7.8 | 6.7 | 48.3 | 35,500 | 10,751 | 19,600 | 3,989 | 2,349 |
| Waterloo-Cedar Falls, IA | 7,541 | 32,145,689 | 3.9 | 7.9 | 3.1 | 4.0 | 6.0 | 7.6 | 66.3 | 894 | 364 | 295 | 1,739 | 1,465 |
| Watertown-Fort Drum, NY | 5,641 | 24,270,939 | 5.8 | 5.5 | 1.8 | 7.2 | 5.7 | 3.2 | 67.8 | 739 | 370 | 279 | 2,469 | 1,419 |
| Wausau-Weston, WI | 6,471 | 28,685,366 | 5.1 | 7.6 | 1.7 | 5.0 | 14.3 | 2.6 | 62.4 | 983 | 577 | 263 | 1,613 | 1,489 |
| Weirton-Steubenville, WV-OH | 4,386 | 14,172,355 | 7.0 | 10.7 | 1.9 | 4.2 | 8.4 | 6.2 | 59.6 | 409 | 170 | 155 | 1,314 | 930 |
| Wenatchee, WA | 5,674 | 30,515,741 | 4.2 | 5.5 | 1.9 | 5.6 | 8.1 | 28.8 | 44.1 | 700 | 303 | 218 | 1,840 | 1,109 |
| Wheeling, WV-OH | 5,775 | 19,162,949 | 8.4 | 8.1 | 3.0 | 4.5 | 7.0 | 11.4 | 55.5 | 534 | 146 | 266 | 1,885 | 1,396 |
| Wichita, KS | 26,450 | 98,085,385 | 6.4 | 10.8 | 4.3 | 3.2 | 4.7 | 5.0 | 63.6 | 2,644 | 1,205 | 868 | 1,362 | 1,003 |
| Wichita Falls, TX | 6,575 | 24,216,273 | 5.5 | 12.2 | 8.9 | 2.7 | 9.6 | 5.4 | 54.0 | 517 | 195 | 221 | 1,467 | 1,204 |
| Williamsport, PA | 3,735 | 16,093,813 | 7.9 | 7.7 | 1.8 | 4.2 | 1.8 | 6.8 | 69.3 | 505 | 219 | 183 | 1,606 | 1,169 |
| Wilmington, NC | 15,220 | 62,968,999 | 3.0 | 6.9 | 2.4 | 1.0 | 45.3 | 2.9 | 34.3 | 2,227 | 381 | 481 | 1,663 | 1,146 |
| Winchester, VA-WV | 5,700 | 19,768,233 | 6.5 | 14.6 | 4.0 | 1.8 | 1.5 | 4.5 | 65.2 | 489 | 174 | 239 | 1,731 | 1,250 |
| Winston-Salem, NC | 23,586 | 79,865,777 | 4.7 | 9.2 | 4.1 | 1.3 | 8.1 | 4.4 | 65.8 | 1,988 | 946 | 794 | 1,191 | 897 |
| Worcester, MA-CT | 28,812 | 140,488,525 | 4.6 | 9.2 | 5.9 | 3.5 | 2.6 | 3.8 | 68.6 | 3,856 | 1,574 | 1,843 | 1,956 | 1,906 |
| Yakima, WA | 9,179 | 43,173,358 | 6.3 | 9.8 | 3.4 | 2.4 | 1.2 | 5.9 | 69.1 | 1,164 | 708 | 278 | 1,114 | 636 |
| York-Hanover, PA | 11,962 | 52,520,313 | 6.6 | 13.8 | 1.8 | 2.1 | 5.9 | 3.5 | 63.6 | 1,957 | 681 | 912 | 2,047 | 1,668 |
| Youngstown-Warren-Boardman, OH-PA | 21,031 | 78,440,663 | 6.9 | 9.9 | 3.3 | 3.3 | 8.0 | 5.3 | 62.1 | 2,015 | 897 | 797 | 1,471 | 991 |
| Yuba City, CA | 6,944 | 37,149,999 | 6.6 | 8.3 | 2.0 | 1.9 | 11.3 | 3.0 | 64.0 | 1,033 | 686 | 204 | 1,179 | 919 |
| Yuma, AZ | 7,776 | 27,599,788 | 11.8 | 12.2 | 3.2 | 2.3 | 2.4 | 7.4 | 59.2 | 762 | 338 | 337 | 1,608 | 1,179 |

1. Based on the resident population estimated as of July 1 of the year shown.

| Area name | Total (mil dol) | Per capita[1] (dollars) | Education | Health and hospitals | Police protection | Public welfare | Highways | Total (mil dol) | Per capita[1] (dollars) | Federal civilian | Federal military | State and local | Number of returns | Mean adjusted gross income | Mean income tax |
|---|---|---|---|---|---|---|---|---|---|---|---|---|---|---|---|
| | 185 | 186 | 187 | 188 | 189 | 190 | 191 | 192 | 193 | 194 | 195 | 196 | 197 | 198 | 199 |
| Texarkana, TX-AR | 514.4 | 3,439 | 55.9 | 2.9 | 5.8 | 1.3 | 4.6 | 447.7 | 2,994 | 3,558 | 397 | 9,209 | 62,160 | 53,796 | 5,248 |
| The Villages, FL | 307.7 | 2,461 | 27.6 | 1.1 | 7.3 | 0.5 | 7.7 | 579.9 | 4,640 | 1,646 | 216 | 3,191 | 61,940 | 89,940 | 11,552 |
| Toledo, OH | 3,180.8 | 4,936 | 41.9 | 5.9 | 8.0 | 4.8 | 3.6 | 2,963.6 | 4,599 | 2,452 | 1,746 | 44,377 | 307,340 | 60,225 | 6,437 |
| Topeka, KS | 1,071.6 | 4,599 | 53.5 | 1.6 | 6.2 | 0.1 | 7.4 | 1,320.2 | 5,666 | 3,646 | 941 | 23,335 | 109,090 | 59,402 | 5,918 |
| Trenton-Princeton, NJ | 2,256.8 | 6,130 | 55.2 | 1.0 | 5.6 | 3.2 | 1.4 | 1,839.7 | 4,997 | 2,259 | 720 | 38,019 | 178,080 | 98,318 | 14,989 |
| Tucson, AZ | 3,167.0 | 3,086 | 39.3 | 2.2 | 12.8 | 1.8 | 6.9 | 5,952.8 | 5,800 | 13,077 | 8,268 | 69,102 | 458,940 | 62,820 | 6,738 |
| Tulsa, OK | 3,004.1 | 3,030 | 47.3 | 5.0 | 6.9 | 0.7 | 6.7 | 4,200.9 | 4,238 | 4,903 | 3,649 | 49,796 | 428,160 | 69,641 | 8,014 |
| Tuscaloosa, AL | 1,269.6 | 5,060 | 32.1 | 41.0 | 4.8 | 0.0 | 5.0 | 1,146.1 | 4,567 | 2,066 | 1,023 | 26,125 | 99,700 | 57,282 | 5,868 |
| Twin Falls, ID | 418.0 | 3,829 | 55.1 | 2.2 | 4.5 | 1.0 | 5.8 | 408.5 | 3,742 | 437 | 346 | 5,995 | 47,200 | 49,101 | 4,292 |
| Tyler, TX | 736.2 | 3,240 | 58.0 | 4.9 | 5.9 | 0.1 | 9.1 | 1,714.9 | 7,548 | 651 | 475 | 13,240 | 101,990 | 68,107 | 8,213 |
| Urban Honolulu, HI | 1,981.5 | 2,009 | 0.0 | 2.2 | 14.5 | 1.1 | 5.4 | 6,208.2 | 6,294 | 31,577 | 52,560 | 66,184 | 486,000 | 70,504 | 7,898 |
| Utica-Rome, NY | 1,774.6 | 6,074 | 54.6 | 2.2 | 2.9 | 10.4 | 5.4 | 1,252.8 | 4,288 | 2,459 | 504 | 28,269 | 133,500 | 55,247 | 5,265 |
| Valdosta, GA | 850.3 | 5,850 | 33.0 | 44.0 | 4.0 | 0.0 | 2.8 | 547.4 | 3,766 | 1,218 | 4,945 | 11,833 | 57,090 | 50,408 | 4,906 |
| Vallejo, CA | 2,261.9 | 5,100 | 38.2 | 8.0 | 9.9 | 6.9 | 4.7 | 2,036.9 | 4,593 | 3,630 | 7,342 | 22,032 | 212,820 | 71,309 | 7,587 |
| Victoria, TX | 550.0 | 5,522 | 37.5 | 32.8 | 5.6 | 0.0 | 3.4 | 534.8 | 5,369 | 213 | 200 | 6,500 | 44,680 | 63,953 | 7,394 |
| Vineland-Bridgeton, NJ | 984.1 | 6,499 | 63.7 | 2.4 | 3.2 | 1.9 | 4.6 | 340.3 | 2,247 | 621 | 280 | 11,801 | 65,950 | 50,790 | 4,546 |
| Virginia Beach-Norfolk-Newport News, VA-NC | 8,517.6 | 4,837 | 42.8 | 6.4 | 5.1 | 3.2 | 3.5 | 8,791.4 | 4,992 | 59,072 | 83,360 | 107,018 | 830,640 | 64,530 | 6,815 |
| Visalia, CA | 3,763.9 | 8,141 | 41.8 | 26.6 | 3.2 | 6.7 | 3.0 | 1,853.2 | 4,009 | 1,079 | 681 | 32,442 | 182,200 | 48,280 | 4,224 |
| Waco, TX | 1,017.8 | 3,784 | 52.9 | 6.2 | 6.8 | 0.8 | 2.1 | 1,287.1 | 4,785 | 3,191 | 632 | 16,053 | 115,950 | 56,907 | 5,839 |
| Walla Walla, WA | 265.3 | 4,384 | 46.0 | 3.5 | 5.4 | 0.1 | 9.6 | 233.0 | 3,850 | 1,334 | 146 | 4,364 | 26,460 | 63,775 | 6,547 |
| Warner Robins, GA | 569.0 | 3,158 | 58.9 | 3.6 | 6.7 | 0.0 | 3.4 | 157.1 | 872 | 16,074 | 3,855 | 11,757 | 82,260 | 56,300 | 5,015 |
| Washington-Arlington-Alexandria, DC-VA-MD-WV | 43,681.9 | 7,033 | 40.5 | 3.3 | 5.2 | 10.7 | 3.0 | 56,275.1 | 9,060 | 379,809 | 66,047 | 340,122 | 3,132,560 | 100,005 | 14,676 |
| Frederick-Gaithersburg-Rockville, MD Div 23224 | 8,887.0 | 6,850 | 42.7 | 1.8 | 3.9 | 0.9 | 3.1 | 13,431.9 | 10,353 | 51,132 | 9,917 | 55,962 | 665,600 | 109,143 | 16,878 |
| Washington-Arlington-Alexandria, DC-VA-MD-WV Div 47894 | 34,794.8 | 7,081 | 40.0 | 3.7 | 5.5 | 13.2 | 3.0 | 42,843.2 | 8,719 | 328,677 | 56,130 | 284,160 | 2,466,960 | 97,539 | 14,082 |
| Waterloo-Cedar Falls, IA | 867.4 | 5,122 | 53.7 | 6.5 | 4.7 | 1.0 | 5.9 | 655.4 | 3,870 | 627 | 615 | 13,460 | 75,700 | 64,126 | 6,414 |
| Watertown-Fort Drum, NY | 718.8 | 6,352 | 52.7 | 3.0 | 2.3 | 8.1 | 6.8 | 408.3 | 3,608 | 3,240 | 15,314 | 8,295 | 50,080 | 51,658 | 4,348 |
| Wausau-Weston, WI | 1,065.5 | 6,529 | 37.8 | 8.6 | 3.6 | 26.0 | 8.2 | 547.1 | 3,352 | 446 | 422 | 9,292 | 82,430 | 63,827 | 6,863 |
| Weirton-Steubenville, WV-OH | 710.0 | 6,014 | 69.1 | 3.6 | 3.3 | 1.8 | 2.8 | 298.3 | 2,527 | 264 | 397 | 5,781 | 54,570 | 52,807 | 4,913 |
| Wenatchee, WA | 619.1 | 5,232 | 46.3 | 12.2 | 4.6 | 0.2 | 6.5 | 1,381.8 | 11,678 | 864 | 300 | 8,722 | 58,380 | 66,432 | 7,238 |
| Wheeling, WV-OH | 517.7 | 3,665 | 46.5 | 0.8 | 5.6 | 2.8 | 5.3 | 494.0 | 3,497 | 601 | 490 | 8,577 | 64,630 | 63,271 | 7,216 |
| Wichita, KS | 2,608.4 | 4,092 | 51.4 | 3.3 | 5.7 | 0.1 | 3.6 | 5,720.1 | 8,973 | 5,210 | 5,222 | 38,020 | 293,120 | 65,915 | 7,191 |
| Wichita Falls, TX | 568.1 | 3,763 | 51.7 | 9.2 | 6.3 | 0.8 | 4.3 | 695.8 | 4,610 | 1,921 | 6,596 | 10,343 | 62,700 | 58,975 | 6,051 |
| Williamsport, PA | 510.6 | 4,479 | 56.7 | 0.0 | 2.8 | 2.6 | 0.6 | 983.6 | 8,627 | 360 | 278 | 8,491 | 54,010 | 54,649 | 5,197 |
| Wilmington, NC | 2,146.0 | 7,414 | 23.7 | 51.7 | 4.3 | 2.5 | 1.1 | 1,816.5 | 6,276 | 1,147 | 877 | 23,084 | 135,560 | 73,403 | 9,320 |
| Winchester, VA-WV | 494.1 | 3,580 | 56.3 | 0.6 | 5.2 | 3.1 | 2.3 | 509.0 | 3,688 | 2,391 | 494 | 8,087 | 66,270 | 64,850 | 6,817 |
| Winston-Salem, NC | 2,150.2 | 3,225 | 50.0 | 3.2 | 8.3 | 4.7 | 4.0 | 1,936.0 | 2,904 | 1,977 | 1,495 | 29,310 | 302,430 | 61,195 | 6,531 |
| Worcester, MA-CT | 4,150.9 | 4,406 | 62.3 | 0.6 | 4.1 | 0.3 | 4.6 | 2,690.1 | 2,855 | 3,263 | 2,133 | 58,753 | 467,870 | 74,866 | 9,046 |
| Yakima, WA | 1,133.6 | 4,536 | 60.9 | 0.9 | 4.8 | 0.2 | 2.9 | 803.9 | 3,217 | 1,283 | 727 | 17,096 | 111,590 | 55,801 | 5,369 |
| York-Hanover, PA | 1,967.7 | 4,417 | 49.6 | 3.5 | 3.5 | 10.0 | | 2,310.6 | 5,187 | 4,360 | 1,310 | 15,615 | 227,080 | 66,519 | 7,030 |
| Youngstown-Warren-Boardman, OH-PA | 2,147.6 | 3,966 | 54.3 | 3.7 | 6.6 | 4.1 | 3.3 | 1,548.0 | 2,858 | 1,955 | 1,354 | 27,296 | 260,700 | 52,023 | 5,008 |
| Yuba City, CA | 1,093.9 | 6,331 | 47.0 | 6.3 | 5.3 | 8.8 | 2.0 | 998.8 | 5,781 | 1,607 | 4,161 | 10,539 | 72,130 | 54,222 | 4,661 |
| Yuma, AZ | 662.2 | 3,161 | 48.5 | 1.4 | 7.5 | 2.1 | 4.6 | 438.5 | 2,093 | 3,689 | 4,585 | 10,755 | 87,250 | 45,251 | 3,565 |

1. Based on the resident population estimated as of July 1 of the year shown.

PART D.

# Cities of 25,000 or More

(For explanation of symbols, see page viii)

Page

| | |
|---|---|
| 897 | City Highlights and Rankings |
| 907 | City Column Headings |
| 910 | Table D |
| 910 | **AL**(Alabaster)—**AR**(Hot Springs) |
| 922 | **AR**(Jacksonville)—**CA**(Davis) |
| 934 | **CA**(Delano)—**CA**(Menifee) |
| 946 | **CA**(Menlo Park)—**CA**(San Dimas) |
| 958 | **CA**(San Francisco)—**CO**(Broomfield) |
| 970 | **CO**(Castle Rock)—**FL**(Daytona Beach) |
| 982 | **FL**(Deerfield Beach)—**FL**(Winter Springs) |
| 994 | **GA**(Albany)—**IL**(Buffalo Grove) |
| 1006 | **IL**(Burbank)—**IL**(Vernon Hills) |
| 1018 | **IL**(Waukegan)—**IA**(West Des Moines) |
| 1030 | **KS**(Dodge City)—**MA**(Barnstable Town) |
| 1042 | **MA**(Beverly)—**MI**(Novi) |
| 1054 | **MI**(Oak Park)—**MS**(Tupelo) |
| 1066 | **MO**(Ballwin)—**NJ**(Fort Lee) |
| 1078 | **NJ**(Garfield)—**NY**(White Plains) |
| 1090 | **NY**(Yonkers)—**OH**(Grove City) |
| 1102 | **OH**(Hamilton)—**OR**(Portland) |
| 1114 | **OR**(Redmond)—**TN**(Collierville) |
| 1126 | **TN**(Columbia)—**TX**(Hurst) |
| 1138 | **TX**(Irving)—**UT**(Holladay) |
| 1150 | **UT**(Kaysville)—**WA**(Olympia) |
| 1162 | **WA**(Pasco)—**WY**(Laramie) |

# Cities of
# 25,000 or More

(For explanation of symbols, see page viii)

| | Page |
|---|---|
| City Highlights and Rankings | 807 |
| City Column Headings | 807 |
| Table D | 810 |
| AL(Alabaster)—AR(Hot Springs) | 810 |
| AR(Jacksonville)—CA(Davis) | 822 |
| CA(Delano)—CA(Menifee) | 834 |
| CA(Menlo Park)—CA(San Dimas) | 846 |
| CA(San Francisco)—CO(Broomfield) | 958 |
| CO(Castle Rock)—FL(Daytona Beach) | 970 |
| FL(Deerfield Beach)—FL(Winter Springs) | 982 |
| GA(Albany)—IL(Buffalo Grove) | 994 |
| IL(Burbank)—IL(Vernon Hills) | 1006 |
| IL(Waukegan)—IA(West Des Moines) | 1018 |
| KS(Dodge City)—MA(Barnstable Town) | 1030 |
| MA(Beverly)—MI(Novi) | 1042 |
| MI(Oak Park)—MS(Tupelo) | 1054 |
| MO(Ballwin)—NJ(Fort Lee) | 1066 |
| NJ(Garfield)—NY(White Plains) | 1078 |
| NY(Yonkers)—OH(Grove City) | 1090 |
| OH(Hamilton)—OR(Portland) | 102? |
| OR(Redmond)—TN(Collierville) | 1114 |
| TN(Columbia)—TX(Hurst) | 1126 |
| TX(Irving)—UT(Holladay) | 1138 |
| UT(Kaysville)—WA(Olympia) | 1150 |
| WA(Pasco)—WY(Laramie) | 1162 |

# City Highlights and Rankings

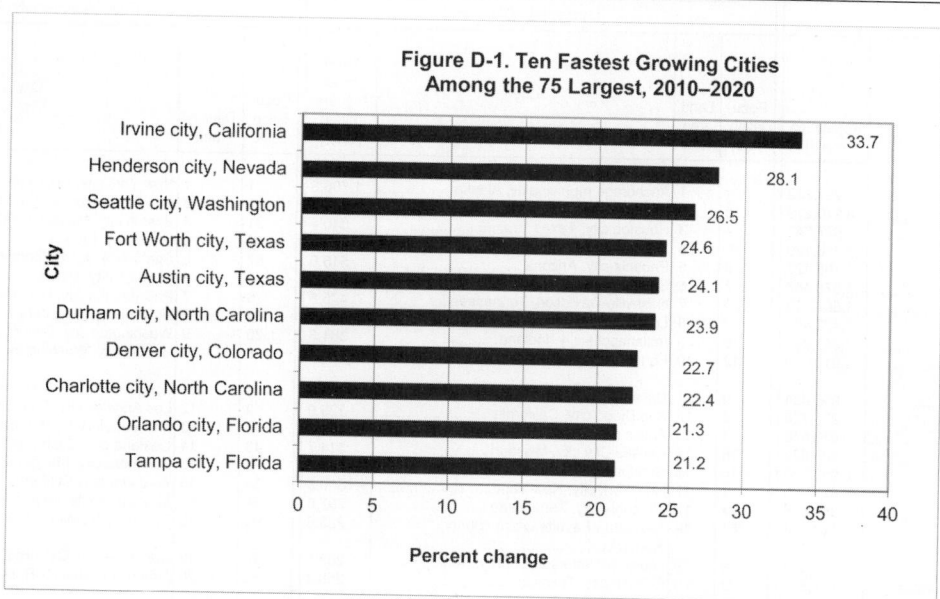

**Figure D-1. Ten Fastest Growing Cities Among the 75 Largest, 2010–2020**

| City | Percent change |
|------|------|
| Irvine city, California | 33.7 |
| Henderson city, Nevada | 28.1 |
| Seattle city, Washington | 26.5 |
| Fort Worth city, Texas | 24.6 |
| Austin city, Texas | 24.1 |
| Durham city, North Carolina | 23.9 |
| Denver city, Colorado | 22.7 |
| Charlotte city, North Carolina | 22.4 |
| Orlando city, Florida | 21.3 |
| Tampa city, Florida | 21.2 |

In 2020, 10 cities had more than 1 million residents, led by New York City with over 8.2 million people, Los Angeles with nearly 4.0 million people, and Chicago with nearly 2.7 million people. California had 15 cities among the nation's 75 most populous, as well as 3 among the top 15 (Los Angeles, San Diego, and San Jose,). Texas had 9 cities in the top 75, and had 5 among the top 15 (Houston, San Antonio, Dallas, Austin, and Fort Worth).

Among the largest cities, 52 had growth rates equaling 5 percent or higher from 2010 to 2020 and 36 had growth rates at 10 percent or higher. Many of these cities were in the south. Irvine, CA had the highest growth rate at 33.7 percent. Texas had two cities in the top ten (Austin and Fort Worth), as did Florida (Orlando and Tampa) and North Carolina (Charlotte and Durham). Forty-eight of the largest cities equaled or exceeded the U.S. growth rate of 6.7 percent.

Among the 75 largest cities, ten lost population between 2010 and 2020. Detroit and St. Louis each lost 6.8 percent of their populations while Cleveland and Baltimore lost 4.6 percent or more.

New Orleans, with a 13.3 percent increase, was among the top twenty-five cities for population growth. Though New Orleans lost more than half of its population after Hurricane Katrina in 2005, it has grown by more than two-thirds between 2006 and 2020. It was the twenty second fastest growing city from 2010 to 2020, but its 2020 population is still nearly 20 percent below its pre-Katrina total.

Among all cities of 25,000 or more, 345 cities had unemployment rates of 10 percent or more, significantly higher than in 2019 when only 10 cities had an unemployment rate of 10 percent or more. Eight of the twenty-five cities with the highest unemployment rates were located in California, seven were in Michigan, three were in New Jersey, two were in Texas, two were in Pennsylvania, while Arizona, Indiana, and Massachusetts each had one. Among the largest cities, Detroit, MI had the highest unemployment rate at 22.4 percent followed by Newark, NJ at 14.7 percent. Two cities in Nebraska had the lowest unemployment rates: Lincoln, NE at 4.3 percent and Omaha, NE at 5.2 percent. Among all cities, only 26 had unemployment rates of 4.0 percent or lower. Three of the ten cities with the lowest unemployment rates were in California, three were in Utah, three were

in Alabama, two were in Kansas, and one each were in Indiana and Idaho. No unemployment data are available for Hawaiian cities, but Honolulu county had an unemployment rate of 10.2 percent.

While the 2010 census provided updated counts of the population and basic demographic characteristics, updated information on social and economic characteristics is now obtained through the ongoing American Community Survey (ACS). For states, counties, and metropolitan areas, the official intercensal estimates are provided by age, sex, race, and Hispanic origin through the Population Estimates Program. Estimates of these demographic characteristics are not included for cities through the Population Estimates Program, but are available through the ACS. This book includes ACS 1-year estimates for 2019, including data on education, income, housing characteristics, and more.

Among the largest cities, San Francisco, San Jose, and Irvine were the top three for both median income and median housing value. Seattle, Washington DC, and San Diego also ranked in the top ten for both median income and housing value. Plano, TX ranked fifth for median household income, but its median housing value ranked only 22nd among the large cities. Los Angeles and New York ranked 7th and 8th for median housing value, but their median household incomes ranked 27th and 22nd among large cities. San Francisco tops the ranking with an estimated median housing value of $1,217,458. Detroit and Cleveland had the lowest median housing values among the large cities in 2019, at $58,902 and $71,108 respectively. They also had the lowest median incomes among the largest cities.

Among the 75 largest cities in 2019, there were four cities where 10 percent or more of residents had moved from another state or county within the previous year. This may result from population growth: Irvine, CA had one of the highest growth rates in recent years as well as one of the highest proportions of people who had moved there in the past year. Irvine has a strong technological economy as well as a large university. Other cities have large shifts because of student or military population groups: East Lansing, MI—site of Michigan State University—lost 1.9 percent of its population from 2010 to 2020 but 27.9 percent of its residents had moved there within the past year, reflecting the 62.2 percent of residents who were between the ages of 18 and 24.

# 75 Largest Cities by 2020 Population
## Selected rankings

| | Population, 2020 | | | Land area, 2020 | | | | | Population density, 2020 | |
|---|---|---|---|---|---|---|---|---|---|---|
| Popu-lation rank | City | Population [col 2] | Popu-lation rank | Land area rank | City | Land area (square miles) [col 1] | Popu-lation rank | Density rank | City | Density (per square kilo-meter) [col 4] |
| 1 | New York city, New York | 8,253,213 | 71 | 1 | Anchorage municipality, Alaska | 1,706.8 | 1 | 1 | New York city, New York | 27,464.9 |
| 2 | Los Angeles city, California | 3,970,219 | 13 | 2 | Jacksonville city, Florida | 747.3 | 17 | 2 | San Francisco city, California | 18,477.7 |
| 3 | Chicago city, Illinois | 2,677,643 | 4 | 3 | Houston city, Texas | 640.4 | 21 | 3 | Boston city, Massachusetts | 14,317.4 |
| 4 | Houston city, Texas | 2,316,120 | 26 | 4 | Oklahoma City city, Oklahoma | 606.2 | 42 | 4 | Miami city, Florida | 13,097.9 |
| 5 | Phoenix city, Arizona | 1,708,127 | 5 | 5 | Phoenix city, Arizona | 518.0 | 57 | 5 | Santa Ana city, California | 12,135.6 |
| 6 | Philadelphia city, Pennsylvania | 1,578,487 | 7 | 6 | San Antonio city, Texas | 498.8 | 3 | 6 | Chicago city, Illinois | 11,759.5 |
| 7 | San Antonio city, Texas | 1,567,118 | 23 | 7 | Nashville-Davidson, Tennessee | 475.8 | 6 | 7 | Philadelphia city, Pennsylvania | 11,744.7 |
| 8 | San Diego city, California | 1,422,420 | 2 | 8 | Los Angeles city, California | 469.5 | 74 | 8 | Newark city, New Jersey | 11,722.8 |
| 9 | Dallas city, Texas | 1,343,266 | 16 | 9 | Indianapolis city, Indiana | 361.6 | 20 | 9 | Washington city, District of Columbia | 11,666.4 |
| 10 | San Jose city, California | 1,013,616 | 12 | 10 | Fort Worth city, Texas | 347.3 | 18 | 10 | Seattle city, Washington | 9,185.1 |
| 11 | Austin city, Texas | 995,484 | 9 | 11 | Dallas city, Texas | 339.6 | 43 | 11 | Long Beach city, California | 8,968.1 |
| 12 | Fort Worth city, Texas | 927,720 | 8 | 12 | San Diego city, California | 325.9 | 2 | 12 | Los Angeles city, California | 8,456.3 |
| 13 | Jacksonville city, Florida | 920,570 | 11 | 13 | Austin city, Texas | 319.9 | 45 | 13 | Minneapolis city, Minnesota | 8,020.6 |
| 14 | Columbus city, Ohio | 903,852 | 38 | 14 | Kansas City city, Missouri | 314.7 | 46 | 14 | Oakland city, California | 7,600.9 |
| 15 | Charlotte city, North Carolina | 900,350 | 15 | 15 | Charlotte city, North Carolina | 308.3 | 31 | 15 | Baltimore city, Maryland | 7,245.1 |
| 16 | Indianapolis city, Indiana | 877,903 | 1 | 16 | New York city, New York | 300.5 | 55 | 16 | Anaheim city, California | 7,031.3 |
| 17 | San Francisco city, California | 866,606 | 28 | 17 | Memphis city, Tennessee | 297.0 | 30 | 17 | Milwaukee city, Wisconsin | 6,123.4 |
| 18 | Seattle city, Washington | 769,714 | 61 | 18 | Lexington-Fayette urban county, Kentucky | 283.6 | 63 | 18 | St. Paul city, Minnesota | 5,898.4 |
| 19 | Denver city, Colorado | 735,538 | 29 | 19 | Louisville/Jefferson County, Kentuck | 263.5 | 10 | 19 | San Jose city, California | 5,684.9 |
| 20 | Washington city, District of Columbia | 712,816 | 22 | 20 | El Paso city, Texas | 258.4 | 56 | 20 | Urban Honolulu CDP, Hawaii | 5,645.5 |
| 21 | Boston city, Massachusetts | 691,531 | 44 | 21 | Virginia Beach city, Virginia | 244.7 | 75 | 21 | Chula Vista city, California | 5,503.6 |
| 22 | El Paso city, Texas | 681,534 | 33 | 22 | Tucson city, Arizona | 241.0 | 65 | 22 | Pittsburgh city, Pennsylvania | 5,401.2 |
| 23 | Nashville-Davidson, Tennessee | 671,295 | 3 | 23 | Chicago city, Illinois | 227.7 | 36 | 23 | Sacramento city, California | 5,201.2 |
| 24 | Detroit city, Michigan | 665,369 | 14 | 24 | Columbus city, Ohio | 220.0 | 62 | 24 | Stockton city, California | 5,027.6 |
| 25 | Las Vegas city, Nevada | 662,368 | 48 | 25 | Tulsa city, Oklahoma | 197.5 | 27 | 25 | Portland city, Oregon | 4,919.5 |
| 26 | Oklahoma City city, Oklahoma | 662,314 | 39 | 26 | Colorado Springs city, Colorado | 195.4 | 54 | 26 | Cleveland city, Ohio | 4,872.4 |
| 27 | Portland city, Oregon | 656,751 | 32 | 27 | Albuquerque city, New Mexico | 187.3 | 67 | 27 | St. Louis city, Missouri | 4,824.1 |
| 28 | Memphis city, Tennessee | 649,705 | 10 | 28 | San Jose city, California | 178.3 | 19 | 28 | Denver city, Colorado | 4,804.3 |
| 29 | Louisville/Jefferson County, Kentucky | 618,338 | 51 | 29 | New Orleans city, Louisiana | 169.5 | 24 | 29 | Detroit city, Michigan | 4,797.2 |
| 30 | Milwaukee city, Wisconsin | 589,067 | 60 | 30 | Corpus Christi city, Texas | 162.2 | 25 | 30 | Las Vegas city, Nevada | 4,671.1 |
| 31 | Baltimore city, Maryland | 586,131 | 50 | 31 | Wichita city, Kansas | 162.0 | 34 | 31 | Fresno city, California | 4,603.0 |
| 32 | Albuquerque city, New Mexico | 562,540 | 52 | 32 | Aurora city, Colorado | 160.1 | 8 | 32 | San Diego city, California | 4,364.6 |
| 33 | Tucson city, Arizona | 553,571 | 19 | 33 | Denver city, Colorado | 153.1 | 73 | 33 | Irvine city, California | 4,324.7 |
| 34 | Fresno city, California | 530,267 | 53 | 34 | Bakersfield city, California | 149.8 | 49 | 34 | Arlington city, Texas | 4,159.2 |
| 35 | Mesa city, Arizona | 528,159 | 41 | 35 | Raleigh city, North Carolina | 147.1 | 14 | 35 | Columbus city, Ohio | 4,108.4 |
| 36 | Sacramento city, California | 512,838 | 25 | 36 | Las Vegas city, Nevada | 141.8 | 58 | 36 | Riverside city, California | 4,073.7 |
| 37 | Atlanta city, Georgia | 512,550 | 40 | 37 | Omaha city, Nebraska | 141.6 | 68 | 37 | Plano city, Texas | 4,062.7 |
| 38 | Kansas City city, Missouri | 497,159 | 24 | 38 | Detroit city, Michigan | 138.7 | 9 | 38 | Dallas city, Texas | 3,955.4 |
| 39 | Colorado Springs city, Colorado | 482,131 | 35 | 38 | Mesa city, Arizona | 138.7 | 64 | 39 | Cincinnati city, Ohio | 3,914.5 |
| 40 | Omaha city, Nebraska | 478,393 | 37 | 40 | Atlanta city, Georgia | 135.3 | 35 | 40 | Mesa city, Arizona | 3,807.9 |
| 41 | Raleigh city, North Carolina | 474,414 | 6 | 41 | Philadelphia city, Pennsylvania | 134.4 | 37 | 41 | Atlanta city, Georgia | 3,788.2 |
| 42 | Miami city, Florida | 471,525 | 27 | 42 | Portland city, Oregon | 133.5 | 4 | 42 | Houston city, Texas | 3,616.7 |
| 43 | Long Beach city, California | 454,681 | 66 | 43 | Greensboro city, North Carolina | 129.6 | 47 | 43 | Tampa city, Florida | 3,575.4 |
| 44 | Virginia Beach city, Virginia | 451,231 | 34 | 44 | Fresno city, California | 115.2 | 40 | 44 | Omaha city, Nebraska | 3,378.5 |
| 45 | Minneapolis city, Minnesota | 433,111 | 47 | 45 | Tampa city, Florida | 114.0 | 5 | 45 | Phoenix city, Arizona | 3,297.5 |
| 46 | Oakland city, California | 424,891 | 72 | 46 | Durham city, North Carolina | 112.8 | 41 | 46 | Raleigh city, North Carolina | 3,225.1 |
| 47 | Tampa city, Florida | 407,599 | 70 | 47 | Orlando city, Florida | 110.6 | 7 | 47 | San Antonio city, Texas | 3,141.8 |
| 48 | Tulsa city, Oklahoma | 403,166 | 59 | 48 | Henderson city, Nevada | 106.2 | 11 | 48 | Austin city, Texas | 3,111.9 |
| 49 | Arlington city, Texas | 398,864 | 36 | 49 | Sacramento city, California | 98.6 | 59 | 49 | Henderson city, Nevada | 3,099.5 |
| 50 | Wichita city, Kansas | 391,731 | 69 | 50 | Lincoln city, Nebraska | 97.7 | 32 | 50 | Albuquerque city, New Mexico | 3,003.4 |
| 51 | New Orleans city, Louisiana | 389,476 | 30 | 51 | Milwaukee city, Wisconsin | 96.2 | 69 | 51 | Lincoln city, Nebraska | 2,973.4 |
| 52 | Aurora city, Colorado | 387,377 | 49 | 52 | Arlington city, Texas | 95.9 | 15 | 52 | Charlotte city, North Carolina | 2,920.4 |
| 53 | Bakersfield city, California | 385,725 | 18 | 53 | Seattle city, Washington | 83.8 | 12 | 53 | Fort Worth city, Texas | 2,671.2 |
| 54 | Cleveland city, Ohio | 378,589 | 58 | 54 | Riverside city, California | 81.2 | 22 | 54 | El Paso city, Texas | 2,637.5 |
| 55 | Anaheim city, California | 353,676 | 31 | 55 | Baltimore city, Maryland | 80.9 | 70 | 55 | Orlando city, Florida | 2,617.2 |
| 56 | Urban Honolulu CDP, Hawaii | 341,555 | 64 | 56 | Cincinnati city, Ohio | 77.8 | 53 | 56 | Bakersfield city, California | 2,574.9 |
| 57 | Santa Ana city, California | 331,301 | 54 | 57 | Cleveland city, Ohio | 77.7 | 72 | 57 | Durham city, North Carolina | 2,534.5 |
| 58 | Riverside city, California | 330,786 | 68 | 58 | Plano city, Texas | 71.7 | 39 | 58 | Colorado Springs city, Colorado | 2,467.4 |
| 59 | Henderson city, Nevada | 329,172 | 73 | 59 | Irvine city, California | 65.6 | 16 | 59 | Indianapolis city, Indiana | 2,427.8 |
| 60 | Corpus Christi city, Texas | 327,248 | 62 | 60 | Stockton city, California | 62.2 | 52 | 60 | Aurora city, Colorado | 2,419.6 |
| 61 | Lexington-Fayette urban county, Kentucky | 324,735 | 67 | 61 | St. Louis city, Missouri | 61.7 | 50 | 61 | Wichita city, Kansas | 2,418.1 |
| 62 | Stockton city, California | 312,716 | 20 | 62 | Washington city, District of Columbia | 61.1 | 29 | 62 | Louisville/Jefferson County, Kentuck | 2,346.6 |
| 63 | St. Paul city, Minnesota | 306,717 | 56 | 63 | Urban Honolulu CDP, Hawaii | 60.5 | 66 | 63 | Greensboro city, North Carolina | 2,298.4 |
| 64 | Cincinnati city, Ohio | 304,548 | 46 | 64 | Oakland city, California | 55.9 | 51 | 64 | New Orleans city, Louisiana | 2,297.8 |
| 65 | Pittsburgh city, Pennsylvania | 299,226 | 65 | 65 | Pittsburgh city, Pennsylvania | 55.4 | 33 | 65 | Tucson city, Arizona | 2,297.0 |
| 66 | Greensboro city, North Carolina | 297,878 | 45 | 66 | Minneapolis city, Minnesota | 54.0 | 28 | 66 | Memphis city, Tennessee | 2,187.6 |
| 67 | St. Louis city, Missouri | 297,645 | 63 | 67 | St. Paul city, Minnesota | 52.0 | 48 | 67 | Tulsa city, Oklahoma | 2,041.3 |
| 68 | Plano city, Texas | 291,296 | 43 | 68 | Long Beach city, California | 50.7 | 60 | 68 | Corpus Christi city, Texas | 2,017.6 |
| 69 | Lincoln city, Nebraska | 290,505 | 55 | 69 | Anaheim city, California | 50.3 | 44 | 69 | Virginia Beach city, Virginia | 1,844.0 |
| 70 | Orlando city, Florida | 289,457 | 75 | 70 | Chula Vista city, California | 49.6 | 38 | 70 | Kansas City city, Missouri | 1,579.8 |
| 71 | Anchorage municipality, Alaska | 287,095 | 21 | 71 | Boston city, Massachusetts | 48.3 | 23 | 71 | Nashville-Davidson, Tennessee | 1,410.9 |
| 72 | Durham city, North Carolina | 285,897 | 17 | 72 | San Francisco city, California | 46.9 | 13 | 72 | Jacksonville city, Florida | 1,231.9 |
| 73 | Irvine city, California | 283,700 | 42 | 73 | Miami city, Florida | 36.0 | 61 | 73 | Lexington-Fayette urban county, Kentucky | 1,145.0 |
| 74 | Newark city, New Jersey | 282,520 | 57 | 74 | Santa Ana city, California | 27.3 | 26 | 74 | Oklahoma City city, Oklahoma | 1,092.6 |
| 75 | Chula Vista city, California | 272,979 | 74 | 75 | Newark city, New Jersey | 24.1 | 71 | 75 | Anchorage municipality, Alaska | 168.2 |

# 75 Largest Cities by 2020 Population
## Selected rankings

| Percent population change, 2010–2020 | | | | Percent White alone, 2019 | | | | Percent Black alone, 2019 | | | |
|---|---|---|---|---|---|---|---|---|---|---|---|
| Population rank | Percent change rank | City | Percent change [col 26] | Population rank | White rank | City | Percent White [col 5] | Population rank | Black rank | City | Percent Black [col 6] |
| 73 | 1 | Irvine city, California | 33.7 | 60 | 1 | Corpus Christi city, Texas | 89.6 | 24 | 1 | Detroit city, Michigan | 77.9 |
| 59 | 2 | Henderson city, Nevada | 28.1 | 69 | 2 | Lincoln city, Nebraska | 84.2 | 28 | 2 | Memphis city, Tennessee | 63.0 |
| 18 | 3 | Seattle city, Washington | 26.5 | 35 | 3 | Mesa city, Arizona | 81.4 | 31 | 3 | Baltimore city, Maryland | 62.6 |
| 12 | 4 | Fort Worth city, Texas | 24.6 | 22 | 4 | El Paso city, Texas | 78.8 | 51 | 4 | New Orleans city, Louisiana | 58.8 |
| 11 | 5 | Austin city, Texas | 24.1 | 39 | 5 | Colorado Springs city, Colorado | 78.5 | 74 | 5 | Newark city, New Jersey | 49.9 |
| 72 | 6 | Durham city, North Carolina | 23.9 | 7 | 6 | San Antonio city, Texas | 78.3 | 54 | 6 | Cleveland city, Ohio | 49.8 |
| 19 | 7 | Denver city, Colorado | 22.7 | 27 | 7 | Portland city, Oregon | 77.3 | 37 | 7 | Atlanta city, Georgia | 49.1 |
| 15 | 8 | Charlotte city, North Carolina | 22.4 | 19 | 8 | Denver city, Colorado | 77.2 | 20 | 8 | Washington city, District of Columbia | 45.4 |
| 70 | 9 | Orlando city, Florida | 21.3 | 42 | 9 | Miami city, Florida | 77.1 | 67 | 9 | St. Louis city, Missouri | 45.2 |
| 47 | 10 | Tampa city, Florida | 21.2 | 40 | 10 | Omaha city, Nebraska | 76.9 | 66 | 10 | Greensboro city, North Carolina | 42.9 |
| 35 | 11 | Mesa city, Arizona | 20.1 | 5 | 11 | Phoenix city, Arizona | 76.7 | 6 | 11 | Philadelphia city, Pennsylvania | 41.5 |
| 37 | 12 | Atlanta city, Georgia | 20.0 | 50 | 12 | Wichita city, Kansas | 74.5 | 64 | 12 | Cincinnati city, Ohio | 41.2 |
| 52 | 13 | Aurora city, Colorado | 19.2 | 32 | 13 | Albuquerque city, New Mexico | 74.3 | 30 | 13 | Milwaukee city, Wisconsin | 38.4 |
| 20 | 14 | Washington city, District of Columbia | 18.5 | 61 | 14 | Lexington-Fayette urban county, Kentucky | 73.5 | 72 | 14 | Durham city, North Carolina | 37.7 |
| 5 | 15 | Phoenix city, Arizona | 18.1 | 11 | 15 | Austin city, Texas | 73.0 | 15 | 15 | Charlotte city, North Carolina | 35.4 |
| 7 | 15 | San Antonio city, Texas | 18.1 | 59 | 16 | Henderson city, Nevada | 72.7 | 13 | 16 | Jacksonville city, Florida | 31.2 |
| 42 | 17 | Miami city, Florida | 18.0 | 33 | 17 | Tucson city, Arizona | 71.9 | 14 | 17 | Columbus city, Ohio | 29.2 |
| 41 | 18 | Raleigh city, North Carolina | 17.4 | 29 | 18 | Louisville/Jefferson County, Kentucky | 70.0 | 3 | 18 | Chicago city, Illinois | 29.0 |
| 39 | 19 | Colorado Springs city, Colorado | 15.5 | 26 | 19 | Oklahoma City city, Oklahoma | 69.0 | 41 | 18 | Raleigh city, North Carolina | 29.0 |
| 14 | 20 | Columbus city, Ohio | 14.6 | 53 | 20 | Bakersfield city, California | 67.5 | 16 | 20 | Indianapolis city, Indiana | 28.8 |
| 26 | 21 | Oklahoma City city, Oklahoma | 14.1 | 70 | 21 | Orlando city, Florida | 67.1 | 23 | 21 | Nashville-Davidson, Tennessee | 27.1 |
| 25 | 22 | Las Vegas city, Nevada | 13.3 | 48 | 22 | Tulsa city, Oklahoma | 66.5 | 38 | 22 | Kansas City city, Missouri | 26.1 |
| 51 | 22 | New Orleans city, Louisiana | 13.3 | 18 | 23 | Seattle city, Washington | 66.3 | 21 | 23 | Boston city, Massachusetts | 24.9 |
| 45 | 24 | Minneapolis city, Minnesota | 13.2 | 65 | 24 | Pittsburgh city, Pennsylvania | 66.2 | 46 | 23 | Oakland city, California | 24.9 |
| 27 | 25 | Portland city, Oregon | 12.5 | 44 | 25 | Virginia Beach city, Virginia | 65.6 | 1 | 25 | New York city, New York | 24.7 |
| 69 | 26 | Lincoln city, Nebraska | 12.3 | 8 | 26 | San Diego city, California | 65.4 | 9 | 26 | Dallas city, Texas | 24.6 |
| 9 | 27 | Dallas city, Texas | 12.2 | 55 | 27 | Anaheim city, California | 65.3 | 29 | 27 | Louisville/Jefferson County, Kentucky | 24.0 |
| 68 | 28 | Plano city, Texas | 12.1 | 12 | 28 | Fort Worth city, Texas | 64.5 | 49 | 28 | Arlington city, Texas | 23.9 |
| 13 | 29 | Jacksonville city, Florida | 12.0 | 45 | 29 | Minneapolis city, Minnesota | 64.3 | 65 | 29 | Pittsburgh city, Pennsylvania | 23.7 |
| 21 | 30 | Boston city, Massachusetts | 11.9 | 68 | 30 | Plano city, Texas | 64.1 | 47 | 30 | Tampa city, Florida | 23.6 |
| 75 | 30 | Chula Vista city, California | 11.9 | 47 | 31 | Tampa city, Florida | 63.7 | 4 | 31 | Houston city, Texas | 23.1 |
| 23 | 32 | Nashville-Davidson, Tennessee | 11.2 | 23 | 32 | Nashville-Davidson, Tennessee | 62.8 | 70 | 32 | Orlando city, Florida | 21.5 |
| 53 | 33 | Bakersfield city, California | 10.9 | 38 | 32 | Kansas City city, Missouri | 62.8 | 45 | 33 | Minneapolis city, Minnesota | 19.3 |
| 66 | 34 | Greensboro city, North Carolina | 10.8 | 75 | 34 | Chula Vista city, California | 62.7 | 44 | 34 | Virginia Beach city, Virginia | 18.9 |
| 4 | 35 | Houston city, Texas | 10.7 | 71 | 35 | Anchorage municipality, Alaska | 62.6 | 12 | 35 | Fort Worth city, Texas | 18.8 |
| 36 | 36 | Sacramento city, California | 10.0 | 9 | 36 | Dallas city, Texas | 62.2 | 52 | 36 | Aurora city, Colorado | 17.1 |
| 61 | 37 | Lexington-Fayette urban county, Kentucky | 9.8 | 25 | 37 | Las Vegas city, Nevada | 61.8 | 63 | 37 | St. Paul city, Minnesota | 16.5 |
| 8 | 38 | San Diego city, California | 9.3 | 52 | 38 | Aurora city, Colorado | 60.6 | 42 | 38 | Miami city, Florida | 15.7 |
| 49 | 39 | Arlington city, Texas | 9.2 | 16 | 39 | Indianapolis city, Indiana | 60.5 | 61 | 39 | Lexington-Fayette urban county, Kentucky | 14.8 |
| 58 | 40 | Riverside city, California | 8.8 | 34 | 40 | Fresno city, California | 60.4 | 48 | 40 | Tulsa city, Oklahoma | 14.5 |
| 46 | 41 | Oakland city, California | 8.7 | 58 | 41 | Riverside city, California | 58.2 | 26 | 41 | Oklahoma City city, Oklahoma | 13.8 |
| 38 | 42 | Kansas City city, Missouri | 8.1 | 41 | 42 | Raleigh city, North Carolina | 58.0 | 40 | 42 | Omaha city, Nebraska | 12.9 |
| 17 | 43 | San Francisco city, California | 7.6 | 14 | 43 | Columbus city, Ohio | 57.9 | 25 | 43 | Las Vegas city, Nevada | 12.0 |
| 63 | 43 | St. Paul city, Minnesota | 7.6 | 63 | 44 | St. Paul city, Minnesota | 57.5 | 36 | 44 | Sacramento city, California | 11.7 |
| 60 | 45 | Corpus Christi city, Texas | 7.2 | 13 | 45 | Jacksonville city, Florida | 56.8 | 43 | 45 | Long Beach city, California | 11.3 |
| 16 | 46 | Indianapolis city, Indiana | 7.0 | 4 | 46 | Houston city, Texas | 54.5 | 50 | 46 | Wichita city, Kansas | 10.5 |
| 62 | 46 | Stockton city, California | 7.0 | 21 | 47 | Boston city, Massachusetts | 53.2 | 62 | 47 | Stockton city, California | 10.1 |
| 34 | 48 | Fresno city, California | 6.7 | 49 | 48 | Arlington city, Texas | 52.5 | 68 | 48 | Plano city, Texas | 9.6 |
| 10 | 49 | San Jose city, California | 6.4 | 43 | 49 | Long Beach city, California | 52.3 | 2 | 49 | Los Angeles city, California | 8.7 |
| 22 | 50 | El Paso city, Texas | 5.2 | 64 | 49 | Cincinnati city, Ohio | 52.3 | 19 | 50 | Denver city, Colorado | 8.6 |
| 55 | 50 | Anaheim city, California | 5.2 | 2 | 51 | Los Angeles city, California | 52.1 | 11 | 51 | Austin city, Texas | 7.8 |
| 33 | 52 | Tucson city, Arizona | 5.1 | 3 | 52 | Chicago city, Illinois | 50.8 | 18 | 52 | Seattle city, Washington | 7.4 |
| 2 | 53 | Los Angeles city, California | 4.7 | 72 | 53 | Durham city, North Carolina | 50.4 | 5 | 53 | Phoenix city, Arizona | 7.3 |
| 40 | 54 | Omaha city, Nebraska | 4.2 | 73 | 54 | Irvine city, California | 47.8 | 7 | 53 | San Antonio city, Texas | 7.3 |
| 29 | 55 | Louisville/Jefferson County, Kentucky | 3.8 | 15 | 55 | Charlotte city, North Carolina | 46.9 | 34 | 55 | Fresno city, California | 7.1 |
| 6 | 56 | Philadelphia city, Pennsylvania | 3.4 | 67 | 56 | St. Louis city, Missouri | 46.8 | 53 | 56 | Bakersfield city, California | 6.9 |
| 32 | 57 | Albuquerque city, New Mexico | 3.0 | 17 | 57 | San Francisco city, California | 45.2 | 39 | 57 | Colorado Springs city, Colorado | 6.6 |
| 44 | 57 | Virginia Beach city, Virginia | 3.0 | 30 | 58 | Milwaukee city, Wisconsin | 44.8 | 59 | 58 | Henderson city, Nevada | 6.2 |
| 48 | 59 | Tulsa city, Oklahoma | 2.9 | 62 | 59 | Stockton city, California | 44.7 | 8 | 59 | San Diego city, California | 6.1 |
| 64 | 60 | Cincinnati city, Ohio | 2.5 | 36 | 60 | Sacramento city, California | 44.2 | 27 | 60 | Portland city, Oregon | 5.6 |
| 50 | 61 | Wichita city, Kansas | 2.4 | 66 | 61 | Greensboro city, North Carolina | 43.1 | 35 | 60 | Mesa city, Arizona | 5.6 |
| 57 | 62 | Santa Ana city, California | 2.0 | 10 | 62 | San Jose city, California | 42.5 | 17 | 62 | San Francisco city, California | 5.5 |
| 74 | 62 | Newark city, New Jersey | 2.0 | 20 | 62 | Washington city, District of Columbia | 42.5 | 58 | 63 | Riverside city, California | 5.3 |
| 56 | 64 | Urban Honolulu CDP, Hawaii | 1.1 | 1 | 64 | New York city, New York | 42.4 | 71 | 63 | Anchorage municipality, Alaska | 5.3 |
| 1 | 65 | New York city, New York | 1.0 | 37 | 65 | Atlanta city, Georgia | 42.3 | 33 | 65 | Tucson city, Arizona | 5.2 |
| 28 | 66 | Memphis city, Tennessee | -0.3 | 6 | 66 | Philadelphia city, Pennsylvania | 39.0 | 75 | 66 | Chula Vista city, California | 5.1 |
| 3 | 67 | Chicago city, Illinois | -0.7 | 54 | 67 | Cleveland city, Ohio | 38.9 | 69 | 67 | Lincoln city, Nebraska | 4.4 |
| 30 | 68 | Milwaukee city, Wisconsin | -0.9 | 46 | 68 | Oakland city, California | 34.5 | 60 | 68 | Corpus Christi city, Texas | 4.2 |
| 43 | 69 | Long Beach city, California | -1.6 | 51 | 69 | New Orleans city, Louisiana | 34.0 | 22 | 69 | El Paso city, Texas | 3.4 |
| 71 | 70 | Anchorage municipality, Alaska | -1.8 | 74 | 70 | Newark city, New Jersey | 33.4 | 55 | 69 | Anaheim city, California | 3.4 |
| 65 | 71 | Pittsburgh city, Pennsylvania | -2.0 | 28 | 71 | Memphis city, Tennessee | 30.4 | 32 | 71 | Albuquerque city, New Mexico | 3.2 |
| 54 | 72 | Cleveland city, Ohio | -4.6 | 31 | 72 | Baltimore city, Maryland | 30.3 | 10 | 72 | San Jose city, California | 3.0 |
| 31 | 73 | Baltimore city, Maryland | -5.6 | 57 | 73 | Santa Ana city, California | 28.9 | 56 | 73 | Urban Honolulu CDP, Hawaii | 2.6 |
| 24 | 74 | Detroit city, Michigan | -6.8 | 56 | 74 | Urban Honolulu CDP, Hawaii | 17.0 | 73 | 74 | Irvine city, California | 2.4 |
| 67 | 74 | St. Louis city, Missouri | -6.8 | 24 | 75 | Detroit city, Michigan | 14.7 | 57 | 75 | Santa Ana city, California | 1.0 |

# 75 Largest Cities by 2020 Population
## Selected rankings

### Percent American Indian, Alaska Native alone, 2019

| Population rank | American Indian, Alaska Native rank | City | Percent American Indian, Alaska Native [col 7] |
|---|---|---|---|
| 71 | 1 | Anchorage municipality, Alaska | 8.9 |
| 32 | 2 | Albuquerque city, New Mexico | 4.9 |
| 48 | 3 | Tulsa city, Oklahoma | 4.7 |
| 33 | 4 | Tucson city, Arizona | 3.8 |
| 26 | 5 | Oklahoma City city, Oklahoma | 3.3 |
| 35 | 6 | Mesa city, Arizona | 3.2 |
| 5 | 7 | Phoenix city, Arizona | 2.1 |
| 46 | 8 | Oakland city, California | 1.3 |
| 25 | 9 | Las Vegas city, Nevada | 1.2 |
| 45 | 9 | Minneapolis city, Minnesota | 1.2 |
| 53 | 11 | Bakersfield city, California | 1.1 |
| 34 | 12 | Fresno city, California | 1.0 |
| 39 | 12 | Colorado Springs city, Colorado | 1.0 |
| 43 | 12 | Long Beach city, California | 1.0 |
| 50 | 12 | Wichita city, Kansas | 1.0 |
| 69 | 12 | Lincoln city, Nebraska | 1.0 |
| 7 | 17 | San Antonio city, Texas | 0.9 |
| 11 | 17 | Austin city, Texas | 0.9 |
| 36 | 17 | Sacramento city, California | 0.9 |
| 58 | 17 | Riverside city, California | 0.9 |
| 2 | 21 | Los Angeles city, California | 0.8 |
| 27 | 21 | Portland city, Oregon | 0.8 |
| 30 | 21 | Milwaukee city, Wisconsin | 0.8 |
| 52 | 21 | Aurora city, Colorado | 0.8 |
| 18 | 25 | Seattle city, Washington | 0.7 |
| 22 | 25 | El Paso city, Texas | 0.7 |
| 24 | 25 | Detroit city, Michigan | 0.7 |
| 40 | 25 | Omaha city, Nebraska | 0.7 |
| 41 | 25 | Raleigh city, North Carolina | 0.7 |
| 54 | 25 | Cleveland city, Ohio | 0.7 |
| 55 | 25 | Anaheim city, California | 0.7 |
| 19 | 32 | Denver city, Colorado | 0.6 |
| 62 | 32 | Stockton city, California | 0.6 |
| 63 | 32 | St. Paul city, Minnesota | 0.6 |
| 66 | 32 | Greensboro city, North Carolina | 0.6 |
| 6 | 36 | Philadelphia city, Pennsylvania | 0.5 |
| 8 | 36 | San Diego city, California | 0.5 |
| 9 | 36 | Dallas city, Texas | 0.5 |
| 10 | 36 | San Jose city, California | 0.5 |
| 12 | 36 | Fort Worth city, Texas | 0.5 |
| 1 | 41 | New York city, New York | 0.4 |
| 3 | 41 | Chicago city, Illinois | 0.4 |
| 4 | 41 | Houston city, Texas | 0.4 |
| 13 | 41 | Jacksonville city, Florida | 0.4 |
| 15 | 41 | Charlotte city, North Carolina | 0.4 |
| 17 | 41 | San Francisco city, California | 0.4 |
| 31 | 41 | Baltimore city, Maryland | 0.4 |
| 57 | 41 | Santa Ana city, California | 0.4 |
| 59 | 41 | Henderson city, Nevada | 0.4 |
| 67 | 41 | St. Louis city, Missouri | 0.4 |
| 14 | 51 | Columbus city, Ohio | 0.3 |
| 16 | 51 | Indianapolis city, Indiana | 0.3 |
| 20 | 51 | Washington city, District of Columbia | 0.3 |
| 21 | 51 | Boston city, Massachusetts | 0.3 |
| 23 | 51 | Nashville-Davidson, Tennessee | 0.3 |
| 44 | 51 | Virginia Beach city, Virginia | 0.3 |
| 51 | 51 | New Orleans city, Louisiana | 0.3 |
| 56 | 51 | Urban Honolulu CDP, Hawaii | 0.3 |
| 65 | 51 | Pittsburgh city, Pennsylvania | 0.3 |
| 70 | 51 | Orlando city, Florida | 0.3 |
| 75 | 51 | Chula Vista city, California | 0.3 |
| 37 | 62 | Atlanta city, Georgia | 0.2 |
| 38 | 62 | Kansas City city, Missouri | 0.2 |
| 60 | 62 | Corpus Christi city, Texas | 0.2 |
| 61 | 62 | Lexington-Fayette urban county, Kentucky | 0.2 |
| 68 | 62 | Plano city, Texas | 0.2 |
| 74 | 62 | Newark city, New Jersey | 0.2 |
| 28 | 68 | Memphis city, Tennessee | 0.1 |
| 29 | 68 | Louisville/Jefferson County, Kentuck | 0.1 |
| 42 | 68 | Miami city, Florida | 0.1 |
| 47 | 68 | Tampa city, Florida | 0.1 |
| 49 | 68 | Arlington city, Texas | 0.1 |
| 64 | 68 | Cincinnati city, Ohio | 0.1 |
| 72 | 68 | Durham city, North Carolina | 0.1 |
| 73 | 68 | Irvine city, California | 0.1 |

### Percent Asian and Pacific Islander alone, 2019

| Population rank | Asian and Pacific Islander rank | City | Percent Asian and Pacific Islander [col 8] |
|---|---|---|---|
| 56 | 1 | Urban Honolulu CDP, Hawaii | 52.0 |
| 73 | 2 | Irvine city, California | 43.0 |
| 10 | 3 | San Jose city, California | 37.7 |
| 17 | 4 | San Francisco city, California | 34.9 |
| 62 | 5 | Stockton city, California | 22.8 |
| 68 | 6 | Plano city, Texas | 22.3 |
| 63 | 7 | St. Paul city, Minnesota | 19.5 |
| 36 | 8 | Sacramento city, California | 18.4 |
| 55 | 9 | Anaheim city, California | 18.2 |
| 75 | 10 | Chula Vista city, California | 18.1 |
| 8 | 11 | San Diego city, California | 17.0 |
| 18 | 12 | Seattle city, Washington | 16.6 |
| 1 | 13 | New York city, New York | 14.4 |
| 46 | 14 | Oakland city, California | 14.3 |
| 34 | 15 | Fresno city, California | 13.8 |
| 43 | 16 | Long Beach city, California | 12.0 |
| 2 | 17 | Los Angeles city, California | 11.7 |
| 57 | 18 | Santa Ana city, California | 11.0 |
| 21 | 19 | Boston city, Massachusetts | 9.7 |
| 59 | 20 | Henderson city, Nevada | 9.2 |
| 71 | 21 | Anchorage municipality, Alaska | 9.0 |
| 11 | 22 | Austin city, Texas | 8.2 |
| 27 | 22 | Portland city, Oregon | 8.2 |
| 58 | 24 | Riverside city, California | 7.8 |
| 53 | 25 | Bakersfield city, California | 7.7 |
| 6 | 26 | Philadelphia city, Pennsylvania | 7.6 |
| 44 | 27 | Virginia Beach city, Virginia | 7.2 |
| 25 | 28 | Las Vegas city, Nevada | 7.1 |
| 3 | 29 | Chicago city, Illinois | 7.0 |
| 49 | 29 | Arlington city, Texas | 7.0 |
| 4 | 31 | Houston city, Texas | 6.5 |
| 15 | 32 | Charlotte city, North Carolina | 6.2 |
| 52 | 32 | Aurora city, Colorado | 6.2 |
| 14 | 34 | Columbus city, Ohio | 5.9 |
| 66 | 35 | Greensboro city, North Carolina | 5.8 |
| 65 | 36 | Pittsburgh city, Pennsylvania | 5.7 |
| 37 | 37 | Atlanta city, Georgia | 5.3 |
| 47 | 38 | Tampa city, Florida | 5.0 |
| 50 | 38 | Wichita city, Kansas | 5.0 |
| 69 | 38 | Lincoln city, Nebraska | 5.0 |
| 72 | 38 | Durham city, North Carolina | 5.0 |
| 12 | 42 | Fort Worth city, Texas | 4.9 |
| 13 | 42 | Jacksonville city, Florida | 4.9 |
| 45 | 44 | Minneapolis city, Minnesota | 4.8 |
| 26 | 45 | Oklahoma City city, Oklahoma | 4.5 |
| 30 | 46 | Milwaukee city, Wisconsin | 4.3 |
| 61 | 47 | Lexington-Fayette urban county, Kentucky | 4.2 |
| 20 | 48 | Washington city, District of Columbia | 4.1 |
| 40 | 48 | Omaha city, Nebraska | 4.1 |
| 5 | 50 | Phoenix city, Arizona | 4.0 |
| 16 | 50 | Indianapolis city, Indiana | 4.0 |
| 41 | 50 | Raleigh city, North Carolina | 4.0 |
| 23 | 53 | Nashville-Davidson, Tennessee | 3.9 |
| 9 | 54 | Dallas city, Texas | 3.7 |
| 19 | 55 | Denver city, Colorado | 3.6 |
| 67 | 56 | St. Louis city, Missouri | 3.4 |
| 33 | 57 | Tucson city, Arizona | 3.2 |
| 70 | 57 | Orlando city, Florida | 3.2 |
| 32 | 59 | Albuquerque city, New Mexico | 3.1 |
| 48 | 59 | Tulsa city, Oklahoma | 3.1 |
| 39 | 61 | Colorado Springs city, Colorado | 3.0 |
| 7 | 62 | San Antonio city, Texas | 2.9 |
| 51 | 63 | New Orleans city, Louisiana | 2.8 |
| 64 | 63 | Cincinnati city, Ohio | 2.8 |
| 31 | 65 | Baltimore city, Maryland | 2.7 |
| 29 | 66 | Louisville/Jefferson County, Kentuck | 2.6 |
| 35 | 66 | Mesa city, Arizona | 2.6 |
| 54 | 68 | Cleveland city, Ohio | 2.5 |
| 38 | 69 | Kansas City city, Missouri | 2.4 |
| 60 | 70 | Corpus Christi city, Texas | 2.3 |
| 28 | 71 | Memphis city, Tennessee | 2.1 |
| 24 | 72 | Detroit city, Michigan | 1.7 |
| 74 | 73 | Newark city, New Jersey | 1.5 |
| 22 | 74 | El Paso city, Texas | 1.4 |
| 42 | 75 | Miami city, Florida | 1.3 |

### Percent Hispanic or Latino,[1] 2019

| Population rank | Hispanic or Latino rank | City | Percent Hispanic or Latino [col 12] |
|---|---|---|---|
| 22 | 1 | El Paso city, Texas | 82.1 |
| 57 | 2 | Santa Ana city, California | 77.9 |
| 42 | 3 | Miami city, Florida | 71.3 |
| 7 | 4 | San Antonio city, Texas | 64.5 |
| 60 | 5 | Corpus Christi city, Texas | 63.4 |
| 75 | 6 | Chula Vista city, California | 58.0 |
| 58 | 7 | Riverside city, California | 57.2 |
| 55 | 8 | Anaheim city, California | 53.4 |
| 53 | 9 | Bakersfield city, California | 52.4 |
| 34 | 10 | Fresno city, California | 49.9 |
| 32 | 11 | Albuquerque city, New Mexico | 49.3 |
| 2 | 12 | Los Angeles city, California | 48.2 |
| 4 | 13 | Houston city, Texas | 45.8 |
| 33 | 14 | Tucson city, Arizona | 45.4 |
| 43 | 15 | Long Beach city, California | 44.8 |
| 62 | 16 | Stockton city, California | 43.5 |
| 5 | 17 | Phoenix city, Arizona | 43.2 |
| 9 | 18 | Dallas city, Texas | 41.2 |
| 70 | 19 | Orlando city, Florida | 37.0 |
| 74 | 20 | Newark city, New Jersey | 36.9 |
| 12 | 21 | Fort Worth city, Texas | 36.4 |
| 25 | 22 | Las Vegas city, Nevada | 35.5 |
| 11 | 23 | Austin city, Texas | 32.5 |
| 36 | 24 | Sacramento city, California | 30.9 |
| 10 | 25 | San Jose city, California | 30.6 |
| 8 | 26 | San Diego city, California | 30.3 |
| 52 | 27 | Aurora city, Colorado | 29.7 |
| 49 | 28 | Arlington city, Texas | 29.6 |
| 19 | 29 | Denver city, Colorado | 29.3 |
| 1 | 30 | New York city, New York | 29.1 |
| 3 | 31 | Chicago city, Illinois | 28.8 |
| 47 | 32 | Tampa city, Florida | 28.4 |
| 46 | 33 | Oakland city, California | 26.8 |
| 35 | 34 | Mesa city, Arizona | 26.2 |
| 26 | 35 | Oklahoma City city, Oklahoma | 20.7 |
| 21 | 36 | Boston city, Massachusetts | 19.7 |
| 30 | 37 | Milwaukee city, Wisconsin | 19.2 |
| 39 | 38 | Colorado Springs city, Colorado | 18.8 |
| 48 | 39 | Tulsa city, Oklahoma | 17.2 |
| 50 | 40 | Wichita city, Kansas | 16.5 |
| 59 | 41 | Henderson city, Nevada | 16.1 |
| 6 | 42 | Philadelphia city, Pennsylvania | 15.2 |
| 17 | 42 | San Francisco city, California | 15.2 |
| 15 | 44 | Charlotte city, North Carolina | 15.0 |
| 40 | 45 | Omaha city, Nebraska | 13.9 |
| 68 | 46 | Plano city, Texas | 13.6 |
| 72 | 47 | Durham city, North Carolina | 13.1 |
| 54 | 48 | Cleveland city, Ohio | 12.7 |
| 41 | 49 | Raleigh city, North Carolina | 12.5 |
| 20 | 50 | Washington city, District of Columbia | 11.3 |
| 38 | 51 | Kansas City city, Missouri | 11.2 |
| 16 | 52 | Indianapolis city, Indiana | 11.1 |
| 13 | 53 | Jacksonville city, Florida | 10.7 |
| 23 | 54 | Nashville-Davidson, Tennessee | 10.6 |
| 73 | 55 | Irvine city, California | 10.4 |
| 45 | 56 | Minneapolis city, Minnesota | 9.8 |
| 27 | 57 | Portland city, Oregon | 9.7 |
| 71 | 58 | Anchorage municipality, Alaska | 9.4 |
| 66 | 59 | Greensboro city, North Carolina | 9.3 |
| 63 | 60 | St. Paul city, Minnesota | 8.7 |
| 44 | 61 | Virginia Beach city, Virginia | 8.5 |
| 24 | 62 | Detroit city, Michigan | 8.3 |
| 69 | 63 | Lincoln city, Nebraska | 8.1 |
| 28 | 64 | Memphis city, Tennessee | 7.7 |
| 56 | 64 | Urban Honolulu CDP, Hawaii | 7.7 |
| 61 | 66 | Lexington-Fayette urban county, Kentucky | 7.4 |
| 14 | 67 | Columbus city, Ohio | 7.0 |
| 18 | 68 | Seattle city, Washington | 6.6 |
| 29 | 69 | Louisville/Jefferson County, Kentuck | 6.0 |
| 31 | 70 | Baltimore city, Maryland | 5.7 |
| 51 | 71 | New Orleans city, Louisiana | 5.5 |
| 37 | 72 | Atlanta city, Georgia | 4.9 |
| 67 | 73 | St. Louis city, Missouri | 4.0 |
| 65 | 74 | Pittsburgh city, Pennsylvania | 3.7 |
| 64 | 75 | Cincinnati city, Ohio | 3.5 |

# 75 Largest Cities by 2020 Population
## Selected rankings

### Percent under 18 years old, 2019

| Population rank | Under 18 years old rank | City | Percent under 18 years old [col 14] |
|---|---|---|---|
| 53 | 1 | Bakersfield city, California | 29.8 |
| 34 | 2 | Fresno city, California | 27.9 |
| 62 | 3 | Stockton city, California | 27.1 |
| 12 | 4 | Fort Worth city, Texas | 26.9 |
| 22 | 5 | El Paso city, Texas | 26.3 |
| 57 | 6 | Santa Ana city, California | 25.6 |
| 30 | 7 | Milwaukee city, Wisconsin | 25.3 |
| 40 | 7 | Omaha city, Nebraska | 25.3 |
| 26 | 9 | Oklahoma City city, Oklahoma | 25.2 |
| 50 | 9 | Wichita city, Kansas | 25.2 |
| 5 | 11 | Phoenix city, Arizona | 25.0 |
| 49 | 11 | Arlington city, Texas | 25.0 |
| 52 | 13 | Aurora city, Colorado | 24.9 |
| 24 | 14 | Detroit city, Michigan | 24.8 |
| 35 | 14 | Mesa city, Arizona | 24.8 |
| 63 | 14 | St. Paul city, Minnesota | 24.8 |
| 4 | 17 | Houston city, Texas | 24.7 |
| 9 | 18 | Dallas city, Texas | 24.6 |
| 16 | 18 | Indianapolis city, Indiana | 24.6 |
| 48 | 18 | Tulsa city, Oklahoma | 24.6 |
| 74 | 18 | Newark city, New Jersey | 24.6 |
| 7 | 22 | San Antonio city, Texas | 24.4 |
| 25 | 22 | Las Vegas city, Nevada | 24.4 |
| 60 | 22 | Corpus Christi city, Texas | 24.4 |
| 58 | 25 | Riverside city, California | 24.3 |
| 28 | 26 | Memphis city, Tennessee | 24.0 |
| 71 | 26 | Anchorage municipality, Alaska | 24.0 |
| 75 | 26 | Chula Vista city, California | 24.0 |
| 15 | 29 | Charlotte city, North Carolina | 23.5 |
| 13 | 30 | Jacksonville city, Florida | 22.6 |
| 36 | 31 | Sacramento city, California | 22.5 |
| 38 | 31 | Kansas City city, Missouri | 22.5 |
| 70 | 33 | Orlando city, Florida | 22.3 |
| 14 | 34 | Columbus city, Ohio | 22.1 |
| 10 | 35 | San Jose city, California | 22.0 |
| 29 | 35 | Louisville/Jefferson County, Kentuck | 22.0 |
| 44 | 35 | Virginia Beach city, Virginia | 22.0 |
| 55 | 35 | Anaheim city, California | 22.0 |
| 66 | 35 | Greensboro city, North Carolina | 22.0 |
| 47 | 40 | Tampa city, Florida | 21.9 |
| 69 | 40 | Lincoln city, Nebraska | 21.9 |
| 54 | 42 | Cleveland city, Ohio | 21.8 |
| 32 | 43 | Albuquerque city, New Mexico | 21.7 |
| 6 | 44 | Philadelphia city, Pennsylvania | 21.6 |
| 39 | 44 | Colorado Springs city, Colorado | 21.6 |
| 43 | 46 | Long Beach city, California | 21.5 |
| 68 | 46 | Plano city, Texas | 21.5 |
| 73 | 46 | Irvine city, California | 21.5 |
| 41 | 49 | Raleigh city, North Carolina | 20.9 |
| 64 | 49 | Cincinnati city, Ohio | 20.9 |
| 61 | 51 | Lexington-Fayette urban county, Kentucky | 20.8 |
| 1 | 52 | New York city, New York | 20.6 |
| 3 | 52 | Chicago city, Illinois | 20.6 |
| 72 | 52 | Durham city, North Carolina | 20.6 |
| 23 | 55 | Nashville-Davidson, Tennessee | 20.5 |
| 45 | 56 | Minneapolis city, Minnesota | 20.3 |
| 2 | 57 | Los Angeles city, California | 20.2 |
| 31 | 58 | Baltimore city, Maryland | 20.1 |
| 33 | 59 | Tucson city, Arizona | 20.0 |
| 46 | 59 | Oakland city, California | 20.0 |
| 51 | 61 | New Orleans city, Louisiana | 19.8 |
| 8 | 62 | San Diego city, California | 19.7 |
| 11 | 63 | Austin city, Texas | 19.5 |
| 59 | 64 | Henderson city, Nevada | 19.4 |
| 19 | 65 | Denver city, Colorado | 19.1 |
| 67 | 66 | St. Louis city, Missouri | 18.7 |
| 20 | 67 | Washington city, District of Columbia | 18.1 |
| 27 | 68 | Portland city, Oregon | 17.7 |
| 42 | 69 | Miami city, Florida | 17.1 |
| 37 | 70 | Atlanta city, Georgia | 17.0 |
| 56 | 71 | Urban Honolulu CDP, Hawaii | 16.6 |
| 21 | 72 | Boston city, Massachusetts | 15.4 |
| 18 | 73 | Seattle city, Washington | 14.7 |
| 65 | 73 | Pittsburgh city, Pennsylvania | 14.7 |
| 17 | 75 | San Francisco city, California | 13.4 |

### Percent 65 years old and over, 2019

| Population rank | 65 years old and over rank | City | Percent 65 years old and over [col 20] |
|---|---|---|---|
| 59 | 1 | Henderson city, Nevada | 21.3 |
| 56 | 2 | Urban Honolulu CDP, Hawaii | 21.1 |
| 42 | 3 | Miami city, Florida | 17.7 |
| 35 | 4 | Mesa city, Arizona | 16.7 |
| 32 | 5 | Albuquerque city, New Mexico | 16.4 |
| 17 | 6 | San Francisco city, California | 16.0 |
| 29 | 7 | Louisville/Jefferson County, Kentuck | 15.9 |
| 51 | 8 | New Orleans city, Louisiana | 15.5 |
| 1 | 9 | New York city, New York | 15.4 |
| 33 | 10 | Tucson city, Arizona | 15.3 |
| 65 | 11 | Pittsburgh city, Pennsylvania | 15.2 |
| 39 | 12 | Colorado Springs city, Colorado | 15.0 |
| 48 | 12 | Tulsa city, Oklahoma | 15.0 |
| 25 | 14 | Las Vegas city, Nevada | 14.9 |
| 44 | 15 | Virginia Beach city, Virginia | 14.7 |
| 24 | 16 | Detroit city, Michigan | 14.5 |
| 31 | 16 | Baltimore city, Maryland | 14.5 |
| 50 | 16 | Wichita city, Kansas | 14.5 |
| 60 | 16 | Corpus Christi city, Texas | 14.5 |
| 13 | 20 | Jacksonville city, Florida | 14.4 |
| 67 | 20 | St. Louis city, Missouri | 14.4 |
| 36 | 22 | Sacramento city, California | 14.2 |
| 54 | 22 | Cleveland city, Ohio | 14.2 |
| 28 | 24 | Memphis city, Tennessee | 14.1 |
| 6 | 25 | Philadelphia city, Pennsylvania | 14.0 |
| 68 | 25 | Plano city, Texas | 14.0 |
| 69 | 27 | Lincoln city, Nebraska | 13.9 |
| 46 | 28 | Oakland city, California | 13.8 |
| 61 | 29 | Lexington-Fayette urban county, Kent | 13.7 |
| 66 | 29 | Greensboro city, North Carolina | 13.7 |
| 10 | 31 | San Jose city, California | 13.6 |
| 27 | 31 | Portland city, Oregon | 13.6 |
| 62 | 31 | Stockton city, California | 13.6 |
| 3 | 34 | Chicago city, Illinois | 13.5 |
| 38 | 34 | Kansas City city, Missouri | 13.5 |
| 40 | 34 | Omaha city, Nebraska | 13.5 |
| 8 | 37 | San Diego city, California | 13.4 |
| 64 | 38 | Cincinnati city, Ohio | 13.2 |
| 2 | 39 | Los Angeles city, California | 13.1 |
| 26 | 39 | Oklahoma City city, Oklahoma | 13.1 |
| 22 | 41 | El Paso city, Texas | 13.0 |
| 47 | 41 | Tampa city, Florida | 13.0 |
| 16 | 43 | Indianapolis city, Indiana | 12.9 |
| 7 | 44 | San Antonio city, Texas | 12.7 |
| 43 | 45 | Long Beach city, California | 12.5 |
| 72 | 45 | Durham city, North Carolina | 12.5 |
| 20 | 47 | Washington city, District of Columbia | 12.4 |
| 23 | 47 | Nashville-Davidson, Tennessee | 12.4 |
| 37 | 49 | Atlanta city, Georgia | 12.3 |
| 75 | 50 | Chula Vista city, California | 12.2 |
| 18 | 51 | Seattle city, Washington | 12.0 |
| 21 | 51 | Boston city, Massachusetts | 12.0 |
| 19 | 53 | Denver city, Colorado | 11.9 |
| 34 | 54 | Fresno city, California | 11.8 |
| 55 | 54 | Anaheim city, California | 11.8 |
| 41 | 56 | Raleigh city, North Carolina | 11.7 |
| 49 | 57 | Arlington city, Texas | 11.6 |
| 52 | 57 | Aurora city, Colorado | 11.6 |
| 63 | 59 | St. Paul city, Minnesota | 11.4 |
| 71 | 59 | Anchorage municipality, Alaska | 11.4 |
| 4 | 61 | Houston city, Texas | 11.3 |
| 5 | 62 | Phoenix city, Arizona | 11.1 |
| 30 | 63 | Milwaukee city, Wisconsin | 10.9 |
| 9 | 64 | Dallas city, Texas | 10.7 |
| 74 | 64 | Newark city, New Jersey | 10.7 |
| 14 | 66 | Columbus city, Ohio | 10.6 |
| 15 | 66 | Charlotte city, North Carolina | 10.6 |
| 53 | 68 | Bakersfield city, California | 10.5 |
| 58 | 68 | Riverside city, California | 10.5 |
| 57 | 70 | Santa Ana city, California | 10.4 |
| 45 | 71 | Minneapolis city, Minnesota | 10.3 |
| 73 | 72 | Irvine city, California | 10.2 |
| 12 | 73 | Fort Worth city, Texas | 10.0 |
| 70 | 73 | Orlando city, Florida | 10.0 |
| 11 | 75 | Austin city, Texas | 9.5 |

### Percent high school graduate or less, 2019

| Population rank | Percent high school graduate or less rank | City | Percent high school graduate or less [col 40] |
|---|---|---|---|
| 74 | 1 | Newark city, New Jersey | 63.4 |
| 57 | 2 | Santa Ana city, California | 61.6 |
| 62 | 3 | Stockton city, California | 53.3 |
| 24 | 4 | Detroit city, Michigan | 51.3 |
| 54 | 5 | Cleveland city, Ohio | 49.9 |
| 42 | 6 | Miami city, Florida | 48.2 |
| 58 | 7 | Riverside city, California | 46.7 |
| 6 | 8 | Philadelphia city, Pennsylvania | 46.6 |
| 55 | 9 | Anaheim city, California | 46.0 |
| 60 | 10 | Corpus Christi city, Texas | 45.7 |
| 34 | 11 | Fresno city, California | 45.6 |
| 30 | 12 | Milwaukee city, Wisconsin | 45.5 |
| 28 | 13 | Memphis city, Tennessee | 45.4 |
| 53 | 14 | Bakersfield city, California | 44.2 |
| 25 | 15 | Las Vegas city, Nevada | 43.9 |
| 22 | 16 | El Paso city, Texas | 43.8 |
| 7 | 17 | San Antonio city, Texas | 43.6 |
| 9 | 18 | Dallas city, Texas | 43.5 |
| 16 | 19 | Indianapolis city, Indiana | 43.1 |
| 4 | 20 | Houston city, Texas | 42.9 |
| 31 | 21 | Baltimore city, Maryland | 42.6 |
| 5 | 22 | Phoenix city, Arizona | 41.9 |
| 12 | 22 | Fort Worth city, Texas | 41.9 |
| 1 | 24 | New York city, New York | 41.2 |
| 2 | 25 | Los Angeles city, California | 40.6 |
| 26 | 26 | Oklahoma City city, Oklahoma | 39.7 |
| 13 | 27 | Jacksonville city, Florida | 39.2 |
| 33 | 28 | Tucson city, Arizona | 38.9 |
| 43 | 29 | Long Beach city, California | 38.3 |
| 29 | 30 | Louisville/Jefferson County, Kentucky | 38.1 |
| 50 | 30 | Wichita city, Kansas | 38.1 |
| 49 | 32 | Arlington city, Texas | 38.0 |
| 48 | 33 | Tulsa city, Oklahoma | 37.4 |
| 75 | 34 | Chula Vista city, California | 37.3 |
| 47 | 35 | Tampa city, Florida | 37.1 |
| 52 | 36 | Aurora city, Colorado | 36.5 |
| 3 | 37 | Chicago city, Illinois | 36.2 |
| 36 | 38 | Sacramento city, California | 35.5 |
| 51 | 38 | New Orleans city, Louisiana | 35.5 |
| 38 | 40 | Kansas City city, Missouri | 35.2 |
| 67 | 41 | St. Louis city, Missouri | 35.1 |
| 64 | 42 | Cincinnati city, Ohio | 34.8 |
| 14 | 43 | Columbus city, Ohio | 34.2 |
| 66 | 44 | Greensboro city, North Carolina | 34.1 |
| 32 | 45 | Albuquerque city, New Mexico | 33.7 |
| 35 | 46 | Mesa city, Arizona | 32.9 |
| 56 | 46 | Urban Honolulu CDP, Hawaii | 32.9 |
| 63 | 48 | St. Paul city, Minnesota | 32.7 |
| 70 | 49 | Orlando city, Florida | 32.2 |
| 40 | 50 | Omaha city, Nebraska | 32.1 |
| 10 | 51 | San Jose city, California | 31.1 |
| 46 | 52 | Oakland city, California | 30.9 |
| 23 | 53 | Nashville-Davidson, Tennessee | 30.7 |
| 65 | 54 | Pittsburgh city, Pennsylvania | 30.6 |
| 21 | 55 | Boston city, Massachusetts | 30.5 |
| 71 | 56 | Anchorage municipality, Alaska | 30.2 |
| 59 | 57 | Henderson city, Nevada | 29.9 |
| 61 | 58 | Lexington-Fayette urban county, Kentucky | 28.7 |
| 72 | 59 | Durham city, North Carolina | 27.9 |
| 69 | 60 | Lincoln city, Nebraska | 27.4 |
| 15 | 61 | Charlotte city, North Carolina | 26.7 |
| 8 | 62 | San Diego city, California | 26.1 |
| 19 | 63 | Denver city, Colorado | 25.8 |
| 39 | 64 | Colorado Springs city, Colorado | 25.6 |
| 41 | 65 | Raleigh city, North Carolina | 25.1 |
| 37 | 66 | Atlanta city, Georgia | 24.6 |
| 44 | 66 | Virginia Beach city, Virginia | 24.6 |
| 11 | 68 | Austin city, Texas | 24.3 |
| 20 | 69 | Washington city, District of Columbia | 24.0 |
| 17 | 70 | San Francisco city, California | 23.5 |
| 45 | 71 | Minneapolis city, Minnesota | 23.4 |
| 27 | 72 | Portland city, Oregon | 21.6 |
| 68 | 73 | Plano city, Texas | 20.1 |
| 18 | 74 | Seattle city, Washington | 14.9 |
| 73 | 75 | Irvine city, California | 9.8 |

# 75 Largest Cities by 2020 Population
## Selected rankings

### Percent college graduates (bachelor's degree or more), 2019

| Population rank | Percent college graduate rank | City | Percent college graduates [col 41] |
|---|---|---|---|
| 73 | 1 | Irvine city, California | 69.5 |
| 18 | 2 | Seattle city, Washington | 65.0 |
| 20 | 3 | Washington city, District of Columbia | 59.7 |
| 17 | 4 | San Francisco city, California | 59.2 |
| 68 | 5 | Plano city, Texas | 57.3 |
| 37 | 6 | Atlanta city, Georgia | 56.5 |
| 11 | 7 | Austin city, Texas | 55.0 |
| 19 | 8 | Denver city, Colorado | 53.1 |
| 27 | 9 | Portland city, Oregon | 52.8 |
| 45 | 10 | Minneapolis city, Minnesota | 52.2 |
| 21 | 11 | Boston city, Massachusetts | 51.7 |
| 72 | 12 | Durham city, North Carolina | 51.1 |
| 41 | 13 | Raleigh city, North Carolina | 50.3 |
| 46 | 14 | Oakland city, California | 48.9 |
| 8 | 15 | San Diego city, California | 47.4 |
| 65 | 16 | Pittsburgh city, Pennsylvania | 47.0 |
| 23 | 17 | Nashville-Davidson, Tennessee | 45.2 |
| 10 | 18 | San Jose city, California | 45.0 |
| 61 | 18 | Lexington-Fayette urban county, Kentucky | 45.0 |
| 15 | 20 | Charlotte city, North Carolina | 44.9 |
| 63 | 21 | St. Paul city, Minnesota | 43.6 |
| 3 | 22 | Chicago city, Illinois | 41.3 |
| 47 | 23 | Tampa city, Florida | 41.0 |
| 70 | 24 | Orlando city, Florida | 40.9 |
| 64 | 25 | Cincinnati city, Ohio | 40.6 |
| 69 | 26 | Lincoln city, Nebraska | 40.4 |
| 39 | 27 | Colorado Springs city, Colorado | 40.3 |
| 40 | 28 | Omaha city, Nebraska | 39.7 |
| 51 | 28 | New Orleans city, Louisiana | 39.7 |
| 1 | 30 | New York city, New York | 39.2 |
| 14 | 31 | Columbus city, Ohio | 39.1 |
| 66 | 32 | Greensboro city, North Carolina | 38.7 |
| 56 | 33 | Urban Honolulu CDP, Hawaii | 38.6 |
| 67 | 33 | St. Louis city, Missouri | 38.6 |
| 44 | 35 | Virginia Beach city, Virginia | 38.2 |
| 71 | 36 | Anchorage municipality, Alaska | 36.3 |
| 32 | 37 | Albuquerque city, New Mexico | 36.2 |
| 38 | 37 | Kansas City city, Missouri | 36.2 |
| 2 | 39 | Los Angeles city, California | 35.9 |
| 59 | 40 | Henderson city, Nevada | 35.3 |
| 9 | 41 | Dallas city, Texas | 34.7 |
| 4 | 42 | Houston city, Texas | 34.0 |
| 36 | 43 | Sacramento city, California | 33.7 |
| 48 | 44 | Tulsa city, Oklahoma | 33.6 |
| 49 | 44 | Arlington city, Texas | 33.6 |
| 31 | 46 | Baltimore city, Maryland | 33.3 |
| 43 | 47 | Long Beach city, California | 32.9 |
| 29 | 48 | Louisville/Jefferson County, Kentuck.. | 31.9 |
| 42 | 49 | Miami city, Florida | 31.4 |
| 12 | 50 | Fort Worth city, Texas | 31.3 |
| 26 | 51 | Oklahoma City city, Oklahoma | 31.1 |
| 52 | 51 | Aurora city, Colorado | 31.1 |
| 6 | 53 | Philadelphia city, Pennsylvania | 31.0 |
| 16 | 54 | Indianapolis city, Indiana | 30.9 |
| 50 | 55 | Wichita city, Kansas | 30.8 |
| 35 | 56 | Mesa city, Arizona | 29.2 |
| 75 | 56 | Chula Vista city, California | 29.2 |
| 13 | 58 | Jacksonville city, Florida | 29.1 |
| 5 | 59 | Phoenix city, Arizona | 28.6 |
| 33 | 60 | Tucson city, Arizona | 27.3 |
| 7 | 61 | San Antonio city, Texas | 26.0 |
| 55 | 62 | Anaheim city, California | 25.9 |
| 30 | 63 | Milwaukee city, Wisconsin | 25.7 |
| 28 | 64 | Memphis city, Tennessee | 25.6 |
| 25 | 65 | Las Vegas city, Nevada | 25.1 |
| 22 | 66 | El Paso city, Texas | 24.3 |
| 60 | 67 | Corpus Christi city, Texas | 23.7 |
| 53 | 68 | Bakersfield city, California | 22.4 |
| 58 | 69 | Riverside city, California | 22.1 |
| 34 | 70 | Fresno city, California | 21.8 |
| 62 | 71 | Stockton city, California | 20.0 |
| 54 | 72 | Cleveland city, Ohio | 18.6 |
| 57 | 73 | Santa Ana city, California | 17.6 |
| 24 | 74 | Detroit city, Michigan | 16.7 |
| 74 | 75 | Newark city, New Jersey | 14.1 |

### Percent married couple family households, 2019

| Population rank | Married-couple households rank | City | Percent married couple households [col 31] |
|---|---|---|---|
| 57 | 1 | Santa Ana city, California | 56.8 |
| 68 | 2 | Plano city, Texas | 56.5 |
| 75 | 3 | Chula Vista city, California | 54.7 |
| 10 | 4 | San Jose city, California | 53.3 |
| 53 | 5 | Bakersfield city, California | 52.1 |
| 59 | 6 | Henderson city, Nevada | 51.6 |
| 44 | 7 | Virginia Beach city, Virginia | 51.4 |
| 55 | 8 | Anaheim city, California | 50.1 |
| 58 | 9 | Riverside city, California | 49.0 |
| 12 | 10 | Fort Worth city, Texas | 48.6 |
| 39 | 11 | Colorado Springs city, Colorado | 48.0 |
| 73 | 12 | Irvine city, California | 47.7 |
| 71 | 13 | Anchorage municipality, Alaska | 47.2 |
| 35 | 14 | Mesa city, Arizona | 47.1 |
| 8 | 15 | San Diego city, California | 46.4 |
| 62 | 16 | Stockton city, California | 46.1 |
| 49 | 17 | Arlington city, Texas | 45.3 |
| 69 | 18 | Lincoln city, Nebraska | 44.5 |
| 52 | 19 | Aurora city, Colorado | 44.2 |
| 60 | 20 | Corpus Christi city, Texas | 44.1 |
| 22 | 21 | El Paso city, Texas | 43.4 |
| 5 | 22 | Phoenix city, Arizona | 42.6 |
| 40 | 22 | Omaha city, Nebraska | 42.6 |
| 26 | 24 | Oklahoma City city, Oklahoma | 41.7 |
| 50 | 24 | Wichita city, Kansas | 41.7 |
| 25 | 26 | Las Vegas city, Nevada | 41.3 |
| 13 | 27 | Jacksonville city, Florida | 39.7 |
| 61 | 27 | Lexington-Fayette urban county, Kent | 39.7 |
| 34 | 29 | Fresno city, California | 39.6 |
| 7 | 30 | San Antonio city, Texas | 39.4 |
| 56 | 30 | Urban Honolulu CDP, Hawaii | 39.4 |
| 43 | 32 | Long Beach city, California | 38.8 |
| 29 | 33 | Louisville/Jefferson County, Kentucky | 38.7 |
| 2 | 34 | Los Angeles city, California | 38.5 |
| 15 | 35 | Charlotte city, North Carolina | 38.4 |
| 41 | 36 | Raleigh city, North Carolina | 38.3 |
| 4 | 37 | Houston city, Texas | 38.0 |
| 48 | 38 | Tulsa city, Oklahoma | 37.7 |
| 36 | 39 | Sacramento city, California | 37.5 |
| 72 | 40 | Durham city, North Carolina | 37.3 |
| 11 | 41 | Austin city, Texas | 37.1 |
| 27 | 42 | Portland city, Oregon | 36.7 |
| 1 | 43 | New York city, New York | 36.4 |
| 32 | 43 | Albuquerque city, New Mexico | 36.4 |
| 46 | 45 | Oakland city, California | 36.3 |
| 63 | 45 | St. Paul city, Minnesota | 36.3 |
| 18 | 47 | Seattle city, Washington | 36.1 |
| 23 | 48 | Nashville-Davidson, Tennessee | 35.9 |
| 38 | 49 | Kansas City city, Missouri | 35.7 |
| 66 | 50 | Greensboro city, North Carolina | 35.6 |
| 47 | 51 | Tampa city, Florida | 35.5 |
| 9 | 52 | Dallas city, Texas | 35.0 |
| 16 | 53 | Indianapolis city, Indiana | 34.9 |
| 17 | 53 | San Francisco city, California | 34.9 |
| 70 | 55 | Orlando city, Florida | 34.3 |
| 19 | 56 | Denver city, Colorado | 34.1 |
| 14 | 57 | Columbus city, Ohio | 33.5 |
| 42 | 58 | Miami city, Florida | 33.1 |
| 3 | 59 | Chicago city, Illinois | 32.5 |
| 33 | 60 | Tucson city, Arizona | 32.3 |
| 45 | 61 | Minneapolis city, Minnesota | 31.1 |
| 74 | 62 | Newark city, New Jersey | 30.9 |
| 28 | 63 | Memphis city, Tennessee | 29.0 |
| 21 | 64 | Boston city, Massachusetts | 28.3 |
| 6 | 65 | Philadelphia city, Pennsylvania | 28.1 |
| 20 | 66 | Washington city, District of Columbia | 26.9 |
| 30 | 67 | Milwaukee city, Wisconsin | 26.5 |
| 51 | 67 | New Orleans city, Louisiana | 26.5 |
| 64 | 69 | Cincinnati city, Ohio | 26.0 |
| 37 | 70 | Atlanta city, Georgia | 25.8 |
| 65 | 71 | Pittsburgh city, Pennsylvania | 25.7 |
| 67 | 72 | St. Louis city, Missouri | 24.7 |
| 31 | 73 | Baltimore city, Maryland | 23.4 |
| 54 | 74 | Cleveland city, Ohio | 19.7 |
| 24 | 75 | Detroit city, Michigan | 19.0 |

### Percent of households composed of one person, 2019

| Population rank | One-person household rank | City | Percent one-person households [col 33] |
|---|---|---|---|
| 67 | 1 | St. Louis city, Missouri | 47.5 |
| 51 | 2 | New Orleans city, Louisiana | 46.8 |
| 20 | 3 | Washington city, District of Columbia | 44.8 |
| 54 | 4 | Cleveland city, Ohio | 44.3 |
| 37 | 5 | Atlanta city, Georgia | 44.1 |
| 65 | 6 | Pittsburgh city, Pennsylvania | 43.6 |
| 24 | 7 | Detroit city, Michigan | 42.0 |
| 64 | 8 | Cincinnati city, Ohio | 41.6 |
| 45 | 9 | Minneapolis city, Minnesota | 40.8 |
| 31 | 10 | Baltimore city, Maryland | 40.6 |
| 16 | 11 | Indianapolis city, Indiana | 40.1 |
| 28 | 12 | Memphis city, Tennessee | 39.0 |
| 3 | 13 | Chicago city, Illinois | 38.5 |
| 19 | 14 | Denver city, Colorado | 38.2 |
| 18 | 15 | Seattle city, Washington | 37.8 |
| 38 | 16 | Kansas City city, Missouri | 37.6 |
| 30 | 17 | Milwaukee city, Wisconsin | 37.4 |
| 33 | 18 | Tucson city, Arizona | 37.3 |
| 9 | 19 | Dallas city, Texas | 36.6 |
| 17 | 19 | San Francisco city, California | 36.6 |
| 21 | 19 | Boston city, Massachusetts | 36.6 |
| 23 | 22 | Nashville-Davidson, Tennessee | 36.4 |
| 46 | 23 | Oakland city, California | 36.3 |
| 6 | 24 | Philadelphia city, Pennsylvania | 36.1 |
| 14 | 24 | Columbus city, Ohio | 36.1 |
| 48 | 26 | Tulsa city, Oklahoma | 35.6 |
| 47 | 27 | Tampa city, Florida | 35.5 |
| 32 | 28 | Albuquerque city, New Mexico | 35.4 |
| 63 | 28 | St. Paul city, Minnesota | 35.4 |
| 11 | 30 | Austin city, Texas | 35.2 |
| 27 | 31 | Portland city, Oregon | 35.0 |
| 29 | 32 | Louisville/Jefferson County, Kentuck.. | 34.6 |
| 56 | 32 | Urban Honolulu CDP, Hawaii | 34.6 |
| 15 | 34 | Charlotte city, North Carolina | 34.5 |
| 50 | 35 | Wichita city, Kansas | 34.2 |
| 42 | 36 | Miami city, Florida | 34.0 |
| 66 | 37 | Greensboro city, North Carolina | 33.9 |
| 70 | 38 | Orlando city, Florida | 33.6 |
| 74 | 38 | Newark city, New Jersey | 33.6 |
| 41 | 40 | Raleigh city, North Carolina | 33.3 |
| 61 | 40 | Lexington-Fayette urban county, Kentucky | 33.3 |
| 1 | 42 | New York city, New York | 32.9 |
| 4 | 43 | Houston city, Texas | 32.8 |
| 72 | 43 | Durham city, North Carolina | 32.8 |
| 13 | 45 | Jacksonville city, Florida | 32.4 |
| 40 | 46 | Omaha city, Nebraska | 32.3 |
| 26 | 47 | Oklahoma City city, Oklahoma | 32.1 |
| 36 | 48 | Sacramento city, California | 31.6 |
| 25 | 49 | Las Vegas city, Nevada | 31.0 |
| 7 | 50 | San Antonio city, Texas | 30.6 |
| 69 | 50 | Lincoln city, Nebraska | 30.6 |
| 43 | 52 | Long Beach city, California | 30.4 |
| 2 | 53 | Los Angeles city, California | 30.2 |
| 71 | 54 | Anchorage municipality, Alaska | 28.2 |
| 5 | 55 | Phoenix city, Arizona | 27.3 |
| 8 | 55 | San Diego city, California | 27.3 |
| 22 | 57 | El Paso city, Texas | 26.8 |
| 52 | 57 | Aurora city, Colorado | 26.8 |
| 39 | 59 | Colorado Springs city, Colorado | 26.7 |
| 59 | 60 | Henderson city, Nevada | 26.6 |
| 12 | 61 | Fort Worth city, Texas | 26.5 |
| 49 | 61 | Arlington city, Texas | 26.5 |
| 60 | 61 | Corpus Christi city, Texas | 26.5 |
| 34 | 64 | Fresno city, California | 26.2 |
| 35 | 65 | Mesa city, Arizona | 26.1 |
| 44 | 66 | Virginia Beach city, Virginia | 24.8 |
| 73 | 66 | Irvine city, California | 24.8 |
| 62 | 68 | Stockton city, California | 24.4 |
| 68 | 69 | Plano city, Texas | 23.9 |
| 53 | 70 | Bakersfield city, California | 21.2 |
| 10 | 71 | San Jose city, California | 20.9 |
| 55 | 72 | Anaheim city, California | 19.9 |
| 58 | 73 | Riverside city, California | 19.8 |
| 75 | 74 | Chula Vista city, California | 16.7 |
| 57 | 75 | Santa Ana city, California | 12.0 |

# 75 Largest Cities by 2020 Population
## Selected rankings

| Median household income, 2019 | | | | Median value of owner-occupied housing units, 2019 | | | | Percent with commutes of 30 minutes or more, 2019 | | | |
|---|---|---|---|---|---|---|---|---|---|---|---|
| Population rank | Median income rank | City | Median income (dollars) [col 42] | Population rank | Median value rank | City | Median value (dollars) [col 53] | Population rank | Commutes of 30 minutes or more rank | City | Commutes of 30 minutes or more [col 56] |
| 17 | 1 | San Francisco city, California | 123,859 | 17 | 1 | San Francisco city, California | 1,217,458 | 1 | 1 | New York city, New York | 70.8 |
| 10 | 2 | San Jose city, California | 115,893 | 10 | 2 | San Jose city, California | 999,943 | 3 | 2 | Chicago city, Illinois | 63.1 |
| 73 | 3 | Irvine city, California | 111,574 | 73 | 3 | Irvine city, California | 933,638 | 74 | 3 | Newark city, New Jersey | 59.0 |
| 18 | 4 | Seattle city, Washington | 102,486 | 46 | 4 | Oakland city, California | 807,615 | 17 | 4 | San Francisco city, California | 58.2 |
| 68 | 5 | Plano city, Texas | 93,321 | 18 | 5 | Seattle city, Washington | 767,024 | 20 | 5 | Washington city, District of Columbia | 57.4 |
| 20 | 6 | Washington city, District of Columbia | 92,266 | 56 | 6 | Urban Honolulu CDP, Hawaii | 754,725 | 6 | 6 | Philadelphia city, Pennsylvania | 56.4 |
| 75 | 7 | Chula Vista city, California | 87,876 | 2 | 7 | Los Angeles city, California | 697,166 | 46 | 7 | Oakland city, California | 55.5 |
| 8 | 8 | San Diego city, California | 85,507 | 1 | 8 | New York city, New York | 680,773 | 2 | 8 | Los Angeles city, California | 55.0 |
| 71 | 9 | Anchorage municipality, Alaska | 82,716 | 8 | 9 | San Diego city, California | 658,394 | 21 | 9 | Boston city, Massachusetts | 54.8 |
| 46 | 10 | Oakland city, California | 82,018 | 20 | 10 | Washington city, District of Columbia | 646,504 | 52 | 10 | Aurora city, Colorado | 50.8 |
| 59 | 11 | Henderson city, Nevada | 81,505 | 55 | 11 | Anaheim city, California | 630,470 | 58 | 11 | Riverside city, California | 50.1 |
| 44 | 12 | Virginia Beach city, Virginia | 79,054 | 21 | 12 | Boston city, Massachusetts | 627,022 | 10 | 12 | San Jose city, California | 49.9 |
| 21 | 13 | Boston city, Massachusetts | 79,018 | 43 | 13 | Long Beach city, California | 614,401 | 42 | 13 | Miami city, Florida | 49.1 |
| 27 | 14 | Portland city, Oregon | 76,231 | 57 | 14 | Santa Ana city, California | 558,410 | 31 | 14 | Baltimore city, Maryland | 48.8 |
| 55 | 15 | Anaheim city, California | 76,075 | 75 | 15 | Chula Vista city, California | 546,403 | 43 | 15 | Long Beach city, California | 48.4 |
| 19 | 16 | Denver city, Colorado | 75,646 | 19 | 16 | Denver city, Colorado | 447,451 | 18 | 16 | Seattle city, Washington | 47.4 |
| 11 | 17 | Austin city, Texas | 75,413 | 27 | 17 | Portland city, Oregon | 445,150 | 4 | 17 | Houston city, Texas | 47.1 |
| 56 | 18 | Urban Honolulu CDP, Hawaii | 72,943 | 58 | 18 | Riverside city, California | 411,043 | 55 | 18 | Anaheim city, California | 46.7 |
| 58 | 19 | Riverside city, California | 71,967 | 36 | 19 | Sacramento city, California | 380,592 | 75 | 18 | Chula Vista city, California | 46.7 |
| 39 | 20 | Colorado Springs city, Colorado | 70,527 | 11 | 20 | Austin city, Texas | 378,259 | 70 | 20 | Orlando city, Florida | 44.2 |
| 57 | 21 | Santa Ana city, California | 70,084 | 59 | 21 | Henderson city, Nevada | 371,465 | 9 | 21 | Dallas city, Texas | 42.9 |
| 1 | 22 | New York city, New York | 69,407 | 68 | 22 | Plano city, Texas | 360,983 | 19 | 22 | Denver city, Colorado | 42.2 |
| 41 | 23 | Raleigh city, North Carolina | 69,333 | 37 | 23 | Atlanta city, Georgia | 359,502 | 5 | 23 | Phoenix city, Arizona | 41.5 |
| 52 | 24 | Aurora city, Colorado | 69,235 | 42 | 24 | Miami city, Florida | 358,512 | 25 | 24 | Las Vegas city, Nevada | 40.9 |
| 36 | 25 | Sacramento city, California | 69,134 | 52 | 25 | Aurora city, Colorado | 342,967 | 68 | 24 | Plano city, Texas | 40.9 |
| 43 | 26 | Long Beach city, California | 67,804 | 39 | 26 | Colorado Springs city, Colorado | 318,167 | 37 | 26 | Atlanta city, Georgia | 40.1 |
| 2 | 27 | Los Angeles city, California | 67,418 | 62 | 27 | Stockton city, California | 317,693 | 12 | 27 | Fort Worth city, Texas | 39.7 |
| 37 | 28 | Atlanta city, Georgia | 66,657 | 71 | 28 | Anchorage municipality, Alaska | 314,555 | 27 | 28 | Portland city, Oregon | 39.3 |
| 45 | 29 | Minneapolis city, Minnesota | 65,889 | 25 | 29 | Las Vegas city, Nevada | 305,937 | 49 | 29 | Arlington city, Texas | 39.2 |
| 72 | 30 | Durham city, North Carolina | 65,534 | 44 | 30 | Virginia Beach city, Virginia | 296,153 | 35 | 30 | Mesa city, Arizona | 38.2 |
| 12 | 31 | Fort Worth city, Texas | 65,356 | 23 | 31 | Nashville-Davidson, Tennessee | 287,288 | 57 | 31 | Santa Ana city, California | 37.9 |
| 35 | 32 | Mesa city, Arizona | 63,836 | 45 | 32 | Minneapolis city, Minnesota | 282,204 | 62 | 32 | Stockton city, California | 37.6 |
| 15 | 33 | Charlotte city, North Carolina | 63,483 | 34 | 33 | Fresno city, California | 276,621 | 11 | 33 | Austin city, Texas | 36.8 |
| 23 | 34 | Nashville-Davidson, Tennessee | 63,462 | 3 | 34 | Chicago city, Illinois | 275,239 | 36 | 33 | Sacramento city, California | 36.8 |
| 63 | 35 | St. Paul city, Minnesota | 63,174 | 41 | 35 | Raleigh city, North Carolina | 274,185 | 23 | 35 | Nashville-Davidson, Tennessee | 35.8 |
| 53 | 36 | Bakersfield city, California | 62,402 | 74 | 36 | Newark city, New Jersey | 270,596 | 15 | 36 | Charlotte city, North Carolina | 35.7 |
| 3 | 37 | Chicago city, Illinois | 61,811 | 5 | 37 | Phoenix city, Arizona | 266,625 | 56 | 36 | Urban Honolulu CDP, Hawaii | 35.7 |
| 49 | 38 | Arlington city, Texas | 61,716 | 47 | 38 | Tampa city, Florida | 265,654 | 8 | 38 | San Diego city, California | 35.6 |
| 40 | 39 | Omaha city, Nebraska | 61,305 | 53 | 39 | Bakersfield city, California | 264,461 | 13 | 38 | Jacksonville city, Florida | 35.6 |
| 5 | 40 | Phoenix city, Arizona | 60,931 | 70 | 40 | Orlando city, Florida | 260,832 | 59 | 40 | Henderson city, Nevada | 35.2 |
| 62 | 41 | Stockton city, California | 59,504 | 35 | 41 | Mesa city, Arizona | 259,307 | 65 | 40 | Pittsburgh city, Pennsylvania | 35.2 |
| 69 | 42 | Lincoln city, Nebraska | 59,228 | 72 | 42 | Durham city, North Carolina | 253,513 | 47 | 42 | Tampa city, Florida | 34.9 |
| 70 | 43 | Orlando city, Florida | 58,819 | 15 | 43 | Charlotte city, North Carolina | 252,131 | 7 | 43 | San Antonio city, Texas | 34.3 |
| 25 | 44 | Las Vegas city, Nevada | 58,713 | 51 | 44 | New Orleans city, Louisiana | 242,911 | 24 | 44 | Detroit city, Michigan | 33.9 |
| 61 | 45 | Lexington-Fayette urban county, Kentucky | 58,356 | 63 | 45 | St. Paul city, Minnesota | 231,486 | 63 | 45 | St. Paul city, Minnesota | 33.2 |
| 47 | 46 | Tampa city, Florida | 57,709 | 9 | 46 | Dallas city, Texas | 231,387 | 16 | 46 | Indianapolis city, Indiana | 32.2 |
| 14 | 47 | Columbus city, Ohio | 57,118 | 49 | 47 | Arlington city, Texas | 213,810 | 44 | 47 | Virginia Beach city, Virginia | 31.8 |
| 13 | 48 | Jacksonville city, Florida | 56,975 | 32 | 48 | Albuquerque city, New Mexico | 211,770 | 41 | 48 | Raleigh city, North Carolina | 31.6 |
| 32 | 49 | Albuquerque city, New Mexico | 55,567 | 61 | 49 | Lexington-Fayette urban county, Kentucky | 211,430 | 67 | 49 | St. Louis city, Missouri | 31.1 |
| 60 | 50 | Corpus Christi city, Texas | 55,564 | 12 | 50 | Fort Worth city, Texas | 209,430 | 45 | 50 | Minneapolis city, Minnesota | 30.6 |
| 26 | 51 | Oklahoma City city, Oklahoma | 55,492 | 13 | 51 | Jacksonville city, Florida | 200,164 | 64 | 51 | Cincinnati city, Ohio | 29.6 |
| 9 | 52 | Dallas city, Texas | 55,332 | 4 | 52 | Houston city, Texas | 195,804 | 33 | 52 | Tucson city, Arizona | 28.5 |
| 38 | 53 | Kansas City city, Missouri | 55,259 | 69 | 53 | Lincoln city, Nebraska | 189,411 | 51 | 53 | New Orleans city, Louisiana | 28.3 |
| 50 | 54 | Wichita city, Kansas | 55,056 | 6 | 54 | Philadelphia city, Pennsylvania | 183,191 | 30 | 54 | Milwaukee city, Wisconsin | 28.0 |
| 29 | 55 | Louisville/Jefferson County, Kentuck | 54,853 | 31 | 55 | Baltimore city, Maryland | 179,135 | 14 | 55 | Columbus city, Ohio | 27.7 |
| 65 | 56 | Pittsburgh city, Pennsylvania | 53,799 | 40 | 56 | Omaha city, Nebraska | 175,784 | 72 | 55 | Durham city, North Carolina | 27.7 |
| 7 | 57 | San Antonio city, Texas | 53,751 | 14 | 57 | Columbus city, Ohio | 173,257 | 32 | 57 | Albuquerque city, New Mexico | 27.6 |
| 34 | 58 | Fresno city, California | 53,161 | 33 | 58 | Tucson city, Arizona | 172,731 | 29 | 58 | Louisville/Jefferson County, Kentuck | 27.5 |
| 4 | 59 | Houston city, Texas | 52,450 | 29 | 59 | Louisville/Jefferson County, Kentuck | 172,095 | 54 | 59 | Cleveland city, Ohio | 27.5 |
| 31 | 60 | Baltimore city, Maryland | 50,177 | 7 | 60 | San Antonio city, Texas | 171,104 | 22 | 60 | El Paso city, Texas | 26.9 |
| 66 | 61 | Greensboro city, North Carolina | 49,748 | 64 | 61 | Cincinnati city, Ohio | 170,774 | 73 | 60 | Irvine city, California | 26.9 |
| 16 | 62 | Indianapolis city, Indiana | 49,661 | 66 | 62 | Greensboro city, North Carolina | 169,553 | 28 | 62 | Memphis city, Tennessee | 26.5 |
| 48 | 63 | Tulsa city, Oklahoma | 49,158 | 38 | 63 | Kansas City city, Missouri | 168,405 | 38 | 63 | Kansas City city, Missouri | 25.8 |
| 22 | 64 | El Paso city, Texas | 48,542 | 26 | 64 | Oklahoma City city, Oklahoma | 165,744 | 34 | 64 | Fresno city, California | 25.7 |
| 6 | 65 | Philadelphia city, Pennsylvania | 47,474 | 60 | 65 | Corpus Christi city, Texas | 157,143 | 66 | 64 | Greensboro city, North Carolina | 25.7 |
| 67 | 66 | St. Louis city, Missouri | 47,176 | 16 | 66 | Indianapolis city, Indiana | 155,399 | 26 | 66 | Oklahoma City city, Oklahoma | 25.2 |
| 64 | 67 | Cincinnati city, Ohio | 46,260 | 48 | 67 | Tulsa city, Oklahoma | 152,656 | 39 | 67 | Colorado Springs city, Colorado | 24.7 |
| 51 | 68 | New Orleans city, Louisiana | 45,615 | 67 | 68 | St. Louis city, Missouri | 150,711 | 53 | 68 | Bakersfield city, California | 23.9 |
| 33 | 69 | Tucson city, Arizona | 44,365 | 65 | 69 | Pittsburgh city, Pennsylvania | 149,165 | 61 | 69 | Lexington-Fayette urban county, Kentucky | 21.4 |
| 30 | 70 | Milwaukee city, Wisconsin | 44,192 | 50 | 70 | Wichita city, Kansas | 146,684 | 40 | 70 | Omaha city, Nebraska | 19.3 |
| 28 | 71 | Memphis city, Tennessee | 43,794 | 30 | 71 | Milwaukee city, Wisconsin | 133,625 | 60 | 71 | Corpus Christi city, Texas | 15.9 |
| 42 | 72 | Miami city, Florida | 42,966 | 22 | 72 | El Paso city, Texas | 133,606 | 71 | 72 | Anchorage municipality, Alaska | 15.0 |
| 74 | 73 | Newark city, New Jersey | 40,235 | 28 | 73 | Memphis city, Tennessee | 115,914 | 48 | 73 | Tulsa city, Oklahoma | 14.4 |
| 24 | 74 | Detroit city, Michigan | 33,965 | 54 | 74 | Cleveland city, Ohio | 71,108 | 50 | 74 | Wichita city, Kansas | 14.1 |
| 54 | 75 | Cleveland city, Ohio | 32,053 | 24 | 75 | Detroit city, Michigan | 58,902 | 69 | 75 | Lincoln city, Nebraska | 13.7 |

# 75 Largest Cities by 2020 Population
## Selected rankings

### Median Non-Family Income, 2019

| Population rank | Poverty rate rank | City | Poverty rate [col 46] |
|---|---|---|---|
| 17 | 1 | San Francisco city, California | 105,235 |
| 20 | 2 | Washington city, District of Columbia | 76,529 |
| 73 | 3 | Irvine city, California | 72,947 |
| 18 | 4 | Seattle city, Washington | 70,847 |
| 10 | 5 | San Jose city, California | 68,749 |
| 21 | 6 | Boston city, Massachusetts | 62,028 |
| 75 | 7 | Chula Vista city, California | 61,964 |
| 8 | 8 | San Diego city, California | 61,890 |
| 46 | 9 | Oakland city, California | 61,028 |
| 19 | 10 | Denver city, Colorado | 59,201 |
| 55 | 11 | Anaheim city, California | 59,091 |
| 11 | 12 | Austin city, Texas | 57,150 |
| 68 | 13 | Plano city, Texas | 55,693 |
| 71 | 14 | Anchorage municipality, Alaska | 54,962 |
| 27 | 15 | Portland city, Oregon | 54,538 |
| 37 | 16 | Atlanta city, Georgia | 53,912 |
| 44 | 17 | Virginia Beach city, Virginia | 53,550 |
| 57 | 18 | Santa Ana city, California | 53,251 |
| 59 | 19 | Henderson city, Nevada | 52,206 |
| 1 | 20 | New York city, New York | 51,293 |
| 43 | 21 | Long Beach city, California | 50,726 |
| 36 | 22 | Sacramento city, California | 50,648 |
| 2 | 23 | Los Angeles city, California | 50,644 |
| 52 | 24 | Aurora city, Colorado | 49,762 |
| 41 | 25 | Raleigh city, North Carolina | 49,736 |
| 45 | 26 | Minneapolis city, Minnesota | 49,618 |
| 56 | 27 | Urban Honolulu CDP, Hawaii | 49,247 |
| 23 | 28 | Nashville-Davidson, Tennessee | 48,784 |
| 70 | 29 | Orlando city, Florida | 47,996 |
| 15 | 30 | Charlotte city, North Carolina | 47,819 |
| 58 | 31 | Riverside city, California | 47,105 |
| 72 | 32 | Durham city, North Carolina | 45,846 |
| 9 | 33 | Dallas city, Texas | 45,844 |
| 3 | 34 | Chicago city, Illinois | 45,744 |
| 35 | 35 | Mesa city, Arizona | 44,906 |
| 14 | 36 | Columbus city, Ohio | 44,679 |
| 63 | 37 | St. Paul city, Minnesota | 44,358 |
| 39 | 38 | Colorado Springs city, Colorado | 44,102 |
| 47 | 39 | Tampa city, Florida | 42,446 |
| 40 | 40 | Omaha city, Nebraska | 42,132 |
| 5 | 41 | Phoenix city, Arizona | 41,932 |
| 49 | 42 | Arlington city, Texas | 41,324 |
| 12 | 43 | Fort Worth city, Texas | 41,150 |
| 4 | 44 | Houston city, Texas | 40,770 |
| 32 | 45 | Albuquerque city, New Mexico | 40,576 |
| 38 | 46 | Kansas City city, Missouri | 40,284 |
| 61 | 47 | Lexington-Fayette urban county, Kentucky | 40,100 |
| 13 | 48 | Jacksonville city, Florida | 39,641 |
| 65 | 49 | Pittsburgh city, Pennsylvania | 39,582 |
| 7 | 50 | San Antonio city, Texas | 38,848 |
| 26 | 51 | Oklahoma City city, Oklahoma | 38,531 |
| 31 | 52 | Baltimore city, Maryland | 37,905 |
| 69 | 53 | Lincoln city, Nebraska | 37,123 |
| 25 | 54 | Las Vegas city, Nevada | 36,839 |
| 29 | 55 | Louisville/Jefferson County, Kentucky | 36,826 |
| 53 | 56 | Bakersfield city, California | 36,127 |
| 6 | 57 | Philadelphia city, Pennsylvania | 35,626 |
| 66 | 58 | Greensboro city, North Carolina | 35,269 |
| 16 | 59 | Indianapolis city, Indiana | 35,104 |
| 62 | 60 | Stockton city, California | 34,420 |
| 67 | 61 | St. Louis city, Missouri | 34,287 |
| 50 | 62 | Wichita city, Kansas | 34,169 |
| 30 | 63 | Milwaukee city, Wisconsin | 33,635 |
| 34 | 64 | Fresno city, California | 33,308 |
| 48 | 65 | Tulsa city, Oklahoma | 32,920 |
| 33 | 66 | Tucson city, Arizona | 32,409 |
| 64 | 67 | Cincinnati city, Ohio | 31,843 |
| 42 | 68 | Miami city, Florida | 31,209 |
| 28 | 69 | Memphis city, Tennessee | 30,951 |
| 51 | 70 | New Orleans city, Louisiana | 30,271 |
| 22 | 71 | El Paso city, Texas | 30,026 |
| 60 | 72 | Corpus Christi city, Texas | 29,869 |
| 74 | 73 | Newark city, New Jersey | 29,331 |
| 24 | 74 | Detroit city, Michigan | 26,607 |
| 54 | 75 | Cleveland city, Ohio | 23,463 |

### Unemployment rate, 2020

| Population rank | Unemployment rate rank | City | Unemployment rate [col 64] |
|---|---|---|---|
| 24 | 1 | Detroit city, Michigan | 22.4 |
| 74 | 2 | Newark city, New Jersey | 14.7 |
| 25 | 3 | Las Vegas city, Nevada | 14.1 |
| 54 | 4 | Cleveland city, Ohio | 13.5 |
| 43 | 5 | Long Beach city, California | 13.3 |
| 2 | 6 | Los Angeles city, California | 12.9 |
| 62 | 7 | Stockton city, California | 12.8 |
| 6 | 8 | Philadelphia city, Pennsylvania | 12.4 |
| 1 | 9 | New York city, New York | 12.3 |
| 51 | 10 | New Orleans city, Louisiana | 12.2 |
| 59 | 10 | Henderson city, Nevada | 12.2 |
| 3 | 12 | Chicago city, Illinois | 12.0 |
| 28 | 13 | Memphis city, Tennessee | 11.3 |
| 53 | 14 | Bakersfield city, California | 11.2 |
| 70 | 14 | Orlando city, Florida | 11.2 |
| 34 | 16 | Fresno city, California | 10.7 |
| 46 | 17 | Oakland city, California | 10.5 |
| 55 | 17 | Anaheim city, California | 10.5 |
| 75 | 19 | Chula Vista city, California | 10.4 |
| 36 | 20 | Sacramento city, California | 9.6 |
| 65 | 21 | Pittsburgh city, Pennsylvania | 9.5 |
| 21 | 22 | Boston city, Massachusetts | 9.2 |
| 50 | 22 | Wichita city, Kansas | 9.2 |
| 66 | 22 | Greensboro city, North Carolina | 9.2 |
| 30 | 25 | Milwaukee city, Wisconsin | 9.1 |
| 57 | 25 | Santa Ana city, California | 9.1 |
| 8 | 27 | San Diego city, California | 9.0 |
| 52 | 27 | Aurora city, Colorado | 9.0 |
| 58 | 27 | Riverside city, California | 9.0 |
| 64 | 27 | Cincinnati city, Ohio | 9.0 |
| 60 | 31 | Corpus Christi city, Texas | 8.9 |
| 4 | 32 | Houston city, Texas | 8.8 |
| 31 | 32 | Baltimore city, Maryland | 8.8 |
| 27 | 34 | Portland city, Oregon | 8.7 |
| 33 | 35 | Tucson city, Arizona | 8.5 |
| 37 | 35 | Atlanta city, Georgia | 8.5 |
| 42 | 35 | Miami city, Florida | 8.5 |
| 67 | 35 | St. Louis city, Missouri | 8.5 |
| 32 | 39 | Albuquerque city, New Mexico | 8.3 |
| 19 | 40 | Denver city, Colorado | 8.2 |
| 10 | 41 | San Jose city, California | 8.1 |
| 5 | 42 | Phoenix city, Arizona | 8.0 |
| 15 | 42 | Charlotte city, North Carolina | 8.0 |
| 20 | 42 | Washington city, District of Columbia | 8.0 |
| 23 | 42 | Nashville-Davidson, Tennessee | 8.0 |
| 47 | 42 | Tampa city, Florida | 8.0 |
| 9 | 47 | Dallas city, Texas | 7.9 |
| 14 | 47 | Columbus city, Ohio | 7.9 |
| 16 | 47 | Indianapolis city, Indiana | 7.9 |
| 22 | 47 | El Paso city, Texas | 7.9 |
| 17 | 51 | San Francisco city, California | 7.8 |
| 49 | 52 | Arlington city, Texas | 7.7 |
| 12 | 53 | Fort Worth city, Texas | 7.6 |
| 63 | 53 | St. Paul city, Minnesota | 7.6 |
| 7 | 55 | San Antonio city, Texas | 7.5 |
| 45 | 55 | Minneapolis city, Minnesota | 7.5 |
| 39 | 57 | Colorado Springs city, Colorado | 7.4 |
| 71 | 57 | Anchorage municipality, Alaska | 7.4 |
| 41 | 59 | Raleigh city, North Carolina | 7.3 |
| 35 | 60 | Mesa city, Arizona | 7.2 |
| 38 | 60 | Kansas City city, Missouri | 7.2 |
| 48 | 62 | Tulsa city, Oklahoma | 7.1 |
| 73 | 62 | Irvine city, California | 7.1 |
| 18 | 64 | Seattle city, Washington | 6.9 |
| 13 | 65 | Jacksonville city, Florida | 6.8 |
| 29 | 65 | Louisville/Jefferson County, Kentucky | 6.8 |
| 72 | 67 | Durham city, North Carolina | 6.6 |
| 26 | 68 | Oklahoma City city, Oklahoma | 6.5 |
| 68 | 69 | Plano city, Texas | 6.3 |
| 11 | 70 | Austin city, Texas | 6.2 |
| 44 | 70 | Virginia Beach city, Virginia | 6.2 |
| 61 | 72 | Lexington-Fayette urban county, Kentucky | 5.7 |
| 40 | 73 | Omaha city, Nebraska | 5.2 |
| 69 | 74 | Lincoln city, Nebraska | 4.3 |
| 56 | 75 | Urban Honolulu CDP, Hawaii | 0.0 |

### Percent change in civilian labor force, 2019-2020

| Population rank | Percent change rank | City | Percent change [col 62] |
|---|---|---|---|
| 24 | 1 | Detroit city, Michigan | 5.1 |
| 74 | 2 | Newark city, New Jersey | 3.4 |
| 50 | 3 | Wichita city, Kansas | 2.5 |
| 14 | 4 | Columbus city, Ohio | 1.7 |
| 51 | 4 | New Orleans city, Louisiana | 1.7 |
| 40 | 6 | Omaha city, Nebraska | 1.6 |
| 62 | 7 | Stockton city, California | 1.4 |
| 64 | 8 | Cincinnati city, Ohio | 1.2 |
| 45 | 9 | Minneapolis city, Minnesota | 1.1 |
| 52 | 9 | Aurora city, Colorado | 1.1 |
| 5 | 11 | Phoenix city, Arizona | 1.0 |
| 12 | 11 | Fort Worth city, Texas | 1.0 |
| 71 | 11 | Anchorage municipality, Alaska | 1.0 |
| 39 | 14 | Colorado Springs city, Colorado | 0.9 |
| 27 | 15 | Portland city, Oregon | 0.8 |
| 28 | 15 | Memphis city, Tennessee | 0.8 |
| 35 | 15 | Mesa city, Arizona | 0.8 |
| 37 | 15 | Atlanta city, Georgia | 0.8 |
| 26 | 19 | Oklahoma City city, Oklahoma | 0.7 |
| 47 | 19 | Tampa city, Florida | 0.7 |
| 19 | 21 | Denver city, Colorado | 0.6 |
| 16 | 22 | Indianapolis city, Indiana | 0.5 |
| 6 | 23 | Philadelphia city, Pennsylvania | 0.3 |
| 22 | 24 | El Paso city, Texas | 0.2 |
| 48 | 24 | Tulsa city, Oklahoma | 0.2 |
| 63 | 26 | St. Paul city, Minnesota | 0.1 |
| 56 | 27 | Urban Honolulu CDP, Hawaii | 0.0 |
| 69 | 27 | Lincoln city, Nebraska | 0.0 |
| 20 | 29 | Washington city, District of Columbia | -0.1 |
| 36 | 29 | Sacramento city, California | -0.1 |
| 30 | 31 | Milwaukee city, Wisconsin | -0.3 |
| 33 | 31 | Tucson city, Arizona | -0.3 |
| 34 | 31 | Fresno city, California | -0.3 |
| 38 | 31 | Kansas City city, Missouri | -0.3 |
| 49 | 31 | Arlington city, Texas | -0.3 |
| 18 | 36 | Seattle city, Washington | -0.4 |
| 9 | 37 | Dallas city, Texas | -0.5 |
| 11 | 37 | Austin city, Texas | -0.5 |
| 58 | 37 | Riverside city, California | -0.5 |
| 67 | 37 | St. Louis city, Missouri | -0.5 |
| 7 | 41 | San Antonio city, Texas | -0.7 |
| 32 | 42 | Albuquerque city, New Mexico | -0.8 |
| 72 | 42 | Durham city, North Carolina | -0.8 |
| 53 | 44 | Bakersfield city, California | -1.0 |
| 3 | 45 | Chicago city, Illinois | -1.1 |
| 75 | 45 | Chula Vista city, California | -1.1 |
| 54 | 47 | Cleveland city, Ohio | -1.2 |
| 13 | 48 | Jacksonville city, Florida | -1.3 |
| 15 | 48 | Charlotte city, North Carolina | -1.3 |
| 46 | 50 | Oakland city, California | -1.5 |
| 4 | 51 | Houston city, Texas | -1.6 |
| 59 | 52 | Henderson city, Nevada | -1.7 |
| 44 | 53 | Virginia Beach city, Virginia | -1.8 |
| 65 | 53 | Pittsburgh city, Pennsylvania | -1.8 |
| 68 | 53 | Plano city, Texas | -1.8 |
| 23 | 56 | Nashville-Davidson, Tennessee | -1.9 |
| 25 | 57 | Las Vegas city, Nevada | -2.0 |
| 61 | 57 | Lexington-Fayette urban county, Kentucky | -2.0 |
| 60 | 59 | Corpus Christi city, Texas | -2.1 |
| 10 | 60 | San Jose city, California | -2.4 |
| 66 | 61 | Greensboro city, North Carolina | -2.6 |
| 41 | 62 | Raleigh city, North Carolina | -2.8 |
| 55 | 62 | Anaheim city, California | -2.8 |
| 70 | 64 | Orlando city, Florida | -2.9 |
| 29 | 65 | Louisville/Jefferson County, Kentucky | -3.0 |
| 31 | 66 | Baltimore city, Maryland | -3.3 |
| 2 | 67 | Los Angeles city, California | -3.4 |
| 8 | 67 | San Diego city, California | -3.4 |
| 21 | 69 | Boston city, Massachusetts | -3.5 |
| 43 | 70 | Long Beach city, California | -3.8 |
| 1 | 71 | New York city, New York | -3.9 |
| 57 | 71 | Santa Ana city, California | -3.9 |
| 73 | 73 | Irvine city, California | -4.0 |
| 17 | 74 | San Francisco city, California | -4.7 |
| 42 | 75 | Miami city, Florida | -5.2 |

# 75 Largest Cities by 2020 Population
## Selected rankings

### Per capita local government taxes, 2017

| Population rank | Local taxes rank | City | Local per capita taxes (dollars) [col 121] |
|---|---|---|---|
| 20 | 1 | Washington city, District of Columbia ... | 10,729 |
| 1 | 2 | New York city, New York | 6,555 |
| 17 | 3 | San Francisco city, California | 4,255 |
| 21 | 4 | Boston city, Massachusetts | 3,439 |
| 31 | 5 | Baltimore city, Maryland | 2,438 |
| 6 | 6 | Philadelphia city, Pennsylvania | 2,346 |
| 23 | 7 | Nashville-Davidson, Tennessee | 2,329 |
| 44 | 8 | Virginia Beach city, Virginia | 2,107 |
| 71 | 9 | Anchorage municipality, Alaska | 2,070 |
| 18 | 10 | Seattle city, Washington | 1,966 |
| 46 | 11 | Oakland city, California | 1,915 |
| 19 | 12 | Denver city, Colorado | 1,893 |
| 67 | 13 | St. Louis city, Missouri | 1,875 |
| 64 | 14 | Cincinnati city, Ohio | 1,810 |
| 38 | 15 | Kansas City city, Missouri | 1,513 |
| 51 | 16 | New Orleans city, Louisiana | 1,416 |
| 65 | 17 | Pittsburgh city, Pennsylvania | 1,385 |
| 27 | 18 | Portland city, Oregon | 1,371 |
| 3 | 19 | Chicago city, Illinois | 1,314 |
| 37 | 20 | Atlanta city, Georgia | 1,300 |
| 61 | 21 | Lexington-Fayette urban county, Kentucky | 1,277 |
| 74 | 22 | Newark city, New Jersey | 1,234 |
| 54 | 23 | Cleveland city, Ohio | 1,225 |
| 10 | 24 | San Jose city, California | 1,200 |
| 42 | 25 | Miami city, Florida | 1,196 |
| 2 | 26 | Los Angeles city, California | 1,184 |
| 45 | 27 | Minneapolis city, Minnesota | 1,120 |
| 13 | 28 | Jacksonville city, Florida | 1,114 |
| 24 | 29 | Detroit city, Michigan | 1,108 |
| 4 | 30 | Houston city, Texas | 1,080 |
| 14 | 31 | Columbus city, Ohio | 1,074 |
| 55 | 32 | Anaheim city, California | 1,052 |
| 68 | 33 | Plano city, Texas | 1,050 |
| 8 | 34 | San Diego city, California | 1,001 |
| 15 | 35 | Charlotte city, North Carolina | 999 |
| 26 | 35 | Oklahoma City city, Oklahoma | 999 |
| 9 | 37 | Dallas city, Texas | 990 |
| 48 | 38 | Tulsa city, Oklahoma | 967 |
| 70 | 39 | Orlando city, Florida | 954 |
| 43 | 40 | Long Beach city, California | 939 |
| 11 | 41 | Austin city, Texas | 937 |
| 36 | 42 | Sacramento city, California | 913 |
| 40 | 43 | Omaha city, Nebraska | 898 |
| 12 | 44 | Fort Worth city, Texas | 875 |
| 52 | 45 | Aurora city, Colorado | 857 |
| 29 | 46 | Louisville/Jefferson County, Kentuck.... | 838 |
| 47 | 47 | Tampa city, Florida | 791 |
| 72 | 48 | Durham city, North Carolina | 783 |
| 73 | 49 | Irvine city, California | 777 |
| 5 | 50 | Phoenix city, Arizona | 767 |
| 41 | 51 | Raleigh city, North Carolina | 765 |
| 62 | 52 | Stockton city, California | 732 |
| 60 | 53 | Corpus Christi city, Texas | 722 |
| 49 | 54 | Arlington city, Texas | 714 |
| 66 | 55 | Greensboro city, North Carolina | 709 |
| 39 | 56 | Colorado Springs city, Colorado | 707 |
| 28 | 57 | Memphis city, Tennessee | 696 |
| 32 | 58 | Albuquerque city, New Mexico | 690 |
| 58 | 59 | Riverside city, California | 686 |
| 22 | 60 | El Paso city, Texas | 673 |
| 57 | 61 | Santa Ana city, California | 634 |
| 7 | 62 | San Antonio city, Texas | 626 |
| 34 | 63 | Fresno city, California | 613 |
| 69 | 64 | Lincoln city, Nebraska | 600 |
| 33 | 65 | Tucson city, Arizona | 578 |
| 63 | 66 | St. Paul city, Minnesota | 574 |
| 75 | 67 | Chula Vista city, California | 508 |
| 16 | 68 | Indianapolis city, Indiana | 492 |
| 59 | 68 | Henderson city, Nevada | 492 |
| 30 | 70 | Milwaukee city, Wisconsin | 482 |
| 50 | 71 | Wichita city, Kansas | 471 |
| 53 | 72 | Bakersfield city, California | 462 |
| 35 | 73 | Mesa city, Arizona | 439 |
| 25 | 74 | Las Vegas city, Nevada | 354 |
| 56 | 75 | Urban Honolulu CDP, Hawaii | 0 |

### Per capita city government debt outstanding, 2017

| Population rank | Debt rank | City | Debt per capita (dollars) [col 138] |
|---|---|---|---|
| 23 | 1 | Nashville-Davidson, Tennessee | 21,246 |
| 20 | 2 | Washington city, District of Columbia ... | 20,747 |
| 17 | 3 | San Francisco city, California | 19,138 |
| 1 | 4 | New York city, New York | 16,666 |
| 37 | 5 | Atlanta city, Georgia | 14,652 |
| 13 | 6 | Jacksonville city, Florida | 9,830 |
| 19 | 7 | Denver city, Colorado | 8,885 |
| 3 | 8 | Chicago city, Illinois | 8,683 |
| 9 | 9 | Dallas city, Texas | 8,129 |
| 63 | 10 | St. Paul city, Minnesota | 7,809 |
| 7 | 11 | San Antonio city, Texas | 7,751 |
| 38 | 12 | Kansas City city, Missouri | 7,339 |
| 2 | 13 | Los Angeles city, California | 6,949 |
| 18 | 14 | Seattle city, Washington | 6,929 |
| 45 | 15 | Minneapolis city, Minnesota | 6,898 |
| 50 | 16 | Wichita city, Kansas | 6,380 |
| 54 | 17 | Cleveland city, Ohio | 6,253 |
| 67 | 18 | St. Louis city, Missouri | 6,173 |
| 51 | 19 | New Orleans city, Louisiana | 6,166 |
| 4 | 20 | Houston city, Texas | 6,083 |
| 11 | 21 | Austin city, Texas | 5,669 |
| 58 | 22 | Riverside city, California | 5,477 |
| 39 | 23 | Colorado Springs city, Colorado | 5,335 |
| 71 | 24 | Anchorage municipality, Alaska | 5,331 |
| 46 | 25 | Oakland city, California | 5,255 |
| 55 | 26 | Anaheim city, California | 5,203 |
| 60 | 27 | Corpus Christi city, Texas | 5,145 |
| 43 | 28 | Long Beach city, California | 5,073 |
| 14 | 29 | Columbus city, Ohio | 4,871 |
| 27 | 30 | Portland city, Oregon | 4,863 |
| 69 | 31 | Lincoln city, Nebraska | 4,811 |
| 15 | 32 | Charlotte city, North Carolina | 4,736 |
| 5 | 33 | Phoenix city, Arizona | 4,729 |
| 24 | 34 | Detroit city, Michigan | 4,726 |
| 10 | 35 | San Jose city, California | 4,618 |
| 36 | 36 | Sacramento city, California | 4,559 |
| 70 | 37 | Orlando city, Florida | 4,406 |
| 47 | 38 | Tampa city, Florida | 4,284 |
| 16 | 39 | Indianapolis city, Indiana | 4,137 |
| 73 | 40 | Irvine city, California | 3,866 |
| 41 | 41 | Raleigh city, North Carolina | 3,608 |
| 48 | 42 | Tulsa city, Oklahoma | 3,512 |
| 22 | 43 | El Paso city, Texas | 3,339 |
| 44 | 44 | Virginia Beach city, Virginia | 3,299 |
| 6 | 45 | Philadelphia city, Pennsylvania | 3,162 |
| 28 | 46 | Memphis city, Tennessee | 3,082 |
| 35 | 47 | Mesa city, Arizona | 3,073 |
| 40 | 48 | Omaha city, Nebraska | 2,912 |
| 26 | 49 | Oklahoma City city, Oklahoma | 2,857 |
| 61 | 50 | Lexington-Fayette urban county, Kentucky | 2,856 |
| 64 | 51 | Cincinnati city, Ohio | 2,846 |
| 21 | 52 | Boston city, Massachusetts | 2,806 |
| 32 | 53 | Albuquerque city, New Mexico | 2,800 |
| 29 | 54 | Louisville/Jefferson County, Kentuck.... | 2,783 |
| 49 | 55 | Arlington city, Texas | 2,643 |
| 30 | 56 | Milwaukee city, Wisconsin | 2,538 |
| 75 | 57 | Chula Vista city, California | 2,487 |
| 34 | 58 | Fresno city, California | 2,296 |
| 52 | 59 | Aurora city, Colorado | 2,203 |
| 8 | 60 | San Diego city, California | 1,968 |
| 62 | 61 | Stockton city, California | 1,899 |
| 33 | 62 | Tucson city, Arizona | 1,858 |
| 53 | 63 | Bakersfield city, California | 1,751 |
| 66 | 64 | Greensboro city, North Carolina | 1,704 |
| 12 | 65 | Fort Worth city, Texas | 1,694 |
| 65 | 66 | Pittsburgh city, Pennsylvania | 1,570 |
| 74 | 67 | Newark city, New Jersey | 1,543 |
| 42 | 68 | Miami city, Florida | 1,429 |
| 31 | 69 | Baltimore city, Maryland | 1,341 |
| 25 | 70 | Las Vegas city, Nevada | 1,255 |
| 72 | 71 | Durham city, North Carolina | 1,232 |
| 59 | 72 | Henderson city, Nevada | 772 |
| 57 | 73 | Santa Ana city, California | 671 |
| 68 | 74 | Plano city, Texas | 471 |
| 56 | 75 | Urban Honolulu CDP, Hawaii | 0 |

### Violent crime rate, 2019 (violent crimes known to police)

| Population rank | Violent crime rate rank | City | Violent crimes (per 100,000 population) [col 36] |
|---|---|---|---|
| 24 | 1 | Detroit city, Michigan | 1,965 |
| 67 | 2 | St. Louis city, Missouri | 1,927 |
| 28 | 3 | Memphis city, Tennessee | 1,901 |
| 31 | 4 | Baltimore city, Maryland | 1,859 |
| 54 | 5 | Cleveland city, Ohio | 1,517 |
| 38 | 6 | Kansas City city, Missouri | 1,431 |
| 62 | 7 | Stockton city, California | 1,397 |
| 32 | 8 | Albuquerque city, New Mexico | 1,352 |
| 30 | 9 | Milwaukee city, Wisconsin | 1,332 |
| 46 | 10 | Oakland city, California | 1,272 |
| 71 | 11 | Anchorage municipality, Alaska | 1,245 |
| 51 | 12 | New Orleans city, Louisiana | 1,145 |
| 50 | 13 | Wichita city, Kansas | 1,141 |
| 4 | 14 | Houston city, Texas | 1,072 |
| 48 | 15 | Tulsa city, Oklahoma | 987 |
| 20 | 16 | Washington city, District of Columbia .. | 977 |
| 3 | 17 | Chicago city, Illinois | 943 |
| 45 | 18 | Minneapolis city, Minnesota | 926 |
| 9 | 19 | Dallas city, Texas | 863 |
| 64 | 20 | Cincinnati city, Ohio | 845 |
| 66 | 21 | Greensboro city, North Carolina | 819 |
| 60 | 22 | Corpus Christi city, Texas | 794 |
| 19 | 23 | Denver city, Colorado | 749 |
| 15 | 24 | Charlotte city, North Carolina | 739 |
| 70 | 25 | Orlando city, Florida | 738 |
| 52 | 26 | Aurora city, Colorado | 735 |
| 2 | 27 | Los Angeles city, California | 732 |
| 72 | 28 | Durham city, North Carolina | 730 |
| 26 | 29 | Oklahoma City city, Oklahoma | 722 |
| 7 | 30 | San Antonio city, Texas | 708 |
| 5 | 31 | Phoenix city, Arizona | 699 |
| 33 | 32 | Tucson city, Arizona | 688 |
| 17 | 33 | San Francisco city, California | 670 |
| 13 | 34 | Jacksonville city, Florida | 647 |
| 36 | 35 | Sacramento city, California | 627 |
| 74 | 36 | Newark city, New Jersey | 619 |
| 40 | 37 | Omaha city, Nebraska | 613 |
| 42 | 38 | Miami city, Florida | 593 |
| 18 | 39 | Seattle city, Washington | 585 |
| 39 | 39 | Colorado Springs city, Colorado | 585 |
| 1 | 41 | New York city, New York | 571 |
| 63 | 42 | St. Paul city, Minnesota | 565 |
| 27 | 43 | Portland city, Oregon | 545 |
| 25 | 44 | Las Vegas city, Nevada | 531 |
| 49 | 45 | Arlington city, Texas | 511 |
| 43 | 46 | Long Beach city, California | 506 |
| 58 | 46 | Riverside city, California | 506 |
| 14 | 48 | Columbus city, Ohio | 503 |
| 53 | 49 | Bakersfield city, California | 455 |
| 12 | 50 | Fort Worth city, Texas | 444 |
| 10 | 51 | San Jose city, California | 438 |
| 57 | 52 | Santa Ana city, California | 435 |
| 47 | 53 | Tampa city, Florida | 405 |
| 11 | 54 | Austin city, Texas | 401 |
| 69 | 55 | Lincoln city, Nebraska | 383 |
| 35 | 56 | Mesa city, Arizona | 377 |
| 8 | 57 | San Diego city, California | 362 |
| 22 | 58 | El Paso city, Texas | 353 |
| 75 | 59 | Chula Vista city, California | 328 |
| 55 | 60 | Anaheim city, California | 316 |
| 61 | 61 | Lexington-Fayette urban county, Kentucky | 297 |
| 56 | 62 | Urban Honolulu CDP, Hawaii | 271 |
| 41 | 63 | Raleigh city, North Carolina | 256 |
| 59 | 64 | Henderson city, Nevada | 171 |
| 68 | 65 | Plano city, Texas | 148 |
| 44 | 66 | Virginia Beach city, Virginia | 129 |
| 21 | 67 | Boston city, Massachusetts | 121 |
| 73 | 68 | Irvine city, California | 64 |
| 6 | | Philadelphia city, Pennsylvania | NA |
| 16 | | Indianapolis city, Indiana | NA |
| 23 | | Nashville-Davidson, Tennessee | NA |
| 29 | | Louisville/Jefferson County, Kentuck... | NA |
| 34 | | Fresno city, California | NA |
| 37 | | Atlanta city, Georgia | NA |
| 65 | | Pittsburgh city, Pennsylvania | NA |

# 75 Largest Cities by 2020 Population
## Selected rankings

Property crime rate, 2019 (property crimes known to police) | Full-time equivalent local government employees, 2017 | Local government employee payroll for March 2017

| Population rank | Property crime rate rank | City | Property crimes (per 100,000 population) [col 38] | Population rank | Government employees rank | City | Government employees [col 108] | Population rank | Government employee payroll rank | City | Government payroll (thousands of dollars) [col 109] |
|---|---|---|---|---|---|---|---|---|---|---|---|
| 46 | 1 | Oakland city, California | 6,421 | 1 | 1 | New York city, New York | 434,443 | 1 | 1 | New York city, New York | 2,716,087,055 |
| 67 | 2 | St. Louis city, Missouri | 6,183 | 2 | 2 | Los Angeles city, California | 52,287 | 2 | 2 | Los Angeles city, California | 452,879,430 |
| 28 | 3 | Memphis city, Tennessee | 6,128 | 20 | 3 | Washington city, District of Columbia | 40,352 | 20 | 3 | Washington city, District of Columbia | 276,215,832 |
| 17 | 4 | San Francisco city, California | 5,506 | 17 | 4 | San Francisco city, California | 33,660 | 17 | 4 | San Francisco city, California | 260,920,281 |
| 50 | 5 | Wichita city, Kansas | 5,322 | 3 | 5 | Chicago city, Illinois | 31,798 | 3 | 5 | Chicago city, Illinois | 220,385,761 |
| 48 | 6 | Tulsa city, Oklahoma | 5,311 | 6 | 6 | Philadelphia city, Pennsylvania | 30,557 | 6 | 6 | Philadelphia city, Pennsylvania | 177,705,357 |
| 51 | 7 | New Orleans city, Louisiana | 5,293 | 31 | 7 | Baltimore city, Maryland | 25,611 | 31 | 7 | Baltimore city, Maryland | 138,264,675 |
| 27 | 8 | Portland city, Oregon | 5,203 | 4 | 8 | Houston city, Texas | 21,657 | 21 | 8 | Boston city, Massachusetts | 132,820,934 |
| 70 | 9 | Orlando city, Florida | 4,827 | 23 | 9 | Nashville-Davidson, Tennessee | 21,283 | 4 | 9 | Houston city, Texas | 108,741,222 |
| 45 | 10 | Minneapolis city, Minnesota | 4,517 | 21 | 10 | Boston city, Massachusetts | 19,672 | 23 | 10 | Nashville-Davidson, Tennessee | 101,098,567 |
| 18 | 11 | Seattle city, Washington | 4,496 | 44 | 11 | Virginia Beach city, Virginia | 17,638 | 5 | 11 | Phoenix city, Arizona | 98,028,340 |
| 54 | 12 | Cleveland city, Ohio | 4,467 | 7 | 12 | San Antonio city, Texas | 15,585 | 18 | 12 | Seattle city, Washington | 94,722,910 |
| 7 | 13 | San Antonio city, Texas | 4,324 | 9 | 13 | Dallas city, Texas | 14,767 | 7 | 13 | San Antonio city, Texas | 91,157,442 |
| 4 | 14 | Houston city, Texas | 4,319 | 19 | 14 | Denver city, Colorado | 14,162 | 19 | 14 | Denver city, Colorado | 86,682,654 |
| 31 | 15 | Baltimore city, Maryland | 4,311 | 11 | 15 | Austin city, Texas | 13,610 | 9 | 15 | Dallas city, Texas | 85,321,347 |
| 24 | 16 | Detroit city, Michigan | 4,303 | 16 | 16 | Indianapolis city, Indiana | 13,456 | 11 | 16 | Austin city, Texas | 84,706,545 |
| 64 | 16 | Cincinnati city, Ohio | 4,303 | 5 | 17 | Phoenix city, Arizona | 12,939 | 8 | 17 | San Diego city, California | 75,006,131 |
| 71 | 18 | Anchorage municipality, Alaska | 4,261 | 18 | 18 | Seattle city, Washington | 11,912 | 44 | 18 | Virginia Beach city, Virginia | 70,561,307 |
| 20 | 19 | Washington city, District of Columbia | 4,246 | 8 | 19 | San Diego city, California | 10,832 | 16 | 19 | Indianapolis city, Indiana | 61,603,859 |
| 53 | 20 | Bakersfield city, California | 4,142 | 28 | 20 | Memphis city, Tennessee | 10,740 | 71 | 20 | Anchorage municipality, Alaska | 52,804,925 |
| 26 | 21 | Oklahoma City city, Oklahoma | 4,092 | 71 | 21 | Anchorage municipality, Alaska | 9,255 | 14 | 21 | Columbus city, Ohio | 52,605,234 |
| 62 | 22 | Stockton city, California | 3,944 | 13 | 22 | Jacksonville city, Florida | 9,087 | 28 | 22 | Memphis city, Tennessee | 51,032,329 |
| 15 | 23 | Charlotte city, North Carolina | 3,926 | 24 | 23 | Detroit city, Michigan | 8,795 | 27 | 23 | Portland city, Oregon | 42,536,596 |
| 38 | 24 | Kansas City city, Missouri | 3,856 | 14 | 24 | Columbus city, Ohio | 8,528 | 13 | 24 | Jacksonville city, Florida | 42,343,407 |
| 72 | 25 | Durham city, North Carolina | 3,808 | 29 | 25 | Louisville/Jefferson County, Kentucky | 8,320 | 15 | 25 | Charlotte city, North Carolina | 40,618,050 |
| 19 | 26 | Denver city, Colorado | 3,744 | 37 | 26 | Atlanta city, Georgia | 8,317 | 37 | 26 | Atlanta city, Georgia | 39,867,373 |
| 11 | 27 | Austin city, Texas | 3,711 | 15 | 27 | Charlotte city, North Carolina | 8,026 | 43 | 27 | Long Beach city, California | 39,697,152 |
| 66 | 28 | Greensboro city, North Carolina | 3,689 | 54 | 28 | Cleveland city, Ohio | 7,046 | 12 | 28 | Fort Worth city, Texas | 38,281,122 |
| 42 | 29 | Miami city, Florida | 3,668 | 51 | 29 | New Orleans city, Louisiana | 6,939 | 32 | 29 | Albuquerque city, New Mexico | 37,997,192 |
| 39 | 30 | Colorado Springs city, Colorado | 3,667 | 32 | 30 | Albuquerque city, New Mexico | 6,938 | 67 | 30 | St. Louis city, Missouri | 37,801,592 |
| 40 | 31 | Omaha city, Nebraska | 3,644 | 12 | 31 | Fort Worth city, Texas | 6,703 | 30 | 31 | Milwaukee city, Wisconsin | 37,635,419 |
| 63 | 32 | St. Paul city, Minnesota | 3,612 | 67 | 32 | St. Louis city, Missouri | 6,658 | 46 | 32 | Oakland city, California | 36,522,164 |
| 60 | 33 | Corpus Christi city, Texas | 3,446 | 22 | 33 | El Paso city, Texas | 6,522 | 29 | 33 | Louisville/Jefferson County, Kentucky | 36,302,120 |
| 9 | 34 | Dallas city, Texas | 3,321 | 30 | 34 | Milwaukee city, Wisconsin | 6,507 | 54 | 34 | Cleveland city, Ohio | 35,836,530 |
| 5 | 35 | Phoenix city, Arizona | 3,315 | 38 | 35 | Kansas City city, Missouri | 6,334 | 51 | 35 | New Orleans city, Louisiana | 33,679,953 |
| 13 | 36 | Jacksonville city, Florida | 3,309 | 27 | 36 | Portland city, Oregon | 6,225 | 24 | 36 | Detroit city, Michigan | 32,957,680 |
| 14 | 37 | Columbus city, Ohio | 3,308 | 64 | 37 | Cincinnati city, Ohio | 5,538 | 36 | 37 | Sacramento city, California | 31,089,039 |
| 33 | 38 | Tucson city, Arizona | 3,272 | 43 | 38 | Long Beach city, California | 5,477 | 64 | 38 | Cincinnati city, Ohio | 30,765,986 |
| 36 | 39 | Sacramento city, California | 3,182 | 33 | 39 | Tucson city, Arizona | 4,956 | 45 | 39 | Minneapolis city, Minnesota | 29,451,317 |
| 56 | 40 | Urban Honolulu CDP, Hawaii | 3,002 | 45 | 40 | Minneapolis city, Minnesota | 4,811 | 38 | 40 | Kansas City city, Missouri | 28,199,434 |
| 61 | 41 | Lexington-Fayette urban county, Kentucky | 2,998 | 41 | 41 | Raleigh city, North Carolina | 4,767 | 39 | 41 | Colorado Springs city, Colorado | 27,500,431 |
| 3 | 42 | Chicago city, Illinois | 2,983 | 26 | 42 | Oklahoma City city, Oklahoma | 4,494 | 26 | 42 | Oklahoma City city, Oklahoma | 27,224,626 |
| 58 | 43 | Riverside city, California | 2,938 | 47 | 43 | Tampa, Florida | 4,198 | 33 | 43 | Tucson city, Arizona | 25,663,881 |
| 52 | 44 | Aurora city, Colorado | 2,918 | 36 | 44 | Sacramento city, California | 4,106 | 22 | 44 | El Paso city, Texas | 25,566,174 |
| 49 | 45 | Arlington city, Texas | 2,807 | 46 | 45 | Oakland city, California | 4,077 | 47 | 45 | Tampa city, Florida | 25,536,216 |
| 25 | 46 | Las Vegas city, Nevada | 2,772 | 39 | 46 | Colorado Springs city, Colorado | 4,068 | 35 | 46 | Mesa city, Arizona | 22,765,005 |
| 69 | 47 | Lincoln city, Nebraska | 2,751 | 42 | 47 | Miami city, Florida | 3,951 | 10 | 47 | San Jose city, California | 22,735,265 |
| 12 | 48 | Fort Worth city, Texas | 2,688 | 48 | 48 | Tulsa city, Oklahoma | 3,773 | 42 | 48 | Miami city, Florida | 21,993,811 |
| 30 | 49 | Milwaukee city, Wisconsin | 2,555 | 61 | 49 | Lexington-Fayette urban county, Kentucky | 3,756 | 34 | 49 | Fresno city, California | 20,907,203 |
| 10 | 50 | San Jose city, California | 2,420 | 74 | 50 | Newark city, New Jersey | 3,735 | 41 | 50 | Raleigh city, North Carolina | 20,560,599 |
| 43 | 51 | Long Beach city, California | 2,414 | 35 | 51 | Mesa city, Arizona | 3,715 | 25 | 51 | Las Vegas city, Nevada | 20,161,448 |
| 2 | 52 | Los Angeles city, California | 2,383 | 34 | 52 | Fresno city, California | 3,451 | 58 | 52 | Riverside city, California | 19,847,957 |
| 55 | 53 | Anaheim city, California | 2,333 | 10 | 53 | San Jose city, California | 3,403 | 55 | 53 | Anaheim city, California | 19,689,571 |
| 57 | 54 | Santa Ana city, California | 2,040 | 65 | 54 | Pittsburgh city, Pennsylvania | 3,293 | 48 | 54 | Tulsa city, Oklahoma | 19,022,256 |
| 8 | 55 | San Diego city, California | 1,883 | 66 | 55 | Greensboro city, North Carolina | 3,249 | 61 | 55 | Lexington-Fayette urban county, Kentucky | 18,866,699 |
| 35 | 56 | Mesa city, Arizona | 1,869 | 63 | 56 | St. Paul city, Minnesota | 3,186 | 63 | 56 | St. Paul city, Minnesota | 18,549,813 |
| 41 | 57 | Raleigh city, North Carolina | 1,783 | 70 | 57 | Orlando city, Florida | 3,145 | 40 | 57 | Omaha city, Nebraska | 18,004,814 |
| 44 | 58 | Virginia Beach city, Virginia | 1,761 | 60 | 58 | Corpus Christi city, Texas | 2,997 | 70 | 58 | Orlando city, Florida | 17,935,004 |
| 59 | 59 | Henderson city, Nevada | 1,748 | 50 | 59 | Wichita city, Kansas | 2,961 | 52 | 59 | Aurora city, Colorado | 17,630,608 |
| 68 | 60 | Plano city, Texas | 1,683 | 40 | 60 | Omaha city, Nebraska | 2,910 | 65 | 60 | Pittsburgh city, Pennsylvania | 17,441,558 |
| 74 | 61 | Newark city, New Jersey | 1,641 | 52 | 61 | Aurora city, Colorado | 2,891 | 59 | 61 | Henderson city, Nevada | 16,130,341 |
| 47 | 62 | Tampa city, Florida | 1,629 | 25 | 62 | Las Vegas city, Nevada | 2,867 | 74 | 62 | Newark city, New Jersey | 15,962,971 |
| 22 | 63 | El Paso city, Texas | 1,511 | 72 | 63 | Durham city, North Carolina | 2,861 | 69 | 63 | Lincoln city, Nebraska | 14,836,123 |
| 1 | 64 | New York city, New York | 1,460 | 69 | 64 | Lincoln city, Nebraska | 2,642 | 50 | 64 | Wichita city, Kansas | 14,226,614 |
| 75 | 65 | Chula Vista city, California | 1,386 | 49 | 65 | Arlington city, Texas | 2,558 | 49 | 65 | Arlington city, Texas | 13,968,721 |
| 21 | 66 | Boston city, Massachusetts | 1,308 | 55 | 66 | Anaheim city, California | 2,310 | 60 | 66 | Corpus Christi city, Texas | 13,665,741 |
| 73 | 67 | Irvine city, California | 1,306 | 59 | 66 | Henderson city, Nevada | 2,310 | 66 | 67 | Greensboro city, North Carolina | 13,016,201 |
| 6 | | Philadelphia city, Pennsylvania | NA | 68 | 68 | Plano city, Texas | 2,302 | 68 | 68 | Plano city, Texas | 12,930,111 |
| 16 | | Indianapolis city, Indiana | NA | 58 | 69 | Riverside city, California | 2,170 | 72 | 69 | Durham city, North Carolina | 12,296,321 |
| 23 | | Nashville-Davidson, Tennessee | NA | 62 | 70 | Stockton city, California | 1,598 | 62 | 70 | Stockton city, California | 11,211,014 |
| 29 | | Louisville/Jefferson County, Kentucky | NA | 53 | 71 | Bakersfield city, California | 1,490 | 73 | 71 | Irvine city, California | 10,354,500 |
| 32 | | Albuquerque city, New Mexico | NA | 57 | 72 | Santa Ana city, California | 1,094 | 53 | 72 | Bakersfield city, California | 9,653,361 |
| 34 | | Fresno city, California | NA | 73 | 73 | Irvine city, California | 1,045 | 57 | 73 | Santa Ana city, California | 9,302,161 |
| 37 | | Atlanta city, Georgia | NA | 75 | 74 | Chula Vista city, California | 1,035 | 75 | 74 | Chula Vista city, California | 7,321,651 |
| 65 | | Pittsburgh city, Pennsylvania | NA | 56 | 75 | Urban Honolulu CDP, Hawaii | NA | 56 | 75 | Urban Honolulu CDP, Hawaii | NA |

## Table D. Cities — **Land Area and Population**

| STATE Place code | City | Land area[1] (sq. mi) | Population, 2020 | | | Race 2019 | | | | | | |
|---|---|---|---|---|---|---|---|---|---|---|---|---|
| | | | Total persons 2020 | Rank | Per square mile | White | Black or African American | American Indian, Alaskan Native | Asian | Hawaiian Pacific Islander | Some other race | Two or more races (percent) |
| | | | | | | | Race alone[2] (percent) | | | | | |
| | | 1 | 2 | 3 | 4 | 5 | 6 | 7 | 8 | 9 | 10 | 11 |

1. Dry land or land partially or temporarily covered by water.    2. Hispanic or Latino persons may be of any race.

## Table D. Cities — **Population**

| City | Percent Hispanic or Latino[1], 2019 | Percent foreign born 2019 | Age of population (percent), 2019 | | | | | | | Median age, 2019 | Percent female, 2019 | Population | | | |
|---|---|---|---|---|---|---|---|---|---|---|---|---|---|---|---|
| | | | Under 18 years | 18 to 24 years | 25 to 34 years | 35 to 44 years | 45 to 54 years | 55 to 64 years | 65 years and over | | | Census counts | | Percent change | |
| | | | | | | | | | | | | 2000 | 2010 | 2000-2010 | 2001-2020 |
| | 12 | 13 | 14 | 15 | 16 | 17 | 18 | 19 | 20 | 21 | 22 | 23 | 24 | 25 | 26 |

1. May be of any race.

## Table D. Cities — **Households, Group Quarters, Crime, and Education**

| City | Households, 2019 | | | | | | | Persons in group quarters, 2019 | Serious crimes known to police[1], 2019 | | | | Educational attainment, 2019 | | |
|---|---|---|---|---|---|---|---|---|---|---|---|---|---|---|---|
| | | | Percent | | | | | | Violent | | Prorerty | | | Attainment[3] (percent) | |
| | Number | Persons per household | Family | Married couple family | Female family | Non-family | One person | | Number | Rate[2] | Number | Rate[2] | Population age 25 and over | High school graduate or less | Bachelor's degree or more |
| | 27 | 28 | 29 | 30 | 31 | 32 | 33 | 34 | 35 | 36 | 37 | 38 | 39 | 40 | 41 |

1. Data for serious crimes have not been adjusted for underreporting. This may affect comparability between geographic areas and over time.    2. Per 100,000 population estimated by the FBI.    3. Persons 25 years old and over.

## Table D. Cities — **Income, Poverty, and Housing**

| City | Money income, 2019 | | | | | Median earnings, 2019 | | | Housing units, 2019 | | | | |
|---|---|---|---|---|---|---|---|---|---|---|---|---|---|
| | Households | | | Median family income | Median non-family income | All persons | Men | Women | Total | Occupied | Percent owner occupied | Median value[1] (dollars) | Median gross rent (dollars) |
| | Median income | Percent with income less than $20,000 | Percent with income of $200,000 or more | | | | | | | | | | |
| | 42 | 43 | 44 | 45 | 46 | 47 | 48 | 49 | 50 | 51 | 52 | 53 | 54 |

1. Based on population estimated by the American Community Survey.

## Table D. Cities — **Commuting, Computer Access, Migration, Labor Force, and Employment**

| City | Commuting[1], 2019 | | Computer access[2], 2019 | | Migration, 2019 | | Civilian labor force, 2020 | | | | Civilian Employment[4], 2019 | | | |
|---|---|---|---|---|---|---|---|---|---|---|---|---|---|---|
| | Percent | | Percent | | | | | | Unemployment | | Population age 16 and older | | Population age 16 to 64 | |
| | Drove alone | With commutes of 30 minutes or more | With a computer in the house | With Internet access | Percent who lived in the same house one year ago | Percent who lived in another state or county one year ago | Total | Percent change 2019-2020 | Total | Rate[3] | Number | Percent in labor force | Number | Percent who worked full-year full-time |
| | 55 | 56 | 57 | 58 | 59 | 60 | 61 | 62 | 63 | 64 | 65 | 66 | 67 | 68 |

1. Employed persons.   2. Households.   3. Percent of civilian labor force.   4. Persons 16 years old and over.

## Table D. Cities — **Construction, Wholesale Trade, and Retail Trade**

| City | Value of residential construction authorized by building permits, 2020 | | | Wholesale trade[1], 2017 | | | | Retail trade[2], 2017 | | | |
|---|---|---|---|---|---|---|---|---|---|---|---|
| | New construction ($1,000) | Number of housing units | Percent single family | Number of establishments | Number of employees | Sales (mil dol) | Annual payroll (mil dol) | Number of establishments | Number of employees | Sales (mil dol) | Annual payroll (mil dol) |
| | 69 | 70 | 71 | 72 | 73 | 74 | 75 | 76 | 77 | 78 | 79 |

1. Merchant wholesalers except manufacturers' sales branches and offices.   2. Establishments with payroll.

## Table D. Cities — **Real Estate, Professional Services, and Manufacturing**

| City | Real estate and rental and leasing, 2017 | | | | Professional, scientific, and technical services[1], 2017 | | | | Manufacturing, 2017 | | | |
|---|---|---|---|---|---|---|---|---|---|---|---|---|
| | Number of establishments | Number of employees | Receipts (mil dol) | Annual payroll (mil dol) | Number of establishments | Number of employees | Receipts (mil dol) | Annual payroll (mil dol) | Number of establishments | Number of employees | Receipts (mil dol) | Annual payroll (mil dol) |
| | 80 | 81 | 82 | 83 | 84 | 85 | 86 | 87 | 88 | 89 | 90 | 91 |

1. Establishments subject to federal tax.

## Table D. Cities — **Accommodation and Food Services, Arts, Entertainment, and Recreation, and Health Care and Social Assistance**

| City | Accommodation and food services, 2017 | | | | Arts, entertainment, and recreation[1], 2017 | | | | Health care and social assistance[1], 2017 | | | |
|---|---|---|---|---|---|---|---|---|---|---|---|---|
| | Number of establishments | Number of employees | Receipts (mil dol) | Annual payroll (mil dol) | Number of establishments | Number of employees | Receipts (mil dol) | Annual payroll (mil dol) | Number of establishments | Number of employees | Receipts (mil dol) | Annual payroll (mil dol) |
| | 92 | 93 | 94 | 95 | 96 | 97 | 98 | 99 | 100 | 101 | 102 | 103 |

1. Establishments subject to federal tax.

## Table D. Cities — **Other Services and Government Employment and Payroll**

| City | Other services[1] | | | | Government employment and payroll, 2017 | | | | | | | | | |
| | | | | | Full-time equivalent employees | March payroll | | | | | | | | |
| | | | | | | Total (dollars) | Perent of total for: | | | | | | | |
| | Number of establish-ments | Number of employees | Receipts (mil dol) | Annual payroll (mil dol) | | | Admin-istrative, judicial, and legal | Police and corrections | Fire protection | Highways and trans-portation | Health and welfare | Natural resources and utilities | Education and libraries |
| | 104 | 105 | 106 | 107 | 108 | 109 | 110 | 111 | 112 | 113 | 114 | 115 | 116 |

1. Establishments subject to federal tax.

## Table D. Cities — **City Government Finances**

| City | City government finances, 2017 | | | | | | | | | | |
| | General revenue | | | | | | | | General expenditure | | |
| | Intergovernmental | | | Taxes | | | | | | | |
| | | | | | Per capita[1] (dollars) | | | | Per capita[1] (dollars) | |
| | Total (mil dol) | Total (mil dol) | Percent from state government | Total (mil dol) | Total | Property | Sales and gross receipts | Total (mil dol) | Total | Capital outlays |
| | 117 | 118 | 119 | 120 | 121 | 122 | 123 | 124 | 125 | 126 |

1. Based on population estimated as of July 1 of the year shown.

## Table D. Cities — **City Government Finances**

| City | City government finances, 2017 (cont.) | | | | | | | | | |
| | General expenditure (cont.) | | | | | | | | | |
| | Percent of total for: | | | | | | | | | |
| | Public welfare | Highways | Parking facilities | Education | Health and hospitals | Police protection | Sewerage and sanitation | Parks and recreation | Housing and community development | Interest on debt |
| | 127 | 128 | 129 | 130 | 131 | 132 | 133 | 134 | 135 | 136 |

## Table D. Cities — **City Government Finances, City Government Employment, and Climate**

| City | City government finances, 2017 (cont.) | | | Climate[2] | | | | | | | |
| | Debt outstanding | | Debt issued during year | Average daily temperature | | | | Annual precipitation (inches) | Heating degree days | Cooling degree days |
| | | | | Mean | | Limits | | | | |
| | Total (mil dol) | Per capita[1] (dollars) | | January | July | January[3] | July[4] | | | |
| | 137 | 138 | 139 | 140 | 141 | 142 | 143 | 144 | 145 | 146 |

1. Based on the population estimated as of July 1 of the year shown.   2. Represents normal values based on the 30-year period, 1971±2000.   3. Average daily minimum.   4. Average daily maximum.

| STATE Place code | | City | Land area[1] (sq. mi) | Total persons 2019 | Rank | Per square mile | White | Black or African American | American Indian, Alaskan Native | Asian | Hawaiian Pacific Islander | Some other race | Two or more races (percent) |
|---|---|---|---|---|---|---|---|---|---|---|---|---|
| | | | 1 | 2 | 3 | 4 | 5 | 6 | 7 | 8 | 9 | 10 | 11 |
| 00 | 00000 | United States ............ | 3,533,043.7 | 329,484,123 | X | 93.3 | 72.0 | 12.8 | 0.9 | 5.7 | 0.2 | 5.0 | 3.4 |
| 01 | 00000 | ALABAMA .............. | 50,646.6 | 4,921,532 | X | 97.2 | 67.8 | 26.9 | 0.5 | 1.3 | 0 | 1.5 | 1.9 |
| 01 | 00820 | Alabaster.............. | 25.3 | 33,701 | 1,144 | 1,332 | D | D | D | D | D | D | D |
| 01 | 03076 | Auburn............... | 60.7 | 68,343 | 538 | 1,126 | 67.7 | 20.7 | 0.4 | 9.9 | 0.0 | 0.1 | 1.2 |
| 01 | 05980 | Bessemer............. | 40.3 | 26,043 | 1,385 | 646 | D | D | D | D | D | D | D |
| 01 | 07000 | Birmingham.......... | 147.0 | 206,950 | 113 | 1,408 | 27.9 | 67.6 | 0.6 | 1.7 | 0.0 | 0.9 | 1.3 |
| 01 | 20104 | Decatur.............. | 54.4 | 54,381 | 712 | 1,000 | D | D | D | D | D | D | D |
| 01 | 21184 | Dothan............... | 89.8 | 69,414 | 529 | 773 | 58.8 | 35.2 | 0.1 | 1.2 | 0.0 | 2.5 | 2.0 |
| 01 | 24184 | Enterprise............ | 31.0 | 28,997 | 1,287 | 935 | 68.4 | 22.4 | 0.6 | 1.8 | 0.0 | 1.8 | 5.1 |
| 01 | 26896 | Florence.............. | 26.5 | 41,166 | 932 | 1,553 | 71.8 | 22.4 | 0.5 | 0.2 | 0.0 | 1.5 | 3.5 |
| 01 | 28696 | Gadsden.............. | 37.4 | 34,816 | 1,108 | 931 | 54.0 | 35.9 | 0.7 | 1.8 | 0.0 | 6.0 | 1.6 |
| 01 | 35800 | Homewood............ | 8.3 | 25,174 | 1,401 | 3,033 | 79.4 | 11.6 | 0.0 | 5.3 | 0.0 | 1.7 | 2.0 |
| 01 | 35896 | Hoover ............... | 48.1 | 85,959 | 390 | 1,787 | 71.6 | 19.8 | 0.0 | 4.3 | 0.0 | 2.3 | 2.0 |
| 01 | 37000 | Huntsville ............ | 218.1 | 202,964 | 116 | 931 | 59.6 | 32.5 | 0.3 | 1.7 | 0.1 | 3.5 | 2.2 |
| 01 | 45784 | Madison .............. | 30.4 | 52,654 | 741 | 1,732 | 82.4 | 9.6 | 0.0 | 5.4 | 0.0 | 0.5 | 2.1 |
| 01 | 50000 | Mobile ............... | 139.5 | 187,746 | 135 | 1,346 | 39.1 | 55.5 | 0.3 | 1.8 | 0.0 | 0.7 | 2.6 |
| 01 | 51000 | Montgomery .......... | 159.9 | 196,268 | 130 | 1,227 | 31.7 | 60.7 | 0.3 | 3.5 | 0.0 | 1.7 | 2.1 |
| 01 | 57048 | Opelika .............. | 59.1 | 31,135 | 1,214 | 527 | D | D | D | D | D | D | D |
| 01 | 59472 | Phenix City .......... | 28.1 | 36,392 | 1,060 | 1,295 | 47.7 | 43.1 | 0.2 | 1.5 | 0.4 | 4.6 | 2.5 |
| 01 | 62328 | Prattville ............ | 35.4 | 36,500 | 1,054 | 1,031 | D | D | D | D | D | D | D |
| 01 | 77256 | Tuscaloosa ........... | 61.9 | 102,819 | 301 | 1,661 | 50.3 | 45.5 | 0.0 | 2.8 | 0.0 | 0.2 | 1.2 |
| 01 | 78552 | Vestavia Hills ........ | 20.0 | 34,317 | 1,123 | 1,716 | D | D | D | D | D | D | D |
| 02 | 00000 | ALASKA ............... | 571,016.9 | 731,158 | X | 1 | 64.2 | 3.1 | 15.8 | 6.0 | 1.4 | 1.7 | 7.9 |
| 02 | 03000 | Anchorage............ | 1,706.8 | 287,095 | 71 | 168 | 62.6 | 5.3 | 8.9 | 9.0 | 2.8 | 2.3 | 9.1 |
| 02 | 24230 | Fairbanks ............ | 31.7 | 30,696 | 1,230 | 968 | 73.8 | 6.7 | 4.6 | 4.0 | 0.4 | 2.2 | 8.3 |
| 02 | 36400 | Juneau .............. | 2,704.0 | 31,849 | 1,199 | 12 | 66.9 | 0.5 | 11.5 | 6.3 | 1.3 | 2.9 | 10.5 |
| 04 | 00000 | ARIZONA .............. | 113,653.1 | 7,421,401 | X | 65 | 78.3 | 4.7 | 4.6 | 3.3 | 0.2 | 5.0 | 3.9 |
| 04 | 02830 | Apache Junction ....... | 35.1 | 43,729 | 884 | 1,246 | 89.5 | 1.4 | 0.0 | 2.5 | 0.0 | 4.6 | 2.0 |
| 04 | 04720 | Avondale ............. | 47.3 | 88,914 | 379 | 1,880 | 70.4 | 11.3 | 1.2 | 1.9 | 0.7 | 10.6 | 3.9 |
| 04 | 07940 | Buckeye .............. | 393.0 | 85,224 | 397 | 217 | 77.4 | 7.7 | 0.9 | 0.6 | 0.2 | 9.4 | 3.7 |
| 04 | 08220 | Bullhead City.......... | 59.4 | 41,733 | 921 | 703 | 89.8 | 1.1 | 1.4 | 0.1 | 0.0 | 2.4 | 5.3 |
| 04 | 10530 | Casa Grande.......... | 110.9 | 61,183 | 611 | 552 | 78.6 | 2.4 | 3.9 | 1.5 | 0.1 | 4.6 | 9.0 |
| 04 | 12000 | Chandler ............. | 65.3 | 265,398 | 79 | 4,064 | 69.2 | 5.5 | 2.4 | 11.7 | 0.0 | 5.5 | 5.6 |
| 04 | 22220 | El Mirage ............ | 9.9 | 36,481 | 1,055 | 3,685 | 84.2 | 6.9 | 0.8 | 1.3 | 0.5 | 3.7 | 2.5 |
| 04 | 23620 | Flagstaff ............. | 66.0 | 73,939 | 486 | 1,120 | 78.0 | 1.7 | 5.1 | 3.5 | 0.4 | 4.5 | 6.8 |
| 04 | 23760 | Florence ............. | 62.6 | 28,049 | 1,311 | 448 | 74.3 | 7.8 | 4.3 | 1.0 | 1.3 | 4.1 | 7.2 |
| 04 | 27400 | Gilbert ............... | 68.6 | 257,658 | 87 | 3,756 | 83.7 | 3.3 | 1.2 | 5.1 | 0.0 | 2.6 | 4.1 |
| 04 | 27820 | Glendale.............. | 61.6 | 255,307 | 88 | 4,145 | 75.3 | 8.1 | 2.7 | 3.9 | 0.0 | 6.7 | 3.3 |
| 04 | 28380 | Goodyear ............ | 191.3 | 90,622 | 368 | 474 | 75.1 | 8.3 | 1.6 | 4.6 | 0.0 | 6.5 | 3.9 |
| 04 | 37620 | Kingman ............. | 37.5 | 31,815 | 1,201 | 848 | 86.5 | 1.7 | 1.1 | 3.2 | 0.0 | 3.6 | 4.0 |
| 04 | 39370 | Lake Havasu City...... | 46.3 | 57,246 | 673 | 1,236 | 93.3 | 0.6 | 0.9 | 0.5 | 0.2 | 3.2 | 1.2 |
| 04 | 44270 | Marana ............... | 121.1 | 51,306 | 760 | 424 | 85.4 | 4.3 | 1.0 | 3.0 | 0.0 | 2.6 | 3.6 |
| 04 | 44410 | Maricopa ............. | 42.5 | 54,539 | 708 | 1,283 | 69.5 | 15.4 | 1.4 | 3.4 | 0.5 | 5.1 | 4.8 |
| 04 | 46000 | Mesa ................ | 138.7 | 528,159 | 35 | 3,808 | 81.4 | 5.6 | 3.2 | 2.6 | 0.4 | 3.8 | 3.1 |
| 04 | 51600 | Oro Valley ........... | 34.9 | 46,678 | 831 | 1,338 | 85.7 | 2.2 | 0.6 | 5.8 | 0.0 | 1.4 | 4.3 |
| 04 | 54050 | Peoria ............... | 176.1 | 179,872 | 143 | 1,021 | 81.5 | 3.6 | 0.5 | 5.0 | 0.1 | 4.0 | 5.3 |
| 04 | 55000 | Phoenix .............. | 518.0 | 1,708,127 | 5 | 3,298 | 76.7 | 7.3 | 2.1 | 4.0 | 0.3 | 5.8 | 3.8 |
| 04 | 57380 | Prescott.............. | 45.0 | 44,837 | 860 | 996 | 92.2 | 0.8 | 1.3 | 0.9 | 0.0 | 2.1 | 2.7 |
| 04 | 57450 | Prescott Valley........ | 40.5 | 47,901 | 809 | 1,183 | 91.8 | 0.2 | 2.3 | 1.5 | 0.0 | 0.7 | 3.5 |
| 04 | 58150 | Queen Creek.......... | 40.3 | 60,097 | 626 | 1,491 | 83.4 | 2.3 | 0.2 | 5.2 | 0.0 | 4.5 | 4.4 |
| 04 | 62140 | Sahuarita............. | 31.7 | 32,327 | 1,182 | 1,020 | 89.2 | 2.9 | 1.3 | 0.9 | 0.0 | 0.7 | 5.0 |
| 04 | 63470 | San Luis ............. | 34.0 | 36,242 | 1,063 | 1,066 | 95.4 | 0.4 | 0.7 | 0.0 | 0.0 | 3.1 | 0.4 |
| 04 | 65000 | Scottsdale............ | 184.0 | 262,647 | 83 | 1,427 | 88.4 | 2.0 | 1.4 | 5.0 | 0.1 | 0.5 | 2.5 |
| 04 | 66820 | Sierra Vista .......... | 152.3 | 43,948 | 878 | 289 | 76.1 | 10.1 | 0.1 | 5.3 | 0.3 | 1.7 | 6.3 |
| 04 | 71510 | Surprise ............. | 110.3 | 147,965 | 179 | 1,342 | 85.8 | 5.7 | 1.0 | 2.2 | 0.2 | 2.8 | 2.4 |
| 04 | 73000 | Tempe ............... | 39.9 | 200,402 | 118 | 5,023 | 73.8 | 7.5 | 1.8 | 9.3 | 0.1 | 4.1 | 3.4 |
| 04 | 77000 | Tucson ............... | 241.0 | 553,571 | 33 | 2,297 | 71.9 | 5.2 | 3.8 | 3.2 | 0.3 | 9.2 | 6.4 |
| 04 | 85540 | Yuma................ | 120.7 | 99,811 | 318 | 827 | 85.3 | 2.4 | 1.7 | 2.0 | 0.2 | 4.3 | 4.1 |
| 05 | 00000 | ARKANSAS ........... | 52,037.5 | 3,030,522 | X | 58 | 76.7 | 15.5 | 0.6 | 1.5 | 0.4 | 2.5 | 2.8 |
| 05 | 04840 | Bella Vista ........... | 45.3 | 29,423 | 1,272 | 650 | D | D | D | D | D | D | D |
| 05 | 05290 | Benton............... | 22.9 | 37,214 | 1,030 | 1,625 | 79.8 | 14.0 | 0.2 | 1.6 | 0.0 | 0.1 | 4.3 |
| 05 | 05320 | Bentonville........... | 33.8 | 57,537 | 670 | 1,702 | 77.0 | 4.7 | 0.3 | 12.9 | 0.0 | 0.3 | 4.6 |
| 05 | 15190 | Conway .............. | 46.1 | 68,100 | 541 | 1,477 | 76.3 | 17.9 | 0.0 | 2.0 | 0.0 | 0.8 | 3.1 |
| 05 | 23290 | Fayetteville .......... | 54.1 | 89,576 | 375 | 1,656 | 81.7 | 6.9 | 1.5 | 3.5 | 0.0 | 3.1 | 3.3 |
| 05 | 24550 | Fort Smith ........... | 64.0 | 87,764 | 385 | 1,371 | 63.5 | 9.6 | 1.1 | 5.9 | 0.0 | 12.6 | 7.4 |
| 05 | 33400 | Hot Springs .......... | 37.5 | 38,939 | 995 | 1,038 | 74.7 | 18.0 | 1.2 | 1.6 | 0.0 | 1.0 | 3.6 |

1. Dry land or land partially or temporarily covered by water.   2. Hispanic or Latino persons may be of any race.

# Table D. Cities — **Population**

| City | Percent Hispanic or Latino[1], 2019 | Percent foreign born, 2019 | Age of population (percent), 2019 | | | | | | | Median age, 2019 | Percent female, 2019 | Population | | | |
|---|---|---|---|---|---|---|---|---|---|---|---|---|---|---|---|
| | | | | | | | | | | | | Census counts | | Percent change | |
| | | | Under 18 years | 18 to 24 years | 25 to 34 years | 35 to 44 years | 45 to 54 years | 55 to 64 years | 65 years and over | | | 2000 | 2010 | 2000-2010 | 2001-2020 |
| | 12 | 13 | 14 | 15 | 16 | 17 | 18 | 19 | 20 | 21 | 22 | 23 | 24 | 25 | 26 |
| United States | 18.4 | 13.7 | 22.2 | 9.3 | 13.9 | 12.8 | 12.4 | 12.9 | 16.5 | 38.5 | 50.8 | 281,421,906 | 308,758,105 | 9.7 | 6.7 |
| ALABAMA | 4.5 | 3.6 | 22.1 | 9.3 | 13.0 | 12.4 | 12.4 | 13.3 | 17.4 | 39.4 | 51.7 | 4,447,100 | 4,780,118 | 7.5 | 3.0 |
| Alabaster | 7.1 | 5.6 | 26.1 | 8.4 | 10.7 | 15.6 | 13.9 | 12.3 | 13.0 | 37.7 | 52.0 | 22,619 | 31,095 | 37.5 | 8.4 |
| Auburn | 1.4 | 11.6 | 19.4 | 30.1 | 13.5 | 13.7 | 9.3 | 6.0 | 7.9 | 25.2 | 49.2 | 42,987 | 53,426 | 24.3 | 27.9 |
| Bessemer | 0.0 | 0.0 | 20.2 | 11.4 | 10.1 | 12.3 | 10.3 | 13.6 | 22.1 | 41.5 | 54.3 | 29,672 | 27,439 | -7.5 | -5.1 |
| Birmingham | 4.2 | 4.1 | 19.0 | 9.9 | 19.0 | 11.9 | 10.9 | 13.8 | 15.5 | 36.7 | 52.9 | 242,820 | 212,013 | -12.7 | -2.4 |
| Decatur | 14.0 | 8.8 | 21.7 | 10.0 | 8.1 | 14.0 | 17.7 | 12.1 | 16.3 | 42.2 | 49.2 | 53,929 | 55,779 | 3.4 | -2.5 |
| Dothan | 3.5 | 4.6 | 21.8 | 7.4 | 12.9 | 13.3 | 12.2 | 13.3 | 19.2 | 41.0 | 51.8 | 57,737 | 66,002 | 14.3 | 5.2 |
| Enterprise | 8.9 | 3.0 | 26.1 | 7.7 | 15.3 | 15.3 | 8.2 | 9.3 | 18.1 | 36.0 | 51.0 | 21,178 | 26,519 | 25.2 | 9.3 |
| Florence | 4.7 | 2.0 | 18.1 | 12.7 | 14.2 | 11.7 | 12.7 | 12.8 | 17.8 | 39.2 | 54.1 | 36,264 | 39,601 | 9.2 | 4.0 |
| Gadsden | 5.9 | 4.9 | 21.1 | 7.1 | 13.6 | 14.3 | 12.3 | 12.8 | 18.9 | 41.1 | 51.7 | 38,978 | 36,930 | -5.3 | -5.7 |
| Homewood | 12.2 | 8.5 | 27.3 | 16.7 | 13.0 | 16.2 | 9.7 | 7.5 | 9.6 | 31.1 | 54.1 | 25,043 | 25,157 | 0.5 | 0.1 |
| Hoover | 6.6 | 6.9 | 24.8 | 7.1 | 14.2 | 14.7 | 13.5 | 9.3 | 16.4 | 36.9 | 52.5 | 62,742 | 80,477 | 28.3 | 6.8 |
| Huntsville | 7.3 | 6.6 | 19.9 | 12.0 | 16.4 | 10.9 | 11.3 | 12.6 | 16.8 | 35.8 | 52.0 | 158,216 | 180,370 | 14.0 | 12.5 |
| Madison | 2.9 | 5.5 | 25.8 | 5.3 | 9.3 | 12.7 | 18.5 | 16.7 | 11.7 | 41.2 | 53.0 | 29,329 | 43,193 | 47.3 | 21.9 |
| Mobile | 2.2 | 2.8 | 21.7 | 10.7 | 14.2 | 11.9 | 11.4 | 13.4 | 16.7 | 37.5 | 53.4 | 198,915 | 194,665 | -2.1 | -3.6 |
| Montgomery | 4.2 | 4.6 | 24.4 | 9.7 | 15.0 | 13.0 | 11.1 | 11.7 | 15.2 | 35.6 | 52.8 | 201,568 | 205,488 | 1.9 | -4.5 |
| Opelika | 13.3 | 3.5 | 13.7 | 12.6 | 15.1 | 11.0 | 12.0 | 15.5 | 20.0 | 41.9 | 51.4 | 23,498 | 26,441 | 12.5 | 17.8 |
| Phenix City | 6.9 | 6.6 | 24.4 | 7.7 | 16.8 | 17.0 | 9.8 | 11.5 | 12.7 | 35.3 | 47.3 | 28,265 | 32,810 | 16.1 | 10.9 |
| Prattville | 3.9 | 6.1 | 22.0 | 11.7 | 11.7 | 14.0 | 12.6 | 11.6 | 16.4 | 38.1 | 51.3 | 24,303 | 34,280 | 41.1 | 6.5 |
| Tuscaloosa | 2.9 | 4.1 | 15.9 | 23.8 | 14.1 | 10.3 | 11.4 | 10.9 | 13.6 | 31.0 | 50.7 | 77,906 | 90,228 | 15.8 | 14.0 |
| Vestavia Hills | 2.0 | 8.3 | 26.6 | 5.7 | 9.0 | 15.0 | 17.0 | 9.7 | 17.0 | 38.8 | 52.5 | 24,476 | 33,826 | 38.2 | 1.5 |
| ALASKA | 7.2 | 8.0 | 24.6 | 9.3 | 16.1 | 12.7 | 12.0 | 13.0 | 12.4 | 35.0 | 48.0 | 626,932 | 710,246 | 13.3 | 2.9 |
| Anchorage | 9.4 | 10.8 | 24.0 | 9.8 | 17.3 | 13.1 | 11.8 | 12.6 | 11.4 | 34.2 | 49.8 | 260,283 | 292,252 | 12.3 | -1.8 |
| Fairbanks | 6.1 | 6.7 | 23.5 | 18.5 | 17.2 | 11.0 | 7.9 | 8.4 | 13.5 | 29.0 | 45.7 | 30,224 | 31,551 | 4.4 | -2.7 |
| Juneau | 7.0 | 9.6 | 21.0 | 5.3 | 18.7 | 12.7 | 13.0 | 14.6 | 14.6 | 37.9 | 50.9 | 30,711 | 31,276 | 1.8 | 1.8 |
| ARIZONA | 31.7 | 13.4 | 22.5 | 9.5 | 13.8 | 12.3 | 11.7 | 12.1 | 18.0 | 38.3 | 50.3 | 5,130,632 | 6,392,292 | 24.6 | 16.1 |
| Apache Junction | 18.1 | 6.6 | 12.2 | 5.6 | 14.1 | 8.0 | 10.9 | 13.7 | 35.4 | 54.3 | 48.4 | 31,814 | 35,709 | 12.2 | 22.5 |
| Avondale | 59.0 | 15.8 | 29.5 | 11.1 | 19.4 | 11.3 | 11.4 | 9.3 | 8.1 | 29.5 | 51.6 | 35,883 | 76,068 | 112.0 | 16.9 |
| Buckeye | 47.6 | 8.3 | 27.4 | 7.3 | 15.6 | 16.2 | 11.6 | 10.4 | 11.4 | 34.8 | 47.3 | 6,537 | 50,829 | 677.6 | 67.7 |
| Bullhead City | 23.7 | 6.9 | 20.4 | 5.1 | 12.3 | 9.6 | 9.0 | 17.8 | 25.8 | 46.5 | 49.6 | 33,769 | 39,545 | 17.1 | 5.5 |
| Casa Grande | 46.3 | 11.1 | 25.3 | 11.2 | 14.3 | 10.1 | 9.5 | 8.8 | 20.8 | 34.4 | 50.5 | 25,224 | 48,571 | 92.6 | 26.0 |
| Chandler | 19.1 | 16.2 | 24.6 | 7.8 | 13.8 | 14.7 | 14.0 | 12.1 | 12.9 | 37.8 | 51.0 | 176,581 | 236,158 | 33.7 | 12.4 |
| El Mirage | 46.6 | 14.3 | 28.2 | 10.1 | 17.1 | 14.4 | 10.5 | 8.5 | 11.3 | 32.4 | 53.4 | 7,609 | 31,797 | 317.9 | 14.7 |
| Flagstaff | 21.6 | 5.7 | 16.1 | 32.6 | 14.1 | 9.9 | 9.2 | 9.5 | 8.7 | 25.8 | 48.3 | 52,894 | 66,018 | 24.8 | 12.0 |
| Florence | 41.8 | 11.6 | 7.9 | 6.0 | 22.8 | 23.1 | 8.5 | 13.0 | 18.6 | 39.8 | 23.2 | 17,054 | 25,424 | 49.1 | 10.3 |
| Gilbert | 14.6 | 9.0 | 29.6 | 8.9 | 11.8 | 15.8 | 14.3 | 9.3 | 10.4 | 34.8 | 51.1 | 109,697 | 208,439 | 90.0 | 23.6 |
| Glendale | 41.7 | 17.6 | 24.4 | 10.5 | 14.2 | 11.9 | 13.8 | 11.5 | 13.7 | 35.7 | 50.1 | 218,812 | 226,134 | 3.3 | 12.9 |
| Goodyear | 32.4 | 12.2 | 25.1 | 9.6 | 11.1 | 13.2 | 14.0 | 12.8 | 14.2 | 37.8 | 52.4 | 18,911 | 65,249 | 245.0 | 38.9 |
| Kingman | 12.5 | 8.3 | 12.8 | 5.7 | 12.5 | 6.4 | 11.1 | 20.1 | 31.4 | 55.8 | 48.4 | 20,069 | 28,071 | 39.9 | 13.3 |
| Lake Havasu City | 14.9 | 6.9 | 14.9 | 7.0 | 8.1 | 9.4 | 12.2 | 15.1 | 33.4 | 54.1 | 49.9 | 41,938 | 52,533 | 25.3 | 9.0 |
| Marana | 28.3 | 11.5 | 26.0 | 4.7 | 10.6 | 12.7 | 9.6 | 11.4 | 24.8 | 42.6 | 49.9 | 13,556 | 34,541 | 154.8 | 48.5 |
| Maricopa | 25.5 | 9.9 | 23.7 | 8.2 | 13.0 | 15.2 | 12.6 | 11.5 | 15.9 | 37.6 | 50.9 | 1,040 | 43,493 | 4,082.0 | 25.4 |
| Mesa | 26.2 | 10.4 | 24.8 | 8.6 | 15.5 | 12.5 | 11.2 | 10.7 | 16.7 | 35.9 | 50.4 | 396,375 | 439,941 | 11.0 | 20.1 |
| Oro Valley | 11.9 | 12.9 | 18.8 | 7.2 | 6.8 | 10.3 | 11.6 | 11.5 | 33.8 | 50.1 | 51.6 | 29,700 | 41,027 | 38.1 | 13.8 |
| Peoria | 17.9 | 11.1 | 20.9 | 6.8 | 12.2 | 13.7 | 12.7 | 13.7 | 20.1 | 42.5 | 53.0 | 108,364 | 154,069 | 42.2 | 16.7 |
| Phoenix | 43.2 | 20.1 | 25.0 | 9.6 | 16.3 | 14.0 | 12.5 | 11.5 | 11.1 | 34.4 | 50.1 | 1,321,045 | 1,446,682 | 9.5 | 18.1 |
| Prescott | 6.7 | 5.3 | 11.9 | 5.6 | 8.3 | 7.2 | 8.9 | 20.4 | 37.9 | 59.8 | 50.0 | 33,938 | 39,662 | 16.9 | 13.0 |
| Prescott Valley | 20.5 | 9.7 | 21.1 | 9.2 | 14.2 | 8.1 | 9.7 | 12.1 | 25.6 | 40.0 | 52.3 | 23,535 | 38,858 | 65.1 | 23.3 |
| Queen Creek | 15.9 | 5.0 | 31.0 | 9.1 | 7.9 | 17.6 | 14.6 | 10.5 | 9.2 | 35.7 | 49.4 | 4,316 | 26,732 | 519.4 | 124.8 |
| Sahuarita | 24.9 | 7.7 | 22.4 | 7.8 | 15.0 | 7.1 | 13.4 | 8.7 | 25.5 | 41.9 | 49.1 | 3,242 | 26,001 | 702.0 | 24.3 |
| San Luis | 97.5 | 49.6 | 25.4 | 14.2 | 14.5 | 12.8 | 14.3 | 9.0 | 9.8 | 32.2 | 50.6 | 15,322 | 27,922 | 82.2 | 29.8 |
| Scottsdale | 8.1 | 10.6 | 13.8 | 6.8 | 13.9 | 9.2 | 13.4 | 15.4 | 27.6 | 50.1 | 51.6 | 202,705 | 217,486 | 7.3 | 20.8 |
| Sierra Vista | 23.9 | 7.9 | 22.8 | 11.4 | 13.7 | 13.8 | 9.2 | 9.6 | 19.5 | 36.0 | 49.6 | 37,775 | 45,258 | 19.8 | -2.9 |
| Surprise | 22.6 | 7.1 | 27.8 | 6.0 | 10.7 | 13.2 | 11.2 | 9.8 | 21.4 | 39.8 | 52.4 | 30,848 | 117,436 | 280.7 | 26.0 |
| Tempe | 20.8 | 15.3 | 12.9 | 22.8 | 24.4 | 10.6 | 9.7 | 8.3 | 11.3 | 30.0 | 47.4 | 158,625 | 161,772 | 2.0 | 23.9 |
| Tucson | 45.4 | 16.3 | 20.0 | 15.2 | 15.6 | 12.2 | 10.7 | 11.1 | 15.3 | 34.5 | 49.9 | 486,699 | 526,510 | 8.2 | 5.1 |
| Yuma | 60.1 | 15.0 | 28.2 | 12.0 | 16.9 | 10.5 | 8.6 | 9.5 | 14.4 | 30.6 | 47.5 | 77,515 | 90,672 | 17.0 | 10.1 |
| ARKANSAS | 7.7 | 5.1 | 23.2 | 9.4 | 12.8 | 12.5 | 12.0 | 12.8 | 17.4 | 38.8 | 51.1 | 2,673,400 | 2,916,029 | 9.1 | 3.9 |
| Bella Vista | 6.3 | 4.1 | 19.1 | 2.7 | 10.6 | 13.6 | 6.8 | 11.7 | 35.5 | 50.9 | 47.0 | 16,582 | 26,510 | 59.9 | 11.0 |
| Benton | 1.6 | 1.2 | 28.8 | 10.1 | 12.8 | 14.7 | 10.9 | 9.5 | 13.2 | 33.3 | 51.8 | 21,906 | 30,765 | 40.4 | 21.0 |
| Bentonville | 11.1 | 15.2 | 26.4 | 8.1 | 19.1 | 16.7 | 12.3 | 9.4 | 8.1 | 32.6 | 49.4 | 19,730 | 35,344 | 79.1 | 62.8 |
| Conway | 3.5 | 2.3 | 21.7 | 22.9 | 13.0 | 13.2 | 8.4 | 9.0 | 11.8 | 28.7 | 53.9 | 43,167 | 58,872 | 36.4 | 15.7 |
| Fayetteville | 6.0 | 6.1 | 16.9 | 26.3 | 18.7 | 11.7 | 8.2 | 8.7 | 9.5 | 27.9 | 47.9 | 58,047 | 73,584 | 26.8 | 21.7 |
| Fort Smith | 19.9 | 12.3 | 23.3 | 10.2 | 14.2 | 12.3 | 12.0 | 12.6 | 15.4 | 37.3 | 51.0 | 80,268 | 86,266 | 7.5 | 1.7 |
| Hot Springs | 9.7 | 8.0 | 18.8 | 10.2 | 9.2 | 14.5 | 11.9 | 14.8 | 20.5 | 42.3 | 53.7 | 35,750 | 37,937 | 6.1 | 2.6 |

1.  May be of any race.

# Households, Group Quarters, Crime, and Education

| City | Households, 2019 | | | | | | | Persons in group quarters, 2019 | Serious crimes known to police[1], 2019 | | | | Educational attainment, 2019 | | |
|---|---|---|---|---|---|---|---|---|---|---|---|---|---|---|---|
| | | | | Percent | | | | | Violent | | Prorerty | | | Attainment[3] (percent) | |
| | Number | Persons per household | Family | Married couple family | Female family | Non-family | One person | | Number | Rate[2] | Number | Rate[2] | Population age 25 and over | High school graduate or less | Bachelor's degree or more |
| | 27 | 28 | 29 | 30 | 31 | 32 | 33 | 34 | 35 | 36 | 37 | 38 | 39 | 40 | 41 |
| United States ..................... | 122,802,852 | 2.61 | 64.8 | 47.5 | 17.3 | 35.2 | 28.3 | 8,084,362 | NA | NA | NA | NA | 224,898,568 | 38.3 | 33.1 |
| ALABAMA ......................... | 1,897,576 | 2.52 | 65.2 | 47.0 | 18.2 | 34.8 | 29.8 | 116,625 | NA | NA | NA | NA | 3,360,058 | 43.8 | 26.3 |
| Alabaster................... | 11,940 | 2.78 | 70.7 | 60.5 | 10.1 | 29.3 | 25.2 | 283 | NA | NA | NA | NA | 21,909 | 32.1 | 35.8 |
| Auburn ..................... | 26,156 | 2.32 | 50.7 | 41.1 | 9.6 | 49.3 | 35.0 | 5,579 | NA | NA | NA | NA | 33,444 | 13.2 | 66.7 |
| Bessemer.................. | 10,163 | 2.53 | 65.4 | 27.4 | 38.0 | 34.6 | 28.6 | 816 | NA | NA | NA | NA | 18,111 | 55.2 | 14.1 |
| Birmingham............... | 93,300 | 2.15 | 50.2 | 26.3 | 23.9 | 49.8 | 41.2 | 9,588 | NA | NA | NA | NA | 149,277 | 40.7 | 29.5 |
| Decatur..................... | 21,969 | 2.43 | 63.0 | 45.0 | 18.1 | 37.0 | 29.1 | 845 | NA | NA | NA | NA | 37,108 | 49.1 | 25.2 |
| Dothan ..................... | 27,103 | 2.49 | 61.3 | 40.6 | 20.6 | 38.7 | 35.6 | 1,118 | NA | NA | NA | NA | 48,558 | 42.0 | 26.1 |
| Enterprise................. | 10,705 | 2.60 | 64.3 | 45.3 | 19.0 | 35.7 | 31.5 | 236 | NA | NA | NA | NA | 18,598 | 24.0 | 30.6 |
| Florence................... | 18,133 | 2.16 | 50.9 | 35.8 | 15.1 | 49.1 | 39.7 | 1,645 | NA | NA | NA | NA | 28,224 | 44.7 | 27.0 |
| Gadsden................... | 14,506 | 2.35 | 56.9 | 30.2 | 26.6 | 43.1 | 39.0 | 883 | NA | NA | NA | NA | 25,142 | 58.5 | 11.1 |
| Homewood................ | 8,716 | 2.67 | 62.8 | 47.1 | 15.7 | 37.2 | 30.2 | 2,122 | NA | NA | NA | NA | 14,214 | 17.5 | 62.4 |
| Hoover ..................... | 32,461 | 2.63 | 72.9 | 56.8 | 16.1 | 27.1 | 22.6 | 356 | 114 | 133 | 1,922 | 2,243 | 58,462 | 16.9 | 56.9 |
| Huntsville.................. | 88,930 | 2.17 | 58.7 | 39.8 | 18.8 | 41.3 | 34.5 | 8,916 | NA | NA | NA | NA | 137,274 | 25.3 | 44.7 |
| Madison.................... | 21,037 | 2.52 | 77.4 | 66.1 | 11.3 | 22.6 | 21.6 | 265 | NA | NA | NA | NA | 36,643 | 9.3 | 64.4 |
| Mobile...................... | 77,628 | 2.37 | 53.2 | 33.7 | 19.5 | 46.8 | 41.6 | 4,882 | NA | NA | NA | NA | 127,563 | 40.5 | 31.6 |
| Montgomery.............. | 78,225 | 2.45 | 57.6 | 33.1 | 24.5 | 42.4 | 36.5 | 6,566 | NA | NA | NA | NA | 130,848 | 35.3 | 34.6 |
| Opelika..................... | 13,690 | 2.22 | 59.9 | 48.3 | 11.6 | 40.1 | 36.1 | 456 | NA | NA | NA | NA | 22,770 | 35.9 | 33.6 |
| Phenix City............... | 16,428 | 2.33 | 65.9 | 42.1 | 23.8 | 34.1 | 30.0 | 552 | NA | NA | NA | NA | 26,386 | 37.0 | 26.7 |
| Prattville................... | 13,984 | 2.47 | 65.9 | 48.4 | 17.5 | 34.1 | 28.0 | 505 | NA | NA | NA | NA | 23,293 | 35.0 | 36.6 |
| Tuscaloosa................ | 36,779 | 2.50 | 54.9 | 34.0 | 20.8 | 45.1 | 35.1 | 9,117 | NA | NA | NA | NA | 60,968 | 39.1 | 36.3 |
| Vestavia Hills............ | 12,855 | 2.68 | 79.8 | 68.4 | 11.4 | 20.2 | 18.4 | 161 | NA | NA | NA | NA | 23,406 | 12.8 | 67.2 |
| ALASKA........................... | 252,199 | 2.79 | 64.7 | 49.4 | 15.3 | 35.3 | 27.4 | 26,673 | NA | NA | NA | NA | 484,058 | 35.2 | 30.2 |
| Anchorage................. | 105,198 | 2.67 | 61.6 | 47.2 | 14.4 | 38.4 | 28.2 | 6,660 | 3,581 | 1,245 | 12,261 | 4,261 | 190,461 | 30.2 | 36.3 |
| Fairbanks.................. | 11,684 | 2.41 | 61.3 | 41.9 | 19.3 | 38.7 | 32.1 | 2,767 | 247 | 784 | 1,353 | 4,296 | 17,920 | 30.1 | 32.9 |
| Juneau ..................... | 12,937 | 2.43 | 58.9 | 44.3 | 14.6 | 41.1 | 35.6 | 498 | 289 | 909 | 1,292 | 4,062 | 23,555 | 28.8 | 37.3 |
| ARIZONA......................... | 2,670,441 | 2.67 | 65.2 | 47.6 | 17.5 | 34.8 | 27.3 | 161,893 | NA | NA | NA | NA | 4,944,540 | 36.1 | 30.2 |
| Apache Junction ........ | 19,877 | 2.13 | 54.8 | 45.9 | 8.9 | 45.2 | 34.9 | 190 | 95 | 223 | NA | NA | 35,002 | 46.7 | 15.4 |
| Avondale................... | 25,339 | 3.46 | 76.0 | 51.5 | 24.4 | 24.0 | 17.4 | 174 | 251 | 288 | 3,095 | 3,553 | 52,292 | 45.5 | 20.5 |
| Buckeye.................... | 21,691 | 3.40 | 83.3 | 66.9 | 16.3 | 16.7 | 11.6 | 5,903 | 107 | 137 | 1,265 | 1,624 | 51,939 | 40.0 | 23.3 |
| Bullhead City............. | 16,684 | 2.44 | 72.8 | 48.3 | 24.5 | 27.2 | 21.9 | 210 | 129 | 318 | 1,352 | 3,336 | 30,459 | 50.5 | 9.2 |
| Casa Grande............. | 19,593 | 2.98 | 71.3 | 49.6 | 21.7 | 28.7 | 20.2 | 228 | 274 | 469 | 1,336 | 2,289 | 37,277 | 51.1 | 12.9 |
| Chandler................... | 94,035 | 2.77 | 69.4 | 53.4 | 15.9 | 30.6 | 24.1 | 917 | 593 | 228 | 5,382 | 2,071 | 176,377 | 25.2 | 45.4 |
| El Mirage.................. | 10,986 | 3.25 | 78.6 | 45.8 | 32.8 | 21.4 | 12.4 | 10 | 84 | 232 | 982 | 2,714 | 22,080 | 53.2 | 13.9 |
| Flagstaff................... | 25,609 | 2.38 | 48.8 | 39.3 | 9.5 | 51.2 | 29.0 | 14,079 | 367 | 489 | 2,371 | 3,161 | 38,488 | 18.4 | 51.3 |
| Florence................... | 4,352 | 2.56 | 65.9 | 47.6 | 18.3 | 34.1 | 31.0 | 16,288 | 44 | 167 | 74 | 280 | 23,591 | 52.8 | 11.3 |
| Gilbert...................... | 81,235 | 3.12 | 78.0 | 63.1 | 14.9 | 22.0 | 17.4 | 508 | 245 | 97 | 3,050 | 1,203 | 156,439 | 19.2 | 46.8 |
| Glendale................... | 81,065 | 3.06 | 69.2 | 45.1 | 24.1 | 30.8 | 24.5 | 4,023 | 863 | 340 | 8,083 | 3,183 | 164,206 | 47.0 | 20.5 |
| Goodyear.................. | 26,158 | 3.14 | 77.2 | 66.7 | 10.5 | 22.8 | 17.6 | 4,750 | 172 | 202 | 2,045 | 2,397 | 56,708 | 32.5 | 29.4 |
| Kingman................... | 13,811 | 2.13 | 69.7 | 56.7 | 13.0 | 30.3 | 27.6 | 1,665 | 100 | 327 | 1,342 | 4,386 | 25,281 | 41.5 | 22.6 |
| Lake Havasu City........ | 25,194 | 2.21 | 69.8 | 59.0 | 10.8 | 30.2 | 23.4 | 274 | 87 | 157 | 670 | 1,209 | 43,668 | 38.9 | 17.7 |
| Marana..................... | 18,474 | 2.62 | 74.6 | 65.4 | 9.1 | 25.4 | 22.6 | 643 | 35 | 72 | 1,132 | 2,319 | 33,949 | 20.8 | 44.8 |
| Maricopa................... | 16,839 | 3.10 | 79.8 | 67.6 | 12.2 | 20.2 | 15.5 | 0 | 85 | 167 | 604 | 1,187 | 35,493 | 39.4 | 21.9 |
| Mesa........................ | 187,111 | 2.75 | 64.4 | 47.1 | 17.3 | 35.6 | 26.1 | 4,101 | 1,953 | 377 | 9,683 | 1,869 | 344,567 | 32.9 | 29.2 |
| Oro Valley................. | 19,711 | 2.33 | 66.3 | 56.3 | 10.0 | 33.7 | 28.8 | 70 | 22 | 48 | 588 | 1,279 | 34,111 | 16.5 | 52.9 |
| Peoria...................... | 66,318 | 2.64 | 71.8 | 57.7 | 14.1 | 28.2 | 24.2 | 1,109 | 407 | 233 | 3,273 | 1,875 | 127,276 | 28.2 | 34.0 |
| Phoenix.................... | 586,878 | 2.83 | 64.2 | 42.6 | 21.6 | 35.8 | 27.3 | 19,499 | 11,803 | 699 | 55,974 | 3,315 | 1,099,238 | 41.9 | 28.6 |
| Prescott.................... | 20,503 | 2.04 | 57.4 | 50.0 | 7.4 | 42.6 | 34.8 | 2,490 | 214 | 489 | 684 | 1,562 | 36,570 | 22.6 | 44.1 |
| Prescott Valley........... | 18,683 | 2.48 | 62.0 | 47.5 | 14.5 | 38.0 | 31.6 | 198 | 93 | 199 | 474 | 1,015 | 32,447 | 36.2 | 25.5 |
| Queen Creek............. | 14,088 | 3.58 | 87.5 | 75.5 | 12.0 | 12.5 | 9.8 | 39 | NA | NA | NA | NA | 30,271 | 27.4 | 38.2 |
| Sahuarita.................. | 10,412 | 3.01 | 81.9 | 74.8 | 7.1 | 18.1 | 14.7 | 116 | 43 | 139 | 345 | 1,115 | 21,910 | 28.4 | 38.8 |
| San Luis................... | 8,792 | 3.75 | 88.9 | 51.5 | 37.4 | 11.1 | 7.3 | 1,786 | 33 | 97 | 420 | 1,228 | 21,010 | 68.5 | 7.2 |
| Scottsdale................. | 123,791 | 2.07 | 53.2 | 43.8 | 9.3 | 46.8 | 37.9 | 1,673 | 415 | 159 | 5,099 | 1,958 | 204,968 | 15.2 | 60.6 |
| Sierra Vista............... | 17,061 | 2.39 | 63.9 | 50.4 | 13.4 | 36.1 | 28.6 | 2,345 | 116 | 262 | 987 | 2,227 | 28,317 | 29.9 | 33.0 |
| Surprise.................... | 49,127 | 2.88 | 76.3 | 59.8 | 16.5 | 23.7 | 18.9 | 377 | 138 | 98 | 2,083 | 1,478 | 93,866 | 31.2 | 32.3 |
| Tempe...................... | 76,669 | 2.40 | 47.4 | 31.1 | 16.3 | 52.6 | 31.7 | 12,165 | 889 | 452 | 7,420 | 3,776 | 125,911 | 15.5 | 50.3 |
| Tucson...................... | 217,993 | 2.39 | 52.6 | 32.3 | 20.3 | 47.4 | 37.3 | 27,838 | 3,775 | 688 | 17,943 | 3,272 | 355,457 | 38.9 | 27.3 |
| Yuma........................ | 34,971 | 2.71 | 71.5 | 45.9 | 25.6 | 28.5 | 23.3 | 3,502 | 376 | 381 | 2,302 | 2,331 | 58,860 | 44.1 | 16.6 |
| ARKANSAS ..................... | 1,163,647 | 2.52 | 64.9 | 47.5 | 17.4 | 35.1 | 29.5 | 84,011 | NA | NA | NA | NA | 2,036,456 | 47.4 | 23.3 |
| Bella Vista................ | 12,662 | 2.27 | 68.9 | 62.2 | 6.6 | 31.1 | 26.4 | 133 | 49 | 169 | 188 | 650 | 22,556 | 30.6 | 32.5 |
| Benton...................... | 12,464 | 2.92 | 70.0 | 45.7 | 24.3 | 30.0 | 27.7 | 430 | 212 | 570 | 1,328 | 3,574 | 22,488 | 44.8 | 32.7 |
| Bentonville ............... | 19,859 | 2.72 | 67.5 | 54.7 | 12.8 | 32.5 | 25.0 | 893 | 119 | 223 | 734 | 1,374 | 35,988 | 25.3 | 51.4 |
| Conway..................... | 26,319 | 2.38 | 55.5 | 34.5 | 21.0 | 44.5 | 29.6 | 5,085 | 324 | 481 | 1,842 | 2,736 | 37,460 | 30.2 | 39.8 |
| Fayetteville............... | 36,705 | 2.19 | 45.5 | 33.8 | 11.7 | 54.5 | 35.1 | 7,056 | 396 | 447 | 3,966 | 4,481 | 49,749 | 28.2 | 48.6 |
| Fort Smith................. | 36,210 | 2.39 | 61.7 | 41.4 | 20.3 | 38.3 | 31.9 | 1,456 | 863 | 980 | 5,127 | 5,823 | 58,414 | 44.4 | 25.8 |
| Hot Springs............... | 16,571 | 2.24 | 54.8 | 33.6 | 21.2 | 45.2 | 42.5 | 1,695 | 241 | 647 | 2,674 | 7,176 | 27,545 | 40.6 | 28.8 |

1. Data for serious crimes have not been adjusted for underreporting. This may affect comparability between geographic areas and over time.  2. Per 100,000 population estimated by the FBI.  3. Persons 25 years old and over.

# Table D. Cities — Income, Poverty, and Housing

| City | Money income, 2019 — Households | | | | | Median earnings, 2019 | | | Housing units, 2019 | | | | |
|---|---|---|---|---|---|---|---|---|---|---|---|---|---|
| | Median income | Percent with income less than $20,000 | Percent with income of $200,000 or more | Median family income | Median non-family income | All persons | Men | Women | Total | Occupied | Percent owner occupied | Median value[1] (dollars) | Median gross rent (dollars) |
| | 42 | 43 | 44 | 45 | 46 | 47 | 48 | 49 | 50 | 51 | 52 | 53 | 54 |
| United States | 65,712 | 13.8 | 8.5 | 80,944 | 39,871 | 36,519 | 42,101 | 30,944 | 139,686,209 | 122,802,852 | 64.1 | 240,463 | 1,097 |
| ALABAMA | 51,734 | 19.3 | 4.3 | 66,171 | 29,458 | 31,962 | 40,190 | 25,958 | 2,284,922 | 1,897,576 | 68.8 | 154,047 | 807 |
| Alabaster | 78,945 | 6.8 | 6.4 | 90,197 | 38,038 | 37,408 | 45,636 | 30,291 | 12,547 | 11,940 | 86.6 | 182,558 | 1,113 |
| Auburn | 46,031 | 30.1 | 8.1 | 91,250 | 18,684 | 24,009 | 35,051 | 19,489 | 28,630 | 26,156 | 50.1 | 304,007 | 770 |
| Bessemer | 32,180 | 31.0 | 2.3 | 41,318 | 18,454 | 20,929 | 20,849 | 20,953 | 13,198 | 10,163 | 40.8 | 101,788 | 738 |
| Birmingham | 36,753 | 29.8 | 3.0 | 47,490 | 26,958 | 30,014 | 31,506 | 26,256 | 113,421 | 93,300 | 40.5 | 98,814 | 839 |
| Decatur | 55,067 | 13.8 | 4.6 | 56,766 | 38,429 | 32,065 | 42,981 | 25,575 | 24,755 | 21,969 | 70.2 | 136,863 | 710 |
| Dothan | 47,411 | 22.1 | 4.9 | 62,582 | 28,919 | 32,123 | 44,539 | 25,052 | 32,728 | 27,103 | 58.6 | 162,242 | 734 |
| Enterprise | 66,013 | 18.4 | 2.0 | 83,683 | 56,343 | 39,239 | 50,723 | 31,200 | 12,276 | 10,705 | 57.3 | 205,436 | 919 |
| Florence | 42,309 | 26.9 | 3.2 | 58,693 | 28,181 | 22,546 | 32,709 | 18,722 | 21,341 | 18,133 | 43.9 | 144,278 | 612 |
| Gadsden | 30,844 | 30.0 | 1.2 | 45,880 | 21,496 | 23,692 | 26,973 | 21,653 | 16,960 | 14,506 | 56.4 | 83,954 | 672 |
| Homewood | 105,049 | 8.6 | 9.8 | 109,902 | 66,695 | 45,594 | 66,320 | 30,279 | 11,076 | 8,716 | 61.7 | 371,886 | 1,131 |
| Hoover | 98,689 | 5.5 | 17.0 | 115,290 | 52,392 | 51,175 | 60,360 | 42,485 | 35,345 | 32,461 | 69.0 | 321,961 | 1,153 |
| Huntsville | 54,342 | 18.2 | 6.1 | 72,343 | 37,111 | 31,733 | 42,715 | 22,612 | 95,947 | 88,930 | 54.8 | 185,227 | 846 |
| Madison | 105,604 | 7.1 | 13.5 | 121,115 | 58,144 | 51,219 | 73,646 | 41,900 | 22,237 | 21,037 | 73.1 | 260,211 | 1,038 |
| Mobile | 46,788 | 21.9 | 3.9 | 59,629 | 31,197 | 30,569 | 37,309 | 26,024 | 92,545 | 77,628 | 52.1 | 129,661 | 876 |
| Montgomery | 51,074 | 17.8 | 3.8 | 61,846 | 38,076 | 30,693 | 35,272 | 26,779 | 93,166 | 78,225 | 54.2 | 124,965 | 875 |
| Opelika | 48,876 | 18.8 | 4.6 | 73,811 | 25,997 | 26,873 | 30,668 | 23,786 | 14,291 | 13,690 | 73.5 | 148,092 | 752 |
| Phenix City | 50,762 | 23.9 | 3.2 | 65,540 | 22,566 | 31,290 | 45,107 | 20,942 | 18,981 | 16,428 | 54.0 | 137,886 | 871 |
| Prattville | 61,558 | 14.5 | 1.2 | 68,842 | 51,747 | 37,062 | 40,248 | 34,724 | 14,750 | 13,984 | 59.4 | 162,892 | 1,106 |
| Tuscaloosa | 40,934 | 24.8 | 7.4 | 63,012 | 27,649 | 22,408 | 26,760 | 20,614 | 50,870 | 36,779 | 46.5 | 200,231 | 886 |
| Vestavia Hills | 114,929 | 4.7 | 27.1 | 128,810 | 44,949 | 60,130 | 62,028 | 56,233 | 13,855 | 12,855 | 75.3 | 428,086 | 1,382 |
| ALASKA | 75,463 | 10.8 | 9.3 | 91,971 | 47,508 | 40,334 | 46,001 | 34,950 | 319,867 | 252,199 | 64.7 | 281,159 | 1,201 |
| Anchorage | 82,716 | 9.2 | 10.2 | 104,660 | 54,962 | 41,868 | 47,712 | 36,960 | 119,272 | 105,198 | 61.0 | 314,555 | 1,230 |
| Fairbanks | 59,914 | 13.1 | 5.8 | 62,285 | 42,333 | 31,646 | 34,422 | 30,113 | 14,459 | 11,684 | 37.4 | 249,315 | 1,296 |
| Juneau | 83,017 | 9.4 | 8.3 | 103,868 | 52,231 | 42,029 | 47,854 | 37,429 | 13,851 | 12,937 | 64.8 | 342,835 | 1,445 |
| ARIZONA | 62,055 | 13.2 | 6.6 | 74,468 | 40,198 | 34,397 | 38,996 | 30,308 | 3,076,048 | 2,670,441 | 65.3 | 255,934 | 1,101 |
| Apache Junction | 54,210 | 18.5 | 3.9 | 69,281 | 29,712 | 41,511 | 47,236 | 29,605 | 24,475 | 19,877 | 75.2 | 145,911 | 935 |
| Avondale | 71,296 | 7.8 | 5.0 | 76,813 | 41,935 | 31,820 | 34,387 | 30,762 | 27,957 | 25,339 | 58.8 | 244,213 | 1,300 |
| Buckeye | 83,348 | 6.8 | 6.7 | 84,337 | 61,672 | 45,900 | 32,180 | 30,762 | 23,952 | 21,691 | 84.8 | 248,660 | 1,516 |
| Bullhead City | 46,066 | 17.1 | 1.0 | 45,323 | 32,197 | 25,429 | 25,113 | 26,092 | 21,610 | 16,684 | 58.0 | 173,911 | 902 |
| Casa Grande | 58,884 | 12.8 | 2.9 | 63,802 | 30,577 | 28,657 | 38,233 | 22,733 | 23,502 | 19,593 | 66.7 | 177,235 | 1,007 |
| Chandler | 83,709 | 6.4 | 11.9 | 100,916 | 60,754 | 46,635 | 53,912 | 40,889 | 101,859 | 94,035 | 67.3 | 342,232 | 1,374 |
| El Mirage | 65,735 | 10.1 | 1.0 | 60,259 | 32,421 | 31,288 | 33,614 | 29,311 | 11,730 | 10,986 | 63.8 | 198,131 | 1,359 |
| Flagstaff | 58,900 | 18.2 | 5.7 | 89,234 | 35,792 | 22,520 | 25,961 | 27,109 | 29,713 | 25,609 | 47.3 | 395,296 | 1,387 |
| Florence | 42,120 | 10.7 | 3.6 | 46,060 | 29,467 | 27,109 | 26,553 | 27,204 | 6,268 | 4,352 | 71.3 | 186,847 | 942 |
| Gilbert | 102,793 | 4.8 | 12.6 | 111,715 | 59,239 | 47,767 | 60,867 | 40,226 | 84,390 | 81,235 | 75.7 | 372,469 | 1,603 |
| Glendale | 57,137 | 16.2 | 4.1 | 64,914 | 31,087 | 31,388 | 35,183 | 27,533 | 87,188 | 81,065 | 58.2 | 241,094 | 1,057 |
| Goodyear | 85,147 | 6.5 | 6.7 | 87,534 | 52,303 | 35,910 | 45,483 | 26,672 | 29,434 | 26,158 | 75.2 | 330,458 | 1,564 |
| Kingman | 56,859 | 12.0 | 3.1 | 62,406 | 29,235 | 32,223 | 35,525 | 31,944 | 15,202 | 13,811 | 69.6 | 191,503 | 850 |
| Lake Havasu City | 56,646 | 13.1 | 3.8 | 63,750 | 36,413 | 31,099 | 39,159 | 23,895 | 33,463 | 25,194 | 73.1 | 273,109 | 1,017 |
| Marana | 76,187 | 5.3 | 5.8 | 87,759 | 41,987 | 41,004 | 57,304 | 32,286 | 19,693 | 18,474 | 82.9 | 262,236 | 1,146 |
| Maricopa | 79,149 | 5.7 | 4.5 | 80,605 | 58,078 | 37,321 | 41,265 | 32,520 | 18,867 | 16,839 | 83.0 | 225,928 | 1,470 |
| Mesa | 63,836 | 10.5 | 6.2 | 74,359 | 44,906 | 36,190 | 40,796 | 31,476 | 215,150 | 187,111 | 63.0 | 259,307 | 1,128 |
| Oro Valley | 86,403 | 8.0 | 11.6 | 101,580 | 51,012 | 42,493 | 57,864 | 36,003 | 22,639 | 19,711 | 76.6 | 331,911 | 1,361 |
| Peoria | 77,368 | 11.8 | 7.1 | 92,962 | 39,203 | 42,113 | 48,832 | 36,763 | 73,751 | 66,318 | 75.0 | 307,836 | 1,332 |
| Phoenix | 60,931 | 12.8 | 7.1 | 68,785 | 41,932 | 34,758 | 37,451 | 30,642 | 637,511 | 586,878 | 54.5 | 266,625 | 1,107 |
| Prescott | 65,494 | 16.2 | 8.0 | 91,038 | 44,934 | 30,729 | 37,487 | 28,302 | 24,027 | 20,503 | 70.6 | 403,839 | 1,249 |
| Prescott Valley | 50,775 | 16.9 | 2.3 | 65,327 | 30,897 | 24,524 | 26,297 | 21,167 | 21,374 | 18,683 | 63.8 | 275,938 | 1,054 |
| Queen Creek | 118,184 | 3.1 | 13.5 | 118,142 | 76,368 | 47,505 | 51,166 | 45,315 | 14,928 | 14,088 | 89.7 | 381,860 | 1,613 |
| Sahuarita | 84,054 | 6.8 | 8.3 | 92,868 | 36,344 | 30,963 | 38,980 | 26,056 | 11,973 | 10,412 | 82.5 | 274,499 | 1,396 |
| San Luis | 41,812 | 18.7 | 1.4 | 40,551 | 44,229 | 21,556 | 25,521 | 17,917 | 9,012 | 8,792 | 63.6 | 121,923 | 658 |
| Scottsdale | 86,097 | 8.9 | 19.5 | 122,948 | 55,779 | 52,772 | 62,385 | 45,666 | 145,936 | 123,791 | 67.2 | 534,251 | 1,471 |
| Sierra Vista | 60,989 | 9.4 | 5.4 | 77,998 | 44,416 | 33,709 | 44,489 | 30,625 | 19,235 | 17,061 | 58.0 | 177,090 | 916 |
| Surprise | 76,405 | 7.0 | 4.2 | 81,527 | 49,486 | 39,268 | 50,378 | 31,071 | 55,634 | 49,127 | 78.5 | 269,894 | 1,523 |
| Tempe | 66,297 | 14.2 | 6.6 | 90,714 | 46,782 | 32,023 | 35,286 | 30,815 | 85,556 | 76,669 | 40.2 | 323,612 | 1,237 |
| Tucson | 44,365 | 20.2 | 2.3 | 57,623 | 32,409 | 26,562 | 29,606 | 23,422 | 243,941 | 217,993 | 49.3 | 172,731 | 846 |
| Yuma | 47,249 | 20.3 | 2.2 | 54,543 | 30,677 | 29,975 | 32,313 | 20,700 | 40,676 | 34,971 | 60.6 | 163,874 | 924 |
| ARKANSAS | 48,952 | 19.1 | 3.8 | 62,387 | 27,607 | 30,818 | 35,650 | 26,457 | 1,389,159 | 1,163,647 | 65.5 | 136,238 | 742 |
| Bella Vista | 74,033 | 8.8 | 8.2 | 81,561 | 39,096 | 42,209 | 50,816 | 35,427 | 14,616 | 12,662 | 88.8 | 177,466 | 850 |
| Benton | 63,141 | 13.0 | 6.3 | 72,056 | 34,434 | 35,075 | 40,828 | 31,824 | 13,661 | 12,464 | 61.2 | 172,712 | 841 |
| Bentonville | 82,362 | 11.6 | 13.6 | 116,444 | 48,346 | 42,079 | 62,225 | 30,409 | 21,575 | 19,859 | 58.2 | 256,940 | 973 |
| Conway | 46,805 | 20.9 | 5.9 | 75,928 | 31,784 | 27,983 | 30,232 | 26,291 | 27,511 | 26,319 | 43.6 | 202,501 | 828 |
| Fayetteville | 44,859 | 23.6 | 6.3 | 75,956 | 27,238 | 25,986 | 28,258 | 23,406 | 39,785 | 36,705 | 37.5 | 215,983 | 848 |
| Fort Smith | 43,097 | 19.6 | 2.8 | 53,899 | 26,620 | 26,685 | 29,063 | 25,325 | 40,894 | 36,210 | 47.6 | 137,968 | 719 |
| Hot Springs | 37,830 | 29.7 | 3.3 | 50,179 | 19,610 | 25,801 | 26,414 | 25,147 | 21,375 | 16,571 | 56.6 | 129,534 | 682 |

1. Based on population estimated by the American Community Survey.

# Table D. Cities — Commuting, Computer Access, Migration, Labor Force, and Employment

| City | Commuting[1], 2019 Percent | | Computer access[2], 2019 Percent | | Migration, 2019 | | Civilian labor force, 2020 | | Unemployment | | Civilian Employment[4], 2019 Population age 16 and older | | Population age 16 to 64 | |
|---|---|---|---|---|---|---|---|---|---|---|---|---|---|---|
| | Drove alone | With commutes of 30 minutes or more | With a computer in the house | With Internet access | Percent who lived in the same house one year ago | Percent who lived in another state or county one year ago | Total | Percent change 2019-2020 | Total | Rate[3] | Number | Percent in labor force | Number | Percent who worked full-year full-time |
| | 55 | 56 | 57 | 58 | 59 | 60 | 61 | 62 | 63 | 64 | 65 | 66 | 67 | 68 |
| United States ...................... | 75.9 | 39.6 | 92.9 | 85.8 | 86.3 | 6.1 | 160,611,064 | -1.5 | 12,933,704 | 8 | 263,534,161 | 63.1 | 209,460,133 | 52.9 |
| ALABAMA ........................... | 85.2 | 35.0 | 89.4 | 80.9 | 86.6 | 5.5 | 2,230,118 | -0.5 | 131,056 | 6 | 3,937,453 | 57.5 | 3,083,141 | 50.4 |
| Alabaster.............................. | 86.1 | 55.1 | 94.6 | 90.3 | 81.4 | 8.2 | 17,995 | -2.2 | 673 | 4 | 25,972 | 68.5 | 21,631 | 55.8 |
| Auburn ................................. | 77.9 | 16.2 | 98.2 | 90.5 | 69.6 | 16.7 | 30,475 | -1.3 | 1,499 | 5 | 54,474 | 62.1 | 49,210 | 41.0 |
| Bessemer ............................ | 77.8 | 32.3 | 90.4 | 80.0 | 87.7 | 2.2 | 10,096 | 1.8 | 1,006 | 10 | 21,828 | 52.0 | 15,978 | 38.8 |
| Birmingham ......................... | 80.6 | 27.2 | 86.2 | 75.7 | 75.5 | 7.8 | 94,011 | 0.9 | 8,117 | 9 | 173,416 | 59.3 | 140,895 | 47.3 |
| Decatur ............................... | 89.9 | 21.6 | 88.9 | 73.3 | 91.8 | 3.9 | 27,683 | 1.6 | 1,443 | 5 | 44,484 | 61.5 | 35,606 | 55.0 |
| Dothan ................................ | 86.3 | 19.9 | 91.2 | 84.0 | 88.7 | 6.1 | 30,280 | 0.7 | 1,784 | 6 | 55,351 | 53.9 | 42,187 | 46.8 |
| Enterprise ........................... | 89.1 | 23.7 | 94.8 | 93.0 | 79.6 | 12.3 | 11,245 | 0.4 | 552 | 5 | 21,262 | 54.1 | 16,182 | 51.5 |
| Florence .............................. | 85.7 | 14.5 | 93.5 | 75.8 | 65.5 | 13.5 | 18,530 | -0.3 | 1,133 | 6 | 34,548 | 53.9 | 27,303 | 38.7 |
| Gadsden .............................. | 92.6 | 22.5 | 81.6 | 70.7 | 84.3 | 7.7 | 12,867 | -3.2 | 1,300 | 10 | 28,154 | 56.0 | 21,535 | 43.8 |
| Homewood ........................... | 80.6 | 15.4 | 97.1 | 94.8 | 84.3 | 5.7 | 13,881 | -3.0 | 453 | 3 | 19,202 | 71.3 | 16,766 | 54.3 |
| Hoover ................................. | 89.0 | 29.6 | 97.5 | 95.1 | 83.4 | 7.9 | 45,137 | -1.7 | 1,782 | 4 | 66,774 | 66.6 | 52,673 | 62.9 |
| Huntsville ............................ | 84.1 | 19.7 | 94.2 | 88.2 | 77.0 | 8.3 | 100,661 | 1.6 | 5,402 | 5 | 164,486 | 63.5 | 130,557 | 53.1 |
| Madison ............................... | 87.4 | 24.6 | NA | NA | 86.2 | 6.0 | 26,699 | 0.7 | 937 | 4 | 40,741 | 66.2 | 34,496 | 60.1 |
| Mobile ................................. | 83.5 | 24.2 | 89.6 | 76.6 | 85.7 | 3.3 | 87,660 | 1.5 | 7,950 | 9 | 151,726 | 59.0 | 120,196 | 51.0 |
| Montgomery ........................ | 84.2 | 17.9 | 90.6 | 82.9 | 77.4 | 8.7 | 93,486 | 1.5 | 7,718 | 8 | 154,371 | 62.3 | 124,223 | 54.5 |
| Opelika ............................... | 77.6 | 31.3 | 85.1 | 78.5 | NA | NA | 14,566 | 1.1 | 1,019 | 7 | 27,087 | 65.4 | 20,897 | 54.4 |
| Phenix City ......................... | 86.1 | 18.9 | 93.4 | 75.6 | 81.8 | 13.2 | 14,964 | -2.1 | 723 | 5 | 30,097 | 60.1 | 25,154 | 54.8 |
| Prattville ............................. | 87.0 | 27.7 | 97.0 | 90.6 | 84.4 | 12.6 | 17,076 | -0.9 | 830 | 5 | 28,914 | 60.6 | 23,146 | 59.0 |
| Tuscaloosa .......................... | 83.3 | 11.3 | 88.6 | 81.9 | 76.3 | 12.8 | 47,193 | -2.1 | 3,629 | 8 | 86,556 | 56.6 | 72,810 | 41.8 |
| Vestavia Hills ...................... | 88.2 | 24.0 | 94.7 | 91.3 | 91.4 | 3.3 | 17,614 | -3.1 | 554 | 3 | 26,558 | 64.1 | 20,680 | 56.4 |
| ALASKA .............................. | 70.0 | 18.7 | 95.3 | 87.6 | 84.4 | 7.5 | 347,414 | -0.1 | 27,195 | 8 | 569,699 | 63.0 | 479,111 | 48.9 |
| Anchorage ........................... | 80.0 | 15.0 | 97.6 | 92.7 | 83.6 | 7.0 | 148,392 | 1.0 | 10,971 | 7 | 225,646 | 67.0 | 192,830 | 53.2 |
| Fairbanks ............................ | 66.2 | 3.9 | 91.3 | 84.9 | 72.1 | 19.8 | 12,085 | 0.8 | 969 | 8 | 24,168 | 46.6 | 20,010 | 50.7 |
| Juneau ................................ | 72.4 | 12.6 | 93.7 | 87.5 | 86.2 | 4.7 | 16,374 | -3.7 | 1,081 | 7 | 25,737 | 60.6 | 21,075 | 54.1 |
| ARIZONA ............................. | 75.6 | 38.8 | 94.6 | 86.8 | 83.9 | 5.9 | 3,570,220 | 0.5 | 282,070 | 8 | 5,820,607 | 59.8 | 4,513,366 | 51.1 |
| Apache Junction .................. | 77.5 | 60.6 | 96.3 | 87.6 | 79.4 | 17.1 | 16,158 | 1.6 | 1,565 | 10 | 37,791 | 48.0 | 22,725 | 57.9 |
| Avondale ............................. | 75.3 | 42.3 | 96.0 | 89.4 | 84.2 | 3.5 | 46,857 | 1.6 | 3,533 | 8 | 64,945 | 73.4 | 57,804 | 60.4 |
| Buckeye ............................... | 76.3 | 51.3 | 97.8 | 94.3 | 90.4 | 4.0 | 33,480 | 6.3 | 2,812 | 8 | 60,100 | 60.4 | 51,017 | 48.2 |
| Bullhead City ....................... | 74.2 | 20.6 | 92.9 | 86.0 | 87.0 | 7.9 | 17,711 | 1.7 | 1,920 | 11 | 33,379 | 49.1 | 22,848 | 40.8 |
| Casa Grande ....................... | 76.5 | 27.7 | 93.4 | 87.2 | 80.4 | 6.1 | 26,282 | 0.6 | 2,018 | 8 | 45,272 | 58.0 | 33,071 | 57.6 |
| Chandler .............................. | 79.2 | 37.7 | 98.3 | 95.8 | 84.5 | 6.8 | 154,167 | 0.4 | 10,169 | 7 | 203,854 | 69.9 | 170,247 | 60.2 |
| El Mirage ............................ | 78.6 | 57.3 | 95.7 | 92.1 | 89.9 | 1.5 | 17,206 | -0.2 | 1,387 | 8 | 26,343 | 72.3 | 22,310 | 57.9 |
| Flagstaff .............................. | 65.6 | 9.0 | 96.6 | 87.0 | 65.1 | 21.4 | 42,674 | -3.0 | 3,109 | 7 | 64,412 | 67.2 | 57,885 | 36.5 |
| Florence .............................. | NA | 49.1 | NA | NA | 76.1 | 15.6 | 4,193 | 10.2 | 364 | 9 | 25,358 | 12.4 | 20,250 | 10.1 |
| Gilbert ................................. | 78.1 | 46.3 | 98.8 | 96.7 | 83.6 | 5.2 | 144,352 | 0.7 | 8,598 | 6 | 188,667 | 70.4 | 162,334 | 58.0 |
| Glendale .............................. | 73.8 | 51.5 | 93.6 | 80.1 | 83.0 | 4.3 | 128,835 | 0.2 | 10,193 | 8 | 196,850 | 63.3 | 162,287 | 50.6 |
| Goodyear ............................. | 75.0 | 47.0 | 98.2 | 95.5 | 84.9 | 6.0 | 42,723 | 3.4 | 3,013 | 7 | 67,796 | 58.6 | 55,459 | 49.9 |
| Kingman .............................. | 79.4 | 20.7 | 94.6 | 89.9 | 80.9 | 7.4 | 14,274 | 1.6 | 1,289 | 9 | 27,661 | 44.5 | 17,922 | 51.5 |
| Lake Havasu City ................ | 78.3 | 19.0 | 94.4 | 84.2 | 83.0 | 6.9 | 25,715 | 1.4 | 2,463 | 10 | 48,639 | 48.8 | 29,970 | 49.5 |
| Marana ................................ | 75.0 | 49.8 | 96.4 | 90.3 | 91.8 | 3.9 | 24,993 | 1.8 | 1,437 | 6 | 37,279 | 54.2 | 25,100 | 57.2 |
| Maricopa ............................. | 72.4 | 70.7 | 97.5 | 96.8 | 83.3 | 13.0 | 27,183 | 3.4 | 2,157 | 8 | 41,176 | 65.2 | 32,912 | 56.9 |
| Mesa ................................... | 75.5 | 38.2 | 95.1 | 89.6 | 83.2 | 4.7 | 264,292 | 0.8 | 19,117 | 7 | 402,029 | 65.9 | 315,785 | 56.2 |
| Oro Valley ........................... | 80.2 | 33.4 | 97.1 | 91.9 | 87.7 | 4.3 | 20,182 | -1.1 | 1,313 | 7 | 38,101 | 49.0 | 22,532 | 49.9 |
| Peoria .................................. | 77.8 | 49.0 | 96.4 | 91.5 | 87.0 | 3.9 | 94,187 | 1.1 | 6,615 | 7 | 143,650 | 59.7 | 108,216 | 54.7 |
| Phoenix ............................... | 75.3 | 41.5 | 96.0 | 84.7 | 84.0 | 4.0 | 876,925 | 1.0 | 69,885 | 8 | 1,307,644 | 67.6 | 1,120,737 | 55.0 |
| Prescott ............................... | 75.1 | 10.6 | 91.6 | 87.9 | 79.7 | 11.0 | 18,658 | 0.5 | 1,388 | 7 | 39,488 | 41.4 | 22,715 | 39.4 |
| Prescott Valley .................... | 82.6 | 17.2 | 95.1 | 92.4 | 89.1 | 5.4 | 22,636 | -0.2 | 1,478 | 7 | 38,156 | 53.7 | 26,262 | 38.0 |
| Queen Creek ....................... | 77.0 | 57.4 | NA | NA | 89.8 | 7.4 | 25,457 | 14.7 | 1,550 | 6 | 37,179 | 71.2 | 32,505 | 56.8 |
| Sahuarita ............................ | 78.5 | 56.9 | 95.2 | 93.4 | 82.4 | 6.1 | 13,408 | 1.6 | 857 | 6 | 25,363 | 52.7 | 17,345 | 54.6 |
| San Luis .............................. | 76.2 | 44.0 | 91.3 | 74.9 | 88.4 | 4.2 | 17,108 | -5.3 | 5,459 | 32 | 26,713 | 56.7 | 23,306 | 39.2 |
| Scottsdale ........................... | 69.9 | 28.2 | 96.2 | 93.5 | 85.1 | 5.7 | 150,955 | 0.0 | 9,966 | 7 | 225,750 | 62.8 | 154,646 | 60.1 |
| Sierra Vista ......................... | 80.7 | 10.3 | 95.0 | 87.3 | 79.2 | 13.5 | 19,059 | 0.4 | 1,213 | 6 | 34,325 | 44.1 | 25,943 | 56.2 |
| Surprise .............................. | 75.6 | 47.5 | 97.4 | 95.2 | 81.0 | 5.3 | 61,154 | 1.4 | 4,592 | 8 | 108,464 | 53.7 | 78,211 | 51.1 |
| Tempe ................................. | 70.6 | 21.0 | 98.8 | 93.6 | 68.4 | 11.9 | 121,257 | 1.0 | 8,651 | 7 | 173,914 | 68.9 | 151,788 | 46.8 |
| Tucson ................................. | 75.1 | 28.5 | 94.0 | 86.9 | 78.4 | 6.8 | 267,222 | -0.3 | 22,668 | 9 | 450,072 | 60.9 | 366,129 | 44.5 |
| Yuma ................................... | 77.6 | 13.4 | 94.1 | 83.5 | 82.9 | 6.4 | 46,373 | 0.0 | 6,251 | 14 | 73,047 | 58.9 | 58,889 | 47.9 |
| ARKANSAS ......................... | 82.4 | 27.9 | 89.5 | 79.4 | 85.6 | 6.2 | 1,354,296 | -0.6 | 81,952 | 6 | 2,401,299 | 58.0 | 1,877,062 | 50.8 |
| Bella Vista ........................... | 89.3 | 40.4 | 94.3 | 91.9 | 90.1 | 5.7 | 12,327 | 0.7 | 686 | 6 | 24,306 | 48.6 | 14,064 | 59.3 |
| Benton ................................. | 84.1 | 34.4 | 91.7 | 90.7 | 91.3 | 4.8 | 17,396 | -1.1 | 914 | 5 | 26,723 | 70.0 | 21,875 | 61.9 |
| Bentonville .......................... | 80.5 | 19.6 | 96.1 | 91.6 | 81.1 | 8.5 | 28,822 | 6.9 | 1,138 | 4 | 42,021 | 71.5 | 37,601 | 61.0 |
| Conway ............................... | 82.9 | 29.6 | 96.4 | 94.4 | 87.5 | 5.4 | 34,022 | -0.1 | 1,900 | 6 | 54,770 | 63.4 | 46,768 | 54.1 |
| Fayetteville ......................... | 80.9 | 19.4 | 96.6 | 84.9 | 68.1 | 15.1 | 48,871 | 1.6 | 2,613 | 5 | 73,671 | 63.6 | 65,347 | 42.9 |
| Fort Smith ........................... | 80.3 | 11.8 | 86.1 | 74.8 | 81.9 | 6.7 | 39,025 | -0.9 | 2,385 | 6 | 70,379 | 61.3 | 56,854 | 50.6 |
| Hot Springs ......................... | 83.3 | 13.6 | 86.5 | 76.3 | 82.7 | 10.7 | 15,574 | 5.4 | 1,487 | 10 | 32,643 | 52.2 | 24,680 | 45.3 |

1. Employed persons.  2. Households.  3. Percent of civilian labor force.  4. Persons 16 years old and over.

| City | Value of residential construction authorized by building permits, 2020 | | | Wholesale trade[1], 2017 | | | | Retail trade[2], 2017 | | | |
|---|---|---|---|---|---|---|---|---|---|---|---|
| | New construction ($1,000) | Number of housing units | Percent single family | Number of establishments | Number of employees | Sales (mil dol) | Annual payroll (mil dol) | Number of establishments | Number of employees | Sales (mil dol) | Annual payroll (mil dol) |
| | 69 | 70 | 71 | 72 | 73 | 74 | 75 | 76 | 77 | 78 | 79 |
| United States | 307,209,905 | 1,471,141 | 67 | 352,065 | 5,156,359 | 5,700,967 | 337,909 | 1,064,087 | 15,938,821 | 4,949,602 | 441,796 |
| ALABAMA | 4,358,692 | 19,982 | 88 | 4,410 | 63,468 | 66,906 | 3,461 | 17,958 | 230,069 | 63,740 | 5,819 |
| Alabaster | 48,989 | 152 | 100 | 42 | 981 | 529 | 46 | 96 | 2,013 | 451 | 49 |
| Auburn | 302,574 | 1,521 | 42 | 37 | 477 | 123 | 19 | 211 | 3,609 | 891 | 84 |
| Bessemer | 1,226 | 11 | 100 | 59 | 1,163 | 560 | 64 | 172 | 2,610 | 757 | 70 |
| Birmingham | 31,288 | 208 | 41 | 472 | 8,998 | 8,242 | 525 | 906 | 11,179 | 3,084 | 296 |
| Decatur | 21,469 | 100 | 92 | 78 | 1,316 | 1,092 | 71 | D | D | D | D |
| Dothan | 64,019 | 231 | 100 | D | D | D | D | 461 | 7,378 | 2,098 | 194 |
| Enterprise | 35,991 | 259 | 61 | 9 | 197 | 39 | 9 | 164 | 2,150 | 689 | 58 |
| Florence | 8,257 | 63 | 100 | 46 | 945 | 432 | 43 | 283 | 4,404 | 1,067 | 107 |
| Gadsden | 0 | 0 | 0 | 30 | 283 | 198 | 14 | 214 | 3,079 | 819 | 72 |
| Homewood | 22,199 | 44 | 100 | 67 | 910 | 584 | 59 | 208 | 3,108 | 724 | 84 |
| Hoover | 173,076 | 399 | 100 | 80 | 706 | 1,970 | 90 | 398 | 7,989 | 2,569 | 228 |
| Huntsville | 134,674 | 1,789 | 95 | 210 | 3,744 | 4,340 | 233 | 952 | 15,387 | 4,333 | 422 |
| Madison | 141,275 | 438 | 100 | 48 | 776 | 463 | 57 | 137 | 2,050 | 500 | 54 |
| Mobile | 21,487 | 119 | 100 | 355 | 4,062 | 2,318 | 216 | 987 | 15,065 | 4,117 | 400 |
| Montgomery | 43,311 | 297 | 96 | 250 | 4,664 | 3,773 | 243 | 869 | 12,352 | 3,349 | 318 |
| Opelika | 96,584 | 409 | 95 | 36 | 500 | 333 | 29 | 199 | 3,063 | 848 | 70 |
| Phenix City | 18,941 | 139 | 40 | 9 | 59 | 84 | 3 | 115 | 1,691 | 406 | 38 |
| Prattville | 105,453 | 332 | 100 | D | D | D | 6 | 156 | 2,930 | 688 | 66 |
| Tuscaloosa | 192,764 | 1,024 | 40 | 84 | 873 | 439 | 48 | 470 | 7,117 | 1,962 | 175 |
| Vestavia Hills | 47,621 | 112 | 100 | 34 | 178 | 163 | 13 | 145 | 2,084 | 515 | 60 |
| ALASKA | 381,356 | 1,420 | 72 | 643 | 7,529 | 5,557 | 458 | 2,480 | 34,498 | 10,385 | 1,095 |
| Anchorage | 271,706 | 970 | 76 | 338 | 5,162 | 3,807 | 314 | 849 | 15,445 | 4,903 | 508 |
| Fairbanks | 0 | 0 | 0 | 44 | 492 | 369 | 28 | 195 | 3,783 | 1,191 | 124 |
| Juneau | 10,110 | 46 | 87 | 33 | 278 | 203 | 16 | 148 | 1,790 | 465 | 53 |
| ARIZONA | 13,798,359 | 60,342 | 70 | 5,508 | 79,541 | 71,206 | 4,827 | 17,918 | 324,912 | 104,366 | 9,506 |
| Apache Junction | 21,234 | 153 | 86 | 13 | 116 | 48 | 7 | 102 | 1,835 | 462 | 47 |
| Avondale | 201,521 | 903 | 61 | 12 | 208 | 70 | 11 | 120 | 4,077 | 1,844 | 146 |
| Buckeye | 943,153 | 3,341 | 100 | 17 | 150 | 87 | 8 | 65 | 1,422 | 477 | 38 |
| Bullhead City | 65,301 | 305 | 100 | 11 | 75 | 36 | 3 | 109 | 2,192 | 630 | 60 |
| Casa Grande | 245,515 | 1,067 | 100 | 25 | 451 | 233 | 23 | 182 | 3,417 | 973 | 87 |
| Chandler | 444,323 | 1,901 | 43 | 273 | 4,128 | 4,529 | 311 | 760 | 15,938 | 4,481 | 440 |
| El Mirage | 10,783 | 44 | 100 | 4 | 40 | 21 | 2 | 31 | 548 | 141 | 16 |
| Flagstaff | 138,056 | 588 | 48 | 74 | 682 | 326 | 32 | 328 | 5,763 | 1,633 | 148 |
| Florence | 112,487 | 399 | 100 | NA | NA | NA | NA | 12 | 172 | 47 | 4 |
| Gilbert | 474,942 | 2,630 | 47 | 160 | 1,175 | 781 | 70 | 526 | 11,863 | 3,953 | 372 |
| Glendale | 148,580 | 548 | 100 | 122 | 1,462 | 1,081 | 68 | 673 | 14,394 | 5,251 | 406 |
| Goodyear | 704,186 | 2,760 | 70 | 30 | 559 | 1,022 | 29 | 149 | 4,398 | 1,445 | 121 |
| Kingman | 60,025 | 407 | 91 | 17 | 318 | 137 | 16 | 137 | 2,654 | 964 | 72 |
| Lake Havasu City | 78,550 | 420 | 81 | 45 | 546 | 132 | 24 | 236 | 3,146 | 1,011 | 87 |
| Marana | 291,647 | 913 | 100 | 29 | 244 | 85 | 9 | 179 | 3,671 | 1,068 | 98 |
| Maricopa | 371,172 | 1,514 | 100 | 6 | 15 | 3 | 1 | 34 | 926 | 284 | 24 |
| Mesa | 974,510 | 4,484 | 63 | 334 | 4,852 | 2,657 | 314 | 1,283 | 23,696 | 6,961 | 671 |
| Oro Valley | 121,996 | 385 | 100 | 24 | 93 | 51 | 5 | 87 | 2,182 | 558 | 55 |
| Peoria | 241,686 | 1,554 | 95 | 63 | 413 | 218 | 18 | 400 | 10,761 | 3,7960 | 349 |
| Phoenix | 1,974,060 | 11,647 | 40 | 1,823 | 36,641 | 36,110 | 2,318 | 3,753 | 73,994 | 27,062 | 2,313 |
| Prescott | 126,628 | 421 | 92 | 54 | 685 | 445 | 41 | 271 | 4,206 | 1,214 | 123 |
| Prescott Valley | 151,591 | 633 | 81 | 27 | 462 | 389 | 20 | 106 | 1,920 | 504 | 50 |
| Queen Creek | 745,217 | 2,001 | 99 | 14 | 18 | 6 | 1 | 81 | 1,257 | 277 | 26 |
| Sahuarita | 185,195 | 578 | 100 | 9 | 40 | 14 | 2 | 31 | 918 | 246 | 23 |
| San Luis | 55,869 | 366 | 100 | D | D | D | D | 33 | 653 | 162 | 15 |
| Scottsdale | 686,557 | 3,086 | 16 | 481 | 4,119 | 2,903 | 251 | 1,358 | 21,051 | 7,973 | 727 |
| Sierra Vista | 16,632 | 97 | 100 | 9 | 67 | 14 | 3 | 148 | 2,714 | 742 | 70 |
| Surprise | 769,226 | 2,515 | 98 | 20 | 504 | 257 | 19 | 200 | 5,435 | 1,803 | 154 |
| Tempe | 237,515 | 1,339 | 9 | 379 | 6,993 | 7,053 | 455 | 771 | 16,650 | 5,905 | 553 |
| Tucson | 370,987 | 1,755 | 59 | 410 | 4,469 | 2,699 | 218 | 1,812 | 32,262 | 8,460 | 887 |
| Yuma | 116,985 | 658 | 88 | 74 | 1,287 | 1,029 | 69 | 313 | 6,228 | 1,686 | 160 |
| ARKANSAS | 2,353,747 | 12,493 | 71 | 2,812 | 36,528 | 29,242 | 1,883 | 10,925 | 142,857 | 40,174 | 3,665 |
| Bella Vista | 108,090 | 416 | 100 | 12 | 49 | 16 | 3 | 26 | 386 | 73 | 9 |
| Benton | 52,606 | 226 | 99 | 26 | 187 | 171 | 7 | 145 | 2,224 | 808 | 63 |
| Bentonville | 238,649 | 1,168 | 41 | 97 | 1,137 | 996 | 86 | 141 | 3,347 | 1,386 | 112 |
| Conway | 108,209 | 782 | 45 | 52 | 595 | 506 | 29 | 296 | 5,085 | 1,532 | 132 |
| Fayetteville | 213,071 | 866 | 67 | 64 | 761 | 732 | 35 | 428 | 7,523 | 1,993 | 187 |
| Fort Smith | 55,824 | 313 | 43 | 165 | 2,568 | 1,524 | 133 | 515 | 7,673 | 2,045 | 187 |
| Hot Springs | 18,872 | 88 | 90 | 42 | 317 | 116 | 13 | 354 | 4,926 | 1,311 | 124 |

1. Merchant wholesalers except manufacturers' sales branches and offices.   2. Establishments with payroll.

# Table D. Cities — Real Estate, Professional Services, and Manufacturing

| City | Real estate and rental and leasing, 2017 | | | | Professional, scientific, and technical services[1], 2017 | | | | Manufacturing, 2017 | | | |
|---|---|---|---|---|---|---|---|---|---|---|---|---|
| | Number of establishments | Number of employees | Receipts (mil dol) | Annual payroll (mil dol) | Number of establishments | Number of employees | Receipts (mil dol) | Annual payroll (mil dol) | Number of establishments | Number of employees | Receipts (mil dol) | Annual payroll (mil dol) |
| | 80 | 81 | 82 | 83 | 84 | 85 | 86 | 87 | 88 | 89 | 90 | 91 |
| United States | 410,820 | 2,194,885 | 674,147 | 113,410 | 913,624 | 9,015,366 | 1,844,781 | 729,371 | 291,586 | 11,522,039 | 5,548,797 | 670,678 |
| ALABAMA | 4,362 | 22,778 | 5,785 | 968 | 9,496 | 100,857 | 20,681 | 7,188 | 4,133 | 245,725 | 137,441 | 12,916 |
| Alabaster | 20 | 79 | 30 | 4 | 62 | 455 | 58 | 16 | NA | NA | NA | NA |
| Auburn | 86 | 344 | 85 | 14 | 158 | 820 | 108 | 39 | NA | NA | NA | NA |
| Bessemer | 25 | 134 | 30 | 6 | 53 | 577 | 106 | 28 | NA | NA | NA | NA |
| Birmingham | 303 | 3,308 | 647 | 173 | 909 | 11,420 | 2,375 | 912 | NA | NA | NA | NA |
| Decatur | 69 | 297 | 72 | 12 | 133 | 1,716 | 439 | 110 | NA | NA | NA | NA |
| Dothan | D | D | D | D | 197 | 1,044 | 151 | 54 | NA | NA | NA | NA |
| Enterprise | 31 | 144 | 20 | 5 | 52 | 728 | 135 | 54 | NA | NA | NA | NA |
| Florence | D | D | D | D | 126 | 645 | 65 | 25 | NA | NA | NA | NA |
| Gadsden | 40 | 141 | 49 | 5 | 91 | 454 | 55 | 20 | NA | NA | NA | NA |
| Homewood | 71 | 473 | 114 | 25 | 176 | 1,551 | 362 | 120 | NA | NA | NA | NA |
| Hoover | 119 | 559 | 401 | 38 | 326 | 2,571 | 404 | 155 | NA | NA | NA | NA |
| Huntsville | 321 | 1,509 | 465 | 61 | 1,146 | 33,606 | 9,011 | 3,054 | NA | NA | NA | NA |
| Madison | 57 | 247 | 86 | 10 | 130 | 1,793 | 669 | 138 | NA | NA | NA | NA |
| Mobile | 319 | 1,869 | 424 | 72 | 707 | 8,917 | 1,349 | 537 | NA | NA | NA | NA |
| Montgomery | 248 | 1,556 | 334 | 67 | 580 | 5,887 | 1,079 | 395 | NA | NA | NA | NA |
| Opelika | 44 | 216 | 33 | 8 | 80 | 421 | 51 | 16 | NA | NA | NA | NA |
| Phenix City | D | D | D | D | 45 | 174 | 20 | 7 | NA | NA | NA | NA |
| Prattville | 46 | 124 | 28 | 4 | 46 | 257 | 27 | 10 | NA | NA | NA | NA |
| Tuscaloosa | 143 | 1,386 | 193 | 48 | D | D | D | D | NA | NA | NA | NA |
| Vestavia Hills | 88 | 466 | 804 | 47 | D | D | D | D | NA | NA | NA | NA |
| ALASKA | 965 | 4,622 | 1,188 | 223 | 1,959 | 17,869 | 3,236 | 1,269 | 535 | 12,515 | 8,617 | 585 |
| Anchorage | 437 | 2,519 | 681 | 124 | 1,188 | 14,219 | 2,583 | 1,045 | NA | NA | NA | NA |
| Fairbanks | 92 | 537 | 124 | 32 | 116 | 984 | 252 | 67 | NA | NA | NA | NA |
| Juneau | 64 | 195 | 60 | 8 | 91 | 471 | 75 | 30 | NA | NA | NA | NA |
| ARIZONA | 9,937 | 48,305 | 12,992 | 2,267 | 18,100 | 146,970 | 24,119 | 10,563 | 4,343 | 142,809 | 56,636 | 9,149 |
| Apache Junction | 38 | 130 | 32 | 3 | 20 | 54 | 6 | 2 | NA | NA | NA | NA |
| Avondale | 50 | 149 | 34 | 6 | 68 | 357 | 26 | 10 | NA | NA | NA | NA |
| Buckeye | 41 | 100 | 16 | 3 | 63 | 165 | 20 | 6 | NA | NA | NA | NA |
| Bullhead City | 62 | 268 | 36 | 6 | 39 | 198 | 28 | 11 | NA | NA | NA | NA |
| Casa Grande | 53 | 246 | 47 | 8 | 52 | 249 | 25 | 8 | NA | NA | NA | NA |
| Chandler | 459 | 1,494 | 506 | 81 | 801 | 13,130 | 1,022 | 1,570 | NA | NA | NA | NA |
| El Mirage | 16 | 55 | 11 | 2 | 10 | 47 | 6 | 2 | NA | NA | NA | NA |
| Flagstaff | 160 | 657 | 148 | 22 | 273 | 1,419 | 188 | 69 | NA | NA | NA | NA |
| Florence | 7 | 35 | 4 | 1 | 15 | 29 | 5 | 2 | NA | NA | NA | NA |
| Gilbert | 491 | 1,220 | 398 | 60 | 739 | 4,785 | 977 | 330 | NA | NA | NA | NA |
| Glendale | 239 | 1,017 | 269 | 38 | 339 | 2,031 | 241 | 135 | NA | NA | NA | NA |
| Goodyear | 98 | 1,004 | 190 | 24 | 134 | 504 | 69 | 25 | NA | NA | NA | NA |
| Kingman | 34 | 94 | 22 | 3 | 51 | 269 | 24 | 10 | NA | NA | NA | NA |
| Lake Havasu City | 113 | 318 | 83 | 10 | 96 | 308 | 45 | 12 | NA | NA | NA | NA |
| Marana | 45 | 154 | 44 | 6 | 78 | 221 | 30 | 9 | NA | NA | NA | NA |
| Maricopa | 19 | 34 | 12 | 1 | 48 | 207 | 12 | 7 | NA | NA | NA | NA |
| Mesa | 669 | 2,481 | 609 | 96 | 1,168 | 6,599 | 949 | 362 | NA | NA | NA | NA |
| Oro Valley | 77 | 131 | 35 | 6 | 144 | 397 | 60 | 22 | NA | NA | NA | NA |
| Peoria | 212 | 688 | 177 | 30 | 306 | 871 | 128 | 41 | NA | NA | NA | NA |
| Phoenix | 2,336 | 16,218 | 4,524 | 876 | 5,288 | 53,825 | 9,297 | 3,734 | NA | NA | NA | NA |
| Prescott | 159 | 349 | 83 | 14 | 229 | 845 | 104 | 42 | NA | NA | NA | NA |
| Prescott Valley | 43 | 111 | 34 | 5 | 62 | 350 | 39 | 15 | NA | NA | NA | NA |
| Queen Creek | 55 | 101 | 16 | 3 | 98 | 225 | 34 | 11 | NA | NA | NA | NA |
| Sahuarita | 18 | 23 | 6 | 1 | 11 | 31 | 4 | 2 | NA | NA | NA | NA |
| San Luis | D | D | D | D | 10 | 36 | 2 | 1 | NA | NA | NA | NA |
| Scottsdale | 1,223 | 6,842 | 2,478 | 388 | 2,332 | 17,664 | 3,929 | 1,445 | NA | NA | NA | NA |
| Sierra Vista | 56 | 283 | 40 | 8 | 97 | 1,507 | 240 | 106 | NA | NA | NA | NA |
| Surprise | 89 | 221 | 61 | 8 | D | D | D | D | NA | NA | NA | NA |
| Tempe | 469 | 2,844 | 922 | 152 | 1,574 | 12,594 | 1,977 | 797 | NA | NA | NA | NA |
| Tucson | 739 | 3,980 | 762 | 148 | 170 | 786 | 97 | 37 | NA | NA | NA | NA |
| Yuma | 145 | 612 | 107 | 18 | | | | | | | | |
| ARKANSAS | 3,115 | 12,776 | 2,588 | 473 | 5,914 | 38,088 | 5,565 | 2,071 | 2,571 | 152,696 | 58,364 | 7,025 |
| Bella Vista | 22 | 94 | 10 | 2 | 38 | 97 | 22 | 4 | NA | NA | NA | NA |
| Benton | 35 | 117 | 30 | 5 | 69 | 385 | 61 | 18 | NA | NA | NA | NA |
| Bentonville | 88 | 358 | 101 | 16 | 264 | 4,158 | 748 | 313 | NA | NA | NA | NA |
| Conway | 115 | 486 | 93 | 16 | 184 | 717 | 93 | 32 | NA | NA | NA | NA |
| Fayetteville | 180 | 667 | 158 | 25 | 413 | 2,501 | 340 | 142 | NA | NA | NA | NA |
| Fort Smith | 143 | 1,137 | 186 | 45 | D | D | D | 34 | NA | NA | NA | NA |
| Hot Springs | 72 | 375 | 58 | 13 | 148 | 798 | 81 | | NA | NA | NA | NA |

1. Establishments subject to federal tax.

## Table D. Cities — Accommodation and Food Services, Arts, Entertainment, and Recreation, and Health Care and Social Assistance

| City | Accommodation and food services, 2017 | | | | Arts, entertainment, and recreation[1], 2017 | | | | Health care and social assistance[1], 2017 | | | |
|---|---|---|---|---|---|---|---|---|---|---|---|---|
| | Number of establishments | Number of employees | Receipts (mil dol) | Annual payroll (mil dol) | Number of establishments | Number of employees | Receipts (mil dol) | Annual payroll (mil dol) | Number of establishments | Number of employees | Receipts (mil dol) | Annual payroll (mil dol) |
| | 92 | 93 | 94 | 95 | 96 | 97 | 98 | 99 | 100 | 101 | 102 | 103 |
| United States | 726,081 | 14,002,624 | 938,237 | 264,604 | 142,938 | 2,390,279 | 265,620 | 82,256 | 892,245 | 20,506,502 | 2,527,903 | 990,056 |
| ALABAMA | 9,085 | 183,590 | 10,528 | 2,762 | 1,168 | 17,939 | 1,158 | 333 | 10,636 | 258,399 | 31,239 | 11,949 |
| Alabaster | 54 | 1,373 | 74 | 21 | 11 | 261 | 18 | 8 | 82 | 2,649 | 322 | 120 |
| Auburn | 226 | 4,598 | 228 | 59 | 15 | 256 | 13 | 4 | 132 | 1,631 | 142 | 48 |
| Bessemer | 91 | 1,807 | 90 | 24 | 8 | 103 | 6 | 2 | 87 | 2,594 | 258 | 98 |
| Birmingham | 608 | 13,704 | 775 | 232 | 83 | 1,873 | 137 | 39 | 730 | 40,064 | 7,423 | 2,528 |
| Decatur | 149 | 3,322 | 160 | 49 | 21 | 264 | 12 | 4 | 285 | 4,439 | 441 | 184 |
| Dothan | 239 | 5,102 | 274 | 72 | D | D | D | 5 | 312 | 9,741 | 1,176 | 492 |
| Enterprise | 91 | 1,560 | 79 | 21 | D | D | D | D | 87 | 1,547 | 151 | 59 |
| Florence | 136 | 3,898 | 174 | 54 | 16 | 250 | 7 | 2 | 208 | 4,768 | 556 | 193 |
| Gadsden | 115 | 2,641 | 136 | 38 | 8 | 168 | 7 | 3 | 211 | 5,388 | 632 | 237 |
| Homewood | 139 | 2,629 | 159 | 44 | 13 | 160 | 7 | 2 | 208 | 6,546 | 955 | 367 |
| Hoover | 220 | 5,140 | 355 | 96 | 33 | 683 | 40 | 13 | 258 | 3,618 | 367 | 171 |
| Huntsville | 580 | 13,530 | 716 | 207 | 88 | 1,943 | 79 | 32 | 773 | 21,639 | 3,100 | 1,103 |
| Madison | 120 | 2,157 | 109 | 29 | 14 | 242 | 9 | 3 | 147 | 2,469 | 262 | 100 |
| Mobile | 510 | 11,756 | 626 | 176 | 73 | 1,465 | 59 | 19 | 627 | 18,940 | 2,506 | 928 |
| Montgomery | 492 | 11,361 | 842 | 182 | 53 | 1,283 | 66 | 25 | 649 | 18,128 | 2,104 | 881 |
| Opelika | 112 | 2,796 | 134 | 39 | D | D | D | D | 102 | 5,116 | 536 | 221 |
| Phenix City | 78 | 1,429 | 76 | 19 | 6 | 36 | 2 | 1 | 62 | 1,407 | 156 | 52 |
| Prattville | D | D | D | D | D | D | D | D | 86 | 1,447 | 124 | 50 |
| Tuscaloosa | 332 | 8,245 | 443 | 123 | 35 | 388 | 23 | 6 | 303 | 9,957 | 1,176 | 489 |
| Vestavia Hills | 83 | 1,836 | 86 | 24 | 21 | 758 | 65 | 18 | 156 | 2,826 | 403 | 158 |
| ALASKA | 2,214 | 28,853 | 2,537 | 742 | 575 | 5,433 | 449 | 104 | 2,679 | 52,125 | 7,856 | 3,048 |
| Anchorage | 798 | 15,618 | 1,225 | 386 | 158 | 2,795 | 199 | 50 | 1,291 | 27,333 | 4,483 | 1,673 |
| Fairbanks | 155 | 2,470 | 175 | 51 | 47 | 508 | 48 | 7 | 216 | 5,826 | 778 | 335 |
| Juneau | 124 | 1,390 | 113 | 31 | 38 | 593 | 24 | 7 | 133 | 2,288 | 310 | 134 |
| ARIZONA | 13,079 | 299,628 | 19,849 | 5,636 | 2,041 | 47,060 | 4,812 | 1,582 | 18,816 | 379,015 | 48,130 | 18,409 |
| Apache Junction | 48 | 869 | 37 | 11 | 9 | 140 | 15 | 3 | 76 | 990 | 81 | 35 |
| Avondale | 113 | 2,255 | 134 | 36 | 10 | D | 59 | D | 124 | 1,519 | 167 | 58 |
| Buckeye | 49 | 1,101 | 60 | 16 | 9 | 154 | 5 | 2 | 38 | 385 | 63 | 25 |
| Bullhead City | 89 | 1,286 | 75 | 20 | 7 | 49 | 3 | 1 | 120 | 1,774 | 226 | 79 |
| Casa Grande | 102 | 2,531 | 123 | 37 | 10 | 113 | 7 | 2 | 188 | 3,444 | 356 | 159 |
| Chandler | 556 | 14,304 | 1,060 | 286 | 89 | 1,974 | 293 | 42 | 839 | 14,927 | 1,833 | 622 |
| El Mirage | 12 | 183 | 10 | 3 | NA | NA | NA | NA | 10 | 88 | 6 | 3 |
| Flagstaff | 329 | 7,249 | 460 | 129 | 56 | 874 | 48 | 18 | 346 | 6,592 | 904 | 314 |
| Florence | 18 | 311 | 14 | 4 | D | D | D | D | 24 | 601 | 81 | 35 |
| Gilbert | 420 | 9,804 | 507 | 150 | 69 | 1,767 | 81 | 27 | 827 | 11,909 | 1,574 | 555 |
| Glendale | 420 | 9,413 | 585 | 152 | 52 | 2,386 | 760 | 345 | 713 | 14,020 | 1,912 | 658 |
| Goodyear | 138 | 4,035 | 224 | 66 | 16 | 553 | 27 | 8 | 176 | 5,066 | 638 | 216 |
| Kingman | 105 | 1,927 | 118 | 32 | 7 | 42 | 3 | 1 | 126 | 3,496 | 442 | 172 |
| Lake Havasu City | 139 | 2,824 | 166 | 49 | 15 | 210 | 13 | 4 | 191 | 2,571 | 342 | 115 |
| Marana | 116 | 2,646 | 190 | 49 | 19 | 370 | 21 | 8 | 69 | 1,220 | 110 | 58 |
| Maricopa | 41 | 726 | 39 | 11 | 5 | D | 5 | D | 45 | 475 | 46 | 19 |
| Mesa | 841 | 17,855 | 979 | 290 | 113 | 2,023 | 140 | 47 | 1,432 | 31,274 | 3,631 | 1,366 |
| Oro Valley | 74 | 1,916 | 136 | 37 | 20 | 415 | 26 | 10 | 149 | 2,038 | 261 | 97 |
| Peoria | 293 | 7,172 | 411 | 124 | 39 | 702 | 53 | 15 | 501 | 7,793 | 830 | 347 |
| Phoenix | 2,958 | 69,330 | 4,789 | 1,382 | 416 | 10,741 | 1,300 | 485 | 4,267 | 107,483 | 15,036 | 5,882 |
| Prescott | 176 | 2,921 | 163 | 51 | 26 | 472 | 23 | 8 | 365 | 7,049 | 845 | 326 |
| Prescott Valley | 76 | 1,329 | 75 | 23 | 9 | 203 | 6 | 2 | 139 | 2,746 | 331 | 102 |
| Queen Creek | 48 | 1,114 | 59 | 19 | D | D | D | D | 89 | 587 | 58 | 21 |
| Sahuarita | 28 | 527 | 26 | 7 | NA | NA | NA | NA | 23 | 111 | 12 | 5 |
| San Luis | 15 | 178 | 10 | 3 | NA | NA | NA | NA | 27 | 388 | 51 | 21 |
| Scottsdale | 886 | 26,231 | 1,873 | 564 | 220 | 5,835 | 479 | 150 | 1,726 | 25,334 | 3,726 | 1,363 |
| Sierra Vista | 97 | 1,783 | 96 | 30 | D | D | D | D | 167 | 2,918 | 280 | 111 |
| Surprise | 178 | 4,625 | 282 | 75 | 14 | 452 | 20 | 7 | 256 | 3,111 | 317 | 122 |
| Tempe | 633 | 14,895 | 876 | 256 | 82 | 1,551 | 102 | 32 | 633 | 11,539 | 1,178 | 478 |
| Tucson | 1,271 | 27,909 | 1,588 | 472 | 165 | 2,921 | 153 | 47 | 1,822 | 44,082 | 5,236 | 2,119 |
| Yuma | 244 | 5,753 | 307 | 86 | D | D | D | D | 328 | 7,066 | 939 | 314 |
| ARKANSAS | 5,977 | 106,280 | 5,485 | 1,573 | 787 | 9,919 | 1,057 | 218 | 7,878 | 182,126 | 19,845 | 7,961 |
| Bella Vista | 22 | 282 | 17 | 5 | 6 | 73 | 3 | 1 | 29 | 435 | 39 | 15 |
| Benton | 73 | 1,281 | 75 | 21 | D | D | D | D | 114 | 2,776 | 273 | 102 |
| Bentonville | 159 | 3,352 | 210 | 57 | 20 | 579 | 89 | 17 | 156 | 2,936 | 250 | 114 |
| Conway | 180 | 3,851 | 193 | 59 | 16 | 224 | 11 | 4 | 253 | 5,012 | 488 | 198 |
| Fayetteville | 367 | 7,796 | 393 | 116 | 48 | 700 | 54 | 14 | 375 | 9,954 | 1,212 | 566 |
| Fort Smith | 273 | 5,742 | 292 | 87 | 28 | 288 | 22 | 7 | 374 | 10,671 | 1,377 | 530 |
| Hot Springs | 218 | 4,441 | 233 | 69 | 37 | 1,445 | 183 | 31 | 256 | 7,527 | 982 | 344 |

1. Establishments subject to federal tax.

# Table D. Cities — Other Services and Government Employment and Payroll

| City | Other services[1] Number of establish-ments | Number of employees | Receipts (mil dol) | Annual payroll (mil dol) | Government employment and payroll, 2017 Full-time equivalent employees | March payroll Total (dollars) | Percent of total for: Admin-istrative, judicial, and legal | Police and corrections | Fire protection | Highways and trans-portation | Health and welfare | Natural resources and utilities | Education and libraries |
|---|---|---|---|---|---|---|---|---|---|---|---|---|---|
| | 104 | 105 | 106 | 107 | 108 | 109 | 110 | 111 | 112 | 113 | 114 | 115 | 116 |
| United States | 560,845 | 3,696,831 | 544,128 | 133,751 | X | X | X | X | X | X | X | X | X |
| ALABAMA | 6,149 | 38,476 | 5,844 | 1,303 | X | X | X | X | X | X | X | X | X |
| Alabaster | D | D | D | D | 217 | 1,013,772 | 0.0 | 41.6 | 31.3 | 0.8 | 0.0 | 22.9 | 3.4 |
| Auburn | 95 | 501 | 48 | 13 | 559 | 2,135,112 | 15.3 | 25.8 | 16.5 | 9.6 | 0.9 | 24.3 | 4.1 |
| Bessemer | 50 | 547 | 99 | 26 | 585 | 2,499,951 | 9.9 | 30.6 | 19.8 | 1.9 | 0.8 | 34.1 | 1.3 |
| Birmingham | 407 | 4,323 | 1,642 | 192 | 4,994 | 23,334,276 | 10.4 | 26.6 | 16.6 | 3.9 | 0.8 | 30.1 | 4.0 |
| Decatur | D | D | D | D | 671 | 2,952,604 | 7.3 | 21.2 | 16.9 | 3.8 | 1.4 | 45.3 | 0.0 |
| Dothan | 174 | 888 | 102 | 25 | 1,008 | 4,064,982 | 12.2 | 22.8 | 19.5 | 2.7 | 0.7 | 34.1 | 0.0 |
| Enterprise | D | D | D | D | 578 | 1,587,934 | 5.2 | 17.1 | 8.9 | 3.7 | 44.1 | 14.3 | 0.7 |
| Florence | D | D | D | 12 | 818 | 3,265,818 | 5.8 | 17.1 | 11.8 | 6.4 | 0.9 | 52.0 | 0.0 |
| Gadsden | D | D | D | D | 693 | 2,605,616 | 8.9 | 20.9 | 22.0 | 8.5 | 0.8 | 28.5 | 3.1 |
| Homewood | 84 | 1,625 | 239 | 65 | 372 | 1,292,236 | 8.0 | 27.9 | 20.4 | 1.1 | 0.0 | 19.4 | 6.2 |
| Hoover | 115 | 640 | 67 | 21 | 698 | 3,829,422 | 11.1 | 31.1 | 26.2 | 5.3 | 0.5 | 8.5 | 8.4 |
| Huntsville | 336 | 3,146 | 413 | 117 | 2,677 | 13,501,024 | 8.7 | 20.9 | 15.5 | 7.7 | 1.6 | 26.7 | 0.0 |
| Madison | 57 | 511 | 67 | 17 | 388 | 1,635,565 | 14.8 | 30.4 | 20.9 | 6.1 | 0.2 | 15.3 | 0.0 |
| Mobile | 381 | 3,075 | 322 | 86 | 2,711 | 10,723,431 | 10.9 | 28.1 | 19.2 | 4.4 | 0.8 | 27.7 | 3.5 |
| Montgomery | 412 | 2,611 | 402 | 107 | 2,677 | 10,755,405 | 8.1 | 24.1 | 18.4 | 6.6 | 0.2 | 28.9 | 1.9 |
| Opelika | D | D | D | D | 388 | 1,689,321 | 9.6 | 27.2 | 17.1 | 3.0 | 0.5 | 39.3 | 2.2 |
| Phenix City | 50 | 248 | 26 | 8 | 433 | 1,405,644 | 9.7 | 27.0 | 19.5 | 7.3 | 0.4 | 31.1 | 1.2 |
| Prattville | D | D | D | D | 330 | 1,254,033 | 11.7 | 32.8 | 30.7 | 2.7 | 0.0 | 21.2 | 0.0 |
| Tuscaloosa | 144 | 1,117 | 122 | 36 | 1,303 | 5,456,667 | 12.2 | 28.7 | 21.6 | 13.3 | 0.3 | 16.9 | 0.0 |
| Vestavia Hills | 66 | 381 | 47 | 14 | 276 | 1,301,919 | 7.8 | 27.1 | 43.8 | 1.9 | 0.0 | 7.0 | 5.3 |
| ALASKA | 1,357 | 7,271 | 1,351 | 270 | X | X | X | X | X | X | X | X | X |
| Anchorage | 571 | 3,928 | 887 | 151 | 9,255 | 52,804,925 | 4.8 | 9.0 | 7.0 | 4.8 | 1.7 | 10.0 | 60.7 |
| Fairbanks | 110 | 568 | 84 | 21 | 204 | 1,433,298 | 8.8 | 26.3 | 28.8 | 33.1 | 0.0 | 0.0 | 0.0 |
| Juneau | 76 | 393 | 62 | 17 | 1,761 | 11,406,328 | 5.4 | 5.8 | 2.9 | 7.2 | 30.8 | 7.3 | 40.6 |
| ARIZONA | 9,271 | 68,350 | 8,728 | 2,306 | X | X | X | X | X | X | X | X | X |
| Apache Junction | 43 | 217 | 21 | 6 | 243 | 1,192,738 | 27.1 | 39.3 | 0.0 | 8.9 | 3.0 | 12.5 | 8.1 |
| Avondale | 72 | 453 | 52 | 14 | 505 | 3,064,120 | 15.5 | 33.9 | 18.1 | 3.2 | 2.8 | 13.1 | 1.9 |
| Buckeye | 21 | 84 | 8 | 2 | 437 | 2,212,861 | 26.8 | 19.3 | 29.1 | 3.5 | 1.7 | 9.8 | 2.0 |
| Bullhead City | 50 | 270 | 23 | 7 | 300 | 1,339,366 | 21.2 | 44.1 | 0.0 | 14.0 | 1.8 | 13.0 | 0.0 |
| Casa Grande | 71 | 436 | 35 | 11 | 401 | 2,083,036 | 21.3 | 30.1 | 18.8 | 5.8 | 0.4 | 16.6 | 3.3 |
| Chandler | 356 | 2,389 | 285 | 77 | 1,718 | 10,949,371 | 15.7 | 33.7 | 16.5 | 7.1 | 4.8 | 15.3 | 2.8 |
| El Mirage | D | D | D | D | 160 | 934,594 | 15.7 | 36.3 | 17.4 | 4.5 | 0.0 | 8.3 | 0.0 |
| Flagstaff | 166 | 932 | 89 | 28 | 786 | 3,843,359 | 18.9 | 21.9 | 13.2 | 4.0 | 3.2 | 21.7 | 5.2 |
| Florence | D | D | D | D | 176 | 806,986 | 19.4 | 24.2 | 22.3 | 8.2 | 0.0 | 20.2 | 2.5 |
| Gilbert | 332 | 1,816 | 185 | 54 | 1,200 | 7,163,116 | 17.5 | 33.5 | 18.7 | 2.9 | 0.1 | 18.8 | 0.0 |
| Glendale | 285 | 1,476 | 161 | 43 | 1,656 | 9,914,002 | 11.9 | 34.4 | 20.1 | 4.8 | 2.1 | 20.0 | 2.2 |
| Goodyear | 66 | 384 | 34 | 11 | 483 | 3,126,807 | 20.7 | 27.4 | 25.5 | 7.2 | 0.0 | 14.5 | 0.0 |
| Kingman | 51 | 331 | 29 | 9 | 328 | 1,497,786 | 17.9 | 22.6 | 22.4 | 10.9 | 0.0 | 24.1 | 0.0 |
| Lake Havasu City | 147 | 598 | 61 | 16 | 503 | 2,453,216 | 17.0 | 26.8 | 20.7 | 7.9 | 0.0 | 23.6 | 0.0 |
| Marana | D | D | D | D | 357 | 1,812,592 | 27.3 | 34.1 | 0.0 | 17.1 | 1.7 | 14.3 | 1.7 |
| Maricopa | D | D | D | D | 313 | 1,580,935 | 9.7 | 31.5 | 29.8 | 3.3 | 0.7 | 17.7 | 1.3 |
| Mesa | 631 | 4,133 | 436 | 130 | 3,715 | 22,765,005 | 16.0 | 35.4 | 16.4 | 3.8 | 3.0 | 19.0 | 0.0 |
| Oro Valley | 58 | 526 | 44 | 13 | 385 | 1,998,940 | 22.1 | 41.7 | 0.0 | 11.6 | 0.0 | 17.9 | 2.0 |
| Peoria | 232 | 1,393 | 137 | 41 | 1,301 | 7,486,848 | 23.2 | 25.0 | 19.2 | 6.4 | 0.0 | 21.0 | 1.5 |
| Phoenix | 2,204 | 20,997 | 2,768 | 778 | 12,939 | 98,028,340 | 11.5 | 34.9 | 19.2 | 9.7 | 4.4 | 17.4 | 3.8 |
| Prescott | 130 | 620 | 69 | 19 | 499 | 2,456,719 | 13.2 | 24.4 | 17.9 | 14.1 | 2.8 | 20.5 | 7.3 |
| Prescott Valley | 82 | 350 | 33 | 10 | 237 | 1,085,810 | 17.3 | 43.8 | 0.0 | 9.2 | 4.8 | 10.0 | 0.0 |
| Queen Creek | 41 | 180 | 24 | 6 | 245 | 1,271,018 | 22.7 | 0.0 | 16.1 | 7.9 | 22.5 | 25.3 | 0.0 |
| Sahuarita | 20 | 139 | 9 | 4 | 139 | 750,836 | 39.8 | 41.0 | 0.0 | 9.5 | 0.0 | 9.7 | 0.0 |
| San Luis | D | D | D | D | 350 | 1,191,616 | 38.6 | 20.9 | 12.5 | 4.5 | 0.0 | 20.3 | 3.3 |
| Scottsdale | 743 | 6,676 | 690 | 214 | 2,358 | 13,918,114 | 21.5 | 30.6 | 15.2 | 5.1 | 1.8 | 17.6 | 2.4 |
| Sierra Vista | 67 | 338 | 34 | 8 | 363 | 1,670,914 | 13.4 | 29.1 | 18.3 | 8.9 | 2.7 | 14.3 | 0.0 |
| Surprise | 101 | 964 | 84 | 25 | 871 | 5,256,566 | 16.8 | 25.0 | 21.5 | 5.7 | 6.9 | 18.8 | 1.4 |
| Tempe | 378 | 3,909 | 1,415 | 245 | 1,813 | 10,766,295 | 16.8 | 31.8 | 14.1 | 5.5 | 9.6 | 17.4 | 0.0 |
| Tucson | 935 | 6,299 | 838 | 197 | 4,956 | 25,663,881 | 13.1 | 31.8 | 18.1 | 9.1 | 3.0 | 19.6 | 0.0 |
| Yuma | 152 | 819 | 76 | 23 | 970 | 4,031,433 | 16.1 | 29.4 | 15.9 | 6.7 | 4.4 | 24.7 | 0.0 |
| ARKANSAS | 3,910 | 21,132 | 2,874 | 655 | X | X | X | X | X | X | X | X | X |
| Bella Vista | 22 | 206 | 26 | 8 | 138 | 573,617 | 6.8 | 29.1 | 40.2 | 12.3 | 6.7 | 0.0 | 2.8 |
| Benton | D | D | D | D | 231 | 838,938 | 11.8 | 38.5 | 33.7 | 5.0 | 0.0 | 7.5 | 0.0 |
| Bentonville | D | D | D | 21 | 515 | 2,182,740 | 12.4 | 23.9 | 18.5 | 3.0 | 0.0 | 33.9 | 2.9 |
| Conway | 106 | 566 | 52 | 15 | 517 | 1,734,322 | 12.3 | 31.7 | 27.5 | 7.8 | 0.0 | 17.9 | 0.0 |
| Fayetteville | 164 | 874 | 364 | 28 | 703 | 2,692,445 | 14.4 | 26.1 | 17.5 | 11.7 | 2.7 | 20.8 | 0.0 |
| Fort Smith | D | D | 83 | D | 927 | 3,748,677 | 9.6 | 22.4 | 20.8 | 11.9 | 1.7 | 32.7 | 0.0 |
| Hot Springs | 115 | 520 | 42 | 14 | 635 | 2,314,170 | 11.8 | 25.0 | 16.9 | 9.6 | 3.4 | 25.8 | 5.2 |

1. Establishments subject to federal tax.

## Table D. Cities — City Government Finances

City government finances, 2017

| City | General revenue Total (mil dol) [117] | Intergovernmental Total (mil dol) [118] | Intergovernmental Percent from state government [119] | Taxes Total (mil dol) [120] | Taxes — Per capita (dollars) Total [121] | Taxes — Per capita (dollars) Property [122] | Taxes — Per capita (dollars) Sales and gross receipts [123] | General expenditure Total (mil dol) [124] | General expenditure — Per capita (dollars) Total [125] | General expenditure — Per capita (dollars) Capital outlays [126] |
|---|---|---|---|---|---|---|---|---|---|---|
| United States | X | X | X | X | X | X | X | X | X | X |
| **ALABAMA** | X | X | X | X | X | X | X | X | X | X |
| Alabaster | 39.5 | 0.8 | 51.1 | 29.4 | 883 | 108 | 775 | 27 | 833 | X |
| Auburn | 122.1 | 1.2 | 41.6 | 91.7 | 1,435 | 372 | 1,062 | 89 | 1,384 | 48 |
| Bessemer | 64.1 | 4.8 | 86.0 | 48.3 | 1,813 | 297 | 1,274 | 74 | 2,791 | 296 |
| Birmingham | 517.2 | 49.6 | 71.5 | 413.2 | 1,954 | 280 | 1,249 | 531 | 2,512 | 379 |
| Decatur | 125.1 | 15.5 | 63.0 | 68.0 | 1,250 | 235 | 967 | 94 | 1,723 | 433 |
| Dothan | 112.3 | 5.2 | 29.5 | 78.1 | 1,144 | 56 | 1,088 | 127 | 1,858 | 414 |
| Enterprise | 37.0 | 0.8 | 98.1 | 29.5 | 1,050 | 94 | 956 | 26 | 913 | 387 |
| Florence | 72.6 | 3.6 | 91.0 | 51.0 | 1,273 | 307 | 962 | 64 | 1,588 | 49 |
| Gadsden | 66.4 | 2.8 | 25.7 | 50.5 | 1,419 | 73 | 869 | 58 | 1,631 | 154 |
| Homewood | 56.0 | 0.9 | 100.0 | 51.9 | 2,037 | 718 | 1,319 | 45 | 1,778 | 64 |
| Hoover | 126.2 | 8.8 | 62.7 | 103.6 | 1,216 | 131 | 1,050 | 129 | 1,516 | 351 |
| Huntsville | 440.2 | 35.8 | 93.6 | 307.8 | 1,575 | 296 | 1,279 | 414 | 2,117 | 620 |
| Madison | 53.5 | 1.3 | 100.0 | 41.5 | 843 | 276 | 567 | 53 | 1,081 | 269 |
| Mobile | 401.3 | 18.8 | 47.7 | 285.1 | 1,499 | 104 | 1,395 | 358 | 1,883 | 362 |
| Montgomery | 270.3 | 30.4 | 77.7 | 196.9 | 984 | 148 | 836 | 225 | 1,126 | 69 |
| Opelika | 74.7 | 1.5 | 51.6 | 57.3 | 1,892 | 321 | 1,236 | 54 | 1,769 | 228 |
| Phenix City | 47.6 | 1.5 | 100.0 | 33.8 | 938 | 164 | 774 | 50 | 1,395 | 233 |
| Prattville | 45.9 | 1.3 | 38.6 | 35.3 | 994 | 79 | 915 | 38 | 1,069 | 190 |
| Tuscaloosa | 191.2 | 52.2 | 36.2 | 89.3 | 896 | 138 | 758 | 157 | 1,576 | 355 |
| Vestavia Hills | 36.2 | 0.0 | 0.0 | 35.4 | 1,029 | 448 | 577 | 19 | 542 | 0 |
| **ALASKA** | X | X | X | X | X | X | X | X | X | X |
| Anchorage | 1,432.3 | 620.4 | 93.7 | 609.3 | 2,070 | 1,797 | 273 | 1,395 | 4,741 | 734 |
| Fairbanks | 60.0 | 20.9 | 99.4 | 24.3 | 767 | 472 | 296 | 49 | 1,558 | 484 |
| Juneau | 341.6 | 89.2 | 91.3 | 102.9 | 3,206 | 1,520 | 1,686 | 327 | 10,198 | 1,922 |
| **ARIZONA** | X | X | X | X | X | X | X | X | X | X |
| Apache Junction | 24.7 | 10.1 | 100.0 | 13.1 | 323 | 0 | 323 | 38 | 923 | 0 |
| Avondale | 112.5 | 33.6 | 96.8 | 53.4 | 630 | 80 | 550 | 103 | 1,216 | 200 |
| Buckeye | 80.9 | 19.3 | 96.6 | 45.1 | 659 | 88 | 571 | 68 | 993 | 33 |
| Bullhead City | 54.2 | 17.2 | 100.0 | 14.4 | 358 | 0 | 358 | 53 | 1,327 | 205 |
| Casa Grande | 83.0 | 23.6 | 90.5 | 35.3 | 633 | 134 | 500 | 79 | 1,426 | 154 |
| Chandler | 375.8 | 99.9 | 91.8 | 175.2 | 691 | 117 | 574 | 449 | 1,771 | 404 |
| El Mirage | 37.5 | 11.3 | 100.0 | 12.0 | 339 | 110 | 229 | 41 | 1,168 | 105 |
| Flagstaff | 146.3 | 37.1 | 86.3 | 66.6 | 924 | 167 | 757 | 134 | 1,854 | 273 |
| Florence | 28.0 | 12.0 | 90.8 | 7.6 | 294 | 82 | 200 | 23 | 878 | 57 |
| Gilbert | 317.7 | 88.6 | 94.1 | 164.7 | 679 | 98 | 577 | 261 | 1,074 | 161 |
| Glendale | 390.7 | 98.4 | 88.8 | 200.2 | 808 | 105 | 703 | 401 | 1,620 | 269 |
| Goodyear | 152.2 | 31.4 | 100.0 | 85.3 | 1,066 | 257 | 809 | 136 | 1,705 | 357 |
| Kingman | 49.9 | 13.2 | 91.4 | 19.3 | 656 | 0 | 656 | 41 | 1,402 | 179 |
| Lake Havasu City | 105.1 | 27.6 | 80.7 | 32.6 | 599 | 219 | 380 | 89 | 1,633 | 130 |
| Marana | 98.7 | 44.7 | 34.4 | 47.8 | 1,062 | 11 | 1,046 | 87 | 1,936 | 787 |
| Maricopa | 56.0 | 22.4 | 87.0 | 27.5 | 573 | 295 | 278 | 66 | 1,368 | 411 |
| Mesa | 651.7 | 206.3 | 77.4 | 219.6 | 439 | 69 | 370 | 699 | 1,398 | 157 |
| Oro Valley | 47.3 | 19.0 | 100.0 | 21.3 | 479 | 0 | 479 | 48 | 1,081 | 117 |
| Peoria | 266.7 | 67.5 | 100.0 | 114.9 | 683 | 129 | 554 | 254 | 1,508 | 168 |
| Phoenix | 3,254.8 | 900.6 | 75.3 | 1,253.3 | 767 | 201 | 563 | 2,422 | 1,483 | 157 |
| Prescott | 85.6 | 21.9 | 75.6 | 37.4 | 877 | 40 | 817 | 92 | 2,146 | 485 |
| Prescott Valley | 57.3 | 18.3 | 88.6 | 25.1 | 564 | 35 | 529 | 49 | 1,111 | 172 |
| Queen Creek | 88.5 | 11.2 | 100.0 | 44.5 | 1,029 | 140 | 889 | 58 | 1,335 | 137 |
| Sahuarita | 26.2 | 12.3 | 98.6 | 9.3 | 312 | 16 | 296 | 26 | 883 | 123 |
| San Luis | 27.7 | 10.5 | 99.4 | 8.8 | 270 | 0 | 270 | 31 | 955 | 74 |
| Scottsdale | 495.2 | 97.2 | 90.8 | 280.4 | 1,115 | 256 | 860 | 497 | 1,975 | 371 |
| Sierra Vista | 58.2 | 17.4 | 99.8 | 22.4 | 521 | 10 | 511 | 60 | 1,388 | 187 |
| Surprise | 158.0 | 43.3 | 100.0 | 71.4 | 530 | 78 | 452 | 196 | 1,454 | 345 |
| Tempe | 392.5 | 81.3 | 75.2 | 206.6 | 1,112 | 247 | 865 | 352 | 1,892 | 261 |
| Tucson | 745.4 | 264.8 | 75.1 | 313.1 | 578 | 94 | 485 | 721 | 1,332 | 185 |
| Yuma | 223.8 | 34.0 | 93.8 | 155.5 | 1,613 | 1,118 | 495 | 133 | 1,380 | 141 |
| **ARKANSAS** | X | X | X | X | X | X | X | X | X | X |
| Bella Vista | 14.4 | 10.4 | 33.0 | 0.1 | 5 | 0 | 5 | 16 | 549 | 42 |
| Benton | 32.2 | 3.2 | 89.3 | 20.6 | 574 | 60 | 513 | 36 | 1,013 | 354 |
| Bentonville | 84.0 | 14.9 | 51.0 | 38.9 | 788 | 157 | 631 | 84 | 1,694 | 699 |
| Conway | 113.6 | 9.1 | 67.7 | 37.8 | 574 | 81 | 493 | 120 | 1,822 | 548 |
| Fayetteville | 139.3 | 27.3 | 43.7 | 68.5 | 801 | 115 | 685 | 130 | 1,516 | 452 |
| Fort Smith | 173.9 | 40.9 | 49.7 | 65.8 | 749 | 155 | 594 | 154 | 1,754 | 639 |
| Hot Springs | 73.3 | 6.4 | 56.4 | 32.3 | 838 | 0 | 838 | 83 | 2,162 | 544 |

1. Based on population estimated as of July 1 of the year shown.

City government finances, 2017 (cont.)

General expenditure (cont.)

Percent of total for:

| City | Public welfare | Highways | Parking facilities | Education | Health and hospitals | Police protection | Sewerage and sanitation | Parks and recreation | Housing and community development | Interest on debt |
|---|---|---|---|---|---|---|---|---|---|---|
| | 127 | 128 | 129 | 130 | 131 | 132 | 133 | 134 | 135 | 136 |
| United States ............... | X | X | X | X | X | X | X | X | X | X |
| ALABAMA ..................... | X | X | X | X | X | 20.3 | 20.0 | 8.7 | 0.0 | 9.8 |
| Alabaster ..................... | 0.0 | 7.5 | 0.0 | 0.0 | 0.0 | 15.0 | 12.7 | 6.2 | 0.5 | 10.2 |
| Auburn ........................ | 0.2 | 5.5 | 0.5 | 0.0 | 2.6 | 16.5 | 5.8 | 2.7 | 1.5 | 3.5 |
| Bessemer ..................... | 0.0 | 6.1 | 0.0 | 0.0 | 0.5 | 18.5 | 0.0 | 5.4 | 2.1 | 3.2 |
| Birmingham .................. | 0.0 | 14.6 | 1.0 | 0.0 | 0.0 | 13.9 | 34.8 | 8.7 | 3.0 | 7.1 |
| Decatur ....................... | 0.2 | 5.2 | 0.0 | 0.0 | 0.9 | 13.5 | 24.7 | 17.5 | 0.3 | 0.0 |
| Dothan ........................ | 0.2 | 6.5 | 0.0 | 0.0 | 0.7 | 18.4 | 15.5 | 8.1 | 0.0 | 0.0 |
| Enterprise .................... | 0.8 | 5.0 | 0.1 | 0.0 | 0.0 | 15.7 | 16.9 | 11.6 | 0.0 | 2.7 |
| Florence ...................... | 0.0 | 8.1 | 0.0 | 0.0 | 1.1 | 17.1 | 14.9 | 9.4 | 0.7 | 3.3 |
| Gadsden ...................... | 0.0 | 5.6 | 0.0 | 0.0 | 0.0 | 23.7 | 4.7 | 8.4 | 0.0 | 5.5 |
| Homewood .................... | 0.0 | 8.7 | 0.0 | 0.0 | 0.0 | 18.6 | 7.1 | 16.6 | 0.0 | 2.5 |
| Hoover ........................ | 0.0 | 9.5 | 0.0 | 0.0 | 0.1 | 11.3 | 5.4 | 9.4 | 1.3 | 7.8 |
| Huntsville .................... | 0.0 | 8.9 | 0.4 | 0.0 | 0.8 | 12.2 | 19.2 | 6.8 | 0.0 | 14.7 |
| Madison ....................... | 0.0 | 15.5 | 0.0 | 0.0 | 0.0 | 12.5 | 20.1 | 8.0 | 1.4 | 5.1 |
| Mobile ......................... | 0.0 | 4.0 | 1.9 | 0.0 | 2.0 | 20.3 | 8.2 | 12.4 | 0.7 | 5.5 |
| Montgomery .................. | 0.0 | 8.1 | 0.3 | 0.0 | 0.0 | 17.6 | 10.2 | 10.6 | 0.6 | 9.1 |
| Opelika ....................... | 0.0 | 8.9 | 0.0 | 0.0 | 0.7 | 14.2 | 16.2 | 11.9 | 0.0 | 5.3 |
| Phenix City .................. | 0.0 | 4.1 | 0.0 | 0.0 | 0.3 | 18.2 | 21.4 | 4.8 | 0.0 | 0.0 |
| Prattville ..................... | 0.0 | 5.9 | 0.0 | 0.0 | 0.0 | 19.4 | 10.3 | 6.2 | 1.6 | 2.2 |
| Tuscaloosa ................... | 0.0 | 16.1 | 0.0 | 0.0 | 0.2 | 32.1 | 4.1 | 4.1 | 0.0 | 12.2 |
| Vestavia Hills ............... | 0.0 | 0.0 | 0.0 | 0.0 | 0.0 | 0.0 | | | | |
| ALASKA ....................... | X | X | X | X | X | X | X | X | X | X |
| Anchorage .................... | 0.1 | 8.5 | 0.0 | 49.8 | 1.5 | 9.5 | 5.2 | 2.5 | 0.9 | 2.1 |
| Fairbanks ..................... | 0.0 | 19.7 | 0.0 | 0.0 | 0.0 | 13.5 | 0.0 | 0.0 | 0.0 | 0.4 |
| Juneau ........................ | 0.0 | 4.0 | 0.1 | 22.6 | 31.8 | 4.3 | 4.0 | 4.5 | 0.9 | 2.3 |
| ARIZONA ...................... | X | X | X | X | X | X | X | X | X | X |
| Apache Junction ............ | 0.0 | 23.7 | 0.0 | 0.0 | 0.0 | 33.4 | 0.0 | 11.3 | 4.9 | 0.9 |
| Avondale ..................... | 3.4 | 4.8 | 0.0 | 0.0 | 3.4 | 33.2 | 18.0 | 4.4 | 1.5 | 3.2 |
| Buckeye ....................... | 0.0 | 5.5 | 0.0 | 0.0 | 0.0 | 24.0 | 25.7 | 3.0 | 0.0 | 4.8 |
| Bullhead City ................ | 0.0 | 7.5 | 0.0 | 0.0 | 4.3 | 24.9 | 23.3 | 5.7 | 0.3 | 3.7 |
| Casa Grande ................. | 0.0 | 12.1 | 0.0 | 0.0 | 0.6 | 21.0 | 13.6 | 12.9 | 0.0 | 2.5 |
| Chandler ...................... | 0.0 | 6.6 | 0.0 | 0.0 | 0.0 | 25.6 | 34.8 | 5.8 | 1.9 | 2.9 |
| El Mirage ..................... | 0.0 | 8.4 | 0.0 | 0.0 | 1.5 | 32.0 | 26.6 | 3.9 | 0.3 | 1.4 |
| Flagstaff ...................... | 0.0 | 13.2 | 0.0 | 0.0 | 0.0 | 15.5 | 18.2 | 6.1 | 6.8 | 1.8 |
| Florence ...................... | 0.0 | 12.6 | 0.0 | 0.0 | 0.0 | 12.8 | 13.1 | 8.2 | 0.0 | 12.3 |
| Gilbert ........................ | 0.0 | 13.0 | 0.0 | 0.0 | 0.0 | 21.0 | 15.2 | 8.8 | 0.5 | 4.4 |
| Glendale ...................... | 0.0 | 10.6 | 0.0 | 0.0 | 0.0 | 23.2 | 13.2 | 6.2 | 5.4 | 6.8 |
| Goodyear ..................... | 0.0 | 16.2 | 0.0 | 0.0 | 0.0 | 15.8 | 10.4 | 4.8 | 1.2 | 9.4 |
| Kingman ....................... | 0.0 | 12.6 | 0.0 | 0.0 | 0.0 | 26.7 | 15.7 | 10.6 | 0.0 | 3.7 |
| Lake Havasu City ........... | 0.0 | 6.8 | 0.0 | 0.0 | 0.0 | 16.8 | 19.9 | 4.2 | 0.5 | 9.2 |
| Marana ........................ | 0.0 | 41.8 | 0.0 | 0.0 | 0.0 | 17.7 | 1.3 | 6.7 | 0.7 | 4.5 |
| Maricopa ...................... | 0.0 | 13.3 | 0.0 | 0.0 | 0.0 | 17.7 | 0.1 | 8.7 | 0.0 | 3.1 |
| Mesa .......................... | 0.0 | 6.8 | 0.0 | 0.0 | 0.0 | 25.6 | 11.8 | 8.4 | 3.9 | 9.3 |
| Oro Valley .................... | 0.0 | 28.3 | 0.0 | 0.0 | 0.0 | 43.1 | 2.2 | 18.8 | 0.0 | 4.0 |
| Peoria ......................... | 0.0 | 17.1 | 0.0 | 0.0 | 0.0 | 18.9 | 13.7 | 11.3 | 5.6 | 5.1 |
| Phoenix ....................... | 0.0 | 3.8 | 0.1 | 1.3 | 0.0 | 18.7 | 12.1 | 6.7 | 0.1 | 11.9 |
| Prescott ...................... | 0.0 | 23.0 | 0.1 | 0.0 | 0.0 | 16.3 | 23.3 | 9.0 | 0.0 | 3.0 |
| Prescott Valley ............. | 0.0 | 16.1 | 0.0 | 0.0 | 0.0 | 20.5 | 12.2 | 7.7 | 0.0 | 5.9 |
| Queen Creek ................. | 0.0 | 10.1 | 0.0 | 0.0 | 0.0 | 0.0 | 18.3 | 7.1 | 0.8 | 6.7 |
| Sahuarita ..................... | 0.0 | 17.7 | 0.0 | 0.0 | 0.0 | 27.0 | 12.9 | 6.7 | 0.0 | 6.6 |
| San Luis ...................... | 0.7 | 11.9 | 0.0 | 0.0 | 0.0 | 15.8 | 11.1 | 7.1 | 0.6 | 17.5 |
| Scottsdale .................... | 0.0 | 5.6 | 0.0 | 0.0 | 0.0 | 19.7 | 12.4 | 8.5 | 1.3 | 8.5 |
| Sierra Vista .................. | 0.0 | 24.3 | 0.0 | 0.0 | 0.0 | 21.1 | 16.1 | 6.1 | 2.3 | 1.2 |
| Surprise ...................... | 0.0 | 12.9 | 0.0 | 0.0 | 0.0 | 14.1 | 13.9 | 11.3 | 6.3 | 1.7 |
| Tempe ......................... | 0.0 | 4.7 | 0.0 | 0.0 | 0.0 | 27.4 | 13.2 | 7.5 | 5.4 | 3.5 |
| Tucson ........................ | 0.0 | 13.6 | 0.7 | 0.0 | 0.0 | 28.6 | 7.2 | 5.4 | 9.2 | 1.5 |
| Yuma .......................... | 0.0 | 9.7 | 0.0 | 0.0 | 0.6 | 22.2 | 13.2 | 14.6 | 1.3 | 3.6 |
| ARKANSAS .................... | X | X | X | X | X | X | X | X | X | X |
| Bella Vista ................... | 0.0 | 26.5 | 0.0 | 0.0 | 0.0 | 20.9 | 0.0 | 9.6 | 0.0 | 0.0 |
| Benton ........................ | 0.0 | 26.8 | 0.0 | 0.0 | 1.3 | 16.6 | 12.5 | 9.0 | 0.2 | 9.3 |
| Bentonville ................... | 0.0 | 20.0 | 0.0 | 0.0 | 0.0 | 10.9 | 16.0 | 10.4 | 0.6 | 2.4 |
| Conway ....................... | 0.0 | 14.5 | 0.0 | 0.0 | 0.3 | 10.2 | 16.0 | 5.2 | 0.5 | 10.3 |
| Fayetteville .................. | 0.0 | 14.9 | 1.0 | 0.0 | 0.6 | 11.0 | 23.3 | 8.9 | 1.4 | 3.1 |
| Fort Smith ................... | 0.0 | 22.8 | 0.1 | 0.0 | 0.1 | 9.4 | 21.0 | 2.4 | 0.3 | 11.6 |
| Hot Springs .................. | 0.0 | 4.2 | 0.1 | 0.0 | 0.8 | 13.7 | 38.9 | 2.2 | | 4.2 |

# Table D. Cities — City Government Finances, City Government Employment, and Climate

| City | City government finances, 2017 (cont.) Debt outstanding | | | Climate² Average daily temperature — Mean | | Limits | | Annual precipitation (inches) | Heating degree days | Cooling degree days |
| | Total (mil dol) | Per capita¹ (dollars) | Debt issued during year | January | July | January³ | July⁴ | | | |
| | 137 | 138 | 139 | 140 | 141 | 142 | 143 | 144 | 145 | 146 |
| United States | X | X | X | X | X | X | X | X | X | X |
| ALABAMA | X | X | X | X | X | X | X | X | X | X |
| Alabaster | 113.1 | 3,400 | 6.6 | NA | NA | NA | NA | NA | NA | NA |
| Auburn | 305.1 | 4,773 | 23.0 | 44.7 | 79.9 | 34.2 | 89.7 | 52.63 | 2,507 | 1,932 |
| Bessemer | 152.1 | 5,708 | 0.0 | 42.9 | 81.0 | 30.8 | 93.6 | 59.38 | 2,766 | 1,943 |
| Birmingham | 1,451.7 | 6,863 | 539.5 | 42.6 | 80.2 | 32.3 | 90.6 | 53.99 | 2,823 | 1,881 |
| Decatur | 200.5 | 3,683 | 21.6 | 38.9 | 79.2 | 29.1 | 90.3 | 55.31 | 3,469 | 1,609 |
| Dothan | 122.5 | 1,794 | 49.6 | 47.7 | 81.3 | 36.2 | 93.3 | 56.61 | 2,058 | 2,264 |
| Enterprise | 87.7 | 3,120 | 2.0 | NA | NA | NA | NA | NA | NA | NA |
| Florence | 99.0 | 2,470 | 0.0 | 39.9 | 80.2 | 30.7 | 90.6 | 55.80 | 3,236 | 1,789 |
| Gadsden | 93.1 | 2,618 | 0.0 | 40.3 | 79.8 | 29.9 | 90.5 | 56.10 | 3,220 | 1,716 |
| Homewood | 65.6 | 2,574 | 2.7 | 42.6 | 80.2 | 32.3 | 90.6 | 53.99 | 2,823 | 1,881 |
| Hoover | 141.6 | 1,661 | 68.6 | 42.9 | 81.0 | 30.8 | 93.6 | 59.38 | 2,766 | 1,943 |
| Huntsville | 983.4 | 5,032 | 137.4 | 39.8 | 79.5 | 30.7 | 89.4 | 57.51 | 3,262 | 1,671 |
| Madison | 276.7 | 5,622 | 72.8 | 46.6 | 81.8 | 35.5 | 92.7 | 54.77 | 2,194 | 2,252 |
| Mobile | 483.5 | 2,542 | 81.6 | 50.1 | 81.5 | 39.5 | 91.2 | 66.29 | 1,681 | 2,539 |
| Montgomery | 356.0 | 1,780 | 31.2 | 46.6 | 81.8 | 35.5 | 92.7 | 54.77 | 2,194 | 2,252 |
| Opelika | 204.8 | 6,756 | 16.4 | NA | NA | NA | NA | NA | NA | NA |
| Phenix City | 121.9 | 3,383 | 15.2 | 46.8 | 82.0 | 36.6 | 91.7 | 48.57 | 2,154 | 2,296 |
| Prattville | 60.5 | 1,707 | 7.4 | NA | NA | NA | NA | NA | NA | NA |
| Tuscaloosa | 191.8 | 1,925 | 33.4 | 42.9 | 80.4 | 32.5 | 90.8 | 54.99 | 2,787 | 1,893 |
| Vestavia Hills | 46.5 | 1,352 | 0.0 | NA | NA | NA | NA | NA | NA | NA |
| ALASKA | X | X | X | X | X | X | X | X | X | X |
| Anchorage | 1,569.1 | 5,331 | 55.9 | 15.8 | 58.4 | 9.3 | 65.3 | 16.08 | 10,470 | 3 |
| Fairbanks | 4.8 | 152 | 0.4 | -9.7 | 62.4 | -19.0 | 73.0 | 10.34 | 13,980 | 74 |
| Juneau | 149.8 | 4,667 | 20.3 | 25.7 | 56.8 | 20.7 | 64.3 | 58.33 | 8,574 | 0 |
| ARIZONA | X | X | X | X | X | X | X | X | X | X |
| Apache Junction | 6.1 | 149 | 0.0 | 52.7 | 89.5 | 40.0 | 104.3 | 12.29 | 1,542 | 3,443 |
| Avondale | 60.1 | 709 | 0.0 | 54.7 | 93.5 | 41.3 | 107.6 | 9.03 | 1,173 | 4,166 |
| Buckeye | 89.2 | 1,302 | 0.0 | NA | NA | NA | NA | NA | NA | NA |
| Bullhead City | 42.5 | 1,057 | 0.0 | 54.4 | 95.6 | 43.3 | 111.7 | 5.84 | 1,164 | 4,508 |
| Casa Grande | 90.7 | 1,629 | 0.0 | 52.4 | 90.4 | 37.3 | 105.1 | 9.22 | 1,572 | 3,554 |
| Chandler | 318.9 | 1,257 | 39.1 | 54.3 | 91.3 | 41.5 | 105.7 | 9.23 | 1,271 | 3,798 |
| El Mirage | 25.6 | 724 | 0.0 | NA | NA | NA | NA | NA | NA | NA |
| Flagstaff | 79.8 | 1,108 | 0.0 | 29.7 | 66.1 | 16.5 | 82.2 | 22.91 | 6,999 | 126 |
| Florence | 70.6 | 2,728 | 5.5 | NA | NA | NA | NA | NA | NA | NA |
| Gilbert | 407.8 | 1,680 | 121.7 | 54.3 | 91.3 | 41.5 | 105.7 | 9.23 | 1,271 | 3,798 |
| Glendale | 804.4 | 3,246 | 0.0 | 52.5 | 90.6 | 39.2 | 104.2 | 7.78 | 1,535 | 3,488 |
| Goodyear | 269.2 | 3,367 | 0.0 | NA | NA | NA | NA | NA | NA | NA |
| Kingman | 36.0 | 1,224 | 0.0 | NA | NA | NA | NA | NA | NA | NA |
| Lake Havasu City | 259.3 | 4,768 | 0.0 | 53.9 | 95.2 | 42.9 | 107.5 | 6.25 | 1,230 | 4,523 |
| Marana | 104.3 | 2,320 | 61.8 | NA | NA | NA | NA | NA | NA | NA |
| Maricopa | 41.4 | 863 | 0.0 | NA | NA | NA | NA | NA | NA | NA |
| Mesa | 1,536.5 | 3,073 | 293.9 | 54.3 | 91.3 | 41.5 | 105.7 | 9.23 | 1,271 | 3,798 |
| Oro Valley | 47.5 | 1,066 | 0.0 | 50.6 | 86.3 | 34.6 | 100.7 | 12.40 | 1,831 | 2,810 |
| Peoria | 274.9 | 1,634 | 0.0 | 54.7 | 93.5 | 41.3 | 107.6 | 9.03 | 1,173 | 4,166 |
| Phoenix | 7,724.4 | 4,729 | 1,130.0 | 54.2 | 92.8 | 43.4 | 104.2 | 8.29 | 1,125 | 4,189 |
| Prescott | 98.9 | 2,318 | 1.0 | 37.1 | 73.4 | 23.3 | 88.3 | 19.19 | 4,849 | 742 |
| Prescott Valley | 68.4 | 1,538 | 0.0 | NA | NA | NA | NA | NA | NA | NA |
| Queen Creek | 154.7 | 3,580 | 0.0 | NA | NA | NA | NA | NA | NA | NA |
| Sahuarita | 46.0 | 1,545 | 9.9 | NA | NA | NA | NA | NA | NA | NA |
| San Luis | 197.1 | 6,021 | 0.0 | NA | NA | NA | NA | NA | NA | NA |
| Scottsdale | 1,218.6 | 4,847 | 247.0 | 54.2 | 92.8 | 43.4 | 104.2 | 8.29 | 1,125 | 4,189 |
| Sierra Vista | 19.9 | 464 | 0.0 | 47.7 | 79.1 | 33.7 | 92.6 | 14.02 | 2,369 | 1,739 |
| Surprise | 73.8 | 547 | 0.0 | 54.7 | 93.5 | 41.3 | 107.6 | 9.03 | 1,173 | 4,166 |
| Tempe | 446.0 | 2,401 | 0.0 | 54.1 | 89.9 | 40.1 | 103.6 | 9.36 | 1,390 | 3,655 |
| Tucson | 1,005.8 | 1,858 | 133.1 | 54.0 | 88.5 | 41.9 | 100.5 | 12.00 | 1,333 | 3,501 |
| Yuma | 24.2 | 251 | 0.0 | 58.1 | 94.1 | 46.2 | 107.3 | 3.01 | 782 | 4,540 |
| ARKANSAS | X | X | X | X | X | X | X | X | X | X |
| Bella Vista | 0.0 | 0 | 0.0 | NA | NA | NA | NA | NA | NA | NA |
| Benton | 98.6 | 2,741 | 2.0 | NA | NA | NA | NA | NA | NA | NA |
| Bentonville | 59.1 | 1,196 | 0.0 | NA | NA | NA | NA | NA | NA | NA |
| Conway | 438.3 | 6,646 | 31.0 | NA | NA | NA | NA | NA | NA | NA |
| Fayetteville | 107.0 | 1,250 | 17.3 | 38.3 | 82.1 | 28.1 | 92.4 | 48.67 | 3,320 | 1,961 |
| Fort Smith | 745.0 | 8,482 | 178.0 | 34.3 | 78.9 | 24.2 | 89.1 | 46.02 | 4,166 | 1,439 |
| Hot Springs | 100.4 | 2,609 | 9.9 | 38.0 | 82.2 | 27.8 | 92.9 | 43.87 | 3,437 | 1,929 |
| | | | | 40.2 | 82.2 | 29.6 | 94.3 | 57.69 | 3,133 | 1,993 |

1. Based on the population estimated as of July 1 of the year shown. 2. Represents normal values based on the 30-year period, 1971±2000. 3. Average daily minimum. 4. Average daily maximum.

# Table D. Cities — Land Area and Population

| | | Population, 2020 | | | | Race 2019 Race alone[2] (percent) | | | | | | |
|---|---|---|---|---|---|---|---|---|---|---|---|---|
| STATE Place code | City | Land area[1] (sq. mi) | Total persons 2019 | Rank | Per square mile | White | Black or African American | American Indian, Alaskan Native | Asian | Hawaiian Pacific Islander | Some other race | Two or more races (percent) |
| | | 1 | 2 | 3 | 4 | 5 | 6 | 7 | 8 | 9 | 10 | 11 |
| | ARKANSAS— Cont'd | | | | | D | D | D | D | D | D | D |
| 05 34750 | Jacksonville | 28.6 | 28,361 | 1,303 | 992 | 70.4 | 22.8 | 0.2 | 1.9 | 0.0 | 0.6 | 4.0 |
| 05 35710 | Jonesboro | 80.2 | 79,859 | 440 | 996 | 51.7 | 42.3 | 0.4 | 3.1 | 0.1 | 0.9 | 1.5 |
| 05 41000 | Little Rock | 120.0 | 197,866 | 123 | 1,649 | 49.2 | 41.8 | 0.2 | 1.4 | 0.0 | 3.8 | 3.6 |
| 05 50450 | North Little Rock | 52.3 | 66,276 | 569 | 1,267 | 90.8 | 2.6 | 1.3 | 0.0 | 0.0 | 2.7 | 2.7 |
| 05 53390 | Paragould | 31.9 | 29,257 | 1,281 | 917 | D | D | D | D | D | D | D |
| 05 55310 | Pine Bluff | 44.2 | 40,435 | 953 | 915 | 88.5 | 1.5 | 0.9 | 2.4 | 2.8 | 1.9 | 2.0 |
| 05 60410 | Rogers | 38.9 | 70,522 | 516 | 1,813 | 84.3 | 8.1 | 0.7 | 1.8 | 0.0 | 1.0 | 4.1 |
| 05 61670 | Russellville | 28.3 | 29,364 | 1,275 | 1,038 | D | D | D | D | D | D | D |
| 05 63800 | Sherwood | 20.7 | 31,653 | 1,205 | 1,529 | 68.3 | 3.4 | 0.0 | 1.7 | 7.8 | 16.5 | 2.3 |
| 05 66080 | Springdale | 47.0 | 81,115 | 429 | 1,726 | D | D | D | D | D | D | D |
| 05 68810 | Texarkana | 42.0 | 29,556 | 1,268 | 704 | D | D | D | D | D | D | D |
| 05 74540 | West Memphis | 28.8 | 24,158 | 1,421 | 839 | D | D | D | D | D | 13.7 | 5.0 |
| 06 00000 | CALIFORNIA | 155,854.0 | 39,368,078 | X | 253 | 59.4 | 5.8 | 0.8 | 14.8 | 0.4 | 13.7 | 5.0 |
| 06 00296 | Adelanto | 52.9 | 34,537 | 1,117 | 653 | 63.7 | 19.0 | 0.7 | 2.5 | 0.0 | 8.3 | 5.8 |
| 06 00562 | Alameda | 10.4 | 78,841 | 449 | 7,581 | 45.0 | 8.3 | 0.5 | 32.5 | 0.1 | 5.2 | 8.3 |
| 06 00884 | Alhambra | 7.6 | 83,261 | 415 | 10,955 | 21.3 | 2.8 | 0.6 | 54.1 | 0.0 | 17.7 | 3.5 |
| 06 00947 | Aliso Viejo | 6.9 | 50,412 | 773 | 7,306 | 69.0 | 4.2 | 0.1 | 16.2 | 0.2 | 3.2 | 7.1 |
| 06 02000 | Anaheim | 50.3 | 353,676 | 55 | 7,031 | 65.3 | 3.4 | 0.7 | 18.2 | 0.7 | 8.4 | 3.2 |
| 06 02252 | Antioch | 29.2 | 111,605 | 269 | 3,822 | 36.6 | 21.1 | 0.5 | 16.4 | 0.7 | 17.2 | 7.4 |
| 06 02364 | Apple Valley | 77.0 | 74,061 | 481 | 962 | 78.0 | 5.8 | 1.0 | 1.6 | 0.7 | 7.6 | 5.3 |
| 06 02462 | Arcadia | 10.9 | 56,323 | 682 | 5,167 | 30.8 | 0.7 | 0.3 | 62.4 | 0.2 | 3.1 | 2.5 |
| 06 03064 | Atascadero | 26.1 | 30,254 | 1,243 | 1,159 | 90.9 | 0.4 | 2.3 | 2.0 | 0.0 | 1.4 | 3.0 |
| 06 03162 | Atwater | 6.6 | 30,296 | 1,239 | 4,590 | 51.8 | 2.1 | 2.1 | 8.0 | 0.1 | 29.6 | 6.2 |
| 06 03386 | Azusa | 9.7 | 49,238 | 789 | 5,076 | 63.1 | 3.2 | 0.1 | 19.0 | 0.0 | 9.7 | 4.9 |
| 06 03526 | Bakersfield | 149.8 | 385,725 | 53 | 2,575 | 67.5 | 6.9 | 1.1 | 7.7 | 0.2 | 12.2 | 4.3 |
| 06 03666 | Baldwin Park | 6.6 | 74,480 | 477 | 11,285 | 54.7 | 0.7 | 1.2 | 20.9 | 0.0 | 15.2 | 7.3 |
| 06 03820 | Banning | 23.2 | 30,380 | 1,236 | 1,310 | 62.1 | 5.2 | 3.6 | 5.5 | 0.0 | 18.8 | 4.8 |
| 06 04758 | Beaumont | 30.3 | 51,449 | 755 | 1,698 | 56.1 | 15.1 | 1.5 | 12.1 | 0.0 | 11.3 | 3.9 |
| 06 04870 | Bell | 2.5 | 34,734 | 1,111 | 13,894 | D | D | D | D | D | D | D |
| 06 04982 | Bellflower | 6.1 | 75,162 | 474 | 12,322 | 31.1 | 17.2 | 0.1 | 8.6 | 0.3 | 39.1 | 3.6 |
| 06 04996 | Bell Gardens | 2.5 | 40,954 | 939 | 16,382 | 85.3 | 3.7 | 0.0 | 0.2 | 0.1 | 9.7 | 1.1 |
| 06 05108 | Belmont | 4.6 | 26,533 | 1,372 | 5,768 | 58.1 | 2.2 | 0.0 | 25.5 | 4.0 | 3.4 | 6.8 |
| 06 05290 | Benicia | 12.8 | 27,988 | 1,314 | 2,187 | 71.9 | 1.6 | 0.0 | 16.2 | 0.0 | 3.2 | 7.1 |
| 06 06000 | Berkeley | 10.4 | 122,726 | 230 | 11,801 | 60.0 | 6.2 | 0.5 | 22.4 | 0.3 | 3.2 | 7.4 |
| 06 06308 | Beverly Hills | 5.7 | 33,001 | 1,163 | 5,790 | D | D | D | D | D | D | D |
| 06 08100 | Brea | 12.2 | 44,484 | 864 | 3,646 | 63.6 | 0.7 | 0.0 | 26.0 | 0.1 | 5.5 | 4.1 |
| 06 08142 | Brentwood | 14.9 | 65,017 | 581 | 4,364 | 59.8 | 11.9 | 0.8 | 16.3 | 0.9 | 6.7 | 3.7 |
| 06 08786 | Buena Park | 10.5 | 81,269 | 425 | 7,740 | 54.0 | 1.1 | 0.1 | 33.8 | 0.1 | 8.1 | 2.9 |
| 06 08954 | Burbank | 17.3 | 101,866 | 308 | 5,888 | 71.3 | 2.9 | 0.7 | 13.7 | 0.0 | 4.6 | 6.7 |
| 06 09066 | Burlingame | 4.4 | 29,557 | 1,267 | 6,718 | 62.6 | 0.3 | 0.0 | 26.9 | 0.1 | 6.4 | 3.7 |
| 06 09710 | Calexico | 8.6 | 39,547 | 981 | 4,599 | D | D | D | D | D | D | D |
| 06 10046 | Camarillo | 19.7 | 69,521 | 527 | 3,529 | 77.7 | 2.0 | 0.8 | 10.6 | 0.1 | 3.6 | 5.2 |
| 06 10345 | Campbell | 6.1 | 42,427 | 909 | 6,955 | 64.7 | 0.6 | 0.0 | 22.4 | 0.0 | 7.0 | 5.4 |
| 06 11194 | Carlsbad | 37.8 | 115,599 | 253 | 3,058 | 83.8 | 0.4 | 0.0 | 10.7 | 0.0 | 0.4 | 4.7 |
| 06 11530 | Carson | 18.7 | 89,860 | 371 | 4,805 | 23.0 | 28.0 | 0.2 | 28.2 | 1.4 | 14.0 | 5.3 |
| 06 12048 | Cathedral City | 22.5 | 55,678 | 691 | 2,475 | 73.1 | 1.8 | 1.3 | 7.8 | 0.5 | 14.0 | 1.5 |
| 06 12524 | Ceres | 9.5 | 48,603 | 802 | 5,116 | 70.6 | 1.3 | 1.4 | 8.9 | 0.0 | 11.0 | 6.7 |
| 06 12552 | Cerritos | 8.7 | 48,624 | 801 | 5,589 | 16.8 | 10.1 | 0.2 | 57.5 | 1.2 | 8.5 | 5.6 |
| 06 13014 | Chico | 34.1 | 102,867 | 300 | 3,017 | 81.5 | 2.6 | 0.5 | 4.0 | 0.3 | 6.6 | 4.5 |
| 06 13210 | Chino | 29.6 | 92,290 | 364 | 3,118 | 67.4 | 3.9 | 0.3 | 15.0 | 1.2 | 7.9 | 4.4 |
| 06 13214 | Chino Hills | 44.7 | 84,586 | 403 | 1,892 | 41.1 | 2.2 | 0.4 | 43.2 | 0.2 | 7.2 | 5.7 |
| 06 13392 | Chula Vista | 49.6 | 272,979 | 75 | 5,504 | 62.7 | 5.1 | 0.3 | 18.1 | 0.4 | 8.3 | 5.0 |
| 06 13588 | Citrus Heights | 14.2 | 87,691 | 387 | 6,175 | 82.2 | 3.4 | 1.8 | 3.4 | 0.0 | 3.2 | 6.1 |
| 06 13756 | Claremont | 13.3 | 35,096 | 1,101 | 2,639 | 60.7 | 6.2 | 0.3 | 16.3 | 0.0 | 9.1 | 7.5 |
| 06 14218 | Clovis | 25.4 | 117,662 | 244 | 4,632 | 70.0 | 3.0 | 1.7 | 10.2 | 0.7 | 8.3 | 6.1 |
| 06 14260 | Coachella | 30.1 | 46,564 | 835 | 1,547 | D | D | D | D | D | D | D |
| 06 14890 | Colton | 15.6 | 54,661 | 706 | 3,504 | 57.0 | 8.4 | 0.9 | 3.4 | 0.0 | 28.6 | 1.7 |
| 06 15044 | Compton | 10.0 | 93,967 | 352 | 9,397 | 34.0 | 22.7 | 0.2 | 0.4 | 0.5 | 40.1 | 2.2 |
| 06 16000 | Concord | 30.6 | 128,544 | 215 | 4,201 | 59.8 | 4.1 | 0.7 | 13.7 | 0.1 | 13.9 | 7.7 |
| 06 16350 | Corona | 39.9 | 170,996 | 155 | 4,286 | 53.9 | 5.2 | 0.6 | 13.0 | 0.2 | 22.9 | 4.3 |
| 06 16532 | Costa Mesa | 15.8 | 112,366 | 264 | 7,112 | 57.9 | 0.9 | 0.9 | 10.9 | 0.5 | 24.6 | 4.4 |
| 06 16742 | Covina | 7.0 | 47,259 | 825 | 6,751 | 69.3 | 3.8 | 0.1 | 14.4 | 0.4 | 6.3 | 5.7 |
| 06 17568 | Culver City | 5.1 | 38,471 | 1,005 | 7,543 | 66.7 | 6.8 | 0.2 | 18.1 | 0.0 | 1.8 | 6.5 |
| 06 17610 | Cupertino | 11.3 | 58,598 | 650 | 5,186 | D | D | D | D | D | D | D |
| 06 17750 | Cypress | 6.6 | 48,982 | 794 | 7,422 | 41.3 | 5.2 | 0.6 | 37.7 | 0.1 | 8.8 | 6.3 |
| 06 17918 | Daly City | 7.6 | 105,868 | 291 | 13,930 | 23.0 | 2.5 | 0.8 | 60.0 | 0.5 | 9.8 | 3.4 |
| 06 17946 | Dana Point | 6.5 | 33,436 | 1,149 | 5,144 | D | D | D | D | D | D | D |
| 06 17988 | Danville | 18.1 | 44,883 | 859 | 2,480 | 79.0 | 1.7 | 0.2 | 14.8 | 0.0 | 0.4 | 3.9 |
| 06 18100 | Davis | 10.0 | 68,805 | 534 | 6,881 | 61.6 | 3.2 | 1.6 | 27.2 | 0.5 | 1.8 | 4.1 |

1. Dry land or land partially or temporarily covered by water.  2. Hispanic or Latino persons may be of any race.

# Table D. Cities — Population

| City | Percent Hispanic or Latino[1], 2019 | Percent foreign born, 2019 | Under 18 years | 18 to 24 years | 25 to 34 years | 35 to 44 years | 45 to 54 years | 55 to 64 years | 65 years and over | Median age, 2019 | Percent female, 2019 | Census counts 2000 | Census counts 2010 | Percent change 2000-2010 | Percent change 2001-2020 |
|---|---|---|---|---|---|---|---|---|---|---|---|---|---|---|---|
| | 12 | 13 | 14 | 15 | 16 | 17 | 18 | 19 | 20 | 21 | 22 | 23 | 24 | 25 | 26 |
| **ARKANSAS— Cont'd** | | | | | | | | | | | | | | | |
| Jacksonville | 10.9 | 6.6 | 24.4 | 12.8 | 18.7 | 6.5 | 8.4 | 13.7 | 15.4 | 32.1 | 50.0 | 29,916 | 28,368 | -5.2 | 0.0 |
| Jonesboro | 6.3 | 5.9 | 22.6 | 14.6 | 13.3 | 14.0 | 12.0 | 10.9 | 12.5 | 34.6 | 52.4 | 55,515 | 67,291 | 21.2 | 18.7 |
| Little Rock | 6.7 | 6.9 | 22.5 | 9.2 | 15.3 | 13.3 | 12.9 | 12.2 | 14.6 | 37.4 | 52.6 | 183,133 | 193,534 | 5.7 | 2.2 |
| North Little Rock | 6.2 | 3.6 | 25.3 | 7.9 | 16.6 | 11.5 | 9.4 | 13.6 | 15.6 | 35.2 | 51.8 | 60,433 | 62,333 | 3.1 | 6.3 |
| Paragould | 4.5 | 0.7 | 24.6 | 9.7 | 16.4 | 15.3 | 9.9 | 11.8 | 12.3 | 34.6 | 53.3 | 22,017 | 26,315 | 19.5 | 11.2 |
| Pine Bluff | 0.2 | 0.9 | 19.5 | 13.9 | 13.9 | 10.6 | 10.4 | 14.3 | 17.3 | 37.1 | 54.2 | 55,085 | 49,060 | -10.9 | -17.6 |
| Rogers | 29.8 | 22.2 | 27.6 | 9.7 | 16.5 | 12.6 | 13.3 | 11.1 | 9.1 | 32.2 | 52.8 | 38,829 | 56,104 | 44.5 | 25.7 |
| Russellville | 17.4 | 15.3 | 20.4 | 20.3 | 12.3 | 6.6 | 13.4 | 12.6 | 14.4 | 31.7 | 57.3 | 23,682 | 28,106 | 18.7 | 4.5 |
| Sherwood | 3.7 | 1.8 | 19.7 | 8.7 | 11.6 | 15.6 | 7.9 | 15.3 | 21.3 | 39.7 | 58.6 | 21,511 | 29,609 | 37.6 | 6.9 |
| Springdale | 38.8 | 23.1 | 29.7 | 11.5 | 13.4 | 14.7 | 11.0 | 8.2 | 11.6 | 31.1 | 49.8 | 45,798 | 70,808 | 54.6 | 14.6 |
| Texarkana | 4.1 | 3.3 | 21.7 | 6.6 | 13.2 | 12.4 | 16.7 | 13.7 | 15.7 | 41.9 | 51.5 | 26,448 | 29,888 | 13.0 | -1.1 |
| West Memphis | 1.8 | 0.8 | 24.3 | 12.2 | 12.3 | 14.6 | 8.9 | 13.8 | 14.0 | 35.7 | 57.2 | 27,666 | 26,245 | -5.1 | -8.0 |
| **CALIFORNIA** | 39.4 | 26.7 | 22.5 | 9.3 | 15.3 | 13.4 | 12.6 | 12.1 | 14.8 | 37.0 | 50.3 | 33,871,648 | 37,254,522 | 10.0 | 5.7 |
| Adelanto | 64.8 | 17.1 | 28.9 | 16.6 | 16.2 | 13.8 | 12.2 | 7.4 | 4.9 | 26.2 | 42.9 | 18,130 | 31,760 | 75.2 | 8.7 |
| Alameda | 11.5 | 24.2 | 20.6 | 6.8 | 15.2 | 15.8 | 12.1 | 12.7 | 16.9 | 39.8 | 49.4 | 72,259 | 73,812 | 2.1 | 6.8 |
| Alhambra | 31.4 | 49.9 | 16.8 | 9.1 | 15.6 | 12.4 | 14.2 | 13.3 | 18.6 | 41.8 | 50.2 | 85,804 | 83,073 | -3.2 | 0.2 |
| Aliso Viejo | 15.6 | 25.8 | 23.9 | 6.8 | 11.0 | 14.9 | 20.2 | 13.8 | 9.4 | 40.8 | 50.2 | 40,166 | 47,668 | 18.7 | 5.8 |
| Anaheim | 53.4 | 35.7 | 22.0 | 10.5 | 17.1 | 13.2 | 13.4 | 12.0 | 11.8 | 35.2 | 50.5 | 328,014 | 336,332 | 2.5 | 5.2 |
| Antioch | 31.5 | 23.2 | 21.4 | 8.9 | 15.5 | 11.0 | 13.0 | 15.0 | 15.2 | 38.8 | 54.6 | 90,532 | 102,745 | 13.5 | 8.6 |
| Apple Valley | 40.8 | 13.3 | 27.1 | 8.8 | 12.9 | 10.4 | 9.1 | 14.5 | 17.1 | 36.3 | 51.2 | 54,239 | 69,146 | 27.5 | 7.1 |
| Arcadia | 12.8 | 48.6 | 22.2 | 6.6 | 10.6 | 12.3 | 14.2 | 13.9 | 20.0 | 42.7 | 50.8 | 53,054 | 56,242 | 6.0 | 0.1 |
| Atascadero | 23.4 | 8.3 | 23.5 | 7.6 | 14.8 | 16.5 | 10.8 | 14.9 | 11.8 | 36.8 | 52.6 | 26,411 | 28,306 | 7.2 | 6.9 |
| Atwater | 51.5 | 20.6 | 32.0 | 8.2 | 16.2 | 15.0 | 11.2 | 8.0 | 9.4 | 29.8 | 50.1 | 23,113 | 28,552 | 23.5 | 6.1 |
| Azusa | 56.5 | 33.0 | 22.5 | 18.0 | 13.2 | 14.4 | 9.0 | 11.8 | 11.0 | 31.8 | 53.4 | 44,712 | 46,574 | 4.2 | 5.7 |
| Bakersfield | 52.4 | 20.3 | 29.8 | 9.6 | 16.2 | 12.9 | 11.2 | 9.7 | 10.5 | 31.4 | 49.8 | 247,057 | 347,783 | 40.8 | 10.9 |
| Baldwin Park | 75.1 | 43.3 | 22.7 | 10.6 | 14.7 | 13.4 | 13.0 | 12.6 | 13.0 | 37.0 | 51.6 | 75,837 | 75,397 | -0.6 | -1.2 |
| Banning | 53.4 | 18.6 | 25.0 | 9.0 | 12.3 | 9.3 | 13.2 | 7.6 | 23.5 | 41.1 | 55.4 | 23,562 | 28,684 | 21.7 | 5.9 |
| Beaumont | 40.4 | 17.6 | 30.6 | 8.1 | 9.0 | 15.1 | 12.4 | 9.7 | 15.1 | 37.0 | 52.7 | 11,384 | 36,932 | 224.4 | 39.3 |
| Bell | 94.4 | 35.4 | 29.0 | 13.5 | 13.8 | 13.7 | 9.6 | 9.8 | 10.6 | 28.9 | 48.2 | 36,664 | 35,469 | -3.3 | -2.1 |
| Bellflower | 60.7 | 30.2 | 21.8 | 9.4 | 15.3 | 13.8 | 12.5 | 13.8 | 13.3 | 37.1 | 51.5 | 72,878 | 76,625 | 5.1 | -1.9 |
| Bell Gardens | 94.0 | 40.9 | 29.4 | 11.1 | 14.8 | 14.8 | 12.0 | 9.4 | 8.5 | 30.3 | 52.4 | 44,054 | 42,058 | -4.5 | -2.6 |
| Belmont | 12.0 | 27.9 | 22.8 | 4.6 | 14.7 | 15.1 | 13.6 | 10.9 | 18.3 | 40.8 | 50.7 | 25,123 | 25,900 | 3.1 | 2.4 |
| Benicia | 14.7 | 14.4 | 18.7 | 6.3 | 8.2 | 14.7 | 12.3 | 17.7 | 22.0 | 46.6 | 52.7 | 26,865 | 27,035 | 0.6 | 3.5 |
| Berkeley | 12.3 | 20.7 | 11.8 | 25.5 | 18.4 | 10.0 | 9.5 | 9.3 | 15.6 | 31.5 | 50.6 | 102,743 | 112,443 | 9.4 | 9.1 |
| Beverly Hills | 5.2 | 34.9 | 19.6 | 3.9 | 7.0 | 12.2 | 15.7 | 16.3 | 25.3 | 48.8 | 52.9 | 33,784 | 33,923 | 0.4 | -2.7 |
| Brea | 36.0 | 27.1 | 21.3 | 9.0 | 10.6 | 11.9 | 14.0 | 18.3 | 14.9 | 41.8 | 56.1 | 35,410 | 39,195 | 10.7 | 13.5 |
| Brentwood | 21.0 | 17.9 | 23.2 | 7.7 | 10.8 | 14.4 | 9.7 | 16.9 | 17.3 | 41.8 | 51.5 | 23,302 | 51,627 | 121.6 | 25.9 |
| Buena Park | 36.4 | 41.7 | 17.4 | 8.4 | 15.7 | 14.5 | 11.8 | 14.5 | 17.7 | 41.1 | 51.7 | 78,282 | 80,456 | 2.8 | 1.0 |
| Burbank | 23.1 | 30.2 | 17.7 | 6.1 | 18.5 | 12.4 | 13.1 | 14.1 | 18.2 | 40.6 | 53.5 | 100,316 | 103,360 | 3.0 | -1.4 |
| Burlingame | 17.5 | 24.9 | 25.7 | 3.4 | 12.9 | 18.4 | 16.4 | 10.1 | 13.1 | 40.3 | 50.2 | 28,158 | 28,815 | 2.3 | 2.6 |
| Calexico | 98.5 | 39.2 | 31.8 | 11.0 | 13.3 | 11.5 | 9.6 | 9.1 | 13.7 | 30.0 | 49.4 | 27,109 | 38,573 | 42.3 | 2.5 |
| Camarillo | 28.0 | 18.6 | 19.0 | 8.7 | 11.5 | 12.2 | 12.7 | 13.9 | 21.9 | 43.7 | 52.7 | 57,077 | 65,175 | 14.2 | 6.7 |
| Campbell | 21.6 | 25.5 | 26.4 | 3.9 | 18.2 | 18.0 | 10.5 | 12.0 | 10.9 | 35.6 | 51.2 | 38,138 | 40,578 | 6.4 | 4.6 |
| Carlsbad | 10.7 | 14.1 | 21.2 | 5.9 | 10.6 | 12.8 | 15.7 | 15.2 | 18.5 | 44.6 | 51.1 | 78,247 | 105,349 | 34.6 | 9.7 |
| Carson | 34.2 | 38.9 | 16.1 | 10.9 | 16.4 | 7.8 | 13.8 | 16.9 | 18.1 | 43.8 | 49.3 | 89,730 | 91,711 | 2.2 | -2.0 |
| Cathedral City | 57.5 | 38.1 | 20.1 | 6.0 | 12.3 | 12.2 | 12.3 | 15.4 | 21.7 | 44.6 | 50.3 | 42,647 | 51,233 | 20.1 | 8.7 |
| Ceres | 64.3 | 25.9 | 27.7 | 8.1 | 19.6 | 13.3 | 9.3 | 10.7 | 11.4 | 31.3 | 46.4 | 34,609 | 45,878 | 32.6 | 5.9 |
| Cerritos | 15.2 | 45.7 | 16.3 | 6.6 | 13.7 | 11.2 | 11.4 | 14.5 | 26.4 | 47.4 | 49.6 | 51,488 | 49,043 | -4.7 | -0.9 |
| Chico | 20.6 | 6.4 | 20.1 | 21.1 | 14.7 | 11.0 | 9.6 | 9.4 | 14.0 | 30.2 | 52.6 | 59,954 | 89,013 | 48.5 | 15.6 |
| Chino | 54.7 | 27.1 | 18.2 | 8.7 | 17.8 | 15.3 | 14.5 | 13.1 | 12.5 | 38.6 | 47.7 | 67,168 | 78,074 | 16.2 | 18.2 |
| Chino Hills | 21.9 | 32.9 | 20.7 | 9.2 | 11.8 | 14.6 | 14.9 | 15.0 | 13.8 | 40.9 | 50.7 | 66,787 | 74,792 | 12.0 | 13.1 |
| Chula Vista | 58.0 | 30.3 | 24.0 | 9.1 | 14.5 | 14.7 | 14.6 | 10.8 | 12.2 | 37.4 | 50.3 | 173,556 | 243,912 | 40.5 | 11.9 |
| Citrus Heights | 19.5 | 13.1 | 17.7 | 8.2 | 17.9 | 11.8 | 9.9 | 16.3 | 18.3 | 39.5 | 52.5 | 85,071 | 83,118 | -2.3 | 5.5 |
| Claremont | 26.7 | 22.0 | 18.7 | 16.2 | 11.6 | 10.6 | 12.6 | 12.0 | 18.3 | 40.2 | 52.4 | 33,998 | 34,867 | 2.6 | 0.7 |
| Clovis | 33.7 | 11.5 | 29.9 | 5.9 | 13.1 | 13.9 | 11.5 | 12.4 | 13.4 | 35.6 | 49.8 | 68,468 | 95,987 | 40.2 | 22.6 |
| Coachella | 95.6 | 34.3 | 21.6 | 14.9 | 19.3 | 13.9 | 8.7 | 11.4 | 10.3 | 33.2 | 49.9 | 22,724 | 40,712 | 79.2 | 14.4 |
| Colton | 69.7 | 20.8 | 24.8 | 12.1 | 12.7 | 16.4 | 14.2 | 9.0 | 10.9 | 35.2 | 53.2 | 47,662 | 52,118 | 9.3 | 4.9 |
| Compton | 73.1 | 31.8 | 27.6 | 9.5 | 14.9 | 13.5 | 12.0 | 11.0 | 11.5 | 33.5 | 50.4 | 93,493 | 96,373 | 3.1 | -2.5 |
| Concord | 32.7 | 29.7 | 20.5 | 9.5 | 16.9 | 12.0 | 13.0 | 13.9 | 14.2 | 37.3 | 49.9 | 121,780 | 122,162 | 0.3 | 5.2 |
| Corona | 51.2 | 22.7 | 25.1 | 8.4 | 16.0 | 15.4 | 12.6 | 13.5 | 9.1 | 35.3 | 50.2 | 124,966 | 152,283 | 21.9 | 12.3 |
| Costa Mesa | 35.2 | 25.3 | 23.1 | 6.9 | 20.2 | 15.5 | 13.2 | 8.7 | 12.5 | 34.9 | 49.1 | 108,724 | 110,082 | 1.2 | 2.1 |
| Covina | 56.3 | 27.4 | 21.4 | 7.0 | 14.0 | 17.5 | 12.5 | 13.0 | 14.5 | 38.9 | 47.6 | 46,837 | 47,797 | 2.0 | -1.1 |
| Culver City | 19.3 | 26.7 | 19.9 | 3.2 | 15.9 | 17.1 | 16.5 | 13.0 | 14.4 | 41.3 | 53.4 | 38,816 | 38,833 | 0.0 | -0.9 |
| Cupertino | 1.8 | 51.8 | 23.3 | 8.3 | 8.2 | 15.0 | 17.3 | 11.4 | 16.5 | 42.7 | 47.3 | 50,546 | 58,547 | 15.8 | 0.1 |
| Cypress | 19.0 | 31.6 | 24.6 | 6.8 | 7.8 | 17.8 | 14.2 | 12.0 | 16.7 | 41.2 | 53.0 | 46,229 | 47,858 | 3.5 | 2.3 |
| Daly City | 25.4 | 56.1 | 16.5 | 9.5 | 15.3 | 16.0 | 11.0 | 14.3 | 17.4 | 39.7 | 52.0 | 103,621 | 101,322 | -2.2 | 4.5 |
| Dana Point | 16.4 | 17.8 | 12.8 | 5.4 | 6.3 | 9.7 | 17.8 | 19.2 | 28.8 | 53.9 | 52.4 | 35,110 | 33,291 | -5.2 | 0.4 |
| Danville | 3.8 | 20.2 | 24.5 | 6.1 | 5.8 | 13.0 | 15.3 | 17.7 | 17.7 | 45.4 | 49.3 | 41,715 | 41,871 | 0.4 | 7.2 |
| Davis | 15.1 | 23.5 | 11.6 | 36.3 | 13.5 | 8.8 | 10.2 | 8.4 | 11.1 | 25.8 | 53.6 | 60,308 | 65,675 | 8.9 | 4.8 |

1. May be of any race.

# Table D. Cities — Households, Group Quarters, Crime, and Education

| City | Households, 2019 Number | Persons per household | Percent Family | Percent Married couple family | Percent Female family | Percent Non-family | Percent One person | Persons in group quarters, 2019 | Serious crimes known to police[1], 2019 — Violent Number | Violent Rate[2] | Property Number | Property Rate[2] | Population age 25 and over | Attainment[3] High school graduate or less | Attainment[3] Bachelor's degree or more |
|---|---|---|---|---|---|---|---|---|---|---|---|---|---|---|---|
| (item) | 27 | 28 | 29 | 30 | 31 | 32 | 33 | 34 | 35 | 36 | 37 | 38 | 39 | 40 | 41 |
| **ARKANSAS— Cont'd** | | | | | | | | | | | | | | | |
| Jacksonville | 11,948 | 2.29 | 58.6 | 34.8 | 23.7 | 41.4 | 35.3 | 857 | 271 | 959 | 1,484 | 5,249 | 17,715 | 36.4 | 21.7 |
| Jonesboro | 29,688 | 2.52 | 66.1 | 45.2 | 20.9 | 33.9 | 27.4 | 3,711 | 537 | 686 | 2,982 | 3,810 | 49,187 | 39.8 | 29.6 |
| Little Rock | 80,063 | 2.42 | 56.9 | 36.5 | 20.5 | 43.1 | 36.8 | 3,833 | 3,009 | 1,517 | 12,145 | 6,122 | 134,842 | 29.9 | 42.7 |
| North Little Rock | 27,903 | 2.33 | 52.8 | 36.2 | 16.6 | 47.2 | 43.2 | 756 | 562 | 844 | 2,479 | 3,722 | 44,059 | 40.5 | 28.7 |
| Paragould | 11,443 | 2.50 | 65.1 | 44.5 | 20.5 | 34.9 | 26.3 | 684 | 307 | 1,050 | 1,378 | 4,712 | 19,273 | 52.2 | 21.4 |
| Pine Bluff | 19,009 | 1.94 | 52.6 | 25.7 | 26.9 | 47.4 | 38.5 | 4,580 | 644 | 1,552 | NA | NA | 27,601 | 45.9 | 19.4 |
| Rogers | 24,297 | 2.80 | 68.0 | 51.7 | 16.3 | 32.0 | 25.2 | 540 | 330 | 477 | 1,968 | 2,845 | 43,049 | 45.4 | 34.1 |
| Russellville | 11,143 | 2.33 | 52.0 | 39.0 | 13.0 | 48.0 | 40.3 | 3,246 | 140 | 475 | 897 | 3,046 | 17,311 | 29.9 | 27.8 |
| Sherwood | 12,282 | 2.55 | 66.4 | 40.9 | 25.5 | 33.6 | 28.8 | 93 | 195 | 620 | 928 | 2,952 | 22,509 | 54.7 | 31.9 |
| Springdale | 26,443 | 3.01 | 73.6 | 49.1 | 24.6 | 26.4 | 19.2 | 978 | NA | NA | NA | NA | 23,995 | 49.1 | 17.7 |
| Texarkana | 11,743 | 2.72 | 64.5 | 43.5 | 21.0 | 35.5 | 29.7 | 1,528 | 183 | 611 | 1,130 | 3,770 | 14,533 | 54.0 | 17.3 |
| West Memphis | 9,787 | 2.29 | 59.2 | 27.1 | 32.0 | 40.8 | 30.3 | 494 | 485 | 1,984 | 1,225 | 5,012 |  |  |  |
| **CALIFORNIA** | 13,157,873 | 2.94 | 68.2 | 49.3 | 18.9 | 31.8 | 24.1 | 826,521 | NA | NA | NA | NA | 26,937,872 | 36.6 | 35.0 |
| Adelanto | 7,775 | 4.01 | 78.6 | 53.8 | 24.8 | 21.4 | 19.0 | 2,882 | 276 | 800 | 459 | 1,331 | 18,565 | 58.0 | 9.3 |
| Alameda | 30,788 | 2.49 | 59.0 | 45.2 | 13.7 | 41.0 | 32.0 | 881 | 162 | 205 | 2,579 | 3,268 | 56,365 | 21.5 | 57.2 |
| Alhambra | 28,777 | 2.88 | 67.9 | 47.4 | 20.5 | 32.1 | 22.9 | 746 | 161 | 190 | 1,749 | 2,062 | 62,028 | 40.1 | 38.0 |
| Aliso Viejo | 18,827 | 2.68 | 67.9 | 53.7 | 14.3 | 32.1 | 23.3 | 404 | 27 | 52 | 433 | 829 | 35,268 | 13.1 | 54.7 |
| Anaheim | 105,301 | 3.28 | 73.1 | 50.1 | 23.0 | 26.9 | 19.9 | 4,880 | 1,120 | 316 | 8,258 | 2,333 | 236,324 | 46.0 | 25.9 |
| Antioch | 36,138 | 3.07 | 68.7 | 45.8 | 22.8 | 31.3 | 25.3 | 556 | 648 | 575 | 3,199 | 2,840 | 77,737 | 43.0 | 20.4 |
| Apple Valley | 23,774 | 3.06 | 73.0 | 54.7 | 18.3 | 27.0 | 21.7 | 610 | 339 | 458 | 1,105 | 1,492 | 47,086 | 43.3 | 18.3 |
| Arcadia | 19,705 | 2.91 | 80.8 | 59.4 | 21.4 | 19.2 | 18.2 | 643 | 84 | 143 | 1,348 | 2,289 | 41,217 | 22.5 | 57.0 |
| Atascadero | 10,868 | 2.74 | 68.8 | 51.5 | 17.3 | 31.2 | 25.6 | 334 | 65 | 213 | 493 | 1,612 | 20,722 | 39.0 | 29.5 |
| Atwater | 9,137 | 3.22 | 79.6 | 47.6 | 32.0 | 20.4 | 15.0 | 172 | 206 | 695 | 1,031 | 3,479 | 17,689 | 47.4 | 16.1 |
| Azusa | 13,846 | 3.19 | 72.0 | 44.2 | 27.8 | 28.0 | 20.4 | 5,845 | 144 | 286 | 959 | 1,903 | 29,698 | 44.0 | 30.0 |
| Bakersfield | 116,558 | 3.26 | 74.9 | 52.1 | 22.8 | 25.1 | 21.2 | 4,089 | 1,766 | 455 | 16,074 | 4,142 | 232,621 | 44.2 | 22.4 |
| Baldwin Park | 19,350 | 3.87 | 81.6 | 57.3 | 24.2 | 18.4 | 15.3 | 431 | 188 | 248 | 1,110 | 1,463 | 50,227 | 63.9 | 13.4 |
| Banning | 10,558 | 2.84 | 71.1 | 41.4 | 29.7 | 28.9 | 26.4 | 1,229 | 136 | 432 | 532 | 1,692 | 20,598 | 50.3 | 15.0 |
| Beaumont | 15,171 | 3.35 | 79.2 | 58.5 | 20.7 | 20.8 | 18.0 | 165 | 92 | 180 | 922 | 1,808 | 31,302 | 35.5 | 32.1 |
| Bell | 8,450 | 4.10 | 88.9 | 53.1 | 35.8 | 11.1 | 10.7 | 855 | 168 | 470 | 441 | 1,233 | 20,424 | 71.4 | 7.8 |
| Bellflower | 23,369 | 3.24 | 72.9 | 42.6 | 30.3 | 27.1 | 21.8 | 686 | 327 | 424 | 1,506 | 1,951 | 52,548 | 47.5 | 17.0 |
| Bell Gardens | 10,905 | 3.81 | 77.6 | 34.2 | 43.4 | 22.4 | 14.5 | 424 | 147 | 347 | 630 | 1,487 | 25,012 | 76.4 | 6.3 |
| Belmont | 10,157 | 2.60 | 74.6 | 63.0 | 11.6 | 25.4 | 19.0 | 507 | 45 | 165 | 392 | 1,437 | 19,551 | 13.8 | 62.1 |
| Benicia | 11,313 | 2.49 | 70.0 | 54.1 | 15.9 | 30.0 | 24.0 | 30 | 16 | 56 | 377 | 1,324 | 21,175 | 17.9 | 44.3 |
| Berkeley | 44,551 | 2.44 | 44.6 | 35.0 | 9.6 | 55.4 | 33.9 | 12,758 | 618 | 503 | 6,256 | 5,095 | 76,114 | 9.5 | 76.5 |
| Beverly Hills | 15,665 | 2.15 | 55.9 | 44.1 | 11.8 | 44.1 | 41.9 | 54 | 103 | 301 | 1,499 | 4,382 | 25,868 | 11.9 | 67.0 |
| Brea | 15,162 | 2.85 | 73.4 | 56.3 | 17.0 | 26.6 | 21.5 | 129 | 72 | 163 | 1,450 | 3,284 | 30,169 | 22.0 | 46.6 |
| Brentwood | 21,626 | 2.97 | 76.8 | 59.2 | 17.5 | 23.2 | 17.3 | 225 | 166 | 254 | 1,335 | 2,039 | 44,551 | 22.8 | 39.8 |
| Buena Park | 25,362 | 3.19 | 84.3 | 68.2 | 16.1 | 15.7 | 11.9 | 939 | 246 | 298 | 2,061 | 2,494 | 60,671 | 35.8 | 32.5 |
| Burbank | 43,167 | 2.36 | 54.4 | 41.1 | 13.2 | 45.6 | 35.6 | 569 | 189 | 182 | 2,617 | 2,523 | 78,172 | 20.8 | 45.4 |
| Burlingame | 12,691 | 2.41 | 58.8 | 54.5 | 4.3 | 41.2 | 33.5 | 366 | 74 | 241 | 1,185 | 3,863 | 21,883 | 10.8 | 71.2 |
| Calexico | 9,972 | 3.98 | 72.0 | 49.0 | 23.0 | 28.0 | 28.0 | 85 | 94 | 233 | 894 | 2,217 | 22,753 | 52.0 | 24.0 |
| Camarillo | 27,274 | 2.54 | 70.6 | 50.6 | 11.9 | 37.4 | 29.9 | 483 | 58 | 83 | 817 | 1,173 | 50,531 | 19.8 | 46.0 |
| Campbell | 15,277 | 2.72 | 71.7 | 53.5 | 18.1 | 28.3 | 20.0 | 301 | 89 | 208 | 1,357 | 3,178 | 29,129 | 18.1 | 55.7 |
| Carlsbad | 44,966 | 2.55 | 69.3 | 57.0 | 12.3 | 30.7 | 23.5 | 548 | 240 | 205 | 2,135 | 1,821 | 84,026 | 11.6 | 61.1 |
| Carson | 25,573 | 3.54 | 80.6 | 58.0 | 22.6 | 19.4 | 16.1 | 883 | 444 | 483 | 2,004 | 2,180 | 66,788 | 37.2 | 29.8 |
| Cathedral City | 21,701 | 2.52 | 57.4 | 41.7 | 15.7 | 42.6 | 36.9 | 384 | 144 | 260 | 602 | 1,088 | 40,612 | 48.8 | 26.5 |
| Ceres | 13,910 | 3.45 | 76.7 | 48.5 | 28.2 | 23.3 | 19.7 | 691 | 209 | 425 | 1,158 | 2,357 | 31,280 | 53.0 | 13.3 |
| Cerritos | 15,051 | 3.30 | 88.7 | 68.8 | 20.0 | 11.3 | 9.0 | 146 | 127 | 251 | 1,634 | 3,227 | 38,422 | 21.2 | 51.3 |
| Chico | 37,040 | 2.69 | 54.5 | 38.1 | 16.4 | 45.5 | 30.2 | 3,825 | 557 | 581 | 2,406 | 2,511 | 60,740 | 25.2 | 40.8 |
| Chino | 26,079 | 3.35 | 81.8 | 61.1 | 20.6 | 18.2 | 14.5 | 6,995 | 329 | 352 | 2,224 | 2,382 | 69,001 | 40.4 | 27.1 |
| Chino Hills | 26,856 | 3.12 | 82.8 | 67.4 | 15.4 | 17.2 | 13.9 | 109 | 95 | 112 | 891 | 1,053 | 58,807 | 17.1 | 52.4 |
| Chula Vista | 82,916 | 3.29 | 78.9 | 54.7 | 24.2 | 21.1 | 16.7 | 1,943 | 904 | 328 | 3,816 | 1,386 | 183,522 | 37.3 | 29.2 |
| Citrus Heights | 36,460 | 2.39 | 58.0 | 37.5 | 20.6 | 42.0 | 33.1 | 732 | 313 | 354 | 2,248 | 2,540 | 65,085 | 34.1 | 17.9 |
| Claremont | 11,514 | 2.78 | 67.6 | 58.0 | 9.6 | 32.4 | 26.5 | 4,242 | 51 | 139 | 774 | 2,110 | 23,636 | 22.3 | 51.2 |
| Clovis | 39,269 | 2.90 | 74.2 | 52.8 | 21.4 | 25.8 | 23.0 | 562 | 243 | 213 | 2,276 | 1,994 | 73,577 | 23.8 | 35.4 |
| Coachella | 11,643 | 3.93 | 84.4 | 55.6 | 28.8 | 15.6 | 11.6 | 0 | 126 | 271 | 894 | 1,923 | 29,056 | 72.3 | 2.9 |
| Colton | 17,298 | 3.15 | 77.2 | 52.8 | 24.4 | 22.8 | 15.6 | 320 | 215 | 390 | 1,517 | 2,755 | 34,620 | 49.6 | 16.2 |
| Compton | 21,757 | 4.37 | 88.2 | 48.4 | 39.8 | 11.8 | 11.0 | 544 | 1,104 | 1,142 | 2,346 | 2,428 | 60,107 | 69.5 | 8.5 |
| Concord | 45,071 | 2.84 | 69.0 | 55.0 | 14.0 | 31.0 | 26.1 | 1,182 | 541 | 414 | 4,560 | 3,491 | 90,544 | 29.5 | 36.9 |
| Corona | 47,793 | 3.54 | 83.5 | 61.1 | 22.4 | 16.5 | 12.6 | 634 | 291 | 170 | 3,482 | 2,038 | 112,966 | 38.9 | 27.8 |
| Costa Mesa | 38,762 | 2.87 | 57.6 | 42.0 | 15.6 | 42.4 | 26.5 | 1,828 | 312 | 274 | 3,739 | 3,278 | 79,196 | 32.3 | 38.5 |
| Covina | 15,428 | 3.05 | 68.0 | 41.2 | 26.9 | 32.0 | 21.9 | 456 | 171 | 356 | 1,189 | 2,478 | 33,975 | 45.5 | 25.3 |
| Culver City | 16,996 | 2.28 | 53.0 | 44.8 | 8.2 | 47.0 | 35.2 | 360 | 182 | 464 | 1,647 | 4,196 | 30,149 | 10.4 | 69.1 |
| Cupertino | 20,820 | 2.83 | 75.2 | 65.5 | 9.7 | 24.8 | 21.4 | 433 | 55 | 91 | 1,018 | 1,687 | 40,535 | 8.2 | 78.5 |
| Cypress | 15,209 | 3.21 | 80.2 | 62.2 | 18.0 | 19.8 | 15.9 | 159 | 52 | 106 | 596 | 1,214 | 33,592 | 24.2 | 44.7 |
| Daly City | 31,289 | 3.37 | 75.6 | 49.8 | 25.8 | 24.4 | 14.9 | 867 | 232 | 215 | 1,868 | 1,734 | 78,656 | 34.7 | 37.4 |
| Dana Point | 15,525 | 2.15 | 58.8 | 45.1 | 13.7 | 41.2 | 35.0 | 133 | 52 | 154 | 399 | 1,181 | 27,476 | 24.8 | 52.3 |
| Danville | 15,827 | 2.80 | 79.3 | 65.4 | 13.9 | 20.7 | 16.5 | 186 | 23 | 51 | 296 | 658 | 30,883 | 8.1 | 72.0 |
| Davis | 26,367 | 2.55 | 45.9 | 36.1 | 9.8 | 54.1 | 23.5 | 2,136 | 107 | 153 | 2,190 | 3,139 | 36,186 | 14.9 | 66.4 |

1. Data for serious crimes have not been adjusted for underreporting. This may affect comparability between geographic areas and over time.   2. Per 100,000 population estimated by the FBI.   3. Persons 25 years old and over.

# Table D. Cities — Income, Poverty, and Housing

| City | Money income, 2019 Households | | | Median family income | Median non-family income | Median earnings, 2019 | | | Housing units, 2019 | | | | |
|---|---|---|---|---|---|---|---|---|---|---|---|---|---|
| | Median income | Percent with income less than $20,000 | Percent with income of $200,000 or more | | | All persons | Men | Women | Total | Occupied | Percent owner occupied | Median value[1] (dollars) | Median gross rent (dollars) |
| | 42 | 43 | 44 | 45 | 46 | 47 | 48 | 49 | 50 | 51 | 52 | 53 | 54 |
| ARKANSAS— Cont'd | | | | | | | | | | | | | |
| Jacksonville | 39,013 | 27.6 | 2.7 | 52,209 | 26,620 | 31,688 | 30,825 | 32,411 | 14,220 | 11,948 | 45.5 | 123,190 | 772 |
| Jonesboro | 42,939 | 21.7 | 5.2 | 63,617 | 27,862 | 25,515 | 31,221 | 20,917 | 33,769 | 29,688 | 53.4 | 166,038 | 731 |
| Little Rock | 54,878 | 17.0 | 6.0 | 69,104 | 39,495 | 34,627 | 36,682 | 31,804 | 97,530 | 80,063 | 55.9 | 171,655 | 875 |
| North Little Rock | 40,870 | 22.5 | 2.0 | 59,099 | 31,341 | 30,789 | 35,069 | 26,984 | 31,875 | 27,903 | 53.5 | 138,765 | 796 |
| Paragould | 42,957 | 19.0 | 0.3 | 59,653 | 23,482 | 27,946 | 28,345 | 27,463 | 11,831 | 11,443 | 62.9 | 131,636 | 728 |
| Pine Bluff | 35,955 | 32.8 | 3.3 | 47,711 | 24,565 | 25,307 | 30,992 | 21,983 | 23,157 | 19,009 | 59.0 | 73,252 | 721 |
| Rogers | 65,823 | 11.1 | 9.5 | 78,791 | 43,965 | 36,914 | 42,850 | 30,222 | 26,325 | 24,297 | 59.4 | 214,076 | 878 |
| Russellville | 35,016 | 25.6 | 2.0 | 68,476 | 22,071 | 25,424 | 34,805 | 15,614 | 13,277 | 11,143 | 56.4 | 134,850 | 682 |
| Sherwood | 66,339 | 6.6 | 3.9 | 83,839 | 55,285 | 40,407 | 44,164 | 36,554 | 13,798 | 12,282 | 78.4 | 156,296 | 769 |
| Springdale | 47,960 | 15.5 | 5.1 | 53,714 | 25,382 | 26,014 | 26,671 | 25,404 | 28,115 | 26,443 | 51.5 | 163,133 | 780 |
| Texarkana | 51,095 | 20.0 | 1.7 | 71,267 | 21,157 | 32,046 | 35,986 | 30,484 | 14,529 | 11,743 | 63.5 | 108,910 | 704 |
| West Memphis | 33,010 | 30.5 | 3.7 | 40,473 | 24,602 | 26,651 | 30,664 | 26,182 | 11,098 | 9,787 | 42.7 | 110,111 | 692 |
| CALIFORNIA | 80,440 | 11.4 | 13.7 | 91,377 | 51,676 | 39,549 | 44,242 | 32,406 | 14,367,012 | 13,157,873 | 54.9 | 568,488 | 1,614 |
| Adelanto | 57,272 | 12.5 | 2.3 | 60,696 | 50,363 | 25,213 | 30,223 | 22,944 | 8,321 | 7,775 | 52.3 | 238,192 | 1,075 |
| Alameda | 109,545 | 9.2 | 24.6 | 132,316 | 70,088 | 61,672 | 61,809 | 61,581 | 33,241 | 30,788 | 46.3 | 971,233 | 2,042 |
| Alhambra | 74,132 | 10.9 | 10.9 | 81,476 | 52,952 | 39,430 | 40,143 | 38,699 | 30,471 | 28,777 | 44.6 | 666,972 | 1,608 |
| Aliso Viejo | 113,800 | 5.7 | 18.4 | 135,213 | 87,400 | 57,582 | 66,320 | 51,491 | 19,925 | 18,827 | 54.4 | 661,560 | 2,309 |
| Anaheim | 76,075 | 9.1 | 9.6 | 81,006 | 59,091 | 35,043 | 40,922 | 30,683 | 110,647 | 105,301 | 46.1 | 630,470 | 1,733 |
| Antioch | 72,412 | 10.3 | 10.8 | 89,108 | 47,076 | 37,714 | 40,486 | 34,685 | 36,663 | 36,138 | 63.4 | 468,828 | 1,789 |
| Apple Valley | 51,314 | 9.2 | 2.8 | 56,205 | 28,818 | 30,568 | 32,406 | 26,226 | 27,510 | 23,774 | 64.8 | 280,184 | 985 |
| Arcadia | 95,048 | 11.2 | 18.8 | 104,705 | 42,167 | 50,527 | 56,440 | 43,992 | 21,482 | 19,705 | 61.5 | 1,017,470 | 1,855 |
| Atascadero | 74,207 | 9.8 | 9.9 | 82,864 | 50,179 | 33,649 | 51,806 | 22,392 | 11,204 | 10,868 | 60.7 | 499,973 | 1,614 |
| Atwater | 63,162 | 14.0 | 4.1 | 63,897 | 33,931 | 33,575 | 37,284 | 28,493 | 9,496 | 9,137 | 48.5 | 267,392 | 1,077 |
| Azusa | 71,120 | 9.7 | 5.4 | 72,857 | 53,095 | 30,022 | 32,148 | 23,772 | 14,516 | 13,846 | 51.2 | 488,911 | 1,459 |
| Bakersfield | 62,402 | 13.2 | 7.1 | 72,247 | 36,127 | 33,297 | 40,668 | 27,240 | 124,863 | 116,558 | 59.4 | 264,461 | 1,105 |
| Baldwin Park | 66,997 | 14.1 | 4.1 | 71,833 | 28,967 | 30,157 | 32,135 | 26,482 | 19,880 | 19,350 | 57.2 | 462,939 | 1,510 |
| Banning | 45,050 | 14.5 | 3.7 | 48,453 | 32,582 | 26,158 | 31,510 | 21,853 | 11,575 | 10,558 | 63.8 | 259,254 | 1,114 |
| Beaumont | 88,763 | 6.9 | 7.6 | 102,714 | 40,879 | 46,405 | 65,351 | 31,604 | 15,578 | 15,171 | 82.3 | 354,171 | 1,273 |
| Bell | 55,045 | 15.6 | 1.7 | 53,469 | 11,932 | 27,023 | 29,853 | 22,578 | 8,450 | 8,450 | 34.0 | 443,089 | 1,276 |
| Bellflower | 67,658 | 15.2 | 3.5 | 75,892 | 34,209 | 33,647 | 37,349 | 30,383 | 25,114 | 23,369 | 37.2 | 522,810 | 1,466 |
| Bell Gardens | 38,484 | 21.5 | 1.0 | 36,317 | 25,071 | 21,586 | 26,427 | 18,357 | 10,930 | 10,905 | 17.4 | 496,465 | 1,332 |
| Belmont | 177,372 | 9.2 | 43.7 | 210,761 | 60,112 | 92,386 | 111,066 | 73,566 | 10,701 | 10,157 | 63.4 | 1,600,262 | 2,497 |
| Benicia | 110,849 | 6.5 | 18.3 | 119,921 | 79,057 | 56,782 | 67,094 | 51,456 | 11,497 | 11,313 | 73.6 | 623,135 | 1,775 |
| Berkeley | 95,360 | 15.2 | 22.0 | 159,002 | 54,761 | 41,458 | 50,222 | 35,407 | 47,994 | 44,551 | 41.4 | 1,168,312 | 1,837 |
| Beverly Hills | 106,540 | 12.9 | 26.8 | 175,704 | 54,375 | 65,955 | 100,035 | 50,186 | 18,887 | 15,665 | 40.1 | 2 | 2,258 |
| Brea | 100,063 | 8.0 | 14.0 | 109,186 | 50,345 | 45,898 | 50,756 | 42,570 | 15,855 | 15,162 | 64.9 | 689,604 | 2,016 |
| Brentwood | 120,899 | 5.7 | 22.3 | 142,625 | 57,057 | 54,543 | 71,515 | 40,341 | 22,574 | 21,626 | 78.6 | 648,896 | 1,926 |
| Buena Park | 87,378 | 5.8 | 10.2 | 85,839 | 68,445 | 41,383 | 50,061 | 33,253 | 26,509 | 25,362 | 61.6 | 616,443 | 1,676 |
| Burbank | 73,613 | 16.3 | 13.1 | 101,138 | 49,826 | 46,498 | 60,603 | 40,413 | 46,254 | 43,167 | 40.6 | 826,065 | 1,728 |
| Burlingame | 148,168 | 4.1 | 37.9 | 215,216 | 79,039 | 97,152 | 122,093 | 78,589 | 13,270 | 12,691 | 49.7 | 2 | 2,343 |
| Calexico | 39,840 | 29.7 | 1.1 | 62,929 | 12,721 | 23,527 | 26,279 | 14,770 | 12,124 | 9,972 | 53.1 | 227,206 | 858 |
| Camarillo | 90,290 | 7.7 | 11.4 | 102,419 | 75,423 | 46,551 | 61,994 | 33,925 | 28,540 | 27,274 | 63.7 | 621,307 | 2,054 |
| Campbell | 131,594 | 6.0 | 30.3 | 147,023 | 71,986 | 71,980 | 100,179 | 56,351 | 16,909 | 15,277 | 41.5 | 1,180,431 | 2,464 |
| Carlsbad | 123,409 | 4.3 | 26.5 | 151,104 | 76,211 | 56,661 | 75,263 | 46,349 | 51,631 | 44,966 | 64.9 | 886,473 | 2,052 |
| Carson | 84,502 | 7.4 | 11.9 | 95,131 | 42,740 | 39,009 | 40,589 | 35,884 | 26,986 | 25,573 | 70.3 | 550,278 | 1,602 |
| Cathedral City | 41,972 | 26.3 | 5.4 | 59,757 | 24,039 | 29,197 | 30,225 | 26,330 | 26,236 | 21,701 | 63.8 | 322,036 | 1,281 |
| Ceres | 64,350 | 10.8 | 5.1 | 64,752 | 40,120 | 40,026 | 41,146 | 27,415 | 14,292 | 13,910 | 60.8 | 325,437 | 1,453 |
| Cerritos | 111,617 | 7.2 | 16.0 | 114,592 | 46,760 | 43,734 | 51,255 | 37,475 | 15,642 | 15,051 | 72.9 | 734,430 | 2,289 |
| Chico | 59,876 | 17.5 | 10.1 | 83,285 | 36,059 | 29,545 | 25,915 | 21,792 | 40,176 | 37,040 | 45.6 | 378,094 | 1,128 |
| Chino | 87,090 | 8.5 | 6.7 | 90,081 | 44,857 | 36,221 | 40,488 | 32,266 | 28,420 | 26,079 | 63.0 | 490,863 | 1,563 |
| Chino Hills | 103,473 | 3.0 | 18.7 | 112,768 | 66,452 | 53,912 | 61,234 | 45,015 | 28,005 | 26,856 | 74.3 | 690,266 | 2,152 |
| Chula Vista | 87,876 | 9.1 | 9.9 | 90,305 | 61,964 | 41,008 | 50,351 | 31,894 | 91,083 | 82,916 | 60.0 | 546,403 | 1,715 |
| Citrus Heights | 61,898 | 11.0 | 3.8 | 68,927 | 41,968 | 34,800 | 40,663 | 30,055 | 37,854 | 36,460 | 60.0 | 337,915 | 1,403 |
| Claremont | 101,826 | 5.2 | 16.3 | 127,787 | 50,942 | 36,833 | 36,750 | 36,880 | 12,190 | 11,514 | 65.1 | 709,511 | 1,402 |
| Clovis | 89,398 | 9.1 | 12.7 | 100,172 | 41,654 | 46,113 | 53,099 | 40,335 | 41,255 | 39,269 | 69.1 | 360,368 | 1,224 |
| Coachella | 47,203 | 13.8 | 3.3 | 52,603 | 35,298 | 26,351 | 28,495 | 24,165 | 11,770 | 11,643 | 71.6 | 239,652 | 926 |
| Colton | 60,922 | 14.3 | 5.5 | 61,834 | 43,153 | 34,072 | 37,445 | 29,556 | 18,173 | 17,298 | 54.1 | 321,466 | 1,326 |
| Compton | 62,200 | 10.6 | 3.2 | 62,238 | 28,801 | 27,210 | 29,472 | 26,183 | 22,411 | 21,757 | 62.4 | 432,514 | 1,315 |
| Concord | 102,477 | 11.1 | 14.8 | 116,708 | 60,822 | 45,308 | 52,481 | 36,737 | 46,435 | 45,071 | 58.3 | 604,343 | 1,982 |
| Corona | 86,790 | 6.3 | 12.7 | 98,273 | 50,985 | 41,511 | 48,799 | 35,671 | 50,282 | 47,793 | 61.4 | 519,803 | 1,726 |
| Costa Mesa | 91,857 | 9.2 | 14.1 | 101,696 | 81,161 | 41,073 | 48,682 | 32,239 | 41,378 | 38,762 | 38.4 | 826,043 | 1,946 |
| Covina | 64,949 | 12.8 | 7.1 | 77,325 | 38,947 | 36,031 | 36,520 | 35,568 | 16,377 | 15,428 | 51.0 | 599,272 | 1,602 |
| Culver City | 123,839 | 5.0 | 23.6 | 172,566 | 67,369 | 71,180 | 87,264 | 51,399 | 18,044 | 16,996 | 51.4 | 1,019,878 | 2,139 |
| Cupertino | 186,419 | 11.0 | 47.8 | 243,336 | 45,024 | 121,562 | 151,107 | 94,725 | 22,373 | 20,820 | 64.5 | 2 | 3,163 |
| Cypress | 96,342 | 6.5 | 17.6 | 115,815 | 44,094 | 44,930 | 46,285 | 44,075 | 15,656 | 15,209 | 69.6 | 684,172 | 1,928 |
| Daly City | 101,834 | 6.2 | 18.0 | 114,914 | 60,477 | 42,952 | 50,502 | 37,135 | 32,638 | 31,289 | 59.3 | 916,501 | 2,371 |
| Dana Point | 90,157 | 12.7 | 19.4 | 95,571 | 61,554 | 45,529 | 50,018 | 41,129 | 20,587 | 15,525 | 70.3 | 874,354 | 2,119 |
| Danville | 193,072 | 2.8 | 48.1 | 210,567 | 92,250 | 77,139 | 105,173 | 60,926 | 16,391 | 15,827 | 82.9 | 1,210,691 | 2,427 |
| Davis | 60,619 | 21.1 | 12.1 | 119,368 | 41,500 | 27,234 | 37,262 | 20,991 | 27,747 | 26,367 | 37.4 | 674,617 | 1,720 |

1. Based on population estimated by the American Community Survey.

# Table D. Cities — Commuting, Computer Access, Migration, Labor Force, and Employment

Column groups: Commuting[1], 2019 (Percent) = items 55–56; Computer access[2], 2019 (Percent) = items 57–58; Migration, 2019 = items 59–60; Civilian labor force, 2020 = items 61–64 (with Unemployment = 63–64); Civilian Employment[4], 2019 = items 65–68 (Population age 16 and older = 65–66; Population age 16 to 64 = 67–68).

| City | Drove alone [55] | With commutes of 30 minutes or more [56] | With a computer in the house [57] | With Internet access [58] | % lived in same house one year ago [59] | % lived in another state or county one year ago [60] | Labor force Total [61] | Percent change 2019-2020 [62] | Unemployment Total [63] | Rate[3] [64] | Pop 16+ Number [65] | Percent in labor force [66] | Pop 16-64 Number [67] | % worked full-year full-time [68] |
|---|---|---|---|---|---|---|---|---|---|---|---|---|---|---|
| **ARKANSAS— Cont'd** | | | | | | | | | | | | | | |
| Jacksonville | 86.4 | 19.9 | 91.8 | 84.2 | 85.2 | 8.5 | 11,726 | -0.3 | 1,027 | 9 | 22,116 | 49.6 | 17,754 | 49.4 |
| Jonesboro | 83.7 | 15.6 | 93.3 | 87.0 | 78.8 | 9.7 | 39,437 | 0.6 | 2,196 | 6 | 62,703 | 62.5 | 52,924 | 46.3 |
| Little Rock | 79.7 | 13.1 | 92.5 | 85.7 | 86.4 | 4.4 | 97,292 | -0.6 | 7,350 | 8 | 156,450 | 65.4 | 127,660 | 57.2 |
| North Little Rock | 86.2 | 18.7 | 91.8 | 78.6 | 87.9 | 4.8 | 30,182 | 0.1 | 2,603 | 9 | 50,258 | 62.4 | 39,975 | 55.1 |
| Paragould | NA | 24.7 | 93.4 | 83.3 | 75.3 | 8.9 | 12,431 | -2.2 | 843 | 7 | 23,164 | 59.7 | 19,548 | 46.8 |
| Pine Bluff | 85.2 | 17.3 | 82.4 | 68.0 | 85.2 | 8.5 | 16,421 | -0.7 | 1,534 | 9 | 34,257 | 54.9 | 27,101 | 44.3 |
| Rogers | 83.7 | 7.3 | 93.6 | 87.4 | 89.6 | 4.2 | 36,240 | 1.3 | 1,586 | 4 | 51,986 | 69.4 | 45,703 | 59.8 |
| Russellville | 80.3 | 14.9 | 87.2 | 78.7 | 81.0 | 12.2 | 12,767 | -4.4 | 690 | 5 | 23,928 | 56.9 | 19,714 | 44.7 |
| Sherwood | 85.6 | 33.8 | 91.6 | 88.2 | 82.7 | 4.6 | 16,039 | -0.8 | 1,005 | 6 | 26,308 | 54.2 | 19,624 | 55.0 |
| Springdale | 74.4 | 24.0 | 93.4 | 81.0 | 88.3 | 4.6 | 39,503 | -0.1 | 1,715 | 4 | 58,603 | 68.5 | 49,275 | 52.1 |
| Texarkana | 83.0 | 16.2 | 86.9 | 78.4 | 83.4 | 3.5 | 13,394 | -1.6 | 1,028 | 8 | 26,812 | 60.2 | 21,555 | 54.7 |
| West Memphis | 84.7 | 26.9 | 88.1 | 81.2 | 78.2 | 3.2 | 10,555 | -0.2 | 1,031 | 10 | 17,633 | 62.3 | 14,438 | 45.8 |
| **CALIFORNIA** | 73.5 | 45.3 | 95.1 | 89.3 | 88.0 | 4.6 | 18,821,167 | -3.0 | 1,908,089 | 10 | 31,617,786 | 63.6 | 25,782,788 | 50.6 |
| Adelanto | 84.8 | 49.5 | NA | NA | 89.7 | 3.1 | 9,969 | 2.3 | 1,594 | 16 | 25,795 | 47.8 | 24,126 | 24.8 |
| Alameda | 54.3 | 57.5 | 97.7 | 92.5 | 83.7 | 7.9 | 39,387 | -4.8 | 3,349 | 9 | 63,509 | 66.4 | 50,419 | 59.6 |
| Alhambra | 79.3 | 51.6 | 95.3 | 92.0 | 93.6 | 2.2 | 44,973 | -3.4 | 5,824 | 13 | 70,780 | 65.5 | 55,237 | 55.1 |
| Aliso Viejo | 76.9 | 39.4 | 99.2 | 93.9 | 87.0 | 5.0 | 28,496 | -6.8 | 2,132 | 8 | 40,251 | 76.2 | 35,449 | 54.6 |
| Anaheim | 78.5 | 46.7 | 95.7 | 91.4 | 88.2 | 3.8 | 167,067 | -2.8 | 17,527 | 11 | 281,221 | 68.8 | 239,938 | 55.7 |
| Antioch | 65.5 | 64.2 | 97.1 | 94.3 | 87.1 | 8.3 | 50,102 | -1.4 | 6,022 | 12 | 90,801 | 63.3 | 73,840 | 46.6 |
| Apple Valley | 83.0 | 38.3 | 95.3 | 88.9 | 89.6 | 2.6 | 28,876 | -1.3 | 2,752 | 10 | 56,662 | 52.7 | 44,095 | 38.1 |
| Arcadia | 73.9 | 51.1 | 97.0 | 92.8 | 92.6 | 2.1 | 28,100 | -6.3 | 2,819 | 10 | 46,870 | 56.2 | 35,271 | 46.4 |
| Atascadero | 71.4 | 17.8 | 95.2 | 90.7 | 87.0 | 1.9 | 14,434 | -6.5 | 1,053 | 7 | 23,975 | 64.4 | 20,412 | 47.5 |
| Atwater | 84.8 | 34.0 | 97.0 | 91.1 | 89.5 | 4.5 | 11,827 | -1.2 | 1,428 | 12 | 21,381 | 69.0 | 18,592 | 50.5 |
| Azusa | 70.6 | 53.1 | 97.1 | 80.7 | 88.0 | 3.5 | 24,245 | -4.3 | 2,817 | 12 | 39,767 | 65.9 | 34,257 | 47.3 |
| Bakersfield | 84.0 | 23.9 | 94.8 | 88.4 | 86.4 | 3.3 | 177,183 | -1.0 | 19,825 | 11 | 280,234 | 61.9 | 239,917 | 41.1 |
| Baldwin Park | 76.3 | 52.0 | 93.2 | 88.1 | 96.3 | 0.6 | 33,601 | -3.2 | 4,593 | 14 | 59,505 | 64.9 | 49,691 | 52.3 |
| Banning | 82.1 | 54.2 | 91.8 | 82.0 | 90.8 | 3.1 | 11,381 | 1.1 | 1,390 | 12 | 24,637 | 49.6 | 17,303 | 39.3 |
| Beaumont | 80.6 | 57.8 | 96.7 | 86.5 | 81.3 | 13.1 | 23,633 | 2.7 | 2,040 | 9 | 37,037 | 61.9 | 29,304 | 51.3 |
| Bell | 79.3 | 57.4 | 94.5 | 81.2 | NA | NA | | | | | | | | |
| Bellflower | 81.2 | 51.9 | 94.6 | 83.2 | NA | NA | 35,619 | -2.7 | 5,186 | 15 | 62,430 | 66.1 | 52,265 | 54.5 |
| Bell Gardens | 79.7 | 63.6 | 94.0 | 85.7 | NA | NA | 17,510 | -3.5 | 2,313 | 13 | 31,011 | 65.4 | 27,424 | 46.1 |
| Belmont | 72.6 | 54.7 | 92.4 | 90.5 | 88.1 | 7.2 | 15,616 | -7.5 | 836 | 5 | 21,352 | 72.0 | 16,435 | 62.4 |
| Benicia | 77.1 | 48.5 | 97.1 | 95.3 | 94.3 | 1.9 | 14,158 | -4.9 | 1,029 | 7 | 23,701 | 63.2 | 17,481 | 56.0 |
| Berkeley | 32.9 | 51.6 | 97.2 | 94.2 | 73.0 | 14.4 | 58,979 | -6.1 | 3,949 | 7 | 109,524 | 61.6 | 90,653 | 38.1 |
| Beverly Hills | 68.6 | 38.0 | 93.6 | 92.5 | 93.9 | 2.3 | 17,219 | -6.4 | 1,771 | 10 | 28,211 | 58.0 | 19,657 | 46.9 |
| Brea | 82.6 | 50.4 | 97.7 | 96.2 | 86.9 | 7.7 | 22,107 | -4.8 | 1,905 | 9 | 35,251 | 69.4 | 28,805 | 57.4 |
| Brentwood | 73.9 | 63.3 | 97.5 | 95.3 | 89.6 | 4.8 | 28,622 | -3.0 | 2,567 | 9 | 51,178 | 64.0 | 40,022 | 50.7 |
| Buena Park | 84.2 | 48.4 | 96.7 | 93.9 | 91.6 | 3.9 | 39,218 | -2.9 | 4,238 | 11 | 68,593 | 67.3 | 54,145 | 57.1 |
| Burbank | 74.0 | 45.2 | 92.2 | 86.9 | 87.8 | 5.6 | 57,304 | -3.4 | 8,091 | 14 | 86,657 | 66.1 | 68,008 | 54.2 |
| Burlingame | 67.4 | 59.1 | 95.8 | 94.7 | 84.2 | 9.7 | 17,600 | -4.9 | 1,025 | 6 | 23,837 | 69.2 | 19,787 | 57.3 |
| Calexico | 79.5 | 30.8 | 84.6 | 79.4 | NA | NA | 17,763 | -4.1 | 6,061 | 34 | 28,393 | 58.5 | 22,950 | 36.8 |
| Camarillo | 81.8 | 29.4 | 96.2 | 94.7 | 89.5 | 5.4 | 33,502 | -2.2 | 2,507 | 8 | 58,152 | 60.3 | 42,829 | 53.0 |
| Campbell | 79.8 | 48.2 | 97.7 | 93.9 | 83.1 | 3.9 | 25,577 | -4.5 | 1,645 | 6 | 31,650 | 73.0 | 27,076 | 53. |
| Carlsbad | 78.3 | 48.1 | 98.0 | 94.4 | 87.7 | 4.0 | 53,006 | -4.4 | 4,348 | 8 | 94,786 | 64.7 | 73,394 | 54. |
| Carson | 81.5 | 37.5 | 95.8 | 92.5 | 90.3 | 3.5 | 44,749 | -3.4 | 6,068 | 14 | 78,969 | 62.6 | 62,398 | 54. |
| Cathedral City | 84.2 | 14.0 | 91.0 | 87.4 | 92.7 | 4.2 | 26,859 | 1.9 | 3,095 | 12 | 45,556 | 57.2 | 33,648 | 46. |
| Ceres | NA | 41.1 | 94.0 | 87.6 | 90.8 | 5.1 | 21,224 | -1.7 | 2,439 | 12 | 36,901 | 61.9 | 31,357 | 45. |
| Cerritos | 81.4 | 59.5 | 95.3 | 93.9 | 92.3 | 3.2 | 24,044 | -5.6 | 2,622 | 11 | 42,280 | 58.9 | 29,099 | 55. |
| Chico | 78.2 | 11.8 | 96.8 | 90.8 | 73.3 | 9.5 | 50,819 | 7.1 | 4,045 | 9 | 85,279 | 64.8 | 70,808 | 49 |
| Chino | 84.6 | 56.0 | 95.0 | 89.2 | 87.1 | 9.3 | 44,623 | 2.4 | 3,938 | 8 | 79,558 | 59.0 | 67,752 | 52 |
| Chino Hills | 82.4 | 57.9 | NA | NA | 85.3 | 10.0 | 45,165 | -1.1 | 3,286 | 7 | 68,635 | 66.2 | 57,031 | 51 |
| Chula Vista | 80.2 | 46.7 | 94.9 | 92.2 | 92.0 | 2.8 | 122,226 | -1.1 | 12,721 | 10 | 217,132 | 65.0 | 183,526 | 50 |
| Citrus Heights | 80.8 | 36.0 | 96.2 | 88.9 | 83.6 | 7.2 | 42,938 | -1.8 | 3,846 | 9 | 73,750 | 65.0 | 57,711 | |
| Claremont | 61.5 | 39.4 | 93.7 | 91.0 | 86.2 | 5.5 | 16,030 | -7.3 | 1,460 | 9 | 30,505 | 62.3 | 23,864 | 37 |
| Clovis | 84.7 | 30.7 | 93.5 | 90.5 | 87.0 | 3.1 | 55,057 | 0.6 | 4,383 | 8 | 83,255 | 66.3 | 67,879 | 51 |
| Coachella | 72.4 | 31.8 | NA | NA | NA | NA | 20,028 | -0.3 | 3,162 | 16 | 36,371 | 67.2 | 31,662 | 44 |
| Colton | 83.8 | 32.5 | 97.6 | 90.0 | 91.1 | 5.6 | 24,710 | -0.1 | 2,444 | 10 | 43,355 | 59.4 | 37,381 | 45 |
| Compton | 78.0 | 52.0 | 94.3 | 83.1 | NA | NA | 38,648 | -1.7 | 6,383 | 17 | 72,207 | 64.2 | 61,178 | 47 |
| Concord | 70.1 | 53.0 | 97.9 | 94.4 | 85.1 | 5.9 | 63,259 | -4.2 | 5,509 | 9 | 105,758 | 68.3 | 87,339 | 56 |
| Corona | 84.0 | 59.4 | 96.7 | 86.6 | 88.9 | 5.2 | 84,198 | -0.6 | 6,942 | 8 | 132,113 | 66.5 | 116,663 | 56 |
| Costa Mesa | 76.8 | 27.1 | 97.6 | 95.2 | 83.6 | 5.1 | 62,869 | -5.0 | 5,103 | 9 | 89,571 | 74.4 | 75,405 | 55 |
| Covina | 83.1 | 58.3 | 92.3 | 80.1 | 91.3 | 3.2 | 23,138 | -4.0 | 3,069 | 13 | 38,652 | 67.1 | 31,763 | 56 |
| Culver City | 77.7 | 56.0 | 96.0 | 82.1 | 91.2 | 2.2 | 21,915 | -5.2 | 2,372 | 11 | 32,133 | 76.5 | 26,485 | 64 |
| Cupertino | 84.3 | 48.4 | 95.6 | 92.5 | 86.5 | 7.3 | 27,962 | -6.1 | 1,363 | 5 | 47,134 | 58.5 | 37,382 | 56 |
| Cypress | 77.2 | 53.3 | 96.8 | 94.1 | 91.6 | 2.7 | 24,286 | -3.4 | 2,270 | 9 | 37,998 | 62.7 | 29,808 | 48 |
| Daly City | 59.3 | 53.6 | 96.6 | 90.0 | 92.6 | 4.4 | 63,625 | -3.1 | 6,134 | 10 | 91,363 | 69.0 | 72,857 | 59 |
| Dana Point | 69.8 | 39.9 | 95.5 | 90.7 | 83.7 | 4.9 | 17,840 | -5.2 | 1,399 | 8 | 29,783 | 57.8 | 20,123 | 49 |
| Danville | 65.5 | 44.4 | NA | NA | 89.9 | 3.8 | 19,430 | -6.5 | 1,219 | 6 | 35,832 | 66.3 | 27,978 | 49 |
| Davis | 57.3 | 27.3 | 97.9 | 91.3 | 66.4 | 15.3 | 34,265 | -4.1 | 1,789 | 5 | 62,556 | 55.6 | 54,842 | 3 |

1. Employed persons.  2. Households.  3. Percent of civilian labor force.  4. Persons 16 years old and over.

# Table D. Cities — Construction, Wholesale Trade, and Retail Trade

| City | Value of residential construction authorized by building permits, 2020 | | | Wholesale trade[1], 2017 | | | | Retail trade[2], 2017 | | | |
|---|---|---|---|---|---|---|---|---|---|---|---|
| | New construction ($1,000) | Number of housing units | Percent single family | Number of establishments | Number of employees | Sales (mil dol) | Annual payroll (mil dol) | Number of establishments | Number of employees | Sales (mil dol) | Annual payroll (mil dol) |
| | 69 | 70 | 71 | 72 | 73 | 74 | 75 | 76 | 77 | 78 | 79 |
| **ARKANSAS— Cont'd** | | | | | | | | | | | |
| Jacksonville | 10,059 | 50 | 100 | 14 | 136 | 43 | 6 | 93 | 1,375 | 407 | 37 |
| Jonesboro | 89,663 | 755 | 62 | 91 | 1,129 | 728 | 54 | 408 | 6,221 | 1,705 | 155 |
| Little Rock | 165,807 | 833 | 50 | 376 | 7,115 | 4,385 | 398 | 1,065 | 15,813 | 4,575 | 434 |
| North Little Rock | 19,737 | 104 | 84 | 162 | 3,705 | 3,898 | 221 | 396 | 6,015 | 1,608 | 147 |
| Paragould | 20,041 | 146 | 81 | 20 | 176 | 80 | 7 | 126 | 1,785 | 451 | 46 |
| Pine Bluff | 1,592 | 10 | 100 | 39 | 319 | 171 | 15 | 233 | 2,714 | 646 | 66 |
| Rogers | 125,486 | 725 | 69 | 50 | 702 | 807 | 42 | 341 | 6,215 | 1,401 | 147 |
| Russellville | 7,385 | 76 | 66 | 43 | 507 | 359 | 23 | 215 | 2,860 | 799 | 71 |
| Sherwood | 46,141 | 248 | 100 | 30 | 275 | 155 | 14 | 99 | 1,817 | 806 | 66 |
| Springdale | 65,740 | 259 | 100 | 133 | 1,486 | 1,332 | 83 | 268 | 4,261 | 1,386 | 132 |
| Texarkana | 4,987 | 21 | 100 | D | D | D | D | 120 | 1,253 | 366 | 31 |
| West Memphis | 1,858 | 20 | 90 | 31 | 868 | 1,362 | 37 | 99 | 1,597 | 512 | 38 |
| **CALIFORNIA** | 2,299 | 8 | 100 | 52,861 | 735,916 | 806,283 | 55,597 | 108,233 | 1,723,278 | 594,861 | 54,229 |
| Adelanto | 17,576 | 163 | 100 | 5 | 102 | 41 | 5 | 18 | 222 | 68 | 7 |
| Alameda | 94,549 | 348 | 23 | 60 | 1,746 | 2,267 | 183 | 180 | 2,415 | 655 | 81 |
| Alhambra | 9,420 | 54 | 39 | 255 | 977 | 924 | 36 | 251 | 4,511 | 1,938 | 145 |
| Aliso Viejo | 0 | 0 | 0 | 74 | 687 | 557 | 65 | 72 | 1,546 | 389 | 45 |
| Anaheim | 92,625 | 449 | 19 | 792 | 10,365 | 7,472 | 605 | 841 | 12,856 | 4,282 | 404 |
| Antioch | 133,636 | 692 | 43 | 27 | 491 | 288 | 25 | 222 | 4,012 | 1,120 | 128 |
| Apple Valley | 18,100 | 110 | 100 | 18 | 108 | 27 | 3 | 118 | 2,452 | 685 | 62 |
| Arcadia | 68,151 | 165 | 56 | 309 | 1,348 | 645 | 63 | 298 | 5,034 | 979 | 117 |
| Atascadero | 12,243 | 44 | 100 | 23 | 128 | 39 | 6 | 117 | 1,224 | 350 | 35 |
| Atwater | 505 | 3 | 100 | 6 | 76 | 59 | 3 | 70 | 1,253 | 281 | 28 |
| Azusa | 16,685 | 98 | 25 | 90 | 1,064 | 905 | 64 | 101 | 1,378 | 492 | 44 |
| Bakersfield | 443,798 | 1,738 | 84 | 283 | 3,708 | 2,800 | 219 | 989 | 20,027 | 6,146 | 576 |
| Baldwin Park | 555 | 2 | 100 | 147 | 829 | 421 | 39 | 133 | 2,330 | 967 | 68 |
| Banning | 142,215 | 476 | 100 | 10 | 247 | 64 | 8 | 61 | 757 | 236 | 22 |
| Beaumont | 68,764 | 271 | 100 | 8 | 30 | 7 | 1 | 78 | 1,574 | 495 | 43 |
| Bell | 9,013 | 65 | 14 | 82 | 1,874 | 1,505 | 124 | 59 | 566 | 158 | 14 |
| Bellflower | 3,218 | 42 | 100 | 35 | 240 | 112 | 10 | 183 | 1,962 | 570 | 62 |
| Bell Gardens | 1,878 | 13 | 46 | 41 | 617 | 233 | 26 | 74 | 1,153 | 277 | 30 |
| Belmont | 4,031 | 8 | 100 | 17 | 61 | 31 | 4 | 50 | 782 | 342 | 37 |
| Benicia | 400 | 1 | 100 | 61 | 1,149 | 695 | 72 | 64 | 1,129 | 275 | 33 |
| Berkeley | 60,592 | 339 | 24 | 108 | 1,225 | 1,093 | 89 | 483 | 6,425 | 1,586 | 215 |
| Beverly Hills | 42,220 | 37 | 49 | 133 | 929 | 696 | 61 | 485 | 6,976 | 3,840 | 355 |
| Brea | 7,134 | 25 | 100 | 327 | 5,016 | 4,197 | 326 | 315 | 5,969 | 1,175 | 140 |
| Brentwood | 57,191 | 274 | 100 | 13 | 62 | 84 | 4 | 155 | 2,722 | 761 | 73 |
| Buena Park | 12,133 | 54 | 83 | 190 | 4,392 | 3,968 | 275 | 228 | 4,460 | 1,863 | 151 |
| Burbank | 11,125 | 129 | 98 | 199 | 3,680 | 1,938 | 279 | 418 | 7,965 | 2,730 | 235 |
| Burlingame | 61,823 | 147 | 13 | 137 | 1,350 | 1,228 | 85 | 165 | 2,368 | 947 | 113 |
| Calexico | 1,927 | 11 | 100 | 53 | 309 | 417 | 9 | 159 | 2,461 | 466 | 48 |
| Camarillo | 21,372 | 98 | 74 | 137 | 2,467 | 2,329 | 211 | 330 | 5,612 | 1,368 | 136 |
| Campbell | 5,103 | 36 | 100 | 77 | 754 | 715 | 83 | D | D | D | 88 |
| Carlsbad | 105,424 | 457 | 49 | 281 | 5,854 | 3,784 | 497 | 487 | 8,305 | 3,111 | 271 |
| Carson | 13,430 | 108 | 4 | 346 | 6,630 | 5,045 | 360 | 221 | 4,827 | 2,085 | 196 |
| Cathedral City | 52,588 | 251 | 99 | 21 | 89 | 35 | 4 | 134 | 2,455 | 1,073 | 93 |
| Ceres | 1,000 | 5 | 100 | 23 | 482 | 335 | 30 | 102 | 1,795 | 447 | 42 |
| Cerritos | 0 | 0 | 0 | 237 | 5,003 | 3,917 | 328 | 251 | 7,796 | 4,022 | 281 |
| Chico | 87,454 | 497 | 60 | 81 | 1,187 | 585 | 60 | 393 | 7,340 | 2,240 | 215 |
| Chino | 109,515 | 440 | 100 | 526 | 7,891 | 6,621 | 424 | 298 | 4,951 | 1,424 | 132 |
| Chino Hills | 12,220 | 44 | 100 | 126 | 428 | 267 | 20 | 165 | 2,535 | 698 | 66 |
| Chula Vista | 247,698 | 1,100 | 20 | 289 | 2,134 | 2,323 | 108 | 624 | 12,160 | 3,156 | 337 |
| Citrus Heights | 4,509 | 22 | 100 | 15 | 67 | 38 | 4 | 246 | 4,525 | 1,137 | 113 |
| Claremont | 0 | 0 | 0 | 25 | 351 | 142 | 18 | 86 | 1,268 | 440 | 41 |
| Clovis | 332,346 | 1,152 | 91 | 45 | 216 | 134 | 10 | 306 | 6,679 | 2,027 | 184 |
| Coachella | 30,800 | 137 | 100 | 18 | 272 | 125 | 19 | 60 | 1,000 | 313 | 26 |
| Colton | 17,163 | 87 | 100 | 41 | 758 | 758 | 40 | 112 | 2,051 | 662 | 66 |
| Compton | 1,658 | 21 | 100 | 122 | 2,107 | 1,457 | 112 | 160 | 2,414 | 574 | 54 |
| Concord | 74,488 | 280 | 19 | 151 | 1,391 | 738 | 88 | 420 | 8,211 | 3,164 | 288 |
| Corona | 64,416 | 253 | 100 | 305 | 7,018 | 13,069 | 483 | 441 | 8,810 | 3,106 | 269 |
| Costa Mesa | 52,022 | 262 | 24 | 262 | 3,735 | 6,243 | 281 | 766 | 15,676 | 4,558 | 486 |
| Covina | 1,690 | 6 | 100 | 70 | 535 | 284 | 24 | 173 | 3,350 | 912 | 102 |
| Culver City | 6,657 | 68 | 100 | 109 | 1,363 | 947 | 88 | 301 | 5,726 | 1,905 | 182 |
| Cupertino | 22,175 | 49 | 96 | 61 | 791 | 722 | 81 | D | D | D | 117 |
| Cypress | 991 | 4 | 50 | 112 | 3,556 | 4,376 | 277 | 90 | 1,569 | 643 | 49 |
| Daly City | 19,235 | 57 | 88 | 28 | 148 | 55 | 6 | 190 | 4,048 | 1,151 | 117 |
| Dana Point | 36,633 | 40 | 55 | 58 | 159 | 167 | 9 | 97 | 1,027 | 380 | 32 |
| Danville | 20,367 | 32 | 100 | 34 | 152 | 235 | 11 | 121 | 1,633 | 597 | 52 |
| Davis | 42,577 | 130 | 14 | 20 | 213 | 153 | 14 | 122 | 2,136 | 595 | 57 |

1. Merchant wholesalers except manufacturers' sales branches and offices.    2. Establishments with payroll.

Items 69—79

# Table D. Cities — Real Estate, Professional Services, and Manufacturing

| City | Real estate and rental and leasing, 2017 | | | | Professional, scientific, and technical services[1], 2017 | | | | Manufacturing, 2017 | | | |
|---|---|---|---|---|---|---|---|---|---|---|---|---|
| | Number of establishments | Number of employees | Receipts (mil dol) | Annual payroll (mil dol) | Number of establishments | Number of employees | Receipts (mil dol) | Annual payroll (mil dol) | Number of establishments | Number of employees | Receipts (mil dol) | Annual payroll (mil dol) |
| | 80 | 81 | 82 | 83 | 84 | 85 | 86 | 87 | 88 | 89 | 90 | 91 |
| **ARKANSAS— Cont'd** | | | | | | | | | | | | |
| Jacksonville | 43 | 194 | 32 | 6 | 41 | 335 | 47 | 19 | NA | NA | NA | NA |
| Jonesboro | 117 | 483 | 118 | 19 | 183 | 1,321 | 166 | 55 | NA | NA | NA | NA |
| Little Rock | 432 | 2,196 | 536 | 109 | 1,245 | 9,394 | 1,631 | 577 | NA | NA | NA | NA |
| North Little Rock | 109 | 495 | 111 | 18 | D | D | D | D | NA | NA | NA | NA |
| Paragould | 31 | 70 | 12 | 2 | 48 | 336 | 32 | 13 | NA | NA | NA | NA |
| Pine Bluff | 49 | 173 | 28 | 5 | D | D | 17 | D | NA | NA | NA | NA |
| Rogers | 98 | 410 | 115 | 20 | 238 | 2,763 | 373 | 150 | NA | NA | NA | NA |
| Russellville | 67 | 175 | 32 | 5 | 121 | 701 | 78 | 42 | NA | NA | NA | NA |
| Sherwood | 43 | 126 | 31 | 4 | 39 | 163 | 17 | 6 | NA | NA | NA | NA |
| Springdale | 71 | 290 | 61 | 11 | 149 | 988 | 121 | 49 | NA | NA | NA | NA |
| Texarkana | 38 | 147 | 35 | 6 | D | D | D | D | NA | NA | NA | NA |
| West Memphis | 31 | 173 | 27 | 6 | 31 | 154 | 13 | 4 | NA | NA | NA | NA |
| **CALIFORNIA** | 57,434 | 314,273 | 110,822 | 18,275 | 127,023 | 1,241,452 | 300,626 | 118,664 | 37,887 | 1,160,890 | 510,859 | 76,483 |
| Adelanto | 9 | 15 | 3 | 1 | 4 | 79 | 10 | 5 | NA | NA | NA | NA |
| Alameda | 90 | 349 | 98 | 14 | 248 | 2,483 | 797 | 263 | NA | NA | NA | NA |
| Alhambra | 140 | 469 | 106 | 23 | 281 | 1,058 | 166 | 48 | NA | NA | NA | NA |
| Aliso Viejo | 86 | 1,883 | 415 | 135 | 307 | 4,789 | 987 | 482 | NA | NA | NA | NA |
| Anaheim | 446 | 3,853 | 970 | 169 | D | D | D | D | NA | NA | NA | NA |
| Antioch | 65 | 395 | 280 | 24 | 90 | 529 | 76 | 29 | NA | NA | NA | NA |
| Apple Valley | 44 | 137 | 25 | 4 | 77 | 301 | 29 | 11 | NA | NA | NA | NA |
| Arcadia | 275 | 890 | 197 | 33 | 314 | 1,023 | 159 | 55 | NA | NA | NA | NA |
| Atascadero | D | D | D | D | 75 | 285 | 37 | 13 | NA | NA | NA | NA |
| Atwater | 12 | 37 | 9 | 1 | 11 | 64 | 9 | 3 | NA | NA | NA | NA |
| Azusa | 30 | 65 | 13 | 3 | 42 | 422 | 80 | 34 | NA | NA | NA | NA |
| Bakersfield | 377 | 2,017 | 388 | 80 | 790 | 5,270 | 779 | 301 | NA | NA | NA | NA |
| Baldwin Park | 23 | 75 | 17 | 4 | 40 | 196 | 19 | 7 | NA | NA | NA | NA |
| Banning | 25 | 89 | 15 | 3 | 14 | 122 | 9 | 3 | NA | NA | NA | NA |
| Beaumont | 28 | 93 | 17 | 3 | 33 | 77 | 9 | 3 | NA | NA | NA | NA |
| Bell | D | D | D | D | 26 | 244 | 40 | 13 | NA | NA | NA | NA |
| Bellflower | 71 | 286 | 67 | 16 | 63 | 218 | 21 | 6 | NA | NA | NA | NA |
| Bell Gardens | D | D | D | D | 9 | 128 | 15 | 6 | NA | NA | NA | NA |
| Belmont | 39 | D | 78 | D | 102 | 429 | 67 | 47 | NA | NA | NA | NA |
| Benicia | 44 | 180 | 83 | 11 | 111 | 964 | 202 | 72 | NA | NA | NA | NA |
| Berkeley | 216 | 904 | 273 | 53 | 663 | 4,639 | 995 | 432 | NA | NA | NA | NA |
| Beverly Hills | 604 | 3,494 | 1,251 | 202 | 1,053 | 6,182 | 1,863 | 595 | NA | NA | NA | NA |
| Brea | 131 | 633 | 175 | 33 | 312 | 2,532 | 440 | 169 | NA | NA | NA | NA |
| Brentwood | 58 | 154 | 60 | 7 | 100 | 399 | 67 | 23 | NA | NA | NA | NA |
| Buena Park | 78 | 309 | 80 | 14 | 141 | 758 | 131 | 38 | NA | NA | NA | NA |
| Burbank | 297 | 3,111 | 2,179 | 222 | 665 | 8,516 | 1,623 | 643 | NA | NA | NA | NA |
| Burlingame | 165 | 830 | 426 | 58 | 307 | 2,963 | 773 | 284 | NA | NA | NA | NA |
| Calexico | 34 | 118 | 16 | 3 | 41 | 115 | 10 | 3 | NA | NA | NA | NA |
| Camarillo | 115 | 598 | 158 | 33 | 339 | 2,751 | 590 | 216 | NA | NA | NA | NA |
| Campbell | 99 | 376 | 104 | 21 | 327 | 3,576 | 899 | 383 | NA | NA | NA | NA |
| Carlsbad | 394 | 1,345 | 426 | 82 | 1,089 | 8,040 | 1,597 | 624 | NA | NA | NA | NA |
| Carson | 80 | 1,227 | 577 | 93 | 115 | 855 | 162 | 55 | NA | NA | NA | NA |
| Cathedral City | 38 | 160 | 47 | 6 | 43 | 95 | 13 | 4 | NA | NA | NA | NA |
| Ceres | 33 | 133 | 51 | 6 | 30 | 433 | 53 | 24 | NA | NA | NA | NA |
| Cerritos | 101 | 410 | 185 | 30 | 206 | 1,389 | 337 | 121 | NA | NA | NA | NA |
| Chico | 165 | 705 | 127 | 23 | 303 | 1,949 | 423 | 106 | NA | NA | NA | NA |
| Chino | 82 | 417 | 140 | 19 | 164 | 940 | 154 | 49 | NA | NA | NA | NA |
| Chino Hills | 97 | 223 | 70 | 10 | 194 | 644 | 127 | 36 | NA | NA | NA | NA |
| Chula Vista | 310 | 931 | 286 | 37 | 455 | 2,438 | 349 | 113 | NA | NA | NA | NA |
| Citrus Heights | 83 | 320 | 129 | 10 | 104 | 649 | 85 | 32 | NA | NA | NA | NA |
| Claremont | D | D | D | D | 152 | 575 | 129 | 41 | NA | NA | NA | NA |
| Clovis | 98 | 344 | 102 | 14 | 186 | 1,595 | 174 | 69 | NA | NA | NA | NA |
| Coachella | 21 | 65 | 14 | 2 | 7 | 39 | 7 | 2 | NA | NA | NA | NA |
| Colton | 40 | 240 | 58 | 9 | 45 | 1,800 | 119 | 50 | NA | NA | NA | NA |
| Compton | 21 | 198 | 44 | 11 | 18 | 325 | 26 | 9 | NA | NA | NA | NA |
| Concord | 182 | 1,414 | 455 | 72 | 298 | 4,687 | 1,045 | 400 | NA | NA | NA | NA |
| Corona | 203 | 765 | 321 | 46 | 384 | 3,434 | 407 | 179 | NA | NA | NA | NA |
| Costa Mesa | 347 | 3,889 | 994 | 184 | 812 | 8,989 | 2,232 | 772 | NA | NA | NA | NA |
| Covina | 99 | 589 | 94 | 29 | 183 | 934 | 123 | 48 | NA | NA | NA | NA |
| Culver City | 128 | 1,475 | 261 | 74 | 459 | 5,140 | 917 | 364 | NA | NA | NA | NA |
| Cupertino | 122 | 696 | 188 | 34 | 413 | 2,420 | 528 | 251 | NA | NA | NA | NA |
| Cypress | 83 | 236 | 74 | 13 | 146 | 1,553 | 376 | 107 | NA | NA | NA | NA |
| Daly City | 57 | 367 | 239 | 17 | 81 | 437 | 45 | 16 | NA | NA | NA | NA |
| Dana Point | 90 | 191 | 81 | 14 | 204 | 606 | 139 | 46 | NA | NA | NA | NA |
| Danville | 128 | 439 | 170 | 30 | 212 | 827 | 223 | 73 | NA | NA | NA | NA |
| Davis | 130 | 586 | 76 | 22 | 238.0 | 1,627 | 298.0 | 126.0 | NA | NA | NA | NA |

1. Establishments subject to federal tax.

# Table D. Cities — Accommodation and Food Services, Arts, Entertainment, and Recreation, and Health Care and Social Assistance

| City | Accommodation and food services, 2017 | | | | Arts, entertainment, and recreation,[1] 2017 | | | | Health care and social assistance,[1] 2017 | | | |
|---|---|---|---|---|---|---|---|---|---|---|---|---|
| | Number of establishments | Number of employees | Receipts (mil dol) | Annual payroll (mil dol) | Number of establishments | Number of employees | Receipts (mil dol) | Annual payroll (mil dol) | Number of establishments | Number of employees | Receipts (mil dol) | Annual payroll (mil dol) |
| | 92 | 93 | 94 | 95 | 96 | 97 | 98 | 99 | 100 | 101 | 102 | 103 |
| **ARKANSAS— Cont'd** | | | | | | | | | | | | |
| Jacksonville | 58 | 835 | 41 | 11 | 4 | 35 | 2 | 1 | 61 | 1,765 | 125 | 53 |
| Jonesboro | D | D | D | D | 20 | 354 | 13 | 5 | 325 | 9,650 | 1,153 | 462 |
| Little Rock | 660 | 13,362 | 755 | 222 | 92 | 1,681 | 155 | 37 | 1,049 | 37,064 | 5,684 | 2,241 |
| North Little Rock | 237 | 4,761 | 234 | 70 | D | D | D | 9 | 245 | 5,693 | 655 | 240 |
| Paragould | D | D | D | D | D | D | D | D | 94 | 1,840 | 183 | 67 |
| Pine Bluff | 99 | 1,584 | 81 | 20 | D | D | D | D | 167 | 3,541 | 373 | 155 |
| Rogers | 194 | 5,028 | 275 | 85 | 17 | 362 | 19 | 7 | 229 | 4,669 | 704 | 230 |
| Russellville | 127 | 2,624 | 127 | 37 | 12 | D | 1 | D | 160 | 3,425 | 312 | 115 |
| Sherwood | 42 | 833 | 45 | 12 | D | D | D | 0 | 62 | 1,095 | 134 | 52 |
| Springdale | 144 | 2,633 | 147 | 40 | D | D | D | D | 165 | 4,952 | 543 | 223 |
| Texarkana | 67 | 1,347 | 57 | 16 | D | D | D | D | D | D | D | D |
| West Memphis | 65 | 1,181 | 62 | 18 | D | D | D | D | 90 | 1,380 | 96 | 41 |
| **CALIFORNIA** | 89,596 | 1,739,010 | 133,717 | 37,677 | 26,390 | 349,752 | 53,107 | 17,209 | 113,390 | 2,043,117 | 311,312 | 119,501 |
| Adelanto | 17 | 228 | 13 | 3 | NA | NA | NA | NA | 16 | 199 | 20 | 10 |
| Alameda | 237 | 3,611 | 241 | 70 | 42 | 1,149 | 376 | 244 | 228 | 4,027 | 430 | 169 |
| Alhambra | 256 | 4,275 | 255 | 79 | 24 | 250 | 18 | 5 | 302 | 3,588 | 504 | 155 |
| Aliso Viejo | 90 | 1,689 | 121 | 37 | 25 | 491 | 40 | 11 | 140 | 2,692 | 462 | 168 |
| Anaheim | 799 | 27,327 | 2,536 | 694 | 94 | 34,259 | 4,595 | 1,388 | 970 | 20,597 | 2,892 | 1,118 |
| Antioch | 139 | 2,028 | 132 | 37 | 28 | 375 | 22 | 6 | 261 | 5,922 | 1,060 | 429 |
| Apple Valley | 95 | 1,565 | 87 | 24 | 8 | 117 | 9 | 2 | 208 | 3,512 | 618 | 198 |
| Arcadia | 224 | 4,170 | 266 | 78 | 71 | 2,277 | 249 | 56 | 441 | 5,310 | 707 | 255 |
| Atascadero | 78 | 1,087 | 72 | 19 | 12 | 180 | 6 | 2 | 119 | 3,123 | 354 | 191 |
| Atwater | 39 | 536 | 36 | 9 | D | D | D | D | 29 | 621 | 37 | 18 |
| Azusa | 95 | 1,319 | 78 | 23 | 6 | 84 | 3 | 1 | 60 | 698 | 48 | 21 |
| Bakersfield | 799 | 15,215 | 962 | 267 | 79 | 1,580 | 94 | 28 | 1,210 | 23,324 | 3,792 | 1,273 |
| Baldwin Park | 103 | 1,504 | 111 | 27 | NA | NA | NA | NA | 86 | 5,226 | 958 | 368 |
| Banning | 57 | 730 | 49 | 13 | NA | NA | NA | NA | 59 | 1,619 | 164 | 73 |
| Beaumont | 58 | 966 | 61 | 18 | 5 | 62 | 3 | 1 | 50 | 686 | 56 | 22 |
| Bell | 63 | 863 | 67 | 17 | NA | NA | NA | NA | 38 | 531 | 45 | 20 |
| Bellflower | 136 | 1,855 | 124 | 32 | 17 | 181 | 13 | 3 | 176 | 2,584 | 368 | 113 |
| Bell Gardens | 66 | 2,578 | 208 | 70 | NA | NA | NA | NA | 39 | 1,012 | 79 | 38 |
| Belmont | 68 | 824 | 79 | 21 | 5 | 57 | 5 | 1 | 71 | 650 | 72 | 27 |
| Benicia | 68 | 1,055 | 66 | 19 | 16 | 137 | 7 | 3 | 64 | 320 | 42 | 14 |
| Berkeley | 538 | 8,416 | 617 | 201 | 115 | 1,619 | 252 | 62 | 491 | 8,402 | 1,341 | 552 |
| Beverly Hills | 238 | 9,436 | 953 | 320 | 1,036 | 5,821 | 3,621 | 1,061 | 1,209 | 6,751 | 1,493 | 433 |
| Brea | 180 | 4,673 | 304 | 94 | D | D | D | 3 | 187 | 2,463 | 264 | 117 |
| Brentwood | 128 | 2,151 | 143 | 41 | 11 | 270 | 15 | 4 | 150 | 1,392 | 139 | 51 |
| Buena Park | 231 | 4,492 | 297 | 79 | D | D | D | D | 220 | 2,215 | 184 | 75 |
| Burbank | 353 | 8,431 | 747 | 182 | 447 | 1,793 | 503 | 156 | 629 | 9,383 | 1,426 | 479 |
| Burlingame | 161 | 5,155 | 587 | 176 | 24 | 433 | 21 | 7 | 211 | 4,729 | 937 | 350 |
| Calexico | 62 | 896 | 51 | 13 | NA | NA | NA | NA | 25 | 360 | 43 | 16 |
| Camarillo | 177 | 3,362 | 218 | 62 | 33 | 474 | 22 | 9 | 274 | 4,745 | 404 | 172 |
| Campbell | 153 | 3,485 | 246 | 79 | 20 | 195 | 27 | 4 | 221 | 3,215 | 420 | 156 |
| Carlsbad | 314 | 9,150 | 734 | 226 | 86 | 3,603 | 367 | 79 | 450 | 7,009 | 652 | 266 |
| Carson | 186 | 3,440 | 239 | 64 | 19 | 561 | 104 | 37 | 197 | 2,260 | 210 | 84 |
| Cathedral City | 109 | 1,777 | 111 | 31 | 12 | 261 | 16 | 5 | 70 | 765 | 87 | 42 |
| Ceres | 75 | 1,067 | 68 | 17 | 6 | 67 | 4 | 1 | 56 | 677 | 74 | 29 |
| Cerritos | 208 | 4,794 | 320 | 92 | 12 | 169 | 15 | 4 | 245 | 3,003 | 362 | 136 |
| Chico | 292 | 5,709 | 315 | 94 | 26 | 540 | 24 | 9 | 443 | 9,129 | 1,188 | 457 |
| Chino | 190 | 3,425 | 180 | 54 | 28 | 815 | 45 | 12 | 231 | 2,926 | 326 | 115 |
| Chino Hills | 178 | 3,343 | 190 | 55 | 23 | 389 | 27 | 8 | 166 | 1,722 | 202 | 74 |
| Chula Vista | 431 | 7,956 | 524 | 150 | 60 | 1,968 | 113 | 33 | 613 | 10,464 | 1,533 | 580 |
| Citrus Heights | D | D | D | D | 15 | 590 | 34 | 12 | 166 | 2,475 | 206 | 85 |
| Claremont | 121 | 1,918 | 128 | 37 | 18 | 469 | 27 | 12 | 182 | 2,377 | 244 | 93 |
| Clovis | 247 | 4,677 | 272 | 76 | 29 | 580 | 49 | 9 | 272 | 5,829 | 679 | 241 |
| Coachella | 54 | 751 | 44 | 11 | D | D | D | D | 22 | 204 | 20 | 9 |
| Colton | 93 | 1,304 | 87 | 21 | NA | NA | NA | NA | 120 | 5,057 | 943 | 289 |
| Compton | D | D | D | D | D | D | D | 1 | 91 | 1,038 | 76 | 32 |
| Concord | 299 | 5,364 | 379 | 104 | 49 | 1,046 | 67 | 18 | 432 | 8,766 | 1,422 | 485 |
| Corona | 346 | 6,667 | 412 | 117 | 43 | 487 | 40 | 10 | 424 | 5,741 | 630 | 230 |
| Costa Mesa | 471 | 10,378 | 771 | 226 | 73 | 1,894 | 515 | 262 | 454 | 5,715 | 570 | 209 |
| Covina | 143 | 2,147 | 139 | 37 | 13 | 128 | 8 | 2 | 242 | 4,622 | 419 | 178 |
| Culver City | 247 | 4,719 | 367 | 101 | 184 | 572 | 267 | 105 | 236 | 4,667 | 691 | 198 |
| Cupertino | 192 | 3,440 | 259 | 83 | 28 | 538 | 30 | 10 | 233 | 2,482 | 288 | 117 |
| Cypress | 129 | 2,082 | 145 | 40 | D | D | D | D | 132 | 1,424 | 256 | 58 |
| Daly City | 198 | 3,416 | 271 | 76 | 14 | 553 | 42 | 13 | 263 | 4,620 | 722 | 256 |
| Dana Point | 112 | 3,795 | 448 | 127 | D | D | D | 7 | 129 | 1,248 | 125 | 48 |
| Danville | 114 | 2,217 | 134 | 41 | 33 | 497 | 39 | 14 | 168 | 1,796 | 205 | 81 |
| Davis | 188 | 3,392 | 187 | 58 | 25 | 289 | 14 | 5 | 172 | 2,626 | 368 | 131 |

1. Establishments subject to federal tax.

Items 92—103

| City | Other services[1] Number of establishments | Other services[1] Number of employees | Other services[1] Receipts (mil dol) | Other services[1] Annual payroll (mil dol) | Full-time equivalent employees | March payroll Total (dollars) | Percent of total for: Administrative, judicial, and legal | Police and corrections | Fire protection | Highways and transportation | Health and welfare | Natural resources and utilities | Education and libraries |
|---|---|---|---|---|---|---|---|---|---|---|---|---|---|
| | 104 | 105 | 106 | 107 | 108 | 109 | 110 | 111 | 112 | 113 | 114 | 115 | 116 |
| **ARKANSAS— Cont'd** | | | | | | | | | | | | | |
| Jacksonville | 31 | 125 | 14 | 4 | 356 | 1,336,592 | 12.7 | 28.5 | 22.3 | 3.9 | 0.0 | 25.8 | 0.0 |
| Jonesboro | 113 | 724 | 73 | 19 | 703 | 2,662,047 | 8.1 | 23.5 | 12.7 | 9.5 | 1.2 | 45.0 | 0.0 |
| Little Rock | 501 | 3,692 | 746 | 152 | 3,175 | 14,264,399 | 7.6 | 26.2 | 17.4 | 11.3 | 12.9 | 18.7 | 0.0 |
| North Little Rock | 147 | 908 | 96 | 28 | 990 | 3,992,647 | 7.7 | 24.2 | 20.1 | 6.6 | 1.3 | 34.0 | 2.6 |
| Paragould | 32 | 146 | 14 | 4 | 285 | 1,104,826 | 4.7 | 15.3 | 11.2 | 7.6 | 0.0 | 34.1 | 0.0 |
| Pine Bluff | 57 | 340 | 33 | 9 | 359 | 1,141,415 | 11.2 | 44.7 | 24.8 | 13.3 | 2.4 | 3.5 | 0.0 |
| Rogers | 87 | 574 | 63 | 20 | 520 | 2,122,890 | 10.2 | 28.2 | 27.6 | 5.1 | 1.2 | 20.4 | 4.1 |
| Russellville | 74 | 416 | 33 | 12 | 262 | 988,177 | 6.5 | 23.9 | 24.5 | 7.1 | 0.0 | 27.4 | 0.0 |
| Sherwood | 54 | 247 | 31 | 7 | 235 | 824,999 | 18.5 | 44.3 | 0.0 | 6.5 | 0.0 | 25.7 | 0.0 |
| Springdale | 119 | 939 | 106 | 32 | 629 | 2,484,976 | 9.5 | 31.9 | 23.1 | 6.1 | 0.3 | 19.8 | 3.5 |
| Texarkana | 33 | 163 | 21 | 5 | 216 | 992,096 | 9.2 | 49.5 | 27.6 | 3.3 | 0.0 | 1.7 | 0.0 |
| West Memphis | 34 | 225 | 23 | 6 | 380 | 1,313,327 | 9.2 | 23.8 | 21.2 | 5.8 | 0.8 | 25.0 | 1.2 |
| **CALIFORNIA** | 62,302 | 437,372 | 64,254 | 15,638 | X | X | X | X | X | X | X | X | X |
| Adelanto | D | D | 9 | D | 130 | 635,090 | 18.0 | 62.9 | 0.0 | 6.8 | 0.0 | 11.6 | 0.0 |
| Alameda | 155 | 1,493 | 207 | 62 | 569 | 4,984,205 | 8.7 | 25.4 | 26.0 | 10.9 | 5.0 | 20.7 | 3.3 |
| Alhambra | 125 | 503 | 65 | 17 | 417 | 3,094,298 | 8.1 | 35.3 | 29.1 | 3.4 | 3.7 | 16.1 | 4.1 |
| Aliso Viejo | 63 | 472 | 78 | 27 | 23 | 150,158 | 88.7 | 0.0 | 0.0 | 4.0 | 0.0 | 7.2 | 0.0 |
| Anaheim | 430 | 3,672 | 407 | 131 | 2,310 | 19,689,571 | 11.0 | 34.3 | 16.7 | 2.4 | 1.2 | 28.3 | 2.0 |
| Antioch | 98 | 554 | 58 | 17 | 307 | 2,742,360 | 9.7 | 58.4 | 0.0 | 7.2 | 5.7 | 16.1 | 0.0 |
| Apple Valley | 48 | 310 | 20 | 8 | 91 | 500,285 | 44.5 | 0.6 | 0.0 | 8.9 | 4.7 | 31.2 | 0.0 |
| Arcadia | 119 | 555 | 57 | 17 | 327 | 2,702,545 | 12.3 | 35.0 | 26.1 | 6.3 | 0.0 | 10.5 | 5.5 |
| Atascadero | 47 | 193 | 20 | 6 | 133 | 836,599 | 14.1 | 33.9 | 21.9 | 3.5 | 6.0 | 12.6 | 0.0 |
| Atwater | 21 | 117 | 11 | 2 | 77 | 468,259 | 21.2 | 53.5 | 0.0 | 8.3 | 0.0 | 16.2 | 0.0 |
| Azusa | 78 | 402 | 39 | 11 | 291 | 2,124,946 | 13.2 | 41.8 | 0.0 | 3.8 | 2.9 | 34.4 | 2.7 |
| Bakersfield | 454 | 3,674 | 363 | 114 | 1,490 | 9,653,365 | 6.6 | 42.2 | 18.5 | 7.2 | 0.6 | 16.9 | 0.0 |
| Baldwin Park | 50 | 137 | 17 | 4 | 234 | 1,398,737 | 11.8 | 58.8 | 0.0 | 10.2 | 3.0 | 10.9 | 0.0 |
| Banning | 33 | 203 | 18 | 5 | 174 | 1,229,485 | 21.3 | 29.0 | 0.0 | 11.9 | 0.0 | 34.9 | 0.0 |
| Beaumont | D | D | D | D | 128 | 830,132 | 3.7 | 57.6 | 0.0 | 13.2 | 4.5 | 12.1 | 0.0 |
| Bell | 33 | 154 | 15 | 5 | 112 | 614,180 | 15.4 | 57.0 | 0.0 | 0.0 | 13.8 | 13.8 | 0.0 |
| Bellflower | 125 | 539 | 61 | 18 | 146 | 689,645 | 32.6 | 10.7 | 0.0 | 28.8 | 9.0 | 19.0 | 0.0 |
| Bell Gardens | D | D | D | D | 191 | 1,176,849 | 11.3 | 58.6 | 0.0 | 13.1 | 3.7 | 11.8 | 0.0 |
| Belmont | 57 | 317 | 39 | 12 | 169 | 1,475,781 | 12.6 | 30.9 | 20.2 | 9.8 | 5.6 | 21.0 | 0.0 |
| Benicia | 79 | 749 | 120 | 41 | 310 | 2,026,006 | 13.0 | 23.5 | 18.0 | 3.4 | 0.0 | 29.9 | 7.0 |
| Berkeley | 326 | 2,358 | 337 | 118 | 1,668 | 15,534,530 | 10.4 | 41.0 | 11.9 | 4.6 | 13.5 | 10.3 | 4.7 |
| Beverly Hills | 404 | 2,441 | 611 | 88 | 568 | 5,900,462 | 0.0 | 44.0 | 27.2 | 2.1 | 2.8 | 17.9 | 6.0 |
| Brea | 103 | 768 | 67 | 21 | 466 | 3,431,286 | 14.7 | 42.8 | 17.4 | 2.4 | 1.3 | 14.0 | 0.0 |
| Brentwood | 96 | 528 | 57 | 15 | 290 | 2,391,198 | 16.9 | 35.1 | 0.0 | 11.2 | 5.8 | 27.1 | 0.0 |
| Buena Park | 96 | 615 | 130 | 29 | 296 | 2,162,651 | 10.3 | 57.0 | 0.0 | 10.5 | 1.8 | 12.2 | 2.8 |
| Burbank | 281 | 2,509 | 325 | 95 | 1,299 | 11,787,918 | 11.7 | 21.9 | 14.1 | 5.4 | 3.5 | 34.6 | 10.6 |
| Burlingame | 141 | 866 | 132 | 39 | 225 | 1,879,269 | 7.4 | 34.5 | 16.3 | 5.8 | 0.0 | 25.6 | 3.9 |
| Calexico | D | D | D | D | 123 | 666,071 | 6.0 | 26.1 | 26.7 | 4.9 | 7.5 | 22.8 | 1.3 |
| Camarillo | 134 | 1,027 | 96 | 29 | 140 | 1,011,277 | 33.6 | 1.4 | 0.0 | 31.3 | 0.7 | 25.3 | 0.0 |
| Campbell | 183 | 1,409 | 147 | 50 | 211 | 1,800,103 | 21.1 | 40.8 | 0.0 | 4.7 | 1.3 | 11.7 | 0.0 |
| Carlsbad | 230 | 1,896 | 297 | 67 | 801 | 4,595,548 | 14.4 | 15.2 | 22.6 | 6.9 | 6.8 | 13.1 | 8.9 |
| Carson | 127 | 1,854 | 238 | 73 | 533 | 2,931,854 | 28.7 | 6.7 | 0.0 | 15.1 | 7.8 | 25.7 | 0.0 |
| Cathedral City | 101 | 725 | 70 | 22 | 173 | 1,659,811 | 17.2 | 47.0 | 24.8 | 6.1 | 0.0 | 0.5 | 0.0 |
| Ceres | 48 | 472 | 138 | 29 | 176 | 1,162,295 | 11.2 | 37.0 | 25.2 | 8.1 | 0.0 | 9.9 | 0.0 |
| Cerritos | 71 | 1,113 | 66 | 36 | 340 | 1,728,051 | 29.0 | 5.6 | 0.0 | 1.7 | 2.6 | 16.5 | 11.0 |
| Chico | 178 | 2,840 | 159 | 52 | 354 | 2,577,844 | 9.9 | 44.5 | 22.6 | 7.6 | 0.4 | 7.3 | 0.0 |
| Chino | 163 | 1,715 | 122 | 44 | 473 | 3,264,265 | 13.5 | 49.5 | 0.0 | 3.6 | 3.9 | 16.2 | 0.0 |
| Chino Hills | 72 | 433 | 25 | 8 | 190 | 1,055,324 | 42.7 | 0.0 | 0.0 | 5.5 | 1.9 | 39.5 | 0.0 |
| Chula Vista | 275 | 1,852 | 139 | 44 | 1,035 | 7,321,658 | 10.7 | 37.4 | 18.4 | 18.0 | 3.8 | 6.5 | 2.8 |
| Citrus Heights | 90 | 562 | 94 | 21 | 207 | 1,592,325 | 17.1 | 67.1 | 0.0 | 5.1 | 2.9 | 1.8 | 0.0 |
| Claremont | 50 | 214 | 25 | 7 | 202 | 1,271,964 | 23.8 | 43.5 | 0.0 | 4.6 | 0.0 | 28.2 | 0.0 |
| Clovis | 145 | 1,047 | 94 | 27 | 560 | 3,609,832 | 9.9 | 34.4 | 16.0 | 6.0 | 9.6 | 15.0 | 0.0 |
| Coachella | 16 | 59 | 6 | 2 | 84 | 642,472 | 15.2 | 4.5 | 0.0 | 6.5 | 11.2 | 45.1 | 0.0 |
| Colton | 57 | 500 | 69 | 15 | 316 | 2,313,222 | 14.2 | 30.5 | 18.4 | 3.1 | 0.4 | 30.3 | 0.8 |
| Compton | D | D | D | D | 329 | 2,108,515 | 20.1 | 6.2 | 37.0 | 8.5 | 4.0 | 13.5 | 0.0 |
| Concord | 251 | 1,589 | 200 | 63 | 475 | 3,242,924 | 14.9 | 60.4 | 0.0 | 9.1 | 1.3 | 10.8 | 0.0 |
| Corona | 240 | 1,634 | 154 | 44 | 670 | 5,204,685 | 11.4 | 38.4 | 24.9 | 6.0 | 0.2 | 15.5 | 1.9 |
| Costa Mesa | 386 | 2,850 | 324 | 89 | 530 | 4,404,296 | 15.6 | 41.6 | 24.5 | 3.9 | 0.0 | 7.6 | 0.0 |
| Covina | 141 | 681 | 75 | 21 | 177 | 1,447,454 | 16.4 | 61.5 | 0.0 | 2.4 | 0.7 | 12.6 | 2.8 |
| Culver City | 167 | 1,521 | 213 | 57 | 686 | 5,240,715 | 10.1 | 24.2 | 21.8 | 25.1 | 6.8 | 10.6 | 0.0 |
| Cupertino | 79 | 470 | 79 | 15 | 164 | 1,353,372 | 34.4 | 0.0 | 0.0 | 21.5 | 0.0 | 26.3 | 0.0 |
| Cypress | 86 | 733 | 96 | 38 | 165 | 1,155,562 | 23.2 | 47.9 | 0.0 | 2.6 | 1.0 | 12.6 | 0.0 |
| Daly City | 98 | 442 | 45 | 12 | 539 | 4,366,533 | 8.6 | 29.3 | 24.4 | 3.1 | 2.8 | 20.4 | 2.4 |
| Dana Point | 76 | 378 | 42 | 11 | 72 | 562,904 | 20.2 | 0.0 | 0.0 | 25.2 | 33.7 | 14.3 | 0.0 |
| Danville | 77 | 374 | 39 | 10 | 117 | 658,459 | 35.3 | 4.1 | 0.0 | 26.7 | 0.0 | 26.5 | 0.0 |
| Davis | 97 | 532 | 58 | 16 | 303 | 2,032,567 | 11.2 | 34.4 | 18.5 | 6.2 | 0.0 | 29.8 | 0.0 |

1. Establishments subject to federal tax.

| City | City government finances, 2017 | | | | | | | | | |
|---|---|---|---|---|---|---|---|---|---|---|
| | General revenue | | | | | | | General expenditure | | |
| | | Intergovernmental | | Taxes | | | | | Per capita[1] (dollars) | |
| | | | | | | Per capita[1] (dollars) | | | | |
| | Total (mil dol) | Total (mil dol) | Percent from state government | Total (mil dol) | Total | Property | Sales and gross receipts | Total (mil dol) | Total | Capital outlays |
| | 117 | 118 | 119 | 120 | 121 | 122 | 123 | 124 | 125 | 126 |
| **ARKANSAS— Cont'd** | | | | | | | | | | |
| Jacksonville | 32.6 | 10.9 | 44.6 | 10.7 | 377 | 29 | 348 | 34 | 1,205 | 148 |
| Jonesboro | 79.1 | 22.7 | 27.0 | 28.1 | 370 | 65 | 305 | 76 | 995 | 130 |
| Little Rock | 461.4 | 90.4 | 24.4 | 186.2 | 939 | 276 | 663 | 513 | 2,589 | 710 |
| North Little Rock | 104.1 | 20.4 | 26.0 | 52.2 | 793 | 287 | 506 | 109 | 1,661 | 285 |
| Paragould | 33.8 | 6.0 | 39.4 | 7.2 | 251 | 34 | 218 | 33 | 1,169 | 142 |
| Pine Bluff | 47.5 | 14.3 | 39.1 | 20.6 | 478 | 93 | 386 | 49 | 1,129 | 102 |
| Rogers | 93.3 | 23.7 | 40.6 | 46.3 | 694 | 79 | 615 | 87 | 1,309 | 540 |
| Russellville | 35.1 | 7.4 | 33.9 | 17.6 | 605 | 33 | 573 | 40 | 1,354 | 529 |
| Sherwood | 25.6 | 9.3 | 26.2 | 8.8 | 284 | 21 | 263 | 25 | 809 | 94 |
| Springdale | 92.0 | 27.1 | 36.9 | 39.1 | 490 | 88 | 402 | 81 | 1,016 | 253 |
| Texarkana | 32.0 | 6.6 | 59.3 | 15.3 | 510 | 127 | 384 | 29 | 964 | 24 |
| West Memphis | 31.4 | 7.6 | 32.3 | 15.2 | 611 | 54 | 557 | 33 | 1,309 | 191 |
| **CALIFORNIA** | X | X | X | X | X | X | X | X | X | X |
| Adelanto | 19.0 | 0.3 | 84.0 | 11.8 | 347 | 191 | 147 | 20 | 577 | 70 |
| Alameda | 173.1 | 14.1 | 38.1 | 96.2 | 1,217 | 633 | 405 | 128 | 1,622 | 63 |
| Alhambra | 98.6 | 3.5 | 86.4 | 56.5 | 665 | 295 | 364 | 103 | 1,212 | 144 |
| Aliso Viejo | 24.2 | 2.6 | 58.7 | 17.5 | 344 | 140 | 197 | 34 | 657 | 95 |
| Anaheim | 796.8 | 150.7 | 10.4 | 368.6 | 1,052 | 279 | 768 | 705 | 2,013 | 241 |
| Antioch | 81.7 | 6.2 | 56.5 | 52.0 | 466 | 202 | 260 | 75 | 672 | 95 |
| Apple Valley | 65.8 | 11.5 | 85.7 | 27.8 | 381 | 155 | 185 | 73 | 1,005 | 111 |
| Arcadia | 82.7 | 4.7 | 38.9 | 51.7 | 886 | 379 | 495 | 76 | 1,309 | 143 |
| Atascadero | 30.9 | 2.4 | 40.5 | 20.0 | 660 | 294 | 328 | 32 | 1,065 | 306 |
| Atwater | 35.4 | 2.6 | 68.1 | 11.5 | 393 | 142 | 150 | 46 | 1,574 | 573 |
| Azusa | 57.5 | 2.5 | 81.1 | 31.6 | 634 | 193 | 374 | 52 | 1,038 | 37 |
| Bakersfield | 444.4 | 118.4 | 11.1 | 174.7 | 462 | 215 | 244 | 468 | 1,238 | 485 |
| Baldwin Park | 47.8 | 10.4 | 77.3 | 27.6 | 364 | 154 | 207 | 50 | 653 | 52 |
| Banning | 30.6 | 6.8 | 72.7 | 10.7 | 345 | 152 | 188 | 30 | 954 | 248 |
| Beaumont | 82.2 | 5.5 | 7.7 | 18.3 | 390 | 166 | 218 | 96 | 2,046 | 957 |
| Bell | 33.2 | 4.4 | 37.1 | 20.2 | 569 | 364 | 202 | 37 | 1,024 | 248 |
| Bellflower | 39.3 | 4.0 | 54.2 | 28.9 | 373 | 138 | 232 | 38 | 493 | 78 |
| Bell Gardens | 43.7 | 2.1 | 58.8 | 26.6 | 627 | 130 | 496 | 39 | 912 | 86 |
| Belmont | 66.3 | 2.5 | 54.7 | 30.2 | 1,114 | 656 | 399 | 50 | 1,863 | 64 |
| Benicia | 51.9 | 2.1 | 75.2 | 36.4 | 1,287 | 597 | 481 | 63 | 2,231 | 715 |
| Berkeley | 381.9 | 35.9 | 58.7 | 186.7 | 1,528 | 738 | 539 | 365 | 2,986 | 172 |
| Beverly Hills | 340.2 | 3.5 | 32.8 | 214.1 | 6,251 | 1,804 | 4,270 | 309 | 9,012 | 1,169 |
| Brea | 75.4 | 5.8 | 21.8 | 44.8 | 1,053 | 420 | 613 | 92 | 2,153 | 615 |
| Brentwood | 141.4 | 5.2 | 48.2 | 46.8 | 752 | 311 | 253 | 106 | 1,704 | 231 |
| Buena Park | 88.2 | 4.1 | 85.8 | 63.1 | 764 | 322 | 424 | 107 | 1,300 | 279 |
| Burbank | 249.2 | 21.0 | 38.4 | 141.7 | 1,360 | 596 | 756 | 319 | 3,060 | 0 |
| Burlingame | 112.1 | 1.3 | 99.3 | 66.5 | 2,173 | 561 | 1,378 | 82 | 2,693 | 481 |
| Calexico | 34.3 | 5.5 | 30.7 | 14.8 | 369 | 131 | 237 | 30 | 754 | 106 |
| Camarillo | 85.5 | 9.2 | 34.0 | 43.9 | 652 | 247 | 397 | 107 | 1,586 | 278 |
| Campbell | 56.2 | 3.6 | 34.8 | 41.3 | 969 | 333 | 614 | 69 | 1,619 | 167 |
| Carlsbad | 226.5 | 19.6 | 44.8 | 141.1 | 1,233 | 593 | 627 | 213 | 1,863 | 385 |
| Carson | 166.8 | 42.6 | 6.6 | 97.5 | 1,058 | 462 | 592 | 242 | 2,627 | 429 |
| Cathedral City | 60.3 | 11.5 | 14.7 | 33.6 | 617 | 139 | 461 | 62 | 1,132 | 141 |
| Ceres | 56.3 | 6.5 | 45.0 | 16.9 | 350 | 123 | 225 | 56 | 1,157 | 94 |
| Cerritos | 65.1 | 1.7 | 97.2 | 52.3 | 1,032 | 218 | 795 | 80 | 1,570 | 181 |
| Chico | 111.0 | 16.6 | 33.2 | 68.5 | 736 | 270 | 461 | 103 | 1,105 | 310 |
| Chino | 169.7 | 15.9 | 97.5 | 64.5 | 719 | 325 | 358 | 130 | 1,448 | 148 |
| Chino Hills | 118.1 | 3.8 | 49.9 | 47.2 | 589 | 157 | 426 | 104 | 1,299 | 209 |
| Chula Vista | 494.5 | 23.3 | 41.6 | 136.6 | 508 | 208 | 259 | 298 | 1,108 | 224 |
| Citrus Heights | 53.7 | 5.7 | 55.1 | 36.1 | 411 | 143 | 264 | 60 | 681 | 143 |
| Claremont | 42.5 | 2.4 | 48.7 | 24.9 | 696 | 265 | 425 | 48 | 1,346 | 128 |
| Clovis | 162.5 | 15.3 | 30.7 | 59.0 | 540 | 224 | 310 | 140 | 1,281 | 213 |
| Coachella | 41.1 | 9.3 | 95.3 | 19.3 | 426 | 128 | 256 | 55 | 1,203 | 535 |
| Colton | 70.4 | 5.9 | 44.7 | 26.1 | 478 | 219 | 257 | 82 | 1,490 | 134 |
| Compton | 83.0 | 15.1 | 17.1 | 44.7 | 461 | 184 | 275 | 123 | 1,262 | 118 |
| Concord | 193.1 | 37.8 | 21.5 | 100.4 | 775 | 259 | 401 | 160 | 1,238 | 106 |
| Corona | 238.9 | 18.6 | 60.0 | 114.1 | 682 | 313 | 344 | 301 | 1,796 | 709 |
| Costa Mesa | 139.3 | 7.5 | 94.0 | 117.8 | 1,036 | 331 | 698 | 131 | 1,150 | 145 |
| Covina | 52.1 | 2.1 | 87.3 | 31.4 | 651 | 238 | 407 | 47 | 981 | 145 |
| Culver City | 192.9 | 53.9 | 10.5 | 88.9 | 2,279 | 322 | 1,627 | 167 | 4,268 | 531 |
| Cupertino | 107.6 | 2.6 | 70.2 | 65.5 | 1,080 | 344 | 705 | 88 | 1,448 | 86 |
| Cypress | 45.6 | 3.9 | 42.7 | 32.2 | 659 | 280 | 362 | 47 | 962 | 245 |
| Daly City | 115.4 | 7.9 | 55.4 | 66.5 | 623 | 315 | 305 | 116 | 1,088 | 126 |
| Dana Point | 40.3 | 1.8 | 97.3 | 35.2 | 1,044 | 341 | 655 | 55 | 1,636 | 132 |
| Danville | 37.0 | 2.0 | 100.0 | 25.2 | 564 | 298 | 251 | 40 | 897 | 248 |
| Davis | 118.3 | 15.5 | 24.7 | 57.2 | 832 | 318 | 297 | 102 | 1,488 | 159 |

1. Based on population estimated as of July 1 of the year shown.

# City Government Finances

City government finances, 2017 (cont.)

General expenditure (cont.)

Percent of total for:

| City | Public welfare | Highways | Parking facilities | Education | Health and hospitals | Police protection | Sewerage and sanitation | Parks and recreation | Housing and community development | Interest on debt |
|---|---|---|---|---|---|---|---|---|---|---|
| | 127 | 128 | 129 | 130 | 131 | 132 | 133 | 134 | 135 | 136 |
| ARKANSAS— Cont'd | | | | | | | | | | |
| Jacksonville | 0.0 | 5.1 | 0.0 | 0.0 | 7.9 | 21.1 | 15.9 | 11.5 | 0.7 | 3.2 |
| Jonesboro | 0.0 | 12.3 | 0.0 | 0.0 | 1.3 | 15.8 | 14.6 | 6.8 | 0.8 | 14.4 |
| Little Rock | 0.0 | 8.2 | 0.3 | 0.0 | 7.4 | 13.9 | 12.9 | 8.4 | 1.4 | 4.2 |
| North Little Rock | 0.1 | 8.1 | 0.0 | 0.0 | 1.0 | 18.4 | 22.3 | 6.8 | 0.6 | 3.9 |
| Paragould | 0.3 | 10.2 | 0.0 | 0.0 | 0.5 | 11.4 | 20.1 | 9.3 | 0.0 | 0.8 |
| Pine Bluff | 0.7 | 8.3 | 0.0 | 0.0 | 0.1 | 30.4 | 17.3 | 5.6 | 4.7 | 1.9 |
| Rogers | 0.0 | 29.7 | 0.0 | 0.0 | 0.5 | 11.6 | 8.8 | 13.2 | 0.0 | 4.3 |
| Russellville | 0.0 | 12.4 | 0.0 | 0.0 | 0.9 | 11.4 | 34.0 | 15.4 | 0.0 | 1.9 |
| Sherwood | 0.0 | 13.7 | 0.0 | 0.0 | 1.3 | 27.2 | 14.8 | 17.6 | 0.0 | 0.7 |
| Springdale | 0.0 | 14.6 | 0.0 | 0.0 | 0.9 | 19.7 | 16.2 | 11.5 | 0.9 | 6.2 |
| Texarkana | 0.0 | 5.5 | 0.0 | 0.0 | 1.3 | 24.4 | 22.5 | 1.2 | 0.9 | 2.9 |
| West Memphis | 0.0 | 15.4 | 0.0 | 0.0 | 1.8 | 23.2 | 14.6 | 5.8 | 0.9 | 2.5 |
| CALIFORNIA | X | X | X | X | X | X | X | 3.1 | 9.6 | 10.7 |
| Adelanto | 0.0 | 9.4 | 0.0 | 0.0 | 0.0 | 31.7 | 3.3 | 5.5 | 19.5 | 3.1 |
| Alameda | 0.0 | 4.2 | 0.0 | 0.0 | 3.1 | 24.1 | 11.2 | 7.3 | 6.0 | 2.1 |
| Alhambra | 0.0 | 6.1 | 2.0 | 0.0 | 1.6 | 26.2 | 0.0 | 9.3 | 0.4 | 3.8 |
| Aliso Viejo | 0.0 | 11.6 | 0.0 | 0.0 | 0.4 | 46.3 | 7.7 | 19.9 | 16.0 | 11.5 |
| Anaheim | 0.0 | 3.6 | 0.0 | 0.0 | 2.5 | 21.0 | 6.7 | 7.7 | 5.4 | 0.5 |
| Antioch | 0.0 | 17.3 | 0.0 | 0.0 | 2.6 | 47.4 | 20.2 | 6.0 | 8.4 | 1.3 |
| Apple Valley | 0.0 | 5.8 | 0.0 | 0.0 | 4.6 | 36.1 | 2.3 | 4.3 | 0.4 | 0.6 |
| Arcadia | 0.0 | 9.3 | 0.0 | 0.0 | 0.7 | 27.3 | 6.1 | 11.3 | 19.3 | 3.8 |
| Atascadero | 0.0 | 7.0 | 0.0 | 0.0 | 0.0 | 21.4 | 51.4 | 1.0 | 1.4 | 8.9 |
| Atwater | 0.0 | 4.4 | 0.0 | 0.0 | 1.1 | 14.1 | 10.1 | 6.9 | 1.5 | 5.3 |
| Azusa | 0.0 | 3.7 | 0.0 | 0.0 | 1.1 | 35.5 | 10.1 | 6.9 | 1.5 | 5.3 |
| Bakersfield | 0.0 | 7.0 | 0.0 | 0.0 | 0.4 | 17.8 | 13.6 | 7.6 | 3.6 | 1.6 |
| Baldwin Park | 0.0 | 15.5 | 0.0 | 0.0 | 0.7 | 37.1 | 0.6 | 8.4 | 14.7 | 1.3 |
| Banning | 0.0 | 12.7 | 0.0 | 0.0 | 0.5 | 25.3 | 20.2 | 3.8 | 13.3 | 1.2 |
| Beaumont | 0.0 | 1.3 | 0.0 | 0.0 | 0.6 | 9.8 | 8.1 | 2.5 | 0.0 | 11.9 |
| Bell | 0.0 | 11.9 | 0.0 | 0.0 | 0.0 | 25.9 | 4.5 | 5.9 | 20.2 | 9.6 |
| Bellflower | 0.0 | 32.1 | 0.0 | 0.0 | 0.0 | 31.8 | 0.0 | 6.1 | 5.9 | 3.5 |
| Bell Gardens | 0.0 | 13.6 | 0.0 | 0.0 | 0.0 | 40.5 | 7.2 | 14.5 | 2.2 | 2.5 |
| Belmont | 0.0 | 5.0 | 0.0 | 0.0 | 0.0 | 22.9 | 20.2 | 9.4 | 0.5 | 4.5 |
| Benicia | 0.0 | 6.3 | 0.0 | 0.0 | 0.0 | 15.1 | 10.1 | 8.7 | 0.1 | 2.0 |
| Berkeley | 0.0 | 10.3 | 2.8 | 0.0 | 10.7 | 18.0 | 13.8 | 5.3 | 3.4 | 1.9 |
| Beverly Hills | 0.0 | 5.8 | 16.7 | 0.0 | 6.2 | 23.1 | 8.0 | 11.0 | 5.0 | 1.5 |
| Brea | 0.0 | 18.2 | 0.0 | 0.0 | 5.8 | 24.6 | 6.5 | 11.6 | 9.2 | 1.4 |
| Brentwood | 0.0 | 12.6 | 0.0 | 0.0 | 0.3 | 20.7 | 21.1 | 13.3 | 6.8 | 6.1 |
| Buena Park | 0.0 | 14.4 | 0.0 | 0.0 | 0.7 | 22.4 | 3.3 | 6.0 | 15.4 | 4.1 |
| Burbank | 0.0 | 5.6 | 0.2 | 0.1 | 4.2 | 16.5 | 7.5 | 3.2 | 3.5 | 1.6 |
| Burlingame | 0.0 | 9.6 | 0.9 | 0.0 | 0.0 | 18.5 | 18.4 | 11.1 | 0.0 | 5.1 |
| Calexico | 0.0 | 8.2 | 0.0 | 0.0 | 0.5 | 24.4 | 22.5 | 1.9 | 8.3 | 4.8 |
| Camarillo | 0.0 | 21.5 | 0.0 | 0.0 | 0.7 | 30.2 | 13.3 | 1.1 | 4.4 | 3.1 |
| Campbell | 0.0 | 14.0 | 0.0 | 0.0 | 0.0 | 24.3 | 0.1 | 10.9 | 2.2 | 1.0 |
| Carlsbad | 0.0 | 12.5 | 0.0 | 0.0 | 8.3 | 16.1 | 8.2 | 13.1 | 5.6 | 0.3 |
| Carson | 0.0 | 7.1 | 0.0 | 0.0 | 0.1 | 17.2 | 0.0 | 8.4 | 18.7 | 4.9 |
| Cathedral City | 0.0 | 12.3 | 0.0 | 0.0 | 5.5 | 24.4 | 1.1 | 2.0 | 8.0 | 11.1 |
| Ceres | 0.0 | 8.6 | 0.0 | 0.0 | 0.9 | 22.2 | 13.0 | 3.7 | 8.0 | 1.6 |
| Cerritos | 0.0 | 10.8 | 0.0 | 0.0 | 0.0 | 33.0 | 6.4 | 12.8 | 5.1 | 0.1 |
| Chico | 0.0 | 21.2 | 0.6 | 0.0 | 1.2 | 23.0 | 5.7 | 2.4 | 11.7 | 5.3 |
| Chino | 0.0 | 7.6 | 0.0 | 0.0 | 0.0 | 26.4 | 24.2 | 5.1 | 12.2 | 2.2 |
| Chino Hills | 0.0 | 14.0 | 0.0 | 0.0 | 0.4 | 25.5 | 12.3 | 7.2 | 0.4 | 3.9 |
| Chula Vista | 0.0 | 9.3 | 0.2 | 0.0 | 1.0 | 20.9 | 9.3 | 19.9 | 1.8 | 2.2 |
| Citrus Heights | 0.0 | 22.5 | 0.0 | 0.0 | 1.2 | 34.7 | 1.3 | 1.3 | 11.0 | 2.0 |
| Claremont | 0.0 | 16.2 | 0.0 | 0.0 | 1.6 | 24.2 | 14.6 | 7.8 | 4.8 | 1.2 |
| Clovis | 0.0 | 9.4 | 0.0 | 0.0 | 1.3 | 22.3 | 25.8 | 4.9 | 9.4 | 6.7 |
| Coachella | 0.0 | 33.4 | 0.0 | 0.0 | 0.5 | 15.0 | 14.2 | 15.2 | 0.3 | 2.1 |
| Colton | 0.0 | 8.0 | 0.0 | 0.0 | 1.0 | 32.5 | 24.8 | 4.0 | 7.2 | 3.3 |
| Compton | 0.0 | 3.4 | 0.0 | 0.0 | 0.2 | 19.5 | 9.3 | 0.8 | 13.3 | 1.9 |
| Concord | 0.0 | 19.3 | 0.0 | 0.0 | 0.0 | 33.3 | 15.8 | 7.5 | 2.3 | 1.6 |
| Corona | 0.0 | 14.5 | 0.0 | 0.0 | 0.0 | 15.7 | 19.2 | 1.7 | 4.3 | 1.4 |
| Costa Mesa | 0.0 | 8.3 | 0.0 | 0.0 | 0.2 | 34.1 | 0.0 | 7.3 | 1.4 | 0.8 |
| Covina | 0.0 | 9.3 | 0.3 | 0.0 | 0.3 | 33.5 | 4.0 | 6.7 | 8.4 | 4.0 |
| Culver City | 0.0 | 5.5 | 1.2 | 0.0 | 4.4 | 22.4 | 13.6 | 5.0 | 10.2 | 0.5 |
| Cupertino | 0.0 | 8.7 | 0.0 | 0.0 | 0.0 | 26.3 | 3.4 | 17.0 | 8.9 | 1.2 |
| Cypress | 0.0 | 22.2 | 0.0 | 0.0 | 1.2 | 33.2 | 3.1 | 14.1 | 8.2 | 0.1 |
| Daly City | 0.0 | 9.3 | 0.0 | 0.0 | 0.0 | 28.3 | 15.2 | 7.9 | 5.8 | 0.5 |
| Dana Point | 0.0 | 16.2 | 0.0 | 0.0 | 0.6 | 44.0 | 0.0 | 13.7 | 0.0 | 1.4 |
| Danville | 0.0 | 18.2 | 0.0 | 0.0 | 1.8 | 19.2 | 0.0 | 21.5 | 0.0 | 1.5 |
| Davis | 0.0 | 8.1 | 0.0 | 0.0 | 0.0 | 18.8 | 21.6 | 18.7 | 1.8 | 0.9 |

# City Government Finances, City Government Employment, and Climate

Table columns group as: **City government finances, 2017 (cont.)** — Debt outstanding [Total (mil dol), Per capita (dollars)[1]], Debt issued during year. **Climate[2]** — Average daily temperature [Mean: January, July; Limits: January[3], July[4]], Annual precipitation (inches), Heating degree days, Cooling degree days.

| City | Total (mil dol) | Per capita (dollars)[1] | Debt issued during year | Mean January | Mean July | Limits January[3] | Limits July[4] | Annual precipitation (inches) | Heating degree days | Cooling degree days |
|---|---|---|---|---|---|---|---|---|---|---|
| | 137 | 138 | 139 | 140 | 141 | 142 | 143 | 144 | 145 | 146 |
| **ARKANSAS— Cont'd** | | | | | | | | | | |
| Jacksonville | 43.5 | 1,526 | 16.1 | 38.2 | 79.9 | 27.4 | 91.1 | 50.56 | 3,470 | 1,699 |
| Jonesboro | 272.8 | 3,592 | 0.4 | 35.6 | 81.6 | 25.8 | 92.3 | 46.18 | 3,737 | 1,858 |
| Little Rock | 592.8 | 2,989 | 52.7 | 40.1 | 82.4 | 30.8 | 92.8 | 50.93 | 3,084 | 2,086 |
| North Little Rock | 146.2 | 2,220 | 21.3 | 40.1 | 82.4 | 30.8 | 92.8 | 50.93 | 3,084 | 2,086 |
| Paragould | 14.3 | 501 | 5.6 | NA | NA | NA | NA | NA | NA | NA |
| Pine Bluff | 23.4 | 543 | 0.0 | 40.8 | 82.4 | 31.5 | 92.4 | 52.48 | 2,935 | 2,099 |
| Rogers | 117.6 | 1,766 | 26.6 | 32.9 | 77.5 | 22.0 | 88.8 | 46.92 | 4,483 | 1,269 |
| Russellville | 47.5 | 1,628 | 0.0 | 34.3 | 78.9 | 24.2 | 89.1 | 46.02 | 4,166 | 1,439 |
| Sherwood | 12.2 | 392 | 0.0 | NA | NA | NA | NA | NA | NA | NA |
| Springdale | 198.8 | 2,491 | 75.5 | NA | NA | NA | NA | NA | NA | NA |
| Texarkana | 32.4 | 1,078 | 0.0 | 44.3 | 82.7 | 35.6 | 92.7 | 47.38 | 2,421 | 2,280 |
| West Memphis | 19.0 | 764 | 0.9 | 37.5 | 81.5 | 28.5 | 90.9 | 52.80 | 3,417 | 1,903 |
| **CALIFORNIA** | X | X | X | X | X | X | X | X | X | X |
| Adelanto | 27.8 | 819 | 1.0 | NA | NA | NA | NA | NA | NA | NA |
| Alameda | 142.2 | 1,799 | 12.9 | 50.9 | 64.9 | 44.7 | 72.7 | 22.94 | 2,400 | 377 |
| Alhambra | 56.1 | 661 | 8.4 | 56.3 | 75.6 | 42.6 | 89.0 | 18.56 | 1,295 | 1,575 |
| Aliso Viejo | 31.2 | 612 | 0.0 | 56.7 | 72.4 | 47.2 | 82.3 | 14.03 | 1,465 | 1,183 |
| Anaheim | 1,823.7 | 5,203 | 321.8 | 56.9 | 73.2 | 45.2 | 84.0 | 11.23 | 1,286 | 1,294 |
| Antioch | 30.1 | 270 | 0.2 | 45.7 | 74.4 | 37.8 | 90.7 | 13.33 | 2,714 | 1,179 |
| Apple Valley | 16.8 | 230 | 0.0 | 45.5 | 80.0 | 31.4 | 99.1 | 6.20 | 2,929 | 1,735 |
| Arcadia | 11.8 | 202 | 0.0 | 56.3 | 75.6 | 42.6 | 89.0 | 18.56 | 1,295 | 1,575 |
| Atascadero | 40.7 | 1,343 | 0.0 | 47.3 | 71.6 | 33.1 | 91.3 | 14.71 | 2,932 | 785 |
| Atwater | 77.1 | 2,635 | 0.0 | NA | NA | NA | NA | NA | NA | NA |
| Azusa | 171.8 | 3,449 | 0.0 | 54.6 | 73.8 | 41.5 | 88.7 | 16.96 | 1,727 | 1,191 |
| Bakersfield | 661.7 | 1,751 | 23.2 | 47.8 | 83.1 | 39.3 | 96.9 | 6.49 | 2,120 | 2,286 |
| Baldwin Park | 13.9 | 183 | 1.6 | 56.3 | 75.6 | 42.6 | 89.0 | 18.56 | 1,295 | 1,575 |
| Banning | 63.2 | 2,028 | 0.2 | NA | NA | NA | NA | NA | NA | NA |
| Beaumont | 216.6 | 4,624 | 0.1 | NA | NA | NA | NA | NA | NA | NA |
| Bell | 69.0 | 1,938 | 0.9 | 58.8 | 76.6 | 47.9 | 88.9 | 14.44 | 949 | 1,837 |
| Bellflower | 36.5 | 472 | 0.0 | 57.0 | 73.8 | 46.0 | 82.9 | 12.94 | 1,211 | 1,186 |
| Bell Gardens | 20.6 | 485 | 0.0 | 58.8 | 76.6 | 47.9 | 88.9 | 14.44 | 949 | 1,837 |
| Belmont | 66.6 | 2,462 | 0.0 | 48.1 | 69.7 | 36.4 | 88.2 | 28.71 | 2,769 | 569 |
| Benicia | 59.7 | 2,114 | 11.0 | 46.3 | 71.2 | 38.8 | 87.4 | 19.58 | 2,757 | 786 |
| Berkeley | 201.8 | 1,652 | 49.0 | 50.0 | 62.8 | 43.6 | 70.4 | 25.40 | 2,857 | 142 |
| Beverly Hills | 214.1 | 6,252 | 0.0 | 57.9 | 69.5 | 49.4 | 76.9 | 18.68 | 1,379 | 893 |
| Brea | 75.2 | 1,766 | 0.0 | 56.9 | 73.2 | 45.2 | 84.0 | 11.23 | 1,286 | 1,294 |
| Brentwood | 177.3 | 2,848 | 0.3 | NA | NA | NA | NA | NA | NA | NA |
| Buena Park | 96.2 | 1,165 | 9.8 | 57.0 | 73.8 | 46.0 | 82.9 | 12.94 | 1,211 | 1,186 |
| Burbank | 281.5 | 2,702 | 0.5 | 54.8 | 75.5 | 42.0 | 88.9 | 17.49 | 1,575 | 1,455 |
| Burlingame | 88.6 | 2,897 | 17.6 | 50.0 | 62.7 | 42.9 | 70.5 | 23.35 | 2,720 | 184 |
| Calexico | 63.6 | 1,586 | 0.0 | 55.8 | 91.4 | 41.3 | 107.0 | 2.96 | 1,080 | 3,952 |
| Camarillo | 82.4 | 1,222 | 41.6 | 55.7 | 66.0 | 45.3 | 74.0 | 13.61 | 1,961 | 389 |
| Campbell | 10.4 | 243 | 8.2 | 48.7 | 70.3 | 38.8 | 85.4 | 22.64 | 2,641 | 613 |
| Carlsbad | 37.0 | 324 | 0.2 | 54.7 | 67.6 | 45.4 | 72.1 | 11.13 | 2,009 | 505 |
| Carson | 204.7 | 2,220 | 33.8 | 56.3 | 69.4 | 46.2 | 77.6 | 14.79 | 1,526 | 742 |
| Cathedral City | 195.3 | 3,589 | 0.0 | 57.3 | 92.1 | 44.2 | 108.2 | 5.23 | 951 | 4,224 |
| Ceres | 22.5 | 466 | 0.0 | 47.2 | 77.7 | 40.1 | 93.6 | 13.12 | 2,358 | 1,570 |
| Cerritos | 11.0 | 217 | 0.0 | 57.0 | 73.8 | 46.0 | 82.9 | 12.94 | 1,211 | 1,186 |
| Chico | 128.4 | 1,379 | 0.5 | 44.5 | 76.9 | 35.2 | 93.0 | 26.23 | 2,945 | 1,334 |
| Chino | 69.6 | 777 | 0.0 | 54.6 | 73.8 | 41.5 | 88.7 | 16.96 | 1,727 | 1,191 |
| Chino Hills | 115.1 | 1,436 | 0.0 | 56.9 | 73.2 | 45.2 | 84.0 | 11.23 | 1,286 | 1,294 |
| Chula Vista | 669.1 | 2,487 | 0.0 | 57.3 | 70.1 | 46.1 | 76.1 | 9.95 | 1,321 | 862 |
| Citrus Heights | 26.6 | 303 | 0.0 | 46.9 | 77.7 | 39.2 | 94.8 | 24.61 | 2,532 | 1,528 |
| Claremont | 13.7 | 383 | 0.0 | 54.6 | 73.8 | 41.5 | 88.7 | 16.96 | 1,727 | 1,191 |
| Clovis | 177.8 | 1,627 | 3.7 | 46.0 | 81.4 | 38.4 | 96.6 | 11.23 | 2,447 | 1,963 |
| Coachella | 93.3 | 2,060 | 25.5 | NA | NA | NA | NA | NA | NA | NA |
| Colton | 94.4 | 1,726 | 0.3 | 54.4 | 79.6 | 41.8 | 96.0 | 16.43 | 1,599 | 1,937 |
| Compton | 103.5 | 1,067 | 28.7 | 57.0 | 73.8 | 46.0 | 82.9 | 12.94 | 1,211 | 1,186 |
| Concord | 76.2 | 588 | 0.0 | 46.3 | 71.2 | 38.8 | 87.4 | 19.58 | 2,757 | 786 |
| Corona | 218.3 | 1,305 | 26.7 | 54.7 | 75.9 | 41.5 | 92.0 | 12.00 | 1,599 | 1,534 |
| Costa Mesa | 25.0 | 220 | 0.0 | 55.9 | 67.3 | 48.2 | 71.4 | 11.65 | 1,719 | 543 |
| Covina | 64.5 | 1,338 | 0.0 | 54.6 | 73.8 | 41.5 | 88.7 | 16.96 | 1,727 | 1,191 |
| Culver City | 15.0 | 384 | 0.0 | 56.7 | 70.8 | 46.1 | 80.0 | 13.32 | 1,344 | 959 |
| Cupertino | 33.7 | 556 | 0.0 | 48.7 | 70.3 | 38.8 | 85.4 | 22.64 | 2,641 | 613 |
| Cypress | 1.8 | 37 | 0.0 | 57.0 | 73.8 | 46.0 | 82.9 | 12.94 | 1,211 | 1,186 |
| Daly City | 29.2 | 273 | 0.0 | 50.6 | 57.3 | 44.6 | 61.1 | 19.77 | 3,665 | 17 |
| Dana Point | 18.8 | 558 | 0.0 | 55.4 | 68.7 | 43.9 | 77.3 | 13.56 | 1,756 | 666 |
| Danville | 10.4 | 232 | 0.0 | 47.5 | 72.4 | 39.3 | 85.2 | 23.96 | 3,267 | 983 |
| Davis | 154.1 | 2,239 | 32.1 | 45.2 | 74.3 | 37.1 | 92.7 | 19.05 | 2,853 | 1,127 |

1. Based on the population estimated as of July 1 of the year shown.
maximum.
2. Represents normal values based on the 30-year period, 1971±2000.   3. Average daily minimum.   4. Average daily

# Table D. Cities — Land Area and Population

| STATE Place code | City | Land area[1] (sq. mi) | Population, 2020 Total persons 2019 | Rank | Per square mile | Race 2019 — Race alone[2] (percent) White | Black or African American | American Indian, Alaskan Native | Asian | Hawaiian Pacific Islander | Some other race | Two or more races (percent) |
|---|---|---|---|---|---|---|---|---|---|---|---|---|
| | | 1 | 2 | 3 | 4 | 5 | 6 | 7 | 8 | 9 | 10 | 11 |
| | **CALIFORNIA— Cont'd** | | | | | | | | | | | |
| 06 18394 | Delano | 14.7 | 53,041 | 738 | 3,608 | 78.0 | 2.7 | 0.5 | 7.7 | 0.0 | 9.1 | 2.0 |
| 06 18996 | Desert Hot Springs | 30.3 | 29,611 | 1,265 | 977 | 66.2 | 4.8 | 2.2 | 2.7 | 0.0 | 18.7 | 5.4 |
| 06 19192 | Diamond Bar | 14.9 | 54,880 | 698 | 3,683 | 27.9 | 3.2 | 0.1 | 59.3 | 0.1 | 6.8 | 2.6 |
| 06 19766 | Downey | 12.4 | 109,202 | 280 | 8,807 | 60.4 | 2.7 | 0.1 | 6.4 | 0.0 | 26.8 | 3.4 |
| 06 20018 | Dublin | 15.2 | 66,936 | 554 | 4,404 | 37.4 | 3.9 | 0.3 | 50.4 | 0.4 | 2.0 | 5.5 |
| 06 20956 | East Palo Alto | 2.5 | 29,059 | 1,285 | 11,624 | 40.8 | 12.6 | 1.7 | 3.2 | 0.3 | 39.6 | 2.1 |
| 06 21712 | El Cajon | 14.5 | 102,401 | 303 | 7,062 | 78.4 | 5.0 | 0.1 | 3.9 | 0.3 | 7.8 | 4.5 |
| 06 21782 | El Centro | 11.8 | 43,601 | 886 | 3,695 | 44.0 | 1.9 | 1.5 | 1.0 | 0.0 | 48.4 | 3.3 |
| 06 22020 | Elk Grove | 42.0 | 177,302 | 147 | 4,222 | 42.7 | 9.3 | 0.7 | 34.1 | 2.0 | 4.3 | 7.0 |
| 06 22230 | El Monte | 9.6 | 112,344 | 266 | 11,703 | 27.9 | 0.6 | 1.4 | 32.4 | 1.6 | 32.1 | 3.9 |
| 06 22300 | El Paso de Robles (Paso Robles) | 19.7 | 31,521 | 1,209 | 1,600 | 80.6 | 0.7 | 0.8 | 3.4 | 0.0 | 8.2 | 6.1 |
| 06 22678 | Encinitas | 19.1 | 62,671 | 599 | 3,281 | 86.5 | 1.4 | 0.1 | 5.1 | 0.0 | 0.8 | 5.3 |
| 06 22804 | Escondido | 37.3 | 149,943 | 173 | 4,020 | 77.9 | 1.6 | 0.3 | 5.8 | 0.6 | 8.5 | 5.3 |
| 06 23042 | Eureka | 9.5 | 26,652 | 1,367 | 2,806 | 73.8 | 2.5 | 1.9 | 4.6 | 0.3 | 6.3 | 10.7 |
| 06 23182 | Fairfield | 41.6 | 117,580 | 245 | 2,826 | 51.9 | 14.5 | 0.3 | 16.0 | 2.0 | 6.7 | 8.6 |
| 06 24638 | Folsom | 27.9 | 82,726 | 420 | 2,965 | 68.2 | 5.5 | 0.0 | 20.5 | 0.5 | 2.3 | 3.1 |
| 06 24680 | Fontana | 43.1 | 216,173 | 103 | 5,016 | 33.5 | 8.2 | 1.2 | 6.1 | 0.1 | 44.3 | 6.6 |
| 06 25338 | Foster City | 3.8 | 32,891 | 1,165 | 8,656 | 42.4 | 0.9 | 0.0 | 50.1 | 0.0 | 0.5 | 6.1 |
| 06 25380 | Fountain Valley | 9.1 | 55,405 | 693 | 6,089 | 49.0 | 0.6 | 0.4 | 40.8 | 0.2 | 2.2 | 6.7 |
| 06 26000 | Fremont | 78.3 | 234,569 | 95 | 2,996 | 23.9 | 3.1 | 0.6 | 61.4 | 0.5 | 6.0 | 4.4 |
| 06 27000 | Fresno | 115.2 | 530,267 | 34 | 4,603 | 60.4 | 7.1 | 1.0 | 13.8 | 0.1 | 13.2 | 4.9 |
| 06 28000 | Fullerton | 22.4 | 139,921 | 195 | 6,247 | 64.0 | 1.4 | 0.4 | 23.1 | 0.0 | 6.1 | 6.1 |
| 06 28168 | Gardena | 5.8 | 58,498 | 652 | 10,086 | 23.1 | 26.1 | 1.4 | 22.5 | 0.1 | 20.7 | 6.1 |
| 06 29000 | Garden Grove | 18.0 | 171,366 | 154 | 9,520 | 42.6 | 0.5 | 0.4 | 43.9 | 0.2 | 9.7 | 2.7 |
| 06 29504 | Gilroy | 16.5 | 54,831 | 702 | 3,323 | 73.0 | 2.1 | 0.5 | 10.3 | 0.0 | 7.9 | 6.1 |
| 06 30000 | Glendale | 30.5 | 197,747 | 124 | 6,484 | 75.5 | 1.6 | 0.1 | 16.2 | 0.2 | 3.5 | 3.0 |
| 06 30014 | Glendora | 19.5 | 50,517 | 771 | 2,591 | 70.8 | 3.2 | 0.2 | 12.8 | 0.4 | 7.0 | 5.6 |
| 06 30378 | Goleta | 7.9 | 32,629 | 1,172 | 4,130 | 60.4 | 5.7 | 0.3 | 14.4 | 0.0 | 11.3 | 7.8 |
| 06 31960 | Hanford | 17.4 | 58,352 | 657 | 3,354 | 81.2 | 4.6 | 2.2 | 1.6 | 0.0 | 5.3 | 5.1 |
| 06 32548 | Hawthorne | 6.1 | 83,775 | 410 | 13,734 | 59.5 | 21.5 | 0.3 | 5.4 | 0.2 | 9.9 | 3.2 |
| 06 33000 | Hayward | 45.8 | 157,966 | 160 | 3,449 | 29.4 | 9.5 | 0.8 | 29.5 | 2.6 | 20.5 | 7.6 |
| 06 33182 | Hemet | 29.3 | 85,161 | 399 | 2,907 | 76.7 | 7.5 | 0.7 | 2.1 | 0.2 | 8.3 | 3.4 |
| 06 33434 | Hesperia | 72.7 | 96,911 | 329 | 1,333 | 79.1 | 3.0 | 0.9 | 2.7 | 0.0 | 14.2 | 5.8 |
| 06 33588 | Highland | 18.6 | 55,696 | 690 | 2,994 | 61.6 | 11.4 | 0.8 | 5.3 | 0.0 | 9.1 | 6.7 |
| 06 34120 | Hollister | 8.0 | 41,805 | 919 | 5,226 | 77.8 | 1.3 | 1.1 | 4.5 | 0.0 | 11.5 | 6.3 |
| 06 36000 | Huntington Beach | 27.0 | 198,246 | 122 | 7,342 | 67.1 | 1.2 | 0.5 | 12.7 | 0.9 | 27.9 | 6.1 |
| 06 36056 | Huntington Park | 3.0 | 56,714 | 675 | 18,905 | 68.6 | 0.0 | 1.3 | 0.5 | 0.7 | 3.2 | 1.0 |
| 06 36294 | Imperial Beach | 4.3 | 27,325 | 1,339 | 6,355 | 78.7 | 4.7 | 1.0 | 10.2 | 0.0 | 15.6 | 2.2 |
| 06 36448 | Indio | 33.2 | 91,919 | 365 | 2,769 | 72.1 | 6.3 | 1.0 | 2.0 | 0.1 | 19.5 | 3.1 |
| 06 36546 | Inglewood | 9.1 | 107,203 | 286 | 11,781 | 33.6 | 40.3 | 0.3 | 2.4 | 0.2 | 1.2 | 5.4 |
| 06 36770 | Irvine | 65.6 | 283,700 | 73 | 4,325 | 47.8 | 2.4 | 0.1 | 43.0 | 0.0 | 13.9 | 3.9 |
| 06 39220 | Laguna Hills | 6.5 | 31,197 | 1,212 | 4,800 | 64.4 | 1.0 | 0.0 | 16.8 | 0.3 | 6.3 | 6.2 |
| 06 39248 | Laguna Niguel | 14.7 | 64,569 | 586 | 4,392 | 75.4 | 2.2 | 0.2 | 9.5 | 0.0 | 25.8 | 2.6 |
| 06 39290 | La Habra | 7.6 | 62,285 | 601 | 8,195 | 61.6 | 2.0 | 0.1 | 7.9 | 0.6 | 21.5 | 7.9 |
| 06 39486 | Lake Elsinore | 38.2 | 65,867 | 572 | 1,724 | 53.6 | 7.6 | 0.3 | 8.5 | 0.2 | 6.8 | 4.3 |
| 06 39496 | Lake Forest | 16.7 | 85,329 | 395 | 5,110 | 66.8 | 0.8 | 0.3 | 20.7 | 0.2 | 10.7 | 5.0 |
| 06 39892 | Lakewood | 9.4 | 78,119 | 453 | 8,311 | 47.3 | 15.0 | 0.5 | 21.3 | 0.8 | 8.6 | 7.7 |
| 06 40004 | La Mesa | 9.1 | 59,724 | 631 | 6,563 | 67.0 | 6.9 | 1.2 | 7.8 | 0.1 | 16.5 | 5.7 |
| 06 40032 | La Mirada | 7.8 | 47,384 | 822 | 6,075 | 49.0 | 0.8 | 0.6 | 27.3 | 0.0 | 6.2 | 3.8 |
| 06 40130 | Lancaster | 94.3 | 155,822 | 163 | 1,652 | 60.2 | 25.4 | 0.3 | 4.1 | 0.1 | 38.3 | 2.0 |
| 06 40340 | La Puente | 3.5 | 39,215 | 990 | 11,204 | 39.6 | 0.3 | 2.5 | 17.2 | 0.0 | 10.7 | 1.2 |
| 06 40354 | La Quinta | 35.3 | 42,271 | 912 | 1,198 | 83.3 | 2.5 | 0.0 | 2.3 | 0.0 | 11.5 | 2.9 |
| 06 40830 | La Verne | 8.4 | 32,160 | 1,190 | 3,829 | 75.4 | 3.2 | 1.2 | 5.8 | 0.7 | 17.1 | 3.3 |
| 06 40886 | Lawndale | 2.0 | 31,935 | 1,196 | 15,968 | 63.8 | 9.1 | 0.0 | 6.0 | 0.0 | 15.2 | 5.4 |
| 06 41124 | Lemon Grove | 3.9 | 26,528 | 1,373 | 6,802 | 55.7 | 12.4 | 0.4 | 11.0 | 0.1 | 2.2 | 4.0 |
| 06 41474 | Lincoln | 24.2 | 48,996 | 793 | 2,025 | 82.7 | 1.1 | 0.5 | 9.3 | 0.4 | 5.0 | 8.4 |
| 06 41992 | Livermore | 26.4 | 91,617 | 366 | 3,470 | 71.5 | 2.0 | 0.5 | 12.3 | 0.0 | 2.4 | 18.9 |
| 06 42202 | Lodi | 13.6 | 68,186 | 540 | 5,014 | 64.4 | 1.5 | 0.1 | 2.7 | 0.7 | 26.3 | 6.7 |
| 06 42524 | Lompoc | 11.6 | 42,253 | 913 | 3,643 | 58.8 | 3.6 | 1.2 | 12.0 | 0.6 | 18.0 | 4.8 |
| 06 43000 | Long Beach | 50.7 | 454,681 | 43 | 8,968 | 52.3 | D | D | D | D | D | D |
| 06 43280 | Los Altos | 6.5 | 30,281 | 1,240 | 4,659 | D | D | D | 11.7 | 0.1 | 22.6 | 4.0 |
| 06 44000 | Los Angeles | 469.5 | 3,970,219 | 2 | 8,456 | 52.1 | 8.7 | 2.5 | 1.7 | 0.0 | 15.6 | 1.3 |
| 06 44028 | Los Banos | 10.0 | 40,571 | 950 | 4,057 | 70.0 | 3.0 | 0.0 | 14.3 | 0.0 | 1.3 | 5.0 |
| 06 44112 | Los Gatos | 11.6 | 30,266 | 1,242 | 2,609 | 76.4 | 1.4 | 0.1 | 11.7 | 0.0 | 18.8 | 1.1 |
| 06 44574 | Lynwood | 4.8 | 68,508 | 536 | 14,273 | 67.5 | 12.2 | 0.1 | 0.3 | 0.0 | 37.1 | 1.2 |
| 06 45022 | Madera | 16.5 | 66,723 | 561 | 4,044 | 54.9 | 3.0 | 0.3 | 2.1 | 1.4 | D | D |
| 06 45400 | Manhattan Beach | 3.9 | 34,298 | 1,125 | 8,794 | D | D | D | 11.0 | 1.4 | 3.7 | 8.6 |
| 06 45484 | Manteca | 21.4 | 84,445 | 408 | 3,946 | 68.5 | 5.9 | 1.0 | 9.6 | 0.0 | 3.5 | 11.5 |
| 06 46114 | Martinez | 12.6 | 38,027 | 1,021 | 3,018 | 71.7 | 3.6 | 0.0 | D | D | D | D |
| 06 46492 | Maywood | 1.2 | 26,595 | 1,368 | 22,163 | D | D | D | 6.4 | 1.4 | 19.1 | 6.0 |
| 06 46842 | Menifee | 46.5 | 97,675 | 326 | 2,101 | 59.2 | 7.6 | 0.4 | 6.4 | 1.4 | 19.1 | 6.0 |

1. Dry land or land partially or temporarily covered by water.   2. Hispanic or Latino persons may be of any race.

# Table D. Cities — **Population**

| City | Percent Hispanic or Latino[1], 2019 | Percent foreign born, 2019 | Age of population (percent), 2019 | | | | | | | Median age, 2019 | Percent female, 2019 | Population | | | |
|---|---|---|---|---|---|---|---|---|---|---|---|---|---|---|---|
| | | | | | | | | | | | | Census counts | | Percent change | |
| | | | Under 18 years | 18 to 24 years | 25 to 34 years | 35 to 44 years | 45 to 54 years | 55 to 64 years | 65 years and over | | | 2000 | 2010 | 2000-2010 | 2001-2020 |
| | 12 | 13 | 14 | 15 | 16 | 17 | 18 | 19 | 20 | 21 | 22 | 23 | 24 | 25 | 26 |
| **CALIFORNIA— Cont'd** | | | | | | | | | | | | | | | |
| Delano | 79.5 | 34.1 | 24.6 | 12.1 | 17.7 | 14.8 | 12.5 | 10.6 | 7.8 | 31.4 | 46.5 | 38,824 | 53,048 | 36.6 | 0.0 |
| Desert Hot Springs | 58.9 | 26.6 | 20.1 | 9.8 | 16.1 | 18.7 | 8.0 | 13.1 | 14.4 | 38.0 | 50.8 | 16,582 | 27,063 | 63.2 | 9.4 |
| Diamond Bar | 20.4 | 46.4 | 18.7 | 6.0 | 11.2 | 11.2 | 13.4 | 17.7 | 21.7 | 47.3 | 53.6 | 56,287 | 55,566 | -1.3 | -1.2 |
| Downey | 78.7 | 34.6 | 22.8 | 10.3 | 14.7 | 12.2 | 16.9 | 10.8 | 12.4 | 36.6 | 50.2 | 107,323 | 111,775 | 4.1 | -2.3 |
| Dublin | 8.7 | 40.9 | 25.7 | 4.2 | 16.7 | 19.7 | 12.4 | 10.9 | 10.4 | 36.9 | 50.0 | 29,973 | 46,036 | 53.6 | 45.4 |
| East Palo Alto | 76.0 | 39.0 | 28.7 | 8.0 | 20.4 | 14.6 | 8.8 | 11.2 | 8.2 | 31.6 | 47.2 | 29,506 | 28,185 | -4.5 | 3.1 |
| El Cajon | 27.5 | 31.0 | 23.8 | 8.8 | 17.5 | 14.3 | 10.3 | 12.8 | 12.5 | 34.9 | 49.5 | 94,869 | 99,589 | 5.0 | 2.8 |
| El Centro | 88.0 | 31.5 | 26.6 | 10.0 | 10.8 | 11.3 | 11.6 | 13.0 | 16.7 | 37.4 | 49.8 | 37,835 | 42,686 | 12.8 | 2.1 |
| Elk Grove | 15.8 | 27.8 | 24.0 | 7.7 | 13.1 | 13.0 | 16.3 | 13.8 | 12.1 | 38.6 | 51.1 | 59,984 | 152,988 | 155.0 | 15.9 |
| El Monte | 60.0 | 51.7 | 20.4 | 10.6 | 14.6 | 15.1 | 13.8 | 9.7 | 15.8 | 38.2 | 52.0 | 115,965 | 113,580 | -2.1 | -1.1 |
| El Paso de Robles (Paso Robles) | 38.1 | 20.7 | 23.5 | 12.0 | 10.5 | 12.4 | 13.1 | 13.1 | 15.3 | 38.1 | 50.1 | 24,297 | 29,744 | 22.4 | 6.0 |
| Encinitas | 14.8 | 9.1 | 22.2 | 6.0 | 10.5 | 13.4 | 13.2 | 13.1 | 21.6 | 42.7 | 52.3 | 58,014 | 59,500 | 2.6 | 5.3 |
| Escondido | 52.1 | 26.9 | 23.3 | 9.2 | 13.8 | 12.7 | 12.2 | 13.4 | 15.5 | 36.9 | 49.8 | 133,559 | 143,990 | 7.8 | 4.1 |
| Eureka | 16.4 | 6.2 | 17.3 | 11.6 | 18.8 | 15.3 | 11.2 | 9.9 | 15.8 | 36.0 | 48.3 | 26,128 | 27,196 | 4.1 | -2.0 |
| Fairfield | 27.6 | 23.8 | 24.1 | 8.8 | 16.7 | 14.2 | 12.4 | 10.8 | 13.0 | 35.3 | 51.4 | 96,178 | 105,453 | 9.6 | 11.5 |
| Folsom | 9.3 | 19.7 | 22.8 | 6.9 | 13.2 | 15.4 | 17.2 | 11.8 | 12.6 | 40.1 | 48.9 | 51,884 | 72,147 | 39.1 | 14.7 |
| Fontana | 70.6 | 26.9 | 28.5 | 11.1 | 13.8 | 15.7 | 12.5 | 8.6 | 9.8 | 32.2 | 49.2 | 128,929 | 196,457 | 52.4 | 10.0 |
| Foster City | 5.2 | 43.4 | 20.8 | 4.6 | 15.6 | 12.8 | 13.7 | 12.3 | 20.3 | 42.3 | 50.2 | 28,803 | 30,563 | 6.1 | 7.6 |
| Fountain Valley | 13.5 | 32.4 | 20.1 | 8.2 | 11.7 | 13.5 | 11.9 | 16.3 | 18.3 | 41.5 | 51.7 | 54,978 | 55,360 | 0.7 | 0.1 |
| Fremont | 12.4 | 48.8 | 22.6 | 5.8 | 15.4 | 16.1 | 14.8 | 12.7 | 12.5 | 38.9 | 50.6 | 203,413 | 214,065 | 5.2 | 9.6 |
| Fresno | 49.9 | 19.9 | 27.9 | 10.5 | 16.6 | 13.7 | 9.9 | 9.7 | 11.8 | 31.8 | 50.6 | 427,652 | 497,059 | 16.2 | 6.7 |
| Fullerton | 37.9 | 29.5 | 22.9 | 11.3 | 15.2 | 13.3 | 12.5 | 10.8 | 13.9 | 35.4 | 50.0 | 126,003 | 135,183 | 7.3 | 3.5 |
| Gardena | 40.7 | 37.9 | 17.6 | 7.5 | 15.5 | 13.2 | 13.3 | 15.8 | 17.1 | 40.9 | 51.6 | 57,746 | 58,829 | 1.9 | -0.6 |
| Garden Grove | 36.0 | 44.9 | 20.4 | 8.5 | 15.4 | 11.1 | 14.4 | 13.1 | 17.1 | 40.1 | 51.1 | 165,196 | 170,960 | 3.5 | 0.2 |
| Gilroy | 65.4 | 29.1 | 31.9 | 9.2 | 11.5 | 15.6 | 11.7 | 11.2 | 8.8 | 33.2 | 50.6 | 41,464 | 48,969 | 18.1 | 12.0 |
| Glendale | 20.3 | 51.8 | 19.0 | 6.9 | 14.9 | 14.5 | 13.3 | 13.2 | 18.2 | 40.8 | 54.0 | 194,973 | 191,682 | -1.7 | 3.2 |
| Glendora | 35.7 | 22.1 | 22.2 | 7.8 | 14.0 | 11.7 | 13.5 | 13.9 | 16.8 | 40.5 | 50.3 | 49,415 | 50,247 | 1.7 | 0.5 |
| Goleta | 23.6 | 25.1 | 23.3 | 15.0 | 15.4 | 12.5 | 8.9 | 11.7 | 13.2 | 32.5 | 48.5 | 55,204 | 30,096 | -45.5 | 8.4 |
| Hanford | 52.8 | 21.8 | 28.0 | 9.1 | 14.3 | 13.7 | 12.8 | 9.4 | 12.7 | 33.7 | 48.3 | 41,686 | 54,457 | 30.6 | 7.2 |
| Hawthorne | 59.1 | 39.8 | 27.2 | 8.9 | 18.3 | 11.7 | 15.0 | 10.3 | 8.6 | 32.2 | 51.4 | 84,112 | 84,171 | 0.1 | -0.5 |
| Hayward | 39.2 | 38.3 | 19.7 | 10.3 | 17.4 | 14.8 | 11.6 | 13.6 | 12.6 | 36.5 | 48.9 | 140,030 | 144,319 | 3.1 | 9.5 |
| Hemet | 49.7 | 17.4 | 24.4 | 8.4 | 10.9 | 12.9 | 7.7 | 12.3 | 23.4 | 40.0 | 54.1 | 58,812 | 78,695 | 33.8 | 8.2 |
| Hesperia | 57.0 | 18.5 | 29.4 | 9.4 | 12.7 | 13.9 | 11.5 | 12.8 | 10.3 | 34.2 | 51.6 | 62,582 | 90,096 | 44.0 | 7.6 |
| Highland | 58.8 | 19.1 | 34.9 | 11.0 | 15.9 | 9.2 | 10.8 | 9.7 | 8.4 | 27.0 | 52.1 | 44,605 | 53,110 | 19.1 | 4.9 |
| Hollister | 69.4 | 20.8 | 29.1 | 8.1 | 14.7 | 16.6 | 11.4 | 9.1 | 11.0 | 33.7 | 49.1 | 34,413 | 35,221 | 2.3 | 18.7 |
| Huntington Beach | 21.9 | 16.7 | 18.1 | 7.9 | 15.5 | 11.2 | 13.3 | 15.0 | 19.0 | 42.5 | 50.3 | 189,594 | 190,987 | 0.7 | 3.8 |
| Huntington Park | 97.9 | 49.5 | 27.2 | 14.4 | 13.5 | 12.2 | 15.0 | 8.2 | 9.5 | 30.6 | 48.3 | 61,348 | 58,127 | -5.3 | -2.4 |
| Imperial Beach | 58.2 | 20.9 | 23.2 | 15.4 | 12.0 | 14.8 | 11.0 | 9.9 | 13.6 | 34.2 | 47.1 | 26,992 | 26,329 | -2.5 | 3.8 |
| Indio | 57.2 | 16.7 | 20.6 | 10.4 | 12.5 | 10.5 | 11.2 | 12.1 | 22.8 | 42.5 | 49.8 | 49,116 | 79,225 | 61.3 | 16.0 |
| Inglewood | 50.7 | 29.7 | 22.4 | 9.9 | 15.7 | 14.5 | 12.9 | 12.7 | 12.1 | 36.8 | 49.4 | 112,580 | 109,605 | -2.6 | -2.2 |
| Irvine | 10.4 | 40.5 | 21.5 | 13.6 | 18.2 | 14.0 | 12.9 | 9.7 | 10.2 | 33.0 | 49.8 | 143,072 | 212,125 | 48.3 | 33.7 |
| Laguna Hills | 21.4 | 32.4 | 17.6 | 9.3 | 12.2 | 9.2 | 19.4 | 14.0 | 18.3 | 46.2 | 50.8 | 31,178 | 30,679 | -1.6 | 1.7 |
| Laguna Niguel | 17.2 | 20.7 | 20.3 | 7.4 | 12.4 | 11.1 | 13.4 | 16.6 | 18.9 | 44.1 | 52.5 | 61,891 | 62,993 | 1.8 | 2.5 |
| La Habra | 59.0 | 28.2 | 22.1 | 10.9 | 15.1 | 11.0 | 14.5 | 12.0 | 14.4 | 36.5 | 51.8 | 58,974 | 61,447 | 4.2 | 1.4 |
| Lake Elsinore | 47.8 | 23.0 | 31.5 | 8.8 | 15.2 | 12.1 | 14.4 | 9.7 | 8.2 | 31.3 | 50.8 | 28,928 | 53,301 | 84.3 | 23.6 |
| Lake Forest | 21.1 | 24.4 | 23.9 | 6.3 | 12.5 | 14.6 | 15.7 | 13.8 | 13.1 | 39.9 | 50.9 | 58,707 | 77,445 | 31.9 | 10.2 |
| Lakewood | 28.7 | 20.8 | 21.5 | 11.3 | 12.6 | 14.0 | 15.0 | 12.8 | 12.7 | 38.1 | 52.5 | 79,345 | 80,086 | 0.9 | -2.5 |
| La Mesa | 27.4 | 18.0 | 21.1 | 7.8 | 19.3 | 16.3 | 12.7 | 9.0 | 13.9 | 36.3 | 52.0 | 54,749 | 57,014 | 4.1 | 4.8 |
| La Mirada | 41.0 | 28.7 | 18.6 | 14.0 | 10.6 | 13.8 | 10.3 | 11.5 | 21.2 | 40.2 | 50.6 | 46,783 | 48,527 | 3.7 | -2.4 |
| Lancaster | 42.2 | 10.2 | 28.8 | 8.4 | 14.8 | 14.0 | 11.4 | 11.6 | 11.0 | 33.5 | 51.4 | 118,718 | 156,653 | 32.0 | -0.5 |
| La Puente | 78.2 | 42.8 | 22.7 | 10.3 | 15.1 | 8.1 | 17.0 | 14.3 | 12.5 | 39.0 | 47.7 | 41,063 | 39,832 | -3.0 | -1.5 |
| La Quinta | 31.6 | 13.8 | 13.3 | 7.5 | 13.4 | 7.8 | 9.2 | 15.9 | 33.0 | 53.7 | 46.7 | 23,694 | 37,462 | 58.1 | 12.8 |
| La Verne | 42.2 | 12.3 | 19.0 | 6.4 | 8.6 | 10.4 | 9.9 | 18.9 | 26.8 | 50.5 | 53.4 | 31,638 | 30,917 | -2.3 | 4.0 |
| Lawndale | 70.5 | 34.2 | 23.0 | 10.0 | 18.3 | 11.2 | 15.9 | 10.9 | 10.6 | 34.4 | 47.8 | 31,711 | 32,769 | 3.3 | -2.5 |
| Lemon Grove | 39.7 | 20.5 | 21.9 | 4.4 | 15.9 | 11.9 | 13.9 | 14.9 | 17.1 | 41.8 | 49.5 | 24,918 | 25,318 | 1.6 | 4.8 |
| Lincoln | 21.3 | 17.7 | 22.9 | 6.0 | 9.4 | 12.4 | 9.4 | 12.7 | 27.2 | 44.4 | 52.3 | 11,205 | 42,922 | 283.1 | 14.2 |
| Livermore | 22.8 | 18.0 | 24.0 | 6.3 | 12.4 | 14.8 | 14.9 | 11.4 | 16.3 | 39.6 | 51.1 | 73,345 | 81,497 | 11.1 | 12.4 |
| Lodi | 31.9 | 19.2 | 29.5 | 6.5 | 14.0 | 13.8 | 13.2 | 9.1 | 14.0 | 35.1 | 47.1 | 56,999 | 62,121 | 9.0 | 9.8 |
| Lompoc | 62.7 | 26.0 | 22.7 | 14.0 | 14.1 | 10.8 | 13.7 | 11.7 | 13.1 | 34.2 | 43.7 | 41,103 | 42,436 | 3.2 | -0.4 |
| Long Beach | 44.8 | 24.5 | 21.5 | 9.8 | 17.2 | 14.4 | 12.8 | 11.8 | 12.5 | 36.0 | 50.1 | 461,522 | 462,189 | 0.1 | -1.6 |
| Los Altos | 4.1 | 34.4 | 27.5 | 3.0 | 5.1 | 13.9 | 19.0 | 11.5 | 20.2 | 45.8 | 52.1 | 27,693 | 29,075 | 5.0 | 4.1 |
| Los Angeles | 48.2 | 36.4 | 20.2 | 10.0 | 18.3 | 14.4 | 13.1 | 10.9 | 13.1 | 35.9 | 50.5 | 3,694,820 | 3,793,206 | 2.7 | 4.7 |
| Los Banos | 71.9 | 30.8 | 33.9 | 9.3 | 13.4 | 10.5 | 13.7 | 7.5 | 11.7 | 29.7 | 51.4 | 25,869 | 35,967 | 39.0 | 12.8 |
| Los Gatos | 6.3 | 23.8 | 24.4 | 4.1 | 7.2 | 13.8 | 15.4 | 15.4 | 19.6 | 45.2 | 49.5 | 28,592 | 29,717 | 3.9 | 1.8 |
| Lynwood | 85.1 | 34.6 | 27.6 | 11.9 | 17.3 | 13.7 | 11.7 | 9.1 | 8.8 | 30.6 | 51.9 | 69,845 | 69,758 | -0.1 | -1.8 |
| Madera | 81.3 | 30.4 | 35.1 | 11.3 | 14.5 | 13.2 | 10.4 | 8.2 | 7.2 | 27.8 | 52.3 | 43,207 | 61,909 | 43.3 | 7.8 |
| Manhattan Beach | 6.6 | 15.6 | 27.6 | 2.1 | 7.3 | 12.5 | 16.0 | 15.2 | 19.4 | 45.2 | 53.8 | 33,852 | 35,130 | 3.8 | -2.4 |
| Manteca | 37.3 | 18.3 | 25.6 | 7.9 | 12.3 | 14.2 | 13.2 | 12.0 | 14.7 | 38.7 | 51.7 | 49,258 | 67,338 | 36.7 | 25.4 |
| Martinez | 15.3 | 15.3 | 21.3 | 4.7 | 12.6 | 15.7 | 12.4 | 16.8 | 16.4 | 42.6 | 51.1 | 35,866 | 36,037 | 0.5 | 5.5 |
| Maywood | 99.1 | 51.5 | 29.0 | 12.0 | 13.3 | 10.4 | 13.9 | 9.6 | 11.7 | 31.3 | 49.4 | 28,083 | 27,380 | -2.5 | -2.9 |
| Menifee | 42.0 | 14.1 | 27.2 | 7.2 | 12.5 | 14.1 | 9.8 | 11.1 | 18.1 | 37.7 | 50.5 | NA | 77,384 | ********** | 26.2 |

1. May be of any race.

Items 12—26

# Table D. Cities — Households, Group Quarters, Crime, and Education

| City | Households, 2019 Number | Persons per household | Percent Family | Married couple family | Female family | Non-family | One person | Persons in group quarters, 2019 | Violent Number | Violent Rate[2] | Property Number | Property Rate[2] | Population age 25 and over | High school graduate or less | Bachelor's degree or more |
|---|---|---|---|---|---|---|---|---|---|---|---|---|---|---|---|
| | 27 | 28 | 29 | 30 | 31 | 32 | 33 | 34 | 35 | 36 | 37 | 38 | 39 | 40 | 41 |
| **CALIFORNIA— Cont'd** | | | | | | | | | | | | | | | |
| Delano | 13,021 | 3.72 | 87.1 | 63.2 | 23.9 | 12.9 | 11.2 | 5,076 | 205 | 387 | 1,118 | 2,109 | 33,872 | 66.5 | 11.1 |
| Desert Hot Springs | 11,146 | 2.58 | 49.8 | 29.3 | 20.5 | 50.2 | 42.0 | 155 | 255 | 876 | 680 | 2,336 | 20,261 | 52.4 | 10.2 |
| Diamond Bar | 18,855 | 2.94 | 79.0 | 66.0 | 12.9 | 21.0 | 17.7 | 200 | 72 | 128 | 908 | 1,611 | 41,947 | 22.5 | 52.7 |
| Downey | 32,073 | 3.44 | 80.0 | 52.3 | 27.7 | 20.0 | 15.3 | 766 | 352 | 313 | 2,473 | 2,202 | 74,450 | 45.5 | 23.9 |
| Dublin | 21,642 | 2.92 | 76.8 | 66.4 | 10.3 | 23.2 | 17.2 | 1,538 | 107 | 162 | 1,259 | 1,905 | 45,476 | 13.6 | 68.5 |
| East Palo Alto | 7,601 | 3.84 | 76.2 | 50.0 | 26.2 | 23.8 | 18.4 | 154 | 144 | 485 | 533 | 1,795 | 18,534 | 60.5 | 22.6 |
| El Cajon | 32,139 | 3.12 | 73.3 | 50.3 | 23.0 | 26.7 | 20.5 | 2,405 | 552 | 532 | 2,046 | 1,973 | 69,222 | 39.3 | 23.5 |
| El Centro | 13,063 | 3.34 | 78.4 | 50.2 | 28.2 | 21.6 | 18.5 | 430 | 143 | 323 | 1,184 | 2,673 | 27,949 | 48.3 | 21.6 |
| Elk Grove | 57,048 | 3.05 | 76.1 | 59.5 | 16.6 | 23.9 | 20.1 | 834 | 372 | 212 | 2,599 | 1,481 | 119,270 | 28.6 | 36.7 |
| El Monte | 31,535 | 3.62 | 77.2 | 49.0 | 28.3 | 22.8 | 17.6 | 1,473 | 304 | 262 | 1,787 | 1,543 | 79,675 | 66.2 | 13.5 |
| El Paso de Robles (Paso Robles) | 11,192 | 2.86 | 73.6 | 60.7 | 13.0 | 26.4 | 22.7 | 116 | 53 | 163 | 646 | 1,986 | 20,740 | 29.6 | 28.1 |
| Encinitas | 23,478 | 2.66 | 69.4 | 58.6 | 10.8 | 30.6 | 22.0 | 343 | 86 | 136 | 605 | 955 | 45,018 | 11.7 | 67.9 |
| Escondido | 51,364 | 2.90 | 68.2 | 46.9 | 21.3 | 31.8 | 25.4 | 2,417 | 536 | 350 | 2,373 | 1,549 | 102,358 | 40.5 | 26.9 |
| Eureka | 11,023 | 2.31 | 46.0 | 31.2 | 14.7 | 54.0 | 35.2 | 1,291 | 195 | 723 | 1,350 | 5,005 | 18,993 | 25.3 | 34.2 |
| Fairfield | 37,310 | 3.10 | 74.7 | 48.4 | 26.3 | 25.3 | 18.1 | 1,596 | 442 | 373 | 3,782 | 3,195 | 78,520 | 32.5 | 31.8 |
| Folsom | 27,286 | 2.79 | 71.2 | 62.9 | 8.3 | 28.8 | 20.6 | 5,125 | 82 | 103 | 1,484 | 1,857 | 57,110 | 17.6 | 53.1 |
| Fontana | 56,457 | 3.79 | 84.5 | 60.7 | 23.8 | 15.5 | 12.4 | 584 | 739 | 342 | 3,094 | 1,433 | 129,706 | 52.0 | 19.6 |
| Foster City | 12,708 | 2.65 | 71.7 | 63.1 | 8.7 | 28.3 | 20.4 | 183 | 39 | 113 | 425 | 1,227 | 25,291 | 9.7 | 70.4 |
| Fountain Valley | 17,848 | 3.07 | 76.0 | 60.1 | 15.9 | 24.0 | 16.6 | 509 | 53 | 95 | 1,163 | 2,082 | 39,683 | 30.2 | 41.4 |
| Fremont | 77,848 | 3.08 | 81.1 | 69.1 | 12.0 | 18.9 | 14.4 | 1,638 | 400 | 166 | 4,523 | 1,878 | 172,527 | 24.4 | 58.3 |
| Fresno | 172,815 | 3.02 | 67.2 | 39.6 | 27.5 | 32.8 | 26.2 | 10,015 | NA | NA | NA | NA | 327,864 | 45.6 | 21.8 |
| Fullerton | 44,365 | 3.05 | 74.7 | 56.9 | 17.8 | 25.3 | 17.4 | 3,399 | 373 | 266 | 3,157 | 2,252 | 91,186 | 27.9 | 44.0 |
| Gardena | 21,006 | 2.79 | 65.5 | 44.7 | 20.7 | 34.5 | 28.8 | 753 | 324 | 542 | 1,202 | 2,009 | 44,440 | 41.0 | 28.2 |
| Garden Grove | 47,378 | 3.58 | 78.6 | 54.7 | 23.9 | 21.4 | 17.1 | 2,084 | 505 | 292 | 3,909 | 2,262 | 122,099 | 49.6 | 24.7 |
| Gilroy | 15,324 | 3.83 | 75.3 | 48.9 | 26.3 | 24.7 | 16.4 | 394 | 251 | 418 | 1,479 | 2,461 | 34,805 | 46.5 | 26.9 |
| Glendale | 74,197 | 2.66 | 64.8 | 45.4 | 19.4 | 35.2 | 28.2 | 1,631 | 231 | 114 | 3,305 | 1,631 | 147,676 | 37.3 | 40.5 |
| Glendora | 16,553 | 3.05 | 73.3 | 55.6 | 17.7 | 26.7 | 20.4 | 1,018 | 155 | 297 | 1,501 | 2,875 | 36,063 | 26.8 | 38.3 |
| Goleta | 10,896 | 2.81 | 61.2 | 48.1 | 13.1 | 38.8 | 29.1 | 248 | 40 | 129 | 509 | 1,646 | 19,080 | 21.6 | 47.8 |
| Hanford | 19,310 | 2.96 | 78.5 | 53.2 | 25.3 | 21.5 | 14.7 | 486 | 257 | 449 | 1,242 | 2,170 | 36,316 | 47.1 | 20.8 |
| Hawthorne | 27,489 | 3.11 | 69.1 | 40.2 | 28.8 | 30.9 | 25.1 | 666 | 636 | 728 | 1,485 | 1,701 | 54,970 | 46.4 | 24.4 |
| Hayward | 47,826 | 3.26 | 72.0 | 50.2 | 21.7 | 28.0 | 19.3 | 3,476 | 552 | 342 | 4,886 | 3,024 | 111,419 | 43.7 | 27.9 |
| Hemet | 28,412 | 2.98 | 67.6 | 43.3 | 24.3 | 32.4 | 26.0 | 753 | 343 | 398 | 2,877 | 3,342 | 57,327 | 56.6 | 10.4 |
| Hesperia | 27,185 | 3.52 | 76.6 | 50.3 | 26.3 | 23.4 | 18.5 | 165 | 462 | 482 | 1,664 | 1,735 | 58,563 | 55.7 | 14.8 |
| Highland | 15,433 | 3.58 | 78.6 | 47.7 | 30.9 | 21.4 | 17.5 | 136 | 362 | 650 | 906 | 1,627 | 29,979 | 48.2 | 26.5 |
| Hollister | 11,963 | 3.43 | 83.9 | 45.7 | 38.2 | 16.1 | 10.5 | 111 | 114 | 282 | 311 | 770 | 25,821 | 50.4 | 16.1 |
| Huntington Beach | 75,273 | 2.63 | 65.7 | 50.3 | 15.5 | 34.3 | 25.3 | 1,082 | 386 | 191 | 4,144 | 2,053 | 147,580 | 26.2 | 41.0 |
| Huntington Park | 14,741 | 3.89 | 81.7 | 38.4 | 43.3 | 18.3 | 13.4 | 140 | 458 | 787 | 1,595 | 2,741 | 33,595 | 71.9 | 7.9 |
| Imperial Beach | 9,734 | 2.70 | 65.9 | 38.4 | 27.4 | 34.1 | 24.2 | 1,146 | 72 | 261 | 331 | 1,200 | 16,849 | 46.7 | 20.6 |
| Indio | 31,760 | 2.86 | 76.3 | 54.8 | 21.6 | 23.7 | 22.0 | 866 | 530 | 571 | 1,880 | 2,026 | 63,334 | 40.0 | 25.9 |
| Inglewood | 34,440 | 3.06 | 67.4 | 36.7 | 30.7 | 32.6 | 25.7 | 2,648 | 671 | 613 | 2,321 | 2,122 | 73,221 | 47.0 | 23.8 |
| Irvine | 104,524 | 2.68 | 62.6 | 47.7 | 14.9 | 37.4 | 24.8 | 7,779 | 188 | 64 | 3,823 | 1,306 | 186,687 | 9.8 | 69.5 |
| Laguna Hills | 11,033 | 2.78 | 66.4 | 57.0 | 9.4 | 33.6 | 25.1 | 498 | 32 | 103 | 405 | 1,301 | 22,809 | 22.2 | 52.4 |
| Laguna Niguel | 24,719 | 2.67 | 71.7 | 57.6 | 14.0 | 28.3 | 20.2 | 379 | 43 | 64 | 607 | 910 | 48,025 | 14.6 | 55.1 |
| La Habra | 18,879 | 3.18 | 73.6 | 53.9 | 19.7 | 26.4 | 20.0 | 502 | 140 | 224 | 1,122 | 1,798 | 40,536 | 44.3 | 26.2 |
| Lake Elsinore | 18,642 | 3.72 | 78.1 | 59.4 | 18.8 | 21.9 | 15.8 | 17 | 173 | 246 | 2,005 | 2,854 | 41,316 | 44.2 | 23.0 |
| Lake Forest | 28,460 | 2.99 | 76.0 | 65.5 | 10.5 | 24.0 | 15.9 | 557 | 66 | 76 | 684 | 789 | 59,734 | 17.7 | 50.9 |
| Lakewood | 25,198 | 3.14 | 79.1 | 57.6 | 21.5 | 20.9 | 17.3 | 139 | 222 | 277 | 1,846 | 2,303 | 53,290 | 29.1 | 32.6 |
| La Mesa | 21,898 | 2.67 | 60.6 | 43.8 | 16.8 | 39.4 | 30.2 | 743 | 174 | 291 | 1,000 | 1,670 | 42,128 | 25.9 | 39.3 |
| La Mirada | 14,240 | 3.17 | 82.7 | 63.8 | 19.0 | 17.3 | 13.3 | 2,964 | 86 | 177 | 701 | 1,439 | 32,452 | 34.0 | 34.6 |
| Lancaster | 49,220 | 3.06 | 73.0 | 45.9 | 27.1 | 27.0 | 21.9 | 7,169 | 1,359 | 853 | 3,346 | 2,100 | 98,924 | 47.1 | 18.4 |
| La Puente | 9,737 | 4.06 | 87.4 | 55.2 | 32.2 | 12.6 | 8.1 | 98 | 151 | 378 | 449 | 1,125 | 26,530 | 62.2 | 8.1 |
| La Quinta | 17,942 | 2.32 | 66.4 | 57.2 | 9.2 | 33.6 | 25.0 | 50 | 68 | 162 | 1,032 | 2,455 | 33,090 | 26.5 | 41.7 |
| La Verne | 12,250 | 2.56 | 70.7 | 52.7 | 17.9 | 29.3 | 23.0 | 651 | 51 | 158 | 627 | 1,939 | 23,845 | 29.1 | 32.7 |
| Lawndale | 10,267 | 3.15 | 76.9 | 46.4 | 30.5 | 23.1 | 15.6 | 92 | 125 | 382 | 411 | 1,255 | 21,683 | 53.0 | 20.1 |
| Lemon Grove | 9,384 | 2.84 | 67.9 | 41.7 | 26.2 | 32.1 | 19.0 | 126 | 167 | 615 | 552 | 2,031 | 19,750 | 39.8 | 21.4 |
| Lincoln | 18,375 | 2.62 | 71.9 | 61.4 | 10.5 | 28.1 | 22.6 | 152 | 32 | 66 | 468 | 962 | 34,350 | 28.3 | 34.4 |
| Livermore | 30,545 | 2.94 | 75.1 | 64.4 | 10.7 | 24.9 | 18.7 | 448 | 193 | 211 | 1,554 | 1,700 | 62,929 | 20.9 | 47.0 |
| Lodi | 22,733 | 2.94 | 69.6 | 53.3 | 16.3 | 30.4 | 24.0 | 681 | 237 | 351 | 1,658 | 2,452 | 43,274 | 46.3 | 18.3 |
| Lompoc | 12,119 | 3.26 | 72.3 | 51.9 | 20.4 | 27.7 | 24.2 | 3,294 | 291 | 680 | 1,057 | 2,469 | 27,133 | 55.6 | 11.1 |
| Long Beach | 165,689 | 2.74 | 59.9 | 38.8 | 21.1 | 40.1 | 30.4 | 9,107 | 2,369 | 506 | 11,297 | 2,414 | 317,851 | 38.3 | 32.9 |
| Los Altos | 10,454 | 2.85 | 80.0 | 74.8 | 5.2 | 20.0 | 18.7 | 259 | 21 | 68 | 310 | 1,009 | 20,940 | 3.0 | 85.6 |
| Los Angeles | 1,398,900 | 2.78 | 59.3 | 38.5 | 20.8 | 40.7 | 30.2 | 93,912 | 29,400 | 732 | 95,704 | 2,383 | 2,777,962 | 40.6 | 35.9 |
| Los Banos | 11,124 | 3.68 | 91.1 | 66.8 | 24.2 | 8.9 | 8.9 | 144 | 143 | 352 | 886 | 2,182 | 23,309 | 56.3 | 14.9 |
| Los Gatos | 11,562 | 2.59 | 68.1 | 57.3 | 10.7 | 31.9 | 27.6 | 330 | 14 | 45 | 419 | 1,361 | 21,598 | 6.8 | 73.1 |
| Lynwood | 15,142 | 4.39 | 85.9 | 56.2 | 29.7 | 14.1 | 11.2 | 3,400 | 454 | 643 | 1,419 | 2,009 | 42,281 | 72.5 | 8.2 |
| Madera | 16,625 | 3.95 | 87.9 | 61.6 | 26.2 | 12.1 | 7.8 | 264 | 334 | 504 | 1,301 | 1,964 | 35,274 | 60.8 | 7.5 |
| Manhattan Beach | 14,139 | 2.49 | 68.9 | 58.7 | 10.2 | 31.1 | 27.0 | 34 | 55 | 155 | 851 | 2,392 | 24,744 | 5.5 | 79.0 |
| Manteca | 26,047 | 3.16 | 81.9 | 62.4 | 19.5 | 18.1 | 14.9 | 623 | 199 | 238 | 1,848 | 2,213 | 55,182 | 40.7 | 20.7 |
| Martinez | 15,462 | 2.44 | 58.3 | 49.7 | 8.7 | 41.7 | 30.6 | 568 | 83 | 215 | 570 | 1,473 | 28,333 | 20.4 | 43.6 |
| Maywood | 6,464 | 4.15 | 84.2 | 55.5 | 28.7 | 15.8 | 13.5 | 137 | 90 | 330 | 296 | 1,085 | 15,899 | 79.7 | 7.9 |
| Menifee | 29,356 | 3.22 | 74.6 | 56.2 | 18.4 | 25.4 | 19.7 | 164 | 145 | 153 | 1,892 | 2,000 | 62,182 | 41.7 | 20.9 |

1. Data for serious crimes have not been adjusted for underreporting. This may affect comparability between geographic areas and over time.    2. Per 100,000 population estimated by the FBI.    3. Persons 25 years old and over.

| City | Money income, 2019 — Households — Median income | Percent with income less than $20,000 | Percent with income of $200,000 or more | Median family income | Median non-family income | Median earnings, 2019 — All persons | Men | Women | Housing units, 2019 — Total | Occupied | Percent owner occupied | Median value¹ (dollars) | Median gross rent (dollars) |
|---|---|---|---|---|---|---|---|---|---|---|---|---|---|
| | 42 | 43 | 44 | 45 | 46 | 47 | 48 | 49 | 50 | 51 | 52 | 53 | 54 |
| **CALIFORNIA— Cont'd** | | | | | | | | | | | | | |
| Delano | 45,849 | 18.8 | 1.7 | 47,239 | 20,466 | 20,114 | 22,326 | 17,013 | 13,505 | 13,021 | 60.7 | 214,333 | 963 |
| Desert Hot Springs | 36,776 | 32.1 | 1.3 | 47,453 | 20,224 | 28,209 | 29,087 | 25,534 | 13,196 | 11,146 | 49.6 | 211,322 | 1,083 |
| Diamond Bar | 98,166 | 7.4 | 14.1 | 108,908 | 42,993 | 48,018 | 55,554 | 40,687 | 19,393 | 18,855 | 77.3 | 677,205 | 2,288 |
| Downey | 74,590 | 5.7 | 8.0 | 78,808 | 37,065 | 35,423 | 38,549 | 31,741 | 34,011 | 32,073 | 46.5 | 619,785 | 1,670 |
| Dublin | 151,813 | 3.1 | 38.2 | 178,420 | 98,817 | 91,873 | 110,387 | 65,257 | 22,534 | 21,642 | 65.7 | 996,226 | 2,777 |
| East Palo Alto | 72,998 | 7.0 | 17.4 | 81,132 | 0 | 33,476 | 40,066 | 30,915 | 8,567 | 7,601 | 45.2 | 1,051,193 | 1,878 |
| El Cajon | 63,236 | 12.3 | 6.9 | 70,428 | 42,018 | 29,997 | 35,257 | 25,605 | 33,385 | 32,139 | 41.1 | 484,502 | 1,464 |
| El Centro | 54,080 | 26.4 | 6.4 | 57,514 | 14,414 | 27,800 | 42,087 | 21,533 | 15,039 | 13,063 | 51.8 | 221,711 | 791 |
| Elk Grove | 94,971 | 8.0 | 13.5 | 107,970 | 58,782 | 48,095 | 51,484 | 42,973 | 58,532 | 57,048 | 75.4 | 462,329 | 1,625 |
| El Monte | 50,829 | 15.4 | 2.1 | 55,400 | 30,759 | 27,121 | 31,787 | 23,081 | 32,740 | 31,535 | 37.6 | 558,136 | 1,385 |
| El Paso de Robles (Paso Robles) | 83,825 | 6.0 | 4.7 | 97,641 | 43,205 | 31,661 | 40,341 | 27,006 | 11,764 | 11,192 | 54.7 | 478,326 | 1,672 |
| Encinitas | 119,040 | 5.3 | 28.7 | 152,462 | 76,877 | 60,785 | 71,713 | 55,754 | 25,542 | 23,478 | 67.6 | 1,114,070 | 2,271 |
| Escondido | 58,157 | 17.5 | 7.1 | 71,771 | 27,405 | 32,937 | 38,151 | 27,006 | 53,773 | 51,364 | 49.0 | 502,553 | 1,531 |
| Eureka | 55,393 | 19.9 | 7.9 | 66,848 | 35,408 | 35,458 | 29,662 | 38,614 | 11,961 | 11,023 | 38.7 | 315,998 | 949 |
| Fairfield | 89,693 | 7.1 | 9.2 | 95,783 | 63,037 | 40,850 | 46,147 | 33,975 | 39,867 | 37,310 | 59.3 | 452,365 | 1,808 |
| Folsom | 119,824 | 7.0 | 21.7 | 150,685 | 68,800 | 65,829 | 80,607 | 51,328 | 28,225 | 27,286 | 67.7 | 594,535 | 1,826 |
| Fontana | 80,800 | 8.5 | 6.4 | 82,062 | 46,191 | 36,597 | 41,066 | 31,200 | 57,905 | 56,457 | 66.2 | 417,960 | 1,501 |
| Foster City | 162,500 | 5.6 | 43.9 | 200,232 | 105,446 | 103,044 | 120,735 | 75,582 | 13,161 | 12,708 | 49.1 | 1,514,414 | 3,257 |
| Fountain Valley | 102,370 | 6.4 | 19.8 | 113,117 | 66,281 | 44,328 | 54,764 | 37,360 | 18,595 | 17,848 | 69.9 | 818,167 | 2,198 |
| Fremont | 144,118 | 5.5 | 31.0 | 156,659 | 73,750 | 75,542 | 95,219 | 59,181 | 81,652 | 77,848 | 58.9 | 1,086,682 | 2,569 |
| Fresno | 53,161 | 19.4 | 3.8 | 58,658 | 33,308 | 30,840 | 35,205 | 25,633 | 180,639 | 172,815 | 46.3 | 276,621 | 1,058 |
| Fullerton | 93,983 | 7.6 | 11.9 | 103,273 | 65,662 | 45,639 | 51,715 | 35,632 | 46,039 | 44,365 | 52.1 | 684,525 | 1,795 |
| Gardena | 60,297 | 14.8 | 5.8 | 73,633 | 36,769 | 33,228 | 36,710 | 31,041 | 22,271 | 21,006 | 44.2 | 561,347 | 1,455 |
| Garden Grove | 72,240 | 12.4 | 8.2 | 80,440 | 40,075 | 34,222 | 39,189 | 29,879 | 49,069 | 47,378 | 54.2 | 603,052 | 1,795 |
| Gilroy | 102,286 | 5.3 | 21.5 | 117,677 | 74,672 | 40,983 | 47,149 | 30,934 | 16,579 | 15,324 | 57.0 | 719,538 | 2,131 |
| Glendale | 68,392 | 17.2 | 9.0 | 79,530 | 42,884 | 41,643 | 45,805 | 38,092 | 80,104 | 74,197 | 32.4 | 821,811 | 1,723 |
| Glendora | 85,432 | 9.3 | 15.5 | 99,988 | 38,675 | 40,970 | 51,244 | 32,003 | 17,834 | 16,553 | 63.8 | 676,790 | 1,860 |
| Goleta | 96,695 | 15.5 | 12.9 | 115,173 | 57,122 | 35,873 | 42,310 | 28,856 | 11,228 | 10,896 | 49.2 | 803,996 | 1,962 |
| Hanford | 61,578 | 12.6 | 3.0 | 62,492 | 50,313 | 36,746 | 42,150 | 25,556 | 20,114 | 19,310 | 57.5 | 243,423 | 891 |
| Hawthorne | 60,361 | 14.2 | 3.9 | 61,117 | 51,321 | 35,132 | 37,473 | 27,067 | 28,410 | 27,489 | 28.7 | 639,771 | 1,404 |
| Hayward | 96,886 | 9.0 | 14.5 | 104,648 | 64,818 | 48,362 | 51,611 | 41,933 | 51,060 | 47,826 | 53.9 | 645,934 | 1,958 |
| Hemet | 39,653 | 20.1 | 2.5 | 45,696 | 25,104 | 31,794 | 32,111 | 31,470 | 31,952 | 28,412 | 62.6 | 231,867 | 1,000 |
| Hesperia | 50,271 | 14.2 | 1.3 | 51,743 | 32,073 | 27,266 | 38,475 | 19,271 | 29,136 | 27,185 | 56.9 | 266,985 | 1,312 |
| Highland | 71,450 | 13.1 | 8.8 | 76,312 | 0 | 35,708 | 36,515 | 32,764 | 16,394 | 15,433 | 67.5 | 363,596 | 1,156 |
| Hollister | 75,206 | 12.0 | 6.3 | 77,322 | 43,456 | 32,087 | 40,320 | 25,418 | 12,184 | 11,963 | 59.1 | 555,171 | 1,669 |
| Huntington Beach | 102,423 | 7.4 | 16.4 | 118,062 | 71,008 | 46,628 | 56,712 | 40,478 | 81,414 | 75,273 | 54.8 | 837,615 | 1,949 |
| Huntington Park | 47,672 | 14.1 | 1.3 | 46,030 | 32,117 | 26,516 | 30,274 | 21,155 | 15,109 | 14,741 | 21.5 | 472,891 | 1,185 |
| Imperial Beach | 52,656 | 11.1 | 3.7 | 61,840 | 32,424 | 31,035 | 35,161 | 25,253 | 10,545 | 9,734 | 32.0 | 642,329 | 1,608 |
| Indio | 74,774 | 10.9 | 6.7 | 81,299 | 38,920 | 31,133 | 35,625 | 25,369 | 39,359 | 31,760 | 78.3 | 323,254 | 1,246 |
| Inglewood | 61,780 | 11.6 | 4.4 | 61,868 | 48,034 | 31,924 | 32,441 | 30,753 | 37,193 | 34,440 | 34.0 | 604,975 | 1,472 |
| Irvine | 111,574 | 13.6 | 23.3 | 137,419 | 72,947 | 61,605 | 76,809 | 48,050 | 114,217 | 104,524 | 43.6 | 933,638 | 2,480 |
| Laguna Hills | 104,050 | 6.9 | 19.5 | 108,750 | 101,777 | 50,721 | 60,909 | 40,095 | 11,750 | 11,033 | 74.4 | 754,087 | 2,209 |
| Laguna Niguel | 120,405 | 6.6 | 27.9 | 143,868 | 61,917 | 53,400 | 64,076 | 50,205 | 27,409 | 24,719 | 68.8 | 887,164 | 2,287 |
| La Habra | 82,151 | 6.7 | 4.5 | 82,709 | 64,639 | 31,386 | 37,572 | 26,742 | 19,200 | 18,879 | 56.9 | 539,207 | 1,614 |
| Lake Elsinore | 77,090 | 7.1 | 4.9 | 80,353 | 65,625 | 38,152 | 46,916 | 28,955 | 19,946 | 18,642 | 72.6 | 383,703 | 1,594 |
| Lake Forest | 118,888 | 5.1 | 21.8 | 136,208 | 65,558 | 54,050 | 71,742 | 41,967 | 30,326 | 28,460 | 71.3 | 730,235 | 2,165 |
| Lakewood | 97,508 | 8.1 | 11.9 | 103,716 | 50,890 | 48,109 | 56,679 | 38,407 | 25,789 | 25,198 | 72.3 | 612,136 | 2,123 |
| La Mesa | 66,901 | 13.1 | 9.1 | 77,222 | 47,721 | 36,061 | 40,184 | 32,164 | 23,194 | 21,898 | 43.4 | 582,266 | 1,723 |
| La Mirada | 97,387 | 10.2 | 9.8 | 104,334 | 32,284 | 42,306 | 48,481 | 33,675 | 14,538 | 14,240 | 74.8 | 630,003 | 1,879 |
| Lancaster | 56,672 | 16.3 | 3.0 | 65,235 | 31,319 | 40,962 | 43,502 | 36,351 | 52,755 | 49,249 | 53.3 | 284,124 | 1,317 |
| La Puente | 69,992 | 8.0 | 2.4 | 72,906 | 55,404 | 28,609 | 30,329 | 26,689 | 10,203 | 9,737 | 54.0 | 465,881 | 1,575 |
| La Quinta | 90,125 | 11.3 | 10.8 | 110,805 | 56,199 | 38,129 | 38,440 | 37,945 | 28,005 | 17,942 | 72.6 | 440,649 | 1,450 |
| La Verne | 81,228 | 9.8 | 14.2 | 105,544 | 29,591 | 50,722 | 64,334 | 30,155 | 12,890 | 12,250 | 71.2 | 627,371 | 1,549 |
| Lawndale | 71,952 | 6.1 | 1.1 | 72,025 | 41,990 | 35,343 | 40,157 | 25,524 | 10,666 | 10,267 | 36.3 | 588,401 | 1,609 |
| Lemon Grove | 58,471 | 16.5 | 3.2 | 71,053 | 36,040 | 36,527 | 47,063 | 30,071 | 9,579 | 9,384 | 52.8 | 477,411 | 1,398 |
| Lincoln | 94,240 | 9.2 | 15.3 | 112,401 | 45,763 | 48,803 | 64,290 | 36,390 | 18,979 | 18,375 | 77.7 | 498,465 | 2,024 |
| Livermore | 138,225 | 3.5 | 27.4 | 152,682 | 86,196 | 62,107 | 81,825 | 51,394 | 31,367 | 30,545 | 73.2 | 852,520 | 2,417 |
| Lodi | 67,484 | 12.1 | 8.6 | 83,252 | 33,909 | 35,174 | 37,484 | 31,024 | 24,445 | 22,733 | 55.0 | 361,133 | 1,192 |
| Lompoc | 58,408 | 9.1 | 3.5 | 69,522 | 40,785 | 24,736 | 26,465 | 22,778 | 12,781 | 12,119 | 54.5 | 325,404 | 1,219 |
| Long Beach | 67,804 | 13.5 | 8.8 | 77,901 | 50,726 | 35,843 | 40,888 | 30,382 | 176,189 | 165,689 | 39.2 | 614,401 | 1,460 |
| Los Altos | 250 | 4.7 | 65.5 | 250 | 96,031 | 151,581 | 183,688 | 92,957 | 10,592 | 10,454 | 79.4 | 2 | 3,013 |
| Los Angeles | 67,418 | 15.9 | 11.5 | 76,277 | 50,644 | 34,669 | 37,159 | 30,987 | 1,532,364 | 1,398,900 | 36.5 | 697,166 | 1,554 |
| Los Banos | 76,363 | 4.9 | 4.6 | 81,103 | 0 | 31,705 | 40,181 | 25,204 | 11,797 | 11,124 | 57.8 | 341,955 | 1,411 |
| Los Gatos | 164,115 | 3.5 | 43.0 | 206,923 | 111,561 | 102,369 | 130,781 | 71,891 | 13,130 | 11,562 | 55.9 | 1,890,086 | 2,415 |
| Lynwood | 53,966 | 12.5 | 4.3 | 56,150 | 32,972 | 28,128 | 30,808 | 25,970 | 15,422 | 15,142 | 50.5 | 455,254 | 1,310 |
| Madera | 50,364 | 9.3 | 2.1 | 50,750 | 27,194 | 23,272 | 29,912 | 18,603 | 17,942 | 16,625 | 47.3 | 271,372 | 1,038 |
| Manhattan Beach | 151,508 | 5.8 | 37.7 | 179,038 | 84,848 | 100,976 | 140,046 | 69,201 | 14,948 | 14,139 | 66.0 | 2 | 2,949 |
| Manteca | 82,265 | 8.6 | 8.6 | 85,206 | 57,045 | 43,970 | 53,829 | 33,892 | 27,147 | 26,047 | 74.7 | 434,992 | 1,649 |
| Martinez | 107,543 | 6.0 | 20.8 | 127,551 | 75,118 | 60,443 | 66,649 | 51,016 | 16,059 | 15,462 | 65.3 | 623,955 | 1,565 |
| Maywood | 52,358 | 17.0 | 2.3 | 57,542 | 26,250 | 27,558 | 30,198 | 26,181 | 6,551 | 6,464 | 23.6 | 434,485 | 1,222 |
| Menifee | 77,033 | 11.8 | 5.3 | 86,867 | 34,872 | 39,552 | 50,496 | 31,456 | 32,054 | 29,356 | 78.4 | 370,210 | 1,848 |

1. Based on population estimated by the American Community Survey.

Items 42—54

# Table D. Cities — Commuting, Computer Access, Migration, Labor Force, and Employment

| City | Commuting[1], 2019 | | Computer access[2], 2019 | | Migration, 2019 | | Civilian labor force, 2020 | | | | Civilian Employment[4], 2019 | | | |
|---|---|---|---|---|---|---|---|---|---|---|---|---|---|---|
| | Percent | | Percent | | Percent who lived in the same house one year ago | Percent who lived in another state or county one year ago | Total | Percent change 2019-2020 | Unemployment | | Population age 16 and older | | Population age 16 to 64 | |
| | Drove alone | With commutes of 30 minutes or more | With a computer in the house | With Internet access | | | | | Total | Rate[3] | Number | Percent in labor force | Number | Percent who worked full-year full-time |
| | 55 | 56 | 57 | 58 | 59 | 60 | 61 | 62 | 63 | 64 | 65 | 66 | 67 | 68 |
| CALIFORNIA— Cont'd | | | | | | | | | | | | | | |
| Delano | 69.4 | 13.3 | 83.6 | 75.6 | 92.5 | 5.2 | 21,155 | -4.5 | 5,160 | 24 | 41,530 | 57.3 | 37,374 | 31.7 |
| Desert Hot Springs | 79.4 | 42.3 | 87.8 | 82.4 | 90.5 | 3.5 | 12,199 | 4.7 | 1,913 | 16 | 23,627 | 57.1 | 19,472 | 47.5 |
| Diamond Bar | 78.0 | 60.3 | 97.4 | 92.8 | 91.7 | 4.1 | 28,749 | -5.7 | 3,048 | 11 | 46,408 | 63.2 | 34,301 | 52.2 |
| Downey | 78.1 | 54.2 | 97.0 | 92.3 | 95.3 | 0.7 | 55,160 | -4.1 | 6,948 | 13 | 89,027 | 66.6 | 75,211 | 53.2 |
| Dublin | 67.3 | 58.3 | 98.7 | 96.8 | 83.7 | 8.7 | 31,361 | -3.6 | 2,080 | 7 | 49,571 | 67.5 | 42,846 | 58.9 |
| East Palo Alto | 74.1 | 33.0 | NA | NA | 92.0 | 3.7 | 14,489 | -4.2 | 1,293 | 9 | 22,243 | 71.3 | 19,852 | 53.4 |
| El Cajon | 74.7 | 36.4 | 96.4 | 91.7 | 83.2 | 4.4 | 44,684 | -0.9 | 5,403 | 12 | 80,379 | 65.7 | 67,577 | 47.9 |
| El Centro | 81.1 | 14.8 | 91.6 | 86.9 | 87.3 | 6.3 | 18,211 | -1.1 | 3,440 | 19 | 33,437 | 60.8 | 26,095 | 43.3 |
| Elk Grove | 77.7 | 59.4 | 97.3 | 93.9 | 89.7 | 2.4 | 82,213 | -1.1 | 6,640 | 8 | 139,873 | 64.7 | 118,718 | 51.6 |
| El Monte | 77.6 | 55.2 | 90.9 | 82.2 | 93.0 | 1.0 | 50,780 | -2.1 | 6,966 | 14 | 95,090 | 61.1 | 76,842 | 50.8 |
| El Paso de Robles (Paso Robles) | 84.3 | 24.8 | 98.0 | 93.3 | 85.6 | 6.3 | 15,473 | -5.3 | 1,229 | 8 | 25,072 | 73.4 | 20,139 | 57.6 |
| Encinitas | 73.1 | 36.4 | 95.7 | 93.0 | 88.5 | 3.5 | 31,631 | -5.0 | 2,229 | 7 | 49,920 | 61.7 | 36,379 | 53.6 |
| Escondido | 81.9 | 48.5 | 95.3 | 91.1 | 88.6 | 3.2 | 66,994 | -3.7 | 5,971 | 9 | 119,187 | 62.4 | 95,724 | 51.0 |
| Eureka | 70.4 | 13.2 | 95.3 | 87.5 | 78.4 | 12.3 | 12,044 | -5.0 | 1,035 | 9 | 22,268 | 64.6 | 18,037 | 46.5 |
| Fairfield | 74.9 | 50.8 | 98.1 | 95.1 | 84.6 | 7.0 | 51,908 | -3.3 | 4,811 | 9 | 91,938 | 66.6 | 76,699 | 53.8 |
| Folsom | 76.6 | 31.3 | 98.7 | 94.5 | 80.7 | 10.4 | 37,223 | -1.0 | 2,266 | 6 | 65,249 | 64.2 | 54,971 | 53.2 |
| Fontana | 77.2 | 47.0 | 96.5 | 88.7 | 88.3 | 3.4 | 99,508 | -0.1 | 9,179 | 9 | 160,322 | 65.4 | 139,309 | 53.2 |
| Foster City | 65.1 | 54.5 | NA | NA | 78.8 | 11.3 | 18,660 | -7.7 | 944 | 5 | 27,905 | 62.8 | 21,038 | 59.4 |
| Fountain Valley | 77.4 | 41.4 | 96.1 | 93.5 | 89.6 | 1.0 | 27,181 | -4.3 | 2,539 | 9 | 45,573 | 63.2 | 35,472 | 51.2 |
| Fremont | 67.5 | 59.3 | 97.7 | 94.6 | 91.1 | 4.9 | 115,508 | -4.3 | 7,750 | 7 | 192,852 | 67.3 | 162,623 | 61.7 |
| Fresno | 76.9 | 25.7 | 92.7 | 83.8 | 87.0 | 3.3 | 233,824 | -0.3 | 25,088 | 11 | 399,279 | 61.2 | 336,668 | 44.0 |
| Fullerton | 79.4 | 48.9 | 97.9 | 93.4 | 83.7 | 4.0 | 67,757 | -3.9 | 6,535 | 10 | 110,293 | 66.0 | 90,967 | 50.6 |
| Gardena | 77.6 | 34.1 | 90.9 | 85.7 | 89.7 | 3.8 | 29,709 | -2.3 | 4,252 | 14 | 50,333 | 63.7 | 40,162 | 51.9 |
| Garden Grove | 77.4 | 45.1 | 95.5 | 89.6 | 92.6 | 1.9 | 80,500 | -1.7 | 9,428 | 12 | 140,767 | 62.7 | 111,470 | 51.0 |
| Gilroy | 70.7 | 56.1 | 98.3 | 92.0 | 84.9 | 3.5 | 30,205 | -1.1 | 2,582 | 9 | 42,188 | 67.6 | 36,969 | 52.6 |
| Glendale | 79.5 | 43.7 | 91.0 | 84.9 | 91.3 | 2.1 | 100,345 | -3.4 | 13,263 | 13 | 165,206 | 64.6 | 128,882 | 55.5 |
| Glendora | 75.3 | 54.4 | 96.1 | 84.4 | 84.2 | 0.8 | 24,612 | -6.1 | 2,591 | 11 | 41,526 | 61.7 | 32,866 | 49.8 |
| Goleta | 68.4 | 15.0 | 98.5 | 92.5 | 85.5 | 6.2 | 17,109 | -0.3 | 1,072 | 6 | 24,340 | 69.8 | 20,273 | 52.8 |
| Hanford | 81.0 | 31.3 | 93.1 | 82.3 | 89.1 | 4.7 | 24,656 | -1.7 | 2,640 | 11 | 43,410 | 55.2 | 36,110 | 48.3 |
| Hawthorne | 72.3 | 44.8 | 95.3 | 90.4 | 89.1 | 4.4 | 44,151 | -1.7 | 6,645 | 15 | 64,934 | 68.2 | 57,504 | 51.8 |
| Hayward | 70.6 | 60.3 | 95.1 | 88.5 | 90.8 | 5.0 | 75,907 | -2.7 | 7,905 | 10 | 132,335 | 67.8 | 112,350 | 57.9 |
| Hemet | 68.2 | 59.6 | 90.0 | 81.7 | 82.2 | 3.9 | 30,167 | 1.5 | 3,935 | 13 | 66,268 | 44.0 | 46,298 | 40.3 |
| Hesperia | 79.3 | 55.5 | 96.3 | 94.1 | 95.1 | 2.5 | 36,169 | 0.4 | 4,010 | 11 | 71,214 | 55.1 | 61,396 | 40.0 |
| Highland | 86.5 | 39.6 | 95.1 | 81.7 | 89.4 | 2.0 | 24,780 | 0.1 | 2,478 | 10 | 37,324 | 67.2 | 32,647 | 53.2 |
| Hollister | 82.4 | 53.4 | 96.8 | 90.3 | 88.5 | 6.6 | 20,333 | 1.1 | 2,210 | 11 | 30,223 | 69.9 | 25,709 | 45.4 |
| Huntington Beach | 77.3 | 51.4 | 98.1 | 91.8 | 87.5 | 5.3 | 103,899 | -4.9 | 8,945 | 9 | 168,288 | 66.8 | 130,390 | 55.5 |
| Huntington Park | 64.7 | 53.2 | 90.8 | 81.5 | NA | NA | 25,226 | -5.3 | 3,009 | 12 | 43,625 | 67.9 | 38,160 | 50.7 |
| Imperial Beach | 76.5 | 44.5 | 88.1 | 81.7 | 90.1 | 3.6 | 11,926 | -0.8 | 1,447 | 12 | 21,897 | 62.8 | 18,160 | 44.9 |
| Indio | 81.6 | 36.2 | 96.8 | 93.0 | 89.7 | 4.5 | 41,618 | 1.1 | 4,932 | 12 | 75,947 | 54.4 | 55,060 | 45.2 |
| Inglewood | 67.9 | 55.0 | 94.6 | 86.9 | 89.4 | 1.7 | 52,289 | -1.1 | 8,589 | 16 | 87,026 | 68.0 | 73,921 | 51.9 |
| Irvine | 71.8 | 26.9 | 98.5 | 95.3 | 76.1 | 9.8 | 142,426 | -4.0 | 10,126 | 7 | 231,951 | 64.5 | 202,655 | 46.2 |
| Laguna Hills | 79.6 | 32.7 | NA | NA | 90.7 | 3.0 | 16,318 | -4.5 | 1,220 | 8 | 26,649 | 64.4 | 20,935 | 49.0 |
| Laguna Niguel | 77.8 | 38.8 | NA | NA | 87.7 | 3.2 | 33,961 | -4.9 | 2,602 | 8 | 54,330 | 64.3 | 41,741 | 51.6 |
| La Habra | 79.0 | 50.5 | 93.2 | 90.1 | 92.9 | 5.1 | 29,183 | -6.0 | 2,751 | 9 | 48,556 | 69.6 | 39,867 | 52.8 |
| Lake Elsinore | 77.2 | 70.1 | 97.7 | 90.0 | 94.2 | 3.8 | 31,266 | 1.6 | 3,144 | 10 | 50,494 | 62.3 | 44,786 | 52.8 |
| Lake Forest | 78.3 | 40.1 | 96.3 | 95.2 | 91.8 | 1.1 | 46,535 | -5.7 | 3,255 | 7 | 67,824 | 71.0 | 56,588 | 58.0 |
| Lakewood | 82.0 | 52.6 | 96.8 | 90.3 | 93.9 | 1.2 | 41,124 | -4.6 | 5,023 | 12 | 63,499 | 65.6 | 53,438 | 50.3 |
| La Mesa | 73.4 | 37.4 | 97.4 | 95.3 | 83.4 | 4.4 | 29,533 | -3.4 | 2,783 | 9 | 48,123 | 70.0 | 39,913 | 54.1 |
| La Mirada | 82.1 | 48.9 | 96.7 | 88.6 | 91.5 | 4.9 | 23,147 | -4.9 | 2,750 | 12 | 40,540 | 61.1 | 30,317 | 47.8 |
| Lancaster | 81.3 | 36.6 | 89.1 | 85.1 | 92.3 | 2.1 | 62,898 | -3.7 | 9,479 | 15 | 118,925 | 52.9 | 101,604 | 47.0 |
| La Puente | 75.2 | 51.8 | 91.0 | 85.0 | NA | NA | 18,375 | -3.8 | 2,381 | 13 | 31,712 | 63.8 | 26,780 | 50.7 |
| La Quinta | 83.4 | 33.8 | 97.1 | 91.9 | 85.3 | 5.7 | 19,267 | 0.6 | 1,984 | 10 | 36,948 | 53.4 | 23,182 | 52.1 |
| La Verne | 75.2 | 52.2 | 95.3 | 91.5 | 86.0 | 6.2 | 15,138 | -6.1 | 1,596 | 11 | 26,808 | 52.8 | 18,242 | 44.6 |
| Lawndale | 76.1 | 42.2 | NA | NA | NA | NA | 16,357 | -1.9 | 2,402 | 15 | 25,779 | 72.5 | 22,353 | 54.1 |
| Lemon Grove | 77.3 | 49.0 | 92.6 | 86.6 | 85.8 | 6.2 | 12,281 | -1.6 | 1,435 | 12 | 21,296 | 57.1 | 16,721 | 47.8 |
| Lincoln | 82.6 | 42.3 | 98.8 | 91.6 | 91.0 | 4.6 | 18,817 | -2.3 | 1,444 | 8 | 38,323 | 51.8 | 25,191 | 55.2 |
| Livermore | 77.0 | 41.1 | 97.8 | 95.0 | 89.6 | 4.3 | 45,995 | -5.1 | 3,316 | 7 | 70,561 | 68.1 | 55,900 | 58.1 |
| Lodi | 81.6 | 28.5 | 94.4 | 90.9 | 85.6 | 3.1 | 29,953 | 0.0 | 3,019 | 10 | 50,143 | 64.1 | 40,649 | 52.9 |
| Lompoc | 62.2 | 43.2 | 90.4 | 83.6 | 81.2 | 3.6 | 17,601 | 1.7 | 1,880 | 11 | 34,005 | 57.6 | 28,398 | 46.1 |
| Long Beach | 78.4 | 48.4 | 95.9 | 88.3 | 88.0 | 3.0 | 230,669 | -3.8 | 30,691 | 13 | 373,834 | 67.0 | 315,796 | 52.8 |
| Los Altos | 74.5 | 31.5 | NA | NA | 87.9 | 2.4 | 13,839 | -6.7 | 582 | 4 | 22,890 | 59.4 | 16,806 | 54.5 |
| Los Angeles | 69.7 | 55.0 | 94.4 | 87.2 | 88.8 | 3.1 | 2,009,709 | -3.4 | 259,331 | 13 | 3,261,493 | 67.0 | 2,741,212 | 50.8 |
| Los Banos | NA | 61.2 | 94.4 | 92.9 | NA | NA | 17,315 | 1.6 | 2,645 | 15 | 29,225 | 57.8 | 24,422 | 50.3 |
| Los Gatos | 80.2 | 51.3 | 92.3 | 89.5 | 86.1 | 3.6 | 15,343 | -5.2 | 907 | 6 | 23,818 | 59.9 | 17,891 | 53.5 |
| Lynwood | 71.9 | 47.7 | 97.0 | 65.9 | NA | NA | 27,780 | -2.8 | 4,039 | 15 | 52,564 | 59.5 | 46,431 | 45.3 |
| Madera | 67.9 | 45.9 | 94.2 | 88.3 | 89.1 | 3.7 | 27,647 | -0.6 | 3,020 | 11 | 45,169 | 65.2 | 40,423 | 41.9 |
| Manhattan Beach | 76.5 | 54.1 | 98.8 | 92.0 | 89.3 | 3.5 | 18,113 | -8.8 | 1,323 | 7 | 27,079 | 58.3 | 20,271 | 52.0 |
| Manteca | 77.2 | 49.5 | 95.9 | 90.2 | 95.0 | 2.0 | 38,416 | 2.0 | 3,805 | 10 | 64,126 | 58.3 | 51,905 | 45.4 |
| Martinez | 75.2 | 50.1 | NA | NA | 87.6 | 1.7 | 19,603 | -5.1 | 1,522 | 8 | 31,273 | 72.5 | 24,982 | 59.4 |
| Maywood | 66.6 | 47.0 | 85.5 | 73.2 | NA | NA | 11,735 | -5.5 | 1,353 | 12 | 20,127 | 60.7 | 16,978 | 53.3 |
| Menifee | 80.0 | 58.5 | 94.3 | 91.3 | 87.2 | 5.9 | 40,447 | 2.2 | 4,095 | 10 | 72,215 | 54.3 | 55,091 | 45.4 |

1. Employed persons.  2. Households.  3. Percent of civilian labor force.  4. Persons 16 years old and over.

# Table D. Cities — Construction, Wholesale Trade, and Retail Trade

| City | Value of residential construction authorized by building permits, 2020 | | | Wholesale trade[1], 2017 | | | | Retail trade[2], 2017 | | | |
| --- | --- | --- | --- | --- | --- | --- | --- | --- | --- | --- | --- |
| | New construction ($1,000) | Number of housing units | Percent single family | Number of establishments | Number of employees | Sales (mil dol) | Annual payroll (mil dol) | Number of establishments | Number of employees | Sales (mil dol) | Annual payroll (mil dol) |
| | 69 | 70 | 71 | 72 | 73 | 74 | 75 | 76 | 77 | 78 | 79 |
| CALIFORNIA— Cont'd | | | | | | | | | | | |
| Delano | 15,066 | 85 | 98 | 12 | 133 | 179 | 9 | 92 | 1,683 | 416 | 40 |
| Desert Hot Springs | 23,518 | 77 | 100 | 4 | 9 | 1 | 0 | 44 | 588 | 161 | 16 |
| Diamond Bar | 7,699 | 21 | 100 | 289 | 1,205 | 2,133 | 56 | 144 | 1,000 | 425 | 32 |
| Downey | 7,462 | 57 | 100 | 86 | 921 | 429 | 59 | 294 | 5,829 | 1,712 | 166 |
| Dublin | 175,274 | 496 | 29 | 63 | 477 | 202 | 27 | 189 | 4,330 | 1,758 | 167 |
| East Palo Alto | 388 | 4 | 100 | 8 | 38 | 7 | 3 | 22 | 858 | 306 | 28 |
| El Cajon | 9,561 | 52 | 100 | 124 | 1,324 | 661 | 69 | 456 | 7,318 | 2,314 | 209 |
| El Centro | 16,747 | 118 | 24 | 57 | 424 | 328 | 20 | 194 | 4,454 | 1,005 | 108 |
| Elk Grove | 157,074 | 760 | 100 | D | D | D | 24 | 279 | 7,241 | 2,562 | 232 |
| El Monte | 11,298 | 51 | 61 | 312 | 2,338 | 1,051 | 99 | 308 | 3,677 | 1,567 | 147 |
| El Paso de Robles (Paso Robles) | 6,778 | 42 | 67 | 42 | 352 | 148 | 15 | 143 | 2,245 | 670 | 64 |
| Encinitas | 23,048 | 107 | 100 | 87 | 432 | 411 | 32 | 284 | 4,402 | 1,263 | 128 |
| Escondido | 48,086 | 254 | 95 | 132 | 998 | 525 | 56 | 511 | 10,528 | 3,319 | 317 |
| Eureka | 3,250 | 31 | 16 | 47 | 448 | 228 | 23 | 222 | 3,418 | 1,008 | 102 |
| Fairfield | 187,193 | 861 | 45 | 77 | 1,377 | 1,082 | 91 | 313 | 5,691 | 1,838 | 168 |
| Folsom | 52,456 | 559 | 89 | 32 | 240 | 1,485 | 37 | 288 | 6,285 | 2,036 | 190 |
| Fontana | 308,025 | 1,143 | 74 | 184 | 3,468 | 5,602 | 176 | 337 | 7,334 | 2,995 | 233 |
| Foster City | 11,823 | 32 | 31 | 35 | 586 | 563 | 56 | 31 | 788 | 316 | 28 |
| Fountain Valley | 6,260 | 39 | 100 | 102 | 1,474 | 9,573 | 128 | 213 | 3,139 | 1,297 | 97 |
| Fremont | 91,600 | 286 | 39 | 454 | 9,727 | 15,445 | 963 | 447 | 8,938 | 4,109 | 356 |
| Fresno | 414,404 | 1,576 | 77 | 510 | 8,150 | 6,495 | 453 | 1,537 | 25,256 | 7,266 | 698 |
| Fullerton | 8,767 | 42 | 45 | 291 | 2,839 | 2,215 | 158 | 428 | 5,476 | 1,767 | 153 |
| Gardena | 9,979 | 63 | 100 | 172 | 1,688 | 956 | 86 | 172 | 2,534 | 954 | 80 |
| Garden Grove | 24,637 | 237 | 100 | 227 | 2,364 | 1,249 | 111 | 444 | 5,988 | 2,237 | 179 |
| Gilroy | 51,719 | 239 | 33 | 47 | 593 | 473 | 43 | 302 | 5,409 | 1,789 | 151 |
| Glendale | 21,986 | 194 | 75 | 241 | 1,913 | 1,308 | 148 | 726 | 12,197 | 3,900 | 384 |
| Glendora | 4,734 | 28 | 100 | 48 | 405 | 219 | 27 | 132 | 2,743 | 867 | 83 |
| Goleta | 40,613 | 197 | 55 | 58 | 1,147 | 1,597 | 90 | 123 | 2,414 | 727 | 73 |
| Hanford | 45,899 | 163 | 100 | 27 | 386 | 249 | 24 | 160 | 2,778 | 848 | 77 |
| Hawthorne | 25,553 | 36 | 86 | 72 | 1,354 | 555 | 61 | 154 | 3,841 | 1,668 | 123 |
| Hayward | 140,189 | 538 | 60 | 442 | 6,836 | 5,408 | 407 | 412 | 6,899 | 2,267 | 216 |
| Hemet | 35,637 | 142 | 82 | 18 | 82 | 82 | 3 | 212 | 4,534 | 1,203 | 125 |
| Hesperia | 61,711 | 396 | 43 | 42 | 194 | 169 | 7 | 179 | 2,602 | 691 | 67 |
| Highland | 13,038 | 48 | 100 | 11 | 81 | 20 | 3 | 57 | 764 | 232 | 22 |
| Hollister | 81,992 | 322 | 100 | 20 | 167 | 72 | 7 | 79 | 1,267 | 394 | 41 |
| Huntington Beach | 36,926 | 124 | 84 | 372 | 4,541 | 3,572 | 295 | 624 | 10,253 | 3,560 | 316 |
| Huntington Park | 120 | 3 | 100 | 60 | 775 | 340 | 31 | 184 | 2,301 | 643 | 60 |
| Imperial Beach | 3,357 | 30 | 100 | 6 | 12 | 5 | 0 | 35 | 251 | 52 | 6 |
| Indio | 71,723 | 494 | 80 | 44 | 514 | 287 | 25 | 169 | 3,206 | 1,167 | 107 |
| Inglewood | 36,334 | 147 | 70 | 77 | 615 | 317 | 28 | 239 | 4,197 | 1,296 | 124 |
| Irvine | 465,813 | 2,040 | 51 | 1,095 | 19,409 | 35,191 | 1,684 | 700 | 11,702 | 5,000 | 437 |
| Laguna Hills | 3,135 | 3 | 100 | 105 | 568 | 312 | 38 | 171 | 1,819 | 486 | 51 |
| Laguna Niguel | 0 | 0 | 0 | 77 | 486 | 260 | 28 | 157 | 2,890 | 1,217 | 101 |
| La Habra | 3,614 | 12 | 100 | 60 | 332 | 191 | 16 | 183 | 3,876 | 1,080 | 105 |
| Lake Elsinore | 83,792 | 447 | 100 | 35 | 207 | 85 | 10 | 193 | 3,314 | 959 | 90 |
| Lake Forest | 123,693 | 412 | 100 | 216 | 3,427 | 4,293 | 284 | 214 | 3,951 | 1,219 | 140 |
| Lakewood | 1,735 | 13 | 85 | 26 | 118 | 64 | 6 | 215 | 5,494 | 1,305 | 132 |
| La Mesa | 24,342 | 268 | 14 | 21 | 220 | 54 | 9 | 224 | 4,099 | 1,297 | 125 |
| La Mirada | 6,875 | 32 | 13 | 130 | 3,395 | 2,139 | 236 | 91 | 1,547 | 419 | 46 |
| Lancaster | 73,169 | 348 | 85 | 63 | 642 | 644 | 34 | 310 | 5,519 | 1,815 | 167 |
| La Puente | 14,853 | 106 | 30 | 22 | 101 | 24 | 3 | 102 | 1,258 | 308 | 31 |
| La Quinta | 56,198 | 190 | 100 | 24 | 44 | 30 | 2 | 90 | 2,405 | 670 | 66 |
| La Verne | 3,669 | 17 | 100 | 58 | 545 | 485 | 33 | 80 | 1,224 | 303 | 30 |
| Lawndale | 6,365 | 23 | 100 | 16 | 113 | 51 | 7 | 83 | 751 | 228 | 22 |
| Lemon Grove | 4,439 | 22 | 100 | 12 | 59 | 34 | 2 | 78 | 1,629 | 619 | 51 |
| Lincoln | 170,148 | 546 | 100 | 18 | 140 | 75 | 8 | 68 | 1,220 | 340 | 33 |
| Livermore | 29,105 | 99 | 57 | 164 | 2,753 | 2,262 | 189 | 366 | 6,829 | 2,295 | 249 |
| Lodi | 79,703 | 423 | 31 | 40 | 280 | 844 | 17 | 214 | 3,991 | 1,271 | 112 |
| Lompoc | 0 | 0 | 0 | 9 | 53 | 17 | 2 | 108 | 1,553 | 409 | 40 |
| Long Beach | 108,271 | 840 | 21 | 309 | 4,118 | 9,447 | 311 | 921 | 13,603 | 4,095 | 383 |
| Los Altos | 36,363 | 49 | 100 | 15 | 87 | 68 | 7 | D | D | D | 35 |
| Los Angeles | 2,810,052 | 12,292 | 15 | 8,196 | 79,204 | 58,688 | 4,522 | 11,728 | 151,931 | 50,764 | 4,828 |
| Los Banos | 51,841 | 104 | 100 | 8 | 110 | 100 | 4 | 80 | 1,473 | 357 | 37 |
| Los Gatos | 30,263 | 104 | 45 | 21 | D | 597 | D | D | D | D | 74 |
| Lynwood | 553 | 5 | 100 | 39 | 487 | 329 | 31 | 102 | 1,250 | 369 | 31 |
| Madera | 58,972 | 323 | 63 | 27 | 363 | 298 | 23 | 156 | 2,282 | 644 | 60 |
| Manhattan Beach | 36,372 | 61 | 95 | 40 | 364 | 1,336 | 64 | 175 | 2,822 | 777 | 85 |
| Manteca | 216,705 | 662 | 98 | 29 | 462 | 516 | 27 | 193 | 3,614 | 1,082 | 106 |
| Martinez | 1,938 | 18 | 33 | 23 | 212 | 61 | 10 | 69 | 1,060 | 328 | 35 |
| Maywood | 1,867 | 12 | 83 | 16 | 652 | 426 | 31 | 45 | 508 | 125 | 12 |
| Menifee | 444,229 | 1,452 | 97 | 15 | 58 | 26 | 2 | 109 | 2,109 | 546 | 53 |

1. Merchant wholesalers except manufacturers' sales branches and offices.  2. Establishments with payroll.

Items 69—79

# Real Estate, Professional Services, and Manufacturing

| City | Real estate and rental and leasing, 2017 | | | | Professional, scientific, and technical services[1], 2017 | | | | Manufacturing, 2017 | | | |
|---|---|---|---|---|---|---|---|---|---|---|---|---|
| | Number of establish-ments | Number of employees | Receipts (mil dol) | Annual payroll (mil dol) | Number of establish-ments | Number of employees | Receipts (mil dol) | Annual payroll (mil dol) | Number of establish-ments | Number of employees | Receipts (mil dol) | Annual payroll (mil dol) |
| | 80 | 81 | 82 | 83 | 84 | 85 | 86 | 87 | 88 | 89 | 90 | 91 |
| CALIFORNIA— Cont'd | | | | | | | | | | | | |
| Delano | 17 | 46 | 13 | 2 | D | D | 8 | D | NA | NA | NA | NA |
| Desert Hot Springs | 18 | 43 | 13 | 2 | 11 | 282 | 7 | 5 | NA | NA | NA | NA |
| Diamond Bar | 145 | D | 73 | D | 273 | 1,194 | 245 | 68 | NA | NA | NA | NA |
| Downey | 165 | 990 | 203 | 41 | 162 | 665 | 84 | 25 | NA | NA | NA | NA |
| Dublin | 60 | 251 | 111 | 14 | 276 | 1,972 | 496 | 188 | NA | NA | NA | NA |
| East Palo Alto | 18 | 76 | 94 | 12 | 24 | 469 | 184 | 62 | NA | NA | NA | NA |
| El Cajon | 144 | 920 | 238 | 40 | 208 | 1,187 | 172 | 58 | NA | NA | NA | NA |
| El Centro | 50 | 229 | 57 | 7 | 91 | 494 | 59 | 24 | NA | NA | NA | NA |
| Elk Grove | 136 | 413 | 102 | 14 | 241 | 1,053 | 186 | 55 | NA | NA | NA | NA |
| El Monte | 75 | 361 | 79 | 19 | 129 | 588 | 102 | 34 | NA | NA | NA | NA |
| El Paso de Robles (Paso Robles) | 63 | 284 | 62 | 10 | 82 | 368 | 49 | 19 | NA | NA | NA | NA |
| Encinitas | 243 | 647 | 164 | 39 | 621 | 2,235 | 500 | 170 | NA | NA | NA | NA |
| Escondido | 205 | 825 | 196 | 35 | 339 | 1,776 | 277 | 100 | NA | NA | NA | NA |
| Eureka | 49 | 257 | 70 | 10 | 91 | 789 | 111 | 46 | NA | NA | NA | NA |
| Fairfield | 109 | 499 | 163 | 21 | 163 | 909 | 139 | 49 | NA | NA | NA | NA |
| Folsom | 123 | 782 | 185 | 48 | 411 | 10,954 | 1,130 | 1,592 | NA | NA | NA | NA |
| Fontana | 112 | 570 | 222 | 31 | 131 | 474 | 69 | 22 | NA | NA | NA | NA |
| Foster City | 64 | 651 | 323 | 55 | 139 | 1,523 | 391 | 175 | NA | NA | NA | NA |
| Fountain Valley | 126 | 407 | 111 | 18 | 233 | 2,226 | 509 | 163 | NA | NA | NA | NA |
| Fremont | 284 | 1,174 | 526 | 68 | 1,251 | 13,598 | 3,166 | 1,414 | NA | NA | NA | NA |
| Fresno | 558 | 3,657 | 784 | 158 | 1,150 | 8,428 | 1,270 | 499 | NA | NA | NA | NA |
| Fullerton | 247 | 917 | 208 | 32 | 448 | 2,968 | 443 | 200 | NA | NA | NA | NA |
| Gardena | 55 | 284 | 49 | 10 | 79 | 283 | 33 | 12 | NA | NA | NA | NA |
| Garden Grove | 167 | 818 | 243 | 34 | 275 | 1,724 | 190 | 74 | NA | NA | NA | NA |
| Gilroy | 46 | 171 | 62 | 8 | 73 | 470 | 88 | 30 | NA | NA | NA | NA |
| Glendale | 401 | 2,913 | 1,696 | 234 | 1,086 | 6,902 | 1,733 | 556 | NA | NA | NA | NA |
| Glendora | 87 | 295 | 70 | 12 | 141 | 934 | 119 | 44 | NA | NA | NA | NA |
| Goleta | 47 | 288 | 57 | 11 | D | D | D | D | NA | NA | NA | NA |
| Hanford | 63 | 250 | 54 | 9 | 61 | 368 | 39 | 16 | NA | NA | NA | NA |
| Hawthorne | 79 | 342 | 46 | 12 | 73 | 606 | 112 | 27 | NA | NA | NA | NA |
| Hayward | 188 | 1,450 | 411 | 77 | 319 | 5,858 | 2,297 | 933 | NA | NA | NA | NA |
| Hemet | 84 | 416 | 95 | 19 | 58 | 379 | 31 | 13 | NA | NA | NA | NA |
| Hesperia | 52 | 227 | 59 | 9 | 57 | 238 | 25 | 8 | NA | NA | NA | NA |
| Highland | 14 | 31 | 9 | 1 | 34 | 155 | 25 | 6 | NA | NA | NA | NA |
| Hollister | 37 | 96 | 25 | 4 | 48 | 179 | 24 | 8 | NA | NA | NA | NA |
| Huntington Beach | 403 | 1,436 | 477 | 71 | 837 | 4,436 | 1,053 | 307 | NA | NA | NA | NA |
| Huntington Park | 28 | 146 | 26 | 5 | 33 | 194 | 18 | 6 | NA | NA | NA | NA |
| Imperial Beach | 19 | 64 | 13 | 2 | 35 | 115 | 14 | 6 | NA | NA | NA | NA |
| Indio | 65 | 353 | 70 | 13 | 79 | 403 | 54 | 18 | NA | NA | NA | NA |
| Inglewood | 76 | 736 | 327 | 33 | 86 | 604 | 79 | 22 | NA | NA | NA | NA |
| Irvine | 945 | 12,200 | 4,969 | 1,209 | 3,259 | 47,066 | 9,986 | 4,614 | NA | NA | NA | NA |
| Laguna Hills | 96 | 594 | 74 | 27 | 367 | 1,995 | 405 | 142 | NA | NA | NA | NA |
| Laguna Niguel | 159 | 481 | 269 | 35 | 354 | 922 | 176 | 52 | NA | NA | NA | NA |
| La Habra | 70 | 255 | 59 | 9 | 94 | 461 | 55 | 19 | NA | NA | NA | NA |
| Lake Elsinore | 40 | 232 | 81 | 10 | 67 | 397 | 37 | 12 | NA | NA | NA | NA |
| Lake Forest | 129 | 970 | 202 | 52 | 465 | 3,203 | 629 | 229 | NA | NA | NA | NA |
| Lakewood | 54 | 300 | 47 | 9 | 97 | 531 | 40 | 15 | NA | NA | NA | NA |
| La Mesa | 151 | 824 | 148 | 30 | 252 | 1,091 | 165 | 62 | NA | NA | NA | NA |
| La Mirada | 56 | 585 | 110 | 39 | 74 | 292 | 56 | 21 | NA | NA | NA | NA |
| Lancaster | 125 | 533 | 154 | 20 | 145 | 1,179 | 146 | 55 | NA | NA | NA | NA |
| La Puente | 23 | 72 | 15 | 3 | 98 | 436 | 14 | 6 | NA | NA | NA | NA |
| La Quinta | 99 | 270 | 73 | 13 | 98 | 337 | 80 | 26 | NA | NA | NA | NA |
| La Verne | 43 | 154 | 37 | 7 | 73 | 402 | 61 | 21 | NA | NA | NA | NA |
| Lawndale | 23 | 285 | 71 | 12 | 38 | 110 | 51 | 9 | NA | NA | NA | NA |
| Lemon Grove | 27 | 75 | 23 | 2 | 24 | 156 | 12 | 4 | NA | NA | NA | NA |
| Lincoln | 42 | 101 | 21 | 4 | 58 | 478 | 41 | 19 | NA | NA | NA | NA |
| Livermore | 110 | 716 | 261 | 52 | 250 | 12,580 | 3,355 | 1,138 | NA | NA | NA | NA |
| Lodi | 65 | 441 | 43 | 16 | 99 | 418 | 53 | 18 | NA | NA | NA | NA |
| Lompoc | 38 | 217 | 29 | 6 | 41 | 272 | 38 | 14 | NA | NA | NA | NA |
| Long Beach | 571 | 3,040 | 1,216 | 151 | 1,123 | 8,457 | 1,589 | 574 | NA | NA | NA | NA |
| Los Altos | 110 | 446 | 217 | 32 | 277 | 4,124 | 1,079 | 614 | NA | NA | NA | NA |
| Los Angeles | 7,059 | 44,587 | 19,767 | 2,717 | 16,741 | 155,056 | 38,075 | 13,649 | NA | NA | NA | NA |
| Los Banos | 25 | 58 | 14 | 2 | 22 | 114 | 15 | 5 | NA | NA | NA | NA |
| Los Gatos | 140 | 571 | 212 | 36 | 270 | 1,786 | 338 | 147 | NA | NA | NA | NA |
| Lynwood | 13 | 129 | 12 | 4 | 26 | 92 | 11 | 4 | NA | NA | NA | NA |
| Madera | 47 | 177 | 31 | 6 | 38 | 152 | 15 | 5 | NA | NA | NA | NA |
| Manhattan Beach | 171 | 546 | 183 | 31 | 407 | 3,584 | 732 | 339 | NA | NA | NA | NA |
| Manteca | 63 | 269 | 57 | 13 | 61 | 274 | 31 | 11 | NA | NA | NA | NA |
| Martinez | 46 | 204 | 52 | 9 | 91 | 402 | 64 | 24 | NA | NA | NA | NA |
| Maywood | 5 | 11 | 1 | 0 | 4 | 10 | 0 | 0 | NA | NA | NA | NA |
| Menifee | 55 | 204 | 44 | 6 | 65 | 175 | 21 | 6 | NA | NA | NA | NA |

1. Establishments subject to federal tax.

# Table D. Cities — Accommodation and Food Services, Arts, Entertainment, and Recreation, and Health Care and Social Assistance

| City | Accommodation and food services, 2017 | | | | Arts, entertainment, and recreation[1], 2017 | | | | Health care and social assistance[1], 2017 | | | |
|---|---|---|---|---|---|---|---|---|---|---|---|---|
| | Number of establishments | Number of employees | Receipts (mil dol) | Annual payroll (mil dol) | Number of establishments | Number of employees | Receipts (mil dol) | Annual payroll (mil dol) | Number of establishments | Number of employees | Receipts (mil dol) | Annual payroll (mil dol) |
| | 92 | 93 | 94 | 95 | 96 | 97 | 98 | 99 | 100 | 101 | 102 | 103 |
| **CALIFORNIA— Cont'd** | | | | | | | | | | | | |
| Delano | 64 | 922 | 52 | 15 | D | D | D | D | 74 | 1,507 | 170 | 67 |
| Desert Hot Springs | 47 | 470 | 32 | 8 | D | D | D | D | 37 | 497 | 29 | 14 |
| Diamond Bar | 141 | 1,950 | 121 | 35 | 11 | 185 | 14 | 4 | 208 | 1,587 | 155 | 63 |
| Downey | 292 | 5,840 | 398 | 112 | 20 | 262 | 29 | 6 | 355 | 12,312 | 1,942 | 801 |
| Dublin | 183 | 3,288 | 265 | 74 | D | D | D | D | 184 | 1,850 | 258 | 100 |
| East Palo Alto | 23 | 565 | 76 | 27 | D | D | D | D | 35 | 483 | 49 | 26 |
| El Cajon | 229 | 3,532 | 201 | 52 | 15 | 333 | 17 | 5 | 287 | 6,238 | 521 | 231 |
| El Centro | 123 | 2,311 | 139 | 41 | 11 | 104 | 4 | 1 | 153 | 2,915 | 305 | 117 |
| Elk Grove | 276 | 5,122 | 310 | 86 | 27 | 546 | 32 | 8 | 384 | 4,044 | 593 | 190 |
| El Monte | 204 | 2,109 | 153 | 35 | 9 | D | 7 | D | 216 | 3,203 | 265 | 107 |
| El Paso de Robles (Paso Robles) | 133 | 2,514 | 174 | 50 | 14 | 325 | 16 | 5 | 88 | 1,154 | 72 | 31 |
| Encinitas | 230 | 4,455 | 285 | 90 | 70 | 853 | 66 | 21 | 438 | 5,156 | 758 | 267 |
| Escondido | 276 | 4,875 | 317 | 91 | 34 | 708 | 37 | 13 | 426 | 8,246 | 1,249 | 441 |
| Eureka | 147 | 2,034 | 124 | 36 | 17 | 150 | 8 | 2 | 153 | 3,477 | 528 | 194 |
| Fairfield | 210 | 3,507 | 221 | 64 | 22 | 580 | 22 | 8 | 283 | 7,141 | 1,195 | 414 |
| Folsom | 231 | 4,867 | 279 | 84 | 36 | 605 | 32 | 10 | 292 | 4,017 | 827 | 264 |
| Fontana | 270 | 4,779 | 310 | 80 | 18 | 370 | 27 | 6 | 207 | 8,480 | 1,981 | 689 |
| Foster City | 72 | 1,394 | 115 | 33 | 5 | 232 | 9 | 4 | 101 | 1,249 | 170 | 69 |
| Fountain Valley | 193 | 3,109 | 195 | 55 | D | D | D | 9 | 430 | 7,277 | 1,113 | 380 |
| Fremont | 464 | 7,787 | 620 | 165 | 52 | 966 | 60 | 20 | 732 | 12,827 | 2,152 | 874 |
| Fresno | 1,018 | 20,209 | 1,209 | 343 | 123 | 3,647 | 171 | 59 | 1,734 | 35,901 | 5,348 | 2,131 |
| Fullerton | 365 | 7,803 | 416 | 142 | 40 | 762 | 46 | 16 | 449 | 8,509 | 1,320 | 582 |
| Gardena | 235 | 3,065 | 223 | 58 | 13 | 1,260 | 91 | 39 | 144 | 2,797 | 314 | 118 |
| Garden Grove | 484 | 8,175 | 674 | 159 | D | D | D | D | 509 | 6,152 | 746 | 242 |
| Gilroy | 155 | 2,711 | 189 | 52 | D | D | D | D | D | D | D | D |
| Glendale | 489 | 8,105 | 640 | 171 | 236 | 1,382 | 190 | 60 | 1,167 | 16,964 | 2,013 | 749 |
| Glendora | 118 | 1,711 | 105 | 30 | D | D | D | D | 223 | 3,712 | 422 | 162 |
| Goleta | 107 | 2,812 | 317 | 73 | 14 | 372 | 17 | 6 | 132 | 1,728 | 274 | 87 |
| Hanford | D | D | D | 29 | D | D | D | D | 165 | 4,184 | 576 | 199 |
| Hawthorne | 129 | 1,955 | 148 | 35 | D | D | D | D | 172 | 2,666 | 305 | 135 |
| Hayward | 326 | 4,417 | 317 | 85 | 25 | D | 37 | D | 331 | 7,052 | 754 | 292 |
| Hemet | 149 | 2,374 | 145 | 41 | D | D | D | D | 222 | 4,585 | 548 | 195 |
| Hesperia | 121 | 2,192 | 124 | 32 | 12 | 77 | 4 | 1 | 104 | 1,050 | 91 | 37 |
| Highland | 60 | 1,038 | 65 | 18 | D | D | D | D | 58 | 1,684 | 108 | 37 |
| Hollister | 63 | 946 | 58 | 17 | 11 | 141 | 7 | 2 | 77 | 1,168 | 168 | 73 |
| Huntington Beach | 546 | 12,331 | 815 | 237 | 113 | 1,459 | 119 | 29 | 751 | 6,662 | 832 | 310 |
| Huntington Park | 128 | 1,772 | 117 | 31 | 3 | 69 | 7 | 2 | 124 | 1,823 | 235 | 70 |
| Imperial Beach | 45 | 431 | 29 | 8 | 4 | 9 | 1 | 0 | 25 | 436 | 37 | 16 |
| Indio | 140 | 3,549 | 320 | 87 | 14 | 532 | 207 | 16 | 131 | 2,200 | 245 | 93 |
| Inglewood | 186 | 2,872 | 220 | 56 | 27 | 1,699 | 209 | 38 | 264 | 5,128 | 685 | 222 |
| Irvine | 736 | 16,000 | 1,177 | 334 | 154 | 2,714 | 259 | 73 | 1,294 | 15,404 | 2,106 | 772 |
| Laguna Hills | 90 | 1,956 | 130 | 38 | D | D | D | D | 374 | 5,753 | 825 | 286 |
| Laguna Niguel | 122 | 1,796 | 121 | 34 | D | D | D | D | 241 | 1,891 | 195 | 76 |
| La Habra | 145 | 2,128 | 139 | 38 | D | D | D | D | 130 | 1,265 | 119 | 49 |
| Lake Elsinore | 108 | 1,654 | 103 | 29 | 16 | 509 | 30 | 10 | 80 | 519 | 57 | 22 |
| Lake Forest | 230 | 3,814 | 264 | 68 | 43 | 452 | 27 | 7 | 252 | 2,526 | 261 | 92 |
| Lakewood | D | D | D | D | 14 | 283 | 25 | 5 | 203 | 2,719 | 399 | 142 |
| La Mesa | 176 | 3,565 | 203 | 67 | 21 | 546 | 22 | 10 | 351 | 8,597 | 1,447 | 538 |
| La Mirada | 107 | 1,407 | 113 | 25 | 14 | 224 | 17 | 5 | 108 | 1,430 | 134 | 54 |
| Lancaster | 239 | 3,950 | 250 | 64 | 19 | 274 | 26 | 7 | 423 | 10,408 | 1,470 | 537 |
| La Puente | 90 | 1,106 | 75 | 19 | 4 | 15 | 2 | 0 | 63 | 472 | 40 | 14 |
| La Quinta | D | D | D | D | 23 | 1,049 | 87 | 33 | 104 | 841 | 77 | 33 |
| La Verne | 74 | 1,443 | 87 | 25 | D | D | D | D | 79 | 927 | 82 | 33 |
| Lawndale | 52 | 700 | 51 | 14 | 7 | 14 | 7 | 1 | 79 | 459 | 47 | 16 |
| Lemon Grove | 55 | 907 | 57 | 15 | 3 | 18 | 1 | 0 | 50 | 969 | 77 | 31 |
| Lincoln | 63 | 942 | 58 | 16 | 11 | 190 | 13 | 5 | 87 | 949 | 165 | 47 |
| Livermore | 219 | 3,592 | 264 | 78 | 38 | 1,079 | 103 | 24 | 214 | 2,551 | 363 | 137 |
| Lodi | 144 | 2,288 | 144 | 41 | 14 | 352 | 18 | 6 | 195 | 3,498 | 499 | 177 |
| Lompoc | 90 | 1,224 | 84 | 23 | 5 | 46 | 3 | 1 | 84 | 1,582 | 193 | 71 |
| Long Beach | 993 | 20,426 | 1,509 | 408 | 145 | 2,242 | 188 | 60 | 1,263 | 29,063 | 4,153 | 1,682 |
| Los Altos | 88 | 1,323 | 124 | 35 | 20 | D | 10 | D | 172 | 1,962 | 288 | 118 |
| Los Angeles | 9,300 | 179,751 | 14,224 | 3,971 | 10,363 | 49,334 | 14,957 | 5,541 | 12,429 | 226,723 | 34,621 | 12,523 |
| Los Banos | 73 | 1,044 | 60 | 16 | D | D | D | D | 54 | 774 | 116 | 40 |
| Los Gatos | 127 | 2,550 | 170 | 59 | 24 | 762 | 57 | 18 | 339 | 3,386 | 525 | 193 |
| Lynwood | 87 | 1,400 | 105 | 24 | NA | NA | NA | NA | 135 | 3,667 | 750 | 212 |
| Madera | 98 | 1,339 | 87 | 22 | 8 | 78 | 4 | 1 | 135 | 2,349 | 242 | 82 |
| Manhattan Beach | 151 | 4,437 | 319 | 103 | 96 | 428 | 53 | 20 | 237 | 1,416 | 222 | 87 |
| Manteca | 144 | 2,779 | 163 | 46 | D | D | D | D | 167 | 4,019 | 576 | 203 |
| Martinez | 90 | 999 | 70 | 18 | 7 | 35 | 2 | 1 | 97 | 5,436 | 867 | 363 |
| Maywood | 41 | 462 | 33 | 9 | NA | NA | NA | NA | 26 | 336 | 35 | 11 |
| Menifee | 95 | 2,379 | 139 | 41 | 12 | 174 | 10 | 3 | 119 | 2,141 | 181 | 71 |

1. Establishments subject to federal tax.

# Table D. Cities — Other Services and Government Employment and Payroll

Other services[1] covers items 104–107. Government employment and payroll, 2017 — March payroll — Percent of total for: covers items 108–116.

| City | Number of establishments | Number of employees | Receipts (mil dol) | Annual payroll (mil dol) | Full-time equivalent employees | Total (dollars) | Administrative, judicial, and legal | Police and corrections | Fire protection | Highways and transportation | Health and welfare | Natural resources and utilities | Education and libraries |
|---|---|---|---|---|---|---|---|---|---|---|---|---|---|
| | 104 | 105 | 106 | 107 | 108 | 109 | 110 | 111 | 112 | 113 | 114 | 115 | 116 |
| **CALIFORNIA— Cont'd** | | | | | | | | | | | | | |
| Delano | 20 | 69 | 7 | 2 | 282 | 1,386,031 | 7.7 | 60.6 | 0.0 | 4.4 | 2.3 | 21.2 | 0.0 |
| Desert Hot Springs | 13 | 52 | 5 | 1 | 70 | 371,024 | 28.6 | 58.7 | 0.0 | 0.0 | 12.7 | 0.0 | 0.0 |
| Diamond Bar | 81 | 369 | 76 | 18 | 77 | 445,656 | 41.9 | 0.0 | 0.0 | 14.0 | 13.8 | 30.3 | 0.0 |
| Downey | 162 | 895 | 89 | 24 | 545 | 3,855,947 | 8.6 | 41.9 | 25.6 | 4.8 | 1.5 | 12.6 | 2.5 |
| Dublin | 107 | 671 | 93 | 27 | 114 | 848,938 | 28.9 | 3.1 | 0.6 | 8.7 | 16.8 | 40.9 | 0.0 |
| East Palo Alto | 17 | 98 | 13 | 4 | 100 | 889,558 | 28.0 | 52.0 | 0.0 | 6.1 | 3.3 | 5.2 | 0.0 |
| El Cajon | 213 | 1,678 | 200 | 57 | 404 | 2,768,459 | 10.9 | 47.1 | 17.9 | 5.9 | 2.1 | 12.1 | 0.0 |
| El Centro | 72 | 324 | 38 | 10 | 1,192 | 5,563,406 | 3.0 | 7.7 | 4.3 | 1.4 | 75.7 | 5.3 | 0.6 |
| Elk Grove | 180 | 1,265 | 166 | 41 | 326 | 2,978,726 | 22.2 | 73.7 | 0.0 | 2.6 | 0.0 | 1.0 | 0.0 |
| El Monte | 129 | 497 | 61 | 13 | 348 | 2,600,084 | 8.9 | 60.6 | 0.0 | 18.7 | 5.2 | 6.6 | 0.0 |
| El Paso de Robles (Paso Robles) | 62 | 300 | 32 | 9 | 184 | 1,253,520 | 11.0 | 27.5 | 15.4 | 4.3 | 3.4 | 23.8 | 3.7 |
| Encinitas | 183 | 1,141 | 118 | 39 | 196 | 1,339,981 | 35.6 | 0.0 | 3.7 | 7.1 | 21.5 | 25.4 | 0.0 |
| Escondido | 279 | 1,541 | 180 | 51 | 875 | 5,261,351 | 15.1 | 30.9 | 15.9 | 6.8 | 0.7 | 24.6 | 2.7 |
| Eureka | 90 | 552 | 61 | 18 | 311 | 1,656,604 | 9.1 | 31.6 | 18.9 | 8.5 | 4.5 | 19.6 | 0.0 |
| Fairfield | 150 | 1,131 | 283 | 53 | 599 | 3,818,932 | 10.2 | 37.1 | 15.8 | 12.4 | 2.9 | 11.2 | 0.0 |
| Folsom | 129 | 974 | 75 | 25 | 456 | 3,094,854 | 10.4 | 26.3 | 21.1 | 6.2 | 5.5 | 26.9 | 1.6 |
| Fontana | 212 | 1,257 | 151 | 34 | 772 | 5,316,586 | 10.6 | 51.7 | 0.0 | 7.4 | 6.0 | 21.1 | 0.0 |
| Foster City | 31 | 148 | 15 | 5 | 207 | 1,825,425 | 19.7 | 32.2 | 19.4 | 5.7 | 0.0 | 22.9 | 0.0 |
| Fountain Valley | 102 | 739 | 65 | 24 | 227 | 1,874,378 | 11.0 | 39.3 | 27.2 | 8.0 | 0.0 | 11.7 | 0.0 |
| Fremont | 350 | 2,386 | 406 | 87 | 957 | 9,501,136 | 16.3 | 31.8 | 22.4 | 8.3 | 8.5 | 8.5 | 0.0 |
| Fresno | 661 | 6,346 | 746 | 209 | 3,451 | 20,907,203 | 6.8 | 36.3 | 15.6 | 19.6 | 0.0 | 16.7 | 0.0 |
| Fullerton | 238 | 2,771 | 269 | 79 | 694 | 4,622,092 | 7.9 | 39.7 | 19.5 | 7.1 | 4.2 | 10.1 | 3.5 |
| Gardena | 117 | 625 | 75 | 20 | 454 | 2,464,867 | 7.9 | 42.6 | 0.0 | 28.2 | 6.3 | 8.7 | 0.0 |
| Garden Grove | 271 | 1,572 | 124 | 38 | 691 | 5,437,306 | 10.6 | 42.6 | 15.3 | 7.1 | 4.6 | 17.8 | 0.0 |
| Gilroy | 111 | 890 | 109 | 30 | 300 | 2,453,398 | 12.5 | 37.2 | 18.1 | 7.2 | 2.7 | 17.5 | 0.0 |
| Glendale | 410 | 1,865 | 232 | 60 | 1,735 | 12,571,242 | 12.4 | 21.4 | 18.8 | 6.5 | 3.9 | 26.6 | 2.9 |
| Glendora | 92 | 554 | 141 | 31 | 348 | 1,891,522 | 10.6 | 43.7 | 0.0 | 11.9 | 6.9 | 19.8 | 7.2 |
| Goleta | D | D | D | 65 | 62 | 536,263 | 73.6 | 0.0 | 0.0 | 14.1 | 0.0 | 12.3 | 0.0 |
| Hanford | 69 | 311 | 33 | 8 | 281 | 1,593,527 | 8.0 | 32.3 | 15.1 | 8.8 | 0.7 | 23.6 | 0.0 |
| Hawthorne | 119 | 683 | 68 | 18 | 299 | 2,186,109 | 15.5 | 56.8 | 0.0 | 7.3 | 6.5 | 8.6 | 0.0 |
| Hayward | 259 | 2,031 | 296 | 74 | 818 | 8,184,527 | 10.0 | 41.9 | 21.2 | 2.7 | 1.4 | 10.8 | 2.8 |
| Hemet | 95 | 461 | 43 | 12 | 427 | 3,039,134 | 11.5 | 33.8 | 20.3 | 7.4 | 2.2 | 11.7 | 3.6 |
| Hesperia | 101 | 601 | 73 | 20 | 120 | 670,338 | 31.5 | 0.0 | 0.0 | 15.9 | 14.5 | 0.0 | 0.0 |
| Highland | 35 | 155 | 16 | 4 | 41 | 318,191 | 66.6 | 0.0 | 0.0 | 21.6 | 0.6 | 1.5 | 0.0 |
| Hollister | D | D | D | 8 | 168 | 1,323,462 | 9.0 | 22.7 | 31.6 | 10.8 | 0.7 | 10.9 | 0.0 |
| Huntington Beach | 411 | 2,525 | 252 | 75 | 1,039 | 8,450,607 | 10.6 | 36.4 | 21.4 | 3.7 | 1.8 | 15.1 | 2.7 |
| Huntington Park | D | D | D | D | 164 | 1,072,674 | 15.5 | 61.0 | 0.0 | 5.9 | 7.4 | 4.5 | 0.0 |
| Imperial Beach | 32 | 138 | 11 | 3 | 96 | 610,849 | 21.7 | 0.0 | 23.5 | 5.7 | 24.4 | 16.3 | 0.0 |
| Indio | 85 | 664 | 62 | 19 | 242 | 1,977,846 | 5.8 | 55.4 | 0.0 | 8.3 | 5.9 | 19.1 | 0.0 |
| Inglewood | 142 | 1,308 | 153 | 42 | 533 | 2,366,678 | 24.9 | 19.9 | 0.0 | 3.7 | 11.2 | 17.9 | 4.5 |
| Irvine | 410 | 3,869 | 610 | 146 | 1,045 | 10,354,501 | 15.2 | 43.5 | 0.0 | 10.4 | 5.8 | 18.3 | 0.0 |
| Laguna Hills | 86 | 458 | 61 | 15 | 43 | 291,744 | 40.0 | 0.0 | 0.0 | 14.4 | 18.8 | 26.2 | 0.0 |
| Laguna Niguel | 109 | 482 | 49 | 14 | 88 | 587,478 | 23.1 | 2.6 | 0.0 | 13.7 | 17.8 | 40.1 | 0.0 |
| La Habra | 112 | 559 | 56 | 16 | 333 | 1,951,632 | 11.4 | 46.4 | 0.0 | 0.0 | 0.0 | 0.0 | 0.0 |
| Lake Elsinore | 72 | 330 | 40 | 12 | 87 | 501,292 | 31.3 | 0.0 | 0.0 | 27.7 | 4.7 | 19.5 | 0.0 |
| Lake Forest | 140 | 881 | 91 | 27 | 84 | 557,618 | 58.8 | 1.2 | 0.0 | 5.1 | 13.9 | 20.9 | 0.0 |
| Lakewood | 82 | 607 | 79 | 18 | 266 | 1,877,960 | 24.0 | 6.4 | 0.0 | 14.6 | 2.1 | 42.8 | 0.0 |
| La Mesa | 130 | 814 | 76 | 23 | 254 | 1,896,004 | 11.5 | 43.4 | 25.0 | 3.7 | 1.0 | 9.9 | 0.0 |
| La Mirada | 48 | 166 | 13 | 4 | 153 | 780,641 | 15.8 | 8.6 | 0.0 | 21.2 | 0.0 | 50.2 | 0.0 |
| Lancaster | 170 | 840 | 92 | 26 | 392 | 2,055,333 | 23.1 | 3.9 | 0.0 | 25.0 | 8.5 | 35.4 | 0.0 |
| La Puente | 44 | 188 | 15 | 4 | 42 | 185,307 | 35.3 | 0.0 | 0.0 | 2.3 | 21.6 | 32.2 | 0.0 |
| La Quinta | 50 | 299 | 45 | 8 | 91 | 505,915 | 26.4 | 0.0 | 0.0 | 19.0 | 0.0 | 11.6 | 0.0 |
| La Verne | D | D | D | D | 182 | 1,379,456 | 14.7 | 37.1 | 26.8 | 3.4 | 0.0 | 13.2 | 0.0 |
| Lawndale | D | D | D | D | 70 | 429,954 | 31.3 | 0.0 | 0.0 | 28.0 | 5.8 | 17.5 | 0.0 |
| Lemon Grove | 55 | 174 | 15 | 4 | 52 | 391,345 | 15.0 | 0.0 | 50.2 | 14.0 | 7.5 | 5.9 | 1.8 |
| Lincoln | D | D | D | 6 | 214 | 1,281,023 | 18.7 | 28.4 | 11.2 | 10.5 | 9.3 | 10.6 | 6.3 |
| Livermore | 148 | 828 | 107 | 30 | 429 | 3,798,311 | 15.9 | 36.7 | 0.0 | 16.7 | 5.5 | 16.2 | 2.7 |
| Lodi | 123 | 606 | 72 | 21 | 495 | 3,158,229 | 10.3 | 27.2 | 16.3 | 7.0 | 1.8 | 29.1 | 2.5 |
| Lompoc | 51 | 351 | 25 | 8 | 380 | 2,298,679 | 13.3 | 22.5 | 12.5 | 4.9 | 0.0 | 32.5 | 1.5 |
| Long Beach | 699 | 5,912 | 628 | 180 | 5,477 | 39,697,152 | 7.1 | 26.3 | 14.1 | 21.8 | 5.4 | 12.7 | 0.0 |
| Los Altos | 78 | 726 | 1,803 | 59 | 139 | 1,212,302 | 9.9 | 34.6 | 0.0 | 13.3 | 6.0 | 32.0 | 0.0 |
| Los Angeles | 7,114 | 53,813 | 8,662 | 1,932 | 52,287 | 452,879,430 | 10.9 | 25.7 | 10.2 | 11.1 | 2.7 | 30.9 | 1.2 |
| Los Banos | 29 | 175 | 13 | 4 | 159 | 915,005 | 11.4 | 40.1 | 15.6 | 5.1 | 2.4 | 19.2 | 0.0 |
| Los Gatos | 112 | 715 | 78 | 21 | 161 | 1,481,593 | 16.6 | 44.7 | 0.0 | 10.6 | 7.3 | 6.5 | 7.7 |
| Lynwood | D | D | D | D | 250 | 1,095,158 | 42.8 | 2.5 | 0.0 | 20.8 | 5.6 | 28.3 | 0.0 |
| Madera | 55 | 239 | 25 | 7 | 294 | 1,462,912 | 16.2 | 35.5 | 0.0 | 9.9 | 12.4 | 22.5 | 0.0 |
| Manhattan Beach | 90 | 705 | 131 | 26 | 351 | 2,662,902 | 14.7 | 46.2 | 8.1 | 8.8 | 8.8 | 11.7 | 0.0 |
| Manteca | 90 | 440 | 51 | 12 | 366 | 2,635,747 | 13.0 | 35.1 | 16.5 | 6.2 | 1.2 | 25.5 | 0.0 |
| Martinez | 62 | 248 | 39 | 11 | 59 | 491,580 | 13.6 | 76.6 | 0.0 | 2.7 | 3.6 | 3.1 | 0.0 |
| Maywood | D | D | D | D | 23 | 76,222 | 44.2 | 0.0 | 0.0 | 0.0 | 6.9 | 36.6 | 0.0 |
| Menifee | 54 | 314 | 39 | 10 | 83 | 474,674 | 40.9 | 0.0 | 0.0 | 24.1 | 0.0 | 20.8 | 0.0 |

1. Establishments subject to federal tax.

## Table D. Cities — City Government Finances

| City | General revenue Total (mil dol) | Intergovernmental Total (mil dol) | Intergovernmental Percent from state government | Taxes Total (mil dol) | Taxes Per capita[1] Total | Taxes Per capita[1] Property | Taxes Per capita[1] Sales and gross receipts | General expenditure Total (mil dol) | General expenditure Per capita[1] Total | General expenditure Per capita[1] Capital outlays |
|---|---|---|---|---|---|---|---|---|---|---|
| | 117 | 118 | 119 | 120 | 121 | 122 | 123 | 124 | 125 | 126 |
| **CALIFORNIA— Cont'd** | | | | | | | | | | |
| Delano................. | 62.7 | 7.4 | 53.0 | 23.6 | 446 | 184 | 259 | 43 | 805 | 70 |
| Desert Hot Springs........... | 27.6 | 1.9 | 99.5 | 17.3 | 603 | 277 | 307 | 29 | 1,026 | 233 |
| Diamond Bar...................... | 30.6 | 3.0 | 64.6 | 21.5 | 381 | 190 | 182 | 45 | 794 | 349 |
| Downey............................ | 111.2 | 19.1 | 37.6 | 63.9 | 568 | 235 | 331 | 108 | 957 | 62 |
| Dublin.............................. | 120.9 | 3.3 | 82.5 | 99.2 | 1,632 | 608 | 695 | 86 | 1,416 | 109 |
| East Palo Alto ................. | 39.8 | 5.0 | 39.0 | 23.9 | 805 | 387 | 378 | 29 | 981 | 159 |
| El Cajon.......................... | 105.2 | 6.5 | 47.2 | 68.4 | 661 | 215 | 443 | 111 | 1,070 | 173 |
| El Centro........................ | 191.6 | 8.4 | 11.5 | 26.3 | 596 | 236 | 351 | 216 | 4,890 | 918 |
| Elk Grove........................ | 226.4 | 24.4 | 53.8 | 69.4 | 405 | 146 | 254 | 112 | 654 | 122 |
| El Monte.......................... | 98.8 | 9.8 | 38.9 | 76.4 | 662 | 301 | 346 | 102 | 880 | 144 |
| El Paso de Robles (Paso Robles)................... | 65.9 | 3.1 | 51.7 | 39.3 | 1,238 | 450 | 781 | 53 | 1,677 | 326 |
| Encinitas......................... | 88.6 | 5.2 | 31.5 | 61.8 | 982 | 695 | 279 | 126 | 2,010 | 420 |
| Escondido ...................... | 175.6 | 12.6 | 60.9 | 89.9 | 595 | 263 | 320 | 181 | 1,195 | 122 |
| Eureka............................ | 75.2 | 30.1 | 12.4 | 26.2 | 965 | 158 | 803 | 59 | 2,158 | 302 |
| Fairfield.......................... | 170.2 | 42.3 | 10.2 | 105.8 | 913 | 367 | 541 | 135 | 1,167 | 236 |
| Folsom........................... | 133.3 | 13.5 | 49.9 | 71.0 | 913 | 413 | 406 | 118 | 1,510 | 228 |
| Fontana.......................... | 297.6 | 16.5 | 39.6 | 168.6 | 799 | 514 | 268 | 224 | 1,063 | 269 |
| Foster City...................... | 83.7 | 1.2 | 100.0 | 41.1 | 1,198 | 794 | 363 | 47 | 1,366 | 53 |
| Fountain Valley................ | 60.2 | 5.5 | 35.7 | 40.1 | 715 | 330 | 371 | 69 | 1,233 | 219 |
| Fremont........................... | 311.6 | 20.1 | 58.8 | 244.3 | 1,042 | 381 | 652 | 283 | 1,208 | 128 |
| Fresno............................ | 767.4 | 154.7 | 39.0 | 322.1 | 613 | 244 | 366 | 441 | 838 | 4 |
| Fullerton......................... | 171.6 | 30.6 | 86.1 | 88.9 | 636 | 367 | 241 | 170 | 1,217 | 287 |
| Gardena.......................... | 106.7 | 46.4 | 11.0 | 47.9 | 800 | 213 | 433 | 82 | 1,363 | 161 |
| Garden Grove .................. | 219.2 | 9.9 | 56.9 | 120.8 | 697 | 362 | 313 | 248 | 1,431 | 418 |
| Gilroy............................. | 82.8 | 3.1 | 67.7 | 41.8 | 727 | 250 | 468 | 91 | 1,575 | 375 |
| Glendale.......................... | 379.8 | 20.7 | 57.5 | 182.9 | 906 | 378 | 523 | 322 | 1,596 | 90 |
| Glendora ......................... | 42.2 | 4.8 | 51.6 | 29.4 | 566 | 211 | 286 | 36 | 684 | 66 |
| Goleta ............................ | 44.5 | 2.4 | 47.4 | 25.3 | 812 | 196 | 611 | 41 | 1,326 | 419 |
| Hanford........................... | 52.2 | 3.7 | 56.8 | 28.6 | 507 | 177 | 237 | 58 | 1,031 | 125 |
| Hawthorne....................... | 99.1 | 7.4 | 54.1 | 65.9 | 755 | 232 | 519 | 106 | 1,219 | 103 |
| Hayward.......................... | 250.4 | 23.7 | 56.5 | 165.7 | 1,034 | 356 | 606 | 226 | 1,412 | 262 |
| Hemet............................. | 55.4 | 4.7 | 66.2 | 37.3 | 440 | 168 | 268 | 65 | 762 | 134 |
| Hesperia.......................... | 66.9 | 5.5 | 46.8 | 48.4 | 512 | 278 | 232 | 98 | 1,033 | 326 |
| Highland.......................... | 28.0 | 3.8 | 51.8 | 21.1 | 383 | 210 | 94 | 48 | 861 | 214 |
| Hollister.......................... | 56.8 | 7.1 | 19.6 | 20.7 | 539 | 112 | 326 | 39 | 1,010 | 13 |
| Huntington Beach............. | 276.3 | 14.8 | 68.9 | 178.5 | 888 | 418 | 464 | 265 | 1,320 | 135 |
| Huntington Park ............... | 62.4 | 8.7 | 52.6 | 37.0 | 633 | 311 | 310 | 54 | 925 | 74 |
| Imperial Beach................. | 26.3 | 2.6 | 93.2 | 11.7 | 428 | 204 | 204 | 27 | 982 | 88 |
| Indio.............................. | 96.2 | 10.1 | 49.7 | 59.5 | 666 | 255 | 406 | 106 | 1,183 | 132 |
| Inglewood........................ | 191.4 | 30.2 | 9.2 | 120.3 | 1,095 | 441 | 571 | 240 | 2,184 | 279 |
| Irvine............................. | 307.9 | 12.6 | 60.1 | 214.3 | 777 | 225 | 343 | 305 | 1,106 | 84 |
| Laguna Hills..................... | 24.2 | 1.8 | 61.8 | 20.1 | 634 | 309 | 314 | 31 | 978 | 258 |
| Laguna Niguel................... | 46.6 | 2.5 | 69.8 | 38.3 | 581 | 314 | 258 | 59 | 890 | 195 |
| La Habra ......................... | 70.0 | 12.2 | 59.7 | 37.1 | 610 | 261 | 345 | 90 | 1,472 | 318 |
| Lake Elsinore ................... | 71.6 | 6.0 | 94.4 | 30.1 | 455 | 116 | 278 | 73 | 1,099 | 172 |
| Lake Forest...................... | 90.9 | 4.9 | 40.8 | 45.4 | 541 | 204 | 328 | 96 | 1,139 | 355 |
| Lakewood........................ | 60.0 | 7.4 | 86.5 | 39.8 | 495 | 185 | 305 | 57 | 712 | 89 |
| La Mesa ......................... | 64.2 | 4.4 | 76.6 | 42.1 | 705 | 240 | 461 | 57 | 1,207 | 186 |
| La Mirada........................ | 54.3 | 6.4 | 17.5 | 36.2 | 742 | 225 | 512 | 72 | 1,169 | 438 |
| Lancaster........................ | 116.7 | 9.7 | 57.7 | 85.9 | 539 | 237 | 240 | 57 | 937 | 185 |
| La Puente........................ | 18.8 | 1.7 | 78.1 | 12.8 | 319 | 149 | 168 | 150 | 455 | 83 |
| La Quinta........................ | 79.8 | 6.9 | 22.5 | 54.9 | 1,332 | 790 | 528 | 18 | 1,555 | 228 |
| La Verne ......................... | 41.5 | 1.9 | 86.0 | 22.6 | 700 | 315 | 381 | 64 | 1,295 | 58 |
| Lawndale......................... | 18.7 | 3.5 | 23.9 | 13.6 | 413 | 139 | 252 | 42 | 892 | 217 |
| Lemon Grove ................... | 24.9 | 3.4 | 96.0 | 13.4 | 498 | 177 | 317 | 34 | 1,272 | 196 |
| Lincoln........................... | 90.3 | 4.5 | 56.1 | 20.4 | 429 | 215 | 205 | 57 | 1,192 | 150 |
| Livermore........................ | 172.3 | 7.9 | 46.6 | 94.0 | 1,043 | 456 | 546 | 204 | 2,260 | 338 |
| Lodi............................... | 102.0 | 17.6 | 62.1 | 42.1 | 640 | 222 | 409 | 118 | 1,789 | 210 |
| Lompoc........................... | 74.0 | 8.9 | 33.8 | 20.2 | 468 | 173 | 292 | 67 | 1,549 | 185 |
| Long Beach...................... | 1,548.9 | 258.1 | 27.0 | 438.3 | 939 | 415 | 516 | 1,648 | 3,531 | 933 |
| Los Altos........................ | 48.7 | 2.3 | 54.4 | 34.1 | 1,112 | 645 | 461 | 47 | 1,535 | 191 |
| Los Angeles..................... | 10,825.2 | 1,131.0 | 26.5 | 4,708.7 | 1,184 | 490 | 639 | 9,618 | 2,419 | 399 |
| Los Banos ....................... | 42.4 | 2.4 | 87.9 | 18.6 | 479 | 146 | 155 | 34 | 878 | 150 |
| Los Gatos........................ | 41.5 | 1.8 | 97.4 | 33.2 | 1,073 | 456 | 596 | 51 | 1,653 | 437 |
| Lynwood.......................... | 49.1 | 6.3 | 92.4 | 27.8 | 393 | 176 | 216 | 48 | 671 | 126 |
| Madera............................ | 58.4 | 10.0 | 35.2 | 23.8 | 366 | 145 | 219 | 62 | 947 | 233 |
| Manhattan Beach............. | 88.6 | 4.2 | 40.2 | 53.7 | 1,503 | 790 | 691 | 86 | 2,415 | 205 |
| Manteca.......................... | 112.8 | 7.4 | 56.0 | 44.4 | 561 | 165 | 249 | 114 | 1,435 | 392 |
| Martinez.......................... | 29.9 | 1.6 | 99.2 | 25.1 | 654 | 346 | 290 | 27 | 711 | 54 |
| Maywood......................... | 11.1 | 1.2 | 50.2 | 8.7 | 318 | 113 | 161 | 12 | 454 | 27 |
| Menifee........................... | 49.7 | 4.5 | 76.4 | 31.6 | 350 | 136 | 208 | 89 | 983 | 190 |

1. Based on population estimated as of July 1 of the year shown.

| City | | | | City government finances, 2017 (cont.) | | | | | |
|---|---|---|---|---|---|---|---|---|---|
| | | | | General expenditure (cont.) | | | | | |
| | | | | Percent of total for: | | | | | |
| | Public welfare | Highways | Parking facilities | Education | Health and hospitals | Police protection | Sewerage and sanitation | Parks and recreation | Housing and community development | Interest on debt |
| | 127 | 128 | 129 | 130 | 131 | 132 | 133 | 134 | 135 | 136 |
| CALIFORNIA— Cont'd | | | | | | | | | | |
| Delano | 0.0 | 7.5 | 0.0 | 0.0 | 0.0 | 24.8 | 11.0 | 5.0 | 2.3 | 4.1 |
| Desert Hot Springs | 0.0 | 18.2 | 0.0 | 0.0 | 1.3 | 21.9 | 0.0 | 3.3 | 10.6 | 15.3 |
| Diamond Bar | 0.0 | 14.6 | 0.0 | 0.0 | 0.5 | 14.2 | 1.3 | 13.8 | 1.1 | 1.0 |
| Downey | 0.0 | 4.2 | 0.0 | 0.0 | 4.8 | 33.4 | 2.9 | 13.2 | 5.2 | 1.1 |
| Dublin | 0.0 | 8.8 | 0.0 | 0.0 | 0.7 | 20.1 | 4.5 | 14.7 | 2.4 | 0.2 |
| East Palo Alto | 0.0 | 15.9 | 0.0 | 0.0 | 0.0 | 34.0 | 8.6 | 4.3 | 7.2 | 0.0 |
| El Cajon | 0.0 | 7.0 | 0.0 | 0.0 | 2.8 | 31.1 | 16.8 | 4.8 | 10.0 | 2.9 |
| El Centro | 0.0 | 3.2 | 0.0 | 0.0 | 75.2 | 5.1 | 3.9 | 2.4 | 1.3 | 2.3 |
| Elk Grove | 0.0 | 17.2 | 0.0 | 0.0 | 0.0 | 31.8 | 14.5 | 0.6 | 5.0 | 0.7 |
| El Monte | 0.0 | 6.9 | 0.1 | 0.0 | 0.4 | 25.0 | 1.7 | 7.4 | 11.4 | 3.4 |
| El Paso de Robles (Paso Robles) | 0.0 | 13.9 | 0.0 | 0.0 | 0.0 | 19.8 | 11.1 | 9.4 | 0.7 | 6.0 |
| Encinitas | 0.0 | 7.4 | 0.0 | 0.0 | 1.4 | 20.7 | 8.2 | 8.7 | 1.6 | 2.7 |
| Escondido | 0.0 | 6.3 | 0.1 | 0.0 | 0.8 | 23.1 | 23.6 | 5.0 | 0.9 | 3.5 |
| Eureka | 0.0 | 6.4 | 0.2 | 0.0 | 0.1 | 21.4 | 18.5 | 5.5 | 11.8 | 5.4 |
| Fairfield | 0.0 | 25.5 | 0.0 | 0.0 | 0.2 | 27.4 | 0.1 | 10.7 | 11.6 | 1.6 |
| Folsom | 0.0 | 7.1 | 0.0 | 0.0 | 0.9 | 19.2 | 10.4 | 11.4 | 5.8 | 4.2 |
| Fontana | 0.0 | 13.0 | 0.0 | 0.0 | 0.0 | 26.6 | 7.9 | 6.8 | 13.0 | 7.3 |
| Foster City | 0.0 | 3.9 | 0.0 | 0.0 | 0.0 | 24.3 | 15.9 | 17.2 | 1.2 | 0.0 |
| Fountain Valley | 0.0 | 17.8 | 0.0 | 0.0 | 5.3 | 23.9 | 7.7 | 6.2 | 8.3 | 1.5 |
| Fremont | 0.1 | 14.7 | 0.0 | 0.0 | 6.2 | 25.6 | 12.5 | 5.4 | 1.7 | 2.1 |
| Fresno | 0.0 | 9.2 | 0.9 | 0.0 | 1.5 | 31.3 | 14.3 | 5.5 | 1.8 | 10.7 |
| Fullerton | 0.0 | 28.7 | 0.0 | 0.0 | 0.8 | 26.7 | 7.8 | 5.4 | 1.0 | 4.9 |
| Gardena | 0.0 | 11.0 | 0.0 | 0.0 | 1.0 | 39.1 | 2.3 | 10.4 | 3.0 | 1.6 |
| Garden Grove | 0.0 | 5.9 | 0.0 | 0.0 | 2.2 | 22.2 | 2.5 | 2.3 | 43.4 | 3.3 |
| Gilroy | 0.0 | 3.9 | 0.0 | 0.0 | 0.1 | 23.4 | 10.7 | 1.8 | 25.0 | 4.4 |
| Glendale | 0.0 | 3.5 | 2.3 | 0.0 | 1.8 | 23.9 | 14.6 | 5.2 | 10.9 | 1.9 |
| Glendora | 0.0 | 9.6 | 0.0 | 0.0 | 0.4 | 44.6 | 0.0 | 10.3 | 9.3 | 1.5 |
| Goleta | 0.0 | 32.2 | 0.0 | 0.0 | 0.9 | 18.7 | 0.0 | 7.5 | 9.8 | 2.8 |
| Hanford | 0.0 | 11.0 | 0.0 | 0.0 | 1.5 | 20.3 | 26.9 | 6.5 | 1.7 | 1.4 |
| Hawthorne | 0.0 | 12.8 | 0.0 | 0.0 | 0.0 | 33.4 | 0.8 | 3.4 | 12.7 | 3.4 |
| Hayward | 0.0 | 19.0 | 0.0 | 0.1 | 0.9 | 30.5 | 10.1 | 0.5 | 0.7 | 2.6 |
| Hemet | 0.0 | 15.4 | 0.0 | 0.0 | 0.5 | 33.1 | 5.7 | 2.4 | 10.2 | 0.7 |
| Hesperia | 0.0 | 21.6 | 0.0 | 0.0 | 1.5 | 15.9 | 0.0 | 0.0 | 21.9 | 10.4 |
| Highland | 0.0 | 26.0 | 0.0 | 0.0 | 6.8 | 36.4 | 0.0 | 2.3 | 7.1 | 0.0 |
| Hollister | 0.0 | 6.0 | 0.0 | 0.0 | 1.4 | 17.1 | 17.9 | 2.6 | 0.0 | 8.6 |
| Huntington Beach | 0.0 | 11.0 | 0.8 | 0.0 | 2.3 | 31.7 | 7.5 | 7.0 | 3.2 | 2.2 |
| Huntington Park | 0.0 | 7.2 | 1.5 | 0.0 | 1.3 | 32.4 | 0.5 | 3.4 | 4.5 | 33.0 |
| Imperial Beach | 0.0 | 8.0 | 0.0 | 0.0 | 1.1 | 32.3 | 15.4 | 2.1 | 5.4 | 5.7 |
| Indio | 0.0 | 9.0 | 0.0 | 0.0 | 4.0 | 23.1 | 0.0 | 4.8 | 12.8 | 5.3 |
| Inglewood | 0.0 | 6.2 | 2.6 | 0.0 | 0.6 | 25.7 | 7.7 | 6.9 | 16.9 | 4.4 |
| Irvine | 0.0 | 16.5 | 0.0 | 0.2 | 1.9 | 23.9 | 0.0 | 20.8 | 2.7 | 12.5 |
| Laguna Hills | 0.0 | 20.6 | 0.0 | 0.0 | 0.8 | 25.2 | 0.0 | 25.9 | 0.2 | 1.5 |
| Laguna Niguel | 0.0 | 20.3 | 0.0 | 0.0 | 1.2 | 40.5 | 0.0 | 15.2 | 0.1 | 0.0 |
| La Habra | 0.0 | 21.3 | 0.0 | 0.0 | 1.9 | 20.6 | 7.7 | 10.6 | 6.1 | 1.7 |
| Lake Elsinore | 0.0 | 1.5 | 0.0 | 0.0 | 1.1 | 16.1 | 0.0 | 22.1 | 11.9 | 15.6 |
| Lake Forest | 0.0 | 31.5 | 0.0 | 0.0 | 1.2 | 16.7 | 0.0 | 15.6 | 1.0 | 0.6 |
| Lakewood | 0.0 | 17.7 | 0.0 | 0.0 | 0.7 | 21.4 | 8.8 | 23.0 | 4.5 | 0.2 |
| La Mesa | 0.0 | 17.1 | 0.3 | 0.0 | 0.8 | 25.0 | 21.1 | 5.5 | 3.7 | 1.7 |
| La Mirada | 0.0 | 43.0 | 0.0 | 0.0 | 0.0 | 14.8 | 0.0 | 22.7 | 10.0 | 5.2 |
| Lancaster | 0.0 | 23.2 | 0.0 | 0.0 | 0.4 | 19.9 | 0.0 | 8.1 | 2.9 | 5.9 |
| La Puente | 0.0 | 15.6 | 0.0 | 0.0 | 0.0 | 33.5 | 2.9 | 17.7 | 8.9 | 4.0 |
| La Quinta | 0.0 | 12.4 | 0.0 | 0.0 | 2.3 | 23.1 | 0.0 | 10.8 | 3.5 | 14.6 |
| La Verne | 0.0 | 6.2 | 0.0 | 0.0 | 8.0 | 31.9 | 8.4 | 9.0 | 20.0 | 2.0 |
| Lawndale | 0.0 | 9.8 | 0.0 | 0.0 | 0.2 | 37.4 | 0.0 | 9.2 | 9.4 | 3.6 |
| Lemon Grove | 0.0 | 7.0 | 0.0 | 0.0 | 0.6 | 31.7 | 13.1 | 0.0 | 4.0 | 3.3 |
| Lincoln | 0.0 | 13.6 | 0.0 | 0.0 | 0.0 | 9.5 | 23.9 | 14.9 | 4.6 | 2.6 |
| Livermore | 0.0 | 12.0 | 0.0 | 0.9 | 0.4 | 15.8 | 10.9 | 1.6 | 1.3 | 1.9 |
| Lodi | 0.0 | 7.4 | 0.0 | 0.0 | 0.4 | 16.9 | 11.8 | 5.0 | 7.7 | 2.5 |
| Lompoc | 0.0 | 9.1 | 0.0 | 0.0 | 0.7 | 16.4 | 22.6 | 6.8 | 4.8 | 5.5 |
| Long Beach | 0.0 | 6.8 | 0.0 | 0.0 | 4.4 | 14.4 | 5.2 | 6.3 | 0.7 | 2.8 |
| Los Altos | 0.0 | 14.1 | 0.0 | 0.0 | 0.0 | 23.1 | 14.6 | 6.0 | 2.5 | 0.1 |
| Los Angeles | 0.0 | 5.7 | 0.4 | 0.0 | 3.9 | 22.8 | 12.1 | 4.8 | 1.9 | 5.6 |
| Los Banos | 0.0 | 9.6 | 0.0 | 0.0 | 0.0 | 24.1 | 30.3 | 15.8 | 1.2 | 0.4 |
| Los Gatos | 0.0 | 13.5 | 0.0 | 0.0 | 0.4 | 26.0 | 0.0 | 6.4 | 11.3 | 2.0 |
| Lynwood | 0.0 | 20.1 | 0.0 | 0.0 | 5.7 | 23.8 | 0.0 | 9.2 | 3.7 | 2.0 |
| Madera | 0.0 | 25.7 | 0.1 | 0.0 | 1.2 | 19.2 | 19.5 | 8.8 | 0.0 | 4.8 |
| Manhattan Beach | 0.0 | 7.9 | 4.5 | 0.0 | 2.8 | 31.7 | 7.5 | 15.0 | 8.0 | 0.9 |
| Manteca | 0.0 | 17.2 | 0.0 | 0.0 | 0.3 | 39.8 | 0.2 | 13.3 | 4.3 | 1.6 |
| Martinez | 0.0 | 9.8 | 0.8 | 0.0 | 0.6 | 30.0 | 0.0 | 3.3 | 3.2 | 4.4 |
| Maywood | 0.0 | 18.8 | 0.0 | 0.0 | 0.6 | 26.9 | 0.0 | 0.8 | | 1.6 |
| Menifee | 0.0 | 28.4 | | | 0.8 | | | | | 1.0 |

# Table D. Cities — City Government Finances, City Government Employment, and Climate

| City | City government finances, 2017 (cont.) | | | Climate[2] | | | | | | |
|---|---|---|---|---|---|---|---|---|---|---|
| | Debt outstanding | | | Average daily temperature | | | | | | |
| | | | | Mean | | Limits | | | | |
| | Total (mil dol) | Per capita[1] (dollars) | Debt issued during year | January | July | January[3] | July[4] | Annual precipitation (inches) | Heating degree days | Cooling degree days |
| | 137 | 138 | 139 | 140 | 141 | 142 | 143 | 144 | 145 | 146 |
| CALIFORNIA— Cont'd | | | | | | | | | | |
| Delano | 65.2 | 1,233 | 1.3 | 46.6 | 81.1 | 36.5 | 99.0 | 7.34 | 2,434 | 1,990 |
| Desert Hot Springs | 75.9 | 2,650 | 35.4 | NA | NA | NA | NA | NA | NA | NA |
| Diamond Bar | 9.6 | 171 | 0.0 | 54.6 | 73.8 | 41.5 | 88.7 | 16.96 | 1,727 | 1,191 |
| Downey | 28.4 | 253 | 0.7 | 57.0 | 73.8 | 46.0 | 82.9 | 12.94 | 1,211 | 1,186 |
| Dublin | 10.4 | 171 | 5.5 | 47.2 | 72.0 | 37.4 | 89.1 | 14.82 | 2,755 | 858 |
| East Palo Alto | 27.1 | 914 | 0.1 | 49.0 | 68.0 | 40.4 | 78.8 | 15.71 | 2,584 | 452 |
| El Cajon | 61.5 | 594 | 0.0 | 54.9 | 74.7 | 41.6 | 87.0 | 11.96 | 1,560 | 1,371 |
| El Centro | 157.7 | 3,576 | 0.0 | 55.8 | 91.4 | 41.3 | 107.0 | 2.96 | 1,080 | 3,852 |
| Elk Grove | 14.4 | 84 | 4.9 | 46.3 | 75.4 | 38.8 | 92.4 | 17.93 | 2,666 | 1,248 |
| El Monte | 72.6 | 629 | 32.0 | 56.3 | 75.6 | 42.6 | 89.0 | 18.56 | 1,295 | 1,575 |
| El Paso de Robles (Paso Robles) | 87.9 | 2,767 | 0.0 | NA | NA | NA | NA | NA | NA | NA |
| Encinitas | 60.8 | 967 | 0.0 | 55.5 | 75.1 | 42.5 | 88.6 | 15.10 | 1,464 | 1,436 |
| Escondido | 242.8 | 1,606 | 3.2 | 55.5 | 75.1 | 42.5 | 88.6 | 15.10 | 1,464 | 1,436 |
| Eureka | 60.6 | 2,234 | 0.0 | 47.9 | 58.1 | 40.8 | 63.3 | 38.10 | 4,430 | 7 |
| Fairfield | 162.8 | 1,406 | 40.3 | 46.1 | 72.6 | 37.5 | 88.8 | 23.46 | 2,649 | 975 |
| Folsom | 219.5 | 2,821 | 0.0 | 46.9 | 77.7 | 39.2 | 94.8 | 24.61 | 2,532 | 1,528 |
| Fontana | 390.2 | 1,848 | 0.0 | 56.6 | 78.3 | 45.3 | 95.0 | 14.77 | 1,364 | 1,901 |
| Foster City | 0.0 | 0 | 0.0 | 48.4 | 68.0 | 39.1 | 80.8 | 20.16 | 2,764 | 422 |
| Fountain Valley | 44.7 | 797 | 2.8 | 58.0 | 72.9 | 46.6 | 87.7 | 13.84 | 1,153 | 1,299 |
| Fremont | 204.3 | 871 | 85.2 | 49.8 | 68.0 | 42.0 | 78.3 | 14.85 | 2,367 | 530 |
| Fresno | 1,206.3 | 2,296 | 272.4 | 46.0 | 81.4 | 38.4 | 96.6 | 11.23 | 2,447 | 1,963 |
| Fullerton | 106.7 | 763 | 2.5 | 56.9 | 73.2 | 45.2 | 84.0 | 11.23 | 1,286 | 1,294 |
| Gardena | 21.9 | 365 | 1.6 | 56.3 | 69.4 | 46.2 | 77.6 | 14.79 | 1,526 | 742 |
| Garden Grove | 127.7 | 736 | 16.0 | 58.0 | 72.9 | 46.6 | 87.7 | 13.84 | 1,153 | 1,299 |
| Gilroy | 84.8 | 1,474 | 0.0 | 49.7 | 72.0 | 39.4 | 88.3 | 20.60 | 2,278 | 913 |
| Glendale | 382.6 | 1,896 | 0.0 | 54.8 | 75.5 | 42.0 | 88.9 | 17.49 | 1,575 | 1,455 |
| Glendora | 34.3 | 660 | 0.0 | 54.6 | 73.8 | 41.5 | 88.7 | 16.96 | 1,727 | 1,191 |
| Goleta | 14.9 | 479 | 0.0 | 53.1 | 67.0 | 40.8 | 76.7 | 16.93 | 2,121 | 482 |
| Hanford | 41.6 | 737 | 0.0 | 44.7 | 79.6 | 35.7 | 95.9 | 8.58 | 2,749 | 1,724 |
| Hawthorne | 70.2 | 804 | 0.0 | 57.1 | 69.3 | 48.6 | 75.3 | 13.15 | 1,274 | 679 |
| Hayward | 211.4 | 1,318 | 0.9 | 49.7 | 64.6 | 41.7 | 75.2 | 26.30 | 2,810 | 261 |
| Hemet | 10.1 | 119 | 0.0 | 52.4 | 79.9 | 38.4 | 97.8 | 12.55 | 1,914 | 1,903 |
| Hesperia | 185.2 | 1,961 | 0.0 | 45.5 | 80.0 | 31.4 | 99.1 | 6.20 | 2,929 | 1,735 |
| Highland | 11.2 | 204 | 0.0 | 54.4 | 79.6 | 41.8 | 96.0 | 16.43 | 1,599 | 1,937 |
| Hollister | 69.0 | 1,797 | 1.5 | 49.5 | 66.6 | 37.8 | 80.9 | 13.61 | 2,724 | 405 |
| Huntington Beach | 142.1 | 707 | 5.8 | 55.9 | 67.3 | 48.2 | 71.4 | 11.65 | 1,719 | 543 |
| Huntington Park | 259.7 | 4,441 | 13.7 | 58.3 | 74.2 | 48.5 | 83.8 | 15.14 | 928 | 1,506 |
| Imperial Beach | 37.8 | 1,385 | 0.0 | 57.3 | 70.1 | 46.1 | 76.1 | 9.95 | 1,321 | 862 |
| Indio | 181.1 | 2,025 | 16.3 | 56.8 | 92.8 | 42.0 | 107.1 | 3.15 | 903 | 4,388 |
| Inglewood | 236.9 | 2,155 | 0.0 | 57.1 | 69.3 | 48.6 | 75.3 | 13.15 | 1,274 | 679 |
| Irvine | 1,066.9 | 3,866 | 166.0 | 54.5 | 72.1 | 41.4 | 83.8 | 13.87 | 1,794 | 1,102 |
| Laguna Hills | 8.7 | 276 | 0.0 | 56.7 | 72.4 | 47.2 | 82.3 | 14.03 | 1,465 | 1,183 |
| Laguna Niguel | 0.3 | 4 | 0.0 | 55.4 | 68.7 | 43.9 | 77.3 | 13.56 | 1,756 | 666 |
| La Habra | 112.3 | 1,845 | 1.2 | 56.9 | 73.2 | 45.2 | 84.0 | 11.23 | 1,286 | 1,294 |
| Lake Elsinore | 246.2 | 3,727 | 0.0 | 52.2 | 79.6 | 38.3 | 98.1 | 12.09 | 1,924 | 1,874 |
| Lake Forest | 14.4 | 171 | 0.0 | 56.7 | 72.4 | 47.2 | 82.3 | 14.03 | 1,465 | 1,183 |
| Lakewood | 2.0 | 25 | 0.0 | 57.0 | 73.8 | 46.0 | 82.9 | 12.94 | 1,211 | 1,186 |
| La Mesa | 43.6 | 730 | 1.9 | 57.1 | 73.0 | 45.7 | 83.6 | 13.75 | 1,313 | 1,261 |
| La Mirada | 70.8 | 1,450 | 1.0 | 58.8 | 76.6 | 47.9 | 88.9 | 14.44 | 949 | 1,837 |
| Lancaster | 233.7 | 1,466 | 80.5 | 43.9 | 80.8 | 31.0 | 95.5 | 7.40 | 3,241 | 1,733 |
| La Puente | 33.7 | 842 | 11.7 | 58.8 | 76.6 | 47.9 | 88.9 | 14.44 | 949 | 1,837 |
| La Quinta | 211.7 | 5,140 | 37.8 | NA | NA | NA | NA | NA | NA | NA |
| La Verne | 10.6 | 327 | 0.0 | 54.6 | 73.8 | 41.5 | 88.7 | 16.96 | 1,727 | 1,191 |
| Lawndale | 19.6 | 597 | 0.0 | 57.1 | 69.3 | 48.6 | 75.3 | 13.15 | 1,274 | 679 |
| Lemon Grove | 27.7 | 1,028 | 0.0 | NA | NA | NA | NA | NA | NA | NA |
| Lincoln | 27.9 | 588 | 27.6 | NA | NA | NA | NA | NA | NA | NA |
| Livermore | 124.3 | 1,379 | 25.1 | 47.2 | 72.0 | 37.4 | 89.1 | 14.82 | 2,755 | 858 |
| Lodi | 152.6 | 2,321 | 0.0 | 46.1 | 73.8 | 37.5 | 91.1 | 18.22 | 2,710 | 1,057 |
| Lompoc | 41.1 | 950 | 0.0 | 53.7 | 64.5 | 41.4 | 75.4 | 15.85 | 2,250 | 322 |
| Long Beach | 2,367.5 | 5,073 | 171.7 | 57.0 | 73.8 | 46.0 | 82.9 | 12.94 | 1,211 | 1,186 |
| Los Altos | 1.4 | 47 | 0.0 | 49.0 | 68.0 | 40.4 | 78.8 | 15.71 | 2,584 | 452 |
| Los Angeles | 27,626.9 | 6,949 | 3,433.3 | 58.3 | 74.2 | 48.5 | 83.8 | 15.14 | 928 | 1,506 |
| Los Banos | 2.0 | 52 | 0.0 | 45.9 | 78.1 | 36.8 | 94.6 | 9.95 | 2,570 | 1,547 |
| Los Gatos | 24.6 | 794 | 0.0 | 48.7 | 70.3 | 38.8 | 85.4 | 22.64 | 2,641 | 613 |
| Lynwood | 40.3 | 571 | 7.7 | 58.3 | 74.2 | 48.5 | 83.8 | 15.14 | 928 | 1,506 |
| Madera | 94.5 | 1,455 | 0.0 | 45.7 | 79.6 | 37.2 | 96.5 | 11.94 | 2,670 | 1,706 |
| Manhattan Beach | 25.0 | 700 | 5.9 | 57.1 | 69.3 | 48.6 | 75.3 | 13.15 | 1,274 | 679 |
| Manteca | 69.7 | 882 | 0.0 | 46.0 | 77.3 | 38.1 | 93.8 | 13.84 | 2,563 | 1,456 |
| Martinez | 33.7 | 877 | 0.0 | 46.3 | 71.2 | 38.8 | 87.4 | 19.58 | 2,757 | 786 |
| Maywood | 2.8 | 104 | 0.0 | 58.3 | 74.2 | 48.5 | 83.8 | 15.14 | 928 | 1,506 |
| Menifee | 18.3 | 203 | 0.0 | NA | NA | NA | NA | NA | NA | NA |

1. Based on the population estimated as of July 1 of the year shown.    2. Represents normal values based on the 30-year period, 1971±2000.    3. Average daily minimum.    4. Average daily maximum.

| STATE Place code | City | Land area[1] (sq. mi) | Population, 2020 | | | Race 2019 Race alone[2] (percent) | | | | | | |
|---|---|---|---|---|---|---|---|---|---|---|---|---|
| | | | Total persons 2019 | Rank | Per square mile | White | Black or African American | American Indian, Alaskan Native | Asian | Hawaiian Pacific Islander | Some other race | Two or more races (percent) |
| | | 1 | 2 | 3 | 4 | 5 | 6 | 7 | 8 | 9 | 10 | 11 |
| | CALIFORNIA— Cont'd | | | | | | | | | | | |
| 06 46870 | Menlo Park | 10.0 | 34,934 | 1,106 | 3,493 | 66.6 | 5.3 | 0.0 | 16.5 | 0.5 | 4.8 | 6.2 |
| 06 46898 | Merced | 23.3 | 88,235 | 381 | 3,787 | 38.0 | 3.3 | 0.8 | 12.9 | 0.0 | 40.9 | 4.2 |
| 06 47766 | Milpitas | 13.5 | 75,867 | 471 | 5,620 | 15.5 | 3.5 | 0.9 | 65.7 | 0.0 | 8.0 | 6.5 |
| 06 48256 | Mission Viejo | 17.7 | 94,242 | 351 | 5,324 | 74.4 | 2.0 | 0.1 | 11.2 | 0.1 | 6.0 | 6.1 |
| 06 48354 | Modesto | 43.0 | 215,666 | 104 | 5,016 | 78.6 | 4.9 | 0.8 | 6.0 | 0.6 | 5.0 | 4.0 |
| 06 48648 | Monrovia | 13.6 | 37,771 | 1,025 | 2,777 | 66.9 | 8.9 | 0.1 | 15.1 | 0.7 | 4.0 | 4.4 |
| 06 48788 | Montclair | 5.5 | 40,600 | 948 | 7,382 | 58.4 | 3.9 | 1.0 | 10.1 | 0.0 | 23.8 | 2.7 |
| 06 48816 | Montebello | 8.3 | 61,531 | 608 | 7,413 | 56.9 | 0.8 | 1.2 | 11.0 | 0.1 | 26.7 | 3.2 |
| 06 48872 | Monterey | 8.6 | 28,511 | 1,298 | 3,315 | 79.2 | 4.6 | 0.0 | 6.9 | 0.3 | 4.2 | 4.7 |
| 06 48914 | Monterey Park | 7.7 | 59,344 | 634 | 7,707 | 18.8 | 1.0 | 0.1 | 65.9 | 0.0 | 12.0 | 2.2 |
| 06 49138 | Moorpark | 12.3 | 36,202 | 1,068 | 2,943 | 79.6 | 1.3 | 1.5 | 4.0 | 0.1 | 6.5 | 6.9 |
| 06 49270 | Moreno Valley | 51.3 | 212,349 | 105 | 4,139 | 34.5 | 18.8 | 0.5 | 6.4 | 0.1 | 37.2 | 2.5 |
| 06 49278 | Morgan Hill | 12.9 | 45,501 | 852 | 3,527 | 74.3 | 1.3 | 0.2 | 13.0 | 0.0 | 3.4 | 7.8 |
| 06 49670 | Mountain View | 12.0 | 80,030 | 437 | 6,669 | 55.2 | 1.6 | 0.3 | 30.9 | 0.0 | 7.6 | 4.4 |
| 06 50076 | Murrieta | 33.6 | 116,522 | 251 | 3,468 | 73.2 | 6.1 | 0.4 | 5.8 | 0.3 | 7.2 | 7.0 |
| 06 50258 | Napa | 18.0 | 77,032 | 460 | 4,280 | 76.9 | 1.0 | 0.3 | 2.3 | 0.1 | 15.4 | 4.0 |
| 06 50398 | National City | 7.3 | 60,932 | 614 | 8,347 | 68.6 | 5.1 | 0.4 | 16.8 | 0.2 | 5.9 | 3.1 |
| 06 50916 | Newark | 13.9 | 49,699 | 779 | 3,576 | 30.9 | 2.5 | 0.6 | 37.6 | 1.0 | 23.2 | 4.3 |
| 06 51182 | Newport Beach | 23.8 | 86,834 | 389 | 3,649 | 83.3 | 0.7 | 0.8 | 9.0 | 0.2 | 2.8 | 3.3 |
| 06 51560 | Norco | 13.9 | 27,892 | 1,320 | 2,007 | 75.3 | 3.5 | 0.3 | 7.8 | 0.0 | 9.2 | 3.9 |
| 06 52526 | Norwalk | 9.7 | 102,635 | 302 | 10,581 | 37.2 | 5.4 | 0.7 | 12.4 | 0.9 | 40.0 | 3.3 |
| 06 52582 | Novato | 27.5 | 53,351 | 726 | 1,940 | 79.4 | 6.3 | 0.0 | 6.7 | 0.0 | 2.4 | 5.2 |
| 06 53000 | Oakland | 55.9 | 424,891 | 46 | 7,601 | 34.5 | 24.9 | 1.3 | 14.3 | 0.4 | 18.3 | 6.3 |
| 06 53070 | Oakley | 15.9 | 42,710 | 900 | 2,686 | 50.7 | 9.9 | 0.2 | 4.6 | 3.3 | 22.2 | 9.0 |
| 06 53322 | Oceanside | 41.3 | 174,648 | 149 | 4,229 | 76.4 | 4.2 | 1.3 | 4.8 | 0.8 | 4.8 | 7.7 |
| 06 53896 | Ontario | 50.0 | 183,393 | 142 | 3,668 | 62.1 | 4.4 | 0.5 | 6.8 | 0.7 | 19.7 | 5.7 |
| 06 53980 | Orange | 25.7 | 138,192 | 200 | 5,377 | 74.0 | 1.3 | 0.6 | 12.1 | 1.4 | 6.3 | 4.1 |
| 06 54652 | Oxnard | 26.5 | 207,945 | 112 | 7,847 | 76.2 | 2.5 | 1.1 | 8.7 | 0.0 | 7.5 | 3.9 |
| 06 54806 | Pacifica | 12.6 | 38,013 | 1,023 | 3,017 | 53.8 | 5.3 | 0.7 | 26.2 | 0.7 | 5.5 | 7.8 |
| 06 55156 | Palmdale | 106.1 | 150,498 | 172 | 1,419 | 59.8 | 15.9 | 3.8 | 4.6 | 0.6 | 12.9 | 2.3 |
| 06 55184 | Palm Desert | 26.8 | 53,524 | 724 | 1,997 | 78.2 | 2.3 | 0.1 | 4.5 | 0.1 | 11.9 | 3.0 |
| 06 55254 | Palm Springs | 94.5 | 48,929 | 796 | 518 | 79.9 | 7.9 | 1.6 | 3.1 | 0.2 | 4.6 | 2.8 |
| 06 55282 | Palo Alto | 24.1 | 67,008 | 553 | 2,780 | 55.0 | 2.6 | 0.1 | 34.1 | 0.1 | 1.8 | 6.3 |
| 06 55520 | Paradise | 18.3 | 4,329 | 1,429 | 237 | NA | NA | NA | NA | NA | NA | NA |
| 06 55618 | Paramount | 4.7 | 53,018 | 739 | 11,280 | 31.9 | 8.7 | 1.1 | 2.5 | 0.0 | 53.8 | 1.9 |
| 06 56000 | Pasadena | 23.0 | 141,045 | 194 | 6,132 | 49.5 | 5.8 | 0.2 | 17.4 | 0.0 | 21.3 | 5.7 |
| 06 56700 | Perris | 31.5 | 79,011 | 447 | 2,508 | 29.2 | 11.1 | 0.0 | 5.5 | 0.1 | 53.2 | 1.0 |
| 06 56784 | Petaluma | 14.4 | 60,632 | 620 | 4,211 | 74.6 | 2.3 | 0.2 | 4.7 | 0.1 | 12.1 | 6.1 |
| 06 56924 | Pico Rivera | 8.3 | 61,338 | 610 | 7,390 | 43.0 | 1.5 | 0.7 | 2.8 | 0.0 | 50.0 | 2.0 |
| 06 57456 | Pittsburg | 17.7 | 72,622 | 496 | 4,103 | 37.0 | 12.9 | 1.0 | 14.6 | 0.3 | 25.1 | 9.1 |
| 06 57526 | Placentia | 6.6 | 51,519 | 754 | 7,806 | 68.3 | 1.4 | 0.2 | 18.4 | 0.3 | 6.4 | 5.1 |
| 06 57764 | Pleasant Hill | 7.1 | 34,781 | 1,110 | 4,899 | 69.8 | 3.0 | 0.8 | 15.8 | 0.0 | 5.7 | 4.9 |
| 06 57792 | Pleasanton | 24.1 | 80,247 | 435 | 3,330 | 56.1 | 1.0 | 0.4 | 33.6 | 0.0 | 4.0 | 4.9 |
| 06 58072 | Pomona | 23.0 | 149,212 | 175 | 6,488 | 47.4 | 5.4 | 2.0 | 10.7 | 0.2 | 30.4 | 3.9 |
| 06 58240 | Porterville | 18.6 | 58,949 | 644 | 3,169 | 75.6 | 0.2 | 0.9 | 4.2 | 0.0 | 13.1 | 6.0 |
| 06 58520 | Poway | 39.1 | 49,246 | 788 | 1,260 | 62.3 | 2.2 | 0.4 | 15.3 | 0.0 | 9.9 | 9.9 |
| 06 59444 | Rancho Cordova | 34.6 | 76,166 | 469 | 2,201 | 53.7 | 10.6 | 0.4 | 14.1 | 0.8 | 7.8 | 12.6 |
| 06 59451 | Rancho Cucamonga | 40.1 | 178,849 | 146 | 4,460 | 59.2 | 8.1 | 0.5 | 14.1 | 0.0 | 10.6 | 7.5 |
| 06 59514 | Rancho Palos Verdes | 13.5 | 40,855 | 942 | 3,026 | 49.6 | 5.3 | 0.8 | 30.3 | 0.3 | 8.2 | 5.6 |
| 06 59587 | Rancho Santa Margarita | 12.9 | 47,569 | 817 | 3,688 | 76.8 | 1.0 | 1.1 | 11.9 | 0.4 | 2.9 | 5.9 |
| 06 59920 | Redding | 59.6 | 92,456 | 361 | 1,551 | 84.0 | 1.6 | 2.8 | 4.6 | 0.0 | 2.6 | 4.3 |
| 06 59962 | Redlands | 36.0 | 72,271 | 499 | 2,008 | 77.9 | 3.6 | 0.2 | 10.5 | 0.0 | 3.2 | 4.6 |
| 06 60018 | Redondo Beach | 6.2 | 65,533 | 575 | 10,570 | 69.8 | 2.0 | 0.4 | 15.9 | 0.0 | 1.8 | 10.1 |
| 06 60102 | Redwood City | 19.3 | 84,466 | 407 | 4,377 | 59.0 | 1.4 | 1.0 | 15.5 | 2.0 | 15.8 | 5.3 |
| 06 60466 | Rialto | 24.1 | 104,132 | 295 | 4,321 | 65.4 | 16.7 | 0.7 | 2.3 | 0.0 | 10.4 | 4.4 |
| 06 60620 | Richmond | 30.1 | 110,038 | 274 | 3,656 | 24.4 | 18.2 | 0.9 | 12.9 | 0.1 | 37.8 | 5.8 |
| 06 60704 | Ridgecrest | 20.9 | 29,135 | 1,283 | 1,394 | 80.0 | 6.9 | 0.0 | 3.5 | 0.0 | 3.3 | 6.3 |
| 06 62000 | Riverside | 81.2 | 330,786 | 58 | 4,074 | 58.2 | 5.3 | 0.9 | 7.8 | 0.4 | 22.6 | 4.8 |
| 06 62364 | Rocklin | 19.8 | 70,159 | 519 | 3,543 | 79.7 | 1.0 | 0.3 | 11.6 | 0.2 | 1.2 | 6.0 |
| 06 62546 | Rohnert Park | 7.3 | 42,975 | 896 | 5,887 | 80.4 | 0.9 | 0.0 | 7.7 | 0.0 | 3.4 | 7.5 |
| 06 62896 | Rosemead | 5.2 | 53,088 | 735 | 10,209 | 15.3 | 1.3 | 1.6 | 58.2 | 0.0 | 21.7 | 1.8 |
| 06 62938 | Roseville | 44.1 | 144,725 | 185 | 3,282 | 80.1 | 2.9 | 0.5 | 10.2 | 0.1 | 0.8 | 5.3 |
| 06 64000 | Sacramento | 98.6 | 512,838 | 36 | 5,201 | 44.2 | 11.7 | 0.9 | 18.4 | 1.9 | 14.8 | 8.2 |
| 06 64224 | Salinas | 23.5 | 154,868 | 164 | 6,590 | 28.3 | 1.8 | 0.3 | 5.3 | 0.0 | 61.7 | 2.5 |
| 06 65000 | San Bernardino | 62.1 | 217,491 | 102 | 3,502 | 57.0 | 15.7 | 0.3 | 4.7 | 0.4 | 17.2 | 4.7 |
| 06 65028 | San Bruno | 5.5 | 44,051 | 873 | 8,009 | 41.2 | 0.5 | 0.1 | 28.5 | 5.3 | 19.9 | 4.4 |
| 06 65042 | San Buenaventura (Ventura) | 21.9 | 107,110 | 287 | 4,891 | 85.3 | 2.0 | 0.8 | 2.7 | 0.0 | 3.6 | 5.7 |
| 06 65070 | San Carlos | 5.4 | 29,641 | 1,262 | 5,489 | 67.9 | 1.6 | 0.2 | 18.2 | 0.0 | 6.2 | 5.9 |
| 06 65084 | San Clemente | 18.4 | 64,551 | 587 | 3,508 | 83.3 | 1.7 | 0.3 | 5.3 | 0.1 | 2.4 | 6.9 |
| 06 66000 | San Diego | 325.9 | 1,422,420 | 8 | 4,365 | 65.4 | 6.1 | 0.5 | 17.0 | 0.5 | 4.9 | 5.6 |
| 06 66070 | San Dimas | 15.0 | 33,165 | 1,157 | 2,211 | 72.9 | 1.4 | 0.8 | 17.6 | 0.0 | 4.2 | 3.1 |

1. Dry land or land partially or temporarily covered by water.   2. Hispanic or Latino persons may be of any race.

# Table D. Cities — Population

| City | Percent Hispanic or Latino[1], 2019 | Percent foreign born, 2019 | Age of population (percent), 2019 | | | | | | | Median age, 2019 | Percent female, 2019 | Population — Census counts | | Population — Percent change | |
|---|---|---|---|---|---|---|---|---|---|---|---|---|---|---|---|
| | | | Under 18 years | 18 to 24 years | 25 to 34 years | 35 to 44 years | 45 to 54 years | 55 to 64 years | 65 years and over | | | 2000 | 2010 | 2000-2010 | 2010-2020 |
| | 12 | 13 | 14 | 15 | 16 | 17 | 18 | 19 | 20 | 21 | 22 | 23 | 24 | 25 | 26 |
| CALIFORNIA— Cont'd | | | | | | | | | | | | | | | |
| Menlo Park | 11.6 | 26.3 | 25.3 | 5.8 | 14.2 | 16.5 | 12.1 | 13.8 | 12.2 | 37.5 | 49.0 | 30,785 | 32,014 | 4.0 | 9.1 |
| Merced | 57.8 | 26.0 | 27.9 | 14.2 | 15.4 | 14.7 | 9.6 | 9.2 | 9.1 | 30.1 | 50.5 | 63,893 | 78,959 | 23.6 | 11.7 |
| Milpitas | 12.9 | 51.0 | 16.5 | 7.2 | 21.2 | 15.5 | 12.3 | 13.1 | 14.2 | 37.7 | 47.3 | 62,698 | 66,824 | 6.6 | 13.5 |
| Mission Viejo | 16.8 | 21.7 | 18.5 | 6.8 | 10.4 | 9.8 | 14.6 | 16.8 | 23.0 | 47.9 | 51.4 | 93,102 | 93,101 | 0.0 | 1.2 |
| Modesto | 41.9 | 17.5 | 25.4 | 8.3 | 15.1 | 11.5 | 12.1 | 12.4 | 15.0 | 35.8 | 50.9 | 188,856 | 203,057 | 7.5 | 6.2 |
| Monrovia | 39.2 | 23.3 | 19.8 | 4.2 | 14.0 | 18.5 | 11.5 | 17.5 | 14.4 | 41.0 | 52.9 | 36,929 | 36,600 | -0.9 | 3.2 |
| Montclair | 65.9 | 29.3 | 22.2 | 11.9 | 17.2 | 13.9 | 11.0 | 10.4 | 13.4 | 34.1 | 51.0 | 33,049 | 36,649 | 10.9 | 10.8 |
| Montebello | 78.5 | 37.8 | 20.7 | 8.1 | 16.9 | 10.6 | 12.6 | 13.7 | 17.4 | 39.3 | 51.9 | 62,150 | 62,483 | 0.5 | -1.5 |
| Monterey | 10.9 | 15.0 | 9.9 | 11.2 | 19.6 | 11.2 | 11.3 | 17.2 | 19.1 | 43.9 | 54.1 | 29,674 | 27,774 | -6.4 | 2.7 |
| Monterey Park | 28.0 | 51.7 | 17.0 | 6.7 | 15.6 | 11.1 | 11.6 | 16.6 | 21.4 | 44.6 | 53.0 | 60,051 | 60,271 | 0.4 | -1.5 |
| Moorpark | 30.9 | 14.2 | 23.7 | 6.2 | 15.9 | 11.6 | 12.2 | 14.1 | 16.2 | 39.5 | 52.4 | 31,415 | 34,594 | 10.1 | 4.6 |
| Moreno Valley | 58.3 | 24.5 | 28.0 | 12.1 | 15.9 | 13.6 | 12.1 | 9.9 | 8.3 | 31.0 | 50.5 | 142,381 | 193,330 | 35.8 | 9.8 |
| Morgan Hill | 34.8 | 18.3 | 24.8 | 6.1 | 12.3 | 14.0 | 12.6 | 14.7 | 15.5 | 40.6 | 48.8 | 33,556 | 37,973 | 13.2 | 19.8 |
| Mountain View | 19.5 | 41.8 | 18.2 | 6.8 | 25.2 | 15.8 | 13.1 | 9.4 | 11.4 | 34.9 | 49.0 | 70,708 | 73,565 | 4.0 | 8.8 |
| Murrieta | 34.5 | 11.9 | 27.5 | 8.5 | 16.0 | 12.0 | 13.4 | 10.0 | 12.6 | 33.1 | 51.6 | 44,282 | 103,726 | 134.2 | 12.3 |
| Napa | 41.7 | 19.4 | 20.9 | 6.6 | 14.0 | 13.5 | 12.6 | 13.0 | 19.3 | 40.5 | 51.9 | 72,585 | 77,094 | 6.2 | -0.1 |
| National City | 65.1 | 36.4 | 20.5 | 14.1 | 19.8 | 12.5 | 10.7 | 9.7 | 12.6 | 33.1 | 48.4 | 54,260 | 58,557 | 7.9 | 4.1 |
| Newark | 34.3 | 42.9 | 20.2 | 10.2 | 18.0 | 15.0 | 13.6 | 11.8 | 11.3 | 35.7 | 50.4 | 42,471 | 42,573 | 0.2 | 16.7 |
| Newport Beach | 10.5 | 14.9 | 18.5 | 6.7 | 11.9 | 11.2 | 13.8 | 16.9 | 21.1 | 46.2 | 51.5 | 70,032 | 85,211 | 21.7 | 1.9 |
| Norco | 31.1 | 9.0 | 16.1 | 9.5 | 9.9 | 15.0 | 18.1 | 13.0 | 18.5 | 44.6 | 47.3 | 24,157 | 27,216 | 12.7 | 2.5 |
| Norwalk | 70.3 | 32.6 | 23.7 | 8.6 | 16.9 | 13.9 | 11.4 | 12.4 | 13.0 | 35.4 | 49.2 | 103,298 | 105,549 | 2.2 | -2.8 |
| Novato | 13.7 | 16.5 | 18.1 | 9.4 | 5.6 | 10.2 | 14.3 | 16.8 | 25.7 | 50.6 | 49.9 | 47,630 | 51,873 | 8.9 | 2.8 |
| Oakland | 26.8 | 25.1 | 20.0 | 7.3 | 19.3 | 16.8 | 12.5 | 10.3 | 13.8 | 36.9 | 52.5 | 399,484 | 390,781 | -2.2 | 8.7 |
| Oakley | 34.3 | 10.8 | 31.0 | 9.9 | 14.9 | 11.6 | 15.3 | 9.5 | 7.9 | 32.4 | 50.5 | 25,619 | 35,421 | 38.3 | 20.6 |
| Oceanside | 41.4 | 18.4 | 22.0 | 8.3 | 15.7 | 13.3 | 10.2 | 14.1 | 16.4 | 37.5 | 52.6 | 161,029 | 167,560 | 4.1 | 4.2 |
| Ontario | 71.4 | 30.1 | 26.7 | 9.3 | 19.5 | 12.8 | 11.9 | 11.0 | 8.8 | 32.1 | 51.3 | 158,007 | 163,931 | 3.7 | 11.9 |
| Orange | 43.2 | 21.4 | 22.0 | 10.3 | 15.4 | 13.2 | 13.0 | 12.6 | 13.5 | 36.6 | 47.2 | 128,821 | 136,778 | 6.2 | 1.0 |
| Oxnard | 73.3 | 33.5 | 26.2 | 10.2 | 14.7 | 12.9 | 13.4 | 11.8 | 10.9 | 34.1 | 49.4 | 170,358 | 198,066 | 16.3 | 5.0 |
| Pacifica | 19.5 | 20.9 | 20.5 | 7.3 | 14.4 | 13.2 | 12.6 | 13.7 | 18.3 | 40.9 | 49.6 | 38,390 | 37,341 | -2.7 | -1.5 |
| Palmdale | 60.4 | 20.9 | 29.1 | 11.3 | 12.9 | 13.4 | 12.6 | 11.0 | 9.6 | 32.0 | 50.9 | 116,670 | 152,731 | 30.9 | 10.5 |
| Palm Desert | 23.8 | 17.1 | 11.3 | 3.9 | 11.0 | 10.7 | 10.2 | 13.0 | 39.8 | 56.9 | 49.8 | 41,155 | 48,449 | 17.7 | 9.8 |
| Palm Springs | 22.8 | 15.2 | 10.2 | 5.1 | 8.9 | 8.4 | 14.5 | 20.4 | 32.6 | 56.4 | 40.4 | 42,807 | 44,554 | 4.1 | 10.5 |
| Palo Alto | 5.4 | 37.9 | 24.1 | 5.9 | 11.9 | 12.1 | 14.3 | 11.0 | 20.5 | 41.3 | 52.1 | 58,598 | 64,347 | 9.8 | 4.1 |
| Paradise | NA | NA | NA | NA | NA | NA | NA | NA | NA | NA | NA | 26,408 | 26,209 | -0.8 | -83.5 |
| Paramount | 81.9 | 33.3 | 29.4 | 11.0 | 16.7 | 13.7 | 12.5 | 8.9 | 7.7 | 30.8 | 49.6 | 55,266 | 54,107 | -2.1 | -2.0 |
| Pasadena | 34.3 | 31.0 | 16.3 | 6.5 | 21.1 | 16.0 | 12.5 | 11.7 | 15.9 | 38.5 | 52.2 | 133,936 | 137,085 | 2.4 | 2.9 |
| Perris | 78.7 | 31.2 | 27.7 | 12.4 | 15.1 | 14.1 | 14.2 | 9.0 | 7.4 | 30.7 | 52.9 | 36,189 | 68,562 | 89.5 | 15.2 |
| Petaluma | 21.5 | 15.4 | 18.1 | 7.4 | 13.2 | 13.3 | 10.3 | 15.3 | 22.4 | 43.6 | 49.6 | 54,548 | 57,977 | 6.3 | 4.6 |
| Pico Rivera | 91.2 | 31.0 | 20.3 | 10.0 | 16.7 | 11.3 | 13.9 | 13.3 | 14.5 | 37.3 | 54.6 | 63,428 | 62,964 | -0.7 | -2.6 |
| Pittsburg | 46.0 | 32.4 | 23.5 | 9.1 | 16.6 | 14.9 | 10.8 | 12.6 | 12.6 | 35.6 | 51.2 | 56,769 | 63,266 | 11.4 | 14.8 |
| Placentia | 32.9 | 26.5 | 24.7 | 7.9 | 12.7 | 13.0 | 14.8 | 12.5 | 14.4 | 38.1 | 50.6 | 46,488 | 50,902 | 9.5 | 1.2 |
| Pleasant Hill | 22.1 | 17.6 | 21.3 | 7.9 | 12.7 | 13.4 | 13.6 | 15.4 | 15.7 | 41.1 | 52.2 | 32,837 | 33,075 | 0.7 | 5.2 |
| Pleasanton | 11.4 | 33.6 | 23.9 | 5.4 | 9.6 | 15.3 | 15.3 | 14.3 | 16.4 | 42.1 | 50.6 | 63,654 | 70,274 | 10.4 | 14.2 |
| Pomona | 72.1 | 32.0 | 23.9 | 12.0 | 15.0 | 14.5 | 11.6 | 10.8 | 12.3 | 34.3 | 50.7 | 149,473 | 149,235 | -0.2 | 0.0 |
| Porterville | 61.7 | 19.8 | 26.8 | 7.6 | 11.5 | 15.1 | 13.0 | 10.6 | 15.4 | 37.8 | 50.8 | 39,615 | 58,095 | 46.6 | 1.5 |
| Poway | 15.5 | 16.7 | 19.2 | 6.8 | 13.8 | 10.7 | 11.7 | 19.3 | 18.6 | 44.4 | 48.6 | 48,044 | 47,805 | -0.5 | 3.0 |
| Rancho Cordova | 23.1 | 25.0 | 28.3 | 8.3 | 16.2 | 14.0 | 10.1 | 10.5 | 12.5 | 33.4 | 52.1 | 55,060 | 64,805 | 17.7 | 17.5 |
| Rancho Cucamonga | 40.0 | 21.2 | 23.1 | 7.9 | 16.2 | 12.4 | 14.5 | 12.2 | 13.7 | 36.9 | 51.1 | 127,743 | 165,373 | 29.5 | 8.1 |
| Rancho Palos Verdes | 15.9 | 26.1 | 19.8 | 6.5 | 5.0 | 10.7 | 13.6 | 21.8 | 22.5 | 51.3 | 50.8 | 41,145 | 41,688 | 1.3 | -2.0 |
| Rancho Santa Margarita | 21.0 | 20.6 | 23.1 | 7.8 | 10.3 | 14.1 | 18.2 | 14.8 | 11.7 | 40.2 | 49.6 | 47,214 | 47,855 | 1.4 | -0.6 |
| Redding | 10.1 | 6.5 | 22.6 | 8.3 | 14.8 | 11.9 | 10.2 | 13.7 | 18.4 | 37.4 | 50.1 | 80,865 | 89,877 | 11.1 | 2.9 |
| Redlands | 26.8 | 12.9 | 18.8 | 9.5 | 15.3 | 11.5 | 11.4 | 16.6 | 16.9 | 39.6 | 51.5 | 63,591 | 68,699 | 8.0 | 5.2 |
| Redondo Beach | 14.2 | 18.8 | 24.6 | 5.0 | 13.7 | 18.5 | 13.1 | 13.3 | 12.0 | 38.7 | 51.0 | 63,261 | 66,870 | 5.7 | -2.0 |
| Redwood City | 30.1 | 34.7 | 20.3 | 6.3 | 17.2 | 16.1 | 13.2 | 13.2 | 13.7 | 38.3 | 52.2 | 75,402 | 76,935 | 2.0 | 9.8 |
| Rialto | 71.5 | 25.8 | 28.2 | 11.8 | 17.6 | 12.2 | 11.2 | 9.3 | 9.7 | 29.8 | 51.4 | 91,873 | 99,301 | 8.1 | 4.9 |
| Richmond | 48.1 | 35.4 | 23.6 | 8.6 | 16.0 | 14.8 | 14.6 | 10.2 | 12.1 | 36.5 | 49.9 | 99,216 | 103,337 | 4.2 | 6.5 |
| Ridgecrest | 16.9 | 2.7 | 21.9 | 9.7 | 17.3 | 14.3 | 11.7 | 13.9 | 11.2 | 35.4 | 53.9 | 24,927 | 27,616 | 10.8 | 5.5 |
| Riverside | 57.2 | 23.4 | 24.3 | 14.4 | 16.3 | 12.5 | 12.6 | 9.4 | 10.5 | 31.1 | 49.1 | 255,166 | 303,905 | 19.1 | 8.8 |
| Rocklin | 10.8 | 11.7 | 27.0 | 8.1 | 10.4 | 14.1 | 13.8 | 9.9 | 16.7 | 39.3 | 51.6 | 36,330 | 57,135 | 57.3 | 22.8 |
| Rohnert Park | 17.6 | 11.6 | 16.6 | 12.3 | 14.2 | 11.7 | 11.5 | 13.6 | 20.2 | 40.7 | 50.3 | 42,236 | 41,163 | -2.5 | 4.4 |
| Rosemead | 34.1 | 53.8 | 20.8 | 9.4 | 13.2 | 12.9 | 14.9 | 12.4 | 16.5 | 40.5 | 50.4 | 53,505 | 53,753 | 0.5 | -1.2 |
| Roseville | 14.5 | 14.3 | 22.9 | 7.3 | 11.8 | 15.4 | 12.9 | 12.3 | 17.4 | 40.0 | 52.7 | 79,921 | 119,277 | 49.2 | 21.3 |
| Sacramento | 30.9 | 20.8 | 22.5 | 9.5 | 18.2 | 14.1 | 11.3 | 10.1 | 14.2 | 34.8 | 50.4 | 407,018 | 466,369 | 14.6 | 10.0 |
| Salinas | 82.2 | 35.4 | 32.1 | 9.7 | 16.0 | 14.5 | 12.2 | 7.4 | 8.1 | 30.3 | 48.8 | 151,060 | 150,560 | -0.3 | 2.9 |
| San Bernardino | 64.9 | 22.4 | 27.7 | 11.9 | 16.1 | 13.3 | 11.7 | 9.4 | 9.9 | 31.1 | 49.7 | 185,401 | 210,309 | 13.4 | 3.4 |
| San Bruno | 31.0 | 39.3 | 18.2 | 11.1 | 15.5 | 12.7 | 15.3 | 12.0 | 15.1 | 39.2 | 48.7 | 40,165 | 41,143 | 2.4 | 7.1 |
| San Buenaventura (Ventura) | 38.9 | 10.9 | 19.0 | 8.1 | 16.9 | 11.7 | 11.6 | 14.7 | 18.0 | 40.1 | 51.7 | 100,916 | 107,212 | 6.2 | -0.1 |
| San Carlos | 13.0 | 26.4 | 23.8 | 2.6 | 10.2 | 17.6 | 13.5 | 13.5 | 18.8 | 42.6 | 52.4 | 27,718 | 28,268 | 2.0 | 4.9 |
| San Clemente | 11.2 | 10.3 | 23.4 | 6.1 | 11.8 | 11.8 | 13.4 | 15.2 | 18.2 | 42.1 | 50.6 | 49,936 | 63,500 | 27.2 | 1.7 |
| San Diego | 30.3 | 25.2 | 19.7 | 10.8 | 18.9 | 14.4 | 12.0 | 10.9 | 13.4 | 35.4 | 49.6 | 1,223,400 | 1,301,927 | 6.4 | 9.3 |
| San Dimas | 26.6 | 23.8 | 22.3 | 9.4 | 12.5 | 12.1 | 11.6 | 12.0 | 20.1 | 40.4 | 52.4 | 34,980 | 33,373 | -4.6 | -0.6 |

1. May be of any race.

| City | Households, 2019 | | | | | | | | Serious crimes known to police[1], 2019 | | | | Educational attainment, 2019 | | |
|---|---|---|---|---|---|---|---|---|---|---|---|---|---|---|---|
| | | | Percent | | | | | Persons in group quarters, 2019 | Violent | | Property | | | Attainment[3] (percent) | |
| | Number | Persons per household | Family | Married couple family | Female family | Non-family | One person | | Number | Rate[2] | Number | Rate[2] | Population age 25 and over | High school graduate or less | Bachelor's degree or more |
| | 27 | 28 | 29 | 30 | 31 | 32 | 33 | 34 | 35 | 36 | 37 | 38 | 39 | 40 | 41 |
| CALIFORNIA— Cont'd | | | | | | | | | | | | | | | |
| Menlo Park | 12,664 | 2.64 | 64.8 | 53.1 | 11.7 | 35.2 | 23.0 | 1,247 | 50 | 143 | 774 | 2,220 | 23,916 | 11.0 | 76.3 |
| Merced | 25,005 | 3.30 | 71.9 | 40.6 | 31.3 | 28.1 | 20.5 | 1,062 | 540 | 644 | 2,342 | 2,793 | 48,448 | 51.9 | 14.6 |
| Milpitas | 26,052 | 3.13 | 79.4 | 62.5 | 16.8 | 20.6 | 12.6 | 2,747 | 102 | 124 | 2,193 | 2,663 | 64,249 | 24.8 | 55.3 |
| Mission Viejo | 34,017 | 2.73 | 74.7 | 64.3 | 10.4 | 25.3 | 22.2 | 1,572 | 65 | 68 | 866 | 907 | 70,524 | 18.5 | 49.8 |
| Modesto | 71,325 | 2.98 | 70.7 | 45.9 | 24.8 | 29.3 | 22.7 | 2,628 | 1,758 | 812 | 7,183 | 3,317 | 142,499 | 48.5 | 17.8 |
| Monrovia | 13,604 | 2.66 | 64.7 | 46.6 | 18.0 | 35.3 | 27.5 | 191 | 43 | 117 | 845 | 2,301 | 27,602 | 30.1 | 39.8 |
| Montclair | 11,406 | 3.49 | 75.8 | 49.0 | 26.9 | 24.2 | 19.7 | 274 | 234 | 588 | 1,469 | 3,692 | 26,411 | 50.0 | 19.4 |
| Montebello | 18,356 | 3.36 | 81.5 | 48.6 | 32.8 | 18.5 | 15.9 | 327 | 178 | 284 | 1,448 | 2,311 | 44,133 | 49.2 | 19.8 |
| Monterey | 12,417 | 2.07 | 48.0 | 37.6 | 10.4 | 52.0 | 38.6 | 2,425 | 78 | 275 | 975 | 3,441 | 22,210 | 17.6 | 56.0 |
| Monterey Park | 19,594 | 3.03 | 73.3 | 43.8 | 29.5 | 26.7 | 18.6 | 249 | 126 | 209 | 1,355 | 2,242 | 45,508 | 43.5 | 33.8 |
| Moorpark | 12,075 | 3.01 | 77.2 | 63.9 | 13.3 | 22.8 | 17.6 | 0 | 26 | 71 | 228 | 619 | 25,472 | 24.2 | 41.6 |
| Moreno Valley | 52,220 | 4.07 | 83.9 | 52.2 | 31.7 | 16.1 | 10.9 | 776 | 847 | 401 | 5,534 | 2,623 | 127,582 | 52.1 | 14.3 |
| Morgan Hill | 14,500 | 3.14 | 82.8 | 67.2 | 15.6 | 17.2 | 11.0 | 398 | 52 | 113 | 703 | 1,524 | 31,741 | 28.5 | 38.7 |
| Mountain View | 35,456 | 2.32 | 51.0 | 41.7 | 9.2 | 49.0 | 34.1 | 291 | 165 | 195 | 2,463 | 2,911 | 62,026 | 14.8 | 71.9 |
| Murrieta | 33,792 | 3.42 | 76.0 | 59.5 | 16.5 | 24.0 | 16.3 | 720 | 77 | 66 | 1,512 | 1,299 | 74,352 | 29.3 | 29.8 |
| Napa | 28,819 | 2.67 | 64.2 | 48.8 | 15.4 | 35.8 | 28.7 | 1,041 | 280 | 352 | 1,232 | 1,549 | 56,644 | 34.5 | 35.3 |
| National City | 15,566 | 3.43 | 76.5 | 50.7 | 25.8 | 23.5 | 19.0 | 7,970 | 364 | 589 | 1,123 | 1,817 | 40,156 | 52.6 | 13.6 |
| Newark | 15,086 | 3.25 | 78.3 | 63.1 | 15.2 | 21.7 | 13.0 | 140 | 85 | 174 | 1,399 | 2,858 | 34,208 | 34.0 | 44.8 |
| Newport Beach | 37,035 | 2.27 | 58.1 | 45.3 | 12.8 | 41.9 | 32.1 | 520 | 135 | 158 | 1,764 | 2,067 | 63,200 | 10.6 | 63.9 |
| Norco | 6,981 | 3.33 | 78.8 | 66.1 | 12.6 | 21.2 | 16.8 | 3,328 | 49 | 185 | 669 | 2,519 | 19,797 | 42.0 | 22.7 |
| Norwalk | 26,257 | 3.85 | 83.9 | 51.5 | 32.4 | 16.1 | 12.3 | 2,873 | 432 | 411 | 1,571 | 1,495 | 70,347 | 50.3 | 22.1 |
| Novato | 22,283 | 2.47 | 68.0 | 55.7 | 12.3 | 32.0 | 25.7 | 570 | 148 | 264 | 890 | 1,585 | 40,251 | 20.5 | 49.9 |
| Oakland | 168,413 | 2.54 | 52.6 | 36.3 | 16.2 | 47.4 | 36.3 | 5,933 | 5,520 | 1,272 | 27,868 | 6,421 | 315,220 | 30.9 | 48.9 |
| Oakley | 11,247 | 3.77 | 84.7 | 68.5 | 16.2 | 15.3 | 11.2 | 101 | 51 | 119 | 497 | 1,155 | 25,133 | 44.0 | 21.0 |
| Oceanside | 60,235 | 2.90 | 68.7 | 53.3 | 15.4 | 31.3 | 23.5 | 786 | 712 | 402 | 3,736 | 2,109 | 122,368 | 34.7 | 30.1 |
| Ontario | 49,894 | 3.69 | 79.1 | 51.9 | 27.2 | 20.9 | 15.0 | 772 | 659 | 359 | 4,290 | 2,340 | 118,368 | 50.6 | 17.6 |
| Orange | 43,492 | 3.01 | 74.4 | 59.7 | 14.6 | 25.6 | 19.7 | 7,777 | 180 | 129 | 2,184 | 1,562 | 93,854 | 31.6 | 39.5 |
| Oxnard | 50,739 | 4.09 | 81.1 | 55.2 | 25.9 | 18.9 | 13.4 | 1,500 | 724 | 343 | 4,242 | 2,007 | 132,858 | 57.2 | 19.7 |
| Pacifica | 12,524 | 3.06 | 72.6 | 52.5 | 20.2 | 27.4 | 18.6 | 194 | 79 | 203 | 533 | 1,369 | 27,818 | 20.6 | 46.9 |
| Palmdale | 41,538 | 3.73 | 79.7 | 55.3 | 24.5 | 20.3 | 16.3 | 229 | 658 | 419 | 2,230 | 1,419 | 92,436 | 54.0 | 16.5 |
| Palm Desert | 26,404 | 2.01 | 51.9 | 41.6 | 10.3 | 48.1 | 38.8 | 299 | 136 | 253 | 1,848 | 3,436 | 45,142 | 24.0 | 42.2 |
| Palm Springs | 25,276 | 1.90 | 39.8 | 29.9 | 9.9 | 60.2 | 45.8 | 452 | 267 | 547 | 1,966 | 4,025 | 41,091 | 23.9 | 44.1 |
| Palo Alto | 25,168 | 2.57 | 67.8 | 56.8 | 11.0 | 32.2 | 24.8 | 590 | 86 | 128 | 1,983 | 2,962 | 45,742 | 6.1 | 84.6 |
| Paradise | NA | NA | NA | NA | NA | NA | NA | NA | 22 | 82 | 202 | 752 | NA | NA | NA |
| Paramount | 14,348 | 3.74 | 85.6 | 50.3 | 35.3 | 14.4 | 12.4 | 328 | 347 | 638 | 1,246 | 2,290 | 32,143 | 62.3 | 9.5 |
| Pasadena | 56,668 | 2.42 | 55.2 | 41.7 | 13.5 | 44.8 | 33.8 | 4,032 | 613 | 432 | 2,866 | 2,020 | 108,899 | 24.6 | 56.8 |
| Perris | 17,582 | 4.49 | 86.2 | 58.8 | 27.3 | 13.8 | 11.7 | 300 | 262 | 325 | 2,071 | 2,573 | 47,476 | 63.0 | 11.0 |
| Petaluma | 24,177 | 2.47 | 64.1 | 48.0 | 16.2 | 35.9 | 30.6 | 771 | 190 | 304 | 789 | 1,264 | 45,091 | 25.2 | 42.7 |
| Pico Rivera | 17,385 | 3.54 | 78.6 | 54.4 | 24.2 | 21.4 | 15.8 | 443 | 234 | 372 | 1,052 | 1,673 | 43,249 | 59.0 | 14.5 |
| Pittsburg | 21,273 | 3.39 | 73.9 | 48.0 | 25.9 | 26.1 | 19.9 | 395 | 446 | 606 | 1,660 | 2,254 | 48,909 | 51.7 | 22.4 |
| Placentia | 16,311 | 3.11 | 79.7 | 62.3 | 17.5 | 20.3 | 16.7 | 469 | 192 | 371 | 889 | 1,718 | 34,552 | 26.0 | 43.7 |
| Pleasant Hill | 13,647 | 2.52 | 69.8 | 54.4 | 15.4 | 30.2 | 22.7 | 423 | 88 | 251 | 1,484 | 4,225 | 24,669 | 19.5 | 55.6 |
| Pleasanton | 28,994 | 2.81 | 79.4 | 69.8 | 9.6 | 20.6 | 17.7 | 316 | 112 | 133 | 1,632 | 1,942 | 57,891 | 13.6 | 66.4 |
| Pomona | 40,579 | 3.62 | 78.1 | 49.4 | 28.7 | 21.9 | 13.7 | 4,852 | 940 | 615 | 4,208 | 2,754 | 97,239 | 52.3 | 18.6 |
| Porterville | 20,347 | 2.87 | 79.3 | 44.6 | 34.7 | 20.7 | 17.7 | 1,276 | 192 | 319 | 1,297 | 2,154 | 39,059 | 45.5 | 15.8 |
| Poway | 16,357 | 2.98 | 80.1 | 62.4 | 17.7 | 19.9 | 16.3 | 626 | 51 | 102 | 490 | 981 | 36,534 | 19.6 | 48.8 |
| Rancho Cordova | 24,907 | 3.00 | 66.3 | 45.8 | 20.6 | 33.7 | 23.2 | 457 | 225 | 297 | 1,422 | 1,874 | 47,602 | 34.3 | 29.1 |
| Rancho Cucamonga | 59,497 | 2.93 | 71.1 | 52.4 | 18.7 | 28.9 | 23.7 | 3,433 | 500 | 279 | 3,456 | 1,928 | 122,525 | 26.8 | 36.1 |
| Rancho Palos Verdes | 14,840 | 2.76 | 72.7 | 65.5 | 7.2 | 27.3 | 22.6 | 507 | 29 | 69 | 313 | 746 | 30,590 | 14.2 | 63.6 |
| Rancho Santa Margarita | 16,183 | 2.96 | 76.2 | 64.2 | 11.9 | 23.8 | 18.1 | 0 | 30 | 62 | 296 | 612 | 33,093 | 17.7 | 46.9 |
| Redding | 37,757 | 2.39 | 58.5 | 41.6 | 17.0 | 41.5 | 28.8 | 2,441 | NA | NA | NA | NA | 63,972 | 28.9 | 24.1 |
| Redlands | 26,115 | 2.63 | 63.4 | 50.0 | 13.3 | 36.6 | 29.4 | 2,724 | 257 | 357 | 2,108 | 2,930 | 51,258 | 22.1 | 53.0 |
| Redondo Beach | 26,360 | 2.53 | 57.3 | 49.2 | 8.1 | 42.7 | 31.5 | 187 | 160 | 237 | 1,370 | 2,030 | 47,043 | 9.2 | 64.3 |
| Redwood City | 32,199 | 2.62 | 61.4 | 48.8 | 12.6 | 38.6 | 26.7 | 1,484 | 189 | 216 | 1,343 | 1,536 | 63,084 | 24.1 | 54.1 |
| Rialto | 24,751 | 4.16 | 85.6 | 52.1 | 33.5 | 14.4 | 11.8 | 548 | 595 | 572 | 3,149 | 3,029 | 62,166 | 56.7 | 12.6 |
| Richmond | 36,634 | 2.98 | 69.6 | 45.2 | 24.5 | 30.4 | 23.0 | 1,443 | 1,034 | 932 | 4,188 | 3,773 | 74,943 | 43.4 | 31.0 |
| Ridgecrest | 11,228 | 2.56 | 61.3 | 32.8 | 28.5 | 38.7 | 23.3 | 202 | 142 | 488 | 401 | 1,378 | 19,811 | 28.7 | 28.0 |
| Riverside | 87,189 | 3.62 | 73.2 | 49.0 | 24.2 | 26.8 | 19.8 | 15,591 | 1,686 | 506 | 9,790 | 2,938 | 203,308 | 46.7 | 22.1 |
| Rocklin | 23,612 | 2.90 | 73.9 | 62.3 | 11.6 | 26.1 | 20.1 | 404 | 67 | 98 | 983 | 1,434 | 44,610 | 18.7 | 49.3 |
| Rohnert Park | 17,906 | 2.41 | 58.0 | 39.3 | 18.7 | 42.0 | 27.3 | 80 | 267 | 605 | 786 | 1,781 | 30,806 | 32.0 | 34.7 |
| Rosemead | 14,269 | 3.75 | 85.1 | 58.5 | 26.6 | 14.9 | 12.3 | 598 | 175 | 321 | 942 | 1,729 | 37,747 | 57.8 | 24.0 |
| Roseville | 53,093 | 2.65 | 71.1 | 56.7 | 14.3 | 28.9 | 23.6 | 953 | 259 | 183 | 3,154 | 2,225 | 98,745 | 21.1 | 43.6 |
| Sacramento | 191,911 | 2.63 | 57.9 | 37.5 | 20.5 | 42.1 | 31.6 | 9,750 | 3,223 | 627 | 16,354 | 3,182 | 349,270 | 35.5 | 33.7 |
| Salinas | 39,862 | 3.86 | 77.2 | 52.5 | 24.7 | 22.8 | 17.6 | 1,459 | 782 | 498 | 3,532 | 2,250 | 90,462 | 64.5 | 11.6 |
| San Bernardino | 60,308 | 3.39 | 74.4 | 41.6 | 32.7 | 25.6 | 20.6 | 11,295 | 2,858 | 1,319 | 9,081 | 4,190 | 130,304 | 60.2 | 12.8 |
| San Bruno | 14,965 | 2.83 | 70.9 | 51.1 | 19.8 | 29.1 | 22.9 | 489 | 138 | 319 | 1,139 | 2,631 | 30,240 | 28.7 | 39.5 |
| San Buenaventura (Ventura) | 40,513 | 2.64 | 66.0 | 47.8 | 18.3 | 34.0 | 25.6 | 2,016 | 458 | 410 | 2,969 | 2,660 | 79,604 | 27.3 | 37.0 |
| San Carlos | 11,439 | 2.63 | 65.3 | 59.8 | 5.5 | 34.7 | 27.7 | 145 | NA | NA | NA | NA | 22,213 | 17.5 | 62.2 |
| San Clemente | 23,182 | 2.77 | 74.4 | 62.7 | 11.7 | 25.6 | 20.2 | 334 | 87 | 134 | 918 | 1,412 | 45,489 | 17.2 | 56.6 |
| San Diego | 512,530 | 2.71 | 60.8 | 46.4 | 14.4 | 39.2 | 27.3 | 36,752 | 5,215 | 362 | 27,141 | 1,883 | 989,949 | 26.1 | 47.4 |
| San Dimas | 12,665 | 2.58 | 62.5 | 38.4 | 24.1 | 37.5 | 31.9 | 872 | 78 | 229 | 833 | 2,446 | 22,956 | 28.3 | 37.7 |

1. Data for serious crimes have not been adjusted for underreporting. This may affect comparability between geographic areas and over time.　　2. Per 100,000 population estimated by the FBI.　　3. Persons 25 years old and over.

## Table D. Cities — Income, Poverty, and Housing

| City | Money income, 2019 — Households — Median income | Percent with income less than $20,000 | Percent with income of $200,000 or more | Median family income | Median non-family income | Median earnings, 2019 — All persons | Men | Women | Housing units, 2019 — Total | Occupied | Percent owner occupied | Median value[1] (dollars) | Median gross rent (dollars) |
|---|---|---|---|---|---|---|---|---|---|---|---|---|---|
| | 42 | 43 | 44 | 45 | 46 | 47 | 48 | 49 | 50 | 51 | 52 | 53 | 54 |
| **CALIFORNIA— Cont'd** | | | | | | | | | | | | | |
| Menlo Park | 180,947 | 5.4 | 47.4 | 222,024 | 129,551 | 82,375 | 120,683 | 61,469 | 14,311 | 12,664 | 53.0 | 2 | 2,595 |
| Merced | 50,309 | 17.5 | 4.7 | 57,904 | 32,309 | 27,546 | 29,318 | 23,347 | 26,025 | 25,005 | 44.8 | 268,105 | 1,111 |
| Milpitas | 140,124 | 4.8 | 33.8 | 142,827 | 117,633 | 65,140 | 80,058 | 53,556 | 27,376 | 26,052 | 64.3 | 978,055 | 2,614 |
| Mission Viejo | 119,798 | 7.6 | 20.6 | 135,386 | 60,228 | 52,354 | 60,359 | 45,960 | 35,818 | 34,017 | 78.9 | 748,940 | 2,293 |
| Modesto | 62,067 | 12.0 | 4.2 | 71,836 | 42,113 | 36,575 | 42,028 | 31,399 | 74,272 | 71,325 | 54.9 | 324,064 | 1,295 |
| Monrovia | 82,474 | 11.4 | 11.4 | 88,191 | 34,849 | 41,493 | 49,678 | 36,986 | 14,705 | 13,604 | 49.7 | 716,390 | 1,718 |
| Montclair | 67,857 | 12.7 | 3.3 | 75,921 | 0 | 27,576 | 26,566 | 30,883 | 11,705 | 11,406 | 54.7 | 420,305 | 1,438 |
| Montebello | 57,397 | 14.4 | 3.4 | 68,130 | 23,465 | 28,849 | 32,030 | 26,147 | 19,271 | 18,356 | 49.8 | 550,633 | 1,387 |
| Monterey | 95,800 | 7.5 | 14.1 | 124,782 | 78,293 | 47,685 | 51,196 | 45,379 | 14,034 | 12,417 | 40.9 | 792,758 | 2,056 |
| Monterey Park | 71,774 | 16.0 | 9.8 | 80,748 | 48,616 | 37,470 | 45,872 | 29,188 | 21,082 | 19,594 | 52.5 | 651,783 | 1,617 |
| Moorpark | 116,633 | 4.9 | 15.1 | 125,820 | 81,422 | 50,786 | 51,839 | 47,911 | 12,316 | 12,075 | 74.9 | 654,975 | 2,277 |
| Moreno Valley | 65,449 | 10.1 | 4.0 | 65,757 | 44,803 | 30,440 | 35,374 | 27,112 | 55,424 | 52,220 | 59.8 | 350,743 | 1,636 |
| Morgan Hill | 134,453 | 2.3 | 32.9 | 154,272 | 78,069 | 60,595 | 69,256 | 55,893 | 14,780 | 14,500 | 75.8 | 934,830 | 1,956 |
| Mountain View | 147,915 | 5.3 | 36.7 | 160,957 | 127,159 | 86,698 | 98,089 | 71,610 | 39,795 | 35,456 | 39.9 | 1,645,797 | 2,525 |
| Murrieta | 100,080 | 6.5 | 14.1 | 105,613 | 72,401 | 42,251 | 56,164 | 25,369 | 37,349 | 33,792 | 65.9 | 441,418 | 1,900 |
| Napa | 88,310 | 7.7 | 16.5 | 99,300 | 57,238 | 45,883 | 50,151 | 40,443 | 31,800 | 28,819 | 61.2 | 663,791 | 1,877 |
| National City | 52,035 | 17.9 | 1.5 | 55,598 | 30,289 | 27,416 | 29,762 | 25,767 | 16,577 | 15,566 | 32.4 | 458,591 | 1,251 |
| Newark | 138,346 | 3.8 | 27.0 | 141,674 | 110,263 | 60,977 | 71,739 | 47,810 | 15,697 | 15,086 | 66.8 | 888,007 | 2,390 |
| Newport Beach | 128,294 | 6.9 | 37.4 | 171,201 | 90,013 | 74,620 | 100,326 | 53,938 | 43,702 | 37,035 | 57.9 | 2 | 2,306 |
| Norco | 116,114 | 9.1 | 15.5 | 130,000 | 33,679 | 40,543 | 50,786 | 29,554 | 7,477 | 6,981 | 88.1 | 607,148 | 1,660 |
| Norwalk | 80,513 | 9.1 | 3.9 | 81,153 | 29,625 | 35,404 | 40,282 | 30,134 | 26,902 | 26,257 | 67.0 | 491,595 | 1,797 |
| Novato | 108,615 | 7.5 | 22.0 | 140,548 | 64,966 | 53,011 | 62,016 | 41,066 | 22,650 | 22,283 | 69.6 | 863,986 | 2,050 |
| Oakland | 82,018 | 15.1 | 16.1 | 99,659 | 61,028 | 50,098 | 51,850 | 45,763 | 182,574 | 168,413 | 41.3 | 807,615 | 1,600 |
| Oakley | 124,895 | 5.0 | 19.9 | 126,741 | 0 | 51,448 | 80,591 | 34,730 | 11,481 | 11,247 | 75.0 | 512,784 | 2,196 |
| Oceanside | 77,226 | 9.0 | 7.8 | 92,902 | 47,778 | 34,904 | 42,250 | 28,299 | 66,334 | 60,235 | 61.8 | 538,967 | 1,815 |
| Ontario | 75,266 | 9.4 | 4.8 | 75,610 | 59,082 | 31,776 | 36,311 | 26,501 | 52,886 | 49,894 | 56.2 | 435,495 | 1,577 |
| Orange | 95,827 | 6.3 | 14.4 | 111,061 | 63,423 | 42,056 | 44,688 | 38,120 | 45,530 | 43,492 | 57.7 | 679,754 | 1,815 |
| Oxnard | 78,981 | 8.0 | 7.8 | 79,105 | 50,324 | 31,098 | 34,304 | 26,367 | 55,451 | 50,739 | 55.7 | 497,265 | 1,539 |
| Pacifica | 142,485 | 6.5 | 28.6 | 156,281 | 54,407 | 58,274 | 55,855 | 60,617 | 12,736 | 12,524 | 74.0 | 1,066,695 | 2,384 |
| Palmdale | 71,774 | 10.6 | 4.8 | 78,269 | 40,028 | 34,511 | 41,164 | 26,898 | 43,382 | 41,538 | 66.3 | 324,958 | 1,411 |
| Palm Desert | 62,453 | 17.6 | 10.0 | 78,205 | 43,398 | 36,700 | 39,883 | 31,024 | 40,153 | 26,404 | 62.2 | 367,277 | 1,392 |
| Palm Springs | 65,320 | 18.3 | 9.7 | 82,091 | 48,250 | 36,847 | 40,494 | 31,872 | 41,934 | 25,276 | 60.5 | 427,901 | 1,180 |
| Palo Alto | 160,360 | 5.4 | 41.2 | 208,198 | 115,969 | 95,275 | 138,315 | 66,408 | 28,541 | 25,168 | 55.4 | 2 | 2,703 |
| Paradise | NA | NA | NA | NA | NA | NA | NA | NA | NA | NA | NA | NA | NA |
| Paramount | 49,755 | 15.6 | 0.3 | 58,202 | 21,613 | 25,765 | 29,187 | 22,522 | 14,988 | 14,348 | 39.0 | 386,507 | 1,445 |
| Pasadena | 92,053 | 11.8 | 13.6 | 118,176 | 66,116 | 51,080 | 55,910 | 48,714 | 62,830 | 56,668 | 38.0 | 889,160 | 1,842 |
| Perris | 70,714 | 10.4 | 1.9 | 74,899 | 0 | 31,117 | 32,473 | 29,469 | 18,906 | 17,582 | 65.2 | 340,801 | 1,300 |
| Petaluma | 92,129 | 7.1 | 16.2 | 123,694 | 54,884 | 45,857 | 52,016 | 40,773 | 24,772 | 24,177 | 68.2 | 675,639 | 2,020 |
| Pico Rivera | 69,199 | 9.3 | 6.4 | 74,166 | 45,303 | 32,163 | 39,079 | 24,482 | 17,954 | 17,385 | 66.4 | 508,680 | 1,636 |
| Pittsburg | 82,960 | 12.9 | 10.1 | 83,198 | 50,565 | 41,139 | 46,579 | 33,608 | 21,953 | 21,273 | 54.2 | 453,730 | 1,833 |
| Placentia | 102,212 | 8.0 | 16.7 | 117,833 | 45,223 | 51,688 | 60,849 | 47,968 | 16,565 | 16,311 | 68.9 | 692,705 | 1,807 |
| Pleasant Hill | 132,698 | 5.5 | 23.8 | 150,183 | 100,941 | 70,780 | 84,523 | 56,181 | 14,474 | 13,647 | 65.8 | 827,419 | 2,107 |
| Pleasanton | 156,063 | 5.6 | 36.2 | 176,993 | 53,447 | 90,433 | 116,660 | 56,397 | 30,391 | 28,994 | 65.4 | 1,087,682 | 2,534 |
| Pomona | 67,202 | 10.7 | 4.3 | 68,757 | 39,096 | 30,390 | 32,952 | 25,077 | 43,096 | 40,579 | 52.5 | 447,628 | 1,449 |
| Porterville | 61,860 | 16.5 | 2.4 | 60,452 | 23,831 | 27,500 | 31,364 | 24,250 | 20,899 | 20,347 | 60.8 | 212,739 | 1,063 |
| Poway | 126,182 | 6.2 | 21.6 | 142,821 | 50,536 | 48,241 | 59,010 | 37,403 | 16,754 | 16,357 | 77.2 | 736,443 | 1,863 |
| Rancho Cordova | 71,655 | 12.4 | 5.2 | 79,102 | 49,247 | 38,073 | 40,164 | 36,124 | 25,977 | 24,907 | 51.2 | 345,172 | 1,378 |
| Rancho Cucamonga | 92,773 | 8.9 | 12.3 | 102,168 | 60,292 | 42,406 | 58,528 | 33,322 | 61,742 | 59,497 | 64.0 | 536,502 | 1,777 |
| Rancho Palos Verdes | 150,117 | 6.3 | 37.8 | 183,015 | 90,854 | 61,558 | 87,419 | 45,314 | 15,952 | 14,840 | 78.1 | 1,245,812 | 2,736 |
| Rancho Santa Margarita | 138,053 | 4.9 | 20.8 | 158,930 | 76,189 | 57,306 | 64,729 | 52,297 | 16,558 | 16,183 | 69.3 | 700,769 | 2,138 |
| Redding | 63,165 | 15.6 | 4.5 | 76,341 | 39,185 | 33,571 | 36,766 | 30,526 | 40,532 | 37,757 | 55.2 | 288,071 | 1,135 |
| Redlands | 72,410 | 11.4 | 12.0 | 101,623 | 41,651 | 41,512 | 50,496 | 36,697 | 28,283 | 26,115 | 63.4 | 420,757 | 1,473 |
| Redondo Beach | 109,019 | 2.7 | 24.4 | 150,504 | 82,328 | 69,447 | 82,331 | 58,695 | 28,137 | 26,360 | 50.3 | 1,028,602 | 2,153 |
| Redwood City | 138,913 | 7.0 | 34.4 | 153,592 | 96,663 | 70,229 | 74,590 | 66,317 | 33,563 | 32,199 | 45.5 | 1,570,310 | 2,760 |
| Rialto | 70,188 | 10.0 | 3.9 | 69,502 | 43,562 | 31,396 | 35,215 | 29,919 | 25,631 | 24,751 | 61.4 | 352,476 | 1,252 |
| Richmond | 72,130 | 11.1 | 7.4 | 76,010 | 56,458 | 36,769 | 41,870 | 31,522 | 39,563 | 36,634 | 44.1 | 572,408 | 1,679 |
| Ridgecrest | 80,166 | 15.0 | 3.8 | 81,239 | 65,714 | 40,966 | 65,991 | 27,599 | 13,319 | 11,228 | 53.7 | 184,424 | 852 |
| Riverside | 71,967 | 10.5 | 6.7 | 82,148 | 47,105 | 32,431 | 36,906 | 28,589 | 91,011 | 87,189 | 53.2 | 411,043 | 1,504 |
| Rocklin | 98,514 | 4.6 | 14.5 | 115,045 | 52,014 | 51,115 | 67,560 | 39,567 | 24,416 | 23,612 | 68.4 | 557,650 | 1,787 |
| Rohnert Park | 82,622 | 7.0 | 12.5 | 92,319 | 61,658 | 41,829 | 47,460 | 30,894 | 18,773 | 17,906 | 56.7 | 571,365 | 1,792 |
| Rosemead | 69,870 | 11.1 | 4.1 | 70,911 | 38,857 | 31,314 | 35,267 | 28,118 | 14,954 | 14,269 | 45.8 | 580,221 | 1,398 |
| Roseville | 101,101 | 8.3 | 15.5 | 120,719 | 61,585 | 53,098 | 68,620 | 48,308 | 56,494 | 53,093 | 68.2 | 487,140 | 1,766 |
| Sacramento | 69,134 | 13.3 | 7.6 | 80,654 | 50,648 | 37,448 | 40,567 | 35,086 | 200,079 | 191,911 | 48.7 | 380,592 | 1,370 |
| Salinas | 69,117 | 12.6 | 4.9 | 69,089 | 50,854 | 27,747 | 31,917 | 23,799 | 41,388 | 39,862 | 45.2 | 478,788 | 1,578 |
| San Bernardino | 49,721 | 17.6 | 3.1 | 52,220 | 34,824 | 26,871 | 31,713 | 22,522 | 63,523 | 60,308 | 48.2 | 284,978 | 1,103 |
| San Bruno | 118,426 | 3.3 | 16.3 | 133,129 | 71,588 | 53,341 | 60,929 | 49,572 | 16,075 | 14,965 | 62.7 | 1,027,465 | 2,432 |
| San Buenaventura (Ventura) | 84,110 | 9.8 | 11.0 | 96,948 | 58,768 | 40,882 | 45,866 | 36,830 | 44,793 | 40,513 | 51.3 | 638,911 | 1,687 |
| San Carlos | 187,754 | 4.2 | 48.1 | 250 | 86,436 | 97,005 | 122,237 | 64,322 | 11,883 | 11,439 | 69.8 | 1,851,474 | 2,334 |
| San Clemente | 127,650 | 7.1 | 31.8 | 154,015 | 57,809 | 59,556 | 80,910 | 51,233 | 26,423 | 23,182 | 72.4 | 946,880 | 2,040 |
| San Diego | 85,507 | 11.0 | 14.1 | 100,470 | 61,890 | 41,778 | 48,933 | 36,013 | 556,735 | 512,530 | 46.5 | 658,394 | 1,806 |
| San Dimas | 74,977 | 15.5 | 7.2 | 101,363 | 39,208 | 46,070 | 50,335 | 45,412 | 13,077 | 12,665 | 63.8 | 607,225 | 1,736 |

1. Based on population estimated by the American Community Survey.

# Table D. Cities — Commuting, Computer Access, Migration, Labor Force, and Employment

| City | Commuting[1], 2019 Percent Drove alone | Commuting[1], 2019 Percent With commutes of 30 minutes or more | Computer access[2], 2019 Percent With a computer in the house | Computer access[2], 2019 Percent With Internet access | Migration, 2019 Percent who lived in the same house one year ago | Migration, 2019 Percent who lived in another state or county one year ago | Civilian labor force, 2020 Total | Civilian labor force, 2020 Percent change 2019-2020 | Unemployment Total | Unemployment Rate[3] | Civilian Employment[4], 2019 Population age 16 and older Number | Population age 16 and older Percent in labor force | Population age 16 to 64 Number | Population age 16 to 64 Percent who worked full-year full-time |
|---|---|---|---|---|---|---|---|---|---|---|---|---|---|---|
| | 55 | 56 | 57 | 58 | 59 | 60 | 61 | 62 | 63 | 64 | 65 | 66 | 67 | 68 |
| **CALIFORNIA— Cont'd** | | | | | | | | | | | | | | |
| Menlo Park | 56.2 | 34.4 | 98.1 | 95.3 | 77.3 | 16.1 | 18,604 | -7.4 | 774 | 4 | 26,581 | 69.2 | 22,332 | 56.6 |
| Merced | 74.9 | 28.5 | 94.4 | 92.3 | 84.4 | 6.0 | 34,065 | -0.7 | 4,053 | 12 | 62,380 | 63.7 | 54,803 | 37.6 |
| Milpitas | 81.7 | 50.3 | 97.7 | 96.2 | 81.1 | 5.2 | 44,160 | 2.2 | 3,294 | 8 | 71,684 | 70.7 | 59,710 | 62.9 |
| Mission Viejo | 78.8 | 42.5 | 98.1 | 96.5 | 90.7 | 2.8 | 47,058 | -5.6 | 3,776 | 8 | 79,209 | 61.1 | 57,466 | 52.5 |
| Modesto | 84.6 | 33.3 | 95.5 | 90.9 | 89.6 | 2.9 | 94,682 | -0.4 | 10,284 | 11 | 166,924 | 59.2 | 134,714 | 45.8 |
| Monrovia | 86.7 | 58.8 | 94.4 | 87.1 | NA | NA | 19,797 | -5.7 | 2,137 | 11 | 30,606 | 66.4 | 25,372 | 54.4 |
| Montclair | 75.9 | 48.3 | 96.7 | 85.9 | 88.3 | 8.2 | 19,549 | 1.4 | 1,771 | 9 | 32,258 | 60.7 | 26,879 | 49.1 |
| Montebello | 76.6 | 53.0 | 88.5 | 76.7 | 94.1 | 0.6 | 28,022 | -3.6 | 3,859 | 14 | 50,736 | 61.1 | 39,973 | 49.6 |
| Monterey | 70.0 | 24.9 | 94.3 | 86.5 | 73.7 | 17.0 | 14,973 | -2.8 | 1,276 | 9 | 25,870 | 61.2 | 20,390 | 60.6 |
| Monterey Park | 74.0 | 53.9 | 94.5 | 86.8 | 90.8 | 2.2 | 29,029 | -1.8 | 4,312 | 15 | 50,538 | 58.2 | 37,795 | 52.8 |
| Moorpark | 79.5 | 42.3 | 96.7 | 94.9 | 91.1 | 1.0 | 18,819 | -3.8 | 1,413 | 8 | 28,809 | 68.3 | 22,900 | 57.3 |
| Moreno Valley | 78.6 | 55.3 | 96.9 | 91.2 | 90.7 | 5.2 | 97,122 | 1.8 | 10,126 | 10 | 159,534 | 63.0 | 141,756 | 48.5 |
| Morgan Hill | 74.3 | 61.5 | 97.5 | 91.0 | 89.7 | 3.0 | 24,172 | -0.8 | 1,743 | 7 | 35,760 | 67.3 | 28,660 | 58.5 |
| Mountain View | 63.9 | 32.9 | 97.0 | 91.8 | 80.9 | 11.2 | 49,883 | -5.8 | 1,977 | 4 | 69,402 | 73.5 | 59,951 | 63.5 |
| Murrieta | 81.6 | 47.5 | 97.4 | 95.9 | 85.2 | 7.0 | 56,247 | 0.3 | 4,969 | 9 | 87,541 | 63.3 | 72,931 | 48.1 |
| Napa | 76.8 | 31.4 | 96.5 | 89.7 | 88.9 | 6.2 | 39,469 | -6.7 | 3,686 | 9 | 63,712 | 65.0 | 48,640 | 55.4 |
| National City | 67.3 | 46.4 | 88.3 | 79.5 | 89.3 | 5.2 | 24,740 | -0.4 | 3,127 | 13 | 50,004 | 55.0 | 42,253 | 58.9 |
| Newark | 74.4 | 64.7 | 98.5 | 92.9 | 81.1 | 8.4 | 24,650 | -2.4 | 2,017 | 8 | 41,207 | 68.8 | 35,665 | 59.9 |
| Newport Beach | 76.5 | 30.7 | 97.7 | 97.0 | 81.4 | 4.4 | 42,177 | -6.6 | 2,864 | 7 | 71,123 | 63.8 | 53,329 | 57.3 |
| Norco | 77.2 | 52.6 | 96.9 | 95.5 | 87.3 | 10.6 | 11,499 | -1.7 | 940 | 8 | 22,625 | 57.9 | 17,706 | 47.2 |
| Norwalk | 82.1 | 51.8 | 97.1 | 92.3 | 94.8 | 1.6 | 48,795 | -3.5 | 6,608 | 14 | 81,998 | 63.0 | 68,450 | 51.9 |
| Novato | 69.3 | 50.7 | 95.8 | 94.0 | 87.1 | 2.8 | 27,753 | -6.6 | 1,966 | 11 | 47,063 | 61.1 | 32,810 | 47.6 |
| Oakland | 51.9 | 55.5 | 90.9 | 83.3 | 87.6 | 5.4 | 211,675 | -1.5 | 22,288 | 11 | 353,995 | 68.3 | 294,100 | 55.2 |
| Oakley | 75.9 | 59.9 | 98.5 | 93.2 | 94.8 | 2.1 | 19,461 | -2.4 | 1,930 | 10 | 30,652 | 68.5 | 27,306 | 50.7 |
| Oceanside | 74.2 | 43.1 | 95.9 | 93.4 | 83.7 | 6.1 | 80,268 | -2.7 | 8,022 | 10 | 140,972 | 65.6 | 112,164 | 56.1 |
| Ontario | 79.9 | 42.5 | 97.3 | 89.8 | 90.5 | 6.6 | 90,095 | 1.4 | 8,009 | 9 | 140,404 | 68.5 | 124,053 | 51.5 |
| Orange | 76.3 | 42.8 | 95.8 | 93.2 | 86.7 | 3.1 | 68,739 | -5.2 | 5,619 | 8 | 112,756 | 64.2 | 93,983 | 51.8 |
| Oxnard | 80.1 | 22.2 | 93.6 | 83.7 | 92.9 | 1.4 | 97,327 | -3.5 | 9,073 | 9 | 159,129 | 67.5 | 136,450 | 48.6 |
| Pacifica | 73.8 | 50.7 | 96.6 | 94.5 | 89.8 | 6.0 | 23,124 | -5.5 | 1,683 | 7 | 31,587 | 66.4 | 24,539 | 54.6 |
| Palmdale | 81.5 | 53.9 | 94.1 | 89.2 | 92.6 | 1.3 | 62,610 | -3.2 | 9,554 | 15 | 114,789 | 61.4 | 99,835 | 46.8 |
| Palm Desert | 75.2 | 19.4 | 96.1 | 90.4 | 82.6 | 6.8 | 25,071 | 0.7 | 2,720 | 11 | 47,969 | 49.1 | 26,753 | 47.9 |
| Palm Springs | 68.4 | 29.5 | 91.1 | 86.2 | 84.8 | 6.6 | 23,273 | 1.1 | 2,480 | 11 | 44,380 | 48.0 | 28,583 | 38.8 |
| Palo Alto | 61.2 | 29.1 | 97.7 | 95.8 | 83.8 | 8.0 | 33,312 | -7.0 | 1,324 | 4 | 51,152 | 62.5 | 37,722 | 53.2 |
| Paradise | NA | NA | NA | NA | NA | NA | 1,773 | -83.4 | 152 | 9 | NA | NA | NA | NA |
| Paramount | 85.7 | 48.8 | 95.6 | 77.9 | 90.9 | 1.2 | 24,033 | -2.9 | 3,411 | 14 | 39,762 | 64.3 | 35,585 | 45.0 |
| Pasadena | 71.1 | 45.4 | 95.7 | 90.1 | 86.3 | 4.6 | 74,974 | -5.6 | 7,781 | 10 | 119,835 | 66.9 | 97,357 | 57.0 |
| Perris | 83.8 | 59.6 | 94.4 | 88.0 | 95.5 | 2.5 | 31,257 | 0.6 | 3,516 | 11 | 58,618 | 63.1 | 52,743 | 54.1 |
| Petaluma | 75.4 | 47.6 | 95.1 | 93.9 | 89.5 | 5.3 | 31,730 | -6.9 | 2,310 | 7 | 51,044 | 63.2 | 37,473 | 53.4 |
| Pico Rivera | 81.2 | 47.1 | 94.2 | 83.6 | NA | NA | 29,117 | -4.5 | 3,860 | 13 | 50,527 | 67.8 | 41,509 | 56.2 |
| Pittsburg | 69.4 | 58.3 | 93.8 | 91.5 | 91.0 | 3.5 | 33,855 | -1.6 | 3,859 | 11 | 57,425 | 61.4 | 48,309 | 47.8 |
| Placentia | 79.8 | 52.8 | 97.5 | 94.2 | 90.7 | 2.7 | 24,641 | -5.0 | 2,154 | 9 | 39,830 | 67.8 | 32,453 | 58.0 |
| Pleasant Hill | 68.2 | 48.4 | NA | NA | 88.4 | 8.5 | 17,181 | -5.2 | 1,280 | 8 | 28,485 | 68.1 | 23,003 | 53.9 |
| Pleasanton | 66.2 | 50.8 | 98.4 | 96.7 | 86.4 | 8.8 | 38,416 | -6.7 | 2,407 | 6 | 64,944 | 64.4 | 51,531 | 54.3 |
| Pomona | 73.9 | 51.5 | 95.3 | 85.1 | 91.5 | 3.8 | 65,971 | -3.5 | 8,815 | 13 | 120,664 | 63.8 | 102,077 | 48.4 |
| Porterville | 65.4 | 30.8 | 92.2 | 85.3 | NA | NA | 24,932 | -3.1 | 3,871 | 16 | 46,059 | 63.7 | 36,900 | 44.3 |
| Poway | 77.6 | 36.8 | 98.0 | 91.9 | 91.4 | 3.5 | 24,130 | -6.0 | 1,582 | 7 | 40,673 | 64.4 | 31,503 | 49.4 |
| Rancho Cordova | 76.2 | 42.9 | 96.0 | 92.6 | 83.4 | 5.8 | 35,049 | -0.9 | 3,225 | 9 | 54,828 | 67.3 | 45,422 | 53.5 |
| Rancho Cucamonga | 81.4 | 46.0 | 96.5 | 91.6 | 87.1 | 3.5 | 95,202 | -1.5 | 7,337 | 8 | 140,836 | 65.9 | 116,555 | 52.3 |
| Rancho Palos Verdes | 77.8 | 60.2 | 96.7 | 93.7 | 92.9 | 2.3 | 18,163 | -7.9 | 1,537 | 9 | 34,634 | 61.7 | 25,298 | 55.7 |
| Rancho Santa Margarita | 79.4 | 49.5 | 97.3 | 95.9 | 91.3 | 2.0 | 26,096 | -6.2 | 1,882 | 7 | 38,598 | 75.3 | 33,000 | 62.1 |
| Redding | 82.7 | 11.0 | 91.8 | 83.3 | 83.4 | 5.6 | 39,347 | -0.9 | 3,345 | 9 | 74,179 | 60.2 | 57,111 | 44.2 |
| Redlands | 80.9 | 31.1 | 94.0 | 86.2 | 89.7 | 4.8 | 35,091 | -1.9 | 2,630 | 8 | 59,601 | 62.1 | 47,504 | 50.7 |
| Redondo Beach | 77.5 | 42.5 | NA | NA | 84.4 | 6.4 | 38,747 | -6.7 | 3,753 | 10 | 51,601 | 75.5 | 43,607 | 58.3 |
| Redwood City | 72.9 | 39.2 | 95.7 | 90.4 | 83.2 | 9.4 | 48,018 | -6.9 | 2,635 | 6 | 70,197 | 71.0 | 58,432 | 60.7 |
| Rialto | 84.4 | 46.4 | 96.5 | 86.9 | 88.3 | 3.2 | 44,994 | 0.2 | 4,794 | 11 | 77,382 | 66.6 | 67,350 | 49.1 |
| Richmond | 68.0 | 51.9 | 94.0 | 89.5 | 89.3 | 4.9 | 52,175 | -1.5 | 5,818 | 11 | 86,778 | 67.6 | 73,351 | 50.9 |
| Ridgecrest | 85.4 | 8.9 | 89.1 | 81.6 | NA | NA | 13,125 | -5.0 | 672 | 5 | 23,359 | 65.4 | 20,123 | 49.7 |
| Riverside | 78.1 | 50.1 | 96.1 | 90.5 | 88.1 | 5.0 | 154,164 | -0.5 | 13,878 | 9 | 260,605 | 63.9 | 225,673 | 49.9 |
| Rocklin | 77.4 | 43.4 | 97.3 | 95.6 | 84.9 | 9.4 | 33,237 | -0.8 | 2,344 | 7 | 53,039 | 61.1 | 41,578 | 50.5 |
| Rohnert Park | 76.0 | 41.3 | 94.5 | 90.9 | 81.0 | 3.3 | 22,819 | -4.7 | 1,966 | 9 | 36,712 | 68.2 | 27,974 | 50.9 |
| Rosemead | 85.3 | 50.6 | 95.7 | 84.5 | 96.8 | 0.4 | 25,441 | 0.2 | 4,118 | 16 | 44,779 | 58.4 | 35,869 | 53.2 |
| Roseville | 80.5 | 43.8 | 96.0 | 91.9 | 87.6 | 8.2 | 68,659 | -1.0 | 5,091 | 7 | 112,968 | 65.1 | 88,331 | 56.5 |
| Sacramento | 74.2 | 36.8 | 94.7 | 89.6 | 82.2 | 6.5 | 236,881 | -0.1 | 22,835 | 10 | 411,100 | 65.4 | 338,257 | 51.6 |
| Salinas | 74.1 | 29.0 | 98.1 | 90.0 | 92.4 | 2.0 | 77,545 | -4.3 | 9,862 | 13 | 111,095 | 63.9 | 98,460 | 47.5 |
| San Bernardino | 79.3 | 35.5 | 92.5 | 80.2 | 89.7 | 3.4 | 85,804 | 1.3 | 10,402 | 12 | 163,136 | 60.7 | 141,776 | 45.0 |
| San Bruno | 61.8 | 44.7 | 97.2 | 91.7 | 89.4 | 5.7 | 25,950 | -4.6 | 2,115 | 8 | 36,330 | 70.2 | 29,849 | 60.5 |
| San Buenaventura (Ventura) | 80.2 | 32.9 | 89.5 | 86.8 | 89.0 | 3.1 | 53,931 | -4.1 | 4,618 | 9 | 90,201 | 64.3 | 70,514 | 51.3 |
| San Carlos | 70.5 | 46.1 | 97.8 | 95.4 | 90.4 | 2.8 | 16,569 | -7.7 | 834 | 5 | 23,646 | 72.7 | 17,972 | 65.6 |
| San Clemente | 76.5 | 44.8 | 98.8 | 96.5 | 88.2 | 7.3 | 30,289 | -5.2 | 2,405 | 8 | 51,317 | 65.1 | 39,538 | 51.7 |
| San Diego | 73.5 | 35.6 | 96.2 | 92.3 | 85.6 | 5.6 | 696,559 | -3.4 | 62,557 | 9 | 1,174,438 | 65.8 | 983,318 | 53.3 |
| San Dimas | 78.4 | 53.9 | 89.0 | 70.4 | 89.6 | 6.0 | 16,886 | -6.2 | 1,796 | 11 | 26,757 | 57.9 | 19,987 | 47.7 |

1. Employed persons.  2. Households.  3. Percent of civilian labor force.  4. Persons 16 years old and over.

# Table D. Cities — Construction, Wholesale Trade, and Retail Trade

| City | Value of residential construction authorized by building permits, 2020 | | | Wholesale trade[1], 2017 | | | | Retail trade[2], 2017 | | | |
|---|---|---|---|---|---|---|---|---|---|---|---|
| | New construction ($1,000) | Number of housing units | Percent single family | Number of establishments | Number of employees | Sales (mil dol) | Annual payroll (mil dol) | Number of establishments | Number of employees | Sales (mil dol) | Annual payroll (mil dol) |
| | 69 | 70 | 71 | 72 | 73 | 74 | 75 | 76 | 77 | 78 | 79 |
| CALIFORNIA— Cont'd | | | | | | | | | | | |
| Menlo Park | 19,914 | 12 | 100 | 32 | 745 | 1,787 | 70 | 113 | 1,798 | 486 | 92 |
| Merced | 222,655 | 744 | 97 | 37 | 643 | 1,091 | 37 | 225 | 4,134 | 1,157 | 109 |
| Milpitas | 66,807 | 409 | 13 | 160 | 8,560 | 9,498 | 1,381 | 295 | 5,471 | 1,352 | 144 |
| Mission Viejo | 10,673 | 39 | 36 | 102 | 530 | 375 | 39 | 336 | 6,221 | 1,874 | 189 |
| Modesto | 59,884 | 217 | 98 | 100 | 2,035 | 2,779 | 133 | 691 | 11,082 | 2,558 | 275 |
| Monrovia | 3,992 | 26 | 92 | 86 | 855 | 491 | 57 | 105 | 2,918 | 1,121 | 100 |
| Montclair | 0 | 0 | 0 | 90 | 695 | 324 | 32 | 217 | 4,086 | 1,242 | 125 |
| Montebello | 2,193 | 29 | 93 | 109 | 1,709 | 1,098 | 97 | 209 | 3,885 | 1,053 | 98 |
| Monterey | 6,200 | 14 | 29 | 35 | 335 | 1,006 | 24 | 200 | 2,232 | 496 | 65 |
| Monterey Park | 3,036 | 12 | 83 | 185 | 1,227 | 977 | 47 | 194 | 1,787 | 619 | 52 |
| Moorpark | 328 | 4 | 100 | 55 | 1,072 | 602 | 69 | 60 | 900 | 246 | 22 |
| Moreno Valley | 78,071 | 462 | 40 | D | D | D | D | 326 | 9,094 | 2,942 | 269 |
| Morgan Hill | 18,398 | 41 | 100 | 56 | 2,108 | 1,222 | 145 | D | D | D | 67 |
| Mountain View | 107,152 | 536 | 16 | 92 | 1,867 | 8,099 | 192 | 210 | 4,054 | 1,305 | 142 |
| Murrieta | 60,551 | 223 | 61 | 106 | 968 | 855 | 62 | 228 | 4,930 | 1,667 | 151 |
| Napa | 143,512 | 742 | 15 | 75 | 505 | 371 | 32 | 295 | 4,518 | 1,318 | 140 |
| National City | 53,957 | 305 | 2 | 93 | 1,571 | 985 | 74 | 295 | 6,265 | 1,759 | 181 |
| Newark | 134,066 | 461 | 64 | 55 | 1,167 | 868 | 69 | 161 | 4,149 | 883 | 114 |
| Newport Beach | 129,875 | 185 | 74 | 224 | 1,915 | 1,571 | 136 | 472 | 7,370 | 3,007 | 281 |
| Norco | 1,266 | 3 | 100 | 28 | 534 | 362 | 35 | 108 | 1,527 | 558 | 44 |
| Norwalk | 1,438 | 6 | 100 | 96 | 1,431 | 1,304 | 70 | 172 | 3,479 | 1,391 | 110 |
| Novato | 11,735 | 38 | 71 | 67 | 627 | 526 | 51 | 162 | 2,656 | 883 | 94 |
| Oakland | 206,913 | 830 | 12 | 341 | 4,811 | 4,229 | 300 | 995 | 11,810 | 3,804 | 425 |
| Oakley | 96,407 | 456 | 72 | 7 | D | 24 | D | 37 | 403 | 143 | 13 |
| Oceanside | 36,494 | 223 | 15 | 144 | 1,563 | 1,039 | 93 | 398 | 6,639 | 1,758 | 174 |
| Ontario | 147,574 | 739 | 87 | 775 | 13,440 | 11,719 | 705 | 590 | 14,963 | 7,008 | 504 |
| Orange | 5,687 | 40 | 100 | 334 | 4,361 | 6,802 | 267 | 538 | 8,816 | 2,639 | 274 |
| Oxnard | 88,708 | 339 | 11 | 211 | 3,685 | 7,401 | 245 | 437 | 8,471 | 3,151 | 275 |
| Pacifica | 600 | 1 | 100 | 10 | 29 | 28 | 2 | 58 | 681 | 172 | 19 |
| Palmdale | 55,357 | 237 | 100 | 30 | 147 | 67 | 5 | 308 | 7,566 | 1,834 | 191 |
| Palm Desert | 30,685 | 49 | 100 | 77 | 569 | 271 | 28 | 453 | 6,849 | 2,287 | 191 |
| Palm Springs | 43,714 | 132 | 89 | 32 | 182 | 76 | 7 | 208 | 3,280 | 973 | 94 |
| Palo Alto | 112,528 | 206 | 44 | 87 | 4,217 | 26,497 | 1,840 | 323 | 5,426 | 1,940 | 237 |
| Paradise | 195,482 | 1,035 | 80 | 7 | 49 | 5 | 1 | 67 | 872 | 186 | 22 |
| Paramount | 0 | 0 | 0 | 219 | 2,167 | 1,178 | 94 | 123 | 1,632 | 419 | 46 |
| Pasadena | 78,806 | 336 | 7 | 167 | 1,143 | 1,084 | 79 | 559 | 9,940 | 3,730 | 335 |
| Perris | 27,963 | 170 | 85 | 29 | 348 | 364 | 20 | 103 | 3,071 | 1,320 | 107 |
| Petaluma | 5,445 | 21 | 100 | 103 | 2,461 | 1,855 | 232 | 273 | 3,900 | 1,259 | 135 |
| Pico Rivera | 6,535 | 60 | 65 | 86 | 2,223 | 1,684 | 130 | 117 | 2,477 | 573 | 64 |
| Pittsburg | 42,209 | 185 | 55 | 33 | 771 | 331 | 42 | 106 | 2,234 | 665 | 67 |
| Placentia | 2,092 | 11 | 100 | 106 | 1,078 | 625 | 71 | 99 | 1,337 | 506 | 46 |
| Pleasant Hill | 4,927 | 16 | 88 | 25 | 139 | 98 | 9 | 135 | 3,149 | 733 | 78 |
| Pleasanton | 46,929 | 119 | 61 | 129 | 3,227 | 2,020 | 171 | 314 | 6,397 | 1,865 | 207 |
| Pomona | 61,636 | 480 | 28 | 316 | 3,279 | 2,922 | 174 | 330 | 4,329 | 1,305 | 119 |
| Porterville | 39,962 | 172 | 65 | 24 | 207 | 567 | 14 | 151 | 2,652 | 620 | 66 |
| Poway | 14,985 | 76 | 33 | 95 | 1,987 | 1,181 | 149 | 140 | 3,001 | 1,122 | 100 |
| Rancho Cordova | 164,436 | 572 | 100 | 165 | 2,261 | 1,573 | 166 | 282 | 3,851 | 1,105 | 125 |
| Rancho Cucamonga | 65,777 | 349 | 28 | 377 | 4,598 | 4,372 | 247 | 476 | 8,566 | 2,281 | 232 |
| Rancho Palos Verdes | 2,513 | 3 | 100 | 35 | 95 | 58 | 5 | 50 | 482 | 162 | 14 |
| Rancho Santa Margarita | 0 | 0 | 0 | 65 | 488 | 260 | 31 | 73 | 1,967 | 644 | 64 |
| Redding | 62,747 | 214 | 98 | 110 | 1,223 | 761 | 67 | 442 | 7,306 | 2,252 | 219 |
| Redlands | 17,958 | 79 | 100 | 53 | 622 | 403 | 29 | 213 | 4,753 | 1,711 | 140 |
| Redondo Beach | 16,339 | 46 | 74 | 73 | 425 | 391 | 29 | 261 | 4,055 | 963 | 105 |
| Redwood City | 38,377 | 145 | 38 | 69 | 2,137 | 1,359 | 269 | 222 | 4,818 | 2,089 | 196 |
| Rialto | 11,251 | 48 | 88 | 54 | 1,109 | 1,195 | 57 | 150 | 3,299 | 1,219 | 91 |
| Richmond | 24,138 | 140 | 98 | 100 | 1,730 | 1,934 | 115 | 228 | 4,845 | 1,815 | 175 |
| Ridgecrest | 13,191 | 73 | 100 | 5 | 29 | 20 | 1 | 78 | 1,333 | 301 | 32 |
| Riverside | 51,403 | 344 | 52 | 324 | 5,890 | 5,222 | 347 | 867 | 15,315 | 4,981 | 463 |
| Rocklin | 164,684 | 424 | 100 | 70 | 1,816 | 1,080 | 106 | 171 | 2,627 | 973 | 98 |
| Rohnert Park | 45,697 | 245 | 85 | 34 | 698 | 348 | 51 | 110 | 2,176 | 716 | 65 |
| Rosemead | 9,772 | 64 | 94 | 129 | 593 | 295 | 20 | 177 | 1,807 | 473 | 48 |
| Roseville | 376,086 | 1,636 | 77 | 98 | 1,165 | 1,403 | 82 | 613 | 14,350 | 4,869 | 463 |
| Sacramento | 726,086 | 3,584 | 26 | 454 | 8,275 | 9,020 | 477 | 1,198 | 19,969 | 5,282 | 550 |
| Salinas | 24,630 | 205 | 47 | 134 | 2,729 | 2,879 | 178 | 443 | 7,831 | 2,664 | 227 |
| San Bernardino | 5,746 | 29 | 86 | 135 | 3,386 | 3,480 | 159 | 515 | 13,661 | 5,216 | 418 |
| San Bruno | 8,756 | 17 | 100 | 31 | 366 | 477 | 31 | 138 | 1,990 | 7,807 | 63 |
| San Buenaventura (Ventura) | 44,884 | 363 | 2 | 162 | 1,753 | 1,029 | 102 | 497 | 7,449 | 2,104 | 217 |
| San Carlos | 8,729 | 32 | 88 | 94 | 943 | 612 | 65 | 137 | 3,217 | 615 | 75 |
| San Clemente | 7,819 | 24 | 83 | 163 | 1,672 | 1,989 | 156 | 239 | 3,320 | 807 | 87 |
| San Diego | 925,481 | 4,682 | 10 | 1,866 | 22,288 | 22,459 | 1,652 | 4,240 | 66,413 | 20,732 | 2,025 |
| San Dimas | 6,756 | 18 | 100 | 100 | 571 | 314 | 30 | 128 | 1,816 | 587 | 51 |

1. Merchant wholesalers except manufacturers' sales branches and offices.  2. Establishments with payroll.

| City | Real estate and rental and leasing, 2017 | | | | Professional, scientific, and technical services[1], 2017 | | | | Manufacturing, 2017 | | | |
|---|---|---|---|---|---|---|---|---|---|---|---|---|
| | Number of establish-ments | Number of employees | Receipts (mil dol) | Annual payroll (mil dol) | Number of establish-ments | Number of employees | Receipts (mil dol) | Annual payroll (mil dol) | Number of establish-ments | Number of employees | Receipts (mil dol) | Annual payroll (mil dol) |
| | 80 | 81 | 82 | 83 | 84 | 85 | 86 | 87 | 88 | 89 | 90 | 91 |
| CALIFORNIA— Cont'd | | | | | | | | | | | | |
| Menlo Park | 80 | 320 | 161 | 27 | 346 | 6,125 | 1,996 | 866 | NA | NA | NA | NA |
| Merced | 69 | 359 | 73 | 13 | 90 | 462 | 46 | 17 | NA | NA | NA | NA |
| Milpitas | 93 | 762 | 198 | 41 | 343 | 7,472 | 2,426 | 975 | NA | NA | NA | NA |
| Mission Viejo | 170 | 750 | 283 | 50 | 496 | 2,675 | 626 | 252 | NA | NA | NA | NA |
| Modesto | 213 | 952 | 237 | 38 | 377 | 3,116 | 403 | 177 | NA | NA | NA | NA |
| Monrovia | 70 | 298 | 55 | 15 | 173 | 1,927 | 417 | 160 | NA | NA | NA | NA |
| Montclair | 32 | 226 | 63 | 10 | 36 | 224 | 32 | 9 | NA | NA | NA | NA |
| Montebello | 63 | 412 | 101 | 18 | 71 | 325 | 35 | 11 | NA | NA | NA | NA |
| Monterey | 95 | 411 | 134 | 17 | 240 | 4,470 | 493 | 203 | NA | NA | NA | NA |
| Monterey Park | 112 | 279 | 77 | 12 | 190 | 1,597 | 212 | 96 | NA | NA | NA | NA |
| Moorpark | 40 | 185 | 27 | 7 | 114 | 388 | 74 | 24 | NA | NA | NA | NA |
| Moreno Valley | 88 | 336 | 100 | 13 | 87 | 538 | 56 | 19 | NA | NA | NA | NA |
| Morgan Hill | 74 | 255 | 73 | 11 | 130 | 858 | 132 | 53 | NA | NA | NA | NA |
| Mountain View | 145 | 812 | 209 | 42 | 537 | 13,889 | 4,427 | 1,851 | NA | NA | NA | NA |
| Murrieta | 167 | 603 | 142 | 24 | 282 | 1,317 | 216 | 74 | NA | NA | NA | NA |
| Napa | 120 | 569 | 167 | 28 | D | D | D | D | NA | NA | NA | NA |
| National City | 55 | 306 | 86 | 14 | 54 | 585 | 75 | 29 | NA | NA | NA | NA |
| Newark | 51 | 247 | 106 | 13 | 157 | 2,442 | 956 | 288 | NA | NA | NA | NA |
| Newport Beach | 841 | 4,996 | 2,407 | 416 | 1,348 | 9,092 | 2,133 | 805 | NA | NA | NA | NA |
| Norco | 37 | 89 | 25 | 3 | 79 | 1,338 | 128 | 61 | NA | NA | NA | NA |
| Norwalk | 55 | 204 | 50 | 9 | 68 | 1,383 | 374 | 160 | NA | NA | NA | NA |
| Novato | 81 | 391 | 96 | 20 | 247 | 2,294 | 448 | 189 | NA | NA | NA | NA |
| Oakland | 544 | 3,196 | 934 | 159 | 1,628 | 14,453 | 3,276 | 1,341 | NA | NA | NA | NA |
| Oakley | 21 | 61 | 18 | 4 | 23 | 121 | 16 | 5 | NA | NA | NA | NA |
| Oceanside | 187 | 678 | 208 | 29 | 350 | 1,814 | 255 | 95 | NA | NA | NA | NA |
| Ontario | 218 | 1,786 | 666 | 97 | D | D | D | D | NA | NA | NA | NA |
| Orange | 300 | 2,196 | 657 | 110 | 744 | 9,814 | 1,166 | 477 | NA | NA | NA | NA |
| Oxnard | 168 | 627 | 189 | 26 | 257 | 2,222 | 395 | 138 | NA | NA | NA | NA |
| Pacifica | 30 | 52 | 18 | 3 | 83 | 236 | 44 | 11 | NA | NA | NA | NA |
| Palmdale | 112 | 373 | 90 | 13 | 157 | 3,894 | 750 | 323 | NA | NA | NA | NA |
| Palm Desert | 165 | 1,056 | 209 | 56 | 296 | 1,079 | 185 | 58 | NA | NA | NA | NA |
| Palm Springs | 151 | 676 | 175 | 31 | 190 | 639 | 131 | 40 | NA | NA | NA | NA |
| Palo Alto | 209 | 1,820 | 1,059 | 139 | 904 | 17,354 | 6,632 | 2,694 | NA | NA | NA | NA |
| Paradise | 30 | 215 | 45 | 7 | 36 | 158 | 19 | 5 | NA | NA | NA | NA |
| Paramount | 28 | 190 | 49 | 9 | 37 | 370 | 27 | 13 | NA | NA | NA | NA |
| Pasadena | 434 | 2,153 | 569 | 109 | 1,352 | 17,626 | 5,095 | 1,867 | NA | NA | NA | NA |
| Perris | 29 | 130 | 32 | 10 | 39 | 121 | 17 | 5 | NA | NA | NA | NA |
| Petaluma | 72 | 360 | 106 | 16 | D | D | D | D | NA | NA | NA | NA |
| Pico Rivera | 43 | 215 | 95 | 13 | 33 | 197 | 33 | 10 | NA | NA | NA | NA |
| Pittsburg | 34 | 219 | 57 | 12 | 49 | 283 | 49 | 15 | NA | NA | NA | NA |
| Placentia | 70 | 388 | 88 | 16 | 121 | 465 | 63 | 23 | NA | NA | NA | NA |
| Pleasant Hill | 55 | 340 | 94 | 24 | 160 | 840 | 148 | 61 | NA | NA | NA | NA |
| Pleasanton | 218 | 1,075 | 401 | 74 | 634 | 8,558 | 2,014 | 1,050 | NA | NA | NA | NA |
| Pomona | 102 | 652 | 157 | 29 | 132 | 1,085 | 156 | 52 | NA | NA | NA | NA |
| Porterville | 25 | 90 | 14 | 3 | 44 | 184 | 20 | 6 | NA | NA | NA | NA |
| Poway | 92 | 343 | 104 | 14 | 265 | 1,479 | 264 | 103 | NA | NA | NA | NA |
| Rancho Cordova | 90 | 447 | 119 | 30 | 243 | 5,602 | 1,174 | 492 | NA | NA | NA | NA |
| Rancho Cucamonga | 256 | 1,003 | 356 | 48 | 491 | 3,294 | 506 | 175 | NA | NA | NA | NA |
| Rancho Palos Verdes | 76 | 172 | 69 | 9 | 157 | 642 | 106 | 43 | NA | NA | NA | NA |
| Rancho Santa Margarita | 75 | 249 | 169 | 35 | 221 | 1,731 | 221 | 67 | NA | NA | NA | NA |
| Redding | 165 | 672 | 145 | 25 | 288 | 1,748 | 231 | 102 | NA | NA | NA | NA |
| Redlands | 98 | 477 | 117 | 16 | 245 | 1,199 | 190 | 66 | NA | NA | NA | NA |
| Redondo Beach | 168 | 888 | 206 | 51 | 476 | 7,910 | 1,427 | 846 | NA | NA | NA | NA |
| Redwood City | 163 | 858 | 327 | 48 | 477 | 10,934 | 3,716 | 1,590 | NA | NA | NA | NA |
| Rialto | 49 | 211 | 49 | 7 | 39 | 266 | 23 | 8 | NA | NA | NA | NA |
| Richmond | 78 | 395 | 128 | 20 | 172 | 2,147 | 225 | 201 | NA | NA | NA | NA |
| Ridgecrest | 26 | 104 | 23 | 3 | 56 | 848 | 145 | 56 | NA | NA | NA | NA |
| Riverside | 370 | 1,809 | 431 | 76 | 674 | 5,250 | 792 | 296 | NA | NA | NA | NA |
| Rocklin | 127 | 352 | 92 | 16 | 211 | 1,853 | 293 | 108 | NA | NA | NA | NA |
| Rohnert Park | 55 | 232 | 74 | 10 | 75 | 459 | 69 | 30 | NA | NA | NA | NA |
| Rosemead | 55 | 156 | 41 | 9 | 100 | 358 | 40 | 12 | NA | NA | NA | NA |
| Roseville | 337 | 2,177 | 608 | 126 | 560 | 5,821 | 961 | 400 | NA | NA | NA | NA |
| Sacramento | 634 | 3,406 | 924 | 176 | 1,847 | 17,260 | 4,127 | 1,489 | NA | NA | NA | NA |
| Salinas | 116 | 542 | 176 | 27 | D | D | D | D | NA | NA | NA | NA |
| San Bernardino | 118 | 523 | 117 | 20 | 220 | 1,925 | 239 | 116 | NA | NA | NA | NA |
| San Bruno | 44 | 164 | 104 | 10 | 77 | 707 | 148 | 82 | NA | NA | NA | NA |
| San Buenaventura (Ventura) | 217 | 996 | 248 | 46 | 497 | 3,948 | 723 | 331 | NA | NA | NA | NA |
| San Carlos | 61 | 341 | 112 | 25 | 243 | 2,208 | 579 | 242 | NA | NA | NA | NA |
| San Clemente | 154 | 887 | 245 | 35 | 448 | 1,530 | 339 | 97 | NA | NA | NA | NA |
| San Diego | 3,293 | 18,066 | 7,825 | 1,125 | 8,428 | 115,825 | 32,670 | 12,607 | NA | NA | NA | NA |
| San Dimas | 61 | 518 | 77 | 18 | 148 | 2,068 | 353 | 130 | NA | NA | NA | NA |

1. Establishments subject to federal tax.

# Table D. Cities — Accommodation and Food Services, Arts, Entertainment, and Recreation, and Health Care and Social Assistance

| City | Accommodation and food services, 2017 | | | | Arts, entertainment, and recreation[1], 2017 | | | | Health care and social assistance[1], 2017 | | | |
|---|---|---|---|---|---|---|---|---|---|---|---|---|
| | Number of establishments | Number of employees | Receipts (mil dol) | Annual payroll (mil dol) | Number of establishments | Number of employees | Receipts (mil dol) | Annual payroll (mil dol) | Number of establishments | Number of employees | Receipts (mil dol) | Annual payroll (mil dol) |
| | 92 | 93 | 94 | 95 | 96 | 97 | 98 | 99 | 100 | 101 | 102 | 103 |
| **CALIFORNIA— Cont'd** | | | | | | | | | | | | |
| Menlo Park | 97 | 1,690 | 128 | 40 | 20 | 243 | 23 | 9 | 163 | 2,418 | 270 | 128 |
| Merced | 138 | 2,513 | 148 | 38 | 20 | 221 | 12 | 4 | 288 | 5,102 | 689 | 263 |
| Milpitas | 314 | 5,598 | 501 | 122 | 26 | 584 | 44 | 12 | 285 | 3,405 | 496 | 175 |
| Mission Viejo | 214 | 3,880 | 230 | 67 | 39 | 661 | 51 | 14 | 508 | 7,651 | 1,219 | 492 |
| Modesto | 430 | 8,484 | 483 | 140 | 47 | 858 | 59 | 17 | 683 | 16,649 | 2,968 | 1,110 |
| Monrovia | 119 | 2,024 | 129 | 38 | 23 | 182 | 19 | 4 | 104 | 1,651 | 173 | 68 |
| Montclair | 93 | 1,627 | 93 | 28 | D | D | D | D | 126 | 2,517 | 214 | 76 |
| Montebello | 142 | 2,355 | 136 | 39 | 7 | 182 | 16 | 5 | 194 | 3,248 | 476 | 164 |
| Monterey | 234 | 6,544 | 567 | 164 | 39 | 994 | 163 | 52 | 307 | 6,055 | 1,131 | 426 |
| Monterey Park | 242 | 3,327 | 209 | 63 | 12 | 132 | 13 | 3 | 282 | 4,342 | 704 | 227 |
| Moorpark | 54 | 817 | 50 | 15 | 15 | 45 | 5 | 2 | 67 | 692 | 67 | 27 |
| Moreno Valley | 257 | 4,746 | 294 | 79 | 22 | 310 | 22 | 5 | 248 | 6,376 | 1,035 | 401 |
| Morgan Hill | 128 | 1,904 | 142 | 39 | 15 | 368 | 16 | 7 | D | D | D | 54 |
| Mountain View | 366 | 6,514 | 486 | 185 | 24 | 838 | 92 | 22 | 324 | 7,574 | 2,048 | 774 |
| Murrieta | 195 | 3,737 | 213 | 63 | 37 | 548 | 31 | 9 | 340 | 5,494 | 735 | 282 |
| Napa | 229 | 5,716 | 553 | 148 | 30 | 451 | 58 | 9 | 306 | 5,637 | 914 | 335 |
| National City | 177 | 2,997 | 215 | 55 | 12 | 139 | 7 | 2 | 177 | 3,842 | 442 | 137 |
| Newark | 176 | 2,471 | 201 | 51 | 4 | 52 | 4 | 1 | 104 | 773 | 80 | 28 |
| Newport Beach | 411 | 13,281 | 1,228 | 345 | 131 | 1,708 | 218 | 68 | 1,036 | 13,306 | 2,612 | 873 |
| Norco | D | D | D | D | 9 | 117 | 13 | 3 | 47 | 557 | 53 | 20 |
| Norwalk | 148 | 2,106 | 141 | 37 | 11 | 209 | 15 | 4 | 160 | 4,090 | 460 | 281 |
| Novato | 127 | 2,081 | 139 | 42 | 27 | 438 | 33 | 11 | 194 | 2,670 | 431 | 175 |
| Oakland | 1,108 | 17,173 | 1,293 | 373 | 164 | 3,257 | 815 | 353 | 1,279 | 30,875 | 5,002 | 2,206 |
| Oakley | D | D | D | D | 3 | D | 7 | | 39 | 248 | 21 | 8 |
| Oceanside | 359 | 7,350 | 587 | 144 | 47 | 1,100 | 70 | 25 | 363 | 5,870 | 761 | 297 |
| Ontario | 364 | 7,906 | 529 | 147 | 30 | 872 | 74 | 19 | 268 | 6,152 | 1,061 | 344 |
| Orange | 450 | 8,572 | 591 | 173 | 53 | 731 | 60 | 15 | 687 | 20,941 | 3,790 | 1,363 |
| Oxnard | 312 | 6,273 | 416 | 114 | 36 | 398 | 32 | 8 | 455 | 7,068 | 847 | 348 |
| Pacifica | 72 | 1,008 | 68 | 19 | D | D | D | 2 | 45 | 563 | 52 | 24 |
| Palmdale | 235 | 4,743 | 300 | 85 | 17 | 520 | 24 | 9 | 261 | 2,704 | 333 | 109 |
| Palm Desert | 195 | 6,563 | 498 | 154 | 48 | 2,602 | 216 | 73 | 286 | 3,891 | 353 | 138 |
| Palm Springs | 276 | 6,406 | 503 | 146 | 49 | 1,363 | 264 | 47 | 283 | 5,879 | 1,027 | 364 |
| Palo Alto | 321 | 7,457 | 710 | 216 | 57 | 1,219 | 100 | 37 | 412 | 27,393 | 7,346 | 2,683 |
| Paradise | D | D | D | D | 13 | 124 | 5 | 2 | 114 | 2,715 | 340 | 132 |
| Paramount | 69 | 830 | 60 | 16 | 4 | 42 | 3 | 1 | 70 | 1,603 | 153 | 69 |
| Pasadena | 557 | 11,645 | 884 | 253 | 249 | 2,311 | 390 | 101 | 1,104 | 19,293 | 2,798 | 1,132 |
| Perris | 77 | 1,310 | 93 | 23 | D | D | D | 4 | 57 | 829 | 81 | 33 |
| Petaluma | 180 | 2,830 | 176 | 55 | 41 | 631 | 55 | 12 | 193 | 3,342 | 405 | 176 |
| Pico Rivera | 115 | 2,022 | 141 | 36 | 5 | 98 | 7 | 2 | 83 | 1,570 | 144 | 55 |
| Pittsburg | 94 | 1,514 | 104 | 29 | 5 | 80 | 4 | 1 | 85 | 1,059 | 98 | 41 |
| Placentia | 106 | 1,479 | 95 | 26 | D | D | D | 4 | 137 | 1,375 | 183 | 61 |
| Pleasant Hill | 114 | 2,104 | 183 | 45 | 18 | 446 | 23 | 11 | 157 | 2,631 | 442 | 117 |
| Pleasanton | 252 | 4,537 | 382 | 102 | 44 | 1,109 | 96 | 27 | 338 | 7,384 | 1,571 | 548 |
| Pomona | 237 | 4,464 | 259 | 79 | 18 | 394 | 83 | 16 | 322 | 9,677 | 1,541 | 540 |
| Porterville | 92 | 1,422 | 84 | 22 | D | D | D | D | 154 | 3,293 | 340 | 140 |
| Poway | 112 | 1,542 | 90 | 25 | 31 | 452 | 37 | 13 | 191 | 2,727 | 418 | 144 |
| Rancho Cordova | 165 | 3,165 | 218 | 57 | 24 | 491 | 32 | 9 | 133 | 5,321 | 851 | 325 |
| Rancho Cucamonga | 402 | 8,913 | 543 | 161 | 51 | 658 | 56 | 15 | 535 | 6,858 | 719 | 278 |
| Rancho Palos Verdes | 40 | 1,532 | 162 | 49 | 20 | 328 | 30 | 11 | 132 | 914 | 118 | 39 |
| Rancho Santa Margarita | 78 | 1,861 | 107 | 32 | 26 | 365 | 36 | 10 | 97 | 900 | 95 | 35 |
| Redding | 260 | 4,767 | 284 | 82 | 35 | 494 | 28 | 9 | 565 | 10,007 | 1,419 | 526 |
| Redlands | 185 | 3,318 | 191 | 56 | 27 | 602 | 26 | 8 | 379 | 7,221 | 953 | 366 |
| Redondo Beach | 231 | 4,891 | 322 | 105 | D | D | D | D | 260 | 1,792 | 222 | 87 |
| Redwood City | 244 | 3,724 | 293 | 87 | 38 | 931 | 76 | 20 | 346 | 6,563 | 1,107 | 565 |
| Rialto | 113 | 1,999 | 125 | 33 | 8 | 123 | 18 | 2 | 96 | 936 | 92 | 33 |
| Richmond | 138 | 1,408 | 110 | 31 | 35 | 1,417 | 94 | 27 | 150 | 4,940 | 677 | 284 |
| Ridgecrest | 68 | 1,012 | 69 | 18 | D | D | D | 1 | 67 | 1,315 | 159 | 58 |
| Riverside | 628 | 12,935 | 771 | 225 | 69 | 1,186 | 89 | 27 | 969 | 21,881 | 3,070 | 1,115 |
| Rocklin | 132 | 1,950 | 116 | 32 | 18 | 409 | 23 | 7 | 192 | 1,663 | 161 | 59 |
| Rohnert Park | 102 | 4,092 | 768 | 120 | 13 | 255 | 20 | 5 | 93 | 1,001 | 139 | 51 |
| Rosemead | 177 | 2,780 | 183 | 50 | D | D | D | 3 | 147 | 2,155 | 203 | 79 |
| Roseville | 405 | 8,923 | 557 | 168 | 58 | 1,894 | 116 | 35 | 583 | 15,615 | 3,210 | 1,328 |
| Sacramento | 1,210 | 24,804 | 1,614 | 466 | 148 | 3,862 | 485 | 208 | 1,351 | 46,565 | 9,154 | 3,550 |
| Salinas | 300 | 4,775 | 318 | 88 | 26 | 451 | 35 | 9 | 345 | 7,217 | 1,304 | 493 |
| San Bernardino | 362 | 6,183 | 393 | 107 | 22 | 467 | 38 | 11 | 452 | 12,967 | 1,868 | 631 |
| San Bruno | 117 | 1,655 | 134 | 37 | D | D | D | D | 107 | 1,131 | 207 | 59 |
| San Buenaventura (Ventura) | 353 | 6,364 | 430 | 126 | 72 | 718 | 60 | 15 | 583 | 11,196 | 1,544 | 576 |
| San Carlos | 105 | 1,698 | 128 | 42 | 21 | 145 | 13 | 4 | 113 | 1,787 | 249 | 111 |
| San Clemente | 174 | 2,918 | 194 | 53 | 48 | 444 | 49 | 17 | 249 | 2,870 | 339 | 132 |
| San Diego | 4,076 | 98,278 | 7,986 | 2,291 | 629 | 18,611 | 1,676 | 529 | 4,712 | 97,758 | 16,501 | 6,428 |
| San Dimas | 89 | 1,800 | 107 | 32 | 18 | 244 | 34 | 8 | 141 | 2,450 | 242 | 87 |

1. Establishments subject to federal tax.

## Table D. Cities — Other Services and Government Employment and Payroll

| City | Other services[1] | | | | Government employment and payroll, 2017 | | | | | | | | |
| | | | | | | March payroll | | | | | | | |
| | | | | | | | Percent of total for: | | | | | | |
| | Number of establishments | Number of employees | Receipts (mil dol) | Annual payroll (mil dol) | Full-time equivalent employees | Total (dollars) | Administrative, judicial, and legal | Police and corrections | Fire protection | Highways and transportation | Health and welfare | Natural resources and utilities | Education and libraries |
| | 104 | 105 | 106 | 107 | 108 | 109 | 110 | 111 | 112 | 113 | 114 | 115 | 116 |
| **CALIFORNIA— Cont'd** | | | | | | | | | | | | | |
| Menlo Park | 93 | 584 | 92 | 22 | 293 | 2,219,306 | 10.4 | 33.8 | 0.0 | 5.7 | 1.0 | 8.5 | 5.5 |
| Merced | 77 | 433 | 43 | 12 | 448 | 2,854,185 | 10.0 | 35.2 | 18.2 | 2.5 | 1.4 | 22.1 | 0.0 |
| Milpitas | 133 | 1,011 | 157 | 48 | 375 | 3,499,150 | 15.0 | 37.9 | 22.9 | 7.2 | 1.2 | 9.3 | 0.0 |
| Mission Viejo | 215 | 1,253 | 135 | 42 | 196 | 1,146,333 | 30.3 | 0.0 | 0.0 | 15.1 | 7.1 | 23.9 | 17.8 |
| Modesto | 309 | 1,761 | 183 | 51 | 1,148 | 8,001,264 | 13.2 | 29.7 | 18.7 | 9.0 | 0.8 | 25.7 | 0.0 |
| Monrovia | 85 | 575 | 103 | 20 | 280 | 1,931,644 | 18.1 | 34.8 | 23.5 | 1.6 | 3.3 | 14.1 | 3.7 |
| Montclair | D | D | 40 | D | 278 | 1,588,376 | 24.9 | 33.6 | 17.3 | 1.8 | 13.6 | 5.5 | 0.0 |
| Montebello | 106 | 571 | 59 | 17 | 392 | 2,384,162 | 7.9 | 36.5 | 21.6 | 29.5 | 0.0 | 2.7 | 0.0 |
| Monterey | 99 | 897 | 172 | 27 | 505 | 3,608,460 | 10.0 | 15.7 | 24.8 | 5.7 | 0.6 | 13.7 | 3.4 |
| Monterey Park | 87 | 378 | 40 | 9 | 333 | 2,214,654 | 9.7 | 35.3 | 26.2 | 4.9 | 2.6 | 16.0 | 5.3 |
| Moorpark | 33 | 148 | 18 | 5 | 74 | 567,743 | 35.0 | 0.0 | 0.0 | 14.4 | 20.1 | 30.5 | 0.0 |
| Moreno Valley | 132 | 572 | 51 | 14 | 412 | 2,054,931 | 30.4 | 4.2 | 2.5 | 17.0 | 20.1 | 17.5 | 3.7 |
| Morgan Hill | 93 | 388 | 42 | 12 | 224 | 1,286,104 | 12.5 | 25.6 | 0.0 | 4.5 | 0.4 | 11.8 | 0.0 |
| Mountain View | 152 | 901 | 318 | 88 | 671 | 5,866,700 | 11.3 | 25.8 | 15.6 | 12.6 | 6.2 | 15.5 | 4.4 |
| Murrieta | 186 | 1,259 | 123 | 36 | 316 | 2,365,710 | 12.9 | 42.9 | 23.5 | 5.3 | 0.3 | 5.1 | 3.5 |
| Napa | 173 | 1,100 | 144 | 38 | 449 | 4,001,327 | 10.5 | 30.6 | 19.6 | 10.0 | 3.1 | 21.2 | 0.0 |
| National City | 124 | 687 | 63 | 20 | 405 | 2,542,406 | 8.6 | 41.8 | 20.2 | 3.0 | 4.4 | 9.9 | 7.7 |
| Newark | 92 | 587 | 61 | 19 | 176 | 878,820 | 14.6 | 51.4 | 0.0 | 5.5 | 1.7 | 15.2 | 0.0 |
| Newport Beach | 322 | 2,215 | 340 | 71 | 784 | 8,009,294 | 13.2 | 35.6 | 19.4 | 8.3 | 1.1 | 10.9 | 4.5 |
| Norco | 72 | 412 | 39 | 12 | 89 | 348,976 | 36.5 | 0.0 | 0.0 | 0.0 | 1.6 | 35.4 | 0.0 |
| Norwalk | 89 | 419 | 54 | 13 | 339 | 1,819,034 | 18.9 | 5.9 | 0.0 | 43.1 | 10.1 | 12.8 | 0.0 |
| Novato | 113 | 881 | 248 | 44 | 208 | 1,580,041 | 7.8 | 47.8 | 0.0 | 4.1 | 8.2 | 13.5 | 0.0 |
| Oakland | 945 | 7,498 | 1,500 | 345 | 4,077 | 36,522,164 | 10.8 | 32.3 | 18.6 | 18.4 | 6.4 | 5.6 | 3.0 |
| Oakley | D | D | D | D | 75 | 669,513 | 13.2 | 61.2 | 0.0 | 10.3 | 4.3 | 6.0 | 0.0 |
| Oceanside | 231 | 1,596 | 156 | 49 | 1,010 | 7,245,519 | 14.3 | 40.5 | 19.0 | 3.1 | 2.9 | 16.1 | 2.4 |
| Ontario | 231 | 2,627 | 346 | 114 | 1,177 | 10,470,584 | 9.1 | 39.3 | 21.9 | 0.0 | 1.2 | 16.1 | 2.3 |
| Orange | 353 | 2,549 | 283 | 86 | 650 | 5,729,395 | 7.6 | 37.2 | 26.7 | 7.7 | 3.1 | 11.4 | 3.7 |
| Oxnard | 227 | 1,132 | 131 | 36 | 1,254 | 8,375,994 | 8.4 | 32.1 | 18.0 | 4.2 | 7.7 | 22.5 | 1.9 |
| Pacifica | 52 | 157 | 17 | 4 | 189 | 1,420,456 | 8.7 | 25.9 | 18.2 | 10.1 | 0.0 | 28.7 | 0.0 |
| Palmdale | 121 | 664 | 48 | 15 | 259 | 1,371,327 | 21.8 | 4.9 | 0.0 | 23.4 | 8.3 | 26.4 | 0.0 |
| Palm Desert | 189 | 1,388 | 115 | 36 | 115 | 977,448 | 44.6 | 0.0 | 0.0 | 25.9 | 3.5 | 4.6 | 0.0 |
| Palm Springs | 114 | 868 | 78 | 21 | 452 | 3,754,400 | 16.6 | 29.9 | 19.3 | 18.5 | 1.5 | 3.9 | 2.1 |
| Palo Alto | 185 | 1,962 | 970 | 137 | 1,032 | 9,194,693 | 18.4 | 13.2 | 14.3 | 3.1 | 0.8 | 34.4 | 3.7 |
| Paradise | 46 | 185 | 15 | 4 | 63 | 354,529 | 17.4 | 56.4 | 0.6 | 11.5 | 5.0 | 0.0 | 0.0 |
| Paramount | 66 | 480 | 68 | 20 | 165 | 844,493 | 21.3 | 19.0 | 0.0 | 23.3 | 10.7 | 24.9 | 0.0 |
| Pasadena | 448 | 2,941 | 338 | 88 | 1,903 | 14,743,235 | 14.4 | 23.4 | 13.4 | 4.8 | 4.4 | 30.0 | 3.2 |
| Perris | 39 | 166 | 30 | 5 | 259 | 1,448,110 | 17.9 | 0.0 | 0.0 | 6.2 | 9.0 | 33.0 | 0.0 |
| Petaluma | 139 | 899 | 90 | 28 | 319 | 2,597,477 | 11.3 | 31.4 | 22.1 | 13.6 | 0.9 | 19.4 | 0.0 |
| Pico Rivera | 67 | 518 | 70 | 19 | 266 | 1,253,039 | 21.6 | 0.0 | 0.0 | 31.8 | 12.8 | 33.8 | 0.0 |
| Pittsburg | 80 | 681 | 63 | 22 | 292 | 2,210,692 | 11.4 | 47.3 | 0.0 | 8.1 | 2.4 | 20.8 | 0.0 |
| Placentia | D | D | D | 13 | 135 | 872,042 | 14.9 | 63.3 | 0.0 | 14.1 | 0.3 | 7.3 | 0.0 |
| Pleasant Hill | 75 | 513 | 49 | 16 | 129 | 1,083,557 | 21.9 | 52.8 | 0.0 | 22.8 | 2.3 | 0.0 | 0.0 |
| Pleasanton | 169 | 1,401 | 203 | 63 | 532 | 4,798,853 | 11.6 | 23.7 | 32.0 | 5.5 | 1.2 | 18.0 | 4.1 |
| Pomona | 163 | 968 | 111 | 34 | 654 | 3,272,255 | 8.4 | 51.9 | 0.0 | 5.9 | 6.4 | 15.2 | 2.9 |
| Porterville | 43 | 281 | 26 | 9 | 564 | 2,144,104 | 8.1 | 22.4 | 11.8 | 4.1 | 3.0 | 45.3 | 2.2 |
| Poway | 108 | 540 | 51 | 15 | 247 | 1,657,178 | 20.5 | 0.0 | 32.3 | 3.2 | 0.0 | 23.0 | 0.0 |
| Rancho Cordova | 118 | 761 | 102 | 32 | 79 | 502,805 | 38.0 | 0.0 | 0.0 | 24.4 | 3.0 | 0.0 | 0.0 |
| Rancho Cucamonga | 282 | 1,496 | 162 | 48 | 625 | 3,766,895 | 15.6 | 0.0 | 36.6 | 11.6 | 0.0 | 26.4 | 4.9 |
| Rancho Palos Verdes | 40 | 100 | 13 | 3 | 92 | 602,068 | 100.0 | 0.0 | 0.0 | 0.0 | 0.0 | 0.0 | 0.0 |
| Rancho Santa Margarita | 71 | 346 | 36 | 10 | 33 | 223,919 | 44.4 | 2.7 | 0.0 | 21.3 | 18.6 | 13.0 | 0.0 |
| Redding | 206 | 1,272 | 168 | 39 | 789 | 5,878,787 | 9.6 | 20.4 | 13.4 | 6.2 | 1.1 | 39.5 | 0.0 |
| Redlands | 143 | 970 | 97 | 31 | 493 | 3,384,099 | 9.8 | 30.4 | 20.7 | 4.5 | 0.4 | 24.3 | 3.4 |
| Redondo Beach | 158 | 762 | 76 | 22 | 505 | 3,899,259 | 11.8 | 37.3 | 21.3 | 8.1 | 5.9 | 9.4 | 3.4 |
| Redwood City | 163 | 1,846 | 1,298 | 196 | 585 | 5,213,786 | 13.5 | 23.5 | 18.2 | 7.9 | 8.8 | 16.4 | 7.2 |
| Rialto | 81 | 388 | 45 | 12 | 375 | 3,064,779 | 11.6 | 40.5 | 27.8 | 7.5 | 1.9 | 4.9 | 0.0 |
| Richmond | 133 | 719 | 102 | 30 | 737 | 6,812,090 | 9.5 | 42.1 | 17.8 | 1.6 | 1.0 | 9.0 | 3.2 |
| Ridgecrest | 34 | 149 | 16 | 4 | 107 | 593,787 | 13.3 | 50.8 | 0.0 | 14.7 | 4.7 | 14.5 | 0.0 |
| Riverside | 456 | 2,928 | 363 | 92 | 2,170 | 19,847,957 | 12.0 | 34.9 | 11.5 | 5.4 | 1.1 | 23.8 | 1.3 |
| Rocklin | 90 | 483 | 55 | 16 | 253 | 1,929,938 | 16.9 | 36.8 | 18.9 | 5.5 | 1.5 | 4.7 | 0.0 |
| Rohnert Park | 71 | 418 | 47 | 14 | 192 | 1,356,881 | 17.9 | 57.4 | 0.0 | 0.0 | 1.6 | 23.1 | 0.0 |
| Rosemead | 100 | 300 | 27 | 7 | 85 | 486,146 | 31.5 | 13.4 | 0.0 | 10.6 | 6.6 | 37.9 | 0.0 |
| Roseville | 250 | 2,574 | 331 | 99 | 1,320 | 9,827,383 | 15.4 | 18.1 | 15.2 | 2.8 | 0.7 | 40.5 | 2.0 |
| Sacramento | 1,017 | 8,213 | 1,322 | 396 | 4,106 | 31,089,039 | 8.1 | 36.2 | 19.8 | 10.7 | 3.7 | 19.8 | 0.0 |
| Salinas | 214 | 1,236 | 166 | 43 | 543 | 3,681,676 | 11.2 | 41.1 | 19.5 | 14.4 | 2.8 | 4.0 | 4.9 |
| San Bernardino | 226 | 1,537 | 130 | 38 | 1,186 | 7,552,176 | 7.2 | 42.0 | 17.0 | 0.3 | 0.0 | 26.4 | 1.1 |
| San Bruno | 95 | 670 | 96 | 25 | 252 | 2,135,604 | 10.4 | 36.3 | 20.5 | 2.1 | 2.2 | 15.9 | 3.3 |
| San Buenaventura (Ventura) | 262 | 1,266 | 148 | 38 | 693 | 4,722,938 | 13.8 | 29.9 | 21.1 | 8.7 | 1.1 | 23.0 | 0.0 |
| San Carlos | 97 | 963 | 99 | 34 | 80 | 649,694 | 35.3 | 0.0 | 1.8 | 10.4 | 6.6 | 29.9 | 0.0 |
| San Clemente | 129 | 745 | 82 | 25 | 216 | 921,067 | 22.0 | 0.0 | 0.0 | 24.0 | 5.9 | 44.8 | 0.0 |
| San Diego | 2,801 | 23,453 | 2,828 | 780 | 10,832 | 75,006,131 | 16.1 | 26.5 | 14.4 | 3.1 | 0.0 | 39.6 | 2.9 |
| San Dimas | 60 | 924 | 72 | 14 | 137 | 641,242 | 26.1 | 0.0 | 0.0 | 17.1 | 0.0 | 24.7 | 0.0 |

1. Establishments subject to federal tax.

# Table D. Cities — City Government Finances

| City | City government finances, 2017 | | | | | | | | | |
| | General revenue | | | | | | | General expenditure | | |
| | Intergovernmental | | | Taxes | | | | | Per capita[1] (dollars) | |
| | | | | | | Per capita[1] (dollars) | | | | |
| | Total (mil dol) | Total (mil dol) | Percent from state government | Total (mil dol) | Total | Property | Sales and gross receipts | Total (mil dol) | Total | Capital outlays |
| | 117 | 118 | 119 | 120 | 121 | 122 | 123 | 124 | 125 | 126 |
| **CALIFORNIA— Cont'd** | | | | | | | | | | |
| Menlo Park | 74.1 | 3.2 | 58.0 | 56.4 | 1,650 | 790 | 842 | 71 | 2,076 | 363 |
| Merced | 93.5 | 7.8 | 72.2 | 38.6 | 467 | 169 | 291 | 107 | 1,296 | 274 |
| Milpitas | 177.5 | 4.8 | 40.4 | 99.9 | 1,282 | 606 | 667 | 133 | 1,713 | 410 |
| Mission Viejo | 83.9 | 4.0 | 59.8 | 59.8 | 626 | 359 | 260 | 101 | 1,060 | 131 |
| Modesto | 285.6 | 70.5 | 69.9 | 107.5 | 505 | 158 | 347 | 214 | 1,005 | 131 |
| Monrovia | 66.4 | 8.2 | 16.0 | 42.1 | 1,142 | 706 | 418 | 73 | 1,986 | 229 |
| Montclair | 63.7 | 3.6 | 40.7 | 43.7 | 1,117 | 579 | 518 | 62 | 1,579 | 455 |
| Montebello | 132.5 | 36.4 | 14.0 | 71.6 | 1,140 | 451 | 652 | 82 | 1,312 | 178 |
| Monterey | 143.5 | 7.0 | 74.5 | 68.4 | 2,407 | 471 | 1,778 | 136 | 4,790 | 363 |
| Monterey Park | 70.1 | 7.6 | 76.0 | 42.6 | 701 | 421 | 266 | 66 | 1,083 | 51 |
| Moorpark | 34.9 | 2.8 | 27.0 | 15.6 | 426 | 246 | 172 | 39 | 1,053 | 85 |
| Moreno Valley | 149.0 | 14.7 | 73.5 | 100.5 | 486 | 201 | 250 | 212 | 1,026 | 177 |
| Morgan Hill | 71.1 | 1.4 | 96.9 | 34.6 | 769 | 232 | 369 | 75 | 1,674 | 425 |
| Mountain View | 300.9 | 4.1 | 65.6 | 148.3 | 1,827 | 1,064 | 661 | 273 | 3,360 | 740 |
| Murrieta | 70.0 | 5.8 | 59.5 | 48.4 | 428 | 182 | 241 | 96 | 849 | 299 |
| Napa | 139.8 | 17.2 | 22.1 | 75.2 | 951 | 389 | 546 | 138 | 1,747 | 158 |
| National City | 92.6 | 18.8 | 28.8 | 54.2 | 888 | 297 | 588 | 101 | 1,646 | 378 |
| Newark | 72.2 | 3.5 | 46.7 | 50.2 | 1,059 | 347 | 661 | 56 | 1,177 | 85 |
| Newport Beach | 259.8 | 10.8 | 20.0 | 174.0 | 2,029 | 1,095 | 900 | 296 | 3,452 | 1,350 |
| Norco | 32.4 | 1.4 | 72.7 | 15.0 | 559 | 151 | 349 | 56 | 2,070 | 522 |
| Norwalk | 101.3 | 33.2 | 24.6 | 56.4 | 535 | 217 | 304 | 74 | 698 | 64 |
| Novato | 54.3 | 3.2 | 58.5 | 38.1 | 685 | 346 | 319 | 59 | 1,055 | 220 |
| Oakland | 1,639.3 | 137.8 | 49.4 | 812.7 | 1,915 | 893 | 747 | 1,319 | 3,109 | 273 |
| Oakley | 34.6 | 1.2 | 100.0 | 18.0 | 432 | 134 | 133 | 32 | 762 | 250 |
| Oceanside | 270.0 | 23.3 | 44.3 | 108.3 | 616 | 344 | 259 | 278 | 1,579 | 262 |
| Ontario | 416.0 | 25.1 | 35.3 | 233.8 | 1,333 | 384 | 944 | 459 | 2,616 | 619 |
| Orange | 147.1 | 9.9 | 56.4 | 105.8 | 754 | 324 | 424 | 171 | 1,217 | 330 |
| Oxnard | 374.7 | 21.2 | 48.0 | 132.9 | 636 | 395 | 237 | 301 | 1,442 | 24 |
| Pacifica | 52.4 | 3.6 | 45.5 | 27.0 | 692 | 377 | 266 | 50 | 1,289 | 119 |
| Palmdale | 171.9 | 15.6 | 66.9 | 131.3 | 839 | 545 | 278 | 101 | 647 | 100 |
| Palm Desert | 125.3 | 2.9 | 50.5 | 55.5 | 1,052 | 192 | 829 | 128 | 2,424 | 451 |
| Palm Springs | 207.4 | 13.3 | 24.7 | 108.5 | 2,258 | 704 | 1,266 | 184 | 3,834 | 753 |
| Palo Alto | 274.2 | 6.8 | 36.9 | 127.2 | 1,899 | 654 | 1,133 | 338 | 5,054 | 1,472 |
| Paradise | 19.4 | 5.8 | 56.6 | 12.5 | 471 | 267 | 201 | 20 | 736 | 20 |
| Paramount | 33.7 | 5.9 | 70.5 | 24.7 | 452 | 142 | 308 | 36 | 656 | 100 |
| Pasadena | 403.5 | 42.8 | 27.6 | 217.6 | 1,535 | 669 | 763 | 323 | 2,281 | 0 |
| Perris | 60.3 | 4.6 | 79.9 | 29.4 | 379 | 153 | 222 | 88 | 1,129 | 150 |
| Petaluma | 103.1 | 16.4 | 31.3 | 42.1 | 694 | 241 | 416 | 91 | 1,491 | 229 |
| Pico Rivera | 59.3 | 12.6 | 42.1 | 37.1 | 588 | 165 | 281 | 77 | 1,214 | 345 |
| Pittsburg | 141.1 | 50.7 | 8.2 | 40.8 | 567 | 253 | 290 | 110 | 1,520 | 198 |
| Placentia | 41.0 | 3.1 | 50.8 | 28.4 | 546 | 264 | 276 | 56 | 1,081 | 129 |
| Pleasant Hill | 29.5 | 3.0 | 58.6 | 23.6 | 677 | 174 | 495 | 42 | 1,208 | 220 |
| Pleasanton | 171.1 | 6.6 | 67.3 | 113.8 | 1,374 | 734 | 512 | 131 | 1,578 | 0 |
| Pomona | 180.3 | 22.1 | 26.2 | 100.2 | 657 | 316 | 328 | 172 | 1,127 | 154 |
| Porterville | 67.1 | 14.5 | 44.9 | 25.5 | 427 | 109 | 256 | 56 | 936 | 180 |
| Poway | 116.0 | 1.9 | 91.9 | 89.0 | 1,786 | 1,379 | 370 | 142 | 2,840 | 930 |
| Rancho Cordova | 91.1 | 11.0 | 34.5 | 66.5 | 907 | 197 | 509 | 107 | 1,462 | 309 |
| Rancho Cucamonga | 210.2 | 17.9 | 47.5 | 156.2 | 883 | 583 | 276 | 194 | 1,095 | 233 |
| Rancho Palos Verdes | 36.1 | 1.7 | 79.9 | 29.9 | 711 | 305 | 371 | 33 | 773 | 72 |
| Rancho Santa Margarita | 20.4 | 1.6 | 73.4 | 16.6 | 342 | 138 | 196 | 20 | 404 | 63 |
| Redding | 156.2 | 21.3 | 32.8 | 56.4 | 616 | 254 | 339 | 147 | 1,600 | 127 |
| Redlands | 115.4 | 6.2 | 73.8 | 55.0 | 771 | 369 | 389 | 111 | 1,553 | 311 |
| Redondo Beach | 123.4 | 6.9 | 79.8 | 64.0 | 948 | 471 | 440 | 123 | 1,815 | 106 |
| Redwood City | 210.3 | 12.6 | 40.6 | 110.9 | 1,284 | 678 | 591 | 221 | 2,559 | 262 |
| Rialto | 191.8 | 13.8 | 91.1 | 68.3 | 662 | 316 | 342 | 127 | 1,227 | 1 |
| Richmond | 293.7 | 44.7 | 30.8 | 153.3 | 1,395 | 587 | 692 | 296 | 2,693 | 389 |
| Ridgecrest | 27.3 | 4.2 | 45.0 | 12.5 | 434 | 69 | 271 | 21 | 740 | 141 |
| Riverside | 451.5 | 47.0 | 68.3 | 224.3 | 686 | 263 | 409 | 404 | 1,237 | 92 |
| Rocklin | 73.2 | 4.2 | 62.4 | 41.4 | 641 | 236 | 395 | 66 | 1,020 | 66 |
| Rohnert Park | 80.3 | 3.5 | 67.8 | 28.8 | 674 | 177 | 492 | 67 | 1,557 | 291 |
| Rosemead | 27.5 | 2.5 | 75.1 | 20.9 | 385 | 160 | 223 | 50 | 927 | 295 |
| Roseville | 382.4 | 49.9 | 36.9 | 108.1 | 800 | 335 | 457 | 251 | 1,859 | 129 |
| Sacramento | 1,022.4 | 59.0 | 46.8 | 457.1 | 913 | 358 | 530 | 776 | 1,550 | 8 |
| Salinas | 172.1 | 18.9 | 50.3 | 127.9 | 817 | 203 | 613 | 161 | 1,026 | 131 |
| San Bernardino | 204.8 | 9.3 | 64.2 | 128.3 | 592 | 146 | 442 | 195 | 900 | 97 |
| San Bruno | 79.9 | 1.5 | 85.8 | 32.7 | 757 | 299 | 410 | 71 | 1,641 | 133 |
| San Buenaventura (Ventura) | 160.9 | 7.0 | 77.9 | 91.3 | 830 | 331 | 474 | 172 | 1,562 | 156 |
| San Carlos | 67.6 | 1.3 | 72.8 | 37.8 | 1,244 | 457 | 704 | 56 | 1,841 | 159 |
| San Clemente | 86.1 | 3.8 | 68.1 | 50.7 | 780 | 468 | 289 | 96 | 1,478 | 316 |
| San Diego | 3,177.9 | 331.0 | 12.6 | 1,414.4 | 1,001 | 419 | 576 | 2,575 | 1,823 | 217 |
| San Dimas | 28.8 | 1.5 | 84.5 | 20.9 | 612 | 238 | 368 | 40 | 1,179 | 110 |

1. Based on population estimated as of July 1 of the year shown.

# Table D. Cities — City Government Finances

General expenditure (cont.)

Percent of total for:

| City | Public welfare | Highways | Parking facilities | Education | Health and hospitals | Police protection | Sewerage and sanitation | Parks and recreation | Housing and community development | Interest on debt |
|---|---|---|---|---|---|---|---|---|---|---|
| | 127 | 128 | 129 | 130 | 131 | 132 | 133 | 134 | 135 | 136 |
| CALIFORNIA— Cont'd | | | | | | | | | | |
| Menlo Park | 0.0 | 21.2 | 1.1 | 0.0 | 0.0 | 21.9 | 0.0 | 21.5 | 3.9 | 4.9 |
| Merced | 0.0 | 14.8 | 0.1 | 0.0 | 1.2 | 19.3 | 25.2 | 4.8 | 7.4 | 1.1 |
| Milpitas | 0.0 | 10.8 | 0.0 | 0.0 | 11.9 | 20.3 | 6.2 | 3.7 | 19.2 | 5.3 |
| Mission Viejo | 0.0 | 13.8 | 0.0 | 0.6 | 3.2 | 19.9 | 0.0 | 14.2 | 5.5 | 1.6 |
| Modesto | 0.0 | 4.4 | 0.6 | 0.0 | 0.5 | 25.4 | 31.1 | 6.2 | 3.4 | 3.6 |
| Monrovia | 0.0 | 4.0 | 0.0 | 0.0 | 0.5 | 33.1 | 1.9 | 3.4 | 11.1 | 10.8 |
| Montclair | 0.0 | 6.1 | 0.0 | 0.0 | 5.7 | 15.9 | 10.6 | 5.9 | 34.0 | 6.5 |
| Montebello | 0.0 | 5.7 | 0.0 | 0.0 | 0.0 | 26.0 | 3.9 | 8.0 | 13.9 | 6.0 |
| Monterey | 0.0 | 14.2 | 4.6 | 0.0 | 0.0 | 11.5 | 1.8 | 9.7 | 5.8 | 5.2 |
| Monterey Park | 0.0 | 7.6 | 0.0 | 0.0 | 15.8 | 28.3 | 12.6 | 6.3 | 2.6 | 8.4 |
| Moorpark | 0.0 | 21.2 | 0.0 | 0.0 | 0.7 | 19.3 | 0.7 | 16.7 | 12.7 | 3.1 |
| Moreno Valley | 0.0 | 12.6 | 0.0 | 0.0 | 1.2 | 38.0 | 0.0 | 4.4 | 4.0 | 2.4 |
| Morgan Hill | 0.0 | 9.8 | 0.0 | 0.0 | 0.0 | 20.7 | 9.0 | 11.8 | 25.4 | 3.9 |
| Mountain View | 0.0 | 3.4 | 0.2 | 1.8 | 0.0 | 11.6 | 12.0 | 16.6 | 1.2 | 0.6 |
| Murrieta | 0.0 | 28.5 | 0.0 | 0.0 | 0.5 | 27.8 | 0.0 | 10.9 | 9.1 | 2.8 |
| Napa | 0.0 | 11.7 | 0.1 | 0.0 | 4.7 | 18.8 | 16.7 | 7.2 | 13.3 | 0.1 |
| National City | 0.0 | 2.7 | 0.0 | 0.0 | 1.2 | 24.6 | 7.3 | 1.6 | 36.4 | 2.5 |
| Newark | 0.0 | 16.2 | 0.0 | 0.0 | 0.4 | 31.7 | 0.1 | 10.2 | 0.8 | 1.4 |
| Newport Beach | 0.0 | 10.8 | 0.0 | 0.0 | 2.8 | 17.5 | 3.1 | 8.2 | 2.3 | 2.5 |
| Norco | 0.0 | 7.5 | 0.0 | 0.0 | 3.0 | 19.6 | 13.5 | 4.2 | 10.9 | 5.6 |
| Norwalk | 0.0 | 14.6 | 0.5 | 0.0 | 8.7 | 18.9 | 1.3 | 12.1 | 14.9 | 4.3 |
| Novato | 0.0 | 11.6 | 0.0 | 0.0 | 2.1 | 24.9 | 0.3 | 20.3 | 5.9 | 6.1 |
| Oakland | 0.0 | 3.4 | 0.6 | 0.0 | 6.1 | 21.0 | 4.4 | 3.0 | 5.2 | 9.5 |
| Oakley | 0.0 | 31.4 | 0.0 | 0.0 | 0.7 | 25.7 | 0.0 | 7.7 | 10.6 | 5.1 |
| Oceanside | 0.0 | 10.5 | 0.3 | 0.0 | 0.2 | 21.3 | 15.8 | 10.4 | 12.8 | 2.1 |
| Ontario | 0.0 | 22.0 | 0.0 | 0.0 | 0.3 | 20.5 | 9.4 | 11.1 | 4.1 | 4.6 |
| Orange | 0.0 | 13.0 | 0.0 | 0.0 | 7.6 | 24.8 | 3.7 | 14.2 | 8.0 | 1.4 |
| Oxnard | 0.0 | 4.7 | 0.0 | 0.0 | 0.0 | 22.1 | 21.5 | 6.7 | 3.4 | 8.0 |
| Pacifica | 0.0 | 8.9 | 1.0 | 0.2 | 1.1 | 19.9 | 20.6 | 10.9 | 0.2 | 6.5 |
| Palmdale | 0.0 | 15.1 | 0.0 | 0.0 | 0.7 | 23.6 | 3.3 | 14.7 | 13.5 | 5.3 |
| Palm Desert | 0.0 | 12.7 | 0.0 | 0.0 | 0.8 | 15.7 | 0.0 | 16.1 | 20.1 | 9.0 |
| Palm Springs | 0.0 | 12.7 | 1.5 | 0.0 | 1.4 | 13.7 | 2.9 | 14.8 | 3.6 | 4.9 |
| Palo Alto | 0.0 | 2.9 | 0.8 | 0.0 | 0.0 | 12.5 | 28.6 | 8.8 | 0.5 | 1.4 |
| Paradise | 0.0 | 11.5 | 0.0 | 0.0 | 0.9 | 22.2 | 0.0 | 0.0 | 0.0 | 3.6 |
| Paramount | 0.0 | 17.9 | 0.0 | 0.0 | 0.0 | 29.7 | 0.0 | 14.5 | 11.3 | 0.0 |
| Pasadena | 0.0 | 5.7 | 3.4 | 0.0 | 4.6 | 22.5 | 5.5 | 8.6 | 7.3 | 6.2 |
| Perris | 0.0 | 10.1 | 0.0 | 0.0 | 0.9 | 33.8 | 3.5 | 3.6 | 11.9 | 8.0 |
| Petaluma | 0.0 | 13.6 | 0.0 | 0.0 | 0.5 | 20.2 | 16.1 | 6.0 | 9.3 | 8.1 |
| Pico Rivera | 0.0 | 33.9 | 0.0 | 0.0 | 0.0 | 16.0 | 0.0 | 12.0 | 8.8 | 0.9 |
| Pittsburg | 0.0 | 12.2 | 0.0 | 0.0 | 0.5 | 23.3 | 2.1 | 5.2 | 25.3 | 14.7 |
| Placentia | 0.0 | 14.0 | 0.7 | 0.0 | 1.0 | 22.1 | 5.6 | 3.9 | 3.2 | 1.3 |
| Pleasant Hill | 0.0 | 37.9 | 0.0 | 0.0 | 0.9 | 26.3 | 0.3 | 0.0 | 6.0 | 0.1 |
| Pleasanton | 0.0 | 7.3 | 0.0 | 0.0 | 0.0 | 21.0 | 10.4 | 18.9 | 7.2 | 0.0 |
| Pomona | 0.0 | 15.7 | 0.0 | 0.0 | 0.4 | 30.7 | 7.5 | 1.9 | 11.5 | 10.0 |
| Porterville | 0.0 | 7.6 | 0.0 | 0.0 | 0.2 | 21.5 | 23.9 | 10.4 | 0.6 | 6.2 |
| Poway | 0.0 | 7.7 | 0.0 | 0.0 | 0.3 | 15.9 | 8.0 | 5.7 | 32.2 | 4.1 |
| Rancho Cordova | 0.0 | 24.3 | 0.0 | 0.0 | 0.5 | 35.5 | 2.9 | 0.0 | 7.1 | 0.5 |
| Rancho Cucamonga | 0.0 | 13.3 | 0.0 | 0.0 | 1.9 | 18.7 | 1.0 | 7.8 | 12.1 | 7.0 |
| Rancho Palos Verdes | 0.0 | 16.0 | 0.0 | 0.0 | 0.3 | 17.8 | 0.7 | 13.3 | 1.6 | 3.2 |
| Rancho Santa Margarita | 0.0 | 21.3 | 0.0 | 0.0 | 4.4 | 45.1 | 0.0 | 6.4 | 0.0 | 1.7 |
| Redding | 0.0 | 4.3 | 0.0 | 0.0 | 0.5 | 19.5 | 32.8 | 4.4 | 6.9 | 3.7 |
| Redlands | 0.0 | 11.1 | 0.0 | 0.0 | 4.6 | 22.8 | 19.3 | 6.2 | 3.4 | 2.1 |
| Redondo Beach | 0.0 | 6.2 | 0.0 | 0.0 | 4.4 | 29.2 | 6.6 | 7.5 | 5.5 | 0.7 |
| Redwood City | 0.0 | 10.8 | 1.2 | 0.0 | 0.7 | 17.7 | 12.3 | 7.4 | 3.0 | 2.2 |
| Rialto | 0.0 | 4.4 | 0.0 | 0.0 | 3.2 | 25.3 | 10.1 | 4.7 | 0.3 | 15.2 |
| Richmond | 0.0 | 10.4 | 0.0 | 0.0 | 0.2 | 22.6 | 4.5 | 1.7 | 16.8 | 5.9 |
| Ridgecrest | 0.0 | 9.0 | 0.0 | 0.0 | 1.7 | 32.6 | 5.6 | 8.9 | 15.6 | 7.8 |
| Riverside | 0.0 | 3.9 | 0.9 | 0.0 | 1.2 | 23.6 | 19.6 | 6.8 | 1.3 | 11.8 |
| Rocklin | 0.0 | 10.5 | 0.0 | 0.0 | 0.3 | 20.6 | 0.0 | 12.2 | 0.8 | 3.6 |
| Rohnert Park | 0.0 | 6.6 | 0.0 | 0.0 | 0.8 | 25.7 | 19.2 | 6.3 | 15.4 | 3.0 |
| Rosemead | 0.0 | 3.1 | 0.0 | 0.0 | 0.5 | 31.9 | 0.0 | 8.9 | 10.3 | 3.7 |
| Roseville | 2.2 | 5.0 | 0.0 | 0.0 | 0.5 | 13.8 | 23.6 | 7.9 | 5.2 | 6.3 |
| Sacramento | 0.7 | 3.4 | 2.3 | 0.0 | 2.5 | 20.6 | 13.9 | 7.9 | 3.4 | 11.6 |
| Salinas | 0.0 | 9.6 | 0.9 | 0.0 | 1.2 | 30.1 | 3.4 | 3.9 | 8.4 | 2.4 |
| San Bernardino | 0.0 | 10.7 | 0.0 | 0.0 | 1.0 | 32.4 | 12.1 | 4.1 | 8.4 | 3.8 |
| San Bruno | 0.0 | 4.1 | 0.0 | 0.0 | 1.7 | 20.1 | 16.5 | 10.1 | 0.0 | 1.6 |
| San Buenaventura (Ventura) | 0.0 | 9.0 | 0.5 | 0.0 | 0.3 | 21.4 | 10.6 | 9.4 | 2.4 | 2.5 |
| San Carlos | 0.0 | 15.4 | 0.1 | 0.0 | 0.0 | 16.5 | 20.6 | 8.0 | 4.7 | 0.3 |
| San Clemente | 0.0 | 7.5 | 0.2 | 0.0 | 1.2 | 14.5 | 14.4 | 24.3 | 2.2 | 1.0 |
| San Diego | 0.0 | 7.2 | 0.1 | 0.0 | 1.9 | 12.4 | 20.0 | 11.3 | 14.7 | 6.4 |
| San Dimas | 0.0 | 15.7 | 0.0 | 0.0 | 0.4 | 31.3 | 0.1 | 8.9 | 4.8 | 1.7 |

# Table D. Cities — City Government Finances, City Government Employment, and Climate

| City | City government finances, 2017 (cont.) | | | Climate[2] | | | | | | |
| | Debt outstanding | | Debt issued during year | Average daily temperature | | | | Annual precipitation (inches) | Heating degree days | Cooling degree days |
| | Total (mil dol) | Per capita[1] (dollars) | | Mean | | Limits | | | | |
| | | | | January | July | January[3] | July[4] | | | |
| | 137 | 138 | 139 | 140 | 141 | 142 | 143 | 144 | 145 | 146 |
|---|---|---|---|---|---|---|---|---|---|---|
| **CALIFORNIA— Cont'd** | | | | | | | | | | |
| Menlo Park | 78.5 | 2,297 | 0.0 | 49.0 | 68.0 | 40.4 | 78.8 | 15.71 | 2,584 | 452 |
| Merced | 49.6 | 600 | 0.0 | 46.3 | 78.6 | 37.5 | 96.5 | 12.50 | 2,602 | 1,578 |
| Milpitas | 125.6 | 1,613 | 0.0 | 50.5 | 70.9 | 41.7 | 84.3 | 15.08 | 2,171 | 811 |
| Mission Viejo | 38.0 | 398 | 0.0 | 56.7 | 72.4 | 47.2 | 82.3 | 14.03 | 1,465 | 1,183 |
| Modesto | 516.5 | 2,425 | 140.3 | 47.2 | 77.7 | 40.1 | 93.6 | 13.12 | 2,358 | 1,570 |
| Monrovia | 149.8 | 4,066 | 27.7 | 56.1 | 75.3 | 44.3 | 89.4 | 21.09 | 1,398 | 1,558 |
| Montclair | 86.9 | 2,221 | 0.0 | 54.6 | 73.8 | 41.5 | 88.7 | 16.96 | 1,727 | 1,191 |
| Montebello | 130.2 | 2,073 | 55.4 | 58.8 | 76.6 | 47.9 | 88.9 | 14.44 | 949 | 1,837 |
| Monterey | 97.5 | 3,432 | 3.8 | 51.6 | 60.2 | 43.4 | 68.1 | 20.35 | 3,092 | 74 |
| Monterey Park | 90.6 | 1,494 | 3.8 | 56.3 | 75.6 | 42.6 | 89.0 | 18.56 | 1,295 | 1,575 |
| Moorpark | 29.0 | 793 | 0.0 | 54.7 | 68.3 | 41.2 | 80.7 | 18.41 | 1,911 | 602 |
| Moreno Valley | 144.4 | 698 | 24.7 | 54.2 | 77.4 | 42.0 | 93.5 | 10.67 | 1,674 | 1,697 |
| Morgan Hill | 90.4 | 2,010 | 0.0 | 43.5 | 70.7 | 37.5 | 78.2 | 23.73 | 4,566 | 747 |
| Mountain View | 44.0 | 541 | 0.0 | 49.0 | 68.0 | 40.4 | 78.8 | 15.71 | 2,584 | 452 |
| Murrieta | 52.1 | 461 | 0.0 | 52.2 | 79.6 | 38.3 | 98.1 | 12.09 | 1,924 | 1,874 |
| Napa | 56.9 | 719 | 12.5 | 47.9 | 68.6 | 39.2 | 82.6 | 26.46 | 2,689 | 529 |
| National City | 60.4 | 988 | 1.3 | 57.3 | 70.1 | 46.1 | 76.1 | 9.95 | 1,321 | 862 |
| Newark | 18.3 | 385 | 0.0 | 49.8 | 68.0 | 42.0 | 78.3 | 14.85 | 2,367 | 530 |
| Newport Beach | 226.9 | 2,646 | 0.0 | 55.9 | 67.3 | 48.2 | 71.4 | 11.65 | 1,719 | 543 |
| Norco | 149.1 | 5,552 | 0.0 | NA | NA | NA | NA | NA | NA | NA |
| Norwalk | 119.3 | 1,131 | 1.0 | 57.0 | 73.8 | 46.0 | 82.9 | 12.94 | 1,211 | 1,186 |
| Novato | 66.4 | 1,193 | 0.0 | 48.8 | 67.7 | 41.3 | 80.9 | 34.29 | 2,621 | 451 |
| Oakland | 2,230.3 | 5,255 | 51.8 | 50.9 | 64.9 | 44.7 | 72.7 | 22.94 | 2,400 | 377 |
| Oakley | 61.8 | 1,484 | 20.1 | 45.7 | 74.4 | 37.8 | 90.7 | 13.33 | 2,714 | 1,179 |
| Oceanside | 171.6 | 975 | 0.0 | 54.7 | 67.6 | 45.4 | 72.1 | 11.13 | 2,009 | 505 |
| Ontario | 250.9 | 1,431 | 2.1 | 54.6 | 73.8 | 41.5 | 88.7 | 16.96 | 1,727 | 1,191 |
| Orange | 107.7 | 768 | 0.0 | 58.0 | 72.9 | 46.6 | 82.7 | 13.84 | 1,153 | 1,299 |
| Oxnard | 442.7 | 2,120 | 0.0 | 55.6 | 65.9 | 45.5 | 72.7 | 15.62 | 1,936 | 403 |
| Pacifica | 72.2 | 1,852 | 36.8 | 49.4 | 62.8 | 42.9 | 71.1 | 20.11 | 2,862 | 142 |
| Palmdale | 184.4 | 1,178 | 18.1 | 46.6 | 81.7 | 34.3 | 97.5 | 7.36 | 2,704 | 1,998 |
| Palm Desert | 416.8 | 7,895 | 245.7 | 56.8 | 92.8 | 42.0 | 107.1 | 3.15 | 903 | 4,388 |
| Palm Springs | 211.4 | 4,400 | 11.5 | 57.3 | 92.1 | 44.2 | 108.2 | 5.23 | 951 | 4,224 |
| Palo Alto | 132.9 | 1,984 | 3.2 | 49.0 | 68.0 | 40.4 | 78.8 | 15.71 | 2,584 | 452 |
| Paradise | 15.6 | 586 | 0.2 | 45.7 | 77.8 | 37.7 | 91.7 | 56.20 | 3,145 | 1,464 |
| Paramount | 5.0 | 92 | 0.3 | 57.0 | 73.8 | 46.0 | 82.9 | 12.94 | 1,211 | 1,186 |
| Pasadena | 946.5 | 6,677 | 197.8 | 56.1 | 75.3 | 44.3 | 89.4 | 21.09 | 1,398 | 1,558 |
| Perris | 311.0 | 4,010 | 5.4 | 51.2 | 78.3 | 36.1 | 97.8 | 11.40 | 2,123 | 1,710 |
| Petaluma | 136.4 | 2,247 | 23.4 | 48.4 | 67.3 | 38.9 | 82.7 | 25.85 | 2,741 | 385 |
| Pico Rivera | 69.4 | 1,099 | 0.0 | 58.3 | 74.2 | 48.5 | 83.8 | 15.14 | 928 | 1,506 |
| Pittsburg | 367.7 | 5,107 | 36.4 | 45.7 | 74.4 | 37.8 | 90.7 | 13.33 | 2,714 | 1,179 |
| Placentia | 14.3 | 275 | 0.0 | 58.0 | 72.9 | 46.6 | 82.7 | 13.84 | 1,153 | 1,299 |
| Pleasant Hill | 11.9 | 342 | 0.0 | 46.3 | 71.2 | 38.8 | 87.4 | 19.58 | 2,757 | 786 |
| Pleasanton | 65.9 | 795 | 11.6 | 47.2 | 72.0 | 37.4 | 89.1 | 14.82 | 2,755 | 858 |
| Pomona | 275.6 | 1,808 | 71.9 | 54.6 | 73.8 | 41.5 | 88.7 | 16.96 | 1,727 | 1,191 |
| Porterville | 78.6 | 1,315 | 0.0 | 48.7 | 82.8 | 39.4 | 98.1 | 11.49 | 2,053 | 2,246 |
| Poway | 159.2 | 3,197 | 0.2 | 55.3 | 70.9 | 43.5 | 80.8 | 11.97 | 1,808 | 979 |
| Rancho Cordova | 17.8 | 243 | 0.0 | 48.2 | 77.4 | 41.3 | 93.8 | 19.87 | 2,226 | 1,597 |
| Rancho Cucamonga | 298.0 | 1,684 | 56.9 | 56.6 | 78.3 | 45.3 | 95.0 | 14.77 | 1,364 | 1,901 |
| Rancho Palos Verdes | 25.9 | 615 | 0.0 | 56.3 | 69.4 | 46.2 | 77.6 | 14.79 | 1,526 | 742 |
| Rancho Santa Margarita | 9.6 | 198 | 0.0 | 56.7 | 72.4 | 47.2 | 82.3 | 14.03 | 1,465 | 1,183 |
| Redding | 231.1 | 2,525 | 4.2 | 45.5 | 81.3 | 35.5 | 98.5 | 33.52 | 2,961 | 1,741 |
| Redlands | 61.2 | 858 | 0.0 | 52.9 | 78.0 | 40.4 | 94.4 | 13.62 | 1,904 | 1,714 |
| Redondo Beach | 32.0 | 475 | 0.0 | 57.1 | 69.3 | 48.6 | 75.3 | 13.15 | 1,274 | 679 |
| Redwood City | 96.0 | 1,111 | 6.3 | 48.4 | 68.0 | 39.1 | 80.8 | 20.16 | 2,764 | 422 |
| Rialto | 312.1 | 3,023 | 0.0 | 54.4 | 79.6 | 41.8 | 96.0 | 16.43 | 1,599 | 1,937 |
| Richmond | 627.7 | 5,714 | 13.1 | 50.0 | 62.7 | 42.9 | 70.5 | 23.35 | 2,720 | 184 |
| Ridgecrest | 32.3 | 1,126 | 0.0 | NA | NA | NA | NA | NA | NA | NA |
| Riverside | 1,790.3 | 5,477 | 7.4 | 55.3 | 78.7 | 42.7 | 94.1 | 10.22 | 1,475 | 1,863 |
| Rocklin | 70.5 | 1,090 | 17.9 | 46.9 | 77.7 | 39.2 | 94.8 | 24.61 | 2,532 | 1,528 |
| Rohnert Park | 61.4 | 1,438 | 0.4 | 48.7 | 67.6 | 39.5 | 82.2 | 31.01 | 2,694 | 526 |
| Rosemead | 45.3 | 836 | 0.0 | 56.3 | 75.6 | 42.6 | 89.0 | 18.56 | 1,295 | 1,575 |
| Roseville | 996.9 | 7,380 | 165.0 | 46.9 | 77.7 | 39.2 | 94.8 | 24.61 | 2,532 | 1,528 |
| Sacramento | 2,283.1 | 4,559 | 57.5 | 46.3 | 75.4 | 38.8 | 92.4 | 17.93 | 2,666 | 1,248 |
| Salinas | 57.8 | 370 | 4.9 | 51.2 | 63.2 | 41.3 | 71.3 | 12.91 | 2,770 | 210 |
| San Bernardino | 220.1 | 1,015 | 104.5 | 54.4 | 79.6 | 41.8 | 96.0 | 16.43 | 1,599 | 1,937 |
| San Bruno | 29.8 | 691 | 1.9 | 49.4 | 62.8 | 42.9 | 71.1 | 20.11 | 2,862 | 142 |
| San Buenaventura (Ventura) | 151.2 | 1,373 | 0.0 | 55.6 | 65.9 | 45.5 | 72.7 | 15.62 | 1,936 | 403 |
| San Carlos | 3.9 | 130 | 0.0 | 48.4 | 68.0 | 39.1 | 80.8 | 20.16 | 2,764 | 422 |
| San Clemente | 82.5 | 1,269 | 65.3 | 55.4 | 68.7 | 43.9 | 77.3 | 13.56 | 1,756 | 666 |
| San Diego | 2,779.5 | 1,968 | 40.6 | 57.8 | 70.9 | 49.7 | 75.8 | 10.77 | 1,063 | 866 |
| San Dimas | 15.2 | 446 | 0.1 | 54.6 | 73.8 | 41.5 | 88.7 | 16.96 | 1,727 | 1,191 |

1. Based on the population estimated as of July 1 of the year shown.    2. Represents normal values based on the 30-year period, 1971±2000.    3. Average daily minimum.    4. Average daily maximum.

# Table D. Cities — Land Area and Population

| STATE Place code | City | Land area[1] (sq. mi) | Population, 2020 Total persons 2019 | Rank | Per square mile | White | Black or African American | American Indian, Alaskan Native | Asian | Hawaiian Pacific Islander | Some other race | Two or more races (percent) |
|---|---|---|---|---|---|---|---|---|---|---|---|---|
| | | 1 | 2 | 3 | 4 | 5 | 6 | 7 | 8 | 9 | 10 | 11 |
| | CALIFORNIA— Cont'd | | | | | | | | | | | |
| 06 67000 | San Francisco | 46.9 | 866,606 | 17 | 18,478 | 45.2 | 5.5 | 0.4 | 34.9 | 0.4 | 7.9 | 5.7 |
| 06 67042 | San Gabriel | 4.1 | 39,470 | 983 | 9,627 | 14.7 | 0.3 | 0.0 | 64.4 | 0.7 | 15.5 | 4.4 |
| 06 67112 | San Jacinto | 26.0 | 51,122 | 763 | 1,966 | 63.9 | 3.7 | 0.6 | 5.8 | 0.0 | 23.0 | 3.0 |
| 06 68000 | San Jose | 178.3 | 1,013,616 | 10 | 5,685 | 42.5 | 3.0 | 0.5 | 37.7 | 0.4 | 10.8 | 5.1 |
| 06 68028 | San Juan Capistrano | 14.4 | 35,810 | 1,079 | 2,487 | D | D | D | D | D | D | D |
| 06 68084 | San Leandro | 13.3 | 89,161 | 378 | 6,704 | 33.4 | 11.9 | 0.4 | 34.2 | 2.1 | 13.6 | 4.4 |
| 06 68154 | San Luis Obispo | 13.3 | 47,590 | 815 | 3,578 | 83.8 | 2.6 | 0.6 | 5.2 | 0.2 | 3.9 | 3.7 |
| 06 68196 | San Marcos | 24.4 | 97,477 | 327 | 3,995 | 77.0 | 2.3 | 0.4 | 9.4 | 0.0 | 5.5 | 5.5 |
| 06 68252 | San Mateo | 12.1 | 102,398 | 304 | 8,463 | 49.8 | 1.7 | 0.0 | 25.7 | 1.6 | 12.9 | 8.4 |
| 06 68294 | San Pablo | 2.6 | 30,779 | 1,227 | 11,838 | 16.9 | 11.8 | 0.4 | 13.5 | 0.0 | 53.9 | 3.4 |
| 06 68364 | San Rafael | 16.6 | 58,528 | 651 | 3,526 | 67.3 | 1.2 | 2.3 | 10.5 | 0.1 | 15.3 | 3.3 |
| 06 68378 | San Ramon | 18.7 | 82,865 | 418 | 4,431 | 39.5 | 0.8 | 0.1 | 53.0 | 0.0 | 0.9 | 5.7 |
| 06 69000 | Santa Ana | 27.3 | 331,301 | 57 | 12,136 | 28.9 | 1.0 | 0.4 | 11.0 | 0.4 | 56.1 | 2.2 |
| 06 69070 | Santa Barbara | 19.5 | 90,254 | 369 | 4,628 | 78.5 | 1.0 | 0.6 | 3.7 | 0.0 | 12.8 | 3.3 |
| 06 69084 | Santa Clara | 18.3 | 131,055 | 212 | 7,162 | 34.7 | 2.9 | 0.6 | 46.5 | 1.9 | 8.1 | 5.3 |
| 06 69088 | Santa Clarita | 70.8 | 209,990 | 108 | 2,966 | 72.6 | 4.9 | 0.6 | 10.4 | 0.3 | 7.2 | 4.0 |
| 06 69112 | Santa Cruz | 12.7 | 64,317 | 589 | 5,064 | 71.7 | 2.3 | 0.7 | 10.2 | 0.0 | 7.2 | 8.0 |
| 06 69196 | Santa Maria | 22.8 | 105,851 | 292 | 4,643 | 85.6 | 2.1 | 1.5 | 3.0 | 0.0 | 6.5 | 1.3 |
| 06 70000 | Santa Monica | 8.4 | 90,024 | 370 | 10,717 | 78.6 | 4.2 | 0.8 | 9.3 | 0.0 | 2.4 | 4.8 |
| 06 70042 | Santa Paula | 5.5 | 30,242 | 1,245 | 5,499 | 93.8 | 0.1 | 1.2 | 0.2 | 0.0 | 1.2 | 3.6 |
| 06 70098 | Santa Rosa | 42.5 | 174,613 | 150 | 4,109 | 61.5 | 2.6 | 1.0 | 6.3 | 0.6 | 21.9 | 5.9 |
| 06 70224 | Santee | 16.5 | 57,770 | 667 | 3,501 | 80.7 | 3.8 | 1.0 | 4.3 | 0.0 | 2.7 | 7.6 |
| 06 70280 | Saratoga | 12.8 | 30,525 | 1,233 | 2,385 | D | D | D | D | D | D | D |
| 06 70742 | Seaside | 8.9 | 34,115 | 1,130 | 3,833 | 52.4 | 6.7 | 0.2 | 16.7 | 0.2 | 17.8 | 6.1 |
| 06 72016 | Simi Valley | 41.5 | 125,168 | 223 | 3,016 | 74.4 | 1.2 | 0.4 | 12.6 | 0.7 | 6.4 | 4.4 |
| 06 72520 | Soledad | 4.5 | 25,247 | 1,399 | 5,610 | 27.3 | 10.1 | 1.9 | 1.9 | 1.0 | 56.8 | 1.2 |
| 06 73080 | South Gate | 7.2 | 93,134 | 358 | 12,935 | 70.1 | 0.6 | 0.0 | 0.1 | 0.1 | 28.7 | 0.4 |
| 06 73220 | South Pasadena | 3.4 | 24,995 | 1,406 | 7,352 | 58.9 | 5.7 | 0.0 | 28.8 | 0.0 | 0.6 | 6.0 |
| 06 73262 | South San Francisco | 9.2 | 66,558 | 563 | 7,235 | 34.2 | 1.9 | 0.7 | 46.4 | 2.3 | 11.1 | 3.5 |
| 06 73962 | Stanton | 3.1 | 37,904 | 1,024 | 12,227 | 43.7 | 0.6 | 0.0 | 30.4 | 0.0 | 23.5 | 1.8 |
| 06 75000 | Stockton | 62.2 | 312,716 | 62 | 5,028 | 44.7 | 10.1 | 0.6 | 22.8 | 0.8 | 8.6 | 12.6 |
| 06 75630 | Suisun City | 4.0 | 29,305 | 1,279 | 7,326 | 42.9 | 29.6 | 0.0 | 10.5 | 1.4 | 5.8 | 9.8 |
| 06 77000 | Sunnyvale | 22.1 | 151,746 | 170 | 6,866 | 39.8 | 1.1 | 0.4 | 44.5 | 0.1 | 8.1 | 6.0 |
| 06 78120 | Temecula | 37.3 | 113,878 | 257 | 3,053 | 66.2 | 2.6 | 0.5 | 14.1 | 0.2 | 8.3 | 8.2 |
| 06 78148 | Temple City | 4.0 | 35,225 | 1,095 | 8,806 | 20.4 | 0.4 | 0.0 | 62.7 | 0.9 | 10.2 | 5.4 |
| 06 78582 | Thousand Oaks | 55.3 | 126,310 | 219 | 2,284 | 82.1 | 2.2 | 0.2 | 8.5 | 0.1 | 3.8 | 3.2 |
| 06 80000 | Torrance | 20.5 | 141,553 | 190 | 6,905 | 42.7 | 3.5 | 0.8 | 37.7 | 0.2 | 6.2 | 8.8 |
| 06 80238 | Tracy | 25.9 | 95,025 | 342 | 3,669 | 61.2 | 4.5 | 0.4 | 13.0 | 1.0 | 10.6 | 9.3 |
| 06 80644 | Tulare | 20.4 | 66,797 | 559 | 3,274 | 84.1 | 1.1 | 1.6 | 1.4 | 0.4 | 8.0 | 3.5 |
| 06 80812 | Turlock | 16.9 | 73,059 | 492 | 4,323 | 82.3 | 2.4 | 0.1 | 5.5 | 0.6 | 3.3 | 5.8 |
| 06 80854 | Tustin | 11.2 | 79,815 | 441 | 7,126 | 59.0 | 1.9 | 0.7 | 21.2 | 0.0 | 12.5 | 4.6 |
| 06 80994 | Twentynine Palms | 58.7 | 27,433 | 1,334 | 467 | 71.2 | 7.7 | 0.5 | 3.3 | 6.2 | 5.0 | 6.1 |
| 06 81204 | Union City | 19.2 | 74,501 | 476 | 3,880 | 21.2 | 4.7 | 0.5 | 53.1 | 1.4 | 10.2 | 8.9 |
| 06 81344 | Upland | 15.6 | 77,850 | 454 | 4,990 | 58.3 | 9.1 | 0.5 | 8.8 | 0.2 | 16.5 | 6.6 |
| 06 81554 | Vacaville | 29.9 | 101,332 | 314 | 3,389 | 63.3 | 8.4 | 1.5 | 11.1 | 1.0 | 5.9 | 8.8 |
| 06 81666 | Vallejo | 30.4 | 120,747 | 236 | 3,972 | 36.9 | 19.8 | 0.4 | 23.6 | 0.2 | 12.3 | 6.9 |
| 06 82590 | Victorville | 73.7 | 124,982 | 224 | 1,696 | 61.2 | 18.5 | 0.9 | 2.0 | 0.0 | 11.7 | 5.7 |
| 06 82954 | Visalia | 37.9 | 135,007 | 203 | 3,562 | 56.7 | 3.0 | 1.3 | 5.9 | 0.1 | 28.8 | 4.2 |
| 06 82996 | Vista | 18.7 | 100,636 | 315 | 5,382 | 86.4 | 1.5 | 1.5 | 3.1 | 1.5 | 3.0 | 3.0 |
| 06 83332 | Walnut | 9.0 | 29,324 | 1,277 | 3,258 | 21.5 | 4.3 | 0.0 | 64.7 | 2.4 | 4.4 | 2.7 |
| 06 83346 | Walnut Creek | 19.8 | 69,936 | 524 | 3,532 | 72.0 | 3.4 | 0.6 | 15.3 | 1.3 | 0.6 | 6.8 |
| 06 83542 | Wasco | 9.4 | 28,734 | 1,294 | 3,057 | 77.1 | 4.4 | 1.2 | 1.4 | 0.0 | 13.8 | 2.2 |
| 06 83668 | Watsonville | 6.7 | 51,673 | 750 | 7,712 | 66.3 | 4.2 | 0.8 | 4.0 | 0.0 | 23.7 | 1.0 |
| 06 84200 | West Covina | 16.0 | 103,656 | 297 | 6,479 | 35.6 | 4.5 | 0.6 | 23.7 | 0.0 | 28.7 | 6.9 |
| 06 84410 | West Hollywood | 1.9 | 35,111 | 1,098 | 18,480 | 76.8 | 3.4 | 0.7 | 4.8 | 0.8 | 2.5 | 10.9 |
| 06 84550 | Westminster | 10.0 | 89,740 | 374 | 8,974 | 34.8 | 2.8 | 0.2 | 46.6 | 0.6 | 11.4 | 3.6 |
| 06 84816 | West Sacramento | 21.5 | 53,680 | 722 | 2,497 | 63.2 | 4.0 | 0.1 | 13.2 | 1.0 | 7.5 | 11.0 |
| 06 85292 | Whittier | 14.6 | 83,194 | 416 | 5,698 | 52.3 | 0.5 | 1.1 | 4.7 | 0.0 | 37.5 | 3.9 |
| 06 85446 | Wildomar | 23.7 | 36,574 | 1,051 | 1,543 | 61.2 | 4.3 | 2.0 | 7.5 | 0.3 | 20.1 | 4.6 |
| 06 85922 | Windsor | 7.4 | 27,322 | 1,340 | 3,692 | 75.3 | 1.1 | 4.5 | 1.7 | 1.3 | 11.3 | 4.8 |
| 06 86328 | Woodland | 15.3 | 60,446 | 622 | 3,951 | 85.4 | 2.7 | 1.6 | 3.8 | 0.3 | 3.9 | 2.8 |
| 06 86832 | Yorba Linda | 20.0 | 68,027 | 542 | 3,401 | 72.0 | 1.7 | 0.2 | 21.3 | 0.1 | 1.9 | 2.8 |
| 06 86972 | Yuba City | 15.0 | 66,444 | 565 | 4,430 | 62.9 | 2.6 | 1.0 | 19.4 | 0.2 | 5.9 | 8.1 |
| 06 87042 | Yucaipa | 28.3 | 54,833 | 701 | 1,938 | 81.4 | 5.7 | 0.2 | 2.9 | 0.0 | 5.4 | 4.3 |
| 08 00000 | COLORADO | 103,637.5 | 5,807,719 | X | 56 | 83.7 | 4.2 | 1.0 | 3.3 | | 3.6 | 4.0 |
| 08 03455 | Arvada | 38.9 | 121,936 | 231 | 3,135 | 91.5 | 0.7 | 1.2 | 1.9 | 0.0 | 1.8 | 2.8 |
| 08 04000 | Aurora | 160.1 | 387,377 | 52 | 2,420 | 60.6 | 17.1 | 0.8 | 6.2 | 0.4 | 9.3 | 5.5 |
| 08 07850 | Boulder | 26.3 | 107,645 | 285 | 4,093 | 87.7 | 1.3 | 0.2 | 7.0 | 0.1 | 0.9 | 2.8 |
| 08 08675 | Brighton | 21.2 | 40,783 | 945 | 1,924 | 87.3 | 2.2 | 1.6 | 2.3 | 0.0 | 3.9 | 2.6 |
| 08 09280 | Broomfield | 33.0 | 72,236 | 500 | 2,189 | 86.0 | 1.4 | 0.4 | 7.2 | 0.0 | 1.9 | 3.1 |

1. Dry land or land partially or temporarily covered by water.  2. Hispanic or Latino persons may be of any race.

# Table D. Cities — Population

| City | Percent Hispanic or Latino[1], 2019 | Percent foreign born, 2019 | Age of population (percent), 2019 | | | | | | | Median age, 2019 | Percent female, 2019 | Population | | | |
| | | | Under 18 years | 18 to 24 years | 25 to 34 years | 35 to 44 years | 45 to 54 years | 55 to 64 years | 65 years and over | | | Census counts | | Percent change | |
| | | | | | | | | | | | | 2000 | 2010 | 2000-2010 | 2001-2020 |
| | 12 | 13 | 14 | 15 | 16 | 17 | 18 | 19 | 20 | 21 | 22 | 23 | 24 | 25 | 26 |
| CALIFORNIA— Cont'd | | | | | | | | | | | | | | | |
| San Francisco | 15.2 | 33.7 | 13.4 | 7.0 | 23.1 | 16.1 | 12.8 | 11.4 | 16.0 | 38.2 | 49.2 | 776,733 | 805,184 | 3.7 | 7.6 |
| San Gabriel | 24.3 | 58.7 | 16.5 | 6.4 | 17.6 | 10.1 | 15.1 | 14.1 | 20.3 | 44.2 | 53.1 | 39,804 | 39,644 | -0.4 | -0.4 |
| San Jacinto | 63.1 | 21.6 | 25.4 | 11.2 | 17.4 | 11.8 | 10.9 | 10.0 | 13.2 | 31.5 | 49.3 | 23,779 | 44,200 | 85.9 | 15.7 |
| San Jose | 30.6 | 41.1 | 22.0 | 8.4 | 16.0 | 14.6 | 13.6 | 11.9 | 13.6 | 37.4 | 49.4 | 894,943 | 952,354 | 6.4 | 6.4 |
| San Juan Capistrano | 42.1 | 24.4 | 20.5 | 6.4 | 15.6 | 9.6 | 12.5 | 17.2 | 18.2 | 41.6 | 52.1 | 33,826 | 34,411 | 1.7 | 4.1 |
| San Leandro | 27.6 | 36.6 | 17.5 | 4.4 | 19.5 | 12.0 | 14.6 | 16.4 | 15.6 | 41.1 | 50.3 | 79,452 | 84,963 | 6.9 | 4.9 |
| San Luis Obispo | 17.0 | 8.0 | 11.3 | 34.4 | 13.6 | 8.2 | 7.1 | 10.3 | 15.1 | 27.4 | 47.3 | 44,174 | 45,141 | 2.2 | 5.4 |
| San Marcos | 37.2 | 23.1 | 26.2 | 9.5 | 11.2 | 14.8 | 13.8 | 9.8 | 14.7 | 37.3 | 52.4 | 54,977 | 83,638 | 52.1 | 16.5 |
| San Mateo | 23.7 | 36.8 | 19.9 | 7.1 | 17.4 | 15.5 | 13.3 | 11.8 | 15.1 | 37.2 | 50.7 | 92,482 | 97,130 | 5.0 | 5.4 |
| San Pablo | 62.3 | 41.1 | 29.5 | 9.4 | 14.6 | 12.0 | 14.3 | 10.1 | 10.0 | 33.0 | 45.8 | 30,215 | 29,505 | -2.3 | 4.3 |
| San Rafael | 31.6 | 27.5 | 19.2 | 8.7 | 10.6 | 12.0 | 14.2 | 13.7 | 21.7 | 44.6 | 50.4 | 56,063 | 57,693 | 2.9 | 1.4 |
| San Ramon | 4.0 | 40.5 | 26.8 | 6.7 | 7.5 | 15.7 | 22.7 | 9.0 | 11.5 | 41.8 | 48.5 | 44,722 | 71,404 | 59.7 | 16.1 |
| Santa Ana | 77.9 | 38.6 | 25.6 | 10.7 | 16.5 | 14.7 | 12.2 | 9.9 | 10.4 | 33.2 | 48.4 | 337,977 | 324,778 | -3.9 | 2.0 |
| Santa Barbara | 37.7 | 21.9 | 17.7 | 12.5 | 14.8 | 13.3 | 11.7 | 11.0 | 18.9 | 37.6 | 50.5 | 92,325 | 88,383 | -4.3 | 2.1 |
| Santa Clara | 16.3 | 47.8 | 17.3 | 11.9 | 23.5 | 13.8 | 12.4 | 11.0 | 10.2 | 33.7 | 47.0 | 102,361 | 116,328 | 13.6 | 12.7 |
| Santa Clarita | 34.6 | 20.6 | 25.4 | 8.3 | 11.8 | 12.6 | 14.7 | 13.8 | 13.4 | 38.5 | 49.6 | 151,088 | 208,778 | 38.2 | 0.6 |
| Santa Cruz | 25.6 | 14.1 | 13.7 | 32.1 | 14.1 | 10.8 | 10.4 | 8.8 | 10.1 | 26.7 | 50.3 | 54,593 | 59,948 | 9.8 | 7.3 |
| Santa Maria | 80.7 | 34.8 | 32.5 | 11.2 | 16.2 | 13.1 | 8.8 | 7.3 | 10.8 | 28.7 | 49.2 | 77,423 | 99,596 | 28.6 | 6.3 |
| Santa Monica | 17.5 | 23.8 | 13.3 | 5.7 | 18.4 | 14.9 | 13.5 | 12.2 | 21.9 | 41.6 | 49.0 | 84,084 | 89,742 | 6.7 | 0.3 |
| Santa Paula | 82.3 | 32.5 | 27.1 | 10.2 | 15.5 | 15.5 | 10.5 | 9.2 | 12.0 | 33.7 | 46.5 | 28,598 | 29,648 | 3.7 | 2.0 |
| Santa Rosa | 35.0 | 20.7 | 22.8 | 7.7 | 14.3 | 12.9 | 14.0 | 11.8 | 16.6 | 39.2 | 52.3 | 147,595 | 175,015 | 18.6 | -0.2 |
| Santee | 16.9 | 11.0 | 23.5 | 8.7 | 13.5 | 14.2 | 11.3 | 13.1 | 15.6 | 38.2 | 50.7 | 52,975 | 53,413 | 0.8 | 8.2 |
| Saratoga | 4.1 | 41.9 | 19.8 | 4.1 | 8.1 | 10.3 | 14.4 | 14.5 | 28.9 | 50.2 | 53.9 | 29,843 | 30,034 | 0.6 | 1.6 |
| Seaside | 34.2 | 28.1 | 22.2 | 11.5 | 17.3 | 10.5 | 9.0 | 15.5 | 14.1 | 32.9 | 48.5 | 31,696 | 33,024 | 4.2 | 3.3 |
| Simi Valley | 24.4 | 20.1 | 19.4 | 9.0 | 13.6 | 12.2 | 15.6 | 14.5 | 15.8 | 41.6 | 51.9 | 111,351 | 124,247 | 11.6 | 0.7 |
| Soledad | 78.4 | 35.3 | 17.2 | 13.2 | 16.5 | 15.7 | 20.2 | 11.3 | 5.9 | 36.3 | 32.4 | 11,263 | 25,738 | 128.5 | -1.9 |
| South Gate | 95.6 | 42.7 | 25.7 | 10.6 | 17.2 | 11.7 | 12.7 | 10.4 | 11.7 | 32.3 | 50.8 | 96,375 | 94,396 | -2.1 | -1.3 |
| South Pasadena | 15.7 | 21.7 | 20.0 | 7.1 | 10.4 | 14.3 | 18.8 | 13.5 | 15.8 | 43.9 | 51.0 | 24,292 | 25,605 | 5.4 | -2.4 |
| South San Francisco | 28.4 | 42.2 | 12.7 | 8.9 | 17.2 | 13.1 | 16.2 | 15.1 | 16.8 | 43.4 | 51.4 | 60,552 | 63,613 | 5.1 | 4.6 |
| Stanton | 51.8 | 43.1 | 27.5 | 7.4 | 13.6 | 14.2 | 12.3 | 11.6 | 13.4 | 35.6 | 52.9 | 37,403 | 37,829 | 1.1 | 0.2 |
| Stockton | 43.5 | 25.7 | 27.1 | 10.1 | 14.0 | 13.7 | 11.2 | 10.4 | 13.6 | 34.0 | 50.4 | 243,771 | 292,348 | 19.9 | 7.0 |
| Suisun City | 29.2 | 15.8 | 28.4 | 10.3 | 12.4 | 14.3 | 10.3 | 12.5 | 11.8 | 34.4 | 46.2 | 26,118 | 28,116 | 7.6 | 4.2 |
| Sunnyvale | 18.3 | 47.1 | 21.5 | 6.3 | 21.1 | 17.6 | 12.2 | 11.0 | 10.4 | 35.6 | 49.7 | 131,760 | 140,492 | 6.6 | 8.0 |
| Temecula | 24.9 | 17.9 | 28.7 | 7.3 | 13.0 | 16.2 | 10.9 | 12.9 | 11.1 | 35.7 | 49.7 | 57,716 | 99,970 | 73.2 | 13.9 |
| Temple City | 21.6 | 51.2 | 19.3 | 8.2 | 12.5 | 11.4 | 16.8 | 16.8 | 15.1 | 43.8 | 53.2 | 33,377 | 35,557 | 6.5 | -0.9 |
| Thousand Oaks | 19.8 | 18.5 | 23.8 | 8.9 | 9.5 | 12.0 | 13.0 | 13.0 | 19.9 | 40.8 | 49.5 | 117,005 | 126,490 | 8.1 | -0.1 |
| Torrance | 19.0 | 30.9 | 19.9 | 4.2 | 15.3 | 13.6 | 13.5 | 15.8 | 17.8 | 42.5 | 51.6 | 137,946 | 145,309 | 5.3 | -2.6 |
| Tracy | 44.3 | 24.7 | 24.6 | 11.8 | 15.9 | 14.5 | 12.9 | 11.4 | 8.9 | 33.6 | 51.9 | 56,929 | 83,426 | 46.5 | 13.9 |
| Tulare | 67.0 | 14.8 | 39.5 | 10.0 | 15.0 | 12.4 | 10.3 | 4.9 | 7.9 | 25.6 | 51.2 | 43,994 | 59,328 | 34.9 | 12.6 |
| Turlock | 38.7 | 20.3 | 27.0 | 11.6 | 14.6 | 11.8 | 11.7 | 10.2 | 13.3 | 32.8 | 54.3 | 55,810 | 68,632 | 23.0 | 6.5 |
| Tustin | 40.2 | 31.3 | 24.2 | 11.7 | 14.5 | 14.0 | 14.2 | 10.3 | 11.2 | 34.7 | 52.3 | 67,504 | 75,299 | 11.5 | 6.0 |
| Twentynine Palms | 24.0 | 5.7 | 29.6 | 25.4 | 17.3 | 12.7 | 4.8 | 5.6 | 4.6 | 22.9 | 44.8 | 14,764 | 25,048 | 69.7 | 9.5 |
| Union City | 18.2 | 44.9 | 21.2 | 5.4 | 14.1 | 16.5 | 11.9 | 13.0 | 18.0 | 39.7 | 46.5 | 66,869 | 69,547 | 4.0 | 7.1 |
| Upland | 44.7 | 15.5 | 20.5 | 9.9 | 16.6 | 12.1 | 15.0 | 11.0 | 14.8 | 38.0 | 53.2 | 68,393 | 73,719 | 7.8 | 5.6 |
| Vacaville | 26.9 | 13.4 | 22.6 | 7.8 | 15.5 | 14.3 | 12.6 | 13.2 | 13.9 | 37.6 | 48.2 | 88,625 | 92,422 | 4.3 | 9.6 |
| Vallejo | 29.6 | 26.7 | 20.4 | 9.3 | 14.5 | 11.6 | 12.9 | 15.2 | 16.1 | 39.7 | 52.3 | 116,760 | 115,908 | -0.7 | 4.2 |
| Victorville | 56.7 | 21.0 | 32.4 | 9.1 | 15.5 | 14.1 | 9.9 | 9.6 | 9.4 | 30.6 | 49.9 | 64,029 | 115,895 | 81.0 | 7.8 |
| Visalia | 54.5 | 15.5 | 27.1 | 9.5 | 16.0 | 12.5 | 11.6 | 11.3 | 12.0 | 32.8 | 51.2 | 91,565 | 124,526 | 36.0 | 8.4 |
| Vista | 56.1 | 28.5 | 24.5 | 9.6 | 17.7 | 13.5 | 14.6 | 9.7 | 10.4 | 34.0 | 49.3 | 89,857 | 93,157 | 3.7 | 8.0 |
| Walnut | 18.1 | 49.3 | 22.8 | 6.2 | 9.9 | 12.6 | 12.1 | 12.0 | 24.5 | 44.0 | 54.0 | 30,004 | 29,166 | -2.8 | 0.5 |
| Walnut Creek | 7.1 | 21.4 | 17.0 | 5.6 | 13.7 | 13.6 | 10.2 | 12.1 | 27.8 | 45.1 | 54.7 | 64,296 | 64,145 | -0.2 | 9.0 |
| Wasco | 84.5 | 28.1 | 23.3 | 12.5 | 22.1 | 12.1 | 12.9 | 11.0 | 6.2 | 30.6 | 42.2 | 21,263 | 25,549 | 20.2 | 12.5 |
| Watsonville | 81.9 | 34.4 | 27.7 | 11.8 | 12.6 | 12.9 | 11.8 | 10.5 | 12.7 | 31.7 | 51.7 | 44,265 | 51,186 | 15.6 | 1.0 |
| West Covina | 56.8 | 35.8 | 19.7 | 9.1 | 16.0 | 12.3 | 14.0 | 12.8 | 16.2 | 39.0 | 52.6 | 105,080 | 106,109 | 1.0 | -2.3 |
| West Hollywood | 11.5 | 25.6 | 4.7 | 4.5 | 34.2 | 21.0 | 10.0 | 11.5 | 14.0 | 37.0 | 46.7 | 35,716 | 34,335 | -3.9 | 2.3 |
| Westminster | 26.2 | 44.2 | 21.0 | 8.7 | 14.5 | 11.6 | 15.9 | 12.6 | 15.8 | 39.6 | 51.4 | 88,207 | 89,511 | 1.5 | 0.3 |
| West Sacramento | 32.2 | 22.2 | 26.1 | 10.4 | 13.8 | 14.6 | 8.6 | 13.2 | 13.3 | 34.9 | 53.0 | 31,615 | 48,744 | 54.2 | 10.1 |
| Whittier | 68.9 | 15.4 | 24.5 | 10.3 | 12.1 | 15.3 | 11.4 | 11.9 | 14.5 | 37.4 | 50.6 | 83,680 | 85,313 | 2.0 | -2.5 |
| Wildomar | 42.7 | 20.2 | 28.6 | 10.4 | 17.0 | 9.6 | 14.5 | 9.4 | 10.6 | 31.5 | 50.8 | 14,064 | 32,211 | 129.0 | 13.5 |
| Windsor | 35.6 | 12.2 | 25.7 | 6.0 | 10.7 | 11.1 | 16.1 | 14.4 | 16.1 | 42.9 | 48.4 | 22,744 | 26,792 | 17.8 | 2.0 |
| Woodland | 45.0 | 16.5 | 23.4 | 9.3 | 13.5 | 15.3 | 13.0 | 9.8 | 15.8 | 37.6 | 48.3 | 49,151 | 55,557 | 13.0 | 8.8 |
| Yorba Linda | 16.1 | 19.8 | 21.3 | 5.6 | 11.1 | 11.4 | 16.8 | 14.8 | 19.0 | 45.4 | 51.7 | 58,918 | 64,176 | 8.9 | 6.0 |
| Yuba City | 31.3 | 24.0 | 22.3 | 10.9 | 13.7 | 11.5 | 12.3 | 12.6 | 16.7 | 37.3 | 50.0 | 36,758 | 65,634 | 78.6 | 1.2 |
| Yucaipa | 34.5 | 10.6 | 24.0 | 9.2 | 15.0 | 8.5 | 13.9 | 12.5 | 16.9 | 38.0 | 52.9 | 41,207 | 51,279 | 24.4 | 6.9 |
| COLORADO | 21.8 | 9.5 | 21.8 | 9.2 | 15.8 | 14.0 | 12.2 | 12.4 | 14.7 | 37.1 | 49.6 | 4,301,261 | 5,029,319 | 16.9 | 15.5 |
| Arvada | 15.3 | 4.8 | 21.5 | 6.8 | 14.7 | 13.6 | 13.5 | 13.0 | 16.9 | 40.2 | 50.6 | 102,153 | 106,759 | 4.5 | 14.2 |
| Aurora | 29.7 | 20.9 | 24.9 | 9.2 | 17.1 | 14.4 | 11.9 | 10.9 | 11.6 | 34.3 | 49.8 | 276,393 | 324,976 | 17.6 | 19.2 |
| Boulder | 10.7 | 8.8 | 12.6 | 29.3 | 17.7 | 9.0 | 10.9 | 8.2 | 12.3 | 28.9 | 47.3 | 94,673 | 97,613 | 3.1 | 10.3 |
| Brighton | 35.1 | 8.0 | 28.8 | 6.6 | 18.4 | 15.6 | 13.0 | 6.8 | 10.9 | 33.5 | 44.0 | 20,905 | 33,840 | 61.9 | 20.5 |
| Broomfield | 12.7 | 9.8 | 22.1 | 7.6 | 15.2 | 14.8 | 14.4 | 12.2 | 13.7 | 38.9 | 49.5 | 38,272 | 55,848 | 45.9 | 29.3 |

1. May be of any race.

# Table D. Cities — Households, Group Quarters, Crime, and Education

| City | Households, 2019 | | | | | | | Persons in group quarters, 2019 | Serious crimes known to police[1], 2019 | | | | Educational attainment, 2019 | | |
|---|---|---|---|---|---|---|---|---|---|---|---|---|---|---|---|
| | | | | Percent | | | | | Violent | | Property | | | Attainment[3] (percent) | |
| | Number | Persons per household | Family | Married couple family | Female family | Non-family | One person | | Number | Rate[2] | Number | Rate[2] | Population age 25 and over | High school graduate or less | Bachelor's degree or more |
| | 27 | 28 | 29 | 30 | 31 | 32 | 33 | 34 | 35 | 36 | 37 | 38 | 39 | 40 | 41 |
| **CALIFORNIA— Cont'd** | | | | | | | | | | | | | | | |
| San Francisco | 365,851 | 2.36 | 45.9 | 34.9 | 11.0 | 54.1 | 36.6 | 19,965 | 5,933 | 670 | 48,780 | 5,506 | 701,279 | 23.5 | 59.2 |
| San Gabriel | 12,570 | 3.13 | 74.5 | 51.5 | 23.0 | 25.5 | 18.7 | 560 | 98 | 242 | 715 | 1,769 | 30,774 | 46.0 | 33.6 |
| San Jacinto | 14,784 | 3.32 | 72.3 | 50.7 | 21.7 | 27.7 | 20.0 | 166 | 108 | 218 | 1,637 | 3,310 | 31,204 | 50.6 | 13.1 |
| San Jose | 327,894 | 3.07 | 71.1 | 53.3 | 17.8 | 28.9 | 20.9 | 14,172 | 4,559 | 438 | 25,164 | 2,420 | 711,040 | 31.1 | 45.0 |
| San Juan Capistrano | 12,114 | 2.95 | 67.3 | 58.1 | 9.2 | 32.7 | 22.1 | 144 | 63 | 174 | 297 | 820 | 26,238 | 34.3 | 40.7 |
| San Leandro | 31,691 | 2.79 | 62.1 | 42.5 | 19.6 | 37.9 | 32.0 | 482 | 454 | 503 | 4,105 | 4,546 | 69,343 | 34.9 | 38.0 |
| San Luis Obispo | 19,457 | 2.40 | 36.7 | 28.6 | 8.1 | 63.3 | 31.0 | 680 | 192 | 402 | 1,738 | 3,641 | 25,781 | 16.7 | 54.6 |
| San Marcos | 30,852 | 3.10 | 70.1 | 52.5 | 17.6 | 29.9 | 24.5 | 979 | 193 | 196 | 955 | 969 | 62,156 | 31.3 | 39.9 |
| San Mateo | 39,564 | 2.60 | 60.5 | 48.3 | 12.2 | 39.5 | 29.9 | 1,386 | 266 | 251 | 2,217 | 2,091 | 76,245 | 21.9 | 57.1 |
| San Pablo | 9,214 | 3.31 | 73.8 | 44.5 | 29.3 | 26.2 | 22.2 | 446 | 194 | 619 | 1,009 | 3,220 | 18,924 | 63.0 | 11.4 |
| San Rafael | 24,133 | 2.33 | 56.4 | 46.2 | 10.3 | 43.6 | 35.8 | 2,146 | 230 | 391 | 1,686 | 2,866 | 42,155 | 27.0 | 49.0 |
| San Ramon | 25,697 | 2.95 | 77.5 | 71.2 | 6.4 | 22.5 | 20.1 | 104 | 58 | 76 | 1,099 | 1,439 | 50,486 | 12.2 | 71.7 |
| Santa Ana | 79,704 | 4.11 | 82.4 | 56.8 | 25.6 | 17.6 | 12.0 | 5,011 | 1,453 | 435 | 6,808 | 2,040 | 211,611 | 61.6 | 17.6 |
| Santa Barbara | 38,674 | 2.33 | 46.3 | 33.7 | 12.6 | 53.7 | 40.1 | 1,404 | 400 | 436 | 2,118 | 2,309 | 63,728 | 26.1 | 51.6 |
| Santa Clara | 46,608 | 2.68 | 66.9 | 53.5 | 13.3 | 33.1 | 23.0 | 5,644 | 214 | 163 | 4,748 | 3,620 | 92,418 | 18.6 | 61.7 |
| Santa Clarita | 69,975 | 3.01 | 77.0 | 62.2 | 14.8 | 23.0 | 19.1 | 2,058 | 277 | 127 | 2,059 | 944 | 141,289 | 28.3 | 36.9 |
| Santa Cruz | 22,166 | 2.43 | 48.4 | 33.1 | 15.3 | 51.6 | 34.2 | 10,679 | 389 | 596 | 2,932 | 4,493 | 35,029 | 17.7 | 54.9 |
| Santa Maria | 27,079 | 3.92 | 78.0 | 41.5 | 36.5 | 22.0 | 16.5 | 1,141 | NA | NA | NA | NA | 60,389 | 64.6 | 13.7 |
| Santa Monica | 45,096 | 1.97 | 44.6 | 34.7 | 9.9 | 55.4 | 45.4 | 1,507 | 664 | 725 | 3,964 | 4,327 | 73,163 | 16.7 | 64.3 |
| Santa Paula | 7,755 | 3.83 | 73.0 | 52.2 | 20.7 | 27.0 | 22.8 | 134 | 91 | 302 | 404 | 1,342 | 18,691 | 62.2 | 13.6 |
| Santa Rosa | 65,590 | 2.64 | 63.9 | 44.5 | 19.5 | 36.1 | 25.9 | 3,304 | 857 | 482 | 2,874 | 1,616 | 122,855 | 35.1 | 33.6 |
| Santee | 19,190 | 2.97 | 72.2 | 52.8 | 19.4 | 27.8 | 19.4 | 1,124 | 98 | 167 | 806 | 1,373 | 39,382 | 35.7 | 29.1 |
| Saratoga | 10,519 | 2.85 | 81.3 | 72.4 | 8.9 | 18.7 | 14.5 | 209 | 16 | 52 | 240 | 783 | 22,943 | 5.8 | 75.4 |
| Seaside | 10,883 | 2.98 | 74.5 | 54.9 | 19.6 | 25.5 | 19.3 | 1,351 | 78 | 229 | 439 | 1,290 | 22,390 | 48.0 | 24.2 |
| Simi Valley | 42,117 | 2.96 | 72.6 | 53.1 | 19.5 | 27.4 | 20.4 | 778 | 146 | 116 | 1,188 | 943 | 89,975 | 30.1 | 33.3 |
| Soledad | 3,536 | 4.89 | 93.2 | 63.4 | 29.9 | 6.8 | 6.8 | 8,717 | 53 | 204 | 136 | 523 | 18,085 | 70.1 | 4.7 |
| South Gate | 23,559 | 3.96 | 86.6 | 53.2 | 33.5 | 13.4 | 10.0 | 66 | 622 | 659 | 2,754 | 2,916 | 59,534 | 69.2 | 8.7 |
| South Pasadena | 10,052 | 2.51 | 66.1 | 49.8 | 16.3 | 33.9 | 27.2 | 145 | 29 | 113 | 566 | 2,210 | 18,463 | 10.4 | 66.4 |
| South San Francisco | 22,355 | 3.00 | 70.4 | 51.4 | 19.0 | 29.6 | 16.4 | 790 | 166 | 243 | 1,484 | 2,174 | 53,153 | 31.4 | 41.3 |
| Stanton | 11,436 | 3.31 | 65.9 | 40.8 | 25.1 | 34.1 | 27.6 | 240 | 95 | 248 | 526 | 1,374 | 24,818 | 54.5 | 19.0 |
| Stockton | 96,149 | 3.18 | 70.5 | 46.1 | 24.5 | 29.5 | 24.4 | 6,571 | 4,380 | 1,397 | 12,367 | 3,944 | 196,661 | 53.3 | 20.0 |
| Suisun City | 9,057 | 3.26 | 71.8 | 45.6 | 26.2 | 28.2 | 25.0 | 128 | 114 | 381 | 664 | 2,219 | 18,174 | 41.0 | 24.8 |
| Sunnyvale | 55,424 | 2.73 | 65.4 | 54.2 | 11.2 | 34.6 | 23.2 | 1,129 | 259 | 167 | 3,380 | 2,183 | 110,317 | 15.2 | 65.4 |
| Temecula | 35,638 | 3.22 | 75.8 | 62.5 | 13.2 | 24.2 | 19.1 | 65 | 166 | 142 | 2,626 | 2,252 | 73,442 | 21.2 | 39.7 |
| Temple City | 11,571 | 3.05 | 83.3 | 62.3 | 21.1 | 16.7 | 14.3 | 553 | 64 | 177 | 331 | 915 | 25,989 | 34.3 | 39.9 |
| Thousand Oaks | 44,618 | 2.80 | 70.4 | 58.6 | 11.8 | 29.6 | 24.8 | 1,951 | 88 | 69 | 1,392 | 1,089 | 85,327 | 17.5 | 54.8 |
| Torrance | 55,483 | 2.56 | 66.3 | 52.3 | 13.9 | 33.7 | 27.2 | 1,511 | 280 | 193 | 2,853 | 1,965 | 108,950 | 22.8 | 53.2 |
| Tracy | 26,771 | 3.53 | 79.7 | 59.1 | 20.6 | 20.3 | 13.7 | 263 | 166 | 179 | 1,843 | 1,984 | 60,295 | 43.4 | 20.3 |
| Tulare | 18,170 | 3.59 | 81.7 | 54.7 | 27.0 | 18.3 | 11.4 | 345 | 261 | 401 | 1,469 | 2,255 | 33,105 | 55.8 | 8.8 |
| Turlock | 25,718 | 2.84 | 69.9 | 47.4 | 22.6 | 30.1 | 22.6 | 610 | 426 | 575 | 2,323 | 3,134 | 45,253 | 37.2 | 25.0 |
| Tustin | 24,125 | 3.24 | 77.4 | 56.0 | 21.4 | 22.6 | 15.6 | 1,228 | 141 | 175 | 2,118 | 2,636 | 50,890 | 30.9 | 43.0 |
| Twentynine Palms | 8,414 | 2.73 | 61.3 | 44.8 | 16.4 | 38.7 | 33.9 | 3,123 | 117 | 440 | 265 | 997 | 11,752 | 25.4 | 21.2 |
| Union City | 22,469 | 3.27 | 84.3 | 69.0 | 15.3 | 15.7 | 11.4 | 733 | 277 | 368 | 1,718 | 2,285 | 54,419 | 29.8 | 48.1 |
| Upland | 26,201 | 2.92 | 73.2 | 50.5 | 22.8 | 26.8 | 18.2 | 670 | 298 | 385 | 1,821 | 2,353 | 53,676 | 30.3 | 32.1 |
| Vacaville | 32,560 | 2.86 | 72.8 | 58.6 | 14.3 | 27.2 | 21.1 | 7,618 | 331 | 327 | 2,442 | 2,414 | 70,015 | 35.1 | 24.6 |
| Vallejo | 42,200 | 2.85 | 64.1 | 42.1 | 22.1 | 35.9 | 28.4 | 1,521 | 1,037 | 845 | 4,941 | 4,028 | 85,563 | 37.8 | 26.5 |
| Victorville | 32,188 | 3.67 | 79.8 | 45.1 | 34.7 | 20.2 | 16.5 | 4,327 | 988 | 803 | 2,271 | 1,845 | 71,601 | 57.8 | 11.3 |
| Visalia | 45,878 | 2.90 | 75.1 | 53.5 | 21.6 | 24.9 | 19.5 | 1,458 | 586 | 434 | 3,900 | 2,890 | 85,368 | 39.4 | 17.5 |
| Vista | 31,075 | 3.20 | 72.2 | 51.4 | 20.8 | 27.8 | 20.5 | 2,095 | 354 | 346 | 1,438 | 1,407 | 66,980 | 44.7 | 22.4 |
| Walnut | 8,392 | 3.53 | 87.0 | 66.0 | 21.0 | 13.0 | 10.4 | 43 | 38 | 126 | 386 | 1,282 | 21,105 | 23.1 | 52.2 |
| Walnut Creek | 31,558 | 2.19 | 57.4 | 47.2 | 10.2 | 42.6 | 33.1 | 961 | 120 | 170 | 2,496 | 3,538 | 54,302 | 10.1 | 65.3 |
| Wasco | 6,358 | 3.60 | 80.2 | 55.3 | 24.9 | 19.8 | 19.8 | 5,797 | NA | NA | NA | NA | 18,450 | 69.5 | 6.2 |
| Watsonville | 15,384 | 3.47 | 75.2 | 48.6 | 26.6 | 24.8 | 18.9 | 497 | 328 | 604 | 971 | 1,789 | 32,619 | 54.9 | 20.2 |
| West Covina | 30,682 | 3.40 | 73.0 | 50.4 | 22.6 | 27.0 | 20.3 | 885 | 260 | 245 | 2,369 | 2,228 | 74,843 | 41.1 | 28.4 |
| West Hollywood | 22,511 | 1.61 | 25.9 | 18.9 | 7.0 | 74.1 | 57.5 | 177 | 294 | 791 | 1,898 | 5,106 | 33,099 | 16.2 | 62.7 |
| Westminster | 26,984 | 3.35 | 78.7 | 56.4 | 22.3 | 21.3 | 16.8 | 272 | 259 | 284 | 2,611 | 2,867 | 63,771 | 47.9 | 26.6 |
| West Sacramento | 18,355 | 2.89 | 69.0 | 44.2 | 24.8 | 31.0 | 19.2 | 398 | 212 | 390 | 1,603 | 2,948 | 34,012 | 40.3 | 30.0 |
| Whittier | 26,884 | 3.10 | 71.8 | 53.5 | 18.3 | 28.2 | 23.5 | 1,776 | 213 | 247 | 1,901 | 2,206 | 55,511 | 41.7 | 26.8 |
| Wildomar | 9,916 | 3.74 | 82.2 | 60.6 | 21.6 | 17.8 | 14.0 | 103 | 61 | 161 | 637 | 1,679 | 22,758 | 50.9 | 15.3 |
| Windsor | 9,300 | 2.91 | 78.8 | 55.6 | 23.2 | 21.2 | 17.7 | 51 | 58 | 207 | 257 | 918 | 18,519 | 28.2 | 32.2 |
| Woodland | 20,605 | 2.87 | 75.2 | 56.0 | 19.3 | 24.8 | 20.4 | 1,338 | 224 | 366 | 1,488 | 2,432 | 40,751 | 35.3 | 29.9 |
| Yorba Linda | 22,676 | 2.97 | 81.0 | 69.4 | 11.6 | 19.0 | 16.3 | 235 | 42 | 62 | 627 | 919 | 49,449 | 18.1 | 52.4 |
| Yuba City | 24,393 | 2.72 | 67.9 | 45.1 | 22.8 | 32.1 | 25.3 | 742 | 241 | 359 | 1,823 | 2,714 | 44,800 | 43.5 | 22.5 |
| Yucaipa | 17,706 | 3.03 | 70.3 | 52.4 | 17.9 | 29.7 | 26.7 | 305 | 200 | 371 | 641 | 1,188 | 35,982 | 38.8 | 25.5 |
| **COLORADO** | 2,235,103 | 2.52 | 63.6 | 50.0 | 13.6 | 36.4 | 27.1 | 119,141 | NA | NA | NA | NA | 3,974,943 | 28.6 | 42.7 |
| Arvada | 48,353 | 2.49 | 67.3 | 51.8 | 15.6 | 32.7 | 27.0 | 549 | 266 | 217 | 3,642 | 2,978 | 86,793 | 28.1 | 44.8 |
| Aurora | 136,443 | 2.77 | 64.6 | 44.2 | 20.3 | 35.4 | 26.8 | 1,794 | 2,799 | 735 | 11,106 | 2,918 | 250,080 | 36.5 | 31.1 |
| Boulder | 42,462 | 2.22 | 40.3 | 34.7 | 5.6 | 59.7 | 34.6 | 11,374 | 278 | 256 | 3,284 | 3,026 | 61,347 | 8.1 | 78.8 |
| Brighton | 12,965 | 3.03 | 69.8 | 57.3 | 12.5 | 30.2 | 23.5 | 1,794 | 138 | 326 | 1,119 | 2,647 | 26,602 | 46.3 | 22.0 |
| Broomfield | 28,062 | 2.50 | 66.8 | 56.4 | 10.4 | 33.2 | 22.1 | 350 | 75 | 106 | 2,046 | 2,890 | 49,543 | 17.2 | 56.9 |

1. Data for serious crimes have not been adjusted for underreporting. This may affect comparability between geographic areas and over time.  2. Per 100,000 population estimated by the FBI.  3. Persons 25 years old and over.

# Table D. Cities — Income, Poverty, and Housing

| City | Money income, 2019 — Households: Median income | Percent with income less than $20,000 | Percent with income of $200,000 or more | Median family income | Median non-family income | Median earnings, 2019: All persons | Men | Women | Housing units, 2019: Total | Occupied | Percent owner occupied | Median value[1] (dollars) | Median gross rent (dollars) |
|---|---|---|---|---|---|---|---|---|---|---|---|---|---|
| | 42 | 43 | 44 | 45 | 46 | 47 | 48 | 49 | 50 | 51 | 52 | 53 | 54 |
| **CALIFORNIA— Cont'd** | | | | | | | | | | | | | |
| San Francisco | 123,859 | 11.8 | 30.7 | 140,707 | 105,235 | 67,926 | 79,284 | 58,478 | 406,399 | 365,851 | 37.1 | 1,217,458 | 1,959 |
| San Gabriel | 75,082 | 14.0 | 8.6 | 77,113 | 41,899 | 40,273 | 37,988 | 40,643 | 13,306 | 12,570 | 44.2 | 714,711 | 1,543 |
| San Jacinto | 53,100 | 21.8 | 3.5 | 62,650 | 37,435 | 31,635 | 35,173 | 25,923 | 17,370 | 14,784 | 71.1 | 292,329 | 1,368 |
| San Jose | 115,893 | 8.0 | 25.9 | 130,829 | 68,749 | 51,580 | 61,169 | 42,315 | 343,234 | 327,894 | 55.2 | 999,943 | 2,223 |
| San Juan Capistrano | 102,958 | 3.2 | 21.3 | 122,194 | 66,650 | 35,682 | 32,224 | 36,141 | 13,563 | 12,114 | 70.8 | 813,961 | 2,113 |
| San Leandro | 95,072 | 7.6 | 13.2 | 110,093 | 47,925 | 47,072 | 50,487 | 45,202 | 33,469 | 31,691 | 55.2 | 654,181 | 1,878 |
| San Luis Obispo | 57,631 | 22.7 | 10.8 | 116,335 | 39,418 | 25,882 | 31,038 | 18,293 | 21,176 | 19,457 | 41.1 | 717,677 | 1,722 |
| San Marcos | 80,814 | 11.1 | 13.6 | 102,595 | 46,472 | 39,706 | 46,078 | 35,376 | 33,095 | 30,852 | 64.6 | 628,825 | 1,779 |
| San Mateo | 149,312 | 8.0 | 34.7 | 180,092 | 82,730 | 70,304 | 64,234 | 71,470 | 41,973 | 39,564 | 49.9 | 1,282,071 | 2,638 |
| San Pablo | 66,823 | 9.9 | 1.4 | 68,345 | 44,792 | 36,056 | 42,327 | 30,692 | 10,166 | 9,214 | 34.5 | 416,519 | 1,629 |
| San Rafael | 85,927 | 10.1 | 22.6 | 106,117 | 52,872 | 40,993 | 47,364 | 35,492 | 25,262 | 24,133 | 50.9 | 1,007,201 | 1,951 |
| San Ramon | 189,185 | 4.1 | 47.6 | 207,593 | 85,144 | 86,334 | 121,129 | 68,074 | 26,697 | 25,697 | 72.4 | 1,089,702 | 2,387 |
| Santa Ana | 70,084 | 8.4 | 6.1 | 70,231 | 53,251 | 28,877 | 32,089 | 24,253 | 82,990 | 79,704 | 43.3 | 558,410 | 1,632 |
| Santa Barbara | 73,974 | 11.4 | 18.5 | 125,123 | 60,223 | 40,224 | 46,728 | 30,526 | 41,215 | 38,674 | 37.7 | 1,163,924 | 1,800 |
| Santa Clara | 147,507 | 2.9 | 31.8 | 166,087 | 100,462 | 65,287 | 87,442 | 50,053 | 50,936 | 46,608 | 41.3 | 1,249,105 | 2,535 |
| Santa Clarita | 97,904 | 7.4 | 14.5 | 112,994 | 51,156 | 48,422 | 60,282 | 35,657 | 72,966 | 69,975 | 72.8 | 584,417 | 2,044 |
| Santa Cruz | 90,855 | 14.9 | 16.2 | 135,244 | 41,481 | 28,162 | 33,341 | 21,905 | 24,029 | 22,166 | 39.6 | 945,551 | 1,979 |
| Santa Maria | 69,393 | 9.8 | 4.1 | 62,939 | 37,287 | 28,495 | 29,926 | 26,445 | 29,084 | 27,079 | 50.1 | 373,195 | 1,371 |
| Santa Monica | 92,490 | 16.4 | 21.9 | 129,007 | 68,255 | 67,463 | 75,952 | 60,014 | 50,563 | 45,096 | 28.2 | 1,590,768 | 1,918 |
| Santa Paula | 71,538 | 10.4 | 1.7 | 82,224 | 36,723 | 30,877 | 35,089 | 25,727 | 8,487 | 7,755 | 48.0 | 436,893 | 1,461 |
| Santa Rosa | 78,574 | 6.3 | 8.3 | 88,362 | 58,820 | 38,984 | 41,821 | 35,739 | 69,406 | 65,590 | 54.7 | 603,056 | 1,728 |
| Santee | 82,945 | 9.9 | 9.7 | 87,825 | 51,180 | 37,104 | 40,590 | 33,663 | 20,342 | 19,190 | 75.9 | 527,530 | 1,706 |
| Saratoga | 207,989 | 4.8 | 52.9 | 248,778 | 79,653 | 102,192 | 150,119 | 81,588 | 11,310 | 10,519 | 83.6 | 2 | 3,241 |
| Seaside | 63,422 | 11.3 | 6.8 | 59,931 | 52,793 | 30,846 | 34,842 | 25,741 | 12,222 | 10,883 | 44.0 | 596,067 | 1,863 |
| Simi Valley | 100,927 | 6.6 | 15.4 | 116,952 | 50,863 | 48,010 | 60,000 | 39,046 | 44,520 | 42,117 | 71.7 | 622,613 | 2,217 |
| Soledad | 81,716 | 2.2 | 8.1 | 82,691 | 0 | 22,642 | 19,817 | 30,140 | 3,536 | 3,536 | 61.0 | 443,833 | 1,531 |
| South Gate | 53,736 | 11.0 | 2.3 | 52,528 | 40,440 | 27,289 | 30,091 | 23,657 | 24,349 | 23,559 | 41.5 | 475,295 | 1,193 |
| South Pasadena | 110,441 | 7.1 | 22.1 | 140,124 | 79,484 | 71,222 | 75,329 | 66,890 | 10,331 | 10,052 | 48.1 | 1,149,148 | 2,102 |
| South San Francisco | 120,573 | 5.8 | 23.0 | 123,353 | 98,291 | 51,199 | 52,297 | 50,237 | 23,307 | 22,355 | 60.8 | 973,448 | 2,417 |
| Stanton | 50,757 | 21.9 | 5.8 | 69,596 | 23,036 | 35,135 | 41,385 | 25,168 | 11,736 | 11,436 | 47.1 | 454,820 | 1,687 |
| Stockton | 59,504 | 14.6 | 5.7 | 66,861 | 34,420 | 31,982 | 35,528 | 27,546 | 104,720 | 96,149 | 49.8 | 317,693 | 1,184 |
| Suisun City | 88,326 | 5.1 | 4.8 | 97,642 | 77,429 | 46,675 | 50,602 | 40,946 | 9,745 | 9,057 | 58.1 | 420,437 | 1,749 |
| Sunnyvale | 151,475 | 4.8 | 35.7 | 168,286 | 118,662 | 79,939 | 101,951 | 61,869 | 59,151 | 55,424 | 44.2 | 1,567,085 | 2,682 |
| Temecula | 95,918 | 7.5 | 12.2 | 104,134 | 51,049 | 45,977 | 55,607 | 32,073 | 37,042 | 35,638 | 62.1 | 485,644 | 1,893 |
| Temple City | 91,959 | 7.1 | 14.3 | 98,185 | 53,038 | 41,732 | 51,254 | 35,066 | 11,958 | 11,571 | 60.4 | 752,886 | 1,754 |
| Thousand Oaks | 113,562 | 8.5 | 23.7 | 145,870 | 63,147 | 50,373 | 62,014 | 37,034 | 47,610 | 44,618 | 67.7 | 793,077 | 2,184 |
| Torrance | 100,107 | 7.1 | 17.0 | 125,231 | 61,778 | 56,961 | 65,319 | 49,040 | 59,079 | 55,483 | 54.4 | 835,928 | 1,876 |
| Tracy | 97,292 | 5.7 | 10.5 | 100,560 | 62,074 | 40,761 | 47,602 | 32,056 | 28,246 | 26,771 | 61.7 | 487,156 | 1,851 |
| Tulare | 60,081 | 12.6 | 2.2 | 60,760 | 35,627 | 33,285 | 37,680 | 26,834 | 18,731 | 18,170 | 57.0 | 257,910 | 1,091 |
| Turlock | 58,263 | 12.9 | 4.5 | 65,922 | 33,475 | 33,720 | 46,504 | 27,139 | 26,517 | 25,718 | 57.5 | 337,723 | 1,210 |
| Tustin | 93,304 | 6.8 | 17.5 | 103,738 | 56,376 | 42,151 | 50,107 | 34,090 | 25,632 | 24,125 | 54.3 | 720,145 | 2,047 |
| Twentynine Palms | 43,979 | 17.0 | 0.2 | 47,202 | 31,849 | 30,227 | 30,871 | 26,647 | 9,640 | 8,414 | 33.3 | 178,867 | 1,037 |
| Union City | 128,108 | 6.7 | 22.4 | 131,248 | 66,197 | 60,008 | 64,519 | 47,747 | 23,648 | 22,469 | 69.1 | 866,100 | 2,290 |
| Upland | 82,426 | 6.3 | 10.7 | 91,608 | 41,729 | 37,156 | 44,598 | 31,762 | 27,462 | 26,201 | 55.2 | 581,455 | 1,600 |
| Vacaville | 91,302 | 8.7 | 11.1 | 108,714 | 42,246 | 50,026 | 59,179 | 42,055 | 34,817 | 32,560 | 61.8 | 464,256 | 1,775 |
| Vallejo | 71,265 | 12.1 | 8.8 | 86,459 | 48,192 | 40,072 | 41,275 | 37,663 | 45,159 | 42,200 | 57.7 | 441,061 | 1,680 |
| Victorville | 60,391 | 14.0 | 0.4 | 60,918 | 38,342 | 28,161 | 31,597 | 23,738 | 33,991 | 32,188 | 51.7 | 258,866 | 1,231 |
| Visalia | 59,911 | 11.5 | 3.0 | 62,399 | 41,384 | 31,743 | 33,563 | 30,005 | 47,900 | 45,878 | 57.1 | 263,999 | 1,147 |
| Vista | 78,656 | 6.0 | 7.6 | 82,059 | 47,170 | 35,318 | 37,494 | 30,707 | 32,469 | 31,075 | 45.7 | 540,499 | 1,731 |
| Walnut | 114,382 | 8.2 | 19.6 | 122,729 | 0 | 42,336 | 42,327 | 42,370 | 8,392 | 8,392 | 83.4 | 819,337 | 2,361 |
| Walnut Creek | 120,238 | 5.8 | 23.4 | 141,285 | 80,216 | 71,093 | 90,920 | 57,287 | 33,971 | 31,558 | 63.6 | 847,256 | 2,179 |
| Wasco | 47,335 | 22.3 | 2.3 | 46,301 | 23,240 | 25,349 | 31,973 | 16,558 | 7,603 | 6,358 | 63.2 | 189,343 | 758 |
| Watsonville | 59,739 | 18.4 | 8.2 | 71,292 | 34,290 | 31,588 | 34,463 | 27,946 | 15,480 | 15,384 | 40.8 | 570,117 | 1,477 |
| West Covina | 82,135 | 9.9 | 10.2 | 98,567 | 35,471 | 35,999 | 41,229 | 31,730 | 31,230 | 30,682 | 58.0 | 618,993 | 1,747 |
| West Hollywood | 77,876 | 16.9 | 15.6 | 114,050 | 65,643 | 54,679 | 58,363 | 52,538 | 24,471 | 22,511 | 17.1 | 810,673 | 1,934 |
| Westminster | 56,378 | 18.3 | 9.8 | 65,325 | 32,755 | 31,990 | 36,026 | 29,075 | 28,112 | 26,984 | 47.2 | 651,314 | 1,689 |
| West Sacramento | 69,413 | 11.6 | 9.4 | 92,163 | 53,662 | 41,268 | 42,388 | 36,975 | 18,956 | 18,355 | 59.5 | 367,216 | 1,234 |
| Whittier | 76,333 | 10.3 | 7.2 | 90,962 | 35,406 | 40,187 | 41,245 | 36,837 | 27,798 | 26,884 | 60.9 | 627,166 | 1,527 |
| Wildomar | 69,397 | 13.8 | 7.4 | 73,971 | 38,924 | 30,771 | 42,062 | 22,773 | 10,227 | 9,916 | 71.3 | 406,899 | 1,555 |
| Windsor | 126,422 | 6.2 | 19.9 | 131,490 | 45,120 | 47,446 | 60,903 | 32,161 | 9,553 | 9,300 | 74.2 | 640,775 | 1,721 |
| Woodland | 80,598 | 9.9 | 10.3 | 81,827 | 56,715 | 41,510 | 49,044 | 35,538 | 21,330 | 20,605 | 57.7 | 412,654 | 1,317 |
| Yorba Linda | 122,962 | 6.1 | 26.9 | 136,319 | 49,454 | 51,486 | 70,252 | 40,436 | 23,938 | 22,676 | 83.8 | 882,646 | 2,306 |
| Yuba City | 61,773 | 11.9 | 5.5 | 71,100 | 40,182 | 30,257 | 31,583 | 28,033 | 24,998 | 24,393 | 54.5 | 303,902 | 1,193 |
| Yucaipa | 74,798 | 16.3 | 7.4 | 89,944 | 32,042 | 39,745 | 46,715 | 26,798 | 19,382 | 17,706 | 68.5 | 349,218 | 1,247 |
| **COLORADO** | 77,127 | 10.2 | 10.3 | 95,164 | 50,092 | 40,646 | 48,082 | 32,371 | 2,464,109 | 2,235,103 | 65.9 | 394,589 | 1,369 |
| Arvada | 89,523 | 6.5 | 10.9 | 115,950 | 51,347 | 47,774 | 55,651 | 40,477 | 49,888 | 48,353 | 71.3 | 443,839 | 1,433 |
| Aurora | 69,235 | 8.6 | 6.1 | 81,550 | 49,762 | 37,286 | 41,963 | 32,163 | 141,909 | 136,443 | 62.0 | 342,967 | 1,420 |
| Boulder | 74,900 | 18.2 | 17.2 | 127,023 | 48,993 | 30,537 | 32,571 | 25,795 | 45,954 | 42,462 | 48.0 | 794,981 | 1,658 |
| Brighton | 76,938 | 7.9 | 5.8 | 85,389 | 61,319 | 41,589 | 43,788 | 36,318 | 12,965 | 12,965 | 66.2 | 352,396 | 1,551 |
| Broomfield | 111,400 | 3.0 | 18.9 | 136,125 | 73,938 | 53,042 | 70,238 | 42,373 | 29,677 | 28,062 | 62.8 | 481,004 | 1,786 |

1. Based on population estimated by the American Community Survey.

## Table D. Cities — Commuting, Computer Access, Migration, Labor Force, and Employment

| City | Commuting¹ 2019 — Drove alone | Commuting — With commutes of 30 minutes or more | Computer access² 2019 — With a computer in the house | Computer access — With Internet access | Migration 2019 — Percent who lived in the same house one year ago | Migration — Percent who lived in another state or county one year ago | Civilian labor force 2020 — Total | Percent change 2019-2020 | Unemployment — Total | Unemployment — Rate³ | Civilian Employment⁴ 2019 — Population age 16 and older — Number | Percent in labor force | Population age 16 to 64 — Number | Percent who worked full-year full-time |
|---|---|---|---|---|---|---|---|---|---|---|---|---|---|---|
| | 55 | 56 | 57 | 58 | 59 | 60 | 61 | 62 | 63 | 64 | 65 | 66 | 67 | 68 |
| CALIFORNIA— Cont'd | | | | | | | | | | | | | | |
| San Francisco | 29.1 | 58.2 | 95.1 | 89.2 | 85.5 | 7.0 | 556,056 | -4.7 | 43,525 | 8 | 774,619 | 72.3 | 633,155 | 61.3 |
| San Gabriel | 76.3 | 52.0 | 95.1 | 91.6 | 95.9 | 1.5 | 20,960 | -2.9 | 2,791 | 13 | 34,072 | 61.8 | 25,988 | 59.8 |
| San Jacinto | 84.0 | 64.1 | 94.0 | 77.1 | NA | NA | 19,319 | 1.6 | 2,317 | 12 | 38,420 | 54.1 | 31,906 | 48.3 |
| San Jose | 75.3 | 49.9 | 95.9 | 92.4 | 88.6 | 4.2 | 542,031 | -2.4 | 43,772 | 8 | 821,088 | 68.5 | 682,591 | 56.7 |
| San Juan Capistrano | 75.2 | 31.5 | 99.0 | 92.6 | 86.2 | 2.4 | 16,389 | -5.3 | 1,251 | 8 | 30,248 | 61.1 | 23,717 | 49.1 |
| San Leandro | 62.8 | 56.2 | 95.4 | 91.3 | 91.3 | 3.7 | 45,201 | -2.9 | 4,879 | 11 | 75,001 | 70.4 | 61,165 | 62.9 |
| San Luis Obispo | 69.3 | 16.5 | 96.4 | 91.8 | 67.8 | 11.9 | 24,140 | -5.1 | 1,849 | 8 | 42,752 | 63.9 | 35,574 | 38.6 |
| San Marcos | 79.3 | 39.6 | 97.0 | 92.2 | 90.2 | 1.5 | 40,932 | -4.1 | 3,427 | 8 | 73,694 | 64.1 | 59,514 | 53.5 |
| San Mateo | 66.6 | 50.2 | 96.3 | 93.3 | 85.7 | 7.2 | 61,304 | -6.5 | 3,757 | 6 | 85,998 | 72.7 | 70,203 | 61.9 |
| San Pablo | 81.5 | 48.3 | 95.0 | 90.8 | 96.0 | 1.0 | 13,635 | -2.0 | 1,556 | 11 | 23,518 | 63.8 | 20,420 | 51.9 |
| San Rafael | 59.7 | 50.8 | 95.1 | 91.0 | 86.8 | 6.0 | 30,233 | -7.0 | 2,036 | 7 | 48,111 | 66.1 | 35,457 | 53.0 |
| San Ramon | 66.0 | 55.3 | 98.8 | 97.4 | 84.5 | 9.4 | 37,538 | -5.4 | 2,504 | 7 | 58,059 | 69.7 | 49,315 | 54.7 |
| Santa Ana | 76.2 | 37.9 | 96.2 | 83.8 | 91.3 | 2.4 | 152,228 | -3.9 | 13,867 | 9 | 257,721 | 66.9 | 223,291 | 49.5 |
| Santa Barbara | 66.8 | 8.7 | 96.2 | 92.9 | 75.1 | 6.9 | 50,616 | 0.2 | 3,528 | 7 | 77,360 | 69.4 | 60,103 | 52.0 |
| Santa Clara | 77.7 | 37.6 | 97.8 | 95.3 | 78.4 | 10.4 | 69,751 | -2.9 | 4,110 | 6 | 109,932 | 72.1 | 96,680 | 58.6 |
| Santa Clarita | 81.0 | 55.9 | 97.2 | 94.0 | 89.0 | 2.1 | 110,304 | -2.8 | 12,954 | 12 | 166,307 | 64.9 | 137,727 | 52.6 |
| Santa Cruz | 60.8 | 23.7 | 98.1 | 91.8 | 71.9 | 15.1 | 31,226 | -6.1 | 2,424 | 7 | 56,817 | 64.0 | 50,304 | 36.2 |
| Santa Maria | 65.6 | 28.9 | 92.2 | 89.8 | 87.2 | 3.6 | 49,483 | -0.4 | 4,997 | 10 | 75,551 | 66.6 | 63,960 | 51.8 |
| Santa Monica | 64.7 | 40.1 | 94.7 | 90.1 | 88.1 | 4.2 | 52,912 | -6.6 | 5,563 | 11 | 79,513 | 65.1 | 59,726 | 56.4 |
| Santa Paula | 71.3 | 53.6 | 85.7 | 80.2 | NA | NA | 13,353 | -4.0 | 1,591 | 12 | 22,401 | 65.0 | 18,822 | 37.8 |
| Santa Rosa | 78.2 | 25.6 | 96.6 | 94.1 | 86.3 | 3.6 | 86,291 | -4.5 | 7,149 | 8 | 141,162 | 67.6 | 111,826 | 51.4 |
| Santee | 83.4 | 43.7 | 92.6 | 90.0 | 84.8 | 4.2 | 28,144 | -4.1 | 2,273 | 8 | 45,934 | 61.3 | 36,855 | 48.3 |
| Saratoga | 79.1 | 59.7 | NA | NA | 90.1 | 5.7 | 13,951 | -6.2 | 677 | 5 | 25,500 | 54.7 | 16,799 | 50.9 |
| Seaside | 68.8 | 27.4 | 93.7 | 88.1 | 86.3 | 8.5 | 17,205 | -1.8 | 1,682 | 10 | 27,498 | 60.4 | 22,752 | 44.7 |
| Simi Valley | 79.5 | 50.7 | 94.2 | 91.9 | 91.9 | 3.5 | 65,693 | -2.7 | 5,439 | 8 | 104,594 | 66.8 | 84,770 | 57.0 |
| Soledad | 66.6 | 35.2 | NA | NA | NA | NA | 7,813 | -4.4 | 1,087 | 14 | 21,878 | 37.8 | 20,343 | 36.9 |
| South Gate | 70.5 | 54.6 | 92.9 | 83.6 | NA | NA | 41,280 | -3.6 | 5,639 | 14 | 72,400 | 63.0 | 61,424 | 48.6 |
| South Pasadena | 68.9 | 59.2 | NA | NA | 91.7 | 5.7 | 13,958 | -6.8 | 1,375 | 10 | 20,874 | 71.3 | 16,861 | 66.2 |
| South San Francisco | 61.4 | 49.2 | 93.5 | 91.8 | 91.8 | 3.5 | 38,626 | -3.4 | 3,435 | 9 | 60,326 | 71.9 | 48,920 | 67.1 |
| Stanton | 72.7 | 49.9 | 90.1 | 83.2 | NA | NA | 18,094 | -2.5 | 1,942 | 11 | 28,412 | 63.9 | 23,289 | 53.0 |
| Stockton | 79.4 | 37.6 | 93.5 | 83.5 | 91.1 | 2.6 | 132,424 | 1.4 | 16,897 | 13 | 237,714 | 58.5 | 195,167 | 46.4 |
| Suisun City | 78.0 | 56.4 | NA | NA | NA | NA | 13,963 | -3.2 | 1,361 | 10 | 22,146 | 70.5 | 18,658 | 56.2 |
| Sunnyvale | 71.8 | 30.9 | 97.0 | 94.8 | 81.5 | 6.9 | 84,100 | -4.4 | 4,287 | 5 | 122,833 | 73.2 | 106,975 | 62.3 |
| Temecula | 77.6 | 52.4 | 97.6 | 95.5 | 82.0 | 8.5 | 55,114 | -0.6 | 4,856 | 9 | 86,219 | 64.8 | 73,512 | 50.8 |
| Temple City | 79.5 | 50.3 | 95.6 | 89.8 | 94.3 | 0.7 | 17,763 | -3.3 | 2,270 | 13 | 29,871 | 65.2 | 24,467 | 53.7 |
| Thousand Oaks | 80.6 | 30.5 | 94.5 | 90.0 | 90.5 | 4.6 | 62,369 | -3.7 | 4,823 | 8 | 100,189 | 64.1 | 75,002 | 51.2 |
| Torrance | 80.7 | 41.6 | 96.8 | 94.4 | 86.7 | 2.8 | 73,717 | -6.1 | 7,622 | 10 | 116,689 | 66.4 | 91,199 | 59.1 |
| Tracy | 79.2 | 61.0 | 98.3 | 95.6 | 88.2 | 3.1 | 45,412 | 3.9 | 4,391 | 10 | 73,883 | 66.1 | 65,440 | 48.1 |
| Tulare | 85.8 | 27.6 | 95.0 | 88.1 | NA | NA | 27,942 | 0.0 | 2,941 | 11 | 42,207 | 62.8 | 37,059 | 49.9 |
| Turlock | 88.3 | 30.1 | 95.3 | 83.2 | 87.9 | 6.9 | 33,343 | -1.5 | 3,013 | 9 | 56,414 | 60.1 | 46,632 | 48.4 |
| Tustin | 81.7 | 35.7 | 96.5 | 92.2 | 86.1 | 4.0 | 40,726 | -4.9 | 3,397 | 8 | 61,831 | 71.0 | 52,983 | 53.2 |
| Twentynine Palms | 68.1 | 10.6 | 98.4 | 92.2 | 72.6 | 20.2 | 6,790 | -3.2 | 628 | 9 | 18,887 | 35.7 | 17,682 | 61.5 |
| Union City | 72.6 | 68.5 | 99.1 | 92.4 | 88.2 | 6.3 | 35,332 | -4.6 | 3,087 | 9 | 59,619 | 65.3 | 46,279 | 59.9 |
| Upland | 75.5 | 45.0 | 95.6 | 88.6 | 86.2 | 5.4 | 39,463 | -0.4 | 3,418 | 7 | 63,877 | 67.3 | 52,448 | 51.1 |
| Vacaville | 82.0 | 41.3 | 96.2 | 93.8 | 87.5 | 8.1 | 44,373 | -3.7 | 3,619 | 8 | 79,805 | 57.0 | 65,797 | 51.5 |
| Vallejo | 71.6 | 60.4 | 93.6 | 93.6 | 87.1 | 6.4 | 55,632 | -2.1 | 6,267 | 11 | 99,543 | 64.3 | 79,911 | 48.5 |
| Victorville | 73.2 | 54.1 | 92.8 | 88.4 | 89.6 | 5.3 | 46,575 | 1.0 | 5,755 | 12 | 86,326 | 54.4 | 74,851 | 40.8 |
| Visalia | 84.6 | 25.2 | 92.2 | 84.5 | 89.7 | 1.9 | 61,189 | -0.7 | 6,059 | 10 | 102,713 | 60.1 | 86,501 | 46.5 |
| Vista | 91.0 | 41.1 | 97.6 | 92.3 | 86.9 | 2.7 | 44,037 | -2.4 | 4,328 | 10 | 80,362 | 68.6 | 69,773 | 55.8 |
| Walnut | 76.2 | 62.5 | NA | NA | 92.8 | 2.0 | 15,110 | -6.1 | 1,538 | 10 | 24,348 | 56.9 | 17,067 | 50.2 |
| Walnut Creek | 59.9 | 54.8 | 96.0 | 91.8 | 83.5 | 9.7 | 32,596 | -5.4 | 2,141 | 9 | 59,483 | 59.3 | 39,975 | 51.4 |
| Wasco | 76.1 | 28.3 | 75.0 | 68.2 | 76.6 | 14.8 | 8,813 | -1.8 | 1,361 | 15 | 22,257 | 44.1 | 20,481 | 27.3 |
| Watsonville | 65.9 | 48.7 | 92.1 | 84.5 | 94.6 | 1.9 | 24,726 | -5.6 | 3,655 | 15 | 40,814 | 68.8 | 33,966 | 50.4 |
| West Covina | 80.8 | 61.6 | 94.7 | 87.2 | 90.7 | 1.2 | 50,858 | -4.0 | 6,647 | 13 | 86,424 | 65.7 | 69,446 | 56.8 |
| West Hollywood | 67.4 | 59.8 | NA | NA | 80.2 | 6.5 | 26,294 | -3.8 | 3,586 | 14 | 34,748 | 80.3 | 29,644 | 62.6 |
| Westminster | 75.7 | 49.7 | 95.0 | 89.2 | NA | NA | 41,399 | -1.4 | 4,885 | 12 | 73,611 | 61.6 | 59,251 | 48.5 |
| West Sacramento | 74.0 | 28.5 | 94.4 | 89.6 | 91.4 | 3.8 | 25,082 | -2.1 | 2,189 | 9 | 40,377 | 66.6 | 33,256 | 49.4 |
| Whittier | 82.4 | 52.6 | 88.7 | 82.8 | 88.6 | 1.6 | 41,746 | -5.2 | 5,025 | 12 | 66,383 | 61.1 | 54,042 | 53.5 |
| Wildomar | 80.2 | 48.8 | 97.0 | 91.7 | 89.0 | 2.9 | 17,579 | -0.8 | 1,626 | 9 | 27,813 | 61.8 | 23,868 | 49.7 |
| Windsor | 79.9 | 22.0 | 98.1 | 92.1 | NA | NA | 13,304 | -7.2 | 1,006 | 8 | 21,222 | 70.4 | 16,852 | 53.5 |
| Woodland | 83.1 | 35.9 | 92.4 | 90.5 | 88.0 | 3.2 | 29,701 | -3.1 | 2,623 | 9 | 47,588 | 62.4 | 38,030 | 53.1 |
| Yorba Linda | 82.9 | 52.3 | NA | NA | 89.9 | 4.8 | 33,091 | -5.7 | 2,367 | 7 | 55,244 | 66.0 | 42,400 | 51.1 |
| Yuba City | 75.5 | 42.0 | 87.1 | 81.5 | 77.4 | 9.8 | 31,567 | -0.6 | 3,574 | 11 | 53,448 | 59.4 | 42,249 | 48. |
| Yucaipa | 86.7 | 46.8 | 95.3 | 89.1 | 86.4 | 7.1 | 24,970 | -1.2 | 1,970 | 8 | 42,422 | 60.2 | 33,329 | 49. |
| COLORADO | 74.2 | 38.6 | 95.7 | 90.6 | 82.4 | 9.7 | 3,122,237 | -0.8 | 226,764 | 7 | 4,645,780 | 67.8 | 3,800,402 | 55. |
| Arvada | 77.8 | 43.9 | 96.2 | 91.4 | 87.1 | 5.5 | 68,750 | -1.3 | 4,886 | 7 | 98,900 | 70.9 | 78,464 | 59. |
| Aurora | 75.9 | 50.8 | 96.0 | 92.6 | 80.6 | 10.9 | 203,038 | 1.1 | 18,260 | 9 | 294,342 | 72.0 | 250,434 | 60. |
| Boulder | 49.6 | 21.6 | 97.5 | 91.3 | 65.7 | 19.0 | 63,536 | -3.8 | 3,792 | 6 | 94,698 | 66.3 | 81,706 | 40 |
| Brighton | 75.7 | 48.0 | 93.7 | 87.3 | 86.9 | 6.1 | 20,639 | -1.1 | 1,590 | 8 | 30,683 | 70.1 | 26,198 | 61 |
| Broomfield | 73.5 | 52.0 | 98.7 | 94.1 | 77.3 | 14.1 | 40,226 | -0.7 | 2,670 | 7 | 57,311 | 73.7 | 47,636 | 61 |

1. Employed persons.    2. Households.    3. Percent of civilian labor force.    4. Persons 16 years old and over.

## Table D. Cities — Construction, Wholesale Trade, and Retail Trade

| City | Value of residential construction authorized by building permits, 2020 | | | Wholesale trade[1], 2017 | | | | Retail trade[2], 2017 | | | |
|---|---|---|---|---|---|---|---|---|---|---|---|
| | New construction ($1,000) | Number of housing units | Percent single family | Number of establishments | Number of employees | Sales (mil dol) | Annual payroll (mil dol) | Number of establishments | Number of employees | Sales (mil dol) | Annual payroll (mil dol) |
| | 69 | 70 | 71 | 72 | 73 | 74 | 75 | 76 | 77 | 78 | 79 |
| **CALIFORNIA— Cont'd** | | | | | | | | | | | |
| San Francisco | 786,074 | 2,004 | 1 | 963 | 14,014 | 24,707 | 1,246 | 3,396 | 49,883 | 19,362 | 2,119 |
| San Gabriel | 36,934 | 199 | 23 | 126 | 513 | 305 | 18 | 218 | 1,627 | 481 | 41 |
| San Jacinto | 22,129 | 126 | 100 | 15 | 71 | 32 | 3 | 68 | 1,092 | 319 | 31 |
| San Jose | 196,373 | 1,369 | 23 | 1,055 | 35,492 | 71,189 | 5,866 | 2,204 | 43,442 | 15,756 | 1,897 |
| San Juan Capistrano | 44,932 | 155 | 52 | 62 | 323 | 300 | 21 | 126 | 1,975 | 789 | 78 |
| San Leandro | 2,975 | 20 | 90 | 234 | 3,337 | 2,576 | 207 | 301 | 7,348 | 5,522 | 239 |
| San Luis Obispo | 79,544 | 440 | 66 | 74 | 933 | 531 | 55 | 342 | 5,770 | 1,883 | 187 |
| San Marcos | 47,366 | 317 | 74 | 134 | 1,139 | 566 | 62 | 268 | 4,415 | 1,414 | 144 |
| San Mateo | 8,910 | 41 | 74 | 83 | 2,791 | 4,211 | 403 | 349 | 6,371 | 2,035 | 211 |
| San Pablo | 1,508 | 9 | 34 | 10 | 61 | 28 | 3 | 78 | 1,024 | 248 | 28 |
| San Rafael | 4,645 | 30 | 100 | 112 | 936 | 896 | 69 | 318 | 5,235 | 1,765 | 201 |
| San Ramon | 25,175 | 70 | 43 | 97 | 1,608 | 2,704 | 129 | 109 | 1,975 | 667 | 68 |
| Santa Ana | 173,290 | 899 | 8 | 540 | 6,092 | 7,321 | 353 | 841 | 12,828 | 3,828 | 386 |
| Santa Barbara | 19,204 | 118 | 12 | 112 | 2,017 | 1,716 | 243 | 544 | 6,220 | 1,658 | 181 |
| Santa Clara | 126,424 | 612 | 21 | 344 | 9,949 | 9,994 | 1,558 | 335 | 4,802 | 2,060 | 200 |
| Santa Clarita | 296,367 | 1,471 | 76 | 219 | 3,121 | 9,702 | 186 | 555 | 10,464 | 3,539 | 327 |
| Santa Cruz | 13,404 | 54 | 74 | 62 | 851 | 758 | 86 | 268 | 4,052 | 985 | 113 |
| Santa Maria | 47,006 | 461 | 26 | 105 | 1,347 | 803 | 73 | 357 | 6,048 | 1,762 | 174 |
| Santa Monica | 63,886 | 153 | 45 | 156 | 2,220 | 1,318 | 148 | 672 | 10,704 | 4,607 | 477 |
| Santa Paula | 61,603 | 160 | 100 | 15 | 283 | 161 | 9 | 51 | 693 | 202 | 19 |
| Santa Rosa | 214,023 | 628 | 70 | 156 | 1,909 | 1,641 | 129 | 679 | 11,842 | 3,622 | 387 |
| Santee | 21,000 | 180 | 44 | 41 | 368 | 490 | 21 | 146 | 3,306 | 893 | 89 |
| Saratoga | 48,839 | 67 | 100 | 14 | 33 | 22 | 3 | D | D | D | 8 |
| Seaside | 0 | 0 | 0 | 12 | 35 | 13 | 1 | 91 | 1,585 | 711 | 62 |
| Simi Valley | 44,218 | 155 | 100 | 149 | 1,332 | 749 | 78 | 410 | 6,234 | 2,026 | 193 |
| Soledad | 6,604 | 29 | 86 | 7 | 53 | 29 | 4 | 29 | 255 | 108 | 8 |
| South Gate | 14,795 | 145 | 57 | 54 | 795 | 748 | 40 | 161 | 3,177 | 893 | 82 |
| South Pasadena | 2,655 | 13 | 100 | 35 | 203 | 196 | 13 | 56 | 831 | 244 | 25 |
| South San Francisco | 500 | 3 | 100 | 289 | 4,212 | 4,457 | 430 | 213 | 3,774 | 1,200 | 135 |
| Stanton | 27,431 | 146 | 99 | 31 | 298 | 87 | 14 | 96 | 1,347 | 514 | 41 |
| Stockton | 176,942 | 736 | 70 | 228 | 5,449 | 5,720 | 299 | 702 | 11,974 | 3,947 | 340 |
| Suisun City | 4,266 | 20 | 45 | D | D | D | D | 32 | 604 | 154 | 16 |
| Sunnyvale | 71,928 | 430 | 39 | 163 | 8,817 | 7,504 | 1,699 | 256 | 4,782 | 1,871 | 174 |
| Temecula | 50,247 | 324 | 81 | 163 | 2,442 | 2,600 | 139 | 460 | 9,426 | 3,164 | 285 |
| Temple City | 9,641 | 51 | 100 | 74 | 247 | 138 | 9 | 103 | 1,188 | 308 | 29 |
| Thousand Oaks | 30,928 | 196 | 26 | 191 | 3,365 | 11,379 | 442 | 531 | 9,044 | 3,434 | 319 |
| Torrance | 8,825 | 69 | 39 | 620 | 5,573 | 7,454 | 380 | 709 | 13,168 | 4,686 | 413 |
| Tracy | 189,625 | 869 | 75 | 71 | 2,418 | 3,129 | 200 | 230 | 4,578 | 1,454 | 138 |
| Tulare | 57,740 | 321 | 94 | 54 | 726 | 1,566 | 44 | 187 | 3,241 | 897 | 82 |
| Turlock | 28,055 | 140 | 74 | 64 | 891 | 472 | 48 | 239 | 4,311 | 1,110 | 114 |
| Tustin | 12,835 | 55 | 64 | 185 | 2,558 | 1,571 | 197 | 297 | 6,742 | 2,749 | 224 |
| Twentynine Palms | 304 | 5 | 100 | 4 | 18 | 9 | 1 | 32 | 286 | 86 | 8 |
| Union City | 529 | 3 | 100 | 143 | 4,672 | 4,431 | 367 | 112 | 2,177 | 792 | 73 |
| Upland | 14,183 | 60 | 75 | 119 | 754 | 283 | 34 | 253 | 4,112 | 1,142 | 113 |
| Vacaville | 146,165 | 661 | 63 | 40 | 333 | 319 | 18 | 328 | 6,309 | 1,799 | 175 |
| Vallejo | 2,856 | 27 | 100 | 30 | 592 | 395 | 47 | 237 | 4,529 | 1,381 | 147 |
| Victorville | 153,322 | 504 | 100 | 44 | 281 | 165 | 11 | 314 | 6,756 | 2,034 | 193 |
| Visalia | 157,284 | 607 | 99 | 138 | 1,934 | 3,799 | 112 | 423 | 7,865 | 2,199 | 204 |
| Vista | 85,068 | 500 | 52 | 193 | 3,641 | 2,108 | 230 | 258 | 4,526 | 1,469 | 141 |
| Walnut | 4,992 | 8 | 100 | 240 | 884 | 562 | 35 | 125 | 1,172 | 355 | 28 |
| Walnut Creek | 51,434 | 210 | 10 | 69 | 700 | 2,061 | 71 | 301 | 6,960 | 2,504 | 282 |
| Wasco | 4,352 | 34 | 100 | 4 | 20 | 7 | 1 | 36 | 450 | 112 | 11 |
| Watsonville | 10,729 | 117 | 51 | 63 | 1,053 | 1,208 | 64 | 150 | 2,378 | 2,046 | 69 |
| West Covina | 8,565 | 32 | 56 | 91 | 234 | 119 | 8 | 284 | 5,745 | 1,801 | 155 |
| West Hollywood | 24,645 | 108 | 7 | 100 | 620 | 504 | 37 | 306 | 3,710 | 1,424 | 136 |
| Westminster | 40,263 | 271 | 21 | 113 | 535 | 452 | 25 | 397 | 5,144 | 1,708 | 150 |
| West Sacramento | 84,979 | 589 | 16 | 144 | 4,360 | 7,098 | 248 | 157 | 2,413 | 741 | 72 |
| Whittier | 71,445 | 327 | 100 | 63 | 399 | 170 | 18 | 215 | 3,587 | 897 | 94 |
| Wildomar | 11,274 | 50 | 100 | 7 | 51 | 30 | 2 | 34 | 424 | 151 | 13 |
| Windsor | 592 | 5 | 100 | 23 | 525 | 481 | 35 | 61 | 1,179 | 342 | 38 |
| Woodland | 111,181 | 339 | 100 | 71 | 1,253 | 1,042 | 68 | 152 | 2,362 | 805 | 71 |
| Yorba Linda | 34,496 | 63 | 100 | 131 | 1,361 | 904 | 105 | 133 | 1,732 | 643 | 58 |
| Yuba City | 18,755 | 54 | 100 | 38 | 378 | 236 | 22 | 243 | 4,174 | 1,180 | 114 |
| Yucaipa | 10,118 | 27 | 100 | 20 | 106 | 27 | 3 | 99 | 1,081 | 298 | 29 |
| **COLORADO** | 45,289 | 186 | 83 | 5,953 | 80,365 | 83,316 | 5,376 | 19,056 | 279,982 | 84,931 | 8,420 |
| Arvada | 81,937 | 295 | 100 | 93 | 1,033 | 549 | 57 | 258 | 3,970 | 1,351 | 121 |
| Aurora | 591,713 | 2,725 | 76 | 229 | 6,441 | 10,716 | 389 | 973 | 17,442 | 4,945 | 472 |
| Boulder | 88,563 | 316 | 13 | 210 | 3,228 | 1,785 | 282 | 593 | 9,518 | 2,647 | 294 |
| Brighton | 153,718 | 827 | 78 | 25 | 712 | 513 | 45 | 106 | 2,210 | 842 | 73 |
| Broomfield | 128,723 | 427 | 52 | 70 | 845 | 430 | 56 | 286 | 5,338 | 1,281 | 140 |

1. Merchant wholesalers except manufacturers' sales branches and offices.  2. Establishments with payroll.

# Table D. Cities — Real Estate, Professional Services, and Manufacturing

| City | Real estate and rental and leasing, 2017 | | | | Professional, scientific, and technical services[1], 2017 | | | | Manufacturing, 2017 | | | |
|---|---|---|---|---|---|---|---|---|---|---|---|---|
| | Number of establishments | Number of employees | Receipts (mil dol) | Annual payroll (mil dol) | Number of establishments | Number of employees | Receipts (mil dol) | Annual payroll (mil dol) | Number of establishments | Number of employees | Receipts (mil dol) | Annual payroll (mil dol) |
| | 80 | 81 | 82 | 83 | 84 | 85 | 86 | 87 | 88 | 89 | 90 | 91 |
| CALIFORNIA— Cont'd | | | | | | | | | | | | |
| San Francisco.................. | 2,132 | 19,351 | 9,399 | 1,564 | 7,193 | 117,838 | 41,005 | 14,436 | NA | NA | NA | NA |
| San Gabriel..................... | 92 | 194 | 41 | 6 | 160 | 905 | 95 | 28 | NA | NA | NA | NA |
| San Jacinto.................... | D | D | D | D | 23 | 113 | 14 | 5 | NA | NA | NA | NA |
| San Jose....................... | 1,139 | 6,062 | 2,777 | 410 | 3,276 | 45,002 | 10,651 | 5,087 | NA | NA | NA | NA |
| San Juan Capistrano ........... | 104 | 323 | 79 | 23 | 234 | 1,209 | 216 | 93 | NA | NA | NA | NA |
| San Leandro.................... | 137 | 834 | 188 | 41 | 131 | 686 | 157 | 53 | NA | NA | NA | NA |
| San Luis Obispo................. | 142 | 876 | 174 | 40 | 358 | 2,507 | 464 | 173 | NA | NA | NA | NA |
| San Marcos..................... | 148 | 494 | 130 | 25 | 263 | 1,510 | 247 | 87 | NA | NA | NA | NA |
| San Mateo...................... | 199 | 1,107 | 514 | 79 | 577 | 6,189 | 1,419 | 591 | NA | NA | NA | NA |
| San Pablo...................... | 17 | 51 | 15 | 2 | 13 | 110 | 12 | 4 | NA | NA | NA | NA |
| San Rafael..................... | 170 | 948 | 966 | 55 | 518 | 2,619 | 565 | 212 | NA | NA | NA | NA |
| San Ramon...................... | 128 | 878 | 452 | 66 | 566 | 5,263 | 1,313 | 510 | NA | NA | NA | NA |
| Santa Ana...................... | 315 | 2,503 | 854 | 183 | 1,067 | 10,446 | 1,926 | 726 | NA | NA | NA | NA |
| Santa Barbara.................. | 353 | 1,362 | 340 | 66 | 670 | 4,249 | 1,065 | 340 | NA | NA | NA | NA |
| Santa Clara.................... | 204 | 2,162 | 670 | 144 | 1,000 | 21,327 | 7,151 | 2,851 | NA | NA | NA | NA |
| Santa Clarita.................. | 285 | 1,204 | 315 | 62 | 630 | 3,375 | 594 | 206 | NA | NA | NA | NA |
| Santa Cruz..................... | 83 | 267 | 100 | 12 | 308 | 1,607 | 256 | 148 | NA | NA | NA | NA |
| Santa Maria.................... | 88 | 403 | 96 | 16 | 154 | 983 | 137 | 53 | NA | NA | NA | NA |
| Santa Monica................... | 523 | 3,086 | 969 | 231 | 1,259 | 10,777 | 2,915 | 1,105 | NA | NA | NA | NA |
| Santa Paula.................... | 23 | 135 | 42 | 10 | D | D | D | D | NA | NA | NA | NA |
| Santa Rosa..................... | 246 | 1,106 | 307 | 54 | 617 | 3,918 | 604 | 259 | NA | NA | NA | NA |
| Santee......................... | 56 | 189 | 53 | 8 | 93 | 411 | 52 | 18 | NA | NA | NA | NA |
| Saratoga....................... | D | D | D | D | 178 | 554 | 116 | 42 | NA | NA | NA | NA |
| Seaside........................ | 14 | 99 | 16 | 4 | D | D | D | D | NA | NA | NA | NA |
| Simi Valley.................... | 142 | 508 | 161 | 24 | 377 | 2,085 | 358 | 134 | NA | NA | NA | NA |
| Soledad........................ | 13 | 46 | 10 | 1 | NA | NA | NA | NA | NA | NA | NA | NA |
| South Gate..................... | 54 | 160 | 53 | 8 | 35 | 664 | 63 | 19 | NA | NA | NA | NA |
| South Pasadena................. | D | D | D | D | D | D | D | D | NA | NA | NA | NA |
| South San Francisco............ | 97 | 917 | 204 | 42 | 239 | 9,444 | 4,596 | 2,331 | NA | NA | NA | NA |
| Stanton........................ | 25 | 93 | 39 | 4 | 44 | 274 | 35 | 12 | NA | NA | NA | NA |
| Stockton....................... | 215 | 1,352 | 283 | 53 | 359 | 2,836 | 347 | 154 | NA | NA | NA | NA |
| Suisun City.................... | 9 | 30 | 7 | 2 | 19 | 74 | 7 | 3 | NA | NA | NA | NA |
| Sunnyvale...................... | 198 | 1,109 | 429 | 73 | 835 | 26,102 | 8,382 | 3,284 | NA | NA | NA | NA |
| Temecula....................... | 224 | 710 | 204 | 30 | 430 | 2,149 | 361 | 132 | NA | NA | NA | NA |
| Temple City.................... | 68 | 153 | 45 | 8 | 77 | 258 | 28 | 9 | NA | NA | NA | NA |
| Thousand Oaks.................. | 311 | 1,224 | 353 | 69 | D | D | D | D | NA | NA | NA | NA |
| Torrance....................... | 401 | 1,907 | 466 | 83 | D | D | D | D | NA | NA | NA | NA |
| Tracy.......................... | 93 | 216 | 80 | 10 | 140 | 580 | 83 | 31 | NA | NA | NA | NA |
| Tulare......................... | 52 | 219 | 34 | 6 | 41 | 285 | 43 | 19 | NA | NA | NA | NA |
| Turlock........................ | 64 | 230 | 55 | 8 | 101 | 749 | 82 | 32 | NA | NA | NA | NA |
| Tustin......................... | 196 | 1,123 | 258 | 61 | 626 | 3,688 | 658 | 248 | NA | NA | NA | NA |
| Twentynine Palms............... | 15 | 166 | 35 | 3 | 12 | 90 | 10 | 5 | NA | NA | NA | NA |
| Union City .................... | 48 | 175 | 64 | 10 | 117 | 523 | 95 | 39 | NA | NA | NA | NA |
| Upland......................... | 120 | 1,096 | 371 | 71 | 230 | 1,530 | 218 | 77 | NA | NA | NA | NA |
| Vacaville...................... | 107 | 485 | 139 | 19 | 133 | 1,077 | 183 | 73 | NA | NA | NA | NA |
| Vallejo........................ | 77 | 272 | 70 | 11 | 105 | 526 | 67 | 26 | NA | NA | NA | NA |
| Victorville.................... | 101 | 451 | 111 | 13 | 90 | 699 | 66 | 21 | NA | NA | NA | NA |
| Visalia........................ | 154 | 923 | 172 | 34 | 256 | 1,457 | 226 | 78 | NA | NA | NA | NA |
| Vista.......................... | 155 | 724 | 251 | 33 | 267 | 1,561 | 266 | 100 | NA | NA | NA | NA |
| Walnut......................... | 75 | 164 | 54 | 8 | 153 | 444 | 75 | 23 | NA | NA | NA | NA |
| Walnut Creek .................. | 238 | 1,482 | 424 | 107 | 724 | 6,893 | 1,425 | 620 | NA | NA | NA | NA |
| Wasco.......................... | 15 | 46 | 7 | 1 | 5 | 21 | 4 | 1 | NA | NA | NA | NA |
| Watsonville.................... | 57 | 255 | 51 | 9 | 78 | 470 | 63 | 30 | NA | NA | NA | NA |
| West Covina.................... | 107 | 453 | 81 | 18 | 155 | 710 | 110 | 36 | NA | NA | NA | NA |
| West Hollywood................. | 185 | 1,104 | 373 | 60 | 446 | 3,140 | 845 | 296 | NA | NA | NA | NA |
| Westminster.................... | 124 | 406 | 107 | 15 | 162 | 755 | 186 | 37 | NA | NA | NA | NA |
| West Sacramento ............... | 94 | 765 | 161 | 35 | 93 | 1,262 | 361 | 109 | NA | NA | NA | NA |
| Whittier....................... | D | D | D | D | 176 | 1,057 | 111 | 40 | NA | NA | NA | NA |
| Wildomar....................... | 20 | 154 | 27 | 5 | 40 | 94 | 12 | 3 | NA | NA | NA | NA |
| Windsor........................ | 25 | 494 | 53 | 15 | 55 | 244 | 47 | 12 | NA | NA | NA | NA |
| Woodland....................... | 53 | 241 | 75 | 11 | 85 | 446 | 64 | 23 | NA | NA | NA | NA |
| Yorba Linda.................... | 135 | 387 | 125 | 18 | 278 | 906 | 178 | 55 | NA | NA | NA | NA |
| Yuba City ..................... | 66 | 306 | 61 | 10 | 97 | 456 | 53 | 19 | NA | NA | NA | NA |
| Yucaipa........................ | 53 | 164 | 32 | 5 | 68 | 198 | 23 | 8 | NA | NA | NA | NA |
| COLORADO ...................... | 12,087 | 47,642 | 12,757 | 2,388 | 26,404 | 192,964 | 39,655 | 15,801 | 5,111 | 121,372 | 50,809 | 7,357 |
| Arvada......................... | 158 | 347 | 71 | 15 | 413 | 1,662 | 256 | 95 | NA | NA | NA | NA |
| Aurora......................... | 403 | 1,857 | 430 | 85 | 733 | 7,726 | 1,384 | 672 | NA | NA | NA | NA |
| Boulder........................ | 393 | 1,608 | 383 | 73 | 1,524 | 16,400 | 3,245 | 1,490 | NA | NA | NA | NA |
| Brighton....................... | 50 | 143 | 24 | 6 | 64 | 415 | 65 | 23 | NA | NA | NA | NA |
| Broomfield..................... | 150 | 425 | 170 | 23 | 404 | 4,147 | 945 | 366 | NA | NA | NA | NA |

1. Establishments subject to federal tax.

# Table D. Cities — Accommodation and Food Services, Arts, Entertainment, and Recreation, and Health Care and Social Assistance

| City | Accommodation and food services, 2017 | | | | Arts, entertainment, and recreation[1], 2017 | | | | Health care and social assistance[1], 2017 | | | |
|---|---|---|---|---|---|---|---|---|---|---|---|---|
| | Number of establish-ments | Number of employees | Receipts (mil dol) | Annual payroll (mil dol) | Number of establish-ments | Number of employees | Receipts (mil dol) | Annual payroll (mil dol) | Number of establish-ments | Number of employees | Receipts (mil dol) | Annual payroll (mil dol) |
| | 92 | 93 | 94 | 95 | 96 | 97 | 98 | 99 | 100 | 101 | 102 | 103 |
| **CALIFORNIA— Cont'd** | | | | | | | | | | | | |
| San Francisco | 4,559 | 89,123 | 9,426 | 2,717 | 616 | 15,360 | 2,541 | 852 | 3,330 | 69,425 | 13,052 | 5,236 |
| San Gabriel | 184 | 2,381 | 147 | 43 | 10 | 127 | 12 | 5 | 231 | 3,617 | 397 | 163 |
| San Jacinto | 48 | 629 | 41 | 10 | NA | NA | NA | NA | 36 | 283 | 31 | 12 |
| San Jose | 2,215 | 39,154 | 2,851 | 825 | 251 | 8,417 | 751 | 313 | 2,617 | 38,211 | 5,678 | 2,323 |
| San Juan Capistrano | 76 | 1,463 | 99 | 29 | 20 | D | D | D | 151 | 1,867 | 214 | 76 |
| San Leandro | 233 | 3,441 | 258 | 70 | 25 | 311 | 15 | 5 | 279 | 8,061 | 1,510 | 600 |
| San Luis Obispo | 237 | 5,150 | 310 | 93 | 23 | 237 | 18 | 4 | 389 | 6,129 | 872 | 333 |
| San Marcos | 211 | 3,820 | 226 | 67 | 40 | 818 | 68 | 21 | 201 | 3,183 | 328 | 150 |
| San Mateo | 371 | 5,655 | 483 | 141 | NA | NA | NA | NA | 553 | 7,558 | 1,191 | 442 |
| San Pablo | 71 | 878 | 64 | 16 | 80 | 602 | 100 | 25 | 57 | 1,357 | 130 | 55 |
| San Rafael | 228 | 3,418 | 253 | 78 | 33 | 502 | 34 | 11 | 365 | 6,843 | 777 | 428 |
| San Ramon | 158 | 2,702 | 208 | 55 | | | | | 360 | 4,323 | 722 | 271 |
| Santa Ana | 602 | 9,481 | 657 | 177 | 58 | 1,140 | 141 | 32 | 876 | 14,624 | 1,920 | 660 |
| Santa Barbara | 487 | 9,711 | 727 | 225 | 117 | 1,434 | 134 | 38 | 590 | 9,902 | 1,780 | 563 |
| Santa Clara | 455 | 7,951 | 773 | 217 | 46 | 2,126 | 605 | 243 | 304 | 9,674 | 1,906 | 917 |
| Santa Clarita | 384 | 7,810 | 505 | 150 | 177 | 1,201 | 109 | 35 | 619 | 8,461 | 1,123 | 420 |
| Santa Cruz | 279 | 5,496 | 385 | 114 | 41 | 2,033 | 95 | 38 | 263 | 3,569 | 409 | 160 |
| Santa Maria | 197 | 3,198 | 224 | 61 | 23 | 340 | 19 | 6 | 335 | 6,512 | 1,087 | 349 |
| Santa Monica | 503 | 13,766 | 1,261 | 386 | 772 | 3,131 | 792 | 329 | 949 | 10,936 | 1,966 | 691 |
| Santa Paula | 43 | 723 | 40 | 12 | D | D | D | D | 43 | 390 | 59 | 19 |
| Santa Rosa | 422 | 6,895 | 455 | 127 | 81 | 1,419 | 126 | 38 | 766 | 14,989 | 2,266 | 1,022 |
| Santee | 133 | 2,420 | 148 | 44 | D | D | D | D | 120 | 956 | 144 | 35 |
| Saratoga | 57 | 628 | 59 | 16 | 16 | 258 | 35 | 8 | 116 | 1,411 | 151 | 67 |
| Seaside | 80 | 1,451 | 97 | 27 | NA | NA | NA | NA | 21 | 290 | 22 | 8 |
| Simi Valley | 247 | 4,200 | 275 | 75 | 73 | 634 | 42 | 15 | 375 | 4,312 | 526 | 202 |
| Soledad | 22 | 302 | 25 | 6 | NA | NA | NA | NA | 14 | 159 | 18 | 6 |
| South Gate | 127 | 2,090 | 138 | 38 | 9 | 204 | 21 | 5 | 83 | 895 | 96 | 34 |
| South Pasadena | 66 | 1,057 | 62 | 19 | 68 | 296 | 50 | 20 | 128 | 731 | 80 | 30 |
| South San Francisco | 262 | 5,006 | 539 | 138 | 22 | 295 | 28 | 9 | 203 | 3,351 | 458 | 277 |
| Stanton | 84 | 1,072 | 74 | 18 | D | D | D | D | 41 | 801 | 44 | 20 |
| Stockton | 492 | 7,809 | 504 | 136 | 50 | 989 | 63 | 23 | 761 | 19,134 | 2,218 | 897 |
| Suisun City | 39 | 609 | 39 | 11 | D | D | D | D | 23 | 144 | 11 | 4 |
| Sunnyvale | 431 | 6,502 | 600 | 171 | 48 | 958 | 69 | 20 | 475 | 5,651 | 757 | 286 |
| Temecula | 332 | 7,648 | 456 | 141 | 51 | 741 | 53 | 15 | 422 | 4,413 | 531 | 198 |
| Temple City | 74 | 1,338 | 79 | 23 | 4 | 6 | 1 | 0 | 111 | 1,140 | 95 | 38 |
| Thousand Oaks | 335 | 7,485 | 478 | 145 | 123 | 1,455 | 113 | 37 | 723 | 8,242 | 1,322 | 437 |
| Torrance | 480 | 10,197 | 696 | 198 | 88 | 1,430 | 112 | 34 | 1,112 | 17,983 | 2,766 | 1,089 |
| Tracy | 185 | 3,065 | 196 | 54 | 21 | 418 | 21 | 8 | 197 | 2,677 | 414 | 132 |
| Tulare | 89 | 1,527 | 99 | 27 | D | D | D | D | 115 | 1,602 | 145 | 62 |
| Turlock | 173 | 3,395 | 205 | 58 | 17 | 325 | 19 | 5 | 213 | 4,512 | 579 | 211 |
| Tustin | 293 | 5,630 | 363 | 103 | 40 | 541 | 35 | 10 | 472 | 4,110 | 565 | 207 |
| Twentynine Palms | 48 | 773 | 61 | 14 | 4 | 46 | 2 | 1 | 15 | 236 | 36 | 15 |
| Union City | 128 | 2,158 | 148 | 42 | D | D | D | D | 165 | 2,453 | 372 | 153 |
| Upland | 185 | 3,288 | 188 | 55 | 32 | 484 | 33 | 9 | 387 | 6,621 | 888 | 295 |
| Vacaville | 198 | 3,994 | 251 | 72 | 24 | 436 | 36 | 6 | 215 | 5,747 | 1,105 | 466 |
| Vallejo | 190 | 2,998 | 204 | 55 | 28 | 1,760 | 106 | 31 | 291 | 8,968 | 1,283 | 698 |
| Victorville | 190 | 3,915 | 217 | 63 | 15 | 343 | 17 | 5 | 257 | 6,432 | 764 | 248 |
| Visalia | 271 | 4,990 | 317 | 88 | 29 | 589 | 33 | 9 | 467 | 10,168 | 1,327 | 480 |
| Vista | 200 | 2,926 | 191 | 53 | 28 | 360 | 33 | 10 | 257 | 4,034 | 387 | 166 |
| Walnut | 62 | 747 | 53 | 14 | 17 | 41 | 8 | 3 | 98 | 632 | 60 | 25 |
| Walnut Creek | 235 | 4,912 | 372 | 112 | 60 | 655 | 254 | 20 | 533 | 13,584 | 3,012 | 1,265 |
| Wasco | D | D | D | D | NA | NA | NA | NA | 18 | 427 | 45 | 16 |
| Watsonville | 105 | 1,508 | 105 | 28 | 10 | D | 8 | D | 190 | 2,927 | 395 | 146 |
| West Covina | 231 | 3,955 | 248 | 67 | 18 | 437 | 33 | 9 | 345 | 7,319 | 1,041 | 354 |
| West Hollywood | 258 | 10,190 | 968 | 288 | D | D | D | D | 302 | 14,861 | 3,599 | 1,374 |
| Westminster | 277 | 3,077 | 202 | 48 | D | D | D | D | 301 | 2,899 | 384 | 120 |
| West Sacramento | D | D | D | D | 11 | 737 | 27 | 7 | 92 | 1,445 | 155 | 62 |
| Whittier | 215 | 3,604 | 225 | 67 | 13 | 486 | 23 | 8 | 319 | 10,435 | 1,332 | 472 |
| Wildomar | 47 | 581 | 37 | 11 | 13 | 161 | 11 | 3 | 66 | 1,695 | 203 | 85 |
| Windsor | 48 | 994 | 59 | 17 | D | D | D | D | 45 | 603 | 53 | 23 |
| Woodland | D | D | D | D | 20 | 288 | 12 | 4 | 130 | 2,586 | 360 | 136 |
| Yorba Linda | 113 | 1,908 | 119 | 33 | D | D | D | D | 211 | 1,644 | 157 | 67 |
| Yuba City | 129 | 2,191 | 146 | 38 | 17 | 283 | 14 | 4 | 216 | 3,527 | 463 | 175 |
| Yucaipa | 75 | 1,374 | 71 | 22 | 18 | 145 | 8 | 2 | 98 | 1,310 | 115 | 41 |
| **COLORADO** | 14,121 | 290,915 | 19,456 | 5,804 | 3,028 | 58,229 | 5,681 | 1,935 | 16,659 | 320,813 | 40,056 | 15,720 |
| Arvada | 208 | 4,129 | 224 | 69 | 44 | 672 | 37 | 14 | 281 | 4,503 | 371 | 168 |
| Aurora | 680 | 13,457 | 813 | 241 | 69 | D | 71 | D | 791 | 26,587 | 3,301 | 1,371 |
| Boulder | 433 | 10,606 | 681 | 210 | 148 | 1,525 | 108 | 33 | 671 | 8,626 | 1,072 | 445 |
| Brighton | 85 | 1,534 | 94 | 26 | 4 | D | 1 | D | 88 | 1,851 | 273 | 98 |
| Broomfield | 178 | 3,689 | 261 | 78 | 35 | 398 | 22 | 7 | 197 | 2,472 | 288 | 117 |

1. Establishments subject to federal tax.

# Table D. Cities — Other Services and Government Employment and Payroll

| City | Other services[1] — Number of establishments (104) | Number of employees (105) | Receipts (mil dol) (106) | Annual payroll (mil dol) (107) | Government — Full-time equivalent employees (108) | Total (dollars) (109) | Percent of total for: Administrative, judicial, and legal (110) | Police and corrections (111) | Fire protection (112) | Highways and transportation (113) | Health and welfare (114) | Natural resources and utilities (115) | Education and libraries (116) |
|---|---|---|---|---|---|---|---|---|---|---|---|---|---|
| CALIFORNIA— Cont'd | | | | | | | | | | | | | |
| San Francisco | 2,470 | 25,451 | 5,265 | 1,113 | 33,660 | 260,920,281 | 11.9 | 17.2 | 8.2 | 23.9 | 23.9 | 11.9 | 1.7 |
| San Gabriel | 141 | 477 | 50 | 14 | 191 | 1,632,525 | 19.5 | 37.5 | 23.1 | 10.7 | 0.0 | 4.2 | 0.0 |
| San Jacinto | 38 | 131 | 13 | 3 | 43 | 215,583 | 42.8 | 0.0 | 0.0 | 5.1 | 0.0 | 34.1 | 0.0 |
| San Jose | 1,569 | 10,237 | 1,297 | 374 | 3,403 | 22,735,265 | 8.9 | 30.6 | 16.9 | 9.7 | 9.5 | 14.9 | 5.1 |
| San Juan Capistrano | 73 | 336 | 42 | 11 | 94 | 671,052 | 28.9 | 0.0 | 0.0 | 18.8 | 17.0 | 35.3 | 0.0 |
| San Leandro | 201 | 1,346 | 176 | 52 | 421 | 3,019,956 | 12.5 | 39.1 | 0.0 | 9.3 | 6.0 | 17.4 | 5.4 |
| San Luis Obispo | 139 | 837 | 87 | 25 | 480 | 3,327,790 | 11.0 | 24.3 | 16.6 | 4.2 | 6.6 | 21.9 | 0.0 |
| San Marcos | 178 | 1,179 | 98 | 32 | 298 | 2,077,947 | 15.8 | 0.0 | 40.1 | 13.9 | 5.5 | 17.3 | 0.0 |
| San Mateo | 245 | 1,601 | 149 | 51 | 501 | 4,613,844 | 9.4 | 37.3 | 24.2 | 0.0 | 2.7 | 20.7 | 5.9 |
| San Pablo | D | D | D | D | 79 | 471,449 | 49.3 | 1.6 | 0.0 | 13.4 | 12.2 | 22.1 | 0.0 |
| San Rafael | 258 | 1,899 | 252 | 79 | 486 | 3,267,504 | 7.4 | 29.5 | 27.2 | 2.5 | 3.9 | 8.1 | 5.0 |
| San Ramon | 148 | 980 | 127 | 39 | 294 | 2,297,560 | 12.7 | 38.3 | 0.0 | 33.0 | 0.0 | 12.5 | 0.0 |
| Santa Ana | 469 | 3,298 | 405 | 107 | 1,094 | 9,302,161 | 14.8 | 60.9 | 0.0 | 8.8 | 7.8 | 5.2 | 2.4 |
| Santa Barbara | 306 | 2,304 | 312 | 76 | 1,161 | 7,949,171 | 14.5 | 23.5 | 15.4 | 13.5 | 1.6 | 20.9 | 2.9 |
| Santa Clara | 292 | 2,126 | 263 | 79 | 1,095 | 11,426,450 | 10.1 | 24.7 | 20.0 | 6.2 | 0.5 | 28.7 | 3.6 |
| Santa Clarita | 333 | 2,013 | 227 | 61 | 509 | 2,841,241 | 22.0 | 0.0 | 0.0 | 10.5 | 0.5 | 34.8 | 0.2 |
| Santa Cruz | 138 | 1,169 | 140 | 54 | 1,069 | 6,006,317 | 13.1 | 18.8 | 13.5 | 4.5 | 1.8 | 33.3 | 9.0 |
| Santa Maria | 165 | 1,181 | 139 | 38 | 614 | 4,153,499 | 8.6 | 38.2 | 16.7 | 5.8 | 1.8 | 24.2 | 2.8 |
| Santa Monica | 482 | 3,854 | 573 | 151 | 2,218 | 17,156,068 | 15.4 | 22.9 | 10.8 | 21.6 | 2.9 | 13.1 | 3.4 |
| Santa Paula | 45 | 261 | 16 | 5 | 114 | 769,663 | 12.5 | 41.5 | 18.2 | 4.2 | 4.4 | 14.0 | 0.0 |
| Santa Rosa | 348 | 2,603 | 311 | 97 | 1,260 | 9,471,990 | 10.7 | 25.7 | 17.1 | 10.1 | 3.6 | 21.9 | 0.0 |
| Santee | 109 | 602 | 70 | 19 | 155 | 1,209,222 | 28.3 | 0.6 | 51.2 | 9.9 | 1.2 | 8.9 | 0.0 |
| Saratoga | 44 | 108 | 19 | 4 | 63 | 533,502 | 27.2 | 0.0 | 0.0 | 25.1 | 17.6 | 20.4 | 0.0 |
| Seaside | 54 | 251 | 31 | 9 | 159 | 1,184,921 | 8.9 | 38.8 | 23.1 | 14.9 | 2.1 | 9.4 | 0.0 |
| Simi Valley | 190 | 1,043 | 145 | 27 | 564 | 3,481,556 | 15.4 | 32.5 | 0.0 | 21.3 | 13.6 | 17.2 | 0.0 |
| Soledad | 5 | 14 | 2 | 0 | 47 | 321,518 | 21.2 | 60.5 | 0.0 | 1.9 | 0.0 | 16.4 | 0.0 |
| South Gate | 74 | 269 | 37 | 7 | 335 | 2,078,258 | 10.7 | 47.2 | 0.0 | 15.4 | 7.1 | 19.8 | 0.0 |
| South Pasadena | 41 | 168 | 23 | 6 | 183 | 1,136,265 | 11.5 | 35.7 | 20.8 | 5.1 | 8.3 | 7.4 | 9.4 |
| South San Francisco | 129 | 1,495 | 204 | 65 | 544 | 4,433,064 | 10.8 | 26.5 | 24.1 | 2.6 | 0.4 | 24.4 | 5.5 |
| Stanton | D | D | D | D | 46 | 243,289 | 36.7 | 0.0 | 0.0 | 19.1 | 21.5 | 22.6 | 0.0 |
| Stockton | 364 | 2,813 | 254 | 80 | 1,598 | 11,211,014 | 9.4 | 45.5 | 17.6 | 4.4 | 3.4 | 13.9 | 3.0 |
| Suisun City | D | D | D | D | 98 | 561,250 | 17.9 | 44.2 | 4.2 | 11.9 | 5.5 | 16.3 | 0.0 |
| Sunnyvale | 197 | 1,086 | 129 | 40 | 849 | 8,013,441 | 9.5 | 44.6 | 7.9 | 9.3 | 8.7 | 16.5 | 3.5 |
| Temecula | 255 | 1,450 | 134 | 41 | 213 | 1,366,140 | 29.5 | 0.0 | 0.9 | 26.1 | 17.0 | 26.3 | 0.0 |
| Temple City | 63 | 191 | 23 | 6 | 65 | 335,559 | 28.2 | 0.0 | 0.0 | 0.0 | 21.0 | 38.3 | 0.0 |
| Thousand Oaks | 259 | 1,238 | 153 | 41 | 418 | 2,687,417 | 30.3 | 1.3 | 0.0 | 26.3 | 0.6 | 22.9 | 9.9 |
| Torrance | 307 | 2,220 | 247 | 71 | 1,383 | 11,106,855 | 10.5 | 31.9 | 18.8 | 15.6 | 4.4 | 10.4 | 3.1 |
| Tracy | 124 | 598 | 62 | 18 | 484 | 4,115,746 | 12.7 | 30.5 | 22.4 | 11.2 | 0.0 | 13.9 | 0.0 |
| Tulare | 53 | 302 | 46 | 12 | 366 | 2,119,156 | 10.1 | 34.1 | 16.1 | 2.4 | 2.1 | 24.7 | 1.8 |
| Turlock | 116 | 1,104 | 140 | 55 | 388 | 2,061,200 | 13.3 | 36.8 | 14.9 | 9.8 | 1.1 | 24.1 | 0.0 |
| Tustin | 166 | 1,415 | 142 | 43 | 309 | 3,042,614 | 10.1 | 58.6 | 0.0 | 13.9 | 6.2 | 11.1 | 0.0 |
| Twentynine Palms | 18 | 95 | 9 | 3 | 29 | 180,695 | 29.6 | 0.0 | 0.0 | 16.9 | 6.4 | 29.2 | 0.0 |
| Union City | 94 | 880 | 139 | 35 | 265 | 2,232,787 | 13.7 | 48.9 | 0.0 | 17.3 | 8.0 | 12.1 | 0.0 |
| Upland | 189 | 1,091 | 103 | 33 | 364 | 2,707,926 | 7.5 | 39.1 | 22.6 | 6.7 | 5.1 | 11.5 | 3.2 |
| Vacaville | 143 | 883 | 86 | 24 | 588 | 4,471,583 | 12.8 | 33.2 | 21.7 | 10.8 | 3.5 | 18.1 | 0.0 |
| Vallejo | 132 | 711 | 77 | 24 | 686 | 6,718,823 | 9.5 | 48.0 | 15.1 | 6.7 | 0.9 | 19.7 | 0.0 |
| Victorville | 122 | 585 | 57 | 15 | 381 | 2,014,212 | 29.2 | 0.0 | 0.0 | 26.2 | 2.1 | 32.7 | 2.0 |
| Visalia | 182 | 1,184 | 128 | 37 | 659 | 3,951,120 | 12.2 | 35.5 | 16.8 | 5.7 | 4.2 | 19.4 | 0.0 |
| Vista | 141 | 834 | 93 | 29 | 300 | 2,120,925 | 26.4 | 0.0 | 39.2 | 4.1 | 2.1 | 13.5 | 0.0 |
| Walnut | 63 | 328 | 37 | 11 | 54 | 314,058 | 52.2 | 0.0 | 0.0 | 3.0 | 0.0 | 44.8 | 0.0 |
| Walnut Creek | 222 | 3,127 | 333 | 153 | 414 | 3,171,980 | 17.0 | 38.0 | 0.0 | 4.8 | 0.6 | 21.8 | 0.0 |
| Wasco | 11 | 37 | 4 | 1 | 61 | 297,732 | 32.8 | 0.0 | 0.0 | 23.1 | 9.1 | 29.1 | 0.0 |
| Watsonville | 83 | 391 | 43 | 12 | 440 | 3,980,296 | 8.4 | 25.4 | 14.9 | 6.0 | 3.9 | 29.8 | 10.4 |
| West Covina | 109 | 518 | 50 | 13 | 399 | 2,822,262 | 11.7 | 43.5 | 31.8 | 5.3 | 3.0 | 4.6 | 0.0 |
| West Hollywood | 241 | 1,936 | 610 | 91 | 240 | 2,150,704 | 28.3 | 1.7 | 0.6 | 4.7 | 13.0 | 9.8 | 0.0 |
| Westminster | 147 | 852 | 115 | 26 | 244 | 1,811,637 | 10.3 | 61.2 | 0.0 | 7.1 | 2.8 | 11.3 | 0.0 |
| West Sacramento | 115 | 985 | 137 | 45 | 424 | 3,883,022 | 14.1 | 25.4 | 25.1 | 1.9 | 6.5 | 14.5 | 0.0 |
| Whittier | 131 | 820 | 70 | 20 | 496 | 3,040,149 | 10.2 | 45.5 | 0.0 | 4.8 | 5.7 | 20.6 | 6.0 |
| Wildomar | 39 | 243 | 19 | 7 | 16 | 94,814 | 32.9 | 0.0 | 0.0 | 7.0 | 0.0 | 0.0 | 0.0 |
| Windsor | 48 | 174 | 19 | 6 | 101 | 704,741 | 24.5 | 0.0 | 0.0 | 8.4 | 4.4 | 43.8 | 0.0 |
| Woodland | 96 | 490 | 48 | 14 | 317 | 2,079,464 | 9.5 | 30.7 | 19.4 | 5.7 | 3.0 | 24.0 | 2.8 |
| Yorba Linda | 96 | 582 | 54 | 18 | 123 | 752,679 | 20.8 | 0.0 | 0.0 | 20.3 | 7.3 | 25.2 | 19.6 |
| Yuba City | 109 | 557 | 60 | 18 | 300 | 1,946,414 | 9.0 | 33.8 | 24.7 | 2.6 | 0.0 | 19.1 | 0.0 |
| Yucaipa | 58 | 301 | 28 | 8 | 58 | 349,100 | 32.6 | 1.0 | 0.0 | 27.3 | 7.7 | 23.9 | 0.0 |
| COLORADO | 11,667 | 74,317 | 10,974 | 2,785 | X | X | X | X | X | X | X | X | X |
| Arvada | 202 | 1,187 | 137 | 36 | 406 | 2,487,923 | 10.4 | 56.5 | 0.0 | 9.1 | 2.7 | 21.3 | 0.0 |
| Aurora | 526 | 3,512 | 442 | 135 | 2,891 | 17,630,608 | 14.5 | 37.8 | 14.5 | 7.0 | 2.4 | 21.1 | 1.4 |
| Boulder | 387 | 2,873 | 422 | 123 | 1,112 | 7,993,053 | 16.7 | 23.2 | 11.2 | 5.4 | 2.1 | 25.5 | 4.7 |
| Brighton | 59 | 290 | 34 | 10 | 308 | 1,473,623 | 20.4 | 31.6 | 0.0 | 6.9 | 7.8 | 26.1 | 0.0 |
| Broomfield | 145 | 732 | 158 | 36 | 765 | 4,448,301 | 17.0 | 35.1 | 0.0 | 2.2 | 18.5 | 18.7 | 3.6 |

1. Establishments subject to federal tax.

# Table D. Cities — **City Government Finances**

| City | City government finances, 2017 | | | | | | | | | |
| | General revenue | | | | | | | General expenditure | | |
| | | Intergovernmental | | Taxes | | | | | | |
| | | | | | Per capita[1] (dollars) | | | | Per capita[1] (dollars) | |
| | Total (mil dol) | Total (mil dol) | Percent from state government | Total (mil dol) | Total | Property | Sales and gross receipts | Total (mil dol) | Total | Capital outlays |
| | 117 | 118 | 119 | 120 | 121 | 122 | 123 | 124 | 125 | 126 |
| **CALIFORNIA— Cont'd** | | | | | | | | | | |
| San Francisco | 9,978.8 | 2,985.9 | 47.3 | 3,736.2 | 4,255 | 2,354 | 1,434 | 10,297 | 11,728 | 921 |
| San Gabriel | 42.8 | 2.6 | 78.2 | 30.3 | 754 | 400 | 349 | 49 | 1,223 | 193 |
| San Jacinto | 28.5 | 2.6 | 76.5 | 16.9 | 351 | 207 | 140 | 29 | 593 | 42 |
| San Jose | 2,270.4 | 98.5 | 51.3 | 1,238.6 | 1,200 | 578 | 522 | 1,559 | 1,510 | 112 |
| San Juan Capistrano | 48.2 | 3.9 | 64.4 | 31.3 | 871 | 483 | 368 | 41 | 1,140 | 183 |
| San Leandro | 157.1 | 14.0 | 29.2 | 99.5 | 1,100 | 317 | 723 | 151 | 1,674 | 144 |
| San Luis Obispo | 103.5 | 6.4 | 29.3 | 60.5 | 1,279 | 319 | 953 | 105 | 2,225 | 372 |
| San Marcos | 130.3 | 5.0 | 63.0 | 77.1 | 806 | 479 | 321 | 154 | 1,612 | 288 |
| San Mateo | 207.2 | 7.8 | 62.0 | 124.9 | 1,197 | 526 | 554 | 177 | 1,698 | 256 |
| San Pablo | 52.2 | 2.4 | 100.0 | 39.5 | 1,271 | 102 | 1,166 | 51 | 1,631 | 394 |
| San Rafael | 103.5 | 6.4 | 43.4 | 68.8 | 1,170 | 327 | 756 | 102 | 1,741 | 51 |
| San Ramon | 90.1 | 19.4 | 20.1 | 45.4 | 599 | 294 | 255 | 85 | 1,125 | 164 |
| Santa Ana | 403.8 | 93.2 | 23.6 | 210.9 | 634 | 246 | 380 | 471 | 1,416 | 212 |
| Santa Barbara | 270.3 | 18.5 | 35.9 | 108.1 | 1,181 | 578 | 594 | 237 | 2,584 | 222 |
| Santa Clara | 494.1 | 28.3 | 15.1 | 146.7 | 1,158 | 402 | 710 | 408 | 3,217 | 304 |
| Santa Clarita | 191.7 | 46.0 | 48.5 | 103.3 | 483 | 194 | 265 | 188 | 878 | 180 |
| Santa Cruz | 154.6 | 5.8 | 49.8 | 76.9 | 1,187 | 380 | 754 | 157 | 2,429 | 335 |
| Santa Maria | 144.0 | 21.0 | 26.1 | 63.6 | 598 | 165 | 341 | 140 | 1,318 | 239 |
| Santa Monica | 690.5 | 116.4 | 22.6 | 338.1 | 3,685 | 813 | 2,718 | 502 | 5,468 | 194 |
| Santa Paula | 30.7 | 2.5 | 63.2 | 10.5 | 350 | 209 | 139 | 32 | 1,068 | 160 |
| Santa Rosa | 302.6 | 56.3 | 79.2 | 116.7 | 643 | 238 | 384 | 318 | 1,753 | 301 |
| Santee | 53.5 | 4.7 | 38.6 | 35.8 | 619 | 283 | 330 | 75 | 1,295 | 192 |
| Saratoga | 24.9 | 2.1 | 45.3 | 19.1 | 620 | 399 | 180 | 33 | 1,070 | 192 |
| Seaside | 32.8 | 2.7 | 81.0 | 25.7 | 760 | 198 | 560 | 32 | 945 | 207 |
| Simi Valley | 113.4 | 15.7 | 38.8 | 68.4 | 543 | 279 | 258 | 107 | 849 | 135 |
| Soledad | 21.0 | 2.7 | 31.0 | 8.1 | 309 | 104 | 147 | 18 | 688 | 48 |
| South Gate | 90.0 | 12.9 | 43.5 | 47.0 | 495 | 195 | 299 | 76 | 798 | 135 |
| South Pasadena | 34.7 | 2.7 | 43.7 | 23.0 | 894 | 512 | 351 | 29 | 1,134 | 0 |
| South San Francisco | 146.8 | 4.5 | 80.0 | 88.6 | 1,319 | 462 | 801 | 145 | 2,156 | 262 |
| Stanton | 26.8 | 3.0 | 81.9 | 18.4 | 479 | 150 | 327 | 37 | 967 | 228 |
| Stockton | 401.2 | 37.4 | 30.9 | 227.1 | 732 | 211 | 519 | 320 | 1,033 | 16 |
| Suisun City | 20.0 | 4.3 | 20.6 | 8.7 | 296 | 150 | 127 | 22 | 735 | 122 |
| Sunnyvale | 373.0 | 27.0 | 70.5 | 141.5 | 923 | 462 | 450 | 290 | 1,893 | 92 |
| Temecula | 118.3 | 9.4 | 30.8 | 78.1 | 685 | 193 | 486 | 180 | 1,580 | 527 |
| Temple City | 19.2 | 3.2 | 75.3 | 11.6 | 321 | 199 | 116 | 18 | 502 | 15 |
| Thousand Oaks | 151.8 | 12.3 | 30.0 | 88.0 | 687 | 269 | 392 | 178 | 1,391 | 218 |
| Torrance | 287.9 | 51.9 | 8.4 | 181.7 | 1,247 | 346 | 878 | 240 | 1,643 | 2 |
| Tracy | 182.9 | 29.4 | 12.1 | 70.7 | 780 | 276 | 500 | 151 | 1,667 | 399 |
| Tulare | 101.5 | 10.5 | 36.7 | 34.8 | 549 | 161 | 385 | 103 | 1,628 | 464 |
| Turlock | 85.6 | 7.1 | 56.0 | 37.4 | 510 | 154 | 293 | 102 | 1,389 | 436 |
| Tustin | 140.3 | 4.2 | 80.5 | 51.4 | 642 | 229 | 404 | 113 | 1,409 | 334 |
| Twentynine Palms | 11.9 | 1.2 | 100.0 | 9.1 | 344 | 180 | 163 | 13 | 491 | 101 |
| Union City | 109.2 | 21.9 | 54.3 | 64.3 | 854 | 393 | 358 | 125 | 1,660 | 319 |
| Upland | 81.3 | 3.1 | 88.4 | 41.9 | 546 | 260 | 259 | 79 | 1,025 | 64 |
| Vacaville | 199.7 | 29.9 | 16.0 | 87.9 | 883 | 391 | 486 | 192 | 1,929 | 553 |
| Vallejo | 191.9 | 33.5 | 23.0 | 90.3 | 743 | 248 | 473 | 172 | 1,415 | 128 |
| Victorville | 151.4 | 33.7 | 18.0 | 60.1 | 492 | 220 | 270 | 212 | 1,738 | 191 |
| Visalia | 227.5 | 55.7 | 38.4 | 81.8 | 618 | 167 | 448 | 157 | 1,185 | 239 |
| Vista | 138.9 | 8.9 | 71.4 | 71.6 | 711 | 329 | 376 | 176 | 1,747 | 208 |
| Walnut | 18.7 | 3.0 | 29.6 | 10.6 | 354 | 196 | 151 | 23 | 769 | 29 |
| Walnut Creek | 107.6 | 2.6 | 76.0 | 63.3 | 908 | 317 | 527 | 111 | 1,589 | 128 |
| Wasco | 20.1 | 4.1 | 16.6 | 7.0 | 260 | 116 | 136 | 18 | 657 | 79 |
| Watsonville | 84.7 | 4.7 | 41.1 | 35.4 | 656 | 407 | 246 | 91 | 1,691 | 117 |
| West Covina | 108.8 | 7.5 | 52.1 | 67.2 | 628 | 340 | 283 | 120 | 1,126 | 314 |
| West Hollywood | 141.9 | 3.4 | 63.9 | 78.3 | 2,129 | 696 | 1,407 | 157 | 4,274 | 682 |
| Westminster | 78.1 | 6.6 | 38.4 | 61.0 | 669 | 309 | 346 | 130 | 1,426 | 529 |
| West Sacramento | 123.0 | 7.0 | 62.1 | 60.7 | 1,138 | 546 | 419 | 153 | 2,864 | 838 |
| Whittier | 85.8 | 4.6 | 63.1 | 49.1 | 569 | 249 | 311 | 85 | 984 | 119 |
| Wildomar | 15.6 | 4.0 | 83.6 | 9.2 | 252 | 108 | 137 | 18 | 498 | 38 |
| Windsor | 30.7 | 1.7 | 43.6 | 16.0 | 583 | 254 | 324 | 36 | 1,312 | 280 |
| Woodland | 119.7 | 3.5 | 62.5 | 53.4 | 891 | 260 | 622 | 108 | 1,794 | 417 |
| Yorba Linda | 63.3 | 4.8 | 31.7 | 37.8 | 557 | 341 | 186 | 68 | 997 | 284 |
| Yuba City | 68.5 | 7.7 | 28.7 | 31.7 | 476 | 174 | 299 | 65 | 979 | 162 |
| Yucaipa | 28.8 | 2.4 | 95.0 | 18.7 | 350 | 271 | 61 | 40 | 738 | 131 |
| **COLORADO** | X | X | X | X | X | X | X | X | X | X |
| Arvada | 5.5 | 0.0 | ********** | 5.5 | 46 | 46 | 0 | 0 | 0 | 0 |
| Aurora | 505.2 | 45.7 | 47.5 | 315.2 | 857 | 99 | 758 | 494 | 1,343 | 228 |
| Boulder | 359.7 | 29.1 | 75.3 | 205.8 | 1,934 | 365 | 1,568 | 346 | 3,254 | 629 |
| Brighton | 62.8 | 7.1 | 100.0 | 40.5 | 1,005 | 191 | 814 | 56 | 1,396 | 256 |
| Broomfield | 248.1 | 15.6 | 98.1 | 133.9 | 1,962 | 670 | 1,156 | 146 | 2,134 | 227 |

1. Based on population estimated as of July 1 of the year shown.

# Table D. Cities — City Government Finances

| City | City government finances, 2017 (cont.) | | | | | | | | | |
| | General expenditure (cont.) | | | | | | | | | |
| | Percent of total for: | | | | | | | | | |
| | Public welfare | Highways | Parking facilities | Education | Health and hospitals | Police protection | Sewerage and sanitation | Parks and recreation | Housing and community development | Interest on debt |
| | 127 | 128 | 129 | 130 | 131 | 132 | 133 | 134 | 135 | 136 |
|---|---|---|---|---|---|---|---|---|---|---|
| **CALIFORNIA— Cont'd** | | | | | | | | | | |
| San Francisco | 13.9 | 1.1 | 1.7 | 2.1 | 31.7 | 5.1 | 4.1 | 2.5 | 4.0 | 5.2 |
| San Gabriel | 0.0 | 9.1 | 0.0 | 0.0 | 0.0 | 28.5 | 2.5 | 12.9 | 2.7 | 0.9 |
| San Jacinto | 0.0 | 22.1 | 0.0 | 0.0 | 0.0 | 36.4 | 3.0 | 1.1 | 2.5 | 1.5 |
| San Jose | 0.0 | 4.1 | 0.7 | 0.0 | 0.2 | 16.5 | 22.2 | 10.1 | 2.8 | 14.8 |
| San Juan Capistrano | 0.0 | 7.1 | 0.0 | 0.0 | 1.7 | 22.2 | 9.0 | 10.9 | 9.7 | 6.9 |
| San Leandro | 0.0 | 5.6 | 0.2 | 0.0 | 0.1 | 21.8 | 7.3 | 5.9 | 7.1 | 3.7 |
| San Luis Obispo | 0.0 | 13.1 | 3.1 | 0.0 | 0.0 | 16.3 | 10.2 | 9.9 | 2.2 | 1.9 |
| San Marcos | 0.0 | 13.8 | 0.0 | 0.0 | 0.0 | 22.7 | 0.0 | 8.9 | 12.4 | 9.3 |
| San Mateo | 0.0 | 3.0 | 1.2 | 0.0 | 0.3 | 21.6 | 27.9 | 8.8 | 0.2 | 4.5 |
| San Pablo | 0.0 | 21.4 | 0.0 | 0.0 | 0.3 | 33.0 | 0.0 | 7.5 | 5.6 | 3.3 |
| San Rafael | 0.0 | 10.6 | 3.5 | 0.0 | 7.3 | 23.6 | 0.0 | 11.2 | 1.2 | 0.5 |
| San Ramon | 0.0 | 13.0 | 0.0 | 0.0 | 0.0 | 23.4 | 0.4 | 22.7 | 6.6 | 4.5 |
| Santa Ana | 0.0 | 8.2 | 1.3 | 0.0 | 1.2 | 28.1 | 6.4 | 7.6 | 12.8 | 2.5 |
| Santa Barbara | 0.0 | 4.9 | 3.2 | 0.0 | 0.3 | 16.4 | 22.3 | 10.4 | 1.6 | 3.8 |
| Santa Clara | 0.0 | 6.8 | 0.4 | 0.0 | 0.3 | 15.3 | 14.6 | 23.9 | 0.5 | 5.8 |
| Santa Clarita | 0.0 | 37.9 | 0.0 | 0.0 | 0.0 | 12.4 | 0.0 | 12.3 | 7.1 | 1.6 |
| Santa Cruz | 0.0 | 5.8 | 4.6 | 0.0 | 0.0 | 17.9 | 21.9 | 11.5 | 8.3 | 2.5 |
| Santa Maria | 0.0 | 6.0 | 0.0 | 0.0 | 0.0 | 21.0 | 31.1 | 9.6 | 0.4 | 0.9 |
| Santa Monica | 0.0 | 2.7 | 2.9 | 0.0 | 4.7 | 17.1 | 8.2 | 10.7 | 8.9 | 3.1 |
| Santa Paula | 0.0 | 8.1 | 0.0 | 0.0 | 0.5 | 20.2 | 16.6 | 14.0 | 0.0 | 12.2 |
| Santa Rosa | 0.0 | 10.5 | 1.5 | 0.0 | 1.0 | 16.6 | 19.8 | 6.1 | 5.4 | 3.4 |
| Santee | 0.0 | 8.1 | 0.0 | 0.2 | 4.8 | 36.2 | 0.0 | 6.1 | 10.9 | 3.3 |
| Saratoga | 0.0 | 19.0 | 0.0 | 0.0 | 3.3 | 31.4 | 0.0 | 15.1 | 0.6 | 1.1 |
| Seaside | 0.0 | 10.3 | 0.0 | 0.0 | 0.7 | 35.3 | 0.0 | 10.4 | 5.9 | 1.8 |
| Simi Valley | 0.0 | 9.8 | 0.0 | 0.0 | 0.8 | 30.4 | 18.7 | 0.6 | 6.5 | 1.7 |
| Soledad | 0.0 | 14.6 | 0.0 | 0.0 | 0.0 | 19.8 | 28.7 | 3.8 | 1.0 | 6.6 |
| South Gate | 0.0 | 24.7 | 0.0 | 0.0 | 0.0 | 32.1 | 5.0 | 9.2 | 10.6 | 5.2 |
| South Pasadena | 0.0 | 5.3 | 0.1 | 0.0 | 2.7 | 28.5 | 3.1 | 13.5 | 0.0 | 0.2 |
| South San Francisco | 0.0 | 15.1 | 0.7 | 0.0 | 7.5 | 18.2 | 15.0 | 10.3 | 3.8 | 0.8 |
| Stanton | 0.0 | 4.9 | 0.0 | 0.0 | 0.9 | 27.7 | 1.5 | 5.8 | 25.9 | 8.8 |
| Stockton | 0.0 | 3.3 | 1.5 | 0.0 | 0.9 | 38.5 | 14.4 | 7.2 | 1.0 | 5.7 |
| Suisun City | 0.0 | 12.1 | 0.0 | 0.0 | 1.2 | 28.0 | 0.0 | 9.5 | 26.4 | 0.4 |
| Sunnyvale | 0.0 | 4.5 | 0.1 | 0.0 | 0.4 | 19.4 | 35.2 | 7.6 | 1.1 | 0.9 |
| Temecula | 0.0 | 14.1 | 0.0 | 0.0 | 0.3 | 14.7 | 4.3 | 3.8 | 1.6 | 3.1 |
| Temple City | 0.0 | 8.5 | 0.8 | 0.0 | 1.9 | 25.2 | 0.0 | 13.0 | 2.4 | 1.5 |
| Thousand Oaks | 0.0 | 21.0 | 0.0 | 0.0 | 0.2 | 31.6 | 9.9 | 5.1 | 2.3 | 0.6 |
| Torrance | 0.0 | 8.2 | 0.0 | 0.0 | 5.3 | 32.3 | 6.1 | 8.9 | 2.8 | 2.0 |
| Tracy | 0.0 | 7.8 | 0.0 | 0.0 | 0.6 | 15.7 | 21.9 | 11.8 | 1.7 | 2.7 |
| Tulare | 0.0 | 19.8 | 0.0 | 0.0 | 1.7 | 15.0 | 25.3 | 4.4 | 0.7 | 11.8 |
| Turlock | 0.0 | 8.8 | 0.0 | 0.0 | 0.5 | 28.3 | 12.8 | 3.8 | 11.5 | 6.7 |
| Tustin | 0.0 | 25.6 | 0.0 | 0.0 | 0.0 | 20.9 | 0.0 | 3.5 | 5.7 | 3.4 |
| Twentynine Palms | 0.0 | 15.0 | 0.0 | 0.0 | 4.7 | 28.1 | 0.0 | 19.0 | 4.4 | 3.6 |
| Union City | 0.0 | 7.4 | 0.0 | 0.0 | 0.4 | 21.3 | 0.5 | 12.1 | 11.8 | 6.3 |
| Upland | 0.0 | 9.1 | 0.0 | 0.0 | 0.1 | 24.1 | 22.1 | 2.9 | 10.4 | 0.0 |
| Vacaville | 0.0 | 10.4 | 0.0 | 0.0 | 6.1 | 18.0 | 23.3 | 6.0 | 15.5 | 1.1 |
| Vallejo | 0.0 | 11.0 | 0.5 | 0.0 | 1.9 | 24.0 | 13.4 | 2.7 | 15.6 | 1.9 |
| Victorville | 0.0 | 4.0 | 0.0 | 0.0 | 0.3 | 10.8 | 20.0 | 4.2 | 9.1 | 9.8 |
| Visalia | 0.0 | 4.6 | 0.0 | 1.8 | 0.0 | 21.2 | 42.6 | 8.6 | 2.1 | 1.0 |
| Vista | 0.0 | 3.6 | 0.0 | 0.0 | 2.4 | 36.7 | 26.7 | 5.0 | 1.8 | 5.5 |
| Walnut | 0.0 | 19.9 | 0.0 | 0.0 | 1.0 | 30.8 | 0.1 | 15.2 | 3.1 | 5.3 |
| Walnut Creek | 0.0 | 10.5 | 6.2 | 0.0 | 0.7 | 23.2 | 0.0 | 24.3 | 5.2 | 0.0 |
| Wasco | 0.0 | 14.5 | 0.0 | 0.0 | 0.9 | 19.7 | 23.3 | 0.0 | 5.8 | 0.4 |
| Watsonville | 0.0 | 9.3 | 0.2 | 0.0 | 1.2 | 20.9 | 21.2 | 4.9 | 2.5 | 0.4 |
| West Covina | 0.0 | 11.3 | 0.3 | 0.0 | 1.8 | 28.7 | 0.0 | 5.1 | 20.1 | 3.1 |
| West Hollywood | 0.0 | 10.9 | 1.6 | 0.0 | 4.0 | 25.4 | 1.7 | 4.6 | 5.9 | 7.4 |
| Westminster | 0.0 | 7.0 | 0.0 | 0.0 | 1.4 | 20.6 | 0.0 | 3.0 | 33.2 | 5.0 |
| West Sacramento | 0.0 | 17.1 | 0.0 | 0.0 | 0.7 | 12.6 | 19.4 | 6.2 | 8.9 | 5.5 |
| Whittier | 0.0 | 10.3 | 0.4 | 0.0 | 0.0 | 38.4 | 10.6 | 14.2 | 2.3 | 2.3 |
| Wildomar | 0.0 | 18.3 | 0.0 | 0.0 | 2.7 | 29.0 | 0.0 | 0.0 | 8.1 | 1.0 |
| Windsor | 0.0 | 12.2 | 0.0 | 0.0 | 0.7 | 19.5 | 21.6 | 7.4 | 7.5 | 5.4 |
| Woodland | 0.2 | 10.6 | 0.0 | 0.0 | 0.6 | 18.1 | 11.1 | 4.9 | 1.6 | 1.2 |
| Yorba Linda | 0.0 | 33.4 | 0.0 | 0.0 | 0.0 | 15.8 | 0.0 | 19.1 | 1.0 | 1.6 |
| Yuba City | 0.0 | 11.2 | 0.0 | 0.0 | 1.1 | 22.0 | 20.6 | 6.8 | 3.7 | 0.0 |
| Yucaipa | 0.0 | 23.5 | 0.0 | 0.0 | 0.0 | 19.5 | 0.3 | 14.0 | 9.1 | |
| **COLORADO** | X | X | X | X | X | X | X | X | X | X |
| Arvada | ********** | ********** | ********** | ********** | ********** | ********** | ********** | ********** | ********** | ********** |
| Aurora | 0.0 | 12.1 | 0.0 | 0.0 | 0.0 | 21.4 | 15.1 | 11.6 | 3.6 | 1.7 |
| Boulder | 0.0 | 18.1 | 2.6 | 0.0 | 0.0 | 11.5 | 7.9 | 16.8 | 4.7 | 2.1 |
| Brighton | 0.0 | 15.7 | 0.0 | 0.0 | 0.5 | 17.4 | 7.8 | 13.5 | 0.0 | 4.6 |
| Broomfield | 9.9 | 7.0 | 0.0 | 0.0 | 1.7 | 12.3 | 14.4 | 15.5 | 0.7 | 8.1 |

# City Government Finances, City Government Employment, and Climate

| City | City government finances, 2017 (cont.) | | | Climate[2] | | | | | | |
| | Debt outstanding | | Debt issued during year | Average daily temperature | | | | Annual precipitation (inches) | Heating degree days | Cooling degree days |
| | Total (mil dol) | Per capita[1] (dollars) | | Mean | | Limits | | | | |
| | | | | January | July | January[3] | July[4] | | | |
| | 137 | 138 | 139 | 140 | 141 | 142 | 143 | 144 | 145 | 146 |
|---|---|---|---|---|---|---|---|---|---|---|
| CALIFORNIA— Cont'd | | | | | | | | | | |
| San Francisco | 16,803.8 | 19,138 | 3,535.7 | 52.3 | 61.3 | 46.4 | 68.2 | 22.28 | 2,597 | 163 |
| San Gabriel | 10.8 | 267 | 0.0 | 56.3 | 75.6 | 42.6 | 89.0 | 18.56 | 1,295 | 1,575 |
| San Jacinto | 26.2 | 545 | 0.0 | NA | NA | NA | NA | NA | NA | NA |
| San Jose | 4,767.7 | 4,618 | 645.5 | 50.5 | 70.9 | 41.7 | 84.3 | 15.08 | 2,171 | 811 |
| San Juan Capistrano | 89.7 | 2,500 | 32.1 | 55.4 | 68.7 | 43.9 | 77.3 | 13.56 | 1,756 | 666 |
| San Leandro | 100.9 | 1,115 | 19.5 | 50.0 | 62.8 | 43.6 | 70.4 | 25.40 | 2,857 | 142 |
| San Luis Obispo | 66.7 | 1,408 | 0.0 | 53.3 | 66.5 | 41.9 | 80.3 | 24.36 | 2,138 | 476 |
| San Marcos | 291.6 | 3,049 | 0.0 | 56.4 | 71.6 | 45.1 | 82.2 | 13.69 | 1,514 | 1,047 |
| San Mateo | 223.2 | 2,138 | 0.7 | 48.4 | 68.0 | 39.1 | 80.8 | 20.16 | 2,764 | 422 |
| San Pablo | 46.4 | 1,492 | 0.3 | 48.8 | 67.7 | 41.3 | 80.9 | 34.29 | 2,621 | 451 |
| San Rafael | 28.2 | 479 | 0.0 | 48.8 | 67.7 | 41.3 | 80.9 | 34.29 | 2,621 | 451 |
| San Ramon | 89.4 | 1,179 | 0.0 | 47.2 | 72.0 | 37.4 | 89.1 | 14.82 | 2,755 | 858 |
| Santa Ana | 223.3 | 671 | 2.1 | 58.0 | 72.9 | 46.6 | 87.7 | 13.84 | 1,153 | 1,299 |
| Santa Barbara | 211.0 | 2,306 | 52.1 | 53.1 | 67.0 | 40.8 | 76.7 | 16.93 | 2,121 | 482 |
| Santa Clara | 704.2 | 5,558 | 4.0 | 50.5 | 70.9 | 41.7 | 84.3 | 15.08 | 2,171 | 811 |
| Santa Clarita | 57.5 | 269 | 0.0 | 50.3 | 74.1 | 36.1 | 94.2 | 13.96 | 2,502 | 1,139 |
| Santa Cruz | 124.5 | 1,921 | 23.6 | 50.6 | 63.7 | 40.2 | 74.8 | 30.67 | 2,836 | 162 |
| Santa Maria | 53.7 | 505 | 0.0 | 51.6 | 63.5 | 39.3 | 73.5 | 14.01 | 2,783 | 121 |
| Santa Monica | 345.2 | 3,762 | 0.0 | 57.0 | 65.5 | 50.2 | 68.8 | 13.27 | 1,810 | 429 |
| Santa Paula | 124.9 | 4,144 | 0.0 | 54.7 | 68.3 | 41.2 | 80.7 | 18.41 | 1,911 | 602 |
| Santa Rosa | 404.9 | 2,230 | 112.5 | 48.7 | 67.6 | 39.5 | 82.2 | 31.01 | 2,694 | 526 |
| Santee | 58.1 | 1,004 | 44.6 | 57.1 | 73.0 | 45.7 | 83.6 | 13.75 | 1,313 | 1,261 |
| Saratoga | 9.6 | 311 | 0.0 | 50.5 | 70.9 | 41.7 | 84.3 | 15.08 | 2,171 | 811 |
| Seaside | 13.2 | 391 | 0.0 | 51.6 | 60.2 | 43.4 | 68.1 | 20.35 | 3,092 | 74 |
| Simi Valley | 180.0 | 1,427 | 19.9 | 53.7 | 76.0 | 39.5 | 95.0 | 17.79 | 1,822 | 1,485 |
| Soledad | 38.2 | 1,459 | 0.0 | NA | NA | NA | NA | NA | NA | NA |
| South Gate | 84.8 | 894 | 0.3 | 58.3 | 74.2 | 48.5 | 83.8 | 15.14 | 928 | 1,506 |
| South Pasadena | 55.4 | 2,155 | 0.0 | NA | NA | NA | NA | NA | NA | NA |
| South San Francisco | 48.4 | 720 | 0.0 | 49.4 | 62.8 | 42.9 | 71.1 | 20.11 | 2,862 | 142 |
| Stanton | 80.6 | 2,101 | 0.0 | 58.0 | 72.9 | 46.6 | 87.7 | 13.84 | 1,153 | 1,299 |
| Stockton | 589.0 | 1,899 | 35.1 | 46.0 | 77.3 | 38.1 | 93.8 | 13.84 | 2,563 | 1,456 |
| Suisun City | 4.4 | 149 | 2.0 | 46.1 | 72.6 | 37.5 | 88.8 | 23.46 | 2,649 | 975 |
| Sunnyvale | 72.2 | 471 | 3.9 | 50.5 | 70.9 | 41.7 | 84.3 | 15.08 | 2,171 | 811 |
| Temecula | 151.3 | 1,328 | 66.3 | 51.2 | 78.3 | 36.1 | 97.8 | 11.40 | 2,123 | 1,710 |
| Temple City | 6.4 | 177 | 0.0 | 56.3 | 75.6 | 42.6 | 89.0 | 18.56 | 1,295 | 1,575 |
| Thousand Oaks | 74.2 | 578 | 0.3 | 53.7 | 76.0 | 39.5 | 95.0 | 17.79 | 1,822 | 1,485 |
| Torrance | 82.3 | 564 | 5.6 | 56.3 | 69.4 | 46.2 | 77.6 | 14.79 | 1,526 | 742 |
| Tracy | 91.6 | 1,010 | 0.0 | 47.1 | 76.4 | 38.5 | 92.5 | 12.51 | 2,421 | 1,470 |
| Tulare | 262.9 | 4,141 | 58.3 | 45.8 | 79.3 | 37.4 | 93.8 | 11.03 | 2,588 | 1,685 |
| Turlock | 104.2 | 1,423 | 0.0 | 46.4 | 77.6 | 39.0 | 93.3 | 12.43 | 2,519 | 1,506 |
| Tustin | 164.0 | 2,047 | 16.9 | 54.5 | 72.1 | 41.4 | 83.8 | 13.87 | 1,794 | 1,102 |
| Twentynine Palms | 11.5 | 435 | 0.0 | 50.0 | 88.4 | 36.1 | 105.8 | 4.57 | 1,910 | 3,064 |
| Union City | 163.3 | 2,170 | 4.7 | 49.8 | 68.0 | 42.0 | 78.3 | 14.85 | 2,367 | 530 |
| Upland | 8.4 | 110 | 0.0 | 54.6 | 73.8 | 41.5 | 88.7 | 16.96 | 1,727 | 1,191 |
| Vacaville | 52.6 | 528 | 0.0 | 47.2 | 77.3 | 38.8 | 95.8 | 24.55 | 2,410 | 1,498 |
| Vallejo | 126.1 | 1,038 | 0.4 | 46.3 | 71.2 | 38.8 | 87.4 | 19.58 | 2,757 | 786 |
| Victorville | 412.3 | 3,380 | 1.0 | 45.5 | 80.0 | 31.4 | 99.1 | 6.20 | 2,929 | 1,735 |
| Visalia | 27.1 | 205 | 0.0 | 45.8 | 79.3 | 37.4 | 93.8 | 11.03 | 2,588 | 1,685 |
| Vista | 222.8 | 2,212 | 5.3 | 56.4 | 71.6 | 45.1 | 82.2 | 13.69 | 1,514 | 1,047 |
| Walnut | 28.7 | 956 | 0.0 | 54.6 | 73.8 | 41.5 | 88.7 | 16.96 | 1,727 | 1,191 |
| Walnut Creek | 0.9 | 14 | 0.2 | 47.5 | 72.4 | 39.3 | 85.2 | 23.96 | 3,267 | 983 |
| Wasco | 1.5 | 56 | 0.0 | NA | NA | NA | NA | NA | NA | NA |
| Watsonville | 4.9 | 91 | 0.0 | 49.7 | 62.4 | 38.7 | 72.0 | 23.25 | 3,080 | 123 |
| West Covina | 81.7 | 764 | 15.4 | 56.3 | 75.6 | 42.6 | 89.0 | 18.56 | 1,295 | 1,575 |
| West Hollywood | 155.2 | 4,217 | 0.0 | 58.3 | 74.2 | 48.5 | 83.8 | 15.14 | 928 | 1,506 |
| Westminster | 126.7 | 1,390 | 77.5 | 58.0 | 72.9 | 46.6 | 82.7 | 13.84 | 1,153 | 1,299 |
| West Sacramento | 334.7 | 6,279 | 0.4 | 46.3 | 75.4 | 38.8 | 92.4 | 17.93 | 2,666 | 1,248 |
| Whittier | 61.6 | 714 | 0.0 | 56.3 | 75.6 | 42.6 | 89.0 | 18.56 | 1,295 | 1,575 |
| Wildomar | 0.0 | 0 | 0.0 | NA | NA | NA | NA | NA | NA | NA |
| Windsor | 15.2 | 554 | 0.0 | NA | NA | NA | NA | NA | NA | NA |
| Woodland | 381.6 | 6,368 | 60.2 | 45.7 | 76.4 | 37.6 | 94.0 | 20.78 | 2,683 | 1,417 |
| Yorba Linda | 20.0 | 294 | 0.1 | 56.9 | 73.2 | 45.2 | 84.0 | 11.23 | 1,286 | 1,294 |
| Yuba City | 100.6 | 1,512 | 15.8 | 46.3 | 78.9 | 37.8 | 96.3 | 22.07 | 2,488 | 1,687 |
| Yucaipa | 38.1 | 712 | 0.0 | 52.9 | 78.0 | 40.4 | 94.4 | 13.62 | 1,904 | 1,714 |
| COLORADO | X | X | X | X | X | X | X | X | X | X |
| Arvada | 71.8 | 604 | 0.0 | 31.2 | 71.5 | 15.6 | 88.3 | 18.17 | 5,988 | 496 |
| Aurora | 809.7 | 2,203 | 471.7 | 29.2 | 73.4 | 15.2 | 88.0 | 15.81 | 6,128 | 696 |
| Boulder | 125.5 | 1,179 | 0.0 | 32.5 | 71.6 | 19.2 | 87.2 | 19.93 | 5,687 | 552 |
| Brighton | 61.4 | 1,522 | 0.0 | NA | NA | NA | NA | NA | NA | NA |
| Broomfield | 261.2 | 3,826 | 0.0 | 32.5 | 71.6 | 19.2 | 87.2 | 19.93 | 5,687 | 552 |

1. Based on the population estimated as of July 1 of the year shown. maximum.    2. Represents normal values based on the 30-year period, 1971±2000.    3. Average daily minimum.    4. Average daily maximum.

Items 137—146

# Table D. Cities — Land Area and Population

| STATE Place code | City | Land area[1] (sq. mi) | Population, 2020 Total persons 2019 | Rank | Per square mile | Race 2019 — Race alone[2] (percent) White | Black or African American | American Indian, Alaskan Native | Asian | Hawaiian Pacific Islander | Some other race | Two or more races (percent) |
|---|---|---|---|---|---|---|---|---|---|---|---|---|
| | | 1 | 2 | 3 | 4 | 5 | 6 | 7 | 8 | 9 | 10 | 11 |
| | **COLORADO— Cont'd** | | | | | | | | | | | |
| 08 12415 | Castle Rock | 34.3 | 70,567 | 515 | 2,057 | 90.6 | 1.3 | 0.6 | 2.3 | 0.0 | 2.4 | 2.8 |
| 08 12815 | Centennial | 29.7 | 110,156 | 273 | 3,709 | 83.3 | 3.4 | 0.0 | 6.4 | 0.0 | 3.2 | 3.6 |
| 08 16000 | Colorado Springs | 195.4 | 482,131 | 39 | 2,467 | 78.5 | 6.6 | 1.0 | 3.0 | 0.4 | 4.6 | 5.9 |
| 08 16495 | Commerce City | 36.0 | 59,317 | 637 | 1,648 | 77.2 | 6.0 | 0.7 | 3.8 | 0.2 | 7.9 | 4.2 |
| 08 20000 | Denver | 153.1 | 735,538 | 19 | 4,804 | 77.2 | 8.6 | 0.6 | 3.6 | 0.1 | 5.3 | 4.5 |
| 08 24785 | Englewood | 6.6 | 35,099 | 1,100 | 5,318 | 83.8 | 2.3 | 1.0 | 2.5 | 0.0 | 5.1 | 5.2 |
| 08 27425 | Fort Collins | 57.2 | 168,234 | 156 | 2,941 | 87.9 | 1.4 | 1.2 | 3.1 | 0.1 | 1.4 | 4.9 |
| 08 27865 | Fountain | 22.5 | 31,017 | 1,217 | 1,379 | 70.5 | 8.7 | 2.9 | 3.9 | 1.3 | 0.9 | 11.8 |
| 08 31660 | Grand Junction | 39.6 | 63,447 | 595 | 1,602 | 93.3 | 1.1 | 0.7 | 1.3 | 0.0 | 0.6 | 3.1 |
| 08 32155 | Greeley | 48.9 | 108,958 | 282 | 2,228 | 84.8 | 3.3 | 1.5 | 1.3 | 0.4 | 4.6 | 4.0 |
| 08 43000 | Lakewood | 43.5 | 157,429 | 162 | 3,619 | 85.9 | 1.2 | 1.2 | 4.3 | 0.0 | 3.1 | 4.3 |
| 08 45255 | Littleton | 12.6 | 46,000 | 847 | 3,651 | 88.1 | 3.6 | 0.3 | 3.1 | 0.0 | 1.6 | 3.4 |
| 08 45970 | Longmont | 28.8 | 98,445 | 325 | 3,418 | 91.9 | 2.3 | 0.3 | 1.7 | 0.0 | 2.2 | 1.8 |
| 08 46465 | Loveland | 34.4 | 83,517 | 412 | 2,428 | 93.5 | 0.6 | 0.7 | 1.7 | 0.1 | 1.3 | 2.2 |
| 08 54330 | Northglenn | 7.4 | 38,845 | 997 | 5,249 | 82.4 | 3.4 | 0.9 | 1.8 | 0.6 | 3.1 | 7.8 |
| 08 57630 | Parker | 22.3 | 58,673 | 648 | 2,631 | 86.0 | 1.3 | 0.4 | 5.8 | 0.0 | 3.5 | 3.0 |
| 08 62000 | Pueblo | 55.4 | 112,751 | 261 | 2,035 | 77.9 | 2.6 | 4.6 | 0.6 | 0.0 | 8.5 | 5.9 |
| 08 77290 | Thornton | 35.9 | 148,113 | 176 | 4,126 | 79.3 | 2.3 | 1.2 | 5.2 | 0.0 | 6.3 | 5.6 |
| 08 83835 | Westminster | 31.6 | 115,046 | 256 | 3,641 | 84.8 | 1.2 | 0.5 | 6.0 | 0.1 | 2.4 | 5.1 |
| 08 84440 | Wheat Ridge | 9.3 | 31,857 | 1,198 | 3,426 | D | D | D | D | D | D | D |
| 09 00000 | **CONNECTICUT** | 4,842.7 | 3,557,006 | X | 735 | 74.6 | 11.1 | 0.3 | 4.7 | 0.0 | 5.6 | 3.7 |
| 09 08000 | Bridgeport | 16.1 | 143,525 | 188 | 8,915 | 39.5 | 33.1 | 0.1 | 4.3 | 0.2 | 19.0 | 3.9 |
| 09 08420 | Bristol | 26.4 | 59,734 | 630 | 2,263 | 81.5 | 7.5 | 0.0 | 2.1 | 0.0 | 2.8 | 6.1 |
| 09 18430 | Danbury | 42.0 | 84,317 | 409 | 2,008 | 58.0 | 12.2 | 0.2 | 6.1 | 0.0 | 19.4 | 4.1 |
| 09 37000 | Hartford | 17.4 | 121,535 | 232 | 6,985 | 27.6 | 37.5 | 0.3 | 1.4 | 0.0 | 21.1 | 12.1 |
| 09 46450 | Meriden | 23.7 | 58,981 | 642 | 2,489 | 60.4 | 14.6 | 0.2 | 3.1 | 0.0 | 12.7 | 9.1 |
| 09 47290 | Middletown | 41.0 | 46,013 | 846 | 1,122 | 70.3 | 17.8 | 0.1 | 2.5 | 0.0 | 5.2 | 4.1 |
| 09 47500 | Milford | 21.9 | 53,259 | 730 | 2,432 | 84.5 | 3.2 | 0.0 | 9.6 | 0.0 | 1.4 | 1.3 |
| 09 49880 | Naugatuck | 16.3 | 30,905 | 1,221 | 1,896 | D | D | D | D | D | D | D |
| 09 50370 | New Britain | 13.4 | 72,198 | 502 | 5,388 | 54.7 | 13.6 | 0.7 | 2.7 | 0.0 | 20.1 | 8.1 |
| 09 52000 | New Haven | 18.7 | 130,801 | 213 | 6,995 | 44.4 | 33.4 | 0.1 | 5.4 | 0.0 | 11.6 | 5.0 |
| 09 52280 | New London | 5.6 | 26,870 | 1,357 | 4,798 | 62.1 | 10.2 | 0.0 | 4.2 | 0.0 | 17.2 | 6.3 |
| 09 55990 | Norwalk | 22.9 | 88,655 | 380 | 3,871 | 66.3 | 16.9 | 0.3 | 8.0 | 0.0 | 2.9 | 5.4 |
| 09 56200 | Norwich | 28.1 | 38,619 | 1,001 | 1,374 | 70.0 | 14.2 | 2.2 | 4.8 | 0.0 | 4.0 | 4.9 |
| 09 68100 | Shelton | 30.6 | 40,957 | 938 | 1,339 | 89.8 | 1.5 | 0.1 | 4.5 | 0.0 | 2.4 | 1.6 |
| 09 73000 | Stamford | 37.6 | 131,397 | 211 | 3,495 | 64.2 | 16.3 | 0.0 | 8.0 | 0.0 | 6.9 | 4.6 |
| 09 76500 | Torrington | 39.8 | 33,870 | 1,138 | 851 | D | D | D | D | D | D | D |
| 09 80000 | Waterbury | 28.5 | 106,826 | 289 | 3,748 | 61.8 | 21.6 | 0.2 | 1.9 | 0.0 | 10.9 | 3.6 |
| 09 82800 | West Haven | 10.8 | 54,367 | 714 | 5,034 | 67.3 | 15.4 | 1.4 | 2.4 | 0.0 | 5.8 | 7.8 |
| 10 00000 | **DELAWARE** | 1,948.5 | 986,809 | X | 506 | 67.7 | 22.5 | 0.4 | 3.8 | 0.0 | 2.4 | 3.1 |
| 10 21200 | Dover | 23.7 | 38,199 | 1,014 | 1,612 | 43.0 | 49.1 | 1.3 | 1.7 | 0.0 | 3.1 | 1.9 |
| 10 50670 | Newark | 9.4 | 33,849 | 1,139 | 3,601 | 75.0 | 10.8 | 0.3 | 7.1 | 0.0 | 2.6 | 4.2 |
| 10 77580 | Wilmington | 10.9 | 70,376 | 518 | 6,457 | 37.0 | 55.3 | 0.2 | 1.2 | 0.0 | 4.6 | 1.7 |
| 11 00000 | **DISTRICT OF COLUMBIA** | 61.1 | 712,816 | X | 11,666 | 42.5 | 45.4 | 0.3 | 4.1 | 0.0 | 4.4 | 3.3 |
| 11 50000 | Washington | 61.1 | 712,816 | 20 | 11,666 | 42.5 | 45.4 | 0.3 | 4.1 | 0.0 | 3.4 | 2.9 |
| 12 00000 | **FLORIDA** | 53,647.9 | 21,733,312 | X | 405 | 74.5 | 16.0 | 0.3 | 2.8 | 0.1 | 3.4 | 2.9 |
| 12 00950 | Altamonte Springs | 9.1 | 43,799 | 883 | 4,813 | 66.0 | 16.1 | 1.4 | 2.8 | 0.0 | 10.3 | 3.4 |
| 12 01700 | Apopka | 34.6 | 53,870 | 719 | 1,557 | 54.2 | 33.6 | 0.0 | 1.4 | 0.0 | 5.6 | 5.2 |
| 12 02681 | Aventura | 2.6 | 36,553 | 1,052 | 14,059 | D | D | D | D | D | D | D |
| 12 07300 | Boca Raton | 29.2 | 99,893 | 316 | 3,421 | 86.0 | 5.2 | 0.2 | 4.3 | 0.0 | 1.8 | 2.5 |
| 12 07525 | Bonita Springs | 38.4 | 60,830 | 617 | 1,584 | 80.1 | 6.4 | 0.0 | 0.8 | 0.1 | 10.8 | 1.8 |
| 12 07875 | Boynton Beach | 16.2 | 80,108 | 436 | 4,945 | 62.6 | 32.8 | 0.0 | 2.0 | 0.0 | 0.9 | 1.7 |
| 12 07950 | Bradenton | 14.3 | 59,737 | 629 | 4,177 | 75.6 | 19.4 | 0.1 | 1.1 | 0.0 | 1.8 | 2.1 |
| 12 10275 | Cape Coral | 106.0 | 200,972 | 117 | 1,896 | 88.4 | 5.8 | 0.0 | 1.8 | 0.0 | 2.0 | 1.9 |
| 12 11050 | Casselberry | 7.0 | 29,413 | 1,273 | 4,202 | D | D | D | D | D | D | 4.1 |
| 12 12875 | Clearwater | 26.1 | 116,552 | 249 | 4,466 | 79.3 | 10.2 | 0.6 | 4.3 | 0.0 | 1.5 | D |
| 12 12925 | Clermont | 18.0 | 39,850 | 967 | 2,214 | D | D | D | D | D | D | 2.1 |
| 12 13275 | Coconut Creek | 11.2 | 61,128 | 612 | 5,458 | 81.5 | 12.8 | 0.0 | 2.9 | 0.0 | 1.8 | 1.5 |
| 12 14125 | Cooper City | 8.0 | 35,700 | 1,083 | 4,463 | 74.6 | 10.4 | 0.4 | 11.3 | 0.0 | 0.9 | 2.6 |
| 12 14250 | Coral Gables | 12.9 | 49,130 | 791 | 3,809 | 92.7 | 1.5 | 0.0 | 2.2 | 0.0 | 1.6 | 3.4 |
| 12 14400 | Coral Springs | 22.9 | 133,519 | 207 | 5,831 | 64.9 | 25.7 | 0.4 | 4.1 | 0.0 | 1.9 | 4.4 |
| 12 15968 | Cutler Bay | 9.9 | 43,203 | 891 | 4,364 | 76.1 | 13.5 | 0.4 | 3.6 | 0.0 | 5.1 | 3.5 |
| 12 16335 | Dania Beach | 7.8 | 32,207 | 1,187 | 4,129 | 68.0 | 21.7 | 0.0 | 1.7 | 0.0 | 10.5 | 3.0 |
| 12 16475 | Davie | 34.9 | 106,621 | 290 | 3,055 | 68.4 | 7.5 | 0.9 | 9.7 | 0.0 | 3.3 | 4.7 |
| 12 16525 | Daytona Beach | 65.6 | 70,143 | 520 | 1,069 | 56.7 | 32.6 | 0.2 | 2.5 | 0.0 | | |

1. Dry land or land partially or temporarily covered by water.   2. Hispanic or Latino persons may be of any race.

# Table D. Cities — **Population**

| City | Percent Hispanic or Latino[1], 2019 | Percent foreign born, 2019 | Age of population (percent), 2019 | | | | | | | Median age, 2019 | Percent female, 2019 | Population | | | |
| | | | Under 18 years | 18 to 24 years | 25 to 34 years | 35 to 44 years | 45 to 54 years | 55 to 64 years | 65 years and over | | | Census counts | | Percent change | |
| | | | | | | | | | | | | 2000 | 2010 | 2000-2010 | 2001-2020 |
| | 12 | 13 | 14 | 15 | 16 | 17 | 18 | 19 | 20 | 21 | 22 | 23 | 24 | 25 | 26 |
| **COLORADO— Cont'd** | | | | | | | | | | | | | | | |
| Castle Rock | 8.8 | 5.8 | 27.5 | 7.0 | 11.9 | 16.4 | 14.7 | 10.9 | 11.7 | 37.0 | 51.2 | 20,224 | 48,238 | 138.5 | 46.3 |
| Centennial | 9.6 | 10.7 | 23.3 | 6.6 | 10.7 | 16.0 | 12.6 | 14.7 | 16.1 | 40.8 | 49.7 | NA | 100,391 | NA | 9.7 |
| Colorado Springs | 18.8 | 7.8 | 21.6 | 10.9 | 16.6 | 12.9 | 11.4 | 11.6 | 15.0 | 35.6 | 49.9 | 360,890 | 417,450 | 15.7 | 15.5 |
| Commerce City | 48.5 | 23.8 | 30.4 | 6.0 | 15.2 | 14.5 | 15.2 | 11.6 | 7.0 | 33.6 | 49.4 | 20,991 | 45,849 | 118.4 | 29.4 |
| Denver | 29.3 | 13.1 | 19.1 | 8.0 | 23.7 | 16.2 | 11.5 | 9.7 | 11.9 | 34.7 | 49.9 | 554,636 | 599,454 | 8.1 | 22.7 |
| Englewood | 20.4 | 8.6 | 19.8 | 7.5 | 18.3 | 15.5 | 11.4 | 14.0 | 13.5 | 37.2 | 49.4 | 31,727 | 30,260 | -4.6 | 16.0 |
| Fort Collins | 10.5 | 6.3 | 16.3 | 22.1 | 18.5 | 10.9 | 10.2 | 10.4 | 11.6 | 30.6 | 48.8 | 118,652 | 144,861 | 22.1 | 16.1 |
| Fountain | 17.9 | 5.1 | 34.6 | 6.6 | 18.8 | 16.1 | 11.5 | 7.6 | 4.8 | 30.9 | 46.7 | 15,197 | 25,899 | 70.4 | 19.8 |
| Grand Junction | 18.2 | 3.6 | 19.0 | 13.1 | 13.5 | 11.1 | 9.4 | 12.9 | 21.1 | 38.4 | 51.9 | 41,986 | 59,034 | 40.6 | 7.5 |
| Greeley | 39.4 | 12.6 | 22.5 | 14.9 | 15.9 | 12.0 | 11.9 | 11.2 | 11.7 | 33.1 | 49.4 | 76,930 | 92,953 | 20.8 | 17.2 |
| Lakewood | 21.6 | 8.3 | 18.0 | 8.8 | 18.9 | 15.3 | 11.4 | 10.6 | 17.1 | 37.6 | 49.7 | 144,126 | 142,592 | -1.1 | 10.4 |
| Littleton | 10.2 | 6.1 | 17.8 | 6.9 | 18.4 | 11.4 | 13.1 | 13.2 | 19.2 | 40.9 | 49.8 | 40,340 | 41,624 | 3.2 | 10.5 |
| Longmont | 23.8 | 11.6 | 21.1 | 9.1 | 12.8 | 14.6 | 12.3 | 14.8 | 15.3 | 40.2 | 51.7 | 71,093 | 86,330 | 21.4 | 14.0 |
| Loveland | 14.9 | 2.2 | 22.5 | 6.7 | 14.7 | 15.5 | 10.0 | 10.8 | 19.9 | 38.1 | 51.3 | 50,608 | 67,005 | 32.4 | 24.6 |
| Northglenn | 35.3 | 8.6 | 20.4 | 12.8 | 16.8 | 16.5 | 9.2 | 11.1 | 13.1 | 35.0 | 46.9 | 31,575 | 35,690 | 13.0 | 8.8 |
| Parker | 11.4 | 5.4 | 26.5 | 8.6 | 14.3 | 15.9 | 16.2 | 9.7 | 8.9 | 35.4 | 47.7 | 23,558 | 45,352 | 92.5 | 29.4 |
| Pueblo | 50.0 | 4.0 | 23.6 | 8.9 | 13.9 | 13.4 | 10.5 | 12.1 | 17.5 | 37.3 | 49.5 | 102,121 | 106,539 | 4.3 | 5.8 |
| Thornton | 37.6 | 15.8 | 27.0 | 7.8 | 17.2 | 13.0 | 13.3 | 10.6 | 11.0 | 34.0 | 49.8 | 82,384 | 118,859 | 44.3 | 24.6 |
| Westminster | 20.6 | 8.7 | 22.9 | 7.4 | 17.9 | 15.2 | 12.5 | 11.8 | 12.3 | 36.2 | 50.3 | 100,940 | 106,142 | 5.2 | 8.4 |
| Wheat Ridge | 26.2 | 4.3 | 20.1 | 7.0 | 15.1 | 13.4 | 9.2 | 16.7 | 18.6 | 40.2 | 50.3 | 32,913 | 30,179 | -8.3 | 5.6 |
| **CONNECTICUT** | 16.9 | 14.8 | 20.4 | 9.6 | 12.5 | 12.0 | 13.5 | 14.5 | 17.6 | 41.2 | 51.1 | 3,405,565 | 3,574,151 | 5.0 | -0.5 |
| Bridgeport | 43.1 | 31.4 | 24.5 | 12.4 | 14.8 | 13.4 | 11.6 | 10.9 | 12.4 | 33.9 | 52.4 | 139,529 | 144,247 | 3.4 | -0.5 |
| Bristol | 13.4 | 11.6 | 20.2 | 7.5 | 16.0 | 15.8 | 13.0 | 11.0 | 16.6 | 37.2 | 52.5 | 60,062 | 60,491 | 0.7 | -1.3 |
| Danbury | 29.2 | 35.8 | 17.0 | 10.4 | 14.1 | 15.0 | 11.1 | 15.9 | 16.6 | 40.4 | 50.8 | 74,848 | 80,834 | 8.0 | 4.3 |
| Hartford | 44.9 | 20.1 | 24.0 | 14.2 | 15.2 | 13.8 | 11.4 | 10.6 | 10.8 | 32.5 | 52.1 | 121,578 | 124,756 | 2.6 | -2.6 |
| Meriden | 34.9 | 13.0 | 22.4 | 8.6 | 18.5 | 13.1 | 12.0 | 10.5 | 14.8 | 35.8 | 53.8 | 58,244 | 60,825 | 4.4 | -3.0 |
| Middletown | 15.1 | 7.9 | 16.9 | 15.9 | 13.6 | 8.4 | 14.8 | 14.4 | 16.0 | 39.4 | 49.3 | 43,167 | 47,650 | 10.4 | -3.4 |
| Milford | 2.8 | 13.1 | 16.2 | 7.2 | 16.2 | 11.1 | 14.0 | 15.6 | 19.6 | 43.9 | 53.5 | 50,594 | 52,756 | 4.2 | 3.8 |
| Naugatuck | 12.8 | 9.9 | 22.9 | 8.7 | 11.8 | 7.7 | 16.2 | 19.0 | 13.6 | 43.0 | 47.9 | 30,989 | 31,871 | 2.8 | -3.0 |
| New Britain | 46.5 | 17.0 | 20.5 | 14.4 | 16.3 | 11.4 | 11.0 | 10.8 | 15.6 | 33.9 | 51.1 | 71,538 | 73,203 | 2.3 | -1.4 |
| New Haven | 32.1 | 19.6 | 20.3 | 16.0 | 18.4 | 12.5 | 12.3 | 9.2 | 11.2 | 32.0 | 51.4 | 123,626 | 129,864 | 5.0 | 0.7 |
| New London | 35.5 | 22.3 | 18.0 | 20.7 | 12.7 | 13.5 | 8.9 | 16.3 | 9.8 | 33.6 | 53.4 | 25,671 | 27,620 | 7.6 | -2.7 |
| Norwalk | 24.8 | 30.6 | 17.6 | 9.7 | 14.0 | 15.6 | 13.5 | 14.1 | 15.5 | 40.6 | 49.6 | 82,951 | 85,609 | 3.2 | 3.6 |
| Norwich | 18.7 | 14.2 | 18.1 | 10.5 | 15.1 | 12.5 | 11.8 | 14.8 | 17.2 | 39.7 | 52.8 | 36,117 | 40,503 | 12.1 | -4.7 |
| Shelton | 10.1 | 9.5 | 15.6 | 5.8 | 12.0 | 11.4 | 14.5 | 15.6 | 25.1 | 49.9 | 52.0 | 38,101 | 39,563 | 3.8 | 3.5 |
| Stamford | 28.0 | 33.9 | 19.7 | 8.3 | 18.1 | 11.9 | 13.5 | 13.4 | 15.1 | 37.4 | 51.5 | 117,083 | 122,633 | 4.7 | 7.1 |
| Torrington | 8.7 | 8.6 | 17.0 | 9.0 | 10.1 | 10.0 | 14.9 | 17.0 | 22.0 | 48.3 | 45.8 | 35,202 | 36,395 | 3.4 | -6.9 |
| Waterbury | 37.4 | 14.9 | 23.0 | 12.3 | 15.0 | 11.3 | 12.3 | 12.5 | 13.6 | 34.9 | 48.0 | 107,271 | 110,305 | 2.8 | -3.2 |
| West Haven | 17.9 | 16.6 | 17.5 | 12.4 | 12.0 | 11.5 | 15.1 | 17.7 | 13.8 | 42.2 | 50.8 | 52,360 | 55,564 | 6.1 | -2.2 |
| **DELAWARE** | 9.6 | 10.0 | 20.9 | 8.5 | 13.2 | 11.7 | 12.2 | 14.1 | 19.5 | 41.4 | 51.7 | 783,600 | 897,947 | 14.6 | 9.9 |
| Dover | 9.1 | 11.8 | 19.3 | 15.7 | 15.1 | 8.6 | 11.4 | 13.1 | 16.8 | 34.9 | 53.6 | 32,135 | 35,795 | 11.4 | 6.7 |
| Newark | 8.9 | 16.2 | 11.3 | 40.2 | 14.5 | 4.6 | 5.8 | 9.3 | 14.3 | 24.4 | 53.6 | 28,547 | 31,776 | 11.3 | 6.5 |
| Wilmington | 12.9 | 6.4 | 21.6 | 6.5 | 20.0 | 12.4 | 12.8 | 13.4 | 13.2 | 36.5 | 52.9 | 72,664 | 70,941 | -2.4 | -0.8 |
| **DISTRICT OF COLUMBIA** | 11.3 | 12.1 | 18.1 | 10.3 | 23.3 | 15.4 | 10.6 | 9.9 | 12.4 | 34.3 | 52.6 | 572,059 | 601,767 | 5.2 | 18.5 |
| Washington | 11.3 | 12.1 | 18.1 | 10.3 | 23.3 | 15.4 | 10.6 | 9.9 | 12.4 | 34.3 | 52.6 | 572,059 | 601,767 | 5.2 | 18.5 |
| **FLORIDA** | 26.4 | 21.1 | 19.7 | 8.2 | 12.9 | 12.2 | 12.6 | 13.5 | 20.9 | 42.4 | 51.1 | 15,982,378 | 18,804,589 | 17.7 | 15.6 |
| Altamonte Springs | 29.6 | 26.6 | 17.5 | 7.7 | 21.0 | 15.3 | 11.6 | 12.0 | 14.9 | 36.5 | 47.4 | 41,200 | 41,573 | 0.9 | 5.4 |
| Apopka | 18.9 | 12.5 | 23.5 | 7.2 | 14.9 | 11.3 | 16.5 | 10.7 | 15.8 | 38.3 | 49.0 | 26,642 | 42,235 | 58.5 | 27.5 |
| Aventura | 45.8 | 52.2 | 15.0 | 6.9 | 6.3 | 11.7 | 16.9 | 15.7 | 27.6 | 51.4 | 53.8 | 25,267 | 35,762 | 41.5 | 2.2 |
| Boca Raton | 12.9 | 21.2 | 17.9 | 10.6 | 9.0 | 10.0 | 12.1 | 14.9 | 25.5 | 47.6 | 53.7 | 74,764 | 84,409 | 12.9 | 18.3 |
| Bonita Springs | 17.9 | 17.0 | 11.3 | 3.3 | 7.0 | 6.0 | 12.8 | 15.4 | 44.1 | 61.6 | 49.1 | 32,797 | 43,936 | 34.0 | 38.5 |
| Boynton Beach | 15.1 | 27.9 | 15.2 | 7.5 | 15.3 | 17.4 | 9.4 | 13.8 | 21.5 | 42.7 | 48.8 | 60,389 | 68,297 | 13.1 | 17.3 |
| Bradenton | 18.1 | 12.1 | 19.8 | 8.3 | 17.5 | 8.4 | 10.1 | 9.8 | 26.1 | 39.8 | 52.5 | 49,504 | 49,221 | -0.6 | 21.4 |
| Cape Coral | 22.1 | 14.0 | 18.2 | 6.8 | 11.2 | 12.2 | 12.5 | 15.0 | 24.0 | 46.4 | 49.7 | 102,286 | 154,307 | 50.9 | 30.2 |
| Casselberry | 31.8 | 17.9 | 18.8 | 12.2 | 20.9 | 7.7 | 14.9 | 10.5 | 14.8 | 33.9 | 54.8 | 22,629 | 26,030 | 15.0 | 13.0 |
| Clearwater | 17.8 | 18.5 | 17.4 | 5.7 | 13.1 | 11.5 | 14.0 | 14.6 | 23.7 | 47.1 | 53.1 | 108,787 | 109,091 | 0.3 | 6.8 |
| Clermont | 31.6 | 26.2 | 20.3 | 6.6 | 9.5 | 14.0 | 14.2 | 11.0 | 24.4 | 44.1 | 51.0 | 9,333 | 28,811 | 208.7 | 38.3 |
| Coconut Creek | 22.2 | 27.1 | 18.9 | 7.0 | 14.3 | 13.2 | 17.1 | 12.9 | 16.6 | 42.9 | 51.4 | 43,566 | 53,046 | 21.8 | 15.2 |
| Cooper City | 27.4 | 25.8 | 25.4 | 6.7 | 8.2 | 15.5 | 13.5 | 16.6 | 14.2 | 40.7 | 51.7 | 27,939 | 28,534 | 2.1 | 25.1 |
| Coral Gables | 56.8 | 41.9 | 16.0 | 15.5 | 11.1 | 8.8 | 14.9 | 12.0 | 21.7 | 44.4 | 55.4 | 42,249 | 46,744 | 10.6 | 5.1 |
| Coral Springs | 33.5 | 32.6 | 27.3 | 9.3 | 11.6 | 13.8 | 12.6 | 13.3 | 12.1 | 36.4 | 52.2 | 117,549 | 122,591 | 4.3 | 8.9 |
| Cutler Bay | 64.8 | 40.8 | 24.5 | 6.7 | 14.9 | 12.3 | 14.4 | 11.4 | 15.9 | 38.1 | 50.2 | NA | 40,289 | ********** | 7.2 |
| Dania Beach | 36.5 | 33.0 | 22.6 | 5.2 | 11.7 | 17.8 | 12.0 | 14.7 | 15.9 | 40.9 | 51.2 | 20,061 | 29,689 | 48.0 | 8.5 |
| Davie | 44.2 | 33.1 | 22.4 | 12.6 | 14.7 | 12.2 | 15.4 | 11.8 | 11.0 | 35.2 | 49.5 | 75,720 | 91,950 | 21.4 | 16.0 |
| Daytona Beach | 10.1 | 9.4 | 14.1 | 13.0 | 14.2 | 10.9 | 12.7 | 12.6 | 22.5 | 42.7 | 52.2 | 64,112 | 61,587 | -3.9 | 13.9 |

1. May be of any race.

# Table D. Cities — Households, Group Quarters, Crime, and Education

| City | Households, 2019 Number | Persons per household | Percent Family | Married couple family | Female family | Non-family | One person | Persons in group quarters, 2019 | Violent Number | Violent Rate[2] | Property Number | Property Rate[2] | Population age 25 and over | High school graduate or less | Bachelor's degree or more |
|---|---|---|---|---|---|---|---|---|---|---|---|---|---|---|---|
| | 27 | 28 | 29 | 30 | 31 | 32 | 33 | 34 | 35 | 36 | 37 | 38 | 39 | 40 | 41 |
| **COLORADO— Cont'd** | | | | | | | | | | | | | | | |
| Castle Rock | 23,803 | 2.85 | 77.4 | 65.1 | 12.3 | 22.6 | 16.5 | 703 | 30 | 45 | 962 | 1,431 | 44,875 | 18.0 | 47.7 |
| Centennial | 40,151 | 2.70 | 75.4 | 63.4 | 11.9 | 24.6 | 20.7 | 2,545 | NA | NA | NA | NA | 77,688 | 16.4 | 56.9 |
| Colorado Springs | 188,837 | 2.49 | 64.3 | 48.0 | 16.3 | 35.7 | 26.7 | 8,014 | 2,806 | 585 | 17,587 | 3,667 | 322,804 | 25.6 | 40.3 |
| Commerce City | 18,407 | 3.26 | 79.2 | 59.6 | 19.6 | 20.8 | 16.5 | 362 | 272 | 452 | 1,845 | 3,065 | 38,374 | 45.1 | 22.9 |
| Denver | 318,445 | 2.24 | 47.3 | 34.1 | 13.1 | 52.7 | 38.2 | 14,176 | 5,459 | 749 | 27,288 | 3,744 | 530,543 | 25.8 | 53.1 |
| Englewood | 14,948 | 2.31 | 53.0 | 37.4 | 15.6 | 47.0 | 35.0 | 341 | 72 | 204 | 1,843 | 5,225 | 25,397 | 33.3 | 35.6 |
| Fort Collins | 70,831 | 2.28 | 51.9 | 39.9 | 11.9 | 48.1 | 26.4 | 8,514 | 371 | 217 | 3,713 | 2,173 | 104,896 | 18.2 | 57.5 |
| Fountain | 9,289 | 3.31 | 68.7 | 56.8 | 11.9 | 31.3 | 22.5 | 0 | 127 | 409 | 665 | 2,142 | 18,065 | 35.8 | 19.6 |
| Grand Junction | 27,300 | 2.24 | 56.6 | 42.5 | 14.0 | 43.4 | 31.9 | 2,409 | 235 | 367 | 2,463 | 3,852 | 43,227 | 33.5 | 34.9 |
| Greeley | 38,299 | 2.67 | 65.4 | 49.7 | 15.7 | 34.6 | 26.0 | 6,543 | 386 | 353 | 2,542 | 2,327 | 68,017 | 40.8 | 26.2 |
| Lakewood | 67,857 | 2.29 | 56.1 | 38.9 | 17.2 | 43.9 | 33.0 | 2,253 | NA | NA | NA | NA | 115,684 | 29.7 | 44.9 |
| Littleton | 21,840 | 2.19 | 56.6 | 44.0 | 12.6 | 43.4 | 35.2 | 584 | 43 | 88 | 1,372 | 2,810 | 36,427 | 21.4 | 54.8 |
| Longmont | 39,984 | 2.40 | 62.4 | 46.9 | 15.5 | 37.6 | 30.9 | 660 | 422 | 431 | 2,548 | 2,602 | 67,502 | 24.9 | 49.6 |
| Loveland | 33,393 | 2.35 | 60.9 | 49.9 | 11.0 | 39.1 | 28.8 | 382 | 222 | 282 | 1,571 | 1,992 | 55,856 | 28.9 | 38.4 |
| Northglenn | 13,678 | 2.83 | 60.9 | 44.5 | 16.5 | 39.1 | 29.7 | 136 | 180 | 457 | 1,229 | 3,118 | 25,930 | 45.7 | 19.2 |
| Parker | 20,856 | 2.76 | 74.5 | 64.9 | 9.6 | 25.5 | 18.5 | 40 | 55 | 96 | 826 | 1,448 | 37,470 | 23.2 | 48.8 |
| Pueblo | 44,945 | 2.43 | 59.7 | 37.8 | 21.9 | 40.3 | 33.2 | 3,354 | NA | NA | 4,801 | 4,272 | 75,797 | 40.7 | 20.2 |
| Thornton | 46,900 | 3.01 | 72.5 | 57.8 | 14.6 | 27.5 | 20.9 | 481 | 388 | 273 | 4,303 | 3,027 | 92,221 | 42.5 | 28.0 |
| Westminster | 43,890 | 2.57 | 62.3 | 47.7 | 14.6 | 37.7 | 28.3 | 416 | 316 | 276 | 3,713 | 3,246 | 78,947 | 29.3 | 41.3 |
| Wheat Ridge | 13,302 | 2.31 | 59.4 | 44.3 | 15.1 | 40.6 | 33.0 | 633 | NA | NA | NA | NA | 22,845 | 33.3 | 37.8 |
| **CONNECTICUT** | 1,377,166 | 2.51 | 64.3 | 47.2 | 17.2 | 35.7 | 29.0 | 110,867 | NA | NA | NA | NA | 2,496,420 | 36.1 | 39.8 |
| Bridgeport | 50,103 | 2.80 | 60.9 | 30.5 | 30.4 | 39.1 | 31.4 | 4,270 | 843 | 582 | 2,465 | 1,701 | 91,157 | 60.3 | 17.6 |
| Bristol | 25,417 | 2.33 | 58.6 | 41.2 | 17.4 | 41.4 | 33.9 | 788 | 53 | 88 | 822 | 1,371 | 43,343 | 36.7 | 28.9 |
| Danbury | 32,310 | 2.52 | 62.0 | 46.0 | 15.9 | 38.0 | 32.9 | 3,396 | 97 | 114 | 1,045 | 1,227 | 61,501 | 43.2 | 31.5 |
| Hartford | 48,168 | 2.38 | 52.1 | 17.7 | 34.4 | 47.9 | 39.9 | 7,368 | 1,049 | 858 | 3,424 | 2,801 | 75,488 | 58.2 | 14.8 |
| Meriden | 22,430 | 2.60 | 59.3 | 35.8 | 23.5 | 40.7 | 33.1 | 1,009 | 157 | 264 | 1,020 | 1,718 | 41,019 | 48.2 | 21.7 |
| Middletown | 20,728 | 2.03 | 48.4 | 31.7 | 16.7 | 51.6 | 46.2 | 4,170 | 38 | 83 | 613 | 1,334 | 31,102 | 34.1 | 39.8 |
| Milford | 22,040 | 2.39 | 61.7 | 49.5 | 12.2 | 38.3 | 33.8 | 446 | NA | NA | NA | NA | 40,718 | 28.6 | 45.5 |
| Naugatuck | 11,425 | 2.70 | 66.7 | 39.8 | 26.9 | 33.3 | 31.3 | 280 | 30 | 96 | 441 | 1,413 | 21,264 | 51.6 | 23.2 |
| New Britain | 29,359 | 2.37 | 51.0 | 26.4 | 24.7 | 49.0 | 38.9 | 2,783 | 306 | 423 | 1,195 | 1,652 | 47,185 | 55.9 | 18.4 |
| New Haven | 46,982 | 2.57 | 56.8 | 27.7 | 29.0 | 43.2 | 34.4 | 9,404 | 1,168 | 895 | 4,958 | 3,799 | 82,937 | 48.6 | 32.4 |
| New London | 10,278 | 2.23 | 62.6 | 36.8 | 25.8 | 37.4 | 32.4 | 3,962 | 79 | 294 | 461 | 1,717 | 16,456 | 41.6 | 27.4 |
| Norwalk | 33,616 | 2.62 | 64.9 | 45.9 | 19.0 | 35.1 | 26.3 | 634 | 183 | 205 | 1,247 | 1,394 | 64,588 | 32.8 | 46.3 |
| Norwich | 16,506 | 2.32 | 60.3 | 35.1 | 25.2 | 39.7 | 32.5 | 470 | 149 | 382 | 533 | 1,368 | 27,683 | 44.5 | 21.2 |
| Shelton | 16,628 | 2.44 | 68.4 | 52.7 | 15.7 | 31.6 | 25.3 | 488 | 22 | 53 | 304 | 736 | 32,307 | 30.4 | 46.4 |
| Stamford | 49,707 | 2.58 | 62.3 | 47.2 | 15.1 | 37.7 | 27.6 | 1,175 | 264 | 202 | 1,803 | 1,380 | 93,336 | 29.8 | 53.3 |
| Torrington | 14,542 | 2.29 | 57.1 | 42.9 | 14.2 | 42.9 | 36.2 | 712 | 36 | 106 | 372 | 1,095 | 25,206 | 45.8 | 26.1 |
| Waterbury | 42,113 | 2.51 | 56.8 | 28.6 | 28.2 | 43.2 | 37.3 | 1,954 | 325 | 301 | 3,263 | 3,027 | 69,692 | 59.8 | 16.5 |
| West Haven | 20,819 | 2.46 | 61.1 | 41.6 | 19.5 | 38.9 | 33.7 | 3,326 | 85 | 155 | 1,091 | 1,991 | 38,284 | 54.1 | 20.4 |
| **DELAWARE** | 376,239 | 2.52 | 64.5 | 47.3 | 17.2 | 35.5 | 28.8 | 24,895 | NA | NA | NA | NA | 687,311 | 39.9 | 33.2 |
| Dover | 15,946 | 2.14 | 55.5 | 28.4 | 27.1 | 44.5 | 39.5 | 4,016 | 334 | 871 | 2,057 | 5,362 | 24,805 | 40.2 | 28.1 |
| Newark | 11,072 | 2.36 | 50.2 | 38.5 | 11.7 | 49.8 | 26.4 | 7,358 | 85 | 250 | 708 | 2,085 | 16,248 | 27.5 | 54.6 |
| Wilmington | 31,754 | 2.10 | 42.7 | 18.7 | 24.0 | 57.3 | 50.2 | 3,356 | 1,058 | 1,498 | 3,150 | 4,460 | 50,453 | 51.6 | 29.0 |
| **DISTRICT OF COLUMBIA** | 291,570 | 2.29 | 42.9 | 26.9 | 16.0 | 57.1 | 44.8 | 38,858 | NA | NA | NA | NA | 505,145 | 24.0 | 59.7 |
| Washington | 291,570 | 2.29 | 42.9 | 26.9 | 16.0 | 57.1 | 44.8 | 38,858 | 6,896 | 977 | 29,965 | 4,246 | 505,145 | 24.0 | 59.7 |
| **FLORIDA** | 7,905,832 | 2.66 | 64.3 | 46.4 | 17.9 | 35.7 | 28.6 | 429,975 | NA | NA | NA | NA | 15,484,502 | 40.0 | 30.7 |
| Altamonte Springs | 19,963 | 2.19 | 51.2 | 39.2 | 12.0 | 48.8 | 38.0 | 330 | 107 | 240 | 1,408 | 3,158 | 33,055 | 27.4 | 43.0 |
| Apopka | 18,989 | 2.80 | 65.7 | 48.4 | 17.4 | 34.3 | 30.7 | 222 | 206 | 374 | 1,931 | 3,506 | 36,994 | 36.2 | 29.0 |
| Aventura | 16,324 | 2.26 | 59.8 | 46.3 | 13.5 | 40.2 | 35.8 | 125 | 69 | 180 | 1,815 | 4,744 | 28,893 | 20.9 | 51.9 |
| Boca Raton | 40,845 | 2.35 | 56.8 | 41.8 | 15.0 | 43.2 | 33.1 | 3,927 | 186 | 184 | 2,183 | 2,158 | 71,434 | 18.9 | 56.1 |
| Bonita Springs | 26,076 | 2.28 | 67.8 | 60.3 | 7.5 | 32.2 | 27.8 | 77 | NA | NA | NA | NA | 50,934 | 31.1 | 41.8 |
| Boynton Beach | 31,863 | 2.45 | 54.9 | 40.8 | 14.0 | 45.1 | 35.7 | 589 | 566 | 713 | 2,618 | 3,299 | 60,863 | 36.4 | 32.9 |
| Bradenton | 21,716 | 2.67 | 58.3 | 38.4 | 19.9 | 41.7 | 35.0 | 1,498 | 360 | 612 | 1,474 | 2,508 | 42,753 | 38.9 | 26.7 |
| Cape Coral | 67,176 | 2.89 | 72.1 | 56.8 | 15.2 | 27.9 | 21.7 | 683 | 226 | 116 | 2,176 | 1,121 | 145,756 | 43.3 | 24.6 |
| Casselberry | 12,254 | 2.34 | 56.9 | 40.7 | 16.2 | 43.1 | 27.5 | 103 | 90 | 308 | 936 | 3,201 | 19,836 | 33.3 | 32.3 |
| Clearwater | 48,641 | 2.35 | 55.3 | 42.6 | 12.7 | 44.7 | 34.5 | 2,711 | 469 | 399 | 2,810 | 2,392 | 89,925 | 32.6 | 36.1 |
| Clermont | 14,160 | 2.71 | 70.8 | 55.4 | 15.4 | 29.2 | 27.2 | 215 | 79 | 209 | 1,793 | 1,617 | 28,267 | 34.7 | 30.2 |
| Coconut Creek | 22,784 | 2.68 | 65.9 | 46.8 | 19.1 | 34.1 | 26.0 | 266 | 82 | 131 | 1,010 | 976 | 45,410 | 31.7 | 35.6 |
| Cooper City | 12,147 | 2.95 | 83.0 | 66.0 | 17.0 | 17.0 | 13.6 | 9 | 26 | 70 | 360 | 1,343 | 24,330 | 19.4 | 50.8 |
| Coral Gables | 19,472 | 2.31 | 59.6 | 48.8 | 10.8 | 40.4 | 34.3 | 4,771 | 50 | 97 | 1,343 | 2,606 | 34,034 | 16.1 | 66.2 |
| Coral Springs | 41,240 | 3.24 | 77.5 | 55.6 | 21.8 | 22.5 | 17.7 | 297 | 195 | 144 | 2,309 | 1,711 | 84,769 | 30.1 | 38.7 |
| Cutler Bay | 13,404 | 3.24 | 78.6 | 52.1 | 26.5 | 21.4 | 14.6 | 322 | 94 | 207 | 1,328 | 2,922 | 30,099 | 47.0 | 25.9 |
| Dania Beach | 12,114 | 2.65 | 63.4 | 39.4 | 24.0 | 36.6 | 29.0 | 170 | 221 | 678 | 1,006 | 3,087 | 23,293 | 28.8 | 34.1 |
| Davie | 35,708 | 2.93 | 69.1 | 43.5 | 25.6 | 30.9 | 22.4 | 1,642 | 276 | 254 | 2,796 | 2,577 | 69,153 | 33.4 | 37.3 |
| Daytona Beach | 28,839 | 2.27 | 48.3 | 29.6 | 18.7 | 51.7 | 39.9 | 3,875 | 794 | 1,137 | 2,833 | 4,057 | 50,471 | 45.5 | 20.6 |

1. Data for serious crimes have not been adjusted for underreporting. This may affect comparability between geographic areas and over time.
2. Per 100,000 population estimated by the FBI.
3. Persons 25 years old and over.

| City | Money income, 2019 Households | | | | | Median earnings, 2019 | | | Housing units, 2019 | | | | |
|---|---|---|---|---|---|---|---|---|---|---|---|---|---|
| | Median income | Percent with income less than $20,000 | Percent with income of $200,000 or more | Median family income | Median non-family income | All persons | Men | Women | Total | Occupied | Percent owner occupied | Median value[1] (dollars) | Median gross rent (dollars) |
| | 42 | 43 | 44 | 45 | 46 | 47 | 48 | 49 | 50 | 51 | 52 | 53 | 54 |
| COLORADO— Cont'd | | | | | | | | | | | | | |
| Castle Rock | 115,543 | 4.6 | 15.8 | 126,461 | 57,885 | 45,834 | 57,002 | 35,456 | 24,870 | 23,803 | 79.0 | 471,211 | 1,619 |
| Centennial | 111,257 | 2.9 | 18.9 | 127,933 | 66,666 | 53,321 | 65,079 | 44,663 | 41,432 | 40,151 | 84.9 | 492,660 | 1,733 |
| Colorado Springs | 70,527 | 10.3 | 6.4 | 84,717 | 44,102 | 36,114 | 41,655 | 30,376 | 199,168 | 188,837 | 59.9 | 318,167 | 1,212 |
| Commerce City | 87,133 | 7.5 | 8.8 | 94,978 | 56,126 | 39,789 | 49,597 | 31,462 | 18,849 | 18,407 | 78.0 | 349,484 | 1,225 |
| Denver | 75,646 | 11.8 | 11.7 | 98,247 | 59,201 | 47,146 | 52,385 | 40,928 | 338,346 | 318,445 | 49.8 | 447,451 | 1,433 |
| Englewood | 70,040 | 14.1 | 5.8 | 87,130 | 46,830 | 40,874 | 45,922 | 35,388 | 15,405 | 14,948 | 51.6 | 398,965 | 1,227 |
| Fort Collins | 70,474 | 14.2 | 8.2 | 107,654 | 45,202 | 32,034 | 39,685 | 26,725 | 72,603 | 70,831 | 54.8 | 428,933 | 1,403 |
| Fountain | 70,314 | 9.5 | 8.5 | 73,493 | 40,055 | 41,347 | 46,248 | 26,777 | 9,578 | 9,289 | 57.1 | 275,797 | 1,501 |
| Grand Junction | 59,325 | 15.8 | 5.4 | 78,773 | 35,884 | 28,836 | 36,145 | 21,650 | 29,035 | 27,300 | 60.8 | 270,491 | 959 |
| Greeley | 61,492 | 12.7 | 4.3 | 79,406 | 34,363 | 31,872 | 36,414 | 27,294 | 40,504 | 38,299 | 57.9 | 305,088 | 1,129 |
| Lakewood | 70,178 | 9.7 | 8.9 | 91,331 | 54,680 | 45,305 | 50,964 | 37,169 | 70,722 | 67,857 | 56.8 | 432,230 | 1,424 |
| Littleton | 81,004 | 8.2 | 11.3 | 101,842 | 50,213 | 46,823 | 51,839 | 40,621 | 23,354 | 21,840 | 58.0 | 457,033 | 1,336 |
| Longmont | 82,974 | 10.4 | 8.2 | 96,610 | 49,530 | 37,298 | 46,722 | 31,308 | 41,696 | 39,984 | 60.3 | 436,678 | 1,546 |
| Loveland | 71,740 | 9.9 | 4.4 | 84,071 | 48,108 | 40,213 | 46,634 | 31,241 | 41,696 | 39,984 | 63.8 | 351,609 | 1,447 |
| Northglenn | 60,502 | 3.8 | 4.0 | 66,052 | 46,431 | 32,624 | 37,139 | 26,724 | 33,990 | 33,393 | 52.9 | 352,048 | 1,474 |
| Parker | 106,382 | 5.1 | 14.7 | 123,601 | 49,695 | 51,950 | 72,162 | 34,196 | 22,081 | 20,856 | 76.2 | 461,492 | 1,744 |
| Pueblo | 43,148 | 24.1 | 2.6 | 55,436 | 29,330 | 30,144 | 32,091 | 26,589 | 48,427 | 44,945 | 59.8 | 159,799 | 808 |
| Thornton | 80,833 | 6.1 | 6.0 | 91,389 | 59,482 | 41,071 | 47,720 | 30,933 | 48,592 | 46,900 | 74.8 | 376,863 | 1,601 |
| Westminster | 81,322 | 6.0 | 10.9 | 102,377 | 55,486 | 43,797 | 52,287 | 37,229 | 45,621 | 43,890 | 65.3 | 388,130 | 1,490 |
| Wheat Ridge | 67,150 | 12.8 | 7.6 | 78,209 | 40,414 | 41,931 | 51,198 | 34,630 | 14,122 | 13,302 | 51.9 | 473,037 | 1,226 |
| CONNECTICUT | 78,833 | 12.1 | 13.2 | 101,272 | 45,001 | 42,196 | 50,790 | 36,085 | 1,524,959 | 1,377,166 | 65.0 | 280,690 | 1,177 |
| Bridgeport | 44,035 | 24.1 | 2.7 | 51,687 | 26,692 | 26,253 | 31,139 | 21,833 | 57,610 | 50,103 | 40.8 | 194,809 | 1,130 |
| Bristol | 69,538 | 11.1 | 5.6 | 87,088 | 46,863 | 44,054 | 49,878 | 36,212 | 27,178 | 25,417 | 64.7 | 199,430 | 1,059 |
| Danbury | 72,226 | 12.0 | 7.4 | 92,778 | 45,229 | 37,731 | 41,194 | 35,916 | 35,922 | 32,310 | 54.3 | 296,396 | 1,494 |
| Hartford | 36,762 | 30.7 | 1.5 | 50,515 | 24,560 | 27,335 | 32,278 | 25,211 | 57,267 | 48,168 | 25.8 | 173,466 | 995 |
| Meriden | 53,471 | 20.1 | 6.6 | 78,180 | 24,889 | 35,579 | 41,014 | 31,883 | 24,814 | 22,430 | 55.8 | 169,079 | 1,058 |
| Middletown | 52,641 | 21.9 | 6.4 | 95,561 | 32,399 | 32,166 | 41,622 | 25,190 | 22,929 | 20,728 | 53.0 | 248,695 | 1,093 |
| Milford | 94,845 | 8.9 | 14.7 | 119,402 | 50,100 | 52,091 | 65,072 | 44,795 | 23,781 | 22,040 | 72.2 | 326,427 | 1,613 |
| Naugatuck | 70,698 | 15.8 | 12.5 | 82,238 | 46,106 | 41,175 | 41,209 | 41,158 | 12,524 | 11,425 | 71.5 | 180,677 | 1,134 |
| New Britain | 50,031 | 19.4 | 3.0 | 60,151 | 38,588 | 32,348 | 41,313 | 25,713 | 32,902 | 29,359 | 39.9 | 165,769 | 978 |
| New Haven | 41,956 | 26.1 | 4.3 | 51,312 | 32,464 | 27,297 | 30,445 | 24,853 | 54,983 | 46,982 | 27.4 | 198,220 | 1,198 |
| New London | 57,799 | 13.0 | 2.9 | 60,302 | 42,077 | 26,735 | 30,727 | 23,803 | 11,348 | 10,278 | 40.1 | 186,457 | 1,015 |
| Norwalk | 89,881 | 8.9 | 17.0 | 97,064 | 70,707 | 50,237 | 52,120 | 48,795 | 38,043 | 33,616 | 54.8 | 457,962 | 1,759 |
| Norwich | 57,679 | 16.4 | 5.8 | 78,971 | 31,714 | 35,134 | 36,792 | 31,042 | 19,562 | 16,506 | 46.3 | 169,266 | 1,070 |
| Shelton | 95,833 | 8.0 | 18.2 | 122,101 | 49,144 | 52,122 | 67,095 | 49,818 | 16,983 | 16,628 | 80.3 | 358,992 | 1,247 |
| Stamford | 100,713 | 8.1 | 20.4 | 122,035 | 76,303 | 50,272 | 60,107 | 42,948 | 54,408 | 49,707 | 49.1 | 565,118 | 1,898 |
| Torrington | 61,789 | 11.2 | 4.6 | 80,000 | 41,358 | 34,080 | 37,382 | 28,924 | 17,181 | 14,542 | 67.4 | 158,797 | 881 |
| Waterbury | 42,754 | 25.6 | 2.1 | 52,038 | 32,291 | 31,985 | 37,932 | 28,044 | 49,367 | 42,113 | 37.6 | 132,264 | 956 |
| West Haven | 58,974 | 12.0 | 3.3 | 75,657 | 40,840 | 40,457 | 46,900 | 27,613 | 23,431 | 20,819 | 54.0 | 200,694 | 1,135 |
| DELAWARE | 70,176 | 12.3 | 7.4 | 87,148 | 44,071 | 37,430 | 42,673 | 32,229 | 443,764 | 376,239 | 70.3 | 261,707 | 1,116 |
| Dover | 46,460 | 23.0 | 2.8 | 52,146 | 30,577 | 25,014 | 31,907 | 17,215 | 18,256 | 15,946 | 43.6 | 183,940 | 1,059 |
| Newark | 64,781 | 18.8 | 9.5 | 87,813 | 43,417 | 12,212 | 25,716 | 6,998 | 12,371 | 11,072 | 46.1 | 300,000 | 1,243 |
| Wilmington | 47,722 | 28.4 | 3.4 | 57,909 | 39,955 | 35,529 | 44,486 | 30,428 | 35,447 | 31,754 | 37.1 | 174,412 | 1,021 |
| DISTRICT OF COLUMBIA | 92,266 | 14.4 | 20.7 | 130,291 | 76,529 | 61,960 | 69,852 | 56,071 | 322,814 | 291,570 | 41.5 | 646,504 | 1,603 |
| Washington | 92,266 | 14.4 | 20.7 | 130,291 | 76,529 | 61,960 | 69,852 | 56,071 | 322,814 | 291,570 | 41.5 | 646,504 | 1,603 |
| FLORIDA | 59,227 | 14.3 | 6.7 | 71,348 | 37,292 | 31,876 | 36,280 | 28,958 | 9,674,053 | 7,905,832 | 66.2 | 245,097 | 1,238 |
| Altamonte Springs | 54,973 | 7.2 | 3.3 | 74,346 | 45,106 | 31,275 | 35,110 | 25,549 | 22,431 | 19,963 | 40.4 | 204,658 | 1,198 |
| Apopka | 65,071 | 9.7 | 4.3 | 77,729 | 40,991 | 36,969 | 36,586 | 37,414 | 20,741 | 18,989 | 79.6 | 259,000 | 1,444 |
| Aventura | 69,038 | 13.8 | 14.9 | 75,743 | 43,825 | 42,328 | 44,192 | 40,836 | 27,948 | 16,324 | 67.4 | 337,828 | 1,883 |
| Boca Raton | 88,828 | 11.7 | 19.7 | 111,790 | 52,476 | 41,058 | 50,877 | 33,926 | 49,313 | 40,845 | 66.1 | 504,911 | 1,871 |
| Bonita Springs | 82,343 | 4.3 | 13.4 | 102,791 | 49,584 | 32,633 | 32,068 | 33,314 | 42,061 | 26,076 | 83.8 | 367,254 | 1,541 |
| Boynton Beach | 65,182 | 13.6 | 5.0 | 66,892 | 52,851 | 32,328 | 38,756 | 29,691 | 38,085 | 31,863 | 59.4 | 238,707 | 1,489 |
| Bradenton | 51,046 | 15.4 | 3.6 | 65,686 | 35,857 | 25,030 | 32,193 | 21,762 | 28,471 | 21,716 | 60.8 | 189,970 | 1,152 |
| Cape Coral | 68,282 | 9.2 | 6.3 | 75,192 | 42,227 | 33,635 | 35,506 | 32,327 | 85,097 | 67,176 | 80.0 | 249,554 | 1,365 |
| Casselberry | 61,459 | 20.6 | 1.8 | 65,983 | 37,182 | 31,591 | 36,666 | 26,812 | 13,119 | 12,254 | 59.0 | 217,512 | 1,237 |
| Clearwater | 52,431 | 15.0 | 5.2 | 80,093 | 35,623 | 34,974 | 41,613 | 28,519 | 60,504 | 48,641 | 60.3 | 224,704 | 1,180 |
| Clermont | 64,283 | 13.9 | 2.8 | 82,521 | 30,828 | 31,822 | 32,902 | 24,176 | 15,591 | 14,160 | 68.9 | 291,749 | 1,335 |
| Coconut Creek | 74,096 | 11.0 | 3.7 | 79,418 | 47,268 | 41,236 | 48,339 | 38,145 | 26,750 | 22,784 | 69.3 | 253,674 | 1,732 |
| Cooper City | 114,375 | 9.2 | 13.1 | 126,462 | 54,438 | 47,354 | 66,087 | 41,173 | 12,848 | 12,117 | 86.5 | 415,043 | 1,508 |
| Coral Gables | 98,343 | 10.1 | 24.7 | 157,649 | 49,196 | 55,165 | 75,296 | 40,628 | 22,396 | 19,472 | 59.0 | 841,986 | 1,712 |
| Coral Springs | 79,115 | 9.4 | 10.8 | 89,283 | 42,232 | 35,989 | 39,830 | 31,625 | 43,887 | 41,240 | 62.1 | 401,669 | 1,636 |
| Cutler Bay | 70,336 | 10.5 | 9.7 | 75,444 | 41,145 | 32,465 | 40,788 | 30,249 | 13,903 | 13,404 | 74.2 | 321,225 | 1,514 |
| Dania Beach | 51,591 | 12.8 | 5.7 | 69,578 | 34,572 | 32,442 | 35,665 | 28,743 | 16,189 | 12,114 | 53.0 | 250,521 | 1,458 |
| Davie | 71,112 | 9.7 | 10.3 | 75,957 | 48,846 | 34,517 | 40,007 | 30,884 | 37,920 | 35,708 | 63.9 | 345,014 | 1,583 |
| Daytona Beach | 40,439 | 25.4 | 1.8 | 55,704 | 29,707 | 26,937 | 27,703 | 26,379 | 35,808 | 28,839 | 46.9 | 168,684 | 981 |

1. Based on population estimated by the American Community Survey.

# Table D. Cities — Commuting, Computer Access, Migration, Labor Force, and Employment

| City | Commuting, 2019 | | Computer access, 2019 | | Migration, 2019 | | Civilian labor force, 2020 | | | | Civilian Employment, 2019 | | | |
|---|---|---|---|---|---|---|---|---|---|---|---|---|---|---|
| | | | | | | | | | Unemployment | | Population age 16 and older | | Population age 16 to 64 | |
| | Drove alone | With commutes of 30 minutes or more | With a computer in the house | With Internet access | Percent who lived in the same house one year ago | Percent who lived in another state or county one year ago | Total | Percent change 2019-2020 | Total | Rate | Number | Percent in labor force | Number | Percent who worked full-year full-time |
| | 55 | 56 | 57 | 58 | 59 | 60 | 61 | 62 | 63 | 64 | 65 | 66 | 67 | 68 |
| **COLORADO— Cont'd** | | | | | | | | | | | | | | |
| Castle Rock | 81.8 | 46.6 | 95.7 | 92.6 | 81.7 | 7.9 | 36,713 | 2.3 | 2,145 | 6 | 52,731 | 72.7 | 44,722 | 56.2 |
| Centennial | 76.8 | 41.2 | 99.1 | 96.6 | 85.8 | 8.3 | 63,023 | -2.9 | 3,792 | 6 | 88,128 | 69.1 | 70,301 | 57.9 |
| Colorado Springs | 76.9 | 24.7 | 96.2 | 92.6 | 78.6 | 10.2 | 240,003 | 0.9 | 17,843 | 7 | 385,499 | 65.6 | 313,940 | 53.9 |
| Commerce City | 76.2 | 48.2 | 93.7 | 88.3 | 86.5 | 7.9 | 30,733 | 1.5 | 2,417 | 8 | 44,823 | 74.8 | 40,590 | 59.5 |
| Denver | 67.2 | 42.2 | 95.3 | 90.8 | 81.1 | 10.8 | 423,824 | 0.6 | 34,900 | 8 | 602,155 | 73.3 | 515,403 | 60.5 |
| Englewood | 75.8 | 35.0 | 93.2 | 89.8 | 86.2 | 10.0 | 21,337 | -0.5 | 1,711 | 8 | 28,662 | 73.5 | 23,940 | 57.2 |
| Fort Collins | 71.4 | 21.7 | 98.7 | 91.9 | 73.4 | 13.4 | 100,922 | -1.7 | 6,359 | 6 | 146,275 | 69.9 | 126,580 | 46.1 |
| Fountain | 87.8 | 30.7 | NA | NA | 76.1 | 13.2 | 13,107 | 0.3 | 1,002 | 8 | 21,595 | 61.1 | 20,115 | 47.9 |
| Grand Junction | 80.3 | 7.4 | 93.5 | 88.4 | 71.5 | 9.8 | 30,877 | -1.3 | 2,372 | 8 | 52,175 | 62.8 | 38,747 | 49.1 |
| Greeley | 80.9 | 29.4 | 95.8 | 85.6 | 82.8 | 8.6 | 52,974 | -3.1 | 4,252 | 8 | 87,301 | 65.7 | 74,641 | 48.4 |
| Lakewood | 77.5 | 46.5 | 95.7 | 91.0 | 79.2 | 13.0 | 88,260 | -0.3 | 7,215 | 8 | 132,036 | 68.1 | 105,104 | 60.1 |
| Littleton | 77.0 | 42.6 | 97.6 | 90.8 | 82.9 | 12.7 | 26,154 | -2.1 | 1,791 | 7 | 40,599 | 71.6 | 31,325 | 61.7 |
| Longmont | 78.3 | 37.6 | 96.9 | 92.0 | 81.0 | 9.6 | 54,346 | -1.1 | 3,670 | 7 | 78,971 | 71.4 | 64,190 | 54.4 |
| Loveland | 80.3 | 37.4 | 96.2 | 91.6 | 83.4 | 8.1 | 43,188 | -1.1 | 2,932 | 7 | 62,032 | 67.0 | 46,358 | 55.9 |
| Northglenn | 89.4 | 49.3 | 93.3 | 80.5 | 79.8 | 7.3 | 21,393 | -1.2 | 1,871 | 9 | 31,887 | 67.9 | 26,785 | 59.0 |
| Parker | 76.3 | 40.3 | NA | NA | 83.9 | 10.3 | 33,174 | 0.9 | 2,094 | 6 | 44,314 | 75.7 | 39,188 | 55.7 |
| Pueblo | 79.4 | 17.7 | 87.2 | 79.9 | 80.6 | 7.8 | 49,877 | 2.1 | 4,566 | 9 | 88,612 | 54.9 | 68,899 | 43.9 |
| Thornton | 79.5 | 54.6 | 97.7 | 93.8 | 81.9 | 6.7 | 78,219 | -0.5 | 5,817 | 7 | 107,124 | 72.5 | 91,515 | 56.4 |
| Westminster | 79.4 | 46.5 | 96.1 | 92.2 | 83.5 | 9.5 | 65,593 | -1.5 | 5,200 | 8 | 90,256 | 75.8 | 76,313 | 61.8 |
| Wheat Ridge | 72.1 | 46.8 | 92.8 | 80.8 | 86.5 | 11.0 | 17,196 | -0.4 | 1,587 | 9 | 25,369 | 66.9 | 19,555 | 57.3 |
| **CONNECTICUT** | 78.3 | 35.9 | 93.1 | 88.1 | 88.0 | 5.2 | 1,872,631 | -2.1 | 148,010 | 8 | 2,926,854 | 65.8 | 2,297,822 | 51.7 |
| Bridgeport | 65.4 | 38.5 | 88.3 | 81.4 | 83.7 | 5.8 | 68,466 | -1.0 | 8,141 | 12 | 112,095 | 64.2 | 94,167 | 43.6 |
| Bristol | 83.8 | 33.2 | 92.1 | 86.8 | 85.6 | 3.3 | 33,032 | -1.2 | 2,838 | 9 | 49,125 | 73.1 | 39,200 | 60.0 |
| Danbury | 79.7 | 39.7 | 93.8 | 89.5 | 83.4 | 8.1 | 45,871 | -3.4 | 3,340 | 7 | 71,464 | 70.1 | 57,376 | 54.0 |
| Hartford | 65.8 | 19.6 | 90.8 | 80.0 | 82.8 | 6.6 | 54,172 | 1.4 | 7,192 | 13 | 96,704 | 64.9 | 83,567 | 43.6 |
| Meriden | 79.0 | 30.3 | 89.1 | 80.5 | 85.1 | 2.4 | 32,583 | 0.6 | 2,879 | 9 | 47,448 | 66.9 | 38,636 | 50.4 |
| Middletown | 77.6 | 27.8 | 90.3 | 80.5 | 82.7 | 12.7 | 25,905 | -1.6 | 1,889 | 7 | 39,705 | 68.1 | 32,284 | 49.0 |
| Milford | 82.2 | 31.4 | 93.9 | 90.4 | 89.2 | 5.4 | 29,780 | -3.0 | 2,242 | 8 | 46,192 | 68.2 | 35,758 | 56.8 |
| Naugatuck | 79.2 | 49.9 | NA | NA | NA | NA | 16,997 | -2.2 | 1,436 | 8 | 25,253 | 69.1 | 21,020 | 49.3 |
| New Britain | 83.9 | 33.3 | 89.6 | 79.7 | 84.1 | 5.6 | 37,174 | 0.4 | 4,017 | 11 | 59,079 | 65.6 | 47,794 | 51.7 |
| New Haven | 61.7 | 24.1 | 89.8 | 85.6 | 81.7 | 7.4 | 65,634 | 0.7 | 5,757 | 9 | 106,654 | 65.8 | 92,016 | 39.8 |
| New London | 67.3 | 22.4 | 90.9 | 86.0 | 80.7 | 10.5 | 11,932 | -1.4 | 1,583 | 13 | 22,664 | 63.2 | 20,019 | 49.0 |
| Norwalk | 76.5 | 38.6 | 96.2 | 92.1 | 91.5 | 3.9 | 49,697 | -2.8 | 4,034 | 8 | 74,799 | 66.5 | 61,009 | 58.4 |
| Norwich | 87.6 | 25.5 | 87.5 | 82.0 | 85.5 | 5.7 | 20,228 | -0.3 | 2,835 | 14 | 32,397 | 66.8 | 25,720 | 50.2 |
| Shelton | 85.7 | 32.4 | NA | NA | 89.6 | 3.5 | 21,449 | -3.1 | 1,680 | 8 | 35,389 | 63.7 | 25,059 | 52.1 |
| Stamford | 67.7 | 35.2 | 94.9 | 91.6 | 80.8 | 6.6 | 68,870 | -2.6 | 5,594 | 8 | 106,569 | 72.8 | 87,040 | 56.0 |
| Torrington | 91.2 | 40.0 | 91.9 | 85.1 | 89.3 | 4.7 | 18,889 | -1.6 | 1,576 | 8 | 29,044 | 58.4 | 21,539 | 50.3 |
| Waterbury | 78.0 | 34.8 | 86.7 | 75.4 | 88.2 | 4.3 | 50,504 | -0.3 | 5,883 | 12 | 84,547 | 62.8 | 69,894 | 43.0 |
| West Haven | 76.4 | 27.8 | 93.1 | 91.0 | 86.7 | 6.7 | 30,697 | 0.4 | 2,641 | 9 | 45,587 | 61.5 | 38,060 | 42.6 |
| **DELAWARE** | 80.7 | 34.9 | 94.1 | 88.1 | 87.4 | 5.8 | 484,358 | -0.6 | 37,900 | 8 | 792,119 | 61.7 | 602,481 | 53.5 |
| Dover | 75.1 | 30.1 | 91.4 | 86.0 | 73.6 | 15.9 | 16,134 | 1.4 | 1,864 | 12 | 31,476 | 58.1 | 25,064 | 44.3 |
| Newark | 64.9 | 28.6 | 97.3 | 94.2 | 74.2 | 13.3 | 16,064 | -2.7 | 937 | 6 | 30,300 | 54.7 | 25,517 | 30.7 |
| Wilmington | 67.6 | 25.8 | 92.3 | 80.6 | 80.7 | 6.7 | 34,433 | 1.5 | 4,052 | 12 | 57,296 | 62.5 | 48,010 | 48.3 |
| **DISTRICT OF COLUMBIA** | 33.0 | 57.4 | 94.4 | 86.8 | 80.8 | 9.9 | 409,734 | -0.1 | 32,895 | 8 | 587,819 | 70.4 | 500,282 | 58.5 |
| Washington | 33.0 | 57.4 | 94.4 | 86.8 | 80.8 | 9.9 | 409,734 | -0.1 | 32,895 | 8 | 587,819 | 70.4 | 500,282 | 58.5 |
| **FLORIDA** | 78.1 | 43.9 | 94.5 | 86.4 | 85.0 | 6.8 | 10,114,329 | -2.2 | 781,491 | 8 | 17,719,854 | 58.9 | 13,221,656 | 52.8 |
| Altamonte Springs | 80.2 | 43.8 | 96.9 | 92.9 | 77.6 | 9.0 | 24,416 | -6.4 | 2,159 | 9 | 37,440 | 71.2 | 30,844 | 61.7 |
| Apopka | 81.1 | 62.1 | 89.3 | 81.9 | 87.8 | 5.5 | 26,501 | -7.5 | 1,949 | 7 | 43,322 | 67.4 | 34,855 | 57.6 |
| Aventura | 73.9 | 51.9 | 95.1 | 85.4 | 90.1 | 5.1 | 15,746 | -8.7 | 1,080 | 7 | 31,927 | 60.7 | 21,731 | 63.2 |
| Boca Raton | 71.3 | 28.8 | 95.7 | 90.6 | 80.1 | 9.7 | 51,430 | -3.3 | 3,213 | 6 | 84,649 | 58.3 | 59,185 | 45.5 |
| Bonita Springs | 75.0 | 31.1 | 98.1 | 94.1 | 90.2 | 4.4 | 25,462 | -0.2 | 1,523 | 6 | 53,141 | 42.8 | 26,820 | 50.5 |
| Boynton Beach | 80.1 | 36.4 | 93.1 | 84.8 | 86.4 | 5.7 | 39,088 | -0.3 | 3,687 | 9 | 67,380 | 68.4 | 50,472 | 63.1 |
| Bradenton | 71.5 | 36.8 | 93.5 | 88.2 | 77.4 | 13.4 | 26,324 | -0.4 | 1,952 | 8 | 48,934 | 59.0 | 33,424 | 52.2 |
| Cape Coral | 80.8 | 44.7 | 97.7 | 94.6 | 85.6 | 8.3 | 91,630 | 0.0 | 6,848 | 8 | 163,534 | 58.0 | 116,868 | 53.7 |
| Casselberry | 74.0 | 34.5 | 95.6 | 92.8 | 84.2 | 7.1 | 14,797 | -7.1 | 1,243 | 8 | 24,024 | 64.1 | 19,752 | 56.1 |
| Clearwater | 77.9 | 36.7 | 94.3 | 85.3 | 82.8 | 5.4 | 57,850 | -1.5 | 4,075 | 7 | 97,849 | 57.9 | 70,097 | 52.1 |
| Clermont | 84.2 | 64.6 | NA | NA | 80.1 | 14.5 | 17,028 | -0.4 | 1,560 | 9 | 31,834 | 59.7 | 22,389 | 58. |
| Coconut Creek | 76.1 | 46.9 | 94.7 | 90.6 | 86.7 | 6.5 | 32,194 | -3.4 | 2,404 | 8 | 51,183 | 68.5 | 41,023 | 63. |
| Cooper City | 84.2 | 45.1 | NA | NA | 90.8 | 2.3 | 19,491 | -4.8 | 1,141 | 6 | 28,017 | 66.7 | 22,943 | 55. |
| Coral Gables | 67.3 | 45.3 | 97.2 | 88.9 | 82.7 | 6.4 | 23,873 | -10.9 | 1,032 | 4 | 42,335 | 61.1 | 31,567 | 53. |
| Coral Springs | 76.9 | 47.9 | 97.5 | 93.7 | 86.8 | 3.4 | 72,328 | -2.8 | 5,638 | 8 | 102,158 | 70.7 | 86,005 | 54. |
| Cutler Bay | 80.8 | 76.8 | 95.1 | 87.3 | 95.7 | 0.5 | 21,570 | -8.8 | 1,422 | 7 | 34,451 | 65.9 | 27,510 | 57. |
| Dania Beach | 78.0 | 43.6 | 97.6 | 93.1 | 79.2 | 5.7 | 16,975 | 1.2 | 2,015 | 12 | 25,269 | 63.5 | 20,123 | 56. |
| Davie | 76.9 | 39.6 | 98.0 | 95.1 | 85.3 | 5.2 | 57,642 | -3.7 | 4,057 | 7 | 85,049 | 71.4 | 73,373 | 55. |
| Daytona Beach | 72.2 | 23.9 | 93.1 | 78.7 | 83.1 | 10.3 | 31,318 | 0.0 | 3,293 | 11 | 59,899 | 56.9 | 44,319 | 46. |

1. Employed persons.  2. Households.  3. Percent of civilian labor force.  4. Persons 16 years old and over.

# Table D. Cities — Construction, Wholesale Trade, and Retail Trade

| City | Value of residential construction authorized by building permits, 2020 | | | Wholesale trade[1], 2017 | | | | Retail trade[2], 2017 | | | |
|---|---|---|---|---|---|---|---|---|---|---|---|
| | New construction ($1,000) | Number of housing units | Percent single family | Number of establishments | Number of employees | Sales (mil dol) | Annual payroll (mil dol) | Number of establishments | Number of employees | Sales (mil dol) | Annual payroll (mil dol) |
| | 69 | 70 | 71 | 72 | 73 | 74 | 75 | 76 | 77 | 78 | 79 |
| **COLORADO— Cont'd** | | | | | | | | | | | |
| Castle Rock | 296,858 | 1,090 | 95 | D | D | D | 10 | 206 | 3,790 | 1,012 | 98 |
| Centennial | 19,754 | 10 | 100 | 185 | 3,450 | 14,662 | 362 | 293 | 5,719 | 2,403 | 213 |
| Colorado Springs | NA | NA | NA | 347 | 3,869 | 2,521 | 261 | 1,698 | 26,651 | 8,162 | 778 |
| Commerce City | 259,898 | 1,142 | 78 | 142 | 3,380 | 2,446 | 206 | 131 | 1,949 | 647 | 58 |
| Denver | 831,523 | 5,059 | 23 | 1,191 | 20,941 | 20,106 | 1,389 | 2,471 | 32,753 | 9,219 | 1,048 |
| Englewood | 51,792 | 303 | 27 | 105 | 1,508 | 1,577 | 115 | 206 | 3,574 | 1,150 | 153 |
| Fort Collins | 118,698 | 627 | 72 | 129 | 2,126 | 1,815 | 179 | 606 | 10,749 | 3,052 | 305 |
| Fountain | NA | NA | NA | NA | NA | NA | NA | 58 | 1,731 | 564 | 46 |
| Grand Junction | NA | NA | NA | 147 | 1,390 | 777 | 72 | 433 | 6,851 | 1,990 | 196 |
| Greeley | 36,427 | 272 | 31 | 88 | 833 | 622 | 46 | 311 | 5,520 | 1,815 | 161 |
| Lakewood | 238,894 | 1,237 | 14 | 140 | 920 | 647 | 63 | 677 | 11,479 | 3,123 | 315 |
| Littleton | 10,535 | 40 | 88 | 63 | 1,245 | 1,121 | 90 | 228 | 4,331 | 2,123 | 174 |
| Longmont | 185,161 | 855 | 36 | 75 | 654 | 448 | 44 | 292 | 5,383 | 1,649 | 164 |
| Loveland | 159,637 | 613 | 79 | 77 | 1,489 | 1,339 | 111 | 336 | 6,144 | 1,880 | 166 |
| Northglenn | 0 | 0 | 0 | 18 | 123 | 60 | 6 | 120 | 2,346 | 776 | 82 |
| Parker | 187,383 | 657 | 61 | 35 | 199 | 120 | 11 | 193 | 3,741 | 1,278 | 107 |
| Pueblo | NA | NA | NA | 59 | 647 | 293 | 35 | 383 | 6,341 | 1,744 | 168 |
| Thornton | 232,421 | 867 | 94 | 34 | 182 | 83 | 14 | 220 | 5,547 | 1,623 | 169 |
| Westminster | 120,983 | 690 | 13 | 66 | 905 | 716 | 62 | 325 | 6,752 | 1,822 | 180 |
| Wheat Ridge | 80,783 | 411 | 36 | 69 | 711 | 299 | 48 | 177 | 2,271 | 791 | 81 |
| **CONNECTICUT** | 700 | 3 | 100 | 3,504 | 62,298 | 102,897 | 5,174 | 12,391 | 186,297 | 55,405 | 5,561 |
| Bridgeport | 5,706 | 71 | 35 | D | D | D | D | 284 | 3,179 | 940 | 112 |
| Bristol | 13,969 | 94 | 64 | 39 | 434 | 360 | 22 | 161 | 2,904 | 878 | 84 |
| Danbury | 24,470 | 188 | 47 | 90 | 1,302 | 998 | 93 | 444 | 8,143 | 2,510 | 251 |
| Hartford | 1,345 | 10 | 100 | 94 | 1,962 | 1,354 | 104 | 371 | 3,334 | 1,374 | 116 |
| Meriden | 1,193 | 10 | 100 | 39 | 337 | 179 | 19 | 227 | 3,049 | 752 | 75 |
| Middletown | 3,178 | 20 | 100 | 45 | 1,190 | 642 | 71 | 124 | 1,633 | 573 | 55 |
| Milford | NA | NA | NA | 95 | 1,239 | 815 | 77 | 312 | 6,036 | 1,781 | 169 |
| Naugatuck | 2,553 | 14 | 100 | 18 | 469 | 188 | 30 | 60 | 1,139 | 414 | 36 |
| New Britain | 1,305 | 12 | 33 | 41 | 378 | 228 | 26 | 162 | 1,797 | 735 | 57 |
| New Haven | 38,123 | 738 | 2 | 69 | 2,269 | 1,991 | 661 | 344 | 3,671 | 1,273 | 118 |
| New London | 6,201 | 37 | 95 | 16 | 726 | 528 | 53 | 105 | 1,538 | 513 | 49 |
| Norwalk | 42,126 | 121 | 12 | 131 | 2,254 | 2,365 | 172 | 345 | 6,332 | 2,094 | 237 |
| Norwich | 3,516 | 68 | 12 | 18 | 419 | 248 | 27 | 128 | 1,912 | 533 | 55 |
| Shelton | 36,897 | 316 | 19 | 60 | 1,926 | 2,110 | 198 | 91 | 1,817 | 604 | 60 |
| Stamford | 65,329 | 312 | 13 | 236 | 5,349 | 41,815 | 654 | 460 | 6,250 | 2,043 | 221 |
| Torrington | 966 | 6 | 100 | 24 | 190 | 63 | 11 | 151 | 2,666 | 830 | 82 |
| Waterbury | 1,700 | 20 | 90 | 68 | 770 | 531 | 48 | 432 | 5,994 | 1,692 | 163 |
| West Haven | 1,260 | 11 | 100 | 55 | 1,291 | 728 | 78 | 111 | 1,295 | 318 | 37 |
| **DELAWARE** | 624,094 | 4,523 | 88 | 962 | 9,315 | 7,225 | 523 | 3,648 | 58,201 | 17,668 | 1,558 |
| Dover | 10,878 | 48 | 100 | 46 | 232 | 174 | 16 | 258 | 4,087 | 1,071 | 91 |
| Newark | 0 | 0 | 0 | 42 | 346 | 126 | 17 | 147 | 2,924 | 987 | 92 |
| Wilmington | 36,112 | 205 | 7 | 129 | 1,078 | 1,287 | 69 | 287 | 3,360 | 1,129 | 101 |
| **DISTRICT OF COLUMBIA** | 886,594 | 7,370 | 2 | 321 | 3,851 | 3,387 | 293 | 1,743 | 23,133 | 5,534 | 673 |
| Washington | 886,594 | 7,370 | 2 | 321 | 3,851 | 3,387 | 293 | 1,743 | 23,133 | 5,534 | 673 |
| **FLORIDA** | 17,191 | 110 | 93 | 27,230 | 283,295 | 288,642 | 16,241 | 74,496 | 1,086,052 | 333,135 | 29,686 |
| Altamonte Springs | 4,158 | 3 | 100 | 67 | 568 | 349 | 34 | 317 | 6,186 | 1,896 | 161 |
| Apopka | 156,742 | 743 | 63 | 55 | 985 | 372 | 44 | 155 | 2,489 | 831 | 68 |
| Aventura | 0 | 0 | 0 | 137 | 586 | 1,797 | 40 | 399 | 8,877 | 2,093 | 241 |
| Boca Raton | 84,304 | 122 | 100 | 436 | 5,801 | 8,008 | 458 | 742 | 10,317 | 2,874 | 355 |
| Bonita Springs | 148,433 | 481 | 99 | 48 | 357 | 192 | 18 | 195 | 2,501 | 700 | 67 |
| Boynton Beach | 4,990 | 21 | 91 | 113 | 928 | 593 | 49 | 332 | 5,615 | 1,378 | 132 |
| Bradenton | 23,376 | 152 | 49 | 31 | 317 | 169 | 18 | 244 | 3,834 | 1,064 | 95 |
| Cape Coral | 619,344 | 3,527 | 76 | 118 | 494 | 317 | 24 | 455 | 6,550 | 1,794 | 165 |
| Casselberry | 27,860 | 276 | 8 | 29 | 296 | 103 | 16 | 149 | 2,571 | 506 | 56 |
| Clearwater | 28,330 | 122 | 71 | 160 | 1,850 | 855 | 109 | 682 | 10,337 | 3,134 | 301 |
| Clermont | 99,603 | 489 | 98 | 22 | 85 | 63 | 3 | 160 | 3,289 | 1,103 | 86 |
| Coconut Creek | 567 | 4 | 100 | 53 | 304 | 258 | 16 | 169 | 3,510 | 1,568 | 134 |
| Cooper City | 2,350 | 8 | 100 | 49 | 159 | 120 | 8 | 96 | 1,753 | 442 | 52 |
| Coral Gables | 44,038 | 33 | 100 | 199 | 1,657 | 7,635 | 181 | 326 | 4,954 | 1,902 | 185 |
| Coral Springs | 12,350 | 60 | 100 | 238 | 2,010 | 1,252 | 118 | 470 | 8,318 | 2,490 | 229 |
| Cutler Bay | 699 | 3 | 100 | 27 | 113 | 35 | 3 | 157 | 2,524 | 796 | 64 |
| Dania Beach | 7,601 | 38 | 100 | 129 | 1,005 | 630 | 49 | 155 | 1,846 | 537 | 57 |
| Davie | 18,924 | 133 | 42 | 268 | 1,753 | 988 | 84 | 443 | 7,614 | 2,973 | 233 |
| Daytona Beach | 309,311 | 1,524 | 50 | 74 | 679 | 374 | 31 | 519 | 7,165 | 2,114 | 193 |

1. Merchant wholesalers except manufacturers' sales branches and offices.  2. Establishments with payroll.

# Table D. Cities — Real Estate, Professional Services, and Manufacturing

| City | Real estate and rental and leasing, 2017 | | | | Professional, scientific, and technical services[1], 2017 | | | | Manufacturing, 2017 | | | |
|---|---|---|---|---|---|---|---|---|---|---|---|---|
| | Number of establishments | Number of employees | Receipts (mil dol) | Annual payroll (mil dol) | Number of establishments | Number of employees | Receipts (mil dol) | Annual payroll (mil dol) | Number of establishments | Number of employees | Receipts (mil dol) | Annual payroll (mil dol) |
| | 80 | 81 | 82 | 83 | 84 | 85 | 86 | 87 | 88 | 89 | 90 | 91 |
| COLORADO— Cont'd | | | | | | | | | | | | |
| Castle Rock | 122 | 324 | 79 | 13 | 258 | 828 | 163 | 54 | NA | NA | NA | NA |
| Centennial | 356 | 1,100 | 426 | 73 | 744 | 4,758 | 1,074 | 377 | NA | NA | NA | NA |
| Colorado Springs | 1,092 | 3,867 | 843 | 161 | 2,097 | 20,579 | 4,152 | 1,668 | NA | NA | NA | NA |
| Commerce City | 60 | 672 | 152 | 34 | D | D | D | D | NA | NA | NA | NA |
| Denver | 2,033 | 11,307 | 3,867 | 710 | 5,133 | 50,003 | 12,643 | 4,747 | NA | NA | NA | NA |
| Englewood | 96 | 430 | 112 | 24 | D | D | D | D | NA | NA | NA | NA |
| Fort Collins | 420 | 1,369 | 368 | 57 | 943 | 5,898 | 1,130 | 487 | NA | NA | NA | NA |
| Fountain | 12 | 41 | 8 | 1 | 23 | 93 | 7 | 2 | NA | NA | NA | NA |
| Grand Junction | 222 | 727 | 180 | 31 | 359 | 3,120 | 491 | 198 | NA | NA | NA | NA |
| Greeley | 142 | 754 | 122 | 32 | 213 | 1,068 | 154 | 53 | NA | NA | NA | NA |
| Lakewood | 295 | 1,151 | 329 | 52 | 871 | 7,165 | 1,286 | 521 | NA | NA | NA | NA |
| Littleton | 116 | 376 | 70 | 14 | 379 | 2,009 | 372 | 144 | NA | NA | NA | NA |
| Longmont | 136 | 393 | 93 | 16 | 402 | 5,245 | 1,025 | 524 | NA | NA | NA | NA |
| Loveland | 146 | 433 | 122 | 23 | 301 | 2,032 | 330 | 115 | NA | NA | NA | NA |
| Northglenn | 35 | 71 | 23 | 3 | 64 | 307 | 31 | 14 | NA | NA | NA | NA |
| Parker | 121 | 223 | 51 | 10 | 247 | 802 | 145 | 46 | NA | NA | NA | NA |
| Pueblo | 110 | 567 | 102 | 17 | D | D | D | D | NA | NA | NA | NA |
| Thornton | 121 | 434 | 114 | 19 | 182 | 1,160 | 237 | 85 | NA | NA | NA | NA |
| Westminster | 198 | 645 | 135 | 28 | D | D | D | D | NA | NA | NA | NA |
| Wheat Ridge | 69 | 188 | 40 | 8 | 201 | 1,360 | 189 | 84 | NA | NA | NA | NA |
| CONNECTICUT | 3,459 | 20,224 | 6,691 | 1,115 | 9,184 | 108,479 | 21,600 | 9,710 | 3,986 | 156,822 | 57,882 | 11,022 |
| Bridgeport | 92 | 518 | 119 | 25 | 197 | 1,285 | 270 | 114 | NA | NA | NA | NA |
| Bristol | 33 | 101 | 28 | 4 | 71 | 322 | 67 | 19 | NA | NA | NA | NA |
| Danbury | 90 | 1,951 | 477 | 143 | 222 | 2,269 | 378 | 163 | NA | NA | NA | NA |
| Hartford | 169 | 1,270 | 308 | 65 | 410 | 8,800 | 2,164 | 825 | NA | NA | NA | NA |
| Meriden | 45 | 491 | 61 | 22 | 64 | 841 | 154 | 69 | NA | NA | NA | NA |
| Middletown | 45 | 301 | 56 | 15 | 101 | 1,022 | 228 | 83 | NA | NA | NA | NA |
| Milford | 58 | 264 | 1,109 | 16 | 181 | 1,189 | 232 | 97 | NA | NA | NA | NA |
| Naugatuck | D | D | D | D | 35 | 210 | 33 | 11 | NA | NA | NA | NA |
| New Britain | 38 | 174 | 41 | 7 | 74 | 601 | 85 | 43 | NA | NA | NA | NA |
| New Haven | 132 | 809 | 200 | 34 | 356 | 3,485 | 781 | 336 | NA | NA | NA | NA |
| New London | 26 | 139 | 28 | 6 | 86 | 642 | 130 | 56 | NA | NA | NA | NA |
| Norwalk | 110 | 393 | 114 | 21 | 325 | 5,218 | 1,430 | 519 | NA | NA | NA | NA |
| Norwich | 35 | 126 | 25 | 4 | 56 | 878 | 122 | 57 | NA | NA | NA | NA |
| Shelton | 47 | 331 | 161 | 26 | 154 | 2,421 | 733 | 204 | NA | NA | NA | NA |
| Stamford | 246 | 1,605 | 680 | 146 | 743 | 16,984 | 4,348 | 1,736 | NA | NA | NA | NA |
| Torrington | 22 | 87 | 14 | 3 | 56 | 377 | 51 | 20 | NA | NA | NA | NA |
| Waterbury | 87 | 344 | 93 | 15 | 121 | 742 | 100 | 39 | NA | NA | NA | NA |
| West Haven | 39 | 110 | 39 | 5 | 60 | 315 | 40 | 14 | NA | NA | NA | NA |
| DELAWARE | 1,286 | 6,096 | 5,164 | 288 | 2,991 | 29,716 | 7,003 | 2,942 | 558 | 27,536 | 16,689 | 1,544 |
| Dover | 80 | 334 | 148 | 13 | 189 | 1,347 | 237 | 106 | NA | NA | NA | NA |
| Newark | 52 | 219 | 544 | 9 | 140 | 1,560 | 268 | 104 | NA | NA | NA | NA |
| Wilmington | 202 | 833 | 833 | 52 | 686 | 9,685 | 3,225 | 1,471 | 113 | 1,092 | 277 | 50 |
| DISTRICT OF COLUMBIA | 1,350 | 11,000 | 4,525 | 883 | 5,812 | 108,016 | 37,245 | 13,126 | NA | NA | NA | NA |
| Washington | 1,350 | 11,000 | 4,525 | 883 | 5,812 | 108,016 | 37,245 | 13,126 | NA | NA | NA | NA |
| FLORIDA | 37,660 | 176,886 | 49,175 | 8,184 | 79,224 | 513,798 | 89,601 | 34,673 | 13,471 | 296,389 | 106,342 | 16,576 |
| Altamonte Springs | 116 | 982 | 186 | 40 | 275 | 2,978 | 211 | 211 | NA | NA | NA | NA |
| Apopka | 52 | 139 | 24 | 5 | 115 | 400 | 59 | 19 | NA | NA | NA | NA |
| Aventura | 285 | 907 | 235 | 45 | 440 | 1,156 | 300 | 86 | NA | NA | NA | NA |
| Boca Raton | 578 | 4,591 | 958 | 238 | 1,769 | 13,203 | 2,821 | 945 | NA | NA | NA | NA |
| Bonita Springs | 147 | 539 | 129 | 26 | 219 | 1,097 | 188 | 69 | NA | NA | NA | NA |
| Boynton Beach | 114 | 455 | 98 | 17 | 328 | 1,364 | 184 | 54 | NA | NA | NA | NA |
| Bradenton | 97 | 269 | 74 | 10 | 228 | 1,315 | 180 | 78 | NA | NA | NA | NA |
| Cape Coral | 413 | 667 | 143 | 25 | 456 | 1,409 | 186 | 61 | NA | NA | NA | NA |
| Casselberry | 51 | 218 | 51 | 9 | 85 | 350 | 42 | 15 | NA | NA | NA | NA |
| Clearwater | 330 | 1,187 | 262 | 50 | 726 | 5,378 | 785 | 333 | NA | NA | NA | NA |
| Clermont | 99 | 247 | 59 | 9 | 107 | 342 | 42 | 15 | NA | NA | NA | NA |
| Coconut Creek | 67 | 294 | 155 | 15 | 198 | 1,234 | 184 | 62 | NA | NA | NA | NA |
| Cooper City | D | D | D | D | 208 | 671 | 94 | 36 | NA | NA | NA | NA |
| Coral Gables | 450 | 1,736 | 533 | 103 | 1,708 | 10,129 | 2,185 | 787 | NA | NA | NA | NA |
| Coral Springs | 267 | 1,222 | 301 | 45 | D | D | D | D | NA | NA | NA | NA |
| Cutler Bay | 34 | 59 | 12 | 2 | 85 | 196 | 18 | 5 | NA | NA | NA | NA |
| Dania Beach | 72 | 578 | 118 | 32 | 145 | 600 | 113 | 37 | NA | NA | NA | NA |
| Davie | 223 | 1,049 | 277 | 48 | 597 | 2,084 | 348 | 104 | NA | NA | NA | NA |
| Daytona Beach | 166 | 825 | 235 | 33 | 309 | 2,256 | 271 | 111 | NA | NA | NA | NA |

1. Establishments subject to federal tax.

# Table D. Cities — Accommodation and Food Services, Arts, Entertainment, and Recreation, and Health Care and Social Assistance

| City | Accommodation and food services, 2017 | | | | Arts, entertainment, and recreation[1], 2017 | | | | Health care and social assistance[1], 2017 | | | |
|---|---|---|---|---|---|---|---|---|---|---|---|---|
| | Number of establishments | Number of employees | Receipts (mil dol) | Annual payroll (mil dol) | Number of establishments | Number of employees | Receipts (mil dol) | Annual payroll (mil dol) | Number of establishments | Number of employees | Receipts (mil dol) | Annual payroll (mil dol) |
| | 92 | 93 | 94 | 95 | 96 | 97 | 98 | 99 | 100 | 101 | 102 | 103 |
| COLORADO— Cont'd | | | | | | | | | | | | |
| Castle Rock | 116 | 2,307 | 138 | 42 | D | D | D | D | 180 | 2,166 | 258 | 90 |
| Centennial | 245 | 6,353 | 364 | 141 | 48 | 897 | 53 | 15 | 450 | 6,621 | 668 | 271 |
| Colorado Springs | 1,111 | 26,615 | 1,682 | 479 | 219 | 3,178 | 202 | 58 | 1,850 | 35,415 | 4,064 | 1,616 |
| Commerce City | 78 | 1,207 | 77 | 21 | D | D | D | D | 44 | 630 | 60 | 23 |
| Denver | 2,354 | 55,545 | 4,477 | 1,297 | 473 | 11,883 | 1,656 | 661 | 2,302 | 56,165 | 8,008 | 3,078 |
| Englewood | 117 | 1,934 | 109 | 34 | D | D | D | D | 232 | 8,097 | 1,140 | 360 |
| Fort Collins | 483 | 10,162 | 561 | 175 | 105 | 1,269 | 77 | 22 | 713 | 11,970 | 1,293 | 564 |
| Fountain | 54 | 1,187 | 65 | 18 | D | D | D | D | D | D | D | D |
| Grand Junction | 247 | 5,101 | 282 | 91 | 39 | 524 | 27 | 9 | 367 | 10,487 | 1,348 | 498 |
| Greeley | 220 | 4,284 | 215 | 64 | 32 | 275 | 17 | 4 | 334 | 6,318 | 902 | 348 |
| Lakewood | 413 | 8,801 | 555 | 167 | 67 | 815 | 48 | 15 | 665 | 13,975 | 1,654 | 635 |
| Littleton | 152 | 2,885 | 161 | 49 | 39 | 666 | 43 | 13 | 275 | 4,712 | 737 | 253 |
| Longmont | 239 | 4,154 | 222 | 71 | 45 | 248 | 18 | 5 | 312 | 4,766 | 490 | 197 |
| Loveland | 224 | 4,916 | 277 | 82 | 41 | 246 | 15 | 4 | 275 | 5,868 | 840 | 299 |
| Northglenn | 57 | 1,668 | 97 | 32 | 11 | D | 16 | D | 79 | 868 | 68 | 28 |
| Parker | 137 | 2,732 | 161 | 47 | 22 | 755 | 36 | 11 | 216 | 3,788 | 597 | 196 |
| Pueblo | 273 | 4,962 | 251 | 74 | 34 | 674 | 92 | 19 | 369 | 12,940 | 1,374 | 589 |
| Thornton | 189 | 4,025 | 249 | 72 | 31 | D | 15 | D | 209 | 3,899 | 497 | 198 |
| Westminster | 258 | 6,014 | 384 | 123 | 47 | 880 | 49 | 16 | 304 | 6,846 | 828 | 314 |
| Wheat Ridge | 91 | 1,515 | 89 | 27 | 19 | 119 | 9 | 3 | 206 | 5,239 | 656 | 277 |
| CONNECTICUT | 8,762 | 146,456 | 10,792 | 3,070 | 1,744 | 30,320 | 2,406 | 785 | 11,065 | 295,083 | 35,302 | 14,069 |
| Bridgeport | 239 | 2,581 | 175 | 48 | 35 | 838 | 48 | 14 | 350 | 15,223 | 1,773 | 766 |
| Bristol | 104 | 1,421 | 93 | 24 | 16 | 403 | 38 | 11 | 172 | 4,618 | 404 | 178 |
| Danbury | 232 | 3,958 | 273 | 82 | 29 | 467 | 26 | 8 | 286 | 9,809 | 1,368 | 637 |
| Hartford | 340 | 5,544 | 390 | 111 | 35 | 1,477 | 108 | 32 | 471 | 24,473 | 3,631 | 1,576 |
| Meriden | 119 | 1,316 | 81 | 20 | 7 | 483 | 15 | 7 | 148 | 5,315 | 548 | 222 |
| Middletown | 126 | 1,535 | 106 | 33 | 20 | 334 | 15 | 6 | 183 | 9,759 | 1,171 | 593 |
| Milford | 194 | 3,121 | 195 | 54 | 33 | 522 | 33 | 11 | 189 | 3,429 | 372 | 174 |
| Naugatuck | 49 | 546 | 31 | 8 | 3 | 120 | 4 | 1 | 51 | 1,326 | 94 | 44 |
| New Britain | 103 | 1,280 | 86 | 23 | 15 | 397 | 19 | 9 | 169 | 8,866 | 807 | 362 |
| New Haven | 366 | 4,686 | 353 | 102 | 44 | 639 | 49 | 16 | 422 | 21,715 | 3,572 | 1,229 |
| New London | 99 | 1,383 | 86 | 27 | 17 | 146 | 10 | 3 | 132 | 4,300 | 567 | 238 |
| Norwalk | 297 | 3,778 | 304 | 85 | 73 | 1,559 | 167 | 64 | 298 | 6,606 | 889 | 345 |
| Norwich | 82 | 1,487 | 88 | 27 | 11 | 113 | 8 | 2 | 192 | 5,301 | 619 | 261 |
| Shelton | 117 | 1,752 | 120 | 33 | 19 | 246 | 18 | 5 | 133 | 3,368 | 446 | 221 |
| Stamford | 410 | 6,333 | 520 | 156 | 79 | 1,789 | 232 | 58 | 475 | 10,082 | 3,737 | 608 |
| Torrington | 92 | 1,010 | 63 | 18 | 17 | 229 | 11 | 3 | 156 | 4,105 | 403 | 174 |
| Waterbury | 237 | 3,084 | 180 | 50 | 15 | 347 | 25 | 7 | 346 | 11,499 | 1,277 | 530 |
| West Haven | 114 | 1,463 | 92 | 26 | NA | NA | NA | NA | 82 | 3,470 | 638 | 253 |
| DELAWARE | 2,141 | 39,950 | 2,554 | 714 | 440 | 8,466 | 854 | 224 | 2,648 | 68,348 | 8,855 | 3,721 |
| Dover | 135 | 4,329 | 334 | 77 | D | D | D | D | 240 | 7,734 | 993 | 389 |
| Newark | 140 | 3,456 | 195 | 60 | 21 | 271 | 15 | 5 | 162 | 10,649 | 1,831 | 809 |
| Wilmington | 222 | 3,407 | 235 | 68 | 55 | 741 | 53 | 18 | 373 | 9,984 | 1,173 | 469 |
| DISTRICT OF COLUMBIA | 2,733 | 72,890 | 6,740 | 2,060 | 380 | 9,817 | 1,737 | 830 | 2,203 | 74,134 | 11,400 | 4,376 |
| Washington | 2,733 | 72,890 | 6,740 | 2,060 | 380 | 9,817 | 1,737 | 830 | 2,203 | 74,134 | 11,400 | 4,376 |
| FLORIDA | 42,071 | 955,006 | 67,950 | 18,462 | 8,883 | 208,733 | 23,435 | 6,639 | 61,554 | 1,127,155 | 155,284 | 55,472 |
| Altamonte Springs | 156 | 3,722 | 233 | 70 | 23 | 229 | 22 | 4 | 243 | 5,741 | 885 | 289 |
| Apopka | 76 | 1,484 | 82 | 23 | 15 | 60 | 9 | 2 | 112 | 1,499 | 173 | 80 |
| Aventura | 111 | 3,330 | 287 | 83 | 32 | 262 | 25 | 6 | 295 | 3,544 | 708 | 227 |
| Boca Raton | 439 | 12,240 | 950 | 274 | 120 | 2,287 | 268 | 73 | 849 | 11,616 | 1,850 | 687 |
| Bonita Springs | 135 | 3,366 | 213 | 67 | 48 | 1,813 | 227 | 50 | 164 | 1,572 | 192 | 59 |
| Boynton Beach | 207 | 4,508 | 254 | 78 | 38 | 765 | 53 | 20 | 363 | 6,057 | 864 | 306 |
| Bradenton | 124 | 2,490 | 141 | 41 | 29 | 309 | 20 | 7 | 315 | 9,649 | 1,163 | 424 |
| Cape Coral | 238 | 4,856 | 265 | 80 | 56 | 420 | 26 | 7 | 321 | 5,242 | 678 | 254 |
| Casselberry | 80 | 1,261 | 68 | 19 | 18 | 179 | 11 | 2 | 66 | 558 | 64 | 20 |
| Clearwater | 397 | 9,686 | 746 | 195 | 86 | 1,711 | 134 | 35 | 578 | 11,955 | 1,653 | 587 |
| Clermont | 107 | 3,032 | 167 | 51 | 27 | 402 | 25 | 8 | 194 | 3,912 | 476 | 170 |
| Coconut Creek | 91 | 1,585 | 89 | 25 | 25 | 1,914 | 369 | 66 | 117 | 1,598 | 1,090 | 66 |
| Cooper City | 61 | 810 | 43 | 13 | 15 | 72 | 3 | 1 | 151 | 892 | 81 | 30 |
| Coral Gables | 267 | 5,925 | 443 | 138 | 79 | 954 | 96 | 33 | 522 | 6,673 | 1,494 | 473 |
| Coral Springs | 318 | 5,280 | 317 | 91 | D | D | D | D | 505 | 3,192 | 406 | 140 |
| Cutler Bay | 71 | 1,297 | 84 | 22 | 9 | 64 | 4 | 1 | 94 | 1,220 | 174 | 42 |
| Dania Beach | 84 | 1,750 | 155 | 34 | 23 | 521 | 51 | 16 | 57 | 519 | 50 | 17 |
| Davie | 257 | 7,377 | 941 | 190 | 69 | 696 | 56 | 16 | 317 | 3,315 | 289 | 107 |
| Daytona Beach | 301 | 6,918 | 449 | 119 | 47 | 2,501 | 439 | 69 | 305 | 11,086 | 1,518 | 503 |

1. Establishments subject to federal tax.

# Table D. Cities — Other Services and Government Employment and Payroll

Government employment and payroll, 2017 — March payroll. "Percent of total for" columns (110–116).

| City | Other services¹ No. of establishments (104) | No. of employees (105) | Receipts (mil dol) (106) | Annual payroll (mil dol) (107) | Full-time equivalent employees (108) | Total (dollars) (109) | Admin., judicial, and legal (110) | Police and corrections (111) | Fire protection (112) | Highways and transportation (113) | Health and welfare (114) | Natural resources and utilities (115) | Education and libraries (116) |
|---|---|---|---|---|---|---|---|---|---|---|---|---|---|
| **COLORADO— Cont'd** | | | | | | | | | | | | | |
| Castle Rock | 133 | 636 | 64 | 21 | 543 | 3,305,326 | 16.7 | 18.9 | 20.8 | 6.4 | 0.0 | 27.4 | 0.0 |
| Centennial | 259 | 1,657 | 253 | 71 | 73 | 420,544 | 100.0 | 0.0 | 0.0 | 0.0 | 0.0 | 0.0 | 0.0 |
| Colorado Springs | 995 | 8,956 | 2,133 | 392 | 4,068 | 27,500,431 | 5.4 | 23.6 | 13.1 | 5.4 | 0.0 | 50.4 | 0.0 |
| Commerce City | 106 | 722 | 90 | 26 | 475 | 2,708,700 | 22.4 | 41.2 | 0.0 | 11.4 | 8.1 | 16.3 | 2.6 |
| Denver | 1,874 | 15,112 | 2,404 | 600 | 14,162 | 86,682,654 | 14.4 | 30.4 | 10.0 | 11.2 | 10.9 | 18.3 | 1.3 |
| Englewood | 137 | 862 | 101 | 36 | 441 | 2,466,812 | 14.6 | 27.1 | 0.0 | 5.3 | 2.3 | 38.5 | 0.0 |
| Fort Collins | 358 | 2,052 | 394 | 66 | 1,708 | 9,440,627 | 28.8 | 23.4 | 0.0 | 10.5 | 0.4 | 32.6 | 0.0 |
| Fountain | 24 | 148 | 15 | 5 | 221 | 1,182,395 | 15.6 | 23.1 | 12.8 | 4.1 | 3.0 | 32.4 | 0.0 |
| Grand Junction | 251 | 1,478 | 170 | 51 | 654 | 3,742,842 | 11.3 | 34.8 | 20.3 | 5.9 | 2.2 | 21.7 | 0.0 |
| Greeley | 170 | 1,085 | 126 | 33 | 966 | 4,912,944 | 13.1 | 27.7 | 16.1 | 10.6 | 0.7 | 27.1 | 0.0 |
| Lakewood | 352 | 2,080 | 226 | 68 | 1,077 | 6,308,895 | 17.4 | 49.7 | 0.0 | 3.9 | 1.6 | 16.9 | 3.9 |
| Littleton | 146 | 965 | 124 | 35 | 462 | 3,320,170 | 11.3 | 21.6 | 46.6 | 4.9 | 4.3 | 5.3 | 5.3 |
| Longmont | 202 | 1,224 | 109 | 35 | 940 | 5,985,157 | 12.9 | 27.6 | 12.3 | 3.8 | 3.9 | 29.2 | 3.1 |
| Loveland | 154 | 1,173 | 127 | 42 | 706 | 3,643,996 | 23.8 | 24.4 | 13.4 | 6.1 | 0.0 | 25.5 | 0.0 |
| Northglenn | 66 | 411 | 50 | 13 | 264 | 1,420,139 | 20.9 | 39.4 | 0.0 | 7.1 | 0.0 | 29.4 | 0.0 |
| Parker | 158 | 992 | 94 | 30 | 418 | 2,100,948 | 21.8 | 33.8 | 0.0 | 10.9 | 0.5 | 27.2 | 0.0 |
| Pueblo | 179 | 1,056 | 98 | 28 | 893 | 5,127,375 | 9.0 | 34.5 | 18.8 | 8.4 | 4.4 | 26.6 | 0.0 |
| Thornton | 138 | 1,196 | 245 | 45 | 882 | 5,877,074 | 20.3 | 28.2 | 10.4 | 10.5 | 3.6 | 23.8 | 2.5 |
| Westminster | 191 | 1,227 | 110 | 39 | 1,038 | 6,205,205 | 18.9 | 27.9 | 15.4 | 4.0 | 0.0 | 25.8 | 0.0 |
| Wheat Ridge | 123 | 585 | 78 | 23 | 353 | 1,288,582 | 18.3 | 41.7 | 0.0 | 10.6 | 0.0 | 29.4 | 0.0 |
| **CONNECTICUT** | 7,507 | 46,490 | 6,056 | 1,646 | X | X | X | X | X | X | X | X | X |
| Bridgeport | 211 | 1,092 | 122 | 34 | 4,257 | 22,084,090 | 4.1 | 24.4 | 10.9 | 1.7 | 1.1 | 3.8 | 53.0 |
| Bristol | 98 | 458 | 53 | 14 | 1,484 | 8,479,799 | 3.6 | 13.0 | 7.6 | 2.4 | 1.5 | 6.1 | 62.8 |
| Danbury | 198 | 1,205 | 177 | 40 | 1,853 | 11,285,779 | 3.3 | 10.4 | 8.2 | 4.2 | 1.1 | 3.0 | 68.3 |
| Hartford | 293 | 2,187 | 415 | 89 | 5,025 | 26,287,752 | 2.7 | 8.4 | 5.0 | 0.3 | 4.3 | 1.5 | 77.7 |
| Meriden | 101 | 477 | 44 | 14 | 1,672 | 9,671,134 | 4.0 | 12.7 | 7.8 | 2.3 | 1.5 | 4.8 | 66.5 |
| Middletown | 106 | 513 | 73 | 19 | 1,538 | 8,328,920 | 3.1 | 12.3 | 8.7 | 2.3 | 5.9 | 5.5 | 59.0 |
| Milford | 157 | 1,010 | 87 | 29 | 1,561 | 7,897,551 | 5.0 | 10.0 | 9.4 | 2.5 | 3.1 | 4.2 | 62.7 |
| Naugatuck | 43 | 163 | 21 | 5 | 871 | 4,716,756 | 3.4 | 12.2 | 5.5 | 1.8 | 0.1 | 1.5 | 74.3 |
| New Britain | 97 | 434 | 58 | 14 | 2,015 | 12,477,375 | 2.9 | 10.7 | 9.6 | 3.0 | 1.5 | 5.1 | 64.5 |
| New Haven | 256 | 1,708 | 266 | 70 | 5,852 | 26,236,826 | 3.6 | 13.4 | 8.2 | 2.1 | 3.1 | 2.3 | 65.6 |
| New London | 69 | 439 | 62 | 14 | 777 | 3,886,300 | 4.7 | 16.1 | 10.9 | 1.2 | 1.5 | 4.6 | 57.8 |
| Norwalk | 231 | 1,432 | 190 | 51 | 3,068 | 15,640,214 | 3.7 | 13.1 | 8.8 | 3.2 | 1.1 | 1.8 | 66.9 |
| Norwich | 76 | 518 | 49 | 12 | 1,217 | 6,728,759 | 2.7 | 10.6 | 6.3 | 0.3 | 2.2 | 4.4 | 52.8 |
| Shelton | 73 | 514 | 36 | 13 | 923 | 5,479,085 | 4.1 | 9.8 | 6.7 | 3.0 | 1.9 | 2.7 | 75.5 |
| Stamford | 340 | 1,882 | 264 | 66 | 3,071 | 21,670,742 | 4.7 | 9.0 | 9.2 | 0.0 | 0.0 | 0.0 | 77.1 |
| Torrington | 83 | 414 | 53 | 13 | 1,032 | 5,822,835 | 3.5 | 13.4 | 7.1 | 2.7 | 2.0 | 2.5 | 65.9 |
| Waterbury | 159 | 1,251 | 117 | 32 | 4,072 | 21,297,073 | 4.6 | 10.6 | 6.3 | 1.4 | 2.3 | 4.7 | 71.2 |
| West Haven | 88 | 427 | 54 | 14 | 1,336 | 8,036,488 | 4.6 | 16.4 | 4.3 | 4.4 | 2.4 | 0.5 | 67.3 |
| **DELAWARE** | 1,608 | 10,252 | 1,221 | 339 | X | X | X | X | X | X | X | X | X |
| Dover | 91 | 627 | 62 | 18 | 360 | 1,823,318 | 20.1 | 46.0 | 1.1 | 5.2 | 0.4 | 23.3 | 3.9 |
| Newark | 65 | 503 | 66 | 22 | 293 | 1,596,221 | 12.5 | 38.5 | 0.4 | 6.2 | 1.0 | 27.7 | 0.0 |
| Wilmington | 226 | 1,638 | 308 | 64 | 1,168 | 6,098,980 | 18.3 | 40.8 | 17.3 | 2.0 | 4.4 | 11.5 | 0.0 |
| **DISTRICT OF COLUMBIA** | 3,502 | 73,756 | 24,226 | 5,459 | X | X | X | X | X | X | X | X | X |
| Washington | 3,502 | 73,756 | 24,226 | 5,459 | 40,352 | 276,215,832 | 14.5 | 18.1 | 5.3 | 2.1 | 17.4 | 9.3 | 22.4 |
| **FLORIDA** | 38,932 | 222,512 | 29,617 | 6,946 | X | X | X | X | X | X | X | X | X |
| Altamonte Springs | 125 | 592 | 48 | 16 | 461 | 2,055,913 | 21.8 | 29.0 | 0.0 | 2.0 | 0.0 | 38.0 | 1.0 |
| Apopka | 69 | 335 | 29 | 8 | 405 | 2,056,784 | 13.9 | 34.9 | 25.1 | 3.2 | 0.0 | 15.0 | 0.0 |
| Aventura | 112 | 1,287 | 154 | 39 | 346 | 1,474,105 | 18.1 | 28.6 | 27.1 | 8.9 | 0.5 | 15.7 | 0.0 |
| Boca Raton | 492 | 2,666 | 397 | 93 | 1,616 | 8,328,920 | 12.4 | 23.5 | 25.3 | 3.1 | 0.1 | 27.9 | 2.4 |
| Bonita Springs | 137 | 632 | 79 | 20 | 60 | 279,533 | 45.7 | 0.0 | 0.0 | 15.4 | 13.0 | 22.2 | 0.0 |
| Boynton Beach | 201 | 1,520 | 125 | 41 | 747 | 4,245,982 | 11.5 | 28.4 | 26.7 | 0.2 | 0.3 | 24.1 | 2.5 |
| Bradenton | 101 | 400 | 58 | 13 | 507 | 2,184,357 | 11.9 | 33.9 | 19.1 | 3.1 | 0.5 | 24.0 | 0.0 |
| Cape Coral | 280 | 1,099 | 93 | 28 | 1,377 | 7,666,811 | 18.1 | 24.1 | 18.5 | 3.6 | 5.3 | 27.0 | 0.0 |
| Casselberry | 71 | 212 | 18 | 5 | 167 | 747,193 | 11.6 | 38.1 | 0.0 | 4.6 | 1.4 | 32.5 | 0.0 |
| Clearwater | 284 | 1,164 | 144 | 36 | 1,649 | 7,771,554 | 13.0 | 26.2 | 16.5 | 4.0 | 0.6 | 29.1 | 3.1 |
| Clermont | 81 | 434 | 40 | 12 | 277 | 1,070,366 | 12.9 | 27.5 | 23.7 | 9.3 | 0.0 | 23.5 | 0.0 |
| Coconut Creek | 106 | 705 | 63 | 21 | 370 | 2,242,383 | 21.0 | 42.8 | 1.5 | 3.4 | 0.0 | 20.8 | 0.0 |
| Cooper City | 66 | 495 | 66 | 15 | 136 | 578,526 | 17.3 | 0.0 | 0.0 | 0.0 | 0.0 | 48.9 | 0.0 |
| Coral Gables | 229 | 1,124 | 158 | 37 | 829 | 4,496,506 | 21.0 | 28.6 | 24.5 | 4.0 | 0.0 | 13.5 | 0.0 |
| Coral Springs | 294 | 1,159 | 119 | 30 | 919 | 5,422,682 | 13.9 | 39.6 | 26.9 | 1.9 | 0.0 | 11.2 | 0.0 |
| Cutler Bay | D | D | D | D | 53 | 224,074 | 39.6 | 1.1 | 0.0 | 9.7 | 0.0 | 35.7 | 0.0 |
| Dania Beach | 112 | 586 | 77 | 20 | 128 | 624,214 | 23.2 | 0.0 | 0.0 | 4.1 | 8.4 | 24.5 | 39.8 |
| Davie | 315 | 1,530 | 220 | 48 | 667 | 4,500,059 | 10.5 | 42.7 | 25.9 | 4.5 | 0.9 | 10.1 | 0.0 |
| Daytona Beach | 173 | 1,297 | 391 | 70 | 964 | 4,588,181 | 12.0 | 39.1 | 11.5 | 5.1 | 1.4 | 26.4 | 0.0 |

1. Establishments subject to federal tax.

## Table D. Cities — City Government Finances

| City | City government finances, 2017 | | | | | | | | | |
| --- | --- | --- | --- | --- | --- | --- | --- | --- | --- | --- |
| | General revenue | | | | | | | General expenditure | | |
| | | Intergovernmental | | Taxes | | | | | Per capita[1] (dollars) | |
| | | | Percent from state government | | | Per capita[1] (dollars) | | | | |
| | Total (mil dol) | Total (mil dol) | | Total (mil dol) | Total | Property | Sales and gross receipts | Total (mil dol) | Total | Capital outlays |
| | 117 | 118 | 119 | 120 | 121 | 122 | 123 | 124 | 125 | 126 |
| **COLORADO— Cont'd** | | | | | | | | | | |
| Castle Rock | 111.2 | 4.7 | 100.0 | 56.6 | 908 | 103 | 805 | 90 | 1,438 | 176 |
| Centennial | 89.5 | 11.6 | 0.0 | 63.4 | 574 | 133 | 434 | 75 | 676 | 164 |
| Colorado Springs | 591.9 | 59.6 | 43.2 | 328.7 | 707 | 75 | 631 | 481 | 1,033 | 192 |
| Commerce City | 77.5 | 2.5 | 73.9 | 63.6 | 1,138 | 50 | 1,088 | 65 | 1,162 | 0 |
| Denver | 3,390.9 | 279.3 | 90.4 | 1,334.7 | 1,893 | 580 | 1,313 | 3,003 | 4,260 | 770 |
| Englewood | 69.6 | 4.1 | 48.3 | 39.3 | 1,140 | 139 | 1,001 | 61 | 1,775 | 4 |
| Fort Collins | 333.0 | 50.6 | 39.7 | 176.4 | 1,065 | 149 | 906 | 271 | 1,638 | 177 |
| Fountain | 20.3 | 1.5 | 8.0 | 17.1 | 575 | 68 | 507 | 29 | 962 | 104 |
| Grand Junction | 118.1 | 16.4 | 40.4 | 58.3 | 940 | 163 | 778 | 128 | 2,066 | 240 |
| Greeley | 159.2 | 15.9 | 53.7 | 100.0 | 936 | 167 | 769 | 147 | 1,380 | 343 |
| Lakewood | 221.3 | 27.3 | 33.1 | 138.8 | 896 | 142 | 695 | 213 | 1,371 | 173 |
| Littleton | 101.2 | 17.9 | 100.0 | 44.7 | 935 | 141 | 793 | 88 | 1,833 | 105 |
| Longmont | 173.3 | 25.8 | 25.7 | 91.7 | 969 | 185 | 784 | 194 | 2,054 | 524 |
| Loveland | 146.6 | 8.5 | 95.8 | 85.0 | 1,105 | 326 | 780 | 137 | 1,783 | 287 |
| Northglenn | 41.8 | 3.8 | 100.0 | 27.4 | 703 | 80 | 623 | 35 | 900 | 202 |
| Parker | 79.5 | 9.0 | 30.8 | 51.2 | 943 | 35 | 909 | 82 | 1,514 | 474 |
| Pueblo | 158.4 | 18.4 | 74.1 | 91.1 | 820 | 178 | 643 | 144 | 1,297 | 103 |
| Thornton | 208.1 | 28.6 | 84.5 | 125.0 | 914 | 102 | 812 | 178 | 1,303 | 288 |
| Westminster | 243.5 | 29.5 | 16.8 | 131.9 | 1,168 | 141 | 1,027 | 245 | 2,170 | 712 |
| Wheat Ridge | 34.1 | 1.8 | 63.8 | 28.8 | 919 | 26 | 893 | 45 | 1,440 | 262 |
| **CONNECTICUT** | X | X | X | X | X | X | X | X | X | X |
| Bridgeport | 786.6 | 412.0 | 97.2 | 318.0 | 2,188 | 2,150 | 38 | 814 | 5,597 | 834 |
| Bristol | 236.3 | 76.2 | 98.8 | 144.8 | 2,408 | 2,352 | 39 | 223 | 3,716 | 424 |
| Danbury | 313.9 | 77.3 | 98.3 | 207.3 | 2,449 | 2,405 | 44 | 284 | 3,357 | 249 |
| Hartford | 843.8 | 535.8 | 87.8 | 267.1 | 2,173 | 2,108 | 55 | 801 | 6,521 | 381 |
| Meriden | 264.9 | 120.4 | 98.7 | 122.5 | 2,054 | 2,054 | 0 | 265 | 4,449 | 547 |
| Middletown | 211.6 | 45.0 | 95.5 | 111.0 | 2,389 | 2,374 | 15 | 230 | 4,958 | 313 |
| Milford | 230.1 | 27.6 | 97.1 | 182.3 | 3,452 | 3,433 | 2 | 229 | 4,327 | 259 |
| Naugatuck | 125.3 | 40.5 | 89.5 | 77.1 | 2,460 | 2,436 | 24 | 134 | 4,261 | 14 |
| New Britain | 242.9 | 120.2 | 99.1 | 121.9 | 1,679 | 1,679 | 0 | 177 | 2,434 | 20 |
| New Haven | 846.0 | 506.3 | 91.8 | 267.9 | 2,051 | 1,932 | 104 | 872 | 6,675 | 998 |
| New London | 130.0 | 63.2 | 92.6 | 49.6 | 1,839 | 1,824 | 15 | 133 | 4,918 | 110 |
| Norwalk | 402.6 | 51.9 | 98.5 | 311.3 | 3,524 | 3,411 | 64 | 383 | 4,335 | 353 |
| Norwich | 224.3 | 127.5 | 94.6 | 79.9 | 2,034 | 2,019 | 0 | 171 | 4,355 | 209 |
| Shelton | 132.3 | 19.1 | 94.1 | 103.5 | 2,520 | 2,469 | 19 | 115 | 2,798 | 75 |
| Stamford | 621.5 | 69.0 | 99.3 | 524.4 | 4,037 | 3,802 | 173 | 543 | 4,180 | 36 |
| Torrington | 146.8 | 40.0 | 97.1 | 90.9 | 2,643 | 2,625 | 11 | 131 | 3,798 | 82 |
| Waterbury | 531.0 | 249.6 | 98.6 | 242.3 | 2,238 | 2,207 | 19 | 577 | 5,325 | 358 |
| West Haven | 199.3 | 76.8 | 99.3 | 99.4 | 1,813 | 1,780 | 33 | 198 | 3,602 | 135 |
| **DELAWARE** | X | X | X | X | X | X | X | X | X | X |
| Dover | 38.6 | 5.3 | 41.6 | 16.8 | 449 | 335 | 70 | 48 | 1,272 | 137 |
| Newark | 25.0 | 1.5 | 86.4 | 9.8 | 292 | 186 | 63 | 38 | 1,135 | 133 |
| Wilmington | 217.8 | 54.1 | 36.6 | 118.8 | 1,681 | 575 | 97 | 184 | 2,604 | 186 |
| **DISTRICT OF COLUMBIA** | X | X | X | X | X | X | X | X | X | X |
| Washington | 13,057.2 | 4,055.8 | 0.0 | 7,455.9 | 10,729 | 3,500 | 2,877 | 13,938 | 20,057 | 2,232 |
| **FLORIDA** | X | X | X | X | X | X | X | X | X | X |
| Altamonte Springs | 56.5 | 5.7 | 97.3 | 26.0 | 588 | 231 | 358 | 66 | 1,490 | 618 |
| Apopka | 57.7 | 10.4 | 93.8 | 20.9 | 399 | 151 | 249 | 57 | 1,087 | 68 |
| Aventura | 54.7 | 13.1 | 89.2 | 31.8 | 845 | 399 | 444 | 49 | 1,296 | 148 |
| Boca Raton | 253.8 | 34.3 | 33.5 | 133.4 | 1,358 | 796 | 561 | 239 | 2,431 | 185 |
| Bonita Springs | 28.0 | 6.0 | 92.4 | 14.7 | 261 | 117 | 144 | 28 | 491 | 219 |
| Boynton Beach | 139.5 | 11.3 | 86.3 | 69.1 | 889 | 628 | 261 | 134 | 1,729 | 194 |
| Bradenton | 78.5 | 9.4 | 71.4 | 30.7 | 541 | 319 | 219 | 65 | 1,138 | 56 |
| Cape Coral | 295.7 | 50.1 | 91.6 | 112.6 | 613 | 416 | 197 | 287 | 1,560 | 106 |
| Casselberry | 36.9 | 3.8 | 91.7 | 12.7 | 449 | 122 | 327 | 42 | 1,494 | 434 |
| Clearwater | 247.2 | 31.1 | 48.2 | 88.7 | 768 | 380 | 389 | 229 | 1,986 | 143 |
| Clermont | 47.9 | 4.9 | 67.0 | 21.0 | 596 | 250 | 346 | 50 | 1,416 | 373 |
| Coconut Creek | 76.0 | 7.2 | 90.4 | 34.1 | 558 | 322 | 236 | 69 | 1,136 | 151 |
| Cooper City | 42.0 | 4.5 | 73.2 | 23.5 | 659 | 415 | 244 | 42 | 1,171 | 48 |
| Coral Gables | 172.9 | 6.4 | 89.0 | 105.2 | 2,077 | 1,411 | 666 | 185 | 3,659 | 687 |
| Coral Springs | 155.2 | 15.5 | 85.5 | 87.8 | 657 | 312 | 327 | 156 | 1,168 | 151 |
| Cutler Bay | 23.1 | 8.5 | 55.8 | 12.0 | 271 | 104 | 166 | 22 | 504 | 7 |
| Dania Beach | 58.6 | 3.7 | 84.9 | 29.5 | 919 | 555 | 364 | 60 | 1,867 | 698 |
| Davie | 123.5 | 15.1 | 80.3 | 76.2 | 725 | 425 | 300 | 129 | 1,230 | 98 |
| Daytona Beach | 147.0 | 15.6 | 54.6 | 50.3 | 739 | 378 | 361 | 130 | 1,916 | 201 |

1. Based on population estimated as of July 1 of the year shown.

City government finances, 2017 (cont.)

General expenditure (cont.)

Percent of total for:

| City | Public welfare | Highways | Parking facilities | Education | Health and hospitals | Police protection | Sewerage and sanitation | Parks and recreation | Housing and community development | Interest on debt |
|---|---|---|---|---|---|---|---|---|---|---|
| | 127 | 128 | 129 | 130 | 131 | 132 | 133 | 134 | 135 | 136 |
| **COLORADO— Cont'd** | | | | | | | | | | |
| Castle Rock | 0.0 | 17.0 | 0.0 | 0.0 | 0.0 | 12.1 | 16.7 | 16.9 | 0.0 | 2.8 |
| Centennial | 0.0 | 35.8 | 0.0 | 0.0 | 0.0 | 28.9 | 0.0 | 3.0 | 0.0 | 0.2 |
| Colorado Springs | 0.0 | 17.0 | 0.6 | 0.0 | 0.3 | 19.2 | 9.5 | 6.0 | 2.2 | 4.3 |
| Commerce City | 0.0 | 11.8 | 0.0 | 0.0 | 0.0 | 24.1 | 0.0 | 16.8 | 5.0 | 15.8 |
| Denver | 4.5 | 4.9 | 0.6 | 0.0 | 2.1 | 7.9 | 4.2 | 8.0 | 4.0 | 8.2 |
| Englewood | 0.0 | 6.2 | 0.0 | 0.0 | 0.0 | 20.6 | 22.3 | 13.3 | 3.5 | 4.7 |
| Fort Collins | 1.6 | 25.5 | 0.9 | 0.0 | 0.7 | 16.0 | 5.2 | 14.9 | 2.3 | 2.4 |
| Fountain | 0.0 | 8.8 | 0.0 | 0.0 | 0.0 | 22.5 | 0.0 | 3.9 | 10.4 | 0.0 |
| Grand Junction | 0.0 | 12.6 | 0.3 | 0.0 | 0.0 | 22.9 | 12.5 | 12.1 | 1.3 | 0.0 |
| Greeley | 0.0 | 19.1 | 0.1 | 0.0 | 0.0 | 15.2 | 9.2 | 13.7 | 5.7 | 1.4 |
| Lakewood | 0.0 | 10.8 | 0.0 | 0.0 | 0.9 | 27.9 | 2.2 | 12.2 | 1.2 | 1.2 |
| Littleton | 0.5 | 7.0 | 0.0 | 0.0 | 0.0 | 13.9 | 10.4 | 4.8 | 0.3 | 2.1 |
| Longmont | 0.6 | 12.6 | 0.1 | 0.0 | 0.0 | 9.6 | 15.6 | 13.7 | 4.0 | 2.3 |
| Loveland | 0.0 | 13.9 | 0.0 | 0.0 | 0.0 | 17.3 | 12.0 | 14.0 | 6.5 | 0.2 |
| Northglenn | 0.0 | 15.2 | 0.0 | 0.0 | 0.0 | 25.0 | 20.5 | 10.7 | 6.5 | 0.0 |
| Parker | 0.0 | 27.2 | 0.0 | 0.0 | 0.0 | 18.6 | 1.4 | 32.7 | 2.2 | 4.1 |
| Pueblo | 0.0 | 8.2 | 0.9 | 0.0 | 0.0 | 23.4 | 12.7 | 5.6 | 9.3 | 1.9 |
| Thornton | 0.0 | 10.3 | 0.0 | 0.0 | 0.3 | 19.2 | 12.5 | 15.7 | 0.1 | 2.8 |
| Westminster | 0.0 | 3.8 | 0.0 | 0.0 | 0.0 | 9.4 | 5.3 | 9.6 | 28.6 | 1.2 |
| Wheat Ridge | 0.0 | 9.8 | 0.0 | 0.0 | 0.0 | 22.4 | 0.0 | 27.6 | 3.1 | 0.1 |
| **CONNECTICUT** | X | X | X | X | X | X | X | X | X | X |
| Bridgeport | 0.2 | 4.1 | 0.0 | 44.6 | 0.3 | 13.6 | 5.2 | 0.9 | 0.5 | 4.7 |
| Bristol | 0.0 | 6.5 | 0.0 | 55.6 | 3.4 | 7.0 | 3.4 | 1.3 | 0.7 | 1.0 |
| Danbury | 0.2 | 3.5 | 0.4 | 53.8 | 1.9 | 6.0 | 3.6 | 2.1 | 0.4 | 2.1 |
| Hartford | 1.7 | 2.2 | 0.1 | 56.0 | 1.8 | 6.9 | 0.8 | 0.7 | 8.9 | 2.5 |
| Meriden | 0.0 | 2.9 | 0.1 | 58.7 | 2.1 | 5.2 | 5.4 | 1.2 | 0.0 | 2.0 |
| Middletown | 0.2 | 3.0 | 0.5 | 40.8 | 0.5 | 6.6 | 16.7 | 1.2 | 0.2 | 2.3 |
| Milford | 0.6 | 2.4 | 0.0 | 53.6 | 1.3 | 7.3 | 5.1 | 0.6 | 0.3 | 1.9 |
| Naugatuck | 0.1 | 3.2 | 0.0 | 54.1 | 0.0 | 5.2 | 3.4 | 0.1 | 0.2 | 3.1 |
| New Britain | 0.0 | 0.0 | 0.0 | 94.6 | 0.0 | 0.0 | 0.0 | 0.0 | 0.0 | 5.4 |
| New Haven | 0.0 | 2.4 | 2.7 | 49.9 | 0.4 | 4.7 | 0.8 | 0.7 | 1.3 | 2.7 |
| New London | 0.0 | 7.6 | 0.4 | 54.1 | 0.2 | 7.4 | 4.2 | 3.1 | 1.1 | 0.4 |
| Norwalk | 0.1 | 4.4 | 1.6 | 54.8 | 1.2 | 5.6 | 3.3 | 1.2 | 0.3 | 2.5 |
| Norwich | 1.9 | 9.6 | 0.0 | 55.4 | 1.8 | 13.4 | 5.8 | 1.0 | 0.5 | 0.0 |
| Shelton | 0.3 | 5.1 | 0.0 | 67.2 | 1.8 | 5.5 | 2.0 | 1.6 | 0.8 | 1.2 |
| Stamford | 0.3 | 2.5 | 0.7 | 57.5 | 1.5 | 11.4 | 4.9 | 0.7 | 2.2 | 3.7 |
| Torrington | 0.0 | 6.6 | 0.0 | 60.3 | 5.7 | 7.5 | 3.9 | 1.4 | 0.1 | 0.9 |
| Waterbury | 0.7 | 4.1 | 0.0 | 54.8 | 0.7 | 5.8 | 2.0 | 0.4 | 0.5 | 4.7 |
| West Haven | 0.2 | 3.6 | 0.0 | 53.3 | 0.9 | 8.3 | 6.3 | 1.1 | 0.4 | 1.7 |
| **DELAWARE** | X | X | X | X | X | X | X | X | X | X |
| Dover | 0.2 | 6.5 | 0.0 | 0.0 | 0.0 | 41.5 | 22.4 | 1.5 | 0.7 | 0.3 |
| Newark | 0.0 | 5.5 | 4.3 | 0.0 | 0.0 | 31.7 | 19.8 | 7.2 | 0.5 | 0.1 |
| Wilmington | 0.0 | 4.1 | 3.8 | 0.0 | 0.0 | 35.0 | 11.2 | 5.2 | 2.8 | 2.7 |
| **DISTRICT OF COLUMBIA** | X | X | X | X | X | X | X | X | X | X |
| Washington | 28.2 | 3.3 | 0.2 | 20.9 | 4.7 | 4.5 | 4.9 | 1.9 | 5.5 | 4.2 |
| **FLORIDA** | X | X | X | X | X | X | X | X | X | X |
| Altamonte Springs | 0.0 | 27.8 | 0.0 | 0.0 | 0.0 | 16.2 | 15.4 | 8.4 | 0.0 | 0.0 |
| Apopka | 0.0 | 5.9 | 0.0 | 0.0 | 9.8 | 28.5 | 18.3 | 6.4 | 0.0 | 0.5 |
| Aventura | 0.0 | 10.9 | 0.0 | 17.1 | 0.0 | 38.7 | 0.0 | 7.7 | 0.0 | 1.5 |
| Boca Raton | 0.0 | 4.6 | 0.0 | 0.0 | 0.0 | 18.6 | 13.9 | 16.0 | 0.3 | 2.1 |
| Bonita Springs | 0.0 | 15.9 | 0.0 | 0.0 | 0.4 | 6.1 | 0.0 | 10.9 | 0.0 | 2.5 |
| Boynton Beach | 0.0 | 1.0 | 0.0 | 0.0 | 0.0 | 23.0 | 22.7 | 6.7 | 0.2 | 0.8 |
| Bradenton | 0.0 | 5.4 | 0.6 | 0.0 | 0.0 | 22.5 | 24.9 | 4.6 | 8.1 | 2.0 |
| Cape Coral | 0.0 | 8.7 | 0.0 | 8.0 | 0.0 | 12.2 | 8.7 | 7.1 | 0.6 | 3.4 |
| Casselberry | 0.0 | 12.7 | 0.0 | 0.0 | 0.0 | 13.8 | 24.0 | 13.2 | 0.0 | 0.8 |
| Clearwater | 0.0 | 5.2 | 1.8 | 0.0 | 3.1 | 17.4 | 26.1 | 14.0 | 0.2 | 0.7 |
| Clermont | 0.0 | 3.1 | 0.0 | 0.0 | 0.0 | 32.6 | 20.9 | 10.1 | 0.0 | 1.4 |
| Coconut Creek | 0.0 | 2.9 | 0.0 | 0.0 | 0.0 | 24.7 | 1.3 | 10.0 | 0.6 | 0.6 |
| Cooper City | 0.0 | 3.7 | 0.3 | 0.0 | 0.0 | 29.0 | 18.4 | 8.3 | 0.0 | 0.1 |
| Coral Gables | 0.0 | 3.2 | 3.9 | 0.0 | 0.0 | 22.8 | 10.6 | 13.8 | 0.0 | 1.2 |
| Coral Springs | 0.0 | 3.9 | 0.0 | 0.0 | 5.9 | 31.6 | 8.8 | 11.4 | 0.0 | 1.2 |
| Cutler Bay | 0.0 | 0.0 | 0.0 | 0.0 | 0.0 | 40.1 | 0.0 | 9.6 | 0.0 | 2.4 |
| Dania Beach | 0.0 | 4.5 | 1.0 | 0.0 | 0.0 | 21.6 | 11.6 | 7.7 | 0.0 | 0.6 |
| Davie | 0.0 | 5.8 | 0.0 | 0.0 | 17.1 | 32.8 | 0.0 | 4.5 | 0.8 | 1.1 |
| Daytona Beach | 0.3 | 4.9 | 0.2 | 0.0 | 0.0 | 23.8 | 16.6 | 11.3 | 1.1 | 1.9 |

# City Government Finances, City Government Employment, and Climate

Columns 137–139: City government finances, 2017 (cont.) — Debt outstanding.
Columns 140–146: Climate[2] — Average daily temperature (Mean; Limits), Annual precipitation, Heating degree days, Cooling degree days.

| City | Total (mil dol) | Per capita[1] (dollars) | Debt issued during year | Mean January | Mean July | Limits January[3] | Limits July[4] | Annual precipitation (inches) | Heating degree days | Cooling degree days |
|---|---|---|---|---|---|---|---|---|---|---|
| | 137 | 138 | 139 | 140 | 141 | 142 | 143 | 144 | 145 | 146 |
| COLORADO— Cont'd | | | | | | | | | | |
| Castle Rock | 93.4 | 1,499 | 0.0 | NA | NA | NA | NA | NA | NA | NA |
| Centennial | 2.9 | 26 | 0.0 | NA | NA | NA | NA | NA | NA | NA |
| Colorado Springs | 2,481.6 | 5,335 | 127.9 | 28.1 | 69.6 | 14.5 | 84.4 | 17.40 | 6,480 | 404 |
| Commerce City | 312.5 | 5,588 | 66.8 | NA | NA | NA | NA | NA | NA | NA |
| Denver | 6,263.5 | 8,885 | 1,361.6 | 31.2 | 71.5 | 15.6 | 88.3 | 18.17 | 5,988 | 496 |
| Englewood | 55.5 | 1,610 | 0.6 | 28.2 | 70.2 | 12.7 | 85.8 | 17.06 | 6,773 | 435 |
| Fort Collins | 143.2 | 865 | 0.0 | 28.2 | 71.5 | 14.5 | 86.2 | 13.98 | 6,256 | 524 |
| Fountain | 49.4 | 1,662 | 7.2 | NA | NA | NA | NA | NA | NA | NA |
| Grand Junction | 92.6 | 1,494 | 1.6 | 27.4 | 77.5 | 16.8 | 91.9 | 9.06 | 5,489 | 1,098 |
| Greeley | 158.8 | 1,487 | 58.2 | 27.8 | 74.0 | 15.6 | 88.7 | 14.22 | 5,980 | 759 |
| Lakewood | 37.1 | 239 | 0.0 | 28.2 | 70.2 | 12.7 | 85.8 | 17.06 | 6,773 | 435 |
| Littleton | 42.8 | 895 | 0.0 | 28.2 | 70.2 | 12.7 | 85.8 | 17.06 | 6,773 | 435 |
| Longmont | 191.2 | 2,020 | 0.0 | 27.1 | 72.2 | 12.0 | 88.9 | 14.15 | 6,415 | 587 |
| Loveland | 2.5 | 32 | 0.0 | 28.2 | 71.5 | 14.5 | 86.2 | 13.98 | 6,256 | 524 |
| Northglenn | 0.2 | 6 | 0.0 | 30.0 | 72.0 | 16.2 | 87.9 | 13.25 | 6,074 | 590 |
| Parker | 65.0 | 1,198 | 0.0 | NA | NA | NA | NA | NA | NA | NA |
| Pueblo | 120.5 | 1,085 | 0.0 | 30.8 | 77.0 | 14.7 | 93.8 | 12.60 | 5,346 | 997 |
| Thornton | 112.8 | 825 | 0.0 | 30.0 | 72.0 | 16.2 | 87.9 | 13.25 | 6,074 | 590 |
| Westminster | 397.6 | 3,519 | 78.8 | 29.2 | 73.4 | 15.2 | 88.0 | 15.81 | 6,128 | 696 |
| Wheat Ridge | 0.1 | 3 | 0.0 | 31.2 | 71.5 | 15.6 | 88.3 | 18.17 | 5,988 | 496 |
| CONNECTICUT | X | X | X | X | X | X | X | X | X | X |
| Bridgeport | 686.8 | 4,725 | 62.8 | 29.9 | 74.0 | 22.9 | 81.9 | 44.15 | 5,466 | 789 |
| Bristol | 89.1 | 1,482 | 25.3 | 23.4 | 70.5 | 12.6 | 83.3 | 51.03 | 6,825 | 395 |
| Danbury | 232.3 | 2,745 | 29.6 | 26.5 | 72.5 | 17.6 | 83.9 | 51.77 | 6,159 | 597 |
| Hartford | 619.6 | 5,042 | 0.0 | 25.9 | 73.6 | 16.3 | 83.8 | 44.29 | 6,121 | 654 |
| Meriden | 157.0 | 2,633 | 0.0 | 28.3 | 73.4 | 20.3 | 84.2 | 52.35 | 5,791 | 669 |
| Middletown | 107.0 | 2,302 | 0.0 | 29.9 | 74.0 | 22.9 | 81.9 | 44.15 | 5,466 | 789 |
| Milford | 102.6 | 1,942 | 0.0 | 29.9 | 74.0 | 22.9 | 81.9 | 44.15 | 5,466 | 789 |
| Naugatuck | 30.8 | 982 | 0.0 | 29.9 | 74.0 | 22.9 | 81.9 | 44.15 | 5,466 | 789 |
| New Britain | 189.1 | 2,604 | 0.0 | 28.3 | 73.4 | 20.3 | 84.2 | 52.35 | 5,791 | 669 |
| New Haven | 605.0 | 4,630 | 125.2 | 25.9 | 72.5 | 16.9 | 82.8 | 52.73 | 6,271 | 558 |
| New London | 44.8 | 1,663 | 0.0 | 28.9 | 71.8 | 20.0 | 80.7 | 48.72 | 5,799 | 511 |
| Norwalk | 261.1 | 2,955 | 38.2 | 27.8 | 73.4 | 18.8 | 84.2 | 48.38 | 5,854 | 652 |
| Norwich | 67.6 | 1,721 | 13.8 | 27.6 | 73.2 | 17.3 | 83.8 | 52.78 | 5,916 | 627 |
| Shelton | 48.1 | 1,171 | 4.0 | 29.9 | 74.0 | 22.9 | 81.9 | 44.15 | 5,466 | 789 |
| Stamford | 559.7 | 4,309 | 47.7 | 28.7 | 73.5 | 19.2 | 85.4 | 52.79 | 5,582 | 692 |
| Torrington | 16.2 | 472 | 0.0 | 23.6 | 69.5 | 13.9 | 80.7 | 54.59 | 6,839 | 323 |
| Waterbury | 448.9 | 4,145 | 0.0 | 25.9 | 72.5 | 16.9 | 82.8 | 52.73 | 6,271 | 558 |
| West Haven | 30.7 | 560 | 0.0 | 25.9 | 72.5 | 16.9 | 82.8 | 52.73 | 6,271 | 558 |
| DELAWARE | X | X | X | X | X | X | X | X | X | X |
| Dover | 36.9 | 987 | 0.0 | 35.3 | 77.8 | 26.9 | 87.4 | 46.28 | 4,212 | 1,262 |
| Newark | 9.9 | 296 | 0.0 | 32.5 | 76.4 | 23.5 | 87.6 | 45.35 | 4,746 | 1,047 |
| Wilmington | 367.9 | 5,207 | 59.7 | 31.5 | 76.6 | 23.7 | 86.0 | 42.81 | 4,888 | 1,125 |
| DISTRICT OF COLUMBIA | X | X | X | X | X | X | X | X | X | X |
| Washington | 14,417.4 | 20,747 | 2,739.2 | 34.9 | 79.2 | 27.3 | 88.3 | 39.35 | 4,055 | 1,531 |
| FLORIDA | X | X | X | X | X | X | X | X | X | X |
| Altamonte Springs | 0.0 | 0 | 0.0 | 58.7 | 81.5 | 47.0 | 91.9 | 51.31 | 799 | 3,017 |
| Apopka | 24.8 | 475 | 1.8 | 58.7 | 81.5 | 47.0 | 91.9 | 51.31 | 799 | 3,017 |
| Aventura | 20.2 | 536 | 0.0 | 67.9 | 82.7 | 62.6 | 87.0 | 46.60 | 141 | 4,090 |
| Boca Raton | 68.8 | 700 | 0.0 | 67.2 | 83.3 | 57.8 | 91.8 | 57.27 | 219 | 4,241 |
| Bonita Springs | 23.0 | 409 | 0.0 | 64.3 | 82.0 | 53.4 | 91.2 | 51.90 | 316 | 3,646 |
| Boynton Beach | 108.1 | 1,391 | 0.0 | 66.2 | 82.5 | 57.3 | 90.1 | 61.39 | 246 | 3,999 |
| Bradenton | 43.3 | 764 | 14.3 | 61.6 | 81.9 | 50.9 | 91.3 | 54.12 | 538 | 3,327 |
| Cape Coral | 820.4 | 4,466 | 1.0 | 62.7 | 81.3 | 50.3 | 91.3 | 50.07 | 427 | 3,287 |
| Casselberry | 21.6 | 762 | 0.0 | NA | NA | NA | NA | NA | NA | NA |
| Clearwater | 205.6 | 1,780 | 0.0 | 61.3 | 82.5 | 52.4 | 89.7 | 44.77 | 591 | 3,482 |
| Clermont | 22.8 | 648 | 10.6 | NA | NA | NA | NA | NA | NA | NA |
| Coconut Creek | 6.3 | 103 | 0.0 | 67.2 | 83.3 | 57.8 | 91.8 | 57.27 | 219 | 4,241 |
| Cooper City | 0.7 | 18 | 0.0 | 67.5 | 82.6 | 59.2 | 89.8 | 64.19 | 167 | 4,120 |
| Coral Gables | 100.2 | 1,979 | 23.8 | 67.9 | 82.7 | 62.6 | 87.0 | 46.60 | 141 | 4,090 |
| Coral Springs | 121.0 | 906 | 42.1 | 67.2 | 83.3 | 57.8 | 91.8 | 57.27 | 219 | 4,241 |
| Cutler Bay | 15.7 | 353 | 0.0 | NA | NA | NA | NA | NA | NA | NA |
| Dania Beach | 29.6 | 922 | 0.0 | NA | NA | NA | NA | NA | NA | NA |
| Davie | 186.7 | 1,776 | 0.4 | 66.2 | 82.5 | 57.3 | 90.1 | 61.39 | 246 | 3,999 |
| Daytona Beach | 138.9 | 2,040 | 0.8 | 57.1 | 81.2 | 44.5 | 91.2 | 57.03 | 954 | 2,819 |

1. Based on the population estimated as of July 1 of the year shown. 2. Represents normal values based on the 30-year period, 1971±2000. 3. Average daily minimum. 4. Average daily maximum.

# Table D. Cities — Land Area and Population

| STATE | Place code | City | Land area[1] (sq. mi) | Population, 2020 Total persons 2019 | Rank | Per square mile | Race 2019 — Race alone[2] (percent) White | Black or African American | American Indian, Alaskan Native | Asian | Hawaiian Pacific Islander | Some other race | Two or more races (percent) |
|---|---|---|---|---|---|---|---|---|---|---|---|---|---|
| | | | 1 | 2 | 3 | 4 | 5 | 6 | 7 | 8 | 9 | 10 | 11 |
| | | **FLORIDA— Cont'd** | | | | | | | | | | | |
| 12 | 16725 | Deerfield Beach | 14.9 | 81,290 | 424 | 5,456 | 65.5 | 24.2 | 0.3 | 3.0 | 0.0 | 4.0 | 3.1 |
| 12 | 16875 | DeLand | 19.0 | 36,019 | 1,073 | 1,896 | 70.0 | 18.6 | 0.0 | 1.2 | 0.0 | 8.5 | 1.7 |
| 12 | 17100 | Delray Beach | 15.9 | 69,577 | 526 | 4,376 | 62.3 | 31.2 | 0.0 | 1.7 | 0.0 | 2.1 | 2.7 |
| 12 | 17200 | Deltona | 37.3 | 94,312 | 350 | 2,529 | 65.4 | 14.0 | 0.4 | 2.3 | 0.0 | 13.8 | 4.0 |
| 12 | 17935 | Doral | 13.8 | 67,792 | 544 | 4,913 | 92.9 | 0.7 | 0.0 | 2.2 | 0.0 | 2.5 | 1.8 |
| 12 | 18575 | Dunedin | 10.4 | 36,464 | 1,057 | 3,506 | D | D | D | D | D | D | D |
| 12 | 24000 | Fort Lauderdale | 34.6 | 184,245 | 140 | 5,325 | 63.4 | 31.1 | 0.2 | 1.7 | 0.2 | 1.5 | 2.0 |
| 12 | 24125 | Fort Myers | 39.8 | 91,086 | 367 | 2,289 | 65.2 | 20.8 | 0.0 | 1.9 | 0.0 | 10.2 | 2.0 |
| 12 | 24300 | Fort Pierce | 23.8 | 46,343 | 838 | 1,947 | D | D | D | D | D | D | D |
| 12 | 25175 | Gainesville | 63.2 | 135,004 | 204 | 2,136 | 64.5 | 22.6 | 0.2 | 5.9 | 0.9 | 2.0 | 4.0 |
| 12 | 27322 | Greenacres | 5.9 | 41,091 | 933 | 6,965 | 63.1 | 23.2 | 0.0 | 7.0 | 0.0 | 6.2 | 0.6 |
| 12 | 28452 | Hallandale Beach | 4.2 | 39,724 | 972 | 9,458 | D | D | D | D | D | D | D |
| 12 | 30000 | Hialeah | 21.6 | 232,027 | 97 | 10,742 | 92.2 | 1.8 | 0.0 | 0.7 | 0.0 | 4.1 | 1.1 |
| 12 | 32000 | Hollywood | 27.3 | 154,611 | 165 | 5,663 | 64.4 | 22.0 | 0.0 | 1.7 | 0.1 | 6.3 | 5.4 |
| 12 | 32275 | Homestead | 15.1 | 69,292 | 531 | 4,589 | 63.2 | 22.6 | 2.3 | 0.5 | 0.0 | 6.1 | 5.3 |
| 12 | 35000 | Jacksonville | 747.3 | 920,570 | 13 | 1,232 | 56.8 | 31.2 | 0.4 | 4.9 | 0.0 | 2.9 | 3.8 |
| 12 | 35875 | Jupiter | 21.6 | 65,834 | 574 | 3,048 | 88.9 | 1.1 | 0.7 | 2.9 | 0.0 | 1.6 | 4.8 |
| 12 | 36950 | Kissimmee | 21.5 | 72,858 | 494 | 3,389 | 61.6 | 11.8 | 0.4 | 3.0 | 0.2 | 21.6 | 1.5 |
| 12 | 38250 | Lakeland | 66.2 | 115,283 | 255 | 1,741 | 67.8 | 21.1 | 1.2 | 1.7 | 0.2 | 5.8 | 2.3 |
| 12 | 39075 | Lake Worth | NA | 38,701 | 1,000 | NA | 58.7 | 25.0 | 3.2 | 0.2 | 0.0 | 8.2 | 4.6 |
| 12 | 39425 | Largo | 18.5 | 84,681 | 402 | 4,577 | 85.4 | 6.9 | 0.0 | 3.6 | 0.2 | 1.4 | 2.7 |
| 12 | 39525 | Lauderdale Lakes | 3.7 | 36,769 | 1,044 | 9,938 | 9.9 | 79.7 | 0.0 | 2.9 | 0.0 | 0.0 | 7.5 |
| 12 | 39550 | Lauderhill | 8.5 | 71,582 | 508 | 8,421 | D | D | D | D | D | D | D |
| 12 | 43125 | Margate | 8.8 | 58,734 | 647 | 6,674 | 58.9 | 33.2 | 0.0 | 4.5 | 0.0 | 1.9 | 1.5 |
| 12 | 43975 | Melbourne | 44.1 | 83,376 | 414 | 1,891 | 84.9 | 7.2 | 0.1 | 3.5 | 0.0 | 0.8 | 3.5 |
| 12 | 45000 | Miami | 36.0 | 471,525 | 42 | 13,098 | 77.1 | 15.7 | 0.1 | 1.3 | 0.0 | 4.0 | 1.8 |
| 12 | 45025 | Miami Beach | 7.7 | 87,726 | 386 | 11,393 | 72.1 | 5.3 | 0.0 | 4.1 | 0.0 | 15.9 | 2.6 |
| 12 | 45060 | Miami Gardens | 18.2 | 109,131 | 281 | 5,996 | 27.7 | 65.7 | 0.3 | 0.5 | 0.0 | 4.2 | 1.6 |
| 12 | 45100 | Miami Lakes | 5.7 | 31,567 | 1,207 | 5,538 | 92.2 | 2.4 | 0.0 | 1.5 | 0.0 | 3.2 | 0.6 |
| 12 | 45975 | Miramar | 28.9 | 141,150 | 193 | 4,884 | 43.4 | 47.1 | 0.2 | 4.8 | 0.0 | 2.2 | 2.3 |
| 12 | 49425 | North Lauderdale | 4.6 | 44,203 | 869 | 9,609 | 27.5 | 55.7 | 0.9 | 3.1 | 0.0 | 4.8 | 7.9 |
| 12 | 49450 | North Miami | 8.4 | 62,247 | 602 | 7,410 | 36.8 | 57.6 | 0.0 | 0.4 | 0.0 | 2.9 | 2.3 |
| 12 | 49475 | North Miami Beach | 4.8 | 42,444 | 907 | 8,843 | D | D | D | D | D | D | D |
| 12 | 49675 | North Port | 99.4 | 73,738 | 488 | 742 | 89.1 | 4.3 | 0.0 | 0.0 | 0.0 | 2.0 | 4.6 |
| 12 | 50575 | Oakland Park | 7.5 | 45,070 | 856 | 6,009 | 68.0 | 23.9 | 0.2 | 0.9 | 0.0 | 4.7 | 2.3 |
| 12 | 50750 | Ocala | 47.3 | 61,669 | 607 | 1,304 | 74.3 | 21.3 | 0.2 | 2.4 | 0.0 | 1.1 | 0.7 |
| 12 | 51075 | Ocoee | 15.6 | 49,168 | 790 | 3,152 | 45.7 | 20.0 | 1.3 | 7.8 | 0.0 | 21.8 | 3.5 |
| 12 | 53000 | Orlando | 110.6 | 289,457 | 70 | 2,617 | 67.1 | 21.5 | 0.3 | 3.2 | 0.0 | 4.7 | 3.2 |
| 12 | 53150 | Ormond Beach | 34.7 | 44,043 | 874 | 1,269 | D | D | D | D | D | D | D |
| 12 | 53575 | Oviedo | 15.5 | 41,652 | 923 | 2,687 | 78.8 | 6.3 | 0.0 | 9.3 | 0.0 | 3.9 | 1.6 |
| 12 | 54000 | Palm Bay | 86.4 | 116,932 | 247 | 1,353 | 72.8 | 14.5 | 0.1 | 1.4 | 0.0 | 4.0 | 7.2 |
| 12 | 54075 | Palm Beach Gardens | 58.7 | 58,210 | 661 | 992 | 83.4 | 9.2 | 0.0 | 4.2 | 0.2 | 1.9 | 1.1 |
| 12 | 54200 | Palm Coast | 95.4 | 92,350 | 363 | 968 | 80.5 | 9.4 | 0.2 | 3.5 | 0.9 | 2.9 | 2.6 |
| 12 | 54700 | Panama City | 35.1 | 33,734 | 1,142 | 961 | 73.2 | 14.1 | 0.7 | 3.2 | 1.9 | 4.0 | 2.9 |
| 12 | 55775 | Pembroke Pines | 32.7 | 174,414 | 151 | 5,334 | 64.0 | 23.1 | 0.2 | 4.3 | 0.3 | 2.8 | 5.2 |
| 12 | 55925 | Pensacola | 22.7 | 53,142 | 733 | 2,341 | 66.8 | 26.3 | 0.0 | 2.5 | 0.0 | 0.4 | 4.1 |
| 12 | 56975 | Pinellas Park | 16.1 | 54,854 | 699 | 3,407 | 78.6 | 3.9 | 0.0 | 11.0 | 0.1 | 2.5 | 4.0 |
| 12 | 57425 | Plantation | 21.8 | 96,351 | 330 | 4,420 | 64.1 | 26.0 | 0.0 | 4.8 | 0.0 | 1.9 | 3.3 |
| 12 | 57550 | Plant City | 27.6 | 39,602 | 977 | 1,435 | 77.7 | 14.0 | 1.1 | 0.2 | 0.0 | 5.5 | 1.5 |
| 12 | 58050 | Pompano Beach | 24.0 | 112,360 | 265 | 4,682 | 66.2 | 26.9 | 0.1 | 1.0 | 0.0 | 1.8 | 4.0 |
| 12 | 58575 | Port Orange | 26.8 | 65,425 | 577 | 2,441 | 89.8 | 4.7 | 0.0 | 3.3 | 0.0 | 1.2 | 1.0 |
| 12 | 58715 | Port St. Lucie | 119.2 | 209,715 | 110 | 1,759 | 73.8 | 15.0 | 0.4 | 2.9 | 0.0 | 3.7 | 4.1 |
| 12 | 60975 | Riviera Beach | 8.3 | 35,658 | 1,085 | 4,296 | 32.8 | 57.1 | 0.0 | 4.5 | 0.2 | 1.1 | 4.3 |
| 12 | 62100 | Royal Palm Beach | 11.3 | 40,851 | 943 | 3,615 | D | D | D | D | D | D | D |
| 12 | 62625 | St. Cloud | 25.5 | 56,590 | 679 | 2,219 | 66.3 | 13.8 | 0.6 | 0.6 | 0.0 | 15.5 | 3.2 |
| 12 | 63000 | St. Petersburg | 61.8 | 267,802 | 78 | 4,333 | 73.4 | 19.4 | 0.2 | 2.3 | 0.0 | 1.0 | 3.8 |
| 12 | 63650 | Sanford | 23.6 | 61,508 | 609 | 2,606 | 48.9 | 32.9 | 0.3 | 3.0 | 0.0 | 12.8 | 2.1 |
| 12 | 64175 | Sarasota | 14.7 | 58,859 | 646 | 4,004 | 76.9 | 16.7 | 0.1 | 2.4 | 0.0 | 1.1 | 2.9 |
| 12 | 69700 | Sunrise | 16.2 | 94,851 | 345 | 5,855 | 54.5 | 32.1 | 0.2 | 4.8 | 0.0 | 3.8 | 4.6 |
| 12 | 70600 | Tallahassee | 100.9 | 196,326 | 129 | 1,946 | 52.8 | 38.8 | 0.2 | 4.8 | 0.0 | 0.9 | 2.5 |
| 12 | 70675 | Tamarac | 11.6 | 67,112 | 551 | 5,786 | 59.5 | 31.0 | 0.7 | 1.3 | 0.1 | 2.7 | 4.7 |
| 12 | 71000 | Tampa | 114.0 | 407,599 | 47 | 3,575 | 63.7 | 23.6 | 0.1 | 5.0 | 0.2 | 3.5 | 3.9 |
| 12 | 71900 | Titusville | 29.2 | 46,619 | 833 | 1,597 | 74.8 | 17.0 | 0.1 | 1.6 | 0.0 | 5.3 | 1.5 |
| 12 | 75812 | Wellington | 45.0 | 65,454 | 576 | 1,455 | 82.9 | 6.8 | 0.0 | 3.5 | 0.0 | 2.7 | 0.6 |
| 12 | 76582 | Weston | 24.6 | 71,086 | 512 | 2,890 | 88.4 | 2.1 | 0.2 | 6.0 | 0.0 | 2.9 | 2.9 |
| 12 | 76600 | West Palm Beach | 53.8 | 112,782 | 260 | 2,096 | 57.0 | 36.0 | 0.0 | 1.2 | 0.0 | 3.5 | 4.1 |
| 12 | 78250 | Winter Garden | 16.3 | 46,646 | 832 | 2,862 | 73.3 | 14.6 | 0.0 | 4.5 | 0.0 | 2.8 | 3.4 |
| 12 | 78275 | Winter Haven | 32.5 | 46,578 | 834 | 1,433 | 67.1 | 24.4 | 0.5 | 1.8 | 0.0 | 2.4 | 3.7 |
| 12 | 78300 | Winter Park | 8.8 | 30,667 | 1,231 | 3,485 | 86.2 | 5.7 | 0.0 | 2.0 | 0.1 | 3.4 | 3.1 |
| 12 | 78325 | Winter Springs | 14.9 | 37,423 | 1,027 | 2,512 | 80.8 | 6.1 | 0.0 | 6.7 | 0.0 | — | — |

1. Dry land or land partially or temporarily covered by water.    2. Hispanic or Latino persons may be of any race.

# Table D. Cities — Population

| City | Percent Hispanic or Latino[1], 2019 | Percent foreign born, 2019 | Age of population (percent), 2019 | | | | | | | Median age, 2019 | Percent female, 2019 | Population | | | |
|---|---|---|---|---|---|---|---|---|---|---|---|---|---|---|---|
| | | | Under 18 years | 18 to 24 years | 25 to 34 years | 35 to 44 years | 45 to 54 years | 55 to 64 years | 65 years and over | | | Census counts | | Percent change | |
| | | | | | | | | | | | | 2000 | 2010 | 2000-2010 | 2001-2020 |
| | 12 | 13 | 14 | 15 | 16 | 17 | 18 | 19 | 20 | 21 | 22 | 23 | 24 | 25 | 26 |
| FLORIDA— Cont'd | | | | | | | | | | | | | | | |
| Deerfield Beach | 15.0 | 33.3 | 18.0 | 4.6 | 13.6 | 11.6 | 15.0 | 13.0 | 24.3 | 46.4 | 54.6 | 64,583 | 75,096 | 16.3 | 8.2 |
| DeLand | 14.2 | 8.7 | 17.1 | 14.6 | 11.6 | 13.7 | 9.7 | 14.5 | 18.8 | 40.8 | 56.6 | 20,904 | 26,872 | 28.5 | 34.0 |
| Delray Beach | 11.7 | 27.7 | 13.4 | 9.1 | 15.7 | 9.5 | 10.1 | 15.9 | 26.2 | 47.7 | 53.2 | 60,020 | 60,611 | 1.0 | 14.8 |
| Deltona | 34.1 | 8.3 | 24.9 | 8.4 | 13.5 | 12.2 | 12.4 | 10.6 | 18.0 | 37.0 | 51.0 | 69,543 | 85,133 | 22.4 | 10.8 |
| Doral | 88.1 | 72.0 | 24.9 | 8.7 | 12.6 | 15.8 | 18.5 | 11.5 | 8.0 | 38.8 | 53.2 | 20,438 | 45,704 | 123.6 | 48.3 |
| Dunedin | 2.0 | 5.6 | 10.2 | 5.4 | 11.0 | 9.4 | 12.6 | 17.2 | 34.3 | 56.5 | 52.0 | 35,691 | 35,353 | -0.9 | 3.1 |
| Fort Lauderdale | 18.1 | 24.6 | 18.1 | 5.2 | 18.0 | 13.4 | 11.9 | 14.4 | 19.1 | 41.1 | 48.6 | 152,397 | 165,998 | 8.9 | 11.0 |
| Fort Myers | 25.1 | 18.2 | 21.8 | 7.6 | 15.9 | 12.7 | 10.5 | 9.7 | 21.7 | 37.9 | 49.7 | 48,208 | 62,308 | 29.2 | 46.2 |
| Fort Pierce | 22.3 | 16.3 | 21.1 | 11.8 | 11.9 | 11.7 | 12.9 | 11.9 | 18.6 | 39.7 | 50.4 | 37,516 | 41,933 | 11.8 | 10.5 |
| Gainesville | 13.2 | 11.0 | 14.6 | 31.5 | 17.0 | 10.5 | 7.9 | 7.6 | 11.0 | 27.0 | 51.2 | 95,447 | 124,492 | 30.4 | 8.4 |
| Greenacres | 42.5 | 44.4 | 17.5 | 10.9 | 12.2 | 12.9 | 13.3 | 14.1 | 19.0 | 41.8 | 54.7 | 27,569 | 37,560 | 36.2 | 9.4 |
| Hallandale Beach | 30.1 | 51.2 | 19.2 | 5.8 | 14.9 | 11.7 | 9.5 | 13.8 | 25.2 | 43.7 | 53.1 | 34,282 | 37,113 | 8.3 | 7.0 |
| Hialeah | 96.0 | 75.7 | 17.8 | 8.9 | 11.1 | 12.4 | 15.7 | 13.0 | 21.1 | 44.9 | 52.0 | 226,419 | 224,697 | -0.8 | 3.3 |
| Hollywood | 36.8 | 37.7 | 21.4 | 7.4 | 12.3 | 14.7 | 13.8 | 14.9 | 15.6 | 41.4 | 50.8 | 139,357 | 140,694 | 1.0 | 9.9 |
| Homestead | 70.6 | 42.5 | 35.1 | 6.3 | 14.1 | 14.5 | 14.1 | 8.9 | 6.9 | 31.5 | 49.1 | 31,909 | 60,753 | 90.4 | 14.1 |
| Jacksonville | 10.7 | 11.7 | 22.6 | 8.9 | 16.5 | 12.8 | 12.2 | 12.6 | 14.4 | 36.1 | 51.7 | 735,617 | 821,758 | 11.7 | 12.0 |
| Jupiter | 15.8 | 11.2 | 20.0 | 7.3 | 9.5 | 11.4 | 13.3 | 14.2 | 24.2 | 46.4 | 48.0 | 39,328 | 55,222 | 40.4 | 19.2 |
| Kissimmee | 71.9 | 28.2 | 25.7 | 8.3 | 13.4 | 15.7 | 13.1 | 11.7 | 12.0 | 35.6 | 46.9 | 47,814 | 59,546 | 24.5 | 22.4 |
| Lakeland | 16.6 | 10.8 | 19.9 | 11.4 | 12.9 | 12.4 | 10.5 | 10.6 | 22.4 | 39.7 | 50.4 | 78,452 | 97,287 | 24.0 | 18.5 |
| Lake Worth | 36.3 | 42.6 | 22.7 | 8.0 | 14.4 | 16.5 | 12.6 | 10.4 | 15.4 | 37.7 | 44.7 | 35,133 | 34,890 | -0.7 | 10.9 |
| Largo | 14.5 | 14.5 | 15.3 | 9.4 | 13.0 | 8.4 | 13.5 | 16.4 | 24.0 | 48.9 | 51.3 | 69,371 | 79,338 | 14.4 | 6.7 |
| Lauderdale Lakes | 4.2 | 49.5 | 25.3 | 5.8 | 11.9 | 11.1 | 11.4 | 16.8 | 17.6 | 39.8 | 52.1 | 31,705 | 32,773 | 3.4 | 12.2 |
| Lauderhill | 5.9 | 35.7 | 26.8 | 7.2 | 14.6 | 11.9 | 14.2 | 10.5 | 14.9 | 36.2 | 56.4 | 57,585 | 66,936 | 16.2 | 6.9 |
| Margate | 33.1 | 39.0 | 12.7 | 7.0 | 16.4 | 7.7 | 16.9 | 15.2 | 24.1 | 50.1 | 54.2 | 53,909 | 53,116 | -1.5 | 10.6 |
| Melbourne | 15.2 | 10.4 | 16.2 | 10.7 | 13.3 | 10.7 | 12.0 | 13.9 | 23.2 | 43.0 | 48.7 | 71,382 | 76,221 | 6.8 | 9.4 |
| Miami | 71.3 | 58.4 | 17.1 | 6.2 | 18.2 | 14.3 | 14.0 | 12.5 | 17.7 | 40.7 | 50.5 | 362,470 | 399,506 | 10.2 | 18.0 |
| Miami Beach | 61.1 | 62.6 | 16.4 | 6.3 | 15.4 | 19.5 | 15.1 | 11.3 | 16.0 | 40.5 | 51.8 | 87,933 | 87,380 | -0.6 | 0.4 |
| Miami Gardens | 31.0 | 37.2 | 23.9 | 9.6 | 15.3 | 11.3 | 12.0 | 12.7 | 15.1 | 36.3 | 54.3 | NA | 107,173 | NA | 1.8 |
| Miami Lakes | 86.1 | 55.3 | 19.6 | 7.1 | 11.8 | 10.7 | 15.4 | 17.5 | 18.0 | 45.5 | 55.8 | 22,676 | 29,373 | 29.5 | 7.5 |
| Miramar | 37.9 | 38.8 | 23.6 | 9.7 | 15.7 | 16.3 | 12.4 | 10.7 | 11.4 | 36.1 | 52.9 | 72,739 | 121,958 | 67.7 | 15.7 |
| North Lauderdale | 31.6 | 46.2 | 25.0 | 9.8 | 14.7 | 16.6 | 16.9 | 7.9 | 9.2 | 35.3 | 47.8 | 32,264 | 41,089 | 27.4 | 7.6 |
| North Miami | 32.7 | 49.8 | 25.2 | 7.9 | 15.3 | 18.1 | 9.7 | 11.6 | 12.2 | 36.2 | 49.1 | 59,880 | 60,169 | 0.5 | 3.5 |
| North Miami Beach | 39.8 | 51.6 | 19.1 | 8.4 | 19.5 | 10.4 | 15.0 | 14.7 | 13.0 | 39.2 | 47.0 | 40,786 | 40,855 | 0.2 | 3.9 |
| North Port | 10.9 | 6.1 | 23.2 | 6.7 | 5.5 | 10.6 | 9.5 | 12.1 | 32.4 | 47.9 | 51.9 | 22,797 | 57,320 | 151.4 | 28.6 |
| Oakland Park | 43.0 | 40.5 | 16.5 | 6.4 | 15.2 | 17.1 | 14.8 | 16.2 | 13.8 | 42.4 | 45.6 | 30,966 | 41,299 | 33.4 | 9.1 |
| Ocala | 16.7 | 9.3 | 21.2 | 9.2 | 17.1 | 12.1 | 9.9 | 13.6 | 17.0 | 35.9 | 47.9 | 45,943 | 56,559 | 23.1 | 9.0 |
| Ocoee | 37.9 | 24.8 | 28.0 | 8.8 | 15.3 | 14.6 | 13.3 | 10.4 | 9.5 | 33.6 | 57.4 | 24,391 | 35,730 | 46.5 | 37.6 |
| Orlando | 37.0 | 19.9 | 22.3 | 7.1 | 21.6 | 17.0 | 12.1 | 10.0 | 10.0 | 34.6 | 52.2 | 185,951 | 238,723 | 28.4 | 21.3 |
| Ormond Beach | 10.4 | 6.3 | 22.0 | 7.1 | 10.0 | 9.5 | 11.7 | 12.5 | 27.2 | 46.3 | 50.3 | 36,301 | 39,429 | 8.6 | 11.7 |
| Oviedo | 15.2 | 15.1 | 26.4 | 5.6 | 13.4 | 14.5 | 14.9 | 13.5 | 11.8 | 37.9 | 51.5 | 26,316 | 33,475 | 27.2 | 24.4 |
| Palm Bay | 17.6 | 13.2 | 20.4 | 7.1 | 12.7 | 11.8 | 10.9 | 16.6 | 20.5 | 43.3 | 52.4 | 79,413 | 104,006 | 31.0 | 12.4 |
| Palm Beach Gardens | 10.6 | 17.4 | 18.0 | 4.5 | 11.3 | 9.5 | 11.4 | 13.2 | 32.1 | 50.7 | 52.5 | 35,058 | 49,906 | 42.4 | 16.6 |
| Palm Coast | 13.4 | 16.4 | 15.6 | 7.3 | 9.2 | 10.1 | 11.8 | 15.7 | 30.2 | 51.6 | 52.8 | 32,732 | 75,203 | 129.8 | 22.8 |
| Panama City | 9.0 | 6.4 | 22.3 | 8.7 | 14.2 | 10.4 | 13.7 | 11.5 | 19.2 | 36.0 | 47.4 | 36,417 | 34,632 | -4.9 | -2.6 |
| Pembroke Pines | 47.9 | 37.8 | 22.6 | 7.6 | 12.9 | 15.0 | 13.2 | 11.5 | 17.3 | 40.7 | 52.8 | 137,427 | 154,898 | 12.7 | 12.6 |
| Pensacola | 2.8 | 4.0 | 24.5 | 6.2 | 15.3 | 11.5 | 7.8 | 15.8 | 18.8 | 37.5 | 53.1 | 56,255 | 52,015 | -7.5 | 2.2 |
| Pinellas Park | 10.3 | 15.7 | 16.5 | 5.5 | 12.6 | 14.7 | 10.8 | 15.9 | 24.0 | 46.1 | 51.2 | 45,658 | 49,568 | 8.6 | 10.7 |
| Plantation | 24.9 | 27.9 | 19.1 | 7.3 | 15.6 | 11.6 | 14.3 | 12.7 | 19.4 | 41.6 | 53.3 | 82,934 | 84,883 | 2.4 | 13.5 |
| Plant City | 28.6 | 14.4 | 18.2 | 5.3 | 20.4 | 11.1 | 13.6 | 16.9 | 14.4 | 41.7 | 54.9 | 29,915 | 34,712 | 16.0 | 14.1 |
| Pompano Beach | 25.5 | 31.4 | 16.2 | 6.9 | 12.5 | 12.7 | 13.4 | 17.0 | 21.2 | 46.3 | 47.0 | 78,191 | 99,752 | 27.6 | 12.6 |
| Port Orange | 7.7 | 11.7 | 20.3 | 6.4 | 9.7 | 10.4 | 12.9 | 14.6 | 25.6 | 47.3 | 50.3 | 45,823 | 56,621 | 23.6 | 15.5 |
| Port St. Lucie | 21.8 | 18.2 | 22.6 | 5.9 | 12.1 | 13.0 | 11.5 | 11.8 | 23.0 | 42.3 | 52.5 | 88,769 | 164,200 | 85.0 | 27.7 |
| Riviera Beach | 9.3 | 18.0 | 25.7 | 6.3 | 14.4 | 10.3 | 11.3 | 14.6 | 17.4 | 38.5 | 46.5 | 29,884 | 32,540 | 8.9 | 9.6 |
| Royal Palm Beach | 22.7 | 19.4 | 19.6 | 11.6 | 9.3 | 10.8 | 13.8 | 17.0 | 17.9 | 44.4 | 57.6 | 21,523 | 34,196 | 58.9 | 19.5 |
| St. Cloud | 51.0 | 18.3 | 23.6 | 6.0 | 12.2 | 12.4 | 11.4 | 12.8 | 21.6 | 41.9 | 55.3 | 20,074 | 37,793 | 88.3 | 49.7 |
| St. Petersburg | 8.7 | 9.4 | 16.1 | 8.1 | 14.9 | 12.4 | 12.1 | 16.0 | 20.4 | 43.8 | 50.4 | 248,232 | 245,173 | -1.2 | 9.2 |
| Sanford | 28.4 | 16.9 | 23.6 | 7.6 | 14.0 | 17.5 | 12.5 | 13.3 | 11.3 | 37.7 | 52.0 | 38,291 | 53,925 | 40.8 | 14.1 |
| Sarasota | 19.4 | 17.4 | 13.8 | 8.1 | 13.2 | 10.3 | 10.3 | 15.2 | 29.0 | 49.2 | 52.6 | 52,715 | 52,106 | -1.2 | 13.0 |
| Sunrise | 34.8 | 44.0 | 15.2 | 8.4 | 14.2 | 12.9 | 12.7 | 16.6 | 19.8 | 44.6 | 53.1 | 85,779 | 84,297 | -1.7 | 12.5 |
| Tallahassee | 6.5 | 8.0 | 17.1 | 28.7 | 16.0 | 10.3 | 9.5 | 7.8 | 10.7 | 26.8 | 52.8 | 150,624 | 181,049 | 20.2 | 8.4 |
| Tamarac | 34.9 | 39.9 | 19.0 | 7.5 | 9.0 | 8.9 | 13.4 | 14.2 | 28.1 | 49.5 | 53.7 | 55,588 | 60,781 | 9.3 | 10.4 |
| Tampa | 28.4 | 19.2 | 21.9 | 9.8 | 16.6 | 14.1 | 12.7 | 11.8 | 13.0 | 36.0 | 50.6 | 303,447 | 336,249 | 10.8 | 21.2 |
| Titusville | 6.4 | 5.2 | 21.3 | 5.9 | 16.4 | 10.6 | 13.2 | 13.1 | 19.5 | 40.3 | 54.0 | 40,670 | 43,687 | 7.4 | 6.7 |
| Wellington | 30.7 | 22.1 | 23.5 | 7.0 | 8.2 | 11.8 | 15.3 | 14.5 | 19.7 | 44.6 | 53.8 | 38,216 | 56,697 | 48.4 | 15.4 |
| Weston | 47.4 | 43.7 | 26.2 | 9.5 | 5.9 | 14.8 | 20.6 | 12.3 | 10.7 | 42.3 | 50.5 | 49,286 | 65,427 | 32.7 | 8.6 |
| West Palm Beach | 25.2 | 26.1 | 21.1 | 9.8 | 17.0 | 11.6 | 12.0 | 10.6 | 17.8 | 36.9 | 54.0 | 82,103 | 100,670 | 22.6 | 12.0 |
| Winter Garden | 29.2 | 19.3 | 23.0 | 5.2 | 19.2 | 12.8 | 9.5 | 16.9 | 13.5 | 37.4 | 51.4 | 14,351 | 34,764 | 142.2 | 34.2 |
| Winter Haven | 13.5 | 9.3 | 22.0 | 6.4 | 11.3 | 5.9 | 13.9 | 12.9 | 27.7 | 48.7 | 55.9 | 26,487 | 34,643 | 30.8 | 34.5 |
| Winter Park | 11.9 | 8.7 | 13.1 | 9.7 | 12.2 | 9.2 | 14.0 | 16.7 | 25.2 | 49.0 | 52.8 | 24,090 | 27,724 | 15.1 | 10.6 |
| Winter Springs | 18.0 | 10.4 | 20.2 | 5.3 | 14.7 | 9.8 | 13.0 | 13.8 | 23.2 | 45.0 | 52.5 | 31,666 | 33,306 | 5.2 | 12.4 |

1. May be of any race.

## Table D. Cities — Households, Group Quarters, Crime, and Education

| City | Households, 2019 | | | | | | | Persons in group quarters, 2019 | Serious crimes known to police[1], 2019 | | | | Educational attainment, 2019 | | |
|---|---|---|---|---|---|---|---|---|---|---|---|---|---|---|---|
| | | | | Percent | | | | | Violent | | Prorerty | | | Attainment[3] (percent) | |
| | Number | Persons per household | Family | Married couple family | Female family | Non-family | One person | | Number | Rate[2] | Number | Rate[2] | Population age 25 and over | High school graduate or less | Bachelor's degree or more |
| | 27 | 28 | 29 | 30 | 31 | 32 | 33 | 34 | 35 | 36 | 37 | 38 | 39 | 40 | 41 |

FLORIDA— Cont'd

| City | 27 | 28 | 29 | 30 | 31 | 32 | 33 | 34 | 35 | 36 | 37 | 38 | 39 | 40 | 41 |
|---|---|---|---|---|---|---|---|---|---|---|---|---|---|---|---|
| Deerfield Beach | 33,439 | 2.38 | 56.7 | 37.3 | 19.4 | 43.3 | 37.3 | 1,523 | 351 | 430 | 2,161 | 2,648 | 62,808 | 40.2 | 31.5 |
| DeLand | 13,350 | 2.39 | 54.4 | 37.7 | 16.7 | 45.6 | 36.4 | 2,910 | 198 | 574 | 1,126 | 3,267 | 23,800 | 30.9 | 29.1 |
| Delray Beach | 28,548 | 2.39 | 52.6 | 36.8 | 15.8 | 47.4 | 38.5 | 1,247 | 336 | 477 | 2,418 | 3,429 | 53,817 | 34.6 | 38.8 |
| Deltona | 31,783 | 2.91 | 75.3 | 51.2 | 24.1 | 24.7 | 18.9 | 233 | NA | NA | NA | NA | 61,843 | 48.7 | 16.0 |
| Doral | 19,722 | 3.33 | 81.6 | 63.4 | 18.2 | 18.4 | 14.9 | 0 | 62 | 97 | 1,475 | 2,299 | 43,696 | 32.2 | 51.1 |
| Dunedin | 17,284 | 2.08 | 48.3 | 39.3 | 9.0 | 51.7 | 45.3 | 671 | 39 | 106 | 472 | 1,285 | 30,865 | 39.9 | 32.4 |
| Fort Lauderdale | 76,384 | 2.34 | 49.6 | 34.2 | 15.4 | 50.4 | 38.5 | 3,625 | 1,098 | 594 | 9,082 | 4,915 | 139,941 | 36.1 | 38.5 |
| Fort Myers | 30,754 | 2.69 | 59.2 | 40.8 | 18.5 | 40.8 | 34.9 | 4,333 | 486 | 571 | 1,952 | 2,293 | 61,478 | 46.2 | 28.7 |
| Fort Pierce | 16,132 | 2.81 | 53.4 | 28.7 | 24.7 | 46.6 | 35.0 | 747 | 259 | 556 | 1,342 | 2,880 | 30,927 | 52.9 | 18.0 |
| Gainesville | 51,990 | 2.32 | 39.2 | 25.1 | 14.1 | 60.8 | 41.6 | 13,504 | 928 | 687 | 4,712 | 3,488 | 72,323 | 25.5 | 46.1 |
| Greenacres | 14,706 | 2.79 | 66.0 | 42.7 | 23.3 | 34.0 | 22.2 | 151 | 117 | 282 | 740 | 1,782 | 29,411 | 54.4 | 23.2 |
| Hallandale Beach | 17,716 | 2.25 | 58.3 | 33.5 | 24.8 | 41.7 | 37.3 | 57 | 178 | 442 | 1,244 | 3,087 | 29,902 | 40.5 | 36.7 |
| Hialeah | 75,989 | 3.05 | 71.9 | 42.1 | 29.8 | 28.1 | 21.1 | 1,973 | 503 | 209 | 4,893 | 2,033 | 171,020 | 62.5 | 18.6 |
| Hollywood | 55,172 | 2.78 | 65.7 | 43.8 | 21.9 | 34.3 | 28.3 | 1,248 | 456 | 291 | 4,159 | 2,655 | 110,296 | 40.1 | 29.7 |
| Homestead | 20,320 | 3.40 | 77.7 | 40.3 | 37.3 | 22.3 | 19.9 | 512 | 684 | 953 | 2,272 | 3,166 | 40,725 | 48.6 | 24.2 |
| Jacksonville | 349,171 | 2.54 | 60.1 | 39.7 | 20.4 | 39.9 | 32.4 | 23,155 | 5,886 | 647 | 30,088 | 3,309 | 624,232 | 39.2 | 29.1 |
| Jupiter | 26,959 | 2.43 | 64.1 | 51.6 | 12.5 | 35.9 | 30.2 | 367 | 118 | 176 | 1,035 | 1,547 | 47,801 | 23.9 | 50.3 |
| Kissimmee | 22,709 | 3.18 | 58.4 | 36.1 | 22.3 | 41.6 | 32.7 | 510 | 368 | 487 | 1,899 | 2,514 | 47,953 | 48.3 | 21.9 |
| Lakeland | 40,236 | 2.67 | 59.5 | 40.4 | 19.1 | 40.5 | 34.3 | 4,911 | 350 | 312 | 3,230 | 2,878 | 77,016 | 43.4 | 24.0 |
| Lake Worth | 13,397 | 2.81 | 51.6 | 25.6 | 26.0 | 48.4 | 29.7 | 844 | 409 | 1,057 | 1,201 | 3,103 | 26,710 | 44.8 | 24.4 |
| Largo | 33,687 | 2.49 | 51.6 | 35.8 | 15.8 | 48.4 | 34.4 | 1,089 | 329 | 384 | 2,348 | 2,739 | 63,980 | 41.3 | 25.8 |
| Lauderdale Lakes | 14,716 | 2.44 | 47.6 | 24.0 | 23.6 | 52.4 | 48.2 | 322 | 329 | 894 | 965 | 2,623 | 24,925 | 62.5 | 9.3 |
| Lauderhill | 24,628 | 2.90 | 62.1 | 27.6 | 34.5 | 37.9 | 33.6 | 530 | 571 | 785 | 2,131 | 2,929 | 47,493 | 53.0 | 22.7 |
| Margate | 24,488 | 2.39 | 55.7 | 37.8 | 17.9 | 44.3 | 37.7 | 260 | 99 | 167 | 935 | 1,575 | 47,209 | 46.7 | 20.8 |
| Melbourne | 31,680 | 2.54 | 51.4 | 39.6 | 11.8 | 48.6 | 40.1 | 2,560 | 579 | 692 | 2,802 | 3,349 | 60,637 | 41.1 | 26.0 |
| Miami | 189,084 | 2.43 | 57.5 | 33.1 | 24.3 | 42.5 | 34.0 | 8,603 | 2,850 | 593 | 17,624 | 3,668 | 358,957 | 48.2 | 31.4 |
| Miami Beach | 40,084 | 2.20 | 52.5 | 39.5 | 12.9 | 47.5 | 37.9 | 727 | 852 | 924 | 6,977 | 7,568 | 68,784 | 33.0 | 45.3 |
| Miami Gardens | 32,171 | 3.38 | 70.3 | 33.2 | 37.2 | 29.7 | 26.6 | 1,190 | 841 | 739 | 4,311 | 3,789 | 73,124 | 58.5 | 12.7 |
| Miami Lakes | 12,177 | 2.57 | 67.3 | 45.7 | 21.6 | 32.7 | 28.1 | 39 | 33 | 103 | 600 | 1,880 | 23,010 | 23.9 | 41.5 |
| Miramar | 39,853 | 3.54 | 77.9 | 48.0 | 29.9 | 22.1 | 13.7 | 160 | 310 | 216 | 1,947 | 1,358 | 94,134 | 47.1 | 24.8 |
| North Lauderdale | 13,478 | 3.28 | 72.0 | 26.4 | 45.6 | 28.0 | 23.6 | 50 | 230 | 513 | 817 | 1,823 | 28,882 | 53.8 | 18.6 |
| North Miami | 20,769 | 2.95 | 67.0 | 39.2 | 27.7 | 33.0 | 30.0 | 1,472 | 507 | 798 | 2,425 | 3,816 | 42,034 | 60.4 | 16.0 |
| North Miami Beach | 14,629 | 2.93 | 68.0 | 38.8 | 29.3 | 32.0 | 27.6 | 245 | 279 | 603 | 1,566 | 3,382 | 31,228 | 49.3 | 19.9 |
| North Port | 26,005 | 2.71 | 69.6 | 53.5 | 16.2 | 30.4 | 19.9 | 127 | 95 | 135 | 967 | 1,378 | 49,590 | 38.9 | 24.2 |
| Oakland Park | 17,690 | 2.54 | 47.8 | 29.9 | 17.9 | 52.2 | 39.5 | 260 | 294 | 641 | 1,770 | 3,860 | 34,825 | 45.6 | 27.2 |
| Ocala | 23,792 | 2.43 | 58.5 | 40.4 | 18.1 | 41.5 | 38.1 | 2,984 | 493 | 809 | 2,548 | 4,182 | 42,331 | 44.3 | 22.9 |
| Ocoee | 13,703 | 3.51 | 78.3 | 55.9 | 22.4 | 21.7 | 19.5 | 216 | 155 | 313 | 1,333 | 2,696 | 30,479 | 39.0 | 28.6 |
| Orlando | 109,454 | 2.61 | 53.8 | 34.3 | 19.5 | 46.2 | 33.6 | 2,003 | 2,157 | 738 | 14,100 | 4,827 | 202,964 | 32.2 | 40.9 |
| Ormond Beach | 16,716 | 2.59 | 69.6 | 49.0 | 20.6 | 30.4 | 23.3 | 440 | 133 | 302 | 1,119 | 2,543 | 31,044 | 36.7 | 36.4 |
| Oviedo | 14,239 | 2.93 | 83.7 | 65.4 | 18.3 | 16.3 | 8.9 | 104 | 56 | 131 | 292 | 684 | 28,473 | 14.1 | 58.3 |
| Palm Bay | 38,504 | 2.99 | 66.3 | 49.7 | 16.6 | 33.7 | 27.5 | 591 | 396 | 343 | 1,981 | 1,715 | 83,749 | 43.4 | 20.4 |
| Palm Beach Gardens | 24,667 | 2.32 | 62.7 | 51.0 | 11.7 | 37.3 | 31.7 | 537 | 78 | 136 | 1,295 | 2,263 | 44,710 | 23.2 | 52.6 |
| Palm Coast | 34,103 | 2.63 | 70.9 | 57.6 | 13.4 | 29.1 | 22.9 | 180 | NA | NA | NA | NA | 69,299 | 45.1 | 23.6 |
| Panama City | 13,937 | 2.44 | 56.5 | 30.1 | 26.4 | 43.5 | 36.2 | 665 | 228 | 613 | 1,914 | 5,145 | 23,917 | 45.1 | 24.9 |
| Pembroke Pines | 57,399 | 3.01 | 69.4 | 46.9 | 22.4 | 30.6 | 26.1 | 663 | 355 | 203 | 3,344 | 1,915 | 121,211 | 40.8 | 34.6 |
| Pensacola | 22,688 | 2.32 | 55.8 | 37.6 | 18.2 | 44.2 | 36.4 | 385 | 312 | 591 | 1,888 | 3,576 | 36,654 | 29.1 | 37.9 |
| Pinellas Park | 20,801 | 2.54 | 57.2 | 39.0 | 18.2 | 42.8 | 35.1 | 859 | 206 | 384 | 2,038 | 3,803 | 41,823 | 52.8 | 21.3 |
| Plantation | 36,203 | 2.60 | 62.9 | 43.1 | 19.8 | 37.1 | 28.1 | 332 | 211 | 221 | 2,578 | 2,700 | 69,636 | 26.2 | 39.8 |
| Plant City | 15,775 | 2.51 | 56.9 | 43.5 | 13.4 | 43.1 | 37.9 | 113 | 162 | 408 | 1,073 | 2,701 | 30,389 | 50.7 | 18.5 |
| Pompano Beach | 44,297 | 2.44 | 52.0 | 35.2 | 16.8 | 48.0 | 39.1 | 4,222 | 900 | 793 | 4,277 | 3,767 | 86,151 | 45.1 | 29.6 |
| Port Orange | 26,736 | 2.42 | 62.6 | 49.0 | 13.6 | 37.4 | 31.9 | 221 | 30 | 46 | 1,178 | 1,805 | 47,501 | 36.0 | 25.0 |
| Port St. Lucie | 68,484 | 2.94 | 74.0 | 56.3 | 17.7 | 26.0 | 21.5 | 733 | 293 | 147 | 1,696 | 850 | 144,282 | 44.3 | 23.6 |
| Riviera Beach | 12,193 | 2.87 | 63.8 | 35.3 | 28.5 | 36.2 | 31.1 | 430 | 346 | 985 | 1,466 | 4,173 | 24,087 | 45.3 | 24.5 |
| Royal Palm Beach | 13,633 | 2.94 | 72.6 | 52.0 | 20.6 | 27.4 | 22.1 | 281 | 57 | 140 | 655 | 1,605 | 27,795 | 34.3 | 36.4 |
| St. Cloud | 16,282 | 3.33 | 67.9 | 38.6 | 29.2 | 32.1 | 30.7 | 427 | 106 | 187 | 636 | 1,124 | 38,417 | 47.1 | 17.0 |
| St. Petersburg | 110,354 | 2.35 | 51.0 | 38.3 | 12.7 | 49.0 | 39.1 | 6,350 | 1,594 | 595 | 8,592 | 3,210 | 200,995 | 30.9 | 38.9 |
| Sanford | 22,523 | 2.67 | 67.4 | 40.0 | 27.4 | 32.6 | 26.4 | 1,327 | 497 | 817 | 2,127 | 3,496 | 42,234 | 49.0 | 18.9 |
| Sarasota | 25,135 | 2.18 | 48.6 | 36.8 | 11.7 | 51.4 | 41.8 | 3,577 | 319 | 546 | 1,857 | 3,176 | 45,537 | 31.6 | 40.1 |
| Sunrise | 39,427 | 2.40 | 56.1 | 37.2 | 18.8 | 43.9 | 38.6 | 595 | 191 | 197 | 2,014 | 2,078 | 72,741 | 42.2 | 26.4 |
| Tallahassee | 76,684 | 2.33 | 45.9 | 24.9 | 21.0 | 54.1 | 34.7 | 15,481 | 1,359 | 697 | 7,763 | 3,979 | 105,464 | 24.3 | 48.1 |
| Tamarac | 26,124 | 2.54 | 57.1 | 34.4 | 22.6 | 42.9 | 36.5 | 324 | 186 | 278 | 1,381 | 2,067 | 49,039 | 43.8 | 21.1 |
| Tampa | 153,918 | 2.52 | 53.9 | 35.5 | 18.4 | 46.1 | 35.5 | 11,586 | 1,622 | 405 | 6,523 | 1,629 | 272,772 | 37.1 | 41.0 |
| Titusville | 18,627 | 2.48 | 52.3 | 38.9 | 13.3 | 47.7 | 41.9 | 447 | 377 | 804 | 1,492 | 3,184 | 33,896 | 38.6 | 24.7 |
| Wellington | 22,506 | 2.91 | 76.0 | 61.5 | 14.5 | 24.0 | 19.9 | 2 | 65 | 98 | 787 | 1,186 | 45,435 | 26.9 | 41.0 |
| Weston | 20,420 | 3.49 | 88.9 | 77.2 | 11.7 | 11.1 | 9.4 | 0 | 33 | 46 | 362 | 503 | 45,745 | 19.6 | 62.7 |
| West Palm Beach | 41,678 | 2.61 | 56.9 | 35.6 | 21.4 | 43.1 | 33.9 | 3,374 | 859 | 762 | 4,231 | 3,751 | 77,374 | 38.0 | 32.9 |
| Winter Garden | 15,785 | 2.89 | 72.2 | 50.2 | 21.9 | 27.8 | 20.8 | 486 | 181 | 387 | 1,112 | 2,379 | 33,055 | 36.3 | 32.4 |
| Winter Haven | 17,073 | 2.60 | 54.4 | 33.0 | 21.3 | 45.6 | 40.3 | 574 | 187 | 423 | 1,126 | 2,547 | 32,192 | 41.0 | 23.6 |
| Winter Park | 13,761 | 2.11 | 52.1 | 40.8 | 11.3 | 47.9 | 37.1 | 1,772 | 70 | 222 | 759 | 2,410 | 23,813 | 18.0 | 60.7 |
| Winter Springs | 15,220 | 2.45 | 68.4 | 48.3 | 20.1 | 31.6 | 27.4 | 0 | 46 | 122 | 249 | 658 | 27,812 | 21.4 | 45.6 |

1. Data for serious crimes have not been adjusted for underreporting. This may affect comparability between geographic areas and over time.    2. Per 100,000 population estimated by the FBI.    3. Persons 25 years old and over.

## Table D. Cities — Income, Poverty, and Housing

| City | Money income, 2019 — Households — Median income | Percent with income less than $20,000 | Percent with income of $200,000 or more | Median family income | Median non-family income | Median earnings, 2019 — All persons | Men | Women | Housing units, 2019 — Total | Occupied | Percent owner occupied | Median value[1] (dollars) | Median gross rent (dollars) |
|---|---|---|---|---|---|---|---|---|---|---|---|---|---|
| | 42 | 43 | 44 | 45 | 46 | 47 | 48 | 49 | 50 | 51 | 52 | 53 | 54 |
| **FLORIDA— Cont'd** | | | | | | | | | | | | | |
| Deerfield Beach | 47,563 | 18.4 | 5.5 | 60,666 | 35,325 | 31,504 | 37,384 | 26,769 | 42,694 | 33,439 | 62.8 | 229,150 | 1,347 |
| DeLand | 60,745 | 20.1 | 3.3 | 80,423 | 22,981 | 34,526 | 37,682 | 31,488 | 13,985 | 13,350 | 58.7 | 243,236 | 908 |
| Delray Beach | 66,766 | 13.2 | 12.1 | 78,073 | 45,830 | 34,971 | 41,295 | 26,764 | 37,213 | 28,548 | 68.2 | 294,661 | 1,675 |
| Deltona | 56,505 | 12.1 | 4.1 | 62,398 | 28,762 | 31,480 | 36,084 | 28,679 | 33,992 | 31,783 | 81.6 | 179,723 | 1,204 |
| Doral | 77,704 | 11.7 | 7.6 | 79,369 | 49,034 | 33,656 | 45,055 | 23,221 | 21,696 | 19,722 | 44.4 | 433,401 | 2,011 |
| Dunedin | 56,220 | 11.8 | 2.9 | 72,136 | 35,992 | 30,633 | 27,028 | 32,386 | 21,168 | 17,284 | 65.5 | 246,891 | 1,148 |
| Fort Lauderdale | 67,950 | 15.1 | 12.1 | 79,697 | 57,169 | 38,254 | 45,701 | 31,122 | 95,905 | 76,348 | 54.5 | 359,231 | 1,413 |
| Fort Myers | 53,246 | 17.5 | 7.9 | 65,109 | 40,311 | 27,636 | 32,399 | 23,346 | 40,820 | 30,754 | 52.9 | 263,723 | 1,107 |
| Fort Pierce | 40,881 | 22.4 | 1.9 | 47,808 | 26,069 | 25,368 | 30,068 | 16,200 | 21,340 | 16,132 | 48.1 | 147,509 | 967 |
| Gainesville | 39,201 | 25.4 | 3.4 | 66,519 | 29,790 | 23,622 | 25,863 | 21,807 | 63,079 | 51,990 | 36.3 | 187,242 | 965 |
| Greenacres | 50,945 | 15.0 | 3.2 | 55,198 | 36,036 | 25,422 | 26,384 | 24,529 | 15,773 | 14,706 | 61.8 | 190,868 | 1,445 |
| Hallandale Beach | 38,955 | 28.5 | 2.8 | 49,094 | 23,964 | 30,171 | 32,158 | 25,674 | 27,677 | 17,716 | 49.4 | 225,354 | 1,370 |
| Hialeah | 40,725 | 28.2 | 1.3 | 47,644 | 14,801 | 27,543 | 30,902 | 23,887 | 79,525 | 75,989 | 44.6 | 290,944 | 1,263 |
| Hollywood | 55,835 | 12.3 | 6.7 | 67,459 | 35,844 | 31,002 | 32,837 | 27,500 | 66,826 | 55,172 | 54.5 | 292,267 | 1,349 |
| Homestead | 52,907 | 21.1 | 1.0 | 52,335 | 42,249 | 28,915 | 31,808 | 24,100 | 21,006 | 20,320 | 45.9 | 242,795 | 1,245 |
| Jacksonville | 56,975 | 14.2 | 4.9 | 68,135 | 39,641 | 32,853 | 40,027 | 30,325 | 396,542 | 349,171 | 56.7 | 200,164 | 1,099 |
| Jupiter | 92,233 | 10.8 | 16.3 | 118,243 | 53,551 | 37,626 | 50,858 | 31,289 | 34,723 | 26,959 | 73.3 | 434,347 | 1,776 |
| Kissimmee | 39,661 | 22.6 | 0.8 | 42,698 | 30,655 | 25,344 | 29,672 | 21,700 | 30,624 | 22,709 | 37.2 | 208,693 | 1,142 |
| Lakeland | 47,902 | 17.5 | 3.7 | 61,200 | 30,434 | 30,905 | 32,487 | 28,734 | 49,998 | 40,236 | 54.5 | 180,535 | 1,010 |
| Lake Worth | 52,869 | 18.1 | 5.6 | 56,423 | 42,372 | 25,607 | 25,654 | 25,476 | 15,919 | 13,397 | 42.3 | 211,099 | 1,213 |
| Largo | 47,427 | 15.8 | 2.3 | 60,823 | 35,887 | 29,734 | 35,296 | 26,120 | 45,879 | 33,687 | 62.4 | 114,603 | 1,197 |
| Lauderdale Lakes | 31,819 | 37.3 | 0.0 | 44,744 | 14,422 | 21,864 | 23,891 | 21,439 | 18,099 | 14,716 | 54.9 | 162,204 | 1,092 |
| Lauderhill | 41,707 | 18.5 | 1.8 | 52,268 | 29,122 | 29,823 | 31,090 | 28,591 | 28,467 | 24,628 | 48.8 | 191,218 | 1,297 |
| Margate | 45,439 | 21.6 | 3.4 | 56,111 | 30,295 | 26,912 | 25,492 | 28,609 | 28,175 | 24,488 | 74.5 | 220,702 | 1,351 |
| Melbourne | 47,933 | 18.5 | 4.3 | 64,784 | 37,244 | 28,609 | 31,622 | 25,918 | 38,651 | 31,680 | 59.5 | 199,362 | 1,115 |
| Miami | 42,966 | 25.8 | 6.0 | 48,649 | 31,209 | 30,244 | 32,200 | 26,133 | 221,040 | 189,084 | 29.6 | 358,512 | 1,298 |
| Miami Beach | 52,816 | 19.3 | 13.0 | 59,004 | 41,733 | 31,317 | 36,853 | 25,092 | 65,746 | 40,084 | 32.1 | 452,836 | 1,435 |
| Miami Gardens | 43,196 | 15.2 | 0.9 | 49,244 | 28,836 | 26,147 | 27,472 | 25,212 | 34,378 | 32,171 | 61.1 | 241,072 | 1,386 |
| Miami Lakes | 75,275 | 5.3 | 13.0 | 87,098 | 53,633 | 50,893 | 41,860 | 52,480 | 12,863 | 12,177 | 65.8 | 434,850 | 1,720 |
| Miramar | 75,079 | 8.0 | 9.5 | 81,508 | 48,778 | 36,519 | 37,709 | 35,589 | 42,700 | 39,853 | 71.6 | 336,514 | 1,475 |
| North Lauderdale | 53,991 | 15.0 | 0.7 | 53,383 | 45,481 | 30,513 | 29,045 | 31,039 | 14,461 | 13,478 | 53.2 | 190,966 | 1,555 |
| North Miami | 40,160 | 19.1 | 4.0 | 42,465 | 25,511 | 26,095 | 27,517 | 25,279 | 21,941 | 20,769 | 41.6 | 244,350 | 1,177 |
| North Miami Beach | 45,995 | 17.1 | 3.0 | 48,302 | 26,256 | 29,310 | 31,513 | 23,069 | 17,552 | 14,629 | 50.9 | 247,977 | 1,223 |
| North Port | 56,226 | 10.3 | 2.8 | 71,071 | 45,042 | 35,353 | 41,422 | 29,254 | 30,862 | 26,005 | 83.8 | 235,495 | 1,309 |
| Oakland Park | 54,030 | 15.3 | 3.0 | 63,155 | 43,216 | 30,228 | 32,069 | 26,188 | 19,889 | 17,690 | 46.4 | 249,967 | 1,325 |
| Ocala | 46,765 | 18.7 | 5.1 | 60,877 | 31,323 | 31,251 | 33,725 | 28,509 | 26,512 | 23,792 | 44.1 | 180,820 | 1,026 |
| Ocoee | 67,282 | 8.5 | 3.1 | 72,708 | 39,818 | 31,258 | 41,667 | 26,187 | 14,023 | 13,703 | 66.3 | 263,342 | 1,437 |
| Orlando | 58,819 | 12.9 | 6.9 | 66,307 | 47,996 | 33,579 | 37,682 | 30,120 | 138,456 | 109,454 | 37.7 | 260,832 | 1,307 |
| Ormond Beach | 59,901 | 10.7 | 4.4 | 66,707 | 37,564 | 31,713 | 34,612 | 31,006 | 19,108 | 16,716 | 70.7 | 250,820 | 1,242 |
| Oviedo | 99,178 | 6.2 | 17.1 | 108,262 | 64,015 | 51,285 | 65,661 | 45,690 | 14,808 | 14,239 | 76.8 | 337,351 | 1,756 |
| Palm Bay | 54,113 | 9.2 | 2.4 | 58,392 | 37,548 | 29,801 | 35,105 | 26,716 | 44,469 | 38,504 | 79.6 | 184,585 | 1,128 |
| Palm Beach Gardens | 101,001 | 7.7 | 24.8 | 136,551 | 59,082 | 47,260 | 61,676 | 40,505 | 30,066 | 24,667 | 76.9 | 423,672 | 1,844 |
| Palm Coast | 60,158 | 10.1 | 3.3 | 66,465 | 45,766 | 31,193 | 36,745 | 27,110 | 38,767 | 34,103 | 76.5 | 237,736 | 1,469 |
| Panama City | 50,913 | 15.8 | 3.4 | 56,651 | 35,266 | 28,643 | 35,465 | 22,056 | 15,792 | 13,937 | 58.1 | 160,015 | 994 |
| Pembroke Pines | 67,232 | 15.7 | 5.6 | 79,466 | 35,544 | 38,414 | 40,412 | 36,405 | 62,089 | 57,399 | 69.8 | 324,210 | 1,497 |
| Pensacola | 57,526 | 16.1 | 9.3 | 83,736 | 37,837 | 36,181 | 37,135 | 35,056 | 25,899 | 22,688 | 58.4 | 226,387 | 926 |
| Pinellas Park | 52,747 | 15.2 | 1.2 | 56,976 | 36,288 | 32,115 | 34,224 | 30,968 | 25,725 | 20,801 | 68.7 | 172,590 | 1,116 |
| Plantation | 71,977 | 8.9 | 6.5 | 84,371 | 51,796 | 36,672 | 41,292 | 34,780 | 39,459 | 36,203 | 60.9 | 350,845 | 1,750 |
| Plant City | 54,028 | 17.0 | 2.6 | 75,521 | 27,807 | 32,125 | 36,170 | 30,562 | 16,636 | 15,775 | 61.5 | 193,982 | 1,002 |
| Pompano Beach | 53,663 | 17.5 | 6.2 | 59,821 | 40,780 | 31,327 | 32,620 | 28,425 | 56,146 | 44,297 | 52.6 | 245,433 | 1,343 |
| Port Orange | 54,941 | 13.8 | 2.1 | 64,493 | 32,010 | 31,114 | 35,578 | 26,927 | 30,569 | 26,736 | 72.4 | 218,075 | 1,140 |
| Port St. Lucie | 68,041 | 9.9 | 2.6 | 75,062 | 42,742 | 32,232 | 40,034 | 30,608 | 77,484 | 68,484 | 80.1 | 244,879 | 1,413 |
| Riviera Beach | 48,075 | 13.7 | 2.2 | 54,373 | 38,170 | 28,802 | 31,421 | 26,182 | 16,095 | 12,193 | 47.5 | 219,089 | 1,189 |
| Royal Palm Beach | 80,933 | 5.3 | 8.8 | 93,991 | 56,117 | 40,237 | 45,759 | 32,459 | 14,243 | 13,633 | 84.1 | 288,106 | 1,687 |
| St. Cloud | 51,535 | 14.8 | 0.2 | 63,976 | 26,431 | 30,262 | 30,676 | 28,292 | 19,859 | 16,282 | 76.5 | 222,760 | 1,162 |
| St. Petersburg | 61,631 | 11.8 | 8.0 | 86,074 | 44,861 | 37,501 | 39,772 | 36,658 | 143,046 | 110,354 | 63.5 | 242,679 | 1,178 |
| Sanford | 48,484 | 21.9 | 5.2 | 53,491 | 33,250 | 31,840 | 33,556 | 28,660 | 24,537 | 22,523 | 50.6 | 205,579 | 1,168 |
| Sarasota | 62,374 | 15.6 | 8.5 | 76,255 | 40,496 | 28,859 | 31,152 | 25,974 | 33,429 | 25,135 | 53.8 | 336,754 | 1,158 |
| Sunrise | 53,321 | 19.3 | 1.5 | 71,250 | 30,347 | 30,952 | 35,564 | 26,657 | 43,749 | 39,427 | 69.9 | 260,403 | 1,493 |
| Tallahassee | 46,964 | 22.2 | 5.1 | 62,800 | 32,531 | 25,575 | 26,933 | 23,804 | 88,043 | 76,684 | 38.8 | 219,656 | 1,024 |
| Tamarac | 44,400 | 22.4 | 3.1 | 46,241 | 35,906 | 23,886 | 25,980 | 22,079 | 29,841 | 26,124 | 70.6 | 179,566 | 1,432 |
| Tampa | 57,709 | 17.3 | 10.0 | 71,151 | 42,446 | 35,963 | 43,480 | 30,324 | 174,379 | 153,918 | 52.0 | 265,654 | 1,202 |
| Titusville | 48,122 | 13.6 | 1.6 | 65,490 | 36,149 | 36,160 | 37,888 | 35,494 | 24,259 | 18,627 | 70.4 | 166,578 | 1,008 |
| Wellington | 85,886 | 5.1 | 13.9 | 95,257 | 56,000 | 42,303 | 50,180 | 35,742 | 26,116 | 22,506 | 75.1 | 396,889 | 2,100 |
| Weston | 125,105 | 4.7 | 29.9 | 130,283 | 71,641 | 56,212 | 74,644 | 39,407 | 23,130 | 20,420 | 73.2 | 494,688 | 2,528 |
| West Palm Beach | 51,089 | 17.6 | 5.7 | 61,673 | 45,739 | 29,714 | 32,872 | 27,811 | 52,996 | 41,678 | 50.6 | 267,832 | 1,451 |
| Winter Garden | 86,940 | 5.2 | 12.5 | 92,235 | 39,059 | 45,250 | 46,449 | 42,033 | 16,233 | 15,785 | 74.8 | 344,180 | 1,339 |
| Winter Haven | 52,316 | 17.1 | 4.7 | 63,985 | 32,071 | 28,450 | 31,551 | 26,962 | 20,961 | 17,073 | 63.3 | 168,935 | 946 |
| Winter Park | 81,343 | 10.6 | 19.1 | 132,338 | 46,285 | 45,842 | 61,828 | 40,386 | 15,561 | 13,761 | 64.3 | 496,892 | 1,358 |
| Winter Springs | 73,162 | 7.2 | 6.0 | 91,775 | 53,670 | 41,234 | 45,244 | 39,916 | 15,519 | 15,220 | 76.5 | 276,893 | 1,370 |

1. Based on population estimated by the American Community Survey.

# Table D. Cities — Commuting, Computer Access, Migration, Labor Force, and Employment

| City | Commuting[1], 2019 Percent Drove alone | With commutes of 30 minutes or more | Computer access[2], 2019 Percent With a computer in the house | With Internet access | Migration, 2019 Percent who lived in the same house one year ago | Percent who lived in another state or county one year ago | Civilian labor force, 2020 Total | Percent change 2019-2020 | Unemployment Total | Rate[3] | Civilian Employment[4], 2019 Population age 16 and older Number | Percent in labor force | Population age 16 to 64 Number | Percent who worked full-year full-time |
|---|---|---|---|---|---|---|---|---|---|---|---|---|---|---|
| | 55 | 56 | 57 | 58 | 59 | 60 | 61 | 62 | 63 | 64 | 65 | 66 | 67 | 68 |
| FLORIDA— Cont'd | | | | | | | | | | | | | | |
| Deerfield Beach | 69.4 | 38.2 | 92.8 | 87.4 | 79.0 | 10.0 | 40,241 | -1.6 | 3,593 | 9 | 68,178 | 59.1 | 48,508 | 46.9 |
| DeLand | 71.7 | 39.0 | 94.5 | 87.5 | 74.1 | 14.6 | 14,495 | 0.7 | 1,138 | 8 | 29,723 | 50.7 | 23,166 | 46.7 |
| Delray Beach | 79.9 | 25.6 | 92.7 | 84.8 | 83.6 | 6.2 | 35,877 | -1.3 | 3,172 | 9 | 61,000 | 61.8 | 42,815 | 56.7 |
| Deltona | 85.2 | 59.2 | 96.9 | 89.7 | 90.4 | 4.6 | 43,590 | -1.6 | 3,548 | 8 | 72,143 | 59.2 | 55,462 | 56.1 |
| Doral | 74.0 | 38.4 | 99.2 | 93.4 | 75.7 | 9.7 | 31,922 | -0.4 | 2,075 | 7 | 51,550 | 62.5 | 46,268 | 53.3 |
| Dunedin | 82.0 | 31.0 | 90.1 | 81.6 | 87.1 | 3.8 | 17,803 | -2.6 | 1,149 | 7 | 33,155 | 57.1 | 20,613 | 52.2 |
| Fort Lauderdale | 72.8 | 33.0 | 95.4 | 85.4 | 81.5 | 5.5 | 97,653 | -0.6 | 10,110 | 10 | 152,980 | 66.9 | 118,218 | 56.2 |
| Fort Myers | 72.0 | 32.4 | 90.8 | 85.7 | 79.3 | 11.5 | 37,005 | 4.0 | 3,168 | 9 | 69,026 | 51.8 | 50,084 | 46.4 |
| Fort Pierce | 79.9 | 36.9 | 90.7 | 77.2 | 93.0 | 2.8 | 18,733 | -1.3 | 1,897 | 10 | 37,685 | 59.5 | 29,090 | 48.1 |
| Gainesville | 66.3 | 16.4 | 96.8 | 86.6 | 68.6 | 16.5 | 67,014 | -2.5 | 3,929 | 6 | 116,084 | 59.3 | 101,284 | 35.7 |
| Greenacres | 85.4 | 55.3 | 89.2 | 81.2 | 91.4 | 2.7 | 21,361 | -2.6 | 1,566 | 7 | 34,592 | 65.5 | 26,795 | 53.9 |
| Hallandale Beach | 74.0 | 57.9 | 92.6 | 80.8 | 81.0 | 7.4 | 18,536 | -0.1 | 2,113 | 11 | 32,559 | 59.2 | 22,513 | 53.5 |
| Hialeah | 76.3 | 54.9 | 86.1 | 73.6 | 93.0 | 1.6 | 105,947 | -7.1 | 8,791 | 8 | 195,707 | 58.9 | 146,525 | 57.4 |
| Hollywood | 79.8 | 52.4 | 96.3 | 89.4 | 85.4 | 5.6 | 80,853 | -1.2 | 7,767 | 10 | 125,400 | 67.3 | 101,297 | 53.9 |
| Homestead | 64.1 | 64.6 | 97.6 | 92.3 | 83.9 | 2.2 | 30,032 | -6.8 | 2,433 | 8 | 47,797 | 69.5 | 43,001 | 52.3 |
| Jacksonville | 80.4 | 35.6 | 93.6 | 87.3 | 81.5 | 6.5 | 458,459 | -1.3 | 31,251 | 7 | 724,870 | 64.2 | 594,040 | 56.2 |
| Jupiter | 85.6 | 25.0 | 96.9 | 92.1 | 86.6 | 5.4 | 33,358 | -4.2 | 1,779 | 5 | 53,990 | 63.6 | 38,098 | 51.8 |
| Kissimmee | 67.2 | 57.7 | 92.5 | 79.1 | 92.5 | 4.2 | 38,395 | -0.8 | 5,733 | 15 | 55,260 | 64.1 | 46,522 | 48.7 |
| Lakeland | 82.6 | 25.6 | 89.8 | 77.3 | 81.2 | 8.0 | 47,851 | 1.4 | 3,790 | 8 | 91,923 | 54.0 | 66,821 | 47.7 |
| Lake Worth | 59.7 | 44.7 | 89.8 | 72.8 | 83.4 | 6.5 | 19,206 | -1.9 | 1,535 | 8 | 30,474 | 71.4 | 24,538 | 55.9 |
| Largo | 76.6 | 28.1 | 93.3 | 83.2 | 81.3 | 7.1 | 41,015 | -2.1 | 2,924 | 7 | 73,195 | 58.8 | 52,781 | 53.4 |
| Lauderdale Lakes | 72.2 | 50.9 | 91.4 | 77.1 | NA | NA | 17,788 | 0.2 | 2,062 | 12 | 28,353 | 66.4 | 21,986 | 58.5 |
| Lauderhill | 79.5 | 50.6 | 89.9 | 81.4 | 88.5 | 3.7 | 36,282 | 0.6 | 4,383 | 12 | 54,085 | 69.5 | 43,386 | 58.1 |
| Margate | 79.7 | 45.2 | 95.2 | 89.6 | 89.5 | 3.0 | 30,274 | -1.7 | 2,764 | 9 | 52,615 | 63.3 | 38,431 | 53.1 |
| Melbourne | 83.3 | 21.4 | 90.6 | 86.6 | 82.3 | 6.2 | 39,724 | -1.2 | 2,762 | 7 | 70,465 | 56.2 | 51,213 | 49.0 |
| Miami | 68.2 | 49.1 | 91.8 | 70.3 | 86.2 | 3.9 | 221,368 | -5.2 | 18,716 | 9 | 395,648 | 65.0 | 312,889 | 59.1 |
| Miami Beach | 45.4 | 43.6 | 94.2 | 80.0 | 72.3 | 12.9 | 51,913 | -8.2 | 3,829 | 7 | 75,873 | 67.0 | 61,645 | 59.8 |
| Miami Gardens | 82.3 | 65.3 | 92.3 | 71.7 | 90.2 | 1.2 | 49,917 | -5.2 | 5,540 | 11 | 87,335 | 61.1 | 70,684 | 53.8 |
| Miami Lakes | 87.5 | 71.0 | NA | NA | NA | NA | 15,206 | -8.1 | 847 | 6 | 26,254 | 66.9 | 20,622 | 59.8 |
| Miramar | 86.8 | 45.5 | 98.6 | 85.1 | 94.8 | 3.2 | 75,880 | -2.0 | 6,492 | 9 | 111,707 | 72.7 | 95,582 | 64.3 |
| North Lauderdale | 75.3 | 49.2 | 96.6 | 92.4 | NA | NA | 23,491 | -1.6 | 2,271 | 10 | 34,306 | 76.8 | 30,237 | 52.3 |
| North Miami | 74.9 | 65.9 | 94.9 | 73.8 | 89.2 | 3.3 | 30,488 | -2.0 | 3,523 | 12 | 48,734 | 66.9 | 41,084 | 54.6 |
| North Miami Beach | 81.2 | 64.0 | 93.2 | 71.1 | NA | NA | 21,534 | -9.0 | 2,175 | 10 | 35,710 | 72.3 | 30,131 | 62.9 |
| North Port | 82.7 | 57.7 | 97.7 | 91.9 | 83.1 | 4.5 | 29,578 | -1.1 | 2,091 | 7 | 55,223 | 43.4 | 32,292 | 46.7 |
| Oakland Park | 75.3 | 38.2 | 93.9 | 84.9 | 83.2 | 6.8 | 26,154 | -1.5 | 2,501 | 10 | 38,364 | 72.6 | 32,145 | 60.4 |
| Ocala | 76.5 | 20.7 | 90.1 | 82.0 | 75.2 | 9.6 | 25,621 | 0.2 | 1,881 | 7 | 48,898 | 61.5 | 38,584 | 51.8 |
| Ocoee | 74.0 | 54.7 | 96.7 | 92.0 | 89.1 | 6.6 | 26,011 | -4.7 | 2,302 | 9 | 36,026 | 74.6 | 31,448 | 57.8 |
| Orlando | 76.7 | 44.2 | 95.7 | 90.6 | 80.7 | 7.2 | 164,262 | -2.9 | 18,391 | 11 | 230,143 | 72.4 | 201,408 | 58.7 |
| Ormond Beach | 79.1 | 22.0 | 95.3 | 88.2 | 86.8 | 4.3 | 19,334 | -2.9 | 1,294 | 7 | 35,220 | 55.1 | 23,312 | 52.3 |
| Oviedo | 78.2 | 52.6 | NA | NA | 87.6 | 9.3 | 20,988 | -8.1 | 1,253 | 6 | 32,419 | 66.2 | 27,498 | 52.1 |
| Palm Bay | 86.2 | 47.3 | 92.5 | 87.2 | 87.4 | 5.5 | 54,897 | -0.2 | 3,899 | 7 | 94,344 | 57.0 | 70,611 | 50.2 |
| Palm Beach Gardens | 75.3 | 20.3 | 96.9 | 93.4 | 86.1 | 5.9 | 29,023 | -1.3 | 1,810 | 6 | 48,807 | 54.5 | 30,284 | 58.2 |
| Palm Coast | 79.8 | 39.8 | 93.9 | 74.9 | 88.8 | 6.7 | 36,809 | -0.9 | 2,838 | 8 | 77,319 | 53.2 | 50,183 | 53.0 |
| Panama City | 79.4 | 28.3 | 96.2 | 82.6 | 64.5 | 15.0 | 15,405 | -1.8 | 1,029 | 7 | 27,354 | 59.6 | 20,690 | 46.5 |
| Pembroke Pines | 85.3 | 52.7 | 96.0 | 84.0 | 92.0 | 3.7 | 90,733 | -2.1 | 7,209 | 8 | 138,131 | 64.1 | 108,065 | 60.1 |
| Pensacola | 74.2 | 19.4 | 93.5 | 85.0 | 84.2 | 6.6 | 27,432 | -1.2 | 1,682 | 6 | 40,597 | 63.1 | 30,617 | 56.7 |
| Pinellas Park | 80.1 | 27.8 | 94.6 | 89.5 | 84.2 | 4.8 | 26,376 | -0.8 | 1,898 | 7 | 45,471 | 59.6 | 32,615 | 51.5 |
| Plantation | 79.9 | 40.6 | 97.8 | 94.5 | 81.2 | 5.8 | 53,070 | -2.8 | 3,963 | 8 | 79,609 | 69.3 | 61,237 | 61.1 |
| Plant City | 85.9 | 42.6 | 93.7 | 88.4 | 87.6 | 3.2 | 19,724 | -1.3 | 1,212 | 6 | 32,797 | 64.6 | 27,059 | 55.6 |
| Pompano Beach | 70.0 | 41.3 | 87.4 | 79.6 | 79.8 | 5.6 | 52,846 | -1.1 | 5,272 | 10 | 95,438 | 61.2 | 71,682 | 51.9 |
| Port Orange | 81.5 | 20.8 | 95.5 | 87.6 | 87.3 | 4.9 | 31,066 | -2.7 | 2,052 | 7 | 52,803 | 57.2 | 36,184 | 47.9 |
| Port St. Lucie | 78.4 | 44.1 | 97.4 | 93.9 | 90.5 | 6.1 | 95,465 | 1.3 | 6,981 | 7 | 160,740 | 61.9 | 114,219 | 59.9 |
| Riviera Beach | 78.9 | 31.7 | 89.7 | 82.0 | 80.0 | 4.8 | 16,647 | 1.8 | 1,822 | 11 | 27,334 | 60.1 | 21,174 | 49.3 |
| Royal Palm Beach | 78.1 | 50.6 | NA | NA | NA | NA | 21,264 | -2.2 | 1,510 | 7 | 33,598 | 65.9 | 26,362 | 55.8 |
| St. Cloud | 90.2 | 66.8 | 93.8 | 86.9 | 87.7 | 8.1 | 25,710 | -3.3 | 2,801 | 11 | 43,495 | 56.4 | 31,729 | 51.4 |
| St. Petersburg | 74.3 | 32.9 | 93.9 | 88.9 | 84.8 | 7.1 | 140,332 | -1.1 | 10,843 | 8 | 228,080 | 64.5 | 173,984 | 58.5 |
| Sanford | 77.0 | 31.0 | 95.8 | 84.9 | 74.0 | 17.6 | 27,642 | -3.3 | 2,841 | 10 | 48,288 | 66.1 | 41,333 | 50.0 |
| Sarasota | 74.4 | 29.6 | 89.5 | 79.5 | 82.2 | 8.1 | 27,633 | -2.7 | 1,966 | 7 | 51,187 | 52.9 | 34,290 | 49.0 |
| Sunrise | 77.7 | 42.5 | 96.1 | 87.1 | 90.8 | 1.5 | 51,816 | -2.3 | 4,594 | 9 | 82,021 | 68.4 | 63,171 | 56.1 |
| Tallahassee | 76.8 | 16.8 | 96.7 | 89.9 | 72.1 | 12.5 | 99,507 | -2.6 | 6,452 | 7 | 164,335 | 66.9 | 143,528 | 46.1 |
| Tamarac | 73.8 | 56.1 | 92.6 | 80.8 | 82.6 | 2.3 | 34,413 | -0.7 | 3,198 | 9 | 56,367 | 58.3 | 37,633 | 50.2 |
| Tampa | 76.4 | 34.9 | 95.2 | 89.5 | 79.5 | 8.1 | 205,176 | 0.7 | 16,386 | 8 | 320,621 | 64.3 | 268,792 | 55.3 |
| Titusville | 88.7 | 29.3 | 93.6 | 90.1 | 85.2 | 11.7 | 21,311 | -0.7 | 1,628 | 8 | 37,786 | 60.0 | 28,690 | 54.2 |
| Wellington | 76.5 | 49.9 | 97.8 | 95.0 | 79.6 | 5.7 | 33,326 | -3.9 | 1,962 | 6 | 51,804 | 61.1 | 38,935 | 48.7 |
| Weston | 80.6 | 49.8 | NA | NA | 83.5 | 8.2 | 35,153 | -4.7 | 2,083 | 6 | 54,605 | 65.0 | 47,018 | 53.6 |
| West Palm Beach | 75.1 | 29.6 | 96.2 | 86.0 | 81.7 | 7.9 | 58,593 | -1.5 | 4,895 | 8 | 91,478 | 63.2 | 71,500 | 54.2 |
| Winter Garden | 74.6 | 59.3 | 97.5 | 91.8 | NA | NA | 23,508 | -1.5 | 2,717 | 12 | 36,220 | 72.7 | 30,025 | 66.5 |
| Winter Haven | 82.4 | 39.1 | 89.6 | 77.6 | 82.7 | 8.9 | 19,075 | 0.5 | 1,946 | 10 | 36,209 | 55.2 | 23,759 | 55.9 |
| Winter Park | 80.9 | 36.9 | 94.7 | 87.1 | 79.4 | 12.0 | 13,804 | -9.1 | 879 | 6 | 27,452 | 55.5 | 19,674 | 50.9 |
| Winter Springs | 82.2 | 53.0 | 95.0 | 91.1 | 85.9 | 8.7 | 18,857 | -8.1 | 1,288 | 7 | 30,550 | 65.0 | 21,909 | 53.7 |

1. Employed persons.   2. Households.   3. Percent of civilian labor force.   4. Persons 16 years old and over.

# Table D. Cities — Construction, Wholesale Trade, and Retail Trade

| City | Value of residential construction authorized by building permits, 2020 | | | Wholesale trade[1], 2017 | | | | Retail trade[2], 2017 | | | |
|---|---|---|---|---|---|---|---|---|---|---|---|
| | New construction ($1,000) | Number of housing units | Percent single family | Number of establishments | Number of employees | Sales (mil dol) | Annual payroll (mil dol) | Number of establishments | Number of employees | Sales (mil dol) | Annual payroll (mil dol) |
| | 69 | 70 | 71 | 72 | 73 | 74 | 75 | 76 | 77 | 78 | 79 |
| **FLORIDA— Cont'd** | | | | | | | | | | | |
| Deerfield Beach | 45,820 | 300 | 100 | 213 | 3,650 | 13,809 | 207 | 315 | 3,668 | 1,150 | 109 |
| DeLand | 154,479 | 521 | 100 | 48 | 401 | 270 | 29 | 164 | 2,872 | 864 | 80 |
| Delray Beach | 65,519 | 131 | 83 | 129 | 1,019 | 777 | 59 | 401 | 5,475 | 2,172 | 184 |
| Deltona | 125,230 | 446 | 99 | 13 | 27 | 11 | 1 | 114 | 1,930 | 474 | 45 |
| Doral | 75,559 | 499 | 21 | 1,480 | 14,646 | 30,047 | 926 | 546 | 7,535 | 2,565 | 245 |
| Dunedin | 22,961 | 64 | 94 | 26 | 256 | 111 | 9 | 134 | 1,462 | 456 | 41 |
| Fort Lauderdale | 126,657 | 537 | 29 | 576 | 5,970 | 5,648 | 350 | 1,120 | 13,335 | 5,092 | 434 |
| Fort Myers | 536,255 | 3,105 | 21 | 168 | 2,615 | 1,367 | 127 | 717 | 11,773 | 3,652 | 323 |
| Fort Pierce | 56,443 | 235 | 100 | 40 | 463 | 285 | 22 | 246 | 3,345 | 1,352 | 104 |
| Gainesville | 50,840 | 1,055 | 14 | 109 | 1,224 | 1,219 | 63 | 549 | 9,697 | 2,663 | 238 |
| Greenacres | 711 | 4 | 100 | 18 | 39 | 24 | 1 | 105 | 1,684 | 616 | 55 |
| Hallandale Beach | 72,960 | 263 | 2 | 87 | 436 | 296 | 21 | 154 | 1,816 | 457 | 45 |
| Hialeah | 118,941 | 1,023 | 8 | 534 | 4,518 | 1,676 | 166 | 1,000 | 11,423 | 2,891 | 264 |
| Hollywood | 126,983 | 563 | 31 | 293 | 2,731 | 1,914 | 155 | 605 | 7,898 | 2,329 | 225 |
| Homestead | 100,799 | 562 | 100 | 43 | 307 | 183 | 12 | 182 | 2,610 | 728 | 64 |
| Jacksonville | 1,212,711 | 9,034 | 61 | 1,099 | 22,090 | 20,467 | 1,376 | 3,145 | 50,554 | 16,704 | 1,410 |
| Jupiter | 54,893 | 80 | 100 | 120 | 679 | 422 | 45 | 318 | 3,528 | 842 | 96 |
| Kissimmee | 64,215 | 262 | 100 | 58 | 305 | 104 | 11 | 370 | 5,535 | 1,276 | 130 |
| Lakeland | 225,842 | 1,224 | 61 | 156 | 3,245 | 10,452 | 172 | 581 | 10,502 | 3,221 | 288 |
| Lake Worth | 29,319 | 264 | 9 | 55 | 548 | 218 | 32 | 145 | 1,202 | 365 | 37 |
| Largo | 21,022 | 85 | 100 | 99 | 1,300 | 2,338 | 64 | 350 | 5,260 | 1,511 | 143 |
| Lauderdale Lakes | 0 | 0 | 0 | 20 | 145 | 44 | 5 | 83 | 1,183 | 294 | 28 |
| Lauderhill | 0 | 0 | 0 | 39 | 154 | 36 | 4 | 175 | 1,564 | 414 | 41 |
| Margate | 4,132 | 28 | 100 | 52 | 268 | 149 | 13 | 189 | 2,820 | 949 | 90 |
| Melbourne | 103,647 | 483 | 46 | 113 | 1,504 | 717 | 91 | 449 | 6,935 | 2,085 | 196 |
| Miami | 681,445 | 3,583 | 2 | 1,436 | 9,090 | 13,044 | 512 | 2,624 | 32,942 | 10,233 | 877 |
| Miami Beach | 31,990 | 33 | 100 | 155 | 519 | 322 | 28 | 551 | 6,909 | 1,823 | 191 |
| Miami Gardens | 71,467 | 679 | 2 | 182 | 3,849 | 2,085 | 207 | 307 | 5,021 | 2,133 | 162 |
| Miami Lakes | 53,478 | 244 | 100 | 101 | 1,402 | 902 | 85 | 85 | 1,357 | 567 | 46 |
| Miramar | 59,367 | 684 | 7 | 199 | 3,278 | 3,317 | 201 | 201 | 3,383 | 1,006 | 87 |
| North Lauderdale | 561 | 8 | 0 | 6 | 18 | 3 | 1 | 65 | 1,234 | 307 | 32 |
| North Miami | 22,390 | 195 | 4 | 71 | 411 | 261 | 22 | 198 | 2,544 | 808 | 70 |
| North Miami Beach | 286 | 1 | 100 | 73 | 236 | 80 | 9 | 190 | 2,670 | 1,084 | 78 |
| North Port | 390,783 | 1,620 | 96 | 24 | 194 | 61 | 9 | 99 | 1,951 | 467 | 47 |
| Oakland Park | 3,126 | 23 | 78 | 108 | 931 | 426 | 70 | 242 | 2,121 | 723 | 64 |
| Ocala | 41,207 | 212 | 82 | 158 | 2,741 | 1,837 | 146 | 629 | 10,971 | 3,462 | 313 |
| Ocoee | 136,969 | 437 | 90 | 25 | 660 | 816 | 46 | 139 | 2,224 | 515 | 51 |
| Orlando | 880,078 | 4,601 | 21 | 661 | 10,635 | 7,905 | 631 | 1,639 | 30,854 | 9,404 | 803 |
| Ormond Beach | 47,212 | 138 | 97 | 48 | 355 | 127 | 17 | 219 | 3,105 | 779 | 76 |
| Oviedo | 5,478 | 14 | 100 | 30 | 201 | 99 | 10 | 146 | 2,127 | 439 | 48 |
| Palm Bay | 334,135 | 1,278 | 100 | 33 | 230 | 121 | 14 | 192 | 3,537 | 950 | 89 |
| Palm Beach Gardens | 124,922 | 319 | 79 | 72 | 550 | 455 | 40 | 374 | 6,904 | 1,703 | 183 |
| Palm Coast | 346,688 | 1,482 | 84 | 35 | 187 | 83 | 10 | 169 | 3,177 | 871 | 80 |
| Panama City | 108,623 | 698 | 19 | 62 | 854 | 328 | 36 | 305 | 4,724 | 1,423 | 134 |
| Pembroke Pines | 10,396 | 44 | 100 | 186 | 506 | 306 | 26 | 577 | 12,099 | 3,744 | 328 |
| Pensacola | 27,345 | 92 | 100 | 72 | 638 | 277 | 34 | 381 | 6,068 | 1,400 | 141 |
| Pinellas Park | 6,780 | 43 | 100 | 195 | 2,598 | 1,287 | 131 | 312 | 4,188 | 1,450 | 126 |
| Plantation | 94,121 | 861 | 2 | 118 | 586 | 387 | 46 | 333 | 5,054 | 1,263 | 123 |
| Plant City | 21,609 | 132 | 100 | 71 | 2,397 | 3,681 | 139 | 183 | 3,768 | 1,837 | 111 |
| Pompano Beach | 37,825 | 334 | 25 | 456 | 7,001 | 5,731 | 387 | 635 | 8,230 | 3,493 | 292 |
| Port Orange | 66,803 | 192 | 100 | 34 | 284 | 186 | 14 | 186 | 3,186 | 778 | 74 |
| Port St. Lucie | 757,289 | 4,085 | 91 | 104 | 614 | 218 | 27 | 371 | 6,893 | 1,891 | 179 |
| Riviera Beach | 84,108 | 368 | 100 | 97 | 2,769 | 2,008 | 176 | 126 | 1,376 | 483 | 47 |
| Royal Palm Beach | 24,769 | 77 | 94 | 24 | 166 | 121 | 9 | 161 | 3,605 | 1,090 | 96 |
| St. Cloud | 365,276 | 1,097 | 100 | 11 | 90 | 23 | 4 | 132 | 2,369 | 612 | 59 |
| St. Petersburg | 143,153 | 589 | 60 | 184 | 2,380 | 1,304 | 159 | 911 | 15,594 | 5,652 | 450 |
| Sanford | 77,600 | 356 | 81 | 97 | 1,476 | 640 | 76 | 364 | 6,148 | 2,137 | 185 |
| Sarasota | 70,406 | 164 | 65 | 80 | 678 | 375 | 41 | 451 | 4,662 | 1,307 | 140 |
| Sunrise | 0 | 0 | 0 | 257 | 2,861 | 2,753 | 175 | 557 | 11,840 | 2,875 | 273 |
| Tallahassee | 224,175 | 1,580 | 30 | 153 | 2,117 | 1,745 | 148 | 877 | 13,577 | 3,392 | 332 |
| Tamarac | 37,593 | 224 | 34 | 71 | 1,522 | 1,113 | 91 | 126 | 2,934 | 732 | 110 |
| Tampa | 802,880 | 3,333 | 39 | 642 | 10,930 | 9,610 | 639 | 1,867 | 25,775 | 7,847 | 758 |
| Titusville | 21,392 | 130 | 100 | 20 | 105 | 33 | 4 | 176 | 2,893 | 751 | 74 |
| Wellington | 20,458 | 28 | 100 | 85 | 311 | 237 | 18 | 260 | 3,485 | 711 | 85 |
| Weston | 3,284 | 14 | 100 | 234 | 1,705 | 3,365 | 118 | 138 | 1,834 | 816 | 54 |
| West Palm Beach | 228,332 | 511 | 18 | 144 | 1,668 | 2,116 | 84 | 595 | 9,297 | 3,434 | 290 |
| Winter Garden | 132,341 | 243 | 100 | 57 | 637 | 1,305 | 31 | 202 | 3,678 | 995 | 89 |
| Winter Haven | 211,719 | 887 | 100 | 37 | 481 | 282 | 19 | 219 | 3,493 | 1,083 | 96 |
| Winter Park | 69,701 | 104 | 94 | 47 | 443 | 254 | 20 | 250 | 2,871 | 979 | 94 |
| Winter Springs | 27,976 | 173 | 59 | 27 | 128 | 54 | 8 | 49 | 424 | 117 | 12 |

1. Merchant wholesalers except manufacturers' sales branches and offices.    2. Establishments with payroll.

# Table D. Cities — Real Estate, Professional Services, and Manufacturing

| City | Real estate and rental and leasing, 2017 | | | | Professional, scientific, and technical services[1], 2017 | | | | Manufacturing, 2017 | | | |
|---|---|---|---|---|---|---|---|---|---|---|---|---|
| | Number of establishments | Number of employees | Receipts (mil dol) | Annual payroll (mil dol) | Number of establishments | Number of employees | Receipts (mil dol) | Annual payroll (mil dol) | Number of establishments | Number of employees | Receipts (mil dol) | Annual payroll (mil dol) |
| | 80 | 81 | 82 | 83 | 84 | 85 | 86 | 87 | 88 | 89 | 90 | 91 |
| FLORIDA— Cont'd | | | | | | | | | | | | |
| Deerfield Beach | 139 | 944 | 250 | 49 | 403 | 2,764 | 432 | 177 | NA | NA | NA | NA |
| DeLand | 62 | 240 | 46 | 8 | 127 | 664 | 88 | 29 | NA | NA | NA | NA |
| Delray Beach | 216 | 539 | 129 | 24 | 488 | 2,350 | 474 | 144 | NA | NA | NA | NA |
| Deltona | 25 | 31 | 7 | 1 | 57 | 178 | 23 | 7 | NA | NA | NA | NA |
| Doral | 360 | 2,441 | 431 | 103 | 691 | 4,103 | 766 | 227 | NA | NA | NA | NA |
| Dunedin | D | D | D | D | 177 | 544 | 89 | 30 | NA | NA | NA | NA |
| Fort Lauderdale | 823 | 3,876 | 1,797 | 227 | 2,491 | 16,701 | 3,569 | 1,301 | NA | NA | NA | NA |
| Fort Myers | 248 | 1,123 | 418 | 48 | 563 | 5,444 | 1,166 | 590 | NA | NA | NA | NA |
| Fort Pierce | 60 | 493 | 80 | 19 | 149 | 944 | 145 | 49 | NA | NA | NA | NA |
| Gainesville | 228 | 1,375 | 230 | 49 | 477 | 3,104 | 397 | 161 | NA | NA | NA | NA |
| Greenacres | 31 | 283 | 41 | 12 | 110 | 367 | 55 | 17 | | | | |
| Hallandale Beach | 123 | 519 | 114 | 27 | 186 | 973 | 196 | 53 | NA | NA | NA | NA |
| Hialeah | 240 | 702 | 153 | 23 | 383 | 1,282 | 151 | 43 | NA | NA | NA | NA |
| Hollywood | 346 | 2,367 | 270 | 94 | D | D | D | D | NA | NA | NA | NA |
| Homestead | 48 | 144 | 25 | 5 | 86 | 526 | 35 | 13 | NA | NA | NA | NA |
| Jacksonville | 1,333 | 8,415 | 3,041 | 472 | 3,058 | 35,882 | 6,387 | 2,534 | NA | NA | NA | NA |
| Jupiter | 205 | 1,015 | 148 | 48 | 538 | 3,335 | 582 | 213 | NA | NA | NA | NA |
| Kissimmee | 190 | 1,028 | 191 | 37 | 196 | 889 | 90 | 32 | NA | NA | NA | NA |
| Lakeland | 207 | 1,135 | 283 | 50 | 374 | 2,330 | 318 | 121 | NA | NA | NA | NA |
| Lake Worth | 51 | 89 | 23 | 4 | 120 | 665 | 102 | 40 | NA | NA | NA | NA |
| Largo | 126 | 511 | 133 | 21 | 244 | 2,878 | 482 | 160 | | | | |
| Lauderdale Lakes | 28 | 432 | 389 | 16 | 27 | 48 | 6 | 2 | NA | NA | NA | NA |
| Lauderhill | 50 | 285 | 60 | 10 | 96 | 264 | 39 | 12 | NA | NA | NA | NA |
| Margate | 56 | 201 | 44 | 5 | 147 | 729 | 90 | 27 | NA | NA | NA | NA |
| Melbourne | 197 | 682 | 163 | 25 | 468 | 8,212 | 1,901 | 704 | NA | NA | NA | NA |
| Miami | 1,541 | 6,240 | 2,578 | 380 | 4,279 | 28,985 | 7,077 | 2,695 | NA | NA | NA | NA |
| Miami Beach | 564 | 1,981 | 588 | 91 | 767 | 2,353 | 505 | 144 | NA | NA | NA | NA |
| Miami Gardens | 75 | 453 | 94 | 18 | 95 | 315 | 35 | 15 | NA | NA | NA | NA |
| Miami Lakes | 116 | 492 | 145 | 29 | 309 | 1,628 | 323 | 100 | NA | NA | NA | NA |
| Miramar | 135 | 560 | 171 | 26 | 339 | 1,882 | 364 | 130 | NA | NA | NA | NA |
| North Lauderdale | 18 | 57 | 20 | 2 | 37 | 585 | 30 | 16 | | | | |
| North Miami | 120 | 402 | 74 | 17 | 192 | 597 | 98 | 33 | NA | NA | NA | NA |
| North Miami Beach | 91 | 190 | 38 | 7 | 223 | 833 | 132 | 45 | NA | NA | NA | NA |
| North Port | D | D | D | D | 56 | 151 | 17 | 6 | NA | NA | NA | NA |
| Oakland Park | 110 | 450 | 96 | 25 | 225 | 1,069 | 217 | 82 | NA | NA | NA | NA |
| Ocala | 194 | 954 | 218 | 30 | 400 | 1,912 | 240 | 86 | NA | NA | NA | NA |
| Ocoee | 41 | 327 | 53 | 14 | 102 | 612 | 126 | 35 | NA | NA | NA | NA |
| Orlando | 987 | 8,964 | 3,542 | 536 | 2,332 | 22,012 | 4,147 | 1,678 | NA | NA | NA | NA |
| Ormond Beach | 95 | D | 71 | D | 191 | 1,069 | 181 | 55 | NA | NA | NA | NA |
| Oviedo | D | D | D | 6 | 149 | 745 | 93 | 40 | NA | NA | NA | NA |
| Palm Bay | 67 | 237 | 45 | 9 | 128 | 607 | 93 | 32 | | | | |
| Palm Beach Gardens | 170 | 765 | 628 | 51 | 543 | 3,060 | 612 | 231 | NA | NA | NA | NA |
| Palm Coast | D | D | D | D | 137 | 342 | 45 | 15 | NA | NA | NA | NA |
| Panama City | 98 | 497 | 114 | 17 | 222 | 1,556 | 261 | 99 | NA | NA | NA | NA |
| Pembroke Pines | 233 | 863 | 238 | 43 | 640 | 5,270 | 798 | 327 | NA | NA | NA | NA |
| Pensacola | 149 | 768 | 169 | 25 | 484 | 1,096 | 132 | 65 | NA | NA | NA | NA |
| Pinellas Park | 73 | 454 | 104 | 19 | 161 | 6,356 | 778 | 566 | NA | NA | NA | NA |
| Plantation | 254 | 2,278 | 632 | 105 | 743 | 419 | 52 | 20 | NA | NA | NA | NA |
| Plant City | 46 | 202 | 42 | 8 | 103 | 2,179 | 422 | 123 | NA | NA | NA | NA |
| Pompano Beach | 279 | 1,140 | 351 | 53 | 523 | 850 | 89 | 43 | NA | NA | NA | NA |
| Port Orange | 111 | 396 | 69 | 13 | 132 | 1,097 | 135 | 47 | | | | |
| Port St. Lucie | 185 | 385 | 98 | 13 | 294 | 319 | 69 | 19 | NA | NA | NA | NA |
| Riviera Beach | 50 | 271 | 54 | 17 | 80 | 467 | 52 | 18 | NA | NA | NA | NA |
| Royal Palm Beach | 37 | 62 | 32 | 2 | 119 | 192 | 21 | 7 | NA | NA | NA | NA |
| St. Cloud | 52 | 124 | 25 | 4 | 64 | 12,571 | 1,948 | 765 | NA | NA | NA | NA |
| St. Petersburg | 467 | 2,667 | 476 | 107 | 1,312 | 1,222 | 152 | 58 | NA | NA | NA | NA |
| Sanford | 92 | 443 | 176 | 19 | 145 | 4,895 | 1,011 | 390 | NA | NA | NA | NA |
| Sarasota | 262 | 708 | 171 | 31 | 656 | 2,528 | 420 | 165 | NA | NA | NA | NA |
| Sunrise | 137 | 763 | 233 | 32 | 396 | 9,135 | 1,673 | 687 | NA | NA | NA | NA |
| Tallahassee | 376 | 2,652 | 413 | 90 | 1,194 | 683 | 86 | 33 | NA | NA | NA | NA |
| Tamarac | 74 | 217 | 58 | 8 | 203 | | | | | | | |
| Tampa | 1,110 | 5,982 | 2,106 | 322 | 2,971 | 41,022 | 8,563 | 3,659 | NA | NA | NA | NA |
| Titusville | 39 | 141 | 25 | 4 | 102 | 521 | 83 | 26 | NA | NA | NA | NA |
| Wellington | 168 | 1,070 | 240 | 60 | 383 | 915 | 166 | 51 | NA | NA | NA | NA |
| Weston | 252 | 472 | 142 | 23 | D | D | D | D | NA | NA | NA | NA |
| West Palm Beach | 333 | 1,462 | 398 | 82 | 1,190 | 7,989 | 1,762 | 734 | NA | NA | NA | NA |
| Winter Garden | 103 | 334 | 72 | 14 | 165 | 598 | 93 | 37 | NA | NA | NA | NA |
| Winter Haven | 60 | 346 | 90 | 13 | 124 | 670 | 77 | 32 | NA | NA | NA | NA |
| Winter Park | 171 | 604 | 164 | 30 | 525 | 3,559 | 637 | 238 | NA | NA | NA | NA |
| Winter Springs | 55 | 98 | 17 | 4 | 116 | 659 | 100 | 37 | NA | NA | NA | NA |

1. Establishments subject to federal tax.

# Table D. Cities — Accommodation and Food Services, Arts, Entertainment, and Recreation, and Health Care and Social Assistance

| City | Accommodation and food services, 2017 | | | | Arts, entertainment, and recreation[1], 2017 | | | | Health care and social assistance[1], 2017 | | | |
|---|---|---|---|---|---|---|---|---|---|---|---|---|
| | Number of establishments | Number of employees | Receipts (mil dol) | Annual payroll (mil dol) | Number of establishments | Number of employees | Receipts (mil dol) | Annual payroll (mil dol) | Number of establishments | Number of employees | Receipts (mil dol) | Annual payroll (mil dol) |
| | 92 | 93 | 94 | 95 | 96 | 97 | 98 | 99 | 100 | 101 | 102 | 103 |
| FLORIDA— Cont'd | | | | | | | | | | | | |
| Deerfield Beach | 179 | 3,367 | 223 | 60 | 43 | 436 | 38 | 10 | 198 | 4,009 | 340 | 125 |
| DeLand | 123 | 2,147 | 105 | 30 | 21 | 284 | 12 | 4 | 141 | 3,703 | 395 | 144 |
| Delray Beach | 259 | 6,536 | 552 | 139 | 56 | 772 | 67 | 22 | 460 | 7,196 | 1,128 | 348 |
| Deltona | 47 | 688 | 38 | 10 | 12 | 240 | 6 | 2 | 90 | 724 | 72 | 25 |
| Doral | 316 | 6,946 | 507 | 144 | 80 | 496 | 57 | 13 | 231 | 3,692 | 416 | 147 |
| Dunedin | D | D | D | D | 22 | 209 | 14 | 4 | 132 | 2,413 | 257 | 104 |
| Fort Lauderdale | 736 | 18,456 | 1,535 | 419 | 187 | 1,572 | 202 | 49 | 945 | 15,261 | 1,951 | 757 |
| Fort Myers | 353 | 8,181 | 472 | 146 | 51 | 1,870 | 124 | 36 | 456 | 17,839 | 2,375 | 1,100 |
| Fort Pierce | 135 | 2,597 | 134 | 39 | D | D | D | D | 202 | 3,879 | 584 | 191 |
| Gainesville | 421 | 9,768 | 516 | 151 | 58 | 721 | 35 | 12 | 447 | 18,947 | 3,260 | 1,183 |
| Greenacres | 68 | 1,086 | 66 | 18 | 12 | 127 | 6 | 2 | 110 | 807 | 106 | 35 |
| Hallandale Beach | 87 | 2,464 | 160 | 53 | 40 | 1,176 | 302 | 43 | 153 | 1,006 | 108 | 38 |
| Hialeah | 377 | 6,741 | 432 | 106 | 44 | D | 81 | D | 796 | 10,440 | 1,383 | 438 |
| Hollywood | 381 | 6,057 | 520 | 138 | 95 | 811 | 75 | 17 | 550 | 12,593 | 1,653 | 697 |
| Homestead | 101 | 2,025 | 139 | 35 | D | D | D | D | 182 | 3,867 | 402 | 122 |
| Jacksonville | 1,998 | 41,498 | 2,489 | 687 | 294 | 6,482 | 776 | 371 | 2,601 | 68,647 | 10,016 | 3,671 |
| Jupiter | 181 | 4,756 | 290 | 86 | 65 | 1,505 | 127 | 39 | 376 | 4,974 | 696 | 262 |
| Kissimmee | 236 | 5,337 | 409 | 99 | 23 | 1,201 | 121 | 34 | 310 | 6,050 | 988 | 326 |
| Lakeland | 300 | 7,573 | 417 | 120 | 37 | 703 | 37 | 12 | 427 | 13,942 | 1,897 | 678 |
| Lake Worth | 88 | 1,110 | 66 | 18 | 21 | 107 | 9 | 2 | 102 | 1,362 | 137 | 55 |
| Largo | 181 | 2,931 | 167 | 46 | 31 | 277 | 18 | 4 | 316 | 10,599 | 1,191 | 505 |
| Lauderdale Lakes | 35 | 596 | 32 | 9 | D | D | D | D | 96 | 2,515 | 309 | 106 |
| Lauderhill | 83 | 1,073 | 71 | 16 | 18 | 112 | 13 | 2 | 154 | 1,894 | 145 | 59 |
| Margate | 93 | 1,221 | 80 | 21 | 18 | 323 | 20 | 5 | 168 | 2,519 | 445 | 130 |
| Melbourne | 265 | 5,501 | 340 | 98 | 36 | 508 | 31 | 11 | 450 | 12,706 | 1,741 | 588 |
| Miami | 1,460 | 35,124 | 2,754 | 743 | 318 | 7,230 | 1,318 | 490 | 1,858 | 41,128 | 6,460 | 2,229 |
| Miami Beach | 691 | 28,900 | 3,170 | 848 | 144 | 1,182 | 170 | 41 | 426 | 7,303 | 1,067 | 430 |
| Miami Gardens | 104 | 1,440 | 103 | 25 | D | D | D | D | 193 | 2,883 | 365 | 175 |
| Miami Lakes | 75 | 1,902 | 133 | 39 | 23 | 144 | 30 | 6 | 235 | 3,060 | 348 | 117 |
| Miramar | 124 | 2,084 | 141 | 36 | 39 | 252 | 19 | 5 | 290 | 3,714 | 567 | 217 |
| North Lauderdale | 27 | 497 | 29 | 7 | D | D | D | 0 | 34 | 311 | 22 | 7 |
| North Miami | 111 | 1,874 | 111 | 28 | 20 | 208 | 28 | 5 | 167 | 4,020 | 284 | 139 |
| North Miami Beach | 112 | 1,775 | 126 | 32 | 17 | 85 | 8 | 2 | 222 | 4,407 | 581 | 209 |
| North Port | 53 | 1,031 | 56 | 16 | 5 | 37 | 1 | 1 | 80 | 717 | 83 | 27 |
| Oakland Park | 114 | 2,283 | 134 | 42 | 26 | 231 | 14 | 3 | 197 | 1,963 | 229 | 88 |
| Ocala | 301 | 6,783 | 385 | 106 | 29 | 383 | 22 | 6 | 583 | 12,284 | 1,658 | 583 |
| Ocoee | 80 | 1,191 | 69 | 18 | 22 | 113 | 11 | 3 | 127 | 3,085 | 575 | 142 |
| Orlando | 1,129 | 33,869 | 2,950 | 703 | 242 | 28,497 | 3,501 | 819 | 1,144 | 38,074 | 6,331 | 2,136 |
| Ormond Beach | 147 | 2,991 | 175 | 50 | 31 | 396 | 23 | 10 | 234 | 3,595 | 454 | 200 |
| Oviedo | D | D | D | D | 20 | 189 | 12 | 3 | 128 | 1,279 | 153 | 51 |
| Palm Bay | 118 | 1,954 | 111 | 29 | 16 | 92 | 5 | 1 | 151 | 2,335 | 223 | 82 |
| Palm Beach Gardens | 165 | 4,965 | 292 | 104 | 62 | 1,313 | 310 | 59 | 409 | 4,790 | 789 | 238 |
| Palm Coast | 110 | 2,182 | 119 | 33 | D | D | D | 7 | 194 | 3,467 | 473 | 156 |
| Panama City | 166 | 3,534 | 192 | 59 | 14 | 152 | 7 | 3 | 354 | 8,121 | 1,059 | 403 |
| Pembroke Pines | 324 | 7,639 | 462 | 132 | 76 | 536 | 53 | 9 | 638 | 9,390 | 1,297 | 497 |
| Pensacola | 195 | 5,482 | 286 | 92 | 34 | 736 | 30 | 10 | 456 | 14,133 | 1,962 | 776 |
| Pinellas Park | 133 | 2,501 | 182 | 45 | 22 | 232 | 13 | 3 | 183 | 3,189 | 296 | 115 |
| Plantation | 212 | 3,869 | 295 | 76 | 42 | 555 | 38 | 13 | 611 | 7,486 | 1,095 | 369 |
| Plant City | 84 | 1,927 | 100 | 29 | 10 | 714 | 14 | 3 | 134 | 2,237 | 371 | 104 |
| Pompano Beach | 279 | 4,425 | 379 | 92 | 73 | 1,567 | 141 | 25 | 335 | 5,819 | 590 | 206 |
| Port Orange | 126 | 2,714 | 142 | 45 | 20 | 393 | 11 | 5 | 182 | 1,915 | 195 | 75 |
| Port St. Lucie | 239 | 4,613 | 253 | 69 | 48 | 379 | 28 | 7 | 488 | 7,881 | 1,058 | 343 |
| Riviera Beach | 37 | 886 | 62 | 18 | 5 | 361 | 24 | 4 | 72 | 3,543 | 518 | 197 |
| Royal Palm Beach | 105 | 2,109 | 125 | 36 | 11 | 98 | 8 | 2 | 154 | 1,544 | 191 | 76 |
| St. Cloud | 82 | 1,475 | 85 | 25 | D | D | D | 3 | 98 | 1,643 | 198 | 65 |
| St. Petersburg | 568 | 10,140 | 658 | 179 | 116 | 2,201 | 307 | 164 | 1,021 | 20,818 | 3,081 | 1,081 |
| Sanford | 150 | 3,039 | 169 | 49 | 15 | 262 | 8 | 3 | 144 | 2,690 | 378 | 142 |
| Sarasota | 272 | 6,120 | 447 | 121 | 70 | 1,573 | 128 | 38 | 470 | 11,452 | 1,675 | 560 |
| Sunrise | 203 | 4,158 | 257 | 73 | 38 | D | 146 | D | 289 | 7,051 | 1,006 | 528 |
| Tallahassee | 647 | 14,587 | 774 | 212 | 84 | 1,268 | 96 | 25 | 680 | 17,885 | 2,343 | 917 |
| Tamarac | 67 | 1,053 | 70 | 17 | 21 | 269 | 17 | 5 | 230 | 3,849 | 548 | 158 |
| Tampa | 1,130 | 28,032 | 2,072 | 557 | 213 | 10,239 | 1,388 | 512 | 1,526 | 38,844 | 6,864 | 2,269 |
| Titusville | 102 | 2,018 | 106 | 29 | 12 | 177 | 11 | 3 | 147 | 2,868 | 332 | 121 |
| Wellington | 120 | 2,830 | 141 | 43 | 81 | 956 | 133 | 25 | 264 | 2,789 | 390 | 140 |
| Weston | 118 | 2,530 | 183 | 49 | 32 | 685 | 50 | 14 | 291 | 3,950 | 735 | 294 |
| West Palm Beach | 389 | 8,290 | 568 | 163 | 81 | 1,945 | 287 | 68 | 621 | 13,961 | 1,840 | 640 |
| Winter Garden | 102 | 2,328 | 131 | 40 | 35 | 271 | 27 | 5 | 106 | 1,272 | 119 | 46 |
| Winter Haven | 117 | 2,369 | 131 | 35 | 18 | 1,971 | 171 | 36 | 191 | 5,761 | 644 | 288 |
| Winter Park | 163 | 4,692 | 258 | 86 | 50 | 937 | 85 | 25 | 357 | 5,942 | 771 | 278 |
| Winter Springs | 29 | 263 | 15 | 4 | 19 | 145 | 10 | 4 | 62 | 323 | 21 | 9 |

1. Establishments subject to federal tax.

# Table D. Cities — Other Services and Government Employment and Payroll

| | Other services[1] | | | | Government employment and payroll, 2017 | | | | | | | | |
| | | | | | Full-time equivalent employees | March payroll | | | | | | | |
| | | | | | | | | | Percent of total for: | | | | |
| City | Number of establish-ments | Number of employees | Receipts (mil dol) | Annual payroll (mil dol) | | Total (dollars) | Admin-istrative, judicial, and legal | Police and corrections | Fire protection | Highways and trans-portation | Health and welfare | Natural resources and utilities | Education and libraries |
| | 104 | 105 | 106 | 107 | 108 | 109 | 110 | 111 | 112 | 113 | 114 | 115 | 116 |
| **FLORIDA— Cont'd** | | | | | | | | | | | | | |
| Deerfield Beach | 231 | 892 | 107 | 26 | 330 | 1,569,268 | 23.0 | 0.0 | 0.0 | 6.8 | 1.7 | 60.9 | 0.0 |
| DeLand | 94 | 422 | 37 | 10 | 370 | 1,617,265 | 12.3 | 22.7 | 13.3 | 4.8 | 7.2 | 32.6 | 0.0 |
| Delray Beach | 272 | 1,035 | 128 | 28 | 819 | 4,654,723 | 13.3 | 31.2 | 29.7 | 2.8 | 0.5 | 18.2 | 0.0 |
| Deltona | D | D | D | D | 308 | 1,299,004 | 19.7 | 0.0 | 34.0 | 4.6 | 0.0 | 32.3 | 0.0 |
| Doral | 202 | 1,917 | 349 | 75 | 368 | 1,893,403 | 13.6 | 55.1 | 0.0 | 6.7 | 0.0 | 8.2 | 0.0 |
| Dunedin | 78 | 323 | 29 | 8 | 361 | 1,657,266 | 8.3 | 0.0 | 21.7 | 5.8 | 3.6 | 40.4 | 5.7 |
| Fort Lauderdale | 793 | 5,081 | 662 | 189 | 2,525 | 15,897,679 | 11.4 | 31.2 | 21.8 | 1.5 | 6.9 | 19.5 | 0.0 |
| Fort Myers | 302 | 1,850 | 253 | 59 | 819 | 3,768,648 | 14.5 | 35.0 | 19.2 | 1.8 | 1.2 | 24.7 | 0.0 |
| Fort Pierce | 94 | 405 | 31 | 11 | 594 | 2,513,558 | 8.1 | 25.4 | 0.0 | 1.8 | 0.4 | 43.1 | 0.0 |
| Gainesville | 274 | 2,770 | 449 | 90 | 2,166 | 10,629,345 | 18.4 | 17.1 | 9.0 | 12.5 | 1.5 | 35.1 | 0.0 |
| Greenacres | 66 | 291 | 29 | 8 | 186 | 1,059,161 | 15.1 | 34.8 | 33.0 | 2.2 | 0.0 | 2.2 | 0.0 |
| Hallandale Beach | 147 | 683 | 77 | 19 | 514 | 3,159,677 | 15.0 | 34.5 | 21.0 | 4.5 | 3.7 | 19.1 | 0.0 |
| Hialeah | 442 | 1,790 | 188 | 44 | 1,278 | 6,831,692 | 5.6 | 31.9 | 32.4 | 3.6 | 4.8 | 19.3 | 0.9 |
| Hollywood | 351 | 1,913 | 246 | 68 | 1,367 | 7,734,587 | 10.9 | 35.3 | 22.5 | 3.5 | 2.3 | 17.9 | 0.0 |
| Homestead | 71 | 318 | 28 | 7 | 402 | 2,558,562 | 11.3 | 39.9 | 0.0 | 3.3 | 1.3 | 34.7 | 0.0 |
| Jacksonville | 1,614 | 11,019 | 1,531 | 388 | 9,087 | 42,343,407 | 12.8 | 31.9 | 15.2 | 6.4 | 1.9 | 17.7 | 2.2 |
| Jupiter | 219 | 1,339 | 174 | 44 | 375 | 2,229,268 | 12.5 | 45.1 | 0.0 | 3.8 | 0.0 | 20.2 | 0.0 |
| Kissimmee | 144 | 758 | 87 | 22 | 897 | 4,628,057 | 6.0 | 20.8 | 13.6 | 3.6 | 0.2 | 51.6 | 0.0 |
| Lakeland | 196 | 1,294 | 145 | 37 | 2,180 | 11,663,514 | 11.3 | 17.1 | 10.6 | 4.3 | 2.7 | 47.2 | 1.4 |
| Lake Worth | 86 | 334 | 33 | 10 | 318 | 1,453,714 | 26.2 | 0.0 | 0.0 | 0.0 | 0.0 | 63.7 | 1.2 |
| Largo | 186 | 1,040 | 113 | 33 | 846 | 4,172,007 | 10.7 | 28.2 | 21.0 | 5.0 | 6.0 | 25.3 | 3.0 |
| Lauderdale Lakes | 47 | 187 | 24 | 6 | 81 | 311,550 | 38.5 | 0.0 | 0.0 | 5.3 | 19.3 | 16.6 | 0.0 |
| Lauderhill | 106 | 387 | 39 | 10 | 483 | 3,060,756 | 11.9 | 35.9 | 22.8 | 4.3 | 0.0 | 21.3 | 0.0 |
| Margate | 128 | 411 | 48 | 14 | 478 | 3,000,546 | 7.0 | 33.3 | 30.5 | 0.3 | 1.5 | 7.3 | 0.0 |
| Melbourne | 205 | 1,002 | 81 | 27 | 876 | 3,893,511 | 12.8 | 26.3 | 21.8 | 12.8 | 0.7 | 22.8 | 0.0 |
| Miami | 1,195 | 7,218 | 1,344 | 223 | 3,951 | 21,993,811 | 8.7 | 37.7 | 30.2 | 1.1 | 0.9 | 10.1 | 0.0 |
| Miami Beach | 305 | 2,250 | 209 | 50 | 1,868 | 11,648,570 | 17.6 | 25.1 | 20.0 | 2.2 | 3.5 | 14.7 | 0.0 |
| Miami Gardens | 73 | 226 | 26 | 6 | 482 | 3,036,273 | 10.1 | 66.9 | 0.0 | 4.2 | 0.7 | 5.9 | 0.0 |
| Miami Lakes | 60 | 613 | 64 | 18 | 46 | 234,782 | 44.8 | 0.0 | 0.0 | 0.0 | 0.0 | 28.4 | 0.0 |
| Miramar | 129 | 986 | 106 | 31 | 868 | 5,621,208 | 15.4 | 36.4 | 23.5 | 2.2 | 2.7 | 15.9 | 0.0 |
| North Lauderdale | 33 | 160 | 25 | 5 | 160 | 737,670 | 14.1 | 0.0 | 34.5 | 6.4 | 9.5 | 25.3 | 0.0 |
| North Miami | 118 | 550 | 49 | 14 | 398 | 1,769,654 | 18.8 | 49.8 | 0.0 | 2.7 | 1.4 | 25.8 | 1.4 |
| North Miami Beach | 96 | 482 | 42 | 10 | 464 | 2,379,780 | 13.4 | 37.8 | 7.0 | 9.8 | 0.5 | 18.4 | 0.0 |
| North Port | 57 | 225 | 32 | 9 | 535 | 2,621,666 | 13.9 | 24.6 | 20.6 | 4.3 | 1.0 | 27.3 | 2.0 |
| Oakland Park | 194 | 668 | 98 | 21 | 246 | 1,355,506 | 20.1 | 0.0 | 34.6 | 6.5 | 0.6 | 35.7 | 0.0 |
| Ocala | 243 | 1,427 | 143 | 43 | 958 | 4,546,983 | 11.3 | 25.4 | 15.8 | 7.4 | 8.7 | 21.6 | 0.0 |
| Ocoee | D | D | D | D | 342 | 1,635,673 | 11.3 | 31.2 | 19.8 | 1.6 | 1.1 | 14.2 | 0.0 |
| Orlando | 729 | 6,858 | 720 | 205 | 3,145 | 17,935,004 | 15.1 | 34.3 | 25.7 | 6.6 | 0.0 | 25.5 | 0.0 |
| Ormond Beach | 108 | 536 | 45 | 13 | 357 | 1,444,897 | 17.0 | 25.9 | 17.2 | 2.1 | 0.0 | 27.0 | 0.0 |
| Oviedo | D | D | D | D | 287 | 1,289,540 | 12.1 | 28.9 | 25.3 | 10.0 | 3.1 | 22.5 | 0.0 |
| Palm Bay | 101 | 441 | 44 | 14 | 770 | 3,203,187 | 10.8 | 28.6 | 22.0 | 10.0 | 3.1 | 22.5 | 0.0 |
| Palm Beach Gardens | 184 | 1,754 | 202 | 77 | 559 | 3,461,550 | 10.2 | 36.5 | 29.6 | 7.0 | 2.1 | 11.6 | 0.0 |
| Palm Coast | 89 | 407 | 33 | 9 | 399 | 1,699,178 | 26.6 | 0.0 | 18.5 | 15.6 | 0.0 | 37.3 | 0.0 |
| Panama City | 106 | 677 | 54 | 16 | 550 | 1,043,351 | 11.7 | 24.2 | 13.2 | 27.6 | 3.7 | 18.9 | 0.0 |
| Pembroke Pines | 255 | 1,522 | 147 | 43 | 656 | 3,958,045 | 11.6 | 46.1 | 35.1 | 0.0 | 0.4 | 4.4 | 0.0 |
| Pensacola | 149 | 1,205 | 104 | 32 | 715 | 2,886,335 | 11.8 | 26.2 | 16.1 | 12.5 | 3.9 | 23.3 | 0.0 |
| Pinellas Park | 182 | 1,202 | 131 | 43 | 493 | 2,244,205 | 13.6 | 31.7 | 19.5 | 4.0 | 5.6 | 17.8 | 3.9 |
| Plantation | 195 | 875 | 124 | 28 | 808 | 4,245,516 | 9.8 | 41.1 | 3.4 | 5.2 | 10.2 | 17.7 | 1.0 |
| Plant City | 67 | 314 | 34 | 11 | 361 | 1,349,864 | 10.5 | 31.6 | 14.7 | 10.8 | 0.4 | 23.7 | 3.0 |
| Pompano Beach | 396 | 2,075 | 303 | 71 | 776 | 4,045,084 | 17.8 | 0.0 | 32.1 | 14.3 | 1.5 | 22.6 | 0.0 |
| Port Orange | 104 | 386 | 34 | 8 | 406 | 1,668,896 | 15.4 | 26.7 | 15.6 | 3.8 | 4.4 | 27.5 | 0.0 |
| Port St. Lucie | 234 | 828 | 80 | 23 | 1,002 | 4,813,985 | 17.2 | 37.8 | 0.0 | 6.1 | 3.3 | 22.2 | 0.0 |
| Riviera Beach | 90 | 648 | 76 | 22 | 492 | 2,387,288 | 16.7 | 31.0 | 21.2 | 5.6 | 2.0 | 15.3 | 1.1 |
| Royal Palm Beach | 90 | 488 | 47 | 14 | 124 | 571,750 | 27.0 | 0.0 | 0.0 | 29.0 | 10.1 | 29.6 | 0.0 |
| St. Cloud | D | D | D | D | 423 | 1,704,067 | 14.6 | 30.3 | 15.2 | 7.6 | 2.5 | 24.2 | 0.0 |
| St. Petersburg | 531 | 3,176 | 402 | 109 | 3,230 | 17,086,876 | 13.0 | 29.4 | 13.2 | 0.8 | 1.5 | 31.2 | 1.6 |
| Sanford | 125 | 655 | 57 | 18 | 565 | 2,407,854 | 7.2 | 25.6 | 17.2 | 16.4 | 0.0 | 24.5 | 0.0 |
| Sarasota | 258 | 1,373 | 247 | 41 | 704 | 3,740,313 | 13.4 | 37.7 | 0.0 | 3.5 | 1.2 | 27.0 | 0.0 |
| Sunrise | 148 | 760 | 146 | 31 | 1,004 | 6,575,439 | 9.0 | 31.4 | 20.5 | 1.2 | 9.1 | 28.1 | 0.0 |
| Tallahassee | 524 | 4,124 | 693 | 185 | 3,177 | 16,131,062 | 17.5 | 17.2 | 11.0 | 8.8 | 1.6 | 36.8 | 0.0 |
| Tamarac | 93 | 274 | 30 | 9 | 363 | 1,876,467 | 25.5 | 0.0 | 24.3 | 8.4 | 1.2 | 29.3 | 0.0 |
| Tampa | 1,043 | 9,738 | 1,446 | 278 | 4,198 | 25,536,216 | 11.1 | 34.9 | 18.1 | 2.0 | 4.0 | 24.4 | 0.0 |
| Titusville | 80 | 406 | 34 | 11 | 462 | 1,914,251 | 19.1 | 30.0 | 15.9 | 4.2 | 0.8 | 24.8 | 0.0 |
| Wellington | 129 | 582 | 119 | 19 | 334 | 1,637,284 | 28.0 | 0.0 | 0.0 | 2.6 | 2.1 | 33.3 | 0.0 |
| Weston | 123 | 483 | 42 | 13 | 11 | 69,116 | 100.0 | 0.0 | 0.0 | 0.0 | 0.0 | 0.0 | 0.0 |
| West Palm Beach | 383 | 2,254 | 377 | 80 | 1,432 | 7,321,395 | 15.2 | 31.5 | 18.1 | 1.4 | 1.0 | 21.8 | 2.1 |
| Winter Garden | 90 | 581 | 53 | 17 | 290 | 1,318,942 | 12.7 | 36.5 | 18.3 | 0.0 | 4.8 | 19.1 | 0.0 |
| Winter Haven | 81 | 317 | 33 | 10 | 458 | 1,781,107 | 14.2 | 26.9 | 18.4 | 5.2 | 0.0 | 28.0 | 2.4 |
| Winter Park | 159 | 878 | 98 | 27 | 524 | 2,722,952 | 10.4 | 22.3 | 21.4 | 4.0 | 0.6 | 34.6 | 0.0 |
| Winter Springs | 39 | 99 | 11 | 3 | 171 | 748,571 | 18.0 | 44.8 | 0.0 | 10.4 | 2.7 | 24.1 | 0.0 |

1. Establishments subject to federal tax.

**City Government Finances**

| City | General revenue | | | | | | | General expenditure | | |
|---|---|---|---|---|---|---|---|---|---|---|
| | | Intergovernmental | | Taxes | | | | | | |
| | | | | | | Per capita[1] (dollars) | | | | Per capita[1] (dollars) | |
| | Total (mil dol) | Total (mil dol) | Percent from state government | Total (mil dol) | Total | Property | Sales and gross receipts | Total (mil dol) | Total | Capital outlays |
| | 117 | 118 | 119 | 120 | 121 | 122 | 123 | 124 | 125 | 126 |
| **FLORIDA— Cont'd** | | | | | | | | | | |
| Deerfield Beach | 134.4 | 17.0 | 63.1 | 60.8 | 755 | 467 | 285 | 140 | 1,733 | 74 |
| DeLand | 49.6 | 6.4 | 83.5 | 20.1 | 623 | 289 | 323 | 43 | 1,336 | 89 |
| Delray Beach | 183.4 | 31.3 | 25.7 | 95.6 | 1,391 | 1,053 | 336 | 142 | 2,062 | 155 |
| Deltona | 59.1 | 10.4 | 90.8 | 28.5 | 315 | 152 | 163 | 56 | 615 | 100 |
| Doral | 72.5 | 8.3 | 84.1 | 52.7 | 875 | 298 | 577 | 74 | 1,228 | 466 |
| Dunedin | 49.5 | 5.0 | 81.6 | 20.9 | 574 | 226 | 347 | 62 | 1,686 | 97 |
| Fort Lauderdale | 444.4 | 48.0 | 44.3 | 205.9 | 1,145 | 663 | 482 | 401 | 2,228 | 92 |
| Fort Myers | 180.2 | 13.7 | 71.5 | 79.6 | 994 | 526 | 468 | 184 | 2,298 | 177 |
| Fort Pierce | 74.1 | 8.0 | 55.9 | 26.0 | 571 | 387 | 184 | 78 | 1,718 | 49 |
| Gainesville | 209.3 | 36.9 | 50.5 | 62.1 | 470 | 221 | 250 | 227 | 1,717 | 263 |
| Greenacres | 26.6 | 5.2 | 89.5 | 16.6 | 408 | 203 | 204 | 30 | 725 | 122 |
| Hallandale Beach | 90.6 | 10.6 | 35.9 | 35.3 | 888 | 598 | 290 | 111 | 2,798 | 525 |
| Hialeah | 254.1 | 63.3 | 73.3 | 101.2 | 428 | 199 | 229 | 241 | 1,021 | 37 |
| Hollywood | 316.5 | 29.1 | 64.7 | 146.5 | 955 | 680 | 275 | 283 | 1,844 | 165 |
| Homestead | 84.6 | 16.7 | 62.7 | 24.5 | 355 | 173 | 182 | 113 | 1,632 | 354 |
| Jacksonville | 2,264.4 | 558.5 | 39.1 | 993.4 | 1,114 | 625 | 488 | 1,680 | 1,884 | 121 |
| Jupiter | 68.3 | 9.7 | 73.5 | 42.2 | 651 | 379 | 272 | 57 | 872 | 73 |
| Kissimmee | 104.6 | 42.0 | 43.7 | 30.8 | 430 | 165 | 266 | 101 | 1,407 | 248 |
| Lakeland | 230.0 | 26.2 | 67.3 | 57.0 | 529 | 289 | 239 | 231 | 2,141 | 340 |
| Lake Worth | 57.1 | 6.7 | 79.6 | 16.1 | 422 | 236 | 186 | 62 | 1,627 | 107 |
| Largo | 121.7 | 13.3 | 83.9 | 46.2 | 545 | 228 | 315 | 168 | 1,986 | 657 |
| Lauderdale Lakes | 35.3 | 7.4 | 63.7 | 15.9 | 438 | 221 | 217 | 29 | 790 | 28 |
| Lauderhill | 84.8 | 11.1 | 79.7 | 34.6 | 481 | 252 | 229 | 79 | 1,093 | 28 |
| Margate | 78.0 | 7.5 | 82.4 | 35.1 | 602 | 364 | 238 | 75 | 1,289 | 123 |
| Melbourne | 172.2 | 57.3 | 84.5 | 54.4 | 665 | 352 | 313 | 138 | 1,688 | 62 |
| Miami | 929.9 | 173.7 | 37.2 | 546.2 | 1,196 | 739 | 457 | 894 | 1,959 | 193 |
| Miami Beach | 565.7 | 66.3 | 17.5 | 347.5 | 3,834 | 1,627 | 2,207 | 646 | 7,129 | 1,755 |
| Miami Gardens | 92.6 | 16.5 | 78.9 | 52.3 | 466 | 244 | 221 | 84 | 746 | 42 |
| Miami Lakes | 24.6 | 6.1 | 78.3 | 15.0 | 489 | 195 | 294 | 23 | 756 | 144 |
| Miramar | 169.4 | 19.1 | 71.6 | 84.5 | 602 | 369 | 233 | 195 | 1,394 | 225 |
| North Lauderdale | 43.1 | 5.5 | 91.8 | 16.4 | 371 | 184 | 187 | 40 | 906 | 25 |
| North Miami | 92.9 | 10.1 | 76.4 | 32.2 | 520 | 302 | 218 | 89 | 1,438 | 26 |
| North Miami Beach | 72.7 | 6.7 | 80.4 | 29.1 | 679 | 328 | 351 | 90 | 2,107 | 174 |
| North Port | 104.1 | 10.3 | 72.6 | 31.9 | 481 | 152 | 328 | 97 | 1,457 | 288 |
| Oakland Park | 68.5 | 8.8 | 49.2 | 27.6 | 614 | 336 | 278 | 63 | 1,410 | 99 |
| Ocala | 136.2 | 15.3 | 69.7 | 44.3 | 750 | 433 | 318 | 130 | 2,199 | 252 |
| Ocoee | 58.7 | 12.4 | 98.0 | 18.6 | 399 | 237 | 161 | 52 | 1,109 | 123 |
| Orlando | 873.9 | 254.3 | 23.9 | 268.8 | 954 | 523 | 431 | 851 | 3,020 | 169 |
| Ormond Beach | 58.4 | 10.5 | 44.0 | 25.7 | 603 | 295 | 305 | 57 | 1,340 | 215 |
| Oviedo | 48.1 | 10.2 | 41.0 | 22.1 | 542 | 274 | 268 | 50 | 1,228 | 133 |
| Palm Bay | 92.1 | 15.6 | 88.0 | 47.0 | 419 | 220 | 199 | 88 | 780 | 104 |
| Palm Beach Gardens | 89.6 | 7.3 | 87.4 | 66.7 | 1,183 | 922 | 261 | 93 | 1,655 | 215 |
| Palm Coast | 64.3 | 10.2 | 51.6 | 25.4 | 295 | 197 | 98 | 89 | 1,028 | 205 |
| Panama City | 70.8 | 10.1 | 89.8 | 33.0 | 888 | 305 | 583 | 61 | 1,634 | 156 |
| Pembroke Pines | 281.9 | 67.5 | 90.0 | 106.7 | 625 | 359 | 266 | 368 | 2,157 | 318 |
| Pensacola | 133.3 | 44.9 | 20.4 | 44.2 | 841 | 261 | 580 | 133 | 2,539 | 460 |
| Pinellas Park | 76.0 | 10.6 | 77.2 | 34.8 | 658 | 297 | 361 | 83 | 1,563 | 164 |
| Plantation | 133.9 | 11.4 | 83.9 | 78.3 | 835 | 493 | 341 | 118 | 1,256 | 100 |
| Plant City | 58.3 | 9.5 | 70.5 | 24.4 | 623 | 220 | 325 | 56 | 1,437 | 304 |
| Pompano Beach | 186.5 | 21.7 | 67.8 | 101.5 | 919 | 569 | 351 | 210 | 1,899 | 300 |
| Port Orange | 68.6 | 7.5 | 90.5 | 25.6 | 406 | 193 | 213 | 69 | 1,097 | 61 |
| Port St. Lucie | 265.6 | 31.7 | 71.6 | 95.4 | 504 | 348 | 157 | 254 | 1,341 | 219 |
| Riviera Beach | 99.2 | 9.6 | 43.5 | 56.7 | 1,633 | 1,286 | 347 | 124 | 3,583 | 557 |
| Royal Palm Beach | 56.2 | 4.7 | 94.8 | 15.1 | 392 | 111 | 281 | 28 | 717 | 120 |
| St. Cloud | 82.9 | 7.0 | 65.0 | 18.1 | 351 | 141 | 210 | 68 | 1,310 | 106 |
| St. Petersburg | 428.7 | 81.9 | 37.9 | 161.8 | 616 | 367 | 249 | 438 | 1,669 | 153 |
| Sanford | 139.1 | 36.8 | 23.1 | 38.3 | 643 | 321 | 322 | 101 | 1,700 | 265 |
| Sarasota | 158.5 | 23.4 | 57.3 | 63.7 | 1,114 | 498 | 616 | 152 | 2,650 | 420 |
| Sunrise | 218.9 | 14.3 | 73.2 | 76.1 | 807 | 404 | 402 | 196 | 2,081 | 261 |
| Tallahassee | 381.8 | 68.3 | 56.9 | 103.4 | 540 | 204 | 336 | 351 | 1,830 | 159 |
| Tamarac | 105.1 | 13.6 | 50.8 | 40.3 | 609 | 312 | 297 | 102 | 1,537 | 128 |
| Tampa | 700.5 | 94.9 | 65.7 | 309.4 | 791 | 369 | 422 | 733 | 1,873 | 273 |
| Titusville | 63.8 | 9.9 | 51.9 | 24.7 | 535 | 271 | 254 | 51 | 1,108 | 39 |
| Wellington | 69.4 | 8.9 | 72.0 | 34.4 | 530 | 255 | 275 | 75 | 1,159 | 266 |
| Weston | 85.2 | 6.2 | 100.0 | 35.3 | 498 | 242 | 256 | 107 | 1,512 | 53 |
| West Palm Beach | 266.9 | 46.7 | 31.3 | 130.3 | 1,181 | 762 | 419 | 272 | 2,467 | 138 |
| Winter Garden | 55.1 | 10.7 | 76.4 | 24.1 | 552 | 229 | 323 | 58 | 1,332 | 312 |
| Winter Haven | 67.9 | 6.7 | 85.5 | 23.4 | 572 | 250 | 322 | 68 | 1,675 | 231 |
| Winter Park | 67.6 | 9.4 | 66.3 | 31.8 | 1,031 | 620 | 411 | 79 | 2,555 | 327 |
| Winter Springs | 34.6 | 4.4 | 79.0 | 15.7 | 429 | 123 | 306 | 29 | 787 | 171 |

1. Based on population estimated as of July 1 of the year shown.

| City | City government finances, 2017 (cont.) | | | | | | | | | |
|---|---|---|---|---|---|---|---|---|---|---|
| | General expenditure (cont.) | | | | | | | | | |
| | Percent of total for: | | | | | | | | | |
| | Public welfare | Highways | Parking facilities | Education | Health and hospitals | Police protection | Sewerage and sanitation | Parks and recreation | Housing and community development | Interest on debt |
| | 127 | 128 | 129 | 130 | 131 | 132 | 133 | 134 | 135 | 136 |
| FLORIDA— Cont'd | | | | | | | | | | |
| Deerfield Beach ................... | 1.6 | 3.3 | 0.4 | 0.0 | 0.0 | 16.8 | 26.2 | 4.8 | 1.4 | 1.5 |
| DeLand ........................ | 0.0 | 6.9 | 0.0 | 0.0 | 0.0 | 18.0 | 24.6 | 8.0 | 0.0 | 0.7 |
| Delray Beach ..................... | 0.0 | 3.2 | 0.8 | 0.0 | 0.0 | 25.4 | 3.6 | 12.7 | 0.2 | 1.7 |
| Deltona............................ | 0.0 | 11.8 | 0.0 | 0.0 | 0.0 | 18.2 | 18.6 | 6.9 | 1.7 | 1.5 |
| Doral.............................. | 0.0 | 22.3 | 0.0 | 0.0 | 0.0 | 27.2 | 0.0 | 24.2 | 0.0 | 1.3 |
| Dunedin........................... | 0.0 | 4.7 | 0.0 | 0.0 | 0.0 | 6.7 | 22.6 | 14.8 | 0.0 | 1.0 |
| Fort Lauderdale................... | 0.0 | 4.2 | 4.1 | 0.0 | 0.0 | 24.8 | 4.7 | 8.0 | 2.4 | 3.3 |
| Fort Myers........................ | 0.0 | 5.5 | 0.6 | 0.0 | 0.0 | 20.4 | 25.0 | 7.9 | 0.4 | 2.3 |
| Fort Pierce ....................... | 0.0 | 3.5 | 0.0 | 0.0 | 0.0 | 17.0 | 21.1 | 17.2 | 4.2 | 6.3 |
| Gainesville ....................... | 0.5 | 10.9 | 0.2 | 0.0 | 0.2 | 17.6 | 21.3 | 7.3 | 9.7 | 3.2 |
| Greenacres ....................... | 0.0 | 5.7 | 0.0 | 0.0 | 20.1 | 31.0 | 4.0 | 17.4 | 0.0 | 0.4 |
| Hallandale Beach................. | 0.0 | 2.1 | 0.0 | 0.0 | 0.0 | 22.7 | 21.7 | 16.8 | 0.0 | 2.0 |
| Hialeah........................... | 0.0 | 3.9 | 0.0 | 2.2 | 0.0 | 21.7 | 28.2 | 5.7 | 0.8 | 2.0 |
| Hollywood ........................ | 0.0 | 3.3 | 2.5 | 0.0 | 0.0 | 24.4 | 19.5 | 4.9 | 0.7 | 3.3 |
| Homestead........................ | 0.7 | 5.8 | 0.0 | 0.0 | 0.0 | 30.9 | 21.1 | 5.6 | 1.4 | 1.2 |
| Jacksonville...................... | 0.7 | 2.8 | 0.2 | 0.0 | 4.9 | 21.0 | 9.8 | 6.7 | 1.1 | 6.7 |
| Jupiter............................ | 0.0 | 12.1 | 0.0 | 0.0 | 0.0 | 34.8 | 0.0 | 4.9 | 0.0 | 1.6 |
| Kissimmee ....................... | 0.0 | 17.0 | 0.0 | 3.9 | 0.0 | 23.8 | 7.7 | 9.8 | 0.1 | 1.8 |
| Lakeland.......................... | 0.0 | 7.2 | 0.5 | 0.0 | 0.0 | 16.7 | 14.3 | 23.6 | 1.0 | 0.3 |
| Lake Worth........................ | 0.0 | 3.9 | 0.5 | 0.0 | 0.0 | 24.7 | 29.5 | 6.9 | 0.0 | 0.0 |
| Largo.............................. | 0.0 | 1.3 | 0.0 | 0.0 | 0.0 | 12.9 | 45.7 | 7.0 | 1.0 | 0.2 |
| Lauderdale Lakes ................ | 0.0 | 1.8 | 0.0 | 0.0 | 0.0 | 22.4 | 3.9 | 4.6 | 0.0 | 3.2 |
| Lauderhill........................ | 0.0 | 1.2 | 0.0 | 0.0 | 5.7 | 23.3 | 6.9 | 8.9 | 1.6 | 3.8 |
| Margate........................... | 0.0 | 2.0 | 0.0 | 0.0 | 0.4 | 23.3 | 13.2 | 5.4 | 0.8 | 2.0 |
| Melbourne........................ | 0.0 | 9.1 | 0.0 | 0.0 | 0.0 | 13.9 | 19.1 | 6.9 | 0.9 | 0.3 |
| Miami............................. | 0.3 | 1.6 | 3.9 | 0.0 | 3.0 | 26.4 | 4.2 | 10.3 | 3.3 | 4.2 |
| Miami Beach...................... | 0.2 | 1.8 | 7.2 | 0.0 | 0.0 | 15.9 | 11.2 | 29.8 | 0.7 | 4.9 |
| Miami Gardens.................... | 0.0 | 5.4 | 0.0 | 0.0 | 0.0 | 41.0 | 0.0 | 8.1 | 1.7 | 8.9 |
| Miami Lakes...................... | 0.0 | 14.9 | 0.0 | 0.0 | 0.0 | 29.8 | 0.0 | 18.0 | 0.0 | 2.9 |
| Miramar........................... | 0.0 | 1.5 | 0.0 | 0.0 | 4.7 | 28.6 | 6.4 | 10.4 | 1.0 | 3.3 |
| North Lauderdale ................ | 0.0 | 4.0 | 0.0 | 0.0 | 0.0 | 23.2 | 19.3 | 12.8 | 0.8 | 0.3 |
| North Miami...................... | 0.0 | 6.6 | 0.0 | 0.0 | 0.0 | 26.8 | 21.1 | 11.5 | 0.6 | 1.4 |
| North Miami Beach .............. | 0.0 | 2.5 | 0.0 | 0.0 | 0.0 | 26.2 | 20.3 | 3.6 | 0.0 | 1.0 |
| North Port........................ | 0.3 | 22.8 | 0.0 | 0.0 | 4.2 | 15.6 | 20.8 | 2.8 | 0.0 | 0.7 |
| Oakland Park ..................... | 0.0 | 7.0 | 0.0 | 0.0 | 0.0 | 23.0 | 17.8 | 8.3 | 0.0 | 1.4 |
| Ocala.............................. | 0.0 | 7.8 | 2.7 | 0.0 | 0.0 | 20.0 | 22.4 | 7.4 | 1.3 | 0.5 |
| Ocoee............................. | 0.0 | 3.2 | 0.0 | 0.0 | 0.0 | 20.0 | 18.1 | 3.8 | 0.0 | 1.5 |
| Orlando........................... | 0.0 | 2.2 | 1.6 | 0.0 | 0.0 | 17.5 | 12.6 | 12.6 | 1.1 | 5.9 |
| Ormond Beach.................... | 0.0 | 7.8 | 0.0 | 0.0 | 0.0 | 13.0 | 25.8 | 12.5 | 0.0 | 0.5 |
| Oviedo............................ | 0.0 | 21.1 | 0.0 | 0.0 | 8.6 | 16.8 | 12.3 | 9.1 | 0.0 | 1.3 |
| Palm Bay.......................... | 0.0 | 19.1 | 0.0 | 0.0 | 0.0 | 20.8 | 21.1 | 4.4 | 1.6 | 5.1 |
| Palm Beach Gardens............ | 0.0 | 4.0 | 0.0 | 0.0 | 0.0 | 24.7 | 0.0 | 9.7 | 0.0 | 0.6 |
| Palm Coast ....................... | 0.0 | 15.4 | 0.0 | 0.0 | 0.0 | 3.1 | 39.9 | 8.0 | 0.0 | 1.0 |
| Panama City ..................... | 0.0 | 6.3 | 0.0 | 0.0 | 0.4 | 17.5 | 12.1 | 10.5 | 0.6 | 0.5 |
| Pembroke Pines.................. | 0.4 | 1.5 | 0.0 | 0.0 | 0.0 | 15.3 | 13.3 | 12.1 | 2.4 | 3.9 |
| Pensacola......................... | 0.0 | 5.4 | 0.3 | 0.0 | 0.0 | 18.9 | 5.5 | 7.4 | 12.4 | 6.2 |
| Pinellas Park..................... | 0.0 | 6.6 | 0.0 | 0.0 | 3.9 | 17.5 | 24.4 | 6.8 | 0.0 | 0.2 |
| Plantation........................ | 0.0 | 2.3 | 0.0 | 0.0 | 6.9 | 30.9 | 17.0 | 13.5 | 0.9 | 0.6 |
| Plant City........................ | 0.0 | 4.9 | 0.0 | 0.0 | 0.0 | 16.5 | 37.6 | 8.3 | 1.9 | 0.5 |
| Pompano Beach .................. | 0.0 | 2.2 | 0.6 | 0.0 | 6.8 | 19.8 | 14.2 | 6.1 | 1.4 | 0.8 |
| Port Orange ...................... | 0.0 | 9.7 | 0.0 | 0.0 | 0.0 | 17.3 | 34.6 | 7.4 | 0.4 | 3.0 |
| Port St. Lucie .................... | 0.0 | 34.0 | 0.0 | 0.0 | 0.5 | 16.0 | 17.3 | 6.4 | 0.1 | 9.3 |
| Riviera Beach..................... | 3.1 | 2.6 | 0.0 | 0.0 | 0.0 | 14.1 | 11.5 | 9.2 | 0.0 | 3.3 |
| Royal Palm Beach ............... | 0.0 | 9.6 | 0.0 | 0.0 | 0.0 | 27.6 | 4.2 | 18.8 | 0.0 | 1.3 |
| St. Cloud ......................... | 0.0 | 5.3 | 0.0 | 0.0 | 3.8 | 16.3 | 28.3 | 5.5 | 0.0 | 4.7 |
| St. Petersburg.................... | 0.0 | 5.0 | 1.3 | 0.0 | 2.8 | 23.5 | 19.5 | 14.9 | 1.3 | 0.5 |
| Sanford ........................... | 0.0 | 6.8 | 0.0 | 0.0 | 7.3 | 15.1 | 25.4 | 6.9 | 0.5 | 0.6 |
| Sarasota........................... | 0.7 | 9.8 | 1.2 | 0.0 | 0.0 | 23.0 | 24.8 | 13.3 | 2.7 | 3.0 |
| Sunrise........................... | 0.0 | 1.6 | 8.3 | 0.0 | 0.0 | 22.3 | 22.3 | 7.1 | 0.4 | 1.1 |
| Tallahassee....................... | 0.0 | 15.8 | 0.0 | 0.0 | 0.0 | 17.2 | 18.0 | 7.0 | 1.0 | 2.0 |
| Tamarac........................... | 0.0 | 6.6 | 0.0 | 0.0 | 0.0 | 15.0 | 16.6 | 7.2 | 0.6 | 1.2 |
| Tampa............................. | 0.0 | 4.6 | 2.1 | 0.0 | 0.0 | 20.9 | 25.8 | 9.6 | 2.3 | 2.0 |
| Titusville......................... | 0.0 | 10.3 | 0.0 | 0.0 | 0.0 | 21.5 | 15.5 | 2.0 | 1.6 | 0.4 |
| Wellington........................ | 0.0 | 12.4 | 0.0 | 0.0 | 0.0 | 11.2 | 9.2 | 20.1 | 0.0 | 0.1 |
| Weston............................ | 0.0 | 0.4 | 0.0 | 0.0 | 8.6 | 10.7 | 33.8 | 9.3 | 0.0 | 2.7 |
| West Palm Beach ................ | 0.0 | 6.1 | 2.1 | 0.0 | 0.6 | 21.4 | 24.5 | 7.3 | 2.5 | 2.8 |
| Winter Garden.................... | 2.1 | 9.0 | 0.0 | 0.0 | 0.0 | 18.8 | 24.9 | 4.1 | 0.0 | 1.2 |
| Winter Haven ..................... | 0.2 | 2.6 | 0.0 | 0.0 | 0.0 | 17.3 | 25.8 | 14.0 | 0.0 | 0.8 |
| Winter Park ....................... | 0.0 | 1.9 | 0.0 | 0.0 | 0.0 | 16.7 | 12.9 | 11.2 | 0.8 | 0.8 |
| Winter Springs ................... | 0.0 | 17.0 | 0.0 | 0.0 | 0.0 | 25.3 | 25.2 | 7.4 | 0.0 | 0.7 |

# Table D. Cities — City Government Finances, City Government Employment, and Climate

| City | City government finances, 2017 (cont.) Debt outstanding | | | Climate[2] Average daily temperature | | | | Annual precipitation (inches) | Heating degree days | Cooling degree days |
| | Total (mil dol) | Per capita[1] (dollars) | Debt issued during year | Mean January | Mean July | Limits January[3] | Limits July[4] | | | |
| | 137 | 138 | 139 | 140 | 141 | 142 | 143 | 144 | 145 | 146 |
|---|---|---|---|---|---|---|---|---|---|---|
| **FLORIDA— Cont'd** | | | | | | | | | | |
| Deerfield Beach | 64.4 | 800 | 0.0 | 67.2 | 83.3 | 57.8 | 91.8 | 57.27 | 219 | 4,241 |
| DeLand | 9.3 | 288 | 0.0 | NA | NA | NA | NA | NA | NA | NA |
| Delray Beach | 68.6 | 998 | 0.0 | 66.2 | 82.5 | 57.3 | 90.1 | 61.39 | 246 | 3,999 |
| Deltona | 166.8 | 1,841 | 39.2 | 57.1 | 81.2 | 44.5 | 91.2 | 57.03 | 954 | 2,819 |
| Doral | 21.5 | 356 | 0.0 | NA | NA | NA | NA | NA | NA | NA |
| Dunedin | 35.2 | 964 | 0.0 | 60.9 | 82.5 | 50.2 | 91.3 | 52.42 | 623 | 3,414 |
| Fort Lauderdale | 719.8 | 4,001 | 158.9 | 67.5 | 82.6 | 59.2 | 89.8 | 64.19 | 167 | 4,120 |
| Fort Myers | 342.2 | 4,277 | 52.1 | 64.9 | 83.0 | 54.5 | 91.7 | 54.19 | 302 | 3,957 |
| Fort Pierce | 151.0 | 3,315 | 0.0 | 62.6 | 81.7 | 50.7 | 91.5 | 53.50 | 477 | 3,430 |
| Gainesville | 1,090.1 | 8,256 | 18.6 | 54.4 | 81.1 | 41.8 | 92.4 | 49.56 | 1,249 | 2,608 |
| Greenacres | 2.4 | 60 | 0.0 | 66.2 | 82.5 | 57.3 | 90.1 | 61.39 | 246 | 3,999 |
| Hallandale Beach | 110.6 | 2,780 | 102.8 | 68.1 | 83.7 | 59.6 | 90.9 | 58.53 | 149 | 4,361 |
| Hialeah | 155.7 | 659 | 46.1 | 67.9 | 82.7 | 62.6 | 87.0 | 46.60 | 141 | 4,090 |
| Hollywood | 329.1 | 2,145 | 103.3 | 67.5 | 82.6 | 59.2 | 89.8 | 64.19 | 167 | 4,120 |
| Homestead | 27.4 | 397 | 4.6 | 67.0 | 81.8 | 56.2 | 90.6 | 55.55 | 238 | 3,923 |
| Jacksonville | 8,768.9 | 9,830 | 476.4 | 54.5 | 82.5 | 42.6 | 92.7 | 51.88 | 1,222 | 2,810 |
| Jupiter | 40.7 | 627 | 0.0 | 66.2 | 82.5 | 57.3 | 90.1 | 61.39 | 246 | 3,999 |
| Kissimmee | 305.3 | 4,273 | 216.5 | 59.7 | 81.8 | 47.7 | 91.6 | 48.01 | 694 | 3,111 |
| Lakeland | 988.0 | 9,173 | 149.3 | 62.5 | 84.0 | 51.1 | 94.6 | 49.13 | 487 | 3,886 |
| Lake Worth | 65.7 | 1,726 | 16.8 | 65.1 | 81.1 | 52.5 | 91.3 | 58.44 | 273 | 3,438 |
| Largo | 36.2 | 428 | 21.9 | 61.3 | 82.5 | 52.4 | 89.7 | 44.77 | 591 | 3,482 |
| Lauderdale Lakes | 20.5 | 566 | 0.0 | 67.2 | 83.3 | 57.8 | 91.8 | 57.27 | 219 | 4,241 |
| Lauderhill | 99.6 | 1,385 | 12.7 | 67.5 | 82.6 | 59.2 | 89.8 | 64.19 | 167 | 4,120 |
| Margate | 26.3 | 451 | 0.0 | 67.2 | 83.3 | 57.8 | 91.8 | 57.27 | 219 | 4,241 |
| Melbourne | 97.8 | 1,195 | 38.6 | 60.9 | 81.2 | 50.0 | 90.5 | 48.29 | 595 | 3,186 |
| Miami | 652.3 | 1,429 | 67.9 | 68.1 | 83.7 | 59.6 | 90.9 | 58.53 | 149 | 4,361 |
| Miami Beach | 982.5 | 10,839 | 575.8 | 68.1 | 83.7 | 59.6 | 90.9 | 58.53 | 149 | 4,361 |
| Miami Gardens | 160.8 | 1,433 | 12.4 | NA | NA | NA | NA | NA | NA | NA |
| Miami Lakes | 8.6 | 281 | 0.2 | NA | NA | NA | NA | NA | NA | NA |
| Miramar | 213.8 | 1,525 | 2.2 | 68.1 | 83.7 | 59.6 | 90.9 | 58.53 | 149 | 4,361 |
| North Lauderdale | 1.7 | 38 | 0.0 | 67.5 | 82.6 | 59.2 | 89.8 | 64.19 | 167 | 4,120 |
| North Miami | 18.7 | 302 | 0.0 | 68.1 | 83.7 | 59.6 | 90.9 | 58.53 | 149 | 4,361 |
| North Miami Beach | 99.0 | 2,310 | 6.2 | 68.1 | 83.7 | 59.6 | 90.9 | 58.53 | 149 | 4,361 |
| North Port | 70.2 | 1,059 | 0.4 | NA | NA | NA | NA | NA | NA | NA |
| Oakland Park | 56.6 | 1,261 | 0.7 | 67.5 | 82.6 | 59.2 | 89.8 | 64.19 | 167 | 4,120 |
| Ocala | 173.3 | 2,932 | 0.0 | 58.1 | 81.7 | 45.7 | 92.2 | 49.68 | 902 | 2,971 |
| Ocoee | 38.8 | 833 | 0.0 | NA | NA | NA | NA | NA | NA | NA |
| Orlando | 1,241.5 | 4,406 | 131.3 | 60.9 | 82.4 | 49.9 | 92.2 | 48.35 | 580 | 3,428 |
| Ormond Beach | 45.0 | 1,053 | 0.0 | 60.9 | 82.4 | 49.9 | 92.2 | 48.35 | 580 | 3,428 |
| Oviedo | 74.1 | 1,817 | 2.2 | 58.7 | 81.5 | 47.0 | 91.9 | 51.31 | 799 | 3,017 |
| Palm Bay | 134.5 | 1,199 | 28.6 | 60.9 | 81.2 | 50.0 | 90.5 | 48.29 | 595 | 3,186 |
| Palm Beach Gardens | 12.7 | 226 | 0.0 | 66.2 | 82.5 | 57.3 | 90.1 | 61.39 | 246 | 3,999 |
| Palm Coast | 166.2 | 1,930 | 45.9 | 57.4 | 82.8 | 46.4 | 92.0 | 49.79 | 909 | 3,193 |
| Panama City | 53.6 | 1,443 | 0.0 | 50.3 | 80.0 | 38.7 | 89.0 | 64.76 | 1,810 | 2,174 |
| Pembroke Pines | 327.5 | 1,918 | 7.6 | 67.5 | 82.6 | 59.2 | 89.8 | 64.19 | 167 | 4,120 |
| Pensacola | 203.0 | 3,865 | 26.8 | 52.0 | 82.6 | 42.7 | 90.7 | 64.28 | 1,498 | 2,650 |
| Pinellas Park | 19.2 | 362 | 0.0 | 61.7 | 83.4 | 54.0 | 90.2 | 49.58 | 548 | 3,718 |
| Plantation | 43.9 | 467 | 0.0 | 67.5 | 82.6 | 59.2 | 89.8 | 64.19 | 167 | 4,120 |
| Plant City | 36.9 | 940 | 0.0 | 61.1 | 81.5 | 49.8 | 90.8 | 51.17 | 625 | 3,261 |
| Pompano Beach | 104.8 | 950 | 0.0 | 67.2 | 83.3 | 57.8 | 91.8 | 57.27 | 219 | 4,241 |
| Port Orange | 101.9 | 1,616 | 0.0 | 57.1 | 81.2 | 44.5 | 91.2 | 57.03 | 954 | 2,819 |
| Port St. Lucie | 806.9 | 4,266 | 372.1 | 64.5 | 81.8 | 54.7 | 89.5 | 59.53 | 315 | 3,600 |
| Riviera Beach | 176.6 | 5,089 | 43.2 | 66.2 | 82.5 | 57.3 | 90.1 | 61.39 | 246 | 3,999 |
| Royal Palm Beach | 0.0 | 0 | 0.0 | NA | NA | NA | NA | NA | NA | NA |
| St. Cloud | 96.6 | 1,868 | 0.0 | NA | NA | NA | NA | NA | NA | NA |
| St. Petersburg | 653.5 | 2,489 | 139.7 | 61.7 | 83.4 | 54.0 | 90.2 | 49.58 | 548 | 3,718 |
| Sanford | 72.0 | 1,209 | 8.4 | 58.7 | 81.5 | 47.0 | 91.9 | 51.31 | 799 | 3,017 |
| Sarasota | 109.2 | 1,910 | 2.9 | 61.7 | 83.4 | 54.0 | 90.2 | 49.58 | 548 | 3,718 |
| Sunrise | 258.1 | 2,737 | 0.0 | 67.5 | 82.6 | 59.2 | 89.8 | 64.19 | 167 | 4,120 |
| Tallahassee | 1,413.0 | 7,373 | 257.0 | 51.8 | 82.4 | 39.7 | 92.0 | 63.21 | 1,604 | 2,551 |
| Tamarac | 52.3 | 791 | 0.0 | 67.2 | 83.3 | 57.8 | 91.8 | 57.27 | 219 | 4,241 |
| Tampa | 1,675.3 | 4,284 | 315.8 | 61.3 | 82.5 | 52.4 | 89.7 | 44.77 | 591 | 3,482 |
| Titusville | 37.7 | 817 | 0.4 | 59.9 | 82.4 | 49.5 | 91.4 | 52.79 | 677 | 3,300 |
| Wellington | 5.5 | 84 | 3.2 | 66.2 | 82.5 | 57.3 | 90.1 | 61.39 | 246 | 3,999 |
| Weston | 68.3 | 963 | 6.2 | 67.5 | 82.6 | 59.2 | 89.8 | 64.19 | 167 | 4,120 |
| West Palm Beach | 452.1 | 4,098 | 120.2 | 66.2 | 82.5 | 57.3 | 90.1 | 61.39 | 246 | 3,999 |
| Winter Garden | 36.2 | 828 | 19.8 | NA | NA | NA | NA | NA | NA | NA |
| Winter Haven | 73.8 | 1,807 | 0.0 | 62.3 | 82.3 | 51.0 | 92.5 | 50.22 | 538 | 3,551 |
| Winter Park | 145.9 | 4,726 | 0.0 | NA | NA | NA | NA | NA | NA | NA |
| Winter Springs | 22.1 | 602 | 0.0 | 60.9 | 82.4 | 49.9 | 92.2 | 48.35 | 580 | 3,428 |

1. Based on the population estimated as of July 1 of the year shown. 2. Represents normal values based on the 30-year period, 1971±2000. 3. Average daily minimum. 4. Average daily maximum.

# Table D. Cities — Land Area and Population

| STATE Place code | City | Land area[1] (sq. mi) | Total persons 2019 | Rank | Per square mile | White | Black or African American | American Indian, Alaskan Native | Asian | Hawaiian Pacific Islander | Some other race | Two or more races (percent) |
|---|---|---|---|---|---|---|---|---|---|---|---|---|
| | | 1 | 2 | 3 | 4 | 5 | 6 | 7 | 8 | 9 | 10 | 11 |
| 13 00000 | GEORGIA | 57,716.3 | 10,710,017 | X | 186 | 57.8 | 31.9 | 0.4 | 4.1 | 0.1 | 3.0 | 2.7 |
| 13 01052 | Albany | 55.1 | 70,839 | 513 | 1,286 | D | D | D | D | D | 0.8 | 3.9 |
| 13 01696 | Alpharetta | 26.9 | 67,897 | 543 | 2,524 | 60.4 | 14.4 | 0.0 | 20.6 | 0.0 | 1.4 | 3.1 |
| 13 03436 | Athens-Clarke County | 116.3 | 126,391 | 218 | 1,087 | 63.1 | 28.0 | 0.2 | 5.3 | 0.1 | 0.9 | 2.1 |
| 13 04000 | Atlanta | 135.3 | 512,550 | 37 | 3,788 | 42.3 | 49.1 | 0.1 | 1.8 | 0.0 | 2.6 | 3.3 |
| 13 04200 | Augusta-Richmond County | 302.3 | 197,468 | 125 | 653 | 34.7 | 57.5 | 0.5 | 2.9 | 0.1 | 3.4 | 3.9 |
| 13 19000 | Columbus | 216.5 | 196,442 | 127 | 907 | 42.6 | 46.7 | 0.7 | 3.2 | 0.0 | 4.9 | 4.8 |
| 13 21380 | Dalton | 21.1 | 33,368 | 1,151 | 1,581 | 82.6 | 3.8 | D | D | D | D | D |
| 13 23900 | Douglasville | 22.8 | 34,036 | 1,134 | 1,493 | D | D | D | D | D | 1.6 | 4.0 |
| 13 24600 | Duluth | 10.2 | 29,676 | 1,259 | 2,909 | 44.3 | 20.4 | 0.0 | 29.7 | 0.0 | 3.0 | 1.3 |
| 13 24768 | Dunwoody | 13.0 | 49,326 | 786 | 3,794 | 62.3 | 12.0 | 0.6 | 20.8 | 0.0 | 1.7 | 1.6 |
| 13 25720 | East Point | 14.7 | 35,042 | 1,104 | 2,384 | 14.4 | 81.2 | 0.8 | 0.3 | 0.0 | 1.7 | 1.6 |
| 13 31908 | Gainesville | 33.4 | 44,213 | 868 | 1,324 | 77.3 | 10.3 | 0.1 | 3.7 | 0.0 | 3.5 | 5.1 |
| 13 38964 | Hinesville | 18.3 | 34,085 | 1,132 | 1,863 | 42.8 | 42.2 | 0.3 | 2.4 | 0.0 | 3.0 | 8.9 |
| 13 42425 | Johns Creek | 30.8 | 85,192 | 398 | 2,766 | 61.0 | 11.9 | 0.1 | 21.7 | 0.0 | 1.8 | 3.5 |
| 13 43192 | Kennesaw | 9.7 | 34,712 | 1,113 | 3,579 | D | D | D | D | D | D | D |
| 13 44340 | LaGrange | 42.1 | 30,741 | 1,228 | 730 | 38.4 | 43.6 | 0.0 | 4.7 | 0.0 | 9.7 | 3.6 |
| 13 45488 | Lawrenceville | 13.6 | 30,842 | 1,226 | 2,268 | D | D | D | D | D | 15.0 | 2.7 |
| 13 49008 | Macon-Bibb County | 249.4 | 152,737 | 167 | 612 | 39.3 | 54.4 | 0.0 | 2.2 | 0.1 | 1.4 | 2.7 |
| 13 49756 | Marietta | 23.4 | 60,786 | 618 | 2,598 | 45.4 | 31.4 | 1.7 | 3.0 | 0.1 | 15.0 | 3.3 |
| 13 51670 | Milton | 38.5 | 40,037 | 961 | 1,040 | D | D | D | D | D | D | D |
| 13 55020 | Newnan | 19.4 | 42,885 | 898 | 2,211 | D | D | D | D | D | D | 3.8 |
| 13 59724 | Peachtree City | 25.1 | 36,763 | 1,046 | 1,465 | 75.5 | 11.0 | 0.0 | 7.8 | 0.0 | 1.8 | 3.8 |
| 13 66668 | Rome | 31.7 | 36,764 | 1,045 | 1,160 | 58.2 | 25.1 | 0.4 | 1.1 | 0.0 | 7.2 | 8.4 |
| 13 67284 | Roswell | 40.7 | 95,434 | 336 | 2,345 | 72.0 | 15.0 | 0.0 | 7.9 | 0.0 | 3.8 | 2.3 |
| 13 68516 | Sandy Springs | 37.7 | 109,928 | 276 | 2,916 | 63.6 | 22.4 | 0.0 | 7.9 | 0.0 | 2.0 | 1.4 |
| 13 69000 | Savannah | 106.8 | 143,632 | 187 | 1,345 | 40.5 | 52.6 | 1.4 | 3.1 | 0.0 | 2.0 | 1.6 |
| 13 71492 | Smyrna | 15.6 | 56,443 | 681 | 3,618 | 46.9 | 28.2 | 0.2 | 8.0 | 0.0 | 13.9 | 6.1 |
| 13 73256 | Statesboro | 15.0 | 33,204 | 1,156 | 2,214 | 47.6 | 41.5 | 0.2 | 1.9 | 0.0 | 2.8 | 6.1 |
| 13 73704 | Stockbridge | 13.7 | 30,125 | 1,249 | 2,199 | D | D | D | D | D | 3.2 | 0.4 |
| 13 78800 | Valdosta | 36.0 | 56,668 | 676 | 1,574 | 36.4 | 57.3 | 1.3 | 0.7 | 0.7 | 0.0 | 4.4 |
| 13 80508 | Warner Robins | 37.8 | 78,601 | 450 | 2,079 | 48.7 | 39.7 | 0.1 | 7.2 | 0.0 | 0.0 | 4.4 |
| 15 00000 | HAWAII | 6,422.4 | 1,407,006 | X | 219 | 24.1 | 1.9 | 0.4 | 38.7 | 10.8 | 1.7 | 22.4 |
| 15 06290 | East Honolulu CDP | 23.0 | NA | 1,433 | NA | 24.0 | 0.1 | 0.0 | 52.9 | 2.2 | 0.2 | 20.7 |
| 15 14650 | Hilo CDP | 53.6 | NA | 1,437 | NA | 19.9 | 0.5 | 0.0 | 28.6 | 13.3 | 0.7 | 37.0 |
| 15 22700 | Kahului CDP | 14.4 | NA | 1,436 | NA | 6.4 | 0.3 | 0.3 | 59.3 | 17.7 | 0.0 | 15.5 |
| 15 23150 | Kailua CDP (Honolulu County) | 7.8 | NA | 1,431 | NA | 48.3 | 1.2 | 0.3 | 23.3 | 4.2 | 1.5 | 21.2 |
| 15 28250 | Kaneohe CDP | 6.5 | NA | 1,430 | NA | 24.3 | 0.7 | 0.3 | 39.2 | 7.7 | 0.5 | 27.2 |
| 15 51050 | Mililani Town CDP | 4.0 | NA | 1,439 | NA | 10.9 | 1.6 | 0.2 | 51.8 | 2.0 | 0.3 | 33.2 |
| 15 62600 | Pearl City CDP | 9.1 | NA | 1,440 | NA | 12.6 | 1.9 | 0.2 | 56.0 | 3.7 | 2.9 | 22.7 |
| 15 71550 | Urban Honolulu CDP | 60.5 | 341,555 | 56 | 5,646 | 17.0 | 2.6 | 0.3 | 52.0 | 9.0 | 1.5 | 17.5 |
| 15 79700 | Waipahu CDP | 2.7 | NA | 1,434 | NA | 4.7 | 0.0 | 0.2 | 69.8 | 13.1 | 0.1 | 12.1 |
| 16 00000 | IDAHO | 82,645.1 | 1,826,913 | X | 22 | 89.4 | 0.7 | 1.3 | 1.5 | 0.1 | 3.8 | 3.3 |
| 16 08830 | Boise City | 84.0 | 229,776 | 98 | 2,735 | 89.3 | 1.0 | 0.9 | 2.5 | 0.1 | 2.5 | 3.6 |
| 16 12250 | Caldwell | 22.9 | 60,674 | 619 | 2,650 | 62.5 | 0.3 | 0.2 | 2.0 | 0.0 | 30.3 | 4.7 |
| 16 16750 | Coeur d'Alene | 16.1 | 53,354 | 725 | 3,314 | 93.0 | 0.1 | 1.4 | 1.7 | 0.0 | 0.4 | 3.4 |
| 16 39700 | Idaho Falls | 24.0 | 64,142 | 590 | 2,673 | 87.2 | 0.5 | 0.8 | 2.5 | D | 3.5 | 5.6 |
| 16 46540 | Lewiston | 17.3 | 33,039 | 1,161 | 1,910 | D | D | D | D | D | D | D |
| 16 52120 | Meridian | 35.1 | 121,182 | 234 | 3,453 | 88.6 | 0.9 | 0.9 | 1.0 | 0.1 | 10.2 | 6.0 |
| 16 56260 | Nampa | 33.5 | 103,215 | 298 | 3,081 | 80.9 | 1.2 | 2.4 | 3.2 | 0.0 | 1.2 | 3.3 |
| 16 64090 | Pocatello | 33.4 | 57,029 | 674 | 1,708 | 88.6 | 0.6 | 0.0 | 1.6 | 0.0 | 0.9 | 4.7 |
| 16 64810 | Post Falls | 15.2 | 38,933 | 996 | 2,561 | 92.2 | D | D | D | D | D | D |
| 16 67420 | Rexburg | 10.0 | 29,658 | 1,261 | 2,966 | D | D | D | D | D | D | D |
| 16 82810 | Twin Falls | 19.4 | 51,411 | 757 | 2,650 | 90.2 | 0.8 | 0.1 | 3.0 | 0.0 | 1.2 | 4.8 |
| 17 00000 | ILLINOIS | 55,513.7 | 12,587,530 | X | 227 | 71.4 | 14.1 | 0.3 | 5.7 | 0.0 | 5.8 | 2.8 |
| 17 00243 | Addison | 9.8 | 36,237 | 1,064 | 3,698 | 70.7 | 4.7 | 0.0 | 11.5 | 0.0 | 12.2 | 0.9 |
| 17 00685 | Algonquin | 12.3 | 30,730 | 1,229 | 2,498 | 88.6 | 0.1 | 0.0 | 3.7 | 0.0 | 2.8 | 4.7 |
| 17 01114 | Alton | 15.7 | 26,049 | 1,384 | 1,659 | D | D | D | D | D | D | D |
| 17 02154 | Arlington Heights | 16.6 | 73,999 | 485 | 4,458 | 80.3 | 2.7 | 0.3 | 11.5 | 0.0 | 3.1 | 2.1 |
| 17 03012 | Aurora | 45.0 | 196,383 | 128 | 4,364 | 64.9 | 10.8 | 0.6 | 9.7 | 0.1 | 8.7 | 5.2 |
| 17 04013 | Bartlett | 15.7 | 40,312 | 955 | 2,568 | 67.1 | 3.9 | 1.5 | 16.8 | 0.0 | 4.7 | 6.1 |
| 17 04078 | Batavia | 10.6 | 26,442 | 1,375 | 2,495 | D | D | D | D | D | D | D |
| 17 04845 | Belleville | 23.2 | 40,536 | 951 | 1,747 | D | D | D | D | D | D | D |
| 17 05092 | Belvidere | 12.1 | 24,731 | 1,411 | 2,044 | D | D | D | D | D | D | D |
| 17 05573 | Berwyn | 3.9 | 53,701 | 720 | 13,770 | 33.8 | 10.4 | 0.2 | 2.0 | 0.0 | 51.4 | 2.2 |
| 17 06613 | Bloomington | 27.1 | 77,132 | 458 | 2,846 | 77.3 | 11.4 | 0.2 | 7.6 | 0.1 | 0.6 | 2.8 |
| 17 07133 | Bolingbrook | 24.8 | 74,025 | 483 | 2,985 | 54.2 | 15.7 | 0.7 | 11.7 | 0.0 | 15.0 | 2.8 |
| 17 09447 | Buffalo Grove | 9.6 | 40,161 | 959 | 4,183 | 74.0 | 1.9 | 0.0 | 19.5 | 0.0 | 0.4 | 4.2 |

1. Dry land or land partially or temporarily covered by water.  2. Hispanic or Latino persons may be of any race.

# Table D. Cities — Population

| City | Percent Hispanic or Latino[1], 2019 | Percent foreign born, 2019 | Age of population (percent), 2019 | | | | | | | Median age, 2019 | Percent female, 2019 | Population | | | |
|---|---|---|---|---|---|---|---|---|---|---|---|---|---|---|---|
| | | | Under 18 years | 18 to 24 years | 25 to 34 years | 35 to 44 years | 45 to 54 years | 55 to 64 years | 65 years and over | | | Census counts | | Percent change | |
| | | | | | | | | | | | | 2000 | 2010 | 2000-2010 | 2001-2020 |
| | 12 | 13 | 14 | 15 | 16 | 17 | 18 | 19 | 20 | 21 | 22 | 23 | 24 | 25 | 26 |
| GEORGIA | 9.8 | 10.3 | 23.6 | 9.7 | 13.7 | 13.2 | 13.2 | 12.2 | 14.3 | 37.2 | 51.3 | 8,186,453 | 9,688,737 | 18.4 | 10.5 |
| Albany | 2.4 | 2.0 | 24.7 | 14.4 | 12.1 | 12.9 | 9.6 | 11.2 | 15.1 | 34.1 | 54.7 | 76,939 | 77,430 | 0.6 | -8.5 |
| Alpharetta | 10.9 | 24.4 | 22.5 | 7.1 | 11.1 | 14.1 | 18.0 | 15.4 | 11.8 | 41.6 | 50.2 | 34,854 | 57,383 | 64.6 | 18.3 |
| Athens-Clarke County | 11.0 | 10.2 | 17.0 | 27.1 | 15.6 | 9.5 | 10.7 | 8.9 | 11.2 | 28.5 | 52.2 | 100,266 | 115,370 | 15.1 | 9.6 |
| Atlanta | 4.9 | 8.4 | 17.0 | 13.4 | 22.2 | 13.3 | 11.7 | 10.1 | 12.3 | 33.3 | 52.2 | 416,474 | 427,042 | 2.5 | 20.0 |
| Augusta-Richmond County | 4.8 | 4.0 | 22.8 | 11.3 | 16.6 | 11.1 | 11.0 | 12.4 | 14.7 | 34.2 | 51.5 | 195,182 | 200,594 | 2.8 | 0.8 |
| Columbus | 7.7 | 6.3 | 24.7 | 10.2 | 15.3 | 12.9 | 11.7 | 11.5 | 13.7 | 34.9 | 51.8 | 186,291 | 190,570 | 2.3 | 3.1 |
| Dalton | 42.9 | 24.2 | 28.0 | 11.1 | 13.3 | 13.4 | 10.5 | 11.2 | 12.5 | 33.9 | 54.3 | 27,912 | 33,104 | 18.6 | 0.8 |
| Douglasville | 8.9 | 9.8 | 20.0 | 11.7 | 13.8 | 14.8 | 11.0 | 12.4 | 16.3 | 37.2 | 58.4 | 20,065 | 29,892 | 49.0 | 13.9 |
| Duluth | 6.7 | 34.5 | 18.6 | 5.7 | 16.5 | 13.7 | 18.9 | 14.0 | 12.6 | 41.9 | 51.0 | 22,122 | 26,672 | 20.6 | 11.3 |
| Dunwoody | 9.6 | 20.6 | 24.5 | 5.9 | 19.1 | 14.3 | 12.6 | 10.2 | 13.4 | 35.4 | 52.2 | 32,808 | 46,428 | 41.5 | 6.2 |
| East Point | 4.1 | 4.9 | 24.4 | 7.1 | 15.3 | 19.8 | 14.2 | 8.5 | 10.6 | 36.2 | 54.1 | 39,595 | 33,452 | -15.5 | 4.8 |
| Gainesville | 48.0 | 20.5 | 29.0 | 13.7 | 14.3 | 11.6 | 8.6 | 9.9 | 12.8 | 28.4 | 53.7 | 25,578 | 34,045 | 33.1 | 29.9 |
| Hinesville | 15.7 | 8.4 | 28.8 | 7.5 | 19.4 | 13.7 | 10.5 | 7.5 | 12.6 | 31.9 | 50.4 | 30,392 | 33,309 | 9.6 | 2.3 |
| Johns Creek | 8.4 | 27.1 | 27.4 | 7.6 | 6.5 | 14.7 | 21.1 | 13.4 | 9.3 | 41.3 | 51.6 | NA | 76,639 | NA | 11.2 |
| Kennesaw | 5.5 | 17.3 | 29.5 | 8.8 | 16.6 | 15.5 | 10.6 | 9.4 | 9.6 | 30.7 | 51.1 | 21,675 | 30,618 | 41.3 | 13.4 |
| LaGrange | 4.3 | 3.5 | 25.2 | 8.4 | 22.0 | 10.6 | 9.9 | 11.1 | 12.8 | 33.0 | 53.0 | 25,998 | 29,365 | 13.0 | 4.7 |
| Lawrenceville | 24.2 | 24.9 | 21.2 | 10.8 | 25.5 | 8.7 | 10.4 | 11.2 | 12.2 | 32.4 | 51.5 | 22,397 | 27,215 | 21.5 | 13.3 |
| Macon-Bibb County | 3.5 | 3.6 | 24.2 | 10.2 | 13.7 | 11.6 | 11.7 | 12.5 | 16.0 | 36.5 | 53.1 | NA | 155,844 | NA | -2.0 |
| Marietta | 18.2 | 15.0 | 19.3 | 10.9 | 14.3 | 17.2 | 10.4 | 11.7 | 16.2 | 38.5 | 51.0 | 58,748 | 56,369 | -4.0 | 7.8 |
| Milton | 4.5 | 15.3 | 24.8 | 2.0 | 7.3 | 18.7 | 20.4 | 16.5 | 10.3 | 42.7 | 52.5 | NA | 32,828 | NA | 22.0 |
| Newnan | 7.1 | 10.4 | 24.8 | 8.2 | 12.9 | 14.4 | 13.4 | 12.0 | 14.4 | 40.2 | 52.8 | 16,242 | 32,874 | 102.4 | 30.5 |
| Peachtree City | 3.2 | 14.6 | 25.9 | 6.4 | 5.5 | 14.0 | 15.6 | 12.6 | 20.1 | 43.4 | 52.0 | 31,580 | 34,409 | 9.0 | 6.8 |
| Rome | 17.9 | 10.6 | 20.4 | 11.2 | 10.2 | 12.8 | 16.4 | 9.9 | 19.2 | 40.0 | 50.3 | 34,980 | 36,417 | 4.1 | 1.0 |
| Roswell | 14.3 | 19.3 | 23.0 | 7.6 | 12.2 | 14.0 | 14.0 | 14.1 | 15.1 | 39.3 | 50.4 | 79,334 | 88,333 | 11.3 | 8.0 |
| Sandy Springs | 10.3 | 17.7 | 18.6 | 7.6 | 19.6 | 17.2 | 12.6 | 11.5 | 12.8 | 37.3 | 52.8 | 85,781 | 93,828 | 9.4 | 17.2 |
| Savannah | 5.4 | 5.2 | 19.0 | 14.4 | 19.4 | 9.7 | 10.4 | 12.1 | 14.9 | 33.4 | 53.4 | 131,510 | 136,919 | 4.1 | 4.9 |
| Smyrna | 19.9 | 20.4 | 19.6 | 5.4 | 23.6 | 15.0 | 14.6 | 10.2 | 11.5 | 36.3 | 51.6 | 40,999 | 50,864 | 24.1 | 11.0 |
| Statesboro | 5.6 | 4.1 | 11.0 | 49.8 | 14.6 | 4.2 | 7.0 | 6.6 | 6.8 | 22.3 | 52.5 | 22,698 | 28,370 | 25.0 | 17.0 |
| Stockbridge | 11.8 | 9.5 | 27.1 | 7.3 | 8.6 | 15.3 | 19.8 | 8.4 | 13.5 | 38.9 | 58.1 | 9,853 | 26,369 | 167.6 | 14.2 |
| Valdosta | 6.0 | 2.6 | 21.7 | 19.7 | 18.6 | 11.0 | 7.5 | 8.2 | 13.4 | 28.1 | 54.8 | 43,724 | 54,753 | 25.2 | 3.5 |
| Warner Robins | 5.5 | 9.4 | 27.3 | 10.6 | 15.3 | 13.9 | 12.4 | 8.9 | 11.7 | 32.8 | 50.3 | 48,804 | 69,859 | 43.1 | 12.5 |
| HAWAII | 10.7 | 19.3 | 21.2 | 8.4 | 13.8 | 13.1 | 11.9 | 12.6 | 19.0 | 39.6 | 50.0 | 1,211,537 | 1,360,304 | 12.3 | 3.4 |
| East Honolulu CDP | 4.9 | 13.2 | 18.1 | 7.2 | 8.7 | 9.7 | 13.9 | 14.5 | 28.0 | 51.0 | 50.6 | NA | NA | NA | NA |
| Hilo CDP | 12.8 | 7.8 | 19.5 | 7.9 | 10.7 | 11.2 | 10.6 | 15.9 | 24.2 | 45.5 | 51.4 | 40,759 | NA | NA | NA |
| Kahului CDP | 4.5 | 49.1 | 18.7 | 8.8 | 11.6 | 18.6 | 11.5 | 12.3 | 18.6 | 40.4 | 46.0 | 20,146 | NA | NA | NA |
| Kailua CDP (Honolulu County) | 7.0 | 11.0 | 18.9 | 7.3 | 10.0 | 14.0 | 10.7 | 14.4 | 24.8 | 44.9 | 50.3 | 36,513 | NA | NA | NA |
| Kaneohe CDP | 9.0 | 8.7 | 18.9 | 6.9 | 14.0 | 10.0 | 13.0 | 17.1 | 20.0 | 45.2 | 50.3 | 34,970 | NA | NA | NA |
| Mililani Town CDP | 11.5 | 10.1 | 20.2 | 5.8 | 14.5 | 13.3 | 11.4 | 14.9 | 19.9 | 42.4 | 50.2 | 28,608 | NA | NA | NA |
| Pearl City CDP | 9.7 | 16.3 | 18.6 | 8.9 | 13.7 | 11.8 | 11.0 | 11.2 | 24.6 | 41.8 | 50.9 | 30,976 | NA | NA | NA |
| Urban Honolulu CDP | 7.7 | 27.2 | 16.6 | 8.5 | 15.6 | 13.3 | 12.2 | 12.8 | 21.1 | 41.4 | 50.1 | NA | 337,721 | NA | 1.1 |
| Waipahu CDP | 5.5 | 41.6 | 21.5 | 7.7 | 11.2 | 12.0 | 15.2 | 12.0 | 20.4 | 42.2 | 50.4 | 33,108 | NA | NA | NA |
| IDAHO | 12.8 | 5.8 | 25.1 | 9.3 | 13.1 | 12.6 | 11.3 | 12.3 | 16.2 | 36.9 | 49.7 | 1,293,953 | 1,567,658 | 21.2 | 16.5 |
| Boise City | 10.4 | 6.0 | 19.4 | 11.3 | 16.2 | 13.5 | 12.2 | 12.6 | 14.8 | 37.6 | 49.0 | 185,787 | 209,382 | 12.7 | 9.7 |
| Caldwell | 45.1 | 12.9 | 31.2 | 10.1 | 14.7 | 10.5 | 13.2 | 8.4 | 11.9 | 30.8 | 49.0 | 25,967 | 46,353 | 78.5 | 30.9 |
| Coeur d'Alene | 6.3 | 2.7 | 23.0 | 9.0 | 14.5 | 13.5 | 13.0 | 12.1 | 14.8 | 37.4 | 48.1 | 34,514 | 44,161 | 28.0 | 20.8 |
| Idaho Falls | 16.0 | 6.0 | 29.9 | 8.3 | 15.0 | 12.2 | 10.6 | 11.0 | 13.1 | 33.0 | 49.9 | 50,730 | 57,833 | 14.0 | 10.9 |
| Lewiston | 4.7 | 3.7 | 21.0 | 9.7 | 13.1 | 15.1 | 11.2 | 12.7 | 17.3 | 38.6 | 53.5 | 30,904 | 31,893 | 3.2 | 3.6 |
| Meridian | 6.1 | 4.9 | 30.4 | 6.9 | 12.0 | 16.1 | 13.2 | 9.3 | 12.1 | 35.8 | 50.6 | 34,919 | 76,983 | 120.5 | 57.4 |
| Nampa | 24.5 | 7.4 | 26.3 | 11.6 | 15.4 | 13.2 | 8.7 | 9.7 | 15.1 | 33.0 | 50.2 | 51,867 | 81,892 | 57.9 | 26.0 |
| Pocatello | 9.0 | 3.0 | 25.8 | 11.5 | 15.8 | 13.2 | 9.4 | 10.4 | 13.8 | 32.8 | 47.6 | 51,466 | 54,235 | 5.4 | 5.2 |
| Post Falls | 6.0 | 1.5 | 28.1 | 8.3 | 15.5 | 12.7 | 10.5 | 11.1 | 13.8 | 33.8 | 54.0 | 17,247 | 27,785 | 61.1 | 40.1 |
| Rexburg | 8.9 | 6.6 | 18.1 | 45.7 | 19.2 | 4.0 | 4.4 | 2.6 | 6.0 | 23.6 | 47.9 | 17,257 | 25,497 | 47.7 | 16.3 |
| Twin Falls | 18.2 | 8.5 | 27.4 | 9.5 | 15.5 | 12.7 | 8.4 | 10.7 | 15.8 | 33.7 | 51.4 | 34,469 | 44,313 | 28.6 | 16.0 |
| ILLINOIS | 17.5 | 13.9 | 22.2 | 9.2 | 13.8 | 13.0 | 12.6 | 13.1 | 16.1 | 38.6 | 50.9 | 12,419,293 | 12,831,572 | 3.3 | -1.9 |
| Addison | 37.9 | 30.3 | 22.4 | 6.3 | 17.3 | 9.9 | 13.9 | 13.5 | 16.7 | 39.1 | 50.1 | 35,914 | 37,084 | 3.3 | -2.3 |
| Algonquin | 14.8 | 11.7 | 21.9 | 8.9 | 14.5 | 9.7 | 15.3 | 13.6 | 16.0 | 39.7 | 50.0 | 23,276 | 30,065 | 29.2 | 2.2 |
| Alton | 0.8 | 1.3 | 18.4 | 7.9 | 13.2 | 11.5 | 15.6 | 13.9 | 19.6 | 44.2 | 47.0 | 30,496 | 27,936 | -8.4 | -6.8 |
| Arlington Heights | 8.3 | 21.2 | 22.5 | 4.1 | 9.8 | 13.9 | 16.2 | 14.4 | 19.0 | 44.6 | 49.3 | 76,031 | 75,185 | -1.1 | -1.6 |
| Aurora | 42.6 | 26.0 | 26.8 | 9.7 | 12.7 | 17.4 | 12.8 | 10.6 | 9.9 | 35.3 | 49.9 | 142,990 | 197,964 | 38.4 | -0.8 |
| Bartlett | 10.8 | 19.9 | 23.8 | 8.2 | 9.8 | 11.6 | 15.5 | 13.4 | 17.7 | 42.8 | 53.0 | 36,706 | 41,235 | 12.3 | -2.2 |
| Batavia | 4.8 | 6.1 | 24.5 | 8.5 | 12.2 | 11.2 | 13.2 | 11.8 | 18.6 | 37.7 | 52.4 | 23,866 | 26,236 | 9.9 | 0.8 |
| Belleville | 6.9 | 3.6 | 24.8 | 11.7 | 14.9 | 13.8 | 9.8 | 12.1 | 12.8 | 33.9 | 51.6 | 41,410 | 44,239 | 6.8 | -8.4 |
| Belvidere | 36.4 | 9.7 | 25.3 | 11.1 | 15.7 | 7.8 | 15.1 | 8.8 | 16.2 | 33.5 | 52.8 | 20,820 | 25,641 | 23.2 | -3.5 |
| Berwyn | 65.7 | 27.0 | 29.2 | 8.7 | 14.3 | 15.9 | 10.4 | 13.1 | 8.3 | 33.7 | 50.0 | 54,016 | 56,650 | 4.9 | -5.2 |
| Bloomington | 6.3 | 11.3 | 23.8 | 11.1 | 14.4 | 14.0 | 12.7 | 11.1 | 12.9 | 35.4 | 51.0 | 64,808 | 76,683 | 18.3 | 0.6 |
| Bolingbrook | 28.0 | 22.9 | 23.8 | 10.6 | 12.0 | 12.6 | 16.8 | 12.9 | 11.3 | 37.9 | 50.7 | 56,321 | 73,362 | 30.3 | 0.9 |
| Buffalo Grove | 16.0 | 34.7 | 25.6 | 4.7 | 10.6 | 16.6 | 14.5 | 14.8 | 13.2 | 40.2 | 51.9 | 42,909 | 41,509 | -3.3 | -3.2 |

1. May be of any race.

# Table D. Cities — Households, Group Quarters, Crime, and Education

| City | Households, 2019 | | | | | | | Persons in group quarters, 2019 | Serious crimes known to police[1], 2019 | | | | Educational attainment, 2019 | | |
|---|---|---|---|---|---|---|---|---|---|---|---|---|---|---|---|
| | | | | Percent | | | | | Violent | | Property | | | Attainment[3] (percent) | |
| | Number | Persons per household | Family | Married couple family | Female family | Non-family | One person | | Number | Rate[2] | Number | Rate[2] | Population age 25 and over | High school graduate or less | Bachelor's degree or more |
| | 27 | 28 | 29 | 30 | 31 | 32 | 33 | 34 | 35 | 36 | 37 | 38 | 39 | 40 | 41 |
| GEORGIA | 3,852,714 | 2.69 | 66.3 | 46.8 | 19.5 | 33.7 | 28.0 | 266,840 | NA | NA | NA | NA | 7,080,222 | 39.5 | 32.5 |
| Albany | 25,640 | 2.59 | 53.9 | 26.9 | 27.0 | 46.1 | 43.6 | 4,282 | 790 | 1,053 | 3,452 | 4,603 | 43,010 | 48.8 | 17.3 |
| Alpharetta | 25,391 | 2.63 | 71.5 | 62.5 | 9.0 | 28.5 | 21.4 | 355 | 75 | 111 | 991 | 1,470 | 47,305 | 9.7 | 75.0 |
| Athens-Clarke County | 51,640 | 2.22 | 45.7 | 29.6 | 16.2 | 54.3 | 32.2 | 12,290 | NA | NA | NA | NA | 70,838 | 24.8 | 48.3 |
| Atlanta | 223,736 | 2.09 | 44.1 | 25.8 | 18.3 | 55.9 | 44.1 | 38,468 | NA | NA | NA | NA | 352,740 | 24.6 | 56.5 |
| Augusta-Richmond County | 66,838 | 2.80 | 62.1 | 32.7 | 29.4 | 37.9 | 34.5 | 10,444 | NA | NA | NA | NA | 129,927 | 47.9 | 20.2 |
| Columbus | 73,134 | 2.59 | 62.5 | 38.5 | 24.0 | 37.5 | 33.9 | 6,115 | NA | NA | NA | NA | 127,592 | 37.5 | 24.7 |
| Dalton | 11,038 | 2.91 | 66.5 | 39.9 | 26.6 | 33.5 | 26.7 | 1,596 | NA | NA | NA | NA | 20,497 | 54.5 | 25.0 |
| Douglasville | 12,237 | 2.71 | 57.5 | 39.1 | 18.5 | 42.5 | 33.9 | 775 | 182 | 526 | 2,013 | 5,816 | 23,210 | 34.3 | 42.7 |
| Duluth | 11,180 | 2.64 | 71.3 | 61.7 | 9.6 | 28.7 | 18.5 | 45 | NA | NA | NA | NA | 22,417 | 19.0 | 50.9 |
| Dunwoody | 20,881 | 2.36 | 58.8 | 50.5 | 8.3 | 41.2 | 38.6 | 30 | 64 | 128 | 1,915 | 3,840 | 34,297 | 9.5 | 72.1 |
| East Point | 16,103 | 2.13 | 48.9 | 20.3 | 28.6 | 51.1 | 44.1 | 521 | NA | NA | NA | NA | 23,866 | 32.2 | 39.4 |
| Gainesville | 13,053 | 3.23 | 69.4 | 52.0 | 17.4 | 30.6 | 28.5 | 1,068 | NA | NA | NA | NA | 24,755 | 45.2 | 31.1 |
| Hinesville | 13,808 | 2.40 | 64.7 | 46.9 | 17.7 | 35.3 | 31.9 | 202 | NA | NA | NA | NA | 21,221 | 28.0 | 21.1 |
| Johns Creek | 28,638 | 2.94 | 81.3 | 69.0 | 12.3 | 18.7 | 17.0 | 312 | NA | NA | NA | NA | 54,999 | 11.8 | 69.6 |
| Kennesaw | 12,325 | 2.76 | 64.5 | 44.3 | 20.2 | 35.5 | 27.2 | 98 | 48 | 139 | 484 | 1,397 | 21,008 | 29.6 | 41.2 |
| LaGrange | 11,692 | 2.52 | 50.0 | 29.6 | 20.5 | 50.0 | 42.3 | 788 | NA | NA | NA | NA | 20,119 | 55.9 | 18.2 |
| Lawrenceville | 11,756 | 2.60 | 72.0 | 46.6 | 25.5 | 28.0 | 25.4 | 334 | NA | NA | NA | NA | 20,987 | 37.5 | 22.3 |
| Macon-Bibb County | 56,726 | 2.60 | 59.2 | 35.7 | 23.5 | 40.8 | 35.0 | 5,867 | NA | NA | NA | NA | 100,417 | 43.7 | 24.9 |
| Marietta | 26,146 | 2.21 | 54.7 | 37.7 | 17.1 | 45.3 | 39.2 | 3,034 | 262 | 427 | 1,930 | 3,147 | 42,471 | 28.8 | 40.9 |
| Milton | 15,795 | 2.51 | 68.9 | 64.7 | 4.1 | 31.1 | 27.0 | 5 | 15 | 37 | 291 | 726 | 28,955 | 8.3 | 77.9 |
| Newnan | 16,911 | 2.43 | 59.8 | 34.1 | 25.7 | 40.2 | 37.9 | 470 | 222 | 545 | 1,090 | 2,677 | 27,883 | 30.2 | 30.7 |
| Peachtree City | 13,266 | 2.72 | 79.6 | 67.7 | 11.9 | 20.4 | 19.7 | 211 | NA | NA | NA | NA | 24,545 | 13.2 | 58.4 |
| Rome | 15,381 | 2.28 | 53.3 | 37.3 | 16.1 | 46.7 | 42.6 | 1,566 | NA | NA | NA | NA | 25,123 | 46.7 | 30.0 |
| Roswell | 35,944 | 2.62 | 71.0 | 57.8 | 13.3 | 29.0 | 24.1 | 445 | NA | NA | NA | NA | 65,754 | 16.8 | 61.3 |
| Sandy Springs | 52,820 | 2.06 | 49.0 | 38.5 | 10.4 | 51.0 | 42.3 | 526 | 143 | 129 | 1,972 | 1,780 | 80,801 | 13.4 | 70.5 |
| Savannah | 53,371 | 2.51 | 55.3 | 30.3 | 25.0 | 44.7 | 35.2 | 10,657 | NA | NA | NA | NA | 96,160 | 38.0 | 27.8 |
| Smyrna | 26,081 | 2.17 | 52.7 | 37.7 | 15.0 | 47.3 | 40.8 | 172 | NA | NA | NA | NA | 42,507 | 24.4 | 55.7 |
| Statesboro | 11,024 | 2.34 | 38.4 | 16.1 | 22.3 | 61.6 | 31.3 | 7,169 | NA | NA | NA | NA | 12,927 | 39.2 | 28.3 |
| Stockbridge | 10,903 | 2.74 | 68.8 | 38.9 | 29.9 | 31.2 | 27.1 | 9 | NA | NA | NA | NA | 19,629 | 37.5 | 22.9 |
| Valdosta | 21,403 | 2.53 | 53.5 | 29.6 | 24.0 | 46.5 | 35.0 | 2,391 | NA | NA | NA | NA | 33,097 | 45.2 | 25.1 |
| Warner Robins | 29,742 | 2.59 | 64.7 | 37.6 | 27.2 | 35.3 | 26.4 | 324 | 420 | 548 | 3,609 | 4,710 | 48,103 | 41.4 | 27.8 |
| HAWAII | 465,299 | 2.95 | 68.0 | 49.9 | 18.1 | 32.0 | 25.2 | 42,918 | NA | NA | NA | NA | 996,668 | 35.1 | 33.6 |
| East Honolulu CDP | 16,253 | 2.80 | 73.9 | 59.8 | 14.2 | 26.1 | 20.7 | 235 | NA | NA | NA | NA | 34,200 | 17.0 | 59.8 |
| Hilo CDP | 14,988 | 2.66 | 66.3 | 43.6 | 22.7 | 33.7 | 28.0 | 1,200 | NA | NA | NA | NA | 29,857 | 31.4 | 33.9 |
| Kahului CDP | 8,140 | 3.43 | 75.3 | 53.1 | 22.3 | 24.7 | 23.9 | 1,508 | NA | NA | NA | NA | 21,334 | 49.8 | 17.6 |
| Kailua CDP (Honolulu County) | 11,995 | 3.01 | 77.0 | 64.9 | 12.1 | 23.0 | 17.7 | 170 | NA | NA | NA | NA | 26,793 | 22.6 | 54.2 |
| Kaneohe CDP | 9,498 | 3.10 | 72.8 | 50.6 | 22.2 | 27.2 | 23.0 | 949 | NA | NA | NA | NA | 22,521 | 31.4 | 36.2 |
| Mililani Town CDP | 9,150 | 3.10 | 79.3 | 64.0 | 15.3 | 20.7 | 17.6 | 0 | NA | NA | NA | NA | 20,986 | 25.7 | 33.8 |
| Pearl City CDP | 13,458 | 3.17 | 74.4 | 55.3 | 19.1 | 25.6 | 18.3 | 1,501 | NA | NA | NA | NA | 32,007 | 35.0 | 31.8 |
| Urban Honolulu CDP | 131,873 | 2.52 | 58.1 | 39.4 | 18.7 | 41.9 | 34.6 | 12,441 | 2,638 | 271 | 29,263 | 3,002 | 258,567 | 32.9 | 38.6 |
| Waipahu CDP | 9,359 | 4.43 | 81.3 | 51.2 | 30.1 | 18.7 | 13.5 | 1,467 | NA | NA | NA | NA | 30,352 | 57.6 | 14.9 |
| IDAHO | 655,859 | 2.68 | 68.5 | 54.5 | 14.0 | 31.5 | 24.6 | 30,331 | NA | NA | NA | NA | 1,170,997 | 34.6 | 28.7 |
| Boise City | 95,359 | 2.36 | 57.6 | 45.3 | 12.3 | 42.4 | 30.7 | 3,488 | 649 | 281 | 3,653 | 1,579 | 158,492 | 25.5 | 42.2 |
| Caldwell | 18,642 | 3.07 | 70.8 | 52.0 | 18.7 | 29.2 | 24.1 | 1,309 | 203 | 350 | 997 | 1,721 | 34,338 | 56.3 | 11.9 |
| Coeur d'Alene | 21,533 | 2.37 | 52.3 | 43.0 | 9.3 | 47.7 | 36.4 | 1,303 | 132 | 253 | 712 | 1,363 | 35,648 | 34.2 | 26.8 |
| Idaho Falls | 22,430 | 2.76 | 64.4 | 49.0 | 15.4 | 35.6 | 30.9 | 940 | 172 | 277 | 940 | 1,514 | 38,900 | 33.5 | 32.2 |
| Lewiston | 13,290 | 2.33 | 62.4 | 47.4 | 15.0 | 37.6 | 26.0 | 710 | 42 | 128 | 955 | 2,900 | 21,994 | 37.7 | 24.9 |
| Meridian | 40,194 | 2.83 | 76.4 | 58.2 | 18.1 | 23.6 | 20.1 | 326 | 177 | 159 | 1,188 | 1,068 | 71,572 | 19.4 | 39.0 |
| Nampa | 34,798 | 2.80 | 70.7 | 45.8 | 24.8 | 29.3 | 21.9 | 1,926 | 306 | 312 | 1,767 | 1,799 | 61,623 | 41.4 | 21.2 |
| Pocatello | 20,958 | 2.62 | 60.2 | 47.8 | 12.4 | 39.8 | 29.6 | 1,811 | 233 | 412 | 1,326 | 2,346 | 35,487 | 29.1 | 34.4 |
| Post Falls | 13,119 | 2.74 | 70.0 | 48.3 | 21.7 | 30.0 | 23.2 | 259 | 68 | 191 | 585 | 1,641 | 23,058 | 25.2 | 23.3 |
| Rexburg | 8,989 | 2.73 | 70.3 | 69.2 | 1.1 | 29.7 | 13.7 | 283 | 6 | 21 | 121 | 416 | 8,980 | 11.7 | 35.8 |
| Twin Falls | 19,143 | 2.59 | 63.5 | 52.6 | 10.9 | 36.5 | 29.6 | 627 | 201 | 398 | 1,034 | 2,049 | 31,642 | 40.7 | 18.1 |
| ILLINOIS | 4,866,006 | 2.54 | 62.9 | 46.3 | 16.6 | 37.1 | 30.8 | 296,807 | NA | NA | NA | NA | 8,694,694 | 36.1 | 35.8 |
| Addison | 12,916 | 2.79 | 75.1 | 50.9 | 24.2 | 24.9 | 21.4 | 160 | 73 | 199 | 423 | 1,153 | 25,808 | 46.2 | 22.8 |
| Algonquin | 11,134 | 2.74 | 72.9 | 56.0 | 16.9 | 27.1 | 22.4 | 0 | 23 | 74 | 303 | 977 | 21,087 | 26.0 | 46.9 |
| Alton | 12,001 | 2.13 | 50.4 | 30.3 | 20.1 | 49.6 | 40.5 | 608 | 268 | 1,017 | 1,101 | 4,177 | 19,325 | 45.8 | 19.8 |
| Arlington Heights | 28,748 | 2.47 | 64.5 | 58.6 | 5.9 | 35.5 | 28.6 | 806 | 33 | 44 | 600 | 797 | 52,736 | 22.0 | 57.6 |
| Aurora | 63,972 | 3.08 | 74.4 | 53.4 | 21.0 | 25.6 | 20.1 | 1,771 | 516 | 258 | 2,014 | 1,008 | 126,144 | 41.5 | 34.3 |
| Bartlett | 14,174 | 2.85 | 76.2 | 59.6 | 16.6 | 23.8 | 22.2 | 93 | 23 | 56 | 188 | 460 | 27,552 | 21.9 | 46.6 |
| Batavia | 9,052 | 2.58 | 62.3 | 56.1 | 6.2 | 37.7 | 28.7 | 92 | 33 | 125 | 337 | 1,280 | 15,732 | 21.3 | 44.7 |
| Belleville | 17,175 | 2.31 | 54.7 | 33.9 | 20.8 | 45.3 | 39.1 | 1,199 | 227 | 555 | 1,093 | 2,670 | 25,979 | 32.1 | 23.3 |
| Belvidere | 9,343 | 2.76 | 67.3 | 36.6 | 30.7 | 32.7 | 26.8 | 275 | 64 | 255 | 258 | 1,026 | 16,576 | 65.2 | 8.7 |
| Berwyn | 17,467 | 3.10 | 75.8 | 45.4 | 30.4 | 24.2 | 20.6 | 218 | 114 | 208 | 595 | 1,088 | 33,774 | 47.6 | 28.0 |
| Bloomington | 31,564 | 2.38 | 62.0 | 46.5 | 15.5 | 38.0 | 32.8 | 2,194 | 387 | 495 | 1,096 | 1,403 | 50,316 | 26.3 | 49.1 |
| Bolingbrook | 22,890 | 3.19 | 75.6 | 54.7 | 21.0 | 24.4 | 19.2 | 387 | 123 | 163 | 650 | 862 | 48,071 | 38.0 | 34.6 |
| Buffalo Grove | 15,509 | 2.77 | 75.1 | 68.1 | 7.0 | 24.9 | 22.2 | 160 | 11 | 27 | 288 | 706 | 29,990 | 15.8 | 63.2 |

1. Data for serious crimes have not been adjusted for underreporting. This may affect comparability between geographic areas and over time.  2. Per 100,000 population estimated by the FBI.  3. Persons 25 years old and over.

# Table D. Cities — Income, Poverty, and Housing

| City | Money income, 2019 — Households — Median income | Percent with income less than $20,000 | Percent with income of $200,000 or more | Median family income | Median non-family income | Median earnings, 2019 — All persons | Men | Women | Housing units, 2019 — Total | Occupied | Percent owner occupied | Median value[1] (dollars) | Median gross rent (dollars) |
|---|---|---|---|---|---|---|---|---|---|---|---|---|---|
| | 42 | 43 | 44 | 45 | 46 | 47 | 48 | 49 | 50 | 51 | 52 | 53 | 54 |
| GEORGIA | 61,980 | 14.7 | 7.3 | 74,833 | 37,881 | 35,019 | 40,395 | 30,412 | 4,378,350 | 3,852,714 | 64.1 | 202,451 | 1,049 |
| Albany | 31,397 | 37.4 | 1.0 | 40,036 | 20,454 | 25,836 | 28,864 | 24,432 | 32,971 | 25,640 | 41.0 | 108,405 | 747 |
| Alpharetta | 136,047 | 5.3 | 28.1 | 168,507 | 77,931 | 67,147 | 101,086 | 48,056 | 27,779 | 25,391 | 71.5 | 469,094 | 1,502 |
| Athens-Clarke County | 39,451 | 26.3 | 4.7 | 66,639 | 30,040 | 20,366 | 22,638 | 17,347 | 53,163 | 51,640 | 36.1 | 195,470 | 849 |
| Atlanta | 66,657 | 18.8 | 14.9 | 95,081 | 53,912 | 47,270 | 54,951 | 41,473 | 253,597 | 223,736 | 46.6 | 359,502 | 1,257 |
| Augusta-Richmond County | 44,914 | 21.0 | 3.0 | 52,530 | 31,011 | 26,897 | 30,724 | 21,986 | 87,800 | 66,838 | 50.9 | 123,949 | 923 |
| Columbus | 46,934 | 20.6 | 5.2 | 60,420 | 33,262 | 31,061 | 36,618 | 26,027 | 85,251 | 73,134 | 49.1 | 139,420 | 885 |
| Dalton | 49,328 | 18.7 | 8.8 | 58,758 | 30,580 | 30,150 | 30,785 | 26,564 | 13,159 | 11,038 | 48.6 | 158,753 | 691 |
| Douglasville | 74,864 | 10.6 | 5.3 | 98,882 | 49,614 | 39,413 | 44,895 | 38,272 | 13,848 | 12,237 | 55.2 | 203,777 | 1,045 |
| Duluth | 85,451 | 10.1 | 13.3 | 108,017 | 61,858 | 47,346 | 51,912 | 42,441 | 11,466 | 11,180 | 61.4 | 319,626 | 1,426 |
| Dunwoody | 84,226 | 4.9 | 18.2 | 120,976 | 45,642 | 50,751 | 67,038 | 45,909 | 22,189 | 20,881 | 55.2 | 454,548 | 1,458 |
| East Point | 46,404 | 19.5 | 3.0 | 54,445 | 41,613 | 37,271 | 40,873 | 35,973 | 19,361 | 16,103 | 40.9 | 166,948 | 1,108 |
| Gainesville | 62,436 | 11.1 | 5.7 | 72,445 | 38,384 | 30,178 | 30,965 | 28,497 | 16,001 | 13,053 | 52.5 | 225,312 | 1,032 |
| Hinesville | 53,354 | 12.9 | 0.9 | 54,699 | 42,588 | 32,020 | 41,229 | 23,257 | 16,317 | 13,808 | 52.1 | 132,575 | 1,017 |
| Johns Creek | 151,773 | 3.5 | 30.8 | 159,997 | 94,368 | 74,016 | 102,817 | 54,735 | 31,296 | 28,638 | 77.1 | 437,203 | 1,755 |
| Kennesaw | 72,794 | 9.0 | 4.8 | 75,879 | 36,710 | 36,426 | 36,827 | 36,047 | 13,143 | 12,325 | 75.1 | 229,263 | 1,393 |
| LaGrange | 34,522 | 35.0 | 2.1 | 50,493 | 16,523 | 23,386 | 22,474 | 23,765 | 13,355 | 11,692 | 38.4 | 161,092 | 804 |
| Lawrenceville | 58,923 | 10.9 | 3.2 | 65,456 | 0 | 27,732 | 24,734 | 28,615 | 12,430 | 11,756 | 49.3 | 211,072 | 1,117 |
| Macon-Bibb County | 42,140 | 23.2 | 4.0 | 55,526 | 31,197 | 29,688 | 32,367 | 25,882 | 70,014 | 56,726 | 51.1 | 128,002 | 857 |
| Marietta | 61,497 | 12.7 | 7.0 | 75,824 | 42,176 | 36,727 | 37,871 | 35,279 | 28,581 | 26,146 | 43.5 | 318,301 | 1,167 |
| Milton | 119,672 | 4.0 | 31.5 | 155,515 | 88,261 | 54,593 | 76,195 | 43,590 | 15,888 | 15,795 | 80.8 | 539,232 | 1,359 |
| Newnan | 61,704 | 16.0 | 3.9 | 86,899 | 35,529 | 37,618 | 49,541 | 32,054 | 17,066 | 16,911 | 55.9 | 222,300 | 953 |
| Peachtree City | 93,271 | 6.3 | 20.7 | 119,792 | 49,034 | 54,900 | 84,727 | 32,388 | 14,252 | 13,266 | 79.5 | 362,966 | 1,382 |
| Rome | 37,299 | 25.3 | 3.9 | 59,090 | 22,663 | 24,744 | 29,447 | 23,837 | 17,376 | 15,381 | 49.1 | 141,667 | 760 |
| Roswell | 109,805 | 4.3 | 21.6 | 130,548 | 72,515 | 50,111 | 63,355 | 42,605 | 38,237 | 35,944 | 68.6 | 422,333 | 1,279 |
| Sandy Springs | 83,111 | 8.7 | 17.7 | 119,949 | 65,311 | 52,780 | 65,189 | 47,414 | 58,128 | 52,820 | 45.9 | 481,590 | 1,457 |
| Savannah | 45,533 | 23.3 | 3.7 | 57,578 | 29,977 | 25,204 | 29,323 | 22,489 | 62,968 | 53,371 | 43.6 | 164,067 | 1,016 |
| Smyrna | 82,784 | 8.0 | 14.6 | 115,225 | 67,676 | 55,038 | 52,908 | 56,046 | 27,419 | 26,081 | 52.8 | 327,368 | 1,358 |
| Statesboro | 36,909 | 26.2 | 2.0 | 56,276 | 24,920 | 11,251 | 9,842 | 13,479 | 12,520 | 11,024 | 25.0 | 121,827 | 844 |
| Stockbridge | 66,231 | 5.6 | 5.4 | 68,587 | 35,417 | 42,431 | 42,156 | 42,624 | 11,157 | 10,903 | 50.5 | 192,269 | 1,176 |
| Valdosta | 38,603 | 23.8 | 1.8 | 47,430 | 29,284 | 23,499 | 30,641 | 17,006 | 25,289 | 21,403 | 39.6 | 104,300 | 832 |
| Warner Robins | 51,289 | 17.9 | 2.1 | 56,127 | 35,289 | 32,065 | 37,588 | 25,868 | 32,536 | 29,742 | 47.1 | 135,829 | 922 |
| HAWAII | 83,102 | 10.3 | 10.9 | 96,462 | 51,343 | 40,240 | 44,411 | 36,077 | 550,328 | 465,299 | 60.2 | 669,204 | 1,651 |
| East Honolulu CDP | 143,412 | 3.5 | 27.1 | 152,308 | 82,243 | 60,464 | 61,620 | 54,224 | 18,144 | 16,253 | 81.7 | 1,001,849 | 2,478 |
| Hilo CDP | 66,813 | 11.8 | 5.6 | 81,196 | 45,268 | 35,453 | 36,173 | 34,764 | 17,465 | 14,988 | 62.7 | 348,194 | 1,153 |
| Kahului CDP | 64,480 | 12.5 | 5.2 | 90,619 | 30,580 | 31,940 | 36,553 | 30,166 | 8,606 | 8,140 | 56.6 | 652,259 | 1,424 |
| Kailua CDP (Honolulu County) | 119,647 | 7.8 | 24.0 | 138,021 | 56,837 | 51,280 | 67,443 | 39,045 | 12,789 | 11,995 | 78.4 | 989,565 | 2,870 |
| Kaneohe CDP | 111,304 | 2.7 | 16.5 | 126,902 | 71,242 | 46,748 | 50,205 | 42,349 | 10,223 | 9,498 | 75.3 | 841,545 | 2,051 |
| Mililani Town CDP | 105,281 | 2.4 | 10.7 | 117,482 | 48,929 | 45,148 | 49,022 | 39,816 | 9,473 | 9,150 | 83.3 | 698,028 | 1,940 |
| Pearl City CDP | 102,909 | 5.6 | 15.8 | 118,776 | 54,963 | 42,205 | 43,192 | 41,653 | 14,214 | 13,458 | 72.0 | 689,417 | 2,130 |
| Urban Honolulu CDP | 72,943 | 13.2 | 10.1 | 89,777 | 49,247 | 40,331 | 44,556 | 36,598 | 154,649 | 131,873 | 46.2 | 754,725 | 1,506 |
| Waipahu CDP | 73,271 | 11.4 | 8.8 | 76,250 | 23,629 | 31,792 | 32,978 | 30,378 | 9,658 | 9,359 | 55.9 | 685,809 | 1,338 |
| IDAHO | 60,999 | 12.7 | 5.0 | 72,365 | 35,157 | 31,111 | 37,742 | 24,562 | 751,113 | 655,859 | 71.6 | 255,188 | 880 |
| Boise City | 65,463 | 12.1 | 8.1 | 83,821 | 42,615 | 33,073 | 36,888 | 29,200 | 102,182 | 95,359 | 61.1 | 310,878 | 1,043 |
| Caldwell | 51,651 | 21.6 | 1.3 | 60,613 | 22,649 | 25,252 | 30,357 | 21,980 | 19,271 | 18,642 | 68.0 | 201,437 | 736 |
| Coeur d'Alene | 59,805 | 10.6 | 5.9 | 82,323 | 38,808 | 30,870 | 35,339 | 24,133 | 23,334 | 21,533 | 46.8 | 294,823 | 991 |
| Idaho Falls | 53,855 | 16.1 | 5.2 | 63,732 | 30,628 | 28,453 | 36,903 | 21,006 | 24,906 | 22,430 | 65.6 | 198,160 | 784 |
| Lewiston | 74,512 | 16.5 | 4.4 | 91,599 | 34,617 | 40,991 | 50,999 | 36,101 | 13,833 | 13,290 | 78.7 | 214,920 | 675 |
| Meridian | 75,515 | 7.2 | 5.9 | 87,477 | 46,728 | 35,927 | 42,381 | 27,261 | 41,043 | 40,194 | 78.3 | 331,568 | 1,246 |
| Nampa | 57,352 | 9.5 | 1.1 | 60,562 | 36,805 | 30,139 | 35,983 | 23,446 | 35,330 | 34,798 | 68.2 | 213,976 | 1,010 |
| Pocatello | 50,846 | 18.7 | 2.2 | 64,192 | 31,798 | 26,287 | 32,334 | 21,597 | 22,865 | 20,958 | 60.8 | 157,155 | 624 |
| Post Falls | 62,730 | 14.5 | 3.0 | 68,289 | 35,915 | 30,816 | 37,530 | 22,427 | 13,764 | 13,119 | 69.4 | 282,254 | 979 |
| Rexburg | 26,798 | 39.9 | 0.3 | 30,417 | 24,055 | 9,850 | 10,466 | 9,212 | 11,387 | 8,989 | 18.8 | 246,148 | 750 |
| Twin Falls | 58,348 | 13.7 | 4.2 | 71,334 | 30,361 | 37,369 | 46,804 | 26,973 | 20,107 | 19,143 | 63.9 | 191,061 | 782 |
| ILLINOIS | 69,187 | 13.4 | 9.0 | 87,771 | 40,860 | 39,055 | 46,333 | 31,736 | 5,388,210 | 4,866,006 | 66.0 | 209,066 | 1,020 |
| Addison | 70,038 | 4.0 | 4.5 | 72,986 | 49,477 | 37,667 | 46,841 | 31,201 | 13,214 | 12,916 | 68.6 | 261,079 | 1,135 |
| Algonquin | 97,298 | 8.3 | 15.8 | 122,058 | 50,700 | 43,635 | 60,959 | 30,621 | 11,400 | 11,134 | 85.2 | 280,768 | 1,773 |
| Alton | 43,262 | 23.0 | 2.6 | 47,827 | 31,456 | 25,043 | 32,558 | 21,905 | 13,582 | 12,001 | 63.4 | 72,335 | 738 |
| Arlington Heights | 102,628 | 9.0 | 15.5 | 126,479 | 57,394 | 60,010 | 70,869 | 49,525 | 31,424 | 28,748 | 71.3 | 374,135 | 1,489 |
| Aurora | 73,071 | 9.0 | 9.9 | 86,453 | 45,157 | 33,855 | 41,324 | 26,097 | 69,856 | 63,972 | 67.1 | 211,416 | 1,145 |
| Bartlett | 110,817 | 4.3 | 10.7 | 125,259 | 66,150 | 51,404 | 60,876 | 42,694 | 14,174 | 14,174 | 84.8 | 289,425 | 1,351 |
| Batavia | 89,310 | 11.2 | 11.2 | 121,175 | 27,474 | 36,340 | 51,646 | 28,736 | 9,647 | 9,052 | 76.2 | 308,063 | 1,311 |
| Belleville | 46,260 | 17.7 | 4.0 | 66,186 | 30,953 | 30,991 | 29,554 | 32,632 | 19,567 | 17,175 | 58.9 | 105,830 | 667 |
| Belvidere | 51,672 | 9.9 | 3.7 | 59,730 | 30,982 | 26,055 | 40,011 | 23,462 | 9,920 | 9,343 | 80.4 | 131,816 | 831 |
| Berwyn | 56,118 | 9.8 | 6.3 | 73,047 | 39,065 | 36,119 | 36,802 | 35,222 | 19,592 | 17,467 | 59.6 | 246,430 | 999 |
| Bloomington | 67,874 | 18.9 | 7.8 | 93,793 | 37,156 | 41,712 | 48,853 | 33,898 | 35,168 | 31,564 | 58.3 | 173,909 | 794 |
| Bolingbrook | 94,198 | 7.1 | 9.7 | 102,416 | 56,851 | 40,380 | 46,524 | 32,079 | 23,422 | 22,890 | 79.7 | 245,657 | 1,330 |
| Buffalo Grove | 118,074 | 4.0 | 21.3 | 134,338 | 54,781 | 51,896 | 62,089 | 40,102 | 16,267 | 15,509 | 83.1 | 329,606 | 1,716 |

1. Based on population estimated by the American Community Survey.

# Table D. Cities — Commuting, Computer Access, Migration, Labor Force, and Employment

| City | Commuting[1], 2019 — Drove alone (55) | Commuting[1], 2019 — With commutes of 30 minutes or more (56) | Computer access[2], 2019 — With a computer in the house (57) | Computer access[2], 2019 — With Internet access (58) | Migration, 2019 — Percent who lived in the same house one year ago (59) | Migration, 2019 — Percent who lived in another state or county one year ago (60) | Civilian labor force, 2020 — Total (61) | Civilian labor force, 2020 — Percent change 2019-2020 (62) | Unemployment — Total (63) | Unemployment — Rate[3] (64) | Civilian Employment[4] — Pop. age 16 and older — Number (65) | Civilian Employment[4] — Pop. age 16 and older — Percent in labor force (66) | Civilian Employment[4] — Pop. age 16 to 64 — Number (67) | Civilian Employment[4] — Pop. age 16 to 64 — Percent who worked full-year full-time (68) |
|---|---|---|---|---|---|---|---|---|---|---|---|---|---|---|
| GEORGIA | 78.4 | 43.5 | 92.6 | 84.4 | 86.0 | 7.9 | 5,072,155 | -0.7 | 330,964 | 7 | 8,394,269 | 62.6 | 6,871,077 | 53.7 |
| Albany | 75.2 | 17.6 | 82.4 | 71.8 | 73.1 | 9.2 | 30,641 | -1.9 | 2,733 | 9 | 55,083 | 55.9 | 44,422 | 44.5 |
| Alpharetta | 78.2 | 45.0 | 96.7 | 93.1 | 86.0 | 6.7 | 35,836 | -2.3 | 1,756 | 5 | 54,414 | 71.1 | 46,455 | 58.6 |
| Athens-Clarke County | 72.2 | 20.5 | 93.4 | 85.6 | 67.8 | 19.0 | 58,206 | -1.5 | 3,753 | 6 | 106,472 | 61.0 | 92,280 | 39.3 |
| Atlanta | 66.4 | 40.1 | 94.1 | 87.2 | 82.9 | 9.0 | 263,851 | 0.8 | 22,321 | 9 | 428,133 | 64.5 | 365,991 | 51.9 |
| Augusta-Richmond County | 76.5 | 22.0 | 92.0 | 80.8 | 83.2 | 9.1 | 85,112 | 0.0 | 6,463 | 8 | 157,483 | 54.6 | 128,471 | 48.3 |
| Columbus | 83.4 | 17.3 | 91.1 | 83.0 | 89.9 | 5.4 | 77,258 | -0.3 | 6,049 | 8 | 152,154 | 55.1 | 125,355 | 48.6 |
| Dalton | 74.9 | 10.1 | 83.1 | 79.0 |  |  | 13,665 | -0.4 | 1,051 | 8 | 25,950 | 62.9 | 21,747 | 60.4 |
| Douglasville | 89.2 | 60.2 | NA | NA | 80.0 | 17.2 | 16,990 | -1.9 | 1,389 | 8 | 27,910 | 67.6 | 22,360 | 60.9 |
| Duluth | 74.1 | 58.0 | 92.8 | 88.1 | 78.9 | 13.6 | 15,714 | -2.3 | 974 | 6 | 25,014 | 68.4 | 21,290 | 62.8 |
| Dunwoody | 67.5 | 38.8 | 96.7 | 95.4 | 87.7 | 7.9 | 26,392 | -4.0 | 1,233 | 5 | 38,856 | 72.5 | 32,251 | 61.4 |
| East Point | 67.4 | 57.1 | 92.6 | 83.4 | 79.4 | 8.2 | 17,249 | 3.3 | 2,273 | 13 | 26,825 | 76.3 | 23,126 | 65.9 |
| Gainesville | 75.5 | 33.7 | 94.1 | 91.2 | 82.3 | 10.6 | 20,365 | 3.3 | 1,062 | 5 | 32,323 | 61.7 | 26,778 | 54.4 |
| Hinesville | 86.0 | 25.9 | 97.0 | 94.5 | 80.6 | 14.1 | 15,194 | -0.2 | 900 | 6 | 25,000 | 49.9 | 20,813 | 48.0 |
| Johns Creek | 69.3 | 56.3 | NA | NA | 88.8 | 5.8 | 43,505 | -3.3 | 2,199 | 5 | 64,793 | 70.8 | 56,889 | 55.0 |
| Kennesaw | 84.0 | 52.5 | NA | NA | 81.3 | 13.1 | 19,386 | -3.4 | 1,093 | 6 | 24,517 | 76.0 | 21,256 | 64.7 |
| LaGrange | NA | 18.2 | 72.4 | 67.9 | 84.0 | 5.8 | 15,507 | 1.9 | 1,379 | 9 | 23,192 | 57.7 | 19,316 | 50.3 |
| Lawrenceville | 62.1 | 74.3 | 96.8 | 80.2 | NA | NA | 14,589 | 1.2 | 1,038 | 7 | 24,510 | 75.5 | 20,736 | 62.4 |
| Macon-Bibb County | 82.7 | 23.6 | 88.1 | 80.4 | 82.6 | 8.3 | 67,421 | -0.7 | 5,050 | 8 | 120,536 | 58.7 | 96,050 | 48.2 |
| Marietta | 76.9 | 46.4 | 98.1 | 88.3 | 79.5 | 8.1 | 34,556 | -2.9 | 2,035 | 6 | 50,616 | 65.8 | 40,746 | 57.1 |
| Milton | 73.6 | 43.1 | NA | NA | NA | NA | 19,556 | -3.1 | 913 | 5 | 31,313 | 75.1 | 27,254 | 58.5 |
| Newnan | 81.4 | 46.7 | 94.3 | 88.4 | 84.6 | 7.8 | 18,978 | 3.0 | 1,446 | 8 | 32,449 | 64.8 | 26,453 | 60.7 |
| Peachtree City | 69.9 | 41.8 | 96.1 | 89.0 | 90.9 | 4.6 | 17,900 | -3.6 | 672 | 4 | 28,632 | 50.9 | 21,349 | 45.1 |
| Rome | 81.1 | 29.7 | 86.7 | 81.6 | 84.5 | 5.5 | 15,219 | -1.3 | 1,145 | 8 | 30,455 | 55.0 | 23,423 | 50.2 |
| Roswell | 74.0 | 49.0 | 97.0 | 95.0 | 84.1 | 8.3 | 51,852 | -3.6 | 2,490 | 5 | 75,227 | 71.3 | 60,966 | 63.3 |
| Sandy Springs | 76.4 | 38.9 | 98.1 | 93.9 | 77.2 | 12.0 | 63,787 | -2.9 | 3,254 | 5 | 91,088 | 74.6 | 77,062 | 64.9 |
| Savannah | 68.8 | 21.2 | 91.9 | 83.3 | 78.1 | 12.1 | 67,905 | 0.8 | 6,149 | 9 | 119,317 | 63.0 | 97,735 | 51.7 |
| Smyrna | 80.9 | 44.4 | 96.4 | 93.8 | 78.7 | 11.8 | 34,199 | -2.8 | 2,054 | 6 | 46,251 | 77.3 | 39,711 | 68.3 |
| Statesboro | 79.3 | 20.0 | 96.1 | 76.3 | 56.3 | 21.4 | 14,193 | 3.2 | 1,213 | 9 | 29,596 | 61.9 | 27,369 | 25.1 |
| Stockbridge | 73.2 | 44.1 | 96.3 | 81.2 | NA | NA | 14,522 | 0.4 | 1,335 | 9 | 22,532 | 69.4 | 18,502 | 59.4 |
| Valdosta | 74.0 | 19.6 | 88.4 | 64.4 | 89.1 | 5.1 | 25,252 | 0.8 | 1,811 | 7 | 45,729 | 62.5 | 38,163 | 42.1 |
| Warner Robins | 86.5 | 29.7 | 91.5 | 86.0 | 83.6 | 7.8 | 33,751 | 1.4 | 1,949 | 6 | 58,801 | 66.2 | 49,751 | 56.2 |
| HAWAII | 69.3 | 44.0 | 93.0 | 87.6 | 87.1 | 5.1 | 648,191 | -2.5 | 75,395 | 12 | 1,147,569 | 60.9 | 878,099 | 57.4 |
| East Honolulu CDP | 64.4 | 62.9 | 96.4 | 92.3 | 90.8 | 3.3 | NA | NA | NA | NA | 38,442 | 61.7 | 25,630 | 61.5 |
| Hilo CDP | 68.2 | 17.7 | 89.1 | 84.0 | 92.4 | 2.5 | NA | NA | NA | NA | 34,398 | 55.9 | 24,459 | 45.2 |
| Kahului CDP | 81.0 | 19.0 | 80.1 | 75.6 | NA | NA | NA | NA | NA | NA | 24,103 | 66.2 | 18,639 | 62.2 |
| Kailua CDP (Honolulu County) | 74.6 | 56.2 | 95.0 | 91.3 | 85.6 | 6.9 | NA | NA | NA | NA | 29,937 | 58.5 | 20,960 | 57.8 |
| Kaneohe CDP | 67.2 | 52.1 | 95.3 | 92.4 | 93.8 | 1.9 | NA | NA | NA | NA | 25,523 | 65.4 | 19,447 | 58.3 |
| Mililani Town CDP | 83.6 | 60.8 | 97.9 | 95.7 | NA | NA | NA | NA | NA | NA | 23,162 | 65.2 | 17,512 | 59.8 |
| Pearl City CDP | 75.9 | 50.7 | 95.1 | 89.6 | 87.1 | 2.2 | NA | NA | NA | NA | 36,976 | 57.3 | 26,091 | 67.6 |
| Urban Honolulu CDP | 59.2 | 35.7 | 92.1 | 86.3 | 86.1 | 5.7 | NA | NA | NA | NA | 294,745 | 63.2 | 221,899 | 57.7 |
| Waipahu CDP | 64.9 | 59.4 | 86.4 | 72.2 | NA | NA | NA | NA | NA | NA | 35,312 | 57.6 | 26,559 | 56.9 |
| IDAHO | 78.3 | 25.5 | 94.7 | 88.4 | 83.8 | 7.6 | 892,151 | 1.2 | 47,786 | 5 | 1,386,665 | 63.3 | 1,098,048 | 50.3 |
| Boise City | 77.6 | 13.3 | 97.0 | 90.6 | 80.9 | 8.0 | 131,390 | -0.7 | 7,365 | 6 | 188,728 | 69.5 | 154,900 | 51.8 |
| Caldwell | 82.2 | 39.6 | 94.0 | 89.9 | 79.0 | 8.4 | 27,365 | 2.1 | 1,647 | 6 | 41,943 | 63.2 | 34,984 | 48.3 |
| Coeur d'Alene | 80.7 | 17.0 | 94.9 | 86.0 | 80.4 | 5.5 | 26,515 | 1.7 | 1,951 | 7 | 40,893 | 68.9 | 33,129 | 55.7 |
| Idaho Falls | 75.0 | 14.8 | 95.0 | 89.4 | 78.0 | 9.2 | 31,344 | 2.5 | 1,364 | 4 | 46,450 | 66.0 | 38,236 | 53.6 |
| Lewiston | 87.2 | 14.6 | 93.1 | 85.3 | 83.8 | 7.6 | 17,545 | -1.3 | 844 | 5 | 25,600 | 68.1 | 20,128 | 62.9 |
| Meridian | 77.9 | 31.4 | 98.3 | 94.2 | 85.0 | 6.0 | 57,811 | 5.4 | 2,979 | 5 | 83,111 | 68.9 | 69,267 | 53.1 |
| Nampa | 81.0 | 36.3 | 95.1 | 91.3 | 82.2 | 8.1 | 45,092 | 2.0 | 2,722 | 6 | 75,491 | 65.5 | 60,484 | 55.1 |
| Pocatello | 78.3 | 17.4 | 94.2 | 89.1 | 80.1 | 9.5 | 28,686 | -0.5 | 1,407 | 5 | 42,936 | 63.8 | 35,132 | 45.4 |
| Post Falls | 84.3 | 21.3 | 97.2 | 91.5 | 88.0 | 5.3 | 18,962 | 3.3 | 1,263 | 7 | 26,813 | 66.2 | 21,807 | 52.7 |
| Rexburg | 74.5 | 11.4 | 98.1 | 68.3 | 55.0 | 25.2 | 16,292 | 2.5 | 429 | 3 | 20,931 | 69.4 | 19,432 | 25.8 |
| Twin Falls | 78.1 | 16.6 | 92.9 | 85.3 | 78.2 | 9.2 | 24,128 | 0.1 | 1,386 | 6 | 36,778 | 64.1 | 28,854 | 58.1 |
| ILLINOIS | 72.4 | 45.7 | 92.2 | 85.4 | 87.9 | 4.6 | 6,249,147 | -3.1 | 591,615 | 10 | 10,184,783 | 64.9 | 8,139,422 | 54.2 |
| Addison | 83.1 | 45.1 | 92.8 | 84.7 | 92.5 | 4.0 | 18,352 | -3.9 | 1,687 | 9 | 29,314 | 65.4 | 23,274 | 61.9 |
| Algonquin | 79.3 | 58.2 | 94.9 | 90.7 | 90.4 | 2.8 | 16,619 | -3.8 | 1,429 | 9 | 24,709 | 75.8 | 19,816 | 59.8 |
| Alton | 83.3 | 34.6 | 88.6 | 79.4 | 86.5 | 7.8 | 11,358 | -1.8 | 1,206 | 11 | 21,877 | 59.6 | 16,748 | 44.3 |
| Arlington Heights | 78.7 | 45.7 | 94.1 | 88.1 | 89.3 | 3.6 | 38,578 | -4.9 | 2,889 | 8 | 57,559 | 66.5 | 43,901 | 62.4 |
| Aurora | 76.3 | 44.9 | 94.4 | 86.4 | 87.2 | 6.2 | 97,080 | -5.0 | 8,693 | 9 | 151,205 | 73.2 | 131,510 | 58.6 |
| Bartlett | 79.1 | 59.5 | 93.0 | 88.3 | 92.8 | 3.9 | 21,941 | -4.4 | 1,832 | 8 | 31,953 | 69.3 | 24,775 | 61.4 |
| Batavia | 79.3 | 37.5 | 91.1 | 84.5 | 91.0 | 5.0 | 13,268 | -4.8 | 1,002 | 8 | 18,589 | 68.8 | 14,219 | 52.0 |
| Belleville | 84.4 | 40.6 | 93.2 | 86.9 | 80.0 | 7.0 | 22,011 | -1.1 | 2,199 | 10 | 31,699 | 65.9 | 26,455 | 55.6 |
| Belvidere | 74.3 | 32.8 | 91.0 | 84.2 | NA | NA | 11,559 | -1.8 | 1,558 | 14 | 20,918 | 63.6 | 16,684 | 53.5 |
| Berwyn | 71.4 | 67.4 | 95.0 | 93.0 | 87.5 | 1.3 | 26,272 | -1.7 | 3,136 | 12 | 40,286 | 70.3 | 35,761 | 55.8 |
| Bloomington | 83.5 | 14.9 | 93.1 | 82.5 | 84.7 | 7.8 | 38,673 | -3.1 | 2,902 | 8 | 60,955 | 60.8 | 50,981 | 51.6 |
| Bolingbrook | 80.1 | 51.6 | 95.4 | 90.2 | 90.0 | 4.7 | 38,969 | -3.6 | 3,781 | 10 | 57,647 | 71.2 | 49,386 | 61.7 |
| Buffalo Grove | 77.4 | 55.7 | 97.8 | 96.4 | 93.8 | 3.1 | 23,522 | -4.3 | 1,715 | 7 | 32,926 | 76.4 | 27,253 | 61.0 |

1. Employed persons.  2. Households.  3. Percent of civilian labor force.  4. Persons 16 years old and over.

| City | Value of residential construction authorized by building permits, 2020 | | | Wholesale trade[1], 2017 | | | | Retail trade[2], 2017 | | | |
|---|---|---|---|---|---|---|---|---|---|---|---|
| | New construction ($1,000) | Number of housing units | Percent single family | Number of establishments | Number of employees | Sales (mil dol) | Annual payroll (mil dol) | Number of establishments | Number of employees | Sales (mil dol) | Annual payroll (mil dol) |
| | 69 | 70 | 71 | 72 | 73 | 74 | 75 | 76 | 77 | 78 | 79 |
| GEORGIA | 7,980 | 38 | 100 | 10,832 | 165,088 | 188,899 | 10,594 | 34,100 | 485,505 | 148,625 | 12,561 |
| Albany | 25,013 | 165 | 10 | 103 | 1,804 | 990 | 85 | 407 | 5,724 | 1,503 | 130 |
| Alpharetta | 96,236 | 367 | 62 | 173 | 3,956 | 7,862 | 392 | 406 | 7,575 | 2,103 | 201 |
| Athens-Clarke County | 99,497 | 635 | 46 | 94 | 2,042 | 2,980 | 96 | 513 | 8,212 | 1,982 | 196 |
| Atlanta | 236,998 | 1,674 | 22 | 671 | 11,052 | 16,286 | 798 | 1,991 | 28,468 | 7,416 | 782 |
| Augusta-Richmond County | NA | NA | NA | D | D | D | D | 752 | 10,886 | 2,800 | 262 |
| Columbus | 155,057 | 1,342 | 67 | 152 | 1,978 | 1,683 | 109 | 794 | 11,547 | 2,979 | 269 |
| Dalton | NA | NA | NA | 124 | 2,737 | 814 | 115 | 230 | 2,748 | 780 | 66 |
| Douglasville | 115,038 | 395 | 39 | 26 | 879 | 1,693 | 76 | 243 | 4,655 | 1,232 | 104 |
| Duluth | 25,350 | 79 | 66 | 172 | 3,427 | 3,026 | 290 | 191 | 3,017 | 1,131 | 98 |
| Dunwoody | 21,592 | 53 | 100 | 53 | 1,559 | 7,628 | 140 | 242 | 5,137 | 991 | 124 |
| East Point | 10,843 | 112 | 100 | 27 | 768 | 897 | 49 | 108 | 1,686 | 400 | 36 |
| Gainesville | 62,269 | 445 | 28 | 97 | 1,874 | 5,940 | 104 | 306 | 5,089 | 1,587 | 153 |
| Hinesville | 33,994 | 138 | 100 | 6 | 28 | 15 | 1 | 123 | 1,765 | 473 | 44 |
| Johns Creek | 61,523 | 143 | 100 | 73 | 395 | 804 | 38 | 160 | 2,089 | 478 | 46 |
| Kennesaw | 13,778 | 56 | 82 | 93 | 1,646 | 1,691 | 114 | 250 | 4,989 | 1,187 | 116 |
| LaGrange | 75,121 | 488 | 16 | 33 | 387 | 261 | 18 | 172 | 2,453 | 651 | 64 |
| Lawrenceville | 17,325 | 114 | 100 | 105 | 3,086 | 1,982 | 283 | 292 | 4,128 | 1,163 | 113 |
| Macon-Bibb County | 27,542 | 157 | 100 | 178 | 2,913 | 1,944 | 149 | D | D | D | D |
| Marietta | 115,138 | 465 | 95 | 237 | 3,882 | 2,644 | 238 | 428 | 5,858 | 2,100 | 193 |
| Milton | 87,054 | 119 | 100 | 31 | 150 | 180 | 14 | 70 | 892 | 199 | 20 |
| Newnan | 51,860 | 190 | 85 | 37 | 513 | 668 | 30 | 220 | 4,095 | 1,128 | 95 |
| Peachtree City | 69,282 | 233 | 93 | 79 | 1,377 | 1,204 | 104 | 170 | 3,335 | 825 | 75 |
| Rome | NA | NA | NA | 48 | 899 | 617 | 46 | 295 | 3,723 | 1,056 | 95 |
| Roswell | 100,717 | 265 | 100 | 169 | 2,445 | 2,483 | 208 | 340 | 5,875 | 2,680 | 204 |
| Sandy Springs | 81,959 | 152 | 91 | 153 | 3,193 | 4,651 | 365 | 271 | 4,407 | 1,527 | 146 |
| Savannah | 73,330 | 349 | 100 | 173 | 2,139 | 2,080 | 121 | 853 | 12,466 | 3,130 | 309 |
| Smyrna | 51,570 | 220 | 100 | 75 | 2,883 | 3,780 | 314 | 206 | 3,557 | 1,605 | 114 |
| Statesboro | 8,484 | 136 | 100 | 20 | 130 | 160 | 7 | 180 | 2,793 | 725 | 65 |
| Stockbridge | 32,364 | 135 | 100 | 7 | 58 | 32 | 5 | 97 | 1,738 | 442 | 42 |
| Valdosta | 103,267 | 496 | 100 | 71 | 820 | 742 | 40 | 387 | 5,792 | 1,613 | 135 |
| Warner Robins | 84,204 | 492 | 81 | 20 | 210 | 88 | 9 | 314 | 4,805 | 1,236 | 116 |
| HAWAII | 337,836 | 870 | 100 | 1,423 | 16,829 | 11,343 | 859 | 4,644 | 72,908 | 21,659 | 2,185 |
| East Honolulu CDP | NA | NA | NA | 29 | 61 | 20 | 3 | 48 | 894 | 358 | 30 |
| Hilo CDP | NA | NA | NA | 61 | 800 | 386 | 35 | 213 | 4,395 | 1,294 | 126 |
| Kahului CDP | NA | NA | NA | 57 | 836 | 631 | 43 | 191 | 4,485 | 1,527 | 148 |
| Kailua CDP (Honolulu County) | NA | NA | NA | 19 | 57 | 23 | 3 | 96 | 1,571 | 407 | 46 |
| Kaneohe CDP | NA | NA | NA | 15 | 78 | 17 | 3 | 104 | 1,743 | 563 | 55 |
| Mililani Town CDP | NA | NA | NA | 5 | 10 | 2 | 0 | 32 | 1,062 | 255 | 27 |
| Pearl City CDP | NA | NA | NA | 24 | 372 | 290 | 21 | 67 | 2,136 | 721 | 58 |
| Urban Honolulu CDP | NA | NA | NA | 674 | 8,497 | 6,625 | 455 | 1,763 | 26,810 | 8,040 | 826 |
| Waipahu CDP | NA | NA | NA | 42 | 624 | 278 | 33 | 98 | 1,585 | 505 | 52 |
| IDAHO | 0 | 0 | 0 | 1,870 | 23,878 | 23,737 | 1,260 | 6,133 | 82,312 | 24,936 | 2,312 |
| Boise City | 275,522 | 1,125 | 67 | 383 | 5,346 | 6,486 | 339 | 920 | 15,803 | 5,043 | 498 |
| Caldwell | 134,338 | 995 | 91 | 34 | 388 | 188 | 19 | 110 | 1,626 | 552 | 53 |
| Coeur d'Alene | 93,740 | 495 | 60 | 45 | 485 | 462 | 23 | 289 | 4,457 | 1,428 | 128 |
| Idaho Falls | 54,683 | 324 | 100 | 114 | 1,243 | 2,099 | 61 | 353 | 5,664 | 1,651 | 145 |
| Lewiston | 14,334 | 68 | 100 | 43 | 489 | 433 | 21 | 194 | 2,478 | 703 | 69 |
| Meridian | 540,458 | 2,530 | 76 | 104 | 2,882 | 2,544 | 168 | 310 | 5,788 | 1,535 | 160 |
| Nampa | 227,263 | 1,738 | 74 | 89 | 1,098 | 1,044 | 55 | 310 | 5,758 | 2,074 | 168 |
| Pocatello | 37,677 | 555 | 17 | 67 | 704 | 542 | 30 | 226 | 3,127 | 944 | 82 |
| Post Falls | 190,024 | 1,364 | 34 | 24 | 365 | 257 | 24 | 109 | 2,133 | 816 | 69 |
| Rexburg | 72,393 | 396 | 29 | 25 | 456 | 252 | 18 | 167 | 1,590 | 448 | 41 |
| Twin Falls | 103,530 | 681 | 77 | 84 | 904 | 540 | 42 | 307 | 4,694 | 1,424 | 127 |
| ILLINOIS | 965 | 6 | 100 | 15,409 | 272,391 | 311,141 | 19,072 | 38,189 | 629,878 | 173,474 | 16,247 |
| Addison | 7,436 | 27 | 89 | 193 | 4,874 | 2,721 | 346 | 108 | 1,738 | 847 | 71 |
| Algonquin | 22,305 | 119 | 24 | 17 | 108 | 280 | 7 | 151 | 3,470 | 717 | 70 |
| Alton | 0 | 0 | 0 | 21 | 192 | 123 | 10 | 141 | 2,175 | 421 | 47 |
| Arlington Heights | 19,445 | 42 | 100 | 155 | 2,334 | 2,838 | 205 | 219 | 4,147 | 1,107 | 112 |
| Aurora | 29,941 | 93 | 98 | 164 | 4,319 | 21,448 | 299 | 568 | 9,559 | 2,173 | 213 |
| Bartlett | 5,038 | 13 | 100 | 62 | 1,409 | 1,245 | 88 | 45 | 383 | 103 | 9 |
| Batavia | 16,489 | 77 | 31 | 70 | 1,431 | 1,568 | 83 | 91 | 1,890 | 431 | 44 |
| Belleville | 8,748 | 43 | 100 | 41 | 517 | 258 | 27 | 178 | 2,788 | 695 | 72 |
| Belvidere | 707 | 6 | 100 | 11 | 89 | 41 | 3 | 72 | 1,083 | 337 | 28 |
| Berwyn | 728 | 5 | 100 | 10 | 67 | 43 | 3 | 102 | 1,273 | 259 | 29 |
| Bloomington | 22,459 | 148 | 43 | 72 | 1,180 | 6,396 | 101 | 323 | 5,767 | 1,311 | 126 |
| Bolingbrook | 10,668 | 61 | 100 | 117 | 5,995 | 5,967 | 422 | 216 | 5,336 | 1,315 | 134 |
| Buffalo Grove | 12,489 | 39 | 100 | 125 | 2,881 | 3,499 | 222 | 101 | 1,282 | 480 | 47 |

1. Merchant wholesalers except manufacturers' sales branches and offices.   2. Establishments with payroll.

| City | Real estate and rental and leasing, 2017 | | | | Professional, scientific, and technical services[1], 2017 | | | | Manufacturing, 2017 | | | |
|---|---|---|---|---|---|---|---|---|---|---|---|---|
| | Number of establish-ments | Number of employees | Receipts (mil dol) | Annual payroll (mil dol) | Number of establish-ments | Number of employees | Receipts (mil dol) | Annual payroll (mil dol) | Number of establish-ments | Number of employees | Receipts (mil dol) | Annual payroll (mil dol) |
| | 80 | 81 | 82 | 83 | 84 | 85 | 86 | 87 | 88 | 89 | 90 | 91 |
| GEORGIA | 12,426 | 63,935 | 23,009 | 3,795 | 29,737 | 266,523 | 55,149 | 20,189 | 7,510 | 368,836 | 169,059 | 19,072 |
| Albany | 109 | 451 | 103 | 16 | 191 | 1,391 | 186 | 71 | NA | NA | NA | NA |
| Alpharetta | 226 | 1,549 | 967 | 114 | 907 | 15,625 | 4,113 | 1,456 | NA | NA | NA | NA |
| Athens-Clarke County | 202 | 1,392 | 243 | 54 | 321 | 1,530 | 194 | 72 | NA | NA | NA | NA |
| Atlanta | 1,461 | 13,207 | 6,525 | 1,012 | 3,987 | 68,768 | 18,728 | 6,781 | NA | NA | NA | NA |
| Augusta-Richmond County | D | D | D | D | 465 | 4,947 | 878 | 283 | NA | NA | NA | NA |
| Columbus | 242 | 1,314 | 343 | 55 | 330 | 2,613 | 312 | 122 | NA | NA | NA | NA |
| Dalton | 32 | 115 | 27 | 4 | 136 | 1,108 | 134 | 66 | NA | NA | NA | NA |
| Douglasville | 53 | 171 | 63 | 7 | 107 | 552 | 67 | 26 | NA | NA | NA | NA |
| Duluth | 111 | 487 | 223 | 28 | 305 | 4,196 | 816 | 306 | NA | NA | NA | NA |
| Dunwoody | 167 | 699 | 270 | 53 | 523 | 7,645 | 1,287 | 589 | NA | NA | NA | NA |
| East Point | 38 | 322 | 63 | 11 | 55 | 466 | 38 | 16 | NA | NA | NA | NA |
| Gainesville | 88 | 261 | 76 | 11 | 177 | 846 | 133 | 43 | NA | NA | NA | NA |
| Hinesville | 41 | 136 | 26 | 4 | 33 | 220 | 23 | 8 | NA | NA | NA | NA |
| Johns Creek | 129 | 291 | 768 | 30 | 638 | 2,987 | 740 | 202 | NA | NA | NA | NA |
| Kennesaw | 65 | 471 | 133 | 18 | 138 | 975 | 154 | 50 | NA | NA | NA | NA |
| LaGrange | 52 | 152 | 45 | 6 | 69 | 307 | 42 | 17 | NA | NA | NA | NA |
| Lawrenceville | 68 | 399 | 127 | 21 | 219 | 1,194 | 163 | 63 | NA | NA | NA | NA |
| Macon-Bibb County | 203 | 1,050 | 209 | 40 | 356 | 2,611 | 415 | 141 | NA | NA | NA | NA |
| Marietta | 190 | 958 | 327 | 57 | 537 | 5,093 | 1,259 | 319 | NA | NA | NA | NA |
| Milton | 56 | 132 | 41 | 8 | 244 | 1,356 | 314 | 97 | NA | NA | NA | NA |
| Newnan | 81 | 212 | 61 | 9 | 103 | 781 | 114 | 44 | NA | NA | NA | NA |
| Peachtree City | 101 | 225 | 64 | 12 | D | D | D | D | NA | NA | NA | NA |
| Rome | 58 | 207 | 44 | 8 | 139 | 706 | 109 | 27 | NA | NA | NA | NA |
| Roswell | 227 | 1,071 | 260 | 55 | 764 | 5,455 | 1,198 | 455 | NA | NA | NA | NA |
| Sandy Springs | 393 | 2,815 | 845 | 153 | 1,016 | 15,440 | 3,777 | 1,410 | NA | NA | NA | NA |
| Savannah | 273 | 1,551 | 344 | 56 | 523 | 4,117 | 562 | 217 | NA | NA | NA | NA |
| Smyrna | 126 | 665 | 139 | 31 | 292 | 3,710 | 741 | 313 | NA | NA | NA | NA |
| Statesboro | 55 | 253 | 43 | 8 | 91 | 468 | 53 | 20 | NA | NA | NA | NA |
| Stockbridge | 34 | 141 | 33 | 6 | 61 | 324 | 86 | 17 | NA | NA | NA | NA |
| Valdosta | 109 | 552 | 108 | 19 | 176 | 903 | 146 | 49 | NA | NA | NA | NA |
| Warner Robins | 75 | 318 | 86 | 11 | 195 | 2,855 | 455 | 179 | NA | NA | NA | NA |
| HAWAII | 2,069 | 13,101 | 4,409 | 675 | 3,380 | 22,668 | 3,799 | 1,466 | 783 | 11,850 | 6,056 | 559 |
| East Honolulu CDP | 56 | 122 | 28 | 6 | 101 | 300 | 44 | 17 | NA | NA | NA | NA |
| Hilo CDP | 87 | 337 | 72 | 12 | 98 | 602 | 86 | 34 | NA | NA | NA | NA |
| Kahului CDP | 47 | 379 | 93 | 16 | 47 | 259 | 30 | 11 | NA | NA | NA | NA |
| Kailua CDP (Honolulu County) | 55 | 173 | 46 | 11 | 101 | 313 | 48 | 18 | NA | NA | NA | NA |
| Kaneohe CDP | 31 | 145 | 27 | 5 | 51 | 190 | 33 | 12 | NA | NA | NA | NA |
| Mililani Town CDP | 10 | 36 | 12 | 2 | 24 | 70 | 5 | 2 | NA | NA | NA | NA |
| Pearl City CDP | 21 | 95 | 14 | 4 | 45 | 313 | 76 | 19 | NA | NA | NA | NA |
| Urban Honolulu CDP | 858 | 6,154 | 1,903 | 338 | 1,725 | 13,894 | 2,558 | 980 | NA | NA | NA | NA |
| Waipahu CDP | 29 | 152 | 29 | 6 | 17 | 549 | 36 | 18 | NA | NA | NA | NA |
| IDAHO | 2,644 | 7,521 | 1,635 | 277 | 4,686 | 33,246 | 5,046 | 1,921 | 1,877 | 58,746 | 20,263 | 3,486 |
| Boise City | 555 | 2,156 | 476 | 85 | 1,298 | 9,518 | 1,498 | 577 | NA | NA | NA | NA |
| Caldwell | 41 | 215 | 32 | 6 | 43 | 212 | 26 | 8 | NA | NA | NA | NA |
| Coeur d'Alene | 153 | 456 | 128 | 18 | 265 | 1,505 | 182 | 83 | NA | NA | NA | NA |
| Idaho Falls | 105 | 366 | 108 | 15 | 305 | 7,969 | 1,447 | 578 | NA | NA | NA | NA |
| Lewiston | D | D | D | D | 71 | 518 | 51 | 21 | NA | NA | NA | NA |
| Meridian | 177 | 527 | 165 | 25 | 268 | 3,002 | 522 | 182 | NA | NA | NA | NA |
| Nampa | 87 | 230 | 47 | 9 | 152 | 885 | 106 | 41 | NA | NA | NA | NA |
| Pocatello | 78 | D | 45 | D | D | D | D | D | NA | NA | NA | NA |
| Post Falls | 54 | 129 | 23 | 5 | 65 | 549 | 136 | 30 | NA | NA | NA | NA |
| Rexburg | 45 | 230 | 28 | 4 | D | D | D | D | NA | NA | NA | NA |
| Twin Falls | 104 | 278 | 59 | 9 | D | D | D | D | NA | NA | NA | NA |
| ILLINOIS | 13,589 | 82,763 | 32,355 | 4,825 | 38,805 | 404,450 | 87,119 | 33,951 | 13,162 | 527,862 | 241,484 | 30,116 |
| Addison | 48 | 167 | 37 | 7 | 90 | 480 | 102 | 28 | NA | NA | NA | NA |
| Algonquin | 30 | 55 | 14 | 3 | 97 | 258 | 35 | 12 | NA | NA | NA | NA |
| Alton | 32 | 126 | 26 | 5 | 57 | 303 | 40 | 18 | NA | NA | NA | NA |
| Arlington Heights | 118 | 3,547 | 322 | 93 | 480 | 3,446 | 420 | 303 | NA | NA | NA | NA |
| Aurora | 140 | 544 | 160 | 21 | 494 | 2,379 | 440 | 147 | NA | NA | NA | NA |
| Bartlett | 26 | 161 | 43 | 11 | 144 | 444 | 111 | 28 | NA | NA | NA | NA |
| Batavia | 26 | 77 | 20 | 4 | 139 | 555 | 116 | 36 | NA | NA | NA | NA |
| Belleville | 50 | 233 | 37 | 8 | 150 | 1,025 | 151 | 68 | NA | NA | NA | NA |
| Belvidere | 14 | 45 | 8 | 1 | 28 | 240 | 50 | 25 | NA | NA | NA | NA |
| Berwyn | 30 | 60 | 15 | 2 | 80 | 278 | 49 | 14 | NA | NA | NA | NA |
| Bloomington | 106 | 455 | 99 | 16 | 232 | 1,584 | 247 | 102 | NA | NA | NA | NA |
| Bolingbrook | 48 | 229 | 65 | 9 | 190 | 1,455 | 169 | 66 | NA | NA | NA | NA |
| Buffalo Grove | 57 | 269 | 88 | 18 | 308 | 2,571 | 469 | 215 | NA | NA | NA | NA |

1. Establishments subject to federal tax.

## Table D. Cities — Accommodation and Food Services, Arts, Entertainment, and Recreation, and Health Care and Social Assistance

| City | Accommodation and food services, 2017 | | | | Arts, entertainment, and recreation[1], 2017 | | | | Health care and social assistance[1], 2017 | | | |
|---|---|---|---|---|---|---|---|---|---|---|---|---|
| | Number of establishments | Number of employees | Receipts (mil dol) | Annual payroll (mil dol) | Number of establishments | Number of employees | Receipts (mil dol) | Annual payroll (mil dol) | Number of establishments | Number of employees | Receipts (mil dol) | Annual payroll (mil dol) |
| | 92 | 93 | 94 | 95 | 96 | 97 | 98 | 99 | 100 | 101 | 102 | 103 |
| GEORGIA | 21,201 | 426,884 | 26,010 | 7,000 | 3,452 | 50,796 | 5,245 | 1,752 | 25,260 | 521,146 | 68,760 | 25,772 |
| Albany | D | D | D | D | 23 | 258 | 15 | 5 | 295 | 8,386 | 1,149 | 454 |
| Alpharetta | 350 | 7,192 | 468 | 129 | 53 | 1,536 | 159 | 29 | 478 | 5,378 | 741 | 287 |
| Athens-Clarke County | 360 | 7,878 | 396 | 110 | 66 | 959 | 51 | 15 | 456 | 10,263 | 1,667 | 621 |
| Atlanta | 1,896 | 54,165 | 4,271 | 1,202 | 469 | 9,271 | 1,414 | 407 | 1,798 | 43,378 | 8,249 | 2,531 |
| Augusta-Richmond County | 461 | 10,227 | 558 | 148 | 53 | 1,298 | 180 | 40 | 605 | 24,637 | 3,882 | 1,370 |
| Columbus | 474 | 10,818 | 575 | 163 | 50 | 1,047 | 56 | 18 | 666 | 15,924 | 1,829 | 702 |
| Dalton | 134 | 2,741 | 170 | 45 | 15 | 275 | 13 | 5 | 164 | D | 639 | D |
| Douglasville | 152 | 3,226 | 173 | 47 | 16 | D | 17 | D | 146 | 2,754 | 421 | 145 |
| Duluth | 186 | 2,349 | 137 | 34 | 27 | 310 | 28 | 12 | D | D | D | D |
| Dunwoody | 149 | 3,804 | 235 | 74 | 25 | 527 | 33 | 14 | 230 | 3,485 | 336 | 122 |
| East Point | 98 | 2,035 | 153 | 40 | 13 | 12 | 30 | 2 | 90 | 1,815 | 233 | 80 |
| Gainesville | 157 | 3,290 | 181 | 51 | 24 | D | 20 | D | 309 | 11,054 | 1,540 | 597 |
| Hinesville | 91 | 1,757 | 88 | 21 | 3 | 15 | 1 | 0 | 69 | 1,137 | 106 | 45 |
| Johns Creek | 162 | 2,402 | 133 | 38 | 43 | 1,131 | 82 | 32 | 216 | 2,864 | 419 | 150 |
| Kennesaw | 168 | 3,962 | 189 | 58 | 24 | 467 | 21 | 5 | 111 | 1,687 | 142 | 65 |
| LaGrange | 90 | 1,742 | 94 | 25 | D | D | D | D | D | D | D | D |
| Lawrenceville | 130 | 2,173 | 123 | 33 | 24 | 492 | 31 | 7 | 281 | 8,469 | 1,282 | 483 |
| Macon-Bibb County | D | D | D | D | 39 | 630 | 31 | 9 | 619 | 16,894 | 2,123 | 756 |
| Marietta | 269 | 4,571 | 283 | 80 | 49 | 929 | 112 | 28 | 434 | 12,131 | 2,325 | 852 |
| Milton | 58 | 925 | 51 | 14 | 18 | 225 | 17 | 5 | 45 | 412 | 50 | 20 |
| Newnan | 118 | 3,262 | 167 | 49 | 20 | 251 | 14 | 3 | 124 | 2,881 | 696 | 150 |
| Peachtree City | 120 | 2,937 | 168 | 51 | 35 | 488 | 72 | 13 | 146 | 1,473 | 161 | 57 |
| Rome | 156 | 3,639 | 195 | 55 | 20 | 177 | 10 | 3 | 268 | 7,906 | 1,104 | 379 |
| Roswell | 276 | 5,642 | 320 | 95 | 70 | 634 | 45 | 15 | 432 | 6,894 | 757 | 299 |
| Sandy Springs | 297 | 5,436 | 382 | 105 | 60 | 1,153 | 101 | 29 | 689 | 22,581 | 3,658 | 1,411 |
| Savannah | 638 | 14,866 | 1,078 | 272 | 75 | 872 | 57 | 17 | 580 | 19,038 | 2,348 | 899 |
| Smyrna | 188 | 3,573 | 229 | 63 | D | D | D | D | 233 | 2,981 | 303 | 120 |
| Statesboro | 137 | 3,679 | 163 | 44 | D | D | D | D | 167 | 3,121 | 329 | 119 |
| Stockbridge | 71 | 1,258 | 73 | 20 | 10 | D | 11 | D | 151 | 2,390 | 286 | 104 |
| Valdosta | 204 | 4,588 | 241 | 64 | 17 | 213 | 8 | 2 | 362 | 7,379 | 856 | 310 |
| Warner Robins | 201 | 4,703 | 221 | 61 | D | D | D | D | 227 | 5,750 | 624 | 222 |
| HAWAII | 3,865 | 112,743 | 12,102 | 3,297 | 510 | 11,912 | 1,026 | 314 | 3,677 | 73,551 | 9,786 | 3,998 |
| East Honolulu CDP | 66 | 1,189 | 74 | 22 | 23 | 526 | 45 | 15 | 82 | 990 | 106 | 39 |
| Hilo CDP | 153 | 2,711 | 191 | 50 | 12 | 132 | 8 | 3 | 233 | 4,621 | 473 | 248 |
| Kahului CDP | D | D | D | D | D | D | D | D | 97 | 1,757 | 187 | 83 |
| Kailua CDP (Honolulu County) | 106 | 1,784 | 112 | 32 | 15 | 205 | 16 | 5 | 147 | 2,022 | 316 | 127 |
| Kaneohe CDP | 87 | 1,483 | 97 | 25 | D | D | D | D | 109 | 2,232 | 204 | 97 |
| Mililani Town CDP | 45 | 1,002 | 66 | 17 | 5 | 238 | 15 | 4 | 30 | 306 | 32 | 16 |
| Pearl City CDP | 79 | 1,608 | 101 | 28 | D | D | D | D | 71 | 909 | 94 | 39 |
| Urban Honolulu CDP | 1,575 | 45,827 | 5,302 | 1,318 | 124 | 2,727 | 241 | 73 | 1,529 | 34,983 | 5,293 | 2,084 |
| Waipahu CDP | 63 | 1,183 | 82 | 22 | NA | NA | NA | NA | 92 | 1,074 | 120 | 46 |
| IDAHO | 3,856 | 65,463 | 3,598 | 1,006 | 804 | 9,644 | 602 | 174 | 5,310 | 98,100 | 10,469 | 4,292 |
| Boise City | 663 | 13,262 | 724 | 217 | 115 | 2,511 | 115 | 36 | 979 | 26,996 | 3,415 | 1,484 |
| Caldwell | 68 | 1,358 | 67 | 17 | 12 | 220 | 8 | 3 | 126 | 1,977 | 233 | 81 |
| Coeur d'Alene | 210 | 4,714 | 261 | 82 | 45 | 286 | 27 | 7 | 293 | 7,752 | 826 | 389 |
| Idaho Falls | 184 | 3,892 | 188 | 58 | 25 | 571 | 15 | 5 | 455 | 7,585 | 998 | 328 |
| Lewiston | D | D | D | D | D | D | D | D | 139 | 3,288 | 309 | 124 |
| Meridian | 220 | 5,201 | 266 | 77 | 42 | 610 | 33 | 10 | 355 | 6,486 | 638 | 248 |
| Nampa | 180 | 3,676 | 186 | 52 | 21 | 555 | 12 | 4 | 219 | 4,737 | 371 | 176 |
| Pocatello | 146 | 2,608 | 134 | 35 | 21 | 296 | 9 | 3 | 314 | 5,003 | 586 | 199 |
| Post Falls | D | D | D | D | 12 | 94 | 6 | 2 | 123 | 1,982 | 184 | 77 |
| Rexburg | 60 | 1,141 | 48 | 12 | 8 | 84 | 3 | 1 | D | D | D | D |
| Twin Falls | 165 | 3,406 | 167 | 47 | 32 | 234 | 12 | 3 | 321 | 5,857 | 636 | 225 |
| ILLINOIS | 29,025 | 536,245 | 35,314 | 10,189 | 5,218 | 91,647 | 10,067 | 2,830 | 34,235 | 817,733 | 97,626 | 38,089 |
| Addison | 87 | 1,503 | 91 | 26 | 10 | 429 | 32 | 12 | 80 | 1,252 | 105 | 57 |
| Algonquin | 93 | 1,984 | 104 | 34 | 17 | 407 | 20 | 6 | 102 | 799 | 75 | 33 |
| Alton | 96 | 1,734 | 87 | 26 | D | D | D | D | 125 | 4,265 | 420 | 173 |
| Arlington Heights | 176 | 3,295 | 211 | 59 | 42 | 594 | 44 | 15 | 414 | 10,686 | 1,348 | 562 |
| Aurora | 287 | 4,738 | 282 | 76 | 57 | 1,398 | 172 | 36 | 390 | 9,379 | 1,366 | 526 |
| Bartlett | 42 | 807 | 41 | 12 | 7 | 44 | 4 | 2 | 63 | 787 | 81 | 34 |
| Batavia | 62 | 1,197 | 66 | 19 | 10 | 145 | 6 | 2 | 69 | 928 | 85 | 32 |
| Belleville | D | D | D | D | 14 | 412 | 11 | 4 | 190 | 5,448 | 670 | 281 |
| Belvidere | 40 | 578 | 32 | 8 | D | D | D | D | 46 | 800 | 73 | 28 |
| Berwyn | 111 | 1,977 | 114 | 34 | 11 | 74 | 3 | 1 | 146 | 4,445 | 452 | 209 |
| Bloomington | 248 | 4,975 | 239 | 74 | 38 | 1,163 | 47 | 14 | 243 | 5,071 | 637 | 257 |
| Bolingbrook | 150 | 3,362 | 195 | 56 | 17 | 363 | 25 | 10 | 175 | 2,527 | 266 | 103 |
| Buffalo Grove | D | D | D | D | 25 | 459 | 26 | 7 | 172 | 1,810 | 211 | 92 |

1. Establishments subject to federal tax.

| City | Other services[1] — Number of establishments (104) | Number of employees (105) | Receipts (mil dol) (106) | Annual payroll (mil dol) (107) | Government employment and payroll, 2017 — Full-time equivalent employees (108) | March payroll — Total (dollars) (109) | Percent of total for: Administrative, judicial, and legal (110) | Police and corrections (111) | Fire protection (112) | Highways and transportation (113) | Health and welfare (114) | Natural resources and utilities (115) | Education and libraries (116) |
|---|---|---|---|---|---|---|---|---|---|---|---|---|---|
| GEORGIA | 14,976 | 97,721 | 14,343 | 3,348 | X | X | X | X | X | X | X | X | X |
| Albany | 123 | 851 | 77 | 24 | 1,093 | 4,003,560 | 13.8 | 19.3 | 17.5 | 7.0 | 1.1 | 33.3 | 0.0 |
| Alpharetta | 179 | 1,439 | 1,864 | 58 | 429 | 2,208,992 | 16.6 | 30.6 | 22.6 | 10.4 | 2.7 | 7.5 | 0.0 |
| Athens-Clarke County | 196 | 1,403 | 276 | 48 | 1,814 | 7,225,052 | 17.1 | 30.3 | 11.4 | 6.0 | 0.5 | 21.1 | 2.5 |
| Atlanta | 1,345 | 15,249 | 3,022 | 623 | 8,317 | 39,867,373 | 17.0 | 31.7 | 11.9 | 11.0 | 1.8 | 25.6 | 0.0 |
| Augusta-Richmond County | 287 | 1,789 | 219 | 59 | 2,658 | 9,166,187 | 21.2 | 32.9 | 13.3 | 7.7 | 2.3 | 17.6 | 2.5 |
| Columbus | D | D | D | 62 | 3,208 | 11,274,331 | 11.9 | 33.3 | 13.6 | 6.3 | 3.3 | 19.7 | 2.6 |
| Dalton | 76 | 624 | 90 | 24 | 606 | 2,675,375 | 3.3 | 15.1 | 13.5 | 10.4 | 0.4 | 28.8 | 0.0 |
| Douglasville | D | D | D | D | 245 | 935,953 | 16.1 | 54.4 | 0.0 | 7.1 | 0.0 | 15.9 | 0.0 |
| Duluth | 162 | 852 | 90 | 25 | 156 | 752,142 | 21.0 | 55.6 | 0.0 | 0.0 | 5.4 | 18.0 | 0.0 |
| Dunwoody | 108 | 557 | 70 | 23 | 87 | 436,150 | 14.3 | 85.7 | 0.0 | 0.0 | 0.0 | 0.0 | 0.0 |
| East Point | D | D | D | 12 | 477 | 1,946,198 | 19.1 | 37.4 | 16.3 | 2.8 | 0.0 | 24.4 | 0.0 |
| Gainesville | 125 | 613 | 73 | 17 | 662 | 2,501,132 | 10.0 | 16.9 | 17.1 | 7.8 | 1.7 | 42.3 | 0.0 |
| Hinesville | 47 | 355 | 26 | 7 | 198 | 737,647 | 9.3 | 50.7 | 24.2 | 0.0 | 3.7 | 2.3 | 0.0 |
| Johns Creek | D | D | D | 27 | 9 | 49,581 | 86.8 | 13.2 | 0.0 | 0.0 | 0.0 | 0.0 | 0.0 |
| Kennesaw | 121 | 1,099 | 146 | 54 | 204 | 787,343 | 15.1 | 36.8 | 0.0 | 6.2 | 0.0 | 25.0 | 0.0 |
| LaGrange | 62 | 286 | 36 | 9 | 402 | 1,914,641 | 4.8 | 28.3 | 13.5 | 4.2 | 0.5 | 28.0 | 0.0 |
| Lawrenceville | 153 | 1,135 | 159 | 52 | 296 | 1,211,161 | 17.4 | 38.5 | 0.0 | 6.0 | 0.0 | 24.9 | 0.0 |
| Macon-Bibb County | D | D | D | D | 1,723 | 6,541,305 | 22.3 | 35.2 | 21.2 | 3.4 | 0.6 | 5.7 | 2.3 |
| Marietta | 248 | 1,352 | 165 | 50 | 691 | 3,014,616 | 14.4 | 24.7 | 22.2 | 4.6 | 1.0 | 23.3 | 0.0 |
| Milton | 55 | 362 | 23 | 8 | 124 | 630,689 | 0.0 | 29.3 | 47.9 | 8.2 | 0.0 | 2.1 | 0.0 |
| Newnan | D | D | D | 11 | 357 | 1,574,402 | 7.1 | 24.9 | 14.7 | 5.0 | 0.5 | 21.7 | 0.8 |
| Peachtree City | 107 | 968 | 70 | 25 | 276 | 1,143,737 | 9.7 | 26.0 | 28.3 | 11.0 | 0.0 | 4.5 | 3.9 |
| Rome | D | D | D | D | 600 | 2,163,223 | 8.1 | 18.6 | 28.7 | 12.7 | 1.2 | 26.4 | 0.0 |
| Roswell | D | D | D | 56 | 765 | 3,218,687 | 15.4 | 31.3 | 13.3 | 8.0 | 0.0 | 29.5 | 0.0 |
| Sandy Springs | 251 | 1,914 | 275 | 80 | 296 | 1,627,788 | 9.1 | 55.2 | 32.9 | 0.0 | 0.0 | 2.8 | 0.1 |
| Savannah | 287 | 1,976 | 228 | 67 | 2,671 | 10,853,045 | 7.1 | 32.7 | 14.8 | 9.4 | 3.1 | 24.2 | 0.1 |
| Smyrna | 131 | 598 | 78 | 22 | 412 | 1,616,640 | 9.7 | 36.1 | 22.3 | 5.4 | 2.9 | 14.9 | 2.8 |
| Statesboro | D | D | D | D | 274 | 940,840 | 12.4 | 28.4 | 12.4 | 12.1 | 0.0 | 32.0 | 0.0 |
| Stockbridge | 56 | 251 | 33 | 9 | 75 | 274,091 | 32.7 | 0.0 | 0.0 | 21.7 | 0.0 | 33.2 | 0.0 |
| Valdosta | 99 | 458 | 42 | 11 | 564 | 2,010,839 | 9.1 | 33.7 | 18.6 | 5.9 | 4.1 | 26.6 | 0.0 |
| Warner Robins | 112 | 658 | 65 | 18 | 576 | 1,911,076 | 8.8 | 29.3 | 21.8 | 7.2 | 1.2 | 24.8 | 0.0 |
| HAWAII | 2,912 | 20,219 | 2,561 | 652 | X | X | X | X | X | X | X | X | X |
| East Honolulu CDP | 60 | 290 | 40 | 9 | NA | NA | NA | NA | NA | NA | NA | NA | NA |
| Hilo CDP | 89 | 392 | 77 | 15 | NA | NA | NA | NA | NA | NA | NA | NA | NA |
| Kahului CDP | 56 | 659 | 64 | 21 | NA | NA | NA | NA | NA | NA | NA | NA | NA |
| Kailua CDP (Honolulu County) | 74 | 518 | 42 | 11 | NA | NA | NA | NA | NA | NA | NA | NA | NA |
| Kaneohe CDP | 73 | 325 | 38 | 11 | NA | NA | NA | NA | NA | NA | NA | NA | NA |
| Mililani Town CDP | 32 | 486 | 25 | 8 | NA | NA | NA | NA | NA | NA | NA | NA | NA |
| Pearl City CDP | 69 | 329 | 47 | 10 | NA | NA | NA | NA | NA | NA | NA | NA | NA |
| Urban Honolulu CDP | 1,333 | 11,192 | 1,493 | 375 | NA | NA | NA | NA | NA | NA | NA | NA | NA |
| Waipahu CDP | 72 | 459 | 42 | 11 | NA | NA | NA | NA | NA | NA | NA | NA | NA |
| IDAHO | 2,749 | 14,016 | 1,574 | 435 | X | X | X | X | X | X | X | X | X |
| Boise City | 531 | 3,282 | 427 | 116 | 1,769 | 9,161,227 | 20.4 | 25.7 | 23.8 | 5.2 | 0.7 | 16.1 | 3.6 |
| Caldwell | 52 | 315 | 29 | 10 | 249 | 1,042,008 | 10.2 | 31.6 | 22.9 | 12.0 | 0.0 | 15.7 | 3.4 |
| Coeur d'Alene | 110 | 653 | 66 | 19 | 354 | 1,949,936 | 16.9 | 30.0 | 21.8 | 7.8 | 0.2 | 19.1 | 3.3 |
| Idaho Falls | 122 | 556 | 64 | 16 | 754 | 3,562,474 | 11.0 | 18.6 | 21.6 | 8.2 | 0.0 | 34.2 | 2.7 |
| Lewiston | D | D | D | D | 314 | 1,471,501 | 14.1 | 22.5 | 25.9 | 7.3 | 0.0 | 16.0 | 3.2 |
| Meridian | 134 | 709 | 72 | 25 | 411 | 2,074,414 | 17.0 | 34.0 | 23.0 | 8.2 | 0.0 | 17.5 | 0.0 |
| Nampa | 124 | 748 | 70 | 21 | 593 | 2,877,951 | 5.7 | 38.0 | 19.0 | 3.3 | 1.5 | 14.5 | 2.5 |
| Pocatello | D | D | 59 | D | 578 | 2,751,585 | 7.3 | 28.4 | 18.9 | 12.9 | 1.5 | 21.7 | 2.5 |
| Post Falls | 59 | 334 | 32 | 10 | 189 | 811,810 | 16.7 | 41.2 | 0.0 | 10.0 | 1.2 | 23.4 | 0.0 |
| Rexburg | 27 | 144 | 10 | 3 | 141 | 586,398 | 21.7 | 31.4 | 3.2 | 11.0 | 4.4 | 16.7 | 0.0 |
| Twin Falls | 115 | 802 | 71 | 23 | 297 | 1,445,319 | 12.5 | 33.2 | 15.8 | 13.0 | 3.3 | 12.8 | 4.2 |
| ILLINOIS | 24,296 | 171,261 | 29,599 | 7,164 | X | X | X | X | X | X | X | X | X |
| Addison | 133 | 892 | 122 | 33 | 288 | 1,938,386 | 9.8 | 48.7 | 0.0 | 3.4 | 8.0 | 18.2 | 6.8 |
| Algonquin | 70 | 324 | 31 | 10 | 147 | 845,145 | 11.3 | 26.1 | 0.0 | 11.5 | 0.3 | 33.9 | 0.0 |
| Alton | 55 | 279 | 24 | 7 | 212 | 1,055,343 | 10.1 | 42.8 | 26.7 | 4.4 | 2.6 | 12.1 | 0.0 |
| Arlington Heights | 201 | 1,262 | 194 | 51 | 570 | 4,103,045 | 12.1 | 28.1 | 24.2 | 9.9 | 2.6 | 7.2 | 15.3 |
| Aurora | 236 | 1,387 | 164 | 39 | 1,439 | 9,828,813 | 9.4 | 32.8 | 18.5 | 7.5 | 8.1 | 14.4 | 5.4 |
| Bartlett | D | D | D | D | 169 | 1,154,138 | 11.7 | 47.0 | 0.0 | 11.2 | 4.4 | 18.2 | 0.0 |
| Batavia | 77 | 642 | 77 | 24 | 174 | 666,982 | 8.5 | 27.3 | 17.7 | 8.7 | 6.8 | 30.9 | 0.0 |
| Belleville | 100 | 629 | 56 | 18 | 327 | 1,705,639 | 5.8 | 36.5 | 25.1 | 6.5 | 0.0 | 16.1 | 4.0 |
| Belvidere | 44 | 210 | 22 | 6 | 136 | 768,644 | 12.9 | 38.7 | 22.9 | 7.6 | 0.0 | 10.9 | 4.1 |
| Berwyn | 71 | 425 | 33 | 9 | 410 | 2,424,924 | 7.2 | 43.9 | 29.7 | 5.6 | 1.1 | 2.4 | 4.9 |
| Bloomington | 176 | 1,706 | 185 | 61 | 789 | 4,402,455 | 5.9 | 27.1 | 19.2 | 17.7 | 0.1 | 20.9 | 4.8 |
| Bolingbrook | 99 | 530 | 52 | 16 | 327 | 2,409,853 | 8.3 | 43.5 | 29.9 | 8.9 | 0.4 | 9.4 | 0.0 |
| Buffalo Grove | 83 | 504 | 62 | 17 | 221 | 1,816,394 | 10.0 | 37.7 | 29.5 | 2.8 | 0.4 | 10.4 | 0.0 |

1. Establishments subject to federal tax.

# Table D. Cities — City Government Finances

| City | City government finances, 2017 | | | | | | | | | |
|---|---|---|---|---|---|---|---|---|---|---|
| | General revenue | | | | | | | General expenditure | | |
| | Intergovernmental | | | Taxes | | | | | | |
| | | | | | Per capita[1] (dollars) | | | | Per capita[1] (dollars) | |
| | Total (mil dol) | Total (mil dol) | Percent from state government | Total (mil dol) | Total | Property | Sales and gross receipts | Total (mil dol) | Total | Capital outlays |
| | 117 | 118 | 119 | 120 | 121 | 122 | 123 | 124 | 125 | 126 |
| GEORGIA | X | X | X | X | X | X | X | X | X | X |
| Albany | 112.3 | 27.8 | 20.8 | 28.0 | 382 | 211 | 172 | 118 | 1,614 | 5 |
| Alpharetta | 116.0 | 31.6 | 21.4 | 52.0 | 790 | 411 | 373 | 115 | 1,749 | 695 |
| Athens-Clarke County | 227.6 | 61.2 | 4.7 | 80.3 | 634 | 424 | 207 | 216 | 1,704 | 241 |
| Atlanta | 1,874.4 | 322.1 | 4.6 | 639.3 | 1,300 | 728 | 572 | 1,608 | 3,270 | 839 |
| Augusta-Richmond County | 393.7 | 123.9 | 3.7 | 123.4 | 627 | 358 | 268 | 326 | 1,654 | 217 |
| Columbus | 331.5 | 90.7 | 8.2 | 141.2 | 728 | 426 | 298 | 329 | 1,698 | 185 |
| Dalton | 88.1 | 17.4 | 1.8 | 14.6 | 434 | 263 | 168 | 76 | 2,252 | 290 |
| Douglasville | 34.3 | 7.6 | 11.4 | 18.4 | 557 | 253 | 268 | 27 | 828 | 27 |
| Duluth | 27.1 | 7.8 | 5.3 | 14.9 | 505 | 277 | 226 | 22 | 741 | 90 |
| Dunwoody | 38.0 | 8.7 | 6.0 | 21.5 | 434 | 137 | 290 | 52 | 1,048 | 363 |
| East Point | 56.9 | 14.2 | 1.4 | 22.4 | 640 | 372 | 268 | 41 | 1,174 | 0 |
| Gainesville | 116.3 | 17.4 | 4.9 | 28.8 | 714 | 396 | 315 | 82 | 2,032 | 241 |
| Hinesville | 30.2 | 4.3 | 10.0 | 13.0 | 395 | 240 | 153 | 30 | 906 | 26 |
| Johns Creek | 57.3 | 22.1 | 9.5 | 31.0 | 368 | 195 | 171 | 79 | 939 | 427 |
| Kennesaw | 33.5 | 7.6 | 3.1 | 17.1 | 499 | 308 | 189 | 39 | 1,151 | 479 |
| LaGrange | 49.5 | 11.9 | 3.3 | 6.2 | 202 | 13 | 184 | 52 | 1,714 | 110 |
| Lawrenceville | 21.7 | 6.1 | 9.8 | 5.7 | 191 | 90 | 93 | 43 | 1,455 | 393 |
| Macon-Bibb County | 136.5 | 57.0 | 3.3 | 39.8 | 261 | 162 | 99 | 138 | 906 | 70 |
| Marietta | 104.6 | 19.1 | 0.6 | 38.9 | 642 | 285 | 349 | 104 | 1,717 | 485 |
| Milton | 31.8 | 11.3 | 3.4 | 17.3 | 445 | 287 | 157 | 33 | 850 | 302 |
| Newnan | 42.7 | 15.2 | 2.3 | 15.1 | 389 | 148 | 227 | 38 | 985 | 156 |
| Peachtree City | 57.1 | 7.7 | 5.1 | 22.9 | 650 | 401 | 246 | 40 | 1,122 | 82 |
| Rome | 73.4 | 20.2 | 3.4 | 23.1 | 635 | 296 | 284 | 92 | 2,541 | 342 |
| Roswell | 102.6 | 28.0 | 4.6 | 44.4 | 469 | 283 | 184 | 131 | 1,385 | 191 |
| Sandy Springs | 117.1 | 31.0 | 4.3 | 70.4 | 660 | 307 | 353 | 208 | 1,951 | 1,186 |
| Savannah | 375.3 | 86.7 | 1.1 | 122.2 | 838 | 451 | 379 | 378 | 2,594 | 304 |
| Smyrna | 72.5 | 12.1 | 4.2 | 34.1 | 606 | 391 | 213 | 50 | 889 | 2 |
| Statesboro | 35.0 | 6.1 | 6.2 | 10.7 | 341 | 145 | 192 | 30 | 948 | 53 |
| Stockbridge | 16.2 | 6.7 | 7.2 | 4.9 | 167 | 4 | 160 | 11 | 375 | 24 |
| Valdosta | 75.0 | 24.2 | 6.6 | 25.8 | 461 | 221 | 236 | 67 | 1,199 | 106 |
| Warner Robins | 68.2 | 8.0 | 12.1 | 31.9 | 423 | 233 | 188 | 70 | 932 | 87 |
| HAWAII | X | X | X | X | X | X | X | X | X | X |
| East Honolulu CDP | NA | NA | NA | NA | NA | NA | NA | NA | NA | NA |
| Hilo CDP | NA | NA | NA | NA | NA | NA | NA | NA | NA | NA |
| Kahului CDP | NA | NA | NA | NA | NA | NA | NA | NA | NA | NA |
| Kailua CDP (Honolulu County) | NA | NA | NA | NA | NA | NA | NA | NA | NA | NA |
| Kaneohe CDP | NA | NA | NA | NA | NA | NA | NA | NA | NA | NA |
| Mililani Town CDP | NA | NA | NA | NA | NA | NA | NA | NA | NA | NA |
| Pearl City CDP | NA | NA | NA | NA | NA | NA | NA | NA | NA | NA |
| Urban Honolulu CDP | NA | NA | NA | NA | NA | NA | NA | NA | NA | NA |
| Waipahu CDP | NA | NA | NA | NA | NA | NA | NA | NA | NA | NA |
| IDAHO | X | X | X | X | X | X | X | X | X | X |
| Boise City | 349.8 | 37.5 | 63.7 | 145.7 | 640 | 564 | 76 | 321 | 1,410 | 270 |
| Caldwell | 48.2 | 7.2 | 57.6 | 17.6 | 322 | 282 | 39 | 46 | 838 | 154 |
| Coeur d'Alene | 59.1 | 10.6 | 56.7 | 26.8 | 529 | 410 | 119 | 59 | 1,162 | 208 |
| Idaho Falls | 87.3 | 18.6 | 87.5 | 33.3 | 538 | 509 | 30 | 79 | 1,280 | 271 |
| Lewiston | 55.4 | 11.6 | 100.0 | 21.2 | 649 | 596 | 53 | 51 | 1,565 | 84 |
| Meridian | 63.0 | 7.8 | 99.7 | 34.5 | 343 | 274 | 69 | 67 | 665 | 177 |
| Nampa | 104.6 | 12.6 | 95.8 | 47.6 | 508 | 461 | 47 | 89 | 950 | 93 |
| Pocatello | 72.2 | 16.3 | 60.3 | 31.9 | 577 | 538 | 39 | 60 | 1,085 | 85 |
| Post Falls | 29.7 | 6.2 | 100.0 | 11.0 | 332 | 332 | 0 | 21 | 617 | 37 |
| Rexburg | 22.7 | 4.8 | 99.7 | 4.1 | 144 | 126 | 18 | 24 | 830 | 103 |
| Twin Falls | 60.0 | 12.5 | 47.2 | 31.1 | 631 | 551 | 81 | 45 | 922 | 176 |
| ILLINOIS | X | X | X | X | X | X | X | X | X | X |
| Addison | 45.9 | 15.4 | 92.6 | 17.2 | 465 | 379 | 86 | 65 | 1,756 | 476 |
| Algonquin | 27.7 | 12.9 | 94.2 | 12.9 | 418 | 203 | 215 | 36 | 1,165 | 353 |
| Alton | 54.8 | 21.1 | 68.7 | 16.9 | 632 | 371 | 261 | 70 | 2,627 | 1,106 |
| Arlington Heights | 183.9 | 38.4 | 97.1 | 131.1 | 1,737 | 1,229 | 507 | 170 | 2,257 | 174 |
| Aurora | 231.6 | 70.3 | 91.8 | 140.7 | 702 | 422 | 280 | 237 | 1,181 | 81 |
| Bartlett | 38.1 | 8.1 | 93.7 | 20.0 | 486 | 354 | 132 | 59 | 1,441 | 374 |
| Batavia | 35.0 | 10.7 | 94.0 | 16.6 | 626 | 284 | 343 | 35 | 1,302 | 226 |
| Belleville | 59.5 | 21.2 | 100.0 | 24.8 | 595 | 473 | 122 | 78 | 1,865 | 675 |
| Belvidere | 21.4 | 7.3 | 100.0 | 9.0 | 358 | 228 | 130 | 24 | 946 | 104 |
| Berwyn | 72.7 | 16.4 | 86.9 | 45.7 | 825 | 601 | 224 | 76 | 1,373 | 57 |
| Bloomington | 136.1 | 29.6 | 98.3 | 70.9 | 907 | 307 | 600 | 131 | 1,681 | 115 |
| Bolingbrook | 105.9 | 26.6 | 90.5 | 54.0 | 720 | 271 | 450 | 104 | 1,392 | 106 |
| Buffalo Grove | 53.0 | 11.7 | 100.0 | 25.7 | 624 | 377 | 247 | 55 | 1,323 | 138 |

1. Based on population estimated as of July 1 of the year shown.

| City | City government finances, 2017 (cont.) General expenditure (cont.) Percent of total for: | | | | | | | | | |
|---|---|---|---|---|---|---|---|---|---|---|
| | Public welfare | Highways | Parking facilities | Education | Health and hospitals | Police protection | Sewerage and sanitation | Parks and recreation | Housing and community development | Interest on debt |
| | 127 | 128 | 129 | 130 | 131 | 132 | 133 | 134 | 135 | 136 |
| GEORGIA ..................... | X | X | X | X | X | X | X | X | X | X |
| Albany.............................. | 0.0 | 5.9 | 0.0 | 0.0 | 0.0 | 14.0 | 42.5 | 4.1 | 0.8 | 1.5 |
| Alpharetta........................ | 0.0 | 19.5 | 0.0 | 0.0 | 0.0 | 17.4 | 2.8 | 11.2 | 0.0 | 3.7 |
| Athens-Clarke County.......... | 0.3 | 5.9 | 0.0 | 0.0 | 0.9 | 14.5 | 10.4 | 5.5 | 0.0 | 7.6 |
| Atlanta............................ | 1.2 | 6.3 | 0.0 | 0.0 | 0.1 | 11.9 | 3.4 | 3.4 | 0.5 | 10.9 |
| Augusta-Richmond County ... | 0.6 | 12.3 | 0.0 | 0.0 | 1.4 | 13.2 | 24.6 | 6.1 | 0.5 | 0.9 |
| Columbus......................... | 0.2 | 10.9 | 0.0 | 0.0 | 9.7 | 15.4 | 8.9 | 5.2 | 1.3 | 1.9 |
| Dalton............................. | 0.0 | 10.6 | 0.0 | 0.0 | 0.0 | 11.8 | 29.6 | 9.2 | 0.0 | 0.8 |
| Douglasville..................... | 0.0 | 5.5 | 0.0 | 0.0 | 0.0 | 35.7 | 7.4 | 13.9 | 0.0 | 4.7 |
| Duluth............................. | 0.0 | 13.9 | 0.0 | 0.0 | 0.0 | 36.4 | 0.0 | 12.0 | 0.0 | 0.9 |
| Dunwoody......................... | 0.0 | 19.7 | 0.0 | 0.0 | 0.0 | 15.2 | 4.2 | 7.8 | 0.0 | 0.2 |
| East Point........................ | 0.0 | 3.4 | 0.0 | 0.0 | 0.0 | 23.3 | 11.1 | 2.8 | 0.0 | 12.0 |
| Gainesville ...................... | 0.1 | 6.6 | 0.0 | 0.0 | 0.0 | 11.0 | 26.1 | 11.3 | 0.0 | 8.7 |
| Hinesville........................ | 0.9 | 5.5 | 0.0 | 0.0 | 0.0 | 22.2 | 28.9 | 1.5 | 2.4 | 2.6 |
| Johns Creek...................... | 0.0 | 26.8 | 0.0 | 0.0 | 0.0 | 11.9 | 0.0 | 30.7 | 0.0 | 0.2 |
| Kennesaw......................... | 0.0 | 11.8 | 0.0 | 0.0 | 0.0 | 17.1 | 4.2 | 7.0 | 15.8 | 3.3 |
| LaGrange......................... | 0.0 | 8.4 | 0.0 | 0.0 | 0.7 | 20.0 | 30.3 | 3.8 | 0.0 | 1.5 |
| Lawrenceville ................... | 0.0 | 6.7 | 0.0 | 0.0 | 0.0 | 24.0 | 11.6 | 0.8 | 1.9 | 0.0 |
| Macon-Bibb County ............ | 0.0 | 3.0 | 0.1 | 0.0 | 0.0 | 19.4 | 7.5 | 1.4 | 7.8 | 1.1 |
| Marietta........................... | 0.3 | 23.3 | 0.0 | 0.0 | 0.0 | 18.9 | 13.4 | 2.1 | 8.6 | 4.6 |
| Milton............................. | 0.0 | 21.9 | 0.0 | 0.0 | 0.0 | 14.8 | 0.0 | 7.2 | 0.0 | 1.1 |
| Newnan............................ | 0.0 | 15.1 | 0.0 | 0.0 | 0.0 | 20.4 | 14.2 | 4.4 | 0.1 | 0.0 |
| Peachtree City ................... | 0.0 | 13.2 | 0.0 | 0.0 | 0.8 | 16.5 | 0.0 | 6.3 | 0.0 | 1.6 |
| Rome .............................. | 0.1 | 5.9 | 0.1 | 0.0 | 0.2 | 8.3 | 36.3 | 11.0 | 0.9 | 3.2 |
| Roswell ........................... | 0.0 | 8.9 | 0.0 | 0.0 | 0.0 | 14.4 | 18.6 | 15.4 | 1.8 | 0.4 |
| Sandy Springs................... | 0.0 | 12.2 | 0.0 | 0.0 | 0.0 | 9.4 | 0.0 | 52.3 | 0.0 | 3.5 |
| Savannah......................... | 3.7 | 11.1 | 1.9 | 0.0 | 0.0 | 18.0 | 18.4 | 7.6 | 0.0 | 1.3 |
| Smyrna............................ | 0.0 | 5.1 | 0.0 | 0.5 | 0.0 | 13.2 | 20.7 | 6.0 | 0.0 | 7.9 |
| Statesboro........................ | 0.0 | 10.1 | 0.0 | 0.0 | 0.2 | 21.8 | 36.8 | 2.0 | 0.0 | 3.3 |
| Stockbridge...................... | 0.0 | 17.3 | 0.0 | 0.0 | 0.0 | 1.2 | 26.5 | 2.6 | 0.0 | 6.3 |
| Valdosta........................... | 0.0 | 12.4 | 0.0 | 0.0 | 0.0 | 24.4 | 22.9 | 0.1 | 0.1 | 5.0 |
| Warner Robins................... | 0.0 | 9.0 | 0.0 | 0.0 | 0.6 | 20.4 | 25.5 | 5.1 | 0.7 | 2.7 |
| HAWAII ........................... | X | X | X | X | X | X | X | X | X | X |
| East Honolulu CDP............. | NA | NA | NA | NA | NA | NA | NA | NA | NA | NA |
| Hilo CDP .......................... | NA | NA | NA | NA | NA | NA | NA | NA | NA | NA |
| Kahului CDP ..................... | NA | NA | NA | NA | NA | NA | NA | NA | NA | NA |
| Kailua CDP (Honolulu County) ........................... | NA | NA | NA | NA | NA | NA | NA | NA | NA | NA |
| Kaneohe CDP..................... | NA | NA | NA | NA | NA | NA | NA | NA | NA | NA |
| Mililani Town CDP.............. | NA | NA | NA | NA | NA | NA | NA | NA | NA | NA |
| Pearl City CDP................... | NA | NA | NA | NA | NA | NA | NA | NA | NA | NA |
| Urban Honolulu CDP ........... | NA | NA | NA | NA | NA | NA | NA | NA | NA | NA |
| Waipahu CDP .................... | NA | NA | NA | NA | NA | NA | NA | NA | NA | NA |
| IDAHO............................. | X | X | X | X | X | X | X | X | X | X |
| Boise City........................ | 0.0 | 0.6 | 0.0 | 0.0 | 0.3 | 16.9 | 29.0 | 10.0 | 1.0 | 0.8 |
| Caldwell .......................... | 0.0 | 10.8 | 0.0 | 0.0 | 0.0 | 18.5 | 25.1 | 7.0 | 0.4 | 1.9 |
| Coeur d'Alene ................... | 0.0 | 9.4 | 0.1 | 0.0 | 0.0 | 22.6 | 24.7 | 5.5 | 0.9 | 1.3 |
| Idaho Falls ...................... | 0.0 | 7.0 | 0.0 | 0.0 | 5.2 | 17.5 | 11.2 | 13.6 | 5.1 | 1.3 |
| Lewiston.......................... | 0.0 | 13.7 | 0.0 | 0.0 | 7.2 | 15.3 | 20.5 | 7.6 | 0.0 | 0.5 |
| Meridian .......................... | 0.0 | 0.0 | 0.0 | 0.0 | 0.0 | 21.6 | 22.7 | 13.0 | 5.1 | 0.0 |
| Nampa............................. | 0.0 | 7.9 | 0.0 | 0.0 | 0.0 | 21.5 | 16.6 | 17.4 | 1.3 | 0.4 |
| Pocatello ......................... | 0.0 | 12.1 | 0.0 | 0.0 | 1.7 | 21.0 | 17.9 | 7.6 | 3.0 | 2.2 |
| Post Falls ........................ | 0.0 | 16.4 | 0.0 | 0.0 | 0.0 | 30.3 | 9.7 | 9.4 | 0.0 | 1.5 |
| Rexburg ........................... | 0.0 | 13.4 | 0.0 | 0.0 | 0.0 | 9.9 | 18.0 | 7.4 | 0.0 | 1.5 |
| Twin Falls........................ | 0.0 | 11.0 | 0.0 | 0.0 | 0.9 | 17.8 | 14.7 | 6.2 | 1.8 | 6.9 |
| ILLINOIS ......................... | X | X | X | X | X | X | X | X | X | X |
| Addison........................... | 0.0 | 11.1 | 0.0 | 0.0 | 0.0 | 38.0 | 11.9 | 0.0 | 0.0 | 2.4 |
| Algonquin......................... | 0.0 | 43.9 | 0.0 | 0.0 | 0.0 | 25.8 | 7.1 | 1.7 | 0.0 | 3.7 |
| Alton............................... | 0.0 | 6.4 | 0.0 | 0.0 | 0.2 | 17.1 | 8.9 | 5.6 | 0.0 | 0.5 |
| Arlington Heights............... | 0.5 | 12.9 | 0.0 | 0.0 | 1.8 | 24.6 | 1.2 | 0.7 | 5.8 | 2.3 |
| Aurora............................. | 4.8 | 11.8 | 1.7 | 0.0 | 0.0 | 32.5 | 0.0 | 3.4 | 0.2 | 3.5 |
| Bartlett............................ | 0.0 | 6.0 | 0.3 | 0.0 | 0.0 | 20.3 | 7.2 | 4.2 | 0.0 | 4.0 |
| Batavia............................ | 0.0 | 14.3 | 0.0 | 0.0 | 0.0 | 25.7 | 13.0 | 0.0 | 0.0 | 3.5 |
| Belleville ......................... | 0.0 | 3.5 | 0.0 | 0.0 | 1.2 | 19.3 | 25.9 | 2.2 | 0.0 | 7.3 |
| Belvidere ......................... | 4.6 | 20.2 | 0.0 | 0.0 | 0.1 | 26.0 | 10.1 | 0.0 | 0.0 | 5.6 |
| Berwyn............................ | 0.0 | 9.6 | 0.1 | 0.0 | 0.0 | 34.6 | 6.3 | 2.4 | 0.3 | 11.3 |
| Bloomington ..................... | 0.0 | 7.0 | 0.1 | 0.0 | 0.0 | 21.0 | 9.8 | 12.3 | 0.6 | 8.0 |
| Bolingbrook...................... | 0.0 | 9.0 | 0.0 | 0.0 | 0.0 | 23.5 | 10.1 | 12.0 | 0.0 | 5.8 |
| Buffalo Grove .................... | 0.0 | 16.7 | 0.0 | 0.0 | 0.0 | 26.3 | 10.0 | 0.0 | 0.0 | 5.9 |

# Table D. Cities — City Government Finances, City Government Employment, and Climate

| City | City government finances, 2017 (cont.) | | | Climate[2] | | | | | | |
| | Debt outstanding | | Debt issued during year | Average daily temperature | | | | Annual precipitation (inches) | Heating degree days | Cooling degree days |
| | Total (mil dol) | Per capita[1] (dollars) | | Mean | | Limits | | | | |
| | | | | January | July | January[3] | July[4] | | | |
| | 137 | 138 | 139 | 140 | 141 | 142 | 143 | 144 | 145 | 146 |
|---|---|---|---|---|---|---|---|---|---|---|
| GEORGIA | X | X | X | X | X | X | X | X | X | X |
| Albany | 29.9 | 408 | 0.0 | 47.5 | 81.4 | 35.1 | 92.5 | 53.40 | 2,106 | 2,264 |
| Alpharetta | 222.0 | 3,377 | 123.4 | 39.5 | 77.2 | 29.1 | 87.5 | 51.82 | 3,490 | 1,327 |
| Athens-Clarke County | 264.0 | 2,082 | 0.0 | 42.2 | 79.8 | 32.9 | 90.2 | 47.83 | 2,861 | 1,785 |
| Atlanta | 7,204.2 | 14,652 | 548.9 | 41.7 | 79.5 | 31.3 | 90.6 | 49.10 | 3,004 | 1,679 |
| Augusta-Richmond County | 604.8 | 3,070 | 66.1 | 44.8 | 80.8 | 33.1 | 92.0 | 44.58 | 2,525 | 1,986 |
| Columbus | 452.8 | 2,336 | 0.0 | 46.8 | 82.0 | 36.6 | 91.7 | 48.57 | 2,154 | 2,296 |
| Dalton | 70.8 | 2,107 | 0.1 | 39.4 | 78.0 | 28.8 | 89.8 | 53.64 | 3,534 | 1,393 |
| Douglasville | 46.5 | 1,410 | 13.6 | NA | NA | NA | NA | NA | NA | NA |
| Duluth | 16.5 | 561 | 0.0 | NA | NA | NA | NA | NA | NA | NA |
| Dunwoody | 1.5 | 29 | 0.0 | NA | NA | NA | NA | NA | NA | NA |
| East Point | 116.1 | 3,317 | 50.9 | 42.7 | 80.0 | 33.5 | 89.4 | 50.20 | 2,827 | 1,810 |
| Gainesville | 779.5 | 19,311 | 0.0 | 36.0 | 72.9 | 24.7 | 84.0 | 58.19 | 4,421 | 752 |
| Hinesville | 31.0 | 940 | 12.7 | 51.6 | 82.6 | 40.7 | 93.3 | 48.32 | 1,551 | 2,539 |
| Johns Creek | 3.9 | 46 | 0.2 | NA | NA | NA | NA | NA | NA | NA |
| Kennesaw | 32.1 | 940 | 0.0 | NA | NA | NA | NA | NA | NA | NA |
| LaGrange | 37.9 | 1,243 | 2.4 | 42.2 | 78.7 | 31.3 | 89.3 | 53.38 | 3,078 | 1,551 |
| Lawrenceville | 0.0 | 0 | 0.0 | NA | NA | NA | NA | NA | NA | NA |
| Macon-Bibb County | 35.9 | 235 | 0.0 | 45.5 | 81.1 | 34.5 | 91.8 | 45.00 | 2,364 | 2,115 |
| Marietta | 115.1 | 1,898 | 0.0 | 39.4 | 77.9 | 28.5 | 89.3 | 54.43 | 3,505 | 1,403 |
| Milton | 9.6 | 247 | 0.7 | NA | NA | NA | NA | NA | NA | NA |
| Newnan | 32.3 | 829 | 0.0 | NA | NA | NA | NA | NA | NA | NA |
| Peachtree City | 13.0 | 370 | 3.2 | 42.6 | 79.4 | 31.8 | 90.5 | 50.10 | 2,958 | 1,679 |
| Rome | 43.9 | 1,207 | 0.0 | 39.4 | 77.5 | 29.1 | 87.7 | 56.16 | 3,510 | 1,360 |
| Roswell | 23.9 | 253 | 0.0 | 39.5 | 77.2 | 29.1 | 87.5 | 51.82 | 3,490 | 1,327 |
| Sandy Springs | 178.3 | 1,672 | 2.8 | NA | NA | NA | NA | NA | NA | NA |
| Savannah | 158.9 | 1,090 | 64.3 | 49.2 | 82.1 | 38.0 | 92.3 | 49.58 | 1,799 | 2,454 |
| Smyrna | 86.0 | 1,528 | 29.1 | 42.7 | 80.0 | 33.5 | 89.4 | 50.20 | 2,827 | 1,810 |
| Statesboro | 15.0 | 480 | 0.0 | NA | NA | NA | NA | NA | NA | NA |
| Stockbridge | 14.5 | 499 | 0.0 | NA | NA | NA | NA | NA | NA | NA |
| Valdosta | 89.7 | 1,601 | 3.9 | 50.0 | 80.9 | 38.0 | 92.0 | 53.06 | 1,782 | 2,319 |
| Warner Robins | 34.5 | 458 | 0.0 | 45.5 | 81.1 | 34.5 | 91.8 | 45.00 | 2,364 | 2,115 |
| HAWAII | X | X | X | X | X | X | X | X | X | X |
| East Honolulu CDP | NA | NA | NA | NA | NA | NA | NA | NA | NA | NA |
| Hilo CDP | NA | NA | NA | NA | NA | NA | NA | NA | NA | NA |
| Kahului CDP | NA | NA | NA | NA | NA | NA | NA | NA | NA | NA |
| Kailua CDP (Honolulu County) | NA | NA | NA | NA | NA | NA | NA | NA | NA | NA |
| Kaneohe CDP | NA | NA | NA | NA | NA | NA | NA | NA | NA | NA |
| Mililani Town CDP | NA | NA | NA | NA | NA | NA | NA | NA | NA | NA |
| Pearl City CDP | NA | NA | NA | NA | NA | NA | NA | NA | NA | NA |
| Urban Honolulu CDP | NA | NA | NA | NA | NA | NA | NA | NA | NA | NA |
| Waipahu CDP | NA | NA | NA | NA | NA | NA | NA | NA | NA | NA |
| IDAHO | X | X | X | X | X | X | X | X | X | X |
| Boise City | 109.3 | 480 | 0.0 | 30.2 | 74.7 | 23.6 | 89.2 | 12.19 | 5,727 | 807 |
| Caldwell | 25.4 | 464 | 2.5 | 29.3 | 68.8 | 19.6 | 85.7 | 10.90 | 6,749 | 410 |
| Coeur d'Alene | 35.7 | 706 | 0.0 | 28.4 | 68.7 | 22.1 | 82.6 | 26.07 | 6,540 | 426 |
| Idaho Falls | 29.3 | 475 | 0.9 | 19.3 | 68.4 | 11.1 | 85.9 | 11.02 | 7,917 | 322 |
| Lewiston | 4.2 | 129 | 0.0 | 33.7 | 73.5 | 28.0 | 87.6 | 12.74 | 5,220 | 792 |
| Meridian | 0.6 | 6 | 0.0 | 29.4 | 71.8 | 22.1 | 89.2 | 9.94 | 5,752 | 579 |
| Nampa | 0.0 | 0 | 0.0 | 28.9 | 73.3 | 20.8 | 90.5 | 11.37 | 5,873 | 692 |
| Pocatello | 40.0 | 723 | 0.0 | 24.4 | 69.2 | 16.3 | 87.5 | 12.58 | 7,109 | 387 |
| Post Falls | 5.9 | 178 | 0.0 | NA | NA | NA | NA | NA | NA | NA |
| Rexburg | 6.9 | 243 | 0.0 | NA | NA | NA | NA | NA | NA | NA |
| Twin Falls | 109.0 | 2,211 | 13.7 | 28.2 | 72.2 | 19.7 | 87.9 | 9.42 | 6,300 | 587 |
| ILLINOIS | X | X | X | X | X | X | X | X | X | X |
| Addison | 59.2 | 1,603 | 17.3 | 22.0 | 73.3 | 14.3 | 83.5 | 36.27 | 6,498 | 830 |
| Algonquin | 29.9 | 968 | 1.8 | NA | NA | NA | NA | NA | NA | NA |
| Alton | 12.1 | 453 | 0.2 | 27.7 | 78.4 | 19.4 | 88.1 | 38.54 | 5,149 | 1,354 |
| Arlington Heights | 66.3 | 878 | 34.5 | 22.0 | 73.3 | 14.3 | 83.5 | 36.27 | 6,498 | 830 |
| Aurora | 1,034.8 | 5,164 | 205.8 | 20.0 | 72.4 | 10.5 | 84.2 | 38.39 | 6,859 | 661 |
| Bartlett | 77.8 | 1,890 | 26.0 | 19.3 | 72.6 | 10.9 | 83.0 | 37.22 | 6,975 | 679 |
| Batavia | 35.9 | 1,356 | 0.0 | NA | NA | NA | NA | NA | NA | NA |
| Belleville | 74.5 | 1,788 | 0.0 | 30.9 | 78.1 | 22.1 | 89.6 | 39.37 | 4,612 | 1,339 |
| Belvidere | 32.4 | 1,288 | 0.7 | NA | NA | NA | NA | NA | NA | NA |
| Berwyn | 264.9 | 4,779 | 30.4 | 25.8 | 75.3 | 17.3 | 86.2 | 40.96 | 5,555 | 1,027 |
| Bloomington | 257.7 | 3,298 | 19.5 | 22.4 | 75.2 | 13.7 | 85.6 | 37.45 | 6,190 | 998 |
| Bolingbrook | 236.7 | 3,156 | 5.1 | 23.1 | 74.8 | 14.2 | 86.8 | 37.94 | 6,053 | 942 |
| Buffalo Grove | 68.8 | 1,671 | 13.2 | 18.4 | 72.1 | 9.6 | 82.3 | 36.56 | 7,149 | 624 |

1. Based on the population estimated as of July 1 of the year shown.   2. Represents normal values based on the 30-year period, 1971±2000.   3. Average daily minimum.   4. Average daily maximum.

| STATE Place code | City | Land area[1] (sq. mi) | Total persons 2019 | Rank | Per square mile | Race 2019 Race alone[2] (percent) White | Black or African American | American Indian, Alaskan Native | Asian | Hawaiian Pacific Islander | Some other race | Two or more races (percent) |
|---|---|---|---|---|---|---|---|---|---|---|---|---|
| | | 1 | 2 | 3 | 4 | 5 | 6 | 7 | 8 | 9 | 10 | 11 |
| | ILLINOIS— Cont'd | | | | | | | | | | | |
| 17 09642 | Burbank | 4.2 | 27,933 | 1,317 | 6,651 | D | D | D | D | D | D | D |
| 17 10487 | Calumet City | 7.2 | 35,514 | 1,090 | 4,933 | D | D | D | D | D | D | D |
| 17 11163 | Carbondale | 17.4 | 25,036 | 1,405 | 1,439 | 60.9 | 25.3 | 1.0 | 7.0 | 1.2 | 1.5 | 3.2 |
| 17 11332 | Carol Stream | 9.1 | 38,821 | 998 | 4,266 | 66.7 | 10.1 | 0.0 | 16.5 | 0.0 | 5.5 | 1.3 |
| 17 11358 | Carpentersville | 7.9 | 36,940 | 1,038 | 4,676 | 55.4 | 3.2 | 2.8 | 5.0 | 0.3 | 30.6 | 2.6 |
| 17 12385 | Champaign | 22.9 | 89,390 | 376 | 3,904 | 64.0 | 16.7 | 0.0 | 14.5 | 0.2 | 0.7 | 3.8 |
| 17 14000 | Chicago | 227.7 | 2,677,643 | 3 | 11,760 | 50.8 | 29.0 | 0.4 | 7.0 | 0.1 | 9.5 | 3.2 |
| 17 14026 | Chicago Heights | 10.3 | 28,990 | 1,288 | 2,815 | 34.4 | 43.5 | 0.0 | 0.0 | 0.0 | 17.4 | 4.7 |
| 17 14351 | Cicero | 5.9 | 79,727 | 442 | 13,513 | 32.1 | 3.9 | 0.1 | 0.6 | 0.0 | 62.6 | 0.7 |
| 17 15599 | Collinsville | 15.1 | 24,217 | 1,420 | 1,604 | 76.2 | 13.9 | 1.6 | 2.4 | 0.0 | 1.2 | 4.6 |
| 17 17887 | Crystal Lake | 18.9 | 39,642 | 975 | 2,098 | 91.5 | 1.4 | 0.2 | 4.1 | 0.0 | 0.7 | 2.1 |
| 17 18563 | Danville | 17.9 | 30,106 | 1,250 | 1,682 | 64.2 | 29.4 | 0.6 | 1.6 | 0.0 | 1.3 | 2.9 |
| 17 18823 | Decatur | 42.9 | 70,007 | 522 | 1,632 | 69.3 | 24.9 | 0.1 | 1.4 | 0.1 | 0.2 | 4.0 |
| 17 19161 | DeKalb | 16.2 | 42,621 | 902 | 2,631 | 70.4 | 12.7 | 0.0 | 4.1 | 0.0 | 5.8 | 7.1 |
| 17 19642 | Des Plaines | 14.2 | 58,179 | 663 | 4,097 | 74.0 | 3.5 | 0.7 | 13.1 | 0.0 | 3.4 | 5.2 |
| 17 20591 | Downers Grove | 14.6 | 48,727 | 800 | 3,338 | 90.4 | 2.5 | 0.0 | 5.3 | 0.0 | 0.5 | 1.3 |
| 17 22255 | East St. Louis | 13.9 | 25,776 | 1,393 | 1,854 | D | D | D | D | D | D | D |
| 17 23074 | Elgin | 38.0 | 110,196 | 272 | 2,900 | 61.6 | 6.6 | 0.3 | 3.3 | 0.0 | 24.6 | 3.6 |
| 17 23256 | Elk Grove Village | 11.6 | 31,831 | 1,200 | 2,744 | 80.0 | 3.4 | 0.0 | 10.7 | 0.0 | 2.0 | 3.9 |
| 17 23620 | Elmhurst | 10.2 | 46,532 | 837 | 4,562 | 87.4 | 0.8 | 0.0 | 7.4 | 0.0 | 1.2 | 3.1 |
| 17 24582 | Evanston | 7.8 | 72,683 | 495 | 9,318 | 67.7 | 16.7 | 0.6 | 8.7 | 0.0 | 2.3 | 4.1 |
| 17 27884 | Freeport | 11.9 | 23,447 | 1,428 | 1,970 | D | D | D | D | D | D | D |
| 17 28326 | Galesburg | 17.7 | 29,803 | 1,258 | 1,684 | 75.5 | 15.1 | 0.2 | 1.6 | 0.0 | 5.9 | 1.7 |
| 17 29730 | Glendale Heights | 5.4 | 33,351 | 1,152 | 6,176 | 54.4 | 15.9 | 0.0 | 19.3 | 0.0 | 6.0 | 4.4 |
| 17 29756 | Glen Ellyn | 6.9 | 27,402 | 1,335 | 3,971 | D | D | D | D | D | D | D |
| 17 29938 | Glenview | 14.0 | 46,823 | 830 | 3,345 | D | D | D | D | D | D | D |
| 17 30926 | Granite City | 19.0 | 27,959 | 1,316 | 1,472 | D | D | D | D | D | D | D |
| 17 32018 | Gurnee | 13.5 | 30,168 | 1,247 | 2,235 | 76.2 | 7.5 | 1.2 | 10.8 | 0.0 | 0.8 | 3.5 |
| 17 32746 | Hanover Park | 6.4 | 37,164 | 1,033 | 5,807 | 54.9 | 7.4 | 0.1 | 17.0 | 0.0 | 14.4 | 6.2 |
| 17 33383 | Harvey | 6.2 | 24,109 | 1,423 | 3,889 | D | D | D | D | D | D | D |
| 17 34722 | Highland Park | 12.2 | 29,427 | 1,271 | 2,412 | 81.6 | 0.2 | 0.0 | 7.5 | 0.0 | 3.9 | 6.7 |
| 17 35411 | Hoffman Estates | 21.1 | 50,495 | 772 | 2,393 | 60.1 | 6.7 | 0.0 | 22.4 | 0.0 | 8.2 | 2.6 |
| 17 38570 | Joliet | 64.5 | 146,673 | 182 | 2,274 | 66.0 | 18.0 | 0.8 | 2.0 | 0.0 | 9.7 | 3.5 |
| 17 38934 | Kankakee | 15.1 | 25,668 | 1,397 | 1,700 | 52.6 | 43.3 | 0.2 | 0.0 | 0.0 | 2.0 | 1.8 |
| 17 41183 | Lake in the Hills | 10.2 | 28,345 | 1,304 | 2,779 | 86.0 | 4.6 | 0.5 | 5.2 | 0.0 | 0.3 | 3.4 |
| 17 42028 | Lansing | 7.5 | 27,059 | 1,348 | 3,608 | 49.6 | 46.0 | 0.0 | 1.3 | 0.0 | 1.0 | 2.0 |
| 17 44407 | Lombard | 10.2 | 43,961 | 877 | 4,310 | 74.0 | 2.1 | 0.0 | 15.3 | 0.0 | 1.2 | 7.4 |
| 17 45694 | McHenry | 14.5 | 27,100 | 1,344 | 1,869 | D | D | D | D | D | D | D |
| 17 48242 | Melrose Park | 4.4 | 24,406 | 1,415 | 5,547 | D | D | D | D | D | D | D |
| 17 49867 | Moline | 16.8 | 41,065 | 934 | 2,444 | 74.7 | 6.3 | 0.4 | 3.0 | 0.0 | 7.8 | 7.7 |
| 17 51089 | Mount Prospect | 10.7 | 53,299 | 728 | 4,981 | 74.1 | 2.9 | 0.9 | 15.2 | 0.2 | 2.1 | 4.6 |
| 17 51349 | Mundelein | 9.6 | 30,945 | 1,219 | 3,223 | 68.7 | 1.3 | 0.0 | 10.3 | 0.0 | 18.9 | 0.9 |
| 17 51622 | Naperville | 39.1 | 147,986 | 178 | 3,785 | 72.0 | 4.4 | 0.1 | 19.4 | 0.0 | 0.9 | 3.3 |
| 17 53000 | Niles | 5.8 | 28,615 | 1,296 | 4,934 | 75.5 | 3.2 | 0.4 | 17.0 | 0.0 | 1.6 | 2.3 |
| 17 53234 | Normal | 18.3 | 54,451 | 711 | 2,976 | 84.7 | 8.0 | 0.0 | 4.7 | 0.0 | 0.0 | 2.5 |
| 17 53481 | Northbrook | 13.2 | 32,657 | 1,170 | 2,474 | D | D | D | D | D | D | D |
| 17 53559 | North Chicago | 8.0 | 29,639 | 1,263 | 3,705 | 47.5 | 29.7 | 0.2 | 3.8 | 0.0 | 15.4 | 3.3 |
| 17 54638 | Oak Forest | 6.0 | 26,884 | 1,356 | 4,481 | 73.9 | 6.3 | 0.0 | 10.0 | 0.0 | 4.8 | 5.1 |
| 17 54820 | Oak Lawn | 8.6 | 54,376 | 713 | 6,323 | 83.7 | 7.0 | 0.0 | 4.5 | 0.1 | 2.9 | 1.8 |
| 17 54885 | Oak Park | 4.7 | 51,852 | 747 | 11,032 | 64.5 | 16.8 | 0.2 | 5.6 | 0.0 | 6.9 | 6.1 |
| 17 55249 | O'Fallon | 15.6 | 29,606 | 1,266 | 1,898 | 81.6 | 8.1 | 0.0 | 2.8 | 0.0 | 0.6 | 6.9 |
| 17 56640 | Orland Park | 22.0 | 57,347 | 671 | 2,607 | 86.0 | 4.8 | 0.0 | 7.2 | 0.0 | 0.8 | 1.2 |
| 17 56887 | Oswego | 14.9 | 36,696 | 1,047 | 2,463 | D | D | D | D | D | D | D |
| 17 57225 | Palatine | 13.6 | 66,830 | 556 | 4,914 | 71.4 | 6.5 | 0.2 | 10.5 | 0.1 | 9.2 | 2.2 |
| 17 57875 | Park Ridge | 7.1 | 36,514 | 1,053 | 5,143 | 85.7 | 2.2 | 0.0 | 10.1 | 0.0 | 0.0 | 2.0 |
| 17 58447 | Pekin | 15.7 | 31,808 | 1,202 | 2,026 | D | D | D | D | D | D | D |
| 17 59000 | Peoria | 48.0 | 109,428 | 278 | 2,280 | 60.8 | 27.6 | 0.1 | 6.4 | 0.0 | 2.4 | 2.8 |
| 17 60287 | Plainfield | 24.7 | 44,542 | 862 | 1,803 | 78.2 | 4.9 | 0.0 | 13.4 | 0.0 | 0.5 | 3.1 |
| 17 62367 | Quincy | 15.8 | 39,589 | 979 | 2,506 | D | D | D | D | D | D | D |
| 17 65000 | Rockford | 64.5 | 144,835 | 183 | 2,246 | 67.7 | 22.3 | 0.1 | 3.8 | 0.0 | 2.2 | 3.8 |
| 17 65078 | Rock Island | 16.9 | 36,870 | 1,040 | 2,182 | 74.2 | 18.0 | 0.7 | 2.7 | 0.0 | 2.6 | 1.8 |
| 17 65442 | Romeoville | 19.1 | 39,770 | 971 | 2,082 | 71.5 | 12.4 | 0.0 | 6.3 | 0.0 | 6.8 | 3.0 |
| 17 66040 | Round Lake Beach | 5.0 | 26,919 | 1,354 | 5,384 | 76.3 | 2.5 | 3.6 | 4.0 | 0.0 | 9.9 | 3.7 |
| 17 66703 | St. Charles | 14.4 | 32,709 | 1,168 | 2,272 | 87.3 | 0.9 | 0.4 | 3.7 | 0.0 | 4.0 | 3.7 |
| 17 68003 | Schaumburg | 19.3 | 72,049 | 504 | 3,733 | 60.9 | 5.7 | 0.0 | 27.4 | 0.0 | 4.0 | 2.0 |
| 17 70122 | Skokie | 10.1 | 62,404 | 600 | 6,179 | 53.3 | 7.8 | 0.1 | 33.8 | 0.0 | 1.1 | 3.8 |
| 17 72000 | Springfield | 61.1 | 113,671 | 258 | 1,860 | 72.3 | 21.3 | 0.1 | 2.8 | 0.0 | 0.2 | 3.3 |
| 17 73157 | Streamwood | 7.8 | 38,777 | 999 | 4,971 | 59.9 | 1.9 | 0.6 | 17.7 | 0.0 | 16.9 | 2.9 |
| 17 75484 | Tinley Park | 16.1 | 55,221 | 695 | 3,430 | D | D | D | D | D | D | D |
| 17 77005 | Urbana | 11.8 | 41,724 | 922 | 3,536 | 62.6 | 16.6 | 0.3 | 15.3 | 0.0 | 2.0 | 3.2 |
| 17 77694 | Vernon Hills | 7.7 | 26,425 | 1,376 | 3,432 | 66.7 | 1.5 | 0.2 | 24.0 | 0.0 | 4.4 | 3.2 |

1. Dry land or land partially or temporarily covered by water.   2. Hispanic or Latino persons may be of any race.

# Table D. Cities — Population

| City | Percent Hispanic or Latino[1], 2019 | Percent foreign born, 2019 | Age of population (percent), 2019 — Under 18 years | 18 to 24 years | 25 to 34 years | 35 to 44 years | 45 to 54 years | 55 to 64 years | 65 years and over | Median age, 2019 | Percent female, 2019 | Census counts 2000 | Census counts 2010 | Percent change 2000-2010 | Percent change 2001-2020 |
|---|---|---|---|---|---|---|---|---|---|---|---|---|---|---|---|
| | 12 | 13 | 14 | 15 | 16 | 17 | 18 | 19 | 20 | 21 | 22 | 23 | 24 | 25 | 26 |
| **ILLINOIS— Cont'd** | | | | | | | | | | | | | | | |
| Burbank | 48.2 | 29.3 | 20.1 | 13.9 | 18.2 | 8.7 | 15.7 | 11.4 | 12.1 | 31.7 | 46.1 | 27,902 | 28,906 | 3.6 | -3.4 |
| Calumet City | 17.5 | 9.0 | 20.3 | 9.0 | 19.4 | 9.8 | 16.4 | 15.2 | 9.9 | 37.1 | 51.7 | 39,071 | 37,159 | -4.9 | -4.4 |
| Carbondale | 3.7 | 6.1 | 9.6 | 40.1 | 16.1 | 6.7 | 12.6 | 7.9 | 7.1 | 25.1 | 46.6 | 20,681 | 26,409 | 27.7 | -5.2 |
| Carol Stream | 13.1 | 23.8 | 23.0 | 7.2 | 12.8 | 12.6 | 13.6 | 15.7 | 15.1 | 41.2 | 53.3 | 40,438 | 39,497 | -2.3 | -1.7 |
| Carpentersville | 52.4 | 24.6 | 31.3 | 10.9 | 12.7 | 10.8 | 17.9 | 9.1 | 7.2 | 29.6 | 49.5 | 30,586 | 37,679 | 23.2 | -2.0 |
| Champaign | 5.8 | 14.6 | 16.1 | 30.5 | 13.8 | 10.7 | 8.6 | 8.9 | 11.4 | 27.2 | 46.9 | 67,518 | 81,245 | 20.3 | 10.0 |
| Chicago | 28.8 | 20.3 | 20.6 | 9.2 | 19.8 | 14.5 | 11.7 | 10.8 | 13.5 | 35.2 | 51.3 | 2,896,016 | 2,695,674 | -6.9 | -0.7 |
| Chicago Heights | 34.5 | 19.8 | 19.9 | 11.9 | 14.1 | 11.3 | 18.9 | 9.5 | 14.3 | 37.4 | 54.5 | 32,776 | 30,393 | -7.3 | -4.6 |
| Cicero | 91.1 | 33.1 | 27.3 | 14.0 | 13.4 | 14.3 | 13.3 | 10.2 | 7.5 | 31.1 | 51.5 | 85,616 | 84,241 | -1.6 | -5.4 |
| Collinsville | 11.5 | 5.6 | 25.6 | 4.5 | 16.1 | 15.7 | 9.8 | 10.7 | 17.6 | 37.9 | 56.3 | 24,707 | 25,591 | 3.6 | -5.4 |
| Crystal Lake | 11.5 | 10.6 | 20.0 | 9.3 | 10.9 | 14.9 | 13.7 | 16.3 | 15.1 | 41.1 | 51.6 | 38,000 | 40,749 | 7.2 | -2.7 |
| Danville | 8.4 | 2.8 | 27.1 | 10.8 | 13.0 | 10.2 | 9.1 | 14.5 | 15.4 | 34.5 | 51.7 | 33,904 | 33,027 | -2.6 | -8.8 |
| Decatur | 2.9 | 3.3 | 19.6 | 9.9 | 14.2 | 12.3 | 9.8 | 13.9 | 20.4 | 39.7 | 52.2 | 81,860 | 76,122 | -7.0 | -8.0 |
| DeKalb | 14.9 | 9.0 | 16.3 | 32.0 | 14.7 | 8.3 | 10.4 | 8.6 | 9.8 | 25.5 | 50.4 | 39,018 | 44,124 | 13.1 | -3.4 |
| Des Plaines | 18.6 | 29.9 | 16.8 | 9.4 | 11.2 | 12.6 | 13.3 | 16.9 | 19.8 | 45.0 | 52.1 | 58,720 | 58,252 | -0.8 | -0.1 |
| Downers Grove | 5.7 | 11.9 | 23.1 | 12.6 | 9.2 | 13.9 | 11.0 | 11.7 | 18.5 | 38.0 | 53.3 | 48,724 | 48,880 | 0.3 | -0.3 |
| East St. Louis | 0.4 | 0.9 | 20.6 | 7.8 | 16.4 | 7.6 | 10.6 | 17.0 | 20.1 | 39.7 | 54.9 | 31,542 | 26,938 | -14.6 | -4.3 |
| Elgin | 47.7 | 22.6 | 26.9 | 9.2 | 14.1 | 12.3 | 11.8 | 12.5 | 13.3 | 34.9 | 51.2 | 94,487 | 108,218 | 14.5 | 1.8 |
| Elk Grove Village | 11.7 | 23.9 | 19.1 | 7.3 | 13.1 | 10.6 | 14.9 | 14.1 | 21.0 | 45.0 | 52.9 | 34,727 | 33,166 | -4.5 | -4.0 |
| Elmhurst | 8.7 | 9.8 | 26.5 | 5.6 | 9.5 | 13.6 | 12.3 | 15.4 | 17.2 | 41.1 | 53.7 | 42,762 | 44,120 | 3.2 | 5.5 |
| Evanston | 7.4 | 18.0 | 19.8 | 13.9 | 16.2 | 10.8 | 11.5 | 11.6 | 16.3 | 35.1 | 53.4 | 74,239 | 74,483 | 0.3 | -2.4 |
| Freeport | 6.5 | 3.3 | 21.1 | 7.1 | 11.1 | 10.3 | 10.6 | 14.6 | 25.1 | 45.3 | 51.6 | 26,443 | 25,676 | -2.9 | -8.7 |
| Galesburg | 10.1 | 6.1 | 18.6 | 13.1 | 12.0 | 11.3 | 10.8 | 12.7 | 21.5 | 39.4 | 48.4 | 33,706 | 32,192 | -4.5 | -7.4 |
| Glendale Heights | 28.5 | 38.7 | 24.4 | 11.9 | 15.5 | 13.4 | 10.7 | 12.6 | 11.5 | 33.5 | 46.5 | 31,765 | 34,326 | 8.1 | -2.8 |
| Glen Ellyn | 6.5 | 7.0 | 22.7 | 3.9 | 10.5 | 15.4 | 12.6 | 14.7 | 20.3 | 43.9 | 47.3 | 26,999 | 27,656 | 2.4 | -0.9 |
| Glenview | 5.1 | 19.5 | 24.0 | 4.9 | 6.0 | 12.6 | 13.6 | 16.5 | 22.5 | 46.9 | 50.0 | 41,847 | 44,726 | 6.9 | 4.7 |
| Granite City | 6.3 | 3.2 | 18.3 | 8.0 | 16.5 | 7.7 | 12.2 | 21.5 | 15.9 | 44.3 | 50.4 | 31,301 | 29,753 | -4.9 | -6.0 |
| Gurnee | 13.1 | 11.7 | 25.6 | 13.1 | 10.4 | 15.8 | 14.7 | 9.2 | 11.1 | 36.1 | 52.2 | 28,834 | 31,231 | 8.3 | -3.4 |
| Hanover Park | 38.6 | 30.1 | 27.0 | 7.9 | 18.1 | 12.0 | 10.0 | 13.5 | 11.5 | 33.1 | 55.1 | 38,278 | 38,215 | -0.2 | -2.8 |
| Harvey | 20.5 | 10.8 | 16.0 | 14.4 | 13.5 | 9.2 | 13.1 | 15.8 | 18.1 | 40.9 | 55.7 | 30,000 | 25,265 | -15.8 | -4.6 |
| Highland Park | 8.1 | 14.6 | 26.3 | 3.4 | 6.8 | 9.9 | 16.5 | 10.5 | 26.5 | 48.0 | 48.1 | 31,365 | 29,745 | -5.2 | -1.1 |
| Hoffman Estates | 19.6 | 27.6 | 24.0 | 7.5 | 12.9 | 12.9 | 17.4 | 14.1 | 11.2 | 38.3 | 50.2 | 49,495 | 51,891 | 4.8 | -2.7 |
| Joliet | 35.4 | 14.6 | 26.1 | 10.4 | 13.1 | 15.5 | 13.4 | 10.5 | 11.0 | 35.4 | 51.8 | 106,221 | 147,307 | 38.7 | -0.4 |
| Kankakee | 17.2 | 8.4 | 25.9 | 10.1 | 10.2 | 13.4 | 13.7 | 13.5 | 13.3 | 37.6 | 52.9 | 27,491 | 27,542 | 0.2 | -6.8 |
| Lake in the Hills | 15.8 | 9.8 | 25.6 | 12.8 | 13.1 | 10.8 | 18.3 | 12.4 | 6.9 | 32.5 | 52.2 | 23,152 | 28,946 | 25.0 | -2.1 |
| Lansing | 29.9 | 11.2 | 31.9 | 7.1 | 12.2 | 12.5 | 10.8 | 12.6 | 12.9 | 34.5 | 52.4 | 28,332 | 28,353 | 0.1 | -4.6 |
| Lombard | 9.1 | 14.8 | 22.1 | 7.5 | 17.7 | 13.1 | 9.9 | 12.8 | 16.8 | 36.6 | 50.5 | 42,322 | 43,462 | 2.7 | 1.1 |
| McHenry | 14.0 | 6.0 | 19.9 | 9.8 | 9.7 | 15.1 | 12.8 | 13.3 | 19.4 | 41.2 | 49.7 | 21,501 | 27,025 | 25.7 | 0.3 |
| Melrose Park | 77.5 | 34.5 | 24.9 | 11.9 | 18.3 | 5.2 | 10.8 | 13.1 | 15.9 | 33.0 | 54.7 | 23,171 | 25,411 | 9.7 | -4.0 |
| Moline | 17.9 | 11.0 | 26.5 | 8.1 | 12.1 | 13.5 | 12.2 | 11.0 | 16.7 | 37.4 | 54.8 | 43,768 | 43,439 | -0.8 | -5.5 |
| Mount Prospect | 10.6 | 30.8 | 18.2 | 4.7 | 15.0 | 14.3 | 15.3 | 13.7 | 18.9 | 43.3 | 50.9 | 56,265 | 55,037 | -2.2 | -3.2 |
| Mundelein | 33.8 | 26.5 | 22.1 | 9.4 | 11.7 | 12.8 | 13.7 | 15.0 | 15.3 | 38.9 | 47.1 | 30,935 | 30,987 | 0.2 | -0.1 |
| Naperville | 7.9 | 20.7 | 24.1 | 8.1 | 11.0 | 12.7 | 17.1 | 14.3 | 12.7 | 40.5 | 50.8 | 128,358 | 142,175 | 10.8 | 4.1 |
| Niles | 7.3 | 38.4 | 14.5 | 9.1 | 10.5 | 9.5 | 13.3 | 16.2 | 26.9 | 50.6 | 50.9 | 30,068 | 29,847 | -0.7 | -4.1 |
| Normal | 5.2 | 4.7 | 14.2 | 35.8 | 10.6 | 9.0 | 9.4 | 9.5 | 11.5 | 25.0 | 53.5 | 45,386 | 52,555 | 15.8 | 3.6 |
| Northbrook | 4.1 | 20.7 | 25.5 | 5.1 | 5.6 | 9.7 | 17.4 | 12.5 | 24.2 | 47.6 | 53.7 | 33,435 | 33,200 | -0.7 | -1.6 |
| North Chicago | 39.6 | 15.4 | 23.8 | 34.5 | 20.8 | 7.8 | 4.6 | 3.8 | 4.8 | 22.6 | 41.7 | 35,918 | 32,594 | -9.3 | -9.1 |
| Oak Forest | 11.5 | 16.2 | 22.2 | 7.2 | 16.4 | 8.5 | 12.3 | 14.6 | 18.8 | 40.2 | 49.7 | 28,051 | 28,001 | -0.2 | -4.0 |
| Oak Lawn | 18.7 | 19.8 | 19.2 | 5.5 | 9.5 | 14.7 | 10.2 | 15.0 | 25.9 | 45.5 | 51.2 | 55,245 | 56,690 | 2.6 | -4.1 |
| Oak Park | 11.7 | 9.9 | 24.3 | 6.5 | 13.0 | 14.7 | 14.6 | 13.0 | 13.9 | 39.0 | 52.3 | 52,524 | 51,878 | -1.2 | -0.1 |
| O'Fallon | 4.8 | 4.0 | 33.8 | 4.9 | 13.2 | 14.6 | 13.0 | 9.3 | 11.3 | 33.7 | 52.1 | 21,910 | 28,767 | 31.3 | 2.9 |
| Orland Park | 4.4 | 17.5 | 20.2 | 7.8 | 11.1 | 10.0 | 14.1 | 15.4 | 21.4 | 45.7 | 50.3 | 51,077 | 56,602 | 10.8 | 1.3 |
| Oswego | 19.3 | 8.5 | 26.7 | 9.0 | 11.6 | 10.2 | 18.9 | 11.2 | 12.5 | 39.4 | 55.0 | 13,326 | 30,452 | 128.5 | 20.5 |
| Palatine | 18.3 | 21.2 | 27.1 | 6.2 | 14.9 | 12.8 | 11.5 | 13.3 | 14.2 | 36.7 | 52.3 | 65,479 | 68,551 | 4.7 | -2.5 |
| Park Ridge | 1.4 | 17.1 | 24.9 | 5.0 | 6.6 | 14.6 | 14.4 | 12.5 | 22.0 | 44.2 | 52.0 | 37,775 | 37,460 | -0.8 | -2.5 |
| Pekin | 2.8 | 0.1 | 20.3 | 6.1 | 14.7 | 13.0 | 9.9 | 16.6 | 19.5 | 41.3 | 52.2 | 33,857 | 34,042 | 0.5 | -6.6 |
| Peoria | 6.7 | 6.4 | 23.2 | 10.9 | 13.7 | 12.2 | 11.2 | 12.1 | 16.8 | 36.5 | 52.0 | 112,936 | 115,141 | 2.0 | -5.0 |
| Plainfield | 7.9 | 12.2 | 29.9 | 7.0 | 9.1 | 13.9 | 18.7 | 10.0 | 11.3 | 37.6 | 48.9 | 13,038 | 39,881 | 205.9 | 11.7 |
| Quincy | 2.2 | 1.7 | 21.7 | 6.7 | 11.9 | 13.1 | 11.4 | 13.8 | 21.4 | 42.6 | 53.5 | 40,366 | 40,717 | 0.9 | -2.8 |
| Rockford | 18.9 | 11.2 | 25.9 | 8.7 | 12.8 | 13.2 | 11.6 | 11.9 | 15.8 | 36.7 | 51.5 | 150,115 | 153,311 | 2.1 | -5.5 |
| Rock Island | 13.6 | 13.7 | 19.7 | 15.8 | 13.4 | 10.5 | 10.1 | 14.8 | 15.8 | 35.8 | 50.9 | 39,684 | 38,979 | -1.8 | -5.4 |
| Romeoville | 32.5 | 18.6 | 23.8 | 12.1 | 9.9 | 14.8 | 16.4 | 10.2 | 12.7 | 38.6 | 50.0 | 21,153 | 39,645 | 87.4 | 0.3 |
| Round Lake Beach | 51.1 | 25.6 | 24.2 | 11.8 | 10.6 | 13.5 | 15.6 | 10.4 | 14.0 | 36.9 | 49.0 | 25,859 | 28,090 | 8.6 | -4.2 |
| St. Charles | 10.2 | 8.6 | 20.3 | 9.2 | 9.4 | 8.8 | 15.5 | 15.5 | 21.3 | 47.7 | 49.9 | 27,896 | 32,289 | 15.7 | 1.3 |
| Schaumburg | 8.3 | 33.0 | 22.0 | 4.9 | 13.5 | 16.5 | 13.7 | 13.3 | 16.1 | 40.4 | 51.8 | 75,386 | 74,233 | -1.5 | -2.9 |
| Skokie | 11.0 | 45.7 | 21.0 | 9.5 | 12.1 | 12.6 | 10.6 | 14.3 | 19.8 | 40.8 | 50.8 | 63,348 | 64,826 | 2.3 | -3.7 |
| Springfield | 3.3 | 4.3 | 22.7 | 8.0 | 13.1 | 12.0 | 11.6 | 13.0 | 19.5 | 40.1 | 53.0 | 111,454 | 116,995 | 5.0 | -2.8 |
| Streamwood | 34.1 | 37.9 | 18.4 | 8.8 | 14.1 | 14.7 | 15.9 | 15.1 | 13.0 | 40.7 | 51.8 | 36,407 | 39,842 | 9.4 | -2.7 |
| Tinley Park | 6.4 | 11.4 | 20.3 | 5.6 | 15.7 | 11.8 | 11.2 | 16.6 | 18.9 | 40.9 | 51.9 | 48,401 | 56,827 | 17.4 | -2.8 |
| Urbana | 8.2 | 15.4 | 13.6 | 34.5 | 19.0 | 7.6 | 8.2 | 7.1 | 9.9 | 25.8 | 55.0 | 36,395 | 42,137 | 15.8 | |
| Vernon Hills | 19.3 | 34.9 | 27.7 | 4.8 | 11.1 | 14.7 | 18.4 | 13.3 | 10.0 | 39.1 | 48.2 | 20,120 | 25,005 | 24.3 | 5.7 |

1. May be of any race.

| City | Households, 2019 Number | Persons per household | Percent Family | Married couple family | Female family | Non-family | One person | Persons in group quarters, 2019 | Violent Number | Violent Rate[2] | Property Number | Property Rate[2] | Population age 25 and over | High school graduate or less | Bachelor's degree or more |
|---|---|---|---|---|---|---|---|---|---|---|---|---|---|---|---|
| | 27 | 28 | 29 | 30 | 31 | 32 | 33 | 34 | 35 | 36 | 37 | 38 | 39 | 40 | 41 |
| **ILLINOIS— Cont'd** | | | | | | | | | | | | | | | |
| Burbank | 8,543 | 3.22 | 73.1 | 53.1 | 20.0 | 26.9 | 25.1 | 226 | 54 | 190 | 244 | 857 | 18,302 | 52.3 | 20.5 |
| Calumet City | 13,963 | 2.57 | 55.8 | 22.8 | 33.0 | 44.2 | 40.2 | 15 | 186 | 515 | 1,040 | 2,879 | 25,406 | 44.1 | 15.8 |
| Carbondale | 11,279 | 1.90 | 27.5 | 18.1 | 9.4 | 72.5 | 49.3 | 2,711 | 208 | 824 | 932 | 3,692 | 12,135 | 21.9 | 37.3 |
| Carol Stream | 14,667 | 2.76 | 75.9 | 57.2 | 18.7 | 24.1 | 20.1 | 327 | 29 | 73 | 412 | 1,040 | 28,486 | 29.5 | 38.4 |
| Carpentersville | 10,602 | 3.51 | 74.6 | 53.7 | 20.9 | 25.4 | 16.3 | 7 | 21 | 56 | 380 | 1,007 | 21,535 | 47.7 | 23.8 |
| Champaign | 35,865 | 2.25 | 43.2 | 25.8 | 17.4 | 56.8 | 41.2 | 8,307 | 827 | 930 | 2,203 | 2,478 | 47,481 | 25.2 | 51.6 |
| Chicago | 1,080,345 | 2.44 | 51.8 | 32.5 | 19.3 | 48.2 | 38.5 | 58,711 | 25,532 | 943 | 80,742 | 2,983 | 1,890,618 | 36.2 | 41.3 |
| Chicago Heights | 9,885 | 2.89 | 74.7 | 39.2 | 35.5 | 25.3 | 22.7 | 769 | 228 | 774 | 576 | 1,954 | 19,990 | 47.5 | 19.0 |
| Cicero | 22,399 | 3.57 | 74.6 | 46.0 | 28.6 | 25.4 | 20.3 | 898 | 258 | 317 | 1,579 | 1,943 | 47,398 | 67.2 | 10.7 |
| Collinsville | 11,144 | 2.15 | 49.8 | 34.4 | 15.4 | 50.2 | 45.0 | 57 | 64 | 261 | 493 | 2,013 | 16,746 | 29.6 | 30.3 |
| Crystal Lake | 13,557 | 2.49 | 70.1 | 57.7 | 12.5 | 29.9 | 25.9 | 398 | 45 | 113 | 491 | 1,229 | 24,199 | 21.6 | 42.4 |
| Danville | 12,065 | 2.37 | 53.1 | 26.9 | 26.3 | 46.9 | 43.8 | 2,422 | 536 | 1,749 | 1,370 | 4,471 | 19,282 | 53.4 | 12.4 |
| Decatur | 29,772 | 2.22 | 50.3 | 36.2 | 14.1 | 49.7 | 43.1 | 3,601 | 375 | 530 | 1,954 | 2,763 | 49,067 | 49.1 | 20.7 |
| DeKalb | 16,706 | 2.36 | 49.2 | 35.6 | 13.6 | 50.8 | 34.3 | 4,135 | 245 | 577 | 1,284 | 3,026 | 22,571 | 22.7 | 49.0 |
| Des Plaines | 24,149 | 2.37 | 58.0 | 43.3 | 14.7 | 42.0 | 37.1 | 998 | 56 | 95 | 650 | 1,101 | 42,918 | 33.2 | 39.6 |
| Downers Grove | 19,091 | 2.54 | 58.5 | 50.4 | 8.1 | 41.5 | 36.3 | 458 | 46 | 93 | 732 | 1,480 | 31,486 | 16.3 | 63.0 |
| East St. Louis | 12,944 | 2.01 | 48.8 | 16.6 | 32.2 | 51.2 | 48.9 | 93 | 268 | 1,020 | 415 | 1,579 | 18,665 | 61.5 | 10.5 |
| Elgin | 36,849 | 2.95 | 66.3 | 46.6 | 19.7 | 33.7 | 28.0 | 2,033 | 223 | 199 | 1,429 | 1,275 | 70,858 | 45.0 | 23.4 |
| Elk Grove Village | 13,325 | 2.49 | 67.4 | 50.9 | 16.5 | 32.6 | 30.9 | 260 | 30 | 93 | 452 | 1,396 | 24,661 | 28.9 | 38.1 |
| Elmhurst | 16,527 | 2.74 | 74.0 | 63.1 | 11.0 | 26.0 | 20.7 | 1,317 | 35 | 75 | 478 | 1,020 | 31,646 | 15.7 | 60.8 |
| Evanston | 27,877 | 2.36 | 55.4 | 43.5 | 11.9 | 44.6 | 35.3 | 7,813 | 115 | 155 | 1,937 | 2,616 | 48,705 | 13.2 | 73.7 |
| Freeport | 11,065 | 2.14 | 53.7 | 36.9 | 16.8 | 46.3 | 42.1 | 709 | 50 | 211 | 498 | 2,099 | 17,511 | 42.2 | 17.1 |
| Galesburg | 11,493 | 2.09 | 55.2 | 28.8 | 26.4 | 44.8 | 41.2 | 4,297 | 138 | 457 | 1,249 | 4,133 | 19,359 | 55.2 | 19.0 |
| Glendale Heights | 10,242 | 3.18 | 73.0 | 57.7 | 15.3 | 27.0 | 19.6 | 37 | 28 | 83 | 373 | 1,101 | 20,774 | 43.3 | 35.5 |
| Glen Ellyn | 11,418 | 2.38 | 67.3 | 55.0 | 12.3 | 32.7 | 31.4 | 46 | 20 | 72 | 244 | 873 | 20,025 | 18.3 | 66.3 |
| Glenview | 17,623 | 2.62 | 76.9 | 66.5 | 10.3 | 23.1 | 20.6 | 757 | 50 | 105 | 393 | 826 | 33,329 | 14.4 | 70.2 |
| Granite City | 11,832 | 2.32 | 70.0 | 43.7 | 26.3 | 30.0 | 28.2 | 271 | 233 | 823 | 655 | 2,313 | 20,429 | 47.3 | 13.3 |
| Gurnee | 10,899 | 2.78 | 66.7 | 53.0 | 13.7 | 33.3 | 26.2 | 44 | 40 | 131 | 994 | 3,260 | 18,595 | 18.4 | 49.4 |
| Hanover Park | 11,890 | 3.52 | 86.6 | 58.6 | 27.9 | 13.4 | 9.7 | 0 | 30 | 80 | 253 | 671 | 27,221 | 46.1 | 26.1 |
| Harvey | 7,048 | 2.75 | 54.0 | 29.3 | 24.6 | 46.0 | 39.2 | 288 | NA | NA | NA | NA | 13,687 | 67.2 | 9.7 |
| Highland Park | 10,657 | 2.74 | 83.1 | 78.1 | 5.0 | 16.9 | 14.0 | 325 | 29 | 98 | 206 | 696 | 20,733 | 12.4 | 71.4 |
| Hoffman Estates | 17,594 | 2.85 | 78.9 | 64.9 | 14.0 | 21.1 | 16.5 | 213 | 65 | 127 | 432 | 845 | 34,458 | 25.8 | 46.3 |
| Joliet | 48,724 | 2.99 | 68.7 | 50.4 | 18.3 | 31.3 | 27.8 | 2,952 | 751 | 507 | 2,160 | 1,458 | 94,436 | 44.2 | 24.3 |
| Kankakee | 9,792 | 2.43 | 55.4 | 34.6 | 20.8 | 44.6 | 38.7 | 1,738 | 236 | 912 | 873 | 3,374 | 16,329 | 61.7 | 12.1 |
| Lake in the Hills | 10,773 | 2.96 | 78.3 | 60.6 | 17.6 | 21.7 | 16.7 | 0 | 19 | 66 | 105 | 364 | 19,650 | 23.4 | 44.4 |
| Lansing | 10,964 | 2.82 | 57.7 | 30.1 | 27.6 | 42.3 | 38.5 | 97 | 104 | 377 | 1,495 | 5,423 | 18,929 | 36.9 | 22.5 |
| Lombard | 17,484 | 2.50 | 60.6 | 50.8 | 9.8 | 39.4 | 31.1 | 608 | 43 | 96 | 949 | 2,124 | 31,219 | 25.3 | 46.0 |
| McHenry | 10,397 | 2.37 | 63.5 | 51.3 | 12.1 | 36.5 | 32.8 | 279 | 40 | 148 | 229 | 848 | 17,538 | 37.9 | 30.6 |
| Melrose Park | 7,711 | 2.96 | 66.6 | 35.5 | 31.1 | 33.4 | 27.7 | 91 | 74 | 298 | 317 | 1,275 | 14,495 | 57.2 | 14.0 |
| Moline | 18,512 | 2.33 | 57.4 | 41.0 | 16.5 | 42.6 | 38.3 | 304 | 219 | 525 | 978 | 2,345 | 28,476 | 38.7 | 30.6 |
| Mount Prospect | 20,895 | 2.38 | 64.8 | 52.5 | 12.3 | 35.2 | 31.1 | 112 | 29 | 54 | 464 | 858 | 38,505 | 29.8 | 45.6 |
| Mundelein | 11,573 | 2.73 | 76.3 | 65.2 | 11.1 | 23.7 | 21.4 | 319 | 23 | 74 | 222 | 710 | 21,856 | 33.3 | 38.4 |
| Naperville | 52,238 | 2.82 | 73.4 | 65.0 | 8.4 | 26.6 | 20.7 | 2,241 | NA | NA | NA | NA | 101,373 | 13.3 | 69.1 |
| Niles | 10,598 | 2.73 | 78.3 | 64.9 | 13.4 | 21.7 | 17.6 | 1,283 | 36 | 124 | 624 | 2,144 | 23,139 | 39.0 | 38.5 |
| Normal | 19,058 | 2.48 | 47.2 | 38.5 | 8.8 | 52.8 | 32.5 | 7,410 | 152 | 276 | 936 | 1,701 | 27,367 | 21.2 | 50.1 |
| Northbrook | 12,761 | 2.49 | 72.5 | 61.8 | 10.6 | 27.5 | 27.2 | 670 | 9 | 27 | 480 | 1,448 | 22,556 | 10.7 | 74.1 |
| North Chicago | 7,630 | 2.91 | 61.1 | 35.2 | 25.9 | 38.9 | 33.8 | 10,140 | NA | NA | NA | NA | 13,511 | 52.0 | 21.5 |
| Oak Forest | 10,101 | 2.68 | 75.0 | 58.2 | 16.8 | 25.0 | 16.9 | 80 | 51 | 187 | 283 | 1,035 | 19,181 | 34.4 | 40.2 |
| Oak Lawn | 22,728 | 2.40 | 60.7 | 43.0 | 17.7 | 39.3 | 36.9 | 475 | 83 | 150 | 826 | 1,492 | 41,429 | 35.4 | 30.1 |
| Oak Park | 21,129 | 2.45 | 62.3 | 47.6 | 14.6 | 37.7 | 32.6 | 562 | 156 | 298 | 1,594 | 3,047 | 36,265 | 12.0 | 68.2 |
| O'Fallon | 11,145 | 2.83 | 77.0 | 57.5 | 19.5 | 23.0 | 23.0 | 0 | 53 | 179 | 423 | 1,425 | 19,312 | 17.8 | 45.5 |
| Orland Park | 22,233 | 2.60 | 68.7 | 56.4 | 12.3 | 31.3 | 29.0 | 409 | 26 | 44 | 948 | 1,620 | 41,969 | 27.6 | 43.9 |
| Oswego | 12,598 | 3.01 | 77.6 | 58.1 | 19.6 | 22.4 | 19.0 | 91 | 23 | 64 | 339 | 945 | 24,430 | 24.6 | 43.9 |
| Palatine | 25,880 | 2.60 | 66.0 | 51.2 | 14.8 | 34.0 | 28.0 | 255 | 42 | 62 | 308 | 453 | 44,986 | 25.6 | 49.8 |
| Park Ridge | 13,894 | 2.65 | 75.2 | 61.6 | 13.6 | 24.8 | 22.1 | 633 | 11 | 30 | 295 | 793 | 26,228 | 17.3 | 61.2 |
| Pekin | 13,216 | 2.26 | 54.2 | 37.7 | 16.5 | 45.8 | 36.0 | 1,626 | 132 | 412 | 618 | 1,929 | 23,199 | 51.5 | 19.5 |
| Peoria | 46,130 | 2.29 | 55.1 | 36.6 | 18.6 | 44.9 | 38.2 | 4,077 | 1,158 | 1,044 | 4,160 | 3,749 | 72,528 | 34.9 | 36.0 |
| Plainfield | 12,247 | 3.27 | 82.0 | 65.5 | 16.4 | 18.0 | 13.9 | 134 | 55 | 123 | 252 | 564 | 25,356 | 24.6 | 49.5 |
| Quincy | 17,189 | 2.25 | 57.4 | 45.0 | 12.3 | 42.6 | 36.5 | 1,271 | 243 | 608 | 1,210 | 3,028 | 28,616 | 44.2 | 22.6 |
| Rockford | 57,463 | 2.43 | 61.8 | 37.5 | 24.2 | 38.2 | 32.3 | 3,930 | 1,711 | 1,174 | 4,848 | 3,327 | 93,983 | 44.5 | 25.5 |
| Rock Island | 16,404 | 2.18 | 52.8 | 35.9 | 16.8 | 47.2 | 34.2 | 2,885 | 139 | 370 | 878 | 2,340 | 24,985 | 43.9 | 20.3 |
| Romeoville | 11,119 | 3.08 | 68.5 | 51.1 | 17.3 | 31.5 | 28.6 | 1,172 | 78 | 197 | 402 | 1,015 | 22,744 | 47.3 | 19.7 |
| Round Lake Beach | 8,506 | 3.09 | 69.9 | 53.5 | 16.3 | 30.1 | 26.0 | 133 | 40 | 147 | 553 | 2,031 | 16,887 | 56.7 | 17.7 |
| St. Charles | 14,427 | 2.32 | 64.1 | 54.9 | 9.2 | 35.9 | 31.8 | 499 | 43 | 130 | 208 | 628 | 23,989 | 27.6 | 46.7 |
| Schaumburg | 30,713 | 2.36 | 58.1 | 46.6 | 11.5 | 41.9 | 36.6 | 356 | 73 | 99 | 1,561 | 2,126 | 53,247 | 23.2 | 54.1 |
| Skokie | 20,060 | 3.08 | 79.6 | 62.6 | 17.0 | 20.4 | 18.4 | 809 | 150 | 238 | 1,870 | 2,964 | 43,562 | 31.9 | 44.9 |
| Springfield | 49,544 | 2.23 | 53.6 | 34.7 | 18.8 | 46.4 | 39.2 | 3,740 | 889 | 777 | 5,080 | 4,441 | 79,081 | 35.9 | 34.0 |
| Streamwood | 13,226 | 2.93 | 71.4 | 52.6 | 18.8 | 28.6 | 25.1 | 231 | 60 | 152 | 415 | 1,050 | 28,353 | 41.6 | 33.7 |
| Tinley Park | 22,998 | 2.62 | 68.5 | 55.0 | 13.5 | 31.5 | 27.1 | 65 | 22 | 39 | 712 | 1,269 | 44,671 | 26.0 | 46.3 |
| Urbana | 16,560 | 2.16 | 41.4 | 25.9 | 15.5 | 58.6 | 41.4 | 6,519 | 124 | 295 | 1,090 | 2,590 | 21,916 | 24.6 | 59.9 |
| Vernon Hills | 10,092 | 2.60 | 79.1 | 64.1 | 15.0 | 20.9 | 17.8 | 46 | 18 | 67 | 435 | 1,620 | 17,746 | 18.3 | 63.6 |

1. Data for serious crimes have not been adjusted for underreporting. This may affect comparability between geographic areas and over time.   2. Per 100,000 population estimated by the FBI.   3. Persons 25 years old and over.

# Table D. Cities — Income, Poverty, and Housing

| City | Money income, 2019 Households | | | Median family income | Median non-family income | Median earnings, 2019 | | | Housing units, 2019 | | | | |
|---|---|---|---|---|---|---|---|---|---|---|---|---|---|
| | Median income | Percent with income less than $20,000 | Percent with income of $200,000 or more | | | All persons | Men | Women | Total | Occupied | Percent owner occupied | Median value[1] (dollars) | Median gross rent (dollars) |
| | 42 | 43 | 44 | 45 | 46 | 47 | 48 | 49 | 50 | 51 | 52 | 53 | 54 |
| ILLINOIS— Cont'd | | | | | | | | | | | | | |
| Burbank | 76,504 | 5.0 | 10.9 | 85,959 | 45,423 | 36,359 | 41,980 | 32,280 | 9,232 | 8,543 | 77.2 | 219,014 | 1,205 |
| Calumet City | 51,506 | 18.0 | 0.5 | 55,947 | 36,054 | 32,014 | 30,543 | 36,474 | 16,479 | 13,963 | 63.4 | 112,500 | 928 |
| Carbondale | 20,801 | 48.2 | 2.6 | 85,771 | 13,575 | 14,785 | 11,615 | 19,303 | 13,997 | 11,279 | 23.2 | 114,956 | 656 |
| Carol Stream | 85,502 | 6.3 | 8.4 | 99,229 | 50,704 | 44,043 | 52,284 | 36,236 | 15,292 | 14,667 | 70.4 | 245,439 | 1,195 |
| Carpentersville | 82,443 | 10.6 | 7.7 | 86,309 | 57,125 | 31,606 | 40,238 | 25,247 | 11,028 | 10,602 | 78.6 | 176,210 | 1,334 |
| Champaign | 47,379 | 24.9 | 5.7 | 77,806 | 33,317 | 27,124 | 30,943 | 25,042 | 42,123 | 35,865 | 43.8 | 154,570 | 985 |
| Chicago | 61,811 | 17.9 | 10.0 | 76,360 | 45,744 | 41,024 | 45,753 | 36,486 | 1,218,078 | 1,080,345 | 44.2 | 275,239 | 1,134 |
| Chicago Heights | 57,386 | 14.4 | 1.2 | 73,524 | 28,502 | 31,861 | 40,476 | 24,981 | 12,635 | 9,885 | 62.3 | 128,688 | 898 |
| Cicero | 51,814 | 11.0 | 1.9 | 56,220 | 30,444 | 30,533 | 33,802 | 25,670 | 24,566 | 22,399 | 52.1 | 179,548 | 986 |
| Collinsville | 54,091 | 6.7 | 4.7 | 68,579 | 46,605 | 40,476 | 46,465 | 27,473 | 11,838 | 11,144 | 60.3 | 125,084 | 849 |
| Crystal Lake | 81,605 | 5.6 | 11.9 | 107,842 | 50,267 | 45,907 | 57,851 | 37,247 | 13,926 | 13,557 | 76.4 | 230,829 | 1,287 |
| Danville | 32,437 | 25.7 | 1.7 | 37,417 | 26,597 | 27,284 | 38,202 | 20,755 | 14,971 | 12,065 | 53.1 | 61,037 | 663 |
| Decatur | 41,440 | 24.8 | 2.7 | 60,603 | 31,631 | 30,913 | 37,023 | 25,797 | 35,769 | 29,772 | 59.7 | 87,790 | 669 |
| DeKalb | 50,130 | 23.6 | 6.7 | 87,347 | 25,058 | 24,455 | 28,850 | 19,120 | 17,493 | 16,706 | 46.7 | 187,849 | 928 |
| Des Plaines | 68,534 | 11.7 | 7.9 | 105,242 | 37,603 | 43,608 | 51,001 | 37,444 | 25,565 | 24,149 | 76.5 | 260,896 | 1,199 |
| Downers Grove | 105,806 | 7.8 | 17.9 | 152,013 | 43,996 | 46,975 | 77,615 | 33,768 | 20,436 | 19,091 | 77.5 | 370,309 | 1,279 |
| East St. Louis | 22,921 | 45.7 | 0.3 | 29,915 | 17,302 | 16,392 | 17,694 | 16,192 | 17,671 | 12,944 | 40.0 | 68,326 | 369 |
| Elgin | 70,192 | 9.5 | 4.9 | 82,372 | 51,880 | 32,248 | 40,755 | 26,809 | 38,632 | 36,849 | 70.0 | 194,467 | 974 |
| Elk Grove Village | 79,042 | 4.7 | 9.8 | 99,946 | 54,134 | 46,124 | 55,538 | 38,944 | 13,783 | 13,325 | 75.2 | 283,259 | 1,254 |
| Elmhurst | 126,841 | 5.1 | 27.1 | 150,721 | 55,197 | 61,013 | 82,459 | 38,658 | 17,437 | 16,527 | 82.9 | 453,929 | 1,404 |
| Evanston | 81,543 | 12.1 | 21.9 | 137,603 | 47,888 | 43,613 | 51,329 | 36,274 | 30,678 | 27,877 | 55.6 | 395,544 | 1,244 |
| Freeport | 40,097 | 21.2 | 0.9 | 54,044 | 24,700 | 26,769 | 32,143 | 22,279 | 12,329 | 11,065 | 64.9 | 69,191 | 700 |
| Galesburg | 31,304 | 34.8 | 0.6 | 41,263 | 24,513 | 21,523 | 32,846 | 16,486 | 13,747 | 11,493 | 57.1 | 68,846 | 593 |
| Glendale Heights | 71,994 | 4.7 | 4.2 | 69,426 | 64,435 | 32,197 | 39,839 | 27,003 | 10,914 | 10,242 | 70.9 | 213,340 | 1,345 |
| Glen Ellyn | 99,521 | 3.0 | 24.3 | 164,889 | 47,885 | 56,455 | 71,469 | 43,555 | 12,208 | 11,418 | 76.2 | 460,793 | 1,135 |
| Glenview | 131,273 | 4.2 | 33.6 | 163,588 | 55,274 | 71,005 | 81,879 | 51,935 | 18,913 | 17,623 | 82.4 | 547,551 | 1,760 |
| Granite City | 55,604 | 24.6 | 1.4 | 59,958 | 30,313 | 30,772 | 36,971 | 23,446 | 13,913 | 11,832 | 59.1 | 88,814 | 744 |
| Gurnee | 106,484 | 6.7 | 16.2 | 140,489 | 55,026 | 51,248 | 71,928 | 35,044 | 11,188 | 10,899 | 71.8 | 279,651 | 1,135 |
| Hanover Park | 80,010 | 5.2 | 4.9 | 83,604 | 30,451 | 35,627 | 41,270 | 29,866 | 12,085 | 11,890 | 73.2 | 223,543 | 1,253 |
| Harvey | 35,516 | 26.7 | 1.7 | 46,826 | 25,096 | 20,917 | 24,602 | 17,415 | 9,427 | 7,048 | 55.0 | 78,663 | 922 |
| Highland Park | 146,479 | 2.2 | 35.5 | 161,826 | 51,260 | 60,896 | 78,691 | 52,256 | 11,219 | 10,657 | 79.0 | 601,157 | 1,618 |
| Hoffman Estates | 95,145 | 2.8 | 10.1 | 108,125 | 62,363 | 48,542 | 54,892 | 40,308 | 19,053 | 17,594 | 77.4 | 302,788 | 1,493 |
| Joliet | 71,284 | 10.0 | 5.9 | 83,853 | 38,982 | 34,447 | 42,326 | 27,391 | 52,516 | 48,724 | 72.5 | 205,618 | 992 |
| Kankakee | 38,773 | 27.2 | 1.5 | 45,435 | 19,554 | 26,745 | 31,385 | 22,523 | 12,301 | 9,792 | 44.3 | 86,544 | 849 |
| Lake in the Hills | 103,036 | 5.5 | 14.2 | 114,840 | 60,578 | 47,677 | 60,797 | 32,378 | 11,056 | 10,773 | 82.9 | 243,627 | 1,291 |
| Lansing | 50,542 | 16.1 | 3.5 | 66,356 | 37,160 | 34,006 | 31,138 | 36,720 | 12,448 | 10,964 | 67.0 | 130,098 | 1,071 |
| Lombard | 80,496 | 6.2 | 7.8 | 106,089 | 50,082 | 44,972 | 52,408 | 34,776 | 18,817 | 17,484 | 68.9 | 275,295 | 1,505 |
| McHenry | 71,180 | 15.0 | 13.0 | 91,068 | 33,200 | 42,244 | 56,439 | 30,832 | 10,469 | 10,397 | 68.8 | 181,790 | 1,110 |
| Melrose Park | 47,123 | 17.4 | 3.3 | 59,303 | 30,905 | 32,511 | 38,024 | 27,175 | 8,649 | 7,711 | 48.8 | 234,780 | 1,076 |
| Moline | 50,308 | 17.6 | 3.2 | 82,550 | 30,154 | 32,127 | 41,308 | 25,169 | 20,414 | 18,512 | 62.5 | 111,851 | 739 |
| Mount Prospect | 89,514 | 10.4 | 11.1 | 107,117 | 44,732 | 50,081 | 62,940 | 35,985 | 22,674 | 20,895 | 67.7 | 339,097 | 1,175 |
| Mundelein | 93,650 | 7.6 | 9.3 | 96,190 | 70,546 | 42,274 | 50,518 | 31,092 | 11,720 | 11,573 | 73.6 | 233,743 | 1,424 |
| Naperville | 140,061 | 5.0 | 28.1 | 153,250 | 75,184 | 61,547 | 81,832 | 45,496 | 54,513 | 52,238 | 74.3 | 422,985 | 1,620 |
| Niles | 83,485 | 8.5 | 10.9 | 99,935 | 46,250 | 36,204 | 50,112 | 29,600 | 10,791 | 10,598 | 82.5 | 294,007 | 1,172 |
| Normal | 58,989 | 22.7 | 5.7 | 103,023 | 28,252 | 20,845 | 20,643 | 21,053 | 21,501 | 19,058 | 54.5 | 169,151 | 859 |
| Northbrook | 128,381 | 4.2 | 33.7 | 165,456 | 50,725 | 71,184 | 91,113 | 53,272 | 13,915 | 12,761 | 85.7 | 582,564 | 1,843 |
| North Chicago | 45,152 | 21.4 | 1.5 | 54,118 | 34,778 | 20,860 | 21,269 | 18,898 | 8,386 | 7,630 | 36.1 | 94,537 | 1,401 |
| Oak Forest | 77,465 | 5.6 | 4.0 | 83,621 | 68,842 | 41,461 | 51,729 | 34,210 | 10,721 | 10,101 | 81.2 | 199,753 | 1,009 |
| Oak Lawn | 63,239 | 16.6 | 5.8 | 84,554 | 34,844 | 40,892 | 48,465 | 32,403 | 24,849 | 22,728 | 83.5 | 212,510 | 1,143 |
| Oak Park | 97,821 | 9.8 | 21.3 | 136,776 | 55,108 | 60,513 | 75,453 | 48,625 | 21,869 | 21,129 | 62.7 | 399,381 | 1,200 |
| O'Fallon | 86,978 | 8.8 | 9.1 | 102,199 | 40,028 | 49,712 | 62,086 | 32,418 | 11,661 | 11,145 | 69.9 | 225,747 | 1,111 |
| Orland Park | 86,635 | 5.1 | 14.1 | 121,344 | 55,689 | 49,699 | 56,737 | 40,851 | 23,766 | 22,233 | 84.0 | 300,677 | 1,267 |
| Oswego | 100,245 | 9.4 | 7.1 | 120,446 | 40,822 | 45,518 | 56,789 | 30,049 | 12,598 | 12,598 | 87.5 | 267,314 | 1,614 |
| Palatine | 83,834 | 8.3 | 13.8 | 111,162 | 60,103 | 50,565 | 56,238 | 43,812 | 27,841 | 25,880 | 68.4 | 304,309 | 1,181 |
| Park Ridge | 116,142 | 6.2 | 23.6 | 136,611 | 70,359 | 68,603 | 85,261 | 50,475 | 15,096 | 13,894 | 84.8 | 428,026 | 1,443 |
| Pekin | 47,596 | 15.1 | 1.5 | 66,183 | 28,915 | 27,705 | 39,280 | 22,128 | 15,448 | 13,216 | 62.8 | 110,874 | 627 |
| Peoria | 53,360 | 20.2 | 6.9 | 72,281 | 30,912 | 31,155 | 36,955 | 26,669 | 53,580 | 46,130 | 58.5 | 124,600 | 759 |
| Plainfield | 133,708 | 3.0 | 19.3 | 134,199 | 74,858 | 52,901 | 79,459 | 41,439 | 12,965 | 12,247 | 92.5 | 320,579 | 1,468 |
| Quincy | 52,182 | 13.0 | 2.2 | 72,211 | 41,519 | 32,372 | 39,000 | 30,300 | 19,265 | 17,189 | 66.4 | 120,157 | 784 |
| Rockford | 46,803 | 23.8 | 3.0 | 55,814 | 28,620 | 26,936 | 30,937 | 23,512 | 64,492 | 57,463 | 51.9 | 98,862 | 824 |
| Rock Island | 50,099 | 21.9 | 1.5 | 67,346 | 35,897 | 30,573 | 31,846 | 27,306 | 18,439 | 16,404 | 56.6 | 109,500 | 746 |
| Romeoville | 79,452 | 8.0 | 4.5 | 91,711 | 42,147 | 36,304 | 43,414 | 29,729 | 12,411 | 11,119 | 84.6 | 193,986 | 1,749 |
| Round Lake Beach | 61,751 | 11.0 | 4.1 | 73,473 | 32,141 | 32,336 | 37,704 | 25,861 | 9,321 | 8,506 | 79.0 | 147,587 | 0 |
| St. Charles | 95,417 | 7.9 | 13.7 | 104,169 | 58,551 | 50,053 | 56,151 | 40,519 | 14,427 | 14,427 | 70.9 | 307,244 | 1,307 |
| Schaumburg | 87,156 | 6.7 | 7.9 | 100,673 | 68,873 | 52,161 | 63,994 | 45,036 | 32,247 | 30,713 | 61.9 | 255,728 | 1,444 |
| Skokie | 76,324 | 11.9 | 9.7 | 86,939 | 54,176 | 32,208 | 39,781 | 27,789 | 22,117 | 20,060 | 74.8 | 328,395 | 1,266 |
| Springfield | 54,484 | 16.9 | 4.2 | 75,437 | 39,291 | 37,950 | 46,435 | 31,843 | 56,010 | 49,544 | 57.2 | 143,172 | 814 |
| Streamwood | 77,389 | 9.3 | 5.7 | 90,206 | 59,359 | 35,259 | 41,199 | 30,546 | 14,084 | 13,226 | 79.6 | 215,833 | 1,687 |
| Tinley Park | 82,770 | 6.8 | 7.1 | 108,614 | 52,494 | 44,720 | 56,010 | 40,644 | 23,881 | 22,998 | 89.3 | 236,297 | 1,169 |
| Urbana | 38,756 | 29.5 | 3.8 | 61,937 | 26,135 | 22,954 | 23,293 | 22,605 | 19,702 | 16,560 | 32.2 | 157,129 | 871 |
| Vernon Hills | 102,457 | 6.5 | 20.2 | 116,028 | 50,510 | 51,180 | 68,207 | 41,438 | 10,761 | 10,092 | 59.5 | 352,008 | 1,464 |

1. Based on population estimated by the American Community Survey.

Items 42—54

# Table D. Cities — Commuting, Computer Access, Migration, Labor Force, and Employment

Column groups: **Commuting[1], 2019** (Percent) = 55, 56 · **Computer access[2], 2019** (Percent) = 57, 58 · **Migration, 2019** = 59, 60 · **Civilian labor force, 2020** = 61, 62 · **Unemployment** = 63, 64 · **Civilian Employment[4], 2019** — Population age 16 and older = 65, 66; Population age 16 to 64 = 67, 68

| City | Drove alone | With commutes of 30 minutes or more | With a computer in the house | With Internet access | Percent who lived in the same house one year ago | Percent who lived in another state or county one year ago | Total | Percent change 2019–2020 | Total | Rate[3] | Number | Percent in labor force | Number | Percent who worked full-year full-time |
|---|---|---|---|---|---|---|---|---|---|---|---|---|---|---|
| | 55 | 56 | 57 | 58 | 59 | 60 | 61 | 62 | 63 | 64 | 65 | 66 | 67 | 68 |
| **ILLINOIS— Cont'd** | | | | | | | | | | | | | | |
| Burbank | 79.3 | 50.1 | 89.0 | 77.4 | NA | NA | 13,490 | -2.7 | 1,462 | 11 | 22,602 | 69.1 | 19,249 | 56.8 |
| Calumet City | 73.5 | 57.2 | 95.1 | 88.7 | NA | NA | 15,838 | -0.2 | 2,433 | 15 | 29,262 | 67.3 | 25,695 | 51.3 |
| Carbondale | 66.4 | 14.6 | 93.6 | 79.3 | 63.6 | 23.7 | 11,352 | -2.2 | 985 | 9 | 22,491 | 58.8 | 20,788 | 24.7 |
| Carol Stream | 75.3 | 46.0 | 97.0 | 95.4 | 88.9 | 1.5 | 22,343 | -4.5 | 1,867 | 8 | 32,190 | 70.7 | 26,020 | 56.7 |
| Carpentersville | 79.6 | 50.8 | NA | NA | 90.3 | 7.9 | 18,285 | -4.7 | 2,073 | 11 | 26,880 | 78.6 | 24,186 | 56.7 |
| Champaign | 64.0 | 7.2 | 94.2 | 84.8 | 74.9 | 13.0 | 46,273 | 1.1 | 3,057 | 7 | 76,171 | 59.5 | 66,025 | 39.1 |
| Chicago | 48.2 | 63.1 | 91.0 | 82.6 | 85.3 | 3.4 | 1,324,384 | -1.1 | 158,943 | 12 | 2,197,345 | 67.1 | 1,834,716 | 54.6 |
| Chicago Heights | 77.4 | 59.4 | 87.9 | 80.8 | NA | NA | 12,254 | -2.1 | 1,616 | 13 | 24,295 | 63.6 | 20,091 | 53.1 |
| Cicero | 70.8 | 47.9 | 88.1 | 85.3 | NA | NA | 34,535 | -2.9 | 3,812 | 11 | 61,635 | 64.0 | 55,570 | 51.9 |
| Collinsville | 84.0 | 38.2 | 92.2 | 89.6 | 87.1 | 7.1 | 12,760 | -2.1 | 1,106 | 9 | 18,241 | 67.3 | 14,011 | 59.1 |
| Crystal Lake | 82.7 | 47.0 | 95.2 | 94.7 | 87.7 | 4.3 | 21,252 | -4.4 | 1,723 | 8 | 27,926 | 72.2 | 22,766 | 58.8 |
| Danville | 84.0 | 6.5 | 83.7 | 77.3 | 88.7 | 2.8 | 12,093 | -0.7 | 1,284 | 11 | 23,223 | 52.0 | 18,442 | 37.4 |
| Decatur | 86.7 | 12.7 | 88.3 | 71.7 | 85.9 | 6.0 | 31,038 | -2.6 | 3,554 | 12 | 57,330 | 53.3 | 43,151 | 48.6 |
| DeKalb | 79.0 | 25.6 | 97.3 | 93.4 | 74.1 | 20.8 | 21,491 | -3.9 | 1,909 | 9 | 36,958 | 69.8 | 32,702 | 45.7 |
| Des Plaines | 78.1 | 49.7 | 91.6 | 84.3 | 89.5 | 2.9 | 31,751 | -2.6 | 3,060 | 10 | 49,260 | 65.1 | 37,774 | 59.3 |
| Downers Grove | 70.7 | 46.3 | 93.9 | 90.5 | 91.3 | 5.0 | 25,657 | -5.3 | 1,815 | 7 | 38,263 | 71.4 | 29,222 | 59.8 |
| East St. Louis | ********** | 28.2 | 62.9 | 38.7 | NA | NA | 8,928 | -0.1 | 1,224 | 14 | 20,903 | 51.1 | 15,657 | 43.6 |
| Elgin | 77.1 | 45.6 | 95.3 | 88.7 | 90.9 | 3.3 | 55,350 | -4.1 | 5,896 | 11 | 84,592 | 66.9 | 69,867 | 55.1 |
| Elk Grove Village | 77.8 | 36.7 | 92.1 | 84.3 | NA | NA | 18,075 | -3.2 | 1,625 | 9 | 27,897 | 65.9 | 20,873 | 59.2 |
| Elmhurst | 69.7 | 46.2 | 95.2 | 92.1 | 90.5 | 3.7 | 22,454 | -4.4 | 1,585 | 7 | 36,177 | 67.0 | 28,178 | 55.1 |
| Evanston | 45.2 | 52.4 | 95.2 | 90.0 | 84.2 | 5.9 | 36,718 | -4.8 | 2,957 | 8 | 60,534 | 62.6 | 48,566 | 47.7 |
| Freeport | 79.1 | 19.7 | 89.7 | 80.9 | NA | NA | 10,435 | -2.2 | 941 | 9 | 19,642 | 55.2 | 13,520 | 52.9 |
| Galesburg | 78.3 | 13.5 | 80.5 | 71.0 | 89.8 | 6.0 | 11,862 | -2.7 | 1,127 | 10 | 23,615 | 44.6 | 17,516 | 33.3 |
| Glendale Heights | 81.2 | 32.9 | NA | NA | 88.9 | 4.1 | 18,193 | -3.8 | 1,698 | 9 | 25,172 | 74.9 | 21,406 | 60.2 |
| Glen Ellyn | 74.7 | 49.3 | 95.6 | 91.5 | 83.8 | 5.6 | 13,323 | -5.2 | 980 | 7 | 21,755 | 67.2 | 16,218 | 51.5 |
| Glenview | 69.1 | 53.6 | 93.1 | 89.9 | 89.2 | 2.3 | 22,006 | -3.8 | 1,738 | 8 | 37,426 | 66.3 | 26,879 | 60.3 |
| Granite City | ********** | 35.4 | 92.4 | 78.9 | 93.0 | 1.6 | 12,884 | -2.2 | 1,195 | 9 | 23,088 | 61.0 | 18,693 | 54.6 |
| Gurnee | 86.4 | 46.5 | 96.9 | 90.1 | 88.7 | 3.8 | 16,888 | -3.6 | 1,294 | 8 | 23,525 | 73.7 | 20,154 | 58.0 |
| Hanover Park | 75.4 | 50.7 | NA | NA | 88.0 | 6.0 | 19,549 | -3.7 | 1,938 | 10 | 31,675 | 70.0 | 26,869 | 58.5 |
| Harvey | 63.7 | 45.2 | 80.6 | 73.5 | NA | NA | 8,028 | 1.1 | 1,401 | 18 | 17,065 | 55.9 | 13,508 | 37.1 |
| Highland Park | 65.5 | 46.4 | 95.6 | 93.8 | 89.8 | 6.6 | 14,900 | -3.8 | 982 | 7 | 23,017 | 59.2 | 15,195 | 53.9 |
| Hoffman Estates | 78.7 | 46.2 | 99.5 | 96.2 | 93.7 | 1.9 | 28,321 | -3.9 | 2,426 | 9 | 39,338 | 73.6 | 33,691 | 62.3 |
| Joliet | 83.6 | 47.7 | 93.9 | 86.7 | 88.4 | 5.0 | 71,896 | -2.8 | 8,077 | 11 | 114,608 | 69.1 | 98,286 | 53.0 |
| Kankakee | 73.8 | 19.6 | 81.7 | 68.0 | 93.8 | 1.5 | 10,877 | -2.8 | 1,289 | 12 | 19,249 | 59.6 | 15,867 | 51.8 |
| Lake in the Hills | 86.4 | 51.7 | NA | NA | 89.3 | 5.4 | 15,369 | -3.9 | 1,398 | 9 | 25,018 | 77.0 | 22,821 | 57.5 |
| Lansing | 80.9 | 46.8 | 94.2 | 85.5 | NA | NA | 13,697 | -1.8 | 1,687 | 12 | 22,926 | 59.5 | 18,913 | 47.4 |
| Lombard | 78.5 | 39.8 | 90.2 | 87.0 | 87.9 | 6.4 | 24,358 | -4.1 | 2,013 | 8 | 35,509 | 64.4 | 28,049 | 55.3 |
| McHenry | 83.6 | 48.4 | NA | NA | 87.1 | 5.6 | 13,580 | -4.0 | 1,109 | 8 | 20,607 | 67.6 | 15,771 | 59.5 |
| Melrose Park | 80.9 | 41.5 | 85.8 | 64.9 | NA | NA | 11,236 | -3.6 | 1,137 | 10 | 18,035 | 63.3 | 14,398 | 58.3 |
| Moline | 82.7 | 19.1 | 93.0 | 86.6 | 85.2 | 3.3 | 21,068 | -4.0 | 2,027 | 10 | 33,396 | 63.5 | 26,119 | 51.2 |
| Mount Prospect | 73.7 | 50.0 | 91.6 | 88.0 | 93.8 | 1.4 | 27,815 | -4.6 | 2,242 | 8 | 41,958 | 68.8 | 32,506 | 60.8 |
| Mundelein | 87.6 | 39.9 | 95.6 | 92.3 | 88.3 | 6.8 | 17,933 | -3.4 | 1,457 | 8 | 25,551 | 70.4 | 20,663 | 57.0 |
| Naperville | 72.4 | 50.7 | 98.7 | 97.7 | 87.7 | 7.5 | 74,588 | -4.8 | 5,194 | 7 | 119,162 | 70.1 | 100,113 | 58.5 |
| Niles | 76.5 | 44.2 | 91.7 | 87.8 | 90.8 | 3.6 | 13,179 | -2.3 | 1,397 | 11 | 26,367 | 58.1 | 18,225 | 47.1 |
| Normal | 76.5 | 10.9 | 96.1 | 77.4 | 74.7 | 14.5 | 27,293 | -3.9 | 1,712 | 6 | 47,686 | 59.6 | 41,375 | 36.1 |
| Northbrook | 73.9 | 41.5 | 98.1 | 93.6 | NA | NA | 15,320 | -4.3 | 1,253 | 8 | 24,900 | 63.7 | 17,058 | 50.8 |
| North Chicago | 41.8 | 17.8 | 91.5 | 87.2 | 63.4 | 29.5 | 8,913 | -3.2 | 896 | 10 | 25,144 | 44.7 | 23,599 | 54.9 |
| Oak Forest | 80.4 | 49.9 | 92.1 | 88.4 | NA | NA | 14,213 | -3.5 | 1,401 | 10 | 21,629 | 69.5 | 16,513 | 60.1 |
| Oak Lawn | 81.5 | 55.7 | 88.8 | 74.6 | NA | NA | 27,055 | -2.9 | 2,822 | 10 | 45,200 | 51.3 | 30,951 | 48.9 |
| Oak Park | 51.2 | 63.5 | 98.7 | 94.4 | 86.5 | 4.2 | 28,371 | -3.7 | 2,298 | 8 | 41,306 | 71.3 | 34,051 | 59.6 |
| O'Fallon | 85.8 | 35.0 | 98.3 | 95.1 | 86.5 | 5.7 | 14,058 | -1.7 | 1,113 | 8 | 21,734 | 69.7 | 18,176 | 64.4 |
| Orland Park | 79.9 | 54.8 | 95.5 | 89.1 | 86.6 | 5.3 | 28,626 | -4.2 | 2,440 | 9 | 48,766 | 63.1 | 36,272 | 53.6 |
| Oswego | 73.9 | 46.9 | NA | NA | NA | NA | 18,856 | -1.5 | 1,479 | 8 | 29,616 | 72.5 | 24,871 | 55.4 |
| Palatine | 79.0 | 46.7 | 96.4 | 93.2 | 89.3 | 4.7 | 36,710 | -4.8 | 2,932 | 8 | 50,400 | 73.6 | 40,796 | 62.0 |
| Park Ridge | 71.6 | 47.7 | 95.2 | 93.1 | 89.1 | 2.6 | 18,348 | -4.6 | 1,482 | 8 | 29,043 | 63.3 | 20,828 | 59.3 |
| Pekin | 79.5 | 30.9 | 82.6 | 71.4 | 83.9 | 4.0 | 14,354 | -3.4 | 1,308 | 8 | 25,801 | 55.9 | 19,674 | 50. |
| Peoria | 82.2 | 10.9 | 90.5 | 82.0 | 87.6 | 4.1 | 50,697 | -1.5 | 5,653 | 11 | 86,417 | 60.4 | 67,895 | 47. |
| Plainfield | 79.4 | 51.1 | 98.2 | 95.3 | 86.1 | 8.7 | 22,016 | -3.6 | 1,782 | 8 | 29,746 | 72.8 | 25,195 | 61. |
| Quincy | 84.5 | 9.9 | 82.5 | 78.3 | 84.4 | 5.6 | 18,699 | -2.9 | 1,279 | 7 | 32,229 | 64.2 | 23,665 | 57. |
| Rockford | 81.8 | 20.9 | 90.5 | 82.1 | 84.6 | 3.7 | 65,041 | -2.2 | 8,719 | 13 | 110,021 | 62.1 | 87,250 | 46. |
| Rock Island | 79.7 | 18.3 | 92.0 | 88.1 | 81.2 | 7.6 | 17,426 | -4.2 | 1,690 | 10 | 31,958 | 63.6 | 25,834 | 51. |
| Romeoville | 78.6 | 49.8 | 94.3 | 87.0 | 90.3 | 5.1 | 19,325 | -1.9 | 2,091 | 11 | 28,372 | 65.8 | 23,855 | 51. |
| Round Lake Beach | 78.6 | 48.0 | 96.4 | 92.9 | 91.8 | 2.8 | 14,649 | -3.3 | 1,565 | 11 | 21,280 | 66.5 | 17,586 | 50. |
| St. Charles | 81.1 | 37.9 | 94.2 | 91.1 | 84.0 | 8.3 | 17,908 | -5.1 | 1,393 | 8 | 28,174 | 63.7 | 20,927 | 57. |
| Schaumburg | 84.2 | 47.6 | 95.0 | 90.3 | 86.3 | 4.2 | 41,347 | -4.4 | 3,452 | 8 | 58,959 | 68.3 | 47,209 | 58. |
| Skokie | 63.9 | 41.6 | 97.0 | 94.0 | 91.7 | 1.1 | 31,359 | -1.9 | 3,394 | 11 | 50,900 | 57.3 | 38,468 | 44. |
| Springfield | 79.6 | 11.2 | 89.8 | 78.5 | 83.2 | 2.1 | 55,342 | -4.1 | 5,003 | 9 | 91,050 | 58.5 | 68,732 | 48 |
| Streamwood | 82.7 | 37.1 | 96.8 | 95.4 | NA | NA | 22,122 | -3.6 | 2,145 | 10 | 33,303 | 74.8 | 28,242 | 59. |
| Tinley Park | 84.6 | 60.2 | 96.3 | 93.1 | 95.1 | 1.8 | 30,187 | -4.0 | 2,656 | 9 | 49,522 | 68.1 | 38,140 | 55. |
| Urbana | 52.4 | 5.7 | 92.4 | 74.9 | 71.6 | 18.8 | 21,788 | 0.1 | 1,405 | 6 | 36,837 | 60.0 | 32,645 | 30 |
| Vernon Hills | 82.6 | 37.1 | 95.5 | 94.1 | 81.5 | 6.7 | 15,107 | -3.6 | 1,098 | 7 | 20,642 | 70.0 | 18,019 | 56 |

1. Employed persons.  2. Households.  3. Percent of civilian labor force.  4. Persons 16 years old and over.

| City | Value of residential construction authorized by building permits, 2020 | | | Wholesale trade[1], 2017 | | | | Retail trade[2], 2017 | | | |
|---|---|---|---|---|---|---|---|---|---|---|---|
| | New construction ($1,000) | Number of housing units | Percent single family | Number of establishments | Number of employees | Sales (mil dol) | Annual payroll (mil dol) | Number of establish-ments | Number of employees | Sales (mil dol) | Annual payroll (mil dol) |
| | 69 | 70 | 71 | 72 | 73 | 74 | 75 | 76 | 77 | 78 | 79 |
| ILLINOIS— Cont'd | | | | | | | | | | | |
| Burbank | 2,472 | 12 | 100 | 9 | 25 | 9 | 1 | 90 | 1,504 | 347 | 31 |
| Calumet City | 0 | 0 | 0 | D | D | D | D | 136 | 2,172 | 429 | 49 |
| Carbondale | 620 | 4 | 100 | 12 | 152 | 46 | 7 | 156 | 2,777 | 716 | 59 |
| Carol Stream | 0 | 0 | 0 | 105 | 3,112 | 5,025 | 210 | 101 | 2,073 | 543 | 54 |
| Carpentersville | 1,105 | 3 | 100 | 11 | 42 | 37 | 3 | 49 | 1,151 | 348 | 30 |
| Champaign | 157,760 | 897 | 7 | 67 | 1,404 | 656 | 65 | 389 | 6,358 | 1,517 | 143 |
| Chicago | 538,025 | 4,133 | 6 | 2,245 | 39,863 | 40,410 | 2,841 | 6,945 | 97,034 | 23,313 | 2,569 |
| Chicago Heights | 175 | 1 | 100 | 36 | 746 | 759 | 42 | 63 | 639 | 162 | 14 |
| Cicero | 681 | 3 | 100 | 34 | 947 | 648 | 51 | 133 | 2,454 | 741 | 62 |
| Collinsville | 1,339 | 6 | 100 | 20 | 206 | 171 | 12 | 80 | 1,588 | 765 | 53 |
| Crystal Lake | 17,652 | 83 | 100 | D | D | D | D | 208 | 3,792 | 1,054 | 101 |
| Danville | 334 | 2 | 100 | 35 | 1,528 | 3,299 | 78 | 151 | 2,587 | 595 | 59 |
| Decatur | 0 | 0 | 0 | 69 | 1,085 | 1,041 | 61 | 285 | 4,159 | 1,198 | 114 |
| DeKalb | 3,468 | 65 | 9 | 21 | 263 | 96 | 12 | 113 | 2,387 | 498 | 51 |
| Des Plaines | 10,391 | 40 | 100 | 125 | 3,192 | 2,652 | 262 | 178 | 3,134 | 886 | 88 |
| Downers Grove | 18,078 | 35 | 100 | 116 | 2,557 | 6,220 | 272 | 228 | 5,096 | 1,599 | 140 |
| East St. Louis | 557 | 14 | 0 | 18 | 199 | 616 | 10 | 52 | 312 | 75 | 7 |
| Elgin | 33,077 | 153 | 100 | 176 | 4,035 | 5,637 | 284 | 208 | 3,703 | 1,186 | 107 |
| Elk Grove Village | 2,620 | 9 | 100 | 426 | 8,055 | 13,689 | 552 | 131 | 2,463 | 1,210 | 86 |
| Elmhurst | 30,700 | 71 | 92 | 126 | 4,038 | 3,055 | 459 | 154 | 2,963 | 1,215 | 90 |
| Evanston | 16,699 | 65 | 8 | 41 | 298 | 447 | 20 | 205 | 4,000 | 1,012 | 100 |
| Freeport | 220 | 2 | 0 | 19 | 218 | 196 | 11 | 110 | 1,563 | 403 | 38 |
| Galesburg | 0 | 0 | 0 | 21 | 450 | 332 | 21 | 142 | 3,326 | 794 | 91 |
| Glendale Heights | 0 | 0 | 0 | 73 | 1,386 | 1,199 | 97 | 85 | 2,212 | 662 | 62 |
| Glen Ellyn | 10,333 | 19 | 100 | 33 | 210 | 278 | 18 | 92 | 1,226 | 324 | 30 |
| Glenview | 26,967 | 51 | 90 | 92 | 1,690 | 1,901 | 133 | 159 | 4,446 | 1,776 | 197 |
| Granite City | 705 | 6 | 100 | 26 | 391 | 441 | 22 | 81 | 1,293 | 361 | 37 |
| Gurnee | 482 | 3 | 100 | 61 | 634 | 504 | 37 | 241 | 5,817 | 1,183 | 117 |
| Hanover Park | 0 | 0 | 0 | 28 | 1,390 | 1,142 | 74 | 66 | 962 | 255 | 24 |
| Harvey | 0 | 0 | 0 | 20 | 274 | 94 | 15 | 60 | 280 | 76 | 6 |
| Highland Park | 26,154 | 77 | 34 | 46 | 160 | 649 | 17 | 159 | 2,623 | 1,229 | 96 |
| Hoffman Estates | 1,018 | 4 | 100 | 68 | 1,016 | 846 | 102 | 118 | 2,755 | 839 | 77 |
| Joliet | 60,443 | 318 | 76 | 91 | 2,124 | 3,045 | 109 | 392 | 8,204 | 2,100 | 193 |
| Kankakee | 0 | 0 | 0 | 30 | 454 | 303 | 25 | 89 | 1,249 | 260 | 27 |
| Lake in the Hills | 500 | 2 | 100 | 19 | 197 | 91 | 15 | 55 | 817 | 264 | 25 |
| Lansing | 0 | 0 | 0 | 25 | 210 | 107 | 9 | 96 | 1,894 | 484 | 48 |
| Lombard | 4,921 | 15 | 100 | 122 | 1,888 | 1,115 | 109 | 236 | 4,109 | 1,050 | 106 |
| McHenry | 12,978 | 98 | 100 | 42 | 1,859 | 775 | 114 | 122 | 2,138 | 526 | 55 |
| Melrose Park | 0 | 0 | 0 | 51 | 787 | 552 | 40 | 102 | 2,641 | 862 | 71 |
| Moline | 3,224 | 40 | 38 | 38 | 711 | 1,001 | 55 | 213 | 4,452 | 1,157 | 111 |
| Mount Prospect | 2,110 | 3 | 100 | 79 | 1,334 | 1,474 | 99 | 151 | 2,723 | 2,642 | 67 |
| Mundelein | 10,277 | 69 | 100 | 55 | 612 | 349 | 37 | 95 | 1,044 | 270 | 27 |
| Naperville | 115,088 | 431 | 74 | 243 | 2,458 | 2,901 | 175 | 460 | 10,909 | 3,928 | 318 |
| Niles | 0 | 0 | 0 | 93 | 2,154 | 1,237 | 196 | 248 | 6,140 | 2,080 | 179 |
| Normal | 15,631 | 81 | 80 | D | D | D | D | 129 | 3,210 | 803 | 72 |
| Northbrook | 25,036 | 44 | 100 | 169 | 3,492 | 5,560 | 411 | 219 | 4,044 | 1,299 | 140 |
| North Chicago | 1,671 | 6 | 100 | 8 | 273 | 202 | 19 | 33 | 125 | 72 | 3 |
| Oak Forest | 150 | 1 | 100 | 23 | 96 | 48 | 5 | 55 | 637 | 179 | 14 |
| Oak Lawn | 3,655 | 17 | 100 | 28 | 80 | 31 | 3 | 172 | 3,707 | 1,250 | 105 |
| Oak Park | 500 | 1 | 100 | D | D | D | 24 | 146 | 1,736 | 358 | 36 |
| O'Fallon | 37,775 | 154 | 70 | 11 | 47 | 22 | 3 | 92 | 2,537 | 922 | 86 |
| Orland Park | 34,875 | 184 | 19 | 52 | 292 | 134 | 18 | 342 | 8,662 | 2,169 | 224 |
| Oswego | 21,500 | 137 | 77 | D | D | D | D | 96 | 2,497 | 519 | 50 |
| Palatine | 3,164 | 7 | 100 | 61 | 401 | 373 | 20 | 179 | 3,328 | 842 | 87 |
| Park Ridge | 13,107 | 20 | 100 | 43 | 192 | 457 | 17 | 72 | 1,245 | 433 | 37 |
| Pekin | 2,079 | 10 | 80 | 18 | 300 | 182 | 16 | 110 | 1,992 | 557 | 54 |
| Peoria | 11,707 | 35 | 100 | 129 | 2,216 | 992 | 107 | 509 | 8,716 | 1,989 | 206 |
| Plainfield | 66,649 | 256 | 100 | 25 | 236 | 276 | 17 | 111 | 2,443 | 611 | 56 |
| Quincy | 13,437 | 62 | 100 | 67 | 1,366 | 649 | 58 | 250 | 4,591 | 1,020 | 106 |
| Rockford | 7,421 | 49 | 25 | 167 | 2,509 | 1,845 | 129 | 517 | 8,093 | 2,239 | 206 |
| Rock Island | 425 | 2 | 100 | 48 | 1,021 | 720 | 56 | 75 | 887 | 169 | 22 |
| Romeoville | 44,358 | 282 | 15 | 68 | 2,516 | 9,743 | 169 | 68 | 1,695 | 414 | 36 |
| Round Lake Beach | 0 | 0 | 0 | 6 | 25 | 7 | 1 | 57 | 1,703 | 376 | 37 |
| St. Charles | 17,797 | 87 | 41 | 113 | 1,307 | 1,411 | 85 | 146 | 3,705 | 1,182 | 110 |
| Schaumburg | 7,635 | 48 | 4 | 263 | 4,993 | 8,377 | 430 | 465 | 12,304 | 3,075 | 323 |
| Skokie | 4,650 | 17 | 100 | 112 | 971 | 617 | 59 | 334 | 6,593 | 1,454 | 171 |
| Springfield | 32,730 | 155 | 39 | 116 | 1,917 | 2,523 | 103 | 570 | 10,751 | 2,646 | 258 |
| Streamwood | 262 | 1 | 100 | 18 | 118 | 87 | 7 | 78 | 1,673 | 480 | 42 |
| Tinley Park | 2,866 | 14 | 100 | 52 | 779 | 282 | 47 | 161 | 3,816 | 1,397 | 109 |
| Urbana | 40,456 | 422 | 5 | 22 | 797 | 821 | 37 | 76 | 1,430 | 390 | 37 |
| Vernon Hills | 20,337 | 33 | 100 | 66 | 3,002 | 2,005 | 256 | 193 | 4,474 | 731 | 91 |

1. Merchant wholesalers except manufacturers' sales branches and offices.  2. Establishments with payroll.

# Table D. Cities — Real Estate, Professional Services, and Manufacturing

| City | Real estate and rental and leasing, 2017 | | | | Professional, scientific, and technical services[1], 2017 | | | | Manufacturing, 2017 | | | |
|---|---|---|---|---|---|---|---|---|---|---|---|---|
| | Number of establishments | Number of employees | Receipts (mil dol) | Annual payroll (mil dol) | Number of establishments | Number of employees | Receipts (mil dol) | Annual payroll (mil dol) | Number of establishments | Number of employees | Receipts (mil dol) | Annual payroll (mil dol) |
| | 80 | 81 | 82 | 83 | 84 | 85 | 86 | 87 | 88 | 89 | 90 | 91 |
| **ILLINOIS— Cont'd** | | | | | | | | | | | | |
| Burbank | 12 | 30 | 7 | 1 | 32 | 96 | 10 | 4 | NA | NA | NA | NA |
| Calumet City | 18 | 126 | 23 | 3 | 22 | 66 | 6 | 2 | NA | NA | NA | NA |
| Carbondale | 53 | 279 | 52 | 7 | 61 | 618 | 66 | 27 | NA | NA | NA | NA |
| Carol Stream | 36 | 278 | 75 | 11 | 96 | 503 | 77 | 29 | NA | NA | NA | NA |
| Carpentersville | 21 | 94 | 19 | 4 | 28 | 67 | 8 | 3 | NA | NA | NA | NA |
| Champaign | 136 | 1,541 | 305 | 68 | 257 | 1,783 | 264 | 110 | NA | NA | NA | NA |
| Chicago | 3,684 | 28,995 | 13,429 | 2,103 | 10,735 | 174,661 | 48,102 | 17,622 | NA | NA | NA | NA |
| Chicago Heights | 14 | 96 | 29 | 5 | 27 | 203 | 37 | 8 | NA | NA | NA | NA |
| Cicero | 28 | 98 | 24 | 4 | 41 | 299 | 24 | 9 | NA | NA | NA | NA |
| Collinsville | 21 | 97 | 19 | 4 | 74 | 520 | 99 | 29 | NA | NA | NA | NA |
| Crystal Lake | 54 | 228 | 39 | 9 | 191 | 947 | 260 | 52 | NA | NA | NA | NA |
| Danville | 33 | 93 | 18 | 3 | 54 | 323 | 52 | 15 | NA | NA | NA | NA |
| Decatur | 72 | 433 | 63 | 13 | 114 | 919 | 105 | 45 | NA | NA | NA | NA |
| DeKalb | 47 | 326 | 70 | 11 | 42 | 146 | 15 | 5 | NA | NA | NA | NA |
| Des Plaines | 82 | 1,248 | 1,382 | 82 | 254 | 2,276 | 420 | 152 | NA | NA | NA | NA |
| Downers Grove | 119 | 816 | 433 | 53 | 347 | 4,485 | 1,016 | 396 | NA | NA | NA | NA |
| East St. Louis | 16 | 77 | 10 | 3 | 7 | 53 | 7 | 3 | NA | NA | NA | NA |
| Elgin | 90 | 474 | 136 | 23 | 276 | 1,678 | 284 | 101 | NA | NA | NA | NA |
| Elk Grove Village | 57 | 630 | 179 | 40 | 164 | 2,308 | 490 | 170 | NA | NA | NA | NA |
| Elmhurst | 85 | 364 | 86 | 19 | 259 | 1,457 | 244 | 78 | NA | NA | NA | NA |
| Evanston | 120 | 408 | 120 | 24 | 410 | 3,350 | 647 | 252 | NA | NA | NA | NA |
| Freeport | 25 | 72 | 8 | 2 | 50 | 282 | 38 | 13 | NA | NA | NA | NA |
| Galesburg | 29 | 102 | 14 | 3 | 49 | 246 | 29 | 11 | NA | NA | NA | NA |
| Glendale Heights | 22 | 103 | 126 | 5 | 46 | 295 | 59 | 19 | NA | NA | NA | NA |
| Glen Ellyn | 55 | 214 | 245 | 16 | 163 | 632 | 149 | 38 | NA | NA | NA | NA |
| Glenview | 106 | 244 | 76 | 14 | 288 | 1,900 | 258 | 141 | NA | NA | NA | NA |
| Granite City | 25 | 145 | 31 | 6 | 39 | 261 | 24 | 12 | NA | NA | NA | NA |
| Gurnee | 34 | 202 | 80 | 11 | 139 | 592 | 75 | 33 | NA | NA | NA | NA |
| Hanover Park | 16 | 51 | 10 | 2 | 36 | 113 | 11 | 4 | NA | NA | NA | NA |
| Harvey | 7 | 63 | 20 | 3 | D | D | 5 | D | NA | NA | NA | NA |
| Highland Park | 78 | 385 | 111 | 25 | 221 | 1,146 | 195 | 81 | NA | NA | NA | NA |
| Hoffman Estates | 51 | 139 | 40 | 8 | 208 | 1,576 | 272 | 105 | NA | NA | NA | NA |
| Joliet | 99 | 464 | 100 | 22 | 223 | 1,370 | 227 | 88 | NA | NA | NA | NA |
| Kankakee | 24 | 71 | 11 | 3 | 49 | 253 | 27 | 12 | NA | NA | NA | NA |
| Lake in the Hills | 9 | 23 | 4 | 1 | 62 | 219 | 35 | 13 | NA | NA | NA | NA |
| Lansing | 24 | 86 | 12 | 4 | 37 | 161 | 18 | 7 | NA | NA | NA | NA |
| Lombard | 82 | 433 | 92 | 23 | 250 | 2,556 | 489 | 202 | NA | NA | NA | NA |
| McHenry | 26 | 102 | 25 | 4 | 73 | 461 | 74 | 27 | NA | NA | NA | NA |
| Melrose Park | 21 | 157 | 33 | 9 | 24 | 77 | 7 | 3 | NA | NA | NA | NA |
| Moline | 62 | 291 | 69 | 11 | 101 | 1,114 | 157 | 50 | NA | NA | NA | NA |
| Mount Prospect | 44 | 168 | 59 | 10 | 175 | 652 | 116 | 42 | NA | NA | NA | NA |
| Mundelein | 23 | 50 | 16 | 2 | 105 | 3,090 | 639 | 249 | NA | NA | NA | NA |
| Naperville | 235 | 931 | 348 | 52 | 1,149 | 9,144 | 1,417 | 842 | NA | NA | NA | NA |
| Niles | 44 | 280 | 51 | 12 | 80 | 286 | 42 | 15 | NA | NA | NA | NA |
| Normal | D | D | D | 8 | 59 | 515 | 87 | 31 | NA | NA | NA | NA |
| Northbrook | 174 | 643 | 211 | 53 | 586 | 5,959 | 1,553 | 669 | NA | NA | NA | NA |
| North Chicago | 6 | 27 | 6 | 1 | 15 | 156 | 20 | 9 | NA | NA | NA | NA |
| Oak Forest | 20 | 81 | 23 | 5 | 43 | 177 | 26 | 8 | NA | NA | NA | NA |
| Oak Lawn | 40 | 146 | 25 | 4 | 90 | 324 | 34 | 15 | NA | NA | NA | NA |
| Oak Park | 86 | 378 | 80 | 17 | 306 | 1,168 | 172 | 69 | NA | NA | NA | NA |
| O'Fallon | 46 | 160 | 33 | 6 | 91 | 1,064 | 187 | 77 | NA | NA | NA | NA |
| Orland Park | 108 | 337 | 139 | 13 | 258 | 1,358 | 175 | 68 | NA | NA | NA | NA |
| Oswego | 33 | 92 | 17 | 3 | 88 | 285 | 46 | 13 | NA | NA | NA | NA |
| Palatine | 83 | 271 | 62 | 12 | 328 | 1,360 | 206 | 79 | NA | NA | NA | NA |
| Park Ridge | 87 | 387 | 64 | 19 | 243 | 792 | 134 | 46 | NA | NA | NA | NA |
| Pekin | 22 | 52 | 9 | 2 | 40 | 238 | 22 | 10 | NA | NA | NA | NA |
| Peoria | 146 | 859 | 151 | 32 | 341 | 4,342 | 672 | 277 | NA | NA | NA | NA |
| Plainfield | 37 | 88 | 29 | 4 | 143 | 924 | 138 | 88 | NA | NA | NA | NA |
| Quincy | 51 | 184 | 31 | 6 | 105 | 685 | 66 | 28 | NA | NA | NA | NA |
| Rockford | 151 | 1,051 | 207 | 47 | 384 | 4,093 | 638 | 232 | NA | NA | NA | NA |
| Rock Island | 30 | 129 | 18 | 3 | 103 | 1,232 | 212 | 83 | NA | NA | NA | NA |
| Romeoville | 25 | 252 | 81 | 19 | 44 | 259 | 34 | 12 | NA | NA | NA | NA |
| Round Lake Beach | 10 | 19 | 6 | 1 | 17 | 52 | 5 | 2 | NA | NA | NA | NA |
| St. Charles | 75 | 579 | 761 | 35 | D | D | D | D | NA | NA | NA | NA |
| Schaumburg | 151 | 1,460 | 560 | 93 | 683 | 12,555 | 2,542 | 1,155 | NA | NA | NA | NA |
| Skokie | 111 | 506 | 109 | 29 | 366 | 10,099 | 1,182 | 430 | NA | NA | NA | NA |
| Springfield | 186 | 762 | 212 | 27 | 391 | 4,850 | 712 | 305 | NA | NA | NA | NA |
| Streamwood | 19 | 212 | 62 | 15 | 75 | 241 | 42 | 9 | NA | NA | NA | NA |
| Tinley Park | 39 | 146 | 27 | 7 | 140 | 866 | 99 | 39 | NA | NA | NA | NA |
| Urbana | 30 | 197 | 30 | 6 | 71 | 560 | 67 | 28 | NA | NA | NA | NA |
| Vernon Hills | 33 | 138 | 67 | 9 | 153 | 1,971 | 437 | 164 | NA | NA | NA | NA |

1. Establishments subject to federal tax.

# Accommodation and Food Services, Arts, Entertainment, and Recreation, and Health Care and Social Assistance

| City | Accommodation and food services, 2017 | | | | Arts, entertainment, and recreation[1], 2017 | | | | Health care and social assistance[1], 2017 | | | |
|---|---|---|---|---|---|---|---|---|---|---|---|---|
| | Number of establishments | Number of employees | Receipts (mil dol) | Annual payroll (mil dol) | Number of establishments | Number of employees | Receipts (mil dol) | Annual payroll (mil dol) | Number of establishments | Number of employees | Receipts (mil dol) | Annual payroll (mil dol) |
| | 92 | 93 | 94 | 95 | 96 | 97 | 98 | 99 | 100 | 101 | 102 | 103 |
| **ILLINOIS— Cont'd** | | | | | | | | | | | | |
| Burbank | 63 | 871 | 54 | 15 | 4 | D | 6 | D | 50 | 683 | 72 | 24 |
| Calumet City | 74 | 1,113 | 69 | 17 | 6 | 58 | 5 | 1 | 58 | 2,316 | 77 | 38 |
| Carbondale | 104 | 2,152 | 79 | 24 | D | D | D | D | 112 | 3,511 | 570 | 162 |
| Carol Stream | 69 | 1,320 | 77 | 22 | 14 | 113 | 10 | 2 | 62 | 998 | 104 | 40 |
| Carpentersville | 43 | 677 | 37 | 10 | 5 | 15 | 1 | 0 | 28 | 246 | 25 | 8 |
| Champaign | 357 | 7,960 | 411 | 122 | 42 | 367 | 22 | 8 | 204 | 3,952 | 451 | 189 |
| Chicago | 6,770 | 142,220 | 12,309 | 3,607 | 1,101 | 25,651 | 3,865 | 1,180 | 6,778 | 186,215 | 22,877 | 8,948 |
| Chicago Heights | 42 | 491 | 32 | 8 | 4 | 83 | 2 | 1 | 72 | 2,582 | 185 | 86 |
| Cicero | 104 | 1,255 | 85 | 22 | 10 | 40 | 5 | 1 | 91 | 1,575 | 120 | 53 |
| Collinsville | 66 | 1,555 | 72 | 22 | 14 | 238 | 30 | 5 | 56 | 400 | 32 | 14 |
| Crystal Lake | 116 | 2,992 | 177 | 51 | 26 | 711 | 47 | 11 | 220 | 2,851 | 302 | 134 |
| Danville | 95 | 1,663 | 74 | 21 | 13 | 152 | 5 | 2 | 114 | 4,311 | 490 | 239 |
| Decatur | 199 | 3,342 | 158 | 49 | 35 | 622 | 24 | 8 | 240 | 7,644 | 941 | 328 |
| DeKalb | 103 | 1,843 | 90 | 26 | 14 | 88 | 7 | 2 | 65 | 2,360 | 291 | 95 |
| Des Plaines | 180 | 3,881 | 319 | 85 | D | D | D | D | 265 | 8,581 | 842 | 275 |
| Downers Grove | 168 | 3,000 | 199 | 56 | 26 | 463 | 20 | 7 | 265 | 5,695 | 945 | 333 |
| East St. Louis | D | D | D | D | NA | NA | NA | NA | 39 | 636 | 39 | 17 |
| Elgin | 180 | 3,329 | 173 | 49 | 26 | 1,037 | 175 | 38 | 294 | 7,370 | 935 | 387 |
| Elk Grove Village | 123 | 2,556 | 159 | 50 | 16 | 92 | 22 | 5 | 147 | 5,466 | 695 | 284 |
| Elmhurst | 128 | 2,135 | 128 | 37 | 30 | 505 | 27 | 7 | 218 | 7,388 | 1,033 | 341 |
| Evanston | 249 | 4,741 | 319 | 103 | 53 | 728 | 47 | 17 | 365 | 10,751 | 1,678 | 804 |
| Freeport | 65 | 1,248 | 55 | 16 | 10 | 279 | 6 | 3 | 90 | 2,408 | 327 | 101 |
| Galesburg | 97 | 1,498 | 73 | 20 | 14 | 146 | 6 | 2 | 122 | 3,375 | 344 | 112 |
| Glendale Heights | 60 | 1,015 | 62 | 17 | 10 | 184 | 22 | 4 | 45 | 871 | 122 | 46 |
| Glen Ellyn | D | D | D | D | 10 | 367 | 12 | 5 | 109 | 1,338 | 210 | 97 |
| Glenview | 170 | 3,400 | 210 | 74 | 41 | 688 | 57 | 24 | 288 | 5,687 | 753 | 268 |
| Granite City | 74 | 1,132 | 57 | 16 | 5 | 67 | 2 | 1 | 81 | 2,083 | 236 | 92 |
| Gurnee | 123 | 3,378 | 167 | 51 | 15 | 1,622 | 150 | 20 | 173 | 2,452 | 206 | 82 |
| Hanover Park | 52 | 481 | 34 | 8 | 3 | D | 4 | D | 41 | 618 | 41 | 17 |
| Harvey | 44 | 684 | 49 | 17 | NA | NA | NA | NA | 52 | 2,960 | 423 | 153 |
| Highland Park | 87 | 1,525 | 90 | 28 | 43 | 721 | 109 | 34 | 163 | 3,022 | 483 | 153 |
| Hoffman Estates | 105 | 2,032 | 138 | 43 | 17 | 173 | 14 | 3 | 237 | 5,751 | 748 | 283 |
| Joliet | 268 | 6,012 | 558 | 131 | 31 | 832 | 67 | 17 | 366 | 8,710 | 964 | 427 |
| Kankakee | 56 | 726 | 38 | 10 | D | D | D | D | 102 | 4,202 | 460 | 222 |
| Lake in the Hills | 29 | 548 | 27 | 8 | 9 | 91 | 6 | 3 | 47 | 402 | 49 | 18 |
| Lansing | 58 | 1,127 | 63 | 16 | 9 | 91 | 5 | 2 | 59 | 635 | 40 | 16 |
| Lombard | 147 | 3,286 | 237 | 71 | D | D | D | D | 172 | 2,853 | 305 | 109 |
| McHenry | D | D | D | D | 8 | 72 | 6 | 2 | 112 | 3,918 | 645 | 214 |
| Melrose Park | 78 | 984 | 69 | 17 | 12 | 217 | 8 | 2 | 100 | 2,851 | 302 | 124 |
| Moline | 149 | 3,051 | 146 | 44 | 16 | 507 | 15 | 6 | 201 | 3,886 | 479 | 166 |
| Mount Prospect | 110 | 1,621 | 103 | 29 | 13 | 276 | 10 | 3 | 134 | 1,161 | 105 | 45 |
| Mundelein | 83 | 1,110 | 65 | 19 | 9 | 106 | 9 | 3 | 61 | 306 | 28 | 10 |
| Naperville | 372 | 9,203 | 576 | 182 | 86 | 2,147 | 287 | 41 | 691 | 12,766 | 1,626 | 673 |
| Niles | 120 | 1,551 | 112 | 28 | 13 | 243 | 13 | 4 | 148 | 3,735 | 377 | 117 |
| Normal | 119 | 3,146 | 155 | 45 | D | D | D | D | 106 | 3,133 | 388 | 131 |
| Northbrook | 99 | 2,121 | 147 | 50 | 40 | 700 | 53 | 22 | 273 | 5,498 | 407 | 188 |
| North Chicago | 31 | 372 | 20 | 5 | 3 | 22 | 1 | 0 | D | D | D | D |
| Oak Forest | D | D | D | D | 7 | 48 | 3 | 1 | 53 | 1,068 | 55 | 25 |
| Oak Lawn | 120 | 2,905 | 239 | 56 | 12 | 142 | 8 | 2 | 240 | 9,338 | 1,689 | 545 |
| Oak Park | 108 | 1,834 | 110 | 33 | 38 | 412 | 28 | 9 | 312 | 5,360 | 569 | 237 |
| O'Fallon | 70 | 1,491 | 82 | 24 | 10 | 421 | 12 | 5 | 85 | 1,828 | 255 | 95 |
| Orland Park | D | D | D | D | 45 | 1,176 | 62 | 20 | 339 | 3,934 | 398 | 167 |
| Oswego | 78 | 1,851 | 95 | 27 | 10 | 117 | 5 | 2 | 82 | 911 | 82 | 30 |
| Palatine | 116 | 1,571 | 102 | 27 | 28 | 583 | 28 | 9 | 161 | 1,761 | 137 | 57 |
| Park Ridge | 70 | 1,121 | 69 | 20 | 21 | 206 | 20 | 7 | 243 | 11,564 | 1,982 | 759 |
| Pekin | 71 | 1,275 | 58 | 17 | 14 | 107 | 4 | 1 | 101 | 1,882 | 170 | 70 |
| Peoria | 315 | 6,046 | 322 | 95 | 51 | 1,391 | 59 | 21 | 439 | 23,741 | 3,175 | 1,337 |
| Plainfield | D | D | D | D | 21 | 496 | 17 | 5 | 161 | 1,449 | 124 | 49 |
| Quincy | 128 | 2,568 | 115 | 34 | 29 | 366 | 13 | 5 | 129 | 5,782 | 989 | 306 |
| Rockford | 342 | 7,811 | 410 | 123 | 58 | 926 | 46 | 13 | 480 | 17,434 | 2,370 | 901 |
| Rock Island | 75 | 1,686 | 125 | 31 | D | D | D | 3 | 114 | 3,488 | 326 | 152 |
| Romeoville | 60 | 1,324 | 71 | 22 | 16 | 625 | 31 | 10 | 40 | 892 | 69 | 32 |
| Round Lake Beach | 43 | 798 | 49 | 13 | 5 | 90 | 6 | 1 | 39 | 414 | 33 | 14 |
| St. Charles | 133 | 3,305 | 258 | 68 | 30 | 374 | 21 | 6 | 184 | 2,229 | 242 | 90 |
| Schaumburg | 305 | 8,473 | 628 | 181 | 43 | 1,320 | 83 | 22 | 330 | 4,671 | 519 | 190 |
| Skokie | 176 | 4,284 | 310 | 88 | 33 | 697 | 64 | 18 | 435 | 8,878 | 827 | 347 |
| Springfield | 420 | 8,603 | 426 | 130 | 67 | 1,124 | 183 | 24 | 401 | 21,292 | 2,965 | 1,018 |
| Streamwood | 74 | 1,063 | 69 | 17 | D | D | D | D | 56 | 979 | 105 | 33 |
| Tinley Park | 122 | 2,781 | 172 | 48 | 19 | 385 | 87 | 14 | 169 | 2,923 | 269 | 102 |
| Urbana | 108 | 1,969 | 106 | 31 | 15 | 141 | 4 | 1 | 75 | 7,531 | 1,405 | 565 |
| Vernon Hills | 97 | 2,048 | 128 | 38 | 21 | 619 | 36 | 11 | 139 | 1,320 | 173 | 62 |

1. Establishments subject to federal tax.

| City | Other services[1] | | | | Government employment and payroll, 2017 | | | | | | | | |
|---|---|---|---|---|---|---|---|---|---|---|---|---|---|
| | | | | | Full-time equivalent employees | March payroll | Percent of total for: | | | | | | |
| | Number of establishments | Number of employees | Receipts (mil dol) | Annual payroll (mil dol) | | Total (dollars) | Administrative, judicial, and legal | Police and corrections | Fire protection | Highways and transportation | Health and welfare | Natural resources and utilities | Education and libraries |
| | 104 | 105 | 106 | 107 | 108 | 109 | 110 | 111 | 112 | 113 | 114 | 115 | 116 |
| **ILLINOIS— Cont'd** | | | | | | | | | | | | | |
| Burbank | D | D | D | D | 119 | 865,103 | 7.4 | 46.2 | 32.6 | 10.9 | 0.0 | 0.0 | 0.0 |
| Calumet City | 46 | 155 | 17 | 5 | 335 | 1,941,229 | 3.7 | 46.0 | 24.2 | 5.8 | 4.4 | 5.3 | 10.6 |
| Carbondale | 48 | 289 | 37 | 5 | 254 | 1,151,535 | 12.4 | 35.5 | 12.5 | 10.7 | 5.0 | 18.9 | 0.0 |
| Carol Stream | 80 | 425 | 47 | 13 | 193 | 1,270,083 | 12.0 | 49.5 | 0.0 | 12.4 | 5.1 | 5.3 | 11.7 |
| Carpentersville | D | D | 11 | D | 185 | 1,271,922 | 7.5 | 39.2 | 25.5 | 10.6 | 5.7 | 10.3 | 0.0 |
| Champaign | 168 | 1,360 | 104 | 33 | 553 | 3,543,612 | 19.7 | 30.0 | 23.2 | 3.7 | 1.6 | 2.7 | 8.3 |
| Chicago | 5,308 | 48,556 | 13,066 | 2,555 | 31,798 | 220,385,761 | 6.4 | 42.8 | 16.1 | 10.9 | 3.0 | 7.7 | 2.2 |
| Chicago Heights | 38 | 175 | 21 | 6 | 270 | 1,550,273 | 6.9 | 46.8 | 27.8 | 6.2 | 0.8 | 5.6 | 2.8 |
| Cicero | 82 | 310 | 35 | 9 | 718 | 3,280,911 | 10.2 | 40.5 | 16.5 | 2.6 | 4.5 | 16.1 | 1.9 |
| Collinsville | 45 | 256 | 28 | 6 | 173 | 1,062,988 | 7.3 | 37.5 | 23.7 | 8.8 | 3.3 | 17.2 | 0.0 |
| Crystal Lake | 134 | 797 | 73 | 24 | 288 | 2,009,773 | 12.4 | 27.0 | 27.6 | 5.9 | 2.3 | 10.0 | 9.4 |
| Danville | 57 | 281 | 25 | 7 | 276 | 1,339,199 | 8.0 | 31.9 | 22.2 | 13.0 | 0.3 | 12.7 | 5.8 |
| Decatur | 121 | 843 | 277 | 31 | 507 | 3,126,016 | 9.1 | 41.0 | 25.8 | 9.0 | 1.6 | 6.8 | 3.9 |
| DeKalb | 58 | 323 | 28 | 8 | 213 | 1,617,137 | 9.4 | 43.9 | 28.1 | 9.0 | 2.7 | 4.3 | 0.0 |
| Des Plaines | 175 | 1,145 | 174 | 48 | 417 | 2,951,169 | 7.2 | 33.7 | 28.6 | 7.9 | 0.5 | 9.4 | 0.0 |
| Downers Grove | 129 | 1,512 | 244 | 87 | 402 | 2,513,313 | 16.2 | 30.8 | 26.2 | 1.6 | 2.8 | 10.9 | 8.4 |
| East St. Louis | 14 | 100 | 9 | 3 | 201 | 932,348 | 14.4 | 50.4 | 29.6 | 3.2 | 0.4 | 0.2 | 1.7 |
| Elgin | 187 | 1,566 | 212 | 70 | 744 | 5,184,943 | 10.3 | 39.1 | 24.6 | 3.4 | 0.0 | 6.5 | 0.0 |
| Elk Grove Village | 125 | 1,240 | 170 | 59 | 314 | 2,010,165 | 11.6 | 35.1 | 30.0 | 3.3 | 6.8 | 13.7 | 22.0 |
| Elmhurst | 116 | 730 | 80 | 24 | 608 | 3,539,277 | 10.6 | 25.1 | 12.8 | 8.0 | 0.0 | 22.9 | 4.8 |
| Evanston | 202 | 1,705 | 315 | 93 | 966 | 6,323,487 | 13.1 | 28.4 | 15.3 | 4.8 | 5.4 | 16.1 | 4.6 |
| Freeport | 49 | 241 | 20 | 6 | 187 | 910,738 | 5.2 | 35.2 | 29.8 | 6.3 | 0.9 | 15.7 | 4.8 |
| Galesburg | 51 | 369 | 120 | 11 | 259 | 1,317,865 | 13.4 | 30.0 | 20.6 | 8.0 | 2.2 | 32.0 | 0.0 |
| Glendale Heights | 33 | 253 | 37 | 10 | 234 | 1,355,867 | 13.8 | 38.4 | 0.0 | 9.3 | 4.9 | 24.2 | 17.6 |
| Glen Ellyn | 64 | 423 | 30 | 12 | 194 | 1,112,684 | 8.7 | 33.3 | 0.2 | 5.1 | 0.4 | 7.4 | 12.1 |
| Glenview | 169 | 969 | 118 | 36 | 360 | 2,839,149 | 3.9 | 24.6 | 27.1 | 5.0 | 1.8 | 14.3 | 0.0 |
| Granite City | 51 | 345 | 40 | 12 | 221 | 1,244,240 | 7.8 | 32.9 | 29.0 | 10.7 | 0.3 | 4.4 | 0.0 |
| Gurnee | 67 | 421 | 45 | 14 | 205 | 1,419,542 | 9.8 | 37.0 | 31.1 | 12.8 | 0.0 | 8.2 | 0.0 |
| Hanover Park | D | D | D | D | 216 | 1,390,130 | 8.8 | 44.4 | 23.2 | 4.2 | 0.0 | 27.3 | 0.0 |
| Harvey | D | D | D | D | 66 | 255,973 | 53.2 | 0.0 | 0.0 | 19.5 | 0.0 | 11.6 | 0.0 |
| Highland Park | 122 | 590 | 65 | 20 | 228 | 1,596,297 | 8.6 | 30.8 | 26.4 | 9.8 | 1.3 | 10.0 | 0.0 |
| Hoffman Estates | 72 | 349 | 30 | 10 | 354 | 2,698,643 | 12.5 | 31.9 | 31.4 | 5.3 | 1.4 | 9.2 | 3.4 |
| Joliet | 187 | 1,167 | 130 | 37 | 892 | 6,910,248 | 6.0 | 39.1 | 27.6 | 5.2 | 2.5 | 19.7 | 4.7 |
| Kankakee | 48 | 226 | 27 | 8 | 234 | 1,312,638 | 8.1 | 36.9 | 25.0 | 0.8 | 2.1 | 13.6 | 0.0 |
| Lake in the Hills | D | D | D | D | 131 | 814,212 | 11.4 | 47.5 | 0.0 | 11.5 | 3.7 | 6.4 | 6.6 |
| Lansing | 46 | 514 | 42 | 15 | 201 | 1,089,482 | 7.3 | 48.7 | 20.2 | 7.5 | 0.0 | 8.4 | 0.0 |
| Lombard | 124 | 1,253 | 188 | 51 | 245 | 1,816,412 | 10.2 | 35.6 | 31.2 | 6.3 | 0.7 | 24.1 | 0.0 |
| McHenry | 76 | 390 | 37 | 11 | 163 | 1,093,776 | 7.1 | 48.9 | 0.0 | 13.6 | 2.6 | 9.1 | 3.0 |
| Melrose Park | D | D | D | D | 265 | 1,554,951 | 10.6 | 37.6 | 29.5 | 5.4 | 0.8 | 18.3 | 6.5 |
| Moline | 91 | 642 | 56 | 18 | 377 | 2,369,164 | 10.8 | 27.4 | 18.3 | 9.7 | 4.3 | 8.5 | 15.7 |
| Mount Prospect | 107 | 735 | 122 | 33 | 418 | 2,902,387 | 7.1 | 30.0 | 23.5 | 3.6 | 2.1 | 7.1 | 9.6 |
| Mundelein | 69 | 322 | 31 | 10 | 191 | 1,191,472 | 8.4 | 42.9 | 18.0 | 7.9 | 0.0 | 20.5 | 9.6 |
| Naperville | 297 | 2,529 | 212 | 84 | 1,105 | 7,205,462 | 9.0 | 28.8 | 17.5 | 14.5 | 5.7 | 10.1 | 5.8 |
| Niles | 88 | 515 | 62 | 18 | 294 | 2,022,452 | 10.0 | 27.9 | 24.3 | 16.5 | 0.0 | 24.5 | 5.8 |
| Normal | 59 | 515 | 42 | 15 | 462 | 2,875,904 | 10.5 | 27.4 | 19.9 | 6.0 | 0.0 | 7.5 | 11.6 |
| Northbrook | 131 | 684 | 115 | 26 | 356 | 2,326,042 | 7.7 | 35.3 | 24.0 | 6.6 | 0.0 | 7.3 | 2.7 |
| North Chicago | 21 | 86 | 11 | 3 | 168 | 1,037,212 | 6.0 | 51.1 | 20.0 | 8.7 | 1.4 | 11.7 | 0.0 |
| Oak Forest | D | D | D | D | 164 | 947,676 | 3.1 | 49.0 | 26.3 | 7.2 | 2.6 | 11.7 | 0.0 |
| Oak Lawn | 108 | 682 | 67 | 18 | 383 | 2,606,591 | 4.2 | 39.0 | 29.5 | 5.7 | 0.8 | 7.1 | 8.8 |
| Oak Park | 130 | 646 | 63 | 17 | 336 | 2,470,159 | 9.5 | 42.0 | 23.8 | 5.3 | 1.2 | 4.4 | 0.0 |
| O'Fallon | 56 | 340 | 29 | 10 | 247 | 1,130,264 | 16.4 | 30.9 | 2.1 | 7.2 | 10.0 | 25.1 | 4.0 |
| Orland Park | 133 | 1,063 | 109 | 36 | 443 | 2,445,732 | 9.3 | 42.6 | 0.0 | 6.9 | 0.0 | 19.7 | 9.3 |
| Oswego | 68 | 353 | 39 | 12 | 109 | 699,543 | 20.4 | 64.0 | 0.0 | 0.0 | 0.0 | 15.5 | 0.0 |
| Palatine | 155 | 765 | 77 | 24 | 340 | 2,512,137 | 9.4 | 39.7 | 32.7 | 3.3 | 1.1 | 4.4 | 0.0 |
| Park Ridge | 97 | 677 | 129 | 30 | 263 | 1,488,872 | 5.8 | 37.5 | 22.9 | 2.2 | 0.0 | 6.4 | 12.6 |
| Pekin | 60 | 287 | 22 | 7 | 260 | 1,406,663 | 4.4 | 32.4 | 24.7 | 8.6 | 2.1 | 6.3 | 19.0 |
| Peoria | 198 | 3,953 | 413 | 192 | 791 | 5,667,304 | 7.3 | 41.8 | 27.6 | 6.1 | 4.0 | 0.4 | 5.1 |
| Plainfield | 78 | 531 | 54 | 17 | 138 | 1,144,565 | 12.2 | 62.6 | 0.0 | 10.6 | 0.0 | 10.1 | 0.0 |
| Quincy | 109 | 525 | 59 | 15 | 352 | 1,613,922 | 7.4 | 26.6 | 22.9 | 15.4 | 1.2 | 14.9 | 5.7 |
| Rockford | 282 | 2,023 | 244 | 62 | 1,103 | 7,086,162 | 5.6 | 33.2 | 36.5 | 5.3 | 7.4 | 4.8 | 4.2 |
| Rock Island | 59 | 388 | 39 | 12 | 368 | 1,968,792 | 6.9 | 32.3 | 18.7 | 14.5 | 4.8 | 16.7 | 4.2 |
| Romeoville | 64 | 450 | 61 | 16 | 281 | 1,619,462 | 9.5 | 41.7 | 17.8 | 4.2 | 0.9 | 22.7 | 0.0 |
| Round Lake Beach | 28 | 130 | 9 | 3 | 66 | 418,314 | 12.0 | 67.5 | 0.0 | 14.4 | 4.9 | 1.2 | 0.0 |
| St. Charles | 113 | 738 | 67 | 23 | 265 | 2,077,030 | 19.1 | 27.9 | 20.1 | 11.8 | 0.0 | 16.2 | 0.0 |
| Schaumburg | 245 | 2,381 | 443 | 133 | 514 | 3,870,281 | 10.7 | 31.8 | 28.6 | 5.7 | 6.8 | 8.8 | 0.0 |
| Skokie | 189 | 1,522 | 174 | 54 | 600 | 3,841,807 | 7.9 | 30.1 | 23.6 | 2.7 | 5.7 | 10.7 | 13.4 |
| Springfield | 355 | 2,754 | 350 | 115 | 1,431 | 9,424,523 | 4.4 | 19.7 | 17.0 | 8.4 | 1.4 | 47.3 | 1.8 |
| Streamwood | 73 | 320 | 30 | 10 | 192 | 1,343,293 | 8.6 | 36.1 | 29.5 | 5.2 | 0.6 | 11.9 | 0.0 |
| Tinley Park | 74 | 397 | 43 | 11 | 361 | 2,034,117 | 10.1 | 44.4 | 14.2 | 7.5 | 0.5 | 9.3 | 9.8 |
| Urbana | 53 | 570 | 614 | 45 | 260 | 1,906,717 | 11.3 | 40.6 | 24.1 | 11.2 | 3.7 | 2.4 | 0.0 |
| Vernon Hills | 65 | 543 | 57 | 18 | 107 | 859,465 | 9.6 | 66.5 | 0.0 | 18.1 | 5.8 | 0.0 | 0.0 |

1. Establishments subject to federal tax.

# Table D. Cities — City Government Finances

ILLINOIS— Cont'd

Column groupings: Items 117–123 fall under **General revenue** (118–119 = *Intergovernmental*; 120–123 = *Taxes*, with 121–123 = *Per capita[1] (dollars)*). Items 124–126 fall under **General expenditure** (125–126 = *Per capita[1] (dollars)*).

| City | Total (mil dol) | Intergovernmental — Total (mil dol) | Intergovernmental — Percent from state government | Taxes — Total (mil dol) | Taxes — Per capita Total | Taxes — Per capita Property | Taxes — Per capita Sales and gross receipts | General expenditure — Total (mil dol) | General expenditure — Per capita Total | General expenditure — Per capita Capital outlays |
|---|---|---|---|---|---|---|---|---|---|---|
| | 117 | 118 | 119 | 120 | 121 | 122 | 123 | 124 | 125 | 126 |
| Burbank | 22.8 | 7.3 | 100.0 | 13.6 | 475 | 209 | 266 | 24 | 823 | 87 |
| Calumet City | 53.3 | 13.3 | 80.0 | 34.4 | 940 | 693 | 247 | 56 | 1,524 | 144 |
| Carbondale | 38.1 | 13.0 | 87.3 | 16.0 | 627 | 98 | 529 | 38 | 1,473 | 109 |
| Carol Stream | 33.4 | 14.2 | 98.0 | 15.0 | 377 | 112 | 265 | 29 | 723 | 3 |
| Carpentersville | 41.7 | 11.9 | 82.1 | 21.4 | 565 | 369 | 195 | 41 | 1,088 | 116 |
| Champaign | 109.6 | 36.0 | 85.2 | 61.2 | 696 | 300 | 396 | 105 | 1,191 | 206 |
| Chicago | 8,260.8 | 1,710.7 | 63.4 | 3,561.4 | 1,314 | 477 | 734 | 7,769 | 2,866 | 464 |
| Chicago Heights | 43.6 | 9.9 | 86.9 | 26.9 | 901 | 651 | 251 | 46 | 1,538 | 99 |
| Cicero | 107.6 | 24.4 | 88.4 | 67.8 | 823 | 581 | 242 | 129 | 1,568 | 206 |
| Collinsville | 33.3 | 17.0 | 95.0 | 8.8 | 355 | 213 | 143 | 33 | 1,321 | 156 |
| Crystal Lake | 55.5 | 17.9 | 99.4 | 24.8 | 618 | 421 | 197 | 54 | 1,341 | 118 |
| Danville | 49.7 | 31.1 | 87.9 | 10.0 | 319 | 204 | 115 | 48 | 1,535 | 60 |
| Decatur | 91.9 | 32.5 | 94.6 | 39.1 | 544 | 217 | 327 | 87 | 1,210 | 6 |
| DeKalb | 55.5 | 17.6 | 93.5 | 29.9 | 699 | 344 | 355 | 57 | 1,341 | 262 |
| Des Plaines | 122.0 | 50.6 | 96.2 | 59.0 | 1,016 | 608 | 408 | 109 | 1,876 | 332 |
| Downers Grove | 71.0 | 28.6 | 94.7 | 30.3 | 612 | 443 | 169 | 65 | 1,313 | 78 |
| East St. Louis | 38.1 | 23.0 | 82.7 | 14.1 | 530 | 363 | 167 | 39 | 1,457 | 0 |
| Elgin | 157.5 | 43.0 | 96.2 | 76.4 | 681 | 452 | 230 | 148 | 1,318 | 182 |
| Elk Grove Village | 94.2 | 16.8 | 97.9 | 51.8 | 1,570 | 785 | 785 | 73 | 2,213 | 136 |
| Elmhurst | 79.1 | 21.2 | 99.0 | 39.7 | 851 | 460 | 391 | 97 | 2,075 | 548 |
| Evanston | 147.8 | 26.7 | 81.6 | 79.3 | 1,063 | 698 | 365 | 172 | 2,298 | 245 |
| Freeport | 32.8 | 13.9 | 83.0 | 8.9 | 371 | 197 | 173 | 32 | 1,338 | 109 |
| Galesburg | 41.3 | 13.6 | 97.6 | 21.3 | 692 | 340 | 352 | 40 | 1,312 | 154 |
| Glendale Heights | 39.5 | 13.6 | 91.2 | 19.4 | 569 | 268 | 302 | 41 | 1,190 | 164 |
| Glen Ellyn | 44.4 | 8.2 | 88.7 | 19.2 | 683 | 437 | 246 | 45 | 1,590 | 322 |
| Glenview | 117.6 | 27.8 | 98.4 | 69.3 | 1,458 | 1,112 | 346 | 102 | 2,146 | 192 |
| Granite City | 47.2 | 12.9 | 92.6 | 21.8 | 762 | 612 | 149 | 46 | 1,603 | 105 |
| Gurnee | 45.8 | 28.0 | 87.5 | 12.2 | 396 | 0 | 396 | 41 | 1,335 | 109 |
| Hanover Park | 45.9 | 11.1 | 97.7 | 25.6 | 674 | 417 | 258 | 43 | 1,140 | 111 |
| Harvey | 32.1 | 7.6 | 99.2 | 18.8 | 756 | 605 | 151 | 43 | 1,708 | 168 |
| Highland Park | 58.8 | 13.4 | 98.5 | 32.7 | 1,101 | 574 | 527 | 59 | 1,976 | 301 |
| Hoffman Estates | 99.5 | 17.6 | 82.4 | 53.6 | 1,042 | 725 | 317 | 79 | 1,537 | 126 |
| Joliet | 228.2 | 72.2 | 97.7 | 88.9 | 601 | 284 | 318 | 192 | 1,297 | 312 |
| Kankakee | 54.7 | 11.1 | 83.2 | 22.9 | 878 | 633 | 245 | 68 | 2,585 | 552 |
| Lake in the Hills | 20.1 | 7.4 | 95.4 | 8.5 | 296 | 213 | 83 | 21 | 726 | 85 |
| Lansing | 38.1 | 9.8 | 88.6 | 20.6 | 740 | 514 | 225 | 35 | 1,267 | 102 |
| Lombard | 99.3 | 29.3 | 90.0 | 19.6 | 447 | 205 | 243 | 109 | 2,485 | 141 |
| McHenry | 30.1 | 12.8 | 98.8 | 6.6 | 247 | 204 | 43 | 53 | 1,983 | 1,081 |
| Melrose Park | 51.3 | 19.7 | 99.9 | 26.0 | 1,033 | 826 | 207 | 57 | 2,268 | 28 |
| Moline | 66.5 | 25.5 | 97.3 | 34.3 | 818 | 466 | 352 | 98 | 2,333 | 669 |
| Mount Prospect | 78.3 | 11.8 | 72.8 | 55.8 | 1,020 | 596 | 425 | 82 | 1,498 | 60 |
| Mundelein | 40.6 | 16.9 | 99.5 | 14.5 | 463 | 410 | 53 | 39 | 1,257 | 109 |
| Naperville | 266.4 | 106.0 | 94.7 | 137.4 | 930 | 659 | 236 | 256 | 1,732 | 133 |
| Niles | 60.3 | 25.7 | 99.5 | 28.2 | 957 | 290 | 667 | 64 | 2,187 | 459 |
| Normal | 81.4 | 18.9 | 99.2 | 43.1 | 791 | 259 | 532 | 93 | 1,700 | 396 |
| Northbrook | 65.3 | 17.7 | 99.2 | 35.8 | 1,074 | 659 | 416 | 74 | 2,219 | 195 |
| North Chicago | 28.9 | 7.2 | 89.4 | 16.4 | 548 | 353 | 195 | 37 | 1,218 | 143 |
| Oak Forest | 26.4 | 5.9 | 90.1 | 16.2 | 587 | 399 | 188 | 31 | 1,129 | 234 |
| Oak Lawn | 77.6 | 20.0 | 96.2 | 39.5 | 705 | 393 | 312 | 67 | 1,199 | 2 |
| Oak Park | 90.3 | 17.8 | 72.6 | 57.3 | 1,099 | 755 | 344 | 95 | 1,825 | 361 |
| O'Fallon | 42.6 | 17.4 | 96.6 | 13.1 | 447 | 239 | 208 | 38 | 1,294 | 187 |
| Orland Park | 88.8 | 30.5 | 98.6 | 29.6 | 506 | 258 | 248 | 93 | 1,594 | 526 |
| Oswego | 24.1 | 14.9 | 98.7 | 4.6 | 131 | 38 | 93 | 24 | 676 | 121 |
| Palatine | 83.0 | 21.8 | 84.1 | 45.1 | 659 | 454 | 205 | 84 | 1,228 | 151 |
| Park Ridge | 53.3 | 8.8 | 99.5 | 39.2 | 1,049 | 667 | 382 | 47 | 1,249 | 84 |
| Pekin | 46.0 | 12.8 | 96.9 | 13.7 | 420 | 206 | 214 | 51 | 1,552 | 347 |
| Peoria | 185.9 | 55.2 | 92.1 | 84.6 | 752 | 323 | 429 | 198 | 1,756 | 250 |
| Plainfield | 42.6 | 10.9 | 93.7 | 18.1 | 412 | 154 | 258 | 38 | 876 | 131 |
| Quincy | 47.4 | 25.0 | 85.7 | 14.8 | 368 | 67 | 300 | 34 | 838 | 4 |
| Rockford | 252.5 | 107.4 | 66.7 | 90.8 | 618 | 401 | 217 | 241 | 1,639 | 286 |
| Rock Island | 63.7 | 18.4 | 90.0 | 24.2 | 640 | 430 | 210 | 74 | 1,945 | 290 |
| Romeoville | 61.5 | 14.1 | 85.7 | 27.2 | 685 | 415 | 271 | 71 | 1,797 | 440 |
| Round Lake Beach | 15.3 | 7.9 | 97.2 | 6.2 | 225 | 91 | 133 | 17 | 627 | 182 |
| St. Charles | 57.2 | 16.9 | 99.2 | 28.5 | 875 | 429 | 446 | 59 | 1,822 | 384 |
| Schaumburg | 168.9 | 48.1 | 91.1 | 67.7 | 915 | 274 | 641 | 197 | 2,658 | 351 |
| Skokie | 115.0 | 29.8 | 98.1 | 64.7 | 1,013 | 491 | 522 | 101 | 1,580 | 187 |
| Springfield | 160.6 | 54.6 | 86.9 | 83.7 | 723 | 243 | 480 | 204 | 1,763 | 454 |
| Streamwood | 33.3 | 10.0 | 98.8 | 19.9 | 500 | 291 | 209 | 29 | 736 | 58 |
| Tinley Park | 68.2 | 21.5 | 98.0 | 38.5 | 681 | 446 | 235 | 58 | 1,018 | 165 |
| Urbana | 46.8 | 15.9 | 76.4 | 23.7 | 553 | 228 | 325 | 44 | 1,022 | 111 |
| Vernon Hills | 31.5 | 17.5 | 100.0 | 10.4 | 399 | 0 | 399 | 29 | 1,098 | 35 |

1. Based on population estimated as of July 1 of the year shown.

| City | | City government finances, 2017 (cont.) | | | | | | | | |
| | | General expenditure (cont.) | | | | | | | | |
| | | Percent of total for: | | | | | | | | |
| | Public welfare | Highways | Parking facilities | Education | Health and hospitals | Police protection | Sewerage and sanitation | Parks and recreation | Housing and community development | Interest on debt |
| | 127 | 128 | 129 | 130 | 131 | 132 | 133 | 134 | 135 | 136 |
| **ILLINOIS— Cont'd** | | | | | | | | | | 14.0 |
| Burbank | 0.0 | 17.5 | 0.0 | 0.0 | 0.0 | 34.2 | 0.0 | 0.0 | 0.2 | 3.8 |
| Calumet City | 2.7 | 9.5 | 0.0 | 0.0 | 0.0 | 27.8 | 7.8 | 0.5 | 9.0 | 8.0 |
| Carbondale | 0.0 | 6.7 | 0.7 | 0.0 | 0.0 | 27.8 | 0.0 | 0.0 | 0.0 | 1.0 |
| Carol Stream | 0.0 | 13.1 | 0.0 | 0.0 | 0.0 | 47.7 | 1.9 | 0.4 | 0.0 | 9.8 |
| Carpentersville | 0.0 | 14.8 | 0.0 | 0.0 | 0.0 | 27.3 | 2.4 | 0.0 | 0.0 | 4.0 |
| Champaign | 4.7 | 16.6 | 3.1 | 0.0 | 1.7 | 22.8 | 5.7 | 0.4 | 3.1 | 12.0 |
| Chicago | 3.9 | 8.6 | 0.1 | 0.0 | 0.0 | 27.9 | 0.0 | 1.0 | 0.0 | 6.3 |
| Chicago Heights | 0.0 | 4.0 | 0.0 | 0.0 | 0.0 | 21.7 | 0.0 | 2.0 | 0.0 | 9.2 |
| Cicero | 4.0 | 14.1 | 0.0 | 0.0 | 1.4 | 25.8 | 9.9 | 0.0 | 0.0 | 5.8 |
| Collinsville | 0.0 | 17.6 | 0.0 | 0.0 | 4.6 | 23.9 | 6.2 | 3.1 | 0.0 | 2.0 |
| Crystal Lake | 0.0 | 15.4 | 0.0 | 0.0 | 0.2 | 0.0 | 12.3 | 4.4 | 0.0 | 0.6 |
| Danville | 0.0 | 11.6 | 0.0 | 0.0 | 0.0 | 30.3 | 2.9 | 1.4 | 0.0 | 7.0 |
| Decatur | 0.0 | 14.6 | 0.0 | 0.0 | 0.0 | 21.5 | 3.9 | 0.0 | 0.4 | 2.2 |
| DeKalb | 0.0 | 7.5 | 0.0 | 0.0 | 0.0 | 21.3 | 0.2 | 0.1 | 1.4 | 0.7 |
| Des Plaines | 0.0 | 18.6 | 0.0 | 0.0 | 0.0 | 25.2 | 1.1 | 0.0 | 0.0 | 4.2 |
| Downers Grove | 1.2 | 12.3 | 2.1 | 0.0 | 0.0 | 20.5 | 1.0 | 1.1 | 5.5 | 3.9 |
| East St. Louis | 0.0 | 12.1 | 0.0 | 0.0 | 0.0 | 30.0 | 3.9 | 12.2 | 0.0 | 2.3 |
| Elgin | 0.7 | 16.8 | 0.0 | 0.0 | 0.0 | 26.9 | 2.6 | 0.0 | 0.0 | 5.0 |
| Elk Grove Village | 0.0 | 11.2 | 0.0 | 0.0 | 0.4 | 17.0 | 25.5 | 1.4 | 0.0 | 3.2 |
| Elmhurst | 0.2 | 16.2 | 2.1 | 0.0 | 1.8 | 22.0 | 5.1 | 6.9 | 0.0 | 8.5 |
| Evanston | 0.0 | 7.9 | 5.0 | 0.0 | 0.0 | 21.4 | 8.4 | 1.2 | 0.0 | 2.9 |
| Freeport | 0.0 | 9.8 | 0.0 | 0.0 | 0.0 | 16.6 | 0.0 | 8.3 | 0.0 | 8.2 |
| Galesburg | 0.0 | 14.3 | 0.5 | 0.0 | 0.0 | 19.2 | 0.6 | 9.6 | 0.0 | 5.8 |
| Glendale Heights | 0.0 | 16.2 | 0.0 | 0.0 | 0.0 | 18.9 | 13.9 | 10.2 | 0.0 | 3.1 |
| Glen Ellyn | 0.0 | 6.5 | 0.6 | 0.0 | 0.0 | 13.1 | 2.4 | 0.0 | 0.0 | 2.4 |
| Glenview | 0.0 | 15.7 | 0.0 | 0.0 | 0.0 | 20.1 | 4.9 | 1.2 | 0.0 | 8.6 |
| Granite City | 0.0 | 11.1 | 0.0 | 0.0 | 0.0 | 31.9 | 0.0 | 0.0 | 0.0 | 5.6 |
| Gurnee | 0.0 | 11.5 | 0.0 | 0.0 | 0.0 | 30.7 | 4.5 | 0.0 | 0.0 | 6.0 |
| Hanover Park | 0.0 | 11.4 | 0.7 | 0.0 | 0.0 | 20.9 | 0.4 | 0.0 | 0.0 | 5.6 |
| Harvey | 0.0 | 7.8 | 0.2 | 0.0 | 0.0 | 28.0 | 7.9 | 0.0 | 0.0 | 2.7 |
| Highland Park | 0.0 | 3.3 | 2.2 | 0.0 | 2.3 | 24.2 | 2.6 | 14.2 | 0.0 | 5.9 |
| Hoffman Estates | 0.0 | 15.8 | 0.0 | 0.0 | 0.0 | 22.2 | 20.0 | 3.7 | 5.8 | 0.6 |
| Joliet | 0.0 | 11.7 | 0.6 | 0.0 | 0.0 | 15.0 | 17.7 | 0.0 | 0.5 | 10.8 |
| Kankakee | 0.0 | 9.7 | 0.0 | 0.0 | 0.0 | 38.7 | 0.0 | 8.1 | 1.8 | 3.4 |
| Lake in the Hills | 0.0 | 12.5 | 0.0 | 0.0 | 0.0 | 34.8 | 0.0 | 0.0 | 0.0 | 12.2 |
| Lansing | 0.0 | 10.1 | 0.0 | 0.0 | 0.0 | 13.3 | 0.1 | 0.0 | 0.0 | 8.0 |
| Lombard | 0.0 | 3.8 | 0.1 | 0.0 | 0.0 | 21.2 | 48.2 | 6.2 | 0.0 | 0.9 |
| McHenry | 0.0 | 14.8 | 0.0 | 0.0 | 0.0 | 24.2 | 3.2 | 3.6 | 0.0 | 18.8 |
| Melrose Park | 0.0 | 7.6 | 0.0 | 0.0 | 0.0 | 16.2 | 17.1 | 4.1 | 2.5 | 7.7 |
| Moline | 0.0 | 7.7 | 0.4 | 0.0 | 0.2 | 20.8 | 5.4 | 0.6 | 0.6 | 15.2 |
| Mount Prospect | 1.9 | 10.0 | 0.0 | 0.0 | 0.0 | 29.2 | 6.7 | 0.0 | 0.0 | 1.5 |
| Mundelein | 0.0 | 30.8 | 0.0 | 0.0 | 0.0 | 25.2 | 2.9 | 4.3 | 0.0 | 4.7 |
| Naperville | 0.0 | 21.6 | 0.0 | 0.0 | 3.0 | 22.0 | 15.4 | 2.5 | 0.0 | 0.4 |
| Niles | 0.0 | 13.4 | 0.0 | 0.0 | 0.0 | 14.1 | 8.4 | 11.5 | 0.0 | 4.2 |
| Normal | 0.0 | 10.8 | 0.2 | 0.0 | 0.0 | 19.8 | 3.2 | 0.0 | 0.9 | 4.8 |
| Northbrook | 0.0 | 14.8 | 0.0 | 0.0 | 0.0 | 27.0 | 0.0 | 0.0 | 0.0 | 14.2 |
| North Chicago | 0.0 | 5.0 | 0.0 | 0.0 | 0.0 | 31.7 | 0.0 | 0.0 | 0.0 | 8.3 |
| Oak Forest | 0.0 | 11.8 | 0.8 | 0.0 | 0.0 | 28.4 | 7.4 | 0.6 | 0.0 | 5.0 |
| Oak Lawn | 0.2 | 9.6 | 0.0 | 0.0 | 0.0 | 21.1 | 3.3 | 0.0 | 10.8 | 10.6 |
| Oak Park | 0.0 | 17.2 | 3.9 | 0.0 | 1.1 | 20.2 | 13.7 | 12.8 | 0.0 | 6.4 |
| O'Fallon | 0.0 | 21.0 | 0.0 | 0.0 | 0.0 | 21.4 | 8.0 | 11.1 | 0.0 | 2.6 |
| Orland Park | 0.0 | 16.6 | 16.7 | 0.0 | 0.0 | 47.9 | 10.0 | 0.0 | 0.0 | 6.6 |
| Oswego | 0.0 | 10.0 | 0.0 | 0.0 | 0.0 | 26.2 | 8.9 | 0.0 | 4.8 | 9.8 |
| Palatine | 0.0 | 12.2 | 0.9 | 0.0 | 0.0 | 18.4 | 8.8 | 1.0 | 0.0 | 2.6 |
| Park Ridge | 0.0 | 15.9 | 0.7 | 0.0 | 0.0 | 20.4 | 8.1 | 0.0 | 0.2 | 2.5 |
| Pekin | 0.0 | 25.5 | 0.0 | 0.0 | 0.0 | 20.0 | 5.2 | 0.0 | 0.1 | 4.2 |
| Peoria | 0.0 | 14.1 | 0.0 | 0.0 | 0.0 | 30.2 | 15.0 | 0.0 | 0.0 | 8.2 |
| Plainfield | 0.0 | 24.7 | 0.0 | 0.0 | 0.6 | 30.2 | 15.0 | 2.7 | 0.2 | 1.3 |
| Quincy | 0.0 | 8.2 | 0.0 | 0.0 | 0.0 | 26.6 | 11.6 | 5.7 | 5.4 | 2.7 |
| Rockford | 6.9 | 20.8 | 0.8 | 0.0 | 0.0 | 21.2 | 5.4 | 8.8 | 0.1 | 5.5 |
| Rock Island | 0.0 | 19.4 | 0.0 | 0.0 | 0.0 | 19.5 | 9.7 | 6.8 | 0.0 | 3.6 |
| Romeoville | 0.0 | 12.0 | 0.8 | 0.0 | 0.0 | 19.2 | 12.6 | 1.3 | 8.3 | 3.7 |
| Round Lake Beach | 0.0 | 38.1 | 0.1 | 0.0 | 0.0 | 37.0 | 0.8 | 0.0 | 0.0 | 6.8 |
| St. Charles | 0.0 | 13.4 | 0.0 | 0.0 | 0.3 | 19.5 | 2.2 | 19.7 | 0.1 | 5.4 |
| Schaumburg | 0.6 | 18.5 | 0.1 | 0.0 | 0.0 | 14.5 | 2.2 | 2.2 | 0.0 | 3.1 |
| Skokie | 0.0 | 21.3 | 0.0 | 0.0 | 2.1 | 22.8 | 4.7 | 1.2 | 3.5 | 2.6 |
| Springfield | 0.0 | 24.9 | 0.5 | 0.0 | 0.0 | 22.3 | 4.6 | 1.2 | 0.0 | 0.7 |
| Streamwood | 0.0 | 9.3 | 0.0 | 0.0 | 0.0 | 38.9 | 0.0 | 1.2 | 0.0 | 5.0 |
| Tinley Park | 0.0 | 14.7 | 0.0 | 0.0 | 0.0 | 30.6 | 10.3 | 0.0 | 0.0 | 0.3 |
| Urbana | 0.0 | 26.4 | 2.1 | 0.0 | 0.0 | 20.5 | 0.2 | 0.0 | 11.8 | 7.8 |
| Vernon Hills | 0.0 | 16.5 | 0.0 | 0.0 | 0.0 | 33.7 | 0.0 | 2.9 | 0.0 | |

# Table D. Cities — City Government Finances, City Government Employment, and Climate

| City | City government finances, 2017 (cont.) | | | Climate[2] | | | | | | |
| | Debt outstanding — Total (mil dol) | Debt outstanding — Per capita[1] (dollars) | Debt issued during year | Avg daily temp — Mean January | Avg daily temp — Mean July | Avg daily temp — Limits January[3] | Avg daily temp — Limits July[4] | Annual precipitation (inches) | Heating degree days | Cooling degree days |
| | 137 | 138 | 139 | 140 | 141 | 142 | 143 | 144 | 145 | 146 |
|---|---|---|---|---|---|---|---|---|---|---|
| **ILLINOIS— Cont'd** | | | | | | | | | | |
| Burbank | 83.9 | 2,920 | 14.1 | 23.5 | 75.5 | 16.2 | 84.7 | 38.35 | 6,083 | 1,001 |
| Calumet City | 44.4 | 1,215 | 6.0 | 22.0 | 74.2 | 14.8 | 83.7 | 38.65 | 6,355 | 866 |
| Carbondale | 72.0 | 2,818 | 2.2 | NA | NA | NA | NA | NA | NA | NA |
| Carol Stream | 7.0 | 175 | 0.0 | 23.1 | 74.8 | 14.2 | 86.8 | 37.94 | 6,053 | 942 |
| Carpentersville | 99.4 | 2,617 | 8.7 | 19.3 | 72.6 | 10.9 | 83.0 | 37.22 | 6,975 | 679 |
| Champaign | 136.1 | 1,550 | 9.0 | 33.7 | 79.0 | 25.0 | 89.5 | 47.93 | 4,183 | 1,501 |
| Chicago | 23,539.6 | 8,683 | 2,428.3 | 25.3 | 75.4 | 18.3 | 84.4 | 38.01 | 5,787 | 994 |
| Chicago Heights | 182.5 | 6,109 | 15.7 | 22.0 | 74.2 | 14.8 | 83.7 | 38.65 | 6,355 | 866 |
| Cicero | 299.4 | 3,633 | 24.5 | 22.0 | 73.3 | 14.3 | 83.5 | 36.27 | 6,498 | 830 |
| Collinsville | 41.3 | 1,666 | 0.0 | NA | NA | NA | NA | NA | NA | NA |
| Crystal Lake | 45.3 | 1,131 | 10.2 | 28.6 | 76.3 | 19.0 | 87.2 | 46.96 | 5,168 | 1,112 |
| Danville | 5.3 | 170 | 0.0 | 25.8 | 75.3 | 17.3 | 86.2 | 40.96 | 5,555 | 1,027 |
| Decatur | 133.5 | 1,859 | 24.3 | 25.8 | 76.2 | 17.1 | 87.8 | 39.74 | 5,458 | 1,142 |
| DeKalb | 33.7 | 788 | 0.0 | 18.5 | 73.1 | 10.3 | 83.6 | 37.38 | 6,979 | 736 |
| Des Plaines | 200.2 | 3,447 | 8.0 | 22.0 | 73.3 | 14.3 | 83.5 | 36.27 | 6,498 | 830 |
| Downers Grove | 75.7 | 1,528 | 8.7 | 23.1 | 74.8 | 14.2 | 86.8 | 37.94 | 6,053 | 942 |
| East St. Louis | 13.3 | 500 | 0.0 | 29.1 | 78.6 | 20.0 | 88.7 | 40.33 | 4,826 | 1,378 |
| Elgin | 87.5 | 781 | 25.0 | 19.3 | 72.6 | 10.9 | 83.0 | 37.22 | 6,975 | 679 |
| Elk Grove Village | 75.6 | 2,294 | 0.0 | 22.0 | 73.3 | 14.3 | 83.5 | 36.27 | 6,498 | 830 |
| Elmhurst | 90.0 | 1,930 | 25.0 | 23.1 | 74.8 | 14.2 | 86.8 | 37.94 | 6,053 | 942 |
| Evanston | 393.9 | 5,279 | 57.4 | 22.0 | 72.9 | 13.7 | 83.2 | 36.80 | 6,630 | 702 |
| Freeport | 50.1 | 2,081 | 5.7 | 17.2 | 71.9 | 9.0 | 82.0 | 34.79 | 7,317 | 611 |
| Galesburg | 105.2 | 3,423 | 31.7 | 21.3 | 74.9 | 13.5 | 84.5 | 37.22 | 6,347 | 941 |
| Glendale Heights | 34.4 | 1,008 | 0.0 | 22.0 | 73.3 | 14.3 | 83.5 | 36.27 | 6,498 | 830 |
| Glen Ellyn | 23.2 | 825 | 0.0 | 21.7 | 74.4 | 12.2 | 85.7 | 38.58 | 6,359 | 888 |
| Glenview | 62.2 | 1,309 | 3.9 | 22.0 | 73.3 | 14.3 | 83.5 | 36.27 | 6,498 | 830 |
| Granite City | 100.4 | 3,500 | 5.9 | 27.7 | 78.4 | 19.4 | 88.1 | 38.54 | 5,149 | 1,354 |
| Gurnee | 52.4 | 1,707 | 0.3 | 19.9 | 72.2 | 12.1 | 82.2 | 35.50 | 6,955 | 634 |
| Hanover Park | 66.9 | 1,759 | 3.2 | 18.4 | 72.1 | 9.6 | 82.3 | 36.56 | 7,149 | 624 |
| Harvey | 63.9 | 2,572 | 10.1 | 22.0 | 74.2 | 14.8 | 83.7 | 38.65 | 6,355 | 866 |
| Highland Park | 53.2 | 1,789 | 11.4 | 22.0 | 72.9 | 13.7 | 83.2 | 36.80 | 6,630 | 702 |
| Hoffman Estates | 205.4 | 3,991 | 17.8 | 18.4 | 72.1 | 9.6 | 82.3 | 36.56 | 7,149 | 624 |
| Joliet | 84.2 | 569 | 12.8 | 21.7 | 73.7 | 13.5 | 84.6 | 36.96 | 6,464 | 809 |
| Kankakee | 171.9 | 6,578 | 1.8 | 21.7 | 74.4 | 12.2 | 85.7 | 38.58 | 6,359 | 888 |
| Lake in the Hills | 16.3 | 568 | 0.3 | NA | NA | NA | NA | NA | NA | NA |
| Lansing | 106.4 | 3,812 | 16.8 | 21.7 | 74.4 | 12.2 | 85.7 | 38.58 | 6,359 | 888 |
| Lombard | 210.5 | 4,798 | 0.0 | 23.1 | 74.8 | 14.2 | 86.8 | 37.94 | 6,053 | 942 |
| McHenry | 38.3 | 1,428 | 19.2 | NA | NA | NA | NA | NA | NA | NA |
| Melrose Park | 271.2 | 10,782 | 13.4 | NA | NA | NA | NA | NA | NA | NA |
| Moline | 182.0 | 4,343 | 37.8 | 21.8 | 76.4 | 13.3 | 85.1 | 35.10 | 6,179 | 1,100 |
| Mount Prospect | 157.4 | 2,878 | 32.4 | 22.0 | 73.3 | 14.3 | 83.5 | 36.27 | 6,498 | 830 |
| Mundelein | 12.6 | 404 | 0.0 | 18.4 | 72.1 | 9.6 | 82.3 | 36.56 | 7,149 | 624 |
| Naperville | 154.5 | 1,047 | 140.5 | 23.1 | 74.8 | 14.2 | 86.8 | 37.94 | 6,053 | 942 |
| Niles | 8.9 | 301 | 0.9 | 22.0 | 73.3 | 14.3 | 83.5 | 36.27 | 6,498 | 830 |
| Normal | 92.3 | 1,695 | 22.4 | 25.8 | 76.2 | 17.1 | 87.8 | 39.74 | 5,458 | 1,142 |
| Northbrook | 100.5 | 3,015 | 0.0 | 22.0 | 72.9 | 13.7 | 83.2 | 36.80 | 6,630 | 702 |
| North Chicago | 121.3 | 4,048 | 12.9 | 20.3 | 71.5 | 12.0 | 81.7 | 34.09 | 7,031 | 613 |
| Oak Forest | 63.8 | 2,310 | 11.3 | 22.0 | 74.2 | 14.8 | 83.7 | 38.65 | 6,355 | 866 |
| Oak Lawn | 71.7 | 1,281 | 14.5 | 23.5 | 75.5 | 16.2 | 84.7 | 38.35 | 6,083 | 1,001 |
| Oak Park | 288.3 | 5,529 | 64.9 | 22.0 | 73.3 | 14.3 | 83.5 | 36.27 | 6,498 | 830 |
| O'Fallon | 54.9 | 1,872 | 9.5 | NA | NA | NA | NA | NA | NA | NA |
| Orland Park | 109.6 | 1,871 | 8.5 | 23.5 | 75.5 | 16.2 | 84.7 | 38.35 | 6,083 | 1,001 |
| Oswego | 49.9 | 1,432 | 27.6 | NA | NA | NA | NA | NA | NA | NA |
| Palatine | 197.8 | 2,887 | 11.0 | 18.4 | 72.1 | 9.6 | 82.3 | 36.56 | 7,149 | 624 |
| Park Ridge | 35.8 | 955 | 12.4 | 22.0 | 73.3 | 14.3 | 83.5 | 36.27 | 6,498 | 830 |
| Pekin | 11.4 | 351 | 0.0 | 24.4 | 75.8 | 15.7 | 87.4 | 35.71 | 5,695 | 1,088 |
| Peoria | 191.9 | 1,704 | 34.9 | 22.5 | 75.1 | 14.3 | 85.7 | 36.03 | 6,097 | 998 |
| Plainfield | 60.4 | 1,377 | 1.1 | NA | NA | NA | NA | NA | NA | NA |
| Quincy | 15.4 | 381 | 4.1 | 24.9 | 76.8 | 16.0 | 88.0 | 35.63 | 5,707 | 1,117 |
| Rockford | 104.8 | 713 | 31.6 | 19.0 | 72.9 | 10.8 | 83.1 | 36.63 | 6,933 | 768 |
| Rock Island | 66.4 | 1,757 | 5.6 | 21.8 | 76.4 | 13.3 | 85.1 | 35.10 | 6,179 | 1,100 |
| Romeoville | 169.3 | 4,273 | 10.2 | NA | NA | NA | NA | NA | NA | NA |
| Round Lake Beach | 20.0 | 727 | 0.0 | 19.9 | 72.2 | 12.1 | 82.2 | 35.50 | 6,955 | 634 |
| St. Charles | 105.9 | 3,254 | 26.4 | 19.3 | 72.6 | 10.9 | 83.0 | 37.22 | 6,975 | 679 |
| Schaumburg | 519.0 | 7,010 | 91.1 | 18.4 | 72.1 | 9.6 | 82.3 | 36.56 | 7,149 | 624 |
| Skokie | 57.5 | 900 | 14.2 | 22.0 | 73.3 | 14.3 | 83.5 | 36.27 | 6,498 | 830 |
| Springfield | 1,436.9 | 12,412 | 75.1 | 25.1 | 76.3 | 17.1 | 86.5 | 35.56 | 5,596 | 1,165 |
| Streamwood | 4.5 | 114 | 0.0 | 19.3 | 72.6 | 10.9 | 83.0 | 37.22 | 6,975 | 679 |
| Tinley Park | 67.8 | 1,199 | 3.1 | 21.8 | 76.4 | 13.3 | 85.1 | 35.10 | 6,179 | 1,100 |
| Urbana | 11.5 | 270 | 0.0 | 20.7 | 73.2 | 12.4 | 83.7 | 34.47 | 6,606 | 774 |
| Vernon Hills | 50.8 | 1,939 | 0.1 | NA | NA | NA | NA | NA | NA | NA |

1. Based on the population estimated as of July 1 of the year shown.  2. Represents normal values based on the 30-year period, 1971±2000.  3. Average daily minimum.  4. Average daily maximum.

## Table D. Cities — **Land Area and Population**

| STATE / Place code | City | Land area[1] (sq. mi) | Population, 2020 | | | Race 2019 — Race alone[2] (percent) | | | | | | |
|---|---|---|---|---|---|---|---|---|---|---|---|---|
| | | | Total persons 2019 | Rank | Per square mile | White | Black or African American | American Indian, Alaskan Native | Asian | Hawaiian Pacific Islander | Some other race | Two or more races (percent) |
| | | 1 | 2 | 3 | 4 | 5 | 6 | 7 | 8 | 9 | 10 | 11 |
| | **ILLINOIS— Cont'd** | | | | | | | | | | | |
| 17 79293 | Waukegan | 24.2 | 85,453 | 394 | 3,531 | 49.1 | 20.6 | 0.4 | 7.1 | 0.0 | 19.3 | 3.6 |
| 17 80060 | West Chicago | 15.4 | 26,593 | 1,369 | 1,727 | 67.0 | 2.4 | 0.0 | 2.9 | 0.0 | 21.6 | 6.3 |
| 17 81048 | Wheaton | 11.3 | 52,451 | 743 | 4,642 | 88.3 | 1.1 | 0.0 | 6.7 | 0.0 | 0.9 | 3.0 |
| 17 81087 | Wheeling | 8.7 | 38,958 | 994 | 4,478 | 54.2 | 0.7 | 1.0 | 20.4 | 0.0 | 21.6 | 2.1 |
| 17 82075 | Wilmette | 5.4 | 26,819 | 1,358 | 4,967 | D | D | D | D | D | D | D |
| 17 83245 | Woodridge | 9.6 | 33,329 | 1,153 | 3,472 | 76.4 | 6.8 | 0.2 | 13.7 | 0.0 | 1.1 | 1.8 |
| 18 00000 | **INDIANA** | 35,826.4 | 6,754,953 | X | 189 | 82.8 | 9.6 | 0.3 | 2.5 | 0.0 | 2.2 | 2.6 |
| 18 01468 | Anderson | 41.6 | 54,631 | 707 | 1,313 | 80.9 | 14.7 | 0.2 | 0.8 | 0.0 | 0.2 | 2.6 |
| 18 05860 | Bloomington | 23.2 | 85,603 | 392 | 3,690 | 82.9 | 3.3 | 0.1 | 10.8 | 0.0 | 0.3 | 2.9 |
| 18 10342 | Carmel | 49.1 | 101,643 | 310 | 2,070 | 80.3 | 4.3 | 0.4 | 11.9 | 0.0 | 3.1 | 4.3 |
| 18 14734 | Columbus | 28.1 | 48,445 | 804 | 1,724 | 79.5 | 1.1 | 0.0 | 12.0 | 0.0 | 1.2 | 2.4 |
| 18 16138 | Crown Point | 18.1 | 30,906 | 1,220 | 1,708 | 85.1 | 8.5 | 0.0 | 2.7 | 0.2 | 1.2 | 3.4 |
| 18 19486 | East Chicago | 14.1 | 27,742 | 1,325 | 1,968 | 24.0 | 43.1 | 0.4 | 0.3 | 0.0 | 28.8 | 2.0 |
| 18 20728 | Elkhart | 27.5 | 52,172 | 745 | 1,897 | 67.5 | 13.3 | 0.0 | 0.3 | 1.0 | 16.8 | 2.8 |
| 18 22000 | Evansville | 47.4 | 118,407 | 242 | 2,498 | 79.5 | 13.4 | 0.0 | 1.7 | 0.0 | 1.6 | 1.4 |
| 18 23278 | Fishers | 35.6 | 97,020 | 328 | 2,725 | 84.8 | 5.9 | 0.0 | 7.2 | 0.0 | 0.6 | 4.6 |
| 18 25000 | Fort Wayne | 110.7 | 272,398 | 76 | 2,461 | 71.5 | 15.4 | 0.4 | 5.4 | 0.0 | 2.8 | 3.4 |
| 18 27000 | Gary | 49.7 | 74,693 | 475 | 1,503 | 16.6 | 78.2 | 0.2 | 0.2 | 0.1 | 1.3 | 3.4 |
| 18 28386 | Goshen | 17.6 | 34,168 | 1,129 | 1,941 | 76.3 | 8.6 | 0.0 | 2.9 | 0.0 | 9.0 | 3.1 |
| 18 29898 | Greenwood | 27.9 | 60,279 | 625 | 2,161 | 79.4 | 5.3 | 0.0 | 9.1 | 0.0 | 3.6 | 2.6 |
| 18 31000 | Hammond | 22.7 | 75,342 | 473 | 3,319 | 41.6 | 25.8 | 0.6 | 0.6 | 0.0 | 29.2 | 2.0 |
| 18 34114 | Hobart | 26.2 | 27,913 | 1,318 | 1,065 | 85.9 | 4.2 | 0.0 | 2.6 | 0.0 | 3.5 | 3.8 |
| 18 36000 | Indianapolis | 361.6 | 877,903 | 16 | 2,428 | 60.5 | 28.8 | 0.3 | 4.0 | 0.0 | 3.0 | 3.3 |
| 18 38358 | Jeffersonville | 34.1 | 48,530 | 803 | 1,423 | D | D | D | D | D | D | D |
| 18 40392 | Kokomo | 36.7 | 58,243 | 659 | 1,587 | 82.8 | 8.1 | 0.0 | 2.8 | 0.0 | 5.4 | 0.9 |
| 18 40788 | Lafayette | 29.4 | 71,484 | 509 | 2,431 | 78.0 | 14.0 | 0.0 | 2.7 | 0.0 | 0.5 | 4.9 |
| 18 42426 | Lawrence | 20.1 | 49,545 | 781 | 2,465 | 63.1 | 27.7 | 0.4 | 0.8 | 0.0 | 2.5 | 5.6 |
| 18 46908 | Marion | 15.7 | 27,674 | 1,330 | 1,763 | 77.9 | 13.2 | 0.3 | 1.8 | 0.0 | 0.7 | 6.0 |
| 18 48528 | Merrillville | 33.2 | 34,916 | 1,107 | 1,052 | 60.5 | 33.1 | 0.0 | 0.4 | 0.2 | 4.1 | 1.8 |
| 18 48798 | Michigan City | 20.4 | 31,054 | 1,216 | 1,522 | 60.8 | 29.3 | 0.0 | 1.4 | 0.0 | 1.5 | 7.0 |
| 18 49932 | Mishawaka | 17.9 | 50,312 | 774 | 2,811 | 85.1 | 5.7 | 0.0 | 3.2 | 0.1 | 2.2 | 3.8 |
| 18 51876 | Muncie | 27.4 | 67,523 | 546 | 2,464 | 85.0 | 9.3 | 0.0 | 1.4 | 0.0 | 0.3 | 4.1 |
| 18 52326 | New Albany | 15.4 | 36,874 | 1,039 | 2,394 | D | D | D | D | D | D | D |
| 18 54180 | Noblesville | 34.3 | 65,835 | 573 | 1,919 | 91.0 | 4.2 | 0.0 | 2.7 | 0.0 | 0.0 | 2.1 |
| 18 60246 | Plainfield | 25.7 | 36,275 | 1,062 | 1,412 | 81.1 | 6.0 | 0.1 | 5.7 | 0.0 | 5.2 | 2.0 |
| 18 61092 | Portage | 25.6 | 37,137 | 1,034 | 1,451 | 84.0 | 12.0 | 0.0 | 0.7 | 0.0 | 2.5 | 0.8 |
| 18 64260 | Richmond | 24.0 | 35,239 | 1,094 | 1,468 | 80.3 | 8.1 | 3.7 | 2.0 | 0.0 | 1.3 | 4.7 |
| 18 68220 | Schererville | 14.9 | 28,593 | 1,297 | 1,919 | 81.1 | 6.9 | 0.0 | 7.3 | 0.0 | 2.1 | 2.6 |
| 18 71000 | South Bend | 42.0 | 101,868 | 307 | 2,425 | 61.1 | 27.6 | 0.7 | 1.2 | 0.0 | 5.9 | 3.5 |
| 18 75428 | Terre Haute | 34.2 | 60,410 | 623 | 1,766 | 82.7 | 10.6 | 0.1 | 2.1 | 0.0 | 0.8 | 3.7 |
| 18 78326 | Valparaiso | 16.4 | 33,944 | 1,136 | 2,070 | D | D | D | D | D | D | D |
| 18 82700 | Westfield | 31.1 | 46,145 | 842 | 1,484 | D | D | D | D | D | D | D |
| 18 82862 | West Lafayette | 13.6 | 51,605 | 751 | 3,795 | 69.4 | 3.2 | 0.7 | 20.6 | 0.1 | 1.0 | 2.2 |
| 19 00000 | **IOWA** | 55,853.7 | 3,163,561 | X | 57 | 89.9 | 4.1 | 0.4 | 2.4 | 0.1 | 1.0 | 2.2 |
| 19 01855 | Ames | 27.6 | 67,033 | 552 | 2,429 | 79.2 | 4.4 | 0.5 | 9.8 | 0.0 | 1.2 | 4.8 |
| 19 02305 | Ankeny | 30.3 | 70,068 | 521 | 2,313 | 93.7 | 1.8 | 0.2 | 1.5 | 0.0 | 0.1 | 2.7 |
| 19 06355 | Bettendorf | 21.3 | 36,665 | 1,048 | 1,721 | 90.0 | 1.2 | 0.1 | 7.4 | 0.0 | 0.5 | 1.1 |
| 19 09550 | Burlington | 14.4 | 24,516 | 1,414 | 1,703 | 83.0 | 5.1 | 0.0 | 1.6 | 0.0 | 0.0 | 9.8 |
| 19 11755 | Cedar Falls | 29.0 | 40,231 | 957 | 1,387 | 91.6 | 1.1 | 0.2 | 4.0 | 0.0 | 1.5 | 4.4 |
| 19 12000 | Cedar Rapids | 72.1 | 134,027 | 205 | 1,859 | 82.6 | 8.6 | 0.5 | 2.4 | 0.0 | 0.8 | 1.3 |
| 19 14430 | Clinton | 35.6 | 25,039 | 1,404 | 703 | 93.0 | 2.6 | 0.5 | 1.9 | 0.0 | 1.0 | 4.1 |
| 19 16860 | Council Bluffs | 43.0 | 62,216 | 603 | 1,447 | D | D | D | D | D | D | D |
| 19 19000 | Davenport | 63.8 | 101,799 | 309 | 1,596 | 79.7 | 12.0 | 0.7 | 2.4 | 0.1 | 1.5 | 2.3 |
| 19 21000 | Des Moines | 88.2 | 212,312 | 106 | 2,407 | 76.4 | 13.6 | 0.3 | 5.8 | 1.0 | 1.3 | 1.9 |
| 19 22395 | Dubuque | 30.9 | 57,781 | 666 | 1,870 | 88.7 | 5.6 | 0.3 | 1.2 | 0.0 | 4.2 | 2.8 |
| 19 28515 | Fort Dodge | 16.0 | 23,906 | 1,424 | 1,494 | 84.0 | 6.8 | 0.5 | 1.7 | 0.1 | 2.2 | 5.0 |
| 19 38595 | Iowa City | 25.6 | 76,608 | 464 | 2,993 | 81.1 | 3.7 | 0.5 | 7.3 | D | D | D |
| 19 49485 | Marion | 17.8 | 40,780 | 946 | 2,291 | D | D | D | D | D | D | D |
| 19 49755 | Marshalltown | 19.2 | 26,736 | 1,362 | 1,393 | D | D | D | D | D | D | D |
| 19 50160 | Mason City | 27.9 | 26,682 | 1,366 | 956 | D | D | D | D | D | 0.0 | 2.0 |
| 19 60465 | Ottumwa | 16.1 | 24,375 | 1,416 | 1,514 | 86.8 | 9.4 | 0.2 | 1.6 | 0.0 | 1.8 | 4.5 |
| 19 73335 | Sioux City | 58.5 | 82,759 | 419 | 1,415 | 84.0 | 4.5 | 2.0 | 3.2 | 0.0 | 0.2 | 1.5 |
| 19 79950 | Urbandale | 22.5 | 44,650 | 861 | 1,984 | 87.5 | 3.1 | 1.1 | 6.6 | 0.0 | 1.7 | 4.6 |
| 19 82425 | Waterloo | 61.6 | 67,292 | 548 | 1,092 | 74.1 | 18.2 | 0.1 | 1.4 | 0.0 | 1.2 | 1.6 |
| 19 83910 | West Des Moines | 47.2 | 69,057 | 533 | 1,463 | 85.0 | 1.5 | 0.0 | 10.6 | 0.0 | 1.2 | 1.6 |

1. Dry land or land partially or temporarily covered by water.   2. Hispanic or Latino persons may be of any race.

# Table D. Cities — Population

| City | Percent Hispanic or Latino[1], 2019 | Percent foreign born, 2019 | Age of population (percent), 2019 | | | | | | | Median age, 2019 | Percent female, 2019 | Population | | | |
|---|---|---|---|---|---|---|---|---|---|---|---|---|---|---|---|
| | | | | | | | | | | | | Census counts | | Percent change | |
| | | | Under 18 years | 18 to 24 years | 25 to 34 years | 35 to 44 years | 45 to 54 years | 55 to 64 years | 65 years and over | | | 2000 | 2010 | 2000-2010 | 2001-2020 |
| | 12 | 13 | 14 | 15 | 16 | 17 | 18 | 19 | 20 | 21 | 22 | 23 | 24 | 25 | 26 |
| **ILLINOIS— Cont'd** | | | | | | | | | | | | | | | |
| Waukegan | 49.7 | 28.2 | 27.4 | 10.6 | 14.7 | 14.8 | 11.5 | 10.5 | 10.4 | 32.8 | 50.8 | 87,901 | 89,119 | 1.4 | -4.1 |
| West Chicago | 51.2 | 25.2 | 22.5 | 14.1 | 10.8 | 14.4 | 14.7 | 11.6 | 12.0 | 36.8 | 53.8 | 23,469 | 27,239 | 16.1 | -2.4 |
| Wheaton | 8.4 | 13.5 | 23.4 | 9.6 | 13.1 | 10.0 | 11.9 | 15.1 | 17.1 | 39.1 | 49.7 | 55,416 | 53,170 | -4.1 | -1.4 |
| Wheeling | 32.8 | 43.3 | 20.5 | 8.1 | 20.0 | 16.4 | 11.1 | 11.2 | 12.6 | 35.8 | 50.6 | 34,496 | 37,644 | 9.1 | 3.5 |
| Wilmette | 4.1 | 20.3 | 31.3 | 2.5 | 3.0 | 14.2 | 14.3 | 14.8 | 19.8 | 44.4 | 52.8 | 27,651 | 27,060 | -2.1 | -0.9 |
| Woodridge | 17.7 | 23.2 | 20.7 | 11.6 | 13.0 | 15.5 | 12.8 | 13.5 | 12.9 | 38.2 | 47.5 | 30,934 | 32,989 | 6.6 | 1.0 |
| **INDIANA** | 7.2 | 5.3 | 23.3 | 9.8 | 13.2 | 12.3 | 12.3 | 13.0 | 16.1 | 38.0 | 50.7 | 6,080,485 | 6,484,050 | 6.6 | 4.2 |
| Anderson | 5.5 | 1.9 | 21.2 | 10.6 | 13.2 | 11.3 | 13.1 | 12.0 | 18.7 | 38.3 | 51.1 | 59,734 | 56,083 | -6.1 | -2.6 |
| Bloomington | 5.0 | 11.2 | 12.6 | 41.4 | 14.5 | 9.8 | 7.2 | 4.4 | 10.0 | 23.7 | 51.0 | 69,291 | 80,299 | 15.9 | 6.6 |
| Carmel | 4.5 | 13.8 | 26.4 | 8.3 | 11.1 | 14.3 | 12.9 | 12.7 | 14.3 | 37.8 | 52.9 | 37,733 | 83,885 | 122.3 | 21.2 |
| Columbus | 5.8 | 15.3 | 24.5 | 7.8 | 17.8 | 12.5 | 9.6 | 11.3 | 16.5 | 35.0 | 51.5 | 39,059 | 44,088 | 12.9 | 9.9 |
| Crown Point | 15.1 | 8.7 | 24.9 | 5.8 | 14.4 | 15.8 | 11.8 | 9.6 | 17.8 | 39.7 | 54.0 | 19,806 | 27,866 | 40.7 | 10.9 |
| East Chicago | 48.7 | 9.4 | 26.4 | 7.2 | 18.4 | 10.7 | 9.3 | 11.1 | 16.9 | 33.9 | 51.5 | 32,414 | 29,698 | -8.4 | -6.6 |
| Elkhart | 34.1 | 17.0 | 25.3 | 8.0 | 17.1 | 11.6 | 9.5 | 13.1 | 15.3 | 34.5 | 50.8 | 51,874 | 51,911 | 0.1 | 0.5 |
| Evansville | 3.5 | 4.1 | 21.9 | 10.4 | 15.4 | 12.8 | 10.6 | 13.2 | 15.6 | 36.7 | 52.3 | 121,582 | 120,069 | -1.1 | -1.4 |
| Fishers | 2.9 | 7.3 | 29.2 | 6.2 | 11.5 | 15.9 | 16.2 | 9.8 | 11.1 | 36.6 | 51.2 | 37,835 | 77,287 | 104.3 | 25.5 |
| Fort Wayne | 9.7 | 9.1 | 24.9 | 10.0 | 14.6 | 12.5 | 11.7 | 11.9 | 14.4 | 35.4 | 52.0 | 205,727 | 253,703 | 23.3 | 7.4 |
| Gary | 7.3 | 1.9 | 25.1 | 9.4 | 13.8 | 10.1 | 8.8 | 14.7 | 18.2 | 36.8 | 55.3 | 102,746 | 80,256 | -21.9 | -6.9 |
| Goshen | 13.3 | 11.1 | 30.5 | 7.8 | 7.1 | 14.6 | 14.8 | 8.4 | 16.9 | 38.0 | 50.7 | 29,383 | 32,572 | 10.9 | 4.9 |
| Greenwood | 8.4 | 10.1 | 27.1 | 8.2 | 15.9 | 13.6 | 11.8 | 9.6 | 13.7 | 34.1 | 52.1 | 36,037 | 51,110 | 41.8 | 17.9 |
| Hammond | 40.9 | 12.8 | 21.8 | 10.2 | 10.4 | 12.5 | 14.9 | 15.4 | 14.7 | 41.2 | 52.6 | 83,048 | 80,828 | -2.7 | -6.8 |
| Hobart | 18.6 | 4.2 | 28.1 | 7.5 | 10.8 | 15.5 | 14.4 | 12.2 | 11.5 | 37.1 | 50.9 | 25,363 | 29,336 | 15.7 | -4.9 |
| Indianapolis | 11.1 | 9.7 | 24.6 | 9.6 | 16.9 | 12.9 | 11.2 | 12.0 | 12.9 | 34.3 | 51.7 | 781,870 | 829,709 | 6.1 | 7.0 |
| Jeffersonville | 4.5 | 2.9 | 22.3 | 9.3 | 15.6 | 13.5 | 9.4 | 12.5 | 17.4 | 37.3 | 51.3 | 27,362 | 45,007 | 64.5 | 7.8 |
| Kokomo | 2.0 | 0.9 | 19.0 | 8.7 | 14.7 | 9.6 | 13.2 | 14.8 | 20.0 | 43.8 | 52.9 | 46,113 | 58,189 | 26.2 | 0.1 |
| Lafayette | 13.7 | 9.4 | 21.6 | 10.9 | 18.7 | 14.0 | 10.3 | 10.9 | 13.5 | 34.4 | 49.5 | 56,397 | 68,864 | 22.1 | 3.8 |
| Lawrence | 12.8 | 4.8 | 24.3 | 7.0 | 20.6 | 14.6 | 10.3 | 12.5 | 10.6 | 34.3 | 53.0 | 38,915 | 45,916 | 18.0 | 7.9 |
| Marion | 7.0 | 2.4 | 18.3 | 17.3 | 14.1 | 8.7 | 9.1 | 11.4 | 21.0 | 35.3 | 56.2 | 31,320 | 29,892 | -4.6 | -7.4 |
| Merrillville | 21.4 | 7.3 | 20.2 | 8.9 | 11.1 | 15.3 | 13.1 | 11.8 | 19.6 | 42.2 | 49.0 | 30,560 | 34,966 | 14.4 | -0.1 |
| Michigan City | 9.8 | 2.2 | 22.9 | 8.3 | 14.1 | 12.0 | 12.3 | 13.8 | 16.6 | 37.7 | 47.0 | 32,900 | 31,553 | -4.1 | -1.6 |
| Mishawaka | 7.4 | 5.8 | 23.7 | 9.7 | 17.3 | 13.2 | 8.3 | 12.4 | 15.4 | 34.2 | 51.4 | 46,557 | 48,292 | 3.7 | 4.2 |
| Muncie | 3.0 | 2.0 | 13.6 | 29.6 | 12.4 | 8.5 | 10.6 | 9.9 | 15.4 | 30.2 | 51.8 | 67,430 | 70,206 | 4.1 | -3.8 |
| New Albany | 6.0 | 1.5 | 22.9 | 9.5 | 14.9 | 13.3 | 11.3 | 10.8 | 17.3 | 37.1 | 49.3 | 37,603 | 36,375 | -3.3 | 1.4 |
| Noblesville | 1.5 | 4.1 | 25.8 | 9.7 | 12.6 | 13.3 | 15.1 | 12.0 | 11.5 | 36.0 | 50.2 | 28,590 | 52,378 | 83.2 | 25.7 |
| Plainfield | 8.4 | 7.2 | 24.9 | 5.6 | 20.2 | 14.7 | 13.1 | 9.4 | 12.0 | 34.6 | 47.9 | 18,396 | 27,700 | 50.6 | 31.0 |
| Portage | 12.7 | 2.4 | 22.5 | 9.5 | 15.5 | 11.5 | 9.7 | 11.9 | 19.4 | 37.4 | 50.4 | 33,496 | 36,826 | 9.9 | 0.8 |
| Richmond | 4.6 | 3.4 | 22.7 | 11.7 | 12.0 | 10.6 | 13.1 | 11.9 | 18.1 | 38.1 | 53.7 | 39,124 | 36,779 | -6.0 | -4.2 |
| Schererville | 12.9 | 11.3 | 21.1 | 8.4 | 12.6 | 12.7 | 15.8 | 14.0 | 15.3 | 41.4 | 46.5 | 24,851 | 29,217 | 17.6 | -2.1 |
| South Bend | 15.3 | 8.6 | 24.9 | 11.1 | 14.5 | 12.1 | 11.8 | 11.9 | 13.8 | 34.7 | 51.6 | 107,789 | 101,239 | -6.1 | 0.6 |
| Terre Haute | 2.4 | 3.8 | 20.8 | 20.2 | 15.2 | 9.6 | 10.6 | 8.2 | 15.4 | 30.4 | 50.9 | 59,614 | 60,791 | 2.0 | -0.6 |
| Valparaiso | 8.9 | 9.0 | 18.9 | 15.5 | 11.0 | 14.5 | 10.1 | 12.1 | 17.8 | 38.0 | 50.2 | 27,428 | 31,740 | 15.7 | 6.9 |
| Westfield | 10.1 | 9.2 | 26.1 | 5.7 | 11.1 | 19.0 | 12.8 | 11.3 | 14.0 | 39.0 | 48.6 | 9,293 | 30,138 | 224.3 | 53.1 |
| West Lafayette | 6.5 | 25.4 | 11.4 | 58.9 | 10.1 | 5.1 | 4.1 | 3.5 | 6.9 | 21.5 | 47.3 | 28,778 | 41,997 | 45.9 | 22.9 |
| **IOWA** | 6.3 | 5.6 | 22.9 | 9.8 | 12.7 | 12.4 | 11.4 | 13.3 | 17.5 | 38.5 | 50.2 | 2,926,324 | 3,046,877 | 4.1 | 3.8 |
| Ames | 3.9 | 10.3 | 11.2 | 38.7 | 18.4 | 8.8 | 6.1 | 6.9 | 10.1 | 25.0 | 46.5 | 50,731 | 59,035 | 16.4 | 13.5 |
| Ankeny | 2.2 | 5.5 | 27.9 | 8.6 | 17.4 | 15.9 | 11.1 | 8.2 | 11.0 | 32.4 | 49.2 | 27,117 | 45,602 | 68.2 | 53.7 |
| Bettendorf | 3.6 | 4.6 | 24.6 | 6.7 | 11.3 | 14.9 | 13.9 | 11.6 | 16.9 | 39.6 | 48.9 | 31,275 | 33,207 | 6.2 | 10.4 |
| Burlington | 1.0 | 3.1 | 18.4 | 11.1 | 9.1 | 10.7 | 10.0 | 15.5 | 25.2 | 46.3 | 56.6 | 26,839 | 25,592 | -4.6 | -4.2 |
| Cedar Falls | 4.3 | 6.5 | 22.7 | 21.1 | 9.0 | 13.8 | 8.4 | 6.1 | 18.8 | 32.0 | 52.5 | 36,145 | 39,284 | 8.7 | 2.4 |
| Cedar Rapids | 5.1 | 7.4 | 22.1 | 9.7 | 15.7 | 12.7 | 11.9 | 12.2 | 15.7 | 36.7 | 51.6 | 120,758 | 126,609 | 4.8 | 5.9 |
| Clinton | 5.1 | 1.0 | 22.1 | 7.1 | 14.8 | 10.3 | 10.6 | 14.9 | 20.1 | 40.8 | 53.3 | 27,772 | 26,873 | -3.2 | -6.8 |
| Council Bluffs | 9.6 | 4.4 | 22.7 | 9.1 | 12.9 | 13.6 | 12.1 | 12.6 | 17.0 | 39.0 | 51.1 | 58,268 | 62,213 | 6.8 | 0.0 |
| Davenport | 9.0 | 4.7 | 22.7 | 8.8 | 14.0 | 14.0 | 12.6 | 12.7 | 15.2 | 38.0 | 51.1 | 98,359 | 99,697 | 1.4 | 2.1 |
| Des Moines | 13.1 | 13.3 | 22.6 | 10.5 | 16.7 | 12.1 | 12.6 | 11.5 | 14.0 | 35.2 | 51.0 | 198,682 | 204,216 | 2.8 | 4.0 |
| Dubuque | 3.5 | 4.7 | 19.6 | 13.1 | 13.6 | 10.2 | 9.5 | 14.9 | 19.1 | 38.4 | 51.1 | 57,686 | 57,605 | -0.1 | 0.3 |
| Fort Dodge | 6.7 | 4.1 | 15.2 | 13.1 | 16.1 | 10.3 | 11.7 | 12.1 | 21.6 | 40.7 | 43.2 | 25,136 | 25,203 | 0.3 | -5.1 |
| Iowa City | 6.5 | 11.0 | 13.2 | 32.3 | 14.3 | 9.6 | 9.2 | 8.7 | 12.7 | 28.4 | 53.0 | 62,220 | 67,963 | 9.2 | 12.7 |
| Marion | 1.2 | 2.7 | 20.9 | 7.5 | 13.4 | 11.3 | 15.4 | 14.0 | 17.4 | 43.0 | 48.6 | 26,294 | 35,225 | 34.0 | 15.8 |
| Marshalltown | 31.6 | 22.4 | 26.3 | 8.8 | 14.2 | 13.4 | 9.9 | 10.0 | 17.5 | 35.3 | 49.4 | 26,009 | 27,558 | 6.0 | -3.0 |
| Mason City | 6.6 | 0.7 | 19.5 | 7.2 | 14.2 | 10.9 | 11.4 | 14.2 | 22.7 | 43.6 | 48.4 | 29,172 | 28,072 | -3.8 | -5.0 |
| Ottumwa | 12.4 | 8.8 | 21.5 | 10.1 | 14.5 | 11.9 | 11.5 | 15.0 | 15.5 | 39.2 | 47.2 | 24,998 | 25,028 | 0.1 | -2.6 |
| Sioux City | 20.5 | 10.5 | 26.2 | 10.2 | 13.5 | 13.3 | 10.0 | 12.6 | 14.3 | 35.1 | 50.1 | 85,013 | 82,685 | -2.7 | 0.1 |
| Urbandale | 4.4 | 9.4 | 27.2 | 4.5 | 13.6 | 15.4 | 12.6 | 12.7 | 14.0 | 37.2 | 50.5 | 29,072 | 39,472 | 35.8 | 13.1 |
| Waterloo | 5.9 | 11.4 | 21.0 | 11.7 | 15.7 | 10.5 | 11.6 | 13.9 | 15.7 | 36.3 | 50.1 | 68,747 | 68,490 | -0.4 | -1.7 |
| West Des Moines | 5.2 | 13.5 | 20.6 | 6.0 | 16.1 | 14.8 | 12.8 | 13.8 | 15.9 | 39.2 | 50.5 | 46,403 | 56,707 | 22.2 | 21.8 |

1. May be of any race.

# Table D. Cities — Households, Group Quarters, Crime, and Education

| City | Households, 2019 | | | | | | | Persons in group quarters, 2019 | Serious crimes known to police[1], 2019 | | | | Educational attainment, 2019 | | |
|---|---|---|---|---|---|---|---|---|---|---|---|---|---|---|---|
| | | | | Percent | | | | | Violent | | Prorerty | | | Attainment[3] (percent) | |
| | Number | Persons per household | Family | Married couple family | Female family | Non-family | One person | | Number | Rate[2] | Number | Rate[2] | Population age 25 and over | High school graduate or less | Bachelor's degree or more |
| | 27 | 28 | 29 | 30 | 31 | 32 | 33 | 34 | 35 | 36 | 37 | 38 | 39 | 40 | 41 |
| **ILLINOIS— Cont'd** | | | | | | | | | | | | | | | |
| Waukegan | 29,100 | 2.89 | 66.5 | 36.4 | 30.1 | 33.5 | 28.9 | 2,207 | 331 | 383 | 1,929 | 2,230 | 53,465 | 49.2 | 17.9 |
| West Chicago | 6,706 | 3.27 | 79.1 | 53.4 | 25.7 | 20.9 | 17.5 | 376 | 38 | 141 | 219 | 811 | 14,139 | 45.1 | 26.6 |
| Wheaton | 19,247 | 2.57 | 70.5 | 62.1 | 8.4 | 29.5 | 25.7 | 3,201 | 44 | 83 | 611 | 1,149 | 35,358 | 12.7 | 67.7 |
| Wheeling | 15,202 | 2.50 | 64.2 | 49.4 | 14.8 | 35.8 | 28.3 | 588 | 29 | 74 | 418 | 1,071 | 27,593 | 33.0 | 45.6 |
| Wilmette | 9,507 | 2.83 | 79.0 | 71.4 | 7.5 | 21.0 | 19.0 | 167 | 9 | 33 | 220 | 806 | 17,916 | 8.6 | 84.3 |
| Woodridge | 13,064 | 2.51 | 64.8 | 51.0 | 13.7 | 35.2 | 29.5 | 75 | 37 | 110 | 301 | 895 | 22,241 | 23.4 | 52.3 |
| **INDIANA** | 2,597,765 | 2.52 | 63.1 | 47.0 | 16.1 | 36.9 | 30.4 | 190,079 | NA | NA | NA | NA | 4,502,015 | 44.3 | 26.9 |
| Anderson | 22,756 | 2.32 | 61.3 | 38.0 | 23.3 | 38.7 | 31.6 | 2,322 | 242 | 441 | 2,096 | 3,818 | 37,596 | 57.1 | 14.6 |
| Bloomington | 30,072 | 2.38 | 37.0 | 27.1 | 9.9 | 63.0 | 41.6 | 15,050 | NA | NA | NA | NA | 39,792 | 23.0 | 57.0 |
| Carmel | 38,677 | 2.63 | 72.6 | 64.7 | 8.0 | 27.4 | 22.9 | 345 | 30 | 31 | 807 | 846 | 66,518 | 9.6 | 75.4 |
| Columbus | 19,035 | 2.56 | 57.8 | 43.8 | 13.9 | 42.2 | 31.4 | 959 | 67 | 140 | 1,439 | 2,998 | 33,631 | 33.9 | 42.9 |
| Crown Point | 12,034 | 2.59 | 66.2 | 52.1 | 14.1 | 33.8 | 29.9 | 1,425 | 7 | 23 | 290 | 956 | 22,631 | 34.6 | 38.5 |
| East Chicago | 11,459 | 2.42 | 56.4 | 21.2 | 35.2 | 43.6 | 39.6 | 98 | 201 | 725 | 1,094 | 3,947 | 18,467 | 65.4 | 9.1 |
| Elkhart | 20,308 | 2.55 | 43.5 | 26.9 | 16.6 | 56.5 | 50.0 | 875 | NA | NA | NA | NA | 35,156 | 59.8 | 15.4 |
| Evansville | 50,692 | 2.23 | 52.7 | 32.7 | 20.0 | 47.3 | 40.9 | 4,389 | 721 | 613 | 4,917 | 4,178 | 79,603 | 49.5 | 20.8 |
| Fishers | 34,846 | 2.76 | 73.3 | 63.3 | 10.0 | 26.7 | 19.4 | 30 | 51 | 53 | 734 | 769 | 62,187 | 12.9 | 64.2 |
| Fort Wayne | 105,414 | 2.45 | 58.8 | 39.4 | 19.4 | 41.2 | 33.4 | 5,094 | 974 | 362 | 7,437 | 2,761 | 171,212 | 39.9 | 27.6 |
| Gary | 29,466 | 2.38 | 58.1 | 21.7 | 36.3 | 41.9 | 36.2 | 740 | 414 | 554 | 2,656 | 3,556 | 46,395 | 53.0 | 14.0 |
| Goshen | 12,235 | 2.74 | 53.1 | 44.9 | 8.2 | 46.9 | 43.6 | 1,415 | NA | NA | NA | NA | 21,563 | 48.4 | 27.5 |
| Greenwood | 22,471 | 2.59 | 65.5 | 48.7 | 16.8 | 34.5 | 26.1 | 644 | 97 | 162 | 1,528 | 2,555 | 38,079 | 45.6 | 28.1 |
| Hammond | 29,402 | 2.53 | 59.5 | 35.2 | 24.3 | 40.5 | 34.5 | 1,044 | 315 | 419 | 2,792 | 3,713 | 51,358 | 59.0 | 13.0 |
| Hobart | 11,086 | 2.63 | 63.1 | 47.9 | 15.3 | 36.9 | 25.7 | 285 | 39 | 140 | 964 | 3,458 | 18,911 | 45.6 | 25.1 |
| Indianapolis | 340,639 | 2.51 | 51.9 | 34.9 | 17.0 | 48.1 | 40.1 | 15,557 | NA | NA | NA | NA | 573,197 | 43.1 | 30.9 |
| Jeffersonville | 18,018 | 2.72 | 69.1 | 40.9 | 28.2 | 30.9 | 24.7 | 973 | 114 | 239 | 1,341 | 2,810 | 34,131 | 40.0 | 22.2 |
| Kokomo | 24,392 | 2.18 | 54.2 | 35.5 | 18.7 | 45.8 | 39.3 | 1,168 | 392 | 678 | 1,493 | 2,581 | 39,209 | 47.4 | 15.0 |
| Lafayette | 33,086 | 2.15 | 49.2 | 33.1 | 16.1 | 50.8 | 41.5 | 1,392 | 352 | 485 | 2,388 | 3,290 | 48,980 | 46.9 | 26.0 |
| Lawrence | 19,833 | 2.49 | 61.4 | 47.4 | 14.0 | 38.6 | 32.5 | 158 | 178 | 360 | 1,087 | 2,199 | 33,955 | 38.6 | 35.9 |
| Marion | 10,936 | 2.00 | 44.2 | 22.9 | 21.3 | 55.8 | 48.4 | 4,682 | 103 | 370 | 870 | 3,126 | 17,100 | 51.9 | 17.8 |
| Merrillville | 14,197 | 2.32 | 61.2 | 38.5 | 22.7 | 38.8 | 35.3 | 375 | 50 | 144 | 580 | 1,670 | 23,564 | 56.0 | 20.1 |
| Michigan City | 13,669 | 2.17 | 52.6 | 22.0 | 30.6 | 47.4 | 37.9 | 3,017 | NA | NA | NA | NA | 22,479 | 56.5 | 13.8 |
| Mishawaka | 21,320 | 2.27 | 52.2 | 33.0 | 19.1 | 47.8 | 43.2 | 1,027 | NA | NA | NA | NA | 32,951 | 46.6 | 25.9 |
| Muncie | 26,772 | 2.27 | 50.9 | 25.8 | 25.2 | 49.1 | 34.9 | 7,126 | NA | NA | NA | NA | 38,598 | 50.3 | 22.6 |
| New Albany | 14,435 | 2.55 | 58.1 | 34.9 | 23.3 | 41.9 | 32.1 | 1,259 | NA | NA | NA | NA | 25,699 | 49.2 | 20.5 |
| Noblesville | 22,515 | 2.78 | 74.7 | 64.5 | 10.3 | 25.3 | 20.7 | 823 | 32 | 50 | 618 | 957 | 40,854 | 19.9 | 57.7 |
| Plainfield | 12,798 | 2.67 | 66.0 | 51.2 | 14.8 | 34.0 | 25.2 | 1,236 | 50 | 142 | 634 | 1,795 | 24,620 | 39.2 | 33.8 |
| Portage | 14,886 | 2.39 | 56.8 | 45.9 | 10.9 | 43.2 | 35.8 | 278 | 73 | 198 | 583 | 1,584 | 24,409 | 55.7 | 15.2 |
| Richmond | 15,465 | 2.18 | 60.5 | 36.3 | 24.2 | 39.5 | 34.7 | 2,511 | NA | NA | NA | NA | 23,793 | 44.4 | 19.0 |
| Schererville | 10,590 | 2.68 | 74.2 | 50.0 | 24.2 | 25.8 | 21.7 | 115 | NA | NA | NA | NA | 20,119 | 30.0 | 32.7 |
| South Bend | 41,196 | 2.40 | 55.1 | 34.2 | 20.9 | 44.9 | 37.2 | 3,654 | 1,357 | 1,331 | 4,296 | 4,214 | 65,683 | 45.1 | 27.5 |
| Terre Haute | 23,148 | 2.30 | 51.1 | 30.8 | 20.3 | 48.9 | 36.4 | 7,546 | NA | NA | NA | NA | 35,889 | 46.1 | 20.7 |
| Valparaiso | 13,360 | 2.24 | 51.2 | 43.5 | 7.7 | 48.8 | 37.0 | 3,321 | NA | NA | NA | NA | 21,819 | 37.7 | 42.7 |
| Westfield | 17,131 | 2.54 | 78.6 | 63.9 | 14.7 | 21.4 | 16.6 | 171 | 16 | 37 | 338 | 782 | 29,828 | 16.4 | 61.3 |
| West Lafayette | 14,196 | 2.50 | 32.6 | 23.9 | 8.7 | 67.4 | 40.1 | 15,277 | 32 | 65 | 357 | 726 | 15,110 | 19.1 | 67.4 |
| **IOWA** | 1,287,221 | 2.38 | 62.8 | 48.9 | 14.0 | 37.2 | 29.6 | 96,609 | NA | NA | NA | NA | 2,123,004 | 38.4 | 29.3 |
| Ames | 26,347 | 2.09 | 41.3 | 34.6 | 6.7 | 58.7 | 32.5 | 11,087 | 135 | 198 | 1,171 | 1,716 | 33,249 | 13.1 | 60.9 |
| Ankeny | 26,150 | 2.55 | 68.3 | 56.3 | 12.0 | 31.7 | 20.3 | 695 | NA | NA | NA | NA | 42,790 | 18.3 | 53.5 |
| Bettendorf | 14,382 | 2.53 | 67.2 | 55.6 | 11.6 | 32.8 | 26.3 | 202 | 46 | 124 | 474 | 1,282 | 25,092 | 18.6 | 53.8 |
| Burlington | 11,798 | 2.22 | 51.2 | 31.6 | 19.6 | 48.8 | 41.2 | 438 | NA | NA | NA | NA | 18,799 | 49.8 | 18.8 |
| Cedar Falls | 15,222 | 2.61 | 57.5 | 51.4 | 6.2 | 42.5 | 27.0 | 3,400 | NA | NA | NA | NA | 24,220 | 22.9 | 48.4 |
| Cedar Rapids | 57,947 | 2.25 | 55.6 | 39.2 | 16.4 | 44.4 | 35.7 | 3,260 | 345 | 257 | 4,470 | 3,336 | 91,093 | 34.6 | 31.7 |
| Clinton | 10,326 | 2.22 | 58.3 | 38.9 | 19.4 | 41.7 | 37.1 | 643 | 77 | 308 | 911 | 3,647 | 16,657 | 53.6 | 18.1 |
| Council Bluffs | 25,053 | 2.40 | 63.1 | 40.3 | 22.8 | 36.9 | 27.5 | 1,970 | 493 | 790 | 3,649 | 5,845 | 42,390 | 41.8 | 22.4 |
| Davenport | 41,123 | 2.39 | 57.7 | 42.3 | 15.4 | 42.3 | 34.9 | 3,133 | 609 | 595 | 3,918 | 3,826 | 69,611 | 38.1 | 26.2 |
| Des Moines | 91,045 | 2.30 | 55.9 | 34.7 | 21.2 | 44.1 | 36.0 | 5,114 | 1,555 | 712 | 8,933 | 4,091 | 143,461 | 41.0 | 27.2 |
| Dubuque | 23,579 | 2.29 | 50.7 | 34.0 | 16.6 | 49.3 | 36.8 | 3,826 | 110 | 190 | 949 | 1,637 | 38,932 | 40.3 | 33.6 |
| Fort Dodge | 10,948 | 1.83 | 41.0 | 30.4 | 10.6 | 59.0 | 48.1 | 2,816 | 101 | 421 | 724 | 3,020 | 16,375 | 45.5 | 18.1 |
| Iowa City | 32,850 | 2.11 | 44.8 | 36.7 | 8.1 | 55.2 | 33.7 | 5,977 | 167 | 216 | 1,252 | 1,618 | 40,976 | 15.6 | 62.1 |
| Marion | 16,421 | 2.44 | 65.1 | 51.8 | 13.2 | 34.9 | 29.1 | 335 | 68 | 167 | 519 | 1,278 | 28,893 | 35.0 | 37.1 |
| Marshalltown | 10,726 | 2.45 | 59.6 | 47.1 | 12.5 | 40.4 | 32.3 | 1,324 | 105 | 389 | 674 | 2,496 | 17,935 | 54.3 | 19.8 |
| Mason City | 12,810 | 2.01 | 52.0 | 36.9 | 15.1 | 48.0 | 41.7 | 765 | 176 | 652 | 789 | 2,925 | 19,476 | 40.4 | 24.6 |
| Ottumwa | 10,433 | 2.22 | 58.7 | 38.6 | 20.2 | 41.3 | 32.8 | 740 | 134 | 547 | 1,009 | 4,120 | 16,328 | 48.8 | 18.1 |
| Sioux City | 31,968 | 2.52 | 64.7 | 44.1 | 20.6 | 35.3 | 26.8 | 2,617 | 366 | 445 | 3,203 | 3,890 | 52,996 | 50.1 | 19.0 |
| Urbandale | 17,881 | 2.63 | 72.1 | 62.3 | 9.8 | 27.9 | 22.5 | 323 | 28 | 63 | 455 | 1,022 | 32,372 | 20.3 | 51.6 |
| Waterloo | 27,914 | 2.38 | 56.5 | 36.5 | 19.9 | 43.7 | 36.2 | 877 | 306 | 452 | 1,682 | 2,484 | 45,316 | 47.7 | 24.1 |
| West Des Moines | 30,464 | 2.21 | 58.5 | 45.6 | 12.9 | 41.5 | 32.2 | 368 | NA | NA | NA | NA | 49,657 | 20.9 | 53.8 |

1. Data for serious crimes have not been adjusted for underreporting. This may affect comparability between geographic areas and over time.  2. Per 100,000 population estimated by the FBI.  3. Persons 25 years old and over.

# Table D. Cities — Income, Poverty, and Housing

| City | Money income, 2019 Households — Median income | Percent with income less than $20,000 | Percent with income of $200,000 or more | Median family income | Median non-family income | Median earnings, 2019 — All persons | Men | Women | Housing units, 2019 — Total | Occupied | Percent owner occupied | Median value[1] (dollars) | Median gross rent (dollars) |
|---|---|---|---|---|---|---|---|---|---|---|---|---|---|
| | 42 | 43 | 44 | 45 | 46 | 47 | 48 | 49 | 50 | 51 | 52 | 53 | 54 |
| **ILLINOIS— Cont'd** | | | | | | | | | | | | | |
| Waukegan | 54,030 | 17.3 | 3.1 | 62,360 | 38,005 | 30,615 | 33,146 | 26,736 | 31,167 | 29,100 | 47.6 | 143,813 | 1,017 |
| West Chicago | 88,378 | 5.5 | 13.9 | 94,742 | 54,714 | 31,200 | 39,383 | 23,522 | 6,874 | 6,706 | 72.2 | 241,862 | 1,110 |
| Wheaton | 107,764 | 4.4 | 18.4 | 133,696 | 64,740 | 52,454 | 75,577 | 37,406 | 21,054 | 19,247 | 77.0 | 377,496 | 1,463 |
| Wheeling | 69,330 | 9.6 | 10.2 | 86,703 | 48,360 | 40,300 | 46,250 | 36,030 | 15,513 | 15,202 | 54.6 | 215,174 | 1,376 |
| Wilmette | 177,786 | 3.2 | 44.3 | 203,767 | 105,928 | 90,811 | 100,004 | 76,003 | 9,869 | 9,507 | 90.8 | 733,412 | 1,378 |
| Woodridge | 78,853 | 7.5 | 12.5 | 101,437 | 50,362 | 42,477 | 50,712 | 39,008 | 13,363 | 13,064 | 60.9 | 286,657 | 1,386 |
| **INDIANA** | 57,603 | 14.3 | 4.7 | 73,876 | 33,994 | 33,756 | 41,268 | 27,480 | 2,921,115 | 2,597,765 | 69.3 | 156,011 | 840 |
| Anderson | 44,279 | 21.3 | 0.3 | 55,811 | 26,784 | 25,668 | 31,177 | 22,118 | 28,290 | 22,756 | 58.0 | 75,486 | 795 |
| Bloomington | 41,017 | 26.0 | 5.0 | 83,431 | 24,307 | 15,501 | 16,223 | 14,384 | 35,311 | 30,072 | 38.8 | 231,339 | 949 |
| Carmel | 113,714 | 7.3 | 26.8 | 153,919 | 56,039 | 56,503 | 91,270 | 40,432 | 41,289 | 38,677 | 73.5 | 370,416 | 1,209 |
| Columbus | 70,869 | 14.3 | 8.5 | 91,125 | 45,865 | 36,569 | 51,683 | 29,299 | 21,604 | 19,035 | 63.0 | 179,899 | 953 |
| Crown Point | 80,783 | 15.6 | 5.1 | 95,388 | 30,417 | 42,032 | 56,497 | 31,739 | 12,266 | 12,034 | 85.5 | 219,543 | 977 |
| East Chicago | 34,871 | 27.6 | 1.2 | 38,902 | 24,246 | 30,087 | 30,588 | 29,680 | 14,640 | 11,459 | 45.9 | 83,777 | 733 |
| Elkhart | 40,466 | 24.6 | 0.9 | 57,627 | 29,819 | 29,192 | 35,597 | 24,262 | 24,726 | 20,308 | 53.8 | 105,861 | 813 |
| Evansville | 42,600 | 22.6 | 1.3 | 51,892 | 30,116 | 28,022 | 32,418 | 21,893 | 58,353 | 50,692 | 51.6 | 97,707 | 804 |
| Fishers | 111,176 | 3.0 | 18.0 | 127,211 | 60,570 | 57,107 | 65,084 | 43,766 | 35,940 | 34,846 | 76.3 | 277,849 | 1,348 |
| Fort Wayne | 49,855 | 15.5 | 2.7 | 61,945 | 31,846 | 31,975 | 40,449 | 26,082 | 116,436 | 105,414 | 63.1 | 124,358 | 764 |
| Gary | 31,341 | 28.9 | 0.9 | 34,471 | 22,438 | 26,986 | 30,664 | 25,458 | 41,170 | 29,466 | 46.3 | 63,753 | 805 |
| Goshen | 44,744 | 20.7 | 2.0 | 63,942 | 25,873 | 29,931 | 40,228 | 25,224 | 13,656 | 12,235 | 57.5 | 135,535 | 844 |
| Greenwood | 65,025 | 9.9 | 4.3 | 72,840 | 38,504 | 37,277 | 49,045 | 30,150 | 25,812 | 22,471 | 64.7 | 179,041 | 1,021 |
| Hammond | 48,333 | 19.1 | 1.5 | 59,007 | 31,658 | 29,272 | 37,389 | 22,769 | 32,053 | 29,402 | 61.6 | 97,475 | 872 |
| Hobart | 68,029 | 10.5 | 1.9 | 104,100 | 46,329 | 42,461 | 46,890 | 31,395 | 12,219 | 11,086 | 81.0 | 163,126 | 1,046 |
| Indianapolis | 49,661 | 18.4 | 3.9 | 64,673 | 35,104 | 33,054 | 36,555 | 30,388 | 384,807 | 340,639 | 53.0 | 155,399 | 908 |
| Jeffersonville | 64,429 | 9.0 | 1.1 | 59,462 | 45,747 | 34,054 | 37,943 | 29,534 | 21,655 | 18,018 | 67.2 | 160,403 | 895 |
| Kokomo | 49,229 | 16.8 | 1.7 | 56,593 | 31,996 | 30,104 | 36,219 | 21,523 | 29,130 | 24,392 | 66.8 | 85,885 | 678 |
| Lafayette | 45,126 | 15.5 | 1.3 | 56,517 | 35,298 | 31,275 | 35,655 | 27,529 | 34,567 | 33,086 | 47.4 | 118,828 | 861 |
| Lawrence | 71,375 | 13.0 | 5.2 | 98,882 | 39,180 | 36,908 | 39,604 | 32,325 | 21,270 | 19,833 | 65.5 | 172,727 | 888 |
| Marion | 37,538 | 23.1 | 2.3 | 52,519 | 30,095 | 21,781 | 30,104 | 16,949 | 13,356 | 10,936 | 59.0 | 76,255 | 644 |
| Merrillville | 54,403 | 17.3 | 2.0 | 74,232 | 38,273 | 37,123 | 52,891 | 26,152 | 14,884 | 14,197 | 61.6 | 150,029 | 1,105 |
| Michigan City | 37,293 | 20.0 | 1.9 | 47,981 | 31,135 | 29,548 | 36,644 | 25,481 | 15,931 | 13,669 | 54.6 | 102,804 | 700 |
| Mishawaka | 43,070 | 21.9 | 2.1 | 59,651 | 28,659 | 31,174 | 36,513 | 27,050 | 23,422 | 21,320 | 53.3 | 110,850 | 847 |
| Muncie | 37,109 | 27.6 | 1.2 | 45,800 | 24,525 | 16,235 | 18,842 | 13,147 | 31,319 | 26,772 | 45.0 | 77,452 | 687 |
| New Albany | 48,317 | 20.8 | 0.7 | 56,717 | 35,589 | 30,059 | 32,530 | 26,788 | 17,732 | 14,435 | 57.5 | 138,458 | 765 |
| Noblesville | 100,051 | 7.5 | 10.3 | 117,494 | 45,504 | 45,954 | 60,554 | 36,217 | 25,169 | 22,515 | 79.4 | 241,702 | 1,037 |
| Plainfield | 65,072 | 7.1 | 8.2 | 80,790 | 46,506 | 37,410 | 42,973 | 31,034 | 13,178 | 12,798 | 58.7 | 218,889 | 995 |
| Portage | 61,512 | 11.1 | 2.4 | 73,105 | 31,784 | 34,179 | 49,559 | 25,505 | 15,966 | 14,886 | 71.4 | 167,537 | 888 |
| Richmond | 40,942 | 18.3 | 1.3 | 44,934 | 34,273 | 27,050 | 29,971 | 21,748 | 18,305 | 15,465 | 52.5 | 85,873 | 689 |
| Schererville | 73,389 | 7.2 | 14.4 | 95,748 | 45,952 | 40,690 | 48,842 | 31,602 | 11,150 | 10,590 | 78.3 | 234,450 | 975 |
| South Bend | 41,599 | 24.4 | 1.9 | 52,573 | 33,008 | 26,590 | 31,780 | 20,815 | 48,551 | 41,196 | 57.3 | 92,178 | 846 |
| Terre Haute | 35,787 | 32.2 | 1.8 | 48,740 | 23,944 | 17,780 | 24,814 | 12,388 | 27,421 | 23,148 | 48.5 | 82,286 | 682 |
| Valparaiso | 64,452 | 15.4 | 8.3 | 81,771 | 42,396 | 33,465 | 42,780 | 28,350 | 14,827 | 13,360 | 58.7 | 218,514 | 879 |
| Westfield | 95,105 | 5.3 | 17.8 | 109,242 | 51,490 | 53,793 | 71,721 | 32,348 | 17,454 | 17,131 | 82.7 | 296,711 | 1,181 |
| West Lafayette | 30,117 | 36.1 | 7.8 | 76,434 | 19,795 | 7,469 | 7,898 | 7,114 | 15,758 | 14,196 | 31.4 | 227,943 | 957 |
| **IOWA** | 61,691 | 13.2 | 5.0 | 78,152 | 36,559 | 35,837 | 42,262 | 30,059 | 1,418,600 | 1,287,221 | 70.5 | 158,856 | 808 |
| Ames | 50,528 | 23.2 | 5.0 | 95,036 | 31,864 | 16,421 | 21,032 | 12,443 | 27,881 | 26,347 | 42.9 | 234,117 | 933 |
| Ankeny | 94,862 | 3.4 | 7.7 | 106,800 | 56,145 | 48,324 | 55,549 | 37,681 | 26,869 | 26,150 | 68.3 | 237,145 | 1,107 |
| Bettendorf | 89,833 | 6.6 | 16.4 | 119,861 | 45,252 | 50,896 | 67,463 | 39,695 | 16,010 | 14,382 | 73.9 | 285,872 | 888 |
| Burlington | 47,987 | 21.9 | 3.0 | 59,565 | 38,638 | 25,914 | 41,269 | 16,857 | 13,186 | 11,798 | 59.8 | 102,629 | 698 |
| Cedar Falls | 71,639 | 8.4 | 10.6 | 100,214 | 44,062 | 27,339 | 32,527 | 18,826 | 16,487 | 15,222 | 68.6 | 219,124 | 990 |
| Cedar Rapids | 56,774 | 13.9 | 4.1 | 71,820 | 40,114 | 37,546 | 41,543 | 33,522 | 62,321 | 57,947 | 64.9 | 146,686 | 758 |
| Clinton | 46,941 | 19.4 | 1.2 | 62,250 | 31,380 | 31,999 | 44,944 | 29,901 | 11,941 | 10,326 | 61.2 | 105,192 | 779 |
| Council Bluffs | 52,508 | 12.1 | 3.4 | 60,709 | 34,774 | 31,810 | 38,595 | 30,004 | 27,407 | 25,053 | 62.2 | 126,914 | 857 |
| Davenport | 57,374 | 16.4 | 3.6 | 72,330 | 34,364 | 37,226 | 42,351 | 31,027 | 45,109 | 41,123 | 61.0 | 137,731 | 752 |
| Des Moines | 53,859 | 16.9 | 3.4 | 65,373 | 40,278 | 33,844 | 38,608 | 30,256 | 97,182 | 91,045 | 60.5 | 150,171 | 872 |
| Dubuque | 54,611 | 17.1 | 3.1 | 71,386 | 39,120 | 30,549 | 38,063 | 25,679 | 26,573 | 23,579 | 63.6 | 146,108 | 827 |
| Fort Dodge | 41,132 | 29.5 | 1.8 | 66,490 | 27,730 | 32,320 | 40,273 | 30,761 | 11,445 | 10,948 | 61.1 | 96,352 | 640 |
| Iowa City | 50,497 | 20.5 | 7.1 | 94,528 | 31,019 | 20,964 | 25,434 | 16,520 | 36,245 | 32,850 | 47.4 | 243,508 | 1,031 |
| Marion | 69,923 | 10.4 | 5.6 | 88,894 | 50,859 | 42,850 | 51,261 | 35,466 | 16,472 | 16,421 | 76.8 | 180,886 | 1,038 |
| Marshalltown | 45,062 | 22.0 | 1.5 | 61,749 | 30,887 | 29,409 | 37,697 | 22,423 | 11,792 | 10,726 | 60.9 | 91,542 | 684 |
| Mason City | 50,064 | 11.6 | 1.6 | 66,272 | 34,647 | 35,377 | 41,617 | 25,694 | 13,652 | 12,810 | 67.6 | 110,366 | 658 |
| Ottumwa | 43,989 | 18.9 | 2.7 | 56,637 | 35,079 | 33,008 | 36,739 | 31,071 | 11,302 | 10,433 | 66.1 | 71,628 | 788 |
| Sioux City | 53,424 | 16.4 | 2.4 | 69,156 | 31,378 | 30,970 | 36,162 | 25,751 | 34,316 | 31,968 | 63.6 | 120,240 | 812 |
| Urbandale | 90,811 | 6.8 | 8.4 | 110,143 | 51,212 | 46,318 | 56,978 | 40,322 | 18,305 | 17,881 | 76.5 | 250,112 | 970 |
| Waterloo | 51,847 | 16.9 | 3.9 | 66,453 | 34,113 | 31,592 | 35,148 | 27,261 | 31,872 | 27,914 | 59.5 | 127,859 | 798 |
| West Des Moines | 82,755 | 7.3 | 11.2 | 115,639 | 55,673 | 48,578 | 58,198 | 36,196 | 32,466 | 30,464 | 63.2 | 239,605 | 1,052 |

1. Based on population estimated by the American Community Survey.

# Table D. Cities — Commuting, Computer Access, Migration, Labor Force, and Employment

| City | Commuting[1], 2019 Percent — Drove alone | With commutes of 30 minutes or more | Computer access[2], 2019 Percent — With a computer in the house | With Internet access | Migration, 2019 — Percent who lived in the same house one year ago | Percent who lived in another state or county one year ago | Civilian labor force, 2020 — Total | Percent change 2019-2020 | Unemployment Total | Rate[3] | Civilian Employment[4], 2019 — Population age 16 and older — Number | Percent in labor force | Population age 16 to 64 — Number | Percent who worked full-year full-time |
|---|---|---|---|---|---|---|---|---|---|---|---|---|---|---|
| | 55 | 56 | 57 | 58 | 59 | 60 | 61 | 62 | 63 | 64 | 65 | 66 | 67 | 68 |
| **ILLINOIS— Cont'd** | | | | | | | | | | | | | | |
| Waukegan | 78.3 | 44.4 | 92.9 | 85.0 | 82.2 | 6.1 | 43,605 | -3.7 | 4,213 | 10 | 65,892 | 64.4 | 56,906 | 52.3 |
| West Chicago | 70.7 | 44.0 | 95.4 | 89.0 | 93.4 | 2.8 | 13,217 | -5.9 | 995 | 8 | 17,992 | 75.5 | 15,320 | 53.3 |
| Wheaton | 74.8 | 36.0 | 96.6 | 94.9 | 87.5 | 7.0 | 26,948 | -5.7 | 1,831 | 7 | 41,863 | 62.7 | 32,866 | 55.9 |
| Wheeling | 77.0 | 37.7 | 96.5 | 91.1 | NA | NA | 21,809 | -4.4 | 1,781 | 8 | 32,299 | 74.7 | 27,430 | 59.3 |
| Wilmette | 49.7 | 58.7 | 97.1 | 95.0 | 90.4 | 2.7 | 12,024 | -5.7 | 797 | 7 | 19,340 | 62.5 | 13,978 | 62.9 |
| Woodridge | 76.4 | 50.1 | 95.0 | 91.3 | 91.1 | 5.6 | 19,304 | -4.1 | 1,568 | 8 | 26,525 | 75.1 | 22,274 | 58.0 |
| **INDIANA** | 81.6 | 32.4 | 91.1 | 83.4 | 86.2 | 6.4 | 3,319,010 | -2.0 | 236,028 | 7 | 5,346,452 | 63.9 | 4,261,980 | 52.8 |
| Anderson | 75.1 | 35.8 | 89.0 | 79.1 | 81.4 | 4.6 | 23,546 | 0.0 | 2,315 | 10 | 44,395 | 57.3 | 34,092 | 45.9 |
| Bloomington | 61.3 | 13.1 | 93.9 | 85.6 | 61.8 | 21.4 | 37,349 | -3.2 | 2,204 | 6 | 77,006 | 57.3 | 68,338 | 28.6 |
| Carmel | 80.0 | 40.4 | 97.0 | 95.8 | 88.6 | 7.4 | 53,678 | 4.4 | 2,428 | 5 | 79,004 | 71.2 | 64,393 | 59.5 |
| Columbus | 87.6 | 16.6 | 93.2 | 88.3 | 80.8 | 8.3 | 25,060 | -2.7 | 1,639 | 7 | 39,392 | 63.9 | 31,177 | 54.9 |
| Crown Point | NA | 38.6 | 93.9 | 92.7 | 88.5 | 2.7 | 15,233 | -2.3 | 1,303 | 9 | 25,204 | 64.1 | 19,409 | 52.0 |
| East Chicago | 83.0 | 36.5 | 76.3 | 59.7 | 77.4 | 14.8 | 9,862 | -1.5 | 1,397 | 14 | 21,476 | 51.9 | 16,782 | 38.7 |
| Elkhart | 76.3 | 20.7 | 91.7 | 86.9 | 81.9 | 7.7 | 26,115 | -1.0 | 2,553 | 10 | 40,831 | 61.9 | 32,752 | 51.7 |
| Evansville | 80.3 | 15.0 | 89.9 | 83.8 | 74.8 | 9.4 | 58,083 | -2.7 | 4,891 | 8 | 93,965 | 64.1 | 75,597 | 50.3 |
| Fishers | 79.0 | 45.5 | 98.6 | 96.7 | 90.1 | 5.8 | 51,952 | -1.4 | 2,277 | 4 | 71,956 | 72.0 | 61,271 | 63.5 |
| Fort Wayne | 81.9 | 22.6 | 94.3 | 89.5 | 83.0 | 4.9 | 129,267 | -0.4 | 10,828 | 8 | 204,347 | 66.4 | 166,501 | 51.8 |
| Gary | 79.8 | 47.1 | 86.4 | 67.1 | 78.6 | 9.8 | 27,009 | 0.2 | 4,316 | 16 | 54,978 | 50.3 | 42,102 | 40.2 |
| Goshen | 76.1 | 19.9 | 89.6 | 83.3 | 87.7 | 4.5 | 17,628 | -0.2 | 1,388 | 8 | 25,269 | 65.2 | 19,373 | 59.2 |
| Greenwood | 87.1 | 41.1 | 93.7 | 85.2 | 80.3 | 12.2 | 30,875 | -0.8 | 1,926 | 6 | 44,066 | 65.8 | 35,998 | 59.4 |
| Hammond | 80.7 | 33.4 | 88.7 | 76.5 | 86.7 | 8.8 | 33,370 | -3.7 | 3,605 | 11 | 60,960 | 60.0 | 49,828 | 47.6 |
| Hobart | 93.6 | 40.0 | 97.9 | 94.9 | 92.1 | 4.1 | 14,516 | -3.1 | 1,629 | 11 | 22,088 | 71.5 | 18,719 | 59.4 |
| Indianapolis | 80.9 | 32.2 | 88.4 | 81.7 | 86.8 | 4.8 | 453,660 | 0.5 | 35,851 | 8 | 679,523 | 66.0 | 567,457 | 52.9 |
| Jeffersonville | 86.7 | 19.8 | 95.2 | 85.3 | 85.3 | 4.0 | 25,309 | 0.2 | 1,872 | 7 | 39,287 | 65.0 | 30,616 | 57.7 |
| Kokomo | 84.0 | 20.1 | 92.5 | 85.1 | 82.7 | 3.8 | 25,662 | 1.1 | 3,085 | 12 | 45,461 | 59.3 | 34,596 | 51.6 |
| Lafayette | 80.9 | 7.3 | 92.3 | 78.5 | 82.8 | 7.8 | 37,093 | -4.5 | 3,127 | 8 | 58,089 | 69.6 | 48,269 | 59.4 |
| Lawrence | 81.6 | 44.3 | 91.2 | 86.1 | 94.7 | 1.9 | 26,728 | -0.9 | 1,745 | 7 | 38,313 | 79.8 | 33,054 | 60.8 |
| Marion | 69.7 | 12.0 | 88.8 | 78.0 | 86.2 | 7.8 | 12,469 | -0.2 | 929 | 8 | 22,180 | 53.4 | 16,591 | 38.0 |
| Merrillville | 84.4 | 48.0 | 88.7 | 85.7 | 90.3 | 3.3 | 16,785 | -2.3 | 1,970 | 12 | 27,543 | 62.6 | 21,034 | 54.3 |
| Michigan City | 79.7 | 23.7 | 92.2 | 75.8 | 81.9 | 7.8 | 12,912 | 1.4 | 1,606 | 12 | 25,867 | 55.7 | 20,449 | 43.7 |
| Mishawaka | 78.7 | 20.3 | 86.8 | 77.8 | 86.2 | 4.0 | 26,191 | -1.6 | 2,202 | 8 | 38,614 | 63.4 | 31,010 | 52.2 |
| Muncie | 68.5 | 21.5 | 91.6 | 82.7 | 66.6 | 12.9 | 30,365 | -2.5 | 2,433 | 8 | 60,527 | 55.8 | 50,071 | 33.3 |
| New Albany | 81.4 | 24.0 | 92.0 | 76.1 | 85.7 | 7.4 | 18,694 | 0.0 | 1,574 | 8 | 30,119 | 60.1 | 23,543 | 45.2 |
| Noblesville | 81.7 | 43.9 | 97.3 | 93.5 | 87.9 | 6.6 | 35,634 | -0.3 | 1,839 | 5 | 48,844 | 72.8 | 41,556 | 56.2 |
| Plainfield | 79.5 | 28.5 | 96.3 | 91.1 | 83.7 | 10.1 | 16,943 | -0.3 | 930 | 6 | 27,677 | 66.9 | 23,413 | 56.7 |
| Portage | NA | 48.1 | 95.2 | 81.9 | 87.4 | 6.8 | 17,710 | -2.6 | 1,792 | 10 | 28,802 | 63.3 | 21,848 | 49.8 |
| Richmond | 76.4 | 12.9 | 89.9 | 86.3 | 77.5 | 5.5 | 14,813 | -1.8 | 1,215 | 8 | 28,915 | 61.5 | 22,365 | 47.5 |
| Schererville | 87.7 | 42.7 | 94.3 | 91.7 | 87.4 | 6.3 | 15,461 | -4.5 | 1,198 | 8 | 23,387 | 66.0 | 19,030 | 58.7 |
| South Bend | 75.4 | 25.0 | 89.2 | 71.6 | 85.1 | 6.1 | 47,817 | -0.6 | 5,020 | 11 | 79,891 | 63.8 | 65,767 | 48.6 |
| Terre Haute | 67.6 | 14.1 | 92.9 | 84.6 | 76.4 | 11.1 | 24,182 | -4.9 | 2,134 | 9 | 49,645 | 53.7 | 40,276 | 34.9 |
| Valparaiso | 82.1 | 34.4 | 92.5 | 83.4 | 74.3 | 11.2 | 16,208 | -4.2 | 1,231 | 8 | 27,373 | 54.7 | 21,435 | 50.4 |
| Westfield | 82.2 | 39.6 | 98.9 | 94.7 | 87.8 | 5.5 | 23,627 | 1.4 | 1,043 | 4 | 32,893 | 71.9 | 26,773 | 62.5 |
| West Lafayette | 50.7 | 5.1 | 95.8 | 87.1 | 62.1 | 25.3 | 21,794 | -4.0 | 732 | 3 | 45,804 | 50.2 | 42,302 | 19.5 |
| **IOWA** | 80.2 | 21.4 | 91.4 | 83.7 | 86.1 | 6.2 | 1,666,420 | -4.2 | 87,655 | 5 | 2,516,898 | 66.8 | 1,963,323 | 57.0 |
| Ames | 69.7 | 13.4 | 96.4 | 70.3 | 64.3 | 17.9 | 39,032 | -4.9 | 1,387 | 4 | 59,653 | 61.5 | 52,965 | 30.2 |
| Ankeny | 89.6 | 26.7 | 96.6 | 88.1 | 86.6 | 6.7 | 37,629 | -1.7 | 1,663 | 4 | 51,133 | 79.1 | 43,736 | 69.6 |
| Bettendorf | 85.9 | 14.4 | 93.4 | 91.0 | 85.1 | 3.7 | 18,068 | -5.2 | 968 | 5 | 28,704 | 69.4 | 22,544 | 60.9 |
| Burlington | 79.1 | 13.5 | 84.5 | 79.8 | 86.6 | 4.4 | 11,605 | -4.5 | 969 | 8 | 22,641 | 61.0 | 15,927 | 52.9 |
| Cedar Falls | 75.0 | 6.9 | 94.6 | 90.3 | 79.2 | 8.3 | 21,591 | -5.3 | 939 | 4 | 34,503 | 69.4 | 26,385 | 49.5 |
| Cedar Rapids | 81.9 | 17.4 | 94.0 | 88.8 | 83.4 | 6.2 | 71,082 | -3.8 | 5,124 | 7 | 107,900 | 68.5 | 86,891 | 54.7 |
| Clinton | 85.1 | 19.0 | 79.8 | 75.8 | 79.7 | 5.0 | 11,410 | -4.6 | 861 | 8 | 18,766 | 61.6 | 14,022 | 53.1 |
| Council Bluffs | 80.6 | 18.3 | 94.2 | 78.1 | 90.3 | 4.9 | 31,071 | -3.7 | 1,843 | 6 | 49,529 | 64.6 | 38,937 | 56.9 |
| Davenport | 84.5 | 14.5 | 87.3 | 80.4 | 83.4 | 6.2 | 49,981 | -4.4 | 3,971 | 8 | 81,049 | 64.2 | 65,575 | 54.0 |
| Des Moines | 78.5 | 16.2 | 89.7 | 80.5 | 83.0 | 7.0 | 111,221 | -4.1 | 8,284 | 7 | 170,751 | 68.6 | 140,784 | 56.2 |
| Dubuque | 77.2 | 10.4 | 93.9 | 84.6 | 84.8 | 5.4 | 32,436 | -3.5 | 2,126 | 7 | 48,100 | 66.8 | 37,061 | 53.0 |
| Fort Dodge | 79.5 | 11.8 | 75.7 | 65.9 | 81.5 | 13.7 | 11,915 | -5.4 | 691 | 6 | 19,818 | 52.5 | 14,891 | 51.5 |
| Iowa City | 54.6 | 14.6 | 96.6 | 89.6 | 70.3 | 13.7 | 41,449 | -5.2 | 2,100 | 5 | 66,228 | 70.1 | 56,693 | 39.1 |
| Marion | 86.4 | 14.6 | 96.2 | 89.6 | 87.6 | 3.8 | 20,617 | -4.5 | 1,109 | 5 | 32,433 | 71.4 | 25,412 | 66.9 |
| Marshalltown | 60.0 | 19.3 | 91.0 | 79.3 | 80.3 | 10.5 | 11,432 | -3.3 | 936 | 8 | 21,423 | 63.8 | 16,594 | 48.8 |
| Mason City | 80.7 | 8.7 | 85.0 | 81.5 | 89.0 | 4.3 | 14,654 | -3.3 | 829 | 6 | 22,030 | 72.1 | 16,012 | 63.2 |
| Ottumwa | 83.8 | 16.6 | 91.0 | 76.2 | 84.0 | 7.2 | 11,877 | -3.7 | 789 | 7 | 19,302 | 69.0 | 15,610 | 62.6 |
| Sioux City | 81.2 | 11.1 | 88.8 | 79.9 | 93.2 | 3.5 | 43,526 | -3.6 | 2,426 | 6 | 63,576 | 69.5 | 51,649 | 57.3 |
| Urbandale | 86.2 | 17.2 | 97.2 | 92.7 | 86.8 | 5.2 | 24,880 | -3.7 | 1,126 | 5 | 35,911 | 71.5 | 29,275 | 61.3 |
| Waterloo | 78.5 | 8.3 | 92.1 | 82.0 | 82.8 | 3.7 | 33,329 | -3.3 | 2,554 | 8 | 55,169 | 67.0 | 44,616 | 54.4 |
| West Des Moines | 80.7 | 11.4 | 96.3 | 90.9 | 86.1 | 8.4 | 39,579 | -2.4 | 1,967 | 5 | 55,627 | 75.8 | 44,870 | 66.4 |

1. Employed persons.  2. Households.  3. Percent of civilian labor force.  4. Persons 16 years old and over.

| City | Value of residential construction authorized by building permits, 2020 | | | Wholesale trade[1], 2017 | | | | Retail trade[2], 2017 | | | |
|---|---|---|---|---|---|---|---|---|---|---|---|
| | New construction ($1,000) | Number of housing units | Percent single family | Number of establishments | Number of employees | Sales (mil dol) | Annual payroll (mil dol) | Number of establishments | Number of employees | Sales (mil dol) | Annual payroll (mil dol) |
| | 69 | 70 | 71 | 72 | 73 | 74 | 75 | 76 | 77 | 78 | 79 |
| **ILLINOIS— Cont'd** | | | | | | | | | | | |
| Waukegan | 309 | 2 | 100 | 74 | 2,909 | 2,928 | 211 | 222 | 3,120 | 746 | 73 |
| West Chicago | 5,350 | 25 | 4 | 74 | 1,591 | 1,063 | 116 | 70 | 1,191 | 338 | 34 |
| Wheaton | 9,922 | 39 | 100 | 42 | 145 | 351 | 11 | 157 | 3,182 | 542 | 58 |
| Wheeling | 0 | 0 | 0 | D | D | D | D | 89 | 1,293 | 359 | 43 |
| Wilmette | 21,001 | 40 | 60 | 27 | 93 | 95 | 6 | 87 | 1,202 | 268 | 32 |
| Woodridge | 6,727 | 23 | 100 | 68 | 2,848 | 2,346 | 199 | 103 | 2,534 | 760 | 63 |
| **INDIANA** | 413,550 | 1,694 | 84 | 6,271 | 96,880 | 87,597 | 5,506 | 21,327 | 336,615 | 102,106 | 8,660 |
| Anderson | 7,910 | 62 | 97 | 39 | 576 | 591 | 31 | 206 | 3,634 | 1,131 | 91 |
| Bloomington | NA | NA | NA | 37 | 401 | 170 | 28 | 361 | 5,762 | 1,659 | 141 |
| Carmel | 179,902 | 592 | 55 | 130 | 1,337 | 1,189 | 111 | 291 | 6,092 | 2,128 | 180 |
| Columbus | NA | NA | NA | 46 | 734 | 563 | 37 | 193 | 3,525 | 967 | 85 |
| Crown Point | 52,516 | 145 | 100 | 34 | 200 | 147 | 11 | 86 | 1,091 | 277 | 26 |
| East Chicago | 3,498 | 28 | 0 | 29 | 400 | 504 | 24 | 53 | 397 | 94 | 9 |
| Elkhart | 22,607 | 138 | 17 | 168 | 3,845 | 3,111 | 201 | 286 | 4,227 | 1,171 | 113 |
| Evansville | 36,617 | 186 | 89 | 211 | 3,167 | 2,197 | 169 | 698 | 11,649 | 3,187 | 294 |
| Fishers | 230,801 | 858 | 68 | 96 | 1,758 | 1,026 | 110 | 206 | 4,188 | 1,469 | 123 |
| Fort Wayne | NA | NA | NA | 372 | 6,095 | 5,212 | 320 | 1,081 | 20,457 | 6,049 | 545 |
| Gary | 1,503 | 8 | 100 | 34 | 624 | 579 | 30 | 173 | 1,393 | 441 | 28 |
| Goshen | 11,335 | 52 | 100 | 27 | 623 | 734 | 32 | 158 | 3,272 | 959 | 88 |
| Greenwood | 101,821 | 475 | 71 | 31 | 1,547 | 1,247 | 108 | 321 | 7,324 | 2,335 | 181 |
| Hammond | 1,762 | 7 | 100 | 73 | 1,109 | 793 | 72 | 210 | 3,522 | 1,101 | 88 |
| Hobart | 9,629 | 34 | 100 | 27 | 299 | 1,026 | 16 | 206 | 4,367 | 1,224 | 108 |
| Indianapolis | 416,711 | 2,112 | 53 | D | D | D | D | 2,711 | 46,121 | 14,347 | 1,257 |
| Jeffersonville | 95,933 | 720 | 37 | 50 | 917 | 1,501 | 52 | 131 | 5,105 | 1,853 | 149 |
| Kokomo | 27,099 | 140 | 79 | 45 | 439 | 284 | 25 | 308 | 4,935 | 1,316 | 113 |
| Lafayette | 48,544 | 436 | 7 | 76 | 1,217 | 1,034 | 65 | 425 | 7,654 | 2,231 | 194 |
| Lawrence | 56,057 | 232 | 44 | 41 | 797 | 751 | 47 | 118 | 1,767 | 463 | 44 |
| Marion | 7,206 | 46 | 100 | 24 | 243 | 113 | 11 | 153 | 2,291 | 631 | 53 |
| Merrillville | 13,754 | 87 | 77 | 32 | 242 | 154 | 14 | 207 | 3,834 | 1,297 | 117 |
| Michigan City | 6,315 | 28 | 100 | 30 | 638 | 420 | 27 | 230 | 3,803 | 776 | 75 |
| Mishawaka | 13,453 | 62 | 100 | 52 | 569 | 543 | 29 | 366 | 7,700 | 2,776 | 199 |
| Muncie | 12,482 | 87 | 37 | 55 | 580 | 372 | 27 | 348 | 5,310 | 1,374 | 125 |
| New Albany | 9,914 | 52 | 100 | 36 | 465 | 261 | 21 | 135 | 2,381 | 607 | 61 |
| Noblesville | 176,692 | 640 | 80 | 81 | 766 | 439 | 51 | 230 | 4,502 | 1,220 | 106 |
| Plainfield | 58,272 | 217 | 98 | 39 | 2,742 | 3,943 | 147 | 136 | 5,233 | 2,517 | 171 |
| Portage | 18,771 | 127 | 100 | 27 | 727 | 586 | 44 | 93 | 2,016 | 567 | 49 |
| Richmond | 1,408 | 7 | 100 | 32 | 542 | 281 | 26 | 198 | 3,313 | 840 | 76 |
| Schererville | 39,553 | 113 | 91 | 25 | 101 | 42 | 5 | 122 | 2,717 | 713 | 67 |
| South Bend | NA | NA | NA | 148 | 2,801 | 2,729 | 160 | 322 | 5,303 | 1,764 | 145 |
| Terre Haute | 6,214 | 38 | 58 | 76 | 1,081 | 435 | 46 | 311 | 4,366 | 987 | 95 |
| Valparaiso | 53,078 | 173 | 86 | 44 | 621 | 302 | 35 | 193 | 3,755 | 996 | 91 |
| Westfield | 371,632 | 1,580 | 64 | 46 | 1,071 | 911 | 85 | 99 | 2,256 | 663 | 57 |
| West Lafayette | 40,912 | 318 | 5 | D | D | D | 1 | 86 | 1,845 | 414 | 37 |
| **IOWA** | 2,701,682 | 12,623 | 68 | 4,426 | 62,658 | 64,163 | 3,389 | 11,479 | 181,416 | 50,063 | 4,519 |
| Ames | 29,194 | 99 | 89 | 33 | 855 | 503 | 23 | 208 | 4,362 | 1,015 | 103 |
| Ankeny | 317,834 | 1,181 | 95 | 46 | 1,458 | 2,347 | 92 | 157 | 4,450 | 1,536 | 125 |
| Bettendorf | 59,053 | 223 | 57 | 48 | 619 | 549 | 32 | 95 | 1,514 | 363 | 37 |
| Burlington | 3,561 | 33 | 58 | 24 | 399 | 554 | 15 | 112 | 1,440 | 293 | 34 |
| Cedar Falls | 32,123 | 114 | 100 | 42 | 920 | 858 | 55 | 168 | 3,444 | 884 | 85 |
| Cedar Rapids | 39,900 | 236 | 55 | 204 | 3,953 | 2,738 | 265 | 479 | 11,763 | 5,379 | 276 |
| Clinton | 4,428 | 28 | 46 | 16 | 120 | 149 | 4 | 119 | 2,023 | 454 | 45 |
| Council Bluffs | 22,991 | 119 | 46 | 57 | 926 | 953 | 53 | 200 | 5,265 | 1,398 | 127 |
| Davenport | 15,870 | 95 | 100 | 162 | 2,646 | 1,330 | 142 | 472 | 9,446 | 2,488 | 248 |
| Des Moines | 83,321 | 441 | 50 | 272 | 5,365 | 4,952 | 289 | 597 | 9,699 | 2,300 | 248 |
| Dubuque | 22,925 | 120 | 33 | 81 | 991 | 1,196 | 51 | 326 | 6,228 | 1,510 | 153 |
| Fort Dodge | 13,650 | 131 | 25 | 37 | 466 | 215 | 29 | 137 | 2,431 | 572 | 58 |
| Iowa City | 44,563 | 159 | 73 | 35 | 508 | 555 | 26 | 216 | 4,292 | 1,167 | 118 |
| Marion | 28,557 | 248 | 75 | 36 | 582 | 222 | 32 | 108 | 1,896 | 463 | 46 |
| Marshalltown | 3,703 | 17 | 88 | 22 | 291 | 195 | 17 | 115 | 1,879 | 410 | 44 |
| Mason City | 15,006 | 141 | 6 | 50 | 630 | 483 | 32 | 165 | 3,105 | 744 | 74 |
| Ottumwa | 345 | 3 | 100 | 21 | 136 | 200 | 8 | 114 | 2,123 | 528 | 52 |
| Sioux City | 60,817 | 521 | 22 | 140 | 2,129 | 1,808 | 113 | 363 | 7,005 | 1,912 | 169 |
| Urbandale | 89,625 | 325 | 76 | 107 | 1,588 | 934 | 103 | 148 | 3,533 | 1,564 | 127 |
| Waterloo | 16,523 | 94 | 49 | 77 | 1,226 | 889 | 60 | 281 | 4,867 | 1,155 | 124 |
| West Des Moines | 161,299 | 718 | 61 | 56 | 623 | 2,252 | 46 | 385 | 8,858 | 1,732 | 192 |

1. Merchant wholesalers except manufacturers' sales branches and offices.  2. Establishments with payroll.

# Table D. Cities — Real Estate, Professional Services, and Manufacturing

| City | Real estate and rental and leasing, 2017 | | | | Professional, scientific, and technical services[1], 2017 | | | | Manufacturing, 2017 | | | |
|---|---|---|---|---|---|---|---|---|---|---|---|---|
| | Number of establishments | Number of employees | Receipts (mil dol) | Annual payroll (mil dol) | Number of establishments | Number of employees | Receipts (mil dol) | Annual payroll (mil dol) | Number of establishments | Number of employees | Receipts (mil dol) | Annual payroll (mil dol) |
| | 80 | 81 | 82 | 83 | 84 | 85 | 86 | 87 | 88 | 89 | 90 | 91 |
| **ILLINOIS— Cont'd** | | | | | | | | | | | | |
| Waukegan | 59 | 268 | 62 | 9 | 161 | 861 | 201 | 63 | NA | NA | NA | NA |
| West Chicago | 19 | 101 | 35 | 7 | 80 | 869 | 309 | 70 | NA | NA | NA | NA |
| Wheaton | 82 | 272 | 65 | 14 | 372 | 1,716 | 310 | 124 | NA | NA | NA | NA |
| Wheeling | 34 | 115 | 39 | 5 | 127 | 882 | 108 | 39 | NA | NA | NA | NA |
| Wilmette | 52 | 333 | 48 | 13 | 168 | 359 | 68 | 23 | NA | NA | NA | NA |
| Woodridge | 37 | 397 | 85 | 18 | D | D | D | D | NA | NA | NA | NA |
| **INDIANA** | 6,686 | 34,736 | 8,392 | 1,520 | 13,061 | 121,821 | 22,391 | 8,538 | 8,064 | 496,083 | 246,672 | 28,131 |
| Anderson | 65 | 303 | 70 | 9 | 87 | 395 | 37 | 13 | NA | NA | NA | NA |
| Bloomington | 158 | 848 | 167 | 31 | 213 | 1,555 | 208 | 81 | NA | NA | NA | NA |
| Carmel | 247 | 1,808 | 847 | 157 | 645 | 7,416 | 1,457 | 584 | NA | NA | NA | NA |
| Columbus | 58 | 216 | 45 | 8 | 128 | 2,145 | 206 | 143 | NA | NA | NA | NA |
| Crown Point | 39 | 191 | 50 | 10 | 104 | 661 | 85 | 33 | NA | NA | NA | NA |
| East Chicago | 13 | 95 | 21 | 6 | 15 | 187 | 27 | 12 | NA | NA | NA | NA |
| Elkhart | 67 | 428 | 94 | 17 | 131 | 885 | 119 | 42 | NA | NA | NA | NA |
| Evansville | 186 | 1,083 | 242 | 40 | 366 | 3,391 | 483 | 193 | NA | NA | NA | NA |
| Fishers | 127 | 468 | 117 | 23 | 396 | 2,434 | 419 | 137 | NA | NA | NA | NA |
| Fort Wayne | 352 | 2,004 | 503 | 82 | 660 | 4,865 | 775 | 269 | NA | NA | NA | NA |
| Gary | 35 | 193 | 31 | 6 | 33 | 157 | 33 | 7 | NA | NA | NA | NA |
| Goshen | 46 | 161 | 28 | 5 | 71 | 434 | 46 | 17 | NA | NA | NA | NA |
| Greenwood | 96 | 339 | 101 | 12 | 117 | 807 | 109 | 38 | NA | NA | NA | NA |
| Hammond | 45 | 201 | 34 | 7 | 87 | 926 | 143 | 64 | NA | NA | NA | NA |
| Hobart | 28 | 124 | 35 | 5 | 48 | 345 | 40 | 16 | NA | NA | NA | NA |
| Indianapolis | D | D | D | D | D | D | D | D | NA | NA | NA | NA |
| Jeffersonville | 61 | 271 | 68 | 12 | 101 | 765 | 117 | 37 | NA | NA | NA | NA |
| Kokomo | 68 | 288 | 61 | 10 | 100 | 626 | 85 | 27 | NA | NA | NA | NA |
| Lafayette | 138 | 821 | 149 | 33 | 190 | 1,584 | 218 | 83 | NA | NA | NA | NA |
| Lawrence | 47 | 277 | 50 | 11 | 112 | 2,460 | 259 | 128 | NA | NA | NA | NA |
| Marion | 37 | 139 | 19 | 4 | 44 | 229 | 27 | 9 | NA | NA | NA | NA |
| Merrillville | 66 | 406 | 77 | 17 | 179 | 1,382 | 182 | 72 | NA | NA | NA | NA |
| Michigan City | 36 | 180 | 37 | 9 | 48 | 265 | 34 | 12 | NA | NA | NA | NA |
| Mishawaka | 71 | 488 | 94 | 17 | 100 | 1,765 | 1,082 | 128 | NA | NA | NA | NA |
| Muncie | 74 | 272 | 57 | 9 | 108 | 1,105 | 170 | 32 | NA | NA | NA | NA |
| New Albany | 45 | 187 | 22 | 5 | 121 | 1,035 | 166 | 54 | NA | NA | NA | NA |
| Noblesville | 77 | 250 | 80 | 8 | 220 | 799 | 132 | 46 | NA | NA | NA | NA |
| Plainfield | 36 | 144 | 39 | 6 | 53 | 671 | 61 | 21 | NA | NA | NA | NA |
| Portage | 32 | 165 | 39 | 7 | 42 | 252 | 35 | 15 | NA | NA | NA | NA |
| Richmond | 41 | 180 | 37 | 7 | 53 | 303 | 29 | 11 | NA | NA | NA | NA |
| Schererville | 41 | 204 | 57 | 7 | 108 | 551 | 85 | 31 | NA | NA | NA | NA |
| South Bend | 93 | 956 | 157 | 38 | 259 | 2,208 | 392 | 135 | NA | NA | NA | NA |
| Terre Haute | 65 | 428 | 79 | 14 | 147 | 969 | 105 | 45 | NA | NA | NA | NA |
| Valparaiso | 67 | 360 | 71 | 12 | 166 | 810 | 144 | 43 | NA | NA | NA | NA |
| Westfield | 41 | 100 | 34 | 4 | 119 | 1,864 | 336 | 127 | NA | NA | NA | NA |
| West Lafayette | 40 | 321 | 41 | 10 | 89 | 814 | 155 | 65 | NA | NA | NA | NA |
| **IOWA** | 3,130 | 13,999 | 2,890 | 576 | 6,460 | 52,607 | 7,937 | 3,172 | 3,489 | 210,722 | 109,728 | 11,374 |
| Ames | 84 | 358 | 60 | 14 | 145 | 912 | 115 | 55 | NA | NA | NA | NA |
| Ankeny | 84 | 236 | 68 | 10 | 115 | 574 | 76 | 34 | NA | NA | NA | NA |
| Bettendorf | 50 | 222 | 35 | 9 | 102 | 493 | 55 | 19 | NA | NA | NA | NA |
| Burlington | 34 | 600 | 95 | 26 | 42 | 248 | 30 | 10 | NA | NA | NA | NA |
| Cedar Falls | 52 | 248 | 52 | 11 | 95 | 2,158 | 139 | 149 | NA | NA | NA | NA |
| Cedar Rapids | 185 | 925 | 223 | 40 | 385 | 5,043 | 763 | 375 | NA | NA | NA | NA |
| Clinton | 26 | 91 | 22 | 4 | 41 | 192 | 19 | 8 | NA | NA | NA | NA |
| Council Bluffs | 64 | 279 | 50 | 10 | 100 | 634 | 91 | 31 | NA | NA | NA | NA |
| Davenport | 125 | 552 | 137 | 21 | 253 | 2,116 | 341 | 123 | NA | NA | NA | NA |
| Des Moines | 239 | 1,505 | 311 | 66 | 638 | 6,647 | 1,235 | 479 | NA | NA | NA | NA |
| Dubuque | 83 | 329 | 81 | 12 | 144 | 1,941 | 306 | 108 | NA | NA | NA | NA |
| Fort Dodge | 37 | 221 | 35 | 6 | D | D | 28 | D | NA | NA | NA | NA |
| Iowa City | 79 | 379 | 78 | 17 | 147 | 961 | 160 | 57 | NA | NA | NA | NA |
| Marion | D | D | D | D | 55 | 385 | 45 | 22 | NA | NA | NA | NA |
| Marshalltown | 24 | 473 | 98 | 17 | 35 | 311 | 55 | 14 | NA | NA | NA | NA |
| Mason City | 49 | 132 | 23 | 3 | 71 | 421 | 54 | 24 | NA | NA | NA | NA |
| Ottumwa | 23 | 119 | 25 | 4 | 30 | 161 | 22 | 7 | NA | NA | NA | NA |
| Sioux City | 87 | 479 | 76 | 15 | 181 | 994 | 146 | 50 | NA | NA | NA | NA |
| Urbandale | 84 | 624 | 156 | 29 | 203 | 3,499 | 747 | 234 | NA | NA | NA | NA |
| Waterloo | 90 | 524 | 90 | 16 | 96 | 807 | 98 | 42 | NA | NA | NA | NA |
| West Des Moines | 164 | 1,550 | 272 | 86 | 424 | 4,496 | 835 | 340 | NA | NA | NA | NA |

1. Establishments subject to federal tax.

## Table D. Cities — Accommodation and Food Services, Arts, Entertainment, and Recreation, and Health Care and Social Assistance

| City | Accommodation and food services, 2017 | | | | Arts, entertainment, and recreation[1], 2017 | | | | Health care and social assistance[1], 2017 | | | |
|---|---|---|---|---|---|---|---|---|---|---|---|---|
| | Number of establishments | Number of employees | Receipts (mil dol) | Annual payroll (mil dol) | Number of establishments | Number of employees | Receipts (mil dol) | Annual payroll (mil dol) | Number of establishments | Number of employees | Receipts (mil dol) | Annual payroll (mil dol) |
| | 92 | 93 | 94 | 95 | 96 | 97 | 98 | 99 | 100 | 101 | 102 | 103 |
| **ILLINOIS— Cont'd** | | | | | | | | | | | | |
| Waukegan | 169 | 2,522 | 164 | 43 | 25 | 401 | 27 | 7 | 154 | 3,066 | 399 | 154 |
| West Chicago | 58 | 679 | 37 | 11 | 10 | 93 | 8 | 2 | 39 | 529 | 49 | 18 |
| Wheaton | D | D | D | D | 37 | 464 | 28 | 9 | 215 | 4,709 | 427 | 195 |
| Wheeling | 76 | 2,017 | 164 | 45 | 12 | 45 | 4 | 1 | 77 | 2,002 | 145 | 61 |
| Wilmette | 54 | 848 | 58 | 19 | 22 | 151 | 20 | 6 | 115 | 829 | 90 | 32 |
| Woodridge | 56 | 1,439 | 71 | 22 | 15 | 254 | 22 | 6 | D | D | D | D |
| **INDIANA** | 13,647 | 280,340 | 15,250 | 4,289 | 2,320 | 37,937 | 3,999 | 1,120 | 16,668 | 436,616 | 51,837 | 19,641 |
| Anderson | 138 | 3,244 | 145 | 44 | D | D | D | D | 191 | 5,536 | 638 | 233 |
| Bloomington | 363 | 7,661 | 394 | 112 | 36 | 688 | 21 | 8 | 356 | 9,307 | 920 | 392 |
| Carmel | 240 | 6,795 | 408 | 131 | 74 | 1,025 | 80 | 26 | 509 | 11,468 | 1,577 | 578 |
| Columbus | 140 | 3,536 | 173 | 50 | 23 | 171 | 8 | 3 | 203 | 5,011 | 598 | 225 |
| Crown Point | 79 | 1,494 | 72 | 20 | 17 | 238 | 8 | 2 | 116 | 3,782 | 513 | 155 |
| East Chicago | 46 | 1,748 | 223 | 49 | NA | NA | NA | NA | 39 | 1,597 | 296 | 78 |
| Elkhart | 185 | 3,470 | 178 | 49 | 19 | 210 | 12 | 4 | 166 | 5,527 | 773 | 253 |
| Evansville | 411 | 9,952 | 581 | 157 | 66 | 1,379 | 81 | 23 | 542 | 18,755 | 2,354 | 832 |
| Fishers | 207 | 4,614 | 240 | 70 | 52 | 831 | 59 | 19 | 280 | 4,402 | 459 | 203 |
| Fort Wayne | 649 | 15,505 | 747 | 226 | 103 | 2,232 | 122 | 40 | 876 | 28,627 | 3,168 | 1,346 |
| Gary | 72 | 959 | 91 | 17 | D | D | D | D | 121 | 3,033 | 259 | 122 |
| Goshen | 100 | 2,019 | 97 | 25 | 10 | 99 | 5 | 2 | 110 | 4,120 | 509 | 186 |
| Greenwood | 163 | 4,298 | 226 | 65 | 17 | 183 | 11 | 4 | 187 | 4,900 | 484 | 199 |
| Hammond | 145 | 2,194 | 121 | 31 | D | D | D | D | 99 | 4,089 | 558 | 202 |
| Hobart | 67 | 1,461 | 74 | 20 | D | D | D | D | 63 | 2,130 | 352 | 100 |
| Indianapolis | 2,138 | 50,367 | 3,120 | 887 | 336 | 8,487 | 1,440 | 535 | 2,700 | 88,524 | 12,566 | 4,847 |
| Jeffersonville | 108 | 2,090 | 102 | 31 | D | D | D | D | 157 | 4,457 | 555 | 215 |
| Kokomo | 183 | 3,933 | 181 | 53 | D | D | D | D | 222 | 5,155 | 580 | 217 |
| Lafayette | 244 | 5,346 | 279 | 79 | 44 | 436 | 17 | 6 | 377 | 10,181 | 1,409 | 515 |
| Lawrence | D | D | D | D | D | D | D | D | 75 | 1,167 | 77 | 35 |
| Marion | 83 | 1,629 | 74 | 21 | D | D | D | D | 139 | 4,305 | 397 | 155 |
| Merrillville | 135 | 3,233 | 182 | 52 | 13 | 192 | 10 | 3 | 266 | 4,964 | 650 | 235 |
| Michigan City | 99 | 3,042 | 279 | 64 | 13 | 126 | 7 | 2 | 97 | 2,617 | 345 | 125 |
| Mishawaka | 204 | 4,544 | 243 | 68 | D | D | D | D | 193 | 5,793 | 734 | 261 |
| Muncie | 177 | 4,511 | 197 | 58 | 32 | 449 | 15 | 5 | 327 | 9,324 | 991 | 385 |
| New Albany | 99 | 1,995 | 100 | 29 | 13 | 231 | 8 | 2 | 182 | 5,619 | 597 | 232 |
| Noblesville | 138 | 3,573 | 175 | 55 | 32 | 347 | 58 | 9 | 179 | 3,425 | 380 | 141 |
| Plainfield | 109 | 2,540 | 130 | 37 | 11 | 109 | 9 | 1 | 78 | 1,144 | 109 | 44 |
| Portage | 86 | 1,955 | 98 | 28 | 10 | 222 | 5 | 2 | 68 | 1,121 | 123 | 43 |
| Richmond | 108 | 2,400 | 119 | 37 | 14 | 70 | 4 | 1 | 136 | 4,518 | 630 | 205 |
| Schererville | 102 | 2,466 | 122 | 37 | 12 | 416 | 18 | 5 | 104 | 2,013 | 134 | 55 |
| South Bend | 245 | 4,627 | 239 | 67 | 35 | 664 | 33 | 11 | 312 | 10,192 | 1,378 | 450 |
| Terre Haute | 235 | 4,807 | 237 | 68 | 26 | 174 | 8 | 3 | 266 | 7,482 | 1,128 | 381 |
| Valparaiso | 128 | 2,707 | 137 | 41 | 13 | 412 | 16 | 6 | 192 | 4,102 | 411 | 161 |
| Westfield | 76 | 1,747 | 86 | 26 | 25 | 324 | 23 | 6 | 77 | 892 | 68 | 29 |
| West Lafayette | 165 | 3,249 | 171 | 44 | 11 | 72 | 3 | 1 | 70 | 2,193 | 196 | 87 |
| **IOWA** | 7,283 | 123,866 | 7,111 | 1,897 | 1,451 | 18,336 | 1,158 | 316 | 8,610 | 216,965 | 22,420 | 9,242 |
| Ames | 205 | 4,628 | 225 | 64 | 27 | 376 | 18 | 7 | 165 | 4,962 | 608 | 244 |
| Ankeny | 137 | 3,252 | 160 | 49 | 22 | 441 | 12 | 4 | 134 | 2,049 | 175 | 74 |
| Bettendorf | 75 | 2,058 | 152 | 32 | 14 | 211 | 6 | 2 | 143 | 3,423 | 358 | 147 |
| Burlington | 86 | 1,848 | 86 | 27 | D | D | D | D | 100 | 1,205 | 78 | 36 |
| Cedar Falls | 125 | 2,838 | 121 | 38 | 19 | 132 | 7 | 2 | 115 | 2,687 | 225 | 79 |
| Cedar Rapids | 417 | 8,001 | 401 | 125 | 60 | 1,504 | 75 | 26 | 467 | 13,259 | 1,612 | 610 |
| Clinton | 81 | 1,583 | 90 | 21 | 13 | 165 | 6 | 2 | 93 | 2,370 | 208 | 85 |
| Council Bluffs | 157 | 5,186 | 575 | 110 | D | D | D | D | 209 | 4,863 | 534 | 203 |
| Davenport | 301 | 6,766 | 383 | 105 | 47 | 863 | 47 | 18 | 331 | 8,939 | 895 | 373 |
| Des Moines | 562 | 9,920 | 567 | 169 | 78 | 1,608 | 100 | 32 | 528 | 22,243 | 2,987 | 1,267 |
| Dubuque | 194 | 4,323 | 244 | 71 | 42 | 1,279 | 116 | 24 | 229 | 7,792 | 882 | 382 |
| Fort Dodge | 84 | 1,361 | 70 | 20 | D | D | D | D | 108 | 2,649 | 280 | 119 |
| Iowa City | 240 | 4,078 | 185 | 57 | 23 | 295 | 11 | 3 | 271 | 16,070 | 2,422 | 886 |
| Marion | 56 | 898 | 40 | 12 | 8 | 127 | 4 | 1 | 91 | 1,037 | 74 | 32 |
| Marshalltown | 70 | 993 | 49 | 14 | D | D | D | D | D | D | D | D |
| Mason City | 86 | 1,472 | 70 | 21 | D | D | D | D | 110 | 2,935 | 312 | 140 |
| Ottumwa | D | D | D | 14 | D | D | D | D | 89 | 2,851 | 254 | 99 |
| Sioux City | D | D | D | D | 41 | 553 | 21 | 6 | 301 | 9,030 | 1,025 | 400 |
| Urbandale | 93 | 1,833 | 108 | 31 | 23 | 644 | 38 | 11 | 110 | 2,848 | 243 | 104 |
| Waterloo | 174 | 3,672 | 233 | 57 | 35 | 578 | 24 | 9 | 238 | 8,396 | 943 | 401 |
| West Des Moines | 246 | 5,390 | 332 | 99 | 38 | 660 | 41 | 15 | 322 | 6,055 | 810 | 351 |

1. Establishments subject to federal tax.

## Table D. Cities — Other Services and Government Employment and Payroll

| City | Other services[1] — Number of establishments | Number of employees | Receipts (mil dol) | Annual payroll (mil dol) | Government employment and payroll, 2017 — March payroll — Full-time equivalent employees | Total (dollars) | Percent of total for: Administrative, judicial, and legal | Police and corrections | Fire protection | Highways and transportation | Health and welfare | Natural resources and utilities | Education and libraries |
|---|---|---|---|---|---|---|---|---|---|---|---|---|---|
|  | 104 | 105 | 106 | 107 | 108 | 109 | 110 | 111 | 112 | 113 | 114 | 115 | 116 |
| **ILLINOIS— Cont'd** | | | | | | | | | | | | | |
| Waukegan | 114 | 478 | 58 | 14 | 514 | 3,531,059 | 9.2 | 37.9 | 27.1 | 2.8 | 0.5 | 7.5 | 5.3 |
| West Chicago | 46 | 251 | 47 | 13 | 115 | 873,891 | 18.2 | 48.2 | 0.0 | 10.8 | 0.7 | 19.0 | 0.0 |
| Wheaton | 118 | 762 | 79 | 20 | 461 | 2,894,226 | 8.6 | 34.0 | 12.6 | 8.3 | 0.0 | 11.5 | 18.4 |
| Wheeling | 83 | 404 | 67 | 14 | 244 | 1,910,542 | 10.0 | 43.3 | 24.9 | 2.4 | 7.2 | 8.2 | 0.0 |
| Wilmette | 77 | 550 | 59 | 18 | 205 | 1,652,110 | 14.0 | 33.5 | 24.5 | 12.1 | 0.9 | 12.5 | 0.0 |
| Woodridge | 55 | 368 | 24 | 12 | 164 | 1,100,802 | 17.3 | 45.2 | 0.0 | 6.2 | 0.0 | 12.3 | 13.6 |
| **INDIANA** | 10,777 | 75,446 | 11,121 | 2,562 | X | X | X | X | X | X | X | X | X |
| Anderson | 86 | 599 | 53 | 14 | 814 | 3,297,848 | 7.5 | 21.4 | 15.7 | 10.1 | 3.5 | 36.0 | 0.0 |
| Bloomington | 150 | 1,205 | 393 | 45 | 775 | 3,297,022 | 15.0 | 21.4 | 16.5 | 13.3 | 4.4 | 24.2 | 0.0 |
| Carmel | 195 | 1,598 | 181 | 58 | 541 | 3,323,712 | 11.6 | 24.8 | 35.4 | 6.7 | 0.0 | 18.1 | 0.0 |
| Columbus | 92 | 659 | 80 | 20 | 449 | 1,866,119 | 6.7 | 23.7 | 24.0 | 7.9 | 3.0 | 33.8 | 0.0 |
| Crown Point | 76 | 506 | 70 | 19 | 206 | 901,061 | 10.4 | 28.3 | 20.4 | 8.1 | 0.0 | 25.4 | 0.0 |
| East Chicago | 37 | 169 | 21 | 4 | 753 | 3,206,403 | 8.4 | 26.2 | 16.2 | 7.5 | 4.5 | 37.1 | 0.0 |
| Elkhart | 130 | 1,151 | 180 | 44 | 571 | 2,684,118 | 5.8 | 28.3 | 25.7 | 9.1 | 0.7 | 14.9 | 0.0 |
| Evansville | 303 | 2,498 | 274 | 76 | 1,284 | 5,829,082 | 2.6 | 29.5 | 23.4 | 14.2 | 1.7 | 9.7 | 0.0 |
| Fishers | 136 | 1,146 | 78 | 24 | 419 | 2,279,967 | 9.1 | 28.9 | 37.3 | 5.7 | 2.0 | 20.5 | 0.0 |
| Fort Wayne | 538 | 4,331 | 472 | 148 | 2,015 | 9,797,922 | 5.7 | 33.5 | 19.2 | 13.2 | 4.7 | 27.0 | 0.0 |
| Gary | 73 | 375 | 50 | 12 | 1,141 | 3,176,242 | 11.7 | 16.3 | 20.2 | 6.4 | 10.1 | 16.0 | 0.0 |
| Goshen | 67 | 463 | 44 | 15 | 252 | 1,147,950 | 9.3 | 31.3 | 23.9 | 6.9 | 1.3 | 12.2 | 0.0 |
| Greenwood | 123 | 848 | 92 | 30 | 297 | 1,295,362 | 18.8 | 27.9 | 28.4 | 9.1 | 0.0 | 19.0 | 0.0 |
| Hammond | 116 | 896 | 99 | 29 | 805 | 4,079,099 | 5.1 | 37.3 | 26.7 | 4.3 | 3.5 | 19.1 | 0.0 |
| Hobart | 65 | 425 | 43 | 13 | 232 | 1,033,616 | 6.4 | 34.5 | 25.9 | 5.8 | 3.9 | 15.3 | 0.0 |
| Indianapolis | 1,518 | 16,940 | 4,075 | 744 | 13,456 | 61,603,859 | 8.7 | 15.6 | 10.9 | 6.1 | 41.6 | 22.2 | 0.0 |
| Jeffersonville | 76 | 731 | 98 | 23 | 344 | 1,489,946 | 7.7 | 29.2 | 30.8 | 7.7 | 0.9 | 16.8 | 0.0 |
| Kokomo | D | D | D | D | 431 | 1,773,018 | 8.0 | 31.6 | 24.1 | 11.7 | 3.4 | 20.5 | 0.0 |
| Lafayette | D | D | D | D | 752 | 3,450,984 | 5.5 | 27.2 | 22.7 | 18.1 | 0.8 | 13.3 | 0.0 |
| Lawrence | 73 | 443 | 59 | 15 | 282 | 1,291,474 | 6.7 | 31.9 | 43.0 | 2.1 | 0.9 | 13.3 | 0.0 |
| Marion | 63 | 316 | 29 | 9 | 239 | 862,860 | 8.3 | 40.0 | 31.7 | 20.1 | 0.0 | 0.0 | 0.0 |
| Merrillville | 81 | 685 | 73 | 22 | 146 | 491,875 | 20.8 | 35.9 | 14.7 | 24.0 | 0.0 | 4.2 | 0.0 |
| Michigan City | 70 | 369 | 38 | 10 | 407 | 1,649,855 | 6.5 | 27.1 | 22.3 | 9.1 | 0.8 | 32.6 | 0.0 |
| Mishawaka | 112 | 709 | 69 | 22 | 503 | 2,309,841 | 6.7 | 25.0 | 27.1 | 4.6 | 1.9 | 32.5 | 0.0 |
| Muncie | 124 | 874 | 104 | 25 | 566 | 2,111,689 | 7.0 | 23.9 | 22.9 | 16.3 | 1.8 | 23.4 | 0.0 |
| New Albany | 76 | 415 | 46 | 12 | 261 | 1,199,198 | 6.4 | 29.6 | 38.6 | 4.7 | 2.2 | 17.6 | 0.0 |
| Noblesville | 102 | 755 | 55 | 18 | 355 | 1,910,485 | 9.9 | 24.2 | 42.0 | 9.6 | 0.7 | 12.7 | 0.0 |
| Plainfield | 59 | 742 | 67 | 29 | 248 | 1,039,451 | 10.1 | 25.1 | 34.3 | 4.5 | 0.0 | 23.0 | 0.0 |
| Portage | 62 | 465 | 51 | 19 | 243 | 1,025,489 | 8.0 | 34.2 | 28.8 | 15.7 | 1.2 | 12.1 | 0.0 |
| Richmond | 64 | 392 | 36 | 10 | 489 | 1,912,770 | 4.8 | 18.8 | 15.3 | 6.8 | 4.7 | 48.6 | 0.0 |
| Schererville | 78 | 676 | 73 | 22 | 200 | 817,686 | 14.8 | 38.0 | 17.9 | 15.1 | 0.0 | 13.8 | 0.0 |
| South Bend | 192 | 1,595 | 218 | 57 | 1,244 | 5,813,124 | 6.4 | 27.3 | 24.2 | 14.0 | 2.6 | 20.8 | 0.0 |
| Terre Haute | 121 | 761 | 91 | 22 | 578 | 2,421,306 | 5.9 | 26.7 | 33.9 | 15.6 | 0.0 | 15.9 | 0.0 |
| Valparaiso | 108 | 745 | 80 | 19 | 284 | 1,314,343 | 5.9 | 21.1 | 26.3 | 6.8 | 0.0 | 39.9 | 0.0 |
| Westfield | 55 | 340 | 34 | 11 | 176 | 992,602 | 19.6 | 28.0 | 32.8 | 8.5 | 1.6 | 9.5 | 0.0 |
| West Lafayette | 40 | 626 | 340 | 31 | 236 | 1,028,818 | 5.8 | 33.3 | 21.8 | 4.2 | 5.4 | 24.8 | 0.0 |
| **IOWA** | 5,975 | 31,306 | 4,187 | 1,022 | X | X | X | X | X | X | X | X | X |
| Ames | 110 | 1,006 | 262 | 33 | 635 | 3,374,874 | 8.8 | 11.7 | 9.7 | 22.4 | 1.9 | 31.1 | 5.3 |
| Ankeny | 98 | 911 | 99 | 24 | 282 | 1,457,235 | 18.7 | 25.1 | 18.9 | 13.1 | 0.0 | 19.5 | 4.8 |
| Bettendorf | 69 | 395 | 35 | 11 | 283 | 1,575,986 | 10.7 | 22.8 | 12.6 | 12.8 | 3.3 | 22.8 | 9.5 |
| Burlington | D | D | D | D | 213 | 1,002,307 | 7.2 | 23.5 | 23.3 | 13.1 | 0.4 | 19.0 | 6.4 |
| Cedar Falls | 59 | 449 | 45 | 14 | 430 | 2,364,435 | 8.8 | 11.4 | 7.2 | 7.5 | 1.8 | 57.7 | 3.0 |
| Cedar Rapids | 248 | 2,058 | 229 | 68 | 1,293 | 7,127,614 | 10.4 | 23.5 | 12.1 | 17.6 | 2.3 | 23.8 | 3.4 |
| Clinton | 57 | 199 | 18 | 5 | 195 | 916,884 | 5.5 | 29.7 | 24.6 | 16.9 | 3.9 | 14.1 | 4.7 |
| Council Bluffs | 107 | 544 | 93 | 17 | 451 | 2,613,092 | 8.2 | 32.9 | 24.9 | 9.2 | 2.7 | 16.5 | 4.5 |
| Davenport | 189 | 1,495 | 142 | 45 | 855 | 4,516,209 | 10.1 | 25.5 | 18.9 | 13.0 | 5.5 | 14.0 | 5.2 |
| Des Moines | 447 | 3,030 | 431 | 116 | 1,933 | 12,154,999 | 7.4 | 25.8 | 16.5 | 12.1 | 5.6 | 26.5 | 3.2 |
| Dubuque | 144 | 873 | 96 | 23 | 648 | 3,329,064 | 12.8 | 20.3 | 16.1 | 22.9 | 4.4 | 13.5 | 4.5 |
| Fort Dodge | D | D | D | D | 193 | 920,989 | 9.5 | 27.0 | 17.8 | 14.0 | 6.6 | 20.0 | 3.7 |
| Iowa City | 118 | 883 | 255 | 38 | 836 | 3,543,459 | 10.2 | 17.4 | 10.4 | 18.2 | 6.2 | 18.5 | 8.6 |
| Marion | 56 | 389 | 37 | 13 | 187 | 1,102,380 | 6.1 | 30.3 | 21.4 | 10.1 | 0.0 | 19.2 | 6.6 |
| Marshalltown | D | D | D | D | 202 | 961,826 | 5.7 | 31.5 | 15.6 | 13.4 | 7.7 | 22.8 | 5.4 |
| Mason City | 72 | 352 | 34 | 10 | 264 | 1,250,056 | 6.8 | 23.1 | 16.1 | 12.2 | 2.7 | 23.4 | 4.4 |
| Ottumwa | 37 | 180 | 15 | 4 | 233 | 968,202 | 9.3 | 23.7 | 15.9 | 11.4 | 2.5 | 30.4 | 3.5 |
| Sioux City | D | D | D | D | 787 | 4,258,995 | 9.0 | 24.9 | 17.8 | 17.5 | 4.8 | 19.1 | 5.1 |
| Urbandale | 112 | 776 | 115 | 35 | 237 | 1,286,100 | 9.6 | 29.1 | 16.1 | 13.3 | 2.2 | 13.9 | 8.0 |
| Waterloo | 117 | 929 | 86 | 23 | 634 | 3,494,655 | 4.9 | 30.4 | 23.1 | 11.5 | 11.2 | 19.3 | 3.7 |
| West Des Moines | 161 | 1,498 | 294 | 80 | 458 | 2,735,559 | 9.7 | 18.5 | 15.1 | 12.7 | | 17.1 | 4.7 |

1. Establishments subject to federal tax.

# Table D. Cities — City Government Finances

| City | General revenue Total (mil dol) | Intergovernmental Total (mil dol) | Intergovernmental Percent from state government | Taxes Total (mil dol) | Taxes Per capita[1] Total | Taxes Per capita[1] Property | Taxes Per capita[1] Sales and gross receipts | General expenditure Total (mil dol) | General expenditure Per capita[1] Total | General expenditure Capital outlays |
|---|---|---|---|---|---|---|---|---|---|---|
| | 117 | 118 | 119 | 120 | 121 | 122 | 123 | 124 | 125 | 126 |
| **ILLINOIS— Cont'd** | | | | | | | | | | |
| Waukegan | 90.9 | 29.2 | 92.3 | 52.4 | 598 | 395 | 202 | 118 | 1,347 | 267 |
| West Chicago | 30.0 | 8.6 | 99.8 | 9.0 | 329 | 164 | 165 | 30 | 1,101 | 36 |
| Wheaton | 59.3 | 14.1 | 100.0 | 35.4 | 663 | 473 | 190 | 53 | 990 | 57 |
| Wheeling | 57.4 | 14.4 | 85.4 | 34.3 | 891 | 611 | 280 | 64 | 1,650 | 281 |
| Wilmette | 48.1 | 7.4 | 99.4 | 29.6 | 1,083 | 603 | 479 | 48 | 1,769 | 192 |
| Woodridge | 33.5 | 11.5 | 91.6 | 16.5 | 490 | 221 | 269 | 28 | 843 | 58 |
| **INDIANA** | X | X | X | X | X | X | X | X | X | X |
| Anderson | 65.9 | 9.6 | 48.7 | 29.9 | 544 | 525 | 19 | 172 | 3,140 | 78 |
| Bloomington | 188.3 | 58.2 | 35.3 | 79.0 | 932 | 909 | 23 | 264 | 3,110 | 107 |
| Carmel | 102.8 | 23.2 | 88.3 | 46.4 | 477 | 431 | 46 | 123 | 1,267 | 54 |
| Columbus | 46.4 | 14.7 | 29.1 | 25.3 | 536 | 501 | 35 | 47 | 992 | 175 |
| Crown Point | 36.5 | 4.4 | 84.2 | 12.6 | 428 | 378 | 50 | 25 | 839 | 99 |
| East Chicago | 83.4 | 31.3 | 97.0 | 37.4 | 1,329 | 1,302 | 27 | 74 | 2,614 | 393 |
| Elkhart | 84.0 | 13.2 | 96.2 | 34.5 | 658 | 645 | 13 | 62 | 1,184 | 55 |
| Evansville | 196.3 | 45.3 | 74.1 | 65.4 | 553 | 511 | 30 | 197 | 1,665 | 283 |
| Fishers | 86.7 | 6.6 | 88.0 | 60.1 | 656 | 366 | 60 | 67 | 728 | 57 |
| Fort Wayne | 240.6 | 46.2 | 100.0 | 183.9 | 694 | 467 | 17 | 183 | 692 | 51 |
| Gary | 125.1 | 22.3 | 84.0 | 39.1 | 516 | 498 | 18 | 133 | 1,757 | 240 |
| Goshen | 38.2 | 4.0 | 92.1 | 17.3 | 517 | 490 | 27 | 34 | 1,012 | 144 |
| Greenwood | 62.6 | 7.3 | 92.8 | 22.4 | 391 | 378 | 13 | 37 | 639 | 41 |
| Hammond | 168.6 | 61.9 | 98.7 | 52.0 | 680 | 641 | 33 | 171 | 2,231 | 365 |
| Hobart | 43.0 | 4.9 | 93.2 | 16.8 | 598 | 559 | 39 | 37 | 1,318 | 311 |
| Indianapolis | 3,008.5 | 919.2 | 61.9 | 424.6 | 492 | 325 | 0 | 2,462 | 2,851 | 95 |
| Jeffersonville | 91.7 | 2.1 | 100.0 | 46.7 | 988 | 709 | 0 | 59 | 1,246 | 332 |
| Kokomo | 82.7 | 29.4 | 45.6 | 37.9 | 655 | 651 | 4 | 60 | 1,044 | 59 |
| Lafayette | 122.9 | 19.0 | 86.6 | 46.8 | 649 | 623 | 15 | 92 | 1,271 | 380 |
| Lawrence | 36.2 | 4.4 | 86.5 | 14.7 | 302 | 289 | 13 | 31 | 631 | 17 |
| Marion | 40.6 | 8.6 | 97.2 | 19.5 | 692 | 688 | 4 | 30 | 1,067 | 72 |
| Merrillville | 15.5 | 0.6 | 100.0 | 14.9 | 427 | 345 | 27 | 12 | 332 | 58 |
| Michigan City | 74.6 | 22.2 | 98.5 | 20.9 | 673 | 643 | 30 | 65 | 2,099 | 353 |
| Mishawaka | 95.6 | 3.7 | 0.0 | 70.8 | 1,446 | 1,245 | 14 | 99 | 2,014 | 760 |
| Muncie | 143.2 | 20.0 | 81.2 | 23.5 | 342 | 321 | 10 | 97 | 1,405 | 349 |
| New Albany | 71.7 | 6.7 | 92.1 | 23.2 | 637 | 602 | 21 | 56 | 1,538 | 228 |
| Noblesville | 88.8 | 18.0 | 88.6 | 43.7 | 706 | 609 | 98 | 65 | 1,050 | 167 |
| Plainfield | 54.8 | 5.2 | 94.5 | 29.2 | 886 | 826 | 60 | 55 | 1,677 | 635 |
| Portage | 46.0 | 1.2 | 100.0 | 26.8 | 730 | 579 | 42 | 31 | 846 | 78 |
| Richmond | 52.0 | 15.0 | 80.1 | 15.9 | 449 | 432 | 16 | 55 | 1,554 | 99 |
| Schererville | 34.0 | 2.7 | 87.8 | 9.0 | 316 | 265 | 51 | 19 | 661 | 42 |
| South Bend | 222.5 | 56.4 | 85.4 | 82.2 | 806 | 780 | 17 | 191 | 1,873 | 278 |
| Terre Haute | 102.6 | 20.5 | 62.6 | 33.5 | 551 | 539 | 10 | 89 | 1,467 | 429 |
| Valparaiso | 55.7 | 4.7 | 93.1 | 26.5 | 792 | 770 | 23 | 46 | 1,364 | 214 |
| Westfield | 28.2 | 4.8 | 76.6 | 18.7 | 474 | 404 | 69 | 49 | 1,236 | 369 |
| West Lafayette | 74.3 | 4.6 | 92.8 | 17.3 | 365 | 359 | 7 | 56 | 1,184 | 581 |
| **IOWA** | X | X | X | X | X | X | X | X | X | X |
| Ames | 279.3 | 24.1 | 60.4 | 39.5 | 595 | 409 | 186 | 247 | 3,729 | 504 |
| Ankeny | 81.8 | 8.7 | 94.3 | 45.1 | 723 | 631 | 92 | 69 | 1,104 | 228 |
| Bettendorf | 69.7 | 11.7 | 60.9 | 35.2 | 984 | 757 | 227 | 68 | 1,908 | 569 |
| Burlington | 41.9 | 7.2 | 72.7 | 20.3 | 815 | 537 | 278 | 37 | 1,472 | 223 |
| Cedar Falls | 83.0 | 14.2 | 55.4 | 34.2 | 834 | 627 | 208 | 73 | 1,788 | 694 |
| Cedar Rapids | 381.2 | 99.4 | 38.3 | 137.5 | 1,037 | 739 | 298 | 388 | 2,925 | 1,067 |
| Clinton | 49.1 | 8.4 | 60.8 | 22.0 | 865 | 633 | 232 | 51 | 2,024 | 360 |
| Council Bluffs | 131.3 | 37.5 | 66.4 | 66.9 | 1,073 | 732 | 341 | 115 | 1,840 | 507 |
| Davenport | 188.2 | 41.4 | 54.1 | 98.4 | 961 | 715 | 247 | 167 | 1,633 | 381 |
| Des Moines | 463.7 | 96.0 | 39.3 | 183.3 | 846 | 663 | 183 | 411 | 1,898 | 251 |
| Dubuque | 147.4 | 46.1 | 38.9 | 53.5 | 918 | 597 | 321 | 125 | 2,144 | 533 |
| Fort Dodge | 58.0 | 12.5 | 53.7 | 21.7 | 891 | 632 | 258 | 60 | 2,459 | 1,168 |
| Iowa City | 146.2 | 39.2 | 60.6 | 64.0 | 845 | 761 | 85 | 128 | 1,693 | 507 |
| Marion | 51.0 | 6.9 | 76.0 | 29.8 | 754 | 590 | 165 | 55 | 1,399 | 438 |
| Marshalltown | 38.7 | 9.3 | 50.5 | 18.2 | 670 | 425 | 245 | 36 | 1,314 | 305 |
| Mason City | 40.6 | 7.3 | 69.5 | 21.7 | 793 | 561 | 232 | 42 | 1,538 | 324 |
| Ottumwa | 44.8 | 5.8 | 59.1 | 19.1 | 781 | 576 | 205 | 40 | 1,648 | 284 |
| Sioux City | 155.2 | 34.8 | 62.7 | 75.8 | 922 | 626 | 295 | 158 | 1,915 | 438 |
| Urbandale | 55.8 | 9.1 | 71.1 | 36.7 | 842 | 758 | 84 | 56 | 1,282 | 536 |
| Waterloo | 144.8 | 46.8 | 41.1 | 65.3 | 964 | 688 | 276 | 130 | 1,916 | 485 |
| West Des Moines | 121.3 | 24.1 | 64.0 | 71.4 | 1,088 | 981 | 108 | 127 | 1,937 | 640 |

1. Based on population estimated as of July 1 of the year shown.

| City | City government finances, 2017 (cont.) General expenditure (cont.) Percent of total for: | | | | | | | | | |
| | Public welfare | Highways | Parking facilities | Education | Health and hospitals | Police protection | Sewerage and sanitation | Parks and recreation | Housing and community development | Interest on debt |
| | 127 | 128 | 129 | 130 | 131 | 132 | 133 | 134 | 135 | 136 |
| **ILLINOIS— Cont'd** | | | | | | | | | | |
| Waukegan | 0.0 | 15.1 | 0.0 | 0.0 | 0.0 | 30.3 | 4.1 | 0.8 | 3.9 | 3.0 |
| West Chicago | 0.0 | 0.0 | 0.2 | 0.0 | 0.0 | 32.9 | 19.8 | 0.0 | 0.0 | 0.5 |
| Wheaton | 0.3 | 18.4 | 1.3 | 0.0 | 0.0 | 26.5 | 7.2 | 1.1 | 0.0 | 1.5 |
| Wheeling | 0.0 | 18.6 | 0.1 | 0.0 | 0.0 | 23.8 | 2.1 | 0.0 | 0.0 | 4.8 |
| Wilmette | 0.0 | 5.9 | 0.8 | 0.0 | 0.5 | 22.1 | 9.2 | 0.0 | 0.0 | 6.1 |
| Woodridge | 0.0 | 10.6 | 0.0 | 0.0 | 0.0 | 35.6 | 4.5 | 0.0 | 0.0 | 7.1 |
| **INDIANA** | X | X | X | X | X | X | X | X | X | X |
| Anderson | 0.0 | 1.7 | 0.0 | 0.0 | 0.0 | 6.8 | 27.9 | 0.9 | 0.0 | 1.3 |
| Bloomington | 0.0 | 3.0 | 0.7 | 0.0 | 0.0 | 5.6 | 14.2 | 3.4 | 51.1 | 6.9 |
| Carmel | 0.0 | 11.8 | 0.0 | 0.0 | 1.1 | 15.1 | 6.0 | 10.0 | 0.0 | 16.4 |
| Columbus | 0.0 | 11.2 | 0.0 | 0.0 | 1.1 | 20.2 | 12.4 | 13.6 | 0.6 | 0.0 |
| Crown Point | 0.0 | 15.2 | 0.0 | 0.0 | 0.3 | 36.3 | 1.3 | 13.2 | 0.1 | 2.4 |
| East Chicago | 0.0 | 10.9 | 0.0 | 0.0 | 0.7 | 16.0 | 8.5 | 2.9 | 2.5 | 4.8 |
| Elkhart | 0.0 | 16.5 | 0.0 | 0.0 | 0.2 | 17.1 | 1.8 | 3.5 | 1.3 | 2.4 |
| Evansville | 0.0 | 4.1 | 0.2 | 0.0 | 2.9 | 17.7 | 15.3 | 5.5 | 1.9 | 12.9 |
| Fishers | 0.0 | 6.3 | 0.0 | 0.0 | 0.0 | 22.1 | 0.1 | 1.9 | 1.6 | 0.0 |
| Fort Wayne | 0.0 | 22.0 | 0.3 | 0.0 | 1.4 | 34.1 | 5.0 | 7.3 | 5.5 | 0.0 |
| Gary | 0.0 | 1.6 | 0.0 | 0.0 | 2.2 | 13.0 | 19.5 | 4.1 | 2.9 | 3.3 |
| Goshen | 0.0 | 8.1 | 2.5 | 0.0 | 0.0 | 16.8 | 11.1 | 6.6 | 0.7 | 3.8 |
| Greenwood | 0.0 | 16.5 | 0.0 | 0.0 | 0.0 | 14.6 | 28.3 | 7.2 | 0.0 | 5.7 |
| Hammond | 0.0 | 6.3 | 0.0 | 0.0 | 0.1 | 14.2 | 21.4 | 6.3 | 2.0 | 4.9 |
| Hobart | 0.0 | 22.8 | 0.0 | 0.0 | 0.0 | 15.6 | 24.2 | 2.3 | 0.0 | 3.0 |
| Indianapolis | 0.0 | 0.0 | 0.0 | 0.0 | 53.8 | 9.0 | 1.8 | 3.4 | 0.0 | 8.3 |
| Jeffersonville | 0.0 | 7.1 | 0.0 | 0.0 | 1.0 | 14.4 | 15.5 | 10.0 | 0.0 | 0.0 |
| Kokomo | 0.0 | 0.0 | 0.0 | 0.0 | 0.0 | 20.5 | 19.8 | 5.6 | 0.0 | 1.7 |
| Lafayette | 0.0 | 25.7 | 0.3 | 0.0 | 0.3 | 17.0 | 2.4 | 5.1 | 1.2 | 6.0 |
| Lawrence | 0.0 | 8.9 | 0.0 | 0.0 | 5.7 | 19.7 | 23.7 | 3.5 | 0.0 | 3.4 |
| Marion | 0.0 | 15.2 | 0.0 | 0.0 | 0.3 | 21.5 | 4.9 | 3.5 | 0.0 | 4.5 |
| Merrillville | 0.0 | 31.2 | 0.0 | 0.0 | 0.3 | 30.2 | 1.3 | 1.2 | 0.0 | 0.0 |
| Michigan City | 0.0 | 4.9 | 0.0 | 0.0 | 0.5 | 22.3 | 24.0 | 5.8 | 1.1 | 1.0 |
| Mishawaka | 0.0 | 18.9 | 0.0 | 0.0 | 0.0 | 13.5 | 0.0 | 4.4 | 22.6 | 0.4 |
| Muncie | 0.0 | 3.9 | 0.3 | 0.0 | 0.6 | 11.8 | 37.9 | 2.0 | 0.4 | 7.3 |
| New Albany | 0.0 | 8.3 | 0.1 | 0.0 | 0.9 | 14.0 | 28.0 | 3.8 | 0.1 | 5.3 |
| Noblesville | 0.0 | 13.6 | 0.1 | 0.0 | 0.0 | 14.2 | 7.3 | 5.6 | 0.0 | 14.4 |
| Plainfield | 0.0 | 2.1 | 0.0 | 0.0 | 0.0 | 10.3 | 9.6 | 7.9 | 0.0 | 9.4 |
| Portage | 0.0 | 10.2 | 0.0 | 0.0 | 0.0 | 19.1 | 28.4 | 3.0 | 0.0 | 8.8 |
| Richmond | 0.0 | 3.4 | 0.1 | 0.0 | 0.0 | 13.1 | 27.7 | 4.4 | 0.0 | 4.9 |
| Schererville | 0.0 | 7.9 | 0.0 | 0.0 | 4.6 | 29.1 | 28.0 | 6.4 | 0.0 | 5.4 |
| South Bend | 0.0 | 5.1 | 0.4 | 0.0 | 3.9 | 16.7 | 12.6 | 6.3 | 0.0 | 3.9 |
| Terre Haute | 0.0 | 4.0 | 0.0 | 0.0 | 1.6 | 13.7 | 12.7 | 3.1 | 4.8 | 6.7 |
| Valparaiso | 0.0 | 20.7 | 0.1 | 0.0 | 0.0 | 8.7 | 7.0 | 6.4 | 0.0 | 7.2 |
| Westfield | 0.0 | 12.0 | 0.0 | 0.0 | 0.3 | 13.0 | 4.8 | 15.1 | 0.0 | 4.5 |
| West Lafayette | 0.2 | 2.9 | 0.0 | 0.0 | 0.0 | 11.8 | 12.5 | 3.6 | 0.3 | 10.7 |
| **IOWA** | X | X | X | X | X | X | X | X | X | X |
| Ames | 0.5 | 1.7 | 0.4 | 0.0 | 71.4 | 3.6 | 5.2 | 1.6 | 0.2 | 2.2 |
| Ankeny | 0.0 | 19.9 | 0.0 | 0.0 | 5.0 | 12.4 | 23.1 | 12.6 | 0.0 | 7.6 |
| Bettendorf | 0.0 | 4.2 | 0.0 | 0.0 | 0.0 | 10.8 | 12.8 | 2.9 | 0.0 | 7.3 |
| Burlington | 0.0 | 7.8 | 0.2 | 0.0 | 3.9 | 15.5 | 18.1 | 6.4 | 0.0 | 3.8 |
| Cedar Falls | 0.0 | 12.3 | 0.2 | 0.0 | 0.4 | 6.8 | 9.3 | 6.4 | 2.0 | 1.2 |
| Cedar Rapids | 0.0 | 5.0 | 0.3 | 0.0 | 0.3 | 11.7 | 15.8 | 2.9 | 2.0 | 3.7 |
| Clinton | 0.0 | 13.4 | 0.0 | 0.0 | 0.6 | 11.2 | 24.8 | 4.2 | 0.5 | 5.7 |
| Council Bluffs | 0.2 | 5.8 | 0.1 | 0.0 | 2.7 | 15.1 | 12.5 | 7.4 | 0.5 | 0.0 |
| Davenport | 0.0 | 10.4 | 0.6 | 0.0 | 0.0 | 15.0 | 12.4 | 4.4 | 4.4 | 5.4 |
| Des Moines | 2.2 | 7.1 | 2.1 | 0.0 | 0.2 | 15.6 | 14.0 | 2.5 | 4.9 | 5.5 |
| Dubuque | 0.4 | 5.4 | 1.3 | 0.0 | 2.2 | 11.7 | 8.3 | 6.6 | 7.2 | 6.9 |
| Fort Dodge | 0.6 | 3.8 | 0.4 | 0.0 | 0.1 | 6.6 | 34.7 | 4.3 | 0.6 | 3.2 |
| Iowa City | 0.2 | 5.1 | 2.9 | 0.0 | 0.6 | 9.8 | 13.3 | 5.6 | 8.8 | 2.0 |
| Marion | 0.0 | 7.2 | 0.0 | 0.0 | 0.1 | 13.1 | 19.2 | 4.6 | 1.5 | 2.7 |
| Marshalltown | 0.1 | 21.5 | 0.1 | 0.0 | 3.7 | 16.6 | 11.7 | 4.5 | 3.7 | 2.9 |
| Mason City | 0.7 | 6.9 | 0.3 | 0.0 | 0.7 | 14.6 | 8.0 | 4.8 | 0.0 | 2.4 |
| Ottumwa | 0.2 | 9.3 | 1.4 | 0.0 | 1.2 | 11.4 | 34.4 | 3.2 | 0.0 | 1.8 |
| Sioux City | 0.0 | 6.1 | 0.8 | 0.0 | 0.4 | 12.5 | 16.2 | 8.5 | 5.6 | 4.2 |
| Urbandale | 0.0 | 9.0 | 0.0 | 0.0 | 0.1 | 14.3 | 4.3 | 9.1 | 1.2 | 2.8 |
| Waterloo | 0.2 | 12.8 | 0.5 | 0.0 | 1.7 | 14.3 | 9.8 | 6.1 | 9.8 | 2.2 |
| West Des Moines | 0.9 | 7.3 | 0.0 | 0.0 | 4.0 | 7.9 | 15.5 | 4.1 | 0.5 | 3.2 |

| City | City government finances, 2017 (cont.) | | | Climate[2] | | | | | | |
|---|---|---|---|---|---|---|---|---|---|---|
| | Debt outstanding | | Debt issued during year | Average daily temperature | | | | Annual precipitation (inches) | Heating degree days | Cooling degree days |
| | Total (mil dol) | Per capita[1] (dollars) | | Mean | | Limits | | | | |
| | | | | January | July | January[3] | July[4] | | | |
| | 137 | 138 | 139 | 140 | 141 | 142 | 143 | 144 | 145 | 146 |
| **ILLINOIS— Cont'd** | | | | | | | | | | |
| Waukegan | 339.5 | 3,875 | 11.8 | 20.3 | 71.5 | 12.0 | 81.7 | 34.09 | 7,031 | 613 |
| West Chicago | 49.4 | 1,814 | 3.2 | NA | NA | NA | NA | NA | NA | NA |
| Wheaton | 23.9 | 448 | 0.0 | 23.1 | 74.8 | 14.2 | 86.8 | 37.94 | 6,053 | 942 |
| Wheeling | 66.3 | 1,721 | 18.0 | 18.4 | 72.1 | 9.6 | 82.3 | 36.56 | 7,149 | 624 |
| Wilmette | 135.7 | 4,960 | 0.3 | 22.0 | 73.3 | 14.3 | 83.5 | 36.27 | 6,498 | 830 |
| Woodridge | 54.5 | 1,622 | 5.9 | 23.1 | 74.8 | 14.2 | 86.8 | 37.94 | 6,053 | 942 |
| **INDIANA** | X | X | X | X | X | X | X | X | X | X |
| Anderson | 189.0 | 3,442 | 63.2 | 25.7 | 74.0 | 18.4 | 83.8 | 39.82 | 5,807 | 872 |
| Bloomington | 350.1 | 4,127 | 44.6 | 27.9 | 75.4 | 19.3 | 86.0 | 44.91 | 5,348 | 1,017 |
| Carmel | 753.7 | 7,738 | 0.0 | 25.3 | 74.2 | 17.0 | 84.5 | 42.85 | 5,901 | 873 |
| Columbus | 14.9 | 316 | 0.0 | 27.9 | 75.9 | 19.1 | 86.4 | 41.94 | 5,367 | 1,059 |
| Crown Point | 27.2 | 921 | 0.0 | NA | NA | NA | NA | NA | NA | NA |
| East Chicago | 98.2 | 3,488 | 0.0 | 23.7 | 74.0 | 15.3 | 84.8 | 38.13 | 6,055 | 887 |
| Elkhart | 47.7 | 910 | 0.0 | 22.8 | 72.1 | 14.3 | 83.3 | 38.56 | 6,487 | 663 |
| Evansville | 745.6 | 6,301 | 0.0 | 33.2 | 79.6 | 24.8 | 90.5 | 45.76 | 4,140 | 1,616 |
| Fishers | 325.6 | 3,554 | 66.3 | 25.3 | 74.2 | 17.0 | 84.5 | 42.85 | 5,901 | 873 |
| Fort Wayne | 497.7 | 1,877 | 24.6 | 23.6 | 73.4 | 16.1 | 84.3 | 36.55 | 6,205 | 830 |
| Gary | 84.3 | 1,113 | 0.0 | 22.2 | 73.5 | 13.9 | 83.9 | 38.02 | 6,497 | 776 |
| Goshen | 85.8 | 2,562 | 0.0 | 24.3 | 73.7 | 17.0 | 84.5 | 36.59 | 6,075 | 826 |
| Greenwood | 58.3 | 1,017 | 0.0 | 25.7 | 74.7 | 18.0 | 84.0 | 40.24 | 5,783 | 942 |
| Hammond | 260.5 | 3,406 | 0.0 | 22.2 | 73.5 | 13.9 | 83.9 | 38.02 | 6,497 | 776 |
| Hobart | 26.8 | 954 | 0.0 | 22.2 | 73.5 | 13.9 | 83.9 | 38.02 | 6,497 | 776 |
| Indianapolis | 3,572.6 | 4,137 | 238.3 | 26.5 | 75.4 | 18.5 | 85.6 | 40.95 | 5,521 | 1,042 |
| Jeffersonville | 115.6 | 2,447 | 3.6 | 31.3 | 75.8 | 21.4 | 88.5 | 45.47 | 4,829 | 1,079 |
| Kokomo | 31.1 | 538 | 11.8 | 22.8 | 73.0 | 15.0 | 84.1 | 41.54 | 6,368 | 771 |
| Lafayette | 148.0 | 2,052 | 0.0 | 23.0 | 73.5 | 14.3 | 84.5 | 36.90 | 6,206 | 842 |
| Lawrence | 26.3 | 541 | 0.0 | 25.7 | 74.7 | 18.0 | 84.0 | 40.24 | 5,783 | 942 |
| Marion | 22.7 | 806 | 0.0 | 24.2 | 73.8 | 16.3 | 84.5 | 39.01 | 6,143 | 819 |
| Merrillville | 31.2 | 896 | 3.2 | 21.1 | 72.7 | 12.1 | 83.6 | 40.04 | 6,642 | 734 |
| Michigan City | 80.9 | 2,608 | 0.0 | 23.4 | 73.0 | 15.7 | 83.1 | 39.70 | 6,294 | 812 |
| Mishawaka | 76.1 | 1,554 | 0.0 | 24.3 | 73.7 | 17.0 | 84.5 | 36.59 | 6,075 | 826 |
| Muncie | 176.4 | 2,562 | 0.0 | 24.4 | 72.5 | 15.9 | 83.9 | 41.23 | 6,215 | 717 |
| New Albany | 74.7 | 2,049 | 0.0 | 31.3 | 75.8 | 21.4 | 88.5 | 45.47 | 4,829 | 1,079 |
| Noblesville | 233.6 | 3,775 | 0.0 | 25.3 | 74.2 | 17.0 | 84.5 | 42.85 | 5,901 | 873 |
| Plainfield | 168.9 | 5,128 | 0.0 | NA | NA | NA | NA | NA | NA | NA |
| Portage | 68.2 | 1,856 | 0.0 | 22.9 | 73.0 | 15.5 | 83.1 | 40.06 | 6,270 | 745 |
| Richmond | 70.1 | 1,979 | 0.0 | 25.7 | 73.1 | 17.2 | 84.6 | 39.55 | 5,942 | 769 |
| Schererville | 46.2 | 1,617 | 0.0 | NA | NA | NA | NA | NA | NA | NA |
| South Bend | 219.6 | 2,152 | 0.3 | 23.4 | 73.0 | 15.7 | 83.1 | 39.70 | 6,294 | 812 |
| Terre Haute | 223.0 | 3,666 | 0.0 | 26.5 | 76.2 | 17.7 | 87.3 | 42.47 | 5,433 | 1,107 |
| Valparaiso | 81.6 | 2,442 | 0.0 | 22.9 | 73.0 | 15.5 | 83.1 | 40.06 | 6,270 | 745 |
| Westfield | 100.3 | 2,540 | 0.0 | NA | NA | NA | NA | NA | NA | NA |
| West Lafayette | 150.3 | 3,165 | 0.0 | 25.2 | 75.5 | 17.2 | 86.3 | 36.32 | 5,732 | 1,024 |
| **IOWA** | X | X | X | X | X | X | X | X | X | X |
| Ames | 221.7 | 3,344 | 0.0 | 18.5 | 73.8 | 9.6 | 84.3 | 34.07 | 6,791 | 830 |
| Ankeny | 148.9 | 2,387 | 0.0 | 18.2 | 74.8 | 8.7 | 85.8 | 33.38 | 6,961 | 881 |
| Bettendorf | 133.4 | 3,728 | 0.0 | 21.1 | 76.2 | 13.2 | 85.4 | 34.11 | 6,246 | 1,072 |
| Burlington | 57.3 | 2,298 | 0.0 | 22.8 | 76.3 | 15.1 | 85.4 | 37.94 | 5,948 | 1,095 |
| Cedar Falls | 67.3 | 1,640 | 0.0 | 16.1 | 73.6 | 6.3 | 85.0 | 33.15 | 7,348 | 758 |
| Cedar Rapids | 489.6 | 3,694 | 0.0 | 19.9 | 74.8 | 11.5 | 85.3 | 36.62 | 6,488 | 910 |
| Clinton | 121.2 | 4,771 | 0.0 | 20.4 | 74.7 | 12.5 | 85.0 | 35.68 | 6,416 | 915 |
| Council Bluffs | 90.8 | 1,455 | 0.0 | 21.1 | 76.2 | 10.4 | 87.7 | 33.25 | 6,323 | 1,057 |
| Davenport | 317.3 | 3,101 | 0.0 | 21.1 | 76.2 | 13.2 | 85.4 | 34.11 | 6,246 | 1,072 |
| Des Moines | 561.4 | 2,591 | 0.0 | 20.4 | 76.1 | 11.7 | 86.0 | 34.72 | 6,436 | 1,052 |
| Dubuque | 271.2 | 4,652 | 0.0 | 17.8 | 75.1 | 8.7 | 85.4 | 33.96 | 6,891 | 908 |
| Fort Dodge | 79.4 | 3,263 | 0.0 | 15.4 | 73.1 | 5.8 | 84.3 | 34.39 | 7,513 | 746 |
| Iowa City | 128.7 | 1,701 | 0.0 | 21.7 | 76.9 | 13.4 | 87.5 | 37.27 | 6,052 | 1,134 |
| Marion | 59.9 | 1,518 | 0.0 | 16.8 | 73.9 | 7.1 | 84.4 | 36.40 | 7,191 | 787 |
| Marshalltown | 58.8 | 2,169 | 7.1 | 16.8 | 73.9 | 7.1 | 84.4 | 36.40 | 7,191 | 787 |
| Mason City | 36.6 | 1,339 | 0.0 | 13.9 | 72.4 | 5.1 | 83.3 | 34.48 | 7,765 | 655 |
| Ottumwa | 46.3 | 1,897 | 0.0 | NA | NA | NA | NA | NA | NA | NA |
| Sioux City | 255.9 | 3,112 | 0.0 | 18.6 | 74.6 | 8.5 | 86.2 | 25.99 | 6,900 | 914 |
| Urbandale | 77.7 | 1,783 | 0.0 | 20.4 | 76.1 | 11.7 | 86.0 | 34.72 | 6,436 | 1,052 |
| Waterloo | 112.0 | 1,653 | 0.0 | 16.1 | 73.6 | 6.3 | 85.0 | 33.15 | 7,348 | 758 |
| West Des Moines | 128.9 | 1,966 | 0.1 | 20.4 | 76.1 | 11.7 | 86.0 | 34.72 | 6,436 | 1,052 |

1. Based on the population estimated as of July 1 of the year shown.   2. Represents normal values based on the 30-year period, 1971±2000.   3. Average daily minimum.   4. Average daily maximum.

| STATE Place code | City | Land area[1] (sq. mi) | Total persons 2019 | Rank | Per square mile | White | Black or African American | American Indian, Alaskan Native | Asian | Hawaiian Pacific Islander | Some other race | Two or more races (percent) |
|---|---|---|---|---|---|---|---|---|---|---|---|---|
| | | 1 | 2 | 3 | 4 | 5 | 6 | 7 | 8 | 9 | 10 | 11 |
| 20 00000 | KANSAS ............ | 81,758.5 | 2,913,805 | X | 36 | 83.6 | 5.7 | 0.8 | 3.0 | 0.1 | 3.0 | 3.7 |
| 20 18250 | Dodge City ........... | 14.7 | 26,687 | 1,365 | 1,815 | D | D | D | D | D | 13.2 | 0.7 |
| 20 25325 | Garden City ........... | 10.9 | 26,003 | 1,387 | 2,386 | 74.6 | 4.6 | 0.9 | 6.0 | 0.0 | D | D |
| 20 33625 | Hutchinson ........... | 24.6 | 40,209 | 958 | 1,635 | 86.0 | 5.0 | 3.2 | 0.7 | 0.1 | 1.7 | 3.3 |
| 20 36000 | Kansas City ........... | 124.7 | 152,727 | 168 | 1,225 | 54.5 | 23.2 | 0.5 | 5.7 | 0.1 | 11.4 | 4.6 |
| 20 38900 | Lawrence ........... | 34.2 | 98,448 | 324 | 2,879 | 79.0 | 4.2 | 2.3 | 7.1 | 0.0 | 0.8 | 6.6 |
| 20 39000 | Leavenworth ........... | 24.2 | 35,934 | 1,075 | 1,485 | 78.2 | 13.2 | 0.6 | 1.9 | 0.0 | 3.2 | 2.9 |
| 20 39075 | Leawood ........... | 15.1 | 34,669 | 1,116 | 2,296 | D | D | D | D | D | D | D |
| 20 39350 | Lenexa ........... | 34.1 | 56,156 | 684 | 1,647 | 81.5 | 6.2 | 0.4 | 5.9 | 0.0 | 3.7 | 2.3 |
| 20 44250 | Manhattan ........... | 19.9 | 54,944 | 697 | 2,761 | 80.2 | 3.4 | 2.4 | 5.9 | 1.8 | 1.1 | 5.2 |
| 20 52575 | Olathe ........... | 61.9 | 141,665 | 189 | 2,289 | 83.7 | 6.9 | 0.0 | 4.2 | 0.0 | 1.3 | 3.8 |
| 20 53775 | Overland Park ........... | 75.2 | 197,381 | 126 | 2,625 | 79.7 | 5.7 | 0.3 | 8.7 | 0.2 | 2.1 | 3.4 |
| 20 62700 | Salina ........... | 25.8 | 46,274 | 839 | 1,794 | 88.8 | 3.2 | 0.3 | 2.9 | 0.2 | 0.2 | 4.4 |
| 20 64500 | Shawnee ........... | 42.0 | 66,298 | 567 | 1,579 | 88.2 | 3.5 | 0.1 | 2.5 | 0.0 | 2.1 | 3.6 |
| 20 71000 | Topeka ........... | 61.4 | 124,558 | 227 | 2,029 | 76.9 | 11.0 | 1.1 | 1.3 | 0.0 | 5.8 | 3.8 |
| 20 79000 | Wichita ........... | 162.0 | 391,731 | 50 | 2,418 | 74.5 | 10.5 | 1.0 | 5.0 | 0.0 | 3.6 | 5.4 |
| 21 00000 | KENTUCKY ........... | 39,491.4 | 4,477,251 | X | 113 | 86.7 | 8.1 | 0.2 | 1.6 | 0.1 | 0.9 | 2.3 |
| 21 08902 | Bowling Green ........... | 40.4 | 70,631 | 514 | 1,748 | 73.7 | 12.1 | 0.7 | 5.4 | 0.0 | 2.9 | 5.2 |
| 21 17848 | Covington ........... | 13.2 | 40,314 | 954 | 3,054 | 80.4 | 9.3 | 0.1 | 1.2 | 0.0 | 4.7 | 4.2 |
| 21 24274 | Elizabethtown ........... | 27.2 | 30,331 | 1,238 | 1,115 | 71.4 | 14.2 | 0.0 | 7.5 | 0.0 | 0.0 | 6.9 |
| 21 27982 | Florence ........... | 10.7 | 33,423 | 1,150 | 3,124 | D | D | D | D | D | D | D |
| 21 28900 | Frankfort ........... | 14.8 | 27,715 | 1,326 | 1,873 | 75.8 | 14.8 | 0.4 | 3.1 | 0.0 | 5.5 | 0.4 |
| 21 30700 | Georgetown ........... | 17.0 | 35,847 | 1,078 | 2,109 | 83.1 | 3.9 | 0.2 | 0.7 | 0.0 | 5.2 | 6.8 |
| 21 35866 | Henderson ........... | 16.1 | 27,974 | 1,315 | 1,738 | 81.1 | 12.4 | 0.3 | 0.4 | 0.0 | 2.8 | 3.0 |
| 21 37918 | Hopkinsville ........... | 31.8 | 30,899 | 1,222 | 972 | D | D | D | D | D | D | D |
| 21 40222 | Jeffersontown ........... | 10.3 | 27,562 | 1,331 | 2,676 | D | D | D | D | D | D | D |
| 21 46027 | Lexington-Fayette ........... | 283.6 | 324,735 | 61 | 1,145 | 73.5 | 14.8 | 0.2 | 4.2 | 0.0 | 2.8 | 4.5 |
| 21 48003 | Louisville/Jefferson County ... | 263.5 | 618,338 | 29 | 2,347 | 70.0 | 24.0 | 0.1 | 2.6 | 0.1 | 0.7 | 2.5 |
| 21 56136 | Nicholasville ........... | 14.6 | 30,888 | 1,223 | 2,116 | D | D | D | D | D | D | D |
| 21 58620 | Owensboro ........... | 20.5 | 60,344 | 624 | 2,944 | 85.1 | 3.2 | 0.4 | 6.8 | 0.1 | 2.3 | 2.1 |
| 21 58836 | Paducah ........... | 20.3 | 24,883 | 1,408 | 1,226 | 75.8 | 15.6 | 2.1 | 2.1 | 0.0 | 1.4 | 3.0 |
| 21 65226 | Richmond ........... | 20.3 | 36,800 | 1,042 | 1,813 | 85.4 | 7.2 | 0.6 | 1.5 | 3.3 | 1.1 | 0.9 |
| 22 00000 | LOUISIANA ........... | 43,204.5 | 4,645,318 | X | 108 | 61.8 | 32.4 | 0.6 | 1.8 | 0.0 | 1.4 | 2.0 |
| 22 00975 | Alexandria ........... | 28.5 | 45,573 | 851 | 1,599 | 33.3 | 61.5 | 0.3 | 1.4 | 0.0 | 1.8 | 1.8 |
| 22 05000 | Baton Rouge ........... | 86.3 | 219,052 | 101 | 2,538 | 41.3 | 51.7 | 0.2 | 3.9 | 0.0 | 1.4 | 1.4 |
| 22 08920 | Bossier City ........... | 43.8 | 68,216 | 539 | 1,557 | 66.5 | 26.2 | 0.0 | 3.7 | 0.0 | 0.9 | 2.8 |
| 22 13960 | Central ........... | 62.3 | 29,450 | 1,270 | 473 | D | D | D | D | D | D | D |
| 22 36255 | Houma ........... | 14.5 | 32,467 | 1,174 | 2,239 | 72.2 | 17.7 | 5.9 | 0.0 | 0.0 | 0.0 | 4.2 |
| 22 39475 | Kenner ........... | 14.9 | 66,288 | 568 | 4,449 | 73.5 | 16.4 | 0.5 | 3.3 | 0.0 | 3.8 | 2.5 |
| 22 40735 | Lafayette ........... | 55.8 | 126,535 | 217 | 2,268 | 65.3 | 29.8 | 0.3 | 2.7 | 0.0 | 0.3 | 1.6 |
| 22 41155 | Lake Charles ........... | 45.7 | 78,433 | 452 | 1,716 | 46.3 | 49.5 | 0.0 | 2.0 | 0.0 | 1.4 | 0.9 |
| 22 51410 | Monroe ........... | 29.7 | 46,998 | 827 | 1,582 | 30.6 | 60.9 | 0.3 | 0.4 | 0.0 | 6.7 | 1.0 |
| 22 54035 | New Iberia ........... | 11.1 | 28,032 | 1,312 | 2,525 | 56.3 | 37.7 | 0.3 | 0.8 | 0.0 | 2.0 | 2.9 |
| 22 55000 | New Orleans ........... | 169.5 | 389,476 | 51 | 2,298 | 34.0 | 58.8 | 0.3 | 2.8 | 0.0 | 2.0 | 2.1 |
| 22 70000 | Shreveport ........... | 107.8 | 184,786 | 137 | 1,714 | 38.1 | 57.9 | 0.4 | 1.4 | 0.2 | 0.3 | 1.6 |
| 22 70805 | Slidell ........... | 15.0 | 27,541 | 1,332 | 1,836 | D | D | D | D | D | D | D |
| 23 00000 | MAINE ........... | 30,844.8 | 1,350,141 | X | 44 | 94.0 | 1.6 | 0.7 | 1.1 | 0.0 | 0.4 | 2.1 |
| 23 02795 | Bangor ........... | 34.3 | 31,998 | 1,195 | 933 | D | D | D | D | D | D | D |
| 23 38740 | Lewiston ........... | 34.1 | 36,226 | 1,066 | 1,062 | 84.4 | 8.6 | 0.0 | 1.3 | 0.0 | 0.6 | 5.1 |
| 23 60545 | Portland ........... | 21.5 | 66,803 | 558 | 3,107 | 85.9 | 7.0 | 0.4 | 3.2 | 0.0 | 0.2 | 3.2 |
| 23 71990 | South Portland ........... | 12.1 | 25,950 | 1,389 | 2,145 | D | D | D | D | D | D | D |
| 24 00000 | MARYLAND ........... | 9,711.1 | 6,055,802 | X | 624 | 54.5 | 30.3 | 0.3 | 6.4 | 0.0 | 5.1 | 3.4 |
| 24 01600 | Annapolis ........... | 7.2 | 39,524 | 982 | 5,489 | 74.7 | 15.4 | 0.0 | 0.7 | 0.0 | 5.7 | 3.5 |
| 24 04000 | Baltimore ........... | 80.9 | 586,131 | 31 | 7,245 | 30.3 | 62.6 | 0.4 | 2.7 | 0.0 | 1.9 | 2.1 |
| 24 08775 | Bowie ........... | 20.4 | 58,212 | 660 | 2,854 | 31.9 | 53.5 | 0.7 | 3.4 | 0.0 | 4.8 | 5.7 |
| 24 18750 | College Park ........... | 5.6 | 32,173 | 1,189 | 5,745 | 56.6 | 18.3 | 0.7 | 14.1 | 0.0 | 8.4 | 2.0 |
| 24 30325 | Frederick ........... | 23.9 | 73,308 | 490 | 3,067 | 74.1 | 16.1 | 0.0 | 5.3 | 0.0 | 2.1 | 6.8 |
| 24 31175 | Gaithersburg ........... | 10.3 | 67,741 | 545 | 6,577 | 50.2 | 17.6 | 0.0 | 21.0 | 0.0 | 4.3 | 6.2 |
| 24 36075 | Hagerstown ........... | 12.2 | 39,918 | 965 | 3,272 | 66.7 | 21.6 | 0.0 | 2.8 | 0.0 | 2.6 | 4.4 |
| 24 45900 | Laurel ........... | 4.8 | 25,548 | 1,398 | 5,323 | 29.4 | 54.2 | 0.0 | 7.4 | 0.0 | 4.6 | 4.4 |
| 24 67675 | Rockville ........... | 13.6 | 69,512 | 528 | 5,111 | 57.4 | 8.1 | 0.2 | 23.6 | 0.0 | 6.3 | 4.4 |
| 24 69925 | Salisbury ........... | 13.8 | 33,017 | 1,162 | 2,393 | 50.0 | 38.6 | 0.5 | 5.7 | 0.0 | 0.3 | 4.8 |
| 25 00000 | MASSACHUSETTS ........... | 7,801.0 | 6,893,574 | X | 884 | 77.0 | 7.9 | 0.3 | 6.9 | 0.0 | 4.3 | 3.6 |
| 25 00840 | Agawam Town ........... | 23.3 | 28,467 | 1,299 | 1,222 | D | D | D | D | D | D | D |
| 25 02690 | Attleboro ........... | 26.8 | 45,444 | 854 | 1,696 | 84.5 | 6.6 | 0.0 | 4.5 | 0.2 | 4.2 | 0.0 |
| 25 03690 | Barnstable Town ........... | 59.9 | 44,535 | 863 | 744 | 84.3 | 10.4 | 0.0 | 2.7 | 0.0 | 0.9 | 1.7 |

1. Dry land or land partially or temporarily covered by water.  2. Hispanic or Latino persons may be of any race.

| City | Percent Hispanic or Latino[1], 2019 | Percent foreign born, 2019 | Age of population (percent), 2019 — Under 18 years | 18 to 24 years | 25 to 34 years | 35 to 44 years | 45 to 54 years | 55 to 64 years | 65 years and over | Median age, 2019 | Percent female, 2019 | Population — Census counts 2000 | 2010 | Percent change 2000-2010 | Percent change 2001-2020 |
|---|---|---|---|---|---|---|---|---|---|---|---|---|---|---|---|
| | 12 | 13 | 14 | 15 | 16 | 17 | 18 | 19 | 20 | 21 | 22 | 23 | 24 | 25 | 26 |
| **KANSAS** | 12.2 | 7.2 | 24.0 | 10.1 | 13.0 | 12.5 | 11.2 | 12.8 | 16.4 | 37.2 | 50.4 | 2,688,418 | 2,853,120 | 6.1 | 2.1 |
| Dodge City | 64.5 | 22.2 | 30.2 | 10.3 | 17.5 | 15.2 | 6.5 | 10.5 | 9.7 | 30.1 | 49.9 | 25,176 | 27,324 | 8.5 | -2.3 |
| Garden City | 47.9 | 25.8 | 31.9 | 10.8 | 12.8 | 10.8 | 10.2 | 11.5 | 12.0 | 31.7 | 49.2 | 28,451 | 26,736 | -6.0 | -2.7 |
| Hutchinson | 14.2 | 3.7 | 22.5 | 11.4 | 9.9 | 11.9 | 12.0 | 10.9 | 21.4 | 37.6 | 50.7 | 40,787 | 42,180 | 3.4 | -4.7 |
| Kansas City | 31.2 | 19.6 | 27.6 | 8.6 | 15.4 | 13.5 | 10.6 | 11.8 | 12.4 | 33.4 | 50.4 | 146,866 | 145,781 | -0.7 | 4.8 |
| Lawrence | 6.9 | 7.5 | 18.3 | 25.8 | 14.6 | 11.3 | 9.6 | 8.3 | 12.2 | 28.4 | 51.1 | 80,098 | 87,944 | 9.8 | 11.9 |
| Leavenworth | 8.0 | 4.0 | 23.8 | 14.3 | 17.2 | 9.0 | 11.1 | 11.9 | 12.6 | 32.5 | 42.2 | 35,420 | 35,245 | -0.5 | 2.0 |
| Leawood | 2.0 | 8.3 | 19.8 | 8.1 | 8.6 | 10.4 | 14.3 | 16.8 | 22.2 | 47.4 | 50.6 | 27,656 | 31,888 | 15.3 | 8.7 |
| Lenexa | 9.0 | 12.1 | 20.4 | 6.7 | 16.4 | 12.8 | 12.6 | 15.7 | 15.4 | 39.4 | 52.0 | 40,238 | 48,214 | 19.8 | 16.5 |
| Manhattan | 8.9 | 8.3 | 14.7 | 37.3 | 15.3 | 8.7 | 7.3 | 6.6 | 10.1 | 24.3 | 50.4 | 44,831 | 52,161 | 16.4 | 5.3 |
| Olathe | 11.7 | 10.5 | 28.2 | 7.9 | 13.9 | 14.5 | 14.0 | 9.5 | 12.0 | 35.0 | 51.1 | 92,962 | 125,900 | 35.4 | 12.5 |
| Overland Park | 6.5 | 11.2 | 22.3 | 8.8 | 14.5 | 14.6 | 12.6 | 12.3 | 14.9 | 37.9 | 50.4 | 149,080 | 173,329 | 16.3 | 13.9 |
| Salina | 12.0 | 5.6 | 22.3 | 9.7 | 13.4 | 11.5 | 10.6 | 12.5 | 20.0 | 37.9 | 49.7 | 45,679 | 47,777 | 4.6 | -3.1 |
| Shawnee | 5.2 | 5.3 | 23.9 | 7.7 | 11.5 | 13.5 | 12.7 | 13.1 | 17.7 | 39.9 | 52.6 | 47,996 | 62,205 | 29.6 | 6.6 |
| Topeka | 16.7 | 4.0 | 22.0 | 10.4 | 13.3 | 11.9 | 10.7 | 13.2 | 18.6 | 39.1 | 52.2 | 122,377 | 127,630 | 4.3 | -2.4 |
| Wichita | 16.5 | 10.2 | 25.2 | 9.5 | 14.9 | 12.7 | 10.8 | 12.4 | 14.5 | 35.3 | 50.8 | 344,284 | 382,424 | 11.1 | 2.4 |
| **KENTUCKY** | 3.8 | 4.4 | 22.4 | 9.3 | 12.9 | 12.5 | 12.6 | 13.4 | 16.9 | 39.2 | 50.8 | 4,041,769 | 4,339,330 | 7.4 | 3.2 |
| Bowling Green | 8.1 | 15.1 | 20.1 | 25.5 | 16.7 | 12.0 | 8.1 | 8.1 | 9.4 | 27.1 | 50.8 | 49,296 | 59,405 | 20.5 | 18.9 |
| Covington | 8.6 | 5.4 | 25.2 | 5.6 | 18.6 | 13.4 | 10.1 | 13.4 | 13.8 | 35.6 | 53.5 | 43,370 | 40,482 | -6.7 | -0.4 |
| Elizabethtown | 5.0 | 6.6 | 19.2 | 7.1 | 13.4 | 15.3 | 16.0 | 11.8 | 17.3 | 41.2 | 47.9 | 22,542 | 28,053 | 24.4 | 8.1 |
| Florence | 1.1 | 7.8 | 23.7 | 8.1 | 14.4 | 11.3 | 13.3 | 10.6 | 18.6 | 37.7 | 45.3 | 23,551 | 29,531 | 25.4 | 13.2 |
| Frankfort | 5.2 | 7.4 | 22.5 | 11.0 | 11.8 | 14.7 | 9.0 | 13.6 | 17.4 | 39.0 | 52.4 | 27,741 | 27,272 | -1.7 | 1.6 |
| Georgetown | 6.4 | 4.6 | 26.1 | 10.9 | 15.4 | 14.5 | 11.8 | 10.0 | 11.2 | 33.3 | 52.3 | 18,080 | 29,137 | 61.2 | 23.0 |
| Henderson | 4.4 | 2.5 | 20.6 | 6.2 | 13.4 | 17.2 | 12.6 | 13.9 | 16.1 | 41.2 | 52.0 | 27,373 | 28,928 | 5.7 | -3.3 |
| Hopkinsville | 4.1 | 2.5 | 22.0 | 8.4 | 13.7 | 10.3 | 12.9 | 12.1 | 20.5 | 40.9 | 48.6 | 30,089 | 31,975 | 6.3 | -3.4 |
| Jeffersontown | 7.7 | 14.0 | 29.1 | 7.0 | 15.2 | 15.0 | 9.1 | 9.5 | 15.1 | 33.7 | 48.6 | 26,633 | 27,813 | 4.4 | -0.9 |
| Lexington-Fayette | 7.4 | 10.2 | 20.8 | 13.7 | 15.4 | 13.5 | 11.2 | 11.7 | 13.7 | 35.0 | 51.0 | 260,512 | 295,875 | 13.6 | 9.8 |
| Louisville/Jefferson County | 6.0 | 8.9 | 22.0 | 8.9 | 15.0 | 12.4 | 12.4 | 13.5 | 15.9 | 37.9 | 51.7 | NA | 595,730 | NA | 3.8 |
| Nicholasville | 3.9 | 2.2 | 32.3 | 8.4 | 13.6 | 14.4 | 10.8 | 9.6 | 10.8 | 32.0 | 55.2 | 19,680 | 28,018 | 42.4 | 10.2 |
| Owensboro | 5.0 | 7.3 | 25.3 | 8.6 | 13.0 | 13.3 | 11.5 | 11.5 | 16.8 | 37.8 | 51.3 | 54,067 | 57,477 | 6.3 | 5.0 |
| Paducah | 4.6 | 1.1 | 21.4 | 7.0 | 8.2 | 14.3 | 16.2 | 15.2 | 17.7 | 44.6 | 52.3 | 26,307 | 25,008 | -4.9 | -0.5 |
| Richmond | 2.4 | 3.0 | 18.0 | 31.4 | 13.4 | 13.1 | 6.5 | 6.8 | 10.8 | 25.2 | 52.1 | 27,152 | 31,320 | 15.4 | 17.5 |
| **LOUISIANA** | 5.4 | 4.2 | 23.3 | 9.1 | 13.8 | 13.1 | 11.7 | 13.0 | 16.0 | 37.7 | 51.2 | 4,468,976 | 4,533,500 | 1.4 | 2.5 |
| Alexandria | 2.5 | 4.9 | 23.7 | 9.9 | 12.2 | 12.4 | 12.4 | 13.6 | 15.9 | 37.9 | 50.9 | 46,342 | 47,901 | 3.4 | -4.9 |
| Baton Rouge | 3.9 | 6.4 | 19.4 | 18.0 | 14.2 | 10.5 | 10.3 | 12.7 | 14.9 | 33.8 | 52.2 | 227,818 | 229,279 | 0.6 | -4.5 |
| Bossier City | 6.7 | 3.8 | 25.3 | 7.7 | 14.8 | 15.5 | 9.6 | 10.4 | 16.8 | 35.4 | 49.6 | 56,461 | 61,596 | 9.1 | 10.7 |
| Central | 5.6 | 2.9 | 31.2 | 7.0 | 8.9 | 18.3 | 10.5 | 8.2 | 15.8 | 38.0 | 47.8 | NA | 27,215 | NA | 8.2 |
| Houma | 1.7 | 1.5 | 24.8 | 5.8 | 11.6 | 12.7 | 10.9 | 16.1 | 18.2 | 38.5 | 53.1 | 32,393 | 33,694 | 4.0 | -3.6 |
| Kenner | 30.3 | 22.2 | 19.2 | 9.3 | 12.1 | 14.9 | 12.3 | 12.9 | 19.4 | 41.2 | 54.9 | 70,517 | 66,663 | -5.5 | -0.6 |
| Lafayette | 3.5 | 3.8 | 20.9 | 12.3 | 12.7 | 13.0 | 10.5 | 12.7 | 17.8 | 37.6 | 53.1 | 110,257 | 121,724 | 10.4 | 4.0 |
| Lake Charles | 2.7 | 3.5 | 21.5 | 9.2 | 15.2 | 11.8 | 12.9 | 11.9 | 17.6 | 37.6 | 53.7 | 71,757 | 72,375 | 0.9 | 8.4 |
| Monroe | 3.4 | 2.2 | 28.0 | 10.6 | 10.8 | 14.2 | 10.4 | 10.4 | 15.7 | 36.1 | 51.4 | 53,107 | 48,927 | -7.9 | -3.9 |
| New Iberia | 3.0 | 1.7 | 26.1 | 8.9 | 15.1 | 11.9 | 13.3 | 11.1 | 13.7 | 35.0 | 52.0 | 32,623 | 30,607 | -6.2 | -8.4 |
| New Orleans | 5.5 | 5.9 | 19.8 | 8.6 | 17.0 | 14.3 | 11.6 | 13.2 | 15.5 | 38.3 | 52.8 | 484,674 | 343,828 | -29.1 | 13.3 |
| Shreveport | 2.9 | 2.0 | 23.7 | 9.5 | 15.2 | 11.2 | 11.3 | 12.4 | 16.6 | 36.1 | 53.2 | 200,145 | 200,141 | 0.4 | -8.0 |
| Slidell | 4.4 | 6.2 | 28.7 | 5.9 | 9.1 | 14.1 | 12.4 | 15.9 | 13.9 | 38.6 | 53.7 | 25,695 | 27,290 | 6.2 | 0.9 |
| **MAINE** | 1.7 | 3.9 | 18.3 | 7.9 | 12.2 | 11.4 | 13.1 | 15.7 | 21.3 | 45.1 | 51.2 | 1,274,923 | 1,328,354 | 4.2 | 1.6 |
| Bangor | 2.1 | 4.5 | 20.9 | 12.3 | 16.4 | 11.9 | 9.3 | 12.0 | 17.0 | 35.2 | 53.8 | 31,473 | 33,031 | 5.0 | -3.1 |
| Lewiston | 3.7 | 8.5 | 24.7 | 12.7 | 12.7 | 10.6 | 10.7 | 11.0 | 17.6 | 35.0 | 51.9 | 35,690 | 36,594 | 2.5 | |
| Portland | 2.2 | 9.5 | 11.9 | 10.5 | 23.7 | 12.2 | 12.0 | 11.7 | 18.0 | 37.2 | 52.9 | 64,249 | 66,190 | 3.0 | 0.9 |
| South Portland | 4.1 | 7.9 | 18.0 | 5.9 | 19.7 | 11.8 | 10.1 | 18.0 | 16.5 | 39.2 | 50.2 | 23,324 | 25,021 | 7.3 | 3.7 |
| **MARYLAND** | 10.6 | 15.4 | 22.0 | 8.8 | 13.6 | 13.0 | 13.2 | 13.6 | 15.9 | 39.0 | 51.6 | 5,296,486 | 5,773,787 | 9.0 | 4.9 |
| Annapolis | 32.4 | 11.9 | 26.9 | 8.1 | 12.8 | 15.7 | 12.8 | 8.3 | 15.5 | 35.9 | 49.4 | 35,838 | 38,334 | 7.0 | 3.1 |
| Baltimore | 5.7 | 7.3 | 20.1 | 9.5 | 18.9 | 13.0 | 11.2 | 12.8 | 14.5 | 35.9 | 53.1 | 651,154 | 620,777 | -4.7 | -5.6 |
| Bowie | 7.2 | 13.9 | 23.4 | 5.6 | 12.6 | 12.2 | 14.5 | 15.9 | 15.7 | 42.3 | 55.6 | 50,269 | 54,822 | 9.1 | 6.2 |
| College Park | 24.0 | 21.5 | 12.7 | 51.4 | 9.6 | 5.8 | 5.6 | 6.2 | 8.7 | 21.4 | 50.4 | 24,657 | 30,410 | 23.3 | 6.2 |
| Frederick | 18.7 | 19.2 | 17.8 | 10.3 | 17.7 | 15.3 | 11.6 | 12.9 | 14.4 | 36.9 | 52.7 | 52,767 | 65,339 | 23.8 | 12.2 |
| Gaithersburg | 20.3 | 37.1 | 22.0 | 8.2 | 13.4 | 15.6 | 16.0 | 10.9 | 14.0 | 38.7 | 53.2 | 52,613 | 59,891 | 13.8 | 13.1 |
| Hagerstown | 9.4 | 8.1 | 23.6 | 7.8 | 15.1 | 12.2 | 11.5 | 12.5 | 17.3 | 37.2 | 53.2 | 36,687 | 39,652 | 8.1 | 0.7 |
| Laurel | 11.2 | 21.1 | 19.3 | 9.3 | 13.2 | 14.1 | 17.2 | 12.2 | 14.7 | 40.8 | 53.1 | 19,960 | 24,909 | 24.8 | 2.6 |
| Rockville | 16.4 | 30.5 | 17.3 | 11.2 | 14.4 | 13.9 | 11.8 | 12.6 | 18.8 | 39.9 | 52.1 | 47,388 | 61,298 | 29.4 | 13.4 |
| Salisbury | 2.4 | 13.1 | 24.6 | 24.6 | 10.8 | 11.1 | 7.9 | 10.8 | 10.4 | 26.2 | 59.2 | 23,743 | 30,268 | 27.5 | 9.1 |
| **MASSACHUSETTS** | 12.4 | 17.3 | 19.6 | 10.0 | 14.4 | 12.3 | 13.0 | 13.7 | 17.0 | 39.7 | 51.5 | 6,349,097 | 6,547,788 | 3.1 | 5.3 |
| Agawam Town | 4.8 | 8.3 | 24.8 | 8.3 | 13.4 | 13.5 | 14.1 | 10.6 | 15.5 | 38.2 | 49.5 | 28,144 | 28,438 | 1.0 | 0.1 |
| Attleboro | 6.7 | 13.0 | 22.0 | 5.0 | 12.9 | 14.4 | 13.0 | 14.3 | 18.3 | 42.2 | 50.9 | 42,068 | 43,570 | 3.6 | 4.3 |
| Barnstable Town | 6.1 | 17.3 | 13.6 | 7.9 | 9.9 | 11.9 | 15.8 | 17.5 | 23.4 | 49.9 | 49.4 | 47,821 | 45,195 | -5.5 | -1.5 |

1. May be of any race.

## Table D. Cities — Households, Group Quarters, Crime, and Education

| City | Households, 2019 — Number | Persons per household | Percent — Family | Percent — Married couple family | Percent — Female family | Percent — Non-family | Percent — One person | Persons in group quarters, 2019 | Serious crimes — Violent — Number | Violent — Rate[2] | Property — Number | Property — Rate[2] | Population age 25 and over | Attainment — High school graduate or less | Attainment — Bachelor's degree or more |
|---|---|---|---|---|---|---|---|---|---|---|---|---|---|---|---|
| | 27 | 28 | 29 | 30 | 31 | 32 | 33 | 34 | 35 | 36 | 37 | 38 | 39 | 40 | 41 |
| KANSAS | 1,138,329 | 2.49 | 64.4 | 50.2 | 14.2 | 35.6 | 29.1 | 80,546 | NA | NA | NA | NA | 1,918,081 | 34.5 | 34.0 |
| Dodge City | 8,840 | 2.91 | 69.4 | 46.7 | 22.7 | 30.6 | 28.9 | 609 | 129 | 472 | 613 | 2,244 | 15,676 | 53.2 | 16.6 |
| Garden City | 9,106 | 2.65 | 59.6 | 36.9 | 22.6 | 40.4 | 39.1 | 613 | 130 | 490 | NA | NA | 14,200 | 52.4 | 21.4 |
| Hutchinson | 15,604 | 2.34 | 59.5 | 44.0 | 15.5 | 40.5 | 34.6 | 2,495 | 153 | 378 | 1,290 | 3,191 | 25,718 | 35.8 | 23.3 |
| Kansas City | 56,873 | 2.67 | 62.2 | 42.1 | 20.1 | 37.8 | 31.0 | 1,228 | NA | NA | NA | NA | 97,492 | 53.3 | 19.3 |
| Lawrence | 39,442 | 2.30 | 48.8 | 37.5 | 11.2 | 51.2 | 29.0 | 7,397 | NA | NA | NA | NA | 54,926 | 24.9 | 51.7 |
| Leavenworth | 12,177 | 2.66 | 58.5 | 44.9 | 13.6 | 41.5 | 32.3 | 3,552 | 19 | 54 | 401 | 1,144 | 22,254 | 41.6 | 27.5 |
| Leawood | 12,964 | 2.67 | 80.7 | 78.5 | 2.2 | 19.3 | 18.3 | 77 | 87 | 155 | 761 | 1,353 | 25,065 | 4.8 | 82.1 |
| Lenexa | 22,329 | 2.47 | 66.1 | 53.4 | 12.7 | 33.9 | 26.8 | 385 | NA | NA | NA | NA | 40,585 | 20.4 | 56.7 |
| Manhattan | 20,061 | 2.43 | 38.6 | 26.2 | 12.4 | 61.4 | 41.8 | 6,456 | 290 | 205 | 1,604 | 1,135 | 26,534 | 17.8 | 49.8 |
| Olathe | 49,971 | 2.78 | 71.5 | 57.2 | 14.3 | 28.5 | 23.6 | 1,517 | NA | NA | NA | NA | 89,789 | 22.8 | 48.5 |
| Overland Park | 79,436 | 2.45 | 65.2 | 51.0 | 14.3 | 34.8 | 28.4 | 1,166 | NA | NA | NA | NA | 134,846 | 15.5 | 61.4 |
| Salina | 18,711 | 2.40 | 56.4 | 44.6 | 11.8 | 43.6 | 35.7 | 1,489 | 213 | 457 | 1,726 | 3,706 | 31,606 | 34.3 | 30.7 |
| Shawnee | 24,678 | 2.65 | 71.2 | 56.7 | 14.5 | 28.8 | 23.7 | 347 | 141 | 213 | NA | NA | 45,006 | 27.0 | 44.6 |
| Topeka | 54,066 | 2.23 | 57.1 | 39.7 | 17.4 | 42.9 | 36.4 | 4,905 | 895 | 712 | 6,271 | 4,991 | 84,788 | 39.3 | 29.4 |
| Wichita | 153,454 | 2.51 | 59.6 | 41.7 | 17.9 | 40.4 | 34.2 | 4,940 | 4,451 | 1,141 | 20,759 | 5,322 | 254,433 | 38.1 | 30.8 |
| KENTUCKY | 1,748,732 | 2.48 | 64.6 | 47.5 | 17.1 | 35.4 | 29.1 | 132,667 | NA | NA | NA | NA | 3,048,442 | 46.0 | 25.1 |
| Bowling Green | 27,178 | 2.33 | 51.2 | 33.6 | 17.6 | 48.8 | 35.4 | 7,108 | 203 | 292 | 3,250 | 4,668 | 38,358 | 39.2 | 33.1 |
| Covington | 17,473 | 2.26 | 47.4 | 26.4 | 21.0 | 52.6 | 44.0 | 888 | 188 | 466 | 1,175 | 2,912 | 27,938 | 44.9 | 27.5 |
| Elizabethtown | 13,512 | 2.16 | 58.3 | 45.5 | 12.8 | 41.7 | 36.2 | 1,155 | 49 | 161 | 452 | 1,488 | 22,328 | 33.4 | 32.9 |
| Florence | 13,480 | 2.43 | 60.5 | 41.3 | 19.2 | 39.5 | 32.5 | 306 | 55 | 167 | 1,415 | 4,308 | 22,486 | 35.6 | 23.0 |
| Frankfort | 12,741 | 2.13 | 55.5 | 33.5 | 22.0 | 44.5 | 39.4 | 1,546 | 81 | 292 | 950 | 3,427 | 19,114 | 39.8 | 30.0 |
| Georgetown | 14,701 | 2.51 | 67.4 | 48.2 | 19.3 | 32.6 | 24.7 | 1,260 | 41 | 117 | 999 | 2,846 | 23,991 | 38.8 | 21.9 |
| Henderson | 13,165 | 2.12 | 61.1 | 40.2 | 20.9 | 38.9 | 35.2 | 1,289 | 65 | 229 | 698 | 2,461 | 21,385 | 57.2 | 16.1 |
| Hopkinsville | 13,523 | 2.13 | 50.1 | 27.7 | 22.4 | 49.9 | 38.8 | 1,908 | 90 | 291 | 1,083 | 3,505 | 21,335 | 45.9 | 24.7 |
| Jeffersontown | 10,342 | 2.66 | 72.5 | 58.6 | 13.9 | 27.5 | 25.9 | 175 | 31 | 111 | 596 | 2,127 | 17,711 | 24.6 | 39.6 |
| Lexington-Fayette | 131,929 | 2.35 | 55.4 | 39.7 | 15.7 | 44.6 | 33.3 | 12,468 | 967 | 297 | 9,776 | 2,998 | 211,577 | 28.7 | 45.0 |
| Louisville/Jefferson County | 252,784 | 2.38 | 57.2 | 38.7 | 18.5 | 42.8 | 34.6 | 14,952 | NA | NA | NA | NA | 426,602 | 38.1 | 31.9 |
| Nicholasville | 10,310 | 2.92 | 70.6 | 46.9 | 23.7 | 29.4 | 25.7 | 298 | 47 | 151 | 758 | 2,430 | 18,029 | 36.9 | 28.4 |
| Owensboro | 25,351 | 2.28 | 58.6 | 35.0 | 23.6 | 41.4 | 37.3 | 2,217 | 135 | 225 | 2,432 | 4,046 | 39,729 | 44.5 | 25.4 |
| Paducah | 12,009 | 1.97 | 44.3 | 28.1 | 16.2 | 55.7 | 44.0 | 1,254 | 71 | 286 | 1,305 | 5,255 | 17,797 | 44.1 | 23.9 |
| Richmond | 13,682 | 2.28 | 48.1 | 30.4 | 17.7 | 51.9 | 35.8 | 4,913 | 75 | 206 | 804 | 2,205 | 18,299 | 40.5 | 32.1 |
| LOUISIANA | 1,741,076 | 2.60 | 64.0 | 42.3 | 21.6 | 36.0 | 30.2 | 128,857 | NA | NA | NA | NA | 3,140,201 | 48.0 | 25.0 |
| Alexandria | 17,256 | 2.55 | 63.1 | 32.2 | 31.0 | 36.9 | 30.7 | 2,118 | 732 | 1,570 | 4,180 | 8,964 | 30,678 | 50.2 | 18.4 |
| Baton Rouge | 82,097 | 2.58 | 51.5 | 28.0 | 23.5 | 48.5 | 37.9 | 8,428 | 2,066 | 936 | 11,673 | 5,290 | 137,948 | 42.4 | 32.1 |
| Bossier City | 26,677 | 2.49 | 65.7 | 38.3 | 27.4 | 34.3 | 25.8 | 1,758 | 546 | 791 | 3,225 | 4,671 | 45,665 | 46.5 | 20.1 |
| Central | 9,638 | 3.04 | 77.9 | 61.0 | 16.9 | 22.1 | 19.2 | 47 | NA | NA | NA | NA | 18,131 | 40.3 | 25.7 |
| Houma | 12,641 | 2.57 | 68.3 | 42.0 | 26.4 | 31.7 | 27.4 | 181 | 209 | 638 | 1,489 | 4,544 | 22,692 | 52.8 | 17.8 |
| Kenner | 26,151 | 2.52 | 64.5 | 44.5 | 20.0 | 35.5 | 30.7 | 464 | 149 | 224 | 1,955 | 2,933 | 47,435 | 44.0 | 28.4 |
| Lafayette | 52,267 | 2.33 | 56.7 | 38.1 | 18.6 | 43.3 | 35.4 | 4,637 | 664 | 524 | 5,454 | 4,305 | 84,260 | 36.4 | 38.9 |
| Lake Charles | 32,614 | 2.32 | 60.0 | 32.3 | 27.7 | 40.0 | 32.3 | 2,723 | 413 | 525 | 2,907 | 3,692 | 54,390 | 45.4 | 26.7 |
| Monroe | 18,719 | 2.33 | 49.5 | 23.4 | 26.0 | 50.5 | 45.9 | 3,613 | 843 | 1,766 | 3,141 | 6,579 | 29,057 | 51.7 | 22.8 |
| New Iberia | 10,260 | 2.74 | 75.8 | 41.6 | 34.2 | 24.2 | 23.3 | 376 | NA | NA | NA | NA | 18,515 | 57.9 | 21.8 |
| New Orleans | 151,753 | 2.50 | 45.7 | 26.5 | 19.2 | 54.3 | 46.8 | 10,788 | 4,516 | 1,145 | 20,879 | 5,293 | 279,405 | 35.5 | 39.7 |
| Shreveport | 73,114 | 2.49 | 58.5 | 33.9 | 24.6 | 41.5 | 35.7 | 5,029 | 1,462 | 780 | 9,189 | 4,899 | 124,813 | 47.3 | 25.2 |
| Slidell | 9,390 | 2.90 | 59.8 | 36.0 | 23.7 | 40.2 | 31.8 | 434 | 75 | 270 | 791 | 2,849 | 18,069 | 42.2 | 25.9 |
| MAINE | 573,618 | 2.28 | 59.8 | 46.8 | 13.0 | 40.2 | 31.2 | 37,206 | NA | NA | NA | NA | 991,152 | 38.2 | 33.2 |
| Bangor | 13,006 | 2.30 | 56.8 | 39.1 | 17.7 | 43.2 | 31.4 | 2,394 | 35 | 110 | 1,168 | 3,665 | 21,536 | 28.9 | 41.4 |
| Lewiston | 14,133 | 2.41 | 60.8 | 38.4 | 22.4 | 39.2 | 28.3 | 2,212 | 107 | 298 | 716 | 1,996 | 22,677 | 47.5 | 24.2 |
| Portland | 30,900 | 2.08 | 40.7 | 30.8 | 9.9 | 59.3 | 38.6 | 2,029 | 161 | 242 | 1,715 | 2,581 | 51,378 | 22.6 | 57.4 |
| South Portland | 11,293 | 2.19 | 50.8 | 41.7 | 9.1 | 49.2 | 34.8 | 837 | 48 | 187 | 563 | 2,192 | 19,419 | 26.2 | 45.5 |
| MARYLAND | 2,226,767 | 2.65 | 65.7 | 47.2 | 18.4 | 34.3 | 28.2 | 140,499 | NA | NA | NA | NA | 4,183,858 | 34.2 | 40.9 |
| Annapolis | 14,376 | 2.69 | 64.5 | 46.0 | 18.4 | 35.5 | 28.5 | 613 | 252 | 642 | 924 | 2,353 | 25,502 | 25.9 | 55.7 |
| Baltimore | 242,694 | 2.34 | 48.6 | 23.4 | 25.2 | 51.4 | 40.6 | 24,636 | 11,101 | 1,859 | 25,748 | 4,311 | 417,550 | 42.6 | 33.3 |
| Bowie | 21,342 | 2.73 | 66.3 | 46.7 | 19.7 | 33.7 | 28.9 | 402 | 59 | 100 | 896 | 1,516 | 41,590 | 27.1 | 46.3 |
| College Park | 7,757 | 2.68 | 41.7 | 30.6 | 11.1 | 58.3 | 42.3 | 11,345 | NA | NA | NA | NA | 11,558 | 30.3 | 42.1 |
| Frederick | 29,293 | 2.41 | 59.0 | 41.4 | 17.6 | 41.0 | 33.3 | 1,705 | 296 | 405 | 1,364 | 1,868 | 51,962 | 35.1 | 40.2 |
| Gaithersburg | 25,589 | 2.64 | 65.5 | 50.2 | 15.4 | 34.5 | 25.4 | 351 | NA | NA | NA | NA | 47,442 | 23.2 | 58.0 |
| Hagerstown | 16,935 | 2.31 | 51.4 | 32.5 | 19.0 | 48.6 | 42.0 | 1,013 | 213 | 529 | 960 | 2,385 | 27,526 | 52.2 | 17.3 |
| Laurel | 9,771 | 2.61 | 58.6 | 40.4 | 18.3 | 41.4 | 35.1 | 171 | 123 | 476 | 887 | 3,436 | 18,305 | 32.7 | 45.2 |
| Rockville | 26,360 | 2.55 | 65.0 | 47.7 | 17.3 | 35.0 | 26.2 | 765 | NA | NA | NA | NA | 48,718 | 21.8 | 62.5 |
| Salisbury | 11,825 | 2.67 | 50.2 | 26.7 | 23.5 | 49.8 | 34.7 | 1,375 | 294 | 887 | 1,422 | 4,292 | 16,752 | 43.3 | 35.9 |
| MASSACHUSETTS | 2,650,680 | 2.51 | 62.8 | 46.7 | 16.1 | 37.2 | 28.6 | 246,195 | NA | NA | NA | NA | 4,850,576 | 32.6 | 45.0 |
| Agawam Town | 10,329 | 2.73 | 65.7 | 48.8 | 16.9 | 34.3 | 29.0 | 443 | 82 | 285 | 376 | 1,308 | 19,140 | 37.0 | 34.5 |
| Attleboro | 17,355 | 2.57 | 63.6 | 48.4 | 15.3 | 36.4 | 27.9 | 589 | 123 | 274 | 522 | 1,161 | 33,031 | 39.7 | 29.1 |
| Barnstable Town | 19,443 | 2.26 | 64.2 | 50.1 | 14.1 | 35.8 | 25.4 | 492 | 178 | 404 | 425 | 965 | 34,881 | 29.1 | 42.7 |

1. Data for serious crimes have not been adjusted for underreporting. This may affect comparability between geographic areas and over time.   2. Per 100,000 population estimated by the FBI.   3. Persons 25 years old and over.

| | Money income, 2019 | | | | | Median earnings, 2019 | | | Housing units, 2019 | | | | |
|---|---|---|---|---|---|---|---|---|---|---|---|---|---|
| City | Households — Median income | Percent with income less than $20,000 | Percent with income of $200,000 or more | Median family income | Median non-family income | All persons | Men | Women | Total | Occupied | Percent owner occupied | Median value[1] (dollars) | Median gross rent (dollars) |
| | 42 | 43 | 44 | 45 | 46 | 47 | 48 | 49 | 50 | 51 | 52 | 53 | 54 |
| KANSAS | 62,087 | 13.6 | 5.9 | 79,006 | 36,347 | 35,266 | 41,524 | 27,519 | 1,288,430 | 1,138,329 | 66.5 | 163,170 | 862 |
| Dodge City | 46,424 | 20.2 | 0.0 | 54,602 | 31,168 | 31,344 | 35,420 | 24,904 | 9,375 | 8,840 | 62.8 | 122,041 | 819 |
| Garden City | 52,072 | 19.6 | 1.6 | 70,573 | 28,886 | 31,820 | 40,147 | 22,210 | 9,839 | 9,106 | 62.8 | 157,043 | 780 |
| Hutchinson | 49,960 | 19.2 | 3.1 | 56,001 | 35,705 | 26,402 | 35,664 | 20,766 | 18,028 | 15,604 | 69.3 | 104,611 | 706 |
| Kansas City | 45,391 | 20.6 | 1.8 | 53,624 | 30,238 | 30,127 | 32,372 | 25,775 | 63,751 | 56,873 | 57.0 | 99,675 | 930 |
| Lawrence | 57,829 | 15.0 | 7.5 | 88,856 | 36,653 | 25,661 | 26,846 | 24,321 | 42,413 | 39,442 | 41.4 | 205,416 | 965 |
| Leavenworth | 58,244 | 20.9 | 1.8 | 80,000 | 24,692 | 27,008 | 36,592 | 20,900 | 14,178 | 12,177 | 45.4 | 145,692 | 790 |
| Leawood | 185,694 | 2.3 | 48.4 | 227,878 | 91,281 | 86,159 | 133,321 | 49,359 | 14,129 | 12,964 | 90.0 | 469,889 | 1,284 |
| Lenexa | 90,414 | 4.3 | 10.2 | 106,785 | 60,870 | 47,881 | 52,253 | 40,086 | 24,519 | 22,329 | 59.1 | 283,026 | 1,209 |
| Manhattan | 48,123 | 22.9 | 1.9 | 77,604 | 31,338 | 18,038 | 17,441 | 18,266 | 25,164 | 20,061 | 37.0 | 217,390 | 849 |
| Olathe | 94,292 | 7.1 | 10.6 | 109,923 | 52,484 | 48,461 | 55,310 | 36,343 | 51,710 | 49,971 | 70.2 | 273,359 | 1,080 |
| Overland Park | 91,518 | 4.6 | 15.0 | 116,721 | 60,393 | 50,853 | 60,049 | 42,007 | 84,858 | 79,436 | 65.0 | 306,333 | 1,196 |
| Salina | 54,307 | 17.9 | 3.7 | 79,633 | 33,057 | 32,903 | 40,730 | 28,022 | 21,226 | 18,711 | 58.4 | 149,120 | 743 |
| Shawnee | 84,909 | 8.7 | 12.4 | 100,082 | 41,885 | 41,057 | 50,693 | 32,173 | 25,834 | 24,678 | 74.0 | 261,705 | 1,008 |
| Topeka | 50,761 | 15.6 | 3.1 | 67,011 | 34,091 | 33,704 | 40,532 | 27,490 | 60,013 | 54,066 | 59.4 | 105,067 | 824 |
| Wichita | 55,056 | 16.5 | 4.5 | 73,745 | 34,169 | 32,360 | 40,120 | 27,251 | 172,738 | 153,454 | 59.2 | 146,684 | 833 |
| KENTUCKY | 52,295 | 18.6 | 4.1 | 66,183 | 30,384 | 32,189 | 38,545 | 27,281 | 2,006,335 | 1,748,732 | 67.0 | 151,696 | 773 |
| Bowling Green | 42,164 | 23.5 | 3.1 | 56,033 | 28,496 | 21,666 | 26,820 | 17,089 | 30,968 | 27,178 | 35.5 | 183,557 | 765 |
| Covington | 44,883 | 25.7 | 3.7 | 56,024 | 37,296 | 35,964 | 39,423 | 31,557 | 20,252 | 17,473 | 46.3 | 126,885 | 690 |
| Elizabethtown | 56,010 | 12.7 | 5.3 | 72,264 | 34,327 | 40,228 | 43,336 | 37,049 | 15,590 | 13,512 | 51.7 | 201,921 | 897 |
| Florence | 61,368 | 13.5 | 1.5 | 90,163 | 35,855 | 32,500 | 38,513 | 27,903 | 14,052 | 13,480 | 63.8 | 159,579 | 941 |
| Frankfort | 51,953 | 17.4 | 3.3 | 74,758 | 31,518 | 31,537 | 37,859 | 24,989 | 13,535 | 12,741 | 50.2 | 132,686 | 820 |
| Georgetown | 66,125 | 10.9 | 3.8 | 68,402 | 50,216 | 35,456 | 51,036 | 25,479 | 15,549 | 14,701 | 61.3 | 184,527 | 880 |
| Henderson | 42,987 | 26.9 | 1.1 | 60,977 | 24,137 | 30,516 | 34,717 | 20,498 | 14,083 | 13,165 | 48.3 | 105,428 | 719 |
| Hopkinsville | 46,402 | 18.1 | 4.4 | 44,630 | 46,164 | 30,545 | 40,561 | 24,063 | 15,167 | 13,523 | 51.1 | 110,521 | 734 |
| Jeffersontown | 64,200 | 7.5 | 2.6 | 83,351 | 49,888 | 38,807 | 41,037 | 35,724 | 10,650 | 10,342 | 61.7 | 190,715 | 1,040 |
| Lexington-Fayette | 58,356 | 14.2 | 6.4 | 82,055 | 40,100 | 31,841 | 35,757 | 30,219 | 144,089 | 131,929 | 53.3 | 211,430 | 932 |
| Louisville/Jefferson County | 54,853 | 16.4 | 5.0 | 71,279 | 36,826 | 35,224 | 40,830 | 30,467 | 282,781 | 252,784 | 59.6 | 172,095 | 878 |
| Nicholasville | 54,903 | 5.4 | 0.0 | 63,272 | 38,943 | 30,669 | 37,682 | 25,933 | 11,080 | 10,310 | 63.9 | 158,586 | 761 |
| Owensboro | 43,367 | 21.5 | 2.9 | 58,209 | 28,718 | 29,946 | 32,662 | 26,399 | 27,310 | 25,351 | 55.0 | 121,641 | 812 |
| Paducah | 36,044 | 25.7 | 6.0 | 53,143 | 24,338 | 31,732 | 34,827 | 26,725 | 14,163 | 12,009 | 47.1 | 87,981 | 649 |
| Richmond | 33,014 | 33.1 | 0.4 | 61,111 | 20,867 | 17,154 | 20,431 | 15,055 | 15,344 | 13,682 | 35.9 | 162,410 | 685 |
| LOUISIANA | 51,073 | 21.4 | 5.0 | 65,105 | 28,907 | 32,102 | 41,101 | 26,538 | 2,089,824 | 1,741,076 | 66.5 | 172,050 | 866 |
| Alexandria | 45,984 | 27.4 | 3.7 | 52,127 | 23,880 | 25,916 | 27,324 | 22,072 | 20,387 | 17,256 | 49.0 | 153,220 | 847 |
| Baton Rouge | 45,819 | 24.5 | 5.0 | 65,340 | 28,353 | 25,503 | 30,325 | 21,898 | 104,145 | 82,097 | 47.7 | 187,524 | 874 |
| Bossier City | 47,339 | 20.9 | 2.6 | 57,180 | 33,672 | 30,752 | 34,020 | 27,100 | 31,386 | 26,677 | 58.1 | 161,132 | 938 |
| Central | 86,535 | 4.5 | 10.7 | 96,234 | 55,679 | 50,036 | 68,667 | 32,077 | 11,060 | 9,638 | 85.0 | 250,258 | 909 |
| Houma | 41,539 | 19.2 | 5.2 | 67,548 | 25,065 | 36,095 | 61,192 | 26,383 | 14,071 | 12,641 | 63.7 | 184,148 | 806 |
| Kenner | 58,283 | 10.9 | 6.2 | 72,158 | 35,960 | 32,413 | 41,614 | 26,015 | 28,369 | 26,151 | 58.2 | 195,684 | 890 |
| Lafayette | 51,477 | 21.1 | 8.6 | 72,145 | 31,845 | 31,815 | 41,788 | 27,187 | 59,431 | 52,267 | 58.3 | 202,580 | 853 |
| Lake Charles | 37,894 | 26.0 | 5.8 | 51,212 | 24,640 | 28,506 | 36,907 | 22,411 | 39,289 | 32,614 | 53.1 | 155,551 | 830 |
| Monroe | 30,313 | 42.5 | 4.7 | 36,143 | 16,856 | 23,733 | 28,463 | 20,607 | 21,532 | 18,719 | 39.8 | 129,432 | 595 |
| New Iberia | 42,467 | 28.5 | 2.1 | 42,133 | 18,222 | 35,863 | 51,591 | 22,272 | 13,004 | 10,260 | 57.3 | 128,827 | 800 |
| New Orleans | 45,615 | 28.3 | 7.7 | 67,322 | 30,271 | 32,032 | 37,244 | 29,877 | 192,236 | 151,753 | 49.6 | 242,911 | 1,010 |
| Shreveport | 45,013 | 27.5 | 4.7 | 57,090 | 25,255 | 30,207 | 35,526 | 24,639 | 91,088 | 73,114 | 57.5 | 151,987 | 830 |
| Slidell | 54,324 | 14.5 | 3.7 | 71,027 | 36,969 | 31,540 | 41,510 | 26,210 | 10,668 | 9,390 | 71.9 | 168,218 | 1,017 |
| MAINE | 58,924 | 14.3 | 5.2 | 76,316 | 35,624 | 35,231 | 40,678 | 30,423 | 750,964 | 573,618 | 72.2 | 200,487 | 870 |
| Bangor | 50,213 | 14.8 | 2.9 | 70,631 | 35,214 | 33,831 | 41,132 | 31,229 | 14,692 | 13,006 | 52.9 | 179,755 | 801 |
| Lewiston | 53,034 | 11.7 | 0.3 | 63,988 | 33,578 | 31,460 | 39,287 | 28,996 | 15,325 | 14,133 | 44.1 | 157,624 | 815 |
| Portland | 61,779 | 15.6 | 8.2 | 88,946 | 48,106 | 38,484 | 40,124 | 37,010 | 34,035 | 30,900 | 46.4 | 334,245 | 1,245 |
| South Portland | 70,492 | 9.9 | 7.9 | 96,125 | 52,990 | 41,887 | 46,914 | 32,291 | 12,044 | 11,293 | 65.2 | 288,439 | 1,278 |
| MARYLAND | 86,738 | 10.0 | 13.6 | 105,679 | 51,990 | 45,891 | 51,361 | 40,552 | 2,470,307 | 2,226,767 | 66.8 | 332,473 | 1,401 |
| Annapolis | 90,375 | 7.9 | 19.9 | 92,057 | 60,086 | 50,426 | 52,782 | 44,657 | 15,466 | 14,376 | 53.7 | 432,906 | 1,804 |
| Baltimore | 50,177 | 21.8 | 6.8 | 65,518 | 37,905 | 37,654 | 40,909 | 34,853 | 293,877 | 242,694 | 47.6 | 179,135 | 1,090 |
| Bowie | 107,592 | 5.8 | 17.7 | 125,843 | 68,502 | 55,528 | 54,080 | 57,185 | 21,902 | 21,342 | 81.1 | 356,646 | 1,694 |
| College Park | 51,528 | 24.4 | 3.8 | 91,278 | 35,163 | 6,371 | 6,842 | 5,834 | 8,715 | 7,757 | 51.2 | 355,932 | 989 |
| Frederick | 86,830 | 7.5 | 9.0 | 105,127 | 47,286 | 40,934 | 43,526 | 35,934 | 31,869 | 29,293 | 58.1 | 274,590 | 1,397 |
| Gaithersburg | 104,567 | 7.5 | 17.5 | 106,377 | 84,921 | 43,650 | 44,015 | 42,947 | 27,451 | 25,589 | 52.5 | 397,274 | 1,788 |
| Hagerstown | 38,752 | 28.7 | 0.9 | 49,697 | 20,561 | 26,420 | 30,684 | 23,150 | 18,878 | 16,935 | 45.1 | 165,503 | 780 |
| Laurel | 85,469 | 4.6 | 8.4 | 117,064 | 62,425 | 45,031 | 46,451 | 42,191 | 10,727 | 9,771 | 45.0 | 282,406 | 1,644 |
| Rockville | 100,677 | 7.1 | 17.2 | 120,193 | 78,425 | 50,503 | 61,899 | 37,499 | 28,028 | 26,360 | 54.5 | 583,138 | 1,883 |
| Salisbury | 50,703 | 14.1 | 3.1 | 66,291 | 33,915 | 25,374 | 28,465 | 21,642 | 13,126 | 11,825 | 28.7 | 195,833 | 1,125 |
| MASSACHUSETTS | 85,843 | 12.2 | 14.6 | 108,348 | 50,338 | 45,737 | 52,201 | 38,781 | 2,928,818 | 2,650,680 | 62.2 | 418,559 | 1,360 |
| Agawam Town | 71,401 | 9.3 | 3.7 | 81,367 | 49,574 | 49,666 | 48,750 | 50,078 | 10,619 | 10,329 | 69.9 | 247,061 | 1,193 |
| Attleboro | 70,850 | 14.1 | 6.4 | 94,408 | 26,859 | 45,533 | 52,424 | 36,942 | 18,921 | 17,355 | 67.3 | 326,834 | 1,170 |
| Barnstable Town | 91,427 | 7.6 | 12.4 | 104,175 | 69,562 | 41,115 | 46,276 | 37,064 | 27,379 | 19,443 | 77.8 | 385,053 | 1,161 |

1. Based on population estimated by the American Community Survey.

# Table D. Cities — Commuting, Computer Access, Migration, Labor Force, and Employment

| City | Commuting[1], 2019 Percent — Drove alone | With commutes of 30 minutes or more | Computer access[2], 2019 Percent — With a computer in the house | With Internet access | Migration, 2019 — Percent who lived in the same house one year ago | Percent who lived in another state or county one year ago | Civilian labor force, 2020 — Total | Percent change 2019-2020 | Unemployment — Total | Rate[3] | Civilian Employment[4], 2019 Population age 16 and older — Number | Percent in labor force | Population age 16 to 64 — Number | Percent who worked full-year full-time |
|---|---|---|---|---|---|---|---|---|---|---|---|---|---|---|
| | 55 | 56 | 57 | 58 | 59 | 60 | 61 | 62 | 63 | 64 | 65 | 66 | 67 | 68 |
| KANSAS | 82.7 | 22.6 | 92.9 | 84.7 | 84.3 | 6.9 | 1,497,003 | 0.7 | 88,008 | 6 | 2,292,476 | 66.0 | 1,814,480 | 56.4 |
| Dodge City | NA | 7.0 | 92.0 | 89.4 | 83.0 | 4.3 | 13,707 | 0.9 | 480 | 4 | 18,859 | 71.8 | 16,299 | 61.6 |
| Garden City | 81.3 | 10.4 | 87.1 | 74.3 | 83.7 | 5.3 | 14,998 | 0.0 | 533 | 4 | 17,714 | 71.7 | 14,731 | 57.3 |
| Hutchinson | 85.9 | 11.4 | 92.0 | 82.7 | 84.0 | 7.8 | 18,864 | 0.3 | 1,153 | 6 | 30,761 | 57.5 | 22,441 | 53.0 |
| Kansas City | 82.6 | 24.1 | 90.6 | 81.9 | 83.6 | 8.3 | 71,695 | 1.4 | 5,610 | 8 | 114,921 | 64.8 | 95,972 | 53.0 |
| Lawrence | 76.8 | 25.1 | 99.2 | 89.5 | 72.2 | 13.8 | 51,937 | -1.4 | 3,202 | 6 | 81,956 | 74.5 | 69,971 | 48.9 |
| Leavenworth | 82.2 | 28.1 | 94.6 | 84.7 | 72.1 | 15.5 | 13,944 | 0.3 | 917 | 7 | 27,930 | 49.8 | 23,395 | 46.6 |
| Leawood | 85.1 | 31.4 | NA | NA | 90.6 | 6.4 | 17,456 | -0.5 | 749 | 4 | 28,543 | 64.3 | 20,849 | 56.9 |
| Lenexa | 84.9 | 18.2 | 98.1 | 95.5 | 82.5 | 8.2 | 32,651 | 1.0 | 1,758 | 5 | 45,507 | 72.1 | 36,920 | 63.3 |
| Manhattan | 70.7 | 13.3 | 95.2 | 81.3 | 69.1 | 18.8 | 27,869 | -4.2 | 1,325 | 5 | 47,898 | 66.9 | 42,330 | 37.5 |
| Olathe | 83.6 | 29.4 | 97.2 | 93.1 | 87.7 | 4.0 | 80,271 | 0.9 | 4,025 | 5 | 105,770 | 74.8 | 88,967 | 63.4 |
| Overland Park | 84.0 | 20.3 | 97.5 | 96.0 | 82.9 | 7.8 | 111,746 | 1.7 | 5,771 | 5 | 157,429 | 72.6 | 128,348 | 60.4 |
| Salina | 85.6 | 9.3 | 88.5 | 79.6 | 83.2 | 8.4 | 25,105 | -2.1 | 1,432 | 6 | 37,026 | 63.1 | 27,719 | 56.6 |
| Shawnee | 85.7 | 26.2 | 93.9 | 91.1 | 85.6 | 6.0 | 36,676 | 0.3 | 2,018 | 6 | 52,219 | 71.0 | 40,597 | 62.8 |
| Topeka | 83.7 | 13.9 | 88.2 | 70.7 | 86.1 | 6.3 | 63,358 | 1.3 | 4,098 | 7 | 100,906 | 65.6 | 77,594 | 59.3 |
| Wichita | 85.2 | 14.1 | 93.3 | 85.1 | 81.3 | 4.4 | 194,215 | 2.5 | 17,819 | 9 | 302,645 | 66.7 | 246,148 | 55.0 |
| KENTUCKY | 82.0 | 31.1 | 90.2 | 82.5 | 85.8 | 6.4 | 2,019,887 | -2.5 | 134,242 | 7 | 3,581,569 | 58.6 | 2,827,010 | 49.6 |
| Bowling Green | 79.8 | 20.3 | 93.8 | 90.6 | 70.0 | 14.1 | 32,896 | -1.3 | 2,274 | 7 | 57,761 | 70.1 | 51,105 | 43.6 |
| Covington | 72.6 | 26.5 | 88.2 | 79.6 | 77.3 | 11.1 | 18,691 | -2.4 | 1,328 | 7 | 31,062 | 64.0 | 25,493 | 55.0 |
| Elizabethtown | 85.6 | 21.0 | 89.9 | 81.9 | 86.9 | 4.1 | 13,651 | -3.0 | 1,024 | 8 | 25,575 | 60.4 | 20,328 | 56.5 |
| Florence | 81.9 | 29.6 | 93.1 | 84.6 | 80.0 | 14.6 | 16,661 | -1.5 | 1,006 | 6 | 26,188 | 69.5 | 20,057 | 64.9 |
| Frankfort | 77.9 | 24.1 | 93.1 | 79.0 | 77.5 | 6.6 | 14,032 | -2.0 | 1,002 | 7 | 23,307 | 57.1 | 18,297 | 51.1 |
| Georgetown | 85.6 | 26.0 | 93.9 | 87.3 | 74.6 | 13.2 | 18,247 | 1.1 | 1,306 | 7 | 28,675 | 70.7 | 24,414 | 61.7 |
| Henderson | 83.9 | 23.2 | 90.4 | 79.5 | 82.2 | 3.4 | 12,611 | -4.9 | 831 | 9 | 23,998 | 62.5 | 19,307 | 53.6 |
| Hopkinsville | NA | 6.0 | 90.8 | 78.4 | 77.7 | 9.3 | 12,134 | -0.4 | 1,139 | 9 | 25,528 | 57.0 | 19,231 | 50.1 |
| Jeffersontown | 86.5 | 29.3 | 97.2 | 93.5 | NA | NA | 15,752 | -4.4 | 859 | 6 | 20,450 | 69.4 | 16,264 | 52.3 |
| Lexington-Fayette | 79.3 | 21.4 | 94.8 | 88.6 | 77.2 | 9.0 | 172,284 | -2.0 | 9,879 | 7 | 262,965 | 66.9 | 218,739 | 53.0 |
| Louisville/Jefferson County | 79.3 | 27.5 | 92.5 | 86.5 | 84.4 | 4.4 | 390,487 | -3.0 | 26,717 | 7 | 497,427 | 64.3 | 399,515 | 51.8 |
| Nicholasville | 80.7 | 41.7 | 94.8 | 90.6 | 84.4 | 9.2 | 15,126 | -1.4 | 974 | 6 | 21,605 | 71.3 | 18,304 | 54.9 |
| Owensboro | 81.2 | 20.5 | 92.2 | 85.8 | 77.5 | 7.9 | 26,064 | -3.3 | 1,750 | 9 | 46,129 | 58.1 | 36,015 | 50.9 |
| Paducah | 78.6 | 13.1 | 90.8 | 83.3 | 91.1 | 3.9 | 10,218 | -1.7 | 889 | 9 | 20,292 | 60.6 | 15,889 | 55.8 |
| Richmond | 76.6 | 32.5 | 91.3 | 85.0 | 69.5 | 14.6 | 18,095 | -3.4 | 1,190 | 7 | 30,332 | 62.7 | 26,416 | 36.2 |
| LOUISIANA | 82.1 | 35.1 | 89.4 | 79.4 | 87.5 | 5.2 | 2,076,643 | -0.9 | 171,405 | 8 | 3,684,053 | 58.6 | 2,941,859 | 48.9 |
| Alexandria | 70.0 | 17.9 | 84.0 | 78.9 | 78.4 | 7.6 | 18,784 | 0.5 | 1,386 | 7 | 37,012 | 56.1 | 29,678 | 47.1 |
| Baton Rouge | 79.3 | 24.0 | 90.8 | 84.8 | 78.3 | 7.3 | 111,951 | -0.1 | 10,159 | 9 | 182,793 | 63.1 | 150,018 | 43.9 |
| Bossier City | 88.6 | 11.8 | 95.0 | 69.0 | 90.8 | 4.0 | 29,864 | -0.9 | 2,082 | 7 | 52,418 | 50.7 | 41,001 | 56.1 |
| Central | 91.3 | 47.0 | 94.3 | 87.2 | NA | NA | 15,449 | -0.9 | 783 | 5 | 21,703 | 61.9 | 17,054 | 56.8 |
| Houma | 80.6 | 15.2 | 84.9 | 76.2 | 86.7 | 5.0 | 13,964 | 0.7 | 1,198 | 9 | 25,453 | 62.1 | 19,520 | 53.9 |
| Kenner | 80.7 | 35.9 | 91.6 | 86.4 | 84.8 | 10.2 | 32,761 | -0.9 | 3,080 | 9 | 54,685 | 67.3 | 41,807 | 64.2 |
| Lafayette | 85.2 | 19.7 | 91.4 | 87.8 | 84.1 | 7.1 | 59,430 | -0.3 | 4,434 | 8 | 102,881 | 64.4 | 80,363 | 51.3 |
| Lake Charles | 80.5 | 21.5 | 88.5 | 78.0 | 82.5 | 6.4 | 38,699 | -5.6 | 4,194 | 11 | 64,038 | 54.3 | 50,264 | 44.9 |
| Monroe | 80.2 | 14.6 | 77.2 | 69.2 | 84.7 | 3.7 | 19,723 | -0.2 | 1,776 | 9 | 35,847 | 49.3 | 28,415 | 35.5 |
| New Iberia | 85.8 | 30.9 | 89.3 | 84.1 | 84.5 | 4.5 | 10,612 | -0.3 | 1,160 | 11 | 21,395 | 54.3 | 17,482 | 49.1 |
| New Orleans | 67.2 | 28.3 | 90.2 | 77.8 | 86.2 | 5.9 | 181,868 | 1.7 | 22,134 | 12 | 320,571 | 63.7 | 259,987 | 48.0 |
| Shreveport | 83.0 | 17.6 | 87.1 | 70.9 | 89.3 | 3.4 | 80,854 | -0.1 | 7,684 | 10 | 147,319 | 54.6 | 116,333 | 46.9 |
| Slidell | 81.1 | 38.6 | 94.5 | 88.8 | 84.9 | 5.6 | 12,132 | -2.0 | 1,045 | 9 | 20,688 | 62.6 | 16,849 | 55.8 |
| MAINE | 78.5 | 32.9 | 92.1 | 84.7 | 87.3 | 5.8 | 676,547 | -2.3 | 36,788 | 5 | 1,128,168 | 62.8 | 842,190 | 52.9 |
| Bangor | 73.1 | 14.3 | 96.1 | 88.6 | 78.8 | 8.3 | 16,523 | -2.1 | 922 | 6 | 26,335 | 60.4 | 20,836 | 51.9 |
| Lewiston | 75.3 | 22.5 | 93.0 | 82.1 | 83.9 | 6.0 | 16,867 | -0.8 | 990 | 6 | 27,858 | 63.6 | 21,476 | 53.8 |
| Portland | 68.6 | 17.8 | 94.4 | 85.4 | 77.9 | 10.7 | 37,642 | -2.3 | 2,368 | 6 | 59,275 | 68.6 | 47,353 | 56.0 |
| South Portland | 78.5 | 17.4 | 97.1 | 88.6 | 89.2 | 3.8 | 14,353 | -2.9 | 822 | 6 | 21,448 | 71.9 | 17,234 | 63.0 |
| MARYLAND | 73.9 | 53.4 | 94.3 | 88.4 | 87.4 | 6.2 | 3,172,796 | -2.7 | 214,509 | 7 | 4,862,424 | 66.8 | 3,902,537 | 56.8 |
| Annapolis | 74.2 | 46.7 | 91.1 | 86.2 | 86.4 | 4.3 | 22,218 | -2.1 | 1,407 | 6 | 29,121 | 75.0 | 23,054 | 67.6 |
| Baltimore | 61.5 | 48.8 | 88.6 | 77.8 | 84.7 | 6.2 | 281,187 | -3.3 | 24,705 | 9 | 485,560 | 62.3 | 399,295 | 49.4 |
| Bowie | 77.2 | 62.1 | 95.2 | 93.0 | 89.5 | 3.8 | 34,283 | -2.5 | 2,490 | 7 | 46,025 | 71.7 | 36,818 | 65.8 |
| College Park | 47.7 | 32.2 | 97.2 | 85.1 | 64.8 | 29.2 | 14,523 | -4.4 | 1,007 | 7 | 28,476 | 49.4 | 25,678 | 21.8 |
| Frederick | 76.0 | 46.4 | 93.7 | 89.8 | 79.4 | 7.8 | 37,667 | -2.9 | 2,513 | 7 | 60,659 | 72.7 | 50,255 | 58.8 |
| Gaithersburg | 64.5 | 52.2 | 98.9 | 92.8 | 81.5 | 8.6 | 36,029 | -2.7 | 2,532 | 7 | 54,567 | 73.1 | 45,079 | 60.6 |
| Hagerstown | 75.7 | 27.7 | 91.4 | 81.8 | 84.2 | 6.2 | 18,617 | -2.2 | 1,672 | 9 | 31,425 | 58.1 | 24,497 | 43.7 |
| Laurel | 69.1 | 59.6 | 98.0 | 94.9 | 78.1 | 11.9 | 15,863 | -0.5 | 1,575 | 10 | 21,186 | 73.7 | 17,416 | 64.5 |
| Rockville | 55.6 | 52.2 | 98.5 | 91.9 | 80.7 | 11.0 | 36,823 | -3.7 | 2,083 | 6 | 57,697 | 67.4 | 44,884 | 60.4 |
| Salisbury | 74.5 | 28.2 | 93.8 | 85.0 | 77.5 | 9.3 | 14,832 | -3.9 | 1,289 | 9 | 26,726 | 62.5 | 23,314 | 40.9 |
| MASSACHUSETTS | 69.6 | 47.8 | 93.5 | 88.3 | 87.5 | 6.0 | 3,658,321 | -4.2 | 324,195 | 9 | 5,707,254 | 67.5 | 4,534,961 | 54.0 |
| Agawam Town | 77.8 | 29.1 | 94.9 | 88.4 | 93.9 | 2.3 | 15,626 | -5.4 | 1,306 | 8 | 22,155 | 71.6 | 17,726 | 57.5 |
| Attleboro | 84.3 | 45.3 | 89.8 | 84.5 | 93.7 | 2.6 | 24,851 | -3.6 | 2,361 | 10 | 36,721 | 64.2 | 28,438 | 54.5 |
| Barnstable Town | 80.7 | 23.7 | 97.4 | 95.8 | 91.3 | 3.4 | 23,212 | -5.9 | 2,380 | 10 | 39,562 | 70.6 | 29,157 | 61.5 |

1. Employed persons.   2. Households.   3. Percent of civilian labor force.   4. Persons 16 years old and over.

| City | Value of residential construction authorized by building permits, 2020 | | | Wholesale trade[1], 2017 | | | | Retail trade[2], 2017 | | | |
|---|---|---|---|---|---|---|---|---|---|---|---|
| | New construction ($1,000) | Number of housing units | Percent single family | Number of establishments | Number of employees | Sales (mil dol) | Annual payroll (mil dol) | Number of establishments | Number of employees | Sales (mil dol) | Annual payroll (mil dol) |
| | 69 | 70 | 71 | 72 | 73 | 74 | 75 | 76 | 77 | 78 | 79 |
| KANSAS | 0 | 0 | 0 | 3,755 | 52,611 | 61,889 | 3,160 | 10,095 | 149,845 | 39,338 | 3,785 |
| Dodge City | 10,776 | 66 | 91 | 44 | 547 | 232 | 27 | 102 | 1,651 | 437 | 39 |
| Garden City | 15,823 | 74 | 100 | 26 | 128 | 143 | 7 | 149 | 2,653 | 638 | 60 |
| Hutchinson | 3,572 | 14 | 86 | 54 | 805 | 1,022 | 34 | 182 | 2,715 | 677 | 63 |
| Kansas City | 36,380 | 201 | 91 | 194 | 5,554 | 7,273 | 392 | 399 | 7,034 | 1,958 | 194 |
| Lawrence | 72,882 | 292 | 66 | 57 | 375 | 166 | 20 | 333 | 6,119 | 1,452 | 141 |
| Leavenworth | 8,471 | 41 | 100 | D | D | D | 1 | 97 | 1,615 | 479 | 43 |
| Leawood | 38,561 | 46 | 100 | 35 | 281 | 431 | 19 | 148 | 2,427 | 432 | 60 |
| Lenexa | 144,051 | 602 | 44 | 312 | 5,800 | 3,663 | 390 | 202 | 4,112 | 1,168 | 125 |
| Manhattan | 41,043 | 206 | 21 | 35 | 610 | 193 | 25 | 238 | 4,998 | 1,044 | 119 |
| Olathe | 305,577 | 1,011 | 56 | 145 | 3,415 | 3,652 | 210 | 331 | 7,589 | 2,551 | 207 |
| Overland Park | 138,373 | 472 | 80 | 228 | 4,117 | 15,131 | 395 | 712 | 13,744 | 3,195 | 367 |
| Salina | 8,420 | 43 | 86 | 73 | 1,047 | 530 | 56 | 229 | 3,997 | 1,090 | 93 |
| Shawnee | 77,852 | 254 | 74 | 63 | 939 | 995 | 62 | 193 | 3,527 | 855 | 88 |
| Topeka | 38,060 | 264 | 42 | 124 | 1,441 | 1,079 | 81 | 525 | 8,881 | 2,180 | 210 |
| Wichita | 194,884 | 836 | 79 | 510 | 7,717 | 7,207 | 477 | 1,467 | 26,178 | 7,133 | 679 |
| KENTUCKY | 132,047 | 817 | 74 | 3,646 | 55,784 | 80,129 | 3,172 | 15,021 | 225,127 | 64,294 | 5,630 |
| Bowling Green | 58,709 | 425 | 69 | 125 | 1,516 | 2,181 | 81 | 460 | 7,420 | 1,863 | 181 |
| Covington | 0 | 0 | 0 | 30 | 1,023 | 313 | 39 | 121 | 1,526 | 403 | 38 |
| Elizabethtown | 33,758 | 310 | 27 | 40 | 489 | 344 | 23 | 255 | 4,735 | 1,235 | 117 |
| Florence | NA | NA | NA | D | D | D | 10 | 301 | 6,867 | 1,901 | 171 |
| Frankfort | 3,217 | 22 | 100 | 43 | 440 | 498 | 26 | 143 | 1,960 | 469 | 44 |
| Georgetown | NA | NA | NA | 14 | 93 | 32 | 4 | 97 | 1,653 | 476 | 40 |
| Henderson | 3,831 | 37 | 41 | 31 | 418 | 230 | 23 | 134 | 1,866 | 631 | 48 |
| Hopkinsville | 6,560 | 55 | 93 | 44 | 620 | 657 | 31 | 168 | 2,320 | 667 | 61 |
| Jeffersontown | 11,815 | 41 | 100 | 152 | 3,127 | 2,997 | 213 | 157 | 2,869 | 1,117 | 95 |
| Lexington-Fayette | 224,453 | 1,437 | 49 | 328 | 4,645 | 3,403 | 259 | 1,211 | 23,102 | 6,342 | 598 |
| Louisville/Jefferson County | 247,806 | 1,223 | 98 | 682 | 10,313 | 6,561 | 619 | 1,987 | 32,443 | 9,078 | 851 |
| Nicholasville | 21,851 | 144 | 83 | 31 | 878 | 1,444 | 42 | 111 | 2,003 | 738 | 58 |
| Owensboro | 31,665 | 338 | 89 | 60 | 853 | 475 | 43 | 329 | 4,867 | 1,295 | 116 |
| Paducah | 4,528 | 24 | 75 | 72 | 1,161 | 4,239 | 53 | 307 | 5,255 | 1,490 | 137 |
| Richmond | 49,450 | 328 | 79 | 23 | 180 | 142 | 6 | 196 | 3,220 | 919 | 79 |
| LOUISIANA | 3,427,774 | 17,283 | 89 | 4,673 | 63,503 | 59,524 | 3,494 | 16,564 | 233,385 | 65,001 | 5,988 |
| Alexandria | 21,860 | 83 | 100 | 76 | 971 | 1,028 | 48 | 340 | 5,180 | 1,479 | 137 |
| Baton Rouge | 88,664 | 306 | 100 | 305 | 3,987 | 2,625 | 233 | 1,211 | 18,219 | 4,602 | 474 |
| Bossier City | 52,916 | 336 | 100 | 74 | 1,354 | 863 | 68 | 368 | 6,775 | 1,860 | 164 |
| Central | 21,734 | 137 | 100 | 9 | 37 | 258 | 5 | 44 | 910 | 218 | 22 |
| Houma | NA | NA | NA | 63 | 552 | 255 | 27 | 157 | 2,072 | 429 | 51 |
| Kenner | 19,446 | 62 | 100 | 94 | 974 | 355 | 53 | 274 | 4,419 | 1,359 | 120 |
| Lafayette | NA | NA | NA | 263 | 3,325 | 1,660 | 176 | 822 | 13,685 | 3,595 | 356 |
| Lake Charles | 69,701 | 528 | 75 | 79 | 978 | 629 | 46 | 469 | 7,026 | 2,120 | 178 |
| Monroe | 29,893 | 184 | 48 | 74 | 1,270 | 1,258 | 59 | 361 | 5,912 | 1,485 | 144 |
| New Iberia | 1,231 | 11 | 27 | 49 | 795 | 562 | 43 | 193 | 2,601 | 745 | 68 |
| New Orleans | 258,232 | 1,360 | 46 | 255 | 3,124 | 2,506 | 187 | 1,327 | 14,795 | 3,499 | 387 |
| Shreveport | 61,215 | 244 | 100 | 283 | 3,848 | 2,686 | 203 | 799 | 12,031 | 3,386 | 320 |
| Slidell | 1,800 | 8 | 100 | D | D | D | 7 | 277 | 4,972 | 1,273 | 119 |
| MAINE | 3,261 | 17 | 100 | 1,309 | 15,527 | 13,952 | 803 | 6,250 | 81,733 | 23,879 | 2,250 |
| Bangor | 13,792 | 81 | 26 | 66 | 888 | 459 | 49 | 283 | 5,699 | 1,801 | 141 |
| Lewiston | 6,550 | 35 | 89 | 39 | 702 | 299 | 34 | 150 | 1,731 | 568 | 46 |
| Portland | 24,517 | 102 | 45 | 173 | 2,580 | 1,805 | 167 | 388 | 5,340 | 2,248 | 176 |
| South Portland | 36,329 | 275 | 6 | 45 | 714 | 2,158 | 43 | 253 | 4,674 | 1,172 | 123 |
| MARYLAND | 326,528 | 1,447 | 52 | 4,598 | 73,388 | 62,763 | 4,872 | 17,911 | 291,814 | 84,966 | 8,241 |
| Annapolis | 20,124 | 48 | 100 | 75 | 621 | 985 | 44 | 423 | 6,433 | 1,516 | 168 |
| Baltimore | 286,444 | 1,621 | 6 | 508 | 7,678 | 9,253 | 509 | 1,912 | 21,064 | 7,175 | 600 |
| Bowie | NA | NA | NA | 18 | 100 | 43 | 7 | 157 | 4,235 | 1,199 | 120 |
| College Park | NA | NA | NA | 7 | 90 | 23 | 6 | 64 | 1,374 | 414 | 40 |
| Frederick | 100,246 | 530 | 67 | 89 | 1,048 | 670 | 59 | 337 | 6,207 | 1,942 | 185 |
| Gaithersburg | 9,427 | 86 | 50 | 62 | 722 | 537 | 56 | 301 | 6,469 | 2,211 | 203 |
| Hagerstown | 9,157 | 82 | 100 | 52 | 503 | 243 | 27 | 211 | 4,439 | 1,247 | 109 |
| Laurel | 7,298 | 27 | 82 | 19 | 610 | 190 | 33 | 144 | 2,544 | 644 | 68 |
| Rockville | 28,767 | 181 | 100 | 83 | 2,003 | 2,010 | 282 | 279 | 4,451 | 1,727 | 155 |
| Salisbury | 15,726 | 92 | 47 | 51 | 642 | 447 | 31 | 218 | 4,500 | 1,277 | 113 |
| MASSACHUSETTS | 4,638 | 21 | 29 | 6,324 | 120,957 | 137,011 | 10,130 | 23,928 | 364,204 | 110,195 | 10,912 |
| Agawam Town | 3,383 | 16 | 100 | 43 | 658 | 451 | 45 | 87 | 967 | 319 | 30 |
| Attleboro | 16,820 | 83 | 98 | 32 | 560 | 505 | 35 | 151 | 3,018 | 870 | 75 |
| Barnstable Town | 40,666 | 57 | 86 | 46 | 602 | 316 | 35 | 355 | 4,187 | 1,338 | 129 |

1. Merchant wholesalers except manufacturers' sales branches and offices.   2. Establishments with payroll.

# Table D. Cities — Real Estate, Professional Services, and Manufacturing

| City | Real estate and rental and leasing, 2017 | | | | Professional, scientific, and technical services[1], 2017 | | | | Manufacturing, 2017 | | | |
|---|---|---|---|---|---|---|---|---|---|---|---|---|
| | Number of establishments | Number of employees | Receipts (mil dol) | Annual payroll (mil dol) | Number of establishments | Number of employees | Receipts (mil dol) | Annual payroll (mil dol) | Number of establishments | Number of employees | Receipts (mil dol) | Annual payroll (mil dol) |
| | 80 | 81 | 82 | 83 | 84 | 85 | 86 | 87 | 88 | 89 | 90 | 91 |
| KANSAS | 3,415 | 14,532 | 3,943 | 586 | 7,201 | 65,648 | 10,972 | 4,174 | 2,760 | 155,968 | 83,418 | 9,032 |
| Dodge City | D | D | D | D | D | D | D | D | NA | NA | NA | NA |
| Garden City | 31 | 115 | 22 | 3 | 45 | 249 | 31 | 11 | NA | NA | NA | NA |
| Hutchinson | 55 | 193 | 33 | 7 | 77 | 551 | 69 | 29 | NA | NA | NA | NA |
| Kansas City | 123 | 652 | 131 | 27 | 178 | 1,464 | 337 | 91 | NA | NA | NA | NA |
| Lawrence | 140 | 691 | 109 | 24 | 259 | 4,867 | 312 | 125 | NA | NA | NA | NA |
| Leavenworth | 37 | 172 | 51 | 5 | 256 | 1,435 | 128 | D | NA | NA | NA | NA |
| Leawood | 104 | 566 | 252 | 26 | 256 | 1,435 | 412 | 151 | NA | NA | NA | NA |
| Lenexa | 133 | 881 | 237 | 45 | 337 | 6,854 | 1,486 | 561 | NA | NA | NA | NA |
| Manhattan | 122 | 483 | 78 | 15 | D | D | D | D | NA | NA | NA | NA |
| Olathe | 156 | 517 | 142 | 25 | 355 | 2,571 | 257 | 98 | NA | NA | NA | NA |
| Overland Park | 439 | 2,325 | 1,053 | 123 | D | D | D | D | NA | NA | NA | NA |
| Salina | 59 | 181 | 41 | 5 | 105 | 1,064 | 93 | 40 | NA | NA | NA | NA |
| Shawnee | 81 | 323 | 88 | 13 | 158 | 752 | 117 | 44 | NA | NA | NA | NA |
| Topeka | 189 | 882 | 160 | 30 | 343 | 3,527 | 553 | 221 | NA | NA | NA | NA |
| Wichita | 527 | 2,896 | 720 | 114 | 1,050 | 10,366 | 1,783 | 643 | 3,699 | 234,010 | 133,416 | 12,769 |
| KENTUCKY | 3,902 | 18,070 | 5,125 | 731 | 8,125 | 67,615 | 10,272 | 3,636 | NA | NA | NA | NA |
| Bowling Green | 116 | 484 | 104 | 16 | 175 | 1,610 | 178 | 60 | NA | NA | NA | NA |
| Covington | 34 | 246 | 60 | 12 | 119 | 1,104 | 212 | 86 | NA | NA | NA | NA |
| Elizabethtown | 62 | 403 | 67 | 11 | 87 | 430 | 50 | 21 | NA | NA | NA | NA |
| Florence | 51 | 198 | 52 | 8 | 105 | 955 | 138 | 56 | NA | NA | NA | NA |
| Frankfort | 29 | 142 | 15 | 4 | 100 | 957 | 110 | 42 | NA | NA | NA | NA |
| Georgetown | 41 | 115 | 25 | 3 | 62 | 324 | 40 | 16 | NA | NA | NA | NA |
| Henderson | 45 | 182 | 22 | 5 | 64 | 382 | 33 | 11 | NA | NA | NA | NA |
| Hopkinsville | 50 | 147 | 26 | 4 | 50 | 305 | 33 | 11 | NA | NA | NA | NA |
| Jeffersontown | 97 | 574 | 170 | 30 | 170 | 3,063 | 448 | 153 | NA | NA | NA | NA |
| Lexington-Fayette | 486 | 2,223 | 581 | 86 | 1,163 | 11,126 | 2,006 | 701 | NA | NA | NA | NA |
| Louisville/Jefferson County | 766 | 4,023 | 1,019 | 177 | 1,697 | 17,890 | 3,089 | 1,147 | NA | NA | NA | NA |
| Nicholasville | 25 | 106 | 16 | 3 | 55 | 515 | 106 | 31 | NA | NA | NA | NA |
| Owensboro | 76 | 453 | 61 | 12 | 128 | 734 | 96 | 37 | NA | NA | NA | NA |
| Paducah | 49 | 341 | 73 | 10 | 126 | 959 | 120 | 45 | NA | NA | NA | NA |
| Richmond | 54 | 197 | 43 | 6 | 82 | 507 | 89 | 24 | NA | NA | NA | NA |
| LOUISIANA | 5,121 | 29,345 | 7,301 | 1,386 | 12,072 | 95,652 | 16,082 | 6,185 | 3,190 | 117,910 | 187,440 | 8,257 |
| Alexandria | 108 | 423 | 84 | 16 | 196 | 1,345 | 168 | 65 | NA | NA | NA | NA |
| Baton Rouge | 384 | 2,005 | 498 | 98 | 1,137 | 14,263 | 2,593 | 1,013 | NA | NA | NA | NA |
| Bossier City | 104 | 546 | 131 | 21 | 126 | 1,093 | 99 | 38 | NA | NA | NA | NA |
| Central | 15 | 30 | 6 | 1 | 27 | 90 | 9 | 4 | NA | NA | NA | NA |
| Houma | 61 | 382 | 122 | 20 | 134 | 1,429 | 155 | 84 | NA | NA | NA | NA |
| Kenner | 99 | 868 | 260 | 37 | 150 | 803 | 145 | 49 | NA | NA | NA | NA |
| Lafayette | 341 | 1,792 | 505 | 95 | 1,018 | 6,933 | 1,268 | 452 | NA | NA | NA | NA |
| Lake Charles | 141 | 642 | 161 | 25 | 303 | 1,789 | 294 | 97 | NA | NA | NA | NA |
| Monroe | 137 | 747 | 184 | 28 | 289 | 1,854 | 280 | 92 | NA | NA | NA | NA |
| New Iberia | 55 | 339 | 69 | 17 | 89 | 426 | 61 | 19 | NA | NA | NA | NA |
| New Orleans | 470 | 2,327 | 574 | 100 | 1,622 | 15,845 | 3,415 | 1,319 | NA | NA | NA | NA |
| Shreveport | 321 | 1,780 | 368 | 69 | 598 | 4,408 | 736 | 249 | NA | NA | NA | NA |
| Slidell | 41 | 235 | 62 | 10 | 131 | 584 | 70 | 22 | NA | NA | NA | NA |
| MAINE | 1,821 | 7,138 | 1,484 | 287 | 3,543 | 23,381 | 3,874 | 1,488 | 1,691 | 48,620 | 15,089 | 2,599 |
| Bangor | 105 | 622 | 225 | 20 | 141 | 993 | 118 | 53 | NA | NA | NA | NA |
| Lewiston | 54 | 226 | 36 | 8 | 78 | 1,014 | 260 | 61 | NA | NA | NA | NA |
| Portland | 303 | 1,577 | 375 | 77 | 677 | 6,052 | 1,133 | 485 | NA | NA | NA | NA |
| South Portland | 55 | 346 | 78 | 17 | 96 | 1,283 | 207 | 84 | NA | NA | NA | NA |
| MARYLAND | 6,811 | 49,157 | 18,087 | 2,852 | 20,974 | 283,999 | 55,405 | 23,645 | 2,967 | 97,992 | 41,777 | 6,343 |
| Annapolis | 101 | 557 | 138 | 36 | 368 | 2,338 | 480 | 212 | NA | NA | NA | NA |
| Baltimore | 748 | 5,107 | 1,607 | 315 | 1,698 | 25,853 | 5,763 | 2,378 | NA | NA | NA | NA |
| Bowie | 38 | 110 | 37 | 4 | 198 | 2,443 | 518 | 196 | NA | NA | NA | NA |
| College Park | 26 | 307 | 60 | 10 | 90 | 547 | 78 | 39 | NA | NA | NA | NA |
| Frederick | 121 | 473 | 134 | 24 | 390 | 6,056 | 879 | 431 | NA | NA | NA | NA |
| Gaithersburg | 96 | 357 | 228 | 23 | 471 | 8,464 | 1,457 | 890 | NA | NA | NA | NA |
| Hagerstown | 62 | 458 | 106 | 16 | 112 | 941 | 109 | 46 | NA | NA | NA | NA |
| Laurel | 50 | 853 | 140 | 36 | 90 | 1,037 | 159 | 79 | NA | NA | NA | NA |
| Rockville | 148 | 1,862 | 1,117 | 143 | 891 | 24,125 | 3,586 | 1,574 | NA | NA | NA | NA |
| Salisbury | 76 | 491 | 85 | 19 | 146 | 1,058 | 146 | 54 | NA | NA | NA | NA |
| MASSACHUSETTS | 7,584 | 52,315 | 17,912 | 3,370 | 21,985 | 310,313 | 78,598 | 32,256 | 6,437 | 231,593 | 82,309 | 15,749 |
| Agawam Town | 24 | 98 | 28 | 6 | 60 | 762 | 112 | 45 | NA | NA | NA | NA |
| Attleboro | 25 | 125 | 26 | 5 | 72 | 298 | 45 | 16 | NA | NA | NA | NA |
| Barnstable Town | 95 | 303 | 96 | 14 | 215 | 999 | 201 | 93 | NA | NA | NA | NA |

1. Establishments subject to federal tax.

# Accommodation and Food Services, Arts, Entertainment, and Recreation, and Health Care and Social Assistance

| City | Accommodation and food services, 2017 | | | | Arts, entertainment, and recreation[1], 2017 | | | | Health care and social assistance[1], 2017 | | | |
|---|---|---|---|---|---|---|---|---|---|---|---|---|
| | Number of establishments | Number of employees | Receipts (mil dol) | Annual payroll (mil dol) | Number of establishments | Number of employees | Receipts (mil dol) | Annual payroll (mil dol) | Number of establishments | Number of employees | Receipts (mil dol) | Annual payroll (mil dol) |
| | 92 | 93 | 94 | 95 | 96 | 97 | 98 | 99 | 100 | 101 | 102 | 103 |
| KANSAS | 6,253 | 118,905 | 5,908 | 1,724 | 1,092 | 17,313 | 1,231 | 323 | 8,104 | 195,941 | 21,440 | 8,522 |
| Dodge City | 88 | 1,503 | 99 | 27 | D | D | D | D | D | D | D | D |
| Garden City | 82 | 1,519 | 82 | 22 | 7 | D | 5 | D | D | D | D | D |
| Hutchinson | 106 | 2,245 | 98 | 30 | 11 | 339 | 13 | 5 | 144 | 3,807 | 447 | 171 |
| Kansas City | 243 | 5,600 | 319 | 91 | 30 | D | 185 | D | 280 | 15,926 | 1,901 | 837 |
| Lawrence | 296 | 6,936 | 313 | 93 | 44 | 652 | 65 | 10 | 283 | 6,166 | 628 | 251 |
| Leavenworth | 61 | 1,044 | 55 | 17 | D | D | D | D | 75 | 2,375 | 283 | 114 |
| Leawood | 68 | 2,908 | 143 | 57 | 17 | 279 | 21 | 8 | 192 | 2,377 | 335 | 123 |
| Lenexa | 122 | 3,094 | 173 | 51 | 32 | 727 | 41 | 14 | 162 | 5,278 | 822 | 260 |
| Manhattan | D | D | D | D | 18 | 233 | 12 | 4 | 212 | 3,714 | 391 | 138 |
| Olathe | 236 | 5,655 | 318 | 92 | 42 | 939 | 30 | 11 | 294 | 6,486 | 880 | 287 |
| Overland Park | 486 | 11,539 | 675 | 212 | 103 | 2,758 | 145 | 48 | 804 | 17,310 | 2,709 | 959 |
| Salina | 135 | 2,800 | 129 | 37 | D | D | D | D | D | D | D | D |
| Shawnee | 120 | 2,614 | 125 | 37 | 22 | 353 | 23 | 5 | 143 | 1,480 | 135 | 56 |
| Topeka | 326 | 6,472 | 325 | 92 | 51 | 976 | 46 | 16 | 454 | 15,708 | 1,920 | 800 |
| Wichita | 1,011 | 21,150 | 1,035 | 306 | 133 | 2,702 | 142 | 41 | 1,336 | 34,913 | 4,017 | 1,599 |
| KENTUCKY | 8,228 | 174,910 | 9,191 | 2,621 | 1,406 | 19,605 | 1,965 | 441 | 11,597 | 268,711 | 32,370 | 12,412 |
| Bowling Green | 259 | 6,307 | 324 | 100 | 33 | 633 | 41 | 11 | 351 | 8,704 | 1,167 | 424 |
| Covington | 126 | 2,329 | 156 | 42 | 13 | 106 | 10 | 2 | 95 | 2,049 | 157 | 72 |
| Elizabethtown | 122 | 3,322 | 171 | 47 | 12 | 100 | 7 | 2 | 225 | 6,032 | 633 | 275 |
| Florence | 168 | 4,045 | 237 | 69 | 23 | 577 | 39 | 10 | 179 | 3,864 | 411 | 171 |
| Frankfort | 87 | 1,870 | 98 | 28 | 10 | 96 | 3 | 1 | 131 | 1,920 | 301 | 93 |
| Georgetown | 100 | 2,138 | 122 | 34 | 11 | 81 | 5 | 2 | 98 | 1,312 | 138 | 55 |
| Henderson | 72 | 1,350 | 69 | 18 | D | D | D | D | 140 | 2,463 | 218 | 90 |
| Hopkinsville | 76 | 1,614 | 80 | 23 | D | D | D | D | 136 | 3,051 | 353 | 127 |
| Jeffersontown | 124 | 3,215 | 181 | 52 | 25 | 451 | 42 | 10 | 138 | 2,322 | 208 | 80 |
| Lexington-Fayette | 849 | 20,035 | 1,112 | 325 | 173 | 3,186 | 349 | 91 | 1,154 | 32,976 | 4,709 | 1,706 |
| Louisville/Jefferson County | 1,397 | 32,702 | 1,911 | 541 | 261 | 5,035 | 789 | 144 | 1,703 | 50,488 | 7,015 | 2,602 |
| Nicholasville | 64 | 1,424 | 63 | 20 | D | D | D | D | 81 | 912 | 76 | 34 |
| Owensboro | 160 | 3,623 | 186 | 55 | 24 | 460 | 26 | 7 | 286 | 9,004 | 1,012 | 442 |
| Paducah | 175 | 3,931 | 187 | 56 | 19 | 309 | 11 | 4 | 213 | 6,571 | 876 | 299 |
| Richmond | 111 | 2,903 | 136 | 42 | 12 | 158 | 5 | 2 | 157 | 2,773 | 285 | 108 |
| LOUISIANA | 9,877 | 215,048 | 14,553 | 3,809 | 1,526 | 25,649 | 2,789 | 845 | 12,685 | 302,408 | 34,618 | 12,913 |
| Alexandria | 179 | 3,685 | 195 | 53 | D | D | D | D | 381 | 8,649 | 1,175 | 414 |
| Baton Rouge | 755 | 18,801 | 1,181 | 318 | 97 | 2,389 | 145 | 42 | 857 | 25,744 | 3,332 | 1,078 |
| Bossier City | 222 | 8,315 | 747 | 160 | 28 | 778 | 76 | 15 | 189 | 4,194 | 433 | 170 |
| Central | D | D | D | D | D | D | D | D | 40 | 585 | 40 | 18 |
| Houma | 99 | 1,395 | 82 | 23 | 8 | 34 | 3 | 1 | 154 | 4,240 | 429 | 182 |
| Kenner | 182 | 3,541 | 253 | 65 | 23 | 972 | 131 | 28 | 173 | 3,184 | 416 | 140 |
| Lafayette | 606 | 13,551 | 899 | 229 | 66 | 1,296 | 68 | 21 | 978 | 22,460 | 2,794 | 991 |
| Lake Charles | 263 | 9,425 | 955 | 222 | 36 | 335 | 22 | 5 | 391 | 11,114 | 1,212 | 468 |
| Monroe | 168 | 3,769 | 193 | 52 | 26 | 483 | 23 | 6 | 420 | 9,475 | 1,027 | 374 |
| New Iberia | 92 | 1,605 | 76 | 21 | D | D | D | D | 171 | 3,524 | 305 | 110 |
| New Orleans | 1,510 | 42,732 | 3,769 | 1,026 | 234 | 6,789 | 657 | 154 | 994 | 26,020 | 3,804 | 1,273 |
| Shreveport | 474 | 11,625 | 754 | 193 | 74 | 1,014 | 63 | 18 | 843 | 28,006 | 4,127 | 1,487 |
| Slidell | 194 | 3,884 | 191 | 55 | 9 | 106 | 7 | 2 | 187 | 5,057 | 585 | 237 |
| MAINE | 4,257 | 55,746 | 4,018 | 1,167 | 887 | 8,027 | 677 | 188 | 4,771 | 112,594 | 11,777 | 5,000 |
| Bangor | 146 | 3,659 | 243 | 66 | 22 | 211 | 42 | 6 | 286 | 11,651 | 1,415 | 630 |
| Lewiston | 90 | 1,258 | 75 | 24 | 14 | 168 | 6 | 3 | 185 | 7,223 | 880 | 384 |
| Portland | 413 | 7,068 | 500 | 159 | 75 | 1,058 | 90 | 26 | 462 | 15,910 | 2,286 | 850 |
| South Portland | 134 | 2,782 | 173 | 52 | D | D | D | D | 136 | 3,289 | 400 | 209 |
| MARYLAND | 12,139 | 237,730 | 16,931 | 4,625 | 2,172 | 43,573 | 4,970 | 1,588 | 16,800 | 384,096 | 48,676 | 19,294 |
| Annapolis | 196 | 5,866 | 383 | 115 | 54 | 838 | 72 | 23 | 204 | 2,005 | 247 | 96 |
| Baltimore | 1,542 | 25,190 | 1,981 | 551 | 203 | 7,484 | 936 | 389 | 1,494 | 77,508 | 12,757 | 4,398 |
| Bowie | 96 | 2,601 | 150 | 47 | D | D | D | D | 265 | 2,585 | 248 | 97 |
| College Park | 120 | 2,022 | 128 | 34 | 10 | 172 | 10 | 4 | 34 | 297 | 25 | 12 |
| Frederick | 265 | 5,127 | 307 | 91 | D | D | D | 14 | 440 | 8,565 | 1,057 | 410 |
| Gaithersburg | 220 | 4,686 | 359 | 98 | 37 | 918 | 60 | 19 | 227 | 4,744 | 589 | 212 |
| Hagerstown | 139 | 2,672 | 142 | 41 | 17 | 362 | 19 | 5 | 210 | 3,935 | 391 | 167 |
| Laurel | 98 | 2,213 | 135 | 38 | D | D | D | D | 131 | 1,517 | 171 | 67 |
| Rockville | 268 | 4,157 | 319 | 87 | 38 | 1,139 | 69 | 28 | 374 | 8,387 | 1,447 | 804 |
| Salisbury | 142 | 3,017 | 157 | 41 | 12 | 111 | 6 | 2 | 206 | 6,774 | 797 | 346 |
| MASSACHUSETTS | 17,773 | 311,058 | 22,893 | 6,857 | 3,477 | 66,160 | 7,037 | 2,360 | 19,349 | 635,012 | 74,024 | 32,038 |
| Agawam Town | 65 | 769 | 39 | 11 | 12 | 1,715 | 92 | 22 | 62 | 1,369 | 103 | 46 |
| Attleboro | 85 | 1,813 | 101 | 34 | 15 | 249 | 8 | 4 | 128 | 4,657 | 474 | 217 |
| Barnstable Town | 189 | 3,118 | 238 | 77 | 55 | 635 | 78 | 25 | 253 | 8,175 | 1,110 | 458 |

1. Establishments subject to federal tax.

# Table D. Cities — Other Services and Government Employment and Payroll

Columns 104–107 under "Other services[1]"; columns 108–116 under "Government employment and payroll, 2017" (March payroll). Columns 110–116 are "Percent of total for."

| City | Number of establishments (104) | Number of employees (105) | Receipts (mil dol) (106) | Annual payroll (mil dol) (107) | Full-time equivalent employees (108) | Total (dollars) (109) | Administrative, judicial, and legal (110) | Police and corrections (111) | Fire protection (112) | Highways and transportation (113) | Health and welfare (114) | Natural resources and utilities (115) | Education and libraries (116) |
|---|---|---|---|---|---|---|---|---|---|---|---|---|---|
| KANSAS | 5,069 | 27,987 | 3,986 | 928 | X | X | X | X | X | X | X | X | X |
| Dodge City | 44 | 243 | 34 | 7 | 237 | 841,843 | 14.1 | 29.4 | 13.0 | 11.1 | 0.7 | 20.1 | 7.2 |
| Garden City | 53 | 243 | 33 | 8 | 322 | 1,330,163 | 19.8 | 27.5 | 12.1 | 6.5 | 0.7 | 32.0 | 0.0 |
| Hutchinson | 78 | 405 | 44 | 10 | 388 | 1,692,937 | 9.1 | 29.1 | 25.0 | 8.2 | 2.0 | 22.6 | 0.0 |
| Kansas City | 190 | 1,349 | 356 | 48 | 2,690 | 15,933,139 | 8.2 | 25.9 | 22.3 | 5.8 | 3.8 | 25.6 | 0.1 |
| Lawrence | 178 | 1,198 | 217 | 41 | 2,243 | 13,193,908 | 3.7 | 9.6 | 8.5 | 2.5 | 59.8 | 13.5 | 0.0 |
| Leavenworth | 59 | 313 | 24 | 8 | 240 | 958,374 | 11.8 | 31.3 | 23.8 | 7.3 | 4.6 | 17.3 | 0.0 |
| Leawood | 77 | 1,039 | 179 | 60 | 260 | 1,468,797 | 17.9 | 29.1 | 24.0 | 12.4 | 1.6 | 11.8 | 0.0 |
| Lenexa | 84 | 722 | 103 | 31 | 428 | 2,250,458 | 15.8 | 31.7 | 22.7 | 11.5 | 2.6 | 10.9 | 0.0 |
| Manhattan | 107 | 1,081 | 290 | 48 | 407 | 1,733,009 | 13.8 | 0.0 | 29.3 | 14.3 | 3.7 | 37.6 | 0.0 |
| Olathe | 197 | 1,292 | 132 | 39 | 697 | 4,432,327 | 16.2 | 28.8 | 19.2 | 8.3 | 0.6 | 27.0 | 0.0 |
| Overland Park | 401 | 2,392 | 262 | 78 | 1,086 | 3,243,332 | 17.6 | 35.9 | 19.4 | 13.1 | 1.0 | 8.8 | 0.0 |
| Salina | 95 | 535 | 99 | 16 | 478 | 2,114,081 | 9.9 | 21.9 | 23.4 | 10.8 | 4.0 | 26.8 | 0.0 |
| Shawnee | 93 | 605 | 47 | 16 | 296 | 1,917,070 | 13.7 | 38.3 | 20.0 | 14.6 | 0.0 | 8.0 | 0.0 |
| Topeka | 312 | 2,171 | 296 | 86 | 1,196 | 5,881,517 | 14.4 | 30.7 | 24.7 | 9.1 | 3.1 | 14.8 | 0.0 |
| Wichita | 659 | 4,761 | 676 | 150 | 2,961 | 14,226,610 | 10.6 | 32.3 | 18.4 | 12.0 | 3.7 | 14.4 | 2.7 |
| KENTUCKY | 5,900 | 38,040 | 4,755 | 1,198 | X | X | X | X | X | X | X | X | X |
| Bowling Green | 179 | 1,128 | 105 | 28 | 681 | 3,194,750 | 7.2 | 22.3 | 20.4 | 5.2 | 1.6 | 26.6 | 0.0 |
| Covington | 71 | 527 | 44 | 14 | 376 | 1,896,750 | 10.8 | 32.1 | 32.5 | 2.6 | 12.6 | 5.9 | 0.0 |
| Elizabethtown | 74 | 458 | 49 | 13 | 317 | 1,236,052 | 12.4 | 23.5 | 20.6 | 9.5 | 0.0 | 25.9 | 0.0 |
| Florence | 74 | 514 | 40 | 13 | 198 | 1,566,875 | 5.7 | 36.0 | 33.0 | 3.7 | 0.0 | 11.0 | 0.0 |
| Frankfort | 76 | 436 | 62 | 16 | 575 | 2,644,113 | 4.0 | 11.4 | 17.3 | 5.4 | 2.3 | 28.9 | 0.0 |
| Georgetown | D | D | D | 7 | 191 | 758,457 | 10.3 | 39.3 | 33.4 | 6.9 | 0.0 | 6.2 | 0.0 |
| Henderson | D | D | D | D | 467 | 1,970,799 | 10.6 | 18.3 | 15.0 | 6.0 | 5.7 | 40.9 | 0.0 |
| Hopkinsville | 54 | 258 | 24 | 6 | 443 | 1,859,036 | 5.8 | 19.4 | 20.4 | 1.6 | 5.4 | 33.0 | 0.0 |
| Jeffersontown | 100 | 787 | 119 | 40 | 128 | 667,386 | 10.4 | 69.3 | 0.0 | 11.8 | 2.6 | 2.0 | 0.0 |
| Lexington-Fayette | 595 | 4,860 | 962 | 165 | 3,756 | 18,866,699 | 10.5 | 30.8 | 20.8 | 5.8 | 7.3 | 14.2 | 3.0 |
| Louisville/Jefferson County | 969 | 8,330 | 1,157 | 272 | 8,320 | 36,302,120 | 9.1 | 28.8 | 6.5 | 12.3 | 17.3 | 18.8 | 2.3 |
| Nicholasville | 59 | 234 | 20 | 6 | 225 | 925,322 | 16.2 | 30.8 | 22.3 | 3.4 | 0.0 | 15.4 | 0.0 |
| Owensboro | D | D | D | D | 776 | 3,786,398 | 6.5 | 16.5 | 12.2 | 5.6 | 2.5 | 41.7 | 0.0 |
| Paducah | D | D | D | D | 531 | 2,391,618 | 5.3 | 17.2 | 15.0 | 8.5 | 6.2 | 37.4 | 0.2 |
| Richmond | 50 | 247 | 25 | 7 | 259 | 421,250 | 9.3 | 26.8 | 23.5 | 5.7 | 16.0 | 10.0 | 0.0 |
| LOUISIANA | 6,531 | 43,472 | 5,702 | 1,572 | X | X | X | X | X | X | X | X | X |
| Alexandria | 92 | 571 | 76 | 19 | 888 | 3,077,029 | 19.1 | 24.1 | 16.5 | 8.4 | 1.3 | 26.5 | 0.0 |
| Baton Rouge | 570 | 4,281 | 680 | 165 | 6,100 | 25,370,879 | 15.8 | 29.2 | 14.6 | 12.1 | 19.0 | 3.3 | 5.8 |
| Bossier City | 105 | 897 | 78 | 25 | 690 | 2,672,349 | 12.1 | 33.0 | 34.0 | 2.8 | 1.1 | 13.0 | 0.0 |
| Central | D | D | D | D | 12 | 21,138 | 55.9 | 44.1 | 0.0 | 0.0 | 0.0 | 0.0 | 3.4 |
| Houma | 57 | 593 | 79 | 34 | 2,640 | 10,714,118 | 9.5 | 14.0 | 2.1 | 2.1 | 60.4 | 8.5 | 0.0 |
| Kenner | 121 | 793 | 110 | 31 | 628 | 2,559,064 | 14.9 | 44.2 | 18.6 | 5.8 | 2.7 | 8.1 | 3.4 |
| Lafayette | 316 | 2,479 | 284 | 79 | 3,190 | 10,047,868 | 16.8 | 32.5 | 12.1 | 8.6 | 3.1 | 20.9 | 3.4 |
| Lake Charles | 139 | 970 | 108 | 30 | 927 | 2,945,183 | 13.5 | 25.6 | 23.5 | 4.3 | 0.0 | 27.3 | 0.0 |
| Monroe | 104 | 708 | 58 | 17 | 1,055 | 3,346,633 | 13.6 | 22.2 | 21.2 | 8.7 | 6.9 | 25.7 | 0.0 |
| New Iberia | 64 | 282 | 35 | 10 | 167 | 549,275 | 13.6 | 0.0 | 41.3 | 11.6 | 8.3 | 25.2 | 2.1 |
| New Orleans | 722 | 5,256 | 751 | 182 | 6,939 | 33,679,953 | 12.6 | 41.0 | 9.3 | 3.1 | 7.9 | 17.4 | 6.0 |
| Shreveport | 330 | 2,326 | 249 | 67 | 2,693 | 9,834,778 | 8.5 | 29.4 | 28.3 | 4.3 | 1.3 | 15.1 | 0.0 |
| Slidell | 105 | 416 | 33 | 10 | 326 | 1,145,559 | 17.2 | 40.7 | 0.0 | 12.9 | 0.0 | 24.0 | 0.0 |
| MAINE | 2,921 | 14,477 | 1,809 | 475 | X | X | X | X | X | X | X | X | X |
| Bangor | 93 | 687 | 83 | 22 | 1,151 | 5,026,690 | 4.0 | 9.1 | 8.6 | 16.7 | 3.1 | 4.8 | 51.5 |
| Lewiston | 78 | 429 | 43 | 14 | 1,348 | 5,468,002 | 3.8 | 8.7 | 7.2 | 4.0 | 0.6 | 3.6 | 69.4 |
| Portland | 286 | 1,905 | 275 | 72 | 2,684 | 12,071,737 | 6.5 | 9.9 | 9.3 | 5.6 | 11.3 | 3.7 | 46.7 |
| South Portland | 84 | 612 | 50 | 16 | 886 | 3,012,021 | 7.2 | 10.5 | 11.2 | 5.0 | 1.1 | 9.1 | 51.8 |
| MARYLAND | 10,355 | 81,469 | 12,118 | 3,519 | X | X | X | X | X | X | X | X | X |
| Annapolis | 202 | 1,802 | 233 | 76 | 671 | 3,640,943 | 12.5 | 28.3 | 24.5 | 16.4 | 0.0 | 13.9 | 0.0 |
| Baltimore | 996 | 9,183 | 1,395 | 360 | 25,611 | 138,264,675 | 6.7 | 17.7 | 8.7 | 3.1 | 3.8 | 8.7 | 49.4 |
| Bowie | 64 | 401 | 33 | 12 | 368 | 1,787,807 | 17.1 | 24.4 | 0.0 | 6.4 | 4.1 | 29.9 | 0.0 |
| College Park | 51 | 282 | 66 | 13 | 113 | 647,321 | 27.1 | 11.9 | 0.0 | 6.8 | 9.6 | 19.4 | 0.0 |
| Frederick | 205 | 1,639 | 233 | 63 | 600 | 2,869,188 | 10.7 | 38.9 | 0.0 | 7.5 | 6.3 | 23.5 | 0.0 |
| Gaithersburg | 151 | 1,236 | 132 | 42 | 347 | 2,140,618 | 25.2 | 25.7 | 0.0 | 10.7 | 8.4 | 22.0 | 0.0 |
| Hagerstown | 106 | 696 | 69 | 20 | 411 | 1,913,537 | 8.7 | 17.2 | 19.2 | 8.0 | 2.4 | 32.8 | 0.0 |
| Laurel | 65 | 487 | 41 | 16 | 207 | 1,129,796 | 17.3 | 48.9 | 0.0 | 5.3 | 0.0 | 14.7 | 0.0 |
| Rockville | 249 | 3,193 | 685 | 273 | 596 | 3,495,551 | 18.6 | 17.9 | 0.0 | 8.5 | 9.3 | 45.2 | 0.0 |
| Salisbury | 108 | 920 | 101 | 29 | 393 | 1,584,284 | 9.7 | 34.4 | 19.1 | 8.4 | 3.2 | 21.1 | 0.0 |
| MASSACHUSETTS | 14,810 | 100,088 | 13,094 | 3,606 | X | X | X | X | X | X | X | X | X |
| Agawam Town | 57 | 385 | 162 | 15 | 1,031 | 4,421,652 | 3.8 | 8.1 | 8.9 | 3.8 | 1.9 | 2.8 | 65.4 |
| Attleboro | 82 | 320 | 36 | 9 | 1,224 | 6,646,168 | 3.6 | 9.7 | 10.1 | 1.4 | 1.5 | 7.5 | 65.7 |
| Barnstable Town | 175 | 1,169 | 122 | 42 | 1,354 | 7,736,084 | 7.5 | 15.0 | 0.0 | 6.0 | 1.8 | 6.2 | 59.3 |

1. Establishments subject to federal tax.

# Table D. Cities — City Government Finances

| City | General revenue Total (mil dol) | Intergovernmental Total (mil dol) | Intergovernmental Percent from state government | Taxes Total (mil dol) | Taxes Per capita[1] (dollars) Total | Taxes Per capita[1] (dollars) Property | Taxes Per capita[1] (dollars) Sales and gross receipts | General expenditure Total (mil dol) | General expenditure Per capita[1] (dollars) Total | General expenditure Per capita[1] (dollars) Capital outlays |
|---|---|---|---|---|---|---|---|---|---|---|
| | 117 | 118 | 119 | 120 | 121 | 122 | 123 | 124 | 125 | 126 |
| KANSAS | X | X | X | X | X | X | X | X | X | X |
| Dodge City | 48.6 | 8.1 | 47.4 | 20.7 | 752 | 400 | 352 | 49 | 1,793 | 438 |
| Garden City | 63.0 | 16.3 | 5.6 | 15.2 | 571 | 246 | 325 | 56 | 2,086 | 588 |
| Hutchinson | 55.4 | 8.0 | 48.2 | 34.4 | 841 | 305 | 536 | 83 | 2,032 | 932 |
| Kansas City | 398.0 | 65.7 | 90.9 | 233.0 | 1,525 | 694 | 824 | 397 | 2,600 | 393 |
| Lawrence | 389.1 | 24.2 | 49.3 | 74.5 | 771 | 369 | 402 | 374 | 3,873 | 598 |
| Leavenworth | 39.6 | 6.6 | 26.9 | 21.9 | 605 | 200 | 405 | 38 | 1,047 | 202 |
| Leawood | 52.6 | 5.4 | 57.2 | 40.3 | 1,163 | 549 | 614 | 55 | 1,576 | 225 |
| Lenexa | 110.3 | 17.8 | 47.3 | 68.0 | 1,269 | 638 | 632 | 134 | 2,507 | 1,242 |
| Manhattan | 100.8 | 9.0 | 35.2 | 66.0 | 1,197 | 538 | 659 | 77 | 1,391 | 228 |
| Olathe | 208.3 | 13.0 | 62.7 | 117.9 | 858 | 300 | 558 | 225 | 1,634 | 444 |
| Overland Park | 303.5 | 66.1 | 20.9 | 125.5 | 656 | 210 | 445 | 230 | 1,202 | 268 |
| Salina | 68.3 | 4.8 | 55.8 | 40.1 | 857 | 312 | 545 | 64 | 1,361 | 104 |
| Shawnee | 77.4 | 19.9 | 55.8 | 46.2 | 704 | 329 | 375 | 71 | 1,078 | 230 |
| Topeka | 196.4 | 25.9 | 75.3 | 123.5 | 976 | 392 | 584 | 194 | 1,532 | 380 |
| Wichita | 638.0 | 175.1 | 49.8 | 184.0 | 471 | 318 | 154 | 576 | 1,475 | 346 |
| KENTUCKY | X | X | X | X | X | X | X | X | X | X |
| Bowling Green | 118.3 | 11.8 | 29.8 | 73.4 | 1,082 | 191 | 103 | 83 | 1,223 | 173 |
| Covington | 73.7 | 16.1 | 15.1 | 45.9 | 1,134 | 179 | 254 | 71 | 1,739 | 553 |
| Elizabethtown | 51.5 | 6.0 | 45.2 | 32.7 | 1,107 | 124 | 348 | 37 | 1,247 | 138 |
| Florence | 47.0 | 3.4 | 91.6 | 33.5 | 1,034 | 244 | 162 | 27 | 847 | 236 |
| Frankfort | 77.8 | 3.5 | 49.5 | 28.2 | 1,021 | 134 | 135 | 64 | 2,309 | 412 |
| Georgetown | 36.5 | 3.8 | 40.2 | 23.1 | 684 | 71 | 160 | 26 | 756 | 35 |
| Henderson | 43.6 | 12.8 | 10.6 | 20.3 | 709 | 313 | 197 | 40 | 1,388 | 269 |
| Hopkinsville | 40.1 | 6.0 | 8.4 | 28.6 | 932 | 173 | 164 | 33 | 1,078 | 57 |
| Jeffersontown | 50.0 | 21.0 | 15.8 | 21.2 | 755 | 140 | 132 | 40 | 1,422 | 119 |
| Lexington-Fayette | 690.2 | 86.8 | 23.5 | 411.2 | 1,277 | 324 | 229 | 590 | 1,832 | 481 |
| Louisville/Jefferson County | 1,127.6 | 192.7 | 37.9 | 645.2 | 838 | 204 | 126 | 1,215 | 1,579 | 151 |
| Nicholasville | 25.4 | 2.4 | 64.4 | 18.4 | 604 | 140 | 203 | 17 | 543 | 57 |
| Owensboro | 112.1 | 20.8 | 17.6 | 42.1 | 707 | 185 | 141 | 92 | 1,537 | 220 |
| Paducah | 68.5 | 14.3 | 52.9 | 36.4 | 1,459 | 253 | 197 | 53 | 2,106 | 561 |
| Richmond | 49.2 | 9.2 | 16.7 | 25.7 | 728 | 98 | 164 | 28 | 780 | 104 |
| LOUISIANA | X | X | X | X | X | X | X | X | X | X |
| Alexandria | 85.5 | 19.2 | 4.6 | 50.0 | 1,064 | 224 | 840 | 104 | 2,204 | 375 |
| Baton Rouge | 991.7 | 113.8 | 42.8 | 550.0 | 2,451 | 919 | 1,532 | 1,138 | 5,072 | 742 |
| Bossier City | 134.8 | 4.4 | 42.7 | 78.9 | 1,152 | 204 | 948 | 95 | 1,393 | 230 |
| Central | 16.5 | 5.1 | 5.8 | 9.6 | 330 | 0 | 330 | 8 | 287 | 251 |
| Houma | 482.7 | 85.3 | 11.7 | 103.7 | 3,121 | 1,587 | 1,534 | 525 | 15,790 | 2,430 |
| Kenner | 102.1 | 51.4 | 4.8 | 31.3 | 467 | 128 | 339 | 107 | 1,601 | 333 |
| Lafayette | 378.0 | 37.9 | 19.2 | 233.7 | 1,847 | 998 | 848 | 400 | 3,158 | 621 |
| Lake Charles | 140.1 | 24.1 | 52.8 | 89.4 | 1,152 | 121 | 1,031 | 122 | 1,573 | 270 |
| Monroe | 134.6 | 22.5 | 14.2 | 84.6 | 1,752 | 237 | 1,515 | 106 | 2,199 | 145 |
| New Iberia | 35.1 | 6.4 | 6.8 | 21.3 | 724 | 152 | 572 | 32 | 1,066 | 169 |
| New Orleans | 1,508.9 | 379.6 | 17.7 | 554.4 | 1,416 | 658 | 747 | 1,359 | 3,472 | 920 |
| Shreveport | 393.6 | 43.1 | 35.0 | 239.8 | 1,249 | 474 | 775 | 378 | 1,970 | 218 |
| Slidell | 38.8 | 0.0 | NA | 31.2 | 1,117 | 255 | 862 | 14 | 491 | 0 |
| MAINE | X | X | X | X | X | X | X | X | X | X |
| Bangor | 147.3 | 41.4 | 94.0 | 62.0 | 1,935 | 1,915 | 20 | 143 | 4,465 | 506 |
| Lewiston | 149.3 | 67.9 | 95.1 | 60.4 | 1,680 | 1,665 | 15 | 143 | 3,987 | 476 |
| Portland | 296.3 | 49.7 | 91.9 | 167.4 | 2,512 | 2,441 | 71 | 331 | 4,964 | 237 |
| South Portland | 100.0 | 18.4 | 90.3 | 68.8 | 2,701 | 2,696 | 5 | 101 | 3,950 | 570 |
| MARYLAND | X | X | X | X | X | X | X | X | X | X |
| Annapolis | 97.1 | 10.0 | 44.7 | 58.7 | 1,495 | 1,150 | 175 | 90 | 2,289 | 20 |
| Baltimore | 3,446.7 | 1,447.1 | 96.3 | 1,488.2 | 2,438 | 1,405 | 323 | 3,479 | 5,698 | 601 |
| Bowie | 52.8 | 3.1 | 86.1 | 42.5 | 726 | 482 | 51 | 50 | 850 | 0 |
| College Park | 20.5 | 1.8 | 80.6 | 14.7 | 456 | 286 | 100 | 16 | 496 | 2 |
| Frederick | 122.6 | 18.9 | 65.5 | 67.1 | 945 | 763 | 55 | 93 | 1,305 | 25 |
| Gaithersburg | 66.7 | 4.5 | 54.7 | 46.5 | 683 | 400 | 117 | 57 | 842 | 19 |
| Hagerstown | 64.4 | 4.0 | 44.0 | 33.9 | 843 | 648 | 130 | 63 | 1,554 | 5 |
| Laurel | 36.1 | 2.3 | 65.0 | 26.2 | 1,018 | 791 | 94 | 35 | 1,339 | 0 |
| Rockville | 111.6 | 8.8 | 51.2 | 60.8 | 896 | 594 | 101 | 105 | 1,545 | 28 |
| Salisbury | 58.8 | 12.6 | 76.7 | 26.4 | 808 | 690 | 58 | 52 | 1,591 | 54 |
| MASSACHUSETTS | X | X | X | X | X | X | X | X | X | X |
| Agawam Town | 101.2 | 30.5 | 96.6 | 61.6 | 2,147 | 2,117 | 30 | 116 | 4,027 | 196 |
| Attleboro | 149.9 | 54.0 | 98.0 | 74.9 | 1,672 | 1,619 | 52 | 133 | 2,976 | 164 |
| Barnstable Town | 185.8 | 28.1 | 84.0 | 126.3 | 2,848 | 2,753 | 95 | 184 | 4,143 | 394 |

1. Based on population estimated as of July 1 of the year shown.

| City | City government finances, 2017 (cont.) | | | | | | | | | |
|---|---|---|---|---|---|---|---|---|---|---|
| | General expenditure (cont.) | | | | | | | | | |
| | Percent of total for: | | | | | | | | | |
| | Public welfare | Highways | Parking facilities | Education | Health and hospitals | Police protection | Sewerage and sanitation | Parks and recreation | Housing and community development | Interest on debt |
| | 127 | 128 | 129 | 130 | 131 | 132 | 133 | 134 | 135 | 136 |
| KANSAS .................... | X | X | X | X | X | X | X | X | X | X |
| Dodge City ................... | 0.0 | 5.9 | 0.0 | 0.0 | 0.9 | 10.4 | 14.8 | 34.1 | 0.4 | 14.3 |
| Garden City ................... | 0.0 | 6.0 | 0.0 | 0.0 | 0.7 | 15.3 | 8.5 | 6.9 | 0.0 | 1.6 |
| Hutchinson ................... | 0.0 | 13.1 | 0.0 | 0.0 | 5.4 | 11.6 | 9.4 | 35.8 | 1.8 | 1.4 |
| Kansas City ................... | 0.0 | 3.1 | 0.1 | 0.0 | 58.3 | 13.0 | 8.8 | 1.8 | 0.8 | 11.0 |
| Lawrence ................... | 0.0 | 2.5 | 0.3 | 0.0 | 0.7 | 5.1 | 12.2 | 2.8 | 0.3 | 2.4 |
| Leavenworth ................... | 0.0 | 16.8 | 0.0 | 0.0 | 0.0 | 17.5 | 14.5 | 7.4 | 8.9 | 2.2 |
| Leawood ................... | 0.0 | 34.8 | 0.0 | 0.0 | 0.0 | 19.5 | 0.0 | 14.7 | 0.0 | 4.3 |
| Lenexa ................... | 0.0 | 55.9 | 0.0 | 0.0 | 0.0 | 11.2 | 1.8 | 4.5 | 0.0 | 9.9 |
| Manhattan ................... | 0.0 | 6.6 | 0.0 | 0.0 | 0.7 | 20.6 | 11.0 | 10.1 | 0.8 | 16.1 |
| Olathe ................... | 0.0 | 23.1 | 0.0 | 0.0 | 0.0 | 11.0 | 6.9 | 12.9 | 0.0 | 15.8 |
| Overland Park ................... | 0.0 | 20.2 | 0.0 | 0.0 | 0.0 | 14.7 | 4.3 | 5.5 | 0.1 | 21.1 |
| Salina ................... | 0.0 | 10.7 | 0.0 | 0.0 | 1.6 | 16.7 | 11.7 | 10.7 | 3.7 | 2.5 |
| Shawnee ................... | 0.0 | 22.8 | 0.0 | 0.0 | 0.0 | 22.3 | 2.4 | 8.7 | 0.4 | 6.9 |
| Topeka ................... | 0.3 | 15.9 | 1.3 | 0.0 | 0.0 | 19.3 | 19.7 | 1.5 | 2.1 | 3.2 |
| Wichita ................... | 0.0 | 14.8 | 0.2 | 0.0 | 0.4 | 14.4 | 11.4 | 5.9 | 0.1 | 18.3 |
| KENTUCKY .................. | X | X | X | X | X | X | X | X | X | X |
| Bowling Green ................... | 0.0 | 14.6 | 0.0 | 0.0 | 0.0 | 13.1 | 14.0 | 12.0 | 6.3 | 4.4 |
| Covington ................... | 0.0 | 16.7 | 1.2 | 0.0 | 0.0 | 13.4 | 2.0 | 9.8 | 21.0 | 2.6 |
| Elizabethtown ................... | 0.0 | 14.6 | 0.0 | 0.0 | 0.0 | 8.7 | 13.4 | 11.4 | 0.0 | 0.0 |
| Florence ................... | 0.0 | 36.3 | 0.0 | 0.0 | 0.0 | 20.0 | 7.7 | 6.5 | 0.0 | 1.9 |
| Frankfort ................... | 0.0 | 4.9 | 0.0 | 0.0 | 0.0 | 10.1 | 19.3 | 5.2 | 0.0 | 0.0 |
| Georgetown ................... | 0.0 | 5.1 | 0.0 | 0.5 | 0.0 | 15.0 | 29.4 | 3.5 | 1.4 | 1.6 |
| Henderson ................... | 0.0 | 6.2 | 0.0 | 0.0 | 0.0 | 11.3 | 6.8 | 3.6 | 17.7 | 3.4 |
| Hopkinsville ................... | 0.0 | 4.7 | 0.0 | 0.0 | 0.0 | 15.6 | 0.0 | 2.6 | 23.2 | 1.7 |
| Jeffersontown ................... | 0.0 | 9.4 | 0.0 | 0.0 | 54.0 | 16.6 | 2.8 | 5.2 | 4.4 | 0.0 |
| Lexington-Fayette ................... | 1.9 | 2.8 | 0.0 | 0.0 | 2.7 | 10.2 | 20.6 | 10.0 | 1.4 | 4.0 |
| Louisville/Jefferson County ... | 1.2 | 6.8 | 2.2 | 4.4 | 8.0 | 15.2 | 2.2 | 4.9 | 4.4 | 10.0 |
| Nicholasville ................... | 0.0 | 12.2 | 0.0 | 0.0 | 0.0 | 26.5 | 19.4 | 0.0 | 2.3 | 2.8 |
| Owensboro ................... | 0.0 | 7.4 | 0.2 | 0.0 | 0.0 | 9.1 | 24.4 | 6.5 | 9.0 | 6.0 |
| Paducah ................... | 0.0 | 21.8 | 0.0 | 0.0 | 0.0 | 12.4 | 5.6 | 5.6 | 16.1 | 1.9 |
| Richmond ................... | 0.0 | 4.9 | 0.0 | 0.0 | 0.0 | 13.1 | 23.0 | 8.1 | 8.3 | 4.0 |
| LOUISIANA .................. | X | X | X | X | X | X | X | X | X | X |
| Alexandria ................... | 0.0 | 8.4 | 0.0 | 0.0 | 0.0 | 16.4 | 12.4 | 6.2 | 4.2 | 8.1 |
| Baton Rouge ................... | 0.5 | 3.9 | 0.1 | 1.0 | 12.8 | 13.7 | 19.6 | 1.2 | 5.9 | 6.9 |
| Bossier City ................... | 0.7 | 10.6 | 0.0 | 0.0 | 0.6 | 20.5 | 11.3 | 5.0 | 4.6 | 9.2 |
| Central ................... | 0.0 | 3.0 | 0.0 | 0.0 | 0.0 | 3.5 | 84.0 | 0.0 | 0.0 | 0.0 |
| Houma ................... | 0.0 | 6.5 | 0.0 | 0.0 | 53.2 | 7.6 | 3.5 | 1.0 | 4.7 | 1.3 |
| Kenner ................... | 0.5 | 6.5 | 0.0 | 0.0 | 0.4 | 15.9 | 19.8 | 9.6 | 11.3 | 0.0 |
| Lafayette ................... | 1.2 | 7.3 | 0.0 | 0.0 | 0.4 | 23.8 | 8.9 | 7.5 | 4.3 | 3.9 |
| Lake Charles ................... | 0.3 | 12.0 | 0.2 | 0.0 | 0.4 | 15.7 | 7.7 | 7.7 | 17.3 | 4.0 |
| Monroe ................... | 0.0 | 23.6 | 0.0 | 0.0 | 0.4 | 16.1 | 12.7 | 8.5 | 11.1 | 0.0 |
| New Iberia ................... | 0.0 | 7.1 | 0.0 | 0.0 | 0.0 | 20.0 | 40.1 | 3.5 | 3.6 | 0.2 |
| New Orleans ................... | 0.5 | 7.1 | 0.0 | 0.0 | 0.0 | | | | | |
| | 0.0 | 4.2 | 0.0 | 0.0 | 3.4 | 6.3 | 10.2 | 6.1 | 18.9 | 6.1 |
| Shreveport ................... | 0.0 | 5.4 | 0.2 | 0.0 | 0.0 | 16.1 | 15.0 | 9.0 | 6.0 | 5.4 |
| Slidell ................... | 0.0 | 0.0 | 0.0 | 0.0 | 0.0 | 0.0 | 39.4 | 11.0 | 0.0 | 0.0 |
| MAINE .................. | X | X | X | X | X | X | X | X | X | X |
| Bangor ................... | 0.0 | 0.0 | 0.5 | 33.2 | 2.2 | 6.2 | 7.3 | 4.9 | 0.8 | 3.3 |
| Lewiston ................... | 0.8 | 3.0 | 0.0 | 50.2 | 0.0 | 4.5 | 8.0 | 0.6 | 1.8 | 2.5 |
| Portland ................... | 11.1 | 5.0 | 0.7 | 33.1 | 1.9 | 5.2 | 7.8 | 2.8 | 1.1 | 5.5 |
| South Portland ................... | 0.5 | 6.1 | 0.0 | 46.4 | 0.0 | 5.1 | 7.0 | 3.6 | 0.0 | 1.7 |
| MARYLAND .................. | X | X | X | X | X | X | X | X | X | X |
| Annapolis ................... | 0.0 | 9.0 | 6.1 | 0.0 | 0.0 | 21.2 | 11.6 | 11.7 | 0.6 | 1.5 |
| Baltimore ................... | 0.0 | 2.8 | 0.7 | 39.7 | 3.4 | 14.3 | 14.6 | 1.1 | 1.5 | 0.9 |
| Bowie ................... | 0.0 | 13.4 | 0.0 | 0.0 | 0.5 | 22.6 | 19.8 | 16.0 | 0.7 | 1.1 |
| College Park ................... | 0.0 | 15.6 | 1.0 | 0.0 | 0.7 | 7.3 | 20.8 | 7.8 | 1.2 | 3.8 |
| Frederick ................... | 0.0 | 11.8 | 4.0 | 0.0 | 0.0 | 32.5 | 12.6 | 8.6 | 0.4 | 5.4 |
| Gaithersburg ................... | 0.0 | 5.7 | 0.0 | 0.0 | 0.8 | 17.4 | 5.1 | 14.3 | 4.6 | 8.5 |
| Hagerstown ................... | 0.0 | 5.7 | 1.1 | 0.0 | 0.0 | 22.5 | 23.9 | 7.1 | 1.4 | 7.1 |
| Laurel ................... | 0.0 | 21.7 | 0.0 | 0.0 | 0.0 | 32.3 | 3.7 | 5.3 | 0.2 | 0.9 |
| Rockville ................... | 0.0 | 15.1 | 0.5 | 0.0 | 0.4 | 10.7 | 15.6 | 22.9 | 1.7 | 5.1 |
| Salisbury ................... | 0.0 | 11.1 | 1.2 | 0.0 | 0.3 | 23.1 | 20.3 | 5.6 | 0.5 | 6.7 |
| MASSACHUSETTS ............ | X | X | X | X | X | X | X | X | X | X |
| Agawam Town ................... | 0.1 | 3.8 | 0.0 | 56.3 | 1.8 | 4.5 | 4.6 | 0.7 | 0.3 | 0.6 |
| Attleboro ................... | 0.7 | 3.2 | 0.0 | 61.6 | 2.2 | 5.9 | 5.6 | 1.4 | 0.4 | 1.2 |
| Barnstable Town ................... | 0.2 | 2.9 | 0.0 | 47.6 | 0.7 | 7.3 | 3.8 | 3.6 | 0.4 | 1.7 |

| City | Debt outstanding — Total (mil dol) [137] | Debt outstanding — Per capita[1] (dollars) [138] | Debt issued during year [139] | Avg daily temp — Mean — January [140] | Avg daily temp — Mean — July [141] | Avg daily temp — Limits — January[3] [142] | Avg daily temp — Limits — July[4] [143] | Annual precipitation (inches) [144] | Heating degree days [145] | Cooling degree days [146] |
|---|---|---|---|---|---|---|---|---|---|---|
| **KANSAS** | X | X | X | X | X | X | X | X | X | X |
| Dodge City | 233.4 | 8,462 | 43.1 | 30.1 | 79.8 | 18.7 | 92.8 | 22.35 | 5,037 | 1,481 |
| Garden City | 41.0 | 1,539 | 2.0 | 28.6 | 77.8 | 14.7 | 92.1 | 18.77 | 5,423 | 1,191 |
| Hutchinson | 78.3 | 1,916 | 31.1 | 28.5 | 79.9 | 17.0 | 92.7 | 30.32 | 5,146 | 1,454 |
| Kansas City | 1,749.6 | 11,455 | 297.3 | 29.1 | 79.0 | 19.9 | 89.4 | 40.17 | 4,847 | 1,406 |
| Lawrence | 388.7 | 4,022 | 73.7 | 29.9 | 80.2 | 20.5 | 90.6 | 39.78 | 4,685 | 1,582 |
| Leavenworth | 29.1 | 803 | 6.6 | 26.6 | 79.1 | 16.4 | 89.8 | 40.94 | 5,331 | 1,356 |
| Leawood | 48.9 | 1,411 | 10.2 | 29.1 | 79.0 | 19.9 | 89.4 | 40.17 | 4,847 | 1,406 |
| Lenexa | 397.5 | 7,423 | 85.3 | 29.1 | 79.0 | 19.9 | 89.4 | 40.17 | 4,847 | 1,406 |
| Manhattan | 373.6 | 6,777 | 22.6 | 27.8 | 79.9 | 16.1 | 92.5 | 34.80 | 5,120 | 1,465 |
| Olathe | 964.4 | 7,012 | 31.1 | 29.1 | 79.0 | 19.9 | 89.4 | 40.17 | 4,847 | 1,406 |
| Overland Park | 480.0 | 2,509 | 31.6 | 29.1 | 79.0 | 19.9 | 89.4 | 40.17 | 4,847 | 1,406 |
| Salina | 135.1 | 2,884 | 0.0 | 29.0 | 81.3 | 18.8 | 93.3 | 32.19 | 4,952 | 1,600 |
| Shawnee | 128.3 | 1,957 | 25.2 | 29.1 | 79.0 | 19.9 | 89.4 | 40.17 | 4,847 | 1,406 |
| Topeka | 392.1 | 3,099 | 74.7 | 27.2 | 78.4 | 17.2 | 89.1 | 35.64 | 5,225 | 1,357 |
| Wichita | 2,489.8 | 6,380 | 153.3 | 30.2 | 81.0 | 20.3 | 92.9 | 30.38 | 4,765 | 1,658 |
| **KENTUCKY** | X | X | X | X | X | X | X | X | X | X |
| Bowling Green | 165.8 | 2,443 | 0.8 | 34.2 | 78.5 | 25.4 | 89.2 | 51.63 | 4,243 | 1,413 |
| Covington | 106.0 | 2,615 | 21.2 | 32.0 | 76.1 | 24.1 | 85.9 | 45.91 | 4,713 | 1,154 |
| Elizabethtown | 54.1 | 1,832 | 4.9 | NA | NA | NA | NA | NA | NA | NA |
| Florence | 18.3 | 566 | 0.0 | NA | NA | NA | NA | NA | NA | NA |
| Frankfort | 60.1 | 2,172 | 5.1 | 30.3 | 75.2 | 20.8 | 86.9 | 43.56 | 5,129 | 994 |
| Georgetown | 443.9 | 13,183 | 0.0 | NA | NA | NA | NA | NA | NA | NA |
| Henderson | 116.1 | 4,051 | 10.2 | 32.6 | 77.6 | 23.6 | 88.4 | 44.77 | 4,374 | 1,344 |
| Hopkinsville | 165.1 | 5,381 | 5.3 | 33.2 | 78.2 | 24.4 | 88.5 | 50.92 | 4,298 | 1,433 |
| Jeffersontown | 102.5 | 3,642 | 0.0 | 33.0 | 78.4 | 24.9 | 87.0 | 44.54 | 4,352 | 1,443 |
| Lexington-Fayette | 919.4 | 2,856 | 199.2 | 31.6 | 75.9 | 22.5 | 86.3 | 46.39 | 4,769 | 1,094 |
| Louisville/Jefferson County | 2,142.7 | 2,783 | 329.7 | NA | NA | NA | NA | NA | NA | NA |
| Nicholasville | 35.8 | 1,175 | 0.0 | NA | NA | NA | NA | NA | NA | NA |
| Owensboro | 450.1 | 7,562 | 32.9 | 33.5 | 79.2 | 24.4 | 90.7 | 46.53 | 4,159 | 1,565 |
| Paducah | 180.3 | 7,222 | 7.1 | 35.2 | 79.9 | 27.2 | 90.8 | 46.04 | 3,893 | 1,635 |
| Richmond | 74.4 | 2,106 | 6.6 | 34.7 | 75.8 | 25.6 | 87.0 | 47.33 | 4,231 | 1,150 |
| **LOUISIANA** | X | X | X | X | X | X | X | X | X | X |
| Alexandria | 201.5 | 4,285 | 12.0 | 48.1 | 83.3 | 38.0 | 92.8 | 61.44 | 1,908 | 2,602 |
| Baton Rouge | 1,319.4 | 5,879 | 0.0 | 50.1 | 81.7 | 40.2 | 90.7 | 63.08 | 1,689 | 2,628 |
| Bossier City | 528.8 | 7,723 | 38.2 | 48.3 | 81.0 | 37.4 | 91.0 | 61.06 | 1,981 | 2,220 |
| Central | 0.0 | 0 | 0.0 | NA | NA | NA | NA | NA | NA | NA |
| Houma | 113.8 | 3,422 | 2.0 | 53.1 | 82.5 | 43.4 | 90.7 | 63.67 | 1,346 | 2,804 |
| Kenner | 91.1 | 1,361 | 8.7 | 52.6 | 82.7 | 43.4 | 91.1 | 64.16 | 1,417 | 2,773 |
| Lafayette | 484.1 | 3,826 | 20.4 | 51.3 | 82.2 | 41.6 | 91.2 | 60.54 | 1,531 | 2,671 |
| Lake Charles | 56.8 | 732 | 0.0 | 50.9 | 82.6 | 41.2 | 91.0 | 57.19 | 1,546 | 2,705 |
| Monroe | 206.8 | 4,285 | 6.1 | 44.6 | 83.0 | 33.5 | 94.1 | 58.04 | 2,399 | 2,311 |
| New Iberia | 1.2 | 41 | 0.0 | 51.3 | 82.3 | 41.4 | 91.1 | 60.89 | 1,544 | 2,680 |
| New Orleans | 2,414.1 | 6,166 | 137.3 | 52.7 | 82.2 | 43.3 | 90.9 | 65.15 | 1,416 | 2,686 |
| Shreveport | 206.4 | 1,075 | 0.0 | 46.4 | 83.4 | 36.5 | 93.3 | 51.30 | 2,251 | 2,405 |
| Slidell | 0.0 | 0 | 0.0 | 50.7 | 82.1 | 40.2 | 91.1 | 62.66 | 1,652 | 2,548 |
| **MAINE** | X | X | X | X | X | X | X | X | X | X |
| Bangor | 134.0 | 4,181 | 4.3 | 18.0 | 69.2 | 8.3 | 79.6 | 39.57 | 7,676 | 313 |
| Lewiston | 125.1 | 3,479 | 11.7 | 20.5 | 71.4 | 11.5 | 81.5 | 45.79 | 7,107 | 465 |
| Portland | 392.1 | 5,883 | 0.0 | 21.7 | 68.7 | 12.5 | 78.8 | 45.83 | 7,318 | 347 |
| South Portland | 56.6 | 2,221 | 12.7 | NA | NA | NA | NA | NA | NA | NA |
| **MARYLAND** | X | X | X | X | X | X | X | X | X | X |
| Annapolis | 161.8 | 4,122 | 17.0 | 32.8 | 77.5 | 23.8 | 87.7 | 44.78 | 4,695 | 1,162 |
| Baltimore | 818.8 | 1,341 | 139.9 | 36.8 | 81.7 | 29.4 | 90.6 | 43.59 | 4,720 | 1,147 |
| Bowie | 13.2 | 225 | 0.1 | 31.8 | 75.2 | 21.2 | 87.1 | 44.66 | 4,970 | 917 |
| College Park | 12.3 | 382 | 0.5 | NA | NA | NA | NA | NA | NA | NA |
| Frederick | 210.7 | 2,970 | 11.2 | 33.3 | 77.9 | 25.1 | 88.9 | 40.64 | 4,430 | 1,272 |
| Gaithersburg | 102.7 | 1,511 | 0.0 | 31.8 | 75.3 | 23.8 | 85.4 | 43.08 | 4,990 | 983 |
| Hagerstown | 109.4 | 2,723 | 10.6 | 29.3 | 75.2 | 20.8 | 86.1 | 39.45 | 5,249 | 902 |
| Laurel | 7.1 | 274 | 0.0 | NA | NA | NA | NA | NA | NA | NA |
| Rockville | 130.8 | 1,929 | 17.5 | 31.8 | 75.3 | 23.8 | 85.4 | 43.08 | 4,990 | 983 |
| Salisbury | 102.0 | 3,127 | 39.6 | NA | NA | NA | NA | NA | NA | NA |
| **MASSACHUSETTS** | X | X | X | X | X | X | X | X | X | X |
| Agawam Town | 25.8 | 899 | 0.0 | NA | NA | NA | NA | NA | NA | NA |
| Attleboro | 60.0 | 1,340 | 0.0 | 27.4 | 72.2 | 17.8 | 83.0 | 48.34 | 6,012 | 558 |
| Barnstable Town | 105.5 | 2,378 | 12.4 | 29.2 | 70.5 | 21.2 | 77.8 | 43.03 | 6,026 | 413 |

1. Based on the population estimated as of July 1 of the year shown.   2. Represents normal values based on the 30-year period, 1971–2000.   3. Average daily minimum.   4. Average daily maximum.

# Table D. Cities — **Land Area and Population**

| STATE Place code | City | Land area[1] (sq. mi) | Total persons 2019 | Rank | Per square mile | White | Black or African American | American Indian, Alaskan Native | Asian | Hawaiian Pacific Islander | Some other race | Two or more races (percent) |
|---|---|---|---|---|---|---|---|---|---|---|---|---|
| | | 1 | 2 | 3 | 4 | 5 | 6 | 7 | 8 | 9 | 10 | 11 |
| | **MASSACHUSETTS—Cont'd** | | | | | | | | | | | |
| 25 05595 | Beverly | 15.1 | 42,242 | 914 | 2,798 | 89.5 | 2.2 | 0.0 | 2.6 | 0.1 | 1.2 | 4.5 |
| 25 07000 | Boston | 48.3 | 691,531 | 21 | 14,317 | 53.2 | 24.9 | 0.3 | 9.7 | 0.1 | 5.4 | 6.3 |
| 25 07740 | Braintree Town | 13.8 | 37,074 | 1,035 | 2,687 | 76.0 | 4.3 | 0.0 | 18.3 | 0.0 | 0.0 | 1.3 |
| 25 09000 | Brockton | 21.3 | 95,770 | 333 | 4,496 | 32.9 | 53.9 | 0.7 | 2.0 | 0.0 | 7.3 | 3.1 |
| 25 11000 | Cambridge | 6.4 | 119,192 | 240 | 18,624 | 62.5 | 11.7 | 0.1 | 17.3 | 0.0 | 2.7 | 5.6 |
| 25 13205 | Chelsea | 2.2 | 39,230 | 989 | 17,832 | 40.6 | 4.3 | 0.0 | 4.3 | 0.0 | 5.1 | 45.6 |
| 25 13660 | Chicopee | 22.9 | 54,825 | 703 | 2,394 | 84.1 | 6.8 | 0.0 | 1.9 | 0.0 | 3.5 | 3.7 |
| 25 21990 | Everett | 3.4 | 46,107 | 845 | 13,561 | 62.5 | 18.2 | 0.0 | 6.7 | 0.0 | 7.9 | 4.7 |
| 25 23000 | Fall River | 33.1 | 89,792 | 372 | 2,713 | 84.1 | 6.7 | 1.3 | 1.6 | 0.0 | 3.5 | 2.9 |
| 25 23875 | Fitchburg | 27.8 | 40,310 | 956 | 1,450 | 83.6 | 3.6 | 0.5 | 1.5 | 0.0 | 4.4 | 6.4 |
| 25 25172 | Franklin Town | 26.6 | 34,711 | 1,114 | 1,305 | D | D | D | D | D | D | D |
| 25 26150 | Gloucester | 26.2 | 30,539 | 1,232 | 1,166 | D | D | D | D | D | D | D |
| 25 29405 | Haverhill | 33.0 | 64,098 | 591 | 1,942 | 74.7 | 4.9 | 0.1 | 1.2 | 0.0 | 17.0 | 1.9 |
| 25 30840 | Holyoke | 21.2 | 39,921 | 964 | 1,883 | 84.2 | 6.1 | 0.0 | 0.3 | 0.0 | 5.2 | 4.2 |
| 25 34550 | Lawrence | 6.9 | 80,007 | 439 | 11,595 | 33.5 | 6.7 | 0.3 | 2.4 | 0.0 | 54.7 | 2.5 |
| 25 35075 | Leominster | 28.8 | 41,548 | 925 | 1,443 | 84.4 | 7.7 | 0.3 | 2.0 | 0.0 | 1.2 | 4.4 |
| 25 37000 | Lowell | 13.6 | 110,904 | 271 | 8,155 | 63.1 | 8.3 | 1.0 | 17.4 | 0.6 | 6.4 | 3.2 |
| 25 37490 | Lynn | 10.7 | 94,539 | 347 | 8,835 | 55.5 | 9.5 | 0.6 | 4.9 | 0.2 | 22.8 | 6.6 |
| 25 37875 | Malden | 5.0 | 59,922 | 627 | 11,984 | 57.9 | 14.7 | 0.0 | 23.5 | 0.0 | 1.6 | 2.1 |
| 25 38715 | Marlborough | 20.9 | 39,290 | 987 | 1,880 | 65.1 | 2.2 | 0.0 | 4.2 | 0.0 | 20.7 | 7.9 |
| 25 39835 | Medford | 8.1 | 60,847 | 616 | 7,512 | 70.5 | 12.0 | 0.6 | 9.2 | 0.0 | 1.3 | 6.4 |
| 25 40115 | Melrose | 4.7 | 27,790 | 1,323 | 5,913 | 85.2 | 3.8 | 0.0 | 8.2 | 0.0 | 0.5 | 2.3 |
| 25 40710 | Methuen Town | 22.2 | 50,910 | 767 | 2,293 | 69.0 | 3.9 | 0.1 | 3.5 | 0.0 | 20.8 | 2.7 |
| 25 45000 | New Bedford | 20.0 | 95,517 | 335 | 4,776 | 61.9 | 9.1 | 0.3 | 2.0 | 0.0 | 21.4 | 5.3 |
| 25 45560 | Newton | 17.8 | 87,803 | 384 | 4,933 | 75.3 | 2.8 | 0.6 | 15.2 | 0.0 | 1.8 | 4.3 |
| 25 46330 | Northampton | 34.2 | 28,362 | 1,302 | 829 | 84.9 | 1.2 | 0.1 | 3.5 | 0.2 | 3.5 | 6.6 |
| 25 52490 | Peabody | 16.2 | 53,063 | 737 | 3,276 | 89.6 | 4.7 | 0.6 | 0.9 | 0.0 | 1.5 | 2.7 |
| 25 53960 | Pittsfield | 40.5 | 41,834 | 917 | 1,033 | 85.0 | 6.7 | 0.0 | 2.0 | 0.0 | 1.7 | 4.6 |
| 25 55745 | Quincy | 16.6 | 94,421 | 348 | 5,688 | 59.6 | 5.1 | 0.3 | 30.4 | 0.1 | 1.6 | 2.8 |
| 25 56585 | Revere | 5.7 | 52,375 | 744 | 9,189 | 79.6 | 6.7 | 0.0 | 3.9 | 0.0 | 7.6 | 2.2 |
| 25 59105 | Salem | 8.3 | 43,581 | 887 | 5,251 | 78.8 | 11.5 | 0.8 | 3.6 | 0.0 | 2.9 | 2.4 |
| 25 62535 | Somerville | 4.1 | 80,935 | 433 | 19,740 | 74.4 | 3.5 | 1.0 | 12.2 | 0.0 | 2.7 | 6.3 |
| 25 67000 | Springfield | 31.9 | 152,646 | 169 | 4,785 | 62.6 | 19.1 | 0.2 | 4.1 | 0.0 | 9.2 | 4.8 |
| 25 69170 | Taunton | 46.7 | 57,675 | 668 | 1,235 | 76.7 | 10.8 | 0.1 | 1.6 | 0.0 | 7.0 | 3.8 |
| 25 72600 | Waltham | 12.7 | 62,069 | 606 | 4,887 | 70.6 | 4.8 | 0.7 | 15.4 | 0.0 | 4.4 | 4.1 |
| 25 73440 | Watertown Town | 4.0 | 36,125 | 1,070 | 9,031 | 82.9 | 1.2 | 0.0 | 12.7 | 0.0 | 0.7 | 2.5 |
| 25 76030 | Westfield | 46.3 | 41,063 | 935 | 887 | D | D | D | D | D | D | D |
| 25 77890 | West Springfield Town | 16.7 | 28,397 | 1,301 | 1,700 | 78.1 | 5.7 | 1.0 | 6.2 | 0.0 | 4.6 | 4.4 |
| 25 78972 | Weymouth Town | 16.8 | 58,104 | 664 | 3,459 | 81.9 | 7.5 | 0.0 | 6.1 | 0.0 | 2.3 | 2.1 |
| 25 81035 | Woburn | 12.7 | 40,135 | 960 | 3,160 | 79.7 | 4.1 | 0.0 | 13.6 | 0.0 | 0.5 | 2.0 |
| 25 82000 | Worcester | 37.4 | 184,570 | 139 | 4,935 | 73.0 | 13.0 | 0.8 | 5.0 | 0.1 | 3.8 | 4.3 |
| 26 00000 | **MICHIGAN** | 56,605.9 | 9,966,555 | X | 176 | 78.2 | 13.7 | 0.6 | 3.3 | | 1.1 | 3.0 |
| 26 01380 | Allen Park | 7.0 | 26,751 | 1,360 | 3,822 | 94.1 | 2.4 | 0.2 | 1.8 | 0.0 | 0.4 | 1.1 |
| 26 03000 | Ann Arbor | 28.2 | 119,280 | 238 | 4,230 | 70.8 | 8.1 | 0.4 | 15.9 | 0.0 | 0.6 | 4.2 |
| 26 05920 | Battle Creek | 42.6 | 50,799 | 768 | 1,193 | 67.5 | 19.4 | 1.1 | 4.8 | 0.0 | 2.1 | 5.1 |
| 26 06020 | Bay City | 10.2 | 32,435 | 1,177 | 3,180 | D | D | D | D | D | D | D |
| 26 12060 | Burton | 23.4 | 28,462 | 1,300 | 1,216 | 88.8 | 3.2 | 0.2 | 1.9 | 0.1 | 1.3 | 4.4 |
| 26 21000 | Dearborn | 24.2 | 93,367 | 355 | 3,858 | 89.4 | 5.9 | 0.1 | 0.7 | 0.0 | 1.2 | 2.8 |
| 26 21020 | Dearborn Heights | 11.7 | 55,027 | 696 | 4,703 | 76.4 | 6.4 | 0.1 | 1.7 | 0.0 | 3.3 | 1.7 |
| 26 22000 | Detroit | 138.7 | 665,369 | 24 | 4,797 | 14.7 | 77.9 | 0.7 | 1.7 | 0.0 | 1.2 | 5.5 |
| 26 24120 | East Lansing | 13.4 | 47,641 | 813 | 3,555 | D | D | D | D | D | D | D |
| 26 24290 | Eastpointe | 5.2 | 31,720 | 1,203 | 6,100 | 51.3 | 35.4 | 0.3 | 1.1 | 0.0 | 6.1 | 5.8 |
| 26 27440 | Farmington Hills | 33.3 | 80,017 | 438 | 2,403 | 61.3 | 20.3 | 0.3 | 16.1 | 0.0 | 0.6 | 2.0 |
| 26 29000 | Flint | 33.4 | 94,968 | 343 | 2,843 | 37.5 | 55.4 | 0.0 | 0.8 | 0.0 | 0.6 | 5.6 |
| 26 31420 | Garden City | 5.9 | 26,216 | 1,380 | 4,443 | D | D | D | D | D | D | D |
| 26 34000 | Grand Rapids | 44.8 | 200,031 | 120 | 4,465 | 65.5 | 18.9 | 0.3 | 2.3 | 0.0 | 6.0 | 7.0 |
| 26 38640 | Holland | 16.7 | 33,140 | 1,158 | 1,984 | 75.1 | 7.7 | 0.7 | 2.0 | 0.0 | 10.3 | 4.2 |
| 26 40680 | Inkster | 6.3 | 24,115 | 1,422 | 3,828 | 17.8 | 76.7 | 0.2 | 0.0 | 0.1 | 1.1 | 4.0 |
| 26 41420 | Jackson | 10.8 | 32,374 | 1,180 | 2,998 | D | D | D | D | D | D | D |
| 26 42160 | Kalamazoo | 24.7 | 76,009 | 470 | 3,077 | 63.4 | 25.3 | 0.4 | 2.8 | 0.0 | 1.2 | 6.8 |
| 26 42820 | Kentwood | 20.9 | 51,719 | 748 | 2,475 | 62.3 | 18.4 | 0.4 | 10.0 | 0.0 | 2.8 | 6.1 |
| 26 46000 | Lansing | 39.1 | 117,540 | 246 | 3,006 | 64.9 | 22.7 | 0.9 | 6.2 | 0.0 | 0.7 | 4.7 |
| 26 47800 | Lincoln Park | 5.8 | 36,066 | 1,072 | 6,218 | 72.2 | 7.8 | 0.6 | 0.1 | 0.0 | 17.3 | 2.0 |
| 26 49000 | Livonia | 35.7 | 93,189 | 356 | 2,610 | 89.5 | 3.1 | 0.0 | 3.1 | 0.0 | 0.1 | 4.2 |
| 26 50560 | Madison Heights | 7.1 | 29,633 | 1,264 | 4,174 | 79.0 | 10.4 | 0.2 | 8.4 | 0.0 | 0.4 | 1.6 |
| 26 53780 | Midland | 34.4 | 41,769 | 920 | 1,214 | 89.9 | 2.6 | 0.4 | 3.3 | 0.0 | 1.1 | 2.6 |
| 26 56020 | Mount Pleasant | 7.7 | 24,274 | 1,419 | 3,153 | 85.7 | 4.5 | 3.2 | 2.1 | 0.0 | 0.3 | 4.4 |
| 26 56320 | Muskegon | 14.1 | 36,462 | 1,058 | 2,586 | 58.1 | 32.4 | 2.0 | 0.4 | 0.2 | 2.5 | 4.4 |
| 26 59440 | Novi | 30.2 | 60,852 | 615 | 2,015 | 67.2 | 8.2 | 0.0 | 21.7 | 0.0 | 0.5 | 2.4 |

1. Dry land or land partially or temporarily covered by water.  2. Hispanic or Latino persons may be of any race.

# Table D. Cities — **Population**

| City | Percent Hispanic or Latino[1], 2019 | Percent foreign born, 2019 | Age of population (percent), 2019 | | | | | | | Median age, 2019 | Percent female, 2019 | Population — Census counts | | Population — Percent change | |
|---|---|---|---|---|---|---|---|---|---|---|---|---|---|---|---|
| | | | Under 18 years | 18 to 24 years | 25 to 34 years | 35 to 44 years | 45 to 54 years | 55 to 64 years | 65 years and over | | | 2000 | 2010 | 2000-2010 | 2001-2020 |
| | 12 | 13 | 14 | 15 | 16 | 17 | 18 | 19 | 20 | 21 | 22 | 23 | 24 | 25 | 26 |
| Beverly | 6.8 | 9.2 | 24.4 | 13.6 | 12.3 | 14.2 | 13.3 | 9.7 | 12.6 | 34.4 | 52.8 | 39,862 | 39,504 | -0.9 | 6.9 |
| Boston | 19.7 | 27.2 | 15.4 | 14.7 | 24.4 | 12.8 | 10.6 | 10.1 | 12.0 | 32.6 | 52.2 | 589,141 | 617,779 | 4.9 | 11.9 |
| Braintree Town | 1.7 | 21.1 | 18.8 | 7.8 | 10.3 | 9.0 | 18.2 | 13.1 | 22.8 | 47.3 | 51.8 | 33,698 | 35,726 | 6.0 | 3.8 |
| Brockton | 11.9 | 32.6 | 25.2 | 10.0 | 13.1 | 11.7 | 14.3 | 12.4 | 13.3 | 36.1 | 51.1 | 94,304 | 93,767 | -0.6 | 2.1 |
| Cambridge | 9.3 | 26.8 | 13.3 | 20.3 | 26.5 | 13.0 | 8.7 | 7.6 | 10.5 | 29.7 | 50.1 | 101,355 | 105,155 | 3.7 | 13.3 |
| Chelsea | 70.5 | 41.8 | 28.9 | 8.3 | 15.9 | 13.2 | 12.1 | 10.5 | 11.0 | 33.3 | 45.9 | 35,080 | 35,181 | 0.3 | 11.5 |
| Chicopee | 19.4 | 9.6 | 20.8 | 11.5 | 13.6 | 11.5 | 10.3 | 12.5 | 19.7 | 37.6 | 48.0 | 54,653 | 55,307 | 1.2 | -0.9 |
| Everett | 21.8 | 45.1 | 17.0 | 8.0 | 15.6 | 17.5 | 15.6 | 10.7 | 15.7 | 40.2 | 48.0 | 38,037 | 41,553 | 9.2 | 11.0 |
| Fall River | 12.0 | 18.7 | 22.5 | 8.3 | 14.3 | 11.8 | 15.2 | 11.3 | 16.6 | 39.2 | 52.0 | 91,938 | 88,865 | -3.3 | 1.0 |
| Fitchburg | 28.0 | 5.5 | 24.5 | 14.4 | 11.0 | 12.0 | 13.0 | 12.3 | 12.9 | 35.2 | 49.4 | 39,102 | 40,325 | 3.1 | 0.0 |
| Franklin Town | 0.5 | 9.9 | 21.2 | 12.8 | 9.0 | 11.8 | 15.9 | 14.4 | 15.0 | 41.4 | 55.2 | 29,560 | 31,633 | 7.0 | 9.7 |
| Gloucester | 1.5 | 5.8 | 17.0 | 7.3 | 10.1 | 10.7 | 11.3 | 23.2 | 20.3 | 50.7 | 51.4 | 30,273 | 28,789 | -4.9 | 6.1 |
| Haverhill | 25.1 | 13.4 | 23.1 | 8.4 | 15.4 | 12.6 | 11.4 | 14.7 | 14.4 | 36.6 | 53.4 | 58,969 | 60,878 | 3.2 | 5.3 |
| Holyoke | 49.0 | 5.4 | 16.3 | 7.5 | 19.5 | 11.2 | 15.0 | 12.8 | 17.7 | 39.0 | 52.7 | 39,838 | 39,881 | 0.1 | 0.1 |
| Lawrence | 82.6 | 45.7 | 23.7 | 10.9 | 17.4 | 12.0 | 13.8 | 11.9 | 10.4 | 34.2 | 50.9 | 72,043 | 76,343 | 6.0 | 4.8 |
| Leominster | 15.7 | 15.0 | 17.6 | 6.8 | 14.2 | 12.8 | 16.8 | 13.0 | 18.9 | 43.1 | 50.3 | 41,303 | 40,762 | -1.3 | 1.9 |
| Lowell | 20.1 | 27.0 | 15.7 | 15.8 | 17.2 | 13.6 | 13.4 | 12.5 | 11.9 | 36.2 | 48.7 | 105,167 | 106,529 | 1.3 | 4.1 |
| Lynn | 46.7 | 37.3 | 26.3 | 9.7 | 13.1 | 13.8 | 12.2 | 11.0 | 13.9 | 35.7 | 48.8 | 89,050 | 90,319 | 1.4 | 4.7 |
| Malden | 11.0 | 40.8 | 17.4 | 9.6 | 24.8 | 11.5 | 10.2 | 11.7 | 14.8 | 33.9 | 48.5 | 56,340 | 59,533 | 5.7 | 0.7 |
| Marlborough | 21.8 | 30.6 | 21.2 | 5.9 | 19.3 | 13.8 | 13.7 | 9.7 | 16.5 | 37.8 | 48.5 | 36,255 | 38,501 | 6.2 | 2.0 |
| Medford | 9.6 | 22.5 | 16.3 | 15.0 | 22.2 | 9.6 | 9.9 | 12.2 | 14.8 | 32.8 | 49.2 | 55,765 | 56,280 | 0.9 | 8.1 |
| Melrose | 1.8 | 11.9 | 19.3 | 7.9 | 10.5 | 14.5 | 16.5 | 12.9 | 18.4 | 44.1 | 56.5 | 27,134 | 26,971 | -0.6 | 3.0 |
| Methuen Town | 31.0 | 23.3 | 19.6 | 8.2 | 12.0 | 13.9 | 13.4 | 14.5 | 18.5 | 42.2 | 50.8 | 43,789 | 47,332 | 8.1 | 7.6 |
| New Bedford | 21.1 | 20.8 | 24.1 | 8.0 | 16.9 | 14.5 | 9.9 | 11.0 | 15.7 | 35.8 | 50.9 | 93,768 | 95,063 | 1.4 | 0.5 |
| Newton | 5.5 | 22.0 | 21.0 | 12.5 | 10.9 | 10.0 | 12.9 | 14.0 | 18.8 | 40.7 | 53.6 | 83,829 | 85,074 | 1.5 | 3.2 |
| Northampton | 9.9 | 8.8 | 11.7 | 17.7 | 11.8 | 10.5 | 12.8 | 14.5 | 21.0 | 43.5 | 64.0 | 28,978 | 28,559 | -1.4 | -0.7 |
| Peabody | 9.3 | 18.1 | 12.7 | 8.6 | 15.1 | 6.6 | 16.5 | 15.0 | 25.5 | 49.7 | 52.0 | 48,129 | 51,270 | 6.5 | 3.5 |
| Pittsfield | 5.6 | 6.1 | 14.0 | 8.9 | 11.2 | 9.1 | 14.0 | 14.9 | 27.8 | 50.2 | 52.9 | 45,793 | 44,743 | -2.3 | -6.5 |
| Quincy | 2.8 | 33.7 | 15.4 | 6.1 | 19.1 | 14.6 | 12.4 | 13.1 | 19.4 | 40.2 | 52.1 | 88,025 | 92,260 | 4.8 | 2.3 |
| Revere | 38.0 | 39.8 | 19.1 | 5.0 | 18.8 | 17.9 | 11.3 | 13.4 | 14.5 | 38.2 | 51.1 | 47,283 | 51,715 | 9.4 | 1.3 |
| Salem | 13.6 | 10.7 | 15.2 | 14.3 | 17.5 | 12.9 | 10.0 | 13.3 | 16.9 | 36.3 | 56.9 | 40,407 | 41,311 | 2.2 | 5.5 |
| Somerville | 12.6 | 26.5 | 13.3 | 12.4 | 32.4 | 15.5 | 7.6 | 9.2 | 9.5 | 32.6 | 53.4 | 77,478 | 75,693 | -2.3 | 6.9 |
| Springfield | 49.0 | 9.4 | 25.8 | 13.0 | 14.9 | 12.4 | 11.1 | 10.7 | 12.1 | 32.1 | 52.6 | 152,082 | 153,132 | 0.7 | -0.3 |
| Taunton | 11.2 | 20.0 | 17.0 | 8.2 | 14.1 | 10.3 | 16.3 | 16.9 | 17.3 | 45.3 | 53.6 | 55,976 | 55,834 | -0.3 | 3.3 |
| Waltham | 15.2 | 27.0 | 12.1 | 20.1 | 21.0 | 13.4 | 8.7 | 11.6 | 13.1 | 33.5 | 52.9 | 59,226 | 60,660 | 2.4 | 2.3 |
| Watertown Town | 4.6 | 22.9 | 16.3 | 5.5 | 23.3 | 18.2 | 9.3 | 14.0 | 16.3 | 36.0 | 51.8 | 32,986 | 31,963 | -3.1 | 13.0 |
| Westfield | 12.1 | 5.3 | 21.0 | 11.3 | 12.0 | 9.4 | 11.5 | 15.5 | 19.2 | 41.0 | 54.4 | 40,072 | 41,093 | 2.5 | -0.1 |
| West Springfield Town | 12.8 | 14.7 | 17.7 | 6.3 | 14.0 | 18.4 | 11.4 | 13.0 | 19.2 | 42.8 | 57.1 | 27,899 | 28,391 | 1.8 | 0.0 |
| Weymouth Town | 5.3 | 15.2 | 16.7 | 9.6 | 13.7 | 12.4 | 11.7 | 16.7 | 19.0 | 42.5 | 54.0 | 53,988 | 53,762 | -0.4 | 8.1 |
| Woburn | 2.6 | 24.7 | 17.4 | 10.2 | 17.2 | 10.3 | 13.2 | 15.2 | 16.5 | 39.3 | 48.1 | 37,258 | 38,908 | 4.4 | 3.2 |
| Worcester | 25.2 | 21.8 | 19.2 | 15.3 | 16.0 | 12.5 | 11.9 | 11.9 | 13.2 | 34.5 | 50.5 | 172,648 | 180,892 | 4.8 | 2.0 |
| MICHIGAN | 5.3 | 7.0 | 21.5 | 9.5 | 13.0 | 11.8 | 12.5 | 14.0 | 17.7 | 39.8 | 50.8 | 9,938,444 | 9,884,112 | -0.5 | 0.8 |
| Allen Park | 4.6 | 2.8 | 22.1 | 4.2 | 13.8 | 14.8 | 13.0 | 16.0 | 16.2 | 41.6 | 56.0 | 29,376 | 28,212 | -4.0 | -5.2 |
| Ann Arbor | 5.3 | 18.5 | 13.3 | 31.3 | 17.4 | 9.7 | 8.4 | 7.9 | 12.0 | 27.6 | 49.9 | 114,024 | 114,008 | 0.0 | 4.6 |
| Battle Creek | 7.2 | 9.6 | 24.0 | 8.3 | 14.0 | 14.3 | 9.5 | 13.0 | 17.0 | 37.0 | 51.3 | 53,364 | 52,388 | -1.8 | -3.0 |
| Bay City | 8.9 | 1.0 | 24.8 | 10.0 | 14.8 | 13.7 | 12.7 | 11.0 | 12.9 | 35.6 | 48.0 | 36,817 | 34,929 | -5.1 | -7.1 |
| Burton | 3.6 | 2.3 | 24.0 | 8.0 | 13.7 | 12.7 | 8.8 | 14.4 | 18.3 | 37.5 | 55.3 | 30,308 | 30,003 |  | -5.1 |
| Dearborn | 3.1 | 27.7 | 30.3 | 11.1 | 15.4 | 11.3 | 10.2 | 9.0 | 12.6 | 29.9 | 50.8 | 97,775 | 98,146 | 0.4 | -4.9 |
| Dearborn Heights | 4.7 | 20.6 | 25.5 | 7.9 | 14.7 | 9.3 | 14.6 | 13.3 | 14.6 | 35.7 | 49.9 | 58,264 | 57,774 | -0.8 | -4.8 |
| Detroit | 8.3 | 6.1 | 24.8 | 9.3 | 15.9 | 11.9 | 11.6 | 11.9 | 14.5 | 35.0 | 52.2 | 951,270 | 713,956 | -24.9 | -6.8 |
| East Lansing | 5.7 | 9.6 | 8.3 | 62.2 | 8.9 | 4.9 | 4.4 | 3.6 | 7.7 | 21.0 | 50.7 | 46,525 | 48,573 | 4.4 | -1.9 |
| Eastpointe | 2.2 | 1.7 | 19.7 | 6.1 | 18.4 | 15.3 | 13.2 | 14.3 | 13.0 | 38.3 | 50.6 | 34,077 | 32,403 | -4.9 | -2.1 |
| Farmington Hills | 1.0 | 21.6 | 20.6 | 7.0 | 15.4 | 13.7 | 12.3 | 12.1 | 18.9 | 40.1 | 49.4 | 82,111 | 79,725 | -2.9 | 0.4 |
| Flint | 4.9 | 2.0 | 24.8 | 9.6 | 14.2 | 11.4 | 12.2 | 14.0 | 13.9 | 36.2 | 52.0 | 124,943 | 102,258 | -18.2 | -7.1 |
| Garden City | 1.3 | 4.5 | 23.8 | 7.2 | 17.3 | 11.9 | 12.2 | 15.6 | 12.1 | 36.6 | 49.3 | 30,047 | 27,636 | -8.0 | -5.1 |
| Grand Rapids | 16.8 | 10.8 | 21.8 | 13.0 | 21.3 | 11.0 | 9.8 | 10.2 | 12.9 | 31.4 | 50.4 | 197,800 | 187,999 | -5.0 | 6.4 |
| Holland | 25.0 | 8.5 | 22.6 | 13.3 | 17.0 | 10.2 | 7.7 | 13.7 | 15.4 | 33.1 | 53.7 | 35,048 | 33,099 | -5.6 | 0.1 |
| Inkster | 1.6 | 5.6 | 26.2 | 9.5 | 18.5 | 9.7 | 15.7 | 8.5 | 11.9 | 31.3 | 55.9 | 30,115 | 25,366 | -15.8 | -4.9 |
| Jackson | 5.7 | 3.3 | 24.0 | 9.8 | 14.5 | 11.1 | 14.4 | 14.1 | 12.1 | 36.8 | 52.4 | 36,316 | 33,474 | -7.8 | -3.3 |
| Kalamazoo | 8.4 | 9.1 | 17.5 | 26.8 | 15.4 | 11.1 | 9.0 | 8.7 | 11.5 | 27.8 | 51.7 | 77,145 | 74,263 | -3.7 | 2.4 |
| Kentwood | 12.8 | 17.9 | 27.0 | 7.2 | 18.5 | 10.9 | 11.6 | 11.8 | 13.0 | 33.8 | 50.1 | 45,255 | 48,698 | 7.6 | 6.2 |
| Lansing | 12.6 | 14.8 | 20.6 | 12.4 | 19.3 | 12.5 | 11.3 | 11.1 | 12.8 | 33.3 | 50.5 | 119,128 | 114,269 | -4.1 | 2.9 |
| Lincoln Park | 27.4 | 11.4 | 23.2 | 8.4 | 15.4 | 10.8 | 12.9 | 12.9 | 15.6 | 37.2 | 48.8 | 40,008 | 38,085 | -4.8 | -5.3 |
| Livonia | 3.2 | 8.2 | 18.9 | 7.6 | 11.2 | 10.1 | 14.6 | 15.0 | 22.6 | 46.4 | 51.8 | 100,545 | 96,857 | -3.7 | -3.8 |
| Madison Heights | 1.6 | 12.9 | 13.5 | 12.9 | 15.6 | 10.9 | 18.7 | 11.5 | 16.9 | 42.7 | 46.3 | 31,101 | 29,694 | -4.5 | -0.2 |
| Midland | 3.1 | 5.5 | 22.2 | 8.7 | 14.3 | 13.3 | 11.3 | 13.1 | 17.2 | 38.5 | 51.9 | 41,685 | 41,872 | 0.4 | -0.2 |
| Mount Pleasant | 6.1 | 5.8 | 12.9 | 40.8 | 13.8 | 8.3 | 4.7 | 8.4 | 11.1 | 22.7 | 55.8 | 25,946 | 26,033 | 0.3 | -6.8 |
| Muskegon | 10.4 | 3.6 | 21.8 | 11.9 | 17.1 | 12.2 | 10.6 | 12.1 | 14.3 | 34.4 | 49.2 | 40,105 | 38,397 | -4.3 | -5.0 |
| Novi | 8.4 | 26.5 | 26.8 | 6.5 | 11.7 | 15.9 | 14.7 | 11.3 | 13.1 | 38.2 | 51.7 | 47,386 | 55,232 | 16.6 | 10.2 |

1. May be of any race.

| City | Households, 2019 Number | Persons per household | Percent Family | Percent Married couple family | Percent Female family | Percent Non-family | Percent One person | Persons in group quarters, 2019 | Serious crimes Violent Number | Violent Rate[2] | Property Number | Property Rate[2] | Population age 25 and over | High school graduate or less | Bachelor's degree or more |
|---|---|---|---|---|---|---|---|---|---|---|---|---|---|---|---|
| | 27 | 28 | 29 | 30 | 31 | 32 | 33 | 34 | 35 | 36 | 37 | 38 | 39 | 40 | 41 |
| **MASSACHUSETTS—Cont'd** | | | | | | | | | | | | | | | |
| Beverly | 14,801 | 2.60 | 65.2 | 48.8 | 16.4 | 34.8 | 27.4 | 3,705 | 47 | 111 | 206 | 487 | 26,149 | 25.4 | 56.4 |
| Boston | 271,553 | 2.39 | 47.1 | 28.3 | 18.8 | 52.9 | 36.6 | 46,369 | 45 | 121 | 486 | 1,308 | 485,350 | 30.5 | 51.7 |
| Braintree Town | 14,570 | 2.51 | 69.3 | 52.9 | 16.3 | 30.7 | 27.2 | 547 | 25 | 257 | 38 | 391 | 27,274 | 32.5 | 42.5 |
| Brockton | 31,440 | 2.98 | 68.0 | 37.1 | 30.9 | 32.0 | 27.0 | 1,986 | 782 | 821 | 1,853 | 1,945 | 62,003 | 55.6 | 16.6 |
| Cambridge | 47,374 | 2.17 | 40.1 | 32.2 | 7.9 | 59.9 | 36.6 | 16,119 | 334 | 279 | 1,983 | 1,654 | 78,861 | 9.9 | 79.7 |
| Chelsea | 11,962 | 3.26 | 56.4 | 31.0 | 25.4 | 43.6 | 32.8 | 743 | 270 | 667 | 594 | 1,467 | 24,907 | 66.0 | 19.0 |
| Chicopee | 23,280 | 2.32 | 57.9 | 32.5 | 25.4 | 42.1 | 34.0 | 1,193 | 340 | 615 | 1,320 | 2,387 | 37,284 | 49.2 | 26.3 |
| Everett | 16,345 | 2.83 | 67.4 | 41.0 | 26.4 | 32.6 | 26.2 | 179 | 245 | 519 | 623 | 1,320 | 34,844 | 52.7 | 20.4 |
| Fall River | 37,899 | 2.33 | 57.1 | 31.3 | 25.8 | 42.9 | 34.9 | 1,411 | 773 | 868 | 1,074 | 1,206 | 61,979 | 58.4 | 15.7 |
| Fitchburg | 15,172 | 2.56 | 59.8 | 38.5 | 21.2 | 40.2 | 28.9 | 1,735 | 217 | 534 | 538 | 1,324 | 24,850 | 46.4 | 19.8 |
| Franklin Town | 12,304 | 2.68 | 72.4 | 63.3 | 9.1 | 27.6 | 23.1 | 1,113 | 2 | 6 | 41 | 124 | 22,496 | 28.6 | 52.2 |
| Gloucester | 12,291 | 2.45 | 67.8 | 51.4 | 16.4 | 32.2 | 21.2 | 362 | 58 | 191 | 164 | 540 | 23,018 | 36.0 | 36.4 |
| Haverhill | 24,279 | 2.60 | 65.4 | 34.8 | 30.6 | 34.6 | 27.0 | 964 | 335 | 524 | 838 | 1,311 | 43,845 | 40.0 | 26.2 |
| Holyoke | 16,427 | 2.37 | 53.3 | 27.0 | 26.3 | 46.7 | 36.0 | 1,220 | 345 | 859 | 1,494 | 3,718 | 30,561 | 53.6 | 24.0 |
| Lawrence | 27,844 | 2.83 | 64.5 | 32.3 | 32.3 | 35.5 | 29.3 | 1,275 | 541 | 674 | 954 | 1,189 | 52,353 | 63.4 | 14.4 |
| Leominster | 17,499 | 2.36 | 70.8 | 49.3 | 21.5 | 29.2 | 23.5 | 338 | 239 | 574 | 599 | 1,439 | 31,541 | 40.6 | 27.7 |
| Lowell | 41,753 | 2.55 | 55.9 | 34.1 | 21.8 | 44.1 | 30.9 | 4,695 | 405 | 363 | 1,652 | 1,483 | 76,130 | 51.3 | 25.7 |
| Lynn | 31,833 | 2.94 | 65.8 | 42.6 | 23.2 | 34.2 | 28.0 | 728 | 465 | 492 | 1,347 | 1,426 | 60,303 | 56.7 | 20.1 |
| Malden | 23,482 | 2.57 | 61.6 | 46.4 | 15.2 | 38.4 | 25.7 | 177 | 162 | 267 | 719 | 1,184 | 44,167 | 36.0 | 42.0 |
| Marlborough | 16,366 | 2.39 | 60.3 | 43.8 | 16.6 | 39.7 | 29.5 | 545 | 149 | 376 | 530 | 1,336 | 28,888 | 41.4 | 35.9 |
| Medford | 21,824 | 2.52 | 53.2 | 38.2 | 15.0 | 46.8 | 30.1 | 2,292 | 116 | 202 | 526 | 915 | 39,385 | 26.7 | 54.2 |
| Melrose | 11,122 | 2.49 | 66.2 | 58.6 | 7.5 | 33.8 | 30.6 | 353 | 32 | 114 | 172 | 612 | 20,397 | 20.8 | 65.4 |
| Methuen Town | 18,670 | 2.69 | 70.6 | 49.0 | 21.6 | 29.4 | 24.1 | 483 | 95 | 187 | 552 | 1,088 | 36,656 | 39.8 | 30.7 |
| New Bedford | 39,216 | 2.38 | 58.7 | 32.3 | 26.4 | 41.3 | 33.6 | 1,887 | 628 | 664 | 2,127 | 2,248 | 64,813 | 56.3 | 16.2 |
| Newton | 30,643 | 2.64 | 69.9 | 62.1 | 7.8 | 30.1 | 22.7 | 7,469 | 49 | 55 | 510 | 575 | 58,816 | 11.0 | 81.1 |
| Northampton | 11,839 | 2.06 | 48.4 | 37.1 | 11.3 | 51.6 | 34.4 | 4,016 | 119 | 414 | 448 | 1,559 | 20,082 | 18.2 | 63.0 |
| Peabody | 22,358 | 2.35 | 63.3 | 46.9 | 16.4 | 36.7 | 31.8 | 576 | 123 | 232 | 435 | 819 | 41,803 | 38.6 | 31.6 |
| Pittsfield | 19,911 | 2.07 | 53.1 | 39.1 | 14.1 | 46.9 | 40.2 | 1,006 | 300 | 710 | 796 | 1,883 | 32,485 | 37.8 | 30.9 |
| Quincy | 42,242 | 2.21 | 51.2 | 38.4 | 12.7 | 48.8 | 39.7 | 1,196 | 375 | 398 | 1,146 | 1,218 | 74,144 | 33.9 | 45.8 |
| Revere | 19,176 | 2.75 | 60.6 | 42.2 | 18.4 | 39.4 | 27.6 | 253 | 173 | 322 | 653 | 1,217 | 40,276 | 50.7 | 26.1 |
| Salem | 18,585 | 2.24 | 55.0 | 35.0 | 20.0 | 45.0 | 33.9 | 1,628 | 108 | 249 | 727 | 1,673 | 30,485 | 26.1 | 51.1 |
| Somerville | 32,120 | 2.44 | 42.3 | 34.3 | 8.0 | 57.7 | 31.1 | 3,145 | 166 | 203 | 1,025 | 1,255 | 60,456 | 24.7 | 63.6 |
| Springfield | 54,922 | 2.69 | 61.8 | 27.3 | 34.5 | 38.2 | 33.4 | 5,753 | 1,397 | 905 | 4,005 | 2,595 | 94,018 | 56.1 | 16.8 |
| Taunton | 23,552 | 2.41 | 62.6 | 39.8 | 22.8 | 37.4 | 29.2 | 656 | 217 | 381 | 345 | 605 | 43,002 | 48.7 | 24.1 |
| Waltham | 23,540 | 2.32 | 54.3 | 42.0 | 12.3 | 45.7 | 30.3 | 7,842 | 102 | 163 | 512 | 816 | 42,316 | 25.8 | 57.3 |
| Watertown Town | 16,419 | 2.18 | 51.0 | 42.1 | 9.0 | 49.0 | 35.2 | 173 | 29 | 80 | 319 | 881 | 28,079 | 15.5 | 71.5 |
| Westfield | 15,403 | 2.47 | 66.2 | 44.9 | 21.4 | 33.8 | 28.1 | 3,204 | 83 | 200 | 398 | 959 | 27,877 | 44.2 | 30.3 |
| West Springfield Town | 13,060 | 2.17 | 53.0 | 37.2 | 15.8 | 47.0 | 39.4 | 147 | 138 | 482 | 969 | 3,385 | 21,648 | 40.1 | 38.7 |
| Weymouth Town | 24,422 | 2.34 | 64.7 | 44.6 | 20.2 | 35.3 | 28.6 | 500 | 170 | 294 | 462 | 800 | 42,524 | 31.7 | 45.5 |
| Woburn | 16,389 | 2.43 | 65.6 | 47.9 | 17.7 | 34.4 | 27.1 | 351 | 70 | 174 | 414 | 1,029 | 29,138 | 33.6 | 43.2 |
| Worcester | 69,125 | 2.47 | 58.1 | 35.1 | 23.0 | 41.9 | 31.7 | 14,883 | 1,165 | 630 | 3,792 | 2,050 | 121,298 | 44.4 | 30.8 |
| **MICHIGAN** | 3,969,880 | 2.46 | 62.9 | 46.0 | 16.9 | 37.1 | 30.5 | 223,499 | NA | NA | NA | NA | 6,894,627 | 37.7 | 30.0 |
| Allen Park | 11,425 | 2.34 | 62.5 | 46.6 | 15.9 | 37.5 | 32.2 | 177 | 48 | 178 | 496 | 1,841 | 19,865 | 34.9 | 27.6 |
| Ann Arbor | 45,827 | 2.31 | 45.2 | 35.2 | 10.0 | 54.8 | 32.0 | 13,901 | 309 | 251 | 2,124 | 1,728 | 66,431 | 9.3 | 76.1 |
| Battle Creek | 21,484 | 2.31 | 51.6 | 29.7 | 21.9 | 48.4 | 39.4 | 1,384 | 568 | 937 | 1,969 | 3,249 | 34,584 | 47.6 | 20.3 |
| Bay City | 13,567 | 2.38 | 53.8 | 32.3 | 21.6 | 46.2 | 37.1 | 437 | 271 | 826 | 654 | 1,994 | 21,324 | 45.9 | 16.8 |
| Burton | 11,773 | 2.42 | 61.9 | 36.3 | 25.6 | 38.1 | 30.9 | 127 | 136 | 477 | NA | NA | 19,424 | 49.8 | 13.9 |
| Dearborn | 30,304 | 3.09 | 65.0 | 49.3 | 15.7 | 35.0 | 27.1 | 352 | 305 | 325 | 2,104 | 2,241 | 55,039 | 38.9 | 30.3 |
| Dearborn Heights | 19,355 | 2.83 | 69.3 | 47.7 | 21.7 | 30.7 | 27.4 | 497 | 227 | 410 | 874 | 1,579 | 36,831 | 47.5 | 21.7 |
| Detroit | 267,139 | 2.46 | 51.9 | 19.0 | 32.9 | 48.1 | 42.0 | 12,937 | 13,040 | 1,965 | 28,550 | 4,303 | 441,647 | 51.3 | 16.7 |
| East Lansing | 14,151 | 2.33 | 32.7 | 23.8 | 8.9 | 67.3 | 42.2 | 16,805 | 88 | 184 | 840 | 1,753 | 14,694 | 5.7 | 76.4 |
| Eastpointe | 12,726 | 2.51 | 70.6 | 38.7 | 31.9 | 29.4 | 26.0 | 169 | 223 | 690 | 828 | 2,560 | 23,808 | 55.4 | 15.3 |
| Farmington Hills | 32,850 | 2.43 | 58.4 | 46.6 | 11.8 | 41.6 | 34.1 | 702 | 87 | 107 | 759 | 934 | 58,335 | 15.6 | 62.0 |
| Flint | 40,121 | 2.32 | 60.1 | 25.8 | 34.3 | 39.9 | 35.1 | 2,440 | 1,284 | 1,349 | 1,986 | 2,086 | 62,742 | 50.1 | 13.5 |
| Garden City | 10,009 | 2.63 | 58.9 | 38.8 | 20.1 | 41.1 | 31.5 | 59 | 96 | 363 | 293 | 1,109 | 18,219 | 47.8 | 12.4 |
| Grand Rapids | 77,128 | 2.50 | 54.0 | 34.1 | 20.0 | 46.0 | 33.7 | 7,814 | 1,286 | 637 | 3,850 | 1,908 | 131,079 | 34.8 | 36.4 |
| Holland | 12,359 | 2.41 | 60.3 | 41.3 | 19.0 | 39.7 | 35.3 | 3,655 | 127 | 381 | 613 | 1,838 | 21,443 | 36.0 | 35.7 |
| Inkster | 9,373 | 2.56 | 58.3 | 21.3 | 37.0 | 41.7 | 34.0 | 327 | 242 | 997 | 595 | 2,452 | 15,600 | 57.8 | 7.4 |
| Jackson | 13,925 | 2.27 | 48.3 | 25.7 | 22.7 | 51.7 | 42.3 | 791 | 360 | 1,108 | 1,344 | 4,135 | 21,488 | 50.2 | 16.9 |
| Kalamazoo | 30,858 | 2.26 | 48.3 | 25.8 | 17.1 | 57.1 | 40.0 | 6,375 | 949 | 1,235 | 3,578 | 4,657 | 42,472 | 36.0 | 31.2 |
| Kentwood | 20,231 | 2.54 | 64.5 | 45.7 | 18.8 | 35.5 | 29.2 | 414 | 162 | 310 | 1,111 | 2,125 | 34,142 | 35.2 | 28.9 |
| Lansing | 51,035 | 2.26 | 46.2 | 26.6 | 19.6 | 53.8 | 44.3 | 1,176 | 1,313 | 1,104 | 3,355 | 2,820 | 78,088 | 35.1 | 28.5 |
| Lincoln Park | 14,742 | 2.46 | 60.2 | 41.2 | 19.0 | 39.8 | 35.7 | 88 | 224 | 616 | 812 | 2,235 | 24,822 | 55.8 | 11.3 |
| Livonia | 37,899 | 2.43 | 69.0 | 57.1 | 11.9 | 31.0 | 27.4 | 1,523 | 167 | 178 | 1,411 | 1,507 | 68,855 | 27.6 | 39.6 |
| Madison Heights | 14,622 | 2.03 | 49.5 | 37.8 | 11.7 | 50.5 | 41.9 | 137 | 53 | 176 | 593 | 1,971 | 21,989 | 42.6 | 25.2 |
| Midland | 17,830 | 2.27 | 62.0 | 49.5 | 12.5 | 38.0 | 32.1 | 1,103 | 56 | 134 | 392 | 938 | 28,729 | 28.0 | 44.8 |
| Mount Pleasant | 8,276 | 2.30 | 42.6 | 27.1 | 15.4 | 57.4 | 38.3 | 5,784 | 48 | 190 | 384 | 1,517 | 11,472 | 23.3 | 46.8 |
| Muskegon | 14,544 | 2.19 | 45.7 | 20.8 | 24.8 | 54.3 | 46.7 | 4,776 | 308 | 828 | 1,293 | 3,478 | 24,225 | 47.5 | 16.0 |
| Novi | 23,192 | 2.61 | 67.7 | 54.1 | 13.5 | 32.3 | 27.3 | 308 | 48 | 78 | 514 | 833 | 40,654 | 13.9 | 57.7 |

1. Data for serious crimes have not been adjusted for underreporting. This may affect comparability between geographic areas and over time.   2. Per 100,000 population estimated by the FBI.   3. Persons 25 years old and over.

# Table D. Cities — Income, Poverty, and Housing

| City | Money income, 2019 — Households | | | | | Median earnings, 2019 | | | Housing units, 2019 | | | | |
|---|---|---|---|---|---|---|---|---|---|---|---|---|---|
| | Median income | Percent with income less than $20,000 | Percent with income of $200,000 or more | Median family income | Median non-family income | All persons | Men | Women | Total | Occupied | Percent owner occupied | Median value[1] (dollars) | Median gross rent (dollars) |
| | 42 | 43 | 44 | 45 | 46 | 47 | 48 | 49 | 50 | 51 | 52 | 53 | 54 |
| **MASSACHUSETTS—Cont'd** | | | | | | | | | | | | | |
| Beverly | 91,647 | 13.1 | 19.3 | 131,402 | 48,750 | 46,183 | 61,853 | 30,445 | 15,491 | 14,801 | 59.9 | 491,173 | 1,415 |
| Boston | 79,018 | 18.4 | 15.9 | 95,885 | 62,028 | 45,046 | 49,279 | 41,690 | 303,791 | 271,553 | 34.7 | 627,022 | 1,735 |
| Braintree Town | 93,621 | 9.1 | 13.5 | 113,285 | 60,593 | 47,211 | 60,320 | 40,322 | 14,858 | 14,570 | 68.9 | 534,683 | 1,456 |
| Brockton | 60,250 | 18.2 | 4.0 | 71,480 | 40,532 | 35,171 | 40,025 | 27,291 | 33,880 | 31,440 | 53.8 | 314,552 | 1,102 |
| Cambridge | 119,540 | 12.0 | 25.5 | 161,246 | 93,314 | 54,381 | 60,539 | 50,882 | 51,621 | 47,374 | 33.1 | 869,922 | 2,354 |
| Chelsea | 60,712 | 20.0 | 4.2 | 60,467 | 36,694 | 35,494 | 36,684 | 31,955 | 12,965 | 11,962 | 31.5 | 386,885 | 1,449 |
| Chicopee | 62,212 | 12.7 | 5.0 | 67,910 | 35,773 | 35,858 | 34,764 | 36,724 | 25,233 | 23,280 | 58.9 | 199,156 | 862 |
| Everett | 66,883 | 16.6 | 6.4 | 74,468 | 27,917 | 35,103 | 36,869 | 31,993 | 17,357 | 16,345 | 45.7 | 470,917 | 1,642 |
| Fall River | 41,346 | 27.9 | 1.9 | 46,791 | 22,994 | 32,022 | 40,878 | 28,262 | 40,999 | 37,899 | 35.8 | 259,613 | 838 |
| Fitchburg | 58,138 | 15.8 | 3.3 | 70,370 | 33,826 | 30,791 | 35,285 | 28,481 | 16,810 | 15,172 | 59.2 | 222,224 | 963 |
| Franklin Town | 120,583 | 6.7 | 20.8 | 145,867 | 60,356 | 51,969 | 73,094 | 34,364 | 12,618 | 12,304 | 78.2 | 435,368 | 1,623 |
| Gloucester | 89,452 | 10.1 | 14.4 | 111,333 | 39,930 | 40,807 | 51,424 | 32,076 | 13,932 | 12,291 | 64.5 | 483,066 | 1,298 |
| Haverhill | 71,434 | 12.3 | 8.2 | 85,762 | 47,254 | 36,285 | 46,657 | 27,494 | 25,908 | 24,279 | 51.0 | 340,256 | 1,357 |
| Holyoke | 42,243 | 22.9 | 4.6 | 58,595 | 31,698 | 35,901 | 38,854 | 31,665 | 18,295 | 16,427 | 40.1 | 230,328 | 867 |
| Lawrence | 50,965 | 21.3 | 2.4 | 60,840 | 27,392 | 30,895 | 36,918 | 26,386 | 29,607 | 27,844 | 33.5 | 313,327 | 1,216 |
| Leominster | 65,904 | 16.8 | 7.9 | 80,687 | 42,426 | 42,830 | 49,289 | 40,149 | 18,093 | 17,499 | 60.8 | 268,793 | 1,061 |
| Lowell | 62,487 | 16.1 | 4.2 | 80,230 | 39,225 | 36,464 | 40,351 | 32,396 | 44,026 | 41,753 | 44.9 | 293,940 | 1,245 |
| Lynn | 61,572 | 19.3 | 5.2 | 71,303 | 27,149 | 32,329 | 36,397 | 29,615 | 33,371 | 31,833 | 49.5 | 380,571 | 1,264 |
| Malden | 72,871 | 13.1 | 9.5 | 85,822 | 58,153 | 48,448 | 50,988 | 46,102 | 24,970 | 23,482 | 47.1 | 486,797 | 1,725 |
| Marlborough | 67,672 | 6.7 | 10.1 | 87,154 | 51,020 | 43,864 | 49,223 | 37,617 | 17,537 | 16,366 | 52.7 | 372,408 | 1,295 |
| Medford | 95,033 | 11.3 | 14.0 | 108,910 | 72,322 | 49,788 | 50,925 | 47,254 | 23,775 | 21,824 | 49.1 | 604,653 | 1,993 |
| Melrose | 119,490 | 8.3 | 25.9 | 165,326 | 50,663 | 70,037 | 86,959 | 60,282 | 11,714 | 11,122 | 68.6 | 653,411 | 1,684 |
| Methuen Town | 78,610 | 12.9 | 9.2 | 95,590 | 48,233 | 42,406 | 54,010 | 33,958 | 19,365 | 18,670 | 70.4 | 356,983 | 1,280 |
| New Bedford | 47,305 | 23.6 | 1.2 | 61,792 | 28,630 | 35,979 | 39,465 | 32,379 | 42,200 | 39,216 | 39.6 | 243,280 | 897 |
| Newton | 150,106 | 6.4 | 37.8 | 195,273 | 72,575 | 61,274 | 81,617 | 47,596 | 33,190 | 30,643 | 72.2 | 944,203 | 2,039 |
| Northampton | 63,332 | 11.3 | 10.0 | 96,958 | 45,592 | 30,242 | 42,597 | 22,190 | 12,217 | 11,839 | 58.1 | 322,844 | 1,175 |
| Peabody | 81,927 | 10.7 | 12.2 | 100,952 | 42,599 | 46,512 | 51,134 | 35,640 | 23,456 | 22,358 | 66.4 | 425,510 | 1,369 |
| Pittsfield | 51,866 | 16.8 | 3.3 | 66,192 | 36,295 | 31,326 | 40,249 | 25,613 | 23,031 | 19,911 | 65.8 | 182,346 | 904 |
| Quincy | 80,959 | 15.5 | 8.4 | 104,449 | 60,297 | 49,124 | 51,882 | 44,471 | 44,034 | 42,242 | 46.4 | 475,736 | 1,515 |
| Revere | 70,269 | 14.3 | 6.4 | 81,253 | 60,610 | 36,734 | 46,287 | 30,446 | 20,672 | 19,176 | 52.5 | 452,717 | 1,525 |
| Salem | 82,528 | 9.3 | 7.9 | 99,190 | 60,218 | 41,963 | 50,508 | 40,221 | 19,803 | 18,585 | 53.5 | 383,434 | 1,448 |
| Somerville | 100,643 | 10.7 | 18.4 | 115,174 | 85,295 | 51,959 | 56,188 | 50,687 | 35,176 | 32,120 | 34.6 | 778,409 | 2,095 |
| Springfield | 44,596 | 27.1 | 1.8 | 55,417 | 22,571 | 31,680 | 35,955 | 30,053 | 60,454 | 54,922 | 47.9 | 165,825 | 936 |
| Taunton | 59,863 | 14.6 | 6.4 | 83,963 | 45,719 | 40,920 | 48,049 | 34,895 | 25,243 | 23,552 | 62.7 | 297,762 | 1,066 |
| Waltham | 101,796 | 6.4 | 15.1 | 108,519 | 83,766 | 43,947 | 51,162 | 39,125 | 24,991 | 23,540 | 46.4 | 610,050 | 1,928 |
| Watertown Town | 99,048 | 8.7 | 14.0 | 131,623 | 83,713 | 61,311 | 61,012 | 61,629 | 17,726 | 16,419 | 49.8 | 644,432 | 2,052 |
| Westfield | 67,041 | 15.4 | 6.2 | 81,201 | 36,321 | 35,368 | 40,693 | 32,189 | 16,686 | 15,403 | 62.9 | 248,215 | 955 |
| West Springfield Town | 56,395 | 15.9 | 6.2 | 72,854 | 43,342 | 41,333 | 48,421 | 36,478 | 13,567 | 13,060 | 56.6 | 261,987 | 898 |
| Weymouth Town | 84,670 | 11.6 | 6.9 | 97,070 | 61,833 | 50,082 | 60,047 | 41,933 | 24,922 | 24,422 | 65.9 | 415,276 | 1,542 |
| Woburn | 88,463 | 10.9 | 12.4 | 104,286 | 66,797 | 51,764 | 70,424 | 36,199 | 17,187 | 16,389 | 56.9 | 549,269 | 1,722 |
| Worcester | 57,092 | 20.2 | 4.7 | 68,612 | 34,656 | 32,251 | 37,625 | 27,800 | 75,585 | 69,125 | 40.7 | 249,641 | 1,133 |
| **MICHIGAN** | 59,584 | 14.4 | 5.7 | 75,703 | 35,755 | 33,328 | 41,137 | 27,339 | 4,629,605 | 3,969,880 | 71.6 | 169,613 | 888 |
| Allen Park | 63,756 | 13.4 | 3.4 | 87,177 | 41,520 | 50,329 | 56,150 | 40,035 | 11,815 | 11,425 | 80.1 | 151,041 | 1,033 |
| Ann Arbor | 72,137 | 14.1 | 13.5 | 124,746 | 52,389 | 27,667 | 31,800 | 20,268 | 51,004 | 45,827 | 46.7 | 370,129 | 1,247 |
| Battle Creek | 40,965 | 20.6 | 3.3 | 61,113 | 26,323 | 26,561 | 30,428 | 24,028 | 25,197 | 21,484 | 59.4 | 88,306 | 787 |
| Bay City | 39,776 | 21.9 | 3.1 | 59,737 | 30,010 | 30,656 | 35,346 | 23,080 | 15,151 | 13,567 | 61.2 | 71,034 | 669 |
| Burton | 45,688 | 15.5 | 2.0 | 61,486 | 28,697 | 30,057 | 40,088 | 24,362 | 13,108 | 11,773 | 74.6 | 104,145 | 840 |
| Dearborn | 53,509 | 18.2 | 4.8 | 63,249 | 40,213 | 31,870 | 35,858 | 25,723 | 33,400 | 30,304 | 66.2 | 157,571 | 997 |
| Dearborn Heights | 51,464 | 12.0 | 3.4 | 62,323 | 31,732 | 31,748 | 41,280 | 23,554 | 20,511 | 19,355 | 75.6 | 132,078 | 1,067 |
| Detroit | 33,965 | 30.7 | 1.8 | 40,529 | 26,607 | 27,046 | 29,461 | 25,905 | 359,623 | 267,139 | 47.8 | 58,902 | 866 |
| East Lansing | 40,794 | 29.3 | 5.5 | 96,435 | 25,988 | 7,289 | 7,795 | 6,933 | 15,673 | 14,151 | 39.1 | 200,985 | 928 |
| Eastpointe | 43,737 | 13.0 | 1.0 | 48,189 | 36,224 | 30,576 | 31,264 | 28,436 | 13,350 | 12,726 | 65.1 | 92,572 | 1,207 |
| Farmington Hills | 86,178 | 11.3 | 11.0 | 119,872 | 58,119 | 52,162 | 65,352 | 41,536 | 34,200 | 32,850 | 64.5 | 282,248 | 1,244 |
| Flint | 32,236 | 34.9 | 1.3 | 37,860 | 21,151 | 23,508 | 26,712 | 20,039 | 53,069 | 40,121 | 59.2 | 34,354 | 735 |
| Garden City | 55,999 | 11.6 | 1.4 | 71,810 | 40,946 | 32,247 | 40,760 | 27,155 | 10,582 | 10,009 | 77.1 | 135,397 | 852 |
| Grand Rapids | 51,817 | 19.5 | 3.3 | 66,472 | 38,228 | 29,996 | 32,171 | 24,573 | 85,549 | 77,128 | 56.5 | 168,265 | 995 |
| Holland | 47,611 | 16.4 | 4.8 | 79,321 | 32,009 | 29,556 | 34,028 | 24,981 | 13,493 | 12,359 | 57.9 | 176,940 | 923 |
| Inkster | 26,534 | 38.9 | 0.2 | 43,265 | 22,066 | 22,398 | 30,983 | 16,763 | 11,589 | 9,373 | 41.9 | 56,437 | 933 |
| Jackson | 33,065 | 33.7 | 2.0 | 41,770 | 20,166 | 21,528 | 27,453 | 16,753 | 16,379 | 13,925 | 51.6 | 75,478 | 762 |
| Kalamazoo | 39,494 | 28.3 | 2.6 | 47,224 | 27,530 | 20,883 | 22,933 | 17,307 | 35,440 | 30,858 | 37.6 | 121,963 | 819 |
| Kentwood | 61,393 | 9.0 | 3.8 | 68,958 | 40,285 | 33,802 | 44,321 | 29,221 | 20,468 | 20,231 | 55.8 | 183,124 | 999 |
| Lansing | 41,066 | 20.1 | 1.3 | 49,595 | 34,499 | 27,415 | 30,082 | 25,930 | 57,126 | 51,035 | 51.8 | 94,165 | 797 |
| Lincoln Park | 43,135 | 19.0 | 0.8 | 59,548 | 30,422 | 31,130 | 36,446 | 26,284 | 15,931 | 14,742 | 69.3 | 88,691 | 795 |
| Livonia | 81,305 | 7.5 | 6.3 | 99,650 | 50,070 | 44,535 | 54,094 | 36,433 | 40,055 | 37,899 | 86.3 | 220,105 | 1,070 |
| Madison Heights | 60,491 | 8.3 | 1.2 | 65,401 | 55,228 | 40,668 | 41,676 | 36,484 | 15,730 | 14,622 | 63.2 | 143,806 | 1,043 |
| Midland | 69,265 | 14.6 | 7.2 | 82,092 | 36,646 | 40,412 | 49,172 | 27,770 | 18,983 | 17,830 | 70.3 | 156,692 | 819 |
| Mount Pleasant | 30,623 | 39.2 | 3.0 | 47,364 | 19,640 | 10,625 | 13,157 | 7,377 | 9,937 | 8,276 | 39.3 | 137,429 | 684 |
| Muskegon | 28,406 | 30.7 | 0.6 | 41,968 | 20,430 | 21,743 | 23,256 | 21,183 | 16,987 | 14,544 | 44.1 | 80,304 | 707 |
| Novi | 101,404 | 4.5 | 17.1 | 121,457 | 59,702 | 60,551 | 83,291 | 37,352 | 23,897 | 23,192 | 70.8 | 333,828 | 1,305 |

1. Based on population estimated by the American Community Survey.

# Table D. Cities — Commuting, Computer Access, Migration, Labor Force, and Employment

| City | Commuting[1], 2019 Percent — Drove alone | With commutes of 30 minutes or more | Computer access[2], 2019 Percent — With a computer in the house | With Internet access | Migration, 2019 — Percent who lived in the same house one year ago | Percent who lived in another state or county one year ago | Civilian labor force, 2020 — Total | Percent change 2019-2020 | Unemployment Total | Rate[3] | Civilian Employment[4], 2019 — Population age 16 and older Number | Percent in labor force | Population age 16 to 64 Number | Percent who worked full-year full-time |
|---|---|---|---|---|---|---|---|---|---|---|---|---|---|---|
| | 55 | 56 | 57 | 58 | 59 | 60 | 61 | 62 | 63 | 64 | 65 | 66 | 67 | 68 |
| **MASSACHUSETTS—Cont'd** | | | | | | | | | | | | | | |
| Beverly | 69.8 | 35.0 | 94.2 | 89.8 | 89.8 | 4.6 | 22,668 | -5.7 | 1,786 | 8 | 32,635 | 72.4 | 27,313 | 56.1 |
| Boston | 36.9 | 54.8 | 92.7 | 86.6 | 81.5 | 9.9 | 385,385 | -3.5 | 35,539 | 9 | 598,649 | 71.1 | 514,993 | 54.4 |
| Braintree Town | 69.8 | 54.7 | 96.5 | 94.5 | 91.7 | 3.2 | 20,517 | -3.2 | 1,906 | 9 | 31,071 | 63.2 | 22,599 | 57.6 |
| Brockton | 75.2 | 46.5 | 93.5 | 80.6 | 85.4 | 6.6 | 48,752 | -1.3 | 6,404 | 13 | 74,643 | 67.7 | 61,919 | 47.7 |
| Cambridge | 26.6 | 46.4 | 96.2 | 90.7 | 74.7 | 15.4 | 68,371 | -6.8 | 3,426 | 5 | 104,303 | 71.6 | 91,840 | 53.5 |
| Chelsea | 50.9 | 61.0 | 89.2 | 82.5 | 88.5 | 4.2 | 20,684 | -1.5 | 2,483 | 12 | 29,329 | 67.3 | 24,962 | 49.4 |
| Chicopee | 88.3 | 20.9 | 89.2 | 81.6 | NA | NA | 27,111 | -4.1 | 2,837 | 11 | 45,862 | 65.9 | 35,000 | 54.4 |
| Everett | 56.2 | 63.0 | 87.3 | 83.4 | 84.1 | 4.9 | 26,745 | -2.3 | 2,885 | 11 | 40,038 | 68.8 | 32,760 | 54.2 |
| Fall River | 83.9 | 34.1 | 84.7 | 72.8 | 86.8 | 4.4 | 40,226 | -2.4 | 5,169 | 13 | 71,752 | 56.2 | 56,933 | 44.7 |
| Fitchburg | 76.1 | 33.4 | 93.2 | 88.5 | 85.5 | 1.8 | 19,697 | -2.5 | 2,359 | 12 | 31,768 | 67.6 | 26,527 | 45.2 |
| Franklin Town | 73.7 | 52.2 | NA | NA | 92.6 | 4.5 | 18,209 | -2.4 | 1,384 | 8 | 28,006 | 71.3 | 22,907 | 58.0 |
| Gloucester | 80.1 | 38.5 | 93.0 | 87.4 | 94.4 | 0.7 | 15,847 | -3.9 | 1,513 | 10 | 25,501 | 68.2 | 19,338 | 48.9 |
| Haverhill | 87.1 | 34.9 | 96.6 | 84.3 | 88.0 | 3.8 | 34,974 | -2.8 | 3,536 | 10 | 50,995 | 75.1 | 41,784 | 54.8 |
| Holyoke | 82.8 | 15.8 | 91.0 | 83.9 | 94.4 | 1.4 | 16,257 | -2.6 | 2,033 | 13 | 34,203 | 58.7 | 27,102 | 45.1 |
| Lawrence | 72.1 | 30.6 | 86.5 | 71.7 | 84.4 | 5.4 | 37,488 | 3.1 | 6,520 | 17 | 62,737 | 70.6 | 54,423 | 48.4 |
| Leominster | 83.6 | 39.7 | 96.7 | 91.2 | 96.1 | 1.5 | 22,193 | -3.7 | 2,212 | 10 | 34,821 | 67.2 | 26,943 | 57.5 |
| Lowell | 77.5 | 39.2 | 90.3 | 85.1 | 83.3 | 6.1 | 56,176 | -2.9 | 5,727 | 10 | 94,766 | 69.9 | 81,536 | 51.8 |
| Lynn | 66.4 | 51.2 | 88.0 | 73.2 | 87.2 | 5.9 | 47,971 | -1.5 | 5,994 | 13 | 72,675 | 64.9 | 59,598 | 47.8 |
| Malden | 57.3 | 74.7 | 93.4 | 87.7 | 86.4 | 10.1 | 34,238 | -1.8 | 3,864 | 11 | 50,487 | 69.6 | 41,509 | 53.9 |
| Marlborough | 76.4 | 47.0 | 92.4 | 89.6 | 80.0 | 6.4 | 22,707 | -5.2 | 1,836 | 8 | 31,790 | 70.7 | 25,274 | 60.3 |
| Medford | 52.2 | 56.9 | 97.0 | 94.9 | 78.8 | 11.2 | 33,418 | -5.0 | 2,624 | 8 | 48,415 | 71.0 | 39,938 | 56.0 |
| Melrose | 57.8 | 72.2 | 93.6 | 90.1 | NA | NA | 16,019 | -5.6 | 1,146 | 7 | 23,100 | 72.6 | 17,931 | 62.8 |
| Methuen Town | 85.1 | 45.2 | 92.7 | 85.1 | 87.2 | 7.2 | 27,175 | -2.6 | 2,822 | 10 | 42,136 | 65.7 | 32,743 | 55.7 |
| New Bedford | 76.4 | 36.4 | 83.9 | 78.5 | 84.2 | 5.0 | 46,998 | -2.0 | 6,181 | 13 | 75,435 | 59.8 | 60,460 | 47.8 |
| Newton | 59.7 | 48.7 | 96.6 | 95.4 | 90.2 | 6.8 | 45,771 | -6.7 | 2,701 | 6 | 72,330 | 65.3 | 55,719 | 49.5 |
| Northampton | 67.9 | 28.0 | 96.7 | 86.8 | 81.6 | 11.2 | 15,813 | -6.0 | 1,128 | 7 | 25,730 | 61.9 | 19,761 | 37.5 |
| Peabody | 81.7 | 41.0 | 91.6 | 87.2 | 89.7 | 4.5 | 28,700 | -4.1 | 2,770 | 10 | 47,470 | 63.5 | 33,926 | 60.3 |
| Pittsfield | 77.0 | 13.9 | 90.2 | 86.1 | 90.1 | 1.6 | 20,185 | -5.9 | 2,188 | 11 | 37,033 | 60.7 | 25,299 | 48.1 |
| Quincy | 58.5 | 63.8 | 93.5 | 89.6 | 87.5 | 7.4 | 55,141 | -1.4 | 6,101 | 11 | 81,637 | 70.6 | 63,344 | 62.9 |
| Revere | 56.5 | 64.7 | 93.2 | 91.1 | 83.9 | 6.1 | 29,725 | 0.0 | 4,068 | 14 | 43,905 | 71.7 | 36,194 | 58.2 |
| Salem | 65.8 | 47.0 | 97.5 | 95.1 | 86.7 | 7.0 | 23,682 | -4.2 | 2,414 | 10 | 37,745 | 69.0 | 30,454 | 52.7 |
| Somerville | 37.7 | 61.7 | 91.7 | 87.8 | 78.7 | 13.1 | 51,020 | -5.6 | 3,325 | 7 | 71,568 | 79.4 | 63,835 | 62.4 |
| Springfield | 76.9 | 27.1 | 86.2 | 74.5 | 91.1 | 1.9 | 63,824 | -2.1 | 8,777 | 14 | 118,003 | 60.0 | 99,363 | 43.0 |
| Taunton | 82.6 | 46.7 | 85.1 | 77.8 | 82.1 | 5.9 | 30,230 | -2.3 | 3,389 | 11 | 49,124 | 67.4 | 39,193 | 54.0 |
| Waltham | 67.0 | 38.6 | 97.6 | 94.9 | 78.4 | 14.0 | 36,553 | -5.9 | 2,585 | 7 | 56,202 | 71.1 | 48,006 | 54.2 |
| Watertown Town | 55.2 | 42.5 | 94.3 | 90.8 | 80.2 | 10.8 | 22,362 | -5.2 | 1,523 | 7 | 30,879 | 77.2 | 25,027 | 63.5 |
| Westfield | 84.5 | 33.6 | 91.5 | 80.5 | 83.5 | 7.1 | 20,504 | -6.1 | 1,693 | 8 | 33,835 | 53.2 | 25,920 | 42.0 |
| West Springfield Town | 88.2 | 18.2 | 90.3 | 82.3 | 88.9 | 3.6 | 14,500 | -4.3 | 1,417 | 10 | 23,851 | 63.0 | 18,385 | 54.7 |
| Weymouth Town | 73.5 | 54.3 | 93.5 | 84.2 | 93.7 | 5.1 | 31,778 | -3.2 | 3,037 | 10 | 48,903 | 68.6 | 37,906 | 59.8 |
| Woburn | 79.4 | 45.5 | 90.3 | 84.9 | 85.8 | 5.2 | 23,199 | -4.3 | 1,965 | 9 | 33,522 | 70.2 | 26,894 | 59.7 |
| Worcester | 68.2 | 37.4 | 91.2 | 82.8 | 83.4 | 6.8 | 91,756 | -2.8 | 9,529 | 10 | 153,995 | 64.6 | 129,483 | 47.9 |
| **MICHIGAN** | 81.8 | 34.1 | 92.3 | 85.3 | 87.1 | 5.5 | 4,840,843 | -2.0 | 478,115 | 10 | 8,095,913 | 61.8 | 6,329,504 | 50.0 |
| Allen Park | 86.2 | 34.7 | 90.2 | 86.6 | NA | NA | 13,750 | -4.8 | 1,269 | 9 | 21,515 | 64.0 | 17,153 | 56.3 |
| Ann Arbor | 59.0 | 21.4 | 98.4 | 92.8 | 69.8 | 18.8 | 63,640 | -3.6 | 3,358 | 5 | 106,048 | 60.5 | 91,618 | 39.7 |
| Battle Creek | 77.7 | 18.0 | 86.1 | 80.0 | 84.0 | 8.4 | 22,634 | -0.1 | 2,561 | 11 | 41,066 | 63.5 | 32,378 | 51.7 |
| Bay City | 86.6 | 21.1 | 88.9 | 84.9 | 83.1 | 4.8 | 15,547 | 0.3 | 1,825 | 12 | 25,776 | 64.3 | 21,553 | 50.6 |
| Burton | 80.7 | 34.6 | 94.4 | 87.6 | NA | NA | 13,245 | -1.6 | 1,486 | 11 | 22,090 | 57.5 | 16,852 | 48.8 |
| Dearborn | 84.0 | 21.6 | 93.4 | 85.0 | 86.5 | 3.9 | 37,769 | -4.7 | 3,502 | 9 | 69,002 | 56.7 | 57,143 | 41.8 |
| Dearborn Heights | 87.8 | 29.3 | 93.5 | 84.7 | 86.3 | 0.5 | 24,650 | -4.0 | 2,555 | 10 | 42,753 | 60.5 | 34,678 | 43.5 |
| Detroit | 68.4 | 33.9 | 88.1 | 74.1 | 85.7 | 3.5 | 265,548 | 5.1 | 59,380 | 22 | 520,143 | 56.5 | 423,045 | 40.7 |
| East Lansing | 52.9 | 13.5 | 94.1 | 84.5 | 53.0 | 27.9 | 22,644 | -3.6 | 1,211 | 5 | 46,364 | 61.4 | 42,531 | 19.6 |
| Eastpointe | 82.5 | 35.5 | 94.6 | 79.6 | 89.5 | 3.5 | 15,630 | 0.3 | 2,638 | 17 | 26,768 | 61.7 | 22,611 | 46.5 |
| Farmington Hills | 85.5 | 47.2 | 94.3 | 91.3 | 85.0 | 6.9 | 42,017 | -7.2 | 2,542 | 6 | 65,869 | 67.7 | 50,670 | 59.9 |
| Flint | 74.5 | 23.6 | 84.6 | 69.5 | 81.3 | 3.4 | 34,874 | 3.4 | 6,669 | 19 | 74,748 | 55.7 | 61,491 | 30.7 |
| Garden City | 87.8 | 33.6 | 91.7 | 84.0 | 92.2 | 0.8 | 14,395 | -4.6 | 1,382 | 10 | 20,543 | 66.6 | 17,362 | 52.5 |
| Grand Rapids | 74.7 | 18.8 | 91.4 | 83.6 | 80.7 | 6.8 | 105,796 | 0.7 | 10,527 | 10 | 161,637 | 68.3 | 135,764 | 49.1 |
| Holland | 77.9 | 14.3 | 93.0 | 86.7 | 82.1 | 9.4 | 16,975 | -1.4 | 1,308 | 8 | 26,811 | 65.6 | 21,645 | 47.5 |
| Inkster | 85.9 | 34.5 | 88.8 | 81.4 | 76.6 | 7.5 | 9,328 | 2.9 | 1,831 | 20 | 18,274 | 63.3 | 15,395 | 46.4 |
| Jackson | 76.5 | 17.9 | 88.9 | 85.1 | 81.5 | 8.0 | 14,298 | 3.3 | 2,072 | 15 | 24,998 | 60.6 | 21,068 | 37.1 |
| Kalamazoo | 75.6 | 17.1 | 93.9 | 81.8 | 71.7 | 13.5 | 37,002 | -1.1 | 3,301 | 9 | 64,607 | 66.4 | 55,816 | 41.6 |
| Kentwood | 89.5 | 13.2 | 95.6 | 93.1 | 85.5 | 3.7 | 29,849 | -1.5 | 2,142 | 7 | 39,151 | 74.1 | 32,428 | 58.6 |
| Lansing | 72.5 | 17.6 | 93.3 | 85.2 | 82.3 | 6.5 | 59,058 | -1.2 | 6,193 | 11 | 95,249 | 67.5 | 80,353 | 45.7 |
| Lincoln Park | 84.0 | 35.5 | 89.5 | 79.4 | 92.8 | 1.4 | 16,972 | -3.8 | 1,828 | 11 | 29,342 | 60.4 | 23,675 | 48.0 |
| Livonia | 86.2 | 34.3 | 93.1 | 90.1 | 90.9 | 3.4 | 50,024 | -6.5 | 3,201 | 6 | 78,143 | 64.1 | 56,951 | 56.2 |
| Madison Heights | 87.7 | 28.4 | 94.4 | 90.7 | 87.0 | 6.7 | 15,442 | -2.5 | 1,992 | 13 | 26,163 | 68.3 | 21,104 | 64.7 |
| Midland | 89.3 | 30.4 | 93.6 | 90.5 | 80.0 | 9.9 | 20,131 | -2.5 | 1,218 | 6 | 33,367 | 61.2 | 26,227 | 51.0 |
| Mount Pleasant | 69.8 | 12.9 | 95.2 | 79.8 | 65.4 | 21.6 | 11,741 | -5.9 | 994 | 9 | 21,978 | 60.6 | 19,232 | 28.6 |
| Muskegon | 77.9 | 21.3 | 83.8 | 77.2 | 75.1 | 13.1 | 14,092 | 3.7 | 2,590 | 18 | 29,768 | 55.3 | 24,532 | 36.9 |
| Novi | 82.8 | 44.0 | 97.8 | 94.8 | 86.3 | 8.1 | 32,157 | -6.5 | 2,034 | 6 | 45,892 | 70.3 | 37,909 | 55.9 |

1. Employed persons.  2. Households.  3. Percent of civilian labor force.  4. Persons 16 years old and over.

# Table D. Cities — Construction, Wholesale Trade, and Retail Trade

| City | Value of residential construction authorized by building permits, 2020 | | | Wholesale trade[1], 2017 | | | | Retail trade[2], 2017 | | | |
|---|---|---|---|---|---|---|---|---|---|---|---|
| | New construction ($1,000) | Number of housing units | Percent single family | Number of establishments | Number of employees | Sales (mil dol) | Annual payroll (mil dol) | Number of establishments | Number of employees | Sales (mil dol) | Annual payroll (mil dol) |
| | 69 | 70 | 71 | 72 | 73 | 74 | 75 | 76 | 77 | 78 | 79 |
| **MASSACHUSETTS—Cont'd** | | | | | | | | | | | |
| Beverly | 11,768 | 20 | 100 | 44 | 364 | 194 | 24 | 141 | 2,006 | 583 | 59 |
| Boston | 1,606,988 | 3,532 | 1 | 512 | 11,960 | 10,101 | 1,476 | 2,191 | 32,451 | 12,080 | 1,104 |
| Braintree Town | 1,275 | 5 | 100 | 48 | 518 | 370 | 40 | 284 | 5,936 | 1,724 | 181 |
| Brockton | 6,585 | 31 | 100 | 69 | 1,455 | 932 | 100 | 327 | 5,070 | 1,350 | 148 |
| Cambridge | 102,488 | 546 | 6 | 82 | 2,489 | 932 | 198 | 387 | 5,773 | 1,287 | 152 |
| Chelsea | 9,155 | 54 | 0 | 84 | 1,512 | 2,290 | 106 | 109 | 1,882 | 492 | 51 |
| Chicopee | 3,614 | 21 | 100 | 42 | 1,458 | 1,400 | 72 | 166 | 2,576 | 841 | 75 |
| Everett | 22,862 | 124 | 0 | 49 | 1,229 | 1,147 | 80 | 125 | 2,009 | 510 | 57 |
| Fall River | 6,660 | 58 | 100 | 69 | 1,486 | 783 | 80 | 276 | 3,329 | 933 | 97 |
| Fitchburg | 5,600 | 28 | 64 | 34 | 349 | 224 | 25 | 122 | 1,643 | 372 | 41 |
| Franklin Town | 40,828 | 248 | 34 | 47 | 1,179 | 1,690 | 81 | 107 | 1,771 | 525 | 55 |
| Gloucester | 47,493 | 258 | 9 | 37 | 368 | 496 | 21 | 126 | 1,550 | 388 | 43 |
| Haverhill | 10,502 | 42 | 76 | 49 | 460 | 295 | 35 | 137 | 2,525 | 808 | 77 |
| Holyoke | 479 | 3 | 100 | 19 | 365 | 98 | 20 | 207 | 3,651 | 694 | 89 |
| Lawrence | 1,704 | 16 | 38 | 53 | 1,319 | 699 | 142 | 227 | 1,729 | 553 | 57 |
| Leominster | 6,447 | 47 | 30 | 39 | 439 | 280 | 26 | 213 | 4,390 | 1,055 | 103 |
| Lowell | 31,935 | 189 | 10 | 48 | 622 | 411 | 53 | 229 | 2,591 | 847 | 78 |
| Lynn | 15,007 | 140 | 22 | 40 | 396 | 192 | 23 | 218 | 2,266 | 770 | 83 |
| Malden | 0 | 0 | 0 | 36 | 623 | 295 | 36 | 129 | 1,184 | 398 | 35 |
| Marlborough | 3,522 | 25 | 100 | 82 | 2,126 | 2,537 | 175 | 213 | 3,293 | 1,030 | 87 |
| Medford | 48,701 | 1,199 | 1 | 41 | 578 | 273 | 39 | 143 | 1,810 | 663 | 62 |
| Melrose | 2,397 | 10 | 40 | 9 | 45 | 23 | 3 | 59 | 777 | 243 | 27 |
| Methuen Town | 17,852 | 53 | 96 | 40 | 647 | 2,882 | 42 | 126 | 2,260 | 600 | 58 |
| New Bedford | 4,211 | 33 | 58 | 91 | 2,012 | 1,205 | 102 | 291 | 2,945 | 827 | 78 |
| Newton | 29,379 | 57 | 65 | 92 | 1,205 | 1,375 | 143 | 320 | 4,964 | 1,182 | 158 |
| Northampton | 7,203 | 29 | 72 | D | D | D | 9 | 159 | 2,149 | 560 | 62 |
| Peabody | 7,164 | 24 | 92 | 54 | 1,250 | 3,286 | 86 | 290 | 5,289 | 1,358 | 162 |
| Pittsfield | 990 | 5 | 100 | 46 | 688 | 270 | 35 | 192 | 3,125 | 937 | 92 |
| Quincy | 36,029 | 153 | 8 | 64 | 1,095 | 1,580 | 66 | 243 | 4,402 | 1,309 | 130 |
| Revere | 6,590 | 40 | 23 | 21 | 191 | 1,137 | 12 | 122 | 2,085 | 547 | 48 |
| Salem | 56,466 | 283 | 11 | 33 | 191 | 97 | 11 | 177 | 2,205 | 509 | 58 |
| Somerville | 21,216 | 103 | 23 | 48 | 637 | 361 | 51 | 206 | 3,555 | 942 | 95 |
| Springfield | 25,766 | 93 | 83 | 95 | 1,405 | 1,188 | 94 | 460 | 5,693 | 1,383 | 155 |
| Taunton | 14,957 | 90 | 79 | 63 | 3,567 | 2,970 | 275 | 195 | 2,861 | 669 | 78 |
| Waltham | 21,780 | 57 | 72 | 93 | 3,195 | 4,481 | 311 | 237 | 3,454 | 1,076 | 109 |
| Watertown Town | 34,889 | 163 | 9 | 33 | 509 | 178 | 34 | 133 | 2,985 | 858 | 97 |
| Westfield | 8,861 | 45 | 100 | 33 | 1,000 | 1,872 | 52 | 134 | 2,127 | 578 | 55 |
| West Springfield Town | 2,099 | 11 | 64 | 57 | 730 | 261 | 47 | 181 | 3,417 | 1,228 | 112 |
| Weymouth Town | 20,250 | 256 | 6 | 49 | 473 | 397 | 29 | 192 | 2,735 | 827 | 92 |
| Woburn | 75,122 | 376 | 22 | 208 | 3,983 | 2,744 | 314 | 182 | 3,582 | 1,649 | 122 |
| Worcester | 10,844 | 70 | 89 | 161 | 2,114 | 939 | 108 | 552 | 7,378 | 2,036 | 207 |
| **MICHIGAN** | 0 | 0 | 0 | 9,173 | 147,939 | 138,546 | 9,136 | 34,201 | 469,987 | 143,437 | 12,482 |
| Allen Park | 0 | 0 | 0 | 12 | 174 | 127 | 9 | 109 | 2,073 | 425 | 42 |
| Ann Arbor | 39,135 | 254 | 38 | 69 | 453 | 346 | 31 | 540 | 8,265 | 1,978 | 213 |
| Battle Creek | 3,088 | 10 | 100 | 33 | 431 | 303 | 23 | 230 | 3,109 | 822 | 74 |
| Bay City | 7,569 | 80 | 30 | 29 | 511 | 213 | 22 | 133 | 1,125 | 273 | 32 |
| Burton | 6,154 | 28 | 100 | D | D | D | 37 | 126 | 1,964 | 486 | 48 |
| Dearborn | 3,309 | 9 | 100 | 138 | 1,765 | 2,291 | 102 | 556 | 6,600 | 2,111 | 166 |
| Dearborn Heights | 2,720 | 10 | 100 | 41 | 123 | 40 | 5 | 214 | 1,846 | 497 | 41 |
| Detroit | 159,190 | 850 | 2 | 377 | 7,476 | 7,245 | 482 | 1,945 | 13,065 | 3,565 | 311 |
| East Lansing | 722 | 6 | 100 | 10 | 30 | 12 | 2 | 85 | 1,588 | 320 | 38 |
| Eastpointe | 16,597 | 52 | 100 | 10 | 39 | 9 | 2 | 111 | 934 | 290 | 27 |
| Farmington Hills | 3,632 | 20 | 60 | 168 | 3,414 | 4,609 | 286 | 270 | 3,983 | 1,540 | 124 |
| Flint | 0 | 0 | 0 | 61 | 948 | 504 | 39 | 331 | 3,025 | 3,231 | 74 |
| Garden City | 508 | 2 | 100 | 17 | 123 | 28 | 5 | 106 | 744 | 296 | 22 |
| Grand Rapids | 70,618 | 305 | 18 | 209 | 4,832 | 3,440 | 269 | 575 | 7,642 | 2,288 | 220 |
| Holland | 2,740 | 10 | 100 | 40 | 618 | 461 | 40 | 156 | 2,218 | 771 | 67 |
| Inkster | 0 | 0 | 0 | 7 | 89 | 57 | 7 | 51 | 291 | 79 | 6 |
| Jackson | 164 | 1 | 100 | 52 | 743 | 576 | 42 | 152 | 1,606 | 384 | 41 |
| Kalamazoo | 5,677 | 23 | 100 | 84 | 1,564 | 712 | 77 | 264 | 2,768 | 778 | 79 |
| Kentwood | 14,229 | 73 | 100 | 133 | 3,588 | 2,042 | 218 | 285 | 4,828 | 1,339 | 135 |
| Lansing | 42,088 | 315 | 9 | 115 | 1,800 | 3,532 | 100 | 450 | 6,257 | 1,733 | 168 |
| Lincoln Park | 0 | 0 | 0 | D | D | D | D | 125 | 1,319 | 345 | 31 |
| Livonia | 16,627 | 56 | 100 | 242 | 4,365 | 5,038 | 261 | 434 | 7,510 | 2,316 | 201 |
| Madison Heights | 1,740 | 15 | 100 | 118 | 2,526 | 1,475 | 166 | 168 | 3,126 | 1,311 | 98 |
| Midland | 6,234 | 52 | 100 | 35 | 612 | 2,621 | 67 | 200 | 3,468 | 989 | 87 |
| Mount Pleasant | 2,702 | 56 | 0 | 22 | 294 | 181 | 13 | 117 | 2,444 | 624 | 59 |
| Muskegon | 6,826 | 43 | 100 | 34 | 473 | 215 | 24 | 117 | 1,699 | 512 | 44 |
| Novi | 45,907 | 321 | 68 | 171 | 3,161 | 6,133 | 254 | 343 | 7,145 | 2,059 | 191 |

1. Merchant wholesalers except manufacturers' sales branches and offices.  2. Establishments with payroll.

| City | Real estate and rental and leasing, 2017 | | | | Professional, scientific, and technical services[1], 2017 | | | | Manufacturing, 2017 | | | |
|---|---|---|---|---|---|---|---|---|---|---|---|---|
| | Number of establishments | Number of employees | Receipts (mil dol) | Annual payroll (mil dol) | Number of establishments | Number of employees | Receipts (mil dol) | Annual payroll (mil dol) | Number of establishments | Number of employees | Receipts (mil dol) | Annual payroll (mil dol) |
| | 80 | 81 | 82 | 83 | 84 | 85 | 86 | 87 | 88 | 89 | 90 | 91 |
| MASSACHUSETTS—Cont'd | | | | | | | | | | | | |
| Beverly | 50 | 314 | 123 | 19 | 174 | 1,961 | 364 | 156 | NA | NA | NA | NA |
| Boston | 1,263 | 13,923 | 5,397 | 1,124 | 3,485 | 86,292 | 25,456 | 9,929 | NA | NA | NA | NA |
| Braintree Town | 83 | 1,659 | 355 | 110 | 201 | 1,919 | 362 | 166 | NA | NA | NA | NA |
| Brockton | 62 | 226 | 53 | 10 | 129 | 672 | 99 | 38 | NA | NA | NA | NA |
| Cambridge | 210 | 1,392 | 725 | 95 | 983 | 33,024 | 11,263 | 4,389 | NA | NA | NA | NA |
| Chelsea | 35 | 178 | 28 | 6 | 36 | 219 | 47 | 22 | NA | NA | NA | NA |
| Chicopee | 41 | 188 | 44 | 7 | 43 | 618 | 48 | 20 | NA | NA | NA | NA |
| Everett | 24 | 76 | 24 | 4 | D | D | D | D | NA | NA | NA | NA |
| Fall River | 75 | 719 | 74 | 17 | 159 | 1,058 | 137 | 48 | NA | NA | NA | NA |
| Fitchburg | 31 | 155 | 29 | 6 | 45 | 303 | 39 | 15 | NA | NA | NA | NA |
| Franklin Town | 35 | 153 | 73 | 10 | 96 | 804 | 160 | 62 | NA | NA | NA | NA |
| Gloucester | 27 | 71 | 14 | 3 | 85 | 292 | 73 | 19 | NA | NA | NA | NA |
| Haverhill | 48 | 325 | 68 | 15 | 84 | 612 | 89 | 33 | NA | NA | NA | NA |
| Holyoke | 42 | 348 | 54 | 15 | 64 | 386 | 66 | 24 | NA | NA | NA | NA |
| Lawrence | 45 | 318 | 52 | 13 | 65 | 390 | 62 | 25 | NA | NA | NA | NA |
| Leominster | 44 | 476 | 101 | 22 | 95 | 577 | 71 | 33 | NA | NA | NA | NA |
| Lowell | 78 | 553 | 92 | 20 | 135 | 1,631 | 297 | 116 | NA | NA | NA | NA |
| Lynn | 46 | 180 | 52 | 9 | 80 | 343 | 41 | 16 | NA | NA | NA | NA |
| Malden | 53 | 374 | 81 | 19 | 58 | 204 | 27 | 10 | NA | NA | NA | NA |
| Marlborough | 53 | 210 | 99 | 12 | 198 | 4,420 | 2,039 | 475 | NA | NA | NA | NA |
| Medford | 46 | 193 | 71 | 12 | 144 | 2,803 | 782 | 278 | NA | NA | NA | NA |
| Melrose | 23 | 95 | 39 | 6 | 77 | 276 | 37 | 17 | NA | NA | NA | NA |
| Methuen Town | 39 | D | 27 | D | 78 | 359 | 70 | 19 | NA | NA | NA | NA |
| New Bedford | 80 | 452 | 87 | 16 | 177 | 1,065 | 135 | 49 | NA | NA | NA | NA |
| Newton | 155 | 3,059 | 2,649 | 302 | 606 | 4,535 | 957 | 487 | NA | NA | NA | NA |
| Northampton | 33 | 92 | 26 | 4 | 122 | 682 | 113 | 41 | NA | NA | NA | NA |
| Peabody | 45 | 285 | 75 | 14 | 115 | 856 | 176 | 66 | NA | NA | NA | NA |
| Pittsfield | 43 | 265 | 37 | 9 | 134 | 2,709 | 570 | 262 | NA | NA | NA | NA |
| Quincy | 107 | 651 | 161 | 40 | 283 | 2,860 | 507 | 231 | NA | NA | NA | NA |
| Revere | 28 | 233 | 45 | 8 | 47 | 153 | 16 | 6 | NA | NA | NA | NA |
| Salem | 42 | 232 | 58 | 10 | 159 | 682 | 107 | 43 | NA | NA | NA | NA |
| Somerville | 79 | 445 | 117 | 23 | 208 | 1,315 | 284 | 107 | NA | NA | NA | NA |
| Springfield | 118 | 780 | 135 | 35 | 270 | 2,349 | 380 | 151 | NA | NA | NA | NA |
| Taunton | 38 | 126 | 39 | 6 | 93 | 567 | 93 | 41 | NA | NA | NA | NA |
| Waltham | 122 | 849 | 309 | 58 | 420 | 17,227 | 3,653 | 2,020 | NA | NA | NA | NA |
| Watertown Town | 41 | 167 | 35 | 8 | 133 | 2,782 | 595 | 297 | NA | NA | NA | NA |
| Westfield | 35 | 150 | 40 | 6 | 64 | 602 | 85 | 41 | NA | NA | NA | NA |
| West Springfield Town | 40 | 158 | 39 | 7 | 73 | 665 | 123 | 51 | NA | NA | NA | NA |
| Weymouth Town | 44 | 127 | 37 | 5 | 99 | 544 | 114 | 41 | NA | NA | NA | NA |
| Woburn | 91 | 1,979 | 299 | 110 | 329 | 6,002 | 1,421 | 582 | NA | NA | NA | NA |
| Worcester | 175 | 762 | 171 | 32 | 419 | 3,958 | 665 | 351 | NA | NA | NA | NA |
| MICHIGAN | 8,467 | 54,808 | 17,783 | 2,374 | 21,832 | 282,246 | 39,437 | 19,670 | 12,418 | 582,365 | 262,495 | 33,349 |
| Allen Park | D | D | D | 4 | 54 | 556 | 160 | 34 | NA | NA | NA | NA |
| Ann Arbor | 167 | 1,864 | 397 | 106 | 646 | 7,279 | 1,434 | 590 | NA | NA | NA | NA |
| Battle Creek | 44 | 271 | 40 | 9 | 100 | 2,176 | 95 | 189 | NA | NA | NA | NA |
| Bay City | 21 | 93 | 10 | 2 | 81 | 588 | 57 | 29 | NA | NA | NA | NA |
| Burton | 20 | 79 | 14 | 2 | 36 | 329 | 25 | 9 | NA | NA | NA | NA |
| Dearborn | D | D | D | D | 271 | 15,810 | 714 | 918 | NA | NA | NA | NA |
| Dearborn Heights | D | D | D | D | 89 | 291 | 29 | 12 | NA | NA | NA | NA |
| Detroit | 268 | 1,832 | 390 | 96 | 764 | 16,904 | 3,975 | 1,398 | NA | NA | NA | NA |
| East Lansing | 44 | 398 | 46 | 14 | 125 | 1,551 | 281 | 104 | NA | NA | NA | NA |
| Eastpointe | 14 | 56 | 17 | 2 | 33 | 182 | 19 | 8 | NA | NA | NA | NA |
| Farmington Hills | 193 | 1,998 | 405 | 117 | 724 | 14,791 | 1,925 | 1,089 | NA | NA | NA | NA |
| Flint | 66 | 519 | 83 | 20 | 123 | 631 | 75 | 35 | NA | NA | NA | NA |
| Garden City | D | D | D | 2 | 24 | 72 | 8 | 2 | NA | NA | NA | NA |
| Grand Rapids | 197 | 1,401 | 227 | 70 | 635 | 8,130 | 1,530 | 545 | NA | NA | NA | NA |
| Holland | 57 | 274 | 65 | 12 | 90 | 2,365 | 434 | 236 | NA | NA | NA | NA |
| Inkster | D | D | D | 3 | 10 | 42 | 5 | 1 | NA | NA | NA | NA |
| Jackson | 36 | 163 | 40 | 6 | 70 | 1,108 | 112 | 58 | NA | NA | NA | NA |
| Kalamazoo | 87 | 639 | 96 | 19 | 195 | 1,686 | 288 | 103 | NA | NA | NA | NA |
| Kentwood | 51 | 188 | 65 | 9 | 115 | 1,431 | 227 | 89 | NA | NA | NA | NA |
| Lansing | 98 | 690 | 128 | 39 | 250 | 2,760 | 521 | 190 | NA | NA | NA | NA |
| Lincoln Park | D | D | D | 5 | 34 | 317 | 37 | 11 | NA | NA | NA | NA |
| Livonia | 96 | 808 | 200 | 39 | 368 | 7,825 | 1,472 | 599 | NA | NA | NA | NA |
| Madison Heights | 40 | 491 | 128 | 30 | 119 | 1,663 | 196 | 120 | NA | NA | NA | NA |
| Midland | 51 | 316 | 66 | 17 | 125 | 2,298 | 132 | 340 | NA | NA | NA | NA |
| Mount Pleasant | 39 | 1,015 | 63 | 27 | 61 | 470 | 45 | 20 | NA | NA | NA | NA |
| Muskegon | 19 | 126 | 22 | 3 | 83 | 768 | 99 | 41 | NA | NA | NA | NA |
| Novi | 79 | 321 | 148 | 14 | 350 | 5,183 | 897 | 405 | NA | NA | NA | NA |

1. Establishments subject to federal tax.

# Accommodation and Food Services, Arts, Entertainment, and Recreation, and Health Care and Social Assistance

| City | Accommodation and food services, 2017 | | | | Arts, entertainment, and recreation[1], 2017 | | | | Health care and social assistance[1], 2017 | | | |
|---|---|---|---|---|---|---|---|---|---|---|---|---|
| | Number of establishments | Number of employees | Receipts (mil dol) | Annual payroll (mil dol) | Number of establishments | Number of employees | Receipts (mil dol) | Annual payroll (mil dol) | Number of establishments | Number of employees | Receipts (mil dol) | Annual payroll (mil dol) |
| | 92 | 93 | 94 | 95 | 96 | 97 | 98 | 99 | 100 | 101 | 102 | 103 |
| **MASSACHUSETTS—Cont'd** | | | | | | | | | | | | |
| Beverly | 119 | 1,823 | 112 | 37 | 28 | 750 | 43 | 15 | 180 | 4,732 | 652 | 282 |
| Boston | 2,456 | 61,705 | 6,073 | 1,769 | 361 | 13,163 | 2,181 | 878 | 1,747 | 133,669 | 21,412 | 9,308 |
| Braintree Town | 123 | 2,746 | 196 | 57 | 21 | 328 | 47 | 8 | 126 | 3,818 | 377 | 144 |
| Brockton | 168 | 2,447 | 165 | 49 | 15 | 337 | 18 | 6 | 294 | 12,717 | 1,445 | 637 |
| Cambridge | 476 | 12,601 | 1,005 | 327 | 82 | 1,284 | 247 | 35 | 361 | 9,551 | 1,323 | 613 |
| Chelsea | 84 | 941 | 72 | 19 | D | D | D | D | 82 | 2,574 | 161 | 83 |
| Chicopee | 101 | 1,665 | 98 | 29 | 9 | 115 | 7 | 2 | 95 | 1,903 | 120 | 58 |
| Everett | 90 | 1,153 | 88 | 24 | 7 | 123 | 6 | 2 | 53 | 1,122 | 178 | 70 |
| Fall River | 189 | 2,538 | 149 | 44 | 19 | 310 | 16 | 6 | 306 | 11,860 | 1,265 | 548 |
| Fitchburg | 81 | 1,032 | 60 | 16 | 7 | 63 | 7 | 3 | 124 | 3,604 | 228 | 104 |
| Franklin Town | 67 | 1,420 | 75 | 23 | D | D | D | D | 73 | 1,100 | 98 | 45 |
| Gloucester | 106 | 1,313 | 105 | 35 | 27 | 336 | 28 | 11 | 87 | 1,783 | 245 | 83 |
| Haverhill | 134 | 1,903 | 121 | 35 | 32 | 743 | 38 | 14 | 143 | 3,560 | 302 | 132 |
| Holyoke | 98 | 1,603 | 100 | 28 | 15 | 197 | 9 | 3 | 143 | 5,845 | 537 | 256 |
| Lawrence | 125 | 1,239 | 85 | 22 | D | D | D | 1 | 190 | 10,384 | 870 | 420 |
| Leominster | 114 | 2,172 | 124 | 37 | 16 | 127 | 7 | 2 | 129 | 3,902 | 467 | 188 |
| Lowell | 207 | 2,972 | 186 | 53 | 22 | 291 | 16 | 5 | 227 | 10,190 | 1,148 | 443 |
| Lynn | 159 | 1,625 | 112 | 30 | 16 | 207 | 9 | 3 | 204 | 7,025 | 520 | 249 |
| Malden | 120 | 1,345 | 86 | 24 | 11 | 111 | 13 | 4 | 118 | 2,984 | 325 | 115 |
| Marlborough | 153 | 2,429 | 183 | 53 | 21 | 367 | 26 | 7 | 145 | 7,591 | 982 | 380 |
| Medford | 128 | 2,288 | 160 | 45 | D | D | D | D | 128 | 3,305 | 422 | 182 |
| Melrose | 43 | 605 | 38 | 12 | 12 | 276 | 11 | 5 | 81 | 2,114 | 258 | 119 |
| Methuen Town | 98 | 1,889 | 126 | 33 | 11 | 340 | 16 | 5 | 116 | 3,894 | 404 | 181 |
| New Bedford | 210 | 2,719 | 158 | 46 | 24 | 361 | 23 | 7 | 226 | 8,588 | 837 | 355 |
| Newton | 235 | 4,069 | 324 | 99 | 79 | 1,401 | 177 | 73 | 490 | 11,068 | 1,291 | 629 |
| Northampton | 105 | 1,806 | 98 | 35 | 28 | 402 | 22 | 7 | 175 | 5,830 | 674 | 321 |
| Peabody | 145 | 2,698 | 190 | 57 | 13 | 334 | 19 | 7 | 170 | 5,803 | 651 | 275 |
| Pittsfield | 146 | 1,822 | 106 | 33 | 30 | 400 | 47 | 12 | 227 | 7,371 | 875 | 382 |
| Quincy | 246 | 3,397 | 244 | 69 | 35 | 892 | 51 | 18 | 287 | 8,035 | 809 | 337 |
| Revere | 105 | 1,463 | 96 | 26 | D | D | D | D | 81 | 1,179 | 97 | 46 |
| Salem | 155 | 2,353 | 163 | 50 | 32 | 688 | 66 | 24 | 134 | 5,978 | 823 | 360 |
| Somerville | 243 | 4,058 | 281 | 94 | 32 | 470 | 84 | 15 | 159 | 4,179 | 561 | 201 |
| Springfield | 252 | 4,603 | 267 | 81 | 20 | 349 | 31 | 8 | 445 | 24,004 | 3,136 | 1,284 |
| Taunton | 118 | 2,103 | 107 | 33 | 13 | 283 | 13 | 4 | 133 | 3,589 | 439 | 165 |
| Waltham | 328 | 4,343 | 345 | 102 | 32 | 581 | 50 | 13 | 219 | 5,244 | 613 | 279 |
| Watertown Town | 98 | 1,354 | 95 | 28 | 25 | 310 | 28 | 7 | 119 | 1,585 | 188 | 74 |
| Westfield | 75 | 1,217 | 62 | 20 | 13 | 324 | 9 | 4 | 96 | 2,421 | 231 | 106 |
| West Springfield Town | 101 | 2,114 | 125 | 38 | 17 | 399 | 37 | 9 | 116 | 3,126 | 221 | 98 |
| Weymouth Town | 105 | 1,443 | 91 | 26 | 17 | 363 | 20 | 9 | 165 | 8,909 | 1,018 | 477 |
| Woburn | 120 | 2,209 | 188 | 47 | 30 | 615 | 37 | 10 | 178 | 4,761 | 443 | 187 |
| Worcester | 487 | 7,323 | 460 | 132 | 44 | 1,661 | 83 | 28 | 648 | 33,563 | 4,889 | 1,813 |
| **MICHIGAN** | 20,696 | 399,032 | 23,056 | 6,563 | 3,469 | 49,733 | 4,998 | 1,750 | 26,977 | 627,808 | 74,195 | 29,310 |
| Allen Park | 72 | 1,394 | 78 | 23 | 6 | 120 | 7 | 2 | 84 | 1,210 | 129 | 55 |
| Ann Arbor | 434 | 10,233 | 625 | 190 | 71 | 849 | 46 | 16 | 372 | 24,404 | 3,919 | 1,641 |
| Battle Creek | D | D | D | D | 19 | 298 | 14 | 5 | 171 | 6,712 | 824 | 327 |
| Bay City | 96 | 1,648 | 78 | 22 | 17 | 248 | 10 | 3 | 150 | 3,562 | 437 | 161 |
| Burton | 57 | 985 | 45 | 12 | 7 | 57 | 6 | 1 | 90 | 1,564 | 68 | 30 |
| Dearborn | 293 | 4,836 | 281 | 77 | 28 | D | 493 | D | 464 | 9,746 | 1,327 | 500 |
| Dearborn Heights | 117 | 1,739 | 98 | 26 | 8 | 73 | 4 | 1 | 133 | 1,287 | 114 | 46 |
| Detroit | 963 | 25,115 | 2,730 | 648 | 92 | 4,307 | 730 | 406 | 982 | 50,181 | 7,485 | 2,858 |
| East Lansing | 136 | 2,704 | 128 | 38 | 10 | 149 | 11 | 3 | 130 | 2,818 | 313 | 134 |
| Eastpointe | 54 | 848 | 41 | 11 | 4 | 19 | 1 | 0 | 94 | 432 | 39 | 16 |
| Farmington Hills | 182 | 3,003 | 176 | 49 | 40 | 675 | 40 | 15 | 421 | 15,018 | 1,341 | 603 |
| Flint | 171 | 2,787 | 142 | 44 | 15 | 547 | 44 | 13 | 201 | 9,385 | 1,279 | 463 |
| Garden City | 45 | 623 | 32 | 8 | D | D | D | D | 81 | 1,990 | 247 | 86 |
| Grand Rapids | 477 | 11,352 | 592 | 197 | 75 | 1,717 | 133 | 39 | 559 | 30,484 | 4,132 | 1,398 |
| Holland | 79 | 2,118 | 102 | 36 | 9 | 238 | 10 | 3 | 135 | 4,238 | 400 | 193 |
| Inkster | 17 | 140 | 10 | 2 | NA | NA | NA | NA | 38 | 691 | 52 | 20 |
| Jackson | 86 | 1,392 | 78 | 20 | 19 | 118 | 9 | 3 | 178 | 7,385 | 869 | 392 |
| Kalamazoo | 228 | 5,225 | 238 | 81 | 41 | 912 | 44 | 16 | 232 | 13,832 | 2,177 | 842 |
| Kentwood | 103 | 2,415 | 129 | 39 | 19 | 549 | 25 | 7 | 147 | 4,146 | 471 | 241 |
| Lansing | 226 | 4,437 | 211 | 68 | 30 | 786 | 108 | 18 | 271 | 14,168 | 1,859 | 723 |
| Lincoln Park | 67 | 1,066 | 58 | 15 | 3 | 27 | 1 | 1 | 60 | 1,048 | 105 | 41 |
| Livonia | 282 | 6,259 | 344 | 102 | 40 | 635 | 39 | 11 | 492 | 11,622 | 1,341 | 587 |
| Madison Heights | 114 | 1,890 | 115 | 32 | 10 | 74 | 4 | 1 | 115 | 2,645 | 325 | 130 |
| Midland | 126 | 2,973 | 155 | 47 | 24 | 394 | 36 | 11 | 249 | 6,583 | 866 | 285 |
| Mount Pleasant | D | D | D | D | D | D | D | D | 159 | 2,200 | 216 | 77 |
| Muskegon | 85 | 1,442 | 73 | 21 | 14 | 260 | 17 | 7 | 118 | 7,161 | 1,115 | 411 |
| Novi | 187 | 4,586 | 265 | 79 | 32 | 847 | 180 | 22 | 296 | 7,623 | 912 | 378 |

1. Establishments subject to federal tax.

## Table D. Cities — Other Services and Government Employment and Payroll

| City | Other services[1] Number of establishments | Number of employees | Receipts (mil dol) | Annual payroll (mil dol) | Government employment and payroll, 2017 — March payroll Full-time equivalent employees | Total (dollars) | Percent of total for: Administrative, judicial, and legal | Police and corrections | Fire protection | Highways and transportation | Health and welfare | Natural resources and utilities | Education and libraries |
|---|---|---|---|---|---|---|---|---|---|---|---|---|---|
| | 104 | 105 | 106 | 107 | 108 | 109 | 110 | 111 | 112 | 113 | 114 | 115 | 116 |
| **MASSACHUSETTS—Cont'd** | | | | | | | | | | | | | |
| Beverly | 100 | 591 | 71 | 20 | 857 | 5,000,348 | 4.2 | 11.7 | 9.4 | 4.7 | 1.9 | 2.7 | 64.5 |
| Boston | 1,852 | 19,411 | 3,451 | 820 | 19,672 | 132,820,934 | 3.5 | 21.3 | 11.1 | 1.9 | 6.9 | 5.2 | 48.7 |
| Braintree Town | 117 | 808 | 95 | 28 | 1,353 | 8,370,821 | 2.4 | 9.0 | 10.9 | 2.5 | 1.8 | 13.7 | 59.5 |
| Brockton | 184 | 1,369 | 127 | 42 | 3,379 | 18,208,042 | 2.5 | 10.9 | 5.1 | 1.3 | 0.7 | 2.6 | 76.1 |
| Cambridge | 284 | 2,154 | 353 | 104 | 7,136 | 36,213,738 | 5.0 | 8.2 | 5.6 | 2.0 | 46.6 | 3.5 | 26.1 |
| Chelsea | 52 | 674 | 51 | 17 | 1,366 | 6,341,623 | 2.8 | 12.7 | 12.0 | 0.5 | 1.4 | 0.7 | 68.5 |
| Chicopee | 96 | 536 | 46 | 14 | 2,238 | 11,063,107 | 2.6 | 7.9 | 7.8 | 0.8 | 0.9 | 8.5 | 69.6 |
| Everett | 76 | 404 | 44 | 15 | 1,308 | 7,801,522 | 3.6 | 12.8 | 9.3 | 1.1 | 0.9 | 1.3 | 64.8 |
| Fall River | 158 | 873 | 84 | 25 | 2,318 | 11,491,808 | 2.3 | 16.8 | 9.0 | 2.1 | 1.1 | 2.1 | 65.4 |
| Fitchburg | 61 | 316 | 36 | 10 | 1,190 | 5,658,922 | 3.8 | 10.3 | 8.1 | 2.9 | 1.5 | 5.6 | 67.2 |
| Franklin Town | D | D | D | 22 | 983 | 5,249,148 | 3.5 | 5.6 | 5.1 | 1.4 | 0.5 | 3.1 | 79.9 |
| Gloucester | 80 | 400 | 36 | 12 | 973 | 4,543,657 | 5.4 | 10.2 | 10.4 | 1.1 | 3.3 | 1.2 | 63.2 |
| Haverhill | 100 | 500 | 48 | 16 | 1,778 | 9,771,010 | 3.6 | 7.9 | 7.8 | 2.3 | 3.5 | 5.6 | 67.9 |
| Holyoke | 56 | 397 | 22 | 8 | 2,086 | 9,796,872 | 2.5 | 12.7 | 7.6 | 1.8 | 1.4 | 14.9 | 58.7 |
| Lawrence | 131 | 812 | 79 | 23 | 497 | 3,463,746 | 14.0 | 35.2 | 24.6 | 16.7 | 5.0 | 3.0 | 1.7 |
| Leominster | 71 | 329 | 29 | 8 | 1,050 | 5,639,777 | 2.0 | 7.4 | 7.8 | 1.3 | 0.6 | 1.9 | 79.0 |
| Lowell | 183 | 906 | 96 | 27 | 3,063 | 17,623,297 | 2.6 | 11.7 | 7.9 | 1.7 | 2.4 | 2.9 | 70.2 |
| Lynn | 142 | 590 | 57 | 19 | 3,251 | 15,409,664 | 1.5 | 7.2 | 7.1 | 0.5 | 1.9 | 1.6 | 78.3 |
| Malden | 119 | 738 | 67 | 18 | 1,320 | 7,778,185 | 2.4 | 8.1 | 5.9 | 0.9 | 0.5 | 1.3 | 78.4 |
| Marlborough | 100 | 984 | 132 | 58 | 1,235 | 6,146,989 | 2.4 | 10.2 | 7.4 | 2.1 | 2.6 | 3.0 | 69.8 |
| Medford | 115 | 1,034 | 307 | 59 | 1,533 | 6,856,790 | 3.8 | 14.8 | 13.1 | 4.6 | 0.5 | 2.3 | 60.8 |
| Melrose | 56 | 343 | 34 | 10 | 705 | 4,004,590 | 4.7 | 11.5 | 9.7 | 0.5 | 3.4 | 3.4 | 59.5 |
| Methuen Town | 67 | 364 | 30 | 10 | 1,188 | 8,629,719 | 2.0 | 11.8 | 7.5 | 1.9 | 1.0 | 3.8 | 71.3 |
| New Bedford | 161 | 1,036 | 109 | 31 | 2,810 | 13,820,091 | 3.1 | 13.6 | 9.2 | 2.0 | 2.1 | 3.7 | 61.4 |
| Newton | 241 | 1,967 | 338 | 79 | 3,271 | 19,625,415 | 3.7 | 7.6 | 6.7 | 3.4 | 1.9 | 3.9 | 71.3 |
| Northampton | 86 | 554 | 60 | 19 | 1,368 | 7,743,025 | 7.0 | 16.4 | 15.1 | 3.4 | 4.9 | 8.4 | 44.6 |
| Peabody | 142 | 679 | 67 | 20 | 1,434 | 7,145,303 | 4.0 | 8.9 | 8.3 | 1.9 | 3.6 | 10.2 | 62.2 |
| Pittsfield | 95 | 712 | 56 | 18 | 1,363 | 6,147,139 | 1.6 | 7.8 | 5.6 | 0.1 | 1.2 | 1.6 | 80.6 |
| Quincy | 214 | 1,482 | 298 | 67 | 2,574 | 15,779,182 | 3.1 | 12.8 | 13.6 | 3.0 | 1.4 | 3.6 | 59.4 |
| Revere | 84 | 364 | 43 | 10 | 1,206 | 7,117,092 | 4.2 | 9.9 | 8.8 | 1.8 | 2.3 | 0.7 | 71.5 |
| Salem | 124 | 709 | 67 | 21 | 1,295 | 6,526,007 | 3.6 | 9.7 | 7.2 | 2.1 | 1.1 | 1.9 | 72.5 |
| Somerville | 132 | 1,436 | 125 | 41 | 1,839 | 11,122,600 | 5.3 | 8.5 | 11.4 | 3.3 | 5.9 | 1.9 | 55.6 |
| Springfield | 227 | 2,584 | 265 | 83 | 5,718 | 28,054,091 | 3.0 | 12.3 | 5.8 | 1.7 | 1.7 | 2.4 | 72.1 |
| Taunton | 103 | 462 | 41 | 12 | 1,684 | 8,439,690 | 2.6 | 11.6 | 9.2 | 1.1 | 6.2 | 4.2 | 63.5 |
| Waltham | 176 | 1,075 | 140 | 52 | 1,717 | 9,886,003 | 4.5 | 13.4 | 12.2 | 5.0 | 1.6 | 2.7 | 57.2 |
| Watertown Town | 101 | 1,116 | 311 | 81 | 644 | 3,642,899 | 6.4 | 19.9 | 15.6 | 1.2 | 1.7 | 3.9 | 51.1 |
| Westfield | 64 | 452 | 37 | 12 | 1,487 | 7,427,763 | 3.8 | 9.6 | 7.8 | 1.7 | 1.4 | 12.9 | 62.0 |
| West Springfield Town | 61 | 467 | 51 | 15 | 991 | 5,114,134 | 2.9 | 11.0 | 8.8 | 5.2 | 0.7 | 0.4 | 69.0 |
| Weymouth Town | 131 | 1,086 | 77 | 26 | 1,301 | 6,282,223 | 0.0 | 0.0 | 0.0 | 0.0 | 0.0 | 0.0 | 100.0 |
| Woburn | 143 | 1,188 | 278 | 48 | 901 | 5,716,920 | 3.6 | 10.0 | 8.0 | 4.0 | 1.7 | 2.3 | 69.6 |
| Worcester | 358 | 2,286 | 232 | 70 | 1,745 | 10,914,232 | 11.0 | 35.4 | 26.7 | 5.4 | 6.3 | 10.3 | 3.4 |
| **MICHIGAN** | 16,545 | 104,291 | 13,205 | 3,415 | X | X | X | X | X | X | X | X | X |
| Allen Park | 57 | 239 | 25 | 5 | 138 | 630,053 | 8.1 | 35.8 | 20.6 | 5.1 | 5.1 | 19.4 | 4.8 |
| Ann Arbor | 235 | 1,855 | 297 | 78 | 949 | 5,739,709 | 12.5 | 35.7 | 10.7 | 11.2 | 2.3 | 11.7 | 0.0 |
| Battle Creek | 96 | 845 | 478 | 45 | 505 | 2,561,641 | 12.1 | 28.3 | 18.5 | 10.7 | 1.5 | 16.6 | 0.0 |
| Bay City | 73 | 458 | 45 | 13 | 268 | 1,253,021 | 17.2 | 24.6 | 10.6 | 5.6 | 1.2 | 36.7 | 0.0 |
| Burton | D | D | D | D | 94 | 426,834 | 23.9 | 46.3 | 2.8 | 10.1 | 2.4 | 10.1 | 0.0 |
| Dearborn | 227 | 1,444 | 137 | 35 | 919 | 3,774,428 | 12.9 | 29.4 | 13.0 | 3.9 | 5.8 | 11.8 | 4.5 |
| Dearborn Heights | 110 | 365 | 34 | 8 | 301 | 1,416,010 | 13.7 | 40.6 | 19.1 | 3.5 | 0.6 | 11.8 | 3.5 |
| Detroit | 785 | 5,867 | 1,029 | 216 | 8,795 | 32,957,680 | 15.3 | 46.9 | 11.2 | 6.3 | 2.8 | 6.9 | 2.0 |
| East Lansing | 57 | 682 | 139 | 35 | 380 | 1,883,723 | 16.9 | 28.6 | 14.7 | 3.7 | 3.3 | 19.2 | 4.0 |
| Eastpointe | 57 | 215 | 25 | 6 | 178 | 824,608 | 14.8 | 39.6 | 19.5 | 6.0 | 0.0 | 11.0 | 3.5 |
| Farmington Hills | 168 | 2,344 | 162 | 74 | 461 | 2,518,181 | 14.5 | 38.0 | 21.3 | 10.6 | 0.5 | 13.1 | 0.0 |
| Flint | 116 | 1,227 | 359 | 45 | 3,499 | 17,528,973 | 1.8 | 3.9 | 4.4 | 1.0 | 84.8 | 3.1 | 0.0 |
| Garden City | D | D | D | D | 87 | 459,448 | 16.9 | 45.3 | 27.3 | 0.0 | 0.0 | 5.7 | 3.8 |
| Grand Rapids | 382 | 2,679 | 344 | 84 | 1,515 | 8,401,875 | 15.7 | 28.8 | 15.5 | 4.6 | 4.4 | 15.1 | 4.6 |
| Holland | 72 | 575 | 66 | 18 | 211 | 1,062,551 | 15.8 | 37.5 | 13.5 | 7.0 | 2.6 | 10.6 | 0.0 |
| Inkster | 19 | 65 | 3 | 1 | 115 | 478,647 | 13.6 | 33.1 | 24.0 | 0.0 | 18.9 | 10.5 | 0.0 |
| Jackson | 74 | 484 | 73 | 15 | 217 | 1,090,462 | 14.9 | 29.2 | 12.6 | 4.0 | 2.6 | 31.8 | 0.0 |
| Kalamazoo | 162 | 1,809 | 247 | 53 | 542 | 2,984,444 | 11.0 | 48.8 | 7.8 | 2.5 | 3.2 | 23.1 | 0.0 |
| Kentwood | 83 | 1,052 | 115 | 38 | 1 | 6,650 | 0.0 | 100.0 | 0.0 | 0.0 | 0.0 | 0.0 | 0.0 |
| Lansing | 266 | 2,122 | 339 | 88 | 1,717 | 10,111,886 | 7.4 | 14.2 | 9.5 | 3.6 | 2.0 | 37.7 | 0.0 |
| Lincoln Park | 75 | 323 | 49 | 11 | 134 | 591,702 | 11.0 | 45.0 | 24.3 | 3.6 | 4.3 | 7.6 | 0.1 |
| Livonia | 245 | 1,811 | 274 | 67 | 677 | 3,317,307 | 15.1 | 30.3 | 17.5 | 7.3 | 2.2 | 11.8 | 4.9 |
| Madison Heights | 88 | 543 | 70 | 21 | 158 | 872,404 | 13.8 | 41.4 | 19.0 | 4.9 | 5.5 | 8.8 | 3.0 |
| Midland | 104 | 845 | 146 | 27 | 384 | 1,874,862 | 12.5 | 15.8 | 14.4 | 16.3 | 3.5 | 24.7 | 7.3 |
| Mount Pleasant | 50 | 317 | 25 | 7 | 125 | 597,030 | 18.6 | 33.6 | 13.7 | 8.7 | 1.2 | 18.2 | 0.0 |
| Muskegon | 58 | 428 | 48 | 11 | 270 | 1,123,408 | 10.3 | 39.2 | 16.9 | 6.3 | 6.9 | 13.7 | 0.0 |
| Novi | 119 | 1,192 | 160 | 49 | 328 | 1,626,861 | 15.0 | 34.0 | 15.0 | 7.2 | 1.6 | 10.4 | 7.2 |

1. Establishments subject to federal tax.

| City | City government finances, 2017 | | | | | | | | | |
|---|---|---|---|---|---|---|---|---|---|---|
| | General revenue | | | | | | | General expenditure | | |
| | Intergovernmental | | | Taxes | | | | | | |
| | | | | | Per capita[1] (dollars) | | | | Per capita[1] (dollars) | |
| | Total (mil dol) | Total (mil dol) | Percent from state government | Total (mil dol) | Total | Property | Sales and gross receipts | Total (mil dol) | Total | Capital outlays |
| | 117 | 118 | 119 | 120 | 121 | 122 | 123 | 124 | 125 | 126 |
| **MASSACHUSETTS—Cont'd** | | | | | | | | | | |
| Beverly | 171.4 | 44.1 | 92.5 | 101.7 | 2,426 | 2,367 | 59 | 162 | 3,869 | 1,021 |
| Boston | 3,800.8 | 945.0 | 91.1 | 2,365.0 | 3,439 | 3,077 | 362 | 3,729 | 5,422 | 605 |
| Braintree Town | 140.0 | 35.0 | 95.9 | 96.0 | 2,578 | 2,438 | 140 | 133 | 3,574 | 177 |
| Brockton | 415.3 | 229.7 | 98.4 | 145.2 | 1,518 | 1,460 | 58 | 376 | 3,927 | 65 |
| Cambridge | 1,207.0 | 357.5 | 73.2 | 431.5 | 3,692 | 3,255 | 437 | 1,103 | 9,436 | 808 |
| Chelsea | 207.0 | 112.4 | 98.5 | 69.4 | 1,722 | 1,626 | 95 | 196 | 4,875 | 717 |
| Chicopee | 221.8 | 88.8 | 98.6 | 86.2 | 1,558 | 1,509 | 49 | 207 | 3,748 | 80 |
| Everett | 202.2 | 89.5 | 97.3 | 102.0 | 2,213 | 2,164 | 48 | 189 | 4,099 | 291 |
| Fall River | 329.3 | 178.6 | 93.9 | 105.0 | 1,176 | 1,114 | 62 | 288 | 3,224 | 161 |
| Fitchburg | 156.4 | 80.9 | 94.7 | 56.0 | 1,374 | 1,299 | 75 | 138 | 3,384 | 257 |
| Franklin Town | 129.9 | 36.9 | 99.8 | 77.5 | 2,336 | 2,270 | 66 | 140 | 4,208 | 376 |
| Gloucester | 129.8 | 21.6 | 95.7 | 85.6 | 2,838 | 2,719 | 120 | 130 | 4,307 | 530 |
| Haverhill | 250.3 | 102.9 | 98.3 | 108.3 | 1,702 | 1,624 | 78 | 230 | 3,616 | 480 |
| Holyoke | 185.1 | 109.1 | 95.9 | 56.9 | 1,415 | 1,355 | 60 | 176 | 4,375 | 34 |
| Lawrence | 335.3 | 249.5 | 84.5 | 74.9 | 937 | 880 | 56 | 326 | 4,080 | 130 |
| Leominster | 149.3 | 63.5 | 95.5 | 72.1 | 1,732 | 1,697 | 35 | 135 | 3,238 | 234 |
| Lowell | 405.0 | 222.8 | 93.0 | 138.2 | 1,236 | 1,190 | 46 | 400 | 3,579 | 212 |
| Lynn | 366.0 | 218.6 | 97.5 | 131.5 | 1,401 | 1,364 | 36 | 388 | 4,127 | 66 |
| Malden | 195.3 | 85.7 | 99.6 | 92.7 | 1,518 | 1,440 | 78 | 168 | 2,746 | 134 |
| Marlborough | 161.2 | 40.1 | 99.6 | 102.9 | 2,587 | 2,496 | 91 | 158 | 3,959 | 89 |
| Medford | 176.5 | 37.5 | 94.6 | 116.7 | 2,020 | 1,936 | 84 | 177 | 3,060 | 81 |
| Melrose | 100.8 | 20.1 | 96.7 | 63.0 | 2,234 | 2,220 | 13 | 107 | 3,808 | 173 |
| Methuen Town | 175.5 | 67.6 | 99.1 | 91.7 | 1,823 | 1,763 | 60 | 156 | 3,094 | 200 |
| New Bedford | 418.3 | 235.2 | 96.1 | 129.2 | 1,359 | 1,281 | 78 | 425 | 4,471 | 1,168 |
| Newton | 456.5 | 57.7 | 89.2 | 344.9 | 3,882 | 3,730 | 152 | 459 | 5,170 | 747 |
| Northampton | 107.4 | 22.5 | 88.9 | 60.9 | 2,131 | 2,038 | 94 | 99 | 3,447 | 136 |
| Peabody | 182.5 | 47.1 | 88.6 | 115.7 | 2,185 | 2,063 | 122 | 197 | 3,714 | 469 |
| Pittsfield | 194.8 | 95.9 | 97.7 | 87.8 | 2,054 | 1,993 | 61 | 216 | 5,054 | 1,134 |
| Quincy | 381.7 | 81.4 | 91.5 | 228.2 | 2,419 | 2,361 | 59 | 445 | 4,721 | 383 |
| Revere | 185.6 | 87.1 | 99.2 | 89.3 | 1,652 | 1,558 | 94 | 196 | 3,631 | 382 |
| Salem | 191.3 | 54.4 | 91.2 | 92.3 | 2,133 | 2,082 | 51 | 173 | 3,992 | 7 |
| Somerville | 266.5 | 70.4 | 93.4 | 155.7 | 1,917 | 1,742 | 175 | 274 | 3,376 | 167 |
| Springfield | 729.9 | 475.9 | 94.2 | 210.9 | 1,368 | 1,283 | 85 | 727 | 4,716 | 336 |
| Taunton | 220.1 | 89.0 | 96.9 | 104.1 | 1,827 | 1,748 | 79 | 221 | 3,885 | 193 |
| Waltham | 260.4 | 33.3 | 96.5 | 194.1 | 3,090 | 2,886 | 204 | 224 | 3,571 | 209 |
| Watertown Town | 138.2 | 18.9 | 97.1 | 100.7 | 2,817 | 2,784 | 33 | 127 | 3,560 | 237 |
| Westfield | 151.2 | 56.9 | 93.5 | 75.3 | 1,820 | 1,784 | 35 | 141 | 3,405 | 513 |
| West Springfield Town | 113.9 | 38.9 | 94.4 | 67.1 | 2,349 | 2,251 | 98 | 102 | 3,564 | 238 |
| Weymouth Town | 183.3 | 49.4 | 97.6 | 110.1 | 1,941 | 1,848 | 93 | 177 | 3,115 | 190 |
| Woburn | 156.6 | 26.9 | 99.2 | 111.4 | 2,758 | 2,599 | 159 | 160 | 3,966 | 428 |
| Worcester | 757.5 | 367.1 | 95.0 | 307.5 | 1,659 | 1,601 | 58 | 808 | 4,360 | 644 |
| **MICHIGAN** | X | X | X | X | X | X | X | X | X | X |
| Allen Park | 60.4 | 8.4 | 61.7 | 22.6 | 830 | 783 | 47 | 59 | 2,148 | 308 |
| Ann Arbor | 258.5 | 61.0 | 37.2 | 102.6 | 844 | 781 | 63 | 256 | 2,105 | 318 |
| Battle Creek | 107.9 | 26.0 | 63.0 | 48.7 | 950 | 606 | 20 | 102 | 1,984 | 308 |
| Bay City | 55.3 | 16.0 | 51.5 | 14.6 | 440 | 423 | 17 | 46 | 1,388 | 287 |
| Burton | 24.2 | 6.9 | 84.1 | 7.7 | 269 | 248 | 21 | 34 | 1,198 | 229 |
| Dearborn | 207.9 | 41.5 | 86.3 | 100.3 | 1,059 | 1,017 | 42 | 230 | 2,423 | 287 |
| Dearborn Heights | 69.1 | 15.6 | 91.2 | 32.7 | 586 | 532 | 54 | 67 | 1,195 | 62 |
| Detroit | 1,894.3 | 608.6 | 59.6 | 747.8 | 1,108 | 290 | 372 | 2,109 | 3,125 | 876 |
| East Lansing | 67.1 | 14.0 | 84.6 | 25.0 | 509 | 464 | 45 | 74 | 1,512 | 312 |
| Eastpointe | 30.5 | 6.9 | 97.7 | 13.0 | 400 | 356 | 44 | 40 | 1,219 | 79 |
| Farmington Hills | 100.0 | 17.0 | 89.4 | 48.6 | 597 | 574 | 23 | 116 | 1,423 | 182 |
| Flint | 614.5 | 175.2 | 89.6 | 35.9 | 372 | 176 | 34 | 578 | 5,987 | 164 |
| Garden City | 31.1 | 7.2 | 74.1 | 12.2 | 456 | 432 | 24 | 39 | 1,459 | 51 |
| Grand Rapids | 361.4 | 81.9 | 48.0 | 154.6 | 776 | 279 | 26 | 372 | 1,866 | 207 |
| Holland | 53.4 | 11.8 | 78.8 | 18.1 | 542 | 518 | 24 | 77 | 2,304 | 1,046 |
| Inkster | 48.7 | 25.5 | 61.8 | 9.8 | 401 | 353 | 48 | 33 | 1,358 | 63 |
| Jackson | 60.4 | 21.9 | 36.7 | 20.5 | 627 | 339 | 12 | 61 | 1,872 | 366 |
| Kalamazoo | 129.0 | 32.1 | 65.7 | 41.4 | 546 | 515 | 32 | 99 | 1,303 | 131 |
| Kentwood | 42.2 | 11.1 | 87.1 | 20.9 | 404 | 366 | 38 | 47 | 912 | 96 |
| Lansing | 238.2 | 55.9 | 49.7 | 82.1 | 697 | 381 | 13 | 218 | 1,854 | 80 |
| Lincoln Park | 37.7 | 9.7 | 84.6 | 13.3 | 362 | 334 | 29 | 37 | 1,000 | 85 |
| Livonia | 121.6 | 18.8 | 92.8 | 58.3 | 619 | 590 | 29 | 111 | 1,178 | 30 |
| Madison Heights | 40.1 | 7.7 | 94.0 | 18.9 | 627 | 601 | 25 | 44 | 1,455 | 211 |
| Midland | 81.6 | 21.3 | 85.8 | 36.4 | 869 | 851 | 18 | 70 | 1,674 | 104 |
| Mount Pleasant | 27.3 | 5.8 | 98.3 | 8.1 | 314 | 297 | 17 | 27 | 1,037 | 51 |
| Muskegon | 56.0 | 12.1 | 74.2 | 18.6 | 490 | 213 | 50 | 56 | 1,484 | 66 |
| Novi | 70.4 | 10.4 | 97.0 | 35.5 | 593 | 552 | 41 | 71 | 1,191 | 312 |

1. Based on population estimated as of July 1 of the year shown.

Items 117—126

# Table D. Cities — City Government Finances

| City | City government finances, 2017 (cont.) | | | | | | | | | |
|---|---|---|---|---|---|---|---|---|---|---|
| | General expenditure (cont.) | | | | | | | | | |
| | Percent of total for: | | | | | | | | | |
| | Public welfare | Highways | Parking facilities | Education | Health and hospitals | Police protection | Sewerage and sanitation | Parks and recreation | Housing and community development | Interest on debt |
| | 127 | 128 | 129 | 130 | 131 | 132 | 133 | 134 | 135 | 136 |
| **MASSACHUSETTS—Cont'd** | | | | | | | | | | |
| Beverly | 0.4 | 4.7 | 0.0 | 60.7 | 0.3 | 4.9 | 5.4 | 1.1 | 0.1 | 1.8 |
| Boston | 0.1 | 3.2 | 0.1 | 37.5 | 7.6 | 10.1 | 6.9 | 2.5 | 2.7 | 1.9 |
| Braintree Town | 0.2 | 5.3 | 0.0 | 63.9 | 0.2 | 7.6 | 1.1 | 2.1 | 0.0 | 0.7 |
| Brockton | 0.3 | 2.4 | 0.1 | 68.3 | 0.3 | 6.1 | 4.8 | 0.5 | 0.0 | 1.9 |
| Cambridge | 0.1 | 1.5 | 0.1 | 20.3 | 53.9 | 3.5 | 3.9 | 1.7 | 0.1 | 1.2 |
| Chelsea | 0.3 | 2.2 | 0.0 | 62.1 | 0.0 | 5.3 | 5.5 | 0.1 | 3.1 | 0.5 |
| Chicopee | 0.4 | 1.4 | 0.0 | 55.2 | 0.3 | 5.5 | 9.9 | 1.7 | 0.5 | 1.8 |
| Everett | 0.2 | 4.4 | 0.0 | 62.2 | 1.2 | 7.0 | 1.9 | 0.4 | 0.6 | 1.3 |
| Fall River | 0.9 | 3.3 | 0.2 | 58.8 | 0.1 | 7.1 | 7.3 | 0.5 | 2.3 | 1.9 |
| Fitchburg | 0.5 | 5.4 | 0.2 | 60.5 | 0.5 | 5.6 | 7.7 | 0.5 | 0.8 | 1.1 |
| Franklin Town | 0.2 | 4.8 | 0.0 | 57.5 | 0.2 | 3.8 | 5.0 | 0.5 | 0.0 | 2.1 |
| Gloucester | 0.4 | 3.5 | 0.0 | 46.5 | 0.3 | 4.9 | 8.0 | 0.2 | 0.8 | 2.5 |
| Haverhill | 0.5 | 3.1 | 0.2 | 58.9 | 0.3 | 4.7 | 6.7 | 1.1 | 0.5 | 1.1 |
| Holyoke | 0.3 | 2.8 | 0.1 | 61.8 | 0.4 | 7.3 | 4.7 | 0.5 | 1.7 | 1.0 |
| Lawrence | 0.3 | 2.5 | 0.1 | 70.3 | 0.0 | 3.9 | 1.2 | 0.0 | 0.3 | 1.2 |
| Leominster | 0.4 | 4.7 | 0.0 | 66.4 | 0.3 | 5.7 | 6.2 | 0.6 | 0.3 | 0.8 |
| Lowell | 0.2 | 2.7 | 0.5 | 58.5 | 0.6 | 6.6 | 5.4 | 0.7 | 1.6 | 1.8 |
| Lynn | 0.3 | 2.1 | 0.2 | 59.7 | 0.0 | 5.3 | 1.4 | 0.1 | 1.1 | 1.4 |
| Malden | 0.2 | 3.1 | 0.0 | 63.0 | 0.6 | 6.8 | 1.7 | 0.5 | 0.0 | 1.8 |
| Marlborough | 0.2 | 6.2 | 0.0 | 58.5 | 0.3 | 4.9 | 10.7 | 0.2 | 0.0 | 2.9 |
| Medford | 0.2 | 2.7 | 0.0 | 46.6 | 0.3 | 7.0 | 11.4 | 0.5 | 0.9 | 1.3 |
| Melrose | 0.4 | 5.7 | 0.0 | 46.2 | 0.8 | 4.2 | 8.1 | 1.6 | 0.1 | 1.8 |
| Methuen Town | 0.3 | 5.9 | 0.0 | 64.3 | 0.5 | 8.2 | 3.9 | 0.1 | 0.3 | 1.0 |
| New Bedford | 0.7 | 0.7 | 0.1 | 66.8 | 1.0 | 5.8 | 4.3 | 0.7 | 0.9 | 1.7 |
| Newton | 0.1 | 3.0 | 0.0 | 59.5 | 0.7 | 4.2 | 8.4 | 1.4 | 1.0 | 1.9 |
| Northampton | 0.9 | 5.6 | 0.6 | 44.4 | 0.3 | 5.9 | 3.3 | 0.3 | 0.3 | 1.7 |
| Peabody | 0.2 | 2.8 | 0.0 | 54.5 | 0.7 | 5.5 | 6.2 | 1.7 | 1.3 | 1.2 |
| Pittsfield | 0.5 | 3.2 | 0.1 | 60.7 | 0.3 | 4.4 | 5.5 | 0.4 | 0.6 | 1.3 |
| Quincy | 0.3 | 2.5 | 0.0 | 44.7 | 0.9 | 6.0 | 8.4 | 0.7 | 1.4 | 1.4 |
| Revere | 0.4 | 2.6 | 0.0 | 56.1 | 0.3 | 4.9 | 9.7 | 0.2 | 0.3 | 1.4 |
| Salem | 0.4 | 3.4 | 0.5 | 48.5 | 0.3 | 5.9 | 4.7 | 0.8 | 0.5 | 1.2 |
| Somerville | 0.2 | 3.4 | 0.0 | 39.3 | 0.9 | 6.1 | 10.4 | 0.8 | 1.3 | 1.4 |
| Springfield | 0.3 | 1.6 | 0.0 | 62.3 | 0.2 | 6.1 | 1.4 | 1.2 | 1.1 | 1.2 |
| Taunton | 3.4 | 2.8 | 0.0 | 52.0 | 0.4 | 6.9 | 4.8 | 0.6 | 0.8 | 1.2 |
| Waltham | 0.1 | 5.6 | 0.2 | 51.3 | 0.3 | 7.4 | 9.4 | 0.8 | 0.6 | 1.2 |
| Watertown Town | 0.2 | 3.0 | 0.0 | 46.7 | 0.5 | 6.9 | 8.4 | 0.5 | 0.0 | 1.1 |
| Westfield | 0.5 | 5.8 | 0.0 | 55.5 | 3.6 | 5.3 | 6.2 | 0.2 | 0.5 | 2.3 |
| West Springfield Town | 0.6 | 4.2 | 0.0 | 58.5 | 0.3 | 6.9 | 4.4 | 0.5 | 1.0 | 2.2 |
| Weymouth Town | 0.4 | 2.4 | 0.0 | 54.9 | 0.3 | 6.4 | 10.8 | 0.3 | 0.5 | 0.7 |
| Woburn | 0.4 | 3.5 | 0.0 | 52.5 | 0.6 | 5.3 | 8.4 | 0.5 | 0.0 | 3.3 |
| Worcester | 0.3 | 3.9 | 0.2 | 53.8 | 0.4 | 5.9 | 6.0 | 0.8 | 1.0 | |
| **MICHIGAN** | X | X | X | X | X | X | X | X | X | X |
| Allen Park | 0.0 | 17.2 | 0.0 | 0.0 | 0.0 | 10.3 | 18.9 | 2.5 | 0.6 | 5.3 |
| Ann Arbor | 0.0 | 13.0 | 8.6 | 0.0 | 0.0 | 10.7 | 19.5 | 5.9 | 8.6 | 2.6 |
| Battle Creek | 0.0 | 13.6 | 1.4 | 0.0 | 0.0 | 19.2 | 18.9 | 5.8 | 1.1 | 1.7 |
| Bay City | 0.0 | 16.2 | 0.0 | 0.0 | 0.0 | 14.2 | 25.0 | 2.2 | 8.3 | 5.2 |
| Burton | 0.0 | 8.9 | 0.0 | 0.0 | 1.0 | 15.9 | 41.6 | 0.5 | 0.0 | 0.9 |
| Dearborn | 0.0 | 7.8 | 0.1 | 0.0 | 0.0 | 16.3 | 33.2 | 7.9 | 1.3 | 1.7 |
| Dearborn Heights | 0.0 | 15.2 | 0.0 | 0.0 | 0.0 | 20.9 | 25.2 | 1.2 | 1.7 | 1.9 |
| Detroit | 0.0 | 4.2 | 0.2 | 0.1 | 1.4 | 12.9 | 17.2 | 26.4 | 5.5 | 7.1 |
| East Lansing | 0.0 | 7.0 | 10.4 | 0.0 | 0.0 | 14.8 | 19.8 | 8.4 | 0.7 | 2.3 |
| Eastpointe | 0.0 | 12.0 | 0.0 | 0.0 | 0.0 | 21.3 | 21.1 | 1.5 | 0.0 | 1.1 |
| Farmington Hills | 0.0 | 25.0 | 0.0 | 0.0 | 0.0 | 16.9 | 14.2 | 8.0 | 1.4 | 1.1 |
| Flint | 0.0 | 1.9 | 0.0 | 0.0 | 78.9 | 4.5 | 5.6 | 0.1 | 0.0 | 0.5 |
| Garden City | 0.0 | 13.6 | 0.0 | 0.0 | 3.2 | 16.8 | 37.4 | 3.1 | 9.9 | 5.5 |
| Grand Rapids | 0.0 | 6.2 | 4.1 | 0.0 | 0.2 | 13.9 | 13.6 | 3.2 | 0.4 | 1.2 |
| Holland | 0.0 | 13.9 | 0.2 | 0.0 | 0.2 | 10.0 | 31.7 | 15.2 | 28.7 | 4.0 |
| Inkster | 0.0 | 6.8 | 0.0 | 0.0 | 0.7 | 12.8 | 14.5 | 1.0 | 10.7 | 3.8 |
| Jackson | 0.0 | 24.9 | 0.5 | 0.0 | 0.0 | 14.4 | 11.0 | 5.8 | 1.8 | 5.6 |
| Kalamazoo | 0.0 | 12.2 | 0.0 | 0.0 | 0.7 | 0.0 | 26.0 | 4.0 | 0.0 | 0.7 |
| Kentwood | 0.0 | 27.1 | 0.0 | 0.0 | 0.0 | 21.9 | 14.2 | 7.4 | 9.4 | 4.0 |
| Lansing | 0.0 | 11.5 | 1.7 | 0.0 | 0.0 | 17.8 | 10.2 | 7.9 | 1.9 | 0.0 |
| Lincoln Park | 0.0 | 12.1 | 0.0 | 0.0 | 0.0 | 26.4 | 19.5 | 1.3 | 1.9 | 0.0 |
| Livonia | 0.0 | 13.8 | 0.0 | 0.0 | 0.0 | 18.1 | 26.7 | 8.1 | 1.7 | 1.2 |
| Madison Heights | 0.0 | 12.4 | 0.2 | 0.0 | 0.0 | 22.8 | 31.3 | 2.4 | 0.0 | 0.7 |
| Midland | 0.0 | 15.9 | 0.0 | 0.0 | 0.0 | 11.2 | 16.7 | 17.5 | 5.0 | 0.8 |
| Mount Pleasant | 0.0 | 13.3 | 0.0 | 0.0 | 0.0 | 16.0 | 19.4 | 11.5 | 0.0 | 0.5 |
| Muskegon | 0.0 | 8.8 | 0.0 | 0.0 | 0.0 | 16.4 | 37.4 | 6.3 | 4.7 | 0.6 |
| Novi | 0.0 | 26.8 | 0.0 | 0.0 | 0.0 | 15.7 | 13.8 | 11.2 | 0.1 | 1.0 |

# Table D. Cities — City Government Finances, City Government Employment, and Climate

| City | Debt outstanding Total (mil dol) | Per capita[1] (dollars) | Debt issued during year | Average daily temperature Mean January | July | Limits January[3] | July[4] | Annual precipitation (inches) | Heating degree days | Cooling degree days |
|---|---|---|---|---|---|---|---|---|---|---|
| | 137 | 138 | 139 | 140 | 141 | 142 | 143 | 144 | 145 | 146 |
| **MASSACHUSETTS—Cont'd** | | | | | | | | | | |
| Beverly | 87.5 | 2,086 | 31.9 | 28.8 | 72.6 | 20.4 | 82.1 | 45.51 | 5,704 | 582 |
| Boston | 1,929.8 | 2,806 | 378.7 | 29.3 | 73.9 | 22.1 | 82.2 | 42.53 | 5,630 | 777 |
| Braintree Town | 121.4 | 3,259 | 8.6 | NA | NA | NA | NA | NA | NA | NA |
| Brockton | 199.6 | 2,086 | 5.0 | 27.9 | 72.1 | 17.8 | 83.2 | 48.25 | 6,008 | 529 |
| Cambridge | 422.1 | 3,612 | 77.4 | 29.3 | 73.9 | 22.1 | 82.2 | 42.53 | 5,630 | 777 |
| Chelsea | 38.0 | 944 | 9.4 | 29.3 | 73.9 | 22.1 | 82.2 | 42.53 | 5,630 | 777 |
| Chicopee | 179.9 | 3,251 | 50.7 | 21.5 | 68.9 | 10.4 | 81.7 | 48.07 | 7,312 | 287 |
| Everett | 85.9 | 1,862 | 14.4 | 29.3 | 73.9 | 22.1 | 82.2 | 42.53 | 5,630 | 777 |
| Fall River | 262.7 | 2,942 | 9.4 | 28.5 | 74.2 | 20.0 | 83.1 | 50.77 | 5,734 | 740 |
| Fitchburg | 72.0 | 1,766 | 1.2 | 24.2 | 71.9 | 15.2 | 81.0 | 49.13 | 6,576 | 548 |
| Franklin Town | 98.8 | 2,978 | 0.0 | NA | NA | NA | NA | NA | NA | NA |
| Gloucester | 159.0 | 5,272 | 22.1 | 28.8 | 72.6 | 20.4 | 82.1 | 45.51 | 5,704 | 582 |
| Haverhill | 109.5 | 1,720 | 1.1 | 25.3 | 72.2 | 15.6 | 83.5 | 46.88 | 6,435 | 550 |
| Holyoke | 86.8 | 2,157 | 5.0 | 21.5 | 68.9 | 10.4 | 81.7 | 48.07 | 7,312 | 287 |
| Lawrence | 142.6 | 1,783 | 46.3 | 24.5 | 71.8 | 14.5 | 82.9 | 44.09 | 6,539 | 510 |
| Leominster | 50.9 | 1,224 | 0.0 | 24.2 | 71.9 | 15.2 | 81.0 | 49.13 | 6,576 | 548 |
| Lowell | 252.2 | 2,257 | 20.4 | 23.6 | 72.4 | 14.1 | 84.5 | 43.14 | 6,575 | 532 |
| Lynn | 63.2 | 673 | 0.0 | 29.3 | 73.9 | 22.1 | 82.2 | 42.53 | 5,630 | 777 |
| Malden | 86.1 | 1,409 | 4.0 | 29.3 | 73.9 | 22.1 | 82.2 | 42.53 | 5,630 | 777 |
| Marlborough | 147.3 | 3,701 | 12.0 | 25.9 | 73.4 | 16.2 | 84.0 | 45.87 | 6,060 | 651 |
| Medford | 56.5 | 978 | 0.0 | 29.3 | 73.9 | 22.1 | 82.2 | 42.53 | 5,630 | 777 |
| Melrose | 60.7 | 2,153 | 0.0 | 29.3 | 73.9 | 22.1 | 82.2 | 42.53 | 5,630 | 777 |
| Methuen Town | 66.5 | 1,321 | 11.9 | 24.5 | 71.8 | 14.5 | 82.9 | 44.09 | 6,539 | 510 |
| New Bedford | 244.0 | 2,566 | 18.4 | 28.5 | 74.2 | 20.0 | 83.1 | 50.77 | 5,734 | 740 |
| Newton | 316.4 | 3,562 | 109.3 | 25.9 | 73.4 | 16.2 | 84.0 | 45.87 | 6,060 | 651 |
| Northampton | 55.5 | 1,942 | 3.4 | 22.3 | 71.2 | 11.2 | 83.2 | 45.57 | 6,856 | 452 |
| Peabody | 88.6 | 1,673 | 10.0 | 28.8 | 72.6 | 20.4 | 82.1 | 45.51 | 5,704 | 582 |
| Pittsfield | 133.3 | 3,116 | 36.1 | 19.9 | 67.6 | 11.2 | 77.5 | 48.71 | 7,689 | 222 |
| Quincy | 257.5 | 2,730 | 43.6 | 26.0 | 71.6 | 18.1 | 81.2 | 51.19 | 6,371 | 558 |
| Revere | 135.9 | 2,513 | 53.5 | 29.3 | 73.9 | 22.1 | 82.2 | 42.53 | 5,630 | 777 |
| Salem | 79.2 | 1,829 | 17.2 | 28.8 | 72.6 | 20.4 | 82.1 | 45.51 | 5,704 | 582 |
| Somerville | 154.7 | 1,905 | 32.2 | 29.3 | 73.9 | 22.1 | 82.2 | 42.53 | 5,630 | 777 |
| Springfield | 200.5 | 1,300 | 71.4 | 25.7 | 73.7 | 17.2 | 84.9 | 46.16 | 6,104 | 759 |
| Taunton | 144.8 | 2,541 | 4.2 | 27.4 | 72.2 | 17.8 | 83.0 | 48.34 | 6,012 | 558 |
| Waltham | 93.7 | 1,491 | 25.7 | 25.4 | 71.5 | 15.7 | 82.7 | 46.95 | 6,370 | 485 |
| Watertown Town | 34.1 | 955 | 0.8 | 29.3 | 73.9 | 22.1 | 82.2 | 42.53 | 5,630 | 777 |
| Westfield | 98.1 | 2,370 | 0.0 | 21.5 | 68.9 | 10.4 | 81.7 | 48.07 | 7,312 | 287 |
| West Springfield Town | 71.4 | 2,498 | 0.0 | NA | NA | NA | NA | NA | NA | NA |
| Weymouth Town | 77.9 | 1,374 | 6.4 | NA | NA | NA | NA | NA | NA | NA |
| Woburn | 74.3 | 1,839 | 9.5 | 25.5 | 71.5 | 15.7 | 82.5 | 48.31 | 6,401 | 472 |
| Worcester | 722.7 | 3,899 | 102.0 | 23.6 | 70.1 | 15.8 | 79.3 | 49.05 | 6,831 | 371 |
| **MICHIGAN** | X | X | X | X | X | X | X | X | X | X |
| Allen Park | 74.2 | 2,724 | 1.5 | 24.5 | 73.5 | 17.8 | 83.4 | 32.89 | 6,422 | 736 |
| Ann Arbor | 313.7 | 2,582 | 38.6 | 23.4 | 72.6 | 16.6 | 83.0 | 35.35 | 6,503 | 691 |
| Battle Creek | 105.2 | 2,053 | 57.0 | 23.1 | 71.0 | 15.3 | 82.5 | 35.15 | 6,742 | 559 |
| Bay City | 81.2 | 2,450 | 10.4 | 21.0 | 71.5 | 13.8 | 81.5 | 31.25 | 7,106 | 545 |
| Burton | 30.2 | 1,055 | 8.6 | 21.3 | 70.6 | 13.3 | 82.0 | 31.61 | 7,005 | 555 |
| Dearborn | 226.7 | 2,394 | 21.6 | 24.7 | 73.7 | 16.1 | 85.7 | 33.58 | 6,224 | 788 |
| Dearborn Heights | 48.2 | 863 | 18.8 | 24.7 | 73.7 | 16.1 | 85.7 | 33.58 | 6,224 | 788 |
| Detroit | 3,188.4 | 4,726 | 982.5 | 24.7 | 73.7 | 16.1 | 85.7 | 33.58 | 6,224 | 788 |
| East Lansing | 62.1 | 1,265 | 17.3 | 21.6 | 70.3 | 13.9 | 82.1 | 31.53 | 7,098 | 558 |
| Eastpointe | 16.6 | 512 | 4.7 | 25.3 | 73.6 | 18.8 | 83.3 | 33.97 | 6,160 | 757 |
| Farmington Hills | 51.9 | 637 | 25.1 | 24.7 | 73.7 | 16.1 | 85.7 | 33.58 | 6,224 | 788 |
| Flint | 145.4 | 1,507 | 0.0 | 21.3 | 70.6 | 13.3 | 82.0 | 31.61 | 7,005 | 555 |
| Garden City | 34.3 | 1,284 | 2.1 | 24.7 | 73.7 | 16.1 | 85.7 | 33.58 | 6,224 | 788 |
| Grand Rapids | 676.1 | 3,395 | 84.7 | 22.4 | 71.4 | 15.6 | 82.3 | 37.13 | 6,896 | 613 |
| Holland | 243.1 | 7,279 | 21.7 | 24.4 | 71.4 | 17.6 | 82.5 | 36.25 | 6,589 | 611 |
| Inkster | 38.2 | 1,556 | 0.3 | 24.5 | 73.5 | 17.8 | 83.4 | 32.89 | 6,422 | 736 |
| Jackson | 45.6 | 1,395 | 18.2 | 22.2 | 71.3 | 14.7 | 82.7 | 30.67 | 6,873 | 570 |
| Kalamazoo | 512.6 | 6,763 | 8.0 | 24.3 | 73.2 | 17.0 | 84.2 | 37.41 | 6,235 | 773 |
| Kentwood | 16.4 | 317 | 0.0 | 22.4 | 71.4 | 15.6 | 82.3 | 37.13 | 6,896 | 613 |
| Lansing | 565.6 | 4,804 | 37.6 | 21.6 | 70.3 | 13.9 | 82.1 | 31.53 | 7,098 | 558 |
| Lincoln Park | 11.9 | 324 | 2.3 | 24.5 | 73.5 | 17.8 | 83.4 | 32.89 | 6,422 | 736 |
| Livonia | 47.6 | 505 | 10.5 | 24.7 | 73.7 | 16.1 | 85.7 | 33.58 | 6,224 | 788 |
| Madison Heights | 23.2 | 767 | 15.3 | 24.7 | 73.7 | 16.1 | 85.7 | 33.58 | 6,224 | 788 |
| Midland | 22.6 | 540 | 0.0 | 22.9 | 72.7 | 16.2 | 83.8 | 30.69 | 6,645 | 679 |
| Mount Pleasant | 7.5 | 288 | 3.2 | 20.7 | 70.6 | 13.5 | 82.2 | 31.57 | 7,329 | 492 |
| Muskegon | 20.1 | 529 | 0.2 | 23.5 | 69.9 | 17.1 | 80.0 | 32.88 | 6,943 | 487 |
| Novi | 23.4 | 390 | 0.0 | 22.1 | 71.0 | 14.3 | 81.7 | 29.28 | 6,989 | 550 |

1. Based on the population estimated as of July 1 of the year shown.  2. Represents normal values based on the 30-year period, 1971±2000.  3. Average daily minimum.  4. Average daily maximum.

# Table D. Cities — Land Area and Population

| STATE Place code | | City | Population, 2020 | | | | Race 2019 Race alone[2] (percent) | | | | | | Two or more races (percent) |
|---|---|---|---|---|---|---|---|---|---|---|---|---|---|
| | | | Land area[1] (sq. mi) | Total persons 2019 | Rank | Per square mile | White | Black or African American | American Indian, Alaskan Native | Asian | Hawaiian Pacific Islander | Some other race | |
| | | | 1 | 2 | 3 | 4 | 5 | 6 | 7 | 8 | 9 | 10 | 11 |
| | | **MICHIGAN— Cont'd** | | | | | | | | | | | |
| 26 | 59920 | Oak Park.................. | 5.1 | 29,293 | 1,280 | 5,744 | 33.2 | 59.2 | 0.1 | 0.8 | 0.0 | 0.0 | 6.6 |
| 26 | 65440 | Pontiac.................. | 19.9 | 58,911 | 645 | 2,960 | 32.6 | 57.3 | 1.2 | 3.2 | 0.0 | 1.5 | 4.2 |
| 26 | 65560 | Portage.................. | 32.3 | 49,787 | 778 | 1,541 | 84.4 | 6.7 | 0.1 | 2.1 | 0.0 | 0.7 | 6.1 |
| 26 | 65820 | Port Huron.............. | 8.1 | 28,631 | 1,295 | 3,535 | 81.3 | 10.7 | 0.1 | 0.8 | 0.0 | 3.3 | 3.8 |
| 26 | 69035 | Rochester Hills........... | 32.8 | 74,111 | 479 | 2,260 | 82.6 | 2.2 | 0.0 | 12.7 | 0.0 | 0.2 | 2.3 |
| 26 | 69800 | Roseville................ | 9.8 | 46,535 | 836 | 4,749 | 66.1 | 24.5 | 0.8 | 2.9 | 0.0 | 0.6 | 5.1 |
| 26 | 70040 | Royal Oak............... | 11.8 | 58,963 | 643 | 4,997 | 90.7 | 3.1 | 0.1 | 4.0 | 0.0 | 0.8 | 1.3 |
| 26 | 70520 | Saginaw................. | 17.1 | 47,823 | 810 | 2,797 | 49.5 | 42.0 | 0.5 | 0.8 | 0.0 | 1.5 | 5.6 |
| 26 | 70760 | St. Clair Shores........ | 11.7 | 58,390 | 655 | 4,991 | 93.0 | 3.7 | 0.0 | 0.9 | 0.0 | 0.1 | 2.3 |
| 26 | 74900 | Southfield.............. | 26.3 | 72,174 | 503 | 2,744 | 27.6 | 62.1 | 0.3 | 2.0 | 0.1 | 2.1 | 5.8 |
| 26 | 74960 | Southgate.............. | 6.9 | 29,135 | 1,284 | 4,223 | 86.0 | 5.8 | 0.0 | 2.5 | 0.0 | 3.6 | 2.1 |
| 26 | 76460 | Sterling Heights........ | 36.4 | 131,709 | 210 | 3,618 | 83.0 | 5.1 | 0.1 | 8.7 | 0.0 | 0.3 | 2.7 |
| 26 | 79000 | Taylor.................. | 23.6 | 60,576 | 621 | 2,567 | 75.7 | 19.4 | 0.9 | 0.8 | 0.0 | 0.6 | 2.6 |
| 26 | 80700 | Troy................... | 33.4 | 83,589 | 411 | 2,503 | 67.6 | 3.2 | 0.0 | 27.2 | 0.0 | 1.0 | 1.0 |
| 26 | 84000 | Warren................. | 34.4 | 132,877 | 208 | 3,863 | 63.0 | 20.5 | 1.4 | 11.6 | 0.0 | 0.9 | 2.7 |
| 26 | 86000 | Westland................ | 20.4 | 81,116 | 428 | 3,976 | 75.6 | 15.8 | 0.5 | 4.5 | 0.0 | 1.2 | 2.4 |
| 26 | 88900 | Wyandotte.............. | 5.3 | 24,707 | 1,412 | 4,662 | D | D | D | D | D | D | D |
| 26 | 88940 | Wyoming................ | 24.7 | 76,628 | 463 | 3,102 | 76.7 | 8.2 | 0.6 | 1.4 | 0.0 | 7.7 | 5.5 |
| 27 | 00000 | **MINNESOTA**.......... | 79,625.9 | 5,657,342 | X | 71 | 82.1 | 6.6 | 1.0 | 5.1 | 0.0 | 1.9 | 3.3 |
| 27 | 01486 | Andover................ | 33.9 | 33,448 | 1,148 | 987 | D | D | D | D | D | D | D |
| 27 | 01900 | Apple Valley........... | 16.9 | 54,847 | 700 | 3,245 | 78.9 | 7.5 | 0.1 | 5.5 | 0.0 | 2.1 | 5.9 |
| 27 | 06382 | Blaine................. | 32.9 | 66,697 | 562 | 2,027 | 77.5 | 4.2 | 0.4 | 10.3 | 0.1 | 1.9 | 5.7 |
| 27 | 06616 | Bloomington............ | 34.7 | 84,583 | 404 | 2,438 | 72.9 | 10.8 | 1.0 | 6.2 | 0.0 | 4.2 | 4.9 |
| 27 | 07948 | Brooklyn Center........ | 8.0 | 30,253 | 1,244 | 3,782 | 46.8 | 28.4 | 1.1 | 20.5 | 0.3 | 1.7 | 1.2 |
| 27 | 07966 | Brooklyn Park.......... | 26.1 | 79,574 | 444 | 3,049 | 46.2 | 24.4 | 0.4 | 21.1 | 0.0 | 0.8 | 7.0 |
| 27 | 08794 | Burnsville.............. | 24.9 | 62,170 | 604 | 2,497 | 69.8 | 15.8 | 0.1 | 4.9 | 0.0 | 6.6 | 2.9 |
| 27 | 13114 | Coon Rapids........... | 22.6 | 62,935 | 598 | 2,785 | 78.3 | 10.6 | 0.9 | 5.8 | 0.0 | 0.4 | 4.1 |
| 27 | 13456 | Cottage Grove........... | 33.6 | 38,170 | 1,018 | 1,136 | 84.5 | 3.7 | 0.0 | 8.2 | 0.2 | 0.7 | 2.7 |
| 27 | 17000 | Duluth................. | 71.7 | 85,773 | 391 | 1,196 | 89.2 | 1.8 | 2.3 | 1.5 | 0.0 | 0.9 | 4.3 |
| 27 | 17288 | Eagan.................. | 31.2 | 65,975 | 570 | 2,115 | 68.8 | 16.0 | 0.7 | 9.3 | 0.0 | 1.6 | 3.6 |
| 27 | 18116 | Eden Prairie............ | 32.5 | 64,894 | 582 | 1,997 | 67.1 | 12.7 | 0.0 | 13.9 | 0.0 | 2.2 | 4.1 |
| 27 | 18188 | Edina.................. | 15.5 | 52,477 | 742 | 3,386 | 87.9 | 1.9 | 0.5 | 6.7 | 0.0 | 0.2 | 2.8 |
| 27 | 22814 | Fridley................ | 10.2 | 27,857 | 1,322 | 2,731 | 67.9 | 16.5 | 0.1 | 6.1 | 0.0 | 4.6 | 4.8 |
| 27 | 31076 | Inver Grove Heights..... | 27.9 | 35,776 | 1,080 | 1,282 | 78.6 | 6.7 | 0.4 | 2.9 | 0.0 | 9.8 | 1.7 |
| 27 | 35180 | Lakeville.............. | 36.3 | 69,583 | 525 | 1,917 | 85.2 | 1.2 | 0.4 | 9.0 | 0.0 | 0.3 | 4.0 |
| 27 | 39878 | Mankato................ | 19.3 | 43,435 | 889 | 2,251 | 87.0 | 6.8 | 0.4 | 3.7 | 0.0 | 0.6 | 1.5 |
| 27 | 40166 | Maple Grove........... | 32.6 | 72,881 | 493 | 2,236 | 86.0 | 5.2 | 0.1 | 4.7 | 0.0 | 0.3 | 3.7 |
| 27 | 40382 | Maplewood............. | 17.0 | 40,653 | 947 | 2,391 | 69.6 | 3.8 | 0.7 | 17.3 | 0.0 | 1.7 | 6.9 |
| 27 | 43000 | Minneapolis............ | 54.0 | 433,111 | 45 | 8,021 | 64.3 | 19.3 | 1.2 | 4.8 | 0.1 | 4.6 | 5.6 |
| 27 | 43252 | Minnetonka ............ | 26.9 | 55,370 | 694 | 2,058 | 81.5 | 4.7 | 0.1 | 10.6 | 0.0 | 0.4 | 2.8 |
| 27 | 43864 | Moorhead.............. | 22.3 | 44,016 | 875 | 1,974 | 88.1 | 6.2 | 1.5 | 2.5 | 0.9 | 0.0 | 0.8 |
| 27 | 47680 | Oakdale................ | 11.0 | 27,773 | 1,324 | 2,525 | 67.1 | 11.9 | 0.3 | 14.1 | 0.0 | 1.7 | 4.9 |
| 27 | 49300 | Owatonna.............. | 14.9 | 25,672 | 1,396 | 1,723 | D | D | D | D | D | D | D |
| 27 | 51730 | Plymouth.............. | 32.7 | 79,635 | 443 | 2,435 | 77.4 | 5.8 | 0.3 | 11.3 | 0.0 | 2.1 | 3.1 |
| 27 | 54214 | Richfield............... | 6.8 | 36,237 | 1,065 | 5,329 | 66.1 | 11.6 | 0.0 | 5.4 | 0.0 | 11.6 | 5.2 |
| 27 | 54880 | Rochester.............. | 55.5 | 119,862 | 237 | 2,160 | 79.0 | 8.0 | 0.3 | 8.0 | 0.4 | 0.7 | 3.7 |
| 27 | 55852 | Roseville.............. | 13.0 | 36,178 | 1,069 | 2,783 | 75.0 | 10.3 | 0.4 | 6.7 | 0.0 | 1.0 | 6.7 |
| 27 | 56896 | St. Cloud.............. | 40.0 | 68,510 | 535 | 1,713 | 75.3 | 17.2 | 0.7 | 1.9 | 0.0 | 1.6 | 3.3 |
| 27 | 57220 | St. Louis Park......... | 10.6 | 48,821 | 798 | 4,606 | 79.5 | 6.7 | 0.4 | 5.9 | 0.0 | 3.5 | 3.9 |
| 27 | 58000 | St. Paul............... | 52.0 | 306,717 | 63 | 5,898 | 57.5 | 16.5 | 0.6 | 19.5 | 0.0 | 1.7 | 4.2 |
| 27 | 58738 | Savage................. | 15.6 | 32,858 | 1,167 | 2,106 | 80.7 | 4.7 | 0.1 | 10.9 | 0.0 | 0.6 | 3.0 |
| 27 | 59350 | Shakopee.............. | 28.4 | 43,020 | 894 | 1,515 | 68.2 | 11.8 | 0.8 | 3.6 | 0.0 | 7.0 | 8.7 |
| 27 | 59998 | Shoreview.............. | 10.8 | 27,092 | 1,345 | 2,509 | D | D | D | D | D | D | D |
| 27 | 71032 | Winona................. | 19.0 | 26,545 | 1,371 | 1,397 | 90.5 | 2.4 | 0.0 | 4.5 | 0.0 | 1.0 | 1.5 |
| 27 | 71428 | Woodbury.............. | 34.9 | 74,255 | 478 | 2,128 | 84.2 | 6.6 | 0.0 | 6.9 | 0.0 | 0.7 | 1.7 |
| 28 | 00000 | **MISSISSIPPI**.......... | 46,925.5 | 2,966,786 | X | 63 | 58.0 | 38.0 | 0.5 | 1.0 | 0.0 | 1.0 | 1.5 |
| 28 | 06220 | Biloxi................. | 42.9 | 46,127 | 843 | 1,075 | 68.0 | 23.8 | 0.0 | 3.7 | 0.0 | 0.5 | 4.0 |
| 28 | 14420 | Clinton................ | 41.9 | 23,873 | 1,425 | 570 | 51.3 | 43.1 | 0.0 | 3.7 | 0.0 | 0.0 | 1.9 |
| 28 | 29180 | Greenville.............. | 26.9 | 28,315 | 1,305 | 1,053 | D | D | D | D | D | D | D |
| 28 | 29700 | Gulfport............... | 55.6 | 71,438 | 510 | 1,285 | 53.1 | 41.8 | 0.0 | 0.5 | 0.0 | 0.5 | 4.0 |
| 28 | 31020 | Hattiesburg............ | 53.4 | 45,806 | 850 | 858 | D | D | D | D | D | D | D |
| 28 | 33700 | Horn Lake.............. | 16.0 | 27,282 | 1,341 | 1,705 | 47.9 | 47.2 | 0.3 | 1.9 | 0.0 | 2.7 | 0.0 |
| 28 | 36000 | Jackson................ | 111.7 | 157,821 | 161 | 1,413 | 17.5 | 80.8 | 0.5 | 0.5 | 0.0 | 0.2 | 0.6 |
| 28 | 46640 | Meridian............... | 53.7 | 36,011 | 1,074 | 671 | 34.8 | 62.7 | 0.1 | 1.1 | 0.0 | 0.4 | 0.9 |
| 28 | 54040 | Olive Branch.......... | 37.2 | 39,646 | 974 | 1,066 | D | D | D | D | D | D | D |
| 28 | 55760 | Pearl.................. | 25.5 | 26,560 | 1,370 | 1,042 | D | D | D | D | D | D | D |
| 28 | 69280 | Southaven.............. | 41.3 | 56,644 | 677 | 1,372 | D | D | D | D | D | D | D |
| 28 | 74840 | Tupelo................. | 64.4 | 38,303 | 1,010 | 595 | D | D | D | D | D | D | D |

1. Dry land or land partially or temporarily covered by water.    2. Hispanic or Latino persons may be of any race.

# Table D. Cities — **Population**

| City | Percent Hispanic or Latino[1], 2019 | Percent foreign born, 2019 | Age of population (percent), 2019 | | | | | | | Median age, 2019 | Percent female, 2019 | Population Census counts | | Percent change | |
|---|---|---|---|---|---|---|---|---|---|---|---|---|---|---|---|
| | | | Under 18 years | 18 to 24 years | 25 to 34 years | 35 to 44 years | 45 to 54 years | 55 to 64 years | 65 years and over | | | 2000 | 2010 | 2000-2010 | 2001-2020 |
| | 12 | 13 | 14 | 15 | 16 | 17 | 18 | 19 | 20 | 21 | 22 | 23 | 24 | 25 | 26 |
| MICHIGAN— Cont'd | | | | | | | | | | | | | | | |
| Oak Park | 0.6 | 8.3 | 22.1 | 11.1 | 17.1 | 10.6 | 10.1 | 14.8 | 14.1 | 34.8 | 53.0 | 29,793 | 29,408 | -1.3 | -0.4 |
| Pontiac | 16.8 | 8.8 | 27.5 | 11.8 | 15.6 | 11.8 | 11.9 | 11.6 | 9.8 | 31.4 | 54.0 | 66,337 | 59,695 | -10.0 | -1.3 |
| Portage | 5.3 | 7.2 | 23.4 | 14.5 | 12.7 | 12.3 | 11.8 | 10.2 | 15.1 | 34.3 | 51.5 | 44,897 | 46,304 | 3.1 | 7.5 |
| Port Huron | 8.6 | 5.3 | 25.0 | 10.1 | 14.5 | 10.5 | 10.0 | 13.4 | 16.4 | 35.2 | 54.1 | 32,338 | 30,206 | -6.6 | -5.2 |
| Rochester Hills | 6.4 | 21.1 | 20.2 | 9.4 | 9.8 | 14.3 | 12.8 | 15.0 | 18.6 | 42.2 | 52.1 | 68,825 | 70,987 | 3.1 | 4.4 |
| Roseville | 2.5 | 5.4 | 20.2 | 7.2 | 19.1 | 10.8 | 13.1 | 14.2 | 15.4 | 37.8 | 51.3 | 48,129 | 47,322 | -1.7 | -1.7 |
| Royal Oak | 4.0 | 9.2 | 14.8 | 7.5 | 24.1 | 13.9 | 9.6 | 12.6 | 17.4 | 37.1 | 51.0 | 60,062 | 57,232 | -4.7 | 3.0 |
| Saginaw | 14.6 | 1.2 | 24.9 | 9.2 | 14.6 | 12.2 | 12.5 | 11.2 | 15.4 | 36.2 | 52.6 | 61,799 | 51,493 | -16.7 | -7.1 |
| St. Clair Shores | 1.5 | 4.1 | 15.5 | 6.0 | 16.6 | 12.5 | 12.1 | 16.5 | 20.8 | 44.1 | 50.9 | 63,096 | 59,765 | -5.3 | -2.3 |
| Southfield | 2.5 | 9.2 | 17.6 | 7.0 | 15.5 | 10.8 | 12.2 | 15.7 | 21.4 | 43.5 | 53.4 | 78,296 | 71,715 | -8.4 | 0.6 |
| Southgate | 8.1 | 3.6 | 16.9 | 7.4 | 11.7 | 10.4 | 14.2 | 16.8 | 22.7 | 48.9 | 51.0 | 30,136 | 30,047 | -0.3 | -3.0 |
| Sterling Heights | 1.5 | 27.3 | 17.7 | 7.3 | 16.2 | 11.3 | 13.5 | 15.2 | 18.9 | 42.2 | 52.3 | 124,471 | 129,675 | 4.2 | 1.6 |
| Taylor | 6.8 | 3.2 | 20.2 | 10.1 | 16.8 | 11.4 | 13.4 | 14.5 | 13.6 | 37.8 | 49.1 | 65,868 | 63,131 | -4.2 | -4.0 |
| Troy | 2.9 | 31.4 | 20.6 | 7.8 | 11.4 | 13.1 | 15.5 | 13.2 | 18.3 | 42.4 | 48.1 | 80,959 | 80,970 | | 3.2 |
| Warren | 3.3 | 14.6 | 21.3 | 8.7 | 14.6 | 12.8 | 13.4 | 13.7 | 15.4 | 38.2 | 50.3 | 138,247 | 134,072 | -3.0 | -0.9 |
| Westland | 5.2 | 8.9 | 19.4 | 8.4 | 16.9 | 9.8 | 14.0 | 14.1 | 17.5 | 40.3 | 53.2 | 86,602 | 84,150 | -2.8 | -3.6 |
| Wyandotte | 4.6 | 2.1 | 20.7 | 4.8 | 16.1 | 10.5 | 13.2 | 14.7 | 19.9 | 42.0 | 53.1 | 28,006 | 25,883 | -7.6 | -4.5 |
| Wyoming | 20.3 | 10.6 | 22.6 | 11.9 | 17.3 | 15.0 | 10.4 | 11.3 | 11.5 | 34.2 | 49.9 | 69,368 | 72,117 | 4.0 | 6.3 |
| MINNESOTA | 5.6 | 8.4 | 23.1 | 8.7 | 13.6 | 13.0 | 11.9 | 13.4 | 16.3 | 38.4 | 50.3 | 4,919,479 | 5,303,933 | 7.8 | 6.7 |
| Andover | 4.0 | 4.6 | 24.5 | 5.3 | 12.9 | 13.4 | 15.3 | 14.7 | 13.9 | 40.3 | 48.9 | 26,588 | 30,588 | 15.0 | 9.4 |
| Apple Valley | 6.1 | 11.6 | 24.0 | 8.4 | 13.1 | 13.9 | 12.2 | 12.5 | 15.9 | 37.6 | 52.4 | 45,527 | 49,092 | 7.8 | 11.7 |
| Blaine | 6.2 | 12.5 | 27.6 | 7.7 | 13.6 | 16.8 | 9.9 | 13.7 | 10.7 | 35.5 | 49.7 | 44,942 | 57,179 | 27.2 | 16.6 |
| Bloomington | 7.6 | 12.6 | 18.4 | 7.6 | 16.2 | 15.0 | 8.7 | 14.7 | 19.3 | 39.6 | 51.8 | 85,172 | 82,893 | -2.7 | 2.0 |
| Brooklyn Center | 9.4 | 25.2 | 23.5 | 6.2 | 19.6 | 10.3 | 13.1 | 12.0 | 15.3 | 35.7 | 48.1 | 29,172 | 30,180 | 3.5 | 0.2 |
| Brooklyn Park | 2.9 | 20.2 | 25.6 | 12.1 | 12.5 | 13.5 | 13.5 | 12.0 | 10.9 | 34.9 | 50.3 | 67,388 | 75,776 | 12.4 | 5.0 |
| Burnsville | 11.6 | 15.5 | 21.6 | 9.9 | 15.3 | 11.3 | 11.4 | 13.4 | 17.1 | 38.2 | 50.6 | 60,220 | 60,286 | 0.1 | 3.1 |
| Coon Rapids | 4.4 | 13.3 | 20.6 | 7.9 | 13.9 | 15.8 | 13.0 | 11.6 | 17.3 | 39.0 | 51.9 | 61,607 | 61,485 | -0.2 | 2.4 |
| Cottage Grove | 2.7 | 7.5 | 27.5 | 6.1 | 13.5 | 17.0 | 13.5 | 10.0 | 12.3 | 36.3 | 47.8 | 30,582 | 34,601 | 13.1 | 10.3 |
| Duluth | 2.6 | 3.4 | 17.2 | 19.3 | 14.4 | 11.3 | 10.5 | 11.7 | 15.6 | 34.5 | 51.3 | 86,918 | 86,266 | -0.8 | -0.6 |
| Eagan | 5.0 | 16.1 | 26.8 | 8.2 | 12.7 | 13.9 | 12.3 | 13.7 | 12.5 | 36.7 | 51.1 | 63,557 | 64,150 | 0.9 | 2.8 |
| Eden Prairie | 4.4 | 16.2 | 29.0 | 4.2 | 10.8 | 16.6 | 12.6 | 12.4 | 14.4 | 38.5 | 51.2 | 54,901 | 60,797 | 10.7 | 6.7 |
| Edina | 3.0 | 12.6 | 24.1 | 7.1 | 9.8 | 14.6 | 11.7 | 12.9 | 19.8 | 42.2 | 51.2 | 47,425 | 47,988 | 1.2 | 9.4 |
| Fridley | 10.0 | 16.6 | 21.4 | 6.5 | 19.1 | 11.8 | 9.3 | 13.6 | 18.3 | 38.0 | 49.9 | 27,449 | 27,222 | -0.8 | 2.3 |
| Inver Grove Heights | 17.6 | 13.5 | 18.8 | 7.5 | 13.5 | 14.6 | 13.8 | 15.4 | 16.3 | 40.9 | 49.5 | 29,751 | 33,986 | 14.2 | 5.3 |
| Lakeville | 3.0 | 7.6 | 28.3 | 7.8 | 11.3 | 13.9 | 15.2 | 15.0 | 8.5 | 36.5 | 49.5 | 43,128 | 55,998 | 29.8 | 24.3 |
| Mankato | 5.6 | 9.3 | 18.0 | 30.4 | 13.6 | 9.4 | 6.6 | 8.1 | 13.8 | 25.7 | 51.4 | 32,427 | 39,862 | 22.9 | 9.0 |
| Maple Grove | 3.4 | 8.7 | 25.1 | 5.0 | 10.5 | 15.1 | 14.8 | 16.4 | 13.0 | 40.8 | 53.9 | 50,365 | 61,548 | 22.2 | 18.4 |
| Maplewood | 6.2 | 10.7 | 19.6 | 6.3 | 15.1 | 13.1 | 8.9 | 18.1 | 18.8 | 41.9 | 48.3 | 34,947 | 38,016 | 8.8 | 6.9 |
| Minneapolis | 9.8 | 14.1 | 20.3 | 12.5 | 22.3 | 14.9 | 9.9 | 9.7 | 10.3 | 32.2 | 49.2 | 382,618 | 382,603 | 0.0 | 13.2 |
| Minnetonka | 2.2 | 14.3 | 21.2 | 4.7 | 13.5 | 13.8 | 12.2 | 14.6 | 19.9 | 40.2 | 51.2 | 51,301 | 49,750 | -3.0 | 11.3 |
| Moorhead | 3.9 | 6.9 | 23.2 | 18.5 | 13.5 | 12.1 | 9.8 | 9.4 | 13.5 | 31.8 | 54.4 | 32,177 | 39,437 | 22.6 | 11.6 |
| Oakdale | 2.2 | 9.1 | 21.2 | 5.7 | 15.9 | 12.3 | 9.4 | 18.4 | 17.0 | 41.7 | 52.2 | 26,653 | 27,364 | 2.7 | 1.5 |
| Owatonna | 7.7 | 4.5 | 25.0 | 8.2 | 12.7 | 12.9 | 12.6 | 11.4 | 17.3 | 38.0 | 50.7 | 22,434 | 25,622 | 14.2 | 0.2 |
| Plymouth | 5.1 | 15.7 | 26.9 | 3.8 | 12.9 | 13.7 | 11.9 | 13.7 | 17.0 | 38.9 | 51.4 | 65,894 | 70,589 | 7.1 | 12.8 |
| Richfield | 20.5 | 14.9 | 22.0 | 7.1 | 18.2 | 15.0 | 9.5 | 12.7 | 15.6 | 36.6 | 49.5 | 34,439 | 35,228 | 2.3 | 2.9 |
| Rochester | 6.1 | 14.1 | 24.1 | 7.4 | 17.1 | 11.8 | 12.0 | 11.6 | 16.0 | 36.3 | 50.8 | 85,806 | 106,825 | 24.5 | 12.2 |
| Roseville | 4.6 | 14.3 | 15.5 | 7.6 | 18.0 | 10.0 | 11.4 | 12.4 | 25.1 | 44.0 | 50.2 | 33,690 | 33,631 | -0.2 | 7.6 |
| St. Cloud | 4.2 | 11.0 | 19.3 | 16.7 | 15.1 | 12.2 | 9.4 | 11.2 | 16.2 | 34.1 | 46.0 | 59,107 | 66,082 | 11.8 | 3.7 |
| St. Louis Park | 8.0 | 8.9 | 17.5 | 6.5 | 23.3 | 13.6 | 10.3 | 10.9 | 18.0 | 36.4 | 54.0 | 44,126 | 45,197 | 2.4 | 8.0 |
| St. Paul | 8.7 | 20.8 | 24.8 | 10.7 | 18.1 | 13.5 | 11.3 | 10.2 | 11.4 | 33.0 | 51.4 | 287,151 | 285,103 | -0.7 | 7.6 |
| Savage | 3.8 | 9.3 | 26.1 | 7.6 | 15.5 | 16.9 | 14.3 | 10.9 | 8.7 | 35.6 | 50.9 | 21,115 | 26,911 | 27.4 | 22.1 |
| Shakopee | 12.8 | 19.6 | 25.8 | 9.3 | 15.5 | 13.5 | 16.6 | 10.9 | 8.4 | 34.8 | 49.5 | 20,568 | 36,993 | 79.9 | 16.3 |
| Shoreview | 2.2 | 10.7 | 22.4 | 3.7 | 11.5 | 13.4 | 9.3 | 15.2 | 24.6 | 44.2 | 52.1 | 25,924 | 25,043 | -3.4 | 8.2 |
| Winona | 4.4 | 6.9 | 12.0 | 27.1 | 16.1 | 10.2 | 8.5 | 11.2 | 14.8 | 30.0 | 53.9 | 27,069 | 27,583 | 1.9 | -3.8 |
| Woodbury | 4.7 | 10.6 | 27.8 | 7.6 | 10.4 | 15.4 | 14.5 | 11.8 | 12.5 | 37.1 | 50.1 | 46,463 | 61,963 | 33.4 | 19.8 |
| MISSISSIPPI | 3.0 | 2.1 | 23.5 | 10.0 | 12.4 | 12.8 | 12.1 | 12.9 | 16.4 | 38.3 | 51.8 | 2,844,658 | 2,968,129 | 4.3 | 0.0 |
| Biloxi | 6.0 | 4.1 | 22.2 | 17.0 | 11.8 | 14.1 | 10.9 | 10.5 | 13.5 | 34.3 | 49.5 | 50,644 | 44,250 | -12.6 | 4.2 |
| Clinton | 2.6 | 3.9 | 23.8 | 8.1 | 15.1 | 12.9 | 8.3 | 14.0 | 17.9 | 37.6 | 58.3 | 23,347 | 25,230 | 8.1 | -5.4 |
| Greenville | NA | NA | 24.6 | 10.0 | 11.7 | 10.6 | 11.2 | 13.9 | 18.0 | 38.7 | 55.6 | 41,633 | 34,396 | -17.4 | -17.7 |
| Gulfport | 7.7 | 3.9 | 24.1 | 11.0 | 15.4 | 12.6 | 12.8 | 11.8 | 12.2 | 34.6 | 50.9 | 71,127 | 67,785 | -4.7 | 5.4 |
| Hattiesburg | 3.8 | 2.9 | 19.2 | 20.7 | 15.1 | 10.7 | 13.6 | 8.0 | 12.8 | 30.9 | 55.6 | 44,779 | 45,732 | 2.1 | 0.2 |
| Horn Lake | 6.6 | 6.1 | 29.3 | 11.3 | 13.9 | 16.6 | 11.7 | 9.2 | 7.9 | 32.0 | 48.8 | 14,099 | 26,068 | 84.9 | 4.7 |
| Jackson | 1.8 | 1.2 | 24.3 | 12.2 | 13.9 | 12.1 | 11.1 | 12.2 | 14.2 | 34.7 | 54.9 | 184,256 | 173,556 | -5.8 | -9.1 |
| Meridian | 2.3 | 1.5 | 23.2 | 8.9 | 14.6 | 13.8 | 11.2 | 12.6 | 15.7 | 37.6 | 54.1 | 39,968 | 41,126 | 2.9 | -12.4 |
| Olive Branch | 6.0 | 3.6 | 23.2 | 7.0 | 13.4 | 12.3 | 13.3 | 11.3 | 19.5 | 39.7 | 52.5 | 21,054 | 33,487 | 59.1 | 18.4 |
| Pearl | 1.3 | 2.9 | 20.5 | 10.1 | 18.5 | 12.8 | 9.3 | 9.6 | 19.1 | 35.2 | 53.3 | 21,961 | 25,675 | 16.9 | 3.4 |
| Southaven | 6.7 | 3.5 | 27.5 | 10.3 | 15.8 | 13.6 | 11.0 | 10.2 | 11.7 | 33.2 | 54.6 | 28,977 | 48,982 | 69.0 | 15.6 |
| Tupelo | 4.8 | 2.3 | 23.9 | 12.2 | 10.1 | 13.9 | 11.2 | 13.7 | 14.9 | 39.4 | 54.0 | 34,211 | 37,622 | 10.0 | 1.8 |

1. May be of any race.

## Table D. Cities — Households, Group Quarters, Crime, and Education

| City | Households 2019 Number | Persons per household | Percent Family | Married couple family | Female family | Non-family | One person | Persons in group quarters, 2019 | Violent Number | Violent Rate[2] | Property Number | Property Rate[2] | Population age 25 and over | High school graduate or less | Bachelor's degree or more |
|---|---|---|---|---|---|---|---|---|---|---|---|---|---|---|---|
| | 27 | 28 | 29 | 30 | 31 | 32 | 33 | 34 | 35 | 36 | 37 | 38 | 39 | 40 | 41 |
| **MICHIGAN— Cont'd** | | | | | | | | | | | | | | | |
| Oak Park | 11,789 | 2.49 | 53.4 | 29.8 | 23.6 | 46.6 | 33.3 | 56 | 72 | 243 | 510 | 1,720 | 19,651 | 25.0 | 41.2 |
| Pontiac | 22,271 | 2.58 | 54.9 | 17.7 | 37.1 | 45.1 | 36.2 | 1,974 | 769 | 1,286 | 1,300 | 2,174 | 36,085 | 53.1 | 15.9 |
| Portage | 19,301 | 2.55 | 62.9 | 49.9 | 13.0 | 37.1 | 26.9 | 165 | 136 | 274 | 1,549 | 3,124 | 30,718 | 19.0 | 47.9 |
| Port Huron | 10,661 | 2.66 | 63.1 | 38.2 | 24.8 | 36.9 | 31.6 | 394 | 218 | 757 | 662 | 2,300 | 18,664 | 44.0 | 20.0 |
| Rochester Hills | 28,261 | 2.59 | 73.5 | 60.3 | 13.1 | 26.5 | 24.4 | 1,270 | 39 | 52 | 461 | 613 | 52,452 | 21.4 | 56.9 |
| Roseville | 19,284 | 2.42 | 63.3 | 34.8 | 28.4 | 36.7 | 30.0 | 279 | 180 | 380 | 1,361 | 2,872 | 34,141 | 46.5 | 13.7 |
| Royal Oak | 29,880 | 1.97 | 46.1 | 39.0 | 7.0 | 53.9 | 43.7 | 282 | 61 | 102 | 439 | 735 | 46,046 | 20.7 | 58.1 |
| Saginaw | 20,154 | 2.32 | 57.9 | 26.3 | 31.6 | 42.1 | 35.9 | 1,437 | 707 | 1,474 | 743 | 1,549 | 31,707 | 52.2 | 10.5 |
| St. Clair Shores | 28,068 | 2.09 | 56.0 | 42.0 | 14.1 | 44.0 | 38.5 | 380 | 107 | 180 | 552 | 930 | 46,266 | 35.6 | 27.8 |
| Southfield | 34,836 | 2.04 | 46.7 | 30.1 | 16.6 | 53.3 | 46.8 | 1,466 | NA | NA | NA | NA | 54,846 | 31.1 | 37.9 |
| Southgate | 12,700 | 2.27 | 56.6 | 42.2 | 14.4 | 43.4 | 35.3 | 95 | NA | NA | NA | NA | 21,939 | 53.0 | 17.0 |
| Sterling Heights | 51,367 | 2.56 | 67.7 | 51.6 | 16.1 | 32.3 | 27.5 | 925 | 167 | 125 | 1,140 | 855 | 99,370 | 43.5 | 28.0 |
| Taylor | 24,231 | 2.49 | 61.6 | 38.2 | 23.4 | 38.4 | 28.9 | 655 | 378 | 620 | 1,386 | 2,275 | 42,460 | 57.3 | 9.6 |
| Troy | 31,410 | 2.67 | 73.2 | 63.6 | 9.6 | 26.8 | 23.8 | 361 | 67 | 79 | 1,042 | 1,230 | 60,212 | 14.4 | 64.3 |
| Warren | 52,359 | 2.53 | 64.1 | 43.4 | 20.6 | 35.9 | 30.7 | 1,399 | 648 | 481 | 2,465 | 1,831 | 93,772 | 47.0 | 19.7 |
| Westland | 35,324 | 2.28 | 56.6 | 35.3 | 21.3 | 43.4 | 37.4 | 838 | 307 | 377 | 1,178 | 1,446 | 58,827 | 45.0 | 21.8 |
| Wyandotte | 10,738 | 2.31 | 58.0 | 38.9 | 19.2 | 42.0 | 36.1 | 70 | 59 | 238 | 354 | 1,426 | 18,503 | 43.5 | 21.3 |
| Wyoming | 28,313 | 2.66 | 67.3 | 47.4 | 19.9 | 32.7 | 23.4 | 475 | 368 | 482 | 1,122 | 1,471 | 49,601 | 42.7 | 24.4 |
| **MINNESOTA** | 2,222,568 | 2.48 | 63.1 | 49.5 | 13.5 | 36.9 | 29.4 | 130,567 | NA | NA | NA | NA | 3,847,212 | 30.8 | 37.3 |
| Andover | 11,515 | 2.87 | 80.9 | 73.1 | 7.9 | 19.1 | 16.7 | 48 | NA | NA | NA | NA | 23,287 | 18.1 | 40.5 |
| Apple Valley | 21,371 | 2.56 | 68.7 | 53.1 | 15.6 | 31.3 | 22.7 | 324 | 51 | 93 | 1,062 | 1,939 | 37,289 | 18.5 | 48.1 |
| Blaine | 23,243 | 2.82 | 71.0 | 56.8 | 14.2 | 29.0 | 23.4 | 136 | 61 | 92 | 1,586 | 2,394 | 42,420 | 33.0 | 36.2 |
| Bloomington | 35,843 | 2.34 | 57.4 | 43.5 | 13.9 | 42.6 | 34.4 | 1,074 | 203 | 236 | 3,079 | 3,584 | 62,821 | 25.9 | 42.1 |
| Brooklyn Center | 10,729 | 2.84 | 61.0 | 39.9 | 21.1 | 39.0 | 26.4 | 196 | 108 | 348 | 1,223 | 3,945 | 21,578 | 46.6 | 22.1 |
| Brooklyn Park | 27,090 | 2.96 | 72.0 | 48.7 | 23.3 | 28.0 | 24.5 | 243 | 299 | 368 | 2,760 | 3,399 | 50,138 | 36.3 | 28.3 |
| Burnsville | 24,463 | 2.49 | 62.0 | 48.2 | 13.8 | 38.0 | 31.3 | 403 | NA | NA | NA | NA | 42,044 | 29.1 | 38.8 |
| Coon Rapids | 24,409 | 2.56 | 67.1 | 47.6 | 19.4 | 32.9 | 25.7 | 420 | NA | NA | NA | NA | 45,077 | 42.2 | 24.1 |
| Cottage Grove | 12,976 | 2.89 | 78.3 | 64.5 | 13.8 | 21.7 | 19.5 | 77 | 29 | 77 | 590 | 1,572 | 24,933 | 25.2 | 36.1 |
| Duluth | 37,374 | 2.13 | 47.5 | 34.8 | 12.7 | 52.5 | 36.9 | 6,040 | 292 | 340 | 3,670 | 4,275 | 54,361 | 27.0 | 38.0 |
| Eagan | 24,515 | 2.69 | 65.0 | 49.3 | 15.7 | 35.0 | 26.7 | 322 | 39 | 58 | 1,294 | 1,936 | 43,162 | 20.6 | 48.2 |
| Eden Prairie | 24,113 | 2.69 | 71.6 | 62.6 | 9.1 | 28.4 | 24.5 | 143 | 52 | 80 | 700 | 1,081 | 43,354 | 14.4 | 61.7 |
| Edina | 21,594 | 2.43 | 61.2 | 53.9 | 7.3 | 38.8 | 34.8 | 288 | 47 | 89 | 983 | 1,852 | 36,354 | 10.4 | 70.1 |
| Fridley | 11,863 | 2.33 | 52.5 | 37.8 | 14.7 | 47.5 | 38.4 | 138 | 88 | 316 | 1,036 | 3,726 | 20,060 | 30.4 | 32.4 |
| Inver Grove Heights | 15,273 | 2.32 | 61.2 | 46.9 | 14.3 | 38.8 | 29.4 | 163 | 78 | 219 | 650 | 1,822 | 26,275 | 30.4 | 40.4 |
| Lakeville | 22,225 | 3.02 | 80.9 | 67.0 | 13.9 | 19.1 | 15.9 | 98 | 45 | 67 | 564 | 839 | 43,020 | 16.8 | 51.3 |
| Mankato | 17,203 | 2.29 | 48.7 | 35.2 | 13.6 | 51.3 | 32.5 | 3,543 | 110 | 256 | 1,462 | 3,404 | 22,136 | 25.6 | 41.6 |
| Maple Grove | 28,583 | 2.54 | 68.5 | 57.3 | 11.2 | 31.5 | 26.6 | 58 | 58 | 79 | 1,107 | 1,513 | 50,710 | 16.6 | 54.8 |
| Maplewood | 16,051 | 2.49 | 59.9 | 40.0 | 19.9 | 40.1 | 31.8 | 908 | 158 | 382 | 2,080 | 5,031 | 30,300 | 38.5 | 33.6 |
| Minneapolis | 181,833 | 2.26 | 45.1 | 31.1 | 14.0 | 54.9 | 40.8 | 18,473 | 3,990 | 926 | 19,469 | 4,517 | 288,610 | 23.4 | 52.2 |
| Minnetonka | 22,646 | 2.37 | 59.3 | 48.8 | 10.5 | 40.7 | 32.3 | 379 | 30 | 55 | 772 | 1,417 | 40,026 | 16.7 | 60.8 |
| Moorhead | 16,831 | 2.39 | 55.4 | 44.4 | 11.0 | 44.6 | 33.5 | 3,288 | NA | NA | NA | NA | 25,373 | 21.9 | 43.3 |
| Oakdale | 11,685 | 2.38 | 60.3 | 44.2 | 16.1 | 39.7 | 34.5 | 186 | 36 | 128 | 719 | 2,559 | 20,420 | 42.6 | 28.7 |
| Owatonna | 10,227 | 2.39 | 64.4 | 50.1 | 14.3 | 35.6 | 32.5 | 519 | 27 | 105 | 403 | 1,563 | 16,656 | 31.7 | 37.2 |
| Plymouth | 30,392 | 2.60 | 73.5 | 61.5 | 12.0 | 26.5 | 21.3 | 890 | 37 | 46 | 938 | 1,164 | 55,238 | 11.9 | 62.0 |
| Richfield | 15,765 | 2.29 | 54.1 | 38.6 | 15.6 | 45.9 | 37.1 | 319 | 73 | 202 | 850 | 2,355 | 25,770 | 27.4 | 43.4 |
| Rochester | 50,479 | 2.31 | 59.5 | 47.7 | 11.8 | 40.5 | 32.4 | 2,497 | 254 | 215 | 2,225 | 1,881 | 81,451 | 25.0 | 49.9 |
| Roseville | 16,271 | 2.15 | 52.9 | 44.9 | 8.0 | 47.1 | 41.7 | 1,471 | 83 | 226 | 2,164 | 5,888 | 28,014 | 25.6 | 49.0 |
| St. Cloud | 27,322 | 2.28 | 52.9 | 34.0 | 18.9 | 47.1 | 34.4 | 4,570 | 298 | 436 | 2,443 | 3,576 | 42,752 | 37.6 | 32.7 |
| St. Louis Park | 23,629 | 2.03 | 48.8 | 39.3 | 9.5 | 51.2 | 40.9 | 667 | 89 | 180 | 1,396 | 2,818 | 37,018 | 17.0 | 58.8 |
| St. Paul | 110,782 | 2.71 | 54.4 | 36.3 | 18.2 | 45.6 | 35.4 | 8,373 | 1,752 | 565 | 11,208 | 3,612 | 198,766 | 32.7 | 43.6 |
| Savage | 10,940 | 2.96 | 83.5 | 65.8 | 17.7 | 16.5 | 13.7 | 2 | 35 | 108 | 404 | 1,249 | 21,455 | 15.5 | 56.6 |
| Shakopee | 15,023 | 2.69 | 64.6 | 46.0 | 18.6 | 35.4 | 27.3 | 1,094 | 81 | 193 | 745 | 1,778 | 26,963 | 34.0 | 33.5 |
| Shoreview | 10,703 | 2.51 | 71.1 | 58.3 | 12.7 | 28.9 | 26.6 | 253 | NA | NA | NA | NA | 20,048 | 18.7 | 55.2 |
| Winona | 10,266 | 2.19 | 45.2 | 33.0 | 12.2 | 54.8 | 38.3 | 3,383 | 30 | 112 | 644 | 2,410 | 15,729 | 31.5 | 33.4 |
| Woodbury | 26,254 | 2.76 | 74.2 | 62.2 | 12.0 | 25.8 | 20.4 | 274 | 45 | 62 | 1,264 | 1,743 | 47,077 | 15.7 | 62.4 |
| **MISSISSIPPI** | 1,100,229 | 2.62 | 65.3 | 43.8 | 21.5 | 34.7 | 30.3 | 93,382 | NA | NA | NA | NA | 1,979,664 | 44.9 | 22.3 |
| Biloxi | 16,611 | 2.61 | 63.4 | 49.4 | 14.1 | 36.6 | 31.0 | 2,924 | 188 | 407 | 2,663 | 5,766 | 28,114 | 35.8 | 26.2 |
| Clinton | 9,528 | 2.51 | 64.3 | 40.0 | 24.3 | 35.7 | 30.3 | 506 | NA | NA | NA | NA | 16,657 | 21.9 | 43.0 |
| Greenville | 11,104 | 2.44 | 61.2 | 30.8 | 30.4 | 38.8 | 34.7 | 505 | NA | NA | NA | NA | 18,076 | 58.3 | 15.7 |
| Gulfport | 25,559 | 2.74 | 61.0 | 35.3 | 25.6 | 39.0 | 32.1 | 1,779 | 236 | 326 | 3,492 | 4,824 | 46,538 | 45.3 | 20.7 |
| Hattiesburg | 18,938 | 2.24 | 49.2 | 26.0 | 23.1 | 50.8 | 41.8 | 3,195 | 106 | 231 | 2,065 | 4,492 | 27,473 | 35.9 | 31.7 |
| Horn Lake | 9,984 | 2.73 | 60.1 | 36.1 | 24.0 | 39.9 | 27.5 | 57 | 31 | 114 | 676 | 2,480 | 16,187 | 50.2 | 18.2 |
| Jackson | 61,590 | 2.49 | 56.9 | 30.3 | 26.7 | 43.1 | 36.5 | 6,923 | NA | NA | 7,411 | 4,538 | 101,868 | 41.0 | 27.0 |
| Meridian | 15,228 | 2.29 | 64.2 | 31.4 | 32.8 | 35.8 | 33.1 | 1,512 | 181 | 491 | 1,475 | 4,000 | 24,702 | 47.7 | 19.9 |
| Olive Branch | 13,681 | 2.85 | 71.2 | 53.9 | 17.3 | 28.8 | 24.0 | 0 | 89 | 230 | 893 | 2,304 | 27,172 | 34.3 | 31.8 |
| Pearl | 10,717 | 2.47 | 58.7 | 33.8 | 24.8 | 41.3 | 36.1 | 63 | NA | NA | NA | NA | 18,378 | 33.7 | 19.1 |
| Southaven | 20,832 | 2.67 | 70.9 | 39.3 | 31.6 | 29.1 | 23.8 | 266 | 92 | 165 | 1,778 | 3,191 | 34,741 | 36.3 | 29.7 |
| Tupelo | 14,576 | 2.58 | 66.0 | 43.7 | 22.3 | 34.0 | 31.0 | 699 | NA | NA | NA | NA | 24,466 | 35.5 | 30.4 |

1. Data for serious crimes have not been adjusted for underreporting. This may affect comparability between geographic areas and over time.  2. Per 100,000 population estimated by the FBI.  3. Persons 25 years old and over.

# Table D. Cities — Income, Poverty, and Housing

Column groups: Money income, 2019 (Households) = columns 42–46; Median earnings, 2019 = columns 47–49; Housing units, 2019 = columns 50–54.

| City | Median income | Percent with income less than $20,000 | Percent with income of $200,000 or more | Median family income | Median non-family income | All persons | Men | Women | Total | Occupied | Percent owner occupied | Median value¹ (dollars) | Median gross rent (dollars) |
|---|---|---|---|---|---|---|---|---|---|---|---|---|---|
| | 42 | 43 | 44 | 45 | 46 | 47 | 48 | 49 | 50 | 51 | 52 | 53 | 54 |
| **MICHIGAN— Cont'd** | | | | | | | | | | | | | |
| Oak Park | 43,377 | 20.8 | 1.2 | 59,485 | 30,551 | 27,349 | 29,229 | 26,387 | 12,451 | 11,789 | 47.6 | 158,964 | 1,189 |
| Pontiac | 34,380 | 30.3 | 0.6 | 39,738 | 24,893 | 23,124 | 26,072 | 21,906 | 25,737 | 22,271 | 39.7 | 80,753 | 827 |
| Portage | 62,998 | 11.6 | 5.2 | 86,299 | 46,209 | 33,565 | 41,871 | 25,375 | 20,728 | 19,301 | 64.1 | 182,426 | 919 |
| Port Huron | 40,912 | 22.1 | 2.8 | 47,136 | 25,531 | 30,122 | 32,231 | 22,500 | 12,273 | 10,661 | 57.9 | 101,055 | 821 |
| Rochester Hills | 101,620 | 5.8 | 14.4 | 118,893 | 50,499 | 50,859 | 66,819 | 30,836 | 30,632 | 28,261 | 79.4 | 308,866 | 1,421 |
| Roseville | 57,564 | 17.2 | 0.5 | 65,813 | 34,243 | 33,237 | 41,379 | 30,991 | 20,052 | 19,284 | 64.3 | 105,440 | 1,026 |
| Royal Oak | 80,017 | 10.0 | 7.9 | 111,257 | 55,146 | 48,236 | 60,385 | 41,944 | 33,579 | 29,880 | 67.2 | 248,335 | 1,083 |
| Saginaw | 29,840 | 28.1 | 1.6 | 32,127 | 21,779 | 20,946 | 21,209 | 20,362 | 23,822 | 20,154 | 59.1 | 41,496 | 712 |
| St. Clair Shores | 64,746 | 10.0 | 3.1 | 81,243 | 40,919 | 41,750 | 51,485 | 35,899 | 29,549 | 28,068 | 84.8 | 161,739 | 884 |
| Southfield | 51,637 | 13.7 | 3.6 | 68,181 | 41,200 | 37,346 | 40,801 | 35,852 | 37,411 | 34,836 | 45.6 | 171,936 | 1,133 |
| Southgate | 60,755 | 9.4 | 1.9 | 74,712 | 41,332 | 36,912 | 41,626 | 34,455 | 13,087 | 12,700 | 67.5 | 131,132 | 959 |
| Sterling Heights | 67,238 | 14.0 | 5.4 | 81,323 | 40,211 | 32,378 | 45,385 | 25,558 | 52,687 | 51,367 | 74.8 | 212,540 | 1,072 |
| Taylor | 51,470 | 14.0 | 2.8 | 59,911 | 38,453 | 31,970 | 37,261 | 27,811 | 25,863 | 24,231 | 67.5 | 116,966 | 875 |
| Troy | 110,909 | 6.7 | 16.5 | 124,809 | 50,328 | 53,336 | 65,665 | 41,478 | 34,268 | 31,410 | 75.3 | 343,506 | 1,301 |
| Warren | 52,739 | 15.0 | 2.8 | 63,352 | 34,566 | 29,977 | 32,628 | 26,447 | 57,681 | 52,359 | 70.0 | 147,139 | 955 |
| Westland | 54,303 | 18.0 | 2.5 | 73,634 | 31,932 | 35,497 | 41,079 | 30,998 | 38,132 | 35,324 | 57.5 | 148,478 | 885 |
| Wyandotte | 56,168 | 15.3 | 2.4 | 71,893 | 39,182 | 38,702 | 40,151 | 37,087 | 11,555 | 10,738 | 74.2 | 134,432 | 831 |
| Wyoming | 57,570 | 12.1 | 1.8 | 66,472 | 39,133 | 32,158 | 38,806 | 26,676 | 29,549 | 28,313 | 63.4 | 155,074 | 898 |
| **MINNESOTA** | 74,593 | 10.7 | 8.5 | 93,584 | 44,136 | 40,977 | 47,704 | 34,418 | 2,477,515 | 2,222,568 | 71.9 | 246,702 | 1,016 |
| Andover | 111,497 | 3.0 | 12.3 | 121,070 | 57,080 | 55,642 | 65,881 | 46,340 | 11,611 | 11,515 | 93.7 | 332,114 | 1,113 |
| Apple Valley | 103,702 | 6.1 | 9.7 | 120,354 | 53,439 | 51,076 | 64,378 | 38,553 | 22,138 | 21,371 | 75.6 | 290,210 | 1,374 |
| Blaine | 86,747 | 8.0 | 7.4 | 103,455 | 49,057 | 43,740 | 55,396 | 32,356 | 23,757 | 23,243 | 84.1 | 256,008 | 1,213 |
| Bloomington | 74,675 | 10.8 | 8.6 | 95,765 | 49,228 | 41,115 | 43,739 | 38,468 | 37,566 | 35,843 | 64.7 | 281,455 | 1,218 |
| Brooklyn Center | 67,321 | 5.2 | 0.5 | 71,354 | 63,236 | 35,559 | 37,063 | 31,478 | 10,908 | 10,729 | 62.3 | 207,176 | 1,115 |
| Brooklyn Park | 81,757 | 6.4 | 9.2 | 93,765 | 50,467 | 40,607 | 46,469 | 31,935 | 27,818 | 27,090 | 68.2 | 248,700 | 1,203 |
| Burnsville | 71,346 | 9.1 | 6.4 | 97,262 | 41,891 | 38,506 | 41,995 | 32,212 | 26,586 | 24,463 | 68.5 | 271,911 | 1,289 |
| Coon Rapids | 72,176 | 7.4 | 4.6 | 84,415 | 48,504 | 37,055 | 45,396 | 31,600 | 24,562 | 24,409 | 69.9 | 236,872 | 1,188 |
| Cottage Grove | 99,948 | 2.4 | 8.6 | 111,436 | 51,572 | 51,424 | 61,218 | 37,833 | 13,621 | 12,976 | 83.7 | 279,495 | 1,313 |
| Duluth | 55,819 | 18.2 | 4.4 | 83,820 | 34,918 | 27,133 | 34,716 | 23,660 | 39,535 | 37,374 | 59.6 | 170,371 | 850 |
| Eagan | 87,029 | 4.7 | 14.5 | 108,088 | 58,663 | 42,403 | 53,922 | 38,874 | 26,284 | 24,515 | 67.0 | 317,734 | 1,342 |
| Eden Prairie | 121,055 | 4.5 | 22.7 | 138,686 | 67,242 | 63,440 | 82,479 | 44,384 | 25,551 | 24,113 | 74.2 | 408,804 | 1,398 |
| Edina | 111,130 | 7.9 | 25.1 | 167,775 | 57,200 | 62,423 | 86,745 | 51,573 | 23,606 | 21,594 | 67.0 | 499,849 | 1,393 |
| Fridley | 72,015 | 8.3 | 3.1 | 79,122 | 58,819 | 44,647 | 50,284 | 35,606 | 12,174 | 11,863 | 63.7 | 223,878 | 983 |
| Inver Grove Heights | 76,571 | 8.7 | 9.4 | 93,200 | 54,166 | 43,759 | 50,587 | 37,263 | 15,525 | 15,273 | 67.6 | 265,269 | 1,199 |
| Lakeville | 117,083 | 4.7 | 17.4 | 129,272 | 41,926 | 50,574 | 65,360 | 41,333 | 22,712 | 22,225 | 87.8 | 344,800 | 1,057 |
| Mankato | 51,408 | 18.8 | 1.6 | 72,449 | 35,689 | 22,290 | 28,959 | 14,731 | 18,993 | 17,203 | 49.1 | 192,103 | 922 |
| Maple Grove | 115,097 | 3.5 | 17.6 | 135,671 | 73,444 | 60,968 | 80,698 | 50,460 | 29,714 | 28,583 | 88.2 | 328,313 | 1,403 |
| Maplewood | 66,993 | 10.8 | 2.9 | 81,842 | 36,986 | 41,109 | 45,857 | 34,761 | 16,427 | 16,051 | 67.0 | 227,697 | 1,171 |
| Minneapolis | 65,889 | 16.1 | 8.9 | 87,727 | 49,618 | 40,141 | 42,829 | 35,981 | 192,708 | 181,833 | 46.7 | 282,204 | 1,068 |
| Minnetonka | 107,730 | 5.5 | 21.5 | 135,489 | 71,698 | 61,166 | 70,659 | 51,451 | 23,148 | 22,646 | 69.5 | 379,941 | 1,559 |
| Moorhead | 64,009 | 12.8 | 3.7 | 82,181 | 37,611 | 36,443 | 40,714 | 32,727 | 18,350 | 16,831 | 65.2 | 226,688 | 904 |
| Oakdale | 78,429 | 11.6 | 4.9 | 96,215 | 42,144 | 41,262 | 41,352 | 41,129 | 11,685 | 11,685 | 72.5 | 243,518 | 1,099 |
| Owatonna | 62,889 | 9.8 | 5.5 | 80,176 | 33,618 | 40,136 | 44,804 | 29,560 | 11,130 | 10,227 | 73.7 | 182,169 | 783 |
| Plymouth | 113,878 | 6.1 | 21.6 | 145,485 | 61,715 | 66,139 | 82,107 | 50,758 | 31,549 | 30,392 | 69.8 | 381,396 | 1,492 |
| Richfield | 64,431 | 11.8 | 4.7 | 90,934 | 44,108 | 41,713 | 46,507 | 36,153 | 16,309 | 15,765 | 60.3 | 250,758 | 1,058 |
| Rochester | 74,527 | 8.9 | 9.8 | 97,298 | 44,118 | 42,116 | 47,999 | 38,973 | 52,640 | 50,479 | 65.7 | 229,848 | 1,030 |
| Roseville | 73,699 | 11.8 | 5.5 | 98,553 | 43,844 | 42,998 | 44,570 | 38,453 | 17,044 | 16,271 | 62.8 | 280,195 | 1,063 |
| St. Cloud | 53,829 | 13.4 | 4.1 | 71,108 | 37,551 | 31,257 | 31,858 | 28,494 | 29,038 | 27,322 | 47.4 | 169,451 | 870 |
| St. Louis Park | 78,560 | 10.3 | 12.2 | 122,634 | 56,442 | 55,816 | 61,002 | 48,761 | 24,454 | 23,629 | 54.1 | 294,246 | 1,276 |
| St. Paul | 63,174 | 15.0 | 6.7 | 80,676 | 44,358 | 35,478 | 37,249 | 32,082 | 121,626 | 110,782 | 52.2 | 231,486 | 1,001 |
| Savage | 136,170 | 3.7 | 24.3 | 149,055 | 66,545 | 55,689 | 71,015 | 48,453 | 11,128 | 10,940 | 87.0 | 349,859 | 1,344 |
| Shakopee | 90,872 | 5.1 | 11.8 | 107,849 | 45,306 | 42,281 | 46,767 | 35,237 | 15,300 | 15,023 | 69.8 | 294,887 | 1,147 |
| Shoreview | 102,362 | 4.7 | 11.8 | 123,138 | 51,784 | 60,301 | 61,620 | 53,167 | 11,451 | 10,703 | 82.7 | 299,794 | 1,089 |
| Winona | 58,618 | 16.6 | 1.5 | 76,460 | 32,176 | 21,934 | 32,317 | 14,598 | 11,224 | 10,266 | 61.3 | 158,265 | 712 |
| Woodbury | 116,517 | 6.0 | 18.7 | 131,599 | 65,648 | 50,960 | 60,705 | 36,962 | 26,735 | 26,254 | 80.3 | 343,826 | 1,608 |
| **MISSISSIPPI** | 45,792 | 22.6 | 3.2 | 58,503 | 25,562 | 29,341 | 34,908 | 24,972 | 1,339,047 | 1,100,229 | 67.3 | 128,239 | 777 |
| Biloxi | 50,488 | 22.4 | 4.4 | 60,203 | 25,676 | 29,677 | 32,182 | 24,978 | 21,885 | 16,611 | 45.6 | 169,984 | 857 |
| Clinton | 56,474 | 16.8 | 5.1 | 65,640 | 41,858 | 35,778 | 32,464 | 35,806 | 10,039 | 9,528 | 62.1 | 178,731 | 958 |
| Greenville | 25,476 | 35.4 | 1.3 | 36,377 | 14,595 | 22,041 | 28,145 | 21,250 | 13,751 | 11,104 | 43.2 | 86,696 | 620 |
| Gulfport | 36,013 | 26.9 | 1.2 | 41,912 | 27,213 | 27,143 | 33,754 | 21,673 | 33,910 | 25,559 | 48.6 | 123,765 | 905 |
| Hattiesburg | 35,238 | 25.5 | 4.6 | 40,442 | 30,950 | 21,764 | 25,142 | 19,467 | 22,321 | 18,938 | 36.7 | 129,987 | 869 |
| Horn Lake | 46,366 | 15.3 | 0.0 | 54,508 | 41,347 | 31,137 | 40,454 | 21,408 | 10,527 | 9,984 | 48.4 | 112,011 | 1,022 |
| Jackson | 38,972 | 24.8 | 2.4 | 50,427 | 25,183 | 25,641 | 30,885 | 21,845 | 75,146 | 61,590 | 47.0 | 91,141 | 858 |
| Meridian | 27,406 | 37.0 | 3.3 | 38,830 | 19,014 | 19,618 | 21,816 | 18,141 | 17,792 | 15,228 | 42.2 | 102,105 | 716 |
| Olive Branch | 80,169 | 11.0 | 6.3 | 89,107 | 55,547 | 41,178 | 46,657 | 36,445 | 14,576 | 13,681 | 82.9 | 187,079 | 1,079 |
| Pearl | 39,833 | 14.1 | 0.7 | 60,562 | 27,230 | 26,768 | 26,161 | 27,109 | 11,518 | 10,717 | 53.3 | 116,883 | 862 |
| Southaven | 57,916 | 11.4 | 1.9 | 61,392 | 45,606 | 40,564 | 45,342 | 36,704 | 21,961 | 20,832 | 60.8 | 173,397 | 1,080 |
| Tupelo | 49,321 | 18.4 | 4.6 | 56,366 | 32,298 | 28,184 | 33,087 | 25,298 | 17,636 | 14,576 | 64.9 | 148,624 | 843 |

1. Based on population estimated by the American Community Survey.

# Table D. Cities — Commuting, Computer Access, Migration, Labor Force, and Employment

| City | Commuting, 2019 Percent | | Computer access, 2019 Percent | | Migration, 2019 | | Civilian labor force, 2020 | | Unemployment | | Civilian Employment, 2019 Population age 16 and older | | Population age 16 to 64 | |
|---|---|---|---|---|---|---|---|---|---|---|---|---|---|---|
| | Drove alone | With commutes of 30 minutes or more | With a computer in the house | With Internet access | Percent who lived in the same house one year ago | Percent who lived in another state or county one year ago | Total | Percent change 2019-2020 | Total | Rate[3] | Number | Percent in labor force | Number | Percent who worked full-year full-time |
| | 55 | 56 | 57 | 58 | 59 | 60 | 61 | 62 | 63 | 64 | 65 | 66 | 67 | 68 |
| **MICHIGAN— Cont'd** | | | | | | | | | | | | | | |
| Oak Park | 76.7 | 26.2 | 95.1 | 92.4 | 93.3 | 2.3 | 14,434 | -1.3 | 2,154 | 15 | 23,642 | 65.2 | 19,481 | 40.0 |
| Pontiac | 74.6 | 30.0 | 88.2 | 79.7 | 75.5 | 9.5 | 26,300 | 2.8 | 5,314 | 20 | 44,668 | 65.0 | 38,822 | 46.3 |
| Portage | 88.5 | 22.7 | 96.7 | 94.3 | 81.9 | 9.6 | 25,094 | -1.5 | 1,661 | 7 | 38,718 | 72.7 | 31,241 | 55.2 |
| Port Huron | 74.5 | 17.6 | 89.1 | 85.4 | 85.0 | 3.9 | 12,793 | -1.0 | 2,031 | 16 | 22,192 | 55.0 | 17,462 | 45.9 |
| Rochester Hills | 83.4 | 39.1 | 96.8 | 94.8 | 88.0 | 4.7 | 38,332 | -5.7 | 2,965 | 8 | 61,077 | 63.7 | 47,243 | 55.2 |
| Roseville | 83.4 | 43.7 | 92.0 | 89.2 | 87.0 | 3.8 | 23,410 | -1.0 | 3,519 | 15 | 38,535 | 64.1 | 31,290 | 51.0 |
| Royal Oak | 86.3 | 39.1 | 96.8 | 91.8 | 87.8 | 5.0 | 36,637 | -7.1 | 2,124 | 6 | 51,248 | 74.5 | 40,929 | 64.7 |
| Saginaw | 78.0 | 19.0 | 88.7 | 83.1 | 90.5 | 2.9 | 19,095 | 1.6 | 3,200 | 17 | 37,035 | 54.6 | 29,619 | 38.5 |
| St. Clair Shores | 88.4 | 43.2 | 93.7 | 90.6 | 93.1 | 3.8 | 30,703 | -3.5 | 3,549 | 12 | 50,892 | 65.2 | 38,647 | 58.0 |
| Southfield | 82.2 | 36.0 | 90.9 | 85.0 | 86.4 | 4.6 | 34,691 | -2.6 | 4,496 | 13 | 61,206 | 58.9 | 45,678 | 50.6 |
| Southgate | 86.5 | 39.6 | 87.4 | 82.4 | 88.2 | 1.9 | 15,226 | -5.3 | 1,285 | 8 | 24,509 | 61.8 | 17,920 | 60.0 |
| Sterling Heights | 86.3 | 41.6 | 94.5 | 91.5 | 91.9 | 2.5 | 66,552 | -3.3 | 7,591 | 11 | 112,533 | 63.2 | 87,516 | 53.2 |
| Taylor | 85.1 | 32.4 | 94.6 | 81.1 | 87.3 | 2.5 | 29,036 | -1.5 | 3,989 | 14 | 49,698 | 63.5 | 41,426 | 54.5 |
| Troy | 85.3 | 41.4 | 98.0 | 95.5 | 91.9 | 3.8 | 42,125 | -5.8 | 3,239 | 8 | 69,038 | 65.0 | 53,621 | 57.1 |
| Warren | 83.3 | 39.2 | 91.9 | 85.1 | 86.0 | 7.3 | 63,799 | -1.4 | 9,038 | 14 | 108,943 | 64.0 | 88,260 | 50.0 |
| Westland | 82.1 | 39.3 | 91.1 | 85.5 | 90.1 | 2.0 | 42,933 | -4.1 | 4,256 | 10 | 67,069 | 65.0 | 52,810 | 57.9 |
| Wyandotte | 85.6 | 38.2 | 94.9 | 81.8 | 89.4 | 3.9 | 13,100 | -4.3 | 1,268 | 10 | 20,092 | 61.1 | 15,147 | 53.1 |
| Wyoming | 77.8 | 18.5 | 94.5 | 89.2 | 81.1 | 8.5 | 43,974 | -1.0 | 3,633 | 8 | 60,021 | 70.3 | 51,297 | 56.5 |
| **MINNESOTA** | 77.6 | 33.3 | 93.6 | 87.6 | 86.8 | 6.7 | 3,094,702 | -0.5 | 191,140 | 6 | 4,479,456 | 69.4 | 3,557,965 | 56.7 |
| Andover | 89.3 | 53.6 | 95.0 | 92.8 | NA | NA | 18,757 | -2.0 | 984 | 5 | 25,833 | 72.7 | 21,219 | 67.0 |
| Apple Valley | 81.4 | 45.2 | 95.5 | 92.5 | 82.8 | 6.4 | 31,701 | 0.5 | 1,936 | 6 | 43,258 | 73.7 | 34,485 | 62.1 |
| Blaine | 78.0 | 42.4 | 95.1 | 93.0 | 85.4 | 5.4 | 37,175 | -0.9 | 2,233 | 6 | 49,616 | 72.4 | 42,586 | 54.8 |
| Bloomington | 81.7 | 29.2 | 93.9 | 86.7 | 83.3 | 7.0 | 46,410 | -0.8 | 3,466 | 8 | 70,671 | 67.9 | 54,277 | 59.1 |
| Brooklyn Center | 76.1 | 34.3 | 93.0 | 88.3 | 87.4 | 9.0 | 15,568 | 0.5 | 1,428 | 9 | 23,935 | 74.0 | 19,230 | 56.8 |
| Brooklyn Park | 77.7 | 35.8 | 98.2 | 93.2 | 92.9 | 4.6 | 42,969 | -0.1 | 3,354 | 8 | 62,472 | 73.6 | 53,711 | 57.8 |
| Burnsville | 75.8 | 32.7 | 95.8 | 89.8 | 82.9 | 11.7 | 35,848 | -0.2 | 2,585 | 7 | 49,570 | 73.9 | 39,072 | 56.3 |
| Coon Rapids | 81.3 | 38.6 | 96.9 | 93.7 | 85.9 | 8.3 | 35,669 | -0.3 | 2,381 | 7 | 51,863 | 72.1 | 40,983 | 59.2 |
| Cottage Grove | 84.7 | 45.2 | 96.6 | 94.8 | 92.2 | 6.2 | 20,804 | -0.7 | 1,190 | 6 | 28,670 | 71.7 | 24,062 | 47.2 |
| Duluth | 76.9 | 12.3 | 92.3 | 86.2 | 86.2 | 10.0 | 45,509 | -0.8 | 3,053 | 7 | 72,318 | 66.3 | 58,973 | 57.0 |
| Eagan | 76.9 | 34.5 | 98.0 | 92.5 | 88.9 | 6.0 | 39,664 | -1.5 | 2,384 | 6 | 50,663 | 76.3 | 42,385 | 60.9 |
| Eden Prairie | 79.5 | 29.8 | 97.8 | 94.5 | 87.4 | 4.1 | 36,489 | -1.1 | 1,886 | 5 | 47,992 | 71.3 | 38,675 | 56.5 |
| Edina | 80.7 | 25.2 | 94.0 | 89.8 | 85.7 | 6.0 | 26,048 | -1.6 | 1,299 | 5 | 41,782 | 65.2 | 31,310 | 63.4 |
| Fridley | 76.2 | 40.7 | 95.8 | 92.7 | 82.7 | 12.4 | 15,037 | 0.6 | 1,217 | 8 | 22,340 | 71.4 | 17,249 | 58.1 |
| Inver Grove Heights | 75.3 | 31.0 | 96.5 | 87.1 | 89.8 | 7.0 | 19,877 | -1.2 | 1,185 | 6 | 30,629 | 67.3 | 24,820 | 58.6 |
| Lakeville | 80.4 | 43.1 | 98.2 | 94.7 | 88.7 | 3.2 | 37,691 | 0.1 | 2,022 | 5 | 50,041 | 78.1 | 44,300 | 39.5 |
| Mankato | 79.8 | 13.1 | 95.0 | 85.7 | 72.0 | 19.6 | 26,307 | -0.4 | 1,489 | 6 | 35,822 | 68.4 | 29,884 | 63.6 |
| Maple Grove | 80.5 | 42.0 | 98.7 | 97.7 | 88.0 | 5.4 | 41,793 | -1.0 | 2,076 | 5 | 56,369 | 75.1 | 46,942 | 53.5 |
| Maplewood | 79.1 | 33.2 | 90.9 | 87.6 | 89.1 | 4.8 | 21,000 | -0.7 | 1,581 | 8 | 34,296 | 62.7 | 26,603 | 52.0 |
| Minneapolis | 60.0 | 30.6 | 95.5 | 87.0 | 76.0 | 9.4 | 247,548 | 1.1 | 18,468 | 8 | 348,761 | 74.9 | 304,384 | 60.4 |
| Minnetonka | 77.5 | 24.6 | 94.7 | 93.5 | 84.1 | 7.1 | 30,560 | -1.7 | 1,613 | 5 | 43,961 | 67.3 | 33,215 | 51.6 |
| Moorhead | 77.1 | 8.2 | 88.8 | 83.7 | 78.4 | 15.1 | 24,838 | 0.5 | 984 | 4 | 34,101 | 70.8 | 28,232 | 57.4 |
| Oakdale | 81.0 | 40.9 | 85.7 | 81.4 | 88.6 | 10.4 | 16,013 | -1.6 | 1,045 | 7 | 22,501 | 69.3 | 17,757 | 59.0 |
| Owatonna | 81.9 | 16.3 | 89.5 | 88.0 | 86.2 | 4.6 | 14,221 | -3.7 | 810 | 6 | 19,115 | 70.1 | 14,805 | 62.3 |
| Plymouth | 82.7 | 37.2 | 96.6 | 93.0 | 86.2 | 4.7 | 44,622 | -1.8 | 2,255 | 5 | 60,521 | 68.6 | 46,964 | 57.4 |
| Richfield | 75.0 | 32.8 | 95.8 | 83.8 | 86.8 | 4.7 | 20,389 | 0.8 | 1,457 | 7 | 29,274 | 70.4 | 23,614 | 63.0 |
| Rochester | 67.6 | 10.1 | 95.5 | 88.7 | 83.1 | 7.6 | 67,911 | 2.7 | 3,844 | 6 | 92,961 | 71.3 | 73,957 | 48.4 |
| Roseville | 74.1 | 27.4 | 93.7 | 87.3 | 84.6 | 8.6 | 19,477 | -1.4 | 1,155 | 6 | 31,010 | 64.3 | 21,844 | 66.0 |
| St. Cloud | 78.4 | 17.7 | 93.0 | 84.5 | 79.0 | 13.4 | 37,609 | 1.1 | 2,956 | 8 | 55,074 | 66.1 | 44,248 | 52.6 |
| St. Louis Park | 78.3 | 22.8 | 92.5 | 88.9 | 79.9 | 8.2 | 30,314 | -2.3 | 1,702 | 6 | 40,788 | 72.5 | 32,053 | 63.2 |
| St. Paul | 68.3 | 33.2 | 93.4 | 88.4 | 82.9 | 10.2 | 160,327 | 0.1 | 12,135 | 8 | 239,516 | 70.6 | 204,514 | 52.6 |
| Savage | 83.0 | 45.5 | NA | NA | 93.7 | 4.7 | 18,958 | 0.8 | 1,104 | 6 | 24,603 | 80.3 | 21,798 | 63.2 |
| Shakopee | 78.4 | 37.3 | 95.6 | 90.6 | 90.1 | 5.6 | 23,621 | -0.5 | 1,532 | 7 | 32,523 | 79.8 | 29,048 | 61.2 |
| Shoreview | 84.1 | 38.5 | 96.2 | 90.4 | 91.7 | 4.3 | 14,973 | -2.6 | 743 | 5 | 21,815 | 59.0 | 15,151 | 59. |
| Winona | 68.7 | 17.1 | 96.3 | 91.5 | 75.2 | 14.2 | 15,055 | -2.2 | 786 | 5 | 22,931 | 68.9 | 19,113 | 43. |
| Woodbury | 77.7 | 41.3 | 96.5 | 93.2 | 88.4 | 8.2 | 40,594 | -0.1 | 2,035 | 5 | 54,826 | 74.2 | 45,721 | 56.2 |
| **MISSISSIPPI** | 84.8 | 33.5 | 87.6 | 76.1 | 87.8 | 5.8 | 1,259,347 | -1.3 | 101,801 | 8 | 2,360,508 | 56.3 | 1,873,704 | 47. |
| Biloxi | 71.8 | 14.4 | 93.6 | 78.0 | 73.8 | 13.3 | 19,771 | 1.7 | 1,991 | 10 | 36,988 | 53.9 | 30,750 | 47. |
| Clinton | 81.2 | 31.2 | 90.1 | 86.8 | 91.7 | 4.9 | 12,281 | -5.9 | 692 | 6 | 19,050 | 62.9 | 14,666 | 47. |
| Greenville | NA | 13.3 | 75.1 | 64.5 | 87.0 | 2.5 | 10,867 | -4.4 | 1,217 | 11 | 21,904 | 45.7 | 16,935 | 32. |
| Gulfport | 87.7 | 22.5 | 89.8 | 78.2 | 76.8 | 7.7 | 29,172 | 1.7 | 3,193 | 11 | 56,322 | 55.4 | 47,574 | 41. |
| Hattiesburg | 84.6 | 13.0 | 91.7 | 81.3 | 79.6 | 9.6 | 21,423 | -0.5 | 1,747 | 8 | 37,868 | 63.4 | 32,031 | 47. |
| Horn Lake | NA | 46.9 | 97.5 | 92.9 | 78.2 | 8.7 | 12,888 | -0.4 | 1,011 | 8 | 20,977 | 65.6 | 18,810 | 49. |
| Jackson | 78.2 | 22.9 | 88.5 | 83.8 | 84.1 | 4.0 | 70,814 | -1.7 | 7,383 | 10 | 125,829 | 60.5 | 103,078 | 44. |
| Meridian | 84.4 | 16.2 | 89.9 | 76.1 | 86.8 | 8.5 | 14,110 | -3.5 | 1,163 | 8 | 28,869 | 53.3 | 23,160 | 47. |
| Olive Branch | 87.2 | 41.2 | 94.2 | 88.3 | 86.7 | 7.4 | 19,992 | -1.4 | 938 | 5 | 31,091 | 62.5 | 23,483 | 61. |
| Pearl | NA | 22.8 | 91.9 | 86.5 | 81.0 | 6.0 | 12,406 | -2.7 | 742 | 6 | 21,502 | 65.4 | 16,445 | 57. |
| Southaven | 86.0 | 33.3 | 94.2 | 87.3 | 79.5 | 7.3 | 27,847 | -1.2 | 1,527 | 6 | 41,889 | 69.4 | 35,357 | 63. |
| Tupelo | 93.1 | 15.9 | 84.9 | 77.4 | 88.3 | 7.0 | 17,992 | 1.4 | 1,666 | 9 | 30,252 | 62.4 | 24,525 | 55 |

1. Employed persons.  2. Households.  3. Percent of civilian labor force.  4. Persons 16 years old and over.

| City | Value of residential construction authorized by building permits, 2020 | | | Wholesale trade[1], 2017 | | | | Retail trade[2], 2017 | | | |
|---|---|---|---|---|---|---|---|---|---|---|---|
| | New construction ($1,000) | Number of housing units | Percent single family | Number of establishments | Number of employees | Sales (mil dol) | Annual payroll (mil dol) | Number of establishments | Number of employees | Sales (mil dol) | Annual payroll (mil dol) |
| | 69 | 70 | 71 | 72 | 73 | 74 | 75 | 76 | 77 | 78 | 79 |
| **MICHIGAN— Cont'd** | | | | | | | | | | | |
| Oak Park | 200 | 1 | 100 | 48 | 588 | 332 | 38 | 114 | 1,166 | 311 | 28 |
| Pontiac | 859 | 6 | 100 | 41 | 774 | 839 | 39 | 197 | 1,966 | 587 | 61 |
| Portage | 28,119 | 135 | 74 | 50 | 1,213 | 640 | 82 | 296 | 5,658 | 1,211 | 120 |
| Port Huron | 177 | 2 | 100 | 22 | 247 | 185 | 12 | 100 | 1,168 | 346 | 33 |
| Rochester Hills | 60,132 | 145 | 100 | 86 | 861 | 625 | 63 | 252 | 5,660 | 1,728 | 148 |
| Roseville | 305 | 2 | 100 | 44 | 780 | 764 | 40 | 233 | 4,877 | 1,364 | 117 |
| Royal Oak | 44,367 | 153 | 100 | 50 | 312 | 727 | 27 | 210 | 2,419 | 666 | 66 |
| Saginaw | 1,191 | 7 | 71 | 30 | 548 | 319 | 29 | 133 | 738 | 165 | 17 |
| St. Clair Shores | 1,185 | 6 | 100 | 33 | 183 | 81 | 8 | 171 | 1,880 | 581 | 54 |
| Southfield | 3,727 | 20 | 100 | 124 | 2,349 | 3,122 | 227 | 362 | 5,298 | 2,018 | 174 |
| Southgate | 22,362 | 96 | 4 | 9 | 29 | 9 | 1 | 123 | 2,586 | 932 | 74 |
| Sterling Heights | 35,531 | 109 | 91 | 153 | 1,877 | 884 | 107 | 458 | 6,921 | 2,224 | 197 |
| Taylor | 0 | 0 | 0 | 61 | 962 | 1,342 | 56 | 290 | 4,978 | 1,557 | 127 |
| Troy | 29,555 | 113 | 96 | 320 | 7,383 | 7,018 | 507 | 582 | 12,256 | 3,595 | 401 |
| Warren | 12,789 | 110 | 20 | 180 | 4,105 | 2,934 | 256 | 463 | 5,697 | 1,784 | 155 |
| Westland | 8,758 | 68 | 100 | 52 | 507 | 170 | 23 | 277 | 4,331 | 1,190 | 103 |
| Wyandotte | 2,083 | 8 | 100 | 7 | 36 | 8 | 1 | 76 | 510 | 154 | 13 |
| Wyoming | 22,612 | 195 | 42 | 176 | 4,245 | 3,809 | 259 | 251 | 4,629 | 1,342 | 135 |
| **MINNESOTA** | 350 | 2 | 100 | 6,397 | 108,895 | 106,477 | 7,091 | 18,827 | 302,886 | 91,994 | 8,118 |
| Andover | 44,754 | 138 | 100 | 17 | 62 | 25 | 3 | 47 | 949 | 209 | 20 |
| Apple Valley | 1,368 | 5 | 100 | 23 | 121 | 42 | 5 | 118 | 3,478 | 976 | 95 |
| Blaine | 87,102 | 265 | 100 | 69 | 1,107 | 559 | 56 | 227 | 4,730 | 1,055 | 110 |
| Bloomington | 113,222 | 739 | 1 | 193 | 5,285 | 6,247 | 435 | 529 | 11,577 | 3,030 | 309 |
| Brooklyn Center | 416 | 2 | 100 | 23 | 597 | 421 | 32 | 64 | 2,053 | 831 | 66 |
| Brooklyn Park | 17,003 | 62 | 100 | 89 | 1,645 | 1,589 | 126 | 153 | 4,081 | 1,293 | 117 |
| Burnsville | 1,087 | 4 | 100 | 151 | 2,570 | 1,257 | 175 | 325 | 6,637 | 2,540 | 206 |
| Coon Rapids | 8,410 | 34 | 100 | 33 | 405 | 219 | 24 | 192 | 4,948 | 1,505 | 132 |
| Cottage Grove | 117,936 | 505 | 76 | 11 | 339 | 423 | 25 | 49 | 1,136 | 300 | 34 |
| Duluth | 27,629 | 195 | 16 | 94 | 1,041 | 495 | 54 | 405 | 6,344 | 1,324 | 145 |
| Eagan | 43,188 | 219 | 7 | 146 | 2,657 | 1,916 | 198 | 265 | 5,074 | 1,114 | 110 |
| Eden Prairie | 40,175 | 145 | 28 | 170 | 4,053 | 13,176 | 305 | 224 | 6,654 | 3,526 | 275 |
| Edina | 167,449 | 551 | 14 | 108 | 934 | 691 | 78 | 277 | 5,064 | 1,062 | 133 |
| Fridley | 59,352 | 381 | 2 | 67 | 1,923 | 1,509 | 134 | 81 | 2,035 | 654 | 60 |
| Inver Grove Heights | 11,990 | 30 | 100 | 18 | 152 | 226 | 10 | 74 | 1,979 | 849 | 65 |
| Lakeville | 261,046 | 1,154 | 69 | 59 | 856 | 623 | 58 | 108 | 2,799 | 825 | 73 |
| Mankato | 56,773 | 304 | 44 | 53 | 915 | 1,595 | 56 | 267 | 5,849 | 1,390 | 137 |
| Maple Grove | 118,972 | 764 | 20 | 97 | 2,632 | 2,243 | 218 | 229 | 5,183 | 1,190 | 124 |
| Maplewood | 2,892 | 13 | 39 | 37 | 379 | 259 | 25 | 215 | 4,384 | 1,455 | 126 |
| Minneapolis | 570,105 | 3,340 | 2 | 466 | 7,907 | 7,533 | 549 | 1,135 | 13,609 | 5,707 | 363 |
| Minnetonka | 38,769 | 159 | 30 | 115 | 1,814 | 2,068 | 134 | 288 | 6,346 | 1,966 | 186 |
| Moorhead | 41,178 | 231 | 56 | 27 | 569 | 327 | 30 | 107 | 2,123 | 535 | 54 |
| Oakdale | 691 | 3 | 100 | 38 | 728 | 445 | 51 | 60 | 1,891 | 453 | 44 |
| Owatonna | 21,218 | 114 | 31 | 16 | 305 | 190 | 16 | 106 | 2,014 | 476 | 49 |
| Plymouth | 93,576 | 280 | 100 | 186 | 5,545 | 5,446 | 397 | 186 | 3,574 | 1,340 | 117 |
| Richfield | 25,967 | 213 | 10 | 12 | 77 | 21 | 3 | 110 | 3,253 | 6,502 | 140 |
| Rochester | 97,425 | 425 | 70 | 72 | 790 | 413 | 44 | 493 | 10,464 | 2,639 | 274 |
| Roseville | 35,127 | 310 | 7 | 78 | 1,157 | 1,190 | 72 | 281 | 6,334 | 1,426 | 153 |
| St. Cloud | 29,680 | 141 | 80 | 84 | 2,399 | 1,443 | 123 | 326 | 6,467 | 1,740 | 166 |
| St. Louis Park | 48,700 | 180 | 3 | 89 | 1,412 | 841 | 101 | 183 | 4,203 | 1,503 | 135 |
| St. Paul | 243,060 | 2,077 | 2 | 246 | 6,243 | 4,224 | 399 | 725 | 9,284 | 1,852 | 225 |
| Savage | 32,322 | 106 | 81 | 49 | 894 | 1,078 | 61 | 70 | 991 | 287 | 34 |
| Shakopee | 60,846 | 165 | 100 | 56 | 1,566 | 1,550 | 111 | 109 | 2,354 | 776 | 73 |
| Shoreview | 38,020 | 208 | 4 | 16 | 567 | 157 | 62 | 32 | 620 | 141 | 14 |
| Winona | 7,594 | 29 | 21 | 27 | 301 | 347 | 14 | 104 | 2,126 | 523 | 51 |
| Woodbury | 172,480 | 891 | 40 | 33 | 303 | 173 | 19 | 255 | 5,566 | 1,180 | 121 |
| **MISSISSIPPI** | 1,428,531 | 7,810 | 92 | 2,347 | 32,051 | 31,127 | 1,617 | 11,525 | 141,410 | 36,921 | 3,385 |
| Biloxi | 46,259 | 261 | 100 | 32 | 226 | 117 | 9 | 205 | 2,578 | 554 | 59 |
| Clinton | 6,313 | 29 | 100 | 14 | 49 | 16 | 2 | 89 | 1,236 | 372 | 36 |
| Greenville | 345 | 2 | 100 | 30 | 336 | 552 | 16 | 166 | 2,238 | 485 | 47 |
| Gulfport | 59,909 | 192 | 100 | 78 | 868 | 466 | 44 | 382 | 5,972 | 1,589 | 145 |
| Hattiesburg | 11,770 | 47 | 100 | 73 | 822 | 564 | 39 | 454 | 6,900 | 1,749 | 164 |
| Horn Lake | 3,293 | 90 | 100 | D | D | D | 6 | 84 | 1,341 | 387 | 35 |
| Jackson | 4,050 | 15 | 80 | 216 | 2,962 | 2,909 | 176 | 624 | 8,303 | 2,631 | 235 |
| Meridian | 1,268 | 7 | 100 | 50 | 1,315 | 1,779 | 57 | 314 | 4,497 | 1,092 | 108 |
| Olive Branch | 41,932 | 284 | 100 | 55 | 1,592 | 1,642 | 84 | 148 | 2,660 | 862 | 70 |
| Pearl | 14,244 | 83 | 100 | 52 | 680 | 422 | 37 | 172 | 2,756 | 561 | 58 |
| Southaven | 69,056 | 474 | 100 | 33 | 1,381 | 3,059 | 72 | 274 | 4,680 | 1,319 | 121 |
| Tupelo | 25,648 | 138 | 100 | 113 | 1,633 | 943 | 77 | 380 | 6,235 | 1,377 | 142 |

1. Merchant wholesalers except manufacturers' sales branches and offices.   2. Establishments with payroll.

# Table D. Cities — Real Estate, Professional Services, and Manufacturing

| City | Real estate and rental and leasing, 2017 | | | | Professional, scientific, and technical services[1], 2017 | | | | Manufacturing, 2017 | | | |
|---|---|---|---|---|---|---|---|---|---|---|---|---|
| | Number of establishments | Number of employees | Receipts (mil dol) | Annual payroll (mil dol) | Number of establishments | Number of employees | Receipts (mil dol) | Annual payroll (mil dol) | Number of establishments | Number of employees | Receipts (mil dol) | Annual payroll (mil dol) |
| | 80 | 81 | 82 | 83 | 84 | 85 | 86 | 87 | 88 | 89 | 90 | 91 |
| MICHIGAN— Cont'd | | | | | | | | | NA | NA | NA | NA |
| Oak Park | 23 | 461 | 30 | 12 | 26 | 250 | 36 | 19 | NA | NA | NA | NA |
| Pontiac | 37 | 198 | 51 | 9 | 48 | 3,042 | 427 | 235 | NA | NA | NA | NA |
| Portage | 56 | 1,375 | 124 | 53 | 136 | 1,453 | 226 | 85 | NA | NA | NA | NA |
| Port Huron | 19 | 708 | 24 | 15 | 68 | 400 | 66 | 30 | NA | NA | NA | NA |
| Rochester Hills | 54 | 318 | 64 | 12 | 236 | 3,533 | 427 | 219 | NA | NA | NA | NA |
| Roseville | 35 | 231 | 51 | 9 | 44 | 212 | 23 | 12 | NA | NA | NA | NA |
| Royal Oak | 81 | 412 | 81 | 18 | 292 | 2,200 | 371 | 155 | NA | NA | NA | NA |
| Saginaw | 20 | 97 | 17 | 3 | 75 | 693 | 80 | 42 | NA | NA | NA | NA |
| St. Clair Shores | 44 | 137 | 30 | 4 | 168 | 766 | 88 | 35 | NA | NA | NA | NA |
| Southfield | 233 | 2,549 | 571 | 153 | 720 | 14,277 | 2,819 | 1,302 | NA | NA | NA | NA |
| Southgate | D | D | D | 2 | 27 | 235 | 21 | 10 | NA | NA | NA | NA |
| Sterling Heights | 105 | 693 | 214 | 29 | 255 | 7,854 | 1,047 | 567 | NA | NA | NA | NA |
| Taylor | 52 | 359 | 106 | 13 | 81 | 791 | 127 | 43 | NA | NA | NA | NA |
| Troy | 181 | 1,354 | 453 | 62 | 902 | 16,472 | 3,026 | 1,260 | NA | NA | NA | NA |
| Warren | 95 | 982 | 264 | 43 | 172 | 21,392 | 263 | 1,262 | NA | NA | NA | NA |
| Westland | D | D | D | 14 | 64 | 285 | 35 | 10 | NA | NA | NA | NA |
| Wyandotte | D | D | D | 1 | 40 | 149 | 19 | 7 | NA | NA | NA | NA |
| Wyoming | 60 | 540 | 205 | 26 | 117 | 822 | 122 | 41 | NA | NA | NA | NA |
| MINNESOTA | 7,218 | 38,077 | 10,433 | 1,915 | 16,689 | 186,597 | 34,697 | 14,712 | 7,198 | 309,097 | 122,014 | 17,926 |
| Andover | 60 | 184 | 26 | 5 | 66 | 157 | 23 | 7 | NA | NA | NA | NA |
| Apple Valley | 67 | 176 | 43 | 7 | 160 | 469 | 81 | 31 | NA | NA | NA | NA |
| Blaine | 80 | 198 | 97 | 9 | D | D | D | D | NA | NA | NA | NA |
| Bloomington | 199 | 4,618 | 1,236 | 276 | 542 | 9,590 | 1,881 | 829 | NA | NA | NA | NA |
| Brooklyn Center | 34 | 148 | 40 | 5 | 60 | 490 | 89 | 46 | NA | NA | NA | NA |
| Brooklyn Park | 70 | 395 | 88 | 19 | 135 | 738 | 129 | 45 | NA | NA | NA | NA |
| Burnsville | 124 | 675 | 160 | 30 | 268 | 1,786 | 329 | 123 | NA | NA | NA | NA |
| Coon Rapids | 70 | 241 | 59 | 8 | 118 | 866 | 128 | 50 | NA | NA | NA | NA |
| Cottage Grove | D | D | D | D | 47 | 103 | 16 | 5 | NA | NA | NA | NA |
| Duluth | 137 | 638 | 118 | 21 | D | D | D | D | NA | NA | NA | NA |
| Eagan | 141 | 770 | 292 | 41 | 346 | 3,219 | 622 | 248 | NA | NA | NA | NA |
| Eden Prairie | 131 | 820 | 225 | 41 | 436 | 4,273 | 886 | 351 | NA | NA | NA | NA |
| Edina | 225 | 1,665 | 632 | 129 | 65 | 282 | 64 | 16 | NA | NA | NA | NA |
| Fridley | 32 | 415 | 50 | 13 | 81 | 445 | 64 | 29 | NA | NA | NA | NA |
| Inver Grove Heights | 34 | 134 | 28 | 5 | 196 | 556 | 92 | 32 | NA | NA | NA | NA |
| Lakeville | 91 | 308 | 75 | 13 | D | D | D | D | NA | NA | NA | NA |
| Mankato | 72 | 561 | 88 | 17 | 314 | 1,435 | 261 | 109 | NA | NA | NA | NA |
| Maple Grove | 117 | 265 | 88 | 14 | 85 | 351 | 41 | 17 | NA | NA | NA | NA |
| Maplewood | D | D | D | D | 2,555 | 36,745 | 8,637 | 3,445 | NA | NA | NA | NA |
| Minneapolis | 794 | 5,194 | 1,613 | 321 | 436 | 4,098 | 982 | 305 | NA | NA | NA | NA |
| Minnetonka | 160 | 1,085 | 377 | 82 | D | D | D | D | NA | NA | NA | NA |
| Moorhead | 33 | 149 | 22 | 4 | 89 | 1,107 | 162 | 82 | NA | NA | NA | NA |
| Oakdale | 32 | 71 | 18 | 4 | 37 | 158 | 32 | 6 | NA | NA | NA | NA |
| Owatonna | 20 | 253 | 17 | 5 | 489 | 7,686 | 1,212 | 462 | NA | NA | NA | NA |
| Plymouth | 167 | 976 | 318 | 64 | 96 | 601 | 71 | 29 | NA | NA | NA | NA |
| Richfield | 32 | 162 | 29 | 6 | D | D | 5,484 | D | NA | NA | NA | NA |
| Rochester | 155 | 716 | 149 | 26 | 187 | 1,787 | 379 | 137 | NA | NA | NA | NA |
| Roseville | 65 | 391 | 119 | 22 | D | D | D | D | NA | NA | NA | NA |
| St. Cloud | 117 | 606 | 150 | 25 | 352 | 3,666 | 591 | 302 | NA | NA | NA | NA |
| St. Louis Park | 148 | 1,005 | 175 | 44 | 939 | 7,440 | 1,669 | 578 | NA | NA | NA | NA |
| St. Paul | 393 | 2,413 | 534 | 113 | 111 | 268 | 47 | 17 | NA | NA | NA | NA |
| Savage | 45 | 155 | 46 | 7 | 96 | 1,003 | 45 | 123 | NA | NA | NA | NA |
| Shakopee | 52 | 151 | 32 | 5 | 100 | 425 | 81 | 26 | NA | NA | NA | NA |
| Shoreview | D | D | D | D | 52 | 232 | 22 | 9 | NA | NA | NA | NA |
| Winona | 28 | 89 | 16 | 2 | 279 | 1,119 | 201 | 77 | NA | NA | NA | NA |
| Woodbury | 98 | 585 | 172 | 39 | | | | | 2,142 | 138,460 | 60,907 | 6,654 |
| MISSISSIPPI | 2,403 | 9,683 | 1,997 | 345 | 4,746 | 30,477 | 4,598 | 1,618 | NA | NA | NA | NA |
| Biloxi | 56 | 231 | 60 | 8 | 138 | 984 | 105 | 35 | NA | NA | NA | NA |
| Clinton | 37 | 117 | 23 | 3 | 56 | 256 | 38 | 12 | NA | NA | NA | NA |
| Greenville | 31 | 113 | 18 | 4 | 59 | 250 | 30 | 10 | NA | NA | NA | NA |
| Gulfport | 111 | 575 | 130 | 19 | 183 | 1,154 | 183 | 60 | NA | NA | NA | NA |
| Hattiesburg | 104 | 507 | 116 | 19 | 163 | 1,519 | 188 | 66 | NA | NA | NA | NA |
| Horn Lake | 23 | 83 | 23 | 3 | 14 | 65 | 7 | 3 | NA | NA | NA | NA |
| Jackson | 222 | 1,037 | 254 | 48 | 576 | 4,964 | 836 | 316 | NA | NA | NA | NA |
| Meridian | 66 | 321 | 68 | 13 | 112 | 870 | 94 | 36 | NA | NA | NA | NA |
| Olive Branch | 27 | 81 | 27 | 4 | 39 | 151 | 13 | 4 | NA | NA | NA | NA |
| Pearl | 37 | 216 | 36 | 9 | 25 | 147 | 29 | 7 | NA | NA | NA | NA |
| Southaven | 50 | 208 | 85 | 10 | 79 | 497 | 81 | 23 | NA | NA | NA | NA |
| Tupelo | 78 | 340 | 75 | 12 | 160 | 983 | 125 | 52 | NA | NA | NA | NA |

1. Establishments subject to federal tax.

# Table D. Cities — Accommodation and Food Services, Arts, Entertainment, and Recreation, and Health Care and Social Assistance

| City | Accommodation and food services, 2017 | | | | Arts, entertainment, and recreation[1], 2017 | | | | Health care and social assistance[1], 2017 | | | |
|---|---|---|---|---|---|---|---|---|---|---|---|---|
| | Number of establishments | Number of employees | Receipts (mil dol) | Annual payroll (mil dol) | Number of establishments | Number of employees | Receipts (mil dol) | Annual payroll (mil dol) | Number of establishments | Number of employees | Receipts (mil dol) | Annual payroll (mil dol) |
| | 92 | 93 | 94 | 95 | 96 | 97 | 98 | 99 | 100 | 101 | 102 | 103 |
| **MICHIGAN— Cont'd** | | | | | | | | | | | | |
| Oak Park | 45 | 551 | 33 | 8 | 4 | 26 | 3 | 1 | 98 | 827 | 156 | 29 |
| Pontiac | 98 | 1,458 | 93 | 25 | 14 | 66 | 11 | 2 | 140 | 6,071 | 774 | 281 |
| Portage | 152 | 3,547 | 166 | 53 | 15 | 288 | 14 | 5 | 198 | 2,769 | 253 | 118 |
| Port Huron | 53 | 967 | 43 | 13 | D | D | D | D | 158 | 4,564 | 552 | 217 |
| Rochester Hills | 154 | 3,301 | 166 | 49 | 28 | 239 | 23 | 7 | 327 | 3,839 | 425 | 185 |
| Roseville | 105 | 2,898 | 142 | 41 | 6 | 98 | 6 | 1 | 143 | 1,479 | 152 | 69 |
| Royal Oak | 177 | 3,633 | 207 | 65 | 27 | 204 | 23 | 5 | 221 | 11,738 | 2,025 | 704 |
| Saginaw | 75 | 1,356 | 61 | 19 | 10 | 204 | 12 | 3 | 173 | 10,734 | 1,423 | 556 |
| St. Clair Shores | D | D | D | D | 23 | 179 | 17 | 4 | 261 | 3,285 | 363 | 172 |
| Southfield | 244 | 4,307 | 273 | 75 | 24 | 337 | 29 | 10 | 719 | 13,913 | 1,673 | 680 |
| Southgate | 86 | 2,135 | 106 | 33 | 8 | 159 | 8 | 2 | 99 | 1,513 | 140 | 59 |
| Sterling Heights | 262 | 5,168 | 263 | 75 | 28 | 295 | 21 | 5 | 453 | 6,039 | 581 | 236 |
| Taylor | 151 | 2,919 | 152 | 40 | 17 | 218 | 21 | 3 | 141 | 3,198 | 344 | 134 |
| Troy | 319 | 7,054 | 468 | 124 | 38 | 626 | 43 | 14 | 499 | 12,559 | 1,619 | 618 |
| Warren | 295 | 5,414 | 310 | 86 | 24 | 301 | 35 | 5 | 447 | 7,385 | 860 | 354 |
| Westland | 153 | 2,911 | 136 | 40 | 15 | 236 | 14 | 3 | 190 | 2,733 | 248 | 100 |
| Wyandotte | 67 | 846 | 43 | 12 | 6 | 31 | 3 | 1 | 58 | 2,832 | 386 | 146 |
| Wyoming | 143 | 2,763 | 139 | 40 | 19 | 665 | 37 | 9 | 149 | 5,085 | 625 | 245 |
| **MINNESOTA** | 12,022 | 239,194 | 14,234 | 4,272 | 3,012 | 48,380 | 4,162 | 1,561 | 17,066 | 473,338 | 50,491 | 21,207 |
| Andover | 34 | 582 | 28 | 8 | 12 | 362 | 13 | 4 | 59 | 717 | 62 | 26 |
| Apple Valley | 81 | 2,022 | 104 | 33 | 20 | D | 12 | D | 147 | 2,011 | 194 | 78 |
| Blaine | D | D | D | D | 25 | 480 | 47 | 13 | 135 | 2,826 | 216 | 86 |
| Bloomington | 286 | 8,771 | 622 | 183 | 52 | 1,329 | 78 | 25 | 296 | 7,142 | 642 | 316 |
| Brooklyn Center | 59 | 1,067 | 71 | 18 | 6 | 115 | 5 | 1 | 141 | 2,781 | 204 | 89 |
| Brooklyn Park | 101 | 1,675 | 95 | 28 | 18 | 159 | 15 | 3 | 176 | 4,282 | 266 | 138 |
| Burnsville | 134 | 2,830 | 167 | 47 | 46 | 1,166 | 39 | 12 | 269 | 6,896 | 747 | 313 |
| Coon Rapids | 115 | 2,968 | 164 | 51 | 22 | 454 | 19 | 6 | 200 | 7,113 | 1,163 | 410 |
| Cottage Grove | 42 | 667 | 37 | 11 | 4 | 17 | 1 | 0 | 63 | 729 | 66 | 30 |
| Duluth | 254 | 6,157 | 318 | 102 | 63 | 950 | 44 | 14 | 470 | 17,433 | 2,210 | 936 |
| Eagan | 176 | 3,519 | 228 | 66 | 28 | D | 461 | D | 206 | 5,045 | 390 | 171 |
| Eden Prairie | 154 | 3,507 | 210 | 68 | 50 | 1,066 | 85 | 25 | 186 | 3,152 | 305 | 145 |
| Edina | 127 | 3,340 | 202 | 68 | 34 | 1,153 | 62 | 22 | 424 | 9,300 | 1,552 | 603 |
| Fridley | D | D | D | D | 8 | 513 | 67 | 15 | 95 | 3,034 | 394 | 186 |
| Inver Grove Heights | 47 | 959 | 51 | 15 | D | D | D | D | 75 | 1,454 | 103 | 50 |
| Lakeville | 73 | 2,040 | 102 | 34 | 17 | 585 | 28 | 9 | 120 | 1,685 | 119 | 51 |
| Mankato | 151 | 4,014 | 172 | 53 | 33 | 564 | 19 | 7 | 228 | 7,128 | 730 | 340 |
| Maple Grove | 122 | 3,673 | 213 | 68 | 25 | 476 | 32 | 10 | 212 | 4,452 | 680 | 223 |
| Maplewood | 98 | 2,028 | 118 | 37 | 20 | 241 | 17 | 4 | 212 | 5,254 | 779 | 260 |
| Minneapolis | 1,298 | 30,932 | 1,999 | 652 | 313 | 6,730 | 910 | 446 | 1,392 | 60,986 | 8,089 | 3,257 |
| Minnetonka | 101 | 2,445 | 166 | 52 | 31 | 788 | 31 | 13 | 233 | 4,100 | 358 | 171 |
| Moorhead | 71 | 1,425 | 74 | 23 | D | D | D | D | 151 | 4,001 | 254 | 109 |
| Oakdale | 44 | 1,459 | 80 | 27 | 15 | 487 | 24 | 6 | 81 | 1,102 | 95 | 31 |
| Owatonna | 72 | 1,448 | 62 | 20 | 12 | 132 | 8 | 2 | 120 | 2,457 | 240 | 106 |
| Plymouth | 127 | 2,954 | 180 | 54 | 34 | 729 | 38 | 10 | 286 | 5,263 | 681 | 304 |
| Richfield | 69 | 1,357 | 78 | 24 | 10 | 111 | 6 | 2 | 124 | 2,753 | 170 | 76 |
| Rochester | 308 | 7,631 | 477 | 147 | 54 | 979 | 42 | 15 | 431 | 20,978 | 2,733 | 929 |
| Roseville | 121 | 3,509 | 201 | 67 | 19 | 289 | 29 | 7 | 202 | 3,857 | 379 | 135 |
| St. Cloud | 180 | 4,324 | 207 | 59 | 41 | 529 | 39 | 10 | 284 | 13,312 | 1,787 | 746 |
| St. Louis Park | 96 | 2,412 | 157 | 49 | 34 | 668 | 45 | 16 | 234 | 9,950 | 1,449 | 634 |
| St. Paul | 674 | 13,422 | 728 | 236 | 152 | 4,543 | 552 | 199 | 1,112 | 45,769 | 4,992 | 2,231 |
| Savage | D | D | D | D | 11 | 226 | 12 | 4 | 67 | 836 | 65 | 29 |
| Shakopee | D | D | D | D | 25 | 949 | 105 | 45 | 89 | 2,666 | 320 | 119 |
| Shoreview | 29 | 571 | 34 | 10 | 13 | 317 | 14 | 4 | 98 | 1,325 | 94 | 46 |
| Winona | 96 | 2,017 | 78 | 23 | 25 | 363 | 12 | 5 | 103 | 2,794 | 218 | 99 |
| Woodbury | 130 | 3,497 | 187 | 58 | 33 | 741 | 33 | 10 | 235 | 3,797 | 593 | 265 |
| **MISSISSIPPI** | 5,651 | 129,836 | 8,181 | 2,134 | 694 | 8,203 | 583 | 156 | 6,391 | 169,010 | 18,752 | 7,533 |
| Biloxi | 156 | 12,235 | 1,272 | 326 | 23 | 771 | 71 | 19 | 150 | 5,455 | 888 | 390 |
| Clinton | 65 | 1,295 | 63 | 15 | 13 | 144 | 12 | 2 | 53 | 1,285 | 71 | 28 |
| Greenville | 78 | 2,140 | 140 | 36 | 11 | 90 | 5 | 2 | 126 | 2,934 | 279 | 112 |
| Gulfport | 205 | 5,964 | 451 | 100 | 20 | 274 | 42 | 5 | 252 | 7,631 | 915 | 391 |
| Hattiesburg | 254 | 6,028 | 308 | 87 | 21 | 355 | 14 | 5 | 229 | 9,360 | 1,232 | 527 |
| Horn Lake | 59 | 1,561 | 84 | 23 | 5 | D | 20 | D | D | D | D | D |
| Jackson | 374 | 8,801 | 434 | 122 | 46 | 813 | 68 | 23 | 624 | 30,160 | 3,812 | 1,733 |
| Meridian | 162 | 3,874 | 186 | 50 | D | D | D | D | 200 | 6,596 | 890 | 319 |
| Olive Branch | 95 | 2,241 | 110 | 29 | D | D | D | D | 74 | 1,183 | 99 | 42 |
| Pearl | 80 | 1,715 | 111 | 27 | 8 | 23 | 3 | 1 | D | D | D | D |
| Southaven | 154 | 3,728 | 193 | 51 | D | D | D | D | 159 | 3,613 | 575 | 190 |
| Tupelo | 182 | 3,932 | 194 | 54 | 25 | 173 | 9 | 2 | 245 | 8,839 | 1,180 | 434 |

1. Establishments subject to federal tax.

# Table D. Cities — Other Services and Government Employment and Payroll

| City | Other services[1] Number of establishments | Number of employees | Receipts (mil dol) | Annual payroll (mil dol) | Government employment and payroll, 2017 Full-time equivalent employees | Total (dollars) | March payroll Percent of total for: Administrative, judicial, and legal | Police and corrections | Fire protection | Highways and transportation | Health and welfare | Natural resources and utilities | Education and libraries |
|---|---|---|---|---|---|---|---|---|---|---|---|---|---|
| | 104 | 105 | 106 | 107 | 108 | 109 | 110 | 111 | 112 | 113 | 114 | 115 | 116 |
| **MICHIGAN— Cont'd** | | | | | | | | | | | | | |
| Oak Park | D | D | D | D | 166 | 785,044 | 29.1 | 51.5 | 0.0 | 11.1 | 1.5 | 3.5 | 3.3 |
| Pontiac | 68 | 577 | 68 | 19 | 81 | 335,718 | 66.2 | 4.7 | 0.0 | 4.4 | 7.5 | 2.5 | 10.9 |
| Portage | 101 | 795 | 69 | 23 | 217 | 1,041,792 | 13.2 | 39.3 | 19.8 | 13.5 | 4.5 | 5.0 | 0.0 |
| Port Huron | 38 | 197 | 23 | 6 | 261 | 1,240,671 | 10.8 | 29.7 | 17.5 | 6.5 | 0.4 | 24.2 | 0.0 |
| Rochester Hills | 111 | 774 | 78 | 27 | 230 | 1,363,246 | 24.3 | 0.4 | 27.8 | 7.3 | 0.0 | 24.8 | 0.0 |
| Roseville | 83 | 411 | 37 | 12 | 262 | 1,316,999 | 19.5 | 36.4 | 18.4 | 4.5 | 4.7 | 5.3 | 4.4 |
| Royal Oak | 138 | 774 | 74 | 23 | 372 | 1,893,874 | 21.0 | 31.8 | 17.9 | 5.8 | 1.3 | 7.0 | 3.8 |
| Saginaw | 60 | 436 | 45 | 12 | 364 | 1,601,557 | 12.6 | 21.9 | 16.7 | 7.3 | 0.9 | 34.1 | 0.0 |
| St. Clair Shores | 117 | 716 | 50 | 15 | 282 | 1,529,888 | 15.0 | 42.8 | 21.6 | 1.7 | 0.7 | 11.6 | 5.1 |
| Southfield | 177 | 1,164 | 216 | 50 | 591 | 3,132,241 | 17.7 | 31.5 | 20.2 | 6.2 | 1.1 | 9.2 | 5.0 |
| Southgate | D | D | D | D | 146 | 681,169 | 15.3 | 37.6 | 21.8 | 7.8 | 0.6 | 9.3 | 2.6 |
| Sterling Heights | 215 | 1,092 | 102 | 33 | 520 | 3,085,789 | 14.4 | 39.5 | 21.3 | 5.3 | 3.0 | 10.2 | 3.8 |
| Taylor | 109 | 1,028 | 133 | 46 | 358 | 1,747,951 | 13.5 | 40.2 | 18.9 | 6.4 | 4.6 | 14.1 | 0.3 |
| Troy | 243 | 2,000 | 646 | 83 | 430 | 2,226,427 | 16.6 | 44.1 | 3.6 | 9.7 | 0.0 | 14.3 | 6.2 |
| Warren | 260 | 1,305 | 191 | 51 | 868 | 4,546,706 | 14.8 | 32.4 | 17.1 | 5.7 | 1.9 | 18.8 | 2.7 |
| Westland | 129 | 851 | 117 | 31 | 332 | 1,685,200 | 18.2 | 35.8 | 25.8 | 3.2 | 4.4 | 9.7 | 0.0 |
| Wyandotte | 50 | 238 | 20 | 7 | 318 | 1,538,335 | 5.8 | 16.0 | 10.7 | 7.7 | 0.0 | 25.4 | 0.0 |
| Wyoming | 119 | 1,167 | 105 | 32 | 403 | 2,036,612 | 15.8 | 32.1 | 9.9 | 11.4 | 3.1 | 21.2 | 0.0 |
| **MINNESOTA** | 11,339 | 77,400 | 10,077 | 2,554 | X | X | X | X | X | X | X | X | X |
| Andover | D | D | D | D | 79 | 409,816 | 15.8 | 0.0 | 14.9 | 21.5 | 0.0 | 30.4 | 0.0 |
| Apple Valley | 66 | 629 | 62 | 19 | 234 | 1,365,809 | 19.0 | 33.2 | 5.4 | 7.5 | 0.0 | 23.9 | 0.0 |
| Blaine | 136 | 1,068 | 134 | 38 | 199 | 1,535,440 | 13.4 | 43.7 | 5.0 | 10.1 | 2.9 | 15.0 | 0.0 |
| Bloomington | 200 | 2,179 | 353 | 91 | 605 | 4,022,432 | 18.0 | 30.1 | 1.7 | 11.3 | 12.4 | 17.3 | 0.0 |
| Brooklyn Center | 27 | 344 | 21 | 8 | 200 | 1,270,549 | 11.0 | 39.2 | 3.3 | 9.7 | 1.6 | 21.9 | 0.0 |
| Brooklyn Park | 93 | 638 | 58 | 19 | 462 | 2,049,529 | 8.1 | 36.7 | 10.5 | 11.0 | 10.1 | 15.7 | 0.0 |
| Burnsville | 133 | 821 | 77 | 31 | 269 | 1,904,343 | 11.2 | 34.9 | 18.8 | 8.2 | 1.4 | 15.0 | 0.0 |
| Coon Rapids | 104 | 896 | 95 | 28 | 255 | 1,606,439 | 14.7 | 33.7 | 14.5 | 7.5 | 2.6 | 15.1 | 0.0 |
| Cottage Grove | 43 | 265 | 17 | 6 | 169 | 951,197 | 9.2 | 35.4 | 12.7 | 9.6 | 8.7 | 19.2 | 0.0 |
| Duluth | 185 | 1,372 | 137 | 35 | 1,059 | 5,770,250 | 10.4 | 20.8 | 15.6 | 19.1 | 1.3 | 21.2 | 3.8 |
| Eagan | 155 | 1,725 | 206 | 67 | 281 | 1,785,151 | 11.9 | 37.7 | 3.8 | 10.0 | 4.4 | 27.5 | 0.0 |
| Eden Prairie | 148 | 1,152 | 220 | 53 | 377 | 2,135,836 | 12.9 | 32.9 | 6.2 | 9.6 | 2.3 | 24.9 | 0.0 |
| Edina | 141 | 1,451 | 203 | 60 | 368 | 2,156,450 | 6.1 | 26.4 | 12.0 | 10.9 | 1.0 | 23.7 | 0.0 |
| Fridley | 59 | 486 | 57 | 14 | 160 | 931,644 | 14.7 | 37.8 | 7.6 | 12.3 | 5.1 | 17.9 | 0.0 |
| Inver Grove Heights | 54 | 405 | 48 | 16 | 195 | 1,026,882 | 11.1 | 32.7 | 12.0 | 13.1 | 7.2 | 24.0 | 0.0 |
| Lakeville | 89 | 599 | 76 | 19 | 226 | 1,415,893 | 9.5 | 34.3 | 3.7 | 8.8 | 3.2 | 18.1 | 0.0 |
| Mankato | 85 | 806 | 78 | 23 | 279 | 1,560,674 | 17.7 | 27.6 | 8.5 | 14.4 | 4.3 | 18.7 | 0.0 |
| Maple Grove | 147 | 1,617 | 115 | 51 | 328 | 1,813,759 | 14.7 | 34.2 | 6.4 | 6.7 | 2.2 | 22.6 | 0.0 |
| Maplewood | 74 | 450 | 59 | 17 | 178 | 1,136,916 | 18.5 | 38.7 | 17.6 | 10.6 | 0.4 | 6.5 | 0.0 |
| Minneapolis | 1,033 | 10,041 | 1,928 | 426 | 4,811 | 29,451,317 | 16.8 | 26.3 | 10.8 | 8.7 | 5.4 | 22.5 | 0.0 |
| Minnetonka | 123 | 837 | 123 | 25 | 294 | 1,683,951 | 17.0 | 28.2 | 7.2 | 12.2 | 1.4 | 22.3 | 0.0 |
| Moorhead | 78 | 386 | 43 | 12 | 341 | 1,807,198 | 8.0 | 22.9 | 12.5 | 9.7 | 1.8 | 40.2 | 0.0 |
| Oakdale | 38 | 138 | 16 | 5 | 111 | 690,728 | 9.6 | 37.3 | 11.0 | 5.1 | 4.1 | 22.6 | 0.0 |
| Owatonna | 63 | 422 | 58 | 12 | 214 | 1,258,086 | 10.3 | 21.6 | 4.1 | 9.9 | 0.5 | 44.1 | 4.6 |
| Plymouth | 131 | 1,101 | 101 | 39 | 315 | 1,920,231 | 12.4 | 35.6 | 5.8 | 4.2 | 2.9 | 28.3 | 0.0 |
| Richfield | 66 | 742 | 125 | 31 | 157 | 885,604 | 14.7 | 33.5 | 0.0 | 13.8 | 5.8 | 23.9 | 0.0 |
| Rochester | 216 | 1,850 | 169 | 53 | 915 | 6,209,034 | 14.5 | 22.7 | 12.7 | 6.8 | 0.0 | 32.3 | 5.6 |
| Roseville | 112 | 1,143 | 300 | 44 | 218 | 1,300,355 | 16.4 | 31.0 | 9.9 | 8.6 | 3.6 | 20.6 | 0.0 |
| St. Cloud | 142 | 1,217 | 128 | 36 | 514 | 2,603,968 | 10.3 | 32.2 | 15.4 | 10.8 | 5.3 | 18.8 | 0.0 |
| St. Louis Park | 143 | 1,088 | 92 | 32 | 276 | 1,991,961 | 10.1 | 28.0 | 11.8 | 8.0 | 13.9 | 19.8 | 0.0 |
| St. Paul | 634 | 5,191 | 859 | 214 | 3,186 | 18,549,813 | 11.3 | 28.4 | 19.3 | 6.9 | 3.5 | 22.3 | 4.2 |
| Savage | 57 | 604 | 38 | 13 | 132 | 726,106 | 15.8 | 36.8 | 2.3 | 13.5 | 0.0 | 19.2 | 0.0 |
| Shakopee | 67 | 383 | 40 | 12 | 219 | 1,326,235 | 15.2 | 28.9 | 4.5 | 3.9 | 0.0 | 36.4 | 0.0 |
| Shoreview | 27 | 152 | 13 | 4 | 123 | 722,231 | 15.2 | 0.0 | 36.9 | 15.9 | 11.6 | 15.9 | 0.0 |
| Winona | 43 | 204 | 27 | 5 | 180 | 907,619 | 9.3 | 26.6 | 14.9 | 10.1 | 3.7 | 23.8 | 5.0 |
| Woodbury | 122 | 998 | 71 | 26 | 211 | 1,311,828 | 7.1 | 20.7 | 13.8 | 10.0 | 0.0 | 38.6 | 5.6 |
| **MISSISSIPPI** | 3,417 | 18,345 | 2,279 | 603 | X | X | X | X | X | X | X | X | X |
| Biloxi | 66 | 326 | 30 | 8 | 637 | 2,469,007 | 10.2 | 27.3 | 33.8 | 9.0 | 2.0 | 12.4 | 0.0 |
| Clinton | 41 | 246 | 31 | 9 | 208 | 675,547 | 9.7 | 34.6 | 25.8 | 5.2 | 0.0 | 19.5 | 0.0 |
| Greenville | D | D | D | D | 308 | 932,163 | 9.8 | 40.2 | 26.7 | 16.1 | 0.0 | 3.9 | 0.0 |
| Gulfport | 114 | 767 | 81 | 24 | 3,674 | 19,047,459 | 1.4 | 4.1 | 3.2 | 0.8 | 88.3 | 1.0 | 0.0 |
| Hattiesburg | 106 | 746 | 74 | 19 | 671 | 2,076,495 | 12.8 | 27.9 | 23.0 | 11.6 | 3.7 | 18.3 | 0.0 |
| Horn Lake | 20 | 121 | 15 | 4 | 188 | 670,233 | 8.3 | 43.2 | 29.3 | 4.5 | 0.0 | 14.0 | 0.0 |
| Jackson | 312 | 2,097 | 287 | 86 | 2,154 | 6,153,172 | 13.1 | 31.3 | 19.3 | 5.2 | 5.2 | 17.5 | 0.0 |
| Meridian | D | D | D | D | 532 | 1,624,537 | 10.4 | 23.6 | 27.2 | 11.8 | 3.7 | 19.2 | 0.0 |
| Olive Branch | D | D | D | D | 404 | 1,656,189 | 13.0 | 24.3 | 31.2 | 3.9 | 0.7 | 21.3 | 0.0 |
| Pearl | 46 | 299 | 38 | 11 | 216 | 903,023 | 8.6 | 31.3 | 29.5 | 10.4 | 2.5 | 17.7 | 0.0 |
| Southaven | 59 | 389 | 32 | 9 | 383 | 1,549,716 | 10.7 | 38.5 | 33.1 | 2.9 | 0.4 | 14.3 | 0.0 |
| Tupelo | D | D | D | D | 514 | 1,914,110 | 12.7 | 24.8 | 21.0 | 9.8 | 0.0 | 29.2 | 0.0 |

1. Establishments subject to federal tax.

# Table D. Cities — City Government Finances

| City | General revenue Total (mil dol) | Intergovernmental Total (mil dol) | Percent from state government | Taxes Total (mil dol) | Per capita[1] (dollars) Total | Per capita[1] (dollars) Property | Sales and gross receipts | General expenditure Total (mil dol) | Per capita[1] (dollars) Total | Capital outlays |
|---|---|---|---|---|---|---|---|---|---|---|
| | 117 | 118 | 119 | 120 | 121 | 122 | 123 | 124 | 125 | 126 |
| **MICHIGAN— Cont'd** | | | | | | | | | | |
| Oak Park | 41.6 | 6.1 | 91.1 | 18.0 | 604 | 586 | 17 | 46 | 1,558 | 27 |
| Pontiac | 52.2 | 16.9 | 93.5 | 27.8 | 463 | 204 | 40 | 48 | 800 | 73 |
| Portage | 57.0 | 11.9 | 96.2 | 24.0 | 493 | 458 | 34 | 52 | 1,063 | 181 |
| Port Huron | 62.9 | 20.0 | 45.2 | 19.0 | 653 | 398 | 26 | 54 | 1,865 | 144 |
| Rochester Hills | 80.4 | 15.2 | 77.0 | 34.2 | 459 | 420 | 39 | 78 | 1,053 | 57 |
| Roseville | 55.0 | 12.7 | 90.4 | 22.3 | 470 | 444 | 26 | 52 | 1,099 | 83 |
| Royal Oak | 107.6 | 13.0 | 81.4 | 51.6 | 869 | 780 | 89 | 108 | 1,819 | 331 |
| Saginaw | 83.1 | 31.9 | 44.3 | 21.2 | 435 | 133 | 30 | 92 | 1,896 | 114 |
| St. Clair Shores | 79.2 | 13.6 | 80.9 | 34.0 | 569 | 524 | 45 | 81 | 1,353 | 243 |
| Southfield | 104.3 | 23.0 | 61.3 | 64.4 | 878 | 809 | 69 | 113 | 1,542 | 265 |
| Southgate | 34.4 | 6.4 | 87.3 | 16.4 | 563 | 540 | 23 | 34 | 1,168 | 3 |
| Sterling Heights | 152.6 | 36.8 | 75.4 | 64.5 | 486 | 467 | 19 | 173 | 1,301 | 207 |
| Taylor | 104.3 | 25.8 | 54.6 | 43.4 | 707 | 657 | 49 | 101 | 1,637 | 20 |
| Troy | 103.7 | 17.4 | 88.7 | 52.7 | 625 | 593 | 33 | 102 | 1,217 | 230 |
| Warren | 191.1 | 49.0 | 89.8 | 97.5 | 721 | 687 | 34 | 168 | 1,241 | 94 |
| Westland | 103.2 | 26.6 | 62.0 | 33.8 | 412 | 396 | 17 | 105 | 1,286 | 27 |
| Wyandotte | 45.2 | 6.3 | 91.7 | 16.3 | 651 | 623 | 28 | 98 | 3,905 | 135 |
| Wyoming | 74.2 | 23.3 | 56.0 | 28.1 | 370 | 330 | 40 | 70 | 916 | 46 |
| **MINNESOTA** | X | X | X | X | X | X | X | X | X | X |
| Andover | 22.2 | 1.6 | 87.0 | 12.5 | 382 | 363 | 19 | 18 | 542 | 121 |
| Apple Valley | 55.9 | 2.7 | 74.9 | 27.5 | 525 | 459 | 66 | 43 | 823 | 188 |
| Blaine | 48.3 | 3.4 | 78.6 | 25.9 | 403 | 348 | 56 | 50 | 784 | 185 |
| Bloomington | 152.9 | 20.6 | 44.3 | 78.0 | 912 | 635 | 277 | 163 | 1,909 | 388 |
| Brooklyn Center | 48.4 | 4.0 | 67.7 | 22.3 | 721 | 632 | 89 | 48 | 1,534 | 276 |
| Brooklyn Park | 82.7 | 7.8 | 71.6 | 49.0 | 610 | 512 | 98 | 73 | 911 | 183 |
| Burnsville | 66.5 | 5.2 | 55.4 | 35.7 | 580 | 523 | 58 | 64 | 1,040 | 278 |
| Coon Rapids | 59.4 | 5.6 | 83.0 | 31.4 | 502 | 407 | 95 | 69 | 1,099 | 321 |
| Cottage Grove | 31.0 | 4.2 | 64.3 | 16.2 | 441 | 405 | 36 | NA | NA | NA |
| Duluth | 223.9 | 79.1 | 70.4 | 63.8 | 742 | 338 | 404 | 228 | 2,649 | 727 |
| Eagan | 63.3 | 4.9 | 87.1 | 34.6 | 520 | 474 | 46 | 67 | 1,013 | 278 |
| Eden Prairie | 64.0 | 2.4 | 59.2 | 40.6 | 632 | 533 | 99 | 74 | 1,154 | 259 |
| Edina | 70.6 | 6.1 | 97.2 | 39.4 | 761 | 605 | 155 | 74 | 1,419 | 407 |
| Fridley | 34.3 | 7.6 | 97.9 | 13.7 | 493 | 441 | 52 | 33 | 1,175 | 323 |
| Inver Grove Heights | 34.9 | 2.8 | 37.2 | 21.4 | 606 | 574 | 0 | 36 | 1,008 | 266 |
| Lakeville | 69.5 | 9.3 | 69.3 | 29.8 | 469 | 410 | 58 | 71 | 1,119 | 591 |
| Mankato | 95.3 | 20.1 | 67.0 | 27.3 | 646 | 416 | 230 | 103 | 2,438 | 789 |
| Maple Grove | 78.0 | 7.7 | 93.7 | 37.4 | 528 | 484 | 43 | 68 | 962 | 282 |
| Maplewood | 49.4 | 5.6 | 98.8 | 24.3 | 597 | 517 | 80 | 37 | 896 | 95 |
| Minneapolis | 1,215.6 | 223.1 | 41.2 | 471.3 | 1,120 | 744 | 376 | 1,582 | 3,759 | 1,427 |
| Minnetonka | 65.4 | 6.1 | 89.3 | 39.4 | 744 | 643 | 101 | 63 | 1,190 | 272 |
| Moorhead | 72.2 | 27.1 | 87.4 | 9.5 | 219 | 216 | 4 | 74 | 1,712 | 743 |
| Oakdale | 23.9 | 1.1 | 88.1 | 11.6 | 416 | 365 | 51 | 24 | 863 | 146 |
| Owatonna | 27.7 | 7.0 | 84.4 | 12.0 | 466 | 436 | 30 | 25 | 956 | 146 |
| Plymouth | 92.1 | 18.8 | 75.5 | 37.3 | 478 | 418 | 60 | 103 | 1,319 | 406 |
| Richfield | 46.3 | 10.3 | 37.8 | 22.2 | 617 | 523 | 94 | 54 | 1,499 | 496 |
| Rochester | 289.2 | 56.0 | 76.0 | 88.5 | 766 | 491 | 274 | 326 | 2,823 | 1,037 |
| Roseville | 32.5 | 3.4 | 68.0 | 20.4 | 566 | 554 | 12 | 37 | 1,036 | 273 |
| St. Cloud | 118.2 | 24.7 | 92.7 | 40.8 | 599 | 360 | 239 | 126 | 1,847 | 821 |
| St. Louis Park | 60.9 | 4.5 | 85.1 | 35.1 | 719 | 568 | 151 | 72 | 1,464 | 437 |
| St. Paul | 658.5 | 197.0 | 57.5 | 174.9 | 574 | 373 | 200 | 811 | 2,660 | 316 |
| Savage | 35.8 | 5.7 | 99.2 | 17.9 | 573 | 531 | 41 | 36 | 1,142 | 438 |
| Shakopee | 34.2 | 3.3 | 97.6 | 21.7 | 532 | 444 | 77 | 65 | 1,598 | 880 |
| Shoreview | 27.5 | 1.4 | 99.9 | 13.4 | 504 | 430 | 74 | 26 | 965 | 116 |
| Winona | 32.3 | 13.1 | 93.8 | 8.9 | 333 | 273 | 60 | 32 | 1,192 | 390 |
| Woodbury | 80.3 | 6.2 | 87.4 | 36.3 | 522 | 454 | 68 | 71 | 1,022 | 329 |
| **MISSISSIPPI** | X | X | X | X | X | X | X | X | X | X |
| Biloxi | 111.4 | 64.0 | 55.6 | 23.0 | 499 | 383 | 116 | 108 | 2,343 | 605 |
| Clinton | 22.8 | 7.3 | 94.7 | 9.4 | 369 | 333 | 36 | 21 | 839 | 121 |
| Greenville | 34.6 | 11.5 | 68.2 | 14.1 | 461 | 383 | 78 | 31 | 1,026 | 158 |
| Gulfport | 597.0 | 45.1 | 93.7 | 32.3 | 449 | 332 | 117 | 587 | 8,156 | 509 |
| Hattiesburg | 80.5 | 32.7 | 79.9 | 28.5 | 620 | 422 | 198 | 87 | 1,883 | 400 |
| Horn Lake | 21.3 | 6.0 | 100.0 | 7.9 | 290 | 262 | 28 | 22 | 806 | 102 |
| Jackson | 219.6 | 57.8 | 68.0 | 80.6 | 483 | 420 | 64 | 201 | 1,206 | 166 |
| Meridian | 44.5 | 15.4 | 99.1 | 18.5 | 488 | 413 | 75 | 38 | 1,012 | 68 |
| Olive Branch | 42.7 | 9.9 | 100.0 | 18.9 | 502 | 452 | 50 | 39 | 1,039 | 114 |
| Pearl | 33.2 | 13.7 | 91.9 | 8.8 | 332 | 288 | 44 | 33 | 1,232 | 83 |
| Southaven | 61.0 | 21.2 | 75.6 | 27.9 | 514 | 453 | 62 | 55 | 1,011 | 207 |
| Tupelo | 63.2 | 32.0 | 85.0 | 16.1 | 421 | 395 | 25 | 74 | 1,949 | 748 |

1. Based on population estimated as of July 1 of the year shown.

| City | Public welfare | Highways | Parking facilities | Education | Health and hospitals | Police protection | Sewerage and sanitation | Parks and recreation | Housing and community development | Interest on debt |
|---|---|---|---|---|---|---|---|---|---|---|
| | 127 | 128 | 129 | 130 | 131 | 132 | 133 | 134 | 135 | 136 |
| **MICHIGAN— Cont'd** | | | | | | | | | | |
| Oak Park | 0.0 | 11.3 | 0.0 | 0.0 | 0.0 | 0.0 | 38.6 | 1.4 | 0.0 | 2.2 |
| Pontiac | 0.0 | 21.3 | 0.0 | 0.0 | 0.0 | 25.5 | 7.2 | 1.6 | 3.2 | 2.3 |
| Portage | 0.3 | 18.0 | 0.0 | 0.0 | 2.8 | 19.0 | 23.9 | 5.9 | 0.4 | 1.5 |
| Port Huron | 0.0 | 12.6 | 0.0 | 0.0 | 0.0 | 16.1 | 18.9 | 8.1 | 13.2 | 3.7 |
| Rochester Hills | 0.0 | 13.1 | 0.0 | 0.0 | 0.0 | 22.6 | 22.3 | 14.1 | 0.0 | 0.5 |
| Roseville | 0.0 | 8.8 | 0.0 | 0.0 | 0.0 | 20.6 | 27.5 | 0.5 | 1.3 | 0.6 |
| Royal Oak | 0.0 | 18.8 | 0.9 | 0.0 | 0.6 | 17.4 | 22.5 | 4.0 | 1.3 | 1.0 |
| Saginaw | 0.0 | 14.4 | 0.0 | 0.0 | 0.0 | 14.6 | 35.1 | 0.4 | 7.3 | 0.6 |
| St. Clair Shores | 0.0 | 7.9 | 0.0 | 0.0 | 0.0 | 24.4 | 28.8 | 10.1 | 2.6 | 1.7 |
| Southfield | 0.0 | 22.2 | 0.0 | 0.0 | 2.6 | 22.9 | 2.3 | 7.5 | 0.7 | 2.1 |
| Southgate | 0.0 | 0.0 | 0.0 | 0.0 | 0.0 | 21.3 | 19.2 | 4.5 | 0.0 | 1.1 |
| Sterling Heights | 0.0 | 20.7 | 0.0 | 0.0 | 0.0 | 17.8 | 28.5 | 2.6 | 0.6 | 1.7 |
| Taylor | 0.0 | 7.7 | 0.0 | 0.1 | 0.0 | 10.7 | 14.7 | 5.7 | 8.0 | 2.5 |
| Troy | 0.0 | 18.9 | 0.0 | 0.0 | 0.0 | 22.6 | 21.9 | 9.7 | 0.0 | 1.4 |
| Warren | 0.0 | 14.3 | 0.0 | 0.0 | 0.3 | 25.5 | 14.1 | 4.2 | 2.4 | 1.1 |
| Westland | 1.2 | 6.4 | 0.0 | 0.0 | 0.8 | 17.5 | 19.6 | 6.1 | 7.4 | 1.2 |
| Wyandotte | 0.0 | 3.3 | 0.0 | 0.0 | 0.0 | 6.1 | 10.5 | 1.5 | 0.0 | 0.5 |
| Wyoming | 0.0 | 12.8 | 0.0 | 0.0 | 0.0 | 17.8 | 20.0 | 7.9 | 12.5 | 1.8 |
| **MINNESOTA** | X | X | X | X | X | X | X | X | X | X |
| Andover | 0.0 | 13.2 | 0.0 | 0.0 | 0.0 | 16.5 | 12.7 | 18.4 | 2.7 | 0.0 |
| Apple Valley | 0.0 | 30.1 | 0.0 | 0.0 | 0.0 | 20.9 | 0.0 | 21.0 | 0.0 | 2.5 |
| Blaine | 0.0 | 30.8 | 0.0 | 0.0 | 0.0 | 20.3 | 18.7 | 4.0 | 0.0 | 1.8 |
| Bloomington | 0.0 | 19.3 | 0.0 | 0.0 | 3.8 | 15.1 | 7.3 | 9.0 | 8.3 | 1.2 |
| Brooklyn Center | 0.3 | 13.3 | 0.0 | 0.0 | 0.0 | 16.9 | 13.0 | 16.3 | 14.3 | 0.0 |
| Brooklyn Park | 0.0 | 24.6 | 0.0 | 0.0 | 0.0 | 28.7 | 0.0 | 16.6 | 0.9 | 2.4 |
| Burnsville | 0.0 | 18.1 | 0.2 | 0.0 | 0.0 | 21.0 | 19.7 | 14.1 | 0.8 | 0.0 |
| Coon Rapids | 0.0 | 25.1 | 0.0 | 0.0 | 0.9 | 15.2 | 12.9 | 23.1 | 2.8 | 2.0 |
| Cottage Grove | NA | NA | NA | NA | NA | NA | NA | NA | NA | NA |
| Duluth | 0.0 | 13.2 | 0.8 | 0.0 | 0.0 | 10.1 | 10.8 | 15.0 | 1.7 | 3.3 |
| Eagan | 0.0 | 26.7 | 0.0 | 0.0 | 0.0 | 19.8 | 9.7 | 15.1 | 0.0 | 1.1 |
| Eden Prairie | 0.0 | 18.2 | 0.0 | 0.0 | 0.0 | 18.7 | 8.1 | 23.1 | 1.2 | 1.7 |
| Edina | 0.0 | 30.3 | 0.0 | 0.0 | 0.0 | 15.5 | 2.2 | 21.4 | 0.1 | 3.5 |
| Fridley | 0.0 | 13.5 | 0.0 | 0.0 | 0.0 | 19.5 | 21.5 | 20.5 | 0.0 | 0.0 |
| Inver Grove Heights | 0.0 | 22.4 | 0.0 | 0.0 | 0.0 | 19.8 | 18.1 | 0.0 | 2.7 | 5.3 |
| Lakeville | 0.0 | 46.2 | 0.0 | 0.0 | 0.0 | 14.5 | 8.9 | 9.2 | 2.1 | 0.0 |
| Mankato | 0.0 | 18.4 | 1.2 | 0.0 | 0.0 | 8.6 | 9.9 | 4.1 | 7.5 | 0.0 |
| Maple Grove | 0.0 | 29.2 | 0.0 | 0.0 | 0.0 | 16.3 | 9.1 | 16.0 | 1.1 | 6.4 |
| Maplewood | 0.0 | 15.8 | 0.0 | 0.0 | 0.1 | 23.6 | 14.8 | 12.1 | 2.1 | 0.0 |
| Minneapolis | 0.0 | 11.8 | 2.8 | 0.0 | 1.6 | 10.1 | 7.8 | 6.8 | 11.9 | 6.6 |
| Minnetonka | 0.0 | 23.7 | 0.0 | 0.0 | 0.6 | 20.8 | 12.9 | 17.4 | 2.9 | 0.3 |
| Moorhead | 0.0 | 47.2 | 0.0 | 0.0 | 0.0 | 12.6 | 12.0 | 7.4 | 0.9 | 0.0 |
| Oakdale | 0.0 | 26.1 | 0.0 | 0.0 | 3.7 | 21.8 | 14.8 | 5.9 | 0.0 | 3.0 |
| Owatonna | 0.0 | 25.5 | 0.0 | 0.0 | 0.0 | 19.2 | 8.1 | 13.1 | 4.2 | 0.0 |
| Plymouth | 0.0 | 26.4 | 0.0 | 0.0 | 0.0 | 12.6 | 8.3 | 22.0 | 5.8 | 0.0 |
| Richfield | 0.0 | 34.8 | 0.0 | 0.0 | 0.1 | 15.3 | 0.0 | 9.4 | 12.9 | 3.1 |
| Rochester | 0.0 | 13.4 | 1.4 | 0.0 | 0.1 | 8.7 | 4.3 | 22.2 | 0.4 | 27.4 |
| Roseville | 0.0 | 7.4 | 0.0 | 0.0 | 0.0 | 19.5 | 15.1 | 16.1 | 12.3 | 3.4 |
| St. Cloud | 0.0 | 13.6 | 1.5 | 0.0 | 0.9 | 13.8 | 12.2 | 29.3 | 2.1 | 0.0 |
| St. Louis Park | 0.0 | 19.3 | 0.0 | 0.0 | 0.0 | 12.8 | 14.5 | 26.5 | 0.9 | 2.2 |
| St. Paul | 0.0 | 10.7 | 0.8 | 0.0 | 1.1 | 12.7 | 9.1 | 9.1 | 17.1 | 5.0 |
| Savage | 0.0 | 44.6 | 0.0 | 0.0 | 0.0 | 15.3 | 12.6 | 7.1 | 0.0 | 5.1 |
| Shakopee | 0.0 | 11.7 | 0.0 | 0.0 | 0.0 | 11.9 | 13.6 | 43.5 | 4.1 | 0.0 |
| Shoreview | 0.0 | 12.0 | 0.0 | 0.0 | 0.0 | 8.2 | 15.8 | 20.4 | 2.2 | 1.4 |
| Winona | 0.0 | 18.4 | 0.0 | 0.0 | 0.0 | 15.2 | 18.1 | 18.5 | 0.0 | 0.5 |
| Woodbury | 0.0 | 40.2 | 0.0 | 0.0 | 3.3 | 15.0 | 9.5 | 13.1 | 0.8 | 2.2 |
| **MISSISSIPPI** | X | X | X | X | X | X | X | X | X | X |
| Biloxi | 0.8 | 9.7 | 0.0 | 0.0 | 0.2 | 13.8 | 23.9 | 4.5 | 1.2 | 2.7 |
| Clinton | 0.0 | 11.5 | 0.0 | 0.0 | 0.0 | 23.0 | 26.1 | 8.1 | 0.0 | 2.7 |
| Greenville | 0.0 | 11.2 | 0.0 | 0.0 | 1.3 | 24.2 | 14.3 | 3.2 | 0.0 | 0.7 |
| Gulfport | 0.0 | 2.1 | 0.0 | 0.0 | 81.8 | 3.2 | 4.1 | 1.8 | 0.3 | 1.7 |
| Hattiesburg | 0.0 | 10.0 | 0.3 | 0.0 | 1.1 | 14.4 | 22.7 | 7.9 | 0.9 | 2.1 |
| Horn Lake | 0.0 | 3.8 | 0.0 | 0.0 | 0.0 | 25.1 | 12.9 | 6.6 | 0.0 | 6.1 |
| Jackson | 1.0 | 8.9 | 0.0 | 0.0 | 0.5 | 17.7 | 18.9 | 8.1 | 0.8 | 4.7 |
| Meridian | 0.0 | 12.6 | 0.2 | 0.0 | 0.0 | 20.9 | 14.9 | 7.4 | 4.0 | 3.6 |
| Olive Branch | 0.0 | 14.4 | 0.0 | 0.0 | 2.5 | 24.6 | 16.3 | 6.8 | 0.0 | 2.4 |
| Pearl | 0.0 | 13.1 | 0.0 | 0.0 | 1.4 | 20.5 | 22.1 | 4.5 | 0.0 | 4.0 |
| Southaven | 0.0 | 12.5 | 0.0 | 0.0 | 0.8 | 20.2 | 9.8 | 9.2 | 0.0 | 2.7 |
| Tupelo | 0.0 | 17.4 | 0.0 | 0.0 | 0.0 | 22.2 | 14.7 | 21.2 | 0.1 | 3.0 |

# Table D. Cities — City Government Finances, City Government Employment, and Climate

| City | City government finances, 2017 (cont.) Debt outstanding | | | Climate[2] Average daily temperature | | | | | | |
| | Total (mil dol) | Per capita[1] (dollars) | Debt issued during year | Mean January | Mean July | Limits January[3] | Limits July[4] | Annual precipitation (inches) | Heating degree days | Cooling degree days |
| | 137 | 138 | 139 | 140 | 141 | 142 | 143 | 144 | 145 | 146 |
|---|---|---|---|---|---|---|---|---|---|---|
| **MICHIGAN— Cont'd** | | | | | | | | | | |
| Oak Park | 40.2 | 1,349 | 0.0 | 24.7 | 73.7 | 16.1 | 85.7 | 33.58 | 6,224 | 788 |
| Pontiac | 22.5 | 373 | 0.0 | 22.9 | 71.9 | 15.9 | 82.3 | 30.03 | 6,680 | 626 |
| Portage | 67.5 | 1,383 | 3.6 | 24.3 | 73.2 | 17.0 | 84.2 | 37.41 | 6,235 | 773 |
| Port Huron | 89.3 | 3,076 | 1.9 | 22.8 | 72.2 | 15.1 | 81.9 | 31.39 | 6,845 | 626 |
| Rochester Hills | 28.1 | 377 | 0.1 | 22.0 | 70.6 | 13.7 | 82.7 | 35.74 | 7,046 | 523 |
| Roseville | 12.0 | 253 | 1.1 | 25.3 | 73.6 | 18.8 | 83.3 | 33.97 | 6,160 | 757 |
| Royal Oak | 221.1 | 3,724 | 140.1 | 24.7 | 73.7 | 16.1 | 85.7 | 33.58 | 6,224 | 788 |
| Saginaw | 62.8 | 1,291 | 26.5 | 21.4 | 71.2 | 14.9 | 81.9 | 31.61 | 7,099 | 548 |
| St. Clair Shores | 24.1 | 404 | 8.6 | 25.3 | 73.6 | 18.8 | 83.3 | 33.97 | 6,160 | 757 |
| Southfield | 76.0 | 1,035 | 0.0 | 24.7 | 73.7 | 16.1 | 85.7 | 33.58 | 6,224 | 788 |
| Southgate | 14.8 | 508 | 1.0 | 24.5 | 73.5 | 17.8 | 83.4 | 32.89 | 6,422 | 736 |
| Sterling Heights | 147.8 | 1,114 | 97.1 | 24.4 | 71.9 | 18.0 | 81.8 | 32.24 | 6,620 | 597 |
| Taylor | 67.1 | 1,092 | 4.2 | 24.5 | 73.5 | 17.8 | 83.4 | 32.89 | 6,422 | 736 |
| Troy | 34.5 | 410 | 0.0 | 22.9 | 71.9 | 15.9 | 82.3 | 30.03 | 6,680 | 626 |
| Warren | 189.7 | 1,404 | 53.8 | 24.7 | 73.7 | 16.1 | 85.7 | 33.58 | 6,680 | 626 |
| Westland | 34.5 | 421 | 8.8 | 24.6 | 73.9 | 17.6 | 84.7 | 32.80 | 6,224 | 788 |
| Wyandotte | 83.9 | 3,351 | 0.5 | 24.5 | 73.5 | 17.8 | 83.4 | 32.89 | 6,167 | 828 |
| Wyoming | 69.6 | 915 | 0.0 | 22.4 | 71.4 | 15.6 | 82.3 | 37.13 | 6,422 | 736 |
| | | | | | | | | | 6,896 | 613 |
| **MINNESOTA** | X | X | X | X | X | X | X | X | X | X |
| Andover | 27.6 | 841 | 4.5 | 10.9 | 70.4 | 1.8 | 80.5 | 31.36 | 8,367 | 500 |
| Apple Valley | 42.3 | 807 | 0.0 | 9.2 | 69.4 | -1.1 | 80.5 | 29.19 | 8,805 | 416 |
| Blaine | 45.2 | 703 | 11.2 | 10.9 | 70.4 | 1.8 | 80.5 | 31.36 | 8,367 | 500 |
| Bloomington | 72.4 | 846 | 13.5 | 13.1 | 73.2 | 4.3 | 83.3 | 29.41 | 7,876 | 699 |
| Brooklyn Center | 99.2 | 3,204 | 38.3 | 13.0 | 71.4 | 2.8 | 82.8 | 30.50 | 7,983 | 587 |
| Brooklyn Park | 70.8 | 881 | 7.9 | 13.0 | 71.4 | 2.8 | 82.8 | 30.50 | 7,983 | 587 |
| Burnsville | 66.2 | 1,078 | 9.3 | 13.8 | 74.0 | 3.4 | 85.8 | 30.44 | 7,549 | 803 |
| Coon Rapids | 95.9 | 1,535 | 9.5 | 13.0 | 71.4 | 2.8 | 82.8 | 30.50 | 7,983 | 587 |
| Cottage Grove | 46.2 | 1,260 | 0.0 | 12.0 | 72.1 | 2.5 | 82.6 | 29.95 | 8,032 | 617 |
| Duluth | 263.8 | 3,068 | 56.1 | 8.4 | 65.5 | -1.2 | 76.3 | 31.00 | 9,724 | 189 |
| Eagan | 64.0 | 961 | 8.1 | 13.1 | 73.2 | 4.3 | 83.3 | 29.41 | 7,876 | 699 |
| Eden Prairie | 83.0 | 1,293 | 4.2 | 10.2 | 71.4 | -0.3 | 82.2 | 28.82 | 8,429 | 567 |
| Edina | 135.3 | 2,612 | 16.4 | 13.8 | 74.0 | 3.4 | 85.8 | 30.44 | 7,549 | 803 |
| Fridley | 13.8 | 495 | 6.0 | 10.9 | 70.4 | 1.8 | 80.5 | 31.36 | 8,367 | 500 |
| Inver Grove Heights | 51.7 | 1,462 | 8.5 | 10.1 | 71.0 | 0.0 | 81.3 | 34.60 | 8,345 | 533 |
| Lakeville | 126.5 | 1,987 | 31.1 | 13.1 | 72.2 | 3.8 | 83.6 | 31.43 | 7,773 | 658 |
| Mankato | 120.9 | 2,866 | 18.5 | 12.5 | 72.1 | 2.4 | 83.4 | 33.42 | 8,029 | 650 |
| Maple Grove | 197.9 | 2,794 | 8.1 | 13.0 | 71.4 | 2.8 | 82.8 | 30.50 | 7,983 | 587 |
| Maplewood | 117.7 | 2,893 | 9.5 | 14.5 | 73.0 | 6.2 | 83.2 | 32.59 | 7,606 | 715 |
| Minneapolis | 2,903.4 | 6,898 | 163.0 | 13.1 | 73.2 | 4.3 | 83.3 | 29.41 | 7,876 | 699 |
| Minnetonka | 136.2 | 2,574 | 29.9 | 13.1 | 73.2 | 4.3 | 83.3 | 29.41 | 7,876 | 699 |
| Moorhead | 381.7 | 8,824 | 128.5 | 3.8 | 69.8 | -7.1 | 81.5 | 21.56 | 9,628 | 478 |
| Oakdale | 23.7 | 846 | 7.8 | 14.5 | 73.0 | 6.2 | 83.2 | 32.59 | 7,606 | 715 |
| Owatonna | 45.8 | 1,780 | 5.7 | NA | NA | NA | NA | NA | NA | NA |
| Plymouth | 75.0 | 960 | 0.0 | 13.0 | 71.4 | 2.8 | 82.8 | 30.50 | 7,983 | 587 |
| Richfield | 116.0 | 3,218 | 14.2 | 13.1 | 73.2 | 4.3 | 83.3 | 29.41 | 7,876 | 699 |
| Rochester | 2,610.2 | 22,582 | 45.1 | 9.8 | 70.0 | 0.0 | 80.9 | 29.10 | 8,703 | 474 |
| Roseville | 31.8 | 882 | 0.0 | 14.5 | 73.0 | 6.2 | 83.2 | 32.59 | 7,606 | 715 |
| St. Cloud | 398.0 | 5,845 | 55.9 | 8.8 | 69.8 | -1.2 | 81.7 | 27.13 | 8,815 | 443 |
| St. Louis Park | 187.1 | 3,830 | 39.4 | 13.0 | 71.4 | 2.8 | 82.8 | 30.50 | 7,983 | 587 |
| St. Paul | 2,381.0 | 7,809 | 126.2 | 14.5 | 73.0 | 6.2 | 83.2 | 32.59 | 7,606 | 715 |
| Savage | 62.5 | 1,999 | 3.3 | NA | NA | NA | NA | NA | NA | NA |
| Shakopee | 93.5 | 2,295 | 29.5 | NA | NA | NA | NA | NA | NA | NA |
| Shoreview | 52.8 | 1,983 | 7.6 | 14.5 | 73.0 | 6.2 | 83.2 | 32.59 | 7,606 | 715 |
| Winona | 3.1 | 114 | 0.0 | 17.6 | 75.8 | 9.2 | 85.3 | 34.20 | 6,839 | 990 |
| Woodbury | 95.4 | 1,372 | 3.6 | 11.5 | 72.1 | 1.9 | 81.9 | 29.92 | 8,104 | 621 |
| **MISSISSIPPI** | X | X | X | X | X | X | X | X | X | X |
| Biloxi | 113.5 | 2,467 | 25.2 | 50.7 | 81.7 | 43.5 | 88.5 | 64.84 | 1,645 | 2,517 |
| Clinton | 34.2 | 1,351 | 10.4 | NA | NA | NA | NA | NA | NA | NA |
| Greenville | 17.0 | 556 | 1.9 | 42.3 | 82.6 | 33.0 | 92.6 | 54.20 | 2,715 | 2,216 |
| Gulfport | 177.0 | 2,460 | 60.2 | 51.6 | 82.6 | 42.6 | 91.3 | 65.20 | 1,514 | 2,679 |
| Hattiesburg | 126.5 | 2,747 | 47.7 | 47.9 | 81.7 | 36.0 | 92.1 | 62.47 | 2,024 | 2,327 |
| Horn Lake | 24.6 | 906 | 0.0 | NA | NA | NA | NA | NA | NA | NA |
| Jackson | 443.6 | 2,659 | 54.1 | 45.0 | 81.4 | 35.0 | 91.4 | 55.95 | 2,401 | 2,264 |
| Meridian | 51.6 | 1,360 | 9.5 | 46.1 | 81.7 | 34.7 | 92.9 | 58.65 | 2,352 | 2,173 |
| Olive Branch | 47.5 | 1,266 | 0.6 | NA | NA | NA | NA | NA | NA | NA |
| Pearl | 81.1 | 3,063 | 16.5 | NA | NA | NA | NA | NA | NA | NA |
| Southaven | 81.9 | 1,513 | 13.9 | 37.9 | 80.4 | 27.8 | 90.3 | 55.06 | 3,442 | 1,749 |
| Tupelo | 97.9 | 2,565 | 31.8 | 40.4 | 80.6 | 30.5 | 91.4 | 55.86 | 3,086 | 1,884 |

1. Based on the population estimated as of July 1 of the year shown. maximum.   2. Represents normal values based on the 30-year period, 1971–2000.   3. Average daily minimum.   4. Average daily maximum.

# Table D. Cities — Land Area and Population

| STATE Place code | City | Land area[1] (sq. mi) | Population, 2020 Total persons 2019 | Rank | Per square mile | Race 2019 — Race alone[2] (percent) White | Black or African American | American Indian, Alaskan Native | Asian | Hawaiian Pacific Islander | Some other race | Two or more races (percent) |
|---|---|---|---|---|---|---|---|---|---|---|---|---|
| | | 1 | 2 | 3 | 4 | 5 | 6 | 7 | 8 | 9 | 10 | 11 |
| 29 00000 | MISSOURI ....................... | 68,745.5 | 6,151,548 | X | 90 | 81.8 | 11.5 | 0.4 | 2.1 | 0.2 | 1.2 | 2.8 |
| 29 03160 | Ballwin.......................... | 9.0 | 30,094 | 1,251 | 3,344 | D | D | D | D | D | D | D |
| 29 06652 | Blue Springs.................. | 22.4 | 56,547 | 680 | 2,524 | 83.1 | 8.3 | 0.2 | 0.9 | 0.0 | 3.9 | 3.6 |
| 29 11242 | Cape Girardeau ............ | 29.1 | 41,588 | 924 | 1,429 | 79.6 | 14.1 | 0.0 | 2.6 | 0.0 | 1.1 | 2.6 |
| 29 13600 | Chesterfield.................. | 31.9 | 47,570 | 816 | 1,491 | 83.5 | 2.4 | 0.0 | 10.6 | 0.0 | 0.2 | 3.4 |
| 29 15670 | Columbia....................... | 66.5 | 124,769 | 226 | 1,876 | 77.9 | 10.6 | 0.3 | 6.4 | 0.1 | 0.6 | 4.1 |
| 29 24778 | Florissant..................... | 12.6 | 50,795 | 769 | 4,031 | D | D | D | D | D | D | D |
| 29 27190 | Gladstone..................... | 8.1 | 27,677 | 1,329 | 3,417 | 83.4 | 1.5 | 1.5 | 0.6 | 2.9 | 1.5 | 8.6 |
| 29 31276 | Hazelwood .................... | 16.0 | 25,042 | 1,403 | 1,565 | D | D | D | D | D | D | D |
| 29 35000 | Independence ............... | 78.0 | 116,774 | 248 | 1,497 | 73.6 | 10.7 | 0.4 | 2.5 | 1.5 | 8.4 | 2.8 |
| 29 37000 | Jefferson City ............... | 36.1 | 41,972 | 916 | 1,163 | 77.7 | 17.1 | 0.2 | 2.6 | 0.0 | 0.6 | 2.0 |
| 29 37592 | Joplin............................ | 38.1 | 50,956 | 766 | 1,337 | 86.7 | 2.3 | 1.8 | 3.1 | 0.1 | 2.0 | 3.9 |
| 29 38000 | Kansas City................... | 314.7 | 497,159 | 38 | 1,580 | 62.8 | 26.1 | 0.2 | 2.4 | 0.2 | 3.9 | 4.4 |
| 29 39044 | Kirkwood ...................... | 9.2 | 27,866 | 1,321 | 3,029 | D | D | D | D | D | D | D |
| 29 41348 | Lee's Summit ................ | 63.9 | 101,467 | 313 | 1,588 | 81.3 | 12.3 | 0.1 | 1.8 | 0.1 | 0.4 | 4.2 |
| 29 42032 | Liberty ......................... | 28.8 | 32,418 | 1,178 | 1,126 | D | D | D | D | D | D | 4.9 |
| 29 46586 | Maryland Heights .......... | 21.9 | 26,895 | 1,355 | 1,228 | 59.2 | 11.4 | 0.0 | 18.6 | 5.3 | 0.6 | 2.9 |
| 29 54074 | O'Fallon ....................... | 29.8 | 89,769 | 373 | 3,012 | 87.2 | 5.3 | 0.1 | 4.0 | 0.0 | 0.4 | 2.4 |
| 29 60788 | Raytown ....................... | 9.9 | 28,816 | 1,293 | 2,911 | 54.0 | 39.4 | 0.4 | 0.7 | 0.0 | 3.0 | 2.8 |
| 29 64082 | St. Charles .................. | 24.8 | 71,761 | 506 | 2,894 | 83.7 | 8.1 | 0.5 | 3.1 | 0.0 | 1.8 | 3.1 |
| 29 64550 | St. Joseph ................... | 44.0 | 74,074 | 480 | 1,684 | 85.9 | 6.1 | 0.9 | 1.5 | 1.9 | 0.6 | 2.9 |
| 29 65000 | St. Louis...................... | 61.7 | 297,645 | 67 | 4,824 | 46.8 | 45.2 | 0.4 | 3.4 | 0.0 | 1.3 | 2.9 |
| 29 65126 | St. Peters .................... | 22.5 | 58,043 | 665 | 2,580 | D | D | D | D | D | D | D |
| 29 70000 | Springfield................... | 82.4 | 168,090 | 157 | 2,040 | 87.7 | 4.4 | 0.8 | 2.1 | 0.0 | 1.5 | 3.5 |
| 29 75220 | University City .............. | 5.9 | 34,049 | 1,133 | 5,771 | 52.6 | 36.8 | 0.4 | 5.8 | 0.1 | 1.5 | 2.7 |
| 29 78442 | Wentzville .................... | 20.1 | 43,109 | 893 | 2,145 | 92.0 | 2.5 | 0.0 | 3.6 | 0.0 | 0.2 | 1.6 |
| 29 79820 | Wildwood ..................... | 66.6 | 35,460 | 1,091 | 532 | D | D | D | D | D | D | 3.4 |
| 30 00000 | MONTANA ..................... | 145,547.7 | 1,080,577 | X | 7 | 88.0 | 0.7 | 6.3 | 0.8 | 0.0 | 0.7 | 3.5 |
| 30 06550 | Billings......................... | 44.8 | 109,736 | 277 | 2,450 | 88.6 | 1.4 | 4.8 | 0.7 | 0.0 | 0.9 | 3.3 |
| 30 08950 | Bozeman....................... | 20.6 | 50,970 | 765 | 2,474 | 91.4 | 2.0 | 1.0 | 1.8 | 0.0 | 0.5 | D |
| 30 11390 | Butte-Silver Bow ........... | 715.8 | 34,465 | 1,118 | 48 | D | D | D | D | D | D | 6.0 |
| 30 32800 | Great Falls ................... | 23.0 | 58,353 | 656 | 2,537 | 85.0 | 1.0 | 6.9 | 0.3 | 0.0 | 0.8 | D |
| 30 35600 | Helena.......................... | 16.8 | 33,737 | 1,141 | 2,008 | D | D | D | D | D | D | 3.5 |
| 30 50200 | Missoula....................... | 34.5 | 76,848 | 461 | 2,228 | 91.4 | 1.2 | 1.4 | 2.2 | 0.0 | 0.3 | 2.8 |
| 31 00000 | NEBRASKA ..................... | 76,816.5 | 1,937,552 | X | 25 | 86.2 | 4.9 | 1.0 | 2.5 | 0.1 | 2.6 | 5.1 |
| 31 03950 | Bellevue........................ | 21.1 | 53,270 | 729 | 2,525 | 85.6 | 3.8 | 1.1 | 0.8 | 0.4 | 3.2 | D |
| 31 17670 | Fremont........................ | 10.7 | 26,227 | 1,379 | 2,451 | D | D | D | D | D | D | 2.9 |
| 31 19595 | Grand Island ................ | 30.1 | 51,004 | 764 | 1,695 | 72.7 | 4.4 | 1.5 | 1.5 | 0.1 | 16.9 | 3.3 |
| 31 25055 | Kearney........................ | 14.5 | 34,235 | 1,127 | 2,361 | 86.2 | 0.6 | 0.7 | 2.0 | 0.0 | 7.2 | 3.6 |
| 31 28000 | Lincoln......................... | 97.7 | 290,505 | 69 | 2,973 | 84.2 | 4.4 | 1.0 | 5.0 | 0.2 | 1.6 | 3.6 |
| 31 37000 | Omaha.......................... | 141.6 | 478,393 | 40 | 3,379 | 76.9 | 12.9 | 0.7 | 4.1 | 0.0 | 1.7 | 4.7 |
| 32 00000 | NEVADA ......................... | 109,860.4 | 3,138,259 | X | 29 | 64.6 | 9.6 | 1.4 | 8.5 | 0.7 | 10.5 | 3.6 |
| 32 09700 | Carson City ................... | 144.5 | 56,034 | 687 | 388 | 85.8 | 2.2 | 2.2 | 2.7 | 0.0 | 3.5 | 5.1 |
| 32 31900 | Henderson .................... | 106.2 | 329,172 | 59 | 3,100 | 72.7 | 6.2 | 0.4 | 9.2 | 0.5 | 5.9 | 5.2 |
| 32 40000 | Las Vegas..................... | 141.8 | 662,368 | 25 | 4,671 | 61.8 | 12.0 | 1.2 | 7.1 | 0.9 | 11.9 | 5.7 |
| 32 51800 | North Las Vegas ........... | 101.3 | 260,098 | 85 | 2,568 | 55.4 | 19.9 | 1.0 | 7.2 | 1.2 | 9.5 | 4.2 |
| 32 60600 | Reno............................. | 108.8 | 259,290 | 86 | 2,383 | 73.6 | 2.7 | 0.9 | 7.5 | 0.5 | 10.6 | 4.4 |
| 32 68400 | Sparks.......................... | 36.4 | 106,900 | 288 | 2,937 | 70.0 | 3.2 | 1.5 | 4.1 | 1.0 | 15.8 | 2.2 |
| 33 00000 | NEW HAMPSHIRE ............... | 8,953.4 | 1,366,275 | X | 153 | 92.6 | 1.6 | 0.1 | 2.6 | 0.1 | 0.8 | 2.9 |
| 33 14200 | Concord ........................ | 64.0 | 43,924 | 879 | 686 | 86.2 | 5.5 | 0.3 | 4.3 | 0.0 | 0.6 | 5.7 |
| 33 18820 | Dover ........................... | 26.7 | 32,443 | 1,176 | 1,215 | 90.3 | 2.1 | 0.0 | 1.2 | 0.0 | 1.8 | 2.7 |
| 33 45140 | Manchester ................... | 33.1 | 112,546 | 262 | 3,400 | 83.9 | 6.3 | 0.3 | 5.0 | 0.0 | 1.7 | 3.6 |
| 33 50260 | Nashua ......................... | 30.8 | 89,167 | 377 | 2,895 | 85.0 | 2.7 | 0.4 | 6.7 | 0.0 | D | D |
| 33 65140 | Rochester...................... | 45.0 | 31,719 | 1,204 | 705 | D | D | D | D | D | 6.4 | 3.0 |
| 34 00000 | NEW JERSEY..................... | 7,354.8 | 8,882,371 | X | 1,208 | 67.1 | 13.6 | 0.2 | 9.6 | 0.0 | 6.4 | 2.9 |
| 34 02080 | Atlantic City ................. | 10.8 | 37,569 | 1,026 | 3,479 | 28.1 | 42.6 | 0.5 | 13.9 | 0.0 | 11.9 | 5.1 |
| 34 03580 | Bayonne ....................... | 5.8 | 65,165 | 580 | 11,235 | 70.9 | 11.1 | 0.2 | 11.3 | 0.0 | 4.9 | 2.6 |
| 34 05170 | Bergenfield ................... | 2.9 | 27,371 | 1,337 | 9,438 | 58.0 | 13.3 | 0.0 | 21.2 | 0.0 | 13.8 | 1.4 |
| 34 07600 | Bridgeton...................... | 6.2 | 23,635 | 1,426 | 3,812 | 48.2 | 35.4 | 0.5 | 0.7 | 0.0 | 34.6 | 3.3 |
| 34 10000 | Camden ........................ | 8.9 | 73,740 | 487 | 8,285 | 17.3 | 41.8 | 0.0 | 3.0 | 0.0 | 7.2 | 4.0 |
| 34 13690 | Clifton.......................... | 11.3 | 85,025 | 400 | 7,524 | 77.8 | 3.9 | 0.0 | 7.1 | 0.0 | 4.1 | 2.3 |
| 34 19390 | East Orange .................. | 3.9 | 64,630 | 585 | 16,572 | 5.5 | 83.5 | 0.5 | 4.0 | 0.2 | 3.2 | 3.0 |
| 34 21000 | Elizabeth ...................... | 12.3 | 128,382 | 216 | 10,438 | 35.0 | 21.2 | 0.2 | 0.9 | 0.0 | 39.7 | 2.0 |
| 34 21480 | Englewood .................... | 4.9 | 28,278 | 1,307 | 5,771 | 58.4 | 21.7 | 0.4 | 6.3 | 0.0 | 11.2 | D |
| 34 22470 | Fair Lawn ..................... | 5.1 | 32,877 | 1,166 | 6,447 | D | D | D | D | D | D | 2.7 |
| 34 24420 | Fort Lee........................ | 2.5 | 38,356 | 1,009 | 15,342 | 44.6 | 1.7 | 0.1 | 47.7 | 0.0 | 3.2 | |

1. Dry land or land partially or temporarily covered by water.    2. Hispanic or Latino persons may be of any race.

# Table D. Cities — Population

| City | Percent Hispanic or Latino[1], 2019 | Percent foreign born, 2019 | Age of population (percent), 2019 | | | | | | | Median age, 2019 | Percent female, 2019 | Population Census counts | | Percent change | |
|---|---|---|---|---|---|---|---|---|---|---|---|---|---|---|---|
| | | | Under 18 years | 18 to 24 years | 25 to 34 years | 35 to 44 years | 45 to 54 years | 55 to 64 years | 65 years and over | | | 2000 | 2010 | 2000-2010 | 2001-2020 |
| | 12 | 13 | 14 | 15 | 16 | 17 | 18 | 19 | 20 | 21 | 22 | 23 | 24 | 25 | 26 |
| MISSOURI | 4.3 | 4.3 | 22.4 | 9.1 | 13.4 | 12.3 | 11.9 | 13.6 | 17.2 | 38.9 | 51.0 | 5,595,211 | 5,988,941 | 7.0 | 2.7 |
| Ballwin | 2.7 | 10.6 | 22.1 | 5.8 | 15.9 | 13.4 | 8.4 | 14.6 | 19.8 | 38.6 | 53.6 | 31,283 | 30,417 | -2.8 | -1.1 |
| Blue Springs | 11.4 | 3.1 | 26.5 | 9.7 | 12.2 | 10.7 | 14.8 | 12.2 | 13.8 | 37.1 | 53.3 | 48,080 | 52,587 | 9.4 | 7.5 |
| Cape Girardeau | 3.9 | 3.3 | 17.5 | 18.1 | 14.8 | 9.9 | 9.8 | 12.8 | 17.2 | 34.8 | 50.2 | 35,349 | 37,966 | 7.4 | 9.5 |
| Chesterfield | 3.0 | 14.4 | 17.9 | 6.4 | 8.7 | 10.2 | 14.4 | 14.5 | 27.9 | 49.5 | 55.1 | 46,802 | 47,483 | 1.5 | 0.2 |
| Columbia | 4.5 | 9.7 | 20.0 | 23.8 | 16.3 | 11.8 | 8.5 | 9.5 | 10.2 | 28.5 | 51.7 | 84,531 | 109,041 | 29.0 | 14.4 |
| Florissant | 0.5 | 1.9 | 23.9 | 7.7 | 15.8 | 12.3 | 12.3 | 14.4 | 13.5 | 36.8 | 52.8 | 50,497 | 52,260 | 3.5 | -2.8 |
| Gladstone | 8.3 | 3.8 | 19.8 | 8.1 | 12.5 | 11.8 | 13.5 | 13.8 | 20.4 | 43.5 | 49.5 | 26,365 | 25,436 | -3.5 | 8.8 |
| Hazelwood | 4.1 | 6.5 | 18.2 | 12.4 | 16.4 | 13.2 | 13.8 | 10.7 | 15.2 | 37.4 | 55.1 | 26,206 | 25,723 | -1.8 | -2.6 |
| Independence | 9.4 | 4.9 | 25.8 | 7.0 | 15.3 | 11.5 | 10.6 | 12.6 | 17.1 | 36.8 | 52.1 | 113,288 | 116,823 | 3.1 | 0.0 |
| Jefferson City | 3.7 | 5.6 | 16.3 | 8.0 | 13.8 | 14.4 | 14.7 | 15.2 | 17.6 | 42.7 | 47.3 | 39,636 | 43,124 | 8.8 | -2.7 |
| Joplin | 4.8 | 3.9 | 22.1 | 11.7 | 16.5 | 11.8 | 9.5 | 11.9 | 16.6 | 34.5 | 49.1 | 45,504 | 50,780 | 11.6 | 0.3 |
| Kansas City | 11.2 | 8.9 | 22.5 | 8.8 | 18.3 | 12.8 | 11.6 | 12.5 | 13.5 | 35.3 | 51.9 | 441,545 | 459,883 | 4.2 | 8.1 |
| Kirkwood | 4.0 | 3.3 | 23.7 | 5.4 | 7.3 | 13.5 | 11.0 | 17.7 | 21.4 | 45.1 | 53.6 | 27,324 | 27,561 | 0.9 | 1.1 |
| Lee's Summit | 2.0 | 3.1 | 27.3 | 6.8 | 10.6 | 15.3 | 13.2 | 12.2 | 14.7 | 38.3 | 52.6 | 70,700 | 91,369 | 29.2 | 11.1 |
| Liberty | 6.4 | 4.3 | 26.7 | 7.2 | 13.3 | 14.5 | 10.9 | 10.6 | 16.9 | 37.0 | 51.9 | 26,232 | 29,242 | 11.5 | 10.9 |
| Maryland Heights | 1.9 | 22.0 | 18.3 | 11.8 | 22.9 | 13.3 | 8.5 | 10.4 | 15.0 | 33.5 | 54.4 | 25,756 | 27,469 | 6.7 | -2.1 |
| O'Fallon | 3.6 | 5.4 | 24.4 | 8.0 | 11.9 | 13.9 | 13.1 | 14.4 | 14.3 | 38.3 | 49.8 | 46,169 | 79,599 | 72.4 | 12.8 |
| Raytown | 0.4 | 4.4 | 22.0 | 6.2 | 12.3 | 10.5 | 12.2 | 13.9 | 22.9 | 44.0 | 52.2 | 30,388 | 29,611 | -2.6 | -2.7 |
| St. Charles | 7.3 | 6.6 | 15.9 | 14.0 | 15.5 | 11.9 | 10.1 | 14.9 | 17.7 | 38.9 | 51.0 | 60,321 | 66,207 | 9.8 | 8.4 |
| St. Joseph | 6.8 | 2.4 | 21.5 | 9.2 | 14.4 | 13.0 | 12.6 | 13.5 | 15.8 | 39.0 | 49.7 | 73,990 | 76,782 | 3.8 | -3.5 |
| St. Louis | 4.2 | 7.1 | 18.7 | 8.8 | 20.4 | 13.3 | 11.4 | 13.0 | 14.4 | 36.4 | 51.7 | 348,189 | 319,308 | -8.3 | -6.8 |
| St. Peters | 1.5 | 2.7 | 22.2 | 5.8 | 11.2 | 16.0 | 12.4 | 15.6 | 16.9 | 42.2 | 54.4 | 51,381 | 52,595 | 2.4 | 10.4 |
| Springfield | 4.0 | 3.4 | 17.9 | 19.1 | 15.1 | 11.5 | 10.0 | 10.9 | 15.5 | 33.6 | 51.4 | 151,580 | 159,333 | 5.1 | 5.5 |
| University City | 2.2 | 11.4 | 17.6 | 11.8 | 18.8 | 10.7 | 5.7 | 13.1 | 22.3 | 37.5 | 54.3 | 37,428 | 35,286 | -5.7 | -3.5 |
| Wentzville | 1.4 | 2.8 | 36.1 | 4.1 | 17.0 | 17.7 | 11.1 | 4.3 | 9.7 | 31.7 | 52.7 | 6,896 | 29,134 | 322.5 | 48.0 |
| Wildwood | 0.5 | 9.9 | 23.3 | 7.9 | 5.3 | 11.9 | 16.2 | 17.8 | 17.6 | 46.0 | 49.6 | 32,884 | 35,301 | 7.4 | 0.5 |
| MONTANA | 3.8 | 2.3 | 21.2 | 9.4 | 12.5 | 12.4 | 11.0 | 14.2 | 19.5 | 40.5 | 49.7 | 902,195 | 989,400 | 9.7 | 9.2 |
| Billings | 6.4 | 2.1 | 23.2 | 8.2 | 14.8 | 13.8 | 10.8 | 12.2 | 17.2 | 37.6 | 50.8 | 89,847 | 104,294 | 16.1 | 5.2 |
| Bozeman | 4.3 | 5.6 | 13.6 | 30.8 | 19.8 | 8.6 | 10.3 | 8.1 | 8.8 | 27.4 | 48.1 | 27,509 | 37,271 | 35.5 | 36.8 |
| Butte-Silver Bow | 3.2 | 3.6 | 19.6 | 11.6 | 12.2 | 12.5 | 10.7 | 14.4 | 19.1 | 40.0 | 48.5 | 33,892 | 33,507 | -0.1 | 2.9 |
| Great Falls | 4.9 | 1.5 | 20.4 | 11.6 | 13.6 | 11.2 | 9.4 | 13.3 | 20.4 | 37.6 | 50.4 | 56,690 | 59,113 | 4.3 | -1.3 |
| Helena | 3.0 | 1.4 | 16.0 | 11.0 | 14.3 | 12.2 | 11.9 | 13.8 | 20.9 | 41.7 | 55.8 | 25,780 | 28,811 | 11.8 | 17.1 |
| Missoula | 3.2 | 2.3 | 17.5 | 16.0 | 17.2 | 13.9 | 10.3 | 10.7 | 14.4 | 34.6 | 50.1 | 57,053 | 67,386 | 18.1 | 14.0 |
| NEBRASKA | 11.3 | 7.4 | 24.6 | 9.7 | 13.3 | 12.6 | 11.2 | 12.5 | 16.1 | 36.8 | 50.0 | 1,711,263 | 1,826,311 | 6.7 | 6.1 |
| Bellevue | 17.6 | 6.7 | 21.6 | 8.4 | 17.9 | 13.4 | 11.5 | 12.1 | 15.0 | 37.0 | 48.7 | 44,382 | 51,525 | 16.1 | 3.4 |
| Fremont | 17.1 | 6.5 | 24.1 | 10.2 | 11.4 | 14.0 | 8.4 | 14.7 | 17.2 | 37.8 | 49.4 | 25,174 | 26,415 | 4.9 | -0.7 |
| Grand Island | 32.8 | 18.0 | 27.3 | 8.4 | 12.8 | 11.7 | 12.9 | 10.6 | 16.3 | 36.2 | 49.6 | 42,940 | 48,678 | 13.4 | 4.8 |
| Kearney | 12.3 | 6.0 | 21.5 | 21.1 | 13.4 | 12.7 | 8.8 | 9.2 | 13.4 | 30.3 | 50.3 | 27,431 | 30,949 | 12.8 | 10.6 |
| Lincoln | 8.1 | 8.6 | 21.9 | 16.3 | 13.9 | 12.8 | 10.0 | 11.2 | 13.9 | 33.5 | 49.8 | 225,581 | 258,791 | 14.7 | 12.3 |
| Omaha | 13.9 | 11.5 | 25.3 | 9.6 | 15.7 | 12.8 | 11.5 | 11.6 | 13.5 | 34.6 | 51.0 | 390,007 | 458,992 | 17.7 | 4.2 |
| NEVADA | 29.2 | 19.8 | 22.4 | 8.2 | 14.6 | 13.3 | 12.9 | 12.4 | 16.2 | 38.4 | 49.8 | 1,998,257 | 2,700,683 | 35.2 | 16.2 |
| Carson City | 24.8 | 14.2 | 20.3 | 7.1 | 13.2 | 12.6 | 11.6 | 14.0 | 21.2 | 41.5 | 49.1 | 52,457 | 55,269 | 5.4 | 1.4 |
| Henderson | 16.1 | 13.2 | 19.4 | 6.1 | 13.7 | 12.5 | 14.4 | 12.7 | 21.3 | 43.7 | 51.5 | 175,381 | 256,995 | 46.5 | 28.1 |
| Las Vegas | 35.5 | 22.6 | 24.4 | 8.0 | 15.0 | 13.0 | 12.7 | 12.0 | 14.9 | 36.8 | 49.4 | 478,434 | 584,500 | 22.2 | 13.3 |
| North Las Vegas | 44.1 | 22.7 | 28.4 | 9.2 | 14.0 | 14.0 | 13.2 | 10.3 | 10.9 | 33.5 | 50.9 | 115,488 | 216,670 | 87.6 | 20.0 |
| Reno | 24.1 | 16.2 | 19.0 | 11.8 | 17.7 | 13.7 | 10.3 | 12.4 | 15.0 | 35.9 | 49.1 | 180,480 | 225,325 | 24.8 | 15.1 |
| Sparks | 34.7 | 14.7 | 25.9 | 8.1 | 11.8 | 13.0 | 14.4 | 11.6 | 15.2 | 38.9 | 49.4 | 66,346 | 91,112 | 37.3 | 17.3 |
| NEW HAMPSHIRE | 4.0 | 6.4 | 18.8 | 9.1 | 12.6 | 11.7 | 13.4 | 15.8 | 18.6 | 43.0 | 50.5 | 1,235,786 | 1,316,457 | 6.5 | 3.8 |
| Concord | 2.4 | 7.5 | 16.8 | 9.1 | 15.4 | 11.9 | 16.1 | 14.7 | 16.0 | 41.0 | 49.9 | 40,687 | 42,686 | 4.9 | 2.9 |
| Dover | 3.5 | 4.1 | 18.2 | 10.5 | 18.8 | 12.0 | 13.9 | 9.7 | 17.0 | 37.9 | 49.2 | 26,884 | 29,999 | 11.6 | 8.1 |
| Manchester | 11.8 | 17.5 | 17.3 | 9.7 | 17.6 | 14.1 | 12.3 | 14.4 | 14.7 | 38.7 | 51.6 | 107,006 | 109,542 | 2.4 | 2.7 |
| Nashua | 12.4 | 14.1 | 17.9 | 10.4 | 16.1 | 12.2 | 13.1 | 13.6 | 16.7 | 39.7 | 47.4 | 86,605 | 86,475 | -0.2 | 3.1 |
| Rochester | 6.0 | 5.4 | 19.0 | 13.1 | 9.6 | 13.4 | 11.0 | 19.0 | 15.0 | 42.8 | 50.0 | 28,461 | 29,789 | 4.7 | 6.5 |
| NEW JERSEY | 20.9 | 23.4 | 21.8 | 8.5 | 12.9 | 12.9 | 13.5 | 13.8 | 16.6 | 40.2 | 51.1 | 8,414,350 | 8,791,959 | 4.5 | 1.0 |
| Atlantic City | 27.6 | 23.6 | 24.3 | 14.8 | 7.3 | 10.6 | 12.9 | 13.7 | 16.4 | 38.0 | 51.8 | 40,517 | 39,552 | -2.4 | -5.0 |
| Bayonne | 29.5 | 29.5 | 22.5 | 6.0 | 16.5 | 15.1 | 11.9 | 15.2 | 12.7 | 38.5 | 54.1 | 61,842 | 63,015 | 1.9 | 3.4 |
| Bergenfield | 37.3 | 34.5 | 20.0 | 12.5 | 11.4 | 12.4 | 16.4 | 12.3 | 15.0 | 38.4 | 49.5 | 26,247 | 26,839 | 2.3 | 2.0 |
| Bridgeton | 50.3 | 26.5 | 28.8 | 7.7 | 21.2 | 15.7 | 12.3 | 7.4 | 6.9 | 31.0 | 41.6 | 22,771 | 25,377 | 11.4 | -6.9 |
| Camden | 53.0 | 11.8 | 32.0 | 10.3 | 13.3 | 14.4 | 9.5 | 10.1 | 10.4 | 30.3 | 55.4 | 79,904 | 76,874 | -3.8 | -4.1 |
| Clifton | 42.9 | 41.5 | 21.9 | 9.6 | 13.3 | 13.2 | 13.1 | 12.8 | 16.2 | 39.2 | 52.1 | 78,672 | 84,175 | 7.0 | 1.0 |
| East Orange | 11.3 | 26.5 | 22.2 | 8.4 | 19.7 | 12.8 | 10.6 | 11.6 | 14.7 | 34.6 | 53.0 | 69,824 | 64,346 | -7.8 | 0.4 |
| Elizabeth | 66.8 | 47.7 | 27.9 | 9.8 | 12.5 | 14.9 | 12.6 | 11.9 | 10.4 | 34.8 | 50.6 | 120,568 | 124,974 | 3.7 | 2.7 |
| Englewood | 29.1 | 28.8 | 15.4 | 8.9 | 13.9 | 12.2 | 18.3 | 13.9 | 17.4 | 44.9 | 62.6 | 26,203 | 27,116 | 3.5 | 4.3 |
| Fair Lawn | 10.3 | 34.3 | 24.4 | 5.0 | 8.3 | 11.3 | 12.6 | 16.7 | 21.7 | 45.4 | 48.1 | 31,637 | 32,379 | 2.3 | 1.5 |
| Fort Lee | 10.0 | 52.8 | 12.5 | 5.4 | 12.9 | 16.0 | 13.5 | 14.8 | 25.0 | 47.7 | 48.1 | 35,461 | 35,426 | -0.1 | 8.3 |

1. May be of any race.

## Table D. Cities — Households, Group Quarters, Crime, and Education

| City | Households, 2019 Number | Persons per household | Percent Family | Married couple family | Female family | Non-family | One person | Persons in group quarters, 2019 | Violent Number | Violent Rate[2] | Property Number | Property Rate[2] | Population age 25 and over | HS graduate or less | Bachelor's degree or more |
|---|---|---|---|---|---|---|---|---|---|---|---|---|---|---|---|
| | 27 | 28 | 29 | 30 | 31 | 32 | 33 | 34 | 35 | 36 | 37 | 38 | 39 | 40 | 41 |
| MISSOURI | 2,458,337 | 2.43 | 62.9 | 47.2 | 15.7 | 37.1 | 30.3 | 174,478 | NA | NA | NA | NA | 4,206,162 | 40.4 | 30.2 |
| Ballwin | 12,206 | 2.47 | 64.0 | 58.0 | 6.0 | 36.0 | 28.5 | 1 | NA | NA | NA | NA | 21,721 | 13.0 | 59.7 |
| Blue Springs | 20,233 | 2.75 | 77.5 | 61.9 | 15.6 | 22.5 | 19.3 | 228 | 99 | 179 | 1,216 | 2,194 | 35,616 | 33.9 | 33.5 |
| Cape Girardeau | 17,159 | 2.19 | 55.6 | 44.6 | 11.0 | 44.4 | 34.1 | 2,899 | 233 | 581 | 1,432 | 3,573 | 26,132 | 32.0 | 39.3 |
| Chesterfield | 20,016 | 2.33 | 69.7 | 65.4 | 4.3 | 30.3 | 28.3 | 856 | NA | NA | NA | NA | 35,987 | 10.3 | 71.0 |
| Columbia | 46,841 | 2.44 | 54.0 | 39.2 | 14.7 | 46.0 | 29.1 | 8,793 | 401 | 321 | 3,243 | 2,594 | 69,232 | 20.9 | 53.5 |
| Florissant | 19,606 | 2.56 | 60.5 | 33.5 | 27.0 | 39.5 | 32.5 | 828 | NA | NA | NA | NA | 34,836 | 39.8 | 22.9 |
| Gladstone | 11,333 | 2.42 | 54.2 | 33.1 | 21.1 | 45.8 | 37.3 | 8 | 89 | 323 | 725 | 2,631 | 19,811 | 40.3 | 27.5 |
| Hazelwood | 10,933 | 2.28 | 59.1 | 36.4 | 22.7 | 40.9 | 37.0 | 175 | 113 | 449 | 695 | 2,764 | 17,428 | 35.8 | 26.8 |
| Independence | 47,329 | 2.44 | 60.5 | 40.6 | 19.8 | 39.5 | 32.3 | 1,268 | 689 | 589 | 5,443 | 4,655 | 78,333 | 47.1 | 18.7 |
| Jefferson City | 17,970 | 2.04 | 55.6 | 44.3 | 11.3 | 44.4 | 39.5 | 5,971 | 119 | 278 | 1,101 | 2,573 | 32,302 | 37.9 | 30.7 |
| Joplin | 20,236 | 2.44 | 58.2 | 34.5 | 23.6 | 41.8 | 29.9 | 1,877 | 307 | 606 | 3,677 | 7,262 | 34,004 | 44.9 | 24.4 |
| Kansas City | 209,768 | 2.32 | 52.2 | 35.7 | 16.5 | 47.8 | 37.6 | 9,481 | 7,099 | 1,431 | 19,124 | 3,856 | 340,306 | 35.2 | 36.2 |
| Kirkwood | 12,157 | 2.26 | 65.7 | 54.1 | 11.6 | 34.3 | 30.9 | 287 | NA | NA | NA | NA | 19,719 | 10.0 | 67.1 |
| Lee's Summit | 36,469 | 2.70 | 77.4 | 62.0 | 15.4 | 22.6 | 18.6 | 663 | 115 | 116 | 2,095 | 2,108 | 65,465 | 21.3 | 50.3 |
| Liberty | 10,951 | 2.83 | 72.5 | 60.9 | 11.6 | 27.5 | 25.5 | 1,099 | NA | NA | NA | NA | 21,221 | 32.1 | 41.7 |
| Maryland Heights | 12,248 | 2.15 | 61.1 | 46.9 | 14.2 | 38.9 | 29.4 | 589 | 56 | 208 | 761 | 2,822 | 18,873 | 23.5 | 53.7 |
| O'Fallon | 32,274 | 2.74 | 73.0 | 61.4 | 11.6 | 27.0 | 22.8 | 364 | 114 | 127 | 1,015 | 1,133 | 59,925 | 26.8 | 40.1 |
| Raytown | 12,170 | 2.34 | 58.8 | 39.0 | 19.8 | 41.2 | 34.3 | 462 | 200 | 691 | 1,294 | 4,473 | 20,835 | 42.5 | 21.0 |
| St. Charles | 29,838 | 2.18 | 56.8 | 47.1 | 9.6 | 43.2 | 34.7 | 6,044 | NA | NA | NA | NA | 49,776 | 31.1 | 40.1 |
| St. Joseph | 29,399 | 2.41 | 56.0 | 40.4 | 15.7 | 44.0 | 38.5 | 4,339 | 431 | 568 | 4,355 | 5,740 | 52,043 | 53.2 | 18.8 |
| St. Louis | 146,779 | 1.98 | 43.5 | 24.7 | 18.8 | 56.5 | 47.5 | 9,925 | 5,792 | 1,927 | 18,582 | 6,183 | 218,004 | 35.1 | 38.6 |
| St. Peters | 23,067 | 2.51 | 65.5 | 55.8 | 9.7 | 34.5 | 28.4 | 236 | 139 | 241 | 1,162 | 2,014 | 41,914 | 26.4 | 41.2 |
| Springfield | 80,693 | 1.94 | 44.9 | 30.0 | 14.9 | 55.1 | 40.5 | 11,083 | 2,571 | 1,519 | 13,188 | 7,793 | 105,719 | 40.4 | 26.2 |
| University City | 17,036 | 1.99 | 41.8 | 26.8 | 15.0 | 58.2 | 46.2 | 303 | NA | NA | NA | NA | 24,131 | 22.5 | 55.8 |
| Wentzville | 13,364 | 3.11 | 77.8 | 62.8 | 15.0 | 22.2 | 17.0 | 165 | 88 | 205 | 450 | 1,049 | 24,987 | 25.5 | 38.5 |
| Wildwood | 13,130 | 2.69 | 81.3 | 72.6 | 8.7 | 18.7 | 15.3 | 113 | NA | NA | NA | NA | 24,384 | 9.8 | 65.4 |
| MONTANA | 437,651 | 2.38 | 61.5 | 48.8 | 12.6 | 38.5 | 30.7 | 28,943 | NA | NA | NA | NA | 741,950 | 34.2 | 33.6 |
| Billings | 48,990 | 2.17 | 55.8 | 41.2 | 14.6 | 44.2 | 34.2 | 3,237 | NA | NA | 4,499 | 4,083 | 75,259 | 35.1 | 35.0 |
| Bozeman | 20,833 | 2.19 | 40.5 | 30.2 | 10.3 | 59.5 | 36.6 | 4,290 | NA | NA | 849 | 1,693 | 27,700 | 16.2 | 62.8 |
| Butte-Silver Bow | 15,025 | 2.16 | 54.1 | 42.8 | 11.3 | 45.9 | 37.9 | 1,490 | NA | NA | NA | NA | 23,410 | 32.1 | 34.7 |
| Great Falls | 24,826 | 2.25 | 57.9 | 35.3 | 22.6 | 42.1 | 34.2 | 1,857 | 302 | 515 | 3,405 | 5,807 | 39,182 | 42.0 | 23.7 |
| Helena | 14,244 | 2.22 | 49.4 | 41.8 | 7.6 | 50.6 | 44.8 | 1,537 | 190 | 579 | 1,400 | 4,268 | 24,193 | 16.1 | 57.2 |
| Missoula | 33,448 | 2.17 | 52.4 | 39.2 | 13.1 | 47.6 | 32.3 | 3,053 | 310 | 411 | 3,082 | 4,086 | 50,221 | 26.1 | 48.8 |
| NEBRASKA | 771,444 | 2.44 | 63.9 | 50.4 | 13.5 | 36.1 | 28.7 | 51,012 | NA | NA | NA | NA | 1,271,770 | 33.6 | 33.2 |
| Bellevue | 21,839 | 2.45 | 63.6 | 47.0 | 16.6 | 36.4 | 25.2 | 132 | 110 | 204 | 961 | 1,784 | 37,446 | 34.5 | 25.6 |
| Fremont | 10,459 | 2.30 | 59.5 | 42.5 | 17.0 | 40.5 | 30.6 | 932 | 49 | 185 | 589 | 2,221 | 16,407 | 44.2 | 19.7 |
| Grand Island | 20,285 | 2.44 | 64.9 | 45.4 | 19.5 | 35.1 | 28.9 | 950 | 236 | 455 | 1,323 | 2,553 | 32,436 | 47.7 | 21.3 |
| Kearney | 12,110 | 2.52 | 57.7 | 42.6 | 15.1 | 42.3 | 34.0 | 1,859 | NA | NA | NA | NA | 18,548 | 29.4 | 40.6 |
| Lincoln | 117,726 | 2.34 | 57.3 | 44.5 | 12.8 | 42.7 | 30.6 | 13,141 | 1,115 | 383 | 8,008 | 2,751 | 178,745 | 27.4 | 40.4 |
| Omaha | 188,824 | 2.47 | 59.3 | 42.6 | 16.7 | 40.7 | 32.3 | 12,017 | 2,883 | 613 | 17,144 | 3,644 | 311,417 | 32.1 | 39.7 |
| NEVADA | 1,143,557 | 2.66 | 63.3 | 44.3 | 19.0 | 36.7 | 28.5 | 37,663 | NA | NA | NA | NA | 2,136,466 | 40.9 | 25.7 |
| Carson City | 22,703 | 2.37 | 63.0 | 42.1 | 20.9 | 37.0 | 32.6 | 2,172 | NA | NA | NA | NA | 40,577 | 38.5 | 21.3 |
| Henderson | 121,285 | 2.63 | 65.7 | 51.6 | 14.1 | 34.3 | 26.6 | 1,474 | 543 | 171 | 5,554 | 1,748 | 238,697 | 29.9 | 35.3 |
| Las Vegas | 235,628 | 2.73 | 61.7 | 41.3 | 20.3 | 38.3 | 31.0 | 8,274 | 8,854 | 531 | 46,197 | 2,772 | 440,508 | 43.9 | 25.1 |
| North Las Vegas | 75,600 | 3.31 | 72.5 | 48.0 | 24.6 | 27.5 | 21.8 | 1,857 | 2,158 | 864 | 5,086 | 2,036 | 157,303 | 51.7 | 17.0 |
| Reno | 110,118 | 2.27 | 54.6 | 39.4 | 15.2 | 45.4 | 32.0 | 5,428 | 1,419 | 558 | 5,344 | 2,101 | 176,934 | 31.2 | 34.7 |
| Sparks | 38,887 | 2.69 | 69.4 | 48.9 | 20.5 | 30.6 | 22.9 | 358 | 450 | 424 | 2,426 | 2,288 | 69,273 | 43.4 | 23.3 |
| NEW HAMPSHIRE | 541,396 | 2.44 | 64.0 | 50.6 | 13.4 | 36.0 | 27.6 | 41,073 | NA | NA | NA | NA | 979,750 | 34.8 | 37.6 |
| Concord | 18,635 | 2.17 | 52.9 | 42.3 | 10.6 | 47.1 | 37.7 | 3,106 | 65 | 149 | 805 | 1,850 | 32,313 | 36.6 | 37.9 |
| Dover | 14,130 | 2.21 | 55.8 | 45.9 | 9.9 | 44.2 | 30.1 | 1,007 | 25 | 78 | 1,086 | 3,389 | 22,956 | 24.7 | 45.1 |
| Manchester | 46,435 | 2.37 | 55.2 | 37.7 | 17.4 | 44.8 | 32.2 | 2,432 | 678 | 601 | 2,678 | 2,372 | 82,293 | 42.5 | 31.7 |
| Nashua | 37,474 | 2.34 | 57.4 | 43.6 | 13.7 | 42.6 | 31.7 | 1,576 | 129 | 144 | 1,151 | 1,285 | 64,062 | 34.8 | 36.8 |
| Rochester | 12,621 | 2.49 | 66.9 | 49.3 | 17.6 | 33.1 | 24.1 | 132 | 94 | 298 | 976 | 3,096 | 21,428 | 46.7 | 22.4 |
| NEW JERSEY | 3,286,264 | 2.65 | 68.2 | 50.7 | 17.5 | 31.8 | 26.4 | 180,487 | NA | NA | NA | NA | 6,191,229 | 36.6 | 41.2 |
| Atlantic City | 16,716 | 2.22 | 48.7 | 21.2 | 27.4 | 51.3 | 45.2 | 606 | 323 | 859 | 1,738 | 4,623 | 23,012 | 65.0 | 17.6 |
| Bayonne | 25,939 | 2.49 | 64.7 | 42.0 | 22.7 | 35.3 | 29.4 | 264 | 145 | 223 | 704 | 1,083 | 46,365 | 38.4 | 39.3 |
| Bergenfield | 9,054 | 3.01 | 77.5 | 55.6 | 21.9 | 22.5 | 16.3 | 56 | 15 | 55 | 75 | 273 | 18,434 | 40.3 | 42.4 |
| Bridgeton | 6,327 | 3.15 | 61.6 | 17.4 | 44.2 | 38.4 | 31.5 | 4,223 | 223 | 917 | 848 | 3,485 | 15,339 | 76.3 | 6.3 |
| Camden | 24,051 | 2.94 | 63.2 | 16.4 | 46.8 | 36.8 | 35.6 | 2,817 | 1,159 | 1,582 | 2,101 | 2,867 | 42,430 | 64.8 | 9.0 |
| Clifton | 28,207 | 2.99 | 72.8 | 49.1 | 23.7 | 27.2 | 22.4 | 788 | 95 | 112 | 1,272 | 1,496 | 58,320 | 49.2 | 32.0 |
| East Orange | 27,960 | 2.27 | 53.2 | 22.4 | 30.8 | 46.8 | 41.8 | 987 | 275 | 428 | 894 | 1,393 | 44,662 | 46.8 | 27.2 |
| Elizabeth | 40,227 | 3.16 | 73.9 | 43.7 | 30.1 | 26.1 | 21.4 | 2,174 | 926 | 719 | 3,898 | 3,028 | 80,534 | 67.4 | 12.4 |
| Englewood | 11,911 | 2.36 | 57.2 | 37.9 | 19.3 | 42.8 | 37.9 | 250 | 71 | 248 | 343 | 1,196 | 21,493 | 41.2 | 42.6 |
| Fair Lawn | 11,639 | 2.81 | 73.7 | 63.9 | 9.8 | 26.3 | 23.6 | 172 | 7 | 21 | 236 | 714 | 23,217 | 17.6 | 66.4 |
| Fort Lee | 18,908 | 2.04 | 54.9 | 42.8 | 12.1 | 45.1 | 39.9 | 7 | 25 | 66 | 323 | 849 | 31,661 | 22.0 | 58.3 |

1. Data for serious crimes have not been adjusted for underreporting. This may affect comparability between geographic areas and over time.   2. Per 100,000 population estimated by the FBI.   3. Persons 25 years old and over.

| City | Money income, 2019 | | | | | Median earnings, 2019 | | | Housing units, 2019 | | | | |
|---|---|---|---|---|---|---|---|---|---|---|---|---|---|
| | Households | | | | | | | | | | | | |
| | Median income | Percent with income less than $20,000 | Percent with income of $200,000 or more | Median family income | Median non-family income | All persons | Men | Women | Total | Occupied | Percent owner occupied | Median value[1] (dollars) | Median gross rent (dollars) |
| | 42 | 43 | 44 | 45 | 46 | 47 | 48 | 49 | 50 | 51 | 52 | 53 | 54 |
| MISSOURI .................. | 57,409 | 15.6 | 5.1 | 73,457 | 33,821 | 34,558 | 40,522 | 29,965 | 2,819,334 | 2,458,337 | 67.1 | 168,044 | 834 |
| Ballwin.......................... | 90,702 | 7.3 | 16.4 | 121,995 | 44,208 | 47,755 | 61,471 | 41,029 | 12,493 | 12,206 | | 284,547 | 1,296 |
| Blue Springs.................. | 74,808 | 7.0 | 5.9 | 84,783 | 35,679 | 40,749 | 49,528 | 32,406 | 21,045 | 20,233 | 77.0 | 184,396 | 1,022 |
| Cape Girardeau ............. | 48,216 | 21.3 | 3.0 | 69,633 | 28,059 | 25,987 | 26,830 | 25,243 | 19,335 | 17,159 | 69.5 | 156,802 | 704 |
| Chesterfield .................. | 122,693 | 2.3 | 24.2 | 151,199 | 49,261 | 58,946 | 95,725 | 44,917 | 21,693 | 20,016 | 55.7 | 385,455 | 1,196 |
| Columbia ...................... | 56,501 | 17.8 | 7.8 | 86,163 | 35,600 | 29,338 | 32,168 | 23,745 | 52,650 | 46,841 | 81.2 | 218,130 | 966 |
| Florissant ..................... | 63,315 | 14.1 | 0.5 | 72,986 | 42,525 | 35,625 | 39,061 | 30,741 | 22,795 | 19,606 | 50.2 | 100,164 | 1,065 |
| Gladstone ..................... | 51,154 | 10.2 | 0.5 | 48,932 | 43,856 | 27,498 | 35,590 | 24,304 | 12,063 | 11,333 | 70.3 | 154,946 | 966 |
| Hazelwood ................... | 55,299 | 10.8 | 1.9 | 62,886 | 30,785 | 32,726 | 41,910 | 30,460 | 11,916 | 10,933 | 63.4 | 115,999 | 743 |
| Independence ............... | 52,325 | 17.6 | 1.5 | 62,926 | 32,284 | 33,771 | 41,095 | 26,882 | 52,927 | 47,329 | 67.2 | 129,070 | 873 |
| Jefferson City ............... | 59,238 | 15.4 | 3.4 | 77,355 | 34,572 | 31,082 | 32,336 | 29,500 | 18,914 | 17,970 | 57.6 | 159,601 | 664 |
| | | | | | | | | | | | 58.3 | | |
| Joplin............................ | 45,334 | 21.1 | 3.5 | 51,134 | 29,081 | 24,736 | 26,976 | 21,693 | 22,614 | 20,236 | 54.5 | 121,036 | 680 |
| Kansas City................... | 55,259 | 17.2 | 5.0 | 72,775 | 40,284 | 36,445 | 40,809 | 31,985 | 238,547 | 209,768 | 52.8 | 168,405 | 979 |
| Kirkwood ...................... | 100,156 | 6.7 | 17.8 | 131,094 | 53,145 | 61,344 | 86,237 | 51,214 | 13,149 | 12,157 | 78.2 | 349,727 | 1,253 |
| Lee's Summit ................ | 106,912 | 6.0 | 11.7 | 120,001 | 46,568 | 54,169 | 63,584 | 45,438 | 38,500 | 36,469 | 77.7 | 245,663 | 1,194 |
| Liberty .......................... | 85,204 | 5.8 | 2.9 | 94,049 | 36,987 | 35,965 | 51,104 | 28,453 | 11,640 | 10,951 | 69.3 | 211,701 | 913 |
| Maryland Heights .......... | 81,253 | 8.9 | 3.4 | 93,331 | 56,433 | 50,858 | 52,783 | 50,527 | 12,767 | 12,248 | 58.2 | 177,525 | 997 |
| O'Fallon........................ | 92,827 | 4.1 | 10.9 | 106,685 | 52,344 | 42,000 | 51,925 | 35,796 | 34,209 | 32,274 | 80.5 | 245,096 | 1,110 |
| Raytown........................ | 54,921 | 16.8 | 1.5 | 69,132 | 34,063 | 33,500 | 37,452 | 30,688 | 12,706 | 12,170 | 68.9 | 107,559 | 883 |
| St. Charles ................... | 72,104 | 8.7 | 8.6 | 86,800 | 48,508 | 40,581 | 47,314 | 32,113 | 31,677 | 29,838 | 61.9 | 218,235 | 979 |
| St. Joseph .................... | 43,614 | 24.1 | 2.8 | 58,652 | 22,211 | 29,240 | 31,752 | 23,996 | 33,844 | 29,399 | 53.9 | 114,363 | 777 |
| St. Louis ...................... | 47,176 | 24.4 | 3.9 | 65,317 | 34,287 | 36,833 | 41,132 | 33,385 | 177,400 | 146,779 | 44.1 | 150,711 | 828 |
| St. Peters .................... | 83,339 | 8.4 | 5.4 | 101,993 | 41,794 | 38,680 | 52,314 | 31,332 | 24,700 | 23,067 | 82.2 | 204,144 | 1,093 |
| Springfield .................... | 35,677 | 25.7 | 1.9 | 50,991 | 25,779 | 26,619 | 28,539 | 24,802 | 81,945 | 80,693 | 38.7 | 124,024 | 733 |
| University City............... | 55,999 | 20.0 | 10.9 | 102,442 | 50,056 | 48,578 | 52,412 | 29,948 | 17,303 | 17,036 | 48.8 | 248,099 | 1,093 |
| Wentzville...................... | 97,537 | 2.8 | 9.8 | 101,985 | 54,941 | 52,626 | 62,566 | 40,618 | 13,585 | 13,364 | 85.1 | 239,326 | 942 |
| Wildwood ...................... | 136,328 | 4.7 | 25.7 | 151,655 | 71,441 | 70,625 | 91,483 | 50,665 | 13,422 | 13,130 | 91.6 | 381,563 | 1,314 |
| MONTANA .................. | 57,153 | 15.2 | 4.3 | 73,014 | 34,919 | 31,313 | 37,976 | 25,500 | 519,938 | 437,651 | 68.9 | 253,605 | 831 |
| Billings.......................... | 58,394 | 15.3 | 5.5 | 71,521 | 38,158 | 35,238 | 40,766 | 27,241 | 51,168 | 48,990 | 62.2 | 232,667 | 877 |
| Bozeman....................... | 64,084 | 15.4 | 5.1 | 93,102 | 48,125 | 22,713 | 25,421 | 18,380 | 22,346 | 20,833 | 41.2 | 405,483 | 1,200 |
| Butte-Silver Bow ........... | 45,536 | 19.1 | 2.6 | 79,544 | 33,050 | 31,526 | 28,784 | 28,453 | 16,914 | 15,025 | 66.8 | 164,698 | 741 |
| Great Falls ................... | 47,175 | 20.0 | 3.1 | 60,510 | 31,178 | 27,266 | 35,517 | 23,465 | 28,582 | 24,826 | 63.9 | 183,882 | 736 |
| Helena.......................... | 60,000 | 17.9 | 8.5 | 102,693 | 38,264 | 35,218 | 37,248 | 30,116 | 16,316 | 14,244 | 57.6 | 278,134 | 832 |
| Missoula........................ | 50,194 | 15.1 | 4.3 | 67,175 | 33,305 | 25,199 | 27,345 | 22,070 | 35,628 | 33,448 | 52.2 | 317,770 | 910 |
| NEBRASKA .................. | 63,229 | 12.2 | 5.4 | 80,062 | 37,791 | 36,077 | 41,536 | 30,661 | 851,167 | 771,444 | 66.3 | 172,736 | 859 |
| Bellevue........................ | 72,164 | 7.7 | 1.5 | 79,404 | 50,831 | 38,991 | 45,772 | 31,961 | 22,859 | 21,839 | 65.2 | 163,990 | 944 |
| Fremont......................... | 53,175 | 11.5 | 2.4 | 73,492 | 32,586 | 30,921 | 40,203 | 25,558 | 11,423 | 10,459 | 51.2 | 148,847 | 828 |
| Grand Island ................. | 52,192 | 15.6 | 2.5 | 62,966 | 31,039 | 33,586 | 36,395 | 31,477 | 21,022 | 20,285 | 60.1 | 143,534 | 727 |
| Kearney......................... | 62,207 | 14.2 | 2.3 | 83,221 | 32,439 | 31,450 | 38,061 | 23,325 | 13,771 | 12,110 | 61.4 | 197,546 | 772 |
| Lincoln.......................... | 59,228 | 12.7 | 5.8 | 78,437 | 37,123 | 32,054 | 36,323 | 29,648 | 124,244 | 117,726 | 57.3 | 189,411 | 857 |
| Omaha .......................... | 61,305 | 13.2 | 5.8 | 77,140 | 42,132 | 36,920 | 41,279 | 31,828 | 203,034 | 188,824 | 58.4 | 175,784 | 940 |
| NEVADA .................. | 63,276 | 13.5 | 6.4 | 76,124 | 41,065 | 35,526 | 39,335 | 31,354 | 1,285,681 | 1,143,557 | 56.6 | 317,791 | 1,168 |
| Carson City................... | 57,270 | 14.9 | 2.7 | 69,901 | 32,407 | 32,183 | 36,187 | 30,306 | 24,348 | 22,703 | 58.9 | 325,206 | 932 |
| Henderson .................... | 81,505 | 9.4 | 12.6 | 101,522 | 52,206 | 42,798 | 51,174 | 40,020 | 132,861 | 121,285 | 65.6 | 371,465 | 1,388 |
| Las Vegas..................... | 58,713 | 16.8 | 5.9 | 72,368 | 36,839 | 35,126 | 36,919 | 31,779 | 258,583 | 235,628 | 52.5 | 305,937 | 1,141 |
| North Las Vegas ........... | 64,460 | 10.8 | 2.9 | 69,927 | 40,405 | 31,678 | 35,071 | 29,344 | 82,145 | 75,600 | 59.4 | 285,175 | 1,276 |
| Reno............................. | 67,082 | 13.1 | 7.2 | 84,533 | 44,707 | 35,866 | 39,951 | 31,228 | 116,026 | 110,118 | 47.8 | 378,570 | 1,149 |
| Sparks.......................... | 73,568 | 9.4 | 3.4 | 84,573 | 41,706 | 36,713 | 40,954 | 32,192 | 40,027 | 38,887 | 61.5 | 352,726 | 1,326 |
| NEW HAMPSHIRE ............. | 77,933 | 10.0 | 9.8 | 97,112 | 45,803 | 40,984 | 48,693 | 33,514 | 642,298 | 541,396 | 71.0 | 281,365 | 1,147 |
| Concord ........................ | 72,311 | 14.5 | 8.0 | 95,411 | 41,859 | 42,204 | 45,630 | 40,727 | 19,736 | 18,635 | 49.5 | 257,296 | 1,108 |
| Dover ............................ | 77,201 | 10.5 | 9.6 | 115,179 | 52,480 | 49,054 | 52,301 | 39,410 | 14,611 | 14,130 | 45.8 | 305,678 | 1,217 |
| Manchester ................... | 64,162 | 13.2 | 4.9 | 77,607 | 50,446 | 37,043 | 47,735 | 31,804 | 49,848 | 46,435 | 48.8 | 253,870 | 1,184 |
| Nashua.......................... | 78,284 | 8.6 | 7.7 | 95,241 | 50,130 | 41,236 | 50,282 | 33,118 | 38,819 | 37,474 | 56.5 | 285,913 | 1,346 |
| Rochester...................... | 58,760 | 11.0 | 2.8 | 67,963 | 38,106 | 29,468 | 36,878 | 21,711 | 13,339 | 12,621 | 63.8 | 202,142 | 990 |
| NEW JERSEY ............... | 85,751 | 10.4 | 15.2 | 105,705 | 49,643 | 45,513 | 52,345 | 37,256 | 3,641,854 | 3,286,264 | 63.3 | 348,845 | 1,376 |
| Atlantic City................... | 26,563 | 40.4 | 0.7 | 31,343 | 23,676 | 24,881 | 26,691 | 21,965 | 20,772 | 16,716 | 30.1 | 152,493 | 1,015 |
| Bayonne........................ | 66,213 | 17.0 | 6.1 | 80,687 | 40,551 | 46,061 | 60,198 | 35,920 | 27,884 | 25,939 | 33.3 | 385,756 | 1,332 |
| Bergenfield.................... | 102,567 | 9.0 | 13.6 | 103,150 | 79,943 | 47,108 | 58,012 | 42,855 | 9,663 | 9,054 | 75.0 | 385,010 | 1,685 |
| Bridgeton....................... | 36,717 | 36.3 | 0.5 | 36,662 | 24,362 | 19,382 | 17,537 | 30,281 | 7,294 | 6,327 | 33.9 | 120,338 | 1,288 |
| Camden......................... | 33,120 | 32.8 | 1.3 | 38,821 | 17,160 | 23,331 | 26,672 | 21,086 | 29,017 | 24,051 | 39.4 | 88,592 | 873 |
| Clifton........................... | 84,804 | 10.4 | 8.7 | 98,219 | 39,979 | 40,844 | 51,878 | 27,243 | 29,706 | 28,207 | 54.8 | 365,435 | 1,402 |
| East Orange .................. | 50,195 | 21.7 | 1.8 | 59,442 | 39,407 | 37,336 | 36,705 | 38,386 | 30,897 | 27,960 | 23.5 | 236,705 | 1,143 |
| Elizabeth ....................... | 50,732 | 14.5 | 3.7 | 53,554 | 31,455 | 31,651 | 35,675 | 27,917 | 43,443 | 40,227 | 25.4 | 290,158 | 1,191 |
| Englewood ..................... | 103,202 | 11.0 | 19.8 | 130,813 | 79,344 | 56,378 | 69,835 | 47,458 | 12,929 | 11,911 | 58.1 | 421,421 | 1,667 |
| Fair Lawn ...................... | 138,719 | 5.7 | 35.3 | 186,758 | 61,563 | 80,405 | 86,010 | 71,707 | 11,976 | 11,639 | 79.6 | 455,149 | 1,450 |
| Fort Lee......................... | 90,464 | 10.5 | 12.7 | 107,042 | 69,167 | 56,736 | 64,858 | 42,385 | 20,670 | 18,908 | 60.6 | 308,678 | 2,166 |

1. Based on population estimated by the American Community Survey.

# Commuting, Computer Access, Migration, Labor Force, and Employment

| City | Commuting[1], 2019 Percent — Drove alone | With commutes of 30 minutes or more | Computer access[2], 2019 Percent — With a computer in the house | With Internet access | Migration, 2019 Percent who lived in the same house one year ago | Percent who lived in another state or county one year ago | Civilian labor force, 2020 Total | Percent change 2019-2020 | Unemployment Total | Rate[3] | Population age 16 and older Number | Percent in labor force | Population age 16 to 64 Number | Percent who worked full-year full-time |
|---|---|---|---|---|---|---|---|---|---|---|---|---|---|---|
| | 55 | 56 | 57 | 58 | 59 | 60 | 61 | 62 | 63 | 64 | 65 | 66 | 67 | 68 |
| MISSOURI | 82.1 | 33.1 | 91.8 | 84.2 | 85.7 | 6.7 | 3,052,700 | -1.0 | 185,538 | 6 | 4,918,035 | 62.2 | 3,860,092 | 54.0 |
| Ballwin | 83.9 | 33.1 | NA | NA | NA | NA | 16,662 | -2.3 | 813 | 5 | 24,313 | 65.6 | 18,347 | 57.7 |
| Blue Springs | 82.9 | 46.7 | 97.7 | 92.6 | 89.1 | 3.2 | 30,564 | -0.6 | 1,839 | 6 | 42,740 | 70.0 | 35,013 | 60.3 |
| Cape Girardeau | 79.9 | 15.4 | 93.8 | 87.8 | 71.4 | 10.8 | 20,385 | 0.8 | 1,151 | 6 | 34,001 | 63.9 | 27,032 | 48.3 |
| Chesterfield | 83.6 | 32.0 | 97.7 | 96.2 | 87.8 | 6.2 | 24,471 | -2.6 | 1,104 | 5 | 40,126 | 61.9 | 26,878 | 54.8 |
| Columbia | 77.6 | 15.0 | 95.3 | 86.0 | 71.4 | 11.0 | 65,869 | -2.1 | 2,821 | 4 | 100,851 | 68.3 | 88,321 | 44.5 |
| Florissant | 85.3 | 35.6 | 95.2 | 90.7 | 81.7 | 3.7 | 27,844 | -1.1 | 2,105 | 8 | 41,241 | 68.5 | 34,380 | 52.1 |
| Gladstone | 71.2 | 22.3 | 95.7 | 89.3 | 77.5 | 8.4 | 14,684 | -0.5 | 1,019 | 7 | 22,734 | 61.3 | 17,122 | 54.7 |
| Hazelwood | 84.9 | 39.1 | 97.3 | 92.6 | NA | NA | 13,701 | -0.6 | 1,050 | 8 | 20,973 | 65.3 | 17,144 | 65.2 |
| Independence | 89.4 | 38.2 | 91.4 | 84.4 | 89.5 | 2.7 | 57,005 | -0.8 | 4,567 | 8 | 89,286 | 65.6 | 69,391 | 59.2 |
| Jefferson City | 85.9 | 12.0 | 92.6 | 81.8 | 83.2 | 11.2 | 20,740 | 0.3 | 1,014 | 5 | 36,798 | 54.6 | 29,303 | 53.4 |
| Joplin | 78.9 | 14.0 | 93.9 | 88.0 | 81.4 | 11.6 | 24,392 | -1.7 | 1,474 | 6 | 40,775 | 66.5 | 32,256 | 51.0 |
| Kansas City | 81.6 | 25.8 | 93.0 | 84.0 | 81.9 | 8.6 | 259,992 | -0.3 | 18,847 | 7 | 394,104 | 69.0 | 327,294 | 58.6 |
| Kirkwood | 82.8 | 28.7 | NA | NA | 90.0 | 4.6 | 15,690 | -2.1 | 695 | 4 | 21,886 | 62.9 | 15,942 | 63.1 |
| Lee's Summit | 86.3 | 45.0 | 98.7 | 93.5 | 89.3 | 4.9 | 54,568 | -1.8 | 2,545 | 5 | 76,986 | 70.0 | 62,378 | 60.9 |
| Liberty | 84.4 | 33.4 | 96.2 | 94.9 | 75.7 | 13.1 | 16,503 | -0.6 | 987 | 6 | 24,298 | 65.6 | 18,889 | 51.9 |
| Maryland Heights | 85.4 | 23.7 | 98.2 | 95.7 | 79.6 | 11.4 | 15,914 | -1.7 | 880 | 6 | 22,653 | 67.8 | 18,612 | 66.6 |
| O'Fallon | 87.6 | 42.4 | 96.9 | 90.9 | 90.4 | 4.4 | 49,475 | -1.5 | 2,608 | 5 | 69,901 | 69.8 | 57,201 | 62.2 |
| Raytown | 90.3 | 32.2 | 91.4 | 84.5 | NA | NA | 14,970 | -0.9 | 1,239 | 8 | 23,105 | 65.2 | 16,467 | 58.5 |
| St. Charles | 84.8 | 26.1 | 95.0 | 87.8 | 82.4 | 7.6 | 39,059 | -1.3 | 2,151 | 6 | 60,180 | 66.1 | 47,586 | 58.3 |
| St. Joseph | 81.0 | 12.9 | 88.2 | 76.8 | 83.3 | 7.0 | 36,888 | -1.8 | 1,911 | 5 | 60,969 | 58.2 | 49,080 | 50.4 |
| St. Louis | 76.2 | 31.1 | 89.3 | 81.7 | 84.5 | 7.9 | 152,292 | -0.5 | 12,958 | 9 | 249,953 | 65.8 | 206,685 | 55.3 |
| St. Peters | 90.2 | 28.1 | 96.5 | 93.5 | 90.7 | 3.9 | 34,373 | 0.0 | 1,779 | 5 | 46,834 | 70.0 | 37,001 | 61.7 |
| Springfield | 80.9 | 19.4 | 83.3 | 70.7 | 81.2 | 7.4 | 86,724 | -0.6 | 5,127 | 6 | 142,351 | 56.2 | 116,270 | 44.5 |
| University City | 74.5 | 18.8 | 90.8 | 87.2 | 84.1 | 8.3 | 18,364 | -1.7 | 1,142 | 6 | 28,741 | 58.9 | 21,122 | 55.5 |
| Wentzville | 89.3 | 50.0 | 96.5 | 92.0 | 88.6 | 4.7 | 22,481 | 0.1 | 1,252 | 6 | 27,867 | 74.9 | 23,816 | 67.8 |
| Wildwood | 86.4 | 59.7 | 98.5 | 97.0 | 88.3 | 6.4 | 18,595 | -2.9 | 793 | 4 | 28,523 | 65.0 | 22,287 | 56.4 |
| MONTANA | 76.2 | 19.3 | 92.4 | 85.0 | 83.5 | 8.2 | 539,883 | 1.2 | 31,788 | 6 | 869,402 | 62.4 | 661,493 | 50.3 |
| Billings | 81.9 | 12.6 | 94.3 | 85.4 | 79.0 | 7.9 | 56,748 | 1.3 | 3,229 | 6 | 87,323 | 68.5 | 68,527 | 56.3 |
| Bozeman | 65.6 | 6.8 | 96.5 | 89.4 | 67.4 | 15.7 | 31,850 | 3.1 | 1,657 | 5 | 44,412 | 71.8 | 40,035 | 37.3 |
| Butte-Silver Bow | 81.7 | 12.8 | 90.2 | 82.9 | 81.1 | 10.4 | 17,295 | 0.6 | 1,142 | 7 | 28,063 | 54.2 | 21,572 | 47.8 |
| Great Falls | 78.5 | 6.1 | 86.8 | 79.1 | 73.5 | 10.1 | 27,895 | -0.3 | 1,701 | 6 | 47,578 | 59.3 | 35,793 | 49.8 |
| Helena | 76.0 | 8.4 | 93.5 | 87.5 | 77.9 | 9.2 | 18,109 | 3.2 | 861 | 5 | 28,320 | 63.7 | 21,386 | 53.3 |
| Missoula | 69.5 | 12.9 | 96.0 | 88.1 | 74.6 | 11.4 | 41,667 | 1.7 | 2,729 | 7 | 63,524 | 72.0 | 52,622 | 45.1 |
| NEBRASKA | 81.8 | 20.8 | 92.7 | 86.6 | 84.9 | 6.6 | 1,035,175 | 0.0 | 43,787 | 4 | 1,509,290 | 69.1 | 1,196,995 | 59.7 |
| Bellevue | 84.6 | 24.2 | 94.1 | 88.1 | 85.1 | 8.2 | 26,823 | -1.1 | 1,235 | 5 | 43,263 | 66.0 | 35,247 | 64.0 |
| Fremont | 86.9 | 25.9 | 93.1 | 80.9 | 82.9 | 6.5 | 14,419 | 3.4 | 533 | 4 | 19,685 | 66.0 | 15,391 | 59.2 |
| Grand Island | 80.4 | 15.9 | 91.0 | 82.7 | 81.2 | 7.9 | 26,648 | 0.6 | 1,586 | 6 | 38,420 | 69.4 | 30,184 | 62.7 |
| Kearney | 85.3 | 11.1 | 95.5 | 89.7 | 73.3 | 13.8 | 19,234 | -0.6 | 776 | 4 | 26,252 | 70.0 | 21,934 | 52.3 |
| Lincoln | 80.4 | 13.7 | 95.9 | 90.0 | 78.5 | 7.5 | 159,825 | 0.0 | 6,894 | 4 | 231,691 | 71.6 | 191,647 | 55.4 |
| Omaha | 80.8 | 19.3 | 93.3 | 88.4 | 84.6 | 5.3 | 247,850 | 1.6 | 12,990 | 5 | 369,885 | 70.6 | 305,354 | 58.0 |
| NEVADA | 77.0 | 35.4 | 94.2 | 85.1 | 82.4 | 6.1 | 1,530,872 | -0.8 | 196,456 | 13 | 2,465,233 | 63.2 | 1,967,014 | 53.5 |
| Carson City | 83.1 | 32.0 | 89.3 | 83.3 | 81.3 | 12.1 | 26,130 | -0.6 | 2,143 | 8 | 45,637 | 57.7 | 33,770 | 53.2 |
| Henderson | 78.9 | 35.2 | 95.5 | 89.8 | 83.6 | 5.9 | 160,414 | -1.7 | 19,550 | 12 | 264,981 | 60.1 | 196,727 | 53.7 |
| Las Vegas | 77.6 | 40.9 | 93.0 | 80.0 | 81.7 | 4.9 | 308,718 | -2.0 | 43,654 | 14 | 510,149 | 62.4 | 413,413 | 52.3 |
| North Las Vegas | 81.4 | 45.6 | 95.0 | 86.5 | 82.3 | 4.2 | 114,937 | -0.1 | 17,019 | 15 | 188,937 | 64.9 | 161,345 | 53.7 |
| Reno | 72.0 | 22.1 | 95.1 | 87.5 | 78.3 | 8.1 | 138,254 | 0.2 | 11,015 | 8 | 212,505 | 69.1 | 174,050 | 54.8 |
| Sparks | 81.5 | 30.8 | 96.3 | 91.5 | 87.0 | 4.4 | 56,813 | -1.0 | 4,439 | 8 | 80,920 | 66.7 | 64,990 | 55.9 |
| NEW HAMPSHIRE | 80.4 | 39.4 | 93.7 | 88.7 | 87.0 | 6.7 | 761,732 | -1.6 | 50,915 | 7 | 1,137,316 | 66.7 | 884,169 | 56.2 |
| Concord | 73.7 | 31.3 | 91.3 | 81.6 | 84.5 | 9.7 | 22,909 | -0.1 | 1,404 | 6 | 37,196 | 67.6 | 30,231 | 58.8 |
| Dover | 77.6 | 31.1 | 92.8 | 87.9 | 79.0 | 14.3 | 18,626 | -0.6 | 1,212 | 7 | 26,770 | 70.1 | 21,310 | 63.6 |
| Manchester | 81.8 | 31.0 | 93.9 | 88.8 | 79.8 | 7.0 | 65,257 | -0.8 | 5,098 | 8 | 95,235 | 70.4 | 78,717 | 57.0 |
| Nashua | 81.6 | 36.6 | 94.7 | 91.8 | 86.5 | 7.7 | 51,108 | -0.3 | 4,094 | 8 | 74,942 | 73.6 | 60,064 | 61.2 |
| Rochester | 82.4 | 43.7 | 92.4 | 86.8 | 89.8 | 5.0 | 17,782 | -1.2 | 1,237 | 7 | 26,515 | 65.0 | 21,798 | 53.6 |
| NEW JERSEY | 71.0 | 49.3 | 93.6 | 88.7 | 89.7 | 5.1 | 4,495,166 | 0.0 | 439,906 | 10 | 7,168,432 | 65.8 | 5,693,357 | 55.2 |
| Atlantic City | 42.1 | 37.5 | 81.7 | 70.7 | 89.1 | 6.9 | 15,995 | 9.2 | 4,113 | 26 | 29,583 | 56.8 | 23,381 | 35.2 |
| Bayonne | 53.0 | 68.1 | 91.8 | 87.6 | 92.9 | 3.6 | 33,192 | 1.8 | 4,193 | 13 | 51,286 | 63.7 | 43,042 | 58.5 |
| Bergenfield | 71.4 | 54.0 | 98.2 | 93.9 | NA | NA | 14,571 | -1.4 | 1,307 | 9 | 22,606 | 63.8 | 18,500 | 52.9 |
| Bridgeton | 62.4 | 53.3 | 79.0 | 68.1 | 87.4 | 6.0 | 8,067 | 0.1 | 821 | 10 | 17,605 | 47.3 | 15,940 | 41.0 |
| Camden | 56.0 | 40.6 | 81.1 | 72.2 | 82.4 | 6.0 | 26,508 | 3.8 | 4,308 | 16 | 52,318 | 54.9 | 44,683 | 37.9 |
| Clifton | 75.4 | 41.7 | 91.0 | 86.3 | 93.2 | 3.8 | 44,846 | 0.5 | 5,020 | 11 | 68,089 | 65.8 | 54,336 | 56.6 |
| East Orange | 65.4 | 53.2 | 90.8 | 79.6 | 85.1 | 5.5 | 30,076 | 2.7 | 4,305 | 14 | 51,865 | 69.0 | 42,399 | 55.8 |
| Elizabeth | 59.0 | 35.3 | 83.9 | 75.7 | 92.2 | 3.6 | 63,532 | 2.2 | 7,672 | 12 | 95,903 | 65.4 | 82,505 | 54.1 |
| Englewood | 63.6 | 45.2 | 95.3 | 85.3 | NA | NA | 14,972 | -1.0 | 1,491 | 10 | 24,623 | 71.9 | 19,678 | 63.5 |
| Fair Lawn | 71.1 | 54.9 | 91.5 | 88.9 | NA | NA | 17,797 | -1.7 | 1,592 | 9 | 26,305 | 65.7 | 19,163 | 59.6 |
| Fort Lee | 52.1 | 61.8 | 96.3 | 92.2 | 91.3 | 3.3 | 19,738 | 0.4 | 1,582 | 8 | 34,419 | 61.9 | 24,784 | 59.8 |

1. Employed persons.   2. Households.   3. Percent of civilian labor force.   4. Persons 16 years old and over.

## Table D. Cities — Construction, Wholesale Trade, and Retail Trade

| City | Value of residential construction authorized by building permits, 2020 | | | Wholesale trade[1], 2017 | | | | Retail trade[2], 2017 | | | |
|---|---|---|---|---|---|---|---|---|---|---|---|
| | New construction ($1,000) | Number of housing units | Percent single family | Number of establishments | Number of employees | Sales (mil dol) | Annual payroll (mil dol) | Number of establishments | Number of employees | Sales (mil dol) | Annual payroll (mil dol) |
| | 69 | 70 | 71 | 72 | 73 | 74 | 75 | 76 | 77 | 78 | 79 |
| MISSOURI | 278 | 1 | 100 | 6,293 | 95,156 | 102,652 | 5,632 | 20,694 | 312,616 | 100,394 | 8,148 |
| Ballwin | 2,490 | 7 | 100 | 12 | 49 | 45 | 3 | 83 | 1,667 | 690 | 54 |
| Blue Springs | 52,120 | 546 | 65 | 39 | 355 | 319 | 24 | 157 | 3,621 | 1,051 | 91 |
| Cape Girardeau | 25,031 | 123 | 55 | 67 | 793 | 550 | 39 | 286 | 4,769 | 1,217 | 116 |
| Chesterfield | NA | NA | NA | 137 | 1,932 | 1,286 | 160 | 380 | 5,835 | 1,219 | 128 |
| Columbia | 146,805 | 522 | 79 | 112 | 1,441 | 745 | 79 | 494 | 10,223 | 2,763 | 249 |
| Florissant | 150 | 2 | 100 | 10 | 33 | 12 | 1 | 157 | 3,510 | 808 | 83 |
| Gladstone | 1,788 | 8 | 100 | 13 | 49 | 31 | 3 | 74 | 1,853 | 562 | 53 |
| Hazelwood | 115 | 1 | 100 | 58 | 1,472 | 1,115 | 84 | 97 | 1,873 | 789 | 68 |
| Independence | 38,198 | 326 | 25 | 73 | 483 | 207 | 23 | 400 | 8,576 | 2,358 | 212 |
| Jefferson City | 14,634 | 56 | 93 | 60 | 926 | 1,273 | 44 | 243 | 5,176 | 1,307 | 128 |
| Joplin | 42,489 | 326 | 55 | 85 | 1,715 | 753 | 73 | 360 | 6,311 | 1,847 | 156 |
| Kansas City | 319,850 | 2,450 | 37 | 590 | 11,714 | 20,250 | 751 | 1,456 | 25,405 | 9,130 | 694 |
| Kirkwood | 29,300 | 75 | 100 | 31 | 172 | 102 | 11 | 110 | 2,386 | 811 | 73 |
| Lee's Summit | 228,894 | 987 | 43 | 104 | 1,522 | 835 | 88 | 288 | 5,910 | 1,681 | 159 |
| Liberty | 22,598 | 71 | 100 | 20 | 206 | 172 | 14 | 96 | 1,881 | 432 | 48 |
| Maryland Heights | 509 | 2 | 100 | 204 | 5,207 | 6,857 | 458 | 115 | 1,536 | 527 | 56 |
| O'Fallon | 150,364 | 1,067 | 39 | 78 | 1,183 | 4,092 | 61 | 227 | 3,670 | 1,162 | 105 |
| Raytown | 670 | 8 | 38 | 18 | 153 | 59 | 9 | 86 | 1,449 | 360 | 37 |
| St. Charles | 106,206 | 406 | 100 | 78 | 1,584 | 3,034 | 98 | 267 | 4,163 | 1,087 | 100 |
| St. Joseph | 21,777 | 194 | 23 | 90 | 1,162 | 1,602 | 62 | 287 | 5,638 | 1,387 | 132 |
| St. Louis | 80,644 | 422 | 29 | 401 | 6,973 | 6,103 | 425 | 878 | 9,928 | 2,857 | 276 |
| St. Peters | 54,264 | 407 | 17 | 73 | 846 | 431 | 45 | 337 | 6,893 | 2,034 | 189 |
| Springfield | 82,910 | 442 | 27 | 320 | 7,221 | 5,038 | 361 | 940 | 16,834 | 4,677 | 450 |
| University City | 800 | 4 | 100 | 20 | 236 | 170 | 17 | 82 | 770 | 163 | 20 |
| Wentzville | 107,004 | 332 | 97 | 35 | 280 | 196 | 18 | 118 | 2,967 | 931 | 83 |
| Wildwood | NA | NA | NA | 33 | 53 | 28 | 3 | 39 | 475 | 105 | 13 |
| MONTANA | 5,374 | 22 | 100 | 1,366 | 13,078 | 11,215 | 677 | 4,754 | 59,032 | 16,936 | 1,623 |
| Billings | 126,613 | 586 | 61 | 247 | 3,468 | 2,559 | 201 | 606 | 9,384 | 2,945 | 277 |
| Bozeman | 142,527 | 945 | 22 | 65 | 830 | 440 | 41 | 363 | 5,443 | 1,458 | 149 |
| Butte-Silver Bow | 13,239 | 141 | 56 | D | D | D | D | 162 | 2,175 | 573 | 55 |
| Great Falls | 7,080 | 45 | 69 | 88 | 970 | 641 | 47 | 325 | 5,075 | 1,364 | 134 |
| Helena | 44,135 | 249 | 17 | 44 | 401 | 322 | 18 | 223 | 3,525 | 1,007 | 99 |
| Missoula | 96,019 | 547 | 42 | 108 | 1,298 | 777 | 71 | 459 | 6,865 | 1,857 | 186 |
| NEBRASKA | 2,320 | 8 | 100 | 2,782 | 35,143 | 39,909 | 1,965 | 7,154 | 109,729 | 31,215 | 2,853 |
| Bellevue | 72,738 | 241 | 94 | 18 | 90 | 149 | 7 | 98 | 2,325 | 792 | 69 |
| Fremont | 42,466 | 376 | 6 | 31 | 592 | 674 | 34 | 115 | 2,209 | 687 | 61 |
| Grand Island | 30,700 | 197 | 38 | 73 | 1,084 | 615 | 62 | 279 | 4,879 | 1,217 | 126 |
| Kearney | 47,444 | 186 | 70 | 49 | 763 | 712 | 38 | 191 | 3,604 | 846 | 88 |
| Lincoln | 303,097 | 1,742 | 55 | 251 | 3,847 | 4,163 | 211 | 951 | 18,008 | 4,681 | 466 |
| Omaha | 349,409 | 3,049 | 55 | 677 | 10,090 | 11,536 | 660 | 1,597 | 33,315 | 9,419 | 897 |
| NEVADA | 91,111 | 200 | 100 | 2,665 | 32,879 | 25,396 | 1,915 | 8,745 | 145,773 | 45,111 | 4,220 |
| Carson City | 57,589 | 263 | 72 | 81 | 620 | 335 | 29 | 220 | 3,141 | 1,171 | 102 |
| Henderson | 421,962 | 2,391 | 95 | 240 | 1,925 | 1,111 | 114 | 768 | 14,462 | 5,280 | 437 |
| Las Vegas | 660,268 | 3,356 | 58 | 409 | 2,708 | 1,938 | 151 | 1,923 | 35,708 | 10,768 | 1,000 |
| North Las Vegas | 883,666 | 4,030 | 75 | 191 | 4,000 | 3,174 | 246 | 344 | 7,295 | 2,094 | 185 |
| Reno | 609,144 | 3,201 | 42 | 306 | 5,519 | 4,685 | 306 | 1,015 | 17,305 | 5,933 | 549 |
| Sparks | 157,634 | 656 | 100 | 226 | 3,757 | 2,272 | 200 | 322 | 5,258 | 1,439 | 150 |
| NEW HAMPSHIRE | 1,241 | 5 | 100 | 1,509 | 22,314 | 20,328 | 1,577 | 6,032 | 96,591 | 30,039 | 2,805 |
| Concord | 22,971 | 109 | 51 | 55 | 1,075 | 1,831 | 60 | 278 | 5,108 | 1,572 | 157 |
| Dover | 20,815 | 102 | 60 | 32 | 500 | 167 | 30 | 91 | 1,369 | 412 | 47 |
| Manchester | 21,059 | 147 | 91 | 195 | 3,163 | 2,758 | 244 | 419 | 7,274 | 2,476 | 238 |
| Nashua | 5,727 | 31 | 81 | 119 | 1,527 | 1,300 | 106 | 464 | 9,984 | 3,301 | 292 |
| Rochester | 11,662 | 71 | 97 | 11 | 167 | 71 | 14 | 131 | 2,897 | 800 | 89 |
| NEW JERSEY | 0 | 0 | 0 | 12,289 | 230,006 | 312,405 | 18,971 | 31,200 | 469,615 | 149,171 | 13,453 |
| Atlantic City | 19,719 | 154 | 5 | 5 | D | 107 | D | 275 | 2,970 | 648 | 63 |
| Bayonne | 91,032 | 878 | 1 | 55 | 905 | 755 | 48 | 178 | 2,011 | 505 | 50 |
| Bergenfield | 2,194 | 6 | 100 | 33 | 234 | 145 | 13 | 84 | 818 | 213 | 22 |
| Bridgeton | 350 | 2 | 100 | 23 | 366 | 130 | 11 | 90 | 594 | 258 | 16 |
| Camden | 36 | 13 | 85 | 50 | 1,214 | 985 | 67 | 190 | 940 | 280 | 22 |
| Clifton | 6,732 | 33 | 52 | 178 | 1,802 | 909 | 100 | 300 | 4,875 | 1,535 | 142 |
| East Orange | 16,186 | 252 | 6 | 18 | 140 | 481 | 7 | 151 | 1,275 | 346 | 30 |
| Elizabeth | 27,941 | 509 | 5 | 106 | 3,450 | 4,803 | 241 | 532 | 7,489 | 1,695 | 155 |
| Englewood | 6,244 | 49 | 8 | 93 | 1,303 | 813 | 81 | 156 | 2,084 | 1,572 | 101 |
| Fair Lawn | 16,044 | 172 | 62 | 56 | 589 | 240 | 24 | 88 | 872 | 340 | 32 |
| Fort Lee | 18,222 | 109 | 18 | 153 | 1,109 | 1,920 | 98 | 135 | 1,340 | 392 | 30 |

1. Merchant wholesalers except manufacturers' sales branches and offices.　2. Establishments with payroll.

# Table D. Cities — Real Estate, Professional Services, and Manufacturing

| City | Real estate and rental and leasing, 2017 | | | | Professional, scientific, and technical services[1], 2017 | | | | Manufacturing, 2017 | | | |
|---|---|---|---|---|---|---|---|---|---|---|---|---|
| | Number of establishments | Number of employees | Receipts (mil dol) | Annual payroll (mil dol) | Number of establishments | Number of employees | Receipts (mil dol) | Annual payroll (mil dol) | Number of establishments | Number of employees | Receipts (mil dol) | Annual payroll (mil dol) |
| | 80 | 81 | 82 | 83 | 84 | 85 | 86 | 87 | 88 | 89 | 90 | 91 |
| MISSOURI | 6,644 | 37,144 | 8,976 | 1,626 | 14,171 | 161,595 | 30,112 | 11,584 | 5,797 | 259,462 | 118,634 | 14,217 |
| Ballwin | 29 | 82 | 19 | 3 | 69 | 111 | 17 | 6 | NA | NA | NA | NA |
| Blue Springs | 70 | 656 | 88 | 23 | 119 | 464 | 62 | 21 | NA | NA | NA | NA |
| Cape Girardeau | 89 | 264 | 62 | 9 | 120 | 864 | 88 | 35 | NA | NA | NA | NA |
| Chesterfield | 129 | 852 | 220 | 39 | 403 | 14,366 | 2,770 | 979 | NA | NA | NA | NA |
| Columbia | 228 | 1,163 | 213 | 38 | 367 | 4,365 | 502 | 226 | NA | NA | NA | NA |
| Florissant | 31 | 112 | 23 | 4 | 57 | 260 | 21 | 8 | NA | NA | NA | NA |
| Gladstone | 47 | 162 | 60 | 8 | D | D | D | D | NA | NA | NA | NA |
| Hazelwood | 34 | 136 | 55 | 6 | 49 | 620 | 112 | 42 | NA | NA | NA | NA |
| Independence | 94 | 426 | 120 | 14 | 224 | 894 | 110 | 49 | NA | NA | NA | NA |
| Jefferson City | 64 | 221 | 57 | 8 | 212 | 1,425 | 210 | 89 | NA | NA | NA | NA |
| Joplin | 85 | 449 | 85 | 14 | 137 | 940 | 123 | 45 | NA | NA | NA | NA |
| Kansas City | 631 | 4,863 | 1,779 | 291 | 1,957 | 32,540 | 7,385 | 2,766 | NA | NA | NA | NA |
| Kirkwood | 49 | 152 | 49 | 5 | 133 | 576 | 111 | 38 | NA | NA | NA | NA |
| Lee's Summit | 143 | 382 | 126 | 20 | 349 | 1,816 | 263 | 101 | NA | NA | NA | NA |
| Liberty | 37 | 161 | 36 | 5 | D | D | D | D | NA | NA | NA | NA |
| Maryland Heights | 66 | 391 | 115 | 21 | 146 | 2,268 | 338 | 143 | NA | NA | NA | NA |
| O'Fallon | 88 | 488 | 105 | 19 | 146 | 591 | 113 | 39 | NA | NA | NA | NA |
| Raytown | 17 | 73 | 10 | 2 | D | D | D | D | NA | NA | NA | NA |
| St. Charles | 96 | 546 | 284 | 31 | 265 | 3,046 | 269 | 99 | NA | NA | NA | NA |
| St. Joseph | 100 | 393 | 65 | 11 | 132 | 1,246 | 233 | 88 | NA | NA | NA | NA |
| St. Louis | 412 | 2,388 | 633 | 109 | 1,114 | 18,010 | 3,685 | 1,389 | NA | NA | NA | NA |
| St. Peters | 84 | 273 | 78 | 11 | D | D | D | D | NA | NA | NA | NA |
| Springfield | 380 | 2,844 | 454 | 97 | 681 | 6,162 | 873 | 357 | NA | NA | NA | NA |
| University City | 46 | 762 | 117 | 27 | 90 | 257 | 34 | 12 | NA | NA | NA | NA |
| Wentzville | D | D | D | D | 48 | 1,695 | 198 | 76 | NA | NA | NA | NA |
| Wildwood | D | D | D | D | 115 | 421 | 54 | 22 | NA | NA | NA | NA |
| MONTANA | 2,036 | 5,951 | 1,145 | 206 | 3,836 | 17,711 | 2,517 | 939 | 1,328 | 17,944 | 10,944 | 936 |
| Billings | 277 | 1,033 | 232 | 44 | 520 | 3,536 | 619 | 218 | NA | NA | NA | NA |
| Bozeman | 219 | 526 | 130 | 20 | 490 | 2,309 | 387 | 140 | NA | NA | NA | NA |
| Butte-Silver Bow | 57 | 156 | 20 | 4 | 107 | 546 | 66 | 29 | NA | NA | NA | NA |
| Great Falls | 103 | 353 | 65 | 12 | 170 | 1,088 | 162 | 58 | NA | NA | NA | NA |
| Helena | 91 | 226 | 58 | 8 | 215 | 1,569 | 222 | 90 | NA | NA | NA | NA |
| Missoula | 196 | 718 | 144 | 26 | 481 | 2,940 | 354 | 149 | NA | NA | NA | NA |
| NEBRASKA | 2,354 | 11,293 | 2,292 | 487 | 4,699 | 39,566 | 6,278 | 2,385 | 1,760 | 93,510 | 53,129 | 4,728 |
| Bellevue | 51 | 189 | 35 | 6 | 90 | 1,172 | 197 | 72 | NA | NA | NA | NA |
| Fremont | 38 | 168 | 31 | 6 | 34 | 168 | 18 | 6 | NA | NA | NA | NA |
| Grand Island | 66 | 249 | 43 | 8 | 91 | 640 | 76 | 33 | NA | NA | NA | NA |
| Kearney | 57 | 144 | 47 | 5 | 85 | 967 | 99 | 44 | NA | NA | NA | NA |
| Lincoln | 391 | 1,957 | 362 | 81 | 835 | 8,293 | 1,432 | 487 | NA | NA | NA | NA |
| Omaha | 803 | 5,999 | 1,242 | 286 | 1,672 | 19,019 | 3,189 | 1,291 | NA | NA | NA | NA |
| NEVADA | 4,684 | 30,562 | 7,359 | 1,312 | 9,018 | 60,168 | 10,816 | 3,951 | 1,828 | 44,182 | 16,408 | 2,517 |
| Carson City | 118 | 361 | 94 | 13 | 246 | 1,126 | 189 | 74 | NA | NA | NA | NA |
| Henderson | 513 | 1,719 | 472 | 80 | 1,020 | 4,993 | 852 | 308 | NA | NA | NA | NA |
| Las Vegas | 1,081 | 5,775 | 1,309 | 284 | 2,610 | 17,301 | 3,413 | 1,206 | NA | NA | NA | NA |
| North Las Vegas | 131 | 839 | 228 | 37 | 153 | 2,526 | 410 | 190 | NA | NA | NA | NA |
| Reno | 540 | 2,960 | 722 | 117 | 1,270 | 8,428 | 1,401 | 532 | NA | NA | NA | NA |
| Sparks | 132 | 817 | 166 | 32 | 136 | 920 | 112 | 41 | NA | NA | NA | NA |
| NEW HAMPSHIRE | 1,523 | 7,920 | 2,000 | 381 | 3,674 | 32,387 | 5,583 | 2,383 | 1,790 | 65,211 | 20,304 | 3,875 |
| Concord | D | D | D | D | 207 | 2,034 | 327 | 161 | NA | NA | NA | NA |
| Dover | 32 | 122 | 25 | 5 | 86 | 469 | 80 | 35 | NA | NA | NA | NA |
| Manchester | 156 | 1,589 | 297 | 72 | 457 | 4,859 | 903 | 426 | NA | NA | NA | NA |
| Nashua | 124 | 505 | 121 | 24 | 306 | 3,316 | 763 | 376 | NA | NA | NA | NA |
| Rochester | 25 | 76 | 17 | 3 | 36 | 295 | 29 | 12 | NA | NA | NA | NA |
| NEW JERSEY | 9,622 | 61,052 | 21,184 | 3,435 | 28,962 | 325,516 | 67,232 | 28,736 | 7,332 | 219,835 | 95,483 | 13,909 |
| Atlantic City | 45 | 452 | 144 | 18 | D | D | D | 68 | NA | NA | NA | NA |
| Bayonne | 37 | 182 | 37 | 8 | 71 | 492 | 64 | 37 | NA | NA | NA | NA |
| Bergenfield | D | D | D | D | 42 | 228 | 32 | 13 | NA | NA | NA | NA |
| Bridgeton | 14 | 123 | 25 | 4 | 29 | 112 | 10 | 3 | NA | NA | NA | NA |
| Camden | 41 | 198 | 46 | 8 | 58 | 1,013 | 585 | 97 | NA | NA | NA | NA |
| Clifton | 107 | 596 | 120 | 26 | 242 | 1,413 | 232 | 82 | NA | NA | NA | NA |
| East Orange | 49 | 248 | 56 | 8 | 47 | 147 | 19 | 5 | NA | NA | NA | NA |
| Elizabeth | 100 | 434 | 230 | 22 | D | D | 84 | D | NA | NA | NA | NA |
| Englewood | 63 | 236 | 94 | 14 | 130 | 649 | 99 | 37 | NA | NA | NA | NA |
| Fair Lawn | D | D | D | D | 209 | 1,168 | 217 | 84 | NA | NA | NA | NA |
| Fort Lee | 135 | 610 | 205 | 28 | 248 | 1,554 | 322 | 134 | NA | NA | NA | NA |

1. Establishments subject to federal tax.

# Table D. Cities — Accommodation and Food Services, Arts, Entertainment, and Recreation, and Health Care and Social Assistance

| City | Accommodation and food services, 2017 | | | | Arts, entertainment, and recreation[1], 2017 | | | | Health care and social assistance[1], 2017 | | | |
|---|---|---|---|---|---|---|---|---|---|---|---|---|
| | Number of establishments | Number of employees | Receipts (mil dol) | Annual payroll (mil dol) | Number of establishments | Number of employees | Receipts (mil dol) | Annual payroll (mil dol) | Number of establishments | Number of employees | Receipts (mil dol) | Annual payroll (mil dol) |
| | 92 | 93 | 94 | 95 | 96 | 97 | 98 | 99 | 100 | 101 | 102 | 103 |
| MISSOURI | 12,896 | 263,644 | 15,082 | 4,275 | 2,274 | 40,484 | 3,953 | 1,554 | 19,097 | 423,057 | 48,193 | 18,682 |
| Ballwin | 52 | 1,128 | 54 | 17 | 14 | 516 | 22 | 9 | 44 | 553 | 34 | 16 |
| Blue Springs | 123 | 2,659 | 137 | 38 | 22 | 231 | 10 | 3 | 144 | 1,868 | 170 | 68 |
| Cape Girardeau | 141 | 3,417 | 163 | 49 | D | D | D | D | 272 | 9,492 | 1,270 | 462 |
| Chesterfield | 199 | 4,217 | 244 | 74 | 47 | 639 | 30 | 10 | 281 | 7,862 | 1,043 | 455 |
| Columbia | 417 | 9,877 | 464 | 136 | 60 | 923 | 38 | 12 | 584 | 18,346 | 2,544 | 871 |
| Florissant | D | D | D | D | D | D | D | D | 211 | 2,589 | 226 | 93 |
| Gladstone | 44 | 946 | 44 | 14 | 4 | D | 2 | D | 76 | 776 | 66 | 27 |
| Hazelwood | 55 | 949 | 48 | 14 | 4 | 6 | 1 | 0 | 111 | 984 | 95 | 35 |
| Independence | 229 | 5,721 | 290 | 86 | 26 | 333 | 23 | 6 | 342 | 6,465 | 838 | 313 |
| Jefferson City | 151 | 3,194 | 154 | 45 | 20 | 426 | 32 | 8 | 254 | 6,697 | 788 | 309 |
| Joplin | 193 | 4,606 | 209 | 65 | 23 | 354 | 11 | 5 | 276 | 10,140 | 1,131 | 456 |
| Kansas City | 1,186 | 30,485 | 2,100 | 594 | 221 | 6,819 | 1,187 | 605 | 1,770 | 45,482 | 6,382 | 2,404 |
| Kirkwood | 71 | 1,877 | 95 | 33 | 21 | 515 | 30 | 10 | 120 | 1,915 | 151 | 63 |
| Lee's Summit | 198 | 4,846 | 230 | 71 | 40 | 423 | 17 | 5 | 330 | 6,882 | 945 | 341 |
| Liberty | 76 | 1,747 | 88 | 26 | 13 | 64 | 4 | 1 | 134 | 3,209 | 335 | 144 |
| Maryland Heights | 96 | 3,063 | 367 | 74 | 20 | 544 | 53 | 10 | 96 | 3,886 | 359 | 158 |
| O'Fallon | 169 | 3,947 | 186 | 58 | 25 | 673 | 26 | 10 | 226 | 3,365 | 301 | 122 |
| Raytown | 51 | 1,169 | 50 | 14 | 4 | 41 | 2 | 1 | 88 | 966 | 68 | 28 |
| St. Charles | 228 | 7,324 | 562 | 138 | 33 | 343 | 18 | 5 | 261 | 5,464 | 555 | 209 |
| St. Joseph | 192 | 3,990 | 195 | 57 | 20 | 566 | 44 | 10 | D | D | D | D |
| St. Louis | 1,002 | 22,623 | 1,685 | 451 | 148 | 4,609 | 617 | 221 | 1,554 | 39,629 | 4,648 | 1,753 |
| St. Peters | 171 | 3,748 | 191 | 58 | 35 | 712 | 32 | 9 | 299 | 4,540 | 439 | 171 |
| Springfield | 702 | 14,999 | 761 | 221 | 75 | 1,334 | 88 | 23 | 768 | 28,280 | 3,730 | 1,390 |
| University City | 85 | 1,439 | 77 | 23 | 12 | D | 24 | D | 169 | 1,682 | 128 | 39 |
| Wentzville | 84 | 1,954 | 98 | 27 | 9 | 113 | 6 | 2 | 88 | 1,292 | 98 | 47 |
| Wildwood | 26 | 664 | 35 | 10 | 9 | 330 | 8 | 2 | 58 | 581 | 52 | 23 |
| MONTANA | 3,568 | 52,415 | 3,126 | 899 | 1,234 | 10,492 | 877 | 180 | 3,716 | 73,254 | 8,448 | 3,304 |
| Billings | 347 | 8,420 | 489 | 141 | 148 | 1,550 | 150 | 26 | 534 | 13,921 | 2,029 | 787 |
| Bozeman | 241 | 4,553 | 273 | 80 | 83 | 462 | 49 | 10 | 305 | 4,646 | 535 | 195 |
| Butte-Silver Bow | 137 | 2,184 | 114 | 36 | 46 | 328 | 24 | 5 | 169 | 3,114 | 296 | 114 |
| Great Falls | 208 | 3,739 | 208 | 60 | 72 | 689 | 59 | 12 | 269 | 6,664 | 792 | 318 |
| Helena | 157 | 3,072 | 164 | 49 | 55 | 756 | 54 | 9 | D | D | D | D |
| Missoula | 293 | 6,180 | 341 | 97 | 84 | 1,299 | 76 | 20 | 453 | 9,689 | 1,151 | 401 |
| NEBRASKA | 4,621 | 76,386 | 3,958 | 1,136 | 913 | 14,879 | 891 | 249 | 5,817 | 135,691 | 16,060 | 6,117 |
| Bellevue | 119 | 2,415 | 118 | 36 | D | D | D | D | 131 | 2,384 | 247 | 93 |
| Fremont | 69 | 1,132 | 55 | 15 | 12 | 414 | 9 | 4 | 106 | 2,629 | 252 | 102 |
| Grand Island | 151 | 2,565 | 140 | 41 | 25 | 464 | 43 | 8 | 182 | 4,408 | 506 | 194 |
| Kearney | 129 | 2,613 | 123 | 37 | 22 | 490 | 17 | 5 | 183 | 3,892 | 662 | 168 |
| Lincoln | 746 | 14,706 | 746 | 212 | 129 | 2,878 | 120 | 36 | 1,012 | 24,568 | 2,870 | 1,094 |
| Omaha | 1,340 | 27,002 | 1,509 | 443 | 244 | 6,172 | 433 | 120 | 1,828 | 51,418 | 6,937 | 2,684 |
| NEVADA | 6,810 | 319,584 | 33,980 | 9,578 | 1,635 | 31,918 | 5,143 | 1,068 | 7,372 | 132,093 | 18,112 | 6,620 |
| Carson City | 170 | 3,036 | 193 | 54 | 51 | 1,019 | 119 | 42 | 215 | 3,865 | 624 | 214 |
| Henderson | 535 | 16,075 | 1,342 | 367 | 197 | 3,474 | 344 | 90 | 879 | 12,658 | 1,848 | 646 |
| Las Vegas | 1,403 | 45,459 | 3,979 | 1,178 | 332 | 6,476 | 1,599 | 307 | 2,077 | 39,266 | 5,199 | 1,914 |
| North Las Vegas | 270 | 7,756 | 629 | 169 | D | D | D | D | 274 | 6,983 | 1,141 | 439 |
| Reno | 821 | 23,651 | 1,802 | 518 | 151 | 3,008 | 341 | 75 | 1,019 | 22,180 | 3,182 | 1,225 |
| Sparks | 210 | 5,393 | 386 | 109 | 46 | 1,317 | 95 | 23 | 182 | 3,102 | 367 | 128 |
| NEW HAMPSHIRE | 3,784 | 59,531 | 3,722 | 1,133 | 815 | 12,442 | 978 | 291 | 3,737 | 94,594 | 11,932 | 4,742 |
| Concord | 131 | 2,877 | 176 | 59 | 25 | 401 | 21 | 7 | 237 | 9,984 | 1,132 | 516 |
| Dover | 110 | 1,905 | 109 | 33 | 17 | 179 | 11 | 4 | 139 | 4,200 | 573 | 204 |
| Manchester | 307 | 6,151 | 373 | 109 | 49 | 1,035 | 84 | 42 | 350 | 14,010 | 1,805 | 767 |
| Nashua | 228 | 4,686 | 272 | 81 | 45 | 592 | 38 | 13 | 370 | 9,314 | 1,277 | 470 |
| Rochester | 73 | 1,164 | 66 | 18 | 8 | 133 | 5 | 2 | 74 | 1,883 | 259 | 101 |
| NEW JERSEY | 21,495 | 318,734 | 23,785 | 6,432 | 3,842 | 65,529 | 5,967 | 1,934 | 28,005 | 613,406 | 74,723 | 30,134 |
| Atlantic City | 199 | 26,558 | 3,006 | 853 | D | D | D | D | 68 | 3,300 | 462 | 176 |
| Bayonne | D | D | D | D | 17 | 300 | 25 | 6 | 170 | 2,708 | 308 | 109 |
| Bergenfield | 41 | 370 | 26 | 6 | D | D | D | D | 77 | 643 | 54 | 20 |
| Bridgeton | 30 | 216 | 14 | 3 | D | D | D | D | 40 | 1,397 | 99 | 42 |
| Camden | 90 | 482 | 43 | 9 | 10 | 357 | 74 | 11 | 161 | 13,231 | 1,959 | 877 |
| Clifton | D | D | D | D | 26 | 314 | 36 | 10 | 442 | 5,344 | 578 | 217 |
| East Orange | 55 | 681 | 37 | 10 | D | D | D | D | 150 | 6,677 | 772 | 355 |
| Elizabeth | 260 | 3,291 | 304 | 74 | 10 | 101 | 6 | 2 | 219 | 6,377 | 630 | 277 |
| Englewood | 59 | 776 | 69 | 18 | 24 | 359 | 29 | 9 | 242 | 5,205 | 1,165 | 338 |
| Fair Lawn | 76 | 891 | 65 | 17 | 15 | 143 | 8 | 2 | 202 | 3,115 | 282 | 117 |
| Fort Lee | 136 | 1,364 | 119 | 26 | 16 | 95 | 7 | 2 | 214 | 1,504 | 200 | 54 |

1. Establishments subject to federal tax.

| City | Other services[1] | | | | Government employment and payroll, 2017 | | | | | | | | |
|---|---|---|---|---|---|---|---|---|---|---|---|---|---|
| | | | | | March payroll | | | | | | | | |
| | | | | | | | Percent of total for: | | | | | | |
| | Number of establishments | Number of employees | Receipts (mil dol) | Annual payroll (mil dol) | Full-time equivalent employees | Total (dollars) | Administrative, judicial, and legal | Police and corrections | Fire protection | Highways and transportation | Health and welfare | Natural resources and utilities | Education and libraries |
| | 104 | 105 | 106 | 107 | 108 | 109 | 110 | 111 | 112 | 113 | 114 | 115 | 116 |
| MISSOURI ..................... | 10,513 | 68,206 | 8,694 | 2,297 | X | X | X | X | X | X | X | X | X |
| Ballwin...................... | 47 | 287 | 22 | 8 | 157 | 685,445 | 12.2 | 43.3 | 0.0 | 16.6 | 0.0 | 22.4 | 0.0 |
| Blue Springs................ | 110 | 540 | 54 | 16 | 291 | 1,550,352 | 12.7 | 51.1 | 0.0 | 8.7 | 6.1 | 19.7 | 0.0 |
| Cape Girardeau............ | 99 | 594 | 58 | 17 | 437 | 1,558,764 | 8.5 | 25.5 | 18.1 | 9.3 | 0.0 | 26.1 | 0.0 |
| Chesterfield................. | 151 | 1,213 | 172 | 42 | 238 | 1,140,711 | 13.9 | 49.2 | 0.0 | 15.7 | 0.0 | 16.3 | 0.0 |
| Columbia.................... | 289 | 1,947 | 197 | 61 | 1,480 | 5,541,903 | 22.8 | 14.8 | 13.4 | 7.9 | 7.4 | 29.8 | 0.0 |
| Florissant................... | 93 | 516 | 49 | 16 | 381 | 1,461,910 | 12.2 | 33.9 | 0.4 | 11.0 | 11.9 | 30.2 | 0.0 |
| Gladstone................... | 54 | 279 | 27 | 8 | 209 | 831,442 | 13.9 | 32.9 | 20.6 | 5.2 | 1.7 | 22.0 | 0.0 |
| Hazelwood.................. | 46 | 279 | 28 | 10 | 126 | 751,106 | 8.0 | 58.1 | 33.9 | 0.0 | 0.0 | 0.0 | 0.0 |
| Independence ............. | 179 | 1,005 | 102 | 31 | 1,031 | 6,155,008 | 6.7 | 24.6 | 16.4 | 3.2 | 3.5 | 44.3 | 0.0 |
| Jefferson City ............. | 192 | 1,151 | 191 | 50 | 370 | 1,669,553 | 10.6 | 27.1 | 18.2 | 11.2 | 0.6 | 21.0 | 0.0 |
| Joplin........................ | 111 | 890 | 72 | 24 | 548 | 1,905,882 | 10.0 | 29.4 | 19.6 | 11.0 | 8.6 | 13.7 | 4.2 |
| Kansas City................. | 860 | 7,708 | 1,753 | 296 | 6,334 | 28,199,434 | 11.8 | 19.0 | 28.8 | 11.7 | 7.4 | 20.5 | 0.0 |
| Kirkwood.................... | 61 | 459 | 47 | 11 | 298 | 1,535,998 | 6.7 | 27.5 | 22.4 | 3.4 | 1.8 | 32.8 | 0.0 |
| Lee's Summit............... | 171 | 1,052 | 100 | 33 | 703 | 3,810,408 | 15.9 | 25.3 | 24.9 | 9.9 | 7.9 | 16.1 | 0.0 |
| Liberty ...................... | 55 | 336 | 30 | 10 | 191 | 864,609 | 20.2 | 33.8 | 25.2 | 7.2 | 1.2 | 7.0 | 0.0 |
| Maryland Heights.......... | 64 | 632 | 73 | 27 | 195 | 1,125,065 | 18.7 | 52.2 | 0.0 | 7.5 | 0.0 | 12.9 | 0.0 |
| O'Fallon..................... | 142 | 1,225 | 114 | 37 | 458 | 2,119,939 | 12.3 | 37.5 | 0.0 | 9.1 | 0.2 | 28.6 | 0.0 |
| Raytown..................... | 48 | 396 | 44 | 12 | 141 | 692,893 | 12.7 | 54.6 | 0.0 | 11.8 | 12.3 | 5.8 | 2.8 |
| St. Charles ................. | 162 | 1,035 | 113 | 33 | 501 | 2,735,489 | 7.7 | 34.1 | 22.1 | 15.6 | 0.0 | 6.7 | 0.0 |
| St. Joseph.................. | 148 | 1,175 | 179 | 48 | 693 | 2,714,225 | 9.4 | 27.2 | 21.4 | 11.5 | 6.6 | 20.9 | 0.0 |
| St. Louis.................... | 625 | 5,377 | 828 | 199 | 6,658 | 37,801,592 | 16.1 | 37.5 | 15.7 | 12.1 | 2.0 | 12.5 | 0.0 |
| St. Peters .................. | 166 | 1,255 | 115 | 34 | 541 | 2,548,634 | 15.4 | 27.5 | 0.0 | 11.9 | 9.0 | 36.2 | 0.0 |
| Springfield.................. | 476 | 4,094 | 449 | 134 | 2,825 | 14,818,819 | 5.5 | 15.9 | 8.3 | 10.0 | 2.8 | 33.9 | 0.0 |
| University City............. | 57 | 513 | 59 | 21 | 286 | 1,206,503 | 5.9 | 47.1 | 16.2 | 8.1 | 0.0 | 13.6 | 7.0 |
| Wentzville................... | 56 | 379 | 38 | 10 | 248 | 1,084,085 | 15.8 | 40.5 | 0.0 | 9.5 | 1.6 | 22.4 | 0.0 |
| Wildwood.................... | 24 | 111 | 6 | 2 | 21 | 145,661 | 74.2 | 0.0 | 0.0 | 11.7 | 0.0 | 0.0 | X |
| MONTANA .................. | 2,409 | 12,077 | 1,609 | 403 | X | X | X | X | X | X | X | X | 2.7 |
| Billings...................... | 302 | 1,927 | 230 | 64 | 907 | 5,144,967 | 10.7 | 22.3 | 17.6 | 14.0 | 1.0 | 18.4 | 4.9 |
| Bozeman.................... | 183 | 1,103 | 193 | 43 | 366 | 1,782,896 | 18.3 | 22.4 | 16.2 | 4.9 | 0.0 | 21.9 | 0.0 |
| Butte-Silver Bow .......... | 65 | 277 | 38 | 8 | 450 | 2,205,916 | 15.6 | 24.3 | 18.9 | 14.3 | 2.0 | 21.5 | 2.9 |
| Great Falls ................. | D | D | D | D | 494 | 2,452,526 | 16.7 | 30.1 | 17.3 | 11.4 | 3.8 | 17.6 | 0.0 |
| Helena....................... | 170 | 1,022 | 147 | 46 | 313 | 1,491,616 | 18.9 | 27.1 | 14.4 | 7.5 | 0.3 | 21.5 | 0.0 |
| Missoula..................... | 252 | 2,039 | 337 | 74 | 561 | 2,768,191 | 12.6 | 29.8 | 24.1 | 9.6 | 1.7 | 17.8 | X |
| NEBRASKA ................. | 4,107 | 21,757 | 3,608 | 701 | X | X | X | X | X | X | X | X | 3.6 |
| Bellevue ..................... | 85 | 474 | 42 | 13 | 330 | 1,482,766 | 6.4 | 48.5 | 24.0 | 8.2 | 1.6 | 5.8 | 2.9 |
| Fremont...................... | 62 | 258 | 26 | 7 | 289 | 1,563,109 | 14.4 | 16.4 | 9.1 | 6.0 | 0.5 | 49.7 | 3.8 |
| Grand Island ............... | 136 | 920 | 86 | 22 | 618 | 3,076,295 | 9.6 | 16.4 | 15.9 | 8.2 | 0.3 | 43.8 | 4.8 |
| Kearney...................... | 98 | 554 | 75 | 13 | 311 | 1,416,471 | 11.1 | 26.4 | 7.2 | 8.3 | 0.0 | 34.7 | 2.7 |
| Lincoln....................... | 678 | 4,126 | 757 | 150 | 2,642 | 14,836,126 | 7.7 | 18.2 | 13.0 | 10.8 | 9.1 | 36.5 | 3.3 |
| Omaha....................... | 1,043 | 7,314 | 1,651 | 250 | 2,910 | 18,004,814 | 5.5 | 37.8 | 29.6 | 8.1 | 1.6 | 10.8 | X |
| NEVADA ..................... | 4,032 | 28,825 | 3,282 | 914 | X | X | X | X | X | X | X | X | 2.1 |
| Carson City ................ | 132 | 618 | 101 | 22 | 650 | 3,550,570 | 21.0 | 28.3 | 16.0 | 8.6 | 5.8 | 14.8 | 0.0 |
| Henderson .................. | 423 | 3,209 | 326 | 92 | 2,310 | 16,130,348 | 16.1 | 32.6 | 16.9 | 1.1 | 1.7 | 21.9 | 0.0 |
| Las Vegas................... | 921 | 6,955 | 777 | 216 | 2,867 | 20,161,448 | 18.7 | 14.2 | 33.0 | 5.9 | 2.9 | 16.3 | 1.2 |
| North Las Vegas .......... | 182 | 2,045 | 202 | 67 | 1,081 | 8,065,862 | 13.5 | 40.2 | 21.3 | 3.6 | 4.3 | 14.1 | 0.0 |
| Reno.......................... | 523 | 3,845 | 453 | 129 | 1,251 | 8,747,437 | 13.9 | 33.3 | 27.4 | 3.9 | 0.0 | 11.5 | 0.0 |
| Sparks........................ | 172 | 1,071 | 133 | 38 | 499 | 3,235,999 | 20.7 | 26.9 | 24.1 | 4.5 | 0.0 | 23.9 | X |
| NEW HAMPSHIRE ............. | 3,011 | 17,864 | 2,096 | 619 | X | X | X | X | X | X | X | X | 3.4 |
| Concord ..................... | 222 | 1,356 | 233 | 56 | 463 | 2,884,048 | 13.0 | 23.4 | 17.7 | 8.5 | 9.1 | 13.8 | 64.8 |
| Dover ........................ | 74 | 467 | 38 | 11 | 847 | 3,683,989 | 5.2 | 10.3 | 8.4 | 1.8 | 1.1 | 5.8 | 51.8 |
| Manchester.................. | 284 | 2,389 | 355 | 99 | 3,155 | 16,041,295 | 3.6 | 12.7 | 9.5 | 10.2 | 2.5 | 7.4 | 58.6 |
| Nashua....................... | 191 | 1,645 | 166 | 55 | 2,699 | 13,478,642 | 3.8 | 13.3 | 8.9 | 3.9 | 1.1 | 8.7 | 71.9 |
| Rochester.................... | 63 | 349 | 37 | 10 | 973 | 4,205,963 | 5.0 | 9.8 | 5.2 | 3.1 | 0.5 | 2.8 | X |
| NEW JERSEY.................... | 19,162 | 113,846 | 14,026 | 3,679 | X | X | X | X | X | X | X | X | X |
| Atlantic City................. | 64 | 844 | 65 | 18 | 1,546 | 9,801,157 | 8.1 | 41.5 | 24.4 | 1.3 | 7.0 | 9.0 | 1.9 |
| Bayonne...................... | 131 | 491 | 35 | 9 | 2,463 | 14,984,419 | 2.7 | 17.0 | 11.6 | 3.4 | 3.0 | 0.9 | 60.5 |
| Bergenfield.................. | 63 | 149 | 18 | 4 | 175 | 1,201,986 | 7.5 | 54.7 | 4.8 | 0.0 | 2.5 | 22.9 | 5.5 |
| Bridgeton..................... | 17 | 86 | 10 | 3 | 18 | 68,880 | 0.0 | 0.0 | 0.0 | 0.0 | 100.0 | 0.0 | 0.0 |
| Camden...................... | 54 | 393 | 42 | 11 | 662 | 3,756,554 | 15.9 | 9.6 | 39.9 | 2.6 | 11.2 | 9.8 | 3.7 |
| Clifton........................ | 209 | 1,103 | 109 | 38 | 564 | 3,939,877 | 6.3 | 35.4 | 29.0 | 0.9 | 2.6 | 4.4 | 66.7 |
| East Orange................ | 69 | 309 | 34 | 9 | 2,811 | 18,542,586 | 3.9 | 14.2 | 7.7 | 1.3 | 3.4 | 2.7 | 1.6 |
| Elizabeth..................... | 208 | 1,671 | 122 | 67 | 1,415 | 9,391,483 | 9.0 | 36.6 | 26.7 | 7.6 | 8.9 | 2.7 | 50.4 |
| Englewood................... | 112 | 652 | 63 | 21 | 731 | 5,467,310 | 3.0 | 22.8 | 11.8 | 2.1 | 2.7 | 4.5 | 5.9 |
| Fair Lawn.................... | 90 | 463 | 37 | 12 | 227 | 1,783,449 | 9.9 | 43.8 | 1.2 | 4.8 | 2.6 | 20.6 | 4.5 |
| Fort Lee...................... | 153 | 663 | 120 | 22 | 306 | 2,105,983 | 8.9 | 50.5 | 3.4 | 9.7 | 7.3 | 3.7 | |

1. Establishments subject to federal tax.

# Table D. Cities — City Government Finances

| City | City government finances, 2017 | | | | | | | | | |
|---|---|---|---|---|---|---|---|---|---|---|
| | General revenue | | | | | | | General expenditure | | |
| | | Intergovernmental | | Taxes | | | | | | |
| | | | | | | Per capita[1] (dollars) | | | Per capita[1] (dollars) | |
| | Total (mil dol) | Total (mil dol) | Percent from state government | Total (mil dol) | Total | Property | Sales and gross receipts | Total (mil dol) | Total | Capital outlays |
| | 117 | 118 | 119 | 120 | 121 | 122 | 123 | 124 | 125 | 126 |
| MISSOURI | X | X | X | X | X | X | X | X | X | X |
| Ballwin | 21.3 | 2.8 | 74.0 | 14.2 | 471 | 0 | 471 | 19 | 637 | 115 |
| Blue Springs | 50.1 | 3.9 | 100.0 | 31.3 | 571 | 112 | 452 | 46 | 845 | 85 |
| Cape Girardeau | 72.6 | 6.5 | 100.0 | 47.6 | 1,214 | 63 | 1,074 | 70 | 1,775 | 755 |
| Chesterfield | 35.2 | 3.3 | 100.0 | 31.9 | 671 | 0 | 671 | 35 | 743 | 210 |
| Columbia | 182.1 | 38.8 | 12.9 | 73.5 | 605 | 99 | 506 | 157 | 1,289 | 337 |
| Florissant | 36.3 | 15.5 | 5.5 | 13.6 | 265 | 13 | 179 | 32 | 627 | 60 |
| Gladstone | 32.3 | 4.0 | 59.0 | 17.1 | 632 | 124 | 508 | 29 | 1,079 | 204 |
| Hazelwood | 34.1 | 1.9 | 49.9 | 29.9 | 1,181 | 319 | 862 | 37 | 1,451 | 99 |
| Independence | 140.0 | 11.0 | 92.8 | 81.4 | 694 | 80 | 614 | 163 | 1,388 | 233 |
| Jefferson City | 63.2 | 6.7 | 100.0 | 36.4 | 848 | 123 | 725 | 63 | 1,471 | 315 |
| Joplin | 135.2 | 59.2 | 7.9 | 49.9 | 988 | 59 | 930 | 120 | 2,380 | 1,063 |
| Kansas City | 1,617.5 | 101.0 | 39.7 | 739.0 | 1,513 | 254 | 698 | 1,711 | 3,504 | 476 |
| Kirkwood | 27.9 | 1.9 | 100.0 | 21.6 | 780 | 166 | 614 | 28 | 993 | 124 |
| Lee's Summit | 160.3 | 22.0 | 13.8 | 84.5 | 870 | 318 | 552 | 143 | 1,472 | 356 |
| Liberty | 43.7 | 1.0 | 1.0 | 27.3 | 865 | 206 | 659 | 57 | 1,812 | 572 |
| Maryland Heights | 39.3 | 11.9 | 21.0 | 22.0 | 815 | 168 | 647 | 61 | 2,275 | 1,209 |
| O'Fallon | 73.2 | 8.9 | 59.3 | 45.8 | 524 | 121 | 402 | 71 | 807 | 280 |
| Raytown | 28.9 | 1.7 | 100.0 | 16.9 | 578 | 74 | 499 | 25 | 849 | 92 |
| St. Charles | 98.3 | 10.3 | 27.9 | 64.5 | 917 | 256 | 661 | 105 | 1,493 | 396 |
| St. Joseph | 119.4 | 10.2 | 81.3 | 66.4 | 872 | 184 | 689 | 116 | 1,520 | 374 |
| St. Louis | 919.8 | 110.5 | 55.9 | 578.0 | 1,875 | 303 | 907 | 1,030 | 3,342 | 243 |
| St. Peters | 89.6 | 7.8 | 28.3 | 46.1 | 806 | 178 | 628 | 103 | 1,801 | 372 |
| Springfield | 348.8 | 57.1 | 68.6 | 173.3 | 1,037 | 137 | 900 | 306 | 1,833 | 334 |
| University City | 36.5 | 2.2 | 100.0 | 26.3 | 765 | 139 | 626 | 36 | 1,034 | 67 |
| Wentzville | 49.6 | 5.6 | 38.8 | 30.5 | 778 | 168 | 611 | 41 | 1,056 | 246 |
| Wildwood | 16.1 | 9.2 | 15.4 | 5.8 | 164 | 0 | 164 | 15 | 415 | 176 |
| MONTANA | X | X | X | X | X | X | X | X | X | X |
| Billings | 163.2 | 26.0 | 60.8 | 46.5 | 424 | 339 | 79 | 153 | 1,391 | 332 |
| Bozeman | 64.9 | 9.1 | 100.0 | 24.7 | 526 | 482 | 43 | 62 | 1,323 | 234 |
| Butte-Silver Bow | 70.0 | 14.8 | 94.8 | 33.0 | 947 | 917 | 30 | 89 | 2,558 | 739 |
| Great Falls | 64.9 | 11.6 | 77.8 | 22.2 | 377 | 334 | 43 | 60 | 1,021 | 120 |
| Helena | 46.1 | 10.1 | 53.2 | 13.1 | 410 | 351 | 58 | 39 | 1,223 | 127 |
| Missoula | 89.1 | 23.0 | 96.4 | 40.4 | 546 | 491 | 55 | 91 | 1,226 | 110 |
| NEBRASKA | X | X | X | X | X | X | X | X | X | X |
| Bellevue | 53.8 | 6.9 | 95.9 | 31.1 | 582 | 298 | 285 | 54 | 1,013 | 147 |
| Fremont | 42.9 | 5.5 | 72.3 | 17.0 | 640 | 193 | 447 | 51 | 1,931 | 366 |
| Grand Island | 76.4 | 5.7 | 100.0 | 34.7 | 679 | 226 | 453 | 86 | 1,685 | 581 |
| Kearney | 47.4 | 7.3 | 100.0 | 18.8 | 560 | 97 | 463 | 57 | 1,683 | 597 |
| Lincoln | 350.3 | 70.6 | 35.0 | 170.9 | 600 | 250 | 350 | 374 | 1,313 | 297 |
| Omaha | 717.8 | 70.6 | 65.3 | 427.8 | 898 | 326 | 572 | 540 | 1,134 | 81 |
| NEVADA | X | X | X | X | X | X | X | X | X | X |
| Carson City | 144.0 | 47.5 | 62.8 | 47.6 | 873 | 505 | 368 | 130 | 2,379 | 315 |
| Henderson | 393.0 | 141.9 | 79.6 | 147.4 | 492 | 253 | 239 | 368 | 1,230 | 97 |
| Las Vegas | 851.6 | 399.8 | 74.1 | 224.7 | 354 | 180 | 173 | 797 | 1,254 | 161 |
| North Las Vegas | 290.7 | 116.1 | 95.7 | 93.9 | 390 | 226 | 164 | 260 | 1,081 | 128 |
| Reno | 392.3 | 90.5 | 88.5 | 146.5 | 594 | 264 | 330 | 349 | 1,413 | 235 |
| Sparks | 129.3 | 41.6 | 94.3 | 44.8 | 448 | 262 | 187 | 118 | 1,175 | 158 |
| NEW HAMPSHIRE | X | X | X | X | X | X | X | X | X | X |
| Concord | 13.7 | 0.0 | NA | 0.0 | 0 | 0 | 0 | 21 | 489 | 0 |
| Dover | 126.3 | 28.7 | 86.1 | 83.3 | 2,637 | 2,596 | 40 | 129 | 4,070 | 976 |
| Manchester | 431.5 | 144.0 | 88.3 | 198.5 | 1,769 | 1,720 | 49 | 377 | 3,356 | 80 |
| Nashua | 331.6 | 71.6 | 99.0 | 215.4 | 2,421 | 2,403 | 17 | 291 | 3,274 | 188 |
| Rochester | 114.2 | 36.3 | 94.0 | 52.6 | 1,695 | 1,512 | 183 | 75 | 2,429 | 232 |
| NEW JERSEY | X | X | X | X | X | X | X | X | X | X |
| Atlantic City | 263.3 | 125.2 | 79.9 | 118.3 | 3,131 | 3,008 | 123 | 214 | 5,653 | 178 |
| Bayonne | 234.7 | 90.2 | 87.3 | 88.4 | 1,352 | 1,310 | 42 | 278 | 4,257 | 70 |
| Bergenfield | 35.9 | 2.6 | 90.5 | 32.3 | 1,177 | 1,136 | 41 | 34 | 1,226 | 133 |
| Bridgeton | 35.6 | 8.1 | 61.2 | 14.5 | 591 | 549 | 42 | 34 | 1,386 | 33 |
| Camden | 262.1 | 21.3 | 3.7 | 48.4 | 656 | 632 | 24 | 181 | 2,451 | 0 |
| Clifton | 122.3 | 14.3 | 80.2 | 89.7 | 1,052 | 1,005 | 48 | 112 | 1,318 | 185 |
| East Orange | 388.8 | 267.5 | 93.5 | 101.3 | 1,575 | 1,535 | 40 | 404 | 6,278 | 121 |
| Elizabeth | 307.8 | 58.5 | 62.1 | 189.2 | 1,476 | 1,323 | 153 | 277 | 2,160 | 245 |
| Englewood | 90.4 | 27.5 | 70.3 | 54.6 | 1,907 | 1,838 | 69 | 147 | 5,129 | 455 |
| Fair Lawn | 49.7 | 4.2 | 97.2 | 43.3 | 1,308 | 1,247 | 61 | 43 | 1,300 | 129 |
| Fort Lee | 83.6 | 9.3 | 24.7 | 68.8 | 1,844 | 1,739 | 105 | 81 | 2,174 | 186 |

1. Based on population estimated as of July 1 of the year shown.

City government finances, 2017 (cont.)

General expenditure (cont.)

Percent of total for:

| City | Public welfare | Highways | Parking facilities | Education | Health and hospitals | Police protection | Sewerage and sanitation | Parks and recreation | Housing and community development | Interest on debt |
|---|---|---|---|---|---|---|---|---|---|---|
| | 127 | 128 | 129 | 130 | 131 | 132 | 133 | 134 | 135 | 136 |
| MISSOURI | X | X | X | X | X | X | X | X | X | X |
| Ballwin | 0.0 | 25.9 | 0.0 | 0.0 | 0.0 | 26.0 | 1.6 | 27.0 | 1.7 | 10.1 |
| Blue Springs | 0.0 | 7.7 | 0.0 | 0.0 | 0.6 | 32.6 | 13.3 | 11.2 | 3.4 | 1.4 |
| Cape Girardeau | 0.0 | 10.7 | 0.0 | 0.0 | 0.0 | 11.8 | 31.8 | 9.5 | 0.0 | 0.0 |
| Chesterfield | 0.0 | 0.0 | 0.0 | 0.0 | 3.4 | 26.6 | 0.0 | 17.8 | 0.9 | 3.2 |
| Columbia | 0.6 | 13.3 | 1.7 | 0.0 | 2.3 | 11.3 | 25.5 | 12.3 | 3.1 | 0.7 |
| Florissant | 0.0 | 12.8 | 0.0 | 0.0 | 0.6 | 36.1 | 0.7 | 15.9 | 5.2 | 0.0 |
| Gladstone | 0.0 | 16.9 | 0.0 | 0.0 | 0.0 | 17.8 | 16.7 | 8.8 | 0.0 | 3.7 |
| Hazelwood | 0.0 | 0.1 | 0.0 | 0.0 | 1.8 | 21.8 | 0.4 | 7.4 | 3.0 | 10.2 |
| Independence | 0.0 | 5.2 | 0.0 | 0.0 | 0.0 | 21.6 | 18.5 | 18.7 | 0.0 | 3.0 |
| Jefferson City | 0.0 | 8.6 | 0.0 | 0.0 | 0.0 | 15.9 | 19.7 | 4.0 | 5.2 | 1.1 |
| Joplin | 0.0 | 21.5 | 0.0 | 0.0 | 2.9 | 8.0 | 19.5 | 4.0 | 5.2 | 4.6 |
| Kansas City | 0.7 | 4.3 | 0.2 | 0.0 | 3.2 | 13.1 | 13.0 | 4.2 | 3.3 | 0.0 |
| Kirkwood | 0.0 | 12.5 | 0.0 | 0.0 | 0.0 | 13.9 | 11.2 | 5.7 | 2.7 | 2.2 |
| Lee's Summit | 0.0 | 11.7 | 0.0 | 0.0 | 0.4 | 12.3 | 43.0 | 7.8 | 8.1 | 0.9 |
| Liberty | 0.0 | 10.2 | 0.0 | 0.0 | 0.0 | 17.6 | 4.6 | 49.2 | 0.5 | 1.7 |
| Maryland Heights | 0.0 | 13.7 | 0.0 | 0.0 | 0.0 | 31.5 | 14.3 | 12.7 | 0.4 | 0.0 |
| O'Fallon | 0.0 | 19.4 | 0.0 | 0.0 | 0.0 | 30.6 | 19.3 | 6.3 | 3.0 | 9.2 |
| Raytown | 0.0 | 6.3 | 0.0 | 0.0 | 1.1 | 18.0 | 13.9 | 4.9 | 0.0 | 8.7 |
| St. Charles | 0.7 | 14.6 | 0.3 | 0.0 | 3.1 | 15.5 | 23.8 | 7.2 | 0.2 | 5.8 |
| St. Joseph | 0.0 | 12.8 | 3.4 | 0.0 | 5.5 | 18.5 | 1.8 | 2.5 | 2.1 | 8.5 |
| St. Louis | 0.0 | 2.4 | 0.9 | 0.0 | 5.5 | 12.6 | 26.8 | 15.8 | 0.2 | 0.0 |
| St. Peters | 0.0 | 13.3 | 0.0 | 0.0 | 0.7 | 24.9 | 11.3 | 11.0 | 1.1 | 3.3 |
| Springfield | 0.5 | 14.1 | 0.1 | 0.0 | 3.0 | 25.1 | 11.3 | 5.6 | 5.9 | 0.2 |
| University City | 0.0 | 18.3 | 0.3 | 0.0 | 0.0 | 18.3 | 21.3 | 10.5 | 0.0 | 6.5 |
| Wentzville | 0.0 | 17.3 | 0.0 | 0.0 | 0.1 | 21.4 | 0.0 | 16.8 | 0.0 | 1.1 |
| Wildwood | 0.0 | 42.0 | 0.0 | 0.0 | 0.0 | X | X | X | X | X |
| MONTANA | X | X | X | X | X | X | X | X | X | X |
| Billings | 0.0 | 22.7 | 1.2 | 0.0 | 0.5 | 14.5 | 22.0 | 4.3 | 3.7 | 1.8 |
| Bozeman | 12.5 | 9.6 | 1.2 | 0.0 | 0.0 | 20.5 | 15.3 | 0.8 | 0.4 | 2.3 |
| Butte-Silver Bow | 0.4 | 7.4 | 0.3 | 0.0 | 5.4 | 9.9 | 30.0 | 3.7 | 3.3 | 1.6 |
| Great Falls | 0.0 | 9.3 | 0.9 | 0.0 | 0.0 | 21.3 | 25.8 | 6.9 | 2.6 | 1.6 |
| Helena | 0.0 | 11.3 | 2.2 | 0.0 | 0.6 | 19.3 | 21.6 | 12.6 | 0.1 | 0.8 |
| Missoula | 0.2 | 9.3 | 2.1 | 0.0 | 2.1 | 19.4 | 7.6 | 8.5 | 3.6 | 2.2 |
| NEBRASKA | X | X | X | X | X | X | X | X | X | X |
| Bellevue | 0.3 | 10.0 | 0.0 | 0.0 | 0.3 | 26.8 | 22.5 | 4.8 | 1.0 | 3.6 |
| Fremont | 0.0 | 29.4 | 0.0 | 0.0 | 0.0 | 11.4 | 24.6 | 11.0 | 0.0 | 5.5 |
| Grand Island | 0.0 | 1.7 | 0.0 | 0.0 | 0.0 | 17.1 | 26.9 | 5.5 | 3.9 | 3.8 |
| Kearney | 0.5 | 23.4 | 0.0 | 0.0 | 0.4 | 10.7 | 19.6 | 20.7 | 1.6 | 1.1 |
| Lincoln | 2.2 | 16.8 | 1.7 | 0.0 | 7.8 | 11.0 | 10.9 | 4.6 | 1.3 | 13.4 |
| Omaha | 0.5 | 14.7 | 0.4 | 0.0 | 0.6 | 25.2 | 10.5 | 5.1 | 0.1 | 9.8 |
| NEVADA | X | X | X | X | X | X | X | X | X | X |
| Carson City | 5.0 | 11.1 | 0.0 | 0.0 | 7.2 | 14.7 | 5.9 | 7.8 | 1.6 | 3.3 |
| Henderson | 0.0 | 3.2 | 0.0 | 0.0 | 0.0 | 22.8 | 8.7 | 12.7 | 4.3 | 2.5 |
| Las Vegas | 0.0 | 12.1 | 1.0 | 0.0 | 0.5 | 18.8 | 7.7 | 7.9 | 2.2 | 4.3 |
| North Las Vegas | 0.0 | 6.4 | 0.0 | 0.0 | 0.0 | 32.5 | 6.9 | 5.5 | 1.7 | 8.5 |
| Reno | 0.0 | 5.0 | 0.2 | 0.0 | 0.0 | 18.2 | 17.5 | 3.3 | 3.2 | 4.6 |
| Sparks | 0.0 | 7.9 | 0.0 | 0.0 | 0.0 | 22.3 | 17.3 | 7.2 | 1.1 | 9.9 |
| NEW HAMPSHIRE | X | X | X | X | X | X | X | X | X | X |
| Concord | 0.0 | 0.0 | 5.0 | 0.0 | 0.0 | 0.0 | 0.0 | 19.2 | 0.0 | 12.0 |
| Dover | 0.6 | 5.5 | 0.6 | 65.0 | 0.0 | 6.4 | 0.7 | 1.6 | 0.0 | 8.3 |
| Manchester | 0.2 | 4.3 | 0.6 | 46.8 | 0.7 | 6.1 | 3.4 | 1.7 | 0.0 | 4.9 |
| Nashua | 0.3 | 5.4 | 0.0 | 54.5 | 0.4 | 9.9 | 6.0 | 2.1 | 0.9 | 5.6 |
| Rochester | 0.2 | 6.2 | 0.0 | 86.1 | 0.0 | 0.0 | 0.0 | 1.0 | 0.0 | 4.0 |
| NEW JERSEY | X | X | X | X | X | X | X | X | X | X |
| Atlantic City | 0.5 | 0.9 | 0.0 | 0.0 | 0.6 | 17.3 | 1.0 | 1.7 | 10.4 | 6.1 |
| Bayonne | 0.0 | 0.8 | 0.4 | 51.9 | 0.3 | 8.3 | 2.0 | 1.9 | 6.2 | 3.1 |
| Bergenfield | 0.0 | 4.7 | 0.0 | 0.0 | 1.5 | 22.1 | 16.9 | 2.2 | 0.2 | 0.3 |
| Bridgeton | 0.0 | 4.4 | 0.0 | 0.0 | 0.6 | 16.3 | 23.8 | 2.2 | 13.0 | 0.5 |
| Camden | 0.0 | 5.1 | 0.0 | 0.0 | 0.0 | 37.9 | 7.0 | 1.7 | 0.0 | 0.1 |
| Clifton | 0.0 | 2.6 | 0.0 | 0.0 | 1.1 | 18.5 | 11.1 | 1.4 | 2.9 | 2.2 |
| East Orange | 0.0 | 1.1 | 0.1 | 62.7 | 1.8 | 6.2 | 2.8 | 1.0 | 5.4 | 0.4 |
| Elizabeth | 0.1 | 4.9 | 1.5 | 0.0 | 1.6 | 16.7 | 12.2 | 3.4 | 9.6 | 1.6 |
| Englewood | 0.0 | 0.8 | 0.0 | 51.8 | 0.5 | 9.0 | 4.4 | 1.0 | 5.8 | 1.0 |
| Fair Lawn | 0.1 | 7.0 | 0.0 | 0.0 | 0.6 | 20.9 | 12.4 | 5.4 | 0.0 | 2.5 |
| Fort Lee | 0.4 | 3.0 | 2.7 | 0.0 | 2.7 | 18.1 | 3.5 | 1.9 | 8.4 | 2.2 |

| City | City government finances, 2017 (cont.) | | | Climate[2] | | | | | | |
| | Debt outstanding | | Debt issued during year | Average daily temperature | | | | Annual precipitation (inches) | Heating degree days | Cooling degree days |
| | Total (mil dol) | Per capita[1] (dollars) | | Mean | | Limits | | | | |
| | | | | January | July | January[3] | July[4] | | | |
| | 137 | 138 | 139 | 140 | 141 | 142 | 143 | 144 | 145 | 146 |
|---|---|---|---|---|---|---|---|---|---|---|
| MISSOURI | X | X | X | X | X | X | X | X | X | X |
| Ballwin | 16.6 | 548 | 0.0 | 27.5 | 78.1 | 17.3 | 89.2 | 38.00 | 5,199 | 1,293 |
| Blue Springs | 123.5 | 2,252 | 0.0 | 24.6 | 76.6 | 14.9 | 87.2 | 41.18 | 5,623 | 1,137 |
| Cape Girardeau | 40.1 | 1,023 | 0.0 | 32.4 | 79.5 | 24.0 | 90.1 | 46.54 | 4,344 | 1,515 |
| Chesterfield | 43.8 | 920 | 0.0 | 27.5 | 78.1 | 17.3 | 89.2 | 38.00 | 5,199 | 1,293 |
| Columbia | 365.2 | 3,005 | 24.7 | 27.8 | 77.4 | 18.2 | 88.6 | 40.28 | 5,177 | 1,246 |
| Florissant | 16.4 | 320 | 0.0 | 29.6 | 80.2 | 21.2 | 89.8 | 38.75 | 4,758 | 1,561 |
| Gladstone | 41.9 | 1,545 | 1.2 | 29.3 | 81.3 | 20.7 | 90.5 | 35.51 | 4,734 | 1,676 |
| Hazelwood | 32.2 | 1,273 | 0.9 | 29.6 | 80.2 | 21.2 | 89.8 | 38.75 | 4,758 | 1,561 |
| Independence | 464.8 | 3,966 | 0.0 | 26.6 | 77.1 | 17.1 | 87.5 | 43.14 | 5,373 | 1,176 |
| Jefferson City | 88.2 | 2,055 | 15.6 | 28.2 | 77.9 | 17.7 | 89.4 | 39.59 | 5,158 | 1,261 |
| Joplin | 37.6 | 744 | 3.3 | 33.1 | 79.9 | 23.7 | 90.4 | 46.07 | 4,253 | 1,555 |
| Kansas City | 3,583.8 | 7,339 | 363.8 | 29.3 | 81.3 | 20.7 | 90.5 | 35.51 | 4,734 | 1,676 |
| Kirkwood | 15.6 | 562 | 1.5 | 29.6 | 80.2 | 21.2 | 89.8 | 38.75 | 4,758 | 1,561 |
| Lee's Summit | 62.7 | 645 | 0.0 | 24.6 | 76.6 | 14.9 | 87.2 | 41.18 | 5,623 | 1,137 |
| Liberty | 0.2 | 5 | 0.0 | 26.6 | 77.1 | 17.1 | 87.5 | 43.14 | 5,373 | 1,176 |
| Maryland Heights | 23.5 | 871 | 0.0 | 29.5 | 80.7 | 21.2 | 90.5 | 38.84 | 4,650 | 1,633 |
| O'Fallon | 207.8 | 2,374 | 47.1 | 28.3 | 79.0 | 19.0 | 90.2 | 38.28 | 5,020 | 1,399 |
| Raytown | 51.5 | 1,757 | 0.0 | 24.6 | 76.6 | 14.9 | 87.2 | 41.18 | 5,623 | 1,137 |
| St. Charles | 209.1 | 2,970 | 3.2 | 27.5 | 78.1 | 17.3 | 89.2 | 38.00 | 5,199 | 1,293 |
| St. Joseph | 726.2 | 9,545 | 15.6 | 26.4 | 78.7 | 15.9 | 89.9 | 35.24 | 5,345 | 1,339 |
| St. Louis | 1,902.6 | 6,173 | 426.3 | 29.5 | 80.7 | 21.2 | 90.5 | 38.84 | 4,650 | 1,633 |
| St. Peters | 126.3 | 2,208 | 24.2 | 28.3 | 79.0 | 19.0 | 90.2 | 38.28 | 5,020 | 1,399 |
| Springfield | 896.0 | 5,361 | 0.0 | 31.7 | 78.5 | 21.8 | 89.9 | 44.97 | 4,602 | 1,366 |
| University City | 3.4 | 98 | 0.0 | 29.5 | 80.7 | 21.2 | 90.5 | 38.84 | 4,650 | 1,633 |
| Wentzville | 62.0 | 1,584 | 0.0 | NA | NA | NA | NA | NA | NA | NA |
| Wildwood | 1.3 | 38 | 0.0 | 27.5 | 78.1 | 17.3 | 89.2 | 38.00 | 5,199 | 1,293 |
| MONTANA | X | X | X | X | X | X | X | X | X | X |
| Billings | 198.4 | 1,807 | 64.0 | 24.0 | 72.0 | 15.1 | 85.8 | 14.77 | 7,006 | 583 |
| Bozeman | 36.7 | 782 | 0.0 | 22.6 | 65.3 | 12.0 | 81.8 | 16.45 | 7,984 | 216 |
| Butte-Silver Bow | 41.6 | 1,194 | 0.0 | 17.6 | 62.7 | 5.4 | 79.8 | 12.78 | 9,399 | 127 |
| Great Falls | 57.9 | 984 | 14.7 | 21.7 | 66.2 | 11.3 | 82.0 | 14.89 | 7,828 | 288 |
| Helena | 23.5 | 733 | 14.2 | 20.2 | 67.8 | 9.9 | 83.4 | 11.32 | 7,975 | 277 |
| Missoula | 66.5 | 898 | 0.0 | 23.5 | 66.9 | 16.2 | 83.6 | 13.82 | 7,622 | 256 |
| NEBRASKA | X | X | X | X | X | X | X | X | X | X |
| Bellevue | 52.5 | 984 | 11.4 | 21.7 | 76.7 | 11.6 | 87.4 | 30.22 | 6,311 | 1,095 |
| Fremont | 71.0 | 2,679 | 0.2 | 21.1 | 76.2 | 10.4 | 87.7 | 29.80 | 6,444 | 1,004 |
| Grand Island | 86.2 | 1,687 | 2.6 | 22.4 | 75.8 | 12.2 | 87.1 | 25.89 | 6,385 | 1,027 |
| Kearney | 55.8 | 1,661 | 14.0 | 22.4 | 74.7 | 11.0 | 85.7 | 25.20 | 6,652 | 852 |
| Lincoln | 1,368.8 | 4,811 | 174.3 | 22.4 | 77.8 | 11.5 | 89.6 | 28.37 | 6,242 | 1,154 |
| Omaha | 1,386.7 | 2,912 | 175.4 | 21.7 | 76.7 | 11.6 | 87.4 | 30.22 | 6,311 | 1,095 |
| NEVADA | X | X | X | X | X | X | X | X | X | X |
| Carson City | 277.0 | 5,079 | 31.3 | 33.7 | 70.0 | 21.7 | 89.2 | 10.36 | 5,661 | 419 |
| Henderson | 230.9 | 772 | 12.7 | 47.0 | 91.2 | 36.8 | 104.1 | 4.49 | 2,239 | 3,214 |
| Las Vegas | 797.5 | 1,255 | 22.9 | 47.0 | 91.2 | 36.8 | 104.1 | 4.49 | 2,239 | 3,214 |
| North Las Vegas | 415.5 | 1,728 | 0.0 | 47.0 | 91.2 | 36.8 | 104.1 | 4.49 | 2,239 | 3,214 |
| Reno | 621.2 | 2,519 | 48.5 | 33.6 | 71.3 | 21.8 | 91.2 | 7.48 | 5,600 | 493 |
| Sparks | 198.9 | 1,988 | 56.3 | 33.6 | 71.3 | 21.8 | 91.2 | 7.48 | 5,600 | 493 |
| NEW HAMPSHIRE | X | X | X | X | X | X | X | X | X | X |
| Concord | 46.1 | 1,066 | 0.0 | 20.1 | 70.0 | 9.7 | 82.9 | 37.60 | 7,478 | 442 |
| Dover | 165.5 | 5,238 | 10.1 | 23.3 | 70.7 | 13.1 | 83.2 | 42.80 | 6,748 | 427 |
| Manchester | 437.7 | 3,899 | 21.0 | 18.8 | 68.4 | 5.2 | 82.1 | 39.82 | 7,742 | 263 |
| Nashua | 189.4 | 2,128 | 13.3 | 22.8 | 70.8 | 12.1 | 82.5 | 45.43 | 6,834 | 445 |
| Rochester | 39.3 | 1,265 | 0.0 | 23.3 | 70.7 | 13.1 | 83.2 | 42.80 | 6,748 | 427 |
| NEW JERSEY | X | X | X | X | X | X | X | X | X | X |
| Atlantic City | 188.2 | 4,979 | 0.0 | 35.2 | 75.2 | 29.0 | 80.6 | 38.37 | 4,480 | 951 |
| Bayonne | 238.9 | 3,657 | 1.7 | 31.3 | 77.2 | 24.4 | 85.2 | 46.25 | 4,843 | 1,220 |
| Bergenfield | 8.9 | 325 | 0.0 | 28.6 | 75.0 | 19.5 | 85.5 | 51.50 | 5,522 | 824 |
| Bridgeton | 14.3 | 584 | 0.0 | NA | NA | NA | NA | NA | NA | NA |
| Camden | 3.2 | 44 | 0.0 | 32.3 | 76.3 | 23.2 | 87.8 | 48.25 | 4,801 | 1,054 |
| Clifton | 84.9 | 995 | 15.8 | 28.6 | 75.0 | 19.5 | 85.5 | 51.50 | 5,522 | 824 |
| East Orange | 78.6 | 1,223 | 0.0 | 31.3 | 77.2 | 24.4 | 85.2 | 46.25 | 4,843 | 1,220 |
| Elizabeth | 183.8 | 1,434 | 49.1 | 29.6 | 74.5 | 19.8 | 85.7 | 50.94 | 5,450 | 787 |
| Englewood | 71.7 | 2,504 | 1.5 | 29.6 | 75.3 | 22.7 | 82.5 | 46.33 | 5,367 | 882 |
| Fair Lawn | 34.7 | 1,049 | 18.9 | 28.6 | 75.0 | 19.5 | 85.5 | 51.50 | 5,522 | 824 |
| Fort Lee | 57.6 | 1,543 | 11.6 | 29.6 | 75.3 | 22.7 | 82.5 | 46.33 | 5,367 | 882 |

1. Based on the population estimated as of July 1 of the year shown.  2. Represents normal values based on the 30-year period, 1971±2000.  3. Average daily minimum.  4. Average daily maximum.

Items 137—146

| STATE Place code | City | Land area¹ (sq. mi) | Total persons 2019 | Rank | Per square mile | White | Black or African American | American Indian, Alaskan Native | Asian | Hawaiian Pacific Islander | Some other race | Two or more races (percent) |
|---|---|---|---|---|---|---|---|---|---|---|---|---|
| | | Population, 2020 | | | | Race 2019 — Race alone² (percent) | | | | | | |
| | | 1 | 2 | 3 | 4 | 5 | 6 | 7 | 8 | 9 | 10 | 11 |
| | **NEW JERSEY—Cont'd** | | | | | | | | | | | |
| 34 25770 | Garfield | 2.1 | 31,641 | 1,206 | 15,067 | 87.9 | 5.6 | 0.0 | 3.9 | 0.0 | 2.5 | 4.0 |
| 34 28680 | Hackensack | 4.2 | 43,981 | 876 | 10,472 | 54.1 | 19.8 | 0.2 | 17.3 | 0.0 | 1.6 | 6.1 |
| 34 32250 | Hoboken | 1.3 | 53,081 | 736 | 40,832 | 71.8 | 5.3 | 0.0 | 15.3 | 0.0 | 7.0 | 5.0 |
| 34 36000 | Jersey City | 14.7 | 262,664 | 82 | 17,868 | 39.6 | 23.1 | 0.5 | 24.9 | 0.0 | 13.6 | 3.4 |
| 34 36510 | Kearny | 8.8 | 40,507 | 952 | 4,603 | 75.2 | 3.8 | 0.3 | 3.6 | 0.0 | 6.1 | 5.6 |
| 34 40350 | Linden | 10.7 | 42,076 | 915 | 3,932 | 56.1 | 26.4 | 0.0 | 5.8 | 0.0 | 14.3 | 4.5 |
| 34 41310 | Long Branch | 5.1 | 30,210 | 1,246 | 5,924 | 62.6 | 18.0 | 0.0 | 0.6 | 0.0 | 1.3 | 4.3 |
| 34 46680 | Millville | 42.0 | 27,025 | 1,351 | 644 | 82.1 | 7.7 | 1.1 | 3.5 | 0.0 | 12.2 | 2.8 |
| 34 51000 | Newark | 24.1 | 282,520 | 74 | 11,723 | 33.4 | 49.9 | 0.2 | 1.5 | 0.0 | 13.0 | 3.2 |
| 34 51210 | New Brunswick | 5.2 | 56,182 | 683 | 10,804 | 61.6 | 13.6 | 0.0 | 8.5 | 0.0 | 13.0 | 2.5 |
| 34 55950 | Paramus | 10.4 | 26,063 | 1,383 | 2,506 | 65.1 | 2.6 | 0.0 | 29.8 | 0.0 | 2.4 | 2.5 |
| 34 56550 | Passaic | 3.1 | 69,340 | 530 | 22,368 | 73.5 | 7.4 | 1.0 | 2.4 | 0.1 | 13.1 | 1.6 |
| 34 57000 | Paterson | 8.4 | 144,801 | 184 | 17,238 | 29.7 | 26.0 | 0.0 | 6.6 | 0.0 | 36.1 | 4.8 |
| 34 58200 | Perth Amboy | 4.7 | 51,309 | 759 | 10,917 | 80.3 | 5.4 | 0.2 | 1.2 | 0.0 | 8.1 | 3.9 |
| 34 59190 | Plainfield | 6.0 | 50,011 | 777 | 8,335 | 14.1 | 37.7 | 0.0 | 1.3 | 0.0 | 43.1 | 2.9 |
| 34 61530 | Rahway | 3.9 | 29,832 | 1,257 | 7,649 | 59.2 | 21.8 | 0.0 | 10.6 | 0.0 | 5.5 | 2.0 |
| 34 65790 | Sayreville | 15.8 | 43,855 | 882 | 2,776 | 63.2 | 15.3 | 0.1 | 18.8 | 0.0 | 0.6 | 3.0 |
| 34 74000 | Trenton | 7.6 | 82,957 | 417 | 10,915 | 29.4 | 50.8 | 0.1 | 2.0 | 0.0 | 14.8 | 4.8 |
| 34 74630 | Union City | 1.3 | 67,137 | 550 | 51,644 | 57.8 | 6.8 | 1.3 | 4.9 | 0.0 | 24.4 | 6.6 |
| 34 76070 | Vineland | 68.4 | 58,615 | 649 | 857 | 64.3 | 20.4 | 0.0 | 0.8 | 0.0 | 8.0 | 4.8 |
| 34 79040 | Westfield | 6.7 | 29,238 | 1,282 | 4,364 | 80.4 | 4.4 | 0.6 | 6.9 | 0.0 | 3.0 | 1.5 |
| 34 79610 | West New York | 1.0 | 53,132 | 734 | 53,132 | 73.9 | 2.3 | 0.0 | 5.0 | 0.0 | 17.3 | 3.5 |
| 35 00000 | **NEW MEXICO** | 121,312.2 | 2,106,319 | X | 17 | 73.9 | 2.3 | 9.5 | 1.7 | 0.0 | 9.0 | 3.5 |
| 35 01780 | Alamogordo | 21.4 | 32,126 | 1,193 | 1,501 | 81.5 | 4.8 | 2.4 | 0.8 | 0.0 | 4.8 | 5.6 |
| 35 02000 | Albuquerque | 187.3 | 562,540 | 32 | 3,003 | 74.3 | 3.2 | 4.9 | 3.4 | 0.0 | 9.8 | 4.3 |
| 35 12150 | Carlsbad | 31.5 | 29,664 | 1,260 | 942 | 85.3 | 1.6 | 0.8 | 1.1 | 0.2 | 6.7 | 7.1 |
| 35 16420 | Clovis | 23.6 | 38,091 | 1,019 | 1,614 | 63.4 | 6.7 | 3.4 | 0.7 | 0.0 | 20.7 | 6.3 |
| 35 25800 | Farmington | 34.5 | 44,156 | 870 | 1,280 | 58.1 | 1.1 | 31.1 | 2.9 | 0.0 | 4.3 | 0.7 |
| 35 32520 | Hobbs | 26.4 | 39,339 | 986 | 1,490 | 85.4 | 5.5 | 1.2 | 2.1 | 0.2 | 9.3 | 3.0 |
| 35 39380 | Las Cruces | 76.9 | 105,096 | 294 | 1,367 | 80.6 | 2.8 | 1.9 | 2.8 | 0.0 | 6.0 | 3.5 |
| 35 63460 | Rio Rancho | 103.4 | 99,885 | 317 | 966 | 80.4 | 3.5 | 3.7 | 1.9 | 0.0 | 7.6 | 1.2 |
| 35 64930 | Roswell | 29.7 | 47,596 | 814 | 1,603 | 86.0 | 2.7 | 0.6 | 1.9 | 0.0 | 6.9 | 3.0 |
| 35 70500 | Santa Fe | 52.2 | 84,996 | 401 | 1,628 | 84.9 | 1.1 | 2.3 | 1.9 | 0.0 | 8.6 | 3.3 |
| 36 00000 | **NEW YORK** | 47,123.8 | 19,336,776 | X | 410 | 63.2 | 15.9 | 0.4 | 8.6 | 0.0 | 2.7 | 9.5 |
| 36 01000 | Albany | 21.4 | 95,429 | 338 | 4,459 | 57.2 | 23.9 | 0.3 | 6.4 | 0.0 | 1.2 | 7.2 |
| 36 03078 | Auburn | 8.3 | 25,915 | 1,391 | 3,122 | 82.3 | 8.1 | 0.0 | 1.3 | 0.0 | 1.8 | 8.4 |
| 36 06607 | Binghamton | 10.5 | 44,137 | 871 | 4,204 | 70.1 | 12.8 | 0.0 | 6.9 | 0.0 | 6.9 | 3.5 |
| 36 11000 | Buffalo | 40.4 | 254,479 | 89 | 6,299 | 45.0 | 37.1 | 0.5 | 7.0 | 0.0 | 0.3 | 7.4 |
| 36 24229 | Elmira | 7.3 | 27,045 | 1,350 | 3,705 | 77.3 | 13.9 | 0.4 | 0.7 | 0.0 | 1.4 | 6.3 |
| 36 27485 | Freeport | 4.6 | 42,705 | 901 | 9,284 | 38.8 | 32.5 | 0.4 | 0.5 | 0.0 | 15.0 | 12.9 |
| 36 29113 | Glen Cove | 6.7 | 27,005 | 1,352 | 4,031 | 83.1 | 3.0 | 0.1 | 4.6 | 0.0 | 8.3 | 1.5 |
| 36 32402 | Harrison | 16.8 | 28,959 | 1,289 | 1,724 | 85.1 | 2.9 | 0.0 | 8.3 | 0.0 | 2.1 | 5.2 |
| 36 33139 | Hempstead | 3.7 | 54,749 | 705 | 14,797 | 24.5 | 43.6 | 0.1 | 2.8 | 0.0 | 24.0 | 5.0 |
| 36 38077 | Ithaca | 5.4 | 30,981 | 1,218 | 5,737 | 68.0 | 3.5 | 0.1 | 17.0 | 0.0 | 3.4 | 8.0 |
| 36 38264 | Jamestown | 8.9 | 28,862 | 1,291 | 3,243 | 86.9 | 2.6 | 0.0 | 0.0 | 0.0 | 6.8 | 3.8 |
| 36 42554 | Lindenhurst | 3.7 | 26,727 | 1,363 | 7,224 | 80.5 | 5.6 | 0.0 | 4.8 | 0.0 | 6.1 | 3.0 |
| 36 43335 | Long Beach | 2.2 | 33,303 | 1,155 | 15,138 | 80.5 | 8.1 | 0.3 | 4.6 | 0.0 | 4.8 | 1.7 |
| 36 47042 | Middletown | 5.3 | 28,023 | 1,313 | 5,287 | 53.6 | 18.1 | 0.0 | 0.0 | 0.0 | 13.6 | 14.7 |
| 36 49121 | Mount Vernon | 4.4 | 66,775 | 560 | 15,176 | 18.5 | 66.6 | 2.4 | 2.1 | D | D | D |
| 36 50034 | Newburgh | 3.8 | 27,902 | 1,319 | 7,343 | D | D | D | D | D | D | D |
| 36 50617 | New Rochelle | 10.3 | 80,973 | 432 | 7,862 | 58.0 | 17.2 | 0.0 | 5.8 | 0.0 | 16.0 | 2.9 |
| 36 51000 | New York | 300.5 | 8,253,213 | 1 | 27,465 | 42.4 | 24.7 | 0.4 | 14.4 | 0.0 | 14.3 | 3.8 |
| 36 51055 | Niagara Falls | 14.1 | 47,392 | 821 | 3,361 | 71.1 | 17.5 | 2.9 | 1.3 | 0.0 | 0.8 | 6.4 |
| 36 53682 | North Tonawanda | 10.1 | 30,050 | 1,252 | 2,975 | D | D | D | D | D | D | D |
| 36 55530 | Ossining | 3.2 | 24,647 | 1,413 | 7,702 | 61.7 | 11.3 | 0.0 | 2.9 | 0.0 | 23.1 | 1.0 |
| 36 59223 | Port Chester | 2.3 | 28,857 | 1,292 | 12,547 | 58.8 | 6.0 | 0.0 | 4.1 | 0.0 | 26.4 | 4.7 |
| 36 59641 | Poughkeepsie | 5.1 | 30,447 | 1,234 | 5,970 | 47.2 | 43.3 | 0.8 | 0.8 | 0.0 | 1.6 | 6.3 |
| 36 63000 | Rochester | 35.8 | 205,225 | 114 | 5,733 | 46.9 | 40.6 | 0.3 | 2.7 | 0.5 | 4.9 | 4.1 |
| 36 63418 | Rome | 74.9 | 32,145 | 1,192 | 429 | 88.3 | 5.8 | 0.1 | 1.6 | 0.0 | 0.6 | 3.6 |
| 36 65255 | Saratoga Springs | 28.1 | 28,287 | 1,306 | 1,007 | 89.7 | 4.0 | 0.0 | 3.8 | 0.4 | 0.3 | 2.1 |
| 36 65508 | Schenectady | 10.8 | 65,231 | 578 | 6,040 | 53.8 | 21.6 | 0.2 | 5.9 | D | 9.0 | 9.2 |
| 36 70420 | Spring Valley | 2.0 | 32,258 | 1,184 | 16,129 | D | D | D | D | D | D | D |
| 36 73000 | Syracuse | 25.1 | 141,229 | 192 | 5,627 | 52.5 | 31.8 | 1.6 | 6.7 | 0.0 | 2.3 | 5.1 |
| 36 75484 | Troy | 10.4 | 48,804 | 799 | 4,693 | 67.7 | 19.5 | 0.0 | 4.8 | 0.0 | 2.6 | 5.4 |
| 36 76540 | Utica | 16.7 | 59,253 | 638 | 3,548 | 57.6 | 15.0 | 0.3 | 11.7 | 0.0 | 3.7 | 11.7 |
| 36 76705 | Valley Stream | 3.5 | 37,196 | 1,031 | 10,627 | 33.8 | 27.2 | 0.0 | 6.2 | 0.0 | 23.0 | 9.7 |
| 36 78608 | Watertown | 9.0 | 24,330 | 1,417 | 2,703 | 86.4 | 6.0 | 0.2 | 3.5 | 0.0 | 0.8 | 3.1 |
| 36 81677 | White Plains | 9.7 | 58,286 | 658 | 6,009 | 58.4 | 11.1 | 0.0 | 8.3 | 0.0 | 20.7 | 1.6 |

1. Dry land or land partially or temporarily covered by water.  2. Hispanic or Latino persons may be of any race.

# Table D. Cities — **Population**

| City | Percent Hispanic or Latino[1], 2019 | Percent foreign born, 2019 | Age of population (percent), 2019 — Under 18 years | 18 to 24 years | 25 to 34 years | 35 to 44 years | 45 to 54 years | 55 to 64 years | 65 years and over | Median age, 2019 | Percent female, 2019 | Population — Census counts 2000 | Census counts 2010 | Percent change 2000-2010 | Percent change 2001-2020 |
|---|---|---|---|---|---|---|---|---|---|---|---|---|---|---|---|
| | 12 | 13 | 14 | 15 | 16 | 17 | 18 | 19 | 20 | 21 | 22 | 23 | 24 | 25 | 26 |
| **NEW JERSEY—Cont'd** | | | | | | | | | | | | | | | |
| Garfield | 39.5 | 43.2 | 25.5 | 5.5 | 15.0 | 14.5 | 12.0 | 12.2 | 15.3 | 38.7 | 48.7 | 29,786 | 30,497 | 2.4 | 3.8 |
| Hackensack | 34.9 | 41.2 | 14.2 | 3.6 | 23.3 | 14.7 | 11.3 | 16.6 | 16.4 | 40.5 | 52.0 | 42,677 | 43,024 | 0.8 | 2.2 |
| Hoboken | 16.6 | 19.3 | 16.7 | 11.4 | 31.9 | 21.0 | 8.5 | 5.6 | 5.1 | 31.4 | 50.2 | 38,577 | 50,020 | 29.7 | 6.1 |
| Jersey City | 25.2 | 44.4 | 18.7 | 7.5 | 26.0 | 15.8 | 11.1 | 9.6 | 11.4 | 34.1 | 49.8 | 240,055 | 247,608 | 3.1 | 6.1 |
| Kearny | 59.2 | 48.2 | 22.2 | 5.5 | 16.9 | 15.5 | 10.0 | 14.4 | 15.5 | 38.2 | 51.2 | 40,513 | 40,711 | 0.5 | -0.5 |
| Linden | 35.4 | 32.3 | 21.0 | 9.3 | 11.4 | 15.7 | 14.4 | 14.0 | 14.2 | 40.6 | 50.0 | 39,394 | 40,533 | 2.9 | 3.8 |
| Long Branch | 27.4 | 25.9 | 24.2 | 11.7 | 13.1 | 16.1 | 9.5 | 12.4 | 12.9 | 35.6 | 54.2 | 31,340 | 30,717 | -2.0 | -1.7 |
| Millville | 17.5 | 5.9 | 22.0 | 7.7 | 10.6 | 13.5 | 14.8 | 11.4 | 20.0 | 42.9 | 55.6 | 26,847 | 28,422 | 5.9 | -4.9 |
| Newark | 36.9 | 35.4 | 24.6 | 10.4 | 15.8 | 13.7 | 13.9 | 10.9 | 10.7 | 34.6 | 50.2 | 273,546 | 276,941 | 1.2 | 2.0 |
| New Brunswick | 48.7 | 32.2 | 26.0 | 25.9 | 12.9 | 11.7 | 9.8 | 5.2 | 8.5 | 23.8 | 48.7 | 48,573 | 54,500 | 12.2 | 3.1 |
| Paramus | 13.5 | 34.4 | 15.9 | 10.1 | 9.2 | 7.2 | 12.0 | 19.4 | 26.1 | 51.9 | 51.2 | 25,737 | 26,342 | 2.4 | -1.1 |
| Passaic | 76.3 | 42.3 | 33.0 | 9.4 | 12.7 | 13.7 | 13.1 | 9.4 | 8.7 | 30.8 | 53.0 | 67,861 | 69,751 | 2.8 | -0.6 |
| Paterson | 60.7 | 46.8 | 27.8 | 9.6 | 16.3 | 12.8 | 10.9 | 11.0 | 11.6 | 32.4 | 50.5 | 149,222 | 146,184 | -2.0 | -0.9 |
| Perth Amboy | 69.3 | 37.4 | 27.5 | 8.5 | 14.6 | 13.4 | 10.7 | 12.3 | 12.9 | 34.3 | 52.6 | 47,303 | 50,827 | 7.4 | 0.9 |
| Plainfield | 43.3 | 42.1 | 25.9 | 6.4 | 17.0 | 14.2 | 13.1 | 10.2 | 13.3 | 35.9 | 48.9 | 47,829 | 49,549 | 3.6 | 0.9 |
| Rahway | 23.7 | 14.2 | 20.5 | 7.1 | 17.0 | 9.8 | 16.0 | 15.0 | 14.7 | 40.8 | 49.6 | 26,500 | 27,320 | 3.1 | 9.2 |
| Sayreville | 17.9 | 30.5 | 21.4 | 8.1 | 19.2 | 12.9 | 15.4 | 10.8 | 12.3 | 35.8 | 51.5 | 40,377 | 42,775 | 5.9 | 2.5 |
| Trenton | 35.9 | 24.7 | 27.4 | 7.7 | 15.9 | 12.9 | 11.8 | 12.6 | 11.6 | 34.5 | 53.2 | 85,403 | 84,962 | -0.5 | -2.4 |
| Union City | 78.7 | 55.4 | 20.6 | 11.6 | 13.8 | 14.1 | 18.1 | 10.0 | 11.9 | 37.8 | 50.3 | 67,088 | 66,467 | -0.9 | 1.0 |
| Vineland | 41.0 | 12.4 | 23.6 | 8.9 | 14.0 | 11.7 | 11.8 | 13.5 | 16.6 | 38.2 | 49.6 | 56,271 | 60,717 | 7.9 | -3.5 |
| Westfield | 11.0 | 13.7 | 28.8 | 6.4 | 4.0 | 14.6 | 17.0 | 14.5 | 14.7 | 42.8 | 53.0 | 29,644 | 30,296 | 2.2 | -3.5 |
| West New York | 80.4 | 55.9 | 25.9 | 5.2 | 16.0 | 19.7 | 10.0 | 8.1 | 15.1 | 36.4 | 46.1 | 45,768 | 49,686 | 8.6 | 6.9 |
| **NEW MEXICO** | 49.3 | 9.6 | 22.6 | 9.4 | 13.3 | 12.5 | 11.2 | 13.0 | 18.0 | 38.6 | 50.6 | 1,819,046 | 2,059,199 | 13.2 | 2.3 |
| Alamogordo | 32.2 | 4.7 | 19.3 | 10.9 | 12.9 | 11.4 | 12.1 | 13.3 | 20.1 | 40.2 | 50.0 | 35,582 | 30,402 | -14.6 | 5.7 |
| Albuquerque | 49.3 | 10.8 | 21.7 | 9.0 | 15.6 | 13.6 | 11.6 | 12.0 | 16.4 | 37.3 | 51.8 | 448,607 | 546,121 | 21.7 | 3.0 |
| Carlsbad | 52.8 | 7.3 | 19.8 | 11.8 | 13.8 | 13.5 | 7.9 | 16.3 | 16.9 | 38.8 | 49.5 | 25,625 | 26,245 | 2.4 | 13.0 |
| Clovis | 45.5 | 5.4 | 27.3 | 11.1 | 15.7 | 14.6 | 7.6 | 10.3 | 13.3 | 32.7 | 48.0 | 32,667 | 37,808 | 15.7 | 0.7 |
| Farmington | 24.4 | 2.9 | 28.5 | 9.1 | 12.4 | 14.3 | 10.0 | 9.4 | 16.3 | 35.0 | 52.8 | 37,844 | 45,947 | 21.4 | -3.9 |
| Hobbs | 61.6 | 17.8 | 32.0 | 11.0 | 13.3 | 14.4 | 10.2 | 10.5 | 8.7 | 28.5 | 46.2 | 28,657 | 34,051 | 18.8 | 15.5 |
| Las Cruces | 58.5 | 10.6 | 21.9 | 16.6 | 17.8 | 8.6 | 9.5 | 10.1 | 15.4 | 30.5 | 50.5 | 74,267 | 97,700 | 31.6 | 7.6 |
| Rio Rancho | 43.7 | 5.9 | 24.1 | 8.2 | 14.6 | 12.8 | 10.9 | 12.7 | 16.6 | 37.9 | 51.3 | 51,765 | 87,389 | 68.8 | 14.3 |
| Roswell | 58.5 | 13.4 | 26.5 | 9.3 | 13.4 | 11.8 | 11.4 | 12.6 | 15.0 | 35.8 | 49.0 | 45,293 | 48,421 | 6.9 | -1.7 |
| Santa Fe | 54.0 | 15.0 | 18.0 | 9.2 | 10.9 | 12.1 | 12.3 | 14.0 | 23.6 | 44.9 | 53.5 | 62,203 | 80,879 | 30.0 | 5.1 |
| **NEW YORK** | 19.3 | 22.4 | 20.7 | 9.1 | 14.7 | 12.5 | 12.7 | 13.4 | 16.9 | 39.2 | 51.4 | 18,976,457 | 19,378,117 | 2.1 | -0.2 |
| Albany | 10.6 | 14.2 | 16.6 | 22.9 | 15.9 | 11.7 | 8.9 | 10.9 | 13.1 | 30.6 | 55.1 | 95,658 | 97,836 | 2.3 | -2.5 |
| Auburn | 3.6 | 1.7 | 21.2 | 6.4 | 15.2 | 16.5 | 12.2 | 12.7 | 15.8 | 38.5 | 48.9 | 28,574 | 27,690 | -3.1 | -6.4 |
| Binghamton | 8.4 | 9.5 | 20.0 | 17.1 | 15.3 | 12.6 | 8.6 | 11.4 | 15.0 | 31.8 | 51.4 | 47,380 | 47,403 | 0.0 | -6.9 |
| Buffalo | 13.8 | 11.4 | 23.3 | 12.3 | 18.1 | 12.0 | 10.5 | 11.4 | 12.3 | 32.7 | 52.4 | 292,648 | 261,346 | -10.7 | -2.6 |
| Elmira | 4.3 | 3.5 | 25.9 | 9.4 | 15.6 | 14.5 | 10.9 | 12.7 | 11.0 | 34.6 | 46.8 | 30,940 | 29,245 | -5.5 | -7.5 |
| Freeport | 45.7 | 32.1 | 26.5 | 7.0 | 14.6 | 13.6 | 12.3 | 9.7 | 16.3 | 36.2 | 52.4 | 43,783 | 42,856 | -2.1 | -0.4 |
| Glen Cove | 28.8 | 27.9 | 17.0 | 9.7 | 8.3 | 10.9 | 13.7 | 13.6 | 26.8 | 47.7 | 54.8 | 26,622 | 26,952 | 1.2 | 0.2 |
| Harrison | 13.6 | 23.3 | 19.8 | 16.3 | 7.9 | 15.3 | 10.2 | 13.8 | 16.7 | 38.3 | 54.8 | 24,154 | 27,469 | 13.7 | 5.4 |
| Hempstead | 47.3 | 37.3 | 26.9 | 9.8 | 15.6 | 16.0 | 9.9 | 9.9 | 11.9 | 33.2 | 51.5 | 56,554 | 54,018 | -4.5 | 1.4 |
| Ithaca | 9.4 | 16.4 | 5.7 | 57.1 | 14.2 | 4.9 | 3.8 | 5.9 | 8.4 | 22.2 | 51.1 | 29,287 | 30,013 | 2.5 | 3.2 |
| Jamestown | 18.0 | 2.1 | 24.0 | 11.4 | 13.0 | 10.9 | 9.8 | 14.2 | 16.6 | 36.6 | 53.5 | 31,730 | 31,160 | -1.8 | -7.4 |
| Lindenhurst | 12.1 | 14.1 | 22.4 | 6.8 | 13.0 | 12.4 | 16.8 | 11.1 | 17.6 | 41.9 | 51.8 | 27,819 | 27,269 | -2.0 | -2.0 |
| Long Beach | 17.3 | 14.5 | 14.1 | 5.1 | 17.0 | 12.5 | 15.1 | 16.5 | 19.7 | 45.5 | 48.9 | 35,462 | 33,333 | -6.0 | -0.1 |
| Middletown | 46.8 | 13.4 | 22.6 | 11.9 | 13.0 | 7.0 | 13.7 | 11.6 | 20.2 | 38.3 | 54.4 | 25,388 | 28,126 | 10.8 | -0.4 |
| Mount Vernon | 12.9 | 28.5 | 16.0 | 8.3 | 14.4 | 11.6 | 14.3 | 14.6 | 20.8 | 44.6 | 52.9 | 68,381 | 67,324 | -1.5 | -0.8 |
| Newburgh | 51.3 | 19.4 | 28.4 | 11.9 | 14.7 | 9.2 | 14.8 | 10.1 | 10.9 | 30.5 | 51.2 | 28,259 | 28,906 | 2.3 | -3.5 |
| New Rochelle | 33.7 | 31.6 | 17.2 | 8.1 | 15.4 | 9.9 | 13.3 | 13.6 | 22.5 | 44.3 | 53.0 | 72,182 | 77,060 | 6.8 | 5.1 |
| New York | 29.1 | 36.2 | 20.6 | 8.3 | 17.8 | 13.7 | 12.3 | 12.0 | 15.4 | 37.2 | 52.3 | 8,008,278 | 8,174,930 | 2.1 | 1.0 |
| Niagara Falls | 4.2 | 3.4 | 20.5 | 7.5 | 15.0 | 13.0 | 10.1 | 16.2 | 17.8 | 39.8 | 52.0 | 55,593 | 50,031 | -10.0 | -5.3 |
| North Tonawanda | 0.1 | 6.7 | 17.5 | 8.2 | 13.8 | 7.4 | 12.6 | 19.5 | 21.1 | 47.7 | 52.9 | 33,262 | 31,574 | -5.1 | -4.8 |
| Ossining | 41.8 | 26.7 | 21.8 | 4.9 | 17.7 | 15.9 | 12.7 | 11.2 | 15.7 | 39.5 | 47.2 | 24,010 | 25,058 | 4.4 | -1.6 |
| Port Chester | 55.8 | 46.0 | 17.9 | 7.4 | 13.3 | 17.0 | 14.7 | 12.6 | 17.1 | 41.9 | 44.0 | 27,867 | 28,922 | 3.8 | -0.2 |
| Poughkeepsie | 10.3 | 10.0 | 16.8 | 9.2 | 13.3 | 12.5 | 12.6 | 14.8 | 20.9 | 44.0 | 51.6 | 29,871 | 30,812 | 3.2 | -1.2 |
| Rochester | 20.3 | 9.0 | 21.7 | 12.8 | 18.3 | 11.6 | 10.6 | 12.1 | 12.8 | 33.3 | 50.4 | 219,773 | 210,645 | -4.2 | -2.6 |
| Rome | 5.2 | 1.9 | 17.7 | 6.6 | 12.3 | 12.1 | 12.4 | 16.3 | 22.6 | 46.3 | 50.7 | 34,950 | 33,715 | -3.5 | -4.7 |
| Saratoga Springs | 4.8 | 7.8 | 9.8 | 16.2 | 16.6 | 10.0 | 11.1 | 11.7 | 24.7 | 40.9 | 51.6 | 26,186 | 26,567 | 1.5 | 6.5 |
| Schenectady | 13.5 | 19.0 | 19.0 | 13.1 | 18.8 | 12.2 | 9.9 | 13.1 | 13.9 | 34.6 | 50.7 | 61,821 | 66,157 | 7.0 | -1.4 |
| Spring Valley | 24.9 | 34.1 | 36.6 | 9.2 | 15.2 | 12.1 | 7.8 | 10.0 | 9.0 | 26.4 | 51.0 | 25,464 | 31,328 | 23.0 | 3.0 |
| Syracuse | 9.6 | 13.1 | 20.0 | 17.6 | 16.0 | 10.2 | 11.0 | 11.6 | 13.6 | 32.2 | 53.9 | 147,306 | 145,047 | -1.5 | -2.6 |
| Troy | 7.9 | 9.6 | 18.1 | 19.5 | 18.9 | 11.0 | 10.6 | 10.1 | 11.9 | 32.4 | 53.0 | 49,170 | 50,162 | 2.0 | -2.7 |
| Utica | 15.7 | 27.9 | 24.3 | 14.3 | 14.1 | 10.4 | 10.2 | 12.0 | 14.8 | 33.4 | 51.3 | 60,651 | 62,245 | 2.6 | -4.8 |
| Valley Stream | 34.9 | 34.4 | 24.9 | 8.2 | 13.5 | 13.6 | 14.3 | 10.8 | 14.7 | 37.5 | 52.0 | 36,368 | 37,385 | 2.8 | -0.5 |
| Watertown | 7.6 | 4.8 | 23.7 | 7.3 | 19.8 | 13.4 | 11.5 | 10.2 | 14.1 | 34.7 | 46.4 | 26,705 | 26,822 | 0.4 | -9.3 |
| White Plains | 31.6 | 31.6 | 19.3 | 9.9 | 15.2 | 13.9 | 12.6 | 11.8 | 17.3 | 39.7 | 51.8 | 53,077 | 56,824 | 7.1 | 2.6 |

1. May be of any race.

| City | Households, 2019 Number (27) | Persons per household (28) | Family (29) | Married couple family (30) | Female family (31) | Non-family (32) | One person (33) | Persons in group quarters, 2019 (34) | Violent Number (35) | Violent Rate[2] (36) | Property Number (37) | Property Rate[2] (38) | Population age 25 and over (39) | High school graduate or less (40) | Bachelor's degree or more (41) |
|---|---|---|---|---|---|---|---|---|---|---|---|---|---|---|---|
| **NEW JERSEY—Cont'd** | | | | | | | | | | | | | | | |
| Garfield | 12,156 | 2.61 | 67.9 | 46.8 | 21.1 | 32.1 | 24.1 | 32 | 51 | 160 | 620 | 1,944 | 21,942 | 59.8 | 22.0 |
| Hackensack | 21,509 | 2.03 | 48.6 | 27.7 | 20.9 | 51.4 | 43.1 | 588 | 103 | 231 | 594 | 1,335 | 36,329 | 36.0 | 46.8 |
| Hoboken | 24,148 | 2.12 | 42.8 | 33.0 | 9.8 | 57.2 | 32.7 | 1,546 | 79 | 147 | 666 | 1,242 | 37,895 | 11.7 | 81.0 |
| Jersey City | 108,939 | 2.38 | 56.8 | 40.5 | 16.3 | 43.2 | 30.7 | 2,507 | 1,393 | 523 | 5,074 | 1,904 | 193,444 | 31.4 | 52.8 |
| Kearny | 13,835 | 2.82 | 77.0 | 58.0 | 19.0 | 23.0 | 22.4 | 1,995 | 54 | 131 | 661 | 1,600 | 29,676 | 51.8 | 23.1 |
| Linden | 14,980 | 2.81 | 73.8 | 47.7 | 26.1 | 26.2 | 22.3 | 353 | 159 | 373 | 1,228 | 2,883 | 29,561 | 48.5 | 24.1 |
| Long Branch | 12,699 | 2.37 | 59.4 | 36.9 | 22.5 | 40.6 | 28.3 | 162 | 109 | 359 | 625 | 2,059 | 19,372 | 42.6 | 31.0 |
| Millville | 10,698 | 2.54 | 65.2 | 45.8 | 19.4 | 34.8 | 27.3 | 247 | 130 | 472 | 1,107 | 4,021 | 19,264 | 54.9 | 18.5 |
| Newark | 102,155 | 2.62 | 61.7 | 30.9 | 30.8 | 38.3 | 33.6 | 14,194 | 1,742 | 619 | 4,618 | 1,641 | 183,362 | 63.4 | 14.1 |
| New Brunswick | 15,888 | 3.02 | 62.2 | 29.9 | 32.3 | 37.8 | 23.4 | 7,664 | 262 | 468 | 1,111 | 1,984 | 26,776 | 62.7 | 22.1 |
| Paramus | 8,023 | 3.05 | 80.9 | 62.9 | 17.9 | 19.1 | 17.3 | 1,803 | 44 | 166 | 919 | 3,473 | 19,429 | 31.3 | 45.7 |
| Passaic | 19,956 | 3.48 | 78.0 | 43.9 | 34.2 | 22.0 | 18.5 | 184 | 369 | 530 | 1,215 | 1,745 | 40,113 | 70.8 | 15.0 |
| Paterson | 43,172 | 3.32 | 75.0 | 38.4 | 36.6 | 25.0 | 20.7 | 2,067 | 1,219 | 841 | 3,132 | 2,162 | 90,893 | 68.1 | 12.3 |
| Perth Amboy | 16,558 | 3.07 | 73.7 | 36.0 | 37.8 | 26.3 | 17.9 | 570 | 147 | 284 | 697 | 1,345 | 32,886 | 66.7 | 12.7 |
| Plainfield | 18,142 | 2.70 | 65.5 | 38.1 | 27.4 | 34.5 | 32.7 | 1,331 | 192 | 380 | 805 | 1,592 | 34,099 | 57.1 | 23.7 |
| Rahway | 13,059 | 2.28 | 56.0 | 39.8 | 16.2 | 44.0 | 37.9 | 133 | 32 | 106 | 157 | 522 | 21,640 | 42.9 | 34.5 |
| Sayreville | 15,390 | 2.87 | 71.7 | 55.4 | 16.2 | 28.3 | 21.8 | 14 | 34 | 76 | 310 | 695 | 31,159 | 30.3 | 41.9 |
| Trenton | 28,246 | 2.78 | 55.7 | 26.6 | 29.1 | 44.3 | 37.0 | 4,791 | 937 | 1,123 | 1,748 | 2,094 | 54,038 | 62.7 | 14.2 |
| Union City | 25,581 | 2.64 | 64.0 | 34.1 | 29.9 | 36.0 | 29.1 | 498 | 193 | 282 | 1,080 | 1,578 | 46,104 | 54.3 | 26.9 |
| Vineland | 22,193 | 2.60 | 67.8 | 38.1 | 29.7 | 32.2 | 29.4 | 1,738 | 211 | 352 | 1,686 | 2,817 | 40,152 | 53.4 | 20.5 |
| Westfield | 10,380 | 2.81 | 75.1 | 69.1 | 6.0 | 24.9 | 24.3 | 325 | 4 | 13 | 141 | 475 | 19,133 | 14.7 | 74.2 |
| West New York | 19,146 | 2.75 | 61.2 | 37.5 | 23.8 | 38.8 | 23.5 | 15 | 128 | 241 | 692 | 1,302 | 36,335 | 55.7 | 28.6 |
| **NEW MEXICO** | 793,420 | 2.59 | 62.6 | 41.8 | 20.9 | 37.4 | 31.3 | 42,907 | NA | NA | NA | NA | 1,425,988 | 40.5 | 27.7 |
| Alamogordo | 13,806 | 2.28 | 55.3 | 38.0 | 17.3 | 44.7 | 37.9 | 556 | 109 | 342 | 796 | 2,501 | 22,332 | 41.2 | 16.4 |
| Albuquerque | 227,179 | 2.44 | 57.1 | 36.4 | 20.7 | 42.9 | 35.4 | 5,659 | 7,596 | 1,352 | NA | NA | 387,982 | 33.7 | 36.2 |
| Carlsbad | 10,587 | 2.76 | 76.8 | 48.0 | 28.8 | 23.2 | 20.5 | 586 | NA | NA | NA | NA | 20,390 | 52.7 | 16.8 |
| Clovis | 15,266 | 2.45 | 57.0 | 34.9 | 22.0 | 43.0 | 37.1 | 563 | 261 | 674 | 1,474 | 3,805 | 23,336 | 45.9 | 19.1 |
| Farmington | 15,647 | 2.74 | 67.5 | 40.5 | 27.0 | 32.5 | 27.8 | 1,495 | NA | NA | NA | NA | 27,695 | 40.8 | 19.4 |
| Hobbs | 12,756 | 2.89 | 71.1 | 47.4 | 23.7 | 28.9 | 20.5 | 2,264 | NA | NA | NA | NA | 22,327 | 46.3 | 15.7 |
| Las Cruces | 40,185 | 2.52 | 57.2 | 35.6 | 21.6 | 42.8 | 30.2 | 2,006 | 514 | 497 | 3,707 | 3,581 | 63,605 | 34.7 | 33.6 |
| Rio Rancho | 35,765 | 2.76 | 77.7 | 53.6 | 24.1 | 22.3 | 18.4 | 508 | 189 | 190 | 1,474 | 1,484 | 67,089 | 29.8 | 32.5 |
| Roswell | 18,339 | 2.50 | 64.3 | 41.3 | 23.0 | 35.7 | 31.5 | 1,970 | 404 | 850 | 1,793 | 3,772 | 30,659 | 45.9 | 18.6 |
| Santa Fe | 35,112 | 2.37 | 55.3 | 35.2 | 20.1 | 44.7 | 36.1 | 1,361 | NA | NA | NA | NA | 61,653 | 32.4 | 39.9 |
| **NEW YORK** | 7,446,812 | 2.54 | 62.3 | 43.3 | 19.0 | 37.7 | 30.5 | 568,613 | NA | NA | NA | NA | 13,664,734 | 38.2 | 37.8 |
| Albany | 42,408 | 2.04 | 42.3 | 24.2 | 18.1 | 57.7 | 42.1 | 9,992 | 736 | 757 | 2,919 | 3,002 | 58,381 | 33.2 | 43.9 |
| Auburn | 10,744 | 2.23 | 51.3 | 28.9 | 22.4 | 48.7 | 37.5 | 2,260 | NA | NA | NA | NA | 18,937 | 44.3 | 20.5 |
| Binghamton | 19,643 | 2.18 | 43.9 | 23.6 | 20.3 | 56.1 | 41.6 | 1,497 | 355 | 798 | 1,907 | 4,288 | 27,918 | 39.7 | 27.5 |
| Buffalo | 109,163 | 2.26 | 51.7 | 25.1 | 26.6 | 48.3 | 38.3 | 9,107 | 2,533 | 991 | 8,298 | 3,245 | 164,432 | 41.2 | 28.7 |
| Elmira | 9,500 | 2.49 | 55.2 | 29.4 | 25.8 | 44.8 | 40.7 | 3,354 | 83 | 308 | 831 | 3,083 | 17,524 | 52.3 | 14.3 |
| Freeport | 13,213 | 3.22 | 72.4 | 47.3 | 25.1 | 27.6 | 25.1 | 454 | 72 | 167 | 428 | 994 | 28,574 | 45.5 | 25.7 |
| Glen Cove | 10,244 | 2.58 | 62.4 | 48.7 | 13.7 | 37.6 | 28.5 | 703 | 5 | 18 | 84 | 309 | 19,904 | 34.7 | 41.2 |
| Harrison | 9,308 | 2.75 | 70.8 | 61.0 | 9.8 | 29.2 | 20.3 | 3,324 | NA | NA | NA | NA | 18,499 | 26.3 | 60.8 |
| Hempstead | 16,943 | 3.20 | 73.6 | 33.7 | 40.0 | 26.4 | 22.7 | 854 | 297 | 536 | 579 | 1,045 | 34,875 | 59.6 | 14.4 |
| Ithaca | 12,248 | 1.91 | 27.3 | 20.0 | 7.3 | 72.7 | 47.2 | 7,428 | 41 | 132 | 840 | 2,699 | 11,455 | 8.3 | 78.4 |
| Jamestown | 13,038 | 2.18 | 50.5 | 28.3 | 22.3 | 49.5 | 42.3 | 611 | 199 | 684 | 863 | 2,965 | 18,764 | 43.4 | 23.7 |
| Lindenhurst | 8,970 | 2.98 | 73.0 | 58.0 | 15.0 | 27.0 | 23.0 | 29 | NA | NA | NA | NA | 18,972 | 40.5 | 27.0 |
| Long Beach | 13,346 | 2.42 | 59.0 | 44.0 | 15.0 | 41.0 | 31.0 | 1,188 | 19 | 57 | 66 | 197 | 27,036 | 25.1 | 50.1 |
| Middletown | 10,945 | 2.54 | 55.6 | 34.5 | 21.0 | 44.4 | 41.8 | 377 | 87 | 313 | 308 | 1,108 | 18,456 | 50.2 | 17.4 |
| Mount Vernon | 26,535 | 2.50 | 57.5 | 32.0 | 25.6 | 42.5 | 38.0 | 912 | 381 | 563 | 842 | 1,245 | 50,930 | 43.7 | 28.9 |
| Newburgh | 10,643 | 2.56 | 59.3 | 28.9 | 30.4 | 40.7 | 35.3 | 903 | 317 | 1,129 | 640 | 2,280 | 16,820 | 45.8 | 21.4 |
| New Rochelle | 32,251 | 2.35 | 63.1 | 43.6 | 19.5 | 36.9 | 34.4 | 2,870 | 98 | 124 | 762 | 965 | 58,715 | 39.2 | 41.0 |
| New York | 3,211,033 | 2.53 | 58.8 | 36.4 | 22.4 | 41.2 | 32.9 | 196,874 | 47,821 | 571 | 122,299 | 1,460 | 5,929,102 | 41.2 | 39.2 |
| Niagara Falls | 23,159 | 2.05 | 47.6 | 25.2 | 22.4 | 52.4 | 43.8 | 357 | 431 | 900 | 1,848 | 3,858 | 34,382 | 39.7 | 24.1 |
| North Tonawanda | 13,584 | 2.22 | 57.5 | 42.9 | 14.6 | 42.5 | 37.2 | 131 | 44 | 146 | 351 | 1,161 | 22,474 | 44.6 | 26.6 |
| Ossining | 10,243 | 2.38 | 58.6 | 43.8 | 14.8 | 41.4 | 37.1 | 1,824 | 20 | 80 | 157 | 628 | 19,220 | 34.4 | 44.2 |
| Port Chester | 9,713 | 3.16 | 70.9 | 53.3 | 17.6 | 29.1 | 21.6 | 231 | 21 | 72 | 200 | 682 | 23,090 | 47.8 | 28.7 |
| Poughkeepsie | 13,895 | 2.13 | 48.9 | 24.0 | 24.9 | 51.1 | 44.0 | 857 | 235 | 772 | 513 | 1,686 | 22,585 | 37.8 | 28.6 |
| Rochester | 87,679 | 2.24 | 44.0 | 20.7 | 23.3 | 56.0 | 44.1 | 9,002 | 1,540 | 748 | 7,142 | 3,471 | 134,679 | 44.1 | 28.2 |
| Rome | 14,338 | 2.08 | 50.1 | 32.4 | 17.7 | 49.9 | 42.8 | 2,370 | 54 | 169 | 524 | 1,637 | 24,328 | 47.7 | 20.2 |
| Saratoga Springs | 12,375 | 2.07 | 52.2 | 45.7 | 6.5 | 47.8 | 30.7 | 2,556 | 64 | 227 | 448 | 1,589 | 20,882 | 15.9 | 61.8 |
| Schenectady | 26,574 | 2.32 | 57.2 | 25.0 | 32.2 | 42.8 | 32.7 | 3,625 | 528 | 806 | 1,770 | 2,702 | 44,326 | 46.7 | 22.4 |
| Spring Valley | 8,379 | 3.50 | 80.9 | 59.0 | 21.9 | 19.1 | 19.1 | 38 | 115 | 355 | 425 | 1,313 | 15,908 | 62.1 | 11.9 |
| Syracuse | 57,114 | 2.23 | 44.9 | 22.8 | 22.0 | 55.1 | 43.4 | 14,700 | 1,129 | 793 | 4,464 | 3,134 | 88,867 | 41.3 | 31.8 |
| Troy | 20,417 | 2.18 | 46.8 | 21.5 | 25.3 | 53.2 | 40.0 | 4,568 | 296 | 601 | 1,365 | 2,770 | 30,706 | 45.1 | 30.2 |
| Utica | 22,071 | 2.59 | 58.9 | 31.1 | 27.8 | 41.1 | 32.8 | 2,582 | 374 | 625 | 1,986 | 3,319 | 36,716 | 53.5 | 19.0 |
| Valley Stream | 10,625 | 3.52 | 86.5 | 62.9 | 23.6 | 13.5 | 9.7 | 36 | NA | NA | NA | NA | 25,038 | 33.1 | 40.2 |
| Watertown | 10,818 | 2.23 | 50.4 | 31.1 | 19.2 | 49.6 | 38.8 | 724 | 154 | 613 | 887 | 3,534 | 17,134 | 49.8 | 18.6 |
| White Plains | 22,740 | 2.51 | 62.2 | 48.5 | 13.7 | 37.8 | 31.4 | 1,104 | NA | NA | NA | NA | 41,163 | 26.2 | 51.8 |

1. Data for serious crimes have not been adjusted for underreporting. This may affect comparability between geographic areas and over time.   2. Per 100,000 population estimated by the FBI.   3. Persons 25 years old and over.

# Table D. Cities — Income, Poverty, and Housing

| City | Money income, 2019 — Households — Median income | Households — Percent with income less than $20,000 | Households — Percent with income of $200,000 or more | Median family income | Median non-family income | Median earnings, 2019 — All persons | Men | Women | Housing units, 2019 — Total | Occupied | Percent owner occupied | Median value[1] (dollars) | Median gross rent (dollars) |
|---|---|---|---|---|---|---|---|---|---|---|---|---|---|
| | 42 | 43 | 44 | 45 | 46 | 47 | 48 | 49 | 50 | 51 | 52 | 53 | 54 |
| **NEW JERSEY—Cont'd** | | | | | | | | | | | | | |
| Garfield | 61,684 | 13.8 | 4.4 | 61,982 | 60,367 | 34,851 | 36,393 | 32,089 | 12,633 | 12,156 | 43.0 | 343,930 | 1,376 |
| Hackensack | 64,639 | 10.2 | 6.9 | 80,146 | 52,492 | 47,266 | 50,086 | 44,400 | 22,424 | 21,509 | 34.1 | 295,026 | 1,552 |
| Hoboken | 143,645 | 7.5 | 34.1 | 208,938 | 120,910 | 90,897 | 95,955 | 83,998 | 27,066 | 24,148 | 31.8 | 742,245 | 2,285 |
| Jersey City | 81,693 | 14.9 | 16.8 | 90,820 | 65,570 | 53,124 | 60,751 | 46,476 | 116,602 | 108,939 | 30.8 | 424,851 | 1,541 |
| Kearny | 80,528 | 9.8 | 7.7 | 83,639 | 75,264 | 44,099 | 50,882 | 30,250 | 14,380 | 13,835 | 48.6 | 345,047 | 1,424 |
| Linden | 74,067 | 6.0 | 5.7 | 87,022 | 36,548 | 38,381 | 43,514 | 31,737 | 16,634 | 14,980 | 56.9 | 290,609 | 1,386 |
| Long Branch | 57,359 | 16.9 | 8.0 | 79,767 | 45,342 | 35,106 | 40,958 | 25,048 | 14,901 | 12,699 | 36.3 | 357,105 | 1,276 |
| Millville | 65,624 | 18.1 | 7.5 | 69,612 | 33,869 | 33,653 | 43,688 | 26,479 | 11,620 | 10,698 | 69.8 | 157,763 | 1,115 |
| Newark | 40,235 | 27.1 | 2.2 | 44,419 | 29,331 | 27,433 | 32,075 | 22,001 | 113,623 | 102,155 | 24.1 | 270,596 | 1,129 |
| New Brunswick | 43,983 | 23.7 | 6.4 | 48,393 | 36,427 | 21,801 | 31,848 | 16,845 | 17,638 | 15,888 | 22.5 | 244,765 | 1,553 |
| Paramus | 132,402 | 8.0 | 26.2 | 154,888 | 41,334 | 61,158 | 63,674 | 51,947 | 8,576 | 8,023 | 75.4 | 634,966 | 3,012 |
| Passaic | 52,158 | 19.4 | 2.5 | 56,016 | 22,985 | 27,500 | 30,910 | 25,744 | 20,622 | 19,956 | 20.1 | 349,381 | 1,160 |
| Paterson | 47,369 | 21.3 | 2.6 | 52,034 | 24,785 | 29,269 | 30,837 | 26,874 | 46,797 | 43,172 | 27.8 | 266,021 | 1,232 |
| Perth Amboy | 46,916 | 21.7 | 4.7 | 45,926 | 0 | 31,646 | 40,297 | 21,870 | 17,047 | 16,558 | 27.7 | 271,947 | 1,376 |
| Plainfield | 49,041 | 17.7 | 6.0 | 53,255 | 32,821 | 27,927 | 29,356 | 24,928 | 19,418 | 18,142 | 39.9 | 280,868 | 1,201 |
| Rahway | 85,313 | 11.4 | 4.8 | 102,342 | 48,958 | 49,722 | 65,852 | 36,946 | 13,578 | 13,059 | 66.4 | 303,545 | 1,238 |
| Sayreville | 94,886 | 4.3 | 15.4 | 121,528 | 66,177 | 51,139 | 66,105 | 35,146 | 15,932 | 15,390 | 64.8 | 324,766 | 1,347 |
| Trenton | 34,000 | 34.6 | 2.2 | 49,720 | 21,013 | 29,165 | 30,510 | 28,097 | 35,409 | 28,246 | 30.7 | 97,574 | 1,002 |
| Union City | 60,140 | 16.4 | 7.0 | 67,648 | 35,506 | 30,632 | 35,454 | 24,688 | 27,109 | 25,581 | 21.4 | 385,543 | 1,246 |
| Vineland | 52,406 | 11.2 | 3.5 | 56,637 | 38,606 | 31,111 | 34,934 | 28,765 | 23,718 | 22,193 | 65.1 | 175,179 | 1,132 |
| Westfield | 155,615 | 7.1 | 41.3 | 192,724 | 57,025 | 82,365 | 111,807 | 66,412 | 11,151 | 10,380 | 77.5 | 758,071 | 1,587 |
| West New York | 48,666 | 19.7 | 5.8 | 41,864 | 44,923 | 31,164 | 35,764 | 25,243 | 21,134 | 19,146 | 15.0 | 369,005 | 1,257 |
| **NEW MEXICO** | 51,945 | 18.9 | 4.5 | 61,826 | 32,958 | 30,276 | 35,175 | 25,617 | 948,470 | 793,420 | 68.1 | 180,936 | 847 |
| Alamogordo | 40,006 | 14.0 | 1.9 | 43,612 | 30,203 | 26,364 | 30,186 | 21,354 | 15,582 | 13,806 | 59.1 | 120,714 | 730 |
| Albuquerque | 55,567 | 17.1 | 4.9 | 66,550 | 40,576 | 32,389 | 37,410 | 29,554 | 247,716 | 227,179 | 60.1 | 211,770 | 905 |
| Carlsbad | 76,442 | 13.4 | 4.7 | 88,982 | 21,787 | 35,922 | 60,145 | 26,436 | 12,540 | 10,587 | 67.9 | 139,465 | 1,160 |
| Clovis | 41,331 | 20.3 | 0.6 | 45,716 | 35,980 | 25,732 | 40,308 | 15,035 | 16,991 | 15,266 | 55.8 | 128,150 | 789 |
| Farmington | 46,146 | 17.8 | 3.5 | 52,732 | 32,390 | 30,427 | 35,200 | 24,553 | 18,247 | 15,647 | 61.1 | 184,708 | 795 |
| Hobbs | 75,826 | 17.4 | 3.4 | 82,618 | 32,117 | 41,987 | 51,967 | 26,688 | 13,143 | 12,756 | 49.5 | 143,389 | 1,001 |
| Las Cruces | 45,130 | 22.7 | 1.6 | 54,827 | 23,514 | 23,638 | 28,548 | 19,845 | 45,249 | 40,185 | 53.5 | 165,320 | 791 |
| Rio Rancho | 74,130 | 5.3 | 5.3 | 77,025 | 47,088 | 35,096 | 41,152 | 31,620 | 37,285 | 35,765 | 80.6 | 192,728 | 1,335 |
| Roswell | 39,598 | 23.7 | 1.3 | 52,183 | 24,398 | 26,215 | 35,218 | 21,566 | 20,469 | 18,339 | 65.4 | 95,228 | 802 |
| Santa Fe | 59,247 | 14.6 | 6.9 | 71,790 | 44,575 | 30,063 | 31,344 | 26,010 | 41,881 | 35,112 | 64.9 | 303,245 | 1,091 |
| **NEW YORK** | 72,108 | 15.0 | 12.1 | 89,475 | 44,681 | 41,492 | 47,239 | 36,555 | 8,404,205 | 7,446,812 | 53.5 | 338,735 | 1,309 |
| Albany | 49,096 | 21.8 | 3.8 | 78,383 | 33,494 | 26,937 | 32,827 | 22,274 | 50,381 | 42,408 | 40.1 | 189,526 | 983 |
| Auburn | 42,468 | 25.5 | 1.0 | 60,817 | 36,239 | 34,014 | 34,053 | 32,491 | 11,987 | 10,744 | 51.3 | 104,266 | 729 |
| Binghamton | 35,949 | 32.9 | 3.1 | 46,392 | 26,118 | 26,217 | 30,624 | 19,832 | 23,541 | 19,643 | 39.4 | 99,276 | 778 |
| Buffalo | 40,843 | 28.0 | 2.9 | 50,258 | 31,097 | 29,954 | 31,207 | 26,867 | 126,735 | 109,163 | 40.6 | 117,539 | 807 |
| Elmira | 40,322 | 27.1 | 1.3 | 53,270 | 17,993 | 25,248 | 28,156 | 21,912 | 12,197 | 9,500 | 46.2 | 77,766 | 682 |
| Freeport | 90,960 | 8.3 | 14.5 | 103,192 | 42,226 | 40,545 | 40,414 | 40,616 | 13,657 | 13,213 | 74.9 | 385,233 | 1,409 |
| Glen Cove | 92,676 | 15.2 | 15.8 | 109,426 | 43,879 | 46,015 | 54,332 | 39,691 | 10,874 | 10,244 | 63.0 | 540,596 | 1,599 |
| Harrison | 136,900 | 2.9 | 38.0 | 201,015 | 73,774 | 67,292 | 82,095 | 30,359 | 9,981 | 9,308 | 71.0 | 893,519 | 2,010 |
| Hempstead | 50,272 | 17.3 | 5.1 | 48,143 | 48,162 | 31,491 | 35,584 | 26,645 | 18,413 | 16,943 | 35.5 | 380,473 | 1,279 |
| Ithaca | 38,584 | 29.3 | 4.9 | 87,292 | 26,809 | 18,045 | 19,666 | 16,717 | 14,284 | 12,248 | 20.8 | 231,004 | 1,278 |
| Jamestown | 36,312 | 31.9 | 2.2 | 50,560 | 20,571 | 27,502 | 33,645 | 25,971 | 14,550 | 13,038 | 49.5 | 61,237 | 626 |
| Lindenhurst | 107,120 | 3.9 | 14.3 | 124,503 | 43,911 | 50,995 | 63,205 | 41,409 | 9,486 | 8,970 | 81.9 | 413,710 | 2,086 |
| Long Beach | 101,976 | 8.3 | 23.9 | 121,470 | 60,601 | 56,201 | 61,747 | 50,481 | 14,821 | 13,346 | 62.4 | 572,786 | 1,858 |
| Middletown | 57,957 | 18.8 | 6.1 | 95,663 | 45,637 | 40,203 | 45,822 | 32,165 | 12,577 | 10,945 | 49.6 | 191,063 | 1,396 |
| Mount Vernon | 59,289 | 23.6 | 9.0 | 82,486 | 30,427 | 41,146 | 41,914 | 40,261 | 27,734 | 26,535 | 42.1 | 398,566 | 1,262 |
| Newburgh | 42,473 | 29.7 | 1.8 | 62,892 | 26,045 | 30,127 | 38,671 | 20,405 | 11,997 | 10,643 | 37.5 | 166,447 | 1,155 |
| New Rochelle | 74,508 | 16.5 | 17.5 | 98,226 | 36,786 | 46,584 | 48,646 | 43,219 | 33,800 | 32,251 | 48.8 | 588,645 | 1,551 |
| New York | 69,407 | 17.8 | 13.0 | 78,113 | 51,293 | 42,326 | 47,057 | 40,205 | 3,546,601 | 3,211,033 | 31.9 | 680,773 | 1,483 |
| Niagara Falls | 39,398 | 28.5 | 1.7 | 57,267 | 28,202 | 29,916 | 32,404 | 26,523 | 27,018 | 23,159 | 50.1 | 83,730 | 608 |
| North Tonawanda | 59,397 | 15.0 | 2.7 | 78,590 | 27,102 | 36,710 | 42,252 | 32,443 | 15,219 | 13,584 | 70.9 | 138,106 | 806 |
| Ossining | 79,336 | 10.0 | 15.9 | 104,300 | 61,553 | 50,353 | 54,833 | 44,198 | 10,942 | 10,243 | 47.2 | 364,243 | 1,856 |
| Port Chester | 87,053 | 11.4 | 15.2 | 86,379 | 70,646 | 32,008 | 35,945 | 26,448 | 10,280 | 9,713 | 52.8 | 449,457 | 1,708 |
| Poughkeepsie | 49,101 | 27.8 | 5.3 | 73,175 | 27,308 | 30,282 | 36,506 | 27,735 | 15,501 | 13,895 | 38.8 | 223,245 | 1,188 |
| Rochester | 37,711 | 26.5 | 2.3 | 47,615 | 30,076 | 30,208 | 32,089 | 27,113 | 100,603 | 87,679 | 35.4 | 86,994 | 847 |
| Rome | 46,944 | 20.1 | 1.4 | 61,076 | 28,348 | 31,110 | 41,049 | 28,659 | 15,977 | 14,338 | 56.8 | 109,560 | 628 |
| Saratoga Springs | 92,648 | 9.2 | 16.6 | 130,000 | 57,138 | 40,658 | 41,103 | 39,856 | 15,102 | 12,375 | 57.4 | 377,014 | 1,357 |
| Schenectady | 40,356 | 22.5 | 2.4 | 40,506 | 34,006 | 28,439 | 34,276 | 24,926 | 30,626 | 26,574 | 39.8 | 107,005 | 939 |
| Spring Valley | 52,137 | 16.7 | 1.0 | 67,911 | NA | 25,653 | 27,222 | 21,317 | 9,470 | 8,379 | 31.1 | 383,139 | 1,296 |
| Syracuse | 39,494 | 29.9 | 2.6 | 49,282 | 33,748 | 27,027 | 30,097 | 25,904 | 71,806 | 57,114 | 39.8 | 99,845 | 806 |
| Troy | 51,813 | 22.2 | 2.3 | 58,827 | 36,244 | 26,247 | 26,903 | 24,867 | 24,226 | 20,417 | 31.3 | 156,944 | 990 |
| Utica | 41,917 | 23.2 | 2.7 | 44,227 | 29,209 | 27,634 | 30,527 | 26,055 | 27,571 | 22,071 | 47.9 | 108,963 | 725 |
| Valley Stream | 107,233 | 5.2 | 12.2 | 112,248 | NA | 41,077 | 44,353 | 32,404 | 11,646 | 10,625 | 80.1 | 486,623 | 1,773 |
| Watertown | 37,010 | 27.8 | 3.1 | 43,893 | 26,548 | 26,926 | 34,208 | 20,501 | 13,748 | 10,818 | 40.7 | 138,773 | 798 |
| White Plains | 103,670 | 15.7 | 24.4 | 130,546 | 45,330 | 46,853 | 62,383 | 40,760 | 25,088 | 22,740 | 47.7 | 623,145 | 1,650 |

1. Based on population estimated by the American Community Survey.

| City | Commuting[1], 2019 | | Computer access[2], 2019 | | Migration, 2019 | | Civilian labor force, 2020 | | | | Civilian Employment[4], 2019 | | | |
|---|---|---|---|---|---|---|---|---|---|---|---|---|---|---|
| | Percent | | Percent | | | | | | Unemployment | | Population age 16 and older | | Population age 16 to 64 | |
| | Drove alone | With commutes of 30 minutes or more | With a computer in the house | With Internet access | Percent who lived in the same house one year ago | Percent who lived in another state or county one year ago | Total | Percent change 2019-2020 | Total | Rate[3] | Number | Percent in labor force | Number | Percent who worked full-year full-time |
| | 55 | 56 | 57 | 58 | 59 | 60 | 61 | 62 | 63 | 64 | 65 | 66 | 67 | 68 |
| NEW JERSEY—Cont'd | | | | | | | | | | | | | | |
| Garfield | 73.1 | 37.9 | 88.3 | 82.5 | 92.6 | 5.6 | 16,450 | 3.3 | 2,403 | 15 | 24,157 | 67.5 | 19,284 | 64.3 |
| Hackensack | 70.2 | 50.6 | 94.0 | 88.9 | 89.4 | 6.9 | 24,183 | 0.2 | 2,758 | 11 | 38,082 | 72.8 | 30,856 | 70.8 |
| Hoboken | 21.6 | 75.0 | 96.4 | 93.8 | 85.4 | 9.6 | 34,415 | -5.7 | 1,626 | 5 | 44,178 | 79.4 | 41,501 | 70.0 |
| Jersey City | 31.6 | 67.6 | 95.1 | 90.3 | 87.3 | 6.3 | 138,858 | -1.3 | 14,449 | 10 | 216,757 | 69.3 | 186,968 | 60.9 |
| Kearny | 65.5 | 57.4 | 94.8 | 90.3 | 89.0 | 4.3 | 20,583 | 1.2 | 2,602 | 13 | 32,627 | 62.9 | 26,247 | 55.0 |
| Linden | 69.7 | 43.6 | 93.7 | 89.6 | 89.0 | 3.9 | 22,004 | 0.9 | 2,446 | 11 | 34,423 | 67.1 | 28,395 | 56.6 |
| Long Branch | 62.4 | 41.0 | 95.7 | 86.4 | NA | NA | 15,995 | -1.0 | 1,396 | 9 | 23,431 | 73.8 | 19,537 | 50.0 |
| Millville | 87.0 | 22.7 | 91.2 | 88.1 | 78.4 | 1.7 | 13,320 | 2.9 | 1,538 | 12 | 21,760 | 59.8 | 16,276 | 44.4 |
| Newark | 52.7 | 59.0 | 89.3 | 69.8 | 88.3 | 5.3 | 119,436 | 3.4 | 17,544 | 15 | 219,790 | 62.3 | 189,657 | 43.4 |
| New Brunswick | 41.4 | 51.4 | 86.5 | 68.9 | 81.2 | 10.2 | 26,450 | -1.6 | 2,105 | 8 | 42,843 | 62.3 | 38,118 | 39.6 |
| Paramus | 71.5 | 43.2 | 93.6 | 90.2 | NA | NA | 12,485 | -1.4 | 1,190 | 10 | 22,629 | 55.6 | 15,772 | 50.8 |
| Passaic | 46.3 | 35.1 | 87.6 | 86.2 | 93.8 | 2.2 | 29,430 | 1.6 | 3,961 | 14 | 48,993 | 69.8 | 42,960 | 54.1 |
| Paterson | 65.2 | 31.5 | 91.0 | 87.2 | 90.7 | 1.3 | 63,873 | 6.1 | 11,829 | 19 | 108,121 | 64.4 | 91,206 | 55.5 |
| Perth Amboy | 64.0 | 42.6 | 92.7 | 74.8 | NA | NA | 25,815 | 4.4 | 4,049 | 16 | 38,385 | 59.4 | 31,729 | 45.9 |
| Plainfield | 62.1 | 46.6 | 90.3 | 85.7 | 85.8 | 10.0 | 26,404 | -0.8 | 2,877 | 11 | 38,543 | 72.2 | 31,868 | 48.8 |
| Rahway | 70.8 | 60.7 | 89.6 | 88.0 | 85.7 | 13.0 | 14,976 | 0.7 | 1,570 | 11 | 24,529 | 67.7 | 20,130 | 61.1 |
| Sayreville | 72.2 | 60.3 | 94.9 | 92.5 | 87.8 | 6.4 | 23,876 | -0.7 | 2,139 | 9 | 35,475 | 70.9 | 30,058 | 59.8 |
| Trenton | 60.1 | 30.1 | 89.3 | 71.3 | 79.4 | 8.6 | 40,540 | 2.4 | 4,411 | 11 | 62,002 | 58.6 | 52,326 | 42.7 |
| Union City | 33.5 | 60.4 | 93.9 | 83.6 | 89.3 | 4.1 | 35,236 | 2.8 | 4,863 | 14 | 55,654 | 71.5 | 47,536 | 59.9 |
| Vineland | 82.8 | 32.7 | 91.8 | 86.3 | 86.2 | 5.9 | 28,466 | 2.3 | 3,115 | 11 | 46,661 | 62.1 | 36,783 | 50.1 |
| Westfield | 65.8 | 56.2 | NA | NA | 92.1 | 7.2 | 13,365 | -4.0 | 775 | 6 | 21,787 | 64.7 | 17,444 | 51.0 |
| West New York | 27.5 | 64.3 | 92.2 | 83.1 | 92.4 | 2.5 | 28,323 | 0.6 | 3,171 | 11 | 40,054 | 67.7 | 32,078 | 64.6 |
| NEW MEXICO | 80.7 | 28.7 | 89.1 | 78.3 | 87.6 | 5.3 | 943,287 | -1.2 | 79,413 | 8 | 1,678,187 | 57.2 | 1,300,457 | 48.3 |
| Alamogordo | 80.7 | 16.2 | 94.1 | 87.0 | 83.8 | 8.5 | 13,258 | 2.3 | 916 | 7 | 26,261 | 48.3 | 19,845 | 43.8 |
| Albuquerque | 82.3 | 27.6 | 93.3 | 84.7 | 84.2 | 3.8 | 278,798 | -0.8 | 23,025 | 8 | 452,509 | 64.3 | 360,631 | 53.0 |
| Carlsbad | NA | 19.9 | 90.9 | 82.3 | NA | NA | 17,006 | -1.4 | 1,220 | 7 | 24,537 | 67.3 | 19,501 | 56.2 |
| Clovis | 79.8 | 17.3 | 86.8 | 84.0 | 77.4 | 14.0 | 17,677 | 3.2 | 1,032 | 6 | 28,447 | 54.6 | 23,409 | 49.1 |
| Farmington | 89.7 | 12.4 | 94.1 | 84.6 | 93.0 | 2.6 | 19,622 | -2.5 | 1,964 | 10 | 33,511 | 50.7 | 26,285 | 46.4 |
| Hobbs | NA | 26.0 | 87.9 | 75.5 | 90.3 | 7.0 | 15,975 | -4.8 | 1,871 | 12 | 27,743 | 62.7 | 24,357 | 55.5 |
| Las Cruces | 74.3 | 15.8 | 91.9 | 80.7 | 80.5 | 7.1 | 47,472 | 0.4 | 3,920 | 8 | 83,263 | 61.5 | 67,325 | 43.6 |
| Rio Rancho | 84.6 | 48.2 | 97.7 | 90.9 | 86.7 | 4.7 | 46,407 | -0.4 | 3,830 | 8 | 77,772 | 65.8 | 61,305 | 57.5 |
| Roswell | 86.4 | 12.7 | 83.1 | 68.0 | 91.3 | 6.4 | 20,836 | 3.7 | 1,770 | 9 | 36,512 | 57.8 | 29,370 | 53.7 |
| Santa Fe | 73.5 | 16.3 | 92.0 | 81.7 | 85.1 | 8.4 | 41,519 | -4.4 | 3,677 | 9 | 71,996 | 62.3 | 52,030 | 49.4 |
| NEW YORK | 52.8 | 52.2 | 91.9 | 85.5 | 89.5 | 4.6 | 9,289,171 | -2.4 | 928,165 | 10 | 15,895,561 | 63.3 | 12,599,593 | 52.6 |
| Albany | 59.9 | 14.4 | 92.0 | 84.9 | 79.4 | 9.1 | 47,125 | 1.2 | 4,437 | 9 | 81,784 | 65.5 | 69,119 | 43.6 |
| Auburn | 81.2 | 28.1 | 88.0 | 81.7 | 79.9 | 7.4 | 11,474 | -0.4 | 1,117 | 10 | 21,225 | 58.2 | 17,103 | 39.7 |
| Binghamton | 73.1 | 15.3 | 89.9 | 79.7 | 72.6 | 8.3 | 17,688 | -0.6 | 1,866 | 11 | 36,705 | 57.4 | 30,049 | 35.2 |
| Buffalo | 67.2 | 20.2 | 89.8 | 80.7 | 84.3 | 4.7 | 110,534 | 2.1 | 14,516 | 13 | 202,599 | 60.1 | 171,073 | 43.3 |
| Elmira | 71.7 | 18.2 | 86.8 | 77.3 | 90.6 | 3.2 | 9,804 | 3.4 | 1,133 | 12 | 20,560 | 54.9 | 17,573 | 35.5 |
| Freeport | 71.1 | 47.6 | 94.7 | 86.4 | 94.7 | 2.3 | 22,978 | 0.8 | 2,555 | 11 | 32,480 | 68.8 | 25,457 | 59.1 |
| Glen Cove | 71.0 | 39.9 | NA | NA | 89.6 | 5.0 | 13,919 | -1.8 | 1,129 | 8 | 22,675 | 61.4 | 15,404 | 59.1 |
| Harrison | 52.6 | 45.7 | NA | NA | 89.9 | 5.8 | 13,407 | 0.3 | 886 | 7 | 24,012 | 59.3 | 19,169 | 45.3 |
| Hempstead | 57.4 | 59.4 | 92.5 | 77.9 | 89.2 | 2.7 | 27,902 | 0.7 | 3,164 | 11 | 41,279 | 66.2 | 34,750 | 53.8 |
| Ithaca | 29.7 | 20.7 | 96.9 | 83.8 | 46.8 | 32.3 | 12,792 | -3.2 | 812 | 6 | 29,405 | 52.9 | 26,803 | 22.8 |
| Jamestown | 72.9 | 13.9 | 84.9 | 72.0 | 82.1 | 6.7 | 11,338 | -2.2 | 1,191 | 11 | 22,782 | 56.3 | 17,946 | 44.8 |
| Lindenhurst | 78.4 | 47.2 | 93.4 | 92.1 | 89.6 | 2.8 | 15,112 | -1.2 | 1,350 | 9 | 21,903 | 68.6 | 17,187 | 57.1 |
| Long Beach | 68.9 | 66.2 | 96.2 | 89.6 | 93.6 | 2.3 | 19,463 | -1.8 | 1,506 | 8 | 29,281 | 68.6 | 22,703 | 61.3 |
| Middletown | 68.8 | 30.8 | 93.9 | 88.0 | 87.4 | 2.1 | 14,075 | 1.1 | 1,418 | 10 | 22,190 | 58.7 | 16,489 | 48.2 |
| Mount Vernon | 48.1 | 61.9 | 91.2 | 72.5 | 90.6 | 5.2 | 33,646 | 1.2 | 4,122 | 12 | 58,244 | 62.1 | 44,273 | 54.9 |
| Newburgh | 62.3 | 30.9 | 90.5 | 87.4 | NA | NA | 12,453 | 0.3 | 1,413 | 11 | 21,322 | 65.4 | 18,249 | 43.5 |
| New Rochelle | 50.4 | 50.9 | 85.8 | 74.0 | 92.4 | 3.7 | 38,574 | -1.4 | 3,470 | 9 | 66,780 | 59.9 | 49,084 | 46.4 |
| New York | 22.1 | 70.8 | 91.4 | 84.2 | 89.6 | 4.5 | 3,909,835 | -3.9 | 480,928 | 12 | 6,803,099 | 63.9 | 5,521,905 | 53.2 |
| Niagara Falls | 68.3 | 24.4 | 86.8 | 79.9 | 85.1 | 6.8 | 21,153 | 1.6 | 2,917 | 14 | 38,605 | 57.8 | 30,088 | 44.4 |
| North Tonawanda | 82.3 | 23.3 | 90.7 | 83.2 | 90.7 | 3.4 | 15,240 | -0.7 | 1,450 | 10 | 25,243 | 65.1 | 18,869 | 57.2 |
| Ossining | 67.9 | 53.2 | NA | NA | 77.6 | 4.9 | 12,892 | -2.5 | 977 | 8 | 20,892 | 66.2 | 16,766 | 62.9 |
| Port Chester | 62.2 | 31.3 | NA | NA | NA | NA | 16,320 | -2.6 | 1,126 | 7 | 26,174 | 69.9 | 20,897 | 54.0 |
| Poughkeepsie | 73.8 | 25.6 | 88.3 | 79.8 | 87.0 | 4.8 | 13,666 | 1.1 | 1,500 | 11 | 25,572 | 57.9 | 19,206 | 47.5 |
| Rochester | 73.0 | 15.6 | 87.4 | 77.8 | 81.8 | 5.3 | 91,366 | 2.2 | 11,686 | 13 | 165,157 | 62.3 | 138,773 | 43.9 |
| Rome | 81.8 | 14.4 | 83.7 | 72.2 | 88.6 | 2.2 | 13,373 | 0.3 | 1,186 | 9 | 26,748 | 51.1 | 19,474 | 49.5 |
| Saratoga Springs | 77.4 | 34.4 | 97.6 | 94.0 | 81.3 | 10.7 | 14,206 | 1.8 | 1,075 | 8 | 25,730 | 63.6 | 18,764 | 53.5 |
| Schenectady | 69.1 | 29.2 | 90.6 | 78.9 | 80.9 | 9.1 | 30,856 | 2.9 | 3,305 | 11 | 53,648 | 61.6 | 44,597 | 45.0 |
| Spring Valley | 66.4 | 22.3 | 83.6 | 75.9 | NA | NA | 14,693 | -1.4 | 1,167 | 8 | 19,611 | 65.0 | 16,959 | 40.1 |
| Syracuse | 64.9 | 15.4 | 85.9 | 74.0 | 75.4 | 9.6 | 59,354 | 2.1 | 6,820 | 12 | 116,670 | 57.2 | 97,271 | 37.3 |
| Troy | 59.8 | 28.2 | 92.0 | 85.7 | 72.3 | 12.5 | 23,212 | 1.3 | 2,209 | 10 | 41,838 | 64.1 | 36,007 | 41.2 |
| Utica | 69.1 | 17.9 | 92.7 | 87.3 | 81.9 | 5.6 | 23,606 | 1.3 | 2,555 | 11 | 46,940 | 55.6 | 38,085 | 40.9 |
| Valley Stream | 59.5 | 64.4 | 97.9 | 95.0 | NA | NA | 19,724 | -0.1 | 1,986 | 10 | 29,424 | 66.8 | 23,925 | 52.0 |
| Watertown | 73.0 | 18.1 | 85.4 | 79.6 | 77.5 | 5.6 | 10,421 | 0.0 | 1,037 | 10 | 19,361 | 50.4 | 15,870 | 39.1 |
| White Plains | 54.6 | 40.4 | 95.1 | 93.6 | 90.9 | 5.1 | 31,712 | -1.8 | 2,365 | 8 | 48,200 | 64.3 | 38,158 | 56.7 |

1. Employed persons.　2. Households.　3. Percent of civilian labor force.　4. Persons 16 years old and over.

# Table D. Cities — Construction, Wholesale Trade, and Retail Trade

| City | Value of residential construction authorized by building permits, 2020 | | | Wholesale trade[1], 2017 | | | | Retail trade[2], 2017 | | | |
|---|---|---|---|---|---|---|---|---|---|---|---|
| | New construction ($1,000) | Number of housing units | Percent single family | Number of establishments | Number of employees | Sales (mil dol) | Annual payroll (mil dol) | Number of establishments | Number of employees | Sales (mil dol) | Annual payroll (mil dol) |
| | 69 | 70 | 71 | 72 | 73 | 74 | 75 | 76 | 77 | 78 | 79 |
| **NEW JERSEY—Cont'd** | | | | | | | | | | | |
| Garfield | 4,176 | 53 | 26 | 41 | 477 | 225 | 24 | 71 | 912 | 270 | 25 |
| Hackensack | 48,689 | 416 | 0 | 181 | 1,562 | 1,675 | 98 | 260 | 3,615 | 990 | 103 |
| Hoboken | 85,448 | 95 | 3 | 34 | 229 | 155 | 15 | 167 | 1,480 | 420 | 38 |
| Jersey City | 516,551 | 3,552 | 4 | 189 | 5,929 | 5,916 | 460 | 818 | 12,319 | 3,302 | 295 |
| Kearny | 41 | 2 | 100 | 49 | 904 | 1,213 | 55 | 98 | 1,838 | 517 | 51 |
| Linden | 13,967 | 158 | 16 | 113 | 1,948 | 1,819 | 115 | 196 | 2,696 | 1,087 | 103 |
| Long Branch | 114,897 | 1,062 | 20 | 28 | 125 | 73 | 6 | 86 | 1,112 | 408 | 37 |
| Millville | 1,156 | 10 | 100 | 25 | 545 | 322 | 36 | 86 | 1,703 | 443 | 42 |
| Newark | 176,240 | 1,494 | 0 | 333 | 6,003 | 6,820 | 440 | 903 | 7,325 | 3,104 | 288 |
| New Brunswick | 36,523 | 209 | 1 | 62 | 840 | 514 | 45 | 123 | 679 | 203 | 17 |
| Paramus | 9,095 | 41 | 100 | 93 | 1,476 | 2,096 | 129 | 586 | 14,840 | 4,147 | 411 |
| Passaic | 13,032 | 105 | 17 | 86 | 1,104 | 474 | 47 | 252 | 2,189 | 567 | 54 |
| Paterson | 42,024 | 590 | 2 | 185 | 2,284 | 1,197 | 120 | 555 | 3,641 | 1,062 | 91 |
| Perth Amboy | 1,643 | 13 | 31 | 39 | 1,152 | 1,425 | 82 | 208 | 1,353 | 467 | 40 |
| Plainfield | 6,495 | 69 | 3 | 21 | D | 39 | D | 124 | 615 | 202 | 20 |
| Rahway | 16,387 | 238 | 6 | 64 | 791 | 1,413 | 55 | 72 | 578 | 232 | 17 |
| Sayreville | 5,427 | 47 | 72 | 59 | 652 | 711 | 41 | 107 | 1,316 | 471 | 37 |
| Trenton | 3,889 | 57 | 0 | 42 | 610 | 656 | 36 | 233 | 1,343 | 363 | 35 |
| Union City | 6,799 | 74 | 4 | 39 | 182 | 103 | 8 | 271 | 1,269 | 323 | 32 |
| Vineland | 5,177 | 72 | 100 | 71 | 1,999 | 1,721 | 98 | 261 | 3,815 | 1,060 | 99 |
| Westfield | 11,593 | 39 | 100 | 15 | D | 22 | D | 109 | 1,352 | 274 | 31 |
| West New York | 37,746 | 409 | 19 | 24 | 166 | 40 | 5 | 215 | 1,287 | 333 | 32 |
| **NEW MEXICO** | 107,347 | 396 | 100 | 1,507 | 16,914 | 11,937 | 835 | 6,335 | 92,557 | 26,404 | 2,511 |
| Alamogordo | NA | NA | NA | 8 | 42 | 12 | 1 | 123 | 2,084 | 534 | 52 |
| Albuquerque | 209,875 | 865 | 100 | 633 | 8,376 | 5,132 | 431 | 1,806 | 32,482 | 9,619 | 934 |
| Carlsbad | 60,333 | 398 | 35 | D | D | D | D | 111 | 1,805 | 564 | 55 |
| Clovis | 19,197 | 111 | 51 | 25 | 277 | 102 | 9 | 163 | 2,240 | 604 | 58 |
| Farmington | 10,780 | 59 | 86 | 91 | 764 | 396 | 45 | 298 | 4,754 | 1,362 | 134 |
| Hobbs | 38,496 | 147 | 100 | 66 | 965 | 473 | 49 | 151 | 2,682 | 894 | 79 |
| Las Cruces | 191,822 | 813 | 91 | 71 | 665 | 355 | 32 | 394 | 7,221 | 1,888 | 174 |
| Rio Rancho | 131,662 | 621 | 97 | 24 | 154 | 92 | 9 | 112 | 2,859 | 801 | 78 |
| Roswell | 7,541 | 44 | 100 | 33 | 335 | 173 | 14 | 179 | 2,985 | 829 | 75 |
| Santa Fe | 117,591 | 577 | 48 | 72 | 695 | 842 | 37 | 710 | 8,126 | 2,327 | 244 |
| **NEW YORK** | 140 | 1 | 100 | 26,900 | 330,990 | 367,972 | 22,510 | 78,260 | 945,360 | 291,725 | 27,815 |
| Albany | 5,710 | 48 | 8 | 98 | 1,373 | 3,169 | 85 | 468 | 6,947 | 1,905 | 198 |
| Auburn | 0 | 0 | 0 | D | D | D | 10 | 114 | 1,893 | 469 | 45 |
| Binghamton | 9,369 | 50 | 0 | 54 | 621 | 255 | 24 | 169 | 1,745 | 483 | 46 |
| Buffalo | 51,603 | 240 | 13 | 233 | 4,611 | 2,814 | 266 | 877 | 8,134 | 1,690 | 183 |
| Elmira | 24,237 | 156 | 0 | 36 | 442 | 191 | 20 | 80 | 1,115 | 313 | 31 |
| Freeport | 2,483 | 7 | 100 | 74 | 589 | 297 | 30 | 186 | 1,904 | 778 | 61 |
| Glen Cove | 690 | 2 | 100 | 49 | 594 | 198 | 12 | 101 | 1,146 | 659 | 43 |
| Harrison | 51,310 | 178 | 16 | 56 | 1,030 | 4,256 | 109 | 53 | 352 | 120 | 12 |
| Hempstead | 960 | 6 | 100 | 34 | 245 | 115 | 14 | 208 | 2,482 | 1,268 | 87 |
| Ithaca | 8,650 | 56 | 0 | 14 | 111 | 31 | 4 | 173 | 3,000 | 749 | 78 |
| Jamestown | 80 | 1 | 100 | 33 | 236 | 110 | 9 | 106 | 1,411 | 433 | 42 |
| Lindenhurst | 547 | 3 | 100 | 41 | 290 | 117 | 15 | 91 | 516 | 175 | 21 |
| Long Beach | 3,800 | 19 | 100 | 28 | 98 | 35 | 6 | 84 | 906 | 325 | 36 |
| Middletown | 15,934 | 66 | 100 | 27 | 210 | 115 | 11 | 115 | 1,509 | 465 | 39 |
| Mount Vernon | 408 | 5 | 20 | 86 | 962 | 665 | 56 | 226 | 2,264 | 559 | 60 |
| Newburgh | 8,019 | 24 | 100 | 34 | 271 | 163 | 13 | 84 | 756 | 251 | 23 |
| New Rochelle | 170,028 | 821 | 1 | 70 | 382 | 338 | 24 | 237 | 3,025 | 1,339 | 106 |
| New York | 517,626 | 4,461 | 0 | 13,813 | 150,295 | 190,586 | 10,285 | 35,488 | 353,094 | 116,310 | 11,295 |
| Niagara Falls | 0 | 0 | 0 | D | D | D | D | 287 | 4,335 | 856 | 81 |
| North Tonawanda | 2,741 | 46 | 15 | D | D | D | 19 | 75 | 1,166 | 256 | 29 |
| Ossining | 0 | 0 | 0 | 12 | D | 86 | D | 75 | 609 | 169 | 19 |
| Port Chester | 0 | 0 | 0 | 39 | 484 | 314 | 31 | 155 | 2,397 | 703 | 70 |
| Poughkeepsie | 19,597 | 171 | 1 | 23 | 219 | 183 | 11 | 125 | 1,649 | 297 | 37 |
| Rochester | 98,987 | 799 | 7 | 242 | 3,356 | 1,616 | 179 | 743 | 7,184 | 1,437 | 178 |
| Rome | 12,305 | 76 | 5 | 19 | 215 | 112 | 9 | 117 | 2,083 | 536 | 54 |
| Saratoga Springs | 43,427 | 105 | 51 | 24 | 554 | 284 | 30 | 146 | 1,947 | 641 | 58 |
| Schenectady | 17,175 | 94 | 0 | 38 | 362 | 350 | 24 | 197 | 1,520 | 415 | 39 |
| Spring Valley | 3,665 | 20 | 30 | 45 | 371 | 496 | 14 | 164 | 1,380 | 553 | 45 |
| Syracuse | 1,075 | 11 | 27 | 120 | 1,855 | 1,947 | 101 | 593 | 7,219 | 1,771 | 184 |
| Troy | 1,951 | 10 | 40 | 26 | 347 | 879 | 22 | 142 | 1,338 | 353 | 36 |
| Utica | 0 | 0 | 0 | 54 | 846 | 476 | 40 | 186 | 2,619 | 619 | 63 |
| Valley Stream | 1,539 | 4 | 100 | 59 | 432 | 358 | 32 | 178 | 2,151 | 728 | 66 |
| Watertown | 0 | 0 | 0 | 18 | 322 | 148 | 16 | 157 | 2,045 | 513 | 51 |
| White Plains | 41,749 | 310 | 1 | 105 | 1,757 | 4,032 | 242 | 396 | 7,271 | 2,112 | 228 |

1.  Merchant wholesalers except manufacturers' sales branches and offices.    2.  Establishments with payroll.

| City | Real estate and rental and leasing, 2017 | | | | Professional, scientific, and technical services[1], 2017 | | | | Manufacturing, 2017 | | | |
|---|---|---|---|---|---|---|---|---|---|---|---|---|
| | Number of establishments | Number of employees | Receipts (mil dol) | Annual payroll (mil dol) | Number of establishments | Number of employees | Receipts (mil dol) | Annual payroll (mil dol) | Number of establishments | Number of employees | Receipts (mil dol) | Annual payroll (mil dol) |
| | 80 | 81 | 82 | 83 | 84 | 85 | 86 | 87 | 88 | 89 | 90 | 91 |
| NEW JERSEY—Cont'd | | | | | | | | | | | | |
| Garfield | 16 | 68 | 15 | 3 | 31 | 209 | 21 | 7 | NA | NA | NA | NA |
| Hackensack | 138 | 615 | 260 | 34 | 411 | 2,870 | 611 | 237 | NA | NA | NA | NA |
| Hoboken | 112 | 529 | 246 | 26 | 225 | 1,286 | 317 | 108 | NA | NA | NA | NA |
| Jersey City | 263 | 1,583 | 499 | 85 | 655 | 8,057 | 2,184 | 793 | NA | NA | NA | NA |
| Kearny | 33 | 238 | 68 | 17 | 54 | 264 | 57 | 14 | NA | NA | NA | NA |
| Linden | 39 | 209 | 81 | 12 | 55 | 934 | 50 | 39 | NA | NA | NA | NA |
| Long Branch | D | D | D | D | 53 | 155 | 27 | 7 | NA | NA | NA | NA |
| Millville | 16 | 78 | 15 | 2 | 37 | 240 | 33 | 9 | NA | NA | NA | NA |
| Newark | 227 | 2,413 | 683 | 96 | 473 | 8,271 | 1,794 | 843 | NA | NA | NA | NA |
| New Brunswick | 58 | 395 | 100 | 19 | 125 | 1,084 | 242 | 87 | NA | NA | NA | NA |
| Paramus | 70 | 470 | 351 | 28 | 206 | 2,412 | 507 | 169 | NA | NA | NA | NA |
| Passaic | 51 | 157 | 32 | 5 | 68 | 350 | 52 | 13 | NA | NA | NA | NA |
| Paterson | 84 | 402 | 67 | 14 | 89 | 368 | 59 | 20 | NA | NA | NA | NA |
| Perth Amboy | 32 | 158 | 50 | 7 | 58 | 211 | 24 | 9 | NA | NA | NA | NA |
| Plainfield | 26 | 84 | 21 | 3 | D | D | D | 28 | NA | NA | NA | NA |
| Rahway | 23 | 197 | 42 | 11 | D | D | D | 25 | D | NA | NA | NA |
| Sayreville | 32 | 130 | 36 | 5 | 88 | 1,149 | 174 | 85 | NA | NA | NA | NA |
| Trenton | 58 | 310 | 86 | 11 | 125 | 840 | 190 | 78 | NA | NA | NA | NA |
| Union City | 54 | 138 | 28 | 4 | 108 | 357 | 60 | 16 | NA | NA | NA | NA |
| Vineland | 57 | 252 | 63 | 12 | 104 | 580 | 64 | 24 | NA | NA | NA | NA |
| Westfield | 30 | 107 | 44 | 4 | 142 | 822 | 169 | 62 | NA | NA | NA | NA |
| West New York | 51 | 234 | 75 | 13 | 79 | 219 | 36 | 12 | NA | NA | NA | NA |
| NEW MEXICO | 2,408 | 9,229 | 2,186 | 376 | 4,728 | 56,695 | 10,082 | 4,190 | 1,332 | 23,235 | 13,724 | 1,287 |
| Alamogordo | 35 | 138 | 20 | 4 | D | D | D | D | NA | NA | NA | NA |
| Albuquerque | 897 | 3,942 | 967 | 158 | D | D | 2,773 | D | NA | NA | NA | NA |
| Carlsbad | 30 | 147 | 43 | 7 | 52 | 386 | 101 | 23 | NA | NA | NA | NA |
| Clovis | 57 | 190 | 32 | 5 | 66 | 302 | 31 | 11 | NA | NA | NA | NA |
| Farmington | 79 | 433 | 124 | 25 | 157 | 897 | 94 | 40 | NA | NA | NA | NA |
| Hobbs | 62 | 408 | 110 | 27 | 55 | 348 | 43 | 17 | NA | NA | NA | NA |
| Las Cruces | D | D | D | 17 | 251 | 2,801 | 331 | 134 | NA | NA | NA | NA |
| Rio Rancho | 63 | 214 | 44 | 9 | 106 | 457 | 58 | 24 | NA | NA | NA | NA |
| Roswell | 64 | 164 | 32 | 5 | 88 | 691 | 124 | 47 | NA | NA | NA | NA |
| Santa Fe | 224 | 703 | 182 | 37 | 516 | 2,060 | 347 | 130 | NA | NA | NA | NA |
| NEW YORK | 34,076 | 193,442 | 70,693 | 11,357 | 61,744 | 668,196 | 172,936 | 60,227 | 15,499 | 411,100 | 155,572 | 23,751 |
| Albany | 143 | 882 | 222 | 39 | 451 | 10,628 | 2,021 | 758 | NA | NA | NA | NA |
| Auburn | 35 | 84 | 22 | 3 | 46 | 258 | 28 | 12 | NA | NA | NA | NA |
| Binghamton | 53 | 217 | 30 | 7 | 115 | 1,485 | 184 | 67 | NA | NA | NA | NA |
| Buffalo | 275 | 2,354 | 483 | 90 | 717 | 14,519 | 2,221 | 894 | NA | NA | NA | NA |
| Elmira | 28 | 102 | 20 | 4 | 46 | 239 | 32 | 10 | NA | NA | NA | NA |
| Freeport | 47 | 233 | 62 | 9 | 123 | 377 | 62 | 22 | NA | NA | NA | NA |
| Glen Cove | 25 | 99 | 22 | 5 | 85 | 310 | 57 | 20 | NA | NA | NA | NA |
| Harrison | 82 | 933 | 309 | 71 | 166 | 2,225 | 972 | 207 | NA | NA | NA | NA |
| Hempstead | 53 | 278 | 73 | 10 | D | D | D | D | NA | NA | NA | NA |
| Ithaca | 55 | 412 | 72 | 14 | 133 | 991 | 132 | 53 | NA | NA | NA | NA |
| Jamestown | 28 | 110 | 20 | 3 | 57 | 280 | 35 | 14 | NA | NA | NA | NA |
| Lindenhurst | 16 | 33 | 10 | 1 | 58 | 161 | 26 | 8 | NA | NA | NA | NA |
| Long Beach | 49 | 129 | 44 | 6 | 105 | 195 | 35 | 9 | NA | NA | NA | NA |
| Middletown | 26 | 95 | 22 | 3 | 45 | 165 | 21 | 8 | NA | NA | NA | NA |
| Mount Vernon | 127 | 549 | 95 | 23 | 101 | 377 | 60 | 21 | NA | NA | NA | NA |
| Newburgh | 18 | 81 | 14 | 3 | 56 | 416 | 51 | 21 | NA | NA | NA | NA |
| New Rochelle | 200 | 692 | 227 | 34 | 215 | 719 | 130 | 48 | NA | NA | NA | NA |
| New York | 20,435 | 127,520 | 53,249 | 8,226 | 28,631 | 406,156 | 126,759 | 42,532 | NA | NA | NA | NA |
| Niagara Falls | 46 | 306 | 54 | 8 | 79 | 391 | 55 | 15 | NA | NA | NA | NA |
| North Tonawanda | 8 | 20 | 3 | 1 | 31 | 151 | 11 | 5 | NA | NA | NA | NA |
| Ossining | 29 | 101 | 35 | 3 | 52 | 166 | 33 | 10 | NA | NA | NA | NA |
| Port Chester | 31 | 107 | 32 | 6 | 72 | 229 | 42 | 11 | NA | NA | NA | NA |
| Poughkeepsie | 47 | 198 | 32 | 7 | 121 | 921 | 131 | 54 | NA | NA | NA | NA |
| Rochester | 286 | 2,481 | 416 | 109 | 637 | 9,109 | 1,513 | 611 | NA | NA | NA | NA |
| Rome | 33 | 97 | 17 | 3 | 85 | 1,444 | 327 | 107 | NA | NA | NA | NA |
| Saratoga Springs | 83 | 380 | 90 | 16 | 184 | 1,294 | 267 | 96 | NA | NA | NA | NA |
| Schenectady | 59 | 264 | 62 | 10 | 125 | 1,234 | 154 | 72 | NA | NA | NA | NA |
| Spring Valley | 104 | 567 | 94 | 21 | 89 | 245 | 36 | 13 | NA | NA | NA | NA |
| Syracuse | 209 | 1,615 | 321 | 78 | 462 | 7,318 | 1,197 | 442 | NA | NA | NA | NA |
| Troy | 40 | 235 | 42 | 8 | 98 | 754 | 102 | 49 | NA | NA | NA | NA |
| Utica | 52 | 177 | 31 | 5 | 137 | 1,238 | 181 | 68 | NA | NA | NA | NA |
| Valley Stream | 52 | 227 | 49 | 9 | 146 | 1,574 | 230 | 93 | NA | NA | NA | NA |
| Watertown | 42 | 260 | 45 | 8 | 50 | 427 | 51 | 21 | NA | NA | NA | NA |
| White Plains | 203 | 771 | 263 | 46 | 690 | 5,859 | 1,448 | 521 | NA | NA | NA | NA |

1. Establishments subject to federal tax.

# Table D. Cities — Accommodation and Food Services, Arts, Entertainment, and Recreation, and Health Care and Social Assistance

| City | Accommodation and food services, 2017 | | | | Arts, entertainment, and recreation[1], 2017 | | | | Health care and social assistance[1], 2017 | | | |
|---|---|---|---|---|---|---|---|---|---|---|---|---|
| | Number of establishments | Number of employees | Receipts (mil dol) | Annual payroll (mil dol) | Number of establishments | Number of employees | Receipts (mil dol) | Annual payroll (mil dol) | Number of establishments | Number of employees | Receipts (mil dol) | Annual payroll (mil dol) |
| | 92 | 93 | 94 | 95 | 96 | 97 | 98 | 99 | 100 | 101 | 102 | 103 |
| **NEW JERSEY—Cont'd** | | | | | | | | | | | | |
| Garfield | 49 | 575 | 46 | 12 | NA | NA | NA | NA | 34 | 501 | 21 | 11 |
| Hackensack | 137 | 2,313 | 157 | 47 | 15 | 448 | 19 | 7 | 373 | 18,335 | 2,906 | 1,139 |
| Hoboken | 260 | 4,079 | 289 | 86 | D | D | D | D | 153 | 2,674 | 432 | 128 |
| Jersey City | 564 | 7,109 | 613 | 164 | 81 | 1,057 | 135 | 30 | 532 | 11,874 | 1,307 | 537 |
| Kearny | 68 | 743 | 51 | 14 | D | D | D | D | 67 | 1,016 | 90 | 40 |
| Linden | D | D | D | D | D | D | D | D | 72 | 1,801 | 192 | 66 |
| Long Branch | 102 | 1,456 | 99 | 28 | 12 | 92 | 11 | 3 | 94 | 5,100 | 1,022 | 422 |
| Millville | 58 | 803 | 45 | 13 | 10 | D | 18 | D | 93 | 1,688 | 133 | 61 |
| Newark | 582 | 9,177 | 862 | 217 | 36 | 2,295 | 288 | 137 | 471 | 18,615 | 2,717 | 1,052 |
| New Brunswick | 182 | 2,307 | 178 | 43 | 13 | 448 | 29 | 9 | 123 | 9,944 | 1,772 | 706 |
| Paramus | 154 | 3,653 | 249 | 70 | 26 | 785 | 67 | 19 | 273 | 6,281 | 754 | 325 |
| Passaic | D | D | D | D | 8 | D | 3 | D | 131 | 2,776 | 322 | 111 |
| Paterson | D | D | D | D | 14 | 186 | 9 | 2 | 238 | 8,526 | 1,049 | 486 |
| Perth Amboy | 111 | 670 | 51 | 11 | 6 | 15 | 2 | 1 | 80 | 1,375 | 107 | 45 |
| Plainfield | D | D | D | D | D | D | D | D | 106 | 1,964 | 145 | 74 |
| Rahway | 59 | 603 | 42 | 11 | 6 | 106 | 3 | 1 | 52 | 1,479 | 182 | 73 |
| Sayreville | D | D | D | D | 12 | 196 | 17 | 5 | 65 | 524 | 42 | 16 |
| Trenton | 142 | 916 | 67 | 17 | 13 | 177 | 12 | 5 | 160 | 8,046 | 790 | 401 |
| Union City | 154 | 1,144 | 77 | 18 | D | D | D | D | 193 | 2,147 | 153 | 65 |
| Vineland | 118 | 2,099 | 106 | 27 | 13 | 90 | 8 | 3 | 219 | 7,064 | 808 | 388 |
| Westfield | D | D | D | D | 21 | 436 | 17 | 8 | 146 | 2,125 | 215 | 97 |
| West New York | 95 | 756 | 49 | 13 | 13 | 53 | 7 | 3 | 117 | 3,153 | 134 | 73 |
| **NEW MEXICO** | 4,392 | 91,601 | 5,526 | 1,583 | 700 | 12,062 | 1,120 | 272 | 5,134 | 127,808 | 13,602 | 5,406 |
| Alamogordo | D | D | D | D | D | D | D | D | 98 | 2,501 | 297 | 104 |
| Albuquerque | 1,399 | 31,062 | 1,779 | 526 | 215 | 3,659 | 254 | 68 | 1,818 | 50,897 | 6,107 | 2,391 |
| Carlsbad | 77 | 1,565 | 114 | 27 | D | D | D | D | 73 | 1,920 | 189 | 77 |
| Clovis | D | D | D | D | 12 | 50 | 3 | 1 | D | D | D | D |
| Farmington | 169 | 3,710 | 184 | 53 | 15 | 179 | 8 | 4 | 237 | 6,045 | 659 | 266 |
| Hobbs | 104 | 2,236 | 194 | 39 | D | D | D | D | D | D | D | D |
| Las Cruces | 283 | 6,408 | 329 | 96 | 24 | 237 | 10 | 4 | 453 | 11,917 | 1,126 | 468 |
| Rio Rancho | 108 | 2,640 | 128 | 38 | D | D | D | D | 174 | 3,483 | 493 | 151 |
| Roswell | 119 | 2,359 | 123 | 35 | 16 | 132 | 8 | 2 | 163 | 3,997 | 407 | 179 |
| Santa Fe | 370 | 8,091 | 569 | 185 | 115 | 1,524 | 141 | 42 | 426 | 7,348 | 1,051 | 382 |
| **NEW YORK** | 54,797 | 824,806 | 66,964 | 19,793 | 13,019 | 185,076 | 29,270 | 8,696 | 58,902 | 1,654,593 | 193,508 | 80,688 |
| Albany | 446 | 6,343 | 406 | 127 | 53 | 1,159 | 56 | 19 | 356 | 23,957 | 3,317 | 1,267 |
| Auburn | 83 | 1,189 | 68 | 18 | 18 | 478 | 15 | 6 | 157 | 3,631 | 335 | 168 |
| Binghamton | 176 | 3,255 | 160 | 48 | 22 | 355 | 20 | 7 | 154 | 7,381 | 745 | 332 |
| Buffalo | 697 | 13,799 | 706 | 229 | 104 | 2,974 | 423 | 162 | 687 | 38,916 | 5,126 | 2,155 |
| Elmira | D | D | D | D | 14 | 238 | 9 | 3 | 127 | 4,622 | 491 | 241 |
| Freeport | 108 | 848 | 65 | 20 | 18 | 54 | 15 | 2 | 159 | 1,902 | 171 | 71 |
| Glen Cove | D | D | D | D | 19 | 396 | 28 | 9 | 112 | 2,819 | 315 | 152 |
| Harrison | D | D | D | D | 27 | 985 | 117 | 46 | 127 | 3,331 | 505 | 200 |
| Hempstead | 134 | 2,710 | 267 | 63 | 8 | 151 | 10 | 4 | 203 | 4,169 | 298 | 134 |
| Ithaca | 204 | 3,447 | 183 | 60 | 25 | 191 | 19 | 5 | 104 | 1,510 | 124 | 55 |
| Jamestown | 79 | 895 | 47 | 14 | 18 | 275 | 19 | 4 | 91 | 2,870 | 250 | 110 |
| Lindenhurst | 68 | 848 | 50 | 14 | D | D | D | D | 49 | 386 | 32 | 12 |
| Long Beach | 96 | 927 | 70 | 19 | 19 | 170 | 11 | 3 | 88 | 1,205 | 150 | 55 |
| Middletown | 72 | 659 | 40 | 13 | D | D | D | D | 92 | 2,064 | 145 | 68 |
| Mount Vernon | 100 | 1,009 | 67 | 17 | 14 | 170 | 14 | 5 | 150 | 3,251 | 280 | 124 |
| Newburgh | 79 | 904 | 58 | 19 | 7 | 45 | 3 | 1 | 112 | 3,323 | 410 | 143 |
| New Rochelle | 213 | 2,172 | 159 | 42 | 41 | 641 | 56 | 16 | 272 | 5,879 | 642 | 267 |
| New York | 24,871 | 396,907 | 38,942 | 11,579 | 6,771 | 96,960 | 20,764 | 6,078 | NA | NA | NA | NA |
| Niagara Falls | 202 | 5,417 | 621 | 124 | 17 | 175 | 11 | 3 | 103 | 2,607 | 235 | 102 |
| North Tonawanda | 55 | 751 | 38 | 13 | 18 | 78 | 7 | 1 | 61 | 1,165 | 68 | 52 |
| Ossining | 50 | 321 | 25 | 7 | 7 | 28 | 4 | 2 | 44 | 900 | 79 | 43 |
| Port Chester | 109 | 1,073 | 98 | 28 | 12 | 375 | 25 | 5 | 49 | 1,233 | 133 | 52 |
| Poughkeepsie | 102 | 1,540 | 78 | 28 | 13 | 227 | 14 | 4 | 145 | 5,520 | 981 | 290 |
| Rochester | 590 | 7,749 | 403 | 134 | 100 | 2,162 | 150 | 48 | 484 | 37,839 | 4,115 | 1,710 |
| Rome | 87 | 1,273 | 59 | 19 | 14 | 63 | 4 | 1 | 108 | 2,528 | 201 | 94 |
| Saratoga Springs | 169 | 3,889 | 321 | 101 | 51 | 1,085 | 200 | 34 | 165 | 4,024 | 500 | 205 |
| Schenectady | 170 | 2,208 | 121 | 37 | 25 | 1,859 | 236 | 58 | 212 | 8,001 | 808 | 370 |
| Spring Valley | D | D | D | D | D | D | D | D | 71 | 4,054 | 185 | 103 |
| Syracuse | 425 | 7,156 | 399 | 133 | 47 | 1,243 | 60 | 21 | 446 | 26,170 | 3,422 | 1,443 |
| Troy | 166 | 2,298 | 133 | 41 | 12 | 239 | 13 | 4 | 187 | 5,567 | 618 | 250 |
| Utica | D | D | D | D | 20 | D | 22 | D | 221 | 10,854 | 1,024 | 469 |
| Valley Stream | 91 | 1,444 | 81 | 24 | 19 | 343 | 21 | 7 | 171 | 5,562 | 314 | 152 |
| Watertown | 107 | 2,156 | 109 | 41 | 16 | 100 | 5 | 2 | 160 | 4,587 | 456 | 211 |
| White Plains | 231 | 3,544 | 288 | 84 | 44 | 778 | 57 | 18 | 391 | 13,062 | 1,958 | 877 |

1. Establishments subject to federal tax.

| City | Other services[1] | | | | Government employment and payroll, 2017 | | | | | | | | |
|---|---|---|---|---|---|---|---|---|---|---|---|---|---|
| | | | | | Full-time equivalent employees | March payroll | Percent of total for: | | | | | | |
| | Number of establishments | Number of employees | Receipts (mil dol) | Annual payroll (mil dol) | | Total (dollars) | Administrative, judicial, and legal | Police and corrections | Fire protection | Highways and transportation | Health and welfare | Natural resources and utilities | Education and libraries |
| | 104 | 105 | 106 | 107 | 108 | 109 | 110 | 111 | 112 | 113 | 114 | 115 | 116 |
| **NEW JERSEY—Cont'd** | | | | | | | | | | | | | |
| Garfield | D | D | D | D | 189 | 1,367,252 | 7.9 | 65.3 | 1.3 | 7.9 | 10.3 | 4.1 | 2.4 |
| Hackensack | 151 | 916 | 119 | 33 | 415 | 3,031,478 | 6.8 | 42.5 | 34.2 | 0.6 | 3.6 | 5.8 | 4.5 |
| Hoboken | 150 | 980 | 79 | 27 | 635 | 3,716,486 | 3.2 | 33.9 | 30.3 | 1.0 | 8.6 | 9.4 | 4.2 |
| Jersey City | 419 | 1,860 | 214 | 50 | 2,982 | 20,957,923 | 8.1 | 44.0 | 27.5 | 3.2 | 7.0 | 6.4 | 2.0 |
| Kearny | D | D | D | D | 325 | 2,322,529 | 4.7 | 46.4 | 33.1 | 8.1 | 1.8 | 1.7 | 1.4 |
| Linden | 105 | 951 | 116 | 38 | 527 | 2,710,901 | 5.9 | 24.0 | 26.3 | 7.6 | 10.6 | 22.3 | 0.0 |
| Long Branch | 63 | 255 | 33 | 7 | 374 | 2,326,716 | 10.2 | 41.5 | 11.0 | 5.3 | 11.1 | 8.6 | 3.1 |
| Millville | 52 | 229 | 19 | 6 | 218 | 1,119,395 | 14.2 | 45.0 | 6.2 | 9.5 | 8.4 | 16.6 | 0.0 |
| Newark | 563 | 3,527 | 505 | 124 | 3,735 | 15,962,973 | 11.5 | 35.4 | 18.0 | 0.6 | 15.6 | 9.9 | 2.9 |
| New Brunswick | 92 | 975 | 259 | 45 | 2,218 | 13,176,046 | 2.4 | 13.9 | 6.4 | 0.7 | 1.0 | 3.6 | 69.5 |
| Paramus | 79 | 669 | 108 | 30 | 348 | 2,063,276 | 9.3 | 52.9 | 1.6 | 0.8 | 6.7 | 19.5 | 6.9 |
| Passaic | 91 | 426 | 46 | 10 | 607 | 3,264,367 | 6.8 | 43.4 | 27.5 | 4.2 | 13.1 | 1.6 | 2.1 |
| Paterson | 232 | 1,348 | 123 | 33 | 1,538 | 9,711,177 | 7.0 | 44.9 | 28.3 | 1.1 | 7.5 | 6.4 | 1.5 |
| Perth Amboy | D | D | D | 18 | 451 | 2,622,482 | 10.6 | 41.9 | 17.2 | 3.2 | 10.1 | 8.5 | 1.7 |
| Plainfield | 87 | 325 | 29 | 9 | 596 | 3,705,643 | 11.8 | 31.9 | 21.7 | 0.6 | 6.8 | 7.2 | 5.4 |
| Rahway | D | D | 27 | D | 286 | 1,963,091 | 7.4 | 40.5 | 21.7 | 9.0 | 8.2 | 6.9 | 4.4 |
| Sayreville | D | D | D | 10 | 279 | 1,964,301 | 15.0 | 52.5 | 0.8 | 3.0 | 0.6 | 15.9 | 4.1 |
| Trenton | 139 | 805 | 112 | 37 | 2,465 | 17,459,250 | 3.1 | 11.0 | 9.2 | 0.9 | 3.3 | 5.4 | 66.0 |
| Union City | 117 | 308 | 39 | 7 | 2,489 | 12,749,167 | 2.2 | 14.6 | 0.2 | 0.0 | 2.9 | 3.8 | 75.1 |
| Vineland | 121 | 731 | 68 | 18 | 683 | 4,105,814 | 10.9 | 28.8 | 5.0 | 1.5 | 11.3 | 38.5 | 2.2 |
| Westfield | 100 | 561 | 55 | 17 | 225 | 1,462,240 | 9.5 | 33.3 | 21.2 | 15.5 | 9.7 | 5.1 | 5.1 |
| West New York | 86 | 262 | 26 | 5 | 1,604 | 9,705,140 | 3.1 | 15.4 | 0.8 | 0.2 | 2.5 | 2.6 | 74.4 |
| **NEW MEXICO** | 2,963 | 17,547 | 2,085 | 558 | X | X | X | X | X | X | X | X | X |
| Alamogordo | 54 | 288 | 21 | 6 | 295 | 867,629 | 16.4 | 29.5 | 8.3 | 3.3 | 8.9 | 20.4 | 4.0 |
| Albuquerque | 1,004 | 6,871 | 800 | 226 | 6,938 | 37,997,192 | 9.8 | 30.8 | 14.7 | 16.1 | 9.2 | 14.1 | 1.6 |
| Carlsbad | 47 | 462 | 65 | 15 | 494 | 2,632,596 | 10.6 | 23.9 | 17.4 | 12.8 | 0.0 | 25.7 | 1.7 |
| Clovis | 71 | 404 | 46 | 10 | 379 | 1,260,854 | 7.8 | 24.0 | 27.7 | 9.2 | 4.8 | 17.3 | 2.4 |
| Farmington | 145 | 959 | 158 | 32 | 954 | 4,000,911 | 9.5 | 22.8 | 11.6 | 5.0 | 2.4 | 39.0 | 3.8 |
| Hobbs | 57 | 500 | 98 | 23 | 479 | 2,228,763 | 9.1 | 31.1 | 16.8 | 7.7 | 3.6 | 23.9 | 2.2 |
| Las Cruces | 176 | 908 | 74 | 23 | 1,470 | 5,706,100 | 9.2 | 24.4 | 13.4 | 5.3 | 4.6 | 21.3 | 1.8 |
| Rio Rancho | 89 | 500 | 45 | 14 | 659 | 2,861,344 | 12.1 | 35.1 | 23.2 | 7.1 | 0.0 | 10.6 | 3.9 |
| Roswell | 64 | 443 | 32 | 9 | 543 | 2,039,709 | 8.2 | 29.5 | 20.0 | 10.0 | 1.5 | 24.5 | 3.3 |
| Santa Fe | 306 | 1,752 | 267 | 72 | 1,444 | 6,129,397 | 10.7 | 17.5 | 12.2 | 10.8 | 5.9 | 32.3 | 2.4 |
| **NEW YORK** | 48,436 | 299,209 | 51,750 | 11,846 | X | X | X | X | X | X | X | X | X |
| Albany | 289 | 1,909 | 366 | 98 | 1,247 | 8,694,425 | 6.8 | 44.9 | 17.0 | 4.1 | 8.7 | 17.4 | 0.0 |
| Auburn | 59 | 254 | 26 | 7 | 302 | 1,533,311 | 12.9 | 28.7 | 25.7 | 7.6 | 4.2 | 20.2 | 0.0 |
| Binghamton | 108 | 652 | 66 | 17 | 554 | 2,898,702 | 5.9 | 31.4 | 28.3 | 6.0 | 6.3 | 15.3 | 0.0 |
| Buffalo | 431 | 3,135 | 311 | 80 | 9,147 | 51,448,265 | 3.0 | 13.1 | 8.5 | 0.6 | 1.9 | 6.0 | 64.6 |
| Elmira | 39 | 216 | 26 | 6 | 303 | 1,797,692 | 5.8 | 37.9 | 27.4 | 6.1 | 4.8 | 16.5 | 0.0 |
| Freeport | 130 | 773 | 71 | 22 | 378 | 2,751,992 | 11.1 | 46.7 | 0.7 | 1.8 | 1.1 | 34.7 | 0.0 |
| Glen Cove | 79 | 292 | 31 | 9 | 229 | 1,410,099 | 9.8 | 44.1 | 1.8 | 7.3 | 7.3 | 17.9 | 0.0 |
| Harrison | 66 | 1,415 | 337 | 56 | 282 | 1,765,849 | 14.2 | 43.2 | 8.0 | 7.9 | 2.8 | 13.9 | 7.8 |
| Hempstead | 125 | 752 | 70 | 20 | 398 | 3,062,661 | 9.0 | 50.1 | 2.5 | 8.5 | 2.1 | 19.5 | 3.9 |
| Ithaca | 86 | 575 | 127 | 19 | 436 | 2,235,422 | 10.2 | 23.5 | 20.2 | 8.2 | 0.6 | 27.6 | 0.0 |
| Jamestown | 64 | 360 | 43 | 10 | 614 | 3,020,128 | 3.6 | 13.4 | 11.8 | 9.4 | 1.4 | 3.1 | 57.2 |
| Lindenhurst | 103 | 345 | 41 | 11 | 64 | 218,306 | 22.1 | 0.0 | 6.6 | 25.3 | 0.0 | 19.0 | 0.0 |
| Long Beach | 104 | 318 | 39 | 7 | 460 | 2,820,272 | 8.6 | 31.5 | 11.6 | 18.0 | 4.8 | 21.8 | 0.0 |
| Middletown | 55 | 301 | 28 | 9 | 244 | 1,425,159 | 14.6 | 45.0 | 10.0 | 5.9 | 1.7 | 22.8 | 0.0 |
| Mount Vernon | 151 | 677 | 106 | 26 | 561 | 3,698,608 | 8.2 | 43.3 | 29.4 | 3.1 | 1.4 | 14.6 | 0.0 |
| Newburgh | 59 | 540 | 59 | 17 | 301 | 1,577,058 | 7.4 | 34.2 | 35.6 | 2.1 | 2.3 | 13.7 | 0.0 |
| New Rochelle | 214 | 1,034 | 128 | 28 | 563 | 3,923,869 | 9.2 | 29.5 | 34.6 | 4.8 | 5.9 | 16.0 | 0.0 |
| New York | 23,119 | 159,346 | 35,688 | 7,493 | 434,443 | 2,716,087,055 | 3.5 | 16.6 | 4.9 | 15.4 | 17.7 | 4.9 | 34.4 |
| Niagara Falls | 73 | 425 | 38 | 10 | 565 | 2,503,768 | 9.9 | 35.5 | 30.5 | 7.3 | 8.2 | 6.5 | 0.0 |
| North Tonawanda | 59 | 186 | 15 | 4 | 287 | 1,463,894 | 6.8 | 28.1 | 17.2 | 18.6 | 0.6 | 25.4 | 0.0 |
| Ossining | 60 | 207 | 26 | 8 | 196 | 1,432,560 | 12.7 | 41.6 | 0.4 | 13.2 | 0.9 | 25.0 | 0.0 |
| Port Chester | 97 | 398 | 74 | 18 | 178 | 1,284,288 | 14.9 | 52.6 | 1.0 | 8.2 | 1.6 | 13.6 | 0.0 |
| Poughkeepsie | 90 | 519 | 59 | 18 | 317 | 1,831,845 | 7.5 | 41.6 | 20.7 | 7.3 | 6.0 | 15.5 | 0.0 |
| Rochester | 376 | 2,559 | 399 | 97 | 9,569 | 52,522,336 | 2.9 | 12.1 | 7.5 | 0.7 | 3.0 | 4.9 | 64.9 |
| Rome | 68 | 412 | 41 | 10 | 335 | 1,649,030 | 9.3 | 27.6 | 32.0 | 13.4 | 2.2 | 12.1 | 0.0 |
| Saratoga Springs | 94 | 496 | 82 | 15 | 356 | 1,861,287 | 9.5 | 32.3 | 19.3 | 23.6 | 4.3 | 7.5 | 0.0 |
| Schenectady | 105 | 776 | 160 | 34 | 500 | 2,823,338 | 0.0 | 44.8 | 24.5 | 7.3 | 10.9 | 10.2 | 0.0 |
| Spring Valley | 62 | 207 | 22 | 6 | 113 | 907,428 | 10.8 | 68.5 | 0.7 | 18.2 | 1.8 | 0.0 | 0.0 |
| Syracuse | 258 | 1,642 | 175 | 50 | 5,554 | 31,911,402 | 1.7 | 12.2 | 7.1 | 4.7 | 3.0 | 4.0 | 67.3 |
| Troy | 81 | 497 | 40 | 15 | 461 | 2,294,977 | 9.2 | 28.7 | 25.0 | 1.6 | 14.8 | 14.3 | 0.0 |
| Utica | 106 | 3,141 | 141 | 55 | 538 | 2,824,997 | 6.7 | 40.3 | 25.9 | 3.4 | 11.4 | 8.8 | 0.0 |
| Valley Stream | D | D | D | D | 211 | 1,315,477 | 13.1 | 0.0 | 0.5 | 17.7 | 0.0 | 36.8 | 6.5 |
| Watertown | 63 | 330 | 47 | 9 | 372 | 1,744,262 | 8.2 | 24.2 | 23.9 | 14.5 | 6.3 | 16.8 | 2.6 |
| White Plains | 243 | 1,590 | 246 | 67 | 1,000 | 6,983,033 | 8.3 | 27.0 | 18.6 | 8.9 | 8.3 | 10.3 | 3.3 |

1. Establishments subject to federal tax.

| City | City government finances, 2017 | | | | | | | | | |
|---|---|---|---|---|---|---|---|---|---|---|
| | General revenue | | | | | | | General expenditure | | |
| | | Intergovernmental | | Taxes | | | | | | |
| | | | | | Per capita[1] (dollars) | | | | Per capita[1] (dollars) | |
| | Total (mil dol) | Total (mil dol) | Percent from state government | Total (mil dol) | Total | Property | Sales and gross receipts | Total (mil dol) | Total | Capital outlays |
| | 117 | 118 | 119 | 120 | 121 | 122 | 123 | 124 | 125 | 126 |
| **NEW JERSEY—Cont'd** | | | | | | | | | | |
| Garfield | 79.0 | 1.8 | 1.1 | 55.4 | 1,739 | 1,729 | 10 | 31 | 984 | 73 |
| Hackensack | 103.5 | 7.1 | 73.1 | 86.4 | 1,940 | 1,856 | 84 | 110 | 2,476 | 369 |
| Hoboken | 136.9 | 25.0 | 55.1 | 69.1 | 1,288 | 1,181 | 106 | 179 | 3,328 | 1,145 |
| Jersey City | 796.7 | 175.1 | 49.5 | 277.6 | 1,055 | 946 | 109 | 623 | 2,365 | 136 |
| Kearny | 78.4 | 24.3 | 100.0 | 46.4 | 1,116 | 1,083 | 33 | 73 | 1,754 | 100 |
| Linden | 107.0 | 25.5 | 81.8 | 71.6 | 1,689 | 1,519 | 169 | 102 | 2,404 | 286 |
| Long Branch | 77.8 | 18.7 | 27.0 | 43.9 | 1,438 | 1,370 | 68 | 81 | 2,633 | 366 |
| Millville | 35.4 | 5.3 | 100.0 | 22.1 | 799 | 745 | 54 | 36 | 1,302 | 0 |
| Newark | 845.4 | 353.0 | 39.2 | 347.1 | 1,234 | 910 | 192 | 891 | 3,168 | 187 |
| New Brunswick | 300.6 | 192.9 | 92.9 | 37.2 | 664 | 603 | 60 | 347 | 6,186 | 237 |
| Paramus | 60.0 | 5.1 | 89.5 | 48.2 | 1,812 | 1,649 | 162 | 65 | 2,443 | 292 |
| Passaic | 124.4 | 37.1 | 40.7 | 67.5 | 963 | 942 | 21 | 125 | 1,778 | 57 |
| Paterson | 340.1 | 108.9 | 54.7 | 175.5 | 1,202 | 1,175 | 27 | 333 | 2,283 | 141 |
| Perth Amboy | 111.5 | 24.1 | 44.3 | 60.7 | 1,171 | 1,124 | 47 | 98 | 1,899 | 110 |
| Plainfield | 124.7 | 23.1 | 1.0 | 95.8 | 1,902 | 1,886 | 16 | 130 | 2,571 | 32 |
| Rahway | 57.5 | 5.0 | 88.8 | 43.0 | 1,450 | 1,334 | 116 | 51 | 1,719 | 88 |
| Sayreville | 56.1 | 12.1 | 80.9 | 36.2 | 814 | 759 | 55 | 57 | 1,280 | 187 |
| Trenton | 467.1 | 382.8 | 92.0 | 54.3 | 651 | 608 | 42 | 512 | 6,131 | 103 |
| Union City | 356.8 | 265.5 | 97.4 | 70.5 | 1,029 | 993 | 36 | 386 | 5,642 | 312 |
| Vineland | 96.1 | 18.4 | 22.9 | 36.1 | 604 | 577 | 26 | 87 | 1,450 | 428 |
| Westfield | 47.1 | 4.1 | 79.0 | 37.2 | 1,244 | 1,160 | 84 | 41 | 1,374 | 117 |
| West New York | 212.3 | 137.3 | 94.9 | 41.5 | 787 | 739 | 48 | 242 | 4,584 | 336 |
| **NEW MEXICO** | X | X | X | X | X | X | X | X | X | X |
| Alamogordo | 48.3 | 14.7 | 76.8 | 19.0 | 605 | 127 | 478 | 40 | 1,260 | 311 |
| Albuquerque | 979.8 | 270.0 | 87.5 | 386.0 | 690 | 261 | 429 | 814 | 1,455 | 218 |
| Carlsbad | 79.2 | 5.3 | 93.6 | 55.1 | 1,908 | 93 | 1,815 | 72 | 2,489 | 862 |
| Clovis | 56.3 | 10.0 | 60.5 | 30.4 | 780 | 52 | 729 | 56 | 1,435 | 463 |
| Farmington | 87.9 | 33.5 | 11.4 | 29.9 | 659 | 42 | 617 | 94 | 2,060 | 126 |
| Hobbs | 127.2 | 50.8 | 98.1 | 54.0 | 1,428 | 66 | 1,362 | 98 | 2,601 | 451 |
| Las Cruces | 180.7 | 17.6 | 55.4 | 123.3 | 1,210 | 148 | 1,062 | 102 | 998 | 174 |
| Rio Rancho | 113.0 | 15.7 | 46.9 | 61.1 | 634 | 203 | 432 | 90 | 931 | 133 |
| Roswell | 66.7 | 42.3 | 94.4 | 7.8 | 163 | 129 | 34 | 65 | 1,351 | 264 |
| Santa Fe | 185.8 | 70.3 | 80.8 | 48.4 | 577 | 134 | 442 | 232 | 2,758 | 332 |
| **NEW YORK** | X | X | X | X | X | X | X | X | X | X |
| Albany | 189.8 | 62.4 | 36.0 | 64.4 | 659 | 578 | 81 | 189 | 1,930 | 240 |
| Auburn | 49.3 | 19.6 | 50.5 | 13.0 | 487 | 442 | 45 | 45 | 1,702 | 184 |
| Binghamton | 111.1 | 50.1 | 34.4 | 38.3 | 850 | 814 | 35 | 126 | 2,788 | 1,159 |
| Buffalo | 1,439.0 | 1,218.1 | 89.2 | 106.4 | 415 | 335 | 80 | 1,483 | 5,785 | 348 |
| Elmira | 33.7 | 15.0 | 40.8 | 13.4 | 492 | 458 | 34 | 39 | 1,437 | 184 |
| Freeport | 66.8 | 1.8 | 82.2 | 45.7 | 1,059 | 985 | 74 | 77 | 1,782 | 73 |
| Glen Cove | 59.7 | 19.7 | 43.5 | 33.4 | 1,226 | 1,101 | 125 | 54 | 1,982 | 213 |
| Harrison | 14.0 | 1.3 | 54.4 | 11.8 | 422 | 378 | 44 | 18 | 656 | 252 |
| Hempstead | 80.6 | 2.0 | 65.7 | 67.4 | 1,218 | 1,168 | 50 | 76 | 1,369 | 58 |
| Ithaca | 66.0 | 10.2 | 63.2 | 38.2 | 1,246 | 715 | 531 | 67 | 2,191 | 286 |
| Jamestown | 76.2 | 35.4 | 57.8 | 16.6 | 562 | 534 | 28 | 83 | 2,803 | 72 |
| Lindenhurst | 14.4 | 2.2 | 83.2 | 8.8 | 328 | 268 | 60 | 13 | 493 | 78 |
| Long Beach | 82.5 | 11.3 | 63.3 | 43.0 | 1,280 | 1,070 | 209 | 90 | 2,679 | 253 |
| Middletown | 44.1 | 14.2 | 24.4 | 20.4 | 731 | 674 | 57 | 48 | 1,711 | 208 |
| Mount Vernon | 115.4 | 14.9 | 64.6 | 85.0 | 1,251 | 874 | 352 | 114 | 1,677 | 63 |
| Newburgh | 55.5 | 19.3 | 46.5 | 23.1 | 820 | 737 | 82 | 60 | 2,114 | 54 |
| New Rochelle | 148.7 | 22.9 | 40.5 | 95.7 | 1,210 | 733 | 477 | 143 | 1,809 | 129 |
| New York | 106,595.1 | 37,653.5 | 81.2 | 55,310.4 | 6,555 | 2,933 | 1,159 | 102,551 | 12,154 | 1,303 |
| Niagara Falls | 113.7 | 62.5 | 73.5 | 42.0 | 870 | 586 | 284 | 125 | 2,594 | 346 |
| North Tonawanda | 43.1 | 17.0 | 33.9 | 18.7 | 615 | 534 | 55 | 45 | 1,465 | 139 |
| Ossining | 40.0 | 10.9 | 5.5 | 23.2 | 923 | 863 | 60 | 45 | 1,803 | 176 |
| Port Chester | 40.0 | 5.6 | 17.6 | 25.4 | 863 | 782 | 82 | 43 | 1,468 | 181 |
| Poughkeepsie | 60.4 | 23.5 | 28.2 | 23.7 | 778 | 682 | 97 | 61 | 2,018 | 141 |
| Rochester | 1,347.3 | 1,042.1 | 79.9 | 121.0 | 586 | 523 | 63 | 1,474 | 7,139 | 748 |
| Rome | 50.6 | 14.9 | 86.9 | 26.2 | 813 | 479 | 334 | 51 | 1,592 | 101 |
| Saratoga Springs | 54.0 | 3.9 | 50.3 | 34.7 | 1,239 | 666 | 573 | 59 | 2,101 | 224 |
| Schenectady | 102.2 | 38.5 | 55.0 | 35.7 | 546 | 465 | 82 | 110 | 1,680 | 257 |
| Spring Valley | 37.0 | 11.4 | 8.0 | 24.5 | 757 | 707 | 50 | 44 | 1,356 | 12 |
| Syracuse | 712.0 | 591.5 | 83.3 | 66.0 | 464 | 412 | 53 | 765 | 5,378 | 223 |
| Troy | 73.4 | 33.1 | 44.6 | 24.8 | 502 | 451 | 51 | 74 | 1,503 | 122 |
| Utica | 90.4 | 35.8 | 72.5 | 43.4 | 719 | 464 | 253 | 83 | 1,378 | 157 |
| Valley Stream | 39.5 | 2.3 | 48.3 | 31.9 | 849 | 773 | 74 | 41 | 1,098 | 140 |
| Watertown | 48.9 | 31.7 | 22.6 | 10.1 | 393 | 345 | 48 | 62 | 2,430 | 315 |
| White Plains | 175.5 | 7.3 | 96.7 | 117.4 | 2,010 | 967 | 1,043 | 185 | 3,167 | 198 |

1. Based on population estimated as of July 1 of the year shown.

# Table D. Cities — City Government Finances

City government finances, 2017 (cont.)

General expenditure (cont.)

Percent of total for:

| City | Public welfare | Highways | Parking facilities | Education | Health and hospitals | Police protection | Sewerage and sanitation | Parks and recreation | Housing and community development | Interest on debt |
|---|---|---|---|---|---|---|---|---|---|---|
| | 127 | 128 | 129 | 130 | 131 | 132 | 133 | 134 | 135 | 136 |
| **NEW JERSEY—Cont'd** | | | | | | | | | | |
| Garfield | 0.0 | 5.0 | 0.0 | 0.0 | 0.9 | 27.2 | 11.5 | 3.5 | 11.9 | 0.0 |
| Hackensack | 0.0 | 2.1 | 0.6 | 0.0 | 0.8 | 15.0 | 11.0 | 9.2 | 4.4 | 1.0 |
| Hoboken | 0.0 | 12.5 | 7.7 | 0.0 | 0.4 | 9.3 | 2.8 | 0.9 | 9.4 | 1.5 |
| Jersey City | 0.0 | 1.4 | 0.4 | 0.0 | 1.5 | 16.6 | 7.8 | 1.8 | 15.0 | 4.1 |
| Kearny | 0.0 | 4.0 | 0.0 | 0.0 | 0.7 | 21.0 | 17.8 | 1.2 | 0.0 | 2.3 |
| Linden | 0.0 | 5.9 | 0.0 | 0.0 | 0.9 | 15.4 | 4.7 | 2.1 | 5.2 | 1.3 |
| Long Branch | 0.0 | 1.9 | 0.2 | 0.0 | 0.8 | 14.7 | 12.3 | 2.4 | 21.7 | 2.7 |
| Millville | 0.0 | 7.0 | 0.0 | 0.0 | 0.6 | 19.1 | 20.9 | 1.7 | 0.5 | 3.0 |
| Newark | 0.3 | 1.8 | 0.5 | 0.0 | 3.4 | 16.1 | 9.6 | 1.8 | 21.3 | 2.0 |
| New Brunswick | 0.0 | 0.7 | 4.5 | 58.2 | 0.3 | 5.8 | 4.4 | 0.5 | 4.4 | 5.5 |
| Paramus | 0.1 | 3.1 | 0.0 | 0.0 | 2.4 | 20.3 | 8.5 | 8.0 | 0.0 | 0.9 |
| Passaic | 0.0 | 2.9 | 0.6 | 0.0 | 2.2 | 17.4 | 9.3 | 1.3 | 21.7 | 0.5 |
| Paterson | 0.2 | 2.9 | 1.9 | 0.0 | 3.7 | 15.7 | 7.3 | 1.9 | 16.4 | 1.3 |
| Perth Amboy | 0.0 | 1.4 | 0.6 | 0.0 | 0.6 | 13.1 | 8.7 | 1.5 | 20.4 | 9.1 |
| Plainfield | 0.0 | 3.5 | 0.0 | 0.0 | 0.7 | 13.1 | 1.0 | 0.7 | 11.7 | 0.0 |
| Rahway | 0.0 | 9.5 | 0.0 | 0.0 | 1.5 | 17.9 | 14.3 | 1.7 | 4.3 | 3.5 |
| Sayreville | 0.0 | 8.1 | 0.0 | 0.0 | 0.2 | 21.8 | 14.8 | 2.3 | 3.5 | 1.2 |
| Trenton | 0.0 | 0.8 | 0.1 | 57.7 | 0.9 | 7.3 | 3.5 | 0.5 | 0.4 | 1.5 |
| Union City | 0.1 | 0.6 | 0.6 | 71.1 | 0.5 | 6.5 | 2.6 | 0.6 | 11.1 | 0.7 |
| Vineland | 0.0 | 29.2 | 0.0 | 0.0 | 4.3 | 16.9 | 13.7 | 0.8 | 11.1 | 0.1 |
| Westfield | 0.0 | 0.7 | 0.6 | 0.0 | 2.0 | 15.8 | 8.4 | 2.1 | 0.0 | 0.3 |
| West New York | 0.0 | 1.0 | 0.8 | 64.6 | 0.6 | 5.8 | 1.7 | 1.1 | 4.5 | 0.6 |
| **NEW MEXICO** | X | X | X | X | X | X | X | X | X | X |
| Alamogordo | 5.3 | 8.8 | 0.0 | 0.0 | 0.0 | 17.3 | 10.2 | 19.5 | 2.9 | 0.0 |
| Albuquerque | 3.5 | 3.4 | 0.5 | 0.0 | 3.9 | 20.8 | 20.1 | 12.8 | 0.2 | 2.0 |
| Carlsbad | 0.0 | 14.5 | 0.0 | 0.0 | 0.5 | 17.3 | 18.5 | 14.6 | 0.0 | 1.1 |
| Clovis | 0.0 | 18.0 | 0.0 | 0.0 | 0.1 | 12.9 | 12.9 | 7.1 | 6.0 | 1.6 |
| Farmington | 0.0 | 11.1 | 0.0 | 0.0 | 1.9 | 18.0 | 10.7 | 14.5 | 0.4 | 0.0 |
| Hobbs | 0.0 | 7.6 | 0.0 | 0.0 | 1.1 | 17.1 | 18.7 | 21.2 | 0.0 | 0.4 |
| Las Cruces | 0.0 | 1.6 | 0.0 | 0.0 | 0.3 | 24.4 | 29.1 | 6.7 | 1.5 | 0.0 |
| Rio Rancho | 1.1 | 16.0 | 0.0 | 0.0 | 0.0 | 20.9 | 15.1 | 7.1 | 0.5 | 0.0 |
| Roswell | 0.0 | 11.4 | 0.0 | 0.0 | 0.0 | 21.1 | 10.8 | 12.5 | 0.0 | 0.6 |
| Santa Fe | 3.4 | 6.5 | 2.4 | 0.0 | 0.0 | 11.2 | 17.1 | 9.4 | 13.3 | 0.1 |
| **NEW YORK** | X | X | X | X | X | X | X | X | X | X |
| Albany | 0.0 | 6.8 | 0.0 | 0.0 | 0.3 | 31.9 | 3.9 | 2.1 | 0.0 | 7.7 |
| Auburn | 0.0 | 7.9 | 0.3 | 0.0 | 0.1 | 14.8 | 19.1 | 3.0 | 1.9 | 5.8 |
| Binghamton | 0.0 | 6.6 | 0.5 | 0.0 | 0.1 | 9.6 | 40.1 | 2.7 | 2.8 | 5.2 |
| Buffalo | 0.0 | 3.8 | 0.2 | 61.6 | 0.1 | 6.3 | 2.8 | 1.2 | 0.0 | 2.5 |
| Elmira | 0.0 | 9.6 | 0.6 | 0.0 | 0.7 | 19.4 | 2.5 | 4.3 | 2.8 | 3.2 |
| Freeport | 0.0 | 6.3 | 0.1 | 0.0 | 0.0 | 24.3 | 4.3 | 4.3 | 0.0 | 10.6 |
| Glen Cove | 7.9 | 10.8 | 0.0 | 0.0 | 1.5 | 27.9 | 5.3 | 10.5 | 0.7 | 7.2 |
| Harrison | 0.0 | 17.3 | 0.0 | 0.0 | 0.0 | 1.6 | 22.9 | 0.0 | 0.0 | 20.8 |
| Hempstead | 0.0 | 3.3 | 0.4 | 0.0 | 0.0 | 30.3 | 4.1 | 4.4 | 0.4 | 4.4 |
| Ithaca | 0.0 | 9.3 | 3.2 | 0.0 | 0.1 | 12.4 | 6.6 | 10.5 | 0.0 | 8.9 |
| Jamestown | 0.0 | 6.2 | 0.1 | 39.5 | 0.1 | 6.8 | 8.3 | 2.1 | 0.0 | 3.0 |
| Lindenhurst | 0.0 | 28.8 | 0.1 | 0.0 | 0.0 | 0.8 | 4.9 | 7.3 | 0.7 | 2.2 |
| Long Beach | 0.0 | 4.3 | 0.0 | 0.2 | 0.1 | 15.5 | 12.1 | 8.6 | 0.5 | 6.7 |
| Middletown | 0.0 | 5.2 | 0.0 | 0.0 | 0.0 | 17.9 | 9.4 | 4.5 | 1.8 | 6.5 |
| Mount Vernon | 0.8 | 2.3 | 0.0 | 0.0 | 0.5 | 21.2 | 5.7 | 5.9 | 3.7 | 2.3 |
| Newburgh | 0.0 | 3.2 | 0.0 | 0.0 | 0.1 | 27.0 | 14.2 | 2.1 | 2.4 | 9.7 |
| New Rochelle | 6.9 | 8.2 | 1.4 | 0.0 | 0.3 | 23.3 | 4.9 | 2.5 | 1.2 | 3.9 |
| New York | 13.8 | 1.9 | 0.0 | 31.0 | 9.9 | 5.5 | 3.6 | 1.3 | 5.7 | 6.1 |
| Niagara Falls | 2.3 | 13.2 | 0.0 | 0.0 | 0.2 | 17.0 | 3.2 | 2.4 | 2.8 | 5.4 |
| North Tonawanda | 0.0 | 12.3 | 0.0 | 0.0 | 0.0 | 13.6 | 12.5 | 4.8 | 0.4 | 2.0 |
| Ossining | 6.9 | 8.3 | 0.1 | 0.0 | 0.3 | 19.2 | 8.3 | 5.8 | 7.7 | 4.1 |
| Port Chester | 0.0 | 9.6 | 0.9 | 0.0 | 0.6 | 21.7 | 8.9 | 3.5 | 0.0 | 6.0 |
| Poughkeepsie | 8.3 | 7.8 | 0.4 | 0.0 | 0.2 | 20.5 | 7.4 | 0.8 | 1.0 | 7.7 |
| Rochester | 0.0 | 2.8 | 1.1 | 58.0 | 0.1 | 6.2 | 3.2 | 1.2 | 0.8 | 1.3 |
| Rome | 0.0 | 9.8 | 0.4 | 0.0 | 0.2 | 12.9 | 10.0 | 2.0 | 2.8 | 7.0 |
| Saratoga Springs | 0.2 | 8.3 | 0.2 | 0.0 | 0.2 | 21.0 | 8.9 | 9.3 | 0.6 | 6.4 |
| Schenectady | 0.0 | 8.4 | 0.6 | 0.0 | 0.1 | 18.5 | 14.4 | 2.0 | 2.7 | 2.8 |
| Spring Valley | 19.5 | 4.0 | 0.1 | 0.0 | 0.0 | 19.1 | 0.1 | 0.6 | 19.5 | 2.3 |
| Syracuse | 0.0 | 3.8 | 0.0 | 60.4 | 0.0 | 6.2 | 2.7 | 1.1 | 0.9 | 2.6 |
| Troy | 0.0 | 6.3 | 0.0 | 0.0 | 0.2 | 25.5 | 7.5 | 3.0 | 2.4 | 7.1 |
| Utica | 7.5 | 5.7 | 0.4 | 0.0 | 1.1 | 18.4 | 3.5 | 3.4 | 1.9 | 4.0 |
| Valley Stream | 0.0 | 14.9 | 0.9 | 0.0 | 0.3 | 0.5 | 13.7 | 10.2 | 0.0 | 4.3 |
| Watertown | 0.0 | 13.9 | 0.1 | 0.0 | 0.2 | 13.2 | 11.7 | 3.9 | 0.5 | 3.2 |
| White Plains | 0.0 | 8.9 | 0.1 | 0.0 | 0.0 | 20.3 | 5.4 | 5.0 | 0.1 | 5.2 |

# Table D. Cities — City Government Finances, City Government Employment, and Climate

| City | City government finances, 2017 (cont.) | | | Climate[2] | | | | | | |
| | Debt outstanding | | Debt issued during year | Average daily temperature | | | | Annual precipitation (inches) | Heating degree days | Cooling degree days |
| | Total (mil dol) | Per capita[1] (dollars) | | Mean | | Limits | | | | |
| | | | | January | July | January[3] | July[4] | | | |
| | 137 | 138 | 139 | 140 | 141 | 142 | 143 | 144 | 145 | 146 |
| **NEW JERSEY—Cont'd** | | | | | | | | | | |
| Garfield | 43.8 | 1,375 | 7.7 | 28.6 | 75.0 | 19.5 | 85.5 | 51.50 | 5,522 | 824 |
| Hackensack | 57.0 | 1,281 | 0.0 | 28.6 | 75.0 | 19.5 | 85.5 | 51.50 | 5,522 | 824 |
| Hoboken | 132.1 | 2,461 | 0.0 | 29.6 | 75.3 | 22.7 | 82.5 | 46.33 | 5,367 | 882 |
| Jersey City | 661.4 | 2,512 | 0.0 | 29.6 | 75.3 | 22.7 | 82.5 | 46.33 | 5,367 | 882 |
| Kearny | 69.8 | 1,678 | 34.0 | 31.3 | 77.2 | 24.4 | 85.2 | 46.25 | 4,843 | 1,220 |
| Linden | 52.9 | 1,249 | 1.0 | 28.5 | 74.0 | 18.2 | 85.8 | 51.61 | 5,595 | 757 |
| Long Branch | 82.5 | 2,699 | 0.0 | 31.7 | 74.1 | 22.8 | 82.6 | 48.63 | 5,168 | 750 |
| Millville | 31.1 | 1,122 | 0.1 | 32.7 | 76.3 | 24.1 | 85.9 | 43.20 | 4,835 | 1,009 |
| Newark | 433.8 | 1,543 | 0.0 | 31.3 | 77.2 | 24.4 | 85.2 | 46.25 | 4,843 | 1,220 |
| New Brunswick | 575.5 | 10,256 | 190.1 | 29.7 | 74.8 | 21.1 | 85.4 | 48.78 | 5,346 | 816 |
| Paramus | 37.7 | 1,416 | 0.0 | 28.6 | 75.0 | 19.5 | 85.5 | 51.50 | 5,522 | 824 |
| Passaic | 5.6 | 80 | 0.0 | 28.6 | 75.0 | 19.5 | 85.5 | 51.50 | 5,522 | 824 |
| Paterson | 82.4 | 564 | 0.0 | 28.6 | 75.0 | 19.5 | 85.5 | 51.50 | 5,522 | 824 |
| Perth Amboy | 175.2 | 3,381 | 4.3 | 29.7 | 74.8 | 21.1 | 85.4 | 48.78 | 5,346 | 816 |
| Plainfield | 36.3 | 720 | 0.0 | 30.0 | 74.9 | 21.5 | 86.6 | 49.63 | 5,266 | 854 |
| Rahway | 79.0 | 2,667 | 20.6 | 29.6 | 74.5 | 19.8 | 85.7 | 50.94 | 5,450 | 787 |
| Sayreville | 51.1 | 1,148 | 7.0 | 29.7 | 74.8 | 21.1 | 85.4 | 48.78 | 5,346 | 816 |
| Trenton | 321.1 | 3,846 | 24.4 | 30.4 | 75.2 | 21.3 | 86.9 | 48.83 | 5,262 | 903 |
| Union City | 108.9 | 1,590 | 10.3 | 29.6 | 75.3 | 22.7 | 82.5 | 46.33 | 5,367 | 882 |
| Vineland | 59.6 | 995 | 29.8 | 26.7 | 70.4 | 16.8 | 81.7 | 53.28 | 6,281 | 438 |
| Westfield | 25.7 | 859 | 0.0 | 30.0 | 74.9 | 21.5 | 86.6 | 49.63 | 5,266 | 854 |
| West New York | 46.5 | 882 | 11.6 | 29.6 | 75.3 | 22.7 | 82.5 | 46.33 | 5,367 | 882 |
| **NEW MEXICO** | X | X | X | X | X | X | X | X | X | X |
| Alamogordo | 63.6 | 2,027 | 17.5 | 42.2 | 79.7 | 28.9 | 93.0 | 13.20 | 31 | 1,715 |
| Albuquerque | 1,567.1 | 2,800 | 129.9 | 35.7 | 78.5 | 23.8 | 92.3 | 9.47 | 4,281 | 1,290 |
| Carlsbad | 52.0 | 1,801 | 0.0 | 42.7 | 81.7 | 27.5 | 95.8 | 14.15 | 2,823 | 2,029 |
| Clovis | 22.0 | 564 | 1.2 | 37.9 | 77.5 | 25.0 | 91.0 | 18.50 | 3,955 | 1,305 |
| Farmington | 1,684.2 | 37,082 | 30.6 | 29.8 | 74.9 | 17.9 | 90.7 | 8.39 | 5,508 | 805 |
| Hobbs | 39.9 | 1,055 | 0.0 | 42.9 | 80.1 | 29.1 | 93.5 | 18.15 | 2,849 | 1,842 |
| Las Cruces | 202.4 | 1,985 | 40.3 | 39.0 | 78.7 | 21.1 | 94.9 | 11.44 | 3,818 | 1,364 |
| Rio Rancho | 228.9 | 2,377 | 61.3 | 33.8 | 73.9 | 19.7 | 90.0 | 9.28 | 4,981 | 773 |
| Roswell | 8.8 | 184 | 0.0 | 40.0 | 80.8 | 24.4 | 94.8 | 13.34 | 3,332 | 1,814 |
| Santa Fe | 259.0 | 3,085 | 46.8 | 29.3 | 69.8 | 15.5 | 85.6 | 14.22 | 6,073 | 414 |
| **NEW YORK** | X | X | X | X | X | X | X | X | X | X |
| Albany | 288.2 | 2,949 | 23.7 | 22.2 | 71.1 | 13.3 | 82.2 | 38.60 | 6,860 | 544 |
| Auburn | 67.1 | 2,518 | 4.9 | 23.7 | 71.2 | 16.0 | 81.5 | 36.98 | 6,694 | 528 |
| Binghamton | 172.1 | 3,820 | 39.5 | 21.7 | 68.7 | 15.0 | 78.1 | 38.65 | 7,237 | 396 |
| Buffalo | 1,263.9 | 4,931 | 157.9 | 24.5 | 70.8 | 17.8 | 79.6 | 40.54 | 6,692 | 548 |
| Elmira | 26.8 | 983 | 0.0 | 23.9 | 70.3 | 15.0 | 82.3 | 34.95 | 6,806 | 446 |
| Freeport | 109.3 | 2,534 | 17.1 | 30.7 | 73.8 | 24.2 | 81.0 | 42.97 | 5,504 | 779 |
| Glen Cove | 57.5 | 2,109 | 5.0 | 31.9 | 74.2 | 25.4 | 82.8 | 46.36 | 5,231 | 839 |
| Harrison | 62.1 | 2,223 | 0.0 | NA | NA | NA | NA | NA | NA | NA |
| Hempstead | 49.2 | 888 | 6.5 | 31.9 | 74.2 | 25.4 | 82.8 | 46.36 | 5,231 | 839 |
| Ithaca | 127.5 | 4,154 | 36.9 | 22.6 | 68.7 | 13.9 | 80.1 | 36.71 | 7,182 | 312 |
| Jamestown | 32.4 | 1,098 | 0.0 | 22.3 | 69.2 | 14.1 | 80.1 | 45.68 | 7,048 | 389 |
| Lindenhurst | 9.0 | 333 | 0.0 | 30.7 | 73.8 | 24.2 | 81.0 | 42.97 | 5,504 | 779 |
| Long Beach | 142.6 | 4,248 | 6.9 | 31.8 | 74.8 | 24.7 | 82.9 | 42.46 | 4,947 | 949 |
| Middletown | 56.3 | 2,016 | 3.5 | 26.5 | 73.0 | 17.5 | 84.0 | 44.00 | 5,820 | 674 |
| Mount Vernon | 29.5 | 435 | 0.0 | 29.7 | 74.2 | 20.1 | 86.0 | 46.46 | 5,400 | 770 |
| Newburgh | 91.0 | 3,226 | 0.0 | 26.6 | 74.3 | 17.1 | 84.9 | 45.79 | 5,813 | 790 |
| New Rochelle | 125.9 | 1,592 | 0.0 | 29.7 | 74.2 | 20.1 | 86.0 | 46.46 | 5,400 | 770 |
| New York | 140,617.8 | 16,666 | 16,835.6 | 32.1 | 76.5 | 26.2 | 84.2 | 49.69 | 4,754 | 1,151 |
| Niagara Falls | 55.1 | 1,142 | 37.5 | 24.2 | 71.4 | 16.8 | 81.8 | 33.93 | 6,752 | 508 |
| North Tonawanda | 13.6 | 448 | 9.1 | 24.2 | 71.4 | 16.8 | 81.8 | 33.93 | 6,752 | 508 |
| Ossining | 33.7 | 1,343 | 6.4 | NA | NA | NA | NA | NA | NA | NA |
| Port Chester | 43.2 | 1,470 | 12.5 | 28.4 | 73.8 | 21.0 | 82.5 | 50.45 | 5,660 | 716 |
| Poughkeepsie | 71.0 | 2,334 | 0.0 | 24.5 | 71.9 | 14.7 | 83.6 | 44.12 | 6,438 | 550 |
| Rochester | 804.2 | 3,896 | 37.5 | 23.9 | 70.7 | 16.6 | 81.4 | 33.98 | 6,728 | 576 |
| Rome | 73.5 | 2,277 | 18.9 | 20.8 | 70.2 | 11.9 | 81.3 | 46.27 | 7,146 | 416 |
| Saratoga Springs | 52.5 | 1,877 | 3.0 | 20.9 | 71.2 | 11.6 | 83.0 | 43.31 | 6,904 | 477 |
| Schenectady | 113.6 | 1,739 | 21.5 | 22.2 | 71.1 | 13.3 | 82.2 | 38.60 | 6,860 | 544 |
| Spring Valley | 11.1 | 343 | 0.0 | 27.3 | 73.1 | 18.2 | 83.8 | 51.01 | 5,809 | 642 |
| Syracuse | 441.8 | 3,105 | 52.8 | 22.7 | 70.9 | 14.0 | 81.7 | 40.05 | 6,803 | 551 |
| Troy | 70.2 | 1,420 | 0.0 | 22.2 | 71.1 | 13.3 | 82.2 | 38.60 | 6,860 | 544 |
| Utica | 68.6 | 1,137 | 2.9 | 22.2 | 70.5 | 12.6 | 83.2 | 41.90 | 6,855 | 441 |
| Valley Stream | 30.3 | 805 | 4.7 | 22.2 | 70.5 | 12.6 | 83.2 | 41.90 | 6,855 | 441 |
| Watertown | 35.9 | 1,398 | 5.1 | 18.6 | 70.2 | 9.1 | 79.4 | 42.57 | 7,517 | 421 |
| White Plains | 160.2 | 2,743 | 38.5 | 29.7 | 74.2 | 20.1 | 86.0 | 46.46 | 5,400 | 770 |

1. Based on the population estimated as of July 1 of the year shown.   2. Represents normal values based on the 30-year period, 1971±2000.   3. Average daily minimum.   4. Average daily maximum.

Items 137—146

# Table D. Cities — Land Area and Population

| STATE | Place code | City | Land area[1] (sq. mi) | Total persons 2019 | Rank | Per square mile | White | Black or African American | American Indian, Alaskan Native | Asian | Hawaiian Pacific Islander | Some other race | Two or more races (percent) |
|---|---|---|---|---|---|---|---|---|---|---|---|---|---|
| | | | 1 | 2 | 3 | 4 | 5 | 6 | 7 | 8 | 9 | 10 | 11 |
| | | **NEW YORK—Cont'd** | | | | | | | | | | | |
| 36 | 84000 | Yonkers................ | 18.0 | 200,040 | 119 | 11,113 | 46.5 | 20.9 | 0.5 | 5.7 | 0.0 | 22.5 | 3.8 |
| 37 | 00000 | NORTH CAROLINA.......... | 48,620.3 | 10,600,823 | X | 218 | 68.1 | 21.5 | 1.2 | 3.0 | 0.1 | 3.4 | 2.8 |
| 37 | 01520 | Apex................ | 22.0 | 64,388 | 588 | 2,927 | 74.7 | 9.5 | 0.2 | 10.6 | 0.1 | 2.8 | 2.2 |
| 37 | 02080 | Asheboro............ | 18.9 | 26,023 | 1,386 | 1,377 | 73.3 | 15.4 | 0.6 | 0.3 | 0.0 | 7.2 | 3.3 |
| 37 | 02140 | Asheville........... | 45.5 | 92,852 | 359 | 2,041 | 82.1 | 12.8 | 1.1 | 1.5 | 0.0 | 0.7 | 1.9 |
| 37 | 09060 | Burlington.......... | 30.1 | 55,599 | 692 | 1,847 | 58.4 | 27.2 | 0.0 | 4.1 | 0.0 | 6.5 | 3.8 |
| 37 | 10740 | Cary................ | 59.2 | 173,587 | 152 | 2,932 | 66.0 | 8.3 | 0.4 | 20.5 | 0.0 | 1.9 | 2.9 |
| 37 | 11800 | Chapel Hill......... | 21.6 | 63,705 | 594 | 2,949 | 71.4 | 9.8 | 0.3 | 14.1 | 0.0 | 1.0 | 3.4 |
| 37 | 12000 | Charlotte........... | 308.3 | 900,350 | 15 | 2,920 | 46.9 | 35.4 | 0.4 | 6.2 | 0.0 | 7.8 | 3.3 |
| 37 | 14100 | Concord............. | 63.5 | 98,500 | 323 | 1,551 | 65.8 | 17.0 | 0.2 | 5.2 | 0.0 | 9.1 | 2.8 |
| 37 | 19000 | Durham.............. | 112.8 | 285,897 | 72 | 2,535 | 50.4 | 37.7 | 0.1 | 5.0 | 0.1 | 3.3 | 3.4 |
| 37 | 22920 | Fayetteville........ | 148.3 | 211,705 | 107 | 1,428 | 44.0 | 41.4 | 1.1 | 3.4 | 0.4 | 3.0 | 6.8 |
| 37 | 25480 | Garner.............. | 16.6 | 32,180 | 1,188 | 1,939 | D | D | D | D | D | D | D |
| 37 | 25580 | Gastonia............ | 51.7 | 77,640 | 455 | 1,502 | 56.6 | 32.7 | 0.2 | 1.7 | 0.0 | 4.9 | 3.8 |
| 37 | 26880 | Goldsboro........... | 28.6 | 34,447 | 1,119 | 1,204 | 35.6 | 57.3 | 0.5 | 2.4 | 0.0 | 0.2 | 4.1 |
| 37 | 28000 | Greensboro.......... | 129.6 | 297,878 | 66 | 2,298 | 43.1 | 42.9 | 0.6 | 5.8 | 0.0 | 3.5 | 4.1 |
| 37 | 28080 | Greenville.......... | 36.6 | 94,779 | 346 | 2,590 | 49.1 | 41.5 | 0.0 | 2.9 | 0.0 | 1.3 | 5.2 |
| 37 | 31060 | Hickory............. | 30.7 | 41,473 | 927 | 1,351 | 66.8 | 19.8 | 0.0 | 3.9 | 0.0 | 6.0 | 3.5 |
| 37 | 31400 | High Point.......... | 56.4 | 113,565 | 259 | 2,014 | 51.0 | 35.2 | 1.0 | 6.9 | 0.1 | 2.2 | 3.7 |
| 37 | 33120 | Huntersville........ | 41.3 | 59,095 | 639 | 1,431 | 82.0 | 9.8 | 0.2 | 3.4 | 0.0 | 1.8 | 2.8 |
| 37 | 33560 | Indian Trail........ | 22.2 | 41,040 | 936 | 1,849 | 86.0 | 6.2 | 0.6 | 1.2 | 0.0 | 3.9 | 2.1 |
| 37 | 34200 | Jacksonville........ | 48.7 | 76,284 | 467 | 1,566 | 69.2 | 16.6 | 0.5 | 3.2 | 0.0 | 1.1 | 9.5 |
| 37 | 35200 | Kannapolis.......... | 32.7 | 52,007 | 746 | 1,590 | 65.8 | 27.2 | 0.0 | 0.4 | 0.0 | 4.3 | 2.2 |
| 37 | 41960 | Matthews............ | 17.1 | 33,718 | 1,143 | 1,972 | 75.4 | 15.9 | 0.0 | 4.3 | 0.0 | 1.5 | 3.0 |
| 37 | 43920 | Monroe.............. | 30.6 | 35,668 | 1,084 | 1,166 | 54.2 | 33.0 | 0.5 | 0.1 | 0.0 | 7.6 | 4.6 |
| 37 | 44220 | Mooresville......... | 24.9 | 39,962 | 963 | 1,605 | 77.8 | 9.1 | 0.2 | 4.6 | 0.0 | 3.8 | 4.6 |
| 37 | 46340 | New Bern............ | 28.4 | 29,940 | 1,255 | 1,054 | D | D | D | D | D | D | D |
| 37 | 55000 | Raleigh............. | 147.1 | 474,414 | 41 | 3,225 | 58.0 | 29.0 | 0.7 | 4.0 | 0.0 | 5.8 | 2.4 |
| 37 | 57500 | Rocky Mount......... | 45.0 | 53,689 | 721 | 1,193 | 30.8 | 64.6 | 0.0 | 0.4 | 0.0 | 2.7 | 1.5 |
| 37 | 58860 | Salisbury........... | 21.8 | 33,821 | 1,140 | 1,551 | 57.7 | 32.3 | 0.0 | 0.8 | 0.0 | 5.1 | 4.0 |
| 37 | 59280 | Sanford............. | 29.3 | 30,361 | 1,237 | 1,036 | D | D | D | D | D | D | D |
| 37 | 67420 | Thomasville......... | 16.8 | 26,713 | 1,364 | 1,590 | D | D | D | D | D | D | D |
| 37 | 70540 | Wake Forest......... | 18.0 | 47,314 | 824 | 2,629 | 80.2 | 15.3 | 0.1 | 0.6 | 0.0 | 1.7 | 2.2 |
| 37 | 74440 | Wilmington.......... | 51.4 | 124,794 | 225 | 2,428 | 75.1 | 18.4 | 0.2 | 0.9 | 0.1 | 2.2 | 3.1 |
| 37 | 74540 | Wilson.............. | 31.0 | 49,479 | 783 | 1,596 | 38.4 | 53.9 | 0.0 | 0.4 | 0.0 | 5.9 | 1.5 |
| 37 | 75000 | Winston-Salem....... | 132.7 | 248,112 | 90 | 1,870 | 56.8 | 35.7 | 0.3 | 2.1 | 0.0 | 2.3 | 2.8 |
| 38 | 00000 | NORTH DAKOTA........ | 68,994.8 | 765,309 | X | 11 | 85.8 | 2.9 | 5.4 | 1.4 | 0.4 | 0.8 | 3.3 |
| 38 | 07200 | Bismarck............ | 34.7 | 74,018 | 484 | 2,133 | 88.3 | 2.4 | 5.2 | 0.9 | 0.2 | 0.2 | 2.8 |
| 38 | 25700 | Fargo............... | 49.8 | 125,209 | 222 | 2,514 | 82.1 | 8.9 | 1.3 | 3.0 | 0.0 | 0.7 | 4.0 |
| 38 | 32060 | Grand Forks......... | 27.9 | 55,950 | 688 | 2,005 | 84.3 | 2.4 | 3.5 | 3.6 | 2.9 | 0.5 | 2.7 |
| 38 | 53380 | Minot............... | 27.3 | 47,428 | 819 | 1,737 | 84.3 | 4.4 | 1.2 | 2.4 | 0.1 | 1.1 | 6.4 |
| 38 | 84780 | West Fargo.......... | 16.1 | 38,194 | 1,015 | 2,372 | D | D | D | D | D | D | D |
| 39 | 00000 | OHIO................ | 40,858.8 | 11,693,217 | X | 286 | 80.9 | 12.6 | 0.2 | 2.3 | 0.0 | 1.0 | 2.9 |
| 39 | 01000 | Akron............... | 61.9 | 195,994 | 131 | 3,166 | 60.5 | 29.7 | 0.3 | 5.5 | 0.0 | 1.2 | 2.9 |
| 39 | 03828 | Barberton........... | 9.0 | 25,815 | 1,392 | 2,868 | D | D | D | D | D | D | D |
| 39 | 04720 | Beavercreek......... | 26.6 | 47,995 | 808 | 1,804 | 85.8 | 4.6 | 0.3 | 3.5 | 0.7 | 0.0 | 5.1 |
| 39 | 07972 | Bowling Green....... | 12.8 | 31,444 | 1,211 | 2,457 | 86.8 | 6.8 | 0.6 | 0.9 | 0.0 | 0.5 | 4.5 |
| 39 | 09680 | Brunswick........... | 13.0 | 35,052 | 1,103 | 2,696 | D | D | D | D | D | D | D |
| 39 | 12000 | Canton.............. | 26.3 | 69,963 | 523 | 2,660 | D | D | D | D | D | D | D |
| 39 | 15000 | Cincinnati.......... | 77.8 | 304,548 | 64 | 3,915 | 52.3 | 41.2 | 0.1 | 2.8 | 0.0 | 0.2 | 3.3 |
| 39 | 16000 | Cleveland........... | 77.7 | 378,589 | 54 | 4,872 | 38.9 | 49.8 | 0.7 | 2.5 | 0.2 | 3.6 | 4.3 |
| 39 | 16014 | Cleveland Heights... | 8.1 | 43,569 | 888 | 5,379 | 54.5 | 38.3 | 0.0 | 4.9 | 0.0 | 0.3 | 2.0 |
| 39 | 18000 | Columbus............ | 220.0 | 903,852 | 14 | 4,108 | 57.9 | 29.2 | 0.3 | 5.9 | 0.0 | 2.7 | 4.0 |
| 39 | 19778 | Cuyahoga Falls...... | 25.8 | 49,005 | 792 | 1,899 | 80.1 | 7.5 | 0.0 | 5.5 | 0.0 | 0.1 | 6.9 |
| 39 | 21000 | Dayton.............. | 55.8 | 139,907 | 196 | 2,507 | 53.0 | 37.5 | 1.2 | 1.4 | 0.0 | 2.4 | 4.4 |
| 39 | 21434 | Delaware............ | 20.0 | 42,607 | 903 | 2,130 | 84.4 | 7.6 | 0.0 | 4.8 | 0.0 | 0.4 | 2.9 |
| 39 | 22694 | Dublin.............. | 24.7 | 49,372 | 784 | 1,999 | D | D | D | D | D | D | D |
| 39 | 25256 | Elyria.............. | 20.5 | 53,873 | 718 | 2,628 | D | D | D | D | D | D | D |
| 39 | 25704 | Euclid.............. | 10.7 | 46,205 | 840 | 4,318 | 79.6 | 9.2 | 0.0 | 4.2 | 0.0 | 0.0 | 6.9 |
| 39 | 25914 | Fairborn............ | 14.6 | 33,990 | 1,135 | 2,328 | 71.1 | 22.7 | 0.2 | 0.8 | 0.0 | 4.0 | 1.2 |
| 39 | 25970 | Fairfield........... | 20.8 | 42,603 | 904 | 2,048 | 91.3 | 1.7 | 0.6 | 2.6 | 0.0 | 2.0 | 1.8 |
| 39 | 27048 | Findlay............. | 19.6 | 40,832 | 944 | 2,083 | 80.2 | 14.5 | 0.3 | 1.4 | 0.2 | 0.9 | 2.5 |
| 39 | 29106 | Gahanna............. | 12.4 | 35,531 | 1,088 | 2,865 | D | D | D | D | D | D | D |
| 39 | 29428 | Garfield Heights.... | 7.2 | 27,272 | 1,342 | 3,788 | D | D | D | D | D | D | D |
| 39 | 31860 | Green............... | 32.0 | 25,678 | 1,395 | 802 | D | D | D | D | D | D | D |
| 39 | 32592 | Grove City.......... | 17.5 | 42,280 | 911 | 2,416 | 89.0 | 3.1 | 0.0 | 4.1 | 0.0 | 1.5 | 2.3 |

1. Dry land or land partially or temporarily covered by water.   2. Hispanic or Latino persons may be of any race.

# Table D. Cities — **Population**

| City | Percent Hispanic or Latino[1], 2019 | Percent foreign born, 2019 | Age of population (percent), 2019 | | | | | | | Median age, 2019 | Percent female, 2019 | Population Census counts | | Percent change | |
|---|---|---|---|---|---|---|---|---|---|---|---|---|---|---|---|
| | | | Under 18 years | 18 to 24 years | 25 to 34 years | 35 to 44 years | 45 to 54 years | 55 to 64 years | 65 years and over | | | 2000 | 2010 | 2000-2010 | 2001-2020 |
| | 12 | 13 | 14 | 15 | 16 | 17 | 18 | 19 | 20 | 21 | 22 | 23 | 24 | 25 | 26 |
| NEW YORK—Cont'd | | | | | | | | | | | | | | | |
| Yonkers | 42.0 | 32.7 | 23.5 | 10.2 | 13.4 | 12.5 | 12.5 | 11.5 | 16.4 | 37.9 | 52.2 | 196,086 | 196,129 | 0.0 | 2.0 |
| NORTH CAROLINA | 9.8 | 8.4 | 21.9 | 9.6 | 13.4 | 12.5 | 13.0 | 13.0 | 16.7 | 39.1 | 51.4 | 8,049,313 | 9,535,762 | 18.5 | 11.2 |
| Apex | 12.2 | 11.9 | 29.4 | 6.3 | 14.7 | 16.6 | 17.3 | 9.6 | 6.1 | 34.8 | 51.7 | 20,212 | 37,802 | 87.0 | 70.3 |
| Asheboro | 29.4 | 10.6 | 24.2 | 10.2 | 16.2 | 12.5 | 9.3 | 8.5 | 18.9 | 34.4 | 53.6 | 21,672 | 25,391 | 17.2 | 2.5 |
| Asheville | 7.4 | 9.2 | 20.0 | 7.9 | 17.2 | 14.7 | 13.6 | 9.4 | 17.2 | 39.0 | 52.6 | 68,889 | 83,359 | 21.0 | 11.4 |
| Burlington | 16.3 | 10.9 | 27.6 | 5.5 | 11.6 | 15.0 | 11.0 | 12.2 | 16.9 | 38.3 | 52.6 | 44,917 | 51,062 | 13.7 | 8.9 |
| Cary | 6.1 | 23.8 | 25.2 | 5.5 | 13.3 | 15.0 | 16.7 | 11.9 | 12.6 | 39.2 | 51.0 | 94,536 | 135,869 | 43.7 | 27.8 |
| Chapel Hill | 5.2 | 16.7 | 13.6 | 33.2 | 13.1 | 8.4 | 11.3 | 6.9 | 13.5 | 26.6 | 53.5 | 48,715 | 57,221 | 17.5 | 11.3 |
| Charlotte | 15.0 | 17.0 | 23.5 | 9.2 | 18.6 | 14.6 | 12.7 | 10.8 | 10.6 | 34.2 | 52.1 | 540,828 | 735,580 | 36.0 | 22.4 |
| Concord | 14.7 | 12.8 | 25.9 | 10.4 | 13.1 | 14.4 | 13.9 | 10.6 | 11.8 | 35.6 | 52.1 | 55,977 | 79,342 | 41.7 | 24.1 |
| Durham | 13.1 | 16.4 | 20.6 | 11.1 | 19.3 | 14.4 | 11.4 | 10.8 | 12.5 | 34.5 | 52.2 | 187,035 | 230,740 | 23.4 | 23.9 |
| Fayetteville | 13.3 | 8.1 | 24.3 | 14.6 | 18.5 | 11.9 | 9.3 | 9.8 | 11.5 | 30.2 | 49.7 | 121,015 | 200,565 | 65.7 | 5.6 |
| Garner | 10.6 | 9.1 | 24.4 | 9.4 | 12.8 | 15.5 | 11.8 | 11.2 | 14.9 | 36.2 | 54.9 | 17,757 | 25,786 | 45.2 | 24.8 |
| Gastonia | 11.5 | 8.6 | 24.3 | 9.0 | 13.0 | 12.9 | 15.0 | 10.2 | 15.5 | 38.2 | 52.9 | 66,277 | 71,729 | 8.2 | 8.2 |
| Goldsboro | 4.7 | 4.0 | 18.0 | 13.5 | 12.6 | 9.7 | 11.5 | 14.6 | 20.1 | 41.0 | 52.8 | 39,043 | 35,413 | -9.3 | -2.7 |
| Greensboro | 9.3 | 14.0 | 22.0 | 12.9 | 15.6 | 12.0 | 12.0 | 11.8 | 13.7 | 34.7 | 54.2 | 223,891 | 268,936 | 20.1 | 10.8 |
| Greenville | 1.8 | 4.5 | 17.4 | 26.5 | 15.6 | 10.3 | 10.0 | 9.3 | 10.9 | 28.1 | 56.1 | 60,476 | 84,630 | 39.9 | 12.0 |
| Hickory | 9.7 | 6.1 | 23.0 | 10.0 | 14.8 | 11.3 | 12.2 | 13.1 | 15.8 | 36.4 | 51.8 | 37,222 | 40,186 | 8.0 | 3.2 |
| High Point | 10.0 | 12.8 | 23.7 | 11.1 | 14.2 | 11.7 | 11.7 | 12.1 | 15.4 | 35.9 | 51.5 | 85,839 | 104,517 | 21.8 | 8.7 |
| Huntersville | 9.6 | 13.0 | 24.9 | 7.4 | 8.7 | 17.0 | 18.8 | 10.5 | 12.7 | 40.9 | 49.7 | 24,960 | 46,924 | 88.0 | 25.9 |
| Indian Trail | 9.1 | 9.4 | 23.8 | 11.0 | 9.2 | 11.3 | 15.9 | 15.4 | 13.5 | 41.7 | 53.9 | 11,905 | 33,618 | 182.4 | 22.1 |
| Jacksonville | 23.3 | 6.3 | 20.7 | 34.5 | 17.6 | 9.0 | 6.7 | 4.8 | 6.8 | 23.3 | 37.6 | 66,715 | 70,285 | 5.4 | 8.5 |
| Kannapolis | 13.3 | 9.2 | 26.7 | 7.8 | 14.6 | 13.3 | 11.6 | 12.2 | 13.8 | 35.6 | 49.6 | 36,910 | 42,618 | 15.5 | 22.0 |
| Matthews | 3.7 | 10.9 | 21.6 | 6.9 | 18.9 | 12.0 | 10.4 | 14.9 | 15.4 | 36.0 | 50.0 | 22,127 | 27,189 | 22.9 | 24.0 |
| Monroe | 26.1 | 12.9 | 27.1 | 11.9 | 8.6 | 13.6 | 13.9 | 11.3 | 13.5 | 37.3 | 51.8 | 26,228 | 32,909 | 25.5 | 8.4 |
| Mooresville | 9.7 | 10.3 | 24.0 | 9.8 | 14.1 | 16.5 | 13.6 | 11.6 | 10.4 | 36.1 | 50.5 | 18,823 | 34,377 | 82.6 | 16.2 |
| New Bern | 6.8 | 6.5 | 20.6 | 5.6 | 12.6 | 7.4 | 11.7 | 13.4 | 28.7 | 47.4 | 55.4 | 23,128 | 29,317 | 26.8 | 2.1 |
| Raleigh | 12.5 | 14.2 | 20.9 | 11.9 | 17.9 | 14.3 | 13.3 | 10.0 | 11.7 | 34.5 | 51.8 | 276,093 | 404,118 | 46.4 | 17.4 |
| Rocky Mount | 3.1 | 3.7 | 17.2 | 12.6 | 9.9 | 11.0 | 11.6 | 17.7 | 20.1 | 44.3 | 51.9 | 55,893 | 57,701 | 3.2 | -7.0 |
| Salisbury | 15.3 | 5.7 | 24.0 | 12.6 | 12.8 | 12.1 | 12.0 | 8.8 | 17.8 | 35.5 | 49.4 | 26,462 | 33,514 | 26.6 | 0.9 |
| Sanford | 25.7 | 9.6 | 28.0 | 9.5 | 14.9 | 8.7 | 11.0 | 12.8 | 15.1 | 33.1 | 47.8 | 23,220 | 28,217 | 21.5 | 7.6 |
| Thomasville | 15.4 | 17.6 | 16.6 | 6.9 | 10.5 | 8.6 | 21.3 | 15.7 | 20.4 | 48.8 | 58.7 | 19,788 | 26,802 | 35.4 | -0.3 |
| Wake Forest | 4.9 | 9.4 | 25.9 | 6.3 | 8.3 | 14.9 | 15.9 | 15.6 | 12.9 | 42.9 | 51.5 | 12,588 | 30,135 | 139.4 | 57.0 |
| Wilmington | 6.8 | 4.9 | 17.5 | 17.2 | 14.5 | 10.3 | 10.2 | 12.4 | 17.9 | 35.9 | 54.7 | 75,838 | 106,465 | 40.4 | 17.2 |
| Wilson | 9.7 | 5.3 | 22.2 | 10.3 | 11.9 | 10.6 | 11.9 | 14.4 | 18.8 | 38.2 | 54.3 | 44,405 | 49,149 | 10.7 | 0.7 |
| Winston-Salem | 16.2 | 9.3 | 25.1 | 11.0 | 13.9 | 13.0 | 12.1 | 11.5 | 13.4 | 35.0 | 52.9 | 185,776 | 229,624 | 23.6 | 8.1 |
| NORTH DAKOTA | 4.0 | 4.1 | 23.2 | 10.6 | 15.4 | 12.6 | 10.0 | 12.3 | 15.8 | 35.5 | 49.0 | 642,200 | 672,575 | 4.7 | 13.8 |
| Bismarck | 2.6 | 3.6 | 22.6 | 10.6 | 14.8 | 13.1 | 9.5 | 13.2 | 16.2 | 35.9 | 50.7 | 55,532 | 61,324 | 10.4 | 20.7 |
| Fargo | 3.1 | 10.2 | 19.1 | 16.0 | 18.3 | 13.7 | 9.5 | 10.4 | 12.9 | 33.4 | 48.7 | 90,599 | 105,612 | 16.6 | 18.6 |
| Grand Forks | 4.1 | 5.2 | 19.5 | 22.3 | 17.4 | 10.0 | 6.6 | 11.5 | 12.8 | 29.5 | 47.8 | 49,321 | 52,921 | 7.3 | 5.7 |
| Minot | 6.9 | 6.5 | 18.4 | 15.2 | 19.2 | 13.0 | 10.5 | 10.1 | 13.6 | 32.6 | 49.8 | 36,567 | 41,095 | 12.4 | 15.4 |
| West Fargo | 0.2 | 2.2 | 26.9 | 8.5 | 20.6 | 13.1 | 9.9 | 9.0 | 11.9 | 31.7 | 51.6 | 14,940 | 25,854 | 73.1 | 47.7 |
| OHIO | 4.0 | 4.8 | 22.0 | 9.1 | 13.2 | 12.0 | 12.4 | 13.7 | 17.5 | 39.6 | 51.0 | 11,353,140 | 11,536,763 | 1.6 | 1.4 |
| Akron | 3.3 | 7.3 | 20.1 | 12.6 | 13.9 | 11.3 | 12.8 | 14.0 | 15.2 | 38.0 | 51.4 | 217,074 | 199,077 | -8.3 | -1.5 |
| Barberton | 0.0 | 0.6 | 21.5 | 3.8 | 22.5 | 7.7 | 6.2 | 16.2 | 22.1 | 38.3 | 53.8 | 27,899 | 26,538 | -4.9 | -2.7 |
| Beavercreek | 0.8 | 5.4 | 21.1 | 5.9 | 14.1 | 11.9 | 14.6 | 14.2 | 18.2 | 41.6 | 51.2 | 37,984 | 45,191 | 19.0 | 6.2 |
| Bowling Green | 6.6 | 2.8 | 12.6 | 40.6 | 13.4 | 6.3 | 7.1 | 9.2 | 10.9 | 23.1 | 52.1 | 29,636 | 30,113 | 1.6 | 4.4 |
| Brunswick | 2.7 | 1.9 | 22.8 | 5.3 | 12.9 | 14.6 | 11.7 | 14.2 | 18.4 | 41.4 | 50.7 | 33,388 | 34,283 | 2.7 | 2.2 |
| Canton | 4.8 | 2.0 | 23.5 | 9.2 | 14.0 | 11.7 | 11.3 | 14.6 | 15.6 | 38.3 | 52.4 | 80,806 | 73,430 | -9.1 | -4.7 |
| Cincinnati | 3.5 | 6.8 | 20.9 | 14.2 | 18.6 | 11.3 | 10.1 | 11.6 | 13.2 | 32.5 | 51.0 | 331,285 | 297,098 | -10.3 | 2.5 |
| Cleveland | 12.7 | 6.9 | 21.8 | 9.9 | 17.7 | 11.3 | 11.4 | 13.8 | 14.2 | 35.5 | 51.9 | 478,403 | 396,831 | -17.1 | -4.6 |
| Cleveland Heights | 2.2 | 8.3 | 18.6 | 7.9 | 19.7 | 12.4 | 9.3 | 12.9 | 19.2 | 37.5 | 51.8 | 49,958 | 46,139 | -7.6 | -5.6 |
| Columbus | 7.0 | 13.3 | 22.1 | 11.5 | 21.0 | 13.0 | 10.8 | 11.0 | 10.6 | 32.4 | 51.0 | 711,470 | 789,007 | 10.9 | 14.6 |
| Cuyahoga Falls | 1.6 | 7.5 | 20.4 | 8.4 | 18.0 | 11.2 | 11.8 | 13.8 | 16.5 | 37.8 | 56.8 | 49,374 | 49,588 | 0.4 | -1.2 |
| Dayton | 5.4 | 4.7 | 22.6 | 15.6 | 15.3 | 11.4 | 10.7 | 11.8 | 12.7 | 33.0 | 50.8 | 166,179 | 141,930 | -14.6 | -1.4 |
| Delaware | 4.9 | 4.0 | 22.0 | 9.7 | 13.4 | 13.7 | 12.1 | 12.1 | 17.0 | 37.4 | 52.2 | 25,243 | 34,800 | 37.9 | 22.4 |
| Dublin | 4.3 | 20.4 | 25.8 | 4.6 | 10.9 | 18.0 | 13.7 | 14.4 | 12.6 | 38.9 | 47.6 | 31,392 | 41,396 | 31.9 | 19.3 |
| Elyria | 9.7 | 3.0 | 22.1 | 10.7 | 11.8 | 12.0 | 14.2 | 12.7 | 16.6 | 40.7 | 53.7 | 55,953 | 54,562 | -2.5 | -1.3 |
| Euclid | 0.5 | 3.2 | 24.9 | 8.1 | 11.7 | 8.9 | 13.8 | 14.8 | 17.8 | 40.9 | 53.4 | 52,717 | 48,901 | -7.2 | -5.5 |
| Fairborn | 4.8 | 6.5 | 23.0 | 14.0 | 21.3 | 10.6 | 7.2 | 7.5 | 16.5 | 30.6 | 53.1 | 32,052 | 32,955 | 2.8 | 3.1 |
| Fairfield | 7.1 | 14.0 | 19.1 | 10.0 | 11.9 | 13.0 | 12.0 | 15.1 | 18.7 | 42.3 | 54.8 | 42,097 | 42,501 | 1.0 | 0.2 |
| Findlay | 7.8 | 3.9 | 19.6 | 10.3 | 17.0 | 10.6 | 11.5 | 13.8 | 17.1 | 39.3 | 51.7 | 38,967 | 41,189 | 5.7 | -0.9 |
| Gahanna | 1.6 | 1.7 | 24.2 | 5.1 | 10.3 | 14.1 | 11.5 | 13.1 | 21.7 | 42.8 | 53.2 | 32,636 | 33,230 | 1.8 | 6.9 |
| Garfield Heights | 3.4 | 2.7 | 26.8 | 7.4 | 9.0 | 15.4 | 10.2 | 16.8 | 14.4 | 39.5 | 54.1 | 30,734 | 28,884 | -6.0 | -5.6 |
| Green | 0.2 | 1.8 | 22.9 | 5.9 | 15.3 | 12.2 | 11.5 | 13.5 | 18.8 | 39.0 | 46.7 | 22,817 | 25,742 | 12.8 | -0.2 |
| Grove City | 3.1 | 3.0 | 20.4 | 5.8 | 15.1 | 14.5 | 14.0 | 11.0 | 19.3 | 40.7 | 49.9 | 27,075 | 35,633 | 31.6 | 18.7 |

1. May be of any race.

## Table D. Cities — Households, Group Quarters, Crime, and Education

| City | Households, 2019 Number | Persons per household | Percent Family | Married couple family | Female family | Non-family | One person | Persons in group quarters, 2019 | Serious crimes known to police[1], 2019 Violent Number | Violent Rate[2] | Property Number | Property Rate[2] | Population age 25 and over | High school graduate or less | Bachelor's degree or more |
|---|---|---|---|---|---|---|---|---|---|---|---|---|---|---|---|
| | 27 | 28 | 29 | 30 | 31 | 32 | 33 | 34 | 35 | 36 | 37 | 38 | 39 | 40 | 41 |
| NEW YORK—Cont'd | | | | | | | | | | | | | | | |
| Yonkers.............. | 74,854 | 2.64 | 64.7 | 40.8 | 23.9 | 35.3 | 31.9 | 2,953 | 708 | 354 | 1,826 | 913 | 132,896 | 41.8 | 35.0 |
| NORTH CAROLINA............. | 4,046,348 | 2.52 | 65.0 | 47.6 | 17.4 | 35.0 | 28.6 | 280,891 | NA | NA | NA | NA | 7,187,077 | 37.0 | 32.3 |
| Apex...................... | 20,971 | 2.82 | 71.8 | 58.3 | 13.5 | 28.2 | 20.3 | 128 | 53 | 94 | 527 | 936 | 38,104 | 13.7 | 62.3 |
| Asheboro............... | 10,248 | 2.45 | 56.7 | 34.5 | 22.1 | 43.3 | 38.4 | 819 | NA | NA | NA | NA | 16,990 | 52.0 | 18.0 |
| Asheville............... | 40,340 | 2.22 | 46.9 | 33.1 | 13.8 | 53.1 | 40.3 | 3,398 | 695 | 742 | 5,923 | 6,325 | 66,935 | 25.5 | 48.9 |
| Burlington............. | 21,455 | 2.47 | 60.6 | 37.8 | 22.8 | 39.4 | 35.3 | 809 | 460 | 850 | 2,490 | 4,602 | 35,938 | 41.1 | 26.6 |
| Cary...................... | 64,728 | 2.64 | 73.6 | 62.1 | 11.5 | 26.4 | 21.8 | 282 | 114 | 66 | 1,560 | 904 | 118,721 | 12.9 | 68.4 |
| Chapel Hill........... | 21,704 | 2.37 | 51.9 | 43.8 | 8.1 | 48.1 | 36.7 | 11,970 | 58 | 94 | 926 | 1,507 | 33,758 | 11.1 | 77.2 |
| Charlotte.............. | 342,448 | 2.55 | 57.1 | 38.4 | 18.7 | 42.9 | 34.5 | 13,670 | 6,982 | 739 | 37,070 | 3,926 | 596,804 | 26.7 | 44.9 |
| Concord................ | 30,660 | 3.09 | 68.5 | 53.0 | 15.6 | 31.5 | 26.6 | 1,453 | 117 | 122 | 1,586 | 1,650 | 61,361 | 33.0 | 40.8 |
| Durham................. | 114,726 | 2.33 | 56.6 | 37.3 | 19.4 | 43.4 | 32.8 | 12,342 | 2,046 | 730 | 10,672 | 3,808 | 190,711 | 27.9 | 51.1 |
| Fayetteville........... | 82,087 | 2.39 | 56.8 | 35.7 | 21.1 | 43.2 | 36.9 | 15,123 | 1,835 | 875 | 7,391 | 3,526 | 129,266 | 34.0 | 26.8 |
| Garner.................. | 11,399 | 2.74 | 62.1 | 44.3 | 17.8 | 37.9 | 28.7 | 205 | 79 | 254 | 1,076 | 3,456 | 20,783 | 29.0 | 36.7 |
| Gastonia............... | 27,796 | 2.72 | 66.1 | 46.4 | 19.6 | 33.9 | 28.5 | 1,704 | NA | NA | NA | NA | 51,520 | 45.5 | 26.8 |
| Goldsboro............. | 16,208 | 2.00 | 47.0 | 26.0 | 20.9 | 53.0 | 45.6 | 1,820 | 267 | 783 | 1,985 | 5,824 | 23,413 | 48.8 | 19.0 |
| Greensboro........... | 118,046 | 2.37 | 58.8 | 35.6 | 23.2 | 41.2 | 33.9 | 17,201 | 2,440 | 819 | 10,994 | 3,689 | 192,908 | 34.1 | 38.7 |
| Greenville............. | 37,402 | 2.32 | 48.4 | 30.3 | 18.2 | 51.6 | 27.5 | 6,723 | 437 | 464 | 2,837 | 3,012 | 52,402 | 35.0 | 34.9 |
| Hickory................. | 15,977 | 2.48 | 58.5 | 39.4 | 19.0 | 41.5 | 34.2 | 1,453 | 165 | 402 | 1,793 | 4,369 | 27,562 | 37.8 | 31.8 |
| High Point............ | 40,877 | 2.62 | 63.1 | 40.7 | 22.4 | 36.9 | 29.2 | 4,675 | 828 | 731 | 3,726 | 3,288 | 72,816 | 34.7 | 33.5 |
| Huntersville.......... | 21,016 | 2.75 | 73.1 | 63.7 | 9.4 | 26.9 | 20.4 | 286 | 79 | 135 | 948 | 1,620 | 39,330 | 17.4 | 57.1 |
| Indian Trail.......... | 13,743 | 2.93 | 81.5 | 67.5 | 14.0 | 18.5 | 16.8 | 0 | NA | NA | NA | NA | 26,255 | 30.7 | 37.6 |
| Jacksonville........... | 21,986 | 2.49 | 70.5 | 48.9 | 21.5 | 29.5 | 25.1 | 17,679 | NA | NA | NA | NA | 32,466 | 37.1 | 22.5 |
| Kannapolis............ | 16,384 | 3.04 | 68.6 | 50.6 | 18.0 | 31.4 | 21.4 | 280 | 132 | 260 | 1,154 | 2,275 | 32,847 | 47.9 | 20.2 |
| Matthews.............. | 12,177 | 2.70 | 70.4 | 61.6 | 8.9 | 29.6 | 28.1 | 248 | 58 | 174 | 1,112 | 3,332 | 23,714 | 22.4 | 50.7 |
| Monroe................. | 11,956 | 2.92 | 72.3 | 48.4 | 23.9 | 27.7 | 27.7 | 588 | 332 | 932 | 1,881 | 5,279 | 21,664 | 45.0 | 20.3 |
| Mooresville........... | 15,730 | 2.47 | 69.8 | 50.4 | 19.3 | 30.2 | 20.9 | 273 | 65 | 167 | 1,050 | 2,695 | 25,886 | 24.7 | 46.0 |
| New Bern.............. | 14,562 | 2.03 | 61.9 | 44.0 | 17.8 | 38.1 | 32.9 | 396 | 125 | 414 | 999 | 3,309 | 22,129 | 33.3 | 28.2 |
| Raleigh................. | 188,412 | 2.41 | 55.6 | 38.3 | 17.4 | 44.4 | 33.3 | 19,964 | 1,222 | 256 | 8,520 | 1,783 | 318,649 | 25.1 | 50.3 |
| Rocky Mount......... | 22,227 | 2.25 | 69.5 | 37.5 | 31.9 | 30.5 | 29.4 | 1,225 | 424 | 788 | 1,627 | 3,023 | 36,008 | 50.3 | 15.7 |
| Salisbury.............. | 11,537 | 2.62 | 67.7 | 42.0 | 25.7 | 32.3 | 28.7 | 3,767 | 249 | 735 | 1,413 | 4,171 | 21,568 | 42.9 | 20.9 |
| Sanford................ | 10,406 | 2.82 | 58.1 | 40.9 | 17.2 | 41.9 | 32.1 | 766 | NA | NA | NA | NA | 18,813 | 41.1 | 26.1 |
| Thomasville........... | 12,064 | 2.18 | 70.3 | 44.7 | 25.6 | 29.7 | 27.1 | 408 | NA | NA | NA | NA | 20,452 | 56.9 | 15.3 |
| Wake Forest.......... | 16,837 | 2.71 | 75.4 | 57.1 | 18.4 | 24.6 | 19.4 | 369 | 24 | 52 | 393 | 852 | 31,195 | 15.5 | 57.2 |
| Wilmington............ | 54,673 | 2.16 | 49.6 | 34.5 | 15.1 | 50.4 | 38.0 | 5,900 | 759 | 608 | 3,593 | 2,880 | 80,774 | 27.8 | 41.9 |
| Wilson.................. | 20,130 | 2.39 | 52.2 | 31.3 | 20.9 | 47.8 | 40.1 | 1,373 | 220 | 446 | 1,545 | 3,131 | 33,411 | 47.5 | 19.5 |
| Winston-Salem....... | 94,884 | 2.50 | 56.6 | 34.1 | 22.6 | 43.4 | 38.0 | 10,328 | NA | NA | NA | NA | 158,491 | 39.3 | 32.4 |
| NORTH DAKOTA ............. | 323,519 | 2.28 | 58.6 | 47.0 | 11.6 | 41.4 | 33.7 | 24,335 | NA | NA | NA | NA | 504,375 | 33.2 | 30.4 |
| Bismarck............... | 30,833 | 2.36 | 58.5 | 45.9 | 12.7 | 41.5 | 35.3 | 1,972 | 228 | 305 | 2,079 | 2,783 | 49,830 | 30.0 | 35.6 |
| Fargo.................... | 57,118 | 2.09 | 47.0 | 37.1 | 9.9 | 53.0 | 41.4 | 4,530 | 574 | 450 | 3,978 | 3,122 | 80,312 | 29.8 | 39.3 |
| Grand Forks........... | 26,391 | 1.98 | 44.5 | 30.5 | 14.0 | 55.5 | 40.2 | 3,537 | 166 | 289 | 1,422 | 2,475 | 32,441 | 22.0 | 39.5 |
| Minot................... | 21,693 | 2.08 | 50.3 | 37.9 | 12.3 | 49.7 | 41.7 | 1,213 | 130 | 270 | 984 | 2,042 | 30,756 | 30.4 | 30.2 |
| West Fargo ........... | 14,582 | 2.54 | 67.6 | 59.9 | 7.7 | 32.4 | 25.0 | 86 | 55 | 144 | 584 | 1,530 | 23,929 | 17.6 | 51.0 |
| OHIO................ | 4,730,340 | 2.40 | 62.2 | 45.0 | 17.2 | 37.8 | 31.1 | 317,286 | NA | NA | NA | NA | 8,049,805 | 41.8 | 29.3 |
| Akron................... | 83,821 | 2.29 | 53.2 | 30.1 | 23.1 | 46.8 | 38.9 | 5,435 | 1,782 | 901 | 6,568 | 3,319 | 132,964 | 45.6 | 22.2 |
| Barberton............. | 11,876 | 2.15 | 55.0 | 26.8 | 28.1 | 45.0 | 34.8 | 372 | 69 | 265 | 689 | 2,648 | 19,382 | 55.5 | 14.9 |
| Beavercreek........... | 18,158 | 2.56 | 68.2 | 59.2 | 9.0 | 31.8 | 23.5 | 458 | 33 | 69 | 1,193 | 2,503 | 34,260 | 16.4 | 57.1 |
| Bowling Green........ | 11,495 | 2.25 | 34.4 | 24.0 | 10.3 | 65.6 | 40.7 | 5,694 | 37 | 117 | 547 | 1,725 | 14,758 | 30.8 | 46.9 |
| Brunswick............. | 13,994 | 2.47 | 63.6 | 50.1 | 13.4 | 36.4 | 29.6 | 288 | 17 | 49 | 189 | 540 | 25,089 | 41.3 | 27.6 |
| Canton................. | 31,981 | 2.12 | 52.9 | 26.0 | 27.0 | 47.1 | 39.2 | 2,532 | 981 | 1,399 | 3,712 | 5,292 | 47,420 | 56.4 | 13.7 |
| Cincinnati............. | 137,582 | 2.11 | 45.7 | 26.0 | 19.7 | 54.3 | 41.6 | 14,129 | 2,562 | 845 | 13,051 | 4,303 | 197,198 | 34.8 | 40.6 |
| Cleveland.............. | 171,632 | 2.14 | 48.3 | 19.7 | 28.6 | 51.7 | 44.3 | 956 | 5,791 | 1,517 | 17,057 | 4,467 | 260,523 | 49.9 | 18.6 |
| Cleveland Heights.............. | 20,155 | 2.14 | 52.7 | 38.9 | 13.8 | 47.3 | 37.1 | 956 | 95 | 215 | 696 | 1,578 | 32,312 | 19.9 | 57.2 |
| Columbus.............. | 368,491 | 2.38 | 52.6 | 33.5 | 19.1 | 47.4 | 36.1 | 23,593 | 4,561 | 503 | 29,974 | 3,308 | 599,181 | 34.2 | 39.1 |
| Cuyahoga Falls.................. | 22,212 | 2.19 | 56.3 | 42.0 | 14.3 | 43.7 | 36.9 | 501 | 50 | 102 | 980 | 1,990 | 34,976 | 35.6 | 35.1 |
| Dayton................. | 56,520 | 2.23 | 49.7 | 23.5 | 26.2 | 50.3 | 43.2 | 14,209 | 1,351 | 962 | 5,673 | 4,040 | 86,820 | 45.6 | 18.0 |
| Delaware............... | 16,268 | 2.58 | 61.4 | 44.2 | 17.2 | 38.6 | 31.9 | 1,926 | 60 | 148 | 477 | 1,174 | 29,996 | 35.8 | 32.4 |
| Dublin.................. | 17,353 | 2.64 | 76.2 | 70.5 | 5.7 | 23.8 | 20.8 | 230 | 26 | 52 | 429 | 864 | 32,072 | 14.1 | 69.1 |
| Elyria.................... | 22,942 | 2.31 | 50.9 | 27.6 | 23.3 | 49.1 | 39.5 | 662 | 137 | 255 | 921 | 1,712 | 36,141 | 47.3 | 14.5 |
| Euclid................... | 20,254 | 2.27 | 51.3 | 24.6 | 26.7 | 48.7 | 43.1 | 639 | 243 | 521 | 1,285 | 2,754 | 31,179 | 39.8 | 18.5 |
| Fairborn............... | 16,125 | 2.13 | 48.8 | 27.5 | 21.3 | 51.2 | 38.3 | 870 | 88 | 261 | 627 | 1,859 | 22,207 | 29.5 | 35.2 |
| Fairfield................ | 18,200 | 2.31 | 66.4 | 45.1 | 21.3 | 33.6 | 32.2 | 551 | 84 | 197 | 879 | 2,062 | 30,135 | 41.6 | 34.5 |
| Findlay................. | 18,582 | 2.11 | 53.1 | 38.2 | 14.9 | 46.9 | 36.7 | 1,940 | 106 | 256 | 902 | 2,181 | 28,898 | 45.7 | 26.9 |
| Gahanna............... | 14,083 | 2.58 | 69.4 | 52.9 | 16.4 | 30.6 | 24.5 | 183 | 44 | 123 | 719 | 2,006 | 25,794 | 26.0 | 46.0 |
| Garfield Heights .......... | 10,816 | 2.50 | 54.8 | 34.0 | 20.8 | 45.2 | 40.6 | 431 | NA | NA | NA | NA | 18,053 | 48.7 | 13.2 |
| Green................... | 10,613 | 2.41 | 66.6 | 55.1 | 11.6 | 33.4 | 28.8 | 188 | NA | NA | NA | NA | 18,342 | 26.8 | 41.3 |
| Grove City............. | 15,515 | 2.63 | 78.0 | 61.0 | 17.1 | 22.0 | 20.2 | 245 | 41 | 97 | 1,193 | 2,812 | 30,368 | 32.4 | 39.7 |

1. Data for serious crimes have not been adjusted for underreporting. This may affect comparability between geographic areas and over time.    2. Per 100,000 population estimated by the FBI.    3. Persons 25 years old and over.

# Table D. Cities — Income, Poverty, and Housing

| City | Money income, 2019 | | | | | Median earnings, 2019 | | | Housing units, 2019 | | | | |
| | Households | | | | | | | | | | | | |
| | Median income | Percent with income less than $20,000 | Percent with income of $200,000 or more | Median family income | Median non-family income | All persons | Men | Women | Total | Occupied | Percent owner occupied | Median value[1] (dollars) | Median gross rent (dollars) |
| | 42 | 43 | 44 | 45 | 46 | 47 | 48 | 49 | 50 | 51 | 52 | 53 | 54 |
| **NEW YORK—Cont'd** | | | | | | | | | | | | | |
| Yonkers | 64,916 | 17.5 | 9.6 | 77,590 | 41,003 | 40,346 | 41,721 | 35,835 | 78,948 | 74,854 | 45.2 | 422,728 | 1,460 |
| **NORTH CAROLINA** | 57,341 | 15.5 | 6.0 | 72,049 | 34,864 | 33,205 | 38,228 | 29,690 | 4,748,148 | 4,046,348 | 65.3 | 193,238 | 931 |
| Apex | 103,978 | 5.0 | 17.0 | 125,750 | 58,030 | 52,862 | 80,918 | 37,624 | 22,330 | 20,971 | 71.8 | 347,519 | 1,406 |
| Asheboro | 41,478 | 26.5 | 0.4 | 46,260 | 33,142 | 27,491 | 33,294 | 21,895 | 11,558 | 10,248 | 45.3 | 127,061 | 685 |
| Asheville | 52,339 | 17.7 | 6.0 | 66,058 | 36,067 | 29,848 | 31,298 | 27,212 | 48,261 | 40,340 | 44.0 | 329,473 | 1,100 |
| Burlington | 50,658 | 19.8 | 4.7 | 59,453 | 32,229 | 32,219 | 40,225 | 27,697 | 25,372 | 21,455 | 58.3 | 148,301 | 795 |
| Cary | 106,304 | 5.7 | 19.9 | 134,294 | 55,488 | 54,470 | 75,418 | 43,251 | 71,182 | 64,728 | 66.5 | 397,868 | 1,300 |
| Chapel Hill | 78,574 | 19.0 | 17.5 | 145,990 | 37,259 | 23,436 | 30,369 | 21,049 | 24,203 | 21,704 | 52.1 | 417,983 | 1,147 |
| Charlotte | 63,483 | 11.9 | 10.3 | 79,910 | 47,819 | 38,528 | 42,450 | 33,680 | 375,538 | 342,448 | 52.4 | 252,131 | 1,174 |
| Concord | 75,578 | 10.7 | 11.5 | 91,766 | 39,441 | 36,349 | 40,762 | 32,201 | 38,380 | 30,660 | 68.8 | 264,369 | 1,082 |
| Durham | 65,534 | 13.0 | 7.9 | 82,642 | 45,846 | 38,598 | 41,079 | 36,181 | 124,001 | 114,726 | 51.6 | 253,513 | 1,127 |
| Fayetteville | 43,789 | 19.9 | 1.9 | 52,248 | 31,834 | 28,072 | 30,570 | 23,614 | 94,480 | 82,087 | 43.6 | 127,988 | 959 |
| Garner | 63,727 | 8.2 | 4.7 | 90,836 | 47,160 | 44,568 | 46,919 | 35,207 | 12,443 | 11,399 | 63.7 | 232,816 | 1,202 |
| Gastonia | 49,172 | 16.7 | 3.5 | 68,544 | 28,804 | 30,622 | 32,926 | 27,409 | 32,422 | 27,796 | 54.2 | 184,290 | 901 |
| Goldsboro | 37,814 | 24.7 | 1.6 | 40,387 | 30,647 | 21,377 | 26,563 | 15,274 | 18,851 | 16,208 | 43.5 | 109,964 | 851 |
| Greensboro | 49,748 | 19.6 | 4.5 | 63,841 | 35,269 | 30,249 | 35,658 | 26,015 | 132,743 | 118,046 | 48.7 | 169,553 | 921 |
| Greenville | 48,169 | 21.5 | 5.7 | 60,630 | 41,147 | 27,653 | 31,395 | 26,437 | 43,097 | 37,402 | 33.4 | 164,986 | 866 |
| Hickory | 47,238 | 19.2 | 4.2 | 63,468 | 29,144 | 27,648 | 30,429 | 25,143 | 18,206 | 15,977 | 48.8 | 168,781 | 757 |
| High Point | 48,348 | 13.9 | 3.0 | 60,932 | 34,194 | 29,148 | 31,714 | 24,315 | 46,840 | 40,877 | 56.4 | 157,769 | 884 |
| Huntersville | 104,814 | 6.0 | 19.5 | 120,504 | 57,095 | 51,754 | 72,828 | 43,797 | 22,161 | 21,016 | 76.8 | 330,866 | 1,379 |
| Indian Trail | 87,425 | 7.4 | 6.4 | 93,272 | 50,098 | 41,352 | 49,098 | 36,363 | 14,455 | 13,743 | 89.7 | 235,578 | 1,639 |
| Jacksonville | 46,374 | 13.8 | 1.3 | 46,713 | 36,083 | 23,546 | 23,763 | 21,323 | 25,710 | 21,986 | 32.2 | 164,835 | 909 |
| Kannapolis | 50,929 | 20.0 | 4.7 | 64,850 | 22,614 | 28,357 | 28,065 | 29,780 | 18,833 | 16,384 | 62.4 | 165,225 | 845 |
| Matthews | 91,619 | 4.8 | 9.2 | 109,809 | 50,659 | 46,819 | 65,589 | 34,675 | 12,437 | 12,177 | 68.3 | 295,917 | 1,355 |
| Monroe | 51,020 | 16.8 | 2.7 | 63,434 | 23,301 | 30,077 | 31,701 | 24,602 | 13,491 | 11,956 | 60.7 | 184,886 | 1,040 |
| Mooresville | 70,885 | 12.4 | 4.5 | 83,927 | 42,044 | 37,838 | 51,359 | 29,427 | 16,654 | 15,730 | 57.6 | 260,514 | 1,130 |
| New Bern | 39,336 | 19.5 | 4.3 | 66,801 | 29,305 | 24,502 | 31,017 | 22,007 | 16,301 | 14,562 | 64.9 | 145,087 | 957 |
| Raleigh | 69,333 | 11.8 | 8.0 | 90,081 | 49,736 | 40,612 | 44,798 | 36,754 | 205,534 | 188,412 | 51.2 | 274,185 | 1,163 |
| Rocky Mount | 39,305 | 24.1 | 3.0 | 53,236 | 19,168 | 26,297 | 31,708 | 25,007 | 27,350 | 22,227 | 50.4 | 120,349 | 795 |
| Salisbury | 42,600 | 19.3 | 3.3 | 55,034 | 26,223 | 24,340 | 31,943 | 20,803 | 14,462 | 11,537 | 48.2 | 149,722 | 858 |
| Sanford | 42,387 | 21.1 | 4.7 | 56,821 | 26,162 | 29,236 | 32,431 | 27,992 | 11,689 | 10,406 | 51.6 | 144,309 | 766 |
| Thomasville | 44,747 | 18.7 | 0.0 | 55,913 | 30,823 | 36,065 | 40,113 | 30,997 | 12,537 | 12,064 | 67.6 | 98,535 | 709 |
| Wake Forest | 98,552 | 5.1 | 9.1 | 106,617 | 60,239 | 50,183 | 62,130 | 36,790 | 17,671 | 16,837 | 76.5 | 332,646 | 1,121 |
| Wilmington | 48,381 | 19.7 | 5.7 | 69,940 | 32,677 | 26,790 | 31,457 | 22,812 | 60,415 | 54,673 | 44.0 | 273,638 | 1,092 |
| Wilson | 32,813 | 33.4 | 3.0 | 59,325 | 19,643 | 30,224 | 35,273 | 24,263 | 22,402 | 20,130 | 45.7 | 154,652 | 698 |
| Winston-Salem | 44,576 | 20.2 | 4.4 | 60,873 | 28,189 | 29,937 | 33,465 | 26,607 | 108,880 | 94,884 | 51.7 | 154,160 | 840 |
| **NORTH DAKOTA** | 64,577 | 13.5 | 6.5 | 87,055 | 38,698 | 37,398 | 45,899 | 30,808 | 379,974 | 323,519 | 61.3 | 205,431 | 804 |
| Bismarck | 65,870 | 13.8 | 7.4 | 89,607 | 37,081 | 40,460 | 46,318 | 31,408 | 35,068 | 30,833 | 64.6 | 251,603 | 856 |
| Fargo | 52,810 | 17.7 | 5.3 | 84,732 | 37,013 | 32,315 | 35,354 | 30,524 | 62,515 | 57,118 | 38.6 | 236,800 | 787 |
| Grand Forks | 52,091 | 17.9 | 3.8 | 80,241 | 32,069 | 29,095 | 35,232 | 25,055 | 27,840 | 26,391 | 42.0 | 212,267 | 815 |
| Minot | 67,222 | 11.2 | 3.6 | 93,005 | 39,951 | 39,682 | 46,010 | 33,369 | 24,043 | 21,693 | 54.7 | 216,421 | 797 |
| West Fargo | 85,648 | 7.8 | 12.9 | 100,778 | 49,366 | 45,397 | 52,337 | 33,389 | 15,069 | 14,582 | 58.6 | 245,810 | 1,092 |
| **OHIO** | 58,642 | 15.1 | 5.4 | 74,911 | 35,147 | 34,852 | 41,358 | 28,819 | 5,232,943 | 4,730,340 | 66.0 | 157,241 | 813 |
| Akron | 41,013 | 22.5 | 2.4 | 52,989 | 30,107 | 27,190 | 31,403 | 22,227 | 96,160 | 83,821 | 51.6 | 86,573 | 770 |
| Barberton | 32,489 | 27.1 | 0.4 | 38,447 | 26,985 | 25,237 | 30,997 | 18,975 | 13,204 | 11,876 | 53.4 | 87,126 | 673 |
| Beavercreek | 96,719 | 4.2 | 12.5 | 118,766 | 60,086 | 53,368 | 70,681 | 41,366 | 19,431 | 18,158 | 73.4 | 209,616 | 1,152 |
| Bowling Green | 41,254 | 29.3 | 5.5 | 61,583 | 28,809 | 12,471 | 14,972 | 9,793 | 11,968 | 11,495 | 38.6 | 179,643 | 726 |
| Brunswick | 71,166 | 10.9 | 3.6 | 81,140 | 48,510 | 42,122 | 52,910 | 31,892 | 14,712 | 13,994 | 71.9 | 177,309 | 934 |
| Canton | 32,125 | 30.0 | 0.5 | 40,692 | 24,619 | 22,604 | 25,327 | 21,189 | 35,252 | 31,981 | 46.0 | 77,890 | 636 |
| Cincinnati | 46,260 | 26.0 | 7.0 | 62,941 | 31,843 | 31,595 | 35,262 | 28,169 | 158,394 | 137,582 | 39.3 | 170,774 | 775 |
| Cleveland | 32,053 | 33.6 | 1.7 | 40,905 | 23,463 | 27,354 | 30,159 | 25,926 | 207,813 | 171,632 | 40.0 | 71,108 | 725 |
| Cleveland Heights | 62,869 | 12.0 | 7.5 | 87,852 | 44,635 | 39,775 | 47,834 | 35,248 | 22,944 | 20,155 | 57.6 | 137,500 | 968 |
| Columbus | 57,118 | 14.5 | 4.2 | 70,936 | 44,679 | 35,013 | 38,898 | 30,219 | 402,520 | 368,491 | 45.5 | 173,257 | 984 |
| Cuyahoga Falls | 56,416 | 14.5 | 3.5 | 70,674 | 38,461 | 36,867 | 42,951 | 30,405 | 23,307 | 22,212 | 58.1 | 143,880 | 857 |
| Dayton | 33,116 | 31.0 | 0.5 | 40,020 | 27,194 | 21,010 | 22,314 | 18,764 | 71,356 | 56,520 | 47.0 | 67,913 | 730 |
| Delaware | 59,291 | 15.7 | 5.1 | 87,832 | 26,735 | 32,677 | 36,957 | 22,485 | 16,983 | 16,268 | 56.0 | 218,496 | 918 |
| Dublin | 121,504 | 3.3 | 28.6 | 142,503 | 60,106 | 66,345 | 96,296 | 45,020 | 17,666 | 17,353 | 75.7 | 375,014 | 1,356 |
| Elyria | 45,399 | 26.3 | 1.8 | 47,873 | 34,823 | 27,862 | 38,650 | 24,965 | 25,337 | 22,942 | 61.6 | 103,322 | 723 |
| Euclid | 45,413 | 23.6 | 1.3 | 62,573 | 27,159 | 33,446 | 36,228 | 31,512 | 22,713 | 20,254 | 46.8 | 85,343 | 805 |
| Fairborn | 44,488 | 16.0 | 1.0 | 52,173 | 39,095 | 30,855 | 41,552 | 27,447 | 17,133 | 16,125 | 45.4 | 121,345 | 841 |
| Fairfield | 56,490 | 14.3 | 3.9 | 73,308 | 33,553 | 37,912 | 43,327 | 34,098 | 19,894 | 18,200 | 58.4 | 170,192 | 947 |
| Findlay | 49,053 | 15.5 | 6.8 | 68,678 | 29,982 | 33,080 | 37,392 | 26,126 | 19,837 | 18,582 | 55.7 | 153,615 | 776 |
| Gahanna | 92,040 | 10.0 | 8.8 | 98,289 | 60,536 | 43,879 | 41,010 | 51,078 | 14,335 | 14,083 | 72.7 | 238,313 | 1,385 |
| Garfield Heights | 45,068 | 20.1 | 1.6 | 60,564 | 29,214 | 30,520 | 40,152 | 22,067 | 12,002 | 10,816 | 62.6 | 82,238 | 863 |
| Green | 73,551 | 9.2 | 12.4 | 93,125 | 50,973 | 40,081 | 50,322 | 32,493 | 11,431 | 10,613 | 69.8 | 237,271 | 928 |
| Grove City | 85,144 | 10.0 | 7.5 | 110,546 | 31,999 | 40,921 | 51,665 | 36,065 | 16,125 | 15,515 | 69.2 | 222,080 | 981 |

1. Based on population estimated by the American Community Survey.

## Table D. Cities — Commuting, Computer Access, Migration, Labor Force, and Employment

| City | Commuting, 2019 — Percent: Drove alone | Commuting, 2019 — Percent: With commutes of 30 minutes or more | Computer access, 2019 — Percent: With a computer in the house | Computer access, 2019 — Percent: With Internet access | Migration, 2019: Percent who lived in the same house one year ago | Migration, 2019: Percent who lived in another state or county one year ago | Civilian labor force, 2020: Total | Civilian labor force, 2020: Percent change 2019–2020 | Unemployment: Total | Unemployment: Rate[3] | Civilian Employment, 2019 — Population age 16 and older: Number | Civilian Employment, 2019 — Population age 16 and older: Percent in labor force | Civilian Employment, 2019 — Population age 16 to 64: Number | Civilian Employment, 2019 — Population age 16 to 64: Percent who worked full-year full-time |
|---|---|---|---|---|---|---|---|---|---|---|---|---|---|---|
| | 55 | 56 | 57 | 58 | 59 | 60 | 61 | 62 | 63 | 64 | 65 | 66 | 67 | 68 |
| **NEW YORK—Cont'd** | | | | | | | | | | | | | | |
| Yonkers | 53.5 | 54.1 | 90.4 | 81.0 | 90.9 | 4.9 | 97,684 | 1.8 | 11,298 | 12 | 157,521 | 64.5 | 124,682 | 50.0 |
| **NORTH CAROLINA** | 80.2 | 35.2 | 92.2 | 84.6 | 85.2 | 7.5 | 4,950,859 | -2.6 | 363,452 | 7 | 8,457,219 | 61.2 | 6,706,284 | 52.2 |
| Apex | 81.7 | 35.4 | NA | NA | 87.9 | 5.4 | 30,785 | 3.9 | 1,503 | 5 | 44,210 | 74.3 | 40,592 | 56.9 |
| Asheboro | 85.4 | 28.3 | 91.0 | 84.8 | 74.1 | 2.7 | 10,720 | -2.9 | 956 | 9 | 20,031 | 51.1 | 15,125 | 45.4 |
| Asheville | 73.9 | 14.3 | 89.0 | 84.9 | 81.1 | 9.7 | 50,704 | -2.8 | 4,875 | 10 | 75,971 | 65.6 | 60,011 | 48.9 |
| Burlington | 86.7 | 27.7 | 92.0 | 79.1 | 83.9 | 9.8 | 25,739 | -1.8 | 2,151 | 8 | 40,047 | 65.5 | 30,942 | 54.9 |
| Cary | 76.8 | 28.8 | 98.8 | 96.1 | 82.0 | 10.5 | 91,078 | -4.1 | 4,816 | 5 | 134,286 | 69.2 | 112,797 | 59.0 |
| Chapel Hill | 56.1 | 22.8 | NA | NA | 65.1 | 23.8 | 31,383 | 0.6 | 1,664 | 5 | 56,201 | 57.5 | 47,637 | 32.8 |
| Charlotte | 74.7 | 35.7 | 95.2 | 90.0 | 83.3 | 7.6 | 492,518 | -1.3 | 39,418 | 8 | 699,598 | 72.5 | 605,334 | 58.8 |
| Concord | 76.7 | 47.4 | 96.3 | 90.1 | 85.0 | 6.5 | 48,863 | -1.0 | 3,643 | 8 | 74,621 | 70.4 | 63,299 | 52.8 |
| Durham | 78.4 | 27.7 | 95.4 | 89.5 | 81.2 | 8.8 | 149,738 | -0.8 | 9,950 | 7 | 225,769 | 68.2 | 190,900 | 54.9 |
| Fayetteville | 77.1 | 20.6 | 91.8 | 86.7 | 75.0 | 12.7 | 76,303 | -1.4 | 7,834 | 10 | 164,710 | 52.4 | 140,262 | 55.4 |
| Garner | 85.6 | 42.2 | NA | NA | 73.4 | 15.9 | 17,218 | -1.1 | 1,182 | 7 | 24,807 | 73.3 | 20,122 | 65.9 |
| Gastonia | 78.6 | 33.4 | 94.8 | 80.7 | 84.6 | 6.8 | 36,861 | -1.5 | 3,415 | 9 | 61,184 | 64.1 | 49,215 | 53.1 |
| Goldsboro | NA | 16.6 | 82.0 | 79.2 | 78.6 | 9.0 | 11,528 | -1.2 | 1,116 | 10 | 28,683 | 42.9 | 21,820 | 45.6 |
| Greensboro | 77.5 | 25.7 | 90.0 | 81.0 | 82.8 | 6.8 | 143,818 | -2.6 | 13,193 | 9 | 237,058 | 65.0 | 196,528 | 48.0 |
| Greenville | 80.5 | 16.6 | 93.6 | 86.5 | 68.8 | 20.7 | 46,971 | -3.3 | 3,477 | 7 | 78,558 | 63.2 | 68,380 | 44.3 |
| Hickory | 78.9 | 15.9 | 91.3 | 86.8 | 81.3 | 12.3 | 20,156 | -2.3 | 1,578 | 8 | 33,021 | 63.7 | 26,544 | 53.1 |
| High Point | 79.4 | 23.7 | 93.2 | 87.7 | 82.4 | 7.3 | 52,973 | -2.7 | 4,929 | 9 | 88,044 | 60.6 | 70,829 | 46.9 |
| Huntersville | 80.5 | 46.3 | 97.1 | 92.5 | 89.5 | 6.1 | 33,478 | -2.5 | 2,029 | 6 | 45,047 | 71.8 | 37,674 | 58.8 |
| Indian Trail | 83.4 | 52.1 | NA | NA | 87.9 | 7.6 | 20,833 | -2.9 | 1,247 | 6 | 32,001 | 71.9 | 26,585 | 55.4 |
| Jacksonville | 55.0 | 15.0 | 97.7 | 91.9 | 58.0 | 27.8 | 19,167 | -3.1 | 1,567 | 8 | 58,652 | 33.2 | 53,727 | 66.0 |
| Kannapolis | 78.3 | 37.1 | 89.3 | 83.0 | 90.4 | 6.3 | 23,848 | -0.6 | 1,964 | 8 | 37,624 | 65.9 | 30,723 | 59.0 |
| Matthews | 80.7 | 47.6 | 96.6 | 91.9 | 89.1 | 5.1 | 17,805 | -3.2 | 1,027 | 6 | 26,745 | 66.1 | 21,645 | 53.4 |
| Monroe | 75.1 | 37.3 | 90.5 | 84.6 | 79.6 | 10.9 | 17,064 | -2.6 | 1,284 | 8 | 27,050 | 64.9 | 22,267 | 51.0 |
| Mooresville | 81.6 | 36.5 | 96.8 | 92.7 | 83.4 | 12.8 | 20,832 | -0.8 | 1,713 | 8 | 30,636 | 69.3 | 26,564 | 56.3 |
| New Bern | 87.2 | 17.6 | 89.3 | 71.2 | 81.2 | 7.2 | 12,517 | -1.9 | 933 | 8 | 24,691 | 57.0 | 16,069 | 55.6 |
| Raleigh | 78.2 | 31.6 | 96.4 | 91.6 | 79.0 | 8.6 | 253,561 | -2.8 | 18,558 | 7 | 385,422 | 69.3 | 330,064 | 55.5 |
| Rocky Mount | 89.5 | 30.4 | 90.3 | 78.3 | 85.5 | 9.1 | 22,677 | -1.9 | 2,707 | 12 | 43,918 | 59.7 | 33,629 | 45.2 |
| Salisbury | 75.6 | 32.0 | 89.1 | 82.9 | 85.7 | 10.4 | 13,822 | -1.1 | 1,413 | 10 | 26,823 | 57.0 | 20,772 | 39.2 |
| Sanford | NA | 31.7 | 92.6 | 59.9 | 87.8 | 5.2 | 12,305 | -0.4 | 1,065 | 9 | 22,354 | 58.9 | 17,802 | 54.3 |
| Thomasville | NA | 36.3 | 93.0 | 87.4 | 92.1 | 5.3 | 12,015 | -2.1 | 1,127 | 9 | 22,852 | 52.8 | 17,399 | 49.0 |
| Wake Forest | 78.8 | 52.8 | NA | NA | 80.7 | 7.6 | 22,866 | -1.4 | 1,386 | 6 | 35,592 | 67.8 | 29,652 | 53.7 |
| Wilmington | 75.5 | 16.0 | 93.8 | 87.9 | 78.9 | 11.2 | 63,516 | -3.2 | 4,869 | 8 | 104,075 | 59.3 | 81,982 | 45.5 |
| Wilson | 81.8 | 26.5 | 76.2 | 66.4 | 86.5 | 8.2 | 20,276 | -1.2 | 2,073 | 10 | 39,270 | 62.1 | 29,952 | 47.6 |
| Winston-Salem | 81.5 | 24.1 | 93.5 | 82.6 | 85.7 | 6.2 | 116,852 | -2.7 | 9,509 | 8 | 192,410 | 62.1 | 159,244 | 48.6 |
| **NORTH DAKOTA** | 82.5 | 15.8 | 92.3 | 83.5 | 82.0 | 8.6 | 406,839 | 0.8 | 20,833 | 5 | 602,982 | 69.1 | 482,805 | 57.7 |
| Bismarck | 83.2 | 8.4 | 91.1 | 85.4 | 84.4 | 6.0 | 38,783 | 3.1 | 1,734 | 5 | 59,658 | 68.4 | 47,601 | 56.2 |
| Fargo | 83.7 | 9.4 | 94.7 | 85.0 | 72.8 | 11.2 | 71,770 | 2.3 | 3,265 | 5 | 101,893 | 76.8 | 85,878 | 56.3 |
| Grand Forks | 80.2 | 11.4 | 95.1 | 87.7 | 74.9 | 12.1 | 30,596 | -1.0 | 1,419 | 5 | 45,709 | 71.6 | 38,602 | 48.8 |
| Minot | 85.3 | 9.2 | 96.3 | 89.3 | 78.4 | 11.3 | 23,465 | 1.8 | 1,404 | 6 | 38,963 | 70.2 | 32,682 | 61.2 |
| West Fargo | 87.3 | 14.5 | 93.1 | 91.9 | 82.4 | 8.6 | 22,127 | 3.3 | 881 | 4 | 27,937 | 78.0 | 23,530 | 70.9 |
| **OHIO** | 82.4 | 32.4 | 91.8 | 84.8 | 85.6 | 5.6 | 5,754,286 | -0.8 | 468,802 | 8 | 9,407,641 | 63.4 | 7,364,093 | 52.9 |
| Akron | 81.2 | 23.9 | 91.7 | 84.5 | 83.2 | 5.7 | 91,563 | 0.6 | 9,232 | 10 | 163,028 | 62.8 | 132,948 | 45.3 |
| Barberton | NA | 22.0 | 86.1 | 80.4 | NA | NA | 12,513 | -0.4 | 1,166 | 9 | 20,759 | 60.3 | 15,012 | 47.3 |
| Beavercreek | 85.4 | 11.6 | 94.8 | 92.8 | 86.8 | 8.5 | 23,559 | -1.1 | 1,292 | 6 | 38,118 | 73.1 | 29,555 | 62.4 |
| Bowling Green | 68.2 | 23.4 | 95.6 | 84.1 | 55.3 | 19.5 | 16,319 | -3.8 | 1,012 | 6 | 27,861 | 63.9 | 24,424 | 31.9 |
| Brunswick | 88.5 | 49.3 | 96.2 | 83.8 | 90.9 | 5.3 | 19,401 | -4.4 | 1,564 | 8 | 28,257 | 54.7 | 21,822 | 62.2 |
| Canton | 84.9 | 23.4 | 86.6 | 71.1 | 84.7 | 3.1 | 31,422 | 0.8 | 3,149 | 10 | 55,084 | 65.7 | 44,083 | 40.9 |
| Cincinnati | 73.2 | 29.6 | 90.9 | 84.3 | 76.3 | 8.8 | 149,129 | 1.2 | 13,467 | 9 | 248,043 | 59.6 | 208,000 | 47.3 |
| Cleveland | 68.5 | 27.5 | 87.5 | 68.3 | 83.8 | 4.1 | 157,399 | -1.2 | 21,230 | 14 | 307,543 | 67.8 | 253,384 | 42.2 |
| Cleveland Heights | 73.7 | 29.0 | 95.0 | 85.2 | 82.6 | 6.5 | 21,944 | -4.3 | 2,032 | 9 | 37,031 | 72.4 | 28,579 | 58.3 |
| Columbus | 78.6 | 27.7 | 94.3 | 88.9 | 77.3 | 6.2 | 483,757 | 1.7 | 38,280 | 8 | 720,275 | 69.8 | 624,932 | 56.9 |
| Cuyahoga Falls | 88.9 | 26.2 | 89.5 | 83.3 | 87.3 | 6.7 | 26,257 | -1.1 | 2,026 | 8 | 40,404 | 69.8 | 32,324 | 56.7 |
| Dayton | 67.5 | 19.1 | 88.4 | 78.4 | 71.9 | 8.3 | 59,991 | 2.0 | 6,576 | 11 | 112,051 | 57.3 | 94,255 | 38.3 |
| Delaware | 78.1 | 51.4 | 93.1 | 86.1 | 85.4 | 8.5 | 22,187 | 2.9 | 1,386 | 6 | 35,490 | 63.5 | 28,022 | 48.3 |
| Dublin | 83.6 | 36.5 | NA | NA | 87.2 | 6.4 | 26,365 | -0.5 | 1,353 | 5 | 35,428 | 75.4 | 29,624 | 61.8 |
| Elyria | 87.3 | 30.1 | 92.9 | 70.1 | 83.3 | 5.2 | 26,591 | -2.7 | 2,902 | 11 | 43,646 | 59.1 | 34,746 | 48.3 |
| Euclid | 73.4 | 27.1 | 88.4 | 81.6 | 88.6 | 2.3 | 22,273 | -1.5 | 2,878 | 13 | 36,384 | 65.2 | 28,117 | 53.7 |
| Fairborn | 83.3 | 23.6 | 93.5 | 91.5 | 77.1 | 13.6 | 17,167 | 0.6 | 1,304 | 8 | 27,853 | 60.6 | 22,046 | 54.6 |
| Fairfield | 87.7 | 35.6 | 90.3 | 82.8 | 82.8 | 8.7 | 24,395 | 0.0 | 1,857 | 8 | 35,215 | 64.9 | 27,243 | 57.8 |
| Findlay | 82.9 | 16.1 | 92.0 | 85.8 | 76.5 | 12.5 | 22,084 | 0.7 | 1,721 | 8 | 34,137 | 61.4 | 27,088 | 56.5 |
| Gahanna | 83.2 | 25.4 | NA | NA | 89.1 | 3.1 | 20,373 | -0.4 | 1,288 | 6 | 28,624 | 67.5 | 20,700 | 60.9 |
| Garfield Heights | 86.1 | 30.4 | 90.4 | 75.7 | 93.0 | 1.9 | 13,483 | -1.4 | 1,751 | 13 | 21,282 | 65.7 | 17,322 | 50.4 |
| Green | 88.6 | 29.8 | 96.7 | 95.1 | 90.8 | 4.3 | 13,653 | -1.6 | 955 | 7 | 20,535 | 69.6 | 15,702 | 54.9 |
| Grove City | 90.1 | 36.2 | 97.4 | 91.3 | NA | NA | 22,541 | -0.1 | 1,373 | 6 | 33,578 | 68.0 | 25,647 | 58.5 |

1. Employed persons.  2. Households.  3. Percent of civilian labor force.  4. Persons 16 years old and over.

# Table D. Cities — Construction, Wholesale Trade, and Retail Trade

| City | Value of residential construction authorized by building permits, 2020 | | | Wholesale trade[1], 2017 | | | | Retail trade[2], 2017 | | | |
|---|---|---|---|---|---|---|---|---|---|---|---|
| | New construction ($1,000) | Number of housing units | Percent single family | Number of establishments | Number of employees | Sales (mil dol) | Annual payroll (mil dol) | Number of establishments | Number of employees | Sales (mil dol) | Annual payroll (mil dol) |
| | 69 | 70 | 71 | 72 | 73 | 74 | 75 | 76 | 77 | 78 | 79 |
| **NEW YORK—Cont'd** | | | | | | | | | | | |
| Yonkers | 48,001 | 490 | 2 | 149 | 1,270 | 844 | 70 | 715 | 10,095 | 2,896 | 290 |
| **NORTH CAROLINA** | 20,082 | 117 | 100 | 9,831 | 154,877 | 132,342 | 10,273 | 34,926 | 496,081 | 141,134 | 12,673 |
| Apex | 318,934 | 1,427 | 100 | 46 | 451 | 359 | 28 | 148 | 3,034 | 961 | 85 |
| Asheboro | 10,153 | 85 | 100 | 33 | 323 | 179 | 12 | 187 | 2,619 | 729 | 65 |
| Asheville | 100,754 | 474 | 84 | 160 | 1,660 | 1,186 | 101 | 872 | 14,168 | 3,502 | 359 |
| Burlington | 94,197 | 851 | 29 | 74 | 1,213 | 487 | 72 | 354 | 5,770 | 1,446 | 135 |
| Cary | 322,326 | 1,515 | 53 | 158 | 3,619 | 6,133 | 362 | 544 | 9,914 | 3,170 | 280 |
| Chapel Hill | 29,136 | 151 | 25 | 25 | 141 | 197 | 10 | 168 | 2,741 | 826 | 80 |
| Charlotte | NA | NA | NA | 1,606 | 27,225 | 22,742 | 1,832 | 2,758 | 46,761 | 13,772 | 1,308 |
| Concord | NA | NA | NA | 109 | 1,785 | 1,935 | 123 | 507 | 10,139 | 2,491 | 240 |
| Durham | 585,352 | 3,832 | 52 | 216 | 5,244 | 4,486 | 539 | 867 | 14,593 | 3,700 | 368 |
| Fayetteville | 63,914 | 291 | 77 | 95 | 1,710 | 668 | 61 | 861 | 14,319 | 3,941 | 360 |
| Garner | 65,095 | 469 | 85 | 66 | 1,411 | 957 | 73 | 156 | 3,136 | 784 | 67 |
| Gastonia | 153,272 | 571 | 85 | 84 | 882 | 353 | 36 | 365 | 5,816 | 1,419 | 138 |
| Goldsboro | 12,590 | 68 | 100 | 61 | 1,302 | 1,093 | 48 | 296 | 4,265 | 1,139 | 97 |
| Greensboro | 220,342 | 1,438 | 46 | 538 | 9,612 | 10,399 | 851 | 1,246 | 20,654 | 5,769 | 539 |
| Greenville | 77,431 | 349 | 91 | 54 | 503 | 370 | 29 | 414 | 6,765 | 1,783 | 170 |
| Hickory | NA | NA | NA | 118 | 4,143 | 3,413 | 234 | 410 | 7,136 | 2,240 | 192 |
| High Point | 60,745 | 512 | 75 | 307 | 4,428 | 4,431 | 257 | 388 | 5,018 | 1,524 | 134 |
| Huntersville | NA | NA | NA | 53 | 924 | 1,347 | 74 | 173 | 3,872 | 1,141 | 98 |
| Indian Trail | NA | NA | NA | 88 | 927 | 501 | 59 | 107 | 2,226 | 741 | 64 |
| Jacksonville | 9,061 | 101 | 17 | 27 | 139 | 70 | 6 | 329 | 6,247 | 1,701 | 153 |
| Kannapolis | NA | NA | NA | 21 | 297 | 226 | 15 | 155 | 1,807 | 528 | 44 |
| Matthews | NA | NA | NA | 54 | 944 | 325 | 59 | 164 | 3,091 | 996 | 85 |
| Monroe | 37,668 | 255 | 100 | 63 | 1,077 | 566 | 59 | 226 | 3,026 | 778 | 70 |
| Mooresville | NA | NA | NA | 105 | 998 | 579 | 66 | 258 | 4,663 | 1,416 | 118 |
| New Bern | 55,334 | 330 | 60 | 28 | 307 | 796 | 17 | 205 | 3,162 | 856 | 78 |
| Raleigh | 705,788 | 3,479 | 33 | 596 | 10,466 | 8,105 | 674 | 1,798 | 29,789 | 9,238 | 846 |
| Rocky Mount | 6,386 | 57 | 100 | 83 | 1,534 | 934 | 69 | 300 | 4,016 | 1,041 | 92 |
| Salisbury | NA | NA | NA | 53 | 844 | 593 | 45 | 200 | 3,193 | 940 | 82 |
| Sanford | 21,838 | 145 | 100 | 33 | 550 | 570 | 25 | 210 | 2,741 | 771 | 67 |
| Thomasville | 10,901 | 63 | 100 | 31 | 711 | 227 | 27 | 131 | 1,350 | 340 | 30 |
| Wake Forest | 154,173 | 860 | 65 | 31 | 296 | 236 | 20 | 801 | 11,925 | 3,747 | 330 |
| Wilmington | NA | NA | NA | 138 | 1,385 | 744 | 85 | D | D | D | D |
| Wilson | 29,227 | 283 | 41 | 80 | 811 | 532 | 43 | 272 | 3,456 | 910 | 84 |
| Winston-Salem | 273,136 | 1,582 | 98 | 236 | 3,954 | 3,454 | 219 | 1,009 | 16,801 | 4,563 | 424 |
| **NORTH DAKOTA** | 0 | 0 | 0 | 1,563 | 20,075 | 23,277 | 1,204 | 3,277 | 49,579 | 19,251 | 1,454 |
| Bismarck | 93,445 | 467 | 51 | 132 | 2,460 | 1,859 | 155 | 388 | 7,549 | 2,074 | 214 |
| Fargo | 190,658 | 1,124 | 40 | 253 | 5,174 | 4,102 | 310 | 539 | 11,419 | 7,916 | 363 |
| Grand Forks | 58,466 | 406 | 32 | 57 | 953 | 631 | 53 | 282 | 5,846 | 1,551 | 150 |
| Minot | 29,623 | 129 | 97 | 79 | 1,305 | 1,706 | 80 | 268 | 5,160 | 1,385 | 144 |
| West Fargo | 104,875 | 373 | 100 | 52 | 546 | 399 | 29 | 101 | 1,531 | 509 | 42 |
| **OHIO** | 0 | 0 | 0 | 11,430 | 193,412 | 172,949 | 11,316 | 35,500 | 588,060 | 174,300 | 14,861 |
| Akron | NA | NA | NA | 229 | 3,602 | 1,875 | 189 | 549 | 7,127 | 1,804 | 179 |
| Barberton | 1,080 | 8 | 100 | 22 | 228 | 237 | 11 | 61 | 718 | 176 | 16 |
| Beavercreek | NA | NA | NA | 25 | 265 | 226 | 16 | 270 | 6,178 | 1,184 | 118 |
| Bowling Green | NA | NA | NA | 12 | 78 | 22 | 3 | 92 | 1,780 | 414 | 39 |
| Brunswick | 2,278 | 8 | 100 | 45 | 401 | 312 | 22 | 83 | 1,487 | 881 | 47 |
| Canton | 3,267 | 28 | 100 | 79 | 1,257 | 1,966 | 67 | 281 | 3,843 | 942 | 92 |
| Cincinnati | 125,827 | 906 | 14 | 309 | 5,562 | 5,483 | 363 | 866 | 15,048 | 3,908 | 371 |
| Cleveland | 14,619 | 117 | 81 | 511 | 9,019 | 5,966 | 515 | 1,136 | 9,648 | 2,487 | 224 |
| Cleveland Heights | 153 | 2 | 0 | 13 | 65 | 14 | 3 | 91 | 1,201 | 284 | 32 |
| Columbus | 703,997 | 6,301 | 13 | 788 | 18,425 | 19,176 | 1,168 | 2,600 | 48,299 | 15,624 | 1,326 |
| Cuyahoga Falls | NA | NA | NA | 44 | 643 | 265 | 32 | 168 | 3,559 | 1,138 | 99 |
| Dayton | 11,742 | 127 | 2 | 141 | 2,405 | 5,432 | 175 | 346 | 3,707 | 858 | 85 |
| Delaware | 125,733 | 453 | 86 | 20 | 191 | 86 | 10 | 112 | 1,940 | 746 | 53 |
| Dublin | 59,402 | 108 | 100 | 90 | 1,815 | 4,741 | 157 | 107 | 2,358 | 1,225 | 83 |
| Elyria | 24,312 | 105 | 100 | 53 | 296 | 164 | 13 | 198 | 3,164 | 804 | 75 |
| Euclid | 0 | 0 | 0 | 37 | 453 | 211 | 26 | 83 | 1,034 | 293 | 27 |
| Fairborn | 25,414 | 132 | 49 | D | D | D | D | 69 | 878 | 308 | 19 |
| Fairfield | 2,081 | 8 | 100 | 75 | 2,299 | 1,816 | 128 | 167 | 4,069 | 1,234 | 114 |
| Findlay | 25,851 | 163 | 19 | 43 | 772 | 895 | 45 | 201 | 3,900 | 1,049 | 94 |
| Gahanna | 21,414 | 101 | 8 | 48 | 904 | 977 | 42 | 91 | 1,468 | 548 | 41 |
| Garfield Heights | 120 | 1 | 100 | 31 | 413 | 132 | 28 | 66 | 876 | 202 | 17 |
| Green | NA | NA | NA | 39 | 645 | 511 | 43 | 62 | 1,869 | 913 | 66 |
| Grove City | 72,594 | 500 | 56 | 29 | 936 | 902 | 58 | 121 | 3,115 | 1,116 | 75 |

1. Merchant wholesalers except manufacturers' sales branches and offices.   2. Establishments with payroll.

| City | Real estate and rental and leasing, 2017 | | | | Professional, scientific, and technical services[1], 2017 | | | | Manufacturing, 2017 | | | |
|---|---|---|---|---|---|---|---|---|---|---|---|---|
| | Number of establish-ments | Number of employees | Receipts (mil dol) | Annual payroll (mil dol) | Number of establish-ments | Number of employees | Receipts (mil dol) | Annual payroll (mil dol) | Number of establish-ments | Number of employees | Receipts (mil dol) | Annual payroll (mil dol) |
| | 80 | 81 | 82 | 83 | 84 | 85 | 86 | 87 | 88 | 89 | 90 | 91 |
| NEW YORK—Cont'd | | | | | | | | | | | | |
| Yonkers | 345 | 1,475 | 360 | 66 | 264 | 1,596 | 321 | 140 | NA | NA | NA | NA |
| NORTH CAROLINA | 12,450 | 56,360 | 14,647 | 2,639 | 24,766 | 221,438 | 40,878 | 16,048 | 8,834 | 422,891 | 200,381 | 21,480 |
| Apex | D | D | D | 5 | 196 | 617 | 99 | 32 | NA | NA | NA | NA |
| Asheboro | 35 | 174 | 29 | 7 | 62 | 334 | 36 | 13 | NA | NA | NA | NA |
| Asheville | 354 | 1,204 | 273 | 49 | 693 | 3,657 | 504 | 213 | NA | NA | NA | NA |
| Burlington | 78 | 289 | 79 | 10 | 94 | 774 | 88 | 38 | NA | NA | NA | NA |
| Cary | 292 | 1,138 | 354 | 54 | 1,049 | 11,172 | 2,079 | 944 | NA | NA | NA | NA |
| Chapel Hill | D | D | D | D | D | D | D | D | NA | NA | NA | NA |
| Charlotte | 1,936 | 11,022 | 4,013 | 694 | 3,655 | 47,392 | 10,078 | 3,864 | NA | NA | NA | NA |
| Concord | 151 | 629 | 153 | 23 | 254 | 993 | 141 | 49 | NA | NA | NA | NA |
| Durham | 337 | 2,013 | 443 | 102 | D | D | D | D | NA | NA | NA | NA |
| Fayetteville | 249 | 1,457 | 351 | 53 | 434 | 5,417 | 653 | 276 | NA | NA | NA | NA |
| Garner | 35 | 145 | 50 | 7 | 87 | 1,302 | 86 | 38 | NA | NA | NA | NA |
| Gastonia | 106 | 470 | 153 | 18 | 154 | 1,040 | 93 | 40 | NA | NA | NA | NA |
| Goldsboro | 48 | 287 | 45 | 9 | 87 | 523 | 60 | 23 | NA | NA | NA | NA |
| Greensboro | 508 | 3,933 | 745 | 203 | 965 | 7,972 | 1,277 | 491 | NA | NA | NA | NA |
| Greenville | 130 | 559 | 117 | 19 | 238 | 1,182 | 149 | 56 | NA | NA | NA | NA |
| Hickory | 110 | 472 | 106 | 19 | 197 | 1,712 | 706 | 100 | NA | NA | NA | NA |
| High Point | 125 | 708 | 105 | 26 | 289 | 2,867 | 470 | 181 | NA | NA | NA | NA |
| Huntersville | 109 | 306 | 97 | 17 | 217 | 1,715 | 270 | 106 | NA | NA | NA | NA |
| Indian Trail | 45 | 111 | 30 | 5 | 72 | 341 | 44 | 14 | NA | NA | NA | NA |
| Jacksonville | 118 | 450 | 92 | 16 | 139 | 943 | 102 | 38 | NA | NA | NA | NA |
| Kannapolis | 45 | 153 | 32 | 6 | 53 | 463 | 43 | 16 | NA | NA | NA | NA |
| Matthews | 63 | 385 | 77 | 15 | 141 | 771 | 106 | 41 | NA | NA | NA | NA |
| Monroe | 39 | 147 | 26 | 5 | 107 | 745 | 94 | 33 | NA | NA | NA | NA |
| Mooresville | 96 | 263 | 86 | 11 | 179 | 1,812 | 243 | 98 | NA | NA | NA | NA |
| New Bern | 59 | 193 | 36 | 6 | 101 | 615 | 96 | 43 | NA | NA | NA | NA |
| Raleigh | 962 | 6,090 | 1,652 | 365 | 2,494 | 33,853 | 7,201 | 2,814 | NA | NA | NA | NA |
| Rocky Mount | 68 | 440 | 75 | 16 | 110 | 730 | 108 | 32 | NA | NA | NA | NA |
| Salisbury | 44 | 134 | 23 | 4 | 90 | 548 | 84 | 26 | NA | NA | NA | NA |
| Sanford | 31 | 91 | 23 | 4 | 71 | 322 | 35 | 14 | NA | NA | NA | NA |
| Thomasville | D | D | D | 3 | 35 | 165 | 21 | 4 | NA | NA | NA | NA |
| Wake Forest | 52 | 158 | 59 | 7 | 154 | 1,530 | 427 | 116 | NA | NA | NA | NA |
| Wilmington | 305 | 1,117 | 286 | 48 | 674 | 4,243 | 677 | 248 | NA | NA | NA | NA |
| Wilson | 71 | 258 | 44 | 7 | 90 | 542 | 65 | 26 | NA | NA | NA | NA |
| Winston-Salem | 351 | 1,687 | 661 | 77 | 694 | 5,556 | 913 | 349 | NA | NA | NA | NA |
| NORTH DAKOTA | 1,100 | 5,440 | 1,280 | 242 | 1,821 | 14,880 | 2,425 | 947 | 701 | 24,214 | 13,605 | 1,266 |
| Bismarck | 140 | 446 | 104 | 16 | 261 | 2,335 | 475 | 166 | NA | NA | NA | NA |
| Fargo | 277 | 1,716 | 293 | 67 | 449 | 5,007 | 727 | 315 | NA | NA | NA | NA |
| Grand Forks | 80 | 569 | 90 | 21 | 119 | 1,301 | 173 | 77 | NA | NA | NA | NA |
| Minot | D | D | D | D | 121 | 737 | 115 | 45 | NA | NA | NA | NA |
| West Fargo | D | D | D | D | 48 | 821 | 188 | 63 | NA | NA | NA | NA |
| OHIO | 10,782 | 62,902 | 20,524 | 2,986 | 23,854 | 250,438 | 43,625 | 17,059 | 13,922 | 652,462 | 306,222 | 36,257 |
| Akron | 165 | 977 | 192 | 38 | 440 | 4,411 | 854 | 305 | NA | NA | NA | NA |
| Barberton | 12 | 37 | 6 | 1 | D | D | D | D | NA | NA | NA | NA |
| Beavercreek | 47 | 171 | 75 | 6 | 229 | 5,170 | 1,038 | 434 | NA | NA | NA | NA |
| Bowling Green | 40 | 168 | 25 | 6 | 49 | 273 | 36 | 15 | NA | NA | NA | NA |
| Brunswick | 29 | 85 | 27 | 3 | 62 | 298 | 34 | 14 | NA | NA | NA | NA |
| Canton | 66 | 307 | 47 | 14 | 127 | 677 | 96 | 31 | NA | NA | NA | NA |
| Cincinnati | 435 | 2,689 | 759 | 156 | 1,115 | 20,598 | 4,138 | 1,629 | NA | NA | NA | NA |
| Cleveland | 395 | 2,997 | 615 | 135 | 1,091 | 20,238 | 4,543 | 1,783 | NA | NA | NA | NA |
| Cleveland Heights | 52 | 226 | 32 | 7 | 95 | 259 | 35 | 12 | NA | NA | NA | NA |
| Columbus | 1,040 | 8,225 | 2,665 | 442 | 2,259 | 30,953 | 6,165 | 2,285 | NA | NA | NA | NA |
| Cuyahoga Falls | 45 | 209 | 25 | 7 | 96 | 578 | 61 | 24 | NA | NA | NA | NA |
| Dayton | 113 | 501 | 136 | 20 | 290 | 3,387 | 594 | 222 | NA | NA | NA | NA |
| Delaware | 34 | 210 | 41 | 10 | 79 | 429 | 85 | 27 | NA | NA | NA | NA |
| Dublin | 113 | 698 | 148 | 40 | 443 | 7,332 | 1,131 | 594 | NA | NA | NA | NA |
| Elyria | 56 | 258 | 51 | 10 | 77 | 440 | 55 | 27 | NA | NA | NA | NA |
| Euclid | 45 | 287 | 44 | 8 | 48 | 326 | 49 | 18 | NA | NA | NA | NA |
| Fairborn | 31 | 105 | 19 | 3 | 73 | 2,138 | 595 | 191 | NA | NA | NA | NA |
| Fairfield | 57 | 340 | 84 | 13 | 81 | 912 | 102 | 35 | NA | NA | NA | NA |
| Findlay | 49 | 419 | 59 | 15 | 86 | 1,241 | 156 | 64 | NA | NA | NA | NA |
| Gahanna | 46 | 401 | 239 | 17 | 141 | 1,278 | 162 | 67 | NA | NA | NA | NA |
| Garfield Heights | 22 | 87 | 19 | 4 | 31 | 328 | 51 | 24 | NA | NA | NA | NA |
| Green | 27 | 195 | 53 | 8 | 73 | 802 | 168 | 54 | NA | NA | NA | NA |
| Grove City | 43 | 152 | 75 | 7 | 51 | 366 | 43 | 14 | NA | NA | NA | NA |

1. Establishments subject to federal tax.

Table D. Cities — **Accommodation and Food Services, Arts, Entertainment, and Recreation, and Health Care and Social Assistance**

| City | Accommodation and food services, 2017 | | | | Arts, entertainment, and recreation[1], 2017 | | | | Health care and social assistance[1], 2017 | | | |
|---|---|---|---|---|---|---|---|---|---|---|---|---|
| | Number of establishments | Number of employees | Receipts (mil dol) | Annual payroll (mil dol) | Number of establishments | Number of employees | Receipts (mil dol) | Annual payroll (mil dol) | Number of establishments | Number of employees | Receipts (mil dol) | Annual payroll (mil dol) |
| | 92 | 93 | 94 | 95 | 96 | 97 | 98 | 99 | 100 | 101 | 102 | 103 |
| NEW YORK—Cont'd | | | | | | | | | | | | |
| Yonkers.................... | 387 | 4,822 | 368 | 102 | 56 | 1,491 | 281 | 55 | 493 | 10,893 | 1,265 | 555 |
| NORTH CAROLINA.............. | 21,437 | 429,125 | 24,913 | 6,866 | 3,868 | 69,027 | 6,354 | 2,057 | 24,080 | 602,444 | 72,732 | 27,627 |
| Apex.................... | 96 | 2,156 | 111 | 33 | 19 | 202 | 16 | 3 | 128 | 1,143 | 130 | 47 |
| Asheboro.................... | 108 | 2,391 | 118 | 32 | 11 | 188 | 6 | 2 | 143 | 3,276 | 289 | 127 |
| Asheville.................... | 646 | 15,437 | 964 | 289 | 126 | 2,342 | 168 | 54 | 739 | 24,227 | 3,351 | 1,329 |
| Burlington.................... | 196 | 4,568 | 230 | 64 | 24 | 492 | 16 | 7 | 247 | 10,615 | 1,590 | 599 |
| Cary.................... | 432 | 9,309 | 541 | 163 | 81 | 2,675 | 107 | 36 | D | D | D | D |
| Chapel Hill.................... | 221 | 4,537 | 268 | 86 | 39 | 471 | 31 | 12 | 274 | 15,270 | 1,943 | 839 |
| Charlotte.................... | 2,214 | 50,810 | 3,396 | 925 | 413 | 12,043 | 1,344 | 614 | 2,288 | 60,916 | 9,331 | 3,379 |
| Concord.................... | 265 | 7,952 | 437 | 118 | 56 | 2,328 | 496 | 131 | 271 | 7,600 | 955 | 418 |
| Durham.................... | 709 | 15,662 | 1,000 | 291 | 106 | 2,162 | 98 | 36 | 823 | 27,001 | 4,718 | 1,480 |
| Fayetteville.................... | 534 | 12,325 | 632 | 175 | 56 | 661 | 33 | 9 | 612 | 18,776 | 2,231 | 887 |
| Garner.................... | 86 | 2,259 | 125 | 32 | 8 | 130 | 7 | 2 | 100 | 1,375 | 150 | 59 |
| Gastonia.................... | 209 | 4,659 | 256 | 71 | 23 | 447 | 15 | 6 | 311 | 9,322 | 1,100 | 440 |
| Goldsboro.................... | 161 | 3,371 | 166 | 46 | 12 | 214 | 7 | 2 | 186 | 6,299 | 696 | 272 |
| Greensboro.................... | 884 | 20,121 | 1,077 | 310 | 122 | 2,541 | 139 | 44 | 890 | 24,398 | 3,050 | 1,153 |
| Greenville.................... | 298 | 7,275 | 364 | 97 | 27 | 396 | 22 | 7 | 386 | 14,873 | 2,070 | 733 |
| Hickory.................... | 193 | 4,176 | 219 | 65 | 27 | 551 | 23 | 9 | 286 | 9,735 | 1,228 | 464 |
| High Point.................... | 236 | 5,662 | 281 | 80 | 31 | 526 | 56 | 7 | 274 | 8,440 | 812 | 370 |
| Huntersville.................... | 121 | 3,005 | 168 | 49 | D | D | D | D | 181 | 3,106 | 595 | 171 |
| Indian Trail.................... | 83 | 1,371 | 65 | 17 | D | D | D | D | D | D | D | D |
| Jacksonville.................... | 246 | 5,373 | 297 | 81 | 15 | 263 | 19 | 5 | 198 | 4,155 | 528 | 173 |
| Kannapolis.................... | 82 | 1,591 | 82 | 22 | D | D | D | D | 63 | 907 | 90 | 36 |
| Matthews.................... | 112 | 2,578 | 135 | 39 | 31 | 684 | 25 | 8 | 144 | 3,779 | 579 | 176 |
| Monroe.................... | 115 | 2,384 | 143 | 38 | 11 | 135 | 8 | 3 | 158 | 4,048 | 459 | 185 |
| Mooresville.................... | 190 | 3,895 | 216 | 60 | 57 | 1,341 | 295 | 78 | 206 | 3,055 | 374 | 137 |
| New Bern.................... | 124 | 2,588 | 142 | 39 | D | D | D | D | 187 | 5,467 | 599 | 244 |
| Raleigh.................... | 1,259 | 29,257 | 1,829 | 510 | 244 | 7,406 | 513 | 184 | 1,635 | 43,361 | 5,535 | 2,256 |
| Rocky Mount.................... | 149 | 3,372 | 171 | 46 | 24 | 397 | 16 | 5 | 219 | 6,098 | 513 | 227 |
| Salisbury.................... | 143 | 3,027 | 168 | 47 | 19 | 300 | 14 | 4 | 187 | 7,950 | 994 | 481 |
| Sanford.................... | 105 | 1,915 | 97 | 26 | 11 | 121 | 6 | 1 | 145 | 2,519 | 256 | 98 |
| Thomasville.................... | 72 | 1,137 | 62 | 16 | 4 | 122 | 3 | 2 | 49 | 1,568 | 167 | 57 |
| Wake Forest.................... | 100 | 2,045 | 109 | 30 | 26 | 278 | 17 | 4 | D | D | D | D |
| Wilmington.................... | 525 | 11,091 | 601 | 174 | 88 | 1,467 | 81 | 25 | 673 | 16,330 | 2,078 | 816 |
| Wilson.................... | 144 | 3,010 | 158 | 43 | D | D | D | D | 206 | 4,782 | 408 | 162 |
| Winston-Salem.................... | 578 | 13,249 | 715 | 207 | 91 | 2,611 | 509 | 82 | 631 | 32,573 | 4,106 | 1,413 |
| NORTH DAKOTA ................. | 2,080 | 36,648 | 2,118 | 620 | 470 | 5,418 | 273 | 85 | 2,057 | 62,455 | 7,298 | 2,913 |
| Bismarck.................... | 197 | 5,353 | 288 | 94 | 42 | 914 | 39 | 14 | 292 | 10,316 | 1,255 | 484 |
| Fargo.................... | 371 | 8,857 | 433 | 138 | 79 | 1,717 | 78 | 27 | 426 | 19,053 | 2,867 | 978 |
| Grand Forks.................... | 182 | 4,165 | 188 | 61 | 45 | 938 | 30 | 9 | 147 | 7,143 | 805 | 370 |
| Minot.................... | 164 | 3,558 | 174 | 56 | 36 | 440 | 25 | 7 | 144 | 4,809 | 594 | 278 |
| West Fargo.................... | 46 | 969 | 45 | 14 | 15 | 62 | 3 | 1 | 66 | 942 | 60 | 28 |
| OHIO.................... | 24,346 | 474,616 | 24,561 | 7,078 | 3,999 | 76,914 | 8,596 | 2,832 | 29,595 | 856,794 | 97,117 | 39,469 |
| Akron.................... | 414 | 6,196 | 312 | 89 | 48 | 1,348 | 104 | 38 | 492 | 31,388 | 3,589 | 1,640 |
| Barberton.................... | 42 | 682 | 29 | 8 | NA | NA | NA | NA | 78 | 2,110 | 213 | 82 |
| Beavercreek.................... | 136 | 4,131 | 200 | 62 | 13 | 221 | 13 | 3 | 154 | 3,520 | 363 | 172 |
| Bowling Green.................... | 106 | 2,628 | 100 | 29 | 10 | 106 | 6 | 2 | 105 | 2,853 | 226 | 89 |
| Brunswick.................... | 69 | 1,309 | 54 | 15 | 13 | 82 | 7 | 2 | 59 | 1,095 | 98 | 46 |
| Canton.................... | 155 | 2,735 | 134 | 36 | 23 | 470 | 49 | 11 | 205 | 10,591 | 1,308 | 504 |
| Cincinnati.................... | 813 | 17,343 | 1,028 | 315 | 146 | 6,496 | 1,328 | 538 | 896 | 55,181 | 7,650 | 3,302 |
| Cleveland.................... | 1,014 | 17,408 | 1,172 | 335 | 127 | 8,202 | 1,361 | 667 | 758 | 79,004 | 11,018 | 5,037 |
| Cleveland Heights.................... | 79 | 1,137 | 66 | 18 | 16 | 58 | 6 | 2 | 99 | 1,182 | 70 | 34 |
| Columbus.................... | 2,072 | 46,502 | 2,801 | 796 | 238 | 6,805 | 748 | 244 | NA | NA | NA | NA |
| Cuyahoga Falls.................... | 138 | 3,002 | 148 | 42 | 15 | 123 | 33 | 3 | 143 | 4,101 | 323 | 146 |
| Dayton.................... | 266 | 5,041 | 257 | 78 | 33 | 1,165 | 175 | 31 | NA | NA | NA | NA |
| Delaware.................... | 91 | 1,709 | 86 | 24 | 12 | 299 | 20 | 4 | 140 | 3,088 | 284 | 118 |
| Dublin.................... | 136 | 3,437 | 194 | 58 | 32 | 745 | 81 | 20 | NA | NA | NA | NA |
| Elyria.................... | 112 | 2,125 | 101 | 28 | D | D | D | D | 144 | 4,257 | 451 | 188 |
| Euclid.................... | 61 | 665 | 41 | 11 | 7 | 125 | 8 | 2 | 98 | 3,581 | 262 | 124 |
| Fairborn.................... | 72 | 1,326 | 73 | 20 | 4 | 103 | 2 | 1 | 53 | 919 | 71 | 30 |
| Fairfield.................... | 103 | 2,241 | 122 | 32 | 10 | 94 | 5 | 1 | NA | NA | NA | NA |
| Findlay.................... | 152 | 3,994 | 182 | 55 | 19 | 309 | 15 | 5 | 159 | 4,521 | 505 | 195 |
| Gahanna.................... | 97 | 2,017 | 110 | 33 | 17 | 299 | 9 | 3 | 170 | 3,722 | 326 | 163 |
| Garfield Heights.................... | 39 | 654 | 35 | 9 | NA | NA | NA | NA | 73 | 2,580 | 293 | 131 |
| Green.................... | 55 | 1,124 | 52 | 14 | D | D | D | D | 89 | 2,108 | 198 | 82 |
| Grove City.................... | 123 | 3,348 | 167 | 48 | 17 | 403 | 14 | 5 | 127 | 1,910 | 271 | 79 |

1. Establishments subject to federal tax.

## Table D. Cities — Other Services and Government Employment and Payroll

| City | Other services[1] | | | | Government employment and payroll, 2017 | | | | | | | | |
|---|---|---|---|---|---|---|---|---|---|---|---|---|---|
| | | | | | Full-time equivalent employees | March payroll | Percent of total for: | | | | | | |
| | Number of establishments | Number of employees | Receipts (mil dol) | Annual payroll (mil dol) | | Total (dollars) | Administrative, judicial, and legal | Police and corrections | Fire protection | Highways and transportation | Health and welfare | Natural resources and utilities | Education and libraries |
| | 104 | 105 | 106 | 107 | 108 | 109 | 110 | 111 | 112 | 113 | 114 | 115 | 116 |
| NEW YORK—Cont'd | | | | | | | | | | | | | |
| Yonkers.................... | 442 | 1,635 | 227 | 51 | 6,946 | 47,004,207 | 4.3 | 17.5 | 11.1 | 1.3 | 2.3 | 7.1 | 55.1 |
| NORTH CAROLINA............ | 15,118 | 93,642 | 13,022 | 3,153 | X | X | X | X | X | X | X | X | X |
| Apex........................ | D | D | D | D | 429 | 2,099,202 | 17.2 | 21.4 | 16.7 | 4.0 | 4.8 | 27.2 | 0.0 |
| Asheboro................... | 45 | 325 | 36 | 9 | 323 | 1,215,465 | 8.0 | 27.6 | 17.9 | 9.0 | 0.0 | 27.8 | 0.0 |
| Asheville................... | 334 | 2,077 | 233 | 67 | 1,209 | 5,026,530 | 11.5 | 19.7 | 22.7 | 5.0 | 2.9 | 15.5 | 0.0 |
| Burlington................. | 105 | 696 | 57 | 18 | 763 | 2,690,756 | 10.6 | 28.9 | 15.3 | 6.7 | 2.8 | 26.4 | 0.0 |
| Cary........................ | 305 | 2,251 | 290 | 104 | 1,388 | 6,713,498 | 13.5 | 19.4 | 17.6 | 12.2 | 0.0 | 25.2 | 0.0 |
| Chapel Hill................. | 102 | 1,334 | 291 | 60 | 886 | 3,497,343 | 12.7 | 26.9 | 15.5 | 18.7 | 2.7 | 8.0 | 3.6 |
| Charlotte .................. | 1,608 | 14,437 | 2,395 | 510 | 8,026 | 40,618,050 | 9.9 | 29.8 | 16.7 | 16.6 | 2.7 | 18.7 | 0.0 |
| Concord.................... | 178 | 1,070 | 125 | 32 | 986 | 3,991,786 | 9.6 | 20.2 | 19.6 | 8.6 | 3.0 | 24.0 | 0.0 |
| Durham..................... | 445 | 5,545 | 1,599 | 316 | 2,861 | 12,296,323 | 9.4 | 29.9 | 17.6 | 5.2 | 2.6 | 22.2 | 0.0 |
| Fayetteville................ | 304 | 1,788 | 175 | 51 | 2,364 | 9,863,073 | 7.3 | 26.2 | 14.5 | 7.1 | 0.9 | 27.2 | 0.0 |
| Garner...................... | D | D | D | D | 186 | 891,895 | 12.2 | 46.9 | 0.0 | 5.3 | 1.2 | 9.3 | 0.0 |
| Gastonia................... | 150 | 931 | 89 | 25 | 901 | 3,649,318 | 15.7 | 23.2 | 15.5 | 9.8 | 1.3 | 28.5 | 0.0 |
| Goldsboro................. | 90 | 596 | 49 | 15 | 437 | 1,702,043 | 11.2 | 24.2 | 18.9 | 7.0 | 0.6 | 29.8 | 0.0 |
| Greensboro................ | 564 | 3,979 | 954 | 139 | 3,249 | 13,016,203 | 8.6 | 26.8 | 20.0 | 7.1 | 1.7 | 20.5 | 2.4 |
| Greenville................. | 138 | 859 | 78 | 23 | 1,200 | 5,562,532 | 7.2 | 18.7 | 9.5 | 7.6 | 4.7 | 35.3 | 2.0 |
| Hickory.................... | 123 | 754 | 71 | 22 | 682 | 2,418,912 | 10.9 | 23.6 | 22.8 | 6.8 | 0.2 | 25.1 | 6.1 |
| High Point................. | 196 | 1,743 | 252 | 54 | 1,550 | 6,874,503 | 10.3 | 32.5 | 13.2 | 5.0 | 2.2 | 31.8 | 3.1 |
| Huntersville .............. | 95 | 525 | 46 | 13 | 161 | 770,414 | 21.0 | 57.5 | 0.0 | 8.9 | 0.0 | 11.0 | 0.0 |
| Indian Trail............... | D | D | D | D | 50 | 166,386 | 56.9 | 0.0 | 0.0 | 9.1 | 0.0 | 9.1 | 0.0 |
| Jacksonville............... | 106 | 702 | 58 | 19 | 564 | 2,057,785 | 15.4 | 29.8 | 16.0 | 4.0 | 0.4 | 25.1 | 0.0 |
| Kannapolis ............... | D | D | D | D | 327 | 1,248,082 | 12.2 | 26.9 | 33.3 | 5.3 | 0.0 | 17.8 | 0.0 |
| Matthews.................. | 85 | 465 | 57 | 15 | 161 | 662,976 | 14.6 | 47.6 | 12.7 | 19.2 | 0.0 | 6.0 | 0.0 |
| Monroe..................... | 90 | 496 | 54 | 18 | 529 | 2,183,062 | 15.7 | 21.3 | 17.2 | 5.6 | 0.0 | 34.5 | 4.8 |
| Mooresville................ | 118 | 653 | 75 | 20 | 461 | 1,929,534 | 12.5 | 25.3 | 20.5 | 3.7 | 0.8 | 23.6 | 0.0 |
| New Bern .................. | 74 | 287 | 33 | 8 | 444 | 1,841,764 | 14.7 | 24.1 | 15.9 | 4.3 | 0.8 | 34.8 | 0.0 |
| Raleigh..................... | 1,106 | 8,160 | 1,090 | 314 | 4,767 | 20,560,599 | 11.1 | 20.7 | 14.3 | 9.8 | 2.2 | 34.5 | 0.0 |
| Rocky Mount.............. | 76 | 458 | 157 | 13 | 947 | 3,736,932 | 14.0 | 24.0 | 17.1 | 5.0 | 0.6 | 34.4 | 0.0 |
| Salisbury.................. | 62 | 330 | 34 | 9 | 421 | 1,754,807 | 15.6 | 18.4 | 19.4 | 9.3 | 0.8 | 24.7 | 0.0 |
| Sanford .................... | 71 | 366 | 27 | 9 | 354 | 1,485,156 | 13.2 | 32.0 | 15.6 | 6.2 | 0.0 | 20.7 | 0.0 |
| Thomasville............... | D | D | D | D | 288 | 1,047,940 | 9.0 | 28.4 | 22.2 | 5.1 | 0.0 | 25.6 | 0.0 |
| Wake Forest............... | 80 | 473 | 47 | 15 | 228 | 1,161,132 | 10.6 | 37.5 | 0.0 | 5.5 | 0.9 | 17.8 | 0.0 |
| Wilmington ............... | 317 | 2,137 | 211 | 61 | 1,005 | 4,472,124 | 15.5 | 34.8 | 21.4 | 8.6 | 1.7 | 16.0 | 0.0 |
| Wilson ..................... | 91 | 565 | 52 | 15 | 774 | 3,369,718 | 16.1 | 18.6 | 12.8 | 3.4 | 0.6 | 39.9 | 0.0 |
| Winston-Salem............ | 418 | 2,661 | 524 | 86 | 2,230 | 8,672,025 | 13.0 | 33.4 | 17.5 | 7.9 | 5.0 | 22.2 | 0.0 |
| NORTH DAKOTA .............. | 1,763 | 9,706 | 1,246 | 318 | X | X | X | X | X | X | X | X | X |
| Bismarck................... | 248 | 1,656 | 241 | 68 | 721 | 3,561,718 | 9.6 | 26.1 | 12.6 | 8.3 | 6.5 | 21.2 | 3.4 |
| Fargo....................... | 307 | 2,393 | 289 | 71 | 982 | 4,940,787 | 10.6 | 21.8 | 13.8 | 19.8 | 13.0 | 13.9 | 3.8 |
| Grand Forks............... | 109 | 656 | 83 | 21 | 547 | 2,676,157 | 12.6 | 21.0 | 16.2 | 14.3 | 8.2 | 19.6 | 3.7 |
| Minot ...................... | 121 | 710 | 87 | 23 | 459 | 2,044,359 | 10.1 | 25.5 | 17.0 | 15.3 | 0.0 | 19.3 | 3.1 |
| West Fargo ................ | 66 | 506 | 56 | 15 | 178 | 824,064 | 32.5 | 39.9 | 0.0 | 4.5 | 0.0 | 18.2 | 4.9 |
| OHIO........................ | 18,425 | 126,378 | 15,268 | 4,044 | X | X | X | X | X | X | X | X | X |
| Akron....................... | 337 | 2,022 | 225 | 58 | 2,074 | 9,349,114 | 18.6 | 29.1 | 22.6 | 3.7 | 1.6 | 18.7 | 0.0 |
| Barberton.................. | D | D | D | D | 258 | 1,204,287 | 17.1 | 23.7 | 20.5 | 6.9 | 9.8 | 22.1 | 0.0 |
| Beavercreek............... | 69 | 460 | 35 | 12 | 154 | 824,107 | 6.7 | 51.2 | 0.0 | 23.2 | 4.2 | 14.6 | 0.0 |
| Bowling Green ............ | 46 | 237 | 17 | 5 | 318 | 1,651,866 | 14.3 | 19.1 | 19.6 | 9.5 | 0.6 | 31.7 | 0.0 |
| Brunswick................. | 62 | 359 | 41 | 13 | 153 | 819,758 | 12.2 | 40.0 | 24.1 | 9.9 | 1.9 | 6.3 | 0.0 |
| Canton ..................... | 125 | 909 | 155 | 35 | 936 | 4,445,774 | 14.7 | 23.0 | 20.0 | 7.9 | 7.6 | 22.3 | 0.0 |
| Cincinnati ................. | 542 | 4,633 | 642 | 164 | 5,538 | 30,765,986 | 7.2 | 25.0 | 18.8 | 5.1 | 8.7 | 27.8 | 0.0 |
| Cleveland .................. | 705 | 5,805 | 757 | 202 | 7,046 | 35,836,530 | 13.7 | 31.1 | 12.4 | 9.6 | 8.0 | 23.2 | 0.0 |
| Cleveland Heights......... | 54 | 308 | 30 | 11 | 424 | 2,142,030 | 9.1 | 30.1 | 26.9 | 4.6 | 2.1 | 17.5 | 0.0 |
| Columbus.................. | 1,251 | 11,396 | 1,777 | 427 | 8,528 | 52,605,234 | 13.2 | 32.1 | 23.7 | 4.3 | 7.2 | 15.7 | 0.0 |
| Cuyahoga Falls............ | 107 | 706 | 48 | 15 | 472 | 2,354,361 | 12.7 | 20.1 | 19.6 | 8.9 | 2.6 | 29.6 | 0.0 |
| Dayton..................... | 197 | 1,667 | 247 | 62 | 1,981 | 10,478,912 | 13.9 | 23.0 | 17.3 | 12.2 | 2.9 | 23.9 | 0.0 |
| Delaware................... | 56 | 273 | 29 | 8 | 302 | 1,895,429 | 23.2 | 22.5 | 30.3 | 7.2 | 0.0 | 14.8 | 0.0 |
| Dublin...................... | 65 | 1,445 | 248 | 102 | 461 | 2,663,324 | 24.2 | 27.9 | 0.0 | 10.6 | 0.0 | 26.1 | 0.0 |
| Elyria....................... | 76 | 382 | 47 | 10 | 481 | 2,392,714 | 14.7 | 27.7 | 15.8 | 6.3 | 3.3 | 26.1 | 0.0 |
| Euclid...................... | 52 | 266 | 27 | 8 | 370 | 1,840,187 | 12.9 | 32.1 | 25.9 | 3.4 | 3.1 | 16.0 | 0.0 |
| Fairborn................... | 39 | 263 | 27 | 8 | 259 | 1,399,619 | 23.2 | 27.0 | 26.0 | 5.6 | 3.1 | 10.6 | 0.0 |
| Fairfield.................... | 86 | 669 | 85 | 21 | 336 | 1,864,962 | 19.6 | 30.1 | 17.5 | 8.7 | 0.0 | 22.0 | 0.0 |
| Findlay..................... | 113 | 837 | 116 | 28 | 314 | 1,518,041 | 15.3 | 27.3 | 23.0 | 11.4 | 0.0 | 22.3 | 0.0 |
| Gahanna................... | 67 | 594 | 70 | 21 | 185 | 1,054,288 | 18.7 | 49.7 | 0.0 | 2.7 | 0.0 | 25.6 | 0.0 |
| Garfield Heights .......... | D | D | D | D | 204 | 1,043,312 | 14.7 | 40.3 | 27.0 | 1.0 | 4.2 | 1.7 | 0.0 |
| Green...................... | D | D | D | 43 | 135 | 745,011 | 17.8 | 0.0 | 50.1 | 29.1 | 0.0 | 3.0 | 0.0 |
| Grove City................. | 65 | 809 | 83 | 24 | 196 | 1,110,362 | 16.8 | 55.0 | 0.0 | 4.9 | 0.0 | 18.2 | 0.0 |

1. Establishments subject to federal tax.

# Table D. Cities — **City Government Finances**

| City | General revenue Intergovernmental Total (mil dol) 117 | Intergovernmental Total (mil dol) 118 | Percent from state government 119 | Taxes Total (mil dol) 120 | Per capita[1] (dollars) Total 121 | Property 122 | Sales and gross receipts 123 | General expenditure Total (mil dol) 124 | Per capita[1] (dollars) Total 125 | Capital outlays 126 |
|---|---|---|---|---|---|---|---|---|---|---|
| **NEW YORK—Cont'd** | | | | | | | | | | |
| Yonkers | 1,011.1 | 527.2 | 98.9 | 401.7 | 2,010 | 1,104 | 589 | 1,168 | 5,843 | 306 |
| **NORTH CAROLINA** | X | X | X | X | X | X | X | X | X | X |
| Apex | 62.9 | 8.4 | 68.1 | 36.5 | 722 | 455 | 267 | 57 | 1,130 | 217 |
| Asheboro | 37.2 | 6.2 | 96.4 | 20.6 | 799 | 613 | 185 | 32 | 1,236 | 31 |
| Asheville | 142.3 | 24.4 | 51.5 | 88.1 | 960 | 639 | 322 | 133 | 1,445 | 226 |
| Burlington | 55.8 | 9.0 | 67.6 | 40.7 | 766 | 523 | 243 | 51 | 957 | 113 |
| Cary | 248.7 | 23.3 | 71.8 | 130.5 | 786 | 542 | 244 | 178 | 1,075 | 232 |
| Chapel Hill | 0.0 | 0.0 | NA | 0.0 | 0 | 0 | 0 | 2 | 37 | 0 |
| Charlotte | 2,085.7 | 502.2 | 46.8 | 858.8 | 999 | 530 | 468 | 1,536 | 1,786 | 485 |
| Concord | 133.7 | 30.0 | 59.0 | 67.6 | 734 | 555 | 178 | 111 | 1,199 | 159 |
| Durham | 376.8 | 58.6 | 61.8 | 211.3 | 783 | 584 | 199 | 309 | 1,144 | 164 |
| Fayetteville | 200.8 | 64.8 | 42.7 | 108.3 | 518 | 344 | 174 | 192 | 919 | 110 |
| Garner | 30.0 | 3.2 | 95.0 | 24.0 | 836 | 620 | 216 | 34 | 1,178 | 283 |
| Gastonia | 88.1 | 15.5 | 79.5 | 45.6 | 597 | 396 | 201 | 91 | 1,198 | 155 |
| Goldsboro | 39.1 | 7.8 | 95.3 | 25.6 | 747 | 458 | 289 | 38 | 1,095 | 95 |
| Greensboro | 385.0 | 52.6 | 74.1 | 206.8 | 709 | 494 | 216 | 415 | 1,423 | 219 |
| Greenville | 151.2 | 24.4 | 86.8 | 49.2 | 535 | 363 | 172 | 138 | 1,502 | 236 |
| Hickory | 67.5 | 10.8 | 73.8 | 37.7 | 925 | 659 | 267 | 58 | 1,431 | 198 |
| High Point | 175.5 | 32.5 | 64.3 | 85.2 | 762 | 550 | 212 | 156 | 1,398 | 79 |
| Huntersville | 42.7 | 6.2 | 80.5 | 29.4 | 521 | 347 | 175 | 40 | 704 | 148 |
| Indian Trail | 14.3 | 2.2 | 37.6 | 9.9 | 254 | 181 | 73 | 17 | 448 | 171 |
| Jacksonville | 70.8 | 9.7 | 74.7 | 35.7 | 490 | 296 | 194 | 60 | 819 | 99 |
| Kannapolis | 48.9 | 7.0 | 65.5 | 34.8 | 713 | 510 | 204 | 49 | 1,009 | 130 |
| Matthews | 25.5 | 4.0 | 71.3 | 17.1 | 532 | 362 | 170 | 20 | 625 | 106 |
| Monroe | 51.7 | 10.3 | 61.3 | 26.7 | 759 | 576 | 183 | 62 | 1,747 | 755 |
| Mooresville | 83.0 | 8.9 | 75.3 | 45.4 | 1,201 | 946 | 255 | 77 | 2,030 | 462 |
| New Bern | 0.0 | 0.0 | NA | 0.0 | 0 | 0 | 0 | 3 | 85 | 0 |
| Raleigh | 724.2 | 109.0 | 54.2 | 356.1 | 765 | 523 | 242 | 552 | 1,186 | 194 |
| Rocky Mount | 82.8 | 17.9 | 58.2 | 35.9 | 658 | 449 | 208 | 78 | 1,436 | 146 |
| Salisbury | 59.4 | 6.1 | 91.3 | 25.2 | 749 | 523 | 226 | 61 | 1,812 | 87 |
| Sanford | 41.0 | 5.6 | 73.0 | 21.5 | 730 | 520 | 210 | 40 | 1,347 | 51 |
| Thomasville | 23.8 | 4.9 | 94.7 | 16.8 | 631 | 443 | 188 | 25 | 955 | 135 |
| Wake Forest | 46.4 | 7.5 | 47.6 | 35.5 | 842 | 582 | 260 | 45 | 1,067 | 291 |
| Wilmington | 150.3 | 20.4 | 74.4 | 95.8 | 793 | 472 | 321 | 151 | 1,245 | 156 |
| Wilson | 81.9 | 10.8 | 91.9 | 30.4 | 619 | 452 | 167 | 70 | 1,430 | 133 |
| Winston-Salem | 324.1 | 51.4 | 57.9 | 171.5 | 701 | 500 | 202 | 413 | 1,689 | 523 |
| **NORTH DAKOTA** | X | X | X | X | X | X | X | X | X | X |
| Bismarck | 128.8 | 25.4 | 55.2 | 44.5 | 608 | 276 | 331 | 162 | 2,211 | 740 |
| Fargo | 278.9 | 77.2 | 92.7 | 91.8 | 750 | 220 | 516 | 373 | 3,048 | 1,949 |
| Grand Forks | 116.8 | 24.1 | 58.9 | 42.2 | 743 | 325 | 419 | 64 | 1,124 | 139 |
| Minot | 138.2 | 63.8 | 29.5 | 40.3 | 839 | 354 | 485 | 120 | 2,487 | 990 |
| West Fargo | 52.2 | 2.9 | 100.0 | 22.6 | 635 | 292 | 343 | 65 | 1,832 | 1,119 |
| **OHIO** | X | X | X | X | X | X | X | X | X | X |
| Akron | 445.0 | 37.7 | 36.9 | 214.3 | 1,082 | 153 | 5 | 416 | 2,100 | 70 |
| Barberton | 35.9 | 6.0 | 87.2 | 15.2 | 582 | 45 | 25 | 35 | 1,342 | 119 |
| Beavercreek | 36.7 | 10.2 | 100.0 | 21.4 | 456 | 438 | 19 | 34 | 714 | 211 |
| Bowling Green | 40.4 | 3.3 | 100.0 | 27.8 | 872 | 86 | 12 | 37 | 1,150 | 122 |
| Brunswick | 34.1 | 4.3 | 63.7 | 22.3 | 640 | 53 | 58 | 25 | 730 | 68 |
| Canton | 122.5 | 20.0 | 100.0 | 66.5 | 934 | 72 | 26 | 164 | 2,305 | 712 |
| Cincinnati | 1,158.1 | 403.9 | 11.7 | 545.9 | 1,810 | 261 | 101 | 929 | 3,079 | 635 |
| Cleveland | 985.2 | 205.9 | 83.4 | 471.8 | 1,225 | 128 | 177 | 1,056 | 2,741 | 374 |
| Cleveland Heights | 55.1 | 5.9 | 54.6 | 38.7 | 869 | 232 | 67 | 55 | 1,225 | 153 |
| Columbus | 1,640.6 | 186.4 | 27.9 | 947.2 | 1,074 | 49 | 74 | 1,750 | 1,984 | 416 |
| Cuyahoga Falls | 71.2 | 1.1 | 55.9 | 44.7 | 908 | 229 | 23 | 64 | 1,299 | 209 |
| Dayton | 314.5 | 46.9 | 63.0 | 144.1 | 1,026 | 116 | 12 | 277 | 1,973 | 164 |
| Delaware | 65.8 | 7.8 | 30.9 | 27.8 | 706 | 43 | 27 | 60 | 1,532 | 319 |
| Dublin | 129.7 | 3.7 | 100.0 | 100.0 | 2,094 | 76 | 133 | 239 | 5,013 | 3,117 |
| Elyria | 66.2 | 5.4 | 80.9 | 38.9 | 723 | 68 | 77 | 62 | 1,161 | 107 |
| Euclid | 101.9 | 43.2 | 97.9 | 30.6 | 649 | 91 | 25 | 132 | 2,806 | 829 |
| Fairborn | 42.7 | 4.9 | 64.7 | 22.4 | 666 | 190 | 37 | 34 | 995 | 150 |
| Fairfield | 64.5 | 6.0 | 100.0 | 38.9 | 913 | 191 | 101 | 54 | 1,262 | 158 |
| Findlay | 53.8 | 5.3 | 100.0 | 29.8 | 718 | 68 | 8 | 49 | 1,181 | 260 |
| Gahanna | 42.6 | 3.4 | 91.9 | 22.6 | 640 | 57 | 48 | 39 | 1,097 | 194 |
| Garfield Heights | 33.7 | 6.0 | 100.0 | 20.0 | 721 | 257 | 29 | 32 | 1,149 | 196 |
| Green | 32.3 | 5.5 | 36.5 | 25.1 | 974 | 59 | 21 | 37 | 1,430 | 167 |
| Grove City | 46.1 | 10.6 | 99.9 | 29.6 | 719 | 64 | 66 | 57 | 1,378 | 606 |

1. Based on population estimated as of July 1 of the year shown.

| City | City government finances, 2017 (cont.) — General expenditure (cont.) — Percent of total for: | | | | | | | | | |
| --- | --- | --- | --- | --- | --- | --- | --- | --- | --- | --- |
| | Public welfare | Highways | Parking facilities | Education | Health and hospitals | Police protection | Sewerage and sanitation | Parks and recreation | Housing and community development | Interest on debt |
| | 127 | 128 | 129 | 130 | 131 | 132 | 133 | 134 | 135 | 136 |
| **NEW YORK—Cont'd** | | | | | | | | | | |
| Yonkers | 0.0 | 0.4 | 0.0 | 51.9 | 0.1 | 8.5 | 1.9 | 0.9 | 0.0 | 3.1 |
| **NORTH CAROLINA** | X | X | X | X | X | X | X | X | X | X |
| Apex | 0.0 | 7.9 | 0.0 | 0.0 | 3.2 | 17.9 | 18.0 | 8.3 | 3.2 | 4.1 |
| Asheboro | 0.0 | 7.7 | 0.0 | 0.0 | 0.1 | 24.1 | 23.4 | 10.0 | 2.9 | 0.2 |
| Asheville | 0.0 | 12.7 | 2.4 | 0.0 | 0.2 | 20.4 | 8.8 | 11.9 | 7.2 | 1.2 |
| Burlington | 0.0 | 5.0 | 0.0 | 0.0 | 0.0 | 34.7 | 5.6 | 16.3 | 1.4 | 0.2 |
| Cary | 0.0 | 14.2 | 0.0 | 0.0 | 0.0 | 13.9 | 5.0 | 14.9 | 2.4 | 4.7 |
| Chapel Hill | 0.0 | 0.0 | 0.0 | 0.0 | 0.0 | 0.0 | 0.0 | 0.0 | 0.0 | 100.0 |
| Charlotte | 0.0 | 8.0 | 0.0 | 0.0 | 0.0 | 17.6 | 17.1 | 7.6 | 4.6 | 7.9 |
| Concord | 0.0 | 6.3 | 0.0 | 0.0 | 0.0 | 20.0 | 18.7 | 7.8 | 11.1 | 2.1 |
| Durham | 0.0 | 7.3 | 0.9 | 0.0 | 0.0 | 28.5 | 17.3 | 6.1 | 9.1 | 3.9 |
| Fayetteville | 0.2 | 8.6 | 0.2 | 0.0 | 0.0 | 27.9 | 8.2 | 7.5 | 2.8 | 6.9 |
| Garner | 0.0 | 13.2 | 0.0 | 0.0 | 0.0 | 23.0 | 5.6 | 24.3 | 3.9 | 3.8 |
| Gastonia | 0.0 | 15.9 | 1.2 | 0.0 | 0.9 | 19.1 | 23.3 | 6.7 | 6.4 | 2.7 |
| Goldsboro | 0.0 | 8.1 | 0.0 | 0.0 | 0.0 | 27.1 | 7.6 | 11.9 | 10.3 | 4.8 |
| Greensboro | 0.0 | 11.8 | 1.1 | 0.0 | 0.1 | 18.6 | 23.3 | 14.9 | 3.7 | 2.2 |
| Greenville | 0.0 | 9.0 | 0.0 | 0.0 | 0.0 | 18.2 | 28.8 | 6.5 | 9.1 | 2.1 |
| Hickory | 0.0 | 6.0 | 0.0 | 0.0 | 0.0 | 20.6 | 26.1 | 6.8 | 7.4 | 0.7 |
| High Point | 0.0 | 8.0 | 0.2 | 0.0 | 0.1 | 17.9 | 15.7 | 13.3 | 7.1 | 5.4 |
| Huntersville | 0.0 | 19.4 | 0.0 | 0.0 | 0.7 | 27.6 | 1.5 | 25.6 | 3.6 | 3.9 |
| Indian Trail | 0.0 | 18.6 | 0.0 | 0.0 | 0.2 | 13.0 | 13.9 | 4.9 | 8.1 | 2.2 |
| Jacksonville | 0.0 | 7.9 | 0.0 | 0.0 | 0.0 | 23.6 | 30.4 | 11.0 | 2.3 | 4.3 |
| Kannapolis | 0.0 | 9.1 | 0.0 | 0.0 | 0.0 | 17.3 | 18.0 | 7.7 | 6.6 | 7.3 |
| Matthews | 0.0 | 16.6 | 0.0 | 0.0 | 3.2 | 30.2 | 7.9 | 10.2 | 2.4 | 1.1 |
| Monroe | 0.0 | 5.2 | 0.0 | 0.0 | 0.1 | 16.6 | 4.0 | 7.4 | 2.6 | 1.2 |
| Mooresville | 0.0 | 6.9 | 0.0 | 0.0 | 0.1 | 13.3 | 14.5 | 15.8 | 9.7 | 8.4 |
| New Bern | 0.0 | 0.0 | 0.0 | 0.0 | 0.0 | 0.0 | 0.0 | 0.0 | 0.0 | 100.0 |
| Raleigh | 0.3 | 9.4 | 1.7 | 0.0 | 0.0 | 18.8 | 16.9 | 16.7 | 3.8 | 5.8 |
| Rocky Mount | 0.6 | 5.0 | 0.0 | 0.0 | 0.0 | 21.2 | 26.3 | 19.7 | 3.9 | 1.3 |
| Salisbury | 0.0 | 8.4 | 0.0 | 0.1 | 0.0 | 11.9 | 18.9 | 3.0 | 4.8 | 7.0 |
| Sanford | 0.0 | 8.0 | 0.0 | 0.0 | 0.1 | 24.1 | 17.6 | 1.9 | 15.3 | 7.4 |
| Thomasville | 0.0 | 8.9 | 0.0 | 0.0 | 0.1 | 25.9 | 10.8 | 6.2 | 3.0 | 3.4 |
| Wake Forest | 0.0 | 5.4 | 0.0 | 0.0 | 0.0 | 20.9 | 7.4 | 28.9 | 6.0 | 1.4 |
| Wilmington | 0.0 | 7.3 | 2.7 | 0.0 | 0.0 | 21.5 | 9.8 | 7.1 | 6.1 | 5.4 |
| Wilson | 0.9 | 5.7 | 0.1 | 0.0 | 0.0 | 18.1 | 12.8 | 9.3 | 7.5 | 1.4 |
| Winston-Salem | 0.0 | 8.4 | 0.4 | 0.0 | 0.0 | 19.2 | 23.8 | 11.5 | 7.6 | 9.9 |
| **NORTH DAKOTA** | X | X | X | X | X | X | X | X | X | X |
| Bismarck | 0.0 | 24.3 | 0.8 | 0.0 | 1.7 | 10.0 | 20.6 | 7.2 | 0.2 | 5.5 |
| Fargo | 0.1 | 29.7 | 0.5 | 0.0 | 2.9 | 5.0 | 5.1 | 4.2 | 0.4 | 6.3 |
| Grand Forks | 0.0 | 11.9 | 0.7 | 0.0 | 3.1 | 16.0 | 23.3 | 3.6 | 3.0 | 0.0 |
| Minot | 0.0 | 49.7 | 0.1 | 0.0 | 2.6 | 8.3 | 8.1 | 2.2 | 0.0 | 0.0 |
| West Fargo | 0.0 | 52.9 | 0.0 | 0.0 | 0.0 | 9.4 | 20.4 | 0.0 | 0.0 | 0.0 |
| **OHIO** | X | X | X | X | X | X | X | X | X | X |
| Akron | 0.2 | 5.9 | 1.2 | 0.1 | 1.1 | 14.3 | 21.2 | 2.3 | 6.0 | 5.2 |
| Barberton | 0.0 | 6.5 | 0.0 | 0.0 | 0.0 | 14.7 | 25.7 | 3.9 | 2.3 | 0.6 |
| Beavercreek | 0.0 | 50.2 | 0.0 | 0.0 | 0.0 | 28.1 | 0.0 | 11.3 | 1.8 | 1.8 |
| Bowling Green | 0.0 | 6.6 | 0.0 | 0.0 | 0.6 | 16.7 | 18.0 | 7.1 | 2.7 | 4.3 |
| Brunswick | 0.4 | 13.4 | 0.0 | 0.0 | 0.0 | 29.5 | 13.3 | 7.6 | 2.6 | 1.2 |
| Canton | 0.0 | 26.1 | 0.4 | 0.0 | 4.2 | 11.6 | 18.8 | 2.0 | 6.3 | 0.1 |
| Cincinnati | 0.0 | 12.4 | 1.4 | 0.6 | 2.3 | 17.0 | 15.4 | 6.0 | 2.6 | 2.8 |
| Cleveland | 0.0 | 3.2 | 0.7 | 0.0 | 3.9 | 22.8 | 4.8 | 3.2 | 8.0 | 6.1 |
| Cleveland Heights | 0.0 | 13.7 | 2.5 | 0.0 | 1.8 | 13.8 | 5.9 | 6.0 | 3.6 | 1.3 |
| Columbus | 0.0 | 10.3 | 0.2 | 0.3 | 3.0 | 17.8 | 16.1 | 8.8 | 1.0 | 6.3 |
| Cuyahoga Falls | 0.0 | 6.6 | 0.0 | 0.0 | 0.0 | 16.3 | 18.9 | 12.9 | 3.1 | 0.4 |
| Dayton | 0.0 | 7.2 | 0.0 | 0.0 | 0.0 | 20.3 | 14.2 | 3.5 | 5.0 | 2.1 |
| Delaware | 0.0 | 7.3 | 0.1 | 0.0 | 0.0 | 12.7 | 12.4 | 4.7 | 0.1 | 0.0 |
| Dublin | 0.0 | 57.2 | 0.0 | 0.0 | 0.2 | 5.9 | 2.7 | 11.2 | 0.0 | 1.6 |
| Elyria | 0.0 | 2.5 | 0.0 | 0.0 | 2.8 | 20.2 | 26.5 | 3.3 | 1.8 | 2.2 |
| Euclid | 0.0 | 1.4 | 0.0 | 0.0 | 0.3 | 8.9 | 71.8 | 2.8 | 0.8 | 0.8 |
| Fairborn | 0.0 | 16.9 | 0.0 | 0.0 | 0.2 | 16.2 | 22.1 | 1.0 | 2.2 | 1.7 |
| Fairfield | 0.0 | 10.7 | 0.0 | 0.0 | 0.0 | 20.2 | 15.6 | 9.9 | 0.0 | 1.8 |
| Findlay | 0.0 | 19.2 | 0.2 | 0.0 | 3.2 | 13.7 | 19.6 | 3.3 | 0.0 | 2.1 |
| Gahanna | 0.0 | 15.4 | 0.2 | 0.0 | 0.7 | 24.9 | 16.8 | 11.0 | 1.2 | 2.5 |
| Garfield Heights | 0.0 | 12.8 | 0.0 | 0.0 | 1.7 | 0.0 | 17.8 | 7.9 | 1.4 | 2.5 |
| Green | 0.0 | 29.6 | 0.0 | 0.0 | 0.8 | 0.0 | 0.0 | 16.4 | 0.0 | 8.0 |
| Grove City | 0.0 | 17.4 | 0.0 | 0.0 | 0.0 | 19.4 | 2.3 | 8.5 | 0.0 | 0.0 |

| City | City government finances, 2017 (cont.) Debt outstanding | | | Climate[2] Average daily temperature | | | | Annual precipitation (inches) | Heating degree days | Cooling degree days |
| | Total (mil dol) | Per capita[1] (dollars) | Debt issued during year | Mean January | Mean July | Limits January[3] | Limits July[4] | | | |
| | 137 | 138 | 139 | 140 | 141 | 142 | 143 | 144 | 145 | 146 |
|---|---|---|---|---|---|---|---|---|---|---|
| **NEW YORK—Cont'd** | | | | | | | | | | |
| Yonkers | 732.1 | 3,663 | 65.8 | 29.7 | 74.2 | 20.1 | 86.0 | 46.46 | 5,400 | 770 |
| **NORTH CAROLINA** | X | X | X | X | X | X | X | X | X | X |
| Apex | 81.0 | 1,604 | 0.0 | NA | NA | NA | NA | NA | NA | NA |
| Asheboro | 5.8 | 225 | 0.0 | NA | NA | NA | NA | NA | NA | NA |
| Asheville | 118.4 | 1,291 | 0.0 | 36.4 | 73.9 | 26.6 | 84.3 | 37.32 | 4,237 | 877 |
| Burlington | 38.2 | 720 | 0.0 | 38.7 | 79.3 | 27.6 | 90.6 | 45.08 | 3,588 | 1,489 |
| Cary | 407.8 | 2,458 | 32.0 | 39.5 | 78.7 | 30.1 | 87.9 | 46.49 | 3,431 | 1,456 |
| Chapel Hill | 59.5 | 1,010 | 10.5 | 38.5 | 79.4 | 27.8 | 88.6 | 48.04 | 3,650 | 1,491 |
| Charlotte | 4,073.0 | 4,736 | 779.3 | 41.7 | 80.3 | 32.1 | 90.1 | 43.51 | 3,162 | 1,681 |
| Concord | 106.6 | 1,156 | 0.0 | 39.4 | 79.2 | 27.9 | 90.3 | 47.30 | 3,463 | 1,540 |
| Durham | 332.3 | 1,232 | 83.4 | 39.7 | 78.8 | 29.6 | 89.1 | 43.05 | 3,465 | 1,521 |
| Fayetteville | 312.5 | 1,495 | 1.9 | 41.7 | 80.4 | 31.1 | 90.4 | 46.78 | 3,097 | 1,721 |
| Garner | 36.5 | 1,268 | 6.2 | NA | NA | NA | NA | NA | NA | NA |
| Gastonia | 78.9 | 1,034 | 5.0 | 42.0 | 79.9 | 31.7 | 89.9 | 49.19 | 3,009 | 1,701 |
| Goldsboro | 57.6 | 1,683 | 26.8 | 43.4 | 81.2 | 33.0 | 91.4 | 49.84 | 2,771 | 1,922 |
| Greensboro | 496.9 | 1,704 | 133.2 | 39.7 | 78.6 | 29.0 | 88.9 | 42.89 | 3,443 | 1,438 |
| Greenville | 303.6 | 3,299 | 22.1 | 42.0 | 78.8 | 31.9 | 88.4 | 49.30 | 3,113 | 1,516 |
| Hickory | 31.2 | 764 | 0.0 | 39.0 | 77.7 | 29.2 | 87.8 | 48.98 | 3,608 | 1,333 |
| High Point | 225.3 | 2,017 | 70.9 | 39.7 | 78.2 | 29.6 | 89.0 | 46.19 | 3,399 | 1,424 |
| Huntersville | 41.1 | 728 | 11.4 | NA | NA | NA | NA | NA | NA | NA |
| Indian Trail | 15.5 | 399 | 0.0 | NA | NA | NA | NA | NA | NA | NA |
| Jacksonville | 92.1 | 1,265 | 0.0 | 44.7 | 80.2 | 33.9 | 89.5 | 54.07 | 2,656 | 1,832 |
| Kannapolis | 96.6 | 1,980 | 0.0 | 39.4 | 79.2 | 27.9 | 90.3 | 47.30 | 3,463 | 1,540 |
| Matthews | 6.4 | 198 | 0.0 | NA | NA | NA | NA | NA | NA | NA |
| Monroe | 74.5 | 2,115 | 20.7 | 41.5 | 79.0 | 31.0 | 89.7 | 48.73 | 3,125 | 1,538 |
| Mooresville | 192.3 | 5,083 | 0.0 | NA | NA | NA | NA | NA | NA | NA |
| New Bern | 56.7 | 1,897 | 0.8 | NA | NA | NA | NA | NA | NA | NA |
| Raleigh | 1,680.4 | 3,608 | 262.9 | 39.1 | 79.4 | 28.1 | 89.9 | 45.70 | 3,514 | 1,550 |
| Rocky Mount | 57.0 | 1,044 | 36.8 | 41.1 | 79.2 | 30.8 | 89.7 | 46.51 | 3,215 | 1,518 |
| Salisbury | 55.6 | 1,650 | 29.7 | 40.2 | 78.7 | 29.5 | 89.5 | 42.86 | 3,356 | 1,466 |
| Sanford | 59.4 | 2,019 | 0.0 | NA | NA | NA | NA | NA | NA | NA |
| Thomasville | 32.4 | 1,216 | 0.0 | NA | NA | NA | NA | NA | NA | NA |
| Wake Forest | 22.5 | 535 | 0.0 | NA | NA | NA | NA | NA | NA | NA |
| Wilmington | 185.7 | 1,537 | 11.8 | 44.8 | 80.1 | 33.3 | 90.0 | 58.44 | 2,606 | 1,791 |
| Wilson | 74.2 | 1,508 | 0.0 | 40.4 | 79.2 | 29.5 | 90.2 | 47.18 | 3,328 | 1,575 |
| Winston-Salem | 778.9 | 3,186 | 88.3 | 39.7 | 78.2 | 29.6 | 89.0 | 46.19 | 3,399 | 1,424 |
| **NORTH DAKOTA** | X | X | X | X | X | X | X | X | X | X |
| Bismarck | 222.9 | 3,048 | 20.3 | 10.2 | 70.4 | -0.6 | 84.5 | 16.84 | 8,802 | 471 |
| Fargo | 795.4 | 6,497 | 169.8 | 6.8 | 70.6 | -2.3 | 82.2 | 21.19 | 9,092 | 533 |
| Grand Forks | 352.1 | 6,204 | 21.4 | 5.3 | 69.4 | -4.3 | 81.9 | 19.60 | 9,489 | 420 |
| Minot | 108.4 | 2,255 | 18.2 | 7.5 | 68.4 | -1.8 | 80.4 | 18.65 | 9,479 | 422 |
| West Fargo | 250.1 | 7,015 | 27.1 | NA | NA | NA | NA | NA | NA | NA |
| **OHIO** | X | X | X | X | X | X | X | X | X | X |
| Akron | 674.3 | 3,404 | 0.0 | 27.2 | 74.1 | 20.1 | 83.9 | 36.07 | 5,752 | 856 |
| Barberton | 4.5 | 172 | 0.0 | 29.1 | 73.6 | 20.3 | 85.0 | 39.16 | 5,348 | 813 |
| Beavercreek | 6.3 | 134 | 0.0 | 27.6 | 73.1 | 19.5 | 83.5 | 40.06 | 5,531 | 768 |
| Bowling Green | 13.3 | 417 | 0.0 | 23.1 | 72.9 | 15.2 | 84.2 | 33.18 | 6,492 | 690 |
| Brunswick | 7.4 | 213 | 0.0 | 25.7 | 71.9 | 18.8 | 81.4 | 38.71 | 6,121 | 702 |
| Canton | 2.4 | 33 | 0.0 | 25.2 | 71.8 | 17.4 | 82.3 | 38.47 | 6,154 | 678 |
| Cincinnati | 858.4 | 2,846 | 0.0 | 30.6 | 76.8 | 22.7 | 86.8 | 39.57 | 4,841 | 1,210 |
| Cleveland | 2,408.9 | 6,253 | 270.6 | 25.7 | 71.9 | 18.8 | 81.4 | 38.71 | 6,121 | 702 |
| Cleveland Heights | 23.5 | 527 | 3.1 | 25.7 | 71.9 | 18.8 | 81.4 | 38.71 | 6,121 | 702 |
| Columbus | 4,294.7 | 4,871 | 1,008.1 | 28.3 | 74.7 | 20.2 | 85.6 | 40.03 | 5,349 | 935 |
| Cuyahoga Falls | 0.6 | 13 | 0.0 | 29.1 | 73.6 | 20.3 | 85.0 | 39.16 | 5,348 | 813 |
| Dayton | 53.5 | 381 | 0.0 | 26.3 | 74.3 | 19.0 | 84.2 | 39.58 | 5,690 | 935 |
| Delaware | 110.0 | 2,794 | 0.0 | 25.1 | 73.0 | 16.6 | 84.6 | 37.58 | 6,178 | 739 |
| Dublin | 158.6 | 3,322 | 32.6 | 28.3 | 74.7 | 20.2 | 85.6 | 40.03 | 5,349 | 935 |
| Elyria | 48.4 | 900 | 0.0 | 27.1 | 73.8 | 19.3 | 85.0 | 38.02 | 5,731 | 818 |
| Euclid | 31.1 | 660 | 1.5 | 23.0 | 68.8 | 14.3 | 80.0 | 47.33 | 6,956 | 372 |
| Fairborn | 12.2 | 361 | 0.0 | 27.9 | 77.0 | 20.6 | 87.2 | 39.41 | 5,343 | 1,214 |
| Fairfield | 16.3 | 382 | 0.0 | 28.7 | 76.6 | 19.9 | 88.1 | 43.36 | 5,261 | 1,135 |
| Findlay | 10.9 | 263 | 0.0 | 24.5 | 73.6 | 17.4 | 83.5 | 36.91 | 6,194 | 809 |
| Gahanna | 23.1 | 652 | 0.0 | 28.3 | 75.1 | 20.3 | 85.3 | 38.52 | 5,492 | 951 |
| Garfield Heights | 19.5 | 700 | 1.4 | 25.7 | 71.9 | 18.8 | 81.4 | 38.71 | 6,121 | 702 |
| Green | 75.5 | 2,929 | 12.5 | NA | NA | NA | NA | NA | NA | NA |
| Grove City | 54.5 | 1,324 | 21.4 | 28.3 | 74.7 | 20.2 | 85.6 | 40.03 | 5,349 | 935 |

1. Based on the population estimated as of July 1 of the year shown.  2. Represents normal values based on the 30-year period, 1971±2000.  3. Average daily minimum.  4. Average daily maximum.

# Table D. Cities — Land Area and Population

| STATE Place code | | City | Land area[1] (sq. mi) | Total persons 2019 | Rank | Per square mile | White | Black or African American | American Indian, Alaskan Native | Asian | Hawaiian Pacific Islander | Some other race | Two or more races (percent) |
|---|---|---|---|---|---|---|---|---|---|---|---|---|---|
| | | | 1 | 2 | 3 | 4 | 5 | 6 | 7 | 8 | 9 | 10 | 11 |
| | | **OHIO—Cont'd** | | | | | | | | | | | |
| 39 | 33012 | Hamilton | 21.5 | 62,148 | 605 | 2,891 | 84.5 | 11.3 | 0.3 | 1.2 | 0.0 | 0.4 | 2.4 |
| 39 | 35476 | Hilliard | 14.3 | 36,797 | 1,043 | 2,573 | D | D | D | D | 0.0 | 0.0 | 5.9 |
| 39 | 36610 | Huber Heights | 22.2 | 38,232 | 1,013 | 1,722 | 78.1 | 14.6 | 0.3 | 1.1 | 0.0 | 0.2 | 10.1 |
| 39 | 39872 | Kent | 9.2 | 29,319 | 1,278 | 3,187 | 78.6 | 7.9 | 0.0 | 3.1 | 0.0 | 1.7 | 2.8 |
| 39 | 40040 | Kettering | 18.7 | 54,773 | 704 | 2,929 | 86.9 | 5.9 | 0.0 | 2.6 | 0.0 | 0.3 | 3.4 |
| 39 | 41664 | Lakewood | 5.5 | 49,350 | 785 | 8,973 | 86.5 | 7.7 | 0.0 | 2.1 | 0.0 | 0.3 | 3.4 |
| 39 | 41720 | Lancaster | 18.8 | 40,932 | 940 | 2,177 | D | D | D | D | D | D | D |
| 39 | 43554 | Lima | 13.6 | 36,440 | 1,059 | 2,679 | 74.0 | 14.5 | 0.9 | 1.4 | 0.0 | 2.1 | 7.0 |
| 39 | 44856 | Lorain | 23.6 | 63,872 | 592 | 2,706 | 70.6 | 22.4 | 0.3 | 0.1 | 0.1 | 1.0 | 5.5 |
| 39 | 47138 | Mansfield | 30.8 | 46,125 | 844 | 1,498 | 84.2 | 7.5 | 0.7 | 0.8 | 0.2 | 0.5 | 6.0 |
| 39 | 47754 | Marion | 13.0 | 35,516 | 1,089 | 2,732 | 76.6 | 5.2 | 0.0 | 14.7 | 0.0 | 0.2 | 3.2 |
| 39 | 48188 | Mason | 19.3 | 34,300 | 1,124 | 1,777 | D | D | D | D | D | D | D |
| 39 | 48244 | Massillon | 19.0 | 32,641 | 1,171 | 1,718 | 91.1 | 5.5 | 0.0 | 1.7 | 0.0 | 0.1 | 1.4 |
| 39 | 48790 | Medina | 11.6 | 25,968 | 1,388 | 2,239 | D | D | D | D | D | D | 4.2 |
| 39 | 49056 | Mentor | 27.8 | 47,090 | 826 | 1,694 | 83.2 | 11.1 | 0.0 | 0.6 | 0.0 | 0.9 | D |
| 39 | 49840 | Middletown | 26.1 | 48,925 | 797 | 1,875 | D | D | D | D | D | D | D |
| 39 | 54040 | Newark | 20.9 | 50,664 | 770 | 2,424 | D | D | D | D | D | D | D |
| 39 | 56882 | North Olmsted | 11.7 | 31,173 | 1,213 | 2,664 | 96.1 | 0.8 | 0.0 | 1.4 | 0.0 | 0.3 | 1.5 |
| 39 | 56966 | North Ridgeville | 23.4 | 35,104 | 1,099 | 1,500 | D | D | D | D | D | D | D |
| 39 | 57008 | North Royalton | 21.4 | 29,952 | 1,254 | 1,400 | D | D | D | D | D | D | 2.7 |
| 39 | 61000 | Parma | 20.0 | 77,558 | 456 | 3,878 | 88.8 | 4.6 | 0.0 | 2.4 | 0.0 | 1.5 | 3.1 |
| 39 | 66390 | Reynoldsburg | 11.1 | 38,374 | 1,008 | 3,457 | 59.0 | 27.0 | 0.0 | 6.3 | 0.0 | 4.6 | 11.9 |
| 39 | 67468 | Riverside | 9.7 | 25,169 | 1,402 | 2,595 | 72.6 | 11.1 | 0.0 | 3.7 | 0.0 | 0.7 | 5.7 |
| 39 | 70380 | Sandusky | 9.6 | 24,322 | 1,418 | 2,534 | 64.7 | 26.4 | 0.0 | 0.0 | 0.0 | 3.3 | D |
| 39 | 71682 | Shaker Heights | 6.3 | 26,754 | 1,359 | 4,247 | D | D | D | D | D | D | D |
| 39 | 74118 | Springfield | 26.0 | 58,456 | 654 | 2,248 | 74.5 | 19.4 | 0.2 | 1.2 | 0.0 | 0.2 | 4.5 |
| 39 | 74944 | Stow | 17.1 | 34,714 | 1,112 | 2,030 | 90.5 | 5.6 | 0.0 | 0.8 | 0.0 | 0.2 | 3.0 |
| 39 | 75098 | Strongsville | 24.6 | 44,481 | 865 | 1,808 | D | D | D | D | D | D | D |
| 39 | 77000 | Toledo | 80.5 | 271,455 | 77 | 3,372 | 61.7 | 28.2 | 0.3 | 1.4 | 0.0 | 2.7 | 5.8 |
| 39 | 77588 | Troy | 12.0 | 26,473 | 1,374 | 2,206 | D | D | D | D | D | D | D |
| 39 | 79002 | Upper Arlington | 9.8 | 35,369 | 1,092 | 3,609 | D | D | D | D | D | D | D |
| 39 | 80892 | Warren | 16.0 | 38,479 | 1,004 | 2,405 | 89.9 | 5.1 | 0.0 | 2.6 | 0.2 | 0.0 | 2.3 |
| 39 | 83342 | Westerville | 12.6 | 41,452 | 928 | 3,290 | 89.8 | 1.3 | 0.3 | 6.3 | 0.0 | 0.6 | 1.6 |
| 39 | 83622 | Westlake | 15.9 | 31,907 | 1,197 | 2,007 | 88.6 | 3.7 | 0.6 | 2.1 | 0.0 | 0.4 | 4.6 |
| 39 | 86548 | Wooster | 17.1 | 26,157 | 1,382 | 1,530 | D | D | D | D | D | D | D |
| 39 | 86772 | Xenia | 13.0 | 27,081 | 1,346 | 2,083 | D | D | D | D | D | D | D |
| 39 | 88000 | Youngstown | 33.9 | 63,828 | 593 | 1,883 | 51.2 | 41.6 | 0.1 | 1.1 | 0.0 | 1.7 | 4.3 |
| 39 | 88084 | Zanesville | 11.8 | 25,193 | 1,400 | 2,135 | D | D | D | D | D | D | D |
| 40 | 00000 | **OKLAHOMA** | 68,595.9 | 3,980,783 | X | 58 | 72.4 | 7.3 | 8.0 | 2.3 | 0.1 | 2.3 | 7.6 |
| 40 | 04450 | Bartlesville | 22.7 | 36,605 | 1,050 | 1,613 | 74.8 | 4.1 | 7.2 | 3.5 | 0.0 | 3.4 | 7.0 |
| 40 | 09050 | Broken Arrow | 63.0 | 111,648 | 268 | 1,772 | 74.7 | 5.1 | 6.0 | 5.3 | 0.2 | 1.1 | 7.7 |
| 40 | 23200 | Edmond | 84.6 | 95,346 | 339 | 1,127 | 81.8 | 5.5 | 2.0 | 4.6 | 0.0 | 0.8 | 5.3 |
| 40 | 23950 | Enid | 73.9 | 49,542 | 782 | 670 | 79.6 | 1.0 | 5.8 | 0.1 | 2.4 | 3.0 | 8.0 |
| 40 | 41850 | Lawton | 81.5 | 93,164 | 357 | 1,143 | 57.4 | 18.5 | 4.6 | 3.2 | 0.2 | 2.5 | 13.6 |
| 40 | 48350 | Midwest City | 24.4 | 57,591 | 669 | 2,360 | 57.0 | 30.7 | 3.4 | 1.5 | 0.1 | 0.6 | 6.6 |
| 40 | 49200 | Moore | 21.9 | 63,102 | 597 | 2,881 | 75.9 | 5.8 | 6.2 | 3.5 | 0.0 | 1.2 | 7.5 |
| 40 | 50050 | Muskogee | 42.9 | 36,831 | 1,041 | 859 | 52.4 | 14.4 | 16.3 | 0.9 | 0.0 | 4.8 | 11.3 |
| 40 | 52500 | Norman | 178.8 | 125,762 | 220 | 703 | 77.8 | 5.2 | 4.0 | 4.6 | 0.1 | 1.9 | 6.4 |
| 40 | 55000 | Oklahoma City | 606.2 | 662,314 | 26 | 1,093 | 69.0 | 13.8 | 3.3 | 4.5 | 0.1 | 3.5 | 5.7 |
| 40 | 56650 | Owasso | 16.9 | 37,241 | 1,029 | 2,204 | 76.3 | 6.1 | 7.1 | 1.5 | 0.0 | 1.3 | 7.7 |
| 40 | 59850 | Ponca City | 18.4 | 23,482 | 1,427 | 1,276 | 74.8 | 2.9 | 10.8 | 0.3 | 0.0 | 4.4 | 6.8 |
| 40 | 66800 | Shawnee | 43.6 | 31,555 | 1,208 | 724 | 71.0 | 5.2 | 16.5 | 0.0 | 0.0 | 2.1 | 7.5 |
| 40 | 70300 | Stillwater | 29.8 | 50,306 | 775 | 1,688 | 74.5 | 4.9 | 3.7 | 7.2 | 0.1 | 3.6 | 7.4 |
| 40 | 75000 | Tulsa | 197.5 | 403,166 | 48 | 2,041 | 66.5 | 14.5 | 4.7 | 3.1 | 0.3 | 3.6 | 4.9 |
| 41 | 00000 | **OREGON** | 95,988.0 | 4,241,507 | X | 44 | 83.5 | 1.8 | 1.2 | 4.6 | 0.3 | 3.6 | 4.9 |
| 41 | 01000 | Albany | 17.7 | 56,129 | 685 | 3,171 | 83.9 | 0.2 | 2.6 | 1.5 | 0.2 | 2.8 | 8.7 |
| 41 | 05350 | Beaverton | 19.6 | 98,957 | 322 | 5,049 | 75.1 | 3.1 | 0.3 | 13.6 | 1.2 | 3.2 | 3.4 |
| 41 | 05800 | Bend | 33.6 | 101,886 | 306 | 3,032 | 92.0 | 0.4 | 0.5 | 2.3 | 0.1 | 1.3 | 3.3 |
| 41 | 15800 | Corvallis | 14.3 | 59,093 | 640 | 4,132 | 79.4 | 2.2 | 1.5 | 9.0 | 0.4 | 3.7 | 3.8 |
| 41 | 23850 | Eugene | 44.2 | 173,236 | 153 | 3,919 | 83.9 | 1.7 | 0.5 | 4.1 | 0.5 | 3.3 | 6.0 |
| 41 | 30550 | Grants Pass | 11.5 | 38,416 | 1,007 | 3,341 | 87.9 | 0.3 | 0.9 | 0.9 | 0.0 | 1.3 | 8.1 |
| 41 | 31250 | Gresham | 23.5 | 110,031 | 275 | 4,682 | 78.7 | 3.9 | 1.1 | 5.2 | 1.6 | 3.6 | 5.9 |
| 41 | 34100 | Hillsboro | 25.7 | 110,985 | 270 | 4,319 | 63.6 | 2.1 | 2.2 | 12.4 | 0.6 | 10.9 | 8.3 |
| 41 | 38500 | Keizer | 7.2 | 39,593 | 978 | 5,499 | 81.9 | 0.0 | 0.2 | 6.6 | 0.4 | 2.1 | 7.0 |
| 41 | 40550 | Lake Oswego | 10.8 | 39,821 | 968 | 3,687 | 83.2 | 1.6 | 0.7 | 6.6 | 0.2 | 0.5 | 4.8 |
| 41 | 45000 | McMinnville | 10.6 | 35,185 | 1,096 | 3,319 | 86.3 | 0.9 | 0.3 | 2.8 | 0.2 | 4.8 | 4.6 |
| 41 | 47000 | Medford | 25.9 | 83,412 | 413 | 3,221 | 90.8 | 1.4 | 0.8 | 1.4 | D | D | D |
| 41 | 55200 | Oregon City | 10.0 | 38,026 | 1,022 | 3,803 | D | D | D | D | D | D | 5.7 |
| 41 | 59000 | Portland | 133.5 | 656,751 | 27 | 4,920 | 77.3 | 5.6 | 0.8 | 8.2 | 0.3 | 2.2 | 5.7 |

1. Dry land or land partially or temporarily covered by water.   2. Hispanic or Latino persons may be of any race.

| City | Percent Hispanic or Latino[1], 2019 | Percent foreign born, 2019 | Age of population (percent), 2019 | | | | | | | Median age, 2019 | Percent female, 2019 | Population Census counts | | Percent change | |
|---|---|---|---|---|---|---|---|---|---|---|---|---|---|---|---|
| | | | Under 18 years | 18 to 24 years | 25 to 34 years | 35 to 44 years | 45 to 54 years | 55 to 64 years | 65 years and over | | | 2000 | 2010 | 2000-2010 | 2001-2020 |
| | 12 | 13 | 14 | 15 | 16 | 17 | 18 | 19 | 20 | 21 | 22 | 23 | 24 | 25 | 26 |
| **OHIO—Cont'd** | | | | | | | | | | | | | | | |
| Hamilton | 5.4 | 4.4 | 23.2 | 9.8 | 18.0 | 11.0 | 9.8 | 14.5 | 13.6 | 34.2 | 51.9 | 60,690 | 62,253 | 2.6 | -0.2 |
| Hilliard | 3.2 | 7.9 | 28.0 | 4.5 | 13.7 | 14.3 | 12.9 | 12.6 | 14.0 | 37.2 | 51.3 | 24,230 | 28,235 | 16.5 | 30.3 |
| Huber Heights | 1.5 | 7.4 | 27.1 | 7.9 | 13.1 | 11.0 | 10.7 | 10.8 | 19.3 | 36.4 | 53.2 | 38,212 | 38,110 | -0.3 | 0.3 |
| Kent | 2.5 | 7.9 | 10.7 | 43.2 | 12.1 | 6.2 | 9.4 | 6.5 | 11.9 | 23.9 | 53.2 | 27,906 | 28,904 | 3.6 | 1.4 |
| Kettering | 3.8 | 7.6 | 23.8 | 7.3 | 15.2 | 13.8 | 10.0 | 12.6 | 17.3 | 37.5 | 50.3 | 57,502 | 56,128 | -2.4 | -2.4 |
| Lakewood | 4.5 | 10.1 | 19.6 | 10.0 | 26.8 | 11.6 | 8.3 | 12.0 | 11.8 | 31.9 | 51.8 | 56,646 | 52,166 | -7.9 | -5.4 |
| Lancaster | 1.0 | 1.0 | 23.5 | 9.1 | 13.7 | 12.3 | 10.8 | 10.8 | 19.7 | 37.6 | 54.3 | 35,335 | 38,764 | 9.7 | 5.6 |
| Lima | 5.5 | 0.6 | 26.3 | 9.9 | 16.8 | 12.6 | 11.2 | 10.0 | 13.1 | 33.5 | 48.3 | 40,081 | 38,620 | -3.6 | -5.6 |
| Lorain | 27.3 | 2.6 | 26.6 | 9.2 | 12.9 | 11.8 | 11.9 | 11.5 | 16.2 | 35.9 | 50.6 | 68,652 | 64,027 | -6.7 | -0.2 |
| Mansfield | 2.4 | 1.4 | 19.8 | 9.6 | 16.2 | 11.0 | 14.5 | 12.8 | 16.2 | 39.0 | 48.0 | 49,346 | 47,811 | -3.1 | -3.5 |
| Marion | 4.7 | 2.1 | 22.2 | 7.9 | 18.9 | 13.4 | 12.0 | 11.4 | 14.2 | 35.6 | 43.6 | 35,318 | 36,831 | 4.3 | -3.6 |
| Mason | 0.0 | 17.0 | 27.8 | 6.0 | 6.3 | 15.8 | 18.0 | 12.4 | 13.6 | 41.1 | 51.0 | 22,016 | 30,860 | 40.2 | 11.1 |
| Massillon | 1.0 | 0.7 | 20.1 | 7.9 | 11.5 | 14.4 | 12.3 | 14.9 | 18.8 | 42.2 | 50.9 | 31,325 | 32,297 | 3.1 | 1.1 |
| Medina | 5.0 | 5.1 | 25.1 | 5.7 | 13.9 | 14.0 | 14.3 | 13.0 | 14.0 | 39.5 | 51.6 | 25,139 | 26,639 | 6.0 | -2.5 |
| Mentor | 2.4 | 4.0 | 19.7 | 6.0 | 12.0 | 11.2 | 12.7 | 15.7 | 22.6 | 45.8 | 47.5 | 50,278 | 47,161 | -6.2 | -0.2 |
| Middletown | 3.5 | 2.6 | 22.7 | 10.0 | 14.5 | 9.9 | 11.1 | 13.2 | 18.6 | 36.3 | 51.4 | 51,605 | 48,691 | -5.6 | 0.5 |
| Newark | 3.0 | 2.6 | 21.1 | 9.8 | 13.6 | 13.1 | 10.7 | 14.0 | 17.7 | 38.4 | 52.3 | 46,279 | 47,551 | 2.7 | 6.5 |
| North Olmsted | 9.0 | 15.7 | 22.3 | 6.9 | 13.2 | 14.6 | 10.6 | 11.4 | 21.0 | 41.5 | 51.7 | 34,113 | 32,753 | -4.0 | -4.8 |
| North Ridgeville | 9.0 | 4.5 | 24.2 | 3.5 | 11.9 | 13.7 | 13.3 | 15.6 | 17.8 | 42.6 | 49.2 | 22,338 | 29,476 | 32.0 | 19.1 |
| North Royalton | 0.6 | 14.5 | 13.8 | 10.4 | 13.7 | 9.8 | 11.9 | 16.8 | 23.5 | 46.5 | 49.3 | 28,648 | 30,470 | 6.4 | -1.7 |
| Parma | 6.6 | 11.2 | 18.9 | 8.8 | 13.9 | 11.3 | 13.0 | 12.2 | 21.9 | 42.4 | 50.6 | 85,655 | 81,618 | -4.7 | -5.0 |
| Reynoldsburg | 6.7 | 16.8 | 20.9 | 5.1 | 14.2 | 15.7 | 12.8 | 15.6 | 15.5 | 41.5 | 52.4 | 32,069 | 35,929 | 12.0 | 6.8 |
| Riverside | 3.1 | 7.9 | 25.0 | 10.2 | 17.4 | 10.4 | 9.5 | 12.1 | 15.4 | 33.4 | 53.0 | 23,545 | 25,175 | 6.9 | 0.0 |
| Sandusky | 6.9 | 1.7 | 26.5 | 9.2 | 17.8 | 5.8 | 12.0 | 15.0 | 13.8 | 32.3 | 55.0 | 27,844 | 25,909 | -6.9 | -6.1 |
| Shaker Heights | 0.8 | 5.6 | 27.0 | 7.8 | 7.4 | 14.4 | 13.5 | 11.6 | 18.3 | 40.0 | 54.5 | 29,405 | 28,395 | -3.4 | -5.8 |
| Springfield | 3.8 | 1.9 | 22.3 | 12.6 | 13.1 | 10.8 | 11.0 | 12.8 | 17.3 | 38.0 | 53.3 | 65,358 | 60,564 | -7.3 | -3.5 |
| Stow | 0.4 | 4.5 | 23.6 | 7.1 | 12.8 | 13.9 | 10.2 | 11.0 | 21.5 | 39.0 | 52.9 | 32,139 | 34,832 | 8.4 | -0.3 |
| Strongsville | 0.8 | 12.4 | 17.8 | 5.3 | 10.7 | 12.0 | 11.5 | 17.5 | 25.3 | 49.2 | 52.4 | 43,858 | 44,750 | 2.0 | -0.6 |
| Toledo | 8.7 | 3.1 | 22.9 | 10.3 | 16.1 | 10.6 | 12.2 | 12.8 | 15.1 | 35.6 | 52.0 | 313,619 | 287,357 | -8.4 | -5.5 |
| Troy | 1.3 | 3.2 | 18.9 | 8.8 | 12.3 | 12.2 | 13.0 | 14.9 | 19.9 | 41.8 | 50.1 | 21,999 | 25,232 | 14.7 | 4.9 |
| Upper Arlington | 1.5 | 8.5 | 24.5 | 3.1 | 11.1 | 14.4 | 15.3 | 15.0 | 16.6 | 42.6 | 52.1 | 33,686 | 33,675 | 0.0 | 5.0 |
| Warren | 4.7 | 1.0 | 18.1 | 9.0 | 13.1 | 13.4 | 13.7 | 12.9 | 19.8 | 43.1 | 48.9 | 46,832 | 41,584 | -11.2 | -7.5 |
| Westerville | 0.7 | 7.5 | 23.9 | 7.7 | 9.7 | 17.0 | 10.4 | 11.1 | 20.1 | 39.8 | 54.3 | 35,318 | 36,268 | 2.7 | 14.3 |
| Westlake | 1.7 | 11.1 | 19.2 | 8.3 | 10.1 | 9.6 | 13.9 | 15.4 | 23.6 | 48.6 | 53.6 | 31,719 | 32,726 | 3.2 | -2.5 |
| Wooster | 1.8 | 2.6 | 18.0 | 18.9 | 12.2 | 10.9 | 11.4 | 11.8 | 16.8 | 35.5 | 50.6 | 24,811 | 26,177 | 5.5 | -0.1 |
| Xenia | 4.5 | 0.9 | 20.9 | 4.2 | 17.3 | 12.1 | 8.2 | 15.6 | 21.8 | 39.7 | 50.5 | 24,164 | 25,643 | 6.1 | 5.6 |
| Youngstown | 8.7 | 3.5 | 20.3 | 14.5 | 10.5 | 10.8 | 10.3 | 14.2 | 19.4 | 39.4 | 48.3 | 82,026 | 66,946 | -18.4 | -4.7 |
| Zanesville | 1.7 | 0.0 | 20.2 | 9.4 | 12.0 | 10.0 | 14.5 | 12.8 | 21.0 | 44.1 | 52.1 | 25,586 | 25,424 | -0.6 | -0.9 |
| **OKLAHOMA** | 11.1 | 6.1 | 24.1 | 9.7 | 13.6 | 12.7 | 11.4 | 12.5 | 16.1 | 37.0 | 50.4 | 3,450,654 | 3,751,582 | 8.7 | 6.1 |
| Bartlesville | 7.2 | 5.9 | 24.8 | 8.6 | 11.8 | 12.1 | 11.8 | 13.0 | 18.0 | 37.9 | 50.8 | 34,748 | 35,726 | 2.8 | 2.5 |
| Broken Arrow | 8.6 | 7.6 | 24.2 | 8.0 | 13.7 | 13.9 | 12.8 | 12.2 | 15.1 | 37.6 | 51.3 | 74,859 | 98,674 | 31.8 | 13.1 |
| Edmond | 6.8 | 5.8 | 23.4 | 10.2 | 10.4 | 12.9 | 13.6 | 14.2 | 15.2 | 39.0 | 51.2 | 68,315 | 81,130 | 18.8 | 17.5 |
| Enid | 15.5 | 5.4 | 25.9 | 10.4 | 14.9 | 13.2 | 9.7 | 11.2 | 14.7 | 34.1 | 50.1 | 47,045 | 49,386 | 5.0 | 0.3 |
| Lawton | 14.6 | 6.1 | 22.5 | 13.7 | 19.2 | 12.8 | 9.2 | 11.6 | 11.1 | 31.5 | 47.4 | 92,757 | 96,862 | 4.4 | -3.8 |
| Midwest City | 2.9 | 2.1 | 25.6 | 9.0 | 12.1 | 13.9 | 12.4 | 10.9 | 16.1 | 37.2 | 53.6 | 54,088 | 54,372 | 0.5 | 5.9 |
| Moore | 8.2 | 5.2 | 21.0 | 10.0 | 19.6 | 9.4 | 11.9 | 11.9 | 16.2 | 34.6 | 52.1 | 41,138 | 55,082 | 33.9 | 14.6 |
| Muskogee | 8.5 | 4.0 | 25.9 | 8.4 | 14.5 | 12.9 | 11.3 | 11.3 | 15.7 | 37.0 | 50.2 | 38,310 | 39,184 | 2.3 | -6.0 |
| Norman | 9.6 | 7.6 | 20.5 | 21.9 | 12.4 | 12.1 | 10.0 | 9.7 | 13.3 | 31.2 | 50.5 | 95,694 | 110,876 | 15.9 | 13.4 |
| Oklahoma City | 20.7 | 12.1 | 25.2 | 9.0 | 16.4 | 13.5 | 10.9 | 11.8 | 13.1 | 34.6 | 50.6 | 506,132 | 580,494 | 14.7 | 14.1 |
| Owasso | 4.5 | 3.5 | 32.4 | 6.6 | 15.5 | 13.0 | 10.9 | 9.3 | 12.3 | 33.4 | 48.5 | 18,502 | 29,813 | 61.1 | 24.9 |
| Ponca City | 9.6 | 1.7 | 24.7 | 7.0 | 13.1 | 12.7 | 11.4 | 10.7 | 20.4 | 37.8 | 53.8 | 25,919 | 25,389 | -2.0 | -7.5 |
| Shawnee | 7.4 | 1.5 | 23.7 | 13.1 | 10.1 | 13.0 | 10.5 | 12.6 | 17.0 | 36.8 | 53.5 | 28,692 | 29,796 | 3.8 | 5.9 |
| Stillwater | 6.0 | 12.8 | 15.0 | 38.0 | 15.0 | 8.0 | 6.1 | 5.6 | 12.2 | 24.0 | 49.9 | 39,065 | 45,717 | 17.0 | 10.0 |
| Tulsa | 17.2 | 11.0 | 24.6 | 9.1 | 15.5 | 12.8 | 11.0 | 12.0 | 15.0 | 35.5 | 51.5 | 393,049 | 391,873 | -0.3 | 2.9 |
| **OREGON** | 13.4 | 9.7 | 20.5 | 8.7 | 14.1 | 13.7 | 12.1 | 12.7 | 18.2 | 39.7 | 50.5 | 3,421,399 | 3,831,083 | 12.0 | 10.7 |
| Albany | 13.0 | 5.4 | 21.2 | 9.0 | 16.2 | 13.4 | 11.3 | 11.4 | 17.4 | 37.8 | 47.9 | 40,852 | 50,147 | 22.8 | 11.9 |
| Beaverton | 11.1 | 17.0 | 15.2 | 9.6 | 19.9 | 14.8 | 13.7 | 11.7 | 15.0 | 38.2 | 47.3 | 76,129 | 89,768 | 17.9 | 10.2 |
| Bend | 10.8 | 5.4 | 21.0 | 7.3 | 14.3 | 16.2 | 13.2 | 12.3 | 15.7 | 39.8 | 48.3 | 52,029 | 76,655 | 47.3 | 32.9 |
| Corvallis | 8.1 | 9.6 | 13.1 | 31.0 | 14.6 | 11.7 | 7.4 | 7.8 | 14.3 | 28.0 | 49.8 | 49,322 | 54,498 | 10.5 | 8.4 |
| Eugene | 10.2 | 6.5 | 15.6 | 18.1 | 13.7 | 13.8 | 10.0 | 10.7 | 18.2 | 36.9 | 51.4 | 137,893 | 156,406 | 13.4 | 10.8 |
| Grants Pass | 7.0 | 2.8 | 21.8 | 8.6 | 13.5 | 13.0 | 10.9 | 10.9 | 21.3 | 39.7 | 53.8 | 23,003 | 35,884 | 56.0 | 7.1 |
| Gresham | 20.9 | 16.5 | 19.3 | 10.1 | 16.6 | 13.3 | 13.0 | 12.3 | 15.4 | 37.7 | 52.1 | 90,205 | 105,607 | 17.1 | 4.2 |
| Hillsboro | 26.3 | 21.1 | 24.1 | 8.5 | 18.4 | 14.2 | 12.2 | 10.8 | 11.8 | 34.4 | 51.5 | 70,186 | 92,002 | 31.1 | 20.6 |
| Keizer | 17.0 | 6.6 | 21.1 | 6.3 | 10.8 | 17.7 | 9.0 | 11.4 | 23.6 | 41.7 | 53.2 | 32,203 | 36,468 | 13.2 | 8.6 |
| Lake Oswego | 2.9 | 13.6 | 24.3 | 4.9 | 6.7 | 10.0 | 16.7 | 11.8 | 25.7 | 47.1 | 51.5 | 35,278 | 36,727 | 4.1 | 8.4 |
| McMinnville | 13.7 | 9.3 | 13.6 | 15.1 | 9.7 | 11.4 | 12.3 | 15.6 | 22.5 | 46.4 | 54.2 | 26,499 | 32,210 | 21.6 | 9.2 |
| Medford | 15.5 | 6.1 | 21.7 | 9.4 | 14.3 | 10.2 | 14.8 | 11.1 | 18.5 | 39.6 | 51.5 | 63,154 | 75,010 | 18.8 | 11.2 |
| Oregon City | 3.1 | 3.5 | 19.9 | 9.5 | 16.1 | 13.9 | 9.9 | 17.1 | 13.6 | 36.1 | 48.8 | 25,754 | 32,630 | 26.7 | 16.5 |
| Portland | 9.7 | 13.4 | 17.7 | 7.6 | 19.4 | 17.6 | 13.0 | 11.1 | 13.6 | 37.5 | 50.3 | 529,121 | 583,810 | 10.3 | 12.5 |

1. May be of any race.

# Households, Group Quarters, Crime, and Education

| City | Households, 2019 Number | Persons per household | Percent Family | Married couple family | Female family | Non-family | One person | Persons in group quarters, 2019 | Violent Number | Violent Rate[2] | Property Number | Property Rate[2] | Population age 25 and over | High school graduate or less | Bachelor's degree or more |
|---|---|---|---|---|---|---|---|---|---|---|---|---|---|---|---|
| | 27 | 28 | 29 | 30 | 31 | 32 | 33 | 34 | 35 | 36 | 37 | 38 | 39 | 40 | 41 |
| **OHIO—Cont'd** | | | | | | | | | | | | | | | |
| Hamilton | 24,268 | 2.48 | 59.7 | 38.5 | 21.2 | 40.3 | 34.5 | 1,899 | 265 | 426 | 2,228 | 3,585 | 41,603 | 58.6 | 15.2 |
| Hilliard | 14,879 | 2.60 | 70.4 | 56.4 | 14.0 | 29.6 | 25.3 | 148 | 26 | 69 | 242 | 644 | 26,201 | 14.9 | 58.0 |
| Huber Heights | 15,402 | 2.68 | 64.1 | 43.8 | 20.3 | 35.9 | 30.1 | 189 | 76 | 199 | 890 | 2,331 | 26,910 | 34.8 | 22.7 |
| Kent | 10,291 | 2.30 | 47.0 | 32.7 | 14.3 | 53.0 | 38.1 | 5,941 | 30 | 101 | 393 | 1,321 | 13,671 | 22.4 | 45.9 |
| Kettering | 24,168 | 2.27 | 58.7 | 46.3 | 12.3 | 41.3 | 33.8 | 541 | 56 | 102 | 794 | 1,444 | 38,253 | 27.0 | 35.3 |
| Lakewood | 23,673 | 2.08 | 43.1 | 29.1 | 14.0 | 56.9 | 44.1 | 337 | 61 | 122 | 643 | 1,291 | 34,987 | 21.7 | 50.2 |
| Lancaster | 16,931 | 2.33 | 53.7 | 35.2 | 18.5 | 46.3 | 35.2 | 1,122 | 110 | 271 | 1,540 | 3,791 | 27,283 | 51.3 | 23.1 |
| Lima | 13,291 | 2.55 | 57.0 | 28.0 | 28.9 | 43.0 | 31.8 | 2,772 | 259 | 707 | 1,493 | 4,073 | 23,369 | 59.2 | 10.8 |
| Lorain | 24,947 | 2.54 | 54.8 | 28.0 | 26.7 | 45.2 | 40.4 | 564 | 289 | 451 | 1,437 | 2,245 | 41,034 | 51.9 | 12.4 |
| Mansfield | 17,993 | 2.23 | 57.8 | 32.9 | 24.9 | 42.2 | 33.9 | 6,494 | 187 | 403 | 1,996 | 4,300 | 32,912 | 55.6 | 15.1 |
| Marion | 13,387 | 2.28 | 53.8 | 27.6 | 26.2 | 46.2 | 40.5 | 5,612 | NA | NA | NA | NA | 25,298 | 58.7 | 10.7 |
| Mason | 11,979 | 2.79 | 78.7 | 67.8 | 10.9 | 21.3 | 18.0 | 410 | 10 | 29 | 218 | 642 | 22,421 | 23.6 | 61.4 |
| Massillon | 13,969 | 2.27 | 57.6 | 39.4 | 18.2 | 42.4 | 34.1 | 859 | 72 | 222 | 743 | 2,291 | 23,463 | 54.0 | 16.4 |
| Medina | 10,656 | 2.38 | 68.7 | 48.7 | 20.1 | 31.3 | 27.2 | 466 | NA | NA | NA | NA | 17,897 | 26.6 | 37.1 |
| Mentor | 20,129 | 2.33 | 62.4 | 50.5 | 11.9 | 37.6 | 32.2 | 371 | 63 | 133 | 703 | 1,487 | 35,110 | 36.9 | 28.2 |
| Middletown | 20,474 | 2.36 | 53.7 | 27.0 | 26.6 | 46.3 | 38.4 | 799 | 206 | 421 | 1,459 | 2,985 | 33,010 | 57.0 | 14.7 |
| Newark | 21,234 | 2.31 | 60.4 | 39.6 | 20.8 | 39.6 | 34.4 | 1,197 | 149 | 296 | 1,676 | 3,329 | 34,808 | 50.0 | 17.0 |
| North Olmsted | 13,455 | 2.30 | 57.7 | 45.8 | 11.9 | 42.3 | 34.0 | 409 | 8 | 25 | 389 | 1,238 | 22,185 | 35.3 | 32.4 |
| North Ridgeville | 12,486 | 2.73 | 69.1 | 61.7 | 7.4 | 30.9 | 21.1 | 255 | 13 | 38 | 112 | 325 | 24,872 | 38.4 | 32.6 |
| North Royalton | 14,588 | 2.04 | 55.9 | 49.7 | 6.2 | 44.1 | 37.2 | 327 | 17 | 56 | 119 | 394 | 22,781 | 30.8 | 43.4 |
| Parma | 34,332 | 2.24 | 59.6 | 41.3 | 18.4 | 40.4 | 33.1 | 1,069 | 112 | 143 | 759 | 969 | 56,457 | 49.6 | 21.4 |
| Reynoldsburg | 14,712 | 2.54 | 64.6 | 43.4 | 21.2 | 35.4 | 26.4 | 36 | 73 | 189 | 988 | 2,561 | 27,709 | 35.6 | 28.9 |
| Riverside | 10,794 | 2.33 | 59.0 | 36.6 | 22.3 | 41.0 | 30.0 | 0 | 35 | 139 | 343 | 1,364 | 16,307 | 43.5 | 20.5 |
| Sandusky | 9,382 | 2.57 | 62.8 | 33.6 | 29.2 | 37.2 | 32.6 | 454 | 74 | 301 | 925 | 3,764 | 15,789 | 51.7 | 16.3 |
| Shaker Heights | 11,124 | 2.41 | 62.5 | 47.5 | 15.0 | 37.5 | 35.5 | 201 | NA | NA | NA | NA | 17,630 | 11.6 | 69.3 |
| Springfield | 23,619 | 2.38 | 63.4 | 33.2 | 30.2 | 36.6 | 32.8 | 2,756 | 292 | 494 | 2,992 | 5,060 | 38,335 | 56.3 | 13.2 |
| Stow | 13,850 | 2.48 | 63.6 | 52.8 | 10.8 | 36.4 | 30.3 | 381 | 15 | 43 | 537 | 1,540 | 24,118 | 29.5 | 46.1 |
| Strongsville | 18,819 | 2.35 | 68.5 | 58.3 | 10.2 | 31.5 | 25.8 | 364 | 18 | 40 | 585 | 1,305 | 34,351 | 27.4 | 45.5 |
| Toledo | 121,022 | 2.19 | 52.6 | 28.3 | 24.2 | 47.4 | 37.8 | 7,788 | 2,604 | 952 | 9,470 | 3,462 | 182,238 | 46.7 | 18.5 |
| Troy | 10,658 | 2.30 | 52.6 | 36.9 | 15.6 | 47.4 | 34.5 | 358 | 23 | 88 | 433 | 1,650 | 17,976 | 41.3 | 22.3 |
| Upper Arlington | 13,471 | 2.61 | 77.6 | 67.8 | 9.8 | 22.4 | 18.1 | 146 | 12 | 34 | 366 | 1,024 | 25,611 | 10.0 | 76.0 |
| Warren | 17,413 | 2.10 | 52.0 | 23.4 | 28.7 | 48.0 | 42.9 | 2,241 | 212 | 558 | 1,478 | 3,888 | 28,218 | 58.0 | 15.8 |
| Westerville | 15,076 | 2.52 | 77.0 | 61.8 | 15.1 | 23.0 | 22.2 | 1,920 | 74 | 181 | 767 | 1,875 | 27,317 | 16.6 | 58.5 |
| Westlake | 14,132 | 2.22 | 63.1 | 48.6 | 14.6 | 36.9 | 33.8 | 682 | NA | NA | NA | NA | 23,233 | 19.6 | 51.9 |
| Wooster | 10,882 | 2.13 | 57.4 | 40.7 | 16.7 | 42.6 | 35.3 | 3,201 | 110 | 413 | 738 | 2,773 | 16,641 | 45.0 | 20.1 |
| Xenia | 10,856 | 2.39 | 59.8 | 40.8 | 19.0 | 40.2 | 30.6 | 1,023 | 74 | 275 | 676 | 2,511 | 20,183 | 45.3 | 19.9 |
| Youngstown | 29,102 | 2.09 | 48.9 | 21.2 | 27.8 | 51.1 | 45.4 | 4,640 | NA | NA | NA | NA | 42,711 | 53.5 | 16.7 |
| Zanesville | 10,739 | 2.29 | 48.1 | 23.3 | 24.9 | 51.9 | 41.2 | 540 | 89 | 351 | 1,071 | 4,226 | 17,704 | 58.9 | 11.6 |
| **OKLAHOMA** | 1,495,151 | 2.57 | 65.3 | 47.7 | 17.5 | 34.7 | 28.6 | 110,059 | NA | NA | NA | NA | 2,619,244 | 43.1 | 26.2 |
| Bartlesville | 14,655 | 2.58 | 66.3 | 51.2 | 15.1 | 33.7 | 29.9 | 645 | 111 | 304 | 1,145 | 3,137 | 25,637 | 39.5 | 30.1 |
| Broken Arrow | 41,730 | 2.64 | 71.5 | 57.5 | 14.0 | 28.5 | 23.9 | 517 | 138 | 125 | 1,977 | 1,789 | 74,991 | 27.8 | 33.5 |
| Edmond | 36,056 | 2.57 | 71.0 | 57.5 | 13.4 | 29.0 | 23.0 | 1,388 | 138 | 146 | 1,599 | 1,689 | 62,418 | 21.3 | 51.4 |
| Enid | 18,995 | 2.52 | 66.7 | 48.3 | 18.4 | 33.3 | 24.7 | 1,632 | 204 | 411 | 1,625 | 3,276 | 31,531 | 43.4 | 23.4 |
| Lawton | 32,174 | 2.61 | 61.0 | 38.2 | 22.8 | 39.0 | 32.4 | 8,958 | 854 | 926 | 2,972 | 3,221 | 59,388 | 42.8 | 20.8 |
| Midwest City | 20,971 | 2.72 | 64.4 | 39.8 | 24.6 | 35.6 | 29.1 | 384 | 156 | 270 | 1,905 | 3,303 | 37,530 | 38.9 | 23.4 |
| Moore | 25,574 | 2.41 | 64.1 | 46.4 | 17.7 | 35.9 | 28.3 | 368 | 110 | 175 | 1,423 | 2,259 | 42,814 | 34.6 | 29.6 |
| Muskogee | 15,286 | 2.33 | 68.5 | 44.2 | 24.3 | 31.5 | 26.6 | 1,451 | 389 | 1,046 | 1,333 | 3,586 | 24,388 | 46.8 | 18.9 |
| Norman | 48,531 | 2.43 | 56.8 | 44.4 | 12.3 | 43.2 | 31.1 | 6,877 | 335 | 268 | 3,267 | 2,612 | 71,909 | 23.0 | 46.9 |
| Oklahoma City | 249,615 | 2.57 | 60.6 | 41.7 | 18.9 | 39.4 | 32.1 | 13,229 | 4,751 | 722 | 26,918 | 4,092 | 431,007 | 39.7 | 31.1 |
| Owasso | 13,721 | 2.70 | 68.9 | 50.9 | 17.9 | 31.1 | 27.2 | 266 | 80 | 212 | 782 | 2,077 | 22,784 | 30.1 | 38.0 |
| Ponca City | 10,077 | 2.36 | 59.1 | 41.7 | 17.4 | 40.9 | 30.7 | 719 | 152 | 637 | 933 | 3,908 | 16,752 | 45.6 | 18.5 |
| Shawnee | 11,792 | 2.53 | 63.3 | 42.9 | 20.3 | 36.7 | 29.2 | 1,604 | 156 | 493 | 1,517 | 4,797 | 19,863 | 42.7 | 26.5 |
| Stillwater | 19,119 | 2.24 | 50.1 | 42.6 | 7.5 | 49.9 | 31.1 | 7,401 | 160 | 314 | 1,086 | 2,129 | 23,637 | 19.0 | 54.3 |
| Tulsa | 163,801 | 2.41 | 56.8 | 37.7 | 19.1 | 43.2 | 35.6 | 6,981 | 3,964 | 987 | 21,336 | 5,311 | 266,253 | 37.4 | 33.6 |
| **OREGON** | 1,649,352 | 2.50 | 61.9 | 47.3 | 14.6 | 38.1 | 28.1 | 89,211 | NA | NA | NA | NA | 2,988,118 | 31.6 | 34.5 |
| Albany | 21,888 | 2.54 | 58.4 | 43.5 | 14.9 | 41.6 | 29.6 | 829 | 70 | 127 | 1,467 | 2,668 | 39,443 | 35.1 | 27.3 |
| Beaverton | 43,659 | 2.25 | 50.6 | 40.1 | 10.4 | 49.4 | 36.1 | 1,015 | 217 | 217 | 1,988 | 1,985 | 74,464 | 24.8 | 45.5 |
| Bend | 39,371 | 2.54 | 60.9 | 47.5 | 13.4 | 39.1 | 28.5 | 574 | 154 | 153 | 1,919 | 1,908 | 72,000 | 21.4 | 46.0 |
| Corvallis | 24,259 | 2.19 | 45.5 | 35.6 | 9.9 | 54.5 | 30.6 | 5,841 | NA | NA | NA | NA | 32,892 | 12.7 | 60.3 |
| Eugene | 73,331 | 2.27 | 50.3 | 36.2 | 14.2 | 49.7 | 34.4 | 6,482 | 675 | 390 | 6,184 | 3,571 | 114,429 | 22.6 | 42.6 |
| Grants Pass | 15,582 | 2.39 | 61.0 | 38.7 | 22.4 | 39.0 | 33.4 | 933 | 121 | 314 | 1,235 | 3,210 | 26,590 | 46.0 | 16.5 |
| Gresham | 40,604 | 2.65 | 65.1 | 44.4 | 20.6 | 34.9 | 24.3 | 1,698 | 490 | 443 | 3,442 | 3,110 | 77,272 | 40.6 | 24.5 |
| Hillsboro | 38,085 | 2.82 | 67.8 | 55.0 | 12.9 | 32.2 | 23.2 | 1,621 | 274 | 248 | 2,164 | 1,958 | 73,531 | 29.7 | 40.9 |
| Keizer | 15,813 | 2.49 | 68.3 | 46.8 | 21.5 | 31.7 | 27.1 | 367 | 61 | 152 | 648 | 1,616 | 28,799 | 33.2 | 29.7 |
| Lake Oswego | 16,820 | 2.36 | 61.8 | 51.6 | 10.2 | 38.2 | 33.8 | 146 | 24 | 60 | 481 | 1,206 | 28,238 | 9.9 | 72.8 |
| McMinnville | 13,207 | 2.50 | 59.6 | 36.9 | 22.7 | 40.4 | 31.6 | 1,782 | 69 | 198 | 758 | 2,170 | 24,802 | 36.3 | 24.9 |
| Medford | 33,047 | 2.46 | 57.4 | 41.1 | 16.3 | 42.6 | 35.2 | 1,811 | NA | NA | NA | NA | 57,249 | 40.4 | 25.8 |
| Oregon City | 14,484 | 2.52 | 61.6 | 43.9 | 17.6 | 38.4 | 24.2 | 841 | 107 | 284 | 644 | 1,707 | 26,341 | 32.6 | 25.7 |
| Portland | 280,176 | 2.27 | 49.6 | 36.7 | 12.8 | 50.4 | 35.0 | 16,637 | 3,606 | 545 | 34,452 | 5,203 | 487,931 | 21.6 | 52.8 |

1. Data for serious crimes have not been adjusted for underreporting. This may affect comparability between geographic areas and over time.   2. Per 100,000 population estimated by the FBI.   3. Persons 25 years old and over.

# Table D. Cities — Income, Poverty, and Housing

| City | Money income, 2019 — Households — Median income | Percent with income less than $20,000 | Percent with income of $200,000 or more | Median family income | Median non-family income | Median earnings, 2019 — All persons | Men | Women | Housing units, 2019 — Total | Occupied | Percent owner occupied | Median value[1] (dollars) | Median gross rent (dollars) |
|---|---|---|---|---|---|---|---|---|---|---|---|---|---|
| | 42 | 43 | 44 | 45 | 46 | 47 | 48 | 49 | 50 | 51 | 52 | 53 | 54 |
| **OHIO—Cont'd** | | | | | | | | | | | | | |
| Hamilton | 52,115 | 17.3 | 0.5 | 65,517 | 31,750 | 30,676 | 36,714 | 26,315 | 27,923 | 24,268 | 54.0 | 115,720 | 822 |
| Hilliard | 103,365 | 4.9 | 11.0 | 120,890 | 64,659 | 56,211 | 80,915 | 45,248 | 15,149 | 14,879 | 69.6 | 292,748 | 1,354 |
| Huber Heights | 63,140 | 8.9 | 3.8 | 75,124 | 39,241 | 30,440 | 36,580 | 26,628 | 16,072 | 15,402 | 74.4 | 117,432 | 1,082 |
| Kent | 30,560 | 34.6 | 1.6 | 67,205 | 17,343 | 15,102 | 16,220 | 12,360 | 13,857 | 10,291 | 38.4 | 150,983 | 803 |
| Kettering | 62,945 | 12.7 | 4.5 | 82,886 | 34,442 | 33,696 | 42,163 | 28,364 | 25,823 | 24,168 | 61.7 | 143,621 | 773 |
| Lakewood | 53,771 | 12.9 | 6.0 | 75,994 | 45,463 | 40,429 | 42,630 | 36,484 | 26,180 | 23,673 | 44.0 | 193,676 | 832 |
| Lancaster | 46,872 | 19.9 | 2.6 | 63,411 | 27,799 | 32,380 | 40,627 | 26,268 | 18,579 | 16,931 | 56.1 | 145,364 | 737 |
| Lima | 38,424 | 23.0 | 0.3 | 46,076 | 27,691 | 22,467 | 21,128 | 25,489 | 15,850 | 13,291 | 52.8 | 68,373 | 681 |
| Lorain | 36,162 | 28.6 | 0.9 | 50,600 | 24,319 | 29,812 | 35,392 | 26,600 | 29,159 | 24,947 | 54.2 | 94,995 | 697 |
| Mansfield | 43,227 | 20.8 | 1.1 | 50,254 | 30,911 | 24,605 | 30,245 | 21,621 | 19,736 | 17,993 | 49.8 | 82,594 | 661 |
| Marion | 41,259 | 22.0 | 0.3 | 41,763 | 36,196 | 26,256 | 30,361 | 22,005 | 15,121 | 13,387 | 49.8 | 80,311 | 740 |
| Mason | 105,506 | 6.3 | 22.5 | 131,289 | 55,390 | 53,677 | 77,295 | 36,637 | 12,116 | 11,979 | 76.4 | 323,789 | 1,421 |
| Massillon | 45,856 | 20.7 | 0.6 | 62,256 | 24,040 | 32,837 | 38,968 | 29,506 | 15,959 | 13,969 | 64.4 | 115,230 | 664 |
| Medina | 63,937 | 14.2 | 5.5 | 81,976 | 38,526 | 47,471 | 52,364 | 40,053 | 11,497 | 10,656 | 65.8 | 197,296 | 848 |
| Mentor | 75,562 | 6.1 | 5.7 | 87,563 | 49,291 | 45,342 | 50,645 | 39,245 | 21,128 | 20,129 | 84.1 | 169,949 | 952 |
| Middletown | 37,389 | 24.1 | 2.4 | 43,689 | 30,214 | 27,470 | 30,980 | 23,693 | 22,340 | 20,474 | 51.5 | 100,110 | 814 |
| Newark | 48,280 | 18.8 | 2.8 | 54,494 | 27,572 | 31,070 | 32,355 | 30,437 | 23,142 | 21,234 | 51.5 | 129,516 | 792 |
| North Olmsted | 63,348 | 12.9 | 1.9 | 77,958 | 41,886 | 35,592 | 41,256 | 27,188 | 13,948 | 13,455 | 69.2 | 164,745 | 988 |
| North Ridgeville | 85,437 | 5.0 | 6.8 | 100,200 | 47,074 | 48,056 | 51,393 | 38,893 | 12,800 | 12,486 | 89.5 | 211,707 | 798 |
| North Royalton | 68,919 | 9.4 | 5.9 | 102,162 | 43,597 | 45,516 | 53,432 | 32,901 | 15,918 | 14,588 | 67.7 | 228,973 | 894 |
| Parma | 57,376 | 12.5 | 1.0 | 68,256 | 40,015 | 37,191 | 42,303 | 30,969 | 36,294 | 34,332 | 71.6 | 119,564 | 885 |
| Reynoldsburg | 70,656 | 2.9 | 3.1 | 84,274 | 42,500 | 31,883 | 35,707 | 28,974 | 15,013 | 14,712 | 60.3 | 197,326 | 1,036 |
| Riverside | 47,337 | 15.9 | 0.4 | 55,210 | 41,111 | 27,497 | 36,483 | 21,413 | 12,673 | 10,794 | 48.7 | 96,269 | 843 |
| Sandusky | 43,807 | 29.6 | 0.9 | 56,855 | 17,151 | 22,259 | 25,625 | 20,755 | 12,554 | 9,382 | 47.7 | 85,850 | 710 |
| Shaker Heights | 89,490 | 9.0 | 21.1 | 136,049 | 42,549 | 56,978 | 70,573 | 47,881 | 12,273 | 11,124 | 60.0 | 258,081 | 895 |
| Springfield | 40,659 | 25.5 | 2.1 | 47,981 | 20,663 | 26,333 | 30,586 | 23,527 | 28,587 | 23,619 | 51.2 | 87,233 | 708 |
| Stow | 63,163 | 10.5 | 7.9 | 94,926 | 36,244 | 40,321 | 45,639 | 36,212 | 14,844 | 13,850 | 70.3 | 185,208 | 998 |
| Strongsville | 85,423 | 5.6 | 10.3 | 98,594 | 50,757 | 50,517 | 60,037 | 36,951 | 19,632 | 18,819 | 80.6 | 212,205 | 1,046 |
| Toledo | 36,709 | 26.3 | 0.6 | 47,113 | 27,055 | 26,871 | 30,491 | 24,947 | 139,224 | 121,022 | 50.5 | 84,622 | 744 |
| Troy | 58,372 | 23.2 | 3.8 | 86,615 | 36,264 | 35,444 | 44,325 | 25,848 | 12,008 | 10,658 | 65.8 | 159,658 | 779 |
| Upper Arlington | 146,702 | 3.3 | 35.7 | 170,625 | 66,327 | 72,685 | 94,017 | 50,664 | 14,514 | 13,471 | 81.1 | 448,768 | 1,280 |
| Warren | 30,471 | 38.0 | 2.4 | 42,612 | 16,959 | 22,275 | 27,170 | 20,417 | 20,492 | 17,413 | 53.5 | 62,750 | 643 |
| Westerville | 94,264 | 9.1 | 8.8 | 118,983 | 51,586 | 53,413 | 56,237 | 51,207 | 16,194 | 15,076 | 80.7 | 259,389 | 1,127 |
| Westlake | 76,127 | 8.4 | 15.4 | 115,266 | 46,816 | 47,014 | 55,493 | 39,112 | 15,487 | 14,132 | 66.7 | 266,269 | 1,204 |
| Wooster | 52,953 | 18.4 | 2.1 | 68,939 | 33,062 | 25,260 | 33,616 | 21,494 | 11,465 | 10,882 | 51.3 | 159,910 | 734 |
| Xenia | 44,226 | 16.6 | 1.3 | 53,681 | 28,571 | 30,371 | 32,130 | 23,910 | 11,764 | 10,856 | 52.3 | 112,858 | 798 |
| Youngstown | 29,143 | 36.2 | 0.6 | 32,806 | 23,959 | 19,083 | 20,439 | 17,995 | 33,665 | 29,102 | 54.4 | 50,184 | 674 |
| Zanesville | 31,400 | 31.3 | 0.9 | 37,345 | 26,486 | 23,957 | 31,021 | 18,753 | 12,900 | 10,739 | 51.3 | 79,327 | 801 |
| **OKLAHOMA** | 54,449 | 16.4 | 4.7 | 68,358 | 32,185 | 32,323 | 40,259 | 26,914 | 1,749,520 | 1,495,151 | 65.5 | 146,951 | 814 |
| Bartlesville | 62,018 | 13.4 | 5.7 | 71,488 | 35,344 | 31,195 | 35,615 | 30,229 | 17,208 | 14,655 | 69.6 | 122,206 | 863 |
| Broken Arrow | 74,290 | 8.0 | 7.5 | 86,991 | 45,678 | 40,945 | 50,322 | 31,860 | 43,831 | 41,730 | 74.0 | 172,128 | 1,007 |
| Edmond | 88,546 | 8.1 | 16.1 | 120,818 | 38,360 | 44,151 | 61,464 | 35,474 | 39,327 | 36,056 | 69.7 | 264,217 | 1,055 |
| Enid | 62,634 | 12.8 | 3.0 | 71,457 | 46,535 | 33,063 | 42,347 | 26,800 | 21,633 | 18,995 | 62.0 | 109,985 | 827 |
| Lawton | 46,886 | 20.9 | 2.2 | 54,188 | 34,749 | 26,062 | 27,273 | 25,048 | 40,665 | 32,174 | 44.8 | 116,670 | 807 |
| Midwest City | 56,983 | 12.9 | 3.3 | 66,238 | 42,840 | 32,258 | 41,080 | 27,128 | 22,754 | 20,971 | 56.2 | 128,224 | 863 |
| Moore | 68,396 | 10.7 | 1.3 | 77,872 | 40,601 | 38,679 | 46,709 | 29,379 | 27,487 | 25,574 | 71.1 | 152,394 | 1,129 |
| Muskogee | 38,929 | 23.4 | 3.0 | 43,705 | 26,262 | 29,079 | 31,676 | 22,195 | 17,739 | 15,286 | 54.0 | 87,949 | 681 |
| Norman | 55,849 | 17.7 | 5.6 | 86,375 | 29,531 | 30,480 | 36,625 | 25,776 | 52,636 | 48,531 | 53.1 | 205,098 | 856 |
| Oklahoma City | 55,492 | 16.0 | 5.2 | 70,386 | 38,531 | 35,040 | 39,582 | 29,988 | 281,616 | 249,615 | 57.5 | 165,744 | 874 |
| Owasso | 71,760 | 6.8 | 8.6 | 88,950 | 49,729 | 42,977 | 50,312 | 32,530 | 14,406 | 13,721 | 65.0 | 173,929 | 1,081 |
| Ponca City | 44,023 | 16.0 | 3.9 | 49,773 | 32,810 | 32,293 | 42,690 | 21,125 | 12,129 | 10,077 | 65.1 | 107,249 | 691 |
| Shawnee | 48,152 | 21.8 | 3.4 | 61,706 | 23,025 | 25,805 | 37,320 | 19,430 | 13,859 | 11,792 | 56.9 | 127,419 | 759 |
| Stillwater | 37,705 | 31.5 | 3.0 | 61,196 | 16,498 | 14,486 | 17,412 | 8,774 | 22,427 | 19,119 | 40.7 | 206,971 | 786 |
| Tulsa | 49,158 | 18.9 | 6.6 | 64,820 | 32,920 | 31,559 | 37,416 | 27,134 | 187,906 | 163,801 | 51.4 | 152,656 | 827 |
| **OREGON** | 67,058 | 12.8 | 7.1 | 82,540 | 41,733 | 35,643 | 41,243 | 30,688 | 1,808,482 | 1,649,352 | 62.9 | 354,551 | 1,185 |
| Albany | 66,154 | 8.8 | 3.2 | 83,232 | 44,543 | 34,185 | 40,248 | 28,924 | 22,113 | 21,888 | 61.3 | 266,116 | 1,033 |
| Beaverton | 73,260 | 11.1 | 8.7 | 101,039 | 53,986 | 44,036 | 46,147 | 41,634 | 46,034 | 43,659 | 46.6 | 428,053 | 1,406 |
| Bend | 69,998 | 11.6 | 8.1 | 87,227 | 40,914 | 35,253 | 41,047 | 31,635 | 44,562 | 39,371 | 62.1 | 426,606 | 1,522 |
| Corvallis | 56,126 | 19.7 | 3.9 | 95,007 | 35,826 | 21,654 | 25,780 | 19,155 | 25,745 | 24,259 | 44.7 | 379,066 | 1,186 |
| Eugene | 51,269 | 19.1 | 5.0 | 84,543 | 30,878 | 28,284 | 31,784 | 24,443 | 76,076 | 73,331 | 46.8 | 329,016 | 1,099 |
| Grants Pass | 43,910 | 19.4 | 1.5 | 46,118 | 35,522 | 24,210 | 30,525 | 21,574 | 16,306 | 15,582 | 45.5 | 265,355 | 1,032 |
| Gresham | 53,892 | 13.2 | 3.0 | 65,998 | 40,385 | 32,055 | 35,199 | 31,087 | 42,870 | 40,604 | 52.8 | 337,819 | 1,289 |
| Hillsboro | 86,038 | 5.9 | 10.3 | 97,515 | 63,147 | 42,062 | 54,873 | 31,996 | 40,299 | 38,085 | 51.8 | 382,348 | 1,500 |
| Keizer | 65,315 | 9.9 | 5.0 | 78,820 | 46,193 | 40,533 | 46,991 | 35,307 | 16,843 | 15,813 | 65.2 | 293,655 | 1,129 |
| Lake Oswego | 117,603 | 6.1 | 27.2 | 160,023 | 53,316 | 68,639 | 97,046 | 51,411 | 17,790 | 16,820 | 74.7 | 693,800 | 1,663 |
| McMinnville | 56,366 | 18.2 | 1.1 | 68,434 | 37,355 | 29,827 | 31,543 | 24,941 | 14,607 | 13,207 | 50.7 | 277,455 | 1,112 |
| Medford | 54,737 | 16.4 | 4.6 | 79,028 | 34,241 | 30,307 | 32,710 | 25,475 | 35,803 | 13,207 | 50.7 | 277,455 | 1,112 |
| Medford | 54,737 | 16.4 | 4.6 | 79,028 | 34,241 | 30,307 | 32,710 | 25,475 | 14,607 | 13,207 | 50.7 | 312,363 | 1,000 |
| Oregon City | 76,656 | 7.1 | 4.5 | 94,754 | 46,944 | 41,220 | 43,332 | 39,440 | 15,355 | 14,484 | 66.2 | 403,483 | 1,277 |
| Portland | 76,231 | 12.8 | 11.0 | 100,589 | 54,538 | 42,060 | 46,246 | 39,623 | 298,837 | 280,176 | 53.8 | 445,150 | 1,312 |

# Table D. Cities — Commuting, Computer Access, Migration, Labor Force, and Employment

| City | Commuting[1], 2019 Percent — Drove alone | With commutes of 30 minutes or more | Computer access[2], 2019 Percent — With a computer in the house | With Internet access | Migration, 2019 — Percent who lived in the same house one year ago | Percent who lived in another state or county one year ago | Civilian labor force, 2020 — Total | Percent change 2019-2020 | Unemployment — Total | Rate[3] | Civilian Employment[4], 2019 — Population age 16 and older — Number | Percent in labor force | Population age 16 to 64 — Number | Percent who worked full-year full-time |
|---|---|---|---|---|---|---|---|---|---|---|---|---|---|---|
| | 55 | 56 | 57 | 58 | 59 | 60 | 61 | 62 | 63 | 64 | 65 | 66 | 67 | 68 |
| **OHIO—Cont'd** | | | | | | | | | | | | | | |
| Hamilton | 77.0 | 37.0 | 92.1 | 86.2 | 81.3 | 6.3 | 28,228 | 0.1 | 2,429 | 9 | 48,691 | 63.1 | 40,218 | 55.7 |
| Hilliard | 90.0 | 33.7 | 97.7 | 94.3 | 84.9 | 5.1 | 19,722 | -0.8 | 1,045 | 5 | 29,085 | 74.9 | 23,636 | 64.5 |
| Huber Heights | 88.2 | 23.9 | 92.0 | 88.1 | 85.9 | 10.5 | 18,404 | 0.7 | 1,605 | 9 | 31,668 | 61.1 | 23,667 | 50.1 |
| Kent | 74.4 | 35.8 | 93.2 | 86.3 | 66.1 | 22.3 | 16,206 | -2.5 | 987 | 6 | 26,964 | 62.4 | 23,422 | 31.7 |
| Kettering | 83.2 | 20.6 | 93.7 | 89.5 | 84.5 | 3.3 | 28,923 | -0.6 | 2,147 | 7 | 43,462 | 66.7 | 33,856 | 59.9 |
| Lakewood | 78.7 | 37.7 | 93.6 | 89.4 | 83.8 | 6.3 | 28,968 | -3.7 | 2,711 | 9 | 40,536 | 73.1 | 34,681 | 58.5 |
| Lancaster | 85.4 | 41.5 | 88.8 | 82.4 | 76.0 | 7.2 | 18,376 | 0.8 | 1,460 | 8 | 32,162 | 57.0 | 24,180 | 55.8 |
| Lima | 82.9 | 19.6 | 91.4 | 86.2 | 88.2 | 5.2 | 14,863 | 2.5 | 1,655 | 11 | 28,290 | 58.7 | 23,485 | 39.6 |
| Lorain | 78.6 | 31.0 | 88.0 | 70.4 | 86.5 | 4.0 | 27,309 | -1.2 | 3,651 | 13 | 48,577 | 53.2 | 38,254 | 42.6 |
| Mansfield | 82.4 | 18.8 | 87.2 | 77.5 | 76.7 | 11.8 | 17,584 | 1.4 | 1,940 | 11 | 38,365 | 52.6 | 30,812 | 37.7 |
| Marion | 82.0 | 28.1 | 90.7 | 82.4 | 81.3 | 5.8 | 13,897 | 0.2 | 1,249 | 9 | 29,175 | 53.5 | 24,043 | 37.2 |
| Mason | 78.4 | 42.8 | NA | NA | 83.9 | 11.2 | 17,306 | -0.5 | 1,065 | 6 | 26,369 | 67.9 | 21,753 | 57.7 |
| Massillon | 88.2 | 30.5 | 91.0 | 84.5 | 91.6 | 2.9 | 15,426 | 0.3 | 1,339 | 9 | 27,012 | 60.7 | 20,889 | 54.0 |
| Medina | 89.4 | 38.8 | 93.9 | 82.6 | 86.9 | 5.8 | 12,728 | -4.4 | 1,116 | 9 | 19,993 | 71.2 | 16,373 | 59.3 |
| Mentor | 90.4 | 29.2 | 94.0 | 89.5 | 92.7 | 1.6 | 25,594 | -5.0 | 1,950 | 8 | 38,992 | 64.1 | 28,298 | 63.9 |
| Middletown | 87.8 | 36.1 | 87.9 | 81.0 | 80.4 | 5.3 | 21,449 | 1.4 | 2,209 | 10 | 39,052 | 55.8 | 29,939 | 47.6 |
| Newark | 85.3 | 36.9 | 91.2 | 85.6 | 82.2 | 4.5 | 24,506 | 1.1 | 1,847 | 8 | 40,506 | 61.8 | 31,611 | 54.4 |
| North Olmsted | 86.2 | 35.7 | 93.0 | 89.2 | 84.8 | 5.8 | 16,700 | -4.6 | 1,484 | 9 | 24,992 | 64.7 | 18,411 | 53.4 |
| North Ridgeville | 78.7 | 40.6 | 94.6 | 81.8 | NA | NA | 17,696 | -3.3 | 1,403 | 8 | 27,175 | 67.1 | 21,044 | 67.7 |
| North Royalton | 85.5 | 46.7 | 93.5 | 91.2 | 85.0 | 2.9 | 16,698 | -5.5 | 1,298 | 8 | 26,146 | 64.4 | 19,076 | 60.2 |
| Parma | 86.0 | 37.0 | 90.7 | 85.9 | 87.5 | 3.9 | 40,697 | -3.7 | 4,129 | 10 | 64,771 | 64.9 | 47,674 | 58.7 |
| Reynoldsburg | 83.9 | 41.9 | 96.2 | 89.7 | 82.4 | 9.7 | 21,442 | 1.1 | 1,669 | 8 | 30,385 | 72.0 | 24,564 | 58.8 |
| Riverside | 86.5 | 12.9 | 94.6 | 90.2 | 91.8 | 3.9 | 10,942 | 0.7 | 986 | 9 | 19,129 | 64.5 | 15,258 | 52.5 |
| Sandusky | 76.5 | 15.4 | 86.9 | 81.2 | 85.5 | 4.3 | 11,759 | 0.9 | 1,596 | 14 | 18,863 | 63.2 | 15,480 | 40.9 |
| Shaker Heights | 76.4 | 39.7 | 92.5 | 83.8 | 83.1 | 7.0 | 13,387 | -5.9 | 1,011 | 8 | 20,641 | 65.7 | 15,683 | 57.7 |
| Springfield | 76.9 | 20.4 | 89.6 | 84.0 | 78.1 | 4.3 | 25,389 | 0.2 | 2,400 | 10 | 48,006 | 57.3 | 37,791 | 41.6 |
| Stow | 91.4 | 39.9 | 95.6 | 92.5 | 86.1 | 9.1 | 18,399 | -1.7 | 1,246 | 7 | 27,131 | 60.1 | 19,655 | 56.9 |
| Strongsville | 82.4 | 44.3 | 93.7 | 90.7 | 87.1 | 4.8 | 23,432 | -5.4 | 1,757 | 8 | 38,041 | 63.7 | 26,740 | 62.7 |
| Toledo | 80.0 | 18.5 | 89.7 | 80.2 | 81.6 | 6.1 | 128,634 | 0.4 | 15,418 | 12 | 216,850 | 60.9 | 175,725 | 44.1 |
| Troy | NA | 22.3 | 91.9 | 90.2 | 89.6 | 3.3 | 13,952 | 0.4 | 1,018 | 7 | 20,881 | 59.3 | 15,928 | 55.7 |
| Upper Arlington | 88.0 | 18.9 | NA | NA | 87.9 | 5.4 | 18,257 | -2.0 | 884 | 5 | 27,535 | 68.9 | 21,681 | 66.3 |
| Warren | 83.7 | 37.2 | 86.3 | 77.2 | 86.8 | 3.8 | 13,689 | 1.1 | 1,750 | 13 | 32,263 | 45.8 | 24,592 | 37.8 |
| Westerville | 79.1 | 32.1 | 96.0 | 92.7 | 86.9 | 6.5 | 22,919 | 1.0 | 1,351 | 6 | 30,860 | 67.4 | 22,834 | 65.3 |
| Westlake | 87.8 | 29.7 | 93.4 | 88.4 | 88.7 | 3.1 | 16,276 | -5.6 | 1,228 | 8 | 27,502 | 66.8 | 19,946 | 56.5 |
| Wooster | 72.0 | 15.3 | 91.8 | 87.4 | 66.2 | 7.2 | 13,821 | -2.1 | 926 | 7 | 21,874 | 64.7 | 17,435 | 47.3 |
| Xenia | 89.1 | 34.0 | 89.5 | 85.1 | 85.6 | 7.3 | 11,707 | 1.0 | 1,029 | 9 | 22,091 | 56.3 | 16,228 | 44.2 |
| Youngstown | 69.9 | 18.4 | 84.1 | 76.8 | 81.0 | 9.3 | 22,768 | 1.2 | 2,935 | 13 | 54,269 | 51.5 | 41,565 | 31.3 |
| Zanesville | 84.0 | 26.1 | 81.1 | 75.2 | 81.3 | 7.2 | 10,046 | 2.6 | 954 | 10 | 20,545 | 47.4 | 15,260 | 32.1 |
| **OKLAHOMA** | 82.7 | 28.4 | 92.1 | 83.0 | 84.0 | 7.0 | 1,848,485 | 0.4 | 113,561 | 6 | 3,110,626 | 60.2 | 2,475,404 | 52.5 |
| Bartlesville | 72.4 | 17.0 | 94.1 | 87.5 | 86.1 | 8.2 | 15,867 | 0.2 | 944 | 6 | 29,991 | 57.4 | 23,053 | 53.5 |
| Broken Arrow | 84.4 | 23.4 | 98.0 | 92.6 | 86.9 | 5.3 | 58,050 | 0.3 | 3,523 | 6 | 87,057 | 69.5 | 70,334 | 60.0 |
| Edmond | 80.6 | 32.2 | 97.2 | 94.6 | 83.9 | 3.9 | 48,305 | -0.4 | 2,299 | 5 | 74,912 | 66.3 | 60,569 | 53.3 |
| Enid | 85.9 | 9.9 | 97.2 | 83.5 | 84.4 | 6.4 | 21,448 | 0.0 | 1,257 | 6 | 38,028 | 61.2 | 30,755 | 58.9 |
| Lawton | 68.8 | 11.8 | 94.0 | 85.8 | 73.0 | 15.7 | 36,055 | 0.3 | 2,563 | 7 | 74,625 | 52.2 | 64,308 | 50.0 |
| Midwest City | 86.4 | 29.6 | 94.4 | 88.5 | 86.7 | 5.6 | 26,734 | 0.2 | 1,919 | 7 | 44,528 | 62.3 | 35,310 | 55.9 |
| Moore | 86.3 | 34.6 | 96.3 | 92.5 | 85.6 | 9.4 | 31,448 | -0.6 | 1,882 | 6 | 50,847 | 68.2 | 41,143 | 63.2 |
| Muskogee | 84.0 | 19.0 | 89.3 | 72.1 | 77.5 | 8.9 | 16,377 | -0.4 | 1,011 | 6 | 28,698 | 52.9 | 22,856 | 44.6 |
| Norman | 78.9 | 32.4 | 96.7 | 88.3 | 76.7 | 10.8 | 63,164 | 0.1 | 3,480 | 6 | 102,344 | 62.6 | 85,786 | 47.7 |
| Oklahoma City | 83.4 | 25.2 | 92.9 | 85.5 | 81.1 | 7.3 | 323,683 | 0.7 | 20,941 | 7 | 507,535 | 65.3 | 421,617 | 55.7 |
| Owasso | 87.7 | 21.9 | 93.1 | 90.5 | 81.2 | 9.0 | 19,255 | -0.6 | 1,078 | 6 | 26,837 | 68.0 | 22,229 | 58.4 |
| Ponca City | 73.8 | 13.8 | 86.1 | 75.6 | 82.5 | 5.8 | 10,043 | 0.2 | 714 | 7 | 19,292 | 55.4 | 14,284 | 53.2 |
| Shawnee | 82.6 | 27.9 | 89.1 | 84.4 | 80.0 | 11.3 | 14,454 | 1.1 | 968 | 7 | 24,666 | 56.6 | 19,313 | 43.6 |
| Stillwater | 79.9 | 12.4 | 95.8 | 82.4 | 68.5 | 18.4 | 24,070 | -0.1 | 1,166 | 5 | 43,261 | 58.0 | 37,121 | 35.3 |
| Tulsa | 80.7 | 14.4 | 93.3 | 85.9 | 80.4 | 6.7 | 196,971 | 0.2 | 14,015 | 7 | 312,854 | 65.4 | 252,580 | 53.4 |
| **OREGON** | 71.5 | 32.9 | 95.0 | 88.8 | 84.4 | 7.1 | 2,104,657 | 0.0 | 159,445 | 8 | 3,451,654 | 62.4 | 2,684,158 | 49.4 |
| Albany | 80.8 | 24.4 | 96.3 | 93.2 | 82.2 | 7.1 | 25,650 | 0.1 | 2,071 | 8 | 46,659 | 62.6 | 36,817 | 52.3 |
| Beaverton | 69.7 | 33.5 | 96.7 | 90.9 | 84.0 | 6.9 | 55,311 | -1.3 | 3,739 | 7 | 85,556 | 71.1 | 70,688 | 57.5 |
| Bend | 67.5 | 15.0 | 96.6 | 93.5 | 81.1 | 10.3 | 54,252 | 1.8 | 4,027 | 7 | 81,637 | 67.7 | 65,870 | 49.6 |
| Corvallis | 61.9 | 19.8 | 96.9 | 94.3 | 69.4 | 19.8 | 29,424 | -4.6 | 1,651 | 6 | 51,796 | 60.6 | 43,365 | 35.5 |
| Eugene | 63.8 | 17.7 | 95.5 | 90.5 | 74.8 | 9.7 | 83,312 | -0.9 | 6,421 | 8 | 148,912 | 62.6 | 117,564 | 43.8 |
| Grants Pass | 72.3 | 14.6 | 88.8 | 84.9 | 85.9 | 6.0 | 16,354 | 1.3 | 1,361 | 8 | 30,682 | 59.7 | 22,563 | 41.2 |
| Gresham | 72.9 | 44.8 | 94.3 | 89.4 | 81.4 | 6.4 | 54,382 | -0.7 | 4,638 | 9 | 91,237 | 64.7 | 74,382 | 47.1 |
| Hillsboro | 74.1 | 29.1 | 98.4 | 93.4 | 81.3 | 9.7 | 57,884 | -1.6 | 3,367 | 6 | 84,719 | 68.3 | 71,890 | 58.8 |
| Keizer | 83.3 | 26.1 | 96.0 | 91.8 | 87.8 | 3.0 | 19,659 | 0.4 | 1,311 | 7 | 32,424 | 55.3 | 23,058 | 50.8 |
| Lake Oswego | 72.8 | 39.9 | 95.4 | 89.9 | 88.6 | 7.1 | 20,825 | -1.5 | 1,206 | 6 | 31,552 | 56.4 | 21,292 | 50.6 |
| McMinnville | 65.6 | 23.3 | 92.5 | 82.4 | 86.4 | 7.2 | 16,855 | -0.4 | 1,247 | 7 | 30,820 | 59.1 | 23,010 | 40.4 |
| Medford | 85.7 | 11.1 | 92.3 | 84.4 | 84.7 | 4.5 | 40,594 | 2.1 | 3,225 | 8 | 67,135 | 59.8 | 51,725 | 44.1 |
| Oregon City | 75.1 | 47.9 | 95.7 | 88.7 | 87.8 | 4.5 | 19,199 | -0.3 | 1,486 | 8 | 30,422 | 71.0 | 25,351 | 55.4 |
| Portland | 56.4 | 39.3 | 96.0 | 91.1 | 82.4 | 7.7 | 378,901 | 0.8 | 32,925 | 9 | 549,540 | 70.6 | 460,851 | 53.6 |

1. Employed persons.  2. Households.  3. Percent of civilian labor force.  4. Persons 16 years old and over.

# Table D. Cities — Construction, Wholesale Trade, and Retail Trade

| City | Value of residential construction authorized by building permits, 2020 | | | Wholesale trade[1], 2017 | | | | Retail trade[2], 2017 | | | |
|---|---|---|---|---|---|---|---|---|---|---|---|
| | New construction ($1,000) | Number of housing units | Percent single family | Number of establishments | Number of employees | Sales (mil dol) | Annual payroll (mil dol) | Number of establishments | Number of employees | Sales (mil dol) | Annual payroll (mil dol) |
| | 69 | 70 | 71 | 72 | 73 | 74 | 75 | 76 | 77 | 78 | 79 |
| OHIO—Cont'd | | | | | | | | | | | |
| Hamilton | 11,092 | 72 | 100 | 40 | 769 | 729 | 50 | 190 | 3,518 | 777 | 74 |
| Hilliard | 37,123 | 114 | 100 | 44 | 724 | 533 | 42 | 87 | 1,613 | 478 | 47 |
| Huber Heights | NA | NA | NA | 23 | 516 | 359 | 32 | 111 | 2,553 | 570 | 56 |
| Kent | 2,017 | 6 | 100 | 10 | 91 | 101 | 7 | 74 | 940 | 342 | 26 |
| Kettering | 4,960 | 16 | 100 | 28 | 196 | 188 | 14 | 150 | 4,391 | 2,074 | 94 |
| Lakewood | 3,456 | 9 | 100 | 20 | 366 | 631 | 41 | 114 | 1,067 | 292 | 26 |
| Lancaster | 18,268 | 143 | 25 | 28 | 231 | 55 | 10 | 188 | 2,858 | 729 | 69 |
| Lima | 5,049 | 51 | 6 | 42 | 528 | 402 | 29 | 109 | 1,374 | 345 | 34 |
| Lorain | 16,728 | 132 | 100 | 25 | 397 | 263 | 22 | 124 | 1,894 | 414 | 41 |
| Mansfield | 11,724 | 104 | 19 | 52 | 677 | 316 | 35 | 172 | 2,288 | 497 | 59 |
| Marion | 5,500 | 9 | 100 | 16 | 202 | 144 | 9 | 71 | 684 | 249 | 19 |
| Mason | 57,670 | 128 | 100 | 55 | 3,326 | 1,473 | 250 | 125 | 2,143 | 507 | 49 |
| Massillon | 11,446 | 128 | 41 | 22 | 391 | 222 | 21 | 111 | 2,561 | 744 | 66 |
| Medina | 4,654 | 34 | 100 | 45 | 463 | 341 | 23 | 108 | 1,677 | 399 | 39 |
| Mentor | 24,156 | 98 | 100 | 88 | 717 | 329 | 38 | 276 | 5,755 | 1,555 | 140 |
| Middletown | 30,418 | 178 | 100 | D | D | D | D | 124 | 3,340 | 1,687 | 94 |
| Newark | NA | NA | NA | 25 | 296 | 138 | 15 | 136 | 1,828 | 557 | 46 |
| North Olmsted | 1,245 | 9 | 0 | 18 | 163 | 251 | 18 | 241 | 5,236 | 1,538 | 134 |
| North Ridgeville | 63,657 | 336 | 100 | 29 | 180 | 130 | 11 | 50 | 592 | 177 | 14 |
| North Royalton | 8,043 | 31 | 100 | 33 | 537 | 202 | 39 | 69 | 606 | 180 | 17 |
| Parma | 3,935 | 25 | 100 | 58 | 1,095 | 589 | 54 | 219 | 3,650 | 1,000 | 89 |
| Reynoldsburg | 7,765 | 150 | 1 | 10 | 36 | 16 | 2 | 118 | 2,455 | 1,137 | 63 |
| Riverside | NA | NA | NA | 8 | 46 | 12 | 2 | 51 | 473 | 133 | 11 |
| Sandusky | 3,871 | 18 | 72 | 21 | 819 | 400 | 56 | 93 | 1,184 | 343 | 30 |
| Shaker Heights | 400 | 1 | 100 | 12 | 25 | 10 | 1 | 44 | 522 | 97 | 11 |
| Springfield | 20,042 | 79 | 87 | 46 | 1,044 | 1,418 | 51 | 224 | 4,171 | 1,044 | 90 |
| Stow | 10,541 | 46 | 57 | 49 | 400 | 411 | 25 | 98 | 2,294 | 587 | 55 |
| Strongsville | 26,709 | 74 | 100 | 71 | 1,897 | 874 | 98 | 240 | 5,320 | 1,191 | 117 |
| Toledo | 22,716 | 354 | 3 | 242 | 3,619 | 2,463 | 168 | 880 | 13,410 | 2,854 | 290 |
| Troy | NA | NA | NA | 20 | 573 | 523 | 34 | 87 | 2,451 | 663 | 56 |
| Upper Arlington | 29,282 | 36 | 78 | 15 | 34 | 11 | 2 | 83 | 1,065 | 222 | 23 |
| Warren | 0 | 0 | 0 | 25 | 448 | 350 | 23 | 142 | 3,681 | 644 | 98 |
| Westerville | 3,238 | 21 | 14 | 50 | 475 | 270 | 30 | 121 | 2,156 | 664 | 63 |
| Westlake | 23,508 | 58 | 88 | 81 | 1,018 | 941 | 62 | 182 | 2,658 | 799 | 79 |
| Wooster | 11,246 | 86 | 17 | 32 | 439 | 361 | 19 | 169 | 2,864 | 758 | 69 |
| Xenia | NA | NA | NA | 11 | 83 | 31 | 5 | 75 | 1,837 | 441 | 44 |
| Youngstown | NA | NA | NA | 83 | 1,223 | 660 | 57 | 162 | 1,441 | 336 | 33 |
| Zanesville | 80 | 1 | 100 | 26 | 245 | 110 | 12 | 168 | 2,937 | 757 | 69 |
| OKLAHOMA | 2,852,519 | 13,733 | 88 | 3,859 | 50,233 | 42,221 | 2,757 | 12,963 | 180,451 | 53,382 | 4,683 |
| Bartlesville | 4,404 | 22 | 100 | 18 | 72 | 34 | 3 | 148 | 2,237 | 657 | 57 |
| Broken Arrow | 185,013 | 1,096 | 77 | 141 | 1,577 | 910 | 90 | 264 | 5,160 | 1,858 | 143 |
| Edmond | 220,369 | 617 | 88 | 113 | 1,196 | 726 | 58 | 355 | 5,653 | 1,551 | 151 |
| Enid | 7,762 | 22 | 100 | 55 | 679 | 1,045 | 35 | 244 | 3,628 | 911 | 94 |
| Lawton | 13,145 | 45 | 100 | D | D | D | D | 336 | 4,833 | 1,250 | 120 |
| Midwest City | 21,832 | 123 | 98 | 11 | 81 | 26 | 3 | 170 | 4,077 | 1,267 | 112 |
| Moore | 40,429 | 174 | 91 | D | D | D | D | 167 | 2,913 | 721 | 69 |
| Muskogee | 5,284 | 25 | 100 | 46 | 594 | 312 | 27 | 204 | 3,027 | 780 | 73 |
| Norman | 174,080 | 795 | 68 | 64 | 943 | 579 | 54 | 434 | 7,673 | 2,234 | 194 |
| Oklahoma City | 847,468 | 3,916 | 97 | 1,026 | 16,717 | 18,804 | 987 | 2,234 | 33,933 | 11,000 | 966 |
| Owasso | 36,147 | 192 | 70 | 19 | 99 | 155 | 5 | 121 | 3,206 | 814 | 70 |
| Ponca City | 3,076 | 9 | 100 | 30 | 142 | 52 | 6 | 121 | 1,598 | 460 | 41 |
| Shawnee | 20,641 | 123 | 100 | 20 | 151 | 70 | 7 | 184 | 2,590 | 682 | 60 |
| Stillwater | 27,335 | 110 | 91 | 26 | 178 | 85 | 7 | 198 | 3,294 | 770 | 74 |
| Tulsa | 165,747 | 831 | 63 | 694 | 11,740 | 8,297 | 728 | 1,664 | 28,067 | 8,171 | 762 |
| OREGON | 40,147 | 368 | 100 | 4,452 | 67,900 | 58,205 | 4,306 | 14,318 | 211,222 | 61,699 | 6,067 |
| Albany | 42,901 | 246 | 76 | 50 | 493 | 437 | 25 | 191 | 3,094 | 845 | 85 |
| Beaverton | 145,578 | 890 | 15 | 163 | 3,933 | 3,366 | 272 | 409 | 9,745 | 4,788 | 311 |
| Bend | 210,799 | 1,076 | 61 | 141 | 1,204 | 891 | 65 | 597 | 8,285 | 2,597 | 250 |
| Corvallis | 15,559 | 71 | 75 | 25 | 213 | 101 | 13 | 217 | 3,632 | 848 | 90 |
| Eugene | 137,289 | 722 | 45 | 251 | 3,325 | 1,806 | 199 | 747 | 11,990 | 3,211 | 363 |
| Grants Pass | 36,632 | 174 | 71 | 36 | 244 | 175 | 12 | 237 | 3,701 | 1,047 | 110 |
| Gresham | 61,996 | 425 | 20 | 53 | 1,756 | 1,099 | 99 | 279 | 4,385 | 1,511 | 128 |
| Hillsboro | 118,596 | 427 | 100 | 109 | 2,182 | 1,450 | 221 | 328 | 6,986 | 2,120 | 201 |
| Keizer | 11,772 | 51 | 51 | 17 | 45 | 41 | 3 | 88 | 1,425 | 295 | 30 |
| Lake Oswego | 35,846 | 64 | 100 | 90 | 1,643 | 1,266 | 200 | 108 | 1,168 | 269 | 32 |
| McMinnville | 37,402 | 160 | 88 | 17 | 143 | 117 | 8 | 146 | 2,296 | 666 | 65 |
| Medford | 75,527 | 285 | 93 | 115 | 1,223 | 854 | 61 | 473 | 8,072 | 2,737 | 239 |
| Oregon City | 23,358 | 86 | 100 | 16 | 239 | 100 | 13 | 123 | 1,919 | 588 | 53 |
| Portland | 308,206 | 1,779 | 33 | 1,103 | 22,237 | 24,079 | 1,518 | 2,703 | 37,084 | 9,801 | 1,113 |

1. Merchant wholesalers except manufacturers' sales branches and offices.  2. Establishments with payroll.

Items 69—79

# Table D. Cities — Real Estate, Professional Services, and Manufacturing

| City | Real estate and rental and leasing, 2017 | | | | Professional, scientific, and technical services[1], 2017 | | | | Manufacturing, 2017 | | | |
|---|---|---|---|---|---|---|---|---|---|---|---|---|
| | Number of establishments | Number of employees | Receipts (mil dol) | Annual payroll (mil dol) | Number of establishments | Number of employees | Receipts (mil dol) | Annual payroll (mil dol) | Number of establishments | Number of employees | Receipts (mil dol) | Annual payroll (mil dol) |
| | 80 | 81 | 82 | 83 | 84 | 85 | 86 | 87 | 88 | 89 | 90 | 91 |
| **OHIO—Cont'd** | | | | | | | | | | | | |
| Hamilton | 39 | 194 | 38 | 7 | 86 | 353 | 36 | 13 | NA | NA | NA | NA |
| Hilliard | 51 | 217 | 44 | 12 | 112 | 796 | 104 | 56 | NA | NA | NA | NA |
| Huber Heights | 26 | 136 | 23 | 6 | 30 | 377 | 31 | 13 | NA | NA | NA | NA |
| Kent | 30 | 153 | 32 | 5 | 37 | 234 | 27 | 11 | NA | NA | NA | NA |
| Kettering | 60 | 324 | 51 | 11 | 108 | 2,721 | 469 | 173 | NA | NA | NA | NA |
| Lakewood | 46 | 160 | 37 | 6 | 125 | 470 | 71 | 28 | NA | NA | NA | NA |
| Lancaster | 49 | 143 | 28 | 5 | 70 | 416 | 43 | 16 | NA | NA | NA | NA |
| Lima | 23 | 114 | 15 | 3 | 71 | 529 | 49 | 21 | NA | NA | NA | NA |
| Lorain | 39 | 143 | 22 | 5 | 52 | 185 | 24 | 8 | NA | NA | NA | NA |
| Mansfield | 55 | 314 | 45 | 10 | 117 | 576 | 66 | 25 | NA | NA | NA | NA |
| Marion | 17 | 50 | 11 | 1 | 33 | 116 | 11 | 4 | NA | NA | NA | NA |
| Mason | 45 | 214 | 71 | 12 | 126 | 784 | 121 | 49 | NA | NA | NA | NA |
| Massillon | 22 | 177 | 45 | 9 | 51 | 242 | 19 | 8 | NA | NA | NA | NA |
| Medina | 41 | 155 | 36 | 6 | 101 | 639 | 73 | 31 | NA | NA | NA | NA |
| Mentor | 52 | 201 | 59 | 7 | 147 | 1,330 | 152 | 66 | NA | NA | NA | NA |
| Middletown | 43 | 153 | 34 | 5 | 70 | 413 | 47 | 20 | NA | NA | NA | NA |
| Newark | 50 | 203 | 34 | 6 | D | D | D | D | NA | NA | NA | NA |
| North Olmsted | 50 | 185 | 50 | 8 | 66 | 284 | 34 | 15 | NA | NA | NA | NA |
| North Ridgeville | 12 | 57 | 12 | 2 | 40 | 212 | 30 | 12 | NA | NA | NA | NA |
| North Royalton | 27 | 177 | 36 | 9 | 79 | 288 | 48 | 15 | NA | NA | NA | NA |
| Parma | 55 | 316 | 82 | 13 | 102 | 632 | 84 | 26 | NA | NA | NA | NA |
| Reynoldsburg | 36 | D | 58 | D | 79 | 498 | 50 | 22 | NA | NA | NA | NA |
| Riverside | 17 | 97 | 28 | 3 | 29 | 492 | 107 | 39 | NA | NA | NA | NA |
| Sandusky | D | D | D | 3 | 51 | 270 | 34 | 14 | NA | NA | NA | NA |
| Shaker Heights | 31 | 75 | 12 | 3 | 102 | 461 | 81 | 37 | NA | NA | NA | NA |
| Springfield | 55 | 222 | 42 | 7 | 85 | 508 | 61 | 22 | NA | NA | NA | NA |
| Stow | D | D | D | D | 79 | 801 | 95 | 45 | NA | NA | NA | NA |
| Strongsville | 54 | 295 | 105 | 12 | 130 | 718 | 114 | 41 | NA | NA | NA | NA |
| Toledo | 247 | 2,232 | 4,130 | 164 | 412 | 4,285 | 762 | 266 | NA | NA | NA | NA |
| Troy | 19 | 104 | 23 | 4 | 51 | 350 | 53 | 22 | NA | NA | NA | NA |
| Upper Arlington | 62 | 385 | 117 | 16 | 153 | 647 | 124 | 45 | NA | NA | NA | NA |
| Warren | 26 | 104 | 16 | 3 | 76 | 322 | 33 | 13 | NA | NA | NA | NA |
| Westerville | 69 | 673 | 115 | 36 | 224 | 2,691 | 405 | 174 | NA | NA | NA | NA |
| Westlake | 80 | 289 | 88 | 16 | 244 | 2,136 | 536 | 180 | NA | NA | NA | NA |
| Wooster | 28 | 114 | 20 | 4 | 68 | 555 | 68 | 28 | NA | NA | NA | NA |
| Xenia | 18 | 54 | 14 | 2 | 21 | 96 | 19 | 6 | NA | NA | NA | NA |
| Youngstown | D | D | D | D | 85 | 724 | 100 | 42 | NA | NA | NA | NA |
| Zanesville | 34 | 180 | 24 | 5 | 64 | 374 | 55 | 18 | NA | NA | NA | NA |
| **OKLAHOMA** | 4,461 | 22,135 | 4,696 | 942 | 9,736 | 70,920 | 11,247 | 4,456 | 3,376 | 123,138 | 61,143 | 6,740 |
| Bartlesville | D | D | D | D | 82 | 1,557 | 138 | 137 | NA | NA | NA | NA |
| Broken Arrow | 97 | 324 | 69 | 11 | 257 | 1,640 | 220 | 93 | NA | NA | NA | NA |
| Edmond | 233 | 561 | 205 | 26 | 510 | 2,554 | 330 | 120 | NA | NA | NA | NA |
| Enid | 83 | 337 | 62 | 11 | 124 | 560 | 71 | 28 | NA | NA | NA | NA |
| Lawton | 122 | 431 | 72 | 12 | 124 | 1,288 | 154 | 58 | NA | NA | NA | NA |
| Midwest City | 65 | 257 | 58 | 8 | 88 | 615 | 67 | 32 | NA | NA | NA | NA |
| Moore | 56 | 248 | 68 | 12 | 76 | 374 | 38 | 13 | NA | NA | NA | NA |
| Muskogee | 43 | 164 | 26 | 5 | 64 | 499 | 41 | 17 | NA | NA | NA | NA |
| Norman | 224 | 1,006 | 179 | 38 | 464 | 2,302 | 311 | 116 | NA | NA | NA | NA |
| Oklahoma City | 1,058 | 6,008 | 1,618 | 324 | 2,577 | 25,695 | 4,406 | 1,799 | NA | NA | NA | NA |
| Owasso | 55 | 190 | 51 | 7 | 74 | 200 | 31 | 9 | NA | NA | NA | NA |
| Ponca City | 27 | 86 | 18 | 2 | 53 | 359 | 42 | 16 | NA | NA | NA | NA |
| Shawnee | 38 | 167 | 22 | 5 | 72 | 577 | 76 | 26 | NA | NA | NA | NA |
| Stillwater | 68 | 344 | 59 | 10 | 108 | 973 | 131 | 47 | NA | NA | NA | NA |
| Tulsa | 773 | 6,322 | 1,105 | 262 | 1,854 | 17,646 | 3,178 | 1,248 | NA | NA | NA | NA |
| **OREGON** | 6,771 | 29,773 | 6,774 | 1,263 | 12,620 | 92,358 | 14,720 | 6,902 | 5,557 | 172,210 | 62,411 | 10,668 |
| Albany | 67 | 226 | 38 | 8 | 114 | 701 | 91 | 31 | NA | NA | NA | NA |
| Beaverton | 281 | 1,150 | 302 | 46 | D | D | D | D | NA | NA | NA | NA |
| Bend | 323 | 808 | 212 | 34 | 638 | 2,520 | 384 | 152 | NA | NA | NA | NA |
| Corvallis | D | D | D | D | 195 | 1,555 | 361 | 117 | NA | NA | NA | NA |
| Eugene | 355 | 1,469 | 337 | 55 | 720 | 4,231 | 543 | 220 | NA | NA | NA | NA |
| Grants Pass | 79 | 365 | 54 | 15 | 93 | 739 | 67 | 24 | NA | NA | NA | NA |
| Gresham | 118 | 442 | 88 | 14 | 123 | 458 | 60 | 22 | NA | NA | NA | NA |
| Hillsboro | 152 | 603 | 171 | 30 | 255 | 9,104 | 772 | 1,274 | NA | NA | NA | NA |
| Keizer | 32 | 151 | 31 | 4 | 54 | 259 | 34 | 10 | NA | NA | NA | NA |
| Lake Oswego | 167 | 909 | 303 | 56 | D | D | D | D | NA | NA | NA | NA |
| McMinnville | 35 | 107 | 21 | 4 | 83 | 338 | 45 | 16 | NA | NA | NA | NA |
| Medford | 191 | 745 | 160 | 27 | 257 | 1,493 | 181 | 66 | NA | NA | NA | NA |
| Oregon City | D | D | D | D | 106 | 515 | 69 | 24 | NA | NA | NA | NA |
| Portland | 1,527 | 10,487 | 2,230 | 508 | 3,934 | 35,617 | 6,946 | 2,806 | NA | NA | NA | NA |

1. Establishments subject to federal tax.

# Accommodation and Food Services, Arts, Entertainment, and Recreation, and Health Care and Social Assistance

| City | Accommodation and food services, 2017 | | | | Arts, entertainment, and recreation,[1] 2017 | | | | Health care and social assistance,[1] 2017 | | | |
|---|---|---|---|---|---|---|---|---|---|---|---|---|
| | Number of establishments | Number of employees | Receipts (mil dol) | Annual payroll (mil dol) | Number of establishments | Number of employees | Receipts (mil dol) | Annual payroll (mil dol) | Number of establishments | Number of employees | Receipts (mil dol) | Annual payroll (mil dol) |
| | 92 | 93 | 94 | 95 | 96 | 97 | 98 | 99 | 100 | 101 | 102 | 103 |
| **OHIO—Cont'd** | | | | | | | | | | | | |
| Hamilton | 141 | 2,958 | 151 | 42 | 16 | 167 | 13 | 3 | 173 | 3,561 | 391 | 157 |
| Hilliard | 78 | 1,651 | 83 | 25 | 20 | 375 | 25 | 8 | 138 | 1,738 | 190 | 73 |
| Huber Heights | 82 | 1,754 | 86 | 27 | 9 | 215 | 4 | 2 | NA | NA | NA | NA |
| Kent | 91 | 1,921 | 82 | 26 | D | D | D | D | 57 | 1,044 | 86 | 31 |
| Kettering | D | D | D | D | 20 | 628 | 22 | 9 | NA | NA | NA | NA |
| Lakewood | 138 | 2,031 | 110 | 33 | 19 | 340 | 14 | 5 | 110 | 1,415 | 112 | 52 |
| Lancaster | 105 | 2,235 | 106 | 32 | 18 | 266 | 12 | 3 | 182 | 5,043 | 604 | 227 |
| Lima | 73 | 1,304 | 69 | 18 | 14 | 205 | 9 | 2 | 162 | 8,181 | 1,133 | 438 |
| Lorain | D | D | D | D | D | D | D | D | 145 | 4,186 | 464 | 209 |
| Mansfield | 104 | 1,960 | 93 | 28 | 12 | 256 | 10 | 3 | 248 | 6,291 | 634 | 251 |
| Marion | 48 | 926 | 46 | 12 | 9 | 51 | 4 | 1 | 141 | 3,495 | 404 | 165 |
| Mason | 105 | 2,460 | 145 | 38 | 20 | 303 | 40 | 9 | 131 | 2,596 | 205 | 88 |
| Massillon | 80 | 1,352 | 66 | 18 | 9 | 82 | 3 | 1 | 87 | 2,931 | 288 | 113 |
| Medina | 61 | 1,182 | 53 | 16 | 7 | 29 | 3 | 1 | 87 | 2,210 | 260 | 105 |
| Mentor | 163 | 3,824 | 186 | 53 | 15 | 132 | 10 | 3 | 178 | 3,056 | 272 | 119 |
| Middletown | 94 | 2,103 | 107 | 31 | 15 | 131 | 6 | 2 | NA | NA | NA | NA |
| Newark | 97 | 1,474 | 82 | 23 | 19 | 220 | 17 | 5 | 159 | 5,008 | 577 | 231 |
| North Olmsted | 120 | 2,547 | 129 | 38 | 9 | 66 | 5 | 1 | 85 | 1,468 | 133 | 46 |
| North Ridgeville | 42 | 765 | 45 | 12 | 9 | 74 | 5 | 1 | 39 | 555 | 42 | 17 |
| North Royalton | D | D | D | D | 10 | 163 | 6 | 2 | 68 | 1,174 | 90 | 39 |
| Parma | 169 | 2,827 | 140 | 40 | 17 | 216 | 9 | 2 | 202 | 5,401 | 417 | 200 |
| Reynoldsburg | 77 | 1,854 | 91 | 27 | 4 | 22 | 1 | 0 | NA | NA | NA | NA |
| Riverside | 56 | 966 | 44 | 12 | 5 | 79 | 5 | 2 | 36 | 749 | 50 | 23 |
| Sandusky | 83 | 1,106 | 99 | 19 | D | D | D | D | 93 | 3,513 | 368 | 153 |
| Shaker Heights | 32 | 400 | 23 | 6 | 8 | 194 | 14 | 6 | 70 | 2,223 | 146 | 61 |
| Springfield | 162 | 3,487 | 162 | 47 | 16 | 286 | 9 | 3 | 229 | 5,888 | 641 | 241 |
| Stow | 97 | 1,951 | 86 | 25 | 18 | 156 | 12 | 2 | 84 | 1,603 | 123 | 49 |
| Strongsville | 124 | 2,812 | 146 | 43 | 22 | 306 | 28 | 7 | 113 | 1,939 | 213 | 80 |
| Toledo | 649 | 11,380 | 583 | 168 | 90 | 4,552 | 389 | 85 | 739 | 24,505 | 3,066 | 1,189 |
| Troy | 75 | 1,897 | 89 | 25 | 8 | 102 | 4 | 2 | 97 | 1,542 | 111 | 48 |
| Upper Arlington | 69 | 1,524 | 77 | 24 | 13 | 269 | 20 | 7 | 97 | 2,137 | 174 | 79 |
| Warren | 69 | 1,061 | 47 | 13 | 14 | 134 | 4 | 2 | 164 | 5,512 | 717 | 243 |
| Westerville | 117 | 2,767 | 134 | 37 | 19 | 163 | 10 | 3 | NA | NA | NA | NA |
| Westlake | 105 | 2,656 | 161 | 48 | 23 | 445 | 32 | 10 | 236 | 4,520 | 458 | 189 |
| Wooster | 92 | 2,023 | 100 | 29 | 11 | 187 | 9 | 3 | 139 | 3,869 | 372 | 162 |
| Xenia | 45 | 1,053 | 48 | 14 | 7 | 106 | 9 | 2 | 87 | 1,586 | 147 | 60 |
| Youngstown | D | D | D | D | D | D | D | D | NA | NA | NA | NA |
| Zanesville | 113 | 2,450 | 120 | 36 | 13 | 177 | 7 | 3 | 131 | 5,352 | 763 | 262 |
| **OKLAHOMA** | 8,397 | 159,826 | 9,251 | 2,470 | 1,152 | 26,523 | 3,470 | 799 | 11,035 | 226,462 | 27,031 | 9,953 |
| Bartlesville | 98 | 1,749 | 85 | 25 | D | D | D | D | 153 | 2,874 | 283 | 117 |
| Broken Arrow | 221 | 4,732 | 235 | 69 | 36 | 554 | 30 | 11 | 244 | 4,451 | 353 | 143 |
| Edmond | 262 | 5,149 | 263 | 75 | 54 | 1,095 | 55 | 18 | 510 | 5,338 | 637 | 224 |
| Enid | 124 | 2,396 | 127 | 32 | 21 | 244 | 12 | 4 | D | D | D | D |
| Lawton | 217 | 4,576 | 266 | 69 | D | D | D | D | 232 | 5,609 | 629 | 261 |
| Midwest City | 133 | 2,792 | 137 | 40 | 11 | 228 | 6 | 3 | 196 | 2,732 | 398 | 128 |
| Moore | 139 | 3,063 | 150 | 41 | 14 | 112 | 7 | 2 | 104 | 924 | 81 | 30 |
| Muskogee | 112 | 2,125 | 96 | 27 | D | D | D | D | 191 | 5,417 | 726 | 295 |
| Norman | 363 | 7,908 | 419 | 120 | D | D | D | D | 486 | 8,811 | 925 | 404 |
| Oklahoma City | 1,597 | 33,987 | 1,959 | 542 | 195 | 5,273 | 932 | 256 | 2,436 | 54,538 | 8,613 | 2,865 |
| Owasso | 101 | 2,601 | 125 | 36 | 9 | 214 | 5 | 1 | 105 | 1,984 | 212 | 75 |
| Ponca City | 63 | 1,113 | 58 | 15 | D | D | D | D | 99 | 1,785 | 154 | 59 |
| Shawnee | 117 | 2,546 | 130 | 36 | 12 | 196 | 11 | 4 | 139 | 2,639 | 229 | 101 |
| Stillwater | 188 | 3,600 | 167 | 45 | 15 | 151 | 5 | 2 | 126 | 3,016 | 272 | 127 |
| Tulsa | 1,205 | 25,731 | 1,503 | 420 | 172 | 3,351 | 178 | 66 | 1,588 | 45,593 | 6,447 | 2,298 |
| **OREGON** | 11,708 | 182,613 | 11,804 | 3,505 | 1,992 | 29,222 | 2,256 | 776 | 13,948 | 263,278 | 33,084 | 13,106 |
| Albany | 133 | 2,191 | 118 | 35 | 16 | 135 | 9 | 2 | 138 | 3,041 | 344 | 138 |
| Beaverton | 341 | 5,608 | 356 | 108 | 76 | 1,544 | 77 | 25 | 473 | 6,216 | 692 | 278 |
| Bend | 382 | 6,525 | 435 | 139 | 87 | 1,770 | 86 | 25 | 533 | 9,929 | 1,388 | 552 |
| Corvallis | 193 | 3,535 | 195 | 56 | 23 | 329 | 10 | 4 | 251 | 5,425 | 686 | 300 |
| Eugene | 576 | 9,516 | 569 | 166 | 96 | 1,793 | 93 | 24 | 829 | 12,879 | 1,218 | 502 |
| Grants Pass | 162 | 2,499 | 140 | 41 | 19 | 348 | 12 | 5 | 253 | 4,939 | 506 | 213 |
| Gresham | 237 | 4,217 | 245 | 73 | 37 | 510 | 30 | 9 | 413 | 5,934 | 634 | 238 |
| Hillsboro | 293 | 5,784 | 403 | 116 | 48 | 952 | 48 | 15 | 350 | 7,724 | 1,088 | 407 |
| Keizer | 74 | 1,187 | 64 | 18 | 16 | 178 | 13 | 4 | 111 | 1,536 | 122 | 53 |
| Lake Oswego | 114 | 2,189 | 140 | 43 | 42 | 429 | 24 | 10 | 240 | 2,419 | 252 | 98 |
| McMinnville | 94 | 1,534 | 88 | 27 | D | D | D | D | 155 | 2,634 | 272 | 109 |
| Medford | 278 | 4,872 | 282 | 85 | 41 | 539 | 38 | 12 | 435 | 12,307 | 1,650 | 640 |
| Oregon City | D | D | D | D | 16 | 145 | 8 | 3 | 157 | 2,617 | 328 | 125 |
| Portland | 2,882 | 47,324 | 3,282 | 999 | 494 | 7,257 | 935 | 352 | 2,906 | 67,025 | 9,250 | 3,545 |

1. Establishments subject to federal tax.

# Table D. Cities — Other Services and Government Employment and Payroll

| City | Other services[1] Number of establish-ments | Number of employees | Receipts (mil dol) | Annual payroll (mil dol) | Government employment and payroll, 2017 Full-time equivalent employees | Total (dollars) | Percent of total for: Admin-istrative, judicial, and legal | Police and corrections | Fire protection | Highways and trans-portation | Health and welfare | Natural resources and utilities | Education and libraries |
|---|---|---|---|---|---|---|---|---|---|---|---|---|---|
| | 104 | 105 | 106 | 107 | 108 | 109 | 110 | 111 | 112 | 113 | 114 | 115 | 116 |
| **OHIO—Cont'd** | | | | | | | | | | | | | |
| Hamilton | 85 | 637 | 65 | 19 | 641 | 3,621,843 | 14.5 | 21.0 | 17.4 | 6.5 | 2.4 | 29.5 | 0.0 |
| Hilliard | 58 | 381 | 45 | 15 | 140 | 748,456 | 19.1 | 53.6 | 0.0 | 3.8 | 0.0 | 18.4 | 0.0 |
| Huber Heights | 61 | 313 | 30 | 8 | 179 | 1,076,657 | 14.1 | 36.4 | 32.1 | 11.3 | 0.0 | 0.1 | 0.0 |
| Kent | 46 | 311 | 29 | 9 | 211 | 1,399,013 | 8.2 | 36.2 | 19.1 | 14.6 | 6.8 | 13.7 | 0.0 |
| Kettering | 87 | 688 | 47 | 16 | 523 | 2,665,641 | 21.1 | 26.8 | 15.4 | 13.8 | 0.8 | 15.2 | 0.0 |
| Lakewood | 71 | 443 | 39 | 12 | 446 | 2,594,885 | 14.0 | 30.6 | 25.2 | 3.6 | 6.1 | 18.6 | 0.0 |
| Lancaster | 78 | 537 | 59 | 16 | 414 | 2,070,670 | 11.8 | 26.0 | 23.6 | 5.5 | 0.8 | 32.0 | 0.0 |
| Lima | 64 | 435 | 36 | 10 | 436 | 1,962,792 | 20.4 | 24.3 | 19.8 | 9.9 | 3.7 | 21.8 | 0.0 |
| Lorain | 74 | 494 | 53 | 11 | 480 | 2,336,209 | 16.9 | 29.7 | 18.4 | 4.8 | 1.3 | 24.5 | 0.0 |
| Mansfield | 107 | 629 | 65 | 16 | 443 | 1,978,647 | 18.0 | 27.5 | 22.9 | 3.0 | 2.3 | 18.7 | 0.0 |
| Marion | 46 | 374 | 24 | 7 | 274 | 1,190,214 | 13.5 | 29.1 | 23.9 | 12.5 | 2.1 | 17.0 | 0.0 |
| Mason | 70 | 642 | 55 | 18 | 272 | 1,438,726 | 19.1 | 23.2 | 24.1 | 7.9 | 0.0 | 20.0 | 0.0 |
| Massillon | 64 | 484 | 50 | 15 | 245 | 1,236,148 | 23.3 | 19.2 | 23.2 | 6.6 | 4.2 | 22.1 | 0.0 |
| Medina | 64 | 406 | 40 | 13 | 222 | 1,012,513 | 22.9 | 28.6 | 5.2 | 5.7 | 3.9 | 25.0 | 0.0 |
| Mentor | 125 | 830 | 65 | 22 | 423 | 2,061,016 | 15.1 | 31.4 | 26.9 | 13.8 | 1.8 | 11.1 | 0.0 |
| Middletown | 81 | 551 | 63 | 16 | 370 | 1,804,982 | 15.7 | 31.1 | 22.6 | 8.3 | 3.4 | 15.5 | 0.0 |
| Newark | 77 | 493 | 57 | 14 | 390 | 1,895,696 | 15.2 | 30.1 | 24.2 | 8.0 | 3.0 | 18.9 | 0.0 |
| North Olmsted | 94 | 682 | 72 | 20 | 263 | 1,477,873 | 11.0 | 37.0 | 18.8 | 8.1 | 4.1 | 17.8 | 0.0 |
| North Ridgeville | 44 | 240 | 23 | 6 | 208 | 1,174,498 | 12.6 | 26.7 | 20.2 | 11.5 | 1.9 | 22.1 | 0.0 |
| North Royalton | 65 | 289 | 28 | 9 | 164 | 1,004,069 | 8.3 | 32.8 | 26.6 | 14.2 | 5.6 | 12.6 | 0.0 |
| Parma | 125 | 822 | 88 | 27 | 594 | 2,839,358 | 17.2 | 31.9 | 23.1 | 15.6 | 1.8 | 3.7 | 0.0 |
| Reynoldsburg | 45 | 238 | 28 | 10 | 144 | 800,384 | 20.4 | 60.7 | 0.0 | 3.6 | 1.3 | 9.5 | 0.0 |
| Riverside | 24 | 110 | 18 | 5 | 85 | 417,592 | 15.6 | 43.5 | 27.4 | 13.5 | 0.0 | 0.0 | 0.0 |
| Sandusky | 43 | 175 | 21 | 4 | 238 | 1,128,593 | 18.9 | 23.6 | 21.7 | 4.9 | 5.4 | 24.5 | 0.0 |
| Shaker Heights | 39 | 122 | 24 | 4 | 393 | 2,183,740 | 15.6 | 30.6 | 18.3 | 3.1 | 3.1 | 14.1 | 0.0 |
| Springfield | 113 | 860 | 136 | 32 | 566 | 2,572,119 | 18.6 | 27.5 | 25.5 | 5.9 | 2.6 | 16.4 | 0.0 |
| Stow | 69 | 385 | 31 | 10 | 256 | 1,403,473 | 24.1 | 28.5 | 25.3 | 9.0 | 0.5 | 11.1 | 0.0 |
| Strongsville | 96 | 870 | 71 | 24 | 374 | 2,259,326 | 6.6 | 41.4 | 20.4 | 16.2 | 4.6 | 9.5 | 0.0 |
| Toledo | 387 | 2,543 | 318 | 73 | 2,644 | 14,297,353 | 12.4 | 29.7 | 29.2 | 8.2 | 1.1 | 15.7 | 0.0 |
| Troy | 60 | 307 | 37 | 8 | 190 | 1,073,928 | 12.4 | 26.5 | 23.2 | 10.8 | 2.0 | 24.0 | 0.0 |
| Upper Arlington | 40 | 293 | 40 | 11 | 223 | 1,554,960 | 16.3 | 28.4 | 31.5 | 3.7 | 2.6 | 15.8 | 0.0 |
| Warren | 69 | 323 | 32 | 9 | 349 | 1,255,656 | 20.8 | 16.0 | 18.3 | 9.6 | 3.0 | 32.2 | 0.0 |
| Westerville | 84 | 1,100 | 363 | 71 | 493 | 2,982,891 | 16.1 | 24.1 | 22.7 | 4.1 | 0.2 | 27.6 | 0.0 |
| Westlake | 80 | 714 | 54 | 19 | 291 | 1,529,306 | 10.5 | 35.2 | 23.7 | 2.4 | 2.7 | 17.9 | 0.0 |
| Wooster | 52 | 335 | 41 | 9 | 1,020 | 4,971,173 | 2.5 | 5.3 | 5.6 | 1.7 | 82.8 | 2.1 | 0.0 |
| Xenia | 39 | 270 | 27 | 7 | 226 | 1,273,227 | 21.5 | 25.0 | 35.1 | 3.1 | 0.4 | 14.9 | 0.0 |
| Youngstown | 99 | 1,165 | 98 | 28 | 803 | 3,326,951 | 12.0 | 27.5 | 20.0 | 5.8 | 3.5 | 27.9 | 0.0 |
| Zanesville | 87 | 712 | 63 | 18 | 302 | 1,219,159 | 11.8 | 32.5 | 19.8 | 6.0 | 0.0 | 22.6 | 0.0 |
| **OKLAHOMA** | 5,565 | 31,947 | 4,934 | 1,084 | X | X | X | X | X | X | X | X | X |
| Bartlesville | 68 | 428 | 41 | 12 | 340 | 1,394,623 | 12.3 | 24.7 | 24.0 | 3.6 | 4.2 | 21.9 | 5.3 |
| Broken Arrow | 167 | 1,188 | 147 | 43 | 727 | 3,786,247 | 11.8 | 30.9 | 30.0 | 6.7 | 0.0 | 16.9 | 0.0 |
| Edmond | 206 | 996 | 99 | 28 | 749 | 4,157,091 | 15.5 | 24.6 | 24.9 | 6.6 | 1.5 | 21.4 | 0.0 |
| Enid | 101 | 530 | 60 | 15 | 463 | 1,922,891 | 13.9 | 26.7 | 24.5 | 7.7 | 0.3 | 14.5 | 2.6 |
| Lawton | D | D | D | D | 872 | 3,517,667 | 14.1 | 30.3 | 18.8 | 7.2 | 2.2 | 20.5 | 1.9 |
| Midwest City | 58 | 338 | 29 | 9 | 468 | 2,438,934 | 15.0 | 26.6 | 24.2 | 3.5 | 3.3 | 19.5 | 0.0 |
| Moore | 74 | 331 | 33 | 9 | 328 | 2,094,680 | 12.1 | 38.6 | 30.4 | 2.3 | 2.5 | 8.6 | 0.0 |
| Muskogee | D | D | D | 13 | 452 | 1,676,180 | 14.9 | 28.8 | 25.0 | 5.3 | 0.0 | 21.3 | 0.0 |
| Norman | 170 | 984 | 237 | 31 | 3,444 | 18,975,140 | 2.4 | 7.4 | 5.5 | 2.4 | 75.8 | 4.7 | 0.0 |
| Oklahoma City | 1,140 | 8,064 | 1,120 | 298 | 4,494 | 27,224,626 | 9.9 | 34.3 | 27.2 | 8.5 | 4.4 | 15.6 | 0.0 |
| Owasso | D | D | D | D | 267 | 1,294,440 | 13.7 | 28.3 | 27.8 | 6.2 | 0.8 | 16.5 | 0.0 |
| Ponca City | 41 | 228 | 22 | 6 | 394 | 1,685,157 | 8.8 | 21.1 | 19.8 | 7.3 | 0.0 | 34.5 | 2.6 |
| Shawnee | 53 | 275 | 29 | 8 | 310 | 1,418,292 | 11.2 | 26.5 | 23.3 | 8.7 | 0.4 | 23.6 | 0.0 |
| Stillwater | 81 | 753 | 237 | 27 | 1,610 | 8,774,509 | 3.4 | 7.3 | 5.3 | 2.0 | 70.8 | 8.7 | 0.8 |
| Tulsa | 900 | 6,379 | 1,439 | 241 | 3,773 | 19,022,256 | 13.6 | 28.1 | 21.7 | 14.1 | 0.3 | 15.1 | 0.0 |
| **OREGON** | 7,414 | 43,543 | 6,289 | 1,555 | X | X | X | X | X | X | X | X | X |
| Albany | 82 | 612 | 40 | 15 | 384 | 2,174,784 | 11.3 | 25.4 | 25.1 | 12.6 | 0.0 | 19.7 | 3.7 |
| Beaverton | 203 | 1,350 | 154 | 46 | 588 | 2,785,440 | 34.7 | 25.0 | 0.0 | 4.9 | 0.0 | 8.8 | 11.5 |
| Bend | 260 | 1,326 | 164 | 45 | 583 | 4,017,372 | 18.8 | 22.1 | 24.0 | 9.8 | 6.4 | 12.1 | 0.0 |
| Corvallis | 118 | 899 | 215 | 37 | 439 | 2,570,510 | 11.3 | 24.1 | 19.5 | 7.7 | 6.8 | 19.3 | 8.2 |
| Eugene | 381 | 2,529 | 428 | 85 | 1,880 | 12,757,120 | 7.9 | 18.3 | 14.0 | 8.0 | 2.6 | 24.7 | 3.3 |
| Grants Pass | 80 | 377 | 34 | 10 | 219 | 1,202,000 | 17.5 | 44.2 | 16.7 | 2.1 | 0.0 | 13.7 | 0.0 |
| Gresham | 145 | 753 | 70 | 23 | 495 | 3,476,105 | 12.8 | 35.9 | 23.6 | 5.2 | 6.4 | 12.3 | 6.6 |
| Hillsboro | 165 | 2,021 | 546 | 132 | 874 | 5,270,127 | 17.5 | 22.8 | 17.1 | 4.6 | 0.0 | 25.0 | 0.0 |
| Keizer | 43 | 176 | 18 | 5 | 91 | 565,772 | 21.9 | 52.4 | 0.0 | 4.6 | 4.2 | 15.4 | 6.2 |
| Lake Oswego | 102 | 526 | 60 | 19 | 344 | 2,197,219 | 16.0 | 24.2 | 20.9 | 3.3 | 0.3 | 16.8 | 4.0 |
| McMinnville | 54 | 289 | 22 | 7 | 259 | 1,535,550 | 14.1 | 21.9 | 18.5 | 5.3 | 1.7 | 32.2 | 0.0 |
| Medford | 163 | 1,169 | 132 | 36 | 485 | 2,896,083 | 13.3 | 33.7 | 22.0 | 15.4 | 0.0 | 10.2 | 5.3 |
| Oregon City | 75 | 404 | 40 | 12 | 187 | 1,094,799 | 19.9 | 35.4 | 0.0 | 13.3 | 0.4 | 17.6 | 0.0 |
| Portland | 1,926 | 13,837 | 2,463 | 557 | 6,225 | 42,536,596 | 20.1 | 23.8 | 14.9 | 10.8 | 1.8 | 26.6 | 0.0 |

1. Establishments subject to federal tax.

# Table D. Cities — **City Government Finances**

| | City government finances, 2017 | | | | | | | | | |
| | General revenue | | | | | | | General expenditure | | |
| City | | Intergovernmental | | Taxes | | | | | Per capita[1] (dollars) | |
| | | | | | | Per capita[1] (dollars) | | | | |
| | Total (mil dol) | Total (mil dol) | Percent from state government | Total (mil dol) | Total | Property | Sales and gross receipts | Total (mil dol) | Total | Capital outlays |
| | 117 | 118 | 119 | 120 | 121 | 122 | 123 | 124 | 125 | 126 |
| **OHIO—Cont'd** | | | | | | | | | | |
| Hamilton | 88.8 | 9.6 | 78.7 | 39.4 | 634 | 169 | 20 | 98 | 1,570 | 288 |
| Hilliard | 46.0 | 3.2 | 56.0 | 30.0 | 833 | 35 | 43 | 25 | 701 | 70 |
| Huber Heights | 36.0 | 5.7 | 40.7 | 21.6 | 567 | 71 | 36 | 38 | 985 | 246 |
| Kent | 37.7 | 7.3 | 43.1 | 19.9 | 663 | 107 | 26 | 35 | 1,155 | 350 |
| Kettering | 86.6 | 7.2 | 89.2 | 60.5 | 1,097 | 169 | 12 | 83 | 1,506 | 205 |
| Lakewood | 68.6 | 8.5 | 70.6 | 40.0 | 796 | 274 | 46 | 64 | 1,282 | 229 |
| Lancaster | 76.4 | 9.7 | 92.0 | 28.7 | 712 | 145 | 1 | 64 | 1,585 | 233 |
| Lima | 71.1 | 29.4 | 68.5 | 19.4 | 523 | 30 | 29 | 77 | 2,075 | 739 |
| Lorain | 47.1 | 13.1 | 24.5 | 27.8 | 436 | 49 | 14 | 51 | 806 | 115 |
| Mansfield | 64.2 | 9.8 | 50.6 | 38.1 | 826 | 46 | 42 | 60 | 1,299 | 61 |
| Marion | 39.4 | 5.3 | 100.0 | 19.4 | 537 | 36 | 13 | 36 | 997 | 92 |
| Mason | 79.8 | 5.1 | 100.0 | 45.2 | 1,361 | 349 | 79 | 80 | 2,398 | 742 |
| Massillon | 44.8 | 4.8 | 100.0 | 21.2 | 655 | 63 | 17 | 41 | 1,275 | 53 |
| Medina | 63.3 | 2.9 | 43.0 | 53.3 | 2,036 | 1,291 | 24 | 43 | 1,659 | 281 |
| Mentor | 74.0 | 6.7 | 98.6 | 54.5 | 1,156 | 106 | 57 | 75 | 1,580 | 0 |
| Middletown | 63.3 | 13.1 | 43.5 | 28.7 | 587 | 113 | 17 | 58 | 1,184 | 289 |
| Newark | 62.7 | 16.0 | 97.4 | 28.9 | 583 | 84 | 2 | 66 | 1,332 | 303 |
| North Olmsted | 47.6 | 14.1 | 9.0 | 24.7 | 780 | 289 | 13 | 44 | 1,401 | 123 |
| North Ridgeville | 35.6 | 3.1 | 100.0 | 18.2 | 546 | 154 | 63 | 42 | 1,258 | 318 |
| North Royalton | 32.3 | 2.0 | 97.8 | 21.0 | 694 | 171 | 8 | 31 | 1,033 | 138 |
| Parma | 85.9 | 12.9 | 100.0 | 59.4 | 751 | 133 | 42 | 86 | 1,081 | 20 |
| Reynoldsburg | 35.7 | 4.4 | 26.9 | 16.1 | 423 | 12 | 10 | 25 | 647 | 28 |
| Riverside | 7.2 | 1.9 | 95.8 | 4.3 | 171 | 75 | 20 | 10 | 392 | 19 |
| Sandusky | 40.6 | 6.0 | 72.8 | 22.3 | 897 | 96 | 408 | 29 | 1,156 | 26 |
| Shaker Heights | 66.1 | 8.6 | 30.5 | 42.2 | 1,542 | 276 | 44 | 71 | 2,584 | 315 |
| Springfield | 66.0 | 14.9 | 78.9 | 35.6 | 602 | 33 | 34 | 87 | 1,467 | 267 |
| Stow | 42.8 | 7.7 | 98.7 | 26.1 | 749 | 204 | 51 | 38 | 1,104 | 197 |
| Strongsville | 75.1 | 5.7 | 42.2 | 46.0 | 1,029 | 208 | 37 | 68 | 1,527 | 260 |
| Toledo | 499.8 | 101.1 | 92.9 | 214.9 | 777 | 42 | 11 | 480 | 1,735 | 249 |
| Troy | 43.5 | 1.6 | 63.6 | 23.5 | 910 | 86 | 30 | 36 | 1,399 | 177 |
| Upper Arlington | 56.5 | 8.9 | 100.0 | 35.9 | 1,015 | 272 | 21 | 47 | 1,338 | 391 |
| Warren | 56.2 | 7.5 | 59.2 | 18.9 | 479 | 32 | 21 | 52 | 1,324 | 124 |
| Westerville | 99.2 | 11.1 | 54.2 | 67.5 | 1,694 | 423 | 57 | 135 | 3,392 | 1,197 |
| Westlake | 77.7 | 8.4 | 88.4 | 44.1 | 1,367 | 407 | 92 | 75 | 2,329 | 967 |
| Wooster | 170.2 | 5.0 | 93.2 | 22.6 | 850 | 86 | 48 | 181 | 6,808 | 293 |
| Xenia | 32.8 | 5.1 | 97.5 | 16.2 | 609 | 26 | 38 | 23 | 855 | 412 |
| Youngstown | 128.6 | 18.3 | 33.3 | 74.6 | 1,155 | 52 | 163 | 102 | 1,584 | 38 |
| Zanesville | 35.8 | 6.3 | 100.0 | 19.7 | 780 | 44 | 31 | 36 | 1,415 | 104 |
| **OKLAHOMA** | X | X | X | X | X | X | X | X | X | X |
| Bartlesville | 44.2 | 2.0 | 69.8 | 25.4 | 698 | 132 | 555 | 45 | 1,242 | 286 |
| Broken Arrow | 116.7 | 12.0 | 92.4 | 68.2 | 629 | 136 | 466 | 111 | 1,021 | 211 |
| Edmond | 124.9 | 8.7 | 89.1 | 72.3 | 789 | 0 | 788 | 107 | 1,165 | 169 |
| Enid | 71.4 | 2.5 | 51.3 | 39.3 | 784 | 11 | 756 | 65 | 1,300 | 182 |
| Lawton | 109.7 | 12.3 | 64.3 | 58.0 | 617 | 40 | 577 | 127 | 1,354 | 402 |
| Midwest City | 57.8 | 4.7 | 14.7 | 14.5 | 253 | 48 | 205 | 63 | 1,106 | 114 |
| Moore | 75.6 | 12.8 | 4.1 | 43.2 | 708 | 99 | 609 | 88 | 1,441 | 190 |
| Muskogee | 53.5 | 6.5 | 37.9 | 30.6 | 809 | 7 | 803 | 56 | 1,475 | 370 |
| Norman | 560.6 | 9.8 | 33.7 | 103.1 | 836 | 119 | 716 | 544 | 4,410 | 588 |
| Oklahoma City | 1,268.2 | 197.6 | 44.4 | 642.5 | 999 | 161 | 839 | 1,026 | 1,596 | 398 |
| Owasso | 48.9 | 2.3 | 40.1 | 33.4 | 919 | 3 | 917 | 43 | 1,180 | 384 |
| Ponca City | 33.4 | 0.8 | 62.2 | 16.8 | 695 | 29 | 664 | 46 | 1,887 | 52 |
| Shawnee | 37.0 | 4.2 | 76.2 | 22.4 | 716 | 2 | 704 | 39 | 1,246 | 219 |
| Stillwater | 55.4 | 2.2 | 63.3 | 32.8 | 656 | 21 | 636 | 43 | 866 | 132 |
| Tulsa | 821.1 | 74.2 | 10.1 | 388.8 | 967 | 184 | 784 | 721 | 1,793 | 554 |
| **OREGON** | X | X | X | X | X | X | X | X | X | X |
| Albany | 76.2 | 9.9 | 51.9 | 38.3 | 715 | 561 | 154 | 85 | 1,582 | 447 |
| Beaverton | 108.9 | 19.8 | 56.1 | 55.9 | 572 | 417 | 155 | 91 | 926 | 69 |
| Bend | 126.6 | 13.2 | 59.6 | 69.3 | 732 | 358 | 374 | 126 | 1,330 | 367 |
| Corvallis | 83.7 | 13.3 | 50.5 | 39.9 | 681 | 484 | 198 | 78 | 1,337 | 137 |
| Eugene | 313.9 | 42.6 | 46.5 | 143.6 | 848 | 683 | 164 | 285 | 1,681 | 228 |
| Grants Pass | 41.1 | 5.0 | 73.1 | 25.4 | 674 | 492 | 182 | 46 | 1,223 | 315 |
| Gresham | 132.1 | 36.2 | 30.6 | 59.4 | 534 | 281 | 253 | 130 | 1,167 | 140 |
| Hillsboro | 189.9 | 12.9 | 87.3 | 101.0 | 940 | 635 | 305 | 165 | 1,538 | 110 |
| Keizer | 20.8 | 3.6 | 89.6 | 8.5 | 216 | 128 | 88 | 19 | 494 | 28 |
| Lake Oswego | 98.4 | 19.2 | 18.5 | 47.5 | 1,210 | 1,011 | 200 | 88 | 2,239 | 446 |
| McMinnville | 41.8 | 5.1 | 66.4 | 19.0 | 556 | 442 | 114 | 49 | 1,441 | 365 |
| Medford | 131.9 | 15.9 | 49.9 | 72.4 | 892 | 517 | 375 | 134 | 1,651 | 339 |
| Oregon City | 57.0 | 7.9 | 100.0 | 23.0 | 631 | 417 | 214 | 39 | 1,074 | 75 |
| Portland | 1,734.3 | 204.2 | 41.5 | 888.0 | 1,371 | 823 | 548 | 1,449 | 2,236 | 419 |

1. Based on population estimated as of July 1 of the year shown.

# City Government Finances

| City | Public welfare 127 | Highways 128 | Parking facilities 129 | Education 130 | Health and hospitals 131 | Police protection 132 | Sewerage and sanitation 133 | Parks and recreation 134 | Housing and community development 135 | Interest on debt 136 |
|---|---|---|---|---|---|---|---|---|---|---|
| **City government finances, 2017 (cont.)** | | | | | | | | | | |
| **General expenditure (cont.)** | | | | | | | | | | |
| **Percent of total for:** | | | | | | | | | | |
| **OHIO—Cont'd** | | | | | | | | | | |
| Hamilton | 0.0 | 14.7 | 0.3 | 0.0 | 4.6 | 15.1 | 20.7 | 3.8 | 2.9 | 4.7 |
| Hilliard | 0.0 | 10.6 | 0.0 | 0.0 | 0.0 | 28.8 | 7.0 | 13.1 | 0.0 | 0.0 |
| Huber Heights | 0.0 | 14.3 | 0.0 | 0.0 | 0.0 | 17.8 | 8.7 | 2.5 | 4.7 | 8.5 |
| Kent | 0.6 | 0.0 | 0.1 | 0.0 | 2.0 | 19.0 | 13.5 | 6.4 | 4.1 | 0.0 |
| Kettering | 0.0 | 12.1 | 0.0 | 0.0 | 2.1 | 20.1 | 0.0 | 18.3 | 2.6 | 0.7 |
| Lakewood | 2.0 | 12.2 | 0.6 | 0.0 | 0.6 | 17.0 | 22.7 | 6.5 | 3.7 | 0.0 |
| Lancaster | 0.0 | 15.8 | 0.0 | 0.0 | 0.8 | 13.6 | 17.9 | 3.6 | 1.9 | 1.5 |
| Lima | 0.0 | 8.9 | 0.1 | 0.0 | 0.0 | 11.1 | 46.5 | 3.2 | 3.3 | 0.0 |
| Lorain | 0.0 | 15.1 | 0.0 | 0.0 | 1.7 | 26.8 | 0.0 | 0.9 | 5.8 | 8.9 |
| Mansfield | 0.8 | 16.2 | 0.0 | 0.4 | 0.0 | 16.4 | 7.9 | 0.7 | 2.8 | 0.4 |
| Marion | 0.0 | 9.0 | 0.0 | 0.0 | 1.3 | 18.7 | 29.4 | 3.6 | 0.0 | 3.3 |
| Mason | 0.0 | 19.4 | 0.0 | 0.0 | 0.0 | 12.7 | 8.3 | 14.4 | 4.9 | 5.0 |
| Massillon | 0.0 | 8.0 | 0.0 | 0.0 | 1.7 | 18.2 | 22.2 | 9.4 | 0.1 | 2.8 |
| Medina | 0.0 | 19.5 | 0.0 | 0.0 | 3.2 | 11.6 | 8.4 | 14.0 | 0.4 | 3.4 |
| Mentor | 0.0 | 26.0 | 0.0 | 0.0 | 0.0 | 19.1 | 0.0 | 13.2 | 0.0 | 1.6 |
| Middletown | 0.6 | 17.0 | 0.0 | 0.0 | 0.8 | 17.8 | 23.6 | 1.3 | 0.8 | 0.0 |
| Newark | 0.0 | 6.7 | 0.0 | 0.0 | 0.1 | 16.2 | 30.9 | 1.2 | 0.0 | 1.6 |
| North Olmsted | 0.0 | 16.4 | 0.0 | 0.0 | 0.0 | 14.8 | 11.1 | 9.6 | 0.0 | 7.6 |
| North Ridgeville | 0.0 | 10.7 | 0.0 | 0.0 | 3.7 | 14.1 | 37.9 | 1.4 | 0.0 | 1.9 |
| North Royalton | 0.8 | 14.9 | 0.0 | 0.0 | 0.0 | 17.9 | 28.9 | 2.1 | 0.0 | 3.2 |
| Parma | 0.0 | 6.9 | 0.0 | 0.0 | 0.4 | 40.1 | 0.5 | 5.2 | 9.0 | 2.0 |
| Reynoldsburg | 0.0 | 4.7 | 0.0 | 0.0 | 1.1 | 38.0 | 30.5 | 6.0 | 0.0 | 0.0 |
| Riverside | 0.0 | 19.1 | 0.0 | 0.0 | 0.0 | 31.8 | 0.0 | 0.5 | 2.0 | 1.5 |
| Sandusky | 0.0 | 7.0 | 0.0 | 0.0 | 0.7 | 17.9 | 25.0 | 2.1 | 1.5 | 2.0 |
| Shaker Heights | 0.0 | 9.8 | 0.0 | 0.0 | 0.7 | 17.7 | 7.9 | 6.2 | 4.5 | 1.1 |
| Springfield | 0.0 | 10.5 | 0.0 | 0.0 | 0.1 | 28.6 | 19.0 | 0.8 | 6.0 | 3.1 |
| Stow | 0.0 | 10.9 | 0.0 | 0.0 | 2.5 | 14.1 | 5.6 | 6.8 | 0.3 | 1.5 |
| Strongsville | 0.0 | 25.9 | 0.0 | 0.0 | 0.5 | 17.7 | 12.1 | 9.1 | 0.0 | 2.8 |
| Toledo | 0.0 | 9.4 | 0.0 | 0.0 | 0.8 | 18.0 | 20.5 | 2.3 | 3.7 | 4.4 |
| Troy | 0.0 | 6.9 | 0.1 | 0.0 | 0.9 | 12.3 | 17.3 | 10.7 | 3.0 | 1.0 |
| Upper Arlington | 0.0 | 26.4 | 0.0 | 0.0 | 0.6 | 16.8 | 6.1 | 11.1 | 0.0 | 0.0 |
| Warren | 0.0 | 10.9 | 0.2 | 0.0 | 1.9 | 12.2 | 20.5 | 1.3 | 4.3 | 0.0 |
| Westerville | 0.0 | 18.0 | 1.3 | 0.0 | 0.0 | 10.2 | 9.9 | 10.0 | 0.0 | 1.6 |
| Westlake | 0.0 | 20.1 | 20.4 | 0.0 | 0.0 | 10.7 | 6.0 | 9.3 | 0.0 | 0.0 |
| Wooster | 0.0 | 2.4 | 0.0 | 0.0 | 74.0 | 4.1 | 6.1 | 1.0 | 0.8 | 0.4 |
| Xenia | 0.0 | 7.7 | 0.2 | 0.0 | 0.0 | 4.9 | 31.4 | 0.9 | 1.2 | 0.2 |
| Youngstown | 0.0 | 6.5 | 0.1 | 0.0 | 1.5 | 22.5 | 27.0 | 2.7 | 6.9 | 0.4 |
| Zanesville | 0.0 | 9.2 | 0.0 | 0.0 | 1.3 | 25.6 | 19.9 | 1.7 | 0.0 | 0.1 |
| **OKLAHOMA** | X | X | X | X | X | X | X | X | X | X |
| Bartlesville | 0.0 | 5.9 | 0.0 | 0.0 | 0.0 | 15.6 | 22.9 | 11.2 | 0.8 | 3.3 |
| Broken Arrow | 0.0 | 14.9 | 0.0 | 0.0 | 0.0 | 21.4 | 9.8 | 7.2 | 1.2 | 5.1 |
| Edmond | 0.7 | 11.6 | 0.0 | 0.0 | 0.7 | 25.2 | 13.5 | 13.0 | 0.0 | 3.3 |
| Enid | 0.3 | 4.0 | 0.0 | 1.8 | 8.3 | 15.6 | 13.8 | 3.0 | 14.0 | 2.3 |
| Lawton | 0.0 | 15.5 | 0.0 | 2.0 | 0.0 | 16.7 | 19.2 | 5.0 | 0.6 | 0.0 |
| Midwest City | 0.0 | 10.7 | 0.0 | 0.0 | 0.7 | 25.6 | 9.9 | 5.3 | 7.0 | 0.0 |
| Moore | 0.2 | 10.3 | 0.0 | 0.0 | 0.0 | 13.6 | 9.2 | 7.5 | 8.7 | 1.7 |
| Muskogee | 0.0 | 9.2 | 0.0 | 0.0 | 0.7 | 16.7 | 7.9 | 11.4 | 0.0 | 0.1 |
| Norman | 0.0 | 5.8 | 0.0 | 0.0 | 65.8 | 5.4 | 8.2 | 3.5 | 1.0 | 2.9 |
| Oklahoma City | 0.0 | 7.5 | 0.4 | 0.0 | 0.5 | 17.9 | 7.3 | 22.5 | 4.9 | 3.5 |
| Owasso | 0.0 | 11.2 | 0.0 | 0.0 | 4.8 | 20.4 | 12.4 | 7.6 | 6.3 | 2.0 |
| Ponca City | 0.0 | 14.0 | 0.0 | 0.0 | 1.7 | 18.4 | 10.7 | 5.5 | 0.4 | 0.9 |
| Shawnee | 0.0 | 15.8 | 0.0 | 0.0 | 0.0 | 22.0 | 8.8 | 10.0 | 5.5 | 0.3 |
| Stillwater | 0.0 | 17.5 | 0.0 | 0.0 | 0.1 | 27.3 | 17.5 | 6.3 | 0.0 | 0.9 |
| Tulsa | 2.5 | 16.9 | 0.5 | 0.0 | 10.6 | 14.5 | 17.3 | 5.7 | 1.0 | 4.7 |
| **OREGON** | X | X | X | X | X | X | X | X | X | X |
| Albany | 0.0 | 4.8 | 0.0 | 0.0 | 0.2 | 27.4 | 14.9 | 7.0 | 3.4 | 3.8 |
| Beaverton | 0.0 | 8.7 | 0.0 | 0.0 | 0.0 | 34.3 | 11.8 | 0.3 | 1.2 | 0.6 |
| Bend | 0.0 | 14.6 | 0.4 | 0.0 | 0.0 | 15.9 | 22.8 | 0.0 | 2.5 | 3.3 |
| Corvallis | 0.0 | 6.4 | 0.2 | 0.0 | 0.0 | 19.0 | 13.0 | 10.2 | 7.4 | 0.0 |
| Eugene | 0.0 | 3.4 | 1.8 | 0.0 | 0.0 | 18.8 | 13.1 | 10.1 | 3.2 | 0.5 |
| Grants Pass | 0.0 | 15.5 | 0.0 | 0.0 | 0.0 | 33.5 | 17.1 | 4.8 | 0.3 | 0.5 |
| Gresham | 0.0 | 8.8 | 0.0 | 0.0 | 0.0 | 22.9 | 20.2 | 2.6 | 1.1 | 2.1 |
| Hillsboro | 0.0 | 9.8 | 0.0 | 0.0 | 0.0 | 18.6 | 23.5 | 14.2 | 1.6 | 1.4 |
| Keizer | 0.0 | 5.0 | 0.0 | 0.0 | 0.0 | 33.3 | 28.1 | 1.9 | 0.0 | 0.0 |
| Lake Oswego | 0.0 | 7.6 | 0.0 | 0.0 | 0.0 | 14.7 | 10.3 | 11.5 | 0.0 | 0.0 |
| McMinnville | 0.0 | 20.4 | 0.0 | 0.0 | 9.5 | 15.2 | 12.4 | 8.3 | 0.4 | 2.3 |
| Medford | 0.0 | 8.3 | 0.3 | 0.0 | 0.0 | 28.3 | 9.2 | 5.4 | 0.7 | 17.0 |
| Oregon City | 0.0 | 9.5 | 1.1 | 0.0 | 0.0 | 22.4 | 20.2 | 9.8 | 1.2 | 3.9 |
| Portland | 0.0 | 12.4 | 0.6 | 0.0 | 0.0 | 14.2 | 16.2 | 9.4 | 9.8 | 7.9 |

# Table D. Cities — City Government Finances, City Government Employment, and Climate

| City | City government finances, 2017 (cont.) Debt outstanding — Total (mil dol) [137] | Per capita[1] (dollars) [138] | Debt issued during year [139] | Climate[2] — Average daily temperature — Mean January [140] | Mean July [141] | Limits January[3] [142] | Limits July[4] [143] | Annual precipitation (inches) [144] | Heating degree days [145] | Cooling degree days [146] |
|---|---|---|---|---|---|---|---|---|---|---|
| **OHIO—Cont'd** | | | | | | | | | | |
| Hamilton | 227.6 | 3,664 | 0.0 | 28.7 | 76.6 | 19.9 | 88.1 | 43.36 | 5,261 | 1,135 |
| Hilliard | 63.9 | 1,774 | 10.4 | NA | NA | NA | NA | NA | NA | NA |
| Huber Heights | 83.3 | 2,188 | 24.7 | 27.9 | 77.0 | 20.6 | 87.2 | 39.41 | 5,343 | 1,214 |
| Kent | 30.0 | 1,002 | 0.0 | 27.2 | 74.1 | 20.1 | 83.9 | 36.07 | 5,752 | 856 |
| Kettering | 14.7 | 267 | 0.0 | 27.9 | 77.0 | 20.6 | 87.2 | 39.41 | 5,343 | 1,214 |
| Lakewood | 79.7 | 1,587 | 41.2 | 25.7 | 71.9 | 18.8 | 81.4 | 38.71 | 6,121 | 702 |
| Lancaster | 0.0 | 0 | 0.0 | 26.5 | 73.1 | 17.8 | 84.4 | 36.55 | 5,887 | 764 |
| Lima | 115.5 | 3,113 | 13.9 | 25.5 | 73.6 | 18.1 | 84.0 | 37.20 | 5,932 | 835 |
| Lorain | 125.1 | 1,961 | 4.7 | 27.1 | 73.8 | 19.3 | 85.0 | 38.02 | 5,731 | 818 |
| Mansfield | 6.9 | 150 | 0.0 | 24.3 | 71.0 | 16.2 | 81.8 | 43.24 | 6,364 | 653 |
| Marion | 24.0 | 666 | 0.0 | 24.5 | 72.7 | 16.0 | 83.7 | 38.35 | 6,300 | 703 |
| Mason | 70.0 | 2,108 | 0.0 | NA | NA | NA | NA | NA | NA | NA |
| Massillon | 0.0 | 0 | 0.0 | 25.2 | 71.8 | 17.4 | 82.3 | 38.47 | 6,154 | 678 |
| Medina | 12.6 | 482 | 0.0 | 23.7 | 71.3 | 16.2 | 82.0 | 38.34 | 6,525 | 558 |
| Mentor | 18.1 | 384 | 0.0 | 23.0 | 68.8 | 14.3 | 80.0 | 47.33 | 6,956 | 372 |
| Middletown | 204.9 | 4,195 | 0.0 | 27.5 | 74.2 | 18.3 | 86.3 | 39.54 | 5,609 | 879 |
| Newark | 20.2 | 409 | 0.0 | 25.8 | 72.7 | 17.3 | 83.8 | 41.62 | 6,084 | 687 |
| North Olmsted | 76.8 | 2,422 | 0.0 | 25.7 | 71.9 | 18.8 | 81.4 | 38.71 | 6,121 | 702 |
| North Ridgeville | 18.1 | 542 | 0.0 | NA | NA | NA | NA | NA | NA | NA |
| North Royalton | 23.6 | 781 | 0.0 | 25.7 | 71.9 | 18.8 | 81.4 | 38.71 | 6,121 | 702 |
| Parma | 24.3 | 307 | 0.0 | 25.7 | 71.9 | 18.8 | 81.4 | 38.71 | 6,121 | 702 |
| Reynoldsburg | 17.8 | 470 | 1.1 | 28.3 | 75.1 | 20.3 | 85.3 | 38.52 | 5,492 | 951 |
| Riverside | 8.8 | 353 | 0.0 | NA | NA | NA | NA | NA | NA | NA |
| Sandusky | 14.4 | 580 | 0.0 | 25.6 | 73.8 | 18.9 | 81.8 | 34.46 | 6,065 | 785 |
| Shaker Heights | 15.2 | 556 | 0.0 | 25.7 | 71.9 | 18.8 | 81.4 | 38.71 | 6,121 | 702 |
| Springfield | 43.3 | 732 | 4.2 | 26.1 | 73.5 | 18.2 | 83.8 | 37.70 | 5,921 | 796 |
| Stow | 6.8 | 195 | 0.0 | 27.2 | 74.1 | 20.1 | 83.9 | 36.07 | 5,752 | 856 |
| Strongsville | 49.2 | 1,100 | 18.4 | 25.7 | 71.9 | 18.8 | 81.4 | 38.71 | 6,121 | 702 |
| Toledo | 422.8 | 1,528 | 0.0 | 27.5 | 77.6 | 21.7 | 87.1 | 33.52 | 5,464 | 702 |
| Troy | 9.8 | 378 | 0.0 | NA | NA | NA | NA | NA | NA | 1,257 |
| Upper Arlington | 68.7 | 1,939 | 15.2 | 28.3 | 75.1 | 20.3 | 85.3 | 38.52 | 5,492 | 951 |
| Warren | 30.6 | 776 | 0.0 | 24.0 | 70.2 | 15.3 | 82.4 | 37.80 | 6,678 | 458 |
| Westerville | 65.2 | 1,637 | 0.0 | 27.7 | 74.4 | 19.7 | 85.4 | 39.35 | 5,434 | 924 |
| Westlake | 90.4 | 2,803 | 0.0 | 27.1 | 73.8 | 19.3 | 85.0 | 38.02 | 5,731 | 818 |
| Wooster | 17.2 | 644 | 0.0 | NA | NA | NA | NA | NA | NA | NA |
| Xenia | 1.2 | 44 | 0.0 | NA | NA | NA | NA | NA | NA | NA |
| Youngstown | 8.6 | 134 | 0.0 | 24.9 | 69.9 | 17.4 | 81.0 | 38.02 | 6,451 | 552 |
| Zanesville | 4.0 | 159 | 2.6 | 24.3 | 68.4 | 16.3 | 78.7 | 36.91 | 6,639 | 373 |
| **OKLAHOMA** | X | X | X | X | X | X | X | X | X | X |
| Bartlesville | 86.5 | 2,378 | 0.0 | 35.4 | 82.2 | 23.7 | 94.5 | 38.99 | 3,743 | 1,894 |
| Broken Arrow | 226.6 | 2,091 | 45.7 | 34.8 | 81.3 | 23.5 | 92.9 | 40.46 | 3,917 | 1,746 |
| Edmond | 129.0 | 1,407 | 41.8 | 36.7 | 82.0 | 26.2 | 93.1 | 35.85 | 3,663 | 1,907 |
| Enid | 65.6 | 1,308 | 15.2 | 33.1 | 82.6 | 21.9 | 94.4 | 34.25 | 4,269 | 1,852 |
| Lawton | 124.2 | 1,322 | 34.4 | 38.2 | 84.2 | 26.4 | 95.7 | 31.64 | 3,326 | 2,199 |
| Midwest City | 76.5 | 1,338 | 0.0 | 36.7 | 82.0 | 26.2 | 93.1 | 35.85 | 3,663 | 1,907 |
| Moore | 95.1 | 1,559 | 1.8 | 36.7 | 82.0 | 26.2 | 93.1 | 35.85 | 3,663 | 1,907 |
| Muskogee | 66.9 | 1,771 | 0.0 | 36.1 | 82.1 | 25.2 | 93.1 | 43.77 | 3,667 | 1,858 |
| Norman | 407.2 | 3,300 | 218.8 | 35.8 | 82.1 | 23.2 | 93.9 | 41.65 | 3,713 | 1,906 |
| Oklahoma City | 1,836.8 | 2,857 | 321.1 | 36.7 | 82.0 | 26.2 | 93.1 | 35.85 | 3,663 | 1,907 |
| Owasso | 42.3 | 1,163 | 21.3 | NA | NA | NA | NA | NA | NA | NA |
| Ponca City | 23.7 | 984 | 0.0 | 33.8 | 82.9 | 23.8 | 94.1 | 36.41 | 4,053 | 1,964 |
| Shawnee | 23.3 | 746 | 4.2 | 37.3 | 83.0 | 25.5 | 94.5 | 40.87 | 3,460 | 2,024 |
| Stillwater | 94.3 | 1,885 | 12.9 | 34.5 | 82.3 | 21.9 | 93.6 | 36.71 | 3,899 | 1,881 |
| Tulsa | 1,412.1 | 3,512 | 296.0 | 37.4 | 81.9 | 27.1 | 92.2 | 45.10 | 3,413 | 1,905 |
| **OREGON** | X | X | X | X | X | X | X | X | X | X |
| Albany | 90.4 | 1,689 | 0.0 | 40.3 | 66.5 | 33.6 | 81.2 | 43.66 | 4,715 | 247 |
| Beaverton | 8.4 | 86 | 0.0 | 40.0 | 66.8 | 33.8 | 79.2 | 39.95 | 4,723 | 287 |
| Bend | 152.2 | 1,609 | 0.0 | 31.2 | 63.5 | 22.6 | 80.7 | 11.73 | 7,042 | 147 |
| Corvallis | 41.0 | 701 | 0.0 | 38.1 | 63.8 | 31.6 | 77.4 | 67.76 | 5,501 | 139 |
| Eugene | 307.6 | 1,816 | 163.6 | 39.8 | 66.2 | 33.0 | 81.5 | 50.90 | 4,786 | 242 |
| Grants Pass | 7.9 | 209 | 0.0 | NA | NA | NA | NA | NA | NA | NA |
| Gresham | 80.5 | 724 | 6.5 | 40.0 | 68.3 | 33.5 | 81.5 | 45.70 | 4,491 | 450 |
| Hillsboro | 92.7 | 863 | 35.2 | 40.5 | 67.6 | 35.1 | 80.4 | 38.19 | 4,532 | 323 |
| Keizer | 15.5 | 396 | 0.0 | 40.3 | 66.8 | 33.5 | 81.5 | 40.00 | 4,784 | 257 |
| Lake Oswego | 183.1 | 4,661 | 0.0 | 41.8 | 69.3 | 35.7 | 82.6 | 46.05 | 4,132 | 475 |
| McMinnville | 33.9 | 991 | 6.1 | 39.6 | 66.6 | 33.0 | 81.9 | 41.66 | 4,815 | 288 |
| Medford | 535.1 | 6,591 | 0.0 | 39.1 | 72.7 | 30.9 | 90.2 | 18.37 | 4,539 | 711 |
| Oregon City | 23.9 | 658 | 0.0 | 41.8 | 69.3 | 35.7 | 82.6 | 46.05 | 4,132 | 475 |
| Portland | 3,150.5 | 4,863 | 631.1 | 41.8 | 69.3 | 35.7 | 82.6 | 46.05 | 4,132 | 475 |

1. Based on the population estimated as of July 1 of the year shown. 2. Represents normal values based on the 30-year period, 1971±2000. 3. Average daily minimum. 4. Average daily maximum.

# Table D. Cities — Land Area and Population

| STATE Place code | City | Land area[1] (sq. mi) 1 | Total persons 2019 2 | Rank 3 | Per square mile 4 | White 5 | Black or African American 6 | American Indian, Alaskan Native 7 | Asian 8 | Hawaiian Pacific Islander 9 | Some other race 10 | Two or more races (percent) 11 |
|---|---|---|---|---|---|---|---|---|---|---|---|---|
| | **OREGON—Cont'd** | | | | | D | D | D | D | D | D | D |
| 41 61200 | Redmond | 18.3 | 34,293 | 1,126 | 1,874 | 78.5 | 1.1 | 0.7 | 3.4 | 0.6 | 8.2 | 7.5 |
| 41 64900 | Salem | 48.8 | 175,891 | 148 | 3,604 | 81.0 | 1.5 | 2.6 | 3.4 | 0.1 | 8.6 | 2.9 |
| 41 69600 | Springfield | 15.8 | 63,209 | 596 | 4,001 | 81.4 | 0.8 | 0.8 | 11.5 | 0.0 | 0.5 | 5.0 |
| 41 73650 | Tigard | 12.7 | 56,049 | 686 | 4,413 | 77.4 | 2.1 | 0.2 | 3.6 | 1.9 | 11.2 | 3.7 |
| 41 74950 | Tualatin | 8.3 | 27,481 | 1,333 | 3,311 | 83.4 | 0.0 | 0.0 | 8.7 | 0.0 | 4.2 | 3.7 |
| 41 80150 | West Linn | 7.4 | 26,747 | 1,361 | 3,615 | 79.6 | 11.4 | 0.2 | 3.5 | 0.0 | 2.6 | 2.6 |
| 42 00000 | **PENNSYLVANIA** | 44,741.7 | 12,783,254 | X | 286 | 68.7 | 17.5 | 0.3 | 3.6 | 0.1 | 7.1 | 2.6 |
| 42 02000 | Allentown | 17.6 | 121,470 | 233 | 6,902 | D | D | D | D | D | D | D |
| 42 02184 | Altoona | 9.8 | 42,989 | 895 | 4,387 | 89.7 | 2.5 | 0.0 | 2.6 | 0.0 | 0.0 | 5.2 |
| 42 06064 | Bethel Park | 11.7 | 32,212 | 1,186 | 2,753 | 64.9 | 13.1 | 0.5 | 3.4 | 0.0 | 14.1 | 3.9 |
| 42 06088 | Bethlehem | 19.1 | 75,466 | 472 | 3,951 | 19.2 | 72.4 | 1.0 | 0.6 | 0.0 | 3.8 | 3.1 |
| 42 13208 | Chester | 4.8 | 33,692 | 1,145 | 7,019 | 66.8 | 20.1 | 0.0 | 1.1 | 0.0 | 8.8 | 3.2 |
| 42 21648 | Easton | 4.3 | 27,364 | 1,338 | 6,364 | 71.0 | 15.5 | 0.4 | 2.3 | 0.0 | 3.5 | 7.3 |
| 42 24000 | Erie | 19.1 | 95,077 | 341 | 4,978 | 36.8 | 47.0 | 1.2 | 4.2 | 0.0 | 5.8 | 4.9 |
| 42 32800 | Harrisburg | 8.1 | 49,322 | 787 | 6,089 | D | D | D | D | D | D | D |
| 42 33408 | Hazleton | 6.0 | 24,809 | 1,410 | 4,135 | 54.9 | 13.5 | 0.1 | 3.9 | 0.2 | 14.2 | 13.2 |
| 42 41216 | Lancaster | 7.2 | 58,990 | 641 | 8,193 | 60.5 | 6.7 | 1.2 | 0.0 | 0.0 | 24.1 | 7.6 |
| 42 42168 | Lebanon | 4.2 | 25,725 | 1,394 | 6,125 | 75.5 | 13.2 | 0.0 | 6.9 | 0.0 | 0.9 | 3.6 |
| 42 50528 | Monroeville | 19.7 | 27,393 | 1,336 | 1,391 | 44.3 | 44.4 | 0.2 | 2.7 | 0.0 | 4.8 | 3.6 |
| 42 54656 | Norristown | 3.5 | 34,323 | 1,122 | 9,807 | 39.0 | 41.5 | 0.5 | 7.6 | 0.1 | 8.1 | 3.3 |
| 42 60000 | Philadelphia | 134.4 | 1,578,487 | 6 | 11,745 | 39.0 | 41.5 | 0.5 | 7.6 | 0.1 | 8.1 | 3.3 |
| 42 61000 | Pittsburgh | 55.4 | 299,226 | 65 | 5,401 | 66.2 | 23.7 | 0.3 | 5.7 | 0.0 | 1.0 | 3.2 |
| 42 61536 | Plum | 28.6 | 26,993 | 1,353 | 944 | D | D | D | D | D | D | D |
| 42 63624 | Reading | 9.8 | 88,217 | 382 | 9,002 | 46.4 | 14.1 | 0.7 | 2.2 | 0.2 | 31.9 | 4.5 |
| 42 69000 | Scranton | 25.3 | 76,532 | 465 | 3,025 | 81.6 | 7.6 | 0.0 | 3.1 | 0.2 | 3.8 | 3.7 |
| 42 73808 | State College | 4.6 | 41,366 | 930 | 8,993 | 83.5 | 2.0 | 0.0 | 11.0 | 0.2 | 0.0 | 3.4 |
| 42 85152 | Wilkes-Barre | 6.8 | 40,597 | 949 | 5,970 | 62.9 | 20.7 | 0.1 | 2.2 | 0.0 | 10.9 | 3.2 |
| 42 85312 | Williamsport | 8.8 | 28,156 | 1,309 | 3,200 | 78.1 | 17.1 | 1.3 | 1.3 | 0.0 | 0.1 | 2.1 |
| 42 87048 | York | 5.3 | 43,907 | 881 | 8,284 | 57.6 | 23.3 | 1.1 | 2.7 | 0.0 | 5.5 | 9.8 |
| 44 00000 | **RHODE ISLAND** | 1,033.9 | 1,057,125 | X | 1,023 | 78.7 | 7.4 | 0.3 | 3.5 | 0.0 | 5.9 | 4.2 |
| 44 19180 | Cranston | 28.3 | 81,142 | 427 | 2,867 | 72.4 | 7.8 | 0.8 | 7.1 | 0.2 | 6.2 | 5.5 |
| 44 22960 | East Providence | 13.3 | 47,337 | 823 | 3,559 | 84.6 | 5.9 | 0.0 | 2.2 | 0.0 | 2.9 | 4.3 |
| 44 54640 | Pawtucket | 8.7 | 71,710 | 507 | 8,243 | 59.5 | 17.7 | 0.4 | 3.1 | 0.0 | 10.7 | 8.5 |
| 44 59000 | Providence | 18.4 | 179,270 | 144 | 9,743 | 52.4 | 21.0 | 0.5 | 5.5 | 0.0 | 13.8 | 6.7 |
| 44 74300 | Warwick | 35.0 | 81,189 | 426 | 2,320 | 90.3 | 3.2 | 0.7 | 2.8 | 0.0 | 0.8 | 2.2 |
| 44 80780 | Woonsocket | 7.7 | 41,545 | 926 | 5,396 | 76.1 | 9.3 | 0.3 | 5.5 | 0.0 | 3.8 | 5.0 |
| 45 00000 | **SOUTH CAROLINA** | 30,063.7 | 5,218,040 | X | 174 | 66.7 | 26.5 | 0.4 | 1.7 | 0.1 | 2.2 | 2.4 |
| 45 00550 | Aiken | 20.9 | 31,121 | 1,215 | 1,489 | D | D | D | D | D | D | D |
| 45 01360 | Anderson | 14.5 | 27,701 | 1,327 | 1,910 | D | D | D | D | D | D | D |
| 45 13330 | Charleston | 114.8 | 139,714 | 198 | 1,217 | 72.8 | 20.7 | 0.1 | 2.5 | 1.0 | 1.1 | 1.9 |
| 45 16000 | Columbia | 136.8 | 132,130 | 209 | 966 | 56.9 | 36.2 | 0.1 | 2.6 | 0.1 | 1.1 | 2.9 |
| 45 25810 | Florence | 23.1 | 38,467 | 1,006 | 1,665 | D | D | D | D | D | D | D |
| 45 29815 | Goose Creek | 41.3 | 44,972 | 857 | 1,089 | 64.4 | 22.2 | 1.0 | 4.9 | 0.0 | 3.7 | 3.8 |
| 45 30850 | Greenville | 29.7 | 72,227 | 501 | 2,432 | 67.7 | 23.4 | 0.1 | 1.5 | 1.9 | 2.5 | 2.8 |
| 45 30985 | Greer | 22.9 | 35,316 | 1,093 | 1,542 | 69.3 | 14.6 | 0.6 | 9.8 | 0.0 | 3.4 | 2.3 |
| 45 34045 | Hilton Head Island | 41.4 | 39,619 | 976 | 957 | 86.4 | 4.7 | 0.1 | 1.1 | 0.0 | 3.7 | 3.9 |
| 45 48535 | Mount Pleasant | 49.5 | 92,799 | 360 | 1,875 | 93.1 | 2.0 | 0.0 | 1.9 | 0.0 | 0.1 | 2.9 |
| 45 49075 | Myrtle Beach | 23.5 | 35,555 | 1,087 | 1,513 | 75.7 | 18.4 | 0.1 | 2.9 | 0.0 | 1.8 | 1.2 |
| 45 50875 | North Charleston | 77.6 | 118,752 | 241 | 1,530 | 39.2 | 43.3 | 0.0 | 2.6 | 2.4 | 9.2 | 3.4 |
| 45 61405 | Rock Hill | 39.0 | 76,318 | 466 | 1,957 | 55.3 | 40.1 | 0.3 | 1.2 | 0.0 | 0.9 | 2.3 |
| 45 68290 | Spartanburg | 20.1 | 37,317 | 1,028 | 1,857 | 43.9 | 49.0 | 0.0 | 2.7 | 0.1 | 0.7 | 3.5 |
| 45 70270 | Summerville | 19.7 | 53,301 | 727 | 2,706 | D | D | D | D | D | D | D |
| 45 70405 | Sumter | 32.6 | 39,439 | 984 | 1,210 | 44.1 | 46.9 | 0.2 | 2.4 | 0.0 | 2.6 | 3.9 |
| 46 00000 | **SOUTH DAKOTA** | 75,809.7 | 892,717 | X | 12 | 84.1 | 2.4 | 8.6 | 1.3 | 0.2 | 0.7 | 2.8 |
| 46 00100 | Aberdeen | 16.5 | 28,180 | 1,308 | 1,708 | D | D | D | D | D | D | D |
| 46 52980 | Rapid City | 54.7 | 78,864 | 448 | 1,442 | 82.1 | 2.5 | 9.3 | 1.3 | 0.0 | 0.0 | 4.7 |
| 46 59020 | Sioux Falls | 79.1 | 187,809 | 134 | 2,374 | 85.3 | 6.7 | 1.4 | 2.2 | 0.0 | 0.6 | 3.8 |
| 47 00000 | **TENNESSEE** | 41,238.0 | 6,886,834 | X | 167 | 77.2 | 16.7 | 0.3 | 1.8 | 0.1 | 1.6 | 2.3 |
| 47 03440 | Bartlett | 32.3 | 59,327 | 635 | 1,837 | 67.2 | 28.2 | 0.2 | 2.1 | 0.0 | 0.0 | 2.3 |
| 47 08280 | Brentwood | 41.1 | 42,876 | 899 | 1,043 | 90.1 | 1.3 | 0.0 | 4.8 | 0.0 | 1.6 | 2.1 |
| 47 08540 | Bristol | 32.7 | 27,142 | 1,343 | 830 | D | D | D | D | D | D | D |
| 47 14000 | Chattanooga | 142.4 | 184,742 | 138 | 1,297 | 61.9 | 29.4 | 0.1 | 3.3 | 0.2 | 2.7 | 2.5 |
| 47 15160 | Clarksville | 99.4 | 161,247 | 158 | 1,622 | 62.4 | 24.6 | 0.6 | 2.2 | 0.4 | 2.3 | 7.4 |
| 47 15400 | Cleveland | 27.1 | 45,948 | 848 | 1,696 | 82.0 | 9.3 | 0.2 | 1.0 | 0.0 | 5.2 | 2.2 |
| 47 16420 | Collierville | 36.3 | 51,441 | 756 | 1,417 | 72.6 | 16.5 | 0.3 | 6.9 | 0.0 | 0.6 | 3.2 |

1. Dry land or land partially or temporarily covered by water.  2. Hispanic or Latino persons may be of any race.

# Table D. Cities — **Population**

| City | Percent Hispanic or Latino[1], 2019 | Percent foreign born, 2019 | Age of population (percent), 2019 | | | | | | | Median age, 2019 | Percent female, 2019 | Population | | | |
|---|---|---|---|---|---|---|---|---|---|---|---|---|---|---|---|
| | | | Under 18 years | 18 to 24 years | 25 to 34 years | 35 to 44 years | 45 to 54 years | 55 to 64 years | 65 years and over | | | Census counts | | Percent change | |
| | | | | | | | | | | | | 2000 | 2010 | 2000-2010 | 2001-2020 |
| | 12 | 13 | 14 | 15 | 16 | 17 | 18 | 19 | 20 | 21 | 22 | 23 | 24 | 25 | 26 |
| **OREGON—Cont'd** | | | | | | | | | | | | | | | |
| Redmond | 11.9 | 4.4 | 23.4 | 9.2 | 11.6 | 14.2 | 14.9 | 9.6 | 17.0 | 36.6 | 56.7 | 13,481 | 26,212 | 94.4 | 30.8 |
| Salem | 23.6 | 10.9 | 24.2 | 11.8 | 14.3 | 13.0 | 11.6 | 11.8 | 13.3 | 34.8 | 49.2 | 136,924 | 154,905 | 13.1 | 13.5 |
| Springfield | 11.2 | 6.5 | 20.6 | 10.2 | 15.8 | 13.9 | 13.1 | 12.0 | 14.4 | 36.5 | 52.6 | 52,864 | 59,407 | 12.4 | 6.4 |
| Tigard | 12.7 | 13.8 | 21.2 | 8.7 | 14.2 | 18.2 | 14.2 | 10.7 | 12.7 | 37.3 | 52.2 | 41,223 | 47,979 | 16.4 | 16.8 |
| Tualatin | 24.3 | 9.2 | 26.0 | 6.4 | 14.2 | 16.6 | 12.5 | 9.2 | 15.1 | 36.5 | 50.2 | 22,791 | 26,112 | 14.6 | 5.2 |
| West Linn | 8.2 | 10.3 | 22.6 | 4.6 | 8.0 | 14.9 | 14.3 | 18.9 | 16.6 | 44.9 | 50.9 | 22,261 | 25,137 | 12.9 | 6.4 |
| **PENNSYLVANIA** | 7.8 | 7.0 | 20.6 | 8.9 | 13.3 | 11.9 | 12.5 | 14.1 | 18.7 | 40.8 | 51.0 | 12,281,054 | 12,702,891 | 3.4 | 0.6 |
| Allentown | 54.7 | 17.6 | 25.8 | 13.6 | 16.5 | 11.0 | 11.3 | 10.6 | 11.2 | 31.2 | 50.8 | 106,632 | 118,018 | 10.7 | 2.9 |
| Altoona | 2.3 | 0.5 | 21.9 | 8.6 | 14.3 | 11.6 | 11.2 | 14.5 | 18.0 | 39.8 | 53.3 | 49,523 | 45,967 | -7.2 | -6.5 |
| Bethel Park | 2.3 | 3.8 | 22.5 | 3.1 | 13.0 | 11.8 | 13.9 | 14.1 | 21.6 | 44.7 | 51.1 | 33,556 | 32,385 | -3.5 | -0.5 |
| Bethlehem | 32.4 | 14.1 | 17.2 | 15.9 | 16.6 | 10.5 | 10.9 | 10.8 | 18.1 | 35.3 | 52.9 | 71,329 | 74,964 | 5.1 | 0.7 |
| Chester | 9.4 | 4.7 | 27.9 | 12.8 | 15.5 | 12.2 | 8.8 | 10.7 | 12.0 | 30.7 | 55.8 | 36,854 | 33,889 | -8.0 | -0.6 |
| Easton | 21.9 | 10.0 | 17.5 | 15.3 | 8.2 | 11.2 | 18.6 | 10.5 | 18.6 | 43.1 | 50.7 | 26,263 | 26,814 | 2.1 | 2.1 |
| Erie | 9.6 | 4.2 | 22.3 | 10.7 | 14.7 | 12.5 | 11.4 | 12.3 | 16.0 | 37.1 | 51.4 | 103,717 | 101,735 | -1.9 | -6.5 |
| Harrisburg | 19.4 | 10.7 | 26.3 | 8.3 | 18.4 | 9.9 | 11.1 | 13.8 | 12.3 | 34.1 | 49.8 | 48,950 | 49,528 | 1.2 | -0.4 |
| Hazleton | 60.5 | 33.2 | 25.3 | 11.2 | 7.3 | 14.2 | 12.3 | 16.1 | 13.6 | 40.5 | 54.4 | 23,329 | 25,416 | 8.9 | -2.4 |
| Lancaster | 35.6 | 9.2 | 19.6 | 13.1 | 19.0 | 11.9 | 13.3 | 12.3 | 10.8 | 33.6 | 52.0 | 56,348 | 59,262 | 5.2 | -0.5 |
| Lebanon | 40.1 | 5.9 | 25.2 | 7.0 | 12.6 | 17.3 | 11.5 | 12.6 | 13.8 | 38.4 | 52.1 | 24,461 | 25,492 | 4.2 | 0.9 |
| Monroeville | 1.3 | 8.7 | 15.1 | 7.5 | 13.0 | 8.5 | 14.2 | 17.7 | 24.0 | 51.1 | 49.6 | 29,349 | 28,348 | -3.4 | -3.4 |
| Norristown | 23.1 | 15.6 | 22.0 | 9.9 | 17.1 | 17.2 | 12.9 | 10.0 | 10.9 | 35.6 | 50.3 | 31,282 | 34,341 | 9.8 | -0.1 |
| Philadelphia | 15.2 | 14.0 | 21.6 | 9.7 | 19.2 | 12.7 | 11.2 | 11.6 | 14.0 | 34.7 | 52.7 | 1,517,550 | 1,525,999 | 0.6 | 3.4 |
| Pittsburgh | 3.7 | 9.0 | 14.7 | 16.2 | 22.4 | 10.8 | 8.9 | 11.7 | 15.2 | 33.3 | 51.3 | 334,563 | 305,306 | -8.7 | -2.0 |
| Plum | 0.4 | 0.9 | 22.0 | 7.7 | 10.5 | 11.3 | 12.9 | 14.2 | 21.4 | 43.8 | 51.6 | 26,940 | 27,110 | 0.6 | -0.4 |
| Reading | 66.1 | 18.1 | 27.4 | 14.2 | 14.2 | 12.7 | 10.0 | 10.9 | 10.6 | 30.4 | 49.4 | 81,207 | 87,996 | 8.4 | 0.3 |
| Scranton | 14.9 | 8.0 | 18.5 | 14.0 | 13.4 | 12.9 | 11.0 | 12.0 | 18.3 | 38.2 | 50.9 | 76,415 | 76,077 | -0.4 | 0.6 |
| State College | 5.4 | 14.8 | 5.0 | 60.3 | 12.1 | 5.8 | 3.8 | 5.1 | 7.9 | 21.7 | 44.6 | 38,420 | 41,980 | 9.3 | -1.5 |
| Wilkes-Barre | 23.6 | 10.0 | 22.3 | 12.3 | 16.6 | 12.7 | 10.0 | 11.3 | 14.8 | 33.8 | 54.0 | 43,123 | 41,510 | -3.7 | -2.2 |
| Williamsport | 1.9 | 3.7 | 20.0 | 17.7 | 14.7 | 10.8 | 8.7 | 13.0 | 15.0 | 32.9 | 52.4 | 30,706 | 29,374 | -4.3 | -4.1 |
| York | 36.1 | 11.5 | 25.5 | 10.3 | 14.8 | 13.7 | 12.2 | 13.2 | 10.3 | 34.7 | 49.2 | 40,862 | 43,807 | 7.2 | 0.2 |
| **RHODE ISLAND** | 16.3 | 13.7 | 19.2 | 10.3 | 14.0 | 11.7 | 12.8 | 14.3 | 17.7 | 40.1 | 51.2 | 1,048,319 | 1,052,970 | 0.4 | 0.4 |
| Cranston | 15.3 | 14.2 | 20.3 | 7.7 | 12.6 | 16.1 | 12.6 | 12.9 | 17.8 | 40.6 | 53.5 | 79,269 | 80,666 | 1.8 | 0.6 |
| East Providence | 7.6 | 16.0 | 17.2 | 4.7 | 16.6 | 13.1 | 9.1 | 16.2 | 23.1 | 41.7 | 53.4 | 48,688 | 47,071 | -3.3 | 0.6 |
| Pawtucket | 26.1 | 24.9 | 21.8 | 8.6 | 18.1 | 10.5 | 14.6 | 13.0 | 13.4 | 36.0 | 48.3 | 72,958 | 71,160 | -2.5 | 0.8 |
| Providence | 44.2 | 29.2 | 20.0 | 16.8 | 17.5 | 13.5 | 10.5 | 9.3 | 12.3 | 31.9 | 51.7 | 173,618 | 177,732 | 2.4 | 0.9 |
| Warwick | 3.0 | 5.6 | 17.9 | 6.7 | 12.8 | 13.1 | 13.1 | 15.6 | 20.8 | 44.7 | 53.6 | 85,808 | 82,687 | -3.6 | -1.8 |
| Woonsocket | 21.1 | 11.8 | 20.9 | 6.3 | 20.4 | 11.8 | 12.1 | 14.7 | 13.8 | 36.5 | 47.9 | 43,224 | 41,202 | -4.7 | 0.8 |
| **SOUTH CAROLINA** | 5.8 | 5.6 | 21.6 | 9.2 | 13.0 | 12.2 | 12.3 | 13.5 | 18.2 | 39.9 | 51.7 | 4,012,012 | 4,625,358 | 15.3 | 12.8 |
| Aiken | 1.2 | 1.8 | 18.3 | 7.1 | 15.9 | 7.0 | 12.1 | 12.1 | 27.5 | 48.7 | 56.4 | 25,337 | 29,609 | 16.9 | 5.1 |
| Anderson | 6.9 | 7.8 | 23.1 | 12.1 | 13.7 | 10.4 | 9.4 | 13.0 | 18.3 | 36.3 | 53.3 | 25,514 | 26,410 | 3.5 | 4.9 |
| Charleston | 3.6 | 5.9 | 19.1 | 10.8 | 22.7 | 12.5 | 9.2 | 10.1 | 15.5 | 34.0 | 53.2 | 96,650 | 120,415 | 24.6 | 16.0 |
| Columbia | 6.9 | 4.7 | 17.0 | 25.9 | 16.1 | 9.5 | 10.5 | 9.9 | 11.0 | 28.6 | 47.8 | 116,278 | 130,550 | 12.3 | 1.2 |
| Florence | 0.9 | 3.4 | 27.9 | 5.0 | 14.2 | 10.7 | 12.7 | 10.7 | 18.7 | 38.7 | 54.6 | 30,248 | 37,886 | 25.3 | 1.5 |
| Goose Creek | 5.3 | 7.9 | 21.7 | 18.1 | 10.2 | 13.8 | 9.6 | 11.7 | 14.9 | 34.9 | 46.6 | 29,208 | 36,428 | 24.7 | 23.5 |
| Greenville | 5.0 | 9.4 | 15.6 | 14.0 | 19.9 | 11.4 | 12.1 | 12.3 | 14.8 | 35.4 | 51.6 | 56,002 | 59,325 | 5.9 | 21.7 |
| Greer | 15.7 | 19.4 | 23.6 | 8.7 | 14.0 | 15.9 | 13.0 | 8.2 | 16.4 | 37.5 | 48.2 | 16,843 | 25,882 | 53.7 | 36.5 |
| Hilton Head Island | 11.2 | 10.6 | 13.9 | 4.5 | 8.1 | 5.9 | 10.0 | 18.5 | 39.0 | 58.8 | 49.3 | 33,862 | 37,097 | 9.6 | 6.8 |
| Mount Pleasant | 1.8 | 5.1 | 21.5 | 6.2 | 9.2 | 16.7 | 13.9 | 13.8 | 18.8 | 43.3 | 51.4 | 47,609 | 68,400 | 43.7 | 35.7 |
| Myrtle Beach | 8.2 | 12.5 | 19.8 | 6.0 | 9.9 | 13.6 | 13.9 | 13.6 | 23.3 | 45.6 | 50.0 | 22,759 | 26,950 | 18.4 | 31.9 |
| North Charleston | 12.0 | 14.6 | 23.1 | 10.1 | 19.5 | 13.4 | 11.0 | 11.5 | 11.5 | 33.4 | 52.0 | 79,641 | 97,591 | 22.5 | 21.7 |
| Rock Hill | 8.0 | 6.8 | 21.0 | 12.5 | 15.6 | 13.4 | 12.4 | 11.1 | 14.0 | 36.1 | 54.6 | 49,765 | 66,867 | 34.4 | 14.1 |
| Spartanburg | 4.0 | 3.8 | 26.9 | 12.0 | 14.7 | 10.3 | 7.4 | 10.9 | 17.9 | 30.8 | 56.1 | 39,673 | 36,717 | -7.5 | 1.6 |
| Summerville | 4.8 | 4.6 | 24.9 | 8.2 | 14.1 | 16.5 | 9.6 | 15.1 | 11.7 | 35.9 | 54.4 | 27,752 | 42,966 | 54.8 | 24.1 |
| Sumter | 6.9 | 4.1 | 28.7 | 7.8 | 13.6 | 15.8 | 10.1 | 9.7 | 14.2 | 34.8 | 50.8 | 39,643 | 40,530 | 2.2 | -2.7 |
| **SOUTH DAKOTA** | 3.7 | 4.1 | 24.3 | 9.2 | 12.9 | 12.0 | 10.9 | 13.2 | 17.4 | 37.7 | 49.2 | 754,844 | 814,198 | 7.9 | 9.6 |
| Aberdeen | 3.6 | 5.4 | 22.2 | 13.5 | 11.6 | 12.4 | 10.9 | 11.6 | 17.8 | 35.8 | 51.6 | 24,658 | 26,112 | 5.9 | 7.9 |
| Rapid City | 4.9 | 2.5 | 22.2 | 8.0 | 13.7 | 12.6 | 9.9 | 14.4 | 19.1 | 39.1 | 49.8 | 59,607 | 68,975 | 15.7 | 14.3 |
| Sioux Falls | 5.5 | 8.8 | 25.2 | 8.9 | 15.3 | 14.0 | 11.9 | 11.0 | 13.6 | 35.3 | 48.9 | 123,975 | 153,976 | 24.2 | 22.0 |
| **TENNESSEE** | 5.7 | 5.5 | 22.1 | 9.1 | 13.7 | 12.4 | 12.8 | 13.2 | 16.7 | 39.0 | 51.3 | 5,689,283 | 6,346,281 | 11.5 | 8.5 |
| Bartlett | 2.5 | 5.9 | 22.2 | 8.5 | 8.6 | 12.4 | 15.6 | 14.1 | 18.6 | 43.4 | 50.8 | 40,543 | 56,938 | 40.4 | 4.2 |
| Brentwood | 5.1 | 8.9 | 29.5 | 7.4 | 4.8 | 12.8 | 17.7 | 12.9 | 14.8 | 42.4 | 49.3 | 23,445 | 37,067 | 58.1 | 15.7 |
| Bristol | 2.1 | 1.3 | 22.3 | 7.0 | 8.1 | 12.6 | 12.6 | 15.7 | 21.7 | 45.0 | 50.5 | 24,821 | 26,765 | 7.8 | 1.4 |
| Chattanooga | 7.5 | 7.7 | 18.3 | 10.3 | 18.1 | 13.2 | 9.4 | 13.5 | 17.3 | 37.6 | 52.8 | 155,554 | 170,319 | 9.5 | 8.5 |
| Clarksville | 11.9 | 6.7 | 27.1 | 12.8 | 20.0 | 14.2 | 9.0 | 8.7 | 8.1 | 29.2 | 49.8 | 103,455 | 132,881 | 28.4 | 21.3 |
| Cleveland | 10.4 | 6.8 | 21.0 | 11.5 | 13.6 | 12.4 | 10.3 | 12.2 | 19.0 | 37.6 | 54.4 | 37,192 | 41,251 | 10.9 | 11.4 |
| Collierville | 1.6 | 8.0 | 26.5 | 6.2 | 14.2 | 10.5 | 16.3 | 11.9 | 14.4 | 38.1 | 54.8 | 31,872 | 45,599 | 43.1 | 12.8 |

1. May be of any race.

## Table D. Cities — Households, Group Quarters, Crime, and Education

| City | Households, 2019 — Number | Persons per household | Family | Married couple family | Female family | Non-family | One person | Persons in group quarters, 2019 | Violent — Number | Violent — Rate[2] | Property — Number | Property — Rate[2] | Population age 25 and over | High school graduate or less | Bachelor's degree or more |
|---|---|---|---|---|---|---|---|---|---|---|---|---|---|---|---|
| | 27 | 28 | 29 | 30 | 31 | 32 | 33 | 34 | 35 | 36 | 37 | 38 | 39 | 40 | 41 |
| **OREGON—Cont'd** | | | | | | | | | | | | | | | |
| Redmond | 11,195 | 2.88 | 65.0 | 48.6 | 16.4 | 35.0 | 24.0 | 225 | 71 | 225 | 1,113 | 3,527 | 21,840 | 34.9 | 21.0 |
| Salem | 61,087 | 2.73 | 64.0 | 45.2 | 18.8 | 36.0 | 28.0 | 7,438 | 670 | 381 | 6,488 | 3,689 | 111,510 | 32.1 | 30.9 |
| Springfield | 25,325 | 2.47 | 63.4 | 42.2 | 21.2 | 36.6 | 27.0 | 730 | 190 | 300 | 1,978 | 3,118 | 43,791 | 36.2 | 21.2 |
| Tigard | 20,073 | 2.75 | 68.5 | 50.9 | 17.6 | 31.5 | 22.5 | 371 | 119 | 214 | 1,410 | 2,535 | 38,888 | 24.4 | 44.6 |
| Tualatin | 10,496 | 2.76 | 62.6 | 50.8 | 11.8 | 37.4 | 30.2 | 166 | 63 | 227 | 765 | 2,753 | 19,694 | 20.7 | 37.5 |
| West Linn | 10,435 | 2.54 | 64.5 | 61.1 | 3.3 | 35.5 | 30.4 | 178 | 19 | 70 | 231 | 857 | 19,464 | 14.5 | 61.7 |
| **PENNSYLVANIA** | 5,119,249 | 2.42 | 63.1 | 46.4 | 16.7 | 36.9 | 30.2 | 421,745 | NA | NA | NA | NA | 9,028,036 | 43.4 | 32.3 |
| Allentown | 40,959 | 2.80 | 66.0 | 34.4 | 31.7 | 34.0 | 26.4 | 6,802 | 471 | 387 | 2,782 | 2,283 | 73,580 | 60.4 | 14.7 |
| Altoona | 19,243 | 2.20 | 60.1 | 30.8 | 29.3 | 39.9 | 32.4 | 1,040 | 264 | 608 | 675 | 1,554 | 30,156 | 56.6 | 20.9 |
| Bethel Park | 13,268 | 2.42 | 67.7 | 59.9 | 7.8 | 32.3 | 28.0 | 293 | NA | NA | NA | NA | 24,068 | 28.3 | 48.2 |
| Bethlehem | 29,213 | 2.39 | 58.7 | 35.8 | 22.8 | 41.3 | 32.2 | 6,111 | NA | NA | NA | NA | 50,850 | 44.8 | 31.3 |
| Chester | 11,008 | 2.77 | 61.9 | 22.3 | 39.6 | 38.1 | 33.0 | 3,545 | 469 | 1,383 | 1,019 | 3,005 | 20,155 | 65.5 | 13.2 |
| Easton | 9,462 | 2.51 | 63.7 | 39.6 | 24.1 | 36.3 | 31.9 | 3,458 | NA | NA | NA | NA | 18,273 | 58.3 | 19.7 |
| Erie | 39,790 | 2.26 | 53.2 | 26.6 | 26.6 | 46.8 | 37.8 | 5,483 | 476 | 497 | 1,817 | 1,896 | 63,928 | 52.1 | 22.3 |
| Harrisburg | 21,399 | 2.25 | 46.8 | 22.2 | 24.6 | 53.2 | 43.7 | 1,128 | 443 | 900 | 1,069 | 2,173 | 32,262 | 50.5 | 25.7 |
| Hazleton | 9,374 | 2.62 | 61.9 | 30.1 | 31.8 | 38.1 | 27.7 | 265 | NA | NA | NA | NA | 15,745 | 63.8 | 8.7 |
| Lancaster | 23,014 | 2.43 | 58.1 | 28.4 | 29.7 | 41.9 | 31.0 | 3,278 | NA | NA | NA | NA | 39,865 | 50.6 | 25.9 |
| Lebanon | 10,458 | 2.44 | 57.7 | 23.4 | 34.3 | 42.3 | 26.5 | 338 | 65 | 250 | 528 | 2,034 | 17,547 | 69.6 | 9.6 |
| Monroeville | 11,711 | 2.29 | 62.1 | 45.9 | 16.2 | 37.9 | 29.1 | 579 | NA | NA | NA | NA | 21,189 | 27.2 | 43.3 |
| Norristown | 14,076 | 2.37 | 60.0 | 21.6 | 36.5 | 42.0 | 36.7 | 941 | 170 | 494 | 543 | 1,577 | 23,394 | 52.9 | 25.8 |
| Philadelphia | 619,505 | 2.48 | 54.7 | 28.1 | 26.6 | 45.3 | 36.1 | 47,589 | NA | NA | NA | NA | 1,088,441 | 46.6 | 31.0 |
| Pittsburgh | 143,276 | 1.94 | 40.6 | 25.7 | 15.0 | 59.4 | 43.6 | 22,120 | NA | NA | NA | NA | 207,348 | 30.6 | 47.0 |
| Plum | 11,147 | 2.42 | 69.3 | 59.7 | 9.6 | 30.7 | 26.0 | 111 | NA | NA | NA | NA | 19,035 | 25.8 | 33.5 |
| Reading | 29,325 | 2.92 | 63.0 | 23.7 | 39.3 | 37.0 | 28.6 | 2,627 | NA | NA | NA | NA | 51,648 | 66.2 | 13.8 |
| Scranton | 30,923 | 2.30 | 60.0 | 38.0 | 22.0 | 40.0 | 33.5 | 5,452 | 1,199 | 1,551 | 1,347 | 1,742 | 51,811 | 53.6 | 24.3 |
| State College | 12,372 | 2.43 | 35.3 | 32.6 | 2.7 | 64.7 | 37.7 | 12,054 | 54 | 92 | 548 | 935 | 14,616 | 15.0 | 71.8 |
| Wilkes-Barre | 16,727 | 2.25 | 56.8 | 32.7 | 24.0 | 43.2 | 35.1 | 3,147 | 242 | 594 | 852 | 2,092 | 26,665 | 53.4 | 16.2 |
| Williamsport | 10,787 | 2.31 | 56.5 | 29.9 | 26.6 | 43.5 | 36.4 | 3,248 | NA | NA | NA | NA | 17,550 | 60.7 | 20.7 |
| York | 17,372 | 2.47 | 51.2 | 23.5 | 27.7 | 48.8 | 35.4 | 1,077 | NA | NA | NA | NA | 28,212 | 66.7 | 13.9 |
| **RHODE ISLAND** | 407,174 | 2.50 | 61.4 | 44.2 | 17.2 | 38.6 | 31.0 | 40,901 | NA | NA | NA | NA | 746,952 | 39.2 | 34.8 |
| Cranston | 31,182 | 2.51 | 61.9 | 44.6 | 17.3 | 38.1 | 31.5 | 3,305 | 127 | 156 | 1,110 | 1,362 | 58,619 | 33.5 | 35.5 |
| East Providence | 19,907 | 2.36 | 58.4 | 44.7 | 13.7 | 41.6 | 37.4 | 595 | 62 | 130 | 498 | 1,046 | 37,177 | 40.3 | 30.2 |
| Pawtucket | 28,452 | 2.52 | 60.8 | 35.7 | 25.1 | 39.2 | 33.5 | 545 | 293 | 407 | 1,640 | 2,277 | 50,197 | 53.1 | 23.2 |
| Providence | 62,046 | 2.65 | 53.1 | 29.8 | 23.3 | 46.9 | 36.0 | 15,498 | 892 | 496 | 5,413 | 3,011 | 113,555 | 46.0 | 33.2 |
| Warwick | 34,833 | 2.31 | 58.7 | 43.4 | 15.2 | 41.3 | 33.3 | 445 | 76 | 94 | 1,239 | 1,534 | 61,103 | 36.2 | 33.3 |
| Woonsocket | 17,100 | 2.40 | 49.2 | 28.2 | 21.0 | 50.8 | 38.2 | 646 | 237 | 568 | 865 | 2,074 | 30,376 | 53.1 | 17.3 |
| **SOUTH CAROLINA** | 1,975,915 | 2.54 | 65.1 | 47.0 | 18.1 | 34.9 | 29.0 | 133,728 | NA | NA | NA | NA | 3,563,204 | 40.2 | 29.6 |
| Aiken | 12,758 | 2.30 | 58.2 | 43.8 | 14.5 | 41.8 | 37.0 | 1,526 | 178 | 576 | 1,499 | 4,848 | 23,024 | 31.1 | 42.8 |
| Anderson | 10,442 | 2.49 | 54.5 | 34.9 | 19.6 | 45.5 | 39.6 | 1,694 | NA | NA | NA | NA | 17,921 | 49.4 | 26.3 |
| Charleston | 58,902 | 2.33 | 54.0 | 43.8 | 10.1 | 46.0 | 31.6 | 6,009 | 516 | 373 | 3,124 | 2,260 | 100,291 | 21.4 | 56.4 |
| Columbia | 45,474 | 2.23 | 48.9 | 32.4 | 16.5 | 51.1 | 40.3 | 30,023 | 1,037 | 775 | 7,027 | 5,252 | 74,991 | 30.8 | 45.9 |
| Florence | 15,499 | 2.44 | 63.0 | 43.1 | 19.9 | 37.0 | 34.4 | 655 | 447 | 1,188 | 2,496 | 6,631 | 25,856 | 32.6 | 35.5 |
| Goose Creek | 14,484 | 2.84 | 72.3 | 52.6 | 19.7 | 27.7 | 21.1 | 2,492 | NA | NA | NA | NA | 26,278 | 32.7 | 27.0 |
| Greenville | 32,250 | 2.05 | 47.8 | 36.3 | 11.6 | 52.2 | 42.6 | 4,366 | 402 | 576 | 2,805 | 4,017 | 49,765 | 24.3 | 52.9 |
| Greer | 12,942 | 2.77 | 64.8 | 52.6 | 12.2 | 35.2 | 29.1 | 199 | 143 | 434 | 1,021 | 3,096 | 24,349 | 36.7 | 32.4 |
| Hilton Head Island | 17,446 | 2.27 | 66.4 | 59.8 | 6.6 | 33.6 | 29.5 | 176 | NA | NA | NA | NA | 32,523 | 21.2 | 57.6 |
| Mount Pleasant | 36,111 | 2.53 | 67.0 | 59.4 | 7.6 | 33.0 | 23.6 | 469 | 105 | 114 | 1,135 | 1,228 | 66,232 | 12.2 | 66.1 |
| Myrtle Beach | 14,792 | 2.34 | 55.6 | 43.3 | 12.3 | 44.4 | 38.6 | 142 | 415 | 1,190 | 3,916 | 11,234 | 25,734 | 38.9 | 31.5 |
| North Charleston | 47,066 | 2.49 | 52.4 | 32.5 | 19.9 | 47.6 | 39.2 | 5,260 | 1,114 | 966 | 6,941 | 6,019 | 81,738 | 44.7 | 25.6 |
| Rock Hill | 32,341 | 2.23 | 56.8 | 34.0 | 22.9 | 43.2 | 33.7 | 2,963 | 498 | 661 | 2,521 | 3,346 | 49,897 | 29.9 | 31.1 |
| Spartanburg | 13,862 | 2.54 | 63.7 | 32.6 | 31.1 | 36.3 | 32.5 | 2,201 | 448 | 1,187 | 2,450 | 6,489 | 22,855 | 47.5 | 30.2 |
| Summerville | 20,815 | 2.70 | 68.2 | 46.2 | 22.0 | 31.8 | 26.4 | 314 | 186 | 352 | 1,662 | 3,143 | 37,771 | 29.1 | 27.4 |
| Sumter | 15,598 | 2.45 | 60.3 | 40.2 | 20.1 | 39.7 | 35.2 | 1,443 | 418 | 1,057 | 1,796 | 4,542 | 25,135 | 38.2 | 27.9 |
| **SOUTH DAKOTA** | 353,799 | 2.40 | 63.3 | 50.2 | 13.1 | 36.7 | 29.9 | 34,018 | NA | NA | NA | NA | 588,029 | 37.9 | 29.7 |
| Aberdeen | 12,867 | 2.28 | 59.1 | 46.4 | 12.7 | 40.9 | 30.5 | 1,090 | 160 | 554 | 629 | 2,179 | 19,600 | 43.0 | 30.9 |
| Rapid City | 31,387 | 2.35 | 55.0 | 42.5 | 12.4 | 45.0 | 38.8 | 3,862 | 540 | 707 | 2,454 | 3,214 | 54,079 | 32.3 | 33.0 |
| Sioux Falls | 76,996 | 2.31 | 60.7 | 47.0 | 13.8 | 39.3 | 31.7 | 6,100 | 897 | 483 | 5,653 | 3,045 | 120,983 | 30.9 | 36.7 |
| **TENNESSEE** | 2,654,737 | 2.51 | 65.1 | 48.0 | 17.1 | 34.9 | 28.7 | 158,077 | NA | NA | NA | NA | 4,693,962 | 43.5 | 28.7 |
| Bartlett | 21,316 | 2.76 | 78.1 | 62.3 | 15.7 | 21.9 | 17.7 | 673 | 188 | 315 | 1,046 | 1,755 | 41,199 | 31.3 | 41.3 |
| Brentwood | 14,010 | 3.05 | 88.0 | 82.2 | 5.8 | 12.0 | 11.8 | 92 | 38 | 88 | 374 | 865 | 26,998 | 10.6 | 75.4 |
| Bristol | 11,201 | 2.36 | 60.2 | 47.7 | 12.5 | 39.8 | 35.8 | 578 | 135 | 502 | 813 | 3,022 | 19,082 | 47.1 | 26.7 |
| Chattanooga | 79,565 | 2.19 | 51.6 | 31.0 | 20.6 | 48.4 | 38.1 | 8,388 | 1,946 | 1,070 | 10,106 | 5,557 | 130,503 | 37.0 | 32.1 |
| Clarksville | 58,985 | 2.62 | 67.1 | 46.7 | 20.4 | 32.9 | 26.0 | 3,363 | 926 | 579 | 4,467 | 2,792 | 95,041 | 33.7 | 28.4 |
| Cleveland | 17,162 | 2.46 | 55.4 | 42.2 | 13.2 | 44.6 | 33.9 | 3,236 | 464 | 1,021 | 2,403 | 5,287 | 30,702 | 43.9 | 26.9 |
| Collierville | 17,190 | 2.96 | 78.7 | 65.6 | 13.1 | 21.3 | 18.1 | 117 | 119 | 232 | 688 | 1,342 | 34,352 | 16.8 | 58.6 |

1. Data for serious crimes have not been adjusted for underreporting. This may affect comparability between geographic areas and over time.
2. Per 100,000 population estimated by the FBI.
3. Persons 25 years old and over.

# Table D. Cities — Income, Poverty, and Housing

| City | Money income, 2019 Households Median income [42] | Percent with income less than $20,000 [43] | Percent with income of $200,000 or more [44] | Median family income [45] | Median non-family income [46] | Median earnings, 2019 All persons [47] | Men [48] | Women [49] | Housing units, 2019 Total [50] | Occupied [51] | Percent owner occupied [52] | Median value[1] (dollars) [53] | Median gross rent (dollars) [54] |
|---|---|---|---|---|---|---|---|---|---|---|---|---|---|
| **OREGON—Cont'd** | | | | | | | | | | | | | |
| Redmond | 68,407 | 11.1 | 2.5 | 88,384 | 47,388 | 32,195 | 35,451 | 31,838 | 11,278 | 11,195 | 65.4 | 326,994 | 1,174 |
| Salem | 61,580 | 16.5 | 3.3 | 72,934 | 35,103 | 31,524 | 36,538 | 26,840 | 64,209 | 61,087 | 56.4 | 286,560 | 1,060 |
| Springfield | 56,802 | 15.5 | 2.0 | 64,375 | 30,994 | 29,263 | 36,324 | 22,108 | 26,476 | 25,325 | 51.3 | 230,916 | 968 |
| Tigard | 87,145 | 11.2 | 10.6 | 108,836 | 53,447 | 42,899 | 52,206 | 30,345 | 21,381 | 20,073 | 65.9 | 230,916 | 968 |
| Tualatin | 85,958 | 10.4 | 9.6 | 107,703 | 46,071 | 48,439 | 55,325 | 44,379 | 11,421 | 10,496 | 52.0 | 462,771 | 1,463 |
| West Linn | 102,049 | 10.5 | 23.8 | 141,351 | 41,758 | 60,421 | 93,283 | 41,958 | 10,835 | 10,435 | 82.5 | 545,733 | 1,561 |
| **PENNSYLVANIA** | 63,463 | 14.3 | 7.3 | 81,075 | 36,748 | 37,000 | 44,105 | 30,813 | 5,732,580 | 5,119,249 | 68.4 | 192,593 | 951 |
| Allentown | 40,835 | 23.1 | 1.0 | 45,223 | 27,279 | 25,369 | 30,688 | 21,106 | 46,003 | 40,959 | 38.5 | 144,016 | 1,057 |
| Altoona | 40,933 | 25.4 | 1.8 | 48,352 | 24,899 | 29,120 | 31,687 | 25,845 | 20,943 | 19,243 | 65.1 | 90,422 | 653 |
| Bethel Park | 85,910 | 7.5 | 7.5 | 104,995 | 39,632 | 54,300 | 63,560 | 44,542 | 13,646 | 13,268 | 77.4 | 208,204 | 1,091 |
| Bethlehem | 60,268 | 16.2 | 4.2 | 69,748 | 36,427 | 30,313 | 32,120 | 26,859 | 31,564 | 29,213 | 50.4 | 175,333 | 1,057 |
| Chester | 39,854 | 26.9 | 0.4 | 43,860 | 25,575 | 26,303 | 29,241 | 25,435 | 13,227 | 11,008 | 37.5 | 84,921 | 918 |
| Easton | 56,029 | 14.9 | 5.5 | 66,411 | 31,207 | 25,445 | 27,172 | 23,333 | 10,484 | 9,462 | 50.8 | 124,608 | 960 |
| Erie | 35,798 | 24.8 | 1.4 | 45,423 | 27,311 | 25,665 | 26,982 | 22,873 | 44,862 | 39,790 | 48.0 | 91,198 | 693 |
| Harrisburg | 37,721 | 23.2 | 1.7 | 36,140 | 35,327 | 30,530 | 35,978 | 25,147 | 25,708 | 21,399 | 40.1 | 81,565 | 828 |
| Hazleton | 36,848 | 24.9 | 1.7 | 39,164 | 28,235 | 27,468 | 32,200 | 26,070 | 10,672 | 9,374 | 43.3 | 114,505 | 855 |
| Lancaster | 55,353 | 21.8 | 1.6 | 59,341 | 35,107 | 30,069 | 32,439 | 25,292 | 23,542 | 23,014 | 41.0 | 144,079 | 955 |
| Lebanon | 35,788 | 21.4 | 2.0 | 36,462 | 27,386 | 28,100 | 30,648 | 25,832 | 11,597 | 10,458 | 38.8 | 114,393 | 857 |
| Monroeville | 82,530 | 12.1 | 6.5 | 103,036 | 39,581 | 44,451 | 58,472 | 34,694 | 13,159 | 11,711 | 63.9 | 154,950 | 1,042 |
| Norristown | 55,222 | 20.0 | 1.5 | 60,521 | 30,892 | 30,221 | 28,865 | 30,947 | 15,207 | 14,076 | 38.7 | 162,491 | 1,048 |
| Philadelphia | 47,474 | 23.6 | 5.7 | 54,978 | 35,626 | 32,355 | 36,131 | 30,601 | 691,653 | 619,505 | 52.3 | 183,191 | 1,079 |
| Pittsburgh | 53,799 | 20.4 | 6.5 | 74,938 | 39,582 | 35,699 | 42,233 | 30,394 | 158,561 | 143,276 | 45.1 | 149,165 | 1,014 |
| Plum | 90,324 | 9.8 | 5.3 | 105,950 | 39,342 | 50,358 | 59,138 | 41,303 | 12,430 | 11,147 | 78.8 | 172,227 | 1,021 |
| Reading | 39,670 | 25.8 | 0.5 | 46,434 | 24,791 | 28,224 | 31,750 | 21,244 | 33,489 | 29,325 | 37.9 | 77,689 | 833 |
| Scranton | 41,169 | 23.9 | 5.0 | 46,637 | 26,559 | 25,902 | 31,542 | 20,613 | 36,634 | 30,923 | 48.3 | 103,616 | 790 |
| State College | 41,516 | 24.6 | 3.9 | 95,988 | 30,095 | 10,012 | 11,289 | 7,231 | 13,634 | 12,372 | 31.4 | 344,948 | 1,011 |
| Wilkes-Barre | 41,219 | 23.2 | 1.7 | 50,060 | 32,652 | 27,975 | 32,206 | 21,527 | 20,127 | 16,727 | 42.7 | 76,816 | 821 |
| Williamsport | 44,824 | 27.0 | 4.1 | 46,477 | 30,275 | 17,123 | 11,973 | 25,302 | 13,640 | 10,787 | 45.6 | 114,558 | 697 |
| York | 34,799 | 26.0 | 0.5 | 45,240 | 26,620 | 25,492 | 27,450 | 22,994 | 19,174 | 17,372 | 34.9 | 73,031 | 823 |
| **RHODE ISLAND** | 71,169 | 13.6 | 7.4 | 89,373 | 40,254 | 39,145 | 45,787 | 34,812 | 470,177 | 407,174 | 61.7 | 282,970 | 1,043 |
| Cranston | 72,930 | 14.7 | 6.4 | 91,116 | 31,721 | 45,516 | 51,396 | 36,674 | 32,471 | 31,182 | 65.9 | 270,574 | 1,101 |
| East Providence | 67,663 | 19.1 | 3.6 | 90,556 | 28,543 | 42,267 | 49,774 | 38,983 | 22,080 | 19,907 | 61.0 | 236,825 | 950 |
| Pawtucket | 54,143 | 19.8 | 1.7 | 67,233 | 35,492 | 32,408 | 37,810 | 30,605 | 31,325 | 28,452 | 45.3 | 217,718 | 935 |
| Providence | 50,097 | 22.6 | 6.1 | 56,064 | 36,136 | 30,634 | 35,382 | 26,133 | 75,963 | 62,046 | 38.0 | 237,178 | 1,059 |
| Warwick | 75,384 | 11.1 | 6.3 | 98,083 | 51,989 | 42,770 | 51,961 | 40,586 | 36,927 | 34,833 | 73.5 | 247,469 | 1,179 |
| Woonsocket | 45,767 | 22.1 | 1.8 | 56,502 | 32,866 | 36,278 | 38,494 | 34,975 | 19,128 | 17,100 | 42.5 | 195,243 | 933 |
| **SOUTH CAROLINA** | 56,227 | 16.4 | 5.4 | 70,537 | 33,610 | 32,373 | 39,893 | 27,579 | 2,351,364 | 1,975,915 | 70.3 | 179,823 | 922 |
| Aiken | 63,974 | 14.8 | 7.3 | 88,321 | 37,223 | 35,666 | 58,410 | 22,422 | 13,763 | 12,758 | 70.7 | 198,023 | 1,162 |
| Anderson | 35,371 | 24.8 | 3.5 | 44,848 | 22,598 | 21,505 | 24,760 | 17,156 | 12,052 | 10,442 | 52.5 | 155,995 | 817 |
| Charleston | 79,359 | 13.3 | 10.1 | 98,992 | 60,014 | 45,716 | 52,039 | 38,127 | 66,893 | 58,902 | 54.6 | 363,566 | 1,378 |
| Columbia | 47,386 | 25.4 | 7.0 | 68,765 | 31,670 | 23,675 | 25,415 | 21,800 | 53,200 | 45,474 | 47.1 | 201,501 | 909 |
| Florence | 51,652 | 22.9 | 5.2 | 70,431 | 35,300 | 32,291 | 41,724 | 28,658 | 18,867 | 15,499 | 61.4 | 158,026 | 841 |
| Goose Creek | 73,505 | 10.6 | 2.6 | 88,743 | 53,005 | 35,675 | 40,531 | 31,666 | 16,061 | 14,484 | 68.3 | 215,516 | 1,314 |
| Greenville | 59,278 | 16.5 | 10.0 | 83,770 | 42,253 | 36,600 | 45,627 | 30,987 | 35,161 | 32,250 | 46.9 | 321,874 | 1,096 |
| Greer | 68,205 | 11.4 | 7.7 | 79,703 | 43,098 | 37,475 | 47,128 | 29,919 | 14,539 | 12,942 | 62.7 | 202,071 | 1,045 |
| Hilton Head Island | 99,246 | 5.8 | 21.7 | 135,687 | 45,503 | 39,078 | 46,883 | 28,359 | 37,066 | 17,446 | 79.8 | 533,919 | 1,399 |
| Mount Pleasant | 121,318 | 6.8 | 24.0 | 152,407 | 61,631 | 60,946 | 90,463 | 44,507 | 39,477 | 36,111 | 79.0 | 514,757 | 1,685 |
| Myrtle Beach | 51,213 | 12.2 | 5.7 | 68,522 | 33,798 | 27,860 | 30,463 | 23,904 | 26,038 | 14,792 | 61.1 | 244,910 | 1,023 |
| North Charleston | 53,470 | 14.9 | 4.1 | 69,278 | 40,983 | 34,255 | 36,945 | 29,662 | 52,834 | 47,066 | 47.0 | 199,986 | 1,093 |
| Rock Hill | 51,879 | 12.4 | 2.8 | 62,847 | 41,140 | 30,791 | 37,108 | 26,965 | 34,067 | 32,341 | 49.4 | 176,775 | 1,058 |
| Spartanburg | 42,688 | 25.5 | 3.4 | 47,323 | 25,183 | 27,400 | 36,110 | 25,962 | 16,126 | 13,862 | 50.2 | 150,419 | 897 |
| Summerville | 61,951 | 8.2 | 4.2 | 76,974 | 43,698 | 37,900 | 45,788 | 32,973 | 22,188 | 20,815 | 65.3 | 223,949 | 1,155 |
| Sumter | 45,296 | 20.8 | 3.0 | 57,002 | 32,750 | 27,362 | 36,874 | 20,934 | 17,673 | 15,598 | 52.0 | 146,130 | 854 |
| **SOUTH DAKOTA** | 59,533 | 12.6 | 4.5 | 76,826 | 35,697 | 32,908 | 40,152 | 28,766 | 401,749 | 353,799 | 67.8 | 185,047 | 769 |
| Aberdeen | 56,169 | 8.7 | 3.4 | 70,655 | 38,893 | 31,859 | 35,770 | 28,154 | 14,051 | 12,867 | 57.7 | 188,532 | 678 |
| Rapid City | 51,296 | 17.3 | 5.5 | 71,931 | 27,503 | 27,579 | 28,666 | 27,054 | 33,517 | 31,387 | 59.2 | 207,329 | 828 |
| Sioux Falls | 61,058 | 10.2 | 5.8 | 79,765 | 39,445 | 38,211 | 42,152 | 33,019 | 80,743 | 76,996 | 59.3 | 218,946 | 849 |
| **TENNESSEE** | 56,071 | 15.8 | 5.4 | 69,993 | 33,992 | 32,604 | 39,010 | 28,699 | 3,028,437 | 2,654,737 | 66.5 | 191,853 | 904 |
| Bartlett | 87,255 | 8.9 | 6.3 | 102,257 | 40,006 | 42,665 | 46,542 | 38,410 | 22,445 | 21,316 | 83.8 | 214,023 | 1,559 |
| Brentwood | 171,089 | 3.9 | 40.9 | 183,940 | 60,420 | 62,400 | 113,068 | 35,721 | 14,212 | 14,010 | 93.5 | 689,177 | 2,055 |
| Bristol | 45,750 | 19.0 | 2.5 | 56,922 | 35,517 | 34,347 | 34,028 | 35,898 | 11,835 | 11,201 | 66.8 | 158,708 | 627 |
| Chattanooga | 46,533 | 19.9 | 5.5 | 58,193 | 35,687 | 30,713 | 33,496 | 26,321 | 89,597 | 79,565 | 49.1 | 198,298 | 907 |
| Clarksville | 51,281 | 13.0 | 1.9 | 62,524 | 40,332 | 31,707 | 40,357 | 24,512 | 64,806 | 58,985 | 50.6 | 158,715 | 923 |
| Cleveland | 46,938 | 21.4 | 4.3 | 72,932 | 34,880 | 24,436 | 32,126 | 19,300 | 18,945 | 17,162 | 52.5 | 175,594 | 779 |
| Collierville | 118,416 | 3.6 | 26.9 | 122,817 | 73,004 | 51,184 | 86,665 | 30,292 | 17,815 | 17,190 | 79.6 | 346,664 | 1,286 |

1. Based on population estimated by the American Community Survey.

# Table D. Cities — Commuting, Computer Access, Migration, Labor Force, and Employment

| City | Commuting[1], 2019 — Drove alone | With commutes of 30 minutes or more | Computer access[2], 2019 — With a computer in the house | With Internet access | Migration, 2019 — Percent who lived in the same house one year ago | Percent who lived in another state or county one year ago | Civilian labor force, 2020 — Total | Percent change 2019-2020 | Unemployment — Total | Rate[3] | Pop. age 16 and older — Number | Percent in labor force | Pop. age 16 to 64 — Number | Percent who worked full-year full-time |
|---|---|---|---|---|---|---|---|---|---|---|---|---|---|---|
| | 55 | 56 | 57 | 58 | 59 | 60 | 61 | 62 | 63 | 64 | 65 | 66 | 67 | 68 |
| **OREGON—Cont'd** | | | | | | | | | | | | | | |
| Redmond | 67.6 | 32.8 | 97.0 | 90.6 | 75.5 | 9.5 | 13,550 | 4.3 | 1,225 | 9 | 25,833 | 65.7 | 20,306 | 49.0 |
| Salem | 78.3 | 34.6 | 95.9 | 90.2 | 84.8 | 7.6 | 81,917 | 1.0 | 5,748 | 7 | 136,643 | 62.0 | 113,368 | 43.3 |
| Springfield | 73.5 | 17.7 | 92.6 | 87.9 | 78.5 | 6.6 | 30,525 | -1.2 | 2,585 | 9 | 51,728 | 69.5 | 42,605 | 48.1 |
| Tigard | 74.7 | 37.0 | 94.9 | 93.1 | 90.5 | 4.9 | 31,484 | -0.1 | 2,090 | 7 | 45,063 | 69.5 | 38,005 | 53.3 |
| Tualatin | 78.0 | 30.4 | 91.9 | 88.9 | 86.0 | 11.9 | 15,158 | -1.0 | 983 | 7 | 22,103 | 67.2 | 17,711 | 57.2 |
| West Linn | 69.2 | 32.6 | 93.4 | 90.6 | NA | NA | 14,348 | -2.4 | 812 | 6 | 21,010 | 65.0 | 16,572 | 51.5 |
| **PENNSYLVANIA** | 75.2 | 39.8 | 90.7 | 85.0 | 87.5 | 5.4 | 6,387,869 | -1.6 | 579,927 | 9 | 10,474,419 | 62.8 | 8,086,201 | 53.6 |
| Allentown | 74.8 | 27.6 | 86.1 | 75.9 | 87.0 | 5.8 | 56,524 | 1.3 | 7,745 | 14 | 93,518 | 63.4 | 79,968 | 46.4 |
| Altoona | 83.8 | 17.4 | 88.8 | 82.2 | 85.2 | 4.5 | 20,038 | 0.6 | 2,114 | 11 | 34,578 | 58.6 | 26,763 | 47.0 |
| Bethel Park | 74.3 | 54.3 | 94.9 | 91.7 | NA | NA | 17,076 | -3.2 | 1,311 | 8 | 25,873 | 67.5 | 18,887 | 67.2 |
| Bethlehem | 72.4 | 24.7 | 91.4 | 86.8 | 80.2 | 11.5 | 37,741 | -1.0 | 3,954 | 11 | 64,669 | 61.2 | 50,928 | 51.0 |
| Chester | 61.8 | 25.3 | 91.6 | 82.6 | 83.2 | 8.7 | 13,562 | 1.0 | 2,025 | 15 | 25,655 | 56.4 | 21,572 | 40.5 |
| Easton | 67.5 | 33.9 | 80.7 | 71.4 | 81.9 | 11.5 | 12,140 | -0.8 | 1,408 | 12 | 22,675 | 57.9 | 17,620 | 52.3 |
| Erie | 71.5 | 13.9 | 86.8 | 78.5 | 80.9 | 5.7 | 43,164 | -0.2 | 5,087 | 12 | 76,445 | 57.3 | 61,176 | 44.7 |
| Harrisburg | 63.2 | 28.3 | 87.9 | 78.0 | 73.1 | 13.2 | 22,441 | 3.3 | 3,201 | 14 | 36,926 | 65.3 | 30,887 | 51.4 |
| Hazleton | 70.1 | 20.1 | 85.4 | 74.4 | 87.4 | 6.6 | 11,924 | 3.6 | 2,006 | 17 | 19,092 | 64.2 | 15,717 | 52.0 |
| Lancaster | 72.0 | 21.2 | 92.2 | 83.9 | 78.9 | 10.8 | 27,671 | 1.8 | 3,416 | 12 | 48,711 | 67.5 | 42,329 | 52.5 |
| Lebanon | 68.7 | 32.0 | 92.3 | 74.9 | 72.2 | 17.6 | 11,976 | 1.5 | 1,524 | 13 | 20,201 | 65.0 | 16,621 | 52.9 |
| Monroeville | 75.6 | 50.6 | 96.2 | 89.0 | 89.0 | 4.2 | 14,120 | -3.0 | 1,202 | 9 | 23,575 | 59.2 | 17,003 | 57.3 |
| Norristown | 75.1 | 44.8 | 91.7 | 76.0 | 83.3 | 6.4 | 18,205 | -0.5 | 1,893 | 10 | 27,668 | 69.1 | 23,935 | 52.8 |
| Philadelphia | 48.8 | 56.4 | 90.7 | 82.8 | 85.3 | 5.1 | 723,938 | 0.3 | 89,529 | 12 | 1,275,366 | 61.9 | 1,054,016 | 47.6 |
| Pittsburgh | 54.1 | 35.2 | 91.3 | 85.9 | 77.6 | 9.8 | 154,697 | -1.8 | 14,744 | 10 | 260,534 | 65.9 | 214,823 | 51.2 |
| Plum | 74.6 | 50.2 | 93.6 | 88.6 | 89.9 | 4.3 | 14,559 | -2.8 | 1,158 | 8 | 21,814 | 65.6 | 16,023 | 64.2 |
| Reading | 68.7 | 28.3 | 85.9 | 76.9 | 73.6 | 14.0 | 35,209 | 1.9 | 5,630 | 16 | 66,729 | 61.6 | 57,354 | 48.1 |
| Scranton | 70.0 | 16.1 | 89.7 | 85.2 | 77.2 | 8.9 | 36,253 | -0.6 | 3,862 | 11 | 64,330 | 57.7 | 50,282 | 42.7 |
| State College | 44.6 | 10.3 | 98.9 | 85.8 | 50.0 | 29.9 | 15,686 | -6.2 | 845 | 5 | 40,328 | 47.1 | 37,002 | 18.6 |
| Wilkes-Barre | 71.8 | 23.0 | 88.5 | 81.8 | 71.0 | 15.6 | 18,807 | 1.7 | 2,550 | 14 | 32,812 | 60.3 | 26,772 | 46.1 |
| Williamsport | 72.1 | 20.5 | 89.4 | 78.8 | 74.7 | 7.5 | 13,131 | -0.7 | 1,394 | 11 | 23,166 | 52.9 | 18,924 | 38.0 |
| York | 67.5 | 26.8 | 86.4 | 74.8 | 81.2 | 10.8 | 18,638 | 4.9 | 3,203 | 17 | 33,512 | 63.0 | 28,982 | 45.9 |
| **RHODE ISLAND** | 79.6 | 35.7 | 92.0 | 86.8 | 88.6 | 5.4 | 541,680 | -2.5 | 50,835 | 9 | 880,978 | 64.9 | 693,823 | 53.3 |
| Cranston | 82.5 | 25.1 | 91.4 | 85.9 | 93.7 | 2.1 | 40,701 | -2.0 | 3,938 | 10 | 66,238 | 61.4 | 51,729 | 53.3 |
| East Providence | 82.5 | 35.7 | 88.3 | 83.6 | 87.4 | 6.9 | 23,809 | -2.2 | 2,309 | 10 | 39,958 | 62.4 | 28,980 | 59.3 |
| Pawtucket | 80.5 | 36.0 | 89.9 | 80.7 | 90.1 | 3.8 | 36,306 | -0.6 | 4,161 | 12 | 58,372 | 69.4 | 48,675 | 58.3 |
| Providence | 65.4 | 30.8 | 90.6 | 85.7 | 83.4 | 6.6 | 85,714 | -0.9 | 9,808 | 11 | 147,854 | 64.8 | 125,650 | 46.4 |
| Warwick | 83.6 | 33.1 | 92.5 | 86.9 | 91.8 | 4.3 | 44,390 | -2.9 | 3,750 | 8 | 68,013 | 68.4 | 51,133 | 58.3 |
| Woonsocket | 80.6 | 47.4 | 96.2 | 82.4 | 84.8 | 4.9 | 18,946 | -0.5 | 2,269 | 12 | 33,831 | 66.7 | 28,071 | 54.0 |
| **SOUTH CAROLINA** | 82.1 | 36.8 | 92.0 | 82.1 | 86.7 | 7.0 | 2,384,590 | 0.4 | 147,183 | 6 | 4,158,317 | 59.5 | 3,222,779 | 52.8 |
| Aiken | 87.4 | 35.0 | 92.4 | 87.3 | 91.5 | 4.6 | 13,400 | 0.8 | 805 | 6 | 25,844 | 51.3 | 17,347 | 51.3 |
| Anderson | 83.6 | 31.0 | 90.3 | 77.1 | 82.3 | 6.0 | 11,656 | 0.2 | 831 | 7 | 22,660 | 56.2 | 17,585 | 44.8 |
| Charleston | 73.6 | 30.7 | 95.5 | 87.0 | 78.1 | 10.7 | 74,909 | -0.3 | 4,827 | 6 | 117,775 | 65.8 | 95,554 | 62.0 |
| Columbia | 61.0 | 14.0 | 95.9 | 84.4 | 63.6 | 25.1 | 58,048 | -1.3 | 3,548 | 6 | 111,779 | 51.9 | 97,309 | 39.4 |
| Florence | 88.9 | 18.3 | 90.7 | 76.4 | 91.9 | 1.7 | 19,725 | 2.6 | 1,145 | 6 | 29,741 | 58.1 | 22,530 | 52.4 |
| Goose Creek | 75.3 | 45.8 | 98.5 | 92.2 | 72.3 | 21.8 | 19,923 | -1.1 | 1,057 | 5 | 35,077 | 57.8 | 28,586 | 56.6 |
| Greenville | 77.5 | 21.5 | 91.6 | 84.1 | 73.0 | 12.0 | 36,497 | 1.4 | 2,121 | 6 | 60,616 | 67.1 | 50,167 | 53.2 |
| Greer | 83.8 | 39.4 | 92.0 | 87.9 | 85.4 | 10.2 | 17,422 | 2.5 | 906 | 5 | 28,707 | 65.8 | 22,789 | 59.9 |
| Hilton Head Island | 69.9 | 14.2 | 95.0 | 89.1 | 87.1 | 5.5 | 17,796 | -1.7 | 979 | 6 | 35,398 | 50.7 | 19,836 | 50.3 |
| Mount Pleasant | 78.9 | 37.0 | 98.3 | 94.5 | 87.5 | 6.7 | 49,331 | -0.7 | 2,117 | 4 | 75,014 | 68.1 | 57,799 | 62.3 |
| Myrtle Beach | 73.6 | 20.7 | 91.3 | 85.2 | 87.8 | 6.7 | 15,880 | 1.3 | 1,771 | 11 | 28,145 | 63.7 | 20,076 | 55.1 |
| North Charleston | 84.0 | 31.1 | 95.3 | 69.3 | 83.3 | 9.2 | 55,867 | 1.5 | 4,288 | 8 | 96,049 | 65.3 | 82,022 | 62.0 |
| Rock Hill | 80.6 | 37.1 | 95.2 | 90.7 | 81.7 | 9.9 | 40,010 | 2.3 | 3,135 | 8 | 61,398 | 69.3 | 50,882 | 58.1 |
| Spartanburg | 78.7 | 29.6 | 85.7 | 74.6 | 82.0 | 7.3 | 17,680 | 3.0 | 1,587 | 9 | 28,041 | 60.5 | 21,330 | 48.2 |
| Summerville | 90.4 | 48.5 | 96.8 | 82.8 | 86.0 | 9.1 | 24,983 | -0.8 | 1,411 | 6 | 43,894 | 70.0 | 37,287 | 54.9 |
| Sumter | 87.8 | 20.2 | 85.6 | 74.5 | 81.5 | 10.1 | 15,626 | -0.2 | 1,149 | 7 | 30,385 | 55.2 | 24,747 | 54.6 |
| **SOUTH DAKOTA** | 79.8 | 16.2 | 91.5 | 84.6 | 84.7 | 8.6 | 463,256 | -0.2 | 21,511 | 5 | 692,113 | 67.5 | 538,314 | 57.9 |
| Aberdeen | 79.1 | 3.6 | 95.3 | 88.4 | 82.6 | 6.3 | 14,978 | -0.5 | 729 | 5 | 24,454 | 73.8 | 19,030 | 64.4 |
| Rapid City | 80.6 | 10.1 | 93.8 | 85.6 | 83.8 | 7.1 | 37,717 | 1.5 | 2,004 | 5 | 62,160 | 63.2 | 47,377 | 51.4 |
| Sioux Falls | 83.6 | 8.9 | 94.3 | 90.8 | 77.6 | 11.0 | 107,672 | 1.1 | 5,006 | 5 | 142,123 | 74.1 | 117,184 | 63.5 |
| **TENNESSEE** | 82.0 | 36.2 | 90.5 | 82.2 | 85.6 | 6.5 | 3,289,426 | -1.7 | 245,532 | 8 | 5,487,875 | 61.4 | 4,348,910 | 52.7 |
| Bartlett | 89.1 | 41.8 | 96.2 | 92.2 | 88.2 | 6.0 | 30,692 | -3.7 | 1,686 | 6 | 47,715 | 66.0 | 36,674 | 60.6 |
| Brentwood | 73.5 | 38.9 | 97.9 | 96.6 | 92.7 | 3.6 | 21,509 | -4.9 | 968 | 5 | 32,481 | 63.2 | 26,130 | 49.4 |
| Bristol | 85.6 | 24.6 | 88.0 | 80.0 | 90.4 | 4.4 | 11,531 | -2.7 | 794 | 7 | 22,005 | 51.7 | 16,155 | 44.6 |
| Chattanooga | 78.3 | 19.4 | 89.2 | 83.8 | 79.8 | 6.2 | 87,497 | -0.2 | 7,452 | 9 | 153,712 | 64.3 | 122,133 | 52.8 |
| Clarksville | 84.3 | 32.3 | 96.5 | 89.5 | 79.1 | 12.7 | 63,468 | 0.1 | 5,400 | 9 | 119,177 | 57.3 | 106,312 | 48.6 |
| Cleveland | 82.4 | 22.5 | 90.1 | 82.9 | 79.3 | 5.3 | 21,724 | 0.9 | 1,629 | 8 | 37,554 | 61.4 | 28,920 | 46.2 |
| Collierville | 84.2 | 39.5 | 98.2 | 96.2 | 88.4 | 5.6 | 25,539 | -3.9 | 1,147 | 5 | 39,511 | 64.7 | 32,167 | 55.9 |

1. Employed persons.  2. Households.  3. Percent of civilian labor force.  4. Persons 16 years old and over.

# Table D. Cities — Construction, Wholesale Trade, and Retail Trade

| City | Value of residential construction authorized by building permits, 2020 | | | Wholesale trade[1], 2017 | | | | Retail trade[2], 2017 | | | |
|---|---|---|---|---|---|---|---|---|---|---|---|
| | New construction ($1,000) | Number of housing units | Percent single family | Number of establishments | Number of employees | Sales (mil dol) | Annual payroll (mil dol) | Number of establishments | Number of employees | Sales (mil dol) | Annual payroll (mil dol) |
| | 69 | 70 | 71 | 72 | 73 | 74 | 75 | 76 | 77 | 78 | 79 |
| OREGON—Cont'd | | | | | | | | | | | |
| Redmond | 147,804 | 703 | 78 | 42 | 253 | 118 | 12 | 126 | 2,090 | 682 | 60 |
| Salem | 219,303 | 1,254 | 38 | 155 | 1,722 | 1,275 | 89 | 621 | 10,944 | 3,409 | 319 |
| Springfield | 31,282 | 166 | 54 | 49 | 1,024 | 520 | 52 | 211 | 3,698 | 970 | 97 |
| Tigard | 94,313 | 632 | 21 | 170 | 2,629 | 2,151 | 202 | 321 | 7,553 | 2,166 | 238 |
| Tualatin | 0 | 0 | 0 | 114 | 1,860 | 1,096 | 122 | 321 | 7,553 | 2,166 | 238 |
| West Linn | 10,735 | 22 | 100 | 20 | 73 | 76 | 5 | 105 | 2,130 | 635 | 68 |
| | | | | | | | | 40 | 568 | 143 | 16 |
| PENNSYLVANIA | 27 | 1 | 100 | 12,071 | 204,256 | 199,204 | 12,959 | 42,514 | 662,560 | 234,836 | 17,354 |
| Allentown | 0 | 0 | 0 | 124 | 2,582 | 2,071 | 244 | 374 | 5,997 | 2,473 | 179 |
| Altoona | 215 | 2 | 100 | 46 | 791 | 520 | 42 | 228 | 4,940 | 1,252 | 113 |
| Bethel Park | 2,779 | 7 | 100 | 37 | 272 | 126 | 16 | 101 | 2,627 | 730 | 73 |
| Bethlehem | 19,630 | 90 | 24 | 73 | 3,434 | 10,488 | 176 | 219 | 3,576 | 1,642 | 91 |
| Chester | 825 | 7 | 71 | 20 | 201 | 188 | 14 | 63 | 341 | 94 | 8 |
| Easton | 7,724 | 73 | 7 | 27 | 401 | 659 | 29 | 136 | 1,859 | 441 | 45 |
| Erie | 394 | 4 | 100 | 102 | 1,741 | 756 | 86 | 339 | 4,889 | 906 | 105 |
| Harrisburg | 8,602 | 70 | 0 | 60 | 1,502 | 2,727 | 86 | 181 | 2,266 | 636 | 53 |
| Hazleton | 1,169 | 38 | 3 | 32 | 482 | 163 | 25 | 110 | 1,055 | 249 | 23 |
| Lancaster | 2,110 | 10 | 100 | 53 | 869 | 515 | 40 | 345 | 5,430 | 1,222 | 135 |
| Lebanon | 0 | 0 | 0 | 17 | 116 | 43 | 6 | 99 | 1,396 | 320 | 34 |
| Monroeville | 4,544 | 25 | 100 | 38 | 317 | 128 | 17 | 268 | 5,533 | 1,798 | 137 |
| Norristown | 6,801 | 42 | 0 | 35 | 399 | 383 | 34 | 90 | 627 | 239 | 22 |
| Philadelphia | 823,775 | 5,665 | 17 | 1,006 | 17,207 | 16,323 | 1,169 | 4,540 | 53,157 | 14,973 | 1,305 |
| Pittsburgh | 92,072 | 620 | 27 | 380 | 7,665 | 7,478 | 531 | 1,121 | 17,233 | 5,116 | 480 |
| Plum | 11,125 | 42 | 100 | 29 | 217 | 109 | 11 | 50 | 572 | 144 | 16 |
| Reading | 0 | 0 | 0 | 51 | 1,282 | 1,249 | 78 | 226 | 3,225 | 828 | 77 |
| Scranton | 749 | 3 | 100 | 76 | 1,235 | 1,911 | 63 | 303 | 3,802 | 1,075 | 102 |
| State College | 1,400 | 2 | 100 | 8 | 51 | 39 | 3 | 115 | 1,654 | 254 | 30 |
| Wilkes-Barre | 1,616 | 11 | 9 | 41 | 313 | 145 | 15 | 242 | 4,820 | 1,328 | 113 |
| Williamsport | 217 | 2 | 100 | 35 | 938 | 658 | 34 | 98 | 1,410 | 314 | 36 |
| York | 8,955 | 57 | 21 | 54 | 901 | 628 | 54 | 128 | 1,482 | 541 | 41 |
| RHODE ISLAND | 3,875 | 22 | 46 | 1,107 | 16,847 | 13,637 | 1,042 | 3,769 | 48,753 | 13,844 | 1,444 |
| Cranston | 6,656 | 59 | 49 | 122 | 2,081 | 1,402 | 138 | 300 | 5,101 | 1,284 | 157 |
| East Providence | 855 | 7 | 100 | 68 | 1,363 | 934 | 93 | 132 | 1,884 | 776 | 64 |
| Pawtucket | 2,892 | 22 | 64 | 64 | 2,278 | 1,082 | 94 | 195 | 1,783 | 502 | 52 |
| Providence | 28,868 | 213 | 9 | 179 | 1,926 | 2,555 | 159 | 649 | 6,659 | 1,480 | 172 |
| Warwick | 12,164 | 60 | 53 | 133 | 1,845 | 1,057 | 126 | 417 | 8,259 | 2,412 | 239 |
| Woonsocket | 4,476 | 35 | 100 | 39 | 658 | 296 | 39 | 131 | 1,372 | 422 | 48 |
| SOUTH CAROLINA | 18,131 | 108 | 44 | 4,360 | 61,421 | 57,337 | 3,538 | 17,700 | 253,384 | 69,980 | 6,206 |
| Aiken | 31,813 | 143 | 96 | 20 | 94 | 125 | 5 | 209 | 3,610 | 866 | 81 |
| Anderson | 16,434 | 134 | 100 | 25 | 202 | 73 | 8 | 243 | 3,951 | 813 | 87 |
| Charleston | 168,450 | 1,281 | 75 | 138 | 2,180 | 2,514 | 159 | 784 | 11,100 | 3,704 | 330 |
| Columbia | 98,020 | 545 | 99 | 140 | 3,118 | 2,507 | 191 | 670 | 11,864 | 3,260 | 287 |
| Florence | 38,000 | 291 | 58 | 55 | 1,006 | 795 | 55 | 365 | 6,543 | 1,671 | 151 |
| Goose Creek | 69,548 | 380 | 100 | D | D | D | D | 83 | 1,944 | 501 | 47 |
| Greenville | 159,090 | 1,146 | 20 | 233 | 3,016 | 6,272 | 207 | 684 | 11,380 | 3,245 | 307 |
| Greer | 278,781 | 1,482 | 78 | 45 | 418 | 223 | 23 | 148 | 3,210 | 995 | 91 |
| Hilton Head Island | 94,621 | 188 | 100 | 55 | 299 | 143 | 13 | 262 | 2,722 | 671 | 72 |
| Mount Pleasant | 309,273 | 1,136 | 58 | 84 | 428 | 206 | 25 | 390 | 6,224 | 1,574 | 158 |
| Myrtle Beach | 167,293 | 505 | 100 | 82 | 795 | 394 | 46 | 668 | 10,778 | 2,579 | 251 |
| North Charleston | 142,507 | 1,396 | 48 | 211 | 3,594 | 2,151 | 205 | 570 | 9,439 | 2,545 | 239 |
| Rock Hill | 59,203 | 219 | 100 | 80 | 1,453 | 1,177 | 88 | 322 | 5,741 | 1,425 | 134 |
| Spartanburg | 8,748 | 37 | 100 | 58 | 432 | 313 | 25 | 344 | 6,234 | 1,424 | 142 |
| Summerville | 79,060 | 432 | 50 | 35 | 409 | 504 | 20 | 223 | 4,891 | 1,222 | 113 |
| Sumter | NA | NA | NA | 29 | 269 | 109 | 13 | 250 | 3,754 | 907 | 85 |
| SOUTH DAKOTA | 0 | 0 | 0 | 1,392 | 16,630 | 17,314 | 887 | 3,884 | 53,134 | 14,674 | 1,372 |
| Aberdeen | 4,815 | 44 | 100 | 41 | 714 | 1,181 | 37 | 160 | 2,746 | 737 | 74 |
| Rapid City | 120,632 | 913 | 28 | 127 | 1,514 | 906 | 78 | 478 | 8,423 | 2,567 | 233 |
| Sioux Falls | 418,567 | 2,767 | 43 | 326 | 5,479 | 4,038 | 322 | 783 | 16,116 | 4,349 | 435 |
| TENNESSEE | 6,066 | 25 | 100 | 5,864 | 100,044 | 111,030 | 5,825 | 22,593 | 322,218 | 101,978 | 8,574 |
| Bartlett | 9,594 | 35 | 100 | 61 | 857 | 466 | 56 | 127 | 1,869 | 667 | 60 |
| Brentwood | 96,149 | 141 | 100 | 67 | 644 | 638 | 52 | 169 | 3,386 | 1,260 | 111 |
| Bristol | 11,500 | 159 | 7 | 41 | 303 | 93 | 13 | 152 | 2,932 | 787 | 73 |
| Chattanooga | 164,146 | 1,038 | 49 | 375 | 5,996 | 3,883 | 364 | 1,084 | 20,026 | 6,224 | 564 |
| Clarksville | 208,955 | 2,248 | 86 | 70 | 915 | 400 | 42 | 491 | 8,241 | 2,098 | 210 |
| Cleveland | 41,480 | 245 | 85 | 46 | 556 | 769 | 42 | 280 | 4,265 | 1,150 | 113 |
| Collierville | 44,705 | 191 | 100 | 43 | 894 | 804 | 49 | 207 | 3,877 | 1,143 | 95 |

1. Merchant wholesalers except manufacturers' sales branches and offices.  2. Establishments with payroll.

# Table D. Cities — Real Estate, Professional Services, and Manufacturing

| City | Real estate and rental and leasing, 2017 | | | | Professional, scientific, and technical services[1], 2017 | | | | Manufacturing, 2017 | | | |
|---|---|---|---|---|---|---|---|---|---|---|---|---|
| | Number of establishments | Number of employees | Receipts (mil dol) | Annual payroll (mil dol) | Number of establishments | Number of employees | Receipts (mil dol) | Annual payroll (mil dol) | Number of establishments | Number of employees | Receipts (mil dol) | Annual payroll (mil dol) |
| | 80 | 81 | 82 | 83 | 84 | 85 | 86 | 87 | 88 | 89 | 90 | 91 |
| **OREGON—Cont'd** | | | | | | | | | | | | |
| Redmond | 54 | 126 | 34 | 4 | 54 | 308 | 39 | 24 | NA | NA | NA | NA |
| Salem | 311 | 1,335 | 300 | 58 | 500 | 2,972 | 402 | 160 | NA | NA | NA | NA |
| Springfield | 71 | 272 | 57 | 11 | 95 | 671 | 61 | 27 | NA | NA | NA | NA |
| Tigard | 132 | 614 | 294 | 45 | 347 | 3,296 | 493 | 215 | NA | NA | NA | NA |
| Tualatin | 78 | 326 | 161 | 16 | 107 | 1,264 | 184 | 61 | NA | NA | NA | NA |
| West Linn | D | D | D | D | 131 | 415 | 68 | 26 | NA | NA | NA | NA |
| **PENNSYLVANIA** | 10,662 | 66,715 | 18,861 | 3,472 | 29,991 | 340,361 | 64,179 | 26,674 | 13,537 | 536,967 | 227,812 | 30,393 |
| Allentown | 102 | 509 | 150 | 23 | 215 | 1,282 | 205 | 90 | NA | NA | NA | NA |
| Altoona | 48 | 213 | 39 | 7 | 84 | 659 | 78 | 35 | NA | NA | NA | NA |
| Bethel Park | 27 | 97 | 33 | 4 | 95 | 504 | 79 | 35 | NA | NA | NA | NA |
| Bethlehem | 70 | 520 | 110 | 23 | 187 | 1,481 | 312 | 107 | NA | NA | NA | NA |
| Chester | 15 | 44 | 10 | 2 | 16 | 262 | 73 | 23 | NA | NA | NA | NA |
| Easton | 24 | 70 | 15 | 3 | 76 | 531 | 57 | 33 | NA | NA | NA | NA |
| Erie | 54 | 284 | 48 | 10 | 181 | 1,735 | 293 | 101 | NA | NA | NA | NA |
| Harrisburg | 57 | 263 | 77 | 13 | 250 | 3,168 | 592 | 256 | NA | NA | NA | NA |
| Hazleton | 15 | 38 | 10 | 1 | 42 | 139 | 17 | 5 | NA | NA | NA | NA |
| Lancaster | 49 | 328 | 54 | 13 | 225 | 2,087 | 308 | 128 | NA | NA | NA | NA |
| Lebanon | 29 | 170 | 26 | 6 | 47 | 325 | 49 | 15 | NA | NA | NA | NA |
| Monroeville | 60 | 409 | 86 | 19 | 89 | 1,291 | 169 | 101 | NA | NA | NA | NA |
| Norristown | 25 | 139 | 24 | 5 | 77 | 491 | 79 | 34 | NA | NA | NA | NA |
| Philadelphia | 1,293 | 10,704 | 3,028 | 617 | 3,123 | 58,180 | 13,016 | 5,290 | NA | NA | NA | NA |
| Pittsburgh | 453 | 4,909 | 1,259 | 234 | 1,569 | 30,512 | 6,465 | 2,643 | NA | NA | NA | NA |
| Plum | 18 | 48 | 9 | 3 | 35 | 502 | 176 | 50 | NA | NA | NA | NA |
| Reading | 137 | D | 121 | D | 125 | 1,973 | 253 | 142 | NA | NA | NA | NA |
| Scranton | 60 | 334 | 60 | 11 | 206 | 1,237 | 176 | 75 | NA | NA | NA | NA |
| State College | D | D | D | D | 85 | 551 | 94 | 35 | NA | NA | NA | NA |
| Wilkes-Barre | 48 | 312 | 91 | 16 | 109 | 1,356 | 196 | 77 | NA | NA | NA | NA |
| Williamsport | 34 | 260 | 44 | 10 | 74 | 661 | 82 | 35 | NA | NA | NA | NA |
| York | 33 | 296 | 57 | 14 | 136 | 1,913 | 273 | 105 | NA | NA | NA | NA |
| **RHODE ISLAND** | 1,110 | 5,287 | 1,351 | 248 | 3,026 | 23,910 | 4,044 | 1,566 | 1,340 | 40,221 | 12,416 | 2,390 |
| Cranston | 83 | 333 | 85 | 16 | 227 | 1,775 | 337 | 103 | NA | NA | NA | NA |
| East Providence | 47 | 231 | 46 | 11 | 130 | 1,099 | 192 | 68 | NA | NA | NA | NA |
| Pawtucket | 64 | 225 | 52 | 11 | 111 | 819 | 117 | 47 | NA | NA | NA | NA |
| Providence | 213 | 1,354 | 354 | 68 | 792 | 6,865 | 1,355 | 552 | NA | NA | NA | NA |
| Warwick | 118 | 939 | 252 | 42 | 353 | 3,072 | 334 | 131 | NA | NA | NA | NA |
| Woonsocket | 35 | 135 | 26 | 6 | 39 | 302 | 71 | 25 | NA | NA | NA | NA |
| **SOUTH CAROLINA** | 5,890 | 26,764 | 6,510 | 1,129 | 10,859 | 100,377 | 17,038 | 6,498 | 3,827 | 225,237 | 138,587 | 12,777 |
| Aiken | 45 | 162 | 32 | 6 | 112 | 1,291 | 206 | 91 | NA | NA | NA | NA |
| Anderson | 55 | 162 | 37 | 5 | 124 | 711 | 76 | 29 | NA | NA | NA | NA |
| Charleston | 434 | 1,617 | 398 | 100 | 855 | 6,668 | 1,194 | 490 | NA | NA | NA | NA |
| Columbia | 254 | 1,548 | 753 | 93 | 825 | 8,579 | 1,684 | 587 | NA | NA | NA | NA |
| Florence | 64 | 304 | 67 | 13 | 140 | 1,111 | 180 | 56 | NA | NA | NA | NA |
| Goose Creek | 27 | 105 | 41 | 4 | 65 | 838 | 144 | 60 | NA | NA | NA | NA |
| Greenville | 333 | 1,878 | 373 | 82 | 954 | 15,341 | 3,704 | 1,296 | NA | NA | NA | NA |
| Greer | 42 | 162 | 36 | 5 | 87 | 638 | 98 | 37 | NA | NA | NA | NA |
| Hilton Head Island | 252 | 1,119 | 276 | 53 | 249 | 982 | 171 | 63 | NA | NA | NA | NA |
| Mount Pleasant | 323 | 818 | 240 | 36 | 530 | 3,244 | 705 | 275 | NA | NA | NA | NA |
| Myrtle Beach | 242 | 1,712 | 318 | 65 | 258 | 1,452 | 211 | 77 | NA | NA | NA | NA |
| North Charleston | 191 | 1,412 | 451 | 61 | 376 | 7,183 | 1,375 | 534 | NA | NA | NA | NA |
| Rock Hill | 106 | 433 | 113 | 20 | 173 | 1,204 | 156 | 63 | NA | NA | NA | NA |
| Spartanburg | 91 | 403 | 193 | 15 | 187 | 1,269 | 198 | 78 | NA | NA | NA | NA |
| Summerville | 84 | 305 | 83 | 11 | 142 | 1,498 | 154 | 54 | NA | NA | NA | NA |
| Sumter | 52 | 271 | 40 | 8 | 100 | 581 | 60 | 19 | NA | NA | NA | NA |
| **SOUTH DAKOTA** | 1,143 | 4,330 | 829 | 155 | 1,967 | 13,063 | 1,763 | 655 | 1,051 | 42,935 | 16,780 | 2,057 |
| Aberdeen | 54 | 280 | 49 | 11 | 54 | 340 | 44 | 16 | NA | NA | NA | NA |
| Rapid City | 193 | 598 | 156 | 20 | 273 | 1,863 | 256 | 91 | NA | NA | NA | NA |
| Sioux Falls | 297 | 1,649 | 357 | 75 | 577 | 5,461 | 764 | 303 | NA | NA | NA | NA |
| **TENNESSEE** | 6,048 | 36,212 | 10,679 | 1,760 | 11,411 | 116,205 | 18,739 | 7,661 | 5,811 | 321,195 | 155,422 | 17,022 |
| Bartlett | 36 | 189 | 175 | 7 | 115 | 2,305 | 164 | 74 | NA | NA | NA | NA |
| Brentwood | 125 | 674 | 241 | 49 | 333 | 4,095 | 796 | 329 | NA | NA | NA | NA |
| Bristol | 36 | 99 | 21 | 4 | 65 | 624 | 72 | 34 | NA | NA | NA | NA |
| Chattanooga | 316 | 1,802 | 581 | 111 | 650 | 7,162 | 1,211 | 456 | NA | NA | NA | NA |
| Clarksville | 162 | 728 | 155 | 23 | 174 | 1,777 | 207 | 78 | NA | NA | NA | NA |
| Cleveland | 59 | 212 | 51 | 8 | 114 | 652 | 77 | 26 | NA | NA | NA | NA |
| Collierville | 48 | 170 | 62 | 8 | 107 | 482 | 60 | 22 | NA | NA | NA | NA |

1. Establishments subject to federal tax.

# Accommodation and Food Services, Arts, Entertainment, and Recreation, and Health Care and Social Assistance

| City | Accommodation and food services, 2017 | | | | Arts, entertainment, and recreation[1], 2017 | | | | Health care and social assistance[1], 2017 | | | |
|---|---|---|---|---|---|---|---|---|---|---|---|---|
| | Number of establishments | Number of employees | Receipts (mil dol) | Annual payroll (mil dol) | Number of establishments | Number of employees | Receipts (mil dol) | Annual payroll (mil dol) | Number of establishments | Number of employees | Receipts (mil dol) | Annual payroll (mil dol) |
| | 92 | 93 | 94 | 95 | 96 | 97 | 98 | 99 | 100 | 101 | 102 | 103 |
| **OREGON—Cont'd** | | | | | | | | | | | | |
| Redmond | 104 | 1,452 | 99 | 27 | 9 | 30 | 3 | 1 | 126 | 1,884 | 225 | 91 |
| Salem | 457 | 7,454 | 432 | 131 | 74 | 984 | 59 | 18 | 724 | 15,658 | 1,963 | 821 |
| Springfield | 178 | 3,149 | 177 | 52 | 20 | 248 | 9 | 3 | 216 | 8,193 | 1,516 | 540 |
| Tigard | 181 | 3,223 | 208 | 62 | 32 | 359 | 26 | 9 | 277 | 3,308 | 368 | 139 |
| Tualatin | 87 | 1,691 | 103 | 31 | D | D | D | D | D | D | D | D |
| West Linn | 53 | 767 | 44 | 14 | 8 | 39 | 3 | 1 | 114 | 870 | 79 | 32 |
| **PENNSYLVANIA** | 28,843 | 481,682 | 28,849 | 7,931 | 4,862 | 101,095 | 10,655 | 3,265 | 37,699 | 1,035,971 | 116,622 | 47,885 |
| Allentown | 278 | 4,016 | 212 | 60 | 38 | 696 | 60 | 16 | 340 | 9,317 | 1,240 | 473 |
| Altoona | 136 | 2,437 | 124 | 34 | 19 | 188 | 9 | 3 | 231 | 7,152 | 761 | 337 |
| Bethel Park | 74 | 1,865 | 98 | 30 | D | D | D | D | 142 | 1,850 | 174 | 80 |
| Bethlehem | 232 | 5,777 | 823 | 159 | 26 | 444 | 32 | 9 | NA | NA | NA | NA |
| Chester | 31 | 452 | 27 | 8 | 4 | 1,605 | 315 | 65 | 75 | 2,020 | 304 | 162 |
| Easton | 114 | 1,558 | 91 | 25 | 11 | 399 | 26 | 9 | 75 | 1,993 | 217 | 77 |
| Erie | 211 | 4,197 | 177 | 51 | 43 | 962 | 45 | 16 | 458 | 16,369 | 1,860 | 806 |
| Harrisburg | 180 | 2,571 | 164 | 46 | 31 | 334 | 30 | 10 | 207 | 8,012 | 992 | 409 |
| Hazleton | 63 | 722 | 40 | 9 | 7 | 16 | 1 | 0 | 103 | 2,938 | 279 | 110 |
| Lancaster | 193 | 4,075 | 247 | 74 | 27 | 572 | 32 | 10 | 201 | 10,120 | 1,535 | 574 |
| Lebanon | D | D | D | D | D | D | D | D | 90 | 2,944 | 355 | 139 |
| Monroeville | 118 | 2,580 | 142 | 39 | 20 | 296 | 13 | 5 | 236 | 6,094 | 704 | 290 |
| Norristown | 68 | 416 | 33 | 7 | 12 | 128 | 11 | 4 | 106 | 2,563 | 227 | 123 |
| Philadelphia | 3,953 | 61,893 | 4,480 | 1,278 | 451 | 15,371 | 2,496 | 980 | 3,992 | 161,976 | 21,516 | 8,285 |
| Pittsburgh | 1,311 | 26,134 | 1,724 | 511 | 209 | 9,309 | 1,741 | 668 | 1,383 | 60,636 | 8,315 | 3,526 |
| Plum | 35 | 588 | 25 | 8 | 7 | 9 | 1 | 0 | 35 | 727 | 46 | 16 |
| Reading | 146 | 1,910 | 123 | 27 | 22 | 769 | 36 | 11 | 159 | 4,881 | 343 | 161 |
| Scranton | 209 | 2,681 | 161 | 44 | D | D | D | D | 337 | 11,865 | 1,274 | 538 |
| State College | 143 | 3,014 | 145 | 40 | 18 | 343 | 16 | 6 | 120 | 3,795 | 569 | 214 |
| Wilkes-Barre | 140 | 2,407 | 126 | 36 | 13 | 511 | 23 | 8 | 146 | 7,076 | 1,092 | 384 |
| Williamsport | 88 | 1,792 | 102 | 35 | 13 | 139 | 19 | 3 | 113 | 5,462 | 847 | 295 |
| York | 95 | 1,533 | 80 | 21 | 17 | 363 | 22 | 7 | 129 | 7,828 | 1,266 | 428 |
| **RHODE ISLAND** | 3,167 | 50,642 | 3,618 | 1,016 | 575 | 8,673 | 740 | 219 | 3,177 | 87,546 | 9,574 | 4,092 |
| Cranston | 209 | 3,249 | 196 | 56 | D | D | D | D | 293 | 5,980 | 473 | 230 |
| East Providence | D | D | D | D | 26 | 546 | 30 | 12 | 160 | 4,521 | 407 | 210 |
| Pawtucket | D | D | D | D | 21 | 448 | 31 | 12 | 160 | 4,131 | 358 | 150 |
| Providence | 642 | 11,268 | 771 | 227 | 60 | 2,110 | 164 | 48 | 577 | 29,944 | 4,427 | 1,785 |
| Warwick | 258 | 5,497 | 335 | 94 | 41 | 746 | 46 | 15 | 357 | 8,635 | 984 | 449 |
| Woonsocket | 85 | 1,200 | 63 | 18 | 7 | 107 | 7 | 2 | 139 | 4,136 | 376 | 166 |
| **SOUTH CAROLINA** | 10,847 | 223,081 | 13,385 | 3,667 | 1,687 | 28,317 | 1,881 | 542 | 10,588 | 244,198 | 29,455 | 11,051 |
| Aiken | 158 | 3,200 | 164 | 48 | 27 | 350 | 30 | 8 | 205 | 4,442 | 463 | 169 |
| Anderson | 155 | 2,959 | 152 | 43 | 13 | 327 | 9 | 3 | 189 | 5,932 | 764 | 255 |
| Charleston | 641 | 16,356 | 1,264 | 350 | 122 | 2,001 | 185 | 50 | 540 | 17,068 | 3,281 | 1,071 |
| Columbia | 573 | 13,902 | 730 | 206 | 60 | 766 | 57 | 15 | 563 | 24,289 | 3,736 | 1,366 |
| Florence | 211 | 4,806 | 249 | 74 | 16 | 330 | 12 | 6 | 267 | 12,687 | 1,818 | 669 |
| Goose Creek | 74 | 1,570 | 75 | 21 | 6 | 103 | 5 | 1 | 52 | 562 | 49 | 19 |
| Greenville | 471 | 12,028 | 737 | 212 | 69 | 1,338 | 100 | 29 | 478 | 16,083 | 1,672 | 736 |
| Greer | 99 | 1,843 | 96 | 25 | 11 | 73 | 4 | 1 | 82 | 1,403 | 142 | 57 |
| Hilton Head Island | 259 | 6,802 | 587 | 172 | 61 | 1,358 | 98 | 30 | 186 | 2,655 | 354 | 111 |
| Mount Pleasant | 254 | 5,290 | 330 | 94 | 81 | 913 | 63 | 17 | D | D | D | D |
| Myrtle Beach | 585 | 15,481 | 1,195 | 289 | 100 | 2,325 | 215 | 48 | 260 | 3,600 | 783 | 211 |
| North Charleston | 363 | 8,051 | 522 | 135 | 24 | 852 | 29 | 10 | 363 | 8,192 | 1,216 | 403 |
| Rock Hill | 199 | 5,221 | 267 | 76 | 28 | 369 | 24 | 6 | 298 | 7,059 | 860 | 283 |
| Spartanburg | 216 | 4,658 | 258 | 74 | 23 | 301 | 11 | 4 | 208 | 9,385 | 1,049 | 578 |
| Summerville | 167 | 3,571 | 191 | 54 | D | D | D | D | 190 | 3,165 | 316 | 119 |
| Sumter | 140 | 2,999 | 148 | 41 | 10 | 223 | 7 | 3 | 162 | 4,648 | 578 | 186 |
| **SOUTH DAKOTA** | 2,495 | 40,704 | 2,316 | 659 | 697 | 6,799 | 541 | 130 | 2,420 | 71,821 | 8,714 | 3,458 |
| Aberdeen | 88 | 1,873 | 96 | 29 | 27 | 270 | 11 | 3 | 102 | 3,072 | 357 | 145 |
| Rapid City | 263 | 6,179 | 348 | 107 | 86 | 659 | 51 | 13 | 365 | 10,448 | 1,443 | 540 |
| Sioux Falls | 466 | 11,856 | 661 | 192 | 159 | 2,761 | 174 | 44 | 560 | 29,192 | 3,928 | 1,597 |
| **TENNESSEE** | 13,518 | 287,534 | 17,182 | 4,839 | 2,824 | 37,967 | 5,037 | 1,720 | 15,891 | 414,598 | 52,089 | 20,220 |
| Bartlett | 98 | 1,613 | 90 | 23 | 16 | 163 | 11 | 5 | 134 | 4,208 | 394 | 145 |
| Brentwood | 115 | 2,905 | 193 | 50 | 64 | 931 | 85 | 30 | 261 | 7,021 | 969 | 440 |
| Bristol | 101 | 2,124 | 96 | 27 | 17 | 254 | 68 | 10 | 134 | 3,617 | 541 | 203 |
| Chattanooga | 730 | 17,030 | 1,023 | 288 | 110 | 2,533 | 150 | 47 | 915 | 27,089 | 3,482 | 1,390 |
| Clarksville | 350 | 7,637 | 374 | 107 | 38 | 407 | 17 | 5 | 335 | 8,076 | 814 | 349 |
| Cleveland | 182 | 3,941 | 202 | 57 | D | D | D | D | 211 | 4,578 | 585 | 201 |
| Collierville | 124 | 2,595 | 143 | 40 | D | D | D | D | 119 | 1,715 | 191 | 75 |

1. Establishments subject to federal tax.

Items 92—103

| City | Other services[1] Number of establishments | Number of employees | Receipts (mil dol) | Annual payroll (mil dol) | Government employment and payroll, 2017 — March payroll Full-time equivalent employees | Total (dollars) | Percent of total for: Administrative, judicial, and legal | Police and corrections | Fire protection | Highways and transportation | Health and welfare | Natural resources and utilities | Education and libraries |
|---|---|---|---|---|---|---|---|---|---|---|---|---|---|
| | 104 | 105 | 106 | 107 | 108 | 109 | 110 | 111 | 112 | 113 | 114 | 115 | 116 |
| **OREGON—Cont'd** | | | | | | | | | | | | | |
| Redmond | 55 | 431 | 53 | 14 | 152 | 837,919 | 12.7 | 34.6 | 0.0 | 20.5 | 5.9 | 18.5 | 0.0 |
| Salem | 322 | 1,811 | 176 | 56 | 1,212 | 7,311,691 | 13.3 | 28.9 | 16.6 | 3.3 | 5.3 | 16.4 | 2.9 |
| Springfield | 96 | 538 | 56 | 15 | 547 | 3,697,459 | 13.5 | 23.1 | 22.1 | 8.5 | 5.0 | 15.3 | 2.1 |
| Tigard | 155 | 1,116 | 227 | 44 | 284 | 1,707,823 | 26.8 | 34.9 | 0.0 | 8.0 | 1.1 | 11.4 | 11.2 |
| Tualatin | 97 | 561 | 62 | 20 | 150 | 925,011 | 22.7 | 37.3 | 0.0 | 2.3 | 4.8 | 12.4 | 11.1 |
| West Linn | 46 | 185 | 14 | 5 | 121 | 640,406 | 23.9 | 20.3 | 0.0 | 12.5 | 0.0 | 21.4 | 12.7 |
| **PENNSYLVANIA** | 26,075 | 161,337 | 22,982 | 5,137 | X | X | X | X | X | X | X | X | X |
| Allentown | 247 | 1,606 | 171 | 52 | 848 | 5,261,837 | 8.4 | 33.2 | 19.8 | 5.0 | 8.9 | 19.7 | 0.0 |
| Altoona | 118 | 613 | 49 | 15 | 237 | 1,025,045 | 9.3 | 37.4 | 29.4 | 17.3 | 4.7 | 2.1 | 0.0 |
| Bethel Park | 85 | 561 | 39 | 15 | 128 | 661,011 | 12.1 | 56.2 | 0.0 | 16.2 | 0.0 | 15.0 | 0.0 |
| Bethlehem | 157 | 1,163 | 111 | 38 | 692 | 3,608,415 | 7.3 | 31.6 | 18.2 | 3.3 | 7.0 | 18.2 | 0.0 |
| Chester | D | D | D | D | 341 | 1,905,322 | 7.9 | 48.0 | 28.7 | 4.3 | 0.2 | 5.0 | 0.0 |
| Easton | 71 | 391 | 39 | 12 | 230 | 1,255,442 | 10.8 | 32.9 | 26.6 | 2.9 | 2.4 | 13.1 | 0.0 |
| Erie | 215 | 1,446 | 154 | 36 | 637 | 3,968,930 | 7.4 | 35.9 | 25.5 | 11.9 | 2.3 | 17.1 | 0.0 |
| Harrisburg | 185 | 1,394 | 250 | 77 | 441 | 2,045,074 | 7.3 | 35.1 | 22.6 | 7.3 | 1.7 | 16.8 | 0.0 |
| Hazleton | 48 | 194 | 14 | 4 | 91 | 483,561 | 10.3 | 46.5 | 20.9 | 13.6 | 6.5 | 0.4 | 0.0 |
| Lancaster | 148 | 1,295 | 109 | 30 | 546 | 3,002,028 | 5.8 | 40.3 | 18.3 | 5.0 | 2.8 | 23.9 | 0.0 |
| Lebanon | 63 | 255 | 29 | 7 | 146 | 754,384 | 5.1 | 39.4 | 16.9 | 5.7 | 2.1 | 30.9 | 0.0 |
| Monroeville | 91 | 599 | 37 | 14 | 143 | 920,606 | 14.4 | 52.5 | 0.0 | 6.2 | 4.2 | 15.3 | 6.2 |
| Norristown | D | D | D | 6 | 164 | 1,032,999 | 11.9 | 61.7 | 14.9 | 5.6 | 0.0 | 1.7 | 0.0 |
| Philadelphia | 2,687 | 20,004 | 2,958 | 786 | 30,557 | 177,705,357 | 16.8 | 37.8 | 10.4 | 3.8 | 12.0 | 15.6 | 1.8 |
| Pittsburgh | 912 | 7,796 | 1,614 | 277 | 3,293 | 17,441,555 | 10.9 | 40.2 | 19.8 | 11.5 | 11.5 | 5.1 | 0.0 |
| Plum | 44 | 232 | 23 | 7 | 67 | 393,790 | 66.5 | 0.0 | 0.0 | 0.0 | 0.0 | 0.0 | 1.3 |
| Reading | 107 | 738 | 91 | 22 | 595 | 3,156,585 | 7.9 | 37.4 | 22.1 | 0.8 | 5.7 | 24.4 | 1.3 |
| Scranton | 148 | 771 | 91 | 22 | 487 | 2,362,715 | 5.0 | 37.1 | 34.4 | 4.4 | 1.2 | 8.8 | 6.9 |
| State College | 53 | 285 | 20 | 8 | 174 | 970,917 | 23.1 | 48.0 | 0.0 | 13.8 | 2.1 | 3.8 | 0.0 |
| Wilkes-Barre | 91 | 540 | 55 | 16 | 282 | 1,551,214 | 9.7 | 36.6 | 28.4 | 2.1 | 5.0 | 9.8 | 0.0 |
| Williamsport | 73 | 728 | 55 | 14 | 205 | 1,066,708 | 6.4 | 32.8 | 21.5 | 33.0 | 1.9 | 4.1 | 0.0 |
| York | 77 | 681 | 117 | 24 | 341 | 1,850,502 | 12.1 | 40.3 | 22.6 | 2.5 | 4.1 | 13.6 | 0.0 |
| **RHODE ISLAND** | 2,318 | 13,882 | 1,773 | 471 | X | X | X | X | X | X | X | X | X |
| Cranston | 181 | 1,300 | 137 | 48 | 2,260 | 12,721,944 | 2.0 | 7.1 | 9.3 | 1.7 | 1.1 | 0.8 | 76.5 |
| East Providence | 103 | 576 | 60 | 18 | 1,176 | 5,613,335 | 3.9 | 9.4 | 8.4 | 3.5 | 0.6 | 5.0 | 68.8 |
| Pawtucket | 140 | 787 | 75 | 22 | 1,646 | 9,798,066 | 3.3 | 7.3 | 10.2 | 0.7 | 0.2 | 3.7 | 73.7 |
| Providence | 421 | 3,336 | 530 | 131 | 4,460 | 27,543,641 | 4.4 | 13.6 | 13.1 | 1.6 | 1.2 | 7.2 | 58.1 |
| Warwick | 229 | 1,598 | 212 | 54 | 2,411 | 13,058,247 | 3.0 | 10.9 | 13.1 | 2.2 | 0.7 | 3.6 | 64.0 |
| Woonsocket | 62 | 242 | 24 | 7 | 1,220 | 6,074,841 | 2.3 | 9.6 | 10.0 | 2.5 | 0.7 | 2.3 | 71.6 |
| **SOUTH CAROLINA** | 7,068 | 46,847 | 5,653 | 1,529 | X | X | X | X | X | X | X | X | X |
| Aiken | 70 | 366 | 40 | 10 | 380 | 1,309,295 | 17.3 | 36.2 | 4.4 | 23.2 | 0.0 | 18.6 | 0.0 |
| Anderson | 72 | 444 | 36 | 12 | 458 | 1,433,486 | 12.4 | 30.6 | 14.0 | 4.1 | 1.1 | 30.3 | 0.0 |
| Charleston | 320 | 2,118 | 347 | 74 | 2,017 | 8,885,603 | 9.7 | 29.2 | 19.7 | 2.8 | 0.5 | 36.0 | 0.0 |
| Columbia | 358 | 3,793 | 407 | 123 | 2,387 | 8,437,005 | 13.1 | 19.3 | 24.7 | 7.8 | 3.1 | 27.1 | 0.0 |
| Florence | 90 | 694 | 74 | 19 | 483 | 1,631,926 | 11.0 | 26.8 | 16.8 | 7.1 | 2.7 | 33.3 | 0.0 |
| Goose Creek | 48 | 282 | 35 | 8 | 261 | 1,051,887 | 14.5 | 35.9 | 21.6 | 3.7 | 0.0 | 20.1 | 0.0 |
| Greenville | 262 | 1,810 | 390 | 61 | 1,090 | 4,394,459 | 9.2 | 17.3 | 14.9 | 5.4 | 1.5 | 42.0 | 0.0 |
| Greer | 61 | 341 | 48 | 12 | 201 | 750,291 | 19.0 | 37.1 | 19.5 | 5.3 | 0.0 | 11.8 | 0.0 |
| Hilton Head Island | 133 | 1,144 | 143 | 43 | 241 | 1,412,148 | 20.9 | 0.0 | 54.8 | 2.2 | 13.1 | 1.7 | 0.0 |
| Mount Pleasant | 190 | 1,122 | 97 | 34 | 576 | 2,297,312 | 20.1 | 29.1 | 22.7 | 2.1 | 0.0 | 23.1 | 2.2 |
| Myrtle Beach | 186 | 1,284 | 161 | 36 | 941 | 3,847,183 | 11.3 | 31.4 | 19.4 | 7.4 | 0.3 | 23.5 | 0.0 |
| North Charleston | 235 | 2,101 | 281 | 88 | 1,139 | 4,423,364 | 11.2 | 38.6 | 22.7 | 5.8 | 0.2 | 14.7 | 0.0 |
| Rock Hill | 118 | 865 | 106 | 31 | 889 | 3,624,179 | 24.2 | 21.2 | 13.9 | 2.5 | 8.3 | 29.0 | 0.0 |
| Spartanburg | 129 | 1,077 | 140 | 33 | 598 | 2,385,343 | 8.6 | 23.3 | 11.2 | 2.7 | 1.9 | 43.7 | 0.0 |
| Summerville | 117 | 669 | 56 | 21 | 419 | 1,532,008 | 8.7 | 31.0 | 21.9 | 7.4 | 0.0 | 21.5 | 0.0 |
| Sumter | 81 | 541 | 50 | 14 | 570 | 1,922,091 | 13.7 | 24.8 | 18.4 | 2.3 | 1.8 | 25.3 | X |
| **SOUTH DAKOTA** | 1,860 | 8,713 | 1,224 | 278 | X | X | X | X | X | X | X | X | X |
| Aberdeen | 65 | 296 | 28 | 8 | 292 | 1,227,627 | 9.9 | 21.1 | 17.4 | 15.0 | 0.0 | 30.7 | 4.5 |
| Rapid City | 257 | 1,486 | 175 | 48 | 972 | 3,985,284 | 6.4 | 19.8 | 19.1 | 15.6 | 0.3 | 30.0 | 3.4 |
| Sioux Falls | 368 | 2,769 | 402 | 97 | 1,310 | 6,656,200 | 14.3 | 23.7 | 16.3 | 10.0 | 8.2 | 19.3 | 4.7 |
| **TENNESSEE** | 8,570 | 61,744 | 8,650 | 2,149 | X | X | X | X | X | X | X | X | X |
| Bartlett | D | D | D | 18 | 565 | 2,426,976 | 15.2 | 29.5 | 20.5 | 6.6 | 1.2 | 19.6 | 0.0 |
| Brentwood | 110 | 904 | 138 | 37 | 273 | 1,548,644 | 19.2 | 27.9 | 30.0 | 6.4 | 0.0 | 11.3 | 5.2 |
| Bristol | 52 | 276 | 32 | 9 | 906 | 3,310,594 | 4.7 | 11.0 | 6.2 | 3.8 | 0.2 | 7.9 | 63.5 |
| Chattanooga | 464 | 3,310 | 498 | 119 | 3,216 | 13,851,575 | 7.3 | 16.4 | 12.1 | 11.9 | 4.6 | 41.4 | 1.8 |
| Clarksville | D | D | D | D | 1,062 | 3,844,522 | 7.4 | 32.3 | 23.3 | 14.7 | 0.6 | 18.7 | 0.0 |
| Cleveland | D | D | D | D | 1,072 | 4,229,059 | 2.1 | 9.9 | 17.4 | 4.8 | 1.2 | 2.7 | 60.1 |
| Collierville | 70 | 509 | 36 | 12 | 501 | 2,125,507 | 17.9 | 29.8 | 19.9 | 6.9 | 0.8 | 19.1 | 0.0 |

1. Establishments subject to federal tax.

# Table D. Cities — **City Government Finances**

| City | General revenue Total (mil dol) 117 | Intergovernmental Total (mil dol) 118 | Intergovernmental Percent from state government 119 | Taxes Total (mil dol) 120 | Taxes Per capita[1] (dollars) Total 121 | Taxes Per capita[1] (dollars) Property 122 | Taxes Per capita[1] (dollars) Sales and gross receipts 123 | General expenditure Total (mil dol) 124 | General expenditure Per capita[1] (dollars) Total 125 | General expenditure Per capita[1] (dollars) Capital outlays 126 |
|---|---|---|---|---|---|---|---|---|---|---|
| **OREGON—Cont'd** | | | | | | | | | | |
| Redmond | 59.0 | 12.1 | 51.4 | 19.3 | 644 | 366 | 278 | 67 | 2,236 | 894 |
| Salem | 313.6 | 73.5 | 61.1 | 127.4 | 751 | 546 | 205 | 285 | 1,682 | 224 |
| Springfield | 149.8 | 9.5 | 99.0 | 44.6 | 715 | 546 | 169 | 101 | 1,619 | 129 |
| Tigard | 63.9 | 10.3 | 61.9 | 31.2 | 585 | 349 | 237 | 52 | 974 | 107 |
| Tualatin | 33.3 | 4.9 | 48.9 | 13.7 | 497 | 358 | 139 | 32 | 1,168 | 71 |
| West Linn | 27.1 | 5.2 | 59.9 | 12.9 | 481 | 324 | 157 | 28 | 1,054 | 281 |
| **PENNSYLVANIA** | X | X | X | X | X | X | X | X | X | X |
| Allentown | 149.9 | 31.8 | 50.9 | 70.6 | 584 | 254 | 117 | 182 | 1,506 | 101 |
| Altoona | 34.7 | 6.4 | 52.1 | 23.7 | 539 | 227 | 82 | 34 | 783 | 69 |
| Bethel Park | 32.3 | 3.2 | 57.4 | 17.9 | 555 | 185 | 68 | 37 | 1,136 | 170 |
| Bethlehem | 96.8 | 23.1 | 92.4 | 42.8 | 566 | 341 | 100 | 97 | 1,288 | 0 |
| Chester | 56.7 | 9.9 | 87.5 | 22.2 | 653 | 273 | 96 | 55 | 1,633 | 14 |
| Easton | 54.0 | 9.3 | 57.4 | 19.1 | 705 | 333 | 71 | 53 | 1,951 | 80 |
| Erie | 120.0 | 23.3 | 35.3 | 53.8 | 554 | 362 | 58 | 108 | 1,114 | 65 |
| Harrisburg | 89.4 | 19.2 | 54.3 | 44.7 | 909 | 353 | 314 | 140 | 2,838 | 1,046 |
| Hazleton | 22.2 | 7.4 | 92.0 | 13.0 | 522 | 187 | 65 | 13 | 535 | 11 |
| Lancaster | 94.4 | 10.8 | 67.2 | 38.0 | 638 | 456 | 68 | 134 | 2,244 | 382 |
| Lebanon | 17.0 | 2.7 | 67.3 | 10.6 | 410 | 153 | 39 | 14 | 548 | 15 |
| Monroeville | 31.3 | 2.7 | 91.5 | 27.3 | 986 | 323 | 355 | 27 | 981 | 37 |
| Norristown | 32.6 | 4.1 | 58.3 | 23.1 | 671 | 320 | 78 | 28 | 807 | 41 |
| Philadelphia | 7,149.5 | 2,318.6 | 66.7 | 3,708.3 | 2,346 | 362 | 419 | 6,861 | 4,341 | 276 |
| Pittsburgh | 628.6 | 115.8 | 64.2 | 417.5 | 1,385 | 470 | 516 | 757 | 2,511 | 54 |
| Plum | 15.5 | 1.7 | 85.0 | 10.8 | 398 | 209 | 42 | 17 | 629 | 70 |
| Reading | 129.9 | 14.0 | 94.7 | 58.8 | 668 | 277 | 91 | 125 | 1,421 | 26 |
| Scranton | 115.7 | 9.2 | 76.7 | 73.5 | 952 | 419 | 133 | 107 | 1,391 | 6 |
| State College | 44.0 | 4.9 | 58.0 | 16.4 | 386 | 174 | 49 | 41 | 975 | 129 |
| Wilkes-Barre | 54.7 | 14.0 | 55.1 | 28.6 | 699 | 267 | 82 | 51 | 1,244 | 133 |
| Williamsport | 37.4 | 15.5 | 69.5 | 19.9 | 698 | 472 | 144 | 27 | 933 | 107 |
| York | 76.8 | 7.8 | 61.0 | 30.2 | 685 | 450 | 128 | 67 | 1,519 | 48 |
| **RHODE ISLAND** | X | X | X | X | X | X | X | X | X | X |
| Cranston | 332.8 | 96.2 | 87.6 | 190.1 | 2,346 | 2,309 | 37 | 306 | 3,777 | 209 |
| East Providence | 121.8 | 53.9 | 96.7 | 45.7 | 966 | 943 | 22 | 175 | 3,686 | 132 |
| Pawtucket | 236.7 | 122.6 | 95.0 | 104.2 | 1,456 | 1,442 | 13 | 239 | 3,343 | 193 |
| Providence | 898.0 | 401.5 | 93.5 | 371.9 | 2,075 | 1,986 | 88 | 791 | 4,411 | 148 |
| Warwick | 337.8 | 55.9 | 96.2 | 240.3 | 2,974 | 2,888 | 86 | 335 | 4,141 | 181 |
| Woonsocket | 176.1 | 86.4 | 95.2 | 65.1 | 1,567 | 1,511 | 56 | 162 | 3,903 | 229 |
| **SOUTH CAROLINA** | X | X | X | X | X | X | X | X | X | X |
| Aiken | 49.4 | 1.9 | 50.1 | 28.8 | 936 | 352 | 391 | 44 | 1,445 | 145 |
| Anderson | 44.7 | 2.7 | 20.5 | 22.0 | 808 | 475 | 333 | 40 | 1,448 | 92 |
| Charleston | 262.8 | 12.4 | 45.3 | 176.6 | 1,302 | 563 | 739 | 177 | 1,304 | 117 |
| Columbia | 262.9 | 66.2 | 40.5 | 98.9 | 738 | 299 | 440 | 309 | 2,306 | 614 |
| Florence | 54.7 | 3.4 | 49.5 | 28.6 | 742 | 94 | 647 | 66 | 1,719 | 684 |
| Goose Creek | 24.8 | 1.2 | 3.3 | 18.7 | 448 | 80 | 367 | 24 | 563 | 71 |
| Greenville | 181.3 | 8.3 | 100.0 | 92.9 | 1,357 | 682 | 675 | 120 | 1,755 | 201 |
| Greer | 37.6 | 3.1 | 26.0 | 22.8 | 735 | 373 | 362 | 31 | 1,002 | 76 |
| Hilton Head Island | 73.0 | 1.8 | 57.9 | 59.3 | 1,476 | 596 | 881 | 63 | 1,572 | 349 |
| Mount Pleasant | 146.0 | 7.3 | 57.7 | 88.5 | 1,014 | 447 | 564 | 146 | 1,670 | 714 |
| Myrtle Beach | 211.7 | 18.1 | 76.5 | 111.3 | 3,409 | 1,004 | 2,405 | 134 | 4,089 | 352 |
| North Charleston | 148.4 | 9.4 | 31.1 | 122.0 | 1,101 | 549 | 497 | 135 | 1,214 | 193 |
| Rock Hill | 105.3 | 11.1 | 22.6 | 45.2 | 616 | 379 | 238 | 102 | 1,394 | 317 |
| Spartanburg | 68.0 | 23.8 | 10.1 | 36.1 | 963 | 487 | 476 | 43 | 1,138 | 63 |
| Summerville | 39.8 | 3.9 | 77.4 | 27.7 | 537 | 246 | 291 | 35 | 684 | 81 |
| Sumter | 61.5 | 6.9 | 23.6 | 28.4 | 715 | 244 | 471 | 48 | 1,209 | 168 |
| **SOUTH DAKOTA** | X | X | X | X | X | X | X | X | X | X |
| Aberdeen | 45.9 | 8.4 | 13.9 | 28.3 | 992 | 296 | 696 | 33 | 1,144 | 0 |
| Rapid City | 146.3 | 13.5 | 53.6 | 85.5 | 1,137 | 212 | 925 | 96 | 1,278 | 104 |
| Sioux Falls | 292.5 | 18.1 | 50.7 | 204.6 | 1,152 | 324 | 828 | 271 | 1,523 | 550 |
| **TENNESSEE** | X | X | X | X | X | X | X | X | X | X |
| Bartlett | 65.4 | 23.9 | 40.6 | 23.8 | 403 | 320 | 84 | 66 | 1,116 | 224 |
| Brentwood | 58.7 | 25.6 | 37.8 | 19.9 | 468 | 273 | 194 | 51 | 1,204 | 285 |
| Bristol | 114.3 | 57.2 | 43.5 | 37.0 | 1,379 | 1,252 | 127 | 75 | 2,804 | 156 |
| Chattanooga | 533.7 | 132.3 | 32.3 | 168.7 | 940 | 760 | 180 | 391 | 2,178 | 187 |
| Clarksville | 158.6 | 54.0 | 69.4 | 42.1 | 275 | 221 | 54 | 123 | 806 | 100 |
| Cleveland | 138.1 | 89.8 | 50.5 | 28.2 | 633 | 543 | 90 | 137 | 3,069 | 837 |
| Collierville | 68.5 | 5.8 | 100.0 | 46.5 | 925 | 510 | 414 | 46 | 908 | 64 |

1. Based on population estimated as of July 1 of the year shown.

# Table D. Cities — City Government Finances

City government finances, 2017 (cont.)

General expenditure (cont.)

Percent of total for:

| City | Public welfare | Highways | Parking facilities | Education | Health and hospitals | Police protection | Sewerage and sanitation | Parks and recreation | Housing and community development | Interest on debt |
|---|---|---|---|---|---|---|---|---|---|---|
| | 127 | 128 | 129 | 130 | 131 | 132 | 133 | 134 | 135 | 136 |
| **OREGON—Cont'd** | | | | | | | | | | |
| Redmond | 0.0 | 9.6 | 0.0 | 0.0 | 0.0 | 11.6 | 17.7 | 6.2 | 12.0 | 4.4 |
| Salem | 0.0 | 13.2 | 0.9 | 0.0 | 0.8 | 14.4 | 14.7 | 2.7 | 12.9 | 7.4 |
| Springfield | 0.0 | 6.6 | 0.0 | 0.0 | 6.3 | 18.9 | 31.1 | 0.0 | 0.8 | 5.8 |
| Tigard | 0.4 | 7.3 | 0.0 | 0.0 | 0.0 | 31.5 | 10.0 | 7.4 | 1.1 | 2.2 |
| Tualatin | 0.0 | 7.5 | 0.0 | 0.0 | 0.0 | 22.6 | 27.0 | 9.1 | 0.3 | 0.7 |
| West Linn | 0.0 | 9.9 | 0.0 | 0.0 | 0.0 | 20.6 | 16.1 | 22.7 | 0.0 | 1.4 |
| **PENNSYLVANIA** | X | X | X | X | X | X | X | X | X | X |
| Allentown | 0.0 | 17.2 | 0.0 | 0.0 | 4.0 | 18.2 | 16.0 | 5.4 | 4.2 | 1.8 |
| Altoona | 0.0 | 21.4 | 0.0 | 0.0 | 0.0 | 26.0 | 0.0 | 1.0 | 11.6 | 2.7 |
| Bethel Park | 0.0 | 14.7 | 0.0 | 0.0 | 0.0 | 19.6 | 30.0 | 6.5 | 0.3 | 2.1 |
| Bethlehem | 0.0 | 7.4 | 0.0 | 0.0 | 3.9 | 14.8 | 20.8 | 3.9 | 2.9 | 8.7 |
| Chester | 0.0 | 5.0 | 0.0 | 0.0 | 1.1 | 31.2 | 2.7 | 5.1 | 1.8 | 0.9 |
| Easton | 0.0 | 6.0 | 0.3 | 0.0 | 0.2 | 19.7 | 20.5 | 3.6 | 3.4 | 4.0 |
| Erie | 0.0 | 13.2 | 0.0 | 0.0 | 0.0 | 16.1 | 16.9 | 2.1 | 5.1 | 3.6 |
| Harrisburg | 0.0 | 8.3 | 0.0 | 0.0 | 0.0 | 13.8 | 6.8 | 0.7 | 48.4 | 5.3 |
| Hazleton | 0.0 | 15.3 | 0.0 | 0.0 | 0.6 | 31.5 | 0.0 | 0.8 | 15.7 | 2.2 |
| Lancaster | 0.0 | 7.1 | 0.0 | 0.0 | 0.0 | 14.6 | 34.5 | 3.7 | 2.9 | 6.0 |
| Lebanon | 0.0 | 8.4 | 0.3 | 0.0 | 1.7 | 29.6 | 5.0 | 3.5 | 1.6 | 0.1 |
| Monroeville | 0.0 | 15.3 | 0.0 | 0.0 | 0.3 | 39.1 | 5.2 | 7.7 | 1.7 | 2.3 |
| Norristown | 0.0 | 7.5 | 0.0 | 0.0 | 2.7 | 30.9 | 10.7 | 1.7 | 3.3 | 3.1 |
| Philadelphia | 2.9 | 3.0 | 0.0 | 2.0 | 24.6 | 9.8 | 7.0 | 1.5 | 5.8 | 3.4 |
| Pittsburgh | 0.0 | 14.1 | 0.0 | 0.0 | 2.5 | 12.5 | 2.3 | 1.3 | 1.7 | 3.2 |
| Plum | 0.0 | 31.5 | 0.0 | 0.0 | 0.0 | 26.9 | 10.1 | 7.0 | 0.0 | 3.4 |
| Reading | 0.0 | 5.6 | 0.0 | 0.0 | 3.6 | 27.8 | 21.0 | 1.2 | 6.1 | 10.0 |
| Scranton | 0.0 | 6.1 | 0.0 | 0.0 | 0.0 | 23.2 | 18.4 | 0.7 | 0.0 | 6.6 |
| State College | 0.0 | 14.3 | 9.0 | 0.0 | 1.0 | 20.9 | 22.3 | 3.7 | 4.1 | 2.0 |
| Wilkes-Barre | 0.0 | 13.3 | 1.7 | 0.0 | 5.9 | 20.0 | 10.1 | 2.8 | 12.8 | 4.5 |
| Williamsport | 0.0 | 14.2 | 0.0 | 0.0 | 0.0 | 30.5 | 0.0 | 7.7 | 5.7 | 3.2 |
| York | 0.0 | 7.0 | 2.6 | 0.0 | 2.3 | 15.9 | 23.9 | 1.7 | 3.1 | 9.2 |
| **RHODE ISLAND** | X | X | X | X | X | X | X | X | X | X |
| Cranston | 0.0 | 6.2 | 0.0 | 59.9 | 1.0 | 8.1 | 7.0 | 1.0 | 0.4 | 1.1 |
| East Providence | 0.0 | 4.6 | 0.0 | 54.0 | 0.2 | 8.9 | 6.7 | 0.5 | 0.6 | 1.0 |
| Pawtucket | 0.0 | 1.1 | 0.0 | 61.6 | 0.0 | 11.3 | 1.7 | 0.7 | 2.0 | 4.2 |
| Providence | 0.0 | 1.2 | 0.0 | 55.7 | 0.0 | 12.0 | 1.5 | 1.4 | 0.4 | 1.2 |
| Warwick | 0.2 | 2.8 | 0.0 | 54.9 | 0.1 | 6.8 | 4.7 | 0.7 | 0.6 | 5.3 |
| Woonsocket | 0.2 | 2.7 | 0.0 | 57.8 | 0.0 | 5.7 | 9.7 | 0.4 | X | X |
| **SOUTH CAROLINA** | X | X | X | X | X | X | X | X | X | X |
| Aiken | 0.0 | 4.8 | 0.0 | 0.0 | 0.0 | 18.7 | 25.4 | 10.5 | 0.4 | 0.0 |
| Anderson | 0.0 | 0.0 | 0.0 | 0.0 | 0.0 | 22.1 | 37.9 | 5.0 | 2.1 | 0.0 |
| Charleston | 0.3 | 4.6 | 6.7 | 0.0 | 0.0 | 27.8 | 5.4 | 9.8 | 2.3 | 0.6 |
| Columbia | 0.0 | 3.3 | 1.2 | 0.0 | 0.6 | 14.2 | 40.0 | 4.5 | 2.7 | 0.0 |
| Florence | 0.0 | 12.2 | 1.2 | 0.0 | 0.0 | 13.1 | 22.0 | 30.8 | 0.3 | 0.0 |
| Goose Creek | 0.0 | 0.0 | 1.6 | 0.0 | 0.0 | 29.2 | 6.1 | 26.2 | 0.0 | 0.0 |
| Greenville | 0.0 | 8.2 | 2.6 | 0.0 | 0.5 | 23.5 | 21.0 | 7.0 | 5.4 | 1.7 |
| Greer | 0.0 | 4.9 | 0.0 | 0.0 | 0.0 | 20.3 | 23.7 | 10.7 | 0.0 | 8.1 |
| Hilton Head Island | 0.0 | 3.4 | 0.0 | 0.0 | 0.0 | 5.9 | 4.4 | 18.2 | 13.3 | 5.7 |
| Mount Pleasant | 0.0 | 9.5 | 0.0 | 0.0 | 5.9 | 8.4 | 10.3 | 5.2 | 1.9 | 0.0 |
| Myrtle Beach | 0.0 | 6.6 | 3.8 | 0.0 | 0.0 | 20.8 | 15.3 | 12.1 | 1.4 | 0.0 |
| North Charleston | 0.0 | 5.0 | 0.6 | 0.0 | 0.0 | 27.0 | 6.4 | 5.4 | 0.3 | 0.0 |
| Rock Hill | 0.0 | 7.9 | 0.0 | 0.0 | 0.5 | 16.0 | 16.6 | 20.2 | 2.1 | 0.0 |
| Spartanburg | 0.0 | 5.7 | 3.0 | 0.0 | 0.0 | 25.5 | 10.7 | 3.7 | 1.4 | 1.2 |
| Summerville | 0.0 | 7.9 | 0.0 | 0.0 | 0.0 | 26.5 | 9.8 | 8.5 | 1.3 | 0.4 |
| Sumter | 0.0 | 2.5 | 0.0 | 0.0 | 0.0 | 25.6 | 20.0 | 7.9 | 6.9 | X |
| **SOUTH DAKOTA** | X | X | X | X | X | X | X | X | X | X |
| Aberdeen | 0.0 | 8.5 | 0.0 | 0.0 | 0.4 | 13.7 | 9.7 | 16.3 | 0.0 | 4.2 |
| Rapid City | 0.1 | 9.1 | 0.4 | 0.0 | 4.6 | 16.9 | 13.0 | 18.4 | 0.4 | 2.7 |
| Sioux Falls | 0.0 | 25.1 | 0.8 | 0.0 | 4.0 | 11.0 | 15.7 | 14.9 | 2.1 | 2.4 |
| **TENNESSEE** | X | X | X | X | X | X | X | X | X | X |
| Bartlett | 0.0 | 18.3 | 0.0 | 0.0 | 0.0 | 21.7 | 14.8 | 10.9 | 0.1 | 0.0 |
| Brentwood | 0.0 | 12.2 | 0.0 | 0.0 | 0.1 | 17.9 | 10.7 | 4.0 | 0.8 | 2.5 |
| Bristol | 0.0 | 5.4 | 0.0 | 57.9 | 0.0 | 9.9 | 7.1 | 5.0 | 0.7 | 0.3 |
| Chattanooga | 1.2 | 5.9 | 0.6 | 2.9 | 0.7 | 18.1 | 21.1 | 6.1 | 2.5 | 3.9 |
| Clarksville | 0.0 | 10.4 | 0.2 | 0.0 | 0.0 | 22.4 | 21.3 | 6.0 | 1.5 | 0.0 |
| Cleveland | 0.0 | 5.3 | 0.2 | 39.4 | 0.5 | 7.6 | 8.8 | 4.4 | 1.0 | 2.0 |
| Collierville | 0.0 | 8.2 | 0.0 | 0.0 | 0.0 | 27.2 | 12.4 | 13.2 | 0.0 | 10.9 |

# Table D. Cities — City Government Finances, City Government Employment, and Climate

| City | City government finances, 2017 (cont.) | | | Climate[2] | | | | | | |
| | Debt outstanding | | Debt issued during year | Average daily temperature | | | | Annual precipitation (inches) | Heating degree days | Cooling degree days |
| | Total (mil dol) | Per capita[1] (dollars) | | Mean | | Limits | | | | |
| | | | | January | July | January[3] | July[4] | | | |
| | 137 | 138 | 139 | 140 | 141 | 142 | 143 | 144 | 145 | 146 |
| **OREGON—Cont'd** | | | | | | | | | | |
| Redmond | 80.6 | 2,686 | 3.9 | NA | NA | NA | NA | NA | NA | NA |
| Salem | 416.0 | 2,452 | 0.0 | 40.3 | 66.8 | 33.5 | 81.5 | 40.00 | NA | NA |
| Springfield | 122.3 | 1,961 | 0.0 | 39.8 | 66.2 | 33.0 | 81.5 | 50.90 | 4,784 | 257 |
| Tigard | 119.4 | 2,237 | 0.0 | 40.0 | 66.8 | 33.8 | 79.2 | 39.95 | 4,786 | 242 |
| Tualatin | 10.0 | 360 | 0.0 | NA | NA | NA | NA | NA | 4,723 | 287 |
| West Linn | 16.5 | 616 | 0.0 | NA | NA | NA | NA | NA | NA | NA |
| **PENNSYLVANIA** | X | X | X | X | X | X | X | X | X | X |
| Allentown | 97.4 | 805 | 3.2 | 27.1 | 73.3 | 19.1 | 83.9 | 45.17 | 5,830 | 787 |
| Altoona | 25.3 | 575 | 1.3 | 26.5 | 71.1 | 18.2 | 81.9 | 42.69 | 6,055 | 546 |
| Bethel Park | 20.2 | 625 | 0.0 | 28.6 | 73.1 | 19.8 | 84.5 | 37.78 | 5,727 | 709 |
| Bethlehem | 195.1 | 2,582 | 3.6 | 27.1 | 73.3 | 19.1 | 83.9 | 45.17 | 5,830 | 787 |
| Chester | 6.7 | 197 | 0.0 | 33.7 | 78.7 | 27.9 | 87.5 | 40.66 | 4,469 | 1,333 |
| Easton | 47.7 | 1,759 | 1.2 | 27.1 | 73.3 | 19.1 | 83.9 | 45.17 | 5,830 | 787 |
| Erie | 178.4 | 1,835 | 14.6 | 26.9 | 72.1 | 20.3 | 80.4 | 42.77 | 6,243 | 620 |
| Harrisburg | 98.9 | 2,009 | 13.0 | 30.3 | 75.9 | 23.1 | 85.7 | 41.45 | 5,201 | 955 |
| Hazleton | 8.4 | 337 | 0.0 | NA | NA | NA | NA | NA | NA | NA |
| Lancaster | 258.7 | 4,348 | 133.0 | 29.1 | 74.4 | 20.7 | 84.7 | 43.47 | 5,448 | 809 |
| Lebanon | 0.2 | 6 | 0.1 | NA | NA | NA | NA | NA | NA | NA |
| Monroeville | 19.3 | 699 | 0.3 | 28.6 | 73.1 | 19.8 | 84.5 | 37.78 | 5,727 | 709 |
| Norristown | 14.4 | 418 | 0.8 | 30.2 | 75.1 | 20.4 | 86.6 | 43.87 | 5,174 | 884 |
| Philadelphia | 4,997.9 | 3,162 | 99.9 | 32.3 | 77.6 | 25.5 | 85.5 | 42.05 | 4,759 | 1,235 |
| Pittsburgh | 473.2 | 1,570 | 27.8 | 27.5 | 72.6 | 19.9 | 82.7 | 37.85 | 5,829 | 726 |
| Plum | 17.0 | 627 | 24.0 | 28.6 | 73.1 | 19.8 | 84.5 | 37.78 | 5,727 | 709 |
| Reading | 257.4 | 2,921 | 25.5 | 27.1 | 73.6 | 19.1 | 83.8 | 45.28 | 5,876 | 723 |
| Scranton | 152.8 | 1,979 | 33.4 | 26.3 | 72.1 | 18.5 | 82.6 | 37.56 | 6,234 | 611 |
| State College | 36.3 | 856 | 6.1 | 25.4 | 71.2 | 18.3 | 80.7 | 39.76 | 6,345 | 538 |
| Wilkes-Barre | 67.4 | 1,646 | 8.8 | 21.5 | 67.7 | 13.2 | 77.4 | 47.89 | 7,466 | 234 |
| Williamsport | 22.5 | 791 | 4.3 | 25.5 | 72.4 | 17.9 | 83.2 | 41.59 | 6,063 | 709 |
| York | 94.3 | 2,140 | 0.0 | 30.0 | 74.6 | 20.9 | 86.5 | 43.00 | 5,233 | 862 |
| **RHODE ISLAND** | X | X | X | X | X | X | X | X | X | X |
| Cranston | 93.9 | 1,159 | 2.2 | 28.7 | 73.3 | 20.3 | 82.6 | 46.45 | 5,754 | 714 |
| East Providence | 86.1 | 1,818 | 0.0 | 28.7 | 73.3 | 20.3 | 82.6 | 46.45 | 5,754 | 714 |
| Pawtucket | 188.3 | 2,630 | 30.5 | 28.7 | 73.3 | 20.3 | 82.6 | 46.45 | 5,754 | 714 |
| Providence | 711.1 | 3,967 | 18.6 | 28.7 | 73.3 | 20.3 | 82.6 | 46.45 | 5,754 | 714 |
| Warwick | 145.6 | 1,801 | 8.5 | 28.7 | 73.3 | 20.3 | 82.6 | 46.45 | 5,754 | 714 |
| Woonsocket | 152.8 | 3,679 | 62.7 | 25.4 | 72.3 | 13.3 | 84.3 | 48.75 | 6,302 | 534 |
| **SOUTH CAROLINA** | X | X | X | X | X | X | X | X | X | X |
| Aiken | 0.0 | 0 | 0.0 | 45.6 | 81.7 | 33.4 | 93.7 | 52.43 | 2,413 | 2,081 |
| Anderson | 179.0 | 6,566 | 0.0 | 41.7 | 79.7 | 31.3 | 90.5 | 46.67 | 3,087 | 1,700 |
| Charleston | 26.3 | 194 | 0.0 | 49.8 | 82.8 | 42.4 | 88.5 | 46.39 | 1,755 | 2,473 |
| Columbia | 595.0 | 4,440 | 211.5 | 47.3 | 83.6 | 36.5 | 95.2 | 47.14 | 2,044 | 2,475 |
| Florence | 191.7 | 4,975 | 51.5 | 45.0 | 81.2 | 35.2 | 90.7 | 44.76 | 2,523 | 2,029 |
| Goose Creek | 26.7 | 638 | 13.3 | 47.9 | 81.7 | 36.9 | 90.9 | 51.53 | 2,005 | 2,306 |
| Greenville | 108.8 | 1,589 | 0.0 | 40.8 | 78.8 | 31.4 | 88.8 | 50.24 | 3,272 | 1,526 |
| Greer | 78.1 | 2,513 | 0.0 | NA | NA | NA | NA | NA | NA | NA |
| Hilton Head Island | 90.1 | 2,243 | 0.0 | 47.9 | 80.5 | 37.3 | 88.2 | 52.52 | 2,128 | 2,012 |
| Mount Pleasant | 226.4 | 2,594 | 119.9 | 47.1 | 81.1 | 37.5 | 88.5 | 49.38 | 2,260 | 2,124 |
| Myrtle Beach | 208.4 | 6,382 | 55.5 | NA | NA | NA | NA | NA | NA | NA |
| North Charleston | 199.6 | 1,801 | 80.0 | 47.9 | 81.7 | 36.9 | 90.9 | 51.53 | 2,005 | 2,306 |
| Rock Hill | 287.3 | 3,914 | 105.4 | 42.2 | 80.1 | 32.5 | 90.1 | 48.32 | 2,934 | 1,721 |
| Spartanburg | 192.7 | 5,149 | 65.6 | 42.1 | 79.3 | 30.1 | 91.1 | 49.95 | 3,080 | 1,591 |
| Summerville | 7.1 | 137 | 0.0 | 49.1 | 81.8 | 38.0 | 91.8 | 48.24 | 1,907 | 2,251 |
| Sumter | 34.8 | 878 | 0.0 | 44.9 | 80.7 | 33.6 | 91.8 | 48.65 | 2,577 | 1,913 |
| **SOUTH DAKOTA** | X | X | X | X | X | X | X | X | X | X |
| Aberdeen | 62.0 | 2,175 | 10.6 | NA | NA | NA | NA | NA | NA | NA |
| Rapid City | 148.0 | 1,967 | 0.0 | 22.3 | 70.2 | 10.3 | 82.7 | 18.45 | 7,623 | 480 |
| Sioux Falls | 380.9 | 2,144 | 38.9 | 14.0 | 73.0 | 2.9 | 85.6 | 24.69 | 7,812 | 747 |
| **TENNESSEE** | X | X | X | X | X | X | X | X | X | X |
| Bartlett | 46.1 | 780 | 7.9 | 37.3 | 79.6 | 27.3 | 89.9 | 55.09 | 3,665 | 1,635 |
| Brentwood | 44.9 | 1,054 | 3.9 | NA | NA | NA | NA | NA | NA | NA |
| Bristol | 2.2 | 82 | 0.0 | NA | NA | NA | NA | NA | NA | NA |
| Chattanooga | 706.5 | 3,935 | 38.4 | 39.4 | 79.6 | 29.9 | 89.8 | 54.52 | 3,427 | 1,608 |
| Clarksville | 795.3 | 5,201 | 0.0 | 35.2 | 79.0 | 25.0 | 90.4 | 51.78 | 4,058 | 1,512 |
| Cleveland | 121.3 | 2,724 | 0.0 | 38.0 | 77.5 | 27.7 | 88.5 | 55.42 | 3,782 | 1,333 |
| Collierville | 123.5 | 2,458 | 0.0 | 37.9 | 81.1 | 28.2 | 91.1 | 53.63 | 3,491 | 1,838 |

1. Based on the population estimated as of July 1 of the year shown.　　2. Represents normal values based on the 30-year period, 1971±2000.　　3. Average daily minimum.　　4. Average daily maximum.

# Table D. Cities — Land Area and Population

| STATE Place code | City | Population, 2020 — Land area[1] (sq. mi) | Total persons 2019 | Rank | Per square mile | Race 2019 — Race alone[2] (percent): White | Black or African American | American Indian, Alaskan Native | Asian | Hawaiian Pacific Islander | Some other race | Two or more races (percent) |
|---|---|---|---|---|---|---|---|---|---|---|---|---|
| | | 1 | 2 | 3 | 4 | 5 | 6 | 7 | 8 | 9 | 10 | 11 |
| | **TENNESSEE—Cont'd** | | | | | | | | | | | |
| 47 16540 | Columbia | 33.1 | 41,826 | 918 | 1,264 | D | D | D | D | D | D | D |
| 47 16920 | Cookeville | 35.8 | 35,060 | 1,102 | 979 | D | D | D | D | D | D | D |
| 47 27740 | Franklin | 42.8 | 85,316 | 396 | 1,993 | 83.2 | 5.6 | 0.1 | 9.0 | 0.0 | 0.4 | 1.7 |
| 47 28540 | Gallatin | 32.2 | 44,335 | 867 | 1,377 | 78.7 | 13.4 | 0.5 | 2.2 | 0.0 | 3.5 | 1.6 |
| 47 28960 | Germantown | 20.0 | 39,186 | 991 | 1,959 | 93.0 | 1.4 | 0.0 | 3.5 | 0.0 | 0.8 | 1.3 |
| 47 33280 | Hendersonville | 31.4 | 59,323 | 636 | 1,889 | 84.0 | 11.6 | 0.2 | 1.7 | 0.1 | 2.1 | 0.3 |
| 47 37640 | Jackson | 58.4 | 67,402 | 547 | 1,154 | 44.3 | 47.4 | 0.5 | 1.6 | 0.0 | 4.0 | 2.3 |
| 47 38320 | Johnson City | 43.3 | 67,142 | 549 | 1,551 | 88.9 | 7.4 | 0.2 | 1.7 | 0.0 | 0.4 | 1.7 |
| 47 39560 | Kingsport | 52.6 | 54,112 | 716 | 1,029 | 88.8 | 4.7 | 0.2 | 1.5 | 0.0 | 1.0 | 3.6 |
| 47 40000 | Knoxville | 98.7 | 190,223 | 133 | 1,927 | 77.7 | 15.3 | 0.9 | 1.6 | 0.0 | 1.0 | 4.2 |
| 47 41200 | La Vergne | 24.7 | 35,646 | 1,086 | 1,443 | 61.6 | 24.1 | 0.0 | 3.1 | 0.0 | 7.0 | 4.2 |
| 47 41520 | Lebanon | 39.7 | 38,086 | 1,020 | 959 | D | D | D | D | D | D | D |
| 47 46380 | Maryville | 17.3 | 29,995 | 1,253 | 1,734 | D | D | D | D | D | D | D |
| 47 48000 | Memphis | 297.0 | 649,705 | 28 | 2,188 | 30.4 | 63.0 | 0.1 | 2.1 | 0.0 | 3.0 | 1.5 |
| 47 50280 | Morristown | 27.6 | 30,275 | 1,241 | 1,097 | 85.4 | 5.3 | 0.0 | 0.5 | 0.3 | 1.4 | 7.2 |
| 47 51560 | Murfreesboro | 62.9 | 150,757 | 171 | 2,397 | 69.8 | 19.7 | 0.3 | 3.9 | 0.1 | 3.4 | 2.3 |
| 47 52004 | Nashville-Davidson | 475.8 | 671,295 | 23 | 1,411 | 62.8 | 27.1 | 0.3 | 3.9 | 0.1 | 3.4 | 2.0 |
| 47 55120 | Oak Ridge | 85.3 | 29,340 | 1,276 | 344 | 84.3 | 6.7 | 1.1 | 3.0 | 1.1 | 1.8 | 2.0 |
| 47 69420 | Smyrna | 33.0 | 53,185 | 732 | 1,612 | D | D | D | D | D | D | D |
| 47 70580 | Spring Hill | 26.8 | 46,152 | 841 | 1,722 | D | D | D | D | D | D | D |
| 48 00000 | **TEXAS** | 261,263.1 | 29,360,759 | X | 112 | 73.4 | 12.3 | 0.5 | 5.0 | 0.1 | 5.9 | 2.9 |
| 48 01000 | Abilene | 106.7 | 124,407 | 228 | 1,166 | 76.5 | 11.3 | 0.1 | 2.2 | 0.0 | 7.6 | 2.3 |
| 48 01924 | Allen | 26.4 | 109,379 | 279 | 4,143 | 68.8 | 8.3 | 0.1 | 18.7 | 0.0 | 2.0 | 2.2 |
| | | | | | | | | | | | | 2.6 |
| 48 03000 | Amarillo | 102.3 | 199,654 | 121 | 1,952 | 81.4 | 7.2 | 0.6 | 5.2 | 0.2 | 2.8 | 3.7 |
| 48 04000 | Arlington | 95.9 | 398,864 | 49 | 4,159 | 52.5 | 23.9 | 0.1 | 7.0 | 0.4 | 12.2 | 4.2 |
| 48 05000 | Austin | 319.9 | 995,484 | 11 | 3,112 | 73.0 | 7.8 | 0.9 | 8.2 | 0.1 | 5.9 | 1.4 |
| 48 06128 | Baytown | 37.0 | 77,288 | 457 | 2,089 | 79.8 | 13.1 | 0.7 | 0.8 | 0.2 | 4.1 | 0.9 |
| 48 07000 | Beaumont | 82.5 | 115,473 | 254 | 1,400 | 47.7 | 45.0 | 0.4 | 3.0 | 0.2 | 2.7 | 4.8 |
| 48 07132 | Bedford | 10.0 | 48,352 | 806 | 4,835 | 76.4 | 10.2 | 0.4 | 5.4 | 1.5 | 1.2 | 4.2 |
| 48 08236 | Big Spring | 18.9 | 28,077 | 1,310 | 1,486 | 86.6 | 3.3 | 0.2 | 1.0 | 0.0 | 3.8 | 0.8 |
| 48 10768 | Brownsville | 131.5 | 183,428 | 141 | 1,395 | 93.8 | 0.4 | 0.1 | 1.0 | 0.0 | 3.8 | 1.5 |
| 48 10912 | Bryan | 54.2 | 88,165 | 383 | 1,627 | 76.9 | 18.6 | 0.1 | 1.5 | 0.4 | 1.2 | 1.4 |
| 48 11428 | Burleson | 28.3 | 49,686 | 780 | 1,756 | D | D | D | D | D | D | D |
| 48 13024 | Carrollton | 36.7 | 139,892 | 197 | 3,812 | 64.7 | 9.6 | 0.6 | 17.2 | 0.4 | 3.2 | 4.3 |
| 48 13492 | Cedar Hill | 35.8 | 47,820 | 811 | 1,336 | 39.7 | 48.3 | 0.0 | 2.3 | 0.0 | 1.8 | 4.0 |
| 48 13552 | Cedar Park | 25.5 | 81,040 | 430 | 3,178 | 80.6 | 3.9 | 0.5 | 9.0 | 0.1 | 1.8 | 3.4 |
| 48 15364 | Cleburne | 35.7 | 31,999 | 1,194 | 896 | D | D | D | D | D | D | D |
| 48 15976 | College Station | 51.2 | 119,260 | 239 | 2,329 | 78.8 | 6.1 | 0.5 | 9.4 | 0.0 | 2.3 | 1.9 |
| 48 16432 | Conroe | 72.0 | 96,214 | 332 | 1,336 | 83.8 | 9.6 | 0.5 | 2.0 | 0.0 | 0.2 | 3.3 |
| 48 16612 | Coppell | 14.4 | 40,958 | 937 | 2,844 | 61.5 | 2.9 | 0.4 | 28.6 | 0.0 | 2.9 | 3.7 |
| 48 16624 | Copperas Cove | 18.0 | 33,498 | 1,147 | 1,861 | 69.0 | 16.1 | 0.4 | 2.3 | 0.0 | 2.0 | 10.9 |
| 48 17000 | Corpus Christi | 162.2 | 327,248 | 60 | 2,018 | 89.6 | 4.2 | 0.2 | 2.3 | 0.0 | 1.4 | 2.3 |
| 48 19000 | Dallas | 339.6 | 1,343,266 | 9 | 3,955 | 62.2 | 24.6 | 0.5 | 3.7 | 0.0 | 6.7 | 2.3 |
| 48 19624 | Deer Park | 10.5 | 32,913 | 1,164 | 3,135 | D | D | D | D | D | D | D |
| 48 19792 | Del Rio | 20.4 | 35,722 | 1,082 | 1,751 | D | D | D | D | D | D | D |
| 48 19972 | Denton | 96.4 | 147,515 | 180 | 1,530 | 77.3 | 10.4 | 1.1 | 4.0 | 0.4 | 5.1 | 1.7 |
| 48 20092 | DeSoto | 21.6 | 52,910 | 740 | 2,450 | 23.5 | 69.0 | 0.2 | 0.7 | 0.0 | 2.9 | 3.5 |
| 48 21628 | Duncanville | 11.2 | 38,189 | 1,016 | 3,410 | 43.6 | 30.2 | 0.2 | 0.7 | 0.0 | 22.5 | 2.9 |
| 48 21892 | Eagle Pass | 9.4 | 29,515 | 1,269 | 3,140 | D | D | D | D | D | D | D |
| 48 22660 | Edinburg | 44.7 | 102,374 | 305 | 2,290 | 87.1 | 2.0 | 0.0 | 2.0 | 0.0 | 7.9 | 0.9 |
| 48 24000 | El Paso | 258.4 | 681,534 | 22 | 2,638 | 78.8 | 3.4 | 0.7 | 1.4 | 0.0 | 13.1 | 2.6 |
| 48 24768 | Euless | 16.1 | 57,332 | 672 | 3,561 | 63.7 | 17.0 | 0.5 | 9.9 | 1.3 | 2.9 | 4.7 |
| 48 25452 | Farmers Branch | 11.9 | 51,373 | 758 | 4,317 | 77.7 | 5.3 | 0.0 | 10.7 | 0.0 | 3.7 | 2.7 |
| 48 26232 | Flower Mound | 42.0 | 81,482 | 423 | 1,940 | 77.3 | 2.9 | 0.1 | 10.8 | 0.0 | 6.1 | 2.8 |
| 48 27000 | Fort Worth | 347.3 | 927,720 | 12 | 2,671 | 64.5 | 18.8 | 0.5 | 4.9 | 0.1 | 8.5 | 2.7 |
| 48 27648 | Friendswood | 20.8 | 40,011 | 962 | 1,924 | 77.1 | 7.9 | 0.2 | 12.7 | 0.0 | 0.9 | 1.2 |
| 48 27684 | Frisco | 68.6 | 209,980 | 109 | 3,061 | 60.7 | 8.4 | 0.4 | 26.0 | 0.1 | 1.1 | 3.2 |
| 48 28068 | Galveston | 41.1 | 50,085 | 776 | 1,219 | 64.7 | 16.4 | 0.6 | 3.0 | 0.1 | 5.1 | 4.0 |
| 48 29000 | Garland | 57.1 | 238,139 | 94 | 4,171 | 54.4 | 7.7 | 0.2 | 9.0 | 0.1 | 16.8 | 2.8 |
| 48 29336 | Georgetown | 57.3 | 85,538 | 393 | 1,493 | 86.5 | 7.7 | 0.2 | 0.8 | 0.1 | 1.9 | 2.2 |
| 48 30464 | Grand Prairie | 72.6 | 195,272 | 132 | 2,690 | 54.4 | 20.6 | 0.2 | 5.8 | 0.0 | 16.8 | 2.2 |
| 48 30644 | Grapevine | 32.1 | 55,780 | 689 | 1,738 | 78.7 | 9.2 | 0.0 | 8.2 | 0.0 | 1.3 | 2.6 |
| 48 30920 | Greenville | 32.3 | 29,374 | 1,274 | 909 | D | D | D | D | D | D | D |
| 48 31928 | Haltom City | 12.3 | 43,728 | 885 | 3,555 | 49.6 | 7.9 | 3.7 | 10.9 | 1.7 | 21.2 | 4.9 |
| 48 32312 | Harker Heights | 15.6 | 33,071 | 1,160 | 2,120 | 59.9 | 25.3 | 0.0 | 1.1 | 0.0 | 3.4 | 3.8 |
| 48 32372 | Harlingen | 40.1 | 65,176 | 579 | 1,625 | 91.0 | 3.7 | 0.4 | 1.1 | 0.0 | 2.3 | 0.9 |
| 48 35000 | Houston | 640.4 | 2,316,120 | 4 | 3,617 | 54.5 | 23.1 | 0.4 | 6.5 | 0.0 | 13.2 | 2.3 |
| 48 35528 | Huntsville | 42.6 | 41,392 | 929 | 972 | D | D | D | D | D | D | D |
| 48 35576 | Hurst | 10.0 | 38,172 | 1,017 | 3,817 | 64.7 | 12.8 | 2.3 | 9.5 | 0.7 | 6.9 | 3.2 |

1. Dry land or land partially or temporarily covered by water.  2. Hispanic or Latino persons may be of any race.

# Table D. Cities — **Population**

| City | Percent Hispanic or Latino[1], 2019 | Percent foreign born, 2019 | Age of population (percent), 2019 | | | | | | | Median age, 2019 | Percent female, 2019 | Population | | | |
|---|---|---|---|---|---|---|---|---|---|---|---|---|---|---|---|
| | | | Under 18 years | 18 to 24 years | 25 to 34 years | 35 to 44 years | 45 to 54 years | 55 to 64 years | 65 years and over | | | Census counts | | Percent change | |
| | | | | | | | | | | | | 2000 | 2010 | 2000-2010 | 2001-2020 |
| | 12 | 13 | 14 | 15 | 16 | 17 | 18 | 19 | 20 | 21 | 22 | 23 | 24 | 25 | 26 |
| **TENNESSEE—Cont'd** | | | | | | | | | | | | | | | |
| Columbia | 8.9 | 3.9 | 23.7 | 8.5 | 17.3 | 11.7 | 13.0 | 11.3 | 14.4 | 36.0 | 55.4 | 33,055 | 34,648 | 4.8 | 20.7 |
| Cookeville | 10.7 | 6.6 | 22.5 | 20.3 | 14.3 | 9.3 | 9.2 | 9.7 | 14.9 | 28.9 | 49.0 | 23,923 | 31,120 | 30.1 | 12.7 |
| Franklin | 4.5 | 13.1 | 25.2 | 7.2 | 13.9 | 16.5 | 12.0 | 12.9 | 12.3 | 37.2 | 54.1 | 41,842 | 62,572 | 49.5 | 36.3 |
| Gallatin | 9.7 | 7.6 | 18.8 | 9.2 | 13.7 | 11.7 | 12.8 | 12.7 | 21.1 | 43.2 | 50.1 | 23,230 | 30,347 | 30.6 | 46.1 |
| Germantown | 2.8 | 7.4 | 24.1 | 5.5 | 6.0 | 16.6 | 11.7 | 16.9 | 19.3 | 42.8 | 49.9 | 37,348 | 38,835 | 4.0 | 0.9 |
| Hendersonville | 3.4 | 6.6 | 24.2 | 7.5 | 12.0 | 14.1 | 15.8 | 12.5 | 13.8 | 40.4 | 52.4 | 40,620 | 51,333 | 26.4 | 15.6 |
| Jackson | 3.4 | 3.7 | 25.1 | 13.3 | 12.1 | 12.0 | 9.0 | 12.6 | 16.0 | 34.8 | 53.4 | 59,643 | 66,844 | 12.1 | 0.8 |
| Johnson City | 6.7 | 5.1 | 17.1 | 18.1 | 15.3 | 12.9 | 9.9 | 11.2 | 15.4 | 34.7 | 51.5 | 55,469 | 63,376 | 14.3 | 5.9 |
| Kingsport | 2.2 | 3.4 | 19.1 | 8.7 | 13.2 | 10.8 | 12.3 | 13.4 | 22.5 | 44.0 | 52.0 | 44,905 | 53,006 | 18.0 | 2.1 |
| Knoxville | 4.3 | 5.4 | 17.5 | 16.6 | 16.8 | 12.2 | 10.9 | 10.8 | 15.2 | 34.2 | 53.3 | 173,890 | 178,137 | 2.4 | 6.8 |
| La Vergne | 17.8 | 15.3 | 28.8 | 11.4 | 16.2 | 13.0 | 14.3 | 10.8 | 5.6 | 31.0 | 51.5 | 18,687 | 32,597 | 74.4 | 9.4 |
| Lebanon | 4.9 | 6.6 | 22.5 | 10.2 | 15.6 | 11.1 | 10.4 | 12.7 | 17.5 | 35.9 | 52.1 | 20,235 | 26,193 | 29.4 | 45.4 |
| Maryville | 3.4 | 3.5 | 26.2 | 10.2 | 12.2 | 10.9 | 14.8 | 10.4 | 15.5 | 37.2 | 54.3 | 23,120 | 27,436 | 18.7 | 9.3 |
| Memphis | 7.7 | 6.1 | 24.0 | 9.8 | 16.9 | 11.8 | 11.3 | 12.0 | 14.1 | 34.6 | 52.5 | 650,100 | 651,874 | 0.3 | -0.3 |
| Morristown | 16.8 | 5.7 | 21.7 | 10.0 | 14.1 | 11.1 | 9.0 | 16.6 | 17.5 | 39.4 | 53.8 | 24,965 | 28,947 | 16.0 | 4.6 |
| Murfreesboro | 7.8 | 8.0 | 22.2 | 16.8 | 15.8 | 12.4 | 12.5 | 9.8 | 10.5 | 31.2 | 49.2 | 68,816 | 109,100 | 58.5 | 38.2 |
| Nashville-Davidson | 10.6 | 13.9 | 20.5 | 9.8 | 20.8 | 14.0 | 11.3 | 11.2 | 12.4 | 34.4 | 51.7 | 545,524 | 603,465 | 10.6 | 11.2 |
| Oak Ridge | 5.6 | 8.2 | 24.7 | 8.0 | 12.0 | 8.0 | 16.0 | 11.0 | 20.3 | 41.4 | 51.7 | 27,387 | 29,328 | 7.1 | 0.0 |
| Smyrna | 13.0 | 15.5 | 26.3 | 7.6 | 13.2 | 17.4 | 12.3 | 13.0 | 10.2 | 36.4 | 51.4 | 25,569 | 40,420 | 58.1 | 31.6 |
| Spring Hill | 11.7 | 8.6 | 31.7 | 7.6 | 7.9 | 20.1 | 13.1 | 9.3 | 10.3 | 35.7 | 46.3 | 7,715 | 29,107 | 277.3 | 58.6 |
| **TEXAS** | 39.7 | 17.1 | 25.5 | 9.7 | 14.6 | 13.7 | 12.3 | 11.3 | 12.9 | 35.1 | 50.4 | 20,851,820 | 25,146,072 | 20.6 | 16.8 |
| Abilene | 27.7 | 6.6 | 23.0 | 14.7 | 16.5 | 12.7 | 8.6 | 10.5 | 14.0 | 34.7 | 49.7 | 115,930 | 117,509 | 1.4 | 5.9 |
| Allen | 12.3 | 19.6 | 26.7 | 9.6 | 7.7 | 17.0 | 16.7 | 12.8 | 9.5 | 39.5 | 49.2 | 43,554 | 84,273 | 93.5 | 29.8 |
| Amarillo | 34.0 | 10.7 | 26.6 | 9.2 | 14.5 | 13.6 | 10.7 | 9.7 | 15.6 | 34.7 | 51.2 | 173,627 | 190,675 | 9.8 | 4.7 |
| Arlington | 29.6 | 19.8 | 25.0 | 11.7 | 15.3 | 13.0 | 12.4 | 11.1 | 11.6 | 33.6 | 51.5 | 332,969 | 365,125 | 9.7 | 9.2 |
| Austin | 32.5 | 19.6 | 19.5 | 10.3 | 22.3 | 16.4 | 12.1 | 9.8 | 9.5 | 33.9 | 49.4 | 656,562 | 801,862 | 22.1 | 24.1 |
| Baytown | 54.0 | 20.0 | 26.1 | 8.7 | 17.7 | 11.7 | 12.1 | 11.7 | 12.1 | 34.1 | 49.8 | 66,430 | 72,075 | 8.5 | 7.2 |
| Beaumont | 18.5 | 9.4 | 24.4 | 9.8 | 16.5 | 12.8 | 9.9 | 12.1 | 14.5 | 34.6 | 52.2 | 113,866 | 117,243 | 3.0 | -1.5 |
| Bedford | 14.7 | 9.3 | 19.4 | 7.5 | 15.4 | 14.4 | 10.3 | 17.0 | 15.9 | 40.0 | 49.8 | 47,152 | 46,990 | -0.3 | 2.9 |
| Big Spring | 44.8 | 15.2 | 21.5 | 10.6 | 18.6 | 17.4 | 10.1 | 11.6 | 10.2 | 34.5 | 47.6 | 25,233 | 27,250 | 8.0 | 3.0 |
| Brownsville | 93.3 | 27.6 | 29.3 | 10.8 | 13.9 | 13.4 | 10.1 | 9.8 | 12.7 | 31.1 | 52.6 | 139,722 | 174,448 | 24.9 | 5.1 |
| Bryan | 39.5 | 15.1 | 25.1 | 11.7 | 18.6 | 13.1 | 11.2 | 9.0 | 11.5 | 31.5 | 51.3 | 65,660 | 76,233 | 16.1 | 15.7 |
| Burleson | 10.4 | 1.9 | 26.9 | 7.7 | 17.8 | 14.1 | 14.2 | 7.6 | 11.7 | 32.7 | 53.5 | 20,976 | 36,876 | 75.8 | 34.7 |
| Carrollton | 30.3 | 28.4 | 20.0 | 9.3 | 17.1 | 14.8 | 12.8 | 14.7 | 11.4 | 37.0 | 50.5 | 109,576 | 119,213 | 8.8 | 17.3 |
| Cedar Hill | 26.6 | 11.5 | 32.3 | 5.5 | 15.6 | 14.7 | 13.0 | 10.0 | 8.9 | 33.7 | 48.0 | 32,093 | 44,935 | 40.0 | 6.4 |
| Cedar Park | 28.2 | 16.3 | 25.8 | 6.6 | 13.9 | 16.8 | 14.4 | 9.8 | 12.8 | 38.2 | 52.7 | 26,049 | 55,113 | 111.6 | 47.0 |
| Cleburne | 30.6 | 4.6 | 28.2 | 6.2 | 14.1 | 10.3 | 12.5 | 13.3 | 15.3 | 36.8 | 48.8 | 26,005 | 29,634 | 14.0 | 8.0 |
| College Station | 15.0 | 11.0 | 16.9 | 38.2 | 13.7 | 10.9 | 6.5 | 6.5 | 7.2 | 23.6 | 49.0 | 67,890 | 94,250 | 38.8 | 26.5 |
| Conroe | 35.9 | 16.5 | 26.1 | 9.2 | 16.7 | 14.3 | 8.2 | 11.9 | 13.6 | 33.0 | 49.8 | 36,811 | 65,368 | 77.6 | 47.2 |
| Coppell | 11.5 | 27.3 | 26.6 | 7.7 | 9.8 | 14.5 | 17.5 | 14.7 | 9.2 | 38.3 | 53.5 | 35,958 | 38,666 | 7.5 | 5.9 |
| Copperas Cove | 16.4 | 8.9 | 25.3 | 10.0 | 18.5 | 11.5 | 14.1 | 8.4 | 12.2 | 33.3 | 53.7 | 29,592 | 32,253 | 9.0 | 3.9 |
| Corpus Christi | 63.4 | 10.0 | 24.4 | 10.2 | 14.9 | 13.2 | 11.3 | 11.7 | 14.5 | 35.3 | 50.5 | 277,454 | 305,202 | 10.0 | 7.2 |
| Dallas | 41.2 | 25.3 | 24.6 | 9.6 | 19.2 | 13.8 | 11.3 | 10.7 | 10.7 | 32.9 | 50.9 | 1,188,580 | 1,197,672 | 0.8 | 12.2 |
| Deer Park | 44.4 | 9.6 | 27.9 | 12.6 | 14.3 | 11.1 | 13.5 | 9.6 | 11.0 | 30.6 | 52.1 | 28,520 | 31,999 | 12.2 | 2.9 |
| Del Rio | 86.9 | 21.1 | 30.0 | 10.0 | 13.6 | 13.0 | 9.8 | 9.3 | 14.3 | 31.9 | 50.5 | 33,867 | 35,921 | 6.1 | -0.6 |
| Denton | 25.3 | 13.8 | 19.1 | 20.1 | 20.5 | 12.3 | 7.7 | 8.8 | 11.4 | 29.5 | 51.3 | 80,537 | 116,329 | 44.4 | 26.8 |
| DeSoto | 16.7 | 11.6 | 22.1 | 7.4 | 8.4 | 15.4 | 13.3 | 14.7 | 18.7 | 42.1 | 54.2 | 37,646 | 49,056 | 30.3 | 7.9 |
| Duncanville | 45.6 | 15.8 | 23.1 | 11.1 | 8.7 | 13.0 | 11.4 | 15.5 | 17.1 | 40.6 | 53.3 | 36,081 | 38,559 | 6.9 | -1.0 |
| Eagle Pass | 95.4 | 29.1 | 24.8 | 17.5 | 10.9 | 10.6 | 11.0 | 10.6 | 14.5 | 32.9 | 46.8 | 22,413 | 26,606 | 18.7 | 10.9 |
| Edinburg | 88.7 | 21.9 | 27.3 | 15.9 | 16.0 | 13.1 | 11.7 | 7.4 | 8.5 | 29.1 | 48.5 | 48,465 | 82,058 | 69.3 | 24.8 |
| El Paso | 82.1 | 21.2 | 26.3 | 11.0 | 15.5 | 12.0 | 11.6 | 10.5 | 13.0 | 32.9 | 50.9 | 563,662 | 648,079 | 15.0 | 5.2 |
| Euless | 14.6 | 20.2 | 22.0 | 8.9 | 20.9 | 13.3 | 11.0 | 13.7 | 10.1 | 34.2 | 51.4 | 46,005 | 51,263 | 11.4 | 11.8 |
| Farmers Branch | 52.2 | 33.8 | 16.9 | 10.7 | 19.1 | 15.6 | 12.4 | 10.5 | 14.9 | 36.2 | 50.9 | 27,508 | 29,139 | 5.9 | 76.3 |
| Flower Mound | 13.7 | 13.6 | 27.7 | 5.7 | 8.3 | 14.2 | 17.9 | 15.6 | 10.7 | 41.2 | 48.6 | 50,702 | 64,658 | 27.5 | 26.0 |
| Fort Worth | 36.4 | 17.6 | 26.9 | 10.1 | 16.5 | 13.9 | 12.3 | 10.2 | 10.0 | 32.8 | 51.4 | 534,694 | 744,800 | 39.3 | 24.6 |
| Friendswood | 16.1 | 15.2 | 26.8 | 11.0 | 8.9 | 10.3 | 15.7 | 11.9 | 15.4 | 38.5 | 51.4 | 29,037 | 35,911 | 23.7 | 11.4 |
| Frisco | 10.7 | 27.2 | 29.9 | 7.1 | 11.4 | 17.9 | 16.2 | 8.9 | 8.6 | 36.0 | 51.5 | 33,714 | 117,159 | 247.5 | 79.2 |
| Galveston | 32.3 | 13.0 | 19.5 | 14.4 | 15.4 | 12.0 | 11.4 | 11.5 | 15.8 | 36.2 | 50.8 | 57,247 | 47,742 | -16.6 | 4.9 |
| Garland | 46.8 | 29.3 | 28.8 | 9.9 | 13.1 | 11.9 | 12.6 | 12.1 | 11.6 | 33.7 | 49.8 | 215,768 | 226,857 | 5.1 | 5.0 |
| Georgetown | 23.1 | 7.2 | 21.0 | 8.8 | 10.0 | 10.9 | 11.5 | 10.7 | 27.1 | 44.6 | 53.8 | 28,339 | 47,491 | 67.6 | 80.1 |
| Grand Prairie | 52.5 | 22.6 | 29.4 | 9.0 | 15.1 | 13.2 | 13.8 | 10.9 | 8.5 | 32.2 | 50.7 | 127,427 | 175,435 | 37.7 | 11.3 |
| Grapevine | 15.0 | 13.2 | 25.5 | 6.5 | 12.8 | 14.5 | 14.4 | 17.3 | 9.2 | 37.9 | 48.3 | 42,059 | 46,336 | 10.2 | 20.4 |
| Greenville | 15.1 | 5.5 | 22.8 | 8.1 | 13.9 | 15.3 | 10.5 | 13.1 | 16.3 | 37.9 | 46.8 | 23,960 | 25,528 | 6.5 | 15.1 |
| Haltom City | 45.5 | 29.7 | 27.4 | 7.0 | 14.9 | 16.0 | 12.6 | 8.0 | 14.2 | 35.4 | 52.7 | 39,018 | 42,566 | 9.1 | 2.7 |
| Harker Heights | 28.1 | 10.1 | 33.6 | 4.1 | 9.5 | 18.7 | 13.6 | 7.6 | 12.8 | 35.5 | 52.4 | 17,308 | 26,730 | 54.4 | 23.7 |
| Harlingen | 84.3 | 14.6 | 30.6 | 11.9 | 13.2 | 9.3 | 12.6 | 7.7 | 14.8 | 30.8 | 51.5 | 57,564 | 64,924 | 12.8 | 0.4 |
| Houston | 45.8 | 28.3 | 24.7 | 9.4 | 18.7 | 14.4 | 11.2 | 10.4 | 11.3 | 33.4 | 50.0 | 1,953,631 | 2,092,919 | 7.1 | 10.7 |
| Huntsville | 19.6 | 12.0 | 13.4 | 30.5 | 12.0 | 12.2 | 15.7 | 6.6 | 9.6 | 29.2 | 38.8 | 35,078 | 38,524 | 9.8 | 7.4 |
| Hurst | 22.6 | 16.4 | 29.3 | 4.1 | 13.5 | 12.7 | 12.2 | 11.6 | 16.7 | 36.5 | 51.1 | 36,273 | 37,344 | 3.0 | 2.2 |

1. May be of any race.

## Table D. Cities — Households, Group Quarters, Crime, and Education

| City | Households, 2019 | | | | | | | Persons in group quarters, 2019 | Serious crimes known to police[1], 2019 | | | | Educational attainment, 2019 | | |
|---|---|---|---|---|---|---|---|---|---|---|---|---|---|---|---|
| | | | Percent | | | | | | Violent | | Prorerty | | | Attainment[3] (percent) | |
| | Number | Persons per household | Family | Married couple family | Female family | Non-family | One person | | Number | Rate[2] | Number | Rate[2] | Population age 25 and over | High school graduate or less | Bachelor's degree or more |
| | 27 | 28 | 29 | 30 | 31 | 32 | 33 | 34 | 35 | 36 | 37 | 38 | 39 | 40 | 41 |
| **TENNESSEE—Cont'd** | | | | | | | | | | | | | | | |
| Columbia | 16,116 | 2.44 | 67.0 | 42.1 | 24.8 | 33.0 | 24.0 | 989 | 257 | 642 | 1,202 | 3,005 | 27,327 | 46.8 | 21.0 |
| Cookeville | 13,519 | 2.37 | 53.6 | 31.3 | 22.3 | 46.4 | 35.9 | 2,714 | 121 | 352 | 1,047 | 3,046 | 19,884 | 41.2 | 33.7 |
| Franklin | 32,690 | 2.53 | 72.4 | 63.8 | 8.6 | 27.6 | 21.3 | 335 | 138 | 165 | 926 | 1,109 | 56,196 | 16.4 | 64.6 |
| Gallatin | 16,980 | 2.45 | 69.4 | 54.2 | 15.3 | 30.6 | 26.0 | 1,232 | 117 | 279 | 451 | 1,076 | 30,916 | 38.9 | 29.9 |
| Germantown | 14,911 | 2.62 | 75.2 | 69.3 | 6.0 | 24.8 | 22.3 | 150 | 36 | 92 | 492 | 1,257 | 27,592 | 10.4 | 67.9 |
| Hendersonville | 21,325 | 2.72 | 69.4 | 56.0 | 13.4 | 30.6 | 27.2 | 193 | 100 | 171 | 652 | 1,117 | 39,689 | 29.4 | 42.5 |
| Jackson | 25,925 | 2.45 | 62.0 | 34.5 | 27.5 | 38.0 | 32.4 | 3,605 | 650 | 971 | 2,473 | 3,696 | 41,378 | 44.4 | 27.3 |
| Johnson City | 30,724 | 2.08 | 51.8 | 36.1 | 15.7 | 48.2 | 37.0 | 3,875 | 286 | 426 | 2,390 | 3,557 | 43,914 | 33.8 | 42.0 |
| Kingsport | 24,638 | 2.14 | 61.4 | 44.4 | 17.0 | 38.6 | 33.8 | 906 | 332 | 612 | 2,829 | 5,218 | 38,798 | 39.8 | 30.6 |
| Knoxville | 83,492 | 2.14 | 48.4 | 31.4 | 17.0 | 51.6 | 40.9 | 8,601 | 1,259 | 667 | 8,207 | 4,350 | 123,677 | 38.1 | 32.9 |
| La Vergne | 10,479 | 3.41 | 83.9 | 59.8 | 24.1 | 16.1 | 13.2 | 12 | 142 | 392 | 650 | 1,794 | 21,344 | 44.5 | 25.2 |
| Lebanon | 13,154 | 2.65 | 68.7 | 48.7 | 20.1 | 31.3 | 24.1 | 1,666 | 149 | 410 | 929 | 2,557 | 24,558 | 44.4 | 27.7 |
| Maryville | 10,610 | 2.63 | 63.0 | 52.1 | 10.8 | 37.0 | 34.5 | 1,828 | 44 | 150 | 463 | 1,574 | 18,913 | 40.2 | 28.1 |
| Memphis | 254,423 | 2.50 | 54.7 | 29.0 | 25.7 | 45.3 | 39.0 | 14,147 | 12,367 | 1,901 | 39,860 | 6,128 | 430,632 | 45.4 | 25.6 |
| Morristown | 12,267 | 2.38 | 58.9 | 37.5 | 21.5 | 41.1 | 36.2 | 948 | 255 | 849 | 1,423 | 4,736 | 20,613 | 52.2 | 21.2 |
| Murfreesboro | 52,530 | 2.72 | 60.4 | 43.3 | 17.1 | 39.6 | 25.4 | 4,288 | 611 | 419 | 4,228 | 2,897 | 89,649 | 27.6 | 34.6 |
| Nashville-Davidson | 279,545 | 2.32 | 52.3 | 35.9 | 16.4 | 47.7 | 36.4 | 21,104 | NA | NA | NA | NA | 465,969 | 30.7 | 45.2 |
| Oak Ridge | 11,364 | 2.52 | 64.7 | 44.9 | 19.8 | 35.3 | 31.0 | 328 | 112 | 385 | 671 | 2,307 | 19,461 | 34.6 | 36.1 |
| Smyrna | 19,188 | 2.67 | 66.6 | 45.1 | 21.4 | 33.4 | 28.6 | 344 | 204 | 391 | 1,464 | 2,803 | 34,079 | 34.9 | 29.1 |
| Spring Hill | 13,466 | 3.13 | 92.7 | 80.0 | 12.8 | 7.3 | 7.3 | 53 | 53 | 122 | 381 | 880 | 25,600 | 23.2 | 45.9 |
| **TEXAS** | 9,985,126 | 2.84 | 68.7 | 49.6 | 19.1 | 31.3 | 25.6 | 600,693 | NA | NA | NA | NA | 18,772,550 | 40.6 | 30.8 |
| Abilene | 43,010 | 2.56 | 64.1 | 46.0 | 18.1 | 35.9 | 29.9 | 13,077 | 458 | 370 | 3,112 | 2,516 | 76,764 | 43.5 | 24.8 |
| Allen | 34,548 | 3.05 | 80.2 | 66.0 | 14.2 | 19.8 | 17.5 | 205 | 88 | 83 | 1,154 | 1,089 | 67,288 | 18.2 | 55.0 |
| Amarillo | 76,207 | 2.57 | 61.7 | 41.7 | 20.0 | 38.3 | 33.1 | 1,862 | 1,447 | 720 | 7,835 | 3,897 | 126,760 | 43.2 | 25.0 |
| Arlington | 138,639 | 2.85 | 67.5 | 45.3 | 22.2 | 32.5 | 26.5 | 4,046 | 2,055 | 511 | 11,291 | 2,807 | 252,612 | 38.0 | 33.6 |
| Austin | 409,903 | 2.33 | 50.9 | 37.1 | 13.8 | 49.1 | 35.2 | 24,623 | 3,953 | 401 | 36,588 | 3,711 | 686,879 | 24.3 | 55.0 |
| Baytown | 26,978 | 2.83 | 69.0 | 47.2 | 21.9 | 31.0 | 25.5 | 594 | 325 | 418 | 2,838 | 3,652 | 50,157 | 52.0 | 14.7 |
| Beaumont | 45,435 | 2.48 | 60.7 | 38.9 | 21.8 | 39.3 | 33.7 | 4,132 | 1,241 | 1,047 | 4,287 | 3,616 | 76,789 | 43.3 | 27.4 |
| Bedford | 20,072 | 2.43 | 60.9 | 42.8 | 18.1 | 39.1 | 31.2 | 374 | 145 | 291 | 1,010 | 2,029 | 35,846 | 27.7 | 33.1 |
| Big Spring | 9,119 | 2.73 | 63.4 | 38.7 | 24.8 | 36.6 | 31.1 | 4,982 | 152 | 538 | 1,049 | 3,710 | 20,260 | 53.1 | 14.0 |
| Brownsville | 55,052 | 3.30 | 80.5 | 49.6 | 30.8 | 19.5 | 17.6 | 1,098 | 776 | 421 | 4,553 | 2,469 | 109,378 | 56.9 | 19.1 |
| Bryan | 31,053 | 2.64 | 61.1 | 34.1 | 27.0 | 38.9 | 33.8 | 4,168 | 370 | 427 | 1,928 | 2,226 | 54,547 | 46.2 | 24.9 |
| Burleson | 16,553 | 2.91 | 76.2 | 51.3 | 24.9 | 23.8 | 18.6 | 77 | 113 | 232 | 726 | 1,489 | 31,623 | 43.4 | 23.8 |
| Carrollton | 48,963 | 2.83 | 68.7 | 54.5 | 14.2 | 31.3 | 25.2 | 681 | NA | NA | NA | NA | 98,480 | 31.6 | 40.8 |
| Cedar Hill | 14,921 | 3.20 | 79.5 | 54.6 | 24.9 | 20.5 | 15.0 | 492 | 99 | 203 | 1,252 | 2,562 | 29,956 | 37.9 | 26.2 |
| Cedar Park | 24,034 | 3.21 | 76.2 | 61.6 | 14.6 | 23.8 | 16.1 | 296 | 57 | 71 | 674 | 841 | 52,352 | 21.7 | 49.4 |
| Cleburne | 11,063 | 2.75 | 69.2 | 48.9 | 20.3 | 30.8 | 28.0 | 851 | 101 | 327 | 588 | 1,905 | 20,511 | 56.3 | 19.7 |
| College Station | 40,281 | 2.63 | 52.8 | 38.3 | 14.4 | 47.2 | 28.4 | 12,062 | 225 | 189 | 2,098 | 1,759 | 52,879 | 19.7 | 62.3 |
| Conroe | 31,236 | 2.86 | 66.6 | 44.8 | 21.8 | 33.4 | 24.8 | 1,594 | NA | NA | NA | NA | 58,949 | 43.3 | 27.5 |
| Coppell | 15,296 | 2.72 | 76.3 | 62.7 | 13.5 | 23.7 | 22.1 | 15 | 25 | 59 | 525 | 1,245 | 27,278 | 10.2 | 69.5 |
| Copperas Cove | 12,248 | 2.64 | 63.5 | 44.0 | 19.5 | 36.5 | 30.6 | 298 | 129 | 395 | 691 | 2,114 | 21,120 | 32.6 | 20.4 |
| Corpus Christi | 119,184 | 2.67 | 67.9 | 44.1 | 23.8 | 32.1 | 26.5 | 8,318 | 2,616 | 794 | 11,347 | 3,446 | 213,562 | 45.7 | 23.7 |
| Dallas | 518,998 | 2.56 | 54.9 | 35.0 | 19.9 | 45.1 | 36.6 | 16,184 | 11,764 | 863 | 45,279 | 3,321 | 882,889 | 43.5 | 34.7 |
| Deer Park | 10,233 | 3.26 | 76.4 | 60.0 | 16.5 | 23.6 | 21.1 | 125 | 48 | 140 | 551 | 1,613 | 19,915 | 45.2 | 17.7 |
| Del Rio | 13,015 | 2.73 | 68.1 | 44.5 | 23.6 | 31.9 | 29.1 | 1,502 | 38 | 106 | 610 | 1,697 | 22,266 | 63.4 | 16.2 |
| Denton | 46,151 | 2.86 | 60.5 | 44.2 | 16.3 | 39.5 | 26.5 | 9,607 | 331 | 234 | 2,665 | 1,883 | 86,135 | 27.5 | 42.7 |
| DeSoto | 20,371 | 2.57 | 63.4 | 44.6 | 18.8 | 36.6 | 32.7 | 636 | 176 | 326 | 1,188 | 2,199 | 37,371 | 37.4 | 23.8 |
| Duncanville | 13,796 | 2.79 | 68.5 | 44.8 | 23.8 | 31.5 | 27.7 | 233 | 174 | 441 | 1,096 | 2,780 | 25,484 | 46.0 | 21.0 |
| Eagle Pass | 9,414 | 3.14 | 88.0 | 55.7 | 32.4 | 12.0 | 12.0 | 162 | 37 | 124 | 687 | 2,302 | 17,125 | 60.7 | 18.3 |
| Edinburg | 30,224 | 3.18 | 68.0 | 42.0 | 26.0 | 32.0 | 21.2 | 5,030 | 281 | 279 | 2,959 | 2,933 | 57,440 | 43.1 | 29.4 |
| El Paso | 223,076 | 3.02 | 69.4 | 43.4 | 26.0 | 30.6 | 26.8 | 7,162 | 2,422 | 353 | 10,378 | 1,511 | 427,304 | 43.8 | 24.3 |
| Euless | 22,951 | 2.49 | 59.3 | 44.5 | 14.9 | 40.7 | 35.5 | 135 | 107 | 184 | 1,186 | 2,040 | 39,482 | 28.9 | 37.4 |
| Farmers Branch | 17,920 | 2.68 | 63.9 | 48.0 | 15.9 | 36.1 | 27.9 | 111 | 63 | 150 | 940 | 2,242 | 34,884 | 45.4 | 38.1 |
| Flower Mound | 27,199 | 3.05 | 83.5 | 72.2 | 11.3 | 16.5 | 14.6 | 191 | 43 | 54 | 730 | 923 | 55,431 | 14.8 | 61.1 |
| Fort Worth | 309,097 | 2.90 | 68.1 | 48.6 | 19.5 | 31.9 | 26.5 | 16,932 | 4,068 | 444 | 24,605 | 2,688 | 575,184 | 41.9 | 31.3 |
| Friendswood | 12,149 | 3.43 | 81.9 | 68.5 | 13.4 | 18.1 | 11.7 | 238 | 26 | 64 | 264 | 648 | 26,078 | 30.9 | 42.3 |
| Frisco | 67,851 | 2.95 | 75.9 | 65.8 | 10.0 | 24.1 | 21.1 | 358 | 160 | 80 | 2,390 | 1,198 | 126,374 | 14.3 | 63.3 |
| Galveston | 20,332 | 2.30 | 57.8 | 32.6 | 25.2 | 42.2 | 35.4 | 3,727 | 243 | 478 | 1,642 | 3,232 | 33,338 | 40.5 | 26.5 |
| Garland | 72,410 | 3.30 | 75.9 | 55.2 | 20.7 | 24.1 | 18.9 | 710 | 738 | 302 | 6,311 | 2,584 | 146,885 | 46.7 | 20.7 |
| Georgetown | 29,882 | 2.60 | 64.9 | 50.3 | 14.6 | 35.1 | 30.2 | 2,027 | 81 | 103 | 730 | 932 | 55,913 | 29.0 | 44.4 |
| Grand Prairie | 62,030 | 3.13 | 72.5 | 50.9 | 21.6 | 27.5 | 23.8 | 407 | 428 | 217 | 4,140 | 2,102 | 119,857 | 47.2 | 26.3 |
| Grapevine | 22,546 | 2.44 | 60.9 | 47.8 | 13.1 | 39.1 | 35.4 | 241 | NA | NA | NA | NA | 37,607 | 17.9 | 56.2 |
| Greenville | 12,086 | 2.32 | 54.9 | 35.0 | 19.8 | 45.1 | 37.3 | 773 | 77 | 269 | 635 | 2,219 | 19,929 | 48.1 | 23.1 |
| Haltom City | 14,084 | 3.11 | 72.8 | 49.6 | 23.2 | 27.2 | 23.4 | 113 | 126 | 283 | 1,171 | 2,627 | 28,799 | 59.7 | 14.3 |
| Harker Heights | 10,265 | 3.14 | 80.3 | 57.7 | 22.5 | 19.7 | 17.5 | 205 | 65 | 200 | 529 | 1,626 | 20,196 | 31.5 | 34.7 |
| Harlingen | 22,901 | 2.76 | 64.1 | 35.9 | 28.2 | 35.9 | 28.4 | 1,905 | 273 | 417 | 3,152 | 4,814 | 52,357 | 54.4 | 18.1 |
| Houston | 876,504 | 2.60 | 59.6 | 38.0 | 21.7 | 40.4 | 32.8 | 34,682 | 25,257 | 1,072 | 101,750 | 4,319 | 1,528,493 | 42.9 | 34.0 |
| Huntsville | 12,719 | 2.50 | 40.8 | 31.0 | 9.9 | 59.2 | 37.1 | 10,446 | NA | NA | NA | NA | 23,692 | 54.2 | 22.2 |
| Hurst | 13,100 | 2.93 | 73.3 | 50.9 | 22.4 | 26.7 | 24.0 | 269 | 82 | 209 | 1,402 | 3,577 | 25,757 | 31.5 | 30.6 |

1. Data for serious crimes have not been adjusted for underreporting. This may affect comparability between geographic areas and over time.   2. Per 100,000 population estimated by the FBI.   3. Persons 25 years old and over.

# Table D. Cities — Income, Poverty, and Housing

| City | Money income, 2019 — Households | | | | | Median earnings, 2019 | | | Housing units, 2019 | | | | |
|---|---|---|---|---|---|---|---|---|---|---|---|---|---|
| | Median income | Percent with income less than $20,000 | Percent with income of $200,000 or more | Median family income | Median non-family income | All persons | Men | Women | Total | Occupied | Percent owner occupied | Median value[1] (dollars) | Median gross rent (dollars) |
| | 42 | 43 | 44 | 45 | 46 | 47 | 48 | 49 | 50 | 51 | 52 | 53 | 54 |
| **TENNESSEE—Cont'd** | | | | | | | | | | | | | |
| Columbia | 60,755 | 10.7 | 3.0 | 67,719 | 42,478 | 32,254 | 36,425 | 30,629 | 17,449 | 16,116 | 65.5 | 176,286 | 895 |
| Cookeville | 38,815 | 24.5 | 1.5 | 59,255 | 21,935 | 25,459 | 24,879 | 26,634 | 15,169 | 13,519 | 41.9 | 198,165 | 755 |
| Franklin | 103,813 | 8.0 | 18.6 | 123,386 | 53,121 | 52,071 | 71,445 | 41,398 | 34,205 | 32,690 | 67.7 | 466,449 | 1,611 |
| Gallatin | 61,737 | 11.1 | 5.6 | 74,992 | 34,567 | 30,920 | 32,860 | 28,247 | 17,726 | 16,980 | 64.6 | 258,279 | 1,015 |
| Germantown | 117,559 | 6.5 | 29.9 | 154,836 | 43,244 | 62,155 | 85,785 | 57,700 | 15,843 | 14,911 | 81.1 | 361,690 | 1,209 |
| Hendersonville | 76,009 | 5.1 | 7.5 | 97,918 | 50,318 | 41,722 | 54,443 | 31,874 | 22,737 | 21,325 | 69.2 | 306,513 | 1,245 |
| Jackson | 46,112 | 22.9 | 2.6 | 52,300 | 37,104 | 27,194 | 30,727 | 25,059 | 29,036 | 25,925 | 51.8 | 117,653 | 898 |
| Johnson City | 42,797 | 25.8 | 6.7 | 71,682 | 22,637 | 26,309 | 27,522 | 24,707 | 33,233 | 30,724 | 47.6 | 170,018 | 754 |
| Kingsport | 48,939 | 21.5 | 3.6 | 65,913 | 26,888 | 29,694 | 40,940 | 21,781 | 27,929 | 24,638 | 66.9 | 157,404 | 639 |
| Knoxville | 41,388 | 23.9 | 3.1 | 59,131 | 30,191 | 28,027 | 30,918 | 25,888 | 93,470 | 83,492 | 45.4 | 155,401 | 871 |
| La Vergne | 72,854 | 4.7 | 3.5 | 80,748 | 58,769 | 34,854 | 39,580 | 30,918 | 11,424 | 10,479 | 64.6 | 200,598 | 1,347 |
| Lebanon | 56,480 | 14.9 | 5.3 | 71,331 | 33,966 | 31,469 | 32,229 | 30,383 | 13,587 | 13,154 | 53.9 | 256,801 | 967 |
| Maryville | 59,586 | 22.2 | 3.0 | 87,538 | 32,275 | 35,896 | 40,922 | 31,444 | 12,377 | 10,610 | 62.1 | 245,635 | 1,004 |
| Memphis | 43,794 | 23.0 | 4.6 | 54,519 | 30,951 | 30,068 | 32,537 | 26,762 | 301,473 | 254,423 | 46.4 | 115,914 | 923 |
| Morristown | 36,191 | 21.8 | 2.9 | 51,140 | 21,562 | 23,728 | 29,663 | 19,849 | 13,148 | 12,267 | 54.8 | 120,533 | 768 |
| Murfreesboro | 68,373 | 8.9 | 5.0 | 79,652 | 43,569 | 33,490 | 34,432 | 31,992 | 59,716 | 52,530 | 53.1 | 273,539 | 1,113 |
| Nashville-Davidson | 63,462 | 11.9 | 7.5 | 80,904 | 48,784 | 37,624 | 41,540 | 35,546 | 314,344 | 279,545 | 53.3 | 287,288 | 1,191 |
| Oak Ridge | 55,303 | 16.3 | 4.6 | 66,059 | 33,434 | 31,155 | 34,904 | 25,331 | 14,172 | 11,364 | 61.5 | 161,015 | 913 |
| Smyrna | 57,566 | 10.9 | 3.7 | 68,958 | 45,387 | 39,837 | 45,605 | 32,917 | 20,561 | 19,188 | 55.9 | 231,891 | 1,167 |
| Spring Hill | 86,697 | 7.1 | 9.0 | 89,006 | 36,987 | 38,990 | 47,394 | 20,011 | 13,466 | 13,466 | 80.4 | 344,725 | 1,390 |
| **TEXAS** | 64,034 | 13.6 | 7.9 | 76,727 | 40,389 | 35,329 | 41,239 | 29,029 | 11,283,892 | 9,985,126 | 61.9 | 200,432 | 1,091 |
| Abilene | 52,997 | 17.3 | 2.8 | 66,653 | 26,953 | 26,893 | 31,389 | 22,025 | 49,516 | 43,010 | 51.8 | 138,378 | 899 |
| Allen | 101,360 | 5.1 | 15.7 | 120,717 | 51,416 | 52,422 | 70,805 | 38,216 | 35,664 | 34,548 | 70.8 | 335,817 | 1,501 |
| Amarillo | 49,802 | 19.2 | 3.4 | 63,909 | 30,542 | 30,572 | 36,737 | 25,520 | 85,889 | 76,207 | 60.8 | 151,671 | 846 |
| Arlington | 61,716 | 10.8 | 5.5 | 73,488 | 41,324 | 33,645 | 38,338 | 30,463 | 148,396 | 138,639 | 54.1 | 213,810 | 1,124 |
| Austin | 75,413 | 11.6 | 11.7 | 98,238 | 57,150 | 42,416 | 49,165 | 37,501 | 442,388 | 409,903 | 44.4 | 378,259 | 1,334 |
| Baytown | 57,765 | 11.2 | 1.9 | 71,944 | 37,411 | 39,187 | 49,583 | 25,831 | 28,802 | 26,978 | 58.1 | 133,857 | 938 |
| Beaumont | 54,488 | 18.4 | 6.0 | 70,794 | 33,433 | 32,274 | 40,526 | 27,732 | 55,245 | 45,435 | 52.1 | 126,216 | 884 |
| Bedford | 70,746 | 9.1 | 5.6 | 88,988 | 43,548 | 41,257 | 48,451 | 35,411 | 20,206 | 20,072 | 54.5 | 246,429 | 1,153 |
| Big Spring | 65,174 | 10.2 | 8.7 | 67,334 | 58,165 | 32,355 | 45,761 | 22,886 | 10,174 | 9,119 | 59.6 | 117,475 | 861 |
| Brownsville | 41,271 | 24.3 | 3.0 | 49,078 | 17,365 | 24,110 | 30,803 | 20,411 | 59,943 | 55,052 | 58.6 | 95,573 | 794 |
| Bryan | 49,769 | 19.6 | 3.0 | 61,322 | 29,455 | 26,931 | 31,780 | 23,523 | 36,121 | 31,053 | 54.2 | 174,005 | 888 |
| Burleson | 77,153 | 8.4 | 2.8 | 86,599 | 44,580 | 40,700 | 45,898 | 35,489 | 17,056 | 16,553 | 62.7 | 219,805 | 1,427 |
| Carrollton | 85,830 | 5.8 | 10.1 | 103,437 | 56,412 | 41,946 | 48,063 | 36,521 | 52,932 | 48,963 | 61.0 | 282,335 | 1,329 |
| Cedar Hill | 71,914 | 9.6 | 5.1 | 75,066 | 56,379 | 38,931 | 46,544 | 35,044 | 15,955 | 14,921 | 67.1 | 203,014 | 1,450 |
| Cedar Park | 101,496 | 4.7 | 12.8 | 106,946 | 56,142 | 42,672 | 56,793 | 31,947 | 24,793 | 24,034 | 71.4 | 334,429 | 1,473 |
| Cleburne | 48,548 | 13.9 | 2.2 | 60,521 | 30,639 | 31,604 | 33,359 | 26,689 | 11,852 | 11,063 | 64.9 | 146,154 | 1,012 |
| College Station | 52,268 | 21.7 | 7.6 | 92,731 | 27,783 | 21,113 | 27,053 | 16,875 | 48,855 | 40,281 | 35.8 | 286,538 | 1,007 |
| Conroe | 69,077 | 6.8 | 8.6 | 72,777 | 46,961 | 36,355 | 41,884 | 30,889 | 35,788 | 31,236 | 51.5 | 212,006 | 1,137 |
| Coppell | 115,581 | 1.7 | 30.1 | 134,250 | 80,926 | 69,323 | 92,181 | 52,377 | 15,296 | 15,296 | 66.1 | 442,222 | 1,540 |
| Copperas Cove | 52,217 | 12.0 | 0.8 | 66,971 | 31,640 | 31,879 | 35,327 | 28,951 | 13,888 | 12,248 | 54.1 | 130,994 | 826 |
| Corpus Christi | 55,564 | 17.4 | 4.1 | 71,505 | 29,869 | 31,987 | 36,877 | 27,281 | 134,604 | 119,184 | 58.1 | 157,143 | 1,037 |
| Dallas | 55,332 | 15.8 | 9.0 | 63,180 | 45,844 | 35,451 | 38,651 | 31,181 | 589,260 | 518,998 | 40.6 | 231,387 | 1,128 |
| Deer Park | 87,086 | 3.7 | 7.8 | 94,118 | 37,676 | 34,697 | 56,657 | 23,350 | 10,467 | 10,233 | 75.5 | 182,198 | 1,054 |
| Del Rio | 40,160 | 30.8 | 0.0 | 48,065 | 18,583 | 23,867 | 45,008 | 16,051 | 14,469 | 13,015 | 58.4 | 135,277 | 780 |
| Denton | 66,644 | 12.8 | 5.3 | 81,792 | 34,408 | 27,188 | 28,709 | 25,763 | 51,628 | 46,151 | 48.0 | 248,981 | 1,117 |
| DeSoto | 61,018 | 13.1 | 3.0 | 77,546 | 37,067 | 36,176 | 42,498 | 34,472 | 22,293 | 20,371 | 61.2 | 221,957 | 1,296 |
| Duncanville | 55,381 | 14.2 | 3.4 | 65,397 | 26,555 | 36,105 | 40,890 | 29,074 | 15,152 | 13,796 | 68.8 | 186,623 | 1,064 |
| Eagle Pass | 39,801 | 28.3 | 0.8 | 42,202 | 0 | 26,527 | 30,430 | 21,236 | 10,134 | 9,414 | 57.6 | 155,888 | 554 |
| Edinburg | 46,000 | 18.8 | 4.6 | 51,681 | 30,572 | 25,435 | 30,285 | 21,424 | 33,635 | 30,224 | 51.5 | 128,470 | 854 |
| El Paso | 48,542 | 19.6 | 2.7 | 54,147 | 30,026 | 26,104 | 30,732 | 21,549 | 250,077 | 223,076 | 57.5 | 133,606 | 847 |
| Euless | 69,847 | 7.9 | 9.4 | 85,245 | 49,504 | 45,097 | 45,980 | 42,250 | 24,521 | 22,951 | 40.9 | 246,285 | 1,198 |
| Farmers Branch | 76,543 | 5.4 | 6.4 | 84,933 | 60,991 | 39,994 | 45,528 | 30,928 | 20,336 | 17,920 | 53.6 | 233,272 | 1,414 |
| Flower Mound | 139,508 | 1.8 | 26.0 | 156,837 | 76,538 | 61,049 | 85,294 | 45,251 | 27,698 | 27,199 | 86.7 | 385,698 | 1,837 |
| Fort Worth | 65,356 | 12.9 | 5.9 | 75,854 | 41,150 | 35,739 | 41,373 | 30,023 | 346,502 | 309,097 | 58.2 | 209,430 | 1,110 |
| Friendswood | 106,485 | 3.8 | 23.7 | 117,687 | 61,362 | 45,956 | 52,356 | 40,560 | 13,355 | 12,149 | 79.5 | 313,346 | 1,581 |
| Frisco | 116,884 | 4.6 | 22.6 | 140,625 | 63,384 | 60,391 | 81,450 | 44,936 | 73,042 | 67,851 | 65.7 | 439,878 | 1,540 |
| Galveston | 46,401 | 26.8 | 3.7 | 64,435 | 31,675 | 27,417 | 31,067 | 23,793 | 30,536 | 20,332 | 40.0 | 234,402 | 1,021 |
| Garland | 62,556 | 9.1 | 4.4 | 70,002 | 44,297 | 31,514 | 33,434 | 27,867 | 76,812 | 72,410 | 64.5 | 190,219 | 1,200 |
| Georgetown | 73,552 | 8.7 | 5.5 | 94,778 | 49,702 | 36,299 | 42,167 | 31,311 | 31,341 | 29,882 | 66.2 | 298,460 | 1,198 |
| Grand Prairie | 66,097 | 8.7 | 4.7 | 71,889 | 46,557 | 37,095 | 41,620 | 28,794 | 67,103 | 62,030 | 59.9 | 208,174 | 1,092 |
| Grapevine | 103,903 | 3.7 | 19.5 | 125,090 | 69,492 | 52,031 | 55,554 | 46,046 | 22,849 | 22,546 | 47.4 | 375,505 | 1,459 |
| Greenville | 55,429 | 17.7 | 3.7 | 60,843 | 29,647 | 37,716 | 40,659 | 31,049 | 12,086 | 12,086 | 48.7 | 119,987 | 999 |
| Haltom City | 56,189 | 13.7 | 3.0 | 60,535 | 35,459 | 30,211 | 36,707 | 24,302 | 16,059 | 14,084 | 55.0 | 141,447 | 999 |
| Harker Heights | 66,439 | 9.8 | 3.0 | 75,615 | 34,529 | 37,757 | 51,933 | 24,863 | 11,933 | 10,265 | 51.9 | 226,894 | 975 |
| Harlingen | 36,868 | 29.7 | 2.9 | 50,332 | 30,697 | 25,763 | 31,139 | 20,240 | 26,908 | 22,901 | 55.2 | 93,854 | 745 |
| Houston | 52,450 | 17.5 | 8.5 | 60,190 | 40,770 | 32,566 | 36,908 | 28,949 | 987,158 | 876,504 | 40.3 | 195,804 | 1,095 |
| Huntsville | 38,659 | 33.8 | 0.4 | 61,107 | 19,644 | 20,414 | 20,672 | 19,494 | 14,433 | 12,719 | 35.9 | 187,834 | 889 |
| Hurst | 64,720 | 11.1 | 7.9 | 80,539 | 34,279 | 39,039 | 44,085 | 31,282 | 14,093 | 13,100 | 62.7 | 226,507 | 1,122 |

1. Based on population estimated by the American Community Survey.

# Table D. Cities — Commuting, Computer Access, Migration, Labor Force, and Employment

| City | Commuting¹, 2019 Percent — Drove alone | Commuting — With commutes of 30 minutes or more | Computer access², 2019 Percent — With a computer in the house | Computer access — With Internet access | Migration, 2019 — Percent who lived in the same house one year ago | Migration — Percent who lived in another state or county one year ago | Civilian labor force, 2020 — Total | Percent change 2019-2020 | Unemployment Total | Rate³ | Civilian Employment⁴, 2019 — Population age 16 and older: Number | Percent in labor force | Population age 16 to 64: Number | Percent who worked full-year full-time |
|---|---|---|---|---|---|---|---|---|---|---|---|---|---|---|
| | 55 | 56 | 57 | 58 | 59 | 60 | 61 | 62 | 63 | 64 | 65 | 66 | 67 | 68 |
| **TENNESSEE—Cont'd** | | | | | | | | | | | | | | |
| Columbia | 82.0 | 40.1 | 97.3 | 90.6 | 92.0 | 4.7 | 19,574 | 1.5 | 1,996 | 10 | 31,798 | 69.2 | 25,978 | 57.4 |
| Cookeville | 79.9 | 19.3 | 81.6 | 78.3 | 72.8 | 17.0 | 14,383 | -0.2 | 1,071 | 7 | 27,582 | 54.7 | 22,424 | 43.0 |
| Franklin | 76.6 | 43.4 | 96.3 | 93.6 | 85.6 | 9.7 | 46,861 | -2.5 | 2,257 | 5 | 64,563 | 72.6 | 54,378 | 60.2 |
| Gallatin | 75.9 | 48.1 | 92.6 | 86.9 | 81.3 | 10.5 | 22,497 | 3.1 | 1,710 | 8 | 35,834 | 59.8 | 26,781 | 53.7 |
| Germantown | 81.5 | 24.7 | 96.1 | 91.9 | 80.3 | 11.0 | 19,420 | -4.4 | 870 | 5 | 30,628 | 69.0 | 23,069 | 66.0 |
| Hendersonville | 79.2 | 42.0 | 96.2 | 92.7 | 90.3 | 7.4 | 32,562 | -2.9 | 2,087 | 6 | 45,842 | 70.9 | 37,816 | 61.5 |
| Jackson | 72.0 | 9.6 | 87.1 | 82.5 | 86.7 | 8.6 | 32,547 | -0.2 | 2,743 | 8 | 50,783 | 60.8 | 40,062 | 51.5 |
| Johnson City | 86.6 | 17.8 | 90.8 | 85.2 | 74.7 | 12.8 | 31,318 | -2.1 | 2,055 | 7 | 57,148 | 63.4 | 46,676 | 54.7 |
| Kingsport | 83.8 | 16.9 | 90.0 | 84.9 | 84.1 | 5.7 | 22,929 | -1.8 | 1,864 | 8 | 44,476 | 56.5 | 32,385 | 52.8 |
| Knoxville | 78.2 | 23.9 | 92.1 | 77.9 | 77.2 | 9.4 | 96,500 | -1.5 | 6,781 | 7 | 158,056 | 62.0 | 129,558 | 50.2 |
| La Vergne | 74.1 | 42.8 | NA | NA | 83.6 | 8.4 | 19,527 | -2.3 | 1,625 | 8 | 26,173 | 76.4 | 24,184 | 59.0 |
| Lebanon | 78.9 | 36.1 | 93.9 | 83.8 | 82.2 | 5.8 | 17,330 | 0.7 | 1,282 | 7 | 28,775 | 63.3 | 22,376 | 56.9 |
| Maryville | 91.4 | 31.1 | 90.6 | 86.0 | 93.0 | 4.3 | 13,974 | -0.8 | 818 | 6 | 23,096 | 58.5 | 18,501 | 47.7 |
| Memphis | 82.6 | 26.5 | 84.9 | 72.3 | 85.7 | 3.3 | 301,072 | 0.8 | 34,094 | 11 | 511,878 | 62.8 | 419,751 | 49.8 |
| Morristown | 92.4 | 27.3 | 84.5 | 80.5 | 84.2 | 9.8 | 11,831 | -0.5 | 922 | 8 | 24,451 | 52.8 | 19,177 | 42.4 |
| Murfreesboro | 80.8 | 45.2 | 96.1 | 93.5 | 77.6 | 9.2 | 83,017 | 0.4 | 5,745 | 7 | 117,564 | 70.8 | 102,086 | 55.8 |
| Nashville-Davidson | 75.5 | 35.8 | 95.4 | 89.0 | 80.1 | 8.5 | 402,058 | -1.9 | 32,271 | 8 | 543,328 | 72.8 | 460,477 | 58.8 |
| Oak Ridge | 82.8 | 33.1 | 93.6 | 90.2 | 83.7 | 12.1 | 14,041 | -2.1 | 922 | 7 | 22,763 | 56.7 | 16,894 | 46.8 |
| Smyrna | 89.9 | 40.4 | 95.7 | 92.2 | 80.0 | 12.7 | 28,669 | -1.3 | 2,119 | 7 | 39,659 | 69.9 | 34,387 | 61.6 |
| Spring Hill | 83.3 | 53.1 | NA | NA | 81.0 | 10.7 | 23,923 | 1.4 | 1,500 | 6 | 30,670 | 69.9 | 26,312 | 55.0 |
| **TEXAS** | 80.1 | 39.9 | 93.6 | 85.7 | 85.0 | 6.5 | 13,983,319 | -0.4 | 1,067,982 | 8 | 22,425,612 | 64.5 | 18,686,885 | 54.0 |
| Abilene | 79.2 | 6.0 | 91.9 | 83.7 | 79.3 | 12.0 | 56,443 | -1.1 | 3,170 | 6 | 97,476 | 56.4 | 80,228 | 51.5 |
| Allen | 79.4 | 46.7 | NA | NA | 85.7 | 6.1 | 57,959 | 0.4 | 3,548 | 6 | 80,500 | 71.6 | 70,493 | 61.2 |
| Amarillo | 84.7 | 16.3 | 94.1 | 87.5 | 83.8 | 8.2 | 100,211 | -0.4 | 5,007 | 5 | 150,689 | 64.9 | 119,803 | 58.3 |
| Arlington | 82.2 | 39.2 | 96.1 | 92.0 | 83.6 | 5.1 | 213,397 | -0.3 | 16,449 | 8 | 311,958 | 67.9 | 265,841 | 52.7 |
| Austin | 71.1 | 36.8 | 94.9 | 88.9 | 77.9 | 9.0 | 589,616 | -0.5 | 36,653 | 6 | 806,581 | 74.1 | 713,142 | 59.6 |
| Baytown | 91.3 | 30.4 | 93.3 | 83.6 | 85.1 | 6.4 | 34,927 | 3.3 | 5,635 | 16 | 58,367 | 61.3 | 49,076 | 52.2 |
| Beaumont | 90.8 | 19.1 | 95.1 | 84.9 | 88.3 | 5.6 | 50,684 | -1.7 | 5,557 | 11 | 90,623 | 61.4 | 73,705 | 53.1 |
| Bedford | 82.8 | 36.4 | 96.3 | 91.0 | 73.6 | 8.1 | 29,102 | -1.3 | 2,199 | 8 | 40,220 | 69.2 | 32,406 | 63.4 |
| Big Spring | NA | 18.7 | 98.7 | 90.8 | 85.8 | 9.2 | 9,768 | -3.9 | 851 | 9 | 23,851 | 51.4 | 20,820 | 44.6 |
| Brownsville | 83.6 | 14.7 | 84.8 | 75.2 | 86.8 | 2.8 | 76,378 | 1.5 | 8,198 | 11 | 135,383 | 58.9 | 112,217 | 47.0 |
| Bryan | 85.0 | 11.8 | 89.2 | 73.0 | 75.7 | 10.4 | 43,171 | -1.6 | 2,519 | 6 | 66,457 | 65.3 | 56,579 | 56.3 |
| Burleson | 83.9 | 52.4 | 97.9 | 93.2 | 84.6 | 8.9 | 25,217 | 0.5 | 1,621 | 6 | 37,275 | 73.5 | 31,624 | 63.5 |
| Carrollton | 83.6 | 40.3 | 98.3 | 90.9 | 86.0 | 9.6 | 82,531 | 0.4 | 5,339 | 7 | 115,235 | 73.9 | 99,327 | 60.9 |
| Cedar Hill | 82.4 | 45.0 | 98.5 | 95.1 | 79.9 | 3.9 | 26,527 | -1.5 | 2,287 | 9 | 33,854 | 73.6 | 29,552 | 61.7 |
| Cedar Park | 71.3 | 42.8 | 97.4 | 95.0 | 77.3 | 16.8 | 43,086 | 0.5 | 2,490 | 6 | 60,355 | 72.9 | 50,448 | 57.1 |
| Cleburne | 83.3 | 41.1 | 94.0 | 88.3 | 87.4 | 8.4 | 13,806 | 0.6 | 968 | 7 | 23,161 | 60.1 | 18,370 | 59.5 |
| College Station | 74.7 | 10.1 | 98.6 | 87.2 | 63.3 | 19.6 | 60,194 | -2.1 | 2,986 | 5 | 100,344 | 60.4 | 91,801 | 36.2 |
| Conroe | 82.7 | 38.0 | 96.2 | 90.7 | 84.9 | 4.8 | 42,725 | 2.0 | 3,458 | 8 | 69,340 | 67.3 | 56,969 | 56.4 |
| Coppell | 80.2 | 38.0 | NA | NA | 85.6 | 7.0 | 23,459 | -2.9 | 1,419 | 6 | 32,068 | 70.6 | 28,265 | 60.0 |
| Copperas Cove | 82.0 | 30.8 | 94.5 | 91.7 | 81.8 | 13.9 | 12,710 | 0.4 | 927 | 7 | 25,232 | 57.4 | 21,258 | 48.4 |
| Corpus Christi | 83.2 | 15.9 | 92.4 | 87.4 | 84.1 | 5.4 | 148,932 | -2.1 | 13,266 | 9 | 255,483 | 61.5 | 208,112 | 53.5 |
| Dallas | 77.4 | 42.9 | 90.4 | 80.7 | 83.8 | 6.2 | 688,543 | -0.5 | 54,077 | 8 | 1,044,712 | 68.4 | 900,674 | 57.4 |
| Deer Park | 89.3 | 36.7 | NA | NA | NA | NA | 16,387 | -3.0 | 1,475 | 9 | 25,448 | 67.7 | 21,778 | 52.1 |
| Del Rio | 85.3 | 8.1 | 84.9 | 48.0 | 78.3 | 8.8 | 16,609 | -1.3 | 1,385 | 8 | 27,550 | 58.4 | 22,230 | 47.3 |
| Denton | 73.7 | 39.8 | 96.7 | 89.7 | 76.5 | 12.3 | 77,013 | 1.3 | 5,290 | 7 | 118,489 | 67.5 | 102,285 | 50.1 |
| DeSoto | 79.2 | 51.2 | 93.8 | 88.1 | 85.2 | 3.3 | 28,455 | -1.4 | 2,585 | 9 | 43,995 | 62.4 | 34,093 | 50.4 |
| Duncanville | 79.0 | 32.9 | 93.6 | 86.8 | 94.0 | 1.7 | 20,009 | -1.9 | 1,691 | 9 | 31,216 | 56.0 | 24,597 | 46.8 |
| Eagle Pass | NA | 22.6 | 77.1 | 59.8 | NA | NA | 13,240 | 4.1 | 2,211 | 17 | 23,317 | 54.3 | 19,002 | 47.5 |
| Edinburg | 78.2 | 20.4 | 90.9 | 80.8 | 88.6 | 5.6 | 47,231 | 2.8 | 4,047 | 9 | 76,262 | 63.3 | 67,618 | 45.9 |
| El Paso | 81.5 | 26.9 | 91.1 | 83.7 | 84.8 | 4.9 | 303,428 | 0.2 | 23,982 | 8 | 521,692 | 60.7 | 432,865 | 50.0 |
| Euless | 86.8 | 31.3 | 96.3 | 93.0 | 81.6 | 4.9 | 33,048 | 0.3 | 2,813 | 9 | 45,562 | 76.7 | 39,763 | 65.7 |
| Farmers Branch | 85.8 | 32.1 | 95.4 | 81.8 | 84.2 | 9.6 | 25,622 | 17.7 | 1,617 | 6 | 41,509 | 74.7 | 34,323 | 67.0 |
| Flower Mound | 75.5 | 49.4 | NA | NA | 83.7 | 8.1 | 43,511 | -0.1 | 2,331 | 5 | 63,841 | 72.4 | 54,957 | 60.7 |
| Fort Worth | 81.7 | 39.7 | 95.5 | 88.8 | 86.1 | 5.6 | 443,950 | 1.0 | 33,883 | 8 | 692,602 | 67.6 | 601,237 | 55.9 |
| Friendswood | 81.1 | 42.6 | 96.5 | 94.4 | 83.3 | 9.2 | 19,455 | -3.1 | 1,258 | 7 | 32,230 | 68.4 | 25,760 | 51.6 |
| Frisco | 77.5 | 44.5 | 99.8 | 97.2 | 82.8 | 13.0 | 109,433 | 4.2 | 6,112 | 6 | 147,312 | 73.3 | 130,006 | 59.7 |
| Galveston | 70.8 | 20.3 | 93.3 | 89.4 | 76.0 | 11.7 | 23,791 | -0.4 | 2,371 | 10 | 41,963 | 61.4 | 33,969 | 46.9 |
| Garland | 77.6 | 56.5 | 95.9 | 90.2 | 91.0 | 2.8 | 124,697 | -1.9 | 9,092 | 7 | 178,975 | 70.0 | 151,155 | 57.2 |
| Georgetown | 72.5 | 34.5 | 95.9 | 91.9 | 83.3 | 9.8 | 33,116 | 4.6 | 2,155 | 7 | 64,084 | 55.9 | 42,492 | 55.3 |
| Grand Prairie | 82.9 | 47.7 | 96.5 | 90.2 | 88.4 | 5.9 | 100,856 | -0.5 | 7,938 | 8 | 144,020 | 71.2 | 127,395 | 61.4 |
| Grapevine | 82.0 | 29.5 | 98.0 | 95.1 | 81.7 | 8.6 | 33,062 | 0.7 | 1,998 | 6 | 43,030 | 81.5 | 37,966 | 63.6 |
| Greenville | 82.5 | 31.2 | 85.7 | 81.0 | 80.4 | 6.9 | 11,896 | 0.5 | 869 | 7 | 22,647 | 59.9 | 17,945 | 54.6 |
| Haltom City | 75.5 | 43.5 | 94.7 | 88.5 | 87.1 | 5.4 | 22,086 | -1.9 | 1,552 | 7 | 33,368 | 66.2 | 27,159 | 58.6 |
| Harker Heights | 82.4 | 21.4 | 94.8 | 87.5 | 62.6 | 17.0 | 12,849 | 0.2 | 872 | 7 | 22,950 | 56.2 | 18,803 | 52.1 |
| Harlingen | 84.1 | 14.8 | 84.0 | 64.4 | 94.9 | 3.2 | 24,742 | 0.7 | 2,345 | 10 | 48,072 | 56.5 | 38,467 | 43.3 |
| Houston | 78.1 | 47.1 | 92.2 | 84.1 | 82.7 | 5.3 | 1,140,188 | -1.6 | 100,644 | 9 | 1,803,388 | 66.6 | 1,541,231 | 53.5 |
| Huntsville | 81.1 | 34.2 | 89.7 | 86.4 | 73.7 | 13.0 | 12,092 | 1.4 | 1,012 | 8 | 36,876 | 50.6 | 32,817 | 26.2 |
| Hurst | 84.2 | 43.2 | 96.3 | 89.0 | 94.8 | 1.5 | 20,194 | -1.5 | 1,490 | 7 | 28,334 | 62.6 | 21,893 | 55.4 |

1. Employed persons.  2. Households.  3. Percent of civilian labor force.  4. Persons 16 years old and over.

# Table D. Cities — Construction, Wholesale Trade, and Retail Trade

| City | Value of residential construction authorized by building permits, 2020 | | | Wholesale trade[1], 2017 | | | | Retail trade[2], 2017 | | | |
|---|---|---|---|---|---|---|---|---|---|---|---|
| | New construction ($1,000) | Number of housing units | Percent single family | Number of establishments | Number of employees | Sales (mil dol) | Annual payroll (mil dol) | Number of establishments | Number of employees | Sales (mil dol) | Annual payroll (mil dol) |
| | 69 | 70 | 71 | 72 | 73 | 74 | 75 | 76 | 77 | 78 | 79 |
| **TENNESSEE—Cont'd** | | | | | | | | | | | |
| Columbia | 74,136 | 636 | 99 | 47 | 540 | 256 | 28 | 214 | 2,597 | 934 | 78 |
| Cookeville | 44,224 | 331 | 70 | 54 | 551 | 312 | 27 | 282 | 4,187 | 1,211 | 113 |
| Franklin | 268,978 | 923 | 44 | 153 | 2,437 | 6,864 | 198 | 481 | 8,792 | 2,878 | 269 |
| Gallatin | 213,325 | 1,178 | 58 | 35 | 530 | 692 | 25 | 145 | 2,411 | 743 | 68 |
| Germantown | NA | NA | NA | 33 | 937 | 1,143 | 86 | 147 | 2,120 | 430 | 49 |
| Hendersonville | 121,074 | 335 | 100 | 53 | 487 | 229 | 33 | 203 | 3,222 | 840 | 76 |
| Jackson | 41,211 | 162 | 100 | 122 | 2,199 | 1,219 | 95 | 439 | 6,871 | 1,987 | 175 |
| Johnson City | 42,441 | 535 | 35 | 83 | 1,527 | 1,063 | 89 | 436 | 7,718 | 2,124 | 188 |
| Kingsport | 23,831 | 100 | 84 | 71 | 989 | 493 | 44 | 317 | 5,487 | 1,464 | 132 |
| Knoxville | 179,372 | 1,149 | 27 | 395 | 6,266 | 3,641 | 357 | 1,264 | 21,652 | 6,319 | 606 |
| La Vergne | 16,300 | 92 | 100 | 62 | 2,827 | 12,427 | 191 | 65 | 898 | 291 | 34 |
| Lebanon | 190,537 | 1,268 | 58 | 57 | 1,265 | 2,091 | 75 | 222 | 2,979 | 984 | 83 |
| Maryville | 36,953 | 208 | 100 | 33 | 334 | 825 | 19 | 173 | 2,686 | 620 | 64 |
| Memphis | NA | NA | NA | 906 | 22,618 | 25,130 | 1,282 | 2,306 | 35,216 | 18,896 | 973 |
| Morristown | 32,740 | 219 | 69 | 40 | 788 | 754 | 38 | 249 | 4,054 | 1,153 | 106 |
| Murfreesboro | 435,709 | 1,762 | 97 | 101 | 974 | 740 | 55 | 575 | 11,809 | 3,739 | 328 |
| Nashville-Davidson | NA | NA | NA | 907 | 18,710 | 17,092 | 1,243 | 2,426 | 36,463 | 10,889 | 1,056 |
| Oak Ridge | 31,038 | 307 | 26 | 27 | 203 | 923 | 11 | 107 | 1,535 | 420 | 38 |
| Smyrna | 127,532 | 763 | 79 | 29 | 860 | 1,464 | 52 | 157 | 2,569 | 610 | 59 |
| Spring Hill | 111,716 | 1,068 | 88 | D | D | D | 8 | 89 | 1,613 | 378 | 39 |
| **TEXAS** | 42,986,432 | 230,503 | 69 | 28,861 | 435,728 | 779,743 | 27,819 | 80,874 | 1,304,540 | 417,232 | 36,628 |
| Abilene | 101,891 | 637 | 59 | D | D | D | D | 503 | 7,962 | 2,322 | 208 |
| Allen | 145,908 | 538 | 100 | D | D | D | 60 | 325 | 6,374 | 1,380 | 130 |
| Amarillo | 164,865 | 845 | 66 | 211 | 3,039 | 3,151 | 173 | 777 | 13,450 | 3,664 | 346 |
| Arlington | 222,170 | 1,447 | 50 | 332 | 5,236 | 3,939 | 326 | 1,196 | 21,315 | 6,710 | 596 |
| Austin | 2,290,427 | 17,690 | 23 | 1,028 | 21,202 | 79,057 | 1,738 | 3,359 | 57,095 | 18,284 | 1,790 |
| Baytown | 66,374 | 283 | 100 | 54 | 612 | 288 | 30 | 287 | 5,915 | 1,982 | 150 |
| Beaumont | 48,823 | 247 | 98 | 178 | 2,276 | 1,897 | 135 | 599 | 9,023 | 2,833 | 260 |
| Bedford | 509 | 3 | 100 | 31 | 140 | 96 | 11 | 128 | 1,982 | 651 | 57 |
| Big Spring | 8,383 | 42 | 95 | 24 | 151 | 163 | 9 | 103 | 1,392 | 484 | 40 |
| Brownsville | 97,487 | 970 | 80 | 184 | 1,317 | 761 | 48 | 513 | 9,075 | 2,244 | 215 |
| Bryan | 139,191 | 772 | 97 | 82 | 1,574 | 1,099 | 89 | 301 | 4,049 | 1,400 | 115 |
| Burleson | 208,033 | 1,381 | 30 | 28 | 228 | 214 | 12 | 181 | 3,925 | 1,370 | 108 |
| Carrollton | 79,031 | 463 | 44 | 404 | 6,804 | 5,736 | 427 | 442 | 6,032 | 2,248 | 227 |
| Cedar Hill | 31,027 | 101 | 98 | 15 | 242 | 141 | 15 | 147 | 3,360 | 625 | 68 |
| Cedar Park | 49,925 | 214 | 100 | 50 | 469 | 234 | 25 | 278 | 4,438 | 1,293 | 121 |
| Cleburne | 69,462 | 326 | 89 | 36 | 417 | 227 | 19 | 139 | 2,469 | 837 | 79 |
| College Station | 160,563 | 1,165 | 42 | 38 | 277 | 152 | 14 | 311 | 6,258 | 1,545 | 135 |
| Conroe | 400,210 | 1,939 | 80 | 123 | 1,557 | 5,416 | 94 | 385 | 7,488 | 2,739 | 236 |
| Coppell | 13,647 | 33 | 100 | 94 | 2,704 | 2,903 | 191 | 89 | 3,727 | 1,822 | 202 |
| Copperas Cove | 39,115 | 328 | 81 | D | D | D | 0 | 76 | 1,169 | 309 | 29 |
| Corpus Christi | 278,237 | 1,190 | 100 | 355 | 5,276 | 2,846 | 280 | 978 | 17,527 | 5,160 | 471 |
| Dallas | 581,654 | 3,691 | 40 | 1,969 | 32,922 | 27,652 | 1,951 | 4,129 | 65,708 | 20,463 | 2,023 |
| Deer Park | 3,968 | 17 | 100 | 45 | 794 | 398 | 55 | 68 | 1,379 | 397 | 36 |
| Del Rio | 13,381 | 64 | 100 | D | D | D | 5 | 140 | 2,078 | 639 | 50 |
| Denton | 510,474 | 2,860 | 34 | 93 | 1,158 | 1,215 | 69 | 445 | 7,546 | 2,281 | 183 |
| DeSoto | 59,138 | 245 | 100 | 31 | 338 | 238 | 17 | 87 | 2,484 | 1,080 | 77 |
| Duncanville | 1,170 | 5 | 100 | 20 | 276 | 107 | 18 | 112 | 1,606 | 618 | 49 |
| Eagle Pass | 29,039 | 139 | 87 | 33 | 165 | 175 | 9 | 148 | 2,641 | 637 | 57 |
| Edinburg | 175,330 | 1,207 | 45 | 79 | 1,692 | 732 | 66 | 220 | 4,634 | 1,421 | 113 |
| El Paso | 389,390 | 1,762 | 89 | 852 | 9,401 | 7,535 | 415 | 1,984 | 34,562 | 9,052 | 815 |
| Euless | 16,211 | 69 | 100 | 40 | 734 | 374 | 55 | 171 | 2,506 | 735 | 65 |
| Farmers Branch | 53,564 | 229 | 100 | 214 | 6,108 | 4,637 | 409 | 192 | 3,002 | 1,118 | 110 |
| Flower Mound | 100,449 | 429 | 100 | 87 | 1,922 | 1,652 | 109 | 155 | 3,358 | 942 | 79 |
| Fort Worth | 1,559,230 | 9,568 | 76 | 699 | 17,277 | 17,429 | 1,216 | 2,246 | 39,925 | 13,691 | 1,123 |
| Friendswood | 46,842 | 156 | 100 | 20 | 127 | 164 | 9 | 90 | 1,292 | 399 | 32 |
| Frisco | 655,193 | 3,179 | 77 | 131 | 2,676 | 21,828 | 198 | 540 | 11,660 | 4,102 | 353 |
| Galveston | 46,377 | 237 | 100 | 34 | 298 | 196 | 10 | 217 | 2,707 | 727 | 72 |
| Garland | 130,780 | 389 | 100 | 182 | 3,339 | 2,040 | 178 | 612 | 9,410 | 2,938 | 247 |
| Georgetown | 673,881 | 3,295 | 67 | 48 | 408 | 283 | 24 | 205 | 3,806 | 1,510 | 123 |
| Grand Prairie | 230,749 | 1,458 | 29 | 295 | 6,961 | 6,161 | 428 | 371 | 5,732 | 2,135 | 170 |
| Grapevine | 23,574 | 322 | 15 | 101 | 2,152 | 3,344 | 161 | 312 | 5,614 | 2,845 | 200 |
| Greenville | 44,428 | 508 | 57 | D | D | D | 8 | 139 | 2,540 | 856 | 75 |
| Haltom City | 35,736 | 123 | 100 | 111 | 1,524 | 1,188 | 79 | 147 | 1,545 | 549 | 52 |
| Harker Heights | 27,897 | 134 | 63 | 3 | D | 2 | D | 57 | 1,246 | 322 | 31 |
| Harlingen | 33,559 | 309 | 76 | 73 | 716 | 1,063 | 29 | 289 | 5,440 | 1,486 | 140 |
| Houston | 2,400,863 | 16,585 | 36 | 4,721 | 81,680 | 364,406 | 5,883 | 9,493 | 159,513 | 53,089 | 4,697 |
| Huntsville | 30,320 | 185 | 63 | 22 | 224 | 178 | 10 | 144 | 2,490 | 843 | 62 |
| Hurst | 15,717 | 112 | 16 | 27 | 173 | 139 | 10 | 274 | 5,553 | 1,350 | 126 |

1. Merchant wholesalers except manufacturers' sales branches and offices.  2. Establishments with payroll.

## Table D. Cities — **Real Estate, Professional Services, and Manufacturing**

| City | Real estate and rental and leasing, 2017 | | | | Professional, scientific, and technical services[1], 2017 | | | | Manufacturing, 2017 | | | |
|---|---|---|---|---|---|---|---|---|---|---|---|---|
| | Number of establishments | Number of employees | Receipts (mil dol) | Annual payroll (mil dol) | Number of establishments | Number of employees | Receipts (mil dol) | Annual payroll (mil dol) | Number of establishments | Number of employees | Receipts (mil dol) | Annual payroll (mil dol) |
| | 80 | 81 | 82 | 83 | 84 | 85 | 86 | 87 | 88 | 89 | 90 | 91 |
| TENNESSEE—Cont'd | | | | | | | | | | | | |
| Columbia | 50 | 199 | 45 | 6 | 59 | 471 | 59 | 26 | NA | NA | NA | NA |
| Cookeville | 53 | 207 | 48 | 7 | 102 | 594 | 71 | 24 | NA | NA | NA | NA |
| Franklin | 163 | 804 | 1,098 | 56 | 480 | 5,922 | 978 | 439 | NA | NA | NA | NA |
| Gallatin | 47 | 618 | 173 | 45 | 63 | 444 | 41 | 17 | NA | NA | NA | NA |
| Germantown | 45 | 325 | 275 | 29 | 110 | 602 | 87 | 35 | NA | NA | NA | NA |
| Hendersonville | 85 | 312 | 97 | 14 | 126 | 2,199 | 246 | 115 | NA | NA | NA | NA |
| Jackson | 105 | 513 | 103 | 19 | 146 | 1,117 | 134 | 55 | NA | NA | NA | NA |
| Johnson City | 117 | 737 | 128 | 25 | 155 | 1,070 | 134 | 51 | NA | NA | NA | NA |
| Kingsport | 60 | 311 | 64 | 9 | 123 | 834 | 106 | 35 | NA | NA | NA | NA |
| Knoxville | 389 | 2,840 | 648 | 126 | 732 | 6,824 | 937 | 403 | NA | NA | NA | NA |
| La Vergne | 17 | 108 | 45 | 7 | 22 | 858 | 26 | 46 | NA | NA | NA | NA |
| Lebanon | 58 | 259 | 84 | 13 | 91 | 478 | 64 | 20 | NA | NA | NA | NA |
| Maryville | 37 | 120 | 31 | 3 | D | D | D | D | NA | NA | NA | NA |
| Memphis | 753 | 7,045 | 1,630 | 367 | 1,243 | 14,880 | 2,337 | 964 | NA | NA | NA | NA |
| Morristown | D | D | D | D | 58 | 233 | 29 | 12 | NA | NA | NA | NA |
| Murfreesboro | 169 | 1,164 | 326 | 52 | D | D | D | D | NA | NA | NA | NA |
| Nashville-Davidson | 1,032 | 8,141 | 2,738 | 459 | 2,094 | 29,524 | 5,524 | 2,329 | NA | NA | NA | NA |
| Oak Ridge | 40 | 134 | 32 | 6 | 153 | 8,104 | 2,164 | 772 | NA | NA | NA | NA |
| Smyrna | 43 | 179 | 81 | 10 | 48 | 730 | 126 | 52 | NA | NA | NA | NA |
| Spring Hill | 26 | 115 | 47 | 5 | 46 | 156 | 19 | 7 | NA | NA | NA | NA |
| TEXAS | 32,290 | 198,712 | 56,443 | 10,560 | 70,033 | 735,744 | 150,903 | 59,488 | 19,844 | 758,722 | 575,218 | 48,004 |
| Abilene | 167 | 776 | 187 | 31 | 256 | 1,600 | 212 | 84 | NA | NA | NA | NA |
| Allen | 106 | 347 | 133 | 17 | 372 | 1,706 | 304 | 120 | NA | NA | NA | NA |
| Amarillo | 289 | 1,246 | 271 | 51 | 478 | 2,665 | 392 | 160 | NA | NA | NA | NA |
| Arlington | 370 | 2,014 | 541 | 97 | 784 | 6,083 | 1,073 | 372 | NA | NA | NA | NA |
| Austin | 2,089 | 14,255 | 4,160 | 887 | 6,026 | 84,531 | 19,151 | 8,356 | NA | NA | NA | NA |
| Baytown | 87 | 607 | 173 | 33 | 107 | 1,685 | 169 | 116 | NA | NA | NA | NA |
| Beaumont | 202 | 1,513 | 388 | 70 | 349 | 4,657 | 848 | 355 | NA | NA | NA | NA |
| Bedford | 70 | 353 | 83 | 26 | 155 | 731 | 139 | 49 | NA | NA | NA | NA |
| Big Spring | 32 | 109 | 32 | 4 | 34 | 155 | 18 | 6 | NA | NA | NA | NA |
| Brownsville | 138 | 621 | 97 | 18 | 271 | 1,956 | 198 | 67 | NA | NA | NA | NA |
| Bryan | 111 | 632 | 106 | 22 | 196 | 1,050 | 144 | 63 | NA | NA | NA | NA |
| Burleson | 50 | 224 | 62 | 16 | 65 | 255 | 33 | 10 | NA | NA | NA | NA |
| Carrollton | 179 | 1,544 | 394 | 87 | 430 | 3,556 | 592 | 204 | NA | NA | NA | NA |
| Cedar Hill | 27 | 92 | 31 | 5 | 58 | 177 | 19 | 6 | NA | NA | NA | NA |
| Cedar Park | 97 | 356 | 113 | 13 | 241 | 1,310 | 198 | 80 | NA | NA | NA | NA |
| Cleburne | 43 | 260 | 29 | 6 | 73 | 298 | 43 | 15 | NA | NA | NA | NA |
| College Station | 149 | 726 | 148 | 22 | 205 | 2,008 | 289 | 120 | NA | NA | NA | NA |
| Conroe | 111 | 515 | 122 | 22 | 227 | 1,007 | 145 | 54 | NA | NA | NA | NA |
| Coppell | 85 | 829 | 218 | 59 | 278 | 5,864 | 949 | 564 | NA | NA | NA | NA |
| Copperas Cove | 32 | 119 | 14 | 4 | 26 | 117 | 12 | 4 | NA | NA | NA | NA |
| Corpus Christi | 413 | 2,647 | 671 | 135 | 786 | 7,070 | 1,153 | 440 | NA | NA | NA | NA |
| Dallas | 2,555 | 24,594 | 8,070 | 1,686 | 5,875 | 83,897 | 19,111 | 7,385 | NA | NA | NA | NA |
| Deer Park | 39 | 476 | 156 | 34 | 73 | 2,057 | 308 | 147 | NA | NA | NA | NA |
| Del Rio | 35 | 105 | 24 | 4 | 35 | 213 | 24 | 6 | NA | NA | NA | NA |
| Denton | 187 | 990 | 198 | 38 | 292 | 1,417 | 160 | 78 | NA | NA | NA | NA |
| DeSoto | 44 | 225 | 50 | 9 | 42 | 240 | 27 | 13 | NA | NA | NA | NA |
| Duncanville | 40 | D | 30 | D | 55 | 286 | 24 | 10 | NA | NA | NA | NA |
| Eagle Pass | D | D | D | D | 34 | 196 | 21 | 8 | NA | NA | NA | NA |
| Edinburg | 69 | 204 | 28 | 5 | 154 | 1,195 | 135 | 41 | NA | NA | NA | NA |
| El Paso | 793 | 3,791 | 833 | 140 | 1,211 | 9,756 | 1,508 | 459 | NA | NA | NA | NA |
| Euless | D | D | D | D | 92 | 839 | 181 | 64 | NA | NA | NA | NA |
| Farmers Branch | 101 | 1,235 | 464 | 83 | 395 | 6,430 | 1,367 | 581 | NA | NA | NA | NA |
| Flower Mound | D | D | D | D | 334 | 1,949 | 334 | 118 | NA | NA | NA | NA |
| Fort Worth | 910 | 5,596 | 1,653 | 273 | 1,891 | 20,570 | 3,490 | 1,424 | NA | NA | NA | NA |
| Friendswood | 48 | 122 | 27 | 4 | 142 | 551 | 81 | 30 | NA | NA | NA | NA |
| Frisco | 251 | 946 | 307 | 45 | 954 | 5,484 | 1,155 | 462 | NA | NA | NA | NA |
| Galveston | 75 | 334 | 67 | 13 | 111 | 606 | 129 | 36 | NA | NA | NA | NA |
| Garland | 175 | 1,385 | 317 | 53 | 298 | 2,261 | 277 | 125 | NA | NA | NA | NA |
| Georgetown | 87 | 319 | 113 | 18 | 203 | 807 | 115 | 44 | NA | NA | NA | NA |
| Grand Prairie | 144 | 1,367 | 331 | 63 | 187 | 1,289 | 242 | 77 | NA | NA | NA | NA |
| Grapevine | 115 | 1,177 | 432 | 68 | 253 | 1,725 | 357 | 133 | NA | NA | NA | NA |
| Greenville | 39 | 167 | 38 | 9 | 45 | 501 | 61 | 23 | NA | NA | NA | NA |
| Haltom City | 31 | 197 | 60 | 10 | 53 | 276 | 39 | 13 | NA | NA | NA | NA |
| Harker Heights | 27 | 89 | 22 | 4 | 27 | 139 | 12 | 5 | NA | NA | NA | NA |
| Harlingen | 113 | 552 | 91 | 15 | 151 | 941 | 111 | 34 | NA | NA | NA | NA |
| Houston | 4,321 | 35,135 | 10,879 | 1,938 | 10,488 | 153,689 | 36,912 | 14,228 | NA | NA | NA | NA |
| Huntsville | 52 | 161 | 38 | 5 | 65 | 412 | 32 | 10 | NA | NA | NA | NA |
| Hurst | 41 | 222 | 59 | 11 | 143 | 747 | 92 | 42 | NA | NA | NA | NA |

1. Establishments subject to federal tax.

# Table D. Cities — Accommodation and Food Services, Arts, Entertainment, and Recreation, and Health Care and Social Assistance

| City | Accommodation and food services, 2017 | | | | Arts, entertainment, and recreation[1], 2017 | | | | Health care and social assistance[1], 2017 | | | |
|---|---|---|---|---|---|---|---|---|---|---|---|---|
| | Number of establishments | Number of employees | Receipts (mil dol) | Annual payroll (mil dol) | Number of establishments | Number of employees | Receipts (mil dol) | Annual payroll (mil dol) | Number of establishments | Number of employees | Receipts (mil dol) | Annual payroll (mil dol) |
| | 92 | 93 | 94 | 95 | 96 | 97 | 98 | 99 | 100 | 101 | 102 | 103 |
| **TENNESSEE—Cont'd** | | | | | | | | | | | | |
| Columbia | 101 | 2,042 | 102.0 | 31.0 | 12 | 106 | 8 | 2 | 181 | 4,691 | 522 | 211 |
| Cookeville | 154 | 3,745 | 190.0 | 56.0 | 17 | 203 | 7 | 2 | 217 | 5,820 | 553 | 234 |
| Franklin | 321 | 8,352 | 540.0 | 154.0 | 176 | 1,379 | 176 | 45 | 387 | 8,667 | 1,093 | 436 |
| Gallatin | 84 | 1,658 | 90.0 | 25.0 | 16 | 181 | 16 | 5 | 120 | 2,937 | 346 | 120 |
| Germantown | 89 | 2,112 | 119.0 | 36.0 | D | D | D | D | 213 | 7,539 | 1,347 | 497 |
| Hendersonville | 123 | 2,797 | 159.0 | 47.0 | D | D | D | D | 188 | 2,791 | 382 | 133 |
| Jackson | 216 | 5,218 | 271.0 | 77.0 | 26 | 330 | 17 | 5 | 320 | 13,468 | 1,475 | 604 |
| Johnson City | 269 | 6,674 | 323.0 | 97.0 | 32 | 261 | 16 | 5 | 334 | 13,174 | 1,722 | 764 |
| Kingsport | 210 | 4,926 | 247.0 | 71.0 | 22 | 383 | 11 | 4 | 277 | 7,487 | 1,030 | 395 |
| Knoxville | 777 | 19,947 | 1,110.0 | 339.0 | 108 | 1,985 | 129 | 35 | 969 | 32,083 | 4,216 | 1,583 |
| La Vergne | 35 | 531 | 32.0 | 8.0 | D | D | D | 0 | 21 | 174 | 20 | 5 |
| Lebanon | 123 | 2,597 | 145.0 | 41.0 | 16 | 124 | 9 | 3 | 170 | 2,573 | 306 | 109 |
| Maryville | 87 | 1,866 | 97.0 | 29.0 | 12 | 85 | 4 | 1 | 162 | 4,624 | 495 | 202 |
| Memphis | 1,399 | 31,459 | 1,858.0 | 530.0 | 148 | 4,111 | 447 | 225 | 1,667 | 55,489 | 7,615 | 2,872 |
| Morristown | D | D | D | D | D | D | D | D | 169 | 4,246 | 416 | 159 |
| Murfreesboro | 397 | 9,538 | 496.0 | 151.0 | 46 | 624 | 29 | 9 | 446 | 11,923 | 1,492 | 579 |
| Nashville-Davidson | 1,982 | 50,360 | 3,955.0 | 1,072.0 | 827 | 9,720 | 2,700 | 951 | 1,912 | 75,503 | 12,112 | 4,645 |
| Oak Ridge | 75 | 1,567 | 83.0 | 24.0 | 11 | 130 | 6 | 2 | 152 | 3,707 | 452 | 165 |
| Smyrna | 123 | 2,887 | 168.0 | 45.0 | 14 | 270 | 13 | 4 | 148 | 2,214 | 317 | 98 |
| Spring Hill | 78 | 1,814 | 81.0 | 27.0 | 11 | 76 | 3 | 1 | 65 | 677 | 62 | 23 |
| **TEXAS** | 57,098 | 1,201,419 | 74,369.0 | 20,611.0 | 7,620 | 153,566 | 14,842 | 4,514 | 69,952 | 1,581,577 | 186,109 | 69,899 |
| Abilene | 295 | 6,704 | 346.0 | 102.0 | D | D | D | D | 382 | 11,855 | 1,312 | 506 |
| Allen | 214 | 4,572 | 273.0 | 78.0 | 35 | 672 | 33 | 10 | 363 | 3,832 | 442 | 156 |
| Amarillo | 530 | 11,165 | 632.0 | 169.0 | 68 | 957 | 51 | 17 | 679 | 17,186 | 2,233 | 796 |
| Arlington | 774 | 19,755 | 1,233.0 | 327.0 | 85 | 5,408 | 797 | 273 | 1,011 | 22,351 | 2,502 | 982 |
| Austin | 2,981 | 73,697 | 5,337.0 | 1,510.0 | 545 | 9,671 | 1,121 | 265 | 3,150 | 71,118 | 8,839 | 3,393 |
| Baytown | 206 | 4,458 | 265.0 | 72.0 | 14 | 194 | 12 | 3 | 226 | 4,800 | 622 | 218 |
| Beaumont | 299 | 6,662 | 382.0 | 102.0 | 43 | 561 | 27 | 8 | 618 | 13,552 | 1,408 | 514 |
| Bedford | 116 | 2,380 | 141.0 | 39.0 | 10 | 95 | 7 | 2 | 232 | 6,185 | 904 | 316 |
| Big Spring | 81 | 1,295 | 72.0 | 19.0 | D | D | D | D | D | D | D | D |
| Brownsville | 342 | 6,555 | 313.0 | 81.0 | 30 | 441 | 19 | 7 | 479 | 17,053 | 966 | 432 |
| Bryan | 175 | 3,446 | 167.0 | 48.0 | 24 | 645 | 27 | 12 | 242 | 4,796 | 667 | 240 |
| Burleson | 109 | 2,547 | 136.0 | 37.0 | 13 | D | 3 | D | 108 | 1,308 | 121 | 43 |
| Carrollton | 351 | 4,655 | 291.0 | 82.0 | 40 | 984 | 74 | 18 | 427 | 6,935 | 837 | 310 |
| Cedar Hill | 91 | 2,145 | 115.0 | 31.0 | D | D | D | D | 100 | 1,309 | 98 | 40 |
| Cedar Park | 185 | 3,925 | 223.0 | 67.0 | 36 | 785 | 48 | 11 | 261 | 3,764 | 462 | 187 |
| Cleburne | 88 | 1,594 | 79.0 | 22.0 | D | D | D | D | 119 | 1,774 | 222 | 77 |
| College Station | 339 | 9,157 | 497.0 | 125.0 | 37 | 698 | 29 | 9 | 180 | 5,705 | 760 | 257 |
| Conroe | 196 | 4,483 | 335.0 | 86.0 | 27 | 542 | 36 | 8 | 278 | 4,808 | 655 | 241 |
| Coppell | 78 | 1,927 | 115.0 | 36.0 | 12 | 293 | 10 | 4 | 173 | 1,685 | 177 | 73 |
| Copperas Cove | 57 | 1,002 | 46.0 | 12.0 | D | D | D | D | 34 | 461 | 29 | 13 |
| Corpus Christi | 827 | 18,028 | 987.0 | 272.0 | 86 | 1,963 | 105 | 36 | 1,071 | 27,875 | 2,870 | 1,119 |
| Dallas | 3,172 | 73,679 | 5,436.0 | 1,525.0 | 433 | 12,428 | 1,469 | 493 | 4,386 | 121,052 | 18,432 | 6,828 |
| Deer Park | 66 | 1,786 | 90.0 | 26.0 | 7 | 10 | 2 | 0 | 52 | 787 | 65 | 26 |
| Del Rio | 82 | 1,767 | 97.0 | 24.0 | D | D | D | 0 | D | D | D | D |
| Denton | 325 | 7,433 | 386.0 | 112.0 | D | D | D | D | 480 | 8,327 | 1,156 | 407 |
| DeSoto | 76 | 1,638 | 81.0 | 23.0 | 7 | 85 | 7 | 2 | 175 | 4,565 | 317 | 140 |
| Duncanville | 67 | 1,788 | 99.0 | 25.0 | D | D | D | D | 131 | 1,970 | 130 | 51 |
| Eagle Pass | 72 | 1,407 | 73.0 | 20.0 | D | D | D | D | 101 | 3,998 | 237 | 96 |
| Edinburg | 142 | 2,768 | 147.0 | 39.0 | D | D | D | D | 332 | 10,504 | 579 | 252 |
| El Paso | 1,448 | 32,400 | 1,566.0 | 436.0 | 148 | 2,663 | 142 | 49 | 1,635 | 45,067 | 4,614 | 1,704 |
| Euless | 96 | 1,756 | 107.0 | 26.0 | 15 | 417 | 24 | 6 | 85 | 1,191 | 87 | 35 |
| Farmers Branch | 106 | 2,964 | 196.0 | 57.0 | 18 | 474 | 32 | 12 | 191 | 5,030 | 667 | 283 |
| Flower Mound | D | D | D | D | 29 | 494 | 25 | 8 | 274 | 3,895 | 497 | 176 |
| Fort Worth | 1,644 | 37,634 | 2,350.0 | 655.0 | 205 | 6,153 | 614 | 148 | 2,113 | 59,067 | 8,753 | 2,973 |
| Friendswood | 66 | 1,136 | 59.0 | 17.0 | 13 | 113 | 8 | 3 | 139 | 1,243 | 113 | 43 |
| Frisco | 434 | 9,494 | 625.0 | 175.0 | 98 | 2,919 | 1,131 | 362 | 647 | 7,966 | 1,132 | 390 |
| Galveston | 240 | 7,318 | 525.0 | 149.0 | 36 | 546 | 47 | 11 | 128 | 8,418 | 824 | 365 |
| Garland | 364 | 6,701 | 390.0 | 107.0 | 34 | 678 | 38 | 9 | 492 | 10,093 | 701 | 287 |
| Georgetown | 127 | 2,822 | 160.0 | 48.0 | 23 | 344 | 26 | 8 | 197 | 4,577 | 429 | 205 |
| Grand Prairie | 249 | 4,808 | 310.0 | 78.0 | 25 | 628 | 107 | 16 | 282 | 4,994 | 327 | 114 |
| Grapevine | 217 | 9,464 | 824.0 | 210.0 | 27 | 736 | 59 | 13 | 254 | 3,640 | 676 | 204 |
| Greenville | 80 | 1,770 | 98.0 | 28.0 | 9 | 84 | 6 | 1 | 130 | 2,922 | 320 | 133 |
| Haltom City | 72 | 905 | 52.0 | 13.0 | D | D | D | D | 43 | 489 | 38 | 17 |
| Harker Heights | 62 | 912 | 34.0 | 9.0 | 8 | 35 | 2 | 0 | D | D | D | D |
| Harlingen | 173 | 4,065 | 217.0 | 62.0 | 20 | D | 11 | 0 | 375 | 15,455 | 1,149 | 464 |
| Houston | 7,240 | 159,256 | 11,039.0 | 3,058.0 | 783 | 25,027 | 3,042 | 1,089 | 8,402 | 220,087 | 32,073 | 11,738 |
| Huntsville | D | D | D | 32.0 | D | D | D | D | 96 | 2,188 | 244 | 88 |
| Hurst | 136 | 4,142 | 168.0 | 46.0 | 17 | 208 | 10 | 3 | 152 | 1,581 | 169 | 70 |

1. Establishments subject to federal tax.

Items 92—103

# Table D. Cities — Other Services and Government Employment and Payroll

| City | Other services[1] | | | | Government employment and payroll, 2017 | | | | | | | | |
|---|---|---|---|---|---|---|---|---|---|---|---|---|---|
| | | | | | | March payroll | | | | | | | |
| | | | | | | | Percent of total for: | | | | | | |
| | Number of establishments | Number of employees | Receipts (mil dol) | Annual payroll (mil dol) | Full-time equivalent employees | Total (dollars) | Administrative, judicial, and legal | Police and corrections | Fire protection | Highways and transportation | Health and welfare | Natural resources and utilities | Education and libraries |
| | 104 | 105 | 106 | 107 | 108 | 109 | 110 | 111 | 112 | 113 | 114 | 115 | 116 |
| TENNESSEE—Cont'd | | | | | | | | | | | | | |
| Columbia | D | D | D | D | 501 | 2,162,900 | 7.1 | 18.0 | 18.8 | 5.3 | 0.0 | 44.4 | 0.0 |
| Cookeville | D | D | D | 13 | 2,932 | 11,250,845 | 1.3 | 1.8 | 1.9 | 1.2 | 87.0 | 5.9 | 0.0 |
| Franklin | 197 | 1,716 | 153 | 56 | 720 | 3,140,176 | 14.1 | 20.1 | 26.1 | 5.7 | 0.0 | 19.6 | 0.0 |
| Gallatin | 75 | 390 | 53 | 14 | 465 | 1,858,965 | 10.3 | 20.0 | 18.0 | 3.7 | 0.0 | 40.0 | 0.0 |
| Germantown | 72 | 575 | 46 | 16 | 428 | 2,182,982 | 10.9 | 33.0 | 23.7 | 9.3 | 6.0 | 15.6 | 0.0 |
| Hendersonville | 102 | 540 | 58 | 17 | 338 | 1,603,915 | 10.1 | 35.4 | 37.5 | 3.5 | 0.4 | 5.0 | 0.0 |
| Jackson | D | D | D | 23 | 808 | 3,216,351 | 9.7 | 35.4 | 25.6 | 9.8 | 0.3 | 16.5 | 0.0 |
| Johnson City | 161 | 920 | 74 | 22 | 2,115 | 8,311,535 | 4.7 | 8.0 | 6.6 | 7.3 | 0.4 | 23.9 | 47.6 |
| Kingsport | 110 | 960 | 107 | 38 | 1,935 | 7,226,954 | 6.5 | 8.5 | 6.8 | 2.7 | 0.4 | 13.3 | 61.0 |
| Knoxville | 493 | 3,430 | 483 | 111 | 2,530 | 12,561,702 | 6.6 | 19.5 | 11.9 | 3.9 | 0.8 | 53.5 | 0.0 |
| La Vergne | 27 | 868 | 114 | 45 | 212 | 793,706 | 12.9 | 37.3 | 19.0 | 3.4 | 0.0 | 16.5 | 3.2 |
| Lebanon | 78 | 1,065 | 134 | 45 | 510 | 1,431,893 | 14.2 | 22.4 | 14.7 | 6.5 | 3.9 | 38.2 | 0.0 |
| Maryville | 72 | 375 | 41 | 11 | 937 | 4,626,698 | 5.0 | 7.6 | 4.0 | 3.9 | 0.0 | 13.9 | 64.8 |
| Memphis | 904 | 8,775 | 2,538 | 395 | 10,740 | 51,032,329 | 4.9 | 26.4 | 16.9 | 5.0 | 1.3 | 41.4 | 1.6 |
| Morristown | 73 | 449 | 34 | 11 | 464 | 2,152,548 | 6.8 | 17.2 | 15.1 | 7.8 | 0.0 | 43.9 | 0.0 |
| Murfreesboro | 231 | 1,510 | 149 | 47 | 2,460 | 9,533,413 | 4.0 | 14.7 | 9.0 | 3.2 | 0.1 | 19.1 | 49.4 |
| Nashville-Davidson | 1,293 | 13,383 | 1,682 | NA | 21,283 | 101,098,567 | 6.3 | 14.3 | 6.5 | 1.1 | 6.4 | 14.5 | 48.8 |
| Oak Ridge | 58 | 268 | 22 | 7 | 1,068 | 4,685,105 | 7.1 | 6.8 | 7.0 | 0.9 | 0.0 | 10.5 | 66.6 |
| Smyrna | D | D | D | D | 398 | 1,846,459 | 18.2 | 26.3 | 24.5 | 1.9 | 0.0 | 25.6 | 0.0 |
| Spring Hill | 42 | 233 | 24 | 6 | 224 | 834,881 | 8.4 | 29.3 | 27.1 | 7.0 | 0.0 | 19.4 | 4.4 |
| TEXAS | 37,506 | 285,777 | 39,008 | 10,468 | X | X | X | X | X | X | X | X | X |
| Abilene | D | D | D | D | 1,201 | 5,368,188 | 10.4 | 31.9 | 22.2 | 4.8 | 5.8 | 18.2 | 1.8 |
| Allen | 118 | 912 | 84 | 29 | 770 | 3,819,781 | 10.4 | 27.7 | 21.8 | 2.6 | 4.9 | 25.2 | 3.8 |
| Amarillo | 376 | 2,926 | 359 | 96 | 1,992 | 8,436,582 | 7.0 | 32.1 | 22.7 | 7.2 | 5.7 | 17.8 | 2.2 |
| Arlington | 476 | 3,291 | 659 | 103 | 2,558 | 13,968,727 | 12.5 | 39.5 | 21.2 | 5.3 | 4.9 | 14.5 | 1.2 |
| Austin | 2,203 | 19,625 | 2,931 | 835 | 13,610 | 84,706,545 | 10.3 | 25.1 | 12.4 | 6.3 | 7.8 | 32.4 | 1.9 |
| Baytown | 116 | 1,119 | 133 | 48 | 871 | 4,767,592 | 12.1 | 29.2 | 25.9 | 2.7 | 3.1 | 19.6 | 2.2 |
| Beaumont | 225 | 2,630 | 238 | 125 | 1,065 | 5,508,111 | 7.0 | 32.3 | 26.9 | 5.2 | 6.6 | 16.1 | 1.5 |
| Bedford | 63 | 508 | 56 | 14 | 365 | 1,912,040 | 12.3 | 26.5 | 26.1 | 6.2 | 0.0 | 13.2 | 3.9 |
| Big Spring | 37 | 286 | 26 | 7 | 244 | 1,004,920 | 14.6 | 24.2 | 30.1 | 6.4 | 2.3 | 20.7 | 0.0 |
| Brownsville | 139 | 625 | 50 | 13 | 1,466 | 6,013,698 | 8.8 | 29.9 | 19.9 | 11.0 | 3.3 | 23.5 | 2.6 |
| Bryan | 150 | 962 | 149 | 32 | 722 | 3,508,034 | 16.8 | 29.1 | 23.3 | 6.0 | 0.9 | 18.3 | 3.7 |
| Burleson | D | D | D | D | 366 | 1,929,487 | 11.7 | 23.3 | 16.5 | 8.0 | 5.1 | 22.9 | 2.2 |
| Carrollton | 234 | 1,673 | 176 | 54 | 880 | 3,924,840 | 15.9 | 22.5 | 26.7 | 8.8 | 4.6 | 11.3 | 3.3 |
| Cedar Hill | D | D | D | D | 344 | 1,659,912 | 11.5 | 28.9 | 23.5 | 4.3 | 5.2 | 20.0 | 2.6 |
| Cedar Park | 161 | 1,034 | 107 | 36 | 441 | 2,303,041 | 13.9 | 30.9 | 21.7 | 5.9 | 0.6 | 13.7 | 2.3 |
| Cleburne | 71 | 442 | 44 | 14 | 342 | 1,450,782 | 7.2 | 26.9 | 22.2 | 3.2 | 5.3 | 23.6 | 1.7 |
| College Station | 111 | 1,071 | 459 | 41 | 902 | 4,439,747 | 21.2 | 23.7 | 20.1 | 5.8 | 0.0 | 24.1 | 0.0 |
| Conroe | 152 | 1,014 | 105 | 31 | 584 | 2,887,973 | 11.2 | 30.4 | 23.3 | 8.9 | 0.4 | 16.8 | 0.0 |
| Coppell | 80 | 1,046 | 137 | 48 | 376 | 1,874,509 | 21.0 | 8.1 | 33.0 | 8.0 | 0.0 | 27.0 | 2.8 |
| Copperas Cove | 52 | 296 | 28 | 8 | 264 | 1,107,609 | 17.2 | 31.6 | 11.6 | 1.3 | 8.3 | 18.8 | 2.2 |
| Corpus Christi | 487 | 3,837 | 576 | 131 | 2,997 | 13,665,749 | 12.9 | 27.4 | 20.1 | 7.2 | 2.6 | 25.0 | 0.9 |
| Dallas | 1,883 | 18,163 | 3,604 | 793 | 14,767 | 85,321,347 | 8.2 | 29.6 | 18.5 | 18.5 | 4.0 | 15.6 | 1.4 |
| Deer Park | D | D | D | D | 372 | 1,670,930 | 18.5 | 32.4 | 5.3 | 2.7 | 0.0 | 32.5 | 2.9 |
| Del Rio | D | D | D | D | 522 | 1,616,515 | 13.5 | 25.6 | 21.8 | 5.8 | 8.5 | 17.9 | 0.0 |
| Denton | 205 | 1,328 | 177 | 51 | 1,496 | 8,533,833 | 14.1 | 16.8 | 16.4 | 5.2 | 1.6 | 40.8 | 2.3 |
| DeSoto | 52 | 214 | 18 | 5 | 358 | 1,779,783 | 8.0 | 33.4 | 23.2 | 1.4 | 0.4 | 13.5 | 2.8 |
| Duncanville | 68 | 445 | 44 | 20 | 278 | 1,360,082 | 13.8 | 29.8 | 21.9 | 5.3 | 0.4 | 16.7 | 3.3 |
| Eagle Pass | 33 | 149 | 10 | 3 | 393 | 1,137,666 | 7.7 | 31.4 | 17.9 | 16.1 | 3.9 | 20.4 | 1.6 |
| Edinburg | 80 | 339 | 26 | 7 | 873 | 3,090,023 | 9.4 | 34.0 | 9.3 | 3.6 | 1.5 | 27.4 | 2.9 |
| El Paso | D | D | D | D | 6,522 | 25,566,174 | 8.6 | 25.6 | 20.7 | 13.2 | 6.2 | 13.5 | 1.7 |
| Euless | 67 | 522 | 89 | 25 | 423 | 2,376,427 | 13.0 | 35.4 | 23.0 | 3.6 | 1.8 | 16.8 | 4.0 |
| Farmers Branch | 85 | 1,144 | 154 | 38 | 385 | 2,278,646 | 11.1 | 24.5 | 24.6 | 3.4 | 0.7 | 19.4 | 0.0 |
| Flower Mound | 110 | 959 | 107 | 32 | 554 | 3,000,454 | 9.0 | 24.5 | 28.6 | 2.9 | 3.5 | 16.7 | 2.5 |
| Fort Worth | 1,001 | 9,263 | 1,155 | 287 | 6,703 | 38,281,122 | 9.2 | 37.3 | 19.2 | 4.6 | 5.5 | 19.7 | 1.9 |
| Friendswood | 59 | 297 | 26 | 9 | 220 | 1,012,376 | 22.9 | 46.4 | 0.5 | 5.2 | 7.5 | 9.5 | 6.1 |
| Frisco | 237 | 1,660 | 195 | 58 | 636 | 3,316,154 | 9.9 | 30.4 | 23.9 | 5.3 | 1.2 | 19.8 | 3.8 |
| Galveston | 116 | 650 | 249 | 21 | 905 | 4,048,763 | 14.8 | 27.8 | 16.8 | 24.0 | 4.1 | 12.2 | 0.0 |
| Garland | 265 | 1,418 | 141 | 42 | 2,118 | 11,691,722 | 10.3 | 24.2 | 16.5 | 3.5 | 2.9 | 28.9 | 2.3 |
| Georgetown | 118 | 654 | 74 | 22 | 665 | 3,544,984 | 15.9 | 18.2 | 21.2 | 3.2 | 2.1 | 27.1 | 3.0 |
| Grand Prairie | 202 | 1,382 | 153 | 44 | 1,359 | 7,454,175 | 10.2 | 33.3 | 24.0 | 3.8 | 6.2 | 12.9 | 1.5 |
| Grapevine | 101 | 1,418 | 276 | 44 | 649 | 3,495,468 | 12.7 | 24.0 | 20.6 | 7.7 | 0.7 | 14.6 | 2.9 |
| Greenville | 41 | 206 | 19 | 5 | 388 | 1,877,528 | 0.0 | 22.0 | 16.8 | 1.6 | 0.2 | 56.7 | 1.4 |
| Haltom City | 81 | 540 | 78 | 21 | 240 | 1,343,974 | 12.6 | 34.3 | 37.7 | 3.6 | 0.0 | 7.5 | 4.3 |
| Harker Heights | 34 | 198 | 14 | 4 | 219 | 967,994 | 17.1 | 30.7 | 21.9 | 3.2 | 0.5 | 17.4 | 3.2 |
| Harlingen | 113 | 698 | 60 | 18 | 765 | 2,944,522 | 7.3 | 29.0 | 19.4 | 8.3 | 1.4 | 21.9 | 1.9 |
| Houston | 4,429 | 40,422 | 5,806 | 1,502 | 21,657 | 108,741,222 | 7.1 | 36.9 | 21.1 | 7.4 | 5.3 | 8.6 | 1.6 |
| Huntsville | 48 | 316 | 27 | 7 | 346 | 1,383,500 | 24.2 | 27.0 | 5.3 | 6.5 | 0.0 | 23.7 | 2.5 |
| Hurst | 85 | 537 | 46 | 15 | 396 | 2,083,352 | 10.9 | 34.9 | 19.8 | 2.4 | 3.0 | 15.5 | 4.2 |

1. Establishments subject to federal tax.

# Table D. Cities — City Government Finances

| City | General revenue Total (mil dol) | Intergovernmental Total (mil dol) | Intergovernmental Percent from state government | Taxes Total (mil dol) | Taxes Per capita[1] (dollars) Total | Per capita[1] Property | Per capita[1] Sales and gross receipts | General expenditure Total (mil dol) | General expenditure Per capita[1] (dollars) Total | Per capita[1] Capital outlays |
|---|---|---|---|---|---|---|---|---|---|---|
| | 117 | 118 | 119 | 120 | 121 | 122 | 123 | 124 | 125 | 126 |
| **TENNESSEE—Cont'd** | | | | | | | | | | |
| Columbia | 61.0 | 16.8 | 9.1 | 16.4 | 426 | 290 | 136 | 52 | 1,361 | 87 |
| Cookeville | 306.3 | 19.4 | 36.1 | 14.3 | 429 | 299 | 130 | 316 | 9,463 | 494 |
| Franklin | 128.0 | 55.8 | 41.4 | 30.3 | 387 | 230 | 157 | 91 | 1,167 | 86 |
| Gallatin | 59.2 | 31.7 | 58.4 | 16.6 | 443 | 326 | 117 | 40 | 1,061 | 103 |
| Germantown | 68.7 | 17.3 | 45.7 | 32.6 | 834 | 740 | 94 | 59 | 1,518 | 122 |
| Hendersonville | 44.9 | 15.1 | 48.4 | 22.4 | 389 | 290 | 99 | 44 | 760 | 115 |
| Jackson | 75.7 | 16.4 | 12.6 | 37.5 | 562 | 469 | 93 | 83 | 1,236 | 168 |
| Johnson City | 251.9 | 136.2 | 41.3 | 59.0 | 892 | 745 | 147 | 201 | 3,035 | 199 |
| Kingsport | 197.0 | 110.3 | 42.0 | 52.8 | 988 | 731 | 257 | 179 | 3,350 | 605 |
| Knoxville | 402.4 | 43.6 | 51.1 | 223.9 | 1,198 | 663 | 535 | 290 | 1,552 | 498 |
| La Vergne | 36.6 | 17.7 | 73.2 | 10.3 | 288 | 212 | 76 | 23 | 637 | 68 |
| Lebanon | 34.6 | 17.8 | 33.3 | 13.1 | 402 | 208 | 194 | 42 | 1,279 | 288 |
| Maryville | 115.4 | 59.0 | 48.8 | 42.8 | 1,471 | 1,335 | 136 | 85 | 2,936 | 212 |
| Memphis | 1,242.1 | 318.8 | 34.2 | 453.0 | 696 | 588 | 108 | 1,049 | 1,611 | 143 |
| Morristown | 63.3 | 19.7 | 28.0 | 15.0 | 506 | 371 | 135 | 62 | 2,074 | 335 |
| Murfreesboro | 224.3 | 109.0 | 62.9 | 100.5 | 738 | 288 | 450 | 260 | 1,909 | 614 |
| Nashville-Davidson | 2,657.7 | 644.9 | 94.4 | 1,600.6 | 2,329 | 1,414 | 914 | 2,894 | 4,212 | 698 |
| Oak Ridge | 149.5 | 89.2 | 33.4 | 43.3 | 1,493 | 919 | 574 | 113 | 3,900 | 218 |
| Smyrna | 66.5 | 22.7 | 25.1 | 15.7 | 314 | 201 | 113 | 53 | 1,067 | 190 |
| Spring Hill | 31.8 | 5.2 | 92.1 | 14.2 | 358 | 135 | 223 | 30 | 759 | 227 |
| **TEXAS** | X | X | X | X | X | X | X | X | X | X |
| Abilene | 150.3 | 11.5 | 24.2 | 99.6 | 815 | 312 | 504 | 186 | 1,518 | 545 |
| Allen | 150.6 | 1.4 | 39.5 | 103.7 | 1,024 | 535 | 490 | 98 | 963 | 145 |
| Amarillo | 285.4 | 34.4 | 43.6 | 151.2 | 757 | 213 | 543 | 289 | 1,449 | 217 |
| Arlington | 514.4 | 28.7 | 23.4 | 283.7 | 714 | 323 | 391 | 463 | 1,166 | 129 |
| Austin | 1,949.7 | 96.4 | 22.6 | 891.1 | 937 | 523 | 413 | 2,010 | 2,113 | 417 |
| Baytown | 135.9 | 8.8 | 74.7 | 60.0 | 782 | 324 | 458 | 131 | 1,704 | 322 |
| Beaumont | 206.3 | 46.5 | 77.5 | 100.5 | 845 | 407 | 438 | 192 | 1,616 | 248 |
| Bedford | 47.4 | 0.7 | 100.0 | 32.5 | 655 | 338 | 317 | 51 | 1,024 | 118 |
| Big Spring | 24.9 | 1.2 | 60.4 | 13.2 | 479 | 236 | 243 | 18 | 662 | 31 |
| Brownsville | 191.7 | 21.5 | 18.3 | 80.9 | 443 | 241 | 202 | 175 | 960 | 98 |
| Bryan | 88.2 | 2.4 | 6.5 | 53.3 | 633 | 353 | 280 | 144 | 1,711 | 173 |
| Burleson | 62.5 | 0.5 | 78.3 | 41.9 | 911 | 452 | 459 | 60 | 1,312 | 376 |
| Carrollton | 156.7 | 2.9 | 82.0 | 117.2 | 862 | 498 | 363 | 151 | 1,112 | 262 |
| Cedar Hill | 61.9 | 0.8 | 9.7 | 42.7 | 876 | 463 | 412 | 57 | 1,160 | 104 |
| Cedar Park | 97.2 | 6.9 | 100.0 | 58.2 | 770 | 348 | 422 | 80 | 1,057 | 259 |
| Cleburne | 51.7 | 2.6 | 36.2 | 29.5 | 979 | 515 | 464 | 53 | 1,768 | 611 |
| College Station | 111.1 | 3.0 | 36.1 | 69.6 | 608 | 283 | 325 | 105 | 914 | 184 |
| Conroe | 108.1 | 10.9 | 10.2 | 78.5 | 926 | 310 | 616 | 119 | 1,402 | 494 |
| Coppell | 91.7 | 3.4 | 2.3 | 75.8 | 1,805 | 864 | 941 | 106 | 2,531 | 963 |
| Copperas Cove | 28.8 | 0.2 | 18.8 | 18.9 | 581 | 446 | 135 | 19 | 593 | 0 |
| Corpus Christi | 448.0 | 30.4 | 41.2 | 234.9 | 722 | 340 | 381 | 436 | 1,339 | 248 |
| Dallas | 3,128.3 | 142.0 | 62.3 | 1,329.3 | 990 | 617 | 373 | 3,236 | 2,410 | 631 |
| Deer Park | 47.0 | 0.2 | 32.7 | 23.2 | 682 | 431 | 251 | 44 | 1,281 | 75 |
| Del Rio | 43.6 | 4.9 | 53.7 | 19.1 | 532 | 240 | 292 | 42 | 1,158 | 177 |
| Denton | 211.0 | 10.9 | 57.7 | 116.4 | 858 | 415 | 443 | 216 | 1,593 | 330 |
| DeSoto | 58.2 | 2.6 | 100.0 | 37.0 | 690 | 453 | 237 | 53 | 982 | 124 |
| Duncanville | 42.1 | 0.6 | 35.1 | 24.4 | 618 | 345 | 274 | 37 | 945 | 15 |
| Eagle Pass | 33.1 | 3.4 | 100.0 | 12.2 | 417 | 249 | 167 | 257 | 8,761 | 209 |
| Edinburg | 82.3 | 6.1 | 80.7 | 48.8 | 512 | 266 | 246 | 83 | 871 | 162 |
| El Paso | 868.3 | 87.0 | 35.9 | 458.5 | 673 | 352 | 320 | 745 | 1,093 | 214 |
| Euless | 82.2 | 0.7 | 46.8 | 56.7 | 1,025 | 267 | 758 | 92 | 1,671 | 322 |
| Farmers Branch | 61.0 | 0.6 | 24.4 | 48.9 | 1,302 | 710 | 592 | 62 | 1,639 | 287 |
| Flower Mound | 93.1 | 2.7 | 21.1 | 73.3 | 956 | 493 | 463 | 79 | 1,026 | 110 |
| Fort Worth | 1,311.3 | 102.9 | 69.5 | 765.5 | 875 | 503 | 372 | 1,300 | 1,486 | 214 |
| Friendswood | 33.0 | 0.7 | 77.3 | 24.3 | 609 | 411 | 199 | 37 | 915 | 238 |
| Frisco | 413.0 | 46.0 | 1.7 | 160.5 | 903 | 547 | 355 | 374 | 2,105 | 1,072 |
| Galveston | 203.8 | 59.1 | 87.4 | 77.9 | 1,540 | 570 | 969 | 166 | 3,284 | 1,001 |
| Garland | 237.9 | 17.5 | 6.5 | 122.7 | 515 | 330 | 185 | 220 | 921 | 62 |
| Georgetown | 89.0 | 4.0 | 99.9 | 52.2 | 738 | 304 | 434 | 100 | 1,421 | 460 |
| Grand Prairie | 281.1 | 34.0 | 13.0 | 152.1 | 784 | 395 | 389 | 244 | 1,255 | 143 |
| Grapevine | 149.9 | 3.1 | 76.6 | 115.1 | 2,132 | 586 | 1,545 | 147 | 2,720 | 716 |
| Greenville | 49.2 | 1.8 | 4.5 | 21.5 | 787 | 390 | 398 | 19 | 679 | 446 |
| Haltom City | 44.7 | 0.2 | 28.2 | 30.5 | 687 | 268 | 419 | 40 | 904 | 130 |
| Harker Heights | 28.2 | 0.6 | 98.9 | 19.1 | 618 | 352 | 265 | 29 | 951 | 247 |
| Harlingen | 87.8 | 7.4 | 39.0 | 45.9 | 704 | 278 | 427 | 84 | 1,283 | 136 |
| Houston | 4,568.5 | 320.9 | 35.7 | 2,501.5 | 1,080 | 623 | 457 | 4,067 | 1,755 | 271 |
| Huntsville | 37.8 | 2.9 | 97.3 | 17.2 | 413 | 140 | 273 | 37 | 876 | 185 |
| Hurst | 57.7 | 1.1 | 44.0 | 40.2 | 1,025 | 359 | 667 | 57 | 1,458 | 184 |

1. Based on population estimated as of July 1 of the year shown.

Items 117—126

| City | Public welfare | Highways | Parking facilities | Education | Health and hospitals | Police protection | Sewerage and sanitation | Parks and recreation | Housing and community development | Interest on debt |
|---|---|---|---|---|---|---|---|---|---|---|
| | 127 | 128 | 129 | 130 | 131 | 132 | 133 | 134 | 135 | 136 |
| **TENNESSEE—Cont'd** | | | | | | | | | | |
| Columbia | 0.1 | 9.8 | 0.0 | 0.0 | 0.4 | 15.0 | 20.8 | 3.7 | 0.4 | 2.0 |
| Cookeville | 0.0 | 1.0 | 0.0 | 0.0 | 88.1 | 2.6 | 1.6 | 1.0 | 0.0 | 0.9 |
| Franklin | 0.0 | 10.8 | 0.0 | 0.0 | 0.2 | 17.0 | 25.9 | 4.7 | 0.3 | 0.0 |
| Gallatin | 0.8 | 3.7 | 0.0 | 0.0 | 0.0 | 20.9 | 20.6 | 10.8 | 0.0 | 1.4 |
| Germantown | 0.0 | 6.5 | 0.0 | 0.0 | 3.4 | 19.1 | 15.0 | 13.1 | 0.0 | 1.5 |
| Hendersonville | 0.0 | 13.7 | 0.0 | 0.0 | 0.0 | 26.9 | 12.2 | 7.0 | 0.0 | 0.9 |
| Jackson | 0.0 | 15.7 | 0.0 | 0.0 | 0.0 | 32.6 | 17.4 | 5.5 | 0.0 | 0.0 |
| Johnson City | 0.0 | 6.3 | 0.0 | 39.3 | 0.0 | 6.8 | 14.8 | 4.3 | 0.4 | 5.8 |
| Kingsport | 0.0 | 3.5 | 0.0 | 44.4 | 0.0 | 6.9 | 5.6 | 4.7 | 0.2 | 0.0 |
| Knoxville | 0.0 | 8.4 | 0.5 | 0.0 | 0.0 | 11.4 | 35.5 | 10.0 | 0.5 | 5.7 |
| La Vergne | 0.5 | 5.2 | 0.0 | 0.0 | 0.0 | 26.7 | 20.6 | 4.0 | 0.0 | 0.0 |
| Lebanon | 0.0 | 13.3 | 0.0 | 0.0 | 0.3 | 24.7 | 12.8 | 5.8 | 0.4 | 0.0 |
| Maryville | 0.0 | 3.7 | 0.0 | 68.1 | 0.3 | 6.2 | 5.5 | 1.8 | 0.0 | 2.7 |
| Memphis | 0.0 | 2.1 | 0.2 | 0.0 | 0.0 | 24.1 | 11.6 | 4.6 | 2.6 | 7.8 |
| Morristown | 0.4 | 10.3 | 0.0 | 0.0 | 0.6 | 13.3 | 24.5 | 3.2 | 0.4 | 3.9 |
| Murfreesboro | 0.0 | 6.7 | 0.0 | 32.7 | 0.0 | 26.2 | 2.4 | 13.8 | 0.4 | 0.0 |
| Nashville-Davidson | 1.7 | 4.7 | 0.1 | 38.6 | 5.6 | 7.4 | 7.1 | 4.5 | 0.0 | 5.2 |
| Oak Ridge | 0.0 | 2.5 | 0.0 | 52.3 | 0.3 | 5.8 | 10.9 | 4.2 | 1.4 | 3.1 |
| Smyrna | 0.4 | 9.3 | 0.0 | 0.0 | 0.0 | 18.6 | 12.1 | 15.6 | 0.0 | 0.0 |
| Spring Hill | 0.0 | 18.4 | 0.0 | 0.0 | 0.0 | 19.3 | 26.8 | 1.6 | 0.0 | 0.0 |
| **TEXAS** | X | X | X | X | X | X | X | X | X | X |
| Abilene | 0.0 | 3.7 | 0.0 | 0.0 | 3.0 | 14.2 | 42.6 | 4.7 | 0.7 | 1.2 |
| Allen | 0.0 | 3.5 | 0.0 | 0.0 | 0.4 | 19.5 | 8.7 | 25.1 | 0.9 | 0.0 |
| Amarillo | 0.0 | 8.1 | 0.0 | 0.0 | 6.6 | 16.1 | 16.6 | 12.0 | 5.1 | 2.4 |
| Arlington | 0.0 | 10.7 | 0.0 | 0.0 | 1.2 | 21.4 | 15.8 | 7.3 | 1.0 | 13.8 |
| Austin | 0.0 | 8.5 | 0.0 | 0.0 | 7.5 | 17.3 | 13.6 | 8.6 | 3.4 | 4.5 |
| Baytown | 0.0 | 10.5 | 0.0 | 0.0 | 2.2 | 25.0 | 8.1 | 9.6 | 0.4 | 3.4 |
| Beaumont | 0.0 | 15.8 | 0.0 | 0.0 | 13.5 | 19.8 | 9.8 | 4.8 | 2.7 | 5.1 |
| Bedford | 0.0 | 5.0 | 0.0 | 0.0 | 0.9 | 25.5 | 13.1 | 6.5 | 0.0 | 4.3 |
| Big Spring | 0.0 | 13.4 | 0.0 | 0.0 | 0.0 | 27.6 | 0.0 | 10.0 | 0.0 | 0.0 |
| Brownsville | 0.7 | 7.6 | 0.3 | 0.0 | 1.1 | 20.6 | 12.2 | 5.3 | 2.3 | 3.3 |
| Bryan | 0.0 | 6.6 | 0.0 | 0.0 | 0.6 | 12.5 | 12.0 | 4.2 | 0.9 | 0.0 |
| Burleson | 0.0 | 18.9 | 0.0 | 0.0 | 0.7 | 18.6 | 12.8 | 11.0 | 0.0 | 5.3 |
| Carrollton | 0.0 | 16.3 | 0.0 | 0.0 | 1.8 | 19.4 | 12.8 | 9.4 | 0.9 | 4.4 |
| Cedar Hill | 0.0 | 11.7 | 0.0 | 0.0 | 0.0 | 18.0 | 13.5 | 5.0 | 0.0 | 10.8 |
| Cedar Park | 0.0 | 17.1 | 0.0 | 0.0 | 0.4 | 12.8 | 16.6 | 4.4 | 0.0 | 8.7 |
| Cleburne | 0.0 | 4.9 | 0.0 | 0.0 | 0.7 | 14.2 | 13.6 | 34.4 | 0.0 | 0.0 |
| College Station | 0.0 | 17.4 | 0.5 | 0.0 | 0.0 | 14.8 | 18.4 | 8.9 | 1.3 | 3.6 |
| Conroe | 0.0 | 15.1 | 0.0 | 0.0 | 0.0 | 26.1 | 14.3 | 7.1 | 0.8 | 3.9 |
| Coppell | 0.0 | 22.7 | 0.0 | 0.0 | 0.8 | 10.4 | 5.8 | 26.1 | 0.0 | 0.0 |
| Copperas Cove | 0.0 | 0.0 | 0.0 | 0.0 | 2.0 | 25.6 | 23.9 | 8.7 | 0.0 | 0.0 |
| Corpus Christi | 0.0 | 12.2 | 0.1 | 0.0 | 1.6 | 19.6 | 19.6 | 8.4 | 2.5 | 5.4 |
| Dallas | 0.3 | 6.8 | 0.0 | 0.0 | 0.9 | 12.7 | 12.1 | 6.0 | 1.2 | 10.7 |
| Deer Park | 0.0 | 3.8 | 0.0 | 0.0 | 0.6 | 17.9 | 14.9 | 15.5 | 0.0 | 3.8 |
| Del Rio | 3.4 | 15.0 | 0.0 | 0.0 | 1.5 | 18.0 | 15.9 | 4.3 | 0.0 | 3.6 |
| Denton | 0.0 | 10.4 | 0.0 | 0.0 | 0.0 | 13.9 | 27.2 | 6.7 | 0.2 | 3.5 |
| DeSoto | 0.0 | 7.5 | 0.0 | 0.0 | 0.0 | 17.0 | 22.9 | 8.1 | 0.0 | 6.3 |
| Duncanville | 0.0 | 7.8 | 0.0 | 0.0 | 1.0 | 21.6 | 20.1 | 9.7 | 0.0 | 0.6 |
| Eagle Pass | 0.0 | 3.2 | 0.0 | 0.0 | 0.0 | 2.7 | 1.6 | 1.3 | 0.0 | 1.1 |
| Edinburg | 0.0 | 4.7 | 0.0 | 0.0 | 2.1 | 18.5 | 25.7 | 14.7 | 1.6 | 2.8 |
| El Paso | 0.1 | 4.3 | 0.0 | 0.0 | 2.7 | 16.1 | 19.2 | 8.0 | 2.2 | 11.8 |
| Euless | 0.0 | 11.9 | 0.0 | 0.0 | 2.5 | 16.8 | 5.0 | 11.5 | 0.0 | 0.0 |
| Farmers Branch | 0.0 | 13.9 | 0.0 | 0.0 | 0.7 | 25.9 | 3.4 | 18.7 | 0.0 | 6.0 |
| Flower Mound | 0.0 | 10.4 | 0.0 | 0.0 | 1.7 | 20.1 | 13.8 | 9.3 | 0.3 | 4.4 |
| Fort Worth | 0.0 | 10.6 | 0.3 | 0.0 | 1.0 | 21.9 | 20.2 | 7.0 | 1.6 | 5.2 |
| Friendswood | 0.0 | 9.4 | 0.0 | 0.0 | 1.1 | 25.3 | 8.4 | 14.4 | 1.2 | 0.0 |
| Frisco | 0.0 | 12.4 | 0.0 | 4.0 | 0.6 | 8.2 | 11.7 | 6.8 | 0.0 | 0.0 |
| Galveston | 0.0 | 8.2 | 0.0 | 0.0 | 0.4 | 9.7 | 23.7 | 15.5 | 2.1 | 3.5 |
| Garland | 0.2 | 7.1 | 0.0 | 0.0 | 1.8 | 21.9 | 15.4 | 6.4 | 5.9 | 6.0 |
| Georgetown | 0.0 | 3.3 | 0.0 | 0.0 | 0.8 | 13.3 | 7.5 | 8.7 | 1.2 | 3.7 |
| Grand Prairie | 0.0 | 5.8 | 0.0 | 0.0 | 1.0 | 21.2 | 16.3 | 8.7 | 13.6 | 3.8 |
| Grapevine | 0.0 | 4.8 | 0.0 | 0.0 | 0.0 | 10.1 | 3.4 | 10.8 | 0.0 | 4.3 |
| Greenville | 0.0 | 26.4 | 0.0 | 0.0 | 0.0 | 0.2 | 4.6 | 0.8 | 0.0 | 8.3 |
| Haltom City | 0.0 | 13.5 | 0.0 | 0.0 | 0.0 | 24.0 | 14.1 | 4.2 | 0.0 | 3.1 |
| Harker Heights | 0.0 | 13.5 | 0.0 | 0.0 | 1.8 | 19.5 | 22.4 | 8.0 | 0.0 | 0.0 |
| Harlingen | 0.0 | 5.2 | 0.0 | 0.0 | 0.8 | 15.6 | 21.9 | 6.7 | 1.8 | 3.6 |
| Houston | 0.0 | 4.5 | 0.3 | 0.0 | 3.0 | 17.9 | 15.1 | 2.6 | 1.6 | 17.3 |
| Huntsville | 0.0 | 8.7 | 0.0 | 0.0 | 0.0 | 19.5 | 39.0 | 3.7 | 0.0 | 1.1 |
| Hurst | 0.0 | 14.8 | 0.0 | 0.0 | 2.2 | 25.6 | 11.9 | 15.1 | 0.0 | 4.0 |

# Table D. Cities — City Government Finances, City Government Employment, and Climate

| City | City government finances, 2017 (cont.) Debt outstanding — Total (mil dol) [137] | Per capita (dollars) [138] | Debt issued during year [139] | Climate — Avg daily temp Mean January [140] | Mean July [141] | Limits January [142] | Limits July [143] | Annual precipitation (inches) [144] | Heating degree days [145] | Cooling degree days [146] |
|---|---|---|---|---|---|---|---|---|---|---|
| **TENNESSEE—Cont'd** | | | | | | | | | | |
| Columbia | 57.4 | 1,496 | 9.0 | 35.6 | 77.2 | 25.0 | 88.5 | 56.13 | 4,183 | 1,267 |
| Cookeville | 91.8 | 2,746 | 0.0 | NA | NA | NA | NA | NA | NA | NA |
| Franklin | 195.9 | 2,503 | 40.2 | 35.1 | 77.4 | 25.2 | 88.9 | 54.33 | 4,199 | 1,294 |
| Gallatin | 68.0 | 1,816 | 0.0 | NA | NA | NA | NA | NA | NA | NA |
| Germantown | 32.0 | 817 | 0.0 | 37.9 | 81.1 | 28.2 | 91.1 | 53.63 | 3,491 | 1,838 |
| Hendersonville | 4.8 | 84 | 0.0 | 36.8 | 79.1 | 27.9 | 88.7 | 48.11 | 3,677 | 1,652 |
| Jackson | 124.2 | 1,861 | 7.3 | 37.1 | 79.6 | 28.2 | 89.4 | 54.86 | 3,649 | 1,648 |
| Johnson City | 382.6 | 5,783 | 83.4 | 34.2 | 74.2 | 24.3 | 84.8 | 41.33 | 4,445 | 956 |
| Kingsport | 241.2 | 4,509 | 30.3 | 35.6 | 76.2 | 26.2 | 86.9 | 44.44 | 4,178 | 1,139 |
| Knoxville | 542.4 | 2,902 | 0.0 | 38.5 | 78.7 | 30.3 | 88.2 | 48.22 | 3,531 | 1,527 |
| La Vergne | 35.7 | 1,003 | 4.3 | NA | NA | NA | NA | NA | NA | NA |
| Lebanon | 69.5 | 2,137 | 13.1 | NA | NA | NA | NA | NA | NA | NA |
| Maryville | 104.2 | 3,587 | 0.0 | NA | NA | NA | NA | NA | NA | NA |
| Memphis | 2,006.0 | 3,082 | 220.9 | 39.9 | 82.5 | 31.3 | 92.1 | 54.65 | 3,041 | 2,187 |
| Morristown | 231.6 | 7,786 | 15.8 | NA | NA | NA | NA | NA | NA | NA |
| Murfreesboro | 368.8 | 2,708 | 51.9 | 35.4 | 78.1 | 25.3 | 89.1 | 54.98 | 4,107 | 1,388 |
| Nashville-Davidson | 14,599.2 | 21,246 | 901.2 | 36.8 | 79.1 | 27.9 | 88.7 | 48.11 | 3,677 | 1,652 |
| Oak Ridge | 125.2 | 4,313 | 0.0 | 36.6 | 77.3 | 27.2 | 88.1 | 55.05 | 3,993 | 1,301 |
| Smyrna | 61.2 | 1,226 | 1.9 | 36.8 | 79.1 | 27.9 | 88.7 | 48.11 | 3,677 | 1,652 |
| Spring Hill | 32.0 | 807 | 0.0 | NA | NA | NA | NA | NA | NA | NA |
| **TEXAS** | X | X | X | X | X | X | X | X | X | X |
| Abilene | 256.2 | 2,096 | 0.0 | 43.5 | 83.5 | 31.8 | 94.8 | 23.78 | 2,659 | 2,386 |
| Allen | 149.6 | 1,478 | 47.0 | 41.8 | 82.4 | 31.1 | 92.7 | 41.01 | 2,843 | 2,060 |
| Amarillo | 366.7 | 1,836 | 0.0 | 35.8 | 78.2 | 22.6 | 91.0 | 19.71 | 4,318 | 1,344 |
| Arlington | 1,049.6 | 2,643 | 0.0 | 44.1 | 85.0 | 34.0 | 95.4 | 34.73 | 2,370 | 2,568 |
| Austin | 5,394.5 | 5,669 | 0.0 | 50.2 | 84.2 | 40.0 | 95.4 | 34.73 | 2,370 | 2,568 |
| Baytown | 204.4 | 2,666 | 53.2 | 51.6 | 83.6 | 41.9 | 95.0 | 33.65 | 1,648 | 2,974 |
| Beaumont | 372.6 | 3,132 | 79.9 | 51.1 | 83.1 | 41.1 | 91.6 | 53.75 | 1,471 | 2,841 |
| Bedford | 49.7 | 1,003 | 0.0 | 44.1 | 85.0 | 34.0 | 92.7 | 57.38 | 1,548 | 2,734 |
| Big Spring | 27.4 | 990 | 8.0 | 42.7 | 82.7 | 29.6 | 95.4 | 34.73 | 2,370 | 2,568 |
| Brownsville | 311.4 | 1,707 | 0.0 | 59.6 | 83.9 | 50.5 | 94.3 | 20.12 | 2,724 | 2,243 |
| Bryan | 440.9 | 5,238 | 118.9 | 50.2 | 84.6 | 39.8 | 92.4 | 27.55 | 644 | 3,874 |
| Burleson | 158.2 | 3,439 | 46.7 | NA | NA | NA | NA | NA | NA | NA |
| Carrollton | 192.5 | 1,415 | 33.8 | 44.1 | 85.0 | 34.0 | 95.4 | 34.73 | 2,370 | 2,568 |
| Cedar Hill | 138.5 | 2,840 | 0.0 | 43.7 | 84.3 | 33.2 | 94.9 | 34.54 | 2,437 | 2,508 |
| Cedar Park | 52.2 | 691 | 0.0 | 47.2 | 83.8 | 35.1 | 95.7 | 36.42 | 1,998 | 2,584 |
| Cleburne | 172.8 | 5,723 | 35.5 | 45.9 | 84.5 | 34.0 | 95.4 | 36.25 | 2,158 | 2,604 |
| College Station | 278.9 | 2,437 | 75.8 | 50.2 | 84.6 | 34.0 | 97.0 | 39.67 | 1,616 | 2,938 |
| Conroe | 265.3 | 3,128 | 25.1 | 50.3 | 83.7 | 39.8 | 95.6 | 49.32 | 1,647 | 2,793 |
| Coppell | 106.6 | 2,538 | 15.1 | 44.1 | 85.0 | 40.0 | 94.3 | 34.73 | 2,370 | 2,568 |
| Copperas Cove | 70.3 | 2,164 | 15.9 | 46.0 | 83.5 | 34.0 | 95.4 | 32.88 | 2,190 | 2,477 |
| Corpus Christi | 1,675.1 | 5,145 | 225.0 | 56.1 | 83.8 | 46.2 | 95.3 | 32.26 | 950 | 3,497 |
| Dallas | 10,912.7 | 8,129 | 1,019.4 | 45.9 | 86.5 | 36.4 | 96.1 | 37.05 | 2,219 | 2,878 |
| Deer Park | 41.5 | 1,222 | 0.0 | 54.3 | 84.5 | 45.2 | 93.6 | 53.96 | 1,174 | 3,179 |
| Del Rio | 38.3 | 1,067 | 0.0 | 51.3 | 85.3 | 39.7 | 96.2 | 18.80 | 1,417 | 3,226 |
| Denton | 569.3 | 4,198 | 0.0 | 42.7 | 83.6 | 32.0 | 94.1 | 37.79 | 2,650 | 2,269 |
| DeSoto | 106.1 | 1,982 | 0.0 | 46.0 | 84.6 | 35.0 | 96.0 | 38.81 | 2,130 | 2,608 |
| Duncanville | 6.5 | 165 | 0.0 | 45.9 | 86.5 | 36.4 | 96.1 | 37.05 | 2,219 | 2,878 |
| Eagle Pass | 63.4 | 2,161 | 4.9 | NA | NA | NA | NA | NA | NA | NA |
| Edinburg | 93.0 | 974 | 0.0 | 58.7 | 85.1 | 48.2 | 95.5 | 22.61 | 719 | 3,898 |
| El Paso | 2,275.1 | 3,339 | 586.9 | 45.1 | 83.3 | 32.9 | 94.5 | 9.43 | 2,543 | 2,254 |
| Euless | 62.5 | 1,132 | 20.8 | 44.1 | 85.0 | 34.0 | 95.4 | 34.73 | 2,370 | 2,568 |
| Farmers Branch | 103.3 | 2,748 | 30.2 | 45.9 | 86.5 | 36.4 | 96.1 | 37.05 | 2,219 | 2,878 |
| Flower Mound | 145.9 | 1,902 | 37.1 | 44.1 | 85.0 | 34.0 | 95.4 | 34.73 | 2,370 | 2,568 |
| Fort Worth | 1,482.3 | 1,694 | 0.0 | 43.3 | 84.5 | 31.4 | 96.6 | 34.01 | 2,509 | 2,466 |
| Friendswood | 67.7 | 1,701 | 36.1 | 54.3 | 84.5 | 45.2 | 93.6 | 53.96 | 1,174 | 3,179 |
| Frisco | 809.1 | 4,552 | 181.3 | 41.8 | 82.4 | 31.1 | 92.7 | 41.01 | 2,843 | 2,060 |
| Galveston | 209.7 | 4,147 | 20.6 | 55.8 | 84.3 | 49.7 | 88.7 | 43.84 | 1,008 | 3,268 |
| Garland | 963.3 | 4,042 | 121.8 | 45.9 | 86.5 | 36.4 | 96.1 | 37.05 | 2,219 | 2,878 |
| Georgetown | 127.0 | 1,797 | 0.0 | 47.2 | 83.8 | 35.1 | 95.7 | 36.42 | 1,998 | 2,584 |
| Grand Prairie | 268.0 | 1,382 | 0.0 | 44.1 | 85.0 | 34.0 | 95.4 | 34.73 | 2,370 | 2,568 |
| Grapevine | 164.5 | 3,048 | 9.4 | 42.4 | 84.0 | 30.8 | 95.5 | 34.66 | 2,649 | 2,340 |
| Greenville | 98.3 | 3,593 | 1.9 | NA | NA | NA | NA | NA | NA | NA |
| Haltom City | 50.7 | 1,141 | 0.0 | 43.0 | 84.1 | 31.4 | 95.7 | 34.12 | 2,608 | 2,358 |
| Harker Heights | 45.8 | 1,482 | 6.5 | NA | NA | NA | NA | NA | NA | NA |
| Harlingen | 99.8 | 1,532 | 23.9 | 58.6 | 84.4 | 48.4 | 94.5 | 28.13 | 737 | 3,736 |
| Houston | 14,092.4 | 6,083 | 362.0 | 54.3 | 84.5 | 45.2 | 93.6 | 53.96 | 1,174 | 3,179 |
| Huntsville | 29.8 | 714 | 0.0 | 48.5 | 83.2 | 39.0 | 93.8 | 48.51 | 1,835 | 2,600 |
| Hurst | 67.0 | 1,711 | 4.7 | 44.1 | 85.0 | 34.0 | 95.4 | 34.73 | 2,370 | 2,568 |

1. Based on the population estimated as of July 1 of the year shown.  2. Represents normal values based on the 30-year period, 1971±2000.  3. Average daily minimum.  4. Average daily maximum.

# Table D. Cities — Land Area and Population

| STATE Place code | City | Land area¹ (sq. mi) | Total persons 2019 | Rank | Per square mile | White | Black or African American | American Indian, Alaskan Native | Asian | Hawaiian Pacific Islander | Some other race | Two or more races (percent) |
|---|---|---|---|---|---|---|---|---|---|---|---|---|
| | | **Population, 2020** | | | | **Race 2019 — Race alone² (percent)** | | | | | | |
| | | 1 | 2 | 3 | 4 | 5 | 6 | 7 | 8 | 9 | 10 | 11 |
| | **TEXAS—Cont'd** | | | | | | | | | | | |
| 48 37000 | Irving | 67.0 | 240,916 | 93 | 3,596 | 45.1 | 14.1 | 0.5 | 22.4 | 0.0 | 15.2 | 2.7 |
| 48 38632 | Keller | 18.4 | 46,939 | 829 | 2,551 | 87.4 | 2.8 | 0.5 | 5.2 | 0.0 | 1.4 | 2.8 |
| 48 39148 | Killeen | 54.8 | 153,991 | 166 | 2,810 | 41.3 | 42.8 | 0.9 | 4.1 | 1.2 | 2.8 | 6.8 |
| 48 39352 | Kingsville | 13.9 | 24,989 | 1,407 | 1,798 | D | D | D | D | D | D | D |
| 48 39952 | Kyle | 31.1 | 51,558 | 753 | 1,658 | 92.2 | 3.1 | 0.4 | 0.4 | 0.0 | 1.8 | 2.1 |
| 48 40588 | Lake Jackson | 19.7 | 27,063 | 1,347 | 1,374 | D | D | D | D | D | D | D |
| 48 41212 | Lancaster | 33.1 | 39,265 | 988 | 1,186 | D | D | 0.0 | 0.2 | 0.0 | 2.5 | 2.7 |
| 48 41440 | La Porte | 18.6 | 35,775 | 1,081 | 1,923 | 89.7 | 5.0 | 0.0 | 0.5 | 0.0 | 2.0 | 0.4 |
| 48 41464 | Laredo | 106.5 | 263,640 | 80 | 2,476 | 96.0 | 0.5 | 0.6 | 0.5 | 0.0 | 0.6 | 4.0 |
| 48 41980 | League City | 51.3 | 108,477 | 283 | 2,115 | 83.4 | 7.4 | 0.4 | 4.2 | 0.0 | 0.5 | 6.5 |
| 48 42016 | Leander | 37.5 | 70,519 | 517 | 1,881 | 81.8 | 5.3 | 0.0 | 5.9 | 0.0 | 3.6 | 3.6 |
| 48 42508 | Lewisville | 37.0 | 112,232 | 267 | 3,033 | 68.5 | 12.6 | 0.8 | 10.9 | 0.0 | 2.0 | 3.0 |
| 48 43012 | Little Elm | 18.1 | 56,625 | 678 | 3,129 | 73.6 | 19.1 | 0.0 | 1.9 | 0.4 | 1.9 | 4.4 |
| 48 43888 | Longview | 55.8 | 81,728 | 421 | 1,465 | 72.8 | 18.9 | 0.1 | 1.9 | 0.0 | 2.5 | 3.5 |
| 48 45000 | Lubbock | 134.6 | 262,611 | 84 | 1,951 | 79.4 | 7.6 | 0.6 | 2.6 | 0.3 | 6.0 | 3.5 |
| 48 45072 | Lufkin | 34.2 | 34,955 | 1,105 | 1,022 | D | D | D | D | D | D | D |
| 48 45384 | McAllen | 62.3 | 143,751 | 186 | 2,307 | 69.2 | 0.6 | 0.1 | 3.6 | 0.0 | 24.1 | 2.4 |
| 48 45744 | McKinney | 67.0 | 208,272 | 111 | 3,109 | 73.6 | 11.7 | 0.5 | 10.1 | 0.0 | 0.8 | 3.3 |
| 48 46452 | Mansfield | 36.6 | 73,094 | 491 | 1,997 | 70.7 | 18.5 | 0.6 | 2.4 | 0.0 | 3.8 | 3.9 |
| 48 47892 | Mesquite | 48.5 | 138,916 | 199 | 2,864 | 65.0 | 20.6 | 0.6 | 3.3 | 0.1 | 15.2 | 3.4 |
| 48 48072 | Midland | 75.5 | 147,069 | 181 | 1,948 | 73.1 | 6.7 | 0.3 | 1.3 | 0.1 | 14.4 | 1.0 |
| 48 48768 | Mission | 36.3 | 84,496 | 406 | 2,328 | 83.5 | 0.1 | 0.0 | 1.0 | 0.0 | 4.1 | 1.9 |
| 48 48804 | Missouri City | 29.0 | 76,643 | 462 | 2,643 | 34.8 | 42.0 | 0.0 | 17.2 | 0.0 | D | D |
| 48 50256 | Nacogdoches | 27.6 | 32,461 | 1,175 | 1,176 | D | D | D | D | D | D | D |
| 48 50820 | New Braunfels | 45.2 | 95,534 | 334 | 2,114 | 92.2 | 3.2 | 0.2 | 1.6 | 0.0 | 1.5 | 1.3 |
| 48 52356 | North Richland Hills | 18.2 | 71,949 | 505 | 3,953 | 82.5 | 8.0 | 0.6 | 3.5 | 0.1 | 2.4 | 2.8 |
| 48 53388 | Odessa | 51.1 | 125,413 | 221 | 2,454 | 74.8 | 8.1 | 1.3 | 1.9 | 0.0 | 11.4 | 2.4 |
| 48 55080 | Paris | 35.2 | 24,872 | 1,409 | 707 | D | D | D | D | D | D | D |
| 48 56000 | Pasadena | 43.7 | 149,440 | 174 | 3,420 | 89.0 | 2.2 | 1.2 | 1.6 | 0.0 | 3.4 | 2.6 |
| 48 56348 | Pearland | 48.7 | 123,562 | 229 | 2,537 | 63.4 | 15.1 | 1.2 | 15.2 | 0.0 | 3.1 | 2.0 |
| 48 57176 | Pflugerville | 25.6 | 66,826 | 557 | 2,610 | 62.9 | 16.8 | 0.2 | 4.9 | 0.0 | 10.0 | 5.2 |
| 48 57200 | Pharr | 23.6 | 79,513 | 445 | 3,369 | D | D | D | D | D | D | D |
| 48 58016 | Plano | 71.7 | 291,296 | 68 | 4,063 | 64.1 | 9.6 | 0.2 | 22.3 | 0.2 | 1.1 | 2.6 |
| 48 58820 | Port Arthur | 77.1 | 53,921 | 717 | 699 | 50.4 | 36.8 | 0.5 | 8.2 | 1.2 | 3.0 | 2.8 |
| 48 61796 | Richardson | 28.6 | 121,112 | 235 | 4,235 | 67.7 | 13.7 | 0.3 | 12.5 | 0.0 | 3.0 | 4.2 |
| 48 62828 | Rockwall | 29.5 | 47,415 | 820 | 1,607 | 83.8 | 4.5 | 1.0 | 3.4 | 0.4 | 2.7 | D |
| 48 63284 | Rosenberg | 37.3 | 39,350 | 985 | 1,055 | D | D | D | D | D | D | D |
| 48 63500 | Round Rock | 37.6 | 137,575 | 201 | 3,659 | 81.3 | 7.9 | 0.0 | 5.2 | 0.0 | 1.5 | 4.0 |
| 48 63572 | Rowlett | 20.7 | 68,388 | 537 | 3,304 | 65.3 | 20.0 | 0.1 | 7.6 | 0.2 | 2.3 | 4.5 |
| 48 64472 | San Angelo | 59.7 | 101,612 | 311 | 1,702 | 86.9 | 4.7 | 0.6 | 1.4 | 0.1 | 4.2 | 2.0 |
| 48 65000 | San Antonio | 498.8 | 1,567,118 | 7 | 3,142 | 78.3 | 7.3 | 0.9 | 2.9 | 0.1 | 6.8 | 3.7 |
| 48 65516 | San Juan | 11.6 | 37,034 | 1,037 | 3,193 | D | D | D | D | D | D | D |
| 48 65600 | San Marcos | 35.6 | 66,350 | 566 | 1,864 | 81.1 | 6.9 | 1.2 | 3.1 | 0.2 | 5.5 | 2.1 |
| 48 66128 | Schertz | 32.1 | 42,440 | 908 | 1,322 | 82.1 | 10.3 | 0.3 | 0.7 | 0.2 | 2.8 | 3.5 |
| 48 66644 | Seguin | 38.3 | 30,165 | 1,248 | 788 | 86.1 | 6.4 | 0.0 | 2.5 | 0.0 | 3.1 | 1.8 |
| 48 67496 | Sherman | 46.1 | 45,136 | 855 | 979 | 86.1 | 6.4 | 0.0 | 2.5 | 0.0 | 3.7 | 2.1 |
| 48 68636 | Socorro | 21.9 | 35,883 | 1,077 | 1,639 | 92.4 | 0.9 | 0.9 | 0.0 | 0.0 | D | D |
| 48 69032 | Southlake | 21.8 | 32,280 | 1,183 | 1,481 | D | D | D | D | D | D | D |
| 48 70808 | Sugar Land | 40.5 | 117,875 | 243 | 2,911 | 50.1 | 8.5 | 0.3 | 35.3 | 0.8 | 1.7 | 3.4 |
| 48 72176 | Temple | 71.2 | 80,761 | 434 | 1,134 | 75.2 | 14.5 | 0.1 | 3.0 | 0.1 | 1.6 | 5.5 |
| 48 72368 | Texarkana | 31.4 | 36,221 | 1,067 | 1,154 | D | D | D | D | D | D | D |
| 48 72392 | Texas City | 66.3 | 51,593 | 752 | 778 | 68.1 | 25.3 | 0.3 | 2.2 | 0.2 | 1.4 | 2.5 |
| 48 72530 | The Colony | 14.0 | 45,498 | 853 | 3,250 | 77.1 | 6.8 | 0.5 | 6.3 | 0.0 | 2.7 | 6.5 |
| 48 74144 | Tyler | 57.5 | 108,222 | 284 | 1,882 | 67.3 | 25.1 | 0.7 | 3.0 | 0.0 | 2.7 | 1.3 |
| 48 75428 | Victoria | 37.4 | 66,838 | 555 | 1,787 | 85.9 | 7.2 | 0.0 | 1.3 | 0.1 | 2.7 | 2.8 |
| 48 76000 | Waco | 88.7 | 141,377 | 191 | 1,594 | 72.7 | 20.3 | 0.3 | 1.9 | 0.0 | 3.4 | 3.5 |
| 48 76816 | Waxahachie | 49.5 | 39,797 | 970 | 804 | 79.9 | 13.8 | 0.7 | 0.1 | 0.0 | 3.4 | 2.0 |
| 48 76864 | Weatherford | 27.1 | 34,698 | 1,115 | 1,280 | 87.2 | 2.3 | 1.2 | 1.5 | 0.0 | 6.0 | 1.9 |
| 48 77272 | Weslaco | 16.4 | 43,273 | 890 | 2,639 | D | D | D | D | D | D | D |
| 48 79000 | Wichita Falls | 72.0 | 105,405 | 293 | 1,464 | 79.1 | 12.1 | 0.6 | 2.2 | 0.1 | 2.5 | 3.3 |
| 48 80356 | Wylie | 22.1 | 54,467 | 710 | 2,465 | 70.4 | 11.0 | 0.3 | 9.6 | 0.1 | 2.9 | 5.7 |
| 49 00000 | UTAH | 82,376.9 | 3,249,879 | X | 40 | 87.3 | 1.2 | 1.1 | 2.4 | 1.0 | 3.8 | 3.2 |
| 49 01310 | American Fork | 11.2 | 33,896 | 1,137 | 3,026 | D | D | D | D | D | D | D |
| 49 07690 | Bountiful | 13.2 | 43,921 | 880 | 3,327 | 91.0 | 0.4 | 0.8 | 3.3 | 1.3 | 1.2 | 2.0 |
| 49 11320 | Cedar City | 35.9 | 36,090 | 1,071 | 1,005 | 93.8 | 0.1 | 1.2 | 1.3 | 0.1 | 0.0 | 3.5 |
| 49 13850 | Clearfield | 7.7 | 32,151 | 1,191 | 4,176 | 85.2 | 5.3 | 0.1 | 1.2 | 0.5 | 3.3 | 4.4 |
| 49 16270 | Cottonwood Heights | 9.2 | 33,327 | 1,154 | 3,623 | D | D | D | D | D | D | D |
| 49 20120 | Draper | 29.9 | 48,963 | 795 | 1,638 | 90.2 | 2.4 | 0.0 | 1.2 | 1.5 | 2.0 | 2.7 |
| 49 36070 | Holladay | 8.5 | 29,885 | 1,256 | 3,516 | D | D | D | D | D | D | D |

1. Dry land or land partially or temporarily covered by water.   2. Hispanic or Latino persons may be of any race.

# Table D. Cities — **Population**

| City | Percent Hispanic or Latino[1], 2019 | Percent foreign born, 2019 | Age of population (percent), 2019 | | | | | | | Median age, 2019 | Percent female, 2019 | Population | | | |
| | | | Under 18 years | 18 to 24 years | 25 to 34 years | 35 to 44 years | 45 to 54 years | 55 to 64 years | 65 years and over | | | Census counts | | Percent change | |
| | | | | | | | | | | | | 2000 | 2010 | 2000-2010 | 2001-2020 |
| | 12 | 13 | 14 | 15 | 16 | 17 | 18 | 19 | 20 | 21 | 22 | 23 | 24 | 25 | 26 |
|---|---|---|---|---|---|---|---|---|---|---|---|---|---|---|---|
| **TEXAS—Cont'd** | | | | | | | | | | | | | | | |
| Irving | 42.6 | 43.7 | 26.5 | 9.8 | 19.3 | 16.0 | 12.6 | 8.2 | 7.7 | 32.0 | 49.3 | 191,615 | 216,285 | 12.9 | 11.4 |
| Keller | 13.9 | 8.1 | 27.0 | 5.4 | 8.8 | 12.8 | 16.6 | 17.5 | 12.0 | 42.2 | 48.3 | 27,345 | 39,627 | 44.9 | 18.5 |
| Killeen | 26.9 | 9.2 | 28.5 | 11.9 | 19.3 | 12.2 | 11.9 | 9.5 | 6.7 | 29.4 | 50.0 | 86,911 | 127,733 | 47.0 | 20.6 |
| Kingsville | 76.3 | 3.9 | 25.0 | 21.0 | 17.1 | 6.7 | 10.5 | 8.1 | 11.6 | 27.5 | 49.2 | 25,575 | 26,453 | 3.4 | -5.5 |
| Kyle | 53.1 | 5.1 | 29.0 | 7.9 | 19.3 | 16.1 | 10.1 | 8.6 | 9.1 | 31.9 | 50.0 | 5,314 | 28,227 | 431.2 | 82.7 |
| Lake Jackson | 25.4 | 10.9 | 23.5 | 9.5 | 14.8 | 14.5 | 11.7 | 11.9 | 14.0 | 37.0 | 47.9 | 26,386 | 26,830 | 1.7 | 0.9 |
| Lancaster | 16.3 | 6.4 | 25.4 | 7.9 | 15.1 | 12.0 | 12.1 | 18.2 | 9.2 | 36.4 | 54.8 | 25,894 | 36,651 | 41.5 | 7.1 |
| La Porte | 38.4 | 8.2 | 22.5 | 7.7 | 16.7 | 14.9 | 12.0 | 13.3 | 12.8 | 37.1 | 46.3 | 31,880 | 33,808 | 6.0 | 5.8 |
| Laredo | 95.4 | 24.6 | 32.2 | 11.5 | 14.0 | 12.9 | 11.2 | 8.5 | 9.8 | 29.2 | 51.1 | 176,576 | 235,781 | 33.5 | 11.8 |
| League City | 21.0 | 11.3 | 25.9 | 6.7 | 16.0 | 15.9 | 11.9 | 11.6 | 12.0 | 35.7 | 48.9 | 45,444 | 83,589 | 83.9 | 29.8 |
| Leander | 14.8 | 9.7 | 31.8 | 7.1 | 13.4 | 21.0 | 11.2 | 8.0 | 7.5 | 34.3 | 50.2 | 7,596 | 27,279 | 259.1 | 158.5 |
| Lewisville | 28.3 | 21.6 | 21.6 | 9.1 | 19.3 | 15.0 | 16.1 | 9.5 | 9.3 | 34.9 | 53.9 | 77,737 | 95,474 | 22.8 | 17.6 |
| Little Elm | 37.4 | 15.8 | 38.1 | 6.2 | 7.9 | 24.3 | 7.5 | 10.0 | 6.1 | 33.5 | 49.6 | 3,646 | 25,877 | 609.7 | 118.8 |
| Longview | 22.7 | 12.4 | 25.3 | 9.8 | 15.8 | 12.6 | 10.8 | 11.2 | 14.6 | 34.4 | 49.6 | 73,344 | 80,419 | 9.6 | 1.6 |
| Lubbock | 37.4 | 6.4 | 23.4 | 18.1 | 15.6 | 12.1 | 9.0 | 10.2 | 11.5 | 29.8 | 51.4 | 199,564 | 229,944 | 15.2 | 14.2 |
| Lufkin | 33.2 | 11.0 | 25.9 | 8.9 | 18.3 | 9.8 | 10.9 | 9.8 | 16.4 | 31.8 | 51.1 | 32,709 | 35,136 | 7.4 | -0.5 |
| McAllen | 84.0 | 24.9 | 26.5 | 9.3 | 13.8 | 14.6 | 10.4 | 10.0 | 15.4 | 35.3 | 50.5 | 106,414 | 131,565 | 23.6 | 9.3 |
| McKinney | 17.8 | 16.8 | 28.8 | 7.0 | 11.1 | 17.3 | 15.1 | 9.4 | 11.4 | 37.2 | 50.6 | 54,369 | 131,154 | 141.2 | 58.8 |
| Mansfield | 13.4 | 11.5 | 29.9 | 8.5 | 11.3 | 13.4 | 15.4 | 10.4 | 11.1 | 35.2 | 52.1 | 28,031 | 56,654 | 102.1 | 29.0 |
| Mesquite | 46.3 | 22.1 | 23.9 | 10.1 | 11.7 | 14.6 | 12.9 | 11.3 | 15.5 | 39.2 | 52.8 | 124,523 | 139,768 | 12.2 | -0.6 |
| Midland | 43.9 | 15.5 | 28.0 | 9.1 | 17.7 | 14.7 | 9.8 | 10.0 | 10.7 | 32.3 | 49.2 | 94,996 | 111,192 | 17.0 | 32.3 |
| Mission | 92.2 | 30.0 | 30.6 | 10.4 | 12.4 | 11.5 | 15.9 | 7.9 | 11.3 | 32.2 | 49.8 | 45,408 | 77,704 | 71.1 | 8.7 |
| Missouri City | 18.3 | 22.3 | 20.8 | 8.7 | 11.4 | 12.3 | 16.1 | 14.2 | 16.5 | 42.3 | 50.7 | 52,913 | 66,653 | 26.0 | 15.0 |
| Nacogdoches | 16.7 | 5.7 | 20.4 | 27.6 | 17.7 | 9.4 | 6.6 | 6.1 | 12.1 | 25.6 | 54.5 | 29,914 | 32,804 | 9.7 | -1.0 |
| New Braunfels | 37.7 | 7.2 | 25.2 | 8.5 | 16.2 | 13.0 | 11.2 | 11.8 | 14.1 | 35.1 | 52.9 | 36,494 | 57,674 | 58.0 | 65.6 |
| North Richland Hills | 14.7 | 5.8 | 23.1 | 6.0 | 14.5 | 11.3 | 12.6 | 16.0 | 16.5 | 40.4 | 52.9 | 55,635 | 63,147 | 13.5 | 13.9 |
| Odessa | 59.0 | 13.2 | 28.6 | 9.8 | 18.4 | 12.3 | 9.3 | 11.3 | 10.4 | 30.9 | 50.8 | 90,943 | 99,878 | 9.8 | 25.6 |
| Paris | 11.3 | 3.6 | 21.7 | 8.4 | 17.3 | 7.8 | 11.2 | 13.4 | 20.1 | 36.8 | 54.2 | 25,898 | 25,262 | -2.5 | -1.5 |
| Pasadena | 73.0 | 26.5 | 29.4 | 9.9 | 13.9 | 11.8 | 13.2 | 10.7 | 11.1 | 32.7 | 51.8 | 141,674 | 149,389 | 5.4 | 0.0 |
| Pearland | 29.3 | 15.1 | 28.5 | 4.7 | 15.2 | 16.3 | 13.7 | 11.0 | 10.6 | 36.1 | 49.2 | 37,640 | 93,159 | 147.5 | 32.6 |
| Pflugerville | 40.3 | 18.6 | 23.5 | 7.7 | 16.5 | 14.9 | 16.7 | 10.9 | 9.8 | 36.6 | 51.4 | 16,335 | 48,370 | 196.1 | 38.2 |
| Pharr | 95.8 | 31.6 | 34.7 | 9.5 | 14.5 | 11.9 | 9.6 | 8.3 | 11.5 | 28.8 | 53.5 | 46,660 | 70,537 | 51.2 | 12.7 |
| Plano | 13.6 | 28.1 | 21.5 | 7.6 | 14.3 | 15.0 | 14.9 | 12.7 | 14.0 | 39.4 | 50.6 | 222,030 | 259,860 | 17.0 | 12.1 |
| Port Arthur | 36.1 | 25.4 | 29.4 | 10.8 | 11.4 | 11.0 | 12.0 | 13.1 | 12.3 | 33.7 | 52.3 | 57,755 | 54,376 | -5.9 | -0.8 |
| Richardson | 19.2 | 21.9 | 22.4 | 11.7 | 15.9 | 12.5 | 12.1 | 11.0 | 14.3 | 35.0 | 52.3 | 91,802 | 99,251 | 8.1 | 22.0 |
| Rockwall | 20.4 | 12.2 | 25.0 | 8.5 | 9.5 | 16.0 | 15.0 | 13.4 | 12.6 | 39.0 | 52.7 | 17,976 | 37,562 | 109.0 | 26.2 |
| Rosenberg | 63.7 | 21.8 | 22.2 | 12.7 | 19.0 | 11.9 | 8.5 | 10.5 | 15.1 | 33.6 | 52.3 | 24,043 | 31,225 | 29.9 | 26.0 |
| Round Rock | 31.9 | 13.6 | 26.6 | 8.9 | 16.6 | 16.3 | 13.2 | 10.4 | 8.1 | 33.7 | 51.0 | 61,136 | 100,019 | 63.6 | 37.5 |
| Rowlett | 16.6 | 11.1 | 23.7 | 7.3 | 13.2 | 9.3 | 18.9 | 15.1 | 12.5 | 40.8 | 50.3 | 44,503 | 56,223 | 26.3 | 21.6 |
| San Angelo | 40.7 | 6.5 | 22.1 | 13.6 | 14.0 | 12.3 | 10.3 | 12.1 | 15.7 | 35.3 | 51.2 | 88,439 | 93,219 | 5.4 | 9.0 |
| San Antonio | 64.5 | 14.1 | 24.4 | 10.3 | 16.3 | 13.8 | 11.8 | 10.7 | 12.7 | 34.4 | 50.7 | 1,144,646 | 1,326,819 | 15.9 | 18.1 |
| San Juan | 98.2 | 21.7 | 34.4 | 11.4 | 14.8 | 14.5 | 9.6 | 7.5 | 7.9 | 26.6 | 53.6 | 26,229 | 33,972 | 29.5 | 9.0 |
| San Marcos | 43.1 | 8.4 | 12.3 | 37.4 | 16.3 | 10.3 | 8.5 | 7.1 | 8.1 | 25.1 | 52.5 | 34,733 | 45,129 | 29.9 | 47.0 |
| Schertz | 29.9 | 3.9 | 28.6 | 7.3 | 13.0 | 16.0 | 12.4 | 10.9 | 11.7 | 36.0 | 51.0 | 18,694 | 31,826 | 70.2 | 33.4 |
| Seguin | 47.4 | 11.1 | 22.5 | 8.8 | 12.3 | 17.0 | 8.5 | 16.2 | 14.7 | 37.8 | 49.4 | 22,011 | 25,606 | 16.3 | 17.8 |
| Sherman | 29.0 | 13.7 | 20.8 | 12.4 | 11.8 | 11.6 | 10.4 | 16.8 | 16.3 | 40.6 | 52.9 | 35,082 | 38,853 | 10.7 | 16.2 |
| Socorro | 98.4 | 32.9 | 23.1 | 11.4 | 10.1 | 15.1 | 13.2 | 11.9 | 15.1 | 38.6 | 52.0 | 27,152 | 32,039 | 18.0 | 12.0 |
| Southlake | 11.9 | 17.6 | 31.5 | 2.8 | 3.8 | 13.0 | 20.3 | 13.8 | 14.9 | 43.5 | 51.6 | 21,519 | 26,573 | 23.5 | 21.5 |
| Sugar Land | 14.0 | 33.6 | 21.9 | 8.9 | 9.8 | 12.6 | 15.3 | 15.1 | 16.2 | 42.7 | 49.8 | 63,328 | 107,700 | 70.1 | 9.4 |
| Temple | 26.8 | 9.3 | 25.6 | 8.9 | 17.4 | 14.4 | 7.8 | 10.2 | 15.8 | 33.9 | 52.7 | 54,514 | 66,076 | 21.2 | 22.2 |
| Texarkana | 12.3 | 5.4 | 23.4 | 8.6 | 13.8 | 10.8 | 15.7 | 11.7 | 16.1 | 37.7 | 54.1 | 34,782 | 36,306 | 4.4 | -0.2 |
| Texas City | 25.2 | 9.2 | 27.8 | 7.0 | 16.2 | 9.7 | 12.2 | 13.8 | 13.4 | 34.6 | 54.1 | 41,521 | 45,105 | 8.6 | 14.4 |
| The Colony | 23.2 | 12.0 | 23.7 | 3.6 | 19.4 | 12.9 | 14.9 | 15.5 | 9.9 | 36.4 | 47.8 | 26,531 | 36,341 | 37.0 | 25.2 |
| Tyler | 23.4 | 11.6 | 21.7 | 13.2 | 14.8 | 10.8 | 10.9 | 11.8 | 16.8 | 35.2 | 52.4 | 83,650 | 96,761 | 15.7 | 11.8 |
| Victoria | 54.3 | 6.2 | 25.0 | 11.4 | 13.5 | 11.7 | 10.7 | 13.1 | 14.7 | 35.1 | 52.9 | 60,603 | 62,619 | 3.3 | 6.7 |
| Waco | 30.0 | 8.8 | 23.5 | 21.2 | 12.5 | 12.1 | 9.9 | 8.9 | 11.9 | 28.9 | 50.9 | 113,726 | 124,787 | 9.7 | 13.3 |
| Waxahachie | 27.8 | 3.5 | 29.8 | 7.2 | 20.5 | 9.5 | 11.2 | 8.7 | 13.1 | 31.7 | 45.7 | 21,426 | 29,536 | 37.9 | 34.7 |
| Weatherford | 21.3 | 12.6 | 23.2 | 10.9 | 14.7 | 7.5 | 13.0 | 14.3 | 16.3 | 39.4 | 46.8 | 19,000 | 25,744 | 35.5 | 34.8 |
| Weslaco | 87.2 | 11.7 | 31.7 | 8.9 | 14.8 | 10.0 | 8.0 | 12.6 | 14.1 | 31.0 | 50.6 | 26,935 | 36,835 | 36.8 | 17.5 |
| Wichita Falls | 21.6 | 6.6 | 22.0 | 15.6 | 15.3 | 11.0 | 11.1 | 11.1 | 14.0 | 32.9 | 47.7 | 104,197 | 104,682 | 0.5 | 0.7 |
| Wylie | 23.2 | 25.1 | 25.0 | 12.7 | 12.4 | 13.1 | 18.2 | 12.1 | 6.6 | 35.0 | 51.5 | 15,132 | 41,706 | 175.6 | 30.6 |
| **UTAH** | 14.4 | 8.6 | 29.0 | 11.4 | 14.7 | 13.8 | 10.3 | 9.4 | 11.4 | 31.2 | 49.8 | 2,233,169 | 2,763,891 | 23.8 | 17.6 |
| American Fork | 10.4 | 7.6 | 30.3 | 10.8 | 17.8 | 13.1 | 9.2 | 9.6 | 9.3 | 28.5 | 51.7 | 21,941 | 26,553 | 21.0 | 27.7 |
| Bountiful | 4.2 | 6.5 | 30.7 | 6.5 | 14.2 | 10.8 | 10.7 | 12.1 | 15.0 | 33.9 | 53.3 | 41,301 | 42,577 | 3.1 | 3.2 |
| Cedar City | 12.6 | 2.0 | 27.6 | 20.6 | 15.4 | 11.3 | 6.5 | 7.8 | 10.8 | 25.5 | 50.3 | 20,527 | 28,867 | 40.6 | 25.0 |
| Clearfield | 11.6 | 6.6 | 29.6 | 11.7 | 20.1 | 15.5 | 8.1 | 7.5 | 7.5 | 29.8 | 52.0 | 25,974 | 29,890 | 15.1 | 7.6 |
| Cottonwood Heights | 3.7 | 5.3 | 19.8 | 9.6 | 11.6 | 15.6 | 13.4 | 10.2 | 19.7 | 39.1 | 51.1 | 27,569 | 33,586 | 21.8 | -0.8 |
| Draper | 8.6 | 6.3 | 36.6 | 4.6 | 13.9 | 16.5 | 13.0 | 8.8 | 6.6 | 31.6 | 47.5 | 25,220 | 42,272 | 67.6 | 15.8 |
| Holladay | 6.5 | 6.8 | 25.9 | 5.2 | 14.6 | 13.9 | 13.1 | 8.8 | 18.5 | 37.0 | 53.1 | 14,561 | 30,127 | 106.9 | -0.8 |

1. May be of any race.

# Table D. Cities — Households, Group Quarters, Crime, and Education

| City | Households, 2019 | | | | | | | Persons in group quarters, 2019 | Serious crimes known to police[1], 2019 — Violent | | Property | | Educational attainment, 2019 | Attainment[3] (percent) | |
|---|---|---|---|---|---|---|---|---|---|---|---|---|---|---|---|
| | Number | Persons per household | Family | Married couple family | Female family | Non-family | One person | | Number | Rate[2] | Number | Rate[2] | Population age 25 and over | High school graduate or less | Bachelor's degree or more |
| | 27 | 28 | 29 | 30 | 31 | 32 | 33 | 34 | 35 | 36 | 37 | 38 | 39 | 40 | 41 |
| **TEXAS—Cont'd** | | | | | | | | | | | | | | | |
| Irving | 83,345 | 2.86 | 68.2 | 47.9 | 20.3 | 31.8 | 25.4 | 1,414 | 617 | 251 | 5,982 | 2,437 | 152,832 | 39.8 | 40.8 |
| Keller | 15,465 | 3.03 | 83.0 | 71.8 | 11.2 | 17.0 | 15.9 | 286 | 25 | 52 | 349 | 721 | 31,943 | 17.2 | 58.2 |
| Killeen | 57,079 | 2.65 | 63.7 | 41.5 | 22.2 | 36.3 | 31.7 | 148 | 583 | 384 | 3,432 | 2,260 | 90,369 | 33.9 | 19.2 |
| Kingsville | 8,987 | 2.75 | 68.1 | 40.3 | 27.8 | 31.9 | 19.7 | 1,792 | 130 | 512 | 753 | 2,964 | 14,334 | 46.4 | 28.1 |
| Kyle | 15,883 | 3.01 | 69.5 | 49.4 | 20.1 | 30.5 | 22.5 | 518 | 68 | 136 | 659 | 1,322 | 30,532 | 41.8 | 30.7 |
| Lake Jackson | 10,602 | 2.56 | 69.5 | 51.0 | 18.4 | 30.5 | 25.4 | 67 | 57 | 206 | 703 | 2,545 | 18,235 | 37.6 | 31.1 |
| Lancaster | 13,309 | 2.91 | 68.1 | 38.9 | 29.2 | 31.9 | 29.3 | 482 | 147 | 369 | 1,014 | 2,548 | 26,160 | 40.0 | 23.7 |
| La Porte | 12,363 | 2.82 | 69.9 | 54.9 | 15.0 | 30.1 | 23.3 | 89 | 83 | 233 | 520 | 1,460 | 24,392 | 50.1 | 14.0 |
| Laredo | 73,066 | 3.57 | 82.5 | 54.0 | 28.6 | 17.5 | 15.4 | 3,380 | 836 | 316 | 4,692 | 1,771 | 148,897 | 52.6 | 20.9 |
| League City | 38,910 | 2.73 | 74.9 | 64.0 | 10.8 | 25.1 | 20.8 | 423 | 121 | 111 | 1,487 | 1,359 | 72,001 | 20.9 | 46.5 |
| Leander | 20,548 | 3.34 | 83.2 | 71.9 | 11.3 | 16.8 | 12.7 | 10 | 67 | 109 | 520 | 848 | 41,901 | 26.1 | 45.8 |
| Lewisville | 41,101 | 2.63 | 59.1 | 40.3 | 18.7 | 40.9 | 33.5 | 416 | 239 | 221 | 2,125 | 1,968 | 75,150 | 35.5 | 35.2 |
| Little Elm | 14,459 | 3.67 | 83.7 | 58.9 | 24.8 | 16.3 | 15.7 | 0 | NA | NA | NA | NA | 29,599 | 43.0 | 32.5 |
| Longview | 29,848 | 2.63 | 62.8 | 43.9 | 18.8 | 37.2 | 27.9 | 4,049 | 374 | 457 | 2,410 | 2,947 | 53,566 | 41.5 | 24.0 |
| Lubbock | 99,104 | 2.49 | 57.9 | 40.0 | 17.9 | 42.1 | 31.8 | 11,619 | 2,613 | 1,008 | 11,940 | 4,606 | 151,294 | 36.4 | 34.0 |
| Lufkin | 12,965 | 2.61 | 71.8 | 48.6 | 23.2 | 28.2 | 25.5 | 1,121 | 184 | 518 | 1,558 | 4,382 | 22,834 | 45.1 | 24.7 |
| McAllen | 47,094 | 3.02 | 74.0 | 47.6 | 26.4 | 26.0 | 21.8 | 1,026 | 140 | 97 | 3,595 | 2,481 | 91,947 | 44.0 | 30.5 |
| McKinney | 68,458 | 2.88 | 75.2 | 59.8 | 15.4 | 24.8 | 19.5 | 1,966 | 287 | 143 | 1,993 | 993 | 127,996 | 25.1 | 47.9 |
| Mansfield | 22,943 | 3.25 | 78.8 | 63.6 | 15.1 | 21.2 | 17.7 | 363 | 65 | 89 | 766 | 1,050 | 46,106 | 21.1 | 46.7 |
| Mesquite | 49,549 | 2.83 | 69.5 | 46.3 | 23.2 | 30.5 | 26.4 | 616 | 685 | 479 | 5,197 | 3,632 | 92,928 | 50.5 | 15.4 |
| Midland | 47,961 | 3.01 | 68.4 | 49.3 | 19.1 | 31.6 | 27.3 | 1,555 | NA | NA | NA | NA | 91,788 | 36.4 | 32.4 |
| Mission | 24,709 | 3.40 | 79.3 | 51.8 | 27.5 | 20.7 | 18.9 | 209 | 105 | 123 | 1,618 | 1,888 | 49,704 | 52.8 | 25.4 |
| Missouri City | 25,225 | 2.98 | 81.1 | 61.8 | 19.3 | 18.9 | 16.4 | 388 | 107 | 141 | 811 | 1,071 | 53,214 | 29.6 | 43.2 |
| Nacogdoches | 12,307 | 2.24 | 50.1 | 27.1 | 23.0 | 49.9 | 36.1 | 5,251 | 111 | 330 | 864 | 2,570 | 17,096 | 40.4 | 29.0 |
| New Braunfels | 33,398 | 2.69 | 72.3 | 54.2 | 18.1 | 27.7 | 23.2 | 947 | 219 | 247 | 1,120 | 1,263 | 60,160 | 35.4 | 37.6 |
| North Richland Hills | 27,070 | 2.60 | 71.9 | 56.3 | 15.5 | 28.1 | 24.1 | 370 | 143 | 199 | 1,279 | 1,781 | 50,099 | 29.1 | 35.0 |
| Odessa | 42,465 | 2.87 | 65.0 | 45.7 | 19.3 | 35.0 | 27.4 | 1,493 | 1,282 | 1,038 | 3,624 | 2,935 | 75,960 | 50.2 | 17.2 |
| Paris | 12,057 | 2.00 | 56.5 | 28.8 | 27.6 | 43.5 | 37.6 | 690 | 252 | 1,017 | 714 | 2,881 | 17,363 | 48.5 | 18.5 |
| Pasadena | 47,706 | 3.15 | 77.0 | 51.3 | 25.7 | 23.0 | 19.6 | 1,023 | 839 | 546 | 3,861 | 2,512 | 91,838 | 60.8 | 15.9 |
| Pearland | 47,106 | 2.78 | 72.7 | 64.0 | 8.7 | 27.3 | 22.9 | 564 | 113 | 90 | 2,071 | 1,641 | 87,811 | 24.7 | 46.6 |
| Pflugerville | 21,971 | 2.95 | 76.4 | 61.7 | 14.7 | 23.6 | 18.7 | 274 | 64 | 96 | 829 | 1,242 | 44,762 | 35.1 | 39.0 |
| Pharr | 23,126 | 3.42 | 85.2 | 51.0 | 34.2 | 14.8 | 13.7 | 47 | 235 | 290 | 1,423 | 1,759 | 44,092 | 64.7 | 13.1 |
| Plano | 107,721 | 2.65 | 70.0 | 56.5 | 13.5 | 30.0 | 23.9 | 768 | 431 | 148 | 4,908 | 1,683 | 202,746 | 20.1 | 57.3 |
| Port Arthur | 18,970 | 2.83 | 68.2 | 41.0 | 27.2 | 31.8 | 30.0 | 657 | 344 | 625 | 1,203 | 2,184 | 32,487 | 63.9 | 9.9 |
| Richardson | 43,322 | 2.73 | 68.9 | 54.1 | 14.8 | 31.1 | 23.1 | 3,219 | 178 | 144 | 2,279 | 1,839 | 79,920 | 19.1 | 54.6 |
| Rockwall | 16,876 | 2.70 | 78.1 | 58.7 | 19.5 | 21.9 | 19.1 | 282 | 50 | 108 | 740 | 1,605 | 30,492 | 24.0 | 41.9 |
| Rosenberg | 13,819 | 2.76 | 63.8 | 36.9 | 26.9 | 36.2 | 32.3 | 213 | 148 | 380 | 684 | 1,757 | 24,919 | 51.4 | 17.1 |
| Round Rock | 42,580 | 3.12 | 70.8 | 54.2 | 16.6 | 29.2 | 23.2 | 417 | 165 | 124 | 2,235 | 1,684 | 86,070 | 28.1 | 37.1 |
| Rowlett | 21,953 | 3.03 | 78.5 | 67.4 | 11.0 | 21.5 | 16.9 | 347 | NA | NA | NA | NA | 46,129 | 26.7 | 35.7 |
| San Angelo | 36,990 | 2.49 | 63.6 | 46.1 | 17.5 | 36.4 | 31.1 | 6,201 | 357 | 353 | 3,173 | 3,139 | 63,153 | 39.9 | 25.6 |
| San Antonio | 512,273 | 2.98 | 62.4 | 39.4 | 23.0 | 37.6 | 30.6 | 21,314 | 11,046 | 708 | 67,422 | 4,324 | 1,010,665 | 43.6 | 26.0 |
| San Juan | 8,788 | 4.20 | 89.6 | 52.7 | 37.0 | 10.4 | 10.4 | 138 | 173 | 461 | 859 | 2,288 | 20,050 | 63.9 | 12.7 |
| San Marcos | 24,286 | 2.40 | 43.3 | 28.9 | 14.5 | 56.7 | 26.7 | 6,554 | 241 | 364 | 1,530 | 2,308 | 32,601 | 34.0 | 37.8 |
| Schertz | 14,798 | 3.07 | 83.7 | 65.8 | 17.9 | 16.3 | 13.4 | 273 | 78 | 184 | 559 | 1,320 | 29,243 | 27.1 | 34.6 |
| Seguin | 11,130 | 2.57 | 63.6 | 43.6 | 20.1 | 36.4 | 28.0 | 1,400 | 84 | 278 | 714 | 2,361 | 20,593 | 62.2 | 20.6 |
| Sherman | 15,756 | 2.70 | 59.7 | 45.5 | 14.2 | 40.3 | 36.2 | 1,536 | 179 | 416 | 1,122 | 2,609 | 29,404 | 43.6 | 18.2 |
| Socorro | 10,643 | 3.23 | 79.9 | 57.0 | 22.9 | 20.1 | 19.8 | 2 | 43 | 123 | 299 | 858 | 22,492 | 63.3 | 8.9 |
| Southlake | 10,346 | 3.15 | 92.4 | 85.6 | 6.7 | 7.6 | 6.5 | 73 | 10 | 30 | 387 | 1,171 | 21,435 | 9.6 | 74.6 |
| Sugar Land | 40,077 | 2.95 | 79.8 | 70.4 | 9.4 | 20.2 | 16.7 | 356 | NA | NA | NA | NA | 81,945 | 20.6 | 58.3 |
| Temple | 28,819 | 2.66 | 62.7 | 41.6 | 21.2 | 37.3 | 29.6 | 1,724 | 217 | 280 | 1,718 | 2,215 | 51,383 | 37.7 | 28.5 |
| Texarkana | 13,650 | 2.52 | 61.8 | 36.0 | 25.8 | 38.2 | 31.6 | 1,892 | 157 | 420 | 1,710 | 4,572 | 24,691 | 43.8 | 16.1 |
| Texas City | 18,651 | 2.59 | 68.7 | 36.4 | 32.3 | 31.3 | 22.2 | 992 | 199 | 401 | 1,400 | 2,819 | 32,660 | 47.7 | 14.7 |
| The Colony | 16,319 | 2.72 | 65.6 | 50.6 | 15.0 | 34.4 | 30.3 | 0 | 122 | 275 | 641 | 1,445 | 32,266 | 27.2 | 41.0 |
| Tyler | 37,504 | 2.71 | 64.5 | 43.7 | 20.8 | 35.5 | 29.2 | 5,288 | 401 | 375 | 3,206 | 3,000 | 69,667 | 31.0 | 32.7 |
| Victoria | 21,806 | 3.02 | 67.4 | 43.7 | 23.8 | 32.6 | 26.6 | 1,145 | 347 | 513 | 1,974 | 2,921 | 42,560 | 47.3 | 19.9 |
| Waco | 50,523 | 2.59 | 57.3 | 37.3 | 20.1 | 42.7 | 34.2 | 8,407 | 799 | 571 | 4,599 | 3,288 | 77,058 | 43.3 | 26.1 |
| Waxahachie | 13,435 | 2.73 | 64.7 | 45.5 | 19.3 | 35.3 | 31.8 | 1,254 | NA | NA | NA | NA | 23,929 | 38.8 | 22.5 |
| Weatherford | 10,879 | 2.96 | 70.2 | 47.8 | 22.3 | 29.8 | 26.4 | 1,342 | 50 | 153 | 482 | 1,476 | 22,102 | 42.8 | 20.7 |
| Weslaco | 12,707 | 3.24 | 82.2 | 51.4 | 30.9 | 17.8 | 16.9 | 454 | 155 | 371 | 1,528 | 3,662 | 24,739 | 46.7 | 21.4 |
| Wichita Falls | 37,332 | 2.46 | 57.3 | 40.7 | 16.7 | 42.7 | 36.5 | 12,948 | 364 | 348 | 3,183 | 3,044 | 65,387 | 48.4 | 22.2 |
| Wylie | 17,086 | 3.07 | 83.0 | 63.2 | 19.8 | 17.0 | 13.9 | 122 | 37 | 70 | 381 | 720 | 32,794 | 32.7 | 34.5 |
| **UTAH** | 1,023,855 | 3.08 | 74.3 | 60.5 | 13.8 | 25.7 | 19.2 | 48,364 | NA | NA | NA | NA | 1,911,592 | 30.1 | 34.8 |
| American Fork | 10,362 | 3.14 | 83.2 | 73.4 | 9.8 | 16.8 | 15.3 | 572 | 18 | 41 | 817 | 1,873 | 19,542 | 20.5 | 41.3 |
| Bountiful | 14,307 | 3.04 | 77.4 | 63.2 | 14.2 | 22.6 | 21.0 | 423 | 36 | 81 | 447 | 1,009 | 27,589 | 19.2 | 43.0 |
| Cedar City | 11,341 | 2.79 | 67.3 | 55.9 | 11.3 | 32.7 | 23.6 | 603 | 93 | 277 | 632 | 1,880 | 16,678 | 21.2 | 37.8 |
| Clearfield | 11,560 | 2.75 | 69.7 | 50.1 | 19.6 | 30.3 | 25.2 | 352 | 56 | 174 | 369 | 1,145 | 18,856 | 35.6 | 21.5 |
| Cottonwood Heights | 12,946 | 2.61 | 62.9 | 50.9 | 12.0 | 37.1 | 26.4 | 49 | 54 | 158 | 806 | 2,358 | 23,901 | 17.8 | 52.3 |
| Draper | 13,147 | 3.53 | 83.6 | 69.8 | 13.8 | 16.4 | 12.2 | 2,934 | 69 | 140 | 906 | 1,845 | 29,011 | 17.4 | 50.2 |
| Holladay | 10,086 | 2.99 | 75.9 | 60.2 | 15.7 | 24.1 | 20.0 | 174 | NA | NA | NA | NA | 20,907 | 13.3 | 54.0 |

1. Data for serious crimes have not been adjusted for underreporting. This may affect comparability between geographic areas and over time.　2. Per 100,000 population estimated by the FBI.　3. Persons 25 years old and over.

| City | Median income (42) | Percent with income less than $20,000 (43) | Percent with income of $200,000 or more (44) | Median family income (45) | Median non-family income (46) | All persons (47) | Men (48) | Women (49) | Total (50) | Occupied (51) | Percent owner occupied (52) | Median value[1] (dollars) (53) | Median gross rent (dollars) (54) |
|---|---|---|---|---|---|---|---|---|---|---|---|---|---|
| **TEXAS—Cont'd** | | | | | | | | | | | | | |
| Irving | 67,263 | 9.0 | 6.8 | 72,951 | 55,643 | 35,403 | 40,372 | 27,321 | 90,335 | 83,345 | 36.0 | 221,749 | 1,225 |
| Keller | 158,770 | 2.3 | 34.3 | 169,052 | 63,651 | 62,149 | 90,023 | 45,540 | 15,648 | 15,465 | 82.2 | 435,720 | 1,566 |
| Killeen | 48,622 | 17.8 | 2.1 | 60,199 | 31,747 | 30,438 | 37,141 | 25,096 | 64,280 | 57,079 | 45.0 | 146,242 | 884 |
| Kingsville | 46,666 | 21.8 | 0.0 | 70,214 | 22,097 | 23,101 | 21,741 | 25,881 | 11,307 | 8,987 | 57.2 | 113,629 | 908 |
| Kyle | 81,297 | 3.1 | 2.8 | 87,843 | 54,483 | 39,131 | 40,811 | 36,457 | 16,782 | 15,883 | 61.5 | 223,442 | 1,496 |
| Lake Jackson | 71,832 | 7.6 | 10.6 | 78,036 | 53,971 | 38,857 | 55,121 | 31,360 | 11,583 | 10,602 | 59.8 | 226,167 | 1,227 |
| Lancaster | 59,983 | 11.1 | 2.7 | 66,419 | 48,217 | 37,875 | 39,943 | 37,498 | 14,491 | 13,309 | 58.8 | 195,698 | 1,284 |
| La Porte | 76,111 | 9.0 | 7.0 | 81,781 | 47,009 | 42,201 | 50,377 | 28,813 | 13,117 | 12,363 | 78.4 | 177,072 | 1,204 |
| Laredo | 57,468 | 17.3 | 4.7 | 62,700 | 24,556 | 27,538 | 35,920 | 21,128 | 80,365 | 73,066 | 60.0 | 148,091 | 860 |
| League City | 115,650 | 2.6 | 16.6 | 130,293 | 58,185 | 58,791 | 67,336 | 50,528 | 40,565 | 38,910 | 76.4 | 267,529 | 1,410 |
| Leander | 117,892 | 1.8 | 15.9 | 129,206 | 56,025 | 48,446 | 60,498 | 35,102 | 20,655 | 20,548 | 77.2 | 332,329 | 1,525 |
| Lewisville | 65,836 | 5.7 | 3.1 | 76,735 | 55,679 | 38,514 | 45,723 | 35,106 | 45,265 | 41,101 | 42.6 | 244,694 | 1,246 |
| Little Elm | 84,484 | 1.1 | 5.1 | 84,116 | 71,814 | 43,065 | 45,197 | 34,017 | 14,709 | 14,459 | 75.1 | 276,197 | 1,829 |
| Longview | 53,987 | 15.5 | 5.0 | 61,603 | 32,228 | 30,894 | 38,277 | 21,787 | 33,728 | 29,848 | 57.1 | 147,189 | 839 |
| Lubbock | 52,254 | 17.3 | 4.2 | 66,551 | 36,453 | 27,888 | 32,204 | 22,663 | 110,644 | 99,104 | 51.5 | 152,755 | 976 |
| Lufkin | 53,681 | 13.4 | 4.6 | 58,733 | 22,786 | 27,063 | 38,087 | 22,099 | 15,402 | 12,965 | 58.8 | 119,712 | 807 |
| McAllen | 46,319 | 20.6 | 5.7 | 56,116 | 30,513 | 27,287 | 30,415 | 24,725 | 53,749 | 47,094 | 60.3 | 129,395 | 786 |
| McKinney | 89,828 | 6.7 | 12.0 | 105,420 | 46,188 | 50,383 | 60,522 | 37,304 | 73,411 | 68,458 | 61.7 | 353,663 | 1,468 |
| Mansfield | 105,182 | 6.7 | 12.5 | 110,927 | 57,832 | 49,114 | 56,525 | 40,191 | 23,848 | 22,943 | 76.2 | 309,808 | 1,379 |
| Mesquite | 61,047 | 10.7 | 1.1 | 71,786 | 39,682 | 32,628 | 37,464 | 27,556 | 54,256 | 49,549 | 66.0 | 172,613 | 1,121 |
| Midland | 84,955 | 9.1 | 13.5 | 106,919 | 49,601 | 40,123 | 52,807 | 30,584 | 50,942 | 47,961 | 67.5 | 270,246 | 1,428 |
| Mission | 54,793 | 18.1 | 5.2 | 60,861 | 29,889 | 28,298 | 39,187 | 19,032 | 28,684 | 24,709 | 72.2 | 131,264 | 760 |
| Missouri City | 86,486 | 2.1 | 11.1 | 90,729 | 0 | 42,118 | 46,557 | 40,059 | 27,005 | 25,225 | 86.7 | 220,882 | 1,706 |
| Nacogdoches | 35,809 | 27.4 | 2.5 | 59,767 | 20,601 | 20,882 | 22,181 | 18,167 | 13,848 | 12,307 | 35.1 | 140,728 | 687 |
| New Braunfels | 81,131 | 8.3 | 7.3 | 93,768 | 41,260 | 40,016 | 48,807 | 29,567 | 36,757 | 33,398 | 63.8 | 242,737 | 1,188 |
| North Richland Hills | 75,830 | 8.0 | 9.9 | 89,265 | 47,694 | 42,662 | 50,664 | 35,123 | 27,744 | 27,070 | 66.4 | 252,054 | 1,244 |
| Odessa | 66,316 | 12.1 | 6.1 | 83,161 | 41,020 | 37,096 | 47,661 | 25,972 | 47,765 | 42,465 | 59.5 | 163,214 | 1,176 |
| Paris | 35,444 | 33.6 | 0.6 | 45,449 | 20,958 | 23,994 | 30,011 | 21,552 | 13,095 | 12,057 | 46.2 | 73,537 | 711 |
| Pasadena | 53,819 | 11.7 | 3.6 | 58,269 | 36,228 | 32,020 | 40,447 | 22,575 | 53,265 | 47,706 | 52.6 | 150,707 | 1,016 |
| Pearland | 106,757 | 3.8 | 10.0 | 127,687 | 70,501 | 61,755 | 73,699 | 52,002 | 47,564 | 47,106 | 80.0 | 269,494 | 1,474 |
| Pflugerville | 90,329 | 6.4 | 7.4 | 106,406 | 54,877 | 46,382 | 55,261 | 41,767 | 22,450 | 21,971 | 73.1 | 261,045 | 1,409 |
| Pharr | 36,342 | 35.0 | 0.5 | 36,241 | 21,227 | 21,880 | 30,718 | 17,888 | 26,385 | 23,126 | 57.3 | 81,971 | 740 |
| Plano | 93,321 | 7.3 | 16.0 | 112,709 | 55,693 | 51,676 | 65,038 | 41,508 | 116,860 | 107,721 | 55.8 | 360,983 | 1,453 |
| Port Arthur | 40,635 | 28.5 | 3.1 | 50,867 | 21,203 | 30,951 | 42,360 | 21,147 | 24,393 | 18,970 | 58.3 | 67,450 | 885 |
| Richardson | 91,118 | 9.6 | 12.7 | 106,522 | 43,426 | 42,555 | 50,645 | 37,924 | 46,978 | 43,322 | 57.4 | 330,684 | 1,531 |
| Rockwall | 86,519 | 6.6 | 14.8 | 100,944 | 65,659 | 42,543 | 68,561 | 42,026 | 17,404 | 16,876 | 77.1 | 284,478 | 1,225 |
| Rosenberg | 48,999 | 15.1 | 0.0 | 68,918 | 32,358 | 29,902 | 32,395 | 23,605 | 14,811 | 13,819 | 42.6 | 184,805 | 1,075 |
| Round Rock | 86,145 | 6.1 | 9.3 | 98,377 | 52,777 | 37,352 | 45,276 | 32,901 | 46,697 | 42,580 | 56.8 | 288,746 | 1,360 |
| Rowlett | 100,465 | 5.4 | 10.2 | 111,991 | 64,399 | 45,059 | 53,912 | 36,860 | 22,389 | 21,953 | 84.4 | 264,026 | 1,473 |
| San Angelo | 54,972 | 14.5 | 3.5 | 77,321 | 29,224 | 30,913 | 35,098 | 27,302 | 41,673 | 36,990 | 64.1 | 149,547 | 969 |
| San Antonio | 53,751 | 16.4 | 4.1 | 63,605 | 38,848 | 30,874 | 33,358 | 26,925 | 561,467 | 512,273 | 53.5 | 171,104 | 1,029 |
| San Juan | 46,237 | 23.9 | 2.3 | 49,531 | 23,412 | 21,002 | 24,821 | 15,048 | 9,261 | 8,788 | 83.2 | 101,978 | 850 |
| San Marcos | 45,155 | 24.9 | 4.8 | 55,782 | 31,325 | 20,695 | 23,105 | 16,497 | 27,187 | 24,286 | 32.9 | 229,542 | 1,097 |
| Schertz | 94,423 | 5.0 | 5.9 | 97,900 | 50,079 | 41,918 | 46,514 | 36,838 | 15,238 | 14,798 | 76.8 | 238,600 | 1,296 |
| Seguin | 49,321 | 10.4 | 0.0 | 60,457 | 42,054 | 25,026 | 25,163 | 24,777 | 12,301 | 11,130 | 70.7 | 178,986 | 899 |
| Sherman | 57,354 | 13.2 | 3.7 | 64,129 | 34,717 | 25,955 | 32,496 | 21,627 | 17,641 | 15,756 | 55.7 | 144,852 | 899 |
| Socorro | 40,709 | 23.6 | 0.0 | 46,365 | 24,030 | 26,762 | 33,230 | 23,006 | 11,062 | 10,643 | 72.1 | 93,283 | 812 |
| Southlake | 215,776 | 6.6 | 56.2 | 241,791 | 78,208 | 126,110 | 180,522 | 25,228 | 10,588 | 10,346 | 94.8 | 746,775 | 0 |
| Sugar Land | 127,598 | 4.3 | 26.2 | 153,485 | 55,096 | 56,864 | 84,273 | 38,446 | 41,472 | 40,077 | 77.4 | 345,059 | 1,777 |
| Temple | 55,792 | 14.6 | 5.6 | 67,236 | 36,868 | 33,872 | 38,469 | 30,194 | 31,200 | 28,819 | 53.3 | 148,206 | 912 |
| Texarkana | 42,515 | 26.1 | 5.5 | 61,250 | 30,774 | 27,773 | 30,345 | 25,544 | 16,039 | 13,650 | 45.9 | 119,903 | 790 |
| Texas City | 48,405 | 16.7 | 3.0 | 48,085 | 46,304 | 31,486 | 35,597 | 29,496 | 20,497 | 18,971 | 47.7 | 134,582 | 975 |
| The Colony | 90,163 | 5.0 | 18.5 | 110,009 | 65,176 | 51,600 | 63,016 | 50,697 | 18,504 | 16,319 | 70.3 | 272,514 | 1,447 |
| Tyler | 52,294 | 15.2 | 4.3 | 65,926 | 31,530 | 30,434 | 36,617 | 24,088 | 43,733 | 37,504 | 51.7 | 164,671 | 1,011 |
| Victoria | 55,968 | 16.3 | 4.6 | 57,173 | 42,405 | 29,675 | 37,251 | 24,213 | 27,414 | 21,806 | 61.8 | 140,250 | 1,005 |
| Waco | 41,481 | 27.2 | 4.2 | 55,310 | 24,274 | 24,662 | 27,376 | 21,021 | 56,686 | 50,523 | 46.9 | 145,539 | 879 |
| Waxahachie | 70,518 | 9.3 | 3.8 | 83,292 | 39,484 | 40,166 | 41,047 | 34,788 | 13,991 | 13,435 | 64.2 | 224,887 | 1,171 |
| Weatherford | 70,288 | 17.5 | 4.1 | 73,685 | 24,156 | 30,691 | 32,374 | 29,167 | 12,502 | 10,879 | 72.1 | 178,602 | 955 |
| Weslaco | 47,511 | 21.6 | 1.1 | 53,291 | 25,401 | 26,332 | 31,197 | 23,065 | 16,731 | 12,707 | 60.7 | 88,414 | 691 |
| Wichita Falls | 48,861 | 19.4 | 2.8 | 67,849 | 26,801 | 24,288 | 26,268 | 21,281 | 44,104 | 37,332 | 56.2 | 111,151 | 770 |
| Wylie | 83,257 | 2.1 | 8.1 | 86,938 | 59,689 | 36,630 | 46,553 | 30,834 | 17,349 | 17,086 | 80.0 | 252,479 | 1,498 |
| **UTAH** | 75,780 | 9.2 | 7.3 | 86,152 | 42,659 | 33,396 | 44,749 | 25,596 | 1,133,543 | 1,023,855 | 70.6 | 330,294 | 1,098 |
| American Fork | 80,308 | 6.2 | 6.5 | 88,320 | 38,505 | 31,597 | 40,065 | 21,406 | 10,486 | 10,362 | 64.1 | 351,386 | 1,333 |
| Bountiful | 90,796 | 3.1 | 14.0 | 100,197 | 50,783 | 37,996 | 52,305 | 30,769 | 14,755 | 14,307 | 74.9 | 371,081 | 1,152 |
| Cedar City | 59,439 | 14.3 | 4.1 | 78,160 | 26,728 | 28,343 | 38,128 | 20,148 | 12,147 | 11,341 | 48.6 | 248,150 | 919 |
| Clearfield | 66,921 | 11.2 | 0.0 | 72,721 | 49,985 | 32,277 | 41,123 | 25,864 | 11,926 | 11,560 | 56.8 | 244,895 | 1,091 |
| Cottonwood Heights | 93,571 | 7.6 | 13.0 | 112,744 | 65,636 | 42,216 | 52,229 | 36,690 | 13,725 | 12,946 | 69.5 | 460,844 | 1,425 |
| Draper | 122,165 | 6.4 | 26.6 | 126,975 | 60,750 | 45,845 | 68,333 | 27,231 | 13,944 | 13,147 | 81.7 | 576,955 | 1,494 |
| Holladay | 97,308 | 6.2 | 21.0 | 128,452 | 53,125 | 39,806 | 51,960 | 31,370 | 11,457 | 10,086 | 79.6 | 535,389 | 1,423 |

1. Based on population estimated by the American Community Survey.

| City | Commuting[1], 2019 Percent — Drove alone | With commutes of 30 minutes or more | Computer access[2], 2019 Percent — With a computer in the house | With Internet access | Migration, 2019 — Percent who lived in the same house one year ago | Percent who lived in another state or county one year ago | Civilian labor force, 2020 — Total | Percent change 2019-2020 | Unemployment Total | Rate[3] | Civilian Employment[4], 2019 — Population age 16 and older Number | Percent in labor force | Population age 16 to 64 Number | Percent who worked full-year full-time |
|---|---|---|---|---|---|---|---|---|---|---|---|---|---|---|
| | 55 | 56 | 57 | 58 | 59 | 60 | 61 | 62 | 63 | 64 | 65 | 66 | 67 | 68 |
| **TEXAS—Cont'd** | | | | | | | | | | | | | | |
| Irving | 82.1 | 32.4 | 97.8 | 88.2 | 81.1 | 8.2 | 131,722 | -1.4 | 9,992 | 8 | 183,419 | 73.1 | 165,071 | 59.6 |
| Keller | 82.7 | 56.9 | NA | NA | 87.8 | 4.4 | 23,970 | -2.7 | 1,277 | 5 | 35,901 | 70.5 | 30,220 | 56.8 |
| Killeen | 80.7 | 24.0 | 93.1 | 84.9 | 74.5 | 10.5 | 57,359 | 1.1 | 4,819 | 8 | 112,945 | 59.3 | 102,822 | 49.7 |
| Kingsville | 80.0 | 25.8 | 84.5 | 78.9 | 77.8 | 13.3 | 10,975 | 0.3 | 940 | 9 | 20,578 | 61.6 | 17,490 | 42.6 |
| Kyle | 84.5 | 55.7 | NA | NA | 89.5 | 8.1 | 25,783 | 1.2 | 1,666 | 7 | 35,226 | 72.2 | 30,838 | 64.0 |
| Lake Jackson | 77.9 | 23.8 | 99.2 | 87.5 | 84.9 | 7.7 | 13,461 | -2.9 | 1,304 | 10 | 21,044 | 69.7 | 17,227 | 51.2 |
| Lancaster | 83.3 | 58.8 | 91.7 | 76.6 | NA | NA | 19,064 | -0.3 | 1,974 | 10 | 30,549 | 69.2 | 26,949 | 53.0 |
| La Porte | 86.8 | 39.0 | 96.0 | 93.4 | 91.2 | 2.6 | 18,043 | -2.6 | 1,719 | 10 | 27,861 | 70.9 | 23,373 | 58.5 |
| Laredo | 81.0 | 20.1 | 93.4 | 79.6 | 85.6 | 3.4 | 111,387 | -2.0 | 9,404 | 8 | 188,202 | 62.6 | 162,224 | 50.3 |
| League City | 86.0 | 49.2 | 98.3 | 95.6 | 86.2 | 9.1 | 56,843 | -1.8 | 3,833 | 7 | 82,933 | 70.4 | 70,122 | 66.5 |
| Leander | 70.7 | 55.0 | NA | NA | 79.7 | 7.8 | 32,632 | 8.3 | 1,817 | 6 | 48,951 | 76.1 | 43,780 | 59.1 |
| Lewisville | 81.9 | 39.1 | 98.4 | 95.4 | 81.2 | 10.2 | 67,112 | 1.9 | 4,900 | 7 | 88,014 | 75.0 | 77,959 | 61.4 |
| Little Elm | 90.8 | 54.5 | NA | NA | 91.5 | 6.4 | 27,443 | 4.1 | 1,793 | 7 | 35,643 | 78.6 | 32,427 | 62.8 |
| Longview | 87.1 | 14.4 | 92.3 | 86.9 | 79.0 | 11.0 | 37,008 | -2.9 | 3,179 | 9 | 63,970 | 61.5 | 51,936 | 55.1 |
| Lubbock | 79.1 | 10.9 | 94.4 | 82.6 | 72.1 | 13.0 | 132,224 | -0.7 | 7,766 | 6 | 204,448 | 65.2 | 174,656 | 48.9 |
| Lufkin | NA | 12.2 | 91.6 | 88.6 | 85.6 | 5.1 | 14,609 | -1.9 | 1,114 | 8 | 27,581 | 64.9 | 21,823 | 59.1 |
| McAllen | 75.9 | 19.2 | 91.8 | 84.6 | 86.2 | 2.6 | 67,442 | 0.8 | 6,103 | 9 | 108,494 | 64.1 | 86,398 | 52.9 |
| McKinney | 81.0 | 42.9 | 96.9 | 89.3 | 80.8 | 8.0 | 104,935 | 2.4 | 6,884 | 7 | 148,541 | 70.0 | 125,895 | 62.0 |
| Mansfield | 79.1 | 49.5 | 96.8 | 93.5 | 86.1 | 8.0 | 38,333 | 0.2 | 2,427 | 6 | 56,233 | 67.3 | 47,948 | 56.2 |
| Mesquite | 83.7 | 56.1 | 95.1 | 87.5 | 88.2 | 2.1 | 74,563 | -1.9 | 5,758 | 8 | 110,747 | 67.3 | 88,938 | 58.9 |
| Midland | 84.4 | 17.7 | 96.5 | 83.5 | 80.9 | 9.6 | 80,745 | -8.8 | 6,261 | 8 | 108,749 | 71.0 | 93,124 | 60.6 |
| Mission | 82.2 | 15.3 | 96.2 | 88.2 | 86.4 | 2.6 | 35,486 | 1.5 | 4,067 | 12 | 61,646 | 61.2 | 52,143 | 43.4 |
| Missouri City | 80.4 | 57.3 | 96.8 | 94.6 | 86.7 | 7.3 | 39,187 | -0.7 | 3,462 | 9 | 62,552 | 65.3 | 50,088 | 52.9 |
| Nacogdoches | 77.1 | 18.5 | 91.1 | 81.0 | 71.5 | 10.9 | 14,063 | -2.1 | 956 | 7 | 26,536 | 64.3 | 22,545 | 42.5 |
| New Braunfels | 81.1 | 34.7 | 94.9 | 91.4 | 85.9 | 9.0 | 44,954 | 3.8 | 2,862 | 6 | 69,827 | 68.4 | 57,024 | 59.1 |
| North Richland Hills | 81.7 | 47.2 | 97.6 | 95.3 | 85.5 | 4.8 | 39,525 | -1.5 | 2,614 | 7 | 56,342 | 66.9 | 44,667 | 58.2 |
| Odessa | 82.5 | 23.1 | 91.6 | 79.5 | 83.2 | 10.5 | 62,629 | -7.4 | 6,139 | 10 | 91,886 | 66.6 | 79,124 | 57.2 |
| Paris | 79.7 | 19.3 | 85.3 | 63.7 | 77.0 | 10.1 | 11,332 | -2.0 | 820 | 7 | 20,541 | 57.7 | 15,546 | 45.3 |
| Pasadena | 80.7 | 38.3 | 93.6 | 90.2 | 79.7 | 4.1 | 66,098 | -1.8 | 7,134 | 11 | 111,022 | 66.3 | 94,239 | 52.6 |
| Pearland | 86.8 | 63.8 | 99.3 | 96.5 | 89.9 | 6.9 | 63,842 | -2.9 | 4,019 | 6 | 98,472 | 70.7 | 84,502 | 68.6 |
| Pflugerville | 75.2 | 45.1 | NA | NA | 85.7 | 6.8 | 37,464 | 0.2 | 2,748 | 7 | 52,456 | 73.0 | 46,043 | 59.4 |
| Pharr | 82.6 | 26.1 | 87.3 | 40.2 | 95.5 | 1.3 | 32,050 | 1.7 | 4,054 | 13 | 55,266 | 51.9 | 46,182 | 40.2 |
| Plano | 79.9 | 40.9 | 98.2 | 95.0 | 86.8 | 7.9 | 161,061 | -1.8 | 10,121 | 6 | 232,799 | 69.3 | 192,779 | 61.2 |
| Port Arthur | 91.6 | 28.3 | 88.6 | 77.9 | 89.6 | 4.7 | 22,024 | 0.5 | 3,458 | 16 | 39,719 | 52.5 | 33,036 | 45.2 |
| Richardson | 77.7 | 42.6 | 99.2 | 96.6 | 86.4 | 5.9 | 65,929 | -1.4 | 4,107 | 6 | 97,017 | 68.3 | 79,663 | 56.8 |
| Rockwall | 79.4 | 56.7 | 98.6 | 97.0 | 88.8 | 7.4 | 23,479 | 0.0 | 1,478 | 6 | 35,584 | 72.3 | 29,803 | 61.8 |
| Rosenberg | 79.1 | 54.0 | 81.5 | 76.2 | 86.6 | 4.1 | 17,884 | -1.1 | 1,510 | 8 | 30,163 | 66.7 | 24,361 | 56.1 |
| Round Rock | 75.5 | 36.5 | 99.1 | 97.2 | 79.8 | 12.1 | 72,388 | 0.5 | 3,882 | 5 | 102,002 | 76.3 | 91,204 | 56.8 |
| Rowlett | 82.3 | 58.5 | NA | NA | 87.4 | 7.4 | 37,449 | -0.2 | 2,407 | 6 | 53,678 | 72.1 | 45,315 | 58.9 |
| San Angelo | 77.1 | 10.1 | 93.1 | 86.7 | 83.3 | 7.4 | 45,006 | -1.5 | 2,912 | 7 | 79,288 | 62.1 | 63,914 | 57.0 |
| San Antonio | 77.4 | 34.3 | 93.3 | 82.4 | 81.9 | 4.7 | 726,954 | -0.7 | 54,168 | 8 | 1,211,043 | 64.7 | 1,013,974 | 53.3 |
| San Juan | NA | 19.5 | 89.4 | 50.3 | NA | NA | 15,870 | 2.7 | 2,175 | 14 | 25,963 | 60.7 | 23,037 | 44.3 |
| San Marcos | 80.1 | 25.4 | 97.7 | 83.6 | 64.9 | 18.4 | 33,991 | 0.9 | 2,613 | 8 | 57,271 | 69.5 | 52,015 | 43.7 |
| Schertz | 86.1 | 60.0 | 98.5 | 94.5 | 80.5 | 12.3 | 19,487 | -0.7 | 1,203 | 6 | 33,872 | 67.0 | 28,531 | 57.3 |
| Seguin | 78.9 | 17.9 | 93.7 | 77.9 | 89.4 | 3.8 | 13,462 | -1.4 | 933 | 7 | 23,971 | 63.4 | 19,568 | 57.4 |
| Sherman | 86.2 | 29.6 | 91.9 | 80.3 | 80.5 | 7.0 | 20,419 | 2.3 | 1,251 | 6 | 35,387 | 67.2 | 28,235 | 52.1 |
| Socorro | 86.5 | 25.8 | 93.9 | 78.8 | NA | NA | 14,644 | 0.9 | 1,326 | 9 | 27,160 | 52.7 | 21,983 | 38.5 |
| Southlake | 72.6 | 61.6 | NA | NA | 87.3 | 9.0 | 14,903 | -2.7 | 734 | 5 | 23,813 | 51.9 | 18,955 | 43.6 |
| Sugar Land | 82.3 | 56.0 | 99.1 | 95.3 | 88.8 | 5.2 | 59,032 | -3.2 | 3,903 | 7 | 96,165 | 65.8 | 76,914 | 54.2 |
| Temple | 82.3 | 24.0 | 93.0 | 86.4 | 81.9 | 9.7 | 35,708 | 0.9 | 2,177 | 6 | 59,854 | 60.1 | 47,422 | 52.2 |
| Texarkana | 82.1 | 15.5 | 83.5 | 74.7 | 89.9 | 4.2 | 15,012 | -3.6 | 1,232 | 8 | 28,753 | 53.1 | 22,919 | 44.8 |
| Texas City | 82.3 | 35.8 | 90.1 | 87.3 | 75.8 | 6.6 | 22,161 | 1.8 | 2,572 | 12 | 37,261 | 62.4 | 30,560 | 51.3 |
| The Colony | 81.1 | 42.4 | NA | NA | 87.6 | 5.3 | 28,531 | 1.7 | 2,085 | 7 | 34,667 | 78.3 | 30,253 | 73.1 |
| Tyler | 85.2 | 21.6 | 94.4 | 80.8 | 87.5 | 6.6 | 51,393 | 1.4 | 3,470 | 7 | 86,220 | 63.7 | 68,276 | 55.1 |
| Victoria | 76.9 | 19.6 | 90.6 | 73.8 | 83.7 | 5.8 | 29,756 | -3.8 | 2,555 | 9 | 51,890 | 63.7 | 42,085 | 51.2 |
| Waco | 83.7 | 9.2 | 92.2 | 82.1 | 78.4 | 10.6 | 61,953 | 1.0 | 4,188 | 7 | 109,806 | 59.6 | 93,215 | 45.1 |
| Waxahachie | 89.0 | 46.5 | 91.5 | 87.4 | 85.5 | 5.0 | 19,234 | 1.8 | 1,299 | 7 | 27,125 | 71.2 | 22,157 | 66.8 |
| Weatherford | NA | 42.5 | 95.7 | 79.0 | 90.1 | 3.8 | 14,940 | 3.6 | 942 | 7 | 26,296 | 54.8 | 20,818 | 46.4 |
| Weslaco | 77.0 | 35.6 | 94.1 | 85.9 | 76.9 | 7.8 | 16,531 | 2.1 | 1,878 | 11 | 29,617 | 59.4 | 23,742 | 46.8 |
| Wichita Falls | 77.0 | 5.7 | 88.0 | 81.7 | 82.4 | 10.2 | 42,796 | -1.7 | 2,959 | 7 | 83,826 | 56.9 | 69,144 | 47.9 |
| Wylie | 83.3 | 59.3 | 98.8 | 94.5 | 89.7 | 3.5 | 28,950 | 1.2 | 1,806 | 6 | 41,791 | 76.5 | 38,319 | 60.3 |
| **UTAH** | 76.1 | 28.1 | 96.9 | 90.6 | 84.2 | 6.8 | 1,632,215 | 1.5 | 76,433 | 5 | 2,378,652 | 69.4 | 2,013,454 | 52.0 |
| American Fork | 72.8 | 25.3 | 98.5 | 93.2 | 84.7 | 5.6 | 15,616 | 2.3 | 692 | 4 | 24,543 | 68.4 | 21,452 | 49.6 |
| Bountiful | 78.9 | 33.2 | 95.4 | 93.9 | 89.4 | 5.4 | 21,024 | 0.2 | 967 | 5 | 31,874 | 65.7 | 25,293 | 54.9 |
| Cedar City | 85.0 | 10.1 | 95.6 | 86.6 | 89.3 | 4.7 | 16,296 | 5.4 | 767 | 5 | 23,761 | 66.8 | 20,268 | 49.0 |
| Clearfield | 79.9 | 26.1 | 95.9 | 86.9 | 79.0 | 11.2 | 15,096 | 1.4 | 808 | 5 | 23,109 | 68.4 | 20,695 | 59.0 |
| Cottonwood Heights | 74.4 | 21.5 | 99.0 | 93.8 | 84.4 | 4.1 | 19,872 | -0.2 | 1,013 | 5 | 27,930 | 70.0 | 21,248 | 60.6 |
| Draper | 73.6 | 29.7 | NA | NA | 87.4 | 6.8 | 23,306 | 0.7 | 1,100 | 5 | 34,269 | 63.2 | 30,994 | 47.2 |
| Holladay | 75.7 | 20.4 | NA | NA | 84.3 | 5.6 | 15,890 | -0.3 | 877 | 6 | 22,941 | 64.7 | 17,342 | 51.9 |

1. Employed persons.   2. Households.   3. Percent of civilian labor force.   4. Persons 16 years old and over.

# Table D. Cities — Construction, Wholesale Trade, and Retail Trade

| City | Value of residential construction authorized by building permits, 2020 | | | Wholesale trade[1], 2017 | | | | Retail trade[2], 2017 | | | |
|---|---|---|---|---|---|---|---|---|---|---|---|
| | New construction ($1,000) | Number of housing units | Percent single family | Number of establishments | Number of employees | Sales (mil dol) | Annual payroll (mil dol) | Number of establishments | Number of employees | Sales (mil dol) | Annual payroll (mil dol) |
| | 69 | 70 | 71 | 72 | 73 | 74 | 75 | 76 | 77 | 78 | 79 |
| **TEXAS—Cont'd** | | | | | | | | | | | |
| Irving | 239,920 | 784 | 80 | 351 | 12,141 | 13,991 | 1,035 | 668 | 12,263 | 5,173 | 380 |
| Keller | 44,233 | 79 | 100 | 25 | 74 | 31 | 4 | 103 | 1,689 | 881 | 41 |
| Killeen | 224,056 | 1,314 | 68 | D | D | D | 4 | 402 | 6,892 | 2,005 | 174 |
| Kingsville | 3,609 | 44 | 55 | D | D | D | D | 88 | 1,310 | 412 | 36 |
| Kyle | 356,601 | 2,042 | 60 | 12 | 118 | 72 | 7 | 68 | 1,872 | 522 | 46 |
| Lake Jackson | 10,276 | 35 | 100 | 7 | 20 | 3 | 1 | 104 | 2,664 | 772 | 71 |
| Lancaster | 15,312 | 71 | 100 | 23 | 900 | 1,135 | 38 | 64 | 1,075 | 279 | 26 |
| La Porte | 89,014 | 549 | 46 | 46 | 1,422 | 700 | 87 | 72 | 576 | 195 | 17 |
| Laredo | 201,512 | 1,447 | 88 | D | D | D | 110 | 795 | 13,447 | 3,357 | 314 |
| League City | 217,326 | 847 | 100 | 61 | 590 | 320 | 32 | 213 | 4,426 | 1,561 | 125 |
| Leander | 443,228 | 2,302 | 88 | 13 | 57 | 23 | 3 | 58 | 835 | 245 | 22 |
| Lewisville | 92,076 | 444 | 81 | 126 | 3,335 | 9,849 | 229 | 438 | 7,375 | 2,595 | 210 |
| Little Elm | 340,255 | 1,498 | 78 | 17 | 114 | 65 | 7 | 52 | 985 | 293 | 24 |
| Longview | 72,065 | 225 | 94 | 165 | 2,129 | 1,040 | 108 | 490 | 7,419 | 2,094 | 201 |
| Lubbock | 547,864 | 3,600 | 51 | 312 | 4,860 | 5,343 | 271 | 889 | 17,441 | 5,351 | 482 |
| Lufkin | 17,144 | 82 | 100 | 45 | 589 | 272 | 31 | 254 | 4,462 | 1,294 | 117 |
| McAllen | 179,983 | 1,096 | 48 | 360 | 3,343 | 2,380 | 135 | 847 | 14,186 | 3,520 | 334 |
| McKinney | 583,210 | 2,302 | 68 | 130 | 1,730 | 1,923 | 122 | 395 | 8,136 | 3,100 | 267 |
| Mansfield | 295,655 | 845 | 100 | 58 | 2,147 | 1,225 | 112 | 160 | 3,527 | 995 | 83 |
| Mesquite | 62,759 | 257 | 100 | 71 | 642 | 622 | 37 | 435 | 7,849 | 2,217 | 221 |
| Midland | 178,729 | 1,289 | 100 | 168 | 2,646 | 4,982 | 184 | 463 | 8,368 | 3,037 | 269 |
| Mission | 50,732 | 381 | 100 | 85 | 613 | 345 | 21 | 208 | 3,852 | 1,219 | 97 |
| Missouri City | 148,470 | 634 | 100 | 71 | 851 | 747 | 51 | 176 | 3,402 | 814 | 80 |
| Nacogdoches | 3,941 | 25 | 92 | 29 | 286 | 148 | 14 | 193 | 2,640 | 800 | 72 |
| New Braunfels | 555,359 | 2,783 | 71 | D | D | D | 119 | 321 | 5,707 | 2,185 | 179 |
| North Richland Hills | 104,491 | 321 | 88 | 37 | 371 | 184 | 18 | 171 | 3,994 | 1,945 | 130 |
| Odessa | 245,144 | 1,188 | 76 | 188 | 2,973 | 2,084 | 196 | 407 | 7,700 | 2,865 | 256 |
| Paris | 4,911 | 29 | 66 | 38 | 250 | 122 | 11 | 173 | 2,417 | 844 | 70 |
| Pasadena | 30,969 | 144 | 100 | 159 | 2,859 | 1,643 | 185 | 419 | 6,655 | 1,721 | 159 |
| Pearland | 124,402 | 531 | 100 | 66 | 546 | 283 | 32 | 300 | 6,635 | 1,756 | 164 |
| Pflugerville | 194,512 | 1,333 | 54 | 48 | 672 | 347 | 39 | 120 | 1,970 | 580 | 51 |
| Pharr | 26,164 | 118 | 93 | 130 | 1,514 | 865 | 64 | 174 | 2,543 | 769 | 67 |
| Plano | 151,069 | 976 | 30 | 402 | 6,240 | 8,209 | 564 | 1,085 | 23,966 | 9,764 | 790 |
| Port Arthur | 80,664 | 592 | 98 | 31 | 577 | 550 | 33 | 181 | 3,235 | 1,015 | 85 |
| Richardson | 37,648 | 45 | 100 | 231 | 5,702 | 12,072 | 570 | 351 | 5,337 | 2,801 | 201 |
| Rockwall | 135,890 | 763 | 51 | 42 | 316 | 374 | 24 | 223 | 4,656 | 1,531 | 142 |
| Rosenberg | 112,725 | 605 | 59 | 25 | 420 | 361 | 26 | 184 | 3,544 | 1,144 | 103 |
| Round Rock | 206,667 | 1,105 | 41 | 98 | 912 | 1,385 | 69 | 426 | 10,564 | 4,355 | 423 |
| Rowlett | 123,612 | 698 | 52 | 33 | 144 | 75 | 8 | 98 | 1,903 | | 45 |
| San Angelo | 143,205 | 686 | 100 | 97 | 1,194 | 701 | 64 | 406 | 6,498 | 1,920 | 187 |
| San Antonio | 1,561,008 | 9,225 | 46 | 1,280 | 25,669 | 16,372 | 1,461 | 4,419 | 81,646 | 24,318 | 2,200 |
| San Juan | 12,276 | 155 | 100 | 14 | 182 | 119 | 7 | 55 | 990 | 431 | 34 |
| San Marcos | 117,548 | 625 | 100 | 26 | 183 | 279 | 15 | 397 | 10,611 | 3,612 | 254 |
| Schertz | 89,246 | 381 | 100 | 52 | 1,602 | 1,250 | 104 | 55 | 1,464 | 427 | 37 |
| Seguin | 152,071 | 792 | 63 | 23 | 322 | 112 | 16 | 124 | 1,959 | 599 | 51 |
| Sherman | 80,777 | 395 | 74 | 40 | 372 | 449 | 19 | 217 | 4,035 | 1,282 | 109 |
| Socorro | 131,486 | 581 | 93 | 15 | 103 | 68 | 5 | 61 | 716 | 197 | 15 |
| Southlake | 74,078 | 64 | 100 | 66 | 731 | 459 | 48 | 216 | 4,760 | 1,439 | 138 |
| Sugar Land | 52,720 | 148 | 100 | 203 | 2,947 | 6,730 | 171 | 546 | 8,629 | 2,637 | 219 |
| Temple | 177,681 | 1,347 | 94 | 62 | 2,208 | 4,015 | 119 | 269 | 4,836 | 1,491 | 129 |
| Texarkana | 10,716 | 47 | 100 | 52 | 579 | 264 | 27 | 292 | 5,078 | 1,328 | 130 |
| Texas City | 168,708 | 1,126 | 69 | 36 | 504 | 270 | 32 | 189 | 2,747 | 731 | 70 |
| The Colony | 55,738 | 239 | 100 | 19 | 339 | 312 | 24 | 76 | 2,811 | 910 | 101 |
| Tyler | 97,717 | 463 | 78 | 121 | 1,503 | 636 | 85 | 634 | 10,870 | 3,294 | 298 |
| Victoria | 30,773 | 132 | 100 | 81 | 1,133 | 862 | 64 | 342 | 5,539 | 1,695 | 155 |
| Waco | 154,496 | 836 | 78 | 148 | 1,864 | 1,023 | 97 | 564 | 8,748 | 2,401 | 229 |
| Waxahachie | 173,646 | 772 | 98 | D | D | D | 21 | 156 | 2,946 | 830 | 79 |
| Weatherford | 136,944 | 877 | 32 | D | D | D | 19 | 196 | 3,561 | 1,446 | 113 |
| Weslaco | 42,156 | 289 | 79 | 29 | 444 | 248 | 16 | 158 | 3,378 | 971 | 84 |
| Wichita Falls | 34,251 | 148 | 88 | 125 | 1,116 | 419 | 52 | 417 | 6,438 | 1,797 | 159 |
| Wylie | 117,871 | 511 | 97 | 19 | 166 | 104 | 10 | 83 | 1,893 | 465 | 41 |
| **UTAH** | 14,797 | 73 | 100 | 3,109 | 46,631 | 41,834 | 2,881 | 9,995 | 153,633 | 50,008 | 4,446 |
| American Fork | 96,983 | 370 | 54 | 32 | 821 | 618 | 44 | 159 | 3,079 | 1,208 | 96 |
| Bountiful | 18,697 | 48 | 60 | D | D | D | 10 | 144 | 2,207 | 980 | 76 |
| Cedar City | 58,704 | 265 | 85 | 23 | 165 | 124 | 8 | 161 | 2,091 | 615 | 51 |
| Clearfield | 24,492 | 234 | 22 | 18 | 245 | 392 | 10 | 66 | 571 | 148 | 12 |
| Cottonwood Heights | 12,441 | 35 | 100 | 41 | 709 | 659 | 83 | 74 | 1,778 | 931 | 55 |
| Draper | 216,076 | 779 | 44 | 68 | 1,995 | 2,304 | 155 | 212 | 4,439 | 2,046 | 158 |
| Holladay | 20,516 | 88 | 19 | 21 | 83 | 28 | 2 | 68 | 609 | 167 | 15 |

1. Merchant wholesalers except manufacturers' sales branches and offices.  2. Establishments with payroll.

| City | Real estate and rental and leasing, 2017 | | | | Professional, scientific, and technical services[1], 2017 | | | | Manufacturing, 2017 | | | |
|---|---|---|---|---|---|---|---|---|---|---|---|---|
| | Number of establishments | Number of employees | Receipts (mil dol) | Annual payroll (mil dol) | Number of establishments | Number of employees | Receipts (mil dol) | Annual payroll (mil dol) | Number of establishments | Number of employees | Receipts (mil dol) | Annual payroll (mil dol) |
| | 80 | 81 | 82 | 83 | 84 | 85 | 86 | 87 | 88 | 89 | 90 | 91 |
| **TEXAS—Cont'd** | | | | | | | | | | | | |
| Irving | 381 | 4,869 | 2,281 | 297 | 1,235 | 25,765 | 5,415 | 2,290 | NA | NA | NA | NA |
| Keller | 62 | 172 | 64 | 8 | 177 | 604 | 113 | 40 | NA | NA | NA | NA |
| Killeen | 152 | 630 | 123 | 21 | 110 | 984 | 110 | 47 | NA | NA | NA | NA |
| Kingsville | 29 | 116 | 29 | 5 | D | D | D | D | NA | NA | NA | NA |
| Kyle | D | D | D | D | 26 | 89 | 11 | 4 | NA | NA | NA | NA |
| Lake Jackson | 31 | 136 | 21 | 4 | 39 | 204 | 29 | 9 | NA | NA | NA | NA |
| Lancaster | 19 | 78 | 22 | 3 | 18 | 62 | 14 | 3 | NA | NA | NA | NA |
| La Porte | 33 | 370 | 162 | 25 | 68 | 2,355 | 375 | 189 | NA | NA | NA | NA |
| Laredo | 247 | 890 | 214 | 29 | 189 | 1,921 | 269 | 110 | NA | NA | NA | NA |
| League City | 86 | 295 | 75 | 13 | 87 | 281 | 47 | 17 | NA | NA | NA | NA |
| Leander | 35 | 63 | 14 | 3 | 248 | 2,665 | 478 | 178 | NA | NA | NA | NA |
| Lewisville | 141 | 746 | 242 | 37 | 65 | 141 | 19 | 7 | NA | NA | NA | NA |
| Little Elm | 19 | 51 | 16 | 2 | 306 | 2,240 | 407 | 132 | NA | NA | NA | NA |
| Longview | 146 | 683 | 164 | 29 | | | | | NA | NA | NA | NA |
| Lubbock | 412 | 1,912 | 398 | 72 | 606 | 4,324 | 586 | 223 | NA | NA | NA | NA |
| Lufkin | 64 | 314 | 94 | 11 | 116 | 659 | 88 | 33 | NA | NA | NA | NA |
| McAllen | 242 | 858 | 196 | 29 | 502 | 3,860 | 392 | 136 | NA | NA | NA | NA |
| McKinney | 214 | 755 | 212 | 36 | 514 | 2,206 | 362 | 155 | NA | NA | NA | NA |
| Mansfield | 58 | 207 | 66 | 11 | 129 | 505 | 77 | 29 | NA | NA | NA | NA |
| Mesquite | 98 | 452 | 121 | 20 | 112 | 736 | 63 | 21 | NA | NA | NA | NA |
| Midland | 293 | 1,719 | 559 | 90 | 443 | 2,948 | 675 | 212 | NA | NA | NA | NA |
| Mission | 73 | 208 | 33 | 6 | 86 | 366 | 42 | 14 | NA | NA | NA | NA |
| Missouri City | 62 | 221 | 61 | 9 | 183 | 781 | 152 | 53 | NA | NA | NA | NA |
| Nacogdoches | 47 | 184 | 27 | 6 | 73 | 367 | 49 | 15 | NA | NA | NA | NA |
| New Braunfels | 149 | 540 | 129 | 23 | 199 | 1,126 | 144 | 50 | NA | NA | NA | NA |
| North Richland Hills | 64 | 501 | 59 | 18 | 151 | 460 | 83 | 26 | NA | NA | NA | NA |
| Odessa | 154 | 1,446 | 589 | 83 | 213 | 1,549 | 257 | 90 | NA | NA | NA | NA |
| Paris | 39 | 161 | 23 | 4 | 50 | 297 | 28 | 12 | NA | NA | NA | NA |
| Pasadena | 134 | 892 | 235 | 42 | 185 | 4,077 | 571 | 292 | NA | NA | NA | NA |
| Pearland | 108 | 380 | 109 | 19 | 236 | 1,286 | 256 | 84 | NA | NA | NA | NA |
| Pflugerville | 39 | 114 | 33 | 5 | 93 | 540 | 84 | 40 | NA | NA | NA | NA |
| Pharr | 46 | 251 | 55 | 8 | 69 | 831 | 50 | 20 | NA | NA | NA | NA |
| Plano | 570 | 5,611 | 1,424 | 355 | 1,750 | 31,565 | 8,237 | 3,203 | NA | NA | NA | NA |
| Port Arthur | 33 | 343 | 58 | 15 | 41 | 365 | 47 | 22 | NA | NA | NA | NA |
| Richardson | 201 | 1,275 | 459 | 61 | D | D | D | D | NA | NA | NA | NA |
| Rockwall | 73 | 235 | 71 | 9 | 198 | 923 | 143 | 57 | NA | NA | NA | NA |
| Rosenberg | 34 | 459 | 38 | 16 | 52 | 194 | 33 | 11 | NA | NA | NA | NA |
| Round Rock | 165 | 555 | 199 | 28 | 444 | 13,529 | 3,172 | 1,287 | NA | NA | NA | NA |
| Rowlett | 22 | 44 | 12 | 2 | 92 | 480 | 77 | 26 | NA | NA | NA | NA |
| San Angelo | 136 | 557 | 113 | 19 | 189 | 1,231 | 162 | 54 | NA | NA | NA | NA |
| San Antonio | 1,783 | 12,132 | 2,900 | 632 | 3,644 | 46,677 | 7,091 | 2,899 | NA | NA | NA | NA |
| San Juan | 9 | 40 | 5 | 2 | 10 | 24 | 2 | 0 | NA | NA | NA | NA |
| San Marcos | 92 | 612 | 137 | 20 | 94 | 799 | 104 | 25 | NA | NA | NA | NA |
| Schertz | 38 | 211 | 42 | 8 | 51 | 466 | 91 | 25 | NA | NA | NA | NA |
| Seguin | 34 | 134 | 26 | 5 | 45 | 255 | 25 | 11 | NA | NA | NA | NA |
| Sherman | 46 | 171 | 50 | 7 | 105 | 400 | 61 | 21 | NA | NA | NA | NA |
| Socorro | 16 | 31 | 4 | 1 | 18 | 121 | 32 | 4 | NA | NA | NA | NA |
| Southlake | 105 | 800 | 105 | 33 | 243 | 1,572 | 278 | 98 | NA | NA | NA | NA |
| Sugar Land | 212 | 516 | 158 | 24 | 643 | 9,908 | 3,584 | 1,038 | NA | NA | NA | NA |
| Temple | 85 | 450 | 64 | 16 | 103 | 652 | 90 | 33 | NA | NA | NA | NA |
| Texarkana | 81 | 395 | 73 | 16 | 123 | 2,248 | 206 | 85 | NA | NA | NA | NA |
| Texas City | 41 | 910 | 209 | 58 | 54 | 671 | 76 | 32 | NA | NA | NA | NA |
| The Colony | 32 | 104 | 49 | 5 | 80 | 480 | 61 | 17 | NA | NA | NA | NA |
| Tyler | 226 | 911 | 200 | 39 | 450 | 3,518 | 734 | 238 | NA | NA | NA | NA |
| Victoria | 107 | 792 | 416 | 39 | 140 | 1,016 | 136 | 43 | NA | NA | NA | NA |
| Waco | 182 | 1,462 | 334 | 68 | 278 | 2,307 | 464 | 146 | NA | NA | NA | NA |
| Waxahachie | 48 | 179 | 46 | 6 | 81 | 402 | 48 | 21 | NA | NA | NA | NA |
| Weatherford | 52 | 230 | 64 | 12 | 106 | 551 | 81 | 26 | NA | NA | NA | NA |
| Weslaco | 41 | 194 | 35 | 6 | 52 | 404 | 41 | 13 | NA | NA | NA | NA |
| Wichita Falls | 150 | 756 | 127 | 25 | 193 | 1,279 | 132 | 52 | NA | NA | NA | NA |
| Wylie | 40 | 123 | 25 | 5 | 64 | 213 | 37 | 12 | NA | NA | NA | NA |
| **UTAH** | 5,577 | 19,457 | 5,297 | 877 | 10,873 | 92,444 | 15,305 | 5,610 | 3,377 | 121,527 | 55,788 | 7,171 |
| American Fork | 67 | 216 | 40 | 6 | 143 | 1,229 | 121 | 47 | NA | NA | NA | NA |
| Bountiful | 93 | 170 | 31 | 7 | 223 | 1,003 | 127 | 47 | NA | NA | NA | NA |
| Cedar City | 83 | 151 | 34 | 5 | 116 | 429 | 52 | 18 | NA | NA | NA | NA |
| Clearfield | 27 | 82 | 38 | 4 | 67 | 1,198 | 150 | 81 | NA | NA | NA | NA |
| Cottonwood Heights | 172 | 600 | 235 | 31 | 228 | 1,497 | 367 | 120 | NA | NA | NA | NA |
| Draper | 142 | 499 | 128 | 21 | 307 | 2,099 | 342 | 111 | NA | NA | NA | NA |
| Holladay | D | D | D | D | 199 | 1,638 | 365 | 196 | NA | NA | NA | NA |

1. Establishments subject to federal tax.

# Accommodation and Food Services, Arts, Entertainment, and Recreation, and Health Care and Social Assistance

| City | Accommodation and food services, 2017 | | | | Arts, entertainment, and recreation[1], 2017 | | | | Health care and social assistance[1], 2017 | | | |
|---|---|---|---|---|---|---|---|---|---|---|---|---|
| | Number of establishments | Number of employees | Receipts (mil dol) | Annual payroll (mil dol) | Number of establishments | Number of employees | Receipts (mil dol) | Annual payroll (mil dol) | Number of establishments | Number of employees | Receipts (mil dol) | Annual payroll (mil dol) |
| | 92 | 93 | 94 | 95 | 96 | 97 | 98 | 99 | 100 | 101 | 102 | 103 |
| **TEXAS—Cont'd** | | | | | | | | | | | | |
| Irving | 656 | 13,735 | 1,081 | 288 | 62 | 773 | 84 | 22 | 676 | 12,494 | 1,927 | 719 |
| Keller | 78 | 1,384 | 72 | 21 | 19 | 199 | 21 | 4 | 153 | 1,670 | 176 | 59 |
| Killeen | 274 | 6,252 | 323 | 85 | D | D | D | D | 211 | 2,898 | 309 | 121 |
| Kingsville | 68 | 1,238 | 66 | 18 | 3 | D | D | 1 | 72 | 1,305 | 104 | 41 |
| Kyle | 59 | 1,337 | 73 | 21 | 8 | 145 | 17 | 3 | 82 | 1,950 | 274 | 104 |
| Lake Jackson | 83 | 1,943 | 114 | 32 | D | D | D | D | 145 | 2,214 | 222 | 96 |
| Lancaster | 46 | 862 | 45 | 12 | 3 | 19 | 1 | 0 | 47 | 1,304 | 103 | 39 |
| La Porte | 68 | 1,391 | 89 | 25 | NA | NA | NA | NA | 44 | 448 | 37 | 15 |
| Laredo | D | D | D | D | D | D | D | D | D | D | D | D |
| League City | 169 | 3,453 | 184 | 55 | D | D | D | D | 200 | 2,907 | 260 | 99 |
| Leander | D | D | D | D | D | D | D | D | 57 | 407 | 38 | 16 |
| Lewisville | 275 | 5,161 | 324 | 86 | 44 | 1,023 | 67 | 19 | 284 | 5,698 | 846 | 346 |
| Little Elm | 56 | 949 | 49 | 14 | 15 | 83 | 6 | 2 | 36 | 294 | 41 | 14 |
| Longview | 274 | 5,630 | 287 | 86 | 30 | 328 | 18 | 6 | 393 | 10,308 | 1,187 | 453 |
| Lubbock | 668 | 15,930 | 904 | 255 | 84 | 1,923 | 87 | 28 | 819 | 23,907 | 2,964 | 1,013 |
| Lufkin | 124 | 2,849 | 158 | 44 | 15 | 184 | 9 | 3 | 263 | 7,245 | 606 | 251 |
| McAllen | 439 | 10,773 | 542 | 152 | 46 | 574 | 52 | 13 | 821 | 18,470 | 1,717 | 607 |
| McKinney | 288 | 6,578 | 374 | 110 | 55 | 1,078 | 64 | 23 | 516 | 7,979 | 995 | 363 |
| Mansfield | 139 | 3,622 | 200 | 55 | 25 | 693 | 42 | 12 | D | D | D | D |
| Mesquite | 236 | 5,700 | 348 | 94 | D | D | D | D | 355 | 8,370 | 615 | 236 |
| Midland | 333 | 7,990 | 609 | 154 | 48 | D | 95 | D | 393 | 7,444 | 882 | 349 |
| Mission | 142 | 2,593 | 137 | 35 | 13 | 266 | 18 | 6 | 306 | 10,015 | 420 | 212 |
| Missouri City | 125 | 1,889 | 109 | 28 | 18 | 261 | 11 | 4 | 230 | 2,249 | 182 | 65 |
| Nacogdoches | 108 | 2,168 | 115 | 35 | 13 | 92 | 6 | 1 | 179 | 3,196 | 364 | 130 |
| New Braunfels | 260 | 6,136 | 345 | 101 | 47 | 1,173 | 82 | 27 | 285 | 5,296 | 614 | 232 |
| North Richland Hills | 145 | 3,074 | 155 | 43 | 21 | 296 | 14 | 4 | 175 | 2,577 | 357 | 129 |
| Odessa | 264 | 6,434 | 449 | 115 | 33 | 483 | 39 | 11 | 290 | 8,055 | 1,011 | 381 |
| Paris | 100 | 1,840 | 87 | 24 | D | D | D | D | 187 | 3,588 | 361 | 135 |
| Pasadena | 238 | 5,473 | 304 | 85 | 17 | 330 | 21 | 4 | 352 | 7,459 | 846 | 325 |
| Pearland | 236 | 5,726 | 338 | 93 | 25 | 499 | 23 | 7 | 372 | 4,937 | 651 | 200 |
| Pflugerville | 93 | 2,267 | 120 | 37 | D | D | D | D | 94 | 1,025 | 92 | 38 |
| Pharr | 109 | 2,209 | 132 | 31 | 8 | 63 | 4 | 1 | 162 | 5,088 | 264 | 100 |
| Plano | 922 | 19,346 | 1,259 | 368 | 117 | 2,399 | 181 | 46 | 1,632 | 29,943 | 4,557 | 1,517 |
| Port Arthur | 117 | 2,263 | 119 | 32 | D | D | D | D | D | D | D | 114 |
| Richardson | 459 | 7,531 | 512 | 139 | 68 | 716 | 45 | 13 | 599 | 8,892 | 960 | 342 |
| Rockwall | 153 | 3,973 | 215 | 62 | 21 | 369 | 25 | 7 | 224 | 2,933 | 445 | 139 |
| Rosenberg | 120 | 2,461 | 149 | 44 | 8 | 37 | 5 | 1 | 76 | 850 | 65 | 24 |
| Round Rock | 333 | 7,804 | 482 | 135 | 42 | 1,317 | 81 | 20 | 364 | 8,980 | 1,326 | 507 |
| Rowlett | 85 | 1,455 | 86 | 24 | 13 | 153 | 11 | 2 | 120 | 2,283 | 359 | 96 |
| San Angelo | 248 | 4,816 | 258 | 73 | 32 | 597 | 26 | 9 | 263 | 7,868 | 815 | 351 |
| San Antonio | 3,546 | 93,890 | 5,970 | 1,652 | D | D | D | D | 4,263 | 113,466 | 13,592 | 4,918 |
| San Juan | 23 | 428 | 27 | 7 | D | D | D | D | 55 | 1,382 | 58 | 32 |
| San Marcos | 256 | 5,899 | 288 | 83 | 22 | 156 | 9 | 3 | 169 | 3,732 | 413 | 167 |
| Schertz | 59 | 1,596 | 81 | 23 | 4 | 162 | 5 | 2 | 69 | 1,042 | 110 | 42 |
| Seguin | 88 | 1,569 | 85 | 24 | D | D | D | D | 101 | 1,914 | 220 | 86 |
| Sherman | 112 | 2,739 | 159 | 46 | D | D | D | D | 213 | 3,898 | 415 | 162 |
| Socorro | 29 | 499 | 21 | 5 | NA | NA | NA | NA | 15 | 293 | 10 | 5 |
| Southlake | 128 | 3,408 | 203 | 63 | 29 | 588 | 38 | 11 | 269 | 2,900 | 459 | 148 |
| Sugar Land | 377 | 8,118 | 526 | 145 | 45 | D | 134 | D | 740 | 12,100 | 1,644 | 582 |
| Temple | 187 | 4,019 | 209 | 60 | 25 | 353 | 20 | 6 | 199 | 21,426 | 3,170 | 1,523 |
| Texarkana | 148 | 3,886 | 180 | 52 | D | D | D | D | 247 | 6,804 | 944 | 339 |
| Texas City | 88 | 1,518 | 81 | 21 | 6 | 50 | 3 | 1 | 103 | 2,264 | 217 | 93 |
| The Colony | 83 | 1,919 | 119 | 34 | 12 | 485 | 31 | 9 | 61 | 632 | 51 | 21 |
| Tyler | 358 | 8,501 | 457 | 140 | 44 | 768 | 47 | 17 | 569 | 20,676 | 2,852 | 1,033 |
| Victoria | 185 | 3,663 | 191 | 56 | D | D | D | D | 277 | 6,721 | 742 | 313 |
| Waco | 367 | 8,862 | 505 | 130 | 53 | 1,382 | 74 | 22 | 454 | 16,024 | 1,651 | 698 |
| Waxahachie | 100 | 2,669 | 141 | 41 | 10 | 138 | 5 | 2 | 116 | 2,078 | 287 | 95 |
| Weatherford | 121 | 2,541 | 140 | 39 | D | D | D | D | 150 | 2,414 | 297 | 103 |
| Weslaco | 91 | 2,073 | 111 | 27 | 8 | 64 | 3 | 1 | 195 | 7,892 | 401 | 187 |
| Wichita Falls | 260 | 5,438 | 265 | 77 | 33 | 440 | 14 | 5 | 365 | 10,507 | 1,108 | 408 |
| Wylie | 49 | 985 | 59 | 16 | D | D | D | D | 69 | 1,074 | 74 | 34 |
| **UTAH** | 5,931 | 118,734 | 7,098 | 2,004 | 1,180 | 28,240 | 1,760 | 553 | 8,254 | 146,365 | 18,708 | 6,703 |
| American Fork | 89 | 1,968 | 103 | 27 | 15 | 122 | 10 | 2 | 152 | 2,603 | 356 | 112 |
| Bountiful | 74 | 1,562 | 74 | 22 | 15 | 98 | 8 | 2 | 223 | 3,291 | 348 | 133 |
| Cedar City | D | D | D | D | 16 | 86 | 5 | 1 | 144 | 2,182 | 182 | 68 |
| Clearfield | D | D | D | D | NA | NA | NA | NA | 61 | 1,672 | 119 | 45 |
| Cottonwood Heights | D | D | D | D | 13 | 165 | 18 | 4 | 147 | 1,559 | 154 | 58 |
| Draper | 117 | 2,066 | 110 | 32 | D | D | D | D | 174 | 1,671 | 210 | 73 |
| Holladay | 55 | 803 | 50 | 13 | D | D | D | D | 126 | 934 | 88 | 35 |

1. Establishments subject to federal tax.

# Table D. Cities — Other Services and Government Employment and Payroll

| City | Other services[1] Number of establishments | Number of employees | Receipts (mil dol) | Annual payroll (mil dol) | Government employment and payroll, 2017 — March payroll Full-time equivalent employees | Total (dollars) | Percent of total for: Administrative, judicial, and legal | Police and corrections | Fire protection | Highways and transportation | Health and welfare | Natural resources and utilities | Education and libraries |
|---|---|---|---|---|---|---|---|---|---|---|---|---|---|
| | 104 | 105 | 106 | 107 | 108 | 109 | 110 | 111 | 112 | 113 | 114 | 115 | 116 |
| **TEXAS—Cont'd** | | | | | | | | | | | | | |
| Irving | 324 | 6,050 | 1,257 | 278 | 1,975 | 10,564,254 | 10.5 | 26.4 | 23.0 | 6.1 | 2.7 | 20.3 | 4.1 |
| Keller | 72 | 434 | 41 | 15 | 321 | 1,555,104 | 14.1 | 30.2 | 22.2 | 2.4 | 5.0 | 21.8 | 3.8 |
| Killeen | 182 | 1,068 | 106 | 28 | 1,201 | 4,971,267 | 12.2 | 33.8 | 25.0 | 4.9 | 1.1 | 16.6 | 1.4 |
| Kingsville | 36 | 155 | 17 | 3 | 283 | 966,742 | 14.4 | 26.1 | 16.5 | 4.7 | 2.8 | 22.8 | 2.8 |
| Kyle | 35 | 221 | 18 | 5 | 184 | 801,895 | 11.6 | 38.9 | 0.0 | 6.8 | 0.8 | 18.8 | 3.6 |
| Lake Jackson | 33 | 229 | 13 | 5 | 230 | 1,227,377 | 12.7 | 28.9 | 0.9 | 5.6 | 0.0 | 33.5 | 0.0 |
| Lancaster | 26 | 170 | 23 | 7 | 227 | 1,083,542 | 7.2 | 31.0 | 33.3 | 3.6 | 0.8 | 18.4 | 2.3 |
| La Porte | D | D | 439 | D | 381 | 1,918,813 | 14.2 | 39.2 | 4.9 | 4.6 | 8.0 | 19.5 | 0.0 |
| Laredo | D | D | D | D | 2,590 | 12,767,862 | 6.2 | 34.8 | 23.4 | 10.1 | 8.4 | 14.7 | 1.3 |
| League City | 117 | 802 | 68 | 20 | 550 | 2,481,831 | 14.5 | 36.3 | 2.2 | 9.6 | 9.0 | 16.5 | 3.8 |
| Leander | 43 | 172 | 22 | 7 | 243 | 1,175,001 | 11.2 | 25.8 | 24.8 | 8.8 | 0.0 | 15.5 | 0.0 |
| Lewisville | 178 | 1,131 | 143 | 41 | 813 | 3,202,868 | 11.3 | 32.7 | 23.1 | 4.7 | 6.3 | 14.4 | 2.1 |
| Little Elm | 29 | 154 | 14 | 4 | 224 | 1,387,532 | 9.0 | 25.4 | 29.4 | 6.9 | 2.1 | 16.6 | 2.0 |
| Longview | 183 | 1,531 | 158 | 51 | 857 | 3,771,794 | 8.9 | 27.7 | 26.8 | 2.0 | 4.5 | 20.6 | 1.6 |
| Lubbock | 405 | 3,610 | 367 | 104 | 2,204 | 10,549,253 | 8.5 | 28.8 | 24.8 | 3.9 | 2.1 | 29.2 | 1.2 |
| Lufkin | 96 | 628 | 89 | 20 | 436 | 1,707,473 | 10.8 | 29.9 | 21.4 | 5.6 | 0.0 | 22.8 | 1.8 |
| McAllen | 195 | 1,286 | 106 | 30 | 1,845 | 6,873,623 | 11.7 | 27.9 | 14.5 | 11.7 | 3.0 | 26.1 | 3.0 |
| McKinney | 193 | 1,398 | 153 | 46 | 1,059 | 6,033,670 | 14.3 | 28.5 | 22.3 | 9.4 | 0.1 | 13.4 | 2.5 |
| Mansfield | 93 | 771 | 78 | 23 | 502 | 2,465,516 | 12.0 | 32.8 | 24.1 | 6.2 | 1.4 | 17.2 | 1.4 |
| Mesquite | 115 | 823 | 101 | 29 | 1,114 | 6,060,197 | 11.2 | 33.9 | 27.2 | 3.7 | 3.8 | 14.6 | 1.7 |
| Midland | 252 | 2,152 | 396 | 88 | 968 | 4,871,703 | 11.3 | 25.9 | 23.9 | 14.0 | 2.7 | 16.7 | 0.0 |
| Mission | 78 | 391 | 38 | 10 | 643 | 2,364,235 | 11.4 | 38.2 | 17.4 | 2.8 | 1.6 | 23.9 | 2.8 |
| Missouri City | D | D | D | D | 315 | 1,632,986 | 9.5 | 42.4 | 22.9 | 9.3 | 0.0 | 4.5 | 0.0 |
| Nacogdoches | D | D | D | D | 312 | 1,270,092 | 11.7 | 30.5 | 23.0 | 6.5 | 1.2 | 17.7 | 1.7 |
| New Braunfels | 147 | 1,096 | 101 | 33 | 827 | 3,664,065 | 7.8 | 24.5 | 26.2 | 5.9 | 1.3 | 26.6 | 2.5 |
| North Richland Hills | 93 | 487 | 61 | 16 | 626 | 2,982,868 | 13.0 | 32.5 | 21.9 | 7.4 | 4.8 | 15.6 | 3.3 |
| Odessa | 193 | 1,687 | 211 | 59 | 943 | 4,456,802 | 14.2 | 28.4 | 22.6 | 8.6 | 2.0 | 16.1 | 0.0 |
| Paris | D | D | D | D | 224 | 1,102,689 | 16.0 | 38.5 | 1.6 | 2.6 | 5.8 | 17.2 | 7.7 |
| Pasadena | 167 | 2,056 | 605 | 109 | 1,024 | 5,195,001 | 10.3 | 46.0 | 2.5 | 8.3 | 5.2 | 17.2 | 2.9 |
| Pearland | 165 | 1,227 | 110 | 35 | 689 | 3,375,508 | 10.7 | 37.0 | 16.5 | 6.6 | 6.6 | 15.5 | 0.0 |
| Pflugerville | 64 | 449 | 58 | 19 | 340 | 1,586,208 | 16.4 | 48.0 | 0.0 | 4.9 | 0.0 | 12.2 | 3.5 |
| Pharr | 64 | 487 | 53 | 13 | 641 | 2,300,354 | 11.6 | 36.5 | 14.0 | 8.4 | 1.0 | 18.4 | 2.2 |
| Plano | 524 | 4,256 | 824 | 155 | 2,302 | 12,930,116 | 10.1 | 29.9 | 23.6 | 2.7 | 0.0 | 13.4 | 4.4 |
| Port Arthur | 52 | 248 | 36 | 10 | 664 | 3,432,181 | 8.3 | 26.8 | 21.8 | 5.6 | 9.0 | 25.1 | 1.7 |
| Richardson | 201 | 1,718 | 237 | 76 | 1,053 | 6,535,292 | 15.4 | 24.9 | 18.9 | 4.9 | 1.8 | 18.8 | 3.4 |
| Rockwall | 92 | 639 | 60 | 19 | 283 | 1,555,754 | 17.3 | 39.5 | 14.2 | 3.1 | 0.0 | 15.8 | 0.0 |
| Rosenberg | 61 | 269 | 36 | 10 | 220 | 935,709 | 12.4 | 48.5 | 20.9 | 5.4 | 1.4 | 5.0 | 0.0 |
| Round Rock | 211 | 1,942 | 197 | 64 | 853 | 4,179,488 | 7.4 | 31.9 | 20.1 | 4.9 | 0.0 | 18.8 | 2.9 |
| Rowlett | 77 | 399 | 46 | 14 | 353 | 1,833,057 | 9.9 | 33.6 | 27.4 | 3.5 | 6.0 | 10.8 | 0.5 |
| San Angelo | 191 | 1,052 | 148 | 35 | 911 | 3,997,824 | 15.6 | 26.1 | 24.2 | 5.3 | 3.8 | 19.3 | 0.0 |
| San Antonio | 2,132 | 16,568 | 1,683 | 497 | 15,585 | 91,157,442 | 7.9 | 22.6 | 14.9 | 3.9 | 5.3 | 41.7 | 1.5 |
| San Juan | 12 | 99 | 11 | 2 | 209 | 675,652 | 0.0 | 34.8 | 11.6 | 1.6 | 4.4 | 30.4 | 3.6 |
| San Marcos | 92 | 529 | 62 | 18 | 604 | 3,071,174 | 14.5 | 27.8 | 14.5 | 5.5 | 4.9 | 23.9 | 2.6 |
| Schertz | D | D | 48 | D | 327 | 1,451,552 | 16.4 | 24.9 | 14.8 | 4.2 | 18.6 | 9.0 | 3.0 |
| Seguin | 51 | 351 | 38 | 11 | 933 | 4,361,087 | 6.2 | 9.2 | 6.8 | 2.3 | 63.3 | 10.6 | 1.3 |
| Sherman | 59 | 387 | 44 | 12 | 422 | 1,937,884 | 11.7 | 24.1 | 24.6 | 5.8 | 2.9 | 25.8 | 1.5 |
| Socorro | 18 | 101 | 8 | 2 | 100 | 322,769 | 27.7 | 45.1 | 0.0 | 16.3 | 0.0 | 5.2 | 0.0 |
| Southlake | 91 | 733 | 58 | 21 | 338 | 1,703,855 | 25.8 | 22.7 | 21.3 | 7.0 | 2.9 | 18.1 | 2.2 |
| Sugar Land | 190 | 1,559 | 129 | 40 | 730 | 4,005,280 | 19.7 | 30.4 | 17.3 | 7.8 | 12.3 | 10.7 | 0.0 |
| Temple | 132 | 876 | 78 | 32 | 809 | 3,363,934 | 13.3 | 28.5 | 21.7 | 6.4 | 2.2 | 23.7 | 2.6 |
| Texarkana | 108 | 788 | 75 | 26 | 529 | 2,087,616 | 14.8 | 23.8 | 17.2 | 9.5 | 3.2 | 18.4 | 1.8 |
| Texas City | 48 | 281 | 38 | 10 | 493 | 2,285,519 | 7.5 | 27.9 | 21.2 | 9.3 | 1.4 | 27.1 | 2.0 |
| The Colony | 53 | 242 | 21 | 6 | 331 | 1,585,838 | 5.5 | 28.7 | 28.8 | 3.9 | 0.0 | 14.9 | 3.3 |
| Tyler | 244 | 1,569 | 173 | 48 | 840 | 4,220,921 | 7.2 | 33.3 | 30.5 | 9.7 | 1.2 | 15.4 | 1.6 |
| Victoria | 128 | 889 | 115 | 33 | 589 | 2,437,506 | 10.7 | 27.7 | 27.2 | 5.4 | 0.0 | 22.2 | 3.3 |
| Waco | 253 | 1,621 | 206 | 53 | 1,545 | 6,661,214 | 9.5 | 30.2 | 17.6 | 4.5 | 6.7 | 25.0 | 2.2 |
| Waxahachie | 61 | 352 | 38 | 11 | 290 | 1,418,072 | 12.4 | 28.0 | 25.4 | 3.4 | 3.0 | 21.1 | 0.0 |
| Weatherford | 64 | 444 | 39 | 14 | 380 | 1,292,771 | 16.2 | 21.4 | 16.6 | 5.4 | 0.4 | 24.3 | 3.0 |
| Weslaco | 61 | 517 | 44 | 16 | 302 | 1,190,869 | 14.3 | 38.2 | 27.2 | 7.7 | 0.0 | 5.7 | 2.6 |
| Wichita Falls | 189 | 970 | 136 | 30 | 1,165 | 5,879,646 | 6.8 | 23.8 | 36.5 | 7.4 | 5.0 | 15.9 | 0.9 |
| Wylie | D | D | D | D | 320 | 1,388,780 | 13.6 | 15.1 | 31.5 | 6.4 | 1.7 | 21.2 | 6.9 |
| **UTAH** | 4,763 | 29,903 | 3,527 | 958 | X | X | X | X | X | X | X | X | X |
| American Fork | 69 | 413 | 42 | 11 | 217 | 914,048 | 15.9 | 22.7 | 17.7 | 2.6 | 5.0 | 23.4 | 4.8 |
| Bountiful | 79 | 565 | 51 | 15 | 205 | 1,038,767 | 13.9 | 29.3 | 0.0 | 14.2 | 0.0 | 42.2 | 0.0 |
| Cedar City | 63 | 222 | 25 | 6 | 208 | 699,101 | 9.9 | 26.3 | 7.2 | 12.9 | 2.1 | 29.8 | 3.2 |
| Clearfield | D | D | D | D | 180 | 685,812 | 19.1 | 29.5 | 17.6 | 2.9 | 0.9 | 22.2 | 0.0 |
| Cottonwood Heights | 34 | 190 | 15 | 5 | 74 | 361,371 | 9.4 | 71.8 | 0.0 | 14.8 | 0.0 | 0.0 | 3.8 |
| Draper | 95 | 693 | 57 | 19 | 186 | 849,667 | 18.8 | 29.5 | 3.8 | 14.1 | 0.6 | 18.3 | 0.0 |
| Holladay | 55 | 260 | 18 | 6 | 21 | 101,357 | 82.9 | 0.0 | 0.0 | 0.0 | 0.0 | 17.1 | 0.0 |

1. Establishments subject to federal tax.

## Table D. Cities — City Government Finances

| City | City government finances, 2017 | | | | | | | | | |
|---|---|---|---|---|---|---|---|---|---|---|
| | General revenue | | | | | | | General expenditure | | |
| | Intergovernmental | | | Taxes | | | | | | |
| | | | | | Per capita[1] (dollars) | | | | Per capita[1] (dollars) | |
| | Total (mil dol) | Total (mil dol) | Percent from state government | Total (mil dol) | Total | Property | Sales and gross receipts | Total (mil dol) | Total | Capital outlays |
| | 117 | 118 | 119 | 120 | 121 | 122 | 123 | 124 | 125 | 126 |
| **TEXAS—Cont'd** | | | | | | | | | | |
| Irving | 325.6 | 15.8 | 80.6 | 241.2 | 1,002 | 517 | 485 | 349 | 1,452 | 254 |
| Keller | 64.1 | 5.9 | 62.8 | 41.7 | 886 | 545 | 340 | 54 | 1,142 | 160 |
| Killeen | 131.7 | 6.4 | 21.1 | 68.8 | 474 | 264 | 210 | 158 | 1,087 | 152 |
| Kingsville | 19.9 | 0.1 | 100.0 | 12.8 | 504 | 255 | 249 | 13 | 511 | 53 |
| Kyle | 21.0 | 1.6 | 24.4 | 12.8 | 297 | 172 | 125 | 25 | 576 | 143 |
| Lake Jackson | 33.2 | 0.0 | 100.0 | 18.1 | 661 | 214 | 447 | 32 | 1,166 | 117 |
| Lancaster | 44.1 | 0.3 | 15.0 | 28.7 | 728 | 398 | 330 | 42 | 1,056 | 33 |
| La Porte | 65.7 | 1.0 | 9.5 | 36.9 | 1,042 | 681 | 361 | 49 | 1,378 | 137 |
| Laredo | 426.9 | 40.9 | 42.8 | 151.4 | 584 | 297 | 287 | 370 | 1,427 | 193 |
| League City | 103.7 | 3.4 | 75.8 | 66.4 | 635 | 368 | 267 | 88 | 844 | 154 |
| Leander | 44.5 | 5.6 | 1.6 | 29.4 | 589 | 395 | 195 | 46 | 923 | 350 |
| Lewisville | 126.6 | 1.5 | 54.3 | 90.7 | 852 | 330 | 522 | 130 | 1,223 | 341 |
| Little Elm | 44.1 | 0.8 | 59.4 | 30.7 | 667 | 353 | 315 | 31 | 667 | 13 |
| Longview | 128.3 | 11.1 | 30.8 | 80.4 | 993 | 405 | 588 | 112 | 1,379 | 135 |
| Lubbock | 322.9 | 37.2 | 12.6 | 168.4 | 662 | 316 | 346 | 366 | 1,440 | 487 |
| Lufkin | 47.9 | 1.5 | 73.5 | 27.8 | 780 | 299 | 481 | 50 | 1,397 | 110 |
| McAllen | 218.2 | 11.1 | 13.2 | 114.7 | 806 | 284 | 522 | 205 | 1,442 | 299 |
| McKinney | 269.3 | 8.6 | 10.8 | 161.9 | 889 | 498 | 391 | 220 | 1,209 | 260 |
| Mansfield | 106.2 | 1.0 | 99.1 | 72.3 | 1,050 | 607 | 443 | 122 | 1,778 | 523 |
| Mesquite | 171.2 | 13.7 | 7.2 | 98.8 | 685 | 287 | 397 | 167 | 1,157 | 118 |
| Midland | 201.0 | 10.0 | 29.8 | 132.1 | 970 | 312 | 658 | 155 | 1,135 | 141 |
| Mission | 79.0 | 8.1 | 45.4 | 40.1 | 480 | 239 | 242 | 70 | 833 | 73 |
| Missouri City | 66.8 | 5.4 | 33.9 | 46.3 | 621 | 421 | 201 | 70 | 941 | 185 |
| Nacogdoches | 35.1 | 1.2 | 69.7 | 18.2 | 547 | 261 | 287 | 35 | 1,059 | 126 |
| New Braunfels | 103.4 | 0.7 | 54.9 | 62.0 | 786 | 275 | 511 | 103 | 1,300 | 268 |
| North Richland Hills | 91.2 | 19.9 | 64.5 | 41.1 | 582 | 213 | 370 | 86 | 1,213 | 338 |
| Odessa | 141.4 | 3.3 | 59.9 | 92.3 | 792 | 228 | 565 | 122 | 1,049 | 93 |
| Paris | 36.3 | 2.1 | 100.0 | 22.0 | 888 | 373 | 515 | 42 | 1,694 | 145 |
| Pasadena | 174.4 | 19.0 | 46.1 | 98.4 | 640 | 274 | 366 | 160 | 1,038 | 275 |
| Pearland | 153.6 | 1.9 | 64.2 | 107.5 | 901 | 543 | 357 | 131 | 1,098 | 188 |
| Pflugerville | 59.3 | 2.9 | 4.1 | 33.7 | 533 | 331 | 202 | 66 | 1,041 | 451 |
| Pharr | 50.0 | 4.0 | 48.7 | 18.6 | 236 | 193 | 41 | NA | NA | NA |
| Plano | 457.1 | 8.4 | 67.3 | 302.0 | 1,050 | 574 | 476 | 370 | 1,285 | 111 |
| Port Arthur | 117.2 | 13.6 | 86.7 | 47.5 | 857 | 363 | 493 | 105 | 1,896 | 171 |
| Richardson | 203.1 | 5.0 | 77.6 | 138.9 | 1,195 | 678 | 517 | 196 | 1,682 | 188 |
| Rockwall | 55.5 | 0.3 | 100.0 | 47.7 | 1,080 | 486 | 594 | 40 | 897 | 0 |
| Rosenberg | 45.0 | 5.4 | 29.3 | 27.4 | 736 | 255 | 480 | 33 | 878 | 159 |
| Round Rock | 171.4 | 1.2 | 49.6 | 135.8 | 1,098 | 340 | 758 | 151 | 1,220 | 285 |
| Rowlett | 64.3 | 2.4 | 7.1 | 40.1 | 637 | 436 | 201 | 63 | 1,000 | 196 |
| San Angelo | 114.5 | 5.2 | 11.1 | 70.9 | 712 | 352 | 359 | 120 | 1,207 | 234 |
| San Antonio | 2,328.5 | 464.5 | 36.9 | 946.1 | 626 | 295 | 322 | 2,274 | 1,505 | 273 |
| San Juan | 20.5 | 1.1 | 30.8 | 11.6 | 316 | 194 | 123 | 22 | 592 | 93 |
| San Marcos | 103.9 | 19.8 | 100.0 | 50.1 | 789 | 282 | 507 | 88 | 1,377 | 240 |
| Schertz | 54.8 | 3.2 | 8.9 | 31.3 | 780 | 406 | 374 | 46 | 1,144 | 255 |
| Seguin | 142.7 | 16.4 | 93.8 | 16.9 | 581 | 273 | 308 | 172 | 5,928 | 1,084 |
| Sherman | 48.9 | 1.3 | 61.2 | 28.9 | 681 | 220 | 461 | 53 | 1,257 | 114 |
| Socorro | 10.3 | 0.2 | 30.2 | 8.8 | 263 | 201 | 62 | 8 | 247 | 0 |
| Southlake | 89.5 | 0.7 | 65.5 | 73.2 | 2,303 | 1,197 | 1,106 | 84 | 2,655 | 496 |
| Sugar Land | 167.8 | 6.0 | 38.5 | 102.0 | 857 | 326 | 532 | 226 | 1,900 | 931 |
| Temple | 108.4 | 4.6 | 95.0 | 58.3 | 784 | 373 | 411 | 101 | 1,363 | 242 |
| Texarkana | 49.3 | 1.5 | 25.8 | 36.4 | 987 | 439 | 548 | 43 | 1,175 | 139 |
| Texas City | 61.0 | 1.7 | 83.3 | 43.5 | 899 | 467 | 432 | 70 | 1,448 | 199 |
| The Colony | 51.5 | 5.0 | 92.3 | 37.6 | 877 | 494 | 383 | 38 | 883 | 49 |
| Tyler | 139.8 | 18.1 | 7.0 | 79.6 | 760 | 174 | 586 | 126 | 1,205 | 228 |
| Victoria | 86.5 | 3.5 | 30.2 | 55.9 | 833 | 355 | 479 | 88 | 1,318 | 298 |
| Waco | 320.4 | 51.1 | 25.6 | 122.7 | 896 | 470 | 427 | 195 | 1,420 | 146 |
| Waxahachie | 51.1 | 0.2 | 100.0 | 36.6 | 1,037 | 533 | 504 | 42 | 1,185 | 71 |
| Weatherford | 37.5 | 1.7 | 48.3 | 24.0 | 780 | 324 | 456 | 43 | 1,403 | 192 |
| Weslaco | 35.3 | 1.8 | 91.3 | 21.6 | 538 | 265 | 273 | 33 | 808 | 80 |
| Wichita Falls | 114.8 | 16.7 | 37.5 | 65.7 | 630 | 331 | 299 | 124 | 1,189 | 131 |
| Wylie | 57.6 | 4.6 | 64.4 | 39.0 | 788 | 548 | 240 | 50 | 1,010 | 109 |
| **UTAH** | X | X | X | X | X | X | X | X | X | X |
| American Fork | 49.1 | 7.5 | 17.8 | 17.6 | 596 | 249 | 346 | 43 | 1,451 | 229 |
| Bountiful | 32.9 | 3.0 | 58.4 | 15.6 | 354 | 84 | 269 | 43 | 978 | 204 |
| Cedar City | 28.5 | 2.3 | 69.2 | 16.4 | 519 | 183 | 335 | 23 | 737 | 97 |
| Clearfield | 34.8 | 1.3 | 73.8 | 14.8 | 474 | 213 | 260 | 30 | 956 | 103 |
| Cottonwood Heights | 23.2 | 1.4 | 97.1 | 16.5 | 484 | 212 | 273 | 27 | 806 | 336 |
| Draper | 117.0 | 8.7 | 40.0 | 60.8 | 1,274 | 570 | 703 | 68 | 1,415 | 555 |
| Holladay | 2.1 | 0.0 | 0.0 | 1.7 | 56 | 2 | 54 | 3 | 94 | 1 |

1. Based on population estimated as of July 1 of the year shown.

## Table D. Cities — City Government Finances

| City | City government finances, 2017 (cont.) General expenditure (cont.) — Percent of total for: | | | | | | | | | |
|---|---|---|---|---|---|---|---|---|---|---|
| | Public welfare | Highways | Parking facilities | Education | Health and hospitals | Police protection | Sewerage and sanitation | Parks and recreation | Housing and community development | Interest on debt |
| | 127 | 128 | 129 | 130 | 131 | 132 | 133 | 134 | 135 | 136 |
| **TEXAS—Cont'd** | | | | | | | | | | |
| Irving | 0.0 | 7.5 | 0.0 | 0.0 | 0.5 | 16.5 | 18.5 | 15.6 | 1.0 | 6.2 |
| Keller | 0.0 | 15.0 | 0.0 | 0.0 | 0.0 | 18.3 | 7.5 | 12.7 | 0.0 | 8.1 |
| Killeen | 0.0 | 17.4 | 0.0 | 0.0 | 0.7 | 18.9 | 17.2 | 5.5 | 1.1 | 7.1 |
| Kingsville | 0.0 | 14.2 | 0.0 | 0.0 | 2.5 | 37.0 | 18.4 | 1.0 | 0.0 | 8.9 |
| Kyle | 0.0 | 2.7 | 0.0 | 0.0 | 0.8 | 18.1 | 24.5 | 10.5 | 0.0 | 10.3 |
| Lake Jackson | 0.0 | 17.3 | 0.0 | 0.0 | 0.0 | 16.7 | 18.1 | 10.7 | 0.0 | 3.1 |
| Lancaster | 0.4 | 1.0 | 0.0 | 0.0 | 0.5 | 16.7 | 21.3 | 9.8 | 0.0 | 8.3 |
| La Porte | 0.0 | 7.6 | 0.0 | 0.0 | 4.7 | 25.7 | 10.8 | 11.9 | 1.7 | 0.0 |
| Laredo | 0.1 | 1.7 | 0.5 | 0.0 | 4.7 | 18.1 | 14.3 | 4.1 | 3.6 | 9.1 |
| League City | 0.0 | 8.9 | 0.0 | 0.0 | 0.2 | 21.1 | 21.7 | 7.7 | 0.2 | 4.4 |
| Leander | 0.0 | 35.9 | 0.0 | 0.0 | 0.0 | 13.7 | 15.5 | 7.5 | 0.0 | 0.0 |
| Lewisville | 0.0 | 13.4 | 0.0 | 0.0 | 0.5 | 19.2 | 12.9 | 7.1 | 5.4 | 8.0 |
| Little Elm | 0.0 | 3.7 | 0.0 | 0.0 | 1.4 | 17.9 | 12.0 | 9.5 | 0.0 | 6.1 |
| Longview | 0.0 | 8.7 | 0.0 | 0.0 | 1.4 | 17.9 | 13.5 | 11.2 | 6.7 | 2.1 |
| Lubbock | 0.0 | 7.3 | 0.0 | 0.0 | 1.3 | 14.6 | 25.0 | 4.2 | 0.7 | 7.9 |
| Lufkin | 0.0 | 11.7 | 0.0 | 0.0 | 1.4 | 19.2 | 26.3 | 6.6 | 0.5 | 4.2 |
| McAllen | 0.8 | 11.1 | 0.5 | 0.0 | 1.3 | 16.6 | 27.7 | 8.8 | 1.0 | 2.0 |
| McKinney | 0.0 | 7.1 | 0.0 | 0.0 | 0.8 | 12.4 | 12.7 | 7.2 | 0.5 | 0.0 |
| Mansfield | 0.0 | 22.3 | 0.0 | 0.0 | 1.8 | 14.6 | 5.8 | 13.6 | 0.0 | 5.7 |
| Mesquite | 0.3 | 5.0 | 0.0 | 0.0 | 0.8 | 24.1 | 11.4 | 8.4 | 9.9 | 3.9 |
| Midland | 0.0 | 4.9 | 0.0 | 0.0 | 3.4 | 17.7 | 15.2 | 12.9 | 1.0 | 1.6 |
| Mission | 0.0 | 5.9 | 0.0 | 0.0 | 0.6 | 20.7 | 14.0 | 5.0 | 1.6 | 17.7 |
| Missouri City | 0.0 | 11.5 | 0.0 | 0.0 | 0.4 | 19.9 | 10.7 | 11.3 | 0.5 | 6.6 |
| Nacogdoches | 0.0 | 6.4 | 0.0 | 0.0 | 1.4 | 23.2 | 22.6 | 4.7 | 0.0 | 0.8 |
| New Braunfels | 0.0 | 18.2 | 0.0 | 0.0 | 1.6 | 14.8 | 15.9 | 7.8 | 0.0 | 4.9 |
| North Richland Hills | 0.0 | 12.3 | 0.0 | 0.0 | 0.0 | 19.9 | 3.2 | 11.5 | 0.4 | 6.7 |
| Odessa | 0.0 | 10.1 | 0.0 | 0.0 | 0.8 | 19.3 | 19.0 | 5.5 | 4.5 | 2.7 |
| Paris | 0.0 | 10.0 | 0.0 | 0.0 | 10.3 | 18.4 | 14.3 | 4.0 | 9.5 | 1.4 |
| Pasadena | 0.0 | 10.7 | 0.0 | 0.0 | 2.3 | 30.4 | 14.2 | 7.9 | 0.0 | 1.7 |
| Pearland | 0.0 | 9.3 | 0.0 | 0.0 | 3.6 | 16.6 | 18.5 | 8.2 | 0.1 | 10.6 |
| Pflugerville | 0.0 | 20.8 | 0.0 | 0.0 | 0.9 | 22.0 | 18.1 | 13.9 | 0.1 | 0.0 |
| Pharr | NA | NA | NA | NA | NA | NA | NA | NA | NA | NA |
| Plano | 0.0 | 4.7 | 0.0 | 0.0 | 1.0 | 17.6 | 17.0 | 12.3 | 0.7 | 3.6 |
| Port Arthur | 2.1 | 10.1 | 0.0 | 0.0 | 1.7 | 21.6 | 14.2 | 4.6 | 2.3 | 2.5 |
| Richardson | 0.0 | 12.3 | 0.0 | 0.0 | 0.9 | 13.6 | 21.5 | 14.9 | 0.0 | 5.9 |
| Rockwall | 0.0 | 6.7 | 0.0 | 0.0 | 0.0 | 26.5 | 11.9 | 10.5 | 0.0 | 12.4 |
| Rosenberg | 0.0 | 16.8 | 0.0 | 0.0 | 0.9 | 25.3 | 18.0 | 3.5 | 0.0 | 4.8 |
| Round Rock | 0.0 | 17.8 | 0.0 | 0.0 | 0.4 | 16.5 | 7.1 | 11.2 | 1.2 | 7.5 |
| Rowlett | 0.0 | 7.5 | 0.0 | 0.0 | 1.5 | 18.2 | 24.0 | 8.3 | 0.3 | 4.4 |
| San Angelo | 0.1 | 5.3 | 0.0 | 0.0 | 2.8 | 16.8 | 9.8 | 14.0 | 2.3 | 3.7 |
| San Antonio | 5.6 | 6.2 | 0.6 | 2.8 | 1.3 | 17.1 | 15.5 | 10.3 | 2.2 | 6.1 |
| San Juan | 0.0 | 7.8 | 0.0 | 0.0 | 0.0 | 18.9 | 24.1 | 7.7 | 0.0 | 3.6 |
| San Marcos | 0.0 | 10.9 | 0.0 | 0.0 | 3.8 | 19.5 | 17.4 | 4.6 | 0.4 | 7.3 |
| Schertz | 0.1 | 4.4 | 0.0 | 0.0 | 12.2 | 11.4 | 17.1 | 18.8 | 0.0 | 4.2 |
| Seguin | 0.0 | 4.9 | 0.0 | 0.0 | 60.4 | 4.2 | 7.0 | 2.5 | 0.0 | 4.7 |
| Sherman | 0.1 | 10.3 | 0.0 | 0.0 | 1.3 | 17.9 | 16.1 | 5.9 | 0.7 | 0.0 |
| Socorro | 0.0 | 15.4 | 0.0 | 0.0 | 4.7 | 33.3 | 0.0 | 9.6 | 0.0 | 10.4 |
| Southlake | 0.0 | 5.1 | 0.0 | 0.0 | 0.0 | 10.9 | 6.2 | 27.9 | 0.0 | 0.0 |
| Sugar Land | 0.0 | 12.9 | 0.0 | 0.0 | 0.5 | 9.7 | 7.6 | 34.4 | 0.0 | 6.6 |
| Temple | 0.0 | 5.2 | 0.0 | 0.0 | 1.0 | 17.8 | 20.7 | 10.4 | 0.0 | 4.0 |
| Texarkana | 0.0 | 12.6 | 0.0 | 0.0 | 2.2 | 20.7 | 19.0 | 5.2 | 2.7 | 4.3 |
| Texas City | 0.0 | 15.9 | 0.0 | 0.0 | 0.5 | 17.9 | 17.0 | 12.8 | 0.4 | 3.4 |
| The Colony | 0.0 | 10.2 | 0.0 | 0.0 | 0.0 | 22.4 | 9.3 | 7.8 | 8.1 | 0.1 |
| Tyler | 0.0 | 10.1 | 0.0 | 0.0 | 0.5 | 21.9 | 21.9 | 3.3 | 0.7 | 4.4 |
| Victoria | 0.1 | 12.8 | 0.0 | 0.0 | 0.0 | 15.6 | 19.5 | 7.8 | 0.7 | 5.6 |
| Waco | 0.0 | 4.6 | 0.0 | 0.0 | 4.7 | 21.4 | 16.2 | 13.0 | 1.9 | 8.0 |
| Waxahachie | 0.0 | 7.5 | 0.0 | 0.0 | 2.3 | 19.1 | 14.2 | 8.4 | 0.0 | 5.6 |
| Weatherford | 0.0 | 13.7 | 0.0 | 0.0 | 2.9 | 18.7 | 16.3 | 5.5 | 0.0 | 0.0 |
| Weslaco | 0.0 | 6.7 | 0.0 | 0.0 | 0.0 | 20.5 | 23.0 | 5.2 | 0.0 | 1.4 |
| Wichita Falls | 0.0 | 7.6 | 0.0 | 0.0 | 4.2 | 19.0 | 14.3 | 6.1 | 4.3 | 6.3 |
| Wylie | 0.0 | 15.2 | 0.0 | 0.0 | 1.0 | 17.3 | 14.5 | 10.4 | 0.0 | |
| **UTAH** | X | X | X | X | X | X | X | X | X | X |
| American Fork | 1.2 | 5.7 | 0.0 | 0.8 | 11.8 | 15.4 | 6.6 | 13.4 | 2.8 | 0.3 |
| Bountiful | 0.0 | 13.7 | 0.0 | 0.0 | 0.0 | 28.7 | 6.9 | 13.3 | 0.8 | 0.0 |
| Cedar City | 0.0 | 12.2 | 0.1 | 0.0 | 0.9 | 19.1 | 9.7 | 18.0 | 7.2 | 0.2 |
| Clearfield | 0.0 | 1.6 | 0.0 | 0.0 | 0.0 | 14.9 | 0.0 | 0.0 | 3.9 | 2.1 |
| Cottonwood Heights | 0.0 | 1.6 | 0.0 | 0.0 | 0.5 | 19.4 | 0.0 | 0.0 | 10.1 | 0.2 |
| Draper | 0.0 | 10.9 | 0.0 | 0.0 | 0.0 | 0.0 | 3.9 | 9.8 | 10.1 | 0.7 |
| Holladay | 0.0 | 0.0 | 9.3 | 0.0 | 1.6 | 0.0 | 0.5 | 0.9 | 4.9 | 0.0 |

# Table D. Cities — City Government Finances, City Government Employment, and Climate

| City | City government finances, 2017 (cont.) Debt outstanding — Total (mil dol) | Per capita[1] (dollars) | Debt issued during year | Climate[2] — Average daily temperature — Mean — January | Mean July | Limits January[3] | Limits July[4] | Annual precipitation (inches) | Heating degree days | Cooling degree days |
|---|---|---|---|---|---|---|---|---|---|---|
| | 137 | 138 | 139 | 140 | 141 | 142 | 143 | 144 | 145 | 146 |
| **TEXAS—Cont'd** | | | | | | | | | | |
| Irving | 668.0 | 2,776 | 304.7 | 45.9 | 86.5 | 36.4 | 96.1 | 37.05 | 2,219 | 2,878 |
| Keller | 84.0 | 1,783 | 0.0 | 43.0 | 84.1 | 31.4 | 95.7 | 34.12 | 2,608 | 2,358 |
| Killeen | 284.6 | 1,961 | 53.0 | 46.0 | 83.5 | 34.0 | 95.3 | 32.88 | 2,190 | 2,477 |
| Kingsville | 27.0 | 1,064 | 0.0 | 55.9 | 84.3 | 43.4 | 95.5 | 29.03 | 1,001 | 3,404 |
| Kyle | 63.8 | 1,479 | 0.0 | NA | NA | NA | NA | NA | NA | NA |
| Lake Jackson | 53.1 | 1,938 | 7.8 | 54.0 | 83.7 | 45.4 | 90.2 | 50.66 | 1,234 | 3,003 |
| Lancaster | 135.2 | 3,431 | 0.0 | 44.4 | 84.0 | 33.3 | 95.7 | 38.69 | 2,380 | 2,452 |
| La Porte | 35.5 | 1,003 | 3.2 | 51.6 | 83.6 | 41.9 | 91.6 | 53.75 | 1,471 | 2,841 |
| Laredo | 524.2 | 2,021 | 0.0 | 55.6 | 88.5 | 43.7 | 101.6 | 21.53 | 931 | 4,213 |
| League City | 210.4 | 2,015 | 0.0 | 52.7 | 82.7 | 43.1 | 91.2 | 51.73 | 1,365 | 2,815 |
| Leander | 127.2 | 2,555 | 0.0 | NA | NA | NA | NA | NA | NA | NA |
| Lewisville | 164.0 | 1,541 | 0.0 | 42.7 | 83.6 | 32.0 | 94.1 | 37.79 | 2,650 | 2,269 |
| Little Elm | 92.5 | 2,012 | 27.8 | NA | NA | NA | NA | NA | NA | NA |
| Longview | 69.9 | 863 | 0.0 | 45.4 | 83.4 | 33.7 | 94.5 | 49.06 | 2,319 | 2,355 |
| Lubbock | 1,242.5 | 4,883 | 135.5 | 38.1 | 79.8 | 24.4 | 91.9 | 18.69 | 3,508 | 1,769 |
| Lufkin | 158.4 | 4,451 | 9.1 | 48.6 | 82.6 | 37.9 | 93.5 | 46.62 | 1,900 | 2,480 |
| McAllen | 253.6 | 1,783 | 52.7 | 58.7 | 85.1 | 48.2 | 95.5 | 22.61 | 719 | 3,898 |
| McKinney | 383.2 | 2,104 | 100.2 | 41.8 | 82.4 | 31.1 | 92.7 | 41.01 | 2,843 | 2,060 |
| Mansfield | 199.3 | 2,894 | 79.2 | 43.7 | 84.3 | 33.2 | 94.9 | 34.54 | 2,437 | 2,508 |
| Mesquite | 145.4 | 1,008 | 0.0 | 45.9 | 86.5 | 36.4 | 96.1 | 37.05 | 2,219 | 2,878 |
| Midland | 78.3 | 575 | 0.0 | 44.5 | 81.8 | 29.5 | 95.6 | 14.84 | 2,479 | 2,241 |
| Mission | 361.0 | 4,320 | 0.0 | 58.8 | 86.3 | 47.5 | 97.7 | 22.13 | 740 | 4,128 |
| Missouri City | 160.9 | 2,156 | 0.0 | 51.8 | 84.1 | 41.6 | 93.7 | 49.34 | 1,475 | 2,950 |
| Nacogdoches | 46.6 | 1,402 | 1.7 | 46.5 | 83.9 | 36.4 | 93.5 | 48.36 | 2,150 | 2,555 |
| New Braunfels | 112.8 | 1,430 | 0.0 | 48.6 | 82.7 | 35.5 | 94.7 | 35.74 | 1,840 | 2,545 |
| North Richland Hills | 136.7 | 1,939 | 8.5 | 44.1 | 85.0 | 34.0 | 95.4 | 34.73 | 2,370 | 2,568 |
| Odessa | 52.9 | 454 | 0.0 | 43.2 | 81.7 | 29.6 | 94.3 | 14.80 | 2,716 | 2,139 |
| Paris | 15.5 | 627 | 0.0 | 40.6 | 83.1 | 29.9 | 94.3 | 47.82 | 2,972 | 2,197 |
| Pasadena | 215.9 | 1,403 | 158.1 | 54.3 | 84.5 | 45.2 | 93.6 | 53.96 | 1,174 | 3,179 |
| Pearland | 231.7 | 1,942 | 0.0 | 54.3 | 84.5 | 45.2 | 93.6 | 53.96 | 1,174 | 3,179 |
| Pflugerville | 222.8 | 3,521 | 69.0 | NA | NA | NA | NA | NA | NA | NA |
| Pharr | 109.7 | 1,395 | 0.0 | 60.1 | 85.9 | 50.3 | 96.1 | 22.96 | 624 | 4,181 |
| Plano | 135.4 | 471 | 0.0 | 44.1 | 85.0 | 34.0 | 95.4 | 34.73 | 2,370 | 2,568 |
| Port Arthur | 57.3 | 1,033 | 0.0 | 52.2 | 82.7 | 42.9 | 91.6 | 59.89 | 1,447 | 2,823 |
| Richardson | 238.9 | 2,056 | 35.4 | 45.9 | 86.5 | 36.4 | 96.1 | 37.05 | 2,219 | 2,878 |
| Rockwall | 125.1 | 2,834 | 0.0 | NA | NA | NA | NA | NA | NA | NA |
| Rosenberg | 17.0 | 456 | 0.0 | NA | NA | NA | NA | NA | NA | NA |
| Round Rock | 247.2 | 1,999 | 0.0 | 47.2 | 83.8 | 35.1 | 95.7 | 36.42 | 1,998 | 2,584 |
| Rowlett | 98.5 | 1,564 | 7.0 | 42.1 | 82.8 | 30.8 | 94.2 | 40.06 | 2,710 | 2,212 |
| San Angelo | 207.7 | 2,084 | 0.0 | 44.9 | 82.4 | 31.8 | 94.4 | 20.91 | 2,396 | 2,383 |
| San Antonio | 11,712.5 | 7,751 | 1,820.6 | 50.3 | 84.3 | 38.6 | 94.6 | 32.92 | 1,573 | 3,038 |
| San Juan | 19.6 | 535 | 0.0 | 60.1 | 85.9 | 50.3 | 96.1 | 22.96 | 624 | 4,181 |
| San Marcos | 70.8 | 1,115 | 0.0 | 49.9 | 84.4 | 38.6 | 95.1 | 37.19 | 1,629 | 2,913 |
| Schertz | 70.3 | 1,752 | 13.9 | NA | NA | NA | NA | NA | NA | NA |
| Seguin | 228.6 | 7,861 | 148.9 | NA | NA | NA | NA | NA | NA | NA |
| Sherman | 175.0 | 4,131 | 75.3 | 41.5 | 82.8 | 32.2 | 92.7 | 42.04 | 2,850 | 2,137 |
| Socorro | 23.4 | 697 | 3.6 | 44.9 | 83.6 | 29.2 | 98.7 | 9.71 | 2,557 | 2,372 |
| Southlake | 171.2 | 5,389 | 26.8 | NA | NA | NA | NA | NA | NA | NA |
| Sugar Land | 469.7 | 3,948 | 88.4 | 51.8 | 84.1 | 41.6 | 93.7 | 49.34 | 1,475 | 2,950 |
| Temple | 49.6 | 667 | 0.0 | 46.1 | 83.7 | 34.9 | 95.0 | 35.81 | 2,191 | 2,551 |
| Texarkana | 55.6 | 1,508 | 0.0 | 41.6 | 82.6 | 30.7 | 93.1 | 51.24 | 2,893 | 2,138 |
| Texas City | 69.9 | 1,444 | 0.0 | 55.8 | 84.3 | 49.7 | 88.7 | 43.84 | 1,008 | 3,268 |
| The Colony | 121.3 | 2,827 | 17.7 | 42.7 | 83.6 | 32.0 | 94.1 | 37.79 | 2,650 | 2,269 |
| Tyler | 55.5 | 530 | 0.0 | 47.5 | 83.4 | 37.7 | 93.6 | 45.27 | 1,958 | 2,521 |
| Victoria | 163.7 | 2,442 | 2.4 | 53.2 | 84.2 | 43.6 | 93.4 | 40.10 | 1,248 | 3,203 |
| Waco | 269.6 | 1,968 | 0.0 | 46.1 | 85.4 | 35.1 | 96.7 | 33.34 | 2,164 | 2,840 |
| Waxahachie | 159.0 | 4,511 | 0.0 | NA | NA | NA | NA | NA | NA | NA |
| Weatherford | 107.3 | 3,493 | 6.1 | NA | NA | NA | NA | NA | NA | NA |
| Weslaco | 80.5 | 2,003 | 0.0 | 58.6 | 84.5 | 47.7 | 95.4 | 25.37 | 755 | 3,791 |
| Wichita Falls | 186.9 | 1,792 | 101.4 | 40.5 | 84.8 | 28.9 | 97.2 | 28.83 | 3,024 | 2,396 |
| Wylie | 89.5 | 1,807 | 35.9 | NA | NA | NA | NA | NA | NA | NA |
| **UTAH** | X | X | X | X | X | X | X | X | X | X |
| American Fork | 0.0 | 0 | 0.0 | NA | NA | NA | NA | NA | NA | NA |
| Bountiful | 0.0 | 0 | 0.0 | 29.1 | 75.8 | 21.6 | 88.4 | 22.40 | 5,937 | 861 |
| Cedar City | 1.6 | 51 | 1.6 | NA | NA | NA | NA | NA | NA | NA |
| Clearfield | 2.4 | 77 | 2.4 | 27.6 | 74.2 | 18.6 | 89.9 | 20.75 | 6,142 | 746 |
| Cottonwood Heights | 4.1 | 120 | 4.1 | NA | NA | NA | NA | NA | NA | NA |
| Draper | 13.5 | 283 | 3.8 | 31.6 | 78.0 | 22.0 | 95.3 | 15.76 | 5,251 | 1,172 |
| Holladay | 0.0 | 0 | 0.0 | NA | NA | NA | NA | NA | NA | NA |

1. Based on the population estimated as of July 1 of the year shown.  2. Represents normal values based on the 30-year period, 1971±2000.  3. Average daily minimum.  4. Average daily maximum.

| STATE Place code | | City | Land area[1] (sq. mi) | Population, 2020 Total persons 2019 | Rank | Per square mile | Race 2019 Race alone[2] (percent) White | Black or African American | American Indian, Alaskan Native | Asian | Hawaiian Pacific Islander | Some other race | Two or more races (percent) |
|---|---|---|---|---|---|---|---|---|---|---|---|---|---|
| | | | 1 | 2 | 3 | 4 | 5 | 6 | 7 | 8 | 9 | 10 | 11 |
| | | UTAH— Cont'd | | | | | | | | | | | |
| 49 | 40360 | Kaysville | 10.5 | 32,603 | 1,173 | 3,105 | D | D | D | D | D | D | D |
| 49 | 43660 | Layton | 22.5 | 79,012 | 446 | 3,512 | 90.5 | 0.9 | 0.6 | 2.6 | 0.2 | 0.8 | 4.4 |
| 49 | 44320 | Lehi | 28.1 | 73,383 | 489 | 2,612 | 95.1 | 0.6 | 0.3 | 1.3 | 0.6 | 0.2 | 1.9 |
| 49 | 45860 | Logan | 17.8 | 51,680 | 749 | 2,903 | 90.2 | 1.7 | 0.4 | 4.5 | 0.0 | 1.2 | 2.0 |
| 49 | 49710 | Midvale | 5.9 | 34,434 | 1,120 | 5,836 | D | D | D | D | D | D | D |
| 49 | 53230 | Murray | 12.3 | 48,355 | 805 | 3,931 | 89.9 | 2.4 | 0.4 | 2.3 | 0.1 | 1.6 | 3.4 |
| 49 | 55980 | Ogden | 27.5 | 87,387 | 388 | 3,178 | 88.4 | 2.1 | 1.6 | 1.2 | 0.2 | 3.0 | 3.5 |
| 49 | 57300 | Orem | 18.6 | 98,970 | 321 | 5,321 | 90.6 | 0.3 | 0.8 | 1.1 | 0.6 | 1.4 | 5.2 |
| 49 | 60930 | Pleasant Grove | 9.2 | 38,241 | 1,012 | 4,157 | 91.5 | 0.2 | 0.0 | 1.5 | 3.1 | 1.4 | 2.3 |
| 49 | 62470 | Provo | 41.7 | 116,295 | 252 | 2,789 | 86.6 | 0.9 | 0.7 | 2.8 | 1.3 | 2.1 | 5.6 |
| 49 | 64340 | Riverton | 12.6 | 44,369 | 866 | 3,521 | 94.0 | 0.6 | 1.0 | 1.5 | 0.0 | 1.1 | 1.8 |
| 49 | 65110 | Roy | 8.1 | 39,559 | 980 | 4,884 | 88.9 | 0.7 | 0.3 | 1.4 | 0.2 | 2.2 | 6.2 |
| 49 | 65330 | St. George | 78.5 | 92,378 | 362 | 1,177 | 89.2 | 0.9 | 1.3 | 0.2 | 0.2 | 4.6 | 3.5 |
| 49 | 67000 | Salt Lake City | 110.3 | 204,087 | 115 | 1,850 | 71.6 | 2.8 | 1.8 | 6.8 | 3.2 | 10.0 | 3.8 |
| 49 | 67440 | Sandy | 24.2 | 94,871 | 344 | 3,920 | 86.7 | 1.4 | 0.1 | 5.6 | 0.1 | 3.5 | 2.6 |
| 49 | 70850 | South Jordan | 22.2 | 78,503 | 451 | 3,536 | 88.6 | 0.2 | 0.5 | 3.4 | 0.9 | 4.3 | 2.3 |
| 49 | 71290 | Spanish Fork | 16.2 | 41,268 | 931 | 2,547 | 94.8 | 0.3 | 0.8 | 0.7 | 0.3 | 0.8 | 2.2 |
| 49 | 72280 | Springville | 14.4 | 33,572 | 1,146 | 2,331 | 95.8 | 0.0 | 0.2 | 0.3 | 1.3 | 0.5 | 1.8 |
| 49 | 75360 | Taylorsville | 10.8 | 59,402 | 633 | 5,500 | 88.5 | 0.2 | 0.6 | 6.2 | 0.1 | 2.6 | 1.8 |
| 49 | 76680 | Tooele | 24.1 | 37,046 | 1,036 | 1,537 | D | D | D | D | D | D | D |
| 49 | 82950 | West Jordan | 32.3 | 116,530 | 250 | 3,608 | 79.5 | 0.8 | 0.3 | 2.2 | 1.8 | 11.4 | 4.1 |
| 49 | 83470 | West Valley City | 35.8 | 133,894 | 206 | 3,740 | 65.8 | 4.2 | 1.0 | 5.1 | 2.6 | 16.7 | 4.7 |
| 50 | 00000 | VERMONT | 9,217.9 | 623,347 | X | 68 | 93.8 | 1.5 | 0.5 | 1.4 | 0.1 | 0.3 | 2.5 |
| 50 | 10675 | Burlington | 10.3 | 42,899 | 897 | 4,165 | 89.9 | 6.0 | 0.0 | 1.5 | 0.0 | 0.2 | 2.4 |
| 51 | 00000 | VIRGINIA | 39,482.1 | 8,590,563 | X | 218 | 67.0 | 19.4 | 0.3 | 6.6 | 0.1 | 2.9 | 3.8 |
| 51 | 01000 | Alexandria | 14.9 | 158,726 | 159 | 10,653 | 63.4 | 20.9 | 0.1 | 5.8 | 0.0 | 4.6 | 5.3 |
| 51 | 07784 | Blacksburg | 19.7 | 44,074 | 872 | 2,237 | 77.4 | 6.9 | 1.6 | 11.9 | 0.2 | 0.6 | 1.5 |
| 51 | 14968 | Charlottesville | 10.2 | 46,950 | 828 | 4,603 | 71.0 | 18.7 | 0.5 | 6.8 | 0.0 | 0.4 | 2.6 |
| 51 | 16000 | Chesapeake | 338.5 | 247,011 | 91 | 730 | 60.1 | 30.0 | 0.2 | 3.5 | 0.0 | 2.3 | 3.8 |
| 51 | 21344 | Danville | 42.8 | 39,869 | 966 | 932 | 43.5 | 49.6 | 0.2 | 1.3 | 0.0 | 5.4 | |
| 51 | 35000 | Hampton | 51.5 | 135,464 | 202 | 2,630 | 40.1 | 50.0 | 0.0 | 2.6 | 0.2 | 2.6 | 4.5 |
| 51 | 35624 | Harrisonburg | 17.3 | 53,204 | 731 | 3,075 | 78.2 | 9.7 | 0.4 | 2.5 | 0.0 | 2.1 | 7.0 |
| 51 | 44984 | Leesburg | 12.4 | 54,488 | 709 | 4,394 | 66.6 | 13.2 | 0.4 | 13.2 | 0.2 | 2.6 | 3.9 |
| 51 | 47672 | Lynchburg | 49.0 | 81,561 | 422 | 1,665 | 63.7 | 28.8 | 0.3 | 1.8 | 0.0 | 2.1 | 3.3 |
| 51 | 48952 | Manassas | 9.8 | 40,869 | 941 | 4,170 | 72.6 | 13.0 | 0.1 | 4.0 | 0.0 | 2.4 | 7.9 |
| 51 | 56000 | Newport News | 69.0 | 179,062 | 145 | 2,595 | 48.1 | 41.4 | 0.3 | 3.1 | 0.2 | 1.7 | 5.1 |
| 51 | 57000 | Norfolk | 53.3 | 242,803 | 92 | 4,555 | 47.1 | 41.2 | 0.3 | 3.7 | 0.0 | 3.1 | 4.6 |
| 51 | 61832 | Petersburg | 22.7 | 30,446 | 1,235 | 1,341 | D | D | D | D | D | D | D |
| 51 | 64000 | Portsmouth | 33.3 | 95,094 | 340 | 2,856 | 39.8 | 54.0 | 0.0 | 1.3 | 0.2 | 1.2 | 3.5 |
| 51 | 67000 | Richmond | 59.9 | 232,226 | 96 | 3,877 | 44.6 | 45.2 | 0.2 | 2.0 | 0.0 | 4.5 | 3.4 |
| 51 | 68000 | Roanoke | 42.5 | 99,058 | 320 | 2,331 | 61.3 | 30.8 | 0.0 | 2.6 | 0.0 | 0.8 | 4.5 |
| 51 | 76432 | Suffolk | 399.2 | 93,913 | 354 | 235 | 50.4 | 42.1 | 0.0 | 1.9 | 0.0 | 1.0 | 4.7 |
| 51 | 82000 | Virginia Beach | 244.7 | 451,231 | 44 | 1,844 | 65.6 | 18.9 | 0.3 | 7.2 | 0.1 | 1.9 | 5.9 |
| 51 | 86720 | Winchester | 9.2 | 27,700 | 1,328 | 3,011 | D | D | D | D | D | D | D |
| 53 | 00000 | WASHINGTON | 66,455.1 | 7,693,612 | X | 116 | 74.2 | 4.0 | 1.4 | 9.0 | 0.7 | 4.8 | 6.0 |
| 53 | 03180 | Auburn | 29.6 | 81,024 | 431 | 2,737 | 54.9 | 8.9 | 3.3 | 11.7 | 0.4 | 10.0 | 10.8 |
| 53 | 05210 | Bellevue | 33.5 | 148,073 | 177 | 4,420 | 55.8 | 1.2 | 0.7 | 37.9 | 0.2 | 1.1 | 3.1 |
| 53 | 05280 | Bellingham | 28.1 | 93,955 | 353 | 3,344 | 84.7 | 1.0 | 1.4 | 4.6 | 0.2 | 2.5 | 5.6 |
| 53 | 07380 | Bothell | 13.6 | 47,524 | 818 | 3,494 | 67.3 | 0.6 | 0.0 | 23.1 | 0.0 | 4.7 | 4.4 |
| 53 | 07695 | Bremerton | 28.4 | 42,322 | 910 | 1,490 | 68.6 | 4.6 | 0.8 | 8.0 | 0.3 | 9.7 | 8.0 |
| 53 | 08850 | Burien | 10.0 | 51,268 | 761 | 5,127 | 50.9 | 8.3 | 0.4 | 13.7 | 0.0 | 18.6 | 8.1 |
| 53 | 17635 | Des Moines | 6.4 | 32,389 | 1,179 | 5,061 | 55.2 | 7.5 | 1.2 | 15.6 | 0.7 | 9.1 | 10.7 |
| 53 | 20750 | Edmonds | 8.9 | 42,515 | 906 | 4,777 | 74.1 | 4.7 | 0.1 | 9.4 | 1.0 | 2.9 | 7.7 |
| 53 | 22640 | Everett | 33.2 | 112,482 | 263 | 3,388 | 68.9 | 7.7 | 1.4 | 11.0 | 0.0 | 4.8 | 6.2 |
| 53 | 23515 | Federal Way | 22.3 | 96,307 | 331 | 4,319 | 41.0 | 21.2 | 0.1 | 13.6 | 3.7 | 11.3 | 9.0 |
| 53 | 33805 | Issaquah | 12.1 | 39,681 | 973 | 3,279 | D | D | D | D | D | D | D |
| 53 | 35275 | Kennewick | 27.4 | 84,515 | 405 | 3,085 | 79.9 | 4.4 | 0.7 | 2.0 | 0.0 | 9.9 | 3.1 |
| 53 | 35415 | Kent | 33.8 | 130,676 | 214 | 3,866 | 41.1 | 11.6 | 0.3 | 24.5 | 2.8 | 11.8 | 8.1 |
| 53 | 35940 | Kirkland | 17.8 | 95,431 | 337 | 5,361 | 70.0 | 1.5 | 0.3 | 17.4 | 0.0 | 3.9 | 6.8 |
| 53 | 36745 | Lacey | 17.2 | 54,178 | 715 | 3,150 | 75.3 | 6.4 | 0.8 | 6.5 | 1.4 | 4.0 | 5.6 |
| 53 | 37900 | Lake Stevens | 9.2 | 34,794 | 1,109 | 3,782 | 83.2 | 3.0 | 1.6 | 3.9 | 0.5 | 1.0 | 6.8 |
| 53 | 38038 | Lakewood | 17.1 | 60,976 | 613 | 3,566 | 56.3 | 12.5 | 0.8 | 9.4 | 2.4 | 8.8 | 9.7 |
| 53 | 40245 | Longview | 14.8 | 38,515 | 1,002 | 2,602 | 85.0 | 0.9 | 0.1 | 2.5 | 0.0 | 1.1 | 10.3 |
| 53 | 40840 | Lynnwood | 7.9 | 39,042 | 993 | 4,942 | 62.2 | 5.7 | 1.0 | 22.7 | 0.4 | 0.0 | 7.9 |
| 53 | 43955 | Marysville | 20.8 | 71,392 | 511 | 3,432 | 79.5 | 2.0 | 2.7 | 5.1 | 0.9 | 4.1 | 5.6 |
| 53 | 47560 | Mount Vernon | 12.3 | 36,336 | 1,061 | 2,954 | 74.2 | 1.2 | 4.2 | 3.5 | 0.0 | 11.4 | 5.6 |
| 53 | 51300 | Olympia | 18.2 | 53,620 | 723 | 2,946 | 80.5 | 1.5 | 2.2 | 5.8 | 1.6 | 0.6 | 7.8 |

1. Dry land or land partially or temporarily covered by water.    2. Hispanic or Latino persons may be of any race.

# Table D. Cities — **Population**

| City | Percent Hispanic or Latino[1], 2019 | Percent foreign born, 2019 | Age of population (percent), 2019 | | | | | | | Median age, 2019 | Percent female, 2019 | Population Census counts | | Percent change | |
| | | | Under 18 years | 18 to 24 years | 25 to 34 years | 35 to 44 years | 45 to 54 years | 55 to 64 years | 65 years and over | | | 2000 | 2010 | 2000-2010 | 2001-2020 |
| | 12 | 13 | 14 | 15 | 16 | 17 | 18 | 19 | 20 | 21 | 22 | 23 | 24 | 25 | 26 |
|---|---|---|---|---|---|---|---|---|---|---|---|---|---|---|---|
| **UTAH— Cont'd** | | | | | | | | | | | | | | | |
| Kaysville | 3.8 | 2.5 | 39.9 | 5.2 | 11.8 | 17.8 | 7.7 | 10.8 | 6.9 | 29.5 | 47.0 | 20,351 | 27,573 | 35.5 | 18.2 |
| Layton | 13.8 | 5.5 | 28.9 | 10.2 | 15.2 | 14.4 | 12.8 | 8.9 | 9.6 | 31.9 | 48.0 | 58,474 | 67,531 | 15.5 | 17.0 |
| Lehi | 9.6 | 7.9 | 43.5 | 5.4 | 12.9 | 17.0 | 9.1 | 6.1 | 5.9 | 26.6 | 49.2 | 19,028 | 47,757 | 151.0 | 53.7 |
| Logan | 13.4 | 10.2 | 22.9 | 28.1 | 19.4 | 9.6 | 6.2 | 5.3 | 8.5 | 24.7 | 52.7 | 42,670 | 48,205 | 13.0 | 7.2 |
| Midvale | 8.0 | 7.0 | 22.3 | 10.3 | 27.5 | 16.9 | 7.5 | 9.5 | 6.0 | 31.3 | 49.2 | 27,029 | 28,000 | 3.6 | 23.0 |
| Murray | 12.9 | 6.3 | 19.7 | 10.2 | 14.4 | 15.0 | 10.7 | 11.1 | 18.9 | 39.4 | 50.0 | 34,024 | 46,680 | 37.2 | 3.6 |
| Ogden | 31.8 | 10.1 | 25.2 | 13.4 | 17.6 | 14.7 | 9.6 | 9.9 | 9.6 | 30.9 | 50.4 | 77,226 | 82,893 | 7.3 | 5.4 |
| Orem | 19.0 | 13.8 | 28.1 | 17.7 | 17.4 | 11.1 | 8.8 | 7.5 | 9.4 | 27.4 | 50.5 | 84,324 | 88,344 | 4.8 | 12.0 |
| Pleasant Grove | 6.1 | 4.2 | 33.0 | 12.3 | 15.9 | 11.0 | 10.1 | 8.9 | 8.8 | 28.7 | 48.9 | 23,468 | 33,547 | 42.9 | 14.0 |
| Provo | 15.7 | 10.0 | 19.5 | 39.3 | 16.0 | 8.1 | 5.2 | 5.6 | 6.2 | 23.6 | 49.8 | 105,166 | 112,485 | 7.0 | 3.4 |
| Riverton | 6.4 | 2.7 | 33.1 | 9.6 | 13.5 | 12.8 | 12.6 | 12.1 | 6.2 | 30.6 | 46.5 | 25,011 | 38,832 | 55.3 | 14.3 |
| Roy | 24.2 | 7.7 | 28.1 | 7.4 | 15.9 | 15.7 | 10.9 | 8.9 | 13.2 | 33.9 | 46.6 | 32,885 | 37,450 | 13.9 | 5.6 |
| St. George | 12.8 | 6.7 | 27.2 | 10.8 | 13.2 | 10.2 | 7.7 | 8.3 | 22.6 | 33.5 | 50.3 | 49,663 | 72,759 | 46.5 | 27.0 |
| Salt Lake City | 19.8 | 17.8 | 18.3 | 13.4 | 21.8 | 13.8 | 11.0 | 10.2 | 11.4 | 32.9 | 48.9 | 181,743 | 186,430 | 2.6 | 9.5 |
| Sandy | 10.2 | 11.7 | 24.8 | 7.6 | 14.1 | 13.6 | 12.7 | 11.4 | 15.7 | 37.0 | 50.5 | 88,418 | 90,176 | 2.0 | 5.2 |
| South Jordan | 9.9 | 5.7 | 33.7 | 7.0 | 14.5 | 13.0 | 10.9 | 9.6 | 11.2 | 31.7 | 52.9 | 29,437 | 50,473 | 71.5 | 55.5 |
| Spanish Fork | 14.5 | 5.6 | 36.2 | 12.3 | 12.5 | 14.5 | 8.5 | 6.9 | 9.2 | 27.0 | 49.6 | 20,246 | 34,764 | 71.7 | 18.7 |
| Springville | 14.9 | 7.2 | 37.4 | 10.0 | 17.2 | 12.4 | 8.2 | 5.9 | 8.9 | 26.0 | 49.8 | 20,424 | 29,563 | 44.7 | 13.6 |
| Taylorsville | 23.6 | 15.6 | 23.7 | 7.2 | 15.9 | 17.5 | 10.2 | 11.4 | 13.9 | 36.5 | 51.8 | 57,439 | 58,691 | 2.2 | 1.2 |
| Tooele | 14.7 | 3.7 | 29.4 | 10.9 | 17.0 | 15.0 | 12.2 | 7.9 | 7.5 | 30.2 | 49.3 | 22,502 | 31,603 | 40.4 | 17.2 |
| West Jordan | 24.4 | 13.4 | 28.9 | 10.7 | 14.8 | 13.9 | 11.5 | 11.8 | 8.4 | 32.3 | 49.3 | 68,336 | 103,605 | 51.6 | 12.5 |
| West Valley City | 39.3 | 22.7 | 30.7 | 9.5 | 15.4 | 14.8 | 9.9 | 10.6 | 9.1 | 30.1 | 49.5 | 108,896 | 129,487 | 18.9 | 3.4 |
| **VERMONT** | 2.0 | 4.7 | 18.2 | 10.4 | 12.3 | 11.3 | 12.5 | 15.2 | 20.1 | 42.8 | 50.5 | 608,827 | 625,727 | 2.8 | -0.4 |
| Burlington | 2.8 | 7.7 | 13.5 | 34.8 | 17.8 | 9.7 | 7.6 | 8.9 | 7.7 | 26.1 | 51.4 | 38,889 | 42,413 | 9.1 | 1.1 |
| **VIRGINIA** | 9.7 | 12.7 | 21.8 | 9.4 | 13.8 | 13.2 | 12.8 | 13.1 | 15.9 | 38.5 | 50.8 | 7,078,515 | 8,001,046 | 13.0 | 7.4 |
| Alexandria | 16.6 | 25.4 | 18.0 | 5.1 | 21.9 | 19.0 | 12.7 | 11.7 | 11.5 | 37.1 | 51.7 | 128,283 | 139,998 | 9.1 | 13.4 |
| Blacksburg | 5.1 | 14.6 | 8.2 | 55.0 | 10.1 | 6.8 | 7.1 | 5.5 | 7.5 | 22.0 | 44.7 | 39,573 | 42,521 | 7.4 | 3.7 |
| Charlottesville | 5.8 | 8.3 | 16.0 | 20.8 | 17.8 | 13.9 | 9.4 | 9.0 | 13.2 | 32.2 | 51.7 | 45,049 | 43,425 | -3.6 | 8.1 |
| Chesapeake | 6.6 | 7.5 | 24.2 | 8.5 | 12.8 | 14.6 | 13.1 | 13.1 | 13.7 | 37.8 | 51.2 | 199,184 | 222,268 | 11.6 | 11.1 |
| Danville | 2.2 | 3.5 | 21.7 | 10.3 | 10.8 | 9.4 | 12.3 | 14.3 | 21.2 | 42.3 | 52.2 | 48,411 | 43,071 | -11.0 | -7.4 |
| Hampton | 6.2 | 4.9 | 21.0 | 11.3 | 16.3 | 11.7 | 10.2 | 13.6 | 15.8 | 35.7 | 51.6 | 146,437 | 137,463 | -6.1 | -1.5 |
| Harrisonburg | 20.7 | 14.7 | 16.8 | 33.6 | 15.9 | 10.1 | 6.9 | 7.5 | 9.2 | 24.8 | 52.7 | 40,468 | 48,902 | 20.8 | 8.8 |
| Leesburg | 16.1 | 20.9 | 28.7 | 6.5 | 13.4 | 16.5 | 15.9 | 10.9 | 8.2 | 36.2 | 51.5 | 28,311 | 42,614 | 50.5 | 27.9 |
| Lynchburg | 4.5 | 4.3 | 19.4 | 23.8 | 15.9 | 9.1 | 7.9 | 9.6 | 14.3 | 28.2 | 52.5 | 65,269 | 75,535 | 15.7 | 8.0 |
| Manassas | 40.9 | 23.4 | 30.1 | 5.9 | 13.9 | 14.6 | 14.7 | 8.8 | 12.0 | 35.0 | 48.8 | 35,135 | 37,799 | 7.6 | 8.1 |
| Newport News | 9.4 | 9.8 | 23.3 | 11.3 | 17.6 | 12.1 | 10.4 | 11.8 | 13.5 | 33.4 | 51.8 | 180,150 | 180,956 | 0.4 | |
| Norfolk | 8.5 | 7.2 | 19.3 | 18.0 | 18.9 | 11.6 | 9.6 | 11.0 | 11.6 | 31.1 | 47.8 | 234,403 | 242,840 | 3.6 | 0.0 |
| Petersburg | 5.1 | 3.8 | 20.7 | 9.9 | 16.3 | 8.6 | 13.2 | 13.6 | 17.8 | 36.8 | 52.9 | 33,740 | 32,441 | -3.9 | -6.1 |
| Portsmouth | 4.7 | 1.8 | 22.7 | 9.9 | 15.5 | 13.5 | 10.7 | 12.6 | 15.1 | 36.7 | 53.0 | 100,565 | 95,531 | -5.0 | -0.5 |
| Richmond | 7.3 | 8.2 | 17.0 | 12.1 | 22.1 | 12.5 | 10.4 | 12.2 | 13.7 | 34.2 | 52.3 | 197,790 | 204,440 | 3.4 | 13.6 |
| Roanoke | 6.7 | 5.2 | 21.9 | 8.8 | 14.8 | 12.4 | 12.1 | 13.2 | 16.8 | 37.8 | 52.3 | 94,911 | 96,910 | 2.1 | 2.2 |
| Suffolk | 4.7 | 4.3 | 23.9 | 7.8 | 13.3 | 13.3 | 13.2 | 13.5 | 15.0 | 37.9 | 51.0 | 63,677 | 84,606 | 32.9 | 11.0 |
| Virginia Beach | 8.5 | 9.3 | 22.0 | 9.0 | 16.2 | 13.6 | 12.0 | 12.5 | 14.7 | 36.6 | 50.9 | 425,257 | 437,885 | 3.0 | 3.0 |
| Winchester | 23.8 | 15.6 | 22.6 | 13.4 | 10.7 | 11.2 | 14.0 | 13.3 | 14.8 | 38.6 | 52.2 | 23,585 | 26,223 | 11.2 | 5.6 |
| **WASHINGTON** | 13.0 | 14.9 | 21.8 | 8.7 | 15.3 | 13.6 | 12.2 | 12.6 | 15.9 | 37.9 | 50.0 | 5,894,121 | 6,724,540 | 14.1 | 14.4 |
| Auburn | 19.9 | 22.3 | 25.3 | 8.6 | 15.2 | 13.0 | 13.9 | 9.6 | 14.3 | 35.4 | 53.8 | 40,314 | 70,299 | 74.4 | 15.3 |
| Bellevue | 7.2 | 41.1 | 21.4 | 6.5 | 20.0 | 14.4 | 13.0 | 10.6 | 14.1 | 36.5 | 49.1 | 109,569 | 127,885 | 16.7 | 15.8 |
| Bellingham | 8.1 | 7.6 | 13.8 | 23.0 | 12.6 | 12.0 | 10.1 | 11.5 | 17.1 | 35.6 | 52.8 | 67,171 | 81,610 | 21.5 | 15.1 |
| Bothell | 10.2 | 25.0 | 22.0 | 6.9 | 16.6 | 16.0 | 12.5 | 11.7 | 14.3 | 37.3 | 50.4 | 30,150 | 39,830 | 32.1 | 19.3 |
| Bremerton | 17.3 | 11.0 | 19.7 | 15.2 | 19.6 | 17.5 | 7.9 | 9.4 | 10.7 | 32.2 | 43.3 | 37,259 | 37,839 | 1.6 | 11.8 |
| Burien | 28.5 | 27.9 | 21.3 | 5.9 | 14.3 | 17.3 | 13.4 | 13.2 | 14.7 | 38.3 | 48.1 | 31,881 | 48,079 | 50.8 | 6.6 |
| Des Moines | 19.5 | 25.5 | 15.9 | 11.3 | 10.5 | 16.2 | 12.3 | 13.6 | 20.4 | 41.7 | 53.0 | 29,267 | 29,685 | 1.4 | 9.1 |
| Edmonds | 7.1 | 16.2 | 19.6 | 6.2 | 11.4 | 15.7 | 14.5 | 12.9 | 19.7 | 42.7 | 50.7 | 39,515 | 39,686 | 0.4 | 7.1 |
| Everett | 14.0 | 18.6 | 21.1 | 8.6 | 19.0 | 12.8 | 11.7 | 14.2 | 12.6 | 36.0 | 51.8 | 91,488 | 102,864 | 12.4 | 9.4 |
| Federal Way | 18.5 | 29.1 | 24.3 | 10.9 | 16.4 | 9.1 | 13.8 | 13.6 | 11.9 | 33.5 | 50.7 | 83,259 | 89,259 | 7.2 | 7.9 |
| Issaquah | 4.8 | 26.7 | 24.9 | 2.7 | 16.3 | 18.9 | 14.2 | 9.2 | 13.8 | 38.9 | 51.3 | 11,212 | 30,436 | 171.5 | 30.4 |
| Kennewick | 25.1 | 12.5 | 25.5 | 8.0 | 15.3 | 13.4 | 10.3 | 12.2 | 15.3 | 35.5 | 51.8 | 54,693 | 73,994 | 35.3 | 14.2 |
| Kent | 15.9 | 35.9 | 23.6 | 9.4 | 16.8 | 14.8 | 11.8 | 13.7 | 9.9 | 35.1 | 50.5 | 79,524 | 118,620 | 49.2 | 10.2 |
| Kirkland | 9.4 | 26.4 | 21.6 | 7.9 | 16.1 | 17.6 | 13.2 | 9.9 | 13.7 | 37.0 | 50.7 | 45,054 | 80,585 | 78.9 | 18.4 |
| Lacey | 17.8 | 13.5 | 19.3 | 11.1 | 18.9 | 14.5 | 9.7 | 12.3 | 14.2 | 35.8 | 53.0 | 31,226 | 42,559 | 36.3 | 27.3 |
| Lake Stevens | 12.7 | 7.5 | 25.7 | 9.5 | 11.7 | 15.2 | 11.3 | 14.6 | 12.1 | 36.7 | 48.9 | 6,361 | 28,270 | 344.4 | 23.1 |
| Lakewood | 20.1 | 14.4 | 20.5 | 13.4 | 14.0 | 12.0 | 9.8 | 13.7 | 16.6 | 36.1 | 50.2 | 58,211 | 57,532 | -1.2 | 6.0 |
| Longview | 11.3 | 3.9 | 23.7 | 6.9 | 13.8 | 13.3 | 9.4 | 11.7 | 21.2 | 38.4 | 52.1 | 34,660 | 36,833 | 6.3 | 4.6 |
| Lynnwood | 7.7 | 27.7 | 12.0 | 8.7 | 16.6 | 12.4 | 11.4 | 14.8 | 24.1 | 45.2 | 48.3 | 33,847 | 35,853 | 5.9 | 8.9 |
| Marysville | 14.7 | 8.5 | 24.5 | 9.5 | 14.3 | 13.6 | 11.0 | 12.6 | 14.7 | 37.1 | 51.7 | 25,315 | 60,022 | 137.1 | 18.9 |
| Mount Vernon | 29.4 | 13.7 | 24.5 | 8.2 | 14.9 | 8.7 | 9.9 | 12.6 | 21.2 | 37.7 | 52.5 | 26,232 | 31,722 | 20.9 | 14.5 |
| Olympia | 9.0 | 8.3 | 18.5 | 6.3 | 16.2 | 16.8 | 11.9 | 12.6 | 17.7 | 39.9 | 50.5 | 42,514 | 46,897 | 10.3 | 14.3 |

1. May be of any race.

## Table D. Cities — Households, Group Quarters, Crime, and Education

| City | Households, 2019 | | | | | | | Persons in group quarters, 2019 | Serious crimes known to police[1], 2019 | | | | Educational attainment, 2019 | | |
| | Number | Persons per household | Percent Family | Married couple family | Female family | Non-family | One person | | Violent Number | Violent Rate[2] | Property Number | Property Rate[2] | Population age 25 and over | High school graduate or less | Bachelor's degree or more |
| | 27 | 28 | 29 | 30 | 31 | 32 | 33 | 34 | 35 | 36 | 37 | 38 | 39 | 40 | 41 |
| **UTAH— Cont'd** | | | | | | | | | | | | | | | |
| Kaysville | 8,718 | 3.72 | 89.9 | 78.7 | 11.2 | 10.1 | 7.4 | 0 | 23 | 70 | 274 | 838 | 17,814 | 9.4 | 51.5 |
| Layton | 25,307 | 3.08 | 78.3 | 62.7 | 15.7 | 21.7 | 16.9 | 157 | 129 | 164 | 1,558 | 1,983 | 47,541 | 28.6 | 36.5 |
| Lehi | 18,411 | 3.78 | 90.1 | 82.0 | 8.1 | 9.9 | 8.1 | 71 | NA | NA | NA | NA | 35,614 | 17.4 | 46.7 |
| Logan | 18,284 | 2.66 | 68.2 | 53.5 | 14.7 | 31.8 | 19.0 | 2,909 | 113 | 217 | 691 | 1,328 | 25,248 | 29.6 | 36.8 |
| Midvale | 14,147 | 2.40 | 56.7 | 43.5 | 13.2 | 43.3 | 29.1 | 128 | NA | NA | NA | NA | 23,004 | 24.1 | 39.7 |
| Murray | 19,937 | 2.44 | 58.9 | 42.2 | 16.7 | 41.1 | 31.6 | 294 | 218 | 439 | 2,890 | 5,822 | 34,264 | 26.9 | 34.9 |
| Ogden | 30,509 | 2.79 | 63.6 | 41.5 | 22.1 | 36.4 | 27.5 | 2,507 | 384 | 437 | 3,211 | 3,654 | 53,899 | 45.2 | 22.1 |
| Orem | 29,380 | 3.27 | 78.4 | 63.4 | 15.0 | 21.6 | 12.2 | 1,681 | 89 | 90 | 1,970 | 1,996 | 53,028 | 22.3 | 41.5 |
| Pleasant Grove | 11,277 | 3.39 | 75.0 | 67.2 | 7.8 | 25.0 | 19.8 | 63 | 39 | 100 | 365 | 934 | 20,949 | 20.6 | 38.0 |
| Provo | 34,454 | 3.05 | 70.2 | 56.2 | 14.0 | 29.8 | 11.9 | 11,675 | 135 | 115 | 1,767 | 1,508 | 48,088 | 20.0 | 45.3 |
| Riverton | 11,728 | 3.78 | 90.5 | 77.9 | 12.6 | 9.5 | 6.7 | 70 | NA | NA | NA | NA | 25,470 | 26.2 | 39.9 |
| Roy | 13,797 | 2.86 | 72.8 | 54.6 | 18.2 | 27.2 | 26.5 | 166 | 55 | 141 | 466 | 1,195 | 25,562 | 37.4 | 22.1 |
| St. George | 30,816 | 2.88 | 75.0 | 59.4 | 15.6 | 25.0 | 17.7 | 822 | 181 | 203 | 1,397 | 1,567 | 55,569 | 26.1 | 30.2 |
| Salt Lake City | 81,839 | 2.37 | 47.1 | 36.5 | 10.7 | 52.9 | 38.2 | 6,819 | 1,442 | 712 | 11,452 | 5,657 | 136,970 | 28.7 | 47.2 |
| Sandy | 31,130 | 3.08 | 78.2 | 65.8 | 12.4 | 21.8 | 16.7 | 389 | 187 | 191 | 2,368 | 2,421 | 65,155 | 21.7 | 44.5 |
| South Jordan | 20,542 | 3.73 | 90.3 | 78.3 | 12.0 | 9.7 | 7.7 | 22 | 73 | 94 | 1,124 | 1,448 | 45,428 | 20.0 | 44.5 |
| Spanish Fork | 11,289 | 3.55 | 82.8 | 70.9 | 11.9 | 17.2 | 15.9 | 880 | 7 | 17 | 332 | 818 | 21,046 | 33.3 | 34.2 |
| Springville | 9,561 | 3.47 | 81.7 | 65.9 | 15.8 | 18.3 | 15.7 | 103 | 19 | 57 | 501 | 1,494 | 17,523 | 27.2 | 35.5 |
| Taylorsville | 21,636 | 2.76 | 69.3 | 51.2 | 18.1 | 30.7 | 26.0 | 152 | NA | NA | NA | NA | 41,287 | 41.6 | 20.8 |
| Tooele | 10,681 | 3.35 | 79.8 | 64.2 | 15.7 | 20.2 | 17.9 | 226 | 73 | 204 | 860 | 2,408 | 21,478 | 44.6 | 17.0 |
| West Jordan | 35,366 | 3.28 | 77.1 | 61.7 | 15.4 | 22.9 | 17.5 | 576 | 264 | 224 | 2,555 | 2,172 | 70,386 | 40.7 | 27.3 |
| West Valley City | 36,842 | 3.66 | 80.5 | 59.8 | 20.7 | 19.5 | 13.0 | 339 | 960 | 699 | 4,378 | 3,189 | 80,868 | 49.9 | 16.1 |
| **VERMONT** | 262,767 | 2.28 | 59.4 | 46.1 | 13.4 | 40.6 | 31.6 | 25,701 | NA | NA | NA | NA | 445,558 | 35.9 | 38.7 |
| Burlington | 16,761 | 2.11 | 38.7 | 27.2 | 11.6 | 61.3 | 35.0 | 7,430 | 174 | 405 | 1,110 | 2,584 | 22,145 | 19.0 | 58.9 |
| **VIRGINIA** | 3,191,847 | 2.60 | 65.6 | 49.8 | 15.8 | 34.4 | 27.7 | 244,897 | NA | NA | NA | NA | 5,872,757 | 33.6 | 39.6 |
| Alexandria | 71,005 | 2.23 | 48.0 | 35.9 | 12.1 | 52.0 | 42.9 | 1,381 | 288 | 177 | 2,517 | 1,551 | 122,536 | 17.5 | 65.9 |
| Blacksburg | 13,340 | 2.62 | 36.9 | 33.2 | 3.7 | 63.1 | 30.5 | 9,257 | 50 | 111 | 275 | 612 | 16,304 | 17.5 | 67.9 |
| Charlottesville | 18,777 | 2.36 | 46.8 | 32.3 | 14.5 | 53.2 | 30.3 | 2,989 | 157 | 324 | 1,124 | 2,320 | 29,876 | 27.1 | 54.5 |
| Chesapeake | 86,878 | 2.76 | 73.6 | 54.3 | 19.3 | 26.4 | 21.4 | 4,951 | 1,112 | 456 | 5,541 | 2,273 | 164,745 | 33.8 | 32.1 |
| Danville | 17,810 | 2.16 | 54.1 | 25.1 | 29.1 | 45.9 | 36.1 | 1,486 | 112 | 279 | 1,467 | 3,650 | 27,259 | 52.0 | 12.7 |
| Hampton | 55,633 | 2.33 | 60.6 | 39.0 | 21.6 | 39.4 | 32.0 | 4,608 | 393 | 295 | 3,994 | 2,999 | 91,041 | 38.4 | 26.3 |
| Harrisonburg | 16,636 | 2.76 | 49.7 | 33.7 | 16.0 | 50.3 | 30.4 | 7,132 | 112 | 206 | 924 | 1,699 | 26,296 | 40.7 | 37.9 |
| Leesburg | 18,108 | 2.95 | 75.0 | 62.8 | 12.2 | 25.0 | 15.6 | 252 | 102 | 184 | 570 | 1,028 | 34,846 | 18.7 | 59.0 |
| Lynchburg | 28,950 | 2.51 | 60.4 | 40.5 | 19.9 | 39.6 | 29.9 | 9,604 | 316 | 383 | 1,802 | 2,184 | 46,645 | 35.8 | 37.2 |
| Manassas | 13,465 | 3.05 | 71.3 | 57.4 | 13.9 | 28.7 | 26.5 | 82 | 103 | 246 | 623 | 1,489 | 26,257 | 43.1 | 31.2 |
| Newport News | 70,918 | 2.41 | 58.5 | 34.3 | 24.2 | 41.5 | 35.3 | 8,574 | 1,056 | 596 | 4,484 | 2,529 | 117,211 | 39.5 | 26.3 |
| Norfolk | 88,387 | 2.45 | 55.8 | 33.3 | 22.5 | 44.2 | 35.2 | 26,123 | 1,325 | 546 | 8,405 | 3,462 | 152,165 | 37.6 | 30.7 |
| Petersburg | 12,915 | 2.38 | 48.1 | 21.6 | 26.5 | 51.9 | 44.4 | 614 | 234 | 748 | 1,034 | 3,306 | 21,779 | 49.9 | 25.2 |
| Portsmouth | 36,089 | 2.49 | 58.9 | 32.3 | 26.6 | 41.1 | 32.3 | 4,385 | 889 | 946 | 5,509 | 5,861 | 63,624 | 41.5 | 21.6 |
| Richmond | 89,878 | 2.44 | 43.4 | 27.6 | 15.8 | 56.6 | 44.1 | 11,140 | 1,068 | 463 | 8,074 | 3,499 | 163,388 | 37.1 | 39.3 |
| Roanoke | 40,810 | 2.39 | 51.8 | 30.4 | 21.4 | 48.2 | 40.0 | 1,598 | 386 | 387 | 4,402 | 4,413 | 68,745 | 47.4 | 23.8 |
| Suffolk | 34,597 | 2.65 | 72.7 | 51.3 | 21.4 | 27.3 | 23.2 | 442 | 270 | 295 | 2,266 | 2,477 | 62,892 | 34.0 | 30.9 |
| Virginia Beach | 175,029 | 2.52 | 67.0 | 51.4 | 15.6 | 33.0 | 24.8 | 8,831 | 581 | 129 | 7,906 | 1,761 | 310,431 | 24.6 | 38.2 |
| Winchester | 9,726 | 2.78 | 67.4 | 42.7 | 24.7 | 32.6 | 26.1 | 1,047 | 80 | 284 | 657 | 2,330 | 17,953 | 42.2 | 32.8 |
| **WASHINGTON** | 2,932,477 | 2.55 | 64.2 | 49.4 | 14.8 | 35.8 | 26.6 | 147,820 | NA | NA | NA | NA | 5,290,324 | 30.4 | 37.0 |
| Auburn | 29,665 | 2.74 | 64.5 | 45.9 | 18.6 | 35.5 | 28.8 | 629 | NA | NA | NA | NA | 54,215 | 39.9 | 29.9 |
| Bellevue | 59,316 | 2.47 | 65.8 | 56.5 | 9.3 | 34.2 | 27.0 | 1,470 | 185 | 123 | 4,263 | 2,838 | 106,863 | 12.2 | 71.2 |
| Bellingham | 38,664 | 2.27 | 47.4 | 32.7 | 14.8 | 52.6 | 31.2 | 4,529 | 251 | 273 | 2,689 | 2,926 | 58,377 | 21.7 | 46.3 |
| Bothell | 18,747 | 2.60 | 64.7 | 58.1 | 6.6 | 35.3 | 21.8 | 736 | 30 | 63 | 964 | 2,027 | 35,196 | 16.8 | 61.5 |
| Bremerton | 16,405 | 2.29 | 51.0 | 37.6 | 13.4 | 49.0 | 36.9 | 3,826 | 168 | 403 | 1,358 | 3,259 | 26,979 | 29.4 | 24.8 |
| Burien | 19,501 | 2.61 | 55.4 | 38.8 | 16.6 | 44.6 | 31.9 | 509 | 180 | 344 | 1,744 | 3,329 | 37,477 | 43.0 | 25.8 |
| Des Moines | 13,299 | 2.39 | 67.6 | 38.2 | 29.5 | 32.4 | 27.9 | 529 | 86 | 263 | 1,007 | 3,079 | 23,564 | 33.9 | 29.1 |
| Edmonds | 17,549 | 2.40 | 67.7 | 56.8 | 10.9 | 32.3 | 26.5 | 393 | 87 | 202 | 909 | 2,107 | 31,608 | 23.7 | 43.8 |
| Everett | 44,219 | 2.42 | 55.2 | 35.6 | 19.6 | 44.8 | 35.5 | 4,568 | 341 | 304 | 3,991 | 3,554 | 78,327 | 36.7 | 26.1 |
| Federal Way | 34,971 | 2.73 | 72.3 | 51.5 | 20.8 | 27.7 | 24.0 | 860 | 330 | 337 | 4,164 | 4,248 | 62,366 | 39.3 | 29.0 |
| Issaquah | 15,931 | 2.46 | 70.5 | 61.5 | 9.0 | 29.5 | 22.7 | 268 | 20 | 49 | 1,138 | 2,799 | 28,628 | 12.4 | 67.5 |
| Kennewick | 31,230 | 2.66 | 62.5 | 41.0 | 21.5 | 37.5 | 34.1 | 1,364 | 253 | 301 | 2,487 | 2,958 | 56,097 | 40.2 | 27.9 |
| Kent | 44,949 | 2.90 | 68.4 | 47.1 | 21.3 | 31.6 | 22.0 | 2,152 | 486 | 371 | 5,200 | 3,969 | 88,712 | 39.8 | 26.8 |
| Kirkland | 36,115 | 2.54 | 65.2 | 55.1 | 10.0 | 34.8 | 24.7 | 1,319 | NA | NA | NA | NA | 65,628 | 15.4 | 62.7 |
| Lacey | 20,493 | 2.52 | 58.6 | 42.3 | 16.3 | 41.4 | 28.6 | 898 | 105 | 203 | 1,436 | 2,771 | 36,569 | 32.0 | 32.0 |
| Lake Stevens | 11,417 | 2.97 | 75.2 | 53.6 | 21.6 | 24.8 | 14.4 | 41 | 44 | 129 | 284 | 833 | 21,973 | 37.5 | 28.1 |
| Lakewood | 25,374 | 2.35 | 58.8 | 37.7 | 21.1 | 41.2 | 34.9 | 1,428 | 461 | 757 | 2,296 | 3,769 | 40,354 | 39.9 | 20.4 |
| Longview | 16,232 | 2.31 | 58.6 | 40.3 | 18.3 | 41.4 | 36.6 | 903 | 101 | 264 | 1,102 | 2,879 | 26,706 | 35.4 | 16.1 |
| Lynnwood | 15,902 | 2.41 | 60.7 | 43.2 | 17.5 | 39.3 | 29.7 | 857 | 113 | 291 | 1,855 | 4,775 | 31,027 | 32.3 | 29.3 |
| Marysville | 24,496 | 2.85 | 70.6 | 53.6 | 17.0 | 29.4 | 20.1 | 471 | 144 | 203 | 1,404 | 1,975 | 46,438 | 39.6 | 21.6 |
| Mount Vernon | 13,155 | 2.68 | 67.0 | 53.3 | 13.6 | 33.0 | 27.7 | 770 | 55 | 152 | 1,140 | 3,143 | 24,224 | 38.1 | 24.7 |
| Olympia | 23,118 | 2.23 | 53.0 | 38.4 | 14.6 | 47.0 | 33.9 | 1,320 | 254 | 477 | 1,854 | 3,479 | 39,805 | 24.8 | 40.8 |

1. Data for serious crimes have not been adjusted for underreporting. This may affect comparability between geographic areas and over time. 2. Per 100,000 population estimated by the FBI. 3. Persons 25 years old and over.

# Table D. Cities — Income, Poverty, and Housing

| City | Money income, 2019 — Households Median income | Percent with income less than $20,000 | Percent with income of $200,000 or more | Median family income | Median non-family income | Median earnings, 2019 All persons | Men | Women | Housing units, 2019 Total | Occupied | Percent owner occupied | Median value[1] (dollars) | Median gross rent (dollars) |
|---|---|---|---|---|---|---|---|---|---|---|---|---|---|
| | 42 | 43 | 44 | 45 | 46 | 47 | 48 | 49 | 50 | 51 | 52 | 53 | 54 |
| **UTAH— Cont'd** | | | | | | | | | | | | | |
| Kaysville | 110,300 | 7.4 | 11.3 | 111,145 | 54,899 | 47,141 | 64,386 | 30,897 | 8,864 | 8,718 | 86.4 | 415,809 | 1,081 |
| Layton | 76,032 | 7.3 | 7.9 | 85,512 | 48,190 | 31,737 | 47,739 | 25,253 | 25,869 | 25,307 | 72.2 | 310,675 | 1,079 |
| Lehi | 105,280 | 2.4 | 10.4 | 110,325 | 67,598 | 50,201 | 67,425 | 28,094 | 18,786 | 18,411 | 82.8 | 407,444 | 1,395 |
| Logan | 45,811 | 20.2 | 1.6 | 56,102 | 28,584 | 16,846 | 22,201 | 12,395 | 19,829 | 18,284 | 41.0 | 248,235 | 794 |
| Midvale | 68,067 | 8.2 | 1.7 | 76,427 | 53,290 | 36,181 | 37,460 | 31,656 | 15,822 | 14,147 | 47.8 | 321,267 | 1,187 |
| Murray | 67,325 | 9.1 | 2.9 | 83,860 | 44,093 | 35,596 | 41,139 | 30,732 | 21,173 | 19,937 | 66.3 | 349,771 | 1,156 |
| Ogden | 60,054 | 11.7 | 3.9 | 69,446 | 34,882 | 31,574 | 37,962 | 27,155 | 33,866 | 30,509 | 58.8 | 210,812 | 911 |
| Orem | 70,465 | 8.8 | 6.2 | 74,606 | 46,243 | 26,573 | 36,980 | 20,291 | 30,913 | 29,380 | 59.4 | 332,310 | 1,106 |
| Pleasant Grove | 80,243 | 8.2 | 9.4 | 99,265 | 36,984 | 35,494 | 49,449 | 19,684 | 11,803 | 11,277 | 68.9 | 353,787 | 1,214 |
| Provo | 53,864 | 17.3 | 4.5 | 58,690 | 40,434 | 15,235 | 17,313 | 11,889 | 36,854 | 34,454 | 41.3 | 335,703 | 913 |
| Riverton | 117,727 | 1.8 | 12.5 | 121,809 | 81,538 | 36,339 | 52,429 | 29,557 | 12,534 | 11,728 | 94.7 | 416,845 | 1,393 |
| Roy | 71,512 | 6.5 | 3.5 | 82,414 | 31,945 | 37,159 | 46,150 | 28,531 | 13,880 | 13,797 | 87.9 | 251,126 | 1,196 |
| St. George | 62,178 | 9.7 | 4.6 | 73,224 | 33,570 | 30,001 | 36,376 | 22,510 | 35,813 | 30,816 | 67.6 | 329,107 | 1,030 |
| Salt Lake City | 63,971 | 16.1 | 9.0 | 87,370 | 46,438 | 32,287 | 37,044 | 29,277 | 89,624 | 81,839 | 46.1 | 378,286 | 1,062 |
| Sandy | 92,032 | 7.7 | 11.7 | 101,967 | 46,504 | 37,850 | 53,916 | 29,074 | 33,386 | 31,130 | 79.8 | 412,776 | 1,422 |
| South Jordan | 108,626 | 2.8 | 15.6 | 115,967 | 65,537 | 42,325 | 61,428 | 30,483 | 21,135 | 20,542 | 82.6 | 475,239 | 1,433 |
| Spanish Fork | 79,076 | 6.3 | 4.5 | 90,109 | 42,129 | 35,290 | 47,763 | 24,434 | 11,452 | 11,289 | 75.0 | 325,265 | 1,061 |
| Springville | 72,384 | 10.0 | 3.2 | 84,786 | 33,019 | 32,659 | 51,222 | 20,762 | 9,805 | 9,561 | 66.3 | 320,467 | 1,148 |
| Taylorsville | 72,139 | 7.0 | 2.5 | 84,070 | 49,314 | 40,109 | 46,912 | 34,025 | 22,639 | 21,636 | 69.5 | 292,585 | 1,123 |
| Tooele | 74,987 | 9.1 | 3.3 | 75,731 | 31,387 | 40,077 | 52,338 | 27,051 | 11,695 | 10,681 | 83.5 | 241,972 | 1,060 |
| West Jordan | 87,006 | 9.6 | 4.2 | 94,911 | 41,884 | 36,563 | 46,875 | 28,160 | 36,892 | 35,366 | 74.9 | 337,554 | 1,206 |
| West Valley City | 73,917 | 8.9 | 3.5 | 79,790 | 41,004 | 32,040 | 38,878 | 26,863 | 39,162 | 36,842 | 68.8 | 273,197 | 1,221 |
| **VERMONT** | 63,001 | 13.2 | 6.0 | 83,458 | 38,034 | 35,474 | 40,185 | 31,360 | 339,412 | 262,767 | 70.9 | 233,242 | 980 |
| Burlington | 55,941 | 21.3 | 6.6 | 105,932 | 35,334 | 21,136 | 25,565 | 15,724 | 17,190 | 16,761 | 34.4 | 325,937 | 1,302 |
| **VIRGINIA** | 76,456 | 11.3 | 11.9 | 93,497 | 46,570 | 40,405 | 46,552 | 33,628 | 3,562,258 | 3,191,847 | 66.1 | 288,833 | 1,254 |
| Alexandria | 103,284 | 5.0 | 23.5 | 138,441 | 90,233 | 62,560 | 69,640 | 60,768 | 76,424 | 71,005 | 44.6 | 572,535 | 1,781 |
| Blacksburg | 46,601 | 31.2 | 8.1 | 125,709 | 21,930 | 11,174 | 9,732 | 12,949 | 16,249 | 13,340 | 36.0 | 319,212 | 1,083 |
| Charlottesville | 56,265 | 19.1 | 12.0 | 91,450 | 40,869 | 29,266 | 36,515 | 22,088 | 20,881 | 18,777 | 43.3 | 335,282 | 1,082 |
| Chesapeake | 77,847 | 8.2 | 7.1 | 92,842 | 42,929 | 40,745 | 47,859 | 35,023 | 92,584 | 86,878 | 69.3 | 290,908 | 1,225 |
| Danville | 35,336 | 27.9 | 2.2 | 46,526 | 24,495 | 24,959 | 30,541 | 22,641 | 21,932 | 17,810 | 47.0 | 82,924 | 677 |
| Hampton | 56,930 | 14.6 | 4.6 | 73,543 | 39,804 | 35,136 | 37,995 | 30,333 | 60,130 | 55,633 | 51.2 | 193,486 | 1,100 |
| Harrisonburg | 49,475 | 16.7 | 4.8 | 58,978 | 39,206 | 20,998 | 23,699 | 15,383 | 18,361 | 16,636 | 38.7 | 222,703 | 915 |
| Leesburg | 122,173 | 6.4 | 25.2 | 128,461 | 76,755 | 60,608 | 76,782 | 50,027 | 18,939 | 18,108 | 76.1 | 459,982 | 1,703 |
| Lynchburg | 54,850 | 16.4 | 3.6 | 65,277 | 37,212 | 25,826 | 30,204 | 21,320 | 32,529 | 28,950 | 47.5 | 170,474 | 886 |
| Manassas | 93,319 | 6.3 | 13.4 | 101,153 | 53,532 | 42,225 | 47,189 | 36,336 | 14,149 | 13,465 | 71.4 | 356,018 | 1,614 |
| Newport News | 53,029 | 17.0 | 3.2 | 61,919 | 38,799 | 31,914 | 38,549 | 27,284 | 77,931 | 70,918 | 48.8 | 186,594 | 1,088 |
| Norfolk | 53,093 | 17.5 | 3.5 | 64,816 | 38,402 | 28,596 | 30,850 | 25,482 | 98,476 | 88,387 | 44.0 | 217,991 | 1,021 |
| Petersburg | 42,771 | 30.9 | 1.3 | 56,817 | 24,355 | 27,688 | 28,004 | 27,549 | 16,309 | 12,915 | 38.1 | 103,719 | 1,002 |
| Portsmouth | 51,195 | 17.7 | 3.1 | 60,820 | 37,155 | 36,103 | 40,642 | 30,778 | 40,906 | 36,089 | 56.0 | 169,568 | 1,042 |
| Richmond | 51,285 | 22.0 | 7.1 | 72,135 | 37,732 | 31,216 | 31,610 | 30,780 | 101,240 | 89,876 | 43.6 | 249,516 | 1,078 |
| Roanoke | 45,838 | 25.1 | 4.5 | 59,962 | 29,059 | 29,994 | 32,090 | 25,844 | 46,941 | 40,810 | 48.3 | 147,272 | 822 |
| Suffolk | 80,481 | 10.3 | 8.4 | 89,753 | 44,729 | 41,543 | 48,897 | 33,095 | 37,954 | 34,597 | 68.3 | 263,536 | 1,241 |
| Virginia Beach | 79,054 | 6.4 | 7.9 | 91,608 | 53,550 | 39,683 | 44,799 | 33,499 | 186,468 | 175,029 | 62.3 | 296,153 | 1,363 |
| Winchester | 69,857 | 8.3 | 12.2 | 77,710 | 47,809 | 28,354 | 29,512 | 27,076 | 11,144 | 9,726 | 47.0 | 272,928 | 1,174 |
| **WASHINGTON** | 78,687 | 10.2 | 11.4 | 94,709 | 50,045 | 41,735 | 50,447 | 34,172 | 3,195,098 | 2,932,477 | 63.1 | 387,631 | 1,359 |
| Auburn | 73,466 | 9.3 | 8.6 | 97,352 | 45,076 | 40,282 | 52,253 | 32,715 | 31,726 | 29,665 | 56.8 | 401,268 | 1,417 |
| Bellevue | 127,402 | 5.4 | 30.7 | 161,989 | 79,206 | 69,410 | 92,402 | 50,862 | 63,685 | 59,316 | 49.7 | 920,858 | 2,119 |
| Bellingham | 58,492 | 16.6 | 5.0 | 87,104 | 34,724 | 30,114 | 35,651 | 24,915 | 41,555 | 38,664 | 49.2 | 432,386 | 1,163 |
| Bothell | 109,592 | 6.3 | 21.4 | 131,859 | 75,924 | 56,000 | 71,482 | 36,435 | 20,233 | 18,747 | 62.1 | 670,584 | 1,914 |
| Bremerton | 63,434 | 14.4 | 3.2 | 82,548 | 42,656 | 33,326 | 36,175 | 31,348 | 17,437 | 16,405 | 45.8 | 322,675 | 1,165 |
| Burien | 80,649 | 7.8 | 7.8 | 94,123 | 53,921 | 38,108 | 45,068 | 31,636 | 20,912 | 19,501 | 60.9 | 433,277 | 1,473 |
| Des Moines | 75,999 | 6.9 | 7.9 | 75,724 | 52,791 | 41,903 | 52,890 | 36,179 | 14,487 | 13,299 | 58.7 | 422,499 | 1,417 |
| Edmonds | 81,064 | 4.5 | 18.0 | 108,176 | 48,924 | 45,681 | 57,984 | 39,849 | 18,816 | 17,549 | 63.9 | 669,368 | 1,530 |
| Everett | 64,183 | 15.4 | 4.6 | 78,086 | 44,295 | 38,734 | 47,646 | 31,461 | 47,724 | 44,219 | 45.9 | 385,448 | 1,346 |
| Federal Way | 64,263 | 10.0 | 6.9 | 77,715 | 41,274 | 34,581 | 42,097 | 29,293 | 37,401 | 34,971 | 51.9 | 372,812 | 1,514 |
| Issaquah | 131,051 | 7.6 | 28.5 | 147,701 | 89,766 | 76,300 | 99,042 | 55,034 | 16,386 | 15,931 | 67.9 | 808,969 | 2,217 |
| Kennewick | 61,983 | 11.7 | 6.2 | 71,009 | 39,343 | 35,016 | 41,480 | 26,070 | 32,849 | 31,230 | 61.6 | 254,829 | 960 |
| Kent | 81,423 | 7.7 | 10.2 | 88,356 | 58,470 | 41,330 | 44,320 | 36,687 | 47,545 | 44,949 | 56.4 | 413,550 | 1,560 |
| Kirkland | 121,588 | 5.6 | 28.7 | 146,195 | 78,699 | 61,214 | 85,070 | 47,272 | 37,739 | 36,115 | 63.0 | 744,402 | 1,932 |
| Lacey | 67,497 | 8.6 | 3.4 | 86,632 | 52,093 | 33,986 | 36,139 | 31,325 | 21,307 | 20,493 | 56.7 | 293,248 | 1,304 |
| Lake Stevens | 95,533 | 4.0 | 8.4 | 99,431 | 61,581 | 47,680 | 65,009 | 36,512 | 11,624 | 11,417 | 75.6 | 417,335 | 1,741 |
| Lakewood | 59,946 | 15.5 | 4.3 | 67,648 | 34,213 | 35,822 | 41,244 | 29,398 | 28,096 | 25,374 | 42.4 | 308,480 | 1,150 |
| Longview | 52,735 | 15.1 | 3.5 | 61,968 | 40,206 | 35,330 | 40,977 | 32,456 | 16,791 | 16,232 | 51.1 | 260,931 | 915 |
| Lynnwood | 77,338 | 15.7 | 9.2 | 87,907 | 35,987 | 44,948 | 43,343 | 48,667 | 16,186 | 15,902 | 53.4 | 518,472 | 1,380 |
| Marysville | 81,335 | 7.9 | 6.0 | 91,638 | 44,179 | 41,156 | 51,256 | 32,236 | 26,158 | 24,496 | 71.6 | 378,259 | 1,351 |
| Mount Vernon | 68,947 | 12.9 | 2.6 | 81,274 | 39,241 | 35,372 | 38,307 | 31,854 | 13,586 | 13,155 | 67.3 | 329,026 | 911 |
| Olympia | 64,549 | 15.9 | 8.3 | 92,717 | 40,085 | 36,685 | 36,877 | 36,442 | 24,607 | 23,118 | 51.7 | 343,281 | 1,224 |

1. Based on population estimated by the American Community Survey.

## Table D. Cities — Commuting, Computer Access, Migration, Labor Force, and Employment

| City | Commuting[1], 2019: Percent: Drove alone | Commuting[1], 2019: Percent: With commutes of 30 minutes or more | Computer access[2], 2019: Percent: With a computer in the house | Computer access[2], 2019: Percent: With Internet access | Migration, 2019: Percent who lived in the same house one year ago | Migration, 2019: Percent who lived in another state or county one year ago | Civilian labor force, 2020: Total | Civilian labor force, 2020: Percent change 2019-2020 | Civilian labor force, 2020: Unemployment: Total | Civilian labor force, 2020: Unemployment: Rate[3] | Civilian Employment[4], 2019: Population age 16 and older: Number | Civilian Employment[4], 2019: Population age 16 and older: Percent in labor force | Civilian Employment[4], 2019: Population age 16 to 64: Number | Civilian Employment[4], 2019: Population age 16 to 64: Percent who worked full-year full-time |
|---|---|---|---|---|---|---|---|---|---|---|---|---|---|---|
| | 55 | 56 | 57 | 58 | 59 | 60 | 61 | 62 | 63 | 64 | 65 | 66 | 67 | 68 |
| UTAH— Cont'd | | | | | | | | | | | | | | |
| Kaysville | 79.9 | 45.6 | NA | NA | NA | NA | 15,028 | 0.6 | 517 | 3 | 20,485 | 71.3 | 18,258 | 51.5 |
| Layton | 78.9 | 33.2 | 97.5 | 94.2 | 85.5 | 5.3 | 39,932 | 1.2 | 1,726 | 4 | 58,605 | 69.0 | 51,153 | 52.0 |
| Lehi | 75.3 | 30.3 | NA | NA | 84.2 | 10.1 | 30,700 | 5.7 | 1,245 | 4 | 41,776 | 70.1 | 37,629 | 52.9 |
| Logan | 65.4 | 11.1 | 97.1 | 86.0 | 67.6 | 17.6 | 28,295 | 0.9 | 853 | 3 | 41,084 | 71.5 | 36,690 | 36.3 |
| Midvale | 78.1 | 21.9 | 97.6 | 94.2 | 78.1 | 4.9 | 21,135 | 2.6 | 1,215 | 6 | 27,143 | 82.5 | 25,087 | 59.8 |
| Murray | 81.6 | 18.6 | 96.4 | 90.4 | 83.3 | 3.4 | 29,445 | -0.1 | 1,574 | 5 | 41,153 | 73.6 | 31,929 | 61.4 |
| Ogden | 74.7 | 23.4 | 94.0 | 89.4 | 81.1 | 6.6 | 42,473 | 1.6 | 2,494 | 6 | 67,724 | 70.5 | 59,265 | 55.3 |
| Orem | 70.7 | 22.6 | 98.1 | 89.8 | 78.5 | 9.1 | 51,370 | 0.5 | 2,134 | 4 | 73,119 | 71.7 | 63,959 | 48.1 |
| Pleasant Grove | 80.6 | 19.7 | 99.1 | 91.1 | 87.5 | 4.5 | 19,000 | -0.4 | 748 | 4 | 26,477 | 72.7 | 23,108 | 49.2 |
| Provo | 61.0 | 16.1 | 98.0 | 72.8 | 63.6 | 14.6 | 66,497 | -0.6 | 2,184 | 3 | 97,270 | 75.0 | 89,997 | 31.1 |
| Riverton | 81.6 | 49.3 | NA | NA | NA | NA | 24,396 | -0.5 | 904 | 4 | 31,935 | 73.2 | 29,164 | 53.3 |
| Roy | 83.2 | 28.0 | 95.9 | 93.3 | 93.1 | 5.4 | 19,735 | 2.3 | 944 | 5 | 29,774 | 71.9 | 24,560 | 64.5 |
| St. George | 70.1 | 11.3 | 94.7 | 90.4 | 83.3 | 7.2 | 40,242 | 3.7 | 2,199 | 6 | 67,876 | 57.1 | 47,651 | 50.7 |
| Salt Lake City | 69.1 | 20.5 | 95.9 | 88.1 | 78.9 | 8.7 | 117,310 | 1.1 | 6,569 | 6 | 167,726 | 72.4 | 144,843 | 53.2 |
| Sandy | 78.6 | 33.6 | 96.3 | 92.6 | 86.6 | 5.7 | 54,322 | -0.4 | 2,529 | 5 | 74,451 | 67.4 | 59,293 | 52.3 |
| South Jordan | 76.6 | 40.8 | 99.2 | 93.5 | 88.4 | 3.9 | 39,232 | 3.1 | 1,674 | 4 | 53,470 | 71.5 | 44,886 | 52.7 |
| Spanish Fork | 85.5 | 22.0 | 97.8 | 92.0 | 88.7 | 4.0 | 18,629 | 2.2 | 708 | 4 | 27,749 | 70.6 | 23,999 | 53.2 |
| Springville | 74.1 | 25.2 | 96.6 | 89.2 | 87.1 | 4.3 | 16,665 | 0.6 | 652 | 4 | 21,631 | 68.8 | 18,665 | 52.0 |
| Taylorsville | 81.1 | 23.9 | 92.7 | 87.3 | 90.2 | 4.7 | 34,061 | -0.2 | 1,728 | 5 | 47,304 | 73.3 | 38,963 | 61.8 |
| Tooele | 77.6 | 55.6 | NA | NA | 90.2 | 6.0 | 17,566 | 2.2 | 910 | 5 | 26,551 | 71.1 | 23,853 | 54.1 |
| West Jordan | 82.1 | 38.2 | 98.9 | 95.5 | 84.5 | 4.2 | 64,338 | 0.4 | 2,874 | 5 | 86,449 | 74.8 | 76,614 | 57.5 |
| West Valley City | 80.2 | 25.5 | 95.3 | 91.7 | 86.7 | 3.8 | 71,265 | -0.1 | 3,897 | 6 | 98,779 | 72.9 | 86,532 | 59.0 |
| VERMONT | 76.2 | 32.3 | 92.1 | 83.2 | 86.9 | 6.1 | 330,058 | -3.6 | 18,413 | 6 | 525,153 | 65.3 | 399,952 | 52.3 |
| Burlington | 51.5 | 20.3 | 99.0 | 81.5 | 65.5 | 17.4 | 23,585 | -2.5 | 1,312 | 6 | 37,869 | 69.1 | 34,552 | 37.1 |
| VIRGINIA | 76.6 | 42.7 | 93.0 | 86.2 | 85.3 | 8.9 | 4,346,644 | -1.5 | 271,407 | 6 | 6,880,704 | 64.0 | 5,522,368 | 56.2 |
| Alexandria | 59.8 | 59.1 | 95.9 | 91.6 | 79.5 | 12.0 | 100,168 | -2.0 | 6,011 | 6 | 133,122 | 76.9 | 114,804 | 68.8 |
| Blacksburg | 59.3 | 9.5 | 98.2 | 96.6 | 66.8 | 16.0 | 19,502 | -5.1 | 836 | 4 | 41,339 | 46.3 | 38,039 | 22.9 |
| Charlottesville | 58.7 | 9.3 | 94.8 | 87.6 | 72.0 | 17.7 | 25,424 | -4.6 | 1,639 | 6 | 40,595 | 66.0 | 34,377 | 44.5 |
| Chesapeake | 85.5 | 38.3 | 96.7 | 91.3 | 86.3 | 9.5 | 122,036 | -1.2 | 7,480 | 6 | 192,317 | 60.8 | 158,803 | 54.9 |
| Danville | 79.0 | 17.0 | 82.8 | 72.6 | 83.4 | 8.9 | 19,250 | 0.1 | 1,871 | 10 | 32,521 | 53.7 | 24,032 | 42.4 |
| Hampton | 83.3 | 28.5 | 93.5 | 83.9 | 81.5 | 12.1 | 64,604 | -0.3 | 5,514 | 9 | 108,949 | 60.3 | 87,665 | 55.6 |
| Harrisonburg | 67.5 | 16.3 | 91.0 | 76.4 | 72.0 | 22.4 | 24,259 | -3.6 | 1,547 | 6 | 44,873 | 61.1 | 40,004 | 38.5 |
| Leesburg | 74.2 | 45.7 | 97.3 | 93.9 | 87.0 | 5.8 | 28,970 | -2.8 | 1,454 | 5 | 39,821 | 77.8 | 35,410 | 65.3 |
| Lynchburg | 73.4 | 13.2 | 96.5 | 92.6 | 79.8 | 11.9 | 36,015 | -1.6 | 2,610 | 7 | 67,750 | 59.9 | 55,995 | 47.6 |
| Manassas | 69.6 | 53.2 | 96.7 | 92.5 | 91.0 | 8.4 | 21,684 | -2.6 | 1,406 | 7 | 29,926 | 71.2 | 25,000 | 60.6 |
| Newport News | 75.4 | 30.9 | 93.7 | 85.1 | 83.7 | 11.2 | 89,715 | 0.3 | 7,786 | 9 | 141,729 | 60.5 | 117,497 | 59.0 |
| Norfolk | 73.4 | 27.6 | 92.4 | 81.4 | 76.3 | 14.4 | 111,825 | -0.5 | 9,751 | 9 | 199,438 | 56.4 | 171,184 | 55.2 |
| Petersburg | 69.5 | 35.3 | 87.3 | 74.3 | 82.4 | 13.8 | 13,459 | 2.5 | 1,877 | 14 | 25,772 | 51.3 | 20,205 | 36.5 |
| Portsmouth | 81.0 | 27.9 | 93.0 | 87.6 | 87.5 | 7.9 | 44,701 | 0.3 | 4,273 | 10 | 75,151 | 53.9 | 60,863 | 51.3 |
| Richmond | 70.0 | 25.0 | 90.6 | 81.7 | 74.7 | 13.2 | 119,962 | 0.6 | 10,592 | 9 | 194,551 | 68.1 | 162,895 | 51.0 |
| Roanoke | 80.9 | 34.4 | 87.2 | 76.9 | 85.1 | 5.4 | 49,174 | -0.5 | 3,786 | 8 | 79,815 | 63.8 | 63,158 | 56.8 |
| Suffolk | 87.0 | 49.3 | 93.0 | 83.9 | 86.4 | 7.9 | 44,546 | -0.9 | 2,903 | 7 | 72,648 | 64.9 | 58,830 | 57.9 |
| Virginia Beach | 82.2 | 31.8 | 96.6 | 92.1 | 83.1 | 9.0 | 230,322 | -1.8 | 14,331 | 6 | 361,953 | 65.1 | 295,694 | 60.1 |
| Winchester | 65.3 | 23.3 | 95.4 | 82.0 | 79.8 | 13.4 | 14,912 | 0.1 | 871 | 6 | 22,452 | 70.0 | 18,309 | 53.9 |
| WASHINGTON | 70.9 | 42.0 | 95.6 | 90.9 | 83.1 | 7.0 | 3,914,869 | 0.0 | 329,087 | 8 | 6,128,444 | 64.1 | 4,920,759 | 52.6 |
| Auburn | 73.9 | 53.5 | 95.7 | 89.3 | 79.9 | 8.5 | 42,657 | -0.3 | 3,952 | 9 | 62,730 | 68.9 | 50,985 | 55.3 |
| Bellevue | 56.0 | 38.3 | 98.9 | 96.8 | 80.3 | 8.2 | 81,351 | -2.2 | 4,792 | 6 | 120,576 | 69.1 | 99,681 | 55.4 |
| Bellingham | 66.1 | 21.4 | 94.3 | 88.8 | 73.1 | 9.4 | 48,267 | -0.1 | 4,473 | 9 | 81,352 | 62.4 | 65,597 | 40.5 |
| Bothell | 70.9 | 56.8 | 95.4 | 93.9 | 82.4 | 11.5 | 26,555 | -0.2 | 1,867 | 7 | 39,551 | 70.5 | 32,462 | 55.4 |
| Bremerton | 57.1 | 31.4 | 94.8 | 91.0 | 73.8 | 16.1 | 18,197 | 0.8 | 1,682 | 9 | 33,737 | 55.3 | 29,306 | 59.7 |
| Burien | 64.0 | 45.2 | 96.4 | 92.2 | 84.4 | 3.7 | 27,964 | 0.2 | 2,789 | 10 | 42,091 | 71.0 | 34,545 | 54.9 |
| Des Moines | 74.6 | 50.2 | 95.2 | 92.0 | 84.5 | 4.2 | 17,413 | 0.9 | 1,740 | 10 | 27,605 | 64.7 | 21,025 | 50.6 |
| Edmonds | 61.5 | 53.1 | 97.2 | 93.2 | 87.3 | 7.1 | 23,243 | -2.0 | 1,664 | 7 | 35,468 | 63.6 | 27,091 | 56.8 |
| Everett | 74.4 | 42.3 | 94.6 | 87.1 | 83.9 | 4.0 | 58,270 | 1.0 | 5,790 | 10 | 90,100 | 64.9 | 76,062 | 53.3 |
| Federal Way | 69.9 | 60.0 | 95.9 | 90.4 | 85.2 | 5.0 | 50,907 | 0.0 | 5,053 | 10 | 75,203 | 67.3 | 63,717 | 51.3 |
| Issaquah | 60.4 | 57.0 | 98.5 | 94.9 | 81.8 | 6.9 | 21,870 | -2.3 | 1,270 | 6 | 30,878 | 67.2 | 25,411 | 61. |
| Kennewick | 78.4 | 19.6 | 95.4 | 84.4 | 85.2 | 8.5 | 42,726 | 0.7 | 3,608 | 8 | 64,835 | 64.9 | 51,943 | 50. |
| Kent | 72.8 | 48.0 | 97.4 | 94.7 | 84.2 | 3.7 | 68,929 | 3.1 | 6,855 | 10 | 104,557 | 71.9 | 91,441 | 57. |
| Kirkland | 68.0 | 43.4 | 98.3 | 94.1 | 82.8 | 7.5 | 54,773 | 1.3 | 3,380 | 6 | 74,731 | 71.4 | 62,007 | 56. |
| Lacey | 75.0 | 33.9 | 94.3 | 85.5 | 79.6 | 8.0 | 23,157 | 3.3 | 2,011 | 9 | 43,998 | 59.1 | 36,542 | 50. |
| Lake Stevens | 81.0 | 57.1 | 97.2 | 95.2 | 89.5 | 3.8 | 18,038 | 1.6 | 1,620 | 9 | 26,570 | 72.3 | 22,465 | 58. |
| Lakewood | 80.1 | 41.9 | 95.4 | 90.2 | 76.7 | 8.2 | 27,241 | 1.0 | 2,750 | 10 | 49,222 | 56.9 | 39,118 | 53. |
| Longview | 85.7 | 29.4 | 91.7 | 85.3 | 78.1 | 13.4 | 16,074 | 0.8 | 1,552 | 10 | 30,266 | 55.2 | 22,107 | 47. |
| Lynnwood | 70.9 | 49.0 | 93.6 | 86.2 | 84.1 | 8.2 | 21,157 | 2.2 | 2,012 | 10 | 35,106 | 61.1 | 25,690 | 53. |
| Marysville | 85.7 | 46.9 | 96.8 | 94.2 | 87.3 | 4.0 | 36,778 | 1.7 | 3,673 | 10 | 55,014 | 64.5 | 44,692 | 58. |
| Mount Vernon | 83.0 | 31.5 | 95.7 | 89.8 | 82.7 | 5.8 | 16,282 | -0.5 | 1,686 | 10 | 28,090 | 59.2 | 20,455 | 53. |
| Olympia | 71.6 | 23.4 | 97.3 | 84.6 | 79.7 | 9.9 | 28,746 | 0.8 | 2,485 | 9 | 44,833 | 62.0 | 35,471 | 48. |

1. Employed persons. 2. Households. 3. Percent of civilian labor force. 4. Persons 16 years old and over.

# Table D. Cities — Construction, Wholesale Trade, and Retail Trade

| City | Value of residential construction authorized by building permits, 2020 | | | Wholesale trade[1], 2017 | | | | Retail trade[2], 2017 | | | |
|---|---|---|---|---|---|---|---|---|---|---|---|
| | New construction ($1,000) | Number of housing units | Percent single family | Number of establishments | Number of employees | Sales (mil dol) | Annual payroll (mil dol) | Number of establishments | Number of employees | Sales (mil dol) | Annual payroll (mil dol) |
| | 69 | 70 | 71 | 72 | 73 | 74 | 75 | 76 | 77 | 78 | 79 |
| **UTAH— Cont'd** | | | | | | | | | | | |
| Kaysville | 43,149 | 106 | 100 | 24 | 250 | 92 | 13 | 84 | 1,019 | 263 | 28 |
| Layton | 171,812 | 734 | 79 | 46 | 338 | 157 | 16 | 294 | 5,222 | 1,407 | 138 |
| Lehi | 298,624 | 1,044 | 100 | 36 | 1,233 | 873 | 77 | 178 | 2,707 | 847 | 73 |
| Logan | 99,377 | 783 | 32 | 71 | 691 | 508 | 35 | 275 | 4,271 | 1,309 | 105 |
| Midvale | 46,357 | 286 | 13 | 52 | 577 | 294 | 30 | 157 | 3,646 | 2,408 | 163 |
| Murray | 36,240 | 132 | 100 | 93 | 868 | 399 | 46 | 340 | 6,394 | 2,171 | 202 |
| Ogden | 53,741 | 353 | 42 | 96 | 1,835 | 1,629 | 101 | 349 | 4,272 | 1,226 | 121 |
| Orem | 46,726 | 155 | 92 | 103 | 1,203 | 593 | 60 | 478 | 7,756 | 2,198 | 213 |
| Pleasant Grove | 56,636 | 153 | 74 | 30 | 135 | 93 | 11 | 97 | 1,144 | 285 | 33 |
| Provo | 69,264 | 357 | 61 | 60 | 1,983 | 817 | 145 | 396 | 4,673 | 1,150 | 118 |
| Riverton | 63,326 | 236 | 61 | 16 | 43 | 23 | 2 | 64 | 1,283 | 331 | 31 |
| Roy | 15,004 | 105 | 83 | D | D | D | D | 71 | 1,056 | 253 | 24 |
| St. George | 240,259 | 1,575 | 88 | 108 | 743 | 354 | 33 | 436 | 6,553 | 2,022 | 189 |
| Salt Lake City | 359,863 | 2,667 | 6 | 601 | 13,791 | 14,431 | 961 | 919 | 15,363 | 5,192 | 483 |
| Sandy | 20,667 | 84 | 55 | 120 | 1,177 | 827 | 62 | 405 | 7,001 | 2,504 | 215 |
| South Jordan | 289,787 | 1,271 | 88 | 41 | 122 | 56 | 5 | 174 | 4,012 | 1,659 | 146 |
| Spanish Fork | 101,982 | 424 | 89 | 23 | 210 | 177 | 9 | 112 | 1,783 | 546 | 46 |
| Springville | 89,655 | 336 | 84 | 23 | 320 | 134 | 17 | 96 | 1,213 | 371 | 36 |
| Taylorsville | 12,428 | 60 | 100 | 21 | 99 | 54 | 3 | 79 | 1,388 | 347 | 34 |
| Tooele | 76,689 | 332 | 87 | 9 | 60 | 31 | 4 | 70 | 1,365 | 400 | 35 |
| West Jordan | 140,278 | 651 | 54 | 73 | 1,496 | 1,706 | 97 | 210 | 4,371 | 1,053 | 106 |
| West Valley City | 131,822 | 644 | 18 | 145 | 2,504 | 2,305 | 155 | 291 | 5,682 | 2,154 | 174 |
| **VERMONT** | 496 | 2 | 100 | 671 | 9,395 | 6,032 | 503 | 3,219 | 38,390 | 10,811 | 1,103 |
| Burlington | 16,994 | 107 | 7 | 43 | 477 | 379 | 36 | 213 | 2,856 | 592 | 77 |
| **VIRGINIA** | 235,295 | 982 | 71 | 5,937 | 86,698 | 98,905 | 5,734 | 27,134 | 429,072 | 120,162 | 11,476 |
| Alexandria | 40,575 | 107 | 100 | 72 | 872 | 652 | 60 | 475 | 7,862 | 2,549 | 259 |
| Blacksburg | 232,994 | 1,233 | 3 | 9 | 310 | 340 | 14 | 96 | 1,468 | 270 | 24 |
| Charlottesville | 36,772 | 190 | 38 | 44 | 451 | 330 | 24 | 281 | 3,282 | 715 | 77 |
| Chesapeake | 313,351 | 1,128 | 100 | 241 | 3,552 | 2,791 | 206 | 785 | 14,925 | 4,182 | 388 |
| Danville | 1,012 | 7 | 100 | 45 | 560 | 256 | 30 | 302 | 4,558 | 1,144 | 104 |
| Hampton | 13,160 | 211 | 100 | 60 | 921 | 425 | 47 | 423 | 6,748 | 1,569 | 169 |
| Harrisonburg | 11,355 | 60 | 65 | 58 | 864 | 439 | 44 | 325 | 5,341 | 1,379 | 136 |
| Leesburg | 44,010 | 264 | 71 | 17 | 277 | 158 | 18 | 279 | 5,980 | 1,766 | 153 |
| Lynchburg | 31,510 | 361 | 43 | 74 | 898 | 479 | 43 | 355 | 6,890 | 2,163 | 165 |
| Manassas | 17,106 | 73 | 100 | 47 | 458 | 228 | 23 | 176 | 2,153 | 962 | 77 |
| Newport News | 20,234 | 160 | 100 | 112 | 1,887 | 1,539 | 130 | 648 | 10,091 | 2,597 | 243 |
| Norfolk | 139,505 | 1,202 | 39 | 187 | 2,714 | 2,297 | 146 | 830 | 11,729 | 2,696 | 293 |
| Petersburg | 7,661 | 63 | 21 | D | D | D | 15 | 141 | 2,454 | 941 | 67 |
| Portsmouth | 31,675 | 280 | 100 | 43 | 659 | 227 | 34 | 261 | 3,060 | 627 | 69 |
| Richmond | 108,842 | 1,023 | 29 | 253 | 3,815 | 3,582 | 212 | 790 | 9,088 | 2,260 | 241 |
| Roanoke | 44,929 | 317 | 16 | 155 | 2,211 | 1,381 | 116 | 533 | 8,966 | 2,604 | 216 |
| Suffolk | 215,656 | 1,108 | 65 | 36 | 1,117 | 1,443 | 79 | 238 | 3,997 | 1,161 | 100 |
| Virginia Beach | 147,855 | 938 | 53 | 351 | 4,770 | 5,597 | 359 | 1,506 | 24,023 | 6,155 | 607 |
| Winchester | 2,840 | 18 | 33 | 36 | 459 | 464 | 22 | 284 | 4,317 | 1,017 | 110 |
| **WASHINGTON** | 9,488,095 | 43,881 | 54 | 7,755 | 120,320 | 112,353 | 7,649 | 21,751 | 347,728 | 160,285 | 11,413 |
| Auburn | 30,198 | 83 | 100 | 180 | 4,529 | 6,417 | 304 | 291 | 5,174 | 1,747 | 165 |
| Bellevue | 154,241 | 333 | 64 | 281 | 4,021 | 7,101 | 446 | 653 | 14,690 | 5,123 | 556 |
| Bellingham | 99,011 | 656 | 19 | 146 | 1,497 | 752 | 74 | 515 | 8,719 | 2,489 | 251 |
| Bothell | 70,102 | 194 | 89 | 54 | 1,754 | 2,375 | 288 | 110 | 1,725 | 469 | 53 |
| Bremerton | 66,949 | 374 | 47 | 22 | 190 | 96 | 10 | 151 | 2,050 | 832 | 82 |
| Burien | 17,568 | 61 | 41 | 23 | 128 | 45 | 6 | 151 | 2,055 | 736 | 72 |
| Des Moines | 53,465 | 222 | 22 | 12 | 237 | 189 | 17 | 40 | 303 | 95 | 11 |
| Edmonds | 40,520 | 234 | 14 | 43 | 176 | 154 | 12 | 132 | 1,902 | 638 | 68 |
| Everett | 72,774 | 560 | 5 | 132 | 2,680 | 2,085 | 143 | 439 | 7,422 | 2,380 | 245 |
| Federal Way | 13,870 | 35 | 100 | 57 | 433 | 442 | 24 | 260 | 4,913 | 1,385 | 143 |
| Issaquah | 22,369 | 103 | 50 | 50 | 532 | 463 | 38 | 160 | 3,736 | 4,809 | 143 |
| Kennewick | 88,829 | 326 | 91 | 78 | 749 | 654 | 39 | 378 | 6,713 | 1,910 | 182 |
| Kent | 118,850 | 377 | 83 | 407 | 10,147 | 11,483 | 655 | 343 | 7,603 | 3,673 | 253 |
| Kirkland | 140,257 | 576 | 73 | 130 | 1,405 | 971 | 93 | 250 | 4,905 | 2,020 | 194 |
| Lacey | 66,808 | 315 | 43 | 30 | 538 | 626 | 31 | 158 | 3,697 | 1,029 | 106 |
| Lake Stevens | 111,189 | 351 | 99 | 13 | 48 | 23 | 2 | 46 | 893 | 247 | 25 |
| Lakewood | 29,424 | 162 | 36 | 68 | 1,157 | 1,297 | 66 | 242 | 3,016 | 799 | 82 |
| Longview | 3,719 | 17 | 77 | 42 | 568 | 343 | 31 | 174 | 3,223 | 992 | 95 |
| Lynnwood | 134,784 | 1,116 | 5 | 84 | 822 | 412 | 58 | 411 | 8,806 | 2,784 | 280 |
| Marysville | 96,151 | 423 | 85 | 38 | 355 | 186 | 20 | 179 | 3,699 | 1,183 | 118 |
| Mount Vernon | 23,245 | 91 | 95 | 40 | 604 | 294 | 31 | 132 | 2,168 | 655 | 69 |
| Olympia | 26,511 | 196 | 20 | 47 | 329 | 266 | 21 | 362 | 5,504 | 1,662 | 162 |

1. Merchant wholesalers except manufacturers' sales branches and offices.  2. Establishments with payroll.

# Table D. Cities — Real Estate, Professional Services, and Manufacturing

| City | Real estate and rental and leasing, 2017 | | | | Professional, scientific, and technical services[1], 2017 | | | | Manufacturing, 2017 | | | |
|---|---|---|---|---|---|---|---|---|---|---|---|---|
| | Number of establishments | Number of employees | Receipts (mil dol) | Annual payroll (mil dol) | Number of establishments | Number of employees | Receipts (mil dol) | Annual payroll (mil dol) | Number of establishments | Number of employees | Receipts (mil dol) | Annual payroll (mil dol) |
| | 80 | 81 | 82 | 83 | 84 | 85 | 86 | 87 | 88 | 89 | 90 | 91 |
| **UTAH— Cont'd** | | | | | | | | | | | | |
| Kaysville | 43 | 75 | 21 | 3 | 130 | 872 | 107 | 51 | NA | NA | NA | NA |
| Layton | 127 | 512 | 119 | 17 | 213 | 1,552 | 201 | 79 | NA | NA | NA | NA |
| Lehi | D | D | D | D | 325 | 3,937 | 848 | 355 | NA | NA | NA | NA |
| Logan | 124 | 361 | 67 | 11 | 189 | 1,868 | 246 | 77 | NA | NA | NA | NA |
| Midvale | 83 | 534 | 145 | 32 | 127 | 3,481 | 252 | 153 | NA | NA | NA | NA |
| Murray | 169 | 1,143 | 233 | 55 | D | D | D | D | NA | NA | NA | NA |
| Ogden | 114 | 381 | 56 | 11 | 275 | 2,689 | 348 | 132 | NA | NA | NA | NA |
| Orem | 210 | 724 | 181 | 28 | D | D | D | D | NA | NA | NA | NA |
| Pleasant Grove | D | D | D | D | 140 | 985 | 178 | 68 | NA | NA | NA | NA |
| Provo | 136 | 556 | 99 | 18 | 415 | 2,798 | 427 | 164 | NA | NA | NA | NA |
| Riverton | D | D | D | D | 111 | 241 | 25 | 8 | NA | NA | NA | NA |
| Roy | 24 | 45 | 10 | 1 | 29 | 176 | 20 | 8 | NA | NA | NA | NA |
| St. George | 262 | 607 | 148 | 19 | 447 | 2,103 | 280 | 83 | NA | NA | NA | NA |
| Salt Lake City | 608 | 3,524 | 1,377 | 200 | 1,616 | 20,088 | 3,814 | 1,472 | NA | NA | NA | NA |
| Sandy | 276 | 1,350 | 389 | 70 | D | D | D | D | NA | NA | NA | NA |
| South Jordan | 173 | 454 | 92 | 16 | 323 | 1,951 | 321 | 136 | NA | NA | NA | NA |
| Spanish Fork | 48 | 81 | 20 | 3 | 108 | 358 | 69 | 17 | NA | NA | NA | NA |
| Springville | 30 | D | 16 | D | 82 | 342 | 52 | 20 | NA | NA | NA | NA |
| Taylorsville | D | D | D | D | 115 | 7,775 | 1,433 | 437 | NA | NA | NA | NA |
| Tooele | 30 | 74 | 15 | 2 | 43 | 190 | 18 | 6 | NA | NA | NA | NA |
| West Jordan | 93 | 211 | 45 | 8 | 160 | 601 | 82 | 26 | NA | NA | NA | NA |
| West Valley City | 94 | 462 | 242 | 23 | 149 | 2,326 | 299 | 120 | NA | NA | NA | NA |
| **VERMONT** | 784 | 2,938 | 652 | 119 | 2,096 | 12,563 | 2,079 | 842 | 1,033 | 28,629 | 8,652 | 1,610 |
| Burlington | 66 | 334 | 81 | 16 | 308 | 3,009 | 526 | 227 | NA | NA | NA | NA |
| **VIRGINIA** | 10,051 | 55,778 | 17,208 | 2,949 | 31,431 | 470,265 | 101,433 | 41,553 | 5,038 | 232,695 | 99,344 | 12,987 |
| Alexandria | 282 | 1,580 | 556 | 84 | 1,211 | 19,170 | 4,414 | 1,910 | NA | NA | NA | NA |
| Blacksburg | 42 | 346 | 70 | 13 | 146 | 1,499 | 248 | 116 | NA | NA | NA | NA |
| Charlottesville | 84 | 555 | 111 | 26 | 334 | 2,763 | 446 | 192 | NA | NA | NA | NA |
| Chesapeake | 307 | 1,421 | 383 | 67 | 519 | 9,596 | 1,369 | 566 | NA | NA | NA | NA |
| Danville | 53 | 237 | 32 | 7 | 64 | 393 | 43 | 18 | NA | NA | NA | NA |
| Hampton | 123 | 784 | 155 | 27 | 290 | 4,585 | 829 | 363 | NA | NA | NA | NA |
| Harrisonburg | 71 | 373 | 99 | 14 | 140 | 1,091 | 127 | 53 | NA | NA | NA | NA |
| Leesburg | 60 | 252 | 89 | 12 | 371 | 3,208 | 551 | 268 | NA | NA | NA | NA |
| Lynchburg | 119 | 458 | 87 | 17 | 192 | 2,919 | 622 | 217 | NA | NA | NA | NA |
| Manassas | 45 | 226 | 61 | 11 | 216 | 1,693 | 296 | 123 | NA | NA | NA | NA |
| Newport News | 280 | 1,419 | 349 | 59 | 363 | 5,003 | 815 | 313 | NA | NA | NA | NA |
| Norfolk | 317 | 3,133 | 609 | 145 | 707 | 10,576 | 2,009 | 728 | NA | NA | NA | NA |
| Petersburg | 37 | 163 | 31 | 5 | 36 | 239 | 28 | 10 | NA | NA | NA | NA |
| Portsmouth | 77 | 388 | 80 | 14 | 144 | 1,450 | 160 | 70 | NA | NA | NA | NA |
| Richmond | 329 | 2,197 | 940 | 106 | 910 | 11,464 | 2,524 | 1,058 | NA | NA | NA | NA |
| Roanoke | 174 | 938 | 193 | 37 | 308 | 2,688 | 421 | 171 | NA | NA | NA | NA |
| Suffolk | 74 | 279 | 54 | 13 | 145 | 923 | 133 | 54 | NA | NA | NA | NA |
| Virginia Beach | 761 | 5,451 | 1,199 | 261 | 1,494 | 21,755 | 4,110 | 1,619 | NA | NA | NA | NA |
| Winchester | 66 | 306 | 98 | 12 | 142 | 995 | 103 | 49 | NA | NA | NA | NA |
| **WASHINGTON** | 11,826 | 53,706 | 15,627 | 2,757 | 22,144 | 214,628 | 42,429 | 17,333 | 7,017 | 263,132 | 140,382 | 17,944 |
| Auburn | 95 | 380 | 98 | 16 | 151 | 707 | 110 | 39 | NA | NA | NA | NA |
| Bellevue | 613 | 3,693 | 1,355 | 270 | 1,361 | 22,379 | 4,638 | 2,102 | NA | NA | NA | NA |
| Bellingham | 234 | 993 | 349 | 43 | 446 | 2,086 | 333 | 119 | NA | NA | NA | NA |
| Bothell | 99 | 249 | 109 | 15 | 273 | 4,074 | 841 | 328 | NA | NA | NA | NA |
| Bremerton | 67 | 242 | 38 | 8 | 75 | 798 | 127 | 52 | NA | NA | NA | NA |
| Burien | 75 | 232 | 73 | 9 | 95 | 484 | 41 | 17 | NA | NA | NA | NA |
| Des Moines | 30 | 71 | 18 | 3 | 31 | 123 | 15 | 5 | NA | NA | NA | NA |
| Edmonds | D | D | D | D | 213 | 1,061 | 159 | 67 | NA | NA | NA | NA |
| Everett | 173 | 1,115 | 227 | 43 | 323 | 3,467 | 494 | 237 | NA | NA | NA | NA |
| Federal Way | 130 | 597 | 209 | 26 | 188 | 1,621 | 195 | 90 | NA | NA | NA | NA |
| Issaquah | 96 | 392 | 161 | 27 | 207 | 1,232 | 183 | 116 | NA | NA | NA | NA |
| Kennewick | 173 | 729 | 177 | 27 | 190 | 1,342 | 212 | 76 | NA | NA | NA | NA |
| Kent | 196 | 985 | 373 | 53 | 246 | 2,534 | 462 | 134 | NA | NA | NA | NA |
| Kirkland | 272 | 1,159 | 882 | 77 | 594 | 6,422 | 1,504 | 659 | NA | NA | NA | NA |
| Lacey | 62 | 241 | 55 | 8 | 82 | 1,729 | 217 | 104 | NA | NA | NA | NA |
| Lake Stevens | 38 | 167 | 47 | 8 | 29 | 85 | 12 | 5 | NA | NA | NA | NA |
| Lakewood | 117 | 686 | 156 | 27 | 106 | 710 | 94 | 36 | NA | NA | NA | NA |
| Longview | 70 | 263 | 53 | 9 | 71 | 448 | 47 | 20 | NA | NA | NA | NA |
| Lynnwood | 103 | 398 | 164 | 18 | 191 | 2,320 | 463 | 170 | NA | NA | NA | NA |
| Marysville | 72 | 240 | 64 | 10 | 67 | 281 | 28 | 10 | NA | NA | NA | NA |
| Mount Vernon | 53 | 148 | 35 | 5 | 103 | 540 | 75 | 24 | NA | NA | NA | NA |
| Olympia | 109 | 425 | 123 | 18 | D | D | D | D | NA | NA | NA | NA |

1. Establishments subject to federal tax.

# Accommodation and Food Services, Arts, Entertainment, and Recreation, and Health Care and Social Assistance

| City | Accommodation and food services, 2017 | | | | Arts, entertainment, and recreation[1], 2017 | | | | Health care and social assistance[1], 2017 | | | |
|---|---|---|---|---|---|---|---|---|---|---|---|---|
| | Number of establishments | Number of employees | Receipts (mil dol) | Annual payroll (mil dol) | Number of establishments | Number of employees | Receipts (mil dol) | Annual payroll (mil dol) | Number of establishments | Number of employees | Receipts (mil dol) | Annual payroll (mil dol) |
| | 92 | 93 | 94 | 95 | 96 | 97 | 98 | 99 | 100 | 101 | 102 | 103 |
| UTAH— Cont'd | | | | | | | | | | | | |
| Kaysville | 24 | 508 | 20 | 6 | 11 | 190 | 10 | 2 | 77 | 668 | 56 | 24 |
| Layton | 166 | 3,673 | 210 | 57 | 19 | 296 | 10 | 3 | 214 | 3,915 | 473 | 167 |
| Lehi | 98 | 2,154 | 129 | 37 | 25 | 674 | 71 | 12 | 114 | 1,058 | 142 | 51 |
| Logan | 128 | 3,069 | 142 | 41 | 28 | 248 | 14 | 4 | 220 | 3,667 | 453 | 141 |
| Midvale | 109 | 1,961 | 113 | 32 | 13 | 451 | 21 | 7 | 81 | 781 | 61 | 26 |
| Murray | 125 | 2,927 | 166 | 51 | 19 | 384 | 22 | 6 | 370 | 10,971 | 1,963 | 654 |
| Ogden | 216 | 3,798 | 189 | 55 | D | D | D | D | 309 | 7,156 | 1,050 | 346 |
| Orem | 186 | 3,680 | 204 | 56 | 57 | 522 | 31 | 8 | 314 | 5,771 | 538 | 193 |
| Pleasant Grove | 34 | 654 | 48 | 13 | 14 | 49 | 15 | 2 | 82 | 660 | 57 | 20 |
| Provo | 202 | 4,386 | 240 | 71 | 47 | 305 | 21 | 6 | 323 | 10,535 | 1,322 | 495 |
| Riverton | 53 | 1,111 | 55 | 17 | 11 | 167 | 6 | 2 | 100 | 1,672 | 206 | 70 |
| Roy | 41 | 674 | 38 | 10 | 4 | D | 1 | D | 56 | 595 | 61 | 27 |
| St. George | 241 | 5,083 | 283 | 77 | 50 | 594 | 35 | 9 | 479 | 7,927 | 1,101 | 347 |
| Salt Lake City | 835 | 19,340 | 1,388 | 379 | 143 | 3,336 | 289 | 76 | 741 | 25,849 | 3,999 | 1,621 |
| Sandy | 213 | 3,892 | 228 | 65 | 47 | 1,217 | 107 | 30 | 315 | 3,567 | 384 | 156 |
| South Jordan | 111 | 2,570 | 134 | 39 | 24 | 434 | 26 | 8 | 174 | 2,436 | 248 | 113 |
| Spanish Fork | 54 | 1,179 | 54 | 14 | 4 | 71 | 3 | 1 | D | D | D | D |
| Springville | 41 | 837 | 39 | 12 | D | D | D | D | 83 | 773 | 67 | 23 |
| Taylorsville | D | D | D | D | 11 | 165 | 8 | 2 | 110 | 1,444 | 116 | 46 |
| Tooele | 51 | 905 | 46 | 12 | 9 | 72 | 4 | 1 | D | D | D | D |
| West Jordan | 133 | 2,622 | 147 | 40 | 13 | 171 | 8 | 2 | 209 | 4,575 | 516 | 142 |
| West Valley City | 226 | 4,126 | 249 | 67 | D | D | D | D | 140 | 4,106 | 550 | 187 |
| VERMONT | 1,979 | 32,891 | 2,020 | 621 | 457 | 7,739 | 517 | 151 | 2,102 | 48,285 | 5,452 | 2,229 |
| Burlington | 172 | 3,588 | 234 | 72 | 37 | 794 | 50 | 18 | 176 | 8,863 | 1,486 | 456 |
| VIRGINIA | 18,199 | 358,010 | 22,075 | 6,227 | 3,071 | 61,703 | 4,447 | 1,283 | 20,522 | 454,501 | 58,793 | 22,722 |
| Alexandria | 402 | 8,518 | 692 | 193 | 75 | 1,283 | 253 | 44 | 426 | 7,386 | 1,056 | 379 |
| Blacksburg | 112 | 2,249 | 115 | 31 | 22 | 202 | 6 | 2 | D | D | D | D |
| Charlottesville | 292 | 5,491 | 330 | 98 | 53 | 750 | 111 | 36 | 158 | 9,293 | 1,743 | 878 |
| Chesapeake | 486 | 10,068 | 531 | 149 | 54 | 1,280 | 59 | 19 | 514 | 10,002 | 1,240 | 504 |
| Danville | 143 | 2,925 | 149 | 41 | 15 | 189 | 8 | 3 | 180 | 5,096 | 489 | 210 |
| Hampton | 256 | 5,902 | 304 | 89 | D | D | D | D | 281 | 7,833 | 1,299 | 455 |
| Harrisonburg | 208 | 5,138 | 259 | 81 | 24 | 234 | 12 | 5 | 185 | 3,080 | 281 | 124 |
| Leesburg | 142 | 2,924 | 240 | 61 | 25 | 364 | 16 | 5 | 182 | 3,129 | 308 | 129 |
| Lynchburg | 270 | 6,023 | 292 | 83 | 35 | 567 | 24 | 8 | 289 | 10,123 | 1,394 | 516 |
| Manassas | 126 | 1,947 | 117 | 33 | 17 | 184 | 10 | 3 | 177 | 3,231 | 376 | 159 |
| Newport News | 411 | 7,516 | 390 | 112 | 52 | 1,025 | 41 | 16 | 452 | 15,710 | 1,843 | 782 |
| Norfolk | 612 | 11,326 | 659 | 182 | 75 | 1,558 | 110 | 36 | 582 | 20,755 | 3,182 | 1,091 |
| Petersburg | 94 | 930 | 51 | 14 | 10 | 72 | 2 | 1 | 126 | 4,361 | 480 | 191 |
| Portsmouth | 163 | 2,414 | 120 | 34 | D | D | D | D | 227 | 7,627 | 1,142 | 386 |
| Richmond | 677 | 14,306 | 794 | 259 | 114 | 2,455 | 179 | 54 | 607 | 27,376 | 4,123 | 1,644 |
| Roanoke | 302 | 7,205 | 372 | 113 | 49 | 545 | 40 | 11 | 316 | 13,827 | 2,050 | 743 |
| Suffolk | 150 | 3,029 | 163 | 45 | 16 | 348 | 13 | 5 | 240 | 4,979 | 593 | 226 |
| Virginia Beach | 1,292 | 25,108 | 1,494 | 419 | 202 | 3,598 | 249 | 67 | 1,162 | 21,729 | 2,767 | 1,053 |
| Winchester | 144 | 3,320 | 160 | 48 | 18 | 258 | 16 | 5 | 269 | 7,080 | 1,125 | 407 |
| WASHINGTON | 17,828 | 289,371 | 21,069 | 6,395 | 3,097 | 65,091 | 6,484 | 2,078 | 21,264 | 438,835 | 56,442 | 22,849 |
| Auburn | 159 | 2,514 | 146 | 46 | D | D | D | D | 215 | 3,917 | 475 | 214 |
| Bellevue | 512 | 10,576 | 919 | 272 | 89 | 2,418 | 192 | 67 | 1,053 | 15,245 | 2,298 | 842 |
| Bellingham | 359 | 6,023 | 348 | 118 | 76 | 1,058 | 41 | 15 | 519 | 9,046 | 1,208 | 416 |
| Bothell | 143 | 2,270 | 141 | 44 | 17 | 414 | 20 | 8 | 174 | 2,487 | 232 | 106 |
| Bremerton | 114 | 2,053 | 130 | 37 | 15 | 370 | 13 | 4 | 138 | 4,816 | 750 | 250 |
| Burien | D | D | D | D | 18 | 278 | 18 | 7 | 194 | 2,908 | 473 | 169 |
| Des Moines | 55 | 831 | 57 | 18 | 6 | 59 | 3 | 2 | 80 | 1,389 | 115 | 48 |
| Edmonds | 133 | 2,032 | 144 | 44 | 18 | 251 | 12 | 4 | 219 | 4,195 | 496 | 208 |
| Everett | 377 | 5,036 | 330 | 100 | 39 | 645 | 42 | 13 | 444 | 14,553 | 1,788 | 737 |
| Federal Way | 242 | 3,796 | 264 | 77 | 37 | 693 | 46 | 15 | 367 | 6,243 | 734 | 289 |
| Issaquah | 130 | 2,324 | 160 | 48 | 25 | 828 | 39 | 18 | 234 | 4,383 | 484 | 180 |
| Kennewick | 214 | 3,894 | 242 | 72 | 28 | 584 | 32 | 11 | 302 | 5,911 | 595 | 266 |
| Kent | 290 | 3,799 | 258 | 73 | 28 | 411 | 41 | 12 | 336 | 4,085 | 369 | 157 |
| Kirkland | 240 | 4,007 | 334 | 99 | 59 | 1,002 | 57 | 23 | 427 | 9,426 | 1,262 | 607 |
| Lacey | 139 | 2,366 | 141 | 43 | D | D | D | D | 148 | 2,832 | 224 | 100 |
| Lake Stevens | 40 | 724 | 44 | 13 | D | D | D | D | 44 | 407 | 40 | 18 |
| Lakewood | 180 | 2,568 | 167 | 50 | 24 | 875 | 36 | 20 | 218 | 5,847 | 607 | 305 |
| Longview | 110 | 1,694 | 91 | 28 | D | D | D | D | 159 | 4,602 | 526 | 217 |
| Lynnwood | 220 | 3,626 | 272 | 77 | 15 | 204 | 14 | 4 | 242 | 2,899 | 281 | 121 |
| Marysville | 123 | 1,795 | 115 | 34 | 14 | 201 | 13 | 3 | 134 | 1,625 | 187 | 82 |
| Mount Vernon | 94 | 1,312 | 77 | 24 | 16 | 359 | 15 | 6 | 151 | 4,991 | 627 | 278 |
| Olympia | 221 | 3,978 | 227 | 75 | 31 | 505 | 26 | 9 | 475 | 10,295 | 1,370 | 536 |

1. Establishments subject to federal tax.

# Table D. Cities — Other Services and Government Employment and Payroll

Columns 104–107: **Other services[1]**.
Columns 108–116: **Government employment and payroll, 2017 — March payroll.** Columns 110–116 are "Percent of total for:".

| City | Number of establishments | Number of employees | Receipts (mil dol) | Annual payroll (mil dol) | Full-time equivalent employees | Total (dollars) | Administrative, judicial, and legal | Police and corrections | Fire protection | Highways and transportation | Health and welfare | Natural resources and utilities | Education and libraries |
|---|---|---|---|---|---|---|---|---|---|---|---|---|---|
| | 104 | 105 | 106 | 107 | 108 | 109 | 110 | 111 | 112 | 113 | 114 | 115 | 116 |
| **UTAH— Cont'd** | | | | | | | | | | | | | |
| Kaysville | 36 | 210 | 22 | 6 | 145 | 558,493 | 10.8 | 23.0 | 8.8 | 7.7 | 0.0 | 39.0 | 0.0 |
| Layton | 123 | 825 | 75 | 20 | 352 | 1,629,092 | 18.5 | 35.0 | 21.5 | 9.4 | 0.1 | 14.0 | 0.0 |
| Lehi | 60 | 527 | 84 | 18 | 387 | 1,583,383 | 14.0 | 19.3 | 14.8 | 5.3 | 0.6 | 36.3 | 4.3 |
| Logan | 115 | 586 | 66 | 18 | 483 | 2,025,945 | 8.9 | 17.1 | 14.8 | 10.0 | 3.6 | 37.3 | 3.2 |
| Midvale | 77 | 447 | 54 | 14 | 78 | 365,027 | 59.3 | 0.0 | 0.0 | 11.8 | 4.3 | 21.6 | 0.0 |
| Murray | 165 | 1,220 | 136 | 43 | 452 | 2,231,784 | 15.7 | 22.8 | 15.9 | 4.1 | 1.5 | 30.0 | 3.0 |
| Ogden | 141 | 1,120 | 125 | 35 | 636 | 2,957,540 | 17.5 | 29.6 | 16.0 | 5.3 | 12.6 | 16.6 | 0.0 |
| Orem | 172 | 1,007 | 93 | 28 | 522 | 2,253,204 | 20.2 | 27.0 | 15.7 | 5.2 | 2.4 | 19.7 | 7.2 |
| Pleasant Grove | 50 | 214 | 25 | 7 | 188 | 671,019 | 18.2 | 26.3 | 13.0 | 6.9 | 0.5 | 26.2 | 4.6 |
| Provo | 141 | 727 | 93 | 21 | 814 | 3,480,203 | 17.3 | 23.0 | 13.0 | 5.3 | 1.4 | 31.9 | 3.9 |
| Riverton | D | D | D | D | 89 | 466,875 | 26.0 | 0.0 | 0.0 | 12.9 | 0.2 | 32.1 | 0.0 |
| Roy | D | D | D | D | 183 | 833,799 | 15.4 | 26.1 | 33.7 | 3.8 | 0.0 | 15.6 | 0.0 |
| St. George | 182 | 1,092 | 108 | 28 | 798 | 3,348,441 | 12.7 | 23.8 | 5.5 | 7.5 | 0.0 | 44.5 | 0.0 |
| Salt Lake City | 699 | 5,916 | 775 | 219 | 3,055 | 15,732,443 | 15.4 | 21.7 | 14.3 | 21.9 | 0.4 | 15.4 | 4.3 |
| Sandy | 169 | 1,079 | 108 | 32 | 514 | 2,541,559 | 23.0 | 27.1 | 17.0 | 9.4 | 0.9 | 18.6 | 0.0 |
| South Jordan | 81 | 505 | 41 | 14 | 373 | 1,630,405 | 22.5 | 19.1 | 20.9 | 3.2 | 2.5 | 16.6 | 0.0 |
| Spanish Fork | D | D | D | D | 344 | 1,474,006 | 26.9 | 14.8 | 2.0 | 4.1 | 4.4 | 17.7 | 2.3 |
| Springville | D | D | D | D | 247 | 1,106,787 | 13.2 | 23.7 | 5.3 | 8.2 | 0.4 | 40.9 | 4.2 |
| Taylorsville | 41 | 272 | 27 | 9 | 42 | 171,501 | 76.3 | 0.0 | 0.0 | 0.0 | 1.2 | 0.0 | 0.0 |
| Tooele | D | D | D | D | 155 | 572,952 | 22.0 | 34.1 | 1.4 | 5.1 | 0.0 | 19.0 | 5.5 |
| West Jordan | 125 | 815 | 124 | 39 | 513 | 2,455,676 | 17.7 | 31.8 | 22.0 | 4.8 | 3.1 | 9.9 | 0.0 |
| West Valley City | 130 | 741 | 92 | 24 | 728 | 3,395,984 | 16.7 | 37.1 | 17.6 | 7.5 | 6.7 | 10.1 | 0.0 |
| **VERMONT** | 1,603 | 7,339 | 922 | 244 | X | X | X | X | X | X | X | X | X |
| Burlington | 128 | 814 | 117 | 28 | 740 | 4,083,314 | 21.2 | 18.4 | 11.7 | 14.6 | 2.6 | 25.3 | 2.7 |
| **VIRGINIA** | 15,700 | 118,537 | 20,344 | 5,188 | X | X | X | X | X | X | X | X | X |
| Alexandria | 611 | 9,983 | 3,295 | 768 | 5,558 | 29,910,334 | 9.3 | 14.2 | 7.3 | 1.9 | 12.9 | 8.3 | 42.1 |
| Blacksburg | 54 | 382 | 248 | 14 | 356 | 1,357,428 | 31.4 | 22.5 | 0.6 | 26.7 | 4.0 | 11.9 | 0.0 |
| Charlottesville | 158 | 1,729 | 580 | 105 | 2,026 | 8,850,434 | 10.6 | 8.3 | 5.6 | 9.5 | 7.6 | 8.1 | 45.4 |
| Chesapeake | 438 | 3,057 | 349 | 111 | 9,189 | 35,172,070 | 5.6 | 11.9 | 5.5 | 2.0 | 5.6 | 4.8 | 61.2 |
| Danville | 89 | 424 | 59 | 11 | 2,182 | 7,837,955 | 10.7 | 13.4 | 6.7 | 4.0 | 5.6 | 10.5 | 44.5 |
| Hampton | 169 | 953 | 112 | 28 | 5,164 | 20,964,935 | 7.4 | 11.7 | 7.5 | 0.9 | 6.1 | 7.1 | 56.6 |
| Harrisonburg | 131 | 715 | 85 | 26 | 1,563 | 6,085,265 | 3.9 | 7.3 | 6.4 | 5.5 | 0.3 | 14.2 | 55.8 |
| Leesburg | 114 | 658 | 81 | 26 | 384 | 2,292,277 | 20.8 | 25.7 | 0.0 | 7.3 | 0.7 | 34.2 | 1.2 |
| Lynchburg | 171 | 1,013 | 107 | 32 | 2,987 | 9,939,533 | 10.6 | 8.8 | 5.9 | 5.0 | 6.0 | 6.7 | 55.9 |
| Manassas | 137 | 902 | 143 | 36 | 1,566 | 8,259,510 | 5.8 | 8.8 | 4.1 | 2.6 | 2.7 | 7.8 | 66.3 |
| Newport News | 283 | 1,731 | 186 | 49 | 7,911 | 33,653,408 | 6.2 | 12.5 | 6.1 | 1.7 | 5.3 | 11.1 | 56.0 |
| Norfolk | 396 | 3,023 | 459 | 106 | 10,850 | 48,264,655 | 7.3 | 13.9 | 4.9 | 3.9 | 8.8 | 7.0 | 51.3 |
| Petersburg | 75 | 576 | 68 | 16 | 1,237 | 4,738,339 | 12.0 | 11.3 | 11.6 | 3.7 | 5.3 | 1.4 | 54.7 |
| Portsmouth | 143 | 1,143 | 141 | 42 | 3,846 | 15,956,245 | 6.4 | 12.4 | 6.7 | 1.7 | 8.2 | 4.8 | 58.0 |
| Richmond | 520 | 4,950 | 609 | 174 | 8,650 | 35,398,423 | 8.7 | 16.8 | 6.2 | 1.8 | 8.0 | 10.1 | 44.9 |
| Roanoke | 252 | 1,629 | 155 | 48 | 3,798 | 15,653,992 | 8.7 | 13.0 | 8.1 | 2.0 | 7.5 | 2.4 | 56.3 |
| Suffolk | 137 | 715 | 65 | 19 | 3,372 | 14,209,808 | 10.1 | 8.4 | 10.9 | 3.5 | 3.3 | 7.4 | 55.5 |
| Virginia Beach | 943 | 5,762 | 919 | 175 | 17,638 | 70,561,307 | 6.4 | 10.4 | 4.2 | 1.5 | 6.9 | 8.9 | 59.0 |
| Winchester | 92 | 536 | 46 | 14 | 1,576 | 6,043,273 | 8.0 | 23.5 | 5.5 | 2.1 | 3.5 | 6.9 | 47.2 |
| **WASHINGTON** | 13,444 | 78,480 | 14,514 | 3,005 | X | X | X | X | X | X | X | X | X |
| Auburn | 162 | 1,180 | 157 | 51 | 454 | 3,059,866 | 23.3 | 34.7 | 0.0 | 16.4 | 4.4 | 18.2 | 0.0 |
| Bellevue | 443 | 3,020 | 336 | 109 | 1,360 | 10,639,183 | 16.4 | 17.5 | 21.4 | 10.1 | 2.5 | 22.5 | 0.0 |
| Bellingham | 276 | 1,771 | 221 | 66 | 825 | 5,436,962 | 14.6 | 24.5 | 27.3 | 6.7 | 0.5 | 13.6 | 3.3 |
| Bothell | 82 | 500 | 40 | 17 | 323 | 2,501,428 | 16.0 | 29.3 | 26.4 | 6.8 | 4.2 | 7.3 | 0.0 |
| Bremerton | 90 | 440 | 49 | 14 | 314 | 2,214,919 | 17.1 | 24.0 | 23.9 | 10.9 | 3.1 | 16.8 | 0.0 |
| Burien | 107 | 598 | 63 | 17 | 77 | 479,562 | 40.5 | 0.0 | 0.0 | 16.2 | 0.0 | 17.4 | 0.0 |
| Des Moines | 35 | 150 | 19 | 5 | 134 | 897,466 | 25.3 | 38.4 | 0.0 | 12.9 | 0.0 | 28.6 | 0.0 |
| Edmonds | 115 | 531 | 52 | 18 | 176 | 1,267,743 | 14.9 | 41.2 | 0.0 | 12.6 | 1.5 | 22.5 | 3.1 |
| Everett | 253 | 1,966 | 241 | 78 | 1,191 | 8,542,797 | 9.5 | 21.4 | 21.8 | 14.5 | 2.7 | 16.4 | 0.0 |
| Federal Way | 154 | 694 | 63 | 23 | 365 | 2,495,462 | 19.6 | 53.5 | 0.0 | 6.7 | 0.0 | 22.0 | 0.0 |
| Issaquah | 106 | 828 | 119 | 31 | 247 | 1,614,855 | 48.4 | 23.8 | 0.0 | 5.7 | 1.8 | 16.0 | 0.0 |
| Kennewick | 169 | 963 | 90 | 29 | 400 | 2,756,636 | 19.0 | 32.2 | 24.4 | 6.7 | 1.1 | 21.8 | 0.0 |
| Kent | 242 | 1,449 | 158 | 50 | 702 | 4,511,920 | 26.0 | 35.1 | 0.0 | 12.2 | 0.0 | 11.2 | 0.0 |
| Kirkland | 237 | 1,286 | 156 | 60 | 605 | 4,373,154 | 28.2 | 22.1 | 23.3 | 2.8 | 3.4 | 25.6 | 0.0 |
| Lacey | 100 | 685 | 91 | 25 | 276 | 1,821,023 | 16.6 | 31.1 | 0.0 | 5.4 | 0.0 | 1.7 | 0.0 |
| Lake Stevens | D | D | 13 | D | 75 | 564,484 | 26.3 | 52.4 | 0.0 | 19.6 | 0.0 | 4.9 | 0.0 |
| Lakewood | 131 | 801 | 92 | 31 | 222 | 1,614,522 | 14.8 | 62.1 | 0.0 | 9.7 | 5.5 | 16.0 | 0.0 |
| Longview | 82 | 547 | 60 | 19 | 329 | 2,272,290 | 14.1 | 25.8 | 21.5 | 13.6 | 0.0 | 19.2 | 5.7 |
| Lynnwood | 153 | 1,066 | 114 | 37 | 397 | 3,100,998 | 18.4 | 30.5 | 20.5 | 2.5 | 0.0 | 16.4 | 0.0 |
| Marysville | 113 | 505 | 53 | 16 | 275 | 1,964,954 | 26.6 | 35.3 | 0.0 | 7.9 | 0.0 | 22.7 | 0.0 |
| Mount Vernon | 83 | 352 | 39 | 12 | 210 | 1,455,600 | 14.6 | 29.7 | 19.8 | 8.6 | 4.7 | 16.0 | 4.5 |
| Olympia | 205 | 1,473 | 237 | 74 | 551 | 3,953,470 | 18.5 | 21.3 | 20.7 | 4.5 | 5.7 | 21.1 | 0.0 |

1. Establishments subject to federal tax.

# Table D. Cities — City Government Finances

| City | General revenue | | | | | | | General expenditure | | |
|---|---|---|---|---|---|---|---|---|---|---|
| | Total (mil dol) | Intergovernmental | | Taxes | | | | Total (mil dol) | Per capita[1] (dollars) | |
| | | Total (mil dol) | Percent from state government | Total (mil dol) | Per capita[1] (dollars) | | | | Total | Capital outlays |
| | | | | | Total | Property | Sales and gross receipts | | | |
| | 117 | 118 | 119 | 120 | 121 | 122 | 123 | 124 | 125 | 126 |
| **UTAH— Cont'd** | | | | | | | | | | |
| Kaysville | 23.3 | 1.2 | 100.0 | 11.2 | 352 | 102 | 250 | 19 | 610 | 53 |
| Layton | 49.3 | 6.8 | 23.0 | 19.7 | 258 | 100 | 158 | 53 | 689 | 143 |
| Lehi | 74.4 | 3.2 | 28.2 | 26.9 | 423 | 144 | 279 | 78 | 1,217 | 413 |
| Logan | 70.4 | 9.9 | 62.4 | 26.1 | 511 | 114 | 391 | 53 | 1,034 | 118 |
| Midvale | 29.7 | 2.2 | 56.1 | 14.1 | 421 | 108 | 313 | 27 | 798 | 289 |
| Murray | 57.2 | 4.6 | 45.2 | 36.8 | 745 | 225 | 520 | 92 | 1,855 | 279 |
| Ogden | 101.2 | 9.9 | 71.6 | 49.5 | 568 | 274 | 294 | 101 | 1,155 | 201 |
| Orem | 86.3 | 6.0 | 88.2 | 45.8 | 469 | 103 | 366 | 72 | 732 | 151 |
| Pleasant Grove | 29.6 | 1.3 | 99.8 | 11.4 | 293 | 104 | 189 | 23 | 600 | 25 |
| Provo | 81.0 | 6.1 | 51.3 | 37.8 | 321 | 119 | 202 | 91 | 776 | 23 |
| Riverton | 21.2 | 1.6 | 82.1 | 9.1 | 209 | 7 | 202 | 17 | 385 | 54 |
| Roy | 25.8 | 1.6 | 91.2 | 13.0 | 332 | 102 | 231 | 21 | 532 | 68 |
| St. George | 108.1 | 6.0 | 58.0 | 51.0 | 604 | 143 | 461 | 86 | 1,022 | 251 |
| Salt Lake City | 639.8 | 16.5 | 11.4 | 264.4 | 1,316 | 846 | 470 | 777 | 3,865 | 1,588 |
| Sandy | 90.3 | 5.1 | 81.7 | 50.5 | 524 | 160 | 364 | 108 | 1,117 | 581 |
| South Jordan | 61.4 | 3.1 | 89.1 | 45.5 | 642 | 327 | 315 | 37 | 523 | 418 |
| Spanish Fork | 34.7 | 1.5 | 76.0 | 11.0 | 279 | 64 | 215 | 27 | 672 | 164 |
| Springville | 33.2 | 1.8 | 93.0 | 19.0 | 573 | 208 | 365 | 41 | 1,229 | 322 |
| Taylorsville | 23.4 | 0.6 | 100.0 | 18.6 | 309 | 133 | 177 | 21 | 348 | 20 |
| Tooele | 24.8 | 1.6 | 84.9 | 15.1 | 437 | 152 | 285 | 19 | 543 | 21 |
| West Jordan | 83.0 | 4.8 | 87.4 | 46.8 | 411 | 155 | 256 | 74 | 652 | 138 |
| West Valley City | 125.9 | 13.4 | 40.4 | 78.2 | 574 | 284 | 290 | 125 | 918 | 18 |
| **VERMONT** | X | X | X | X | X | X | X | X | X | X |
| Burlington | 136.5 | 22.6 | 44.8 | 45.1 | 1,065 | 732 | 332 | 120 | 2,835 | 661 |
| **VIRGINIA** | X | X | X | X | X | X | X | X | X | X |
| Alexandria | 931.2 | 216.1 | 74.5 | 611.7 | 3,842 | 2,970 | 872 | 954 | 5,989 | 792 |
| Blacksburg | 46.5 | 15.2 | 48.5 | 19.9 | 449 | 153 | 297 | 37 | 841 | 161 |
| Charlottesville | 284.7 | 117.4 | 65.7 | 115.9 | 2,442 | 1,469 | 928 | 279 | 5,879 | 627 |
| Chesapeake | 1,034.6 | 456.8 | 82.5 | 465.8 | 1,939 | 1,356 | 564 | 975 | 4,060 | 455 |
| Danville | 227.3 | 124.1 | 77.6 | 56.2 | 1,369 | 721 | 616 | 228 | 5,562 | 202 |
| Hampton | 602.6 | 274.9 | 77.6 | 244.4 | 1,814 | 1,233 | 581 | 586 | 4,348 | 378 |
| Harrisonburg | 203.2 | 81.0 | 85.6 | 79.3 | 1,479 | 722 | 735 | 221 | 4,111 | 1,090 |
| Leesburg | 74.1 | 22.4 | 90.3 | 32.4 | 599 | 274 | 325 | 70 | 1,292 | 277 |
| Lynchburg | 379.8 | 190.6 | 73.1 | 136.8 | 1,699 | 1,007 | 675 | 331 | 4,104 | 978 |
| Manassas | 219.2 | 85.2 | 84.8 | 99.7 | 2,423 | 1,844 | 553 | 210 | 5,092 | 470 |
| Newport News | 904.3 | 403.4 | 73.8 | 368.0 | 2,051 | 1,455 | 581 | 912 | 5,080 | 625 |
| Norfolk | 1,219.1 | 544.7 | 73.0 | 443.6 | 1,814 | 1,104 | 690 | 1,241 | 5,073 | 704 |
| Petersburg | 147.0 | 79.1 | 86.4 | 55.2 | 1,772 | 1,308 | 450 | 145 | 4,669 | 116 |
| Portsmouth | 418.6 | 204.0 | 83.7 | 166.4 | 1,755 | 1,255 | 500 | 491 | 5,177 | 314 |
| Richmond | 1,288.8 | 545.0 | 79.3 | 525.3 | 2,312 | 1,451 | 819 | 1,218 | 5,359 | 668 |
| Roanoke | 488.0 | 259.4 | 83.9 | 184.9 | 1,868 | 1,142 | 726 | 450 | 4,547 | 298 |
| Suffolk | 393.1 | 171.0 | 88.9 | 168.9 | 1,874 | 1,377 | 481 | 396 | 4,399 | 928 |
| Virginia Beach | 1,932.4 | 740.4 | 80.9 | 948.1 | 2,107 | 1,432 | 651 | 1,859 | 4,132 | 485 |
| Winchester | 221.6 | 109.2 | 45.2 | 75.6 | 2,683 | 1,545 | 1,138 | 158 | 5,600 | 561 |
| **WASHINGTON** | X | X | X | X | X | X | X | X | X | X |
| Auburn | 145.3 | 13.7 | 74.7 | 61.1 | 751 | 222 | 477 | 122 | 1,504 | 244 |
| Bellevue | 414.6 | 47.4 | 27.0 | 221.7 | 1,526 | 284 | 1,086 | 403 | 2,773 | 690 |
| Bellingham | 169.4 | 13.8 | 60.9 | 98.1 | 1,101 | 250 | 809 | 139 | 1,564 | 165 |
| Bothell | 89.7 | 10.8 | 90.1 | 44.2 | 968 | 277 | 623 | 78 | 1,701 | 385 |
| Bremerton | 71.9 | 4.6 | 83.1 | 36.5 | 898 | 238 | 628 | 60 | 1,476 | 129 |
| Burien | 41.5 | 4.0 | 68.6 | 27.3 | 525 | 144 | 342 | 31 | 586 | 58 |
| Des Moines | 42.2 | 9.1 | 94.7 | 19.4 | 618 | 152 | 421 | 36 | 1,140 | 332 |
| Edmonds | 67.4 | 8.8 | 78.0 | 36.3 | 859 | 347 | 437 | 60 | 1,432 | 180 |
| Everett | 228.6 | 15.2 | 75.1 | 139.7 | 1,268 | 382 | 829 | 199 | 1,802 | 219 |
| Federal Way | 90.4 | 22.8 | 74.2 | 48.3 | 497 | 108 | 334 | 92 | 948 | 295 |
| Issaquah | 78.4 | 8.7 | 80.9 | 43.2 | 1,146 | 253 | 805 | 68 | 1,812 | 354 |
| Kennewick | 88.9 | 8.2 | 87.6 | 53.9 | 661 | 154 | 473 | 87 | 1,065 | 208 |
| Kent | 197.9 | 23.2 | 69.2 | 101.7 | 772 | 171 | 553 | 169 | 1,286 | 257 |
| Kirkland | 166.4 | 8.7 | 65.8 | 99.7 | 1,120 | 321 | 683 | 151 | 1,700 | 270 |
| Lacey | 78.1 | 6.5 | 61.3 | 34.0 | 685 | 141 | 492 | 56 | 1,132 | 96 |
| Lake Stevens | 19.8 | 2.2 | 99.9 | 14.3 | 435 | 131 | 235 | 18 | 558 | 151 |
| Lakewood | 61.1 | 15.6 | 70.9 | 35.6 | 591 | 123 | 432 | 57 | 943 | 281 |
| Longview | 69.2 | 9.6 | 49.4 | 29.3 | 780 | 248 | 520 | 62 | 1,644 | 158 |
| Lynnwood | 84.4 | 6.9 | 71.5 | 49.1 | 1,283 | 310 | 916 | 77 | 2,012 | 209 |
| Marysville | 82.6 | 4.4 | 83.4 | 45.0 | 653 | 232 | 385 | 76 | 1,099 | 177 |
| Mount Vernon | 47.9 | 2.8 | 92.6 | 23.8 | 677 | 212 | 429 | 49 | 1,379 | 150 |
| Olympia | 126.6 | 11.5 | 35.9 | 68.2 | 1,320 | 288 | 984 | 125 | 2,424 | 259 |

1. Based on population estimated as of July 1 of the year shown.

City government finances, 2017 (cont.)

General expenditure (cont.)

Percent of total for:

| City | Public welfare | Highways | Parking facilities | Education | Health and hospitals | Police protection | Sewerage and sanitation | Parks and recreation | Housing and community development | Interest on debt |
|---|---|---|---|---|---|---|---|---|---|---|
| | 127 | 128 | 129 | 130 | 131 | 132 | 133 | 134 | 135 | 136 |
| **UTAH— Cont'd** | | | | | | | | | | |
| Kaysville | 0.0 | 10.1 | 0.0 | 0.0 | 3.7 | 17.9 | 22.5 | 9.6 | 0.0 | 0.1 |
| Layton | 0.0 | 7.8 | 0.0 | 0.0 | 7.9 | 16.8 | 22.6 | 6.4 | 2.5 | 0.0 |
| Lehi | 0.2 | 8.3 | 0.0 | 0.3 | 0.2 | 10.0 | 14.6 | 20.9 | 0.7 | 5.7 |
| Logan | 1.1 | 7.7 | 0.0 | 0.0 | 4.7 | 14.0 | 17.5 | 14.7 | 4.6 | 0.2 |
| Midvale | 0.0 | 1.8 | 0.0 | 0.0 | 0.5 | 24.1 | 10.8 | 1.5 | 27.0 | 0.6 |
| Murray | 1.2 | 4.8 | 0.0 | 0.0 | 0.6 | 22.5 | 7.6 | 14.2 | 9.0 | 0.6 |
| Ogden | 0.0 | 8.9 | 0.0 | 0.0 | 4.9 | 19.4 | 3.2 | 1.3 | 6.8 | 1.7 |
| Orem | 0.2 | 7.3 | 0.0 | 0.0 | 0.4 | 16.8 | 11.7 | 10.6 | 2.2 | 0.2 |
| Pleasant Grove | 5.5 | 12.8 | 0.0 | 0.0 | 0.5 | 14.9 | 19.3 | 9.0 | 3.3 | 0.6 |
| Provo | 0.0 | 5.6 | 0.0 | 0.0 | 0.5 | 16.5 | 12.7 | 10.4 | 6.4 | 0.3 |
| Riverton | 0.9 | 11.5 | 0.0 | 0.0 | 1.4 | 0.0 | 7.5 | 18.1 | 2.8 | 1.7 |
| Roy | 0.0 | 13.8 | 0.0 | 0.0 | 0.0 | 23.2 | 7.7 | 14.3 | 1.7 | 0.0 |
| St. George | 0.2 | 4.1 | 0.0 | 0.0 | 0.0 | 15.3 | 15.8 | 19.1 | 3.7 | 0.6 |
| Salt Lake City | 0.0 | 2.3 | 0.0 | 0.1 | 0.7 | 12.6 | 2.5 | 5.0 | 2.9 | 1.7 |
| Sandy | 0.0 | 2.0 | 0.0 | 0.0 | 0.5 | 12.4 | 3.5 | 12.6 | 1.2 | 0.2 |
| South Jordan | 0.0 | 0.0 | 0.0 | 0.0 | 0.0 | 0.0 | 0.0 | 0.0 | 0.0 | 1.2 |
| Spanish Fork | 0.0 | 9.0 | 0.0 | 0.0 | 3.2 | 14.6 | 13.9 | 21.4 | 4.9 | 0.2 |
| Springville | 0.8 | 3.6 | 0.0 | 17.0 | 0.0 | 12.8 | 11.2 | 8.0 | 3.4 | 0.1 |
| Taylorsville | 0.0 | 0.0 | 0.0 | 0.0 | 0.0 | 42.5 | 0.0 | 0.3 | 4.8 | 0.0 |
| Tooele | 0.5 | 7.0 | 0.0 | 0.0 | 1.3 | 23.5 | 13.6 | 14.0 | 5.0 | 4.6 |
| West Jordan | 0.0 | 10.1 | 0.0 | 0.0 | 0.0 | 24.2 | 7.6 | 6.9 | 0.4 | 0.3 |
| West Valley City | 4.7 | 12.6 | 0.0 | 0.0 | 2.7 | 19.3 | 3.8 | 8.3 | 17.1 | 1.2 |
| **VERMONT** | X | X | X | X | X | X | X | X | X | X |
| Burlington | 0.0 | 9.6 | 4.6 | 0.0 | 0.1 | 12.9 | 5.9 | 9.7 | 3.9 | 3.9 |
| **VIRGINIA** | X | X | X | X | X | X | X | X | X | X |
| Alexandria | 5.1 | 6.3 | 0.0 | 29.3 | 5.1 | 9.9 | 8.8 | 4.0 | 3.4 | 0.4 |
| Blacksburg | 0.0 | 21.0 | 0.0 | 28.5 | 1.6 | 21.3 | 17.3 | 6.5 | 6.3 | 0.1 |
| Charlottesville | 10.5 | 7.0 | 0.1 | 50.7 | 7.9 | 6.5 | 5.0 | 1.6 | 2.5 | 0.0 |
| Chesapeake | 2.1 | 5.8 | 0.0 | 40.0 | 3.6 | 5.8 | 4.7 | 2.2 | 7.9 | 1.9 |
| Danville | 5.4 | 9.2 | 0.0 | 39.1 | 1.3 | 5.3 | 5.2 | 5.9 | 8.5 | 0.7 |
| Hampton | 5.6 | 3.4 | 0.1 | 52.9 | 0.6 | 4.6 | 9.0 | 2.6 | 0.4 | 2.6 |
| Harrisonburg | 1.6 | 4.6 | 0.1 | 0.0 | 0.0 | 16.9 | 21.4 | 12.1 | 1.1 | 2.5 |
| Leesburg | 0.0 | 23.2 | 0.1 | 44.2 | 0.6 | 6.4 | 7.5 | 2.6 | 3.2 | 0.0 |
| Lynchburg | 8.2 | 3.7 | 0.2 | 55.7 | 1.9 | 7.1 | 11.0 | 0.2 | 2.1 | 0.1 |
| Manassas | 0.7 | 5.4 | 0.0 | | 0.6 | 5.6 | 4.7 | 2.6 | 4.3 | 2.2 |
| Newport News | 4.2 | 2.3 | 0.0 | 43.4 | 0.6 | 5.9 | 4.8 | 4.6 | 7.4 | 2.7 |
| Norfolk | 3.7 | 2.9 | 1.2 | 33.8 | 3.6 | 9.2 | 4.0 | 1.3 | 1.0 | 3.1 |
| Petersburg | 11.6 | 4.6 | 0.0 | 40.0 | 0.9 | 6.6 | 4.7 | 2.7 | 5.6 | 0.7 |
| Portsmouth | 4.6 | 1.3 | 0.0 | 34.2 | 2.2 | 8.5 | 10.1 | 2.2 | 10.6 | 3.5 |
| Richmond | 5.8 | 4.0 | 0.5 | 31.2 | 6.4 | 5.4 | 3.3 | 1.8 | 7.6 | 3.3 |
| Roanoke | 13.1 | 5.4 | 0.6 | 41.2 | 0.5 | 6.3 | 3.7 | 2.5 | 5.0 | 0.0 |
| Suffolk | 3.2 | 10.5 | 0.0 | 44.7 | 0.7 | 5.2 | 8.2 | 7.4 | 1.8 | 0.0 |
| Virginia Beach | 3.2 | 5.9 | 0.5 | 46.1 | 3.6 | 5.2 | 5.2 | 2.1 | 1.1 | 1.2 |
| Winchester | 4.4 | 3.0 | 0.4 | 48.0 | 1.1 | 5.2 | 5.2 | 2.1 | 1.1 | 2.7 |
| **WASHINGTON** | X | X | X | X | X | X | X | X | X | X |
| Auburn | 0.0 | 10.8 | 0.0 | 0.0 | 0.5 | 19.3 | 36.9 | 14.5 | 0.3 | 1.5 |
| Bellevue | 0.0 | 19.7 | 0.0 | 0.0 | 0.0 | 10.1 | 19.9 | 14.7 | 6.0 | 4.0 |
| Bellingham | 0.0 | 10.9 | 1.6 | 0.0 | 6.3 | 14.3 | 17.5 | 8.3 | 2.8 | 2.5 |
| Bothell | 0.4 | 27.1 | 0.0 | 0.0 | 0.2 | 14.8 | 11.6 | 1.5 | 2.1 | 5.4 |
| Bremerton | 0.1 | 9.2 | 1.4 | 0.0 | 0.6 | 18.3 | 21.2 | 12.3 | 1.3 | 2.5 |
| Burien | 0.0 | 11.2 | 0.0 | 0.0 | 0.6 | 36.6 | 6.3 | 8.3 | 2.7 | 4.9 |
| Des Moines | 0.1 | 31.5 | 0.0 | 0.0 | 0.7 | 22.6 | 8.5 | 14.6 | 0.0 | 0.8 |
| Edmonds | 0.0 | 11.1 | 0.0 | 0.0 | 0.1 | 15.7 | 16.9 | 8.0 | 0.0 | 1.5 |
| Everett | 0.2 | 6.9 | 0.2 | 0.0 | 4.9 | 16.7 | 18.9 | 10.5 | 0.8 | 3.4 |
| Federal Way | 0.5 | 19.7 | 0.0 | 0.0 | 1.3 | 24.9 | 4.6 | 9.2 | 1.8 | 0.5 |
| Issaquah | 0.2 | 19.4 | 0.0 | 0.0 | 0.2 | 8.7 | 16.2 | 9.3 | 4.2 | 1.5 |
| Kennewick | 0.0 | 11.0 | 0.0 | 0.0 | 5.4 | 20.4 | 6.2 | 12.4 | 0.2 | 3.3 |
| Kent | 1.3 | 16.0 | 0.0 | 0.0 | 0.3 | 17.6 | 23.8 | 10.1 | 1.1 | 2.0 |
| Kirkland | 0.0 | 12.1 | 0.0 | 0.0 | 0.7 | 13.8 | 23.1 | 5.7 | 0.0 | 1.4 |
| Lacey | 0.0 | 11.2 | 0.0 | 0.0 | 0.5 | 17.0 | 29.0 | 9.9 | 0.0 | 1.1 |
| Lake Stevens | 0.0 | 22.7 | 0.0 | 0.0 | 0.2 | 40.0 | 7.4 | 2.6 | 6.2 | 4.2 |
| Lakewood | 0.7 | 28.9 | 0.0 | 0.0 | 0.4 | 40.0 | 6.2 | 4.1 | 2.6 | 0.9 |
| Longview | 0.0 | 8.0 | 0.0 | 0.0 | 0.3 | 19.3 | 29.8 | 3.8 | 2.2 | 1.5 |
| Lynnwood | 0.0 | 13.9 | 0.0 | 0.0 | 0.2 | 18.8 | 8.7 | 5.3 | 2.2 | 2.9 |
| Marysville | 0.0 | 14.8 | 0.0 | 0.0 | 0.3 | 17.6 | 21.8 | 5.0 | 3.1 | 4.3 |
| Mount Vernon | 0.1 | 13.4 | 0.0 | 0.0 | 0.4 | 14.9 | 33.2 | 2.3 | 1.1 | 3.3 |
| Olympia | 0.0 | 5.2 | 0.0 | 0.0 | 0.0 | 11.5 | 28.6 | 11.1 | 0.2 | 2.9 |

# Table D. Cities — City Government Finances, City Government Employment, and Climate

| City | City government finances, 2017 (cont.) | | | Climate[2] | | | | | | |
| | Debt outstanding | | | Average daily temperature | | | | Annual precipitation (inches) | Heating degree days | Cooling degree days |
| | | | Debt issued during year | Mean | | Limits | | | | |
| | Total (mil dol) | Per capita[1] (dollars) | | January | July | January[3] | July[4] | | | |
| | 137 | 138 | 139 | 140 | 141 | 142 | 143 | 144 | 145 | 146 |
|---|---|---|---|---|---|---|---|---|---|---|
| **UTAH— Cont'd** | | | | | | | | | | |
| Kaysville | 0.0 | 0 | 0.0 | NA | NA | NA | NA | NA | NA | NA |
| Layton | 0.0 | 0 | 0.0 | 27.6 | 74.2 | 18.6 | 89.9 | 20.75 | 6,142 | 746 |
| Lehi | 120.3 | 1,890 | 4.2 | NA | NA | NA | NA | NA | NA | NA |
| Logan | 0.0 | 0 | 0.0 | 21.8 | 71.6 | 12.7 | 88.3 | 17.86 | 7,174 | 522 |
| Midvale | 18.5 | 552 | 18.3 | 30.4 | 78.5 | 22.1 | 90.9 | 26.19 | 5,441 | 1,197 |
| Murray | 10.3 | 209 | 10.0 | 29.2 | 77.0 | 21.3 | 90.6 | 16.50 | 5,631 | 1,066 |
| Ogden | 117.4 | 1,348 | 17.0 | 28.1 | 76.6 | 20.1 | 90.0 | 23.67 | 5,868 | 980 |
| Orem | 3.7 | 38 | 2.9 | 28.6 | 76.5 | 20.3 | 92.3 | 12.84 | 5,564 | 1,016 |
| Pleasant Grove | 15.7 | 404 | 15.7 | NA | NA | NA | NA | NA | NA | NA |
| Provo | 4.6 | 39 | 2.1 | 30.9 | 76.9 | 22.5 | 93.4 | 20.13 | 5,264 | 1,028 |
| Riverton | 427.2 | 9,851 | 0.0 | 31.6 | 78.0 | 22.0 | 95.3 | 15.76 | 5,251 | 1,172 |
| Roy | 1.8 | 45 | 0.0 | 27.6 | 74.2 | 18.6 | 89.9 | 20.75 | 6,142 | 746 |
| St. George | 109.0 | 1,289 | 0.0 | 41.8 | 86.3 | 28.9 | 102.8 | 8.77 | 3,103 | 2,471 |
| Salt Lake City | 1,677.2 | 8,347 | 1,236.3 | 32.0 | 78.1 | 25.4 | 89.0 | 17.75 | 5,095 | 1,190 |
| Sandy | 0.0 | 0 | 0.0 | 30.4 | 78.5 | 22.1 | 90.9 | 26.19 | 5,441 | 1,197 |
| South Jordan | 23.7 | 335 | 23.7 | 31.6 | 78.0 | 22.0 | 95.3 | 15.76 | 5,251 | 1,172 |
| Spanish Fork | 0.0 | 0 | 0.0 | NA | NA | NA | NA | NA | NA | NA |
| Springville | 35.8 | 1,078 | 1.2 | NA | NA | NA | NA | NA | NA | NA |
| Taylorsville | 0.5 | 9 | 0.0 | 29.2 | 77.0 | 21.3 | 90.6 | 16.50 | 5,631 | 1,066 |
| Tooele | 11.4 | 328 | 11.4 | NA | NA | NA | NA | NA | NA | NA |
| West Jordan | 55.3 | 486 | 49.9 | 30.4 | 78.5 | 22.1 | 90.9 | 26.19 | 5,441 | 1,197 |
| West Valley City | 76.2 | 559 | 119.1 | 29.2 | 77.0 | 21.3 | 90.6 | 16.50 | 5,631 | 1,066 |
| **VERMONT** | X | X | X | X | X | X | X | X | X | X |
| Burlington | 185.1 | 4,365 | 37.2 | 18.0 | 70.6 | 9.3 | 81.4 | 36.05 | 7,665 | 489 |
| **VIRGINIA** | X | X | X | X | X | X | X | X | X | X |
| Alexandria | 802.7 | 5,042 | 130.9 | 34.9 | 79.2 | 27.3 | 88.3 | 39.35 | 4,055 | 1,531 |
| Blacksburg | 0.0 | 0 | 0.0 | 30.9 | 71.1 | 20.6 | 82.5 | 42.63 | 5,559 | 533 |
| Charlottesville | 140.6 | 2,963 | 15.2 | 35.5 | 76.9 | 26.2 | 88.0 | 48.87 | 4,103 | 1,212 |
| Chesapeake | 548.2 | 2,283 | 104.3 | 40.1 | 79.1 | 32.3 | 86.8 | 45.74 | 3,368 | 1,612 |
| Danville | 136.3 | 3,321 | 0.0 | 36.6 | 78.8 | 25.8 | 90.0 | 44.98 | 3,970 | 1,418 |
| Hampton | 309.3 | 2,296 | 0.0 | 39.4 | 78.5 | 32.0 | 85.2 | 47.90 | 3,535 | 1,432 |
| Harrisonburg | 6.2 | 115 | 0.0 | 30.5 | 73.5 | 20.4 | 85.3 | 36.12 | 5,333 | 758 |
| Leesburg | 124.3 | 2,299 | 16.8 | 31.5 | 75.2 | 20.8 | 87.1 | 43.21 | 5,031 | 911 |
| Lynchburg | 332.2 | 4,125 | 0.0 | 34.5 | 75.1 | 24.5 | 86.4 | 43.31 | 4,354 | 1,075 |
| Manassas | 110.0 | 2,674 | 0.0 | 31.7 | 75.7 | 21.9 | 87.4 | 41.80 | 4,925 | 1,075 |
| Newport News | 993.6 | 5,537 | 177.7 | 41.2 | 80.3 | 33.8 | 87.9 | 43.53 | 3,179 | 1,682 |
| Norfolk | 1,518.8 | 6,211 | 436.5 | 40.1 | 79.1 | 32.3 | 86.8 | 45.74 | 3,368 | 1,612 |
| Petersburg | 21.8 | 702 | 0.0 | 39.7 | 79.6 | 29.2 | 91.0 | 45.26 | 3,334 | 1,619 |
| Portsmouth | 589.7 | 6,218 | 17.1 | 40.1 | 79.1 | 32.3 | 86.8 | 45.74 | 3,368 | 1,612 |
| Richmond | 1,610.3 | 7,087 | 505.0 | 36.4 | 77.9 | 27.6 | 87.5 | 43.91 | 3,919 | 1,435 |
| Roanoke | 120.4 | 1,217 | 17.9 | 35.8 | 76.2 | 26.6 | 87.5 | 42.49 | 4,284 | 1,134 |
| Suffolk | 628.7 | 6,977 | 169.6 | 39.6 | 78.5 | 30.3 | 88.1 | 48.71 | 3,467 | 1,427 |
| Virginia Beach | 1,484.0 | 3,299 | 35.5 | 40.7 | 78.8 | 32.2 | 86.9 | 44.50 | 3,336 | 1,482 |
| Winchester | 182.7 | 6,485 | 0.0 | NA | NA | NA | NA | NA | NA | NA |
| **WASHINGTON** | X | X | X | X | X | X | X | X | X | X |
| Auburn | 69.3 | 852 | 4.7 | 40.8 | 66.4 | 34.6 | 77.4 | 39.59 | 4,624 | 219 |
| Bellevue | 284.6 | 1,959 | 0.0 | 41.5 | 65.5 | 36.0 | 74.5 | 38.25 | 4,615 | 192 |
| Bellingham | 106.3 | 1,194 | 10.6 | 40.5 | 63.3 | 34.8 | 72.5 | 34.84 | 4,980 | 68 |
| Bothell | 130.1 | 2,846 | 17.9 | 40.8 | 65.2 | 35.2 | 75.0 | 35.96 | 4,756 | 174 |
| Bremerton | 71.1 | 1,749 | 8.4 | 40.1 | 64.6 | 34.7 | 75.2 | 53.96 | 4,994 | 158 |
| Burien | 33.2 | 638 | 7.0 | 40.9 | 65.3 | 35.9 | 75.3 | 37.07 | 4,797 | 173 |
| Des Moines | 17.9 | 570 | 3.7 | 40.9 | 65.3 | 35.9 | 75.3 | 37.07 | 4,797 | 173 |
| Edmonds | 73.3 | 1,737 | 6.8 | 40.8 | 65.2 | 35.2 | 75.0 | 35.96 | 4,756 | 174 |
| Everett | 349.2 | 3,170 | 153.2 | 39.7 | 63.6 | 33.6 | 73.0 | 37.54 | 5,199 | 121 |
| Federal Way | 51.6 | 531 | 7.8 | 41.0 | 65.6 | 35.1 | 76.1 | 38.95 | 4,650 | 167 |
| Issaquah | 39.3 | 1,042 | 3.0 | NA | NA | NA | NA | NA | NA | NA |
| Kennewick | 82.0 | 1,004 | 14.6 | 34.2 | 75.2 | 28.0 | 89.3 | 8.01 | 4,731 | 909 |
| Kent | 263.8 | 2,003 | 142.1 | 40.8 | 66.4 | 34.6 | 77.4 | 39.59 | 4,624 | 219 |
| Kirkland | 46.5 | 522 | 3.0 | 40.8 | 65.2 | 35.2 | 75.0 | 35.96 | 4,756 | 174 |
| Lacey | 37.8 | 762 | 7.8 | 38.1 | 62.8 | 31.8 | 76.1 | 50.79 | 5,531 | 97 |
| Lake Stevens | 18.2 | 554 | 0.9 | NA | NA | NA | NA | NA | NA | NA |
| Lakewood | 12.5 | 207 | 4.6 | 41.0 | 65.6 | 35.1 | 76.1 | 38.95 | 4,650 | 167 |
| Longview | 60.1 | 1,599 | 1.3 | 39.9 | 64.5 | 33.8 | 76.5 | 48.02 | 4,900 | 148 |
| Lynnwood | 85.7 | 2,239 | 3.6 | 40.8 | 65.2 | 35.2 | 75.0 | 35.96 | 4,756 | 174 |
| Marysville | 75.1 | 1,090 | 5.0 | 39.7 | 63.6 | 33.6 | 73.0 | 37.54 | 5,199 | 121 |
| Mount Vernon | 38.6 | 1,096 | 3.2 | 39.9 | 62.3 | 34.1 | 73.0 | 32.70 | 5,197 | 47 |
| Olympia | 98.6 | 1,909 | 7.3 | 38.1 | 62.8 | 31.8 | 76.1 | 50.79 | 5,531 | 97 |

1. Based on the population estimated as of July 1 of the year shown.  2. Represents normal values based on the 30-year period, 1971±2000.  3. Average daily minimum.  4. Average daily maximum.

Items 137—146

# Table D. Cities — Land Area and Population

| STATE code | Place code | City | Land area¹ (sq. mi) 1 | Population, 2020 — Total persons 2019 — 2 | Rank 3 | Per square mile 4 | Race 2019 — Race alone² (percent) — White 5 | Black or African American 6 | American Indian, Alaskan Native 7 | Asian 8 | Hawaiian Pacific Islander 9 | Some other race 10 | Two or more races (percent) 11 |
|---|---|---|---|---|---|---|---|---|---|---|---|---|---|
| | | **WASHINGTON—Cont'd** | | | | | | | | | | | |
| 53 | 53545 | Pasco | 34.0 | 77,095 | 459 | 2,268 | 68.9 | 1.8 | 2.1 | 1.2 | 0.2 | 20.3 | 5.7 |
| 53 | 56625 | Pullman | 10.9 | 34,336 | 1,121 | 3,150 | 77.5 | 3.0 | 0.4 | 12.5 | 0.0 | 1.5 | 5.2 |
| 53 | 56695 | Puyallup | 14.1 | 42,569 | 905 | 3,019 | 86.3 | 3.8 | 0.7 | 3.9 | 0.0 | 0.9 | 4.4 |
| 53 | 57535 | Redmond | 16.6 | 72,507 | 497 | 4,368 | 57.3 | 2.5 | 0.0 | 35.7 | 0.3 | 0.8 | 3.4 |
| 53 | 57745 | Renton | 23.5 | 101,494 | 312 | 4,319 | 53.6 | 9.2 | 0.4 | 25.6 | 0.2 | 4.1 | 6.9 |
| 53 | 58235 | Richland | 39.2 | 59,439 | 632 | 1,516 | 87.3 | 1.1 | 0.4 | 3.1 | 0.0 | 3.2 | 4.8 |
| 53 | 61115 | Sammamish | 20.4 | 65,960 | 571 | 3,233 | 61.1 | 2.5 | 0.7 | 31.4 | 0.1 | 0.7 | 3.5 |
| 53 | 62288 | SeaTac | 10.1 | 28,904 | 1,290 | 2,862 | 34.6 | 26.7 | 0.3 | 8.7 | 3.8 | 20.9 | 5.0 |
| 53 | 63000 | Seattle | 83.8 | 769,714 | 18 | 9,185 | 66.3 | 7.4 | 0.7 | 16.6 | 0.4 | 1.9 | 6.7 |
| 53 | 63960 | Shoreline | 11.6 | 58,197 | 662 | 5,017 | 64.2 | 3.3 | 1.0 | 23.6 | 0.2 | 1.5 | 6.2 |
| 53 | 67000 | Spokane | 68.8 | 222,050 | 99 | 3,228 | 84.6 | 3.0 | 2.3 | 3.0 | 0.6 | 1.1 | 5.4 |
| 53 | 67167 | Spokane Valley | 37.7 | 102,893 | 299 | 2,729 | 89.3 | 0.7 | 0.5 | 2.1 | 1.5 | 1.6 | 4.3 |
| 53 | 70000 | Tacoma | 49.7 | 219,945 | 100 | 4,426 | 63.7 | 11.5 | 1.8 | 6.2 | 1.0 | 5.3 | 10.4 |
| 53 | 73465 | University Place | 8.3 | 34,090 | 1,131 | 4,107 | 66.9 | 8.0 | 0.0 | 11.5 | 2.4 | 0.2 | 11.0 |
| 53 | 74060 | Vancouver | 48.7 | 186,192 | 136 | 3,823 | 78.0 | 3.3 | 0.7 | 5.2 | 2.2 | 3.5 | 7.1 |
| 53 | 75775 | Walla Walla | 13.8 | 33,096 | 1,159 | 2,398 | 79.5 | 0.9 | 1.0 | 2.4 | 0.3 | 9.1 | 6.8 |
| 53 | 77105 | Wenatchee | 10.6 | 34,219 | 1,128 | 3,228 | 62.8 | 3.3 | 1.5 | 0.4 | 0.0 | 27.1 | 5.0 |
| 53 | 80010 | Yakima | 27.8 | 94,322 | 349 | 3,393 | 66.6 | 2.6 | 2.9 | 1.5 | 1.3 | 20.7 | 4.4 |
| 54 | 00000 | **WEST VIRGINIA** | 24,041.1 | 1,784,787 | X | 74 | 93.1 | 3.7 | 0.2 | 0.8 | 0.0 | 0.4 | 1.8 |
| 54 | 14600 | Charleston | 31.5 | 45,879 | 849 | 1,457 | 75.9 | 12.8 | 0.6 | 2.4 | 0.0 | 0.7 | 7.6 |
| 54 | 39460 | Huntington | 16.2 | 44,934 | 858 | 2,774 | 85.7 | 8.0 | 0.0 | 1.6 | 0.0 | 0.1 | 4.5 |
| 54 | 55756 | Morgantown | 10.2 | 30,847 | 1,225 | 3,024 | 87.5 | 4.1 | 0.0 | 3.7 | 0.0 | 2.0 | 2.7 |
| 54 | 62140 | Parkersburg | 11.8 | 29,009 | 1,286 | 2,458 | D | D | D | D | D | D | D |
| 54 | 86452 | Wheeling | 13.8 | 26,283 | 1,378 | 1,905 | D | D | D | D | D | D | 2.4 |
| 55 | 00000 | **WISCONSIN** | 54,167.4 | 5,832,655 | X | 108 | 85.2 | 6.4 | 0.9 | 2.9 | 0.1 | 2.1 | 2.4 |
| 55 | 02375 | Appleton | 24.8 | 74,058 | 482 | 2,986 | 82.7 | 3.9 | 0.4 | 7.3 | 0.0 | 2.8 | 2.9 |
| 55 | 06500 | Beloit | 17.3 | 36,637 | 1,049 | 2,118 | 64.6 | 11.4 | 0.0 | 1.9 | 0.2 | 14.4 | 7.5 |
| 55 | 10025 | Brookfield | 27.3 | 39,164 | 992 | 1,435 | D | D | D | D | D | D | D |
| 55 | 22300 | Eau Claire | 32.9 | 69,087 | 532 | 2,100 | 89.1 | 2.2 | 0.4 | 6.1 | 0.1 | 0.2 | 1.9 |
| 55 | 25950 | Fitchburg | 34.9 | 30,866 | 1,224 | 884 | 76.7 | 8.7 | 0.3 | 7.8 | 0.4 | 4.3 | 1.7 |
| 55 | 26275 | Fond du Lac | 19.2 | 43,178 | 892 | 2,249 | 88.8 | 2.7 | 0.4 | 0.4 | 0.6 | 2.8 | 4.4 |
| 55 | 27300 | Franklin | 34.6 | 35,925 | 1,076 | 1,038 | 77.6 | 6.0 | 0.2 | 4.5 | 0.0 | 5.7 | 6.0 |
| 55 | 31000 | Green Bay | 45.5 | 103,836 | 296 | 2,282 | 75.9 | 4.9 | 4.0 | 4.0 | 0.1 | 5.7 | 5.5 |
| 55 | 31175 | Greenfield | 11.5 | 37,183 | 1,032 | 3,233 | 84.1 | 4.5 | 0.5 | 3.0 | 0.0 | 6.0 | 1.8 |
| 55 | 37825 | Janesville | 34.2 | 64,664 | 584 | 1,891 | 90.6 | 4.1 | 0.1 | 1.5 | 0.0 | 1.6 | 2.1 |
| 55 | 39225 | Kenosha | 28.3 | 99,570 | 319 | 3,518 | 80.3 | 12.0 | 0.7 | 2.0 | 0.5 | 2.1 | 2.4 |
| 55 | 40775 | La Crosse | 21.7 | 51,163 | 762 | 2,358 | 90.2 | 1.9 | 0.6 | 4.8 | 0.2 | 0.4 | 2.0 |
| 55 | 48000 | Madison | 79.6 | 263,094 | 81 | 3,305 | 78.1 | 7.3 | 0.3 | 9.1 | 0.1 | 1.2 | 3.8 |
| 55 | 48500 | Manitowoc | 17.8 | 32,359 | 1,181 | 1,818 | 87.1 | 3.6 | 0.9 | 6.7 | 0.0 | 0.7 | 0.9 |
| 55 | 51000 | Menomonee Falls | 32.9 | 38,243 | 1,011 | 1,162 | D | D | D | D | D | D | D |
| 55 | 53000 | Milwaukee | 96.2 | 589,067 | 30 | 6,123 | 44.8 | 38.4 | 0.8 | 4.3 | 0.0 | 7.9 | 3.9 |
| 55 | 54875 | Mount Pleasant | 33.9 | 27,048 | 1,349 | 798 | D | D | D | D | D | D | D |
| 55 | 55750 | Neenah | 9.4 | 26,363 | 1,377 | 2,805 | 88.5 | 6.5 | 1.4 | 1.4 | 0.0 | 0.1 | 2.2 |
| 55 | 56375 | New Berlin | 36.4 | 39,812 | 969 | 1,094 | D | D | D | D | D | D | D |
| 55 | 58800 | Oak Creek | 28.4 | 36,469 | 1,056 | 1,284 | 85.8 | 2.5 | 0.2 | 7.8 | 0.0 | 0.9 | 2.8 |
| 55 | 60500 | Oshkosh | 27.0 | 66,495 | 564 | 2,463 | 90.3 | 2.8 | 0.4 | 3.5 | 0.0 | 0.4 | 4.4 |
| 55 | 66000 | Racine | 15.5 | 76,237 | 468 | 4,919 | 65.6 | 23.2 | 1.0 | 0.7 | 0.0 | 4.4 | 5.7 |
| 55 | 72975 | Sheboygan | 15.6 | 47,819 | 812 | 3,065 | 76.6 | 1.8 | 1.0 | 13.2 | 0.0 | 3.0 | 4.4 |
| 55 | 77200 | Stevens Point | 17.2 | 25,949 | 1,390 | 1,509 | D | D | D | D | D | D | D |
| 55 | 78600 | Sun Prairie | 12.9 | 35,119 | 1,097 | 2,722 | 82.9 | 4.7 | 1.2 | 6.4 | 0.0 | 0.5 | 4.3 |
| 55 | 78650 | Superior | 36.6 | 26,213 | 1,381 | 716 | 89.9 | 2.8 | 3.7 | 2.4 | 0.0 | 3.2 | 1.8 |
| 55 | 84250 | Waukesha | 25.5 | 72,376 | 498 | 2,838 | 88.3 | 0.6 | 0.6 | 5.2 | 0.2 | 0.3 | 2.5 |
| 55 | 84475 | Wausau | 19.2 | 38,511 | 1,003 | 2,006 | 80.5 | 1.1 | 0.7 | 13.2 | 0.1 | 0.7 | 3.7 |
| 55 | 84675 | Wauwatosa | 13.2 | 48,084 | 807 | 3,643 | 86.0 | 4.9 | 0.2 | 4.5 | 0.1 | 2.0 | 6.3 |
| 55 | 85300 | West Allis | 11.4 | 59,748 | 628 | 5,241 | 81.3 | 6.3 | 0.4 | 3.5 | 0.1 | D | D |
| 55 | 85350 | West Bend | 15.5 | 31,478 | 1,210 | 2,031 | D | D | D | D | D | 1.9 | 2.7 |
| 56 | 00000 | **WYOMING** | 97,088.7 | 582,328 | X | 6 | 90.9 | 1.2 | 2.4 | 0.8 | 0.1 | 1.9 | 2.7 |
| 56 | 13150 | Casper | 26.5 | 58,498 | 653 | 2,208 | 93.7 | 1.1 | 1.4 | 1.1 | 0.0 | 0.9 | 1.8 |
| 56 | 13900 | Cheyenne | 32.3 | 64,742 | 583 | 2,004 | 86.3 | 2.1 | 1.5 | 1.5 | 0.4 | 3.4 | 4.9 |
| 56 | 31855 | Gillette | 23.1 | 32,220 | 1,185 | 1,395 | 86.6 | 1.3 | 0.6 | 0.0 | 0.7 | 5.1 | 5.8 |
| 56 | 45050 | Laramie | 18.4 | 32,661 | 1,169 | 1,775 | 88.6 | 3.1 | 0.2 | 4.3 | 0.0 | 0.9 | 2.9 |

1. Dry land or land partially or temporarily covered by water.    2. Hispanic or Latino persons may be of any race.

| City | Percent Hispanic or Latino[1], 2019 | Percent foreign born, 2019 | Age of population (percent), 2019 | | | | | | | Median age, 2019 | Percent female, 2019 | Census counts | | Percent change | |
|---|---|---|---|---|---|---|---|---|---|---|---|---|---|---|---|
| | | | Under 18 years | 18 to 24 years | 25 to 34 years | 35 to 44 years | 45 to 54 years | 55 to 64 years | 65 years and over | | | 2000 | 2010 | 2000-2010 | 2001-2020 |
| | 12 | 13 | 14 | 15 | 16 | 17 | 18 | 19 | 20 | 21 | 22 | 23 | 24 | 25 | 26 |
| **WASHINGTON—Cont'd** | | | | | | | | | | | | | | | |
| Pasco | 54.3 | 25.4 | 34.0 | 9.6 | 15.0 | 14.4 | 10.9 | 7.9 | 8.2 | 29.7 | 49.2 | 32,066 | 62,161 | 93.9 | 24.0 |
| Pullman | 8.5 | 12.6 | 12.3 | 49.4 | 16.3 | 5.2 | 5.4 | 4.2 | 7.2 | 22.4 | 50.7 | 24,675 | 29,820 | 20.9 | 15.1 |
| Puyallup | 5.8 | 6.4 | 16.1 | 5.7 | 15.8 | 14.8 | 12.7 | 18.2 | 16.7 | 42.9 | 51.1 | 33,011 | 37,244 | 12.8 | 14.3 |
| Redmond | 5.0 | 41.3 | 22.4 | 4.6 | 22.4 | 17.0 | 13.9 | 9.0 | 10.7 | 35.2 | 47.9 | 45,256 | 54,511 | 20.5 | 33.0 |
| Renton | 15.1 | 29.5 | 21.1 | 6.1 | 19.4 | 16.1 | 13.3 | 10.6 | 13.3 | 36.2 | 49.5 | 50,052 | 91,929 | 83.7 | 10.4 |
| Richland | 13.7 | 6.7 | 25.4 | 8.1 | 14.5 | 12.8 | 9.7 | 14.1 | 15.4 | 36.2 | 47.2 | 38,708 | 48,491 | 25.3 | 22.6 |
| Sammamish | 4.1 | 33.9 | 31.9 | 4.0 | 8.9 | 20.4 | 16.0 | 10.0 | 8.8 | 37.6 | 50.7 | 34,104 | 57,447 | 68.4 | 14.8 |
| SeaTac | 25.8 | 39.2 | 24.1 | 5.9 | 19.1 | 18.0 | 10.5 | 7.4 | 15.0 | 35.7 | 50.0 | 25,496 | 26,903 | 5.5 | 7.4 |
| Seattle | 6.6 | 19.6 | 14.7 | 10.3 | 25.5 | 15.5 | 11.6 | 10.3 | 12.0 | 34.7 | 48.8 | 563,374 | 608,660 | 8.0 | 26.5 |
| Shoreline | 5.5 | 24.8 | 17.9 | 8.5 | 13.0 | 14.8 | 14.0 | 11.6 | 20.2 | 41.4 | 50.2 | 53,025 | 53,031 | 0.0 | 9.7 |
| Spokane | 6.2 | 5.9 | 19.7 | 10.3 | 16.9 | 12.6 | 11.7 | 13.1 | 15.6 | 37.0 | 50.6 | 195,629 | 209,459 | 7.1 | 6.0 |
| Spokane Valley | 6.7 | 6.0 | 24.1 | 6.5 | 16.3 | 12.6 | 11.3 | 12.6 | 16.7 | 37.5 | 52.2 | NA | 89,746 | ********** | 14.6 |
| Tacoma | 12.8 | 12.3 | 21.8 | 9.1 | 17.1 | 15.5 | 11.3 | 12.0 | 13.0 | 35.9 | 50.4 | 193,556 | 198,338 | 2.5 | 10.9 |
| University Place | 5.7 | 14.7 | 28.4 | 8.2 | 11.6 | 15.2 | 11.2 | 8.6 | 16.9 | 35.6 | 48.9 | 29,933 | 31,133 | 4.0 | 9.5 |
| Vancouver | 13.7 | 15.5 | 21.4 | 9.2 | 16.3 | 14.5 | 10.3 | 12.4 | 15.9 | 36.9 | 49.8 | 143,560 | 167,167 | 16.4 | 11.4 |
| Walla Walla | 19.2 | 9.9 | 17.5 | 12.6 | 13.5 | 12.4 | 13.3 | 11.5 | 19.2 | 40.2 | 46.2 | 29,686 | 32,461 | 9.3 | 2.0 |
| Wenatchee | 36.9 | 16.6 | 28.4 | 6.9 | 15.6 | 15.4 | 8.2 | 10.8 | 14.7 | 34.4 | 51.1 | 27,856 | 32,821 | 17.8 | 4.3 |
| Yakima | 46.0 | 21.6 | 27.6 | 9.6 | 15.2 | 12.0 | 11.3 | 9.0 | 15.3 | 33.4 | 49.0 | 71,845 | 91,285 | 27.1 | 3.3 |
| **WEST VIRGINIA** | 1.5 | 1.6 | 20.0 | 8.5 | 11.8 | 12.3 | 12.6 | 14.3 | 20.5 | 42.9 | 50.6 | 1,808,344 | 1,853,008 | 2.5 | -3.7 |
| Charleston | 0.7 | 3.9 | 16.9 | 6.0 | 13.8 | 16.5 | 10.7 | 15.2 | 20.9 | 41.8 | 53.7 | 53,421 | 51,280 | -4.0 | -10.5 |
| Huntington | 0.7 | 2.5 | 20.3 | 15.5 | 13.3 | 11.2 | 12.0 | 13.4 | 14.4 | 36.1 | 52.8 | 51,475 | 49,312 | -4.2 | -8.9 |
| Morgantown | 1.6 | 8.1 | 10.0 | 41.2 | 13.8 | 7.6 | 7.8 | 9.4 | 10.3 | 24.5 | 50.7 | 26,809 | 28,497 | 6.3 | 8.2 |
| Parkersburg | NA | NA | 20.4 | 7.7 | 11.2 | 11.4 | 14.2 | 14.1 | 21.0 | 44.0 | 53.2 | 33,099 | 31,284 | -5.5 | -7.3 |
| Wheeling | 1.5 | 2.6 | 21.1 | 8.2 | 12.1 | 9.6 | 11.0 | 16.3 | 21.7 | 44.4 | 51.8 | 31,419 | 28,381 | -9.7 | -7.4 |
| **WISCONSIN** | 7.1 | 5.1 | 21.7 | 9.3 | 12.7 | 12.3 | 12.4 | 14.2 | 17.5 | 39.9 | 50.3 | 5,363,675 | 5,687,285 | 6.0 | 2.6 |
| Appleton | 8.4 | 5.3 | 24.0 | 10.6 | 17.4 | 12.2 | 11.7 | 10.7 | 13.4 | 33.9 | 52.6 | 70,087 | 72,584 | 3.6 | 2.0 |
| Beloit | 22.7 | 11.1 | 27.8 | 13.4 | 10.6 | 11.4 | 12.7 | 11.4 | 12.7 | 33.6 | 54.7 | 35,775 | 37,023 | 3.5 | -1.0 |
| Brookfield | 3.3 | 9.6 | 22.0 | 6.1 | 11.2 | 12.1 | 12.2 | 14.4 | 22.1 | 43.9 | 50.4 | 38,649 | 37,920 | -1.9 | 3.3 |
| Eau Claire | 2.9 | 4.0 | 17.9 | 20.6 | 15.2 | 11.6 | 8.9 | 10.6 | 15.3 | 32.2 | 52.7 | 61,704 | 66,237 | 7.3 | 4.3 |
| Fitchburg | 17.9 | 18.9 | 14.3 | 13.9 | 18.7 | 14.8 | 9.2 | 13.8 | 15.4 | 36.2 | 47.3 | 20,501 | 25,156 | 22.7 | 22.7 |
| Fond du Lac | 6.5 | 3.2 | 22.4 | 10.7 | 11.8 | 13.4 | 11.1 | 12.6 | 18.1 | 38.1 | 56.0 | 42,203 | 43,023 | 1.9 | 0.4 |
| Franklin | 11.2 | 8.1 | 21.1 | 4.1 | 10.9 | 17.6 | 9.7 | 16.4 | 20.3 | 43.0 | 49.7 | 29,494 | 35,462 | 20.2 | 1.3 |
| Green Bay | 16.9 | 9.0 | 23.8 | 10.5 | 14.5 | 11.3 | 12.9 | 12.4 | 14.5 | 36.0 | 50.7 | 102,313 | 103,877 | 1.5 | 0.0 |
| Greenfield | 13.8 | 8.8 | 15.2 | 6.3 | 11.2 | 12.5 | 12.3 | 15.5 | 27.0 | 49.7 | 49.7 | 35,476 | 36,753 | 3.6 | 1.2 |
| Janesville | 4.7 | 3.1 | 21.7 | 8.8 | 14.3 | 12.6 | 12.0 | 14.5 | 16.2 | 39.8 | 51.0 | 59,498 | 63,667 | 7.0 | 1.6 |
| Kenosha | 18.8 | 6.0 | 24.1 | 10.8 | 15.0 | 13.7 | 11.8 | 11.1 | 13.6 | 35.1 | 52.3 | 90,352 | 99,274 | 9.9 | 0.3 |
| La Crosse | 2.4 | 3.0 | 13.2 | 27.1 | 16.3 | 10.5 | 7.5 | 9.7 | 15.7 | 30.1 | 51.1 | 51,818 | 51,338 | -0.9 | -0.3 |
| Madison | 6.8 | 11.9 | 15.5 | 20.2 | 19.5 | 12.7 | 10.2 | 10.4 | 11.4 | 31.7 | 50.7 | 208,054 | 233,413 | 12.2 | 12.7 |
| Manitowoc | 6.9 | 4.0 | 20.2 | 7.3 | 11.1 | 11.2 | 13.9 | 14.6 | 21.7 | 45.1 | 49.3 | 34,053 | 33,699 | -1.0 | -4.0 |
| Menomonee Falls | 4.5 | 2.3 | 19.3 | 6.1 | 12.0 | 12.7 | 12.2 | 16.1 | 21.6 | 44.8 | 53.1 | 32,647 | 35,625 | 9.1 | 7.3 |
| Milwaukee | 19.2 | 10.2 | 25.3 | 11.6 | 18.0 | 12.5 | 11.0 | 10.7 | 10.9 | 31.5 | 51.7 | 596,974 | 594,503 | -0.4 | -0.9 |
| Mount Pleasant | 14.0 | 7.9 | 19.1 | 9.5 | 11.7 | 10.0 | 10.8 | 14.4 | 24.6 | 44.5 | 52.4 | NA | 26,698 | ********** | 1.3 |
| Neenah | 1.4 | 2.3 | 24.6 | 6.5 | 12.7 | 11.2 | 16.7 | 14.2 | 14.1 | 39.8 | 48.7 | 24,507 | 25,500 | 4.1 | 3.4 |
| New Berlin | 2.5 | 6.0 | 16.5 | 8.0 | 11.4 | 10.1 | 14.1 | 17.3 | 22.6 | 49.0 | 51.2 | 38,220 | 39,607 | 3.6 | 0.5 |
| Oak Creek | 5.6 | 9.8 | 21.5 | 5.8 | 10.5 | 16.5 | 10.7 | 18.4 | 16.6 | 41.9 | 50.9 | 28,456 | 34,449 | 21.1 | 5.9 |
| Oshkosh | 3.8 | 4.5 | 18.5 | 14.6 | 16.2 | 11.7 | 10.5 | 11.3 | 17.3 | 35.5 | 49.9 | 62,916 | 66,342 | 5.4 | 0.2 |
| Racine | 19.9 | 5.7 | 25.2 | 10.6 | 11.5 | 13.8 | 12.7 | 12.6 | 13.6 | 37.6 | 51.4 | 81,855 | 78,559 | -4.0 | -3.0 |
| Sheboygan | 11.9 | 10.6 | 23.0 | 10.4 | 15.4 | 11.4 | 12.4 | 12.6 | 14.7 | 36.6 | 49.7 | 50,792 | 49,399 | -2.7 | -3.2 |
| Stevens Point | 1.9 | 2.4 | 14.1 | 29.8 | 11.3 | 8.2 | 11.9 | 9.2 | 15.6 | 28.9 | 50.1 | 24,551 | 26,711 | 8.8 | -2.9 |
| Sun Prairie | 5.1 | 6.4 | 25.1 | 5.8 | 16.8 | 16.2 | 11.4 | 11.7 | 13.0 | 36.2 | 48.5 | 20,369 | 29,539 | 45.0 | 18.9 |
| Superior | 1.9 | 2.6 | 20.3 | 11.3 | 14.0 | 14.5 | 11.8 | 11.7 | 16.4 | 38.7 | 51.6 | 27,368 | 27,229 | -0.5 | -3.7 |
| Waukesha | 14.0 | 7.0 | 19.3 | 10.6 | 13.5 | 13.7 | 15.4 | 12.8 | 14.7 | 39.7 | 54.0 | 64,825 | 71,224 | 9.9 | 1.6 |
| Wausau | 4.2 | 9.4 | 22.2 | 9.6 | 13.0 | 12.3 | 13.8 | 13.5 | 15.6 | 39.6 | 47.3 | 38,426 | 39,299 | 2.3 | -2.0 |
| Wauwatosa | 3.8 | 5.9 | 22.5 | 5.4 | 17.9 | 12.3 | 11.9 | 11.6 | 18.4 | 37.5 | 52.6 | 47,271 | 46,424 | -1.8 | 3.6 |
| West Allis | 15.4 | 5.8 | 20.3 | 6.3 | 17.7 | 13.8 | 10.6 | 14.0 | 17.4 | 38.6 | 50.2 | 61,254 | 60,386 | -1.4 | -1.1 |
| West Bend | 4.6 | 0.3 | 23.7 | 7.6 | 12.9 | 12.6 | 12.8 | 10.7 | 19.8 | 38.7 | 51.2 | 28,152 | 31,186 | 10.8 | 0.9 |
| **WYOMING** | 10.1 | 3.1 | 23.2 | 9.5 | 13.2 | 12.7 | 11.0 | 13.3 | 17.1 | 38.1 | 48.9 | 493,782 | 563,775 | 14.2 | 3.3 |
| Casper | 6.4 | 2.2 | 20.9 | 9.8 | 16.3 | 10.4 | 11.9 | 14.1 | 16.6 | 39.3 | 49.9 | 49,644 | 55,311 | 11.4 | 5.8 |
| Cheyenne | 17.3 | 4.2 | 21.2 | 7.6 | 16.8 | 11.1 | 11.5 | 12.8 | 19.0 | 37.9 | 51.8 | 53,011 | 59,531 | 12.3 | 8.8 |
| Gillette | 11.9 | 3.5 | 26.8 | 9.0 | 19.7 | 13.9 | 7.8 | 11.8 | 11.1 | 31.9 | 48.4 | 19,646 | 31,419 | 59.9 | 2.5 |
| Laramie | 10.5 | 5.2 | 14.5 | 33.3 | 19.1 | 11.4 | 6.2 | 7.7 | 7.8 | 25.8 | 45.4 | 27,204 | 30,781 | 13.1 | 6.1 |

1. May be of any race.

Items 12—26

# Table D. Cities — Households, Group Quarters, Crime, and Education

| City | Households, 2019 Number | Persons per household | Percent Family | Married couple family | Female family | Non-family | One person | Persons in group quarters, 2019 | Violent crimes Number | Violent Rate | Property Number | Property Rate | Population age 25 and over | High school graduate or less | Bachelor's degree or more |
|---|---|---|---|---|---|---|---|---|---|---|---|---|---|---|---|
| | 27 | 28 | 29 | 30 | 31 | 32 | 33 | 34 | 35 | 36 | 37 | 38 | 39 | 40 | 41 |
| WASHINGTON—Cont'd | | | | | | | | | | | | | | | |
| Pasco | 22,747 | 3.27 | 76.9 | 48.0 | 28.9 | 23.1 | 18.3 | 528 | NA | NA | NA | NA | 42,199 | 55.6 | 17.1 |
| Pullman | 12,008 | 2.31 | 40.5 | 27.5 | 13.1 | 59.5 | 40.4 | 6,720 | 42 | 121 | 273 | 789 | 13,198 | 11.5 | 61.2 |
| Puyallup | 18,851 | 2.21 | 57.9 | 42.7 | 15.2 | 42.1 | 31.9 | 765 | 133 | 313 | 2,187 | 5,145 | 33,149 | 35.4 | 24.0 |
| Redmond | 29,881 | 2.40 | 61.4 | 52.9 | 8.6 | 38.6 | 29.9 | 234 | 71 | 102 | 1,846 | 2,656 | 52,472 | 10.6 | 73.3 |
| Renton | 39,908 | 2.53 | 59.3 | 45.4 | 13.9 | 40.7 | 26.7 | 721 | 325 | 314 | 4,182 | 4,042 | 74,004 | 31.1 | 36.6 |
| Richland | 21,582 | 2.69 | 67.9 | 53.5 | 14.5 | 32.1 | 26.8 | 253 | 100 | 171 | 1,199 | 2,049 | 38,760 | 23.2 | 42.3 |
| Sammamish | 21,456 | 3.07 | 84.5 | 73.9 | 10.6 | 15.5 | 12.7 | 108 | 23 | 34 | 404 | 605 | 42,221 | 7.0 | 77.5 |
| SeaTac | 10,393 | 2.70 | 59.9 | 38.1 | 21.8 | 40.1 | 34.8 | 978 | 130 | 440 | 1,098 | 3,718 | 20,335 | 45.0 | 19.9 |
| Seattle | 343,988 | 2.11 | 44.3 | 36.1 | 8.2 | 55.7 | 37.8 | 26,334 | 4,471 | 585 | 34,333 | 4,496 | 565,224 | 14.9 | 65.0 |
| Shoreline | 21,663 | 2.56 | 50.9 | 35.0 | 14.2 | | 23.5 | 1,647 | 95 | 166 | 1,104 | 1,930 | 41,959 | 20.2 | 52.3 |
| Spokane | 95,222 | 2.25 | 52.4 | 36.2 | 16.2 | 47.6 | 35.0 | 7,392 | 1,520 | 690 | 13,048 | 5,919 | 155,305 | 29.9 | 31.5 |
| Spokane Valley | 41,483 | 2.42 | 62.4 | 46.0 | 16.4 | 37.6 | 28.6 | 866 | 311 | 308 | 4,465 | 4,422 | 70,226 | 34.5 | 20.5 |
| Tacoma | 87,016 | 2.44 | 55.5 | 37.2 | 18.3 | 44.5 | 33.4 | 5,163 | 1,848 | 845 | 11,415 | 5,221 | 150,410 | 35.3 | 33.1 |
| University Place | 13,514 | 2.50 | 64.2 | 54.0 | 10.2 | 35.8 | 30.6 | 194 | 88 | 258 | 541 | 1,587 | 21,586 | 22.1 | 48.2 |
| Vancouver | 75,199 | 2.42 | 56.6 | 38.9 | 17.7 | 43.4 | 30.7 | 2,444 | 889 | 480 | 5,998 | 3,242 | 128,028 | 33.2 | 31.7 |
| Walla Walla | 13,325 | 2.17 | 57.0 | 43.3 | 13.7 | 43.0 | 31.5 | 4,048 | 114 | 345 | 1,052 | 3,183 | 23,010 | 30.4 | 30.6 |
| Wenatchee | 12,428 | 2.70 | 66.7 | 41.2 | 25.5 | 33.3 | 30.1 | 863 | 62 | 180 | 638 | 1,849 | 22,233 | 42.4 | 24.6 |
| Yakima | 33,424 | 2.73 | 64.4 | 40.9 | 23.5 | 35.6 | 28.4 | 2,513 | 420 | 446 | 2,723 | 2,892 | 58,803 | 52.1 | 17.3 |
| WEST VIRGINIA | 728,175 | 2.40 | 64.6 | 48.4 | 16.2 | 35.4 | 29.6 | 46,910 | NA | NA | NA | NA | 1,281,086 | 53.1 | 21.1 |
| Charleston | 22,147 | 2.00 | 52.1 | 29.8 | 22.3 | 47.9 | 40.5 | 2,146 | NA | NA | NA | NA | 35,897 | 39.0 | 37.6 |
| Huntington | 19,057 | 2.24 | 49.8 | 34.4 | 15.4 | 50.2 | 41.2 | 3,112 | 330 | 722 | 1,792 | 3,923 | 29,396 | 45.5 | 29.2 |
| Morgantown | 11,488 | 2.23 | 35.9 | 26.9 | 8.9 | 64.1 | 38.4 | 4,967 | 67 | 214 | 489 | 1,563 | 14,929 | 21.3 | 59.3 |
| Parkersburg | 12,480 | 2.32 | 59.4 | 39.7 | 19.7 | 40.6 | 35.5 | 401 | NA | NA | NA | NA | 21,078 | 54.9 | 15.5 |
| Wheeling | 11,563 | 2.26 | 49.9 | 37.4 | 12.5 | 50.1 | 43.1 | 1,092 | NA | NA | NA | NA | 19,228 | 42.8 | 29.4 |
| WISCONSIN | 2,386,623 | 2.38 | 61.6 | 47.8 | 13.8 | 38.4 | 30.5 | 143,849 | NA | NA | NA | NA | 4,015,285 | 37.7 | 31.3 |
| Appleton | 31,049 | 2.35 | 57.4 | 43.6 | 13.8 | 42.6 | 34.7 | 2,555 | 206 | 276 | 1,073 | 1,435 | 49,310 | 31.4 | 34.7 |
| Beloit | 13,108 | 2.69 | 58.1 | 37.8 | 20.3 | 41.9 | 30.8 | 1,621 | 154 | 416 | 1,114 | 3,009 | 21,727 | 52.1 | 14.9 |
| Brookfield | 14,527 | 2.66 | 72.9 | 65.7 | 7.1 | 27.1 | 21.8 | 398 | 31 | 80 | 691 | 1,777 | 28,122 | 19.7 | 59.0 |
| Eau Claire | 27,233 | 2.39 | 53.3 | 39.8 | 13.5 | 46.7 | 30.9 | 3,970 | 191 | 276 | 1,634 | 2,361 | 42,443 | 27.9 | 33.3 |
| Fitchburg | 14,794 | 2.02 | 51.4 | 38.3 | 13.1 | 48.6 | 36.7 | 978 | 93 | 301 | 529 | 1,715 | 22,129 | 24.1 | 55.7 |
| Fond du Lac | 17,922 | 2.28 | 53.6 | 36.8 | 16.8 | 46.4 | 39.4 | 2,428 | 134 | 312 | 797 | 1,855 | 28,951 | 42.9 | 23.2 |
| Franklin | 14,062 | 2.39 | 69.9 | 66.4 | 3.5 | 30.1 | 28.8 | 2,243 | 25 | 70 | 576 | 1,606 | 26,801 | 30.0 | 42.8 |
| Green Bay | 43,004 | 2.34 | 58.4 | 38.7 | 19.7 | 41.6 | 34.1 | 3,733 | 529 | 504 | 1,724 | 1,642 | 68,646 | 43.6 | 23.8 |
| Greenfield | 17,697 | 2.08 | 53.4 | 38.0 | 15.4 | 46.6 | 39.9 | 433 | 64 | 171 | 946 | 2,530 | 29,234 | 43.7 | 26.7 |
| Janesville | 27,301 | 2.33 | 56.8 | 40.4 | 16.5 | 43.2 | 33.6 | 1,013 | 148 | 229 | 1,577 | 2,438 | 44,893 | 44.0 | 22.7 |
| Kenosha | 38,042 | 2.54 | 58.5 | 37.3 | 21.2 | 41.5 | 33.6 | 3,413 | 311 | 310 | 1,582 | 1,578 | 65,080 | 40.7 | 26.8 |
| La Crosse | 21,117 | 2.20 | 42.2 | 28.8 | 13.3 | 57.8 | 38.3 | 4,684 | 122 | 236 | 1,850 | 3,586 | 30,555 | 38.5 | 35.4 |
| Madison | 114,096 | 2.17 | 44.4 | 33.6 | 10.8 | 55.6 | 36.2 | 12,027 | 940 | 360 | 6,464 | 2,474 | 166,784 | 18.0 | 58.1 |
| Manitowoc | 14,799 | 2.15 | 50.5 | 38.1 | 12.4 | 49.5 | 44.6 | 767 | 73 | 225 | 631 | 1,942 | 23,629 | 43.8 | 24.0 |
| Menomonee Falls | 16,865 | 2.24 | 62.1 | 54.7 | 7.4 | 37.9 | 34.1 | 178 | 28 | 74 | 331 | 873 | 28,366 | 26.5 | 46.7 |
| Milwaukee | 232,176 | 2.48 | 52.9 | 26.5 | 26.4 | 47.1 | 37.4 | 15,492 | 7,874 | 1,332 | 15,097 | 2,555 | 372,155 | 45.5 | 25.7 |
| Mount Pleasant | 11,308 | 2.37 | 67.0 | 55.4 | 11.5 | 33.0 | 26.6 | 304 | 40 | 148 | 536 | 1,980 | 19,342 | 33.8 | 30.5 |
| Neenah | 10,985 | 2.38 | 55.7 | 42.5 | 13.2 | 44.3 | 35.5 | 152 | 45 | 172 | 330 | 1,263 | 18,120 | 32.8 | 40.1 |
| New Berlin | 16,493 | 2.40 | 72.7 | 63.7 | 9.0 | 27.3 | 23.6 | 183 | 20 | 50 | 452 | 1,137 | 29,973 | 23.6 | 47.0 |
| Oak Creek | 15,172 | 2.39 | 61.1 | 49.7 | 11.3 | 38.9 | 31.3 | 46 | 40 | 109 | 722 | 1,968 | 26,411 | 33.7 | 31.1 |
| Oshkosh | 26,794 | 2.24 | 53.3 | 39.0 | 14.3 | 46.7 | 36.8 | 7,114 | 190 | 284 | 1,067 | 1,597 | 44,867 | 44.9 | 26.1 |
| Racine | 31,303 | 2.40 | 55.6 | 33.3 | 22.3 | 44.4 | 35.5 | 1,665 | 291 | 377 | 990 | 1,281 | 49,290 | 49.1 | 18.9 |
| Sheboygan | 20,727 | 2.27 | 55.7 | 40.4 | 15.2 | 44.3 | 35.8 | 916 | 172 | 358 | 783 | 1,630 | 31,944 | 43.8 | 22.5 |
| Stevens Point | 11,219 | 2.03 | 38.2 | 26.7 | 11.5 | 61.8 | 41.7 | 3,055 | 55 | 211 | 260 | 996 | 14,532 | 34.2 | 42.0 |
| Sun Prairie | 13,742 | 2.51 | 65.3 | 51.6 | 13.7 | 34.7 | 26.8 | 111 | 33 | 95 | 498 | 1,441 | 23,944 | 20.9 | 52.0 |
| Superior | 11,551 | 2.16 | 57.5 | 38.1 | 19.5 | 42.5 | 33.3 | 1,005 | 72 | 277 | 1,101 | 4,240 | 17,746 | 31.2 | 31.0 |
| Waukesha | 30,364 | 2.29 | 59.5 | 43.8 | 15.7 | 40.5 | 34.0 | 2,765 | 67 | 92 | 690 | 949 | 50,674 | 29.8 | 37.9 |
| Wausau | 17,293 | 2.17 | 52.6 | 36.9 | 15.7 | 47.4 | 38.4 | 955 | 156 | 405 | 521 | 1,353 | 26,282 | 38.3 | 27.9 |
| Wauwatosa | 20,889 | 2.27 | 54.8 | 43.3 | 11.4 | 45.2 | 36.7 | 654 | 57 | 117 | 1,331 | 2,741 | 34,691 | 18.8 | 57.0 |
| West Allis | 27,168 | 2.18 | 51.0 | 38.5 | 12.4 | 49.0 | 39.6 | 744 | 211 | 356 | 1,686 | 2,843 | 43,969 | 38.9 | 27.0 |
| West Bend | 13,020 | 2.39 | 66.1 | 51.5 | 14.6 | 33.9 | 30.6 | 497 | 67 | 212 | 512 | 1,618 | 21,700 | 35.2 | 27.8 |
| WYOMING | 233,128 | 2.42 | 65.6 | 53.0 | 12.6 | 34.4 | 27.4 | 14,240 | NA | NA | NA | NA | 389,847 | 35.6 | 29.1 |
| Casper | 23,543 | 2.22 | 64.4 | 48.6 | 15.8 | 35.6 | 29.3 | 1,213 | 169 | 293 | 1,677 | 2,904 | 37,049 | 40.4 | 21.8 |
| Cheyenne | 28,911 | 2.19 | 59.2 | 46.0 | 13.1 | 40.8 | 36.4 | 957 | 220 | 341 | 1,999 | 3,099 | 45,732 | 33.2 | 31.0 |
| Gillette | 13,027 | 2.49 | 72.4 | 55.5 | 16.9 | 27.6 | 21.1 | 323 | 43 | 135 | 626 | 1,959 | 21,040 | 43.9 | 13.8 |
| Laramie | 13,247 | 2.30 | 46.0 | 37.2 | 8.8 | 54.0 | 23.9 | 2,363 | 58 | 178 | 318 | 973 | 17,182 | 18.5 | 60.4 |

1. Data for serious crimes have not been adjusted for underreporting. This may affect comparability between geographic areas and over time.   2. Per 100,000 population estimated by the FBI.   3. Persons 25 years old and over.

## Table D. Cities — Income, Poverty, and Housing

| City | Money income, 2019 — Households — Median income | Percent with income less than $20,000 | Percent with income of $200,000 or more | Median family income | Median non-family income | Median earnings, 2019 — All persons | Men | Women | Housing units, 2019 — Total | Occupied | Percent owner occupied | Median value[1] (dollars) | Median gross rent (dollars) |
|---|---|---|---|---|---|---|---|---|---|---|---|---|---|
| | 42 | 43 | 44 | 45 | 46 | 47 | 48 | 49 | 50 | 51 | 52 | 53 | 54 |
| **WASHINGTON—Cont'd** | | | | | | | | | | | | | |
| Pasco | 60,377 | 10.7 | 3.9 | 61,501 | 52,000 | 31,639 | 36,556 | 25,844 | 23,617 | 22,747 | 65.2 | 254,739 | 1,035 |
| Pullman | 29,019 | 37.2 | 3.7 | 48,781 | 18,065 | 10,796 | 11,322 | 10,409 | 13,639 | 12,008 | 28.4 | 323,812 | 886 |
| Puyallup | 75,953 | 8.9 | 7.8 | 101,203 | 47,306 | 45,796 | 50,346 | 41,755 | 19,432 | 18,851 | 49.8 | 368,173 | 1,423 |
| Redmond | 142,920 | 5.5 | 26.0 | 170,476 | 105,903 | 90,090 | 120,568 | 50,388 | 31,686 | 29,881 | 50.8 | 810,221 | 1,896 |
| Renton | 80,310 | 7.5 | 12.0 | 99,057 | 57,664 | 48,012 | 51,419 | 43,585 | 41,502 | 39,908 | 55.1 | 461,846 | 1,624 |
| Richland | 81,674 | 7.0 | 11.1 | 98,544 | 56,651 | 47,956 | 59,017 | 36,754 | 22,834 | 21,582 | 64.7 | 309,289 | 1,156 |
| Sammamish | 188,120 | 3.5 | 46.8 | 205,441 | 121,752 | 97,616 | 126,399 | 60,620 | 21,832 | 21,456 | 81.8 | 931,680 | 2,487 |
| SeaTac | 62,915 | 12.9 | 4.6 | 81,075 | 35,356 | 37,111 | 40,787 | 32,724 | 10,989 | 10,393 | 40.5 | 389,575 | 1,531 |
| Seattle | 102,486 | 10.4 | 22.1 | 152,242 | 70,847 | 58,698 | 66,004 | 51,210 | 372,011 | 343,988 | 43.9 | 767,024 | 1,744 |
| Shoreline | 92,866 | 9.2 | 18.1 | 118,704 | 51,268 | 47,253 | 53,650 | 42,610 | 23,705 | 21,663 | 65.1 | 632,701 | 1,673 |
| Spokane | 52,447 | 16.1 | 4.7 | 70,371 | 36,647 | 31,600 | 33,662 | 30,472 | 101,438 | 95,222 | 55.1 | 228,527 | 893 |
| Spokane Valley | 57,513 | 13.1 | 2.1 | 65,247 | 36,202 | 36,067 | 41,020 | 31,047 | 43,376 | 41,483 | 56.4 | 239,943 | 986 |
| Tacoma | 70,411 | 12.3 | 6.2 | 84,114 | 48,130 | 37,773 | 42,807 | 35,005 | 92,394 | 87,016 | 53.9 | 344,457 | 1,273 |
| University Place | 70,598 | 7.2 | 5.5 | 96,675 | 47,119 | 51,236 | 63,966 | 42,524 | 14,994 | 13,514 | 54.1 | 443,820 | 1,384 |
| Vancouver | 66,679 | 9.4 | 4.3 | 75,880 | 54,297 | 36,701 | 42,405 | 30,298 | 79,266 | 75,199 | 48.4 | 339,808 | 1,343 |
| Walla Walla | 53,993 | 19.5 | 6.2 | 69,786 | 23,961 | 33,130 | 35,655 | 27,495 | 14,832 | 13,325 | 62.3 | 257,400 | 840 |
| Wenatchee | 52,897 | 13.2 | 3.2 | 54,837 | 34,754 | 28,927 | 28,066 | 31,198 | 13,307 | 12,428 | 45.8 | 288,701 | 1,007 |
| Yakima | 44,292 | 17.7 | 2.4 | 51,250 | 31,248 | 27,465 | 28,752 | 26,232 | 35,834 | 33,424 | 51.2 | 216,621 | 842 |
| **WEST VIRGINIA** | 48,850 | 20.1 | 3.4 | 60,920 | 26,752 | 30,970 | 37,783 | 25,714 | 894,983 | 728,175 | 73.4 | 124,586 | 727 |
| Charleston | 45,250 | 25.9 | 5.9 | 62,215 | 34,341 | 31,165 | 35,153 | 28,842 | 26,418 | 22,147 | 58.7 | 138,828 | 675 |
| Huntington | 36,966 | 29.3 | 3.2 | 45,863 | 26,889 | 25,294 | 27,777 | 22,158 | 25,706 | 19,057 | 49.2 | 104,599 | 727 |
| Morgantown | 45,000 | 31.2 | 4.8 | 109,573 | 22,018 | 12,706 | 14,181 | 11,345 | 13,541 | 11,488 | 43.9 | 225,255 | 829 |
| Parkersburg | 36,629 | 29.5 | 2.0 | 48,112 | 20,535 | 24,349 | 29,361 | 21,837 | 14,824 | 12,480 | 65.5 | 98,029 | 635 |
| Wheeling | 43,015 | 24.2 | 3.5 | 69,768 | 26,791 | 26,933 | 28,728 | 22,147 | 14,076 | 11,563 | 60.5 | 123,286 | 688 |
| **WISCONSIN** | 64,168 | 12.4 | 5.2 | 81,829 | 38,937 | 37,044 | 43,188 | 31,240 | 2,725,153 | 2,386,623 | 67.2 | 197,221 | 867 |
| Appleton | 60,129 | 12.3 | 3.9 | 75,187 | 38,254 | 34,400 | 41,745 | 28,633 | 32,104 | 31,049 | 65.9 | 165,611 | 781 |
| Beloit | 47,794 | 17.4 | 0.8 | 52,141 | 33,191 | 29,529 | 32,316 | 26,858 | 14,729 | 13,108 | 59.3 | 98,431 | 824 |
| Brookfield | 102,339 | 3.5 | 21.3 | 130,277 | 65,498 | 51,701 | 67,253 | 36,254 | 15,150 | 14,527 | 83.2 | 335,739 | 1,383 |
| Eau Claire | 59,684 | 11.8 | 2.6 | 85,359 | 37,801 | 31,990 | 41,142 | 25,595 | 29,026 | 27,233 | 57.6 | 175,288 | 820 |
| Fitchburg | 76,684 | 13.2 | 11.1 | 111,107 | 48,949 | 41,142 | 50,068 | 39,735 | 15,395 | 14,794 | 43.2 | 334,066 | 1,103 |
| Fond du Lac | 51,749 | 11.4 | 3.1 | 73,332 | 32,451 | 35,122 | 39,068 | 26,370 | 19,344 | 17,922 | 53.2 | 134,696 | 748 |
| Franklin | 92,439 | 7.7 | 7.8 | 114,580 | 42,688 | 52,121 | 62,401 | 42,399 | 15,009 | 14,062 | 77.4 | 269,963 | 1,105 |
| Green Bay | 49,029 | 16.2 | 3.2 | 65,336 | 34,701 | 31,441 | 36,396 | 26,787 | 45,090 | 43,004 | 52.4 | 154,665 | 756 |
| Greenfield | 65,673 | 11.3 | 2.5 | 83,704 | 49,291 | 40,880 | 40,846 | 40,918 | 19,066 | 17,697 | 59.5 | 194,375 | 991 |
| Janesville | 55,784 | 15.9 | 1.8 | 66,707 | 35,720 | 35,427 | 41,195 | 30,023 | 28,629 | 27,301 | 63.0 | 161,170 | 877 |
| Kenosha | 55,594 | 15.4 | 2.3 | 71,508 | 36,473 | 32,388 | 37,592 | 30,725 | 41,197 | 38,042 | 54.3 | 172,714 | 967 |
| La Crosse | 46,419 | 18.0 | 3.5 | 71,174 | 33,943 | 22,094 | 26,320 | 17,288 | 22,854 | 21,117 | 50.4 | 160,669 | 827 |
| Madison | 66,847 | 12.8 | 7.1 | 98,001 | 47,359 | 35,881 | 38,982 | 33,157 | 119,881 | 114,096 | 45.9 | 275,890 | 1,155 |
| Manitowoc | 47,913 | 17.3 | 3.5 | 81,991 | 35,528 | 31,155 | 37,334 | 26,283 | 16,621 | 14,799 | 61.6 | 120,602 | 737 |
| Menomonee Falls | 87,559 | 8.7 | 10.7 | 116,011 | 41,809 | 50,886 | 60,114 | 41,529 | 17,597 | 16,865 | 74.9 | 263,333 | 1,147 |
| Milwaukee | 44,192 | 22.6 | 2.0 | 51,038 | 33,635 | 31,157 | 33,189 | 28,459 | 260,024 | 232,176 | 40.0 | 133,625 | 865 |
| Mount Pleasant | 77,689 | 10.7 | 6.9 | 88,831 | 49,714 | 41,898 | 41,567 | 42,079 | 11,759 | 11,308 | 81.6 | 208,325 | 879 |
| Neenah | 70,375 | 8.9 | 6.9 | 100,000 | 40,193 | 35,934 | 41,218 | 32,076 | 11,661 | 10,985 | 66.1 | 164,669 | 706 |
| New Berlin | 85,857 | 3.4 | 10.5 | 102,123 | 48,941 | 44,419 | 54,329 | 38,659 | 16,817 | 16,493 | 75.9 | 283,361 | 1,247 |
| Oak Creek | 72,703 | 9.4 | 6.5 | 106,201 | 43,033 | 51,015 | 54,972 | 37,221 | 15,503 | 15,172 | 65.6 | 241,701 | 1,060 |
| Oshkosh | 54,925 | 15.4 | 2.6 | 64,819 | 34,661 | 31,764 | 37,413 | 27,202 | 28,873 | 26,794 | 54.5 | 130,965 | 788 |
| Racine | 43,748 | 21.9 | 1.6 | 53,544 | 33,656 | 28,609 | 30,875 | 25,487 | 33,126 | 31,303 | 52.8 | 124,508 | 792 |
| Sheboygan | 52,809 | 11.9 | 1.5 | 65,015 | 34,914 | 35,359 | 41,492 | 28,475 | 21,805 | 20,727 | 62.8 | 131,316 | 717 |
| Stevens Point | 45,625 | 24.4 | 2.2 | 74,911 | 31,450 | 22,968 | 32,158 | 16,708 | 11,724 | 11,219 | 51.7 | 144,648 | 722 |
| Sun Prairie | 86,408 | 5.1 | 5.7 | 108,408 | 55,219 | 48,387 | 62,231 | 43,620 | 14,405 | 13,742 | 65.0 | 274,718 | 1,213 |
| Superior | 49,550 | 16.2 | 2.0 | 67,315 | 31,262 | 32,221 | 41,127 | 26,802 | 12,370 | 11,551 | 62.2 | 124,929 | 693 |
| Waukesha | 72,411 | 12.2 | 5.3 | 95,576 | 40,715 | 40,891 | 49,428 | 35,980 | 31,309 | 30,364 | 58.4 | 234,661 | 994 |
| Wausau | 50,380 | 13.9 | 4.4 | 64,497 | 35,661 | 30,911 | 34,229 | 26,911 | 19,830 | 17,293 | 54.7 | 123,841 | 734 |
| Wauwatosa | 79,223 | 9.3 | 11.7 | 120,647 | 45,598 | 50,527 | 61,600 | 41,890 | 22,334 | 20,889 | 61.6 | 260,086 | 1,147 |
| West Allis | 60,038 | 14.4 | 3.6 | 74,329 | 38,873 | 41,294 | 46,421 | 35,762 | 29,126 | 27,168 | 53.6 | 158,443 | 866 |
| West Bend | 66,148 | 9.3 | 3.4 | 78,690 | 34,243 | 35,929 | 40,935 | 31,054 | 13,050 | 13,020 | 71.9 | 194,970 | 815 |
| **WYOMING** | 65,003 | 13.0 | 5.0 | 79,946 | 37,299 | 34,780 | 46,506 | 26,594 | 280,281 | 233,128 | 71.9 | 235,151 | 822 |
| Casper | 59,269 | 14.9 | 5.2 | 76,020 | 37,438 | 34,240 | 44,049 | 26,532 | 26,307 | 23,543 | 68.8 | 216,406 | 745 |
| Cheyenne | 68,173 | 14.9 | 6.0 | 86,407 | 45,385 | 36,692 | 45,414 | 32,285 | 30,704 | 28,911 | 64.5 | 252,558 | 798 |
| Gillette | 79,836 | 8.4 | 4.3 | 82,842 | 39,032 | 42,644 | 66,690 | 32,894 | 14,746 | 13,027 | 77.2 | 214,293 | 785 |
| Laramie | 48,786 | 18.2 | 4.3 | 71,703 | 35,158 | 22,382 | 25,430 | 20,103 | 15,075 | 13,247 | 44.4 | 237,356 | 922 |

1. Based on population estimated by the American Community Survey.

# Commuting, Computer Access, Migration, Labor Force, and Employment

| City | Commuting[1], 2019 Percent — Drove alone (55) | With commutes of 30 minutes or more (56) | Computer access[2], 2019 Percent — With a computer in the house (57) | With Internet access (58) | Migration, 2019 — Percent who lived in the same house one year ago (59) | Percent who lived in another state or county one year ago (60) | Civilian labor force, 2020 — Total (61) | Percent change 2019-2020 (62) | Unemployment — Total (63) | Rate[3] (64) | Civilian Employment[4], 2019 — Population age 16 and older — Number (65) | Percent in labor force (66) | Population age 16 to 64 — Number (67) | Percent who worked full-year full-time (68) |
|---|---|---|---|---|---|---|---|---|---|---|---|---|---|---|
| WASHINGTON—Cont'd | | | | | | | | | | | | | | |
| Pasco | 80.3 | 19.9 | 95.0 | 91.0 | 84.5 | 7.3 | 34,369 | -1.2 | 3,216 | 9 | 51,994 | 67.1 | 45,826 | 48.4 |
| Pullman | 52.5 | 6.6 | 96.5 | 83.1 | 57.4 | 26.9 | 15,232 | -7.6 | 926 | 6 | 30,695 | 59.2 | 28,224 | 22.4 |
| Puyallup | 78.4 | 52.4 | 94.0 | 89.6 | 82.7 | 6.4 | 23,233 | 1.5 | 2,153 | 9 | 36,242 | 65.3 | 29,151 | 60.2 |
| Redmond | 66.1 | 37.0 | 98.0 | 95.5 | 81.3 | 9.5 | 40,997 | 2.6 | 1,985 | 5 | 57,563 | 71.6 | 49,876 | 59.9 |
| Renton | 74.4 | 56.1 | 94.7 | 92.3 | 81.9 | 6.5 | 58,973 | -0.3 | 5,198 | 9 | 81,649 | 70.3 | 68,103 | 58.9 |
| Richland | 83.7 | 18.2 | 95.1 | 89.3 | 81.3 | 9.7 | 29,968 | 0.5 | 2,334 | 8 | 44,805 | 62.4 | 35,825 | 53.2 |
| Sammamish | 69.8 | 65.6 | NA | NA | 87.6 | 6.3 | 33,179 | -2.8 | 1,796 | 5 | 47,911 | 69.7 | 42,131 | 55.3 |
| SeaTac | 72.0 | 44.2 | 94.6 | 91.0 | 82.1 | 3.1 | 15,715 | 4.0 | 2,111 | 13 | 22,373 | 71.3 | 18,019 | 53.2 |
| Seattle | 44.5 | 47.4 | 96.8 | 92.8 | 76.6 | 8.3 | 474,625 | -0.4 | 32,864 | 7 | 651,639 | 74.9 | 560,901 | 58.4 |
| Shoreline | 70.2 | 55.7 | 97.1 | 91.3 | 86.2 | 4.9 | 31,209 | -0.2 | 2,549 | 8 | 47,435 | 68.6 | 35,928 | 58.7 |
| Spokane | 74.3 | 23.7 | 93.5 | 88.4 | 79.5 | 6.6 | 110,291 | 1.2 | 10,506 | 10 | 182,595 | 61.3 | 147,912 | 46.5 |
| Spokane Valley | 73.2 | 23.2 | 93.6 | 87.6 | 80.4 | 6.5 | 51,639 | 0.9 | 4,625 | 9 | 77,893 | 62.3 | 61,020 | 50.5 |
| Tacoma | 74.4 | 42.9 | 92.9 | 86.6 | 80.5 | 7.9 | 110,832 | 2.0 | 11,904 | 11 | 176,286 | 67.6 | 147,867 | 53.1 |
| University Place | 75.3 | 41.7 | 95.5 | 92.9 | 85.3 | 4.8 | 17,694 | 0.5 | 1,492 | 8 | 25,185 | 59.1 | 19,436 | 56.7 |
| Vancouver | 72.3 | 33.1 | 93.9 | 88.0 | 79.1 | 9.8 | 93,508 | 0.3 | 8,556 | 9 | 149,375 | 64.1 | 120,054 | 52.0 |
| Walla Walla | 76.5 | 10.1 | 92.6 | 87.7 | 76.1 | 13.6 | 15,298 | 0.0 | 1,160 | 8 | 27,728 | 56.4 | 21,416 | 43.3 |
| Wenatchee | 75.9 | 14.3 | 87.3 | 83.1 | 77.5 | 11.4 | 19,209 | -2.0 | 1,692 | 9 | 25,205 | 59.4 | 20,145 | 37.5 |
| Yakima | 82.6 | 16.3 | 91.0 | 85.7 | 78.8 | 5.7 | 48,522 | -0.8 | 4,493 | 9 | 70,394 | 62.2 | 56,062 | 48.9 |
| WEST VIRGINIA | 82.0 | 33.7 | 87.9 | 80.0 | 88.2 | 5.4 | 792,156 | -0.6 | 66,133 | 8 | 1,476,688 | 53.7 | 1,109,288 | 47.0 |
| Charleston | 72.3 | 12.6 | 85.3 | 80.6 | 85.1 | 7.6 | 22,524 | 1.0 | 2,081 | 9 | 39,742 | 55.4 | 30,024 | 49.1 |
| Huntington | 76.9 | 15.4 | 85.6 | 80.6 | 82.2 | 8.1 | 20,003 | 0.7 | 1,765 | 9 | 37,808 | 54.6 | 31,234 | 39.9 |
| Morgantown | 66.7 | 15.5 | 97.4 | 89.5 | 58.3 | 28.9 | 15,009 | -0.7 | 1,051 | 7 | 27,588 | 52.5 | 24,443 | 29.7 |
| Parkersburg | 71.7 | 11.6 | 84.5 | 75.6 | 85.2 | 5.7 | 12,194 | -0.2 | 1,255 | 10 | 23,954 | 53.5 | 17,797 | 40.7 |
| Wheeling | 74.8 | 21.5 | 76.9 | 72.8 | 86.8 | 7.3 | 12,580 | -3.9 | 1,197 | 10 | 22,077 | 54.2 | 16,176 | 44.3 |
| WISCONSIN | 80.8 | 28.3 | 91.7 | 85.7 | 86.8 | 5.6 | 3,065,402 | -1.3 | 192,793 | 6 | 4,707,424 | 65.9 | 3,687,528 | 56.3 |
| Appleton | 81.6 | 20.3 | 93.4 | 87.4 | 84.6 | 7.6 | 39,302 | -1.0 | 2,297 | 6 | 59,239 | 70.0 | 49,148 | 58.8 |
| Beloit | 75.2 | 32.3 | 93.0 | 89.1 | 88.9 | 3.2 | 17,142 | -0.9 | 1,346 | 8 | 28,035 | 63.4 | 23,332 | 49.9 |
| Brookfield | 86.4 | 23.4 | 97.0 | 94.5 | 87.3 | 11.3 | 19,337 | -1.2 | 1,090 | 6 | 31,293 | 62.6 | 22,639 | 54.8 |
| Eau Claire | 77.8 | 9.4 | 96.2 | 93.4 | 75.4 | 13.1 | 39,279 | -1.1 | 2,181 | 6 | 57,703 | 69.9 | 47,152 | 50.2 |
| Fitchburg | 82.7 | 21.7 | 95.8 | 92.4 | 83.3 | 9.9 | 17,084 | 0.2 | 747 | 4 | 26,651 | 70.7 | 21,910 | 62.5 |
| Fond du Lac | 79.0 | 21.5 | 86.3 | 77.9 | 80.6 | 8.2 | 22,557 | -1.3 | 1,473 | 7 | 34,230 | 66.5 | 26,420 | 59.6 |
| Franklin | 87.2 | 40.3 | 90.9 | 88.3 | 85.6 | 1.6 | 17,684 | -2.3 | 1,042 | 6 | 28,886 | 58.1 | 21,624 | 55.1 |
| Green Bay | 79.8 | 15.2 | 90.0 | 84.7 | 83.2 | 4.1 | 53,597 | -1.0 | 3,730 | 7 | 82,154 | 69.7 | 66,967 | 55.9 |
| Greenfield | 85.8 | 23.3 | 88.3 | 83.8 | 90.2 | 2.7 | 19,599 | -1.3 | 1,431 | 7 | 32,444 | 61.0 | 22,386 | 62.9 |
| Janesville | 80.8 | 24.5 | 90.4 | 81.4 | 84.6 | 5.5 | 33,731 | -0.1 | 2,533 | 8 | 52,222 | 64.3 | 41,754 | 52.5 |
| Kenosha | 85.2 | 34.1 | 93.7 | 88.4 | 84.8 | 4.5 | 50,075 | -0.8 | 3,969 | 8 | 78,777 | 67.2 | 65,213 | 56.4 |
| La Crosse | 72.0 | 13.8 | 93.9 | 86.2 | 72.7 | 11.8 | 28,439 | -2.3 | 1,650 | 6 | 45,048 | 65.6 | 37,019 | 43.0 |
| Madison | 63.7 | 21.6 | 96.8 | 92.6 | 75.0 | 9.3 | 155,747 | -0.8 | 7,527 | 5 | 222,949 | 73.0 | 193,245 | 54.8 |
| Manitowoc | 78.4 | 18.4 | 86.7 | 82.7 | 86.7 | 6.8 | 15,863 | -0.4 | 1,168 | 7 | 27,201 | 61.9 | 20,120 | 53.1 |
| Menomonee Falls | 84.7 | 26.3 | 90.3 | 88.6 | 92.1 | 5.3 | 20,429 | -1.1 | 1,160 | 6 | 31,650 | 69.0 | 23,438 | 67.8 |
| Milwaukee | 73.7 | 28.0 | 89.0 | 78.2 | 85.0 | 4.4 | 273,263 | -0.3 | 24,955 | 9 | 455,418 | 63.5 | 391,231 | 49.1 |
| Mount Pleasant | 80.6 | 23.8 | 95.5 | 93.3 | 88.9 | 4.5 | 14,161 | -1.2 | 924 | 7 | 22,258 | 62.5 | 15,592 | 50.9 |
| Neenah | 84.2 | 15.4 | 88.7 | 85.9 | 88.8 | 3.1 | 14,028 | 0.6 | 823 | 6 | 20,228 | 70.6 | 16,522 | 59.1 |
| New Berlin | 90.9 | 30.8 | 95.4 | 89.5 | 90.4 | 7.2 | 21,348 | -1.9 | 1,283 | 6 | 34,014 | 66.7 | 25,031 | 58.2 |
| Oak Creek | 88.0 | 29.5 | 93.9 | 88.6 | 83.6 | 6.7 | 20,498 | -2.2 | 1,218 | 6 | 28,892 | 65.2 | 22,871 | 60.7 |
| Oshkosh | 83.9 | 19.3 | 89.4 | 83.5 | 78.1 | 8.9 | 34,384 | -0.4 | 1,915 | 6 | 56,135 | 62.1 | 44,535 | 48.3 |
| Racine | 81.7 | 30.8 | 90.5 | 83.0 | 86.4 | 4.6 | 34,636 | -1.4 | 3,118 | 9 | 60,090 | 60.8 | 49,619 | 45.9 |
| Sheboygan | 85.0 | 7.4 | 88.6 | 86.7 | 87.8 | 1.2 | 25,049 | -0.3 | 1,738 | 7 | 37,564 | 67.3 | 30,500 | 55.6 |
| Stevens Point | 68.9 | 14.2 | 89.9 | 83.5 | 75.1 | 14.1 | 13,827 | -3.5 | 761 | 6 | 22,615 | 67.6 | 18,577 | 41.2 |
| Sun Prairie | 78.8 | 28.3 | 95.8 | 90.3 | 90.4 | 2.5 | 20,226 | 0.7 | 1,019 | 5 | 26,855 | 74.8 | 22,357 | 60.8 |
| Superior | 78.8 | 16.0 | 90.7 | 83.5 | 88.4 | 5.3 | 14,192 | -1.3 | 1,216 | 9 | 21,845 | 68.1 | 17,589 | 54.5 |
| Waukesha | 75.5 | 28.1 | 94.6 | 90.9 | 86.2 | 8.3 | 40,997 | -2.0 | 2,497 | 6 | 60,994 | 74.3 | 50,346 | 62.3 |
| Wausau | 74.5 | 12.4 | 92.7 | 85.0 | 83.9 | 3.5 | 19,427 | -0.9 | 1,217 | 6 | 31,606 | 66.9 | 25,594 | 53.2 |
| Wauwatosa | 85.9 | 21.8 | 91.2 | 86.0 | 87.2 | 3.6 | 26,842 | -2.3 | 1,566 | 6 | 38,920 | 66.6 | 30,057 | 62.6 |
| West Allis | 81.3 | 26.7 | 88.4 | 81.5 | 82.4 | 4.3 | 33,043 | 0.5 | 2,755 | 8 | 48,893 | 68.8 | 38,459 | 63.5 |
| West Bend | 85.6 | 38.7 | 92.3 | 89.7 | 88.7 | 6.7 | 16,627 | -1.9 | 1,024 | 6 | 24,967 | 71.4 | 18,730 | 61.1 |
| WYOMING | 76.9 | 19.1 | 94.6 | 87.2 | 82.6 | 8.0 | 296,801 | 1.6 | 17,339 | 6 | 458,477 | 64.2 | 359,688 | 54.9 |
| Casper | 83.6 | 7.3 | 94.3 | 84.6 | 84.5 | 5.1 | 29,813 | 3.3 | 2,301 | 8 | 43,391 | 67.6 | 34,526 | 59.0 |
| Cheyenne | 77.9 | 8.3 | 93.4 | 84.5 | 84.8 | 6.5 | 32,921 | 4.9 | 1,751 | 5 | 51,945 | 64.6 | 39,750 | 61.4 |
| Gillette | 84.3 | 23.1 | 98.0 | 94.9 | 83.8 | 7.2 | 15,904 | 1.1 | 1,171 | 7 | 24,648 | 73.9 | 21,015 | 65.3 |
| Laramie | 69.7 | 8.4 | 97.6 | 91.0 | 53.8 | 19.2 | 16,445 | -2.4 | 638 | 4 | 28,480 | 66.1 | 25,912 | 39.3 |

1. Employed persons.  2. Households.  3. Percent of civilian labor force.  4. Persons 16 years old and over.

# Table D. Cities — Construction, Wholesale Trade, and Retail Trade

| City | Value of residential construction authorized by building permits, 2020 | | | Wholesale trade[1], 2017 | | | | Retail trade[2], 2017 | | | |
|---|---|---|---|---|---|---|---|---|---|---|---|
| | New construction ($1,000) | Number of housing units | Percent single family | Number of establishments | Number of employees | Sales (mil dol) | Annual payroll (mil dol) | Number of establishments | Number of employees | Sales (mil dol) | Annual payroll (mil dol) |
| | 69 | 70 | 71 | 72 | 73 | 74 | 75 | 76 | 77 | 78 | 79 |
| **WASHINGTON—Cont'd** | | | | | | | | | | | |
| Pasco | 134,064 | 525 | 100 | 74 | 1,064 | 773 | 63 | 161 | 2,708 | 1,002 | 100 |
| Pullman | 11,641 | 49 | 84 | 10 | 177 | 129 | 8 | 50 | 1,209 | 313 | 29 |
| Puyallup | 41,056 | 102 | 100 | 42 | 571 | 503 | 31 | 216 | 5,482 | 2,234 | 201 |
| Redmond | 232,716 | 2,005 | 4 | 154 | 3,422 | 5,522 | 337 | 249 | 4,497 | 1,295 | 145 |
| Renton | 77,530 | 263 | 60 | 113 | 3,949 | 3,064 | 271 | 297 | 5,735 | 2,118 | 203 |
| Richland | 151,733 | 572 | 68 | 26 | 298 | 124 | 12 | 162 | 2,726 | 834 | 76 |
| Sammamish | 80,058 | 172 | 100 | 43 | 84 | 108 | 5 | 60 | 680 | 214 | 21 |
| SeaTac | 7,893 | 25 | 64 | 31 | 296 | 176 | 18 | 84 | 866 | 225 | 23 |
| Seattle | 748,381 | 5,726 | 4 | 1,041 | 19,093 | 17,790 | 1,341 | 2,502 | 39,883 | 60,444 | 1,782 |
| Shoreline | 43,861 | 182 | 84 | 31 | 124 | 63 | 6 | 128 | 2,759 | 996 | 98 |
| Spokane | 111,757 | 539 | 62 | 276 | 3,988 | 5,883 | 222 | 849 | 13,695 | 3,668 | 399 |
| Spokane Valley | 181,655 | 1,110 | 20 | 212 | 3,496 | 2,794 | 187 | 482 | 8,405 | 2,552 | 264 |
| Tacoma | 170,122 | 932 | 26 | 202 | 2,949 | 2,396 | 177 | 752 | 12,652 | 3,630 | 389 |
| University Place | 20,460 | 134 | 15 | 7 | 57 | 24 | 3 | 45 | 848 | 222 | 26 |
| Vancouver | 249,817 | 1,573 | 33 | 203 | 2,268 | 3,145 | 136 | 588 | 12,125 | 3,802 | 408 |
| Walla Walla | 12,126 | 49 | 67 | 30 | 161 | 85 | 8 | 132 | 1,505 | 354 | 42 |
| Wenatchee | 25,197 | 193 | 14 | 64 | 1,157 | 890 | 59 | 199 | 2,992 | 807 | 87 |
| Yakima | 36,693 | 203 | 71 | 103 | 2,417 | 1,523 | 113 | 344 | 5,794 | 1,960 | 182 |
| **WEST VIRGINIA** | 600 | 24 | 100 | 1,205 | 14,578 | 12,189 | 677 | 5,963 | 82,985 | 23,058 | 2,004 |
| Charleston | 6,835 | 43 | 91 | 101 | 1,097 | 930 | 55 | 302 | 5,318 | 1,349 | 128 |
| Huntington | 545 | 7 | 29 | 65 | 901 | 400 | 47 | 209 | 2,884 | 755 | 70 |
| Morgantown | 5,146 | 12 | 100 | 24 | 218 | 140 | 9 | 233 | 4,421 | 1,231 | 100 |
| Parkersburg | 569 | 2 | 100 | 39 | 357 | 133 | 14 | 183 | 3,194 | 975 | 88 |
| Wheeling | 5,277 | 40 | 3 | 57 | 1,099 | 3,864 | 47 | 138 | 1,595 | 413 | 39 |
| **WISCONSIN** | 550 | 1 | 100 | 5,934 | 104,382 | 81,566 | 6,129 | 18,908 | 317,668 | 91,764 | 8,274 |
| Appleton | 29,743 | 104 | 73 | 96 | 1,479 | 5,973 | 97 | 278 | 4,874 | 1,361 | 123 |
| Beloit | 20,420 | 155 | 15 | 19 | 352 | 231 | 26 | 104 | 1,761 | 540 | 45 |
| Brookfield | 25,824 | 59 | 39 | 109 | 1,689 | 985 | 113 | 271 | 5,303 | 1,066 | 136 |
| Eau Claire | 44,219 | 308 | 25 | 88 | 1,472 | 1,036 | 73 | 318 | 6,459 | 1,563 | 146 |
| Fitchburg | 123,183 | 714 | 16 | 28 | 1,002 | 569 | 57 | 52 | 999 | 269 | 28 |
| Fond du Lac | 9,816 | 26 | 100 | 47 | 858 | 577 | 54 | 197 | 4,289 | 1,237 | 111 |
| Franklin | 35,178 | 97 | 79 | 41 | 409 | 426 | 28 | 79 | 2,312 | 666 | 64 |
| Green Bay | 56,895 | 323 | 22 | 114 | 3,378 | 2,686 | 186 | 313 | 5,673 | 1,640 | 145 |
| Greenfield | 2,935 | 31 | 3 | 13 | 80 | 27 | 5 | 152 | 3,181 | 1,098 | 97 |
| Janesville | 41,367 | 271 | 44 | 71 | 1,740 | 1,274 | 95 | 276 | 6,749 | 1,609 | 199 |
| Kenosha | 6,209 | 26 | 100 | 53 | 897 | 1,334 | 87 | 290 | 9,394 | 4,528 | 273 |
| La Crosse | 5,319 | 27 | 78 | 56 | 1,406 | 2,317 | 77 | 243 | 4,846 | 1,200 | 117 |
| Madison | 377,953 | 2,185 | 19 | 271 | 5,246 | 3,744 | 314 | 946 | 19,081 | 6,336 | 531 |
| Manitowoc | 8,981 | 77 | 20 | 23 | 364 | 204 | 18 | 142 | 2,614 | 585 | 58 |
| Menomonee Falls | 59,881 | 230 | 46 | 90 | 1,597 | 844 | 105 | 124 | 2,867 | 813 | 75 |
| Milwaukee | 21,195 | 112 | 30 | 439 | 11,197 | 6,788 | 825 | D | D | D | D |
| Mount Pleasant | 14,589 | 55 | 75 | 17 | 262 | 120 | 13 | 81 | 1,782 | 574 | 46 |
| Neenah | 26,239 | 176 | 21 | 17 | 561 | 281 | 31 | 83 | 1,904 | 557 | 47 |
| New Berlin | 17,737 | 44 | 64 | 112 | 2,516 | 1,317 | 141 | 99 | 2,076 | 566 | 65 |
| Oak Creek | 22,894 | 87 | 100 | 39 | 2,298 | 1,576 | 162 | 88 | 2,244 | 680 | 53 |
| Oshkosh | 22,999 | 183 | 27 | 50 | 1,091 | 542 | 54 | 259 | 5,404 | 1,338 | 128 |
| Racine | 0 | 0 | 0 | 47 | 411 | 206 | 23 | 240 | 3,126 | 585 | 61 |
| Sheboygan | 21,430 | 195 | 2 | 42 | 889 | 587 | 42 | 171 | 3,226 | 831 | 80 |
| Stevens Point | 11,153 | 182 | 7 | 24 | 372 | 181 | 17 | 113 | 1,947 | 563 | 49 |
| Sun Prairie | 63,250 | 217 | 71 | 32 | 901 | 420 | 47 | 78 | 1,923 | 629 | 50 |
| Superior | 2,260 | 8 | 75 | 38 | 652 | 979 | 34 | 117 | 2,027 | 592 | 58 |
| Waukesha | 10,850 | 47 | 45 | 103 | 2,200 | 1,237 | 160 | 233 | 5,326 | 1,986 | 167 |
| Wausau | 11,712 | 72 | 33 | 47 | 692 | 364 | 41 | 175 | 5,863 | 1,811 | 145 |
| Wauwatosa | 4,379 | 31 | 13 | 56 | 1,110 | 520 | 60 | 291 | 5,783 | 1,144 | 135 |
| West Allis | 919 | 6 | 33 | 99 | 2,156 | 1,409 | 136 | 228 | 4,008 | 1,176 | 120 |
| West Bend | 16,980 | 78 | 59 | 29 | 218 | 96 | 9 | 132 | 2,525 | 771 | 65 |
| **WYOMING** | 703,859 | 2,128 | 83 | 696 | 5,967 | 6,061 | 342 | 2,583 | 29,786 | 9,124 | 853 |
| Casper | 23,659 | 117 | 51 | 92 | 1,039 | 2,443 | 64 | 291 | 4,198 | 1,159 | 120 |
| Cheyenne | 63,347 | 280 | 67 | 95 | 626 | 400 | 35 | 324 | 5,039 | 1,587 | 141 |
| Gillette | 8,749 | 29 | 93 | 60 | 699 | 414 | 47 | 151 | 2,171 | 675 | 65 |
| Laramie | 16,878 | 114 | 56 | 16 | 147 | 134 | 7 | 117 | 1,647 | 456 | 38 |

1. Merchant wholesalers except manufacturers' sales branches and offices.　2. Establishments with payroll.

# Table D. Cities — Real Estate, Professional Services, and Manufacturing

| City | Real estate and rental and leasing, 2017 | | | | Professional, scientific, and technical services[1], 2017 | | | | Manufacturing, 2017 | | | |
|---|---|---|---|---|---|---|---|---|---|---|---|---|
| | Number of establishments | Number of employees | Receipts (mil dol) | Annual payroll (mil dol) | Number of establishments | Number of employees | Receipts (mil dol) | Annual payroll (mil dol) | Number of establishments | Number of employees | Receipts (mil dol) | Annual payroll (mil dol) |
| | 80 | 81 | 82 | 83 | 84 | 85 | 86 | 87 | 88 | 89 | 90 | 91 |
| **WASHINGTON—Cont'd** | | | | | | | | | | | | |
| Pasco | 69 | D | 98 | D | 80 | 505 | 78 | 29 | NA | NA | NA | NA |
| Pullman | 43 | 252 | 35 | 6 | 41 | 273 | 30 | 13 | NA | NA | NA | NA |
| Puyallup | 112 | 431 | 193 | 20 | 118 | 885 | 94 | 49 | NA | NA | NA | NA |
| Redmond | 171 | 1,051 | 371 | 80 | 487 | 11,054 | 2,279 | 975 | NA | NA | NA | NA |
| Renton | 157 | 548 | 223 | 28 | 246 | 1,705 | 252 | 107 | NA | NA | NA | NA |
| Richland | 102 | 298 | 73 | 11 | 203 | 6,935 | 1,635 | 661 | NA | NA | NA | NA |
| Sammamish | D | D | D | D | 255 | 701 | 171 | 65 | NA | NA | NA | NA |
| SeaTac | 58 | 823 | 311 | 37 | 31 | 279 | 48 | 20 | NA | NA | NA | NA |
| Seattle | 2,133 | 12,440 | 3,881 | 780 | 5,375 | 82,213 | 18,758 | 7,649 | NA | NA | NA | NA |
| Shoreline | 84 | 268 | 146 | 13 | 128 | 516 | 66 | 26 | NA | NA | NA | NA |
| Spokane | 342 | 1,885 | 415 | 70 | 786 | 6,065 | 1,007 | 395 | NA | NA | NA | NA |
| Spokane Valley | 147 | 892 | 213 | 35 | 200 | 1,291 | 174 | 66 | NA | NA | NA | NA |
| Tacoma | 301 | 1,573 | 404 | 72 | 529 | 3,905 | 670 | 271 | NA | NA | NA | NA |
| University Place | D | D | D | D | 73 | 243 | 34 | 15 | NA | NA | NA | NA |
| Vancouver | 347 | 1,708 | 394 | 79 | 608 | 4,989 | 898 | 350 | NA | NA | NA | NA |
| Walla Walla | 44 | 154 | 29 | 6 | 81 | 412 | 53 | 20 | NA | NA | NA | NA |
| Wenatchee | 62 | 264 | 52 | 10 | 127 | 812 | 105 | 44 | NA | NA | NA | NA |
| Yakima | 163 | 746 | 133 | 25 | 215 | 1,522 | 236 | 90 | NA | NA | NA | NA |
| **WEST VIRGINIA** | 1,432 | 6,061 | 1,369 | 229 | 2,813 | 22,856 | 3,295 | 1,218 | 1,142 | 48,533 | 24,602 | 2,748 |
| Charleston | 129 | 607 | 176 | 27 | 368 | 3,580 | 685 | 231 | NA | NA | NA | NA |
| Huntington | 72 | 358 | 62 | 12 | 138 | 1,534 | 164 | 62 | NA | NA | NA | NA |
| Morgantown | 73 | 342 | 70 | 11 | 131 | 1,879 | 383 | 121 | NA | NA | NA | NA |
| Parkersburg | 48 | 212 | 52 | 8 | 80 | 524 | 67 | 25 | NA | NA | NA | NA |
| Wheeling | 48 | D | 38 | D | 128 | 1,957 | 260 | 103 | NA | NA | NA | NA |
| **WISCONSIN** | 4,956 | 27,163 | 5,908 | 1,106 | 11,553 | 106,041 | 18,880 | 7,307 | 8,837 | 455,538 | 171,896 | 25,181 |
| Appleton | D | D | D | D | 166 | 1,617 | 240 | 104 | NA | NA | NA | NA |
| Beloit | 19 | 141 | 75 | 10 | 41 | 1,084 | 29 | 102 | NA | NA | NA | NA |
| Brookfield | 111 | 1,494 | 232 | 64 | 314 | 4,513 | 933 | 418 | NA | NA | NA | NA |
| Eau Claire | 96 | 541 | 81 | 18 | 151 | 1,328 | 189 | 75 | NA | NA | NA | NA |
| Fitchburg | 44 | 233 | 52 | 9 | 86 | 567 | 90 | 36 | NA | NA | NA | NA |
| Fond du Lac | 28 | 228 | 32 | 9 | 74 | 1,201 | 130 | 95 | NA | NA | NA | NA |
| Franklin | 29 | 130 | 27 | 4 | 61 | 416 | 72 | 19 | NA | NA | NA | NA |
| Green Bay | 87 | 637 | 132 | 27 | 212 | 2,075 | 365 | 134 | NA | NA | NA | NA |
| Greenfield | 44 | 198 | 53 | 8 | 77 | 707 | 61 | 31 | NA | NA | NA | NA |
| Janesville | 62 | 225 | 41 | 7 | 101 | 837 | 92 | 36 | NA | NA | NA | NA |
| Kenosha | 72 | 375 | 73 | 13 | 136 | 717 | 95 | 32 | NA | NA | NA | NA |
| La Crosse | 87 | 915 | 111 | 30 | 152 | 1,406 | 146 | 74 | NA | NA | NA | NA |
| Madison | 416 | 3,311 | 777 | 153 | 1,074 | 15,405 | 3,578 | 1,441 | NA | NA | NA | NA |
| Manitowoc | D | D | D | D | 53 | 400 | 62 | 20 | NA | NA | NA | NA |
| Menomonee Falls | 23 | 422 | 76 | 38 | 96 | 891 | 113 | 56 | NA | NA | NA | NA |
| Milwaukee | 471 | 3,767 | 1,075 | 194 | 1,148 | 17,420 | 3,547 | 1,477 | NA | NA | NA | NA |
| Mount Pleasant | 27 | 135 | 24 | 5 | 59 | 414 | 50 | 26 | NA | NA | NA | NA |
| Neenah | 24 | 167 | 96 | 8 | 50 | 562 | 53 | 44 | NA | NA | NA | NA |
| New Berlin | 39 | 125 | 46 | 6 | 122 | 2,110 | 400 | 84 | NA | NA | NA | NA |
| Oak Creek | 27 | 358 | 96 | 19 | 48 | 399 | 61 | 23 | NA | NA | NA | NA |
| Oshkosh | 60 | 366 | 66 | 12 | 95 | 1,604 | 705 | 79 | NA | NA | NA | NA |
| Racine | D | D | D | D | 101 | 657 | 73 | 37 | NA | NA | NA | NA |
| Sheboygan | 31 | 112 | 24 | 4 | 85 | 690 | 108 | 33 | NA | NA | NA | NA |
| Stevens Point | 27 | 122 | 19 | 4 | 61 | 451 | 71 | 25 | NA | NA | NA | NA |
| Sun Prairie | 38 | 155 | 37 | 6 | 89 | 516 | 86 | 29 | NA | NA | NA | NA |
| Superior | 31 | D | 22 | D | 58 | 365 | 30 | 15 | NA | NA | NA | NA |
| Waukesha | 58 | 539 | 139 | 16 | 170 | 1,218 | 219 | 83 | NA | NA | NA | NA |
| Wausau | 42 | 208 | 34 | 7 | 121 | 865 | 130 | 56 | NA | NA | NA | NA |
| Wauwatosa | 51 | 222 | 63 | 12 | 240 | 2,106 | 371 | 127 | NA | NA | NA | NA |
| West Allis | 50 | 255 | 76 | 12 | 86 | 863 | 157 | 63 | NA | NA | NA | NA |
| West Bend | D | D | D | D | 58 | 534 | 66 | 29 | NA | NA | NA | NA |
| **WYOMING** | 1,199 | 4,777 | 1,217 | 220 | 2,395 | 9,589 | 1,539 | 532 | 581 | 9,354 | 7,898 | 643 |
| Casper | 138 | 551 | 115 | 23 | 216 | 1,190 | 231 | 74 | NA | NA | NA | NA |
| Cheyenne | 145 | 577 | 167 | 27 | 520 | 2,272 | 338 | 125 | NA | NA | NA | NA |
| Gillette | 72 | 328 | 151 | 21 | 103 | 486 | 65 | 23 | NA | NA | NA | NA |
| Laramie | D | D | D | D | 117 | 656 | 110 | 39 | NA | NA | NA | NA |

1. Establishments subject to federal tax.

# Table D. Cities — Accommodation and Food Services, Arts, Entertainment, and Recreation, and Health Care and Social Assistance

| City | Accommodation and food services, 2017 | | | | Arts, entertainment, and recreation[1], 2017 | | | | Health care and social assistance[1], 2017 | | | |
|---|---|---|---|---|---|---|---|---|---|---|---|---|
| | Number of establishments | Number of employees | Receipts (mil dol) | Annual payroll (mil dol) | Number of establishments | Number of employees | Receipts (mil dol) | Annual payroll (mil dol) | Number of establishments | Number of employees | Receipts (mil dol) | Annual payroll (mil dol) |
| | 92 | 93 | 94 | 95 | 96 | 97 | 98 | 99 | 100 | 101 | 102 | 103 |
| **WASHINGTON—Cont'd** | | | | | | | | | | | | |
| Pasco | 105 | 1,625 | 96 | 27 | D | D | D | D | 126 | 1,987 | 190 | 90 |
| Pullman | 98 | 1,651 | 75 | 23 | D | D | D | D | 72 | 1,467 | 147 | 67 |
| Puyallup | 149 | 3,139 | 192 | 60 | 20 | 640 | 52 | 13 | 213 | 5,725 | 1,021 | 378 |
| Redmond | 295 | 5,942 | 509 | 169 | 37 | 692 | 38 | 14 | 250 | 3,359 | 434 | 164 |
| Renton | 282 | 4,099 | 328 | 92 | 24 | 904 | 487 | 276 | 362 | 8,124 | 1,234 | 560 |
| Richland | 156 | 3,019 | 183 | 56 | 32 | 652 | 24 | 9 | 242 | 5,653 | 819 | 326 |
| Sammamish | 40 | 658 | 45 | 14 | 23 | 463 | 21 | 9 | 99 | 700 | 64 | 27 |
| SeaTac | 113 | 3,690 | 453 | 125 | 3 | 599 | 24 | 8 | 42 | 1,090 | 109 | 31 |
| Seattle | 3,159 | 58,136 | 4,918 | 1,565 | 591 | 12,795 | 1,635 | 605 | 2,912 | 85,733 | 13,242 | 5,229 |
| Shoreline | 102 | 1,139 | 78 | 24 | 31 | 823 | 70 | 23 | 217 | 2,973 | 269 | 114 |
| Spokane | 642 | 12,036 | 703 | 234 | 95 | 1,696 | 82 | 30 | 956 | 29,351 | 3,806 | 1,483 |
| Spokane Valley | 226 | 3,918 | 269 | 76 | 32 | 463 | 26 | 9 | 374 | 7,637 | 779 | 362 |
| Tacoma | 541 | 9,576 | 600 | 194 | 77 | 2,714 | 323 | 84 | 673 | 27,464 | 4,362 | 1,778 |
| University Place | 40 | 697 | 36 | 12 | D | D | D | D | 100 | 1,095 | 85 | 35 |
| Vancouver | 494 | 8,677 | 557 | 168 | 62 | 1,091 | 62 | 18 | 693 | 15,603 | 1,794 | 754 |
| Walla Walla | 112 | 1,801 | 106 | 33 | 16 | 319 | 15 | 6 | 127 | 4,493 | 549 | 238 |
| Wenatchee | 129 | 1,974 | 120 | 38 | 19 | 245 | 10 | 4 | 154 | 5,506 | 855 | 373 |
| Yakima | 244 | 4,118 | 252 | 78 | 40 | 816 | 43 | 18 | 374 | 10,187 | 1,356 | 525 |
| **WEST VIRGINIA** | 3,614 | 68,102 | 4,069 | 1,076 | 775 | 6,681 | 488 | 108 | 4,869 | 132,627 | 15,237 | 5,844 |
| Charleston | 222 | 4,558 | 262 | 75 | 43 | 803 | 53 | 14 | 369 | 12,200 | 2,000 | 744 |
| Huntington | 189 | 3,616 | 166 | 51 | 28 | 561 | 19 | 7 | 135 | 10,407 | 1,409 | 528 |
| Morgantown | 225 | 4,566 | 213 | 60 | 31 | 279 | 15 | 4 | 176 | 5,377 | 533 | 199 |
| Parkersburg | 120 | 2,298 | 103 | 33 | 27 | 348 | 29 | 4 | NA | NA | NA | NA |
| Wheeling | 97 | 2,162 | 199 | 38 | 31 | 358 | 27 | 7 | NA | NA | NA | NA |
| **WISCONSIN** | 14,840 | 244,099 | 13,496 | 3,688 | 2,809 | 47,048 | 3,494 | 1,218 | 15,983 | 417,365 | 48,478 | 19,198 |
| Appleton | 217 | 4,163 | 206 | 59 | 31 | 1,106 | 38 | 14 | 296 | 9,061 | 1,093 | 477 |
| Beloit | 95 | 1,738 | 90 | 25 | D | D | D | D | 63 | 2,108 | 391 | 132 |
| Brookfield | 130 | 3,419 | 206 | 58 | 33 | 709 | 36 | 13 | 388 | 5,085 | 651 | 302 |
| Eau Claire | 229 | 4,861 | 214 | 65 | 41 | 1,034 | 51 | 10 | 269 | 8,306 | 981 | 356 |
| Fitchburg | 45 | 984 | 53 | 15 | 19 | 403 | 14 | 5 | 59 | 2,084 | 190 | 78 |
| Fond du Lac | 116 | 2,144 | 104 | 29 | 19 | 473 | 32 | 6 | 199 | 5,515 | 754 | 265 |
| Franklin | 64 | 1,110 | 54 | 17 | 19 | 196 | 11 | 4 | 115 | 2,156 | 271 | 105 |
| Green Bay | 255 | 5,340 | 267 | 76 | 44 | D | 499 | D | 331 | 16,466 | 2,489 | 838 |
| Greenfield | 70 | 1,987 | 95 | 29 | 14 | 418 | 13 | 4 | 154 | 2,823 | 280 | 115 |
| Janesville | 165 | 3,467 | 168 | 46 | 27 | 436 | 20 | 7 | 186 | 5,385 | 710 | 301 |
| Kenosha | 229 | 4,304 | 205 | 59 | 23 | 495 | 30 | 7 | 345 | 7,213 | 738 | 313 |
| La Crosse | 227 | 4,549 | 213 | 63 | 45 | 912 | 36 | 11 | 186 | 9,192 | 1,145 | 497 |
| Madison | 808 | 17,393 | 960 | 293 | 156 | 2,474 | 159 | 50 | 741 | 39,689 | 5,639 | 2,306 |
| Manitowoc | 99 | 1,559 | 71 | 20 | 15 | 341 | 16 | 5 | 124 | 3,288 | 265 | 123 |
| Menomonee Falls | 72 | 1,570 | 79 | 24 | 22 | 486 | 18 | 6 | 82 | 4,159 | 458 | 189 |
| Milwaukee | 1,229 | 26,656 | 1,947 | 484 | 164 | 6,049 | 830 | 355 | 1,744 | 61,852 | 7,646 | 2,756 |
| Mount Pleasant | 64 | 1,435 | 77 | 20 | 6 | 155 | 7 | 2 | 124 | 2,099 | 254 | 99 |
| Neenah | 87 | 1,283 | 62 | 19 | 17 | 479 | 10 | 5 | 93 | 1,554 | 164 | 72 |
| New Berlin | 63 | 1,410 | 65 | 21 | 21 | 593 | 24 | 7 | 105 | 2,009 | 177 | 87 |
| Oak Creek | 73 | 1,672 | 85 | 24 | 12 | 80 | 3 | 1 | 71 | 714 | 68 | 29 |
| Oshkosh | 200 | 4,194 | 176 | 52 | 34 | 606 | 20 | 8 | 201 | 5,961 | 630 | 270 |
| Racine | D | D | D | D | 35 | 294 | 32 | 8 | 207 | 5,788 | 474 | 194 |
| Sheboygan | 140 | 2,347 | 116 | 35 | 22 | 491 | 20 | 7 | 196 | 4,863 | 612 | 238 |
| Stevens Point | 121 | 1,912 | 83 | 25 | 20 | 611 | 18 | 7 | 133 | 3,890 | 398 | 157 |
| Sun Prairie | 73 | 1,429 | 66 | 20 | D | D | D | D | 75 | 1,486 | 127 | 54 |
| Superior | 100 | 1,528 | 62 | 19 | D | D | D | D | 94 | 1,776 | 128 | 54 |
| Waukesha | 163 | 2,998 | 153 | 45 | 37 | 807 | 23 | 8 | 275 | 6,515 | 934 | 300 |
| Wausau | 117 | 2,122 | 101 | 30 | 27 | 476 | 19 | 6 | 232 | 7,274 | 1,108 | 402 |
| Wauwatosa | 155 | 3,902 | 200 | 63 | 26 | 674 | 33 | 10 | 384 | 12,197 | 1,527 | 570 |
| West Allis | 163 | 2,383 | 122 | 33 | 11 | 345 | 10 | 3 | 233 | 8,225 | 892 | 381 |
| West Bend | 72 | 1,479 | 66 | 18 | 18 | 459 | 14 | 5 | 120 | 1,658 | 129 | 52 |
| **WYOMING** | 1,839 | 27,248 | 1,946 | 556 | 445 | 4,791 | 384 | 107 | 2,001 | 33,540 | 3,872 | 1,657 |
| Casper | 158 | 3,086 | 171 | 52 | 35 | 327 | 20 | 6 | 317 | 5,458 | 732 | 307 |
| Cheyenne | 181 | 3,277 | 201 | 57 | 31 | 448 | 32 | 7 | 322 | 6,534 | 853 | 371 |
| Gillette | 99 | 1,713 | 89 | 28 | 16 | 148 | 6 | 2 | 107 | 2,083 | 275 | 120 |
| Laramie | 95 | 1,617 | 76 | 22 | D | D | D | D | D | D | D | D |

1. Establishments subject to federal tax.

# Table D. Cities — Other Services and Government Employment and Payroll

| City | Other services[1] Number of establishments | Number of employees | Receipts (mil dol) | Annual payroll (mil dol) | Government employment and payroll, 2017 — March payroll Full-time equivalent employees | Total (dollars) | Percent of total for: Administrative, judicial, and legal | Police and corrections | Fire protection | Highways and transportation | Health and welfare | Natural resources and utilities | Education and libraries |
|---|---|---|---|---|---|---|---|---|---|---|---|---|---|
| | 104 | 105 | 106 | 107 | 108 | 109 | 110 | 111 | 112 | 113 | 114 | 115 | 116 |
| **WASHINGTON—Cont'd** | | | | | | | | | | | | | |
| Pasco | D | D | D | D | 296 | 1,995,450 | 11.3 | 31.5 | 27.9 | 6.9 | 0.0 | 20.5 | 0.0 |
| Pullman | 40 | 179 | 16 | 5 | 260 | 1,300,701 | 12.6 | 21.3 | 22.6 | 20.5 | 0.0 | 14.3 | 5.0 |
| Puyallup | 118 | 918 | 102 | 33 | 281 | 1,790,507 | 24.2 | 35.9 | 0.0 | 10.0 | 0.0 | 24.2 | 4.4 |
| Redmond | 162 | 1,043 | 113 | 41 | 663 | 4,945,324 | 26.5 | 18.0 | 28.4 | 2.9 | 0.0 | 16.7 | 0.0 |
| Renton | 189 | 1,039 | 103 | 35 | 925 | 6,233,220 | 9.2 | 38.4 | 20.7 | 6.2 | 0.6 | 14.9 | 0.0 |
| Richland | 89 | 553 | 61 | 19 | 479 | 3,366,979 | 11.0 | 18.5 | 16.9 | 1.6 | 1.3 | 33.2 | 1.7 |
| Sammamish | 37 | 182 | 15 | 5 | 102 | 675,663 | 21.7 | 0.4 | 0.0 | 22.6 | 20.2 | 22.4 | 0.0 |
| SeaTac | 73 | 836 | 111 | 32 | 115 | 755,800 | 40.7 | 0.8 | 0.0 | 24.5 | 3.0 | 16.1 | 0.0 |
| Seattle | 2,080 | 15,523 | 7,433 | 859 | 11,912 | 94,722,910 | 16.9 | 18.6 | 11.2 | 6.8 | 3.5 | 35.8 | 3.4 |
| Shoreline | 89 | 448 | 43 | 14 | 175 | 943,320 | 56.5 | 0.0 | 0.0 | 15.1 | 0.0 | 24.4 | 0.0 |
| Spokane | 458 | 2,736 | 306 | 86 | 2,145 | 13,755,462 | 13.5 | 23.9 | 20.9 | 6.7 | 0.9 | 26.0 | 2.9 |
| Spokane Valley | 207 | 1,477 | 164 | 53 | 90 | 563,427 | 64.4 | 0.0 | 0.0 | 21.1 | 0.0 | 14.5 | 0.0 |
| Tacoma | 453 | 3,393 | 438 | 124 | 3,566 | 27,168,562 | 11.2 | 12.0 | 13.6 | 9.5 | 2.5 | 44.6 | 1.9 |
| University Place | 60 | 262 | 25 | 8 | 48 | 380,073 | 66.6 | 0.0 | 0.0 | 29.5 | 0.0 | 3.9 | 0.0 |
| Vancouver | 404 | 2,443 | 367 | 81 | 1,065 | 6,647,181 | 12.7 | 25.6 | 25.6 | 6.2 | 0.0 | 20.3 | 0.0 |
| Walla Walla | 71 | 324 | 35 | 10 | 302 | 1,791,859 | 16.0 | 27.3 | 22.3 | 9.5 | 0.8 | 18.6 | 3.3 |
| Wenatchee | 109 | 452 | 56 | 15 | 155 | 925,754 | 11.9 | 32.3 | 0.0 | 6.7 | 5.1 | 14.0 | 0.0 |
| Yakima | 177 | 1,101 | 109 | 30 | 689 | 4,349,766 | 14.2 | 31.5 | 20.8 | 9.0 | 1.0 | 17.1 | 0.0 |
| **WEST VIRGINIA** | 2,524 | 14,894 | 1,897 | 465 | X | X | X | X | X | X | X | X | X |
| Charleston | 205 | 1,383 | 187 | 56 | 817 | 3,163,140 | 14.0 | 28.2 | 22.0 | 9.1 | 1.3 | 12.0 | 0.0 |
| Huntington | 86 | 503 | 70 | 16 | 322 | 1,195,342 | 11.3 | 39.6 | 27.3 | 4.2 | 2.0 | 7.7 | 0.0 |
| Morgantown | 81 | 652 | 184 | 22 | 440 | 1,827,342 | 5.9 | 24.3 | 13.4 | 11.1 | 0.0 | 35.1 | 3.6 |
| Parkersburg | 97 | 604 | 63 | 17 | 364 | 1,227,450 | 11.3 | 25.6 | 18.5 | 8.8 | 1.5 | 32.0 | 0.0 |
| Wheeling | 98 | 893 | 94 | 29 | 746 | 2,380,152 | 3.3 | 13.9 | 15.6 | 7.1 | 2.0 | 56.2 | 0.0 |
| **WISCONSIN** | 10,351 | 63,120 | 9,241 | 2,069 | X | X | X | X | X | X | X | X | X |
| Appleton | 139 | 1,292 | 141 | 49 | 648 | 3,496,019 | 10.3 | 25.2 | 17.1 | 7.8 | 4.2 | 15.7 | 6.3 |
| Beloit | 50 | 254 | 19 | 6 | 364 | 1,992,675 | 10.0 | 27.6 | 20.5 | 14.2 | 3.9 | 15.1 | 5.5 |
| Brookfield | 136 | 1,163 | 121 | 38 | 335 | 2,052,842 | 11.7 | 27.8 | 22.6 | 7.6 | 1.4 | 13.1 | 5.8 |
| Eau Claire | 141 | 1,045 | 87 | 25 | 605 | 3,089,358 | 9.0 | 25.3 | 17.6 | 16.3 | 10.8 | 12.9 | 5.0 |
| Fitchburg | 41 | 471 | 46 | 17 | 190 | 974,003 | 18.2 | 37.1 | 14.6 | 5.1 | 3.0 | 7.7 | 7.2 |
| Fond du Lac | 92 | 634 | 59 | 17 | 335 | 1,792,502 | 8.0 | 28.0 | 22.7 | 13.6 | 1.5 | 14.1 | 7.1 |
| Franklin | 52 | 568 | 47 | 17 | 223 | 1,298,124 | 8.0 | 34.4 | 26.4 | 11.5 | 3.2 | 2.1 | 4.6 |
| Green Bay | 164 | 1,003 | 116 | 30 | 887 | 4,787,549 | 6.9 | 33.8 | 24.7 | 10.4 | 1.1 | 19.3 | 0.0 |
| Greenfield | 76 | 451 | 36 | 13 | 243 | 1,425,578 | 10.9 | 36.4 | 25.4 | 13.7 | 2.4 | 5.7 | 4.0 |
| Janesville | 121 | 685 | 57 | 16 | 564 | 2,898,547 | 7.2 | 25.2 | 21.3 | 14.1 | 4.1 | 12.4 | 12.7 |
| Kenosha | 153 | 1,037 | 78 | 25 | 830 | 4,343,227 | 5.8 | 28.2 | 22.2 | 10.6 | 2.7 | 18.1 | 7.6 |
| La Crosse | 121 | 989 | 127 | 36 | 613 | 2,798,236 | 11.4 | 21.5 | 19.2 | 15.4 | 0.0 | 13.8 | 8.5 |
| Madison | 645 | 5,782 | 1,189 | 262 | 2,789 | 13,351,958 | 8.7 | 28.1 | 15.4 | 21.6 | 1.9 | 11.0 | 4.1 |
| Manitowoc | 72 | 310 | 34 | 7 | 353 | 1,846,790 | 5.5 | 22.1 | 17.5 | 11.2 | 0.0 | 37.6 | 5.2 |
| Menomonee Falls | 71 | 672 | 48 | 16 | 91 | 437,223 | 16.8 | 35.8 | 15.3 | 11.3 | 3.3 | 9.5 | 6.8 |
| Milwaukee | 846 | 6,350 | 2,494 | 245 | 6,507 | 37,635,419 | 7.8 | 42.5 | 16.2 | 7.4 | 4.0 | 14.5 | 2.9 |
| Mount Pleasant | D | D | D | D | 146 | 801,664 | 5.6 | 32.8 | 45.0 | 8.2 | 0.5 | 5.4 | 0.0 |
| Neenah | 55 | 370 | 34 | 13 | 261 | 1,435,666 | 7.9 | 23.3 | 28.1 | 15.5 | 3.3 | 15.6 | 6.2 |
| New Berlin | 76 | 834 | 90 | 34 | 252 | 1,462,929 | 11.6 | 38.4 | 18.7 | 10.1 | 0.0 | 11.6 | 4.6 |
| Oak Creek | 50 | 303 | 25 | 8 | 286 | 1,625,666 | 11.4 | 31.8 | 22.3 | 8.7 | 2.1 | 14.7 | 2.8 |
| Oshkosh | 100 | 870 | 105 | 30 | 585 | 4,170,961 | 7.4 | 25.0 | 24.4 | 19.4 | 2.1 | 12.2 | 4.7 |
| Racine | 119 | 786 | 57 | 18 | 782 | 4,263,192 | 6.4 | 32.1 | 20.3 | 9.9 | 5.1 | 20.5 | 3.6 |
| Sheboygan | 100 | 523 | 39 | 12 | 432 | 2,262,569 | 7.3 | 27.6 | 19.5 | 15.5 | 2.1 | 18.2 | 5.8 |
| Stevens Point | 63 | 384 | 52 | 13 | 233 | 1,262,902 | 6.8 | 32.3 | 20.1 | 19.0 | 1.3 | 19.7 | 0.0 |
| Sun Prairie | 64 | 391 | 31 | 10 | 237 | 1,186,258 | 15.9 | 29.6 | 0.0 | 7.8 | 7.8 | 24.1 | 7.1 |
| Superior | D | D | D | D | 265 | 1,368,991 | 6.5 | 28.8 | 17.1 | 8.5 | 2.4 | 21.6 | 5.1 |
| Waukesha | 138 | 957 | 105 | 30 | 575 | 3,333,088 | 10.7 | 28.3 | 19.7 | 16.2 | 0.3 | 16.9 | 5.6 |
| Wausau | 75 | 425 | 54 | 13 | 314 | 1,558,210 | 11.7 | 30.0 | 21.8 | 22.6 | 3.6 | 7.8 | 0.0 |
| Wauwatosa | 104 | 808 | 148 | 30 | 410 | 2,437,615 | 9.3 | 30.2 | 28.9 | 11.1 | 3.8 | 9.4 | 4.0 |
| West Allis | 119 | 828 | 131 | 32 | 542 | 3,010,688 | 9.3 | 29.5 | 21.3 | 9.5 | 8.7 | 10.0 | 3.1 |
| West Bend | 84 | 582 | 42 | 14 | 234 | 1,051,988 | 12.4 | 34.5 | 19.9 | 6.5 | 0.9 | 15.2 | 5.4 |
| **WYOMING** | 1,368 | 6,245 | 931 | 221 | X | X | X | X | X | X | X | X | X |
| Casper | 160 | 826 | 117 | 29 | 496 | 2,420,062 | 12.8 | 27.5 | 16.6 | 7.0 | 1.7 | 28.3 | 0.0 |
| Cheyenne | 176 | 1,042 | 113 | 33 | 638 | 2,712,553 | 11.3 | 26.7 | 19.8 | 12.6 | 2.9 | 20.6 | 0.0 |
| Gillette | 86 | 475 | 78 | 23 | 303 | 1,429,261 | 20.3 | 28.0 | 0.0 | 4.0 | 0.0 | 28.0 | 0.0 |
| Laramie | 70 | 326 | 91 | 10 | 283 | 1,341,348 | 14.9 | 27.3 | 20.5 | 7.5 | 0.0 | 26.7 | 0.0 |

1. Establishments subject to federal tax.

# Table D. Cities — City Government Finances

| City | City government finances, 2017 | | | | | | | | | |
|---|---|---|---|---|---|---|---|---|---|---|
| | General revenue | | | | | | | General expenditure | | |
| | | Intergovernmental | | Taxes | | | | | Per capita[1] (dollars) | |
| | | | Percent from state government | | | Per capita[1] (dollars) | | | | |
| | Total (mil dol) | Total (mil dol) | | Total (mil dol) | Total | Property | Sales and gross receipts | Total (mil dol) | Total | Capital outlays |
| | 117 | 118 | 119 | 120 | 121 | 122 | 123 | 124 | 125 | 126 |
| **WASHINGTON—Cont'd** | | | | | | | | | | |
| Pasco | 83.7 | 5.0 | 79.2 | 37.8 | 519 | 109 | 387 | 74 | 1,013 | 126 |
| Pullman | 53.8 | 19.2 | 15.8 | 19.8 | 590 | 220 | 359 | 32 | 958 | 156 |
| Puyallup | 72.8 | 5.7 | 96.6 | 41.6 | 1,015 | 218 | 746 | 62 | 1,508 | 337 |
| Redmond | 185.4 | 25.0 | 35.3 | 98.5 | 1,518 | 369 | 1,022 | 163 | 2,512 | 562 |
| Renton | 253.4 | 57.8 | 30.8 | 111.4 | 1,092 | 364 | 670 | 199 | 1,948 | 240 |
| Richland | 97.0 | 11.9 | 53.5 | 51.3 | 906 | 298 | 562 | 100 | 1,766 | 163 |
| Sammamish | 88.7 | 3.5 | 66.2 | 49.0 | 755 | 422 | 232 | 47 | 722 | 128 |
| SeaTac | 61.9 | 13.1 | 53.4 | 41.7 | 1,425 | 500 | 882 | 55 | 1,882 | 504 |
| Seattle | 2,791.7 | 256.3 | 67.1 | 1,432.8 | 1,966 | 746 | 1,085 | 2,586 | 3,549 | 749 |
| Shoreline | 54.2 | 9.7 | 62.3 | 35.9 | 635 | 221 | 354 | 49 | 860 | 140 |
| Spokane | 411.8 | 42.1 | 62.2 | 171.7 | 790 | 318 | 446 | 404 | 1,860 | 278 |
| Spokane Valley | 59.5 | 12.0 | 93.4 | 42.9 | 439 | 117 | 298 | 61 | 627 | 191 |
| Tacoma | 584.1 | 62.6 | 36.7 | 224.1 | 1,050 | 349 | 656 | 560 | 2,625 | 341 |
| University Place | 28.5 | 6.8 | 100.0 | 16.0 | 479 | 121 | 321 | 30 | 895 | 256 |
| Vancouver | 284.4 | 29.9 | 49.8 | 143.3 | 794 | 249 | 506 | 239 | 1,326 | 217 |
| Walla Walla | 66.9 | 6.5 | 66.8 | 26.3 | 800 | 217 | 533 | 69 | 2,087 | 401 |
| Wenatchee | 48.3 | 5.8 | 72.5 | 25.7 | 750 | 98 | 620 | 36 | 1,060 | 127 |
| Yakima | 142.6 | 25.8 | 29.2 | 69.0 | 738 | 194 | 525 | 115 | 1,225 | 122 |
| **WEST VIRGINIA** | X | X | X | X | X | X | X | X | X | X |
| Charleston | 151.7 | 9.0 | 41.9 | 84.4 | 1,758 | 299 | 1,459 | 169 | 3,516 | 1,075 |
| Huntington | 70.6 | 6.4 | 38.9 | 30.7 | 657 | 126 | 532 | 73 | 1,562 | 18 |
| Morgantown | 65.2 | 1.6 | 40.9 | 27.9 | 902 | 187 | 715 | 63 | 2,048 | 578 |
| Parkersburg | 40.8 | 1.6 | 12.9 | 15.7 | 521 | 195 | 320 | 40 | 1,314 | 114 |
| Wheeling | 79.8 | 3.6 | 58.0 | 27.5 | 1,025 | 240 | 777 | 75 | 2,809 | 135 |
| **WISCONSIN** | X | X | X | X | X | X | X | X | X | X |
| Appleton | 109.9 | 32.9 | 58.2 | 46.1 | 619 | 554 | 59 | 112 | 1,498 | 325 |
| Beloit | 63.9 | 24.7 | 88.6 | 23.4 | 636 | 590 | 45 | 61 | 1,654 | 228 |
| Brookfield | 62.9 | 7.1 | 60.6 | 42.1 | 1,105 | 984 | 120 | 68 | 1,780 | 121 |
| Eau Claire | 93.8 | 23.9 | 66.4 | 44.5 | 648 | 580 | 63 | 95 | 1,378 | 226 |
| Fitchburg | 39.1 | 5.0 | 84.8 | 27.1 | 918 | 878 | 40 | 38 | 1,281 | 387 |
| Fond du Lac | 61.0 | 15.6 | 68.2 | 27.3 | 640 | 576 | 60 | 58 | 1,358 | 165 |
| Franklin | 38.4 | 4.5 | 76.7 | 25.4 | 702 | 642 | 60 | 38 | 1,049 | 122 |
| Green Bay | 147.7 | 42.2 | 72.8 | 59.3 | 565 | 535 | 28 | 144 | 1,373 | 160 |
| Greenfield | 41.5 | 5.6 | 81.8 | 24.5 | 666 | 620 | 45 | 60 | 1,641 | 626 |
| Janesville | 90.4 | 18.7 | 60.4 | 41.0 | 637 | 576 | 60 | 110 | 1,702 | 522 |
| Kenosha | 146.4 | 31.1 | 73.3 | 81.9 | 821 | 777 | 41 | 130 | 1,302 | 158 |
| La Crosse | 101.1 | 25.7 | 78.4 | 47.6 | 920 | 820 | 81 | 95 | 1,833 | 332 |
| Madison | 481.5 | 105.0 | 66.8 | 246.0 | 962 | 874 | 83 | 423 | 1,654 | 173 |
| Manitowoc | 52.5 | 15.2 | 62.8 | 18.8 | 577 | 532 | 43 | 43 | 1,324 | 171 |
| Menomonee Falls | 51.2 | 4.7 | 93.6 | 31.0 | 828 | 708 | 121 | 58 | 1,547 | 173 |
| Milwaukee | 940.5 | 378.3 | 76.2 | 286.3 | 482 | 454 | 28 | 1,117 | 1,881 | 235 |
| Mount Pleasant | 38.4 | 5.5 | 57.7 | 20.1 | 745 | 694 | 50 | 40 | 1,465 | 219 |
| Neenah | 37.5 | 6.7 | 71.5 | 20.9 | 806 | 770 | 36 | 38 | 1,473 | 193 |
| New Berlin | 48.5 | 4.6 | 85.5 | 26.0 | 654 | 611 | 42 | 59 | 1,474 | 204 |
| Oak Creek | 47.8 | 10.2 | 93.6 | 25.5 | 704 | 593 | 106 | 59 | 1,637 | 456 |
| Oshkosh | 98.4 | 24.3 | 79.3 | 42.5 | 637 | 592 | 42 | 105 | 1,578 | 227 |
| Racine | 141.7 | 50.4 | 80.1 | 55.9 | 725 | 693 | 30 | 145 | 1,882 | 202 |
| Sheboygan | 61.7 | 19.8 | 84.6 | 27.6 | 570 | 491 | 69 | 63 | 1,302 | 129 |
| Stevens Point | 43.4 | 9.9 | 69.1 | 17.1 | 652 | 589 | 57 | 41 | 1,555 | 329 |
| Sun Prairie | 41.0 | 5.3 | 86.9 | 24.2 | 733 | 689 | 40 | 36 | 1,103 | 154 |
| Superior | 56.6 | 22.0 | 89.6 | 15.3 | 582 | 499 | 77 | 48 | 1,821 | 230 |
| Waukesha | 113.0 | 24.2 | 80.1 | 63.7 | 879 | 820 | 50 | 104 | 1,438 | 307 |
| Wausau | 56.4 | 13.8 | 85.6 | 30.0 | 776 | 719 | 47 | 65 | 1,684 | 572 |
| Wauwatosa | 76.9 | 10.3 | 65.6 | 45.7 | 952 | 876 | 68 | 93 | 1,927 | 439 |
| West Allis | 89.6 | 23.4 | 64.6 | 43.8 | 732 | 685 | 32 | 84 | 1,408 | 88 |
| West Bend | 37.9 | 6.8 | 77.3 | 23.6 | 747 | 688 | 52 | 32 | 1,029 | 110 |
| **WYOMING** | X | X | X | X | X | X | X | X | X | X |
| Casper | 109.5 | 52.4 | 65.1 | 11.7 | 202 | 98 | 90 | 121 | 2,092 | 543 |
| Cheyenne | 116.7 | 64.2 | 51.9 | 15.5 | 244 | 102 | 136 | 110 | 1,739 | 488 |
| Gillette | 101.0 | 65.0 | 73.3 | 4.1 | 127 | 99 | 28 | 68 | 2,098 | 414 |
| Laramie | 52.4 | 30.6 | 62.6 | 5.3 | 163 | 72 | 92 | 51 | 1,583 | 544 |

1. Based on population estimated as of July 1 of the year shown.

# Table D. Cities — City Government Finances

| City | City government finances, 2017 (cont.) | | | | | | | | | |
|------|---|---|---|---|---|---|---|---|---|---|
| | General expenditure (cont.) | | | | | | | | | |
| | Percent of total for: | | | | | | | | | |
| | Public welfare | Highways | Parking facilities | Education | Health and hospitals | Police protection | Sewerage and sanitation | Parks and recreation | Housing and community development | Interest on debt |
| | 127 | 128 | 129 | 130 | 131 | 132 | 133 | 134 | 135 | 136 |
| **WASHINGTON—Cont'd** | | | | | | | | | | |
| Pasco | 0.0 | 7.0 | 0.0 | 0.0 | 8.0 | 19.2 | 11.7 | 9.1 | 1.6 | 1.8 |
| Pullman | 0.0 | 7.2 | 0.0 | 0.0 | 6.6 | 18.7 | 18.2 | 9.9 | 0.1 | 0.7 |
| Puyallup | 0.0 | 12.7 | 0.0 | 0.0 | 0.4 | 26.3 | 23.2 | 5.3 | 3.1 | 4.5 |
| Redmond | 0.0 | 14.0 | 0.0 | 0.0 | 4.2 | 12.8 | 18.9 | 7.6 | 1.2 | 1.5 |
| Renton | 0.6 | 11.2 | 0.0 | 0.0 | 0.4 | 12.9 | 26.1 | 6.5 | 1.0 | 3.5 |
| Richland | 0.0 | 9.2 | 0.0 | 0.0 | 3.5 | 11.4 | 14.8 | 5.3 | 4.8 | 1.8 |
| Sammamish | 0.0 | 20.6 | 0.0 | 0.0 | 0.5 | 14.1 | 7.6 | 17.7 | 1.9 | 0.3 |
| SeaTac | 1.3 | 30.4 | 0.0 | 0.0 | 0.7 | 17.6 | 2.5 | 3.3 | 1.6 | 0.4 |
| Seattle | 4.0 | 17.0 | 0.5 | 1.8 | 0.5 | 8.3 | 20.9 | 9.8 | 2.8 | 2.4 |
| Shoreline | 0.0 | 15.8 | 0.0 | 0.0 | 0.7 | 27.0 | 5.6 | 8.8 | 0.0 | 3.5 |
| Spokane | 2.3 | 11.2 | 0.0 | 0.0 | 0.3 | 14.1 | 32.5 | 7.1 | 1.0 | 2.0 |
| Spokane Valley | 0.1 | 27.3 | 0.0 | 0.0 | 0.7 | 30.8 | 2.9 | 4.7 | 3.3 | 0.7 |
| Tacoma | 0.0 | 9.7 | 0.7 | 0.0 | 0.8 | 15.3 | 22.2 | 3.0 | 3.0 | 4.8 |
| University Place | 0.0 | 58.3 | 0.1 | 0.0 | 0.4 | 12.0 | 0.0 | 3.4 | 3.3 | 7.0 |
| Vancouver | 0.0 | 13.8 | 0.7 | 0.0 | 0.4 | 16.9 | 14.0 | 7.1 | 3.0 | 2.8 |
| Walla Walla | 0.0 | 10.5 | 0.0 | 0.0 | 8.3 | 12.5 | 23.0 | 12.6 | 0.0 | 3.0 |
| Wenatchee | 3.2 | 14.7 | 0.0 | 0.0 | 0.6 | 17.7 | 14.7 | 11.7 | 2.4 | 2.7 |
| Yakima | 0.0 | 9.6 | 0.1 | 0.0 | 0.3 | 20.9 | 20.1 | 7.1 | 1.1 | 1.3 |
| **WEST VIRGINIA** | X | X | X | X | X | X | X | X | X | X |
| Charleston | 0.5 | 4.8 | 1.3 | 0.0 | 0.1 | 15.8 | 13.3 | 27.6 | 2.3 | 2.3 |
| Huntington | 0.6 | 4.1 | 1.2 | 0.0 | 0.1 | 17.9 | 26.3 | 5.0 | 3.1 | 0.1 |
| Morgantown | 0.3 | 9.4 | 3.4 | 0.0 | 0.0 | 13.5 | 35.5 | 7.0 | 0.6 | 2.1 |
| Parkersburg | 0.0 | 16.9 | 0.4 | 0.0 | 0.0 | 17.2 | 27.3 | 3.7 | 1.4 | 0.3 |
| Wheeling | 0.1 | 8.8 | 0.8 | 0.0 | 0.0 | 11.2 | 8.4 | 45.3 | 0.7 | 1.5 |
| **WISCONSIN** | X | X | X | X | X | X | X | X | X | X |
| Appleton | 0.6 | 23.6 | 1.6 | 0.0 | 1.0 | 15.9 | 11.9 | 10.3 | 0.4 | 6.6 |
| Beloit | 0.0 | 15.1 | 0.0 | 0.0 | 2.4 | 20.0 | 19.2 | 5.6 | 1.2 | 5.7 |
| Brookfield | 0.0 | 11.6 | 0.0 | 0.0 | 7.4 | 15.6 | 24.3 | 5.1 | 0.0 | 3.2 |
| Eau Claire | 0.2 | 18.1 | 4.4 | 0.0 | 7.8 | 18.6 | 9.0 | 8.6 | 1.6 | 4.2 |
| Fitchburg | 0.1 | 15.7 | 0.0 | 0.0 | 1.3 | 21.2 | 10.1 | 7.1 | 0.0 | 3.5 |
| Fond du Lac | 0.0 | 16.8 | 0.7 | 0.0 | 6.7 | 17.7 | 19.1 | 4.6 | 1.0 | 8.6 |
| Franklin | 0.0 | 15.6 | 0.0 | 0.0 | 5.9 | 25.0 | 18.9 | 1.9 | 0.0 | 2.4 |
| Green Bay | 0.0 | 13.2 | 2.3 | 0.0 | 0.2 | 18.6 | 19.3 | 8.1 | 0.6 | 5.3 |
| Greenfield | 0.0 | 20.6 | 0.4 | 0.0 | 3.8 | 17.2 | 9.0 | 3.7 | 0.0 | 3.1 |
| Janesville | 0.6 | 16.2 | 0.1 | 0.0 | 3.1 | 12.8 | 18.5 | 4.6 | 3.2 | 2.3 |
| Kenosha | 0.0 | 13.5 | 0.0 | 0.0 | 8.9 | 22.3 | 11.8 | 6.5 | 8.1 | 5.8 |
| La Crosse | 0.0 | 14.6 | 7.6 | 0.0 | 0.4 | 15.5 | 9.1 | 11.5 | 4.9 | 2.2 |
| Madison | 0.0 | 13.6 | 2.2 | 0.0 | 3.2 | 17.3 | 10.8 | 9.9 | 10.2 | 5.5 |
| Manitowoc | 0.0 | 18.0 | 0.0 | 0.0 | 0.0 | 18.1 | 11.9 | 5.1 | 0.0 | 5.5 |
| Menomonee Falls | 0.0 | 15.9 | 0.0 | 0.0 | 5.0 | 16.2 | 31.7 | 2.3 | 0.0 | 5.0 |
| Milwaukee | 0.0 | 13.6 | 2.2 | 0.0 | 2.1 | 28.3 | 8.0 | 0.3 | 2.9 | 4.3 |
| Mount Pleasant | 0.0 | 18.5 | 0.0 | 0.0 | 0.6 | 19.4 | 30.5 | 0.7 | 0.3 | 3.3 |
| Neenah | 0.0 | 18.3 | 0.7 | 0.0 | 0.3 | 16.5 | 13.3 | 6.4 | 1.4 | 7.6 |
| New Berlin | 0.0 | 19.1 | 0.0 | 0.0 | 2.8 | 16.8 | 34.4 | 4.4 | 0.0 | 2.6 |
| Oak Creek | 0.0 | 16.6 | 0.0 | 0.0 | 9.4 | 18.3 | 13.4 | 4.1 | 0.0 | 5.1 |
| Oshkosh | 0.0 | 16.0 | 1.1 | 0.0 | 1.8 | 12.8 | 11.2 | 7.1 | 10.9 | 9.8 |
| Racine | 0.0 | 12.9 | 1.0 | 0.0 | 12.1 | 22.5 | 14.4 | 7.1 | 2.3 | 4.5 |
| Sheboygan | 0.0 | 13.4 | 0.7 | 0.0 | 1.9 | 22.5 | 14.2 | 4.5 | 3.6 | 3.3 |
| Stevens Point | 0.0 | 24.3 | 0.1 | 0.0 | 5.0 | 13.4 | 18.9 | 6.5 | 1.1 | 4.1 |
| Sun Prairie | 0.0 | 25.8 | 0.0 | 0.0 | 4.5 | 21.0 | 12.4 | 5.2 | 0.0 | 5.8 |
| Superior | 0.0 | 19.8 | 0.0 | 0.0 | 0.4 | 16.3 | 24.7 | 9.6 | 1.9 | 3.0 |
| Waukesha | 0.0 | 25.5 | 0.7 | 0.0 | 2.6 | 16.7 | 12.4 | 6.6 | 0.3 | 5.7 |
| Wausau | 0.0 | 23.3 | 3.2 | 0.0 | 4.8 | 14.4 | 8.4 | 16.0 | 1.4 | 2.4 |
| Wauwatosa | 0.0 | 15.6 | 0.1 | 0.0 | 11.0 | 16.7 | 8.8 | 1.4 | 0.9 | 3.9 |
| West Allis | 0.3 | 16.9 | 0.1 | 0.0 | 5.1 | 23.5 | 10.8 | 1.0 | 5.6 | 2.7 |
| West Bend | 0.0 | 14.6 | 0.0 | 0.0 | 2.0 | 24.4 | 13.6 | 9.4 | 0.9 | 6.0 |
| **WYOMING** | X | X | X | X | X | X | X | X | X | X |
| Casper | 1.2 | 13.1 | 0.0 | 0.0 | 1.3 | 13.4 | 18.9 | 17.5 | 0.1 | 0.2 |
| Cheyenne | 1.7 | 12.9 | 0.6 | 0.0 | 2.0 | 16.8 | 25.4 | 8.7 | 0.4 | 0.9 |
| Gillette | 0.0 | 12.9 | 0.0 | 0.0 | 0.8 | 11.7 | 10.3 | 4.0 | 1.3 | 0.4 |
| Laramie | 0.0 | 12.5 | 0.0 | 0.0 | 1.7 | 13.3 | 13.5 | 11.5 | 0.6 | 0.7 |

# Table D. Cities — City Government Finances, City Government Employment, and Climate

| City | City government finances, 2017 (cont.) Debt outstanding — Total (mil dol) | Per capita[1] (dollars) | Debt issued during year | Climate[2] Average daily temperature Mean — January | July | Limits — January[3] | July[4] | Annual precipitation (inches) | Heating degree days | Cooling degree days |
|---|---|---|---|---|---|---|---|---|---|---|
| | 137 | 138 | 139 | 140 | 141 | 142 | 143 | 144 | 145 | 146 |
| **WASHINGTON—Cont'd** | | | | | | | | | | |
| Pasco | 67.7 | 930 | 8.6 | 34.2 | 75.2 | 28.0 | 89.3 | 8.01 | 4,731 | 909 |
| Pullman | 9.4 | 279 | 0.0 | NA | NA | NA | NA | NA | NA | NA |
| Puyallup | 67.5 | 1,646 | 8.1 | 39.9 | 64.9 | 32.9 | 77.8 | 40.51 | 4,991 | 153 |
| Redmond | 136.5 | 2,105 | 31.0 | 25.1 | 55.0 | 20.0 | 65.0 | 82.86 | 9,630 | 12 |
| Renton | 144.1 | 1,413 | 8.9 | 40.9 | 65.3 | 35.9 | 75.3 | 37.07 | 4,797 | 173 |
| Richland | 196.5 | 3,469 | 12.1 | 33.0 | 73.2 | 26.0 | 87.9 | 7.55 | 5,133 | 739 |
| Sammamish | 2.8 | 43 | 0.1 | 40.8 | 65.2 | 35.2 | 75.0 | 35.96 | 4,756 | 174 |
| SeaTac | 4.5 | 152 | 0.7 | 40.9 | 65.3 | 35.9 | 75.3 | 37.07 | 4,797 | 173 |
| Seattle | 5,049.0 | 6,929 | 671.4 | 41.5 | 65.5 | 36.0 | 74.5 | 38.25 | 4,615 | 192 |
| Shoreline | 44.9 | 794 | 5.2 | 40.8 | 65.2 | 35.2 | 75.0 | 35.96 | 4,756 | 174 |
| Spokane | 408.0 | 1,877 | 34.7 | 27.3 | 68.6 | 21.7 | 82.5 | 16.67 | 6,820 | 394 |
| Spokane Valley | 20.0 | 205 | 8.4 | NA | NA | NA | NA | NA | NA | NA |
| Tacoma | 1,703.7 | 7,983 | 197.3 | 41.0 | 65.6 | 35.1 | 76.1 | 38.95 | 4,650 | 167 |
| University Place | 42.0 | 1,256 | 19.7 | 41.0 | 65.6 | 35.1 | 76.1 | 38.95 | 4,650 | 167 |
| Vancouver | 180.5 | 1,000 | 9.3 | 39.0 | 65.4 | 32.4 | 77.3 | 41.92 | 4,990 | 197 |
| Walla Walla | 69.8 | 2,124 | 22.3 | 34.7 | 75.3 | 28.8 | 89.9 | 20.88 | 4,882 | 957 |
| Wenatchee | 38.2 | 1,117 | 9.0 | 29.2 | 74.4 | 23.2 | 87.8 | 9.12 | 5,533 | 832 |
| Yakima | 90.6 | 969 | 2.6 | 29.1 | 69.1 | 20.5 | 87.2 | 8.26 | 6,104 | 431 |
| **WEST VIRGINIA** | X | X | X | X | X | X | X | X | X | X |
| Charleston | 97.1 | 2,021 | 0.0 | 33.4 | 73.9 | 24.2 | 84.9 | 44.05 | 4,644 | 978 |
| Huntington | 35.7 | 764 | 1.2 | 32.1 | 76.3 | 23.5 | 87.1 | 41.74 | 4,737 | 1,128 |
| Morgantown | 187.8 | 6,064 | 73.6 | 30.8 | 73.5 | 22.3 | 83.4 | 43.30 | 5,174 | 815 |
| Parkersburg | 61.4 | 2,042 | 1.7 | 30.7 | 75.4 | 22.3 | 85.8 | 40.69 | 5,091 | 1,038 |
| Wheeling | 96.8 | 3,603 | 0.0 | 29.6 | 74.8 | 21.4 | 85.2 | 40.34 | 5,313 | 926 |
| **WISCONSIN** | X | X | X | X | X | X | X | X | X | X |
| Appleton | 167.6 | 2,249 | 57.7 | 16.0 | 71.6 | 7.8 | 81.4 | 30.16 | 7,721 | 572 |
| Beloit | 75.6 | 2,053 | 17.5 | 19.1 | 72.4 | 11.6 | 82.5 | 35.25 | 6,969 | 664 |
| Brookfield | 72.3 | 1,899 | 18.9 | 20.0 | 74.3 | 11.5 | 85.1 | 32.09 | 6,886 | 791 |
| Eau Claire | 158.5 | 2,312 | 27.1 | 11.9 | 71.4 | 2.5 | 82.6 | 32.12 | 8,196 | 554 |
| Fitchburg | 47.4 | 1,604 | 9.7 | NA | NA | NA | NA | NA | NA | NA |
| Fond du Lac | 153.9 | 3,606 | 6.1 | 16.6 | 71.8 | 9.1 | 81.1 | 30.15 | 7,534 | 586 |
| Franklin | 32.6 | 903 | 5.8 | 20.7 | 72.0 | 13.4 | 81.1 | 34.81 | 7,087 | 616 |
| Green Bay | 262.5 | 2,501 | 21.2 | 15.6 | 69.9 | 7.1 | 81.2 | 29.19 | 7,963 | 463 |
| Greenfield | 77.0 | 2,095 | 29.3 | 19.9 | 73.8 | 12.7 | 81.9 | 33.86 | 6,847 | 764 |
| Janesville | 112.9 | 1,756 | 28.7 | 17.7 | 72.1 | 8.6 | 83.8 | 32.78 | 7,238 | 629 |
| Kenosha | 111.1 | 1,113 | 20.7 | 20.8 | 71.3 | 13.2 | 78.7 | 34.74 | 6,999 | 549 |
| La Crosse | 63.5 | 1,227 | 15.7 | 15.9 | 74.0 | 6.3 | 85.2 | 32.36 | 7,340 | 775 |
| Madison | 906.4 | 3,543 | 152.3 | 17.3 | 71.6 | 9.3 | 82.1 | 32.95 | 7,493 | 582 |
| Manitowoc | 84.7 | 2,595 | 7.6 | 18.7 | 69.9 | 10.8 | 79.6 | 30.49 | 7,563 | 425 |
| Menomonee Falls | 94.8 | 2,535 | 19.9 | 16.5 | 69.3 | 8.1 | 80.2 | 33.45 | 7,832 | 407 |
| Milwaukee | 1,507.0 | 2,538 | 492.0 | 20.0 | 74.3 | 11.5 | 85.1 | 32.09 | 6,886 | 791 |
| Mount Pleasant | 39.3 | 1,459 | 8.9 | NA | NA | NA | NA | NA | NA | NA |
| Neenah | 65.0 | 2,506 | 6.5 | NA | NA | NA | NA | NA | NA | NA |
| New Berlin | 50.0 | 1,258 | 7.7 | 19.9 | 73.8 | 12.7 | 81.9 | 33.86 | 6,847 | 764 |
| Oak Creek | 115.1 | 3,173 | 33.9 | 20.7 | 72.0 | 13.4 | 81.1 | 34.81 | 7,087 | 616 |
| Oshkosh | 305.9 | 4,586 | 79.8 | 16.1 | 72.0 | 7.8 | 81.8 | 31.57 | 7,639 | 591 |
| Racine | 179.5 | 2,325 | 29.0 | 20.7 | 71.3 | 13.3 | 78.6 | 35.35 | 7,032 | 567 |
| Sheboygan | 56.0 | 1,158 | 16.3 | 20.9 | 71.4 | 13.2 | 81.4 | 31.90 | 7,056 | 559 |
| Stevens Point | 59.6 | 2,274 | 4.8 | NA | NA | NA | NA | NA | NA | NA |
| Sun Prairie | 69.9 | 2,120 | 21.2 | NA | NA | NA | NA | NA | NA | NA |
| Superior | 44.6 | 1,700 | 2.8 | 12.1 | 66.6 | 3.4 | 76.2 | 30.78 | 9,006 | 241 |
| Waukesha | 301.2 | 4,153 | 80.9 | 19.5 | 73.8 | 11.4 | 84.2 | 34.64 | 6,893 | 784 |
| Wausau | 75.1 | 1,939 | 27.4 | 13.0 | 70.1 | 3.6 | 80.8 | 33.36 | 8,237 | 464 |
| Wauwatosa | 126.1 | 2,627 | 19.4 | 20.0 | 74.3 | 11.5 | 85.1 | 32.09 | 6,886 | 791 |
| West Allis | 72.2 | 1,207 | 22.7 | 19.9 | 73.8 | 12.7 | 81.9 | 33.86 | 6,847 | 764 |
| West Bend | 63.7 | 2,020 | 7.5 | 18.4 | 70.6 | 10.7 | 81.3 | 32.85 | 7,371 | 502 |
| **WYOMING** | X | X | X | X | X | X | X | X | X | X |
| Casper | 15.8 | 273 | 0.0 | 22.3 | 70.0 | 12.2 | 86.8 | 13.03 | 7,571 | 428 |
| Cheyenne | 83.0 | 1,306 | 10.2 | 25.9 | 67.7 | 14.8 | 81.9 | 15.45 | 7,388 | 273 |
| Gillette | 158.6 | 4,925 | 5.4 | NA | NA | NA | NA | NA | NA | NA |
| Laramie | 23.5 | 723 | 0.6 | 20.3 | 62.9 | 7.8 | 79.4 | 11.19 | 9,233 | 75 |

1. Based on the population estimated as of July 1 of the year shown.  2. Represents normal values based on the 30-year period, 1971±2000.  3. Average daily minimum.  4. Average daily maximum.

# Congressional Districts of the 116th Congress

(For explanation of symbols, see page viii)

Page

| Page | |
|---|---|
| 1177 | Congressional District Highlights and Rankings |
| 1184 | Congressional District Column Headings |
| 1186 | Table E |
| 1186 | **AL**(District 1)—**CA**(District 53) |
| 1193 | **CO**(District 1)—**IL**(District 2) |
| 1200 | **IL**(District 3)—**MA**(District 9) |
| 1207 | **MI**(District 1)—**NM**(District 2) |
| 1214 | **NM**(District 3)—**OR**(District 2) |
| 1221 | **OR**(District 3)—**TX**(District 25) |
| 1228 | **TX**(District 26)—**WY**(At Large) |

| Page | |
|---|---|
| 1177 | Congressional District Highlights and Rankings |
| 1184 | Congressional District Column Headings |
| 1186 | Table B |
| 1166 | AL (District 1)—CA (District 53) |
| 1188 | CO (District 1)—IL (District 7) |
| 1200 | IL (District 8)—MA (District 9) |
| 1207 | MI (District 1)—NM (District 3) |
| 1214 | NM (District 3)—OR (District 2) |
| 1221 | OR (District 3)—TX (District 28) |
| 1228 | TX (District 29)—WY (At Large) |

# Congressional District Highlights and Rankings

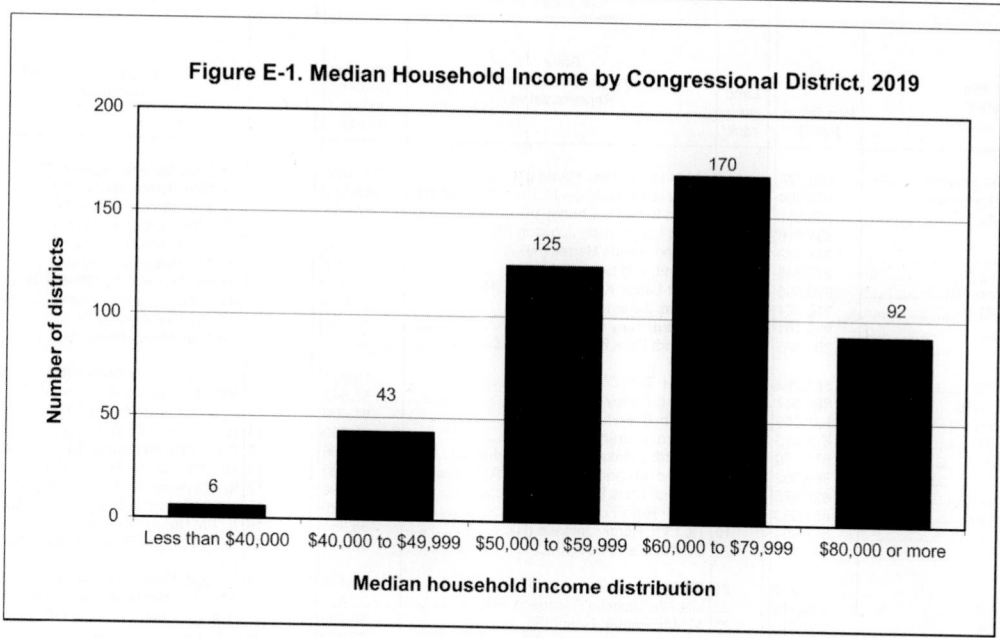

## Figure E-1. Median Household Income by Congressional District, 2019

Number of districts (y-axis)

- Less than $40,000: 6
- $40,000 to $49,999: 43
- $50,000 to $59,999: 125
- $60,000 to $79,999: 170
- $80,000 or more: 92

Median household income distribution

Every 10 years, the Census Bureau conducts a count to reapportion the seats in the U.S. House of Representatives. The House's 435 seats are divided among the 50 states. (The District of Columbia has no representative in Congress, although it has a nonvoting delegate.) The seats are reapportioned according to the population measured on April 1 of the census year in order to account for population changes among the states over the previous decade. The number of districts within a state may change after each decennial census, and the districts' boundaries may change more than once during a decade. The 117th Congress which convened in 2021 was the fourth to reflect the new boundaries based on the 2010 census. The Representatives of the 117th Congress are listed. Most of the data in Table E have been collected or updated for the 116th congress, but some sources were collected for the 115th congress or earlier. Boundary changes between the 114th and 115th congresses occurred in Florida, Minnesota, North Carolina, and Virginia.

As the state with the largest population, California had the most representatives with 53. Texas (36) and New York (27) were second and third largest, respectively. There were 7 states with just 1 representative: Alaska, Delaware, Montana, North Dakota, South Dakota, Vermont, and Wyoming. These states' representatives were considered 'At Large,' as they represented an entire state instead of a specific congressional district within the state.

Because the number of representatives is limited to 435, states with larger population growth add seats, while states with little or no growth lose seats. When the 113th Congress convened in January 2013, eight states had more representatives in congress and 10 states had fewer. Based on the 2010 census, Texas gained 4 seats, Florida gained 2, while Washington, Nevada, Utah, Arizona, Georgia, and South Carolina each gained one seat. New York and Ohio each lost two seats, while Massachusetts, New Jersey, Pennsylvania, Michigan, Illinois, Iowa, Missouri, and Louisiana each lost a seat.

As each decade progresses, population shifts alter the size of districts, leading up to the reapportionment of the next census. After the 2000 census, the population of each congressional district was about 645,000. Based on the 2019 population estimates, the congressional districts average over 750,000 people. Montana's at-large congressional district had over a million people, while Delaware and South Dakota also had at-large seats with above average populations. Texas' 22nd district is the largest apportioned congressional district in the United States with a population over 960,000 in 2019. Rhode Island's two congressional districts have the smallest populations at 530,066 and 529,295. Congressional districts in Nebraska and West Virginia also have smaller populations. For years, Louisiana's 2nd district was the least populous congressional district in the nation due to outmigration after Hurricane Katrina.

While most of the congressional districts had about the same population size, they varied widely in other characteristics. In California's 33rd district, 96.5 percent of the residents were high school graduates, compared with just 57.3 percent in California's 40th district and 62.3 in California's 21st district. These two districts also had the lowest proportion of college graduates. In California's 21st district only 10.7 percent of residents had a college degree compared with 10.1 percent in California's 40th district. Texas' 29th district and 33rd districts also had low rates of college attainment with 11.6 percent and 13.1 percent of their residents, respectively having a bachelor's degree or more.

The highest unemployment rates were found in New York's 15th district and California's 16th and 51st districts, all over 10 percent. Four districts from California, and one district each from Georgia, Michigan, New York, Mississippi, Ohio, and Illinois ranked among the 10 highest unemployment rates. Thirty-one congressional districts had 20 percent or more of their populations living in poverty. New York's 15th district had the highest poverty rate in the nation at 34.4 percent and the lowest median household income at $31,061 in 2019. In California's 18th district, the median household income was $149,375, the highest in the nation. One hundred thirty congressional districts had median household incomes exceeding $75,000 per year, while six districts had median household incomes below $40,000. The U.S. median household income in 2019 was $65,712.

# Congressional Districts of the 116th Congress of the United States
## Selected Rankings

| Population, 2019 | | | Land area, 2019 | | | Population density, 2019 | | |
|---|---|---|---|---|---|---|---|---|
| Population rank | State congressional district Representative | Population [col 2] | Land area rank | State congressional district Representative | Land area (square miles) [col 1] | Density rank | State congressional district Representative | Population density (per square mile) [col 3] |
| 1 | MT At large: Matthew M. Rosendale Sr. (R) | 1,068,778 | 1 | AK At-Large: Don Young (R) | 571,022 | 1 | NY 13th: Adriano Espaillat (D) | 73,946 |
| 2 | DE: At-Large: Lisa Blunt Rochester (D) | 973,764 | 2 | MT At large: Matthew M. Rosendale Sr. (R) | 145,550 | 2 | NY 10th: Jerrold Nadler (D) | 52,167 |
| 3 | TX 22nd: Troy E. Nehls (R) | 960,957 | 3 | WY At-large Liz Cheney (R) | 97,089 | 3 | NY 15th: Ritchie Torres (D) | 50,894 |
| 4 | ID 1st: Russ Fulcher (R) | 934,826 | 4 | SD At-Large Dusty Johnson (R) | 75,810 | 4 | NY 12th: Carolyn B. Maloney (D) | 49,064 |
| 5 | FL 9th: Darren Soto (D) | 931,872 | 5 | NM 2nd: Yvette Herrell (R) | 71,746 | 5 | NY 9th: Yvette D. Clarke (D) | 46,329 |
| 6 | Tx 10th: Michael T. McCaul (R) | 925,348 | 6 | OR 2nd: Cliff Bentz (R) | 69,451 | 6 | NY 7th: Nydia M. Velázquez (D) | 43,296 |
| 7 | TX 26th: Michael C. Burgess (R) | 920,865 | 7 | ND At-Large: Kelly Armstrong (R) | 68,996 | 7 | NY 8th: Hakeem S. Jeffries (D) | 27,335 |
| 8 | TX 31st: John R. Carter (R) | 916,064 | 8 | NE 3rd: Adrian Smith (R) | 67,433 | 8 | NY 14th: Alexandria Ocasio Cortez (D) | 24,374 |
| 9 | TX 3rd: Van Taylor (R) | 913,161 | 9 | TX 23rd: Tony Gonzales (R) | 58,059 | 9 | NY 6th: Grace Meng (D) | 23,975 |
| 10 | TX 8th: Kevin Brady (R) | 895,861 | 10 | NV 2nd: Mark E. Amodei (R) | 55,898 | 10 | CA 12th: Nancy Pelosi (D) | 19,998 |
| 11 | NC 12th: Alma S. Adams (D) | 891,792 | 11 | AZ 1st: Tom O'Halleran (D) | 55,034 | 11 | CA 34th: Jimmy Gomez (D) | 15,313 |
| 12 | NC 2nd: Deborah K. Ross (D) | 888,547 | 12 | KS 1st: Tracey Mann (R) | 52,542 | 12 | NY 5th: Gregory W. Meeks (D) | 14,617 |
| 13 | SD At-Large Dusty Johnson (R) | 884,659 | 13 | NV 4th: Steven Horsford (D) | 51,009 | 13 | PA 3rd: Dwight Evans (D) | 14,036 |
| 14 | FL 16th: Vern Buchanan (R) | 873,875 | 14 | CO 3rd: Lauren Boebert (R) | 49,730 | 14 | NJ 8th: Albio Sires (D) | 14,027 |
| 15 | NC 4th: David E. Price (D) | 873,270 | 15 | NM 3rd: Teresa Leger Fernandez (D) | 44,966 | 15 | CA 37th: Karen Bass (D) | 13,276 |
| 16 | CO 4th: Ken Buck (R) | 868,302 | 16 | ID 2nd: Michael K. Simpson (R) | 43,225 | 16 | MA 7th: Ayanna Presley (D) | 13,060 |
| 17 | OR 1st: Suzanne Bonamici (D) | 858,875 | 17 | UT 2nd: Chris Stewart (R) | 40,196 | 17 | IL 4th: Jesus G. "Chuy" Garcia (D) | 12,905 |
| 18 | VA 10th: Jennifer Wexton (D) | 857,693 | 18 | ID 1st: Russ Fulcher (R) | 39,420 | 18 | CA 40th: Lucille Roybal-Allard (D) | 12,411 |
| 19 | TX 35th: Lloyd Doggett (D) | 857,654 | 19 | TX 13th: Ronny Jackson (R) | 38,351 | 19 | IL 7th: Danny K. Davis (D) | 11,674 |
| 20 | NV 3rd: Susie Lee (D) | 857,197 | 20 | CO 4th: Ken Buck (R) | 38,101 | 20 | DC: At-Large: Eleanor Holmes Norton (D) | 11,546 |
| 21 | AZ 7th: Ruben Gallego (D) | 853,856 | 21 | OK 3rd: Frank D. Lucas (R) | 34,116 | 21 | PA 2nd: Brendan F. Boyle (D) | 11,514 |
| 22 | OR 3rd: Earl Blumenauer (D) | 853,116 | 22 | MN 7th: Michelle Fischbach (R) | 33,429 | 22 | NY 11th: Nicole Malliotakis (R) | 11,357 |
| 23 | CO 1st: Diana DeGette (D) | 852,816 | 23 | AZ 4th: Paul A. Gosar (R) | 33,220 | 23 | CA 43rd: Maxine Waters (D) | 10,378 |
| 24 | ID 2nd: Michael K. Simpson (R) | 852,239 | 24 | CA 8th: Jay Obernolte (R) | 32,893 | 24 | CA 46th: J. Luis Correa (D) | 10,174 |
| 25 | UT 4th: Burgess Owens (R) | 850,432 | 25 | CA 1st: Doug LaMalfa (R) | 28,121 | 25 | NJ 10th: Donald M. Payne, Jr. (D) | 10,046 |
| 26 | AZ 5th: Andy Biggs (R) | 849,917 | 26 | MN 8th: Pete Stauber (R) | 27,909 | 26 | NY 16th: Jamaal Bowman (D) | 9,440 |
| 27 | IA 3rd: Cynthia Axne (D) | 848,170 | 27 | TX 11th: August Pfluger (R) | 27,835 | 27 | CA 44th: Nanette Diaz Barragan (D) | 8,968 |
| 28 | GA 7th Carolyn Bourdeaux (D) | 844,917 | 28 | ME 2nd: Jared F. Golden (D) | 27,558 | 28 | NJ 9th: Bill Pascrell, Jr. (D) | 7,998 |
| 29 | TX 12th: Kay Granger (R) | 844,563 | 29 | TX 19th: Jodey C. Arrington (R) | 25,836 | 29 | CA 13th: Barbara Lee (D) | 7,943 |
| 30 | OR 5th: Kurt Schrader (D) | 844,220 | 30 | MI 1st: Jack Bergman (R) | 25,029 | 30 | CA 29th: Tony Cárdenas (D) | 7,778 |
| 31 | OR 2nd: Cliff Bentz (R) | 841,022 | 31 | WI 7th: Thomas P. Tiffany (R) | 23,039 | 31 | IL 5th: Michael Quigley (D) | 7,735 |
| 32 | CA 42nd: Ken Calvert (R) | 840,562 | 32 | IA 4th: Randy Feenstra (R) | 22,755 | 32 | FL 24th: Frederica S. Wilson (D) | 7,368 |
| 33 | FL 4th: John H. Rutherford (R) | 836,235 | 33 | AR 4th: Bruce Westerman (R) | 22,336 | 33 | CA 38th: Linda T. Sánchez (D) | 6,947 |
| 34 | FL 19th: Byron Donalds (R) | 833,013 | 34 | OK 2nd: Markwayne Mullin (R) | 20,996 | 34 | IL 9th: Janice D. Schakowsky (D) | 6,824 |
| 35 | TX 20th: Joaquin Castro (D) | 832,518 | 35 | UT 3rd: John R. Curtis (R) | 20,074 | 35 | NV 1st: Dina Titus (D) | 6,818 |
| 36 | TX 24rd: Beth Van Duyne (R) | 832,445 | 36 | MO 8th: Jason T. Smith (R) | 19,902 | 36 | FL 27th: Maria Elvira Salazar (R) | 6,629 |
| 37 | FL 14th: Kathy Castor (D) | 831,508 | 37 | UT 1st: Blake D. Moore (R) | 19,557 | 37 | NY 4th: Kathleen M. Rice (D) | 6,595 |
| 38 | TX 21st: Chip Roy (R) | 829,628 | 38 | AR 1st: Eric A. "Rick" Crawford (R) | 19,280 | 38 | CA 53rd: Sara Jacobs (D) | 5,752 |
| 39 | AR 3rd: Steve Womack (R) | 829,149 | 39 | WA 4th: Dan Newsome (R) | 19,247 | 39 | WA 7th: Pramila Jayapal (D) | 5,682 |
| 40 | CO 6th: Jason Crow (D) | 828,201 | 40 | MO 6th: Sam Graves (R) | 18,198 | 40 | CA 32nd: Grace F. Napolitano (D) | 5,640 |
| 41 | TX 18th: Sheila Jackson-Lee (D) | 827,015 | 41 | OR 4th: Peter A. DeFazio (D) | 17,273 | 41 | CA 30th: Brad Sherman (D) | 5,621 |
| 42 | AZ 4th: Paul A. Gosar (R) | 825,763 | 42 | AZ 3rd: Raul M. Grijalva (D) | 15,688 | 42 | WI 4th: Gwen Moore (D) | 5,483 |
| 43 | OK 5th: Stephanie I. Bice (R) | 825,115 | 43 | MS 2nd: Bennie G. Thompson (D) | 15,552 | 43 | VA 8th: Donald S. Beyer, Jr. (D) | 5,446 |
| 44 | VA 1st: Robert J. Wittman (R) | 824,492 | 44 | WA 5th: Cathy McMorris Rodgers (R) | 15,474 | 44 | MN 5th: Ilhan Omar (D) | 5,337 |
| 45 | CO 2nd: Joe Neguse (D) | 824,050 | 45 | NY 21st: Elise M. Stefanik (R) | 15,114 | 45 | AZ 9th: Greg Stanton (D) | 4,964 |
| 46 | SC 1st: Nancy Mace (R) | 821,107 | 46 | IL 15th: Mary E. Miller (R) | 14,696 | 46 | CA 48th: Michelle Steel (R) | 4,922 |
| 47 | OR 4th: Peter A. DeFazio (D) | 820,504 | 47 | LA 5th: Julia Letlow (R) | 14,452 | 47 | TX 7th: Lizzie Fletcher (D) | 4,707 |
| 48 | CO 5th: Doug Lamborn (R) | 820,255 | 48 | MO 4th: Vicky Hartzler (R) | 14,407 | 48 | FL 22nd: Theodore E. Deutch (D) | 4,647 |
| 49 | MA 7th: Ayanna Pressley (D) | 819,035 | 49 | KS 4th: Ron Estes (R) | 14,315 | 49 | TX 9th: Al Green (D) | 4,641 |
| 50 | AZ 9th: Greg Stanton (D) | 818,850 | 50 | KS 2nd: Jake LaTurner (R) | 14,144 | 50 | CA 35th: Norma J. Torres (D) | 4,529 |
| 51 | TX 25th: Roger Williams (R) | 818,807 | 51 | CA 2nd: Jared Huffman (D) | 12,953 | 51 | CO 1st: Diana DeGette (D) | 4,498 |
| 52 | TX 6th: Vacant | 818,442 | 52 | CA 4th: Tom McClintock (R) | 12,837 | 52 | CA 6th: Doris O. Matsui (D) | 4,468 |
| 53 | WA 7th: Pramila Jayapal (D) | 817,787 | 53 | MS 3rd: Michael Guest (R) | 12,755 | 53 | VA 11th: Gerald E. Connolly (D) | 4,266 |
| 54 | NC 7th: David Rouzer (R) | 816,402 | 54 | LA 4th: Mike Johnson (R) | 12,437 | 54 | CA 17th: Ro Khanna (D) | 4,264 |
| 55 | NC 8th: Richard Hudson (R) | 815,055 | 55 | IA 2nd: Mariannette Miller-Meeks(R) | 12,261 | 55 | TX 31st: Colin Z. Allred (D) | 4,183 |
| 56 | FL 7th: Stephanie N. Murphy (D) | 814,980 | 56 | KY 1st: James Comer (R) | 12,084 | 56 | Sylvia R. Garcia (D) | 4,178 |
| 57 | AZ 6th: David Schweikert (R) | 814,971 | 57 | IA 1st: Ashley Hinson (R) | 12,049 | 57 | AZ 7th: Ruben Gallego (D) | 4,171 |
| 58 | OH 3rd: Joyce Beatty (D) | 813,890 | 58 | MN 1st: Jim Hagedorn (R) | 11,974 | 58 | TX 20th: Joaquin Castro (D) | 4,165 |
| 59 | VA 8th: Donald S. Beyer, Jr. (D) | 813,568 | 59 | KY 5th: Harold Rogers (R) | 11,234 | 59 | WA 9th: Adam Smith (D) | 4,097 |
| 60 | FL 11th: Daniel Webster (R) | 813,112 | 60 | WI 3rd: Ron Kind (D) | 11,117 | 60 | FL 23rd: Debbie Wasserman Schultz (D) | 4,083 |
| 61 | TN 4th: Scott DesJarlais (R) | 812,697 | 61 | FL 2nd: Neal P. Dunn (R) | 11,004 | 61 | FL 13th: Charlie Crist (D) | 4,027 |
| 62 | FL 10th: Val Butler Demings (D) | 811,634 | 62 | MS 1st: Trent Kelly (R) | 10,573 | 62 | MI 9th: Andy Levin (D) | 3,912 |
| 63 | FL 12th: Gus M. Bilirakis (R) | 811,308 | 63 | IL 18th: Darin LaHood (R) | 10,515 | 63 | NY 2nd: Andrew R. Garbarino (R) | 3,864 |
| 64 | OK 1st: Kevin Hern (R) | 809,500 | 64 | AL 7th: Terri A. Sewell (R) | 10,155 | 64 | MI 14th: Brenda L. Lawrence (D) | 3,722 |
| 65 | CO 7th: Ed Perlmutter (D) | 808,543 | 65 | AL 2nd: Barry Moore(R) | 10,143 | 65 | MI 13th: Rashida Tlaib (D) | 3,636 |
| 66 | FL 17th: W. Gregory Steube (R) | 804,754 | 66 | TX 4th: Pat Fallon (R) | 10,131 | 66 | OH 3rd: Joyce Beatty (D) | 3,568 |
| 67 | TX 15th: Vicente Gonzalez (D) | 804,562 | 67 | VA 5th: Bob Good (R) | 10,031 | 67 | TX 32nd: Marc A. Veasey (D) | 3,541 |
| 68 | LA 6th: Garret Graves (R) | 803,704 | 68 | CA 23rd: Kevin McCarthy (R) | 9,901 | 68 | TX 18th: Sheila Jackson-Lee (D) | 3,512 |
| 69 | GA 13th: David Scott (D) | 802,943 | 69 | PA 12th: Fred Keller (R) | 9,896 | 69 | CA 39th: Young Kim (R) | 3,495 |
| 70 | VA 7th: Abigail Davis Spanberger (D) | 802,921 | 70 | OK 4th: Tom Cole (R) | 9,777 | 70 | IL 8th: Raja Krishnamoorthi (D) | 3,487 |
| 71 | MO 3rd: Blaine Luetkemeyer (R) | 802,919 | 71 | WV 3rd: Carol D. Miller (R) | 9,746 | 71 | HI 1st: Ed Case (D) | 3,449 |
| 72 | FL 20th: Vacant | 802,463 | 72 | PA 15th: Glen Thompson (R) | 9,735 | 72 | NJ 6th: Frank Pallone Jr. (D) | 3,436 |
| 73 | AZ 3rd: Raul M. Grijalva (D) | 801,531 | 73 | GA 2nd: Sanford D. Bishop Jr. (R) | 9,630 | 73 | CA 31st: Pete Aguilar (D) | 3,428 |
| 74 | FL 15th: C. Scott Franklin (R) | 801,294 | 74 | TX 28th: Henry Cuellar (D) | 9,379 | 74 | PA 5th: Mary Gay Scanlon (D) | 3,400 |
| 75 | TN 7th: Mark E. Green (R) | 800,536 | 75 | VT: At-large Peter Welch (D) | 9,217 | 75 | CA 47th: Alan S. Lowenthal (D) | 3,318 |

# Congressional Districts of the 116th Congress of the United States
## Selected Rankings

### Percent Non-Hispanic White alone, 2019

| Non-Hispanic White alone rank | State congressional district Representative | Percent white [col 11] |
|---|---|---|
| 1 | KY 5th: Harold Rogers (R) | 95.3 |
| 2 | PA 15th: Glen Thompson (R) | 94.6 |
| 3 | OH 6th: Bill Johnson (R) | 94.5 |
| 4 | ME 2nd: Jared F. Golden (D) | 93.4 |
| 5 | WV 3rd: Carol D. Miller (R) | 93.2 |
| 6 | WV 1st: David McKinley (R) | 93.0 |
| 7 | VT: At-large Peter Welch (D) | 92.5 |
| 8 | ME 1st: Chellie Pingree (D) | 92.3 |
| 8 | PA 14th: Guy Reschenthaler (R) | 92.3 |
| 10 | PA 12th: Fred Keller (R) | 91.8 |
| 11 | MN 8th: Pete Stauber (R) | 91.4 |
| 12 | WI 3rd: Ron Kind (D) | 91.3 |
| 13 | NY 27th: Chris Jacobs (R) | 91.2 |
| 13 | PA 13th: John Joyce (R) | 91.2 |
| 13 | WI 7th: Thomas P. Tiffany (R) | 91.2 |
| 16 | MI 4th: John L. Moolenaar (R) | 91.1 |
| 17 | MI 1st: Jack Bergman (R) | 90.8 |
| 17 | OH 16th: Anthony Gonzalez (R) | 90.8 |
| 17 | TN 1st: Diana Harshbarger (R) | 90.8 |
| 20 | IN 6th: Greg Pence (R) | 90.6 |
| 21 | MO 8th: Jason T. Smith (R) | 90.4 |
| 21 | NY 21st: Elise M. Stefanik (R) | 90.4 |
| 23 | IL 15th: Mary E. Miller (R) | 90.3 |
| 24 | OH 7th: Bob Gibbs (R) | 90.2 |
| 25 | WV 2nd: Alexander X. Mooney (R) | 90.1 |
| 26 | NH 1st: Chris Pappas (D) | 89.9 |
| 27 | MO 3rd: Blaine Luetkemeyer (R) | 89.8 |
| 28 | MI 10th: Lisa McClain (R) | 89.7 |
| 29 | IN 7th: Larry Bucshon (R) | 89.6 |
| 30 | KY 4th: Thomas Massie (R) | 89.5 |
| 30 | NH 2nd: Ann M. Kuster (D) | 89.5 |
| 32 | PA 16th: Mike Kelly (R) | 89.2 |
| 33 | VA 9th: Morgan Griffith (R) | 88.6 |
| 34 | WI 6th: Glenn Grothman (R) | 88.4 |
| 35 | IL 18th: Darin LaHood (R) | 88.1 |
| 35 | IN 8th: Trey Hollingsworth (R) | 88.1 |
| 35 | PA 9th: Daniel Meuser (R) | 88.1 |
| 38 | NY 23rd: Tom Reed (R) | 88.0 |
| 39 | OH 5th: Robert E. Latta (R) | 87.7 |
| 39 | OH 14th: David P. Joyce (R) | 87.7 |
| 41 | MO 7th: Bill Long (R) | 87.6 |
| 41 | OH 15th: Vacant | 87.6 |
| 43 | MN 7th: Michelle Fischbach (R) | 87.3 |
| 44 | IA 1st: Ashley Hinson (R) | 87.2 |
| 44 | MI 7th: Tim Walberg (R) | 87.2 |
| 46 | MN 6th: Tom Emmer (R) | 87.0 |
| 46 | PA 17th: Conor Lamb (D) | 87.0 |
| 48 | KY 1st: James Comer (R) | 86.9 |
| 48 | KY 2nd: Brett Guthrie (R) | 86.9 |
| 48 | OH 4th: Jim Jordan (R) | 86.9 |
| 51 | TN 6th: John W. Rose (R) | 86.8 |
| 52 | NY 22nd: Claudia Tenney (R) | 86.7 |
| 53 | IA 4th: Randy Feenstra (R) | 86.6 |
| 53 | MO 6th: Sam Graves (R) | 86.6 |
| 55 | WI 8th: Mike Gallagher (R) | 86.2 |
| 56 | MO 2nd: Ann Wagner (R) | 86.0 |
| 56 | WI 5th: Scott Fitzgerald (R) | 86.0 |
| 58 | MO 4th: Vicky Hartzler (R) | 85.9 |
| 59 | MT At large: Matthew M. Rosendale Sr. (R) | 85.8 |
| 60 | NC 11th: Madison Cawthorn (R) | 85.6 |
| 61 | TN 2nd: Tim Burchett (R) | 85.3 |
| 62 | MA 8th: William R. Keating (D) | 85.1 |
| 63 | OH 8th: Warren Davidson (R) | 84.7 |
| 64 | MN 1st: Jim Hagedorn (R) | 84.6 |
| 64 | OH 12th: Troy Balderson (R) | 84.6 |
| 66 | IA 2nd: Mariannette Miller-Meeks (R) | 84.5 |
| 66 | OH 2nd: Brad R. Wenstrup (R) | 84.5 |
| 68 | PA 11th: Lloyd Smucker (R) | 84.1 |
| 69 | WY At-large: Liz Cheney (R) | 83.7 |
| 70 | ND At-Large: Kelly Armstrong (R) | 83.6 |
| 71 | IN 4th: James R. Baird (R) | 83.5 |
| 71 | NE 3rd: Adrian Smith (R) | 83.5 |
| 73 | OR 4th: Peter A. DeFazio (D) | 83.2 |
| 74 | WA 5th: Cathy McMorris Rodgers (R) | 83.0 |
| 75 | AL 4th: Robert B. Aderholt (R) | 82.9 |

### Percent Black alone, 2019

| Black rank | State congressional district Representative | Percent black [col 5] |
|---|---|---|
| 1 | MS 2nd: Bennie G. Thompson (D) | 67.0 |
| 2 | TN 9th: Steve Cohen (D) | 65.1 |
| 3 | AL 7th: Terri A. Sewell (D) | 62.6 |
| 4 | GA 13th: David Scott (D) | 61.6 |
| 5 | LA 2nd: Troy Carter (D) | 61.5 |
| 6 | GA 4th: Henry C. "Hank" Johnson Jr. (D) | 60.9 |
| 7 | PA 3rd: Dwight Evans (D) | 56.4 |
| 8 | IL 2nd: Robin L. Kelly (D) | 55.9 |
| 8 | MI 14th: Brenda L. Lawrence (D) | 55.9 |
| 10 | GA 5th: Nikema Williams (D) | 55.4 |
| 11 | SC 6th: James E. Clyburn (D) | 54.3 |
| 12 | MI 13th: Rashida Tlaib (D) | 54.1 |
| 13 | FL 20th: Vacant | 52.9 |
| 13 | OH 11th: Vacant | 52.9 |
| 15 | GA 2nd: Sanford D. Bishop Jr. (D) | 52.8 |
| 16 | MD 7th: Kweisi Mfume (D) | 52.5 |
| 17 | MD 4th: Anthony G. Brown (D) | 52.4 |
| 18 | NY 8th: Hakeem S. Jeffries (D) | 51.9 |
| 19 | NJ 10th: Donald M. Payne, Jr. (D) | 51.0 |
| 20 | IL 1st: Bobby L. Rush (D) | 50.5 |
| 21 | NY 5th: Gregory W. Meeks (D) | 49.2 |
| 22 | MO 1st: Cori Bush (D) | 49.0 |
| 23 | VA 3rd: Robert C. "Bobby" Scott (D) | 47.4 |
| 24 | FL 5th: Al Lawson (D) | 47.1 |
| 24 | FL 24th: Frederica S. Wilson (D) | 47.1 |
| 26 | NY 9th: Yvette D. Clarke (D) | 46.7 |
| 27 | DC: At-Large: Eleanor Holmes Norton (D) | 45.4 |
| 28 | NC 1st: G. K. Butterfield (D) | 44.5 |
| 29 | IL 7th: Danny K. Davis (D) | 42.5 |
| 30 | NY 15th: Ritchie Torres (D) | 42.0 |
| 31 | TX 30th: Eddie Bernice Johnson (D) | 41.2 |
| 32 | VA 4th: A. Donald McEachin (D) | 41.1 |
| 33 | MD 5th: Steny H. Hoyer (D) | 39.2 |
| 34 | TX 9th: Al Green (D) | 39.0 |
| 35 | NC 12th: Alma S. Adams (D) | 37.5 |
| 36 | MD 2nd: C. A. Dutch Ruppersberger (D) | 36.9 |
| 37 | MS 3rd: Michael Guest (R) | 35.4 |
| 38 | LA 5th: Julia Letlow (R) | 35.1 |
| 39 | OH 3rd: Joyce Beatty (D) | 34.6 |
| 40 | GA 12th: Rick W. Allen (R) | 34.3 |
| 41 | LA 4th: Mike Johnson (R) | 33.9 |
| 42 | TX 18th: Sheila Jackson-Lee (D) | 33.7 |
| 42 | WI 4th: Gwen Moore (D) | 33.7 |
| 44 | AL 2nd: Barry Moore(R) | 32.2 |
| 45 | NY 16th: Jamaal Bowman (D) | 32.0 |
| 46 | GA 8th: Austin Scott (R) | 31.4 |
| 47 | GA 1st: Earl L. "Buddy" Carter (R) | 30.7 |
| 47 | IN 7th André Carson (D) | 30.7 |
| 49 | FL 10th: Val Butler Demings (D) | 29.5 |
| 49 | NY 13th: Adriano Espaillat (D) | 29.5 |
| 51 | MS 1st: Trent Kelly (R) | 29.0 |
| 52 | SC 7th: Tom Rice (R) | 27.8 |
| 53 | AL 1st: Jerry Carl (R) | 27.4 |
| 54 | AL 3rd: Mike Rogers (R) | 26.3 |
| 55 | SC 5th: Ralph Norman (R) | 26.2 |
| 56 | GA 10th: Jody B. Hice (R) | 26.1 |
| 57 | PA 5th: Mary Gay Scanlon (D) | 25.8 |
| 58 | GA 3rd  A. Drew Ferguson IV (R) | 25.4 |
| 58 | MA 7th: Ayanna Presley (D) | 25.4 |
| 60 | LA 6th: Garret Graves (R) | 25.2 |
| 60 | PA 2nd: Brendan F. Boyle (D) | 25.2 |
| 62 | LA 3rd: Clay Higgins (R) | 25.0 |
| 63 | SC 2nd: Joe Wilson (R) | 24.6 |
| 64 | TN 5th: Jim Cooper (D) | 24.4 |
| 65 | MD 3rd: John P. Sarbanes (D) | 23.5 |
| 65 | MS 4th: Steven M. Palazzo (R) | 23.5 |
| 67 | AR 2nd: J. French Hill (R) | 23.3 |
| 68 | KY 3rd: John A. Yarmuth (D) | 23.0 |
| 68 | NC 8th: Richard Hudson (R) | 23.0 |
| 70 | NC 13th: Ted Budd (R) | 22.8 |
| 71 | NC 4th: David E. Price (D) | 22.6 |
| 72 | DE: At-Large: Lisa Blunt Rochester (D) | 22.5 |
| 73 | TN 8th: David Kustoff (R) | 22.0 |
| 74 | CA 37th: Karen Bass (D) | 21.9 |
| 75 | GA 7th  Carolyn Bourdeaux (D) | 21.5 |

### Percent American Indian, Alaska Native alone, 2019

| American Indian Alaska Native rank | State congressional district Representative | Percent American Indian Alaska Native [col 6] |
|---|---|---|
| 1 | AZ 1st: Tom O'Halleran (D) | 23.0 |
| 2 | NM 3rd: Teresa Leger Fernandez (D) | 18.7 |
| 3 | OK 2nd: Markwayne Mullin (R) | 17.2 |
| 4 | AK At-Large: Don Young (R) | 15.8 |
| 5 | SD At-Large Dusty Johnson (R) | 8.6 |
| 6 | NC 9th: Dan Bishop (R) | 7.6 |
| 7 | OK 3rd: Frank D. Lucas (R) | 6.7 |
| 8 | MT At large: Matthew M. Rosendale Sr. (R) | 6.3 |
| 9 | OK 1st: Kevin Hern (R) | 6.2 |
| 10 | OK 4th: Tom Cole (R) | 5.9 |
| 11 | ND At-Large: Kelly Armstrong (R) | 5.4 |
| 12 | NM 2nd: Yvette Herrell (R) | 5.3 |
| 13 | OK 5th: Stephanie I. Bice (R) | 4.8 |
| 14 | AZ 3rd: Raul M. Grijalva (D) | 4.6 |
| 15 | NM 1st: Melanie A. Stansbury (D) | 4.5 |
| 16 | AZ 9th: Greg Stanton (D) | 3.6 |
| 17 | MN 7th: Michelle Fischbach (R) | 3.3 |
| 18 | MI 1st: Jack Bergman (R) | 2.7 |
| 18 | WA 4th: Dan Newsome (R) | 2.7 |
| 20 | WI 8th: Mike Gallagher (R) | 2.6 |
| 21 | AZ 7th: Ruben Gallego (D) | 2.5 |
| 21 | CA 2nd: Jared Huffman (D) | 2.5 |
| 23 | WY At-large Liz Cheney (R) | 2.4 |
| 24 | MN 8th: Pete Stauber (R) | 2.3 |
| 24 | OR 2nd: Cliff Bentz (R) | 2.3 |
| 26 | AZ 4th: Paul A. Gosar (R) | 2.2 |
| 26 | CO 3rd: Lauren Boebert (R) | 2.2 |
| 26 | NV 2nd: Mark E. Amodei (R) | 2.2 |
| 29 | WA 6th: Derek Kilmer (D) | 2.1 |
| 29 | WI 7th: Thomas P. Tiffany (R) | 2.1 |
| 31 | AZ 6th: David Schweikert (R) | 1.9 |
| 32 | NC 11th: Madison Cawthorn (R) | 1.8 |
| 33 | CA 21st: David G. Valadao (R) | 1.7 |
| 33 | CA 34th: Jimmy Gomez (D) | 1.7 |
| 33 | NV 1st: Dina Titus (D) | 1.7 |
| 33 | WA 2nd: Rick Larsen (D) | 1.7 |
| 33 | WA 5th: Cathy McMorris Rodgers (R) | 1.7 |
| 38 | CA 1st: Doug LaMalfa (R) | 1.6 |
| 38 | CA 25th: Mike Garcia (R) | 1.6 |
| 40 | CA 8th: Jay Obernolte (R) | 1.5 |
| 40 | NE 1st: Jeff Fortenberry (R) | 1.5 |
| 40 | UT 3rd: John R. Curtis (R) | 1.5 |
| 43 | AZ 2nd: Ann Kirkpatrick (D) | 1.4 |
| 43 | CA 6th: Lucy McBath (D) | 1.4 |
| 43 | ID 2nd: Michael K. Simpson (R) | 1.4 |
| 43 | WA 10th: Marilyn Strickland (D) | 1.4 |
| 47 | AZ 8th: Debbie Lesko (R) | 1.3 |
| 47 | CA 36th: Raul Ruiz (D) | 1.3 |
| 47 | KS 2nd: Jake LaTurner (R) | 1.3 |
| 50 | CA 3rd: John Garamendi (D) | 1.2 |
| 50 | CA 4th: Tom McClintock (R) | 1.2 |
| 50 | CA 24th: Salud O. Carbajal (D) | 1.2 |
| 50 | ID 1st: Russ Fulcher (R) | 1.2 |
| 50 | NV 4th: Steven Horsford (D) | 1.2 |
| 50 | NC 8th: Richard Hudson (R) | 1.2 |
| 50 | OR 5th: Kurt Schrader (D) | 1.2 |
| 50 | UT 2nd: Chris Stewart (R) | 1.2 |
| 50 | WA 8th: Kim Schrier (D) | 1.2 |
| 59 | AZ 5th: Andy Biggs (R) | 1.1 |
| 59 | AR 3rd: Steve Womack (R) | 1.1 |
| 59 | CA 16th: Jim Costa (D) | 1.1 |
| 59 | CA 35th: Norma J. Torres (D) | 1.1 |
| 59 | CA 40th: Lucille Roybal-Allard (D) | 1.1 |
| 59 | LA 1st: Steve Scalise (R) | 1.1 |
| 59 | TX 20th: Joaquin Castro (D) | 1.1 |
| 59 | UT 1st: Blake D. Moore (R) | 1.1 |
| 59 | WA 1st: Suzan K. DelBene (D) | 1.1 |
| 68 | AL 1st: Jerry Carl (R) | 1.0 |
| 68 | CA 22nd: Devin Nunes (R) | 1.0 |
| 68 | CA 23rd: Kevin McCarthy (R) | 1.0 |
| 68 | CA 50th: Darrell Issa (R) | 1.0 |
| 68 | CA 51st: Juan Vargas (D) | 1.0 |
| 68 | CO 5th: Doug Lamborn (R) | 1.0 |
| 68 | ME 2nd: Jared F. Golden (D) | 1.0 |
| 68 | MS 3rd: Michael Guest (R) | 1.0 |

# Congressional Districts of the 116th Congress of the United States
## Selected Rankings

### Percent Asian or Pacific Islander alone, 2019

| Asian or Pacific Islander rank | State congressional district Representative | Percent Asian or Pacific Islander [col 7] |
|---|---|---|
| 1 | HI 1st: Ed Case (D) | 58.5 |
| 2 | CA 17th: Ro Khanna (D) | 56.4 |
| 3 | NY 6th: Grace Meng (D) | 40.4 |
| 4 | HI 2nd: Kaiali'i Kahele (D) | 40.1 |
| 5 | CA 27th: Judy Chu (D) | 39.4 |
| 6 | CA 15th: Eric Swalwell (D) | 38.1 |
| 7 | CA 14th Jackie Speier (D) | 37.5 |
| 8 | CA 12th: Nancy Pelosi (D) | 33.2 |
| 9 | CA 39th: Young Kim (R) | 32.9 |
| 10 | CA 19th: Zoe Lofgren (D) | 31.3 |
| 11 | CA 45th: Katie Porter (D) | 26.4 |
| 12 | WA 9th: Adam Smith (D) | 25.0 |
| 13 | CA 18th: Anna G. Eshoo (D) | 24.6 |
| 14 | CA 47th: Alan S. Lowenthal (D) | 23.8 |
| 15 | CA 13th: Barbara Lee (D) | 21.2 |
| 16 | CA 52nd: Scott H. Peters (D) | 20.9 |
| 17 | CA 34th: Jimmy Gomez (D) | 20.4 |
| 18 | CA 48th: Michelle Steel (R) | 20.1 |
| 19 | CA 32nd: Grace F. Napolitano (D) | 19.6 |
| 20 | CA 7th: Ami Bera (D) | 18.9 |
| 20 | VA 11th: Gerald E. Connolly (D) | 18.9 |
| 22 | NY 10th: Jerrold Nadler (D) | 18.7 |
| 23 | NJ 6th: Frank Pallone Jr. (D) | 18.6 |
| 24 | NY 7th: Nydia M. Velázquez (D) | 18.4 |
| 24 | TX 22nd: Troy E. Nehls (R) | 18.4 |
| 26 | TX 3rd: Van Taylor (R) | 17.9 |
| 27 | NY 14th: Alexandria Ocasio Cortez (D) | 17.8 |
| 28 | NJ 12th: Bonnie Watson Coleman (D) | 17.5 |
| 29 | CA 9th: Jerry McNerney (D) | 17.1 |
| 30 | NY 3rd: Thomas R. Souzzi (D) | 16.7 |
| 31 | CA 6th: Doris O. Matsui (D) | 16.6 |
| 32 | CA 33rd: Ted Lieu (D) | 16.4 |
| 32 | CA 38th: Linda T. Sánchez (D) | 16.4 |
| 34 | GA 7th Carolyn Bourdeaux (D) | 15.8 |
| 34 | TX 24th: Beth Van Duyne (R) | 15.8 |
| 36 | VA 10th: Jennifer Wexton (D) | 15.7 |
| 37 | CA 53rd: Sara Jacobs (D) | 15.5 |
| 37 | NY 12th: Carolyn B. Maloney (D) | 15.5 |
| 39 | CA 28th: Adam B. Schiff (D) | 15.4 |
| 39 | NY 11th: Nicole Malliotakis (R) | 15.4 |
| 41 | NY 5th: Gregory W. Meeks (D) | 15.2 |
| 42 | WA 7th: Pramila Jayapal (D) | 15.1 |
| 43 | CA 11th: Mark Desaulnier (D) | 15.0 |
| 44 | IL 9th: Janice D. Schakowsky (D) | 14.7 |
| 45 | NV 3rd: Susie Lee (D) | 14.6 |
| 46 | CA 46th: J. Luis Correa (D) | 14.4 |
| 47 | NJ 9th: Bill Pascrell, Jr. (D) | 13.8 |
| 48 | IL 8th: Raja Krishnamoorthi (D) | 13.5 |
| 49 | MA 5th: Katherine Clark (D) | 13.4 |
| 50 | WA 1st: Suzan K. DelBene (D) | 13.3 |
| 51 | MN 4th: Betty McCollum (D) | 12.9 |
| 52 | CA 43rd: Maxine Waters (D) | 12.6 |
| 53 | CA 30th: Brad Sherman (D) | 12.5 |
| 54 | TX 7th: Lizzie Fletcher (D) | 12.2 |
| 54 | VA 8th: Donald S. Beyer, Jr. (D) | 12.2 |
| 56 | TX 9th: Al Green (D) | 12.1 |
| 57 | CA 3rd: John Garamendi (D) | 12.0 |
| 57 | CA 5th: Mike Thompson (D) | 12.0 |
| 59 | MD 6th: David J. Trone (D) | 11.8 |
| 59 | NJ 7th: Tom Malinowski (D) | 11.8 |
| 61 | GA 6th: Lucy McBath (D) | 11.7 |
| 62 | MI 11th: Haley M. Stevens (D) | 11.2 |
| 63 | IL 10th: Bradley Scott Schneider (D) | 10.9 |
| 64 | MA 7th: Ayanna Presley (D) | 10.8 |
| 65 | WA 8th: Kim Schrier (D) | 10.6 |
| 66 | NJ 5th: Josh Gottheimer (D) | 10.5 |
| 66 | NJ 11th: Mikie Sherill (D) | 10.5 |
| 68 | CA 42nd: Ken Calvert (R) | 10.2 |
| 69 | NV 1st: Dina Titus (D) | 10.0 |
| 70 | WA 2nd: Rick Larsen (D) | 9.9 |
| 71 | NJ 8th: Albio Sires (D) | 9.8 |
| 72 | CA 16th: Jim Costa (D) | 9.5 |
| 72 | IL 6th: Sean Casten (D) | 9.5 |
| 72 | MD 8th: Jamie Raskin (D) | 9.5 |
| 75 | NC 4th: David E. Price (D) | 9.4 |

### Percent Hispanic or Latino,[1] 2019

| Hispanic rank | State congressional district Representative | Percent Hispanic [col 10] |
|---|---|---|
| 1 | CA 40th: Lucille Roybal-Allard (D) | 89.2 |
| 2 | TX 34th: Filemon Vela (D) | 84.5 |
| 3 | TX 15th: Vicente Gonzalez (D) | 82.0 |
| 4 | TX 16th: Veronica Escobar (D) | 81.3 |
| 5 | TX 28th: Henry Cuellar (D) | 78.9 |
| 6 | TX 29th: Sylvia R. Garcia (D) | 78.7 |
| 7 | CA 21st: David G. Valadao (R) | 75.3 |
| 8 | CA 44th: Nanette Diaz Barragan (D) | 71.4 |
| 9 | CA 51st: Juan Vargas (D) | 71.3 |
| 10 | CA 35th: Norma J. Torres (D) | 71.2 |
| 11 | TX 23rd: Tony Gonzales (R) | 69.6 |
| 12 | TX 20th: Joaquin Castro (D) | 68.4 |
| 13 | IL 4th: Jesus G. "Chuy" Garcia (D) | 67.1 |
| 14 | CA 29th: Tony Cárdenas (D) | 67.0 |
| 14 | CA 46th: J. Luis Correa (D) | 67.0 |
| 16 | TX 33rd: Marc A. Veasey (D) | 66.8 |
| 17 | AZ 3rd: Raul M. Grijalva (D) | 65.1 |
| 18 | AZ 7th: Ruben Gallego (D) | 64.0 |
| 18 | NY 15th: Ritchie Torres (D) | 64.0 |
| 20 | CA 41st: Mark Takano (D) | 62.7 |
| 21 | CA 16th: Jim Costa (D) | 62.1 |
| 22 | CA 38th: Linda T. Sánchez (D) | 61.5 |
| 23 | CA 32nd: Grace F. Napolitano (D) | 61.0 |
| 24 | TX 35th: Lloyd Doggett (D) | 60.6 |
| 25 | CA 34th: Jimmy Gomez (D) | 59.8 |
| 26 | NM 2nd: Yvette Herrell (R) | 55.9 |
| 27 | NJ 8th: Albio Sires (D) | 55.4 |
| 28 | NY 13th: Adriano Espaillat (D) | 54.5 |
| 29 | TX 27th: Michael Cloud (R) | 54.3 |
| 30 | CA 20th: Jimmy Panetta (D) | 54.1 |
| 31 | CA 31st: Pete Aguilar (D) | 53.5 |
| 32 | NM 1st: Melanie A. Stansbury (D) | 51.0 |
| 33 | CA 22nd: Devin Nunes (R) | 50.6 |
| 34 | NY 14th: Alexandria Ocasio Cortez (D) | 49.9 |
| 35 | CA 43rd: Maxine Waters (D) | 49.7 |
| 36 | CA 36th: Raul Ruiz (D) | 48.8 |
| 37 | NV 1st: Dina Titus (D) | 47.0 |
| 38 | CA 26th: Julia Brownley (D) | 46.2 |
| 39 | CA 10th: Josh Harder (D) | 45.1 |
| 40 | CA 8th: Jay Obernolte (R) | 44.8 |
| 40 | TX 18th: Sheila Jackson-Lee (D) | 44.8 |
| 42 | CA 42nd: Ken Calvert (R) | 41.7 |
| 43 | NM 3rd: Teresa Leger Fernandez (D) | 40.8 |
| 44 | CA 23rd: Kevin McCarthy (R) | 40.7 |
| 45 | CA 30th: Eddie Bernice Johnson (D) | 40.5 |
| 46 | NJ 9th: Bill Pascrell, Jr. (D) | 40.1 |
| 47 | CA 25th: Mike Garcia (R) | 40.0 |
| 47 | CA 37th: Karen Bass (D) | 40.0 |
| 49 | TX 11th: August Pfluger (R) | 39.9 |
| 50 | WA 4th: Dan Newsome (R) | 39.7 |
| 51 | NY 7th: Nydia M. Velázquez (D) | 39.1 |
| 52 | CA 19th: Zoe Lofgren (D) | 39.0 |
| 53 | CA 9th: Jerry McNerney (D) | 38.4 |
| 54 | TX 9th: Al Green (D) | 38.0 |
| 55 | TX 19th: Jodey C. Arrington (R) | 37.4 |
| 56 | CA 24th: Salud O. Carbajal (D) | 37.0 |
| 57 | CA 47th: Alan S. Lowenthal (D) | 36.0 |
| 58 | CA 53rd: Sara Jacobs (D) | 33.8 |
| 59 | CA 39th: Young Kim (R) | 32.9 |
| 60 | IL 3rd: Marie Newman (D) | 32.3 |
| 61 | CA 50th: Darrell Issa (R) | 32.2 |
| 62 | NV 4th: Steven Horsford (D) | 32.0 |
| 63 | TX 2nd: Dan Crenshaw (R) | 31.6 |
| 64 | Tx 10th: Michael T. McCaul (R) | 30.7 |
| 65 | TX 21st: Chip Roy (R) | 30.5 |
| 66 | TX 7th: Lizzie Fletcher (D) | 30.3 |
| 67 | CA 3rd: John Garamendi (D) | 30.2 |
| 68 | TX 5th: Lance Gooden (R) | 30.1 |
| 69 | CA 6th: Doris O. Matsui (D) | 29.9 |
| 70 | AZ 2nd: Ann Kirkpatrick (D) | 29.4 |
| 71 | CA 30th: Brad Sherman (D) | 29.3 |
| 72 | CA 11th: Mark Desaulnier (D) | 29.1 |
| 73 | CA 5th: Mike Thompson (D) | 28.9 |
| 74 | TX 36th: Brian Babin (R) | 28.5 |
| 75 | TX 13th: Ronny Jackson (R) | 28.4 |

### Percent foreign born, 2019

| Foreign-born rank | State congressional district Representative | Percent foreign born [col 13] |
|---|---|---|
| 1 | FL 25th Mario Diaz-Balart (R) | 57.4 |
| 2 | FL 27th: Maria Elvira Salazar (R) | 55.0 |
| 3 | NY 6th: Grace Meng (D) | 50.4 |
| 4 | CA 17th: Ro Khanna (D) | 50.2 |
| 5 | FL 26th: Carlos A. Gimenez (R) | 48.3 |
| 6 | NJ 8th: Albio Sires (D) | 46.3 |
| 7 | NY 14th: Alexandria Ocasio Cortez (D) | 45.8 |
| 8 | NY 5th: Gregory W. Meeks (D) | 44.4 |
| 9 | CA 34th: Jimmy Gomez (D) | 44.1 |
| 10 | FL 24th: Frederica S. Wilson (D) | 43.9 |
| 11 | CA 29th: Tony Cárdenas (D) | 42.3 |
| 12 | NJ 9th: Bill Pascrell, Jr. (D) | 41.2 |
| 13 | CA 40th: Lucille Roybal-Allard (D) | 39.8 |
| 14 | CA 14th Jackie Speier (D) | 39.1 |
| 15 | CA 28th: Adam B. Schiff (D) | 38.9 |
| 16 | CA 27th: Judy Chu (D) | 38.6 |
| 17 | FL 23rd: Debbie Wasserman Schultz (D) | 37.7 |
| 18 | CA 19th: Zoe Lofgren (D) | 37.0 |
| 19 | NY 9th: Yvette D. Clarke (D) | 36.8 |
| 20 | CA 46th: J. Luis Correa (D) | 36.4 |
| 21 | CA 32nd: Grace F. Napolitano (D) | 36.1 |
| 22 | FL 20th: Vacant | 36.0 |
| 23 | CA 15th: Eric Swalwell (D) | 35.4 |
| 24 | NY 7th: Nydia M. Velázquez (D) | 34.9 |
| 25 | TX 9th: Al Green (D) | 34.6 |
| 26 | NY 13th: Adriano Espaillat (D) | 34.2 |
| 27 | CA 44th: Nanette Diaz Barragan (D) | 34.1 |
| 28 | CA 30th: Brad Sherman (D) | 33.6 |
| 29 | CA 39th: Young Kim (R) | 33.5 |
| 30 | NY 15th: Ritchie Torres (D) | 33.2 |
| 31 | NV 1st: Dina Titus (D) | 32.7 |
| 32 | CA 12th: Nancy Pelosi (D) | 32.2 |
| 32 | CA 43rd: Maxine Waters (D) | 32.2 |
| 34 | IL 4th: Jesus G. "Chuy" Garcia (D) | 32.0 |
| 34 | NY 8th: Hakeem S. Jeffries (D) | 32.0 |
| 36 | CA 37th: Karen Bass (D) | 31.8 |
| 36 | TX 33rd: Marc A. Veasey (D) | 31.8 |
| 38 | MA 7th: Ayanna Presley (D) | 31.7 |
| 39 | WA 9th: Adam Smith (D) | 31.6 |
| 40 | VA 11th: Gerald E. Connolly (D) | 31.5 |
| 41 | CA 18th: Anna G. Eshoo (D) | 31.2 |
| 41 | CA 51st: Juan Vargas (D) | 31.2 |
| 43 | NY 11th: Nicole Malliotakis (R) | 30.9 |
| 44 | CA 38th: Linda T. Sánchez (D) | 30.8 |
| 45 | NY 10th: Jerrold Nadler (D) | 30.6 |
| 45 | TX 7th: Lizzie Fletcher (D) | 30.6 |
| 47 | TX 29th: Sylvia R. Garcia (D) | 30.4 |
| 48 | NJ 10th: Donald M. Payne, Jr. (D) | 30.3 |
| 49 | CA 35th: Norma J. Torres (D) | 30.1 |
| 50 | CA 45th: Katie Porter (D) | 30.0 |
| 51 | NY 16th: Jamaal Bowman (D) | 29.7 |
| 52 | FL 21st: Lois Frankel (D) | 29.5 |
| 53 | NJ 12th: Bonnie Watson Coleman (D) | 29.2 |
| 54 | CA 47th: Alan S. Lowenthal (D) | 29.0 |
| 55 | FL 22nd: Theodore E. Deutch (D) | 28.9 |
| 56 | NJ 6th: Frank Pallone Jr. (D) | 28.8 |
| 57 | IL 8th: Raja Krishnamoorthi (D) | 28.1 |
| 57 | VA 8th: Donald S. Beyer, Jr. (D) | 28.1 |
| 59 | CA 11th: Mark Desaulnier (D) | 27.9 |
| 60 | GA 7th Carolyn Bourdeaux (D) | 27.5 |
| 61 | CA 21st: David G. Valadao (R) | 27.3 |
| 62 | NY 12th: Carolyn B. Maloney (D) | 26.9 |
| 63 | TX 24th: Beth Van Duyne (R) | 26.7 |
| 64 | IL 9th: Janice D. Schakowsky (D) | 26.6 |
| 65 | CA 41st: Mark Takano (D) | 25.8 |
| 66 | CA 13th: Barbara Lee (D) | 25.7 |
| 67 | CA 20th: Jimmy Panetta (D) | 25.1 |
| 68 | FL 10th: Val Butler Demings (D) | 25.0 |
| 69 | MD 8th: Jamie Raskin (D) | 24.9 |
| 70 | CA 16th: Jim Costa (D) | 24.8 |
| 70 | TX 22nd: Troy E. Nehls (R) | 24.8 |
| 72 | MA 5th: Katherine Clark (D) | 24.7 |
| 73 | AZ 7th: Ruben Gallego (D) | 24.5 |
| 73 | CA 48th: Michelle Steel (R) | 24.5 |
| 75 | IL 10th: Bradley Scott Schneider (D) | 24.0 |

# Congressional Districts of the 116th Congress of the United States
## Selected Rankings

| Under 18 years old rank | State congressional district Representative | Percent under 18 years old [col 15 and 16] | 65 years old and over rank | State congressional district Representative | Percent 65 years old and over [col 22 and 23] | College graduates rank | State congressional district Representative | Percent college graduates [col 27] |
|---|---|---|---|---|---|---|---|---|
| 1 | TX 29th: Sylvia R. Garcia (D) | 31.5 | 1 | FL 11th: Daniel Webster (R) | 37.5 | 1 | NY 12th: Carolyn B. Maloney (D) | 72.5 |
| 2 | CA 21st: David G. Valadao (R) | 30.6 | 2 | FL 17th: W. Gregory Steube (R) | 32.9 | 2 | CA 33rd: Ted Lieu (D) | 66.6 |
| 3 | TX 28th: Henry Cuellar (D) | 30.4 | 3 | FL 19th: Byron Donalds (R) | 32.6 | 3 | CA 18th: Anna G. Eshoo (D) | 65.8 |
| 4 | TX 33rd: Marc A. Veasey (D) | 30.3 | 4 | AZ 4th: Paul A. Gosar (R) | 28.0 | 4 | GA 6th: Lucy McBath (D) | 65.1 |
| 5 | UT 1st: Blake D. Moore (R) | 30.1 | 5 | FL 16th: Vern Buchanan (R) | 27.4 | 5 | NY 10th: Jerrold Nadler (D) | 63.3 |
| 5 | UT 4th: Burgess Owens (R) | 30.1 | 6 | FL 21st: Lois Frankel (D) | 26.9 | 5 | VA 8th: Donald S. Beyer, Jr. (D) | 63.3 |
| 7 | TX 15th: Vicente Gonzalez (D) | 29.7 | 7 | FL 18th: Brian J. Mast (R) | 26.4 | 7 | WA 7th: Pramila Jayapal (D) | 62.3 |
| 8 | CA 16th: Jim Costa (D) | 29.6 | 8 | FL 6th: Michael Waltz (R) | 26.0 | 8 | CA 52nd: Scott H. Peters (D) | 61.0 |
| 9 | UT 3rd: John R. Curtis (R) | 29.2 | 9 | FL 8th Bill Posey (R) | 25.9 | 9 | CA 12th: Nancy Pelosi (D) | 60.9 |
| 10 | AZ 7th: Ruben Gallego (D) | 28.8 | 10 | FL 12th: Gus M. Bilirakis (R) | 24.5 | 10 | CA 17th: Ro Khanna (D) | 60.2 |
| 10 | CA 22nd: Devin Nunes (R) | 28.8 | 11 | MI 1st: Jack Bergman (R) | 24.1 | 11 | MA 5th: Katherine Clark (D) | 60.1 |
| 12 | TX 34th: Filemon Vela (D) | 28.6 | 12 | FL 13th: Charlie Crist (D) | 24.0 | 12 | DC: At-Large: Eleanor Holmes Norton (D) | 59.7 |
| 13 | WA 4th: Dan Newsome (R) | 28.5 | 13 | CA 36th: Raul Ruiz (D) | 23.5 | 13 | VA 11th: Gerald E. Connolly (D) | 57.2 |
| 14 | CA 40th: Lucille Roybal-Allard (D) | 28.4 | 14 | NC 11th: Madison Cawthorn (R) | 23.3 | 14 | NJ 11th: Mikie Sherill (D) | 57.1 |
| 15 | CA 8th: Jay Olbernolte (R) | 28.2 | 15 | MA 8th: William R. Keating (D) | 22.8 | 15 | CO 2nd: Joe Neguse (D) | 56.9 |
| 16 | NY 15th: Ritchie Torres (D) | 27.5 | 16 | AZ 8th: Debbie Lesko (R) | 22.4 | 16 | IL 5th: Michael Quigley (D) | 56.8 |
| 17 | TX 22nd: Troy E. Nehls (R) | 27.4 | 16 | FL 22nd: Theodore E. Deutch (D) | 22.4 | 17 | CA 45th: Katie Porter (D) | 56.6 |
| 17 | TX 23rd: Tony Gonzales (R) | 27.4 | 18 | AZ 2nd: Ann Kirkpatrick (D) | 22.3 | 17 | VA 10th: Jennifer Wexton (D) | 56.6 |
| 19 | AZ 3rd: Raul M. Grijalva (D) | 26.9 | 18 | CA 2nd: Jared Huffman (D) | 22.3 | 19 | NY 3rd: Thomas R. Souzzi (D) | 56.2 |
| 19 | GA 7th Carolyn Bourdeaux (D) | 26.9 | 20 | CA 4th: Tom McClintock (R) | 22.2 | 20 | NC 4th: David E. Price (D) | 55.9 |
| 19 | TX 16th: Veronica Escobar (D) | 26.9 | 20 | SC 7th: Tom Rice (R) | 22.2 | 21 | IL 9th: Janice D. Schakowsky (D) | 55.7 |
| 19 | TX 30th: Eddie Bernice Johnson (D) | 26.9 | 22 | PA 14th: Guy Reschenthaler (R) | 22.0 | 22 | NJ 7th: Tom Malinowski (D) | 54.8 |
| 23 | CA 41st: Mark Takano (D) | 26.8 | 23 | WV 3rd: Carol D. Miller (R) | 21.9 | 23 | TX 3rd: Van Taylor (R) | 54.5 |
| 24 | CA 42nd: Ken Calvert (R) | 26.7 | 24 | OR 4th: Peter A. DeFazio (D) | 21.7 | 24 | MD 8th: Jamie Raskin (D) | 54.4 |
| 25 | UT 2nd: Chris Stewart (R) | 26.5 | 25 | ME 2nd: Jared F. Golden (D) | 21.6 | 25 | CA 13th: Barbara Lee (D) | 53.8 |
| 26 | CA 9th: Jerry McNerney (D) | 26.3 | 25 | OR 2nd: Cliff Bentz (R) | 21.6 | 26 | IL 6th: Sean Casten (D) | 53.2 |
| 26 | CA 10th: Josh Harder (D) | 26.3 | 27 | NY 3rd: Thomas R. Souzzi (D) | 21.5 | 27 | CT 4th: James A. Himes (D) | 52.6 |
| 26 | GA 13th: David Scott (D) | 26.3 | 28 | CA 1st: Doug LaMalfa (R) | 21.4 | 28 | MD 3rd: John P. Sarbanes (D) | 52.2 |
| 26 | NE 2nd: Don Bacon (R) | 26.3 | 28 | PA 13th: John Joyce (R) | 21.4 | 29 | CO 1st: Diana DeGette (D) | 52.1 |
| 26 | VA 10th: Jennifer Wexton (D) | 26.3 | 30 | PA 15th: Glen Thompson (R) | 21.3 | 30 | MA 4th: Jake Auchincloss (D) | 52.0 |
| 31 | CA 23rd: Kevin McCarthy (R) | 26.2 | 30 | WA 6th: Derek Kilmer (D) | 21.3 | 31 | MO 2nd: Ann Wagner (R) | 51.8 |
| 32 | TX 31st: John R. Carter (R) | 26.0 | 32 | MN 8th: Pete Stauber (R) | 21.2 | 32 | NJ 5th: Josh Gottheimer (D) | 50.6 |
| 33 | TX 6th: Vacant | 25.9 | 33 | NY 19th: Antonio Delgado (D) | 21.1 | 33 | TX 7th: Lizzie Fletcher (D) | 50.3 |
| 33 | TX 18th: Sheila Jackson-Lee (D) | 25.9 | 33 | VA 5th: Bob Good (R) | 21.1 | 34 | MN 3rd: Dean Philips (D) | 50.2 |
| 35 | CA 35th: Norma J. Torres (D) | 25.8 | 33 | VA 9th: Morgan Griffith (R) | 21.1 | 35 | NY 17th: Mondaire Jones (D) | 49.6 |
| 35 | ID 2nd: Michael K. Simpson (R) | 25.8 | 33 | WI 7th: Thomas P. Tiffany (R) | 21.1 | 36 | CA 14th Jackie Speier (D) | 49.5 |
| 35 | TX 20th: Joaquin Castro (D) | 25.8 | 37 | ME 1st: Chellie Pingree (D) | 20.9 | 37 | MA 8th: Stephen F. Lynch (D) | 49.2 |
| 38 | Tx 10th: Michael T. McCaul (R) | 25.7 | 38 | FL 2nd: Neal P. Dunn (R) | 20.8 | 38 | MI 11th: Haley M. Stevens (D) | 49.1 |
| 39 | CA 25th: Mike Garcia (R) | 25.6 | 38 | TN 1st: Diana Harshbarger (R) | 20.8 | 39 | KS 3rd: Sharice Davids (D) | 48.5 |
| 39 | IN 7th André Carson (D) | 25.6 | 40 | NC 7th: David Rouzer (R) | 20.7 | 39 | PA 4th: Madeleine Dean (D) | 48.5 |
| 39 | TX 26th: Michael C. Burgess (R) | 25.6 | 40 | OH 14th: David P. Joyce (R) | 20.7 | 41 | TX 24rd: Beth Van Duyne (R) | 48.4 |
| 42 | CA 44th: Nanette Diaz Barragan (D) | 25.5 | 40 | OH 16th: Anthony Gonzalez (R) | 20.7 | 42 | MN 5th: Ilhan Omar (D) | 48.3 |
| 42 | TX 11th: August Pfluger (R) | 25.5 | 43 | PA 17th: Conor Lamb (D) | 20.5 | 43 | CA 28th: Adam B. Schiff (D) | 48.1 |
| 44 | AZ 5th: Andy Biggs (R) | 25.4 | 44 | OH 6th: Bill Johnson (R) | 20.4 | 44 | CA 15th: Eric Swalwell (D) | 48.0 |
| 44 | IN 3rd: Jim Banks (R) | 25.4 | 45 | NE 3rd: Adrian Smith (R) | 20.3 | 45 | IN 5th: Victoria Spartz (R) | 47.9 |
| 44 | TX 3rd: Van Taylor (R) | 25.4 | 45 | PA 9th: Daniel Meuser (R) | 20.3 | 46 | GA 5th: Nikema Williams (D) | 47.4 |
| 47 | MN 6th: Tom Emmer (R) | 25.3 | 45 | WV 1st: David McKinley (R) | 20.3 | 46 | MA 6th: Seth Moulton (D) | 47.4 |
| 47 | TX 9th: Al Green (D) | 25.3 | 48 | MO 2nd: Ann Wagner (R) | 20.2 | 46 | TX 21st: Chip Roy (R) | 47.4 |
| 49 | IL 14th: Lauren Underwood (D) | 25.1 | 48 | NJ 3rd: Andy Kim (D) | 20.2 | 49 | CA 49th: Mike Levin (D) | 47.2 |
| 49 | KS 4th: Ron Estes (R) | 25.1 | 48 | PA 16th: Mike Kelly (R) | 20.2 | 50 | MA 7th: Ayanna Presley (D) | 47.1 |
| 49 | TX 36th: Brian Babin (R) | 25.1 | 51 | MN 7th: Michelle Fischbach (R) | 20.1 | 50 | NJ 12th: Bonnie Watson Coleman (D) | 47.1 |
| 49 | WA 8th: Kim Schrier (D) | 25.1 | 52 | AZ 6th: David Schweikert (R) | 20.0 | 52 | PA 6th: Chrissy Houlahan (D) | 46.7 |
| 53 | OK 5th: Stephanie I. Bice (R) | 25.0 | 52 | NY 27th: Chris Jacobs (R) | 20.0 | 53 | CA 27th: Judy Chu (D) | 45.9 |
| 53 | WI 4th: Gwen Moore (D) | 25.0 | 52 | PA 8th: Matt Cartwright (D) | 20.0 | 54 | TX 26th: Michael C. Burgess (R) | 45.6 |
| 55 | CA 6th: Doris O. Matsui (D) | 24.9 | 52 | VT: At-large Peter Welch (D) | 20.0 | 55 | TX 22nd: Troy E. Nehls (R) | 45.4 |
| 55 | CA 51st: Juan Vargas (D) | 24.9 | 56 | NJ 2nd: Jefferson Van Drew (R) | 19.9 | 55 | TX 32nd: Colin Z. Allred (D) | 45.3 |
| 55 | KS 3rd: Sharice Davids (D) | 24.9 | 57 | AR 4th: Bruce Westerman (R) | 19.8 | 56 | WA 1st: Suzan K. DelBene (D) | 45.3 |
| 55 | NC 2nd: Deborah K. Ross (D) | 24.9 | 57 | PA 12th: Fred Keller (R) | 19.8 | 58 | CA 30th: Brad Sherman (D) | 45.2 |
| 55 | OK 1st: Kevin Hern (R) | 24.9 | 59 | AZ 1st: Tom O'Halleran (D) | 19.7 | 59 | GA 7th Carolyn Bourdeaux (D) | 45.0 |
| 55 | PA 2nd: Brendan F. Boyle (D) | 24.9 | 59 | MI 4th: John L. Moolenaar (R) | 19.7 | 60 | IL 7th: Danny K. Davis (D) | 44.9 |
| 55 | TX 8th: Kevin Brady (R) | 24.9 | 61 | CO 3rd: Lauren Boebert (R) | 19.6 | 60 | WI 2nd: Mark Pocan (D) | 44.9 |
| 62 | NV 4th: Steven Horsford (D) | 24.8 | 61 | NY 23rd: Tom Reed (R) | 19.6 | 62 | CA 11th: Mark Desaulnier (D) | 44.9 |
| 62 | TX 13th: Ronny Jackson (R) | 24.8 | 63 | IL 18th: Darin LaHood (R) | 19.5 | 63 | MN 4th: Betty McCollum (D) | 44.7 |
| 64 | CA 31st: Pete Aguilar (D) | 24.7 | 63 | NC 10th: Patrick T. McHenry (R) | 19.5 | 64 | CA 48th: Michelle Steel (R) | 44.6 |
| 64 | TN 9th: Steve Cohen (D) | 24.7 | 63 | OH 13th: Tim Ryan (D) | 19.5 | 64 | TX 2nd: Dan Crenshaw (R) | 44.6 |
| 64 | TX 5th: Lance Gooden (R) | 24.7 | 63 | WV 2nd: Alexander X. Mooney (R) | 19.5 | 64 | WA 9th: Adam Smith (D) | 44.6 |
| 64 | TX 12th: Kay Granger (R) | 24.7 | 67 | DE: At-Large: Lisa Blunt Rochester (D) | 19.4 | 67 | NJ 4th: Christopher H. Smith (R) | 44.3 |
| 68 | AK At-Large: Don Young (R) | 24.6 | 67 | MD 1st: Andrew Harris (R) | 19.4 | 67 | OR 3rd: Earl Blumenauer (D) | 44.3 |
| 68 | LA 3rd: Clay Higgins (R) | 24.6 | 67 | MT At large: Matthew M. Rosendale Sr. (R) | 19.4 | 69 | IL 10th: Bradley Scott Schneider (D) | 44.2 |
| 68 | MI 13th: Rashida Tlaib (D) | 24.6 | 70 | IL 17th: Cheri Bustos (D) | 19.3 | 70 | CO 6th: Jason Crow (D) | 44.0 |
| 68 | TX 27th: Michael Cloud (R) | 24.6 | 70 | MI 10th: Lisa McClain (R) | 19.3 | 71 | NY 4th: Kathleen M. Rice (D) | 43.6 |
| 72 | AR 3rd: Steve Womack (R) | 24.5 | 70 | OK 2nd: Markwayne Mullin (R) | 19.3 | 72 | AZ 6th: David Schweikert (R) | 43.5 |
| 72 | GA 4th: Henry C. "Hank" Johnson Jr. (D) | 24.5 | 73 | IL 15th: Mary E. Miller (R) | 19.2 | 73 | MD 6th: David J. Trone (D) | 43.3 |
| 72 | ID 1st: Russ Fulcher (R) | 24.5 | 73 | NJ 4th: Christopher H. Smith (R) | 19.2 | 74 | CA 39th: Young Kim (R) | 43.1 |
| 72 | IL 11th: Bill Foster (D) | 24.5 | 75 | FL 25th Mario Diaz-Balart (R) | 19.1 | 75 | TN 5th: Jim Cooper (D) | 43.0 |

## Selected Rankings

| | Median value of owner-occupied housing units, 2019 | | | Percent female-headed family households, 2019 | | | Percent of households with one person, 2019 | |
|---|---|---|---|---|---|---|---|---|
| Median value rank | State congressional district Representative | Median value (dollars) [col 43] | Female householder rank | State congressional district Representative | Percent with female householder [col 32] | One-person household rank | State congressional district Representative | Percent one-person households [col 33] |
| 1 | CA 18th: Anna G. Eshoo (D) | 1,588,400 | 1 | NY 15th: Ritchie Torres (D) | 32.7 | 1 | NY 12th: Carolyn B. Maloney (D) | 47.1 |
| 2 | CA 33rd: Ted Lieu (D) | 1,419,000 | 2 | CA 40th: Lucille Roybal-Allard (D) | 25.6 | 2 | DC: At-Large: Eleanor Holmes Norton (D) | 44.8 |
| 3 | CA 12th: Nancy Pelosi (D) | 1,226,000 | 3 | MS 2nd: Bennie G. Thompson (D) | 23.3 | 3 | IL 7th: Danny K. Davis (D) | 43.0 |
| 4 | CA 14th Jackie Speier (D) | 1,161,400 | 4 | PA 2nd: Brendan F. Boyle (D) | 22.6 | 4 | OH 11th: Vacant | 42.7 |
| 5 | NY 12th: Carolyn B. Maloney (D) | 1,141,500 | 5 | MI 13th: Rashida Tlaib (D) | 22.4 | 5 | PA 3rd: Dwight Evans (D) | 42.2 |
| 6 | CA 17th: Ro Khanna (D) | 1,137,100 | 5 | NY 8th: Hakeem S. Jeffries (D) | 22.4 | 6 | GA 5th: Nikema Williams (D) | 41.9 |
| 7 | CA 10th: Jerrold Nadler (D) | 1,121,800 | 7 | NY 5th: Gregory W. Meeks (D) | 22.1 | 7 | NY 10th: Jerrold Nadler (D) | 41.6 |
| 8 | CA 19th: Zoe Lofgren (D) | 894,400 | 8 | IL 2nd: Robin L. Kelly (D) | 22.0 | 8 | MO 1st: Cori Bush (D) | 41.5 |
| 9 | CA 15th: Eric Swalwell (D) | 888,000 | 9 | TN 9th: Steve Cohen (D) | 21.7 | 9 | NY 13th: Adriano Espaillat (D) | 40.7 |
| 10 | CA 28th: Adam B. Schiff (D) | 882,900 | 10 | NY 13th: Adriano Espaillat (D) | 21.6 | 10 | IN 7th André Carson (D) | 40.1 |
| 11 | CA 13th: Barbara Lee (D) | 858,500 | 10 | FL 20th: Vacant | 21.6 | 11 | MN 5th: Ilhan Omar (D) | 39.5 |
| 12 | NY 7th: Nydia M. Velázquez (D) | 849,900 | 12 | AZ 7th: Ruben Gallego (D) | 21.2 | 12 | MI 13th: Rashida Tlaib (D) | 39.4 |
| 13 | CA 48th: Michelle Steel (R) | 840,200 | 13 | CA 44th: Nanette Diaz Barragan (D) | 21.0 | 13 | LA 2nd: Troy Carter (D) | 39.0 |
| 14 | CA 37th: Karen Bass (D) | 828,000 | 14 | AL 7th: Terri A. Sewell (D) | 20.9 | 14 | PA 18th: Michael F. Doyle (D) | 38.8 |
| 15 | CA 45th: Katie Porter (D) | 794,400 | 14 | TX 29th: Sylvia R. Garcia (D) | 20.9 | 15 | TN 9th: Steve Cohen (D) | 38.2 |
| 16 | CA 52nd: Scott H. Peters (D) | 775,300 | 16 | FL 24th: Frederica S. Wilson (D) | 20.8 | 16 | CA 12th: Nancy Pelosi (D) | 38.0 |
| 17 | CA 27th: Judy Chu (D) | 770,100 | 17 | GA 13th: David Scott (D) | 20.7 | 17 | CA 28th: Adam B. Schiff (D) | 37.5 |
| 18 | NY 9th: Yvette D. Clarke (D) | 759,500 | 18 | CA 16th: Jim Costa (D) | 20.4 | 18 | MI 14th: Brenda L. Lawrence (D) | 37.3 |
| 19 | CA 49th: Mike Levin (D) | 748,200 | 19 | TX 34th: Filemon Vela (D) | 20.1 | 19 | NY 26th: Brian Higgins (D) | 37.1 |
| 20 | CA 11th: Mark Desaulnier (D) | 735,200 | 19 | GA 2nd: Sanford D. Bishop Jr. (D) | 20.1 | 20 | FL 13th: Charlie Crist (D) | 37.0 |
| 21 | HI 1st: Ed Case (D) | 728,900 | 21 | NY 9th: Yvette D. Clarke (D) | 19.8 | 21 | CO 1st: Diana DeGette (D) | 36.5 |
| 22 | WA 7th: Pramila Jayapal (D) | 718,500 | 22 | TX 30th: Eddie Bernice Johnson (D) | 19.7 | 21 | MD 7th: Kweisi Mfume (D) | 36.5 |
| 23 | CA 30th: Brad Sherman (D) | 717,000 | 23 | TX 33rd: Marc A. Veasey (D) | 19.6 | 23 | SC 6th: James E. Clyburn (D) | 36.4 |
| 24 | CA 2nd: Jared Huffman (D) | 694,400 | 24 | TX 16th: Veronica Escobar (D) | 19.5 | 23 | WA 7th: Pramila Jayapal (D) | 36.4 |
| 25 | NY 6th: Grace Meng (D) | 694,100 | 24 | LA 2nd: Troy Carter (D) | 19.5 | 25 | IL 13th: Rodney Davis (R) | 36.2 |
| 26 | CA 34th: Jimmy Gomez (D) | 688,100 | 26 | CA 21st: David G. Valadao (R) | 19.3 | 25 | WI 4th: Gwen Moore (D) | 36.2 |
| 27 | CA 39th: Young Kim (R) | 685,300 | 26 | OH 11th: Vacant | 19.3 | 27 | OH 9th: Marcy Kaptur (D) | 36.1 |
| 28 | CA 20th: Jimmy Panetta (D) | 666,600 | 26 | VA 3rd: Robert C. "Bobby" Scott (D) | 19.3 | 28 | OH 13th: Tim Ryan (D) | 35.9 |
| 29 | NY 8th: Hakeem S. Jeffries (D) | 661,900 | 28 | NJ 10th: Donald M. Payne, Jr. (D) | 19.1 | 29 | IL 1st: Bobby L. Rush (D) | 35.5 |
| 30 | DC: At-Large: Eleanor Holmes Norton (D) | 646,500 | 29 | TX 15th: Vicente Gonzalez (D) | 19.1 | 29 | IL 2nd: Robin L. Kelly (D) | 35.5 |
| 31 | NY 3rd: Thomas R. Souzzi (D) | 645,300 | 29 | FL 5th: Al Lawson (D) | 19.1 | 31 | AL 7th: Terri A. Sewell (D) | 35.4 |
| 32 | CA 47th: Alan S. Lowenthal (D) | 640,300 | 32 | CA 51st: Juan Vargas (D) | 19.0 | 32 | NV 1st: Dina Titus (D) | 35.2 |
| 33 | CA 26th: Julia Brownley (D) | 631,200 | 33 | TX 18th: Sheila Jackson-Lee (D) | 18.9 | 33 | TN 5th: Jim Cooper (D) | 34.8 |
| 34 | NY 11th: Nicole Malliotakis (R) | 630,200 | 33 | GA 4th: Henry C. "Hank" Johnson Jr. (D) | 18.9 | 34 | OH 3rd: Joyce Beatty (D) | 34.7 |
| 35 | CA 24th: Salud O. Carbajal (D) | 624,900 | 35 | MI 14th: Brenda L. Lawrence (D) | 18.8 | 35 | CA 13th: Barbara Lee (D) | 34.6 |
| 36 | MA 5th: Katherine Clark (D) | 622,800 | 35 | SC 6th: James E. Clyburn (D) | 18.8 | 35 | MO 5th: Emanuel Cleaver (D) | 34.6 |
| 37 | HI 2nd: Kaiali'l Kahele (D) | 611,600 | 35 | TX 9th: Al Green (D) | 18.8 | 37 | NY 25th: Joseph D. Morelle (D) | 34.5 |
| 38 | CA 43rd: Maxine Waters (D) | 608,300 | 38 | WI 4th: Gwen Moore (D) | 18.7 | 38 | KY 3rd: John A. Yarmuth (D) | 34.3 |
| 39 | NY 14th: Alexandria Ocasio Cortez (D) | 601,000 | 39 | TX 20th: Joaquin Castro (D) | 18.6 | 38 | NC 12th: Alma S. Adams (D) | 34.3 |
| 40 | MA 7th: Ayanna Presley (D) | 589,700 | 40 | FL 26th: Carlos A. Gimenez (R) | 18.5 | 40 | FL 22nd: Theodore E. Deutch (D) | 34.2 |
| 41 | VA 8th: Donald S. Beyer, Jr. (D) | 588,000 | 41 | PA 3rd: Dwight Evans (D) | 18.4 | 41 | AZ 2nd: Ann Kirkpatrick (D) | 34.1 |
| 42 | CA 46th: J. Luis Correa (D) | 586,500 | 41 | GA 12th: Rick W. Allen (R) | 18.4 | 41 | IL 5th: Michael Quigley (D) | 34.1 |
| 43 | CA 53rd: Sara Jacobs (D) | 583,900 | 41 | IL 1st: Bobby L. Rush (D) | 18.4 | 41 | IL 9th: Janice D. Schakowsky (D) | 34.1 |
| 44 | CA 5th: Mike Thompson (D) | 583,600 | 44 | TX 28th: Henry Cuellar (D) | 18.2 | 41 | MA 7th: Ayanna Presley (D) | 34.1 |
| 45 | CA 38th: Linda T. Sánchez (D) | 578,600 | 45 | CA 35th: Norma J. Torres (D) | 18.1 | 41 | NY 15th: Ritchie Torres (D) | 34.1 |
| 46 | CA 29th: Tony Cárdenas (D) | 565,000 | 46 | CA 43rd: Maxine Waters (D) | 17.8 | 46 | CA 33rd: Ted Lieu (D) | 33.9 |
| 47 | CA 32nd: Grace F. Napolitano (D) | 555,700 | 46 | NY 16th: Jamaal Bowman (D) | 17.8 | 46 | FL 5th: Al Lawson (D) | 33.9 |
| 48 | WA 1st: Suzan K. DelBene (D) | 552,700 | 46 | LA 5th: Julia Letlow (R) | 17.8 | 46 | IL 17th: Cheri Bustos (D) | 33.9 |
| 49 | CA 50th: Darrell Issa (R) | 551,200 | 49 | MD 4th: Anthony G. Brown (D) | 17.7 | 46 | NM 1st: Melanie A. Stansbury (D) | 33.9 |
| 50 | NY 5th: Gregory W. Meeks (D) | 545,900 | 50 | CA 29th: Tony Cárdenas (D) | 17.4 | 50 | CA 37th: Karen Bass (D) | 33.7 |
| 51 | WA 9th: Adam Smith (D) | 536,400 | 50 | MO 1st: Cori Bush (D) | 17.4 | 50 | ND At-Large: Kelly Armstrong (R) | 33.7 |
| 52 | NY 4th: Kathleen M. Rice (D) | 535,000 | 50 | TX 35th: Lloyd Doggett (D) | 17.4 | 50 | VA 8th: Donald S. Beyer, Jr. (D) | 33.7 |
| 53 | VA 10th: Jennifer Wexton (D) | 529,100 | 50 | MD 7th: Kweisi Mfume (D) | 17.4 | 53 | GA 2nd: Sanford D. Bishop Jr. (D) | 33.5 |
| 54 | CT 4th: James A. Himes (D) | 528,400 | 54 | NC 1st: G. K. Butterfield (D) | 17.3 | 53 | OH 10th: Michael R. Turner (R) | 33.5 |
| 55 | NY 16th: Jamaal Bowman (D) | 514,200 | 55 | AZ 3rd: Raul M. Grijalva (D) | 17.2 | 55 | MI 9th: Andy Levin (D) | 33.2 |
| 56 | MA 6th: Seth Moulton (D) | 508,900 | 55 | CA 32nd: Grace F. Napolitano (D) | 17.2 | 56 | PA 17th: Conor Lamb (D) | 33.1 |
| 57 | CA 40th: Lucille Roybal-Allard (D) | 500,800 | 55 | LA 4th: Mike Johnson (R) | 17.2 | 56 | TX 21st: Chip Roy (R) | 33.1 |
| 58 | VA 11th: Gerald E. Connolly (D) | 500,100 | 58 | CA 38th: Linda T. Sánchez (D) | 17.1 | 58 | NY 24th: John Katko (R) | 33.0 |
| 59 | CA 25th: Mike Garcia (R) | 496,700 | 59 | CA 41st: Mark Takano (D) | 17.0 | 59 | CT 3rd: Rosa L. DeLauro (D) | 32.9 |
| 60 | NY 17th: Mondaire Jones (D) | 496,500 | 60 | CA 31st: Pete Aguilar (D) | 16.8 | 60 | AZ 9th: Greg Stanton (D) | 32.8 |
| 61 | NY 13th: Adriano Espaillat (D) | 493,900 | 60 | NV 1st: Dina Titus (D) | 16.8 | 60 | FL 8th Bill Posey (R) | 32.8 |
| 62 | CO 2nd: Joe Neguse (D) | 489,700 | 62 | OH 9th: Marcy Kaptur (D) | 16.6 | 60 | MS 3rd: Michael Guest (R) | 32.8 |
| 63 | NY 15th: Ritchie Torres (D) | 479,100 | 63 | AL 2nd: Barry Moore(R) | 16.3 | 60 | RI 1st: David Cicilline (D) | 32.8 |
| 64 | MA 8th: Stephen F. Lynch (D) | 474,100 | 63 | CA 46th: J. Luis Correa (D) | 16.3 | 64 | NY 23rd: Tom Reed (R) | 32.6 |
| 65 | CA 4th: Tom McClintock (R) | 473,100 | 65 | NJ 8th: Albio Sires (D) | 16.2 | 64 | NC 1st: G. K. Butterfield (D) | 32.6 |
| 66 | CA 44th: Nanette Diaz Barragan (D) | 471,900 | 65 | OH 3rd: Joyce Beatty (D) | 16.2 | 66 | VA 4th: A. Donald McEachin (D) | 32.5 |
| 67 | NJ 11th: Mikie Sherill (D) | 467,800 | 65 | GA 5th: Nikema Williams (D) | 16.2 | 67 | IL 12th: Mike Bost (R) | 32.4 |
| 68 | MD 8th: Jamie Raskin (D) | 467,200 | 68 | NY 7th: Nydia M. Velázquez (D) | 16.1 | 68 | CA 34th: Jimmy Gomez (D) | 32.3 |
| 69 | NJ 7th: Tom Malinowski (D) | 461,800 | 69 | MS 4th: Steven M. Palazzo (R) | 16.0 | 68 | MI 5th: Daniel T. Kildee (D) | 32.3 |
| 70 | CA 42nd: Ken Calvert (R) | 455,800 | 70 | NY 14th: Alexandria Ocasio Cortez (D) | 15.9 | 68 | PA 14th: Guy Reschenthaler (R) | 32.3 |
| 71 | MA 4th: Jake Auchincloss (D) | 454,500 | 70 | SC 7th: Tom Rice (R) | 15.9 | 71 | MN 4th: Betty McCollum (D) | 32.1 |
| 72 | CO 1st: Diana DeGette (D) | 449,200 | 70 | LA 3rd: Clay Higgins (R) | 15.9 | 71 | NY 20th: Paul Tonko (D) | 32.1 |
| 73 | WA 8th: Kim Schrier (D) | 442,900 | 73 | FL 25th Mario Diaz-Balart (R) | 15.8 | 73 | NJ 10th: Donald M. Payne, Jr. (D) | 31.9 |
| 74 | NJ 5th: Josh Gottheimer (D) | 431,000 | 74 | TX 23rd: Tony Gonzales (R) | 15.7 | 73 | VA 9th: Morgan Griffith (R) | 31.9 |
| 75 | WA 2nd: Rick Larsen (D) | 430,100 | 74 | MD 2nd: C. A. Dutch Ruppersberger (D) | 15.7 | 75 | ME 1st: Chellie Pingree (D) | 31.8 |

# Congressional Districts of the 116th Congress of the United States
## Selected Rankings

### Median household income, 2019

| Median income rank | State congressional district Representative | Median income (dollars) [col 47] |
|---|---|---|
| 1 | CA 18th: Anna G. Eshoo (D) | 149,375 |
| 2 | CA 17th: Ro Khanna (D) | 147,671 |
| 3 | VA 10th: Jennifer Wexton (D) | 132,226 |
| 4 | CA 12th: Nancy Pelosi (D) | 127,290 |
| 5 | NY 3rd: Thomas R. Souzzi (D) | 126,191 |
| 6 | CA 14th Jackie Speier (D) | 125,980 |
| 7 | CA 15th: Eric Swalwell (D) | 125,018 |
| 8 | NY 12th: Carolyn B. Maloney (D) | 124,502 |
| 9 | NJ 11th: Mikie Sherill (D) | 120,847 |
| 10 | VA 11th: Gerald E. Connolly (D) | 118,099 |
| 11 | CA 33rd: Ted Lieu (D) | 117,012 |
| 12 | NJ 7th: Tom Malinowski (D) | 115,585 |
| 13 | CA 45th: Katie Porter (D) | 115,427 |
| 14 | VA 8th: Donald S. Beyer, Jr. (D) | 111,627 |
| 15 | NY 4th: Kathleen M. Rice (D) | 110,677 |
| 16 | NJ 5th: Josh Gottheimer (D) | 110,329 |
| 17 | MD 8th: Jamie Raskin (D) | 109,016 |
| 18 | NY 2nd: Andrew R. Garbarino (R) | 108,725 |
| 19 | NY 17th: Mondaire Jones (D) | 108,449 |
| 20 | CA 19th: Zoe Lofgren (D) | 107,240 |
| 21 | CA 52nd: Scott H. Peters (D) | 106,524 |
| 22 | MA 5th: Katherine Clark (D) | 106,311 |
| 23 | WA 1st: Suzan K. DelBene (D) | 106,190 |
| 24 | IL 6th: Sean Casten (D) | 105,292 |
| 25 | MA 4th: Jake Auchincloss (D) | 104,857 |
| 26 | CA 11th: Mark Desaulnier (D) | 103,580 |
| 27 | NY 10th: Jerrold Nadler (D) | 103,331 |
| 28 | NY 1st: Lee M. Zeldin (R) | 101,701 |
| 29 | TX 22nd: Troy E. Nehls (R) | 101,658 |
| 30 | CT 4th: James A. Himes (D) | 101,646 |
| 31 | MD 5th: Steny H. Hoyer (D) | 101,298 |
| 32 | WA 7th: Pramila Jayapal (D) | 100,991 |
| 33 | MA 8th: Stephen F. Lynch (D) | 100,690 |
| 34 | CA 48th: Michelle Steel (R) | 100,604 |
| 35 | GA 6th: Lucy McBath (D) | 100,110 |
| 36 | CA 49th: Mike Levin (D) | 100,037 |
| 37 | IL 14th: Lauren Underwood (D) | 100,011 |
| 38 | MN 3rd: Dean Philips (D) | 98,877 |
| 39 | MA 6th: Seth Moulton (D) | 97,115 |
| 40 | CA 39th: Young Kim (R) | 96,431 |
| 41 | TX 26th: Michael C. Burgess (R) | 96,307 |
| 42 | WA 8th: Kim Schrier (D) | 95,968 |
| 43 | TX 3rd: Van Taylor (R) | 95,619 |
| 44 | MD 3rd: John P. Sarbanes (D) | 94,736 |
| 45 | IL 5th: Michael Quigley (D) | 94,144 |
| 46 | PA 1st: Brian K. Fitzpatrick (R) | 93,474 |
| 47 | DC: At-Large: Eleanor Holmes Norton (D) | 92,266 |
| 48 | NY 18th: Sean Patrick Maloney (D) | 91,723 |
| 49 | NJ 6th: Frank Pallone Jr. (D) | 91,654 |
| 50 | CA 26th: Julia Brownley (D) | 91,602 |
| 51 | CA 13th: Barbara Lee (D) | 91,514 |
| 52 | NJ 4th: Christopher H. Smith (R) | 91,212 |
| 53 | PA 4th: Madeleine Dean (D) | 91,030 |
| 54 | CA 27th: Judy Chu (D) | 90,792 |
| 55 | CA 42nd: Ken Calvert (R) | 90,651 |
| 56 | MN 2nd: Angie Craig (D) | 90,531 |
| 57 | WA 9th: Adam Smith (D) | 90,353 |
| 58 | VA 1st: Robert J. Wittman (R) | 90,181 |
| 59 | CA 30th: Brad Sherman (D) | 89,460 |
| 60 | MO 2nd: Ann Wagner (R) | 88,684 |
| 61 | MD 6th: David J. Trone (D) | 88,592 |
| 62 | MI 11th: Haley M. Stevens (D) | 88,253 |
| 63 | MD 4th: Anthony G. Brown (D) | 88,207 |
| 64 | CO 2nd: Joe Neguse (D) | 87,585 |
| 65 | NJ 12th: Bonnie Watson Coleman (D) | 87,559 |
| 66 | CO 6th: Jason Crow (D) | 87,312 |
| 67 | HI 1st: Ed Case (D) | 86,674 |
| 68 | CA 4th: Tom McClintock (R) | 86,374 |
| 69 | CA 5th: Mike Thompson (D) | 85,856 |
| 70 | NJ 3rd: Andy Kim (D) | 85,746 |
| 71 | PA 6th: Chrissy Houlahan (D) | 85,665 |
| 72 | CA 25th: Mike Garcia (R) | 84,670 |
| 73 | IL 10th: Bradley Scott Schneider (D) | 84,608 |
| 74 | MN 6th: Tom Emmer (R) | 84,159 |
| 75 | CO 4th: Ken Buck (R) | 83,609 |

### Percent of persons below 65 years with no health insurance, 2019

| No health insurance rank | State congressional district Representative | Percent with no health insurance [col 59] |
|---|---|---|
| 1 | TX 29th: Sylvia R. Garcia (D) | 32.3 |
| 2 | TX 33rd: Marc A. Veasey (D) | 31.9 |
| 3 | TX 15th: Vicente Gonzalez (D) | 27.2 |
| 3 | TX 34th: Filemon Vela (D) | 27.2 |
| 5 | TX 18th: Sheila Jackson-Lee (D) | 26.2 |
| 6 | TX 28th: Henry Cuellar (D) | 26.1 |
| 7 | TX 9th: Al Green (D) | 25.9 |
| 8 | TX 30th: Eddie Bernice Johnson (D) | 22.5 |
| 9 | TX 16th: Veronica Escobar (D) | 21.9 |
| 10 | TX 5th: Lance Gooden (R) | 21.5 |
| 11 | TX 35th: Lloyd Doggett (D) | 21.0 |
| 12 | AZ 7th: Ruben Gallego (D) | 20.2 |
| 13 | FL 24th: Frederica S. Wilson (D) | 19.4 |
| 14 | TX 36th: Brian Babin (R) | 19.2 |
| 15 | TX 20th: Joaquin Castro (D) | 19.0 |
| 16 | FL 20th: Vacant | 18.6 |
| 16 | NV 1st: Dina Titus (D) | 18.6 |
| 18 | TX 27th: Michael Cloud (R) | 18.5 |
| 19 | OK 2nd: Markwayne Mullin (R) | 18.2 |
| 19 | TX 32nd: Colin Z. Allred (D) | 18.2 |
| 21 | TX 11th: August Pfluger (R) | 17.8 |
| 22 | TX 19th: Jodey C. Arrington (R) | 17.7 |
| 23 | GA 4th: Henry C. "Hank" Johnson Jr. (D) | 17.0 |
| 23 | TX 14th: Randy K. Weber Sr. (R) | 17.0 |
| 25 | TX 7th: Lizzie Fletcher (D) | 16.9 |
| 26 | TX 1st: Louie Gohmert (R) | 16.7 |
| 27 | TX 23rd: Tony Gonzales (R) | 16.6 |
| 28 | TX 13th: Ronny Jackson (R) | 16.5 |
| 29 | FL 25th Mario Diaz-Balart (R) | 16.3 |
| 30 | CA 34th: Jimmy Gomez (D) | 16.2 |
| 31 | NJ 8th: Albio Sires (D) | 16.0 |
| 32 | FL 26th: Carlos A. Gimenez (R) | 15.9 |
| 33 | TX 4th: Pat Fallon (R) | 15.8 |
| 33 | TX 12th: Kay Granger (R) | 15.8 |
| 35 | CA 40th: Lucille Roybal-Allard (D) | 15.6 |
| 36 | TX 2nd: Dan Crenshaw (R) | 15.3 |
| 37 | GA 9th: Andrew S. Clyde (R) | 15.1 |
| 38 | IL 4th: Jesus G. "Chuy" Garcia (D) | 15.0 |
| 39 | GA 8th: Austin Scott (D) | 14.9 |
| 40 | OK 5th: Stephanie I. Bice (R) | 14.7 |
| 41 | FL 21st: Lois Frankel (D) | 14.5 |
| 41 | GA 1st: Earl L. "Buddy" Carter (R) | 14.5 |
| 41 | TX 6th: Vacant | 14.5 |
| 44 | AZ 3rd: Raul M. Grijalva (D) | 14.4 |
| 44 | GA 2nd: Sanford D. Bishop Jr. (D) | 14.4 |
| 46 | GA 13th: David Scott (D) | 14.3 |
| 46 | TN 9th: Steve Cohen (D) | 14.3 |
| 48 | MS 2nd: Bennie G. Thompson (D) | 14.2 |
| 48 | FL 10th: Val Butler Demings (D) | 14.2 |
| 50 | MS 4th: Steven M. Palazzo (R) | 14.1 |
| 51 | GA 7th Carolyn Bourdeaux (D) | 14.0 |
| 52 | Tx 10th: Michael T. McCaul (R) | 13.9 |
| 53 | FL 27th: Maria Elvira Salazar (R) | 13.7 |
| 53 | MO 7th: Bill Long (R) | 13.7 |
| 53 | OK 1st: Kevin Hern (R) | 13.7 |
| 53 | FL 5th: Al Lawson (D) | 13.7 |
| 57 | GA 14th: Marjorie Taylor Greene (R) | 13.6 |
| 57 | NC 12th: Alma S. Adams (D) | 13.6 |
| 57 | TX 24rd: Beth Van Duyne (R) | 13.6 |
| 60 | FL 22nd: Theodore E. Deutch (D) | 13.3 |
| 60 | FL 6th: Michael Waltz (R) | 13.3 |
| 62 | NC 11th: Madison Cawthorn (R) | 13.2 |
| 63 | GA 12th: Rick W. Allen (R) | 13.1 |
| 63 | OK 3rd: Frank D. Lucas (R) | 13.1 |
| 63 | SC 6th: James E. Clyburn (D) | 13.1 |
| 66 | SC 7th: Tom Rice (R) | 13.0 |
| 67 | AZ 1st: Tom O'Halleran (D) | 12.9 |
| 67 | FL 15th: C. Scott Franklin (R) | 12.9 |
| 67 | MO 8th: Jason T. Smith (R) | 12.9 |
| 67 | NJ 9th: Bill Pascrell, Jr. (D) | 12.9 |
| 67 | TX 8th: Kevin Brady (R) | 12.9 |
| 72 | GA 11th: Barry Loudermilk (R) | 12.8 |
| 72 | TX 17th: Pete Sessions (R) | 12.8 |
| 72 | FL 9th: Darren Soto (D) | 12.8 |
| 75 | CA 29th: Tony Cárdenas (D) | 12.7 |

### Percent of persons below the poverty level, 2019

| Poverty rate rank | State congressional district Representative | Poverty rate [col 49] |
|---|---|---|
| 1 | NY 15th: Ritchie Torres (D) | 34.4 |
| 2 | MI 13th: Rashida Tlaib (D) | 26.1 |
| 3 | MS 2nd: Bennie G. Thompson (D) | 25.9 |
| 4 | PA 2nd: Brendan F. Boyle (D) | 25.4 |
| 5 | KY 5th: Harold Rogers (R) | 25.0 |
| 6 | TX 34th: Filemon Vela (D) | 24.7 |
| 7 | GA 2nd: Sanford D. Bishop Jr. (D) | 23.9 |
| 8 | AL 7th: Terri A. Sewell (D) | 23.7 |
| 9 | SC 6th: James E. Clyburn (D) | 23.5 |
| 10 | CA 21st: David G. Valadao (R) | 23.4 |
| 10 | LA 5th: Julia Letlow (R) | 23.4 |
| 12 | AZ 7th: Ruben Gallego (D) | 23.1 |
| 13 | CA 16th: Jim Costa (D) | 23.0 |
| 14 | OH 11th: Vacant | 22.7 |
| 15 | LA 4th: Mike Johnson (R) | 22.5 |
| 16 | TX 15th: Vicente Gonzalez (D) | 22.3 |
| 17 | NY 13th: Adriano Espaillat (D) | 22.2 |
| 17 | TX 29th: Sylvia R. Garcia (D) | 22.2 |
| 19 | CA 34th: Jimmy Gomez (D) | 22.1 |
| 20 | FL 5th: Al Lawson (D) | 21.9 |
| 21 | PA 3rd: Dwight Evans (D) | 21.8 |
| 22 | LA 2nd: Troy Carter (D) | 21.6 |
| 23 | TX 28th: Henry Cuellar (D) | 21.4 |
| 24 | NM 2nd: Yvette Herrell (R) | 20.7 |
| 25 | TN 9th: Steve Cohen (D) | 20.7 |
| 26 | TX 9th: Al Green (D) | 20.6 |
| 27 | NV 1st: Dina Titus (D) | 20.5 |
| 28 | WI 4th: Gwen Moore (D) | 20.3 |
| 29 | TX 33rd: Marc A. Veasey (D) | 20.1 |
| 30 | CA 40th: Lucille Roybal-Allard (D) | 20.0 |
| 30 | OH 9th: Marcy Kaptur (D) | 20.0 |
| 32 | NY 8th: Hakeem S. Jeffries (D) | 19.8 |
| 32 | OK 2nd: Markwayne Mullin (R) | 19.8 |
| 32 | TX 18th: Sheila Jackson-Lee (D) | 19.8 |
| 35 | IL 7th: Danny K. Davis (D) | 19.7 |
| 36 | LA 3rd: Clay Higgins (R) | 19.5 |
| 37 | MS 4th: Steven M. Palazzo (R) | 19.4 |
| 37 | NC 1st: G. K. Butterfield (D) | 19.4 |
| 39 | MI 14th: Brenda L. Lawrence (D) | 19.3 |
| 40 | NY 7th: Nydia M. Velázquez (D) | 19.0 |
| 41 | MS 3rd: Michael Guest (R) | 18.9 |
| 41 | TX 16th: Veronica Escobar (D) | 18.9 |
| 43 | GA 5th: Nikema Williams (D) | 18.7 |
| 43 | TX 30th: Eddie Bernice Johnson (D) | 18.7 |
| 45 | WV 3rd: Carol D. Miller (R) | 18.5 |
| 46 | TX 35th: Lloyd Doggett (D) | 18.4 |
| 47 | AR 1st: Eric A. "Rick" Crawford (R) | 18.3 |
| 47 | GA 12th: Rick W. Allen (R) | 18.3 |
| 49 | AZ 1st: Tom O'Halleran (D) | 18.0 |
| 49 | MO 8th: Jason T. Smith (R) | 18.0 |
| 51 | IL 13th: Rodney Davis (R) | 17.9 |
| 52 | AR 4th: Bruce Westerman (R) | 17.8 |
| 52 | FL 24th: Frederica S. Wilson (D) | 17.8 |
| 52 | OH 3rd: Joyce Beatty (D) | 17.8 |
| 55 | KY 1st: James Comer (R) | 17.7 |
| 56 | OH 13th: Tim Ryan (D) | 17.6 |
| 57 | CA 51st: Juan Vargas (D) | 17.5 |
| 57 | NM 3rd: Teresa Leger Fernandez (D) | 17.5 |
| 59 | MI 5th: Daniel T. Kildee (D) | 17.4 |
| 60 | GA 8th: Austin Scott (R) | 17.3 |
| 61 | VA 9th: Morgan Griffith (R) | 17.2 |
| 62 | IL 2nd: Robin L. Kelly (D) | 17.1 |
| 63 | FL 20th: Vacant | 17.0 |
| 63 | MA 7th: Ayanna Presley (D) | 17.0 |
| 63 | NY 26th: Brian Higgins (D) | 17.0 |
| 66 | AZ 3rd: Raul M. Grijalva (D) | 16.9 |
| 66 | TX 17th: Pete Sessions (R) | 16.9 |
| 68 | TN 1st: Diana Harshbarger (R) | 16.8 |
| 68 | TX 27th: Michael Cloud (R) | 16.8 |
| 70 | TX 1st: Louie Gohmert (R) | 16.7 |
| 71 | CA 8th: Jay Obernolte (R) | 16.6 |
| 71 | IN 7th André Carson (D) | 16.6 |
| 73 | SC 7th: Tom Rice (R) | 16.5 |
| 74 | AL 2nd: Barry Moore (R) | 16.4 |
| 74 | MO 1st: Cori Bush (D) | 16.4 |

## Table E. Congressional Districts 116th Congress — **Land Area and Population Characteristics**

| STATE District | Representative, 117th Congress | Land area,[1] 2019 (sq mi) | Total persons | Per square mile | Race alone (percent) | | | | | | | Non-Hispanic White alone (percent) | Percent female | Percent foreign-born | Percent born in state of residence |
| | | | | | White | Black | American Indian, Alaska Native | Asian and Pacific Islander | Some other race (percent) | Two or more races (percent) | Hispanic or Latino[2] (percent) | | | | |
| | 1 | 2 | 3 | | 4 | 5 | 6 | 7 | 8 | 9 | 10 | 11 | 12 | 13 | 14 |

1. Dry land or land partially or temporarily covered by water.    2. May be of any race.

## Table E. Congressional Districts 116th Congress — **Age and Education**

| STATE District | Population and population characteristics, 2019 (cont.) | | | | | | | | | | Education, 2019 | | |
| | Age (percent) | | | | | | | | | Median age | Total Enrollment[1] | Attainment[2] (percent) | |
| | Under 5 years | 5 to 17 years | 18 to 24 years | 25 to 34 years | 35 to 44 years | 45 to 54 years | 55 to 64 years | 65 to 74 years | 75 years and over | | | High school graduate or more | Bachelor's degree or more |
| | 15 | 16 | 17 | 18 | 19 | 20 | 21 | 22 | 23 | 24 | 25 | 26 | 27 |

1. All persons 3 years old and over enrolled in nursery school through college and graduate or professional school.    2. Persons 25 years old and over.

## Table E. Congressional Districts 116th Congress — **Households and Group Quarters**

| STATE District | Households, 2019 | | | | | | Total in group quarters, 2018 | Group Quarters, 2010 | | | | |
| | Number | Average household size | Family households (percent) | Married couple family (percent) | Female family house-holder[1] | One person households (percent) | | Percent 65 years and over | Persons in correctional institutions | Persons in nursing facilities | Persons in college dormitories | Persons in military quarters |
| | 28 | 29 | 30 | 31 | 32 | 33 | 34 | 35 | 36 | 37 | 38 | 39 |

1. No spouse present.

## Table E. Congressional Districts 116th Congress — **Housing and Money Income**

| STATE District | Housing units, 2019 | | | | | | Money income, 2019 | | |
| | Total | Occupied units | | | | | Per capita income (dollars) | Households | |
| | | Occupied units as a percent of all units | Owner-occupied | | | Renter-occupied | | Median income (dollars) | Percent with income of $100,000 or more |
| | | | Owner-occupied units as a percent of occupied units | Median value[1] (dollars) | Percent valued at $500,000 or more | Median rent[2] | | | |
| | 40 | 41 | 42 | 43 | 44 | 45 | 46 | 47 | 48 |

1. Specified owner-occupied units; $1,000,000 represents $1,000,000    2. Specified renter-occupied units.

## Table E. Congressional Districts 116th Congress — **Poverty, Labor Force, Employment, and Social Security**

| STATE District | Poverty, 2019 | | | Civilian labor force, 2019 | | | Civilian employment,[2] 2019 | | | | Persons under 65 years of age with no health insurance, 2019 (percent) | Social Security beneficiaries, December 2020 | | Supplemental Security Income recipients, December 2019 |
|---|---|---|---|---|---|---|---|---|---|---|---|---|---|---|
| | Persons below poverty level (percent) | Families below poverty level (percent) | Percent of households receiving food stamps in past 12 months | Total | Unemployment | | Total | Percent | | | | Number | Rate[3] | |
| | | | | | Total | Rate[1] | | Management, business, science, and arts occupations | Service, sales, and office | Construction and production | | | | |
| | 49 | 50 | 51 | 52 | 53 | 54 | 55 | 56 | 57 | 58 | 59 | 60 | 61 | 62 |

1. Percent of civilian labor force.    2. Persons 16 years old and over.    3. Per 1,000 resident population estimated in the 2019 American Community Survey.

## Table E. Congressional Districts 116th Congress — **Agriculture**

| STATE District | Agriculture 2017 | | | | | | | | | |
|---|---|---|---|---|---|---|---|---|---|---|
| | Land in farms | | | Value of products sold | | | | Government payments | |
| | Number of farms | Acres | Average size of farm (acres) | Irrigated land (acres) | Total ($1,000) | Average per farm (dollars) | Percent from crops | Percent from livestock and poultry products | Total ($1,000) | Average per farm receiving payments (dollars) |
| | 63 | 64 | 65 | 66 | 67 | 68 | 69 | 70 | 71 | 72 |

## Table E. Congressional Districts 116th Congress — **Nonfarm Employment and Payroll**

| STATE District | Private nonfarm employment and payroll, 2019 | | | | | | | | | | | | | |
|---|---|---|---|---|---|---|---|---|---|---|---|---|---|---|
| | Number of establishments | Employment | | | | | | | | | | | Annual payroll | |
| | | Total | Percent by selected industries | | | | | | | | | | Total (mil dol) | Average per employee (dollars) |
| | | | Manufac-turing | Construc-tion | Wholesale trade | Retail trade | Health care and social assistance | Finance and Insurance | Real estate and rental and leasing | Profes-sional, sci-entific, and technical services | Information | | | |
| | 73 | 74 | 75 | 76 | 77 | 78 | 79 | 80 | 81 | 82 | 83 | | 84 | 85 |

# Table E. Congressional Districts 116th Congress — **Land Area and Population Characteristics**

| STATE District | Representative, 117th Congress | Land area,[1] 2018 (sq mi) | Population and population characteristics, 2019 | | | | | | | | | | | |
|---|---|---|---|---|---|---|---|---|---|---|---|---|---|---|
| | | | Total persons | Per square mile | Race alone (percent) | | | | Some other race (percent) | Two or more races (percent) | Hispanic or Latino[2] (percent) | Non-Hispanic White alone (percent) | Percent female | Percent foreign-born | Percent born in state of residence |
| | | | | | White | Black | American Indian, Alaska Native | Asian and Pacific Islander | | | | | | | |
| | | 1 | 2 | 3 | 4 | 5 | 6 | 7 | 8 | 9 | 10 | 11 | 12 | 13 | 14 |
| UNITED STATES | | 3,533,038.28 | 328,239,523 | 92.906 | 72.0 | 12.8 | 0.9 | 5.9 | 5.0 | 3.4 | 18.4 | 59.953 | 50.8 | 13.7 | 58.0 |
| ALABAMA | | 50,647.13 | 4,903,185 | 96.811 | 67.8 | 26.9 | 0.5 | 1.3 | 1.5 | 1.9 | 4.5 | 65.120 | 51.7 | 3.6 | 69.2 |
| District 1 | Jerry Carl (R) | 6,067.128 | 717,438 | 118.250 | 66.9 | 27.4 | 1.0 | 1.4 | 1.1 | 2.1 | 3.3 | 65.014 | 51.8 | 3.1 | 68.1 |
| District 2 | Barry Moore(R) | 10,143.014 | 674,920 | 66.540 | 62.8 | 32.2 | 0.3 | 1.3 | 1.4 | 1.9 | 3.9 | 60.356 | 51.8 | 2.9 | 68.0 |
| District 3 | Mike Rogers (R) | 7,543.488 | 717,896 | 95.168 | 68.9 | 26.3 | 0.3 | 1.9 | 1.0 | 1.7 | 2.9 | 67.036 | 51.3 | 3.2 | 64.2 |
| District 4 | Robert B. Aderholt (R) | 8,888.913 | 687,453 | 77.338 | 86.8 | 7.4 | 0.5 | 0.6 | 3.0 | 1.8 | 6.9 | 82.947 | 50.9 | 4.1 | 74.2 |
| District 5 | Mo Brooks (R) | 3,678.146 | 735,858 | 200.062 | 75.0 | 18.3 | 0.5 | 1.7 | 1.8 | 2.6 | 5.5 | 71.846 | 51.2 | 4.2 | 59.2 |
| District 6 | Gary J. Palmer (R) | 4,171.636 | 699,605 | 167.705 | 78.6 | 16.0 | 0.4 | 1.6 | 1.8 | 1.7 | 5.1 | 75.736 | 51.9 | 4.4 | 72.3 |
| District 7 | Terri A. Sewell (D) | 10,154.805 | 670,015 | 65.980 | 34.2 | 62.6 | 0.2 | 1.2 | 0.6 | 1.2 | 3.7 | 31.217 | 52.8 | 2.9 | 79.7 |
| ALASKA | | 571,022.379 | 731,545 | 1.281 | 64.2 | 3.1 | 15.8 | 7.4 | 1.7 | 7.9 | 7.2 | 59.834 | 48.0 | 8.0 | 42.9 |
| At Large | Don Young (R) | 571,022.379 | 731,545 | 1.281 | 64.2 | 3.1 | 15.8 | 7.4 | 1.7 | 7.9 | 7.2 | 59.834 | 48.0 | 8.0 | 42.9 |
| ARIZONA | | 113,653.420 | 7,278,717 | 64.043 | 78.3 | 4.7 | 4.6 | 3.5 | 5.0 | 3.9 | 31.7 | 53.982 | 50.3 | 13.4 | 39.9 |
| District 1 | Tom O'Halleran (D) | 55,034.156 | 782,088 | 14.211 | 64.2 | 2.8 | 23.0 | 1.8 | 4.5 | 3.7 | 23.0 | 48.308 | 49.0 | 6.4 | 52.4 |
| District 2 | Ann Kirkpatrick (D) | 7,882.602 | 733,197 | 93.015 | 81.8 | 4.3 | 1.4 | 3.0 | 4.6 | 4.8 | 29.4 | 59.969 | 50.5 | 11.6 | 37.4 |
| District 3 | Raul M. Grijalva (D) | 15,687.885 | 801,531 | 51.092 | 74.5 | 4.4 | 4.6 | 1.8 | 9.9 | 4.8 | 65.1 | 24.714 | 49.8 | 20.9 | 47.7 |
| District 4 | Paul A. Gosar (R) | 33,220.313 | 825,763 | 24.857 | 87.8 | 1.7 | 2.2 | 1.1 | 3.9 | 3.2 | 20.5 | 72.573 | 49.6 | 7.1 | 28.8 |
| District 5 | Andy Biggs (R) | 293.553 | 849,917 | 2,895.276 | 82.9 | 4.0 | 1.1 | 5.4 | 2.7 | 3.9 | 16.6 | 70.433 | 50.9 | 9.7 | 37. |
| District 6 | David Schweikert (R) | 625.097 | 814,971 | 1,303.751 | 82.6 | 3.0 | 1.9 | 5.8 | 2.9 | 3.7 | 17.9 | 68.811 | 50.6 | 14.6 | 31. |
| District 7 | Ruben Gallego (D) | 204.708 | 853,856 | 4,171.092 | 72.9 | 9.9 | 2.5 | 3.0 | 8.3 | 3.4 | 64.0 | 20.217 | 50.5 | 24.5 | 49. |
| District 8 | Debbie Lesko (R) | 540.148 | 798,544 | 1,478.380 | 82.8 | 5.1 | 1.3 | 3.9 | 3.3 | 3.6 | 21.6 | 66.430 | 51.9 | 10.6 | 33. |
| District 9 | Greg Stanton (D) | 164.958 | 818,850 | 4,963.991 | 75.1 | 7.0 | 3.6 | 5.6 | 4.7 | 3.8 | 26.6 | 55.180 | 49.4 | 14.5 | 39. |
| ARKANSAS | | 51,992.846 | 3,017,804 | 58.043 | 76.7 | 15.5 | 0.6 | 1.9 | 2.5 | 2.8 | 7.7 | 71.971 | 51.1 | 5.1 | 60. |
| District 1 | Eric A. "Rick" Crawford (R) | 19,280.440 | 719,048 | 37.294 | 77.4 | 17.6 | 0.3 | 0.8 | 0.9 | 3.1 | 3.4 | 75.365 | 50.5 | 2.0 | 65. |
| District 2 | J. French Hill (R) | 4,976.022 | 767,662 | 154.272 | 70.8 | 23.3 | 0.4 | 1.6 | 1.5 | 2.5 | 5.4 | 67.262 | 51.8 | 4.1 | 66. |
| District 3 | Steve Womack (R) | 5,399.946 | 829,149 | 153.548 | 83.8 | 3.1 | 1.1 | 4.2 | 4.4 | 3.5 | 14.9 | 73.985 | 50.9 | 10.2 | 48. |
| District 4 | Bruce Westerman (R) | 22,336.438 | 701,945 | 31.426 | 74.0 | 19.4 | 0.5 | 0.9 | 3.1 | 1.9 | 6.0 | 71.262 | 51.4 | 3.4 | 63. |
| CALIFORNIA | | 155,858.329 | 39,512,223 | 253.514 | 59.4 | 5.8 | 0.8 | 15.2 | 13.7 | 5.0 | 39.4 | 36.333 | 50.3 | 26.7 | 56. |
| District 1 | Doug LaMalfa (R) | 28,120.782 | 711,905 | 25.316 | 84.8 | 1.6 | 1.6 | 3.3 | 3.3 | 5.4 | 15.3 | 74.657 | 49.8 | 6.3 | 71. |
| District 2 | Jared Huffman (D) | 12,952.633 | 708,434 | 54.694 | 79.9 | 1.8 | 2.5 | 4.1 | 6.5 | 5.1 | 18.2 | 70.139 | 50.6 | 12.7 | 61. |
| District 3 | John Garamendi (D) | 6,184.553 | 755,811 | 122.209 | 68.7 | 6.3 | 1.2 | 12.0 | 5.2 | 6.6 | 30.2 | 45.934 | 50.0 | 18.2 | 63. |
| District 4 | Tom McClintock (R) | 12,836.830 | 757,806 | 59.034 | 84.4 | 1.7 | 1.2 | 6.1 | 1.9 | 4.6 | 13.1 | 74.929 | 50.3 | 9.9 | 65. |
| District 5 | Mike Thompson (D) | 1,730.603 | 726,072 | 419.549 | 62.5 | 6.6 | 0.5 | 12.0 | 12.8 | 5.6 | 28.9 | 49.171 | 51.4 | 19.7 | 61. |
| District 6 | Doris O. Matsui (D) | 175.011 | 781,943 | 4,467.965 | 49.6 | 11.8 | 0.7 | 16.6 | 12.5 | 8.8 | 29.9 | 35.745 | 51.0 | 22.2 | 62. |
| District 7 | Ami Bera (D) | 548.891 | 756,668 | 1,378.540 | 60.4 | 7.2 | 0.6 | 18.9 | 5.3 | 7.6 | 17.4 | 50.859 | 51.5 | 20.7 | 60 |
| District 8 | Jay Olbernolte (R) | 32,893.040 | 723,311 | 21.990 | 73.9 | 8.5 | 1.5 | 3.0 | 7.9 | 5.3 | 44.8 | 40.711 | 49.7 | 14.1 | 65. |
| District 9 | Jerry McNerney (D) | 1,250.958 | 784,956 | 627.484 | 53.0 | 9.5 | 0.7 | 17.1 | 8.7 | 11.0 | 38.4 | 31.570 | 50.2 | 21.9 | 65. |
| District 10 | Josh Harder (D) | 1,820.276 | 764,859 | 420.188 | 76.7 | 3.5 | 0.7 | 8.1 | 5.7 | 5.2 | 45.1 | 39.857 | 50.8 | 20.0 | 68. |
| District 11 | Mark Desaulnier (D) | 489.679 | 765,504 | 1,563.277 | 54.0 | 7.4 | 0.6 | 15.0 | 16.6 | 6.4 | 29.1 | 43.786 | 51.2 | 27.9 | 52 |
| District 12 | Nancy Pelosi (D) | 38.996 | 779,824 | 19,997.538 | 46.9 | 5.8 | 0.4 | 33.2 | 7.7 | 5.9 | 15.2 | 41.503 | 48.8 | 32.2 | 39 |
| District 13 | Barbara Lee (D) | 96.796 | 768,889 | 7,943.396 | 40.6 | 17.5 | 0.9 | 21.2 | 13.2 | 6.5 | 22.3 | 34.276 | 51.4 | 25.7 | 50 |
| District 14 | Jackie Speier (D) | 259.871 | 742,980 | 2,859.034 | 43.6 | 2.5 | 0.5 | 37.5 | 10.6 | 5.3 | 23.1 | 32.814 | 50.8 | 39.1 | 45 |
| District 15 | Eric Swalwell (D) | 598.192 | 782,312 | 1,307.794 | 41.0 | 4.9 | 0.5 | 38.1 | 8.8 | 6.6 | 21.5 | 30.530 | 50.0 | 35.4 | 49 |
| District 16 | Jim Costa (D) | 2,842.073 | 753,152 | 265.001 | 56.0 | 5.3 | 1.1 | 9.5 | 24.9 | 3.2 | 62.1 | 20.936 | 50.3 | 24.8 | 65 |
| District 17 | Ro Khanna (D) | 185.410 | 790,519 | 4,263.627 | 27.6 | 2.3 | 0.6 | 56.4 | 8.5 | 4.7 | 16.1 | 21.127 | 48.5 | 50.2 | 37 |
| District 18 | Anna G. Eshoo (D) | 695.842 | 753,806 | 1,083.301 | 62.6 | 2.4 | 0.2 | 24.6 | 5.0 | 5.2 | 15.8 | 52.832 | 50.9 | 31.2 | 46 |
| District 19 | Zoe Lofgren (D) | 915.895 | 737,535 | 805.262 | 46.9 | 2.8 | 0.5 | 31.3 | 13.0 | 5.5 | 39.0 | 23.680 | 49.1 | 37.0 | 50 |
| District 20 | Jimmy Panetta (D) | 4,875.220 | 741,838 | 152.165 | 58.2 | 2.4 | 0.6 | 6.0 | 28.3 | 4.5 | 54.1 | 35.348 | 49.9 | 25.1 | 58 |
| District 21 | David G. Valadao (R) | 6,730.250 | 729,460 | 108.385 | 71.1 | 2.9 | 1.7 | 3.3 | 18.3 | 2.7 | 75.3 | 16.734 | 47.6 | 27.3 | 64 |
| District 22 | Devin Nunes (R) | 1,165.353 | 768,917 | 659.815 | 66.0 | 3.2 | 1.0 | 8.6 | 16.5 | 4.7 | 50.6 | 35.730 | 50.2 | 17.3 | 70 |
| District 23 | Kevin McCarthy (R) | 9,901.494 | 741,557 | 74.893 | 72.1 | 6.8 | 1.0 | 5.1 | 10.0 | 4.9 | 40.7 | 44.301 | 49.8 | 14.2 | 69 |
| District 24 | Salud O. Carbajal (D) | 6,882.329 | 737,443 | 107.150 | 80.3 | 2.0 | 1.2 | 5.1 | 7.9 | 3.5 | 37.0 | 53.177 | 49.9 | 18.2 | 61 |
| District 25 | Mike Garcia (R) | 1,690.739 | 718,949 | 425.228 | 67.1 | 10.1 | 1.6 | 8.9 | 8.2 | 4.0 | 40.0 | 38.091 | 50.2 | 18.6 | 66 |
| District 26 | Julia Brownley (D) | 938.873 | 725,535 | 772.772 | 81.1 | 2.0 | 0.7 | 6.8 | 4.9 | 4.6 | 46.2 | 42.363 | 50.4 | 21.9 | 59 |
| District 27 | Judy Chu (D) | 699.914 | 712,783 | 1,018.387 | 38.3 | 3.9 | 0.7 | 39.4 | 13.2 | 4.4 | 27.8 | 25.770 | 51.4 | 38.6 | 47 |
| District 28 | Adam B. Schiff (D) | 218.458 | 693,299 | 3,173.603 | 67.2 | 2.7 | 0.5 | 15.4 | 9.4 | 4.8 | 23.9 | 54.111 | 50.9 | 38.9 | 38 |
| District 29 | Tony Cárdenas (D) | 92.264 | 717,659 | 7,778.321 | 58.2 | 4.0 | 0.5 | 7.5 | 26.4 | 3.4 | 67.0 | 19.757 | 50.3 | 42.3 | 47 |
| District 30 | Brad Sherman (D) | 135.929 | 764,062 | 5,621.037 | 61.5 | 4.5 | 0.3 | 12.5 | 16.0 | 5.1 | 29.3 | 50.108 | 51.3 | 33.6 | 45 |
| District 31 | Pete Aguilar (D) | 219.841 | 753,576 | 3,427.823 | 60.5 | 11.2 | 0.4 | 8.6 | 14.2 | 5.2 | 53.5 | 24.711 | 50.9 | 20.8 | 65 |
| District 32 | Grace F. Napolitano (D) | 124.250 | 700,726 | 5,639.646 | 50.3 | 2.6 | 0.8 | 19.6 | 21.8 | 4.8 | 61.0 | 15.185 | 51.4 | 36.1 | 56 |
| District 33 | Ted Lieu (D) | 288.667 | 703,908 | 2,438.478 | 71.4 | 3.2 | 0.3 | 16.4 | 3.1 | 5.7 | 13.5 | 62.678 | 51.3 | 22.7 | 46 |
| District 34 | Jimmy Gomez (D) | 47.676 | 730,042 | 15,312.568 | 34.7 | 5.0 | 1.7 | 20.4 | 34.5 | 3.8 | 59.8 | 12.772 | 49.3 | 44.1 | 43 |
| District 35 | Norma J. Torres (D) | 168.836 | 764,643 | 4,528.910 | 52.1 | 5.3 | 1.1 | 7.8 | 28.8 | 5.0 | 71.2 | 13.700 | 49.7 | 30.1 | 61 |
| District 36 | Raul Ruiz (D) | 5,914.974 | 755,764 | 127.771 | 71.2 | 4.9 | 1.3 | 4.0 | 15.0 | 3.5 | 48.8 | 40.510 | 49.7 | 20.6 | 56 |
| District 37 | Karen Bass (D) | 55.263 | 733,668 | 13,275.935 | 50.2 | 21.9 | 0.5 | 9.3 | 14.2 | 3.9 | 40.0 | 25.733 | 50.9 | 31.8 | 49 |
| District 38 | Linda T. Sánchez (D) | 101.420 | 704,515 | 6,946.510 | 41.4 | 4.8 | 0.9 | 16.4 | 32.8 | 3.7 | 61.5 | 15.082 | 51.0 | 30.8 | 61 |
| District 39 | Young Kim (R) | 205.216 | 717,176 | 3,494.737 | 51.9 | 1.7 | 0.3 | 32.9 | 9.3 | 3.8 | 32.9 | 30.359 | 51.3 | 33.5 | 54 |
| District 40 | Lucille Roybal-Allard (D) | 57.684 | 715,934 | 12,411.310 | 55.7 | 5.1 | 1.1 | 2.0 | 34.4 | 1.7 | 89.2 | 3.239 | 49.6 | 39.8 | 56 |
| District 41 | Mark Takano (D) | 316.727 | 786,719 | 2,483.903 | 46.6 | 9.1 | 0.8 | 6.9 | 33.0 | 3.6 | 62.7 | 19.851 | 50.0 | 25.8 | 6 |
| District 42 | Ken Calvert (R) | 936.203 | 840,562 | 897.842 | 61.1 | 6.2 | 0.5 | 10.2 | 16.5 | 5.5 | 41.7 | 39.164 | 50.7 | 18.1 | 6 |
| District 43 | Maxine Waters (D) | 72.083 | 748,092 | 10,378.203 | 43.3 | 20.2 | 0.9 | 12.6 | 18.2 | 4.8 | 49.7 | 14.847 | 50.8 | 32.2 | 5 |
| District 44 | Nanette Diaz Barragan (D) | 79.963 | 717,140 | 8,968.398 | 45.8 | 14.6 | 0.7 | 6.5 | 28.6 | 3.8 | 71.4 | 6.010 | 50.9 | 34.1 | 5 |
| District 45 | Katie Porter (D) | 330.596 | 791,311 | 2,393.589 | 61.9 | 1.8 | 0.3 | 26.4 | 4.6 | 4.9 | 18.2 | 49.533 | 50.7 | 30.0 | 4 |
| District 46 | J. Luis Correa (D) | 72.212 | 734,651 | 10,173.531 | 50.7 | 2.1 | 0.6 | 14.4 | 29.6 | 2.7 | 67.0 | 15.300 | 49.2 | 36.4 | 5 |
| District 47 | Alan S. Lowenthal (D) | 216.270 | 717,594 | 3,318.047 | 52.5 | 6.9 | 0.4 | 23.8 | 12.1 | 4.2 | 36.0 | 30.372 | 51.0 | 29.0 | 5 |
| District 48 | Michelle Steel (R) | 145.947 | 718,359 | 4,922.054 | 62.1 | 1.4 | 0.4 | 20.1 | 11.2 | 4.9 | 21.5 | 53.240 | 50.9 | 24.5 | 5 |
| District 49 | Mike Levin (D) | 553.500 | 731,366 | 1,321.348 | 80.6 | 2.3 | 0.9 | 7.4 | 3.1 | 5.7 | 27.8 | 58.213 | 50.5 | 17.2 | 5 |
| District 50 | Darrell Issa (R) | 2,789.589 | 758,142 | 271.776 | 78.8 | 2.7 | 1.0 | 6.2 | 5.5 | 5.9 | 32.2 | 54.524 | 49.6 | 19.3 | 5 |
| District 51 | Juan Vargas (D) | 4,791.060 | 740,797 | 154.621 | 68.5 | 6.6 | 1.0 | 8.3 | 11.9 | 3.7 | 71.3 | 12.745 | 48.8 | 31.2 | 5 |
| District 52 | Scott H. Peters (D) | 267.142 | 767,151 | 2,871.697 | 66.4 | 3.0 | 0.3 | 20.9 | 3.2 | 6.1 | 14.4 | 57.033 | 49.3 | 22.8 | 4 |
| District 53 | Sara Jacobs (D) | 136.056 | 782,599 | 5,752.036 | 62.7 | 8.1 | 0.6 | 15.5 | 7.1 | 6.1 | 33.8 | 39.349 | 50.3 | 22.9 | 5 |

1. Dry land or land partially or temporarily covered by water. 2. May be of any race.

# Table E. Congressional Districts 116th Congress — **Age and Education**

|  | Population and population characteristics, 2019 (cont.) | | | | | | | | | | Education, 2019 | | |
| STATE District | Age (percent) | | | | | | | | | | Total Enrollment[1] | Attainment[2] (percent) | |
|  | Under 5 years | 5 to 17 years | 18 to 24 years | 25 to 34 years | 35 to 44 years | 45 to 54 years | 55 to 64 years | 65 to 74 years | 75 years and over | Median age |  | High school graduate or more | Bachelor's degree or more |
|  | 15 | 16 | 17 | 18 | 19 | 20 | 21 | 22 | 23 | 24 | 25 | 26 | 27 |
| UNITED STATES | 5.9 | 16.3 | 9.3 | 13.9 | 12.7 | 12.4 | 12.9 | 9.6 | 6.9 | 38.5 | 80,465,620 | 88.6 | 33.1 |
| ALABAMA | 5.8 | 16.3 | 9.3 | 13.0 | 12.4 | 12.4 | 13.4 | 10.3 | 7.1 | 39.4 | 1,164,195 | 87.1 | 26.3 |
| District 1 | 6.0 | 16.5 | 8.1 | 12.9 | 12.4 | 12.2 | 13.6 | 11.0 | 7.3 | 40.4 | 156,759 | 88.2 | 25.0 |
| District 2 | 6.1 | 16.4 | 9.0 | 13.6 | 12.2 | 12.0 | 12.7 | 10.4 | 7.4 | 39.0 | 159,113 | 86.6 | 23.5 |
| District 3 | 5.4 | 16.3 | 10.6 | 12.7 | 12.5 | 12.5 | 13.4 | 10.1 | 6.6 | 38.9 | 184,664 | 85.1 | 24.0 |
| District 4 | 5.9 | 16.5 | 7.9 | 12.4 | 12.1 | 12.9 | 13.6 | 10.8 | 7.9 | 41.1 | 152,292 | 82.9 | 17.5 |
| District 5 | 5.7 | 15.9 | 9.3 | 12.8 | 12.4 | 13.2 | 13.8 | 10.0 | 7.1 | 39.7 | 172,764 | 89.3 | 33.5 |
| District 6 | 5.7 | 17.1 | 7.9 | 13.2 | 13.7 | 12.8 | 13.2 | 10.0 | 6.6 | 39.9 | 167,806 | 91.0 | 38.5 |
| District 7 | 6.1 | 15.4 | 12.7 | 13.5 | 11.4 | 11.2 | 12.7 | 9.9 | 6.9 | 36.8 | 170,797 | 86.4 | 21.6 |
| ALASKA | 7.0 | 17.6 | 9.3 | 16.2 | 12.7 | 11.9 | 13.0 | 8.4 | 4.0 | 35.0 | 180,074 | 93.6 | 30.2 |
| At Large | 7.0 | 17.6 | 9.3 | 16.2 | 12.7 | 11.9 | 13.0 | 8.4 | 4.0 | 35.0 | 180,074 | 93.6 | 30.2 |
| ARIZONA | 5.9 | 16.6 | 9.5 | 13.8 | 12.4 | 11.8 | 12.1 | 10.3 | 7.6 | 38.3 | 1,773,809 | 87.6 | 30.2 |
| District 1 | 5.8 | 16.9 | 10.4 | 12.2 | 11.5 | 11.1 | 12.3 | 11.7 | 8.0 | 39.0 | 199,375 | 87.9 | 25.2 |
| District 2 | 4.9 | 14.0 | 10.1 | 13.2 | 11.6 | 11.1 | 12.8 | 12.2 | 10.1 | 41.5 | 165,998 | 91.2 | 35.3 |
| District 3 | 7.5 | 19.4 | 12.2 | 15.0 | 12.0 | 11.4 | 10.3 | 7.4 | 4.8 | 32.1 | 223,404 | 77.6 | 16.6 |
| District 4 | 4.5 | 14.1 | 7.0 | 10.3 | 10.7 | 10.2 | 15.2 | 15.9 | 12.1 | 48.7 | 166,902 | 89.1 | 21.0 |
| District 5 | 6.0 | 19.4 | 7.9 | 11.1 | 13.7 | 13.0 | 11.7 | 9.4 | 7.9 | 39.3 | 229,734 | 94.2 | 38.3 |
| District 6 | 4.8 | 14.2 | 7.5 | 13.7 | 11.5 | 13.2 | 15.1 | 11.9 | 8.1 | 43.2 | 163,514 | 91.6 | 43.5 |
| District 7 | 8.1 | 20.7 | 11.5 | 15.8 | 14.6 | 11.5 | 9.5 | 5.3 | 3.2 | 30.6 | 233,429 | 71.3 | 15.8 |
| District 8 | 5.2 | 16.3 | 7.1 | 12.0 | 12.0 | 12.3 | 12.7 | 12.5 | 9.9 | 42.9 | 184,465 | 91.8 | 31.5 |
| District 9 | 5.8 | 14.3 | 12.2 | 20.5 | 13.5 | 11.5 | 10.0 | 7.2 | 5.0 | 33.3 | 206,988 | 90.8 | 41.4 |
| ARKANSAS | 6.1 | 17.1 | 9.4 | 12.8 | 12.5 | 12.0 | 12.8 | 10.0 | 7.3 | 38.8 | 738,272 | 87.5 | 23.3 |
| District 1 | 6.1 | 16.7 | 8.3 | 12.5 | 12.5 | 12.1 | 13.3 | 10.4 | 8.0 | 40.1 | 165,159 | 86.1 | 16.2 |
| District 2 | 6.2 | 16.7 | 9.8 | 13.4 | 12.8 | 12.0 | 12.5 | 9.8 | 6.8 | 38.0 | 192,423 | 90.3 | 29.5 |
| District 3 | 6.4 | 18.1 | 10.3 | 13.9 | 12.9 | 11.7 | 11.6 | 8.8 | 6.3 | 36.2 | 221,185 | 86.2 | 29.0 |
| District 4 | 5.6 | 16.8 | 8.8 | 11.2 | 11.5 | 12.4 | 14.0 | 11.4 | 8.4 | 41.4 | 159,505 | 87.3 | 17.7 |
| CALIFORNIA | 6.0 | 16.5 | 9.3 | 15.3 | 13.4 | 12.6 | 12.1 | 8.6 | 6.2 | 37.0 | 10,305,759 | 84.0 | 35.0 |
| District 1 | 5.4 | 14.8 | 9.1 | 12.8 | 11.7 | 10.7 | 14.1 | 12.8 | 8.6 | 41.1 | 164,456 | 90.5 | 25.8 |
| District 2 | 4.5 | 14.7 | 8.1 | 10.7 | 12.0 | 12.9 | 14.7 | 13.6 | 8.7 | 44.9 | 162,748 | 90.8 | 42.7 |
| District 3 | 6.6 | 16.3 | 12.0 | 14.3 | 12.4 | 11.6 | 11.9 | 8.8 | 6.0 | 35.5 | 215,632 | 86.5 | 27.2 |
| District 4 | 5.0 | 15.5 | 6.6 | 10.3 | 12.7 | 12.5 | 15.2 | 13.2 | 9.0 | 44.9 | 162,351 | 92.9 | 35.2 |
| District 5 | 5.1 | 15.0 | 8.4 | 12.9 | 12.8 | 12.8 | 14.2 | 11.5 | 7.3 | 41.5 | 165,563 | 88.4 | 33.3 |
| District 6 | 7.3 | 17.6 | 9.3 | 17.4 | 14.0 | 10.6 | 10.8 | 7.7 | 5.2 | 33.9 | 214,115 | 86.6 | 29.6 |
| District 7 | 5.3 | 16.8 | 7.6 | 14.1 | 13.3 | 13.3 | 13.6 | 9.5 | 6.4 | 39.1 | 194,162 | 91.0 | 33.7 |
| District 8 | 7.9 | 20.3 | 9.9 | 14.0 | 11.9 | 10.6 | 12.4 | 8.1 | 5.1 | 33.4 | 197,481 | 82.7 | 17.3 |
| District 9 | 6.2 | 20.1 | 9.3 | 13.6 | 12.8 | 11.9 | 11.8 | 8.4 | 5.7 | 35.7 | 224,429 | 81.3 | 22.5 |
| District 10 | 6.8 | 19.5 | 9.7 | 14.3 | 13.1 | 12.0 | 11.7 | 7.5 | 5.4 | 34.8 | 214,342 | 81.7 | 18.0 |
| District 11 | 6.0 | 16.1 | 8.0 | 13.1 | 13.4 | 13.3 | 13.5 | 9.5 | 7.1 | 39.8 | 186,312 | 87.6 | 44.8 |
| District 12 | 4.6 | 9.0 | 6.4 | 24.0 | 16.7 | 12.9 | 11.0 | 8.6 | 6.9 | 38.0 | 138,671 | 89.0 | 60.9 |
| District 13 | 5.6 | 12.8 | 9.7 | 18.9 | 14.9 | 12.4 | 11.0 | 9.1 | 5.7 | 37.0 | 193,916 | 88.6 | 53.8 |
| District 14 | 4.8 | 13.6 | 8.2 | 15.7 | 14.1 | 13.2 | 13.5 | 9.6 | 7.4 | 40.1 | 169,988 | 89.1 | 49.5 |
| District 15 | 5.8 | 17.1 | 6.8 | 13.5 | 15.7 | 14.1 | 12.7 | 8.3 | 6.0 | 39.2 | 194,313 | 90.6 | 48.0 |
| District 16 | 8.0 | 21.6 | 10.6 | 15.5 | 13.1 | 10.4 | 9.5 | 6.6 | 4.5 | 31.0 | 233,273 | 68.3 | 13.2 |
| District 17 | 5.9 | 13.8 | 7.9 | 19.5 | 15.4 | 13.6 | 12.0 | 6.5 | 5.4 | 36.7 | 180,806 | 91.6 | 60.2 |
| District 18 | 5.9 | 17.0 | 7.1 | 13.3 | 14.1 | 14.3 | 12.4 | 8.7 | 7.3 | 39.7 | 201,203 | 95.3 | 65.8 |
| District 19 | 5.5 | 16.5 | 9.1 | 15.5 | 13.9 | 13.2 | 12.1 | 8.4 | 5.8 | 37.4 | 196,855 | 82.4 | 37.1 |
| District 20 | 6.4 | 17.7 | 11.8 | 13.6 | 13.0 | 11.4 | 11.7 | 9.1 | 5.6 | 35.6 | 217,176 | 76.3 | 29.7 |
| District 21 | 8.4 | 22.2 | 11.2 | 16.2 | 12.3 | 11.4 | 9.1 | 5.4 | 3.8 | 29.8 | 226,117 | 62.3 | 10.7 |
| District 22 | 7.6 | 21.2 | 9.1 | 14.8 | 13.3 | 10.7 | 10.5 | 7.7 | 5.0 | 33.0 | 231,328 | 82.9 | 24.5 |
| District 23 | 6.7 | 19.5 | 8.7 | 14.9 | 13.2 | 11.3 | 11.8 | 8.4 | 5.5 | 35.1 | 212,541 | 85.0 | 22.1 |
| District 24 | 5.4 | 14.6 | 15.3 | 12.7 | 11.3 | 10.7 | 12.1 | 10.2 | 7.6 | 36.5 | 211,217 | 85.2 | 35.7 |
| District 25 | 5.8 | 19.8 | 9.2 | 12.5 | 12.9 | 14.0 | 13.3 | 7.4 | 5.3 | 37.2 | 210,171 | 84.8 | 28.2 |
| District 26 | 5.8 | 17.3 | 9.3 | 13.4 | 12.5 | 12.6 | 13.0 | 9.4 | 6.8 | 38.4 | 184,716 | 85.5 | 35.2 |
| District 27 | 5.1 | 13.6 | 7.7 | 14.9 | 13.1 | 14.2 | 13.7 | 9.8 | 8.0 | 41.5 | 175,018 | 87.1 | 45.9 |
| District 28 | 4.4 | 11.0 | 6.4 | 19.7 | 16.0 | 13.0 | 12.9 | 9.2 | 7.2 | 40.0 | 130,422 | 89.3 | 48.1 |
| District 29 | 5.8 | 17.0 | 10.4 | 16.9 | 14.3 | 11.3 | 11.3 | 6.9 | 4.6 | 35.1 | 190,608 | 71.3 | 22.3 |
| District 30 | 5.5 | 13.2 | 7.9 | 15.2 | 13.9 | 14.9 | 13.1 | 8.9 | 7.3 | 40.5 | 176,222 | 89.6 | 45.2 |
| District 31 | 6.3 | 18.4 | 10.1 | 16.4 | 12.6 | 13.0 | 10.8 | 7.9 | 4.5 | 34.2 | 215,605 | 82.3 | 25.9 |
| District 32 | 5.7 | 15.2 | 9.8 | 14.3 | 13.5 | 12.8 | 13.1 | 9.1 | 6.5 | 38.9 | 179,450 | 77.9 | 22.3 |
| District 33 | 5.0 | 13.8 | 8.9 | 13.9 | 12.9 | 13.6 | 13.9 | 9.4 | 8.7 | 41.2 | 172,708 | 96.5 | 66.6 |
| District 34 | 5.4 | 13.7 | 9.8 | 21.0 | 14.5 | 12.7 | 10.1 | 6.9 | 5.8 | 35.0 | 173,279 | 70.6 | 30.3 |
| District 35 | 6.7 | 19.1 | 10.7 | 16.0 | 14.4 | 12.0 | 10.4 | 6.2 | 4.3 | 33.3 | 223,387 | 73.4 | 18.3 |
| District 36 | 4.8 | 15.4 | 8.0 | 12.4 | 11.4 | 11.1 | 13.3 | 12.9 | 10.6 | 43.1 | 168,974 | 82.2 | 24.2 |
| District 37 | 5.1 | 13.6 | 10.6 | 19.8 | 14.2 | 13.0 | 10.8 | 7.2 | 5.5 | 35.7 | 179,890 | 80.9 | 40.2 |
| District 38 | 5.3 | 16.2 | 9.6 | 14.8 | 13.4 | 12.6 | 12.5 | 8.9 | 6.8 | 38.0 | 186,919 | 81.2 | 24.8 |
| District 39 | 5.6 | 15.3 | 8.8 | 13.1 | 12.7 | 13.9 | 14.0 | 9.9 | 6.7 | 40.5 | 185,448 | 89.9 | 43.1 |
| District 40 | 7.4 | 21.0 | 11.2 | 15.1 | 13.2 | 13.0 | 9.8 | 5.6 | 3.6 | 31.2 | 214,216 | 57.3 | 10.1 |
| District 41 | 6.8 | 20.0 | 12.5 | 15.5 | 13.4 | 12.3 | 9.8 | 5.8 | 3.9 | 31.5 | 242,240 | 76.6 | 17.6 |
| District 42 | 7.0 | 19.7 | 8.3 | 14.1 | 13.6 | 13.3 | 11.7 | 7.0 | 5.3 | 35.6 | 238,060 | 88.7 | 26.5 |
| District 43 | 6.9 | 16.5 | 8.5 | 17.4 | 13.4 | 13.0 | 11.6 | 7.0 | 5.6 | 35.4 | 193,839 | 78.4 | 29.7 |
| District 44 | 7.0 | 18.5 | 10.4 | 16.3 | 12.4 | 12.3 | 11.2 | 7.4 | 4.4 | 33.5 | 200,702 | 65.8 | 14.9 |
| District 45 | 5.7 | 15.2 | 9.6 | 13.9 | 12.8 | 14.2 | 12.9 | 8.9 | 6.7 | 39.3 | 224,487 | 94.6 | 56.6 |
| District 46 | 6.4 | 17.3 | 11.1 | 16.9 | 13.9 | 13.0 | 10.7 | 5.8 | 4.9 | 33.8 | 204,594 | 70.8 | 21.1 |
| District 47 | 5.7 | 15.4 | 9.0 | 15.1 | 14.1 | 13.4 | 12.8 | 8.7 | 6.1 | 38.2 | 182,961 | 82.2 | 34.1 |
| District 48 | 5.6 | 14.6 | 7.3 | 14.4 | 12.5 | 13.7 | 13.8 | 10.3 | 7.7 | 41.0 | 167,761 | 89.6 | 44.6 |
| District 49 | 6.3 | 16.1 | 9.0 | 13.5 | 12.5 | 12.9 | 13.2 | 9.7 | 6.6 | 39.0 | 176,074 | 90.0 | 47.2 |
| District 50 | 6.7 | 17.1 | 9.1 | 13.3 | 12.8 | 11.9 | 13.3 | 9.2 | 6.6 | 37.6 | 193,484 | 87.8 | 30.9 |
| District 51 | 6.8 | 18.1 | 10.6 | 16.4 | 13.6 | 11.5 | 10.6 | 7.1 | 5.3 | 33.7 | 205,585 | 73.2 | 15.7 |
| District 52 | 5.4 | 13.9 | 10.4 | 17.7 | 13.1 | 12.7 | 12.2 | 8.3 | 6.3 | 36.8 | 195,249 | 95.3 | 61.0 |
| District 53 | 5.9 | 13.7 | 9.8 | 19.0 | 15.4 | 11.9 | 10.6 | 8.0 | 5.5 | 35.8 | 204,695 | 90.8 | 40.5 |

1. All persons 3 years old and over enrolled in nursery school through college and graduate or professional school.   2. Persons 25 years old and over.

Items 15—27

# Table E. Congressional Districts 116th Congress — **Households and Group Quarters**

| STATE District | Households, 2019 | | | | | | Total in group quarters, 2018 | Group Quarters, 2010 | | | | |
|---|---|---|---|---|---|---|---|---|---|---|---|---|
| | Number | Average household size | Family households (percent) | Married couple family (percent) | Female family householder[1] | One person households (percent) | | Percent 65 years and over | Persons in correctional institutions | Persons in nursing facilities | Persons in college dormitories | Persons in military quarters |
| | 28 | 29 | 30 | 31 | 32 | 33 | 34 | 35 | 36 | 37 | 38 | 39 |
| UNITED STATES | 122,802,852 | 2.61 | 64.8 | 47.5 | 12.3 | 28.3 | 8,084,362 | 18.3 | 2,263,602 | 1,502,264 | 2,521,090 | 338,191 |
| ALABAMA | 1,897,576 | 2.52 | 65.2 | 47.0 | 14.0 | 29.8 | 116,625 | 18.7 | 41,177 | 22,995 | 36,341 | 2,152 |
| District 1 | 268,033 | 2.63 | 64.3 | 46.1 | 13.9 | 31.5 | 12,535 | 19.1 | 5,651 | 2,671 | 2,579 | 0 |
| District 2 | 262,180 | 2.50 | 64.1 | 43.5 | 16.3 | 30.8 | 20,398 | 15.4 | 13,081 | 3,727 | 3,107 | 1,274 |
| District 3 | 279,236 | 2.50 | 66.0 | 49.5 | 12.4 | 28.3 | 20,682 | 15.9 | 7,014 | 3,016 | 6,949 | 0 |
| District 4 | 262,865 | 2.58 | 68.3 | 52.1 | 11.7 | 27.9 | 8,178 | 39.0 | 2,922 | 3,652 | 791 | 0 |
| District 5 | 293,146 | 2.45 | 66.7 | 50.2 | 12.4 | 28.1 | 17,360 | 17.5 | 5,087 | 2,879 | 5,347 | 823 |
| District 6 | 272,470 | 2.53 | 68.5 | 54.4 | 10.9 | 27.3 | 9,614 | 23.8 | 4,884 | 2,733 | 2,830 | 0 |
| District 7 | 259,646 | 2.47 | 58.2 | 32.3 | 20.9 | 35.4 | 27,858 | 15.1 | 2,538 | 4,317 | 14,738 | 55 |
| ALASKA | 252,199 | 2.79 | 64.7 | 49.4 | 10.2 | 27.4 | 26,673 | 5.9 | 4,206 | 1,626 | 1,872 | 5,055 |
| At Large | 252,199 | 2.79 | 64.7 | 49.4 | 10.2 | 27.4 | 26,673 | 5.9 | 4,206 | 1,626 | 1,872 | 5055 |
| ARIZONA | 2,670,441 | 2.67 | 65.2 | 47.6 | 12.0 | 27.3 | 161,893 | 12.2 | 67,767 | 13,819 | 27,987 | 5,172 |
| District 1 | 273,698 | 2.73 | 67.8 | 49.3 | 13.8 | 26.4 | 35,681 | 4.2 | 15,071 | 816 | 7,630 | 0 |
| District 2 | 309,633 | 2.31 | 57.4 | 43.3 | 9.9 | 34.1 | 19,390 | 19.9 | 7,538 | 2,914 | 118 | 3,352 |
| District 3 | 248,829 | 3.11 | 72.3 | 46.5 | 17.2 | 21.8 | 27,775 | 5.3 | 11,711 | 863 | 6,913 | 0 |
| District 4 | 329,367 | 2.42 | 67.0 | 53.1 | 9.1 | 26.1 | 28,323 | 6.9 | 19,869 | 1,431 | 1,213 | 1,164 |
| District 5 | 300,867 | 2.81 | 70.6 | 57.0 | 9.5 | 23.2 | 3,639 | 55.3 | 11 | 859 | 346 | 0 |
| District 6 | 333,370 | 2.43 | 62.3 | 48.4 | 9.2 | 30.2 | 4,810 | 53.4 | 10 | 1,594 | 274 | 0 |
| District 7 | 258,353 | 3.26 | 67.4 | 37.5 | 21.2 | 23.8 | 12,376 | 5.2 | 8,915 | 840 | 1,162 | 656 |
| District 8 | 293,215 | 2.68 | 70.3 | 55.9 | 9.6 | 24.7 | 12,980 | 30.5 | 4,624 | 2,809 | 861 | 0 |
| District 9 | 323,109 | 2.48 | 54.6 | 36.7 | 11.5 | 32.8 | 16,919 | 12.2 | 18 | 1,693 | 9,470 | 619 |
| ARKANSAS | 1,163,647 | 2.52 | 64.9 | 47.5 | 12.9 | 29.5 | 84,011 | 21.7 | 25,844 | 18,532 | 24,144 | 0 |
| District 1 | 283,256 | 2.44 | 64.3 | 46.1 | 13.5 | 30.7 | 26,826 | 20.6 | 15,495 | 5,770 | 2,669 | 616 |
| District 2 | 298,841 | 2.51 | 62.8 | 44.5 | 13.8 | 31.6 | 17,577 | 20.1 | 3,461 | 3,521 | 6,184 | 0 |
| District 3 | 310,011 | 2.62 | 66.6 | 51.3 | 11.4 | 26.0 | 17,746 | 21.3 | 1,599 | 3,663 | 9,499 | 3 |
| District 4 | 271,539 | 2.50 | 65.9 | 48.1 | 13.2 | 29.8 | 21,862 | 24.7 | 5,289 | 5,578 | 5,792 | 0 |
| CALIFORNIA | 13,157,873 | 2.94 | 68.2 | 49.3 | 12.8 | 24.1 | 826,521 | 15.4 | 256,807 | 111,884 | 172,843 | 57,628 |
| District 1 | 271,323 | 2.54 | 63.6 | 47.9 | 9.9 | 27.9 | 23,316 | 14.1 | 11,908 | 2,834 | 2,728 | 5 |
| District 2 | 285,860 | 2.42 | 60.0 | 46.1 | 9.2 | 30.9 | 17,974 | 13.5 | 9,975 | 2,246 | 2,645 | 393 |
| District 3 | 255,155 | 2.87 | 68.8 | 49.7 | 12.9 | 21.7 | 23,375 | 9.1 | 12,154 | 1,728 | 4,981 | 1,321 |
| District 4 | 286,740 | 2.60 | 69.4 | 57.1 | 8.5 | 24.5 | 13,531 | 15.3 | 8,633 | 1,677 | 458 | 0 |
| District 5 | 268,680 | 2.65 | 63.7 | 46.8 | 11.0 | 27.8 | 12,981 | 22.6 | 1,940 | 3,657 | 3,581 | 0 |
| District 6 | 281,104 | 2.74 | 61.0 | 39.4 | 15.4 | 29.1 | 11,208 | 19.4 | 2,254 | 1,789 | 1,493 | 0 |
| District 7 | 272,771 | 2.73 | 67.9 | 50.1 | 12.4 | 24.8 | 11,098 | 23.1 | 6,682 | 1,704 | 22 | 0 |
| District 8 | 228,723 | 3.09 | 71.5 | 49.7 | 14.5 | 23.9 | 15,883 | 10.8 | 6,717 | 1,325 | 117 | 4,748 |
| District 9 | 241,764 | 3.17 | 73.7 | 52.7 | 15.5 | 21.1 | 18,780 | 21.5 | 1,358 | 2,403 | 2,194 | 0 |
| District 10 | 238,915 | 3.17 | 75.6 | 53.5 | 14.4 | 18.6 | 8,473 | 24.1 | 4,848 | 2,580 | 584 | 0 |
| District 11 | 268,492 | 2.82 | 70.3 | 54.0 | 11.6 | 23.2 | 8,265 | 35.3 | 865 | 3,004 | 1,568 | 0 |
| District 12 | 331,958 | 2.30 | 43.9 | 33.4 | 7.8 | 38.0 | 16,186 | 18.3 | 1,573 | 2,810 | 5,993 | 0 |
| District 13 | 294,896 | 2.54 | 53.6 | 38.6 | 10.7 | 34.6 | 20,111 | 13.2 | 1,044 | 2,600 | 11,927 | 661 |
| District 14 | 255,130 | 2.87 | 67.4 | 52.0 | 9.8 | 23.1 | 10,774 | 29.2 | 1,090 | 1,939 | 2,487 | 0 |
| District 15 | 254,639 | 3.03 | 75.9 | 61.9 | 9.4 | 18.8 | 9,529 | 23.6 | 5,669 | 2,136 | 1,076 | 0 |
| District 16 | 217,249 | 3.36 | 75.5 | 46.0 | 20.4 | 19.8 | 22,477 | 11.0 | 12,113 | 2,113 | 1,443 | 0 |
| District 17 | 264,673 | 2.94 | 73.6 | 60.1 | 8.9 | 17.8 | 12,385 | 20.8 | 2,564 | 1,666 | 2,592 | 0 |
| District 18 | 271,858 | 2.71 | 69.0 | 56.7 | 8.2 | 23.2 | 17,486 | 24.9 | 4 | 2,889 | 6,712 | 5 |
| District 19 | 228,533 | 3.17 | 70.9 | 52.2 | 11.7 | 20.9 | 12,034 | 11.7 | 1,448 | 916 | 3,356 | 0 |
| District 20 | 231,132 | 3.08 | 69.4 | 49.4 | 14.5 | 23.2 | 30,679 | 8.9 | 11,516 | 2,072 | 8,411 | 2,504 |
| District 21 | 196,617 | 3.53 | 77.9 | 51.2 | 19.3 | 18.6 | 35,636 | 4.1 | 44,549 | 1,252 | 239 | 2,038 |
| District 22 | 250,996 | 3.03 | 74.5 | 51.9 | 14.7 | 20.1 | 8,199 | 28.5 | 1,349 | 1,709 | 1,566 | 0 |
| District 23 | 247,660 | 2.92 | 70.8 | 49.3 | 15.0 | 23.9 | 18,811 | 9.1 | 17,160 | 2,293 | 560 | 329 |
| District 24 | 257,087 | 2.72 | 62.7 | 47.3 | 10.4 | 27.7 | 36,921 | 7.2 | 11,387 | 1,805 | 15,714 | 475 |
| District 25 | 221,082 | 3.20 | 77.3 | 57.2 | 13.2 | 18.3 | 11,115 | 5.7 | 7,129 | 251 | 1,775 | 0 |
| District 26 | 227,406 | 3.14 | 72.4 | 55.1 | 12.2 | 21.6 | 12,551 | 18.2 | 1,535 | 1,644 | 2,316 | 1,349 |
| District 27 | 243,697 | 2.87 | 69.8 | 51.5 | 12.6 | 23.6 | 13,927 | 25.4 | 331 | 3,659 | 5,809 | 0 |
| District 28 | 295,486 | 2.31 | 51.1 | 37.7 | 8.9 | 37.5 | 11,240 | 32.8 | 183 | 3,356 | 647 | 0 |
| District 29 | 211,871 | 3.36 | 71.7 | 44.6 | 17.4 | 20.8 | 6,133 | 35.7 | 307 | 2,842 | 151 | 0 |
| District 30 | 282,925 | 2.66 | 62.7 | 46.2 | 11.2 | 28.7 | 10,264 | 41.0 | 20 | 2,996 | 2,746 | 0 |
| District 31 | 229,969 | 3.18 | 73.4 | 49.0 | 16.8 | 21.1 | 22,495 | 15.4 | 5,464 | 2,391 | 2,739 | 0 |
| District 32 | 202,442 | 3.40 | 75.1 | 49.4 | 17.2 | 19.1 | 12,513 | 29.1 | 74 | 2,895 | 3,298 | 0 |
| District 33 | 291,655 | 2.33 | 57.0 | 46.7 | 6.3 | 33.9 | 24,594 | 11.2 | 98 | 1,851 | 14,107 | 0 |
| District 34 | 257,349 | 2.73 | 55.3 | 32.9 | 14.5 | 32.3 | 27,227 | 11.7 | 8,722 | 3,270 | 2,438 | 0 |
| District 35 | 201,969 | 3.72 | 81.0 | 54.7 | 18.1 | 13.8 | 13,952 | 11.5 | 6,996 | 2,041 | 1,948 | 0 |
| District 36 | 277,381 | 2.68 | 64.3 | 46.3 | 13.3 | 29.1 | 13,222 | 16.9 | 9,107 | 1,504 | 70 | 0 |
| District 37 | 275,527 | 2.60 | 54.4 | 33.5 | 13.9 | 33.7 | 16,261 | 14.0 | 155 | 2,263 | 7,976 | 0 |
| District 38 | 203,171 | 3.42 | 79.7 | 55.1 | 17.1 | 16.2 | 10,379 | 26.5 | 156 | 2,794 | 3,305 | 0 |
| District 39 | 229,737 | 3.09 | 79.0 | 61.3 | 12.2 | 16.2 | 7,765 | 26.5 | 2 | 1,230 | 2,630 | 0 |
| District 40 | 179,420 | 3.97 | 82.0 | 45.3 | 25.6 | 13.0 | 3,520 | 37.5 | 7 | 1,571 | 5 | 0 |
| District 41 | 195,146 | 3.94 | 79.0 | 52.8 | 17.0 | 15.5 | 17,863 | 11.9 | 1,171 | 1,826 | 7,379 | 0 |
| District 42 | 236,023 | 3.53 | 80.1 | 62.2 | 12.1 | 14.6 | 6,972 | 10.4 | 5,652 | 440 | 161 | 0 |
| District 43 | 246,103 | 2.99 | 67.6 | 42.1 | 17.8 | 24.9 | 11,100 | 27.0 | 120 | 2,478 | 3,293 | 0 |
| District 44 | 191,030 | 3.70 | 78.9 | 48.3 | 21.0 | 17.0 | 9,670 | 16.1 | 2,843 | 1,687 | 587 | 20 |
| District 45 | 281,965 | 2.76 | 69.3 | 56.4 | 8.0 | 22.4 | 13,151 | 22.2 | 534 | 789 | 5,705 | 0 |
| District 46 | 194,502 | 3.69 | 76.9 | 52.6 | 16.3 | 16.6 | 17,716 | 19.0 | 4,906 | 2,887 | 1,967 | 0 |
| District 47 | 243,994 | 2.90 | 65.5 | 45.4 | 13.6 | 26.9 | 10,726 | 31.5 | 235 | 2,931 | 2,040 | 2 |
| District 48 | 264,799 | 2.69 | 64.8 | 49.7 | 10.2 | 26.1 | 5,566 | 30.5 | 13 | 1,555 | 1,249 | 14 |
| District 49 | 253,903 | 2.80 | 71.2 | 57.7 | 10.0 | 22.1 | 20,208 | 6.5 | 782 | 1,612 | 7,910 | 16,563 |
| District 50 | 245,557 | 3.05 | 72.8 | 54.7 | 11.8 | 21.4 | 8,416 | 26.3 | 889 | 1,975 | 757 | 0 |
| District 51 | 207,851 | 3.39 | 74.0 | 47.5 | 19.0 | 21.0 | 35,213 | 4.0 | 17,756 | 1,059 | 0 | 11,846 |
| District 52 | 287,694 | 2.59 | 62.3 | 50.6 | 7.4 | 26.6 | 22,765 | 8.6 | 2,774 | 1,881 | 4,151 | 15,213 |
| District 53 | 276,264 | 2.78 | 62.8 | 46.1 | 11.4 | 25.6 | 13,885 | 28.3 | 46 | 3,059 | 3,237 | 142 |

1. No spouse present.

# Table E. Congressional Districts 116th Congress — **Housing and Money Income**

| STATE District | Housing units, 2019 — Occupied units: Total | Occupied units as a percent of all units | Owner-occupied units as a percent of occupied units | Median value[1] (dollars) | Percent valued at $500,000 or more | Median rent[2] | Money income, 2019 — Households: Per capita income (dollars) | Median income (dollars) | Percent with income of $100,000 or more |
|---|---|---|---|---|---|---|---|---|---|
| | 40 | 41 | 42 | 43 | 44 | 45 | 46 | 47 | 48 |
| UNITED STATES | 139,686,209 | 87.9 | 64.1 | 240,500 | 17.0 | 1,097 | 35,672 | 65,712 | 31.4 |
| ALABAMA | 2,284,922 | 83.0 | 68.8 | 154,000 | 5.1 | 807 | 28,650 | 51,734 | 22.8 |
| District 1 | 345,987 | 77.5 | 69.1 | 160,600 | 5.9 | 872 | 27,463 | 50,663 | 21.2 |
| District 2 | 317,151 | 82.7 | 65.6 | 135,500 | 2.7 | 815 | 26,790 | 50,494 | 20.1 |
| District 3 | 337,424 | 82.8 | 71.6 | 144,700 | 4.6 | 771 | 27,867 | 51,925 | 21.9 |
| District 4 | 319,267 | 82.3 | 73.9 | 132,800 | 3.5 | 661 | 25,211 | 47,531 | 17.7 |
| District 5 | 326,251 | 89.9 | 69.5 | 173,300 | 4.6 | 783 | 32,465 | 59,950 | 28.5 |
| District 6 | 302,129 | 90.2 | 75.9 | 206,000 | 10.8 | 1,026 | 38,316 | 69,072 | 33.7 |
| District 7 | 336,713 | 77.1 | 55.2 | 107,400 | 2.1 | 789 | 21,877 | 38,023 | 15.0 |
| ALASKA | 319,867 | 78.8 | 64.7 | 281,200 | 10.8 | 1,201 | 36,978 | 75,463 | 36.8 |
| At Large | 319,867 | 78.8 | 64.7 | 281,200 | 10.8 | 1,201 | 36,978 | 75,463 | 36.8 |
| ARIZONA | 3,076,048 | 86.8 | 65.3 | 255,900 | 12.3 | 1,101 | 32,173 | 62,055 | 28.1 |
| District 1 | 351,215 | 77.9 | 72.1 | 205,700 | 7.9 | 971 | 26,924 | 56,117 | 23.9 |
| District 2 | 354,389 | 87.4 | 62.5 | 210,000 | 8.1 | 883 | 33,921 | 54,835 | 24.9 |
| District 3 | 280,339 | 88.8 | 63.2 | 183,200 | 2.4 | 951 | 21,947 | 54,583 | 19.2 |
| District 4 | 413,616 | 79.6 | 75.7 | 228,200 | 7.6 | 962 | 29,233 | 55,040 | 20.6 |
| District 5 | 336,750 | 89.3 | 75.4 | 324,400 | 14.7 | 1,331 | 38,747 | 82,540 | 41.2 |
| District 6 | 371,166 | 89.8 | 65.6 | 387,000 | 34.9 | 1,277 | 47,943 | 76,615 | 38.6 |
| District 7 | 284,258 | 90.9 | 47.8 | 208,300 | 3.9 | 991 | 20,229 | 49,066 | 16.5 |
| District 8 | 326,621 | 89.8 | 75.5 | 282,000 | 9.4 | 1,366 | 33,761 | 73,632 | 33.3 |
| District 9 | 357,694 | 90.3 | 48.0 | 310,800 | 16.1 | 1,164 | 36,986 | 65,379 | 30.5 |
| ARKANSAS | 1,389,159 | 83.8 | 65.5 | 136,200 | 3.5 | 742 | 27,274 | 48,952 | 19.4 |
| District 1 | 346,700 | 81.7 | 65.1 | 107,800 | 1.9 | 675 | 24,862 | 43,193 | 15.4 |
| District 2 | 346,110 | 86.3 | 63.6 | 155,600 | 4.0 | 812 | 30,333 | 53,600 | 22.8 |
| District 3 | 348,991 | 88.8 | 62.8 | 170,500 | 5.7 | 787 | 29,507 | 54,310 | 23.5 |
| District 4 | 347,358 | 78.2 | 71.0 | 103,600 | 2.3 | 650 | 23,763 | 43,824 | 15.0 |
| CALIFORNIA | 14,367,012 | 91.6 | 54.9 | 568,500 | 56.9 | 1,614 | 39,393 | 80,440 | 40.5 |
| District 1 | 313,616 | 86.5 | 65.6 | 312,100 | 19.9 | 1,095 | 33,085 | 61,433 | 27.6 |
| District 2 | 321,642 | 88.9 | 62.4 | 694,400 | 65.5 | 1,555 | 52,296 | 80,051 | 41.0 |
| District 3 | 275,348 | 92.7 | 59.9 | 397,600 | 29.2 | 1,419 | 33,812 | 73,191 | 35.7 |
| District 4 | 371,336 | 77.2 | 74.7 | 473,100 | 44.9 | 1,465 | 43,882 | 86,374 | 42.5 |
| District 5 | 293,052 | 91.7 | 61.9 | 583,600 | 63.2 | 1,744 | 42,831 | 85,856 | 41.6 |
| District 6 | 293,219 | 95.9 | 47.9 | 356,100 | 21.3 | 1,321 | 31,749 | 64,687 | 29.9 |
| District 7 | 282,288 | 96.6 | 64.7 | 421,700 | 30.9 | 1,412 | 37,827 | 77,786 | 38.5 |
| District 8 | 304,813 | 75.0 | 61.5 | 270,000 | 10.7 | 1,123 | 24,138 | 56,140 | 23.1 |
| District 9 | 258,370 | 93.6 | 61.2 | 409,900 | 34.5 | 1,257 | 31,161 | 72,237 | 35.8 |
| District 10 | 251,334 | 95.1 | 60.1 | 372,000 | 20.2 | 1,319 | 28,584 | 69,647 | 32.8 |
| District 11 | 281,944 | 95.2 | 62.2 | 735,200 | 73.5 | 1,891 | 53,700 | 103,580 | 52.3 |
| District 12 | 369,976 | 89.7 | 34.7 | 1,226,000 | NA | 1,933 | 77,858 | 127,290 | 58.2 |
| District 13 | 318,117 | 92.7 | 44.2 | 858,500 | 84.3 | 1,752 | 50,721 | 91,514 | 46.5 |
| District 14 | 269,681 | 94.6 | 58.7 | 1,161,400 | 94.6 | 2,453 | 62,192 | 125,980 | 60.6 |
| District 15 | 266,833 | 95.4 | 63.5 | 888,000 | 90.0 | 2,253 | 53,430 | 125,018 | 60.6 |
| District 16 | 231,051 | 94.0 | 46.8 | 262,400 | 8.4 | 1,002 | 21,081 | 50,401 | 20.8 |
| District 17 | 281,828 | 93.9 | 53.6 | 1,137,100 | 91.3 | 2,663 | 63,572 | 147,671 | 67.7 |
| District 18 | 297,048 | 91.5 | 58.8 | 1,588,400 | 95.6 | 2,399 | 81,112 | 149,375 | 64.7 |
| District 19 | 237,857 | 96.1 | 57.5 | 894,400 | 87.3 | 2,075 | 46,572 | 107,240 | 53.4 |
| District 20 | 253,991 | 91.0 | 54.4 | 666,600 | 69.7 | 1,700 | 35,613 | 80,461 | 39.4 |
| District 21 | 209,479 | 93.9 | 51.9 | 214,400 | 4.8 | 895 | 17,875 | 46,037 | 17.0 |
| District 22 | 262,541 | 95.6 | 60.9 | 298,300 | 11.6 | 1,114 | 28,593 | 66,532 | 30.8 |
| District 23 | 275,854 | 89.8 | 62.6 | 264,300 | 7.8 | 1,034 | 29,846 | 63,550 | 31.6 |
| District 24 | 288,483 | 89.1 | 56.2 | 624,900 | 62.2 | 1,650 | 38,707 | 76,308 | 37.8 |
| District 25 | 233,085 | 94.9 | 70.0 | 496,700 | 49.6 | 1,660 | 34,815 | 84,670 | 41.6 |
| District 26 | 247,311 | 92.0 | 61.2 | 631,200 | 70.6 | 1,847 | 39,524 | 91,602 | 45.6 |
| District 27 | 263,223 | 92.6 | 53.2 | 770,100 | 89.1 | 1,719 | 43,216 | 90,792 | 45.2 |
| District 28 | 328,824 | 89.9 | 33.6 | 882,900 | NA | 1,686 | 47,528 | 74,431 | 37.8 |
| District 29 | 224,774 | 94.3 | 41.7 | 565,000 | 63.3 | 1,481 | 25,078 | 60,970 | 27.1 |
| District 30 | 302,277 | 93.6 | 51.4 | 717,000 | 84.6 | 1,826 | 48,228 | 89,460 | 44.9 |
| District 31 | 243,978 | 94.3 | 56.8 | 390,400 | 27.5 | 1,365 | 28,990 | 70,554 | 33.0 |
| District 32 | 209,828 | 96.5 | 57.0 | 555,700 | 60.7 | 1,566 | 26,851 | 71,136 | 33.0 |
| District 33 | 326,929 | 89.2 | 50.3 | 1,419,000 | 94.0 | 2,235 | 77,811 | 117,012 | 56.5 |
| District 34 | 282,548 | 91.1 | 20.1 | 688,100 | 78.8 | 1,335 | 28,489 | 52,043 | 24.3 |
| District 35 | 212,152 | 95.2 | 58.4 | 423,700 | 21.9 | 1,480 | 24,246 | 73,069 | 32.2 |
| District 36 | 371,370 | 74.7 | 69.3 | 310,900 | 15.6 | 1,200 | 32,698 | 58,728 | 27.9 |
| District 37 | 306,724 | 89.8 | 34.8 | 828,000 | 82.4 | 1,593 | 41,792 | 69,573 | 37.1 |
| District 38 | 210,725 | 96.4 | 62.3 | 578,600 | 66.3 | 1,631 | 28,595 | 79,573 | 38.9 |
| District 39 | 238,145 | 96.5 | 66.7 | 685,300 | 81.5 | 1,836 | 39,423 | 96,431 | 48.3 |
| District 40 | 187,465 | 95.7 | 32.9 | 500,800 | 50.1 | 1,311 | 17,920 | 51,774 | 19.7 |
| District 41 | 205,811 | 94.8 | 58.2 | 378,800 | 18.2 | 1,512 | 23,528 | 69,984 | 31.6 |
| District 42 | 252,239 | 93.6 | 72.0 | 455,800 | 36.8 | 1,841 | 34,168 | 90,651 | 46.2 |
| District 43 | 261,400 | 94.1 | 42.4 | 608,300 | 68.5 | 1,542 | 31,681 | 66,670 | 33.2 |
| District 44 | 199,233 | 95.9 | 48.6 | 471,900 | 40.5 | 1,263 | 21,949 | 59,030 | 25.4 |
| District 45 | 301,841 | 93.4 | 62.6 | 794,400 | 84.8 | 2,318 | 52,974 | 115,427 | 57.2 |
| District 46 | 205,086 | 94.8 | 41.9 | 586,500 | 70.6 | 1,693 | 25,765 | 71,800 | 33.6 |
| District 47 | 258,045 | 94.6 | 46.3 | 640,300 | 76.7 | 1,584 | 34,384 | 72,493 | 37.1 |
| District 48 | 289,410 | 91.5 | 55.8 | 840,200 | 82.9 | 2,046 | 53,688 | 100,604 | 50.5 |
| District 49 | 287,475 | 88.3 | 63.5 | 748,200 | 79.3 | 1,972 | 51,004 | 100,037 | 50.0 |
| District 50 | 263,842 | 93.1 | 63.5 | 551,200 | 58.3 | 1,610 | 34,562 | 78,346 | 39.4 |
| District 51 | 230,227 | 90.3 | 43.7 | 385,300 | 26.1 | 1,311 | 21,758 | 52,471 | 21.2 |
| District 52 | 315,338 | 91.2 | 53.5 | 775,300 | 83.4 | 2,098 | 53,927 | 106,524 | 54.1 |
| District 53 | 298,006 | 92.7 | 50.2 | 583,900 | 65.7 | 1,741 | 39,302 | 82,083 | 39.1 |

1. Specified owner-occupied units; $1,000,000 represents $1,000,000    2. Specified renter-occupied units.

# Table E. Congressional Districts 116th Congress — **Poverty, Labor Force, Employment, and Social Security**

| STATE District | Poverty, 2019 | | | Civilian labor force, 2019 | | | Civilian employment,[2] 2019 | | | | Persons under 65 years of age with no health insurance, 2019 (percent) | Social Security beneficiaries, December 2020 | | Supplemental Security Income recipients, December 2019 |
|---|---|---|---|---|---|---|---|---|---|---|---|---|---|---|
| | | | | | Unemployment | | | Percent | | | | | | |
| | Persons below poverty level (percent) | Families below poverty level (percent) | Percent of households receiving food stamps in past 12 months | Total | Total | Rate[1] | Total | Management, business, science, and arts occupations | Service, sales, and office | Construction and production | | Number | Rate[3] | |
| | 49 | 50 | 51 | 52 | 53 | 54 | 55 | 56 | 57 | 58 | 59 | 60 | 61 | 62 |
| UNITED STATES | 12.3 | 8.6 | 10.7 | 167,501,734 | 7,515,579 | 4.5 | 158,758,794 | 39.9 | 38.1 | 22.1 | 9.0 | 63,281,222 | 192.8 | 7,958,723 |
| ALABAMA | 15.5 | 11.2 | 13.0 | 2,281,982 | 111,622 | 4.9 | 2,153,467 | 35.9 | 36.8 | 27.3 | 9.6 | 1,165,990 | 237.8 | 157,314 |
| District 1 | 15.4 | 11.1 | 11.8 | 323,412 | 18,449 | 5.7 | 304,338 | 35.2 | 39.3 | 25.6 | 10.1 | 177,840 | 247.9 | 21,835 |
| District 2 | 16.4 | 13.0 | 13.7 | 307,687 | 13,470 | 4.5 | 285,238 | 33.7 | 39.4 | 26.8 | 10.4 | 166,101 | 246.1 | 26,330 |
| District 3 | 15.0 | 10.8 | 13.2 | 336,667 | 19,545 | 5.9 | 312,851 | 35.1 | 35.0 | 29.9 | 8.5 | 171,420 | 238.8 | 21,485 |
| District 4 | 15.6 | 10.9 | 13.7 | 307,509 | 13,277 | 4.3 | 293,868 | 30.3 | 34.6 | 35.1 | 11.3 | 182,594 | 265.6 | 22,660 |
| District 5 | 13.2 | 9.8 | 10.7 | 359,132 | 15,243 | 4.3 | 342,119 | 40.4 | 35.0 | 24.6 | 9.3 | 160,021 | 217.5 | 16,013 |
| District 6 | 9.5 | 6.3 | 7.2 | 351,845 | 11,571 | 3.3 | 339,524 | 45.3 | 33.9 | 20.9 | 7.2 | 150,795 | 215.5 | 11,424 |
| District 7 | 23.7 | 18.1 | 21.5 | 295,730 | 20,067 | 6.8 | 275,529 | 28.8 | 41.8 | 29.4 | 11.0 | 157,219 | 234.6 | 37,567 |
| ALASKA | 10.1 | 6.6 | 10.3 | 377,728 | 20,650 | 5.8 | 338,011 | 38.3 | 36.8 | 24.8 | 12.1 | 107,982 | 147.6 | 12,424 |
| At Large | 10.1 | 6.6 | 10.3 | 377,728 | 20,650 | 5.8 | 338,011 | 38.3 | 36.8 | 24.8 | 12.1 | 107,982 | 147.6 | 12,424 |
| ARIZONA | 13.5 | 9.5 | 9.4 | 3,506,306 | 176,868 | 5.1 | 3,305,302 | 37.9 | 41.5 | 20.6 | 11.1 | 1,433,237 | 196.9 | 118,853 |
| District 1 | 18.0 | 12.2 | 11.9 | 324,277 | 22,260 | 6.9 | 301,393 | 35.5 | 41.6 | 22.9 | 12.9 | 172,878 | 221.0 | 17,359 |
| District 2 | 13.1 | 8.4 | 10.0 | 352,679 | 16,097 | 4.7 | 325,435 | 41.0 | 42.2 | 16.9 | 8.3 | 187,776 | 256.1 | 14,373 |
| District 3 | 16.9 | 14.2 | 17.4 | 377,960 | 29,039 | 7.7 | 346,135 | 26.6 | 45.2 | 28.1 | 14.4 | 126,777 | 158.2 | 19,398 |
| District 4 | 12.2 | 8.1 | 9.3 | 329,726 | 17,416 | 5.4 | 307,949 | 31.5 | 45.4 | 23.1 | 8.7 | 247,464 | 299.7 | 12,071 |
| District 5 | 6.5 | 4.6 | 3.8 | 427,019 | 16,325 | 3.8 | 410,184 | 45.4 | 38.6 | 16.0 | 7.4 | 152,731 | 179.7 | 7,327 |
| District 6 | 8.6 | 5.9 | 5.1 | 433,738 | 14,938 | 3.4 | 418,800 | 46.4 | 39.1 | 14.5 | 8.9 | 155,381 | 190.7 | 8,151 |
| District 7 | 23.1 | 19.7 | 17.6 | 415,102 | 22,664 | 5.5 | 391,716 | 24.6 | 43.5 | 31.9 | 20.2 | 88,051 | 103.1 | 21,844 |
| District 8 | 9.1 | 6.3 | 6.1 | 373,448 | 16,762 | 4.5 | 353,220 | 40.4 | 40.8 | 18.8 | 6.8 | 192,932 | 241.6 | 8,581 |
| District 9 | 13.9 | 8.8 | 6.9 | 472,357 | 21,367 | 4.5 | 450,470 | 45.3 | 38.9 | 15.8 | 12.0 | 109,247 | 133.4 | 9,749 |
| ARKANSAS | 16.2 | 11.7 | 11.0 | 1,398,200 | 67,142 | 4.8 | 1,325,091 | 33.9 | 37.7 | 28.4 | 9.1 | 707,846 | 234.6 | 103,122 |
| District 1 | 18.3 | 13.4 | 14.3 | 320,478 | 20,270 | 6.3 | 299,282 | 30.3 | 36.9 | 32.8 | 8.5 | 186,499 | 259.4 | 31,622 |
| District 2 | 14.6 | 10.9 | 10.1 | 371,641 | 13,598 | 3.7 | 353,600 | 37.1 | 38.6 | 24.3 | 8.6 | 167,994 | 218.8 | 25,864 |
| District 3 | 14.6 | 9.4 | 6.9 | 399,402 | 19,333 | 4.8 | 379,820 | 36.4 | 36.9 | 26.7 | 10.4 | 165,214 | 199.3 | 17,181 |
| District 4 | 17.8 | 13.7 | 13.1 | 306,679 | 13,941 | 4.6 | 292,389 | 30.4 | 38.3 | 31.3 | 8.4 | 188,139 | 268.0 | 28,455 |
| CALIFORNIA | 11.8 | 8.2 | 8.4 | 20,246,507 | 1,017,660 | 5.1 | 19,078,101 | 40.7 | 38.5 | 20.9 | 7.6 | 6,150,009 | 155.6 | 1,192,888 |
| District 1 | 13.9 | 8.6 | 10.0 | 318,103 | 17,742 | 5.6 | 299,668 | 37.4 | 40.7 | 22.0 | 6.1 | 181,014 | 254.3 | 28,153 |
| District 2 | 9.8 | 5.5 | 6.1 | 361,227 | 16,784 | 4.7 | 343,868 | 44.9 | 38.9 | 16.3 | 5.3 | 159,245 | 224.8 | 16,659 |
| District 3 | 12.8 | 7.9 | 9.7 | 358,722 | 17,005 | 4.8 | 335,367 | 34.5 | 39.7 | 25.7 | 5.6 | 134,501 | 178.0 | 21,601 |
| District 4 | 8.4 | 5.5 | 5.8 | 359,912 | 15,150 | 4.2 | 343,019 | 43.7 | 39.2 | 17.0 | 4.8 | 180,557 | 238.3 | 12,640 |
| District 5 | 7.8 | 5.0 | 5.5 | 387,178 | 14,504 | 3.7 | 372,431 | 38.5 | 39.8 | 21.7 | 5.8 | 144,623 | 199.2 | 14,911 |
| District 6 | 15.5 | 10.8 | 13.2 | 393,968 | 22,511 | 5.7 | 370,793 | 35.8 | 42.2 | 21.9 | 6.3 | 116,125 | 148.5 | 39,172 |
| District 7 | 10.1 | 6.3 | 7.8 | 388,957 | 20,973 | 5.4 | 367,575 | 41.6 | 42.0 | 16.4 | 4.8 | 136,305 | 180.1 | 24,514 |
| District 8 | 16.6 | 12.7 | 16.0 | 301,477 | 24,499 | 8.5 | 264,457 | 29.9 | 40.4 | 29.8 | 9.3 | 131,872 | 182.3 | 27,877 |
| District 9 | 13.4 | 10.0 | 13.8 | 363,985 | 21,960 | 6.0 | 341,905 | 30.7 | 40.0 | 29.3 | 6.1 | 123,952 | 157.9 | 27,160 |
| District 10 | 11.8 | 8.8 | 9.6 | 358,246 | 23,704 | 6.6 | 334,499 | 27.7 | 40.3 | 32.0 | 6.6 | 123,247 | 161.1 | 23,893 |
| District 11 | 8.7 | 6.3 | 5.5 | 396,598 | 16,529 | 4.2 | 380,069 | 45.0 | 38.5 | 16.4 | 5.9 | 131,371 | 171.6 | 18,533 |
| District 12 | 9.5 | 5.1 | 5.5 | 500,103 | 18,920 | 3.8 | 481,017 | 61.8 | 30.7 | 7.5 | 3.7 | 109,027 | 139.8 | 34,372 |
| District 13 | 13.2 | 8.2 | 7.2 | 434,602 | 19,199 | 4.4 | 414,325 | 54.5 | 31.5 | 14.0 | 4.8 | 109,577 | 142.5 | 30,176 |
| District 14 | 6.6 | 3.3 | 3.0 | 430,562 | 13,057 | 3.0 | 417,400 | 50.1 | 36.1 | 13.8 | 4.4 | 117,289 | 157.9 | 12,954 |
| District 15 | 5.3 | 3.4 | 3.7 | 419,208 | 14,370 | 3.4 | 403,744 | 52.2 | 32.7 | 15.1 | 2.9 | 101,404 | 129.6 | 14,503 |
| District 16 | 23.0 | 19.3 | 23.7 | 324,668 | 31,100 | 9.6 | 292,779 | 22.3 | 40.4 | 37.3 | 10.0 | 100,124 | 132.9 | 36,711 |
| District 17 | 5.1 | 2.9 | 2.7 | 459,359 | 16,014 | 3.5 | 442,953 | 62.1 | 25.1 | 12.8 | 2.8 | 86,219 | 109.1 | 12,819 |
| District 18 | 4.6 | 2.3 | 2.5 | 391,030 | 10,338 | 2.6 | 380,274 | 66.0 | 25.2 | 8.8 | 3.7 | 104,574 | 138.7 | 9,718 |
| District 19 | 7.2 | 4.1 | 5.4 | 401,716 | 15,277 | 3.8 | 385,650 | 43.1 | 37.1 | 19.9 | 5.8 | 97,355 | 132.0 | 24,143 |
| District 20 | 12.7 | 8.9 | 8.6 | 368,485 | 17,709 | 4.9 | 345,120 | 33.9 | 37.5 | 28.7 | 8.5 | 118,808 | 160.2 | 13,582 |
| District 21 | 23.4 | 19.9 | 19.6 | 297,865 | 27,163 | 9.3 | 263,952 | 18.8 | 35.6 | 45.6 | 10.4 | 85,935 | 117.8 | 24,888 |
| District 22 | 15.2 | 11.7 | 15.0 | 359,069 | 23,790 | 6.6 | 334,061 | 36.0 | 36.9 | 27.0 | 6.1 | 120,383 | 156.6 | 26,765 |
| District 23 | 15.6 | 12.2 | 15.0 | 333,169 | 26,588 | 8.1 | 302,987 | 36.0 | 37.1 | 26.9 | 6.1 | 122,498 | 165.2 | 27,078 |
| District 24 | 12.0 | 6.0 | 6.5 | 374,343 | 15,488 | 4.2 | 354,158 | 38.7 | 39.2 | 22.1 | 8.7 | 141,625 | 192.0 | 12,382 |
| District 25 | 10.3 | 7.6 | 7.5 | 340,951 | 15,355 | 4.5 | 325,250 | 38.6 | 40.2 | 21.2 | 5.2 | 106,824 | 148.6 | 22,285 |
| District 26 | 8.1 | 5.2 | 7.0 | 372,892 | 15,429 | 4.2 | 354,043 | 36.3 | 40.5 | 23.2 | 10.0 | 126,813 | 174.8 | 13,663 |
| District 27 | 8.8 | 5.8 | 4.0 | 377,614 | 12,522 | 3.3 | 364,788 | 50.0 | 37.2 | 12.8 | 6.2 | 119,702 | 167.9 | 26,682 |
| District 28 | 12.6 | 7.7 | 7.3 | 410,773 | 22,495 | 5.5 | 388,105 | 52.0 | 36.5 | 11.5 | 8.6 | 94,981 | 137.0 | 40,340 |
| District 29 | 15.9 | 12.9 | 12.0 | 385,922 | 19,035 | 4.9 | 366,746 | 29.4 | 43.5 | 27.1 | 12.7 | 84,111 | 117.2 | 30,935 |
| District 30 | 9.5 | 5.8 | 4.3 | 432,801 | 19,675 | 4.5 | 412,751 | 49.4 | 37.0 | 13.6 | 7.0 | 115,790 | 151.5 | 20,662 |
| District 31 | 12.3 | 9.5 | 11.3 | 372,523 | 19,396 | 5.2 | 352,798 | 34.3 | 39.2 | 26.5 | 8.6 | 101,388 | 134.5 | 26,217 |
| District 32 | 11.6 | 8.4 | 9.5 | 361,948 | 18,975 | 5.2 | 342,648 | 30.5 | 42.4 | 27.1 | 9.1 | 110,576 | 157.8 | 24,564 |
| District 33 | 7.7 | 3.7 | 2.5 | 379,185 | 16,442 | 4.3 | 361,761 | 65.1 | 28.3 | 6.6 | 3.9 | 119,006 | 169.1 | 10,448 |
| District 34 | 22.1 | 17.1 | 12.9 | 396,577 | 22,351 | 5.6 | 373,876 | 35.1 | 42.5 | 22.4 | 16.2 | 81,878 | 112.2 | 38,841 |
| District 35 | 12.6 | 10.9 | 12.2 | 377,527 | 17,222 | 4.6 | 360,079 | 24.5 | 40.6 | 34.8 | 9.9 | 89,008 | 116.4 | 21,778 |
| District 36 | 15.5 | 11.7 | 10.8 | 325,423 | 23,328 | 7.2 | 301,378 | 30.7 | 47.3 | 21.9 | 8.3 | 172,689 | 228.5 | 24,663 |
| District 37 | 16.0 | 10.9 | 10.4 | 415,128 | 18,045 | 4.4 | 395,868 | 45.2 | 39.1 | 15.7 | 9.9 | 96,768 | 131.9 | 25,846 |
| District 38 | 8.9 | 6.5 | 6.8 | 361,407 | 16,359 | 4.5 | 344,351 | 33.7 | 42.5 | 23.8 | 8.0 | 115,073 | 163.3 | 24,577 |
| District 39 | 7.4 | 5.0 | 4.3 | 384,566 | 18,026 | 4.7 | 366,268 | 44.6 | 39.2 | 16.1 | 6.7 | 115,420 | 160.9 | 15,493 |
| District 40 | 20.0 | 18.3 | 16.9 | 349,131 | 21,887 | 6.3 | 327,244 | 17.2 | 45.1 | 37.8 | 15.6 | 76,249 | 106.5 | 23,752 |
| District 41 | 11.3 | 8.8 | 10.7 | 378,185 | 21,296 | 5.6 | 356,456 | 25.8 | 41.0 | 33.2 | 11.0 | 96,558 | 122.7 | 23,461 |
| District 42 | 7.8 | 5.9 | 4.7 | 400,648 | 20,144 | 5.1 | 377,243 | 37.2 | 40.5 | 22.3 | 6.9 | 120,544 | 143.4 | 12,986 |
| District 43 | 14.4 | 10.9 | 11.6 | 397,384 | 19,336 | 4.9 | 377,402 | 33.8 | 43.7 | 22.5 | 10.3 | 100,131 | 133.8 | 29,895 |
| District 44 | 14.4 | 12.6 | 16.8 | 355,549 | 25,238 | 7.1 | 329,514 | 21.2 | 45.0 | 33.7 | 12.7 | 92,997 | 129.7 | 32,645 |
| District 45 | 8.9 | 5.5 | 2.6 | 420,359 | 13,605 | 3.2 | 406,531 | 57.3 | 33.0 | 9.6 | 5.1 | 119,243 | 150.7 | 9,921 |
| District 46 | 12.6 | 9.4 | 8.3 | 393,754 | 16,922 | 4.3 | 376,252 | 27.9 | 44.4 | 27.6 | 12.3 | 80,895 | 110.1 | 26,276 |
| District 47 | 13.5 | 9.8 | 9.1 | 381,054 | 17,597 | 4.6 | 362,313 | 40.5 | 39.9 | 19.6 | 8.8 | 102,640 | 143.0 | 30,643 |
| District 48 | 7.9 | 5.2 | 4.2 | 389,541 | 16,714 | 4.3 | 372,363 | 46.3 | 39.1 | 14.6 | 6.2 | 129,027 | 179.6 | 9,709 |
| District 49 | 6.0 | 3.9 | 3.2 | 393,735 | 18,877 | 5.1 | 349,517 | 46.3 | 38.7 | 15.0 | 7.3 | 124,110 | 169.7 | 7,974 |
| District 50 | 11.7 | 8.3 | 7.2 | 372,079 | 16,064 | 4.4 | 351,242 | 36.8 | 41.2 | 22.0 | 7.6 | 129,997 | 171.5 | 15,520 |
| District 51 | 17.5 | 14.4 | 14.3 | 358,998 | 34,023 | 10.0 | 306,358 | 23.2 | 50.0 | 26.8 | 12.2 | 122,384 | 165.2 | 35,423 |
| District 52 | 8.1 | 4.3 | 3.1 | 433,580 | 14,966 | 3.6 | 398,767 | 60.8 | 30.2 | 9.0 | 4.5 | 115,459 | 150.5 | 13,793 |
| District 53 | 9.9 | 5.7 | 6.4 | 444,721 | 25,962 | 6.0 | 404,118 | 46.2 | 38.8 | 15.1 | 7.1 | 112,118 | 143.3 | 18,692 |

1. Percent of civilian labor force. 2. Persons 16 years old and over. 3. Per 1,000 resident population estimated in the 2019 American Community Survey.

# Table E. Congressional Districts 116th Congress — **Agriculture**

| STATE District | Agriculture 2017 | | | | | | | | | |
|---|---|---|---|---|---|---|---|---|---|---|
| | Land in farms | | | | Value of products sold | | | | Government payments | |
| | Number of farms | Acres | Average size of farm (acres) | Irrigated land (acres) | Total ($1,000) | Average per farm (dollars) | Percent from crops | Percent from livestock and poultry products | Total ($1,000) | Average per farm receiving payments (dollars) |
| | 63 | 64 | 65 | 66 | 67 | 68 | 69 | 70 | 71 | 72 |
| UNITED STATES.............. | 2,042,220 | 900,217,576 | 441 | 320,041,858 | 388,522,695 | 190,245 | 49.8 | 50.2 | 8,943,574 | 13,906 |
| ALABAMA.......................... | 40,592 | 8,580,940 | 211 | 2,205,766 | 5,980,595 | 147,334 | 20.3 | 79.7 | 134,654 | 8,892 |
| District 1....................... | 2,928 | 652,956 | 223 | 227,241 | 310,410 | 106,014 | 76.8 | 23.2 | 15,572 | 16,762 |
| District 2....................... | 8,100 | 2,095,530 | 259 | 504,145 | 1,514,029 | 186,917 | 18.4 | 81.6 | 43,478 | 11,103 |
| District 3....................... | 5,232 | 1,104,479 | 211 | 215,369 | 813,111 | 155,411 | 20.1 | 79.9 | 13,810 | 7,955 |
| District 4....................... | 11,374 | 1,639,238 | 144 | 473,554 | 2,172,299 | 190,988 | 8.7 | 91.3 | 24,234 | 6,379 |
| District 5....................... | 5,927 | 965,903 | 163 | 459,776 | 519,220 | 87,602 | 43.1 | 56.9 | 14,741 | 7,456 |
| District 6....................... | 2,580 | 399,443 | 155 | 81,843 | 217,477 | 84,293 | 16.4 | 83.6 | 4,185 | 6,707 |
| District 7....................... | 4,451 | 1,723,391 | 387 | 243,838 | 434,050 | 97,517 | 19.2 | 80.8 | 18,634 | 8,615 |
| ALASKA.......................... | 990 | 849,753 | 858 | 31,877 | 70,459 | 71,171 | 42.1 | 57.9 | 2,091 | 9,293 |
| At Large ....................... | 990 | 849,753 | 858 | 31,877 | 70,459 | 71,171 | 42.1 | 57.9 | 2,091 | 9,293 |
| ARIZONA.......................... | 19,086 | 26,125,819 | 1,369 | 915,647 | 3,852,008 | 201,824 | 54.4 | 45.6 | 22,331 | 29,735 |
| District 1....................... | 13,560 | 20,195,286 | 1,489 | 289,469 | 1,031,964 | 76,104 | 34.3 | 65.7 | 6,836 | 24,768 |
| District 2....................... | 1,323 | 1,040,464 | 786 | 88,140 | 161,280 | 121,905 | 60.0 | 40.0 | 3,212 | 25,291 |
| District 3....................... | 1,257 | 3,078,961 | 2,449 | 219,807 | 1,195,275 | 950,895 | 54.7 | 45.3 | 5,920 | 39,467 |
| District 4....................... | 1,717 | 1,598,936 | 931 | 239,342 | 1,017,973 | 592,879 | 67.6 | 32.4 | 4,311 | 35,628 |
| District 5....................... | 435 | 60,281 | 139 | 22,775 | 126,184 | 290,078 | D | D | 1,053 | 32,906 |
| District 6....................... | 297 | D | D | 21,075 | 113,635 | 382,609 | 92.0 | 8.0 | 239 | 21,727 |
| District 7....................... | 106 | 20,624 | 195 | 16,332 | 102,007 | 962,330 | 84.0 | 16.0 | 383 | 42,556 |
| District 8....................... | 343 | D | D | 18,354 | 86,807 | 253,082 | 71.7 | 28.3 | 193 | 9,190 |
| District 9....................... | 48 | 2,095 | 44 | 353 | 16,884 | 351,750 | D | D | 184 | 46,000 |
| ARKANSAS ..................... | 42,625 | 13,888,929 | 326 | 7,098,672 | 9,651,160 | 226,420 | 37.6 | 62.4 | 321,742 | 38,624 |
| District 1....................... | 13,399 | 7,758,029 | 579 | 5,424,425 | 4,087,818 | 305,084 | 77.5 | 22.5 | 272,375 | 48,500 |
| District 2....................... | 5,351 | 1,039,161 | 194 | 315,834 | 422,992 | 79,049 | 21.3 | 78.7 | 13,321 | 19,195 |
| District 3....................... | 9,329 | 1,669,233 | 179 | 314,685 | 1,928,511 | 206,722 | 2.3 | 97.7 | 4,899 | 10,080 |
| District 4....................... | 14,546 | 3,422,506 | 235 | 1,043,728 | 3,211,839 | 220,806 | 10.0 | 90.0 | 31,147 | 20,304 |
| CALIFORNIA ..................... | 70,521 | 24,522,801 | 348 | 7,857,512 | 45,154,359 | 640,297 | 73.9 | 26.1 | 127,938 | 24,112 |
| District 1....................... | 7,949 | 3,690,258 | 464 | 650,694 | 1,471,521 | 185,120 | 81.2 | 18.8 | 17,554 | 27,300 |
| District 2....................... | 4,847 | 1,983,379 | 409 | 121,953 | 1,206,957 | 249,011 | 56.6 | 43.4 | 2,866 | 14,697 |
| District 3....................... | 5,761 | 2,222,919 | 386 | 1,075,534 | 2,524,704 | 438,241 | 92.6 | 7.4 | 27,813 | 31,356 |
| District 4....................... | 5,025 | 1,537,953 | 306 | 149,086 | 783,366 | 155,894 | 70.6 | 29.4 | 5,706 | 16,444 |
| District 5....................... | 3,616 | 502,424 | 139 | 106,858 | 889,982 | 246,123 | 89.6 | 10.4 | 2,234 | 17,591 |
| District 6....................... | 95 | 7,723 | 81 | 7,171 | 14,672 | 154,442 | 99.6 | 0.4 | 232 | 38,667 |
| District 7....................... | 714 | 134,035 | 188 | 30,374 | 186,365 | 261,015 | 63.7 | 36.3 | 860 | 18,696 |
| District 8....................... | 830 | 405,833 | 489 | 25,014 | 160,823 | 193,763 | 31.2 | 68.8 | 1,370 | 42,813 |
| District 9....................... | 3,034 | 682,651 | 225 | 417,973 | 1,881,357 | 620,091 | 74.8 | 25.2 | 5,136 | 24,226 |
| District 10..................... | 4,282 | 881,704 | 206 | 463,869 | 2,915,414 | 680,853 | 56.4 | 43.6 | 5,996 | 18,855 |
| District 11..................... | 230 | 105,382 | 458 | 19,726 | 36,383 | 158,187 | 77.0 | 23.0 | 332 | 15,091 |
| District 12..................... | 9 | 89 | 10 | 5 | 452 | 50,222 | D | D | D | D |
| District 13..................... | 35 | D | D | 34 | 2,608 | 74,514 | 12.6 | 87.4 | D | D |
| District 14..................... | 121 | 24,065 | 199 | 2,329 | 48,318 | 399,322 | 98.3 | 1.7 | D | D |
| District 15..................... | 409 | 138,780 | 339 | 6,969 | 34,387 | 84,076 | 76.5 | 23.5 | 194 | 7,462 |
| District 16..................... | 3,710 | 1,394,781 | 376 | 800,605 | 4,460,128 | 1,202,191 | 54.7 | 45.3 | 10,377 | 27,094 |
| District 17..................... | 60 | D | D | 541 | 11,233 | 187,217 | 97.5 | 2.5 | D | D |
| District 18..................... | 492 | 63,346 | 129 | 10,962 | 195,998 | 398,370 | 99.3 | 0.7 | 18 | 3,600 |
| District 19..................... | 612 | 268,704 | 439 | 14,521 | 197,717 | 323,067 | 92.1 | 7.9 | 490 | 16,897 |
| District 20..................... | 2,210 | 1,901,631 | 860 | 338,292 | 4,831,405 | 2,186,156 | 98.4 | 1.6 | 1,925 | 12,419 |
| District 21..................... | 4,578 | 2,751,011 | 601 | 1,718,981 | 9,790,565 | 2,138,612 | 62.0 | 38.0 | 18,580 | 24,609 |
| District 22..................... | 3,428 | 771,371 | 225 | 483,801 | 3,124,957 | 911,598 | 60.9 | 39.1 | 8,611 | 28,703 |
| District 23..................... | 2,800 | 1,981,120 | 708 | 419,024 | 2,438,036 | 870,727 | 87.9 | 12.1 | 4,921 | 24,605 |
| District 24..................... | 4,091 | 1,708,370 | 418 | 237,661 | 2,708,747 | 662,123 | 97.4 | 2.6 | 5,168 | 17,227 |
| District 25..................... | 403 | 36,030 | 89 | 7,779 | 53,359 | 132,404 | D | D | 64 | 6,400 |
| District 26..................... | 1,814 | 193,157 | 106 | 69,560 | 1,128,887 | 622,319 | 99.5 | 0.5 | 743 | 16,152 |
| District 27..................... | 124 | 3,062 | 25 | 989 | 24,781 | 199,847 | 99.3 | 0.7 | 43 | 14,333 |
| District 28..................... | 93 | 1,488 | 16 | 67 | 965 | 10,376 | 72.2 | 27.8 | D | D |
| District 29..................... | 75 | 1,341 | 18 | 686 | 12,735 | 169,800 | 98.7 | 1.3 | D | D |
| District 30..................... | 47 | 1,864 | 40 | 1,529 | 15,037 | 319,936 | 99.7 | 0.3 | D | D |
| District 31..................... | 177 | 6,904 | 39 | 2,458 | 19,407 | 109,644 | 99.5 | 0.5 | 124 | 24,800 |
| District 32..................... | 49 | 804 | 16 | 211 | 15,520 | 316,735 | 99.7 | 0.3 | D | D |
| District 33..................... | 101 | 5,819 | 58 | 1,248 | 9,241 | 91,495 | 98.7 | 1.3 | D | D |
| District 34..................... | 12 | 151 | 13 | 13 | 54 | 4,500 | 100.0 | 0.0 | D | D |
| District 35..................... | 168 | 10,073 | 60 | 3,962 | 207,967 | 1,237,899 | 13.0 | 87.0 | 310 | 34,444 |
| District 36..................... | 961 | 185,512 | 193 | 106,983 | 649,142 | 675,486 | 86.7 | 13.3 | 1,502 | 34,930 |
| District 37..................... | 13 | D | D | 31 | 169 | 13,000 | D | D | D | D |
| District 38..................... | 37 | 179 | 5 | 98 | 5,583 | 150,892 | D | D | D | D |
| District 39..................... | 81 | 5,324 | 66 | 649 | 10,590 | 130,741 | 95.7 | 4.3 | 300 | 50,000 |
| District 40..................... | 5 | D | D | 21 | 617 | 123,400 | 100.0 | 0.0 | D | D |
| District 41..................... | 413 | 29,236 | 71 | 9,848 | 57,385 | 138,947 | 87.0 | 13.0 | 16 | 1,778 |
| District 42..................... | 1,178 | 45,409 | 39 | 24,697 | 212,078 | 180,032 | 38.4 | 61.6 | 101 | 11,222 |
| District 43..................... | 27 | 1,709 | 63 | 330 | 8,044 | 297,926 | D | D | 139 | 46,333 |
| District 44..................... | 36 | 188 | 5 | 118 | 21,704 | 602,889 | D | D | D | D |
| District 45..................... | 82 | 26,333 | 321 | 3,066 | 41,076 | 500,927 | 99.9 | 0.1 | 3 | 1,000 |
| District 46..................... | 17 | D | D | 91 | 4,874 | 286,706 | D | D | D | D |
| District 47..................... | 28 | 9,985 | 357 | 214 | 5,939 | 212,107 | 99.3 | 0.7 | D | D |
| District 48..................... | 28 | 2,641 | 94 | 704 | 20,354 | 726,929 | 99.0 | 1.0 | 3 | 1,000 |
| District 49..................... | 813 | 21,797 | 27 | 7,962 | 174,311 | 214,405 | 90.9 | 9.1 | 13 | 2,167 |
| District 50..................... | 3,783 | 153,781 | 41 | 37,897 | 597,859 | 158,038 | 94.4 | 5.6 | 542 | 12,044 |
| District 51..................... | 697 | 563,066 | 808 | 470,449 | 1,908,532 | 2,738,209 | 66.5 | 33.5 | 3,641 | 32,509 |
| District 52..................... | 214 | 7,688 | 36 | 2,821 | 19,376 | 90,542 | 93.3 | 6.7 | D | D |
| District 53..................... | 107 | 3,830 | 36 | 1,054 | 12,679 | 118,495 | 98.8 | 1.2 | D | D |

# Table E. Congressional Districts 116th Congress — Nonfarm Employment and Payroll

| STATE District | Number of establishments | Private nonfarm employment and payroll, 2019 — Employment | Percent by selected industries | | | | | | | | | Annual payroll | |
|---|---|---|---|---|---|---|---|---|---|---|---|---|---|
| | | Total | Manufac-turing | Construc-tion | Wholesale trade | Retail trade | Health care and social assistance | Finance and Insurance | Real estate and rental and leasing | Profes-sional, sci-entific, and technical services | Information | Total (mil dol) | Average per employee (dollars) |
| | 73 | 74 | 75 | 76 | 77 | 78 | 79 | 80 | 81 | 82 | 83 | 84 | 85 |
| UNITED STATES | 7,959,103 | 132,989,428 | 9.1 | 5.3 | 4.6 | 11.8 | 15.7 | 4.9 | 1.7 | 7.0 | 2.7 | 7,428,554 | 55,858 |
| ALABAMA | 100,731 | 1,758,609 | 15.0 | 5.2 | 4.3 | 13.0 | 15.0 | 4.0 | 1.3 | 6.2 | 1.8 | 79,548 | 45,234 |
| District 1 | 16,042 | 246,199 | 11.3 | 7.2 | 4.3 | 15.3 | 13.8 | 3.6 | 1.8 | 4.9 | 1.6 | 10,561 | 42,895 |
| District 2 | 14,142 | 219,624 | 13.0 | 4.3 | 5.0 | 14.7 | 17.2 | 3.3 | 1.2 | 4.5 | 1.4 | 8,833 | 40,218 |
| District 3 | 11,862 | 189,702 | 22.2 | 4.2 | 3.6 | 15.2 | 14.9 | 2.3 | 1.1 | 2.9 | 1.1 | 7,022 | 37,017 |
| District 4 | 12,582 | 197,681 | 28.3 | 4.2 | 3.7 | 14.0 | 15.8 | 3.0 | 1.0 | 2.1 | 1.2 | 7,428 | 37,574 |
| District 5 | 15,364 | 277,067 | 15.1 | 4.7 | 3.5 | 12.7 | 15.0 | 2.5 | 1.1 | 17.3 | 2.0 | 14,110 | 50,928 |
| District 6 | 16,577 | 264,118 | 6.9 | 6.7 | 4.5 | 15.1 | 13.3 | 7.5 | 1.5 | 5.0 | 4.1 | 13,571 | 51,381 |
| District 7 | 13,636 | 298,347 | 16.6 | 5.6 | 5.5 | 8.9 | 18.8 | 5.0 | 1.8 | 4.0 | 1.4 | 15,471 | 51,856 |
| ALASKA | 21,399 | 264,971 | 4.7 | 6.1 | 3.2 | 12.6 | 19.5 | 2.7 | 1.7 | 7.3 | 2.5 | 16,572 | 62,542 |
| At Large | 21,399 | 264,971 | 4.7 | 6.1 | 3.2 | 12.6 | 19.5 | 2.7 | 1.7 | 7.3 | 2.5 | 16,572 | 62,542 |
| ARIZONA | 147,163 | 2,614,641 | 5.9 | 6.6 | 3.8 | 12.5 | 14.8 | 6.8 | 2.0 | 6.2 | 2.2 | 129,708 | 49,608 |
| District 1 | 12,269 | 174,346 | 8.4 | 4.8 | 2.0 | 16.6 | 18.1 | 1.6 | 1.5 | 2.6 | 1.4 | 7,172 | 41,136 |
| District 2 | 15,527 | 223,637 | 2.7 | 6.2 | 1.3 | 16.1 | 22.3 | 4.2 | 2.3 | 6.8 | 2.2 | 9,168 | 40,993 |
| District 3 | 10,146 | 181,789 | 9.7 | 6.7 | 4.6 | 14.5 | 16.2 | 3.0 | 1.6 | 3.6 | 1.9 | 8,475 | 46,619 |
| District 4 | 13,501 | 151,907 | 6.7 | 8.2 | 3.1 | 20.1 | 18.4 | 2.2 | 1.6 | 3.0 | 1.1 | 5,567 | 36,648 |
| District 5 | 16,889 | 221,827 | 7.5 | 8.3 | 2.6 | 17.1 | 15.4 | 6.8 | 2.3 | 5.2 | 1.7 | 10,003 | 45,095 |
| District 6 | 26,422 | 424,012 | 3.3 | 7.5 | 2.6 | 10.9 | 12.6 | 11.8 | 2.5 | 7.2 | 3.4 | 23,935 | 56,449 |
| District 7 | 14,380 | 410,157 | 9.7 | 8.0 | 8.7 | 8.2 | 13.2 | 6.1 | 1.5 | 4.4 | 1.7 | 22,169 | 54,050 |
| District 8 | 12,672 | 167,902 | 3.4 | 5.6 | 1.9 | 21.6 | 23.0 | 3.4 | 1.9 | 3.0 | 1.0 | 6,389 | 38,050 |
| District 9 | 24,618 | 523,470 | 5.8 | 6.5 | 3.9 | 9.6 | 12.6 | 9.9 | 2.9 | 11.1 | 3.5 | 30,670 | 58,590 |
| ARKANSAS | 67,243 | 1,053,453 | 15.4 | 4.7 | 4.6 | 13.1 | 17.7 | 3.8 | 1.3 | 3.8 | 2.0 | 46,260 | 43,912 |
| District 1 | 13,819 | 195,359 | 22.0 | 3.9 | 5.1 | 15.2 | 21.0 | 3.0 | 1.1 | 2.3 | 0.9 | 7,503 | 38,404 |
| District 2 | 19,341 | 309,356 | 7.0 | 5.4 | 5.0 | 13.1 | 21.5 | 6.0 | 1.4 | 4.4 | 3.7 | 14,379 | 46,479 |
| District 3 | 19,972 | 336,093 | 15.7 | 4.6 | 5.0 | 12.2 | 13.6 | 2.6 | 1.4 | 4.9 | 1.7 | 16,469 | 49,001 |
| District 4 | 13,710 | 186,400 | 23.8 | 5.2 | 2.7 | 14.4 | 17.6 | 3.0 | 1.1 | 2.2 | 0.9 | 7,104 | 38,113 |
| CALIFORNIA | 966,224 | 15,516,824 | 7.6 | 5.4 | 5.3 | 10.8 | 13.5 | 4.0 | 2.0 | 8.2 | 4.5 | 1,077,176 | 69,420 |
| District 1 | 15,715 | 179,970 | 6.6 | 6.3 | 3.2 | 17.3 | 22.1 | 3.6 | 1.5 | 4.2 | 1.4 | 7,846 | 43,595 |
| District 2 | 22,060 | 235,505 | 8.5 | 7.2 | 3.8 | 15.2 | 16.3 | 3.5 | 2.4 | 6.4 | 2.9 | 13,820 | 58,683 |
| District 3 | 12,455 | 179,724 | 9.0 | 8.6 | 5.8 | 16.0 | 16.2 | 2.9 | 1.7 | 3.9 | 1.7 | 9,122 | 50,754 |
| District 4 | 19,782 | 244,058 | 3.9 | 9.5 | 2.5 | 14.2 | 14.6 | 5.6 | 2.6 | 6.4 | 2.4 | 12,788 | 52,396 |
| District 5 | 17,663 | 246,618 | 12.1 | 9.1 | 4.0 | 13.2 | 18.6 | 2.5 | 1.5 | 4.0 | 1.3 | 14,518 | 58,869 |
| District 6 | 16,283 | 286,176 | 4.1 | 7.3 | 6.2 | 11.2 | 19.4 | 2.9 | 2.1 | 7.2 | 2.3 | 16,207 | 56,632 |
| District 7 | 14,594 | 228,292 | 4.6 | 7.3 | 2.4 | 13.6 | 15.6 | 9.6 | 1.8 | 10.3 | 2.6 | 14,281 | 62,558 |
| District 8 | 9,882 | 120,770 | 5.2 | 5.3 | 1.5 | 18.5 | 18.2 | 1.6 | 2.5 | 2.9 | 1.3 | 4,705 | 38,957 |
| District 9 | 11,438 | 172,806 | 7.5 | 7.7 | 6.0 | 14.1 | 17.4 | 3.6 | 1.7 | 2.8 | 1.1 | 8,626 | 49,914 |
| District 10 | 12,925 | 205,206 | 12.3 | 6.0 | 5.8 | 16.0 | 16.2 | 2.1 | 1.3 | 3.6 | 1.2 | 9,703 | 47,285 |
| District 11 | 17,614 | 241,380 | 4.4 | 7.7 | 3.5 | 14.1 | 18.0 | 6.8 | 2.2 | 7.3 | 1.8 | 15,653 | 64,846 |
| District 12 | 33,219 | 693,119 | 1.2 | 3.2 | 2.2 | 7.1 | 10.0 | 9.2 | 2.6 | 17.5 | 11.7 | 85,239 | 122,959 |
| District 13 | 19,903 | 316,512 | 6.3 | 6.0 | 4.5 | 9.6 | 17.8 | 3.8 | 1.7 | 9.6 | 3.7 | 23,253 | 73,466 |
| District 14 | 19,898 | 394,424 | 5.9 | 5.0 | 5.2 | 9.3 | 8.9 | 4.0 | 1.9 | 9.7 | 16.6 | 52,123 | 132,149 |
| District 15 | 17,759 | 311,741 | 8.5 | 8.3 | 8.2 | 9.7 | 11.8 | 3.8 | 1.6 | 12.3 | 7.3 | 25,696 | 82,427 |
| District 16 | 9,325 | 160,034 | 16.2 | 6.9 | 6.3 | 12.3 | 19.3 | 1.7 | 1.6 | 2.6 | 1.8 | 7,212 | 45,068 |
| District 17 | 21,779 | 611,335 | 16.2 | 4.8 | 11.1 | 5.6 | 6.1 | 2.1 | 1.2 | 13.6 | 9.1 | 87,294 | 142,793 |
| District 18 | 21,957 | 420,324 | 2.8 | 3.0 | 2.4 | 8.7 | 17.4 | 3.4 | 1.5 | 13.7 | 17.5 | 57,719 | 137,321 |
| District 19 | 14,535 | 232,087 | 7.1 | 8.0 | 4.6 | 9.8 | 10.8 | 2.4 | 2.2 | 11.0 | 6.3 | 18,070 | 77,860 |
| District 20 | 16,685 | 217,036 | 7.9 | 5.8 | 5.3 | 16.1 | 15.4 | 2.2 | 1.7 | 5.7 | 1.4 | 10,920 | 50,316 |
| District 21 | 7,157 | 115,290 | 16.5 | 5.3 | 8.3 | 17.6 | 11.3 | 1.2 | 1.1 | 1.5 | 1.0 | 4,953 | 42,961 |
| District 22 | 15,168 | 227,157 | 7.7 | 6.0 | 3.7 | 16.4 | 18.5 | 4.5 | 1.9 | 4.2 | 1.4 | 9,780 | 43,052 |
| District 23 | 12,550 | 182,731 | 4.9 | 6.7 | 3.7 | 14.0 | 20.1 | 3.0 | 1.7 | 5.3 | 2.2 | 8,733 | 47,794 |
| District 24 | 20,906 | 258,566 | 8.0 | 6.3 | 3.8 | 13.0 | 15.3 | 2.5 | 2.0 | 6.7 | 3.8 | 14,033 | 54,274 |
| District 25 | 13,214 | 170,533 | 9.4 | 7.2 | 3.9 | 15.9 | 13.8 | 2.4 | 1.9 | 8.2 | 1.5 | 8,065 | 47,294 |
| District 26 | 19,472 | 250,866 | 9.5 | 5.6 | 5.9 | 14.7 | 14.7 | 5.1 | 1.8 | 8.8 | 2.3 | 15,034 | 59,928 |
| District 27 | 23,308 | 260,952 | 2.1 | 2.4 | 2.7 | 11.7 | 19.4 | 6.9 | 2.1 | 10.1 | 2.0 | 14,470 | 55,452 |
| District 28 | 27,700 | 335,177 | 4.0 | 3.3 | 2.4 | 10.8 | 19.4 | 2.7 | 2.9 | 6.8 | 9.7 | 23,075 | 68,845 |
| District 29 | 13,149 | 173,805 | 13.6 | 8.8 | 6.7 | 12.0 | 19.8 | 1.8 | 2.7 | 3.4 | 2.4 | 8,874 | 51,059 |
| District 30 | 32,986 | 356,134 | 5.4 | 4.2 | 3.3 | 10.7 | 17.2 | 6.2 | 2.6 | 8.0 | 9.6 | 22,878 | 64,239 |
| District 31 | 14,078 | 260,705 | 7.9 | 5.4 | 4.7 | 14.2 | 22.0 | 3.0 | 1.4 | 3.4 | 2.0 | 12,286 | 47,126 |
| District 32 | 15,914 | 252,841 | 14.9 | 4.8 | 9.8 | 12.8 | 15.6 | 3.0 | 1.6 | 4.4 | 1.4 | 12,659 | 50,066 |
| District 33 | 42,828 | 521,048 | 5.5 | 1.9 | 2.7 | 9.9 | 12.1 | 4.4 | 3.6 | 13.6 | 10.3 | 45,164 | 86,678 |
| District 34 | 24,756 | 352,694 | 4.0 | 1.8 | 8.4 | 6.1 | 14.4 | 7.4 | 2.5 | 11.8 | 2.1 | 25,092 | 71,144 |
| District 35 | 14,544 | 281,652 | 11.0 | 6.9 | 11.7 | 12.6 | 11.3 | 1.8 | 1.4 | 2.1 | 1.4 | 14,019 | 49,775 |
| District 36 | 13,192 | 180,431 | 3.5 | 6.6 | 1.8 | 18.8 | 17.0 | 1.6 | 2.5 | 2.7 | 1.5 | 6,997 | 38,780 |
| District 37 | 22,818 | 279,015 | 2.8 | 1.8 | 2.9 | 9.1 | 13.2 | 4.3 | 3.5 | 11.6 | 12.1 | 22,734 | 81,481 |
| District 38 | 15,347 | 251,822 | 12.7 | 5.6 | 14.3 | 13.4 | 14.8 | 2.0 | 1.6 | 3.7 | 1.4 | 12,253 | 48,659 |
| District 39 | 21,069 | 273,088 | 10.5 | 8.2 | 11.4 | 11.0 | 9.3 | 6.4 | 1.7 | 4.8 | 1.4 | 14,017 | 51,329 |
| District 40 | 11,736 | 217,010 | 19.8 | 2.7 | 16.1 | 11.0 | 11.9 | 1.4 | 1.2 | 1.9 | 0.8 | 10,385 | 47,856 |
| District 41 | 11,011 | 208,119 | 8.4 | 9.2 | 5.3 | 14.5 | 14.2 | 2.0 | 1.4 | 3.0 | 1.2 | 9,223 | 44,316 |
| District 42 | 13,151 | 174,039 | 10.8 | 15.8 | 7.4 | 13.6 | 12.2 | 1.7 | 1.5 | 4.6 | 0.9 | 7,616 | 43,762 |
| District 43 | 15,859 | 311,873 | 10.8 | 2.8 | 6.4 | 9.8 | 12.4 | 2.0 | 2.5 | 6.1 | 2.0 | 17,861 | 57,271 |
| District 44 | 9,117 | 165,933 | 19.2 | 7.2 | 11.1 | 10.2 | 9.7 | 1.0 | 1.3 | 2.0 | 1.4 | 8,488 | 51,153 |
| District 45 | 27,970 | 433,213 | 9.7 | 5.8 | 8.0 | 7.6 | 9.5 | 7.1 | 3.7 | 13.9 | 5.1 | 34,178 | 78,893 |
| District 46 | 18,369 | 399,783 | 10.2 | 7.7 | 6.3 | 7.4 | 14.1 | 4.6 | 1.9 | 5.3 | 1.2 | 20,741 | 51,881 |
| District 47 | 16,411 | 256,701 | 8.0 | 5.4 | 5.1 | 11.2 | 15.7 | 5.2 | 1.9 | 5.6 | 1.7 | 14,242 | 55,479 |
| District 48 | 26,717 | 338,137 | 6.8 | 5.1 | 4.9 | 13.1 | 12.4 | 6.4 | 4.1 | 10.2 | 1.6 | 21,901 | 64,769 |
| District 49 | 22,897 | 294,025 | 11.3 | 5.3 | 5.7 | 11.8 | 12.8 | 2.6 | 2.3 | 13.2 | 2.1 | 18,504 | 62,933 |
| District 50 | 17,074 | 210,793 | 7.9 | 13.1 | 3.5 | 16.4 | 12.8 | 1.9 | 1.7 | 4.3 | 1.2 | 9,366 | 44,431 |
| District 51 | 10,908 | 146,541 | 10.5 | 4.2 | 7.1 | 21.6 | 16.0 | 2.3 | 1.7 | 3.1 | 1.7 | 5,895 | 40,227 |
| District 52 | 30,815 | 550,683 | 7.8 | 5.2 | 3.8 | 8.0 | 7.9 | 6.7 | 2.5 | 17.6 | 4.1 | 40,825 | 74,135 |
| District 53 | 17,223 | 249,260 | 3.0 | 5.4 | 1.7 | 11.7 | 29.5 | 4.3 | 2.4 | 7.2 | 1.0 | 12,848 | 51,544 |

# Table E. Congressional Districts 116th Congress — Land Area and Population Characteristics

| STATE District | Representative, 117th Congress | Population and population characteristics, 2019 | | | | | | | | | | | | |
|---|---|---|---|---|---|---|---|---|---|---|---|---|---|---|
| | | | | | Race alone (percent) | | | | | | | | | |
| | | Land area,[1] 2018 (sq mi) | Total persons | Per square mile | White | Black | American Indian, Alaska Native | Asian and Pacific Islander | Some other race (percent) | Two or more races (percent) | Hispanic or Latino[2] (percent) | Non-Hispanic White alone (percent) | Percent female | Percent foreign-born | Percent born in state of residence |
| | | 1 | 2 | 3 | 4 | 5 | 6 | 7 | 8 | 9 | 10 | 11 | 12 | 13 | 14 |
| COLORADO | | 103,637.062 | 5,758,736 | 55.566 | 83.7 | 4.2 | 1.0 | 3.4 | 3.6 | 4.0 | 21.8 | 67.517 | 49.6 | 9.5 | 42.2 |
| District 1 | Diana DeGette (D) | 189.599 | 852,816 | 4,497.998 | 78.5 | 7.8 | 0.6 | 3.5 | 5.1 | 4.5 | 27.1 | 58.089 | 50.0 | 12.2 | 40.4 |
| District 2 | Joe Neguse (D) | 7,538.163 | 824,050 | 109.317 | 90.3 | 1.0 | 0.7 | 3.7 | 1.3 | 3.1 | 11.0 | 81.371 | 49.4 | 6.9 | 36.6 |
| District 3 | Lauren Boebert (R) | 49,729.680 | 756,569 | 15.214 | 89.5 | 0.9 | 2.2 | 1.2 | 3.4 | 2.7 | 25.0 | 70.217 | 49.6 | 6.2 | 48.4 |
| District 4 | Ken Buck (R) | 38,101.054 | 868,302 | 22.789 | 89 | 1.6 | 0.9 | 2.3 | 2.9 | 3.1 | 22.4 | 71.145 | 49.5 | 8.2 | 47.1 |
| District 5 | Doug Lamborn (R) | 7,265.784 | 820,255 | 112.893 | 80.8 | 6.0 | 1.0 | 2.9 | 3.5 | 5.9 | 16.7 | 70.314 | 48.8 | 6.2 | 32.3 |
| District 6 | Jason Crow (D) | 473.692 | 828,201 | 1,748.396 | 73.2 | 9.8 | 0.7 | 6.5 | 5.5 | 4.4 | 21.0 | 59.582 | 49.9 | 15.8 | 40.3 |
| District 7 | Ed Perlmutter (D) | 339.09 | 808,543 | 2,384.450 | 85.4 | 1.8 | 0.9 | 3.6 | 3.8 | 4.5 | 29.7 | 62.209 | 49.6 | 10.8 | 50.9 |
| CONNECTICUT | | 4,842.374 | 3,565,287 | 736.268 | 74.6 | 11.1 | 0.3 | 4.7 | 5.6 | 3.7 | 16.9 | 65.623 | 51.1 | 14.8 | 53.8 |
| District 1 | John B. Larson (D) | 675.383 | 703,138 | 1,041.095 | 67.8 | 15.3 | 0.3 | 5.7 | 6.3 | 4.6 | 17.2 | 59.555 | 52.0 | 15.2 | 57.8 |
| District 2 | Joe Courtney (D) | 1,988.105 | 701,590 | 352.894 | 86.2 | 4.3 | 0.4 | 3.8 | 2.4 | 3.0 | 9.0 | 80.710 | 49.9 | 7.5 | 55.1 |
| District 3 | Rosa L. DeLauro (D) | 470.426 | 717,989 | 1,526.253 | 71.5 | 15.5 | 0.2 | 4.2 | 5.4 | 3.2 | 16.7 | 63.105 | 51.9 | 13.1 | 59.8 |
| District 4 | James A. Himes (D) | 460.793 | 737,733 | 1,601.007 | 72.3 | 12.4 | 0.1 | 5.6 | 6.1 | 3.6 | 20.8 | 59.773 | 51.3 | 22.8 | 42.0 |
| District 5 | Jahana Haynes (D) | 1,247.667 | 704,837 | 564.924 | 75.5 | 8.0 | 0.2 | 4.2 | 8.0 | 4.1 | 20.3 | 65.349 | 50.3 | 15.2 | 54.6 |
| DELAWARE | | 1,948.554 | 973,764 | 499.737 | 67.7 | 22.5 | 0.4 | 3.8 | 2.4 | 3.1 | 9.6 | 61.309 | 51.7 | 10.0 | 44.0 |
| At Large | Lisa Blunt Rochester (D) | 1,948.554 | 973,764 | 499.737 | 67.7 | 22.5 | 0.4 | 3.8 | 2.4 | 3.1 | 9.6 | 61.309 | 51.7 | 10.0 | 44.0 |
| DISTRICT OF COLUMBIA | | 61.126 | 705,749 | 11,545.807 | 42.5 | 45.4 | 0.3 | 4.1 | 4.4 | 3.3 | 11.3 | 37.287 | 52.6 | 12.1 | 37.2 |
| Delegate District (At Large) | Eleanor Holmes Norton (D) | 61.126 | 705,749 | 11,545.807 | 42.5 | 45.4 | 0.3 | 4.1 | 4.4 | 3.3 | 11.3 | 37.287 | 52.6 | 12.1 | 37.2 |
| FLORIDA | | 53,652.171 | 21,477,737 | 400.314 | 74.5 | 16.0 | 0.3 | 2.9 | 3.4 | 2.9 | 26.4 | 52.978 | 51.1 | 21.1 | 35.8 |
| District 1 | Matt Gaetz (R) | 4,018.264 | 798,305 | 198.669 | 76.4 | 12.9 | 0.4 | 2.5 | 2.3 | 5.5 | 6.9 | 72.726 | 49.7 | 6.0 | 41.2 |
| District 2 | Neal P. Dunn (R) | 11,004.227 | 720,777 | 65.500 | 81.2 | 12.6 | 0.2 | 2.2 | 1.3 | 2.5 | 6.6 | 75.912 | 49.9 | 5.5 | 51.8 |
| District 3 | Kat Cammack (R) | 3,565.652 | 758,939 | 212.847 | 74.7 | 16.4 | 0.3 | 3.5 | 1.8 | 3.4 | 11.4 | 65.948 | 50.6 | 8.0 | 49.7 |
| District 4 | John H. Rutherford (R) | 1,569.706 | 836,235 | 532.734 | 79.5 | 9.9 | 0.4 | 4.7 | 1.9 | 3.6 | 8.7 | 73.206 | 51.2 | 10.6 | 40.4 |
| District 5 | Al Lawson (D) | 3,819.980 | 742,643 | 194.410 | 44.5 | 47.1 | 0.4 | 2.4 | 2.8 | 2.8 | 9.7 | 38.685 | 51.0 | 7.8 | 59.1 |
| District 6 | Michael Waltz (R) | 2,173.681 | 790,455 | 363.648 | 79.7 | 10.9 | 0.2 | 2.0 | 4.7 | 2.5 | 13.4 | 71.827 | 51.8 | 8.5 | 35.0 |
| District 7 | Stephanie N. Murphy (D) | 392.691 | 814,980 | 2,075.392 | 73.9 | 11.6 | 0.4 | 5.2 | 6.0 | 2.9 | 26.7 | 54.840 | 51.7 | 14.8 | 37.2 |
| District 8 | Bill Posey (R) | 1,751.359 | 780,036 | 445.389 | 82.8 | 9.0 | 0.2 | 2.3 | 1.8 | 3.8 | 11.8 | 73.382 | 51.1 | 9.2 | 31.6 |
| District 9 | Darren Soto (D) | 2,312.018 | 931,872 | 403.056 | 74 | 13.1 | 0.4 | 3.1 | 6.4 | 3.0 | 44.0 | 39.177 | 50.4 | 17.8 | 29.5 |
| District 10 | Val Butler Demings (D) | 435.773 | 811,634 | 1,862.516 | 54.2 | 29.5 | 0.2 | 5.2 | 6.9 | 4.1 | 28.5 | 34.428 | 51.2 | 25.0 | 35.6 |
| District 11 | Daniel Webster (R) | 2,408.306 | 813,112 | 337.628 | 86.5 | 7.4 | 0.4 | 1.5 | 1.8 | 2.5 | 11.8 | 77.066 | 51.5 | 7.2 | 30.1 |
| District 12 | Gus M. Bilirakis (R) | 858.308 | 811,308 | 945.241 | 86.3 | 5.5 | 0.3 | 3.1 | 1.3 | 3.5 | 14.0 | 75.043 | 52.0 | 10.8 | 33.7 |
| District 13 | Charlie Crist (D) | 181.684 | 731,658 | 4,027.091 | 78.9 | 12.1 | 0.2 | 3.6 | 1.5 | 3.7 | 10.8 | 70.768 | 51.4 | 11.7 | 34.2 |
| District 14 | Kathy Castor (D) | 275.978 | 831,508 | 3,012.950 | 68.6 | 19.0 | 0.1 | 4.5 | 3.7 | 4.1 | 32.9 | 42.918 | 50.1 | 21.5 | 39.0 |
| District 15 | C. Scott Franklin (R) | 1,088.053 | 801,294 | 736.448 | 73.9 | 14.1 | 0.5 | 3.4 | 5.5 | 2.6 | 23.8 | 56.700 | 51.6 | 14.3 | 39.5 |
| District 16 | Vern Buchanan (R) | 1,294.683 | 873,875 | 674.972 | 82.2 | 9.7 | 0.2 | 2.2 | 3.4 | 2.3 | 17.7 | 68.608 | 52.1 | 13.0 | 31.6 |
| District 17 | W. Gregory Steube (R) | 5,573.904 | 804,754 | 144.379 | 85.7 | 7.6 | 0.4 | 1.3 | 3.2 | 1.8 | 16.8 | 72.564 | 50.5 | 11.0 | 33.1 |
| District 18 | Brian J. Mast (R) | 1,509.866 | 795,742 | 527.028 | 79.6 | 12.8 | 0.3 | 2.4 | 2.0 | 2.9 | 17.5 | 65.214 | 51.3 | 15.9 | 33.7 |
| District 19 | Byron Donalds (R) | 747.502 | 833,013 | 1,114.396 | 85 | 7.8 | 0.1 | 2.0 | 3.6 | 1.5 | 20.5 | 68.746 | 51.0 | 17.4 | 23.6 |
| District 20 | Vacant | 2,159.116 | 802,463 | 371.663 | 38.4 | 52.9 | 0.3 | 2.5 | 2.2 | 3.6 | 26.8 | 16.444 | 51.4 | 36.0 | 42.5 |
| District 21 | Lois Frankel (D) | 256.140 | 786,566 | 3,070.844 | 74.4 | 16.3 | 0.4 | 2.7 | 3.4 | 2.8 | 25.4 | 53.815 | 51.8 | 29.5 | 26.7 |
| District 22 | Theodore E. Deutch (D) | 163.749 | 760,953 | 4,647.070 | 76.8 | 15.1 | 0.1 | 3.6 | 1.9 | 2.5 | 23.2 | 56.352 | 50.7 | 28.9 | 28.1 |
| District 23 | Debbie Wasserman Schultz (D) | 186.826 | 762,858 | 4,083.254 | 72 | 15.2 | 0.3 | 5.0 | 3.9 | 3.5 | 40.1 | 37.778 | 52.2 | 37.7 | 33.5 |
| District 24 | Frederica S. Wilson (D) | 102.431 | 754,731 | 7,368.189 | 43.3 | 47.1 | 0.2 | 1.2 | 5.6 | 2.5 | 42.4 | 10.213 | 50.8 | 43.9 | 41.5 |
| District 25 | Mario Diaz-Balart (R) | 3,504.593 | 796,422 | 227.251 | 90.5 | 4.1 | 0.1 | 1.1 | 2.9 | 1.2 | 76.0 | 19.063 | 51.1 | 57.4 | 25.7 |
| District 26 | Carlos A. Gimenez (R) | 2,184.436 | 780,951 | 357.507 | 81.2 | 11.2 | 0.3 | 1.4 | 3.5 | 2.3 | 72.4 | 15.323 | 50.8 | 48.3 | 35.3 |
| District 27 | Maria Elvira Salazar (R) | 113.245 | 750,653 | 6,628.575 | 85.6 | 5.7 | 0.1 | 2.4 | 4.1 | 2.1 | 70.8 | 21.153 | 51.5 | 55.0 | 28.3 |
| GEORGIA | | 57,716.959 | 10,617,423 | 183.957 | 57.8 | 31.9 | 0.4 | 4.2 | 3.0 | 2.7 | 9.8 | 51.822 | 51.3 | 10.3 | 54.4 |
| District 1 | Earl L. "Buddy" Carter (R) | 8,126.850 | 749,949 | 92.280 | 62.1 | 30.7 | 0.4 | 1.8 | 2.4 | 2.8 | 7.0 | 58.260 | 50.1 | 4.4 | 54.0 |
| District 2 | Sanford D. Bishop Jr. (D) | 9,630.462 | 671,831 | 69.761 | 41.6 | 52.8 | 0.3 | 1.1 | 2.1 | 2.0 | 5.2 | 39.316 | 51.2 | 3.6 | 71.8 |
| District 3 | A. Drew Ferguson IV (R) | 3,836.371 | 750,998 | 195.757 | 68.6 | 25.4 | 0.2 | 2.2 | 1.6 | 2.0 | 6.0 | 64.400 | 51.1 | 5.7 | 61.6 |
| District 4 | Henry C. "Hank" Johnson Jr. (D) | 497.296 | 782,142 | 1,572.790 | 27.3 | 60.9 | 0.3 | 4.8 | 4.2 | 2.5 | 10.2 | 21.883 | 52.7 | 16.6 | 48.4 |
| District 5 | Nikema Williams (D) | 264.983 | 788,996 | 2,977.534 | 34.8 | 55.4 | 0.4 | 5.4 | 1.7 | 2.3 | 6.8 | 30.567 | 51.9 | 9.3 | 50.9 |
| District 6 | Lucy McBath (D) | 298.906 | 742,932 | 2,485.504 | 66.7 | 13.8 | 0.4 | 11.7 | 3.2 | 3.3 | 12.2 | 59.195 | 51.4 | 20.4 | 34.7 |
| District 7 | Carolyn Bourdeaux (D) | 392.875 | 844,773 | 2,150.234 | 51.3 | 21.5 | 0.5 | 15.8 | 7.7 | 3.1 | 20.2 | 40.410 | 51.8 | 27.5 | 36.4 |
| District 8 | Austin Scott (R) | 8,738.899 | 706,237 | 80.815 | 61.4 | 31.4 | 0.7 | 1.8 | 2.3 | 2.5 | 6.9 | 57.530 | 50.8 | 4.7 | 67.2 |
| District 9 | Andrew S. Clyde (R) | 5,212.180 | 771,168 | 147.955 | 87.4 | 6.5 | 0.2 | 1.4 | 2.4 | 2.0 | 13.4 | 76.898 | 51.2 | 7.4 | 59.0 |
| District 10 | Jody B. Hice (R) | 7,098.407 | 757,807 | 106.757 | 67.4 | 26.1 | 0.1 | 2.5 | 1.3 | 2.6 | 5.3 | 64.307 | 50.6 | 5.7 | 64.2 |
| District 11 | Barry Loudermilk (R) | 1,070.095 | 782,704 | 731.434 | 71.2 | 17.8 | 0.5 | 3.1 | 5.0 | 2.4 | 11.5 | 65.103 | 51.0 | 11.9 | 45.5 |
| District 12 | Rick W. Allen (R) | 8,211.462 | 732,810 | 89.242 | 58.1 | 34.3 | 0.1 | 1.7 | 2.3 | 3.4 | 6.5 | 54.308 | 50.3 | 4.4 | 66.6 |
| District 13 | David Scott (D) | 714.330 | 802,943 | 1,124.051 | 29 | 61.6 | 0.2 | 2.7 | 3.5 | 3.0 | 11.9 | 22.398 | 53.0 | 11.6 | 48.7 |
| District 14 | Marjorie Taylor Greene (R) | 3,623.843 | 732,133 | 202.032 | 84.6 | 8.7 | 0.3 | 1.1 | 1.9 | 3.3 | 12.2 | 75.116 | 50.8 | 7.1 | 58.7 |
| HAWAII | | 6,422.480 | 1,415,872 | 220.456 | 24.1 | 1.9 | 0.4 | 49.5 | 1.7 | 22.4 | 10.7 | 21.543 | 50.0 | 19.3 | 52.3 |
| District 1 | Ed Case (D) | 209.014 | 720,786 | 3,448.506 | 16.8 | 2.4 | 0.2 | 58.5 | 1.8 | 20.3 | 9.1 | 14.897 | 50.3 | 23.1 | 52.4 |
| District 2 | Kaiali'i Kahele (D) | 6,213.466 | 695,086 | 111.868 | 31.7 | 1.4 | 0.7 | 40.1 | 1.7 | 24.4 | 12.3 | 28.435 | 49.7 | 15.4 | 52.2 |
| IDAHO | | 82,645.141 | 1,787,065 | 21.623 | 89.4 | 0.7 | 1.3 | 1.6 | 3.8 | 3.3 | 12.8 | 81.588 | 49.7 | 5.8 | 46.1 |
| District 1 | Russ Fulcher (R) | 39,420.388 | 934,826 | 23.714 | 88.0 | 0.8 | 1.2 | 1.3 | 4.4 | 3.6 | 11.5 | 82.721 | 50.0 | 5.1 | 41.4 |
| District 2 | Michael K. Simpson (R) | 43,224.753 | 852,239 | 19.716 | 90.1 | 0.6 | 1.4 | 1.8 | 3.1 | 3.0 | 14.3 | 80.344 | 49.5 | 6.5 | 51.2 |
| ILLINOIS | | 1,338.930 | 12,671,821 | 9,464.140 | 71.4 | 14.1 | 0.3 | 5.7 | 5.8 | 2.8 | 17.5 | 60.694 | 50.9 | 13.9 | 67.4 |
| District 1 | Bobby L. Rush (D) | 258.309 | 711,039 | 2,752.668 | 42 | 50.5 | 0.1 | 2.6 | 2.6 | 2.1 | 11.1 | 34.352 | 53.3 | 8.8 | 76.8 |
| District 2 | Robin L. Kelly (D) | 1,080.621 | 685,695 | 634.538 | 37.9 | 55.9 | 0.1 | 0.7 | 3.1 | 2.3 | 16.0 | 25.534 | 53.7 | 7.5 | 77.3 |

1. Dry land or land partially or temporarily covered by water.    2. May be of any race.

# Table E. Congressional Districts 116th Congress — **Age and Education**

| STATE District | Population and population characteristics, 2019 (cont.) Age (percent) | | | | | | | | | | Education, 2019 | Attainment[2] (percent) | |
| | Under 5 years | 5 to 17 years | 18 to 24 years | 25 to 34 years | 35 to 44 years | 45 to 54 years | 55 to 64 years | 65 to 74 years | 75 years and over | Median age | Total Enrollment[1] | High school graduate or more | Bachelor's degree or more |
| | 15 | 16 | 17 | 18 | 19 | 20 | 21 | 22 | 23 | 24 | 25 | 26 | 27 |
| COLORADO | 5.7 | 16.1 | 9.2 | 15.8 | 14.0 | 12.2 | 12.3 | 9.1 | 5.6 | 37.1 | 1,406,414 | 92.4 | 42.7 |
| District 1 | 5.8 | 13.6 | 7.7 | 22.3 | 16.2 | 11.6 | 10.3 | 7.7 | 4.9 | 35.3 | 183,050 | 91.8 | 52.1 |
| District 2 | 4.5 | 14.7 | 12.4 | 14.0 | 12.9 | 12.8 | 13.2 | 9.9 | 5.6 | 38.0 | 225,741 | 96.3 | 56.9 |
| District 3 | 5.0 | 15.8 | 8.5 | 12.8 | 13.0 | 11.4 | 13.9 | 12.3 | 7.3 | 40.6 | 170,454 | 91.4 | 33.0 |
| District 4 | 6.2 | 18.3 | 8.4 | 13.0 | 14.3 | 13.0 | 12.6 | 8.6 | 5.5 | 38.0 | 222,973 | 90.9 | 37.2 |
| District 5 | 6.1 | 16.2 | 10.1 | 15.7 | 12.7 | 11.6 | 12.7 | 9.2 | 5.6 | 36.3 | 205,491 | 94.7 | 37.8 |
| District 6 | 6.1 | 18.2 | 8.3 | 15.0 | 14.3 | 13.3 | 12.1 | 8.1 | 4.7 | 36.7 | 208,801 | 92.3 | 44.0 |
| District 7 | 6.0 | 16.0 | 8.7 | 17.2 | 14.4 | 12.0 | 11.9 | 8.4 | 5.5 | 36.3 | 189,904 | 89.5 | 36.2 |
| CONNECTICUT | 5.1 | 15.3 | 9.6 | 12.4 | 12.0 | 13.5 | 14.5 | 9.9 | 7.8 | 41.2 | 882,104 | 90.7 | 39.8 |
| District 1 | 5.3 | 15.3 | 8.7 | 13.3 | 12.4 | 13.5 | 13.9 | 9.5 | 8.2 | 40.9 | 172,129 | 90.5 | 38.6 |
| District 2 | 4.4 | 14.4 | 10.9 | 11.8 | 11.4 | 13.2 | 15.3 | 11.0 | 7.6 | 42.3 | 166,070 | 93.1 | 35.8 |
| District 3 | 4.6 | 14.0 | 10.5 | 13.5 | 11.6 | 13.6 | 14.6 | 9.9 | 7.7 | 41.3 | 180,538 | 90.6 | 36.1 |
| District 4 | 5.8 | 17.7 | 8.9 | 11.2 | 12.7 | 14.4 | 13.6 | 8.6 | 7.2 | 40.0 | 199,909 | 90.4 | 52.6 |
| District 5 | 5.1 | 15.4 | 8.8 | 12.6 | 11.9 | 12.8 | 14.8 | 10.5 | 8.2 | 41.9 | 163,458 | 89.0 | 36.1 |
| DELAWARE | 5.6 | 15.4 | 8.5 | 13.2 | 11.7 | 12.2 | 14.1 | 11.5 | 7.9 | 41.4 | 227,551 | 90.3 | 33.2 |
| At Large | 5.6 | 15.4 | 8.5 | 13.2 | 11.7 | 12.2 | 14.1 | 11.5 | 7.9 | 41.4 | 227,551 | 90.3 | 33.2 |
| DISTRICT OF COLUMBIA | 6.4 | 11.7 | 10.3 | 23.3 | 15.5 | 10.6 | 9.9 | 7.2 | 5.2 | 34.3 | 169,418 | 91.9 | 59.7 |
| Delegate District (At Large) | 6.4 | 11.7 | 10.3 | 23.3 | 15.5 | 10.6 | 9.9 | 7.2 | 5.2 | 34.3 | 169,418 | 91.9 | 59.7 |
| FLORIDA | 5.3 | 14.4 | 8.2 | 12.9 | 12.1 | 12.6 | 13.5 | 11.5 | 9.5 | 42.4 | 4,795,224 | 88.4 | 30.7 |
| District 1 | 5.9 | 15.5 | 9.2 | 14.7 | 11.9 | 11.7 | 13.9 | 10.3 | 6.9 | 38.8 | 179,953 | 90.4 | 26.9 |
| District 2 | 5.0 | 14.9 | 9.3 | 11.9 | 11.6 | 12.6 | 14.1 | 12.2 | 8.6 | 42.7 | 162,980 | 87.9 | 24.6 |
| District 3 | 5.3 | 15.0 | 12.3 | 13.3 | 12.1 | 12.0 | 13.0 | 10.1 | 7.0 | 37.9 | 200,824 | 90.8 | 30.1 |
| District 4 | 5.7 | 15.5 | 8.0 | 14.0 | 13.1 | 12.8 | 13.5 | 10.6 | 6.8 | 39.3 | 190,096 | 93.5 | 41.9 |
| District 5 | 6.4 | 15.7 | 12.1 | 16.2 | 12.3 | 11.7 | 12.1 | 8.1 | 5.3 | 34.7 | 189,891 | 85.6 | 20.4 |
| District 6 | 4.5 | 13.2 | 7.3 | 10.9 | 10.5 | 11.9 | 15.7 | 14.7 | 11.3 | 48.2 | 151,766 | 91.3 | 24.8 |
| District 7 | 5.1 | 14.3 | 11.4 | 16.3 | 13.8 | 12.8 | 11.6 | 8.3 | 6.3 | 36.9 | 212,208 | 92.8 | 41.1 |
| District 8 | 4.5 | 13.4 | 7.0 | 11.2 | 10.2 | 11.7 | 16.1 | 13.9 | 12.0 | 48.3 | 157,821 | 92.1 | 30.5 |
| District 9 | 6.3 | 17.3 | 8.7 | 13.4 | 14.3 | 12.2 | 11.5 | 9.5 | 7.0 | 38.0 | 233,494 | 87.1 | 25.6 |
| District 10 | 6.3 | 16.9 | 8.3 | 17.0 | 14.2 | 13.2 | 12.3 | 7.5 | 4.4 | 36.1 | 209,662 | 86.2 | 32.6 |
| District 11 | 4.1 | 11.3 | 5.8 | 8.7 | 8.7 | 10.2 | 13.8 | 19.9 | 17.6 | 56.0 | 126,632 | 89.9 | 23.5 |
| District 12 | 4.9 | 14.6 | 6.4 | 10.6 | 11.9 | 13.1 | 13.8 | 12.9 | 11.6 | 46.5 | 165,000 | 91.1 | 28.4 |
| District 13 | 4.2 | 11.3 | 7.4 | 12.8 | 11.2 | 12.5 | 16.5 | 13.3 | 10.7 | 47.8 | 137,509 | 90.5 | 33.0 |
| District 14 | 6.3 | 15.3 | 9.8 | 16.5 | 13.1 | 13.7 | 11.8 | 8.0 | 5.5 | 36.4 | 201,921 | 89.1 | 38.5 |
| District 15 | 4.8 | 16.8 | 8.2 | 14.1 | 14.1 | 12.7 | 12.9 | 9.5 | 7.0 | 39.4 | 192,995 | 87.6 | 25.8 |
| District 16 | 4.6 | 13.4 | 7.0 | 10.4 | 11.1 | 12.0 | 14.1 | 14.8 | 12.6 | 48.3 | 165,317 | 91.2 | 33.1 |
| District 17 | 4.2 | 12.4 | 6.8 | 9.2 | 9.4 | 10.9 | 14.4 | 17.0 | 15.9 | 52.7 | 142,364 | 88.4 | 23.0 |
| District 18 | 4.5 | 13.8 | 6.9 | 11.0 | 10.8 | 11.9 | 14.7 | 13.5 | 12.9 | 48.1 | 161,032 | 90.2 | 32.3 |
| District 19 | 4.0 | 11.9 | 6.6 | 9.9 | 10.0 | 10.9 | 14.1 | 16.9 | 15.7 | 52.0 | 145,040 | 89.8 | 33.8 |
| District 20 | 6.8 | 17.2 | 8.6 | 14.9 | 13.2 | 12.8 | 12.3 | 7.4 | 6.6 | 36.9 | 207,949 | 83.2 | 20.8 |
| District 21 | 4.7 | 12.9 | 6.8 | 11.4 | 11.7 | 12.4 | 13.2 | 12.9 | 14.0 | 47.0 | 157,673 | 88.8 | 37.1 |
| District 22 | 4.8 | 12.7 | 6.9 | 12.5 | 12.1 | 14.0 | 14.5 | 11.9 | 10.5 | 45.8 | 159,611 | 91.9 | 41.6 |
| District 23 | 5.5 | 15.7 | 8.1 | 12.1 | 13.5 | 14.7 | 13.4 | 9.7 | 7.2 | 41.8 | 192,865 | 92.2 | 40.6 |
| District 24 | 6.7 | 15.4 | 8.4 | 16.1 | 14.2 | 12.1 | 12.4 | 8.5 | 6.2 | 37.3 | 188,088 | 79.7 | 20.4 |
| District 25 | 5.1 | 14.6 | 8.3 | 12.3 | 12.8 | 14.7 | 13.3 | 9.5 | 9.6 | 42.8 | 178,594 | 76.1 | 26.9 |
| District 26 | 6.2 | 16.0 | 8.4 | 12.6 | 13.4 | 15.0 | 12.7 | 9.0 | 6.4 | 40.3 | 194,723 | 81.8 | 28.5 |
| District 27 | 5.3 | 12.2 | 7.9 | 14.8 | 13.3 | 14.6 | 12.9 | 9.1 | 9.8 | 42.1 | 189,216 | 85.1 | 41.1 |
| GEORGIA | 6.0 | 17.5 | 9.7 | 13.8 | 13.2 | 13.1 | 12.3 | 8.8 | 5.6 | 37.2 | 2,771,807 | 87.9 | 32.5 |
| District 1 | 5.8 | 17.4 | 10.1 | 14.5 | 12.7 | 11.9 | 11.9 | 9.7 | 5.9 | 36.6 | 186,030 | 88.8 | 26.3 |
| District 2 | 6.0 | 17.3 | 10.5 | 13.0 | 12.8 | 11.4 | 12.9 | 9.6 | 6.4 | 37.3 | 174,939 | 84.1 | 18.4 |
| District 3 | 5.7 | 18.0 | 9.1 | 12.4 | 12.9 | 13.3 | 12.9 | 9.6 | 6.1 | 38.9 | 186,189 | 88.9 | 26.0 |
| District 4 | 6.3 | 18.2 | 9.0 | 14.3 | 14.0 | 12.4 | 12.4 | 8.0 | 4.4 | 36.5 | 209,227 | 87.7 | 30.7 |
| District 5 | 6.2 | 13.7 | 11.6 | 19.9 | 13.9 | 12.0 | 10.4 | 7.6 | 4.9 | 34.0 | 204,906 | 90.6 | 47.4 |
| District 6 | 5.8 | 17.0 | 7.5 | 13.4 | 14.6 | 14.9 | 13.2 | 8.1 | 5.5 | 36.2 | 194,837 | 94.6 | 65.1 |
| District 7 | 7.0 | 19.9 | 8.7 | 12.7 | 14.7 | 14.7 | 11.4 | 7.0 | 3.9 | 36.2 | 241,520 | 89.1 | 45.0 |
| District 8 | 6.1 | 17.2 | 10.3 | 12.9 | 13.0 | 11.9 | 12.3 | 9.5 | 6.9 | 37.6 | 182,178 | 86.5 | 22.7 |
| District 9 | 5.5 | 16.6 | 9.2 | 11.5 | 11.2 | 13.3 | 13.6 | 11.4 | 7.7 | 41.3 | 180,511 | 84.0 | 24.2 |
| District 10 | 5.7 | 17.9 | 11.1 | 12.4 | 12.3 | 13.6 | 12.0 | 9.3 | 5.9 | 37.7 | 213,703 | 87.8 | 29.6 |
| District 11 | 5.8 | 17.2 | 8.8 | 14.6 | 13.7 | 14.5 | 12.0 | 8.6 | 4.7 | 37.6 | 201,466 | 90.5 | 42.8 |
| District 12 | 5.7 | 17.4 | 11.8 | 14.0 | 11.9 | 12.2 | 12.2 | 9.1 | 5.9 | 36.0 | 190,831 | 85.4 | 22.2 |
| District 13 | 6.7 | 19.6 | 9.6 | 14.2 | 13.7 | 13.4 | 11.9 | 7.2 | 3.8 | 35.0 | 224,837 | 89.7 | 29.6 |
| District 14 | 6.0 | 18.0 | 9.3 | 12.1 | 13.0 | 13.9 | 12.5 | 8.8 | 6.2 | 38.3 | 180,633 | 81.8 | 18.4 |
| HAWAII | 6.0 | 15.2 | 8.4 | 13.8 | 13.1 | 11.9 | 12.6 | 10.8 | 8.2 | 39.6 | 317,419 | 92.4 | 33.6 |
| District 1 | 5.9 | 14.0 | 8.5 | 15.4 | 13.2 | 12.1 | 11.9 | 10.3 | 8.8 | 39.1 | 160,547 | 92.0 | 36.7 |
| District 2 | 6.2 | 16.4 | 8.3 | 12.2 | 12.9 | 11.7 | 13.3 | 11.4 | 7.6 | 40.0 | 156,872 | 92.8 | 30.3 |
| IDAHO | 6.5 | 18.6 | 9.3 | 13.1 | 12.6 | 11.3 | 12.3 | 9.7 | 6.4 | 36.9 | 457,955 | 91.5 | 28.7 |
| District 1 | 6.0 | 18.5 | 8.1 | 12.3 | 13.0 | 11.9 | 12.7 | 10.6 | 6.9 | 38.8 | 226,295 | 91.6 | 27.2 |
| District 2 | 7.0 | 18.8 | 10.7 | 14.0 | 12.3 | 10.7 | 11.8 | 8.7 | 5.9 | 34.6 | 231,660 | 91.3 | 30.6 |
| ILLINOIS | 5.9 | 16.3 | 9.2 | 13.8 | 13.0 | 12.6 | 13.0 | 9.4 | 6.8 | 38.6 | 3,123,418 | 89.8 | 35.8 |
| District 1 | 5.9 | 16.2 | 9.4 | 14.0 | 12.2 | 12.4 | 13.4 | 9.6 | 7.1 | 38.8 | 175,127 | 89.2 | 31.3 |
| District 2 | 6.1 | 16.6 | 10.4 | 13.3 | 11.2 | 13.1 | 13.6 | 9.0 | 6.8 | 37.9 | 170,779 | 89.2 | 23.3 |

1. All persons 3 years old and over enrolled in nursery school through college and graduate or professional school.     2. Persons 25 years old and over.

# Table E. Congressional Districts 116th Congress — **Households and Group Quarters**

| STATE District | Households, 2019 | | | | | | Group Quarters, 2010 | | | | | |
|---|---|---|---|---|---|---|---|---|---|---|---|---|
| | Number | Average household size | Family households (percent) | Married couple family (percent) | Female family householder[1] | One person households (percent) | Total in group quarters, 2018 | Percent 65 years and over | Persons in correctional institutions | Persons in nursing facilities | Persons in college dormitories | Persons in military quarters |
| | 28 | 29 | 30 | 31 | 32 | 33 | 34 | 35 | 36 | 37 | 38 | 39 |
| COLORADO | 2,235,103 | 2.52 | 63.6 | 50.0 | 9.0 | 27.1 | 119,141 | 14.3 | 40,568 | 18,079 | 29,952 | 10,945 |
| District 1 | 368,498 | 2.27 | 49.9 | 36.6 | 8.9 | 36.5 | 14,835 | 12.3 | 3,960 | 2,592 | 4,940 | 0 |
| District 2 | 333,774 | 2.40 | 60.9 | 52.0 | 5.4 | 25.9 | 23,010 | 11.6 | 1,634 | 2,213 | 11,841 | 0 |
| District 3 | 303,788 | 2.44 | 64.4 | 50.8 | 9.0 | 27.7 | 16,056 | 20.4 | 3,982 | 3,460 | 5,041 | 0 |
| District 4 | 318,691 | 2.67 | 71.7 | 58.2 | 8.5 | 22.4 | 17,345 | 12.8 | 13,075 | 2,577 | 3,993 | 0 |
| District 5 | 305,276 | 2.59 | 68.0 | 53.8 | 9.8 | 24.3 | 30,323 | 6.9 | 12,094 | 2,245 | 2,580 | 10,678 |
| District 6 | 301,487 | 2.72 | 68.7 | 52.7 | 11.1 | 24.4 | 7,207 | 32.7 | 3,009 | 2,128 | 0 | 267 |
| District 7 | 303,589 | 2.63 | 64.6 | 48.0 | 10.7 | 26.8 | 10,365 | 28.1 | 2,814 | 2,864 | 1,557 | 0 |
| CONNECTICUT | 1,377,166 | 2.51 | 64.3 | 47.2 | 12.4 | 29.0 | 110,867 | 20.9 | 20,059 | 26,371 | 48,537 | 3,977 |
| District 1 | 280,863 | 2.45 | 62.7 | 44.0 | 13.7 | 30.7 | 16,333 | 33.0 | 1,335 | 6,778 | 5,649 | 0 |
| District 2 | 278,655 | 2.38 | 65.3 | 50.9 | 10.0 | 26.4 | 37,619 | 10.7 | 12,975 | 4,306 | 16,779 | 3,977 |
| District 3 | 279,104 | 2.48 | 60.9 | 42.3 | 14.5 | 32.9 | 26,425 | 17.9 | 902 | 5,257 | 16,566 | 0 |
| District 4 | 263,988 | 2.74 | 69.5 | 52.5 | 11.7 | 24.5 | 13,326 | 31.6 | 1,159 | 4,719 | 5,030 | 0 |
| District 5 | 274,556 | 2.50 | 63.5 | 46.4 | 12.0 | 30.3 | 17,164 | 28.1 | 3,688 | 5,311 | 4,513 | 0 |
| DELAWARE | 376,239 | 2.52 | 64.5 | 47.3 | 11.9 | 28.8 | 24,895 | 16.8 | 6,457 | 4,591 | 10,184 | 283 |
| At Large | 376,239 | 2.52 | 64.5 | 47.3 | 11.9 | 28.8 | 24,895 | 16.8 | 6,457 | 4,591 | 10,184 | 283 |
| DISTRICT OF COLUMBIA | 291,570 | 2.29 | 42.9 | 26.9 | 13.1 | 44.8 | 38,858 | 7.8 | 3,598 | 3,064 | 24,087 | 1,504 |
| Delegate District (At Large) | 291,570 | 2.29 | 42.9 | 26.9 | 13.1 | 44.8 | 38,858 | 7.8 | 3,598 | 3,064 | 24,087 | 1,504 |
| FLORIDA | 7,905,832 | 2.66 | 64.3 | 46.4 | 12.8 | 28.6 | 429,975 | 17.5 | 167,453 | 73,372 | 85,243 | 14,612 |
| District 1 | 293,683 | 2.61 | 63.9 | 47.2 | 12.8 | 29.3 | 31,549 | 8.8 | 10,881 | 2,652 | 4,832 | 8,580 |
| District 2 | 274,771 | 2.46 | 65.3 | 48.8 | 11.9 | 28.5 | 45,429 | 6.9 | 32,671 | 3,280 | 10,370 | 513 |
| District 3 | 279,725 | 2.58 | 63.7 | 45.8 | 13.0 | 28.9 | 36,105 | 9.3 | 21,156 | 3,439 | 9,235 | 0 |
| District 4 | 317,700 | 2.58 | 65.8 | 51.8 | 10.3 | 26.6 | 18,026 | 18.0 | 2,597 | 2,852 | 2,776 | 4,662 |
| District 5 | 281,508 | 2.52 | 56.2 | 30.6 | 19.1 | 33.9 | 34,584 | 10.6 | 10,195 | 2,281 | 1,357 | 0 |
| District 6 | 310,974 | 2.50 | 64.4 | 48.5 | 10.7 | 28.9 | 12,161 | 20.3 | 4,351 | 3,572 | 5,748 | 4 |
| District 7 | 301,756 | 2.63 | 63.8 | 45.9 | 12.7 | 25.8 | 22,582 | 15.3 | 757 | 3,056 | 13,306 | 0 |
| District 8 | 299,585 | 2.58 | 61.3 | 48.9 | 9.2 | 32.8 | 7,587 | 27.2 | 3,105 | 2,612 | 1,259 | 182 |
| District 9 | 293,726 | 3.14 | 69.8 | 48.4 | 15.5 | 24.6 | 8,347 | 24.1 | 3,819 | 1,838 | 245 | 0 |
| District 10 | 262,196 | 3.06 | 67.9 | 47.6 | 15.1 | 24.9 | 8,267 | 16.2 | 3,929 | 2,605 | 315 | 0 |
| District 11 | 340,680 | 2.34 | 66.6 | 54.1 | 8.9 | 27.9 | 17,337 | 15.3 | 16,221 | 2,953 | 312 | 0 |
| District 12 | 315,056 | 2.55 | 65.6 | 49.3 | 11.6 | 28.9 | 8,993 | 40.5 | 1,981 | 2,902 | 1,050 | 0 |
| District 13 | 297,213 | 2.40 | 53.5 | 40.2 | 9.7 | 37.0 | 17,529 | 35.3 | 3,506 | 5,478 | 1,751 | 0 |
| District 14 | 309,771 | 2.64 | 58.4 | 39.6 | 13.4 | 31.6 | 14,011 | 18.8 | 3,483 | 3,007 | 3,375 | 324 |
| District 15 | 279,096 | 2.82 | 65.2 | 46.8 | 12.1 | 27.2 | 12,938 | 15.6 | 854 | 2,303 | 7,847 | 0 |
| District 16 | 332,027 | 2.60 | 64.4 | 51.4 | 9.6 | 29.1 | 9,927 | 38.1 | 2,011 | 3,857 | 1,569 | 0 |
| District 17 | 310,259 | 2.53 | 65.1 | 50.9 | 10.6 | 28.2 | 19,470 | 15.9 | 12,761 | 3,312 | 606 | 2 |
| District 18 | 304,300 | 2.59 | 63.6 | 48.7 | 10.6 | 29.9 | 8,195 | 25.5 | 3,930 | 1,838 | 339 | 9 |
| District 19 | 321,480 | 2.55 | 64.9 | 52.8 | 8.0 | 29.3 | 12,162 | 26.8 | 2,727 | 2,860 | 2,688 | 0 |
| District 20 | 260,511 | 3.02 | 65.2 | 35.5 | 21.6 | 28.4 | 16,814 | 18.4 | 10,719 | 3,611 | 214 | 0 |
| District 21 | 304,170 | 2.57 | 62.0 | 46.2 | 10.1 | 30.2 | 6,182 | 52.8 | 619 | 1,396 | 19 | 0 |
| District 22 | 309,545 | 2.42 | 57.1 | 42.3 | 10.8 | 34.2 | 10,434 | 29.7 | 1,291 | 3,658 | 3,672 | 59 |
| District 23 | 277,260 | 2.74 | 66.4 | 47.0 | 13.5 | 27.7 | 3,617 | 23.7 | 938 | 1,239 | 1,463 | 0 |
| District 24 | 254,252 | 2.91 | 65.9 | 36.3 | 20.8 | 29.3 | 13,731 | 21.9 | 3,004 | 2,733 | 3,132 | 0 |
| District 25 | 252,220 | 3.12 | 75.4 | 51.6 | 15.8 | 18.4 | 8,843 | 12.7 | 5,752 | 717 | 596 | 277 |
| District 26 | 235,037 | 3.26 | 76.9 | 50.7 | 18.5 | 18.5 | 14,378 | 17.9 | 4,186 | 1,100 | 2,657 | 0 |
| District 27 | 287,331 | 2.57 | 62.6 | 42.1 | 15.0 | 29.1 | 10,777 | 24.6 | 9 | 2,221 | 4,510 | 0 |
| GEORGIA | 3,852,714 | 2.69 | 66.3 | 46.8 | 14.7 | 28.0 | 266,840 | 12.5 | 104,012 | 34,738 | 72,288 | 16,072 |
| District 1 | 276,916 | 2.62 | 65.8 | 45.8 | 15.3 | 28.2 | 25,797 | 10.4 | 10,733 | 3,028 | 5,798 | 5,069 |
| District 2 | 244,980 | 2.59 | 62.2 | 37.2 | 20.1 | 33.5 | 36,875 | 8.4 | 17,145 | 3,559 | 7,343 | 5,167 |
| District 3 | 271,059 | 2.72 | 71.0 | 51.8 | 14.6 | 25.4 | 12,525 | 20.6 | 4,777 | 2,837 | 4,020 | 64 |
| District 4 | 271,813 | 2.85 | 65.1 | 40.9 | 18.9 | 29.8 | 7,194 | 17.9 | 4,367 | 1,472 | 653 | 0 |
| District 5 | 328,926 | 2.27 | 47.6 | 26.5 | 16.2 | 41.9 | 43,261 | 5.3 | 6,731 | 2,500 | 22,635 | 78 |
| District 6 | 297,852 | 2.48 | 63.7 | 52.9 | 7.3 | 30.8 | 3,329 | 49.0 | 89 | 1,109 | 620 | 0 |
| District 7 | 273,983 | 3.06 | 77.4 | 60.5 | 11.3 | 18.0 | 5,654 | 18.9 | 4,093 | 1,167 | 0 | 0 |
| District 8 | 263,205 | 2.59 | 66.3 | 45.5 | 15.5 | 28.4 | 24,449 | 13.4 | 13,355 | 4,044 | 5,390 | 401 |
| District 9 | 273,121 | 2.78 | 73.0 | 56.8 | 11.8 | 22.0 | 12,401 | 17.7 | 4,824 | 2,514 | 4,115 | 44 |
| District 10 | 271,451 | 2.68 | 69.5 | 50.6 | 14.6 | 23.2 | 28,978 | 12.0 | 10,678 | 2,840 | 9,699 | 74 |
| District 11 | 299,424 | 2.57 | 65.8 | 50.7 | 10.6 | 27.7 | 12,534 | 14.1 | 3,520 | 1,539 | 4,296 | 36 |
| District 12 | 247,719 | 2.82 | 64.7 | 42.2 | 18.4 | 30.3 | 33,946 | 10.7 | 15,029 | 3,980 | 5,294 | 5,139 |
| District 13 | 273,017 | 2.92 | 69.8 | 43.0 | 20.7 | 25.1 | 5,956 | 21.4 | 3,437 | 1,322 | 0 | 0 |
| District 14 | 259,248 | 2.77 | 70.4 | 52.6 | 12.8 | 25.3 | 13,941 | 21.5 | 5,234 | 2,827 | 2,425 | 0 |
| HAWAII | 465,299 | 2.95 | 68.0 | 49.9 | 12.5 | 25.2 | 42,918 | 12.2 | 5,673 | 5,198 | 7,540 | 12,551 |
| District 1 | 243,440 | 2.87 | 66.1 | 48.1 | 12.4 | 26.9 | 20,927 | 13.3 | 3,581 | 2,672 | 4,641 | 5,332 |
| District 2 | 221,859 | 3.03 | 70.0 | 51.8 | 12.6 | 23.3 | 21,991 | 10.9 | 2,092 | 2,526 | 2,899 | 7,219 |
| IDAHO | 655,859 | 2.68 | 68.5 | 54.5 | 9.3 | 24.6 | 30,331 | 16.9 | 11,275 | 4,820 | 7,223 | 466 |
| District 1 | 341,529 | 2.68 | 71.0 | 55.9 | 9.7 | 22.8 | 19,167 | 15.4 | 7,951 | 2,375 | 4,359 | 0 |
| District 2 | 314,330 | 2.68 | 65.7 | 52.9 | 8.8 | 26.7 | 11,164 | 19.2 | 3,324 | 2,445 | 2,864 | 466 |
| ILLINOIS | 4,866,006 | 2.54 | 62.9 | 46.3 | 11.9 | 30.8 | 296,807 | 22.6 | 70,828 | 81,516 | 92,960 | 12,483 |
| District 1 | 271,729 | 2.57 | 59.7 | 36.0 | 18.4 | 35.5 | 11,646 | 21.7 | 0 | 4,063 | 5,957 | 0 |
| District 2 | 259,880 | 2.59 | 59.9 | 31.9 | 22.0 | 35.5 | 12,064 | 31.1 | 685 | 5,037 | 2,054 | 0 |

1. No spouse present.

# Table E. Congressional Districts 116th Congress — **Housing and Money Income**

| STATE District | Housing units, 2019 | | | | | | Money income, 2019 | | |
| | Total | Occupied units as a percent of all units | Owner-occupied units as a percent of occupied units | Median value[1] (dollars) | Percent valued at $500,000 or more | Median rent[2] | Per capita income (dollars) | Median income (dollars) | Percent with income of $100,000 or more |
| | | | Occupied units | | | | | Households | |
| | | | Owner-occupied | | | Renter-occupied | | | |
| | 40 | 41 | 42 | 43 | 44 | 45 | 46 | 47 | 48 |
|---|---|---|---|---|---|---|---|---|---|
| COLORADO | 2,464,109 | 90.7 | 65.9 | 394,600 | 30.3 | 1,369 | 41,053 | 77,127 | 37.7 |
| District 1 | 389,199 | 94.7 | 52.7 | 449,200 | 40.0 | 1,432 | 48,187 | 77,507 | 39.3 |
| District 2 | 395,016 | 84.5 | 67.1 | 489,700 | 48.0 | 1,539 | 49,522 | 87,585 | 43.4 |
| District 3 | 377,350 | 80.5 | 70.3 | 275,600 | 20.9 | 977 | 34,029 | 59,973 | 25.8 |
| District 4 | 337,783 | 94.3 | 73.1 | 400,000 | 30.0 | 1,285 | 39,293 | 83,609 | 40.7 |
| District 5 | 335,355 | 91.0 | 66.1 | 318,700 | 15.0 | 1,239 | 35,033 | 71,244 | 32.5 |
| District 6 | 313,861 | 96.1 | 68.6 | 427,800 | 32.8 | 1,486 | 42,132 | 87,312 | 43.5 |
| District 7 | 315,545 | 96.2 | 65.4 | 393,600 | 24.4 | 1,421 | 38,358 | 77,164 | 37.7 |
| CONNECTICUT | 1,524,959 | 90.3 | 65.0 | 280,700 | 17.0 | 1,177 | 45,359 | 78,833 | 39.8 |
| District 1 | 306,950 | 91.5 | 64.1 | 237,700 | 6.1 | 1,102 | 40,572 | 75,502 | 37.2 |
| District 2 | 308,690 | 90.3 | 70.7 | 261,600 | 10.0 | 1,136 | 41,827 | 80,280 | 39.1 |
| District 3 | 309,546 | 90.2 | 61.2 | 257,600 | 7.2 | 1,200 | 38,721 | 70,574 | 34.1 |
| District 4 | 290,420 | 90.9 | 64.4 | 528,400 | 52.5 | 1,549 | 63,117 | 101,646 | 51.1 |
| District 5 | 309,353 | 88.8 | 64.8 | 275,200 | 11.0 | 1,082 | 41,827 | 73,616 | 38.2 |
| DELAWARE | 443,764 | 84.8 | 70.3 | 261,700 | 9.2 | 1,116 | 36,858 | 70,176 | 32.9 |
| At Large | 443,764 | 84.8 | 70.3 | 261,700 | 9.2 | 1,116 | 36,858 | 70,176 | 32.9 |
| DISTRICT OF COLUMBIA | 322,814 | 90.3 | 41.5 | 646,500 | 64.2 | 1,603 | 59,808 | 92,266 | 47.5 |
| Delegate District (At Large) | 322,814 | 90.3 | 41.5 | 646,500 | 64.2 | 1,603 | 59,808 | 92,266 | 47.5 |
| FLORIDA | 9,674,053 | 81.7 | 66.2 | 245,100 | 12.0 | 1,238 | 32,887 | 59,227 | 26.2 |
| District 1 | 376,749 | 78.0 | 67.1 | 206,600 | 8.0 | 1,076 | 30,934 | 58,358 | 25.3 |
| District 2 | 357,583 | 76.8 | 73.8 | 170,500 | 4.9 | 932 | 28,380 | 54,087 | 21.7 |
| District 3 | 322,944 | 86.6 | 64.2 | 188,700 | 4.4 | 970 | 28,705 | 56,005 | 24.4 |
| District 4 | 362,998 | 87.5 | 70.3 | 278,900 | 15.9 | 1,274 | 41,202 | 77,026 | 37.4 |
| District 5 | 331,115 | 85.0 | 51.1 | 150,100 | 2.2 | 953 | 22,957 | 43,667 | 14.4 |
| District 6 | 376,360 | 82.6 | 73.5 | 217,600 | 6.7 | 1,114 | 30,512 | 55,281 | 21.9 |
| District 7 | 328,729 | 91.8 | 58.7 | 274,400 | 12.0 | 1,324 | 35,879 | 66,748 | 30.7 |
| District 8 | 371,815 | 80.6 | 77.5 | 231,600 | 8.9 | 1,109 | 33,534 | 58,740 | 25.8 |
| District 9 | 394,918 | 74.4 | 65.0 | 218,900 | 3.9 | 1,291 | 25,110 | 54,878 | 20.6 |
| District 10 | 331,662 | 79.1 | 59.8 | 261,300 | 11.1 | 1,276 | 31,275 | 61,737 | 26.0 |
| District 11 | 408,508 | 83.4 | 80.2 | 189,900 | 4.5 | 992 | 29,981 | 52,749 | 18.6 |
| District 12 | 371,238 | 84.9 | 74.9 | 214,500 | 7.3 | 1,130 | 31,882 | 56,761 | 25.3 |
| District 13 | 395,748 | 75.1 | 67.3 | 222,400 | 13.1 | 1,165 | 36,765 | 56,612 | 25.2 |
| District 14 | 343,593 | 90.2 | 53.6 | 267,200 | 15.3 | 1,184 | 34,292 | 60,022 | 27.7 |
| District 15 | 323,137 | 86.4 | 64.0 | 215,700 | 3.5 | 1,147 | 28,201 | 57,496 | 23.7 |
| District 16 | 431,870 | 76.9 | 75.6 | 265,700 | 14.3 | 1,269 | 38,286 | 68,071 | 31.3 |
| District 17 | 400,916 | 77.4 | 79.1 | 203,900 | 6.5 | 966 | 30,361 | 53,693 | 21.2 |
| District 18 | 375,853 | 81.0 | 76.3 | 284,500 | 18.2 | 1,374 | 40,735 | 68,744 | 33.1 |
| District 19 | 494,699 | 65.0 | 75.4 | 279,400 | 21.4 | 1,289 | 43,471 | 67,598 | 32.0 |
| District 20 | 292,964 | 88.9 | 54.8 | 229,400 | 6.5 | 1,318 | 23,499 | 49,223 | 18.6 |
| District 21 | 373,810 | 81.4 | 72.2 | 288,400 | 16.2 | 1,481 | 40,007 | 65,394 | 31.1 |
| District 22 | 382,914 | 80.8 | 65.2 | 348,900 | 26.3 | 1,565 | 44,819 | 69,452 | 35.4 |
| District 23 | 343,622 | 80.7 | 65.3 | 343,700 | 21.5 | 1,591 | 38,787 | 69,198 | 34.0 |
| District 24 | 288,320 | 88.2 | 45.1 | 259,100 | 11.2 | 1,236 | 23,098 | 44,275 | 15.6 |
| District 25 | 283,400 | 89.0 | 56.2 | 312,600 | 14.2 | 1,423 | 25,977 | 56,270 | 23.9 |
| District 26 | 264,021 | 89.0 | 63.0 | 339,100 | 14.0 | 1,509 | 27,241 | 65,584 | 28.4 |
| District 27 | 344,567 | 83.4 | 45.1 | 424,600 | 39.2 | 1,457 | 41,836 | 60,384 | 31.1 |
| GEORGIA | 4,378,350 | 88.0 | 64.1 | 202,500 | 9.5 | 1,049 | 32,657 | 61,980 | 28.4 |
| District 1 | 332,162 | 83.4 | 62.4 | 168,500 | 7.3 | 981 | 28,612 | 55,542 | 22.3 |
| District 2 | 302,889 | 80.9 | 55.1 | 111,700 | 2.2 | 760 | 23,010 | 39,728 | 14.9 |
| District 3 | 299,566 | 90.5 | 70.3 | 189,500 | 5.6 | 973 | 31,433 | 66,614 | 28.9 |
| District 4 | 289,933 | 93.8 | 61.6 | 190,900 | 3.7 | 1,169 | 28,684 | 60,128 | 24.8 |
| District 5 | 375,962 | 87.5 | 46.7 | 257,900 | 25.2 | 1,175 | 43,210 | 60,247 | 30.7 |
| District 6 | 319,655 | 93.2 | 64.8 | 416,500 | 33.9 | 1,418 | 56,330 | 100,110 | 50.1 |
| District 7 | 288,278 | 95.0 | 68.4 | 288,200 | 11.5 | 1,357 | 35,391 | 80,926 | 40.4 |
| District 8 | 310,103 | 84.9 | 63.8 | 130,900 | 3.4 | 800 | 26,751 | 50,745 | 20.6 |
| District 9 | 340,320 | 80.3 | 75.4 | 210,700 | 8.5 | 863 | 29,707 | 59,728 | 25.6 |
| District 10 | 308,528 | 88.0 | 70.1 | 201,600 | 7.0 | 876 | 29,525 | 61,151 | 27.5 |
| District 11 | 316,334 | 94.7 | 65.0 | 279,500 | 13.8 | 1,269 | 42,154 | 76,640 | 36.7 |
| District 12 | 309,538 | 80.0 | 62.8 | 139,800 | 2.4 | 828 | 25,367 | 50,217 | 19.3 |
| District 13 | 293,828 | 92.9 | 63.3 | 181,700 | 2.9 | 1,141 | 29,028 | 66,203 | 28.8 |
| District 14 | 291,254 | 89.0 | 70.7 | 155,900 | 2.6 | 781 | 25,770 | 56,150 | 21.2 |
| HAWAII | 550,328 | 84.5 | 60.2 | 669,200 | 68.7 | 1,651 | 36,989 | 83,102 | 41.1 |
| District 1 | 273,133 | 89.1 | 56.2 | 728,900 | 75.4 | 1,681 | 39,390 | 86,674 | 43.0 |
| District 2 | 277,195 | 80.0 | 64.6 | 611,600 | 62.3 | 1,602 | 34,500 | 79,985 | 39.1 |
| IDAHO | 751,113 | 87.3 | 71.6 | 255,200 | 11.7 | 880 | 29,606 | 60,999 | 25.2 |
| District 1 | 388,649 | 87.9 | 75.1 | 275,800 | 13.0 | 951 | 29,760 | 62,886 | 25.6 |
| District 2 | 362,464 | 86.7 | 67.7 | 227,500 | 10.2 | 843 | 29,436 | 58,708 | 24.8 |
| ILLINOIS | 5,388,210 | 90.3 | 66.0 | 209,100 | 9.5 | 1,020 | 37,728 | 69,187 | 33.7 |
| District 1 | 307,269 | 88.4 | 60.4 | 199,800 | 4.3 | 970 | 31,023 | 56,680 | 28.0 |
| District 2 | 303,550 | 85.6 | 59.7 | 138,500 | 1.9 | 957 | 26,931 | 51,472 | 22.4 |

1. Specified owner-occupied units; $1,000,000 represents $1,000,000    2. Specified renter-occupied units.

# Table E. Congressional Districts 116th Congress — **Poverty, Labor Force, Employment, and Social Security**

| STATE District | Poverty, 2019 | | | Civilian labor force, 2019 | Unemployment | | Civilian employment,[2] 2019 | Percent | | | Persons under 65 years of age with no health insurance, 2019 (percent) | Social Security beneficiaries, December 2020 | | Supplemental Security Income recipients, December 2019 |
|---|---|---|---|---|---|---|---|---|---|---|---|---|---|---|
| | Persons below poverty level (percent) | Families below poverty level (percent) | Percent of households receiving food stamps in past 12 months | Total | Total | Rate[1] | Total | Management, business, science, and arts occupations | Service, sales, and office | Construction and production | | Number | Rate[3] | |
| | 49 | 50 | 51 | 52 | 53 | 54 | 55 | 56 | 57 | 58 | 59 | 60 | 61 | 62 |
| COLORADO ....................... | 9.3 | 5.8 | 6.9 | 3,182,096 | 115,497 | 3.7 | 3,032,173 | 45.1 | 36.3 | 18.6 | 7.9 | 915,854 | 159.0 | 71,971 |
| District 1 ......................... | 10.8 | 7.0 | 6.4 | 511,950 | 15,285 | 3.0 | 495,986 | 51.7 | 33.4 | 14.9 | 7.9 | 108,255 | 126.9 | 14,736 |
| District 2 ......................... | 9.8 | 3.8 | 4.1 | 466,661 | 14,800 | 3.2 | 451,258 | 52.2 | 33.5 | 14.3 | 5.7 | 132,772 | 161.1 | 4,776 |
| District 3 ......................... | 13.6 | 9.0 | 10.8 | 379,110 | 14,437 | 3.8 | 363,986 | 37.2 | 40.5 | 22.3 | 9.9 | 165,127 | 218.3 | 14,006 |
| District 4 ......................... | 7.6 | 4.9 | 6.5 | 465,100 | 14,809 | 3.2 | 449,312 | 42.9 | 34.3 | 22.8 | 7.3 | 135,705 | 156.3 | 8,935 |
| District 5 ......................... | 8.9 | 5.9 | 8.8 | 430,194 | 21,597 | 5.4 | 380,885 | 44.7 | 38.4 | 16.8 | 8.0 | 141,741 | 172.8 | 10,826 |
| District 6 ......................... | 6.6 | 4.3 | 5.5 | 469,387 | 17,213 | 3.7 | 448,845 | 44.3 | 37.0 | 18.7 | 8.4 | 111,918 | 135.1 | 9,376 |
| District 7 ......................... | 8.5 | 6.3 | 6.4 | 459,694 | 17,356 | 3.8 | 441,901 | 40.4 | 38.5 | 21.2 | 8.5 | 120,336 | 148.8 | 9,316 |
| CONNECTICUT................... | 10.0 | 6.8 | 11.1 | 1,937,195 | 102,571 | 5.3 | 1,823,915 | 44.7 | 37.9 | 17.4 | 5.8 | 695,402 | 195.0 | 66,925 |
| District 1 ......................... | 10.7 | 6.9 | 12.6 | 385,599 | 18,663 | 4.9 | 365,733 | 45.6 | 38.0 | 16.4 | 4.6 | 144,976 | 206.2 | 17,268 |
| District 2 ......................... | 7.5 | 4.4 | 9.4 | 387,257 | 17,988 | 4.8 | 360,442 | 43.8 | 38.0 | 18.3 | 3.9 | 150,216 | 214.1 | 9,175 |
| District 3 ......................... | 10.8 | 7.4 | 12.6 | 395,837 | 24,134 | 6.1 | 371,428 | 42.6 | 38.7 | 18.6 | 5.4 | 139,730 | 194.6 | 15,180 |
| District 4 ......................... | 9.6 | 7.3 | 8.7 | 386,554 | 21,403 | 5.5 | 365,038 | 50.0 | 35.8 | 14.1 | 9.0 | 118,406 | 160.5 | 10,649 |
| District 5 ......................... | 11.4 | 8.0 | 12.3 | 381,948 | 20,383 | 5.3 | 361,274 | 41.6 | 38.9 | 19.6 | 5.9 | 142,074 | 201.6 | 14,653 |
| DELAWARE....................... | 11.3 | 7.6 | 10.1 | 494,432 | 22,631 | 4.6 | 466,061 | 40.5 | 38.8 | 20.6 | 6.5 | 224,617 | 230.7 | 17,143 |
| At Large ......................... | 11.3 | 7.6 | 10.1 | 494,432 | 22,631 | 4.6 | 466,061 | 40.5 | 38.8 | 20.6 | 6.5 | 224,617 | 230.7 | 17,143 |
| DISTRICT OF COLUMBIA.... | 13.5 | 8.8 | 11.3 | 416,359 | 25,916 | 6.3 | 387,826 | 67.8 | 26.2 | 6.0 | 3.4 | 83,647 | 118.5 | 24,920 |
| Delegate District (At Large)... | 13.5 | 8.8 | 11.3 | 416,359 | 25,916 | 6.3 | 387,826 | 67.8 | 26.2 | 6.0 | 3.4 | 83,647 | 118.5 | 24,920 |
| FLORIDA .......................... | 12.7 | 8.7 | 11.8 | 10,500,485 | 471,449 | 4.5 | 9,958,518 | 36.5 | 43.4 | 20.1 | 12.9 | 4,840,275 | 225.4 | 575,272 |
| District 1 ......................... | 12.5 | 8.1 | 10.2 | 394,462 | 18,230 | 4.9 | 350,550 | 36.2 | 42.9 | 20.9 | 12.5 | 181,698 | 227.6 | 17,279 |
| District 2 ......................... | 13.6 | 9.5 | 13.2 | 314,782 | 14,720 | 4.7 | 297,243 | 37.0 | 41.3 | 21.7 | 11.7 | 188,358 | 261.3 | 19,924 |
| District 3 ......................... | 15.6 | 9.4 | 10.4 | 355,434 | 18,613 | 5.3 | 334,880 | 40.4 | 40.1 | 19.6 | 11.1 | 166,231 | 219.0 | 19,845 |
| District 4 ......................... | 8.0 | 5.7 | 5.8 | 445,596 | 19,156 | 4.4 | 412,241 | 45.2 | 39.0 | 15.7 | 9.1 | 170,532 | 203.9 | 11,305 |
| District 5 ......................... | 21.9 | 15.6 | 19.9 | 368,790 | 22,756 | 6.2 | 342,047 | 30.3 | 45.3 | 24.4 | 13.7 | 139,758 | 188.2 | 29,570 |
| District 6 ......................... | 12.7 | 8.9 | 11.1 | 356,976 | 13,468 | 3.8 | 342,619 | 33.1 | 45.4 | 21.5 | 13.3 | 241,083 | 305.0 | 17,486 |
| District 7 ......................... | 11.7 | 8.0 | 10.0 | 445,481 | 17,197 | 3.9 | 427,653 | 44.1 | 39.9 | 16.0 | 10.6 | 132,229 | 162.2 | 13,933 |
| District 8 ......................... | 9.5 | 5.8 | 8.2 | 355,700 | 19,607 | 5.6 | 333,612 | 38.6 | 41.9 | 19.5 | 10.9 | 229,975 | 294.8 | 15,199 |
| District 9 ......................... | 13.4 | 9.4 | 14.1 | 449,569 | 18,303 | 4.1 | 431,037 | 31.0 | 45.4 | 23.6 | 12.8 | 188,015 | 201.8 | 27,683 |
| District 10 ....................... | 12.5 | 9.5 | 13.7 | 450,133 | 19,153 | 4.3 | 430,500 | 35.5 | 45.7 | 18.8 | 14.2 | 126,618 | 156.0 | 22,915 |
| District 11 ....................... | 12.5 | 7.6 | 10.6 | 282,273 | 15,750 | 5.6 | 265,594 | 32.5 | 46.7 | 20.9 | 10.7 | 327,131 | 402.3 | 17,264 |
| District 12 ....................... | 11.0 | 7.7 | 9.4 | 368,017 | 15,910 | 4.3 | 350,973 | 41.4 | 41.5 | 17.1 | 12.2 | 216,818 | 267.2 | 15,927 |
| District 13 ....................... | 11.5 | 5.9 | 8.9 | 373,759 | 19,369 | 5.2 | 352,837 | 40.5 | 43.5 | 15.9 | 11.4 | 188,842 | 258.1 | 18,422 |
| District 14 ....................... | 14.5 | 9.6 | 12.2 | 447,759 | 18,787 | 4.2 | 425,198 | 41.0 | 39.7 | 19.3 | 12.7 | 127,779 | 153.7 | 26,041 |
| District 15 ....................... | 12.9 | 8.4 | 11.9 | 398,925 | 16,633 | 4.2 | 381,220 | 35.8 | 41.2 | 23.0 | 12.9 | 165,743 | 206.8 | 22,558 |
| District 16 ....................... | 9.7 | 6.4 | 7.4 | 411,804 | 21,043 | 5.1 | 389,238 | 36.9 | 43.1 | 19.9 | 11.9 | 232,433 | 266.0 | 12,398 |
| District 17 ....................... | 12.7 | 8.3 | 10.3 | 307,090 | 15,154 | 4.9 | 291,799 | 28.8 | 46.5 | 24.7 | 12.0 | 258,424 | 321.1 | 14,846 |
| District 18 ....................... | 8.6 | 5.4 | 8.6 | 387,082 | 16,684 | 4.3 | 369,451 | 39.0 | 43.0 | 18.0 | 11.5 | 209,254 | 263.0 | 12,248 |
| District 19 ....................... | 10.1 | 6.6 | 6.0 | 358,903 | 13,718 | 3.8 | 344,670 | 34.1 | 46.6 | 19.3 | 11.3 | 240,462 | 288.7 | 12,323 |
| District 20 ....................... | 17.0 | 14.0 | 19.5 | 422,402 | 30,461 | 7.2 | 391,339 | 26.9 | 50.2 | 22.9 | 18.6 | 123,643 | 154.1 | 25,416 |
| District 21 ....................... | 10.4 | 7.1 | 8.0 | 391,261 | 15,965 | 4.1 | 375,018 | 37.8 | 42.7 | 19.5 | 14.5 | 189,515 | 240.9 | 10,520 |
| District 22 ....................... | 11.2 | 7.9 | 6.9 | 409,443 | 18,308 | 4.5 | 390,837 | 40.7 | 42.8 | 16.5 | 13.3 | 153,844 | 202.2 | 11,555 |
| District 23 ....................... | 9.7 | 7.2 | 8.7 | 407,166 | 14,481 | 3.6 | 391,921 | 42.8 | 42.7 | 14.4 | 11.2 | 132,825 | 174.1 | 14,981 |
| District 24 ....................... | 17.8 | 13.4 | 25.8 | 388,553 | 19,249 | 5.0 | 368,904 | 26.9 | 48.7 | 24.4 | 19.4 | 122,372 | 162.1 | 47,736 |
| District 25 ....................... | 14.3 | 11.5 | 21.9 | 400,681 | 11,987 | 3.0 | 388,456 | 29.6 | 42.6 | 27.7 | 16.3 | 138,894 | 174.4 | 44,227 |
| District 26 ....................... | 13.4 | 10.7 | 19.1 | 401,119 | 15,914 | 4.0 | 382,785 | 32.9 | 43.7 | 23.4 | 15.9 | 124,258 | 159.1 | 38,021 |
| District 27 ....................... | 14.8 | 10.6 | 16.8 | 407,325 | 10,833 | 2.7 | 395,896 | 42.6 | 41.4 | 16.1 | 13.7 | 123,541 | 164.6 | 35,650 |
| GEORGIA ......................... | 13.3 | 9.6 | 11.2 | 5,308,730 | 251,981 | 4.8 | 5,002,153 | 39.1 | 37.0 | 23.8 | 13.3 | 1,902,790 | 179.2 | 258,304 |
| District 1 ......................... | 14.8 | 10.8 | 12.3 | 361,149 | 19,623 | 5.7 | 322,341 | 33.5 | 41.6 | 24.9 | 14.5 | 144,703 | 193.0 | 17,960 |
| District 2 ......................... | 23.9 | 19.5 | 22.7 | 300,774 | 27,290 | 9.4 | 262,089 | 29.0 | 41.4 | 29.6 | 14.4 | 145,011 | 215.8 | 30,779 |
| District 3 ......................... | 12.3 | 8.8 | 11.3 | 360,844 | 15,773 | 4.4 | 342,408 | 36.2 | 35.9 | 27.9 | 10.6 | 156,447 | 208.3 | 19,228 |
| District 4 ......................... | 11.2 | 8.4 | 12.2 | 409,520 | 21,214 | 5.2 | 388,167 | 35.2 | 37.9 | 26.9 | 17.0 | 120,792 | 154.4 | 22,027 |
| District 5 ......................... | 18.7 | 13.8 | 15.1 | 420,464 | 21,021 | 5.0 | 398,835 | 52.1 | 33.6 | 14.3 | 11.5 | 105,952 | 134.3 | 25,622 |
| District 6 ......................... | 6.7 | 4.1 | 2.8 | 420,848 | 16,148 | 3.8 | 403,838 | 60.0 | 29.8 | 10.2 | 8.2 | 100,297 | 135.0 | 5,715 |
| District 7 ......................... | 8.2 | 6.3 | 3.8 | 444,428 | 13,186 | 3.0 | 430,668 | 43.4 | 37.3 | 19.3 | 14.0 | 102,100 | 120.9 | 10,206 |
| District 8 ......................... | 17.3 | 13.2 | 15.1 | 333,390 | 17,392 | 5.3 | 309,480 | 32.9 | 39.8 | 27.3 | 14.9 | 148,135 | 209.8 | 23,881 |
| District 9 ......................... | 13.0 | 9.0 | 9.4 | 359,541 | 13,022 | 3.6 | 345,417 | 33.3 | 38.1 | 28.6 | 15.1 | 184,537 | 239.3 | 15,505 |
| District 10 ....................... | 13.6 | 8.8 | 11.5 | 360,723 | 16,927 | 4.7 | 343,102 | 38.7 | 36.5 | 24.8 | 11.9 | 153,341 | 202.3 | 18,192 |
| District 11 ....................... | 9.0 | 6.5 | 4.9 | 425,012 | 12,478 | 2.9 | 411,915 | 46.0 | 35.2 | 18.8 | 12.8 | 122,806 | 156.9 | 9,885 |
| District 12 ....................... | 18.3 | 13.5 | 14.0 | 341,072 | 18,767 | 5.7 | 312,898 | 32.6 | 39.5 | 27.9 | 13.1 | 145,935 | 199.1 | 23,562 |
| District 13 ....................... | 10.5 | 7.2 | 13.0 | 421,859 | 22,044 | 5.2 | 399,416 | 35.6 | 38.8 | 25.6 | 14.3 | 122,628 | 152.7 | 18,522 |
| District 14 ....................... | 11.7 | 8.5 | 11.0 | 349,106 | 17,096 | 4.9 | 331,579 | 28.3 | 36.6 | 35.1 | 13.6 | 150,106 | 205.0 | 17,220 |
| HAWAII............................. | 9.3 | 6.4 | 10.4 | 745,236 | 27,112 | 3.9 | 671,768 | 36.1 | 45.7 | 18.2 | 4.0 | 282,623 | 199.6 | 22,412 |
| District 1 ......................... | 7.6 | 4.6 | 8.5 | 393,013 | 12,538 | 3.4 | 353,173 | 38.4 | 45.0 | 16.6 | 3.6 | 138,192 | 191.7 | 10,927 |
| District 2 ......................... | 11.1 | 8.3 | 12.5 | 352,223 | 14,574 | 4.4 | 318,595 | 33.6 | 46.5 | 19.8 | 4.5 | 144,431 | 207.8 | 11,485 |
| IDAHO.............................. | 11.2 | 7.4 | 8.3 | 881,838 | 29,127 | 3.3 | 848,223 | 35.5 | 39.4 | 25.1 | 10.8 | 370,385 | 207.3 | 30,780 |
| District 1 ......................... | 9.4 | 6.3 | 7.5 | 451,655 | 12,089 | 2.7 | 438,184 | 34.8 | 40.3 | 24.9 | 10.8 | 211,037 | 225.8 | 15,197 |
| District 2 ......................... | 13.1 | 8.6 | 9.2 | 430,183 | 17,038 | 4.0 | 410,039 | 36.3 | 38.4 | 25.2 | 10.7 | 159,348 | 187.0 | 15,583 |
| ILLINOIS ........................... | 11.5 | 7.9 | 11.8 | 6,628,213 | 320,069 | 4.8 | 6,286,647 | 40.7 | 37.3 | 22.0 | 7.3 | 2,274,372 | 179.5 | 259,810 |
| District 1 ......................... | 16.0 | 11.7 | 18.4 | 359,565 | 28,629 | 8.0 | 330,798 | 37.2 | 41.7 | 21.1 | 8.1 | 125,175 | 176.0 | 23,767 |
| District 2 ......................... | 17.1 | 12.3 | 20.3 | 347,386 | 29,494 | 8.5 | 317,623 | 29.5 | 43.4 | 27.1 | 8.8 | 131,795 | 192.2 | 28,231 |

1. Percent of civilian labor force.    2. Persons 16 years old and over.    3. Per 1,000 resident population estimated in the 2019 American Community Survey.

Items 49—62

# Table E. Congressional Districts 116th Congress — **Agriculture**

| STATE District | Agriculture 2017 | | | | | | | | | |
|---|---|---|---|---|---|---|---|---|---|---|
| | Land in farms | | | | Value of products sold | | | | Government payments | |
| | Number of farms | Acres | Average size of farm (acres) | Irrigated land (acres) | Total ($1,000) | Average per farm (dollars) | Percent from crops | Percent from livestock and poultry products | Total ($1,000) | Average per farm receiving payments (dollars) |
| | 63 | 64 | 65 | 66 | 67 | 68 | 69 | 70 | 71 | 72 |
| COLORADO | 38,893 | 31,820,957 | 818 | 5,916,737 | 7,491,702 | 192,623 | 29.9 | 70.1 | 198,697 | 22,206 |
| District 1 | 71 | 6,390 | 90 | 523 | D | D | D | D | D | D |
| District 2 | 3,627 | 953,177 | 263 | 126,497 | 206,343 | 56,891 | 51.3 | 48.7 | 1,035 | 6,273 |
| District 3 | 14,808 | 8,719,448 | 589 | 1,025,088 | 967,155 | 65,313 | 53.4 | 46.6 | 178,446 | 10,711 |
| District 4 | 16,578 | 20,777,185 | 1,253 | 4,667,347 | 6,156,045 | 371,338 | 24.5 | 75.5 | 178,446 | 25,540 |
| District 5 | 3,044 | 1,225,026 | 402 | 47,661 | 71,820 | 23,594 | 32.5 | 67.5 | 1,626 | 11,698 |
| District 6 | 395 | 96,913 | 245 | 38,832 | D | D | D | D | 644 | 9,200 |
| District 7 | 370 | 42,818 | 116 | 10,789 | 47,821 | 129,246 | 90.4 | 9.6 | D | D |
| CONNECTICUT | 5,521 | 381,539 | 69 | 122,074 | 580,114 | 105,074 | 72.4 | 27.6 | 1,850 | 7,551 |
| District 1 | 720 | 41,976 | 58 | 17,349 | 88,995 | 123,604 | 97.0 | 3.0 | 202 | 7,481 |
| District 2 | 2,501 | 171,834 | 69 | 60,637 | 290,026 | 115,964 | 61.6 | 38.4 | 869 | 7,364 |
| District 3 | 560 | 22,760 | 41 | 8,099 | 30,317 | 54,138 | D | D | 322 | 11,500 |
| District 4 | 336 | 49,386 | 147 | 2,989 | 32,355 | 96,295 | 41.5 | 58.5 | 10 | 1,667 |
| District 5 | 1,404 | 95,583 | 68 | 33,000 | 138,421 | 98,590 | D | D | 446 | 6,758 |
| DELAWARE | 2,302 | 525,324 | 228 | 435,085 | 1,465,973 | 636,826 | 22.2 | 77.8 | 15,162 | 18,604 |
| At Large | 2,302 | 525,324 | 228 | 435,085 | 1,465,973 | 636,826 | 22.2 | 77.8 | 15,162 | 18,604 |
| DISTRICT OF COLUMBIA | X | X | X | X | X | X | X | X | X | X |
| Delegate District (At Large) | X | X | X | X | X | X | X | X | X | X |
| FLORIDA | 47,590 | 9,731,731 | 204 | 2,093,330 | 7,357,343 | 154,599 | 77.5 | 22.5 | 59,120 | 14,795 |
| District 1 | 2,929 | 349,886 | 119 | 112,935 | 125,902 | 42,985 | 61.0 | 39.0 | 9,336 | 16,322 |
| District 2 | 7,980 | 1,696,547 | 213 | 398,462 | 809,576 | 101,451 | 60.9 | 39.1 | 19,453 | 20,872 |
| District 3 | 5,415 | 531,791 | 98 | 87,596 | 239,941 | 44,310 | 61.9 | 38.1 | 1,881 | 9,746 |
| District 4 | 809 | 103,274 | 128 | 14,834 | 43,075 | 53,245 | 62.4 | 37.6 | 301 | 5,017 |
| District 5 | 2,815 | 597,271 | 212 | 90,114 | 290,125 | 103,064 | 53.2 | 46.8 | 4,332 | 13,929 |
| District 6 | 2,353 | 274,585 | 117 | 37,953 | 354,142 | 150,507 | 92.4 | 7.6 | 1,507 | 11,773 |
| District 7 | 456 | 41,487 | 91 | 3,212 | 26,224 | 57,509 | 91.2 | 8.8 | 180 | 10,000 |
| District 8 | 1,106 | 367,011 | 332 | 61,950 | 182,637 | 165,133 | 85.7 | 14.3 | 703 | 11,339 |
| District 9 | 1,290 | 816,015 | 633 | 71,832 | 238,917 | 185,207 | 75.6 | 24.4 | 1,739 | 14,372 |
| District 10 | 361 | 18,909 | 52 | 5,215 | 196,875 | 545,360 | 99.6 | 0.4 | 69 | 6,900 |
| District 11 | 4,389 | 445,594 | 102 | 48,445 | 165,219 | 37,644 | 64.8 | 35.2 | 1,994 | 6,391 |
| District 12 | 1,318 | 198,908 | 151 | 16,710 | 68,035 | 51,620 | 25.8 | 74.2 | 944 | 6,695 |
| District 13 | 78 | 198 | 3 | 58 | 877 | 11,244 | D | D | D | D |
| District 14 | 249 | 23,507 | 94 | 3,592 | 34,590 | 138,916 | 84.3 | 15.7 | D | D |
| District 15 | 2,439 | 271,445 | 111 | 53,811 | 357,975 | 146,771 | 82.2 | 17.8 | 1,530 | 9,000 |
| District 16 | 1,625 | 281,092 | 173 | 77,338 | 557,864 | 343,301 | 90.4 | 9.6 | 540 | 9,000 |
| District 17 | 5,060 | 2,148,236 | 425 | 295,998 | 1,085,016 | 214,430 | 60.4 | 39.6 | 9,989 | 14,799 |
| District 18 | 1,327 | 545,846 | 411 | 199,100 | 462,117 | 348,242 | 93.9 | 6.1 | 1,462 | 16,805 |
| District 19 | 461 | 60,374 | 131 | 21,045 | 104,295 | 226,236 | 95.8 | 4.2 | 36 | 3,000 |
| District 20 | 645 | 295,454 | 458 | 235,636 | 507,802 | 787,290 | 98.9 | 1.1 | 93 | 7,154 |
| District 21 | 403 | 28,185 | 70 | 18,733 | 187,957 | 466,395 | 96.6 | 3.4 | D | D |
| District 22 | 95 | 1,077 | 11 | 264 | 4,866 | 51,221 | 95.7 | 4.3 | D | D |
| District 23 | 479 | 3,809 | 8 | 910 | 16,598 | 34,651 | 94.2 | 5.8 | D | D |
| District 24 | 40 | 772 | 19 | 311 | 2,007 | 50,175 | D | D | 1,511 | 23,246 |
| District 25 | 831 | 556,369 | 670 | 189,265 | 474,762 | 571,314 | 93.0 | 7.0 | 1,500 | 32,609 |
| District 26 | 2,298 | 70,917 | 31 | 45,578 | 800,022 | 348,138 | 98.6 | 1.4 | D | D |
| District 27 | 339 | 3,172 | 9 | 2,433 | 19,925 | 58,776 | 97.4 | 2.6 | D | D |
| GEORGIA | 42,439 | 9,953,730 | 235 | 3,628,707 | 9,573,252 | 225,577 | 34.2 | 65.8 | 247,428 | 18,310 |
| District 1 | 2,271 | 444,632 | 196 | 144,495 | 285,752 | 125,827 | 63.5 | 36.5 | 5,441 | 11,985 |
| District 2 | 5,565 | 2,604,471 | 468 | 1,196,536 | 1,835,745 | 329,873 | 61.5 | 38.5 | 96,196 | 31,406 |
| District 3 | 3,547 | 473,893 | 134 | 73,710 | 386,470 | 108,957 | 12.2 | 87.8 | 2,164 | 5,116 |
| District 4 | 236 | 19,841 | 84 | 3,114 | 2,128 | 9,017 | 54.6 | 45.4 | 60 | 4,615 |
| District 5 | 37 | 3,697 | 100 | 101 | 216 | 5,838 | 49.1 | 50.9 | 97 | 8,083 |
| District 6 | 96 | 3,153 | 33 | 323 | 1,399 | 14,573 | 43.5 | 56.5 | D | D |
| District 7 | 191 | 11,824 | 62 | 2,645 | 15,251 | 79,848 | 81.1 | 18.9 | 20 | 2,000 |
| District 8 | 6,410 | 2,070,355 | 323 | 924,328 | 1,416,022 | 220,908 | 64.0 | 36.0 | 68,008 | 22,692 |
| District 9 | 6,825 | 670,548 | 98 | 151,029 | 2,238,033 | 327,917 | 3.3 | 96.7 | 8,570 | 5,343 |
| District 10 | 5,529 | 1,116,120 | 202 | 235,447 | 921,978 | 166,753 | 19.6 | 80.4 | 12,584 | 9,717 |
| District 11 | 969 | 103,747 | 107 | 20,615 | 97,670 | 100,795 | 18.5 | 81.5 | 1,105 | 7,367 |
| District 12 | 6,001 | 1,866,526 | 311 | 733,828 | 1,293,651 | 215,573 | 51.9 | 48.1 | 47,457 | 19,522 |
| District 13 | 313 | 17,735 | 57 | 2,715 | 2,894 | 9,246 | 58.9 | 41.1 | 26 | 1,733 |
| District 14 | 4,449 | 547,188 | 123 | 139,821 | 1,076,043 | 241,862 | 4.5 | 95.5 | 5,700 | 5,449 |
| HAWAII | 7,328 | 1,135,352 | 155 | 84,767 | 563,802 | 76,938 | 74.0 | 26.0 | 8,362 | 12,631 |
| District 1 | 252 | 12,158 | 48 | 3,152 | 32,162 | 127,627 | 91.2 | 8.8 | 53 | 2,524 |
| District 2 | 7,076 | 1,123,194 | 159 | 81,615 | 531,641 | 75,133 | 72.9 | 27.1 | 8,309 | 12,963 |
| IDAHO | 24,996 | 11,691,912 | 468 | 4,576,077 | 7,567,440 | 302,746 | 42.4 | 57.6 | 129,605 | 21,306 |
| District 1 | 12,254 | 4,107,446 | 335 | 1,325,074 | 1,577,768 | 128,755 | 48.8 | 51.2 | 34,226 | 16,969 |
| District 2 | 12,742 | 7,584,466 | 595 | 3,251,003 | 5,989,671 | 470,073 | 40.7 | 59.3 | 95,379 | 23,458 |
| ILLINOIS | 72,651 | 27,006,288 | 372 | 22,701,382 | 17,009,972 | 234,133 | 81.4 | 18.6 | 521,229 | 10,727 |
| District 1 | 186 | 61,309 | 330 | 58,228 | 38,288 | 205,849 | 92.4 | 7.6 | 198 | 3,536 |
| District 2 | 1,119 | 419,020 | 374 | 395,183 | 295,161 | 263,772 | 91.3 | 8.7 | 2,900 | 6,416 |

# Table E. Congressional Districts 116th Congress — **Nonfarm Employment and Payroll**

| STATE District | | Private nonfarm employment and payroll, 2019 | | | | | | | | | | | |
|---|---|---|---|---|---|---|---|---|---|---|---|---|---|
| | | Employment | | | | | | | | | | Annual payroll | |
| | | Total | Percent by selected industries | | | | | | | | | | |
| | Number of establishments | Total | Manufac-turing | Construc-tion | Wholesale trade | Retail trade | Health care and social assistance | Finance and Insurance | Real estate and rental and leasing | Profes-sional, sci-entific, and technical services | Information | Total (mil dol) | Average per employee (dollars) |
| | 73 | 74 | 75 | 76 | 77 | 78 | 79 | 80 | 81 | 82 | 83 | 84 | 85 |
| COLORADO .................. | 174,258 | 2,473,192 | 5.2 | 7.1 | 4.2 | 11.3 | 13.0 | 4.9 | 1.9 | 8.3 | 3.9 | 142,461 | 57,602 |
| District 1 .................... | 31,371 | 525,864 | 3.8 | 5.3 | 5.1 | 8.0 | 12.3 | 5.5 | 2.5 | 10.1 | 3.6 | 35,474 | 67,459 |
| District 2 .................... | 31,428 | 385,720 | 8.6 | 5.9 | 4.4 | 12.5 | 11.7 | 2.7 | 2.3 | 10.5 | 5.2 | 22,743 | 58,963 |
| District 3 .................... | 26,548 | 262,145 | 4.9 | 9.4 | 2.8 | 14.9 | 17.7 | 3.1 | 2.7 | 4.5 | 1.5 | 11,251 | 42,921 |
| District 4 .................... | 21,163 | 253,858 | 9.6 | 11.0 | 3.8 | 14.8 | 13.3 | 3.7 | 1.6 | 6.4 | 3.3 | 13,516 | 53,244 |
| District 5 .................... | 20,900 | 272,339 | 4.1 | 6.3 | 2.3 | 13.0 | 16.2 | 4.7 | 1.7 | 9.2 | 3.6 | 12,913 | 47,415 |
| District 6 .................... | 21,440 | 341,099 | 2.9 | 6.5 | 5.6 | 11.2 | 14.9 | 8.7 | 1.7 | 7.9 | 8.0 | 21,705 | 63,632 |
| District 7 .................... | 20,597 | 292,556 | 5.9 | 11.0 | 5.3 | 13.7 | 12.3 | 5.3 | 1.6 | 8.0 | 1.9 | 14,907 | 50,955 |
| CONNECTICUT.................. | 88,916 | 1,538,341 | 10.4 | 3.8 | 4.7 | 11.7 | 19.2 | 7.0 | 1.3 | 6.9 | 2.7 | 100,305 | 65,203 |
| District 1 .................... | 18,499 | 387,325 | 12.1 | 3.6 | 4.7 | 9.9 | 18.1 | 11.9 | 1.2 | 6.9 | 3.1 | 25,678 | 66,296 |
| District 2 .................... | 14,410 | 215,640 | 13.8 | 3.9 | 4.3 | 15.7 | 17.8 | 2.8 | 0.7 | 5.6 | 1.5 | 10,499 | 48,689 |
| District 3 .................... | 16,364 | 311,379 | 12.0 | 4.0 | 4.8 | 10.7 | 21.3 | 2.8 | 1.4 | 4.2 | 1.5 | 17,770 | 57,069 |
| District 4 .................... | 21,673 | 334,182 | 4.6 | 2.8 | 6.0 | 10.8 | 17.1 | 8.8 | 1.4 | 10.2 | 4.4 | 29,679 | 88,811 |
| District 5 .................... | 17,405 | 260,867 | 11.7 | 5.3 | 3.2 | 14.6 | 23.2 | 4.3 | 1.8 | 5.0 | 2.2 | 14,420 | 55,278 |
| DELAWARE.................... | 26,142 | 413,410 | 7.2 | 5.6 | 4.1 | 13.8 | 17.4 | 10.1 | 1.5 | 7.6 | 1.5 | 23,948 | 57,928 |
| At Large ..................... | 26,142 | 413,410 | 7.2 | 5.6 | 4.1 | 13.8 | 17.4 | 10.1 | 1.5 | 7.6 | 1.5 | 23,948 | 57,928 |
| DISTRICT OF COLUMBIA.... | 23,993 | 528,826 | 0.2 | 1.9 | 0.7 | 4.1 | 13.9 | 3.8 | 2.0 | 19.3 | 4.7 | 44,758 | 84,636 |
| Delegate District (At Large)... | 23,993 | 528,826 | 0.2 | 1.9 | 0.7 | 4.1 | 13.9 | 3.8 | 2.0 | 19.3 | 4.7 | 44,758 | 84,636 |
| FLORIDA ...................... | 574,512 | 8,860,042 | 3.7 | 5.3 | 3.8 | 12.3 | 13.2 | 4.2 | 2.1 | 6.0 | 2.0 | 426,908 | 48,184 |
| District 1 .................... | 18,081 | 227,863 | 3.6 | 6.7 | 2.7 | 16.9 | 15.9 | 5.9 | 2.5 | 7.5 | 1.2 | 9,430 | 41,384 |
| District 2 .................... | 15,298 | 174,249 | 6.7 | 6.2 | 3.1 | 17.4 | 18.4 | 3.2 | 1.9 | 7.3 | 1.9 | 7,191 | 41,271 |
| District 3 .................... | 16,967 | 233,756 | 5.7 | 5.6 | 3.5 | 17.1 | 24.6 | 3.0 | 1.7 | 4.8 | 1.8 | 9,484 | 40,574 |
| District 4 .................... | 25,733 | 388,569 | 4.3 | 6.4 | 4.7 | 12.1 | 14.4 | 9.0 | 2.0 | 6.5 | 1.5 | 18,918 | 48,687 |
| District 5 .................... | 14,843 | 235,836 | 5.8 | 7.7 | 4.7 | 15.0 | 14.7 | 5.8 | 1.8 | 5.4 | 3.1 | 11,281 | 47,836 |
| District 6 .................... | 17,882 | 196,324 | 5.8 | 7.3 | 2.6 | 18.2 | 18.0 | 3.3 | 2.0 | 4.8 | 2.0 | 7,310 | 37,234 |
| District 7 .................... | 25,392 | 369,213 | 2.6 | 7.5 | 3.6 | 12.2 | 15.3 | 8.1 | 2.0 | 10.1 | 4.1 | 19,425 | 52,612 |
| District 8 .................... | 19,189 | 238,227 | 7.4 | 6.6 | 2.2 | 16.2 | 17.9 | 2.7 | 1.9 | 9.2 | 1.7 | 10,859 | 45,581 |
| District 9 .................... | 15,286 | 213,454 | 5.2 | 6.3 | 2.8 | 16.4 | 13.7 | 2.9 | 3.5 | 4.5 | 1.5 | 8,371 | 39,217 |
| District 10 ................... | 24,675 | 528,942 | 4.2 | 4.9 | 4.7 | 11.8 | 7.7 | 1.9 | 3.6 | 6.0 | 2.3 | 24,377 | 46,086 |
| District 11 ................... | 13,205 | 153,778 | 4.7 | 8.7 | 2.0 | 19.3 | 24.8 | 2.3 | 3.3 | 2.8 | 2.0 | 5,631 | 36,616 |
| District 12 ................... | 17,610 | 179,174 | 4.6 | 7.1 | 2.9 | 20.0 | 20.0 | 4.4 | 1.7 | 6.7 | 1.4 | 7,176 | 40,050 |
| District 13 ................... | 22,313 | 326,415 | 8.4 | 4.9 | 3.6 | 12.6 | 18.4 | 7.1 | 2.5 | 6.7 | 3.2 | 16,907 | 51,795 |
| District 14 ................... | 26,104 | 449,583 | 3.0 | 5.1 | 4.5 | 9.4 | 15.2 | 9.5 | 2.3 | 11.0 | 2.9 | 25,054 | 55,728 |
| District 15 ................... | 17,130 | 280,978 | 6.0 | 8.0 | 6.6 | 14.7 | 13.8 | 6.9 | 2.0 | 5.5 | 2.0 | 12,505 | 44,504 |
| District 16 ................... | 22,518 | 247,807 | 5.5 | 8.0 | 3.6 | 18.3 | 17.7 | 3.2 | 2.2 | 6.2 | 1.6 | 10,865 | 43,845 |
| District 17 ................... | 14,575 | 146,167 | 6.2 | 8.9 | 3.2 | 21.0 | 19.7 | 2.6 | 1.9 | 3.5 | 1.3 | 5,505 | 37,659 |
| District 18 ................... | 23,001 | 245,475 | 4.0 | 6.7 | 2.9 | 15.6 | 17.7 | 3.1 | 2.2 | 6.7 | 1.5 | 11,618 | 47,329 |
| District 19 ................... | 27,123 | 327,215 | 2.5 | 10.5 | 3.0 | 17.4 | 16.3 | 2.7 | 3.3 | 5.4 | 1.6 | 14,736 | 45,036 |
| District 20 ................... | 17,902 | 274,121 | 5.8 | 8.2 | 8.3 | 13.1 | 14.9 | 3.2 | 2.0 | 6.1 | 4.3 | 13,930 | 50,816 |
| District 21 ................... | 22,672 | 212,081 | 1.5 | 6.8 | 2.2 | 16.8 | 20.0 | 2.9 | 2.8 | 6.2 | 2.0 | 9,244 | 43,588 |
| District 22 ................... | 34,621 | 376,738 | 2.7 | 5.6 | 4.9 | 13.1 | 13.7 | 5.2 | 3.3 | 10.2 | 3.0 | 19,914 | 52,859 |
| District 23 ................... | 27,311 | 307,760 | 1.8 | 3.9 | 4.4 | 18.7 | 12.4 | 4.5 | 3.2 | 6.7 | 3.3 | 16,052 | 52,158 |
| District 24 ................... | 17,142 | 220,078 | 5.5 | 3.8 | 7.8 | 12.9 | 23.4 | 2.0 | 2.2 | 5.9 | 1.6 | 10,722 | 48,720 |
| District 25 ................... | 27,300 | 354,580 | 5.8 | 6.4 | 11.4 | 11.8 | 9.5 | 3.8 | 2.3 | 4.6 | 3.4 | 17,146 | 48,356 |
| District 26 ................... | 16,262 | 137,960 | 3.0 | 7.3 | 5.1 | 20.1 | 13.6 | 3.8 | 2.8 | 5.2 | 1.1 | 5,210 | 37,767 |
| District 27 ................... | 33,031 | 349,245 | 0.8 | 2.7 | 2.1 | 12.6 | 16.0 | 7.6 | 3.0 | 11.9 | 2.2 | 21,817 | 62,470 |
| GEORGIA ...................... | 239,034 | 4,040,559 | 9.7 | 4.9 | 5.1 | 11.8 | 13.1 | 4.5 | 1.7 | 7.3 | 3.0 | 212,578 | 52,611 |
| District 1 .................... | 16,459 | 247,243 | 11.7 | 5.1 | 3.8 | 14.7 | 13.5 | 2.0 | 1.6 | 3.7 | 1.3 | 10,368 | 41,935 |
| District 2 .................... | 12,860 | 211,592 | 14.3 | 4.2 | 5.3 | 12.3 | 18.4 | 10.1 | 1.3 | 3.9 | 1.7 | 9,237 | 43,657 |
| District 3 .................... | 15,624 | 238,359 | 16.2 | 5.5 | 4.0 | 16.5 | 14.9 | 2.8 | 1.1 | 3.1 | 1.4 | 9,627 | 40,388 |
| District 4 .................... | 11,655 | 155,502 | 11.1 | 8.1 | 5.0 | 16.2 | 16.0 | 1.7 | 1.6 | 4.0 | 1.9 | 6,639 | 42,695 |
| District 5 .................... | 23,250 | 552,415 | 3.5 | 2.4 | 4.1 | 6.5 | 12.3 | 3.5 | 2.3 | 10.3 | 4.6 | 38,424 | 69,556 |
| District 6 .................... | 29,563 | 513,095 | 2.0 | 3.3 | 4.9 | 8.2 | 11.5 | 10.2 | 2.4 | 13.2 | 7.6 | 37,518 | 73,122 |
| District 7 .................... | 26,892 | 373,861 | 7.9 | 7.6 | 10.7 | 13.1 | 9.6 | 4.2 | 1.7 | 9.1 | 3.7 | 20,360 | 54,459 |
| District 8 .................... | 13,486 | 195,227 | 12.9 | 4.3 | 3.5 | 16.4 | 18.2 | 3.7 | 1.1 | 4.2 | 1.1 | 7,231 | 37,037 |
| District 9 .................... | 15,652 | 222,035 | 23.8 | 4.9 | 4.7 | 14.7 | 12.9 | 2.2 | 0.9 | 2.5 | 1.0 | 9,109 | 41,025 |
| District 10 ................... | 14,024 | 171,237 | 13.0 | 5.9 | 4.2 | 16.5 | 16.1 | 2.5 | 1.7 | 3.5 | 1.3 | 6,437 | 37,592 |
| District 11 ................... | 22,414 | 395,204 | 7.3 | 7.0 | 4.9 | 10.9 | 9.8 | 6.0 | 2.8 | 12.6 | 3.3 | 25,187 | 63,731 |
| District 12 ................... | 13,536 | 217,225 | 13.0 | 6.6 | 4.0 | 14.9 | 20.7 | 2.2 | 1.2 | 3.5 | 1.7 | 9,133 | 42,044 |
| District 13 ................... | 11,697 | 184,079 | 7.1 | 7.3 | 7.6 | 15.8 | 16.6 | 1.8 | 1.7 | 2.6 | 1.2 | 8,414 | 45,708 |
| District 14 ................... | 11,020 | 181,112 | 26.8 | 4.3 | 5.0 | 14.3 | 13.8 | 1.9 | 0.9 | 2.3 | 1.0 | 7,183 | 39,660 |
| HAWAII ...................... | 32,889 | 553,206 | 2.2 | 5.6 | 3.4 | 12.9 | 13.2 | 3.9 | 2.6 | 4.2 | 1.5 | 25,839 | 46,707 |
| District 1 .................... | 18,166 | 329,256 | 2.7 | 6.3 | 4.2 | 12.4 | 14.1 | 5.4 | 2.5 | 5.1 | 2.0 | 16,457 | 49,983 |
| District 2 .................... | 14,485 | 191,769 | 1.8 | 5.3 | 2.3 | 16.1 | 13.9 | 1.6 | 3.2 | 3.0 | 1.0 | 7,956 | 41,486 |
| IDAHO.......................... | 50,547 | 616,778 | 10.3 | 7.9 | 5.3 | 13.7 | 16.7 | 3.8 | 1.4 | 6.0 | 2.1 | 26,955 | 43,703 |
| District 1 .................... | 24,629 | 274,657 | 11.0 | 10.4 | 4.8 | 14.5 | 15.8 | 3.9 | 1.5 | 4.8 | 2.2 | 11,294 | 41,119 |
| District 2 .................... | 25,566 | 327,672 | 10.1 | 6.1 | 5.8 | 13.5 | 18.1 | 3.6 | 1.4 | 7.0 | 2.1 | 15,037 | 45,889 |
| ILLINOIS ...................... | 320,417 | 5,530,388 | 9.9 | 3.9 | 5.8 | 10.5 | 14.8 | 6.2 | 1.5 | 7.3 | 2.4 | 325,435 | 58,845 |
| District 1 .................... | 11,898 | 189,555 | 5.7 | 4.6 | 4.2 | 14.9 | 19.4 | 2.2 | 1.5 | 2.9 | 1.2 | 8,598 | 45,360 |
| District 2 .................... | 9,927 | 166,903 | 18.2 | 4.6 | 5.4 | 12.6 | 20.8 | 2.1 | 1.1 | 2.1 | 1.9 | 7,891 | 47,279 |

# Table E. Congressional Districts 116th Congress — **Land Area and Population Characteristics**

| STATE District | Representative, 117th Congress | Land area,[1] 2018 (sq mi) | Total persons | Per square mile | Race alone (percent) White | Black | American Indian, Alaska Native | Asian and Pacific Islander | Some other race (percent) | Two or more races (percent) | Hispanic or Latino[2] (percent) | Non-Hispanic White alone (percent) | Percent female | Percent foreign-born | Percent born in state of residence |
|---|---|---|---|---|---|---|---|---|---|---|---|---|---|---|---|
| | | 1 | 2 | 3 | 4 | 5 | 6 | 7 | 8 | 9 | 10 | 11 | 12 | 13 | 14 |
| ILLINOIS—Cont'd | | | | | | | | | | | | | | | |
| District 3 | Marie Newman (D) | 236.881 | 702,503 | 2,965.637 | 77.3 | 5.3 | 0.5 | 4.7 | 9.6 | 2.6 | 32.3 | 56.622 | 50.0 | 20.8 | 69.8 |
| District 4 | Jesus G. "Chuy" Garcia (D) | 52.437 | 676,674 | 12,904.514 | 58.1 | 4.5 | 0.6 | 4.5 | 28.9 | 3.4 | 67.1 | 23.170 | 49.4 | 32.0 | 53.7 |
| District 5 | Michael Quigley (D) | 95.591 | 739,401 | 7,735.048 | 79.3 | 3.0 | 0.3 | 7.1 | 7.2 | 3.1 | 21.9 | 65.909 | 50.7 | 20.7 | 55.6 |
| District 6 | Sean Casten (D) | 378.735 | 710,626 | 1,876.315 | 82.0 | 2.1 | 0.2 | 9.5 | 3.1 | 3.0 | 11.7 | 74.081 | 50.8 | 15.3 | 65.7 |
| District 7 | Danny K. Davis (D) | 62.339 | 727,761 | 11,674.249 | 39.5 | 42.5 | 0.2 | 7.9 | 7.3 | 2.7 | 17.3 | 30.559 | 51.3 | 14.5 | 61.6 |
| District 8 | Raja Krishnamoorthi (D) | 205.661 | 717,115 | 3,486.879 | 65.1 | 6.4 | 0.2 | 13.5 | 11.2 | 3.4 | 28.0 | 49.864 | 50.4 | 28.1 | 59.5 |
| District 9 | Janice D. Schakowsky (D) | 105.401 | 719,256 | 6,823.996 | 68.4 | 9.5 | 0.2 | 14.7 | 3.5 | 3.6 | 11.2 | 62.112 | 51.2 | 26.6 | 53.4 |
| District 10 | Bradley Scott Schneider (D) | 299.848 | 706,189 | 2,355.157 | 68.8 | 7.1 | 0.6 | 10.9 | 9.5 | 3.0 | 25.1 | 54.421 | 50.4 | 24.0 | 57.6 |
| District 11 | Bill Foster (D) | 280.307 | 721,594 | 2,574.299 | 69.1 | 11.5 | 0.4 | 7.9 | 7.4 | 3.7 | 28.1 | 50.454 | 50.9 | 19.4 | 65.4 |
| District 12 | Mike Bost (R) | 5,008.043 | 679,002 | 135.582 | 77.8 | 16.7 | 0.3 | 1.4 | 0.8 | 3.0 | 3.4 | 75.611 | 50.6 | 2.5 | 67.3 |
| District 13 | Rodney Davis (R) | 5,793.555 | 698,830 | 120.622 | 80.9 | 11.9 | 0.2 | 4.1 | 0.5 | 2.5 | 4.0 | 77.642 | 51.1 | 5.1 | 73.7 |
| District 14 | Lauren Underwood (D) | 1,597.234 | 727,525 | 455.491 | 86.1 | 3.1 | 0.1 | 5.3 | 2.9 | 2.5 | 12.0 | 77.707 | 50.6 | 8.9 | 72.3 |
| District 15 | Mary E. Miller (R) | 14,696.144 | 685,859 | 46.669 | 92.3 | 4.7 | 0.2 | 0.7 | 0.4 | 1.6 | 2.8 | 90.277 | 50.1 | 1.8 | 77.5 |
| District 16 | Adam Kinzinger (R) | 7,916.299 | 694,262 | 87.700 | 89.3 | 3.8 | 0.1 | 1.6 | 2.9 | 2.2 | 10.6 | 82.160 | 50.2 | 4.9 | 77.2 |
| District 17 | Cheri Bustos (D) | 6,930.338 | 666,201 | 96.128 | 81.3 | 12.4 | 0.3 | 1.0 | 2.3 | 2.8 | 9.6 | 74.800 | 50.6 | 4.8 | 72.4 |
| District 18 | Darin LaHood (R) | 10,515.434 | 702,289 | 66.786 | 90.3 | 3.8 | 0.1 | 3.0 | 0.6 | 2.3 | 2.9 | 88.139 | 51.4 | 3.4 | 78.2 |
| INDIANA | | 35,826.035 | 6,732,219 | 187.914 | 82.8 | 9.6 | 0.3 | 2.5 | 2.2 | 2.6 | 7.2 | 78.314 | 50.7 | 5.3 | 67.7 |
| District 1 | Frank J. Mrvan (D) | 1,156.727 | 719,122 | 621.687 | 71.1 | 18.5 | 0.3 | 1.6 | 5.9 | 2.7 | 16.3 | 62.231 | 50.9 | 5.5 | 58.5 |
| District 2 | Jackie Walorski (R) | 3,958.697 | 721,469 | 182.249 | 84.7 | 7.1 | 0.3 | 1.6 | 3.3 | 3.0 | 10.2 | 78.388 | 50.3 | 5.9 | 69.2 |
| District 3 | Jim Banks (R) | 4,180.113 | 753,051 | 180.151 | 86.4 | 6.1 | 0.4 | 2.7 | 1.7 | 2.8 | 6.3 | 82.370 | 50.7 | 5.1 | 71.6 |
| District 4 | James R. Baird (R) | 6,351.772 | 767,105 | 120.770 | 87.0 | 4.6 | 0.2 | 3.3 | 2.6 | 2.3 | 6.3 | 83.454 | 50.1 | 5.8 | 70.1 |
| District 5 | Victoria Spartz (R) | 1,924.969 | 791,257 | 411.049 | 84.3 | 8.6 | 0.2 | 3.8 | 0.8 | 2.3 | 5.1 | 80.389 | 50.7 | 5.9 | 64.1 |
| District 6 | Greg Pence (R) | 6,206.894 | 720,190 | 116.031 | 92.6 | 2.1 | 0.3 | 1.7 | 1.0 | 2.4 | 3.1 | 90.616 | 50.7 | 2.6 | 70.8 |
| District 7 | André Carson (D) | 304.049 | 777,205 | 2,556.183 | 57.9 | 30.7 | 0.3 | 4.2 | 3.3 | 3.5 | 11.5 | 50.474 | 51.6 | 10.1 | 66.4 |
| District 8 | Larry Bucshon (R) | 7,255.476 | 716,924 | 98.811 | 91.2 | 4.5 | 0.1 | 1.3 | 0.8 | 2.1 | 2.7 | 89.649 | 50.3 | 2.7 | 75.5 |
| District 9 | Trey Hollingsworth (R) | 4,487.338 | 765,896 | 170.679 | 90.7 | 3.0 | 0.2 | 2.7 | 1.0 | 2.3 | 3.6 | 88.138 | 50.8 | 4.0 | 63.8 |
| IOWA | | 55,853.421 | 3,155,070 | 56.488 | 89.9 | 4.1 | 0.4 | 2.5 | 1.0 | 2.2 | 6.3 | 85.092 | 50.2 | 5.6 | 69.7 |
| District 1 | Ashley Hinson (R) | 12,048.832 | 774,014 | 64.240 | 90.5 | 4.2 | 0.4 | 1.7 | 0.9 | 2.3 | 4.3 | 87.159 | 50.4 | 4.8 | 73.4 |
| District 2 | Mariannette Miller-Meeks (R) | 12,260.968 | 782,989 | 63.860 | 88.6 | 4.9 | 0.3 | 2.5 | 1.2 | 2.6 | 5.8 | 84.470 | 50.3 | 5.2 | 66.8 |
| District 3 | Cynthia Axne (D) | 8,788.817 | 848,170 | 96.505 | 88.2 | 5.2 | 0.3 | 3.8 | 0.8 | 1.7 | 7.2 | 82.430 | 50.7 | 7.6 | 67.3 |
| District 4 | Randy Feenstra (R) | 22,754.764 | 749,897 | 32.956 | 92.5 | 2.0 | 0.5 | 1.8 | 1.2 | 2.1 | 7.7 | 86.617 | 49.4 | 4.7 | 71.7 |
| KANSAS | | 81,758.594 | 2,913,314 | 35.633 | 83.6 | 5.7 | 0.8 | 3.1 | 3.0 | 3.7 | 12.2 | 75.395 | 50.4 | 7.2 | 59.2 |
| District 1 | Tracey Mann (R) | 52,542.054 | 694,498 | 13.218 | 87.9 | 2.8 | 0.9 | 1.8 | 4.0 | 2.7 | 16.4 | 76.388 | 49.8 | 7.6 | 64.2 |
| District 2 | Jake LaTurner (R) | 14,144.318 | 715,881 | 50.613 | 86.4 | 4.7 | 1.3 | 1.6 | 1.7 | 4.2 | 7.1 | 81.764 | 50.3 | 2.7 | 64.3 |
| District 3 | Sharice Davids (D) | 757.493 | 779,860 | 1,029.528 | 79.0 | 8.4 | 0.5 | 5.2 | 3.6 | 3.5 | 12.5 | 71.237 | 50.9 | 10.9 | 44.4 |
| District 4 | Ron Estes (R) | 14,314.729 | 723,075 | 50.513 | 81.9 | 6.7 | 0.9 | 3.4 | 2.6 | 4.5 | 12.9 | 72.619 | 50.4 | 7.1 | 65.1 |
| KENTUCKY | | 39,491.609 | 4,467,673 | 113.130 | 86.7 | 8.1 | 0.2 | 1.7 | 0.9 | 2.3 | 3.8 | 84.221 | 50.8 | 4.4 | 68.3 |
| District 1 | James Comer (R) | 12,084.391 | 717,704 | 59.391 | 88.7 | 7.0 | 0.4 | 0.8 | 0.8 | 2.3 | 3.1 | 86.918 | 50.2 | 1.8 | 68.0 |
| District 2 | Brett Guthrie (R) | 7,182.258 | 774,897 | 107.890 | 89.1 | 5.2 | 0.2 | 2.2 | 0.8 | 2.5 | 3.4 | 86.923 | 50.8 | 4.2 | 70.7 |
| District 3 | John A. Yarmuth (D) | 319.781 | 742,543 | 2,322.348 | 70.9 | 23.0 | 0.2 | 2.9 | 0.6 | 2.4 | 6.1 | 65.968 | 51.7 | 9.1 | 66.3 |
| District 4 | Thomas Massie (R) | 4,373.899 | 761,936 | 174.201 | 91.9 | 3.2 | 0.0 | 1.6 | 1.2 | 2.1 | 3.4 | 89.457 | 50.4 | 4.1 | 59.9 |
| District 5 | Harold Rogers (R) | 11,234.054 | 689,793 | 61.402 | 96.4 | 1.2 | 0.2 | 0.7 | 0.2 | 1.4 | 1.4 | 95.310 | 50.2 | 1.0 | 77.7 |
| District 6 | Garland "Andy" Barr (R) | 4,297.269 | 780,800 | 181.697 | 83.9 | 8.9 | 0.2 | 2.2 | 1.7 | 3.2 | 4.9 | 81.510 | 51.3 | 5.8 | 67.8 |
| LOUISIANA | | 43,210.209 | 4,648,794 | 107.585 | 61.8 | 32.4 | 0.6 | 1.8 | 1.4 | 2.0 | 5.4 | 58.245 | 51.2 | 4.2 | 77.6 |
| District 1 | Steve Scalise (R) | 4,032.671 | 799,917 | 198.359 | 79.4 | 14.0 | 1.1 | 2.2 | 0.9 | 2.3 | 8.9 | 72.191 | 52.1 | 7.0 | 73.2 |
| District 2 | Troy Carter (D) | 1,271.280 | 788,021 | 619.864 | 31.4 | 61.5 | 0.4 | 2.6 | 2.2 | 1.9 | 6.6 | 27.560 | 51.5 | 5.7 | 77.4 |
| District 3 | Clay Higgins (R) | 6,984.205 | 785,101 | 112.411 | 69.8 | 25.0 | 0.4 | 1.5 | 1.2 | 2.1 | 4.2 | 67.142 | 51.4 | 3.0 | 82.6 |
| District 4 | Mike Johnson (R) | 12,436.610 | 737,674 | 59.315 | 60.8 | 33.9 | 0.9 | 1.1 | 0.7 | 2.5 | 4.4 | 57.557 | 50.5 | 2.5 | 74.1 |
| District 5 | Julia Letlow (D) | 14,452.401 | 734,377 | 50.813 | 60.6 | 35.1 | 0.3 | 0.8 | 1.3 | 1.9 | 2.7 | 59.072 | 51.6 | 2.2 | 79.9 |
| District 6 | Garret Graves (R) | 4,033.062 | 803,704 | 199.279 | 68.5 | 25.2 | 0.4 | 2.3 | 2.2 | 1.4 | 5.2 | 65.638 | 51.6 | 4.6 | 78.3 |
| MAINE | | 30,845.096 | 1,344,212 | 43.579 | 94.0 | 1.6 | 0.7 | 1.1 | 0.4 | 2.1 | 1.7 | 92.822 | 51.2 | 3.9 | 61.6 |
| District 1 | Chellie Pingree (D) | 3,287.206 | 686,731 | 208.910 | 93.6 | 2.1 | 0.5 | 1.5 | 0.4 | 1.9 | 1.9 | 92.294 | 51.6 | 4.8 | 55.2 |
| District 2 | Jared F. Golden (D) | 27,557.890 | 657,481 | 23.858 | 94.4 | 1.1 | 1.0 | 0.8 | 0.4 | 2.3 | 1.5 | 93.373 | 50.8 | 3.0 | 68.3 |
| MARYLAND | | 9,711.202 | 6,045,680 | 622.547 | 54.5 | 30.3 | 0.3 | 6.4 | 5.1 | 3.4 | 10.6 | 49.820 | 51.6 | 15.4 | 47.4 |
| District 1 | Andrew Harris (R) | 3,977.403 | 737,341 | 185.383 | 81.8 | 12.4 | 0.3 | 2.2 | 1.0 | 2.3 | 4.6 | 78.738 | 50.9 | 5.1 | 61.9 |
| District 2 | C. A. Dutch Ruppersberger (D) | 348.651 | 750,702 | 2,153.162 | 50.9 | 36.9 | 0.2 | 5.4 | 3.0 | 3.5 | 7.7 | 47.167 | 52.5 | 11.6 | 61.3 |
| District 3 | John P. Sarbanes (D) | 304.529 | 779,502 | 2,559.697 | 61.7 | 23.5 | 0.5 | 7.4 | 2.8 | 4.1 | 9.4 | 56.127 | 52.1 | 15.4 | 49.0 |
| District 4 | Anthony G. Brown (D) | 297.788 | 758,795 | 2,548.105 | 28.5 | 52.4 | 0.2 | 3.0 | 12.8 | 3.1 | 17.5 | 24.548 | 51.3 | 20.8 | 30.9 |
| District 5 | Steny H. Hoyer (D) | 1,482.643 | 756,743 | 510.401 | 46.8 | 39.2 | 0.3 | 4.2 | 5.7 | 3.9 | 10.4 | 42.819 | 51.8 | 12.6 | 39.9 |
| District 6 | David J. Trone (D) | 1,952.343 | 769,046 | 393.909 | 66.3 | 14.3 | 0.3 | 11.8 | 3.1 | 4.2 | 13.1 | 57.820 | 50.9 | 20.4 | 42.9 |
| District 7 | Kweisi Mfume (D) | 488.212 | 717,158 | 1,468.942 | 35.7 | 52.5 | 0.5 | 7.6 | 1.2 | 2.6 | 4.0 | 33.412 | 52.1 | 11.3 | 61.0 |
| District 8 | Jamie Raskin (D) | 859.631 | 776,393 | 903.170 | 63.5 | 12.7 | 0.3 | 9.5 | 10.5 | 3.5 | 17.7 | 57.347 | 50.9 | 24.9 | 33.6 |
| MASSACHUSETTS | | 7,800.962 | 6,892,503 | 883.545 | 77.0 | 7.9 | 0.3 | 6.9 | 4.3 | 3.6 | 12.4 | 70.311 | 51.5 | 17.3 | 59.4 |
| District 1 | Richard E. Neal (D) | 2,350.130 | 723,831 | 307.996 | 83.8 | 6.6 | 0.2 | 2.2 | 3.6 | 3.6 | 19.3 | 70.605 | 51.4 | 7.4 | 65.9 |
| District 2 | James P. McGovern (D) | 1,627.997 | 759,750 | 466.678 | 83.1 | 5.2 | 0.3 | 6.1 | 2.1 | 3.3 | 10.4 | 75.699 | 51.8 | 13.0 | 62.6 |
| District 3 | Lori Trahan (D) | 757.634 | 771,723 | 1,018.596 | 80.4 | 4.1 | 0.3 | 7.5 | 11.4 | 2.7 | 21.3 | 64.995 | 49.9 | 19.2 | 59.5 |
| District 4 | Jake Auchincloss (D) | 668.286 | 765,466 | 1,145.417 | 83.7 | 3.5 | 0.2 | 7.2 | 2.1 | 3.3 | 5.0 | 80.722 | 51.7 | 15.2 | 58.4 |
| District 5 | Katherine Clark (D) | 265.131 | 768,043 | 2,896.843 | 83.7 | 5.6 | 0.2 | 13.4 | 3.8 | 3.2 | 10.0 | 67.595 | 51.1 | 24.6 | 50.2 |
| District 6 | Seth Moulton (D) | 526.855 | 770,998 | 1,463.397 | 84.9 | 4.1 | 0.2 | 4.5 | 3.7 | 2.6 | 10.5 | 79.259 | 52.2 | 14.6 | 67.7 |
| District 7 | Ayanna Pressley (D) | 62.715 | 819,035 | 13,059.635 | 49.5 | 25.4 | 0.3 | 10.8 | 5.5 | 8.4 | 22.4 | 40.096 | 51.6 | 31.7 | 41.4 |
| District 8 | Stephen F. Lynch (D) | 326.001 | 765,516 | 2,348.201 | 74.7 | 11.4 | 0.3 | 8.3 | 2.6 | 2.7 | 6.1 | 71.172 | 51.6 | 17.9 | 62.1 |
| District 9 | William R. Keating (D) | 1,216.213 | 748,141 | 615.140 | 87.9 | 3.7 | 0.5 | 1.5 | 4.0 | 2.6 | 6.0 | 85.146 | 51.9 | 10.0 | 68.3 |

1. Dry land or land partially or temporarily covered by water.   2. May be of any race.

# Table E. Congressional Districts 116th Congress — Age and Education

| STATE District | Population and population characteristics, 2019 (cont.) | | | | | | | | | | | Education, 2019 | |
| | Age (percent) | | | | | | | | | | Total Enrollment[1] | Attainment[2] (percent) | |
| | Under 5 years | 5 to 17 years | 18 to 24 years | 25 to 34 years | 35 to 44 years | 45 to 54 years | 55 to 64 years | 65 to 74 years | 75 years and over | Median age | | High school graduate or more | Bachelor's degree or more |
| | 15 | 16 | 17 | 18 | 19 | 20 | 21 | 22 | 23 | 24 | 25 | 26 | 27 |
| ILLINOIS—Cont'd | | | | | | | | | | | | | |
| District 3 | 5.9 | 17.0 | 8.2 | 12.7 | 13.7 | 12.5 | 13.5 | 9.7 | 6.7 | 40.0 | 171,970 | 87.5 | 31.2 |
| District 4 | 6.0 | 17.0 | 9.9 | 18.8 | 14.7 | 11.6 | 10.8 | 7.0 | 4.2 | 34.0 | 171,452 | 77.0 | 27.8 |
| District 5 | 6.0 | 13.6 | 8.1 | 20.6 | 15.2 | 12.9 | 10.6 | 7.2 | 5.7 | 35.8 | 164,232 | 92.7 | 56.8 |
| District 6 | 5.6 | 17.0 | 7.8 | 10.6 | 12.7 | 13.8 | 15.4 | 10.1 | 7.1 | 42.1 | 181,930 | 95.5 | 53.2 |
| District 7 | 5.7 | 14.1 | 9.4 | 20.6 | 14.2 | 10.9 | 11.2 | 8.4 | 5.6 | 35.1 | 170,571 | 86.9 | 44.9 |
| District 8 | 6.8 | 16.9 | 8.3 | 14.9 | 13.1 | 13.4 | 12.9 | 8.1 | 5.6 | 37.3 | 175,067 | 88.0 | 35.0 |
| District 9 | 6.1 | 14.7 | 8.1 | 13.9 | 13.2 | 13.2 | 12.7 | 10.0 | 8.2 | 40.3 | 180,417 | 91.7 | 55.7 |
| District 10 | 5.7 | 17.7 | 9.7 | 11.8 | 12.8 | 13.0 | 13.4 | 9.2 | 6.6 | 38.7 | 181,466 | 88.8 | 44.2 |
| District 11 | 5.9 | 18.6 | 9.7 | 12.6 | 14.4 | 13.9 | 12.0 | 7.6 | 5.2 | 37.2 | 190,654 | 88.1 | 37.5 |
| District 12 | 5.8 | 15.7 | 8.6 | 13.1 | 12.1 | 12.3 | 14.1 | 10.5 | 7.7 | 40.4 | 158,106 | 90.3 | 23.9 |
| District 13 | 5.0 | 14.7 | 15.4 | 12.7 | 11.4 | 11.1 | 13.0 | 9.5 | 7.1 | 36.8 | 204,685 | 92.6 | 31.3 |
| District 14 | 5.7 | 19.4 | 7.6 | 10.9 | 13.2 | 14.7 | 13.6 | 8.9 | 5.9 | 39.8 | 194,829 | 94.2 | 42.6 |
| District 15 | 5.8 | 16.2 | 8.1 | 11.7 | 12.3 | 12.1 | 14.5 | 10.7 | 8.5 | 41.4 | 151,842 | 90.4 | 20.9 |
| District 16 | 5.5 | 15.9 | 9.5 | 11.7 | 11.8 | 11.8 | 14.0 | 10.6 | 8.2 | 41.2 | 160,587 | 91.0 | 23.5 |
| District 17 | 5.9 | 16.1 | 9.1 | 11.8 | 12.2 | 11.8 | 14.0 | 11.1 | 8.2 | 40.7 | 148,638 | 89.0 | 20.2 |
| District 18 | 6.2 | 16.8 | 8.2 | 11.9 | 12.4 | 11.9 | 13.1 | 10.9 | 8.6 | 40.7 | 171,066 | 94.0 | 33.6 |
| INDIANA | 6.2 | 17.1 | 9.8 | 13.2 | 12.2 | 12.3 | 13.0 | 9.5 | 6.6 | 38.0 | 1,646,376 | 89.6 | 26.9 |
| District 1 | 5.9 | 16.9 | 8.7 | 12.4 | 12.9 | 12.4 | 13.8 | 10.0 | 6.9 | 39.9 | 168,561 | 90.8 | 23.2 |
| District 2 | 6.1 | 18.1 | 9.6 | 12.5 | 11.4 | 12.4 | 13.0 | 9.6 | 7.2 | 38.2 | 178,187 | 87.1 | 22.5 |
| District 3 | 6.9 | 18.5 | 8.8 | 13.2 | 12.0 | 11.8 | 12.9 | 9.5 | 6.4 | 37.4 | 178,295 | 87.7 | 24.5 |
| District 4 | 5.7 | 16.9 | 12.3 | 13.2 | 11.6 | 12.1 | 12.6 | 9.1 | 6.5 | 36.7 | 206,110 | 90.3 | 26.6 |
| District 5 | 6.0 | 17.9 | 8.2 | 13.1 | 13.4 | 13.3 | 12.7 | 8.9 | 6.5 | 38.3 | 193,455 | 94.3 | 47.9 |
| District 6 | 5.7 | 16.0 | 9.8 | 12.3 | 11.5 | 12.8 | 13.9 | 10.3 | 7.7 | 40.5 | 166,367 | 88.9 | 21.6 |
| District 7 | 7.5 | 18.1 | 9.7 | 16.6 | 13.1 | 11.4 | 11.4 | 7.5 | 4.8 | 33.9 | 193,189 | 84.9 | 25.0 |
| District 8 | 5.9 | 16.4 | 9.4 | 12.5 | 11.9 | 11.8 | 14.1 | 10.5 | 7.6 | 40.3 | 162,674 | 90.2 | 22.1 |
| District 9 | 5.8 | 15.6 | 11.9 | 12.5 | 12.4 | 12.2 | 13.1 | 10.0 | 6.5 | 38.4 | 199,538 | 91.7 | 27.1 |
| IOWA | 6.0 | 16.9 | 9.8 | 12.7 | 12.4 | 11.4 | 13.3 | 9.9 | 7.7 | 38.5 | 783,468 | 92.6 | 29.3 |
| District 1 | 6.0 | 16.6 | 9.9 | 12.1 | 11.8 | 11.6 | 13.7 | 10.2 | 8.2 | 39.2 | 188,647 | 93.4 | 27.6 |
| District 2 | 5.8 | 16.2 | 10.5 | 12.7 | 12.1 | 11.5 | 13.5 | 10.1 | 7.6 | 39.0 | 197,718 | 92.2 | 28.4 |
| District 3 | 6.5 | 17.9 | 8.4 | 13.7 | 13.9 | 12.0 | 12.3 | 8.9 | 6.3 | 37.4 | 204,634 | 92.7 | 35.2 |
| District 4 | 5.8 | 16.6 | 10.6 | 12.2 | 11.8 | 10.3 | 13.8 | 10.4 | 8.7 | 38.9 | 192,469 | 92.2 | 25.4 |
| KANSAS | 6.3 | 17.7 | 10.1 | 13.0 | 12.4 | 11.2 | 12.7 | 9.4 | 7.0 | 37.2 | 753,423 | 91.8 | 34.0 |
| District 1 | 6.3 | 17.1 | 12.2 | 12.5 | 11.8 | 9.9 | 12.5 | 9.5 | 8.0 | 36.3 | 183,362 | 89.7 | 26.1 |
| District 2 | 5.7 | 17.0 | 11.6 | 12.0 | 11.4 | 11.3 | 13.4 | 10.3 | 7.5 | 38.5 | 183,480 | 93.3 | 29.4 |
| District 3 | 6.6 | 18.3 | 8.0 | 13.9 | 14.1 | 12.3 | 12.3 | 8.7 | 5.8 | 37.1 | 197,744 | 93.0 | 48.5 |
| District 4 | 6.6 | 18.5 | 8.9 | 13.3 | 12.3 | 11.2 | 13.0 | 9.2 | 6.9 | 36.9 | 188,837 | 90.8 | 30.0 |
| KENTUCKY | 6.1 | 16.3 | 9.3 | 12.9 | 12.4 | 12.6 | 13.4 | 10.2 | 6.7 | 39.2 | 1,025,269 | 87.2 | 25.1 |
| District 1 | 5.9 | 15.8 | 9.5 | 12.2 | 11.9 | 12.1 | 13.7 | 11.2 | 7.7 | 40.4 | 152,277 | 86.3 | 17.2 |
| District 2 | 6.2 | 17.1 | 9.6 | 12.6 | 12.5 | 12.4 | 13.3 | 9.8 | 6.6 | 38.8 | 183,615 | 87.4 | 21.9 |
| District 3 | 6.1 | 15.7 | 8.6 | 14.8 | 12.7 | 11.9 | 13.4 | 9.9 | 6.8 | 38.3 | 166,854 | 91.3 | 34.4 |
| District 4 | 6.2 | 17.2 | 8.4 | 12.3 | 12.7 | 13.6 | 13.5 | 10.1 | 6.1 | 39.7 | 177,273 | 89.5 | 29.0 |
| District 5 | 6.0 | 16.1 | 8.1 | 12.1 | 12.6 | 13.1 | 13.8 | 11.0 | 7.2 | 41.3 | 147,459 | 78.0 | 13.8 |
| District 6 | 6.3 | 15.9 | 11.6 | 13.4 | 12.6 | 12.4 | 12.7 | 9.3 | 6.0 | 37.2 | 197,791 | 90.0 | 33.3 |
| LOUISIANA | 6.3 | 17.0 | 9.1 | 13.8 | 13.0 | 11.7 | 13.0 | 9.6 | 6.4 | 37.7 | 1,149,360 | 86.0 | 25.0 |
| District 1 | 5.9 | 17.0 | 8.2 | 12.9 | 13.2 | 12.0 | 13.5 | 10.4 | 6.9 | 39.1 | 195,711 | 87.9 | 32.2 |
| District 2 | 6.4 | 15.9 | 8.4 | 15.6 | 13.2 | 11.4 | 13.9 | 9.7 | 5.7 | 37.6 | 195,958 | 84.6 | 25.5 |
| District 3 | 6.7 | 17.9 | 8.6 | 14.0 | 13.0 | 11.7 | 12.9 | 9.0 | 6.1 | 36.8 | 190,827 | 85.5 | 23.3 |
| District 4 | 6.6 | 16.8 | 9.1 | 13.1 | 12.5 | 11.9 | 12.7 | 10.2 | 7.2 | 38.3 | 174,180 | 85.3 | 19.3 |
| District 5 | 6.2 | 17.2 | 9.6 | 12.6 | 13.1 | 11.7 | 13.3 | 9.4 | 6.9 | 38.3 | 182,739 | 83.0 | 18.2 |
| District 6 | 6.1 | 17.3 | 10.8 | 14.7 | 13.5 | 11.7 | 11.6 | 8.9 | 5.5 | 35.7 | 209,945 | 89.1 | 30.2 |
| MAINE | 4.7 | 13.6 | 7.9 | 12.2 | 11.5 | 13.2 | 15.7 | 12.6 | 8.7 | 45.1 | 275,036 | 93.2 | 33.2 |
| District 1 | 4.5 | 13.5 | 7.8 | 13.0 | 11.6 | 13.1 | 15.4 | 12.3 | 8.6 | 44.5 | 137,860 | 94.1 | 40.5 |
| District 2 | 4.8 | 13.8 | 8.1 | 11.4 | 11.2 | 13.1 | 16.0 | 12.9 | 8.7 | 45.7 | 137,176 | 92.3 | 25.4 |
| MARYLAND | 5.9 | 16.1 | 8.8 | 13.6 | 13.0 | 13.2 | 13.5 | 9.3 | 6.5 | 39.0 | 1,509,619 | 90.4 | 40.9 |
| District 1 | 4.9 | 16.0 | 8.8 | 11.5 | 11.1 | 13.3 | 15.0 | 11.3 | 8.1 | 43.0 | 181,018 | 91.2 | 33.1 |
| District 2 | 6.5 | 16.1 | 8.8 | 15.5 | 13.4 | 12.3 | 12.8 | 8.9 | 5.7 | 37.1 | 187,176 | 90.2 | 32.9 |
| District 3 | 6.6 | 16.0 | 9.0 | 15.4 | 13.4 | 12.5 | 12.4 | 8.4 | 6.4 | 36.8 | 198,496 | 92.6 | 52.2 |
| District 4 | 6.8 | 16.4 | 7.7 | 14.6 | 14.0 | 13.0 | 13.4 | 8.6 | 5.6 | 38.0 | 186,376 | 87.5 | 35.6 |
| District 5 | 5.7 | 16.3 | 10.2 | 12.6 | 12.4 | 14.1 | 14.1 | 9.2 | 5.4 | 39.1 | 199,130 | 91.4 | 34.3 |
| District 6 | 5.6 | 16.7 | 8.1 | 12.8 | 13.8 | 13.7 | 13.8 | 9.2 | 6.2 | 39.6 | 189,597 | 90.2 | 43.3 |
| District 7 | 5.2 | 15.2 | 9.3 | 14.3 | 12.7 | 12.5 | 14.4 | 9.2 | 7.1 | 39.2 | 176,948 | 89.2 | 40.1 |
| District 8 | 6.0 | 16.2 | 8.2 | 11.9 | 13.3 | 13.8 | 12.9 | 10.0 | 7.9 | 40.5 | 190,878 | 90.8 | 54.4 |
| MASSACHUSETTS | 5.2 | 14.5 | 10.0 | 14.4 | 12.4 | 13.0 | 13.6 | 9.8 | 7.2 | 39.7 | 1,691,491 | 91.3 | 45.0 |
| District 1 | 5.0 | 15.3 | 9.9 | 12.4 | 11.7 | 12.5 | 14.4 | 11.1 | 7.7 | 41.1 | 175,725 | 89.1 | 30.8 |
| District 2 | 4.9 | 14.2 | 12.4 | 12.5 | 11.4 | 13.8 | 14.0 | 10.0 | 6.7 | 40.3 | 201,417 | 92.5 | 40.7 |
| District 3 | 5.3 | 16.0 | 8.9 | 13.8 | 12.9 | 13.6 | 14.4 | 8.8 | 6.2 | 39.6 | 187,252 | 89.9 | 38.4 |
| District 4 | 5.1 | 16.3 | 9.5 | 11.7 | 12.2 | 14.1 | 14.4 | 9.7 | 7.0 | 41.1 | 193,303 | 93.2 | 52.0 |
| District 5 | 6.0 | 13.7 | 9.6 | 16.1 | 13.2 | 12.6 | 12.6 | 9.4 | 6.6 | 38.2 | 190,395 | 94.3 | 60.1 |
| District 6 | 5.2 | 15.0 | 8.5 | 11.3 | 12.2 | 13.9 | 14.9 | 10.7 | 8.4 | 43.0 | 175,942 | 92.6 | 47.4 |
| District 7 | 5.0 | 12.0 | 15.2 | 22.9 | 13.2 | 10.6 | 9.7 | 6.4 | 5.1 | 32.3 | 242,375 | 86.7 | 47.1 |
| District 8 | 5.0 | 13.7 | 7.7 | 16.7 | 12.7 | 13.1 | 13.9 | 9.6 | 7.6 | 40.1 | 168,932 | 91.9 | 49.2 |
| District 9 | 4.9 | 14.2 | 8.0 | 11.3 | 11.1 | 12.9 | 14.9 | 13.2 | 9.6 | 45.5 | 156,150 | 91.9 | 37.8 |

1. All persons 3 years old and over enrolled in nursery school through college and graduate or professional school.  2. Persons 25 years old and over.

Items 15—27

# Table E. Congressional Districts 116th Congress — **Households and Group Quarters**

| STATE District | Households, 2019 | | | | | | Group Quarters, 2010 | | | | | |
| --- | --- | --- | --- | --- | --- | --- | --- | --- | --- | --- | --- | --- |
| | Number | Average household size | Family households (percent) | Married couple family (percent) | Female family house-holder[1] | One person households (percent) | Total in group quarters, 2018 | Percent 65 years and over | Persons in correctional institutions | Persons in nursing facilities | Persons in college dormitories | Persons in military quarters |
| | 28 | 29 | 30 | 31 | 32 | 33 | 34 | 35 | 36 | 37 | 38 | 39 |
| **ILLINOIS—Cont'd** | | | | | | | | | | | | |
| District 3 | 245,983 | 2.81 | 68.7 | 51.5 | 12.1 | 27.1 | 10,554 | 30.2 | 3,160 | 3,740 | 1,890 | 0 |
| District 4 | 227,057 | 2.96 | 64.2 | 41.3 | 14.5 | 25.9 | 3,694 | 34.4 | 1 | 1,268 | 285 | 0 |
| District 5 | 303,552 | 2.40 | 53.8 | 42.1 | 8.1 | 34.1 | 10,018 | 32.7 | 0 | 3,882 | 4,823 | 0 |
| District 6 | 266,761 | 2.63 | 71.4 | 60.3 | 7.8 | 24.4 | 9,380 | 36.8 | 786 | 4,034 | 3,517 | 16 |
| District 7 | 306,915 | 2.28 | 48.4 | 29.3 | 14.9 | 43.0 | 28,136 | 6.3 | 11,612 | 3,240 | 7,992 | 0 |
| District 8 | 255,684 | 2.78 | 68.1 | 50.8 | 12.0 | 26.4 | 5,366 | 45.7 | 0 | 2,564 | 744 | 0 |
| District 9 | 285,485 | 2.44 | 59.0 | 48.0 | 7.6 | 34.1 | 23,299 | 27.4 | 31 | 8,543 | 8,803 | 12,155 |
| District 10 | 253,447 | 2.70 | 71.1 | 55.5 | 10.9 | 24.7 | 20,645 | 19.5 | 704 | 5,665 | 2,088 | 0 |
| District 11 | 250,397 | 2.85 | 69.2 | 51.9 | 12.1 | 25.6 | 7,043 | 33.9 | 1,085 | 2,891 | 1,495 | 0 |
| District 12 | 282,834 | 2.33 | 61.6 | 43.2 | 12.8 | 32.4 | 19,056 | 22.6 | 10,700 | 5,209 | 3,332 | 307 |
| District 13 | 283,501 | 2.32 | 55.2 | 39.8 | 11.2 | 36.2 | 41,671 | 12.5 | 5,125 | 5,481 | 27,454 | 0 |
| District 14 | 256,715 | 2.81 | 74.0 | 62.1 | 8.2 | 20.9 | 5,012 | 33.0 | 1,364 | 1,562 | 158 | 0 |
| District 15 | 277,166 | 2.38 | 65.5 | 51.2 | 10.1 | 29.5 | 26,946 | 23.2 | 12,642 | 6,376 | 3,900 | 0 |
| District 16 | 278,976 | 2.42 | 63.8 | 49.0 | 9.6 | 29.8 | 18,512 | 26.0 | 6,821 | 5,619 | 6,178 | 0 |
| District 17 | 279,176 | 2.30 | 59.5 | 41.4 | 13.7 | 33.9 | 24,532 | 21.2 | 8,864 | 5,921 | 5,835 | 0 |
| District 18 | 280,748 | 2.43 | 64.8 | 51.7 | 9.1 | 29.7 | 19,233 | 27.7 | 7,248 | 6,421 | 6,455 | 5 |
| **INDIANA** | 2,597,765 | 2.52 | 63.1 | 47.0 | 11.0 | 30.4 | 190,079 | 20.5 | 48,694 | 41,158 | 75,434 | 228 |
| District 1 | 279,194 | 2.52 | 65.5 | 44.8 | 14.0 | 28.5 | 16,703 | 19.0 | 7,380 | 3,116 | 3,194 | 0 |
| District 2 | 268,256 | 2.60 | 63.2 | 47.7 | 11.2 | 31.5 | 23,363 | 20.0 | 5,936 | 4,482 | 8,659 | 0 |
| District 3 | 291,660 | 2.54 | 66.1 | 50.3 | 9.7 | 28.2 | 13,485 | 33.2 | 2,229 | 4,852 | 3,948 | 0 |
| District 4 | 294,928 | 2.50 | 62.6 | 48.0 | 9.6 | 29.9 | 31,256 | 15.3 | 7,233 | 4,934 | 15,651 | 0 |
| District 5 | 313,075 | 2.48 | 66.1 | 53.7 | 8.8 | 28.1 | 15,890 | 25.1 | 4,416 | 4,444 | 5,257 | 178 |
| District 6 | 280,444 | 2.49 | 63.8 | 47.0 | 11.0 | 28.9 | 21,774 | 23.8 | 5,096 | 5,532 | 8,541 | 0 |
| District 7 | 296,794 | 2.57 | 52.1 | 33.3 | 13.9 | 40.1 | 14,357 | 18.3 | 3,216 | 3,152 | 5,162 | 50 |
| District 8 | 286,330 | 2.40 | 64.0 | 48.8 | 10.7 | 30.3 | 28,584 | 19.4 | 10,561 | 5,980 | 10,509 | 0 |
| District 9 | 287,084 | 2.58 | 64.7 | 49.3 | 10.5 | 27.7 | 24,667 | 17.9 | 2,627 | 4,666 | 14,513 | 0 |
| **IOWA** | 1,287,221 | 2.38 | 62.8 | 48.9 | 9.6 | 29.6 | 96,609 | 26.1 | 13,309 | 26,871 | 44,574 | 3 |
| District 1 | 314,430 | 2.38 | 61.8 | 48.5 | 9.3 | 30.5 | 24,804 | 26.1 | 2,415 | 7,519 | 13,934 | 3 |
| District 2 | 316,300 | 2.40 | 63.1 | 49.5 | 9.2 | 29.1 | 23,501 | 21.8 | 5,185 | 5,702 | 11,804 | 0 |
| District 3 | 342,025 | 2.43 | 64.9 | 48.6 | 11.2 | 28.1 | 16,687 | 27.7 | 3,690 | 5,218 | 5,248 | 0 |
| District 4 | 314,466 | 2.28 | 61.4 | 48.8 | 8.4 | 31.0 | 31,617 | 29.1 | 2,019 | 8,432 | 13,588 | 0 |
| **KANSAS** | 1,138,329 | 2.49 | 64.4 | 50.2 | 9.7 | 29.1 | 80,546 | 24.4 | 18,009 | 20,672 | 27,754 | 3,943 |
| District 1 | 272,404 | 2.44 | 62.6 | 50.4 | 8.0 | 30.9 | 30,972 | 23.2 | 5,306 | 7,084 | 10,873 | 3,425 |
| District 2 | 286,494 | 2.40 | 63.2 | 48.6 | 9.7 | 28.6 | 29,061 | 17.4 | 7,436 | 5,478 | 12,652 | 192 |
| District 3 | 297,879 | 2.60 | 67.5 | 53.2 | 10.0 | 26.6 | 6,271 | 49.6 | 1,355 | 3,598 | 544 | 0 |
| District 4 | 281,552 | 2.52 | 64.2 | 48.4 | 11.1 | 30.5 | 14,242 | 29.4 | 3,912 | 4,512 | 3,685 | 326 |
| **KENTUCKY** | 1,748,732 | 2.48 | 64.6 | 47.5 | 11.9 | 29.1 | 132,667 | 19.2 | 41,122 | 26,044 | 36,340 | 5,856 |
| District 1 | 284,068 | 2.43 | 65.2 | 48.8 | 10.7 | 29.2 | 27,512 | 20.4 | 7,869 | 5,548 | 4,101 | 3,843 |
| District 2 | 293,329 | 2.56 | 67.7 | 51.0 | 10.9 | 27.1 | 22,547 | 21.3 | 4,501 | 4,378 | 7,527 | 2,013 |
| District 3 | 307,525 | 2.36 | 58.0 | 40.0 | 13.8 | 34.3 | 15,942 | 30.7 | 2,662 | 4,723 | 3,307 | 0 |
| District 4 | 287,996 | 2.59 | 67.7 | 51.7 | 11.0 | 25.8 | 16,048 | 21.9 | 9,255 | 3,657 | 1,768 | 0 |
| District 5 | 266,136 | 2.50 | 67.9 | 49.3 | 12.9 | 28.0 | 24,577 | 15.1 | 12,451 | 4,570 | 5,039 | 0 |
| District 6 | 309,678 | 2.44 | 61.9 | 44.8 | 12.2 | 30.1 | 26,041 | 12.1 | 4,384 | 3,168 | 14,598 | 0 |
| **LOUISIANA** | 1,741,076 | 2.60 | 64.0 | 42.3 | 16.4 | 30.2 | 128,857 | 16.7 | 60,804 | 24,524 | 24,891 | 2,861 |
| District 1 | 299,885 | 2.62 | 65.9 | 47.5 | 13.8 | 28.1 | 13,516 | 24.4 | 1,182 | 2,957 | 5,469 | 57 |
| District 2 | 296,502 | 2.60 | 55.2 | 30.3 | 19.5 | 39.0 | 17,475 | 12.0 | 11,687 | 3,058 | 4,007 | 138 |
| District 3 | 297,618 | 2.60 | 67.6 | 46.3 | 15.9 | 26.5 | 12,552 | 26.9 | 4,329 | 4,273 | 2,935 | 0 |
| District 4 | 284,771 | 2.51 | 65.0 | 42.4 | 17.2 | 30.1 | 23,813 | 18.4 | 13,410 | 5,425 | 6,333 | 2,524 |
| District 5 | 269,916 | 2.53 | 64.3 | 40.6 | 17.8 | 30.8 | 50,477 | 11.6 | 28,559 | 5,925 | 3,757 | 142 |
| District 6 | 292,384 | 2.71 | 66.1 | 46.7 | 14.3 | 27.0 | 11,024 | 23.4 | 1,637 | 2,886 | 2,390 | 0 |
| **MAINE** | 573,618 | 2.28 | 59.8 | 46.8 | 9.0 | 31.2 | 37,206 | 22.5 | 3,679 | 7,878 | 17,251 | 131 |
| District 1 | 293,012 | 2.28 | 58.9 | 47.1 | 8.5 | 31.8 | 18,341 | 24.5 | 2,708 | 4,128 | 7,433 | 119 |
| District 2 | 280,606 | 2.28 | 60.7 | 46.4 | 9.5 | 30.6 | 18,865 | 20.4 | 971 | 3,750 | 9,818 | 12 |
| **MARYLAND** | 2,226,767 | 2.65 | 65.7 | 47.2 | 13.6 | 28.2 | 140,499 | 19.4 | 35,832 | 28,001 | 48,141 | 7,534 |
| District 1 | 280,274 | 2.57 | 68.2 | 54.0 | 9.8 | 25.6 | 17,255 | 20.6 | 5,648 | 3,465 | 5,803 | 7 |
| District 2 | 281,891 | 2.59 | 63.4 | 42.2 | 15.7 | 30.0 | 19,717 | 17.1 | 7,411 | 3,533 | 3,971 | 2,113 |
| District 3 | 292,767 | 2.60 | 64.6 | 47.6 | 12.8 | 27.7 | 19,429 | 19.9 | 583 | 3,399 | 6,910 | 4,515 |
| District 4 | 274,009 | 2.75 | 65.1 | 41.2 | 17.7 | 29.6 | 5,245 | 42.2 | 1,149 | 2,597 | 410 | 115 |
| District 5 | 261,745 | 2.82 | 70.9 | 51.0 | 14.9 | 24.5 | 19,893 | 11.8 | 1,111 | 2,598 | 14,602 | 387 |
| District 6 | 274,499 | 2.72 | 68.9 | 53.2 | 11.8 | 25.3 | 22,780 | 17.6 | 12,569 | 3,716 | 2,766 | 93 |
| District 7 | 279,322 | 2.48 | 56.7 | 34.7 | 17.4 | 36.5 | 25,718 | 14.8 | 6,305 | 4,677 | 10,895 | 0 |
| District 8 | 282,260 | 2.71 | 67.9 | 54.1 | 9.3 | 25.9 | 10,462 | 37.8 | 1,056 | 4,016 | 2,784 | 304 |
| **MASSACHUSETTS** | 2,650,680 | 2.51 | 62.8 | 46.7 | 11.5 | 28.6 | 246,195 | 17.7 | 24,683 | 43,833 | 135,773 | 498 |
| District 1 | 287,065 | 2.43 | 62.3 | 41.3 | 14.6 | 30.8 | 25,114 | 20.9 | 1,897 | 5,855 | 13,226 | 0 |
| District 2 | 289,018 | 2.48 | 64.2 | 48.7 | 10.7 | 27.1 | 43,933 | 13.9 | 1,576 | 5,572 | 26,184 | 0 |
| District 3 | 287,751 | 2.62 | 66.1 | 48.1 | 12.9 | 26.6 | 18,252 | 19.7 | 6,829 | 3,823 | 4,944 | 0 |
| District 4 | 282,691 | 2.61 | 69.7 | 57.0 | 9.6 | 23.4 | 26,980 | 19.7 | 2,520 | 4,914 | 15,480 | 25 |
| District 5 | 296,859 | 2.50 | 61.9 | 49.6 | 8.4 | 28.1 | 27,207 | 16.5 | 771 | 4,440 | 18,907 | 113 |
| District 6 | 294,824 | 2.55 | 67.3 | 52.7 | 10.5 | 26.5 | 18,729 | 28.9 | 2,014 | 4,919 | 5,965 | 73 |
| District 7 | 302,369 | 2.52 | 49.7 | 29.9 | 14.9 | 34.1 | 55,669 | 4.7 | 1,895 | 2,728 | 42,309 | 26 |
| District 8 | 304,233 | 2.46 | 60.0 | 45.7 | 10.1 | 31.2 | 15,986 | 35.8 | 3,879 | 6,322 | 3,522 | 26 |
| District 9 | 305,870 | 2.40 | 64.9 | 48.2 | 11.4 | 29.1 | 14,325 | 30.2 | 3,302 | 5,260 | 5,236 | 26 |

1. No spouse present.

# Table E. Congressional Districts 116th Congress — **Housing and Money Income**

| STATE District | Housing units, 2019 — Occupied units — Total | Occupied units as a percent of all units | Owner-occupied — Owner-occupied units as a percent of occupied units | Owner-occupied — Median value[1] (dollars) | Owner-occupied — Percent valued at $500,000 or more | Renter-occupied — Median rent[2] | Money income, 2019 — Households — Per capita income (dollars) | Households — Median income (dollars) | Households — Percent with income of $100,000 or more |
|---|---|---|---|---|---|---|---|---|---|
| | 40 | 41 | 42 | 43 | 44 | 45 | 46 | 47 | 48 |
| **ILLINOIS—Cont'd** | | | | | | | | | |
| District 3 | 263,934 | 93.2 | 75.6 | 251,400 | 8.7 | 1,055 | 35,311 | 75,411 | 36.7 |
| District 4 | 248,819 | 91.3 | 45.7 | 264,900 | 11.1 | 1,054 | 28,606 | 59,548 | 27.1 |
| District 5 | 327,459 | 92.7 | 55.1 | 370,800 | 30.8 | 1,396 | 57,264 | 94,144 | 47.5 |
| District 6 | 280,012 | 95.3 | 80.7 | 334,200 | 20.6 | 1,360 | 53,192 | 105,292 | 53.4 |
| District 7 | 358,847 | 85.5 | 41.8 | 293,700 | 21.8 | 1,217 | 47,632 | 64,312 | 35.0 |
| District 8 | 268,928 | 95.1 | 67.7 | 240,300 | 4.2 | 1,271 | 34,575 | 77,991 | 36.1 |
| District 9 | 310,591 | 91.9 | 60.8 | 350,300 | 25.7 | 1,198 | 47,825 | 78,569 | 40.8 |
| District 10 | 270,820 | 93.6 | 69.8 | 274,500 | 23.3 | 1,236 | 45,923 | 84,608 | 43.0 |
| District 11 | 268,547 | 93.2 | 71.7 | 233,100 | 7.0 | 1,224 | 37,080 | 81,598 | 39.7 |
| District 12 | 325,979 | 86.8 | 67.0 | 116,300 | 1.3 | 736 | 30,152 | 53,008 | 22.2 |
| District 13 | 326,650 | 86.8 | 62.9 | 128,000 | 2.2 | 811 | 30,717 | 53,578 | 24.0 |
| District 14 | 270,110 | 95.0 | 81.8 | 274,900 | 7.6 | 1,185 | 43,820 | 100,011 | 50.0 |
| District 15 | 316,888 | 87.5 | 75.8 | 109,400 | 1.8 | 681 | 30,254 | 56,268 | 23.4 |
| District 16 | 303,428 | 91.9 | 73.2 | 146,600 | 1.7 | 844 | 32,354 | 62,868 | 28.0 |
| District 17 | 321,437 | 86.9 | 66.5 | 107,100 | 1.5 | 705 | 27,618 | 50,346 | 19.2 |
| District 18 | 314,942 | 89.1 | 75.3 | 149,400 | 2.4 | 768 | 35,274 | 67,284 | 30.8 |
| **INDIANA** | 2,921,115 | 88.9 | 69.3 | 156,000 | 3.9 | 840 | 30,988 | 57,603 | 24.8 |
| District 1 | 311,836 | 89.5 | 70.5 | 171,100 | 3.5 | 881 | 31,641 | 61,104 | 28.0 |
| District 2 | 310,599 | 86.4 | 72.5 | 141,700 | 3.1 | 812 | 27,938 | 54,788 | 22.3 |
| District 3 | 327,347 | 89.1 | 73.4 | 144,800 | 3.9 | 747 | 29,721 | 57,288 | 22.8 |
| District 4 | 324,018 | 91.0 | 69.0 | 155,100 | 2.2 | 854 | 30,840 | 60,119 | 24.6 |
| District 5 | 338,362 | 92.5 | 71.5 | 219,100 | 10.8 | 1,005 | 43,837 | 76,417 | 38.4 |
| District 6 | 318,656 | 88.0 | 72.0 | 142,400 | 2.8 | 745 | 28,919 | 55,959 | 22.4 |
| District 7 | 338,317 | 87.7 | 51.2 | 136,500 | 2.2 | 877 | 25,074 | 46,118 | 16.1 |
| District 8 | 323,834 | 88.4 | 71.1 | 128,800 | 2.6 | 731 | 29,423 | 54,326 | 21.2 |
| District 9 | 328,146 | 87.5 | 73.3 | 167,500 | 3.3 | 858 | 30,776 | 60,448 | 26.1 |
| **IOWA** | 1,418,600 | 90.7 | 70.5 | 158,900 | 4.0 | 808 | 33,107 | 61,691 | 26.2 |
| District 1 | 346,740 | 90.7 | 73.5 | 156,200 | 3.5 | 760 | 32,576 | 61,542 | 25.6 |
| District 2 | 350,935 | 90.1 | 68.4 | 154,600 | 3.9 | 793 | 32,724 | 59,569 | 25.0 |
| District 3 | 366,626 | 93.3 | 69.2 | 187,800 | 5.5 | 908 | 35,769 | 67,681 | 31.1 |
| District 4 | 354,299 | 88.8 | 71.1 | 134,000 | 3.0 | 733 | 31,042 | 58,270 | 22.8 |
| **KANSAS** | 1,288,430 | 88.4 | 66.5 | 163,200 | 5.0 | 862 | 32,885 | 62,087 | 27.2 |
| District 1 | 323,644 | 84.2 | 66.5 | 128,600 | 2.1 | 748 | 28,195 | 55,188 | 21.2 |
| District 2 | 323,252 | 88.6 | 66.2 | 139,200 | 3.0 | 789 | 29,886 | 56,515 | 23.0 |
| District 3 | 321,749 | 92.6 | 67.3 | 261,800 | 11.2 | 1,080 | 42,430 | 81,792 | 39.1 |
| District 4 | 319,785 | 88.0 | 65.8 | 144,100 | 3.1 | 820 | 30,064 | 58,119 | 24.5 |
| **KENTUCKY** | 2,006,335 | 87.2 | 67.0 | 151,700 | 4.2 | 773 | 29,029 | 52,295 | 21.6 |
| District 1 | 340,891 | 83.3 | 68.5 | 115,700 | 2.4 | 683 | 25,443 | 46,999 | 15.8 |
| District 2 | 334,717 | 87.6 | 68.9 | 158,100 | 3.7 | 766 | 27,629 | 53,496 | 20.8 |
| District 3 | 340,492 | 90.3 | 60.4 | 179,500 | 5.5 | 902 | 33,102 | 57,546 | 24.0 |
| District 4 | 316,822 | 90.9 | 73.2 | 178,500 | 6.1 | 801 | 34,354 | 66,327 | 31.7 |
| District 5 | 332,906 | 79.9 | 71.1 | 87,500 | 1.8 | 578 | 21,750 | 35,636 | 12.1 |
| District 6 | 340,507 | 90.9 | 61.2 | 173,400 | 5.7 | 820 | 31,075 | 55,613 | 23.9 |
| **LOUISIANA** | 2,089,824 | 83.3 | 66.5 | 172,100 | 5.4 | 866 | 28,662 | 51,073 | 23.2 |
| District 1 | 339,426 | 88.4 | 71.3 | 219,900 | 9.8 | 966 | 33,254 | 61,431 | 29.0 |
| District 2 | 356,807 | 83.1 | 55.7 | 172,400 | 6.6 | 935 | 27,049 | 44,124 | 20.2 |
| District 3 | 349,938 | 85.0 | 68.3 | 161,000 | 3.8 | 817 | 28,560 | 51,504 | 23.6 |
| District 4 | 353,356 | 80.6 | 66.3 | 138,500 | 3.4 | 778 | 25,601 | 44,580 | 18.5 |
| District 5 | 340,933 | 79.2 | 64.9 | 124,500 | 3.0 | 721 | 23,668 | 41,257 | 16.4 |
| District 6 | 349,364 | 83.7 | 72.2 | 201,500 | 5.4 | 938 | 33,147 | 65,549 | 30.8 |
| **MAINE** | 750,964 | 76.4 | 72.2 | 200,500 | 7.4 | 870 | 34,078 | 58,924 | 25.3 |
| District 1 | 364,607 | 80.4 | 71.1 | 267,600 | 11.7 | 1,019 | 38,567 | 67,392 | 30.7 |
| District 2 | 386,357 | 72.6 | 73.3 | 152,500 | 3.1 | 755 | 29,390 | 51,202 | 19.7 |
| **MARYLAND** | 2,470,307 | 90.1 | 66.8 | 332,500 | 22.1 | 1,401 | 43,325 | 86,738 | 43.3 |
| District 1 | 344,735 | 81.3 | 76.0 | 291,800 | 15.0 | 1,095 | 39,943 | 80,022 | 38.8 |
| District 2 | 306,946 | 91.8 | 59.6 | 251,900 | 7.2 | 1,337 | 36,303 | 73,004 | 34.2 |
| District 3 | 313,721 | 93.3 | 65.2 | 342,100 | 24.8 | 1,553 | 47,631 | 94,736 | 47.5 |
| District 4 | 288,867 | 94.9 | 62.4 | 345,700 | 20.3 | 1,479 | 41,018 | 88,207 | 42.5 |
| District 5 | 284,047 | 92.1 | 78.5 | 348,700 | 14.8 | 1,550 | 42,548 | 101,298 | 51.0 |
| District 6 | 304,912 | 90.0 | 68.9 | 322,200 | 25.4 | 1,396 | 43,001 | 88,592 | 44.8 |
| District 7 | 327,088 | 85.4 | 57.2 | 285,600 | 25.8 | 1,144 | 39,650 | 63,082 | 33.8 |
| District 8 | 299,991 | 94.1 | 67.6 | 467,200 | 43.6 | 1,733 | 55,730 | 109,016 | 54.1 |
| **MASSACHUSETTS** | 2,928,818 | 90.5 | 62.2 | 418,600 | 37.0 | 1,360 | 46,241 | 85,843 | 43.6 |
| District 1 | 321,418 | 89.3 | 64.4 | 235,400 | 5.8 | 917 | 34,225 | 61,559 | 30.2 |
| District 2 | 308,426 | 93.7 | 64.3 | 309,500 | 14.5 | 1,117 | 38,878 | 77,375 | 39.3 |
| District 3 | 303,327 | 94.9 | 63.0 | 372,700 | 27.0 | 1,254 | 42,260 | 81,879 | 40.8 |
| District 4 | 299,801 | 94.3 | 71.6 | 454,500 | 42.9 | 1,415 | 55,642 | 104,857 | 52.6 |
| District 5 | 312,880 | 94.9 | 58.8 | 622,800 | 67.6 | 1,796 | 57,158 | 106,311 | 53.1 |
| District 6 | 310,190 | 95.0 | 70.3 | 508,900 | 51.3 | 1,414 | 50,100 | 97,115 | 48.4 |
| District 7 | 334,963 | 90.3 | 33.7 | 589,700 | 61.4 | 1,734 | 42,315 | 75,461 | 38.8 |
| District 8 | 325,759 | 93.4 | 63.1 | 474,100 | 45.2 | 1,624 | 52,090 | 100,690 | 50.5 |
| District 9 | 412,054 | 74.2 | 71.3 | 387,600 | 28.6 | 1,049 | 42,957 | 77,167 | 38.2 |

1. Specified owner-occupied units; $1,000,000 represents $1,000,000   2. Specified renter-occupied units.

# Table E. Congressional Districts 116th Congress — Poverty, Labor Force, Employment, and Social Security

| STATE District | Poverty, 2019 | | | Civilian labor force, 2019 | | | Civilian employment,[2] 2019 | | | | Persons under 65 years of age with no health insurance, 2019 (percent) | Social Security beneficiaries, December 2020 | | Supplemental Security Income recipients, December 2019 |
|---|---|---|---|---|---|---|---|---|---|---|---|---|---|---|
| | Persons below poverty level (percent) | Families below poverty level (percent) | Percent of households receiving food stamps in past 12 months | Total | Unemployment Total | Rate[1] | Total | Management, business, science, and arts occupations | Service, sales, and office | Construction and production | | Number | Rate[3] | |
| | 49 | 50 | 51 | 52 | 53 | 54 | 55 | 56 | 57 | 58 | 59 | 60 | 61 | 62 |
| **ILLINOIS—Cont'd** | | | | | | | | | | | | | | |
| District 3 | 9.0 | 6.8 | 8.2 | 352,068 | 20,252 | 5.8 | 331,816 | 36.8 | 38.6 | 24.7 | 8.8 | 119,908 | 170.7 | 12,210 |
| District 4 | 12.9 | 10.7 | 13.6 | 370,541 | 18,039 | 4.9 | 352,458 | 32.9 | 38.3 | 28.7 | 15.0 | 77,168 | 114.0 | 16,649 |
| District 5 | 8.1 | 5.2 | 5.8 | 448,703 | 12,350 | 2.8 | 436,257 | 54.2 | 32.2 | 13.6 | 6.0 | 97,940 | 132.5 | 10,867 |
| District 6 | 4.6 | 3.1 | 3.6 | 388,105 | 12,184 | 3.1 | 375,841 | 52.1 | 34.3 | 13.7 | 4.3 | 123,574 | 173.9 | 5,322 |
| District 7 | 19.7 | 13.8 | 19.3 | 382,507 | 27,171 | 7.1 | 354,730 | 50.9 | 33.8 | 15.3 | 7.9 | 99,677 | 137.0 | 30,926 |
| District 8 | 8.9 | 7.0 | 9.3 | 397,382 | 15,031 | 3.8 | 382,084 | 36.0 | 37.4 | 26.6 | 10.0 | 110,472 | 154.1 | 7,811 |
| District 9 | 9.7 | 5.6 | 8.5 | 374,526 | 13,625 | 3.6 | 360,269 | 54.8 | 32.9 | 12.3 | 7.3 | 125,317 | 174.2 | 16,743 |
| District 10 | 7.8 | 5.2 | 8.5 | 382,960 | 15,343 | 4.1 | 356,872 | 42.6 | 37.3 | 20.1 | 8.7 | 115,604 | 163.7 | 9,740 |
| District 11 | 8.0 | 6.4 | 10.3 | 397,495 | 16,447 | 4.1 | 380,825 | 39.4 | 36.8 | 23.8 | 7.8 | 105,549 | 146.3 | 8,287 |
| District 12 | 14.5 | 10.1 | 15.6 | 335,883 | 17,315 | 5.2 | 313,123 | 34.3 | 41.4 | 24.3 | 6.7 | 150,447 | 221.6 | 19,129 |
| District 13 | 17.9 | 10.8 | 13.6 | 344,208 | 18,122 | 5.3 | 325,649 | 40.5 | 38.7 | 20.8 | 4.5 | 140,510 | 201.1 | 15,232 |
| District 14 | 4.3 | 2.9 | 5.3 | 393,392 | 12,678 | 3.2 | 380,643 | 44.3 | 36.1 | 19.6 | 4.9 | 125,697 | 172.8 | 5,085 |
| District 15 | 11.9 | 8.2 | 13.0 | 326,765 | 12,414 | 3.8 | 313,764 | 33.3 | 36.7 | 30.0 | 6.4 | 160,078 | 233.4 | 13,930 |
| District 16 | 11.2 | 7.2 | 11.0 | 356,483 | 17,822 | 5.0 | 338,357 | 33.2 | 36.8 | 30.0 | 4.6 | 152,731 | 220.0 | 9,122 |
| District 17 | 16.3 | 12.7 | 19.1 | 324,961 | 19,197 | 5.9 | 304,780 | 29.9 | 41.1 | 29.0 | 6.4 | 157,659 | 236.7 | 18,218 |
| District 18 | 10.2 | 6.3 | 8.7 | 345,283 | 13,956 | 4.0 | 330,758 | 42.4 | 36.9 | 20.6 | 4.6 | 155,071 | 220.8 | 8,541 |
| **INDIANA** | 11.9 | 7.8 | 8.5 | 3,419,651 | 142,084 | 4.2 | 3,272,727 | 35.1 | 36.5 | 28.4 | 8.6 | 1,382,024 | 205.3 | 127,242 |
| District 1 | 12.9 | 8.3 | 10.4 | 350,583 | 18,801 | 5.4 | 331,606 | 32.0 | 39.1 | 28.9 | 7.4 | 153,730 | 213.8 | 15,877 |
| District 2 | 12.4 | 9.2 | 8.4 | 354,036 | 12,751 | 3.6 | 341,040 | 31.4 | 37.0 | 31.6 | 11.1 | 151,017 | 209.3 | 12,771 |
| District 3 | 9.8 | 6.8 | 7.6 | 386,608 | 13,383 | 3.5 | 372,618 | 33.5 | 33.5 | 33.0 | 11.5 | 154,492 | 205.2 | 12,780 |
| District 4 | 11.0 | 6.2 | 6.2 | 394,778 | 13,175 | 3.3 | 381,449 | 33.0 | 36.4 | 30.5 | 7.6 | 152,750 | 199.1 | 10,506 |
| District 5 | 8.1 | 5.8 | 5.1 | 424,904 | 14,986 | 3.5 | 409,413 | 49.9 | 33.1 | 17.0 | 6.7 | 147,509 | 186.4 | 10,136 |
| District 6 | 12.8 | 8.9 | 9.1 | 360,775 | 13,882 | 3.9 | 345,948 | 34.0 | 35.9 | 30.1 | 7.8 | 124,728 | 160.5 | 23,604 |
| District 7 | 16.6 | 11.2 | 13.5 | 390,495 | 24,526 | 6.3 | 365,524 | 32.8 | 39.2 | 28.0 | 11.3 | 165,763 | 231.2 | 14,814 |
| District 8 | 12.3 | 8.7 | 8.8 | 360,377 | 17,367 | 4.8 | 342,185 | 31.3 | 37.7 | 31.1 | 7.9 | 161,756 | 211.2 | 12,209 |
| District 9 | 11.5 | 6.5 | 7.5 | 397,095 | 13,213 | 3.3 | 382,944 | 35.5 | 37.6 | 26.9 | 6.6 | 170,279 | 236.4 | 14,545 |
| **IOWA** | 11.2 | 7.3 | 9.5 | 1,682,701 | 62,357 | 3.7 | 1,618,556 | 37.8 | 35.2 | 27.1 | 5.0 | 663,803 | 210.4 | 51,573 |
| District 1 | 11.6 | 7.7 | 9.1 | 415,520 | 15,738 | 3.8 | 399,688 | 35.3 | 35.8 | 28.9 | 4.1 | 171,170 | 221.1 | 13,081 |
| District 2 | 12.9 | 7.3 | 10.4 | 409,578 | 16,770 | 4.1 | 392,326 | 37.5 | 34.7 | 27.9 | 6.0 | 166,364 | 212.5 | 15,163 |
| District 3 | 9.4 | 7.0 | 10.2 | 463,389 | 18,987 | 4.1 | 443,737 | 43.2 | 34.8 | 22.0 | 4.7 | 156,137 | 184.1 | 13,358 |
| District 4 | 11.0 | 7.1 | 8.2 | 394,214 | 10,862 | 2.8 | 382,805 | 34.4 | 35.4 | 30.2 | 5.1 | 170,132 | 226.9 | 9,971 |
| **KANSAS** | 11.4 | 7.5 | 7.0 | 1,533,054 | 56,737 | 3.8 | 1,455,746 | 40.0 | 36.0 | 24.0 | 9.1 | 569,120 | 195.4 | 47,303 |
| District 1 | 13.3 | 8.3 | 5.9 | 364,297 | 12,268 | 3.5 | 339,960 | 35.3 | 36.9 | 27.8 | 9.9 | 140,777 | 202.7 | 9,637 |
| District 2 | 12.3 | 7.9 | 8.0 | 370,441 | 12,849 | 3.5 | 354,073 | 37.1 | 36.6 | 26.3 | 7.8 | 154,584 | 215.9 | 14,498 |
| District 3 | 8.6 | 6.2 | 4.7 | 429,255 | 13,401 | 3.1 | 414,921 | 49.0 | 34.2 | 16.8 | 8.6 | 127,647 | 163.7 | 8,889 |
| District 4 | 11.9 | 7.8 | 9.4 | 369,061 | 18,219 | 5.0 | 346,792 | 36.8 | 36.6 | 26.7 | 10.1 | 146,112 | 202.1 | 14,279 |
| **KENTUCKY** | 16.3 | 11.8 | 11.9 | 2,112,749 | 100,537 | 4.8 | 1,997,773 | 36.2 | 36.4 | 27.4 | 6.4 | 1,009,092 | 225.9 | 167,786 |
| District 1 | 17.7 | 14.0 | 13.3 | 321,342 | 15,460 | 4.9 | 297,611 | 29.8 | 36.5 | 33.7 | 7.3 | 181,782 | 253.3 | 26,068 |
| District 2 | 14.8 | 9.8 | 10.2 | 368,904 | 17,920 | 4.9 | 346,096 | 33.0 | 35.7 | 31.3 | 5.7 | 173,306 | 223.7 | 22,291 |
| District 3 | 14.4 | 10.2 | 9.3 | 389,220 | 17,961 | 4.6 | 370,835 | 40.0 | 36.3 | 23.7 | 6.0 | 150,626 | 202.9 | 24,212 |
| District 4 | 11.4 | 8.2 | 7.6 | 386,757 | 11,487 | 3.0 | 375,038 | 39.6 | 34.4 | 26.0 | 6.0 | 153,625 | 201.6 | 17,436 |
| District 5 | 25.0 | 19.8 | 22.0 | 252,771 | 18,947 | 7.5 | 233,561 | 31.0 | 40.1 | 28.9 | 7.4 | 192,520 | 279.1 | 55,250 |
| District 6 | 15.2 | 9.5 | 10.0 | 393,755 | 18,762 | 4.8 | 374,632 | 40.4 | 36.5 | 23.1 | 6.2 | 157,233 | 201.4 | 22,529 |
| **LOUISIANA** | 19.0 | 14.3 | 14.4 | 2,179,183 | 118,011 | 5.5 | 2,040,325 | 35.4 | 40.7 | 23.9 | 8.8 | 925,400 | 199.1 | 170,026 |
| District 1 | 14.4 | 10.0 | 9.5 | 390,411 | 16,370 | 4.2 | 371,255 | 39.9 | 40.1 | 20.1 | 7.9 | 163,772 | 204.7 | 18,177 |
| District 2 | 21.6 | 15.1 | 18.4 | 393,762 | 29,934 | 7.6 | 362,298 | 34.4 | 43.7 | 21.9 | 10.0 | 146,016 | 185.3 | 39,064 |
| District 3 | 19.5 | 15.2 | 14.0 | 359,562 | 18,666 | 5.2 | 339,969 | 35.1 | 38.8 | 26.1 | 7.4 | 155,418 | 198.0 | 24,519 |
| District 4 | 22.5 | 17.5 | 16.9 | 319,562 | 18,647 | 6.1 | 287,460 | 31.1 | 42.5 | 26.3 | 9.0 | 157,280 | 213.2 | 31,748 |
| District 5 | 23.4 | 19.0 | 18.7 | 306,587 | 16,834 | 5.5 | 287,930 | 30.4 | 43.3 | 26.3 | 9.0 | 160,732 | 218.9 | 37,464 |
| District 6 | 13.6 | 10.0 | 13.6 | 409,399 | 17,560 | 4.3 | 391,413 | 39.2 | 37.2 | 23.7 | 9.5 | 142,182 | 176.9 | 19,054 |
| **MAINE** | 10.9 | 6.5 | 12.3 | 711,235 | 24,471 | 3.5 | 684,413 | 40.1 | 37.8 | 22.2 | 8.0 | 355,433 | 264.4 | 35,947 |
| District 1 | 8.9 | 5.5 | 9.0 | 382,290 | 11,288 | 3.0 | 369,546 | 43.9 | 36.5 | 19.6 | 6.6 | 176,154 | 256.5 | 14,094 |
| District 2 | 13.0 | 7.5 | 15.7 | 328,945 | 13,183 | 4.0 | 314,867 | 35.5 | 39.3 | 25.1 | 9.4 | 179,279 | 272.7 | 21,853 |
| **MARYLAND** | 9.0 | 5.8 | 9.8 | 3,279,799 | 147,356 | 4.5 | 3,098,870 | 48.0 | 35.1 | 16.9 | 5.8 | 1,032,078 | 170.7 | 120,333 |
| District 1 | 9.4 | 5.3 | 9.5 | 383,322 | 12,421 | 3.2 | 369,772 | 43.0 | 37.0 | 20.0 | 4.1 | 166,789 | 226.2 | 11,780 |
| District 2 | 11.9 | 8.3 | 13.1 | 400,286 | 19,753 | 5.0 | 372,671 | 43.0 | 38.2 | 18.8 | 6.0 | 129,449 | 172.4 | 18,205 |
| District 3 | 7.7 | 4.8 | 7.5 | 436,039 | 20,206 | 4.7 | 406,837 | 56.0 | 32.2 | 11.8 | 9.5 | 122,459 | 157.1 | 15,469 |
| District 4 | 7.7 | 5.1 | 8.5 | 431,551 | 21,444 | 5.0 | 405,458 | 41.3 | 38.0 | 20.7 | 4.7 | 109,827 | 144.7 | 11,063 |
| District 5 | 6.6 | 4.2 | 6.2 | 410,648 | 17,513 | 4.3 | 387,982 | 45.3 | 34.9 | 19.8 | 5.6 | 119,884 | 158.4 | 10,023 |
| District 6 | 8.3 | 5.6 | 10.6 | 415,811 | 16,351 | 3.9 | 397,929 | 48.4 | 35.4 | 16.2 | 5.1 | 129,824 | 181.0 | 14,282 |
| District 7 | 14.3 | 9.4 | 17.3 | 370,245 | 22,597 | 6.1 | 346,992 | 49.8 | 35.7 | 14.5 | 7.3 | 124,421 | 160.3 | 31,917 |
| District 8 | 6.8 | 4.3 | 5.3 | 431,897 | 17,071 | 4.0 | 411,229 | 56.6 | 30.0 | 13.4 | 2.9 | 129,425 | 145.1 | 7,594 |
| **MASSACHUSETTS** | 9.4 | 6.0 | 11.2 | 3,857,469 | 151,812 | 3.9 | 3,700,243 | 48.4 | 35.5 | 16.1 | 2.9 | 1,294,623 | 187.8 | 179,322 |
| District 1 | 12.3 | 8.3 | 18.3 | 375,992 | 17,291 | 4.6 | 357,576 | 39.0 | 39.3 | 21.7 | 3.2 | 170,503 | 235.6 | 36,730 |
| District 2 | 9.5 | 5.5 | 13.0 | 417,515 | 17,577 | 4.2 | 399,501 | 45.4 | 35.7 | 18.9 | 2.7 | 142,507 | 187.6 | 19,596 |
| District 3 | 8.1 | 4.9 | 12.0 | 428,488 | 19,304 | 4.5 | 408,701 | 42.4 | 37.2 | 20.4 | 3.6 | 136,691 | 177.1 | 22,933 |
| District 4 | 6.6 | 4.1 | 7.6 | 421,554 | 14,323 | 3.4 | 406,924 | 54.7 | 30.9 | 14.5 | 2.1 | 139,396 | 182.1 | 10,826 |
| District 5 | 7.7 | 4.6 | 5.7 | 445,983 | 14,047 | 3.2 | 431,505 | 60.2 | 29.4 | 10.4 | 2.5 | 121,935 | 158.8 | 10,426 |
| District 6 | 6.6 | 3.9 | 7.6 | 430,589 | 15,291 | 3.6 | 414,986 | 50.2 | 35.5 | 14.3 | 2.2 | 153,863 | 199.6 | 13,363 |
| District 7 | 17.0 | 12.3 | 16.9 | 495,532 | 23,081 | 4.7 | 472,007 | 49.9 | 38.2 | 11.9 | 4.6 | 92,255 | 112.6 | 33,002 |
| District 8 | 7.8 | 5.6 | 9.7 | 449,861 | 15,632 | 3.5 | 433,680 | 51.9 | 34.5 | 13.6 | 2.1 | 138,752 | 181.3 | 14,537 |
| District 9 | 8.9 | 6.1 | 10.0 | 391,955 | 15,266 | 3.9 | 375,363 | 38.4 | 39.7 | 22.0 | 3.2 | 198,721 | 265.6 | 17,909 |

1. Percent of civilian labor force.  2. Persons 16 years old and over.  3. Per 1,000 resident population estimated in the 2019 American Community Survey.

# Table E. Congressional Districts 116th Congress — **Agriculture**

| STATE District | Agriculture 2017 | | | | | | | | | |
|---|---|---|---|---|---|---|---|---|---|---|
| | Land in farms | | | | Value of products sold | | | | Government payments | |
| | Number of farms | Acres | Average size of farm (acres) | Irrigated land (acres) | Total ($1,000) | Average per farm (dollars) | Percent from crops | Percent from livestock and poultry products | Total ($1,000) | Average per farm receiving payments (dollars) |
| | 63 | 64 | 65 | 66 | 67 | 68 | 69 | 70 | 71 | 72 |
| **ILLINOIS—Cont'd** | | | | | | | | | | |
| District 3 | 97 | 7,643 | 79 | 6,731 | 5,512 | 56,825 | 91.1 | 8.9 | 45 | 2,368 |
| District 4 | 7 | D | D | 5 | 149 | 21,286 | D | D | D | D |
| District 5 | 10 | D | D | 21 | 53 | 5,300 | D | D | D | D |
| District 6 | 197 | 15,572 | 79 | 9,818 | 38,691 | 196,401 | 93.6 | 6.4 | 417 | 15,444 |
| District 7 | 16 | 668 | 42 | D | 1,072 | 67,000 | 98.0 | 2.0 | D | D |
| District 8 | 32 | 2,974 | 93 | D | 4,267 | 133,344 | 78.2 | 21.8 | D | D |
| District 9 | 20 | 286 | 14 | 93 | 245 | 12,250 | 91.0 | 9.0 | D | D |
| District 10 | 118 | 10,377 | 88 | 7,981 | 21,179 | 179,483 | 96.9 | 3.1 | 110 | 10,000 |
| District 11 | 121 | 18,409 | 152 | 16,488 | 13,715 | 113,347 | 98.9 | 1.1 | 148 | 3,524 |
| District 12 | 6,820 | 1,946,182 | 285 | 1,466,610 | 805,718 | 118,140 | 84.2 | 15.8 | 41,813 | 10,414 |
| District 13 | 7,766 | 3,135,635 | 404 | 2,669,082 | 1,883,640 | 242,550 | 89.4 | 10.6 | 52,897 | 9,823 |
| District 14 | 2,036 | 618,099 | 304 | 563,973 | 509,435 | 250,214 | 80.5 | 19.5 | 11,944 | 13,792 |
| District 15 | 21,078 | 7,426,442 | 352 | 6,212,563 | 4,259,863 | 202,100 | 79.8 | 20.2 | 170,063 | 11,271 |
| District 16 | 9,983 | 4,128,064 | 414 | 3,788,243 | 2,978,591 | 298,366 | 83.3 | 16.7 | 80,693 | 12,544 |
| District 17 | 9,578 | 3,577,140 | 373 | 2,881,538 | 2,562,993 | 267,592 | 72.6 | 27.4 | 79,988 | 11,942 |
| District 18 | 13,467 | 5,638,365 | 419 | 4,621,479 | 3,591,400 | 266,682 | 82.0 | 18.0 | 79,989 | 8,430 |
| **INDIANA** | 56,649 | 14,969,996 | 264 | 12,345,774 | 11,107,336 | 196,073 | 64.1 | 35.9 | 342,914 | 12,628 |
| District 1 | 1,148 | 340,509 | 297 | 308,926 | 214,590 | 186,925 | 89.0 | 11.0 | 8,942 | 15,259 |
| District 2 | 7,227 | 1,837,512 | 254 | 1,567,768 | 1,626,207 | 225,018 | 55.4 | 44.6 | 51,386 | 14,720 |
| District 3 | 10,574 | 2,108,580 | 199 | 1,764,657 | 1,976,214 | 186,894 | 47.3 | 52.7 | 40,561 | 8,858 |
| District 4 | 8,273 | 3,194,762 | 386 | 2,852,760 | 2,418,273 | 292,309 | 71.6 | 28.4 | 65,287 | 13,445 |
| District 5 | 2,607 | 846,826 | 325 | 787,740 | 567,515 | 217,689 | 87.7 | 12.3 | 14,688 | 10,424 |
| District 6 | 10,193 | 2,588,026 | 254 | 2,073,367 | 1,682,472 | 165,062 | 68.4 | 31.6 | 62,303 | 12,376 |
| District 7 | 163 | 15,719 | 96 | 13,871 | 11,513 | 70,632 | 92.0 | 8.0 | 419 | 13,516 |
| District 8 | 9,750 | 2,730,610 | 280 | 2,135,119 | 1,896,295 | 194,492 | 67.0 | 33.0 | 70,322 | 14,544 |
| District 9 | 6,714 | 1,307,452 | 195 | 841,566 | 714,256 | 106,383 | 60.3 | 39.7 | 29,007 | 12,423 |
| **IOWA** | 86,104 | 30,563,878 | 355 | 24,347,862 | 28,956,455 | 336,296 | 47.8 | 52.2 | 682,995 | 11,146 |
| District 1 | 21,735 | 6,684,219 | 308 | 5,250,747 | 6,067,485 | 279,157 | 50.2 | 49.8 | 156,577 | 9,697 |
| District 2 | 19,439 | 5,982,719 | 308 | 4,032,959 | 4,288,086 | 220,592 | 50.2 | 49.8 | 144,109 | 11,010 |
| District 3 | 11,923 | 4,582,245 | 384 | 3,312,351 | 2,634,857 | 220,989 | 65.9 | 34.1 | 95,406 | 12,633 |
| District 4 | 33,007 | 13,314,695 | 403 | 11,751,805 | 15,966,026 | 483,716 | 43.2 | 56.8 | 286,904 | 11,716 |
| **KANSAS** | 58,569 | 45,759,319 | 781 | 21,837,465 | 18,782,726 | 320,694 | 34.4 | 65.6 | 509,205 | 14,089 |
| District 1 | 27,977 | 29,850,908 | 1,067 | 13,748,148 | 14,276,677 | 510,300 | 28.6 | 71.4 | 368,435 | 17,621 |
| District 2 | 19,374 | 7,619,877 | 393 | 4,210,470 | 2,184,002 | 112,729 | 63.1 | 36.9 | 59,851 | 6,569 |
| District 3 | 1,128 | 202,593 | 180 | 127,288 | 73,657 | 65,299 | 81.5 | 18.5 | 971 | 3,996 |
| District 4 | 10,090 | 8,085,941 | 801 | 3,751,559 | 2,248,390 | 222,833 | 41.9 | 58.1 | 79,947 | 13,596 |
| **KENTUCKY** | 75,966 | 12,961,784 | 171 | 5,474,346 | 5,737,920 | 75,533 | 44.3 | 55.7 | 126,697 | 7,502 |
| District 1 | 21,346 | 4,852,908 | 227 | 2,696,597 | 3,093,230 | 144,909 | 47.9 | 52.1 | 76,262 | 9,436 |
| District 2 | 19,839 | 3,134,196 | 158 | 1,336,434 | 1,157,083 | 58,324 | 51.3 | 48.7 | 36,761 | 7,633 |
| District 3 | 190 | 10,644 | 56 | 2,639 | 5,100 | 26,842 | 85.6 | 14.4 | 56 | 7,000 |
| District 4 | 11,807 | 1,645,348 | 139 | 555,968 | 316,984 | 26,847 | 62.5 | 37.5 | 4,898 | 4,230 |
| District 5 | 11,110 | 1,472,160 | 133 | 350,486 | 244,748 | 22,030 | 34.5 | 65.5 | 3,322 | 2,579 |
| District 6 | 11,674 | 1,846,528 | 158 | 532,222 | 920,776 | 78,874 | 19.5 | 80.5 | 5,398 | 3,512 |
| **LOUISIANA** | 27,386 | 7,997,511 | 292 | 3,314,955 | 3,172,978 | 115,861 | 65.0 | 35.0 | 177,399 | 22,822 |
| District 1 | 1,724 | 380,822 | 221 | 57,206 | 71,150 | 41,270 | 55.9 | 44.1 | 251 | 12,550 |
| District 2 | 412 | 269,457 | 654 | 103,019 | 70,301 | 170,633 | 94.4 | 5.6 | 330 | 10,313 |
| District 3 | 5,682 | 1,812,789 | 319 | 565,038 | 505,329 | 88,935 | 77.2 | 22.8 | 50,916 | 24,562 |
| District 4 | 7,211 | 1,572,422 | 218 | 430,884 | 728,780 | 101,065 | 25.0 | 75.0 | 23,611 | 17,860 |
| District 5 | 9,960 | 3,395,715 | 341 | 1,894,297 | 1,548,447 | 155,467 | 77.7 | 22.3 | 96,718 | 24,149 |
| District 6 | 2,397 | 566,306 | 236 | 264,511 | 248,972 | 103,868 | 71.9 | 28.1 | 5,574 | 17,364 |
| **MAINE** | 7,600 | 1,307,613 | 172 | 360,295 | 666,962 | 87,758 | 61.3 | 38.7 | 8,947 | 10,806 |
| District 1 | 2,501 | 204,079 | 82 | 53,902 | 94,149 | 37,645 | 66.4 | 33.6 | 1,829 | 12,113 |
| District 2 | 5,099 | 1,103,534 | 216 | 306,393 | 572,813 | 112,338 | 60.5 | 39.5 | 7,118 | 10,514 |
| **MARYLAND** | 12,429 | 1,990,122 | 160 | 1,290,212 | 2,472,806 | 198,955 | 38.3 | 61.7 | 44,410 | 12,471 |
| District 1 | 5,298 | 1,147,145 | 217 | 812,238 | 1,915,054 | 361,467 | 33.5 | 66.5 | 32,857 | 14,181 |
| District 2 | 143 | 7,808 | 55 | 2,261 | 7,223 | 50,510 | 71.1 | 28.9 | 111 | 11,100 |
| District 3 | 183 | 14,350 | 78 | 7,991 | 11,926 | 65,169 | 83.0 | 17.0 | 249 | 20,750 |
| District 4 | 129 | 5,728 | 44 | 2,148 | 8,721 | 67,605 | 96.6 | 3.4 | 10 | 2,500 |
| District 5 | 1,878 | 182,049 | 97 | 89,237 | 73,022 | 38,883 | 79.5 | 20.5 | 2,353 | 8,715 |
| District 6 | 2,462 | 330,277 | 134 | 179,485 | 237,611 | 96,511 | 38.3 | 61.7 | 2,972 | 7,375 |
| District 7 | 653 | 79,254 | 121 | 45,853 | 55,821 | 85,484 | 86.0 | 14.0 | 1,092 | 11,258 |
| District 8 | 1,683 | 223,511 | 133 | 150,999 | 163,428 | 97,105 | 53.0 | 47.0 | 4,767 | 10,641 |
| **MASSACHUSETTS** | 7,241 | 491,653 | 68 | 140,922 | 475,185 | 65,624 | 76.5 | 23.5 | 4,004 | 7,583 |
| District 1 | 1,955 | 190,928 | 98 | 45,678 | 73,907 | 37,804 | 61.8 | 38.2 | 1,502 | 8,632 |
| District 2 | 1,721 | 123,085 | 72 | 41,818 | 143,703 | 83,500 | 76.0 | 24.0 | 1,146 | 8,128 |
| District 3 | 842 | 36,665 | 44 | 12,632 | 40,465 | 48,058 | 74.0 | 26.0 | 285 | 6,477 |
| District 4 | 656 | 28,229 | 43 | 9,054 | 32,585 | 49,672 | 87.1 | 12.9 | 272 | 4,772 |
| District 5 | 170 | 6,214 | 37 | 2,032 | 36,345 | 213,794 | 98.2 | 1.8 | 20 | 3,333 |
| District 6 | 417 | 19,362 | 46 | 7,705 | 29,882 | 71,659 | 86.5 | 13.5 | 45 | 5,625 |
| District 7 | 27 | 60 | 2 | 30 | 268 | 9,926 | 90.7 | 9.3 | D | D |
| District 8 | 175 | 5,678 | 32 | 1,976 | 9,289 | 53,080 | 90.8 | 9.2 | 53 | 5,300 |
| District 9 | 1,278 | 81,432 | 64 | 19,997 | 108,740 | 85,086 | 73.7 | 26.3 | 680 | 7,727 |

# Table E. Congressional Districts 116th Congress — Nonfarm Employment and Payroll

Private nonfarm employment and payroll, 2019

| STATE District | Number of establishments (73) | Total (74) | Manufacturing (75) | Construction (76) | Wholesale trade (77) | Retail trade (78) | Health care and social assistance (79) | Finance and Insurance (80) | Real estate and rental and leasing (81) | Professional, scientific, and technical services (82) | Information (83) | Total (mil dol) (84) | Average per employee (dollars) (85) |
|---|---|---|---|---|---|---|---|---|---|---|---|---|---|
| **ILLINOIS—Cont'd** | | | | | | | | | | | | | |
| District 3 | 15,981 | 229,089 | 11.8 | 5.1 | 5.3 | 12.5 | 13.9 | 2.3 | 1.2 | 4.3 | 0.9 | 10,795 | 47,123 |
| District 4 | 11,259 | 138,643 | 15.1 | 3.6 | 7.8 | 16.5 | 14.7 | 2.5 | 1.3 | 3.0 | 1.1 | 5,922 | 42,713 |
| District 5 | 23,538 | 377,631 | 7.0 | 4.0 | 4.9 | 9.6 | 13.4 | 3.6 | 2.8 | 5.1 | 2.1 | 20,981 | 55,558 |
| District 6 | 25,554 | 389,433 | 8.8 | 4.5 | 6.8 | 10.9 | 13.6 | 7.0 | 1.4 | 7.8 | 3.2 | 23,262 | 59,733 |
| District 7 | 30,306 | 866,627 | 2.7 | 1.7 | 3.0 | 4.3 | 12.6 | 15.7 | 2.1 | 18.7 | 5.5 | 79,807 | 92,089 |
| District 8 | 24,114 | 413,529 | 13.5 | 5.8 | 11.7 | 9.3 | 9.3 | 4.9 | 2.0 | 6.7 | 3.6 | 25,832 | 62,468 |
| District 9 | 20,496 | 293,830 | 7.0 | 2.8 | 4.5 | 11.1 | 24.1 | 2.8 | 1.8 | 5.9 | 1.9 | 17,073 | 58,106 |
| District 10 | 21,612 | 360,940 | 11.1 | 3.3 | 12.0 | 10.2 | 13.8 | 6.5 | 1.2 | 9.8 | 1.5 | 28,800 | 79,790 |
| District 11 | 16,951 | 290,260 | 9.4 | 5.1 | 8.5 | 13.4 | 13.9 | 3.0 | 1.2 | 5.1 | 1.4 | 14,694 | 50,622 |
| District 12 | 14,283 | 217,850 | 12.8 | 4.7 | 4.1 | 15.1 | 19.5 | 3.1 | 1.0 | 4.0 | 1.1 | 9,152 | 42,009 |
| District 13 | 15,204 | 242,013 | 8.0 | 4.5 | 3.8 | 13.4 | 22.0 | 4.4 | 1.7 | 4.3 | 2.2 | 10,697 | 44,199 |
| District 14 | 18,187 | 199,312 | 13.5 | 8.4 | 5.6 | 14.1 | 11.6 | 2.9 | 1.3 | 5.8 | 1.0 | 9,225 | 46,283 |
| District 15 | 15,078 | 203,725 | 20.1 | 4.5 | 5.7 | 12.4 | 17.9 | 3.6 | 0.9 | 2.3 | 1.3 | 8,058 | 39,551 |
| District 16 | 15,015 | 233,510 | 20.8 | 5.0 | 4.5 | 13.7 | 15.1 | 3.1 | 0.9 | 2.8 | 1.3 | 10,669 | 45,689 |
| District 17 | 14,257 | 261,461 | 15.7 | 3.7 | 4.1 | 11.0 | 19.4 | 3.6 | 0.8 | 3.7 | 1.4 | 13,761 | 52,631 |
| District 18 | 15,800 | 251,154 | 10.3 | 4.3 | 6.6 | 15.3 | 14.7 | 11.1 | 0.9 | 3.5 | 1.3 | 11,312 | 45,040 |
| **INDIANA** | 148,917 | 2,834,056 | 18.4 | 5.1 | 4.5 | 11.6 | 15.7 | 3.9 | 1.2 | 4.7 | 1.5 | 132,502 | 46,753 |
| District 1 | 15,015 | 245,796 | 14.5 | 6.8 | 3.7 | 14.4 | 19.2 | 2.3 | 1.3 | 3.5 | 1.0 | 11,623 | 47,289 |
| District 2 | 15,822 | 328,475 | 34.5 | 3.6 | 4.8 | 10.7 | 12.6 | 2.4 | 0.8 | 2.6 | 1.3 | 14,414 | 43,883 |
| District 3 | 17,752 | 339,292 | 27.2 | 4.6 | 4.9 | 11.5 | 15.8 | 3.5 | 1.0 | 2.7 | 1.0 | 15,249 | 44,944 |
| District 4 | 15,499 | 275,129 | 22.7 | 5.0 | 5.3 | 10.6 | 15.7 | 8.0 | 1.8 | 9.3 | 2.3 | 11,586 | 42,112 |
| District 5 | 22,342 | 406,882 | 6.6 | 4.7 | 4.2 | 12.3 | 17.3 | 2.9 | 0.8 | 3.7 | 1.1 | 20,895 | 51,353 |
| District 6 | 14,355 | 241,263 | 22.4 | 4.4 | 5.7 | 8.3 | 17.3 | 4.3 | 2.1 | 6.1 | 2.6 | 10,445 | 43,295 |
| District 7 | 15,412 | 356,471 | 9.3 | 6.6 | 4.2 | 12.3 | 17.6 | 3.0 | 1.0 | 3.2 | 1.3 | 20,124 | 56,454 |
| District 8 | 16,222 | 284,062 | 20.8 | 6.1 | 4.2 | 12.3 | 17.6 | 3.0 | 1.0 | 3.2 | 1.3 | 12,636 | 44,484 |
| District 9 | 15,785 | 249,896 | 18.2 | 5.9 | 3.4 | 15.6 | 17.4 | 3.2 | 1.2 | 3.0 | 1.3 | 10,159 | 40,655 |
| **IOWA** | 82,770 | 1,380,747 | 15.8 | 4.7 | 5.0 | 13.6 | 15.8 | 6.8 | 1.1 | 4.3 | 2.1 | 63,671 | 46,113 |
| District 1 | 19,797 | 342,989 | 19.0 | 4.5 | 4.6 | 13.7 | 15.2 | 5.7 | 0.9 | 4.5 | 2.4 | 15,827 | 46,143 |
| District 2 | 18,933 | 319,004 | 19.7 | 4.8 | 3.8 | 14.4 | 18.0 | 3.4 | 0.9 | 3.7 | 2.0 | 13,734 | 43,053 |
| District 3 | 21,674 | 389,239 | 7.9 | 5.1 | 5.4 | 13.5 | 14.1 | 12.3 | 1.6 | 5.8 | 2.4 | 20,156 | 51,784 |
| District 4 | 21,925 | 289,815 | 20.7 | 5.2 | 6.6 | 14.5 | 18.1 | 3.6 | 0.9 | 2.9 | 1.9 | 12,103 | 41,763 |
| **KANSAS** | 74,292 | 1,209,318 | 14.1 | 5.4 | 5.2 | 12.2 | 15.7 | 5.3 | 1.2 | 5.5 | 2.5 | 57,272 | 47,359 |
| District 1 | 19,618 | 238,681 | 18.0 | 4.6 | 5.5 | 14.6 | 17.8 | 3.9 | 1.0 | 3.2 | 2.1 | 8,933 | 37,428 |
| District 2 | 15,697 | 232,195 | 15.1 | 6.0 | 3.4 | 13.3 | 20.0 | 4.6 | 1.1 | 4.4 | 1.4 | 9,634 | 41,493 |
| District 3 | 21,421 | 416,883 | 8.2 | 5.6 | 7.1 | 10.9 | 13.4 | 7.7 | 1.4 | 8.1 | 3.9 | 23,778 | 57,037 |
| District 4 | 17,027 | 288,943 | 20.3 | 5.9 | 3.9 | 12.4 | 15.3 | 3.1 | 1.3 | 4.7 | 1.9 | 13,650 | 47,243 |
| **KENTUCKY** | 91,219 | 1,666,637 | 15.2 | 4.5 | 4.3 | 13.1 | 15.6 | 4.5 | 1.1 | 4.5 | 1.8 | 74,884 | 44,931 |
| District 1 | 14,228 | 217,571 | 23.7 | 4.7 | 3.7 | 14.8 | 15.6 | 3.0 | 0.9 | 2.7 | 1.2 | 8,549 | 39,293 |
| District 2 | 14,128 | 239,285 | 21.1 | 5.3 | 3.3 | 15.0 | 15.9 | 3.6 | 1.0 | 3.2 | 1.2 | 9,671 | 40,418 |
| District 3 | 19,479 | 453,983 | 11.9 | 3.8 | 4.5 | 9.8 | 14.4 | 6.7 | 1.3 | 5.1 | 2.2 | 24,351 | 53,639 |
| District 4 | 14,426 | 261,618 | 13.2 | 4.6 | 5.6 | 13.6 | 13.8 | 5.1 | 1.1 | 3.7 | 1.2 | 12,170 | 46,519 |
| District 5 | 11,064 | 161,637 | 10.9 | 3.1 | 3.0 | 17.7 | 22.4 | 3.4 | 1.0 | 5.0 | 2.5 | 5,713 | 35,345 |
| District 6 | 17,258 | 295,919 | 15.0 | 6.0 | 3.5 | 13.8 | 16.8 | 2.9 | 1.1 | 5.8 | 2.2 | 13,079 | 44,199 |
| **LOUISIANA** | 106,302 | 1,719,561 | 7.1 | 8.5 | 4.3 | 13.0 | 17.8 | 3.7 | 1.8 | 5.7 | 1.3 | 82,452 | 47,950 |
| District 1 | 21,322 | 314,755 | 4.8 | 6.0 | 5.1 | 14.4 | 19.1 | 4.5 | 1.9 | 5.4 | 1.3 | 15,582 | 49,505 |
| District 2 | 16,109 | 296,666 | 9.5 | 6.1 | 3.5 | 10.3 | 12.9 | 3.2 | 1.7 | 6.7 | 1.2 | 15,499 | 52,242 |
| District 3 | 19,830 | 295,726 | 9.1 | 6.2 | 5.0 | 13.9 | 17.8 | 2.9 | 2.8 | 5.8 | 1.2 | 14,351 | 48,528 |
| District 4 | 14,926 | 213,628 | 8.9 | 5.6 | 4.2 | 14.7 | 23.9 | 3.3 | 1.7 | 3.6 | 1.8 | 8,680 | 40,630 |
| District 5 | 15,033 | 205,090 | 7.9 | 5.5 | 3.8 | 15.9 | 26.6 | 4.3 | 1.4 | 3.9 | 1.4 | 7,958 | 38,805 |
| District 6 | 18,520 | 338,615 | 4.7 | 19.4 | 4.1 | 12.7 | 14.1 | 4.2 | 1.5 | 6.9 | 2.1 | 17,212 | 50,831 |
| **MAINE** | 41,843 | 522,191 | 9.7 | 5.3 | 3.6 | 16.0 | 21.6 | 5.4 | 1.4 | 4.4 | 1.9 | 24,129 | 46,208 |
| District 1 | 23,927 | 310,741 | 9.2 | 5.1 | 3.7 | 15.1 | 21.1 | 5.9 | 1.5 | 4.9 | 2.3 | 15,154 | 48,768 |
| District 2 | 17,587 | 206,627 | 10.6 | 5.7 | 3.4 | 17.8 | 22.7 | 4.3 | 1.3 | 3.4 | 2.1 | 8,679 | 42,003 |
| **MARYLAND** | 139,449 | 2,380,865 | 4.2 | 6.8 | 3.8 | 12.1 | 16.7 | 4.2 | 2.0 | 12.2 | 1.1 | 136,778 | 57,449 |
| District 1 | 17,648 | 203,160 | 8.6 | 7.7 | 3.7 | 16.9 | 18.1 | 2.6 | 1.3 | 4.6 | 1.9 | 8,655 | 42,601 |
| District 2 | 16,484 | 348,146 | 6.3 | 7.9 | 7.4 | 13.5 | 11.7 | 4.1 | 1.9 | 11.7 | 2.9 | 20,371 | 58,513 |
| District 3 | 21,351 | 397,298 | 4.6 | 5.2 | 4.0 | 10.7 | 16.2 | 3.7 | 2.0 | 13.2 | 2.2 | 24,125 | 60,722 |
| District 4 | 13,170 | 206,940 | 2.2 | 8.7 | 4.3 | 15.0 | 13.3 | 2.3 | 2.7 | 9.5 | 1.6 | 9,520 | 46,003 |
| District 5 | 14,466 | 221,802 | 2.3 | 13.5 | 2.5 | 14.6 | 14.7 | 1.8 | 2.6 | 14.8 | 2.7 | 11,609 | 52,339 |
| District 6 | 19,249 | 297,814 | 6.2 | 6.6 | 3.3 | 15.0 | 15.4 | 4.6 | 1.4 | 13.5 | 1.4 | 15,843 | 53,198 |
| District 7 | 15,505 | 309,653 | 2.1 | 2.8 | 1.9 | 8.0 | 28.1 | 6.8 | 2.2 | 10.5 | 2.5 | 19,954 | 64,441 |
| District 8 | 20,903 | 338,130 | 2.1 | 6.4 | 2.4 | 9.4 | 18.3 | 5.1 | 2.4 | 15.9 | 3.6 | 22,724 | 67,206 |
| **MASSACHUSETTS** | 181,061 | 3,386,372 | 6.9 | 4.5 | 4.5 | 10.7 | 18.5 | 5.5 | 1.5 | 9.3 | 3.6 | 238,938 | 70,559 |
| District 1 | 16,000 | 258,368 | 11.2 | 4.5 | 3.6 | 12.9 | 23.2 | 4.2 | 1.4 | 4.1 | 1.2 | 12,037 | 46,590 |
| District 2 | 17,009 | 294,505 | 9.2 | 4.1 | 4.3 | 12.9 | 22.6 | 5.2 | 1.2 | 5.6 | 1.4 | 14,463 | 49,110 |
| District 3 | 16,957 | 294,795 | 15.4 | 5.2 | 5.9 | 10.5 | 18.0 | 3.4 | 1.1 | 9.1 | 3.0 | 20,802 | 70,565 |
| District 4 | 21,052 | 348,332 | 7.7 | 4.3 | 7.5 | 12.7 | 16.5 | 3.7 | 1.9 | 5.7 | 2.5 | 21,656 | 62,171 |
| District 5 | 21,194 | 405,822 | 3.9 | 4.7 | 4.4 | 9.1 | 13.7 | 3.7 | 1.6 | 15.1 | 5.6 | 32,139 | 79,196 |
| District 6 | 21,856 | 377,159 | 12.1 | 5.6 | 5.0 | 12.6 | 17.9 | 3.2 | 1.2 | 8.6 | 5.1 | 25,115 | 66,591 |
| District 7 | 17,760 | 540,777 | 1.8 | 1.8 | 2.5 | 6.7 | 21.1 | 6.1 | 1.9 | 11.9 | 4.8 | 49,116 | 90,824 |
| District 8 | 25,289 | 543,429 | 3.4 | 5.0 | 3.7 | 9.4 | 17.1 | 12.4 | 2.1 | 12.7 | 4.4 | 46,447 | 85,470 |
| District 9 | 23,246 | 251,708 | 6.3 | 7.9 | 3.7 | 17.9 | 20.9 | 2.7 | 1.3 | 4.8 | 1.5 | 12,414 | 49,318 |

# Table E. Congressional Districts 116th Congress — Land Area and Population Characteristics

| STATE District | Representative, 117th Congress | Land area,[1] 2018 (sq mi) | Total persons | Per square mile | White | Black | American Indian, Alaska Native | Asian and Pacific Islander | Some other race (percent) | Two or more races (percent) | Hispanic or Latino[2] (percent) | Non-Hispanic White alone (percent) | Percent female | Percent foreign-born | Percent born in state of residence |
|---|---|---|---|---|---|---|---|---|---|---|---|---|---|---|---|
| | | | | | Race alone (percent) | | | | | | | | | | |
| | | 1 | 2 | 3 | 4 | 5 | 6 | 7 | 8 | 9 | 10 | 11 | 12 | 13 | 14 |
| MICHIGAN | | 56,608.218 | 9,986,857 | 176.421 | 78.2 | 13.7 | 0.6 | 3.3 | 1.1 | 3.0 | 5.3 | 74.668 | 50.8 | 7.0 | 76.2 |
| District 1 | Jack Bergman (R) | 25,029.019 | 697,102 | 27.852 | 92.2 | 1.4 | 2.7 | 0.8 | 0.2 | 2.7 | 1.9 | 90.776 | 49.3 | 1.8 | 78.1 |
| District 2 | Bill Huizenga (R) | 3,337.808 | 746,998 | 223.799 | 84.8 | 6.2 | 0.5 | 2.3 | 2.7 | 3.5 | 9.6 | 78.962 | 50.4 | 5.6 | 78.7 |
| District 3 | Peter Meijer (R) | 2,630.835 | 752,287 | 285.950 | 82.8 | 8.3 | 0.3 | 2.1 | 2.1 | 4.4 | 8.0 | 77.982 | 50.4 | 5.6 | 78.5 |
| District 4 | John L. Moolenaar (R) | 8,459.864 | 702,887 | 83.085 | 93.7 | 1.8 | 0.8 | 1.0 | 0.5 | 2.1 | 3.5 | 91.075 | 49.7 | 1.7 | 85.1 |
| District 5 | Daniel T. Kildee (D) | 2,349.077 | 672,466 | 286.268 | 77.2 | 17.5 | 0.5 | 1.0 | 0.5 | 3.4 | 5.1 | 73.437 | 51.7 | 2.2 | 84.3 |
| District 6 | Fred Upton (R) | 3,547.492 | 721,508 | 203.385 | 84.8 | 8.5 | 0.4 | 1.5 | 1.0 | 3.7 | 6.5 | 80.132 | 50.8 | 4.7 | 68.7 |
| District 7 | Tim Walberg (R) | 4,228.171 | 710,064 | 167.936 | 90.9 | 4.0 | 0.5 | 1.2 | 0.4 | 3.0 | 4.5 | 87.197 | 49.8 | 3.3 | 73.7 |
| District 8 | Elissa Slotkin (D) | 1,503.181 | 740,750 | 492.788 | 85.1 | 6.1 | 0.4 | 5.1 | 0.5 | 2.8 | 5.2 | 81.051 | 51.4 | 8.8 | 74.3 |
| District 9 | Andy Levin (D) | 183.597 | 718,223 | 3,911.954 | 76.6 | 13.9 | 0.5 | 5.4 | 0.4 | 3.3 | 2.6 | 74.650 | 51.3 | 11.6 | 76.2 |
| District 10 | Lisa McClain (R) | 4,146.819 | 721,753 | 174.050 | 91.9 | 3.2 | 0.2 | 1.8 | 0.9 | 2.0 | 3.4 | 89.682 | 50.6 | 6.4 | 83.3 |
| District 11 | Haley M. Stevens (D) | 419.109 | 735,677 | 1,755.336 | 80.6 | 4.9 | 0.2 | 11.2 | 0.4 | 2.8 | 4.4 | 77.039 | 50.3 | 15.8 | 68.9 |
| District 12 | Debbie Dingell (D) | 402.931 | 704,912 | 1,749.461 | 77.6 | 11.3 | 0.3 | 5.5 | 1.8 | 3.6 | 6.2 | 73.652 | 51.0 | 12.1 | 68.9 |
| District 13 | Rashida Tlaib (D) | 184.921 | 672,291 | 3,635.558 | 38.5 | 54.1 | 0.7 | 1.2 | 3.0 | 2.4 | 7.7 | 34.181 | 52.1 | 7.3 | 77.4 |
| District 14 | Brenda L. Lawrence (D) | 185.394 | 689,939 | 3,721.474 | 34.1 | 55.9 | 0.4 | 5.7 | 1.4 | 2.5 | 4.7 | 31.434 | 52.0 | 10.9 | 71.8 |
| MINNESOTA | | 79,626.685 | 5,639,632 | 70.826 | 82.1 | 6.6 | 1.0 | 5.1 | 1.9 | 3.3 | 5.6 | 78.940 | 50.3 | 8.4 | 67.7 |
| District 1 | Jim Hagedorn (R) | 11,974.407 | 679,003 | 56.705 | 89.1 | 3.1 | 0.3 | 3.1 | 1.9 | 2.5 | 6.8 | 84.575 | 50.4 | 6.6 | 68.1 |
| District 2 | Angie Craig (D) | 2,437.923 | 717,698 | 294.389 | 82.7 | 5.7 | 0.3 | 5.0 | 2.7 | 3.6 | 6.5 | 79.323 | 50.2 | 8.7 | 67.3 |
| District 3 | Dean Philips (D) | 527.424 | 730,214 | 1,384.491 | 77.3 | 8.3 | 0.4 | 8.8 | 1.2 | 3.9 | 4.0 | 74.813 | 51.2 | 12.5 | 62.4 |
| District 4 | Betty McCollum (D) | 332.601 | 719,873 | 2,164.374 | 70.1 | 10.8 | 0.4 | 12.9 | 1.6 | 4.2 | 6.7 | 65.809 | 51.0 | 14.3 | 61.7 |
| District 5 | Ilhan Omar (D) | 135.734 | 724,373 | 5,336.710 | 67.2 | 16.7 | 0.8 | 5.7 | 4.7 | 4.8 | 9.9 | 63.179 | 50.1 | 14.0 | 56.9 |
| District 6 | Tom Emmer (R) | 2,880.649 | 729,029 | 253.078 | 88.9 | 4.1 | 0.4 | 2.6 | 1.3 | 2.6 | 3.4 | 86.992 | 49.6 | 4.9 | 76.1 |
| District 7 | Michelle Fischbach (R) | 33,429.350 | 668,096 | 19.985 | 90.4 | 1.6 | 3.3 | 1.2 | 1.5 | 2.1 | 5.2 | 87.275 | 50.1 | 3.1 | 71.8 |
| District 8 | Pete Stauber (R) | 27,908.597 | 671,346 | 24.055 | 92.5 | 1.1 | 2.3 | 0.8 | 0.5 | 2.7 | 1.9 | 91.369 | 49.5 | 1.9 | 78.5 |
| MISSISSIPPI | | 46,923.958 | 2,976,149 | 63.425 | 58.0 | 38.0 | 0.5 | 1.0 | 1.0 | 1.5 | 3.0 | 56.298 | 51.8 | 2.1 | 71.5 |
| District 1 | Trent Kelly (R) | 10,573.389 | 769,026 | 72.732 | 66.8 | 29.0 | 0.1 | 1.1 | 1.8 | 1.2 | 3.3 | 65.533 | 51.4 | 2.4 | 63.7 |
| District 2 | Bennie G. Thompson (D) | 15,551.698 | 692,452 | 44.526 | 30.6 | 67.0 | 0.5 | 0.5 | 0.4 | 1.0 | 1.9 | 29.484 | 51.6 | 1.2 | 84.6 |
| District 3 | Michael Guest (R) | 12,754.717 | 738,992 | 57.939 | 60.5 | 35.4 | 1.0 | 1.2 | 1.1 | 0.8 | 2.4 | 59.226 | 52.2 | 2.3 | 77.1 |
| District 4 | Steven M. Palazzo (R) | 8,044.154 | 775,679 | 96.428 | 71.5 | 23.5 | 0.4 | 1.3 | 0.6 | 2.7 | 4.4 | 68.291 | 51.9 | 2.6 | 62.3 |
| MISSOURI | | 68,746.483 | 6,137,428 | 89.276 | 81.8 | 11.5 | 0.4 | 2.3 | 1.2 | 2.8 | 4.3 | 79.095 | 51.0 | 4.3 | 65.9 |
| District 1 | Cori Bush (D) | 225.366 | 727,772 | 3,229.289 | 42.5 | 49.0 | 0.2 | 3.5 | 1.9 | 2.9 | 3.6 | 40.960 | 52.7 | 6.5 | 67.9 |
| District 2 | Ann Wagner (R) | 465.730 | 751,926 | 1,614.511 | 88.7 | 3.7 | 0.2 | 4.8 | 0.4 | 2.2 | 3.1 | 86.018 | 51.9 | 7.5 | 64.9 |
| District 3 | Blaine Luetkemeyer (R) | 6,851.492 | 802,919 | 117.189 | 91.7 | 4.0 | 0.4 | 1.3 | 0.5 | 2.1 | 2.7 | 89.780 | 49.8 | 2.4 | 73.2 |
| District 4 | Vicky Hartzler (R) | 14,406.941 | 775,664 | 53.840 | 88.7 | 4.7 | 0.5 | 2.1 | 0.5 | 3.5 | 3.9 | 85.863 | 50.1 | 3.5 | 63.7 |
| District 5 | Emanuel Cleaver (D) | 2,425.313 | 777,659 | 320.643 | 68.7 | 21.2 | 0.4 | 2.1 | 4.1 | 3.5 | 9.4 | 63.535 | 51.2 | 6.3 | 59.8 |
| District 6 | Sam Graves (R) | 18,198.101 | 777,104 | 42.702 | 89.2 | 4.5 | 0.4 | 1.7 | 0.8 | 3.3 | 4.2 | 86.630 | 50.9 | 2.9 | 65.1 |
| District 7 | Bill Long (R) | 6,271.658 | 787,917 | 125.631 | 91.1 | 1.9 | 0.9 | 1.5 | 1.4 | 3.2 | 3.1 | 87.557 | 50.9 | 3.6 | 60.3 |
| District 8 | Jason T. Smith (R) | 19,901.882 | 736,467 | 37.005 | 92.0 | 4.8 | 0.3 | 1.0 | 0.2 | 1.7 | 1.9 | 90.361 | 50.6 | 1.6 | 72.1 |
| MONTANA | | 145,550.355 | 1,068,778 | 7.343 | 88.0 | 0.7 | 6.3 | 0.8 | 0.7 | 3.4 | 3.8 | 85.757 | 49.7 | 2.3 | 53.4 |
| At Large | Matthew M. Rosendale Sr. (R) | 145,550.355 | 1,068,778 | 7.343 | 88.0 | 0.7 | 6.3 | 0.8 | 0.7 | 3.4 | 3.8 | 85.757 | 49.7 | 2.3 | 53.4 |
| NEBRASKA | | 76,817.875 | 1,934,408 | 25.182 | 86.2 | 4.9 | 1.0 | 2.6 | 2.6 | 2.8 | 11.3 | 78.358 | 50.0 | 7.4 | 64.7 |
| District 1 | Jeff Fortenberry (R) | 8,877.683 | 651,958 | 73.438 | 88.1 | 2.8 | 1.5 | 2.9 | 1.9 | 2.8 | 10.0 | 80.931 | 49.5 | 7.0 | 67.2 |
| District 2 | Don Bacon (R) | 506.882 | 684,882 | 1,351.167 | 80.8 | 9.9 | 0.7 | 3.7 | 1.6 | 3.3 | 11.9 | 71.404 | 50.7 | 9.3 | 59.4 |
| District 3 | Adrian Smith (R) | 67,433.310 | 597,568 | 8.862 | 90.2 | 1.5 | 0.7 | 0.7 | 4.7 | 2.1 | 12.2 | 83.520 | 49.9 | 5.6 | 68.1 |
| NEVADA | | 109,860.456 | 3,080,156 | 28.037 | 64.6 | 9.6 | 1.4 | 9.2 | 10.5 | 4.7 | 29.2 | 47.841 | 49.8 | 19.8 | 27.2 |
| District 1 | Dina Titus (D) | 104.486 | 712,411 | 6,818.244 | 51.4 | 12.1 | 1.7 | 10.0 | 20.7 | 4.2 | 47.0 | 27.914 | 49.6 | 32.7 | 24.3 |
| District 2 | Mark E. Amodei (R) | 55,897.584 | 736,907 | 13.183 | 78.5 | 2.0 | 2.2 | 5.0 | 8.8 | 3.6 | 23.5 | 65.262 | 49.1 | 12.7 | 31.9 |
| District 3 | Susie Lee (D) | 2,848.930 | 857,197 | 300.884 | 64.5 | 8.8 | 0.6 | 14.6 | 5.9 | 5.6 | 17.0 | 54.883 | 50.8 | 18.0 | 23.8 |
| District 4 | Steven Horsford (D) | 51,009.456 | 773,641 | 15.167 | 63.6 | 15.6 | 1.2 | 6.6 | 8.0 | 5.1 | 32.0 | 41.796 | 49.8 | 16.7 | 29.2 |
| NEW HAMPSHIRE | | 8,953.754 | 1,359,711 | 151.859 | 92.6 | 1.6 | 0.1 | 2.7 | 0.8 | 2.2 | 4.0 | 89.656 | 50.5 | 6.4 | 40.6 |
| District 1 | Chris Pappas (D) | 2,464.994 | 686,735 | 278.595 | 92.8 | 1.6 | 0.1 | 2.7 | 1.0 | 1.8 | 4.1 | 89.851 | 51.1 | 7.0 | 39.4 |
| District 2 | Ann M. Kuster (D) | 6,488.760 | 672,976 | 103.714 | 92.3 | 1.6 | 0.2 | 2.7 | 0.6 | 2.6 | 3.8 | 89.457 | 49.9 | 5.8 | 41.9 |
| NEW JERSEY | | 7,354.757 | 8,882,190 | 1,207.680 | 67.1 | 13.6 | 0.2 | 9.6 | 6.4 | 3.0 | 20.9 | 54.321 | 51.1 | 23.4 | 51.5 |
| District 1 | Donald Norcross (D) | 350.255 | 726,825 | 2,075.131 | 68.0 | 17.0 | 0.1 | 5.0 | 6.7 | 3.1 | 14.1 | 62.030 | 51.6 | 9.1 | 55.8 |
| District 2 | Jefferson Van Drew (R) | 2,094.210 | 707,255 | 337.719 | 74.0 | 12.9 | 0.3 | 3.8 | 5.7 | 3.3 | 17.2 | 64.333 | 50.7 | 10.2 | 60.7 |
| District 3 | Andy Kim (D) | 898.276 | 735,981 | 819.326 | 79.1 | 11.5 | 0.2 | 3.8 | 1.9 | 3.6 | 9.1 | 72.885 | 50.7 | 9.3 | 61.3 |
| District 4 | Christopher H. Smith (R) | 691.009 | 748,199 | 1,082.763 | 86.0 | 6.2 | 0.1 | 4.2 | 2.2 | 1.3 | 11.2 | 77.625 | 50.5 | 12.0 | 59.4 |
| District 5 | Josh Gottheimer (D) | 991.593 | 734,764 | 740.994 | 79.1 | 5.5 | 0.1 | 10.5 | 2.0 | 2.9 | 14.6 | 67.705 | 51.2 | 20.7 | 51.3 |
| District 6 | Frank Pallone Jr. (D) | 215.286 | 739,726 | 3,436.015 | 61.0 | 10.8 | 0.1 | 18.6 | 5.7 | 3.8 | 22.4 | 46.528 | 50.7 | 28.8 | 47.8 |
| District 7 | Tom Malinowski (D) | 969.983 | 734,239 | 756.961 | 74.9 | 5.0 | 0.2 | 11.8 | 5.6 | 2.4 | 14.3 | 66.339 | 51.0 | 21.7 | 54.0 |
| District 8 | Albio Sires (D) | 54.634 | 766,357 | 14,027.108 | 57.8 | 10.7 | 0.5 | 9.8 | 17.4 | 3.7 | 55.4 | 24.507 | 50.1 | 46.3 | 35.2 |
| District 9 | Bill Pascrell, Jr. (D) | 95.320 | 762,322 | 7,997.503 | 62.7 | 8.9 | 0.2 | 13.8 | 11.4 | 2.9 | 40.1 | 36.207 | 51.2 | 41.2 | 42.2 |
| District 10 | Donald M. Payne, Jr. (D) | 75.830 | 761,783 | 10,045.932 | 31.1 | 51.0 | 0.3 | 6.4 | 7.3 | 3.9 | 20.0 | 20.772 | 51.9 | 30.3 | 47.6 |
| District 11 | Mikie Sherill (D) | 506.238 | 717,657 | 1,417.628 | 81.7 | 4.6 | 0.1 | 10.5 | 1.1 | 1.9 | 11.5 | 71.944 | 51.5 | 19.0 | 58.1 |
| District 12 | Bonnie Watson Coleman (D) | 412.123 | 747,082 | 1,812.765 | 52.9 | 17.3 | 0.3 | 17.5 | 9.3 | 2.8 | 18.4 | 44.800 | 51.2 | 29.2 | 45.7 |
| NEW MEXICO | | 76,347.142 | 2,096,829 | 27.464 | 73.9 | 2.3 | 9.5 | 1.7 | 9.0 | 3.5 | 49.3 | 36.754 | 50.6 | 9.6 | 53.8 |
| District 1 | Melanie A. Stansbury (D) | 4,600.930 | 691,229 | 150.237 | 74.4 | 2.7 | 4.5 | 2.8 | 11.2 | 4.4 | 51.0 | 38.230 | 51.3 | 10.8 | 54.1 |
| District 2 | Yvette Herrell (R) | 71,746.212 | 705,615 | 9.835 | 81.9 | 2.4 | 5.3 | 1.2 | 7.1 | 2.0 | 55.9 | 35.373 | 49.4 | 11.9 | 50.1 |

1. Dry land or land partially or temporarily covered by water.   2. May be of any race.

# Table E. Congressional Districts 116th Congress — **Age and Education**

| STATE District | Population and population characteristics, 2019 (cont.) | | | | | | | | | | Education, 2019 | | |
| | Age (percent) | | | | | | | | | | | Attainment[2] (percent) | |
| | Under 5 years | 5 to 17 years | 18 to 24 years | 25 to 34 years | 35 to 44 years | 45 to 54 years | 55 to 64 years | 65 to 74 years | 75 years and over | Median age | Total Enrollment[1] | High school graduate or more | Bachelor's degree or more |
| | 15 | 16 | 17 | 18 | 19 | 20 | 21 | 22 | 23 | 24 | 25 | 26 | 27 |
|---|---|---|---|---|---|---|---|---|---|---|---|---|---|
| MICHIGAN | 5.7 | 15.8 | 9.5 | 13.0 | 11.8 | 12.5 | 14.0 | 10.4 | 7.3 | 39.8 | 2,387,441 | 91.4 | 30.0 |
| District 1 | 4.6 | 13.5 | 8.4 | 10.8 | 10.4 | 11.7 | 16.3 | 14.1 | 10.0 | 47.1 | 137,514 | 92.5 | 26.2 |
| District 2 | 6.2 | 17.2 | 10.6 | 13.0 | 12.1 | 11.3 | 13.1 | 9.6 | 6.8 | 37.3 | 185,942 | 91.5 | 27.5 |
| District 3 | 6.1 | 17.1 | 9.0 | 14.5 | 12.7 | 12.1 | 13.0 | 9.3 | 6.1 | 37.4 | 179,720 | 91.8 | 32.1 |
| District 4 | 5.3 | 15.1 | 10.5 | 11.5 | 11.0 | 12.5 | 14.4 | 11.4 | 8.3 | 41.8 | 166,481 | 91.6 | 23.4 |
| District 5 | 5.7 | 15.9 | 8.0 | 12.4 | 11.3 | 12.6 | 15.0 | 11.1 | 8.0 | 41.8 | 146,940 | 90.0 | 20.2 |
| District 6 | 5.9 | 16.3 | 10.9 | 12.1 | 11.9 | 11.7 | 13.6 | 10.6 | 7.2 | 39.1 | 179,059 | 90.6 | 28.5 |
| District 7 | 5.3 | 15.6 | 8.5 | 11.5 | 11.9 | 12.8 | 15.1 | 11.2 | 7.9 | 42.2 | 158,767 | 91.7 | 26.7 |
| District 8 | 5.2 | 15.3 | 12.6 | 12.0 | 12.0 | 12.8 | 14.4 | 9.7 | 6.2 | 39.0 | 209,422 | 94.7 | 42.2 |
| District 9 | 5.2 | 14.1 | 7.9 | 16.1 | 12.2 | 13.4 | 13.8 | 9.7 | 7.6 | 39.9 | 147,984 | 91.1 | 31.3 |
| District 10 | 5.3 | 15.8 | 7.9 | 11.1 | 11.5 | 13.9 | 15.3 | 11.4 | 7.9 | 43.7 | 161,787 | 91.8 | 24.9 |
| District 11 | 5.5 | 15.6 | 7.5 | 12.2 | 12.0 | 14.6 | 15.0 | 10.4 | 7.3 | 42.4 | 176,706 | 95.5 | 49.1 |
| District 12 | 5.5 | 15.0 | 13.4 | 14.5 | 11.7 | 11.9 | 12.5 | 9.6 | 5.9 | 36.4 | 201,262 | 90.9 | 34.9 |
| District 13 | 7.1 | 17.5 | 9.7 | 15.2 | 11.9 | 12.4 | 11.8 | 8.5 | 5.8 | 35.3 | 167,217 | 84.2 | 16.0 |
| District 14 | 6.6 | 16.9 | 8.1 | 15.7 | 11.6 | 12.0 | 12.9 | 9.5 | 6.6 | 37.0 | 168,640 | 89.7 | 35.0 |
| MINNESOTA | 6.2 | 16.9 | 8.7 | 13.6 | 12.9 | 11.9 | 13.5 | 9.5 | 6.9 | 38.4 | 1,386,106 | 93.6 | 37.3 |
| District 1 | 6.3 | 16.8 | 9.7 | 12.8 | 11.8 | 11.4 | 13.1 | 9.9 | 8.0 | 38.3 | 169,054 | 93.0 | 31.1 |
| District 2 | 6.3 | 18.2 | 8.4 | 12.4 | 13.9 | 13.0 | 13.4 | 8.6 | 5.8 | 38.1 | 188,407 | 95.1 | 40.7 |
| District 3 | 6.0 | 17.5 | 6.8 | 12.3 | 13.9 | 13.3 | 14.3 | 9.6 | 6.4 | 39.9 | 180,145 | 95.6 | 50.2 |
| District 4 | 6.3 | 17.0 | 9.1 | 15.2 | 13.1 | 11.5 | 12.7 | 9.0 | 6.1 | 36.5 | 183,096 | 92.7 | 44.7 |
| District 5 | 6.4 | 13.9 | 9.9 | 20.6 | 13.9 | 10.7 | 10.9 | 7.8 | 5.8 | 34.5 | 174,024 | 92.3 | 48.3 |
| District 6 | 6.6 | 18.7 | 9.0 | 12.4 | 13.5 | 12.8 | 13.5 | 8.0 | 5.4 | 37.3 | 189,114 | 94.2 | 32.5 |
| District 7 | 6.2 | 17.4 | 8.6 | 11.3 | 11.5 | 10.8 | 14.2 | 10.9 | 9.2 | 40.5 | 155,734 | 92.4 | 23.4 |
| District 8 | 5.4 | 15.4 | 7.9 | 11.3 | 11.7 | 11.8 | 15.5 | 12.3 | 8.9 | 43.5 | 146,532 | 93.4 | 24.6 |
| MISSISSIPPI | 6.0 | 17.4 | 10.0 | 12.4 | 12.7 | 12.2 | 12.9 | 9.7 | 6.6 | 38.3 | 754,558 | 85.3 | 22.3 |
| District 1 | 6.0 | 17.5 | 10.2 | 12.0 | 12.9 | 12.7 | 12.8 | 9.6 | 6.5 | 38.3 | 196,245 | 84.9 | 21.5 |
| District 2 | 6.2 | 18.2 | 10.2 | 12.7 | 12.8 | 11.0 | 12.7 | 9.6 | 6.5 | 37.4 | 181,139 | 82.4 | 19.0 |
| District 3 | 6.0 | 17.1 | 10.1 | 12.4 | 12.7 | 12.0 | 13.1 | 9.8 | 7.0 | 38.4 | 183,606 | 85.6 | 26.5 |
| District 4 | 5.9 | 17.1 | 9.6 | 12.4 | 12.7 | 12.8 | 13.0 | 9.9 | 6.6 | 39.0 | 193,568 | 88.0 | 22.0 |
| MISSOURI | 6.0 | 16.4 | 9.1 | 13.5 | 12.4 | 11.9 | 13.6 | 9.9 | 7.3 | 38.9 | 1,444,801 | 90.7 | 30.2 |
| District 1 | 6.2 | 15.4 | 9.7 | 17.3 | 12.8 | 11.2 | 12.8 | 8.6 | 6.0 | 36.0 | 182,026 | 90.4 | 33.5 |
| District 2 | 5.4 | 15.5 | 7.2 | 11.6 | 12.1 | 12.6 | 15.4 | 11.2 | 9.0 | 43.3 | 171,242 | 95.4 | 51.8 |
| District 3 | 5.9 | 17.2 | 7.8 | 12.6 | 13.2 | 12.6 | 14.1 | 10.1 | 6.6 | 39.8 | 187,630 | 91.4 | 28.3 |
| District 4 | 5.9 | 16.1 | 12.2 | 12.9 | 11.9 | 11.1 | 12.9 | 9.8 | 7.2 | 37.1 | 190,938 | 90.4 | 26.0 |
| District 5 | 6.5 | 16.2 | 8.4 | 15.7 | 12.5 | 11.7 | 13.2 | 8.9 | 6.9 | 37.5 | 173,326 | 91.3 | 29.8 |
| District 6 | 6.0 | 17.4 | 8.9 | 12.2 | 12.5 | 12.5 | 13.4 | 9.8 | 7.2 | 39.3 | 189,298 | 92.0 | 28.8 |
| District 7 | 5.9 | 16.8 | 10.2 | 13.2 | 11.7 | 11.5 | 12.9 | 10.2 | 7.6 | 38.2 | 188,907 | 89.2 | 24.9 |
| District 8 | 5.8 | 16.7 | 8.4 | 11.9 | 12.0 | 11.9 | 14.4 | 10.8 | 8.1 | 40.7 | 161,434 | 85.1 | 17.8 |
| MONTANA | 5.5 | 15.7 | 9.4 | 12.4 | 12.4 | 10.9 | 14.2 | 11.8 | 7.6 | 40.5 | 242,758 | 94.2 | 33.6 |
| At Large | 5.5 | 15.7 | 9.4 | 12.4 | 12.4 | 10.9 | 14.2 | 11.8 | 7.6 | 40.5 | 242,758 | 94.2 | 33.6 |
| NEBRASKA | 6.7 | 17.9 | 9.7 | 13.4 | 12.6 | 11.1 | 12.5 | 9.2 | 6.9 | 36.8 | 507,156 | 92.0 | 33.2 |
| District 1 | 6.4 | 17.3 | 11.9 | 13.1 | 12.4 | 11.0 | 12.3 | 8.9 | 6.5 | 36.0 | 180,154 | 92.6 | 33.0 |
| District 2 | 7.4 | 18.9 | 8.8 | 14.9 | 13.8 | 11.6 | 11.5 | 8.1 | 5.1 | 35.0 | 187,836 | 92.4 | 42.2 |
| District 3 | 6.2 | 17.4 | 8.3 | 11.8 | 11.4 | 10.8 | 13.9 | 11.0 | 9.3 | 40.2 | 139,166 | 91.0 | 23.7 |
| NEVADA | 6.0 | 16.4 | 8.2 | 14.6 | 13.3 | 12.8 | 12.4 | 9.9 | 6.2 | 38.4 | 707,075 | 86.9 | 25.7 |
| District 1 | 6.2 | 16.6 | 8.9 | 16.0 | 13.5 | 12.7 | 12.1 | 8.8 | 5.5 | 36.4 | 156,576 | 77.1 | 17.6 |
| District 2 | 5.7 | 15.6 | 8.6 | 14.3 | 12.1 | 12.1 | 13.8 | 11.3 | 6.6 | 39.6 | 168,190 | 88.4 | 27.3 |
| District 3 | 5.8 | 15.2 | 7.2 | 14.4 | 14.8 | 14.2 | 11.9 | 10.0 | 6.6 | 39.7 | 189,669 | 94.0 | 34.6 |
| District 4 | 6.3 | 18.5 | 8.3 | 13.7 | 12.8 | 12.3 | 12.0 | 9.7 | 6.2 | 37.3 | 192,640 | 86.1 | 21.1 |
| NEW HAMPSHIRE | 4.7 | 14.1 | 9.1 | 12.5 | 11.7 | 13.4 | 15.7 | 11.2 | 7.4 | 43.0 | 298,926 | 93.3 | 37.6 |
| District 1 | 4.3 | 14.3 | 9.4 | 12.4 | 12.1 | 13.4 | 15.8 | 10.9 | 7.3 | 42.5 | 152,651 | 93.4 | 38.2 |
| District 2 | 5.1 | 13.9 | 8.8 | 12.7 | 11.3 | 13.4 | 15.8 | 11.4 | 7.6 | 43.5 | 146,275 | 93.1 | 36.9 |
| NEW JERSEY | 5.8 | 16.0 | 8.5 | 12.9 | 12.9 | 13.5 | 13.8 | 9.4 | 7.2 | 40.2 | 2,192,217 | 90.3 | 41.2 |
| District 1 | 5.8 | 16.5 | 8.4 | 13.4 | 12.9 | 12.9 | 13.9 | 9.4 | 6.8 | 39.2 | 177,238 | 90.9 | 33.4 |
| District 2 | 4.9 | 15.6 | 8.2 | 11.9 | 11.6 | 12.9 | 15.0 | 11.7 | 8.2 | 42.8 | 162,961 | 88.1 | 27.2 |
| District 3 | 4.9 | 15.3 | 7.6 | 12.0 | 11.7 | 12.7 | 14.2 | 10.5 | 8.7 | 43.5 | 169,784 | 93.7 | 36.0 |
| District 4 | 6.7 | 17.8 | 7.7 | 10.6 | 11.2 | 12.7 | 14.2 | 10.4 | 8.5 | 41.1 | 195,483 | 94.1 | 44.3 |
| District 5 | 4.6 | 16.9 | 8.2 | 9.9 | 11.8 | 14.5 | 15.1 | 10.4 | 8.5 | 43.9 | 187,836 | 93.9 | 50.6 |
| District 6 | 6.0 | 16.2 | 10.2 | 13.9 | 13.3 | 13.5 | 12.9 | 8.3 | 5.6 | 37.7 | 200,240 | 88.7 | 41.1 |
| District 7 | 5.2 | 16.4 | 8.1 | 10.1 | 12.8 | 15.2 | 15.4 | 9.4 | 7.5 | 43.0 | 179,658 | 94.4 | 54.8 |
| District 8 | 7.5 | 14.6 | 8.6 | 19.3 | 16.4 | 12.1 | 10.7 | 6.3 | 4.7 | 35.1 | 179,717 | 81.2 | 35.3 |
| District 9 | 6.6 | 15.5 | 8.3 | 14.7 | 13.5 | 13.1 | 13.1 | 8.7 | 6.5 | 38.9 | 178,088 | 87.2 | 35.6 |
| District 10 | 6.6 | 15.8 | 9.4 | 15.9 | 15.3 | 13.4 | 13.4 | 12.1 | 8.0 | 36.7 | 194,484 | 86.4 | 32.3 |
| District 11 | 4.9 | 16.0 | 8.0 | 10.6 | 12.6 | 15.0 | 14.6 | 10.1 | 8.2 | 43.5 | 174,040 | 95.3 | 57.1 |
| District 12 | 5.2 | 16.0 | 9.0 | 12.5 | 13.0 | 13.6 | 13.8 | 9.5 | 7.2 | 40.4 | 192,688 | 90.2 | 47.1 |
| NEW MEXICO | 5.6 | 17.0 | 9.4 | 13.3 | 12.5 | 11.2 | 13.0 | 10.7 | 7.3 | 38.6 | 516,495 | 85.9 | 27.7 |
| District 1 | 5.3 | 15.7 | 8.7 | 14.5 | 13.0 | 11.7 | 13.3 | 10.6 | 7.1 | 39.1 | 163,642 | 88.5 | 34.5 |
| District 2 | 5.9 | 18.0 | 10.9 | 12.2 | 12.2 | 10.8 | 12.2 | 10.4 | 7.3 | 37.3 | 182,731 | 81.2 | 20.5 |

1. All persons 3 years old and over enrolled in nursery school through college and graduate or professional school.　2. Persons 25 years old and over.

**1208　MI(District 1)—NM(District 2)**

Items 15—27

## Table E. Congressional Districts 116th Congress — Households and Group Quarters

| STATE District | Households, 2019 | | | | | | Total in group quarters, 2018 | Group Quarters, 2010 | | | | |
|---|---|---|---|---|---|---|---|---|---|---|---|---|
| | Number | Average household size | Family households (percent) | Married couple family (percent) | Female family house-holder[1] | One person households (percent) | | Percent 65 years and over | Persons in correctional institutions | Persons in nursing facilities | Persons in college dormitories | Persons in military quarters |
| | 28 | 29 | 30 | 31 | 32 | 33 | 34 | 35 | 36 | 37 | 38 | 39 |
| MICHIGAN | 3,969,880 | 2.46 | 62.9 | 46.0 | 11.8 | 30.5 | 223,499 | 19.8 | 62,083 | 42,473 | 78,033 | 214 |
| District 1 | 298,748 | 2.26 | 61.9 | 49.1 | 8.2 | 31.6 | 22,476 | 20.2 | 12,990 | 4,884 | 5,714 | 129 |
| District 2 | 280,717 | 2.59 | 66.0 | 50.2 | 11.1 | 27.3 | 19,848 | 17.4 | 5,021 | 2,948 | 6,414 | 0 |
| District 3 | 283,800 | 2.59 | 64.8 | 48.7 | 11.0 | 27.2 | 18,157 | 18.1 | 7,012 | 3,695 | 5,151 | 82 |
| District 4 | 272,535 | 2.48 | 65.6 | 51.5 | 9.3 | 27.6 | 26,863 | 13.6 | 9,024 | 3,606 | 13,524 | 0 |
| District 5 | 284,841 | 2.33 | 62.4 | 41.7 | 15.2 | 32.3 | 9,073 | 33.6 | 1,630 | 3,103 | 1,303 | 3 |
| District 6 | 283,902 | 2.49 | 64.9 | 48.7 | 11.2 | 27.7 | 14,562 | 22.3 | 1,476 | 3,141 | 6,864 | 0 |
| District 7 | 280,096 | 2.45 | 65.6 | 50.6 | 10.2 | 29.0 | 24,468 | 15.0 | 14,438 | 2,638 | 4,594 | 0 |
| District 8 | 284,819 | 2.52 | 64.1 | 50.0 | 10.0 | 28.8 | 22,626 | 10.0 | 988 | 1,873 | 16,015 | 0 |
| District 9 | 304,748 | 2.34 | 60.3 | 43.3 | 11.2 | 33.2 | 6,238 | 50.7 | 1,143 | 3,267 | 0 | 0 |
| District 10 | 287,019 | 2.48 | 67.7 | 53.8 | 8.8 | 26.9 | 9,223 | 36.3 | 3,120 | 2,375 | 0 | 0 |
| District 11 | 292,059 | 2.50 | 67.4 | 54.9 | 7.9 | 27.7 | 6,112 | 45.5 | 183 | 2,598 | 1,581 | 0 |
| District 12 | 277,907 | 2.46 | 58.1 | 41.8 | 11.7 | 31.6 | 21,991 | 11.9 | 387 | 2,155 | 13,873 | 0 |
| District 13 | 264,757 | 2.50 | 55.0 | 25.6 | 22.4 | 39.4 | 10,737 | 21.2 | 114 | 2,806 | 2,287 | 0 |
| District 14 | 273,932 | 2.48 | 55.8 | 31.6 | 18.8 | 37.3 | 11,125 | 23.4 | 4,557 | 3,384 | 713 | 0 |
| MINNESOTA | 2,222,568 | 2.48 | 63.1 | 49.5 | 8.8 | 29.4 | 130,567 | 24.9 | 20,397 | 32,989 | 50,444 | 0 |
| District 1 | 273,667 | 2.40 | 63.9 | 50.4 | 8.0 | 29.2 | 22,551 | 23.9 | 4,808 | 5,137 | 9,723 | 0 |
| District 2 | 266,732 | 2.65 | 68.9 | 55.1 | 8.9 | 24.8 | 11,690 | 25.8 | 1,042 | 2,781 | 4,486 | 0 |
| District 3 | 281,550 | 2.58 | 67.9 | 55.2 | 8.5 | 26.2 | 4,737 | 38.8 | 551 | 1,864 | 644 | 0 |
| District 4 | 269,193 | 2.61 | 60.1 | 44.9 | 10.0 | 32.1 | 17,281 | 16.8 | 3,286 | 3,540 | 9,958 | 0 |
| District 5 | 309,153 | 2.27 | 48.8 | 35.0 | 9.9 | 39.5 | 21,886 | 20.9 | 876 | 5,425 | 8,740 | 0 |
| District 6 | 263,748 | 2.71 | 69.8 | 56.6 | 8.4 | 23.1 | 13,720 | 19.1 | 3,328 | 2,246 | 5,885 | 0 |
| District 7 | 275,213 | 2.36 | 64.1 | 51.5 | 8.1 | 29.5 | 17,498 | 40.1 | 1,022 | 7,051 | 6,518 | 0 |
| District 8 | 283,312 | 2.29 | 63.0 | 49.5 | 8.7 | 29.2 | 21,204 | 25.9 | 5,484 | 4,945 | 4,490 | 0 |
| MISSISSIPPI | 1,100,229 | 2.62 | 65.3 | 43.8 | 16.8 | 30.3 | 93,382 | 16 | 34,273 | 16,496 | 26,472 | 3,938 |
| District 1 | 278,795 | 2.69 | 67.3 | 47.7 | 14.5 | 28.3 | 18,350 | 22.9 | 2,943 | 4,381 | 7,786 | 0 |
| District 2 | 251,496 | 2.63 | 63.6 | 35.0 | 23.3 | 31.6 | 30,074 | 11.7 | 14,994 | 4,226 | 8,476 | 8 |
| District 3 | 280,709 | 2.55 | 63.0 | 45.0 | 14.2 | 32.8 | 24,310 | 18.4 | 9,560 | 4,625 | 5,828 | 601 |
| District 4 | 289,229 | 2.61 | 67.0 | 46.3 | 16.0 | 28.5 | 20,648 | 14.6 | 6,776 | 3,264 | 4,382 | 3,329 |
| MISSOURI | 2,458,337 | 2.43 | 62.9 | 47.2 | 11.1 | 30.3 | 174,478 | 22.7 | 41,956 | 44,866 | 52,869 | 10,217 |
| District 1 | 320,959 | 2.21 | 50.6 | 28.3 | 17.4 | 41.5 | 18,991 | 17.6 | 1,344 | 4,741 | 9,050 | 0 |
| District 2 | 307,657 | 2.41 | 67.0 | 55.5 | 8.3 | 28.4 | 10,865 | 55.7 | 2,300 | 7,231 | 1,046 | 0 |
| District 3 | 307,761 | 2.54 | 69.2 | 56.2 | 8.3 | 24.6 | 20,811 | 19.1 | 6,107 | 3,662 | 5,666 | 0 |
| District 4 | 293,159 | 2.52 | 65.2 | 49.7 | 10.8 | 27.5 | 35,610 | 13.4 | 6,015 | 5,555 | 11,262 | 10,217 |
| District 5 | 324,295 | 2.35 | 57.1 | 39.3 | 13.4 | 34.6 | 13,991 | 31.8 | 1,420 | 5,172 | 2,626 | 0 |
| District 6 | 290,828 | 2.58 | 65.5 | 52.2 | 9.2 | 28.5 | 27,445 | 20.5 | 11,433 | 6,573 | 8,326 | 0 |
| District 7 | 322,806 | 2.38 | 63.0 | 48.5 | 9.5 | 28.5 | 20,612 | 22.7 | 2,358 | 4,919 | 9,841 | 0 |
| District 8 | 290,872 | 2.44 | 66.9 | 49.7 | 11.8 | 27.4 | 26,153 | 23.9 | 10,979 | 7,013 | 5,052 | 0 |
| MONTANA | 437,651 | 2.38 | 61.5 | 48.8 | 8.1 | 30.7 | 28,943 | 19.2 | 5,338 | 5,200 | 8,332 | 678 |
| At Large | 437,651 | 2.38 | 61.5 | 48.8 | 8.1 | 30.7 | 28,943 | 19.2 | 5,338 | 5,200 | 8,332 | 678 |
| NEBRASKA | 771,444 | 2.44 | 63.9 | 50.4 | 9.1 | 28.7 | 51,012 | 24.4 | 8,084 | 13,519 | 22,073 | 443 |
| District 1 | 259,127 | 2.44 | 62.5 | 50.2 | 8.0 | 28.4 | 19,580 | 19.1 | 3,362 | 4,124 | 10,719 | 443 |
| District 2 | 262,877 | 2.55 | 64.5 | 48.9 | 10.8 | 28.0 | 13,779 | 17.6 | 2,325 | 2,632 | 4,750 | 0 |
| District 3 | 249,440 | 2.32 | 64.7 | 52.1 | 8.5 | 29.9 | 17,653 | 35.5 | 2,397 | 6,763 | 6,604 | 0 |
| NEVADA | 1,143,557 | 2.66 | 63.3 | 44.3 | 12.8 | 28.5 | 37,663 | 15.2 | 19,891 | 5,005 | 3,336 | 1,022 |
| District 1 | 257,962 | 2.72 | 56.6 | 32.5 | 16.8 | 35.2 | 10,068 | 11.3 | 4,771 | 954 | 1,211 | 0 |
| District 2 | 298,135 | 2.43 | 63.9 | 48.6 | 10.2 | 26.9 | 12,462 | 12.8 | 6,685 | 1,334 | 2,125 | 166 |
| District 3 | 321,058 | 2.66 | 65.1 | 48.5 | 11.1 | 26.3 | 2,330 | 58.1 | 352 | 1,044 | 0 | 0 |
| District 4 | 266,402 | 2.86 | 66.9 | 45.8 | 14.1 | 26.5 | 12,803 | 14.2 | 8,083 | 1,673 | 0 | 856 |
| NEW HAMPSHIRE | 541,396 | 2.44 | 64.0 | 50.6 | 8.8 | 27.6 | 41,073 | 19.5 | 4,851 | 7,767 | 22,820 | 454 |
| District 1 | 274,786 | 2.43 | 64.3 | 50.4 | 9.0 | 27.1 | 18,899 | 21.5 | 1,640 | 3,963 | 10,357 | 454 |
| District 2 | 266,610 | 2.44 | 63.6 | 50.7 | 8.6 | 28.2 | 22,174 | 17.9 | 3,211 | 3,804 | 12,463 | 0 |
| NEW JERSEY | 3,286,264 | 2.65 | 68.2 | 50.7 | 12.5 | 26.4 | 180,487 | 23.7 | 44,468 | 45,512 | 55,483 | 1,452 |
| District 1 | 270,997 | 2.64 | 66.3 | 46.3 | 14.8 | 28.3 | 11,191 | 33.0 | 2,221 | 3,995 | 2,424 | 0 |
| District 2 | 271,508 | 2.52 | 66.8 | 48.3 | 13.3 | 27.8 | 22,924 | 17.9 | 11,704 | 4,360 | 2,376 | 768 |
| District 3 | 284,794 | 2.53 | 68.1 | 52.0 | 11.7 | 27.0 | 14,974 | 26.5 | 6,383 | 4,169 | 0 | 360 |
| District 4 | 275,617 | 2.69 | 67.4 | 54.9 | 8.2 | 27.9 | 7,504 | 50.3 | 1,319 | 4,614 | 1,306 | 254 |
| District 5 | 267,665 | 2.70 | 72.1 | 58.8 | 10.0 | 23.4 | 12,595 | 39.2 | 1,041 | 5,005 | 4,364 | 0 |
| District 6 | 256,290 | 2.80 | 71.3 | 52.7 | 13.1 | 22.5 | 22,792 | 12.3 | 2,976 | 3,169 | 15,266 | 70 |
| District 7 | 271,992 | 2.66 | 72.4 | 60.2 | 8.6 | 23.0 | 11,460 | 37.9 | 2,789 | 3,543 | 129 | 0 |
| District 8 | 287,329 | 2.63 | 62.7 | 40.2 | 16.2 | 27.1 | 9,307 | 22.0 | 3,043 | 2,492 | 1,409 | 0 |
| District 9 | 270,244 | 2.80 | 70.4 | 48.6 | 14.8 | 25.2 | 5,231 | 26.3 | 1,068 | 1,145 | 468 | 0 |
| District 10 | 285,444 | 2.59 | 62.7 | 36.7 | 19.1 | 31.9 | 22,002 | 11.0 | 7,245 | 2,896 | 6,771 | 0 |
| District 11 | 270,984 | 2.59 | 70.3 | 58.6 | 8.2 | 25.2 | 16,592 | 30.3 | 411 | 5,741 | 9,278 | 0 |
| District 12 | 273,400 | 2.65 | 68.9 | 53.1 | 11.0 | 26.9 | 23,915 | 16.9 | 4,268 | 4,383 | 11,692 | 0 |
| NEW MEXICO | 793,420 | 2.59 | 62.6 | 41.8 | 14.7 | 31.3 | 42,907 | 14.3 | 17,907 | 5,567 | 8,478 | 1,789 |
| District 1 | 276,649 | 2.47 | 59.0 | 39.2 | 14.2 | 33.9 | 9,219 | 15.9 | 4,489 | 1,768 | 2,750 | 530 |
| District 2 | 254,925 | 2.68 | 64.8 | 44.7 | 14.4 | 29.4 | 22,077 | 11.7 | 9,874 | 2,334 | 3,613 | 815 |

1. No spouse present.

# Table E. Congressional Districts 116th Congress — Housing and Money Income

| STATE District | Housing units, 2019 | | | | | | Money income, 2019 | | |
| | Total (40) | Occupied units as a percent of all units (41) | Owner-occupied units as a percent of occupied units (42) | Median value¹ (dollars) (43) | Percent valued at $500,000 or more (44) | Median rent² (45) | Per capita income (dollars) (46) | Median income (dollars) (47) | Percent with income of $100,000 or more (48) |
|---|---|---|---|---|---|---|---|---|---|
| MICHIGAN | 4,629,605 | 85.7 | 71.6 | 169,600 | 5.4 | 888 | 32,892 | 59,584 | 26.1 |
| District 1 | 453,433 | 65.9 | 78.0 | 147,000 | 5.4 | 707 | 30,326 | 51,553 | 18.6 |
| District 2 | 324,672 | 86.5 | 73.2 | 172,200 | 3.7 | 874 | 29,265 | 59,356 | 23.5 |
| District 3 | 308,723 | 91.9 | 74.3 | 186,800 | 5.1 | 897 | 33,508 | 64,919 | 28.5 |
| District 4 | 352,752 | 77.3 | 79.3 | 139,800 | 2.2 | 738 | 29,414 | 54,166 | 22.1 |
| District 5 | 331,980 | 85.8 | 70.7 | 112,400 | 1.6 | 755 | 27,715 | 47,655 | 17.5 |
| District 6 | 333,415 | 85.1 | 71.3 | 163,700 | 5.7 | 803 | 30,467 | 56,520 | 23.2 |
| District 7 | 310,670 | 90.2 | 76.1 | 168,000 | 4.2 | 799 | 32,850 | 61,379 | 26.5 |
| District 8 | 305,376 | 93.3 | 74.2 | 243,300 | 8.8 | 938 | 39,350 | 74,841 | 36.8 |
| District 9 | 324,929 | 93.8 | 70.1 | 172,100 | 5.4 | 990 | 35,660 | 62,943 | 26.5 |
| District 10 | 318,805 | 90.0 | 81.1 | 210,300 | 3.9 | 890 | 34,017 | 67,472 | 31.5 |
| District 11 | 309,782 | 94.3 | 77.4 | 267,700 | 12.9 | 1,123 | 47,857 | 88,253 | 44.2 |
| District 12 | 298,737 | 93.0 | 65.4 | 169,700 | 6.2 | 1,030 | 34,920 | 62,253 | 28.3 |
| District 13 | 328,504 | 80.6 | 53.6 | 85,600 | 1.8 | 859 | 22,056 | 39,005 | 13.2 |
| District 14 | 327,827 | 83.6 | 55.4 | 151,900 | 7.0 | 992 | 31,441 | 50,438 | 23.2 |
| MINNESOTA | 2,477,515 | 89.7 | 71.9 | 246,700 | 9.2 | 1,016 | 39,025 | 74,593 | 35.3 |
| District 1 | 297,641 | 91.9 | 74.2 | 190,000 | 5.1 | 802 | 34,310 | 66,330 | 28.4 |
| District 2 | 278,569 | 95.8 | 77.4 | 290,200 | 10.4 | 1,173 | 41,934 | 90,531 | 45.0 |
| District 3 | 294,234 | 95.7 | 74.8 | 333,600 | 20.0 | 1,339 | 51,768 | 98,877 | 49.4 |
| District 4 | 286,906 | 93.8 | 64.2 | 267,300 | 10.2 | 1,085 | 39,620 | 75,665 | 37.5 |
| District 5 | 324,598 | 95.2 | 53.1 | 265,100 | 11.4 | 1,091 | 40,713 | 68,709 | 31.7 |
| District 6 | 277,399 | 95.1 | 79.3 | 266,600 | 7.4 | 1,011 | 38,157 | 84,159 | 41.1 |
| District 7 | 336,498 | 81.8 | 76.9 | 171,500 | 4.5 | 726 | 31,815 | 60,932 | 24.1 |
| District 8 | 381,670 | 74.2 | 77.8 | 190,200 | 5.0 | 787 | 32,481 | 61,659 | 26.0 |
| MISSISSIPPI | 1,339,047 | 82.2 | 67.3 | 128,200 | 3.1 | 777 | 25,301 | 45,792 | 18.1 |
| District 1 | 337,641 | 82.6 | 70.6 | 130,100 | 2.5 | 795 | 26,549 | 50,243 | 19.2 |
| District 2 | 314,726 | 79.9 | 58.9 | 96,800 | 2.0 | 695 | 20,877 | 37,372 | 12.8 |
| District 3 | 338,392 | 83.0 | 71.5 | 137,800 | 4.1 | 776 | 27,397 | 49,863 | 20.6 |
| District 4 | 348,288 | 83.0 | 67.5 | 146,300 | 3.4 | 850 | 26,018 | 47,340 | 19.4 |
| MISSOURI | 2,819,334 | 87.2 | 67.1 | 168,000 | 5.3 | 834 | 31,756 | 57,409 | 24.7 |
| District 1 | 370,556 | 86.6 | 50.8 | 120,600 | 4.4 | 869 | 30,261 | 50,163 | 19.6 |
| District 2 | 323,931 | 95.0 | 79.0 | 261,500 | 12.1 | 1,045 | 50,562 | 88,684 | 43.9 |
| District 3 | 351,928 | 87.4 | 77.6 | 189,500 | 5.4 | 848 | 32,870 | 69,621 | 30.4 |
| District 4 | 358,471 | 81.8 | 69.9 | 160,500 | 4.4 | 802 | 27,917 | 53,237 | 21.5 |
| District 5 | 367,355 | 88.3 | 57.5 | 151,900 | 4.3 | 921 | 30,780 | 55,239 | 22.7 |
| District 6 | 337,469 | 86.2 | 71.5 | 170,400 | 4.0 | 817 | 30,824 | 62,094 | 27.6 |
| District 7 | 362,150 | 89.1 | 62.5 | 152,500 | 3.6 | 745 | 26,853 | 47,679 | 17.3 |
| District 8 | 347,474 | 83.7 | 70.2 | 127,000 | 2.4 | 672 | 24,122 | 45,656 | 15.1 |
| MONTANA | 519,938 | 84.2 | 68.9 | 253,600 | 12.7 | 831 | 32,625 | 57,153 | 24.0 |
| At Large | 519,938 | 84.2 | 68.9 | 253,600 | 12.7 | 831 | 32,625 | 57,153 | 24.0 |
| NEBRASKA | 851,167 | 90.6 | 66.3 | 172,700 | 4.5 | 859 | 33,272 | 63,229 | 27.4 |
| District 1 | 280,254 | 92.5 | 65.2 | 182,000 | 5.0 | 849 | 32,965 | 63,921 | 27.8 |
| District 2 | 280,322 | 93.8 | 63.8 | 199,000 | 5.8 | 965 | 36,505 | 71,277 | 33.0 |
| District 3 | 290,591 | 85.8 | 70.0 | 132,100 | 2.8 | 714 | 29,900 | 55,729 | 20.9 |
| NEVADA | 1,285,681 | 88.9 | 56.6 | 317,800 | 16.4 | 1,168 | 33,575 | 63,276 | 28.4 |
| District 1 | 308,804 | 83.5 | 39.8 | 238,000 | 6.1 | 994 | 24,019 | 44,078 | 14.4 |
| District 2 | 323,746 | 92.1 | 61.4 | 349,100 | 22.2 | 1,124 | 37,731 | 69,972 | 31.7 |
| District 3 | 357,356 | 89.8 | 61.2 | 363,100 | 20.8 | 1,409 | 42,240 | 79,169 | 38.4 |
| District 4 | 295,775 | 90.1 | 62.0 | 290,400 | 11.0 | 1,203 | 28,817 | 62,241 | 26.0 |
| NEW HAMPSHIRE | 642,298 | 84.3 | 71.0 | 281,400 | 11.3 | 1,147 | 41,241 | 77,933 | 37.7 |
| District 1 | 324,309 | 84.7 | 69.6 | 297,400 | 14.2 | 1,191 | 42,619 | 79,996 | 38.8 |
| District 2 | 317,989 | 83.8 | 72.4 | 265,200 | 8.3 | 1,077 | 39,836 | 76,368 | 36.6 |
| NEW JERSEY | 3,641,854 | 90.2 | 63.3 | 348,800 | 25.5 | 1,376 | 44,888 | 85,751 | 43.6 |
| District 1 | 296,659 | 91.3 | 69.3 | 209,500 | 3.5 | 1,134 | 38,237 | 75,244 | 37.9 |
| District 2 | 389,667 | 69.7 | 71.5 | 223,100 | 10.0 | 1,144 | 37,281 | 68,127 | 33.1 |
| District 3 | 322,088 | 88.4 | 78.5 | 275,500 | 10.2 | 1,422 | 43,661 | 85,746 | 41.9 |
| District 4 | 299,932 | 91.9 | 75.0 | 382,700 | 31.6 | 1,398 | 48,021 | 91,212 | 46.6 |
| District 5 | 285,892 | 93.6 | 75.5 | 431,000 | 38.2 | 1,501 | 55,532 | 110,329 | 54.7 |
| District 6 | 275,124 | 93.2 | 61.1 | 365,600 | 21.9 | 1,499 | 41,148 | 91,654 | 46.6 |
| District 7 | 286,289 | 95.0 | 74.5 | 461,800 | 43.6 | 1,597 | 60,729 | 115,585 | 56.2 |
| District 8 | 308,578 | 93.1 | 29.2 | 384,500 | 31.6 | 1,384 | 37,546 | 65,658 | 33.3 |
| District 9 | 285,409 | 94.7 | 47.8 | 392,100 | 27.6 | 1,421 | 39,514 | 81,431 | 41.1 |
| District 10 | 311,698 | 91.6 | 39.9 | 319,200 | 16.4 | 1,231 | 33,299 | 61,975 | 29.5 |
| District 11 | 284,919 | 95.1 | 75.4 | 467,800 | 43.7 | 1,581 | 60,429 | 120,847 | 59.5 |
| District 12 | 295,599 | 92.5 | 64.7 | 361,200 | 26.7 | 1,384 | 44,196 | 87,559 | 44.4 |
| NEW MEXICO | 948,470 | 83.7 | 68.1 | 180,900 | 7.3 | 847 | 28,423 | 51,945 | 21.9 |
| District 1 | 305,362 | 90.6 | 64.4 | 209,500 | 7.3 | 890 | 32,793 | 55,318 | 24.6 |
| District 2 | 320,168 | 79.6 | 68.0 | 145,200 | 3.6 | 760 | 24,058 | 46,817 | 18.0 |

1. Specified owner-occupied units; $1,000,000 represents $1,000,000    2. Specified renter-occupied units.

# Table E. Congressional Districts 116th Congress — Poverty, Labor Force, Employment, and Social Security

| STATE District | Poverty, 2019 Persons below poverty level (percent) | Families below poverty level (percent) | Percent of households receiving food stamps in past 12 months | Civilian labor force, 2019 Total | Unemployment Total | Unemployment Rate[1] | Civilian employment,[2] 2019 Total | Management, business, science, and arts occupations | Service, sales, and office | Construction and production | Persons under 65 years of age with no health insurance, 2019 (percent) | Social Security beneficiaries, December 2020 Number | Rate[3] | Supplemental Security Income recipients, December 2019 |
|---|---|---|---|---|---|---|---|---|---|---|---|---|---|---|
| | 49 | 50 | 51 | 52 | 53 | 54 | 55 | 56 | 57 | 58 | 59 | 60 | 61 | 62 |
| MICHIGAN | 13.0 | 8.8 | 11.7 | 5,010,213 | 250,252 | 5.0 | 4,755,016 | 37.7 | 37.3 | 25.0 | 5.7 | 2,250,141 | 225.3 | 265,956 |
| District 1 | 11.7 | 7.5 | 10.4 | 328,854 | 16,958 | 5.2 | 311,167 | 32.8 | 40.7 | 26.5 | 6.6 | 216,011 | 309.9 | 14,574 |
| District 2 | 10.5 | 7.2 | 9.7 | 385,683 | 14,488 | 3.8 | 370,603 | 32.8 | 36.9 | 30.3 | 5.2 | 161,878 | 216.7 | 15,309 |
| District 3 | 12.2 | 8.3 | 9.2 | 396,781 | 17,844 | 4.5 | 378,444 | 36.1 | 36.1 | 27.7 | 6.2 | 148,252 | 197.1 | 16,038 |
| District 4 | 13.5 | 8.1 | 11.4 | 323,883 | 15,683 | 4.8 | 308,086 | 33.3 | 39.1 | 27.6 | 6.5 | 184,460 | 262.4 | 16,134 |
| District 5 | 17.4 | 12.4 | 19.9 | 312,042 | 23,521 | 7.5 | 288,175 | 31.1 | 40.7 | 28.2 | 5.2 | 178,321 | 265.2 | 28,376 |
| District 6 | 13.9 | 9.6 | 9.4 | 363,084 | 15,740 | 4.3 | 347,023 | 34.7 | 37.6 | 27.7 | 6.7 | 164,277 | 227.7 | 16,757 |
| District 7 | 10.9 | 7.9 | 10.1 | 344,077 | 14,240 | 4.1 | 329,730 | 35.9 | 35.1 | 29.0 | 5.4 | 172,315 | 242.7 | 13,254 |
| District 8 | 9.6 | 4.7 | 7.3 | 399,366 | 17,858 | 4.5 | 381,335 | 45.7 | 37.0 | 17.3 | 4.4 | 144,320 | 194.8 | 11,336 |
| District 9 | 9.3 | 6.5 | 9.7 | 392,093 | 17,710 | 4.5 | 373,863 | 38.7 | 37.8 | 23.6 | 5.6 | 152,038 | 211.7 | 20,406 |
| District 10 | 8.1 | 5.2 | 8.1 | 362,525 | 17,514 | 4.8 | 344,386 | 38.3 | 35.1 | 26.6 | 5.5 | 172,801 | 239.4 | 12,991 |
| District 11 | 5.7 | 3.9 | 4.1 | 397,527 | 11,808 | 3.0 | 385,501 | 53.4 | 32.1 | 14.4 | 3.7 | 146,435 | 199.0 | 8,029 |
| District 12 | 15.2 | 9.1 | 9.8 | 360,440 | 14,292 | 4.0 | 345,730 | 43.2 | 34.4 | 22.4 | 4.3 | 139,641 | 198.1 | 16,157 |
| District 13 | 26.1 | 21.8 | 26.1 | 311,727 | 26,419 | 8.5 | 285,246 | 26.4 | 43.4 | 30.2 | 8.6 | 130,642 | 194.3 | 47,265 |
| District 14 | 19.3 | 15.0 | 19.9 | 332,131 | 26,177 | 7.9 | 305,737 | 39.3 | 38.6 | 22.1 | 6.6 | 138,750 | 201.1 | 29,330 |
| MINNESOTA | 9.0 | 5.2 | 7.4 | 3,112,899 | 98,727 | 3.2 | 3,010,452 | 42.8 | 35.4 | 21.8 | 4.8 | 1,069,913 | 189.7 | 92,761 |
| District 1 | 9.3 | 5.1 | 6.1 | 371,318 | 8,910 | 2.4 | 361,919 | 38.2 | 35.2 | 26.6 | 6.2 | 142,502 | 209.9 | 9,263 |
| District 2 | 5.9 | 3.6 | 4.8 | 412,808 | 13,671 | 3.3 | 398,600 | 43.8 | 36.4 | 19.8 | 3.9 | 121,157 | 168.8 | 7,248 |
| District 3 | 5.2 | 3.0 | 4.5 | 410,408 | 10,152 | 2.5 | 399,997 | 50.7 | 33.6 | 15.7 | 3.0 | 128,889 | 176.5 | 7,455 |
| District 4 | 10.7 | 6.5 | 9.7 | 388,735 | 12,902 | 3.3 | 375,442 | 45.9 | 36.2 | 17.9 | 5.2 | 121,356 | 168.6 | 17,855 |
| District 5 | 13.7 | 8.6 | 10.9 | 430,188 | 17,002 | 4.0 | 412,946 | 50.6 | 35.4 | 14.0 | 6.3 | 98,533 | 136.0 | 21,349 |
| District 6 | 6.3 | 3.5 | 5.6 | 411,769 | 12,463 | 3.0 | 398,498 | 40.0 | 34.5 | 25.5 | 3.9 | 122,080 | 167.5 | 7,276 |
| District 7 | 10.4 | 5.7 | 8.3 | 347,355 | 11,108 | 3.2 | 335,990 | 35.3 | 35.5 | 29.2 | 5.4 | 158,852 | 237.8 | 9,970 |
| District 8 | 10.6 | 6.2 | 8.4 | 340,318 | 12,519 | 3.7 | 327,060 | 34.7 | 36.8 | 28.5 | 5.0 | 176,544 | 263.0 | 12,345 |
| MISSISSIPPI | 19.6 | 14.8 | 13.5 | 1,342,991 | 87,817 | 6.6 | 1,240,752 | 32.8 | 38.6 | 28.7 | 12.9 | 681,219 | 228.9 | 114,080 |
| District 1 | 14.9 | 10.2 | 10.0 | 356,713 | 15,263 | 4.3 | 339,015 | 31.3 | 35.8 | 32.9 | 11.3 | 176,307 | 229.3 | 23,713 |
| District 2 | 25.9 | 20.5 | 20.0 | 292,794 | 27,800 | 9.5 | 264,587 | 28.3 | 43.4 | 28.3 | 14.2 | 163,340 | 235.9 | 41,776 |
| District 3 | 18.9 | 13.9 | 12.0 | 336,328 | 16,946 | 5.1 | 316,982 | 37.2 | 36.2 | 26.6 | 12.2 | 167,574 | 226.8 | 24,559 |
| District 4 | 19.4 | 15.3 | 12.6 | 357,156 | 27,808 | 8.0 | 320,168 | 33.6 | 39.9 | 26.5 | 14.1 | 173,998 | 224.3 | 24,032 |
| MISSOURI | 12.9 | 8.8 | 10.2 | 3,078,235 | 116,062 | 3.8 | 2,942,459 | 38.4 | 37.9 | 23.7 | 9.9 | 1,323,195 | 215.6 | 134,636 |
| District 1 | 16.4 | 12.2 | 15.6 | 384,067 | 19,912 | 5.2 | 363,473 | 40.4 | 40.2 | 19.4 | 9.6 | 133,984 | 184.1 | 27,037 |
| District 2 | 5.1 | 3.1 | 2.7 | 404,759 | 10,926 | 2.7 | 393,602 | 54.1 | 31.6 | 14.3 | 3.9 | 162,747 | 216.4 | 6,096 |
| District 3 | 8.2 | 5.6 | 7.1 | 411,051 | 13,276 | 3.2 | 397,183 | 37.3 | 37.4 | 25.3 | 7.9 | 175,526 | 218.6 | 10,654 |
| District 4 | 14.8 | 9.1 | 10.7 | 381,758 | 14,879 | 4.0 | 352,761 | 35.1 | 39.0 | 25.9 | 11.1 | 168,921 | 217.8 | 15,515 |
| District 5 | 14.3 | 10.5 | 10.4 | 412,068 | 14,693 | 3.6 | 396,442 | 35.7 | 39.6 | 24.7 | 12.0 | 151,040 | 194.2 | 19,394 |
| District 6 | 11.2 | 6.7 | 7.3 | 385,428 | 12,473 | 3.2 | 371,596 | 37.6 | 35.7 | 26.7 | 8.5 | 160,471 | 206.5 | 11,762 |
| District 7 | 16.0 | 10.3 | 12.2 | 371,290 | 14,359 | 3.9 | 356,220 | 33.4 | 42.3 | 24.4 | 13.7 | 182,273 | 231.3 | 17,606 |
| District 8 | 18.0 | 13.8 | 15.4 | 327,814 | 15,544 | 4.8 | 311,182 | 31.3 | 38.5 | 30.2 | 12.9 | 188,233 | 255.6 | 26,572 |
| MONTANA | 12.6 | 7.9 | 8.7 | 545,370 | 22,004 | 4.1 | 520,261 | 39.1 | 38.3 | 22.6 | 8.2 | 244,937 | 229.2 | 17,491 |
| At Large | 12.6 | 7.9 | 8.7 | 545,370 | 22,004 | 4.1 | 520,261 | 39.1 | 38.3 | 22.6 | 8.2 | 244,937 | 229.2 | 17,491 |
| NEBRASKA | 9.9 | 6.2 | 7.6 | 1,049,680 | 34,133 | 3.3 | 1,008,957 | 40.0 | 35.5 | 24.4 | 8.3 | 357,164 | 184.6 | 28,920 |
| District 1 | 9.6 | 5.2 | 7.3 | 360,636 | 10,553 | 3.0 | 346,046 | 38.9 | 36.6 | 24.5 | 7.8 | 117,981 | 181.0 | 8,915 |
| District 2 | 9.5 | 6.5 | 7.8 | 377,866 | 15,274 | 4.1 | 360,275 | 45.4 | 35.4 | 19.3 | 8.4 | 106,622 | 155.7 | 11,417 |
| District 3 | 10.7 | 6.9 | 7.8 | 311,178 | 8,306 | 2.7 | 302,636 | 35.0 | 34.5 | 30.5 | 8.6 | 132,561 | 221.8 | 8,588 |
| NEVADA | 12.5 | 8.7 | 10.5 | 1,568,228 | 79,000 | 5.1 | 1,479,868 | 30.4 | 47.3 | 22.3 | 11.2 | 565,671 | 183.7 | 56,484 |
| District 1 | 20.5 | 16.9 | 16.9 | 351,825 | 21,435 | 6.1 | 330,032 | 19.0 | 55.2 | 25.8 | 18.6 | 114,816 | 161.2 | 18,582 |
| District 2 | 11.0 | 6.1 | 8.9 | 385,754 | 17,123 | 4.5 | 367,060 | 31.8 | 40.2 | 28.0 | 9.4 | 153,943 | 208.9 | 10,399 |
| District 3 | 8.2 | 5.5 | 5.6 | 461,943 | 19,905 | 4.3 | 441,089 | 39.0 | 46.6 | 14.3 | 7.6 | 152,647 | 178.1 | 10,070 |
| District 4 | 11.4 | 8.5 | 12.1 | 368,706 | 20,537 | 5.7 | 341,687 | 28.7 | 48.0 | 23.3 | 10.2 | 144,265 | 186.5 | 17,433 |
| NEW HAMPSHIRE | 7.3 | 4.2 | 6.0 | 759,931 | 22,609 | 3.0 | 735,493 | 42.5 | 35.6 | 21.9 | 6.2 | 317,389 | 233.4 | 17,880 |
| District 1 | 7.7 | 4.7 | 5.9 | 390,533 | 12,384 | 3.2 | 376,508 | 41.6 | 36.5 | 21.9 | 5.6 | 156,800 | 228.3 | 9,183 |
| District 2 | 6.8 | 3.7 | 6.2 | 369,398 | 10,225 | 2.8 | 358,985 | 43.6 | 34.6 | 21.9 | 5.6 | 160,589 | 238.6 | 8,697 |
| NEW JERSEY | 9.2 | 6.2 | 7.4 | 4,731,226 | 220,476 | 4.7 | 4,496,699 | 44.9 | 36.3 | 18.8 | 7.7 | 1,651,408 | 185.9 | 173,226 |
| District 1 | 10.0 | 6.4 | 10.0 | 388,240 | 20,275 | 5.2 | 367,470 | 41.9 | 37.7 | 20.4 | 5.8 | 146,367 | 201.4 | 19,920 |
| District 2 | 10.4 | 7.2 | 9.8 | 355,589 | 20,794 | 5.9 | 333,247 | 34.6 | 43.2 | 22.3 | 7.1 | 171,366 | 242.3 | 18,114 |
| District 3 | 5.9 | 3.7 | 4.8 | 388,413 | 17,830 | 4.7 | 362,070 | 43.4 | 38.6 | 18.0 | 4.3 | 178,334 | 242.3 | 9,967 |
| District 4 | 7.9 | 4.8 | 4.7 | 369,642 | 13,428 | 3.6 | 354,792 | 48.4 | 36.2 | 15.4 | 5.3 | 164,394 | 219.7 | 8,521 |
| District 5 | 5.0 | 3.4 | 3.3 | 390,557 | 13,952 | 3.6 | 376,605 | 52.5 | 33.9 | 13.6 | 4.4 | 142,565 | 194.0 | 8,462 |
| District 6 | 9.4 | 6.7 | 6.4 | 399,481 | 15,355 | 3.8 | 383,985 | 45.6 | 33.4 | 21.0 | 8.2 | 120,065 | 162.3 | 12,326 |
| District 7 | 5.4 | 3.1 | 2.9 | 410,144 | 18,611 | 4.5 | 391,227 | 53.7 | 32.3 | 14.0 | 5.5 | 130,872 | 178.2 | 5,962 |
| District 8 | 14.7 | 10.9 | 13.0 | 425,248 | 16,248 | 3.8 | 408,303 | 36.4 | 36.5 | 27.0 | 16.0 | 90,559 | 118.2 | 24,186 |
| District 9 | 11.5 | 8.6 | 10.8 | 411,590 | 16,273 | 4.0 | 395,067 | 38.7 | 36.8 | 24.5 | 12.9 | 124,167 | 162.9 | 17,721 |
| District 10 | 14.7 | 11.2 | 14.4 | 400,163 | 27,764 | 6.9 | 371,882 | 36.6 | 43.0 | 20.4 | 10.9 | 109,903 | 144.3 | 27,592 |
| District 11 | 3.8 | 2.2 | 2.1 | 399,095 | 16,214 | 4.1 | 382,828 | 56.0 | 32.6 | 11.4 | 3.5 | 140,299 | 195.5 | 5,588 |
| District 12 | 10.9 | 6.9 | 6.3 | 393,064 | 23,732 | 6.0 | 369,223 | 50.1 | 32.9 | 17.0 | 7.7 | 132,517 | 177.4 | 14,867 |
| NEW MEXICO | 18.2 | 13.7 | 16.4 | 972,732 | 52,489 | 5.5 | 907,775 | 37.6 | 41.0 | 21.3 | 9.8 | 453,282 | 216.2 | 60,930 |
| District 1 | 16.3 | 12.9 | 15.4 | 349,161 | 17,548 | 5.1 | 328,194 | 42.5 | 39.9 | 17.6 | 8.6 | 144,320 | 208.8 | 18,142 |
| District 2 | 20.7 | 15.4 | 17.9 | 303,204 | 17,672 | 5.9 | 280,181 | 32.0 | 41.2 | 26.8 | 9.7 | 153,143 | 217.0 | 23,096 |

1. Percent of civilian labor force.    2. Persons 16 years old and over.    3. Per 1,000 resident population estimated in the 2019 American Community Survey.

# Table E. Congressional Districts 116th Congress — **Agriculture**

| STATE District | | Land in farms | | | Value of products sold | | | | Government payments | |
| --- | --- | --- | --- | --- | --- | --- | --- | --- | --- | --- |
| | Number of farms | Acres | Average size of farm (acres) | Irrigated land (acres) | Total ($1,000) | Average per farm (dollars) | Percent from crops | Percent from livestock and poultry products | Total ($1,000) | Average per farm receiving payments (dollars) |
| | 63 | 64 | 65 | 66 | 67 | 68 | 69 | 70 | 71 | 72 |
| MICHIGAN | 47,641 | 9,764,090 | 205 | 7,214,667 | 8,220,936 | 172,560 | 56.5 | 43.5 | 167,189 | 10,892 |
| District 1 | 6,952 | 1,082,209 | 156 | 449,001 | 357,657 | 51,447 | 59.1 | 40.9 | 5,710 | 4,827 |
| District 2 | 3,516 | 578,489 | 165 | 390,567 | 958,662 | 272,657 | 58.1 | 41.9 | 4,619 | 8,118 |
| District 3 | 3,868 | 773,809 | 200 | 586,992 | 846,663 | 218,889 | 40.5 | 59.5 | 14,843 | 12,009 |
| District 4 | 9,471 | 2,085,639 | 220 | 1,557,525 | 1,553,454 | 164,022 | 44.7 | 55.3 | 32,154 | 8,711 |
| District 5 | 2,612 | 607,782 | 233 | 503,507 | 348,992 | 133,611 | 78.8 | 21.2 | 12,854 | 9,578 |
| District 6 | 5,312 | 1,106,709 | 208 | 855,681 | 1,567,911 | 295,164 | 59.4 | 40.6 | 21,268 | 14,925 |
| District 7 | 7,424 | 1,618,131 | 218 | 1,300,677 | 997,521 | 134,364 | 70.7 | 29.3 | 40,056 | 13,185 |
| District 8 | 1,988 | 293,838 | 148 | 219,882 | 175,102 | 88,079 | 64.9 | 35.1 | 5,626 | 18,568 |
| District 9 | 23 | D | D | D | 3,176 | 138,087 | D | D | D | D |
| District 10 | 5,921 | 1,578,233 | 267 | 1,324,773 | 1,367,410 | 230,942 | 56.2 | 43.8 | 29,581 | 11,729 |
| District 11 | 167 | 8,360 | 50 | 4,172 | 12,429 | 74,425 | 98.8 | 1.2 | 413 | 13,767 |
| District 12 | 291 | 25,639 | 88 | 18,446 | 26,613 | 91,454 | D | D | 8 | 1,143 |
| District 13 | 71 | 1,872 | 26 | D | 4,984 | 70,197 | D | D | D | D |
| District 14 | 25 | D | D | 136 | 361 | 14,440 | 56.5 | 43.5 | D | D |
| MINNESOTA | 68,822 | 25,516,982 | 371 | 20,054,132 | 18,395,390 | 267,289 | 55.4 | 44.6 | 394,491 | 9,568 |
| District 1 | 18,080 | 6,432,710 | 356 | 5,514,901 | 6,521,603 | 360,708 | 49.1 | 50.9 | 139,174 | 10,630 |
| District 2 | 4,323 | 1,084,055 | 251 | 862,034 | 968,711 | 224,083 | 56.2 | 43.8 | 24,916 | 10,648 |
| District 3 | 514 | 56,055 | 109 | 37,496 | 59,877 | 116,492 | 84.5 | 15.5 | 374 | 0 |
| District 4 | 306 | 26,745 | 87 | 20,368 | 18,073 | 59,062 | D | D | D | 5,194 |
| District 5 | 24 | 1,241 | 52 | 875 | 5,449 | 227,042 | D | D | 8,131 | D |
| District 6 | 5,231 | 981,166 | 188 | 719,093 | 927,726 | 177,352 | 47.0 | 53.0 | 213,860 | 4,055 |
| District 7 | 30,340 | 14,905,007 | 491 | 11,889,723 | 9,079,273 | 299,251 | 61.9 | 38.1 | 7,721 | 9,993 |
| District 8 | 10,004 | 2,030,003 | 203 | 1,009,642 | 814,678 | 81,435 | 39.1 | 60.9 | 213,785 | 3,575 |
| MISSISSIPPI | 34,988 | 10,415,136 | 298 | 4,174,210 | 6,195,969 | 177,088 | 37.0 | 63.0 | 323,801 | 14,986 |
| District 1 | 9,940 | 2,535,334 | 255 | 815,727 | 741,685 | 74,616 | 51.9 | 48.1 | 38,445 | 7,803 |
| District 2 | 8,921 | 4,904,171 | 550 | 2,870,319 | 2,405,086 | 269,598 | 28.6 | 71.4 | 147,529 | 27,936 |
| District 3 | 9,886 | 2,098,567 | 212 | 369,795 | 2,283,032 | 230,936 | 94.6 | 5.4 | 21,401 | 7,317 |
| District 4 | 6,241 | 877,064 | 141 | 118,369 | 766,165 | 122,763 | 91.4 | 8.6 | 6,410 | 5,658 |
| MISSOURI | 95,320 | 27,781,883 | 291 | 13,486,275 | 10,525,937 | 110,427 | 52.0 | 48.0 | 323,801 | 10,366 |
| District 1 | 60 | 4,038 | 67 | 1,734 | 3,917 | 65,283 | 95.5 | 4.5 | 78 | 6,000 |
| District 2 | 232 | 56,434 | 243 | 15,973 | 19,886 | 85,716 | 95.4 | 4.6 | 223 | 5,718 |
| District 3 | 11,671 | 2,582,240 | 221 | 1,017,877 | 763,114 | 65,385 | 46.9 | 53.1 | 16,790 | 5,510 |
| District 4 | 23,209 | 6,321,590 | 272 | 2,819,491 | 2,390,061 | 102,980 | 41.6 | 58.4 | 57,041 | 8,620 |
| District 5 | 3,599 | 1,113,070 | 309 | 772,182 | 521,467 | 144,892 | 75.8 | 24.2 | 11,009 | 6,817 |
| District 6 | 25,268 | 9,149,092 | 362 | 5,265,587 | 3,473,872 | 137,481 | 37.0 | 63.0 | 148,718 | 10,355 |
| District 7 | 12,945 | 2,313,829 | 179 | 662,898 | 1,412,550 | 109,119 | 90.6 | 9.4 | 6,746 | 5,324 |
| District 8 | 18,336 | 6,241,590 | 340 | 2,930,533 | 1,941,071 | 105,861 | 71.4 | 28.6 | 83,197 | 19,457 |
| MONTANA | 27,048 | 58,122,878 | 2,149 | 9,901,226 | 3,520,623 | 130,162 | 45.0 | 55.0 | 284,244 | 27,014 |
| At Large | 27,048 | 58,122,878 | 2,149 | 9,901,226 | 3,520,623 | 130,162 | 45.0 | 55.0 | 284,244 | 27,014 |
| NEBRASKA | 46,332 | 44,986,821 | 971 | 19,460,222 | 21,983,429 | 474,476 | 42.4 | 57.6 | 639,975 | 20,745 |
| District 1 | 12,307 | 5,386,864 | 438 | 4,463,921 | 5,299,528 | 430,611 | 45.6 | 54.4 | 134,195 | 15,708 |
| District 2 | 731 | 177,616 | 243 | 152,180 | 102,706 | 140,501 | 92.6 | 7.4 | 3,892 | 12,761 |
| District 3 | 33,294 | 39,422,341 | 1,184 | 14,844,121 | 16,581,195 | 498,024 | 41.0 | 59.0 | 501,888 | 22,812 |
| NEVADA | 3,423 | 6,128,153 | 1,790 | 573,785 | 665,758 | 194,495 | 41.5 | 58.5 | 5,049 | 16,183 |
| District 1 | 16 | 674 | 42 | D | 394 | 24,625 | 52.0 | 49.0 | D | D |
| District 2 | 2,440 | 5,359,802 | 2,197 | 464,131 | 456,516 | 187,097 | 43.7 | 56.3 | 3,915 | 16,313 |
| District 3 | 72 | 3,477 | 48 | D | 1,501 | 20,847 | 78.0 | 22.0 | D | D |
| District 4 | 895 | 764,200 | 854 | 108,581 | 207,348 | 231,674 | 36.3 | 63.7 | D | D |
| NEW HAMPSHIRE | 4,123 | 425,393 | 103 | 85,793 | 187,794 | 45,548 | 57.4 | 42.6 | 3,494 | 11,344 |
| District 1 | 1,429 | 114,081 | 80 | 21,437 | 43,425 | 30,388 | 65.6 | 34.4 | 1,243 | 12,186 |
| District 2 | 2,694 | 311,312 | 116 | 64,356 | 144,369 | 53,589 | 54.9 | 45.1 | 2,251 | 10,927 |
| NEW JERSEY | 9,883 | 734,084 | 74 | 411,785 | 1,097,951 | 111,095 | 89.7 | 10.3 | 7,503 | 10,071 |
| District 1 | 323 | 19,769 | 61 | 12,012 | 50,873 | 157,502 | 97.0 | 3.0 | 126 | 6,632 |
| District 2 | 2,470 | 242,801 | 98 | 164,407 | 522,859 | 211,684 | 94.6 | 5.4 | 3,879 | 12,930 |
| District 3 | 950 | 95,629 | 101 | 45,970 | 98,998 | 104,208 | 92.3 | 7.7 | 829 | 12,014 |
| District 4 | 973 | 46,949 | 48 | 25,493 | 103,382 | 106,251 | 84.7 | 15.3 | 431 | 10,512 |
| District 5 | 1,813 | 109,258 | 60 | 46,313 | 102,405 | 56,484 | 69.5 | 30.5 | 885 | 8,349 |
| District 6 | 129 | 6,165 | 48 | 2,383 | 13,972 | 108,310 | D | D | D | D |
| District 7 | 2,409 | 161,144 | 67 | 89,857 | 130,857 | 54,320 | 85.3 | 14.7 | 1,076 | 6,482 |
| District 8 | 3 | D | D | 20 | D | D | D | D | D | D |
| District 9 | 10 | 77 | 8 | 55 | 0 | 0 | D | D | D | D |
| District 10 | 4 | D | D | 5 | 60 | 15,000 | D | D | D | D |
| District 11 | 287 | 11,013 | 38 | 3,451 | 17,851 | 62,199 | 95.0 | 5.0 | 255 | 6,538 |
| District 12 | 512 | 41,235 | 81 | 21,819 | 54,054 | 105,574 | 85.7 | 14.3 | 63,660 | 18,436 |
| NEW MEXICO | 25,044 | 40,659,836 | 1,624 | 806,138 | 2,582,343 | 103,112 | 25.2 | 74.8 | 2,343 | 16,385 |
| District 1 | 2,270 | 1,912,502 | 843 | 20,328 | 62,931 | 27,723 | 30.0 | 70.0 | 25,845 | 19,580 |
| District 2 | 9,762 | 20,343,252 | 2,084 | 390,375 | 1,562,618 | 160,072 | 30.7 | 69.3 | | |

# Table E. Congressional Districts 116th Congress — Nonfarm Employment and Payroll

| STATE District | Number of establishments | Total | Manufacturing | Construction | Wholesale trade | Retail trade | Health care and social assistance | Finance and Insurance | Real estate and rental and leasing | Professional, scientific, and technical services | Information | Total (mil dol) | Average per employee (dollars) |
|---|---|---|---|---|---|---|---|---|---|---|---|---|---|
| | | | | | Private nonfarm employment and payroll, 2019 — Employment — Percent by selected industries | | | | | | | Annual payroll | |
| | 73 | 74 | 75 | 76 | 77 | 78 | 79 | 80 | 81 | 82 | 83 | 84 | 85 |
| MICHIGAN | 222,226 | 3,978,872 | 15.4 | 4.0 | 4.6 | 11.6 | 16.1 | 4.2 | 1.4 | 7.3 | 1.7 | 203,510 | 51,148 |
| District 1 | 19,789 | 212,518 | 13.8 | 6.3 | 2.8 | 17.0 | 19.1 | 3.8 | 1.3 | 3.2 | 1.4 | 8,747 | 41,158 |
| District 2 | 16,788 | 334,668 | 27.6 | 4.5 | 6.1 | 11.8 | 11.3 | 2.5 | 1.0 | 3.3 | 1.0 | 15,378 | 45,949 |
| District 3 | 16,112 | 316,825 | 20.1 | 4.1 | 5.8 | 9.2 | 16.9 | 4.5 | 1.2 | 5.1 | 1.7 | 15,987 | 50,461 |
| District 4 | 13,431 | 189,799 | 16.6 | 6.0 | 3.3 | 16.7 | 16.8 | 3.5 | 1.5 | 3.8 | 1.2 | 8,095 | 42,649 |
| District 5 | 13,270 | 216,173 | 13.3 | 4.1 | 4.7 | 14.4 | 22.6 | 3.0 | 1.3 | 3.6 | 1.2 | 9,529 | 44,079 |
| District 6 | 14,611 | 241,147 | 22.1 | 4.3 | 4.2 | 12.6 | 15.5 | 3.2 | 1.5 | 4.7 | 0.9 | 11,847 | 49,129 |
| District 7 | 12,827 | 207,578 | 22.1 | 3.9 | 3.7 | 13.1 | 14.3 | 4.1 | 0.9 | 4.1 | 1.5 | 9,687 | 46,665 |
| District 8 | 16,210 | 241,163 | 11.7 | 5.0 | 3.1 | 14.0 | 17.9 | 6.6 | 1.6 | 6.3 | 1.8 | 11,000 | 45,612 |
| District 9 | 17,547 | 278,333 | 14.7 | 3.8 | 5.6 | 11.8 | 20.0 | 2.2 | 2.1 | 11.5 | 1.7 | 14,923 | 53,615 |
| District 10 | 15,069 | 219,776 | 29.2 | 5.7 | 3.5 | 14.5 | 12.6 | 2.2 | 1.0 | 6.2 | 0.9 | 9,994 | 45,473 |
| District 11 | 24,313 | 479,556 | 10.2 | 3.7 | 7.4 | 11.5 | 12.2 | 5.2 | 1.4 | 15.3 | 2.3 | 30,209 | 62,994 |
| District 12 | 15,588 | 303,653 | 12.3 | 2.7 | 3.3 | 12.7 | 20.7 | 3.1 | 1.2 | 8.5 | 2.6 | 17,767 | 58,510 |
| District 13 | 10,245 | 220,090 | 13.9 | 2.5 | 4.4 | 8.2 | 21.5 | 1.1 | 1.1 | 2.9 | 1.3 | 12,054 | 54,770 |
| District 14 | 15,621 | 342,997 | 5.5 | 3.2 | 4.1 | 7.1 | 18.9 | 11.7 | 2.8 | 13.1 | 3.2 | 22,516 | 65,645 |
| MINNESOTA | 151,495 | 2,729,420 | 11.6 | 4.7 | 5.3 | 11.0 | 17.6 | 6.1 | 1.4 | 6.7 | 2.2 | 153,756 | 56,333 |
| District 1 | 16,825 | 302,222 | 17.2 | 4.0 | 3.0 | 12.8 | 18.7 | 3.1 | 0.9 | 13.4 | 1.9 | 14,511 | 48,015 |
| District 2 | 17,146 | 281,765 | 12.8 | 5.9 | 5.9 | 13.1 | 13.5 | 5.0 | 1.2 | 4.2 | 2.8 | 14,455 | 51,303 |
| District 3 | 23,626 | 480,326 | 13.0 | 4.3 | 6.9 | 11.0 | 12.5 | 9.0 | 2.3 | 7.9 | 2.5 | 31,653 | 65,899 |
| District 4 | 17,931 | 363,996 | 8.2 | 3.7 | 5.0 | 9.7 | 20.2 | 6.1 | 1.6 | 5.1 | 2.4 | 21,397 | 58,784 |
| District 5 | 21,200 | 524,418 | 6.0 | 3.6 | 4.3 | 6.3 | 19.6 | 9.0 | 1.7 | 9.3 | 3.1 | 36,147 | 68,927 |
| District 6 | 17,285 | 237,465 | 16.1 | 9.8 | 5.2 | 14.6 | 16.5 | 2.9 | 1.1 | 3.2 | 1.2 | 10,963 | 46,165 |
| District 7 | 19,222 | 235,595 | 19.5 | 4.9 | 6.4 | 14.6 | 21.8 | 3.6 | 0.6 | 2.9 | 1.4 | 9,528 | 40,444 |
| District 8 | 17,579 | 217,524 | 9.1 | 5.5 | 2.8 | 15.4 | 25.2 | 3.5 | 1.1 | 3.4 | 1.4 | 9,160 | 42,110 |
| MISSISSIPPI | 59,130 | 958,126 | 15.5 | 4.8 | 4.0 | 14.3 | 18.4 | 3.4 | 1.0 | 3.3 | 1.4 | 37,731 | 39,379 |
| District 1 | 14,732 | 241,840 | 22.7 | 3.5 | 4.6 | 15.0 | 14.9 | 2.7 | 0.9 | 2.3 | 1.1 | 8,954 | 37,026 |
| District 2 | 12,410 | 187,875 | 16.6 | 4.2 | 3.9 | 14.1 | 19.5 | 2.7 | 1.0 | 2.5 | 1.4 | 7,085 | 37,712 |
| District 3 | 17,319 | 271,158 | 10.7 | 5.7 | 4.9 | 13.7 | 21.4 | 4.7 | 1.3 | 4.1 | 2.1 | 11,526 | 42,507 |
| District 4 | 14,272 | 233,110 | 14.5 | 5.8 | 2.3 | 16.0 | 18.1 | 3.1 | 1.1 | 3.7 | 1.0 | 9,392 | 40,289 |
| MISSOURI | 151,816 | 2,547,310 | 10.9 | 5.1 | 5.1 | 12.1 | 16.5 | 5.4 | 1.4 | 6.5 | 2.2 | 125,302 | 49,190 |
| District 1 | 19,858 | 422,074 | 11.2 | 5.0 | 6.0 | 6.9 | 14.8 | 4.2 | 1.4 | 5.9 | 2.3 | 24,981 | 59,187 |
| District 2 | 24,915 | 458,995 | 4.2 | 5.0 | 3.7 | 11.4 | 15.6 | 8.6 | 1.6 | 10.4 | 3.1 | 26,988 | 58,798 |
| District 3 | 17,679 | 242,693 | 15.8 | 7.8 | 4.5 | 15.6 | 14.3 | 3.6 | 1.1 | 3.6 | 2.3 | 9,866 | 40,651 |
| District 4 | 16,056 | 207,184 | 12.1 | 4.8 | 3.4 | 16.7 | 20.1 | 5.4 | 1.3 | 4.3 | 1.4 | 7,785 | 37,577 |
| District 5 | 20,409 | 418,272 | 10.3 | 5.8 | 8.1 | 9.8 | 15.5 | 6.2 | 1.3 | 11.1 | 2.5 | 23,707 | 56,679 |
| District 6 | 16,337 | 216,796 | 14.5 | 5.0 | 4.3 | 16.5 | 17.1 | 4.6 | 1.5 | 3.3 | 1.5 | 8,955 | 41,308 |
| District 7 | 19,395 | 315,900 | 12.7 | 4.1 | 5.1 | 13.7 | 17.3 | 3.5 | 2.0 | 3.9 | 2.3 | 12,762 | 40,399 |
| District 8 | 16,521 | 203,786 | 16.2 | 4.3 | 4.1 | 16.6 | 24.5 | 3.6 | 1.2 | 2.3 | 1.3 | 7,280 | 35,726 |
| MONTANA | 38,959 | 375,176 | 5.4 | 7.4 | 4.1 | 15.5 | 19.2 | 4.4 | 1.6 | 5.1 | 2.3 | 16,051 | 42,784 |
| At Large | 38,959 | 375,176 | 5.4 | 7.4 | 4.1 | 15.5 | 19.2 | 4.4 | 1.6 | 5.1 | 2.3 | 16,051 | 42,784 |
| NEBRASKA | 54,939 | 856,242 | 11.7 | 5.7 | 4.7 | 12.8 | 16.1 | 8.5 | 1.4 | 4.7 | 2.4 | 39,433 | 46,053 |
| District 1 | 17,142 | 251,065 | 14.1 | 5.9 | 3.8 | 13.9 | 16.9 | 7.3 | 1.3 | 5.4 | 2.4 | 10,702 | 42,626 |
| District 2 | 18,692 | 366,092 | 6.9 | 6.3 | 4.7 | 11.6 | 16.0 | 11.3 | 1.9 | 5.6 | 3.2 | 19,043 | 52,016 |
| District 3 | 18,739 | 207,289 | 19.0 | 5.1 | 6.3 | 15.6 | 17.3 | 4.6 | 0.8 | 2.6 | 1.3 | 8,062 | 38,891 |
| NEVADA | 68,567 | 1,261,577 | 3.9 | 7.0 | 3.3 | 11.8 | 10.9 | 3.2 | 2.4 | 4.9 | 1.6 | 58,122 | 46,071 |
| District 1 | 18,323 | 434,480 | 1.2 | 4.8 | 1.7 | 10.8 | 10.9 | 2.0 | 2.4 | 3.9 | 1.2 | 18,808 | 43,289 |
| District 2 | 19,411 | 290,037 | 8.8 | 7.5 | 5.5 | 12.8 | 13.0 | 2.6 | 2.0 | 5.1 | 1.7 | 14,383 | 49,589 |
| District 3 | 19,971 | 333,237 | 3.6 | 8.7 | 3.2 | 12.2 | 9.0 | 5.0 | 3.2 | 5.8 | 2.1 | 15,787 | 47,373 |
| District 4 | 10,386 | 164,188 | 3.7 | 10.0 | 4.0 | 15.1 | 13.2 | 3.7 | 2.0 | 5.2 | 2.0 | 7,178 | 43,719 |
| NEW HAMPSHIRE | 38,494 | 620,164 | 11.0 | 4.8 | 4.0 | 15.8 | 15.6 | 4.7 | 1.3 | 5.7 | 2.6 | 33,048 | 53,289 |
| District 1 | 19,955 | 308,249 | 10.7 | 5.0 | 4.2 | 16.2 | 15.1 | 6.5 | 1.7 | 5.8 | 3.1 | 16,810 | 54,535 |
| District 2 | 18,099 | 274,270 | 12.7 | 5.2 | 4.0 | 17.6 | 18.2 | 2.7 | 1.0 | 5.7 | 2.3 | 13,997 | 51,033 |
| NEW JERSEY | 233,888 | 3,805,357 | 5.9 | 4.3 | 7.3 | 11.9 | 16.4 | 5.2 | 1.6 | 8.8 | 2.3 | 239,862 | 63,033 |
| District 1 | 16,079 | 270,116 | 8.1 | 5.4 | 6.5 | 14.7 | 21.9 | 2.2 | 1.4 | 5.5 | 1.6 | 13,542 | 50,136 |
| District 2 | 17,137 | 234,321 | 6.7 | 6.0 | 4.7 | 15.1 | 17.4 | 2.5 | 1.5 | 3.6 | 0.8 | 10,057 | 42,920 |
| District 3 | 16,863 | 256,915 | 4.7 | 4.2 | 5.5 | 15.8 | 20.8 | 7.5 | 2.0 | 7.2 | 1.8 | 13,190 | 51,340 |
| District 4 | 22,273 | 304,038 | 4.7 | 5.5 | 4.1 | 16.6 | 20.5 | 4.3 | 2.1 | 7.8 | 2.7 | 14,932 | 49,112 |
| District 5 | 22,847 | 299,731 | 5.7 | 4.3 | 8.2 | 14.7 | 21.2 | 3.7 | 1.3 | 6.9 | 1.6 | 18,382 | 61,328 |
| District 6 | 18,859 | 332,169 | 5.3 | 5.1 | 8.1 | 11.6 | 14.5 | 3.2 | 1.8 | 14.1 | 2.8 | 21,173 | 63,740 |
| District 7 | 23,166 | 366,679 | 6.4 | 5.2 | 7.2 | 12.9 | 15.6 | 5.3 | 1.4 | 10.3 | 5.0 | 28,374 | 77,382 |
| District 8 | 16,158 | 255,067 | 5.1 | 3.1 | 6.7 | 10.7 | 14.0 | 16.5 | 1.5 | 4.4 | 2.1 | 18,905 | 74,119 |
| District 9 | 21,467 | 313,905 | 10.1 | 5.2 | 11.2 | 9.5 | 14.3 | 2.6 | 2.5 | 7.4 | 2.7 | 19,061 | 60,721 |
| District 10 | 13,724 | 221,070 | 5.2 | 3.3 | 5.0 | 10.2 | 20.4 | 4.8 | 2.0 | 5.3 | 1.4 | 13,812 | 62,480 |
| District 11 | 25,385 | 437,518 | 5.7 | 3.9 | 8.1 | 9.8 | 14.5 | 5.9 | 1.8 | 14.5 | 2.0 | 34,670 | 79,242 |
| District 12 | 19,041 | 345,886 | 5.7 | 3.0 | 11.7 | 9.2 | 14.1 | 5.3 | 1.4 | 13.5 | 2.1 | 24,123 | 69,742 |
| NEW MEXICO | 43,804 | 644,537 | 4.1 | 6.5 | 3.1 | 14.5 | 19.6 | 3.8 | 1.5 | 9.0 | 1.7 | 28,177 | 43,717 |
| District 1 | 16,343 | 276,712 | 4.7 | 7.1 | 4.0 | 13.0 | 19.1 | 4.4 | 1.5 | 11.7 | 2.4 | 12,488 | 45,131 |
| District 2 | 12,887 | 178,896 | 4.0 | 7.5 | 2.4 | 15.8 | 20.6 | 3.2 | 1.5 | 4.1 | 1.1 | 7,388 | 41,295 |

# Table E. Congressional Districts 116th Congress — **Land Area and Population Characteristics**

| STATE District | Representative, 117th Congress | Land area,[1] 2018 (sq mi) | Total persons | Per square mile | White | Black | American Indian, Alaska Native | Asian and Pacific Islander | Some other race (percent) | Two or more races (percent) | Hispanic or Latino[2] (percent) | Non-Hispanic White alone (percent) | Percent female | Percent foreign-born | Percent born in state of residence |
|---|---|---|---|---|---|---|---|---|---|---|---|---|---|---|---|
| | | 1 | 2 | 3 | 4 | 5 | 6 | 7 | 8 | 9 | 10 | 11 | 12 | 13 | 14 |
| **NEW MEXICO— Cont'd** | | | | | | | | | | | | | | | |
| District 3 | Teresa Leger Fernandez (D) | 44,965.606 | 699,985 | 15.567 | 65.3 | 1.9 | 18.7 | 1.4 | 8.8 | 3.9 | 40.8 | 36.691 | 51.0 | 6.2 | 57.2 |
| **NEW YORK** | | 47,123.586 | 19,453,561 | 412.820 | 63.2 | 15.9 | 0.4 | 8.6 | 8.6 | 3.3 | 19.3 | 55.057 | 51.4 | 22.4 | 63.1 |
| District 1 | Lee M. Zeldin (R) | 650.214 | 713,168 | 1,096.820 | 85.5 | 5.2 | 0.1 | 4.1 | 3.0 | 2.1 | 17.9 | 71.728 | 51.5 | 13.3 | 77.7 |
| District 2 | Andrew R. Garbarino (R) | 180.916 | 698,974 | 3,863.528 | 73.9 | 10.0 | 0.5 | 4.2 | 8.9 | 2.6 | 23.6 | 60.527 | 50.9 | 17.2 | 75.6 |
| District 3 | Thomas R. Souzzi (D) | 254.822 | 725,746 | 2,848.051 | 73.3 | 2.9 | 0.5 | 16.7 | 4.3 | 2.3 | 11.1 | 66.598 | 50.6 | 23.5 | 68.8 |
| District 4 | Kathleen M. Rice (D) | 110.741 | 730,314 | 6,594.793 | 64.5 | 15.2 | 0.2 | 7.3 | 8.7 | 4.1 | 22.4 | 52.900 | 51.1 | 23.8 | 68.6 |
| District 5 | Gregory W. Meeks (D) | 51.927 | 759,001 | 14,616.693 | 17.3 | 49.2 | 0.7 | 15.2 | 13.9 | 3.7 | 20.0 | 10.571 | 52.8 | 44.4 | 48.2 |
| District 6 | Grace Meng (D) | 29.794 | 714,299 | 23,974.592 | 41.3 | 5.4 | 0.4 | 40.4 | 9.2 | 3.3 | 19.9 | 32.383 | 51.7 | 50.4 | 43.6 |
| District 7 | Nydia M. Velázquez (D) | 16.140 | 698,794 | 43,295.787 | 48.5 | 11.7 | 0.6 | 18.4 | 16.6 | 4.4 | 39.1 | 31.738 | 50.7 | 34.9 | 45.6 |
| District 8 | Hakeem S. Jeffries (D) | 28.419 | 776,825 | 27,334.706 | 31.6 | 51.9 | 0.2 | 6.1 | 7.0 | 3.1 | 17.1 | 25.238 | 55.0 | 32.0 | 53.8 |
| District 9 | Yvette D. Clarke (D) | 15.548 | 720,316 | 46,328.531 | 36.2 | 46.7 | 0.3 | 7.3 | 5.8 | 3.7 | 12.1 | 32.138 | 53.4 | 36.8 | 47.8 |
| District 10 | Jerrold Nadler (D) | 14.046 | 732,732 | 52,166.595 | 68.7 | 3.7 | 0.3 | 18.7 | 5.5 | 3.1 | 13.4 | 61.593 | 51.1 | 30.6 | 44.9 |
| District 11 | Nicole Malliotakis (R) | 64.931 | 737,390 | 11,356.517 | 69.4 | 7.7 | 0.2 | 15.4 | 4.9 | 2.4 | 17.3 | 58.276 | 51.4 | 30.9 | 62.3 |
| District 12 | Carolyn B. Maloney (D) | 14.792 | 725,760 | 49,064.359 | 71.0 | 5.4 | 0.1 | 15.5 | 4.7 | 3.4 | 14.4 | 62.434 | 52.4 | 26.9 | 41.1 |
| District 13 | Adriano Espaillat (D) | 10.165 | 751,661 | 73,945.991 | 30.0 | 29.5 | 0.6 | 4.5 | 30.0 | 5.5 | 54.5 | 15.634 | 52.4 | 34.2 | 48.1 |
| District 14 | Alexandria Ocasio Cortez (D) | 28.582 | 696,664 | 24,374.222 | 41.4 | 10.5 | 0.6 | 17.8 | 25.2 | 4.4 | 49.9 | 21.558 | 49.3 | 45.8 | 43.2 |
| District 15 | Ritchie Torres (D) | 14.528 | 739,390 | 50,894.135 | 16.3 | 42.0 | 0.4 | 1.8 | 35.3 | 4.1 | 64.0 | 2.642 | 53.6 | 33.2 | 53.8 |
| District 16 | Jamaal Bowman (D) | 78.380 | 739,893 | 9,439.819 | 44.4 | 32.0 | 0.7 | 5.1 | 13.7 | 4.0 | 28.1 | 34.763 | 52.4 | 29.7 | 55.9 |
| District 17 | Mondaire Jones (D) | 382.598 | 737,355 | 1,927.232 | 69.1 | 11.3 | 0.1 | 6.1 | 10.6 | 2.8 | 22.7 | 59.516 | 51.2 | 23.3 | 61.3 |
| District 18 | Sean Patrick Maloney (D) | 1,354.079 | 718,624 | 530.711 | 77.2 | 10.6 | 0.3 | 3.2 | 5.4 | 3.3 | 18.3 | 66.619 | 50.0 | 11.7 | 70.6 |
| District 19 | Antonio Delgado (D) | 7,937.223 | 701,011 | 88.319 | 87.1 | 4.9 | 0.3 | 1.6 | 3.0 | 3.1 | 8.4 | 82.722 | 50.0 | 6.7 | 74.6 |
| District 20 | Paul Tonko (D) | 1,231.485 | 725,669 | 589.263 | 78.3 | 9.3 | 0.1 | 5.4 | 2.2 | 4.8 | 6.7 | 74.902 | 51.3 | 9.9 | 74.2 |
| District 21 | Elise M. Stefanik (R) | 15,114.265 | 694,835 | 45.972 | 92.5 | 2.8 | 0.5 | 1.2 | 0.7 | 2.2 | 3.4 | 90.405 | 48.8 | 3.3 | 77.1 |
| District 22 | Claudia Tenney (R) | 5,077.273 | 688,391 | 135.583 | 88.7 | 3.9 | 0.2 | 3.0 | 1.0 | 3.3 | 4.0 | 86.700 | 50.4 | 6.0 | 80.8 |
| District 23 | Tom Reed (R) | 7,371.894 | 687,583 | 93.271 | 90.4 | 2.9 | 0.6 | 2.4 | 1.3 | 2.4 | 4.2 | 87.953 | 50.4 | 3.9 | 74.1 |
| District 24 | John Katko (R) | 2,388.575 | 701,841 | 293.833 | 83.3 | 8.5 | 0.5 | 2.8 | 1.2 | 3.8 | 4.7 | 81.102 | 51.1 | 6.3 | 80.0 |
| District 25 | Joseph D. Morelle (D) | 510.145 | 714,657 | 1,400.890 | 75.0 | 15.8 | 0.2 | 3.9 | 2.0 | 3.1 | 9.4 | 69.215 | 51.6 | 9.3 | 72.6 |
| District 26 | Brian Higgins (D) | 219.161 | 703,114 | 3,208.208 | 70.4 | 18.3 | 0.6 | 4.7 | 3.0 | 3.1 | 7.0 | 67.719 | 52.1 | 8.1 | 79.0 |
| District 27 | Chris Jacobs (R) | 3,972.943 | 719,554 | 181.114 | 92.7 | 2.6 | 0.6 | 1.0 | 1.2 | 1.9 | 2.9 | 91.169 | 50.2 | 3.2 | 85.2 |
| **NORTH CAROLINA** | | 48,623.015 | 10,488,084 | 215.702 | 68.1 | 21.5 | 1.2 | 3.1 | 3.4 | 2.8 | 9.8 | 62.472 | 51.4 | 8.4 | 56.0 |
| District 1 | G. K. Butterfield (D) | 5,871.515 | 763,500 | 130.035 | 47.1 | 44.5 | 0.5 | 1.9 | 3.3 | 2.6 | 9.0 | 41.506 | 52.7 | 8.4 | 64.0 |
| District 2 | Deborah K. Ross (D) | 2,697.968 | 888,547 | 329.339 | 71.8 | 19.2 | 0.4 | 2.6 | 3.2 | 2.7 | 9.7 | 66.419 | 50.9 | 7.2 | 54.8 |
| District 3 | Gregory F. Murphy (R) | 7,218.130 | 761,753 | 105.533 | 73.2 | 19.8 | 0.4 | 1.5 | 1.5 | 3.6 | 8.4 | 67.379 | 49.1 | 4.5 | 52.2 |
| District 4 | David E. Price (D) | 732.440 | 873,270 | 1,192.275 | 60.0 | 22.6 | 0.5 | 9.4 | 4.9 | 2.6 | 11.4 | 54.589 | 51.6 | 16.7 | 42.6 |
| District 5 | Virginia Foxx (R) | 3,969.162 | 765,013 | 192.739 | 79.3 | 14.7 | 0.5 | 1.6 | 1.8 | 2.1 | 10.0 | 71.846 | 51.7 | 5.9 | 64.0 |
| District 6 | Kathy E. Manning (D) | 3,910.779 | 791,470 | 202.382 | 70.9 | 19.9 | 0.6 | 1.7 | 3.9 | 3.0 | 10.5 | 65.331 | 52.1 | 7.4 | 63.2 |
| District 7 | David Rouzer (R) | 5,946.891 | 816,402 | 137.282 | 74.6 | 17.9 | 0.9 | 1.0 | 2.8 | 2.9 | 9.7 | 67.978 | 52.3 | 6.1 | 58.6 |
| District 8 | Richard Hudson (R) | 2,954.147 | 815,055 | 275.902 | 64.6 | 23.0 | 1.2 | 3.1 | 3.7 | 4.5 | 11.0 | 58.569 | 50.5 | 7.5 | 53.7 |
| District 9 | Dan Bishop (R) | 3,874.743 | 796,413 | 205.540 | 64.4 | 19.6 | 7.6 | 3.0 | 2.9 | 2.5 | 7.6 | 59.974 | 51.5 | 7.9 | 56.7 |
| District 10 | Patrick T. McHenry (R) | 2,590.31 | 771,791 | 297.953 | 81.2 | 12.6 | 0.5 | 1.9 | 2.0 | 1.9 | 7.1 | 76.549 | 51.7 | 5.2 | 62.7 |
| District 11 | Madison Cawthorn (R) | 6,604.757 | 772,612 | 116.978 | 89.5 | 3.3 | 1.8 | 1.2 | 2.6 | 1.8 | 6.7 | 85.577 | 51.1 | 5.3 | 58.6 |
| District 12 | Alma S. Adams (D) | 420.002 | 891,792 | 2,123.304 | 44.9 | 37.5 | 0.4 | 5.8 | 8.0 | 3.3 | 15.9 | 38.000 | 52.0 | 16.7 | 41.7 |
| District 13 | Ted Budd (R) | 1,832.171 | 780,466 | 425.979 | 68.0 | 22.8 | 0.5 | 3.7 | 2.0 | 3.0 | 8.5 | 62.467 | 51.4 | 8.7 | 58.6 |
| **NORTH DAKOTA** | | 68,995.863 | 762,062 | 11.045 | 85.8 | 2.9 | 5.4 | 1.8 | 0.8 | 3.3 | 4.0 | 83.638 | 49.0 | 4.1 | 61.8 |
| At Large | Kelly Armstrong (R) | 68,995.863 | 762,062 | 11.045 | 85.8 | 2.9 | 5.4 | 1.8 | 0.8 | 3.3 | 4.0 | 83.638 | 49.0 | 4.1 | 61.8 |
| **OHIO** | | 40,858.762 | 11,689,100 | 286.086 | 80.9 | 12.6 | 0.2 | 2.3 | 1.0 | 2.9 | 4.0 | 78.344 | 51.0 | 4.8 | 74.7 |
| District 1 | Steve Chabot (R) | 686.675 | 749,773 | 1,091.889 | 70.7 | 21.4 | 0.0 | 4.1 | 0.7 | 3.0 | 3.4 | 68.298 | 50.7 | 6.8 | 70.3 |
| District 2 | Brad R. Wenstrup (R) | 3,221.371 | 730,151 | 226.658 | 86.2 | 9.2 | 0.1 | 1.4 | 0.5 | 2.6 | 2.4 | 84.454 | 51.6 | 2.9 | 74.4 |
| District 3 | Joyce Beatty (D) | 228.086 | 813,890 | 3,568.347 | 52.8 | 34.6 | 0.2 | 4.5 | 3.2 | 4.7 | 7.2 | 49.205 | 51.3 | 12.9 | 64.8 |
| District 4 | Jim Jordan (R) | 4,665.374 | 712,261 | 152.670 | 89.3 | 5.2 | 0.2 | 1.3 | 0.9 | 3.1 | 3.9 | 86.867 | 50.2 | 2.1 | 82.7 |
| District 5 | Robert E. Latta (R) | 5,626.051 | 721,212 | 128.192 | 91.2 | 3.0 | 0.3 | 1.5 | 1.4 | 2.6 | 6.0 | 87.656 | 50.7 | 2.6 | 79.5 |
| District 6 | Bill Johnson (R) | 7,214.612 | 698,284 | 96.787 | 95.2 | 2.1 | 0.0 | 0.5 | 0.5 | 1.6 | 1.2 | 94.511 | 50.3 | 1.1 | 69.7 |
| District 7 | Bob Gibbs (R) | 3,864.889 | 727,011 | 188.107 | 92.1 | 4.2 | 0.2 | 0.7 | 0.4 | 2.4 | 2.9 | 90.192 | 50.4 | 1.5 | 83.8 |
| District 8 | Warren Davidson (R) | 2,449.408 | 733,811 | 299.587 | 87.5 | 6.8 | 0.1 | 2.4 | 0.8 | 2.3 | 3.7 | 84.665 | 50.9 | 4.4 | 73.9 |
| District 9 | Marcy Kaptur (D) | 463.033 | 697,570 | 1,506.523 | 74.5 | 15.9 | 0.4 | 2.2 | 2.6 | 4.4 | 10.9 | 68.093 | 52.2 | 5.5 | 74.6 |
| District 10 | Michael R. Turner (R) | 1,129.433 | 723,716 | 640.778 | 75.5 | 16.2 | 0.4 | 2.2 | 1.2 | 4.4 | 3.1 | 73.888 | 51.6 | 4.5 | 70.1 |
| District 11 | Vacant | 244.630 | 684,617 | 2,798.582 | 39.8 | 52.9 | 0.4 | 2.4 | 1.5 | 3.0 | 4.8 | 37.158 | 52.1 | 5.4 | 74.2 |
| District 12 | Troy Balderson (R) | 2,271.949 | 788,335 | 346.986 | 86.5 | 5.0 | 0.2 | 4.9 | 0.6 | 2.8 | 2.6 | 84.660 | 50.9 | 6.4 | 71.1 |
| District 13 | Tim Ryan (D) | 893.985 | 704,191 | 787.699 | 81.2 | 12.6 | 0.1 | 2.4 | 0.5 | 3.2 | 3.5 | 78.706 | 51.6 | 4.1 | 76.6 |
| District 14 | David P. Joyce (R) | 1,955.163 | 714,870 | 365.632 | 90.2 | 4.9 | 0.1 | 1.9 | 0.5 | 2.3 | 3.3 | 87.708 | 51.0 | 4.8 | 76.8 |
| District 15 | Vacant | 4,738.806 | 769,664 | 162.417 | 89.3 | 4.7 | 0.2 | 2.8 | 0.5 | 2.4 | 2.4 | 87.600 | 49.4 | 4.2 | 77.1 |
| District 16 | Anthony Gonzalez (R) | 1,205.297 | 719,744 | 597.151 | 92.7 | 1.9 | 0.1 | 2.5 | 0.8 | 1.9 | 2.6 | 90.829 | 50.7 | 6.1 | 77.7 |
| **OKLAHOMA** | | 68,596.53 | 3,956,971 | 57.685 | 72.4 | 7.3 | 8.0 | 2.4 | 2.3 | 7.6 | 11.1 | 64.898 | 50.4 | 6.1 | 60.6 |
| District 1 | Kevin Hern (R) | 1,632.178 | 809,500 | 495.963 | 74.1 | 8.6 | 6.2 | 3.4 | 2.6 | 8.0 | 11.9 | 63.073 | 51.0 | 8.1 | 57.7 |
| District 2 | Markwayne Mullin (R) | 20,996.459 | 747,337 | 35.593 | 65.9 | 3.2 | 17.2 | 0.9 | 1.9 | 11.0 | 5.7 | 63.124 | 50.2 | 2.3 | 61.7 |
| District 3 | Frank D. Lucas (R) | 34,116.351 | 782,091 | 22.924 | 79.8 | 3.6 | 6.7 | 1.6 | 2.3 | 6.0 | 10.6 | 72.697 | 49.9 | 5.3 | 63.9 |
| District 4 | Tom Cole (R) | 9,777.405 | 792,928 | 81.097 | 75.1 | 7.1 | 5.9 | 2.5 | 1.8 | 7.5 | 8.8 | 69.458 | 49.8 | 4.6 | 60.1 |
| District 5 | Stephanie I. Bice (R) | 2,074.048 | 825,115 | 397.828 | 70.1 | 13.2 | 4.8 | 3.5 | 2.9 | 5.5 | 17.6 | 56.522 | 51.1 | 9.7 | 59.7 |
| **OREGON** | | 72,458.362 | 4,217,737 | 58.209 | 83.5 | 1.8 | 1.2 | 4.9 | 3.6 | 4.9 | 13.4 | 74.897 | 50.5 | 9.7 | 45.6 |
| District 1 | Suzanne Bonamici (D) | 3,007.117 | 858,875 | 285.614 | 78.5 | 1.8 | 0.7 | 9.4 | 4.8 | 4.9 | 15.0 | 69.344 | 50.5 | 14.5 | 43.7 |
| District 2 | Cliff Bentz (R) | 69,451.245 | 841,022 | 12.110 | 89.7 | 0.8 | 2.3 | 1.3 | 2.1 | 3.7 | 14.2 | 78.874 | 50.2 | 6.1 | 43.1 |

1. Dry land or land partially or temporarily covered by water.    2. May be of any race.

# Table E. Congressional Districts 116th Congress — **Age and Education**

| STATE District | Population and population characteristics, 2019 (cont.) | | | | | | | | | | Education, 2019 | | |
| | Age (percent) | | | | | | | | | | Total Enrollment[1] | Attainment[2] (percent) | |
| | Under 5 years | 5 to 17 years | 18 to 24 years | 25 to 34 years | 35 to 44 years | 45 to 54 years | 55 to 64 years | 65 to 74 years | 75 years and over | Median age | | High school graduate or more | Bachelor's degree or more |
| | 15 | 16 | 17 | 18 | 19 | 20 | 21 | 22 | 23 | 24 | 25 | 26 | 27 |
|---|---|---|---|---|---|---|---|---|---|---|---|---|---|
| **NEW MEXICO—Cont'd** | | | | | | | | | | | | | |
| District 3 | 5.5 | 17.3 | 8.5 | 13.2 | 12.1 | 11.2 | 13.4 | 11.0 | 7.6 | 39.3 | 170,122 | 87.8 | 27.7 |
| **NEW YORK** | 5.8 | 14.9 | 9.1 | 14.7 | 12.5 | 12.8 | 13.4 | 9.5 | 7.4 | 39.2 | 4,594,259 | 87.6 | 37.8 |
| District 1 | 5.6 | 15.4 | 8.9 | 11.8 | 11.7 | 13.9 | 14.6 | 10.5 | 7.4 | 42.3 | 171,134 | 91.5 | 38.3 |
| District 2 | 5.5 | 15.6 | 9.3 | 12.7 | 11.6 | 14.3 | 14.8 | 8.8 | 7.4 | 40.6 | 169,891 | 90.2 | 32.6 |
| District 3 | 4.5 | 15.1 | 8.3 | 10.0 | 10.8 | 14.3 | 15.5 | 11.0 | 10.5 | 45.9 | 167,531 | 93.5 | 56.2 |
| District 4 | 5.9 | 16.3 | 8.1 | 13.0 | 13.3 | 12.9 | 13.9 | 9.8 | 7.0 | 40.4 | 177,258 | 89.9 | 43.6 |
| District 5 | 5.8 | 15.8 | 9.2 | 13.9 | 12.2 | 13.3 | 13.9 | 9.2 | 6.7 | 39.7 | 191,348 | 83.1 | 25.9 |
| District 6 | 6.1 | 13.7 | 6.4 | 14.2 | 13.8 | 13.8 | 14.7 | 9.7 | 7.7 | 42.0 | 157,029 | 84.3 | 38.2 |
| District 7 | 6.8 | 15.1 | 8.3 | 20.2 | 14.7 | 12.0 | 10.5 | 7.1 | 5.4 | 34.9 | 163,058 | 76.8 | 36.8 |
| District 8 | 6.4 | 14.3 | 8.6 | 18.5 | 13.8 | 11.0 | 11.8 | 8.5 | 7.0 | 36.4 | 177,276 | 86.1 | 35.2 |
| District 9 | 7.0 | 14.6 | 7.5 | 17.5 | 13.6 | 12.1 | 11.9 | 9.5 | 6.2 | 37.2 | 170,370 | 87.3 | 40.6 |
| District 10 | 6.7 | 12.7 | 8.7 | 18.4 | 14.1 | 11.9 | 11.5 | 9.4 | 6.7 | 37.0 | 163,783 | 88.3 | 63.3 |
| District 11 | 5.8 | 16.6 | 7.6 | 13.7 | 12.9 | 13.2 | 13.2 | 9.9 | 7.2 | 39.8 | 181,012 | 86.9 | 38.1 |
| District 12 | 4.6 | 7.4 | 8.7 | 26.4 | 15.4 | 11.2 | 9.9 | 8.6 | 7.8 | 36.2 | 121,604 | 93.6 | 72.5 |
| District 13 | 5.7 | 14.5 | 8.7 | 20.8 | 13.4 | 12.4 | 11.3 | 7.4 | 5.9 | 35.3 | 183,105 | 76.4 | 32.9 |
| District 14 | 6.4 | 13.9 | 7.4 | 18.1 | 14.7 | 13.1 | 11.3 | 8.2 | 7.1 | 37.8 | 154,671 | 79.3 | 28.2 |
| District 15 | 7.7 | 19.8 | 10.6 | 16.1 | 12.5 | 12.2 | 11.1 | 6.0 | 4.3 | 32.1 | 216,271 | 70.4 | 14.6 |
| District 16 | 6.0 | 15.6 | 8.9 | 12.7 | 12.3 | 13.1 | 13.0 | 9.6 | 8.9 | 40.4 | 183,939 | 85.6 | 40.6 |
| District 17 | 6.7 | 17.5 | 8.7 | 11.9 | 12.5 | 12.9 | 13.4 | 8.9 | 7.5 | 39.0 | 191,748 | 90.4 | 49.6 |
| District 18 | 5.6 | 16.9 | 10.1 | 11.4 | 12.1 | 13.8 | 14.2 | 9.2 | 6.7 | 39.7 | 184,758 | 91.2 | 36.2 |
| District 19 | 4.6 | 13.6 | 9.3 | 11.0 | 11.1 | 13.2 | 16.1 | 12.5 | 8.6 | 45.4 | 146,486 | 90.8 | 31.2 |
| District 20 | 5.0 | 14.0 | 11.6 | 14.0 | 11.9 | 12.3 | 13.4 | 10.3 | 7.5 | 39.2 | 174,129 | 92.4 | 40.4 |
| District 21 | 5.2 | 14.8 | 9.8 | 12.7 | 11.5 | 12.6 | 14.8 | 11.0 | 7.7 | 41.5 | 149,700 | 89.0 | 24.7 |
| District 22 | 5.3 | 14.9 | 11.3 | 11.9 | 10.6 | 12.0 | 14.6 | 10.8 | 8.3 | 41.0 | 166,606 | 89.9 | 26.1 |
| District 23 | 5.1 | 14.5 | 11.9 | 11.9 | 10.6 | 11.9 | 14.6 | 11.4 | 8.2 | 41.2 | 164,342 | 90.9 | 28.0 |
| District 24 | 5.5 | 15.3 | 9.9 | 12.7 | 11.5 | 12.7 | 14.6 | 10.2 | 7.6 | 40.4 | 174,546 | 90.6 | 32.0 |
| District 25 | 5.4 | 15.2 | 9.8 | 14.5 | 11.5 | 12.0 | 13.7 | 10.2 | 7.7 | 39.1 | 170,978 | 90.5 | 40.4 |
| District 26 | 5.6 | 14.6 | 10.0 | 15.7 | 11.4 | 11.3 | 14.0 | 10.0 | 7.5 | 38.1 | 164,734 | 91.6 | 31.5 |
| District 27 | 4.9 | 14.9 | 7.9 | 11.6 | 11.4 | 13.6 | 15.7 | 11.5 | 8.5 | 44.4 | 156,952 | 93.5 | 33.1 |
| **NORTH CAROLINA** | 5.7 | 16.2 | 9.6 | 13.3 | 12.5 | 13.0 | 13.0 | 10.0 | 6.7 | 39.1 | 2,556,961 | 88.6 | 32.3 |
| District 1 | 5.5 | 15.0 | 11.5 | 13.8 | 11.9 | 12.0 | 13.1 | 10.1 | 7.0 | 38.4 | 188,711 | 85.1 | 28.9 |
| District 2 | 6.3 | 18.6 | 7.9 | 11.8 | 14.3 | 14.0 | 13.2 | 9.0 | 5.1 | 39.1 | 236,401 | 91.4 | 38.3 |
| District 3 | 5.6 | 15.9 | 11.7 | 13.2 | 11.2 | 11.7 | 12.8 | 10.7 | 7.1 | 37.9 | 180,783 | 89.2 | 25.3 |
| District 4 | 5.4 | 16.3 | 11.6 | 16.3 | 14.3 | 13.6 | 10.5 | 7.3 | 4.6 | 35.2 | 247,472 | 93.0 | 55.9 |
| District 5 | 5.1 | 15.2 | 10.4 | 12.3 | 11.6 | 12.7 | 14.0 | 10.8 | 7.8 | 40.9 | 182,219 | 86.9 | 26.8 |
| District 6 | 5.7 | 16.6 | 8.5 | 11.4 | 12.0 | 13.8 | 14.0 | 10.6 | 7.4 | 41.5 | 184,562 | 85.5 | 25.3 |
| District 7 | 5.2 | 15.2 | 9.3 | 12.0 | 11.7 | 11.8 | 14.2 | 12.8 | 7.9 | 42.3 | 184,407 | 88.9 | 28.2 |
| District 8 | 6.8 | 17.2 | 10.0 | 14.9 | 12.1 | 12.0 | 12.0 | 8.8 | 6.3 | 35.8 | 212,287 | 89.6 | 27.6 |
| District 9 | 5.8 | 18.6 | 8.6 | 11.0 | 13.3 | 14.2 | 13.2 | 9.1 | 6.1 | 39.6 | 208,050 | 89.4 | 35.9 |
| District 10 | 5.4 | 15.5 | 8.0 | 12.3 | 12.0 | 13.5 | 13.8 | 11.5 | 8.0 | 42.1 | 164,797 | 87.5 | 26.2 |
| District 11 | 4.6 | 13.7 | 7.9 | 11.6 | 11.1 | 13.1 | 14.5 | 13.5 | 9.8 | 45.8 | 152,827 | 87.7 | 26.3 |
| District 12 | 6.9 | 16.2 | 9.8 | 18.6 | 15.0 | 12.7 | 10.3 | 6.7 | 3.8 | 34.1 | 226,750 | 88.6 | 41.5 |
| District 13 | 5.5 | 15.9 | 9.8 | 13.8 | 11.3 | 13.5 | 13.0 | 9.9 | 7.2 | 39.4 | 187,695 | 88.3 | 30.5 |
| **NORTH DAKOTA** | 6.6 | 16.5 | 10.6 | 15.4 | 12.6 | 10.1 | 12.3 | 8.8 | 6.9 | 35.5 | 187,040 | 93.5 | 30.4 |
| At Large | 6.6 | 16.5 | 10.6 | 15.4 | 12.6 | 10.1 | 12.3 | 8.8 | 6.9 | 35.5 | 187,040 | 93.5 | 30.4 |
| **OHIO** | 5.9 | 16.1 | 9.1 | 13.2 | 12.0 | 12.4 | 13.8 | 10.2 | 7.3 | 39.6 | 2,771,284 | 90.8 | 29.3 |
| District 1 | 6.1 | 17.4 | 10.1 | 13.9 | 12.1 | 12.4 | 13.4 | 8.5 | 6.1 | 36.7 | 195,369 | 92.0 | 37.9 |
| District 2 | 6.1 | 16.6 | 7.3 | 13.7 | 12.6 | 12.2 | 13.7 | 10.6 | 7.4 | 39.9 | 161,583 | 90.6 | 33.1 |
| District 3 | 7.5 | 17.0 | 11.0 | 18.4 | 12.9 | 10.9 | 11.2 | 6.6 | 4.3 | 32.6 | 221,435 | 88.3 | 29.9 |
| District 4 | 5.7 | 16.2 | 8.7 | 11.8 | 12.3 | 12.7 | 14.2 | 10.7 | 7.7 | 41.4 | 163,081 | 91.1 | 20.0 |
| District 5 | 5.9 | 16.3 | 9.7 | 12.7 | 11.2 | 12.2 | 14.1 | 10.4 | 7.6 | 39.8 | 174,678 | 93.2 | 28.0 |
| District 6 | 5.5 | 15.5 | 8.0 | 11.3 | 11.4 | 12.9 | 15.1 | 11.8 | 8.6 | 43.5 | 147,643 | 89.7 | 17.8 |
| District 7 | 5.9 | 17.5 | 8.0 | 11.4 | 11.8 | 12.4 | 14.1 | 10.5 | 8.4 | 41.4 | 160,737 | 88.0 | 21.1 |
| District 8 | 6.0 | 17.0 | 10.6 | 11.4 | 11.9 | 12.4 | 13.5 | 10.0 | 7.0 | 39.0 | 184,730 | 91.4 | 25.3 |
| District 9 | 6.3 | 16.7 | 9.0 | 14.7 | 11.5 | 11.8 | 13.4 | 10.0 | 6.7 | 37.5 | 167,218 | 87.5 | 23.8 |
| District 10 | 6.0 | 15.6 | 9.9 | 13.7 | 11.5 | 11.8 | 13.4 | 10.4 | 7.8 | 38.7 | 179,609 | 91.2 | 31.0 |
| District 11 | 5.9 | 15.0 | 9.7 | 14.9 | 11.1 | 11.4 | 14.3 | 9.7 | 7.9 | 39.3 | 169,471 | 88.3 | 28.0 |
| District 12 | 5.7 | 17.2 | 8.6 | 12.3 | 13.9 | 13.5 | 12.7 | 9.6 | 6.3 | 39.2 | 197,743 | 93.5 | 42.2 |
| District 13 | 5.2 | 13.4 | 11.0 | 13.1 | 10.9 | 12.7 | 14.2 | 11.2 | 8.3 | 41.7 | 152,866 | 90.6 | 23.7 |
| District 14 | 5.2 | 16.0 | 7.5 | 10.8 | 11.8 | 12.8 | 15.1 | 11.8 | 8.9 | 43.8 | 157,975 | 91.9 | 33.9 |
| District 15 | 5.7 | 15.2 | 9.0 | 15.0 | 13.1 | 13.5 | 13.0 | 9.3 | 6.1 | 38.8 | 179,084 | 91.6 | 34.7 |
| District 16 | 5.2 | 15.7 | 7.7 | 11.7 | 11.6 | 12.7 | 14.7 | 12.0 | 8.7 | 43.5 | 158,062 | 93.6 | 35.5 |
| **OKLAHOMA** | 6.3 | 17.7 | 9.7 | 13.6 | 12.7 | 11.4 | 12.4 | 9.3 | 6.7 | 37.0 | 999,896 | 88.4 | 26.2 |
| District 1 | 6.9 | 18.0 | 8.6 | 14.3 | 12.9 | 11.8 | 12.2 | 8.9 | 6.3 | 36.6 | 204,165 | 89.9 | 32.0 |
| District 2 | 6.0 | 17.1 | 9.1 | 11.8 | 11.7 | 11.9 | 13.3 | 11.1 | 8.2 | 40.3 | 171,028 | 86.3 | 17.4 |
| District 3 | 5.8 | 18.2 | 10.4 | 12.7 | 13.0 | 12.4 | 12.4 | 9.3 | 7.1 | 37.0 | 203,610 | 88.2 | 23.7 |
| District 4 | 5.8 | 17.3 | 11.5 | 13.6 | 12.7 | 11.4 | 12.3 | 9.0 | 6.3 | 36.5 | 210,193 | 90.2 | 25.7 |
| District 5 | 7.2 | 17.8 | 9.1 | 15.3 | 13.0 | 11.1 | 12.0 | 8.5 | 5.8 | 35.3 | 210,900 | 87.4 | 31.6 |
| **OREGON** | 5.3 | 15.1 | 8.7 | 14.1 | 13.7 | 12.1 | 12.8 | 11.0 | 7.1 | 39.7 | 947,627 | 91.4 | 34.5 |
| District 1 | 5.5 | 16.2 | 8.1 | 15.3 | 14.5 | 12.8 | 12.3 | 9.3 | 6.0 | 38.0 | 199,001 | 93.2 | 40.7 |
| District 2 | 5.1 | 15.9 | 7.5 | 12.1 | 12.5 | 12.0 | 13.2 | 12.9 | 8.7 | 42.3 | 178,845 | 89.9 | 27.0 |

1. All persons 3 years old and over enrolled in nursery school through college and graduate or professional school.  2. Persons 25 years old and over.

# Table E. Congressional Districts 116th Congress — **Households and Group Quarters**

| STATE District | Households, 2019 | | | | | | Total in group quarters, 2018 | Group Quarters, 2010 | | | | |
|---|---|---|---|---|---|---|---|---|---|---|---|---|
| | Number | Average household size | Family households (percent) | Married couple family (percent) | Female family house-holder[1] | One person households (percent) | | Percent 65 years and over | Persons in correctional institutions | Persons in nursing facilities | Persons in college dormitories | Persons in military quarters |
| | 28 | 29 | 30 | 31 | 32 | 33 | 34 | 35 | 36 | 37 | 38 | 39 |
| **NEW MEXICO—Cont'd** | | | | | | | | | | | | |
| District 3 | 261,846 | 2.63 | 64.3 | 41.7 | 15.4 | 30.4 | 11,611 | 17.2 | 3,544 | 1,465 | 2,115 | 444 |
| **NEW YORK** | 7,446,812 | 2.54 | 62.3 | 43.3 | 13.9 | 30.5 | 568,613 | 19.7 | 95,306 | 116,558 | 218,960 | 8,100 |
| District 1 | 252,630 | 2.75 | 69.3 | 53.7 | 10.6 | 25.5 | 18,541 | 20.6 | 1,648 | 4,504 | 9,203 | 6 |
| District 2 | 223,418 | 3.10 | 75.2 | 56.7 | 13.2 | 20.6 | 5,404 | 46.7 | 6 | 2,952 | 720 | 4 |
| District 3 | 257,333 | 2.77 | 73.7 | 60.8 | 8.6 | 22.6 | 11,650 | 41.2 | 10 | 5,607 | 3,982 | 4 |
| District 4 | 238,571 | 3.02 | 74.5 | 55.6 | 12.9 | 21.6 | 10,774 | 28.5 | 1,657 | 4,200 | 4,748 | 0 |
| District 5 | 230,322 | 3.23 | 75.2 | 44.9 | 22.1 | 20.5 | 15,237 | 30.2 | 234 | 5,713 | 2,367 | 0 |
| District 6 | 266,127 | 2.66 | 67.4 | 49.1 | 12.0 | 26.7 | 6,962 | 62.0 | 17 | 4,179 | 812 | 0 |
| District 7 | 253,213 | 2.71 | 61.6 | 39.5 | 16.1 | 25.3 | 11,627 | 12.2 | 3,636 | 1,600 | 1,803 | 0 |
| District 8 | 308,982 | 2.46 | 60.1 | 31.4 | 22.4 | 30.0 | 18,258 | 26.0 | 256 | 4,120 | 2,056 | 0 |
| District 9 | 285,637 | 2.48 | 61.2 | 36.1 | 19.8 | 30.7 | 12,554 | 27.0 | 0 | 2,585 | 611 | 0 |
| District 10 | 317,659 | 2.23 | 48.8 | 38.9 | 7.1 | 41.6 | 23,766 | 10.2 | 165 | 2,076 | 16,711 | 0 |
| District 11 | 265,022 | 2.75 | 70.0 | 52.2 | 13.5 | 25.3 | 8,348 | 36.4 | 924 | 3,540 | 1,457 | 60 |
| District 12 | 366,996 | 1.90 | 39.4 | 31.1 | 5.3 | 47.1 | 28,767 | 10.4 | 414 | 3,896 | 16,755 | 0 |
| District 13 | 295,057 | 2.49 | 50.0 | 22.3 | 21.6 | 40.7 | 16,889 | 29.6 | 334 | 5,101 | 1,924 | 0 |
| District 14 | 242,702 | 2.80 | 64.5 | 40.2 | 15.9 | 27.6 | 17,846 | 21.9 | 11,095 | 5,195 | 2,355 | 0 |
| District 15 | 262,833 | 2.72 | 61.1 | 20.5 | 32.7 | 34.1 | 24,832 | 8.9 | 981 | 2,145 | 2,352 | 0 |
| District 16 | 277,231 | 2.61 | 65.0 | 42.1 | 17.8 | 31.6 | 17,691 | 41.8 | 0 | 7,221 | 5,113 | 0 |
| District 17 | 249,845 | 2.87 | 72.6 | 58.9 | 10.0 | 22.9 | 19,599 | 23.3 | 3,245 | 4,756 | 8,571 | 0 |
| District 18 | 253,104 | 2.74 | 69.5 | 53.8 | 11.9 | 24.6 | 24,438 | 14.3 | 8,143 | 3,627 | 7,046 | 4,409 |
| District 19 | 272,241 | 2.44 | 63.3 | 48.2 | 10.6 | 30.2 | 37,411 | 13.1 | 9,986 | 4,771 | 12,644 | 3 |
| District 20 | 302,852 | 2.30 | 59.3 | 41.7 | 12.8 | 32.1 | 30,539 | 16.3 | 1,298 | 4,651 | 18,920 | 0 |
| District 21 | 280,765 | 2.33 | 64.2 | 48.3 | 10.3 | 28.5 | 39,591 | 10.4 | 18,001 | 3,692 | 11,032 | 3,614 |
| District 22 | 277,265 | 2.37 | 61.3 | 45.4 | 11.7 | 31.1 | 31,856 | 18.0 | 6,104 | 6,243 | 17,830 | 0 |
| District 23 | 282,880 | 2.29 | 59.4 | 44.5 | 10.4 | 32.6 | 38,853 | 13.9 | 6,988 | 5,257 | 23,896 | 0 |
| District 24 | 280,843 | 2.40 | 59.5 | 43.0 | 11.9 | 33.0 | 28,771 | 17.7 | 4,484 | 4,882 | 14,806 | 0 |
| District 25 | 294,719 | 2.34 | 57.6 | 40.7 | 12.6 | 34.5 | 26,022 | 20.7 | 1,499 | 4,931 | 14,361 | 0 |
| District 26 | 311,623 | 2.19 | 54.5 | 34.5 | 14.6 | 37.1 | 20,357 | 20.0 | 926 | 3,778 | 11,661 | 0 |
| District 27 | 296,942 | 2.35 | 64.8 | 50.4 | 9.6 | 28.5 | 22,030 | 20.3 | 13,255 | 5,336 | 5,224 | 0 |
| **NORTH CAROLINA** | 4,046,348 | 2.52 | 65.0 | 47.6 | 12.9 | 28.6 | 280,891 | 16.6 | 61,680 | 46,638 | 89,795 | 26,326 |
| District 1 | 305,873 | 2.38 | 59.2 | 37.6 | 17.3 | 32.6 | 36,415 | 15.6 | 13,922 | 6,285 | 10,496 | 594 |
| District 2 | 321,361 | 2.74 | 74.5 | 58.2 | 11.9 | 21.2 | 9,286 | 21.8 | 1,132 | 3,214 | 1,773 | 5,949 |
| District 3 | 296,678 | 2.45 | 67.0 | 51.3 | 12.1 | 26.5 | 35,570 | 6.2 | 6,616 | 2,817 | 8,725 | 19,749 |
| District 4 | 337,926 | 2.49 | 60.1 | 44.9 | 11.2 | 30.6 | 33,226 | 7.0 | 5,518 | 2,406 | 22,784 | 0 |
| District 5 | 305,576 | 2.42 | 62.7 | 47.2 | 11.5 | 31.1 | 25,346 | 18.3 | 2,268 | 3,598 | 10,492 | 0 |
| District 6 | 305,260 | 2.54 | 68.3 | 49.1 | 13.7 | 27.1 | 15,354 | 23.6 | 2,493 | 3,811 | 6,738 | 0 |
| District 7 | 319,329 | 2.50 | 63.6 | 48.0 | 11.6 | 30.0 | 18,054 | 29.2 | 5,511 | 3,495 | 9 | 33 |
| District 8 | 298,572 | 2.65 | 66.1 | 47.7 | 14.6 | 28.3 | 24,173 | 17.1 | 8,325 | 3,469 | 3,449 | 0 |
| District 9 | 285,623 | 2.74 | 70.6 | 52.9 | 13.3 | 26.0 | 13,362 | 45.8 | 50 | 2,736 | 2,155 | 1 |
| District 10 | 298,394 | 2.54 | 65.0 | 47.2 | 12.0 | 28.1 | 15,230 | 30.8 | 2,503 | 4,741 | 4,033 | 0 |
| District 11 | 323,519 | 2.34 | 66.1 | 51.1 | 10.2 | 27.7 | 16,229 | 23.1 | 5,915 | 4,622 | 4,965 | 0 |
| District 12 | 341,716 | 2.57 | 57.1 | 38.1 | 14.5 | 34.3 | 14,378 | 10.1 | 4,648 | 3,049 | 13,560 | 0 |
| District 13 | 306,521 | 2.47 | 66.1 | 46.8 | 13.8 | 27.5 | 24,268 | 28.6 | 2,779 | 2,395 | 616 | 0 |
| **NORTH DAKOTA** | 323,519 | 2.28 | 58.6 | 47.0 | 7.5 | 33.7 | 24,335 | 25.3 | 2,489 | 6,433 | 10,570 | 1,380 |
| At Large | 323,519 | 2.28 | 58.6 | 47.0 | 7.5 | 33.7 | 24,335 | 25.3 | 2,489 | 6,433 | 10,570 | 1,380 |
| **OHIO** | 4,730,340 | 2.40 | 62.2 | 45.0 | 12.3 | 31.1 | 317,286 | 25.3 | 76,590 | 83,019 | 106,042 | 571 |
| District 1 | 300,336 | 2.42 | 62.2 | 43.8 | 13.1 | 30.1 | 24,073 | 17.7 | 6,431 | 4,486 | 7,060 | 0 |
| District 2 | 301,403 | 2.39 | 62.1 | 45.7 | 12.0 | 31.1 | 10,299 | 55.0 | 880 | 5,887 | 712 | 0 |
| District 3 | 318,400 | 2.49 | 55.5 | 33.1 | 16.2 | 34.7 | 22,360 | 10.7 | 2,353 | 2,793 | 11,851 | 0 |
| District 4 | 286,055 | 2.39 | 65.6 | 48.8 | 11.0 | 28.5 | 27,293 | 17.6 | 15,892 | 5,563 | 5,261 | 0 |
| District 5 | 293,869 | 2.39 | 64.4 | 50.6 | 9.5 | 28.7 | 17,746 | 31.9 | 971 | 6,014 | 8,367 | 0 |
| District 6 | 275,138 | 2.46 | 65.5 | 49.9 | 10.3 | 29.6 | 21,825 | 30.8 | 9,881 | 6,354 | 3,595 | 0 |
| District 7 | 284,461 | 2.50 | 67.2 | 51.9 | 10.5 | 27.2 | 15,545 | 35.8 | 795 | 6,223 | 5,129 | 0 |
| District 8 | 280,620 | 2.56 | 67.5 | 50.2 | 11.5 | 27.0 | 16,784 | 24.8 | 1,369 | 4,682 | 8,537 | 0 |
| District 9 | 297,555 | 2.30 | 55.7 | 33.9 | 16.6 | 36.1 | 14,162 | 26.5 | 1,894 | 4,930 | 5,512 | 547 |
| District 10 | 300,821 | 2.31 | 59.3 | 41.3 | 13.1 | 33.5 | 28,582 | 21.5 | 1,997 | 5,845 | 12,267 | 0 |
| District 11 | 307,046 | 2.15 | 51.0 | 26.4 | 19.3 | 42.7 | 23,801 | 21.5 | 3,906 | 6,243 | 7,446 | 0 |
| District 12 | 301,083 | 2.56 | 68.1 | 54.2 | 10.0 | 25.3 | 18,673 | 19.4 | 6,071 | 3,929 | 6,352 | 0 |
| District 13 | 301,973 | 2.26 | 57.5 | 37.4 | 15.3 | 35.9 | 20,981 | 20.6 | 5,134 | 5,394 | 10,204 | 0 |
| District 14 | 289,183 | 2.43 | 65.2 | 52.0 | 9.5 | 29.3 | 11,749 | 40.5 | 2,076 | 5,260 | 1,338 | 24 |
| District 15 | 293,600 | 2.51 | 66.2 | 51.1 | 10.9 | 26.3 | 31,859 | 11.1 | 16,670 | 3,476 | 9,591 | 0 |
| District 16 | 298,797 | 2.37 | 64.3 | 51.8 | 7.9 | 30.0 | 11,554 | 51.7 | 270 | 5,940 | 2,820 | 0 |
| **OKLAHOMA** | 1,495,151 | 2.57 | 65.3 | 47.7 | 12.2 | 28.6 | 110,059 | 24.0 | 40,562 | 21,678 | 30,148 | 7,203 |
| District 1 | 314,493 | 2.54 | 64.0 | 47.0 | 11.7 | 30.3 | 9,382 | 28.8 | 2,371 | 3,656 | 2,981 | 0 |
| District 2 | 285,136 | 2.54 | 67.9 | 49.0 | 13.3 | 26.9 | 23,278 | 19.9 | 9,915 | 5,506 | 4,623 | 0 |
| District 3 | 286,280 | 2.62 | 67.7 | 51.2 | 11.3 | 26.4 | 33,368 | 13.8 | 15,091 | 4,818 | 10,089 | 463 |
| District 4 | 293,347 | 2.61 | 65.3 | 49.5 | 11.2 | 27.8 | 27,017 | 12.1 | 7,991 | 3,797 | 7,490 | 6,740 |
| District 5 | 315,895 | 2.56 | 61.9 | 42.6 | 13.5 | 31.4 | 17,014 | 18.8 | 5,194 | 3,901 | 4,965 | 0 |
| **OREGON** | 1,649,352 | 2.50 | 61.9 | 47.3 | 10.0 | 28.1 | 89,211 | 18 | 22,203 | 11,491 | 23,704 | 178 |
| District 1 | 325,950 | 2.59 | 63.2 | 50.2 | 8.7 | 28.3 | 14,669 | 20.7 | 4,641 | 1,964 | 3,338 | 127 |
| District 2 | 330,768 | 2.48 | 64.2 | 48.9 | 10.1 | 27.0 | 21,640 | 19.0 | 8,762 | 2,660 | 1,775 | 13 |

1. No spouse present.

# Table E. Congressional Districts 116th Congress — Housing and Money Income

| STATE District | Housing units, 2019 Total | Occupied units as a percent of all units | Owner-occupied units as a percent of occupied units | Median value[1] (dollars) | Percent valued at $500,000 or more | Renter-occupied Median rent[2] | Per capita income (dollars) | Median income (dollars) | Percent with income of $100,000 or more |
|---|---|---|---|---|---|---|---|---|---|
| | 40 | 41 | 42 | 43 | 44 | 45 | 46 | 47 | 48 |
| **NEW MEXICO—Cont'd** | | | | | | | | | |
| District 3 | 322,940 | 81.1 | 72.0 | 194,000 | 10.7 | 880 | 28,509 | 53,357 | 22.9 |
| **NEW YORK** | 8,404,205 | 88.6 | 53.5 | 338,700 | 33.0 | 1,309 | 41,857 | 72,108 | 36.7 |
| District 1 | 316,401 | 79.8 | 80.5 | 418,000 | 33.2 | 1,817 | 49,513 | 101,701 | 51.1 |
| District 2 | 236,904 | 94.3 | 82.1 | 422,700 | 26.7 | 1,775 | 42,564 | 108,725 | 54.5 |
| District 3 | 271,289 | 94.9 | 82.0 | 645,300 | 69.4 | 1,935 | 63,276 | 126,191 | 60.2 |
| District 4 | 252,055 | 94.7 | 76.6 | 535,000 | 55.6 | 1,732 | 48,176 | 110,677 | 54.5 |
| District 5 | 246,467 | 93.4 | 57.1 | 545,900 | 58.0 | 1,457 | 31,038 | 76,150 | 37.9 |
| District 6 | 289,510 | 91.9 | 45.6 | 694,100 | 64.5 | 1,630 | 35,979 | 70,529 | 36.2 |
| District 7 | 274,455 | 92.3 | 23.0 | 849,900 | NA | 1,501 | 37,598 | 66,891 | 34.9 |
| District 8 | 337,429 | 91.6 | 32.4 | 661,900 | 73.8 | 1,336 | 34,649 | 59,806 | 30.2 |
| District 9 | 311,376 | 91.7 | 30.1 | 759,500 | 74.2 | 1,460 | 39,445 | 69,754 | 35.7 |
| District 10 | 365,475 | 86.9 | 31.4 | 1,121,800 | 87.4 | 1,882 | 83,387 | 103,331 | 52.1 |
| District 11 | 288,888 | 91.7 | 54.7 | 630,200 | 68.1 | 1,431 | 38,188 | 81,253 | 41.4 |
| District 12 | 445,781 | 82.3 | 27.5 | 1,141,500 | NA | 2,269 | 98,414 | 124,502 | 58.7 |
| District 13 | 320,968 | 91.9 | 10.0 | 493,900 | 49.4 | 1,253 | 30,388 | 46,298 | 21.3 |
| District 14 | 268,752 | 90.3 | 30.2 | 601,000 | 62.1 | 1,583 | 30,468 | 66,749 | 29.9 |
| District 15 | 275,448 | 95.4 | 8.8 | 479,100 | 46.1 | 1,134 | 17,731 | 31,061 | 10.9 |
| District 16 | 288,721 | 96.0 | 50.3 | 514,200 | 51.5 | 1,415 | 47,963 | 74,799 | 38.2 |
| District 17 | 267,068 | 93.6 | 67.9 | 496,500 | 49.3 | 1,663 | 53,010 | 108,449 | 55.4 |
| District 18 | 278,065 | 91.0 | 69.5 | 325,000 | 19.0 | 1,346 | 42,739 | 91,723 | 46.0 |
| District 19 | 370,850 | 73.4 | 72.9 | 214,600 | 7.7 | 955 | 36,492 | 67,004 | 32.0 |
| District 20 | 339,116 | 89.3 | 60.1 | 225,700 | 5.9 | 1,034 | 38,404 | 71,156 | 35.4 |
| District 21 | 379,410 | 74.0 | 70.3 | 154,400 | 5.1 | 826 | 30,382 | 57,320 | 24.2 |
| District 22 | 326,249 | 85.0 | 69.4 | 124,300 | 2.5 | 758 | 30,477 | 56,615 | 23.9 |
| District 23 | 347,510 | 81.4 | 69.4 | 112,900 | 3.6 | 792 | 30,443 | 53,769 | 22.1 |
| District 24 | 322,243 | 87.2 | 68.0 | 140,700 | 2.5 | 839 | 33,469 | 60,899 | 26.9 |
| District 25 | 318,884 | 92.4 | 62.1 | 154,500 | 3.1 | 921 | 35,555 | 61,336 | 28.5 |
| District 26 | 342,286 | 91.0 | 58.6 | 143,400 | 3.9 | 823 | 31,249 | 52,122 | 22.3 |
| District 27 | 322,605 | 92.0 | 76.8 | 167,700 | 4.4 | 822 | 37,159 | 69,186 | 32.1 |
| **NORTH CAROLINA** | 4,748,148 | 85.2 | 65.3 | 193,200 | 8.1 | 931 | 32,021 | 57,341 | 25.3 |
| District 1 | 354,662 | 86.2 | 57.0 | 157,000 | 4.3 | 865 | 27,891 | 47,469 | 19.4 |
| District 2 | 350,361 | 91.7 | 74.7 | 248,900 | 9.8 | 940 | 37,234 | 75,366 | 37.1 |
| District 3 | 391,515 | 75.8 | 66.3 | 168,300 | 5.0 | 878 | 28,635 | 53,545 | 21.9 |
| District 4 | 369,251 | 91.5 | 55.7 | 300,900 | 16.0 | 1,170 | 41,025 | 75,687 | 37.5 |
| District 5 | 372,435 | 82.0 | 67.7 | 161,500 | 4.9 | 792 | 28,563 | 49,376 | 20.0 |
| District 6 | 345,723 | 88.3 | 69.4 | 160,400 | 6.1 | 762 | 29,064 | 54,132 | 22.2 |
| District 7 | 407,061 | 78.4 | 68.5 | 186,900 | 7.8 | 914 | 30,226 | 53,066 | 22.0 |
| District 8 | 351,033 | 85.1 | 63.8 | 172,100 | 4.2 | 929 | 28,755 | 54,507 | 21.8 |
| District 9 | 323,073 | 88.4 | 72.1 | 232,400 | 14.8 | 926 | 37,689 | 66,208 | 33.3 |
| District 10 | 353,757 | 84.3 | 68.5 | 174,300 | 7.2 | 795 | 29,359 | 53,189 | 21.9 |
| District 11 | 407,687 | 79.4 | 71.9 | 186,400 | 7.6 | 791 | 29,406 | 51,884 | 19.5 |
| District 12 | 376,911 | 90.7 | 51.7 | 236,500 | 11.5 | 1,169 | 35,791 | 61,658 | 28.1 |
| District 13 | 344,679 | 88.9 | 63.9 | 172,700 | 6.3 | 872 | 30,160 | 56,718 | 23.3 |
| **NORTH DAKOTA** | 379,974 | 85.1 | 61.3 | 205,400 | 6.4 | 804 | 36,611 | 64,577 | 29.8 |
| At Large | 379,974 | 85.1 | 61.3 | 205,400 | 6.4 | 804 | 36,611 | 64,577 | 29.8 |
| **OHIO** | 5,232,943 | 90.4 | 66.0 | 157,200 | 4.0 | 813 | 32,780 | 58,642 | 25.5 |
| District 1 | 328,980 | 91.3 | 61.1 | 189,300 | 7.9 | 837 | 37,572 | 63,648 | 32.1 |
| District 2 | 329,900 | 91.4 | 67.9 | 169,600 | 7.4 | 799 | 36,628 | 61,220 | 30.0 |
| District 3 | 348,389 | 91.4 | 46.0 | 155,200 | 2.9 | 930 | 27,744 | 51,435 | 19.8 |
| District 4 | 316,213 | 90.5 | 72.8 | 136,800 | 2.3 | 729 | 29,940 | 60,212 | 22.5 |
| District 5 | 314,363 | 93.5 | 72.9 | 150,900 | 2.0 | 748 | 33,826 | 63,702 | 27.2 |
| District 6 | 325,469 | 84.5 | 75.8 | 122,900 | 2.7 | 662 | 27,291 | 52,442 | 20.0 |
| District 7 | 309,087 | 92.0 | 73.5 | 153,300 | 3.2 | 706 | 28,761 | 57,897 | 22.2 |
| District 8 | 306,634 | 91.5 | 69.9 | 163,800 | 2.9 | 829 | 31,721 | 62,845 | 27.0 |
| District 9 | 348,734 | 85.3 | 58.1 | 113,700 | 2.8 | 761 | 27,126 | 45,076 | 16.6 |
| District 10 | 337,202 | 89.2 | 62.4 | 137,200 | 2.7 | 810 | 32,082 | 56,595 | 23.9 |
| District 11 | 360,797 | 85.1 | 48.7 | 101,400 | 3.6 | 768 | 30,077 | 42,207 | 17.4 |
| District 12 | 322,811 | 93.3 | 70.9 | 246,200 | 7.2 | 946 | 40,191 | 76,631 | 37.4 |
| District 13 | 337,621 | 89.4 | 63.4 | 105,400 | 0.9 | 747 | 27,437 | 46,582 | 16.2 |
| District 14 | 312,101 | 92.7 | 76.2 | 190,800 | 5.9 | 874 | 39,068 | 67,698 | 31.8 |
| District 15 | 318,068 | 92.3 | 67.9 | 195,100 | 5.8 | 949 | 36,454 | 69,844 | 33.3 |
| District 16 | 316,574 | 94.4 | 73.1 | 181,300 | 3.4 | 876 | 37,503 | 68,534 | 31.3 |
| **OKLAHOMA** | 1,749,520 | 85.5 | 65.5 | 147,000 | 4.1 | 814 | 29,666 | 54,449 | 22.9 |
| District 1 | 353,147 | 89.1 | 62.4 | 165,600 | 5.2 | 874 | 33,557 | 59,660 | 26.6 |
| District 2 | 359,289 | 79.4 | 71.1 | 112,600 | 2.7 | 681 | 23,964 | 45,207 | 15.9 |
| District 3 | 341,109 | 83.9 | 70.6 | 136,400 | 3.0 | 766 | 27,900 | 54,629 | 21.2 |
| District 4 | 336,727 | 87.1 | 65.7 | 151,300 | 2.8 | 854 | 30,266 | 59,463 | 26.1 |
| District 5 | 359,248 | 87.9 | 58.9 | 161,200 | 7.2 | 853 | 32,108 | 54,928 | 24.3 |
| **OREGON** | 1,808,482 | 91.2 | 62.9 | 354,600 | 23.3 | 1,185 | 35,531 | 67,058 | 31.1 |
| District 1 | 351,429 | 92.7 | 61.4 | 417,800 | 32.2 | 1,386 | 41,663 | 81,473 | 39.7 |
| District 2 | 380,896 | 86.8 | 66.3 | 288,600 | 17.3 | 1,013 | 30,483 | 57,870 | 24.8 |

1. Specified owner-occupied units; $1,000,000 represents $1,000,000    2. Specified renter-occupied units.

# Table E. Congressional Districts 116th Congress — Poverty, Labor Force, Employment, and Social Security

| STATE District | Poverty, 2019 — Persons below poverty level (percent) | Families below poverty level (percent) | Percent of households receiving food stamps in past 12 months | Civilian labor force, 2019 — Total | Unemployment Total | Unemployment Rate[1] | Civilian employment,[2] 2019 — Total | Percent — Management, business, science, and arts occupations | Service, sales, and office | Construction and production | Persons under 65 years of age with no health insurance, 2019 (percent) | Social Security beneficiaries, December 2020 — Number | Rate[3] | Supplemental Security Income recipients, December 2019 |
|---|---|---|---|---|---|---|---|---|---|---|---|---|---|---|
| | 49 | 50 | 51 | 52 | 53 | 54 | 55 | 56 | 57 | 58 | 59 | 60 | 61 | 62 |
| **NEW MEXICO—Cont'd** | | | | | | | | | | | | | | |
| District 3 | 17.5 | 12.7 | 16.0 | 320,367 | 17,269 | 5.5 | 299,400 | 37.6 | 42.1 | 20.4 | 11.1 | 155,819 | 222.6 | 19,692 |
| **NEW YORK** | 13.0 | 9.3 | 14.1 | 10,085,219 | 445,165 | 4.4 | 9,611,029 | 43.1 | 40.0 | 16.9 | 5.1 | 3,680,264 | 189.2 | 601,717 |
| District 1 | 7.0 | 4.5 | 5.6 | 375,470 | 13,194 | 3.5 | 361,475 | 43.6 | 38.1 | 18.2 | 4.7 | 155,786 | 218.4 | 9,322 |
| District 2 | 5.9 | 4.0 | 6.3 | 388,698 | 10,165 | 2.6 | 378,000 | 38.9 | 41.4 | 19.7 | 3.7 | 137,320 | 196.5 | 9,110 |
| District 3 | 5.2 | 2.9 | 3.0 | 381,484 | 12,432 | 3.3 | 368,920 | 53.8 | 35.0 | 11.2 | 3.4 | 149,939 | 206.6 | 6,811 |
| District 4 | 5.7 | 3.9 | 4.9 | 391,689 | 14,208 | 3.6 | 377,304 | 44.0 | 39.2 | 16.8 | 4.9 | 139,945 | 191.6 | 9,079 |
| District 5 | 9.9 | 7.5 | 14.4 | 386,927 | 23,919 | 6.2 | 362,775 | 30.8 | 48.3 | 20.9 | 8.0 | 114,353 | 150.7 | 25,358 |
| District 6 | 10.8 | 8.6 | 11.2 | 368,615 | 13,643 | 3.7 | 354,642 | 40.4 | 42.6 | 17.0 | 8.0 | 122,747 | 171.8 | 20,064 |
| District 7 | 19.0 | 15.6 | 22.3 | 368,138 | 17,679 | 4.8 | 350,257 | 43.8 | 40.9 | 15.3 | 7.3 | 88,949 | 127.3 | 30,070 |
| District 8 | 19.8 | 15.8 | 23.8 | 385,839 | 18,047 | 4.7 | 367,352 | 42.6 | 41.7 | 15.6 | 4.7 | 113,623 | 146.3 | 46,700 |
| District 9 | 14.0 | 10.7 | 19.3 | 374,814 | 17,202 | 4.6 | 357,232 | 46.6 | 40.3 | 13.1 | 6.3 | 104,773 | 145.5 | 28,661 |
| District 10 | 13.6 | 9.4 | 10.8 | 410,114 | 14,769 | 3.6 | 395,345 | 61.9 | 30.4 | 7.7 | 4.1 | 105,372 | 143.8 | 19,903 |
| District 11 | 10.1 | 7.5 | 14.7 | 362,790 | 14,202 | 3.9 | 348,004 | 42.2 | 40.6 | 17.2 | 5.1 | 133,260 | 180.7 | 27,051 |
| District 12 | 10.0 | 5.3 | 6.2 | 462,440 | 13,013 | 2.8 | 449,117 | 68.0 | 27.0 | 5.0 | 3.5 | 107,180 | 147.7 | 13,310 |
| District 13 | 22.2 | 19.4 | 27.3 | 402,307 | 30,667 | 7.6 | 371,482 | 39.8 | 46.4 | 13.8 | 7.7 | 111,504 | 148.3 | 52,325 |
| District 14 | 12.9 | 8.8 | 15.0 | 371,212 | 17,782 | 4.8 | 353,265 | 29.5 | 48.3 | 22.2 | 11.0 | 96,778 | 138.9 | 17,929 |
| District 15 | 34.4 | 29.7 | 44.4 | 326,023 | 34,399 | 10.6 | 291,354 | 19.6 | 59.0 | 21.4 | 8.9 | 97,641 | 132.1 | 68,474 |
| District 16 | 11.7 | 9.4 | 15.4 | 372,009 | 21,199 | 5.7 | 350,760 | 44.8 | 40.1 | 15.1 | 5.3 | 127,579 | 172.4 | 20,853 |
| District 17 | 9.1 | 5.9 | 7.2 | 381,957 | 18,659 | 4.9 | 363,176 | 50.0 | 36.4 | 13.6 | 4.0 | 130,999 | 177.7 | 9,650 |
| District 18 | 9.7 | 6.3 | 8.2 | 370,601 | 10,858 | 3.0 | 355,080 | 42.5 | 39.5 | 18.0 | 3.6 | 139,579 | 194.2 | 11,550 |
| District 19 | 11.2 | 7.8 | 9.1 | 354,838 | 13,150 | 3.7 | 340,626 | 39.2 | 39.3 | 21.5 | 4.4 | 172,271 | 245.7 | 14,602 |
| District 20 | 11.6 | 7.7 | 10.4 | 392,229 | 16,340 | 4.2 | 374,499 | 47.5 | 37.1 | 15.4 | 3.0 | 156,089 | 215.1 | 18,437 |
| District 21 | 14.4 | 9.8 | 13.8 | 329,536 | 13,771 | 4.4 | 301,994 | 35.8 | 40.7 | 23.5 | 4.3 | 177,089 | 254.9 | 18,703 |
| District 22 | 15.0 | 10.1 | 14.7 | 329,589 | 14,265 | 4.3 | 314,831 | 37.0 | 41.5 | 21.5 | 4.2 | 169,370 | 246.0 | 21,125 |
| District 23 | 15.4 | 9.4 | 14.7 | 327,964 | 14,446 | 4.4 | 313,411 | 38.9 | 37.1 | 24.0 | 5.6 | 169,963 | 247.2 | 19,045 |
| District 24 | 14.2 | 10.0 | 13.1 | 359,954 | 13,924 | 3.9 | 344,883 | 38.5 | 41.7 | 19.8 | 3.5 | 160,251 | 228.3 | 20,230 |
| District 25 | 13.0 | 8.6 | 15.0 | 374,265 | 17,260 | 4.6 | 356,893 | 46.3 | 37.5 | 16.2 | 2.7 | 160,557 | 224.7 | 25,918 |
| District 26 | 17.0 | 12.4 | 18.8 | 361,188 | 13,652 | 3.8 | 346,938 | 39.8 | 41.4 | 18.9 | 2.8 | 158,135 | 224.9 | 27,896 |
| District 27 | 8.1 | 5.5 | 8.5 | 374,529 | 12,320 | 3.3 | 361,414 | 41.1 | 36.2 | 22.7 | 2.9 | 179,212 | 249.1 | 9,541 |
| **NORTH CAROLINA** | 13.6 | 9.6 | 11.6 | 5,272,092 | 236,872 | 4.6 | 4,937,837 | 39.3 | 37.0 | 23.7 | 11.1 | 2,183,353 | 208.2 | 227,652 |
| District 1 | 19.4 | 14.3 | 16.6 | 369,166 | 20,431 | 5.5 | 348,499 | 39.1 | 38.0 | 22.9 | 12.7 | 174,371 | 228.4 | 29,375 |
| District 2 | 9.6 | 6.3 | 8.4 | 456,352 | 24,465 | 5.4 | 427,020 | 45.3 | 34.4 | 20.2 | 9.5 | 121,250 | 136.5 | 9,506 |
| District 3 | 12.7 | 10.1 | 12.9 | 377,355 | 18,461 | 5.5 | 317,181 | 35.2 | 39.7 | 25.1 | 10.5 | 174,695 | 229.3 | 18,043 |
| District 4 | 10.3 | 5.8 | 6.9 | 485,724 | 18,389 | 3.8 | 466,889 | 54.5 | 31.6 | 13.8 | 9.1 | 150,341 | 172.2 | 13,346 |
| District 5 | 15.7 | 10.6 | 11.2 | 372,903 | 20,244 | 5.4 | 351,778 | 36.4 | 38.1 | 25.4 | 12.0 | 187,557 | 245.2 | 18,367 |
| District 6 | 15.6 | 12.0 | 13.8 | 380,977 | 15,994 | 4.2 | 364,129 | 33.5 | 38.5 | 28.0 | 11.4 | 155,201 | 196.1 | 19,369 |
| District 7 | 14.4 | 9.6 | 13.7 | 386,163 | 15,121 | 4.0 | 363,854 | 35.1 | 39.3 | 25.6 | 11.2 | 205,381 | 251.6 | 18,894 |
| District 8 | 12.8 | 9.0 | 11.6 | 411,368 | 18,392 | 4.9 | 357,270 | 34.5 | 39.2 | 26.3 | 11.0 | 152,919 | 187.6 | 19,726 |
| District 9 | 14.6 | 10.5 | 13.7 | 383,270 | 16,242 | 4.3 | 364,566 | 43.5 | 34.6 | 21.9 | 8.4 | 153,008 | 192.1 | 18,083 |
| District 10 | 14.0 | 9.8 | 12.5 | 377,710 | 18,769 | 5.0 | 357,807 | 34.9 | 36.1 | 29.0 | 12.0 | 186,983 | 242.3 | 14,875 |
| District 11 | 13.2 | 9.4 | 9.7 | 368,021 | 13,242 | 3.6 | 353,762 | 33.6 | 39.7 | 26.7 | 13.2 | 115,796 | 129.8 | 15,637 |
| District 12 | 11.5 | 8.0 | 8.3 | 513,638 | 20,365 | 4.0 | 492,975 | 41.9 | 38.5 | 19.5 | 13.6 | 183,395 | 235.0 | 15,981 |
| District 13 | 14.4 | 11.1 | 12.5 | 389,445 | 16,757 | 4.3 | 372,107 | 36.0 | 35.6 | 28.4 | 10.6 | 138,461 | 181.7 | 8,304 |
| **NORTH DAKOTA** | 10.6 | 6.5 | 6.9 | 424,480 | 10,920 | 2.6 | 405,699 | 38.3 | 36.1 | 25.6 | 6.8 | 138,461 | 181.7 | 8,304 |
| At Large | 10.6 | 6.5 | 6.9 | 424,480 | 10,920 | 2.6 | 405,699 | 38.3 | 36.1 | 25.6 | 6.8 | 138,461 | 181.7 | 8,304 |
| **OHIO** | 13.1 | 9.2 | 12.0 | 5,976,393 | 272,654 | 4.6 | 5,692,943 | 38.1 | 37.4 | 24.5 | 6.5 | 2,405,217 | 205.8 | 306,163 |
| District 1 | 12.4 | 7.9 | 9.4 | 398,094 | 17,821 | 4.5 | 380,083 | 44.9 | 36.0 | 19.0 | 5.9 | 134,257 | 179.1 | 19,298 |
| District 2 | 12.8 | 9.9 | 11.2 | 367,295 | 14,858 | 4.0 | 352,217 | 42.7 | 35.9 | 21.4 | 5.9 | 151,408 | 207.4 | 20,705 |
| District 3 | 17.8 | 14.9 | 14.3 | 443,867 | 23,307 | 5.3 | 420,334 | 36.6 | 41.7 | 21.7 | 9.8 | 105,903 | 130.1 | 28,620 |
| District 4 | 11.5 | 8.7 | 10.7 | 359,354 | 13,151 | 3.7 | 345,995 | 30.7 | 34.9 | 34.4 | 5.0 | 159,589 | 224.1 | 14,845 |
| District 5 | 9.0 | 5.5 | 7.8 | 383,710 | 14,417 | 3.8 | 368,717 | 35.5 | 34.7 | 29.8 | 4.5 | 157,124 | 217.9 | 10,628 |
| District 6 | 13.4 | 9.7 | 13.6 | 318,122 | 15,199 | 4.8 | 302,868 | 32.2 | 38.1 | 29.7 | 6.3 | 174,982 | 250.6 | 23,184 |
| District 7 | 10.9 | 7.6 | 10.8 | 357,441 | 12,636 | 3.5 | 344,394 | 32.5 | 36.7 | 30.7 | 9.7 | 163,911 | 225.5 | 14,177 |
| District 8 | 11.9 | 8.2 | 10.4 | 368,355 | 15,827 | 4.3 | 351,301 | 33.9 | 37.4 | 28.7 | 6.2 | 150,085 | 204.5 | 14,738 |
| District 9 | 20.0 | 14.9 | 19.4 | 343,404 | 21,623 | 6.3 | 321,521 | 31.8 | 42.2 | 25.9 | 7.4 | 144,308 | 206.9 | 29,873 |
| District 10 | 14.4 | 10.1 | 12.1 | 365,571 | 19,867 | 5.5 | 341,556 | 39.9 | 38.6 | 21.6 | 7.0 | 149,596 | 206.7 | 19,116 |
| District 11 | 22.7 | 16.7 | 23.2 | 342,265 | 32,455 | 9.5 | 309,392 | 38.2 | 42.3 | 19.5 | 5.5 | 136,962 | 200.1 | 39,871 |
| District 12 | 9.2 | 5.9 | 8.3 | 417,539 | 12,658 | 3.0 | 404,182 | 48.1 | 33.6 | 18.3 | 5.5 | 145,002 | 183.9 | 13,556 |
| District 13 | 17.6 | 12.8 | 15.9 | 355,068 | 19,445 | 5.5 | 334,666 | 30.6 | 42.0 | 27.4 | 7.1 | 165,499 | 235.0 | 26,283 |
| District 14 | 8.9 | 6.4 | 8.1 | 370,254 | 13,916 | 3.8 | 355,618 | 41.5 | 36.4 | 22.2 | 6.5 | 163,835 | 229.2 | 9,808 |
| District 15 | 10.6 | 7.0 | 10.2 | 402,128 | 13,699 | 3.4 | 387,992 | 44.5 | 33.2 | 22.3 | 5.1 | 139,169 | 180.8 | 14,261 |
| District 16 | 6.8 | 4.0 | 6.2 | 383,926 | 11,775 | 3.1 | 372,107 | 42.0 | 36.1 | 21.9 | 6.1 | 163,587 | 227.3 | 7,200 |
| **OKLAHOMA** | 15.2 | 10.8 | 12.0 | 1,895,848 | 81,705 | 4.4 | 1,791,228 | 36.2 | 38.4 | 25.4 | 14.2 | 811,064 | 205.0 | 96,502 |
| District 1 | 13.3 | 9.6 | 10.6 | 413,454 | 19,807 | 4.8 | 392,588 | 39.4 | 37.8 | 22.8 | 13.7 | 156,773 | 193.7 | 18,233 |
| District 2 | 19.8 | 15.0 | 15.6 | 318,497 | 15,711 | 4.9 | 302,319 | 30.1 | 39.7 | 30.1 | 18.2 | 187,513 | 250.9 | 26,293 |
| District 3 | 15.1 | 10.5 | 11.4 | 361,784 | 12,241 | 3.4 | 346,432 | 33.2 | 38.0 | 28.8 | 13.1 | 160,344 | 205.0 | 14,375 |
| District 4 | 12.5 | 7.6 | 9.5 | 393,005 | 15,974 | 4.2 | 361,778 | 39.1 | 36.9 | 24.0 | 11.3 | 159,374 | 201.0 | 16,229 |
| District 5 | 15.5 | 11.4 | 12.9 | 409,108 | 17,972 | 4.4 | 388,111 | 37.7 | 39.7 | 22.6 | 14.7 | 147,060 | 178.2 | 21,372 |
| **OREGON** | 11.4 | 6.6 | 13.4 | 2,158,612 | 107,991 | 5.0 | 2,045,338 | 40.4 | 38.2 | 21.4 | 7.1 | 906,127 | 214.8 | 87,980 |
| District 1 | 8.9 | 4.9 | 8.9 | 460,499 | 17,477 | 3.8 | 441,121 | 45.0 | 35.2 | 19.8 | 5.8 | 148,480 | 172.9 | 11,988 |
| District 2 | 13.0 | 8.1 | 16.0 | 395,701 | 22,264 | 5.6 | 372,031 | 35.9 | 40.6 | 23.5 | 8.2 | 217,079 | 258.1 | 18,420 |

1. Percent of civilian labor force.  2. Persons 16 years old and over.  3. Per 1,000 resident population estimated in the 2019 American Community Survey.

# Table E. Congressional Districts 116th Congress — **Agriculture**

| STATE District | Number of farms | Land in farms — Acres | Land in farms — Average size of farm (acres) | Land in farms — Irrigated land (acres) | Value of products sold — Total ($1,000) | Value of products sold — Average per farm (dollars) | Value of products sold — Percent from crops | Value of products sold — Percent from livestock and poultry products | Government payments — Total ($1,000) | Government payments — Average per farm receiving payments (dollars) |
|---|---|---|---|---|---|---|---|---|---|---|
| | 63 | 64 | 65 | 66 | 67 | 68 | 69 | 70 | 71 | 72 |
| **NEW MEXICO—Cont'd** | | | | | | | | | | |
| District 3 | 13,012 | 18,404,082 | 1,414 | 395,435 | 956,793 | 73,532 | 15.9 | 84.1 | 35,472 | 17,825 |
| **NEW YORK** | 33,438 | 6,866,171 | 205 | 3,581,095 | 5,369,212 | 160,572 | 39.3 | 60.7 | 59,106 | 9,162 |
| District 1 | 466 | 25,862 | 55 | 16,618 | 199,192 | 427,451 | 90.1 | 9.9 | 76 | 6,333 |
| District 2 | 42 | 432 | 10 | 120 | 2,985 | 71,071 | 81.0 | 19.0 | D | D |
| District 3 | 78 | 4,488 | 58 | 2,845 | 26,075 | 334,295 | 87.7 | 12.3 | D | D |
| District 4 | 7 | 207 | 30 | 22 | 95 | 13,571 | 100.0 | 0.0 | D | D |
| District 5 | X | X | X | X | X | X | X | X | X | X |
| District 6 | 3 | D | D | D | X | X | X | X | X | X |
| District 7 | 3 | D | D | 3 | 58 | 19,333 | D | D | D | D |
| District 8 | 5 | 6 | 1 | 6 | 34 | 6,800 | 100.0 | 0.0 | D | D |
| District 9 | 11 | 14 | 1 | 14 | 6,641 | 603,727 | D | D | D | D |
| District 10 | 3 | 3 | 1 | 3 | 14 | 4,667 | 100.0 | 0.0 | D | D |
| District 11 | 6 | D | D | D | D | D | D | D | D | D |
| District 12 | X | X | X | X | X | X | X | X | X | X |
| District 13 | 4 | 8 | 2 | 4 | 31 | 7,750 | 100.0 | 0.0 | X | X |
| District 14 | X | X | X | X | X | X | X | X | X | X |
| District 15 | X | X | X | X | X | X | X | X | X | X |
| District 16 | 4 | D | D | 6 | 122 | 30,500 | 100.0 | 0.0 | D | D |
| District 17 | 64 | 3,607 | 56 | 456 | 5,688 | 88,875 | 88.5 | 11.5 | D | D |
| District 18 | 834 | 97,545 | 117 | 41,571 | 101,205 | 121,349 | 76.6 | 23.4 | 472 | 6,841 |
| District 19 | 4,978 | 893,753 | 180 | 386,911 | 453,031 | 91,007 | 47.5 | 52.5 | 4,562 | 5,894 |
| District 20 | 1,144 | 153,570 | 134 | 77,467 | 117,643 | 102,835 | 52.4 | 47.6 | 680 | 4,533 |
| District 21 | 5,867 | 1,422,164 | 242 | 667,428 | 987,044 | 168,237 | 20.5 | 79.5 | 6,091 | 6,945 |
| District 22 | 4,325 | 857,969 | 198 | 394,978 | 472,075 | 109,150 | 25.6 | 74.4 | 4,719 | 5,157 |
| District 23 | 8,113 | 1,647,668 | 203 | 854,912 | 1,065,088 | 131,282 | 35.9 | 64.1 | 15,512 | 9,066 |
| District 24 | 2,543 | 572,957 | 225 | 352,684 | 699,584 | 275,102 | 39.9 | 60.1 | 6,333 | 11,349 |
| District 25 | 392 | 69,264 | 177 | 42,514 | 49,178 | 125,454 | 89.2 | 10.8 | 1,757 | 23,427 |
| District 26 | 80 | 12,569 | 157 | 4,774 | 8,567 | 107,088 | 98.9 | 1.1 | D | D |
| District 27 | 4,466 | 1,104,024 | 247 | 737,743 | 1,174,577 | 263,004 | 42.5 | 57.5 | 18,827 | 14,482 |
| **NORTH CAROLINA** | 46,418 | 8,430,522 | 182 | 4,407,160 | 12,900,674 | 277,924 | 29.0 | 71.0 | 107,565 | 10,746 |
| District 1 | 3,443 | 1,424,052 | 414 | 837,774 | 1,284,852 | 373,178 | 51.6 | 48.4 | 33,934 | 18,676 |
| District 2 | 2,882 | 560,837 | 195 | 320,561 | 728,529 | 252,786 | 61.9 | 38.1 | 3,999 | 5,448 |
| District 3 | 3,031 | 1,258,543 | 415 | 978,038 | 1,722,737 | 568,372 | 40.4 | 59.6 | 27,850 | 18,805 |
| District 4 | 877 | 84,981 | 97 | 31,708 | 50,145 | 57,178 | 65.9 | 34.1 | 519 | 3,437 |
| District 5 | 6,957 | 787,246 | 113 | 262,330 | 1,059,940 | 152,356 | 20.0 | 80.0 | 1,938 | 3,101 |
| District 6 | 5,825 | 740,881 | 127 | 239,786 | 713,084 | 122,418 | 26.2 | 73.8 | 2,135 | 2,986 |
| District 7 | 4,208 | 1,182,179 | 281 | 703,243 | 3,901,879 | 927,253 | 17.0 | 83.0 | 15,606 | 11,052 |
| District 8 | 3,054 | 428,078 | 140 | 177,235 | 617,584 | 202,221 | 20.6 | 79.4 | 3,163 | 6,817 |
| District 9 | 2,964 | 742,381 | 250 | 443,241 | 1,767,017 | 596,160 | 14.6 | 85.4 | 9,531 | 12,065 |
| District 10 | 3,954 | 379,387 | 96 | 134,064 | 370,291 | 93,650 | 23.6 | 76.4 | 3,028 | 4,420 |
| District 11 | 5,778 | 457,986 | 79 | 111,024 | 327,908 | 56,751 | 50.1 | 49.9 | 2,837 | 3,422 |
| District 12 | 174 | 9,981 | 57 | 4,609 | D | D | D | D | 131 | 10,917 |
| District 13 | 3,271 | 373,990 | 114 | 163,547 | D | D | D | D | 2,894 | 9,810 |
| **NORTH DAKOTA** | 26,364 | 39,341,591 | 1,492 | 23,976,011 | 8,234,102 | 312,324 | 81.1 | 18.9 | 467,034 | 22,770 |
| At Large | 26,364 | 39,341,591 | 1,492 | 23,976,011 | 8,234,102 | 312,324 | 81.1 | 18.9 | 467,034 | 22,770 |
| **OHIO** | 77,805 | 13,965,295 | 179 | 10,190,952 | 9,341,225 | 120,059 | 58.1 | 41.9 | 351,125 | 12,301 |
| District 1 | 1,198 | 104,683 | 87 | 72,291 | 61,864 | 51,639 | 92.5 | 7.5 | 2,766 | 17,731 |
| District 2 | 5,788 | 971,188 | 168 | 561,643 | 351,574 | 60,742 | 72.2 | 27.8 | 23,903 | 11,729 |
| District 3 | 112 | 10,036 | 90 | 8,336 | 18,150 | 162,054 | 98.2 | 1.8 | 286 | 11,440 |
| District 4 | 9,520 | 2,277,947 | 239 | 1,977,190 | 1,777,616 | 186,724 | 63.8 | 36.2 | 82,104 | 14,059 |
| District 5 | 10,683 | 2,913,459 | 273 | 2,607,814 | 2,094,949 | 196,101 | 63.3 | 36.7 | 86,759 | 11,594 |
| District 6 | 12,271 | 1,662,328 | 135 | 605,518 | 480,616 | 39,167 | 40.8 | 59.2 | 10,283 | 6,967 |
| District 7 | 10,030 | 1,332,469 | 133 | 848,892 | 1,093,721 | 109,045 | 39.2 | 60.8 | 24,540 | 10,825 |
| District 8 | 6,086 | 1,139,909 | 187 | 972,960 | 1,323,050 | 217,392 | 41.0 | 59.0 | 30,802 | 9,952 |
| District 9 | 522 | 75,694 | 145 | 63,632 | 57,816 | 110,759 | 94.8 | 5.2 | 2,273 | 9,883 |
| District 10 | 1,876 | 386,147 | 206 | 323,254 | 240,924 | 128,424 | 86.6 | 13.4 | 14,157 | 17,521 |
| District 11 | 108 | 1,849 | 17 | 570 | 1,573 | 14,565 | 93.5 | 6.5 | 140 | 12,727 |
| District 12 | 4,556 | 698,970 | 153 | 494,043 | 438,440 | 96,234 | 61.7 | 38.3 | 14,830 | 12,008 |
| District 13 | 1,285 | 99,455 | 77 | 55,420 | 44,477 | 34,612 | 61.7 | 38.3 | 1,129 | 7,428 |
| District 14 | 3,432 | 337,648 | 98 | 187,729 | 217,096 | 63,256 | 75.1 | 24.9 | 2,394 | 6,435 |
| District 15 | 7,093 | 1,612,792 | 227 | 1,164,067 | 763,351 | 107,620 | 82.8 | 17.2 | 48,785 | 17,479 |
| District 16 | 3,245 | 340,721 | 105 | 247,593 | 376,008 | 115,873 | 29.3 | 70.7 | 5,974 | 10,555 |
| **OKLAHOMA** | 78,531 | 34,156,290 | 435 | 7,812,594 | 7,465,512 | 95,065 | 20.3 | 79.7 | 232,018 | 11,248 |
| District 1 | 3,313 | 588,647 | 178 | 182,107 | 110,733 | 33,424 | 49.2 | 50.8 | 2,543 | 6,218 |
| District 2 | 29,177 | 7,890,823 | 270 | 1,392,848 | 2,290,737 | 78,512 | 11.2 | 88.8 | 25,153 | 5,613 |
| District 3 | 28,506 | 19,928,136 | 699 | 5,158,149 | 4,192,913 | 147,089 | 23.7 | 76.3 | 164,878 | 13,484 |
| District 4 | 13,511 | 5,008,747 | 371 | 969,680 | 797,804 | 59,048 | 23.8 | 76.2 | 37,282 | 12,272 |
| District 5 | 4,024 | 739,937 | 184 | 109,810 | 73,324 | 18,222 | 30.4 | 69.6 | 2,162 | 4,590 |
| **OREGON** | 37,616 | 15,962,322 | 424 | 2,965,392 | 5,006,821 | 133,103 | 65.6 | 34.4 | 92,406 | 22,918 |
| District 1 | 5,086 | 342,106 | 67 | 178,334 | 581,142 | 114,263 | 85.9 | 14.1 | 3,347 | 8,973 |
| District 2 | 13,969 | 13,669,551 | 979 | 2,031,807 | 2,404,429 | 172,126 | 55.0 | 45.0 | 81,710 | 28,244 |

# Table E. Congressional Districts 116th Congress — Nonfarm Employment and Payroll

Private nonfarm employment and payroll, 2019

| STATE District | Number of establishments | Total | Manufac- turing | Construc- tion | Wholesale trade | Retail trade | Health care and social assistance | Finance and Insurance | Real estate and rental and leasing | Profes- sional, sci- entific, and technical services | Information | Total (mil dol) | Average per employee (dollars) |
|---|---|---|---|---|---|---|---|---|---|---|---|---|---|
| | 73 | 74 | 75 | 76 | 77 | 78 | 79 | 80 | 81 | 82 | 83 | 84 | 85 |
| **NEW MEXICO—Cont'd** | | | | | | | | | | | | | |
| District 3 | 14,213 | 178,364 | 3.6 | 5.0 | 2.2 | 16.4 | 20.4 | 3.0 | 1.5 | 9.5 | 1.2 | 7,758 | 43,498 |
| **NEW YORK** | 547,351 | 8,597,216 | 4.8 | 4.5 | 4.1 | 10.6 | 20.2 | 6.5 | 2.2 | 7.6 | 3.6 | 603,078 | 70,148 |
| District 1 | 23,463 | 249,934 | 6.9 | 8.6 | 6.1 | 15.9 | 22.8 | 2.9 | 1.0 | 6.4 | 1.4 | 13,922 | 55,703 |
| District 2 | 20,970 | 261,615 | 11.6 | 9.8 | 7.9 | 13.7 | 16.5 | 2.4 | 1.2 | 5.2 | 1.8 | 13,336 | 50,977 |
| District 3 | 30,281 | 396,898 | 3.3 | 4.9 | 5.9 | 10.2 | 24.6 | 7.6 | 2.2 | 8.2 | 1.9 | 26,310 | 66,290 |
| District 4 | 24,962 | 278,426 | 2.7 | 6.1 | 3.0 | 13.9 | 24.4 | 4.8 | 1.6 | 6.8 | 1.5 | 13,814 | 49,616 |
| District 5 | 11,878 | 167,660 | 1.8 | 5.4 | 2.1 | 12.2 | 21.8 | 1.3 | 1.9 | 2.0 | 0.8 | 8,181 | 48,796 |
| District 6 | 19,471 | 188,619 | 2.3 | 6.7 | 3.1 | 12.5 | 41.1 | 3.5 | 2.5 | 3.7 | 1.1 | 7,986 | 42,340 |
| District 7 | 23,283 | 247,727 | 3.7 | 6.9 | 5.1 | 9.9 | 28.2 | 3.6 | 2.9 | 5.3 | 2.1 | 11,464 | 46,277 |
| District 8 | 12,146 | 159,471 | 1.9 | 3.4 | 3.2 | 13.6 | 35.5 | 2.2 | 2.2 | 2.5 | 2.0 | 6,601 | 41,390 |
| District 9 | 13,421 | 133,076 | 1.0 | 3.0 | 1.6 | 11.2 | 46.8 | 1.5 | 3.1 | 3.2 | 2.1 | 5,625 | 42,267 |
| District 10 | 39,903 | 858,092 | 1.1 | 2.5 | 2.4 | 6.4 | 13.1 | 12.7 | 2.9 | 11.6 | 9.0 | 84,634 | 98,630 |
| District 11 | 16,692 | 172,204 | 0.9 | 7.5 | 2.0 | 14.4 | 36.3 | 2.8 | 1.6 | 3.4 | 1.3 | 7,482 | 43,447 |
| District 12 | 68,539 | 1,558,850 | 1.1 | 2.7 | 4.5 | 6.0 | 9.9 | 13.0 | 3.8 | 15.1 | 8.2 | 195,064 | 125,134 |
| District 13 | 10,370 | 160,326 | 0.2 | 1.6 | 0.5 | 9.6 | 48.2 | 1.2 | 3.2 | 1.8 | 1.5 | 10,105 | 63,026 |
| District 14 | 12,621 | 160,824 | 2.7 | 10.3 | 3.6 | 10.1 | 26.7 | 1.9 | 3.0 | 1.8 | 1.5 | 8,374 | 52,069 |
| District 15 | 9,438 | 139,459 | 3.6 | 4.7 | 7.6 | 13.1 | 32.4 | 1.2 | 3.4 | 1.8 | 1.2 | 6,686 | 47,942 |
| District 16 | 16,224 | 179,554 | 3.1 | 10.2 | 2.4 | 15.6 | 22.0 | 2.2 | 3.8 | 3.5 | 1.8 | 8,884 | 49,479 |
| District 17 | 26,624 | 355,401 | 3.4 | 5.9 | 5.4 | 10.6 | 23.6 | 6.0 | 1.9 | 6.9 | 2.1 | 25,109 | 70,650 |
| District 18 | 19,610 | 233,275 | 7.7 | 6.0 | 5.4 | 16.7 | 20.5 | 2.5 | 1.6 | 5.5 | 1.8 | 11,164 | 47,859 |
| District 19 | 16,928 | 171,997 | 7.9 | 5.9 | 3.1 | 16.5 | 22.6 | 3.4 | 1.2 | 3.4 | 1.4 | 7,031 | 40,880 |
| District 20 | 19,143 | 347,157 | 7.3 | 4.3 | 3.9 | 12.3 | 20.2 | 5.5 | 1.4 | 9.3 | 2.6 | 18,414 | 53,043 |
| District 21 | 15,324 | 184,082 | 11.5 | 5.0 | 2.5 | 19.1 | 22.5 | 2.4 | 1.2 | 2.7 | 1.8 | 7,420 | 40,309 |
| District 22 | 14,013 | 225,594 | 13.1 | 3.4 | 3.5 | 14.2 | 21.6 | 4.8 | 0.9 | 4.3 | 1.8 | 9,342 | 41,410 |
| District 23 | 14,157 | 222,945 | 14.7 | 3.6 | 2.8 | 14.2 | 18.5 | 2.5 | 1.1 | 3.0 | 1.5 | 9,505 | 42,633 |
| District 24 | 16,186 | 278,093 | 10.2 | 4.6 | 5.5 | 13.6 | 19.6 | 4.1 | 1.4 | 5.8 | 1.9 | 13,151 | 47,289 |
| District 25 | 17,032 | 357,181 | 10.0 | 3.7 | 4.5 | 11.2 | 20.3 | 3.6 | 1.8 | 7.4 | 2.9 | 17,265 | 48,337 |
| District 26 | 17,388 | 357,561 | 8.1 | 3.1 | 5.2 | 10.8 | 20.2 | 8.4 | 1.5 | 7.1 | 1.6 | 17,062 | 47,717 |
| District 27 | 16,355 | 215,760 | 17.7 | 6.4 | 4.4 | 18.0 | 14.8 | 2.4 | 1.3 | 3.3 | 1.3 | 9,660 | 44,773 |
| **NORTH CAROLINA** | 238,015 | 3,932,620 | 11.5 | 5.6 | 4.9 | 12.9 | 15.7 | 4.9 | 1.5 | 6.0 | 2.3 | 194,776 | 49,528 |
| District 1 | 14,445 | 251,739 | 13.0 | 4.2 | 2.9 | 11.9 | 24.3 | 4.4 | 1.2 | 4.8 | 1.8 | 11,670 | 46,358 |
| District 2 | 16,800 | 238,250 | 11.1 | 6.8 | 7.1 | 14.5 | 17.7 | 2.7 | 1.2 | 4.4 | 1.3 | 11,552 | 48,486 |
| District 3 | 16,349 | 190,290 | 9.6 | 6.2 | 2.7 | 19.9 | 16.4 | 2.9 | 1.8 | 3.6 | 1.3 | 6,660 | 35,001 |
| District 4 | 26,492 | 514,682 | 3.6 | 5.1 | 6.3 | 10.7 | 12.2 | 5.6 | 1.8 | 15.2 | 4.5 | 33,229 | 64,561 |
| District 5 | 16,324 | 266,278 | 13.7 | 4.9 | 4.2 | 14.5 | 21.5 | 2.1 | 0.7 | 2.3 | 1.2 | 11,940 | 44,842 |
| District 6 | 13,935 | 231,384 | 19.5 | 6.1 | 3.9 | 13.1 | 17.2 | 3.1 | 1.7 | 3.8 | 2.3 | 9,303 | 40,204 |
| District 7 | 17,678 | 223,762 | 11.5 | 6.6 | 3.6 | 17.3 | 19.0 | 2.0 | 1.4 | 4.0 | 1.5 | 8,938 | 39,943 |
| District 8 | 14,629 | 215,065 | 11.1 | 5.7 | 2.9 | 18.2 | 14.3 | 3.1 | 1.5 | 3.8 | 1.2 | 8,171 | 37,992 |
| District 9 | 17,596 | 251,622 | 15.2 | 6.9 | 3.9 | 13.6 | 14.3 | 5.8 | 1.5 | 4.4 | 1.2 | 12,110 | 48,128 |
| District 10 | 18,966 | 302,658 | 19.4 | 6.2 | 4.5 | 13.3 | 18.9 | 1.9 | 1.2 | 3.0 | 0.9 | 12,528 | 41,393 |
| District 11 | 16,989 | 218,253 | 19.9 | 5.8 | 3.6 | 16.5 | 16.6 | 2.0 | 1.1 | 2.6 | 1.2 | 8,206 | 37,597 |
| District 12 | 26,092 | 557,598 | 4.5 | 5.1 | 6.2 | 8.6 | 11.2 | 11.6 | 2.0 | 8.2 | 4.1 | 36,974 | 66,309 |
| District 13 | 20,895 | 373,574 | 15.5 | 4.9 | 6.2 | 11.4 | 14.0 | 3.3 | 1.5 | 4.1 | 1.4 | 17,565 | 47,018 |
| **NORTH DAKOTA** | 24,654 | 353,333 | 7.5 | 6.6 | 6.1 | 13.5 | 18.4 | 5.2 | 1.6 | 4.7 | 2.0 | 18,292 | 51,770 |
| At Large | 24,654 | 353,333 | 7.5 | 6.6 | 6.1 | 13.5 | 18.4 | 5.2 | 1.6 | 4.7 | 2.0 | 18,292 | 51,770 |
| **OHIO** | 250,981 | 4,916,956 | 14.0 | 4.1 | 4.8 | 11.3 | 17.6 | 5.3 | 1.3 | 5.4 | 2.0 | 243,907 | 49,605 |
| District 1 | 16,980 | 413,593 | 10.9 | 4.5 | 5.1 | 8.9 | 18.3 | 6.8 | 1.4 | 7.6 | 3.1 | 26,336 | 63,677 |
| District 2 | 15,749 | 273,723 | 9.8 | 4.4 | 5.1 | 13.3 | 16.4 | 5.7 | 1.4 | 7.5 | 2.4 | 13,370 | 48,845 |
| District 3 | 15,620 | 396,892 | 5.7 | 4.6 | 5.2 | 9.9 | 21.0 | 7.5 | 2.0 | 6.2 | 1.8 | 21,835 | 55,014 |
| District 4 | 14,132 | 273,227 | 30.0 | 4.2 | 4.4 | 10.9 | 14.8 | 2.0 | 0.7 | 4.3 | 1.0 | 12,214 | 44,704 |
| District 5 | 16,057 | 303,178 | 22.9 | 4.2 | 3.7 | 12.4 | 14.6 | 3.1 | 0.9 | 4.3 | 1.0 | 13,169 | 43,435 |
| District 6 | 12,729 | 175,427 | 15.0 | 5.4 | 3.4 | 14.8 | 20.6 | 2.6 | 1.2 | 2.4 | 0.8 | 6,872 | 39,171 |
| District 7 | 14,301 | 226,059 | 23.1 | 5.7 | 4.6 | 14.0 | 17.7 | 2.8 | 0.8 | 2.6 | 1.0 | 9,102 | 40,264 |
| District 8 | 13,866 | 253,073 | 20.2 | 4.1 | 7.4 | 12.3 | 12.9 | 5.2 | 0.9 | 3.3 | 1.0 | 11,362 | 44,897 |
| District 9 | 12,916 | 241,993 | 18.6 | 4.2 | 4.6 | 11.3 | 18.2 | 3.7 | 1.4 | 2.9 | 1.4 | 11,919 | 49,253 |
| District 10 | 14,852 | 300,355 | 12.2 | 3.5 | 3.6 | 12.7 | 20.8 | 4.3 | 1.2 | 8.5 | 2.5 | 14,884 | 49,555 |
| District 11 | 17,679 | 413,642 | 8.2 | 3.3 | 4.6 | 6.8 | 29.0 | 5.8 | 1.8 | 7.9 | 1.9 | 25,324 | 61,221 |
| District 12 | 18,148 | 341,966 | 8.5 | 3.5 | 3.3 | 10.8 | 19.0 | 10.9 | 1.1 | 6.4 | 1.9 | 17,362 | 50,771 |
| District 13 | 14,432 | 245,037 | 14.4 | 4.6 | 4.8 | 15.9 | 18.7 | 2.0 | 1.5 | 2.9 | 1.6 | 10,101 | 41,221 |
| District 14 | 19,946 | 333,748 | 18.6 | 4.2 | 7.7 | 11.5 | 12.6 | 6.1 | 1.7 | 5.2 | 1.6 | 17,772 | 53,249 |
| District 15 | 14,377 | 243,829 | 12.3 | 4.1 | 4.7 | 14.0 | 14.8 | 4.4 | 1.9 | 5.0 | 2.5 | 10,866 | 44,564 |
| District 16 | 18,162 | 311,740 | 13.2 | 4.1 | 5.0 | 14.6 | 15.9 | 4.1 | 1.1 | 4.5 | 5.8 | 13,659 | 43,815 |
| **OKLAHOMA** | 93,761 | 1,404,725 | 9.5 | 5.5 | 4.2 | 12.8 | 15.6 | 4.2 | 1.5 | 5.5 | 1.9 | 65,227 | 46,434 |
| District 1 | 21,749 | 385,566 | 11.8 | 5.4 | 4.3 | 11.1 | 14.6 | 3.9 | 1.4 | 6.5 | 2.8 | 20,028 | 51,944 |
| District 2 | 13,017 | 163,558 | 15.1 | 5.0 | 3.4 | 15.6 | 23.2 | 3.5 | 0.8 | 2.7 | 1.2 | 6,025 | 36,840 |
| District 3 | 17,197 | 198,402 | 12.3 | 7.4 | 4.4 | 14.9 | 13.6 | 3.6 | 1.6 | 3.8 | 1.5 | 8,297 | 41,821 |
| District 4 | 16,763 | 211,514 | 8.1 | 5.7 | 2.8 | 16.3 | 17.4 | 4.2 | 1.7 | 4.7 | 1.3 | 8,269 | 39,094 |
| District 5 | 24,536 | 391,788 | 5.7 | 5.3 | 5.5 | 11.9 | 14.9 | 5.3 | 1.8 | 6.5 | 1.9 | 19,325 | 49,325 |
| **OREGON** | 119,074 | 1,643,425 | 11.0 | 6.3 | 4.7 | 12.7 | 15.8 | 4.0 | 1.9 | 5.9 | 2.3 | 87,517 | 53,253 |
| District 1 | 24,008 | 400,589 | 14.3 | 6.3 | 6.0 | 11.4 | 11.6 | 4.0 | 2.2 | 7.0 | 2.9 | 27,179 | 67,849 |
| District 2 | 24,873 | 270,689 | 11.1 | 6.1 | 3.1 | 16.2 | 18.4 | 2.8 | 1.5 | 4.0 | 1.9 | 11,283 | 41,684 |

# Table E. Congressional Districts 116th Congress — **Land Area and Population Characteristics**

Population and population characteristics, 2019

Race alone (percent)

| STATE District | Representative, 117th Congress | Land area,[1] 2018 (sq mi) | Total persons | Per square mile | White | Black | American Indian, Alaska Native | Asian and Pacific Islander | Some other race (percent) | Two or more races (percent) | Hispanic or Latino[2] (percent) | Non-Hispanic White alone (percent) | Percent female | Percent foreign-born | Percent born in state of residence |
|---|---|---|---|---|---|---|---|---|---|---|---|---|---|---|---|
| | | 1 | 2 | 3 | 4 | 5 | 6 | 7 | 8 | 9 | 10 | 11 | 12 | 13 | 14 |
| **OREGON—Cont'd** | | | | | | | | | | | | | | | |
| District 3 | Earl Blumenauer (D) | 1,074.5 | 853,116 | 794.0 | 78.2 | 4.8 | 0.8 | 8.0 | 2.6 | 5.6 | 11.6 | 70.0 | 50.2 | 13.5 | 43.1 |
| District 4 | Peter A. DeFazio (D) | 17,272.9 | 820,504 | 47.5 | 87.9 | 0.9 | 0.9 | 2.5 | 2.7 | 5.0 | 8.5 | 83.2 | 50.8 | 4.7 | 45.9 |
| District 5 | Kurt Schrader (D) | 5,190.2 | 844,220 | 162.7 | 83.8 | 0.9 | 1.2 | 3.1 | 5.5 | 5.4 | 17.7 | 73.5 | 50.7 | 9.5 | 52.2 |
| **PENNSYLVANIA** | | 44,742.3 | 12,801,989 | 286.1 | 79.6 | 11.4 | 0.2 | 3.5 | 2.6 | 2.6 | 7.8 | 75.6 | 51.0 | 7.0 | 71.4 |
| District 1 | Brian K. Fitzpatrick (R) | 638.4 | 713,411 | 1,117.4 | 85.4 | 4.7 | 0.1 | 6.2 | 1.6 | 2.1 | 5.5 | 81.9 | 51.1 | 11.0 | 64.9 |
| District 2 | Brendan F. Boyle (D) | 62.8 | 722,722 | 11,513.8 | 45.0 | 25.2 | 0.4 | 8.8 | 16.6 | 4.0 | 27.9 | 37.3 | 51.9 | 18.6 | 62.6 |
| District 3 | Dwight Evans (D) | 52.8 | 741,654 | 14,036.1 | 33.3 | 56.4 | 0.6 | 5.7 | 1.1 | 3.0 | 4.3 | 31.1 | 53.3 | 8.8 | 65.6 |
| District 4 | Madeleine Dean (D) | 476.9 | 730,701 | 1,532.1 | 79.6 | 9.7 | 0.1 | 6.7 | 1.2 | 2.7 | 5.7 | 75.6 | 51.1 | 9.6 | 72.5 |
| District 5 | Mary Gay Scanlon (D) | 211.8 | 719,973 | 3,399.6 | 63.3 | 25.8 | 0.3 | 7.1 | 1.2 | 2.3 | 4.5 | 60.8 | 52.3 | 12.1 | 69.2 |
| District 6 | Chrissy Houlahan (D) | 912.8 | 735,283 | 805.5 | 78.0 | 7.0 | 0.2 | 4.7 | 7.0 | 3.1 | 16.1 | 71.3 | 50.6 | 9.7 | 61.8 |
| District 7 | Susan Wild (D) | 856.9 | 731,467 | 853.6 | 82.0 | 7.5 | 0.2 | 3.0 | 4.1 | 3.1 | 20.5 | 68.2 | 50.9 | 9.8 | 56.9 |
| District 8 | Matt Cartwright (D) | 2,666.9 | 698,973 | 262.1 | 84.3 | 7.9 | 0.3 | 1.9 | 3.4 | 2.2 | 12.7 | 77.5 | 50.6 | 6.7 | 60.4 |
| District 9 | Daniel Meuser (R) | 3,294.6 | 699,832 | 212.4 | 92.2 | 2.6 | 0.3 | 1.2 | 2.1 | 1.4 | 7.1 | 88.1 | 50.2 | 3.4 | 80.1 |
| District 10 | Scott Perry (R) | 1,080.2 | 744,681 | 689.4 | 78.1 | 11.4 | 0.2 | 4.1 | 2.3 | 3.7 | 9.1 | 73.3 | 50.9 | 7.5 | 68.6 |
| District 11 | Lloyd Smucker (R) | 1,503.0 | 734,038 | 488.4 | 89.1 | 3.7 | 0.1 | 1.8 | 2.1 | 3.2 | 8.9 | 84.1 | 50.9 | 4.2 | 70.5 |
| District 12 | Fred Keller (R) | 9,895.9 | 701,387 | 70.9 | 93.8 | 2.2 | 0.2 | 1.8 | 0.4 | 1.7 | 2.6 | 91.8 | 49.9 | 3.1 | 76.9 |
| District 13 | John Joyce (R) | 6,020.1 | 697,051 | 115.8 | 93.4 | 3.2 | 0.4 | 0.5 | 0.8 | 1.7 | 3.5 | 91.2 | 50.0 | 2.3 | 75.9 |
| District 14 | Guy Reschenthaler (R) | 2,847.5 | 678,915 | 238.4 | 93.4 | 3.0 | 0.0 | 0.9 | 0.4 | 2.2 | 1.5 | 92.3 | 50.8 | 1.8 | 83.5 |
| District 15 | Glen Thompson (R) | 9,734.7 | 672,749 | 69.1 | 95.2 | 2.4 | 0.1 | 0.7 | 0.3 | 1.3 | 1.3 | 94.6 | 49.0 | 1.6 | 84.9 |
| District 16 | Mike Kelly (R) | 3,311.5 | 678,333 | 204.8 | 90.6 | 4.5 | 0.2 | 1.1 | 0.7 | 3.0 | 2.7 | 89.2 | 51.1 | 2.2 | 80.3 |
| District 17 | Conor Lamb (D) | 882.8 | 706,961 | 800.8 | 88.1 | 6.0 | 0.1 | 2.7 | 0.5 | 2.6 | 1.7 | 87.0 | 51.1 | 5.2 | 77.2 |
| District 18 | Michael F. Doyle (D) | 292.5 | 693,858 | 2,371.9 | 73.5 | 18.1 | 0.1 | 4.3 | 0.7 | 3.3 | 2.7 | 71.8 | 52.2 | 6.6 | 75.8 |
| **RHODE ISLAND** | | 1,033.9 | 1,059,361 | 1,024.6 | 78.7 | 7.4 | 0.3 | 3.5 | 5.9 | 4.2 | 16.3 | 70.8 | 51.2 | 13.7 | 56.6 |
| District 1 | David Cicilline (D) | 268.5 | 530,066 | 1,973.9 | 75.9 | 8.6 | 0.4 | 3.2 | 7.3 | 4.7 | 19.1 | 66.1 | 50.4 | 15.6 | 51.7 |
| District 2 | James R. Langevin (D) | 765.4 | 529,295 | 691.6 | 81.5 | 6.2 | 0.3 | 3.8 | 4.4 | 3.7 | 13.5 | 75.5 | 51.9 | 11.7 | 61.6 |
| **SOUTH CAROLINA** | | 30,064.3 | 5,148,714 | 171.3 | 66.7 | 26.5 | 0.4 | 1.8 | 2.2 | 2.4 | 5.8 | 63.5 | 51.7 | 5.6 | 55.0 |
| District 1 | Nancy Mace (R) | 1,549.6 | 821,107 | 529.9 | 72.0 | 19.0 | 0.4 | 2.6 | 2.9 | 3.0 | 6.1 | 69.4 | 51.8 | 7.0 | 40.8 |
| District 2 | Joe Wilson (R) | 3,022.3 | 722,542 | 239.1 | 69.4 | 24.6 | 0.4 | 2.0 | 1.2 | 2.4 | 5.9 | 65.3 | 51.8 | 5.0 | 52.8 |
| District 3 | Jeff Duncan (R) | 5,269.0 | 706,961 | 134.2 | 75.1 | 19.2 | 0.2 | 1.0 | 1.8 | 2.6 | 4.7 | 72.4 | 51.1 | 3.9 | 65.4 |
| District 4 | William R. Timmons IV (R) | 1,299.7 | 754,148 | 580.3 | 73.5 | 17.6 | 0.4 | 2.8 | 3.2 | 2.5 | 9.0 | 68.4 | 51.5 | 8.3 | 55.2 |
| District 5 | Ralph Norman (R) | 5,505.9 | 738,205 | 134.1 | 68.1 | 26.2 | 0.5 | 1.7 | 0.9 | 2.6 | 4.4 | 65.1 | 52.3 | 5.1 | 53.6 |
| District 6 | James E. Clyburn (D) | 8,063.3 | 665,215 | 82.5 | 38.1 | 54.3 | 0.2 | 1.3 | 3.9 | 2.2 | 6.0 | 36.3 | 51.3 | 5.2 | 66.6 |
| District 7 | Tom Rice (R) | 5,354.5 | 740,536 | 138.3 | 67.5 | 27.8 | 0.5 | 1.1 | 1.4 | 1.7 | 4.2 | 64.9 | 52.2 | 4.3 | 53.7 |
| **SOUTH DAKOTA** | | 75,809.6 | 884,659 | 11.7 | 84.1 | 2.4 | 8.6 | 1.5 | 0.7 | 2.8 | 3.7 | 81.5 | 49.2 | 4.1 | 63.6 |
| At Large | Dusty Johnson (R) | 75,809.6 | 884,659 | 11.7 | 84.1 | 2.4 | 8.6 | 1.5 | 0.7 | 2.8 | 3.7 | 81.5 | 49.2 | 4.1 | 63.6 |
| **TENNESSEE** | | 41,232.5 | 6,829,174 | 165.6 | 77.2 | 16.7 | 0.3 | 1.9 | 1.6 | 2.3 | 5.7 | 73.3 | 51.3 | 5.5 | 59.1 |
| District 1 | Diana Harshbarger (R) | 4,143.6 | 725,173 | 175.0 | 94.1 | 2.3 | 0.4 | 0.9 | 0.5 | 1.9 | 4.1 | 90.8 | 51.0 | 3.1 | 61.4 |
| District 2 | Tim Burchett (R) | 2,322.0 | 758,519 | 326.7 | 89.0 | 5.9 | 0.3 | 1.6 | 0.7 | 2.5 | 4.4 | 85.3 | 51.7 | 4.6 | 58.1 |
| District 3 | Chuck Fleischmann (R) | 4,570.3 | 743,225 | 162.6 | 84.4 | 10.5 | 0.3 | 1.5 | 1.2 | 2.1 | 4.3 | 81.6 | 50.6 | 4.2 | 61.2 |
| District 4 | Scott DesJarlais (R) | 5,985.0 | 812,697 | 135.8 | 83.3 | 9.9 | 0.2 | 2.0 | 1.5 | 3.0 | 7.3 | 78.1 | 50.9 | 6.0 | 60.4 |
| District 5 | Jim Cooper (D) | 1,248.3 | 778,094 | 623.3 | 66.5 | 24.4 | 0.3 | 3.5 | 3.1 | 2.2 | 9.7 | 59.7 | 51.7 | 12.3 | 49.8 |
| District 6 | John W. Rose (R) | 6,475.3 | 799,365 | 123.4 | 89.4 | 5.2 | 0.4 | 1.1 | 2.1 | 1.9 | 4.6 | 86.8 | 50.5 | 4.0 | 61.1 |
| District 7 | Mark E. Green (R) | 9,161.2 | 800,536 | 87.4 | 83.4 | 9.7 | 0.4 | 2.2 | 0.9 | 3.5 | 5.4 | 79.4 | 50.7 | 5.0 | 50.7 |
| District 8 | David Kustoff (R) | 6,846.1 | 711,068 | 103.9 | 73.0 | 22.0 | 0.3 | 1.7 | 1.2 | 1.8 | 3.1 | 71.1 | 51.5 | 3.4 | 65.4 |
| District 9 | Steve Cohen (D) | 480.7 | 700,497 | 1,457.3 | 27.6 | 65.1 | 0.1 | 2.4 | 3.2 | 1.6 | 8.1 | 23.0 | 52.9 | 6.4 | 65.6 |
| **TEXAS** | | 222,850.4 | 28,995,881 | 130.1 | 73.4 | 12.3 | 0.5 | 5.1 | 5.9 | 2.9 | 39.7 | 41.1 | 50.4 | 17.1 | 59.5 |
| District 1 | Louie Gohmert (R) | 7,869.1 | 726,094 | 92.3 | 77.5 | 17.2 | 0.4 | 1.1 | 1.2 | 2.5 | 18.0 | 61.6 | 51.1 | 7.2 | 71.6 |
| District 2 | Dan Crenshaw (R) | 309.1 | 787,271 | 2,547.1 | 66.4 | 11.9 | 0.5 | 8.3 | 9.7 | 3.2 | 31.6 | 46.5 | 49.7 | 21.5 | 50.4 |
| District 3 | Van Taylor (R) | 480.9 | 913,161 | 1,898.8 | 67.2 | 9.9 | 0.3 | 17.9 | 1.8 | 2.9 | 14.9 | 54.7 | 50.8 | 22.9 | 42.7 |
| District 4 | Pat Fallon (R) | 10,131.2 | 782,743 | 77.3 | 80.4 | 10.9 | 0.7 | 1.7 | 3.0 | 3.3 | 15.2 | 69.4 | 50.6 | 7.2 | 65.3 |
| District 5 | Lance Gooden (R) | 5,044.4 | 759,749 | 150.6 | 75.3 | 15.1 | 0.5 | 2.8 | 4.0 | 2.2 | 30.1 | 50.3 | 49.8 | 15.5 | 65.1 |
| District 6 | Vacant | 2,148.8 | 818,442 | 380.9 | 64.1 | 21.1 | 0.3 | 5.1 | 6.3 | 3.2 | 25.4 | 45.5 | 51.1 | 13.4 | 58.5 |
| District 7 | Lizzie Fletcher (D) | 162.1 | 762,826 | 4,706.6 | 60.0 | 15.5 | 0.7 | 12.2 | 8.3 | 3.4 | 30.3 | 38.9 | 50.9 | 30.6 | 45.0 |
| District 8 | Kevin Brady (R) | 6,054.8 | 895,861 | 148.0 | 83.0 | 8.4 | 0.3 | 3.0 | 2.5 | 2.7 | 24.2 | 62.4 | 49.2 | 13.2 | 57.6 |
| District 9 | Al Green (D) | 165.8 | 769,335 | 4,640.5 | 34.5 | 39.0 | 0.3 | 12.1 | 11.7 | 2.3 | 38.0 | 10.2 | 50.8 | 34.6 | 50.0 |
| District 10 | Michael T. McCaul (R) | 5,072.2 | 925,348 | 182.4 | 70.1 | 12.7 | 0.6 | 5.1 | 8.3 | 3.3 | 30.7 | 49.8 | 51.0 | 17.8 | 54.7 |
| District 11 | August Pfluger (R) | 27,834.6 | 790,264 | 28.4 | 83.1 | 4.3 | 0.6 | 1.2 | 8.2 | 2.6 | 39.9 | 53.3 | 50.1 | 10.0 | 68.6 |
| District 12 | Kay Granger (R) | 1,441.5 | 844,563 | 585.9 | 78.9 | 9.7 | 0.8 | 3.9 | 3.8 | 3.0 | 23.2 | 60.6 | 51.7 | 11.3 | 60.2 |
| District 13 | Ronny Jackson (R) | 38,350.8 | 714,733 | 18.6 | 85.1 | 5.5 | 0.8 | 2.6 | 3.1 | 2.8 | 28.4 | 61.5 | 50.6 | 9.3 | 66.7 |
| District 14 | Randy K. Weber Sr. (R) | 2,448.3 | 760,530 | 310.6 | 73.1 | 19.3 | 0.2 | 3.3 | 2.0 | 2.1 | 26.3 | 49.7 | 49.5 | 11.1 | 67.3 |
| District 15 | Vicente Gonzalez (D) | 7,804.2 | 804,562 | 103.1 | 83.8 | 1.9 | 0.2 | 1.2 | 10.3 | 2.6 | 82.0 | 14.0 | 50.5 | 21.5 | 65.7 |
| District 16 | Veronica Escobar (D) | 710.9 | 747,648 | 1,051.7 | 79.0 | 3.8 | 0.6 | 1.3 | 12.5 | 2.7 | 81.3 | 12.7 | 50.5 | 21.4 | 59.2 |
| District 17 | Pete Sessions (R) | 7,650.9 | 786,023 | 102.7 | 76.1 | 13.7 | 0.5 | 4.6 | 1.7 | 3.4 | 26.7 | 52.6 | 50.2 | 11.4 | 67.3 |
| District 18 | Sheila Jackson-Lee (D) | 235.5 | 827,015 | 3,511.6 | 51.0 | 33.7 | 0.4 | 4.5 | 8.1 | 2.2 | 44.8 | 15.7 | 50.2 | 23.4 | 58.6 |
| District 19 | Jodey C. Arrington (R) | 25,835.8 | 729,664 | 28.2 | 83.6 | 6.2 | 0.4 | 1.7 | 5.3 | 2.8 | 37.4 | 53.1 | 50.2 | 8.3 | 71.3 |
| District 20 | Joaquin Castro (D) | 199.9 | 832,518 | 4,165.2 | 78.5 | 5.7 | 1.1 | 3.1 | 7.7 | 3.8 | 68.4 | 21.3 | 50.9 | 14.9 | 63.6 |
| District 21 | Chip Roy (R) | 5,921.0 | 829,628 | 140.1 | 84.7 | 4.2 | 0.8 | 3.8 | 3.1 | 3.5 | 30.5 | 59.0 | 50.7 | 10.5 | 56.0 |
| District 22 | Troy E. Nehls (R) | 1,033.7 | 960,957 | 929.7 | 59.7 | 14.5 | 0.6 | 18.4 | 3.2 | 3.7 | 26.8 | 37.8 | 51.0 | 24.8 | 51.8 |
| District 23 | Tony Gonzales (R) | 58,058.9 | 786,712 | 13.6 | 84.3 | 3.5 | 0.8 | 2.2 | 6.7 | 2.5 | 69.6 | 23.9 | 49.9 | 15.1 | 64.7 |
| District 24 | Beth Van Duyne (R) | 263.6 | 832,445 | 3,157.7 | 62.2 | 13.1 | 0.5 | 15.8 | 4.7 | 3.6 | 23.6 | 44.6 | 50.3 | 26.7 | 42.5 |
| District 25 | Roger Williams (R) | 7,622.5 | 818,807 | 107.4 | 82.4 | 7.4 | 0.5 | 4.1 | 2.1 | 3.6 | 18.7 | 67.7 | 49.6 | 8.2 | 57.1 |

1. Dry land or land partially or temporarily covered by water.  2. May be of any race.

| | Population and population characteristics, 2019 (cont.) | | | | | | | | | | Education, 2019 | | |
| STATE District | Age (percent) | | | | | | | | | | Total Enrollment[1] | Attainment[2] (percent) | |
| | Under 5 years | 5 to 17 years | 18 to 24 years | 25 to 34 years | 35 to 44 years | 45 to 54 years | 55 to 64 years | 65 to 74 years | 75 years and over | Median age | | High school graduate or more | Bachelor's degree or more |
| | 15 | 16 | 17 | 18 | 19 | 20 | 21 | 22 | 23 | 24 | 25 | 26 | 27 |
|---|---|---|---|---|---|---|---|---|---|---|---|---|---|
| **OREGON—Cont'd** | | | | | | | | | | | | | |
| District 3 | 5.2 | 13.9 | 8.2 | 17.8 | 16.6 | 13.0 | 11.1 | 9.2 | 5.1 | 37.6 | 190,291 | 91.7 | 44.3 |
| District 4 | 4.8 | 13.5 | 11.1 | 12.5 | 11.8 | 10.8 | 13.9 | 13.1 | 8.6 | 41.6 | 186,206 | 91.9 | 28.1 |
| District 5 | 5.9 | 16.3 | 8.6 | 12.7 | 13.2 | 12.0 | 13.1 | 11.0 | 7.3 | 39.9 | 193,284 | 90.1 | 32.0 |
| **PENNSYLVANIA** | 5.4 | 15.1 | 8.9 | 13.3 | 11.9 | 12.5 | 14.1 | 10.6 | 8.1 | 40.8 | 2,896,469 | 91.0 | 32.3 |
| District 1 | 4.9 | 15.5 | 7.6 | 11.3 | 12.4 | 13.6 | 15.8 | 10.7 | 8.2 | 43.5 | 161,151 | 94.0 | 42.6 |
| District 2 | 7.2 | 17.7 | 7.4 | 16.7 | 13.1 | 12.3 | 12.3 | 7.8 | 5.8 | 35.7 | 178,344 | 81.0 | 22.6 |
| District 3 | 5.6 | 12.5 | 12.1 | 22.2 | 12.0 | 10.2 | 10.8 | 8.5 | 6.0 | 33.6 | 188,150 | 90.4 | 40.6 |
| District 4 | 5.5 | 15.7 | 7.8 | 12.6 | 12.7 | 12.7 | 14.4 | 10.3 | 8.1 | 41.2 | 167,275 | 94.7 | 48.5 |
| District 5 | 6.1 | 16.3 | 9.5 | 13.2 | 12.8 | 12.2 | 13.5 | 9.2 | 7.1 | 38.6 | 183,111 | 92.2 | 39.2 |
| District 6 | 5.8 | 17.4 | 9.3 | 12.1 | 12.1 | 13.0 | 13.6 | 9.4 | 7.0 | 39.3 | 191,570 | 91.4 | 46.7 |
| District 7 | 5.3 | 15.8 | 9.5 | 12.6 | 12.6 | 12.0 | 14.0 | 10.1 | 8.0 | 40.6 | 173,109 | 90.6 | 30.1 |
| District 8 | 5.0 | 14.8 | 8.7 | 12.1 | 11.8 | 12.6 | 14.8 | 11.4 | 8.6 | 43.0 | 152,729 | 91.4 | 24.3 |
| District 9 | 4.9 | 14.6 | 8.6 | 11.8 | 11.2 | 13.6 | 15.2 | 11.6 | 8.7 | 44.2 | 144,954 | 89.3 | 21.3 |
| District 10 | 5.8 | 15.9 | 8.4 | 13.4 | 12.1 | 13.1 | 13.7 | 10.2 | 7.5 | 40.3 | 170,434 | 90.8 | 32.2 |
| District 11 | 6.0 | 16.9 | 8.4 | 13.0 | 11.8 | 11.9 | 13.4 | 10.4 | 8.2 | 39.5 | 163,059 | 87.2 | 28.1 |
| District 12 | 5.0 | 14.6 | 12.1 | 11.8 | 10.6 | 11.9 | 14.1 | 11.1 | 8.7 | 41.1 | 165,741 | 89.7 | 23.8 |
| District 13 | 5.1 | 15.1 | 8.2 | 11.1 | 11.3 | 13.0 | 14.8 | 11.8 | 9.6 | 44.2 | 141,766 | 89.4 | 20.6 |
| District 14 | 4.9 | 14.0 | 7.8 | 11.3 | 11.2 | 13.2 | 15.6 | 12.6 | 9.4 | 45.6 | 134,116 | 92.5 | 27.5 |
| District 15 | 4.7 | 13.7 | 8.8 | 12.1 | 11.0 | 13.2 | 15.3 | 12.1 | 9.2 | 44.8 | 131,992 | 91.4 | 20.6 |
| District 16 | 5.3 | 14.7 | 9.5 | 11.8 | 10.9 | 12.9 | 14.7 | 11.6 | 8.6 | 43.0 | 146,327 | 91.5 | 28.3 |
| District 17 | 5.3 | 14.4 | 6.1 | 13.1 | 12.9 | 12.4 | 15.3 | 11.9 | 8.6 | 43.5 | 142,719 | 95.4 | 42.5 |
| District 18 | 5.2 | 12.5 | 10.7 | 17.1 | 11.3 | 11.0 | 13.5 | 10.2 | 8.7 | 38.5 | 159,922 | 94.5 | 40.0 |
| **RHODE ISLAND** | 5.2 | 14.0 | 10.3 | 14.0 | 11.8 | 12.8 | 14.3 | 10.2 | 7.5 | 40.1 | 253,795 | 89.3 | 34.8 |
| District 1 | 5.7 | 14.4 | 10.7 | 13.8 | 11.1 | 12.9 | 14.2 | 9.8 | 7.3 | 39.3 | 130,826 | 87.1 | 34.2 |
| District 2 | 4.6 | 13.7 | 9.9 | 14.1 | 12.4 | 12.6 | 14.5 | 10.5 | 7.7 | 40.8 | 122,969 | 91.4 | 35.4 |
| **SOUTH CAROLINA** | 5.6 | 15.9 | 9.2 | 13.1 | 12.2 | 12.3 | 13.5 | 11.1 | 7.1 | 39.9 | 1,180,396 | 88.3 | 29.6 |
| District 1 | 5.8 | 15.9 | 7.7 | 14.1 | 13.0 | 11.8 | 13.0 | 11.7 | 7.2 | 39.8 | 182,152 | 92.7 | 40.9 |
| District 2 | 5.5 | 17.1 | 9.1 | 13.2 | 12.2 | 13.1 | 13.1 | 10.3 | 6.6 | 39.0 | 171,064 | 89.9 | 35.1 |
| District 3 | 5.4 | 16.2 | 10.0 | 11.9 | 11.3 | 12.7 | 13.6 | 11.2 | 7.7 | 40.6 | 164,696 | 86.5 | 23.9 |
| District 4 | 6.2 | 16.3 | 8.9 | 14.0 | 12.6 | 12.6 | 12.8 | 9.7 | 6.7 | 38.4 | 176,448 | 87.9 | 33.4 |
| District 5 | 5.7 | 16.9 | 8.1 | 12.2 | 13.1 | 13.2 | 13.4 | 10.5 | 6.8 | 40.1 | 169,805 | 89.0 | 26.3 |
| District 6 | 5.5 | 14.6 | 13.1 | 14.2 | 12.2 | 11.0 | 13.3 | 9.7 | 6.5 | 36.9 | 163,200 | 84.5 | 22.9 |
| District 7 | 5.2 | 14.6 | 8.1 | 11.6 | 11.0 | 12.2 | 15.0 | 14.3 | 7.9 | 44.6 | 153,031 | 86.8 | 22.4 |
| **SOUTH DAKOTA** | 6.7 | 17.7 | 9.2 | 12.9 | 12.0 | 11.0 | 13.2 | 10.3 | 7.2 | 37.7 | 214,461 | 92.1 | 29.7 |
| At Large | 6.7 | 17.7 | 9.2 | 12.9 | 12.0 | 11.0 | 13.2 | 10.3 | 7.2 | 37.7 | 214,461 | 92.1 | 29.7 |
| **TENNESSEE** | 6.0 | 16.1 | 9.1 | 13.7 | 12.5 | 12.8 | 13.1 | 9.9 | 6.7 | 39.0 | 1,572,331 | 88.0 | 28.7 |
| District 1 | 5.2 | 14.4 | 8.5 | 12.0 | 11.0 | 13.7 | 14.4 | 12.0 | 8.8 | 44.0 | 141,551 | 85.7 | 21.9 |
| District 2 | 5.2 | 14.9 | 10.5 | 12.9 | 11.9 | 12.8 | 13.4 | 10.9 | 7.4 | 40.5 | 173,025 | 89.3 | 31.2 |
| District 3 | 5.5 | 15.4 | 8.3 | 12.7 | 12.2 | 13.0 | 14.0 | 11.1 | 7.9 | 41.6 | 160,829 | 87.5 | 26.1 |
| District 4 | 6.2 | 17.3 | 10.1 | 13.3 | 12.5 | 12.9 | 12.7 | 9.1 | 5.9 | 37.8 | 199,471 | 87.2 | 23.8 |
| District 5 | 6.6 | 14.2 | 9.3 | 19.8 | 13.7 | 11.5 | 11.8 | 7.9 | 5.0 | 35.0 | 171,260 | 89.5 | 43.0 |
| District 6 | 5.5 | 16.4 | 8.2 | 12.4 | 12.6 | 13.2 | 13.6 | 10.7 | 7.3 | 40.9 | 179,977 | 87.3 | 25.0 |
| District 7 | 6.3 | 18.0 | 9.1 | 12.2 | 13.5 | 13.2 | 12.5 | 9.2 | 6.0 | 38.2 | 197,876 | 89.7 | 32.1 |
| District 8 | 5.9 | 17.2 | 8.4 | 11.0 | 12.6 | 13.0 | 14.1 | 10.5 | 7.3 | 41.2 | 169,328 | 89.6 | 30.6 |
| District 9 | 7.2 | 17.5 | 9.7 | 16.9 | 11.8 | 11.5 | 12.1 | 8.4 | 4.8 | 34.1 | 179,014 | 86.1 | 24.2 |
| **TEXAS** | 6.8 | 18.7 | 9.7 | 14.6 | 13.7 | 12.3 | 11.3 | 7.8 | 5.1 | 35.1 | 7,763,174 | 84.6 | 30.8 |
| District 1 | 6.4 | 17.8 | 9.9 | 13.0 | 12.1 | 10.9 | 12.2 | 10.1 | 7.6 | 37.4 | 180,920 | 86.0 | 22.1 |
| District 2 | 6.2 | 17.4 | 9.1 | 16.0 | 14.5 | 12.2 | 11.8 | 8.2 | 4.7 | 35.9 | 201,117 | 88.3 | 44.6 |
| District 3 | 5.9 | 19.5 | 8.1 | 12.8 | 16.3 | 14.7 | 11.2 | 7.1 | 4.5 | 37.5 | 249,306 | 94.6 | 54.5 |
| District 4 | 6.2 | 18.3 | 8.1 | 12.0 | 12.7 | 12.7 | 13.4 | 9.7 | 7.1 | 39.1 | 192,021 | 87.6 | 23.6 |
| District 5 | 6.9 | 17.8 | 9.1 | 14.0 | 13.1 | 12.5 | 11.7 | 8.8 | 6.0 | 36.4 | 185,425 | 81.9 | 19.8 |
| District 6 | 6.5 | 19.4 | 10.0 | 13.8 | 13.4 | 13.0 | 11.6 | 7.7 | 4.7 | 35.2 | 233,758 | 89.7 | 32.3 |
| District 7 | 7.5 | 16.4 | 7.5 | 17.6 | 15.6 | 12.3 | 11.3 | 7.3 | 4.5 | 35.6 | 194,074 | 90.7 | 50.3 |
| District 8 | 6.2 | 18.7 | 8.9 | 12.4 | 13.7 | 13.5 | 12.3 | 8.9 | 5.4 | 37.9 | 232,937 | 87.5 | 30.5 |
| District 9 | 7.0 | 18.3 | 9.5 | 15.9 | 14.2 | 12.6 | 11.2 | 6.8 | 4.3 | 34.4 | 209,818 | 79.0 | 26.5 |
| District 10 | 6.4 | 19.3 | 8.3 | 14.9 | 14.8 | 13.2 | 10.8 | 7.3 | 5.0 | 35.7 | 252,011 | 91.4 | 41.4 |
| District 11 | 7.3 | 18.2 | 8.9 | 14.5 | 12.8 | 10.6 | 12.1 | 8.8 | 6.8 | 35.6 | 194,413 | 84.6 | 23.1 |
| District 12 | 6.8 | 17.9 | 9.2 | 14.8 | 14.2 | 12.4 | 11.4 | 8.0 | 5.3 | 35.9 | 218,214 | 89.4 | 31.9 |
| District 13 | 6.7 | 18.1 | 9.6 | 13.1 | 12.7 | 11.6 | 12.1 | 9.1 | 6.8 | 37.0 | 176,770 | 85.1 | 21.3 |
| District 14 | 6.3 | 17.6 | 9.2 | 13.8 | 13.1 | 12.2 | 13.3 | 8.6 | 5.8 | 37.3 | 189,305 | 87.6 | 25.0 |
| District 15 | 8.3 | 21.4 | 11.0 | 13.6 | 13.2 | 11.2 | 9.6 | 6.9 | 4.8 | 31.8 | 239,556 | 72.0 | 21.2 |
| District 16 | 7.2 | 19.7 | 11.4 | 15.2 | 12.1 | 11.7 | 10.4 | 7.1 | 5.4 | 32.4 | 223,596 | 81.2 | 24.5 |
| District 17 | 6.4 | 15.8 | 15.3 | 14.8 | 12.7 | 11.1 | 11.0 | 7.4 | 5.4 | 33.3 | 236,673 | 87.3 | 32.3 |
| District 18 | 7.3 | 18.6 | 10.8 | 17.8 | 13.7 | 11.7 | 10.1 | 6.3 | 3.7 | 32.4 | 215,550 | 78.0 | 23.0 |
| District 19 | 6.6 | 17.7 | 13.4 | 13.7 | 12.7 | 10.1 | 11.3 | 8.3 | 6.1 | 33.9 | 203,243 | 83.4 | 23.9 |
| District 20 | 7.4 | 18.4 | 11.8 | 15.9 | 13.5 | 11.2 | 9.9 | 7.1 | 4.7 | 32.6 | 233,455 | 82.0 | 24.5 |
| District 21 | 5.0 | 13.9 | 10.1 | 15.7 | 13.4 | 12.5 | 12.2 | 10.5 | 6.7 | 38.9 | 196,145 | 93.5 | 47.4 |
| District 22 | 6.7 | 20.7 | 7.9 | 12.1 | 15.6 | 14.0 | 11.5 | 7.4 | 3.8 | 36.7 | 279,860 | 91.2 | 45.4 |
| District 23 | 7.0 | 20.4 | 9.2 | 14.4 | 13.7 | 11.6 | 10.6 | 8.1 | 5.0 | 34.4 | 217,927 | 79.0 | 23.9 |
| District 24 | 6.4 | 15.9 | 8.5 | 18.2 | 15.2 | 12.8 | 12.1 | 6.7 | 4.2 | 35.6 | 195,493 | 91.1 | 48.4 |
| District 25 | 5.7 | 17.6 | 9.3 | 13.2 | 14.0 | 13.4 | 12.6 | 8.7 | 5.4 | 37.8 | 207,000 | 91.3 | 40.6 |

1. All persons 3 years old and over enrolled in nursery school through college and graduate or professional school.   2. Persons 25 years old and over.

| STATE District | Households, 2019 | | | | | | Total in group quarters, 2018 | Group Quarters, 2010 | | | | |
|---|---|---|---|---|---|---|---|---|---|---|---|---|
| | Number | Average household size | Family households (percent) | Married couple family (percent) | Female family house-holder[1] | One person households (percent) | | Percent 65 years and over | Persons in correctional institutions | Persons in nursing facilities | Persons in college dormitories | Persons in military quarters |
| | 28 | 29 | 30 | 31 | 32 | 33 | 34 | 35 | 36 | 37 | 38 | 39 |
| OREGON—Cont'd | | | | | | | | | | | | |
| District 3 | 341,703 | 2.44 | 56.1 | 41.4 | 10.1 | 30.1 | 18,506 | 15.5 | 2,065 | 2,399 | 6,604 | 0 |
| District 4 | 335,886 | 2.39 | 60.9 | 46.1 | 10.1 | 28.6 | 18,431 | 16.5 | 1,354 | 2,067 | 8,825 | 21 |
| District 5 | 315,045 | 2.63 | 65.7 | 50.5 | 11.2 | 26.6 | 15,965 | 19.1 | 5,381 | 2,401 | 3,162 | 17 |
| PENNSYLVANIA | 5,119,249 | 2.42 | 63.1 | 46.4 | 11.8 | 30.2 | 421,745 | 21.0 | 97,820 | 87,775 | 177,332 | 259 |
| District 1 | 272,563 | 2.58 | 69.6 | 56.4 | 8.8 | 24.6 | 10,334 | 9.2 | 10,896 | 2,024 | 7,108 | 14 |
| District 2 | 259,215 | 2.74 | 64.1 | 34.5 | 22.6 | 29.4 | 12,455 | 13.8 | 680 | 4,909 | 23,446 | 0 |
| District 3 | 315,700 | 2.24 | 46.1 | 22.7 | 18.4 | 42.2 | 33,967 | 19.5 | 6,328 | 5,188 | 11,315 | 11 |
| District 4 | 283,530 | 2.51 | 67.8 | 53.8 | 10.3 | 26.5 | 18,679 | 20.4 | 6,485 | 3,922 | 7,205 | 37 |
| District 5 | 267,150 | 2.60 | 65.1 | 45.3 | 15.4 | 30.1 | 25,394 | 11.4 | 14,533 | 5,088 | 23,276 | 0 |
| District 6 | 270,267 | 2.65 | 68.5 | 52.8 | 11.7 | 25.4 | 19,093 | 24.0 | 3,336 | 4,461 | 7,424 | 0 |
| District 7 | 272,657 | 2.59 | 67.9 | 49.4 | 12.8 | 25.4 | 24,082 | 21.2 | 5,890 | 5,118 | 10,160 | 0 |
| District 8 | 275,550 | 2.46 | 64.9 | 45.5 | 13.4 | 29.7 | 22,435 | 46.3 | 1,039 | 4,098 | 1,569 | 0 |
| District 9 | 275,546 | 2.43 | 67.0 | 51.0 | 10.0 | 26.7 | 31,288 | 23.8 | 6,117 | 5,134 | 8,771 | 0 |
| District 10 | 299,825 | 2.40 | 62.0 | 46.7 | 10.6 | 31.1 | 26,302 | 15.5 | 11,878 | 4,486 | 10,876 | 0 |
| District 11 | 278,868 | 2.59 | 69.9 | 54.6 | 10.3 | 23.8 | 12,372 | 22.1 | 7,333 | 6,231 | 9,991 | 6 |
| District 12 | 276,084 | 2.40 | 64.8 | 49.3 | 9.8 | 28.2 | 38,173 | 32.7 | 4,480 | 4,694 | 3,073 | 6 |
| District 13 | 281,524 | 2.39 | 66.2 | 51.4 | 10.1 | 28.2 | 25,148 | 57.8 | 2 | 6,384 | 1,329 | 121 |
| District 14 | 292,836 | 2.25 | 61.3 | 46.8 | 9.8 | 32.3 | 21,495 | 15.0 | 4,663 | 4,194 | 15,905 | 0 |
| District 15 | 280,673 | 2.28 | 62.1 | 49.1 | 8.4 | 31.6 | 33,570 | 21.1 | 1,424 | 5,210 | 13,039 | 63 |
| District 16 | 283,627 | 2.28 | 62.4 | 46.2 | 11.4 | 31.7 | 30,758 | 27.5 | 1,132 | 5,327 | 9,156 | 0 |
| District 17 | 313,101 | 2.23 | 60.2 | 47.8 | 8.9 | 33.1 | 9,759 | 25.7 | 7,876 | 6,346 | 7,783 | 1 |
| District 18 | 320,533 | 2.08 | 50.6 | 35.3 | 11.1 | 38.8 | 26,441 | 30.3 | 3,728 | 4,961 | 5,906 | 0 |
| RHODE ISLAND | 407,174 | 2.50 | 61.4 | 44.2 | 11.9 | 31.0 | 40,901 | 18.5 | 3,783 | 8,420 | 24,687 | 1,385 |
| District 1 | 201,963 | 2.52 | 60.0 | 42.2 | 11.4 | 32.8 | 21,001 | 21.3 | 350 | 4,996 | 13,035 | 1,385 |
| District 2 | 205,211 | 2.48 | 62.8 | 46.1 | 12.3 | 29.3 | 19,900 | 15.5 | 3,433 | 3,424 | 11,652 | 0 |
| SOUTH CAROLINA | 1,975,915 | 2.54 | 65.1 | 47.0 | 13.7 | 29.0 | 133,728 | 13.0 | 41,649 | 19,020 | 46,463 | 19,413 |
| District 1 | 311,963 | 2.61 | 67.4 | 53.8 | 10.1 | 25.9 | 8,425 | 15.4 | 531 | 1,662 | 2,560 | 4,436 |
| District 2 | 280,589 | 2.50 | 66.2 | 48.8 | 12.9 | 28.4 | 19,962 | 12.8 | 1,458 | 2,391 | 823 | 11,567 |
| District 3 | 266,671 | 2.57 | 67.1 | 48.2 | 14.0 | 27.2 | 20,658 | 14.3 | 7,044 | 3,116 | 9,817 | 0 |
| District 4 | 289,536 | 2.54 | 65.9 | 49.5 | 11.4 | 29.3 | 17,944 | 14.5 | 3,776 | 2,832 | 9,668 | 0 |
| District 5 | 286,307 | 2.54 | 66.7 | 48.5 | 13.7 | 28.4 | 12,021 | 20.0 | 5,318 | 2,792 | 3,289 | 611 |
| District 6 | 251,006 | 2.48 | 56.4 | 33.2 | 18.8 | 36.4 | 41,532 | 6.5 | 18,400 | 3,087 | 17,383 | 2,799 |
| District 7 | 289,843 | 2.51 | 64.8 | 44.8 | 15.9 | 28.4 | 13,186 | 22.4 | 5,122 | 3,140 | 2,923 | 0 |
| SOUTH DAKOTA | 353,799 | 2.40 | 63.3 | 50.2 | 8.7 | 29.9 | 34,018 | 21.1 | 6,327 | 7,005 | 10,248 | 597 |
| At Large | 353,799 | 2.40 | 63.3 | 50.2 | 8.7 | 29.9 | 34,018 | 21.1 | 6,327 | 7,005 | 10,248 | 597 |
| TENNESSEE | 2,654,737 | 2.51 | 65.1 | 48.0 | 12.3 | 28.7 | 158,077 | 19.8 | 46,957 | 33,041 | 53,136 | 1,544 |
| District 1 | 301,451 | 2.35 | 65.4 | 49.0 | 11.4 | 29.3 | 17,555 | 26.7 | 4,577 | 4,466 | 4,176 | 0 |
| District 2 | 304,715 | 2.43 | 64.0 | 49.2 | 10.5 | 29.8 | 17,198 | 19.1 | 1,901 | 3,686 | 10,443 | 0 |
| District 3 | 294,614 | 2.45 | 64.6 | 47.9 | 11.2 | 28.9 | 20,043 | 23.8 | 4,913 | 4,071 | 4,884 | 3 |
| District 4 | 296,321 | 2.69 | 69.6 | 52.9 | 11.8 | 24.2 | 15,600 | 21.1 | 3,301 | 3,584 | 6,619 | 0 |
| District 5 | 320,384 | 2.36 | 54.5 | 38.2 | 12.1 | 34.8 | 22,400 | 9.0 | 6,769 | 2,588 | 13,660 | 0 |
| District 6 | 303,648 | 2.59 | 71.0 | 56.0 | 10.1 | 24.0 | 14,157 | 36.4 | 2,385 | 3,730 | 2,283 | 0 |
| District 7 | 292,062 | 2.68 | 70.7 | 57.1 | 9.6 | 24.0 | 18,179 | 21.2 | 8,311 | 3,918 | 2,350 | 1,250 |
| District 8 | 271,504 | 2.55 | 70.6 | 52.2 | 13.2 | 25.4 | 18,376 | 21.3 | 7,494 | 4,158 | 4,748 | 10 |
| District 9 | 270,038 | 2.54 | 55.8 | 29.2 | 21.7 | 38.2 | 14,569 | 14.0 | 7,306 | 2,840 | 3,973 | 281 |
| TEXAS | 9,985,126 | 2.84 | 68.7 | 49.6 | 13.7 | 25.6 | 600,693 | 14.9 | 267,405 | 94,278 | 119,834 | 35,260 |
| District 1 | 259,325 | 2.70 | 69.0 | 49.2 | 14.9 | 25.6 | 27,073 | 17.3 | 8,132 | 5,093 | 9,123 | 0 |
| District 2 | 291,668 | 2.67 | 66.3 | 51.2 | 10.9 | 26.0 | 9,278 | 18.4 | 2,441 | 1,873 | 2,921 | 0 |
| District 3 | 324,907 | 2.79 | 72.3 | 58.9 | 9.8 | 22.4 | 5,629 | 31.9 | 1,061 | 1,326 | 387 | 0 |
| District 4 | 284,355 | 2.69 | 70.8 | 53.7 | 12.3 | 25.3 | 17,772 | 25.8 | 8,850 | 5,078 | 2,237 | 0 |
| District 5 | 261,420 | 2.81 | 69.5 | 48.9 | 13.9 | 25.6 | 24,890 | 14.9 | 17,768 | 3,964 | 938 | 0 |
| District 6 | 280,492 | 2.89 | 72.3 | 52.0 | 14.5 | 23.4 | 7,486 | 32.8 | 689 | 2,419 | 2,137 | 0 |
| District 7 | 301,869 | 2.52 | 62.7 | 46.2 | 11.8 | 30.4 | 1,381 | 58.1 | 3 | 774 | 5 | 0 |
| District 8 | 306,739 | 2.83 | 73.3 | 58.7 | 9.7 | 21.7 | 26,740 | 8.5 | 23,107 | 2,021 | 2,615 | 0 |
| District 9 | 266,557 | 2.87 | 67.0 | 42.0 | 18.8 | 27.3 | 3,027 | 32.7 | 154 | 1,629 | 1,520 | 0 |
| District 10 | 321,723 | 2.83 | 69.7 | 54.7 | 10.3 | 23.6 | 14,564 | 25.7 | 2,453 | 3,156 | 4,344 | 0 |
| District 11 | 280,662 | 2.74 | 67.9 | 52.5 | 10.9 | 27.1 | 20,183 | 18.3 | 9,050 | 3,721 | 3,615 | 1,924 |
| District 12 | 295,728 | 2.80 | 69.2 | 53.0 | 11.6 | 25.4 | 16,443 | 20.3 | 5,807 | 3,540 | 3,405 | 202 |
| District 13 | 259,762 | 2.62 | 66.3 | 49.6 | 11.3 | 29.1 | 33,020 | 11.8 | 17,511 | 4,208 | 2,479 | 5,487 |
| District 14 | 272,741 | 2.67 | 68.8 | 49.2 | 14.7 | 26.3 | 32,749 | 9.5 | 22,748 | 2,861 | 2,869 | 40 |
| District 15 | 237,783 | 3.31 | 77.0 | 52.3 | 19.1 | 19.1 | 17,811 | 14.6 | 8,632 | 2,035 | 1,514 | 0 |
| District 16 | 240,099 | 3.05 | 70.2 | 44.1 | 19.5 | 26.1 | 14,455 | 9.8 | 6,076 | 1,482 | 491 | 5,683 |
| District 17 | 286,608 | 2.63 | 63.9 | 46.1 | 13.1 | 28.1 | 31,380 | 11.1 | 8,355 | 3,792 | 14,831 | 0 |
| District 18 | 293,472 | 2.74 | 62.1 | 36.5 | 18.9 | 31.7 | 22,673 | 3.8 | 12,624 | 783 | 5,661 | 0 |
| District 19 | 263,521 | 2.61 | 64.5 | 48.1 | 12.2 | 28.7 | 40,657 | 10.0 | 19,845 | 4,073 | 10,748 | 621 |
| District 20 | 255,836 | 3.20 | 66.5 | 41.7 | 18.6 | 26.6 | 13,070 | 11.9 | 4 | 2,081 | 4,846 | 9,322 |
| District 21 | 332,269 | 2.45 | 57.6 | 45.5 | 8.5 | 33.1 | 16,551 | 19.9 | 600 | 4,024 | 7,505 | 3,793 |
| District 22 | 310,680 | 3.07 | 78.0 | 64.5 | 9.2 | 18.2 | 7,205 | 21.6 | 4,494 | 1,409 | 4 | 0 |
| District 23 | 242,400 | 3.15 | 73.0 | 51.7 | 15.7 | 23.9 | 22,381 | 7.6 | 16,613 | 1,566 | 193 | 877 |
| District 24 | 325,176 | 2.55 | 62.0 | 47.3 | 9.7 | 31.1 | 3,335 | 65.8 | 53 | 1,913 | 480 | 0 |
| District 25 | 291,833 | 2.68 | 69.2 | 55.7 | 9.7 | 24.2 | 37,027 | 9.6 | 11,967 | 3,274 | 10,926 | 2,822 |

1. No spouse present.

# Table E. Congressional Districts 116th Congress — Housing and Money Income

| STATE District | Housing units, 2019 | | | | | | Money income, 2019 | | |
|---|---|---|---|---|---|---|---|---|---|
| | Total | Occupied units | | | | | Households | | |
| | | Occupied units | Owner-occupied | | | Renter-occupied | | | |
| | | Occupied units as a percent of all units | Owner-occupied units as a percent of occupied units | Median value[1] (dollars) | Percent valued at $500,000 or more | Median rent[2] | Per capita income (dollars) | Median income (dollars) | Percent with income of $100,000 or more |
| | 40 | 41 | 42 | 43 | 44 | 45 | 46 | 47 | 48 |
| OREGON—Cont'd | | | | | | | | | |
| District 3 | 364,943 | 93.6 | 58.4 | 417,100 | 33.2 | 1,301 | 39,967 | 73,091 | 36.1 |
| District 4 | 361,505 | 92.9 | 62.9 | 280,400 | 13.1 | 997 | 30,799 | 55,886 | 23.4 |
| District 5 | 349,709 | 90.1 | 65.7 | 352,900 | 22.0 | 1,177 | 34,437 | 68,757 | 32.0 |
| PENNSYLVANIA | 5,732,580 | 89.3 | 68.4 | 192,600 | 7.7 | 951 | 35,804 | 63,463 | 29.4 |
| District 1 | 285,580 | 95.4 | 76.8 | 340,400 | 20.4 | 1,256 | 46,834 | 93,474 | 46.7 |
| District 2 | 280,330 | 92.5 | 58.0 | 176,000 | 3.1 | 1,021 | 25,261 | 46,248 | 20.3 |
| District 3 | 361,781 | 87.3 | 47.1 | 197,500 | 12.6 | 1,126 | 35,247 | 49,897 | 22.7 |
| District 4 | 299,508 | 94.7 | 71.8 | 323,500 | 16.8 | 1,308 | 48,679 | 91,030 | 45.9 |
| District 5 | 288,409 | 92.6 | 66.7 | 241,700 | 14.9 | 1,110 | 40,492 | 71,880 | 36.1 |
| District 6 | 287,997 | 93.8 | 69.7 | 327,200 | 22.2 | 1,180 | 47,600 | 85,665 | 43.7 |
| District 7 | 296,400 | 92.0 | 66.3 | 223,000 | 5.6 | 1,118 | 34,364 | 69,105 | 32.5 |
| District 8 | 355,934 | 77.4 | 68.9 | 150,500 | 3.0 | 826 | 29,848 | 56,149 | 22.3 |
| District 9 | 316,254 | 87.1 | 76.3 | 165,000 | 3.0 | 833 | 31,203 | 62,078 | 26.6 |
| District 10 | 321,181 | 93.4 | 67.5 | 183,800 | 4.5 | 962 | 35,674 | 67,155 | 30.2 |
| District 11 | 292,232 | 95.4 | 72.4 | 221,700 | 5.5 | 1,019 | 33,862 | 68,811 | 29.5 |
| District 12 | 335,203 | 82.4 | 70.1 | 167,400 | 4.8 | 779 | 29,090 | 55,203 | 22.0 |
| District 13 | 325,103 | 86.6 | 75.2 | 160,500 | 3.8 | 756 | 29,888 | 56,618 | 21.2 |
| District 14 | 326,809 | 89.6 | 74.1 | 156,700 | 4.0 | 735 | 34,120 | 59,165 | 26.2 |
| District 15 | 346,466 | 81.0 | 74.8 | 117,400 | 2.3 | 701 | 28,603 | 53,741 | 20.2 |
| District 16 | 320,106 | 88.6 | 70.8 | 145,600 | 3.4 | 739 | 30,682 | 54,627 | 22.8 |
| District 17 | 337,352 | 92.8 | 71.0 | 188,100 | 6.5 | 860 | 43,660 | 70,857 | 34.6 |
| District 18 | 355,935 | 90.1 | 57.1 | 147,600 | 4.1 | 928 | 37,821 | 58,743 | 26.7 |
| RHODE ISLAND | 470,177 | 86.6 | 61.7 | 283,000 | 13.0 | 1,043 | 37,525 | 71,169 | 33.1 |
| District 1 | 231,244 | 87.3 | 57.5 | 281,300 | 14.7 | 997 | 36,808 | 66,652 | 30.6 |
| District 2 | 238,933 | 85.9 | 65.8 | 284,100 | 11.6 | 1,093 | 38,243 | 74,180 | 35.6 |
| SOUTH CAROLINA | 2,351,364 | 84.0 | 70.3 | 179,800 | 8.2 | 922 | 31,295 | 56,227 | 24.7 |
| District 1 | 373,178 | 83.6 | 72.2 | 291,500 | 21.0 | 1,255 | 41,363 | 77,185 | 37.2 |
| District 2 | 315,303 | 89.0 | 74.6 | 171,800 | 5.8 | 950 | 33,250 | 60,781 | 27.5 |
| District 3 | 318,026 | 83.9 | 72.9 | 156,400 | 5.2 | 757 | 27,671 | 50,815 | 19.6 |
| District 4 | 317,792 | 91.1 | 70.4 | 187,200 | 7.5 | 927 | 32,995 | 60,731 | 26.8 |
| District 5 | 316,971 | 90.3 | 73.4 | 167,500 | 5.6 | 892 | 30,712 | 56,282 | 25.6 |
| District 6 | 308,011 | 81.5 | 57.7 | 114,700 | 4.7 | 870 | 23,416 | 41,128 | 14.2 |
| District 7 | 402,083 | 72.1 | 69.1 | 165,400 | 5.5 | 880 | 27,613 | 49,494 | 19.0 |
| SOUTH DAKOTA | 401,749 | 88.1 | 67.8 | 185,000 | 5.2 | 769 | 31,550 | 59,533 | 25.1 |
| At Large | 401,749 | 88.1 | 67.8 | 185,000 | 5.2 | 769 | 31,550 | 59,533 | 25.1 |
| TENNESSEE | 3,028,437 | 87.7 | 66.5 | 191,900 | 8.6 | 904 | 31,224 | 56,071 | 24.0 |
| District 1 | 357,642 | 84.3 | 71.2 | 152,600 | 4.0 | 718 | 27,902 | 47,478 | 17.7 |
| District 2 | 340,956 | 89.4 | 68.6 | 195,400 | 8.4 | 888 | 32,352 | 57,777 | 25.1 |
| District 3 | 337,758 | 87.2 | 68.9 | 174,500 | 5.9 | 821 | 29,889 | 52,491 | 21.9 |
| District 4 | 332,133 | 89.2 | 68.9 | 205,600 | 5.2 | 886 | 28,010 | 59,461 | 23.6 |
| District 5 | 360,099 | 89.0 | 56.3 | 278,100 | 16.2 | 1,178 | 38,710 | 63,295 | 28.5 |
| District 6 | 340,297 | 89.2 | 73.4 | 223,100 | 7.6 | 862 | 30,329 | 59,421 | 24.7 |
| District 7 | 337,867 | 86.4 | 72.9 | 206,900 | 18.0 | 891 | 34,014 | 62,720 | 29.8 |
| District 8 | 302,703 | 89.7 | 71.9 | 176,500 | 7.3 | 777 | 34,289 | 60,152 | 28.5 |
| District 9 | 318,982 | 84.7 | 45.7 | 115,000 | 2.6 | 937 | 24,997 | 43,708 | 15.5 |
| TEXAS | 11,283,892 | 88.5 | 61.9 | 200,400 | 8.9 | 1,091 | 32,267 | 64,034 | 30.5 |
| District 1 | 314,348 | 82.5 | 67.4 | 142,000 | 4.0 | 842 | 27,758 | 54,396 | 22.0 |
| District 2 | 321,005 | 90.9 | 61.9 | 240,900 | 11.7 | 1,255 | 43,242 | 80,922 | 41.1 |
| District 3 | 349,650 | 92.9 | 60.2 | 365,700 | 19.4 | 1,459 | 45,082 | 95,619 | 48.0 |
| District 4 | 324,709 | 87.6 | 72.5 | 168,900 | 6.7 | 859 | 30,534 | 60,060 | 27.2 |
| District 5 | 301,961 | 86.6 | 64.6 | 171,600 | 5.7 | 964 | 27,372 | 57,026 | 24.0 |
| District 6 | 302,954 | 92.6 | 65.5 | 222,800 | 4.2 | 1,173 | 31,980 | 71,161 | 32.2 |
| District 7 | 338,474 | 89.2 | 47.5 | 270,800 | 26.7 | 1,235 | 50,060 | 73,730 | 37.1 |
| District 8 | 348,336 | 88.1 | 70.9 | 235,000 | 13.8 | 1,157 | 38,409 | 78,615 | 39.9 |
| District 9 | 295,621 | 90.2 | 46.4 | 161,500 | 2.0 | 1,042 | 23,500 | 49,147 | 17.8 |
| District 10 | 359,151 | 89.6 | 66.3 | 267,700 | 15.7 | 1,269 | 38,812 | 80,528 | 39.5 |
| District 11 | 340,997 | 82.3 | 69.5 | 164,700 | 7.3 | 1,051 | 31,864 | 64,070 | 30.3 |
| District 12 | 332,183 | 89.0 | 65.0 | 229,800 | 8.1 | 1,148 | 34,605 | 73,000 | 34.1 |
| District 13 | 309,868 | 83.8 | 67.6 | 128,900 | 3.3 | 799 | 27,487 | 54,004 | 22.6 |
| District 14 | 325,311 | 83.8 | 67.2 | 180,000 | 3.7 | 1,028 | 33,009 | 67,459 | 32.8 |
| District 15 | 275,787 | 86.2 | 67.8 | 113,500 | 2.0 | 797 | 20,909 | 48,113 | 21.2 |
| District 16 | 269,267 | 89.2 | 58.9 | 133,300 | 1.1 | 849 | 22,371 | 49,013 | 18.9 |
| District 17 | 331,951 | 86.3 | 57.7 | 199,300 | 5.0 | 1,010 | 29,501 | 58,929 | 26.3 |
| District 18 | 325,988 | 90.0 | 45.3 | 160,900 | 8.9 | 1,046 | 27,290 | 48,625 | 20.7 |
| District 19 | 306,425 | 86.0 | 62.4 | 127,100 | 2.5 | 891 | 27,005 | 55,278 | 22.9 |
| District 20 | 280,466 | 91.2 | 55.4 | 165,500 | 1.1 | 1,013 | 24,840 | 54,908 | 22.4 |
| District 21 | 370,836 | 89.6 | 58.9 | 325,600 | 21.5 | 1,225 | 44,806 | 73,472 | 36.8 |
| District 22 | 329,806 | 94.2 | 74.6 | 289,200 | 10.8 | 1,369 | 42,142 | 101,658 | 51.2 |
| District 23 | 279,790 | 86.6 | 71.7 | 158,000 | 7.0 | 944 | 27,368 | 59,074 | 26.6 |
| District 24 | 352,117 | 92.3 | 47.2 | 327,700 | 22.0 | 1,284 | 45,897 | 79,667 | 38.4 |
| District 25 | 330,232 | 88.4 | 70.1 | 290,100 | 24.1 | 1,155 | 41,622 | 79,975 | 40.7 |

1. Specified owner-occupied units; $1,000,000 represents $1,000,000.    2. Specified renter-occupied units.

# Table E. Congressional Districts 116th Congress — Poverty, Labor Force, Employment, and Social Security

| STATE District | Poverty, 2019 Persons below poverty level (percent) | Families below poverty level (percent) | Percent of households receiving food stamps in past 12 months | Civilian labor force, 2019 Total | Unemployment Total | Rate[1] | Civilian employment,[2] 2019 Total | Percent Management, business, science, and arts occupations | Service, sales, and office | Construction and production | Persons under 65 years of age with no health insurance, 2019 (percent) | Social Security beneficiaries, December 2020 Number | Rate[3] | Supplemental Security Income recipients, December 2019 |
|---|---|---|---|---|---|---|---|---|---|---|---|---|---|---|
| | 49 | 50 | 51 | 52 | 53 | 54 | 55 | 56 | 57 | 58 | 59 | 60 | 61 | 62 |
| OREGON—Cont'd | | | | | | | | | | | | | | |
| District 3 | 11.6 | 6.9 | 11.8 | 487,922 | 20,759 | 4.3 | 466,910 | 45.3 | 37.1 | 17.5 | 7.0 | 136,413 | 159.9 | 20,396 |
| District 4 | 13.8 | 7.1 | 16.8 | 397,491 | 23,464 | 5.9 | 373,269 | 35.5 | 40.0 | 24.5 | 6.5 | 218,778 | 266.6 | 21,809 |
| District 5 | 9.7 | 5.9 | 13.4 | 416,999 | 24,027 | 5.8 | 392,007 | 38.4 | 38.7 | 22.9 | 7.8 | 185,377 | 219.6 | 15,367 |
| PENNSYLVANIA | 12.0 | 8.2 | 13.5 | 6,592,650 | 295,633 | 4.5 | 6,285,109 | 40.3 | 37.2 | 22.5 | 5.7 | 2,877,728 | 224.8 | 348,636 |
| District 1 | 5.3 | 3.8 | 6.6 | 397,764 | 15,339 | 3.9 | 381,441 | 47.2 | 34.3 | 18.5 | 3.8 | 157,071 | 220.2 | 8,368 |
| District 2 | 25.4 | 21.7 | 32.7 | 341,692 | 28,374 | 8.3 | 313,136 | 30.6 | 46.2 | 23.2 | 9.8 | 114,416 | 158.3 | 45,379 |
| District 3 | 21.8 | 15.0 | 22.9 | 388,544 | 30,636 | 7.9 | 357,238 | 48.8 | 40.0 | 11.1 | 6.4 | 124,104 | 167.3 | 47,893 |
| District 4 | 6.4 | 4.3 | 5.8 | 401,779 | 13,281 | 3.3 | 387,414 | 51.7 | 33.3 | 15.1 | 3.6 | 147,678 | 202.1 | 7,635 |
| District 5 | 11.4 | 8.6 | 14.3 | 375,380 | 21,645 | 5.8 | 353,577 | 45.4 | 38.3 | 16.3 | 5.2 | 135,744 | 188.5 | 20,491 |
| District 6 | 8.5 | 5.7 | 9.2 | 392,301 | 15,765 | 4.0 | 376,085 | 48.6 | 34.2 | 17.2 | 4.8 | 138,390 | 188.2 | 13,184 |
| District 7 | 10.0 | 6.8 | 12.4 | 380,041 | 16,618 | 4.4 | 363,172 | 37.2 | 38.5 | 24.3 | 5.2 | 164,666 | 225.1 | 17,151 |
| District 8 | 14.6 | 11.1 | 16.2 | 347,250 | 16,438 | 4.7 | 330,241 | 32.8 | 41.0 | 26.3 | 6.2 | 177,072 | 253.3 | 20,097 |
| District 9 | 9.8 | 6.7 | 10.6 | 356,873 | 12,828 | 3.6 | 343,281 | 32.6 | 36.4 | 31.0 | 5.4 | 177,094 | 253.1 | 13,354 |
| District 10 | 9.9 | 6.5 | 10.0 | 396,746 | 13,811 | 3.5 | 381,134 | 40.4 | 36.1 | 23.5 | 5.3 | 160,261 | 215.2 | 17,124 |
| District 11 | 10.1 | 6.5 | 7.5 | 385,441 | 10,772 | 2.8 | 374,086 | 37.1 | 35.6 | 27.4 | 9.0 | 161,012 | 219.4 | 11,159 |
| District 12 | 12.7 | 7.9 | 11.6 | 335,666 | 13,114 | 3.9 | 322,062 | 34.4 | 36.1 | 29.5 | 7.9 | 167,994 | 239.5 | 14,350 |
| District 13 | 11.0 | 8.1 | 13.4 | 341,576 | 10,809 | 3.2 | 330,170 | 31.1 | 36.7 | 32.2 | 6.9 | 182,367 | 261.6 | 16,075 |
| District 14 | 12.0 | 8.0 | 15.7 | 336,134 | 16,881 | 5.0 | 318,936 | 36.8 | 38.3 | 24.9 | 4.6 | 188,062 | 277.0 | 20,253 |
| District 15 | 12.3 | 7.7 | 14.1 | 321,117 | 12,525 | 3.9 | 307,918 | 30.5 | 37.6 | 31.8 | 5.6 | 184,416 | 274.1 | 17,858 |
| District 16 | 13.5 | 9.5 | 16.9 | 334,030 | 15,234 | 4.6 | 317,818 | 36.6 | 38.4 | 25.0 | 5.1 | 176,478 | 260.2 | 22,067 |
| District 17 | 7.7 | 4.6 | 9.1 | 384,072 | 11,489 | 3.0 | 372,036 | 47.9 | 35.2 | 17.0 | 3.2 | 170,755 | 241.5 | 13,216 |
| District 18 | 14.1 | 9.1 | 15.2 | 376,244 | 20,074 | 5.3 | 355,364 | 48.3 | 36.8 | 14.9 | 4.3 | 150,148 | 216.4 | 22,982 |
| RHODE ISLAND | 10.8 | 6.6 | 14.5 | 575,681 | 24,199 | 4.2 | 547,914 | 42.2 | 38.8 | 18.9 | 4.0 | 230,018 | 217.1 | 32,124 |
| District 1 | 12.2 | 7.1 | 15.8 | 282,807 | 11,976 | 4.3 | 268,026 | 39.6 | 40.5 | 19.8 | 4.6 | 111,656 | 210.6 | 17,857 |
| District 2 | 9.5 | 6.0 | 13.2 | 292,874 | 12,223 | 4.2 | 279,888 | 44.7 | 37.2 | 18.1 | 3.5 | 118,362 | 223.6 | 14,267 |
| SOUTH CAROLINA | 13.8 | 9.8 | 9.9 | 2,513,088 | 116,037 | 4.7 | 2,359,714 | 36.9 | 38.0 | 25.1 | 10.8 | 1,197,138 | 232.5 | 113,864 |
| District 1 | 8.8 | 6.3 | 5.3 | 424,517 | 13,866 | 3.3 | 401,103 | 43.1 | 37.6 | 19.3 | 10.0 | 171,246 | 208.6 | 8,972 |
| District 2 | 11.6 | 8.5 | 9.6 | 365,549 | 16,506 | 4.7 | 336,153 | 41.9 | 36.7 | 21.5 | 9.4 | 153,423 | 212.3 | 12,078 |
| District 3 | 15.1 | 10.6 | 11.2 | 330,428 | 15,657 | 4.7 | 314,021 | 33.9 | 35.0 | 31.1 | 10.1 | 176,829 | 250.1 | 14,594 |
| District 4 | 11.0 | 7.7 | 7.2 | 382,393 | 15,341 | 4.0 | 366,549 | 38.8 | 35.8 | 25.4 | 10.2 | 159,712 | 211.8 | 15,252 |
| District 5 | 12.4 | 8.7 | 9.6 | 364,717 | 17,331 | 4.8 | 342,402 | 36.0 | 37.6 | 26.4 | 9.8 | 170,316 | 230.7 | 17,196 |
| District 6 | 23.5 | 17.5 | 16.6 | 309,310 | 20,204 | 6.7 | 280,920 | 30.2 | 41.4 | 28.5 | 13.1 | 149,433 | 224.6 | 24,812 |
| District 7 | 16.5 | 11.9 | 11.1 | 336,174 | 17,132 | 5.1 | 318,566 | 31.6 | 43.0 | 25.3 | 13.0 | 216,179 | 291.9 | 20,960 |
| SOUTH DAKOTA | 11.9 | 7.3 | 8.7 | 469,624 | 14,033 | 3.0 | 452,975 | 39.2 | 36.7 | 24.1 | 10.1 | 185,752 | 210.0 | 14,495 |
| At Large | 11.9 | 7.3 | 8.7 | 469,624 | 14,033 | 3.0 | 452,975 | 39.2 | 36.7 | 24.1 | 10.1 | 185,752 | 210.0 | 14,495 |
| TENNESSEE | 13.9 | 10.0 | 11.4 | 3,387,696 | 152,373 | 4.5 | 3,215,401 | 36.2 | 38.2 | 25.6 | 10.1 | 1,496,750 | 219.2 | 172,815 |
| District 1 | 16.8 | 12.4 | 11.7 | 336,976 | 17,056 | 5.1 | 319,499 | 31.7 | 41.3 | 27.1 | 11.5 | 206,264 | 284.4 | 21,739 |
| District 2 | 12.7 | 8.9 | 10.0 | 379,212 | 12,803 | 3.4 | 364,586 | 36.4 | 40.4 | 23.2 | 8.8 | 178,145 | 234.9 | 17,494 |
| District 3 | 15.3 | 11.2 | 12.4 | 356,502 | 20,492 | 5.8 | 335,644 | 37.3 | 37.0 | 25.6 | 10.1 | 182,484 | 245.5 | 20,755 |
| District 4 | 12.9 | 9.4 | 10.5 | 408,205 | 16,526 | 4.1 | 391,084 | 31.2 | 38.3 | 30.6 | 10.1 | 169,017 | 208.0 | 16,980 |
| District 5 | 11.9 | 7.8 | 7.9 | 452,677 | 17,135 | 3.8 | 434,950 | 44.6 | 35.2 | 20.1 | 10.9 | 119,756 | 153.9 | 15,780 |
| District 6 | 12.4 | 9.1 | 10.7 | 384,958 | 13,437 | 3.5 | 370,970 | 34.1 | 38.0 | 27.9 | 8.9 | 192,176 | 240.4 | 16,135 |
| District 7 | 11.0 | 7.4 | 9.4 | 382,876 | 16,006 | 4.3 | 353,848 | 38.6 | 37.8 | 23.6 | 8.0 | 164,133 | 205.0 | 15,318 |
| District 8 | 11.9 | 8.6 | 10.9 | 337,810 | 12,537 | 3.7 | 324,141 | 40.3 | 34.5 | 25.2 | 8.2 | 165,016 | 232.1 | 17,853 |
| District 9 | 20.7 | 16.3 | 20.2 | 348,480 | 26,381 | 7.6 | 320,679 | 29.7 | 42.1 | 28.2 | 14.3 | 119,759 | 171.0 | 30,761 |
| TEXAS | 13.6 | 10.5 | 10.8 | 14,574,146 | 635,077 | 4.4 | 13,830,576 | 37.6 | 38.4 | 23.9 | 18.1 | 4,421,803 | 152.5 | 633,501 |
| District 1 | 16.7 | 12.7 | 12.8 | 336,159 | 16,428 | 4.9 | 319,448 | 32.6 | 39.0 | 28.4 | 16.7 | 157,343 | 216.7 | 20,608 |
| District 2 | 10.1 | 7.4 | 4.9 | 421,559 | 19,549 | 4.6 | 401,885 | 47.4 | 33.4 | 19.1 | 15.3 | 106,293 | 135.0 | 13,266 |
| District 3 | 6.2 | 4.0 | 2.5 | 500,045 | 20,092 | 4.0 | 479,693 | 54.9 | 33.3 | 11.8 | 10.1 | 107,173 | 117.4 | 7,961 |
| District 4 | 13.3 | 10.3 | 10.7 | 363,155 | 13,598 | 3.7 | 349,201 | 33.5 | 39.5 | 26.9 | 15.8 | 165,280 | 211.2 | 18,890 |
| District 5 | 13.0 | 10.2 | 11.3 | 369,257 | 15,117 | 4.1 | 353,882 | 30.3 | 39.4 | 30.2 | 21.5 | 132,837 | 174.8 | 16,338 |
| District 6 | 8.7 | 6.2 | 8.4 | 435,193 | 19,997 | 4.6 | 413,607 | 37.0 | 38.2 | 24.7 | 14.5 | 120,502 | 147.2 | 12,692 |
| District 7 | 10.3 | 7.8 | 6.4 | 427,844 | 17,771 | 4.2 | 409,650 | 49.6 | 32.4 | 18.0 | 16.9 | 93,009 | 121.9 | 11,292 |
| District 8 | 10.1 | 6.9 | 6.7 | 426,815 | 15,427 | 3.6 | 411,153 | 40.6 | 36.4 | 23.0 | 12.9 | 145,471 | 162.4 | 13,150 |
| District 9 | 20.6 | 16.8 | 16.9 | 401,832 | 24,700 | 6.2 | 376,550 | 28.0 | 44.9 | 27.2 | 25.9 | 95,484 | 124.1 | 25,891 |
| District 10 | 8.1 | 6.2 | 6.3 | 493,335 | 20,492 | 4.2 | 472,392 | 46.7 | 32.9 | 20.5 | 13.9 | 126,949 | 137.2 | 10,883 |
| District 11 | 11.6 | 8.0 | 7.1 | 384,698 | 12,521 | 3.3 | 368,704 | 33.0 | 36.3 | 30.7 | 17.8 | 147,916 | 187.2 | 15,001 |
| District 12 | 8.3 | 5.7 | 7.2 | 431,837 | 15,315 | 3.6 | 414,195 | 39.9 | 37.8 | 22.3 | 15.8 | 130,735 | 154.8 | 13,785 |
| District 13 | 13.2 | 10.0 | 10.0 | 344,977 | 10,863 | 3.2 | 326,600 | 30.1 | 39.4 | 30.4 | 16.5 | 134,952 | 188.8 | 13,186 |
| District 14 | 12.4 | 9.8 | 10.5 | 360,567 | 13,812 | 3.8 | 346,422 | 36.5 | 36.8 | 26.8 | 17.0 | 135,094 | 177.6 | 17,596 |
| District 15 | 22.3 | 19.5 | 22.8 | 357,296 | 19,858 | 5.6 | 335,765 | 28.9 | 45.2 | 25.9 | 27.2 | 113,465 | 141.0 | 28,391 |
| District 16 | 18.9 | 16.0 | 20.2 | 359,271 | 20,250 | 6.0 | 320,065 | 32.5 | 44.9 | 22.6 | 21.9 | 122,107 | 163.3 | 25,060 |
| District 17 | 16.9 | 11.2 | 10.4 | 394,005 | 14,286 | 3.6 | 378,350 | 38.8 | 38.3 | 22.9 | 12.8 | 121,850 | 155.0 | 16,235 |
| District 18 | 19.8 | 16.8 | 16.3 | 411,538 | 21,908 | 5.3 | 389,418 | 30.3 | 39.5 | 30.3 | 26.2 | 97,435 | 117.8 | 26,468 |
| District 19 | 15.6 | 10.7 | 11.5 | 353,300 | 9,359 | 2.7 | 338,457 | 32.9 | 40.4 | 26.7 | 17.7 | 126,855 | 173.9 | 15,828 |
| District 20 | 16.3 | 12.6 | 17.5 | 426,864 | 22,022 | 5.3 | 395,947 | 33.1 | 45.6 | 21.3 | 19.0 | 119,235 | 143.2 | 19,879 |
| District 21 | 9.6 | 6.2 | 4.7 | 457,963 | 14,301 | 3.1 | 439,988 | 48.2 | 38.5 | 13.3 | 12.0 | 156,316 | 188.4 | 11,967 |
| District 22 | 5.6 | 4.1 | 5.5 | 492,764 | 16,272 | 3.3 | 475,679 | 52.7 | 32.0 | 15.3 | 12.0 | 115,921 | 120.6 | 12,019 |
| District 23 | 15.5 | 13.9 | 15.8 | 362,607 | 12,280 | 3.5 | 342,932 | 32.1 | 41.6 | 26.3 | 16.6 | 136,359 | 173.3 | 21,329 |
| District 24 | 7.4 | 5.6 | 4.1 | 496,444 | 15,720 | 3.2 | 479,999 | 48.0 | 35.1 | 16.9 | 13.6 | 96,301 | 115.7 | 7,354 |
| District 25 | 9.7 | 6.7 | 6.4 | 423,085 | 16,205 | 3.9 | 396,132 | 47.5 | 32.4 | 20.1 | 12.0 | 142,436 | 174.0 | 10,994 |

1. Percent of civilian labor force.  2. Persons 16 years old and over.  3. Per 1,000 resident population estimated in the 2019 American Community Survey.

# Table E. Congressional Districts 116th Congress — **Agriculture**

| STATE District | Land in farms | | | | Value of products sold | | | | Government payments | |
|---|---|---|---|---|---|---|---|---|---|---|
| | Number of farms | Acres | Average size of farm (acres) | Irrigated land (acres) | Total ($1,000) | Average per farm (dollars) | Percent from crops | Percent from livestock and poultry products | Total ($1,000) | Average per farm receiving payments (dollars) |
| | 63 | 64 | 65 | 66 | 67 | 68 | 69 | 70 | 71 | 72 |
| OREGON—Cont'd | | | | | | | | | | |
| District 3 | 2,807 | 90,555 | 32 | 31,975 | 239,212 | 85,220 | 88.0 | 12.0 | 146 | 3,650 |
| District 4 | 9,064 | 1,280,043 | 141 | 388,197 | 628,301 | 69,318 | 63.0 | 37.0 | 3,910 | 9,799 |
| District 5 | 6,690 | 580,067 | 87 | 335,079 | 1,153,738 | 172,457 | 74.2 | 25.8 | 3,294 | 10,073 |
| PENNSYLVANIA | 53,157 | 7,278,668 | 137 | 3,931,996 | 7,758,885 | 145,962 | 35.8 | 64.2 | 74,182 | 6,823 |
| District 1 | 866 | 78,687 | 91 | 52,691 | 84,647 | 97,745 | 75.2 | 24.8 | 637 | 7,583 |
| District 2 | 22 | 68 | 3 | 23 | 133 | 6,045 | 85.7 | 15.0 | D | D |
| District 3 | 21 | 216 | 10 | 94 | 194 | 9,238 | 77.8 | 21.6 | D | D |
| District 4 | 596 | 37,975 | 64 | 20,987 | 32,564 | 54,638 | 62.9 | 37.1 | 233 | 5,825 |
| District 5 | 61 | 2,385 | 39 | 668 | 9,494 | 155,639 | 97.8 | 2.3 | D | D |
| District 6 | 1,814 | 162,693 | 90 | 103,764 | 727,609 | 401,107 | 79.2 | 20.8 | 1,908 | 10,258 |
| District 7 | 942 | 142,994 | 152 | 110,081 | 118,328 | 125,614 | 74.0 | 26.0 | 2,367 | 15,173 |
| District 8 | 1,236 | 193,249 | 156 | 64,158 | 60,087 | 48,614 | 49.4 | 50.6 | 588 | 3,973 |
| District 9 | 5,166 | 636,513 | 123 | 426,263 | 1,214,767 | 235,147 | 36.4 | 63.6 | 9,756 | 6,668 |
| District 10 | 1,551 | 189,487 | 122 | 134,814 | 211,943 | 136,649 | 46.4 | 53.6 | 2,500 | 8,251 |
| District 11 | 6,578 | 576,434 | 88 | 427,814 | 1,689,756 | 256,880 | 18.6 | 81.4 | 8,468 | 9,645 |
| District 12 | 10,546 | 1,728,148 | 164 | 820,450 | 1,414,932 | 134,168 | 23.0 | 77.0 | 19,379 | 6,302 |
| District 13 | 8,124 | 1,343,863 | 165 | 780,880 | 1,379,471 | 169,802 | 27.3 | 72.7 | 13,151 | 7,331 |
| District 14 | 4,142 | 528,568 | 128 | 205,558 | 140,027 | 33,807 | 55.3 | 44.7 | 2,522 | 5,447 |
| District 15 | 5,867 | 906,075 | 154 | 398,379 | 315,753 | 53,818 | 48.5 | 51.5 | 8,177 | 6,495 |
| District 16 | 4,602 | 665,252 | 145 | 350,735 | 320,968 | 69,745 | 56.2 | 43.8 | 4,262 | 4,491 |
| District 17 | 885 | 75,415 | 85 | 31,530 | 35,258 | 39,840 | 70.1 | 29.9 | D | D |
| District 18 | 138 | 10,646 | 77 | 3,107 | 2,953 | 21,399 | 89.5 | 10.5 | D | D |
| RHODE ISLAND | 1,043 | 56,864 | 55 | 14,302 | 57,998 | 55,607 | 70.5 | 29.5 | 1,037 | 14,205 |
| District 1 | 322 | 14,540 | 45 | 5,779 | 22,877 | 71,047 | 58.4 | 41.6 | 174 | 10,235 |
| District 2 | 721 | 42,324 | 59 | 8,523 | 35,122 | 48,713 | 78.4 | 21.6 | 862 | 15,393 |
| SOUTH CAROLINA | 24,791 | 4,744,913 | 191 | 1,599,887 | 3,008,739 | 121,364 | 36.4 | 63.6 | 55,192 | 10,400 |
| District 1 | 616 | 76,208 | 124 | 11,039 | 35,526 | 57,672 | 91.0 | 9.0 | 166 | 3,689 |
| District 2 | 3,388 | 457,224 | 135 | 143,344 | 497,742 | 146,913 | 31.5 | 68.5 | 6,844 | 14,562 |
| District 3 | 6,839 | 881,416 | 129 | 194,971 | 658,491 | 96,285 | 12.2 | 87.8 | 6,681 | 5,971 |
| District 4 | 1,792 | 108,346 | 60 | 24,522 | 33,573 | 18,735 | 71.8 | 28.2 | 518 | 4,709 |
| District 5 | 4,851 | 840,194 | 173 | 230,826 | 627,665 | 129,389 | 27.0 | 73.0 | 7,124 | 7,916 |
| District 6 | 4,388 | 1,527,163 | 348 | 566,689 | 597,829 | 136,242 | 36.3 | 63.7 | 23,870 | 14,564 |
| District 7 | 2,917 | 854,362 | 293 | 428,496 | 557,913 | 191,263 | 45.2 | 54.8 | 9,989 | 9,755 |
| SOUTH DAKOTA | 29,968 | 43,243,742 | 1,443 | 16,371,543 | 9,721,523 | 324,397 | 53.1 | 46.9 | 419,508 | 19,416 |
| At Large | 29,968 | 43,243,742 | 1,443 | 16,371,543 | 9,721,522 | 324,397 | 53.1 | 46.9 | 419,508 | 19,416 |
| TENNESSEE | 69,983 | 10,874,238 | 155 | 4,566,352 | 3,798,934 | 54,284 | 57.4 | 42.6 | 115,945 | 6,254 |
| District 1 | 10,091 | 895,703 | 89 | 279,460 | 232,281 | 23,019 | 30.1 | 69.9 | 5,109 | 3,355 |
| District 2 | 5,504 | 500,612 | 91 | 153,466 | 180,981 | 32,882 | 60.2 | 39.8 | 5,341 | 3,980 |
| District 3 | 5,878 | 609,260 | 104 | 179,821 | 259,736 | 44,188 | 14.5 | 85.5 | 5,365 | 4,341 |
| District 4 | 11,384 | 1,622,906 | 143 | 534,915 | 727,550 | 63,910 | 36.7 | 63.3 | 12,949 | 4,982 |
| District 5 | 2,099 | 234,406 | 112 | 64,699 | 43,218 | 20,590 | 70.3 | 29.7 | 866 | 2,656 |
| District 6 | 14,561 | 1,977,518 | 136 | 640,720 | 655,337 | 45,006 | 50.9 | 49.1 | 15,680 | 3,678 |
| District 7 | 13,036 | 2,353,365 | 181 | 747,529 | 469,912 | 36,047 | 58.8 | 41.2 | 24,773 | 6,542 |
| District 8 | 7,246 | 2,648,961 | 366 | 1,949,573 | 1,217,343 | 168,002 | 14.0 | 86.0 | 45,431 | 13,218 |
| District 9 | 184 | 31,507 | 171 | 16,169 | 12,575 | 68,342 | 96.9 | 3.1 | 433 | 16,654 |
| TEXAS | 248,416 | 12,703,6,184 | 511 | 17,595,330 | 24,924,041 | 100,332 | 27.7 | 72.3 | 749,231 | 20,984 |
| District 1 | 11,811 | 1,756,575 | 149 | 240,554 | 1,342,954 | 113,704 | 5.0 | 95.0 | 1,334 | 7,058 |
| District 2 | 213 | 19,714 | 93 | 3,901 | 8,856 | 41,577 | 87.4 | 12.6 | 186 | 14,308 |
| District 3 | 1,362 | 137,649 | 101 | 52,404 | 46,263 | 33,967 | 32.2 | 67.8 | 1,166 | 18,219 |
| District 4 | 23,721 | 4,211,086 | 178 | 1,078,170 | 1,306,516 | 55,078 | 18.2 | 81.8 | 39,599 | 10,272 |
| District 5 | 12,604 | 1,947,480 | 155 | 361,393 | 454,125 | 36,030 | 35.1 | 64.9 | 2,243 | 11,049 |
| District 6 | 5,347 | 1,020,043 | 191 | 307,085 | 143,690 | 26,873 | 58.3 | 41.7 | 8,011 | 13,026 |
| District 7 | 59 | 27,971 | 474 | 540 | 687 | 11,644 | 83.7 | 16.3 | D | D |
| District 8 | 9,409 | 1,692,702 | 180 | 199,691 | 341,363 | 36,280 | 40.0 | 60.0 | 1,033 | 8,134 |
| District 9 | 98 | D | 0 | 1,267 | 1,442 | 14,714 | 57.5 | 42.6 | 35 | 5,833 |
| District 10 | 14,829 | 2,529,360 | 171 | 363,629 | 371,304 | 25,039 | 45.5 | 54.5 | 16,927 | 19,569 |
| District 11 | 20,886 | 15,547,323 | 744 | 1,502,355 | 1,239,636 | 59,352 | 34.9 | 65.1 | 55,547 | 14,203 |
| District 12 | 5,962 | 762,005 | 128 | 76,555 | 85,465 | 14,335 | 19.5 | 80.5 | 513 | 5,830 |
| District 13 | 19,229 | 22,289,498 | 1,159 | 3,647,588 | 8,790,318 | 457,139 | 14.9 | 85.1 | 214,954 | 29,277 |
| District 14 | 3,586 | 835,042 | 233 | 106,179 | 103,019 | 28,728 | 50.1 | 49.9 | 15,268 | 44,000 |
| District 15 | 8,984 | 3,792,508 | 422 | 387,448 | 372,617 | 41,476 | 60.5 | 39.5 | 10,272 | 11,834 |
| District 16 | 358 | 83,720 | 234 | 12,446 | 20,176 | 56,358 | 96.8 | 3.2 | D | D |
| District 17 | 15,795 | 3,896,462 | 247 | 774,561 | 1,056,334 | 66,878 | 20.4 | 79.6 | 16,618 | 14,772 |
| District 18 | 118 | D | D | 1,051 | 5,875 | 49,788 | 51.6 | 48.4 | D | D |
| District 19 | 14,252 | 15,108,704 | 1,060 | 4,440,380 | 4,845,855 | 340,012 | 34.6 | 65.4 | 209,918 | 25,706 |
| District 20 | 170 | 13,030 | 77 | 1,924 | 1,380 | 8,118 | 57.5 | 42.5 | 8 | 2,000 |
| District 21 | 8,076 | 2,999,190 | 371 | 62,516 | 96,177 | 11,909 | 24.4 | 75.6 | 3,402 | 7,162 |
| District 22 | 1,787 | 342,738 | 192 | 119,119 | 103,549 | 57,946 | 80.1 | 19.9 | 6,691 | 25,249 |
| District 23 | 9,676 | 27,813,527 | 2,874 | 447,435 | 831,884 | 85,974 | 45.2 | 54.8 | 19,770 | 17,875 |
| District 24 | 165 | 32,948 | 200 | 3,550 | 2,799 | 16,964 | 57.9 | 42.1 | 45 | 5,000 |
| District 25 | 14,207 | 3,944,915 | 278 | 572,672 | 463,442 | 32,621 | 28.1 | 71.9 | 16,336 | 12,615 |

# Table E. Congressional Districts 116th Congress — Nonfarm Employment and Payroll

| STATE District | Number of establishments | Private nonfarm employment and payroll, 2019 — Employment | | | | | | | | | | Annual payroll | |
|---|---|---|---|---|---|---|---|---|---|---|---|---|---|
| | | | Percent by selected industries | | | | | | | | | | |
| | | Total | Manufac-turing | Construc-tion | Wholesale trade | Retail trade | Health care and social assistance | Finance and Insurance | Real estate and rental and leasing | Profes-sional, sci-entific, and technical services | Information | Total (mil dol) | Average per employee (dollars) |
| | 73 | 74 | 75 | 76 | 77 | 78 | 79 | 80 | 81 | 82 | 83 | 84 | 85 |
| OREGON—Cont'd | | | | | | | | | | | | | |
| District 3 | 26,938 | 414,603 | 7.5 | 5.5 | 5.1 | 9.7 | 15.9 | 5.2 | 2.2 | 7.7 | 2.6 | 23,141 | 55,814 |
| District 4 | 20,008 | 251,810 | 12.5 | 6.1 | 3.7 | 15.3 | 19.0 | 3.3 | 1.6 | 4.5 | 1.5 | 11,190 | 44,438 |
| District 5 | 22,707 | 283,005 | 10.7 | 8.5 | 5.0 | 14.5 | 17.7 | 3.4 | 2.0 | 4.6 | 2.0 | 13,395 | 47,331 |
| PENNSYLVANIA | 303,224 | 5,557,885 | 10.2 | 4.6 | 4.5 | 11.7 | 19.5 | 5.3 | 1.2 | 6.0 | 2.0 | 294,727 | 53,029 |
| District 1 | 21,745 | 302,462 | 12.0 | 6.1 | 7.7 | 14.1 | 19.1 | 3.3 | 1.1 | 6.6 | 1.3 | 15,005 | 49,610 |
| District 2 | 10,972 | 189,151 | 7.9 | 3.7 | 4.8 | 12.9 | 32.4 | 1.5 | 1.0 | 2.0 | 0.7 | 8,577 | 45,347 |
| District 3 | 15,968 | 408,473 | 0.8 | 0.9 | 1.0 | 5.2 | 25.3 | 7.7 | 1.6 | 11.6 | 4.5 | 28,442 | 69,629 |
| District 4 | 23,144 | 463,577 | 7.7 | 5.6 | 5.4 | 10.4 | 15.8 | 8.5 | 1.5 | 9.3 | 3.0 | 32,056 | 69,150 |
| District 5 | 16,415 | 321,311 | 5.2 | 4.3 | 3.4 | 10.5 | 19.5 | 5.0 | 1.7 | 3.8 | 2.5 | 18,342 | 57,086 |
| District 6 | 19,065 | 350,020 | 8.1 | 3.8 | 5.6 | 10.5 | 16.9 | 9.8 | 2.0 | 8.3 | 3.1 | 24,139 | 68,964 |
| District 7 | 16,594 | 317,603 | 10.5 | 3.7 | 6.9 | 13.3 | 21.2 | 3.2 | 1.1 | 4.0 | 1.3 | 16,174 | 50,924 |
| District 8 | 15,352 | 263,956 | 11.8 | 3.6 | 4.2 | 14.4 | 19.9 | 3.9 | 0.9 | 4.0 | 1.7 | 10,948 | 41,478 |
| District 9 | 13,550 | 226,852 | 21.2 | 4.1 | 4.5 | 13.1 | 18.4 | 2.3 | 0.7 | 2.6 | 1.4 | 9,903 | 43,653 |
| District 10 | 17,652 | 385,148 | 8.9 | 4.7 | 4.3 | 12.3 | 18.7 | 5.4 | 1.1 | 5.7 | 1.4 | 19,248 | 49,975 |
| District 11 | 16,619 | 295,113 | 16.8 | 8.5 | 6.8 | 13.7 | 15.1 | 3.1 | 1.0 | 4.8 | 0.9 | 13,193 | 44,705 |
| District 12 | 15,111 | 215,549 | 16.9 | 4.3 | 3.0 | 15.0 | 19.3 | 2.7 | 1.1 | 3.4 | 1.6 | 8,679 | 40,266 |
| District 13 | 14,475 | 219,363 | 16.2 | 4.7 | 3.4 | 13.9 | 20.2 | 2.5 | 0.8 | 3.4 | 1.4 | 8,882 | 40,489 |
| District 14 | 16,059 | 249,858 | 11.3 | 8.5 | 5.3 | 13.3 | 16.6 | 2.0 | 1.2 | 4.4 | 1.6 | 11,929 | 47,741 |
| District 15 | 14,985 | 202,468 | 19.3 | 4.0 | 3.0 | 15.3 | 20.2 | 3.1 | 0.7 | 2.6 | 0.9 | 7,914 | 39,087 |
| District 16 | 16,349 | 270,075 | 18.3 | 4.2 | 3.6 | 12.9 | 21.5 | 3.9 | 0.9 | 3.1 | 1.3 | 11,017 | 40,791 |
| District 17 | 19,178 | 335,205 | 8.3 | 6.2 | 5.0 | 13.4 | 17.2 | 3.7 | 1.2 | 6.8 | 1.8 | 17,045 | 50,850 |
| District 18 | 18,722 | 439,573 | 3.8 | 3.9 | 2.3 | 8.2 | 21.5 | 9.7 | 1.4 | 8.3 | 2.7 | 26,741 | 60,834 |
| RHODE ISLAND | 28,801 | 444,948 | 8.9 | 4.5 | 4.5 | 10.8 | 19.6 | 7.1 | 1.2 | 5.2 | 1.5 | 22,726 | 51,077 |
| District 1 | 13,475 | 209,719 | 8.7 | 4.6 | 4.2 | 9.2 | 18.9 | 7.4 | 1.3 | 5.3 | 0.9 | 10,294 | 49,084 |
| District 2 | 15,014 | 229,827 | 9.2 | 4.4 | 4.8 | 12.4 | 20.5 | 6.8 | 1.2 | 4.6 | 2.0 | 12,110 | 52,690 |
| SOUTH CAROLINA | 111,926 | 1,949,406 | 12.8 | 4.6 | 3.9 | 12.9 | 12.8 | 3.9 | 1.4 | 5.1 | 1.9 | 83,500 | 42,834 |
| District 1 | 20,919 | 255,306 | 4.9 | 5.2 | 2.0 | 17.6 | 13.2 | 3.0 | 2.5 | 6.7 | 3.3 | 10,879 | 42,613 |
| District 2 | 13,972 | 218,921 | 10.5 | 5.2 | 3.8 | 17.2 | 13.4 | 2.8 | 1.3 | 4.3 | 2.4 | 9,025 | 41,226 |
| District 3 | 11,555 | 180,934 | 27.6 | 4.4 | 4.8 | 14.1 | 13.1 | 2.3 | 0.8 | 3.2 | 0.8 | 7,585 | 41,924 |
| District 4 | 19,453 | 369,332 | 16.2 | 4.4 | 4.6 | 11.6 | 12.8 | 3.5 | 1.2 | 6.2 | 2.4 | 17,050 | 46,164 |
| District 5 | 12,689 | 205,483 | 19.8 | 5.5 | 4.0 | 14.0 | 11.1 | 4.9 | 1.1 | 5.5 | 1.5 | 8,949 | 43,551 |
| District 6 | 15,581 | 285,469 | 13.6 | 6.0 | 6.1 | 10.3 | 15.4 | 5.3 | 1.6 | 6.4 | 2.2 | 13,779 | 48,268 |
| District 7 | 17,083 | 248,414 | 10.3 | 4.9 | 3.4 | 17.3 | 19.3 | 2.8 | 2.4 | 2.9 | 1.2 | 9,497 | 38,230 |
| SOUTH DAKOTA | 27,108 | 358,943 | 13.0 | 5.7 | 5.1 | 14.6 | 19.1 | 7.0 | 1.2 | 3.5 | 1.9 | 15,767 | 43,925 |
| At Large | 27,108 | 358,943 | 13.0 | 5.7 | 5.1 | 14.6 | 19.1 | 7.0 | 1.2 | 3.5 | 1.9 | 15,767 | 43,925 |
| TENNESSEE | 139,760 | 2,724,545 | 12.5 | 4.4 | 4.5 | 11.7 | 15.6 | 4.6 | 1.3 | 4.5 | 1.7 | 129,929 | 47,688 |
| District 1 | 13,998 | 247,720 | 18.4 | 4.0 | 2.7 | 15.6 | 16.5 | 3.0 | 1.5 | 1.9 | 1.3 | 9,829 | 39,678 |
| District 2 | 16,373 | 302,098 | 8.9 | 4.6 | 5.5 | 13.3 | 16.0 | 4.7 | 1.5 | 4.9 | 1.9 | 13,439 | 44,485 |
| District 3 | 14,730 | 292,232 | 18.9 | 4.6 | 2.8 | 12.2 | 15.3 | 5.6 | 0.9 | 6.1 | 1.4 | 13,518 | 46,257 |
| District 4 | 13,660 | 256,637 | 22.9 | 3.8 | 4.0 | 14.2 | 11.8 | 3.8 | 1.0 | 3.0 | 1.5 | 10,768 | 41,960 |
| District 5 | 21,721 | 484,574 | 5.3 | 5.1 | 5.0 | 8.2 | 16.8 | 5.5 | 1.8 | 5.8 | 2.8 | 28,186 | 58,166 |
| District 6 | 14,206 | 212,358 | 19.7 | 5.3 | 3.8 | 14.4 | 14.6 | 3.2 | 1.1 | 3.5 | 1.3 | 8,435 | 39,722 |
| District 7 | 16,022 | 252,998 | 11.1 | 4.7 | 3.1 | 12.8 | 17.1 | 6.9 | 1.2 | 5.4 | 2.4 | 13,228 | 52,284 |
| District 8 | 14,987 | 256,971 | 13.5 | 4.1 | 4.5 | 12.4 | 21.1 | 4.3 | 1.7 | 4.3 | 1.1 | 11,765 | 45,784 |
| District 9 | 13,325 | 329,767 | 6.9 | 4.1 | 7.9 | 10.0 | 14.1 | 2.8 | 1.5 | 3.4 | 1.3 | 17,061 | 51,737 |
| TEXAS | 609,476 | 11,104,054 | 7.4 | 6.7 | 4.8 | 11.9 | 14.3 | 5.0 | 1.9 | 6.8 | 2.4 | 611,142 | 55,038 |
| District 1 | 16,946 | 261,226 | 13.2 | 5.6 | 4.0 | 13.7 | 18.9 | 3.4 | 1.5 | 4.3 | 1.5 | 11,180 | 42,799 |
| District 2 | 21,317 | 363,457 | 9.3 | 9.6 | 7.3 | 12.3 | 8.5 | 4.8 | 2.2 | 7.1 | 1.6 | 20,286 | 55,815 |
| District 3 | 23,862 | 427,772 | 4.0 | 4.1 | 4.6 | 11.9 | 13.1 | 12.3 | 1.9 | 11.2 | 4.5 | 29,183 | 68,220 |
| District 4 | 14,723 | 207,701 | 19.4 | 5.5 | 3.3 | 15.7 | 19.9 | 3.2 | 1.0 | 3.6 | 1.2 | 8,320 | 40,060 |
| District 5 | 13,018 | 193,021 | 12.8 | 10.4 | 5.0 | 15.7 | 15.2 | 2.1 | 1.9 | 2.6 | 1.1 | 7,665 | 39,712 |
| District 6 | 14,953 | 245,285 | 13.4 | 6.4 | 6.0 | 14.8 | 13.7 | 5.2 | 1.6 | 3.4 | 1.7 | 10,743 | 43,799 |
| District 7 | 25,756 | 462,036 | 2.1 | 5.9 | 3.8 | 10.6 | 10.5 | 6.3 | 3.1 | 11.4 | 2.6 | 36,602 | 79,219 |
| District 8 | 16,200 | 226,535 | 6.5 | 7.1 | 3.9 | 15.1 | 12.8 | 3.3 | 1.9 | 7.2 | 2.0 | 13,872 | 61,234 |
| District 9 | 12,944 | 327,556 | 4.8 | 4.3 | 5.5 | 7.4 | 34.7 | 2.2 | 1.5 | 7.3 | 1.8 | 20,385 | 62,234 |
| District 10 | 21,754 | 339,828 | 7.0 | 7.6 | 5.4 | 14.8 | 11.6 | 3.3 | 2.3 | 8.7 | 4.0 | 18,347 | 53,990 |
| District 11 | 19,860 | 289,194 | 6.7 | 8.8 | 5.8 | 12.7 | 11.9 | 2.8 | 2.5 | 3.4 | 1.3 | 16,481 | 56,989 |
| District 12 | 18,836 | 354,391 | 11.4 | 5.8 | 4.4 | 13.1 | 14.9 | 5.6 | 1.4 | 5.4 | 1.7 | 19,682 | 55,538 |
| District 13 | 16,384 | 223,421 | 14.1 | 6.1 | 4.9 | 14.6 | 17.7 | 4.3 | 1.2 | 2.7 | 1.4 | 9,782 | 43,782 |
| District 14 | 14,130 | 240,462 | 12.4 | 10.2 | 3.1 | 14.4 | 15.5 | 3.2 | 2.4 | 4.7 | 0.9 | 12,281 | 51,071 |
| District 15 | 12,357 | 195,704 | 6.6 | 3.5 | 4.7 | 17.2 | 26.2 | 3.4 | 1.2 | 3.4 | 1.9 | 6,770 | 34,594 |
| District 16 | 14,021 | 232,305 | 6.6 | 4.7 | 4.0 | 14.8 | 19.6 | 3.2 | 1.7 | 3.8 | 3.5 | 7,906 | 34,034 |
| District 17 | 15,304 | 274,166 | 11.9 | 7.7 | 3.7 | 12.6 | 14.1 | 4.7 | 1.6 | 7.7 | 2.1 | 13,106 | 47,802 |
| District 18 | 17,438 | 419,942 | 7.6 | 6.3 | 10.2 | 6.0 | 7.3 | 3.7 | 1.8 | 9.2 | 2.0 | 34,730 | 82,702 |
| District 19 | 16,980 | 247,873 | 6.8 | 6.0 | 5.2 | 14.5 | 19.9 | 4.1 | 1.5 | 3.2 | 2.0 | 10,309 | 41,588 |
| District 20 | 12,113 | 274,899 | 3.4 | 3.2 | 2.3 | 15.5 | 15.6 | 13.6 | 1.6 | 7.1 | 2.0 | 12,401 | 45,110 |
| District 21 | 28,377 | 474,234 | 2.6 | 6.0 | 3.3 | 9.9 | 16.0 | 5.9 | 2.5 | 11.1 | 4.5 | 27,688 | 58,384 |
| District 22 | 18,098 | 238,658 | 5.4 | 6.8 | 3.7 | 19.2 | 16.7 | 3.3 | 1.7 | 5.7 | 1.6 | 10,584 | 44,350 |
| District 23 | 11,945 | 190,529 | 6.0 | 14.9 | 3.1 | 14.7 | 13.4 | 3.5 | 1.8 | 3.0 | 0.9 | 8,378 | 43,974 |
| District 24 | 29,104 | 733,698 | 4.5 | 5.4 | 6.4 | 7.4 | 7.4 | 8.3 | 2.3 | 10.5 | 4.8 | 48,175 | 65,661 |
| District 25 | 19,449 | 237,516 | 6.7 | 6.7 | 3.5 | 13.4 | 16.9 | 4.3 | 2.6 | 7.6 | 3.7 | 13,197 | 55,562 |

# Table E. Congressional Districts 116th Congress — **Land Area and Population Characteristics**

| STATE District | Representative, 117th Congress | Land area,[1] 2018 (sq mi) | Total persons | Per square mile | White | Black | American Indian, Alaska Native | Asian and Pacific Islander | Some other race (percent) | Two or more races (percent) | Hispanic or Latino[2] (percent) | Non-Hispanic White alone (percent) | Percent female | Percent foreign-born | Percent born in state of residence |
|---|---|---|---|---|---|---|---|---|---|---|---|---|---|---|---|
| | | | | | Race alone (percent) | | | | | | | | | | |
| | | 1 | 2 | 3 | 4 | 5 | 6 | 7 | 8 | 9 | 10 | 11 | 12 | 13 | 14 |
| **TEXAS — Cont'd** | | | | | | | | | | | | | | | |
| District 26 | Michael C. Burgess (R) | 907.991 | 920,865 | 1,014.179 | 76.0 | 9.1 | 0.6 | 7.9 | 2.6 | 3.8 | 19.8 | 60.189 | 51.2 | 13.9 | 48.8 |
| District 27 | Michael Cloud (R) | 9,120.849 | 745,526 | 81.739 | 85.7 | 4.9 | 0.2 | 1.6 | 5.5 | 2.0 | 54.3 | 38.157 | 50.2 | 8.9 | 75.3 |
| District 28 | Henry Cuellar (D) | 9,379.036 | 772,410 | 82.355 | 87.7 | 3.9 | 0.5 | 1.0 | 4.6 | 2.4 | 78.9 | 15.711 | 50.5 | 19.7 | 65.7 |
| District 29 | Sylvia R. Garcia (D) | 187.624 | 783,915 | 4,178.117 | 72.2 | 10.5 | 0.3 | 1.8 | 13.5 | 1.6 | 78.7 | 8.483 | 50.2 | 30.4 | 61.0 |
| District 30 | Eddie Bernice Johnson (D) | 356.724 | 792,445 | 2,221.451 | 47.9 | 41.2 | 0.3 | 1.8 | 7.0 | 1.9 | 40.5 | 15.613 | 51.1 | 18.9 | 62.3 |
| District 31 | John R. Carter (R) | 2,154.833 | 916,064 | 425.121 | 74.6 | 12.3 | 0.5 | 6.1 | 2.1 | 4.5 | 25.3 | 54.018 | 50.5 | 11.6 | 49.5 |
| District 32 | Colin Z. Allred (D) | 186.014 | 778,087 | 4,182.949 | 67.4 | 14.9 | 0.5 | 8.3 | 5.2 | 3.7 | 27.0 | 46.487 | 51.1 | 21.8 | 50.6 |
| District 33 | Marc A. Veasey (D) | 212.147 | 751,182 | 3,540.856 | 60.8 | 16.2 | 0.4 | 2.1 | 18.7 | 1.8 | 66.8 | 14.481 | 49.8 | 31.8 | 56.4 |
| District 34 | Filemon Vela (D) | 8,191.582 | 712,596 | 86.991 | 91.7 | 1.4 | 0.3 | 0.5 | 4.6 | 1.5 | 84.5 | 13.382 | 50.3 | 18.6 | 71.6 |
| District 35 | Lloyd Doggett (D) | 594.784 | 857,654 | 1,441.959 | 76.9 | 9.2 | 0.6 | 1.8 | 8.8 | 2.7 | 60.6 | 27.413 | 49.7 | 15.2 | 62.9 |
| District 36 | Brian Babin (R) | 7,125.887 | 758,238 | 106.406 | 82.4 | 9.0 | 0.5 | 1.8 | 3.6 | 2.8 | 28.5 | 58.785 | 50.2 | 10.9 | 68.8 |
| **UTAH** | | 82,376.853 | 3,205,958 | 38.918 | 87.3 | 1.2 | 1.1 | 3.4 | 3.8 | 3.2 | 14.4 | 77.688 | 49.8 | 8.6 | 61.7 |
| District 1 | Blake D. Moore (R) | 19,557.402 | 787,582 | 40.270 | 91.4 | 0.9 | 1.1 | 2.0 | 1.6 | 2.9 | 13.5 | 80.544 | 49.2 | 5.6 | 65.5 |
| District 2 | Chris Stewart (R) | 40,195.721 | 788,484 | 19.616 | 83.4 | 1.5 | 1.2 | 4.7 | 6.3 | 3.0 | 15.8 | 74.992 | 50.1 | 10.0 | 59.0 |
| District 3 | John R. Curtis (R) | 20,073.914 | 779,460 | 38.829 | 90.3 | 0.7 | 1.5 | 2.7 | 1.3 | 3.4 | 10.9 | 81.429 | 49.9 | 7.6 | 59.7 |
| District 4 | Burgess Owens (R) | 2,549.816 | 850,432 | 333.527 | 84.5 | 1.5 | 0.6 | 4.0 | 6.0 | 3.6 | 17.1 | 74.114 | 49.9 | 10.8 | 62.5 |
| **VERMONT** | | 9,217.449 | 623,989 | 67.696 | 93.8 | 1.5 | 0.5 | 1.5 | 0.3 | 2.5 | 2.0 | 92.452 | 50.5 | 4.7 | 48.8 |
| At Large | Peter Welch (D) | 9,217.449 | 623,989 | 67.696 | 93.8 | 1.5 | 0.5 | 1.5 | 0.3 | 2.5 | 2.0 | 92.452 | 50.5 | 4.7 | 48.8 |
| **VIRGINIA** | | 39,482.107 | 8,535,519 | 216.187 | 67.0 | 19.4 | 0.3 | 6.7 | 2.9 | 3.8 | 9.7 | 61.071 | 50.8 | 12.7 | 49.5 |
| District 1 | Robert J. Wittman (R) | 4,211.649 | 824,492 | 195.765 | 72.4 | 16.4 | 0.2 | 3.6 | 3.2 | 4.2 | 11.1 | 65.373 | 50.2 | 10.8 | 48.6 |
| District 2 | Elaine G. Luria (D) | 1,103.765 | 723,927 | 655.871 | 67.2 | 19.3 | 0.3 | 6.3 | 2.2 | 4.9 | 8.3 | 62.472 | 50.3 | 8.7 | 43.0 |
| District 3 | Robert C. "Bobby" Scott (D) | 625.168 | 760,127 | 1,215.876 | 43.3 | 47.4 | 0.1 | 2.7 | 2.0 | 4.4 | 6.8 | 39.922 | 51.9 | 6.6 | 56.2 |
| District 4 | A. Donald McEachin (D) | 3,642.186 | 768,382 | 210.967 | 51.3 | 41.1 | 0.3 | 1.7 | 2.2 | 3.4 | 6.0 | 48.140 | 50.7 | 5.2 | 62.7 |
| District 5 | Bob Good (R) | 10,030.711 | 735,766 | 73.351 | 74.9 | 19.4 | 0.4 | 1.7 | 0.7 | 2.8 | 3.8 | 72.020 | 51.2 | 4.1 | 64.3 |
| District 6 | Ben Cline (R) | 5,929.001 | 755,012 | 127.342 | 82.2 | 11.9 | 0.3 | 1.6 | 1.1 | 2.9 | 5.6 | 78.691 | 51.5 | 5.1 | 65.0 |
| District 7 | Abigail Davis Spanberger (D) | 3,118.270 | 802,921 | 257.489 | 69.0 | 18.0 | 0.2 | 5.5 | 4.0 | 3.3 | 8.0 | 65.886 | 50.9 | 10.7 | 54.5 |
| District 8 | Donald S. Beyer, Jr. (D) | 149.397 | 813,568 | 5,445.678 | 62.9 | 14.1 | 0.3 | 12.2 | 5.8 | 4.6 | 18.5 | 51.024 | 50.7 | 28.1 | 25.1 |
| District 9 | Morgan Griffith (R) | 9,115.487 | 704,078 | 77.240 | 90.8 | 5.5 | 0.2 | 1.6 | 0.4 | 1.5 | 2.8 | 88.636 | 50.1 | 2.7 | 64.5 |
| District 10 | Jennifer Wexton (D) | 1,371.385 | 857,693 | 625.421 | 69.1 | 8.0 | 0.3 | 15.7 | 2.5 | 4.5 | 14.4 | 58.761 | 50.2 | 22.2 | 35.8 |
| District 11 | Gerald E. Connolly (D) | 185.088 | 789,553 | 4,265.825 | 55.5 | 13.9 | 0.5 | 18.9 | 7.0 | 4.4 | 19.1 | 44.570 | 50.8 | 31.5 | 29.7 |
| **WASHINGTON** | | 66,455.117 | 7,614,893 | 114.587 | 74.2 | 4.0 | 1.4 | 9.7 | 4.8 | 6.0 | 13.0 | 67.325 | 50.0 | 14.9 | 46.4 |
| District 1 | Suzan K. DelBene (D) | 6,181.857 | 791,545 | 128.043 | 75.1 | 1.5 | 1.1 | 13.3 | 3.7 | 5.3 | 9.7 | 70.274 | 49.3 | 18.5 | 47.2 |
| District 2 | Rick Larsen (D) | 1,019.034 | 760,064 | 745.867 | 76.1 | 3.3 | 1.7 | 9.9 | 3.5 | 5.4 | 11.1 | 70.185 | 50.2 | 14.1 | 48.1 |
| District 3 | Jaime Herrera Beutler (R) | 9,117.724 | 756,675 | 82.989 | 85.9 | 1.6 | 0.7 | 3.8 | 2.1 | 5.8 | 10.2 | 79.306 | 50.2 | 9.3 | 40.3 |
| District 4 | Dan Newsome (R) | 19,246.625 | 735,797 | 38.230 | 75.2 | 1.4 | 2.7 | 1.7 | 15.2 | 3.8 | 39.7 | 52.749 | 49.8 | 16.7 | 55.9 |
| District 5 | Cathy McMorris Rodgers (R) | 15,473.512 | 734,322 | 47.457 | 87.3 | 1.7 | 1.7 | 3.0 | 1.6 | 4.7 | 8.0 | 75.835 | 50.2 | 6.4 | 47.8 |
| District 6 | Derek Kilmer (D) | 6,902.336 | 726,540 | 105.260 | 80.8 | 4.0 | 2.1 | 4.3 | 1.8 | 7.0 | 8.1 | 65.992 | 49.4 | 18.8 | 38.0 |
| District 7 | Pramila Jayapal (D) | 143.934 | 817,787 | 5,681.680 | 69.7 | 5.0 | 0.5 | 15.1 | 2.9 | 6.8 | 12.0 | 68.204 | 50.4 | 15.5 | 50.7 |
| District 8 | Kim Schrier (D) | 7,359.539 | 770,177 | 104.650 | 72.4 | 3.1 | 1.2 | 10.6 | 6.9 | 5.8 | 12.9 | 43.758 | 49.7 | 31.6 | 37.1 |
| District 9 | Adam Smith (D) | 183.472 | 751,595 | 4,096.511 | 49.2 | 12.0 | 0.7 | 25.0 | 6.5 | 6.7 | 12.6 | 64.144 | 50.0 | 11.3 | 47.8 |
| District 10 | Marilyn Strickland (D) | 827.084 | 770,391 | 931.454 | 71.3 | 6.4 | 1.4 | 8.9 | 3.8 | 8.1 | | | | | |
| **WEST VIRGINIA** | | 24,041.152 | 1,792,147 | 74.545 | 93.1 | 3.7 | 0.2 | 0.8 | 0.4 | 1.8 | 1.5 | 92.039 | 50.6 | 1.6 | 68.5 |
| District 1 | David McKinley (R) | 6,276.065 | 601,811 | 95.890 | 93.6 | 2.5 | 0.3 | 0.9 | 0.7 | 1.8 | 1.2 | 92.990 | 50.1 | 1.7 | 67.8 |
| District 2 | Alexander X. Mooney (R) | 8,019.233 | 623,039 | 77.693 | 91.8 | 4.4 | 0.2 | 0.9 | 0.4 | 2.3 | 2.2 | 90.081 | 50.9 | 2.2 | 61.0 |
| District 3 | Carol D. Miller (R) | 9,745.854 | 567,297 | 58.209 | 93.9 | 4.1 | 0.0 | 0.5 | 0.1 | 1.3 | 1.0 | 93.179 | 50.7 | 0.9 | 77.5 |
| **WISCONSIN** | | 54,167.142 | 5,822,434 | 107.490 | 85.2 | 6.4 | 0.9 | 3.0 | 2.1 | 2.4 | 7.1 | 80.801 | 50.3 | 5.1 | 70.9 |
| District 1 | Brian Steil (R) | 1,727.988 | 721,691 | 417.648 | 86.9 | 6.0 | 0.3 | 2.3 | 2.1 | 2.4 | 10.1 | 79.484 | 50.4 | 5.5 | 66.6 |
| District 2 | Mark Pocan (D) | 4,536.911 | 773,663 | 170.526 | 85.9 | 4.4 | 0.4 | 4.6 | 1.9 | 2.8 | 6.8 | 81.293 | 49.7 | 7.7 | 63.4 |
| District 3 | Ron Kind (D) | 11,117.445 | 723,169 | 65.048 | 93.2 | 1.5 | 0.5 | 2.6 | 0.8 | 1.5 | 2.9 | 91.292 | 49.7 | 2.8 | 71.0 |
| District 4 | Gwen Moore (D) | 128.423 | 704,146 | 5,483.021 | 50.8 | 33.7 | 0.7 | 4.0 | 7.1 | 3.7 | 17.6 | 41.502 | 51.9 | 9.8 | 65.8 |
| District 5 | Scott Fitzgerald (R) | 1,891.004 | 733,314 | 387.791 | 90.5 | 2.4 | 0.3 | 3.2 | 1.5 | 2.2 | 6.5 | 85.975 | 50.7 | 4.5 | 76.4 |
| District 6 | Glenn Grothman (R) | 4,919.256 | 715,828 | 145.516 | 91.6 | 2.0 | 0.5 | 2.8 | 1.2 | 1.9 | 4.8 | 88.363 | 50.0 | 4.1 | 77.9 |
| District 7 | Thomas P. Tiffany (R) | 23,039.292 | 714,544 | 31.014 | 93.2 | 0.8 | 2.1 | 1.8 | 0.3 | 1.9 | 2.6 | 91.239 | 49.6 | 2.2 | 68.0 |
| District 8 | Mike Gallagher (R) | 6,806.823 | 736,079 | 108.138 | 89.0 | 1.7 | 2.6 | 2.5 | 1.9 | 2.3 | 5.6 | 86.224 | 50.1 | 3.6 | 78.5 |
| **WYOMING** | | 97,088.756 | 578,759 | 5.961 | 90.9 | 1.2 | 2.4 | 0.9 | 1.9 | 2.7 | 10.1 | 83.697 | 48.9 | 3.1 | 43.0 |
| At Large | Liz Cheney (R) | 97,088.756 | 578,759 | 5.961 | 90.9 | 1.2 | 2.4 | 0.9 | 1.9 | 2.7 | 10.1 | 83.697 | 48.9 | 3.1 | 43.0 |

1. Dry land or land partially or temporarily covered by water.   2. May be of any race.

# Table E. Congressional Districts 116th Congress — **Age and Education**

| STATE District | Population and population characteristics, 2019 (cont.) | | | | | | | | | | Education, 2019 | | |
| | Age (percent) | | | | | | | | | | | Attainment[2] (percent) | |
| | Under 5 years | 5 to 17 years | 18 to 24 years | 25 to 34 years | 35 to 44 years | 45 to 54 years | 55 to 64 years | 65 to 74 years | 75 years and over | Median age | Total Enrollment[1] | High school graduate or more | Bachelor's degree or more |
| | 15 | 16 | 17 | 18 | 19 | 20 | 21 | 22 | 23 | 24 | 25 | 26 | 27 |
| **TEXAS — Cont'd** | | | | | | | | | | | | | |
| District 26 | 6.2 | 19.4 | 9.1 | 13.5 | 14.7 | 14.6 | 11.7 | 7.0 | 3.8 | 36.3 | 268,390 | 92.7 | 45.6 |
| District 27 | 6.6 | 18.0 | 9.2 | 13.6 | 12.6 | 11.1 | 12.2 | 10.0 | 6.8 | 36.7 | 183,463 | 83.3 | 20.1 |
| District 28 | 8.1 | 22.3 | 10.5 | 13.8 | 12.4 | 11.4 | 9.7 | 6.8 | 5.0 | 31.2 | 232,082 | 73.5 | 20.0 |
| District 29 | 9.7 | 21.8 | 10.3 | 15.3 | 13.3 | 11.3 | 9.4 | 5.7 | 3.0 | 30.3 | 224,318 | 65.5 | 11.6 |
| District 30 | 7.1 | 19.8 | 10.0 | 15.8 | 12.9 | 11.7 | 11.6 | 6.8 | 4.1 | 33.0 | 217,763 | 79.7 | 22.4 |
| District 31 | 7.1 | 18.9 | 9.0 | 14.9 | 15.2 | 12.5 | 10.1 | 7.6 | 4.6 | 35.0 | 246,613 | 92.4 | 36.4 |
| District 32 | 7.1 | 17.4 | 8.9 | 16.2 | 12.8 | 13.0 | 11.8 | 7.7 | 5.0 | 35.3 | 200,686 | 88.4 | 45.3 |
| District 33 | 8.1 | 22.2 | 10.2 | 15.4 | 13.6 | 11.9 | 9.9 | 5.2 | 3.3 | 30.6 | 213,416 | 62.4 | 13.1 |
| District 34 | 7.6 | 21.0 | 11.0 | 13.3 | 11.4 | 11.4 | 9.8 | 8.2 | 6.2 | 32.5 | 200,691 | 69.9 | 15.5 |
| District 35 | 6.3 | 17.5 | 11.4 | 18.3 | 14.6 | 11.6 | 9.8 | 6.4 | 4.1 | 33.0 | 231,938 | 80.1 | 25.2 |
| District 36 | 6.5 | 18.6 | 8.6 | 13.5 | 12.5 | 13.2 | 12.6 | 9.0 | 5.5 | 37.0 | 185,227 | 85.2 | 19.4 |
| **UTAH** | 7.7 | 21.4 | 11.4 | 14.7 | 13.8 | 10.3 | 9.4 | 6.8 | 4.6 | 31.2 | 1,012,792 | 93.0 | 34.8 |
| District 1 | 7.8 | 22.3 | 11.0 | 14.3 | 13.9 | 10.5 | 9.7 | 6.6 | 4.0 | 31.1 | 249,418 | 92.8 | 31.8 |
| District 2 | 6.9 | 19.6 | 10.6 | 14.8 | 13.6 | 10.3 | 10.0 | 8.0 | 6.1 | 33.4 | 231,164 | 92.1 | 32.2 |
| District 3 | 7.8 | 21.4 | 14.8 | 13.6 | 12.4 | 10.0 | 8.8 | 6.7 | 4.4 | 28.7 | 278,053 | 95.1 | 41.9 |
| District 4 | 8.1 | 22.0 | 9.2 | 16.0 | 15.2 | 10.5 | 9.0 | 6.1 | 3.8 | 31.5 | 254,157 | 92.4 | 33.8 |
| **VERMONT** | 4.7 | 13.5 | 10.4 | 12.3 | 11.4 | 12.5 | 15.2 | 12.0 | 8.0 | 42.8 | 144,066 | 93.1 | 38.7 |
| At Large | 4.7 | 13.5 | 10.4 | 12.3 | 11.4 | 12.5 | 15.2 | 12.0 | 8.0 | 42.8 | 144,066 | 93.1 | 38.7 |
| **VIRGINIA** | 5.9 | 15.9 | 9.4 | 13.8 | 13.3 | 12.8 | 13.1 | 9.5 | 6.4 | 38.5 | 2,118,036 | 90.0 | 39.6 |
| District 1 | 6.0 | 17.5 | 8.3 | 12.2 | 13.3 | 13.3 | 13.6 | 9.5 | 6.3 | 39.3 | 205,677 | 91.0 | 37.8 |
| District 2 | 5.8 | 14.8 | 11.0 | 15.6 | 12.5 | 11.5 | 13.3 | 9.0 | 6.6 | 36.9 | 173,613 | 92.6 | 37.7 |
| District 3 | 6.7 | 15.8 | 11.0 | 16.4 | 12.6 | 11.1 | 12.2 | 8.4 | 5.8 | 35.1 | 203,996 | 89.4 | 26.5 |
| District 4 | 5.9 | 14.8 | 9.9 | 15.1 | 13.1 | 12.3 | 13.4 | 9.5 | 6.1 | 37.8 | 185,517 | 87.8 | 30.8 |
| District 5 | 4.7 | 14.7 | 9.7 | 11.8 | 11.0 | 12.5 | 14.7 | 12.7 | 8.4 | 43.2 | 171,845 | 87.2 | 29.0 |
| District 6 | 5.1 | 14.7 | 12.0 | 12.4 | 11.8 | 11.7 | 13.6 | 10.6 | 8.1 | 39.7 | 187,588 | 87.9 | 28.3 |
| District 7 | 5.9 | 16.9 | 7.6 | 13.2 | 13.3 | 13.1 | 13.4 | 10.1 | 6.4 | 39.8 | 189,962 | 91.6 | 40.6 |
| District 8 | 6.9 | 14.5 | 7.2 | 18.7 | 16.2 | 13.0 | 11.1 | 7.5 | 4.9 | 36.5 | 188,711 | 92.0 | 63.3 |
| District 9 | 4.6 | 13.4 | 11.1 | 11.2 | 11.2 | 13.1 | 14.3 | 12.1 | 9.0 | 43.7 | 155,089 | 85.4 | 21.9 |
| District 10 | 6.2 | 20.1 | 7.6 | 10.6 | 14.7 | 15.5 | 12.3 | 7.8 | 5.1 | 38.8 | 241,936 | 93.1 | 56.6 |
| District 11 | 6.3 | 16.6 | 9.2 | 14.2 | 15.1 | 13.6 | 12.4 | 7.7 | 4.9 | 37.0 | 214,102 | 91.7 | 57.2 |
| **WASHINGTON** | 6.0 | 15.9 | 8.7 | 15.3 | 13.5 | 12.2 | 12.6 | 9.6 | 6.2 | 37.9 | 1,764,122 | 91.7 | 37.0 |
| District 1 | 6.3 | 16.9 | 7.2 | 13.7 | 14.7 | 13.9 | 13.0 | 9.0 | 5.3 | 38.8 | 185,170 | 94.4 | 45.3 |
| District 2 | 5.7 | 14.1 | 9.6 | 15.2 | 13.4 | 11.7 | 13.1 | 10.6 | 6.7 | 38.9 | 165,878 | 92.5 | 33.4 |
| District 3 | 5.9 | 16.8 | 7.7 | 13.2 | 12.7 | 12.6 | 13.2 | 11.2 | 6.8 | 39.8 | 169,078 | 91.7 | 27.2 |
| District 4 | 7.5 | 21.0 | 8.8 | 13.6 | 12.6 | 10.6 | 11.3 | 8.8 | 5.6 | 34.4 | 188,586 | 80.5 | 22.2 |
| District 5 | 5.7 | 15.5 | 11.0 | 13.8 | 12.0 | 11.2 | 13.3 | 10.4 | 7.0 | 38.2 | 185,477 | 93.5 | 30.4 |
| District 6 | 5.2 | 14.1 | 8.0 | 13.7 | 12.4 | 11.3 | 14.1 | 13.2 | 8.1 | 41.9 | 148,719 | 94.1 | 33.0 |
| District 7 | 4.6 | 11.2 | 9.9 | 22.5 | 15.4 | 12.2 | 10.9 | 7.9 | 5.4 | 35.8 | 176,303 | 95.1 | 62.3 |
| District 8 | 6.6 | 18.5 | 7.7 | 12.1 | 14.8 | 13.2 | 12.9 | 8.8 | 5.6 | 38.3 | 195,328 | 91.6 | 36.8 |
| District 9 | 6.1 | 14.6 | 7.7 | 19.1 | 13.9 | 12.7 | 11.8 | 8.1 | 5.7 | 36.3 | 168,365 | 90.0 | 44.6 |
| District 10 | 6.3 | 16.3 | 9.3 | 15.3 | 13.4 | 12.0 | 12.6 | 9.2 | 5.8 | 37.0 | 181,218 | 91.5 | 27.9 |
| **WEST VIRGINIA** | 5.1 | 15.0 | 8.5 | 11.8 | 12.3 | 12.7 | 14.3 | 12.1 | 8.4 | 42.9 | 373,698 | 87.1 | 21.1 |
| District 1 | 5.0 | 13.9 | 10.4 | 11.9 | 12.1 | 12.6 | 13.9 | 11.8 | 8.5 | 42.3 | 131,972 | 89.3 | 23.3 |
| District 2 | 5.4 | 15.8 | 7.3 | 12.2 | 12.5 | 12.7 | 14.7 | 11.7 | 7.8 | 42.3 | 127,970 | 88.6 | 22.6 |
| District 3 | 4.9 | 15.1 | 7.7 | 11.2 | 12.2 | 12.6 | 14.4 | 13.1 | 8.8 | 44.2 | 113,756 | 83.3 | 17.0 |
| **WISCONSIN** | 5.6 | 16.0 | 9.3 | 12.6 | 12.3 | 12.3 | 14.1 | 10.3 | 7.2 | 39.9 | 1,384,039 | 92.8 | 31.3 |
| District 1 | 5.6 | 16.9 | 8.1 | 11.6 | 12.9 | 12.9 | 14.8 | 10.2 | 7.0 | 40.9 | 165,178 | 92.2 | 30.5 |
| District 2 | 5.5 | 15.4 | 11.7 | 14.3 | 13.2 | 12.0 | 12.6 | 9.3 | 6.0 | 37.2 | 199,184 | 95.0 | 44.9 |
| District 3 | 5.3 | 15.3 | 12.6 | 11.6 | 11.6 | 11.3 | 14.0 | 10.8 | 7.6 | 39.2 | 185,973 | 93.6 | 27.0 |
| District 4 | 7.0 | 18.0 | 10.8 | 17.2 | 12.5 | 11.2 | 11.1 | 7.3 | 4.8 | 32.8 | 194,895 | 87.0 | 29.3 |
| District 5 | 5.2 | 15.1 | 8.1 | 11.9 | 12.3 | 13.0 | 15.2 | 10.9 | 8.2 | 42.5 | 162,905 | 94.9 | 38.2 |
| District 6 | 5.2 | 15.7 | 8.6 | 11.9 | 11.8 | 12.9 | 14.6 | 11.1 | 8.0 | 42.0 | 159,840 | 93.2 | 28.3 |
| District 7 | 5.4 | 15.9 | 6.6 | 10.5 | 11.6 | 12.7 | 16.2 | 12.2 | 8.9 | 45.0 | 146,325 | 92.4 | 24.4 |
| District 8 | 5.7 | 16.5 | 8.1 | 12.3 | 12.2 | 12.8 | 14.7 | 10.4 | 7.4 | 40.8 | 169,739 | 93.0 | 27.2 |
| **WYOMING** | 5.7 | 17.4 | 9.5 | 13.3 | 12.7 | 11.0 | 13.3 | 10.4 | 6.6 | 38.1 | 146,180 | 94.5 | 29.1 |
| At Large | 5.7 | 17.4 | 9.5 | 13.3 | 12.7 | 11.0 | 13.3 | 10.4 | 6.6 | 38.1 | 146,180 | 94.5 | 29.1 |

1. All persons 3 years old and over enrolled in nursery school through college and graduate or professional school.   2. Persons 25 years old and over.

Items 15—27

# Table E. Congressional Districts 116th Congress — Households and Group Quarters

| STATE District | Households, 2019 | | | | | | Group Quarters, 2010 | | | | | |
| | Number | Average household size | Family households (percent) | Married couple family (percent) | Female family house-holder[1] | One person households (percent) | Total in group quarters, 2018 | Percent 65 years and over | Persons in correctional institutions | Persons in nursing facilities | Persons in college dormitories | Persons in military quarters |
| | 28 | 29 | 30 | 31 | 32 | 33 | 34 | 35 | 36 | 37 | 38 | 39 |
| **TEXAS — Cont'd** | | | | | | | | | | | | |
| District 26 | 306,921 | 2.96 | 74.5 | 60.4 | 9.3 | 20.3 | 11,348 | 14.1 | 1,179 | 1,636 | 6,475 | 0 |
| District 27 | 262,779 | 2.78 | 69.3 | 48.0 | 15.3 | 25.6 | 14,912 | 31.7 | 3,832 | 4,197 | 1,365 | 103 |
| District 28 | 221,494 | 3.46 | 77.9 | 52.2 | 18.2 | 19.1 | 6,763 | 32.3 | 3,252 | 2,888 | 740 | 146 |
| District 29 | 230,537 | 3.39 | 75.8 | 44.8 | 20.9 | 19.7 | 2,203 | 36.1 | 31 | 1,008 | 0 | 0 |
| District 30 | 272,334 | 2.84 | 62.3 | 36.6 | 19.7 | 31.6 | 17,792 | 9.2 | 11,108 | 2,286 | 2,507 | 0 |
| District 31 | 306,854 | 2.93 | 69.9 | 53.7 | 12.3 | 24.4 | 16,034 | 16.1 | 3,442 | 2,183 | 2,041 | 4,084 |
| District 32 | 291,460 | 2.65 | 63.0 | 48.4 | 9.9 | 30.1 | 5,212 | 38.4 | 22 | 2,226 | 2,437 | 0 |
| District 33 | 225,078 | 3.31 | 69.8 | 42.2 | 19.6 | 24.0 | 5,890 | 19.3 | 2,147 | 1,389 | 534 | 0 |
| District 34 | 216,930 | 3.21 | 75.9 | 49.6 | 20.1 | 20.5 | 17,121 | 13.7 | 13,244 | 2,881 | 2,180 | 41 |
| District 35 | 294,971 | 2.84 | 61.8 | 38.1 | 17.4 | 27.8 | 19,614 | 9.9 | 7,971 | 2,563 | 5,761 | 115 |
| District 36 | 268,143 | 2.76 | 71.3 | 51.3 | 12.5 | 24.8 | 17,024 | 16.7 | 12,140 | 2,926 | 0 | 0 |
| **UTAH** | 1,023,855 | 3.08 | 74.3 | 60.5 | 9.1 | 19.2 | 48,364 | 10.4 | 12,666 | 5,854 | 15,666 | 523 |
| District 1 | 252,011 | 3.09 | 77.0 | 62.4 | 9.5 | 18.1 | 8,666 | 10.6 | 1,892 | 1,190 | 3,227 | 488 |
| District 2 | 268,916 | 2.88 | 68.6 | 55.9 | 7.8 | 23.7 | 14,940 | 12.1 | 3,459 | 1,763 | 3,443 | 35 |
| District 3 | 236,363 | 3.23 | 77.8 | 65.0 | 8.7 | 15.5 | 16,540 | 7.5 | 441 | 1,279 | 8,960 | 0 |
| District 4 | 266,565 | 3.16 | 74.4 | 59.3 | 10.4 | 18.9 | 8,218 | 12.4 | 6,874 | 1,622 | 36 | 0 |
| **VERMONT** | 262,767 | 2.28 | 59.4 | 46.1 | 8.8 | 31.6 | 25,701 | 15.0 | 1,592 | 3,588 | 16,895 | 5 |
| At Large | 262,767 | 2.28 | 59.4 | 46.1 | 8.8 | 31.6 | 25,701 | 15.0 | 1,592 | 3,588 | 16,895 | 5 |
| **VIRGINIA** | 3,191,847 | 2.60 | 65.6 | 49.8 | 11.4 | 27.7 | 244,897 | 12.0 | 1,592 | 30,324 | 84,048 | 37,568 |
| District 1 | 285,625 | 2.83 | 73.6 | 59.1 | 10.4 | 20.9 | 16,505 | 13.8 | 4,215 | 2,568 | 6,380 | 3,351 |
| District 2 | 279,179 | 2.47 | 66.3 | 50.9 | 11.9 | 26.5 | 34,794 | 11.2 | 1,685 | 2,534 | 7,086 | 7,370 |
| District 3 | 294,139 | 2.49 | 61.4 | 37.2 | 19.3 | 31.5 | 28,105 | 5.4 | 6,045 | 3,045 | 11,014 | 24,287 |
| District 4 | 288,854 | 2.55 | 60.2 | 41.6 | 14.6 | 32.5 | 31,452 | 10.2 | 19,352 | 2,662 | 2,942 | 316 |
| District 5 | 295,095 | 2.39 | 64.4 | 47.2 | 12.6 | 29.2 | 31,286 | 13.8 | 10,972 | 4,052 | 12,206 | 0 |
| District 6 | 291,647 | 2.47 | 63.5 | 47.3 | 10.8 | 29.9 | 34,839 | 12.6 | 4,623 | 4,349 | 19,615 | 1,398 |
| District 7 | 296,638 | 2.67 | 68.6 | 53.9 | 10.1 | 26.1 | 11,698 | 19.8 | 4,500 | 2,649 | 3,721 | 0 |
| District 8 | 326,597 | 2.47 | 55.5 | 43.3 | 8.6 | 33.7 | 7,433 | 28.6 | 944 | 2,026 | 1,084 | 846 |
| District 9 | 277,983 | 2.41 | 62.3 | 47.5 | 10.5 | 31.9 | 34,858 | 13.1 | 10,091 | 4,288 | 14,085 | 0 |
| District 10 | 283,522 | 3.00 | 78.1 | 66.2 | 8.0 | 17.9 | 5,821 | 18.7 | 1,606 | 907 | 1,068 | 0 |
| District 11 | 272,568 | 2.87 | 69.7 | 55.8 | 8.9 | 23.5 | 8,106 | 13.6 | 1,207 | 1,244 | 4,847 | 0 |
| **WASHINGTON** | 2,932,477 | 2.55 | 64.2 | 49.4 | 9.9 | 26.6 | 147,820 | 18.6 | 31,960 | 22,156 | 35,534 | 12,385 |
| District 1 | 288,473 | 2.71 | 71.4 | 59.4 | 7.3 | 20.9 | 8,815 | 23.7 | 2,676 | 1,313 | 620 | 2,504 |
| District 2 | 296,423 | 2.51 | 63.3 | 47.6 | 11.0 | 26.5 | 14,888 | 20.5 | 1,754 | 2,323 | 3,862 | 14 |
| District 3 | 288,096 | 2.60 | 67.8 | 52.3 | 9.9 | 24.5 | 7,286 | 35.8 | 1,539 | 1,617 | 0 | 4 |
| District 4 | 250,101 | 2.90 | 71.5 | 49.9 | 15.0 | 23.5 | 10,198 | 25.7 | 4,028 | 2,246 | 382 | 513 |
| District 5 | 295,592 | 2.39 | 60.9 | 46.7 | 9.8 | 29.2 | 28,948 | 11.8 | 5,656 | 2,927 | 13,048 | 5,694 |
| District 6 | 294,537 | 2.40 | 64.3 | 49.2 | 10.3 | 27.6 | 19,881 | 16 | 8,063 | 3,211 | 1,022 | 362 |
| District 7 | 366,275 | 2.16 | 47.5 | 38.8 | 5.8 | 36.4 | 25,148 | 13.6 | 1,851 | 2,700 | 10,456 | 0 |
| District 8 | 273,857 | 2.79 | 71.8 | 57.7 | 9.7 | 21.7 | 6,841 | 24.8 | 839 | 1,305 | 2,155 | 0 |
| District 9 | 290,396 | 2.55 | 63.2 | 47.1 | 10.6 | 27.2 | 10,876 | 26.9 | 3,031 | 2,422 | 1,348 | 3,294 |
| District 10 | 288,727 | 2.62 | 66.3 | 48.6 | 11.2 | 25.5 | 14,939 | 17.9 | 2,523 | 2,092 | 2,641 | |
| **WEST VIRGINIA** | 728,175 | 2.40 | 64.6 | 48.4 | 10.9 | 29.6 | 46,910 | 18.9 | 16,591 | 9,748 | 17,113 | 79 |
| District 1 | 239,559 | 2.42 | 62.7 | 48.4 | 8.9 | 29.8 | 21,099 | 15.8 | 6,965 | 3,632 | 10,300 | 0 |
| District 2 | 251,318 | 2.44 | 65.9 | 48.6 | 12.2 | 28.9 | 10,658 | 23.6 | 2,583 | 2,791 | 3,447 | 79 |
| District 3 | 237,298 | 2.33 | 65.3 | 48.3 | 11.5 | 30.3 | 15,153 | 20.1 | 7,043 | 3,325 | 3,366 | 0 |
| **WISCONSIN** | 2,386,623 | 2.38 | 61.6 | 47.8 | 9.3 | 30.5 | 143,849 | 23.1 | 38,102 | 33,808 | 56,773 | 132 |
| District 1 | 284,862 | 2.48 | 65.1 | 51.5 | 9.6 | 28.5 | 14,821 | 21.6 | 6,723 | 3,632 | 2,368 | 0 |
| District 2 | 320,423 | 2.36 | 57.9 | 46.0 | 7.9 | 30.8 | 17,421 | 19.7 | 2,269 | 2,791 | 8,805 | 0 |
| District 3 | 294,067 | 2.35 | 59.6 | 47.3 | 7.5 | 30.5 | 31,648 | 15.5 | 6,311 | 3,325 | 18,297 | 122 |
| District 4 | 279,444 | 2.46 | 54.5 | 30.2 | 18.7 | 36.2 | 16,773 | 17.5 | 2,539 | 3,173 | 9,814 | 0 |
| District 5 | 302,776 | 2.38 | 63.7 | 51.6 | 8.2 | 29.6 | 13,404 | 33.1 | 1,490 | 4,832 | 5,703 | 0 |
| District 6 | 297,909 | 2.33 | 61.9 | 49.7 | 7.7 | 31.3 | 23,177 | 20.2 | 12,174 | 5,078 | 6,097 | 0 |
| District 7 | 303,088 | 2.32 | 65.0 | 52.6 | 7.7 | 29.0 | 12,053 | 42.6 | 3,541 | 5,073 | 1,198 | 10 |
| District 8 | 304,054 | 2.37 | 64.8 | 52.2 | 8.1 | 28.3 | 14,552 | 31.5 | 3,055 | 4,742 | 4,491 | 503 |
| **WYOMING** | 233,128 | 2.42 | 65.6 | 53.0 | 8.1 | 27.4 | 14,240 | 18.4 | 3,576 | 2,450 | 4,443 | 503 |
| At Large | 233,128 | 2.42 | 65.6 | 53.0 | 8.1 | 27.4 | 14,240 | 18.4 | 3,576 | 2,450 | 4,443 | 503 |

1. No spouse present.

# Table E. Congressional Districts 116th Congress — Housing and Money Income

| STATE District | Housing units, 2019 | | | | | | Money income, 2019 | | |
| | Occupied units | | Owner-occupied | | | Renter-occupied | Per capita income (dollars) | Households | |
| | Total | Occupied units as a percent of all units | Owner-occupied units as a percent of occupied units | Median value[1] (dollars) | Percent valued at $500,000 or more | Median rent[2] | | Median income (dollars) | Percent with income of $100,000 or more |
| --- | --- | --- | --- | --- | --- | --- | --- | --- | --- |
| | 40 | 41 | 42 | 43 | 44 | 45 | 46 | 47 | 48 |
| **TEXAS — Cont'd** | | | | | | | | | |
| District 26 | 326,992 | 93.9 | 71.2 | 302,400 | 13.7 | 1,350 | 42,870 | 96,307 | 48.1 |
| District 27 | 326,076 | 80.6 | 66.0 | 151,000 | 4.1 | 1,007 | 26,912 | 55,987 | 24.7 |
| District 28 | 251,114 | 88.2 | 69.7 | 140,900 | 2.0 | 907 | 21,411 | 53,597 | 22.1 |
| District 29 | 255,183 | 90.3 | 52.3 | 133,000 | 2.4 | 978 | 19,124 | 48,300 | 17.5 |
| District 30 | 302,749 | 90.0 | 49.9 | 169,200 | 3.5 | 1,121 | 25,907 | 51,819 | 21.2 |
| District 31 | 334,759 | 91.7 | 61.6 | 259,900 | 7.8 | 1,165 | 34,862 | 75,813 | 36.8 |
| District 32 | 323,786 | 90.0 | 57.2 | 285,900 | 21.8 | 1,276 | 46,866 | 76,464 | 38.8 |
| District 33 | 248,907 | 90.4 | 48.3 | 139,400 | 2.0 | 985 | 18,672 | 45,997 | 13.8 |
| District 34 | 263,749 | 82.2 | 67.3 | 86,500 | 1.4 | 779 | 19,016 | 42,092 | 17.7 |
| District 35 | 322,887 | 91.4 | 51.3 | 184,800 | 2.6 | 1,112 | 25,763 | 53,898 | 21.0 |
| District 36 | 310,457 | 86.4 | 72.4 | 158,600 | 2.4 | 982 | 30,030 | 62,206 | 29.7 |
| **UTAH** | 1,133,543 | 90.3 | 70.6 | 330,300 | 18.3 | 1,098 | 31,771 | 75,780 | 34.7 |
| District 1 | 288,417 | 87.4 | 74.6 | 288,400 | 14.0 | 963 | 30,730 | 73,964 | 32.1 |
| District 2 | 306,004 | 87.9 | 66.2 | 305,700 | 16.3 | 1,041 | 32,217 | 69,852 | 30.9 |
| District 3 | 257,365 | 91.8 | 69.9 | 376,900 | 26.8 | 1,139 | 32,613 | 78,931 | 38.3 |
| District 4 | 281,757 | 94.6 | 72.0 | 350,300 | 17.2 | 1,204 | 31,551 | 80,918 | 37.8 |
| **VERMONT** | 339,412 | 77.4 | 70.9 | 233,200 | 7.4 | 980 | 35,702 | 63,001 | 28.4 |
| At Large | 339,412 | 77.4 | 70.9 | 233,200 | 7.4 | 980 | 35,702 | 63,001 | 28.4 |
| **VIRGINIA** | 3,562,258 | 89.6 | 66.1 | 288,800 | 22.0 | 1,254 | 40,635 | 76,456 | 37.9 |
| District 1 | 322,345 | 88.6 | 77.6 | 332,100 | 16.7 | 1,330 | 39,688 | 90,181 | 44.4 |
| District 2 | 309,423 | 90.2 | 62.7 | 284,000 | 14.1 | 1,293 | 38,025 | 74,704 | 34.0 |
| District 3 | 323,372 | 91.0 | 52.5 | 210,800 | 4.4 | 1,074 | 30,193 | 56,455 | 22.8 |
| District 4 | 320,159 | 90.2 | 62.2 | 221,300 | 8.3 | 1,087 | 33,720 | 60,407 | 28.3 |
| District 5 | 356,946 | 82.7 | 70.7 | 201,800 | 12.7 | 849 | 33,362 | 57,615 | 26.4 |
| District 6 | 333,693 | 87.4 | 66.1 | 200,000 | 5.5 | 869 | 30,459 | 59,939 | 25.0 |
| District 7 | 320,043 | 92.7 | 72.4 | 282,500 | 13.1 | 1,239 | 40,092 | 81,364 | 40.6 |
| District 8 | 345,964 | 94.4 | 51.4 | 588,000 | 59.7 | 1,863 | 61,204 | 111,627 | 55.3 |
| District 9 | 346,711 | 80.2 | 70.4 | 135,600 | 3.4 | 688 | 26,170 | 46,909 | 17.1 |
| District 10 | 299,501 | 94.7 | 78.3 | 529,100 | 53.6 | 1,729 | 56,736 | 132,226 | 63.4 |
| District 11 | 284,101 | 95.9 | 65.2 | 500,100 | 50.0 | 1,895 | 52,077 | 118,099 | 58.6 |
| **WASHINGTON** | 3,195,098 | 91.8 | 63.1 | 387,600 | 33.0 | 1,359 | 41,521 | 78,687 | 38.8 |
| District 1 | 311,102 | 92.7 | 72.7 | 552,700 | 55.8 | 1,685 | 51,190 | 106,190 | 52.9 |
| District 2 | 326,558 | 90.8 | 62.1 | 430,100 | 34.6 | 1,386 | 37,968 | 75,095 | 35.3 |
| District 3 | 311,463 | 92.5 | 67.3 | 338,300 | 17.2 | 1,206 | 35,075 | 70,936 | 32.7 |
| District 4 | 277,182 | 90.2 | 65.9 | 234,100 | 7.4 | 896 | 27,707 | 59,872 | 26.1 |
| District 5 | 322,282 | 91.7 | 62.5 | 255,100 | 9.4 | 917 | 31,736 | 57,837 | 24.8 |
| District 6 | 334,163 | 88.1 | 69.1 | 351,000 | 22.6 | 1,220 | 37,520 | 72,169 | 33.3 |
| District 7 | 395,094 | 92.7 | 48.0 | 718,500 | 76.8 | 1,736 | 62,693 | 100,991 | 50.9 |
| District 8 | 300,884 | 91.0 | 74.4 | 442,900 | 39.0 | 1,454 | 44,140 | 95,968 | 48.1 |
| District 9 | 310,785 | 93.4 | 52.8 | 536,400 | 53.3 | 1,679 | 49,560 | 90,353 | 45.8 |
| District 10 | 305,585 | 94.5 | 61.1 | 335,900 | 14.2 | 1,326 | 34,776 | 75,114 | 34.3 |
| **WEST VIRGINIA** | 894,983 | 81.4 | 73.4 | 124,600 | 2.3 | 727 | 27,446 | 48,850 | 19.1 |
| District 1 | 291,928 | 82.1 | 72.8 | 133,000 | 2.9 | 717 | 28,450 | 51,480 | 21.1 |
| District 2 | 302,268 | 83.1 | 75.0 | 149,900 | 2.1 | 808 | 28,947 | 52,166 | 21.8 |
| District 3 | 300,787 | 78.9 | 72.4 | 94,100 | 1.9 | 688 | 24,733 | 42,553 | 14.3 |
| **WISCONSIN** | 2,725,153 | 87.6 | 67.2 | 197,200 | 5.9 | 867 | 34,568 | 64,168 | 28.1 |
| District 1 | 309,253 | 92.1 | 70.5 | 224,400 | 5.9 | 912 | 34,654 | 68,695 | 31.9 |
| District 2 | 340,644 | 94.1 | 61.5 | 262,400 | 10.0 | 1,051 | 39,318 | 72,036 | 33.8 |
| District 3 | 331,212 | 88.8 | 70.9 | 171,500 | 3.9 | 794 | 31,358 | 59,426 | 23.4 |
| District 4 | 309,678 | 90.2 | 44.4 | 148,300 | 3.7 | 871 | 26,909 | 47,421 | 18.8 |
| District 5 | 321,169 | 94.3 | 70.1 | 252,700 | 9.0 | 977 | 42,435 | 77,386 | 37.6 |
| District 6 | 329,866 | 90.3 | 70.4 | 174,700 | 5.8 | 785 | 34,796 | 63,251 | 26.4 |
| District 7 | 428,533 | 70.7 | 77.7 | 168,100 | 4.6 | 728 | 32,759 | 60,706 | 24.1 |
| District 8 | 354,798 | 85.7 | 71.1 | 180,000 | 3.8 | 771 | 33,671 | 65,346 | 28.0 |
| **WYOMING** | 280,281 | 83.2 | 71.9 | 235,200 | 9.7 | 822 | 34,104 | 65,003 | 27.9 |
| At Large | 280,281 | 83.2 | 71.9 | 235,200 | 9.7 | 822 | 34,104 | 65,003 | 27.9 |

1. Specified owner-occupied units; $1,000,000 represents $1,000,000   2. Specified renter-occupied units.

# Table E. Congressional Districts 116th Congress — Poverty, Labor Force, Employment, and Social Security

| STATE District | Poverty, 2019 | | | Civilian labor force, 2019 | | | Civilian employment,[2] 2019 | | | | Persons under 65 years of age with no health insurance, 2019 (percent) | Social Security beneficiaries, December 2020 | | Supplemental Security Income recipients, December 2019 |
|---|---|---|---|---|---|---|---|---|---|---|---|---|---|---|
| | | | | | Unemployment | | | Percent | | | | | | |
| | Persons below poverty level (percent) | Families below poverty level (percent) | Percent of households receiving food stamps in past 12 months | Total | Total | Rate[1] | Total | Management, business, science, and arts occupations | Service, sales, and office | Construction and production | | Number | Rate[3] | |
| | 49 | 50 | 51 | 52 | 53 | 54 | 55 | 56 | 57 | 58 | 59 | 60 | 61 | 62 |
| TEXAS — Cont'd | | | | | | | | | | | | | | |
| District 26 | 6.1 | 3.6 | 4.0 | 514,228 | 15,681 | 3.1 | 498,053 | 46.7 | 36.7 | 16.5 | 11.2 | 111,612 | 121.2 | 6,795 |
| District 27 | 16.8 | 13.3 | 13.4 | 346,553 | 17,129 | 5.0 | 326,655 | 29.2 | 40.1 | 30.6 | 18.5 | 144,262 | 193.5 | 19,878 |
| District 28 | 21.4 | 17.5 | 19.4 | 343,699 | 20,352 | 6.0 | 320,749 | 29.4 | 43.3 | 27.4 | 26.1 | 121,416 | 157.2 | 28,088 |
| District 29 | 22.2 | 19.3 | 18.5 | 359,214 | 21,462 | 6.0 | 337,546 | 19.2 | 38.9 | 41.9 | 32.3 | 81,790 | 104.3 | 21,088 |
| District 30 | 18.7 | 15.2 | 15.2 | 380,109 | 22,637 | 6.0 | 357,184 | 29.4 | 41.4 | 29.3 | 22.5 | 107,597 | 135.8 | 30,408 |
| District 31 | 8.4 | 6.2 | 7.5 | 489,322 | 26,802 | 5.7 | 443,300 | 43.7 | 38.5 | 17.8 | 11.3 | 139,242 | 152.0 | 12,505 |
| District 32 | 9.9 | 6.9 | 5.9 | 431,920 | 15,980 | 3.7 | 415,690 | 46.3 | 36.1 | 17.6 | 18.2 | 104,633 | 134.5 | 11,171 |
| District 33 | 20.1 | 17.5 | 18.6 | 363,740 | 21,488 | 5.9 | 342,181 | 17.2 | 39.8 | 43.0 | 31.9 | 83,289 | 110.9 | 21,542 |
| District 34 | 24.7 | 21.7 | 24.4 | 299,107 | 14,960 | 5.0 | 282,923 | 26.3 | 48.5 | 25.3 | 27.2 | 123,766 | 173.7 | 36,769 |
| District 35 | 18.4 | 14.5 | 15.3 | 454,589 | 18,311 | 4.1 | 433,582 | 32.0 | 43.4 | 24.7 | 21.0 | 111,015 | 129.4 | 22,625 |
| District 36 | 13.0 | 10.3 | 10.6 | 359,255 | 22,132 | 6.2 | 336,599 | 32.7 | 36.0 | 31.3 | 19.2 | 145,863 | 192.4 | 16,583 |
| UTAH | 8.9 | 6.1 | 5.5 | 1,656,987 | 52,422 | 3.2 | 1,598,530 | 40.4 | 37.3 | 22.3 | 9.5 | 430,247 | 134.2 | 31,519 |
| District 1 | 7.7 | 5.5 | 6.7 | 401,598 | 10,591 | 2.7 | 387,154 | 38.1 | 37.4 | 24.5 | 8.3 | 104,954 | 133.3 | 7,749 |
| District 2 | 10.1 | 6.4 | 5.9 | 401,470 | 10,905 | 2.7 | 389,942 | 39.4 | 37.5 | 23.1 | 11.1 | 127,417 | 161.6 | 8,956 |
| District 3 | 9.8 | 7.0 | 5.1 | 400,567 | 13,997 | 3.5 | 385,883 | 44.2 | 37.8 | 18.0 | 8.7 | 98,582 | 126.5 | 6,273 |
| District 4 | 8.1 | 5.6 | 4.5 | 453,352 | 16,929 | 3.7 | 435,551 | 40.0 | 36.6 | 23.4 | 10.0 | 99,294 | 116.8 | 8,541 |
| VERMONT | 10.2 | 5.8 | 10.0 | 343,455 | 11,552 | 3.4 | 331,247 | 42.9 | 36.1 | 20.9 | 4.5 | 156,005 | 250.0 | 14,947 |
| At Large | 10.2 | 5.8 | 10.0 | 343,455 | 11,552 | 3.4 | 331,247 | 42.9 | 36.1 | 20.9 | 4.5 | 156,005 | 250.0 | 14,947 |
| VIRGINIA | 9.9 | 6.5 | 7.8 | 4,537,998 | 175,528 | 4.0 | 4,229,399 | 45.4 | 36.1 | 18.5 | 7.8 | 1,585,194 | 185.7 | 155,200 |
| District 1 | 7.9 | 5.3 | 6.4 | 435,411 | 16,306 | 3.9 | 406,686 | 43.9 | 36.5 | 19.6 | 7.5 | 149,670 | 181.5 | 8,975 |
| District 2 | 8.5 | 5.5 | 6.2 | 410,270 | 14,606 | 4.1 | 346,272 | 43.9 | 38.3 | 17.8 | 6.4 | 134,997 | 186.5 | 10,425 |
| District 3 | 14.8 | 11.4 | 13.3 | 397,185 | 23,109 | 6.3 | 346,272 | 33.7 | 42.3 | 24.0 | 9.0 | 139,626 | 183.7 | 21,818 |
| District 4 | 12.3 | 7.4 | 11.9 | 403,726 | 20,613 | 5.2 | 373,694 | 39.1 | 39.0 | 21.9 | 7.9 | 149,888 | 195.1 | 22,010 |
| District 5 | 12.9 | 8.6 | 10.2 | 355,629 | 14,177 | 4.0 | 340,364 | 39.0 | 38.0 | 23.1 | 7.7 | 188,007 | 255.5 | 17,466 |
| District 6 | 12.2 | 7.3 | 8.8 | 380,600 | 12,427 | 3.3 | 367,673 | 35.4 | 39.2 | 25.4 | 7.6 | 177,921 | 235.7 | 16,825 |
| District 7 | 7.9 | 5.3 | 5.9 | 426,494 | 16,243 | 3.8 | 408,381 | 43.6 | 38.2 | 18.3 | 7.7 | 156,148 | 194.5 | 11,085 |
| District 8 | 8.4 | 6.3 | 3.6 | 494,246 | 14,815 | 3.1 | 463,471 | 60.2 | 29.0 | 10.9 | 8.8 | 83,914 | 103.1 | 7,597 |
| District 9 | 17.2 | 10.1 | 12.4 | 306,575 | 13,895 | 4.5 | 291,853 | 35.5 | 38.7 | 25.8 | 7.9 | 201,404 | 286.1 | 23,854 |
| District 10 | 3.5 | 2.2 | 3.6 | 466,489 | 13,056 | 2.8 | 451,458 | 57.2 | 30.5 | 12.3 | 6.4 | 111,255 | 129.7 | 5,991 |
| District 11 | 6.3 | 3.9 | 3.5 | 461,373 | 16,281 | 3.6 | 439,086 | 56.5 | 32.0 | 11.4 | 8.7 | 92,364 | 117.0 | 9,154 |
| WASHINGTON | 9.8 | 6.2 | 10.6 | 3,988,571 | 179,353 | 4.6 | 3,749,409 | 43.3 | 35.4 | 21.3 | 6.5 | 1,401,525 | 184.1 | 146,666 |
| District 1 | 6.3 | 4.1 | 6.3 | 421,675 | 14,716 | 3.5 | 406,234 | 51.9 | 29.5 | 18.6 | 5.1 | 123,781 | 156.4 | 7,456 |
| District 2 | 9.0 | 5.0 | 10.1 | 402,232 | 17,406 | 4.4 | 374,100 | 39.3 | 37.4 | 23.3 | 6.4 | 147,566 | 194.1 | 13,702 |
| District 3 | 10.1 | 6.8 | 12.3 | 366,865 | 18,201 | 5.0 | 347,514 | 35.3 | 38.6 | 26.1 | 6.0 | 167,564 | 221.4 | 16,902 |
| District 4 | 15.1 | 12.0 | 16.7 | 342,418 | 20,324 | 5.9 | 321,771 | 32.3 | 34.0 | 33.8 | 12.1 | 132,939 | 180.7 | 16,961 |
| District 5 | 14.0 | 7.4 | 14.8 | 354,871 | 21,230 | 6.1 | 329,581 | 40.0 | 39.9 | 20.2 | 5.8 | 163,876 | 223.2 | 19,686 |
| District 6 | 8.9 | 5.8 | 10.9 | 356,058 | 17,033 | 5.1 | 319,087 | 40.1 | 38.5 | 21.4 | 5.4 | 179,510 | 247.1 | 16,649 |
| District 7 | 8.6 | 3.7 | 7.1 | 515,596 | 18,023 | 3.5 | 496,578 | 60.1 | 30.2 | 9.7 | 5.0 | 110,478 | 135.1 | 12,472 |
| District 8 | 7.9 | 5.3 | 7.3 | 399,987 | 15,372 | 3.9 | 383,569 | 42.1 | 36.0 | 21.9 | 5.9 | 125,857 | 163.4 | 10,546 |
| District 9 | 8.7 | 6.0 | 10.1 | 426,991 | 16,989 | 4.0 | 409,295 | 46.4 | 35.4 | 18.2 | 6.8 | 104,930 | 139.6 | 15,270 |
| District 10 | 9.8 | 6.7 | 12.2 | 401,878 | 20,059 | 5.3 | 361,680 | 35.6 | 38.4 | 26.0 | 6.8 | 145,024 | 188.2 | 17,022 |
| WEST VIRGINIA | 16.0 | 11.1 | 16.9 | 795,104 | 47,424 | 6.0 | 744,981 | 33.3 | 40.6 | 26.1 | 6.6 | 479,303 | 267.4 | 69,208 |
| District 1 | 14.8 | 8.5 | 13.9 | 281,158 | 15,703 | 5.6 | 264,930 | 34.6 | 40.0 | 25.3 | 5.9 | 150,614 | 250.3 | 18,299 |
| District 2 | 15.0 | 11.2 | 15.1 | 287,634 | 14,196 | 5.0 | 271,620 | 32.9 | 40.6 | 26.5 | 6.9 | 160,416 | 257.5 | 19,150 |
| District 3 | 18.5 | 13.7 | 22.0 | 226,312 | 17,525 | 7.8 | 208,431 | 32.1 | 41.5 | 26.5 | 6.9 | 168,273 | 296.6 | 31,759 |
| WISCONSIN | 10.4 | 6.2 | 9.4 | 3,104,192 | 98,873 | 3.2 | 3,002,074 | 37.9 | 35.7 | 26.4 | 5.7 | 1,275,932 | 219.1 | 116,390 |
| District 1 | 8.7 | 5.5 | 10.0 | 374,407 | 15,878 | 4.3 | 357,686 | 37.2 | 36.2 | 26.7 | 5.5 | 157,899 | 218.8 | 13,638 |
| District 2 | 10.1 | 4.7 | 6.7 | 443,291 | 9,626 | 2.2 | 432,620 | 48.8 | 32.5 | 18.7 | 4.4 | 143,946 | 186.1 | 11,338 |
| District 3 | 11.4 | 5.9 | 8.6 | 379,534 | 11,101 | 2.9 | 368,118 | 34.3 | 36.4 | 29.3 | 6.1 | 167,205 | 231.2 | 12,059 |
| District 4 | 20.3 | 15.4 | 21.3 | 348,699 | 17,213 | 4.9 | 331,449 | 36.3 | 39.7 | 23.9 | 8.9 | 114,465 | 162.6 | 38,743 |
| District 5 | 6.4 | 3.7 | 5.8 | 413,081 | 13,107 | 3.2 | 399,974 | 43.2 | 34.3 | 22.5 | 3.6 | 160,893 | 219.4 | 7,764 |
| District 6 | 8.0 | 4.6 | 7.7 | 381,274 | 9,662 | 2.5 | 371,282 | 33.5 | 36.3 | 30.2 | 4.6 | 170,986 | 238.9 | 10,236 |
| District 7 | 9.9 | 6.0 | 9.2 | 365,488 | 10,681 | 2.9 | 354,597 | 33.7 | 34.6 | 31.7 | 6.5 | 193,557 | 270.9 | 11,455 |
| District 8 | 9.0 | 5.6 | 6.9 | 398,418 | 11,605 | 2.9 | 386,543 | 34.0 | 36.5 | 29.5 | 5.9 | 166,981 | 226.9 | 11,157 |
| WYOMING | 10.1 | 6.6 | 4.7 | 297,245 | 10,539 | 3.6 | 283,750 | 36.3 | 36.6 | 27.1 | 12.3 | 118,420 | 204.6 | 6,992 |
| At Large | 10.1 | 6.6 | 4.7 | 297,245 | 10,539 | 3.6 | 283,750 | 36.3 | 36.6 | 27.1 | 12.3 | 118,420 | 204.6 | 6,992 |

1. Percent of civilian labor force.   2. Persons 16 years old and over.   3. Per 1,000 resident population estimated in the 2019 American Community Survey.

# Table E. Congressional Districts 116th Congress — **Agriculture**

| STATE District | Agriculture 2017 | | | | | | | | | |
|---|---|---|---|---|---|---|---|---|---|---|
| | Land in farms | | | | Value of products sold | | | | Government payments | |
| | Number of farms | Acres | Average size of farm (acres) | Irrigated land (acres) | Total ($1,000) | Average per farm (dollars) | Percent from crops | Percent from livestock and poultry products | Total ($1,000) | Average per farm receiving payments (dollars) |
| | 63 | 64 | 65 | 66 | 67 | 68 | 69 | 70 | 71 | 72 |
| TEXAS — Cont'd | | | | | | | | | | |
| District 26 | 3,405 | 379,288 | 111 | 109,449 | 135,031 | 39,657 | 26.8 | 73.2 | 1,967 | 9,106 |
| District 27 | 12,006 | 4,509,855 | 376 | 1,246,336 | 1,211,354 | 100,896 | 56.1 | 43.9 | 54,669 | 28,533 |
| District 28 | 7,866 | 4,740,139 | 603 | 201,075 | 266,700 | 33,905 | 28.7 | 71.3 | 7,328 | 14,092 |
| District 29 | 118 | 1,957 | 17 | 176 | 2,006 | 17,000 | 89.2 | 10.8 | D | D |
| District 30 | 339 | 30,440 | 90 | 8,664 | 24,115 | 71,136 | 94.4 | 5.6 | 343 | 14,913 |
| District 31 | 4,972 | 1,029,430 | 207 | 316,810 | 191,223 | 38,460 | 54.8 | 45.2 | 10,186 | 14,326 |
| District 32 | 268 | 24,824 | 93 | 6,903 | 4,973 | 18,556 | 47.4 | 52.6 | 149 | 49,667 |
| District 33 | 105 | 3,913 | 37 | 1,382 | 595 | 5,667 | 86.1 | 13.8 | D | D |
| District 34 | 8,728 | 4,198,181 | 481 | 742,909 | 891,631 | 102,158 | 47.6 | 52.4 | 22,756 | 21,488 |
| District 35 | 1,119 | 279,641 | 250 | 55,770 | 50,929 | 45,513 | 76.2 | 23.8 | 1,015 | 11,154 |
| District 36 | 6,786 | 987,386 | 146 | 141,443 | 109,757 | 16,174 | 54.5 | 45.5 | 10,717 | 50,552 |
| UTAH | 18,409 | 10,811,604 | 587 | 1,062,894 | 1,838,609 | 99,876 | 30.5 | 69.5 | 27,868 | 12,633 |
| District 1 | 7,638 | 5,450,417 | 714 | 483,806 | 534,894 | 70,031 | 34.5 | 65.5 | 16,242 | 16,034 |
| District 2 | 4,838 | 2,447,277 | 506 | 379,948 | 964,666 | 199,394 | 23.8 | 76.2 | 6,629 | 9,156 |
| District 3 | 3,559 | 2,492,368 | 700 | 126,072 | 194,706 | 54,708 | 40.8 | 59.2 | 3,232 | 11,107 |
| District 4 | 2,374 | 421,542 | 178 | 73,068 | 144,343 | 60,802 | 46.7 | 53.3 | 1,766 | 9,921 |
| VERMONT | 6,808 | 1,193,437 | 175 | 417,925 | 780,968 | 114,713 | 24.0 | 76.0 | 5,698 | 8,355 |
| At Large | 6,808 | 1,193,437 | 175 | 417,925 | 780,968 | 114,713 | 24.0 | 76.0 | 5,698 | 8,355 |
| VIRGINIA | 43,225 | 7,797,979 | 180 | 2,613,010 | 3,960,501 | 91,625 | 34.4 | 65.6 | 60,805 | 10,122 |
| District 1 | 3,054 | 645,919 | 211 | 388,966 | 301,198 | 98,624 | 77.7 | 22.3 | 10,885 | 18,233 |
| District 2 | 610 | 149,494 | 245 | 108,539 | 274,025 | 449,221 | 46.7 | 53.3 | 3,964 | 26,079 |
| District 3 | 333 | 105,974 | 318 | 64,944 | 89,535 | 268,874 | 69.8 | 30.2 | 3,263 | 26,967 |
| District 4 | 1,879 | 581,527 | 309 | 348,136 | 278,687 | 148,317 | 77.8 | 22.2 | 18,864 | 24,123 |
| District 5 | 11,408 | 2,257,513 | 198 | 614,799 | 601,654 | 52,740 | 42.7 | 57.3 | 9,223 | 5,255 |
| District 6 | 8,091 | 1,278,122 | 158 | 392,804 | 1,481,922 | 183,157 | 9.4 | 90.6 | 5,475 | 6,411 |
| District 7 | 3,235 | 576,339 | 178 | 193,179 | 350,186 | 108,249 | 44.3 | 55.7 | 2,518 | 5,097 |
| District 8 | 27 | 1,634 | 61 | 201 | 378 | 14,000 | 92.1 | 7.7 | D | D |
| District 9 | 11,933 | 1,893,432 | 159 | 389,465 | 486,668 | 40,783 | 21.5 | 78.5 | 6,227 | 5,506 |
| District 10 | 2,629 | 305,569 | 116 | 111,745 | 95,792 | 36,437 | 66.0 | 34.0 | 378 | 3,231 |
| District 11 | 26 | 2,456 | 94 | 232 | 457 | 17,577 | 54.9 | 44.9 | D | D |
| WASHINGTON | 35,793 | 14,679,857 | 410 | 4,472,130 | 9,634,461 | 269,172 | 72.5 | 27.5 | 168,990 | 30,692 |
| District 1 | 3,760 | 184,504 | 49 | 92,737 | 609,749 | 162,167 | 50.0 | 50.0 | 2,098 | 6,812 |
| District 2 | 1,672 | 128,023 | 77 | 67,888 | 301,189 | 180,137 | 64.3 | 35.7 | 462 | 4,574 |
| District 3 | 5,744 | 907,479 | 158 | 201,363 | 409,081 | 71,219 | 45.1 | 54.9 | 5,633 | 14,297 |
| District 4 | 9,281 | 7,215,328 | 777 | 2,037,982 | 6,606,543 | 711,835 | 74.9 | 25.1 | 72,231 | 40,488 |
| District 5 | 7,248 | 5,707,399 | 787 | 1,939,306 | 963,198 | 132,892 | 90.6 | 9.4 | 85,143 | 32,081 |
| District 6 | 2,429 | 167,201 | 69 | 29,213 | 93,124 | 38,338 | 38.3 | 61.7 | 326 | 5,175 |
| District 7 | 210 | 2,788 | 13 | 721 | 2,850 | 13,571 | 76.5 | 23.5 | D | D |
| District 8 | 4,031 | 312,692 | 78 | 89,276 | 472,565 | 117,233 | 80.7 | 19.3 | 2,914 | 18,327 |
| District 9 | 73 | 884 | 12 | 150 | 1,013 | 13,877 | 88.6 | 11.4 | D | D |
| District 10 | 1,345 | 53,559 | 40 | 13,494 | 175,150 | 130,223 | 35.2 | 64.8 | 79 | 2,548 |
| WEST VIRGINIA | 23,622 | 3,662,178 | 155 | 736,151 | 754,279 | 31,931 | 20.3 | 79.7 | 9,094 | 4,853 |
| District 1 | 9,112 | 1,317,106 | 145 | 271,231 | 155,397 | 17,054 | 21.1 | 78.9 | D | D |
| District 2 | 8,567 | 1,406,851 | 164 | 298,139 | 437,526 | 51,071 | 16.5 | 83.5 | 3,636 | 4,722 |
| District 3 | 5,943 | 938,221 | 158 | 166,781 | 161,355 | 27,150 | 29.8 | 70.2 | D | D |
| WISCONSIN | 64,793 | 14,318,630 | 221 | 9,234,611 | 11,427,423 | 176,368 | 35.6 | 64.4 | 126,583 | 4,609 |
| District 1 | 2,798 | 574,614 | 205 | 472,422 | 470,615 | 168,197 | D | D | 13,871 | 10,871 |
| District 2 | 9,579 | 2,086,402 | 218 | 1,403,770 | 1,652,129 | 172,474 | 39.3 | 60.7 | 30,032 | 5,925 |
| District 3 | 18,254 | 4,142,195 | 227 | 2,319,178 | 2,783,186 | 152,470 | 40.8 | 59.2 | 31,987 | 3,951 |
| District 4 | 39 | 310 | 8 | 136 | 3,019 | 77,410 | D | D | D | D |
| District 5 | 3,120 | 647,716 | 208 | 495,739 | 686,285 | 219,963 | 39.0 | 61.0 | 5,237 | 4,007 |
| District 6 | 7,975 | 1,753,113 | 220 | 1,333,881 | 1,724,547 | 216,244 | 33.4 | 66.6 | 16,772 | 4,706 |
| District 7 | 15,205 | 3,428,587 | 225 | 1,960,317 | 2,249,708 | 147,958 | 33.3 | 66.7 | 13,222 | 2,949 |
| District 8 | 7,823 | 1,685,693 | 215 | 1,249,168 | 1,857,935 | 237,496 | 22.2 | 77.8 | 15,462 | 4,217 |
| WYOMING | 11,938 | 29,004,884 | 2,430 | 1,544,826 | 1,472,113 | 123,313 | 21.6 | 78.4 | 30,218 | 14,410 |
| At Large | 11,938 | 29,004,884 | 2,430 | 1,544,826 | 1,472,113 | 123,313 | 21.6 | 78.4 | 30,218 | 14,410 |

| STATE District | Number of establishments | Total | Private nonfarm employment and payroll, 2019 — Employment — Percent by selected industries | | | | | | | | | Annual payroll | |
| | | | Manufacturing | Construction | Wholesale trade | Retail trade | Health care and social assistance | Finance and Insurance | Real estate and rental and leasing | Professional, scientific, and technical services | Information | Total (mil dol) | Average per employee (dollars) |
| | 73 | 74 | 75 | 76 | 77 | 78 | 79 | 80 | 81 | 82 | 83 | 84 | 85 |
|---|---|---|---|---|---|---|---|---|---|---|---|---|---|
| **TEXAS — Cont'd** | | | | | | | | | | | | | |
| District 26 | 16,918 | 263,961 | 6.7 | 6.8 | 4.5 | 13.1 | 11.5 | 9.6 | 2.2 | 4.7 | 1.8 | 12,606 | 47,758 |
| District 27 | 15,956 | 250,397 | 9.0 | 9.7 | 4.6 | 14.9 | 18.6 | 2.9 | 2.1 | 4.2 | 1.2 | 11,269 | 45,003 |
| District 28 | 11,532 | 167,121 | 2.5 | 4.7 | 3.7 | 18.3 | 23.2 | 2.9 | 1.3 | 2.4 | 0.9 | 5,590 | 33,446 |
| District 29 | 10,830 | 220,616 | 14.9 | 13.9 | 7.0 | 11.6 | 6.0 | 2.3 | 1.5 | 5.5 | 0.8 | 11,600 | 52,581 |
| District 30 | 13,870 | 341,532 | 5.4 | 3.4 | 5.0 | 8.3 | 15.7 | 6.1 | 3.1 | 10.6 | 4.3 | 25,155 | 73,654 |
| District 31 | 17,026 | 266,486 | 5.7 | 7.4 | 2.8 | 16.0 | 18.2 | 4.6 | 1.4 | 9.3 | 2.4 | 13,312 | 49,956 |
| District 32 | 22,389 | 362,038 | 5.9 | 4.4 | 2.7 | 11.2 | 16.8 | 7.5 | 3.4 | 8.4 | 3.9 | 22,303 | 61,603 |
| District 33 | 12,930 | 296,202 | 15.1 | 9.2 | 10.4 | 8.8 | 12.8 | 1.7 | 1.6 | 3.6 | 1.3 | 14,772 | 49,872 |
| District 34 | 10,931 | 175,838 | 3.8 | 3.0 | 2.7 | 17.2 | 33.1 | 3.2 | 1.5 | 2.5 | 1.6 | 5,386 | 30,630 |
| District 35 | 16,690 | 335,765 | 7.4 | 6.8 | 7.0 | 13.6 | 11.3 | 2.8 | 1.9 | 5.1 | 3.9 | 15,333 | 45,666 |
| District 36 | 13,216 | 261,122 | 16.0 | 13.3 | 3.7 | 10.9 | 10.0 | 1.9 | 1.4 | 8.1 | 0.5 | 15,896 | 60,874 |
| **UTAH** | 83,924 | 1,373,876 | 9.4 | 7.0 | 4.4 | 11.4 | 10.9 | 5.3 | 1.5 | 7.0 | 4.1 | 67,688 | 49,268 |
| District 1 | 19,309 | 248,314 | 17.0 | 7.6 | 3.5 | 14.3 | 12.2 | 3.9 | 1.8 | 6.2 | 1.3 | 10,588 | 42,639 |
| District 2 | 22,827 | 409,136 | 10.0 | 6.2 | 5.4 | 10.0 | 11.3 | 6.0 | 1.4 | 6.5 | 2.6 | 21,659 | 52,938 |
| District 3 | 20,803 | 289,562 | 5.9 | 6.1 | 3.9 | 12.2 | 12.1 | 4.5 | 1.6 | 7.0 | 5.7 | 13,114 | 45,291 |
| District 4 | 20,477 | 342,073 | 8.4 | 9.8 | 5.2 | 12.9 | 11.0 | 6.4 | 1.7 | 8.7 | 6.8 | 17,788 | 52,001 |
| **VERMONT** | 20,829 | 261,196 | 11.8 | 5.4 | 4.2 | 14.5 | 19.0 | 3.5 | 1.2 | 4.6 | 3.1 | 11,885 | 45,502 |
| At Large | 20,829 | 261,196 | 11.8 | 5.4 | 4.2 | 14.5 | 19.0 | 3.5 | 1.2 | 4.6 | 3.1 | 11,885 | 45,502 |
| **VIRGINIA** | 203,467 | 3,455,993 | 7.0 | 5.6 | 3.1 | 12.5 | 13.6 | 4.8 | 1.6 | 14.3 | 2.9 | 197,418 | 57,123 |
| District 1 | 17,614 | 231,936 | 4.8 | 10.8 | 5.5 | 17.3 | 12.7 | 4.0 | 1.3 | 8.6 | 1.2 | 10,218 | 44,055 |
| District 2 | 16,680 | 235,095 | 4.9 | 5.8 | 2.1 | 15.3 | 12.5 | 5.6 | 2.9 | 10.0 | 1.8 | 9,706 | 41,287 |
| District 3 | 16,849 | 348,068 | 12.9 | 4.8 | 3.4 | 12.7 | 18.4 | 3.5 | 2.0 | 7.9 | 2.0 | 17,105 | 49,143 |
| District 4 | 16,149 | 286,985 | 9.1 | 6.8 | 4.9 | 12.5 | 15.2 | 4.7 | 1.8 | 6.5 | 1.0 | 14,811 | 51,610 |
| District 5 | 16,797 | 215,322 | 11.6 | 7.2 | 2.7 | 15.4 | 18.4 | 3.2 | 1.3 | 5.7 | 1.7 | 9,599 | 44,580 |
| District 6 | 18,291 | 320,771 | 13.9 | 5.5 | 3.3 | 12.7 | 16.0 | 4.2 | 1.2 | 3.9 | 1.2 | 13,410 | 41,806 |
| District 7 | 19,547 | 307,670 | 3.1 | 6.2 | 3.9 | 15.1 | 13.6 | 11.9 | 1.8 | 8.4 | 2.9 | 17,003 | 55,262 |
| District 8 | 20,433 | 377,867 | 1.0 | 3.7 | 1.4 | 9.4 | 9.4 | 2.8 | 2.0 | 27.4 | 3.3 | 27,625 | 73,107 |
| District 9 | 14,004 | 209,443 | 19.9 | 4.0 | 3.8 | 16.8 | 17.6 | 2.8 | 1.0 | 3.7 | 1.5 | 8,065 | 38,505 |
| District 10 | 23,961 | 374,272 | 5.1 | 7.4 | 2.5 | 11.2 | 10.8 | 2.8 | 1.3 | 20.5 | 3.7 | 23,697 | 63,315 |
| District 11 | 22,376 | 477,480 | 0.7 | 3.4 | 2.1 | 8.6 | 11.1 | 5.6 | 1.6 | 32.2 | 7.4 | 41,143 | 86,167 |
| **WASHINGTON** | 195,105 | 2,898,378 | 9.3 | 7.2 | 4.7 | 11.7 | 15.3 | 3.8 | 1.9 | 7.5 | 5.2 | 191,592 | 66,103 |
| District 1 | 20,212 | 296,193 | 9.8 | 10.0 | 4.6 | 8.4 | 9.5 | 2.2 | 1.5 | 10.4 | 18.4 | 27,177 | 91,753 |
| District 2 | 20,697 | 283,781 | 22.0 | 8.7 | 3.1 | 15.0 | 13.9 | 4.0 | 1.7 | 4.3 | 2.0 | 15,734 | 55,444 |
| District 3 | 17,239 | 218,636 | 12.4 | 10.0 | 4.2 | 13.2 | 16.4 | 3.5 | 1.6 | 4.9 | 1.7 | 11,103 | 50,785 |
| District 4 | 15,092 | 207,421 | 12.1 | 7.4 | 6.1 | 15.0 | 17.5 | 2.3 | 1.4 | 5.8 | 1.3 | 10,055 | 48,475 |
| District 5 | 17,775 | 245,652 | 9.0 | 6.1 | 5.1 | 13.6 | 22.2 | 5.0 | 1.8 | 4.2 | 2.3 | 11,663 | 47,476 |
| District 6 | 17,804 | 209,174 | 4.6 | 5.9 | 2.3 | 16.3 | 24.8 | 3.4 | 1.8 | 5.4 | 1.1 | 10,055 | 48,069 |
| District 7 | 29,369 | 510,512 | 3.4 | 4.4 | 3.4 | 8.1 | 13.4 | 5.6 | 2.6 | 12.5 | 7.2 | 45,591 | 89,304 |
| District 8 | 16,638 | 184,257 | 9.1 | 10.5 | 7.2 | 14.1 | 13.0 | 1.9 | 1.5 | 3.6 | 2.3 | 10,117 | 54,909 |
| District 9 | 23,884 | 483,168 | 9.6 | 5.7 | 6.6 | 8.8 | 13.5 | 3.6 | 2.0 | 6.8 | 6.6 | 36,207 | 74,936 |
| District 10 | 15,730 | 218,616 | 6.4 | 9.3 | 4.7 | 16.1 | 18.4 | 3.5 | 1.9 | 4.2 | 1.7 | 10,378 | 47,470 |
| **WEST VIRGINIA** | 35,795 | 554,433 | 8.8 | 4.7 | 3.3 | 14.5 | 24.0 | 3.0 | 1.2 | 4.5 | 1.5 | 23,907 | 43,119 |
| District 1 | 12,708 | 207,676 | 9.6 | 4.2 | 3.0 | 14.2 | 23.9 | 2.9 | 1.0 | 4.8 | 1.3 | 9,236 | 44,474 |
| District 2 | 12,006 | 181,491 | 9.6 | 5.6 | 3.7 | 14.3 | 22.2 | 3.4 | 1.3 | 4.6 | 1.6 | 7,895 | 43,502 |
| District 3 | 10,736 | 155,367 | 7.4 | 4.3 | 3.1 | 16.1 | 27.3 | 2.2 | 0.9 | 3.5 | 1.6 | 6,227 | 40,076 |
| **WISCONSIN** | 141,635 | 2,610,712 | 18.3 | 4.6 | 4.8 | 11.9 | 15.7 | 5.6 | 1.1 | 4.4 | 2.2 | 130,985 | 50,172 |
| District 1 | 15,526 | 266,132 | 20.8 | 4.7 | 5.6 | 16.5 | 13.5 | 2.3 | 1.0 | 2.8 | 1.0 | 11,984 | 45,028 |
| District 2 | 19,484 | 380,895 | 11.5 | 5.0 | 4.7 | 12.0 | 16.5 | 7.1 | 1.5 | 8.1 | 5.4 | 21,426 | 56,251 |
| District 3 | 16,916 | 283,202 | 18.8 | 4.3 | 4.1 | 13.7 | 16.9 | 5.3 | 1.0 | 2.6 | 1.9 | 12,513 | 44,185 |
| District 4 | 13,510 | 317,032 | 10.6 | 2.1 | 4.9 | 6.8 | 20.3 | 9.6 | 1.4 | 5.5 | 2.7 | 18,853 | 59,466 |
| District 5 | 21,316 | 411,460 | 16.3 | 5.2 | 6.7 | 11.7 | 15.4 | 5.1 | 1.3 | 5.0 | 1.8 | 22,062 | 53,619 |
| District 6 | 16,682 | 316,241 | 28.0 | 4.5 | 3.6 | 11.4 | 13.9 | 3.5 | 0.9 | 3.4 | 1.1 | 15,364 | 48,583 |
| District 7 | 19,167 | 259,018 | 24.5 | 5.1 | 3.7 | 13.9 | 18.0 | 4.0 | 0.6 | 2.4 | 1.1 | 11,023 | 42,558 |
| District 8 | 18,437 | 332,728 | 22.1 | 5.8 | 4.4 | 12.3 | 13.3 | 5.1 | 0.9 | 3.2 | 1.4 | 15,685 | 47,142 |
| **WYOMING** | 21,578 | 207,016 | 5.1 | 8.8 | 3.6 | 14.5 | 15.8 | 3.4 | 2.2 | 5.1 | 1.9 | 10,423 | 50,347 |
| At Large | 21,578 | 207,016 | 5.1 | 8.8 | 3.6 | 14.5 | 15.8 | 3.4 | 2.2 | 5.1 | 1.9 | 10,423 | 50,347 |

# APPENDIX A
# GEOGRAPHIC CONCEPTS AND CODES

## GEOGRAPHIC AREAS COVERED

*County and City Extra* presents data for states (Table A), states and counties (Table B), metropolitan areas (Table C), cities with populations of 25,000 or more in 2010 (Table D), and congressional districts (Table E).

## STATES AND COUNTIES

Data are presented for each of the 50 states, the District of Columbia, and the United States as a whole. The states are arranged alphabetically and counties in Table B are arranged alphabetically within each state. Data are presented for 3,143 counties and county equivalents.

## County equivalents

In Louisiana, the primary divisions of the state are known as parishes rather than counties. In Alaska, the county equivalents are the organized boroughs, together with the census areas that were developed for general statistical purposes by the state of Alaska and the U.S. Census Bureau. Four states—Maryland, Missouri, Nevada, and Virginia—have one or more incorporated places that are legally independent of any county and thus constitute primary divisions of their states. Within each state, independent cities are listed alphabetically following the list of counties. The District of Columbia is not divided into counties or county equivalents—data for the entire district are presented as a county equivalent. New York City contains five counties: Bronx, Kings, New York, Queens, and Richmond.

## County changes since the 2010 census

- The independent city of Bedford, Virginia changed to town status and was added to Bedford county, effective July 1, 2013.
- Wade Hampton Census Area, AK (02-270) changed its name and FIPS code to Kusilvak Census Area (02-158), effective July 1, 2015.
- Shannon County, SD (46-113) changed its name and FIPS code to Oglala Lakota County (46-102), effective May 1, 2015.
- Petersburg Borough, AK was created from part of Petersburg Census Area and part of Hoonah-Angoon Census Area. Petersburg Borough retains the FIPS code 02-195, formerly used by Petersburg Census Area, effective January 3, 2013.
- Prince of Wales-Hyder Census Area added part of the former Petersburg Census Area, effective January 3, 2013.
- Valdez-Cordova Census Area, AK (02-261) was split to form two new census areas, effective January 2, 2019.
- Chugach Census Area, AK (02-063) was created from part of the former Valdez-Cordova Census Area (02-261), effective January 2, 2019.
- Copper River Census Area, AK (02-066) was created from part of the former Valdez-Cordova Census Area (02-261), effective January 2, 2019.

## County changes since the 2000 census

- Broomfield County, CO, was created from parts of Adams, Boulder, Jefferson, and Weld Counties, effective November 15, 2001. The boundaries of Broomfield County reflect the boundaries of Broomfield city legally in effect on that date.
- Clifton Forge city, VA, formerly an independent city, became a town within Alleghany County, effective July 1, 2001.
- Effective June 20, 2007, the Skagway-Hoonah-Angoon Census Area in Alaska was divided into the Skagway Municipality and the Hoonah-Angoon Census Area.
- In May and June, 2008, the Wrangell-Petersburg and Prince of Wales-Outer Ketchikan Census Areas were dissolved and replaced by Wrangell City and Borough, Petersburg Census Area, and Prince of Wales Census Area. Some territory from the Prince of Wales Outer Ketchikan Census Area became part of the existing Ketchikan Gateway Borough.

## METROPOLITAN AREAS

Table C presents data for 384 metropolitan statistical areas and 31 metropolitan divisions, which are located within the 11 largest metropolitan statistical areas. The metropolitan statistical areas are listed alphabetically, and the metropolitan divisions are listed alphabetically under the metropolitan statistical area of which they are components.

The U.S. Office of Management and Budget (OMB) defines metropolitan and micropolitan statistical areas according to published standards. The major purpose of defining these areas is to enable all U.S. government agencies to use the same geographic definitions in tabulating and publishing data. The general concept of a metropolitan or micropolitan statistical area is that of a core area containing a substantial population nucleus, together with adjacent communities that have a high degree of economic and social integration with the core.

New delineations of these Core Based Statistical Areas (CBSAs) based on the 2010 census were released in February 2013 and updated in July 2015, August 2017, and September 2018. Table C in this book uses these 2018 delineations for metropolitan areas and metropolitan divisions. Micropolitan areas are not included in Table C. Many of the data items in Table C were released under the old scheme but have been aggregated from county data to the newly defined metropolitan areas. This results in a higher level of data suppression for those items. The September 2018 changes were far more substantial than those of previous years. Users should be aware of these changes. A complete list is included as Appendix D.

Appendix B lists the metropolitan areas and metropolitan divisions with their component counties and 2010 census populations and their 2020 estimated populations. Appendix C lists the metropolitan and micropolitan areas, together with their 2010 census populations and their 2020 estimated populations. Appendix C includes one new micropolitan area from an updated OMB Bulletin from March, 2020.

Standard definitions of metropolitan areas were first issued in 1949 by the Bureau of the Budget (the predecessor of OMB), under the designation "standard metropolitan area" (SMA). The term was changed to "standard metropolitan statistical area" (SMSA) in 1959, and to "metropolitan statistical area" (MSA) in 1983. The term "metropolitan area" (MA) was adopted in 1990 and referred collectively to metropolitan statistical areas (MSAs), consolidated metropolitan statistical areas (CMSAs), and primary metropolitan statistical areas (PMSAs). The term "core based statistical area" (CBSA) became effective in 2000 and refers collectively to metropolitan and micropolitan statistical areas.

The 2010 standards provide that each CBSA must contain at least one urban area of 10,000 or more population. Each metropolitan statistical area must have at least one urbanized area of 50,000 or more inhabitants. Each micropolitan statistical area must have at least one urban cluster of at least 10,000 but less than 50,000 people.

Under the standards, A metro area contains a core urban area of 50,000 or more population, and a micro area contains an urban core of at least 10,000 (but less than 50,000) population. Each metro or micro area consists of one or more counties and includes the counties containing the core urban area, as well as any adjacent counties that have a high degree of social and economic integration (as measured by commuting to work) with the urban core.

If specified criteria are met, a metropolitan statistical area containing a single core with a population of 2.5 million or more may be subdivided to form smaller groupings of counties referred to as "metropolitan divisions."

As of September 14, 2018, there were 384 metropolitan statistical areas and 542 micropolitan statistical areas in the United States. Table C includes the 384 metropolitan statistical areas and the 31 metropolitan divisions. The metropolitan areas and metropolitan divisions are listed in Appendix B with their 2010 census population counts. The metropolitan areas, metropolitan divisions, and micropolitan areas are listed in Appendix C with their 2010 census populations and their 2020 estimated populations. Appendix D lists the many changes made to metropolitan areas with the 2018 delineations. A new delineation in March, 2020 established Bluffton, IN as a new micropolitan area.

The largest city in each metropolitan or micropolitan statistical area is designated a "principal city." Additional cities qualify if specified requirements are met concerning population size and employment. The title of each metropolitan or micropolitan statistical area consists of the names of up to three of its principal cities and the name of each state into which the metropolitan or micropolitan statistical area extends. Titles of metropolitan divisions also typically are based on principal city names, but in certain cases consist of county names. The principal city need not be an incorporated place if it meets the requirements of population size and employment. Usually such a principal city is a census designated place in decennial census data, but it is not included in most other data sources and is not in Table D (cities) in this volume.

In view of the importance of cities and towns in New England, the 2010 standards also provide for a set of geographic areas that are defined using cities and towns in the six New England states. These New England city and town areas (NECTAs) are not included in this volume.

Appendix B lists the 384 metropolitan statistical areas, together with their component metropolitan divisions, where appropriate, the component counties of each area, and their 2010 census populations. Appendix C provides the same information for the 384 metropolitan areas delineated in September 2018 and it also includes the 543 micropolitan statistical areas delineated in March 2020. Maps showing the metropolitan and micropolitan areas within each state can be found at https://www.census.gov/geographies/reference-maps/2018/geo/cbsa.html

## CITIES.

Table D presents data for 1,437 cities with 2010 census populations of 25,000 or more. Corresponding data for states are also provided. The states are arranged alphabetically and the cities are ordered alphabetically within each state.

As used in this volume, the term *city* refers to places that have been incorporated as cities, boroughs, towns, or villages under the laws of their respective states. Towns in the New England states and New York are treated as minor civil divisions (MCDs) and are not included in the cities database. For Hawaii, data for the census designated places (CDPs) are included in the cities table, since the Census Bureau does not recognize any incorporated places in Hawaii. CDPs are delineated by the Census Bureau, in cooperation with states and localities, as statistical counterparts of incorporated places for purposes of the decennial census. CDPs comprise densely settled concentrations of population that are identifiable by name but are not legally incorporated as places.

Appendix E lists the 1,437 cities followed by the county where each city is located. If a city includes portions of more than one county, the population in each part is specified.

A consolidated city is an incorporated place that has combined its government functions with a county or subcounty entity but contains one or more other semi-independent incorporated places that continue to function as local governments within the consolidated government. Each consolidated city contains a core city, the area of a consolidated city not included in another separately incorporated place. The census geographic term for this core is the "balance" of the consolidated city. Thus the "balance" is essentially the core city of the consolidated government. This volume includes the consolidated city data where possible, but some data sources include numbers only for the "balance" and others do not specify which entity is represented.

Consolidated cities included in this volume are Milford, CT; Athens-Clarke County, GA; Augusta-Richmond County, GA; Indianapolis, IN; Louisville-Jefferson County, KY; Butte-Silver Bow, MT; and Nashville-Davidson, TN.

Appendix E lists these seven consolidated cities, followed by the component places and their 2010 census populations.

On January 1, 2014, Macon, Georgia consolidated with Bibb county. A small portion of Macon that was in Jones county was de-annexed. Data from years prior to 2015 represent the smaller Macon city rather than the consolidated Macon-Bibb County.

## CONGRESSIONAL DISTRICTS

The congressional districts shown in this volume are the districts used for the election of the 116th Congress, which convened in January 2019. These are the districts that were established following the 2010 Census and are based on population data from that census. Data are shown for the 435 regular districts plus the District of Columbia, which has a non-voting delegate, but no representative. Corresponding data for each state also are included. States are listed alphabetically and districts numerically within each state. A map showing congressional districts for the 113th Congress is included in Appendix D.

## GEOGRAPHIC CODES

Tables A, B, C, and D provide, in one or more columns at the beginning of the table, a geographic code or codes for each area.

In Table B (states and counties), a five-digit state and county code is given for each state and county. The first two digits indicate the state; the remaining three represent the county. Within each state, the counties are listed in order, beginning with 001, with even numbers usually omitted. Independent cities follow the counties and begin with the number 510. In the second column of Table B, a five-digit core based statistical area (CBSA) code is given for those counties that are within metropolitan and micropolitan areas. In Table A, a two-digit state code is provided. The state code is a sequential numbering, with some gaps, of the states and the District of Columbia in alphabetical order from Alabama (01) to Wyoming (56).

These codes have been established by the U.S. government as Federal Information Processing Standards and are often referred to as *FIPS codes*. They are used by U.S. government agencies and many other organizations for data presentation. The codes are provided in this volume for use in matching the data given here with other data sources in which counties are identified by FIPS code. The metro area codes will also enable the user to identify the metro area of which a county is a component. Table C (metropolitan areas) provides the same metro area codes for each metropolitan area, as well as metropolitan division codes where appropriate.

Table D (cities) provides, in the first column, a seven-digit state and place code. The first two digits identify the state and are the same as the FIPS codes described above. The remaining five digits are the place FIPS codes established by the U.S. government.

## INDEPENDENT CITIES

The following independent cities are not included in any county; their data are presented separately in this volume.

### MARYLAND
Baltimore (separate from Baltimore County)

### MISSOURI
St. Louis (separate from St. Louis County)

### NEVADA
Carson City

### VIRGINIA

| | |
|---|---|
| Alexandria | Manassas |
| Bristol | Manassas Park |
| Buena Vista | Martinsville |
| Charlottesville | Newport News |
| Chesapeake | Norfolk |
| Colonial Heights | Norton |
| Covington | Petersburg |
| Danville | Poquoson |
| Emporia | Portsmouth |
| Fairfax | Radford |
| Falls Church | Richmond |
| Franklin | Roanoke |
| Fredericksburg | Salem |
| Galax | Staunton |
| Hampton | Suffolk |
| Harrisonburg | Virginia Beach |
| Hopewell | Waynesboro |
| Lexington | Williamsburg |
| Lynchburg | Winchester |

## COUNTY TYPE

Table B (states and counties) provides, in the third column, a *county type* code that identifies each county by its metropolitan/nonmetropolitan status and its size. These are the "rural-urban continuum codes" developed by the Economic Research Service of the U.S. Department of Agriculture.

The 2013 rural-urban continuum codes form a classification scheme that distinguishes metropolitan counties by size and nonmetropolitan counties by degree of urbanization and proximity to metro areas. The standard OMB metro and nonmetro categories have been subdivided into three metro and six nonmetro categories, resulting in a nine-part county codification. This scheme was originally developed in 1974. The codes were updated in 1983, 1993, and 2003 and slightly revised in 1988. The 1988 revision was first published in 1990. This scheme allows researchers to break county data into finer residential groups, beyond metro and nonmetro, particularly for the analysis of trends in nonmetro areas that are related to population density and metro influence. The 2013 and 2003 rural-urban continuum codes are not directly comparable with the codes from previous years because of the new methodology used in developing the 2013 metropolitan areas.

**Metropolitan counties**
1. Counties in metro areas of 1 million population or more.
2. Counties in metro areas of 250,000 to 1 million population.
3. Counties in metro areas of fewer than 250,000 population.

**Nonmetropolitan counties**
4. Urban population of 20,000 or more, adjacent to a metro area.
5. Urban population of 20,000 or more, not adjacent to a metro area.
6. Urban population of 2,500 to 19,999, adjacent to a metro area.
7. Urban population of 2,500 to 19,999, not adjacent to a metro area.
8. Completely rural or less than 2,500 urban population, adjacent to a metro area.
9. Completely rural or less than 2,500 urban population, not adjacent to a metro area.

# APPENDIX B
# METROPOLITAN STATISTICAL AREAS, METROPOLITAN DIVISIONS, AND COMPONENTS
## (as defined September 2018)

| State/County FIPS code | Core based statistical area | Title and Geographic Components | 2010 census population | 2020 estimated population |
|---|---|---|---|---|
| 10180 | | Abilene, TX Metro area | 165,252 | 173,185 |
| 10180 | 48059 | Callahan County | 13,545 | 14,110 |
| 10180 | 48253 | Jones County | 20,192 | 19,875 |
| 10180 | 48441 | Taylor County | 131,515 | 139,200 |
| 10420 | | Akron, OH Metro area | 703,196 | 701,449 |
| 10420 | 39133 | Portage County | 161,424 | 162,583 |
| 10420 | 39153 | Summit County | 541,772 | 538,866 |
| 10500 | | Albany, GA Metro area | 154,033 | 145,206 |
| 10500 | 13095 | Dougherty County | 94,564 | 86,477 |
| 10500 | 13177 | Lee County | 28,295 | 30,234 |
| 10500 | 13273 | Terrell County | 9,507 | 8,523 |
| 10500 | 13321 | Worth County | 21,667 | 19,972 |
| 10540 | | Albany-Lebanon, OR Metro area | 116,681 | 131,054 |
| 10540 | 41043 | Linn County | 116,681 | 131,054 |
| 10580 | | Albany-Schenectady-Troy, NY Metro area.. | 870,713 | 878,550 |
| 10580 | 36001 | Albany County | 304,208 | 303,654 |
| 10580 | 36083 | Rensselaer County | 159,433 | 158,108 |
| 10580 | 36091 | Saratoga County | 219,598 | 230,298 |
| 10580 | 36093 | Schenectady County | 154,751 | 155,358 |
| 10580 | 36095 | Schoharie County | 32,723 | 31,132 |
| 10740 | | Albuquerque, NM Metro area | 887,063 | 923,630 |
| 10740 | 35001 | Bernalillo County | 662,477 | 681,666 |
| 10740 | 35043 | Sandoval County | 131,621 | 148,904 |
| 10740 | 35057 | Torrance County | 16,380 | 15,486 |
| 10740 | 35061 | Valencia County | 76,585 | 77,574 |
| 10780 | | Alexandria, LA Metro area | 153,918 | 150,821 |
| 10780 | 22043 | Grant Parish | 22,309 | 22,254 |
| 10780 | 22079 | Rapides Parish | 131,609 | 128,567 |
| 10900 | | Allentown-Bethlehem-Easton, PA-NJ Metro area | 821,273 | 846,399 |
| 10900 | 34041 | Warren County | 108,642 | 105,624 |
| 10900 | 42025 | Carbon County | 65,244 | 64,081 |
| 10900 | 42077 | Lehigh County | 349,675 | 370,802 |
| 10900 | 42095 | Northampton County | 297,712 | 305,892 |
| 11020 | | Altoona, PA Metro area | 127,117 | 121,007 |
| 11020 | 42013 | Blair County | 127,117 | 121,007 |
| 11100 | | Amarillo, TX Metro area | 251,935 | 265,761 |
| 11100 | 48011 | Armstrong County | 1,901 | 1,869 |
| 11100 | 48065 | Carson County | 6,184 | 5,854 |
| 11100 | 48359 | Oldham County | 2,052 | 2,135 |
| 11100 | 48375 | Potter County | 121,078 | 116,004 |
| 11100 | 48381 | Randall County | 120,720 | 139,899 |
| 11180 | | Ames, IA Metro area | 115,850 | 124,514 |
| 11180 | 19015 | Boone County | 26,308 | 26,277 |
| 11180 | 19169 | Story County | 89,542 | 98,237 |
| 11260 | | Anchorage, AK Metro area | 380,821 | 397,308 |
| 11260 | 02020 | Anchorage Municipality | 291,836 | 287,095 |
| 11260 | 02170 | Matanuska-Susitna Borough | 88,985 | 110,213 |
| 11460 | | Ann Arbor, MI Metro area | 345,163 | 366,473 |
| 11460 | 26161 | Washtenaw County | 345,163 | 366,473 |
| 11500 | | Anniston-Oxford, AL Metro area | 118,526 | 113,469 |
| 11500 | 01015 | Calhoun County | 118,526 | 113,469 |
| 11540 | | Appleton, WI Metro area | 225,667 | 238,975 |
| 11540 | 55015 | Calumet County | 48,981 | 50,209 |
| 11540 | 55087 | Outagamie County | 176,686 | 188,766 |
| 11700 | | Asheville, NC Metro area | 424,863 | 466,634 |
| 11700 | 37021 | Buncombe County | 238,330 | 263,477 |
| 11700 | 37087 | Haywood County | 59,032 | 62,972 |
| 11700 | 37089 | Henderson County | 106,719 | 118,445 |
| 11700 | 37115 | Madison County | 20,782 | 21,740 |

| State/County FIPS code | Core based statistical area | Title and Geographic Components | 2010 census population | 2020 estimated population |
|---|---|---|---|---|
| 12020 | | Athens-Clarke County, GA Metro area | 192,567 | 214,759 |
| 12020 | 13059 | Clarke County | 116,688 | 127,795 |
| 12020 | 13195 | Madison County | 28,167 | 30,457 |
| 12020 | 13219 | Oconee County | 32,831 | 41,124 |
| 12020 | 13221 | Oglethorpe County | 14,881 | 15,383 |
| 12060 | | Atlanta-Sandy Springs-Alpharetta, GA Metro area | 5,286,718 | 6,087,762 |
| 12060 | 13013 | Barrow County | 69,356 | 85,588 |
| 12060 | 13015 | Bartow County | 100,092 | 109,426 |
| 12060 | 13035 | Butts County | 23,695 | 25,426 |
| 12060 | 13045 | Carroll County | 110,570 | 121,633 |
| 12060 | 13057 | Cherokee County | 214,381 | 265,274 |
| 12060 | 13063 | Clayton County | 259,630 | 292,646 |
| 12060 | 13067 | Cobb County | 688,065 | 762,944 |
| 12060 | 13077 | Coweta County | 127,369 | 150,849 |
| 12060 | 13085 | Dawson County | 22,382 | 27,113 |
| 12060 | 13089 | DeKalb County | 691,961 | 762,009 |
| 12060 | 13097 | Douglas County | 132,282 | 147,988 |
| 12060 | 13113 | Fayette County | 106,560 | 115,821 |
| 12060 | 13117 | Forsyth County | 175,484 | 250,847 |
| 12060 | 13121 | Fulton County | 920,445 | 1,077,402 |
| 12060 | 13135 | Gwinnett County | 805,286 | 942,627 |
| 12060 | 13143 | Haralson County | 28,775 | 30,383 |
| 12060 | 13149 | Heard County | 11,829 | 11,973 |
| 12060 | 13151 | Henry County | 203,777 | 239,139 |
| 12060 | 13159 | Jasper County | 13,894 | 14,483 |
| 12060 | 13171 | Lamar County | 18,311 | 19,261 |
| 12060 | 13199 | Meriwether County | 21,978 | 21,164 |
| 12060 | 13211 | Morgan County | 17,871 | 19,636 |
| 12060 | 13217 | Newton County | 99,967 | 113,295 |
| 12060 | 13223 | Paulding County | 142,397 | 173,359 |
| 12060 | 13227 | Pickens County | 29,404 | 33,127 |
| 12060 | 13231 | Pike County | 17,873 | 19,121 |
| 12060 | 13247 | Rockdale County | 85,169 | 90,939 |
| 12060 | 13255 | Spalding County | 64,109 | 67,414 |
| 12060 | 13297 | Walton County | 83,806 | 96,875 |
| 12100 | | Atlantic City-Hammonton, NJ Metro area.... | 274,525 | 262,945 |
| 12100 | 34001 | Atlantic County | 274,525 | 262,945 |
| 12220 | | Auburn-Opelika, AL Metro area | 140,287 | 166,831 |
| 12220 | 01081 | Lee County | 140,287 | 166,831 |
| 12260 | | Augusta-Richmond County, GA-SC Metro area | 564,893 | 614,312 |
| 12260 | 13033 | Burke County | 23,326 | 22,648 |
| 12260 | 13073 | Columbia County | 124,016 | 160,377 |
| 12260 | 13181 | Lincoln County | 7,996 | 8,031 |
| 12260 | 13189 | McDuffie County | 21,867 | 21,162 |
| 12260 | 13245 | Richmond County | 200,594 | 202,079 |
| 12260 | 45003 | Aiken County | 160,129 | 172,895 |
| 12260 | 45037 | Edgefield County | 26,965 | 27,120 |
| 12420 | | Austin-Round Rock-Georgetown, TX Metro area | 1,716,323 | 2,295,303 |
| 12420 | 48021 | Bastrop County | 74,217 | 91,601 |
| 12420 | 48055 | Caldwell County | 38,055 | 43,979 |
| 12420 | 48209 | Hays County | 157,103 | 241,365 |
| 12420 | 48453 | Travis County | 1,024,444 | 1,300,503 |
| 12420 | 48491 | Williamson County | 422,504 | 617,855 |
| 12540 | | Bakersfield, CA Metro area | 839,621 | 901,362 |
| 12540 | 06029 | Kern County | 839,621 | 901,362 |
| 12580 | | Baltimore-Columbia-Towson, MD Metro area | 2,710,598 | 2,800,189 |
| 12580 | 24003 | Anne Arundel County | 537,631 | 582,777 |
| 12580 | 24005 | Baltimore County | 805,324 | 826,017 |
| 12580 | 24013 | Carroll County | 167,141 | 169,092 |
| 12580 | 24025 | Harford County | 244,824 | 256,805 |
| 12580 | 24027 | Howard County | 287,123 | 328,200 |
| 12580 | 24035 | Queen Anne's County | 47,785 | 51,167 |
| 12580 | 24510 | Baltimore city | 620,770 | 586,131 |
| 12620 | | Bangor, ME Metro area | 153,931 | 151,655 |
| 12620 | 23019 | Penobscot County | 153,931 | 151,655 |

| State/County FIPS code | Core based statistical area | Title and Geographic Components | 2010 census population | 2020 estimated population |
|---|---|---|---|---|
| 12700 | | Barnstable Town, MA Metro area | 215,880 | 213,164 |
| 12700 | 25001 | Barnstable County | 215,880 | 213,164 |
| 12940 | | Baton Rouge, LA Metro area | 825,917 | 858,571 |
| 12940 | 22005 | Ascension Parish | 107,215 | 128,665 |
| 12940 | 22007 | Assumption Parish | 23,416 | 21,621 |
| 12940 | 22033 | East Baton Rouge Parish | 440,525 | 439,729 |
| 12940 | 22037 | East Feliciana Parish | 20,260 | 18,882 |
| 12940 | 22047 | Iberville Parish | 33,404 | 32,070 |
| 12940 | 22063 | Livingston Parish | 127,674 | 143,737 |
| 12940 | 22077 | Pointe Coupee Parish | 22,808 | 21,529 |
| 12940 | 22091 | St. Helena Parish | 11,205 | 10,081 |
| 12940 | 22121 | West Baton Rouge Parish | 23,785 | 26,792 |
| 12940 | 22125 | West Feliciana Parish | 15,625 | 15,465 |
| 12980 | | Battle Creek, MI Metro area | 136,150 | 133,580 |
| 12980 | 26025 | Calhoun County | 136,150 | 133,580 |
| 13020 | | Bay City, MI Metro area | 107,773 | 102,387 |
| 13020 | 26017 | Bay County | 107,773 | 102,387 |
| 13140 | | Beaumont-Port Arthur, TX Metro area | 388,749 | 391,310 |
| 13140 | 48199 | Hardin County | 54,635 | 58,305 |
| 13140 | 48245 | Jefferson County | 252,277 | 250,127 |
| 13140 | 48361 | Orange County | 81,837 | 82,878 |
| 13220 | | Beckley, WV Metro area | 124,914 | 114,982 |
| 13220 | 54019 | Fayette County | 46,049 | 42,062 |
| 13220 | 54081 | Raleigh County | 78,865 | 72,920 |
| 13380 | | Bellingham, WA Metro area | 201,146 | 231,016 |
| 13380 | 53073 | Whatcom County | 201,146 | 231,016 |
| 13460 | | Bend, OR Metro area | 157,728 | 201,769 |
| 13460 | 41017 | Deschutes County | 157,728 | 201,769 |
| 13740 | | Billings, MT Metro area | 167,165 | 183,799 |
| 13740 | 30009 | Carbon County | 10,078 | 10,921 |
| 13740 | 30095 | Stillwater County | 9,096 | 9,888 |
| 13740 | 30111 | Yellowstone County | 147,991 | 162,990 |
| 13780 | | Binghamton, NY Metro area | 251,724 | 237,324 |
| 13780 | 36007 | Broome County | 200,675 | 189,420 |
| 13780 | 36107 | Tioga County | 51,049 | 47,904 |
| 13820 | | Birmingham-Hoover, AL Metro area | 1,061,039 | 1,091,921 |
| 13820 | 01007 | Bibb County | 22,915 | 22,136 |
| 13820 | 01009 | Blount County | 57,322 | 57,879 |
| 13820 | 01021 | Chilton County | 43,632 | 44,397 |
| 13820 | 01073 | Jefferson County | 658,567 | 655,342 |
| 13820 | 01115 | St. Clair County | 83,350 | 90,739 |
| 13820 | 01117 | Shelby County | 195,253 | 221,428 |
| 13900 | | Bismarck, ND Metro area | 110,625 | 129,641 |
| 13900 | 38015 | Burleigh County | 81,308 | 96,212 |
| 13900 | 38059 | Morton County | 27,469 | 31,503 |
| 13900 | 38065 | Oliver County | 1,848 | 1,926 |
| 13980 | | Blacksburg-Christiansburg, VA Metro area | 162,960 | 167,244 |
| 13980 | 51071 | Giles County | 17,286 | 16,663 |
| 13980 | 51121 | Montgomery County | 94,422 | 98,391 |
| 13980 | 51155 | Pulaski County | 34,857 | 33,935 |
| 13980 | 51750 | Radford city | 16,395 | 18,255 |
| 14010 | | Bloomington, IL Metro area | 169,577 | 171,256 |
| 14010 | 17113 | McLean County | 169,577 | 171,256 |
| 14020 | | Bloomington, IN Metro area | 159,535 | 169,052 |
| 14020 | 18105 | Monroe County | 137,962 | 148,219 |
| 14020 | 18119 | Owen County | 21,573 | 20,833 |
| 14100 | | Bloomsburg-Berwick, PA Metro area | 85,555 | 82,884 |
| 14100 | 42037 | Columbia County | 67,299 | 64,842 |
| 14100 | 42093 | Montour County | 18,256 | 18,042 |
| 14260 | | Boise City, ID Metro area | 616,566 | 770,353 |
| 14260 | 16001 | Ada County | 392,372 | 494,399 |
| 14260 | 16015 | Boise County | 7,027 | 8,065 |
| 14260 | 16027 | Canyon County | 188,922 | 237,053 |
| 14260 | 16045 | Gem County | 16,719 | 18,703 |
| 14260 | 16073 | Owyhee County | 11,526 | 12,133 |
| 14460 | | Boston-Cambridge-Newton, MA-NH Metro area | 4,552,595 | 4,878,211 |
| 14460 | | Boston, MA Div 14454 | 1,888,025 | 2,034,729 |
| 14460 | 25021 | Norfolk County | 670,910 | 709,409 |
| 14460 | 25023 | Plymouth County | 494,932 | 523,738 |
| 14460 | 25025 | Suffolk County | 722,183 | 801,582 |
| 14460 | | Cambridge-Newton-Framingham, MA Div 15764 | 2,246,215 | 2,400,642 |
| 14460 | 25009 | Essex County | 743,082 | 791,263 |
| 14460 | 25017 | Middlesex County | 1,503,133 | 1,609,379 |
| 14460 | | Rockingham County-Strafford County, NH Div 40484 | 418,355 | 442,840 |
| 14460 | 33015 | Rockingham County | 295,204 | 311,307 |
| 14460 | 33017 | Strafford County | 123,151 | 131,533 |
| 14500 | | Boulder, CO Metro area | 294,560 | 327,171 |
| 14500 | 08013 | Boulder County | 294,560 | 327,171 |
| 14540 | | Bowling Green, KY Metro area | 158,613 | 180,751 |
| 14540 | 21003 | Allen County | 19,958 | 21,303 |
| 14540 | 21031 | Butler County | 12,697 | 12,703 |
| 14540 | 21061 | Edmonson County | 12,177 | 12,235 |
| 14540 | 21227 | Warren County | 113,781 | 134,510 |
| 14740 | | Bremerton-Silverdale-Port Orchard, WA Metro area | 251,143 | 272,787 |
| 14740 | 53035 | Kitsap County | 251,143 | 272,787 |
| 14860 | | Bridgeport-Stamford-Norwalk, CT Metro area | 916,904 | 942,426 |
| 14860 | 09001 | Fairfield County | 916,904 | 942,426 |
| 15180 | | Brownsville-Harlingen, TX Metro area | 406,215 | 424,180 |
| 15180 | 48061 | Cameron County | 406,215 | 424,180 |
| 15260 | | Brunswick, GA Metro area | 112,385 | 119,157 |
| 15260 | 13025 | Brantley County | 18,428 | 19,202 |
| 15260 | 13127 | Glynn County | 79,627 | 85,568 |
| 15260 | 13191 | McIntosh County | 14,330 | 14,387 |
| 15380 | | Buffalo-Cheektowaga, NY Metro area | 1,135,614 | 1,125,637 |
| 15380 | 36029 | Erie County | 919,134 | 917,241 |
| 15380 | 36063 | Niagara County | 216,480 | 208,396 |
| 15500 | | Burlington, NC Metro area | 151,155 | 171,346 |
| 15500 | 37001 | Alamance County | 151,155 | 171,346 |
| 15540 | | Burlington-South Burlington, VT Metro area | 211,264 | 221,160 |
| 15540 | 50007 | Chittenden County | 156,535 | 164,306 |
| 15540 | 50011 | Franklin County | 47,759 | 49,685 |
| 15540 | 50013 | Grand Isle County | 6,970 | 7,169 |
| 15680 | | California-Lexington Park, MD Metro area | 105,144 | 114,687 |
| 15680 | 24037 | St. Mary's County | 105,144 | 114,687 |
| 15940 | | Canton-Massillon, OH Metro area | 404,425 | 396,669 |
| 15940 | 39019 | Carroll County | 28,834 | 26,897 |
| 15940 | 39151 | Stark County | 375,591 | 369,772 |
| 15980 | | Cape Coral-Fort Myers, FL Metro area | 618,755 | 790,767 |
| 15980 | 12071 | Lee County | 618,755 | 790,767 |
| 16020 | | Cape Girardeau, MO-IL Metro area | 96,270 | 97,120 |
| 16020 | 17003 | Alexander County | 8,238 | 5,497 |
| 16020 | 29017 | Bollinger County | 12,358 | 12,111 |
| 16020 | 29031 | Cape Girardeau County | 75,674 | 79,512 |

| State/ County FIPS code | Core based statistical area | Title and Geographic Components | 2010 census population | 2020 estimated population | State/ County FIPS code | Core based statistical area | Title and Geographic Components | 2010 census population | 2020 estimated population |
|---|---|---|---|---|---|---|---|---|---|
| 16060 | | Carbondale-Marion, IL Metro area .............. | 139,148 | 135,448 | 16980 | | Elgin, IL Div 20994 .................................. | 735,288 | 766,139 |
| 16060 | 17077 | Jackson County ........................ | 60,206 | 56,675 | 16980 | 17037 | DeKalb County ........................ | 105,162 | 104,491 |
| 16060 | 17087 | Johnson County ........................ | 12,577 | 12,358 | 16980 | 17089 | Kane County ........................ | 515,322 | 531,010 |
| 16060 | 17199 | Williamson County ........................ | 66,365 | 66,415 | 16980 | 17093 | Kendall County ........................ | 114,804 | 130,638 |
| 16180 | | Carson City, NV Metro area ...................... | 55,269 | 56,034 | 16980 | | Gary, IN Div 23844 .................................. | 708,117 | 705,863 |
| 16180 | 32510 | Carson City ........................ | 55,269 | 56,034 | 16980 | 18073 | Jasper County ........................ | 33,481 | 33,440 |
| | | | | | 16980 | 18089 | Lake County ........................ | 496,108 | 487,536 |
| 16220 | | Casper, WY Metro area............................. | 75,448 | 80,815 | 16980 | 18111 | Newton County ........................ | 14,239 | 13,907 |
| 16220 | 56025 | Natrona County ........................ | 75,448 | 80,815 | 16980 | 18127 | Porter County ........................ | 164,289 | 170,980 |
| 16300 | | Cedar Rapids, IA Metro area...................... | 257,948 | 273,885 | 16980 | | Lake County-Kenosha County, IL-WI Div 29404 | 869,824 | 863,264 |
| 16300 | 19011 | Benton County ........................ | 26,069 | 25,414 | 16980 | 17097 | Lake County ........................ | 703,400 | 693,593 |
| 16300 | 19105 | Jones County ........................ | 20,636 | 20,617 | 16980 | 55059 | Kenosha County ........................ | 166,424 | 169,671 |
| 16300 | 19113 | Linn County ........................ | 211,243 | 227,854 | | | | | |
| 16540 | | Chambersburg-Waynesboro, PA Metro area | 149,631 | 155,637 | 17020 | | Chico, CA Metro area .................................. | 220,005 | 212,744 |
| 16540 | 42055 | Franklin County ........................ | 149,631 | 155,637 | 17020 | 06007 | Butte County ........................ | 220,005 | 212,744 |
| 16580 | | Champaign-Urbana, IL Metro area.............. | 217,806 | 225,547 | 17140 | | Cincinnati, OH-KY-IN Metro area .............. | 2,137,713 | 2,232,907 |
| 16580 | 17019 | Champaign County ........................ | 201,081 | 209,192 | 17140 | 18029 | Dearborn County ........................ | 50,027 | 49,824 |
| 16580 | 17147 | Piatt County ........................ | 16,725 | 16,355 | 17140 | 18047 | Franklin County ........................ | 23,098 | 22,761 |
| | | | | | 17140 | 18115 | Ohio County ........................ | 6,098 | 5,892 |
| 16620 | | Charleston, WV Metro area...................... | 277,985 | 254,145 | 17140 | 18161 | Union County ........................ | 7,516 | 7,119 |
| 16620 | 54005 | Boone County ........................ | 24,625 | 21,055 | 17140 | 21015 | Boone County ........................ | 118,810 | 135,396 |
| 16620 | 54015 | Clay County ........................ | 9,384 | 8,341 | 17140 | 21023 | Bracken County ........................ | 8,486 | 8,286 |
| 16620 | 54035 | Jackson County ........................ | 29,214 | 28,453 | 17140 | 21037 | Campbell County ........................ | 90,338 | 94,020 |
| 16620 | 54039 | Kanawha County ........................ | 193,053 | 176,253 | 17140 | 21077 | Gallatin County ........................ | 8,586 | 8,779 |
| 16620 | 54043 | Lincoln County ........................ | 21,709 | 20,043 | 17140 | 21081 | Grant County ........................ | 24,655 | 25,387 |
| | | | | | 17140 | 21117 | Kenton County ........................ | 159,728 | 167,949 |
| 16700 | | Charleston-North Charleston, SC Metro area | 664,645 | 819,705 | 17140 | 21191 | Pendleton County ........................ | 14,874 | 14,586 |
| 16700 | 45015 | Berkeley County ........................ | 178,373 | 235,987 | 17140 | 39015 | Brown County ........................ | 44,826 | 43,414 |
| 16700 | 45019 | Charleston County ........................ | 350,128 | 417,981 | 17140 | 39017 | Butler County ........................ | 368,136 | 385,648 |
| 16700 | 45035 | Dorchester County ........................ | 136,144 | 165,737 | 17140 | 39025 | Clermont County ........................ | 197,366 | 207,449 |
| | | | | | 17140 | 39061 | Hamilton County ........................ | 802,371 | 817,985 |
| 16740 | | Charlotte-Concord-Gastonia, NC-SC Metro area | 2,243,963 | 2,684,276 | 17140 | 39165 | Warren County ........................ | 212,798 | 238,412 |
| 16740 | 37007 | Anson County ........................ | 26,929 | 24,097 | 17300 | | Clarksville, TN-KY Metro area.................... | 273,942 | 314,364 |
| 16740 | 37025 | Cabarrus County ........................ | 178,121 | 221,479 | 17300 | 21047 | Christian County ........................ | 73,940 | 71,478 |
| 16740 | 37071 | Gaston County ........................ | 206,098 | 226,568 | 17300 | 21221 | Trigg County ........................ | 14,327 | 14,776 |
| 16740 | 37097 | Iredell County ........................ | 159,464 | 185,770 | 17300 | 47125 | Montgomery County ........................ | 172,362 | 214,251 |
| 16740 | 37109 | Lincoln County ........................ | 78,014 | 88,097 | 17300 | 47161 | Stewart County ........................ | 13,313 | 13,859 |
| 16740 | 37119 | Mecklenburg County ........................ | 919,664 | 1,128,945 | | | | | |
| 16740 | 37159 | Rowan County ........................ | 138,464 | 142,495 | 17420 | | Cleveland, TN Metro area........................ | 115,747 | 125,906 |
| 16740 | 37179 | Union County ........................ | 201,332 | 244,562 | 17420 | 47011 | Bradley County ........................ | 98,926 | 109,071 |
| 16740 | 45023 | Chester County ........................ | 33,159 | 32,232 | 17420 | 47139 | Polk County ........................ | 16,821 | 16,835 |
| 16740 | 45057 | Lancaster County ........................ | 76,651 | 100,926 | | | | | |
| 16740 | 45091 | York County ........................ | 226,037 | 289,105 | 17460 | | Cleveland-Elyria, OH Metro area .............. | 2,077,277 | 2,043,807 |
| | | | | | 17460 | 39035 | Cuyahoga County ........................ | 1,280,114 | 1,227,883 |
| 16820 | | Charlottesville, VA Metro area.................... | 201,569 | 219,910 | 17460 | 39055 | Geauga County ........................ | 93,405 | 93,271 |
| 16820 | 51003 | Albemarle County ........................ | 98,998 | 110,652 | 17460 | 39085 | Lake County ........................ | 230,051 | 229,569 |
| 16820 | 51065 | Fluvanna County ........................ | 25,742 | 27,422 | 17460 | 39093 | Lorain County ........................ | 301,374 | 312,172 |
| 16820 | 51079 | Greene County ........................ | 18,389 | 20,131 | 17460 | 39103 | Medina County ........................ | 172,333 | 180,912 |
| 16820 | 51125 | Nelson County ........................ | 15,015 | 14,755 | | | | | |
| 16820 | 51540 | Charlottesville city ........................ | 43,425 | 46,950 | 17660 | | Coeur d'Alene, ID Metro area.................... | 138,466 | 170,628 |
| | | | | | 17660 | 16055 | Kootenai County ........................ | 138,466 | 170,628 |
| 16860 | | Chattanooga, TN-GA Metro area................ | 528,126 | 569,931 | 17780 | | College Station-Bryan, TX Metro area ........ | 228,668 | 268,224 |
| 16860 | 13047 | Catoosa County ........................ | 63,925 | 67,996 | 17780 | 48041 | Brazos County ........................ | 194,861 | 232,555 |
| 16860 | 13083 | Dade County ........................ | 16,643 | 16,057 | 17780 | 48051 | Burleson County ........................ | 17,187 | 18,514 |
| 16860 | 13295 | Walker County ........................ | 68,738 | 70,116 | 17780 | 48395 | Robertson County ........................ | 16,620 | 17,155 |
| 16860 | 47065 | Hamilton County ........................ | 336,477 | 371,662 | | | | | |
| 16860 | 47115 | Marion County ........................ | 28,222 | 28,924 | 17820 | | Colorado Springs, CO Metro area................ | 645,612 | 753,839 |
| 16860 | 47153 | Sequatchie County ........................ | 14,121 | 15,176 | 17820 | 08041 | El Paso County ........................ | 622,253 | 728,310 |
| | | | | | 17820 | 08119 | Teller County ........................ | 23,359 | 25,529 |
| 16940 | | Cheyenne, WY Metro area........................ | 91,885 | 100,595 | | | | | |
| 16940 | 56021 | Laramie County ........................ | 91,885 | 100,595 | 17860 | | Columbia, MO Metro area........................ | 190,398 | 210,094 |
| | | | | | 17860 | 29019 | Boone County ........................ | 162,652 | 182,991 |
| 16980 | | Chicago-Naperville-Elgin, IL-IN-WI Metro area | 9,461,537 | 9,406,638 | 17860 | 29053 | Cooper County ........................ | 17,604 | 17,102 |
| | | | | | 17860 | 29089 | Howard County ........................ | 10,142 | 10,001 |
| 16980 | | Chicago-Naperville-Evanston, IL Div 16984 | 7,148,308 | 7,071,372 | 17900 | | Columbia, SC Metro area........................ | 767,469 | 847,397 |
| 16980 | 17031 | Cook County ........................ | 5,195,026 | 5,108,284 | 17900 | 45017 | Calhoun County ........................ | 15,178 | 14,554 |
| 16980 | 17043 | DuPage County ........................ | 916,741 | 917,481 | 17900 | 45039 | Fairfield County ........................ | 23,959 | 22,059 |
| 16980 | 17063 | Grundy County ........................ | 50,079 | 50,993 | 17900 | 45055 | Kershaw County ........................ | 61,594 | 67,472 |
| 16980 | 17111 | McHenry County ........................ | 308,882 | 305,888 | 17900 | 45063 | Lexington County ........................ | 262,453 | 303,946 |
| 16980 | 17197 | Will County ........................ | 677,580 | 688,726 | 17900 | 45079 | Richland County ........................ | 384,425 | 419,051 |
| | | | | | 17900 | 45081 | Saluda County ........................ | 19,860 | 20,315 |

| State/County FIPS code | Core based statistical area | Title and Geographic Components | 2010 census population | 2020 estimated population | State/County FIPS code | Core based statistical area | Title and Geographic Components | 2010 census population | 2020 estimated population |
|---|---|---|---|---|---|---|---|---|---|
| 17980 | | Columbus, GA-AL Metro area | 308,478 | 322,658 | 19460 | | Decatur, AL Metro area | 153,827 | 152,740 |
| 17980 | 01113 | Russell County | 52,963 | 58,237 | 19460 | 01079 | Lawrence County | 34,337 | 32,857 |
| 17980 | 13053 | Chattahoochee County | 11,263 | 10,551 | 19460 | 01103 | Morgan County | 119,490 | 119,883 |
| 17980 | 13145 | Harris County | 31,994 | 36,080 | | | | | |
| 17980 | 13197 | Marion County | 8,738 | 8,516 | 19500 | | Decatur, IL Metro area | 110,777 | 103,015 |
| 17980 | 13215 | Muscogee County | 190,570 | 196,442 | 19500 | 17115 | Macon County | 110,777 | 103,015 |
| 17980 | 13259 | Stewart County | 6,060 | 6,689 | | | | | |
| 17980 | 13263 | Talbot County | 6,890 | 6,143 | 19660 | | Deltona-Daytona Beach-Ormond Beach, FL Metro area | 590,288 | 679,948 |
| | | | | | 19660 | 12035 | Flagler County | 95,692 | 118,451 |
| 18020 | | Columbus, IN Metro area | 76,783 | 84,447 | 19660 | 12127 | Volusia County | 494,596 | 561,497 |
| 18020 | 18005 | Bartholomew County | 76,783 | 84,447 | | | | | |
| | | | | | 19740 | | Denver-Aurora-Lakewood, CO Metro area | 2,543,608 | 2,991,231 |
| 18140 | | Columbus, OH Metro area | 1,902,008 | 2,138,946 | 19740 | 08001 | Adams County | 441,697 | 519,883 |
| 18140 | 39041 | Delaware County | 174,172 | 213,554 | 19740 | 08005 | Arapahoe County | 572,118 | 657,452 |
| 18140 | 39045 | Fairfield County | 146,194 | 159,709 | 19740 | 08014 | Broomfield County | 55,861 | 72,236 |
| 18140 | 39049 | Franklin County | 1,163,476 | 1,324,624 | 19740 | 08019 | Clear Creek County | 9,073 | 9,586 |
| 18140 | 39073 | Hocking County | 29,369 | 28,095 | 19740 | 08031 | Denver County | 599,825 | 735,538 |
| 18140 | 39089 | Licking County | 166,482 | 178,100 | 19740 | 08035 | Douglas County | 285,465 | 360,750 |
| 18140 | 39097 | Madison County | 43,438 | 44,559 | 19740 | 08039 | Elbert County | 23,088 | 27,313 |
| 18140 | 39117 | Morrow County | 34,825 | 35,411 | 19740 | 08047 | Gilpin County | 5,449 | 6,235 |
| 18140 | 39127 | Perry County | 36,037 | 36,215 | 19740 | 08059 | Jefferson County | 534,829 | 583,283 |
| 18140 | 39129 | Pickaway County | 55,684 | 58,658 | 19740 | 08093 | Park County | 16,203 | 18,955 |
| 18140 | 39159 | Union County | 52,331 | 60,021 | | | | | |
| | | | | | 19780 | | Des Moines-West Des Moines, IA Metro area | 606,474 | 707,915 |
| 18580 | | Corpus Christi, TX Metro area | 405,025 | 430,217 | 19780 | 19049 | Dallas County | 66,139 | 96,963 |
| 18580 | 48355 | Nueces County | 340,223 | 363,148 | 19780 | 19077 | Guthrie County | 10,955 | 10,737 |
| 18580 | 48409 | San Patricio County | 64,802 | 67,069 | 19780 | 19099 | Jasper County | 36,842 | 37,148 |
| | | | | | 19780 | 19121 | Madison County | 15,681 | 16,521 |
| 18700 | | Corvallis, OR Metro area | 85,581 | 93,239 | 19780 | 19153 | Polk County | 430,631 | 494,281 |
| 18700 | 41003 | Benton County | 85,581 | 93,239 | 19780 | 19181 | Warren County | 46,226 | 52,265 |
| | | | | | | | | | |
| 18880 | | Crestview-Fort Walton Beach-Destin, FL Metro area | 235,870 | 289,468 | 19820 | | Detroit-Warren-Dearborn, MI Metro area | 4,296,227 | 4,304,136 |
| 18880 | 12091 | Okaloosa County | 180,824 | 212,820 | 19820 | | Detroit-Dearborn-Livonia, MI Div 19804 | 1,820,473 | 1,740,623 |
| 18880 | 12131 | Walton County | 55,046 | 76,648 | 19820 | 26163 | Wayne County | 1,820,473 | 1,740,623 |
| | | | | | | | | | |
| 19060 | | Cumberland, MD-WV Metro area | 103,272 | 96,779 | 19820 | | Warren-Troy-Farmington Hills, MI Div 47664 | 2,475,754 | 2,563,513 |
| 19060 | 24001 | Allegany County | 75,047 | 70,057 | 19820 | 26087 | Lapeer County | 88,316 | 87,635 |
| 19060 | 54057 | Mineral County | 28,225 | 26,722 | 19820 | 26093 | Livingston County | 180,964 | 192,335 |
| | | | | | 19820 | 26099 | Macomb County | 841,039 | 870,791 |
| 19100 | | Dallas-Fort Worth-Arlington, TX Metro area | 6,366,537 | 7,694,138 | 19820 | 26125 | Oakland County | 1,202,384 | 1,253,459 |
| | | | | | 19820 | 26147 | St. Clair County | 163,051 | 159,293 |
| 19100 | | Dallas-Plano-Irving, TX Div 19124 | 4,228,853 | 5,171,934 | | | | | |
| 19100 | 48085 | Collin County | 781,419 | 1,072,069 | 20020 | | Dothan, AL Metro area | 145,640 | 150,214 |
| 19100 | 48113 | Dallas County | 2,367,430 | 2,635,888 | 20020 | 01061 | Geneva County | 26,781 | 26,411 |
| 19100 | 48121 | Denton County | 662,557 | 919,324 | 20020 | 01067 | Henry County | 17,299 | 17,223 |
| 19100 | 48139 | Ellis County | 149,610 | 191,760 | 20020 | 01069 | Houston County | 101,560 | 106,580 |
| 19100 | 48231 | Hunt County | 86,144 | 99,807 | | | | | |
| 19100 | 48257 | Kaufman County | 103,348 | 143,198 | 20100 | | Dover, DE Metro area | 162,350 | 183,643 |
| 19100 | 48397 | Rockwall County | 78,345 | 109,888 | 20100 | 10001 | Kent County | 162,350 | 183,643 |
| | | | | | | | | | |
| 19100 | | Fort Worth-Arlington-Grapevine, TX Div 23104 | 2,137,684 | 2,522,204 | 20220 | | Dubuque, IA Metro area | 93,643 | 97,590 |
| 19100 | 48251 | Johnson County | 150,956 | 179,575 | 20220 | 19061 | Dubuque County | 93,643 | 97,590 |
| 19100 | 48367 | Parker County | 116,949 | 148,198 | | | | | |
| 19100 | 48439 | Tarrant County | 1,810,664 | 2,123,347 | 20260 | | Duluth, MN-WI Metro area | 290,636 | 288,648 |
| 19100 | 48497 | Wise County | 59,115 | 71,084 | 20260 | 27017 | Carlton County | 35,386 | 35,769 |
| | | | | | 20260 | 27075 | Lake County | 10,862 | 10,639 |
| 19140 | | Dalton, GA Metro area | 142,233 | 143,869 | 20260 | 27137 | St. Louis County | 200,229 | 198,538 |
| 19140 | 13213 | Murray County | 39,628 | 40,032 | 20260 | 55031 | Douglas County | 44,159 | 43,702 |
| 19140 | 13313 | Whitfield County | 102,605 | 103,837 | | | | | |
| | | | | | 20500 | | Durham-Chapel Hill, NC Metro area | 564,193 | 652,542 |
| 19180 | | Danville, IL Metro area | 81,625 | 74,855 | 20500 | 37037 | Chatham County | 63,485 | 75,748 |
| 19180 | 17183 | Vermilion County | 81,625 | 74,855 | 20500 | 37063 | Durham County | 270,001 | 327,306 |
| | | | | | 20500 | 37077 | Granville County | 57,538 | 60,486 |
| 19300 | | Daphne-Fairhope-Foley, AL Metro area | 182,265 | 229,287 | 20500 | 37135 | Orange County | 133,693 | 149,077 |
| 19300 | 01003 | Baldwin County | 182,265 | 229,287 | 20500 | 37145 | Person County | 39,476 | 39,925 |
| | | | | | | | | | |
| 19340 | | Davenport-Moline-Rock Island, IA-IL Metro area | 379,681 | 377,759 | 20700 | | East Stroudsburg, PA Metro area | 169,841 | 170,154 |
| 19340 | 17073 | Henry County | 50,483 | 48,411 | 20700 | 42089 | Monroe County | 169,841 | 170,154 |
| 19340 | 17131 | Mercer County | 16,434 | 15,225 | | | | | |
| 19340 | 17161 | Rock Island County | 147,541 | 140,907 | 20740 | | Eau Claire, WI Metro area | 161,383 | 169,997 |
| 19340 | 19163 | Scott County | 165,223 | 173,216 | 20740 | 55017 | Chippewa County | 62,502 | 64,737 |
| | | | | | 20740 | 55035 | Eau Claire County | 98,881 | 105,260 |
| 19430 | | Dayton-Kettering, OH Metro area | 799,280 | 809,248 | | | | | |
| 19430 | 39057 | Greene County | 161,577 | 170,122 | | | | | |
| 19430 | 39109 | Miami County | 102,503 | 107,516 | | | | | |
| 19430 | 39113 | Montgomery County | 535,200 | 531,610 | | | | | |

| State/County FIPS code | Core based statistical area | Title and Geographic Components | 2010 census population | 2020 estimated population | State/County FIPS code | Core based statistical area | Title and Geographic Components | 2010 census population | 2020 estimated population |
|---|---|---|---|---|---|---|---|---|---|
| 20940 | | El Centro, CA Metro area | 174,524 | 180,267 | 23060 | | Fort Wayne, IN Metro area | 388,626 | 416,565 |
| 20940 | 06025 | Imperial County | 174,524 | 180,267 | 23060 | 18003 | Allen County | 355,339 | 382,187 |
| | | | | | 23060 | 18183 | Whitley County | 33,287 | 34,378 |
| 21060 | | Elizabethtown-Fort Knox, KY Metro area | 148,331 | 154,356 | | | | | |
| 21060 | 21093 | Hardin County | 105,537 | 111,309 | 23420 | | Fresno, CA Metro area | 930,507 | 1,000,918 |
| 21060 | 21123 | Larue County | 14,181 | 14,431 | 23420 | 06019 | Fresno County | 930,507 | 1,000,918 |
| 21060 | 21163 | Meade County | 28,613 | 28,616 | | | | | |
| | | | | | 23460 | | Gadsden, AL Metro area | 104,429 | 102,371 |
| 21140 | | Elkhart-Goshen, IN Metro area | 197,569 | 206,161 | 23460 | 01055 | Etowah County | 104,429 | 102,371 |
| 21140 | 18039 | Elkhart County | 197,569 | 206,161 | | | | | |
| | | | | | 23540 | | Gainesville, FL Metro area | 305,076 | 332,317 |
| 21300 | | Elmira, NY Metro area | 88,847 | 82,622 | 23540 | 12001 | Alachua County | 247,337 | 271,218 |
| 21300 | 36015 | Chemung County | 88,847 | 82,622 | 23540 | 12041 | Gilchrist County | 16,941 | 18,885 |
| | | | | | 23540 | 12075 | Levy County | 40,798 | 42,214 |
| 21340 | | El Paso, TX Metro area | 804,109 | 846,192 | | | | | |
| 21340 | 48141 | El Paso County | 800,633 | 841,286 | 23580 | | Gainesville, GA Metro area | 179,724 | 206,591 |
| 21340 | 48229 | Hudspeth County | 3,476 | 4,906 | 23580 | 13139 | Hall County | 179,724 | 206,591 |
| 21420 | | Enid, OK Metro area | 60,580 | 60,869 | 23900 | | Gettysburg, PA Metro area | 101,428 | 102,742 |
| 21420 | 40047 | Garfield County | 60,580 | 60,869 | 23900 | 42001 | Adams County | 101,428 | 102,742 |
| 21500 | | Erie, PA Metro area | 280,584 | 268,426 | 24020 | | Glens Falls, NY Metro area | 128,946 | 124,362 |
| 21500 | 42049 | Erie County | 280,584 | 268,426 | 24020 | 36113 | Warren County | 65,692 | 63,756 |
| | | | | | 24020 | 36115 | Washington County | 63,254 | 60,606 |
| 21660 | | Eugene-Springfield, OR Metro area | 351,705 | 382,986 | | | | | |
| 21660 | 41039 | Lane County | 351,705 | 382,986 | 24140 | | Goldsboro, NC Metro area | 122,661 | 123,967 |
| | | | | | 24140 | 37191 | Wayne County | 122,661 | 123,967 |
| 21780 | | Evansville, IN-KY Metro area | 311,548 | 315,731 | | | | | |
| 21780 | 18129 | Posey County | 25,912 | 25,275 | 24220 | | Grand Forks, ND-MN Metro area | 98,464 | 100,381 |
| 21780 | 18163 | Vanderburgh County | 179,701 | 182,447 | 24220 | 27119 | Polk County | 31,600 | 30,900 |
| 21780 | 18173 | Warrick County | 59,689 | 63,269 | 24220 | 38035 | Grand Forks County | 66,864 | 69,481 |
| 21780 | 21101 | Henderson County | 46,246 | 44,740 | | | | | |
| | | | | | 24260 | | Grand Island, NE Metro area | 72,744 | 75,325 |
| 21820 | | Fairbanks, AK Metro area | 97,585 | 95,651 | 24260 | 31079 | Hall County | 58,611 | 61,028 |
| 21820 | 02090 | Fairbanks North Star Borough | 97,585 | 95,651 | 24260 | 31093 | Howard County | 6,274 | 6,488 |
| | | | | | 24260 | 31121 | Merrick County | 7,859 | 7,809 |
| 22020 | | Fargo, ND-MN Metro area | 208,777 | 248,594 | | | | | |
| 22020 | 27027 | Clay County | 58,999 | 64,690 | 24300 | | Grand Junction, CO Metro area | 146,733 | 155,603 |
| 22020 | 38017 | Cass County | 149,778 | 183,904 | 24300 | 08077 | Mesa County | 146,733 | 155,603 |
| 22140 | | Farmington, NM Metro area | 130,045 | 123,312 | 24340 | | Grand Rapids-Kentwood, MI Metro area | 993,663 | 1,081,372 |
| 22140 | 35045 | San Juan County | 130,045 | 123,312 | 24340 | 26067 | Ionia County | 63,901 | 64,553 |
| | | | | | 24340 | 26081 | Kent County | 602,625 | 658,708 |
| 22180 | | Fayetteville, NC Metro area | 481,011 | 529,252 | 24340 | 26117 | Montcalm County | 63,342 | 63,476 |
| 22180 | 37051 | Cumberland County | 319,431 | 336,364 | 24340 | 26139 | Ottawa County | 263,795 | 294,635 |
| 22180 | 37085 | Harnett County | 114,691 | 137,058 | | | | | |
| 22180 | 37093 | Hoke County | 46,889 | 55,830 | 24420 | | Grants Pass, OR Metro area | 82,719 | 88,053 |
| | | | | | 24420 | 41033 | Josephine County | 82,719 | 88,053 |
| 22220 | | Fayetteville-Springdale-Rogers, AR Metro area | 440,121 | 548,634 | | | | | |
| 22220 | 05007 | Benton County | 221,348 | 288,774 | 24500 | | Great Falls, MT Metro area | 81,326 | 81,346 |
| 22220 | 05087 | Madison County | 15,723 | 16,644 | 24500 | 30013 | Cascade County | 81,326 | 81,346 |
| 22220 | 05143 | Washington County | 203,050 | 243,216 | | | | | |
| | | | | | 24540 | | Greeley, CO Metro area | 252,827 | 333,983 |
| 22380 | | Flagstaff, AZ Metro area | 134,426 | 142,481 | 24540 | 08123 | Weld County | 252,827 | 333,983 |
| 22380 | 04005 | Coconino County | 134,426 | 142,481 | | | | | |
| | | | | | 24580 | | Green Bay, WI Metro area | 306,241 | 323,379 |
| 22420 | | Flint, MI Metro area | 425,787 | 404,794 | 24580 | 55009 | Brown County | 248,003 | 264,610 |
| 22420 | 26049 | Genesee County | 425,787 | 404,794 | 24580 | 55061 | Kewaunee County | 20,578 | 20,386 |
| | | | | | 24580 | 55083 | Oconto County | 37,660 | 38,383 |
| 22500 | | Florence, SC Metro area | 205,576 | 204,097 | | | | | |
| 22500 | 45031 | Darlington County | 68,611 | 66,509 | 24660 | | Greensboro-High Point, NC Metro area | 723,923 | 776,363 |
| 22500 | 45041 | Florence County | 136,965 | 137,588 | 24660 | 37081 | Guilford County | 488,454 | 540,521 |
| | | | | | 24660 | 37151 | Randolph County | 141,824 | 144,557 |
| 22520 | | Florence-Muscle Shoals, AL Metro area | 147,137 | 148,779 | 24660 | 37157 | Rockingham County | 93,645 | 91,285 |
| 22520 | 01033 | Colbert County | 54,428 | 55,411 | | | | | |
| 22520 | 01077 | Lauderdale County | 92,709 | 93,368 | 24780 | | Greenville, NC Metro area | 168,176 | 182,924 |
| | | | | | 24780 | 37147 | Pitt County | 168,176 | 182,924 |
| 22540 | | Fond du Lac, WI Metro area | 101,623 | 102,902 | | | | | |
| 22540 | 55039 | Fond du Lac County | 101,623 | 102,902 | 24860 | | Greenville-Anderson, SC Metro area | 824,031 | 932,705 |
| | | | | | 24860 | 45007 | Anderson County | 186,922 | 204,353 |
| 22660 | | Fort Collins, CO Metro area | 299,630 | 360,428 | 24860 | 45045 | Greenville County | 451,211 | 532,486 |
| 22660 | 08069 | Larimer County | 299,630 | 360,428 | 24860 | 45059 | Laurens County | 66,513 | 67,883 |
| | | | | | 24860 | 45077 | Pickens County | 119,385 | 127,983 |
| 22900 | | Fort Smith, AR-OK Metro area | 248,240 | 250,434 | | | | | |
| 22900 | 05033 | Crawford County | 61,935 | 63,409 | 25060 | | Gulfport-Biloxi, MS Metro area | 388,591 | 418,963 |
| 22900 | 05047 | Franklin County | 18,137 | 17,897 | 25060 | 28045 | Hancock County | 44,023 | 48,000 |
| 22900 | 05131 | Sebastian County | 125,740 | 127,590 | 25060 | 28047 | Harrison County | 187,109 | 208,801 |
| 22900 | 40135 | Sequoyah County | 42,428 | 41,538 | 25060 | 28059 | Jackson County | 139,669 | 143,802 |
| | | | | | 25060 | 28131 | Stone County | 17,790 | 18,360 |

| State/ County FIPS code | Core based statistical area | Title and Geographic Components | 2010 census population | 2020 estimated population | State/ County FIPS code | Core based statistical area | Title and Geographic Components | 2010 census population | 2020 estimated population |
|---|---|---|---|---|---|---|---|---|---|
| 25180 | | Hagerstown-Martinsburg, MD-WV Metro area ..... | 269,146 | 291,144 | 26620 | | Huntsville, AL Metro area ............................ | 417,593 | 481,681 |
| 25180 | 24043 | Washington County ................................. | 147,417 | 151,146 | 26620 | 01083 | Limestone County ................................... | 82,786 | 102,228 |
| 25180 | 54003 | Berkeley County ..................................... | 104,188 | 122,125 | 26620 | 01089 | Madison County ...................................... | 334,807 | 379,453 |
| 25180 | 54065 | Morgan County ....................................... | 17,541 | 17,873 | | | | | |
| | | | | | 26820 | | Idaho Falls, ID Metro area ........................... | 133,331 | 155,361 |
| 25220 | | Hammond, LA Metro area ........................... | 121,109 | 136,765 | 26820 | 16019 | Bonneville County ................................... | 104,294 | 122,134 |
| 25220 | 22105 | Tangipahoa Parish ................................... | 121,109 | 136,765 | 26820 | 16023 | Butte County .......................................... | 2,893 | 2,646 |
| | | | | | 26820 | 16051 | Jefferson County ..................................... | 26,144 | 30,581 |
| 25260 | | Hanford-Corcoran, CA Metro area ............. | 152,974 | 152,692 | | | | | |
| 25260 | 06031 | Kings County .......................................... | 152,974 | 152,692 | 26900 | | Indianapolis-Carmel-Anderson, IN Metro area ..... | 1,888,075 | 2,091,019 |
| | | | | | 26900 | 18011 | Boone County ......................................... | 56,641 | 69,347 |
| 25420 | | Harrisburg-Carlisle, PA Metro area ............. | 549,444 | 581,943 | 26900 | 18013 | Brown County ......................................... | 15,244 | 15,112 |
| 25420 | 42041 | Cumberland County................................. | 235,387 | 255,857 | 26900 | 18057 | Hamilton County ...................................... | 274,557 | 344,238 |
| 25420 | 42043 | Dauphin County....................................... | 268,126 | 279,874 | 26900 | 18059 | Hancock County ...................................... | 70,060 | 79,553 |
| 25420 | 42099 | Perry County........................................... | 45,931 | 46,212 | 26900 | 18063 | Hendricks County ................................... | 145,456 | 173,251 |
| | | | | | 26900 | 18081 | Johnson County ...................................... | 139,855 | 160,607 |
| 25500 | | Harrisonburg, VA Metro area ..................... | 125,221 | 135,550 | 26900 | 18095 | Madison County ...................................... | 131,637 | 129,681 |
| 25500 | 51165 | Rockingham County ................................ | 76,321 | 82,346 | 26900 | 18097 | Marion County ........................................ | 903,375 | 966,183 |
| 25500 | 51660 | Harrisonburg city..................................... | 48,900 | 53,204 | 26900 | 18109 | Morgan County ....................................... | 68,926 | 70,707 |
| | | | | | 26900 | 18133 | Putnam County ....................................... | 37,945 | 37,469 |
| 25540 | | Hartford-East Hartford-Middletown, CT Metro area ..... | 1,212,471 | 1,201,483 | 26900 | 18145 | Shelby County ........................................ | 44,379 | 44,871 |
| 25540 | 09003 | Hartford County ...................................... | 894,052 | 889,226 | | | | | |
| 25540 | 09007 | Middlesex County ................................... | 165,672 | 161,657 | 26980 | | Iowa City, IA Metro area ............................. | 152,586 | 175,732 |
| 25540 | 09013 | Tolland County ....................................... | 152,747 | 150,600 | 26980 | 19103 | Johnson County ...................................... | 130,882 | 153,740 |
| | | | | | 26980 | 19183 | Washington County ................................. | 21,704 | 21,992 |
| 25620 | | Hattiesburg, MS Metro area ....................... | 162,418 | 169,554 | | | | | |
| 25620 | 28031 | Covington County .................................... | 19,573 | 18,518 | 27060 | | Ithaca, NY Metro area ............................... | 101,592 | 101,058 |
| 25620 | 28035 | Forrest County ....................................... | 74,862 | 75,009 | 27060 | 36109 | Tompkins County .................................... | 101,592 | 101,058 |
| 25620 | 28073 | Lamar County ........................................ | 55,732 | 64,165 | | | | | |
| 25620 | 28111 | Perry County........................................... | 12,251 | 11,862 | 27100 | | Jackson, MI Metro area .............................. | 160,233 | 156,920 |
| | | | | | 27100 | 26075 | Jackson County ...................................... | 160,233 | 156,920 |
| 25860 | | Hickory-Lenoir-Morganton, NC Metro area . | 365,794 | 370,266 | | | | | |
| 25860 | 37003 | Alexander County .................................... | 37,182 | 37,441 | 27140 | | Jackson, MS Metro area ............................. | 587,115 | 589,082 |
| 25860 | 37023 | Burke County ......................................... | 90,837 | 90,418 | 27140 | 28029 | Copiah County ........................................ | 29,448 | 27,933 |
| 25860 | 37027 | Caldwell County ..................................... | 83,058 | 82,100 | 27140 | 28049 | Hinds County ......................................... | 245,364 | 227,966 |
| 25860 | 37035 | Catawba County ..................................... | 154,717 | 160,307 | 27140 | 28051 | Holmes County ....................................... | 19,483 | 16,726 |
| | | | | | 27140 | 28089 | Madison County ...................................... | 95,203 | 106,871 |
| 25940 | | Hilton Head Island-Bluffton, SC Metro area . | 187,010 | 227,244 | 27140 | 28121 | Rankin County ........................................ | 142,052 | 155,975 |
| 25940 | 45013 | Beaufort County ..................................... | 162,219 | 195,656 | 27140 | 28127 | Simpson County ..................................... | 27,500 | 26,629 |
| 25940 | 45053 | Jasper County......................................... | 24,791 | 31,588 | 27140 | 28163 | Yazoo County ........................................ | 28,065 | 26,982 |
| 25980 | | Hinesville, GA Metro area ......................... | 77,929 | 83,175 | 27180 | | Jackson, TN Metro area .............................. | 179,711 | 179,131 |
| 25980 | 13179 | Liberty County........................................ | 63,585 | 63,004 | 27180 | 47023 | Chester County ...................................... | 17,145 | 17,432 |
| 25980 | 13183 | Long County ........................................... | 14,344 | 20,171 | 27180 | 47033 | Crockett County ..................................... | 14,576 | 14,180 |
| | | | | | 27180 | 47053 | Gibson County ....................................... | 49,687 | 49,159 |
| 26140 | | Homosassa Springs, FL Metro area............ | 141,230 | 153,010 | 27180 | 47113 | Madison County ...................................... | 98,303 | 98,360 |
| 26140 | 12017 | Citrus County ......................................... | 141,230 | 153,010 | | | | | |
| | | | | | 27260 | | Jacksonville, FL Metro area ........................ | 1,345,594 | 1,587,892 |
| 26300 | | Hot Springs, AR Metro area ....................... | 95,999 | 99,789 | 27260 | 12003 | Baker County ......................................... | 27,115 | 29,566 |
| 26300 | 05051 | Garland County ...................................... | 95,999 | 99,789 | 27260 | 12019 | Clay County ........................................... | 190,878 | 221,770 |
| | | | | | 27260 | 12031 | Duval County ......................................... | 864,253 | 966,728 |
| 26380 | | Houma-Thibodaux, LA Metro area ............. | 208,185 | 207,455 | 27260 | 12089 | Nassau County ...................................... | 73,310 | 91,113 |
| 26380 | 22057 | Lafourche Parish..................................... | 96,642 | 97,596 | 27260 | 12109 | St. Johns County .................................... | 190,038 | 278,715 |
| 26380 | 22109 | Terrebonne Parish .................................. | 111,543 | 109,859 | | | | | |
| | | | | | 27340 | | Jacksonville, NC Metro area ....................... | 177,801 | 203,943 |
| 26420 | | Houston-The Woodlands-Sugar Land, TX Metro are ..... | 5,920,487 | 7,154,478 | 27340 | 37133 | Onslow County ....................................... | 177,801 | 203,943 |
| 26420 | 48015 | Austin County ........................................ | 28,412 | 29,972 | | | | | |
| 26420 | 48039 | Brazoria County...................................... | 313,117 | 380,518 | 27500 | | Janesville-Beloit, WI Metro area.................. | 160,325 | 163,084 |
| 26420 | 48071 | Chambers County ................................... | 35,107 | 45,590 | 27500 | 55105 | Rock County .......................................... | 160,325 | 163,084 |
| 26420 | 48157 | Fort Bend County.................................... | 584,699 | 839,706 | | | | | |
| 26420 | 48167 | Galveston County ................................... | 291,312 | 345,089 | 27620 | | Jefferson City, MO Metro area ..................... | 149,820 | 150,198 |
| 26420 | 48201 | Harris County ........................................ | 4,093,176 | 4,738,253 | 27620 | 29027 | Callaway County ..................................... | 44,331 | 44,887 |
| 26420 | 48291 | Liberty County........................................ | 75,643 | 91,547 | 27620 | 29051 | Cole County .......................................... | 75,975 | 76,191 |
| 26420 | 48339 | Montgomery County ................................ | 455,747 | 626,351 | 27620 | 29135 | Moniteau County .................................... | 15,605 | 15,585 |
| 26420 | 48473 | Waller County ........................................ | 43,274 | 57,452 | 27620 | 29151 | Osage County ....................................... | 13,909 | 13,535 |
| 26580 | | Huntington-Ashland, WV-KY-OH Metro area ..... | 370,899 | 354,085 | 27740 | | Johnson City, TN Metro area ...................... | 198,757 | 204,540 |
| 26580 | 21019 | Boyd County .......................................... | 49,534 | 46,516 | 27740 | 47019 | Carter County ........................................ | 57,383 | 56,418 |
| 26580 | 21043 | Carter County ........................................ | 27,721 | 26,542 | 27740 | 47171 | Unicoi County ........................................ | 18,311 | 17,755 |
| 26580 | 21089 | Greenup County ..................................... | 36,906 | 34,865 | 27740 | 47179 | Washington County ................................. | 123,063 | 130,367 |
| 26580 | 39087 | Lawrence County.................................... | 62,448 | 59,091 | | | | | |
| 26580 | 54011 | Cabell County ........................................ | 96,295 | 91,589 | 27780 | | Johnstown, PA Metro area .......................... | 143,695 | 128,672 |
| 26580 | 54079 | Putnam County ...................................... | 55,489 | 56,428 | 27780 | 42021 | Cambria County ..................................... | 143,695 | 128,672 |
| 26580 | 54099 | Wayne County ....................................... | 42,506 | 39,054 | 27860 | | Jonesboro, AR Metro area .......................... | 121,019 | 135,528 |
| | | | | | 27860 | 05031 | Craighead County .................................. | 96,443 | 112,245 |
| | | | | | 27860 | 05111 | Poinsett County ..................................... | 24,576 | 23,283 |

| State/ County FIPS code | Core based statistical area | Title and Geographic Components | 2010 census population | 2020 estimated population | State/ County FIPS code | Core based statistical area | Title and Geographic Components | 2010 census population | 2020 estimated population |
|---|---|---|---|---|---|---|---|---|---|
| 27900 | | Joplin, MO Metro area | 175,509 | 180,099 | 29340 | | Lake Charles, LA Metro area | 199,640 | 210,313 |
| 27900 | 29097 | Jasper County | 117,391 | 121,648 | 29340 | 22019 | Calcasieu Parish | 192,772 | 203,310 |
| 27900 | 29145 | Newton County | 58,118 | 58,451 | 29340 | 22023 | Cameron Parish | 6,868 | 7,003 |
| 27980 | | Kahului-Wailuku-Lahaina, HI Metro area | 154,840 | 167,902 | 29420 | | Lake Havasu City-Kingman, AZ Metro area | 200,182 | 217,206 |
| 27980 | 15009 | Maui County | 154,840 | 167,902 | 29420 | 04015 | Mohave County | 200,182 | 217,206 |
| 28020 | | Kalamazoo-Portage, MI Metro area | 250,327 | 265,988 | 29460 | | Lakeland-Winter Haven, FL Metro area | 602,073 | 744,552 |
| 28020 | 26077 | Kalamazoo County | 250,327 | 265,988 | 29460 | 12105 | Polk County | 602,073 | 744,552 |
| 28100 | | Kankakee, IL Metro area | 113,450 | 108,594 | 29540 | | Lancaster, PA Metro area | 519,443 | 546,192 |
| 28100 | 17091 | Kankakee County | 113,450 | 108,594 | 29540 | 42071 | Lancaster County | 519,443 | 546,192 |
| 28140 | | Kansas City, MO-KS Metro area | 2,009,355 | 2,173,212 | 29620 | | Lansing-East Lansing, MI Metro area | 534,684 | 548,248 |
| 28140 | 20091 | Johnson County | 544,181 | 607,220 | 29620 | 26037 | Clinton County | 75,364 | 79,753 |
| 28140 | 20103 | Leavenworth County | 76,220 | 82,246 | 29620 | 26045 | Eaton County | 107,761 | 110,148 |
| 28140 | 20107 | Linn County | 9,656 | 9,654 | 29620 | 26065 | Ingham County | 280,891 | 290,609 |
| 28140 | 20121 | Miami County | 32,781 | 34,334 | 29620 | 26155 | Shiawassee County | 70,668 | 67,738 |
| 28140 | 20209 | Wyandotte County | 157,523 | 165,265 | | | | | |
| 28140 | 29013 | Bates County | 17,047 | 16,242 | 29700 | | Laredo, TX Metro area | 250,304 | 277,681 |
| 28140 | 29025 | Caldwell County | 9,418 | 9,051 | 29700 | 48479 | Webb County | 250,304 | 277,681 |
| 28140 | 29037 | Cass County | 99,500 | 106,806 | | | | | |
| 28140 | 29047 | Clay County | 221,906 | 253,463 | 29740 | | Las Cruces, NM Metro area | 209,217 | 221,262 |
| 28140 | 29049 | Clinton County | 20,743 | 20,553 | 29740 | 35013 | Dona Ana County | 209,217 | 221,262 |
| 28140 | 29095 | Jackson County | 674,166 | 705,925 | | | | | |
| 28140 | 29107 | Lafayette County | 33,369 | 33,006 | 29820 | | Las Vegas-Henderson-Paradise, NV Metro area | 1,951,268 | 2,315,963 |
| 28140 | 29165 | Platte County | 89,329 | 106,532 | 29820 | 32003 | Clark County | 1,951,268 | 2,315,963 |
| 28140 | 29177 | Ray County | 23,516 | 22,915 | | | | | |
| 28420 | | Kennewick-Richland, WA Metro area | 253,328 | 303,501 | 29940 | | Lawrence, KS Metro area | 110,826 | 122,530 |
| 28420 | 53005 | Benton County | 175,168 | 206,426 | 29940 | 20045 | Douglas County | 110,826 | 122,530 |
| 28420 | 53021 | Franklin County | 78,160 | 97,075 | | | | | |
| | | | | | 30020 | | Lawton, OK Metro area | 130,288 | 126,775 |
| 28660 | | Killeen-Temple, TX Metro area | 405,308 | 468,453 | 30020 | 40031 | Comanche County | 124,098 | 121,099 |
| 28660 | 48027 | Bell County | 310,159 | 369,927 | 30020 | 40033 | Cotton County | 6,190 | 5,676 |
| 28660 | 48099 | Coryell County | 75,474 | 76,737 | | | | | |
| 28660 | 48281 | Lampasas County | 19,675 | 21,789 | 30140 | | Lebanon, PA Metro area | 133,597 | 141,663 |
| | | | | | 30140 | 42075 | Lebanon County | 133,597 | 141,663 |
| 28700 | | Kingsport-Bristol, TN-VA Metro area | 309,493 | 308,183 | | | | | |
| 28700 | 47073 | Hawkins County | 56,826 | 56,775 | 30300 | | Lewiston, ID-WA Metro area | 60,893 | 63,575 |
| 28700 | 47163 | Sullivan County | 156,800 | 158,755 | 30300 | 16069 | Nez Perce County | 39,270 | 40,755 |
| 28700 | 51169 | Scott County | 23,168 | 21,629 | 30300 | 53003 | Asotin County | 21,623 | 22,820 |
| 28700 | 51191 | Washington County | 54,961 | 53,695 | | | | | |
| 28700 | 51520 | Bristol city | 17,738 | 17,329 | 30340 | | Lewiston-Auburn, ME Metro area | 107,709 | 108,547 |
| | | | | | 30340 | 23001 | Androscoggin County | 107,709 | 108,547 |
| 28740 | | Kingston, NY Metro area | 182,519 | 177,716 | | | | | |
| 28740 | 36111 | Ulster County | 182,519 | 177,716 | 30460 | | Lexington-Fayette, KY Metro area | 472,103 | 520,391 |
| | | | | | 30460 | 21017 | Bourbon County | 20,008 | 19,901 |
| 28940 | | Knoxville, TN Metro area | 815,025 | 878,124 | 30460 | 21049 | Clark County | 35,603 | 36,463 |
| 28940 | 47001 | Anderson County | 75,082 | 77,558 | 30460 | 21067 | Fayette County | 295,870 | 324,735 |
| 28940 | 47009 | Blount County | 123,098 | 134,751 | 30460 | 21113 | Jessamine County | 48,580 | 54,057 |
| 28940 | 47013 | Campbell County | 40,723 | 39,837 | 30460 | 21209 | Scott County | 47,096 | 58,470 |
| 28940 | 47093 | Knox County | 432,260 | 475,609 | 30460 | 21239 | Woodford County | 24,946 | 26,765 |
| 28940 | 47105 | Loudon County | 48,561 | 54,910 | | | | | |
| 28940 | 47129 | Morgan County | 21,986 | 21,431 | 30620 | | Lima, OH Metro area | 106,313 | 101,980 |
| 28940 | 47145 | Roane County | 54,208 | 53,841 | 30620 | 39003 | Allen County | 106,313 | 101,980 |
| 28940 | 47173 | Union County | 19,107 | 20,187 | | | | | |
| | | | | | 30700 | | Lincoln, NE Metro area | 302,157 | 337,836 |
| 29020 | | Kokomo, IN Metro area | 82,748 | 82,732 | 30700 | 31109 | Lancaster County | 285,407 | 320,650 |
| 29020 | 18067 | Howard County | 82,748 | 82,732 | 30700 | 31159 | Seward County | 16,750 | 17,186 |
| 29100 | | La Crosse-Onalaska, WI-MN Metro area | 133,658 | 137,134 | 30780 | | Little Rock-North Little Rock-Conway, AR Metro | 699,790 | 746,564 |
| 29100 | 27055 | Houston County | 19,020 | 18,632 | 30780 | 05045 | Faulkner County | 113,238 | 126,919 |
| 29100 | 55063 | La Crosse County | 114,638 | 118,502 | 30780 | 05053 | Grant County | 17,842 | 18,449 |
| | | | | | 30780 | 05085 | Lonoke County | 68,382 | 73,921 |
| 29180 | | Lafayette, LA Metro area | 466,733 | 489,759 | 30780 | 05105 | Perry County | 10,444 | 10,327 |
| 29180 | 22001 | Acadia Parish | 61,787 | 61,918 | 30780 | 05119 | Pulaski County | 382,749 | 392,980 |
| 29180 | 22045 | Iberia Parish | 73,087 | 68,991 | 30780 | 05125 | Saline County | 107,135 | 123,968 |
| 29180 | 22055 | Lafayette Parish | 221,778 | 246,518 | | | | | |
| 29180 | 22099 | St. Martin Parish | 52,122 | 52,954 | 30860 | | Logan, UT-ID Metro area | 125,442 | 144,219 |
| 29180 | 22113 | Vermilion Parish | 57,959 | 59,378 | 30860 | 16041 | Franklin County | 12,786 | 14,215 |
| | | | | | 30860 | 49005 | Cache County | 112,656 | 130,004 |
| 29200 | | Lafayette-West Lafayette, IN Metro area | 210,310 | 233,278 | | | | | |
| 29200 | 18007 | Benton County | 8,836 | 8,741 | 30980 | | Longview, TX Metro area | 280,007 | 287,105 |
| 29200 | 18015 | Carroll County | 20,160 | 20,228 | 30980 | 48183 | Gregg County | 121,747 | 124,229 |
| 29200 | 18157 | Tippecanoe County | 172,803 | 196,115 | 30980 | 48203 | Harrison County | 65,640 | 66,386 |
| 29200 | 18171 | Warren County | 8,511 | 8,194 | 30980 | 48401 | Rusk County | 53,309 | 54,324 |
| | | | | | 30980 | 48459 | Upshur County | 39,311 | 42,166 |

| State/County FIPS code | Core based statistical area | Title and Geographic Components | 2010 census population | 2020 estimated population |
|---|---|---|---|---|
| 31020 | | Longview, WA Metro area | 102,408 | 111,371 |
| 31020 | 53015 | Cowlitz County | 102,408 | 111,371 |
| 31080 | | Los Angeles-Long Beach-Anaheim, CA Metro area | 12,828,957 | 13,109,903 |
| 31080 | | Anaheim-Santa Ana-Irvine, CA Div 11244 | 3,008,989 | 3,166,857 |
| 31080 | 06059 | Orange County | 3,008,989 | 3,166,857 |
| 31080 | | Los Angeles-Long Beach-Glendale, CA Div 31084 | 9,819,968 | 9,943,046 |
| 31080 | 06037 | Los Angeles County | 9,819,968 | 9,943,046 |
| 31140 | | Louisville/Jefferson County, KY-IN Metro area | 1,202,686 | 1,268,993 |
| 31140 | 18019 | Clark County | 110,222 | 119,266 |
| 31140 | 18043 | Floyd County | 74,579 | 78,936 |
| 31140 | 18061 | Harrison County | 39,366 | 40,682 |
| 31140 | 18175 | Washington County | 28,254 | 28,213 |
| 31140 | 21029 | Bullitt County | 74,308 | 82,182 |
| 31140 | 21103 | Henry County | 15,413 | 16,067 |
| 31140 | 21111 | Jefferson County | 741,075 | 767,452 |
| 31140 | 21185 | Oldham County | 60,356 | 66,999 |
| 31140 | 21211 | Shelby County | 41,971 | 49,611 |
| 31140 | 21215 | Spencer County | 17,142 | 19,585 |
| 31180 | | Lubbock, TX Metro area | 290,889 | 326,364 |
| 31180 | 48107 | Crosby County | 6,056 | 5,567 |
| 31180 | 48303 | Lubbock County | 278,918 | 314,772 |
| 31180 | 48305 | Lynn County | 5,915 | 6,025 |
| 31340 | | Lynchburg, VA Metro area | 252,654 | 264,386 |
| 31340 | 51009 | Amherst County | 32,354 | 31,667 |
| 31340 | 51011 | Appomattox County | 15,028 | 16,043 |
| 31340 | 51019 | Bedford County | 74,929 | 79,811 |
| 31340 | 51031 | Campbell County | 54,809 | 55,304 |
| 31340 | 51680 | Lynchburg city | 75,534 | 81,561 |
| 31420 | | Macon-Bibb County, GA Metro area | 232,245 | 229,900 |
| 31420 | 13021 | Bibb County | 155,783 | 152,737 |
| 31420 | 13079 | Crawford County | 12,601 | 12,231 |
| 31420 | 13169 | Jones County | 28,668 | 28,787 |
| 31420 | 13207 | Monroe County | 26,167 | 28,042 |
| 31420 | 13289 | Twiggs County | 9,026 | 8,103 |
| 31460 | | Madera, CA Metro area | 150,834 | 157,761 |
| 31460 | 06039 | Madera County | 150,834 | 157,761 |
| 31540 | | Madison, WI Metro area | 605,466 | 670,447 |
| 31540 | 55021 | Columbia County | 56,859 | 57,668 |
| 31540 | 55025 | Dane County | 488,081 | 552,536 |
| 31540 | 55045 | Green County | 36,839 | 36,603 |
| 31540 | 55049 | Iowa County | 23,687 | 23,640 |
| 31700 | | Manchester-Nashua, NH Metro area | 400,706 | 418,735 |
| 31700 | 33011 | Hillsborough County | 400,706 | 418,735 |
| 31740 | | Manhattan, KS Metro area | 127,094 | 130,142 |
| 31740 | 20061 | Geary County | 34,356 | 32,218 |
| 31740 | 20149 | Pottawatomie County | 21,608 | 24,722 |
| 31740 | 20161 | Riley County | 71,130 | 73,202 |
| 31860 | | Mankato, MN Metro area | 96,742 | 102,723 |
| 31860 | 27013 | Blue Earth County | 64,013 | 68,241 |
| 31860 | 27103 | Nicollet County | 32,729 | 34,482 |
| 31900 | | Mansfield, OH Metro area | 124,474 | 120,891 |
| 31900 | 39139 | Richland County | 124,474 | 120,891 |
| 32580 | | McAllen-Edinburg-Mission, TX Metro area | 774,764 | 875,200 |
| 32580 | 48215 | Hidalgo County | 774,764 | 875,200 |
| 32780 | | Medford, OR Metro area | 203,204 | 221,844 |
| 32780 | 41029 | Jackson County | 203,204 | 221,844 |
| 32820 | | Memphis, TN-MS-AR Metro area | 1,316,102 | 1,348,678 |
| 32820 | 05035 | Crittenden County | 50,907 | 47,616 |
| 32820 | 28033 | DeSoto County | 161,267 | 188,275 |
| 32820 | 28093 | Marshall County | 37,151 | 35,301 |
| 32820 | 28137 | Tate County | 28,872 | 28,539 |
| 32820 | 28143 | Tunica County | 10,778 | 9,392 |
| 32820 | 47047 | Fayette County | 38,439 | 41,620 |
| 32820 | 47157 | Shelby County | 927,682 | 936,017 |
| 32820 | 47167 | Tipton County | 61,006 | 61,918 |
| 32900 | | Merced, CA Metro area | 255,796 | 279,252 |
| 32900 | 06047 | Merced County | 255,796 | 279,252 |
| 33100 | | Miami-Fort Lauderdale-Pompano Beach, FL Metro area | 5,566,274 | 6,173,008 |
| 33100 | | Fort Lauderdale-Pompano Beach-Sunrise, FL Div 22744 | 1,748,146 | 1,958,105 |
| 33100 | 12011 | Broward County | 1,748,146 | 1,958,105 |
| 33100 | | Miami-Miami Beach-Kendall, FL Div 33124 | 2,497,993 | 2,707,303 |
| 33100 | 12086 | Miami-Dade County | 2,497,993 | 2,707,303 |
| 33100 | | West Palm Beach-Boca Raton-Boynton Beach, FL Div 48424 | 1,320,135 | 1,507,600 |
| 33100 | 12099 | Palm Beach County | 1,320,135 | 1,507,600 |
| 33140 | | Michigan City-La Porte, IN Metro area | 111,466 | 109,663 |
| 33140 | 18091 | LaPorte County | 111,466 | 109,663 |
| 33220 | | Midland, MI Metro area | 83,621 | 83,441 |
| 33220 | 26111 | Midland County | 83,621 | 83,441 |
| 33260 | | Midland, TX Metro area | 141,671 | 183,679 |
| 33260 | 48317 | Martin County | 4,799 | 5,816 |
| 33260 | 48329 | Midland County | 136,872 | 177,863 |
| 33340 | | Milwaukee-Waukesha, WI Metro area | 1,555,954 | 1,577,676 |
| 33340 | 55079 | Milwaukee County | 947,728 | 945,016 |
| 33340 | 55089 | Ozaukee County | 86,395 | 90,043 |
| 33340 | 55131 | Washington County | 131,885 | 136,445 |
| 33340 | 55133 | Waukesha County | 389,946 | 406,172 |
| 33460 | | Minneapolis-St. Paul-Bloomington, MN Metro area | 3,333,628 | 3,657,477 |
| 33460 | 27003 | Anoka County | 330,858 | 359,921 |
| 33460 | 27019 | Carver County | 91,086 | 106,565 |
| 33460 | 27025 | Chisago County | 53,890 | 56,794 |
| 33460 | 27037 | Dakota County | 398,590 | 431,807 |
| 33460 | 27053 | Hennepin County | 1,152,385 | 1,268,408 |
| 33460 | 27059 | Isanti County | 37,810 | 41,429 |
| 33460 | 27079 | Le Sueur County | 27,701 | 28,741 |
| 33460 | 27095 | Mille Lacs County | 26,097 | 26,146 |
| 33460 | 27123 | Ramsey County | 508,639 | 547,903 |
| 33460 | 27139 | Scott County | 129,908 | 150,689 |
| 33460 | 27141 | Sherburne County | 88,492 | 98,811 |
| 33460 | 27163 | Washington County | 238,109 | 265,476 |
| 33460 | 27171 | Wright County | 124,697 | 140,249 |
| 33460 | 55093 | Pierce County | 41,029 | 42,700 |
| 33460 | 55109 | St. Croix County | 84,337 | 91,838 |
| 33540 | | Missoula, MT Metro area | 109,296 | 121,630 |
| 33540 | 30063 | Missoula County | 109,296 | 121,630 |
| 33660 | | Mobile, AL Metro area | 430,719 | 428,692 |
| 33660 | 01097 | Mobile County | 413,139 | 412,716 |
| 33660 | 01129 | Washington County | 17,580 | 15,976 |
| 33700 | | Modesto, CA Metro area | 514,450 | 550,081 |
| 33700 | 06099 | Stanislaus County | 514,450 | 550,081 |
| 33740 | | Monroe, LA Metro area | 204,487 | 198,836 |
| 33740 | 22067 | Morehouse Parish | 27,979 | 24,227 |
| 33740 | 22073 | Ouachita Parish | 153,734 | 152,439 |
| 33740 | 22111 | Union Parish | 22,774 | 22,170 |
| 33780 | | Monroe, MI Metro area | 152,031 | 150,568 |
| 33780 | 26115 | Monroe County | 152,031 | 150,568 |

| State/County FIPS code | Core based statistical area | Title and Geographic Components | 2010 census population | 2020 estimated population | State/County FIPS code | Core based statistical area | Title and Geographic Components | 2010 census population | 2020 estimated population |
|---|---|---|---|---|---|---|---|---|---|
| 33860 | | Montgomery, AL Metro area.............. | 374,540 | 372,583 | 35620 | | Newark, NJ-PA Div 35084.............. | 2,146,271 | 2,167,853 |
| 33860 | 01001 | Autauga County .............. | 54,597 | 56,145 | 35620 | 34013 | Essex County.............. | 783,891 | 800,501 |
| 33860 | 01051 | Elmore County .............. | 79,272 | 82,158 | 35620 | 34019 | Hunterdon County .............. | 127,361 | 124,797 |
| 33860 | 01085 | Lowndes County .............. | 11,296 | 9,641 | 35620 | 34027 | Morris County .............. | 492,281 | 491,087 |
| 33860 | 01101 | Montgomery County .............. | 229,375 | 224,639 | 35620 | 34037 | Sussex County .............. | 148,936 | 140,002 |
| | | | | | 35620 | 34039 | Union County .............. | 536,464 | 555,394 |
| 34060 | | Morgantown, WV Metro area .............. | 129,702 | 140,199 | 35620 | 42103 | Pike County .............. | 57,338 | 56,072 |
| 34060 | 54061 | Monongalia County .............. | 96,184 | 106,819 | | | | | |
| 34060 | 54077 | Preston County.............. | 33,518 | 33,380 | 35620 | | New Brunswick-Lakewood, NJ Div 35154... | 2,340,425 | 2,384,685 |
| | | | | | 35620 | 34023 | Middlesex County .............. | 810,038 | 822,736 |
| 34100 | | Morristown, TN Metro area.............. | 136,858 | 143,982 | 35620 | 34025 | Monmouth County .............. | 630,362 | 618,381 |
| 34100 | 47057 | Grainger County .............. | 22,656 | 23,565 | 35620 | 34029 | Ocean County .............. | 576,546 | 614,237 |
| 34100 | 47063 | Hamblen County .............. | 62,534 | 65,110 | 35620 | 34035 | Somerset County.............. | 323,479 | 329,331 |
| 34100 | 47089 | Jefferson County .............. | 51,668 | 55,307 | | | | | |
| | | | | | 35620 | | New York-Jersey City-White Plains, NY-NJ Div 35614 | 11,576,585 | 11,746,214 |
| 34580 | | Mount Vernon-Anacortes, WA Metro area .. | 116,892 | 130,789 | 35620 | 34003 | Bergen County .............. | 905,107 | 930,394 |
| 34580 | 53057 | Skagit County .............. | 116,892 | 130,789 | 35620 | 34017 | Hudson County .............. | 634,284 | 671,666 |
| | | | | | 35620 | 34031 | Passaic County .............. | 501,600 | 500,382 |
| 34620 | | Muncie, IN Metro area.............. | 117,670 | 113,454 | 35620 | 36005 | Bronx County .............. | 1,384,580 | 1,401,142 |
| 34620 | 18035 | Delaware County .............. | 117,670 | 113,454 | 35620 | 36047 | Kings County .............. | 2,504,721 | 2,538,934 |
| | | | | | 35620 | 36061 | New York County .............. | 1,586,381 | 1,611,989 |
| 34740 | | Muskegon, MI Metro area .............. | 172,194 | 173,883 | 35620 | 36079 | Putnam County .............. | 99,654 | 98,532 |
| 34740 | 26121 | Muskegon County .............. | 172,194 | 173,883 | 35620 | 36081 | Queens County .............. | 2,230,619 | 2,225,821 |
| | | | | | 35620 | 36085 | Richmond County .............. | 468,730 | 475,327 |
| 34820 | | Myrtle Beach-Conway-North Myrtle Beach, SC-NC Metro area.............. | 376,575 | 514,488 | 35620 | 36087 | Rockland County .............. | 311,691 | 326,225 |
| 34820 | 37019 | Brunswick County .............. | 107,429 | 149,039 | 35620 | 36119 | Westchester County .............. | 949,218 | 965,802 |
| 34820 | 45051 | Horry County.............. | 269,146 | 365,449 | | | | | |
| | | | | | 35660 | | Niles, MI Metro area .............. | 156,808 | 153,025 |
| 34900 | | Napa, CA Metro area .............. | 136,535 | 135,965 | 35660 | 26021 | Berrien County .............. | 156,808 | 153,025 |
| 34900 | 06055 | Napa County.............. | 136,535 | 135,965 | | | | | |
| | | | | | 35840 | | North Port-Sarasota-Bradenton, FL Metro area .............. | 702,312 | 854,684 |
| 34940 | | Naples-Marco Island, FL Metro area.......... | 321,522 | 392,973 | 35840 | 12081 | Manatee County .............. | 322,879 | 411,219 |
| 34940 | 12021 | Collier County .............. | 321,522 | 392,973 | 35840 | 12115 | Sarasota County .............. | 379,433 | 443,465 |
| | | | | | | | | | |
| 34980 | | Nashville-Davidson--Murfreesboro--Franklin, TN Metro area.............. | 1,646,183 | 1,961,232 | 35980 | | Norwich-New London, CT Metro area.......... | 274,070 | 264,999 |
| 34980 | 47015 | Cannon County .............. | 13,813 | 14,847 | 35980 | 09011 | New London County .............. | 274,070 | 264,999 |
| 34980 | 47021 | Cheatham County .............. | 39,110 | 41,101 | | | | | |
| 34980 | 47037 | Davidson County .............. | 626,558 | 694,176 | 36100 | | Ocala, FL Metro area .............. | 331,299 | 373,513 |
| 34980 | 47043 | Dickson County .............. | 49,650 | 54,376 | 36100 | 12083 | Marion County .............. | 331,299 | 373,513 |
| 34980 | 47111 | Macon County .............. | 22,226 | 24,827 | | | | | |
| 34980 | 47119 | Maury County .............. | 80,932 | 99,590 | 36140 | | Ocean City, NJ Metro area.............. | 97,257 | 91,546 |
| 34980 | 47147 | Robertson County .............. | 66,319 | 72,275 | 36140 | 34009 | Cape May County .............. | 97,257 | 91,546 |
| 34980 | 47149 | Rutherford County .............. | 262,588 | 339,261 | | | | | |
| 34980 | 47159 | Smith County .............. | 19,150 | 20,285 | 36220 | | Odessa, TX Metro area.............. | 137,136 | 167,701 |
| 34980 | 47165 | Sumner County .............. | 160,634 | 195,561 | 36220 | 48135 | Ector County .............. | 137,136 | 167,701 |
| 34980 | 47169 | Trousdale County .............. | 7,864 | 11,455 | | | | | |
| 34980 | 47187 | Williamson County .............. | 183,277 | 245,348 | 36260 | | Ogden-Clearfield, UT Metro area.............. | 597,162 | 691,359 |
| 34980 | 47189 | Wilson County .............. | 114,062 | 148,130 | 36260 | 49003 | Box Elder County .............. | 49,983 | 57,007 |
| | | | | | 36260 | 49011 | Davis County .............. | 306,492 | 359,232 |
| 35100 | | New Bern, NC Metro area.............. | 126,808 | 123,198 | 36260 | 49029 | Morgan County .............. | 9,469 | 12,462 |
| 35100 | 37049 | Craven County .............. | 103,498 | 101,233 | 36260 | 49057 | Weber County .............. | 231,218 | 262,658 |
| 35100 | 37103 | Jones County .............. | 10,167 | 9,250 | | | | | |
| 35100 | 37137 | Pamlico County .............. | 13,143 | 12,715 | 36420 | | Oklahoma City, OK Metro area .............. | 1,253,002 | 1,425,375 |
| | | | | | 36420 | 40017 | Canadian County .............. | 115,566 | 153,192 |
| 35300 | | New Haven-Milford, CT Metro area.............. | 862,442 | 851,948 | 36420 | 40027 | Cleveland County .............. | 255,990 | 287,066 |
| 35300 | 09009 | New Haven County .............. | 862,442 | 851,948 | 36420 | 40051 | Grady County.............. | 52,430 | 55,906 |
| | | | | | 36420 | 40081 | Lincoln County .............. | 34,274 | 35,045 |
| 35380 | | New Orleans-Metairie, LA Metro area.......... | 1,189,891 | 1,272,258 | 36420 | 40083 | Logan County .............. | 41,854 | 48,777 |
| 35380 | 22051 | Jefferson Parish .............. | 432,576 | 432,346 | 36420 | 40087 | McClain County .............. | 34,503 | 41,348 |
| 35380 | 22071 | Orleans Parish .............. | 343,828 | 389,476 | 36420 | 40109 | Oklahoma County .............. | 718,385 | 804,041 |
| 35380 | 22075 | Plaquemines Parish.............. | 23,035 | 23,113 | | | | | |
| 35380 | 22087 | St. Bernard Parish .............. | 35,897 | 47,647 | 36500 | | Olympia-Lacey-Tumwater, WA Metro area. | 252,260 | 294,074 |
| 35380 | 22089 | St. Charles Parish .............. | 52,888 | 52,987 | 36500 | 53067 | Thurston County .............. | 252,260 | 294,074 |
| 35380 | 22093 | St. James Parish.............. | 22,101 | 20,727 | | | | | |
| 35380 | 22095 | St. John the Baptist Parish .............. | 45,810 | 42,516 | 36540 | | Omaha-Council Bluffs, NE-IA Metro area ... | 865,347 | 954,270 |
| 35380 | 22103 | St. Tammany Parish .............. | 233,756 | 263,446 | 36540 | 19085 | Harrison County.............. | 14,937 | 13,928 |
| | | | | | 36540 | 19129 | Mills County .............. | 15,059 | 14,766 |
| 35620 | | New York-Newark-Jersey City, NY-NJ-PA Metro area .............. | 18,896,277 | 19,124,359 | 36540 | 19155 | Pottawattamie County.............. | 93,149 | 93,328 |
| | | | | | 36540 | 31025 | Cass County .............. | 25,241 | 26,232 |
| 35620 | | Nassau County-Suffolk County, NY Div 35004 .............. | 2,832,996 | 2,825,607 | 36540 | 31055 | Douglas County .............. | 517,116 | 574,332 |
| | | | | | 36540 | 31153 | Sarpy County .............. | 158,835 | 188,856 |
| 35620 | 36059 | Nassau County .............. | 1,339,880 | 1,351,334 | 36540 | 31155 | Saunders County .............. | 20,778 | 21,927 |
| 35620 | 36103 | Suffolk County .............. | 1,493,116 | 1,474,273 | 36540 | 31177 | Washington County .............. | 20,232 | 20,901 |

| State/County FIPS code | Core based statistical area | Title and Geographic Components | 2010 census population | 2020 estimated population | State/County FIPS code | Core based statistical area | Title and Geographic Components | 2010 census population | 2020 estimated population |
|---|---|---|---|---|---|---|---|---|---|
| 36740 | | Orlando-Kissimmee-Sanford, FL Metro area | 2,134,399 | 2,639,374 | 38300 | | Pittsburgh, PA Metro area | 2,356,294 | 2,309,246 |
| 36740 | 12069 | Lake County | 297,047 | 375,492 | 38300 | 42003 | Allegheny County | 1,223,303 | 1,211,358 |
| 36740 | 12095 | Orange County | 1,145,957 | 1,404,396 | 38300 | 42005 | Armstrong County | 69,059 | 64,162 |
| 36740 | 12097 | Osceola County | 268,685 | 385,315 | 38300 | 42007 | Beaver County | 170,531 | 162,575 |
| 36740 | 12117 | Seminole County | 422,710 | 474,171 | 38300 | 42019 | Butler County | 183,880 | 189,135 |
| | | | | | 38300 | 42051 | Fayette County | 136,601 | 128,126 |
| 36780 | | Oshkosh-Neenah, WI Metro area | 167,000 | 171,631 | 38300 | 42125 | Washington County | 207,849 | 206,803 |
| 36780 | 55139 | Winnebago County | 167,000 | 171,631 | 38300 | 42129 | Westmoreland County | 365,071 | 347,087 |
| 36980 | | Owensboro, KY Metro area | 114,746 | 119,795 | 38340 | | Pittsfield, MA Metro area | 131,274 | 124,571 |
| 36980 | 21059 | Daviess County | 96,641 | 101,978 | 38340 | 25003 | Berkshire County | 131,274 | 124,571 |
| 36980 | 21091 | Hancock County | 8,565 | 8,742 | | | | | |
| 36980 | 21149 | McLean County | 9,540 | 9,075 | 38540 | | Pocatello, ID Metro area | 90,661 | 96,438 |
| | | | | | 38540 | 16005 | Bannock County | 82,842 | 88,795 |
| 37100 | | Oxnard-Thousand Oaks-Ventura, CA Metro area | 823,398 | 841,387 | 38540 | 16077 | Power County | 7,819 | 7,643 |
| 37100 | 06111 | Ventura County | 823,398 | 841,387 | 38860 | | Portland-South Portland, ME Metro area | 514,108 | 543,221 |
| | | | | | 38860 | 23005 | Cumberland County | 281,690 | 298,111 |
| 37340 | | Palm Bay-Melbourne-Titusville, FL Metro area | 543,372 | 608,459 | 38860 | 23023 | Sagadahoc County | 35,287 | 36,044 |
| 37340 | 12009 | Brevard County | 543,372 | 608,459 | 38860 | 23031 | York County | 197,131 | 209,066 |
| 37460 | | Panama City, FL Metro area | 168,850 | 171,322 | 38900 | | Portland-Vancouver-Hillsboro, OR-WA Metro area | 2,226,003 | 2,510,259 |
| 37460 | 12005 | Bay County | 168,850 | 171,322 | 38900 | 41005 | Clackamas County | 375,996 | 421,596 |
| | | | | | 38900 | 41009 | Columbia County | 49,353 | 52,876 |
| 37620 | | Parkersburg-Vienna, WV Metro area | 92,667 | 88,643 | 38900 | 41051 | Multnomah County | 735,146 | 815,637 |
| 37620 | 54105 | Wirt County | 5,714 | 5,705 | 38900 | 41067 | Washington County | 529,862 | 603,514 |
| 37620 | 54107 | Wood County | 86,953 | 82,938 | 38900 | 41071 | Yamhill County | 99,216 | 107,664 |
| | | | | | 38900 | 53011 | Clark County | 425,360 | 496,865 |
| 37860 | | Pensacola-Ferry Pass-Brent, FL Metro area | 448,991 | 511,503 | 38900 | 53059 | Skamania County | 11,070 | 12,107 |
| 37860 | 12033 | Escambia County | 297,620 | 322,364 | | | | | |
| 37860 | 12113 | Santa Rosa County | 151,371 | 189,139 | 38940 | | Port St. Lucie, FL Metro area | 424,107 | 499,274 |
| | | | | | 38940 | 12085 | Martin County | 146,852 | 162,088 |
| 37900 | | Peoria, IL Metro area | 416,253 | 396,781 | 38940 | 12111 | St. Lucie County | 277,255 | 337,186 |
| 37900 | 17057 | Fulton County | 37,071 | 33,690 | | | | | |
| 37900 | 17123 | Marshall County | 12,646 | 11,309 | 39100 | | Poughkeepsie-Newburgh-Middletown, NY Metro area | 670,280 | 678,527 |
| 37900 | 17143 | Peoria County | 186,496 | 177,652 | 39100 | 36027 | Dutchess County | 297,454 | 293,293 |
| 37900 | 17175 | Stark County | 5,992 | 5,262 | 39100 | 36071 | Orange County | 372,826 | 385,234 |
| 37900 | 17179 | Tazewell County | 135,392 | 130,777 | | | | | |
| 37900 | 17203 | Woodford County | 38,656 | 38,091 | 39150 | | Prescott Valley-Prescott, AZ Metro area | 211,017 | 240,226 |
| | | | | | 39150 | 04025 | Yavapai County | 211,017 | 240,226 |
| 37980 | | Philadelphia-Camden-Wilmington, PA-NJ-DE-MD Metro area | 5,965,677 | 6,107,906 | 39300 | | Providence-Warwick, RI-MA Metro area | 1,601,206 | 1,623,890 |
| | | | | | 39300 | 25005 | Bristol County | 548,242 | 566,765 |
| 37980 | | Camden, NJ Div 15804 | 1,251,021 | 1,246,650 | 39300 | 44001 | Bristol County | 49,844 | 48,350 |
| 37980 | 34005 | Burlington County | 448,730 | 446,596 | 39300 | 44003 | Kent County | 166,109 | 164,646 |
| 37980 | 34007 | Camden County | 513,535 | 506,809 | 39300 | 44005 | Newport County | 83,141 | 81,836 |
| 37980 | 34015 | Gloucester County | 288,756 | 293,245 | 39300 | 44007 | Providence County | 626,781 | 636,547 |
| | | | | | 39300 | 44009 | Washington County | 127,089 | 125,746 |
| 37980 | | Montgomery County-Bucks County-Chester County, PA Div 33874 | 1,924,229 | 1,988,615 | 39340 | | Provo-Orem, UT Metro area | 526,885 | 663,181 |
| 37980 | 42017 | Bucks County | 625,256 | 627,987 | 39340 | 49023 | Juab County | 10,246 | 12,122 |
| 37980 | 42029 | Chester County | 499,133 | 526,759 | 39340 | 49049 | Utah County | 516,639 | 651,059 |
| 37980 | 42091 | Montgomery County | 799,840 | 833,869 | | | | | |
| | | | | | 39380 | | Pueblo, CO Metro area | 159,063 | 169,823 |
| 37980 | | Philadelphia, PA Div 37964 | 2,084,769 | 2,145,240 | 39380 | 08101 | Pueblo County | 159,063 | 169,823 |
| 37980 | 42045 | Delaware County | 558,757 | 566,753 | | | | | |
| 37980 | 42101 | Philadelphia County | 1,526,012 | 1,578,487 | 39460 | | Punta Gorda, FL Metro area | 159,967 | 194,711 |
| | | | | | 39460 | 12015 | Charlotte County | 159,967 | 194,711 |
| 37980 | | Wilmington, DE-MD-NJ Div 48864 | 705,658 | 727,401 | 39540 | | Racine, WI Metro area | 195,428 | 195,802 |
| 37980 | 10003 | New Castle County | 538,484 | 561,531 | 39540 | 55101 | Racine County | 195,428 | 195,802 |
| 37980 | 24015 | Cecil County | 101,102 | 103,419 | | | | | |
| 37980 | 34033 | Salem County | 66,072 | 62,451 | 39580 | | Raleigh-Cary, NC Metro area | 1,130,493 | 1,420,376 |
| | | | | | 39580 | 37069 | Franklin County | 60,563 | 71,859 |
| 38060 | | Phoenix-Mesa-Chandler, AZ Metro area | 4,193,129 | 5,059,909 | 39580 | 37101 | Johnston County | 168,878 | 216,246 |
| 38060 | 04013 | Maricopa County | 3,817,365 | 4,579,081 | 39580 | 37183 | Wake County | 901,052 | 1,132,271 |
| 38060 | 04021 | Pinal County | 375,764 | 480,828 | | | | | |
| | | | | | 39660 | | Rapid City, SD Metro area | 126,400 | 144,514 |
| 38220 | | Pine Bluff, AR Metro area | 100,289 | 86,278 | 39660 | 46093 | Meade County | 25,440 | 28,588 |
| 38220 | 05025 | Cleveland County | 8,692 | 7,957 | 39660 | 46103 | Pennington County | 100,960 | 115,926 |
| 38220 | 05069 | Jefferson County | 77,456 | 65,377 | | | | | |
| 38220 | 05079 | Lincoln County | 14,141 | 12,944 | 39740 | | Reading, PA Metro area | 411,570 | 421,017 |
| | | | | | 39740 | 42011 | Berks County | 411,570 | 421,017 |

| State/County FIPS code | Core based statistical area | Title and Geographic Components | 2010 census population | 2020 estimated population | State/County FIPS code | Core based statistical area | Title and Geographic Components | 2010 census population | 2020 estimated population |
|---|---|---|---|---|---|---|---|---|---|
| 39820 | | Redding, CA Metro area | 177,221 | 179,027 | 41140 | | St. Joseph, MO-KS Metro area | 127,319 | 122,556 |
| 39820 | 06089 | Shasta County | 177,221 | 179,027 | 41140 | 20043 | Doniphan County | 7,948 | 7,496 |
| | | | | | 41140 | 29003 | Andrew County | 17,296 | 17,586 |
| 39900 | | Reno, NV Metro area | 425,442 | 481,289 | 41140 | 29021 | Buchanan County | 89,191 | 86,530 |
| 39900 | 32029 | Storey County | 4,013 | 4,207 | 41140 | 29063 | DeKalb County | 12,884 | 10,944 |
| 39900 | 32031 | Washoe County | 421,429 | 477,082 | | | | | |
| | | | | | 41180 | | St. Louis, MO-IL Metro area | 2,787,751 | 2,805,473 |
| 40060 | | Richmond, VA Metro area | 1,186,471 | 1,303,469 | 41180 | 17005 | Bond County | 17,768 | 16,262 |
| 40060 | 51007 | Amelia County | 12,695 | 13,014 | 41180 | 17013 | Calhoun County | 5,089 | 4,616 |
| 40060 | 51036 | Charles City County | 7,256 | 6,821 | 41180 | 17027 | Clinton County | 37,762 | 37,398 |
| 40060 | 51041 | Chesterfield County | 316,240 | 358,245 | 41180 | 17083 | Jersey County | 23,010 | 21,616 |
| 40060 | 51053 | Dinwiddie County | 28,012 | 28,688 | 41180 | 17117 | Macoupin County | 47,763 | 44,567 |
| 40060 | 51075 | Goochland County | 21,692 | 24,431 | 41180 | 17119 | Madison County | 269,298 | 262,635 |
| 40060 | 51085 | Hanover County | 99,850 | 108,262 | 41180 | 17133 | Monroe County | 32,949 | 34,739 |
| 40060 | 51087 | Henrico County | 306,756 | 333,766 | 41180 | 17163 | St. Clair County | 270,078 | 258,046 |
| 40060 | 51097 | King and Queen County | 6,940 | 6,942 | 41180 | 29071 | Franklin County | 101,468 | 104,469 |
| 40060 | 51101 | King William County | 15,927 | 17,641 | 41180 | 29099 | Jefferson County | 218,722 | 226,543 |
| 40060 | 51127 | New Kent County | 18,432 | 23,648 | 41180 | 29113 | Lincoln County | 52,536 | 60,119 |
| 40060 | 51145 | Powhatan County | 28,069 | 30,148 | 41180 | 29183 | St. Charles County | 360,495 | 406,204 |
| 40060 | 51149 | Prince George County | 35,715 | 38,686 | 41180 | 29189 | St. Louis County | 998,985 | 994,020 |
| 40060 | 51183 | Sussex County | 12,070 | 10,925 | 41180 | 29219 | Warren County | 32,539 | 36,594 |
| 40060 | 51570 | Colonial Heights city | 17,410 | 17,205 | 41180 | 29510 | St. Louis city | 319,289 | 297,645 |
| 40060 | 51670 | Hopewell city | 22,591 | 22,375 | | | | | |
| 40060 | 51730 | Petersburg city | 32,441 | 30,446 | 41420 | | Salem, OR Metro area | 390,738 | 436,948 |
| 40060 | 51760 | Richmond city | 204,375 | 232,226 | 41420 | 41047 | Marion County | 315,338 | 349,204 |
| | | | | | 41420 | 41053 | Polk County | 75,400 | 87,744 |
| 40140 | | Riverside-San Bernardino-Ontario, CA Metro area | 4,224,948 | 4,678,371 | | | | | |
| 40140 | 06065 | Riverside County | 2,189,765 | 2,489,188 | 41500 | | Salinas, CA Metro area | 415,059 | 430,906 |
| 40140 | 06071 | San Bernardino County | 2,035,183 | 2,189,183 | 41500 | 06053 | Monterey County | 415,059 | 430,906 |
| | | | | | | | | | |
| 40220 | | Roanoke, VA Metro area | 308,666 | 313,784 | 41540 | | Salisbury, MD-DE Metro area | 373,754 | 423,481 |
| 40220 | 51023 | Botetourt County | 33,152 | 33,633 | 41540 | 10005 | Sussex County | 197,103 | 241,546 |
| 40220 | 51045 | Craig County | 5,175 | 5,077 | 41540 | 24039 | Somerset County | 26,470 | 25,453 |
| 40220 | 51067 | Franklin County | 56,128 | 56,167 | 41540 | 24045 | Wicomico County | 98,733 | 103,990 |
| 40220 | 51161 | Roanoke County | 92,465 | 94,509 | 41540 | 24047 | Worcester County | 51,448 | 52,403 |
| 40220 | 51770 | Roanoke city | 96,910 | 99,058 | | | | | |
| 40220 | 51775 | Salem city | 24,836 | 25,340 | 41620 | | Salt Lake City, UT Metro area | 1,087,808 | 1,240,029 |
| | | | | | 41620 | 49035 | Salt Lake County | 1,029,590 | 1,165,517 |
| 40340 | | Rochester, MN Metro area | 206,888 | 223,062 | 41620 | 49045 | Tooele County | 58,218 | 74,512 |
| 40340 | 27039 | Dodge County | 20,087 | 20,987 | | | | | |
| 40340 | 27045 | Fillmore County | 20,868 | 21,135 | 41660 | | San Angelo, TX Metro area | 112,968 | 122,889 |
| 40340 | 27109 | Olmsted County | 144,268 | 159,298 | 41660 | 48235 | Irion County | 1,597 | 1,564 |
| 40340 | 27157 | Wabasha County | 21,665 | 21,642 | 41660 | 48431 | Sterling County | 1,143 | 1,315 |
| | | | | | 41660 | 48451 | Tom Green County | 110,228 | 120,010 |
| 40380 | | Rochester, NY Metro area | 1,079,704 | 1,067,486 | | | | | |
| 40380 | 36051 | Livingston County | 65,206 | 62,398 | 41700 | | San Antonio-New Braunfels, TX Metro area | 2,142,520 | 2,590,732 |
| 40380 | 36055 | Monroe County | 744,394 | 740,900 | 41700 | 48013 | Atascosa County | 44,923 | 51,724 |
| 40380 | 36069 | Ontario County | 108,099 | 110,091 | 41700 | 48019 | Bandera County | 20,487 | 23,861 |
| 40380 | 36073 | Orleans County | 42,890 | 39,978 | 41700 | 48029 | Bexar County | 1,714,781 | 2,026,823 |
| 40380 | 36117 | Wayne County | 93,751 | 89,339 | 41700 | 48091 | Comal County | 108,520 | 164,812 |
| 40380 | 36123 | Yates County | 25,364 | 24,780 | 41700 | 48187 | Guadalupe County | 131,527 | 170,608 |
| | | | | | 41700 | 48259 | Kendall County | 33,384 | 48,523 |
| 40420 | | Rockford, IL Metro area | 349,431 | 334,072 | 41700 | 48325 | Medina County | 45,993 | 52,358 |
| 40420 | 17007 | Boone County | 54,167 | 52,777 | 41700 | 48493 | Wilson County | 42,905 | 52,023 |
| 40420 | 17201 | Winnebago County | 295,264 | 281,295 | | | | | |
| | | | | | 41740 | | San Diego-Chula Vista-Carlsbad, CA Metro area | 3,095,349 | 3,332,427 |
| 40580 | | Rocky Mount, NC Metro area | 152,368 | 145,688 | 41740 | 06073 | San Diego County | 3,095,349 | 3,332,427 |
| 40580 | 37065 | Edgecombe County | 56,539 | 50,829 | | | | | |
| 40580 | 37127 | Nash County | 95,829 | 94,859 | 41860 | | San Francisco-Oakland-Berkeley, CA Metro area | 4,335,593 | 4,696,902 |
| | | | | | | | | | |
| 40660 | | Rome, GA Metro area | 96,314 | 98,604 | 41860 | | Oakland-Berkeley-Livermore, CA Div 36084 | 2,559,462 | 2,814,656 |
| 40660 | 13115 | Floyd County | 96,314 | 98,604 | 41860 | 06001 | Alameda County | 1,510,258 | 1,662,323 |
| | | | | | 41860 | 06013 | Contra Costa County | 1,049,204 | 1,152,333 |
| 40900 | | Sacramento-Roseville-Folsom, CA Metro area | 2,149,150 | 2,374,749 | | | | | |
| 40900 | 06017 | El Dorado County | 181,058 | 192,925 | 41860 | | San Francisco-San Mateo-Redwood City, CA Div 41884 | 1,523,701 | 1,624,914 |
| 40900 | 06061 | Placer County | 348,502 | 402,950 | 41860 | 06075 | San Francisco County | 805,184 | 866,606 |
| 40900 | 06067 | Sacramento County | 1,418,735 | 1,559,146 | 41860 | 06081 | San Mateo County | 718,517 | 758,308 |
| 40900 | 06113 | Yolo County | 200,855 | 219,728 | | | | | |
| | | | | | 41860 | | San Rafael, CA Div 42034 | 252,430 | 257,332 |
| 40980 | | Saginaw, MI Metro area | 200,169 | 189,868 | 41860 | 06041 | Marin County | 252,430 | 257,332 |
| 40980 | 26145 | Saginaw County | 200,169 | 189,868 | | | | | |
| | | | | | 41940 | | San Jose-Sunnyvale-Santa Clara, CA Metro area | 1,836,951 | 1,971,160 |
| 41060 | | St. Cloud, MN Metro area | 189,093 | 202,996 | 41940 | 06069 | San Benito County | 55,265 | 64,055 |
| 41060 | 27009 | Benton County | 38,451 | 40,958 | 41940 | 06085 | Santa Clara County | 1,781,686 | 1,907,105 |
| 41060 | 27145 | Stearns County | 150,642 | 162,038 | | | | | |
| 41100 | | St. George, UT Metro area | 138,115 | 184,913 | | | | | |
| 41100 | 49053 | Washington County | 138,115 | 184,913 | | | | | |

| State/County FIPS code | Core based statistical area | Title and Geographic Components | 2010 census population | 2020 estimated population |
|---|---|---|---|---|
| 42020 | | San Luis Obispo-Paso Robles, CA Metro area ..... | 269,597 | 282,249 |
| 42020 | 06079 | San Luis Obispo County ......................... | 269,597 | 282,249 |
| 42100 | | Santa Cruz-Watsonville, CA Metro area ..... | 262,350 | 269,925 |
| 42100 | 06087 | Santa Cruz County ..................................... | 262,350 | 269,925 |
| 42140 | | Santa Fe, NM Metro area............................ | 144,232 | 151,946 |
| 42140 | 35049 | Santa Fe County ........................................ | 144,232 | 151,946 |
| 42200 | | Santa Maria-Santa Barbara, CA Metro area | 423,947 | 444,766 |
| 42200 | 06083 | Santa Barbara County ............................. | 423,947 | 444,766 |
| 42220 | | Santa Rosa-Petaluma, CA Metro area....... | 483,861 | 489,819 |
| 42220 | 06097 | Sonoma County ........................................ | 483,861 | 489,819 |
| 42340 | | Savannah, GA Metro area............................ | 347,597 | 395,983 |
| 42340 | 13029 | Bryan County .............................................. | 30,215 | 40,755 |
| 42340 | 13051 | Chatham County.......................................... | 265,127 | 289,463 |
| 42340 | 13103 | Effingham County ........................................ | 52,255 | 65,765 |
| 42540 | | Scranton--Wilkes-Barre, PA Metro area...... | 563,604 | 552,528 |
| 42540 | 42069 | Lackawanna County ..................................... | 214,415 | 208,989 |
| 42540 | 42079 | Luzerne County ........................................... | 320,906 | 316,982 |
| 42540 | 42131 | Wyoming County ......................................... | 28,283 | 26,557 |
| 42660 | | Seattle-Tacoma-Bellevue, WA Metro area.. | 3,439,808 | 4,018,598 |
| 42660 | | Seattle-Bellevue-Kent, WA Div 42644......... | 2,644,586 | 3,104,708 |
| 42660 | 53033 | King County................................................. | 1,931,287 | 2,274,315 |
| 42660 | 53061 | Snohomish County ...................................... | 713,299 | 830,393 |
| 42660 | | Tacoma-Lakewood, WA Div 45104.............. | 795,222 | 913,890 |
| 42660 | 53053 | Pierce County ............................................. | 795,222 | 913,890 |
| 42680 | | Sebastian-Vero Beach, FL Metro area........ | 138,028 | 162,518 |
| 42680 | 12061 | Indian River County .................................... | 138,028 | 162,518 |
| 42700 | | Sebring-Avon Park, FL Metro area.............. | 98,784 | 106,639 |
| 42700 | 12055 | Highlands County ........................................ | 98,784 | 106,639 |
| 43100 | | Sheboygan, WI Metro area ......................... | 115,512 | 115,240 |
| 43100 | 55117 | Sheboygan County ...................................... | 115,512 | 115,240 |
| 43300 | | Sherman-Denison, TX Metro area .............. | 120,877 | 138,318 |
| 43300 | 48181 | Grayson County ........................................... | 120,877 | 138,318 |
| 43340 | | Shreveport-Bossier City, LA Metro area...... | 398,606 | 392,404 |
| 43340 | 22015 | Bossier Parish ............................................ | 117,036 | 127,275 |
| 43340 | 22017 | Caddo Parish .............................................. | 254,914 | 237,479 |
| 43340 | 22031 | De Soto Parish ........................................... | 26,656 | 27,650 |
| 43420 | | Sierra Vista-Douglas, AZ Metro area .......... | 131,359 | 127,450 |
| 43420 | 04003 | Cochise County ........................................... | 131,359 | 127,450 |
| 43580 | | Sioux City, IA-NE-SD Metro area................ | 143,582 | 144,996 |
| 43580 | 19193 | Woodbury County ........................................ | 102,175 | 103,138 |
| 43580 | 31043 | Dakota County ............................................. | 21,006 | 20,070 |
| 43580 | 31051 | Dixon County .............................................. | 6,003 | 5,596 |
| 43580 | 46127 | Union County ............................................... | 14,398 | 16,192 |
| 43620 | | Sioux Falls, SD Metro area ........................ | 228,264 | 273,566 |
| 43620 | 46083 | Lincoln County ............................................ | 44,823 | 63,019 |
| 43620 | 46087 | McCook County........................................... | 5,618 | 5,520 |
| 43620 | 46099 | Minnehaha County....................................... | 169,474 | 196,659 |
| 43620 | 46125 | Turner County.............................................. | 8,349 | 8,368 |
| 43780 | | South Bend-Mishawaka, IN-MI Metro area . | 319,203 | 323,068 |
| 43780 | 18141 | St. Joseph County ...................................... | 266,914 | 271,484 |
| 43780 | 26027 | Cass County ............................................... | 52,289 | 51,584 |
| 43900 | | Spartanburg, SC Metro area ...................... | 284,304 | 326,205 |
| 43900 | 45083 | Spartanburg County..................................... | 284,304 | 326,205 |
| 44060 | | Spokane-Spokane Valley, WA Metro area.. | 514,752 | 574,585 |
| 44060 | 53063 | Spokane County .......................................... | 471,220 | 528,225 |
| 44060 | 53065 | Stevens County ........................................... | 43,532 | 46,360 |

| State/County FIPS code | Core based statistical area | Title and Geographic Components | 2010 census population | 2020 estimated population |
|---|---|---|---|---|
| 44100 | | Springfield, IL Metro area............................ | 210,170 | 205,950 |
| 44100 | 17129 | Menard County ............................................ | 12,705 | 12,068 |
| 44100 | 17167 | Sangamon County ....................................... | 197,465 | 193,882 |
| 44140 | | Springfield, MA Metro area.......................... | 693,059 | 695,654 |
| 44140 | 25011 | Franklin County........................................... | 71,381 | 70,267 |
| 44140 | 25013 | Hampden County ......................................... | 463,615 | 463,986 |
| 44140 | 25015 | Hampshire County ....................................... | 158,063 | 161,401 |
| 44180 | | Springfield, MO Metro area.......................... | 436,756 | 475,220 |
| 44180 | 29043 | Christian County.......................................... | 77,414 | 90,655 |
| 44180 | 29059 | Dallas County ............................................. | 16,769 | 17,219 |
| 44180 | 29077 | Greene County ............................................ | 275,179 | 294,997 |
| 44180 | 29167 | Polk County ................................................ | 31,130 | 32,490 |
| 44180 | 29225 | Webster County........................................... | 36,264 | 39,859 |
| 44220 | | Springfield, OH Metro area.......................... | 138,339 | 133,593 |
| 44220 | 39023 | Clark County ............................................... | 138,339 | 133,593 |
| 44300 | | State College, PA Metro area...................... | 154,005 | 161,496 |
| 44300 | 42027 | Centre County ............................................. | 154,005 | 161,496 |
| 44420 | | Staunton, VA Metro area............................. | 118,496 | 124,475 |
| 44420 | 51015 | Augusta County........................................... | 73,753 | 76,544 |
| 44420 | 51790 | Staunton city .............................................. | 23,745 | 25,190 |
| 44420 | 51820 | Waynesboro city ......................................... | 20,998 | 22,741 |
| 44700 | | Stockton, CA Metro area ............................ | 685,306 | 767,967 |
| 44700 | 06077 | San Joaquin County .................................... | 685,306 | 767,967 |
| 44940 | | Sumter, SC Metro area ............................... | 142,434 | 139,775 |
| 44940 | 45027 | Clarendon County........................................ | 34,949 | 33,415 |
| 44940 | 45085 | Sumter County ............................................ | 107,485 | 106,360 |
| 45060 | | Syracuse, NY Metro area ........................... | 662,624 | 646,038 |
| 45060 | 36053 | Madison County .......................................... | 73,452 | 70,478 |
| 45060 | 36067 | Onondaga County ....................................... | 467,067 | 459,214 |
| 45060 | 36075 | Oswego County ........................................... | 122,105 | 116,346 |
| 45220 | | Tallahassee, FL Metro area ........................ | 368,771 | 389,599 |
| 45220 | 12039 | Gadsden County ......................................... | 47,744 | 45,277 |
| 45220 | 12065 | Jefferson County ......................................... | 14,761 | 14,543 |
| 45220 | 12073 | Leon County ................................................ | 275,483 | 295,460 |
| 45220 | 12129 | Wakulla County........................................... | 30,783 | 34,319 |
| 45300 | | Tampa-St. Petersburg-Clearwater, FL Metro area ........................................... | 2,783,485 | 3,243,963 |
| 45300 | 12053 | Hernando County......................................... | 172,778 | 198,792 |
| 45300 | 12057 | Hillsborough County .................................... | 1,229,202 | 1,497,957 |
| 45300 | 12101 | Pasco County .............................................. | 464,705 | 570,412 |
| 45300 | 12103 | Pinellas County ........................................... | 916,800 | 976,802 |
| 45460 | | Terre Haute, IN Metro area......................... | 189,774 | 185,632 |
| 45460 | 18021 | Clay County ................................................ | 26,888 | 26,246 |
| 45460 | 18121 | Parke County .............................................. | 17,349 | 16,871 |
| 45460 | 18153 | Sullivan County ........................................... | 21,475 | 20,578 |
| 45460 | 18165 | Vermillion County ........................................ | 16,210 | 15,329 |
| 45460 | 18167 | Vigo County ................................................ | 107,852 | 106,608 |
| 45500 | | Texarkana, TX-AR Metro area .................... | 149,194 | 148,838 |
| 45500 | 05081 | Little River County ...................................... | 13,168 | 12,180 |
| 45500 | 05091 | Miller County ............................................... | 43,462 | 43,177 |
| 45500 | 48037 | Bowie County .............................................. | 92,564 | 93,481 |
| 45540 | | The Villages, FL Metro area........................ | 93,420 | 139,018 |
| 45540 | 12119 | Sumter County ............................................ | 93,420 | 139,018 |
| 45780 | | Toledo, OH Metro area................................ | 651,435 | 641,549 |
| 45780 | 39051 | Fulton County ............................................. | 42,698 | 41,889 |
| 45780 | 39095 | Lucas County .............................................. | 441,815 | 428,294 |
| 45780 | 39123 | Ottawa County ............................................ | 41,433 | 40,253 |
| 45780 | 39173 | Wood County .............................................. | 125,489 | 131,113 |
| 45820 | | Topeka, KS Metro area ............................... | 233,860 | 230,878 |
| 45820 | 20085 | Jackson County........................................... | 13,460 | 13,171 |
| 45820 | 20087 | Jefferson County ......................................... | 19,108 | 19,032 |
| 45820 | 20139 | Osage County ............................................. | 16,294 | 15,770 |
| 45820 | 20177 | Shawnee County ......................................... | 177,943 | 175,999 |
| 45820 | 20197 | Wabaunsee County..................................... | 7,055 | 6,906 |

| State/County FIPS code | Core based statistical area | Title and Geographic Components | 2010 census population | 2020 estimated population | State/County FIPS code | Core based statistical area | Title and Geographic Components | 2010 census population | 2020 estimated population |
|---|---|---|---|---|---|---|---|---|---|
| 45940 | | Trenton-Princeton, NJ Metro area............... | 367,485 | 367,239 | 47380 | | Waco, TX Metro area............................. | 252,766 | 277,005 |
| 45940 | 34021 | Mercer County............................. | 367,485 | 367,239 | 47380 | 48145 | Falls County............................. | 17,863 | 17,275 |
| | | | | | 47380 | 48309 | McLennan County............................. | 234,903 | 259,730 |
| 46060 | | Tucson, AZ Metro area............................. | 980,263 | 1,061,175 | | | | | |
| 46060 | 04019 | Pima County............................. | 980,263 | 1,061,175 | 47460 | | Walla Walla, WA Metro area............................. | 58,781 | 61,292 |
| | | | | | 47460 | 53071 | Walla Walla County............................. | 58,781 | 61,292 |
| 46140 | | Tulsa, OK Metro area............................. | 937,523 | 1,006,411 | | | | | |
| 46140 | 40037 | Creek County............................. | 69,992 | 71,485 | 47580 | | Warner Robins, GA Metro area............................. | 167,626 | 188,060 |
| 46140 | 40111 | Okmulgee County............................. | 40,062 | 38,234 | 47580 | 13153 | Houston County............................. | 139,819 | 160,110 |
| 46140 | 40113 | Osage County............................. | 47,473 | 46,642 | 47580 | 13225 | Peach County............................. | 27,807 | 27,950 |
| 46140 | 40117 | Pawnee County............................. | 16,570 | 16,381 | | | | | |
| 46140 | 40131 | Rogers County............................. | 86,914 | 93,155 | 47900 | | Washington-Arlington-Alexandria, DC-VA-MD-WV Metro area............................. | 5,649,688 | 6,324,629 |
| 46140 | 40143 | Tulsa County............................. | 603,430 | 657,589 | | | | | |
| 46140 | 40145 | Wagoner County............................. | 73,082 | 82,925 | 47900 | | Frederick-Gaithersburg-Rockville, MD Div 23224............................. | 1,204,687 | 1,316,977 |
| | | | | | 47900 | 24021 | Frederick County............................. | 233,403 | 265,161 |
| 46220 | | Tuscaloosa, AL Metro area............................. | 239,214 | 253,211 | 47900 | 24031 | Montgomery County............................. | 971,284 | 1,051,816 |
| 46220 | 01063 | Greene County............................. | 9,039 | 7,990 | | | | | |
| 46220 | 01065 | Hale County............................. | 15,762 | 14,670 | 47900 | | Washington-Arlington-Alexandria, DC-VA-MD-WV Div 47894............................. | 4,445,001 | 5,007,652 |
| 46220 | 01107 | Pickens County............................. | 19,746 | 19,793 | 47900 | 11001 | District of Columbia............................. | 601,767 | 712,816 |
| 46220 | 01125 | Tuscaloosa County............................. | 194,667 | 210,758 | 47900 | 24009 | Calvert County............................. | 88,739 | 93,072 |
| | | | | | 47900 | 24017 | Charles County............................. | 146,564 | 164,436 |
| 46300 | | Twin Falls, ID Metro area............................. | 99,596 | 112,989 | 47900 | 24033 | Prince George's County............................. | 864,029 | 909,612 |
| 46300 | 16053 | Jerome County............................. | 22,355 | 24,578 | 47900 | 51013 | Arlington County............................. | 207,696 | 240,119 |
| 46300 | 16083 | Twin Falls County............................. | 77,241 | 88,411 | 47900 | 51043 | Clarke County............................. | 14,025 | 14,622 |
| | | | | | 47900 | 51047 | Culpeper County............................. | 46,688 | 53,569 |
| 46340 | | Tyler, TX Metro area............................. | 209,725 | 235,806 | 47900 | 51059 | Fairfax County............................. | 1,081,703 | 1,150,847 |
| 46340 | 48423 | Smith County............................. | 209,725 | 235,806 | 47900 | 51061 | Fauquier County............................. | 65,228 | 71,361 |
| | | | | | 47900 | 51107 | Loudoun County............................. | 312,348 | 422,784 |
| 46520 | | Urban Honolulu, HI Metro area............................. | 953,206 | 963,826 | 47900 | 51113 | Madison County............................. | 13,306 | 13,312 |
| 46520 | 15003 | Honolulu County............................. | 953,206 | 963,826 | 47900 | 51153 | Prince William County............................. | 402,009 | 475,533 |
| | | | | | 47900 | 51157 | Rappahannock County............................. | 7,506 | 7,260 |
| 46540 | | Utica-Rome, NY Metro area............................. | 299,329 | 288,291 | 47900 | 51177 | Spotsylvania County............................. | 122,453 | 138,449 |
| 46540 | 36043 | Herkimer County............................. | 64,469 | 60,945 | 47900 | 51179 | Stafford County............................. | 128,984 | 156,748 |
| 46540 | 36065 | Oneida County............................. | 234,860 | 227,346 | 47900 | 51187 | Warren County............................. | 37,450 | 40,475 |
| | | | | | 47900 | 51510 | Alexandria city............................. | 139,998 | 158,726 |
| 46660 | | Valdosta, GA Metro area............................. | 139,662 | 148,364 | 47900 | 51600 | Fairfax city............................. | 22,554 | 23,429 |
| 46660 | 13027 | Brooks County............................. | 16,314 | 15,357 | 47900 | 51610 | Falls Church city............................. | 12,244 | 14,631 |
| 46660 | 13101 | Echols County............................. | 4,023 | 4,002 | 47900 | 51630 | Fredericksburg city............................. | 24,118 | 29,492 |
| 46660 | 13173 | Lanier County............................. | 10,077 | 10,737 | 47900 | 51683 | Manassas city............................. | 37,799 | 40,869 |
| 46660 | 13185 | Lowndes County............................. | 109,248 | 118,268 | 47900 | 51685 | Manassas Park city............................. | 14,243 | 18,004 |
| | | | | | 47900 | 54037 | Jefferson County............................. | 53,490 | 57,486 |
| 46700 | | Vallejo, CA Metro area............................. | 413,343 | 446,935 | | | | | |
| 46700 | 06095 | Solano County............................. | 413,343 | 446,935 | 47940 | | Waterloo-Cedar Falls, IA Metro area............................. | 167,819 | 168,314 |
| | | | | | 47940 | 19013 | Black Hawk County............................. | 131,086 | 130,786 |
| 47020 | | Victoria, TX Metro area............................. | 94,003 | 99,562 | 47940 | 19017 | Bremer County............................. | 24,280 | 25,311 |
| 47020 | 48175 | Goliad County............................. | 7,210 | 7,626 | 47940 | 19075 | Grundy County............................. | 12,453 | 12,217 |
| 47020 | 48469 | Victoria County............................. | 86,793 | 91,936 | | | | | |
| | | | | | 48060 | | Watertown-Fort Drum, NY Metro area............................. | 116,232 | 108,095 |
| 47220 | | Vineland-Bridgeton, NJ Metro area............................. | 156,627 | 147,008 | 48060 | 36045 | Jefferson County............................. | 116,232 | 108,095 |
| 47220 | 34011 | Cumberland County............................. | 156,627 | 147,008 | | | | | |
| | | | | | 48140 | | Wausau-Weston, WI Metro area............................. | 162,804 | 163,159 |
| 47260 | | Virginia Beach-Norfolk-Newport News, VA-NC Metro area............................. | 1,713,955 | 1,779,824 | 48140 | 55069 | Lincoln County............................. | 28,743 | 27,566 |
| 47260 | 37029 | Camden County............................. | 9,980 | 10,984 | 48140 | 55073 | Marathon County............................. | 134,061 | 135,593 |
| 47260 | 37053 | Currituck County............................. | 23,547 | 29,052 | | | | | |
| 47260 | 37073 | Gates County............................. | 12,185 | 11,464 | 48260 | | Weirton-Steubenville, WV-OH Metro area.. | 124,455 | 115,184 |
| 47260 | 51073 | Gloucester County............................. | 36,859 | 37,459 | 48260 | 39081 | Jefferson County............................. | 69,716 | 64,939 |
| 47260 | 51093 | Isle of Wight County............................. | 35,277 | 37,725 | 48260 | 54009 | Brooke County............................. | 24,051 | 21,674 |
| 47260 | 51095 | James City County............................. | 67,379 | 77,612 | 48260 | 54029 | Hancock County............................. | 30,688 | 28,571 |
| 47260 | 51115 | Mathews County............................. | 8,976 | 8,766 | | | | | |
| 47260 | 51175 | Southampton County............................. | 18,573 | 17,636 | 48300 | | Wenatchee, WA Metro area............................. | 110,887 | 121,134 |
| 47260 | 51199 | York County............................. | 65,191 | 69,199 | 48300 | 53007 | Chelan County............................. | 72,460 | 77,574 |
| 47260 | 51550 | Chesapeake city............................. | 222,311 | 247,011 | 48300 | 53017 | Douglas County............................. | 38,427 | 43,560 |
| 47260 | 51620 | Franklin city............................. | 8,578 | 7,833 | | | | | |
| 47260 | 51650 | Hampton city............................. | 137,464 | 135,464 | 48540 | | Wheeling, WV-OH Metro area............................. | 147,957 | 137,217 |
| 47260 | 51700 | Newport News city............................. | 180,955 | 179,062 | 48540 | 39013 | Belmont County............................. | 70,400 | 65,932 |
| 47260 | 51710 | Norfolk city............................. | 242,827 | 242,803 | 48540 | 54051 | Marshall County............................. | 33,131 | 30,103 |
| 47260 | 51735 | Poquoson city............................. | 12,159 | 12,257 | 48540 | 54069 | Ohio County............................. | 44,426 | 41,182 |
| 47260 | 51740 | Portsmouth city............................. | 95,526 | 95,094 | | | | | |
| 47260 | 51800 | Suffolk city............................. | 84,565 | 93,913 | 48620 | | Wichita, KS Metro area............................. | 623,061 | 643,768 |
| 47260 | 51810 | Virginia Beach city............................. | 437,903 | 451,231 | 48620 | 20015 | Butler County............................. | 65,884 | 66,992 |
| 47260 | 51830 | Williamsburg city............................. | 13,700 | 15,259 | 48620 | 20079 | Harvey County............................. | 34,684 | 34,291 |
| | | | | | 48620 | 20173 | Sedgwick County............................. | 498,356 | 519,907 |
| 47300 | | Visalia, CA Metro area............................. | 442,182 | 468,680 | 48620 | 20191 | Sumner County............................. | 24,137 | 22,578 |
| 47300 | 06107 | Tulare County............................. | 442,182 | 468,680 | | | | | |

Appendix B

| State/County FIPS code | Core based statistical area | Title and Geographic Components | 2010 census population | 2020 estimated population | State/County FIPS code | Core based statistical area | Title and Geographic Components | 2010 census population | 2020 estimated population |
|---|---|---|---|---|---|---|---|---|---|
| 48660 | | Wichita Falls, TX Metro area | 151,474 | 152,485 | 49340 | | Worcester, MA-CT Metro area | 916,763 | 945,752 |
| 48660 | 48009 | Archer County | 9,061 | 8,730 | 49340 | 09015 | Windham County | 118,380 | 116,540 |
| 48660 | 48077 | Clay County | 10,754 | 10,550 | 49340 | 25027 | Worcester County | 798,383 | 829,212 |
| 48660 | 48485 | Wichita County | 131,659 | 133,205 | | | | | |
| | | | | | 49420 | | Yakima, WA Metro area | 243,240 | 251,879 |
| 48700 | | Williamsport, PA Metro area | 116,102 | 113,209 | 49420 | 53077 | Yakima County | 243,240 | 251,879 |
| 48700 | 42081 | Lycoming County | 116,102 | 113,209 | | | | | |
| | | | | | 49620 | | York-Hanover, PA Metro area | 435,015 | 450,448 |
| 48900 | | Wilmington, NC Metro area | 254,879 | 301,284 | 49620 | 42133 | York County | 435,015 | 450,448 |
| 48900 | 37129 | New Hanover County | 202,683 | 236,613 | | | | | |
| 48900 | 37141 | Pender County | 52,196 | 64,671 | 49660 | | Youngstown-Warren-Boardman, OH-PA Metro area | 565,782 | 531,420 |
| | | | | | 49660 | 39099 | Mahoning County | 238,787 | 226,075 |
| 49020 | | Winchester, VA-WV Metro area | 128,452 | 142,009 | 49660 | 39155 | Trumbull County | 210,332 | 196,800 |
| 49020 | 51069 | Frederick County | 78,269 | 91,119 | 49660 | 42085 | Mercer County | 116,663 | 108,545 |
| 49020 | 51840 | Winchester city | 26,223 | 27,700 | | | | | |
| 49020 | 54027 | Hampshire County | 23,960 | 23,190 | 49700 | | Yuba City, CA Metro area | 166,898 | 176,545 |
| | | | | | 49700 | 06101 | Sutter County | 94,756 | 96,385 |
| 49180 | | Winston-Salem, NC Metro area | 640,503 | 679,731 | 49700 | 06115 | Yuba County | 72,142 | 80,160 |
| 49180 | 37057 | Davidson County | 162,822 | 169,234 | | | | | |
| 49180 | 37059 | Davie County | 41,221 | 43,286 | 49740 | | Yuma, AZ Metro area | 195,750 | 217,824 |
| 49180 | 37067 | Forsyth County | 350,638 | 383,843 | 49740 | 04027 | Yuma County | 195,750 | 217,824 |
| 49180 | 37169 | Stokes County | 47,413 | 45,743 | | | | | |
| 49180 | 37197 | Yadkin County | 38,409 | 37,625 | | | | | |

# APPENDIX C
# CORE BASED STATISTICAL AREAS
## (Metropolitan and Micropolitan),
# METROPOLITAN DIVISIONS, AND COMPONENTS
## (as defined March, 2020)

| Core based statistical area | State/County FIPS code | Title and Geographic Components | 2010 Census Population | 2020 Estimated Population | Core based statistical area | State/County FIPS code | Title and Geographic Components | 2010 Census Population | 2020 Estimated Population |
|---|---|---|---|---|---|---|---|---|---|
| 10100 | | Aberdeen, SD Micro Area...................... | 40,602 | 42,555 | 10780 | | Alexandria, LA Metro Area ................... | 153,922 | 150,821 |
| 10100 | 46013 | Brown County, SD ..................... | 36,531 | 38,738 | 10780 | 22043 | Grant Parish, LA .......................... | 22,309 | 22,254 |
| 10100 | 46045 | Edmunds County, SD ...................... | 4,071 | 3,817 | 10780 | 22079 | Rapides Parish, LA ..................... | 131,613 | 128,567 |
| 10140 | | Aberdeen, WA Micro Area.................. | 72,797 | 75,950 | 10820 | | Alexandria, MN Micro Area................ | 36,009 | 38,328 |
| 10140 | 53027 | Grays Harbor County, WA ............... | 72,797 | 75,950 | 10820 | 27041 | Douglas County, MN...................... | 36,009 | 38,328 |
| 10180 | | Abilene, TX Metro Area | 165,252 | 173,185 | 10860 | | Alice, TX Micro Area...................... | 52,620 | 51,510 |
| 10180 | 48059 | Callahan County, TX.................. | 13,544 | 14,110 | 10860 | 48131 | Duval County, TX...................... | 11,782 | 11,058 |
| 10180 | 48253 | Jones County, TX ..................... | 20,202 | 19,875 | 10860 | 48249 | Jim Wells County, TX.................... | 40,838 | 40,452 |
| 10180 | 48441 | Taylor County, TX.................... | 131,506 | 139,200 | | | | | |
| 10220 | | Ada, OK Micro Area...................... | 37,492 | 38,397 | 10900 | | Allentown-Bethlehem-Easton, PA-NJ Metro Area................ | 821,173 | 846,399 |
| 10220 | 40123 | Pontotoc County, OK..................... | 37,492 | 38,397 | 10900 | 34041 | Warren County, NJ ..................... | 108,692 | 105,624 |
| | | | | | 10900 | 42025 | Carbon County, PA..................... | 65,249 | 64,081 |
| 10300 | | Adrian, MI Micro Area............... | 99,892 | 97,808 | 10900 | 42077 | Lehigh County, PA...................... | 349,497 | 370,802 |
| 10300 | 26091 | Lenawee County, MI.................... | 99,892 | 97,808 | 10900 | 42095 | Northampton County, PA.................. | 297,735 | 305,892 |
| 10420 | | Akron, OH Metro Area................ | 703,200 | 701,449 | 10940 | | Alma, MI Micro Area.................... | 42,476 | 40,283 |
| 10420 | 39133 | Portage County, OH................ | 161,419 | 162,583 | 10940 | 26057 | Gratiot County, MI...................... | 42,476 | 40,283 |
| 10420 | 39153 | Summit County, OH.................. | 541,781 | 538,866 | | | | | |
| | | | | | 10980 | | Alpena, MI Micro Area.................... | 29,598 | 28,238 |
| 10460 | | Alamogordo, NM Micro Area .............. | 63,797 | 67,967 | 10980 | 26007 | Alpena County, MI...................... | 29,598 | 28,238 |
| 10460 | 35035 | Otero County, NM...................... | 63,797 | 67,967 | | | | | |
| | | | | | 11020 | | Altoona, PA Metro Area.................... | 127,089 | 121,007 |
| 10500 | | Albany, GA Metro Area.................. | 153,857 | 145,206 | 11020 | 42013 | Blair County, PA...................... | 127,089 | 121,007 |
| 10500 | 13095 | Dougherty County, GA..................... | 94,565 | 86,477 | | | | | |
| 10500 | 13177 | Lee County, GA ..................... | 28,298 | 30,234 | 11060 | | Altus, OK Micro Area.................... | 26,446 | 24,305 |
| 10500 | 13273 | Terrell County, GA................. | 9,315 | 8,523 | 11060 | 40065 | Jackson County, OK ..................... | 26,446 | 24,305 |
| 10500 | 13321 | Worth County, GA................... | 21,679 | 19,972 | | | | | |
| | | | | | 11100 | | Amarillo, TX Metro Area................ | 251,933 | 265,761 |
| 10540 | | Albany-Lebanon, OR Metro Area ......... | 116,672 | 131,054 | 11100 | 48011 | Armstrong County, TX ................ | 1,901 | 1,869 |
| 10540 | 41043 | Linn County, OR ............................. | 116,672 | 131,054 | 11100 | 48065 | Carson County, TX .................... | 6,182 | 5,854 |
| | | | | | 11100 | 48359 | Oldham County, TX .................... | 2,052 | 2,135 |
| 10580 | | Albany-Schenectady-Troy, NY Metro Area..................... | 870,716 | 878,550 | 11100 | 48375 | Potter County, TX ..................... | 121,073 | 116,004 |
| | | | | | 11100 | 48381 | Randall County, TX.................... | 120,725 | 139,899 |
| 10580 | 36001 | Albany County, NY..................... | 304,204 | 303,654 | | | | | |
| 10580 | 36083 | Rensselaer County, NY .................... | 159,429 | 158,108 | 11140 | | Americus, GA Micro Area.................... | 37,829 | 34,478 |
| 10580 | 36091 | Saratoga County, NY ..................... | 219,607 | 230,298 | 11140 | 13249 | Schley County, GA.................... | 5,010 | 5,196 |
| 10580 | 36093 | Schenectady County, NY............... | 154,727 | 155,358 | 11140 | 13261 | Sumter County, GA .................... | 32,819 | 29,282 |
| 10580 | 36095 | Schoharie County, NY ..................... | 32,749 | 31,132 | | | | | |
| | | | | | 11180 | | Ames, IA Metro Area ..................... | 115,848 | 124,514 |
| 10620 | | Albemarle, NC Micro Area.................. | 60,585 | 63,239 | 11180 | 19015 | Boone County, IA...................... | 26,306 | 26,277 |
| 10620 | 37167 | Stanly County, NC ............................. | 60,585 | 63,239 | 11180 | 19169 | Story County, IA ..................... | 89,542 | 98,237 |
| 10660 | | Albert Lea, MN Micro Area .............. | 31,255 | 30,364 | 11220 | | Amsterdam, NY Micro Area................ | 50,219 | 49,170 |
| 10660 | 27047 | Freeborn County, MN ..................... | 31,255 | 30,364 | 11220 | 36057 | Montgomery County, NY.................. | 50,219 | 49,170 |
| 10700 | | Albertville, AL Micro Area ............... | 93,019 | 96,990 | 11260 | | Anchorage, AK Metro Area................ | 380,821 | 397,308 |
| 10700 | 01095 | Marshall County, AL......................... | 93,019 | 96,990 | 11260 | 02020 | Anchorage Municipality, AK............... | 291,826 | 287,095 |
| | | | | | 11260 | 02170 | Matanuska-Susitna Borough, AK....... | 88,995 | 110,213 |
| 10740 | | Albuquerque, NM Metro Area............... | 887,077 | 923,630 | | | | | |
| 10740 | 35001 | Bernalillo County, NM..................... | 662,564 | 681,666 | 11380 | | Andrews, TX Micro Area.................... | 14,786 | 18,879 |
| 10740 | 35043 | Sandoval County, NM..................... | 131,561 | 148,904 | 11380 | 48003 | Andrews County, TX ..................... | 14,786 | 18,879 |
| 10740 | 35057 | Torrance County, NM..................... | 16,383 | 15,486 | | | | | |
| 10740 | 35061 | Valencia County, NM..................... | 76,569 | 77,574 | 11420 | | Angola, IN Micro Area ..................... | 34,185 | 34,831 |
| | | | | | 11420 | 18151 | Steuben County, IN..................... | 34,185 | 34,831 |
| 10760 | | Alexander City, AL Micro Area .............. | 53,155 | 50,783 | | | | | |
| 10760 | 01037 | Coosa County, AL......................... | 11,539 | 10,650 | 11460 | | Ann Arbor, MI Metro Area.................. | 344,791 | 366,473 |
| 10760 | 01123 | Tallapoosa County, AL..................... | 41,616 | 40,133 | 11460 | 26161 | Washtenaw County, MI...................... | 344,791 | 366,473 |

| Core based statisti-cal area | State/ County FIPS code | Title and Geographic Components | 2010 Census Population | 2020 Estimated Population | Core based statisti-cal area | State/ County FIPS code | Title and Geographic Components | 2010 Census Population | 2020 Estimated Population |
|---|---|---|---|---|---|---|---|---|---|
| 11500 | | Anniston-Oxford, AL Metro Area .......... | 118,572 | 113,469 | 12060 | | Atlanta-Sandy Springs-Alpharetta, GA Metro Area................................................ | 5,286,728 | 6,087,762 |
| 11500 | 01015 | Calhoun County, AL.............................. | 118,572 | 113,469 | 12060 | 13013 | Barrow County, GA.............................. | 69,367 | 85,588 |
| | | | | | 12060 | 13015 | Bartow County, GA............................... | 100,157 | 109,426 |
| 11540 | | Appleton, WI Metro Area ...................... | 225,666 | 238,975 | 12060 | 13035 | Butts County, GA ................................. | 23,655 | 25,426 |
| 11540 | 55015 | Calumet County, WI............................. | 48,971 | 50,209 | 12060 | 13045 | Carroll County, GA .............................. | 110,527 | 121,633 |
| 11540 | 55087 | Outagamie County, WI.......................... | 176,695 | 188,766 | 12060 | 13057 | Cherokee County, GA........................... | 214,346 | 265,274 |
| | | | | | 12060 | 13063 | Clayton County, GA.............................. | 259,424 | 292,646 |
| 11580 | | Arcadia, FL Micro Area......................... | 34,862 | 38,520 | 12060 | 13067 | Cobb County, GA................................. | 688,078 | 762,944 |
| 11580 | 12027 | DeSoto County, FL .............................. | 34,862 | 38,520 | 12060 | 13077 | Coweta County, GA ............................. | 127,317 | 150,849 |
| | | | | | 12060 | 13085 | Dawson County, GA ............................ | 22,330 | 27,113 |
| 11620 | | Ardmore, OK Micro Area ...................... | 56,980 | 58,583 | 12060 | 13089 | DeKalb County, GA.............................. | 691,893 | 762,009 |
| 11620 | 40019 | Carter County, OK ............................... | 47,557 | 48,353 | 12060 | 13097 | Douglas County, GA............................. | 132,403 | 147,988 |
| 11620 | 40085 | Love County, OK.................................. | 9,423 | 10,230 | 12060 | 13113 | Fayette County, GA ............................. | 106,567 | 115,821 |
| | | | | | 12060 | 13117 | Forsyth County, GA.............................. | 175,511 | 250,847 |
| 11660 | | Arkadelphia, AR Micro Area ................. | 22,995 | 22,103 | 12060 | 13121 | Fulton County, GA ............................... | 920,581 | 1,077,402 |
| 11660 | 05019 | Clark County, AR ................................ | 22,995 | 22,103 | 12060 | 13135 | Gwinnett County, GA............................ | 805,321 | 942,627 |
| | | | | | 12060 | 13143 | Haralson County, GA............................ | 28,780 | 30,383 |
| 11700 | | Asheville, NC Metro Area...................... | 424,858 | 466,634 | 12060 | 13149 | Heard County, GA................................ | 11,834 | 11,973 |
| 11700 | 37021 | Buncombe County, NC .......................... | 238,318 | 263,477 | 12060 | 13151 | Henry County, GA................................ | 203,922 | 239,139 |
| 11700 | 37087 | Haywood County, NC ........................... | 59,036 | 62,972 | 12060 | 13159 | Jasper County, GA............................... | 13,900 | 14,483 |
| 11700 | 37089 | Henderson County, NC.......................... | 106,740 | 118,445 | 12060 | 13171 | Lamar County, GA................................ | 18,317 | 19,261 |
| 11700 | 37115 | Madison County, NC ............................ | 20,764 | 21,740 | 12060 | 13199 | Meriwether County, GA......................... | 21,992 | 21,164 |
| | | | | | 12060 | 13211 | Morgan County, GA.............................. | 17,868 | 19,636 |
| 11740 | | Ashland, OH Micro Area........................ | 53,139 | 53,362 | 12060 | 13217 | Newton County, GA.............................. | 99,958 | 113,295 |
| 11740 | 39005 | Ashland County, OH ............................ | 53,139 | 53,362 | 12060 | 13223 | Paulding County, GA ........................... | 142,324 | 173,359 |
| | | | | | 12060 | 13227 | Pickens County, GA ............................ | 29,431 | 33,127 |
| 11780 | | Ashtabula, OH Micro Area..................... | 101,497 | 96,513 | 12060 | 13231 | Pike County, GA .................................. | 17,869 | 19,121 |
| 11780 | 39007 | Ashtabula County, OH .......................... | 101,497 | 96,513 | 12060 | 13247 | Rockdale County, GA ........................... | 85,215 | 90,939 |
| | | | | | 12060 | 13255 | Spalding County, GA ............................ | 64,073 | 67,414 |
| 11820 | | Astoria, OR Micro Area......................... | 37,039 | 40,423 | 12060 | 13297 | Walton County, GA............................... | 83,768 | 96,875 |
| 11820 | 41007 | Clatsop County, OR ............................. | 37,039 | 40,423 | | | | | |
| | | | | | 12100 | | Atlantic City-Hammonton, NJ Metro Area ....................................................... | 274,549 | 262,945 |
| 11860 | | Atchison, KS Micro Area....................... | 16,924 | 16,015 | 12100 | 34001 | Atlantic County, NJ ............................. | 274,549 | 262,945 |
| 11860 | 20005 | Atchison County, KS ............................ | 16,924 | 16,015 | | | | | |
| | | | | | 12120 | | Atmore, AL Micro Area ......................... | 38,319 | 36,281 |
| 11900 | | Athens, OH Micro Area......................... | 64,757 | 65,481 | 12120 | 01053 | Escambia County, AL ........................... | 38,319 | 36,281 |
| 11900 | 39009 | Athens County, OH .............................. | 64,757 | 65,481 | | | | | |
| | | | | | 12140 | | Auburn, IN Micro Area.......................... | 42,223 | 43,670 |
| 11940 | | Athens, TN Micro Area ......................... | 52,266 | 54,208 | 12140 | 18033 | DeKalb County, IN ............................... | 42,223 | 43,670 |
| 11940 | 47107 | McMinn County, TN .............................. | 52,266 | 54,208 | | | | | |
| | | | | | 12180 | | Auburn, NY Micro Area......................... | 80,026 | 76,029 |
| 11980 | | Athens, TX Micro Area ......................... | 78,532 | 83,792 | 12180 | 36011 | Cayuga County, NY ............................. | 80,026 | 76,029 |
| 11980 | 48213 | Henderson County, TX ........................ | 78,532 | 83,792 | | | | | |
| | | | | | 12220 | | Auburn-Opelika, AL Metro Area ............ | 140,247 | 166,831 |
| 12020 | | Athens-Clarke County, GA Metro Area . | 192,541 | 214,759 | 12220 | 01081 | Lee County, AL ................................... | 140,247 | 166,831 |
| 12020 | 13059 | Clarke County, GA............................... | 116,714 | 127,795 | | | | | |
| 12020 | 13195 | Madison County, GA............................ | 28,120 | 30,457 | 12260 | | Augusta-Richmond County, GA-SC Metro Area....................................................... | 564,873 | 614,312 |
| 12020 | 13219 | Oconee County, GA............................. | 32,808 | 41,124 | 12260 | 13033 | Burke County, GA............................... | 23,316 | 22,648 |
| 12020 | 13221 | Oglethorpe County, GA........................ | 14,899 | 15,383 | 12260 | 13073 | Columbia County, GA ......................... | 124,053 | 160,377 |
| | | | | | 12260 | 13181 | Lincoln County, GA.............................. | 7,996 | 8,031 |
| | | | | | 12260 | 13189 | McDuffie County, GA ........................... | 21,875 | 21,162 |
| | | | | | 12260 | 13245 | Richmond County, GA ......................... | 200,549 | 202,079 |
| | | | | | 12260 | 45003 | Aiken County, SC................................ | 160,099 | 172,895 |
| | | | | | 12260 | 45037 | Edgefield County, SC.......................... | 26,985 | 27,120 |
| | | | | | 12300 | | Augusta-Waterville, ME Micro Area....... | 122,151 | 122,955 |
| | | | | | 12300 | 23011 | Kennebec County, ME.......................... | 122,151 | 122,955 |
| | | | | | 12380 | | Austin, MN Micro Area.......................... | 39,163 | 40,150 |
| | | | | | 12380 | 27099 | Mower County, MN ............................. | 39,163 | 40,150 |

| Core based statistical area | State/ County FIPS code | Title and Geographic Components | 2010 Census Population | 2020 Estimated Population | Core based statistical area | State/ County FIPS code | Title and Geographic Components | 2010 Census Population | 2020 Estimated Population |
|---|---|---|---|---|---|---|---|---|---|
| 12420 | | Austin-Round Rock-Georgetown, TX Metro Area | 1,716,289 | 2,295,303 | 13060 | | Bay City, TX Micro Area | 36,702 | 36,725 |
| 12420 | 48021 | Bastrop County, TX | 74,171 | 91,601 | 13060 | 48321 | Matagorda County, TX | 36,702 | 36,725 |
| 12420 | 48055 | Caldwell County, TX | 38,066 | 43,979 | | | | | |
| 12420 | 48209 | Hays County, TX | 157,107 | 241,365 | 13100 | | Beatrice, NE Micro Area | 22,311 | 21,431 |
| 12420 | 48453 | Travis County, TX | 1,024,266 | 1,300,503 | 13100 | 31067 | Gage County, NE | 22,311 | 21,431 |
| 12420 | 48491 | Williamson County, TX | 422,679 | 617,855 | | | | | |
| | | | | | 13140 | | Beaumont-Port Arthur, TX Metro Area | 388,745 | 391,310 |
| 12460 | | Bainbridge, GA Micro Area | 27,842 | 26,457 | 13140 | 48199 | Hardin County, TX | 54,635 | 58,305 |
| 12460 | 13087 | Decatur County, GA | 27,842 | 26,457 | 13140 | 48245 | Jefferson County, TX | 252,273 | 250,127 |
| | | | | | 13140 | 48361 | Orange County, TX | 81,837 | 82,878 |
| 12540 | | Bakersfield, CA Metro Area | 839,631 | 901,362 | | | | | |
| 12540 | 06029 | Kern County, CA | 839,631 | 901,362 | 13180 | | Beaver Dam, WI Micro Area | 88,759 | 87,336 |
| | | | | | 13180 | 55027 | Dodge County, WI | 88,759 | 87,336 |
| 12580 | | Baltimore-Columbia-Towson, MD Metro Area | 2,710,489 | 2,800,189 | | | | | |
| | | | | | 13220 | | Beckley, WV Metro Area | 124,898 | 114,982 |
| 12580 | 24003 | Anne Arundel County, MD | 537,656 | 582,777 | 13220 | 54019 | Fayette County, WV | 46,039 | 42,062 |
| 12580 | 24005 | Baltimore County, MD | 805,029 | 826,017 | 13220 | 54081 | Raleigh County, WV | 78,859 | 72,920 |
| 12580 | 24013 | Carroll County, MD | 167,134 | 169,092 | | | | | |
| 12580 | 24025 | Harford County, MD | 244,826 | 256,805 | 13260 | | Bedford, IN Micro Area | 46,134 | 45,496 |
| 12580 | 24027 | Howard County, MD | 287,085 | 328,200 | 13260 | 18093 | Lawrence County, IN | 46,134 | 45,496 |
| 12580 | 24035 | Queen Anne's County, MD | 47,798 | 51,167 | | | | | |
| 12580 | 24510 | Baltimore city, MD | 620,961 | 586,131 | 13300 | | Beeville, TX Micro Area | 31,861 | 32,513 |
| | | | | | 13300 | 48025 | Bee County, TX | 31,861 | 32,513 |
| 12620 | | Bangor, ME Metro Area | 153,923 | 151,655 | | | | | |
| 12620 | 23019 | Penobscot County, ME | 153,923 | 151,655 | 13340 | | Bellefontaine, OH Micro Area | 45,858 | 45,326 |
| | | | | | 13340 | 39091 | Logan County, OH | 45,858 | 45,326 |
| 12660 | | Baraboo, WI Micro Area | 61,976 | 64,449 | | | | | |
| 12660 | 55111 | Sauk County, WI | 61,976 | 64,449 | 13380 | | Bellingham, WA Metro Area | 201,140 | 231,016 |
| | | | | | 13380 | 53073 | Whatcom County, WA | 201,140 | 231,016 |
| 12680 | | Bardstown, KY Micro Area | 43,437 | 46,450 | | | | | |
| 12680 | 21179 | Nelson County, KY | 43,437 | 46,450 | 13420 | | Bemidji, MN Micro Area | 44,442 | 47,442 |
| | | | | | 13420 | 27007 | Beltrami County, MN | 44,442 | 47,442 |
| 12700 | | Barnstable Town, MA Metro Area | 215,888 | 213,164 | | | | | |
| 12700 | 25001 | Barnstable County, MA | 215,888 | 213,164 | 13460 | | Bend, OR Metro Area | 157,733 | 201,769 |
| | | | | | 13460 | 41017 | Deschutes County, OR | 157,733 | 201,769 |
| 12740 | | Barre, VT Micro Area | 59,534 | 58,328 | | | | | |
| 12740 | 50023 | Washington County, VT | 59,534 | 58,328 | 13500 | | Bennettsville, SC Micro Area | 28,933 | 25,581 |
| | | | | | 13500 | 45069 | Marlboro County, SC | 28,933 | 25,581 |
| 12780 | | Bartlesville, OK Micro Area | 50,976 | 52,222 | | | | | |
| 12780 | 40147 | Washington County, OK | 50,976 | 52,222 | 13540 | | Bennington, VT Micro Area | 37,125 | 35,338 |
| | | | | | 13540 | 50003 | Bennington County, VT | 37,125 | 35,338 |
| 12860 | | Batavia, NY Micro Area | 60,079 | 56,994 | | | | | |
| 12860 | 36037 | Genesee County, NY | 60,079 | 56,994 | 13620 | | Berlin, NH Micro Area | 33,055 | 31,174 |
| | | | | | 13620 | 33007 | Coos County, NH | 33,055 | 31,174 |
| 12900 | | Batesville, AR Micro Area | 53,911 | 55,181 | | | | | |
| 12900 | 05063 | Independence County, AR | 36,647 | 37,757 | 13660 | | Big Rapids, MI Micro Area | 42,798 | 43,907 |
| 12900 | 05135 | Sharp County, AR | 17,264 | 17,424 | 13660 | 26107 | Mecosta County, MI | 42,798 | 43,907 |
| | | | | | 13700 | | Big Spring, TX Micro Area | 35,012 | 36,540 |
| 12940 | | Baton Rouge, LA Metro Area | 825,905 | 858,571 | 13700 | 48227 | Howard County, TX | 35,012 | 36,540 |
| 12940 | 22005 | Ascension Parish, LA | 107,215 | 128,665 | | | | | |
| 12940 | 22007 | Assumption Parish, LA | 23,421 | 21,621 | 13720 | | Big Stone Gap, VA Micro Area | 45,410 | 41,191 |
| 12940 | 22033 | East Baton Rouge Parish, LA | 440,171 | 439,729 | 13720 | 51195 | Wise County, VA | 41,452 | 37,206 |
| 12940 | 22037 | East Feliciana Parish, LA | 20,267 | 18,882 | 13720 | 51720 | Norton city, VA | 3,958 | 3,985 |
| 12940 | 22047 | Iberville Parish, LA | 33,387 | 32,070 | | | | | |
| 12940 | 22063 | Livingston Parish, LA | 128,026 | 143,737 | 13740 | | Billings, MT Metro Area | 167,167 | 183,799 |
| 12940 | 22077 | Pointe Coupee Parish, LA | 22,802 | 21,529 | 13740 | 30009 | Carbon County, MT | 10,078 | 10,921 |
| 12940 | 22091 | St. Helena Parish, LA | 11,203 | 10,081 | 13740 | 30095 | Stillwater County, MT | 9,117 | 9,888 |
| 12940 | 22121 | West Baton Rouge Parish, LA | 23,788 | 26,792 | 13740 | 30111 | Yellowstone County, MT | 147,972 | 162,990 |
| 12940 | 22125 | West Feliciana Parish, LA | 15,625 | 15,465 | | | | | |
| | | | | | 13780 | | Binghamton, NY Metro Area | 251,725 | 237,324 |
| 12980 | | Battle Creek, MI Metro Area | 136,146 | 133,580 | 13780 | 36007 | Broome County, NY | 200,600 | 189,420 |
| 12980 | 26025 | Calhoun County, MI | 136,146 | 133,580 | 13780 | 36107 | Tioga County, NY | 51,125 | 47,904 |
| | | | | | | | | | |
| 13020 | | Bay City, MI Metro Area | 107,771 | 102,387 | | | | | |
| 13020 | 26017 | Bay County, MI | 107,771 | 102,387 | | | | | |

| Core based statistical area | State/County FIPS code | Title and Geographic Components | 2010 Census Population | 2020 Estimated Population | Core based statistical area | State/County FIPS code | Title and Geographic Components | 2010 Census Population | 2020 Estimated Population |
|---|---|---|---|---|---|---|---|---|---|
| 13820 | | Birmingham-Hoover, AL Metro Area ..... | 1,061,024 | 1,091,921 | 14460 | | Boston, MA Division 14454 .................. | 1,887,792 | 2,034,729 |
| 13820 | 01007 | Bibb County, AL........................ | 22,915 | 22,136 | 14460 | 25021 | Norfolk County, MA...................... | 670,850 | 709,409 |
| 13820 | 01009 | Blount County, AL........................ | 57,322 | 57,879 | 14460 | 25023 | Plymouth County, MA ..................... | 494,919 | 523,738 |
| 13820 | 01021 | Chilton County, AL........................ | 43,643 | 44,397 | 14460 | 25025 | Suffolk County, MA ....................... | 722,023 | 801,582 |
| 13820 | 01073 | Jefferson County, AL..................... | 658,466 | 655,342 | | | | | |
| 13820 | 01115 | St. Clair County, AL...................... | 83,593 | 90,739 | 14460 | | Cambridge-Newton-Framingham, MA Division 15764 ................................. | 2,246,244 | 2,400,642 |
| 13820 | 01117 | Shelby County, AL....................... | 195,085 | 221,428 | 14460 | 25009 | Essex County, MA ....................... | 743,159 | 791,263 |
| 13900 | | Bismarck, ND Metro Area.................. | 110,625 | 129,641 | 14460 | 25017 | Middlesex County, MA .................. | 1,503,085 | 1,609,379 |
| 13900 | 38015 | Burleigh County, ND ...................... | 81,308 | 96,212 | | | | | |
| 13900 | 38059 | Morton County, ND ....................... | 27,471 | 31,503 | 14460 | | Rockingham County-Strafford County, NH Division 40484.............................. | 418,366 | 442,840 |
| 13900 | 38065 | Oliver County, ND........................ | 1,846 | 1,926 | 14460 | 33015 | Rockingham County, NH .................. | 295,223 | 311,307 |
| | | | | | 14460 | 33017 | Strafford County, NH...................... | 123,143 | 131,533 |
| 13940 | | Blackfoot, ID Micro Area................. | 45,607 | 47,202 | | | | | |
| 13940 | 16011 | Bingham County, ID........................ | 45,607 | 47,202 | 14500 | | Boulder, CO Metro Area.................. | 294,567 | 327,171 |
| | | | | | 14500 | 08013 | Boulder County, CO...................... | 294,567 | 327,171 |
| 13980 | | Blacksburg-Christiansburg, VA Metro Area.................................................... | 162,958 | 167,244 | 14540 | | Bowling Green, KY Metro Area ............ | 158,599 | 180,751 |
| 13980 | 51071 | Giles County, VA.......................... | 17,286 | 16,663 | 14540 | 21003 | Allen County, KY.......................... | 19,956 | 21,303 |
| 13980 | 51121 | Montgomery County, VA.................. | 94,392 | 98,391 | 14540 | 21031 | Butler County, KY......................... | 12,690 | 12,703 |
| 13980 | 51155 | Pulaski County, VA....................... | 34,872 | 33,935 | 14540 | 21061 | Edmonson County, KY.................... | 12,161 | 12,235 |
| 13980 | 51750 | Radford city, VA........................... | 16,408 | 18,255 | 14540 | 21227 | Warren County, KY........................ | 113,792 | 134,510 |
| 14010 | | Bloomington, IL Metro Area.................. | 169,572 | 171,256 | 14580 | | Bozeman, MT Micro Area.................. | 89,513 | 116,806 |
| 14010 | 17113 | McLean County, IL......................... | 169,572 | 171,256 | 14580 | 30031 | Gallatin County, MT ...................... | 89,513 | 116,806 |
| 14020 | | Bloomington, IN Metro Area ................ | 159,549 | 169,052 | 14620 | | Bradford, PA Micro Area.................. | 43,450 | 40,333 |
| 14020 | 18105 | Monroe County, IN........................ | 137,974 | 148,219 | 14620 | 42083 | McKean County, PA....................... | 43,450 | 40,333 |
| 14020 | 18119 | Owen County, IN.......................... | 21,575 | 20,833 | | | | | |
| | | | | | 14660 | | Brainerd, MN Micro Area ................. | 91,067 | 95,572 |
| 14100 | | Bloomsburg-Berwick, PA Metro Area.... | 85,562 | 82,884 | 14660 | 27021 | Cass County, MN.......................... | 28,567 | 29,928 |
| 14100 | 42037 | Columbia County, PA...................... | 67,295 | 64,842 | 14660 | 27035 | Crow Wing County, MN ................... | 62,500 | 65,644 |
| 14100 | 42093 | Montour County, PA....................... | 18,267 | 18,042 | | | | | |
| | | | | | 14700 | | Branson, MO Micro Area .................. | 51,675 | 56,104 |
| 14140 | | Bluefield, WV-VA Micro Area............... | 114,166 | 105,026 | 14700 | 29213 | Taney County, MO.......................... | 51,675 | 56,104 |
| 14140 | 51021 | Bland County, VA......................... | 6,824 | 6,239 | | | | | |
| 14140 | 51185 | Tazewell County, VA...................... | 45,078 | 40,529 | 14720 | | Breckenridge, CO Micro Area............... | 27,994 | 30,631 |
| 14140 | 54055 | Mercer County, WV....................... | 62,264 | 58,258 | 14720 | 08117 | Summit County, CO ....................... | 27,994 | 30,631 |
| 14160 | | Bluffton, IN Micro Area ..................... | 27,636 | 28,142 | 14740 | | Bremerton-Silverdale-Port Orchard, WA Metro Area........................................ | 251,133 | 272,787 |
| 14160 | 18179 | Wells County, IN........................... | 27,636 | 28,142 | 14740 | 53035 | Kitsap County, WA........................ | 251,133 | 272,787 |
| 14180 | | Blytheville, AR Micro Area ................ | 46,480 | 40,066 | 14780 | | Brenham, TX Micro Area.................... | 33,718 | 35,771 |
| 14180 | 05093 | Mississippi County, AR .................. | 46,480 | 40,066 | 14780 | 48477 | Washington County, TX.................... | 33,718 | 35,771 |
| 14220 | | Bogalusa, LA Micro Area.................... | 47,168 | 45,773 | 14820 | | Brevard, NC Micro Area ..................... | 33,090 | 34,498 |
| 14220 | 22117 | Washington Parish, LA .................... | 47,168 | 45,773 | 14820 | 37175 | Transylvania County, NC .................. | 33,090 | 34,498 |
| 14260 | | Boise City, ID Metro Area.................... | 616,561 | 770,353 | 14860 | | Bridgeport-Stamford-Norwalk, CT Metro Area.................................................... | 916,829 | 942,426 |
| 14260 | 16001 | Ada County, ID............................ | 392,365 | 494,399 | 14860 | 09001 | Fairfield County, CT....................... | 916,829 | 942,426 |
| 14260 | 16015 | Boise County, ID......................... | 7,028 | 8,065 | | | | | |
| 14260 | 16027 | Canyon County, ID........................ | 188,923 | 237,053 | 15020 | | Brookhaven, MS Micro Area................ | 34,869 | 33,936 |
| 14260 | 16045 | Gem County, ID........................... | 16,719 | 18,703 | 15020 | 28085 | Lincoln County, MS....................... | 34,869 | 33,936 |
| 14260 | 16073 | Owyhee County, ID........................ | 11,526 | 12,133 | | | | | |
| | | | | | 15060 | | Brookings, OR Micro Area .................. | 22,364 | 23,305 |
| 14300 | | Bonham, TX Micro Area..................... | 33,915 | 35,913 | 15060 | 41015 | Curry County, OR.......................... | 22,364 | 23,305 |
| 14300 | 48147 | Fannin County, TX........................ | 33,915 | 35,913 | | | | | |
| | | | | | 15100 | | Brookings, SD Micro Area ................. | 31,965 | 35,603 |
| 14380 | | Boone, NC Micro Area....................... | 51,079 | 56,441 | 15100 | 46011 | Brookings County, SD...................... | 31,965 | 35,603 |
| 14380 | 37189 | Watauga County, NC ...................... | 51,079 | 56,441 | | | | | |
| | | | | | 15140 | | Brownsville, TN Micro Area ............... | 18,787 | 17,002 |
| 14420 | | Borger, TX Micro Area...................... | 22,150 | 20,677 | 15140 | 47075 | Haywood County, TN....................... | 18,787 | 17,002 |
| 14420 | 48233 | Hutchinson County, TX.................... | 22,150 | 20,677 | | | | | |
| 14460 | | Boston-Cambridge-Newton, MA-NH Metro Area........................................ | 4,552,402 | 4,878,211 | | | | | |

# CORE BASED STATISTICAL AREAS
## (Metropolitan and Micropolitan),
## METROPOLITAN DIVISIONS, AND COMPONENTS
## (as defined March, 2020)—*Continued*

| Core based statisti-cal area | State/ County FIPS code | Title and Geographic Components | 2010 Census Population | 2020 Estimated Population | Core based statisti-cal area | State/ County FIPS code | Title and Geographic Components | 2010 Census Population | 2020 Estimated Population |
|---|---|---|---|---|---|---|---|---|---|
| 15180 | | Brownsville-Harlingen, TX Metro Area .. | 406,220 | 424,180 | 15940 | | Canton-Massillon, OH Metro Area........ | 404,422 | 396,669 |
| 15180 | 48061 | Cameron County, TX ...................... | 406,220 | 424,180 | 15940 | 39019 | Carroll County, OH........................... | 28,836 | 26,897 |
| | | | | | 15940 | 39151 | Stark County, OH............................ | 375,586 | 369,772 |
| 15220 | | Brownwood, TX Micro Area................ | 38,106 | 37,633 | | | | | |
| 15220 | 48049 | Brown County, TX........................... | 38,106 | 37,633 | 15980 | | Cape Coral-Fort Myers, FL Metro Area . | 618,754 | 790,767 |
| | | | | | 15980 | 12071 | Lee County, FL............................... | 618,754 | 790,767 |
| 15260 | | Brunswick, GA Metro Area ............... | 112,370 | 119,157 | | | | | |
| 15260 | 13025 | Brantley County, GA ....................... | 18,411 | 19,202 | 16020 | | Cape Girardeau, MO-IL Metro Area ...... | 96,275 | 97,120 |
| 15260 | 13127 | Glynn County, GA .......................... | 79,626 | 85,568 | 16020 | 17003 | Alexander County, IL ....................... | 8,238 | 5,497 |
| 15260 | 13191 | McIntosh County, GA....................... | 14,333 | 14,387 | 16020 | 29017 | Bollinger County, MO ...................... | 12,363 | 12,111 |
| | | | | | 16020 | 29031 | Cape Girardeau County, MO ............ | 75,674 | 79,512 |
| 15340 | | Bucyrus-Galion, OH Micro Area ........... | 43,784 | 41,338 | | | | | |
| 15340 | 39033 | Crawford County, OH....................... | 43,784 | 41,338 | 16060 | | Carbondale-Marion, IL Metro Area........ | 139,157 | 135,448 |
| | | | | | 16060 | 17077 | Jackson County, IL .......................... | 60,218 | 56,675 |
| 15380 | | Buffalo-Cheektowaga, NY Metro Area .. | 1,135,509 | 1,125,637 | 16060 | 17087 | Johnson County, IL ......................... | 12,582 | 12,358 |
| 15380 | 36029 | Erie County, NY ............................. | 919,040 | 917,241 | 16060 | 17199 | Williamson County, IL...................... | 66,357 | 66,415 |
| 15380 | 36063 | Niagara County, NY ........................ | 216,469 | 208,396 | | | | | |
| | | | | | 16100 | | Carlsbad-Artesia, NM Micro Area ......... | 53,829 | 58,418 |
| 15420 | | Burley, ID Micro Area ...................... | 43,021 | 45,493 | 16100 | 35015 | Eddy County, NM............................ | 53,829 | 58,418 |
| 15420 | 16031 | Cassia County, ID .......................... | 22,952 | 24,277 | | | | | |
| 15420 | 16067 | Minidoka County, ID........................ | 20,069 | 21,216 | 16140 | | Carroll, IA Micro Area ...................... | 20,816 | 19,914 |
| | | | | | 16140 | 19027 | Carroll County, IA........................... | 20,816 | 19,914 |
| 15460 | | Burlington, IA-IL Micro Area ................ | 47,656 | 45,243 | | | | | |
| 15460 | 17071 | Henderson County, IL ...................... | 7,331 | 6,535 | 16180 | | Carson City, NV Metro Area................ | 55,274 | 56,034 |
| 15460 | 19057 | Des Moines County, IA ................... | 40,325 | 38,708 | 16180 | 32510 | Carson City, NV ............................. | 55,274 | 56,034 |
| | | | | | | | | | |
| 15500 | | Burlington, NC Metro Area .................. | 151,131 | 171,346 | 16220 | | Casper, WY Metro Area ..................... | 75,450 | 80,815 |
| 15500 | 37001 | Alamance County, NC ..................... | 151,131 | 171,346 | 16220 | 56025 | Natrona County, WY ........................ | 75,450 | 80,815 |
| | | | | | | | | | |
| 15540 | | Burlington-South Burlington, VT Metro Area.............................................. | 211,261 | 221,160 | 16260 | | Cedar City, UT Micro Area ................. | 46,163 | 56,814 |
| 15540 | 50007 | Chittenden County, VT..................... | 156,545 | 164,306 | 16260 | 49021 | Iron County, UT ............................. | 46,163 | 56,814 |
| 15540 | 50011 | Franklin County, VT ........................ | 47,746 | 49,685 | | | | | |
| 15540 | 50013 | Grand Isle County, VT ..................... | 6,970 | 7,169 | 16300 | | Cedar Rapids, IA Metro Area .............. | 257,940 | 273,885 |
| | | | | | 16300 | 19011 | Benton County, IA........................... | 26,076 | 25,414 |
| 15580 | | Butte-Silver Bow, MT Micro Area ......... | 34,200 | 35,180 | 16300 | 19105 | Jones County, IA............................ | 20,638 | 20,617 |
| 15580 | 30093 | Silver Bow County, MT ................... | 34,200 | 35,180 | 16300 | 19113 | Linn County, IA.............................. | 211,226 | 227,854 |
| | | | | | | | | | |
| 15620 | | Cadillac, MI Micro Area ..................... | 47,584 | 48,895 | 16340 | | Cedartown, GA Micro Area.................. | 41,475 | 42,840 |
| 15620 | 26113 | Missaukee County, MI ..................... | 14,849 | 15,152 | 16340 | 13233 | Polk County, GA ............................ | 41,475 | 42,840 |
| 15620 | 26165 | Wexford County, MI ........................ | 32,735 | 33,743 | | | | | |
| | | | | | 16380 | | Celina, OH Micro Area...................... | 40,814 | 41,274 |
| 15660 | | Calhoun, GA Micro Area...................... | 55,186 | 58,780 | 16380 | 39107 | Mercer County, OH ......................... | 40,814 | 41,274 |
| 15660 | 13129 | Gordon County, GA ........................ | 55,186 | 58,780 | | | | | |
| | | | | | 16420 | | Central City, KY Micro Area................. | 31,499 | 30,457 |
| 15680 | | California-Lexington Park, MD Metro Area.............................................. | 105,151 | 114,687 | 16420 | 21177 | Muhlenberg County, KY.................... | 31,499 | 30,457 |
| 15680 | 24037 | St. Mary's County, MD..................... | 105,151 | 114,687 | 16460 | | Centralia, IL Micro Area...................... | 39,437 | 37,045 |
| | | | | | 16460 | 17121 | Marion County, IL............................ | 39,437 | 37,045 |
| 15700 | | Cambridge, MD Micro Area ................. | 32,618 | 31,853 | | | | | |
| 15700 | 24019 | Dorchester County, MD .................... | 32,618 | 31,853 | 16500 | | Centralia, WA Micro Area ................... | 75,455 | 82,109 |
| | | | | | 16500 | 53041 | Lewis County, WA........................... | 75,455 | 82,109 |
| 15740 | | Cambridge, OH Micro Area ................. | 40,087 | 38,779 | | | | | |
| 15740 | 39059 | Guernsey County, OH ..................... | 40,087 | 38,779 | 16540 | | Chambersburg-Waynesboro, PA Metro Area.............................................. | 149,618 | 155,637 |
| | | | | | 16540 | 42055 | Franklin County, PA ........................ | 149,618 | 155,637 |
| 15780 | | Camden, AR Micro Area..................... | 31,488 | 28,280 | | | | | |
| 15780 | 05013 | Calhoun County, AR ....................... | 5,368 | 5,113 | 16580 | | Champaign-Urbana, IL Metro Area ....... | 217,810 | 225,547 |
| 15780 | 05103 | Ouachita County, AR ...................... | 26,120 | 23,167 | 16580 | 17019 | Champaign County, IL ..................... | 201,081 | 209,192 |
| | | | | | 16580 | 17147 | Piatt County, IL .............................. | 16,729 | 16,355 |
| 15820 | | Campbellsville, KY Micro Area ............ | 35,770 | 36,702 | | | | | |
| 15820 | 21087 | Green County, KY........................... | 11,258 | 10,995 | 16620 | | Charleston, WV Metro Area................ | 278,009 | 254,145 |
| 15820 | 21217 | Taylor County, KY........................... | 24,512 | 25,707 | 16620 | 54005 | Boone County, WV ......................... | 24,629 | 21,055 |
| | | | | | 16620 | 54015 | Clay County, WV............................ | 9,386 | 8,341 |
| 15860 | | Cañon City, CO Micro Area................. | 46,824 | 47,867 | 16620 | 54035 | Jackson County, WV........................ | 29,211 | 28,453 |
| 15860 | 08043 | Fremont County, CO ....................... | 46,824 | 47,867 | 16620 | 54039 | Kanawha County, WV...................... | 193,063 | 176,253 |
| | | | | | 16620 | 54043 | Lincoln County, WV ........................ | 21,720 | 20,043 |

# CORE BASED STATISTICAL AREAS
## (Metropolitan and Micropolitan),
## METROPOLITAN DIVISIONS, AND COMPONENTS
## (as defined March, 2020)—*Continued*

| Core based statistical area | State/County FIPS code | Title and Geographic Components | 2010 Census Population | 2020 Estimated Population | Core based statistical area | State/County FIPS code | Title and Geographic Components | 2010 Census Population | 2020 Estimated Population |
|---|---|---|---|---|---|---|---|---|---|
| 16660 | | Charleston-Mattoon, IL Micro Area....... | 64,921 | 61,032 | 16980 | | Lake County-Kenosha County, IL-WI Division 29404..................................... | 869,888 | 863,264 |
| 16660 | 17029 | Coles County, IL ................................ | 53,873 | 50,383 | 16980 | 17097 | Lake County, IL ................................. | 703,462 | 693,593 |
| 16660 | 17035 | Cumberland County, IL...................... | 11,048 | 10,649 | 16980 | 55059 | Kenosha County, WI.......................... | 166,426 | 169,671 |
| 16700 | | Charleston-North Charleston, SC Metro Area.................................................. | 664,607 | 819,705 | 17020 | | Chico, CA Metro Area........................ | 220,000 | 212,744 |
| 16700 | 45015 | Berkeley County, SC.......................... | 177,843 | 235,987 | 17020 | 06007 | Butte County, CA.............................. | 220,000 | 212,744 |
| 16700 | 45019 | Charleston County, SC....................... | 350,209 | 417,981 | 17060 | | Chillicothe, OH Micro Area ................ | 78,064 | 76,420 |
| 16700 | 45035 | Dorchester County, SC....................... | 136,555 | 165,737 | 17060 | 39141 | Ross County, OH............................... | 78,064 | 76,420 |
| 16740 | | Charlotte-Concord-Gastonia, NC-SC Metro Area........................................ | 2,243,960 | 2,684,276 | 17140 | | Cincinnati, OH-KY-IN Metro Area.......... | 2,137,667 | 2,232,907 |
| 16740 | 37007 | Anson County, NC............................. | 26,948 | 24,097 | 17140 | 18029 | Dearborn County, IN .......................... | 50,047 | 49,824 |
| 16740 | 37025 | Cabarrus County, NC.......................... | 178,011 | 221,479 | 17140 | 18047 | Franklin County, IN ........................... | 23,087 | 22,761 |
| 16740 | 37071 | Gaston County, NC............................ | 206,086 | 226,568 | 17140 | 18115 | Ohio County, IN ............................... | 6,128 | 5,892 |
| 16740 | 37097 | Iredell County, NC............................. | 159,437 | 185,770 | 17140 | 18161 | Union County, IN............................... | 7,516 | 7,119 |
| 16740 | 37109 | Lincoln County, NC............................ | 78,265 | 88,097 | 17140 | 21015 | Boone County, KY............................. | 118,811 | 135,396 |
| 16740 | 37119 | Mecklenburg County, NC.................... | 919,628 | 1,128,945 | 17140 | 21023 | Bracken County, KY........................... | 8,488 | 8,286 |
| 16740 | 37159 | Rowan County, NC............................ | 138,428 | 142,495 | 17140 | 21037 | Campbell County, KY......................... | 90,336 | 94,020 |
| 16740 | 37179 | Union County, NC............................. | 201,292 | 244,562 | 17140 | 21077 | Gallatin County, KY........................... | 8,589 | 8,779 |
| 16740 | 45023 | Chester County, SC........................... | 33,140 | 32,232 | 17140 | 21081 | Grant County, KY.............................. | 24,662 | 25,387 |
| 16740 | 45057 | Lancaster County, SC......................... | 76,652 | 100,926 | 17140 | 21117 | Kenton County, KY............................ | 159,720 | 167,949 |
| 16740 | 45091 | York County, SC............................... | 226,073 | 289,105 | 17140 | 21191 | Pendleton County, KY........................ | 14,877 | 14,586 |
| | | | | | 17140 | 39015 | Brown County, OH............................ | 44,846 | 43,414 |
| 16820 | | Charlottesville, VA Metro Area ............ | 201,559 | 219,910 | 17140 | 39017 | Butler County, OH............................. | 368,130 | 385,648 |
| 16820 | 51003 | Albemarle County, VA......................... | 98,970 | 110,652 | 17140 | 39025 | Clermont County, OH......................... | 197,363 | 207,449 |
| 16820 | 51065 | Fluvanna County, VA.......................... | 25,691 | 27,422 | 17140 | 39061 | Hamilton County, OH......................... | 802,374 | 817,985 |
| 16820 | 51079 | Greene County, VA............................. | 18,403 | 20,131 | 17140 | 39165 | Warren County, OH........................... | 212,693 | 238,412 |
| 16820 | 51125 | Nelson County, VA............................. | 15,020 | 14,755 | | | | | |
| 16820 | 51540 | Charlottesville city, VA ...................... | 43,475 | 46,950 | 17220 | | Clarksburg, WV Micro Area ................ | 94,196 | 91,937 |
| | | | | | 17220 | 54017 | Doddridge County, WV ...................... | 8,202 | 8,368 |
| 16860 | | Chattanooga, TN-GA Metro Area......... | 528,143 | 569,931 | 17220 | 54033 | Harrison County, WV ......................... | 69,099 | 66,870 |
| 16860 | 13047 | Catoosa County, GA.......................... | 63,942 | 67,996 | 17220 | 54091 | Taylor County, WV............................ | 16,895 | 16,699 |
| 16860 | 13083 | Dade County, GA.............................. | 16,633 | 16,057 | | | | | |
| 16860 | 13295 | Walker County, GA............................ | 68,756 | 70,116 | 17260 | | Clarksdale, MS Micro Area ................. | 26,151 | 21,564 |
| 16860 | 47065 | Hamilton County, TN.......................... | 336,463 | 371,662 | 17260 | 28027 | Coahoma County, MS......................... | 26,151 | 21,564 |
| 16860 | 47115 | Marion County, TN............................ | 28,237 | 28,924 | | | | | |
| 16860 | 47153 | Sequatchie County, TN....................... | 14,112 | 15,176 | 17300 | | Clarksville, TN-KY Metro Area.............. | 273,949 | 314,364 |
| | | | | | 17300 | 21047 | Christian County, KY.......................... | 73,955 | 71,478 |
| 16940 | | Cheyenne, WY Metro Area................... | 91,738 | 100,595 | 17300 | 21221 | Trigg County, KY.............................. | 14,339 | 14,776 |
| 16940 | 56021 | Laramie County, WY.......................... | 91,738 | 100,595 | 17300 | 47125 | Montgomery County, TN...................... | 172,331 | 214,251 |
| | | | | | 17300 | 47161 | Stewart County, TN............................ | 13,324 | 13,859 |
| 16980 | | Chicago-Naperville-Elgin, IL-IN-WI Metro Area........................................ | 9,461,105 | 9,406,638 | 17340 | | Clearlake, CA Micro Area.................... | 64,665 | 64,479 |
| | | | | | 17340 | 06033 | Lake County, CA............................... | 64,665 | 64,479 |
| 16980 | | Chicago-Naperville-Evanston, IL Division 16984 ..................................... | 7,147,982 | 7,071,372 | 17380 | | Cleveland, MS Micro Area................... | 34,145 | 30,142 |
| 16980 | 17031 | Cook County, IL................................ | 5,194,675 | 5,108,284 | 17380 | 28011 | Bolivar County, MS........................... | 34,145 | 30,142 |
| 16980 | 17043 | DuPage County, IL............................ | 916,924 | 917,481 | | | | | |
| 16980 | 17063 | Grundy County, IL............................. | 50,063 | 50,993 | 17420 | | Cleveland, TN Metro Area .................. | 115,788 | 125,906 |
| 16980 | 17111 | McHenry County, IL........................... | 308,760 | 305,888 | 17420 | 47011 | Bradley County, TN............................ | 98,963 | 109,071 |
| 16980 | 17197 | Will County, IL................................. | 677,560 | 688,726 | 17420 | 47139 | Polk County, TN............................... | 16,825 | 16,835 |
| 16980 | | Elgin, IL Division 20994 ...................... | 735,165 | 766,139 | 17460 | | Cleveland-Elyria, OH Metro Area .......... | 2,077,240 | 2,043,807 |
| 16980 | 17037 | DeKalb County, IL............................. | 105,160 | 104,491 | 17460 | 39035 | Cuyahoga County, OH........................ | 1,280,122 | 1,227,883 |
| 16980 | 17089 | Kane County, IL............................... | 515,269 | 531,010 | 17460 | 39055 | Geauga County, OH .......................... | 93,389 | 93,271 |
| 16980 | 17093 | Kendall County, IL............................ | 114,736 | 130,638 | 17460 | 39085 | Lake County, OH .............................. | 230,041 | 229,569 |
| | | | | | 17460 | 39093 | Lorain County, OH............................ | 301,356 | 312,172 |
| 16980 | | Gary, IN Division 23844 ...................... | 708,070 | 705,863 | 17460 | 39103 | Medina County, OH ........................... | 172,332 | 180,912 |
| 16980 | 18073 | Jasper County, IN ............................. | 33,478 | 33,440 | | | | | |
| 16980 | 18089 | Lake County, IN................................ | 496,005 | 487,536 | 17500 | | Clewiston, FL Micro Area ................... | 39,140 | 42,813 |
| 16980 | 18111 | Newton County, IN............................ | 14,244 | 13,907 | 17500 | 12051 | Hendry County, FL............................ | 39,140 | 42,813 |
| 16980 | 18127 | Porter County, IN.............................. | 164,343 | 170,980 | | | | | |
| | | | | | 17540 | | Clinton, IA Micro Area........................ | 49,116 | 46,392 |
| | | | | | 17540 | 19045 | Clinton County, IA............................. | 49,116 | 46,392 |

| Core based statistical area | State/ County FIPS code | Title and Geographic Components | 2010 Census Population | 2020 Estimated Population | Core based statistical area | State/ County FIPS code | Title and Geographic Components | 2010 Census Population | 2020 Estimated Population |
|---|---|---|---|---|---|---|---|---|---|
| 17580 | | Clovis, NM Micro Area | 48,376 | 48,793 | 18220 | | Connersville, IN Micro Area | 24,277 | 22,892 |
| 17580 | 35009 | Curry County, NM | 48,376 | 48,793 | 18220 | 18041 | Fayette County, IN | 24,277 | 22,892 |
| 17660 | | Coeur d'Alene, ID Metro Area | 138,494 | 170,628 | 18260 | | Cookeville, TN Micro Area | 106,042 | 115,359 |
| 17660 | 16055 | Kootenai County, ID | 138,494 | 170,628 | 18260 | 47087 | Jackson County, TN | 11,638 | 11,864 |
| | | | | | 18260 | 47133 | Overton County, TN | 22,083 | 22,566 |
| 17700 | | Coffeyville, KS Micro Area | 35,471 | 31,502 | 18260 | 47141 | Putnam County, TN | 72,321 | 80,929 |
| 17700 | 20125 | Montgomery County, KS | 35,471 | 31,502 | | | | | |
| | | | | | 18300 | | Coos Bay, OR Micro Area | 63,043 | 64,711 |
| 17740 | | Coldwater, MI Micro Area | 45,248 | 43,424 | 18300 | 41011 | Coos County, OR | 63,043 | 64,711 |
| 17740 | 26023 | Branch County, MI | 45,248 | 43,424 | | | | | |
| | | | | | 18380 | | Cordele, GA Micro Area | 23,439 | 22,034 |
| 17780 | | College Station-Bryan, TX Metro Area | 228,660 | 268,224 | 18380 | 13081 | Crisp County, GA | 23,439 | 22,034 |
| 17780 | 48041 | Brazos County, TX | 194,851 | 232,555 | | | | | |
| 17780 | 48051 | Burleson County, TX | 17,187 | 18,514 | 18420 | | Corinth, MS Micro Area | 37,057 | 36,889 |
| 17780 | 48395 | Robertson County, TX | 16,622 | 17,155 | 18420 | 28003 | Alcorn County, MS | 37,057 | 36,889 |
| 17820 | | Colorado Springs, CO Metro Area | 645,613 | 753,839 | 18460 | | Cornelia, GA Micro Area | 43,041 | 46,047 |
| 17820 | 08041 | El Paso County, CO | 622,263 | 728,310 | 18460 | 13137 | Habersham County, GA | 43,041 | 46,047 |
| 17820 | 08119 | Teller County, CO | 23,350 | 25,529 | | | | | |
| | | | | | 18500 | | Corning, NY Micro Area | 98,990 | 94,657 |
| 17860 | | Columbia, MO Metro Area | 190,387 | 210,094 | 18500 | 36101 | Steuben County, NY | 98,990 | 94,657 |
| 17860 | 29019 | Boone County, MO | 162,642 | 182,991 | | | | | |
| 17860 | 29053 | Cooper County, MO | 17,601 | 17,102 | 18580 | | Corpus Christi, TX Metro Area | 405,027 | 430,217 |
| 17860 | 29089 | Howard County, MO | 10,144 | 10,001 | 18580 | 48355 | Nueces County, TX | 340,223 | 363,148 |
| | | | | | 18580 | 48409 | San Patricio County, TX | 64,804 | 67,069 |
| 17900 | | Columbia, SC Metro Area | 767,598 | 847,397 | | | | | |
| 17900 | 45017 | Calhoun County, SC | 15,175 | 14,554 | 18620 | | Corsicana, TX Micro Area | 47,735 | 50,694 |
| 17900 | 45039 | Fairfield County, SC | 23,956 | 22,059 | 18620 | 48349 | Navarro County, TX | 47,735 | 50,694 |
| 17900 | 45055 | Kershaw County, SC | 61,697 | 67,472 | | | | | |
| 17900 | 45063 | Lexington County, SC | 262,391 | 303,946 | 18660 | | Cortland, NY Micro Area | 49,336 | 47,173 |
| 17900 | 45079 | Richland County, SC | 384,504 | 419,051 | 18660 | 36023 | Cortland County, NY | 49,336 | 47,173 |
| 17900 | 45081 | Saluda County, SC | 19,875 | 20,315 | | | | | |
| | | | | | 18700 | | Corvallis, OR Metro Area | 85,579 | 93,239 |
| 17980 | | Columbus, GA-AL Metro Area | 307,788 | 322,658 | 18700 | 41003 | Benton County, OR | 85,579 | 93,239 |
| 17980 | 01113 | Russell County, AL | 52,947 | 58,237 | | | | | |
| 17980 | 13053 | Chattahoochee County, GA | 11,267 | 10,551 | 18740 | | Coshocton, OH Micro Area | 36,901 | 36,449 |
| 17980 | 13145 | Harris County, GA | 32,024 | 36,080 | 18740 | 39031 | Coshocton County, OH | 36,901 | 36,449 |
| 17980 | 13197 | Marion County, GA | 8,742 | 8,516 | | | | | |
| 17980 | 13215 | Muscogee County, GA | 189,885 | 196,442 | 18780 | | Craig, CO Micro Area | 13,795 | 13,144 |
| 17980 | 13259 | Stewart County, GA | 6,058 | 6,689 | 18780 | 08081 | Moffat County, CO | 13,795 | 13,144 |
| 17980 | 13263 | Talbot County, GA | 6,865 | 6,143 | | | | | |
| | | | | | 18820 | | Crawfordsville, IN Micro Area | 38,124 | 38,365 |
| 18020 | | Columbus, IN Metro Area | 76,794 | 84,447 | 18820 | 18107 | Montgomery County, IN | 38,124 | 38,365 |
| 18020 | 18005 | Bartholomew County, IN | 76,794 | 84,447 | | | | | |
| | | | | | 18860 | | Crescent City, CA Micro Area | 28,610 | 27,968 |
| 18060 | | Columbus, MS Micro Area | 59,779 | 58,309 | 18860 | 06015 | Del Norte County, CA | 28,610 | 27,968 |
| 18060 | 28087 | Lowndes County, MS | 59,779 | 58,309 | | | | | |
| | | | | | 18880 | | Crestview-Fort Walton Beach-Destin, FL Metro Area | 235,865 | 289,468 |
| 18100 | | Columbus, NE Micro Area | 32,237 | 33,364 | 18880 | 12091 | Okaloosa County, FL | 180,822 | 212,820 |
| 18100 | 31141 | Platte County, NE | 32,237 | 33,364 | 18880 | 12131 | Walton County, FL | 55,043 | 76,648 |
| 18140 | | Columbus, OH Metro Area | 1,901,974 | 2,138,946 | | | | | |
| 18140 | 39041 | Delaware County, OH | 174,214 | 213,554 | 18900 | | Crossville, TN Micro Area | 56,053 | 61,603 |
| 18140 | 39045 | Fairfield County, OH | 146,156 | 159,709 | 18900 | 47035 | Cumberland County, TN | 56,053 | 61,603 |
| 18140 | 39049 | Franklin County, OH | 1,163,414 | 1,324,624 | | | | | |
| 18140 | 39073 | Hocking County, OH | 29,380 | 28,095 | 18980 | | Cullman, AL Micro Area | 80,406 | 84,515 |
| 18140 | 39089 | Licking County, OH | 166,492 | 178,100 | 18980 | 01043 | Cullman County, AL | 80,406 | 84,515 |
| 18140 | 39097 | Madison County, OH | 43,435 | 44,559 | | | | | |
| 18140 | 39117 | Morrow County, OH | 34,827 | 35,411 | 19000 | | Cullowhee, NC Micro Area | 54,252 | 58,212 |
| 18140 | 39127 | Perry County, OH | 36,058 | 36,215 | 19000 | 37099 | Jackson County, NC | 40,271 | 44,033 |
| 18140 | 39129 | Pickaway County, OH | 55,698 | 58,658 | 19000 | 37173 | Swain County, NC | 13,981 | 14,179 |
| 18140 | 39159 | Union County, OH | 52,300 | 60,021 | | | | | |
| | | | | | 19060 | | Cumberland, MD-WV Metro Area | 103,299 | 96,779 |
| 18180 | | Concord, NH Micro Area | 146,445 | 152,622 | 19060 | 24001 | Allegany County, MD | 75,087 | 70,057 |
| 18180 | 33013 | Merrimack County, NH | 146,445 | 152,622 | 19060 | 54057 | Mineral County, WV | 28,212 | 26,722 |

| Core based statistical area | State/County FIPS code | Title and Geographic Components | 2010 Census Population | 2020 Estimated Population | Core based statistical area | State/County FIPS code | Title and Geographic Components | 2010 Census Population | 2020 Estimated Population |
|---|---|---|---|---|---|---|---|---|---|
| 19100 | | Dallas-Fort Worth-Arlington, TX Metro Area | 6,366,542 | 7,694,138 | 19620 | | Del Rio, TX Micro Area | 48,879 | 49,028 |
| | | | | | 19620 | 48465 | Val Verde County, TX | 48,879 | 49,028 |
| 19100 | | Dallas-Plano-Irving, TX Division 19124 | 4,230,520 | 5,171,934 | 19660 | | Deltona-Daytona Beach-Ormond Beach, FL Metro Area | 590,289 | 679,948 |
| 19100 | 48085 | Collin County, TX | 782,341 | 1,072,069 | | | | | |
| 19100 | 48113 | Dallas County, TX | 2,368,139 | 2,635,888 | 19660 | 12035 | Flagler County, FL | 95,696 | 118,451 |
| 19100 | 48121 | Denton County, TX | 662,614 | 919,324 | 19660 | 12127 | Volusia County, FL | 494,593 | 561,497 |
| 19100 | 48139 | Ellis County, TX | 149,610 | 191,760 | | | | | |
| 19100 | 48231 | Hunt County, TX | 86,129 | 99,807 | 19700 | | Deming, NM Micro Area | 25,095 | 23,905 |
| 19100 | 48257 | Kaufman County, TX | 103,350 | 143,198 | 19700 | 35029 | Luna County, NM | 25,095 | 23,905 |
| 19100 | 48397 | Rockwall County, TX | 78,337 | 109,888 | | | | | |
| | | | | | 19740 | | Denver-Aurora-Lakewood, CO Metro Area | 2,543,482 | 2,991,231 |
| 19100 | | Fort Worth-Arlington-Grapevine, TX Division 23104 | 2,136,022 | 2,522,204 | 19740 | 08001 | Adams County, CO | 441,603 | 519,883 |
| 19100 | 48251 | Johnson County, TX | 150,934 | 179,575 | 19740 | 08005 | Arapahoe County, CO | 572,003 | 657,452 |
| 19100 | 48367 | Parker County, TX | 116,927 | 148,198 | 19740 | 08014 | Broomfield County, CO | 55,889 | 72,236 |
| 19100 | 48439 | Tarrant County, TX | 1,809,034 | 2,123,347 | 19740 | 08019 | Clear Creek County, CO | 9,088 | 9,586 |
| 19100 | 48497 | Wise County, TX | 59,127 | 71,084 | 19740 | 08031 | Denver County, CO | 600,158 | 735,538 |
| | | | | | 19740 | 08035 | Douglas County, CO | 285,465 | 360,750 |
| 19140 | | Dalton, GA Metro Area | 142,227 | 143,869 | 19740 | 08039 | Elbert County, CO | 23,086 | 27,313 |
| 19140 | 13213 | Murray County, GA | 39,628 | 40,032 | 19740 | 08047 | Gilpin County, CO | 5,441 | 6,235 |
| 19140 | 13313 | Whitfield County, GA | 102,599 | 103,837 | 19740 | 08059 | Jefferson County, CO | 534,543 | 583,283 |
| | | | | | 19740 | 08093 | Park County, CO | 16,206 | 18,955 |
| 19180 | | Danville, IL Metro Area | 81,625 | 74,855 | | | | | |
| 19180 | 17183 | Vermilion County, IL | 81,625 | 74,855 | 19760 | | DeRidder, LA Micro Area | 35,654 | 37,881 |
| | | | | | 19760 | 22011 | Beauregard Parish, LA | 35,654 | 37,881 |
| 19220 | | Danville, KY Micro Area | 53,174 | 54,833 | | | | | |
| 19220 | 21021 | Boyle County, KY | 28,432 | 30,367 | 19780 | | Des Moines-West Des Moines, IA Metro Area | 606,475 | 707,915 |
| 19220 | 21137 | Lincoln County, KY | 24,742 | 24,466 | | | | | |
| | | | | | 19780 | 19049 | Dallas County, IA | 66,135 | 96,963 |
| 19260 | | Danville, VA Micro Area | 106,561 | 99,719 | 19780 | 19077 | Guthrie County, IA | 10,954 | 10,737 |
| 19260 | 51143 | Pittsylvania County, VA | 63,506 | 59,850 | 19780 | 19099 | Jasper County, IA | 36,842 | 37,148 |
| 19260 | 51590 | Danville city, VA | 43,055 | 39,869 | 19780 | 19121 | Madison County, IA | 15,679 | 16,521 |
| | | | | | 19780 | 19153 | Polk County, IA | 430,640 | 494,281 |
| 19300 | | Daphne-Fairhope-Foley, AL Metro Area | 182,265 | 229,287 | 19780 | 19181 | Warren County, IA | 46,225 | 52,265 |
| 19300 | 01003 | Baldwin County, AL | 182,265 | 229,287 | | | | | |
| | | | | | 19820 | | Detroit-Warren-Dearborn, MI Metro Area | 4,296,250 | 4,304,136 |
| 19340 | | Davenport-Moline-Rock Island, IA-IL Metro Area | 379,690 | 377,759 | | | | | |
| 19340 | 17073 | Henry County, IL | 50,486 | 48,411 | 19820 | | Detroit-Dearborn-Livonia, MI Division 19804 | 1,820,584 | 1,740,623 |
| 19340 | 17131 | Mercer County, IL | 16,434 | 15,225 | 19820 | 26163 | Wayne County, MI | 1,820,584 | 1,740,623 |
| 19340 | 17161 | Rock Island County, IL | 147,546 | 140,907 | | | | | |
| 19340 | 19163 | Scott County, IA | 165,224 | 173,216 | 19820 | | Warren-Troy-Farmington Hills, MI Division 47664 | 2,475,666 | 2,563,513 |
| | | | | | 19820 | 26087 | Lapeer County, MI | 88,319 | 87,635 |
| 19420 | | Dayton, TN Micro Area | 31,809 | 33,443 | 19820 | 26093 | Livingston County, MI | 180,967 | 192,335 |
| 19420 | 47143 | Rhea County, TN | 31,809 | 33,443 | 19820 | 26099 | Macomb County, MI | 840,978 | 870,791 |
| | | | | | 19820 | 26125 | Oakland County, MI | 1,202,362 | 1,253,459 |
| 19430 | | Dayton-Kettering, OH Metro Area | 799,232 | 809,248 | 19820 | 26147 | St. Clair County, MI | 163,040 | 159,293 |
| 19430 | 39057 | Greene County, OH | 161,573 | 170,122 | | | | | |
| 19430 | 39109 | Miami County, OH | 102,506 | 107,516 | 19860 | | Dickinson, ND Micro Area | 24,982 | 32,997 |
| 19430 | 39113 | Montgomery County, OH | 535,153 | 531,610 | 19860 | 38007 | Billings County, ND | 783 | 890 |
| | | | | | 19860 | 38089 | Stark County, ND | 24,199 | 32,107 |
| 19460 | | Decatur, AL Metro Area | 153,829 | 152,740 | | | | | |
| 19460 | 01079 | Lawrence County, AL | 34,339 | 32,857 | 19940 | | Dixon, IL Micro Area | 36,031 | 33,647 |
| 19460 | 01103 | Morgan County, AL | 119,490 | 119,883 | 19940 | 17103 | Lee County, IL | 36,031 | 33,647 |
| | | | | | 19980 | | Dodge City, KS Micro Area | 33,848 | 33,094 |
| 19500 | | Decatur, IL Metro Area | 110,768 | 103,015 | 19980 | 20057 | Ford County, KS | 33,848 | 33,094 |
| 19500 | 17115 | Macon County, IL | 110,768 | 103,015 | | | | | |
| | | | | | 20020 | | Dothan, AL Metro Area | 145,639 | 150,214 |
| 19540 | | Decatur, IN Micro Area | 34,387 | 35,839 | 20020 | 01061 | Geneva County, AL | 26,790 | 26,411 |
| 19540 | 18001 | Adams County, IN | 34,387 | 35,839 | 20020 | 01067 | Henry County, AL | 17,302 | 17,223 |
| | | | | | 20020 | 01069 | Houston County, AL | 101,547 | 106,580 |
| 19580 | | Defiance, OH Micro Area | 39,037 | 37,778 | | | | | |
| 19580 | 39039 | Defiance County, OH | 39,037 | 37,778 | | | | | |

| Core based statistical area | State/ County FIPS code | Title and Geographic Components | 2010 Census Population | 2020 Estimated Population | Core based statistical area | State/ County FIPS code | Title and Geographic Components | 2010 Census Population | 2020 Estimated Population |
|---|---|---|---|---|---|---|---|---|---|
| 20060 | | Douglas, GA Micro Area | 50,731 | 51,611 | 20900 | | El Campo, TX Micro Area | 41,280 | 41,685 |
| 20060 | 13003 | Atkinson County, GA | 8,375 | 8,393 | 20900 | 48481 | Wharton County, TX | 41,280 | 41,685 |
| 20060 | 13069 | Coffee County, GA | 42,356 | 43,218 | | | | | |
| | | | | | 20940 | | El Centro, CA Metro Area | 174,528 | 180,267 |
| 20100 | | Dover, DE Metro Area | 162,310 | 183,643 | 20940 | 06025 | Imperial County, CA | 174,528 | 180,267 |
| 20100 | 10001 | Kent County, DE | 162,310 | 183,643 | | | | | |
| | | | | | 20980 | | El Dorado, AR Micro Area | 41,639 | 38,219 |
| 20140 | | Dublin, GA Micro Area | 65,299 | 64,001 | 20980 | 05139 | Union County, AR | 41,639 | 38,219 |
| 20140 | 13167 | Johnson County, GA | 9,980 | 9,667 | | | | | |
| 20140 | 13175 | Laurens County, GA | 48,434 | 47,512 | 21020 | | Elizabeth City, NC Micro Area | 54,114 | 54,039 |
| 20140 | 13283 | Treutlen County, GA | 6,885 | 6,822 | 21020 | 37139 | Pasquotank County, NC | 40,661 | 40,372 |
| | | | | | 21020 | 37143 | Perquimans County, NC | 13,453 | 13,667 |
| 20180 | | DuBois, PA Micro Area | 81,642 | 78,612 | | | | | |
| 20180 | 42033 | Clearfield County, PA | 81,642 | 78,612 | 21060 | | Elizabethtown-Fort Knox, KY Metro Area | 148,338 | 154,356 |
| 20220 | | Dubuque, IA Metro Area | 93,653 | 97,590 | 21060 | 21093 | Hardin County, KY | 105,543 | 111,309 |
| 20220 | 19061 | Dubuque County, IA | 93,653 | 97,590 | 21060 | 21123 | Larue County, KY | 14,193 | 14,431 |
| | | | | | 21060 | 21163 | Meade County, KY | 28,602 | 28,616 |
| 20260 | | Duluth, MN-WI Metro Area | 290,637 | 288,648 | | | | | |
| 20260 | 27017 | Carlton County, MN | 35,386 | 35,769 | 21120 | | Elk City, OK Micro Area | 22,119 | 21,468 |
| 20260 | 27075 | Lake County, MN | 10,866 | 10,639 | 21120 | 40009 | Beckham County, OK | 22,119 | 21,468 |
| 20260 | 27137 | St. Louis County, MN | 200,226 | 198,538 | | | | | |
| 20260 | 55031 | Douglas County, WI | 44,159 | 43,702 | 21140 | | Elkhart-Goshen, IN Metro Area | 197,559 | 206,161 |
| | | | | | 21140 | 18039 | Elkhart County, IN | 197,559 | 206,161 |
| 20300 | | Dumas, TX Micro Area | 21,904 | 20,654 | | | | | |
| 20300 | 48341 | Moore County, TX | 21,904 | 20,654 | 21180 | | Elkins, WV Micro Area | 29,405 | 28,387 |
| | | | | | 21180 | 54083 | Randolph County, WV | 29,405 | 28,387 |
| 20340 | | Duncan, OK Micro Area | 45,048 | 43,100 | | | | | |
| 20340 | 40137 | Stephens County, OK | 45,048 | 43,100 | 21220 | | Elko, NV Micro Area | 50,805 | 55,071 |
| | | | | | 21220 | 32007 | Elko County, NV | 48,818 | 53,006 |
| 20420 | | Durango, CO Micro Area | 51,334 | 56,564 | 21220 | 32011 | Eureka County, NV | 1,987 | 2,065 |
| 20420 | 08067 | La Plata County, CO | 51,334 | 56,564 | | | | | |
| | | | | | 21260 | | Ellensburg, WA Micro Area | 40,915 | 49,204 |
| 20460 | | Durant, OK Micro Area | 42,416 | 48,998 | 21260 | 53037 | Kittitas County, WA | 40,915 | 49,204 |
| 20460 | 40013 | Bryan County, OK | 42,416 | 48,998 | | | | | |
| | | | | | 21300 | | Elmira, NY Metro Area | 88,830 | 82,622 |
| 20500 | | Durham-Chapel Hill, NC Metro Area | 564,273 | 652,542 | 21300 | 36015 | Chemung County, NY | 88,830 | 82,622 |
| 20500 | 37037 | Chatham County, NC | 63,505 | 75,748 | | | | | |
| 20500 | 37063 | Durham County, NC | 267,587 | 327,306 | 21340 | | El Paso, TX Metro Area | 804,123 | 846,192 |
| 20500 | 37077 | Granville County, NC | 59,916 | 60,486 | 21340 | 48141 | El Paso County, TX | 800,647 | 841,286 |
| 20500 | 37135 | Orange County, NC | 133,801 | 149,077 | 21340 | 48229 | Hudspeth County, TX | 3,476 | 4,906 |
| 20500 | 37145 | Person County, NC | 39,464 | 39,925 | | | | | |
| | | | | | 21380 | | Emporia, KS Micro Area | 36,480 | 35,631 |
| 20540 | | Dyersburg, TN Micro Area | 38,335 | 36,693 | 21380 | 20017 | Chase County, KS | 2,790 | 2,586 |
| 20540 | 47045 | Dyer County, TN | 38,335 | 36,693 | 21380 | 20111 | Lyon County, KS | 33,690 | 33,045 |
| 20580 | | Eagle Pass, TX Micro Area | 54,258 | 58,378 | 21420 | | Enid, OK Metro Area | 60,580 | 60,869 |
| 20580 | 48323 | Maverick County, TX | 54,258 | 58,378 | 21420 | 40047 | Garfield County, OK | 60,580 | 60,869 |
| 20660 | | Easton, MD Micro Area | 37,782 | 36,972 | 21460 | | Enterprise, AL Micro Area | 49,948 | 53,230 |
| 20660 | 24041 | Talbot County, MD | 37,782 | 36,972 | 21460 | 01031 | Coffee County, AL | 49,948 | 53,230 |
| 20700 | | East Stroudsburg, PA Metro Area | 169,842 | 170,154 | 21500 | | Erie, PA Metro Area | 280,566 | 268,426 |
| 20700 | 42089 | Monroe County, PA | 169,842 | 170,154 | 21500 | 42049 | Erie County, PA | 280,566 | 268,426 |
| 20740 | | Eau Claire, WI Metro Area | 161,151 | 169,997 | 21540 | | Escanaba, MI Micro Area | 37,069 | 35,612 |
| 20740 | 55017 | Chippewa County, WI | 62,415 | 64,737 | 21540 | 26041 | Delta County, MI | 37,069 | 35,612 |
| 20740 | 55035 | Eau Claire County, WI | 98,736 | 105,260 | | | | | |
| | | | | | 21580 | | Espa±ola, NM Micro Area | 40,246 | 38,521 |
| 20780 | | Edwards, CO Micro Area | 52,197 | 54,929 | 21580 | 35039 | Rio Arriba County, NM | 40,246 | 38,521 |
| 20780 | 08037 | Eagle County, CO | 52,197 | 54,929 | | | | | |
| | | | | | 21640 | | Eufaula, AL-GA Micro Area | 29,970 | 26,860 |
| 20820 | | Effingham, IL Micro Area | 34,242 | 34,065 | 21640 | 01005 | Barbour County, AL | 27,457 | 24,589 |
| 20820 | 17049 | Effingham County, IL | 34,242 | 34,065 | 21640 | 13239 | Quitman County, GA | 2,513 | 2,271 |

| Core based statistical area | State/ County FIPS code | Title and Geographic Components | 2010 Census Population | 2020 Estimated Population | Core based statistical area | State/ County FIPS code | Title and Geographic Components | 2010 Census Population | 2020 Estimated Population |
|---|---|---|---|---|---|---|---|---|---|
| 21660 | | Eugene-Springfield, OR Metro Area...... | 351,715 | 382,986 | 22380 | | Flagstaff, AZ Metro Area...................... | 134,421 | 142,481 |
| 21660 | 41039 | Lane County, OR ............................. | 351,715 | 382,986 | 22380 | 04005 | Coconino County, AZ...................... | 134,421 | 142,481 |
| 21700 | | Eureka-Arcata, CA Micro Area............. | 134,623 | 134,977 | 22420 | | Flint, MI Metro Area............................ | 425,790 | 404,794 |
| 21700 | 06023 | Humboldt County, CA ....................... | 134,623 | 134,977 | 22420 | 26049 | Genesee County, MI...................... | 425,790 | 404,794 |
| 21740 | | Evanston, WY Micro Area ................... | 21,118 | 20,215 | 22500 | | Florence, SC Metro Area..................... | 205,566 | 204,097 |
| 21740 | 56041 | Uinta County, WY ............................. | 21,118 | 20,215 | 22500 | 45031 | Darlington County, SC ................... | 68,681 | 66,509 |
| | | | | | 22500 | 45041 | Florence County, SC...................... | 136,885 | 137,588 |
| 21780 | | Evansville, IN-KY Metro Area............... | 311,552 | 315,731 | | | | | |
| 21780 | 18129 | Posey County, IN............................. | 25,910 | 25,275 | 22520 | | Florence-Muscle Shoals, AL Metro Area | 147,137 | 148,779 |
| 21780 | 18163 | Vanderburgh County, IN ................... | 179,703 | 182,447 | 22520 | 01033 | Colbert County, AL........................ | 54,428 | 55,411 |
| 21780 | 18173 | Warrick County, IN........................... | 59,689 | 63,269 | 22520 | 01077 | Lauderdale County, AL ................... | 92,709 | 93,368 |
| 21780 | 21101 | Henderson County, KY ..................... | 46,250 | 44,740 | | | | | |
| | | | | | 22540 | | Fond du Lac, WI Metro Area ................. | 101,633 | 102,902 |
| 21820 | | Fairbanks, AK Metro Area ................... | 97,581 | 95,651 | 22540 | 55039 | Fond du Lac County, WI .................. | 101,633 | 102,902 |
| 21820 | 02090 | Fairbanks North Star Borough, AK .... | 97,581 | 95,651 | | | | | |
| | | | | | 22580 | | Forest City, NC Micro Area.................. | 67,810 | 67,076 |
| 21840 | | Fairfield, IA Micro Area ...................... | 16,843 | 18,347 | 22580 | 37161 | Rutherford County, NC ................... | 67,810 | 67,076 |
| 21840 | 19101 | Jefferson County, IA ........................ | 16,843 | 18,347 | | | | | |
| | | | | | 22620 | | Forrest City, AR Micro Area................. | 28,258 | 24,682 |
| 21860 | | Fairmont, MN Micro Area .................... | 20,840 | 19,484 | 22620 | 05123 | St. Francis County, AR ................... | 28,258 | 24,682 |
| 21860 | 27091 | Martin County, MN ........................... | 20,840 | 19,484 | | | | | |
| | | | | | 22660 | | Fort Collins, CO Metro Area ................. | 299,630 | 360,428 |
| 21900 | | Fairmont, WV Micro Area .................... | 56,418 | 55,962 | 22660 | 08069 | Larimer County, CO....................... | 299,630 | 360,428 |
| 21900 | 54049 | Marion County, WV........................... | 56,418 | 55,962 | | | | | |
| | | | | | 22700 | | Fort Dodge, IA Micro Area ................... | 38,013 | 35,934 |
| 21980 | | Fallon, NV Micro Area ........................ | 24,877 | 25,363 | 22700 | 19187 | Webster County, IA........................ | 38,013 | 35,934 |
| 21980 | 32001 | Churchill County, NV........................ | 24,877 | 25,363 | | | | | |
| | | | | | 22780 | | Fort Leonard Wood, MO Micro Area ..... | 52,274 | 52,709 |
| 22020 | | Fargo, ND-MN Metro Area ................... | 208,777 | 248,594 | 22780 | 29169 | Pulaski County, MO........................ | 52,274 | 52,709 |
| 22020 | 27027 | Clay County, MN.............................. | 58,999 | 64,690 | | | | | |
| 22020 | 38017 | Cass County, ND............................. | 149,778 | 183,904 | 22800 | | Fort Madison-Keokuk, IA-IL-MO Micro Area................................................ | 62,105 | 57,732 |
| 22060 | | Faribault-Northfield, MN Micro Area ..... | 64,142 | 67,084 | 22800 | 17067 | Hancock County, IL........................ | 19,104 | 17,422 |
| 22060 | 27131 | Rice County, MN.............................. | 64,142 | 67,084 | 22800 | 19111 | Lee County, IA.............................. | 35,862 | 33,480 |
| | | | | | 22800 | 29045 | Clark County, MO .......................... | 7,139 | 6,830 |
| 22100 | | Farmington, MO Micro Area ................. | 65,359 | 66,485 | | | | | |
| 22100 | 29187 | St. Francois County, MO................... | 65,359 | 66,485 | 22820 | | Fort Morgan, CO Micro Area ................ | 28,159 | 28,941 |
| | | | | | 22820 | 08087 | Morgan County, CO ....................... | 28,159 | 28,941 |
| 22140 | | Farmington, NM Metro Area ................. | 130,044 | 123,312 | | | | | |
| 22140 | 35045 | San Juan County, NM...................... | 130,044 | 123,312 | 22840 | | Fort Payne, AL Micro Area .................. | 71,109 | 71,658 |
| | | | | | 22840 | 01049 | DeKalb County, AL ........................ | 71,109 | 71,658 |
| 22180 | | Fayetteville, NC Metro Area ................. | 481,061 | 529,252 | | | | | |
| 22180 | 37051 | Cumberland County, NC.................... | 319,431 | 336,364 | 22860 | | Fort Polk South, LA Micro Area............. | 52,334 | 47,894 |
| 22180 | 37085 | Harnett County, NC.......................... | 114,678 | 137,058 | 22860 | 22115 | Vernon Parish, LA......................... | 52,334 | 47,894 |
| 22180 | 37093 | Hoke County, NC............................. | 46,952 | 55,830 | | | | | |
| | | | | | 22900 | | Fort Smith, AR-OK Metro Area.............. | 248,208 | 250,434 |
| 22220 | | Fayetteville-Springdale-Rogers, AR Metro Area............................................ | 440,121 | 548,634 | 22900 | 40135 | Sequoyah County, OK..................... | 42,391 | 41,538 |
| | | | | | 22900 | 05033 | Crawford County, AR...................... | 61,948 | 63,409 |
| 22220 | 05007 | Benton County, AR........................... | 221,339 | 288,774 | 22900 | 05047 | Franklin County, AR....................... | 18,125 | 17,897 |
| 22220 | 05087 | Madison County, AR......................... | 15,717 | 16,644 | 22900 | 05131 | Sebastian County, AR..................... | 125,744 | 127,590 |
| 22220 | 05143 | Washington County, AR .................... | 203,065 | 243,216 | | | | | |
| | | | | | 23060 | | Fort Wayne, IN Metro Area................... | 388,621 | 416,565 |
| 22260 | | Fergus Falls, MN Micro Area................ | 57,303 | 58,741 | 23060 | 18003 | Allen County, IN............................ | 355,329 | 382,187 |
| 22260 | 27111 | Otter Tail County, MN ...................... | 57,303 | 58,741 | 23060 | 18183 | Whitley County, IN ......................... | 33,292 | 34,378 |
| 22280 | | Fernley, NV Micro Area ....................... | 51,980 | 58,319 | 23140 | | Frankfort, IN Micro Area ..................... | 33,224 | 32,206 |
| 22280 | 32019 | Lyon County, NV.............................. | 51,980 | 58,319 | 23140 | 18023 | Clinton County, IN.......................... | 33,224 | 32,206 |
| 22300 | | Findlay, OH Micro Area ....................... | 74,782 | 75,407 | 23180 | | Frankfort, KY Micro Area..................... | 70,706 | 73,951 |
| 22300 | 39063 | Hancock County, OH ....................... | 74,782 | 75,407 | 23180 | 21005 | Anderson County, KY...................... | 21,421 | 22,833 |
| | | | | | 23180 | 21073 | Franklin County, KY ....................... | 49,285 | 51,118 |
| 22340 | | Fitzgerald, GA Micro Area ................... | 17,634 | 16,614 | | | | | |
| 22340 | 13017 | Ben Hill County, GA ........................ | 17,634 | 16,614 | 23240 | | Fredericksburg, TX Micro Area............. | 24,837 | 26,960 |
| | | | | | 23240 | 48171 | Gillespie County, TX ...................... | 24,837 | 26,960 |

| Core based statistical area | State/ County FIPS code | Title and Geographic Components | 2010 Census Population | 2020 Estimated Population | Core based statistical area | State/ County FIPS code | Title and Geographic Components | 2010 Census Population | 2020 Estimated Population |
|---|---|---|---|---|---|---|---|---|---|
| 23300 | | Freeport, IL Micro Area | 47,711 | 43,831 | 24100 | | Gloversville, NY Micro Area | 55,531 | 52,812 |
| 23300 | 17177 | Stephenson County, IL | 47,711 | 43,831 | 24100 | 36035 | Fulton County, NY | 55,531 | 52,812 |
| 23340 | | Fremont, NE Micro Area | 36,691 | 36,222 | 24140 | | Goldsboro, NC Metro Area | 122,623 | 123,967 |
| 23340 | 31053 | Dodge County, NE | 36,691 | 36,222 | 24140 | 37191 | Wayne County, NC | 122,623 | 123,967 |
| 23380 | | Fremont, OH Micro Area | 60,944 | 58,351 | 24180 | | Granbury, TX Micro Area | 51,182 | 63,527 |
| 23380 | 39143 | Sandusky County, OH | 60,944 | 58,351 | 24180 | 48221 | Hood County, TX | 51,182 | 63,527 |
| 23420 | | Fresno, CA Metro Area | 930,450 | 1,000,918 | 24220 | | Grand Forks, ND-MN Metro Area | 98,461 | 100,381 |
| 23420 | 06019 | Fresno County, CA | 930,450 | 1,000,918 | 24220 | 27119 | Polk County, MN | 31,600 | 30,900 |
| 23460 | | Gadsden, AL Metro Area | 104,430 | 102,371 | 24220 | 38035 | Grand Forks County, ND | 66,861 | 69,481 |
| 23460 | 01055 | Etowah County, AL | 104,430 | 102,371 | 24260 | | Grand Island, NE Metro Area | 72,726 | 75,325 |
| 23500 | | Gaffney, SC Micro Area | 55,342 | 57,316 | 24260 | 31079 | Hall County, NE | 58,607 | 61,028 |
| 23500 | 45021 | Cherokee County, SC | 55,342 | 57,316 | 24260 | 31093 | Howard County, NE | 6,274 | 6,488 |
| 23540 | | Gainesville, FL Metro Area | 305,076 | 332,317 | 24260 | 31121 | Merrick County, NE | 7,845 | 7,809 |
| 23540 | 12001 | Alachua County, FL | 247,336 | 271,218 | 24300 | | Grand Junction, CO Metro Area | 146,723 | 155,603 |
| 23540 | 12041 | Gilchrist County, FL | 16,939 | 18,885 | 24300 | 08077 | Mesa County, CO | 146,723 | 155,603 |
| 23540 | 12075 | Levy County, FL | 40,801 | 42,214 | 24330 | | Grand Rapids, MN Micro Area | 45,058 | 45,268 |
| 23580 | | Gainesville, GA Metro Area | 179,684 | 206,591 | 24330 | 27061 | Itasca County, MN | 45,058 | 45,268 |
| 23580 | 13139 | Hall County, GA | 179,684 | 206,591 | 24340 | | Grand Rapids-Kentwood, MI Metro Area | 993,670 | 1,081,372 |
| 23620 | | Gainesville, TX Micro Area | 38,437 | 41,393 | 24340 | 26067 | Ionia County, MI | 63,905 | 64,553 |
| 23620 | 48097 | Cooke County, TX | 38,437 | 41,393 | 24340 | 26081 | Kent County, MI | 602,622 | 658,708 |
| 23660 | | Galesburg, IL Micro Area | 52,919 | 49,053 | 24340 | 26117 | Montcalm County, MI | 63,342 | 63,476 |
| 23660 | 17095 | Knox County, IL | 52,919 | 49,053 | 24340 | 26139 | Ottawa County, MI | 263,801 | 294,635 |
| 23700 | | Gallup, NM Micro Area | 71,492 | 70,824 | 24380 | | Grants, NM Micro Area | 27,213 | 26,354 |
| 23700 | 35031 | McKinley County, NM | 71,492 | 70,824 | 24380 | 35006 | Cibola County, NM | 27,213 | 26,354 |
| 23780 | | Garden City, KS Micro Area | 40,753 | 39,662 | 24420 | | Grants Pass, OR Metro Area | 82,713 | 88,053 |
| 23780 | 20055 | Finney County, KS | 36,776 | 35,917 | 24420 | 41033 | Josephine County, OR | 82,713 | 88,053 |
| 23780 | 20093 | Kearny County, KS | 3,977 | 3,745 | 24460 | | Great Bend, KS Micro Area | 27,674 | 25,658 |
| 23820 | | Gardnerville Ranchos, NV Micro Area | 46,997 | 49,088 | 24460 | 20009 | Barton County, KS | 27,674 | 25,658 |
| 23820 | 32005 | Douglas County, NV | 46,997 | 49,088 | 24500 | | Great Falls, MT Metro Area | 81,327 | 81,346 |
| 23860 | | Georgetown, SC Micro Area | 60,158 | 63,353 | 24500 | 30013 | Cascade County, MT | 81,327 | 81,346 |
| 23860 | 45043 | Georgetown County, SC | 60,158 | 63,353 | 24540 | | Greeley, CO Metro Area | 252,825 | 333,983 |
| 23900 | | Gettysburg, PA Metro Area | 101,407 | 102,742 | 24540 | 08123 | Weld County, CO | 252,825 | 333,983 |
| 23900 | 42001 | Adams County, PA | 101,407 | 102,742 | 24580 | | Green Bay, WI Metro Area | 306,241 | 323,379 |
| 23940 | | Gillette, WY Micro Area | 60,424 | 61,012 | 24580 | 55009 | Brown County, WI | 248,007 | 264,610 |
| 23940 | 56005 | Campbell County, WY | 46,133 | 46,676 | 24580 | 55061 | Kewaunee County, WI | 20,574 | 20,386 |
| 23940 | 56011 | Crook County, WY | 7,083 | 7,593 | 24580 | 55083 | Oconto County, WI | 37,660 | 38,383 |
| 23940 | 56045 | Weston County, WY | 7,208 | 6,743 | 24620 | | Greeneville, TN Micro Area | 68,831 | 69,571 |
| 23980 | | Glasgow, KY Micro Area | 52,272 | 54,358 | 24620 | 47059 | Greene County, TN | 68,831 | 69,571 |
| 23980 | 21009 | Barren County, KY | 42,173 | 44,300 | 24660 | | Greensboro-High Point, NC Metro Area | 723,801 | 776,363 |
| 23980 | 21169 | Metcalfe County, KY | 10,099 | 10,058 | 24660 | 37081 | Guilford County, NC | 488,406 | 540,521 |
| 24020 | | Glens Falls, NY Metro Area | 128,923 | 124,362 | 24660 | 37151 | Randolph County, NC | 141,752 | 144,557 |
| 24020 | 36113 | Warren County, NY | 65,707 | 63,756 | 24660 | 37157 | Rockingham County, NC | 93,643 | 91,285 |
| 24020 | 36115 | Washington County, NY | 63,216 | 60,606 | 24700 | | Greensburg, IN Micro Area | 25,740 | 26,584 |
| 24060 | | Glenwood Springs, CO Micro Area | 73,537 | 78,260 | 24700 | 18031 | Decatur County, IN | 25,740 | 26,584 |
| 24060 | 08045 | Garfield County, CO | 56,389 | 60,366 | 24740 | | Greenville, MS Micro Area | 51,137 | 42,837 |
| 24060 | 08097 | Pitkin County, CO | 17,148 | 17,894 | 24740 | 28151 | Washington County, MS | 51,137 | 42,837 |

# CORE BASED STATISTICAL AREAS
## (Metropolitan and Micropolitan),
## METROPOLITAN DIVISIONS, AND COMPONENTS
## (as defined March, 2020)—*Continued*

| Core based statistical area | State/County FIPS code | Title and Geographic Components | 2010 Census Population | 2020 Estimated Population | Core based statistical area | State/County FIPS code | Title and Geographic Components | 2010 Census Population | 2020 Estimated Population |
|---|---|---|---|---|---|---|---|---|---|
| 24780 | | Greenville, NC Metro Area ..................... | 168,148 | 182,924 | 25540 | | Hartford-East Hartford-Middletown, CT Metro Area.............................. | 1,212,381 | 1,201,483 |
| 24780 | 37147 | Pitt County, NC ................................. | 168,148 | 182,924 | 25540 | 09003 | Hartford County, CT.......................... | 894,014 | 889,226 |
| | | | | | 25540 | 09007 | Middlesex County, CT........................ | 165,676 | 161,657 |
| 24820 | | Greenville, OH Micro Area.................... | 52,959 | 51,205 | 25540 | 09013 | Tolland County, CT............................ | 152,691 | 150,600 |
| 24820 | 39037 | Darke County, OH.............................. | 52,959 | 51,205 | | | | | |
| | | | | | 25580 | | Hastings, NE Micro Area .................... | 31,364 | 31,321 |
| 24860 | | Greenville-Anderson, SC Metro Area .... | 824,112 | 932,705 | 25580 | 31001 | Adams County, NE ............................ | 31,364 | 31,321 |
| 24860 | 45007 | Anderson County, SC ........................ | 187,126 | 204,353 | | | | | |
| 24860 | 45045 | Greenville County, SC........................ | 451,225 | 532,486 | 25620 | | Hattiesburg, MS Metro Area................. | 162,410 | 169,554 |
| 24860 | 45059 | Laurens County, SC.......................... | 66,537 | 67,883 | 25620 | 28031 | Covington County, MS........................ | 19,568 | 18,518 |
| 24860 | 45077 | Pickens County, SC.......................... | 119,224 | 127,983 | 25620 | 28035 | Forrest County, MS............................ | 74,934 | 75,009 |
| | | | | | 25620 | 28073 | Lamar County, MS............................. | 55,658 | 64,165 |
| 24900 | | Greenwood, MS Micro Area ................. | 42,914 | 37,586 | 25620 | 28111 | Perry County, MS.............................. | 12,250 | 11,862 |
| 24900 | 28015 | Carroll County, MS............................ | 10,597 | 9,732 | | | | | |
| 24900 | 28083 | Leflore County, MS............................ | 32,317 | 27,854 | 25700 | | Hays, KS Micro Area ......................... | 28,452 | 28,671 |
| | | | | | 25700 | 20051 | Ellis County, KS ............................... | 28,452 | 28,671 |
| 24940 | | Greenwood, SC Micro Area.................. | 69,661 | 71,074 | | | | | |
| 24940 | 45047 | Greenwood County, SC...................... | 69,661 | 71,074 | 25720 | | Heber, UT Micro Area.......................... | 59,854 | 77,799 |
| | | | | | 25720 | 49043 | Summit County, UT............................ | 36,324 | 42,499 |
| 24980 | | Grenada, MS Micro Area...................... | 21,906 | 20,610 | 25720 | 49051 | Wasatch County, UT.......................... | 23,530 | 35,300 |
| 24980 | 28043 | Grenada County, MS ......................... | 21,906 | 20,610 | | | | | |
| | | | | | 25740 | | Helena, MT Micro Area........................ | 74,801 | 82,589 |
| 25060 | | Gulfport-Biloxi, MS Metro Area.............. | 388,488 | 418,963 | 25740 | 30043 | Jefferson County, MT......................... | 11,406 | 12,360 |
| 25060 | 28045 | Hancock County, MS.......................... | 43,929 | 48,000 | 25740 | 30049 | Lewis and Clark County, MT................ | 63,395 | 70,229 |
| 25060 | 28047 | Harrison County, MS.......................... | 187,105 | 208,801 | | | | | |
| 25060 | 28059 | Jackson County, MS.......................... | 139,668 | 143,802 | 25760 | | Helena-West Helena, AR Micro Area .... | 21,757 | 17,299 |
| 25060 | 28131 | Stone County, MS............................. | 17,786 | 18,360 | 25760 | 05107 | Phillips County, AR ........................... | 21,757 | 17,299 |
| | | | | | | | | | |
| 25100 | | Guymon, OK Micro Area ...................... | 20,640 | 19,997 | 25780 | | Henderson, NC Micro Area .................. | 45,422 | 44,718 |
| 25100 | 40139 | Texas County, OK............................. | 20,640 | 19,997 | 25780 | 37181 | Vance County, NC ............................ | 45,422 | 44,718 |
| | | | | | | | | | |
| 25180 | | Hagerstown-Martinsburg, MD-WV Metro Area.......................................... | 269,140 | 291,144 | 25820 | | Hereford, TX Micro Area...................... | 19,372 | 18,277 |
| 25180 | 24043 | Washington County, MD..................... | 147,430 | 151,146 | 25820 | 48117 | Deaf Smith County, TX ...................... | 19,372 | 18,277 |
| 25180 | 54003 | Berkeley County, WV........................ | 104,169 | 122,125 | | | | | |
| 25180 | 54065 | Morgan County, WV.......................... | 17,541 | 17,873 | 25840 | | Hermiston-Pendleton, OR Micro Area ... | 87,062 | 89,452 |
| | | | | | 25840 | 41049 | Morrow County, OR ........................... | 11,173 | 11,700 |
| 25200 | | Hailey, ID Micro Area.......................... | 22,493 | 24,556 | 25840 | 41059 | Umatilla County, OR.......................... | 75,889 | 77,752 |
| 25200 | 16013 | Blaine County, ID.............................. | 21,376 | 23,426 | | | | | |
| 25200 | 16025 | Camas County, ID............................. | 1,117 | 1,130 | 25860 | | Hickory-Lenoir-Morganton, NC Metro Area .................................................. | 365,497 | 370,266 |
| | | | | | 25860 | 37003 | Alexander County, NC ....................... | 37,198 | 37,441 |
| 25220 | | Hammond, LA Metro Area.................... | 121,097 | 136,765 | 25860 | 37023 | Burke County, NC............................. | 90,912 | 90,418 |
| 25220 | 22105 | Tangipahoa Parish, LA ...................... | 121,097 | 136,765 | 25860 | 37027 | Caldwell County, NC.......................... | 83,029 | 82,100 |
| | | | | | 25860 | 37035 | Catawba County, NC.......................... | 154,358 | 160,307 |
| 25260 | | Hanford-Corcoran, CA Metro Area ........ | 152,982 | 152,692 | | | | | |
| 25260 | 06031 | Kings County, CA.............................. | 152,982 | 152,692 | 25880 | | Hillsdale, MI Micro Area...................... | 46,688 | 45,658 |
| | | | | | 25880 | 26059 | Hillsdale County, MI.......................... | 46,688 | 45,658 |
| 25300 | | Hannibal, MO Micro Area .................... | 38,948 | 38,722 | | | | | |
| 25300 | 29127 | Marion County, MO........................... | 28,781 | 28,423 | 25900 | | Hilo, HI Micro Area ............................ | 185,079 | 203,340 |
| 25300 | 29173 | Ralls County, MO.............................. | 10,167 | 10,299 | 25900 | 15001 | Hawaii County, HI ............................. | 185,079 | 203,340 |
| | | | | | | | | | |
| 25420 | | Harrisburg-Carlisle, PA Metro Area....... | 549,475 | 581,943 | 25940 | | Hilton Head Island-Bluffton, SC Metro Area .................................................. | 187,010 | 227,244 |
| 25420 | 42041 | Cumberland County, PA..................... | 235,406 | 255,857 | 25940 | 45013 | Beaufort County, SC.......................... | 162,233 | 195,656 |
| 25420 | 42043 | Dauphin County, PA.......................... | 268,100 | 279,874 | 25940 | 45053 | Jasper County, SC ........................... | 24,777 | 31,588 |
| 25420 | 42099 | Perry County, PA............................... | 45,969 | 46,212 | | | | | |
| | | | | | 25980 | | Hinesville, GA Metro Area .................. | 77,917 | 83,175 |
| 25460 | | Harrison, AR Micro Area...................... | 45,233 | 45,227 | 25980 | 13179 | Liberty County, GA............................ | 63,453 | 63,004 |
| 25460 | 05009 | Boone County, AR ............................ | 36,903 | 37,625 | 25980 | 13183 | Long County, GA .............................. | 14,464 | 20,171 |
| 25460 | 05101 | Newton County, AR........................... | 8,330 | 7,602 | | | | | |
| | | | | | 26020 | | Hobbs, NM Micro Area........................ | 64,727 | 71,830 |
| 25500 | | Harrisonburg, VA Metro Area ............... | 125,228 | 135,550 | 26020 | 35025 | Lea County, NM ............................... | 64,727 | 71,830 |
| 25500 | 51165 | Rockingham County, VA..................... | 76,314 | 82,346 | | | | | |
| 25500 | 51660 | Harrisonburg city, VA......................... | 48,914 | 53,204 | 26090 | | Holland, MI Micro Area........................ | 111,408 | 118,927 |
| | | | | | 26090 | 26005 | Allegan County, MI............................ | 111,408 | 118,927 |

Appendix C

| Core based statistical area | State/ County FIPS code | Title and Geographic Components | 2010 Census Population | 2020 Estimated Population | Core based statistical area | State/ County FIPS code | Title and Geographic Components | 2010 Census Population | 2020 Estimated Population |
|---|---|---|---|---|---|---|---|---|---|
| 26140 | | Homosassa Springs, FL Metro Area ..... | 141,236 | 153,010 | 26780 | | Hutchinson, MN Micro Area.................. | 36,651 | 35,710 |
| 26140 | 12017 | Citrus County, FL ........................... | 141,236 | 153,010 | 26780 | 27085 | McLeod County, MN ....................... | 36,651 | 35,710 |
| 26220 | | Hood River, OR Micro Area.................. | 22,346 | 23,280 | 26820 | | Idaho Falls, ID Metro Area................... | 133,265 | 155,361 |
| 26220 | 41027 | Hood River County, OR ..................... | 22,346 | 23,280 | 26820 | 16019 | Bonneville County, ID ...................... | 104,234 | 122,134 |
| | | | | | 26820 | 16023 | Butte County, ID.............................. | 2,891 | 2,646 |
| 26260 | | Hope, AR Micro Area........................... | 31,606 | 29,352 | 26820 | 16051 | Jefferson County, ID........................ | 26,140 | 30,581 |
| 26260 | 05057 | Hempstead County, AR ..................... | 22,609 | 21,253 | | | | | |
| 26260 | 05099 | Nevada County, AR ......................... | 8,997 | 8,099 | 26860 | | Indiana, PA Micro Area........................ | 88,880 | 83,664 |
| | | | | | 26860 | 42063 | Indiana County, PA .......................... | 88,880 | 83,664 |
| 26300 | | Hot Springs, AR Metro Area................. | 96,024 | 99,789 | | | | | |
| 26300 | 05051 | Garland County, AR ......................... | 96,024 | 99,789 | 26900 | | Indianapolis-Carmel-Anderson, IN Metro Area........................................... | 1,887,877 | 2,091,019 |
| 26340 | | Houghton, MI Micro Area..................... | 38,784 | 37,245 | 26900 | 18011 | Boone County, IN............................. | 56,640 | 69,347 |
| 26340 | 26061 | Houghton County, MI ....................... | 36,628 | 35,126 | 26900 | 18013 | Brown County, IN............................ | 15,242 | 15,112 |
| 26340 | 26083 | Keweenaw County, MI ..................... | 2,156 | 2,119 | 26900 | 18057 | Hamilton County, IN........................ | 274,569 | 344,238 |
| | | | | | 26900 | 18059 | Hancock County, IN......................... | 70,002 | 79,553 |
| 26380 | | Houma-Thibodaux, LA Metro Area....... | 208,178 | 207,455 | 26900 | 18063 | Hendricks County, IN....................... | 145,448 | 173,251 |
| 26380 | 22057 | Lafourche Parish, LA ....................... | 96,318 | 97,596 | 26900 | 18081 | Johnson County, IN......................... | 139,654 | 160,607 |
| 26380 | 22109 | Terrebonne Parish, LA..................... | 111,860 | 109,859 | 26900 | 18095 | Madison County, IN......................... | 131,636 | 129,681 |
| | | | | | 26900 | 18097 | Marion County, IN........................... | 903,393 | 966,183 |
| 26420 | | Houston-The Woodlands-Sugar Land, TX Metro Area...................................... | 5,920,416 | 7,154,478 | 26900 | 18109 | Morgan County, IN.......................... | 68,894 | 70,707 |
| 26420 | 48015 | Austin County, TX............................. | 28,417 | 29,972 | 26900 | 18133 | Putnam County, IN.......................... | 37,963 | 37,469 |
| 26420 | 48039 | Brazoria County, TX......................... | 313,166 | 380,518 | 26900 | 18145 | Shelby County, IN........................... | 44,436 | 44,871 |
| 26420 | 48071 | Chambers County, TX....................... | 35,096 | 45,590 | | | | | |
| 26420 | 48157 | Fort Bend County, TX....................... | 585,375 | 839,706 | 26940 | | Indianola, MS Micro Area.................... | 29,450 | 24,740 |
| 26420 | 48167 | Galveston County, TX....................... | 291,309 | 345,089 | 26940 | 28133 | Sunflower County, MS ..................... | 29,450 | 24,740 |
| 26420 | 48201 | Harris County, TX............................. | 4,092,459 | 4,738,253 | | | | | |
| 26420 | 48291 | Liberty County, TX........................... | 75,643 | 91,547 | 26980 | | Iowa City, IA Metro Area..................... | 152,586 | 175,732 |
| 26420 | 48339 | Montgomery County, TX.................... | 455,746 | 626,351 | 26980 | 19103 | Johnson County, IA.......................... | 130,882 | 153,740 |
| 26420 | 48473 | Waller County, TX............................ | 43,205 | 57,452 | 26980 | 19183 | Washington County, IA..................... | 21,704 | 21,992 |
| | | | | | 27020 | | Iron Mountain, MI-WI Micro Area.......... | 30,591 | 29,410 |
| 26460 | | Hudson, NY Micro Area....................... | 63,096 | 59,534 | 27020 | 26043 | Dickinson County, MI....................... | 26,168 | 25,112 |
| 26460 | 36021 | Columbia County, NY ....................... | 63,096 | 59,534 | 27020 | 55037 | Florence County, WI........................ | 4,423 | 4,298 |
| 26500 | | Huntingdon, PA Micro Area.................. | 45,913 | 44,590 | 27060 | | Ithaca, NY Metro Area......................... | 101,564 | 101,058 |
| 26500 | 42061 | Huntingdon County, PA .................... | 45,913 | 44,590 | 27060 | 36109 | Tompkins County, NY ...................... | 101,564 | 101,058 |
| 26540 | | Huntington, IN Micro Area................... | 37,124 | 36,395 | 27100 | | Jackson, MI Metro Area....................... | 160,248 | 156,920 |
| 26540 | 18069 | Huntington County, IN...................... | 37,124 | 36,395 | 27100 | 26075 | Jackson County, MI ......................... | 160,248 | 156,920 |
| 26580 | | Huntington-Ashland, WV-KY-OH Metro Area.................................................... | 370,908 | 354,085 | 27140 | | Jackson, MS Metro Area...................... | 586,320 | 589,082 |
| 26580 | 21019 | Boyd County, KY.............................. | 49,542 | 46,516 | 27140 | 28029 | Copiah County, MS .......................... | 29,449 | 27,933 |
| 26580 | 21043 | Carter County, KY............................ | 27,720 | 26,542 | 27140 | 28049 | Hinds County, MS ........................... | 245,285 | 227,966 |
| 26580 | 21089 | Greenup County, KY......................... | 36,910 | 34,865 | 27140 | 28051 | Holmes County, MS ........................ | 19,198 | 16,726 |
| 26580 | 39087 | Lawrence County, OH....................... | 62,450 | 59,091 | 27140 | 28089 | Madison County, MS ....................... | 95,203 | 106,871 |
| 26580 | 54011 | Cabell County, WV........................... | 96,319 | 91,589 | 27140 | 28121 | Rankin County, MS .......................... | 141,617 | 155,975 |
| 26580 | 54079 | Putnam County, WV ......................... | 55,486 | 56,428 | 27140 | 28127 | Simpson County, MS ....................... | 27,503 | 26,629 |
| 26580 | 54099 | Wayne County, WV........................... | 42,481 | 39,054 | 27140 | 28163 | Yazoo County, MS ........................... | 28,065 | 26,982 |
| | | | | | 27160 | | Jackson, OH Micro Area...................... | 33,225 | 32,493 |
| 26620 | | Huntsville, AL Metro Area.................... | 417,593 | 481,681 | 27160 | 39079 | Jackson County, OH......................... | 33,225 | 32,493 |
| 26620 | 01083 | Limestone County, AL....................... | 82,782 | 102,228 | | | | | |
| 26620 | 01089 | Madison County, AL.......................... | 334,811 | 379,453 | 27180 | | Jackson, TN Metro Area...................... | 179,694 | 179,131 |
| | | | | | 27180 | 47023 | Chester County, TN ......................... | 17,131 | 17,432 |
| 26660 | | Huntsville, TX Micro Area.................... | 67,861 | 72,164 | 27180 | 47033 | Crockett County, TN......................... | 14,586 | 14,180 |
| 26660 | 48471 | Walker County, TX............................ | 67,861 | 72,164 | 27180 | 47053 | Gibson County, TN .......................... | 49,683 | 49,159 |
| | | | | | 27180 | 47113 | Madison County, TN ........................ | 98,294 | 98,360 |
| 26700 | | Huron, SD Micro Area ........................ | 19,469 | 20,498 | | | | | |
| 26700 | 46005 | Beadle County, SD ........................... | 17,398 | 18,513 | 27220 | | Jackson, WY-ID Micro Area ................. | 31,464 | 35,998 |
| 26700 | 46073 | Jerauld County, SD.......................... | 2,071 | 1,985 | 27220 | 16081 | Teton County, ID.............................. | 10,170 | 12,501 |
| | | | | | 27220 | 56039 | Teton County, WY............................ | 21,294 | 23,497 |
| 26740 | | Hutchinson, KS Micro Area ................. | 64,511 | 61,793 | | | | | |
| 26740 | 20155 | Reno County, KS ............................. | 64,511 | 61,793 | | | | | |

| Core based statistical area | State/County FIPS code | Title and Geographic Components | 2010 Census Population | 2020 Estimated Population | Core based statistical area | State/County FIPS code | Title and Geographic Components | 2010 Census Population | 2020 Estimated Population |
|---|---|---|---|---|---|---|---|---|---|
| 27260 | | Jacksonville, FL Metro Area | 1,345,596 | 1,587,892 | 27940 | | Juneau, AK Micro Area | 31,275 | 31,849 |
| 27260 | 12003 | Baker County, FL | 27,115 | 29,566 | 27940 | 02110 | Juneau City and Borough, AK | 31,275 | 31,849 |
| 27260 | 12019 | Clay County, FL | 190,865 | 221,770 | | | | | |
| 27260 | 12031 | Duval County, FL | 864,263 | 966,728 | 27980 | | Kahului-Wailuku-Lahaina, HI Metro Area | 154,834 | 167,902 |
| 27260 | 12089 | Nassau County, FL | 73,314 | 91,113 | 27980 | 15009 | Maui County, HI | 154,834 | 167,902 |
| 27260 | 12109 | St. Johns County, FL | 190,039 | 278,715 | | | | | |
| | | | | | 28020 | | Kalamazoo-Portage, MI Metro Area | 250,331 | 265,988 |
| 27300 | | Jacksonville, IL Micro Area | 40,902 | 38,350 | 28020 | 26077 | Kalamazoo County, MI | 250,331 | 265,988 |
| 27300 | 17137 | Morgan County, IL | 35,547 | 33,400 | | | | | |
| 27300 | 17171 | Scott County, IL | 5,355 | 4,950 | 28060 | | Kalispell, MT Micro Area | 90,928 | 105,851 |
| | | | | | 28060 | 30029 | Flathead County, MT | 90,928 | 105,851 |
| 27340 | | Jacksonville, NC Metro Area | 177,772 | 203,943 | | | | | |
| 27340 | 37133 | Onslow County, NC | 177,772 | 203,943 | 28100 | | Kankakee, IL Metro Area | 113,449 | 108,594 |
| | | | | | 28100 | 17091 | Kankakee County, IL | 113,449 | 108,594 |
| 27380 | | Jacksonville, TX Micro Area | 50,845 | 52,875 | | | | | |
| 27380 | 48073 | Cherokee County, TX | 50,845 | 52,875 | 28140 | | Kansas City, MO-KS Metro Area | 2,009,342 | 2,173,212 |
| | | | | | 28140 | 20091 | Johnson County, KS | 544,179 | 607,220 |
| 27420 | | Jamestown, ND Micro Area | 21,100 | 20,498 | 28140 | 20103 | Leavenworth County, KS | 76,227 | 82,246 |
| 27420 | 38093 | Stutsman County, ND | 21,100 | 20,498 | 28140 | 20107 | Linn County, KS | 9,656 | 9,654 |
| | | | | | 28140 | 20121 | Miami County, KS | 32,787 | 34,334 |
| 27460 | | Jamestown-Dunkirk-Fredonia, NY Micro Area | 134,905 | 126,032 | 28140 | 20209 | Wyandotte County, KS | 157,505 | 165,265 |
| 27460 | 36013 | Chautauqua County, NY | 134,905 | 126,032 | 28140 | 29013 | Bates County, MO | 17,049 | 16,242 |
| | | | | | 28140 | 29025 | Caldwell County, MO | 9,424 | 9,051 |
| 27500 | | Janesville-Beloit, WI Metro Area | 160,331 | 163,084 | 28140 | 29037 | Cass County, MO | 99,478 | 106,806 |
| 27500 | 55105 | Rock County, WI | 160,331 | 163,084 | 28140 | 29047 | Clay County, MO | 221,939 | 253,463 |
| | | | | | 28140 | 29049 | Clinton County, MO | 20,743 | 20,553 |
| 27530 | | Jasper, AL Micro Area | 67,023 | 63,143 | 28140 | 29095 | Jackson County, MO | 674,158 | 705,925 |
| 27530 | 01127 | Walker County, AL | 67,023 | 63,143 | 28140 | 29107 | Lafayette County, MO | 33,381 | 33,006 |
| | | | | | 28140 | 29165 | Platte County, MO | 89,322 | 106,532 |
| 27540 | | Jasper, IN Micro Area | 54,734 | 54,920 | 28140 | 29177 | Ray County, MO | 23,494 | 22,915 |
| 27540 | 18037 | Dubois County, IN | 41,889 | 42,542 | | | | | |
| 27540 | 18125 | Pike County, IN | 12,845 | 12,378 | 28180 | | Kapaa, HI Micro Area | 67,091 | 71,851 |
| | | | | | 28180 | 15007 | Kauai County, HI | 67,091 | 71,851 |
| 27600 | | Jefferson, GA Micro Area | 60,485 | 76,199 | | | | | |
| 27600 | 13157 | Jackson County, GA | 60,485 | 76,199 | 28260 | | Kearney, NE Micro Area | 52,591 | 56,766 |
| | | | | | 28260 | 31019 | Buffalo County, NE | 46,102 | 50,114 |
| 27620 | | Jefferson City, MO Metro Area | 149,807 | 150,198 | 28260 | 31099 | Kearney County, NE | 6,489 | 6,652 |
| 27620 | 29027 | Callaway County, MO | 44,332 | 44,887 | | | | | |
| 27620 | 29051 | Cole County, MO | 75,990 | 76,191 | 28300 | | Keene, NH Micro Area | 77,117 | 76,228 |
| 27620 | 29135 | Moniteau County, MO | 15,607 | 15,585 | 28300 | 33005 | Cheshire County, NH | 77,117 | 76,228 |
| 27620 | 29151 | Osage County, MO | 13,878 | 13,535 | | | | | |
| | | | | | 28340 | | Kendallville, IN Micro Area | 47,536 | 47,832 |
| 27660 | | Jennings, LA Micro Area | 31,594 | 31,208 | 28340 | 18113 | Noble County, IN | 47,536 | 47,832 |
| 27660 | 22053 | Jefferson Davis Parish, LA | 31,594 | 31,208 | | | | | |
| | | | | | 28380 | | Kennett, MO Micro Area | 31,953 | 28,878 |
| 27700 | | Jesup, GA Micro Area | 30,099 | 30,023 | 28380 | 29069 | Dunklin County, MO | 31,953 | 28,878 |
| 27700 | 13305 | Wayne County, GA | 30,099 | 30,023 | | | | | |
| | | | | | 28420 | | Kennewick-Richland, WA Metro Area | 253,340 | 303,501 |
| 27740 | | Johnson City, TN Metro Area | 198,716 | 204,540 | 28420 | 53005 | Benton County, WA | 175,177 | 206,426 |
| 27740 | 47019 | Carter County, TN | 57,424 | 56,418 | 28420 | 53021 | Franklin County, WA | 78,163 | 97,075 |
| 27740 | 47171 | Unicoi County, TN | 18,313 | 17,755 | | | | | |
| 27740 | 47179 | Washington County, TN | 122,979 | 130,367 | 28500 | | Kerrville, TX Micro Area | 49,625 | 52,869 |
| | | | | | 28500 | 48265 | Kerr County, TX | 49,625 | 52,869 |
| 27780 | | Johnstown, PA Metro Area | 143,679 | 128,672 | | | | | |
| 27780 | 42021 | Cambria County, PA | 143,679 | 128,672 | 28540 | | Ketchikan, AK Micro Area | 13,477 | 13,747 |
| | | | | | 28540 | 02130 | Ketchikan Gateway Borough, AK | 13,477 | 13,747 |
| 27860 | | Jonesboro, AR Metro Area | 121,026 | 135,528 | | | | | |
| 27860 | 05031 | Craighead County, AR | 96,443 | 112,245 | 28580 | | Key West, FL Micro Area | 73,090 | 73,900 |
| 27860 | 05111 | Poinsett County, AR | 24,583 | 23,283 | 28580 | 12087 | Monroe County, FL | 73,090 | 73,900 |
| | | | | | | | | | |
| 27900 | | Joplin, MO Metro Area | 175,518 | 180,099 | 28620 | | Kill Devil Hills, NC Micro Area | 33,920 | 37,547 |
| 27900 | 29097 | Jasper County, MO | 117,404 | 121,648 | 28620 | 37055 | Dare County, NC | 33,920 | 37,547 |
| 27900 | 29145 | Newton County, MO | 58,114 | 58,451 | | | | | |

| Core based statistical area | State/ County FIPS code | Title and Geographic Components | 2010 Census Population | 2020 Estimated Population | Core based statistical area | State/ County FIPS code | Title and Geographic Components | 2010 Census Population | 2020 Estimated Population |
|---|---|---|---|---|---|---|---|---|---|
| 28660 | | Killeen-Temple, TX Metro Area ............. | 405,300 | 468,453 | 29300 | | LaGrange, GA-AL Micro Area .............. | 101,259 | 103,079 |
| 28660 | 48027 | Bell County, TX................................... | 310,235 | 369,927 | 29300 | 01017 | Chambers County, AL...................... | 34,215 | 32,865 |
| 28660 | 48099 | Coryell County, TX.............................. | 75,388 | 76,737 | 29300 | 13285 | Troup County, GA............................ | 67,044 | 70,214 |
| 28660 | 48281 | Lampasas County, TX ......................... | 19,677 | 21,789 | | | | | |
| | | | | | 29340 | | Lake Charles, LA Metro Area .............. | 199,607 | 210,313 |
| 28700 | | Kingsport-Bristol, TN-VA Metro Area..... | 309,544 | 308,183 | 29340 | 22019 | Calcasieu Parish, LA....................... | 192,768 | 203,310 |
| 28700 | 47073 | Hawkins County, TN............................ | 56,833 | 56,775 | 29340 | 22023 | Cameron Parish, LA......................... | 6,839 | 7,003 |
| 28700 | 47163 | Sullivan County, TN............................ | 156,823 | 158,755 | | | | | |
| 28700 | 51169 | Scott County, VA................................ | 23,177 | 21,629 | 29380 | | Lake City, FL Micro Area .................... | 67,531 | 72,654 |
| 28700 | 51191 | Washington County, VA...................... | 54,876 | 53,695 | 29380 | 12023 | Columbia County, FL........................ | 67,531 | 72,654 |
| 28700 | 51520 | Bristol city, VA.................................. | 17,835 | 17,329 | | | | | |
| | | | | | 29420 | | Lake Havasu City-Kingman, AZ Metro Area ................................................. | | |
| 28740 | | Kingston, NY Metro Area...................... | 182,493 | 177,716 | | | | 200,186 | 217,206 |
| 28740 | 36111 | Ulster County, NY .............................. | 182,493 | 177,716 | 29420 | 04015 | Mohave County, AZ.......................... | 200,186 | 217,206 |
| | | | | | | | | | |
| 28780 | | Kingsville, TX Micro Area .................... | 32,477 | 30,717 | 29460 | | Lakeland-Winter Haven, FL Metro Area | 602,095 | 744,552 |
| 28780 | 48261 | Kenedy County, TX............................. | 416 | 379 | 29460 | 12105 | Polk County, FL .............................. | 602,095 | 744,552 |
| 28780 | 48273 | Kleberg County, TX............................. | 32,061 | 30,338 | | | | | |
| | | | | | 29500 | | Lamesa, TX Micro Area...................... | 13,833 | 12,974 |
| 28820 | | Kinston, NC Micro Area ....................... | 59,495 | 55,720 | 29500 | 48115 | Dawson County, TX.......................... | 13,833 | 12,974 |
| 28820 | 37107 | Lenoir County, NC .............................. | 59,495 | 55,720 | | | | | |
| | | | | | 29540 | | Lancaster, PA Metro Area ................... | 519,445 | 546,192 |
| 28860 | | Kirksville, MO Micro Area .................... | 30,038 | 29,933 | 29540 | 42071 | Lancaster County, PA....................... | 519,445 | 546,192 |
| 28860 | 29001 | Adair County, MO ............................... | 25,607 | 25,399 | | | | | |
| 28860 | 29197 | Schuyler County, MO .......................... | 4,431 | 4,534 | 29620 | | Lansing-East Lansing, MI Metro Area ... | 534,684 | 548,248 |
| | | | | | 29620 | 26037 | Clinton County, MI............................ | 75,382 | 79,753 |
| 28900 | | Klamath Falls, OR Micro Area .............. | 66,380 | 68,739 | 29620 | 26045 | Eaton County, MI............................. | 107,759 | 110,148 |
| 28900 | 41035 | Klamath County, OR............................ | 66,380 | 68,739 | 29620 | 26065 | Ingham County, MI........................... | 280,895 | 290,609 |
| | | | | | 29620 | 26155 | Shiawassee County, MI ..................... | 70,648 | 67,738 |
| 28940 | | Knoxville, TN Metro Area...................... | 814,914 | 878,124 | | | | | |
| 28940 | 47001 | Anderson County, TN ......................... | 75,129 | 77,558 | 29660 | | Laramie, WY Micro Area .................... | 36,299 | 38,950 |
| 28940 | 47009 | Blount County, TN.............................. | 123,010 | 134,751 | 29660 | 56001 | Albany County, WY........................... | 36,299 | 38,950 |
| 28940 | 47013 | Campbell County, TN.......................... | 40,716 | 39,837 | | | | | |
| 28940 | 47093 | Knox County, TN................................ | 432,226 | 475,609 | 29700 | | Laredo, TX Metro Area....................... | 250,304 | 277,681 |
| 28940 | 47105 | Loudon County, TN............................ | 48,556 | 54,910 | 29700 | 48479 | Webb County, TX............................. | 250,304 | 277,681 |
| 28940 | 47129 | Morgan County, TN............................ | 21,987 | 21,431 | | | | | |
| 28940 | 47145 | Roane County, TN ............................. | 54,181 | 53,841 | 29740 | | Las Cruces, NM Metro Area ................ | 209,233 | 221,262 |
| 28940 | 47173 | Union County, TN .............................. | 19,109 | 20,187 | 29740 | 35013 | Do±a Ana County, NM....................... | 209,233 | 221,262 |
| | | | | | | | | | |
| 29020 | | Kokomo, IN Metro Area ....................... | 82,752 | 82,732 | 29780 | | Las Vegas, NM Micro Area ................. | 34,274 | 31,622 |
| 29020 | 18067 | Howard County, IN............................. | 82,752 | 82,732 | 29780 | 35033 | Mora County, NM............................. | 4,881 | 4,478 |
| | | | | | 29780 | 35047 | San Miguel County, NM..................... | 29,393 | 27,144 |
| 29060 | | Laconia, NH Micro Area ...................... | 60,088 | 61,551 | | | | | |
| 29060 | 33001 | Belknap County, NH ........................... | 60,088 | 61,551 | 29820 | | Las Vegas-Henderson-Paradise, NV Metro Area .................................... | 1,951,269 | 2,315,963 |
| 29100 | | La Crosse-Onalaska, WI-MN Metro Area .............................................. | 133,665 | 137,134 | 29820 | 32003 | Clark County, NV............................. | 1,951,269 | 2,315,963 |
| 29100 | 27055 | Houston County, MN.......................... | 19,027 | 18,632 | 29860 | | Laurel, MS Micro Area........................ | 84,823 | 84,325 |
| 29100 | 55063 | La Crosse County, WI....................... | 114,638 | 118,502 | 29860 | 28061 | Jasper County, MS........................... | 17,062 | 16,332 |
| | | | | | 29860 | 28067 | Jones County, MS............................ | 67,761 | 67,993 |
| 29180 | | Lafayette, LA Metro Area..................... | 466,750 | 489,759 | | | | | |
| 29180 | 22001 | Acadia Parish, LA ............................. | 61,773 | 61,918 | 29900 | | Laurinburg, NC Micro Area ................. | 36,157 | 34,637 |
| 29180 | 22045 | Iberia Parish, LA............................... | 73,240 | 68,991 | 29900 | 37165 | Scotland County, NC........................ | 36,157 | 34,637 |
| 29180 | 22055 | Lafayette Parish, LA.......................... | 221,578 | 246,518 | | | | | |
| 29180 | 22099 | St. Martin Parish, LA......................... | 52,160 | 52,954 | 29940 | | Lawrence, KS Metro Area ................... | 110,826 | 122,530 |
| 29180 | 22113 | Vermilion Parish, LA.......................... | 57,999 | 59,378 | 29940 | 20045 | Douglas County, KS.......................... | 110,826 | 122,530 |
| | | | | | | | | | |
| 29200 | | Lafayette-West Lafayette, IN Metro Area.............................................. | 210,297 | 233,278 | 29980 | | Lawrenceburg, TN Micro Area............. | 41,869 | 44,432 |
| 29200 | 18007 | Benton County, IN.............................. | 8,854 | 8,741 | 29980 | 47099 | Lawrence County, TN ...................... | 41,869 | 44,432 |
| 29200 | 18015 | Carroll County, IN.............................. | 20,155 | 20,228 | | | | | |
| 29200 | 18157 | Tippecanoe County, IN ...................... | 172,780 | 196,115 | 30020 | | Lawton, OK Metro Area...................... | 130,291 | 126,775 |
| 29200 | 18171 | Warren County, IN............................. | 8,508 | 8,194 | 30020 | 40031 | Comanche County, OK....................... | 124,098 | 121,099 |
| | | | | | 30020 | 40033 | Cotton County, OK........................... | 6,193 | 5,676 |
| 29260 | | La Grande, OR Micro Area................... | 25,748 | 26,551 | | | | | |
| 29260 | 41061 | Union County, OR.............................. | 25,748 | 26,551 | 30060 | | Lebanon, MO Micro Area .................... | 35,571 | 35,895 |
| | | | | | 30060 | 29105 | Laclede County, MO ......................... | 35,571 | 35,895 |

# CORE BASED STATISTICAL AREAS
## (Metropolitan and Micropolitan),
## METROPOLITAN DIVISIONS, AND COMPONENTS
## (as defined March, 2020)—*Continued*

| Core based statistical area | State/County FIPS code | Title and Geographic Components | 2010 Census Population | 2020 Estimated Population | Core based statistical area | State/County FIPS code | Title and Geographic Components | 2010 Census Population | 2020 Estimated Population |
|---|---|---|---|---|---|---|---|---|---|
| 30100 | | Lebanon, NH-VT Micro Area | 218,466 | 217,783 | 30860 | | Logan, UT-ID Metro Area | 125,442 | 144,219 |
| 30100 | 33009 | Grafton County, NH | 89,118 | 90,691 | 30860 | 16041 | Franklin County, ID | 12,786 | 14,215 |
| 30100 | 33019 | Sullivan County, NH | 43,742 | 43,267 | 30860 | 49005 | Cache County, UT | 112,656 | 130,004 |
| 30100 | 50017 | Orange County, VT | 28,936 | 28,837 | | | | | |
| 30100 | 50027 | Windsor County, VT | 56,670 | 54,988 | 30900 | | Logansport, IN Micro Area | 38,966 | 37,388 |
| | | | | | 30900 | 18017 | Cass County, IN | 38,966 | 37,388 |
| 30140 | | Lebanon, PA Metro Area | 133,568 | 141,663 | | | | | |
| 30140 | 42075 | Lebanon County, PA | 133,568 | 141,663 | 30940 | | London, KY Micro Area | 148,099 | 148,342 |
| | | | | | 30940 | 21051 | Clay County, KY | 21,730 | 19,631 |
| 30220 | | Levelland, TX Micro Area | 22,935 | 22,921 | 30940 | 21121 | Knox County, KY | 31,883 | 31,022 |
| 30220 | 48219 | Hockley County, TX | 22,935 | 22,921 | 30940 | 21125 | Laurel County, KY | 58,849 | 61,238 |
| | | | | | 30940 | 21235 | Whitley County, KY | 35,637 | 36,451 |
| 30260 | | Lewisburg, PA Micro Area | 44,947 | 44,294 | | | | | |
| 30260 | 42119 | Union County, PA | 44,947 | 44,294 | 30980 | | Longview, TX Metro Area | 280,000 | 287,105 |
| | | | | | 30980 | 48183 | Gregg County, TX | 121,730 | 124,229 |
| 30280 | | Lewisburg, TN Micro Area | 30,617 | 35,016 | 30980 | 48203 | Harrison County, TX | 65,631 | 66,386 |
| 30280 | 47117 | Marshall County, TN | 30,617 | 35,016 | 30980 | 48401 | Rusk County, TX | 53,330 | 54,324 |
| | | | | | 30980 | 48459 | Upshur County, TX | 39,309 | 42,166 |
| 30300 | | Lewiston, ID-WA Metro Area | 60,888 | 63,575 | | | | | |
| 30300 | 16069 | Nez Perce County, ID | 39,265 | 40,755 | 31020 | | Longview, WA Metro Area | 102,410 | 111,371 |
| 30300 | 53003 | Asotin County, WA | 21,623 | 22,820 | 31020 | 53015 | Cowlitz County, WA | 102,410 | 111,371 |
| | | | | | | | | | |
| 30340 | | Lewiston-Auburn, ME Metro Area | 107,702 | 108,547 | 31060 | | Los Alamos, NM Micro Area | 17,950 | 19,462 |
| 30340 | 23001 | Androscoggin County, ME | 107,702 | 108,547 | 31060 | 35028 | Los Alamos County, NM | 17,950 | 19,462 |
| | | | | | | | | | |
| 30380 | | Lewistown, PA Micro Area | 46,682 | 46,064 | 31080 | | Los Angeles-Long Beach-Anaheim, CA Metro Area | 12,828,837 | 13,109,903 |
| 30380 | 42087 | Mifflin County, PA | 46,682 | 46,064 | | | | | |
| | | | | | 31080 | | Anaheim-Santa Ana-Irvine, CA Division 11244 | 3,010,232 | 3,166,857 |
| 30420 | | Lexington, NE Micro Area | 26,370 | 25,496 | 31080 | 06059 | Orange County, CA | 3,010,232 | 3,166,857 |
| 30420 | 31047 | Dawson County, NE | 24,326 | 23,510 | | | | | |
| 30420 | 31073 | Gosper County, NE | 2,044 | 1,986 | 31080 | | Los Angeles-Long Beach-Glendale, CA Division 31084 | 9,818,605 | 9,943,046 |
| | | | | | 31080 | 06037 | Los Angeles County, CA | 9,818,605 | 9,943,046 |
| 30460 | | Lexington-Fayette, KY Metro Area | 472,099 | 520,391 | | | | | |
| 30460 | 21017 | Bourbon County, KY | 19,985 | 19,901 | 31140 | | Louisville/Jefferson County, KY-IN Metro Area | 1,202,718 | 1,268,993 |
| 30460 | 21049 | Clark County, KY | 35,613 | 36,463 | 31140 | 18019 | Clark County, IN | 110,232 | 119,266 |
| 30460 | 21067 | Fayette County, KY | 295,803 | 324,735 | 31140 | 18043 | Floyd County, IN | 74,578 | 78,936 |
| 30460 | 21113 | Jessamine County, KY | 48,586 | 54,057 | 31140 | 18061 | Harrison County, IN | 39,364 | 40,682 |
| 30460 | 21209 | Scott County, KY | 47,173 | 58,470 | 31140 | 18175 | Washington County, IN | 28,262 | 28,213 |
| 30460 | 21239 | Woodford County, KY | 24,939 | 26,765 | 31140 | 21029 | Bullitt County, KY | 74,319 | 82,182 |
| | | | | | 31140 | 21103 | Henry County, KY | 15,416 | 16,067 |
| 30580 | | Liberal, KS Micro Area | 22,952 | 21,038 | 31140 | 21111 | Jefferson County, KY | 741,096 | 767,452 |
| 30580 | 20175 | Seward County, KS | 22,952 | 21,038 | 31140 | 21185 | Oldham County, KY | 60,316 | 66,999 |
| | | | | | 31140 | 21211 | Shelby County, KY | 42,074 | 49,611 |
| 30620 | | Lima, OH Metro Area | 106,331 | 101,980 | 31140 | 21215 | Spencer County, KY | 17,061 | 19,585 |
| 30620 | 39003 | Allen County, OH | 106,331 | 101,980 | | | | | |
| | | | | | 31180 | | Lubbock, TX Metro Area | 290,805 | 326,364 |
| 30660 | | Lincoln, IL Micro Area | 30,305 | 28,383 | 31180 | 48107 | Crosby County, TX | 6,059 | 5,567 |
| 30660 | 17107 | Logan County, IL | 30,305 | 28,383 | 31180 | 48303 | Lubbock County, TX | 278,831 | 314,772 |
| | | | | | 31180 | 48305 | Lynn County, TX | 5,915 | 6,025 |
| 30700 | | Lincoln, NE Metro Area | 302,157 | 337,836 | | | | | |
| 30700 | 31109 | Lancaster County, NE | 285,407 | 320,650 | 31220 | | Ludington, MI Micro Area | 28,705 | 29,164 |
| 30700 | 31159 | Seward County, NE | 16,750 | 17,186 | 31220 | 26105 | Mason County, MI | 28,705 | 29,164 |
| | | | | | | | | | |
| 30780 | | Little Rock-North Little Rock-Conway, AR Metro Area | 699,757 | 746,564 | 31260 | | Lufkin, TX Micro Area | 86,771 | 86,796 |
| 30780 | 05045 | Faulkner County, AR | 113,237 | 126,919 | 31260 | 48005 | Angelina County, TX | 86,771 | 86,796 |
| 30780 | 05053 | Grant County, AR | 17,853 | 18,449 | | | | | |
| 30780 | 05085 | Lonoke County, AR | 68,356 | 73,921 | 31300 | | Lumberton, NC Micro Area | 134,168 | 129,999 |
| 30780 | 05105 | Perry County, AR | 10,445 | 10,327 | 31300 | 37155 | Robeson County, NC | 134,168 | 129,999 |
| 30780 | 05119 | Pulaski County, AR | 382,748 | 392,980 | | | | | |
| 30780 | 05125 | Saline County, AR | 107,118 | 123,968 | | | | | |
| | | | | | | | | | |
| 30820 | | Lock Haven, PA Micro Area | 39,238 | 37,957 | | | | | |
| 30820 | 42035 | Clinton County, PA | 39,238 | 37,957 | | | | | |

# CORE BASED STATISTICAL AREAS
## (Metropolitan and Micropolitan),
## METROPOLITAN DIVISIONS, AND COMPONENTS
## (as defined March, 2020)—*Continued*

| Core based statisti-cal area | State/County FIPS code | Title and Geographic Components | 2010 Census Population | 2020 Estimated Population | Core based statisti-cal area | State/County FIPS code | Title and Geographic Components | 2010 Census Population | 2020 Estimated Population |
|---|---|---|---|---|---|---|---|---|---|
| 31340 | | Lynchburg, VA Metro Area .................. | 252,634 | 264,386 | 31980 | | Marion, IN Micro Area........................... | 70,061 | 65,225 |
| 31340 | 51009 | Amherst County, VA........................ | 32,353 | 31,667 | 31980 | 18053 | Grant County, IN............................... | 70,061 | 65,225 |
| 31340 | 51011 | Appomattox County, VA.................... | 14,973 | 16,043 | | | | | |
| 31340 | 51019 | Bedford County, VA......................... | 74,898 | 79,811 | 32000 | | Marion, NC Micro Area.......................... | 44,996 | 45,782 |
| 31340 | 51031 | Campbell County, VA....................... | 54,842 | 55,304 | 32000 | 37111 | McDowell County, NC....................... | 44,996 | 45,782 |
| 31340 | 51680 | Lynchburg city, VA.......................... | 75,568 | 81,561 | | | | | |
| | | | | | 32020 | | Marion, OH Micro Area.......................... | 66,501 | 64,820 |
| 31380 | | Macomb, IL Micro Area....................... | 32,612 | 29,295 | 32020 | 39101 | Marion County, OH........................... | 66,501 | 64,820 |
| 31380 | 17109 | McDonough County, IL ................... | 32,612 | 29,295 | | | | | |
| | | | | | 32100 | | Marquette, MI Micro Area..................... | 67,077 | 65,834 |
| 31420 | | Macon-Bibb County, GA Metro Area .... | 232,293 | 229,900 | 32100 | 26103 | Marquette County, MI ....................... | 67,077 | 65,834 |
| 31420 | 13021 | Bibb County, GA.............................. | 155,547 | 152,737 | | | | | |
| 31420 | 13079 | Crawford County, GA........................ | 12,630 | 12,231 | 32140 | | Marshall, MN Micro Area...................... | 25,857 | 25,271 |
| 31420 | 13169 | Jones County, GA............................ | 28,669 | 28,787 | 32140 | 27083 | Lyon County, MN.............................. | 25,857 | 25,271 |
| 31420 | 13207 | Monroe County, GA.......................... | 26,424 | 28,042 | | | | | |
| 31420 | 13289 | Twiggs County, GA........................... | 9,023 | 8,103 | 32180 | | Marshall, MO Micro Area...................... | 23,370 | 22,858 |
| | | | | | 32180 | 29195 | Saline County, MO............................ | 23,370 | 22,858 |
| 31460 | | Madera, CA Metro Area....................... | 150,865 | 157,761 | | | | | |
| 31460 | 06039 | Madera County, CA ......................... | 150,865 | 157,761 | 32260 | | Marshalltown, IA Micro Area................. | 40,648 | 39,495 |
| | | | | | 32260 | 19127 | Marshall County, IA.......................... | 40,648 | 39,495 |
| 31500 | | Madison, IN Micro Area....................... | 32,428 | 32,110 | | | | | |
| 31500 | 18077 | Jefferson County, IN ....................... | 32,428 | 32,110 | 32280 | | Martin, TN Micro Area ......................... | 35,021 | 33,334 |
| | | | | | 32280 | 47183 | Weakley County, TN......................... | 35,021 | 33,334 |
| 31540 | | Madison, WI Metro Area...................... | 605,435 | 670,447 | | | | | |
| 31540 | 55021 | Columbia County, WI........................ | 56,833 | 57,668 | 32300 | | Martinsville, VA Micro Area .................. | 67,972 | 62,664 |
| 31540 | 55025 | Dane County, WI.............................. | 488,073 | 552,536 | 32300 | 51089 | Henry County, VA............................. | 54,151 | 50,309 |
| 31540 | 55045 | Green County, WI ............................ | 36,842 | 36,603 | 32300 | 51690 | Martinsville city, VA......................... | 13,821 | 12,355 |
| 31540 | 55049 | Iowa County, WI.............................. | 23,687 | 23,640 | | | | | |
| | | | | | 32340 | | Maryville, MO Micro Area .................... | 23,370 | 21,743 |
| 31580 | | Madisonville, KY Micro Area................. | 46,920 | 44,662 | 32340 | 29147 | Nodaway County, MO........................ | 23,370 | 21,743 |
| 31580 | 21107 | Hopkins County, KY ........................ | 46,920 | 44,662 | | | | | |
| | | | | | 32380 | | Mason City, IA Micro Area.................... | 51,749 | 49,462 |
| 31620 | | Magnolia, AR Micro Area..................... | 24,552 | 23,331 | 32380 | 19033 | Cerro Gordo County, IA .................... | 44,151 | 42,103 |
| 31620 | 05027 | Columbia County, AR ....................... | 24,552 | 23,331 | 32380 | 19195 | Worth County, IA.............................. | 7,598 | 7,359 |
| | | | | | | | | | |
| 31660 | | Malone, NY Micro Area ....................... | 51,599 | 49,965 | 32460 | | Mayfield, KY Micro Area....................... | 37,121 | 36,818 |
| 31660 | 36033 | Franklin County, NY......................... | 51,599 | 49,965 | 32460 | 21083 | Graves County, KY........................... | 37,121 | 36,818 |
| | | | | | | | | | |
| 31680 | | Malvern, AR Micro Area....................... | 32,923 | 33,787 | 32500 | | Maysville, KY Micro Area...................... | 17,490 | 17,035 |
| 31680 | 05059 | Hot Spring County, AR...................... | 32,923 | 33,787 | 32500 | 21161 | Mason County, KY............................ | 17,490 | 17,035 |
| | | | | | | | | | |
| 31700 | | Manchester-Nashua, NH Metro Area .... | 400,721 | 418,735 | 32540 | | McAlester, OK Micro Area.................... | 45,837 | 43,679 |
| 31700 | 33011 | Hillsborough County, NH ................. | 400,721 | 418,735 | 32540 | 40121 | Pittsburg County, OK ....................... | 45,837 | 43,679 |
| | | | | | | | | | |
| 31740 | | Manhattan, KS Metro Area.................. | 127,081 | 130,142 | 32580 | | McAllen-Edinburg-Mission, TX Metro Area ........................................ | 774,769 | 875,200 |
| 31740 | 20061 | Geary County, KS............................ | 34,362 | 32,218 | | | | | |
| 31740 | 20149 | Pottawatomie County, KS.................. | 21,604 | 24,722 | 32580 | 48215 | Hidalgo County, TX........................... | 774,769 | 875,200 |
| 31740 | 20161 | Riley County, KS.............................. | 71,115 | 73,202 | | | | | |
| | | | | | 32620 | | McComb, MS Micro Area...................... | 40,404 | 38,997 |
| 31820 | | Manitowoc, WI Micro Area................... | 81,442 | 78,757 | 32620 | 28113 | Pike County, MS .............................. | 40,404 | 38,997 |
| 31820 | 55071 | Manitowoc County, WI...................... | 81,442 | 78,757 | | | | | |
| | | | | | 32660 | | McMinnville, TN Micro Area.................. | 39,839 | 41,605 |
| 31860 | | Mankato, MN Metro Area..................... | 96,740 | 102,723 | 32660 | 47177 | Warren County, TN........................... | 39,839 | 41,605 |
| 31860 | 27013 | Blue Earth County, MN ..................... | 64,013 | 68,241 | | | | | |
| 31860 | 27103 | Nicollet County, MN ......................... | 32,727 | 34,482 | 32700 | | McPherson, KS Micro Area................... | 29,180 | 28,448 |
| | | | | | 32700 | 20113 | McPherson County, KS...................... | 29,180 | 28,448 |
| 31900 | | Mansfield, OH Metro Area.................... | 124,475 | 120,891 | | | | | |
| 31900 | 39139 | Richland County, OH ........................ | 124,475 | 120,891 | 32740 | | Meadville, PA Micro Area..................... | 88,765 | 83,697 |
| | | | | | 32740 | 42039 | Crawford County, PA ........................ | 88,765 | 83,697 |
| 31930 | | Marietta, OH Micro Area...................... | 61,778 | 59,652 | | | | | |
| 31930 | 39167 | Washington County, OH ................... | 61,778 | 59,652 | 32780 | | Medford, OR Metro Area ...................... | 203,206 | 221,844 |
| | | | | | 32780 | 41029 | Jackson County, OR.......................... | 203,206 | 221,844 |
| 31940 | | Marinette, WI-MI Micro Area................ | 65,778 | 62,870 | | | | | |
| 31940 | 26109 | Menominee County, MI...................... | 24,029 | 22,608 | | | | | |
| 31940 | 55075 | Marinette County, WI........................ | 41,749 | 40,262 | | | | | |

| Core based statisti-cal area | State/ County FIPS code | Title and Geographic Components | 2010 Census Population | 2020 Estimated Population | Core based statisti-cal area | State/ County FIPS code | Title and Geographic Components | 2010 Census Population | 2020 Estimated Population |
|---|---|---|---|---|---|---|---|---|---|
| 32820 | | Memphis, TN-MS-AR Metro Area.......... | 1,316,100 | 1,348,678 | 33380 | | Minden, LA Micro Area.......................... | 41,207 | 37,943 |
| 32820 | 28033 | DeSoto County, MS ...................... | 161,252 | 188,275 | 33380 | 22119 | Webster Parish, LA............................ | 41,207 | 37,943 |
| 32820 | 28093 | Marshall County, MS...................... | 37,144 | 35,301 | | | | | |
| 32820 | 28137 | Tate County, MS ........................... | 28,886 | 28,539 | 33420 | | Mineral Wells, TX Micro Area................ | 28,111 | 29,320 |
| 32820 | 28143 | Tunica County, MS ....................... | 10,778 | 9,392 | 33420 | 48363 | Palo Pinto County, TX......................... | 28,111 | 29,320 |
| 32820 | 47047 | Fayette County, TN ...................... | 38,413 | 41,620 | | | | | |
| 32820 | 47157 | Shelby County, TN ....................... | 927,644 | 936,017 | 33460 | | Minneapolis-St. Paul-Bloomington, | | |
| 32820 | 47167 | Tipton County, TN ........................ | 61,081 | 61,918 | | | MN-WI Metro Area................................ | 3,333,633 | 3,657,477 |
| 32820 | 05035 | Crittenden County, AR ................. | 50,902 | 47,616 | 33460 | 27003 | Anoka County, MN ............................ | 330,844 | 359,921 |
| | | | | | 33460 | 27019 | Carver County, MN ........................... | 91,042 | 106,565 |
| 32860 | | Menomonie, WI Micro Area.................. | 43,857 | 45,452 | 33460 | 27025 | Chisago County, MN.......................... | 53,887 | 56,794 |
| 32860 | 55033 | Dunn County, WI............................ | 43,857 | 45,452 | 33460 | 27037 | Dakota County, MN........................... | 398,552 | 431,807 |
| | | | | | 33460 | 27053 | Hennepin County, MN........................ | 1,152,425 | 1,268,408 |
| 32900 | | Merced, CA Metro Area...................... | 255,793 | 279,252 | 33460 | 27059 | Isanti County, MN ............................ | 37,816 | 41,429 |
| 32900 | 06047 | Merced County, CA......................... | 255,793 | 279,252 | 33460 | 27079 | Le Sueur County, MN ....................... | 27,703 | 28,741 |
| | | | | | 33460 | 27095 | Mille Lacs County, MN ..................... | 26,097 | 26,146 |
| 32940 | | Meridian, MS Micro Area..................... | 107,449 | 98,571 | 33460 | 27123 | Ramsey County, MN.......................... | 508,640 | 547,903 |
| 32940 | 28023 | Clarke County, MS......................... | 16,732 | 15,299 | 33460 | 27139 | Scott County, MN............................. | 129,928 | 150,689 |
| 32940 | 28069 | Kemper County, MS ....................... | 10,456 | 9,521 | 33460 | 27141 | Sherburne County, MN....................... | 88,499 | 98,811 |
| 32940 | 28075 | Lauderdale County, MS ................. | 80,261 | 73,751 | 33460 | 27163 | Washington County, MN..................... | 238,136 | 265,476 |
| | | | | | 33460 | 27171 | Wright County, MN............................ | 124,700 | 140,249 |
| 33020 | | Mexico, MO Micro Area........................ | 25,529 | 24,835 | 33460 | 55093 | Pierce County, WI ............................ | 41,019 | 42,700 |
| 33020 | 29007 | Audrain County, MO ........................ | 25,529 | 24,835 | 33460 | 55109 | St. Croix County, WI ......................... | 84,345 | 91,838 |
| 33060 | | Miami, OK Micro Area .......................... | 31,848 | 30,879 | 33500 | | Minot, ND Micro Area .......................... | 69,540 | 76,444 |
| 33060 | 40115 | Ottawa County, OK ......................... | 31,848 | 30,879 | 33500 | 38049 | McHenry County, ND ......................... | 5,395 | 5,695 |
| | | | | | 33500 | 38075 | Renville County, ND.......................... | 2,470 | 2,283 |
| 33100 | | Miami-Fort Lauderdale-Pompano | | | 33500 | 38101 | Ward County, ND ............................. | 61,675 | 68,466 |
| | | Beach, FL Metro Area ......................... | 5,564,635 | 6,173,008 | | | | | |
| | | | | | 33540 | | Missoula, MT Metro Area...................... | 109,299 | 121,630 |
| 33100 | | Fort Lauderdale-Pompano Beach- | | | 33540 | 30063 | Missoula County, MT ........................ | 109,299 | 121,630 |
| | | Sunrise, FL Division 22744................... | 1,748,066 | 1,958,105 | | | | | |
| 33100 | 12011 | Broward County, FL....................... | 1,748,066 | 1,958,105 | 33580 | | Mitchell, SD Micro Area....................... | 22,835 | 23,301 |
| | | | | | 33580 | 46035 | Davison County, SD.......................... | 19,504 | 19,812 |
| 33100 | | Miami-Miami Beach-Kendall, FL Divi- | | | 33580 | 46061 | Hanson County, SD .......................... | 3,331 | 3,489 |
| | | sion 33124 ......................................... | 2,496,435 | 2,707,303 | | | | | |
| 33100 | 12086 | Miami-Dade County, FL.................... | 2,496,435 | 2,707,303 | 33620 | | Moberly, MO Micro Area...................... | 25,414 | 24,409 |
| | | | | | 33620 | 29175 | Randolph County, MO ....................... | 25,414 | 24,409 |
| 33100 | | West Palm Beach-Boca Raton-Boynton | | | | | | | |
| | | Beach, FL Division 48424.................... | 1,320,134 | 1,507,600 | 33660 | | Mobile, AL Metro Area......................... | 430,573 | 428,692 |
| 33100 | 12099 | Palm Beach County, FL ................... | 1,320,134 | 1,507,600 | 33660 | 01097 | Mobile County, AL............................ | 412,992 | 412,716 |
| | | | | | 33660 | 01129 | Washington County, AL ..................... | 17,581 | 15,976 |
| 33140 | | Michigan City-La Porte, IN Metro Area.. | 111,467 | 109,663 | | | | | |
| 33140 | 18091 | LaPorte County, IN ......................... | 111,467 | 109,663 | 33700 | | Modesto, CA Metro Area...................... | 514,453 | 550,081 |
| | | | | | 33700 | 06099 | Stanislaus County, CA....................... | 514,453 | 550,081 |
| 33180 | | Middlesborough, KY Micro Area............ | 28,691 | 25,482 | | | | | |
| 33180 | 21013 | Bell County, KY ............................. | 28,691 | 25,482 | 33740 | | Monroe, LA Metro Area........................ | 204,420 | 198,836 |
| | | | | | 33740 | 22067 | Morehouse Parish, LA....................... | 27,979 | 24,227 |
| 33220 | | Midland, MI Metro Area ....................... | 83,629 | 83,441 | 33740 | 22073 | Ouachita Parish, LA.......................... | 153,720 | 152,439 |
| 33220 | 26111 | Midland County, MI......................... | 83,629 | 83,441 | 33740 | 22111 | Union Parish, LA.............................. | 22,721 | 22,170 |
| | | | | | | | | | |
| 33260 | | Midland, TX Metro Area....................... | 141,671 | 183,679 | 33780 | | Monroe, MI Metro Area........................ | 152,021 | 150,568 |
| 33260 | 48317 | Martin County, TX........................... | 4,799 | 5,816 | 33780 | 26115 | Monroe County, MI ........................... | 152,021 | 150,568 |
| 33260 | 48329 | Midland County, TX......................... | 136,872 | 177,863 | | | | | |
| | | | | | 33860 | | Montgomery, AL Metro Area ................ | 374,536 | 372,583 |
| 33300 | | Milledgeville, GA Micro Area ............... | 55,149 | 53,593 | 33860 | 01001 | Autauga County, AL.......................... | 54,571 | 56,145 |
| 33300 | 13009 | Baldwin County, GA......................... | 45,720 | 45,099 | 33860 | 01051 | Elmore County, AL ........................... | 79,303 | 82,158 |
| 33300 | 13141 | Hancock County, GA ....................... | 9,429 | 8,494 | 33860 | 01085 | Lowndes County, AL......................... | 11,299 | 9,641 |
| | | | | | 33860 | 01101 | Montgomery County, AL ..................... | 229,363 | 224,639 |
| 33340 | | Milwaukee-Waukesha, WI Metro Area .. | 1,555,908 | 1,577,676 | | | | | |
| 33340 | 55079 | Milwaukee County, WI ..................... | 947,735 | 945,016 | 33940 | | Montrose, CO Micro Area.................... | 45,712 | 48,323 |
| 33340 | 55089 | Ozaukee County, WI........................ | 86,395 | 90,043 | 33940 | 08085 | Montrose County, CO ....................... | 41,276 | 43,322 |
| 33340 | 55131 | Washington County, WI .................... | 131,887 | 136,445 | 33940 | 08091 | Ouray County, CO ........................... | 4,436 | 5,001 |
| 33340 | 55133 | Waukesha County, WI ...................... | 389,891 | 406,172 | | | | | |
| | | | | | 33980 | | Morehead City, NC Micro Area.............. | 66,469 | 69,558 |
| | | | | | 33980 | 37031 | Carteret County, NC ......................... | 66,469 | 69,558 |

| Core based statistical area | State/County FIPS code | Title and Geographic Components | 2010 Census Population | 2020 Estimated Population | Core based statistical area | State/County FIPS code | Title and Geographic Components | 2010 Census Population | 2020 Estimated Population |
|---|---|---|---|---|---|---|---|---|---|
| 34020 | | Morgan City, LA Micro Area ................. | 54,650 | 48,330 | 34780 | | Muskogee, OK Micro Area .................... | 70,990 | 67,610 |
| 34020 | 22101 | St. Mary Parish, LA........................ | 54,650 | 48,330 | 34780 | 40101 | Muskogee County, OK...................... | 70,990 | 67,610 |
| 34060 | | Morgantown, WV Metro Area ............... | 129,709 | 140,199 | 34820 | | Myrtle Beach-Conway-North Myrtle | | |
| 34060 | 54061 | Monongalia County, WV ................... | 96,189 | 106,819 | | | Beach, SC-NC Metro Area .................... | 376,722 | 514,488 |
| 34060 | 54077 | Preston County, WV ....................... | 33,520 | 33,380 | 34820 | 37019 | Brunswick County, NC...................... | 107,431 | 149,039 |
| 34100 | | Morristown, TN Metro Area ................. | 136,608 | 143,982 | 34820 | 45051 | Horry County, SC........................... | 269,291 | 365,449 |
| 34100 | 47057 | Grainger County, TN...................... | 22,657 | 23,565 | 34860 | | Nacogdoches, TX Micro Area............... | 64,524 | 64,753 |
| 34100 | 47063 | Hamblen County, TN....................... | 62,544 | 65,110 | 34860 | 48347 | Nacogdoches County, TX.................. | 64,524 | 64,753 |
| 34100 | 47089 | Jefferson County, TN...................... | 51,407 | 55,307 | 34900 | | Napa, CA Metro Area ......................... | 136,484 | 135,965 |
| 34140 | | Moscow, ID Micro Area ...................... | 37,244 | 40,830 | 34900 | 06055 | Napa County, CA........................... | 136,484 | 135,965 |
| 34140 | 16057 | Latah County, ID............................ | 37,244 | 40,830 | 34940 | | Naples-Marco Island, FL Metro Area..... | 321,520 | 392,973 |
| 34180 | | Moses Lake, WA Micro Area ............... | 89,120 | 99,377 | 34940 | 12021 | Collier County, FL.......................... | 321,520 | 392,973 |
| 34180 | 53025 | Grant County, WA.......................... | 89,120 | 99,377 | 34980 | | Nashville-Davidson--Murfreesboro- | | |
| 34220 | | Moultrie, GA Micro Area ..................... | 45,498 | 45,542 | | | -Franklin, TN Metro Area .................... | 1,646,200 | 1,961,232 |
| 34220 | 13071 | Colquitt County, GA........................ | 45,498 | 45,542 | 34980 | 47015 | Cannon County, TN........................ | 13,801 | 14,847 |
| 34260 | | Mountain Home, AR Micro Area........... | 41,513 | 42,242 | 34980 | 47021 | Cheatham County, TN..................... | 39,105 | 41,101 |
| 34260 | 05005 | Baxter County, AR ......................... | 41,513 | 42,242 | 34980 | 47037 | Davidson County, TN...................... | 626,681 | 694,176 |
| 34300 | | Mountain Home, ID Micro Area ............ | 27,038 | 27,448 | 34980 | 47043 | Dickson County, TN........................ | 49,666 | 54,376 |
| 34300 | 16039 | Elmore County, ID.......................... | 27,038 | 27,448 | 34980 | 47111 | Macon County, TN......................... | 22,248 | 24,827 |
| 34340 | | Mount Airy, NC Micro Area................. | 73,673 | 71,683 | 34980 | 47119 | Maury County, TN.......................... | 80,956 | 99,590 |
| 34340 | 37171 | Surry County, NC........................... | 73,673 | 71,683 | 34980 | 47147 | Robertson County, TN..................... | 66,283 | 72,275 |
| 34350 | | Mount Gay-Shamrock, WV Micro Area . | 36,743 | 31,688 | 34980 | 47149 | Rutherford County, TN..................... | 262,604 | 339,261 |
| 34350 | 54045 | Logan County, WV.......................... | 36,743 | 31,688 | 34980 | 47159 | Smith County, TN........................... | 19,166 | 20,285 |
| 34380 | | Mount Pleasant, MI Micro Area ........... | 70,311 | 69,504 | 34980 | 47165 | Sumner County, TN........................ | 160,645 | 195,561 |
| 34380 | 26073 | Isabella County, MI........................ | 70,311 | 69,504 | 34980 | 47169 | Trousdale County, TN..................... | 7,870 | 11,455 |
| 34420 | | Mount Pleasant, TX Micro Area........... | 44,735 | 45,986 | 34980 | 47187 | Williamson County, TN.................... | 183,182 | 245,348 |
| 34420 | 48063 | Camp County, TX .......................... | 12,401 | 13,060 | 34980 | 47189 | Wilson County, TN......................... | 113,993 | 148,130 |
| 34420 | 48449 | Titus County, TX........................... | 32,334 | 32,926 | 35020 | | Natchez, MS-LA Micro Area............... | 53,119 | 49,189 |
| 34460 | | Mount Sterling, KY Micro Area ............ | 44,396 | 47,169 | 35020 | 22029 | Concordia Parish, LA...................... | 20,822 | 18,914 |
| 34460 | 21011 | Bath County, KY........................... | 11,591 | 12,481 | 35020 | 28001 | Adams County, MS......................... | 32,297 | 30,275 |
| 34460 | 21165 | Menifee County, KY........................ | 6,306 | 6,502 | 35060 | | Natchitoches, LA Micro Area................ | 39,566 | 37,655 |
| 34460 | 21173 | Montgomery County, KY.................... | 26,499 | 28,186 | 35060 | 22069 | Natchitoches Parish, LA.................... | 39,566 | 37,655 |
| 34500 | | Mount Vernon, IL Micro Area............... | 38,827 | 37,235 | 35100 | | New Bern, NC Metro Area ................... | 126,802 | 123,198 |
| 34500 | 17081 | Jefferson County, IL....................... | 38,827 | 37,235 | 35100 | 37049 | Craven County, NC........................ | 103,505 | 101,233 |
| 34540 | | Mount Vernon, OH Micro Area ............. | 60,921 | 62,423 | 35100 | 37103 | Jones County, NC.......................... | 10,153 | 9,250 |
| 34540 | 39083 | Knox County, OH........................... | 60,921 | 62,423 | 35100 | 37137 | Pamlico County, NC........................ | 13,144 | 12,715 |
| 34580 | | Mount Vernon-Anacortes, WA Metro Area.............. | 116,901 | 130,789 | 35140 | | Newberry, SC Micro Area.................... | 37,508 | 38,445 |
| 34580 | 53057 | Skagit County, WA......................... | 116,901 | 130,789 | 35140 | 45071 | Newberry County, SC ...................... | 37,508 | 38,445 |
| 34620 | | Muncie, IN Metro Area....................... | 117,671 | 113,454 | 35220 | | New Castle, IN Micro Area .................. | 49,462 | 48,033 |
| 34620 | 18035 | Delaware County, IN....................... | 117,671 | 113,454 | 35220 | 18065 | Henry County, IN........................... | 49,462 | 48,033 |
| 34660 | | Murray, KY Micro Area ...................... | 37,191 | 39,300 | 35260 | | New Castle, PA Micro Area.................. | 91,108 | 85,083 |
| 34660 | 21035 | Calloway County, KY ...................... | 37,191 | 39,300 | 35260 | 42073 | Lawrence County, PA ...................... | 91,108 | 85,083 |
| 34700 | | Muscatine, IA Micro Area ................... | 42,745 | 42,394 | 35300 | | New Haven-Milford, CT Metro Area ...... | 862,477 | 851,948 |
| 34700 | 19139 | Muscatine County, IA...................... | 42,745 | 42,394 | 35300 | 09009 | New Haven County, CT ................... | 862,477 | 851,948 |
| 34740 | | Muskegon, MI Metro Area................... | 172,188 | 173,883 | 35380 | | New Orleans-Metairie, LA Metro Area... | 1,189,866 | 1,272,258 |
| 34740 | 26121 | Muskegon County, MI...................... | 172,188 | 173,883 | 35380 | 22051 | Jefferson Parish, LA....................... | 432,552 | 432,346 |
| | | | | | 35380 | 22071 | Orleans Parish, LA......................... | 343,829 | 389,476 |
| | | | | | 35380 | 22075 | Plaquemines Parish, LA................... | 23,042 | 23,113 |
| | | | | | 35380 | 22087 | St. Bernard Parish, LA..................... | 35,897 | 47,647 |
| | | | | | 35380 | 22089 | St. Charles Parish, LA..................... | 52,780 | 52,987 |
| | | | | | 35380 | 22093 | St. James Parish, LA...................... | 22,102 | 20,727 |
| | | | | | 35380 | 22095 | St. John the Baptist Parish, LA .......... | 45,924 | 42,516 |
| | | | | | 35380 | 22103 | St. Tammany Parish, LA................... | 233,740 | 263,446 |

| Core based statistical area | State/County FIPS code | Title and Geographic Components | 2010 Census Population | 2020 Estimated Population | Core based statistical area | State/County FIPS code | Title and Geographic Components | 2010 Census Population | 2020 Estimated Population |
|---|---|---|---|---|---|---|---|---|---|
| 35420 | | New Philadelphia-Dover, OH Micro Area........... | 92,582 | 91,776 | 35840 | | North Port-Sarasota-Bradenton, FL Metro Area | 702,281 | 854,684 |
| 35420 | 39157 | Tuscarawas County, OH............... | 92,582 | 91,776 | 35840 | 12081 | Manatee County, FL........... | 322,833 | 411,219 |
| | | | | | 35840 | 12115 | Sarasota County, FL........... | 379,448 | 443,465 |
| 35440 | | Newport, OR Micro Area | 46,034 | 50,583 | | | | | |
| 35440 | 41041 | Lincoln County, OR............... | 46,034 | 50,583 | 35860 | | North Vernon, IN Micro Area ............... | 28,525 | 27,515 |
| | | | | | 35860 | 18079 | Jennings County, IN........... | 28,525 | 27,515 |
| 35460 | | Newport, TN Micro Area...... | 35,662 | 36,225 | | | | | |
| 35460 | 47029 | Cocke County, TN............... | 35,662 | 36,225 | 35900 | | North Wilkesboro, NC Micro Area ........ | 69,340 | 68,043 |
| | | | | | 35900 | 37193 | Wilkes County, NC........... | 69,340 | 68,043 |
| 35580 | | New Ulm, MN Micro Area...... | 25,893 | 24,846 | | | | | |
| 35580 | 27015 | Brown County, MN............... | 25,893 | 24,846 | 35940 | | Norwalk, OH Micro Area........... | 59,626 | 57,979 |
| | | | | | 35940 | 39077 | Huron County, OH ........... | 59,626 | 57,979 |
| 35620 | | New York-Newark-Jersey City, NY-NJ-PA Metro Area | 18,897,109 | 19,124,359 | 35980 | | Norwich-New London, CT Metro Area... | 274,055 | 264,999 |
| | | | | | 35980 | 09011 | New London County, CT........... | 274,055 | 264,999 |
| 35620 | | Nassau County-Suffolk County, NY Division 35004 | 2,832,882 | 2,825,607 | 36020 | | Oak Harbor, WA Micro Area............... | 78,506 | 86,014 |
| 35620 | 36059 | Nassau County, NY ............... | 1,339,532 | 1,351,334 | 36020 | 53029 | Island County, WA ........... | 78,506 | 86,014 |
| 35620 | 36103 | Suffolk County, NY............... | 1,493,350 | 1,474,273 | | | | | |
| | | | | | 36100 | | Ocala, FL Metro Area............... | 331,298 | 373,513 |
| 35620 | | Newark, NJ-PA Division 35084 ............ | 2,147,727 | 2,167,853 | 36100 | 12083 | Marion County, FL........... | 331,298 | 373,513 |
| 35620 | 34013 | Essex County, NJ............... | 783,969 | 800,501 | | | | | |
| 35620 | 34019 | Hunterdon County, NJ ............... | 128,349 | 124,797 | 36140 | | Ocean City, NJ Metro Area............... | 97,265 | 91,546 |
| 35620 | 34027 | Morris County, NJ ............... | 492,276 | 491,087 | 36140 | 34009 | Cape May County, NJ........... | 97,265 | 91,546 |
| 35620 | 34037 | Sussex County, NJ............... | 149,265 | 140,002 | | | | | |
| 35620 | 34039 | Union County, NJ............... | 536,499 | 555,394 | 36220 | | Odessa, TX Metro Area............... | 137,130 | 167,701 |
| 35620 | 42103 | Pike County, PA............... | 57,369 | 56,072 | 36220 | 48135 | Ector County, TX........... | 137,130 | 167,701 |
| | | | | | 36260 | | Ogden-Clearfield, UT Metro Area.......... | 597,159 | 691,359 |
| 35620 | | New Brunswick-Lakewood, NJ Division 35154 | 2,340,249 | 2,384,685 | 36260 | 49003 | Box Elder County, UT........... | 49,975 | 57,007 |
| 35620 | 34023 | Middlesex County, NJ ............... | 809,858 | 822,736 | 36260 | 49011 | Davis County, UT........... | 306,479 | 359,232 |
| 35620 | 34025 | Monmouth County, NJ ............... | 630,380 | 618,381 | 36260 | 49029 | Morgan County, UT........... | 9,469 | 12,462 |
| 35620 | 34029 | Ocean County, NJ............... | 576,567 | 614,237 | 36260 | 49057 | Weber County, UT........... | 231,236 | 262,658 |
| 35620 | 34035 | Somerset County, NJ ............... | 323,444 | 329,331 | | | | | |
| | | | | | 36300 | | Ogdensburg-Massena, NY Micro Area.. | 111,944 | 107,185 |
| 35620 | | New York-Jersey City-White Plains, NY-NJ Division 35614 ......... | 11,576,251 | 11,746,214 | 36300 | 36089 | St. Lawrence County, NY........... | 111,944 | 107,185 |
| 35620 | 34003 | Bergen County, NJ............... | 905,116 | 930,394 | 36340 | | Oil City, PA Micro Area............... | 54,984 | 50,328 |
| 35620 | 34017 | Hudson County, NJ............... | 634,266 | 671,666 | 36340 | 42121 | Venango County, PA ........... | 54,984 | 50,328 |
| 35620 | 34031 | Passaic County, NJ............... | 501,226 | 500,382 | | | | | |
| 35620 | 36005 | Bronx County, NY............... | 1,385,108 | 1,401,142 | 36380 | | Okeechobee, FL Micro Area............... | 39,996 | 42,297 |
| 35620 | 36047 | Kings County, NY............... | 2,504,700 | 2,538,934 | 36380 | 12093 | Okeechobee County, FL ........... | 39,996 | 42,297 |
| 35620 | 36061 | New York County, NY............... | 1,585,873 | 1,611,989 | | | | | |
| 35620 | 36079 | Putnam County, NY............... | 99,710 | 98,532 | 36420 | | Oklahoma City, OK Metro Area............ | 1,252,987 | 1,425,375 |
| 35620 | 36081 | Queens County, NY............... | 2,230,722 | 2,225,821 | 36420 | 40017 | Canadian County, OK........... | 115,541 | 153,192 |
| 35620 | 36085 | Richmond County, NY ............... | 468,730 | 475,327 | 36420 | 40027 | Cleveland County, OK........... | 255,755 | 287,066 |
| 35620 | 36087 | Rockland County, NY............... | 311,687 | 326,225 | 36420 | 40051 | Grady County, OK........... | 52,431 | 55,906 |
| 35620 | 36119 | Westchester County, NY............... | 949,113 | 965,802 | 36420 | 40081 | Lincoln County, OK........... | 34,273 | 35,045 |
| | | | | | 36420 | 40083 | Logan County, OK........... | 41,848 | 48,777 |
| 35660 | | Niles, MI Metro Area............... | 156,813 | 153,025 | 36420 | 40087 | McClain County, OK ........... | 34,506 | 41,348 |
| 35660 | 26021 | Berrien County, MI............... | 156,813 | 153,025 | 36420 | 40109 | Oklahoma County, OK........... | 718,633 | 804,041 |
| 35700 | | Nogales, AZ Micro Area ............... | 47,420 | 46,808 | 36460 | | Olean, NY Micro Area............... | 80,317 | 75,863 |
| 35700 | 04023 | Santa Cruz County, AZ............... | 47,420 | 46,808 | 36460 | 36009 | Cattaraugus County, NY ........... | 80,317 | 75,863 |
| 35740 | | Norfolk, NE Micro Area............... | 48,271 | 47,877 | 36500 | | Olympia-Lacey-Tumwater, WA Metro Area | 252,264 | 294,074 |
| 35740 | 31119 | Madison County, NE............... | 34,876 | 34,813 | 36500 | 53067 | Thurston County, WA............... | 252,264 | 294,074 |
| 35740 | 31139 | Pierce County, NE ............... | 7,266 | 7,184 | | | | | |
| 35740 | 31167 | Stanton County, NE ............... | 6,129 | 5,880 | | | | | |
| 35820 | | North Platte, NE Micro Area ............... | 37,590 | 35,568 | | | | | |
| 35820 | 31111 | Lincoln County, NE ............... | 36,288 | 34,347 | | | | | |
| 35820 | 31113 | Logan County, NE............... | 763 | 747 | | | | | |
| 35820 | 31117 | McPherson County, NE ............... | 539 | 474 | | | | | |

| Core based statistical area | State/County FIPS code | Title and Geographic Components | 2010 Census Population | 2020 Estimated Population | Core based statistical area | State/County FIPS code | Title and Geographic Components | 2010 Census Population | 2020 Estimated Population |
|---|---|---|---|---|---|---|---|---|---|
| 36540 | | Omaha-Council Bluffs, NE-IA Metro Area.......... | 865,350 | 954,270 | 37120 | | Ozark, AL Micro Area.......... | 50,251 | 48,959 |
| 36540 | 19085 | Harrison County, IA.......... | 14,928 | 13,928 | 37120 | 01045 | Dale County, AL.......... | 50,251 | 48,959 |
| 36540 | 19129 | Mills County, IA.......... | 15,059 | 14,766 | | | | | |
| 36540 | 19155 | Pottawattamie County, IA | 93,158 | 93,328 | 37140 | | Paducah, KY-IL Micro Area.......... | 98,762 | 96,090 |
| 36540 | 31025 | Cass County, NE.......... | 25,241 | 26,232 | 37140 | 17127 | Massac County, IL.......... | 15,429 | 13,636 |
| 36540 | 31055 | Douglas County, NE.......... | 517,110 | 574,332 | 37140 | 21007 | Ballard County, KY.......... | 8,249 | 7,769 |
| 36540 | 31153 | Sarpy County, NE.......... | 158,840 | 188,856 | 37140 | 21139 | Livingston County, KY.......... | 9,519 | 9,041 |
| 36540 | 31155 | Saunders County, NE.......... | 20,780 | 21,927 | 37140 | 21145 | McCracken County, KY.......... | 65,565 | 65,644 |
| 36540 | 31177 | Washington County, NE.......... | 20,234 | 20,901 | | | | | |
| | | | | | 37220 | | Pahrump, NV Micro Area.......... | 43,946 | 48,054 |
| 36580 | | Oneonta, NY Micro Area.......... | 62,259 | 58,701 | 37220 | 32023 | Nye County, NV.......... | 43,946 | 48,054 |
| 36580 | 36077 | Otsego County, NY.......... | 62,259 | 58,701 | | | | | |
| | | | | | 37260 | | Palatka, FL Micro Area.......... | 74,364 | 74,815 |
| 36620 | | Ontario, OR-ID Micro Area.......... | 53,936 | 55,754 | 37260 | 12107 | Putnam County, FL.......... | 74,364 | 74,815 |
| 36620 | 16075 | Payette County, ID.......... | 22,623 | 24,771 | | | | | |
| 36620 | 41045 | Malheur County, OR.......... | 31,313 | 30,983 | 37300 | | Palestine, TX Micro Area.......... | 58,458 | 57,805 |
| | | | | | 37300 | 48001 | Anderson County, TX.......... | 58,458 | 57,805 |
| 36660 | | Opelousas, LA Micro Area.......... | 83,384 | 81,440 | | | | | |
| 36660 | 22097 | St. Landry Parish, LA.......... | 83,384 | 81,440 | 37340 | | Palm Bay-Melbourne-Titusville, FL Metro Area.......... | 543,376 | 608,459 |
| | | | | | 37340 | 12009 | Brevard County, FL.......... | 543,376 | 608,459 |
| 36700 | | Orangeburg, SC Micro Area.......... | 92,501 | 85,343 | | | | | |
| 36700 | 45075 | Orangeburg County, SC.......... | 92,501 | 85,343 | 37420 | | Pampa, TX Micro Area.......... | 23,464 | 22,471 |
| | | | | | 37420 | 48179 | Gray County, TX.......... | 22,535 | 21,658 |
| 36740 | | Orlando-Kissimmee-Sanford, FL Metro Area.......... | 2,134,411 | 2,639,374 | 37420 | 48393 | Roberts County, TX.......... | 929 | 813 |
| 36740 | 12069 | Lake County, FL.......... | 297,052 | 375,492 | | | | | |
| 36740 | 12095 | Orange County, FL.......... | 1,145,956 | 1,404,396 | 37460 | | Panama City, FL Metro Area.......... | 168,852 | 171,322 |
| 36740 | 12097 | Osceola County, FL.......... | 268,685 | 385,315 | 37460 | 12005 | Bay County, FL.......... | 168,852 | 171,322 |
| 36740 | 12117 | Seminole County, FL.......... | 422,718 | 474,171 | | | | | |
| | | | | | 37500 | | Paragould, AR Micro Area.......... | 42,090 | 45,597 |
| 36780 | | Oshkosh-Neenah, WI Metro Area.......... | 166,994 | 171,631 | 37500 | 05055 | Greene County, AR.......... | 42,090 | 45,597 |
| 36780 | 55139 | Winnebago County, WI.......... | 166,994 | 171,631 | | | | | |
| | | | | | 37540 | | Paris, TN Micro Area.......... | 32,330 | 32,056 |
| 36820 | | Oskaloosa, IA Micro Area.......... | 22,381 | 22,370 | 37540 | 47079 | Henry County, TN.......... | 32,330 | 32,056 |
| 36820 | 19123 | Mahaska County, IA.......... | 22,381 | 22,370 | | | | | |
| | | | | | 37580 | | Paris, TX Micro Area.......... | 49,793 | 49,905 |
| 36830 | | Othello, WA Micro Area.......... | 18,728 | 20,027 | 37580 | 48277 | Lamar County, TX.......... | 49,793 | 49,905 |
| 36830 | 53001 | Adams County, WA.......... | 18,728 | 20,027 | | | | | |
| | | | | | 37620 | | Parkersburg-Vienna, WV Metro Area.... | 92,673 | 88,643 |
| 36837 | | Ottawa, IL Micro Area.......... | 154,908 | 145,590 | 37620 | 54105 | Wirt County, WV.......... | 5,717 | 5,705 |
| 36837 | 17011 | Bureau County, IL.......... | 34,978 | 32,303 | 37620 | 54107 | Wood County, WV.......... | 86,956 | 82,938 |
| 36837 | 17099 | LaSalle County, IL.......... | 113,924 | 107,571 | | | | | |
| 36837 | 17155 | Putnam County, IL.......... | 6,006 | 5,716 | 37660 | | Parsons, KS Micro Area.......... | 21,607 | 19,586 |
| | | | | | 37660 | 20099 | Labette County, KS.......... | 21,607 | 19,586 |
| 36840 | | Ottawa, KS Micro Area.......... | 25,992 | 25,703 | | | | | |
| 36840 | 20059 | Franklin County, KS.......... | 25,992 | 25,703 | 37740 | | Payson, AZ Micro Area.......... | 53,597 | 54,303 |
| | | | | | 37740 | 04007 | Gila County, AZ.......... | 53,597 | 54,303 |
| 36900 | | Ottumwa, IA Micro Area.......... | 35,625 | 34,985 | | | | | |
| 36900 | 19179 | Wapello County, IA.......... | 35,625 | 34,985 | 37770 | | Pearsall, TX Micro Area.......... | 17,217 | 20,379 |
| | | | | | 37770 | 48163 | Frio County, TX.......... | 17,217 | 20,379 |
| 36940 | | Owatonna, MN Micro Area.......... | 36,576 | 36,596 | | | | | |
| 36940 | 27147 | Steele County, MN.......... | 36,576 | 36,596 | 37780 | | Pecos, TX Micro Area.......... | 13,865 | 16,130 |
| | | | | | 37780 | 48301 | Loving County, TX.......... | 82 | 181 |
| 36980 | | Owensboro, KY Metro Area.......... | 114,752 | 119,795 | 37780 | 48389 | Reeves County, TX.......... | 13,783 | 15,949 |
| 36980 | 21059 | Daviess County, KY.......... | 96,656 | 101,978 | | | | | |
| 36980 | 21091 | Hancock County, KY.......... | 8,565 | 8,742 | 37800 | | Pella, IA Micro Area.......... | 33,309 | 33,168 |
| 36980 | 21149 | McLean County, KY.......... | 9,531 | 9,075 | 37800 | 19125 | Marion County, IA.......... | 33,309 | 33,168 |
| | | | | | 37860 | | Pensacola-Ferry Pass-Brent, FL Metro Area.......... | 448,991 | 511,503 |
| 37060 | | Oxford, MS Micro Area.......... | 47,351 | 54,408 | 37860 | 12033 | Escambia County, FL.......... | 297,619 | 322,364 |
| 37060 | 28071 | Lafayette County, MS.......... | 47,351 | 54,408 | 37860 | 12113 | Santa Rosa County, FL.......... | 151,372 | 189,139 |
| 37100 | | Oxnard-Thousand Oaks-Ventura, CA Metro Area.......... | 823,318 | 841,387 | | | | | |
| 37100 | 06111 | Ventura County, CA.......... | 823,318 | 841,387 | | | | | |

Appendix C

| Core based statisti-cal area | State/ County FIPS code | Title and Geographic Components | 2010 Census Population | 2020 Estimated Population | Core based statisti-cal area | State/ County FIPS code | Title and Geographic Components | 2010 Census Population | 2020 Estimated Population |
|---|---|---|---|---|---|---|---|---|---|
| 37900 | | Peoria, IL Metro Area | 416,255 | 396,781 | 38340 | | Pittsfield, MA Metro Area | 131,219 | 124,571 |
| 37900 | 17057 | Fulton County, IL | 37,069 | 33,690 | 38340 | 25003 | Berkshire County, MA | 131,219 | 124,571 |
| 37900 | 17123 | Marshall County, IL | 12,640 | 11,309 | | | | | |
| 37900 | 17143 | Peoria County, IL | 186,494 | 177,652 | 38380 | | Plainview, TX Micro Area | 36,273 | 32,754 |
| 37900 | 17175 | Stark County, IL | 5,994 | 5,262 | 38380 | 48189 | Hale County, TX | 36,273 | 32,754 |
| 37900 | 17179 | Tazewell County, IL | 135,394 | 130,777 | | | | | |
| 37900 | 17203 | Woodford County, IL | 38,664 | 38,091 | 38420 | | Platteville, WI Micro Area | 51,208 | 51,021 |
| | | | | | 38420 | 55043 | Grant County, WI | 51,208 | 51,021 |
| 37940 | | Peru, IN Micro Area | 36,903 | 35,328 | | | | | |
| 37940 | 18103 | Miami County, IN | 36,903 | 35,328 | 38460 | | Plattsburgh, NY Micro Area | 82,128 | 79,778 |
| | | | | | 38460 | 36019 | Clinton County, NY | 82,128 | 79,778 |
| 37980 | | Philadelphia-Camden-Wilmington, PA-NJ-DE-MD Metro Area | 5,965,343 | 6,107,906 | 38500 | | Plymouth, IN Micro Area | 47,051 | 46,108 |
| | | | | | 38500 | 18099 | Marshall County, IN | 47,051 | 46,108 |
| 37980 | | Camden, NJ Division 15804 | 1,250,679 | 1,246,650 | | | | | |
| 37980 | 34005 | Burlington County, NJ | 448,734 | 446,596 | 38540 | | Pocatello, ID Metro Area | 90,656 | 96,438 |
| 37980 | 34007 | Camden County, NJ | 513,657 | 506,809 | 38540 | 16005 | Bannock County, ID | 82,839 | 88,795 |
| 37980 | 34015 | Gloucester County, NJ | 288,288 | 293,245 | 38540 | 16077 | Power County, ID | 7,817 | 7,643 |
| 37980 | | Montgomery County-Bucks County-Chester County, PA Division 33874 | 1,924,009 | 1,988,615 | 38580 | | Point Pleasant, WV-OH Micro Area | 58,258 | 56,137 |
| 37980 | 42017 | Bucks County, PA | 625,249 | 627,987 | 38580 | 39053 | Gallia County, OH | 30,934 | 29,802 |
| 37980 | 42029 | Chester County, PA | 498,886 | 526,759 | 38580 | 54053 | Mason County, WV | 27,324 | 26,335 |
| 37980 | 42091 | Montgomery County, PA | 799,874 | 833,869 | 38620 | | Ponca City, OK Micro Area | 46,562 | 43,274 |
| | | | | | 38620 | 40071 | Kay County, OK | 46,562 | 43,274 |
| 37980 | | Philadelphia, PA Division 37964 | 2,084,985 | 2,145,240 | | | | | |
| 37980 | 42045 | Delaware County, PA | 558,979 | 566,753 | 38700 | | Pontiac, IL Micro Area | 38,950 | 35,414 |
| 37980 | 42101 | Philadelphia County, PA | 1,526,006 | 1,578,487 | 38700 | 17105 | Livingston County, IL | 38,950 | 35,414 |
| 37980 | | Wilmington, DE-MD-NJ Division 48864 | 705,670 | 727,401 | 38740 | | Poplar Bluff, MO Micro Area | 56,894 | 55,478 |
| 37980 | 10003 | New Castle County, DE | 538,479 | 561,531 | 38740 | 29023 | Butler County, MO | 42,794 | 42,178 |
| 37980 | 24015 | Cecil County, MD | 101,108 | 103,419 | 38740 | 29181 | Ripley County, MO | 14,100 | 13,300 |
| 37980 | 34033 | Salem County, NJ | 66,083 | 62,451 | | | | | |
| | | | | | 38780 | | Portales, NM Micro Area | 19,846 | 18,350 |
| 38060 | | Phoenix-Mesa-Chandler, AZ Metro Area | 4,192,887 | 5,059,909 | 38780 | 35041 | Roosevelt County, NM | 19,846 | 18,350 |
| 38060 | 04013 | Maricopa County, AZ | 3,817,117 | 4,579,081 | 38820 | | Port Angeles, WA Micro Area | 71,404 | 78,067 |
| 38060 | 04021 | Pinal County, AZ | 375,770 | 480,828 | 38820 | 53009 | Clallam County, WA | 71,404 | 78,067 |
| 38100 | | Picayune, MS Micro Area | 55,834 | 55,876 | 38860 | | Portland-South Portland, ME Metro Area | 514,098 | 543,221 |
| 38100 | 28109 | Pearl River County, MS | 55,834 | 55,876 | 38860 | 23005 | Cumberland County, ME | 281,674 | 298,111 |
| | | | | | 38860 | 23023 | Sagadahoc County, ME | 35,293 | 36,044 |
| 38180 | | Pierre, SD Micro Area | 19,988 | 20,457 | 38860 | 23031 | York County, ME | 197,131 | 209,066 |
| 38180 | 46065 | Hughes County, SD | 17,022 | 17,336 | | | | | |
| 38180 | 46117 | Stanley County, SD | 2,966 | 3,121 | 38900 | | Portland-Vancouver-Hillsboro, OR-WA Metro Area | 2,226,009 | 2,510,259 |
| | | | | | 38900 | 41005 | Clackamas County, OR | 375,992 | 421,596 |
| 38220 | | Pine Bluff, AR Metro Area | 100,258 | 86,278 | 38900 | 41009 | Columbia County, OR | 49,351 | 52,876 |
| 38220 | 05025 | Cleveland County, AR | 8,689 | 7,957 | 38900 | 41051 | Multnomah County, OR | 735,334 | 815,637 |
| 38220 | 05069 | Jefferson County, AR | 77,435 | 65,377 | 38900 | 41067 | Washington County, OR | 529,710 | 603,514 |
| 38220 | 05079 | Lincoln County, AR | 14,134 | 12,944 | 38900 | 41071 | Yamhill County, OR | 99,193 | 107,664 |
| | | | | | 38900 | 53011 | Clark County, WA | 425,363 | 496,865 |
| 38240 | | Pinehurst-Southern Pines, NC Micro Area | 88,247 | 103,352 | 38900 | 53059 | Skamania County, WA | 11,066 | 12,107 |
| 38240 | 37125 | Moore County, NC | 88,247 | 103,352 | 38920 | | Port Lavaca, TX Micro Area | 21,381 | 21,001 |
| | | | | | 38920 | 48057 | Calhoun County, TX | 21,381 | 21,001 |
| 38260 | | Pittsburg, KS Micro Area | 39,134 | 38,730 | | | | | |
| 38260 | 20037 | Crawford County, KS | 39,134 | 38,730 | 38940 | | Port St. Lucie, FL Metro Area | 424,107 | 499,274 |
| | | | | | 38940 | 12085 | Martin County, FL | 146,318 | 162,088 |
| 38300 | | Pittsburgh, PA Metro Area | 2,356,285 | 2,309,246 | 38940 | 12111 | St. Lucie County, FL | 277,789 | 337,186 |
| 38300 | 42003 | Allegheny County, PA | 1,223,348 | 1,211,358 | | | | | |
| 38300 | 42005 | Armstrong County, PA | 68,941 | 64,162 | 39020 | | Portsmouth, OH Micro Area | 79,499 | 74,347 |
| 38300 | 42007 | Beaver County, PA | 170,539 | 162,575 | 39020 | 39145 | Scioto County, OH | 79,499 | 74,347 |
| 38300 | 42019 | Butler County, PA | 183,862 | 189,135 | | | | | |
| 38300 | 42051 | Fayette County, PA | 136,606 | 128,126 | | | | | |
| 38300 | 42125 | Washington County, PA | 207,820 | 206,803 | | | | | |
| 38300 | 42129 | Westmoreland County, PA | 365,169 | 347,087 | | | | | |

# CORE BASED STATISTICAL AREAS
## (Metropolitan and Micropolitan),
## METROPOLITAN DIVISIONS, AND COMPONENTS
## (as defined March, 2020)—*Continued*

| Core based statisti-cal area | State/ County FIPS code | Title and Geographic Components | 2010 Census Population | 2020 Estimated Population | Core based statisti-cal area | State/ County FIPS code | Title and Geographic Components | 2010 Census Population | 2020 Estimated Population |
|---|---|---|---|---|---|---|---|---|---|
| 39060 | | Pottsville, PA Micro Area | 148,289 | 140,709 | 39860 | | Red Wing, MN Micro Area | 46,183 | 46,318 |
| 39060 | 42107 | Schuylkill County, PA | 148,289 | 140,709 | 39860 | 27049 | Goodhue County, MN | 46,183 | 46,318 |
| 39100 | | Poughkeepsie-Newburgh-Middletown, NY Metro Area | 670,301 | 678,527 | 39900 | | Reno, NV Metro Area | 425,417 | 481,289 |
| 39100 | 36027 | Dutchess County, NY | 297,488 | 293,293 | 39900 | 32029 | Storey County, NV | 4,010 | 4,207 |
| 39100 | 36071 | Orange County, NY | 372,813 | 385,234 | 39900 | 32031 | Washoe County, NV | 421,407 | 477,082 |
| 39150 | | Prescott Valley-Prescott, AZ Metro Area | 211,033 | 240,226 | 39940 | | Rexburg, ID Micro Area | 50,778 | 53,536 |
| 39150 | 04025 | Yavapai County, AZ | 211,033 | 240,226 | 39940 | 16043 | Fremont County, ID | 13,242 | 13,218 |
| | | | | | 39940 | 16065 | Madison County, ID | 37,536 | 40,318 |
| 39220 | | Price, UT Micro Area | 21,403 | 20,760 | 39980 | | Richmond, IN Micro Area | 68,917 | 65,778 |
| 39220 | 49007 | Carbon County, UT | 21,403 | 20,760 | 39980 | 18177 | Wayne County, IN | 68,917 | 65,778 |
| 39260 | | Prineville, OR Micro Area | 20,978 | 25,105 | 40060 | | Richmond, VA Metro Area | 1,186,501 | 1,303,469 |
| 39260 | 41013 | Crook County, OR | 20,978 | 25,105 | 40060 | 51007 | Amelia County, VA | 12,690 | 13,014 |
| | | | | | 40060 | 51036 | Charles City County, VA | 7,256 | 6,821 |
| 39300 | | Providence-Warwick, RI-MA Metro Area | 1,600,852 | 1,623,890 | 40060 | 51041 | Chesterfield County, VA | 316,236 | 358,245 |
| 39300 | 25005 | Bristol County, MA | 548,285 | 566,765 | 40060 | 51053 | Dinwiddie County, VA | 28,001 | 28,688 |
| 39300 | 44001 | Bristol County, RI | 49,875 | 48,350 | 40060 | 51075 | Goochland County, VA | 21,717 | 24,431 |
| 39300 | 44003 | Kent County, RI | 166,158 | 164,646 | 40060 | 51085 | Hanover County, VA | 99,863 | 108,262 |
| 39300 | 44005 | Newport County, RI | 82,888 | 81,836 | 40060 | 51087 | Henrico County, VA | 306,935 | 333,766 |
| 39300 | 44007 | Providence County, RI | 626,667 | 636,547 | 40060 | 51097 | King and Queen County, VA | 6,945 | 6,942 |
| 39300 | 44009 | Washington County, RI | 126,979 | 125,746 | 40060 | 51101 | King William County, VA | 15,935 | 17,641 |
| | | | | | 40060 | 51127 | New Kent County, VA | 18,429 | 23,648 |
| 39340 | | Provo-Orem, UT Metro Area | 526,810 | 663,181 | 40060 | 51145 | Powhatan County, VA | 28,046 | 30,148 |
| 39340 | 49023 | Juab County, UT | 10,246 | 12,122 | 40060 | 51149 | Prince George County, VA | 35,725 | 38,686 |
| 39340 | 49049 | Utah County, UT | 516,564 | 651,059 | 40060 | 51183 | Sussex County, VA | 12,087 | 10,925 |
| | | | | | 40060 | 51570 | Colonial Heights city, VA | 17,411 | 17,205 |
| 39380 | | Pueblo, CO Metro Area | 159,063 | 169,823 | 40060 | 51670 | Hopewell city, VA | 22,591 | 22,375 |
| 39380 | 08101 | Pueblo County, CO | 159,063 | 169,823 | 40060 | 51730 | Petersburg city, VA | 32,420 | 30,446 |
| | | | | | 40060 | 51760 | Richmond city, VA | 204,214 | 232,226 |
| 39420 | | Pullman, WA Micro Area | 44,776 | 49,500 | | | | | |
| 39420 | 53075 | Whitman County, WA | 44,776 | 49,500 | 40080 | | Richmond-Berea, KY Micro Area | 97,588 | 108,374 |
| | | | | | 40080 | 21065 | Estill County, KY | 14,672 | 14,109 |
| 39460 | | Punta Gorda, FL Metro Area | 159,978 | 194,711 | 40080 | 21151 | Madison County, KY | 82,916 | 94,265 |
| 39460 | 12015 | Charlotte County, FL | 159,978 | 194,711 | | | | | |
| | | | | | 40100 | | Rio Grande City-Roma, TX Micro Area | 60,968 | 64,266 |
| 39500 | | Quincy, IL-MO Micro Area | 77,314 | 74,593 | 40100 | 48427 | Starr County, TX | 60,968 | 64,266 |
| 39500 | 17001 | Adams County, IL | 67,103 | 64,783 | | | | | |
| 39500 | 29111 | Lewis County, MO | 10,211 | 9,810 | 40140 | | Riverside-San Bernardino-Ontario, CA Metro Area | 4,224,851 | 4,678,371 |
| | | | | | 40140 | 06065 | Riverside County, CA | 2,189,641 | 2,489,188 |
| 39540 | | Racine, WI Metro Area | 195,408 | 195,802 | 40140 | 06071 | San Bernardino County, CA | 2,035,210 | 2,189,183 |
| 39540 | 55101 | Racine County, WI | 195,408 | 195,802 | | | | | |
| | | | | | 40180 | | Riverton, WY Micro Area | 40,123 | 39,317 |
| 39580 | | Raleigh-Cary, NC Metro Area | 1,130,490 | 1,420,376 | 40180 | 56013 | Fremont County, WY | 40,123 | 39,317 |
| 39580 | 37069 | Franklin County, NC | 60,619 | 71,859 | | | | | |
| 39580 | 37101 | Johnston County, NC | 168,878 | 216,246 | 40220 | | Roanoke, VA Metro Area | 308,707 | 313,784 |
| 39580 | 37183 | Wake County, NC | 900,993 | 1,132,271 | 40220 | 51023 | Botetourt County, VA | 33,148 | 33,633 |
| | | | | | 40220 | 51045 | Craig County, VA | 5,190 | 5,077 |
| 39660 | | Rapid City, SD Metro Area | 126,382 | 144,514 | 40220 | 51067 | Franklin County, VA | 56,159 | 56,167 |
| 39660 | 46093 | Meade County, SD | 25,434 | 28,588 | 40220 | 51161 | Roanoke County, VA | 92,376 | 94,509 |
| 39660 | 46103 | Pennington County, SD | 100,948 | 115,926 | 40220 | 51770 | Roanoke city, VA | 97,032 | 99,058 |
| | | | | | 40220 | 51775 | Salem city, VA | 24,802 | 25,340 |
| 39700 | | Raymondville, TX Micro Area | 22,134 | 21,161 | | | | | |
| 39700 | 48489 | Willacy County, TX | 22,134 | 21,161 | 40260 | | Roanoke Rapids, NC Micro Area | 76,790 | 68,567 |
| | | | | | 40260 | 37083 | Halifax County, NC | 54,691 | 49,479 |
| 39740 | | Reading, PA Metro Area | 411,442 | 421,017 | 40260 | 37131 | Northampton County, NC | 22,099 | 19,088 |
| 39740 | 42011 | Berks County, PA | 411,442 | 421,017 | | | | | |
| | | | | | 40300 | | Rochelle, IL Micro Area | 53,497 | 50,306 |
| 39780 | | Red Bluff, CA Micro Area | 63,463 | 64,494 | 40300 | 17141 | Ogle County, IL | 53,497 | 50,306 |
| 39780 | 06103 | Tehama County, CA | 63,463 | 64,494 | | | | | |
| 39820 | | Redding, CA Metro Area | 177,223 | 179,027 | | | | | |
| 39820 | 06089 | Shasta County, CA | 177,223 | 179,027 | | | | | |

| Core based statistical area | State/County FIPS code | Title and Geographic Components | 2010 Census Population | 2020 Estimated Population |
|---|---|---|---|---|
| 40340 | | Rochester, MN Metro Area | 206,877 | 223,062 |
| 40340 | 27039 | Dodge County, MN | 20,087 | 20,987 |
| 40340 | 27045 | Fillmore County, MN | 20,866 | 21,135 |
| 40340 | 27109 | Olmsted County, MN | 144,248 | 159,298 |
| 40340 | 27157 | Wabasha County, MN | 21,676 | 21,642 |
| 40380 | | Rochester, NY Metro Area | 1,079,671 | 1,067,486 |
| 40380 | 36051 | Livingston County, NY | 65,393 | 62,398 |
| 40380 | 36055 | Monroe County, NY | 744,344 | 740,900 |
| 40380 | 36069 | Ontario County, NY | 107,931 | 110,091 |
| 40380 | 36073 | Orleans County, NY | 42,883 | 39,978 |
| 40380 | 36117 | Wayne County, NY | 93,772 | 89,339 |
| 40380 | 36123 | Yates County, NY | 25,348 | 24,780 |
| 40420 | | Rockford, IL Metro Area | 349,431 | 334,072 |
| 40420 | 17007 | Boone County, IL | 54,165 | 52,777 |
| 40420 | 17201 | Winnebago County, IL | 295,266 | 281,295 |
| 40460 | | Rockingham, NC Micro Area | 46,639 | 44,332 |
| 40460 | 37153 | Richmond County, NC | 46,639 | 44,332 |
| 40530 | | Rockport, TX Micro Area | 23,158 | 23,814 |
| 40530 | 48007 | Aransas County, TX | 23,158 | 23,814 |
| 40540 | | Rock Springs, WY Micro Area | 43,806 | 42,673 |
| 40540 | 56037 | Sweetwater County, WY | 43,806 | 42,673 |
| 40580 | | Rocky Mount, NC Metro Area | 152,392 | 145,688 |
| 40580 | 37065 | Edgecombe County, NC | 56,552 | 50,829 |
| 40580 | 37127 | Nash County, NC | 95,840 | 94,859 |
| 40620 | | Rolla, MO Micro Area | 45,156 | 44,414 |
| 40620 | 29161 | Phelps County, MO | 45,156 | 44,414 |
| 40660 | | Rome, GA Metro Area | 96,317 | 98,604 |
| 40660 | 13115 | Floyd County, GA | 96,317 | 98,604 |
| 40700 | | Roseburg, OR Micro Area | 107,667 | 111,364 |
| 40700 | 41019 | Douglas County, OR | 107,667 | 111,364 |
| 40740 | | Roswell, NM Micro Area | 65,645 | 64,711 |
| 40740 | 35005 | Chaves County, NM | 65,645 | 64,711 |
| 40760 | | Ruidoso, NM Micro Area | 20,497 | 19,939 |
| 40760 | 35027 | Lincoln County, NM | 20,497 | 19,939 |
| 40780 | | Russellville, AR Micro Area | 83,939 | 85,515 |
| 40780 | 05115 | Pope County, AR | 61,754 | 64,334 |
| 40780 | 05149 | Yell County, AR | 22,185 | 21,181 |
| 40820 | | Ruston, LA Micro Area | 46,735 | 46,552 |
| 40820 | 22061 | Lincoln Parish, LA | 46,735 | 46,552 |
| 40860 | | Rutland, VT Micro Area | 61,642 | 57,764 |
| 40860 | 50021 | Rutland County, VT | 61,642 | 57,764 |
| 40900 | | Sacramento-Roseville-Folsom, CA Metro Area | 2,149,127 | 2,374,749 |
| 40900 | 06017 | El Dorado County, CA | 181,058 | 192,925 |
| 40900 | 06061 | Placer County, CA | 348,432 | 402,950 |
| 40900 | 06067 | Sacramento County, CA | 1,418,788 | 1,559,146 |
| 40900 | 06113 | Yolo County, CA | 200,849 | 219,728 |
| 40940 | | Safford, AZ Micro Area | 37,220 | 39,211 |
| 40940 | 04009 | Graham County, AZ | 37,220 | 39,211 |
| 40980 | | Saginaw, MI Metro Area | 200,169 | 189,868 |
| 40980 | 26145 | Saginaw County, MI | 200,169 | 189,868 |
| 41060 | | St. Cloud, MN Metro Area | 189,093 | 202,996 |
| 41060 | 27009 | Benton County, MN | 38,451 | 40,958 |
| 41060 | 27145 | Stearns County, MN | 150,642 | 162,038 |
| 41100 | | St. George, UT Metro Area | 138,115 | 184,913 |
| 41100 | 49053 | Washington County, UT | 138,115 | 184,913 |
| 41140 | | St. Joseph, MO-KS Metro Area | 127,329 | 122,556 |
| 41140 | 20043 | Doniphan County, KS | 7,945 | 7,496 |
| 41140 | 29003 | Andrew County, MO | 17,291 | 17,586 |
| 41140 | 29021 | Buchanan County, MO | 89,201 | 86,530 |
| 41140 | 29063 | DeKalb County, MO | 12,892 | 10,944 |
| 41180 | | St. Louis, MO-IL Metro Area | 2,787,701 | 2,805,473 |
| 41180 | 17005 | Bond County, IL | 17,768 | 16,262 |
| 41180 | 17013 | Calhoun County, IL | 5,089 | 4,616 |
| 41180 | 17027 | Clinton County, IL | 37,762 | 37,398 |
| 41180 | 17083 | Jersey County, IL | 22,985 | 21,616 |
| 41180 | 17117 | Macoupin County, IL | 47,765 | 44,567 |
| 41180 | 17119 | Madison County, IL | 269,282 | 262,635 |
| 41180 | 17133 | Monroe County, IL | 32,957 | 34,739 |
| 41180 | 17163 | St. Clair County, IL | 270,056 | 258,046 |
| 41180 | 29071 | Franklin County, MO | 101,492 | 104,469 |
| 41180 | 29099 | Jefferson County, MO | 218,733 | 226,543 |
| 41180 | 29113 | Lincoln County, MO | 52,566 | 60,119 |
| 41180 | 29183 | St. Charles County, MO | 360,485 | 406,204 |
| 41180 | 29189 | St. Louis County, MO | 998,954 | 994,020 |
| 41180 | 29219 | Warren County, MO | 32,513 | 36,594 |
| 41180 | 29510 | St. Louis city, MO | 319,294 | 297,645 |
| 41220 | | St. Marys, GA Micro Area | 50,513 | 55,388 |
| 41220 | 13039 | Camden County, GA | 50,513 | 55,388 |
| 41260 | | St. Marys, PA Micro Area | 31,946 | 29,607 |
| 41260 | 42047 | Elk County, PA | 31,946 | 29,607 |
| 41400 | | Salem, OH Micro Area | 107,841 | 101,118 |
| 41400 | 39029 | Columbiana County, OH | 107,841 | 101,118 |
| 41420 | | Salem, OR Metro Area | 390,738 | 436,948 |
| 41420 | 41047 | Marion County, OR | 315,335 | 349,204 |
| 41420 | 41053 | Polk County, OR | 75,403 | 87,744 |
| 41460 | | Salina, KS Micro Area | 61,697 | 59,638 |
| 41460 | 20143 | Ottawa County, KS | 6,091 | 5,712 |
| 41460 | 20169 | Saline County, KS | 55,606 | 53,926 |
| 41500 | | Salinas, CA Metro Area | 415,057 | 430,906 |
| 41500 | 06053 | Monterey County, CA | 415,057 | 430,906 |
| 41540 | | Salisbury, MD-DE Metro Area | 373,802 | 423,481 |
| 41540 | 10005 | Sussex County, DE | 197,145 | 241,635 |
| 41540 | 24039 | Somerset County, MD | 26,470 | 25,453 |
| 41540 | 24045 | Wicomico County, MD | 98,733 | 103,990 |
| 41540 | 24047 | Worcester County, MD | 51,454 | 52,403 |
| 41620 | | Salt Lake City, UT Metro Area | 1,087,873 | 1,240,029 |
| 41620 | 49035 | Salt Lake County, UT | 1,029,655 | 1,165,517 |
| 41620 | 49045 | Tooele County, UT | 58,218 | 74,512 |

| Core based statistical area | State/ County FIPS code | Title and Geographic Components | 2010 Census Population | 2020 Estimated Population | Core based statistical area | State/ County FIPS code | Title and Geographic Components | 2010 Census Population | 2020 Estimated Population |
|---|---|---|---|---|---|---|---|---|---|
| 41660 | | San Angelo, TX Metro Area.................. | 112,966 | 122,889 | 42300 | | Sault Ste. Marie, MI Micro Area............. | 38,520 | 36,958 |
| 41660 | 48235 | Irion County, TX........................... | 1,599 | 1,564 | 42300 | 26033 | Chippewa County, MI........................ | 38,520 | 36,958 |
| 41660 | 48431 | Sterling County, TX........................ | 1,143 | 1,315 | | | | | |
| 41660 | 48451 | Tom Green County, TX...................... | 110,224 | 120,010 | 42340 | | Savannah, GA Metro Area ................... | 347,611 | 395,983 |
| | | | | | 42340 | 13029 | Bryan County, GA........................... | 30,233 | 40,755 |
| 41700 | | San Antonio-New Braunfels, TX Metro | | | 42340 | 13051 | Chatham County, GA........................ | 265,128 | 289,463 |
| | | Area....................................... | 2,142,508 | 2,590,732 | 42340 | 13103 | Effingham County, GA....................... | 52,250 | 65,765 |
| 41700 | 48013 | Atascosa County, TX....................... | 44,911 | 51,724 | | | | | |
| 41700 | 48019 | Bandera County, TX ........................ | 20,485 | 23,861 | 42380 | | Sayre, PA Micro Area ....................... | 62,622 | 60,221 |
| 41700 | 48029 | Bexar County, TX........................... | 1,714,773 | 2,026,823 | 42380 | 42015 | Bradford County, PA........................ | 62,622 | 60,221 |
| 41700 | 48091 | Comal County, TX........................... | 108,472 | 164,812 | | | | | |
| 41700 | 48187 | Guadalupe County, TX...................... | 131,533 | 170,608 | 42420 | | Scottsbluff, NE Micro Area ................. | 38,971 | 37,285 |
| 41700 | 48259 | Kendall County, TX......................... | 33,410 | 48,523 | 42420 | 31007 | Banner County, NE ......................... | 690 | 786 |
| 41700 | 48325 | Medina County, TX......................... | 46,006 | 52,358 | 42420 | 31157 | Scotts Bluff County, NE .................... | 36,970 | 35,299 |
| 41700 | 48493 | Wilson County, TX.......................... | 42,918 | 52,023 | 42420 | 31165 | Sioux County, NE........................... | 1,311 | 1,200 |
| 41740 | | San Diego-Chula Vista-Carlsbad, CA | | | 42460 | | Scottsboro, AL Micro Area.................. | 53,227 | 51,582 |
| | | Metro Area................................. | 3,095,313 | 3,332,427 | 42460 | 01071 | Jackson County, AL......................... | 53,227 | 51,582 |
| 41740 | 06073 | San Diego County, CA...................... | 3,095,313 | 3,332,427 | | | | | |
| | | | | | 42500 | | Scottsburg, IN Micro Area .................. | 24,181 | 23,788 |
| 41760 | | Sandpoint, ID Micro Area ................... | 40,877 | 46,817 | 42500 | 18143 | Scott County, IN............................ | 24,181 | 23,788 |
| 41760 | 16017 | Bonner County, ID ......................... | 40,877 | 46,817 | | | | | |
| | | | | | 42540 | | Scranton--Wilkes-Barre, PA Metro Area | 563,631 | 552,528 |
| 41780 | | Sandusky, OH Micro Area ................... | 77,079 | 73,719 | 42540 | 42069 | Lackawanna County, PA..................... | 214,437 | 208,989 |
| 41780 | 39043 | Erie County, OH............................ | 77,079 | 73,719 | 42540 | 42079 | Luzerne County, PA......................... | 320,918 | 316,982 |
| | | | | | 42540 | 42131 | Wyoming County, PA........................ | 28,276 | 26,557 |
| 41820 | | Sanford, NC Micro Area .................... | 57,866 | 62,353 | | | | | |
| 41820 | 37105 | Lee County, NC ............................ | 57,866 | 62,353 | 42620 | | Searcy, AR Micro Area ...................... | 77,076 | 78,729 |
| | | | | | 42620 | 05145 | White County, AR........................... | 77,076 | 78,729 |
| 41860 | | San Francisco-Oakland-Berkeley, CA | | | | | | | |
| | | Metro Area................................. | 4,335,391 | 4,696,902 | 42660 | | Seattle-Tacoma-Bellevue, WA Metro | | |
| | | | | | | | Area....................................... | 3,439,809 | 4,018,598 |
| 41860 | | Oakland-Berkeley-Livermore, CA Divi- | | | | | | | |
| | | sion 36084................................ | 2,559,296 | 2,814,656 | 42660 | | Seattle-Bellevue-Kent, WA Division | | |
| 41860 | 06001 | Alameda County, CA....................... | 1,510,271 | 1,662,323 | | | 42644...................................... | 2,644,584 | 3,104,708 |
| 41860 | 06013 | Contra Costa County, CA .................. | 1,049,025 | 1,152,333 | 42660 | 53033 | King County, WA........................... | 1,931,249 | 2,274,315 |
| | | | | | 42660 | 53061 | Snohomish County, WA.................... | 713,335 | 830,393 |
| 41860 | | San Francisco-San Mateo-Redwood | | | | | | | |
| | | City, CA Division 41884.................... | 1,523,686 | 1,624,914 | 42660 | | Tacoma-Lakewood, WA Division 45104 | 795,225 | 913,890 |
| 41860 | 06075 | San Francisco County, CA................. | 805,235 | 866,606 | 42660 | 53053 | Pierce County, WA.......................... | 795,225 | 913,890 |
| 41860 | 06081 | San Mateo County, CA .................... | 718,451 | 758,308 | | | | | |
| | | | | | 42680 | | Sebastian-Vero Beach, FL Metro Area.. | 138,028 | 162,518 |
| 41860 | | San Rafael, CA Division 42034 ............ | 252,409 | 257,332 | 42680 | 12061 | Indian River County, FL ................... | 138,028 | 162,518 |
| 41860 | 06041 | Marin County, CA.......................... | 252,409 | 257,332 | | | | | |
| | | | | | 42700 | | Sebring-Avon Park, FL Metro Area ....... | 98,786 | 106,639 |
| 41940 | | San Jose-Sunnyvale-Santa Clara, CA | | | 42700 | 12055 | Highlands County, FL ...................... | 98,786 | 106,639 |
| | | Metro Area................................. | 1,836,911 | 1,971,160 | | | | | |
| 41940 | 06069 | San Benito County, CA..................... | 55,269 | 64,055 | 42740 | | Sedalia, MO Micro Area ..................... | 42,201 | 42,490 |
| 41940 | 06085 | Santa Clara County, CA.................... | 1,781,642 | 1,907,105 | 42740 | 29159 | Pettis County, MO.......................... | 42,201 | 42,490 |
| 42020 | | San Luis Obispo-Paso Robles, CA | | | 42780 | | Selinsgrove, PA Micro Area................. | 39,702 | 40,317 |
| | | Metro Area................................. | 269,637 | 282,249 | 42780 | 42109 | Snyder County, PA ......................... | 39,702 | 40,317 |
| 42020 | 06079 | San Luis Obispo County, CA ............. | 269,637 | 282,249 | | | | | |
| | | | | | 42820 | | Selma, AL Micro Area........................ | 43,820 | 36,098 |
| 42100 | | Santa Cruz-Watsonville, CA Metro Area | 262,382 | 269,925 | 42820 | 01047 | Dallas County, AL .......................... | 43,820 | 36,098 |
| 42100 | 06087 | Santa Cruz County, CA..................... | 262,382 | 269,925 | | | | | |
| | | | | | 42860 | | Seneca, SC Micro Area....................... | 74,273 | 80,015 |
| 42140 | | Santa Fe, NM Metro Area.................. | 144,170 | 151,946 | 42860 | 45073 | Oconee County, SC ........................ | 74,273 | 80,015 |
| 42140 | 35049 | Santa Fe County, NM ...................... | 144,170 | 151,946 | | | | | |
| | | | | | 42900 | | Seneca Falls, NY Micro Area ............... | 35,251 | 33,991 |
| 42200 | | Santa Maria-Santa Barbara, CA Metro | | | 42900 | 36099 | Seneca County, NY ........................ | 35,251 | 33,991 |
| | | Area....................................... | 423,895 | 444,766 | | | | | |
| 42200 | 06083 | Santa Barbara County, CA ................ | 423,895 | 444,766 | 42940 | | Sevierville, TN Micro Area................... | 89,889 | 99,244 |
| | | | | | 42940 | 47155 | Sevier County, TN.......................... | 89,889 | 99,244 |
| 42220 | | Santa Rosa-Petaluma, CA Metro Area.. | 483,878 | 489,819 | | | | | |
| 42220 | 06097 | Sonoma County, CA ....................... | 483,878 | 489,819 | | | | | |

Appendix C

| Core based statistical area | State/County FIPS code | Title and Geographic Components | 2010 Census Population | 2020 Estimated Population | Core based statistical area | State/County FIPS code | Title and Geographic Components | 2010 Census Population | 2020 Estimated Population |
|---|---|---|---|---|---|---|---|---|---|
| 42980 | | Seymour, IN Micro Area | 42,376 | 44,222 | 43740 | | Somerset, PA Micro Area | 77,742 | 72,916 |
| 42980 | 18071 | Jackson County, IN | 42,376 | 44,222 | 43740 | 42111 | Somerset County, PA | 77,742 | 72,916 |
| 43020 | | Shawano, WI Micro Area | 46,181 | 45,332 | 43760 | | Sonora, CA Micro Area | 55,365 | 54,515 |
| 43020 | 55078 | Menominee County, WI | 4,232 | 4,546 | 43760 | 06109 | Tuolumne County, CA | 55,365 | 54,515 |
| 43020 | 55115 | Shawano County, WI | 41,949 | 40,786 | 43780 | | South Bend-Mishawaka, IN-MI Metro Area | 319,224 | 323,068 |
| 43060 | | Shawnee, OK Micro Area | 69,442 | 72,998 | 43780 | 18141 | St. Joseph County, IN | 266,931 | 271,484 |
| 43060 | 40125 | Pottawatomie County, OK | 69,442 | 72,998 | 43780 | 26027 | Cass County, MI | 52,293 | 51,584 |
| 43100 | | Sheboygan, WI Metro Area | 115,507 | 115,240 | 43900 | | Spartanburg, SC Metro Area | 284,307 | 326,205 |
| 43100 | 55117 | Sheboygan County, WI | 115,507 | 115,240 | 43900 | 45083 | Spartanburg County, SC | 284,307 | 326,205 |
| 43140 | | Shelby, NC Micro Area | 98,078 | 99,035 | 43940 | | Spearfish, SD Micro Area | 24,097 | 26,221 |
| 43140 | 37045 | Cleveland County, NC | 98,078 | 99,035 | 43940 | 46081 | Lawrence County, SD | 24,097 | 26,221 |
| 43180 | | Shelbyville, TN Micro Area | 45,058 | 50,179 | 43980 | | Spencer, IA Micro Area | 16,667 | 15,976 |
| 43180 | 47003 | Bedford County, TN | 45,058 | 50,179 | 43980 | 19041 | Clay County, IA | 16,667 | 15,976 |
| 43220 | | Shelton, WA Micro Area | 60,699 | 68,224 | 44020 | | Spirit Lake, IA Micro Area | 16,667 | 17,549 |
| 43220 | 53045 | Mason County, WA | 60,699 | 68,224 | 44020 | 19059 | Dickinson County, IA | 16,667 | 17,549 |
| 43260 | | Sheridan, WY Micro Area | 29,116 | 30,863 | 44060 | | Spokane-Spokane Valley, WA Metro Area | 514,752 | 574,585 |
| 43260 | 56033 | Sheridan County, WY | 29,116 | 30,863 | 44060 | 53063 | Spokane County, WA | 471,221 | 528,225 |
| 43300 | | Sherman-Denison, TX Metro Area | 120,877 | 138,318 | 44060 | 53065 | Stevens County, WA | 43,531 | 46,360 |
| 43300 | 48181 | Grayson County, TX | 120,877 | 138,318 | 44100 | | Springfield, IL Metro Area | 210,170 | 205,950 |
| 43320 | | Show Low, AZ Micro Area | 107,449 | 112,112 | 44100 | 17129 | Menard County, IL | 12,705 | 12,068 |
| 43320 | 04017 | Navajo County, AZ | 107,449 | 112,112 | 44100 | 17167 | Sangamon County, IL | 197,465 | 193,882 |
| 43340 | | Shreveport-Bossier City, LA Metro Area | 398,604 | 392,404 | 44140 | | Springfield, MA Metro Area | 692,942 | 695,654 |
| 43340 | 22015 | Bossier Parish, LA | 116,979 | 127,275 | 44140 | 25011 | Franklin County, MA | 71,372 | 70,267 |
| 43340 | 22017 | Caddo Parish, LA | 254,969 | 237,479 | 44140 | 25013 | Hampden County, MA | 463,490 | 463,986 |
| 43340 | 22031 | De Soto Parish, LA | 26,656 | 27,650 | 44140 | 25015 | Hampshire County, MA | 158,080 | 161,401 |
| 43380 | | Sidney, OH Micro Area | 49,423 | 48,337 | 44180 | | Springfield, MO Metro Area | 436,712 | 475,220 |
| 43380 | 39149 | Shelby County, OH | 49,423 | 48,337 | 44180 | 29043 | Christian County, MO | 77,422 | 90,655 |
| 43420 | | Sierra Vista-Douglas, AZ Metro Area | 131,346 | 127,450 | 44180 | 29059 | Dallas County, MO | 16,777 | 17,219 |
| 43420 | 04003 | Cochise County, AZ | 131,346 | 127,450 | 44180 | 29077 | Greene County, MO | 275,174 | 294,997 |
| 43460 | | Sikeston, MO Micro Area | 39,191 | 38,288 | 44180 | 29167 | Polk County, MO | 31,137 | 32,490 |
| 43460 | 29201 | Scott County, MO | 39,191 | 38,288 | 44180 | 29225 | Webster County, MO | 36,202 | 39,859 |
| 43500 | | Silver City, NM Micro Area | 29,514 | 27,007 | 44220 | | Springfield, OH Metro Area | 138,333 | 133,593 |
| 43500 | 35017 | Grant County, NM | 29,514 | 27,007 | 44220 | 39023 | Clark County, OH | 138,333 | 133,593 |
| 43580 | | Sioux City, IA-NE-SD Metro Area | 143,577 | 144,996 | 44260 | | Starkville, MS Micro Area | 57,924 | 59,465 |
| 43580 | 19193 | Woodbury County, IA | 102,172 | 103,138 | 44260 | 28105 | Oktibbeha County, MS | 47,671 | 49,789 |
| 43580 | 31043 | Dakota County, NE | 21,006 | 20,070 | 44260 | 28155 | Webster County, MS | 10,253 | 9,676 |
| 43580 | 31051 | Dixon County, NE | 6,000 | 5,596 | 44300 | | State College, PA Metro Area | 153,990 | 161,496 |
| 43580 | 46127 | Union County, SD | 14,399 | 16,192 | 44300 | 42027 | Centre County, PA | 153,990 | 161,496 |
| 43620 | | Sioux Falls, SD Metro Area | 228,261 | 273,566 | 44340 | | Statesboro, GA Micro Area | 70,217 | 80,839 |
| 43620 | 46083 | Lincoln County, SD | 44,828 | 63,019 | 44340 | 13031 | Bulloch County, GA | 70,217 | 80,839 |
| 43620 | 46087 | McCook County, SD | 5,618 | 5,520 | 44420 | | Staunton, VA Metro Area | 118,502 | 124,475 |
| 43620 | 46099 | Minnehaha County, SD | 169,468 | 196,659 | 44420 | 51015 | Augusta County, VA | 73,750 | 76,544 |
| 43620 | 46125 | Turner County, SD | 8,347 | 8,368 | 44420 | 51790 | Staunton city, VA | 23,746 | 25,190 |
| 43660 | | Snyder, TX Micro Area | 16,921 | 16,662 | 44420 | 51820 | Waynesboro city, VA | 21,006 | 22,741 |
| 43660 | 48415 | Scurry County, TX | 16,921 | 16,662 | 44460 | | Steamboat Springs, CO Micro Area | 23,509 | 25,560 |
| 43700 | | Somerset, KY Micro Area | 63,063 | 65,530 | 44460 | 08107 | Routt County, CO | 23,509 | 25,560 |
| 43700 | 21199 | Pulaski County, KY | 63,063 | 65,530 | | | | | |

# CORE BASED STATISTICAL AREAS
## (Metropolitan and Micropolitan),
## METROPOLITAN DIVISIONS, AND COMPONENTS
## (as defined March, 2020)—*Continued*

| Core based statistical area | State/County FIPS code | Title and Geographic Components | 2010 Census Population | 2020 Estimated Population | Core based statistical area | State/County FIPS code | Title and Geographic Components | 2010 Census Population | 2020 Estimated Population |
|---|---|---|---|---|---|---|---|---|---|
| 44500 | | Stephenville, TX Micro Area | 37,890 | 43,224 | 45340 | | Taos, NM Micro Area | 32,937 | 32,593 |
| 44500 | 48143 | Erath County, TX | 37,890 | 43,224 | 45340 | 35055 | Taos County, NM | 32,937 | 32,593 |
| 44540 | | Sterling, CO Micro Area | 22,709 | 21,974 | 45380 | | Taylorville, IL Micro Area | 34,800 | 32,075 |
| 44540 | 08075 | Logan County, CO | 22,709 | 21,974 | 45380 | 17021 | Christian County, IL | 34,800 | 32,075 |
| 44580 | | Sterling, IL Micro Area | 58,498 | 54,656 | 45460 | | Terre Haute, IN Metro Area | 189,764 | 185,632 |
| 44580 | 17195 | Whiteside County, IL | 58,498 | 54,656 | 45460 | 18021 | Clay County, IN | 26,890 | 26,246 |
| 44620 | | Stevens Point, WI Micro Area | 70,019 | 71,032 | 45460 | 18121 | Parke County, IN | 17,339 | 16,871 |
| 44620 | 55097 | Portage County, WI | 70,019 | 71,032 | 45460 | 18153 | Sullivan County, IN | 21,475 | 20,578 |
| 44660 | | Stillwater, OK Micro Area | 77,350 | 81,755 | 45460 | 18165 | Vermillion County, IN | 16,212 | 15,329 |
| 44660 | 40119 | Payne County, OK | 77,350 | 81,755 | 45460 | 18167 | Vigo County, IN | 107,848 | 106,608 |
| 44700 | | Stockton, CA Metro Area | 685,306 | 767,967 | 45500 | | Texarkana, TX-AR Metro Area | 149,198 | 148,838 |
| 44700 | 06077 | San Joaquin County, CA | 685,306 | 767,967 | 45500 | 48037 | Bowie County, TX | 92,565 | 93,481 |
| 44740 | | Storm Lake, IA Micro Area | 20,260 | 19,772 | 45500 | 05081 | Little River County, AR | 13,171 | 12,180 |
| 44740 | 19021 | Buena Vista County, IA | 20,260 | 19,772 | 45500 | 05091 | Miller County, AR | 43,462 | 43,177 |
| 44780 | | Sturgis, MI Micro Area | 61,295 | 60,848 | 45520 | | The Dalles, OR Micro Area | 25,213 | 26,403 |
| 44780 | 26149 | St. Joseph County, MI | 61,295 | 60,848 | 45520 | 41065 | Wasco County, OR | 25,213 | 26,403 |
| 44860 | | Sulphur Springs, TX Micro Area | 35,161 | 37,170 | 45540 | | The Villages, FL Metro Area | 93,420 | 139,018 |
| 44860 | 48223 | Hopkins County, TX | 35,161 | 37,170 | 45540 | 12119 | Sumter County, FL | 93,420 | 139,018 |
| 44900 | | Summerville, GA Micro Area | 26,015 | 24,843 | 45580 | | Thomaston, GA Micro Area | 27,153 | 26,527 |
| 44900 | 13055 | Chattooga County, GA | 26,015 | 24,843 | 45580 | 13293 | Upson County, GA | 27,153 | 26,527 |
| 44940 | | Sumter, SC Metro Area | 142,427 | 139,775 | 45620 | | Thomasville, GA Micro Area | 44,720 | 44,372 |
| 44940 | 45027 | Clarendon County, SC | 34,971 | 33,415 | 45620 | 13275 | Thomas County, GA | 44,720 | 44,372 |
| 44940 | 45085 | Sumter County, SC | 107,456 | 106,360 | 45660 | | Tiffin, OH Micro Area | 56,745 | 54,938 |
| 44980 | | Sunbury, PA Micro Area | 94,528 | 90,258 | 45660 | 39147 | Seneca County, OH | 56,745 | 54,938 |
| 44980 | 42097 | Northumberland County, PA | 94,528 | 90,258 | 45700 | | Tifton, GA Micro Area | 40,118 | 40,719 |
| 45000 | | Susanville, CA Micro Area | 34,895 | 30,016 | 45700 | 13277 | Tift County, GA | 40,118 | 40,719 |
| 45000 | 06035 | Lassen County, CA | 34,895 | 30,016 | 45740 | | Toccoa, GA Micro Area | 26,175 | 26,107 |
| 45020 | | Sweetwater, TX Micro Area | 15,216 | 14,835 | 45740 | 13257 | Stephens County, GA | 26,175 | 26,107 |
| 45020 | 48353 | Nolan County, TX | 15,216 | 14,835 | 45780 | | Toledo, OH Metro Area | 651,429 | 641,549 |
| 45060 | | Syracuse, NY Metro Area | 662,577 | 646,038 | 45780 | 39051 | Fulton County, OH | 42,698 | 41,889 |
| 45060 | 36053 | Madison County, NY | 73,442 | 70,478 | 45780 | 39095 | Lucas County, OH | 441,815 | 428,294 |
| 45060 | 36067 | Onondaga County, NY | 467,026 | 459,214 | 45780 | 39123 | Ottawa County, OH | 41,428 | 40,253 |
| 45060 | 36075 | Oswego County, NY | 122,109 | 116,346 | 45780 | 39173 | Wood County, OH | 125,488 | 131,113 |
| 45140 | | Tahlequah, OK Micro Area | 46,987 | 49,019 | 45820 | | Topeka, KS Metro Area | 233,870 | 230,878 |
| 45140 | 40021 | Cherokee County, OK | 46,987 | 49,019 | 45820 | 20085 | Jackson County, KS | 13,462 | 13,171 |
| 45180 | | Talladega-Sylacauga, AL Micro Area | 82,291 | 79,985 | 45820 | 20087 | Jefferson County, KS | 19,126 | 19,032 |
| 45180 | 01121 | Talladega County, AL | 82,291 | 79,985 | 45820 | 20139 | Osage County, KS | 16,295 | 15,770 |
| 45220 | | Tallahassee, FL Metro Area | 367,413 | 389,599 | 45820 | 20177 | Shawnee County, KS | 177,934 | 175,999 |
| 45220 | 12039 | Gadsden County, FL | 46,389 | 45,277 | 45820 | 20197 | Wabaunsee County, KS | 7,053 | 6,906 |
| 45220 | 12065 | Jefferson County, FL | 14,761 | 14,543 | 45860 | | Torrington, CT Micro Area | 189,927 | 179,610 |
| 45220 | 12073 | Leon County, FL | 275,487 | 295,460 | 45860 | 09005 | Litchfield County, CT | 189,927 | 179,610 |
| 45220 | 12129 | Wakulla County, FL | 30,776 | 34,319 | 45900 | | Traverse City, MI Micro Area | 143,372 | 151,190 |
| 45300 | | Tampa-St. Petersburg-Clearwater, FL Metro Area | 2,783,243 | 3,243,963 | 45900 | 26019 | Benzie County, MI | 17,525 | 17,852 |
| 45300 | 12053 | Hernando County, FL | 172,778 | 198,792 | 45900 | 26055 | Grand Traverse County, MI | 86,986 | 93,592 |
| 45300 | 12057 | Hillsborough County, FL | 1,229,226 | 1,497,957 | 45900 | 26079 | Kalkaska County, MI | 17,153 | 18,003 |
| 45300 | 12101 | Pasco County, FL | 464,697 | 570,412 | 45900 | 26089 | Leelanau County, MI | 21,708 | 21,743 |
| 45300 | 12103 | Pinellas County, FL | 916,542 | 976,802 | 45940 | | Trenton-Princeton, NJ Metro Area | 366,513 | 367,239 |
| | | | | | 45940 | 34021 | Mercer County, NJ | 366,513 | 367,239 |
| | | | | | 45980 | | Troy, AL Micro Area | 32,899 | 32,966 |
| | | | | | 45980 | 01109 | Pike County, AL | 32,899 | 32,966 |

# CORE BASED STATISTICAL AREAS
## (Metropolitan and Micropolitan),
## METROPOLITAN DIVISIONS, AND COMPONENTS
## (as defined March, 2020)—*Continued*

| Core based statistical area | State/County FIPS code | Title and Geographic Components | 2010 Census Population | 2020 Estimated Population | Core based statistical area | State/County FIPS code | Title and Geographic Components | 2010 Census Population | 2020 Estimated Population |
|---|---|---|---|---|---|---|---|---|---|
| 46020 | | Truckee-Grass Valley, CA Micro Area... | 98,764 | 99,606 | 46700 | | Vallejo, CA Metro Area | 413,344 | 446,935 |
| 46020 | 06057 | Nevada County, CA | 98,764 | 99,606 | 46700 | 06095 | Solano County, CA | 413,344 | 446,935 |
| | | | | | | | | | |
| 46060 | | Tucson, AZ Metro Area | 980,263 | 1,061,175 | 46780 | | Van Wert, OH Micro Area | 28,744 | 28,159 |
| 46060 | 04019 | Pima County, AZ | 980,263 | 1,061,175 | 46780 | 39161 | Van Wert County, OH | 28,744 | 28,159 |
| | | | | | | | | | |
| 46100 | | Tullahoma-Manchester, TN Micro Area. | 100,210 | 106,555 | 46820 | | Vermillion, SD Micro Area | 13,864 | 14,246 |
| 46100 | 47031 | Coffee County, TN | 52,796 | 57,632 | 46820 | 46027 | Clay County, SD | 13,864 | 14,246 |
| 46100 | 47051 | Franklin County, TN | 41,052 | 42,485 | | | | | |
| 46100 | 47127 | Moore County, TN | 6,362 | 6,438 | 46860 | | Vernal, UT Micro Area | 32,588 | 35,970 |
| | | | | | 46860 | 49047 | Uintah County, UT | 32,588 | 35,970 |
| 46140 | | Tulsa, OK Metro Area | 937,478 | 1,006,411 | | | | | |
| 46140 | 40037 | Creek County, OK | 69,967 | 71,485 | 46900 | | Vernon, TX Micro Area | 13,535 | 12,552 |
| 46140 | 40111 | Okmulgee County, OK | 40,069 | 38,234 | 46900 | 48487 | Wilbarger County, TX | 13,535 | 12,552 |
| 46140 | 40113 | Osage County, OK | 47,472 | 46,642 | | | | | |
| 46140 | 40117 | Pawnee County, OK | 16,577 | 16,381 | 46980 | | Vicksburg, MS Micro Area | 48,773 | 44,841 |
| 46140 | 40131 | Rogers County, OK | 86,905 | 93,155 | 46980 | 28149 | Warren County, MS | 48,773 | 44,841 |
| 46140 | 40143 | Tulsa County, OK | 603,403 | 657,589 | | | | | |
| 46140 | 40145 | Wagoner County, OK | 73,085 | 82,925 | 47020 | | Victoria, TX Metro Area | 94,003 | 99,562 |
| | | | | | 47020 | 48175 | Goliad County, TX | 7,210 | 7,626 |
| 46180 | | Tupelo, MS Micro Area | 161,544 | 166,201 | 47020 | 48469 | Victoria County, TX | 86,793 | 91,936 |
| 46180 | 28057 | Itawamba County, MS | 23,401 | 23,261 | | | | | |
| 46180 | 28081 | Lee County, MS | 82,910 | 85,466 | 47080 | | Vidalia, GA Micro Area | 36,346 | 35,985 |
| 46180 | 28115 | Pontotoc County, MS | 29,957 | 32,461 | 47080 | 13209 | Montgomery County, GA | 9,123 | 9,012 |
| 46180 | 28117 | Prentiss County, MS | 25,276 | 25,013 | 47080 | 13279 | Toombs County, GA | 27,223 | 26,973 |
| | | | | | | | | | |
| 46220 | | Tuscaloosa, AL Metro Area | 239,207 | 253,211 | 47180 | | Vincennes, IN Micro Area | 38,440 | 36,522 |
| 46220 | 01063 | Greene County, AL | 9,045 | 7,990 | 47180 | 18083 | Knox County, IN | 38,440 | 36,522 |
| 46220 | 01065 | Hale County, AL | 15,760 | 14,670 | | | | | |
| 46220 | 01107 | Pickens County, AL | 19,746 | 19,793 | 47220 | | Vineland-Bridgeton, NJ Metro Area | 156,898 | 147,008 |
| 46220 | 01125 | Tuscaloosa County, AL | 194,656 | 210,758 | 47220 | 34011 | Cumberland County, NJ | 156,898 | 147,008 |
| | | | | | | | | | |
| 46300 | | Twin Falls, ID Metro Area | 99,604 | 112,989 | 47240 | | Vineyard Haven, MA Micro Area | 16,535 | 17,461 |
| 46300 | 16053 | Jerome County, ID | 22,374 | 24,578 | 47240 | 25007 | Dukes County, MA | 16,535 | 17,461 |
| 46300 | 16083 | Twin Falls County, ID | 77,230 | 88,411 | | | | | |
| | | | | | 47260 | | Virginia Beach-Norfolk-Newport News, VA-NC Metro Area | 1,713,954 | 1,779,824 |
| 46340 | | Tyler, TX Metro Area | 209,714 | 235,806 | 47260 | 37029 | Camden County, NC | 9,980 | 10,984 |
| 46340 | 48423 | Smith County, TX | 209,714 | 235,806 | 47260 | 37053 | Currituck County, NC | 23,547 | 29,052 |
| | | | | | 47260 | 37073 | Gates County, NC | 12,197 | 11,464 |
| 46380 | | Ukiah, CA Micro Area | 87,841 | 86,061 | 47260 | 51073 | Gloucester County, VA | 36,858 | 37,459 |
| 46380 | 06045 | Mendocino County, CA | 87,841 | 86,061 | 47260 | 51093 | Isle of Wight County, VA | 35,270 | 37,725 |
| | | | | | 47260 | 51095 | James City County, VA | 67,009 | 77,612 |
| 46420 | | Union, SC Micro Area | 28,961 | 26,991 | 47260 | 51115 | Mathews County, VA | 8,978 | 8,766 |
| 46420 | 45087 | Union County, SC | 28,961 | 26,991 | 47260 | 51175 | Southampton County, VA | 18,570 | 17,636 |
| | | | | | 47260 | 51199 | York County, VA | 65,464 | 69,199 |
| 46460 | | Union City, TN Micro Area | 31,807 | 30,131 | 47260 | 51550 | Chesapeake city, VA | 222,209 | 247,011 |
| 46460 | 47131 | Obion County, TN | 31,807 | 30,131 | 47260 | 51620 | Franklin city, VA | 8,582 | 7,833 |
| | | | | | 47260 | 51650 | Hampton city, VA | 137,436 | 135,464 |
| 46500 | | Urbana, OH Micro Area | 40,097 | 38,960 | 47260 | 51700 | Newport News city, VA | 180,719 | 179,062 |
| 46500 | 39021 | Champaign County, OH | 40,097 | 38,960 | 47260 | 51710 | Norfolk city, VA | 242,803 | 242,803 |
| | | | | | 47260 | 51735 | Poquoson city, VA | 12,150 | 12,257 |
| 46520 | | Urban Honolulu, HI Metro Area | 953,207 | 963,826 | 47260 | 51740 | Portsmouth city, VA | 95,535 | 95,094 |
| 46520 | 15003 | Honolulu County, HI | 953,207 | 963,826 | 47260 | 51800 | Suffolk city, VA | 84,585 | 93,913 |
| | | | | | 47260 | 51810 | Virginia Beach city, VA | 437,994 | 451,231 |
| 46540 | | Utica-Rome, NY Metro Area | 299,397 | 288,291 | 47260 | 51830 | Williamsburg city, VA | 14,068 | 15,259 |
| 46540 | 36043 | Herkimer County, NY | 64,519 | 60,945 | | | | | |
| 46540 | 36065 | Oneida County, NY | 234,878 | 227,346 | 47300 | | Visalia, CA Metro Area | 442,179 | 468,680 |
| | | | | | 47300 | 06107 | Tulare County, CA | 442,179 | 468,680 |
| 46620 | | Uvalde, TX Micro Area | 26,405 | 26,742 | | | | | |
| 46620 | 48463 | Uvalde County, TX | 26,405 | 26,742 | 47340 | | Wabash, IN Micro Area | 32,888 | 30,784 |
| | | | | | 47340 | 18169 | Wabash County, IN | 32,888 | 30,784 |
| 46660 | | Valdosta, GA Metro Area | 139,588 | 148,364 | | | | | |
| 46660 | 13027 | Brooks County, GA | 16,243 | 15,357 | 47380 | | Waco, TX Metro Area | 252,772 | 277,005 |
| 46660 | 13101 | Echols County, GA | 4,034 | 4,002 | 47380 | 48145 | Falls County, TX | 17,866 | 17,275 |
| 46660 | 13173 | Lanier County, GA | 10,078 | 10,737 | 47380 | 48309 | McLennan County, TX | 234,906 | 259,730 |
| 46660 | 13185 | Lowndes County, GA | 109,233 | 118,268 | | | | | |

| Core based statistical area | State/ County FIPS code | Title and Geographic Components | 2010 Census Population | 2020 Estimated Population | Core based statistical area | State/ County FIPS code | Title and Geographic Components | 2010 Census Population | 2020 Estimated Population |
|---|---|---|---|---|---|---|---|---|---|
| 47420 | | Wahpeton, ND-MN Micro Area............. | 22,897 | 22,317 | 47940 | | Waterloo-Cedar Falls, IA Metro Area .... | 167,819 | 168,314 |
| 47420 | 27167 | Wilkin County, MN ........................... | 6,576 | 6,161 | 47940 | 19013 | Black Hawk County, IA ...................... | 131,090 | 130,786 |
| 47420 | 38077 | Richland County, ND ........................ | 16,321 | 16,156 | 47940 | 19017 | Bremer County, IA ........................... | 24,276 | 25,311 |
| | | | | | 47940 | 19075 | Grundy County, IA ........................... | 12,453 | 12,217 |
| 47460 | | Walla Walla, WA Metro Area............... | 58,781 | 61,292 | | | | | |
| 47460 | 53071 | Walla Walla County, WA.................... | 58,781 | 61,292 | 47980 | | Watertown, SD Micro Area ................... | 33,130 | 34,420 |
| | | | | | 47980 | 46029 | Codington County, SD ...................... | 27,227 | 28,186 |
| 47540 | | Wapakoneta, OH Micro Area............... | 45,949 | 45,680 | 47980 | 46057 | Hamlin County, SD .......................... | 5,903 | 6,234 |
| 47540 | 39011 | Auglaize County, OH ........................ | 45,949 | 45,680 | | | | | |
| | | | | | 48020 | | Watertown-Fort Atkinson, WI Micro Area........ | 83,686 | 85,038 |
| 47580 | | Warner Robins, GA Metro Area............. | 167,595 | 188,060 | 48020 | 55055 | Jefferson County, WI ........................ | 83,686 | 85,038 |
| 47580 | 13153 | Houston County, GA ......................... | 139,900 | 160,110 | | | | | |
| 47580 | 13225 | Peach County, GA ........................... | 27,695 | 27,950 | 48060 | | Watertown-Fort Drum, NY Metro Area .. | 116,229 | 108,095 |
| | | | | | 48060 | 36045 | Jefferson County, NY........................ | 116,229 | 108,095 |
| 47620 | | Warren, PA Micro Area...................... | 41,815 | 38,911 | | | | | |
| 47620 | 42123 | Warren County, PA ........................... | 41,815 | 38,911 | 48100 | | Wauchula, FL Micro Area ................... | 27,731 | 26,822 |
| | | | | | 48100 | 12049 | Hardee County, FL........................... | 27,731 | 26,822 |
| 47660 | | Warrensburg, MO Micro Area............... | 52,595 | 54,219 | | | | | |
| 47660 | 29101 | Johnson County, MO ........................ | 52,595 | 54,219 | 48140 | | Wausau-Weston, WI Metro Area........... | 162,806 | 163,159 |
| | | | | | 48140 | 55069 | Lincoln County, WI ........................... | 28,743 | 27,566 |
| 47700 | | Warsaw, IN Micro Area...................... | 77,358 | 78,988 | 48140 | 55073 | Marathon County, WI......................... | 134,063 | 135,593 |
| 47700 | 18085 | Kosciusko County, IN........................ | 77,358 | 78,988 | | | | | |
| | | | | | 48180 | | Waycross, GA Micro Area .................... | 55,070 | 55,348 |
| 47780 | | Washington, IN Micro Area................. | 31,648 | 33,505 | 48180 | 13229 | Pierce County, GA ........................... | 18,758 | 19,522 |
| 47780 | 18027 | Daviess County, IN .......................... | 31,648 | 33,505 | 48180 | 13299 | Ware County, GA............................. | 36,312 | 35,826 |
| | | | | | | | | | |
| 47820 | | Washington, NC Micro Area ................ | 47,759 | 47,073 | 48220 | | Weatherford, OK Micro Area ................ | 27,469 | 28,648 |
| 47820 | 37013 | Beaufort County, NC......................... | 47,759 | 47,073 | 48220 | 40039 | Custer County, OK........................... | 27,469 | 28,648 |
| | | | | | | | | | |
| 47900 | | Washington-Arlington-Alexandria, DC-VA-MD-WV Metro Area ...................... | 5,649,540 | 6,324,629 | 48260 | | Weirton-Steubenville, WV-OH Metro Area............ | 124,454 | 115,184 |
| | | | | | 48260 | 39081 | Jefferson County, OH ....................... | 69,709 | 64,939 |
| 47900 | | Frederick-Gaithersburg-Rockville, MD Division 23224 | 1,205,162 | 1,316,977 | 48260 | 54009 | Brooke County, WV .......................... | 24,069 | 21,674 |
| 47900 | 24021 | Frederick County, MD ....................... | 233,385 | 265,161 | 48260 | 54029 | Hancock County, WV......................... | 30,676 | 28,571 |
| 47900 | 24031 | Montgomery County, MD.................... | 971,777 | 1,051,816 | | | | | |
| | | | | | 48300 | | Wenatchee, WA Metro Area................. | 110,884 | 121,134 |
| 47900 | | Washington-Arlington-Alexandria, DC-VA-MD-WV Division 47894 ............... | 4,444,378 | 5,007,652 | 48300 | 53007 | Chelan County, WA .......................... | 72,453 | 77,574 |
| 47900 | 11001 | District of Columbia, DC.................... | 601,723 | 712,816 | 48300 | 53017 | Douglas County, WA......................... | 38,431 | 43,560 |
| 47900 | 24009 | Calvert County, MD........................... | 88,737 | 93,072 | | | | | |
| 47900 | 24017 | Charles County, MD.......................... | 146,551 | 164,436 | 48460 | | West Plains, MO Micro Area ................ | 40,400 | 40,262 |
| 47900 | 24033 | Prince George's County, MD............... | 863,420 | 909,612 | 48460 | 29091 | Howell County, MO........................... | 40,400 | 40,262 |
| 47900 | 51013 | Arlington County, VA......................... | 207,627 | 240,119 | | | | | |
| 47900 | 51043 | Clarke County, VA ........................... | 14,034 | 14,622 | 48500 | | West Point, MS Micro Area ................. | 20,634 | 19,352 |
| 47900 | 51047 | Culpeper County, VA......................... | 46,689 | 53,569 | 48500 | 28025 | Clay County, MS ............................. | 20,634 | 19,352 |
| 47900 | 51059 | Fairfax County, VA........................... | 1,081,726 | 1,150,847 | | | | | |
| 47900 | 51061 | Fauquier County, VA ........................ | 65,203 | 71,361 | 48540 | | Wheeling, WV-OH Metro Area .............. | 147,950 | 137,217 |
| 47900 | 51107 | Loudoun County, VA......................... | 312,311 | 422,784 | 48540 | 39013 | Belmont County, OH ........................ | 70,400 | 65,932 |
| 47900 | 51113 | Madison County, VA......................... | 13,308 | 13,312 | 48540 | 54051 | Marshall County, WV ........................ | 33,107 | 30,103 |
| 47900 | 51153 | Prince William County, VA................. | 402,002 | 475,533 | 48540 | 54069 | Ohio County, WV............................. | 44,443 | 41,182 |
| 47900 | 51157 | Rappahannock County, VA................. | 7,373 | 7,260 | | | | | |
| 47900 | 51177 | Spotsylvania County, VA................... | 122,397 | 138,449 | 48580 | | Whitewater, WI Micro Area.................. | 102,228 | 103,953 |
| 47900 | 51179 | Stafford County, VA.......................... | 128,961 | 156,748 | 48580 | 55127 | Walworth County, WI......................... | 102,228 | 103,953 |
| 47900 | 51187 | Warren County, VA........................... | 37,575 | 40,475 | | | | | |
| 47900 | 51510 | Alexandria city, VA.......................... | 139,966 | 158,726 | 48620 | | Wichita, KS Metro Area...................... | 623,061 | 643,768 |
| 47900 | 51600 | Fairfax city, VA............................... | 22,565 | 23,429 | 48620 | 20015 | Butler County, KS ........................... | 65,880 | 66,992 |
| 47900 | 51610 | Falls Church city, VA........................ | 12,332 | 14,631 | 48620 | 20079 | Harvey County, KS .......................... | 34,684 | 34,291 |
| 47900 | 51630 | Fredericksburg city, VA..................... | 24,286 | 29,492 | 48620 | 20173 | Sedgwick County, KS ....................... | 498,365 | 519,907 |
| 47900 | 51683 | Manassas city, VA ........................... | 37,821 | 40,869 | 48620 | 20191 | Sumner County, KS .......................... | 24,132 | 22,578 |
| 47900 | 51685 | Manassas Park city, VA..................... | 14,273 | 18,004 | | | | | |
| 47900 | 54037 | Jefferson County, WV........................ | 53,498 | 57,486 | 48660 | | Wichita Falls, TX Metro Area................ | 151,306 | 152,485 |
| | | | | | 48660 | 48009 | Archer County, TX ........................... | 9,054 | 8,730 |
| 47920 | | Washington Court House, OH Micro Area.............. | 29,030 | 28,579 | 48660 | 48077 | Clay County, TX.............................. | 10,752 | 10,550 |
| 47920 | 39047 | Fayette County, OH ......................... | 29,030 | 28,579 | 48660 | 48485 | Wichita County, TX........................... | 131,500 | 133,205 |

| Core based statisti-cal area | State/ County FIPS code | Title and Geographic Components | 2010 Census Population | 2020 Estimated Population | Core based statisti-cal area | State/ County FIPS code | Title and Geographic Components | 2010 Census Population | 2020 Estimated Population |
|---|---|---|---|---|---|---|---|---|---|
| 48700 | | Williamsport, PA Metro Area ................ | 116,111 | 113,209 | 49260 | | Woodward, OK Micro Area.................. | 24,232 | 23,642 |
| 48700 | 42081 | Lycoming County, PA ...................... | 116,111 | 113,209 | 49260 | 40045 | Ellis County, OK............................ | 4,151 | 3,830 |
| | | | | | 49260 | 40153 | Woodward County, OK ..................... | 20,081 | 19,812 |
| 48780 | | Williston, ND Micro Area...................... | 22,398 | 38,700 | | | | | |
| 48780 | 38105 | Williams County, ND ...................... | 22,398 | 38,700 | 49300 | | Wooster, OH Micro Area ..................... | 114,520 | 115,694 |
| | | | | | 49300 | 39169 | Wayne County, OH.......................... | 114,520 | 115,694 |
| 48820 | | Willmar, MN Micro Area....................... | 42,239 | 43,130 | | | | | |
| 48820 | 27067 | Kandiyohi County, MN ..................... | 42,239 | 43,130 | 49340 | | Worcester, MA-CT Metro Area.............. | 916,980 | 945,752 |
| | | | | | 49340 | 25027 | Worcester County, MA...................... | 798,552 | 829,212 |
| 48900 | | Wilmington, NC Metro Area .................. | 254,884 | 301,284 | 49340 | 09015 | Windham County, CT........................ | 118,428 | 116,540 |
| 48900 | 37129 | New Hanover County, NC.................. | 202,667 | 236,613 | | | | | |
| 48900 | 37141 | Pender County, NC.......................... | 52,217 | 64,671 | 49380 | | Worthington, MN Micro Area .............. | 21,378 | 21,400 |
| | | | | | 49380 | 27105 | Nobles County, MN......................... | 21,378 | 21,400 |
| 48940 | | Wilmington, OH Micro Area ................. | 42,040 | 41,921 | | | | | |
| 48940 | 39027 | Clinton County, OH ........................ | 42,040 | 41,921 | 49420 | | Yakima, WA Metro Area ..................... | 243,231 | 251,879 |
| | | | | | 49420 | 53077 | Yakima County, WA......................... | 243,231 | 251,879 |
| 48980 | | Wilson, NC Micro Area ....................... | 81,234 | 81,979 | | | | | |
| 48980 | 37195 | Wilson County, NC........................... | 81,234 | 81,979 | 49460 | | Yankton, SD Micro Area .................... | 22,438 | 22,742 |
| | | | | | 49460 | 46135 | Yankton County, SD ........................ | 22,438 | 22,742 |
| 49020 | | Winchester, VA-WV Metro Area ........... | 128,472 | 142,009 | | | | | |
| 49020 | 51069 | Frederick County, VA....................... | 78,305 | 91,119 | 49620 | | York-Hanover, PA Metro Area.............. | 434,972 | 450,448 |
| 49020 | 51840 | Winchester city, VA......................... | 26,203 | 27,700 | 49620 | 42133 | York County, PA ............................ | 434,972 | 450,448 |
| 49020 | 54027 | Hampshire County, WV .................... | 23,964 | 23,190 | | | | | |
| | | | | | 49660 | | Youngstown-Warren-Boardman, OH-PA Metro Area ......................... | 565,773 | 531,420 |
| 49060 | | Winfield, KS Micro Area...................... | 36,311 | 34,628 | 49660 | 39099 | Mahoning County, OH ..................... | 238,823 | 226,075 |
| 49060 | 20035 | Cowley County, KS ........................ | 36,311 | 34,628 | 49660 | 39155 | Trumbull County, OH ...................... | 210,312 | 196,800 |
| | | | | | 49660 | 42085 | Mercer County, PA.......................... | 116,638 | 108,545 |
| 49080 | | Winnemucca, NV Micro Area ................ | 16,528 | 16,962 | | | | | |
| 49080 | 32013 | Humboldt County, NV ...................... | 16,528 | 16,962 | 49700 | | Yuba City, CA Metro Area ................... | 166,892 | 176,545 |
| | | | | | 49700 | 06101 | Sutter County, CA .......................... | 94,737 | 96,385 |
| 49100 | | Winona, MN Micro Area ...................... | 51,461 | 50,485 | 49700 | 06115 | Yuba County, CA ........................... | 72,155 | 80,160 |
| 49100 | 27169 | Winona County, MN ........................ | 51,461 | 50,485 | | | | | |
| | | | | | 49740 | | Yuma, AZ Metro Area........................ | 195,751 | 217,824 |
| 49180 | | Winston-Salem, NC Metro Area ........... | 640,595 | 679,731 | 49740 | 04027 | Yuma County, AZ........................... | 195,751 | 217,824 |
| 49180 | 37057 | Davidson County, NC ...................... | 162,878 | 169,234 | | | | | |
| 49180 | 37059 | Davie County, NC .......................... | 41,240 | 43,286 | 49780 | | Zanesville, OH Micro Area.................. | 86,074 | 86,020 |
| 49180 | 37067 | Forsyth County, NC ........................ | 350,670 | 383,843 | 49780 | 39119 | Muskingum County, OH................... | 86,074 | 86,020 |
| 49180 | 37169 | Stokes County, NC ......................... | 47,401 | 45,743 | | | | | |
| 49180 | 37197 | Yadkin County, NC ......................... | 38,406 | 37,625 | 49820 | | Zapata, TX Micro Area ..................... | 14,018 | 14,172 |
| | | | | | 49820 | 48505 | Zapata County, TX........................... | 14,018 | 14,172 |
| 49220 | | Wisconsin Rapids-Marshfield, WI Micro Area........................................ | 74,749 | 72,560 | | | | | |
| 49220 | 55141 | Wood County, WI............................ | 74,749 | 72,560 | | | | | |

# APPENDIX D
# CHANGES TO METROPOLITAN AREAS IN THE DELINEATIONS OF OMB BULLETIN 18-04

A map of the Core Based Statistical Areas from OMB Bulletin 18-04 is available at
https://www2.census.gov/geo/maps/metroarea/us_wall/Sep2018/ CBSA_WallMap_Sep2018.pdf?#

**Albany, GA** lost Baker county

**Albany-Lebanon, OR** has a new name

**Ames, IA** added Boone county

**Atlanta-Sandy Springs-Alpharetta, GA** has a new name

**Baton Rouge, LA** added Assumption parish

**Beaumont-Port Arthur, TX** lost Newton county

**Bend, OR** has a new name

**Billings, MT** added Stillwater county and lost Golden Valley county

**Birmingham-Hoover, AL** lost Walker county

**Bismarck, ND** lost Sioux county

**Blacksburg-Christiansburg, VA** has a new name and lost Floyd county

**Bloomington, IL** lost De Witt county

**Bremerton-Silverdale-Port Orchard, WA** has a new name

**Buffalo-Cheektowaga, NY** has a new name

**Carbondale-Marion, IL** added Johnson county

**Champaign-Urbana, IL** lost Ford county

**Charleston, WV** added Jackson and Lincoln counties

**Charlotte-Concord-Gastonia, NC-SC** added Anson county, NC

**The Chicago-Naperville-Evanston, IL** Division in the Chicago-Naperville-Elgin, IL-IN-WI Metro area has a new name and new CBSA code; this Division lost Kendall county which has been added to the Elgin, IL Division.

**Cincinnati, OH-KY-IN** added Franklin, IN county

**Clarksville, TN-KY** added Stewart, TN county

**Columbus, GA-AL** added Stewart and Talbot, GA counties

**Corpus Christi, TX** lost Aransas county

**The Fort Worth-Arlington, TX** Division in the Dallas-Fort Worth-Arlington, TX Metro area lost Hood and Somervell counties

**Dayton-Kettering, OH** has a new name and a new CBSA code

**Des Moines-West Des Moines, IA** added Jasper county

**Duluth, MN-WI added Lake, MN** county

**Durham-Chapel Hill, NC** added Granville county

**Fayetteville, NC** added Harnett county

**Fayetteville-Springdale-Rogers, AR** lost McDonald, MO county

**Fort Smith, AR-OK added Franklin, AR** county and lost Le Flore, OK county

**Fort Wayne, IN** lost Wells county

**Gainesville, FL** added Levy county

**Grand Island, NE** lost Hamilton county

**Grand Rapids-Kentwood, MI** has a new name and added Ionia county

**Greenville-Anderson, SC** has a new name

**Gulfport-Biloxi, MS** has a new name and added Stone county

**Hagerstown-Martinsburg, MD-WV** added Morgan, WV county

**Hattiesburg, MS** added Covington county

**Hilton Head Island-Bluffton, NC** has a new name

**Huntington-Ashland, WV-KY-OH** added Carter, KY county and lost Lincoln, WV county

**Jackson, MS** added Holmes county

**Jackson, TN** added Gibson county

**Kahului-Wailuku-Lahaina, HI** lost Kalawao county

**Kalamazoo-Portage, MI** lost Van Buren county

**Kingsport-Bristol, TN-VA** has a new name

**Lafayette-West Lafayette, IN** added Warren county

**Lansing-East Lansing, MI** added Shiawassee county

**Longview, TX** added Harrison county

**Louisville/Jefferson County, KY-IN** lost Scott, IN county and Trimble, KY county

**Lynchburg, VA** lost Bedford city but it stayed in the Metro area because it was absorbed by Bedford county

**Macon-Bibb County, GA** has a new name

**Manhattan, KA** added Geary county

**Memphis, TN-MS-AR** lost Benton, MS county

**Miami-Fort Lauderdale-Pompano Beach, FL** has a new name, as do two of its three Divisions: Fort Lauderdale-Pompano Beach-Sunrise Division and West Palm Beach-Boca Raton-Boynton Beach Division.

**Minneapolis-St. Paul-Bloomington, MN-WI** lost Sibley, MN county

**Mobile, Al** added Washington county

**Monroe, LA** added Morehouse county

**Morrison, TN** added Granger county

**Naples-Marco Island, FL** has a new name

**Nashville-Davidson—Murfreesboro-Franklin, TN** lost Hickman county

**The New York-Newark-Jersey City, NY-NJ-PA** Metro Area lost the Dutchess County-Putnam County Division, with Dutchess county moving out of the Metro area to the new Poughkeepsie-Newburgh-Middletown, NY Metro Area. The Newark, NJ-PA Division lost Somerset county. New Brunswick-Lakewood, NJ was established as a new Division, with Middlesex, Monmouth, Ocean, and Somerset counties. The New York-Jersey City-White Plains, NY-NJ

Division lost Middlesex, Monmouth, and Ocean, NJ counties and Orange, NY county and added Putnam, NY county

**Olympia-Lacey-Tumwater, WA** has a new name

**Panama City, FL** lost Gulf county

**Peoria, IL** added Fulton county

**Phoenix-Mesa-Chandler, AZ** has a new name

**Pocatello, ID** added Power county

**Prescott Valley-Prescott, AZ** has a new name

**Poughkeepsie-Newburgh-Middletown, NY** is a new Metropolitan Area, consisting of Dutchess and Orange counties, previously in the New York Metropolitan Area

**Raleigh-Cary, NC** has a new name

**Rapid City, SD** lost Custer county

**Richmond, VA** added King and Queen county

**Sacramento-Roseville-Folsom, CA** has a new name

**St Louis, MO-IL** added the part of Sullivan city that is in Crawford, MO county

**San Angelo, TX** added Sterling county

**San Diego-Chula Vista-Carlsbad, CA** has a new name

**San Francisco-Oakland-Berkeley, CA** has a new name, as do two of its Divisions: Oakland-Berkeley-Livermore, CA and San Francisco-Redwood City-San Mateo Divisions

**San Luis Obispo-Paso Robles, CA** has a new name

**Santa Rosa-Petaluma, CA** has a new name

**Scranton-Wilkes Barre, PA** has a new name

**The Seattle-Bellevue-Kent, WA** Division, in the Seattle-Tacoma-Bellevue, WA Metropolitan area, has a new name

**Sebring-Avon Park, FL** has a new name

**Shreveport-Bossier City, LA** lost Webster parish

**Sioux City, IA-NE-SD** lost Plymouth, IA county

**Spartanburg, SC** lost Union county

**Spokane-Spokane Valley, WA** dopped Pend Oreille county

**Springfield, MA** added Franklin county

**Staunton, VA** has a new name

**Stockton, CA** has a new name

**Sumter, SC** added Clarendon county

**Terre Haute, IN** added Parke county

**Toledo, OH** added Ottawa county

**Trenton-Princeton, NJ** has a new name

**Tuscaloosa, AL** added Greene county

**Twin Falls, ID** formerly a Micropolitan area, became a Metropolitan area consisting of Jerome and Twin Falls counties

**Vallejo, CA** has a new name

**Virginia Beach-Norfolk-Newport News, VA-NC** added Camden, NC county and Southampton County and Franklin City, VA

**Visalia, CA** has a new name

**Walla Walla, WA** lost Columbia county

**Warner Robins, GA** lost Pulaski county

**The Frederick-Gaithersburg-Rockville, MD** Division, in the Washington-Arlington-Alexandria, DC-VA-MD-WV Metropolitan area, has a new name and Division code, and the Washington-Arlington-Alexandria, DC-VA-MD-WV Division added Madison county, VA

**Wausau-Weston, WI** added Lincoln county

**Wichita, KS** lost Kingman county

# APPENDIX E
# CITIES BY COUNTY

The following table is arranged alphabetically by state. Under each state heading are listed all cities with a 2010 census population of 25,000 or more, along with their component counties and the population in each component.

| State code | Place code | County code | Geographic Area Name | 2010 census population |
|---|---|---|---|---|
| 01 | | | **ALABAMA** | 4,779,736 |
| 01 | | 00820 | Alabaster city | 30,352 |
| 01 | 117 | 00820 | Shelby County | 30,352 |
| 01 | | 03076 | Auburn city | 53,380 |
| 01 | 081 | 03076 | Lee County | 53,380 |
| 01 | | 05980 | Bessemer city | 27,456 |
| 01 | 073 | 05980 | Jefferson County | 27,456 |
| 01 | | 07000 | Birmingham city | 212,237 |
| 01 | 073 | 07000 | Jefferson County | 210,609 |
| 01 | 117 | 07000 | Shelby County | 1,628 |
| 01 | | 20104 | Decatur city | 55,683 |
| 01 | 083 | 20104 | Limestone County | 84 |
| 01 | 103 | 20104 | Morgan County | 55,599 |
| 01 | | 21184 | Dothan city | 65,496 |
| 01 | 045 | 21184 | Dale County | 887 |
| 01 | 067 | 21184 | Henry County | 5 |
| 01 | 069 | 21184 | Houston County | 64,604 |
| 01 | | 24184 | Enterprise city | 26,562 |
| 01 | 031 | 24184 | Coffee County | 26,139 |
| 01 | 045 | 24184 | Dale County | 423 |
| 01 | | 26896 | Florence city | 39,319 |
| 01 | 077 | 26896 | Lauderdale County | 39,319 |
| 01 | | 28696 | Gadsden city | 36,856 |
| 01 | 055 | 28696 | Etowah County | 36,856 |
| 01 | | 35800 | Homewood city | 25,167 |
| 01 | 073 | 35800 | Jefferson County | 25,167 |
| 01 | | 35896 | Hoover city | 81,619 |
| 01 | 073 | 35896 | Jefferson County | 58,582 |
| 01 | 117 | 35896 | Shelby County | 23,037 |
| 01 | | 37000 | Huntsville city | 180,105 |
| 01 | 083 | 37000 | Limestone County | 1,521 |
| 01 | 089 | 37000 | Madison County | 178,584 |
| 01 | | 45784 | Madison city | 42,938 |
| 01 | 083 | 45784 | Limestone County | 3,453 |
| 01 | 089 | 45784 | Madison County | 39,485 |
| 01 | | 50000 | Mobile city | 195,111 |
| 01 | 097 | 50000 | Mobile County | 195,111 |
| 01 | | 51000 | Montgomery city | 205,764 |
| 01 | 101 | 51000 | Montgomery County | 205,764 |
| 01 | | 57048 | Opelika city | 26,477 |
| 01 | 081 | 57048 | Lee County | 26,477 |
| 01 | | 59472 | Phenix City city | 32,822 |
| 01 | 081 | 59472 | Lee County | 4,153 |
| 01 | 113 | 59472 | Russell County | 28,669 |
| 01 | | 62328 | Prattville city | 33,960 |
| 01 | 001 | 62328 | Autauga County | 32,168 |
| 01 | 051 | 62328 | Elmore County | 1,792 |
| 01 | | 77256 | Tuscaloosa city | 90,468 |
| 01 | 125 | 77256 | Tuscaloosa County | 90,468 |
| 01 | | 78552 | Vestavia Hills city | 34,033 |
| 01 | 073 | 78552 | Jefferson County | 34,019 |
| 01 | 117 | 78552 | Shelby County | 14 |
| 02 | | | **ALASKA** | 710,231 |
| 02 | | 03000 | Anchorage municipality | 291,826 |
| 02 | 020 | 03000 | Anchorage Municipality | 291,826 |
| 02 | | 24230 | Fairbanks city | 31,535 |
| 02 | 090 | 24230 | Fairbanks North Star Borough | 31,535 |
| 02 | | 36400 | Juneau city and borough | 31,275 |
| 02 | 110 | 36400 | Juneau City and Borough | 31,275 |
| 04 | | | **ARIZONA** | 6,392,017 |
| 04 | | 02830 | Apache Junction city | 35,840 |
| 04 | 013 | 02830 | Maricopa County | 294 |
| 04 | 021 | 02830 | Pinal County | 35,546 |
| 04 | | 04720 | Avondale city | 76,238 |
| 04 | 013 | 04720 | Maricopa County | 76,238 |
| 04 | | 07940 | Buckeye town | 50,876 |
| 04 | 013 | 07940 | Maricopa County | 50,876 |
| 04 | | 08220 | Bullhead City city | 39,540 |
| 04 | 015 | 08220 | Mohave County | 39,540 |
| 04 | | 10530 | Casa Grande city | 48,571 |
| 04 | 021 | 10530 | Pinal County | 48,571 |
| 04 | | 12000 | Chandler city | 236,123 |
| 04 | 013 | 12000 | Maricopa County | 236,123 |
| 04 | | 22220 | El Mirage city | 31,797 |
| 04 | 013 | 22220 | Maricopa County | 31,797 |
| 04 | | 23620 | Flagstaff city | 65,870 |
| 04 | 005 | 23620 | Coconino County | 65,870 |
| 04 | | 23760 | Florence town | 25,536 |
| 04 | 021 | 23760 | Pinal County | 25,536 |
| 04 | | 27400 | Gilbert town | 208,453 |
| 04 | 013 | 27400 | Maricopa County | 208,453 |
| 04 | | 27820 | Glendale city | 226,721 |
| 04 | 013 | 27820 | Maricopa County | 226,721 |
| 04 | | 28380 | Goodyear city | 65,275 |
| 04 | 013 | 28380 | Maricopa County | 65,275 |
| 04 | | 37620 | Kingman city | 28,068 |
| 04 | 015 | 37620 | Mohave County | 28,068 |
| 04 | | 39370 | Lake Havasu City city | 52,527 |
| 04 | 015 | 39370 | Mohave County | 52,527 |
| 04 | | 44270 | Marana town | 34,961 |
| 04 | 019 | 44270 | Pima County | 34,961 |
| 04 | 021 | 44270 | Pinal County | 0 |
| 04 | | 44410 | Maricopa city | 43,482 |
| 04 | 021 | 44410 | Pinal County | 43,482 |

| State code | Place code | County code | Geographic Area Name | 2010 census population | State code | Place code | County code | Geographic Area Name | 2010 census population |
|---|---|---|---|---|---|---|---|---|---|
| 04 | | 46000 | Mesa city | 439,041 | 05 | | 35710 | Jonesboro city | 67,263 |
| 04 | 013 | 46000 | Maricopa County | 439,041 | 05 | 031 | 35710 | Craighead County | 67,263 |
| 04 | | 51600 | Oro Valley town | 41,011 | 05 | | 41000 | Little Rock city | 193,524 |
| 04 | 019 | 51600 | Pima County | 41,011 | 05 | 119 | 41000 | Pulaski County | 193,524 |
| 04 | | 54050 | Peoria city | 154,065 | 05 | | 50450 | North Little Rock city | 62,304 |
| 04 | 013 | 54050 | Maricopa County | 154,058 | 05 | 119 | 50450 | Pulaski County | 62,304 |
| 04 | 025 | 54050 | Yavapai County | 7 | 05 | | 53390 | Paragould city | 26,113 |
| 04 | | 55000 | Phoenix city | 1,445,632 | 05 | 055 | 53390 | Greene County | 26,113 |
| 04 | 013 | 55000 | Maricopa County | 1,445,632 | 05 | | 55310 | Pine Bluff city | 49,083 |
| 04 | | 57380 | Prescott city | 39,843 | 05 | 069 | 55310 | Jefferson County | 49,083 |
| 04 | 025 | 57380 | Yavapai County | 39,843 | 05 | | 60410 | Rogers city | 55,964 |
| 04 | | 57450 | Prescott Valley town | 38,822 | 05 | 007 | 60410 | Benton County | 55,964 |
| 04 | 025 | 57450 | Yavapai County | 38,822 | 05 | | 61670 | Russellville city | 27,920 |
| 04 | | 58150 | Queen Creek town | 26,361 | 05 | 115 | 61670 | Pope County | 27,920 |
| 04 | 013 | 58150 | Maricopa County | 25,912 | 05 | | 63800 | Sherwood city | 29,523 |
| 04 | 021 | 58150 | Pinal County | 449 | 05 | 119 | 63800 | Pulaski County | 29,523 |
| 04 | | 62140 | Sahuarita town | 25,259 | 05 | | 66080 | Springdale city | 69,797 |
| 04 | 019 | 62140 | Pima County | 25,259 | 05 | 007 | 66080 | Benton County | 6,054 |
| 04 | | 63470 | San Luis city | 25,505 | 05 | 143 | 66080 | Washington County | 63,743 |
| 04 | 027 | 63470 | Yuma County | 25,505 | 05 | | 68810 | Texarkana city | 29,919 |
| 04 | | 65000 | Scottsdale city | 217,385 | 05 | 091 | 68810 | Miller County | 29,919 |
| 04 | 013 | 65000 | Maricopa County | 217,385 | 05 | | 74540 | West Memphis city | 26,245 |
| 04 | | 66820 | Sierra Vista city | 43,888 | 05 | 035 | 74540 | Crittenden County | 26,245 |
| 04 | 003 | 66820 | Cochise County | 43,888 | 06 | | | **CALIFORNIA** | 37,253,956 |
| 04 | | 71510 | Surprise city | 117,517 | 06 | | 00296 | Adelanto city | 31,765 |
| 04 | 013 | 71510 | Maricopa County | 117,517 | 06 | 071 | 00296 | San Bernardino County | 31,765 |
| 04 | | 73000 | Tempe city | 161,719 | 06 | | 00562 | Alameda city | 73,812 |
| 04 | 013 | 73000 | Maricopa County | 161,719 | 06 | 001 | 00562 | Alameda County | 73,812 |
| 04 | | 77000 | Tucson city | 520,116 | 06 | | 00884 | Alhambra city | 83,089 |
| 04 | 019 | 77000 | Pima County | 520,116 | 06 | 037 | 00884 | Los Angeles County | 83,089 |
| 04 | | 85540 | Yuma city | 93,064 | 06 | | 00947 | Aliso Viejo city | 47,823 |
| 04 | 027 | 85540 | Yuma County | 93,064 | 06 | 059 | 00947 | Orange County | 47,823 |
| 05 | | | **ARKANSAS** | 2,915,918 | 06 | | 02000 | Anaheim city | 336,265 |
| 05 | | 04840 | Bella Vista town | 26,461 | 06 | 059 | 02000 | Orange County | 336,265 |
| 05 | 007 | 04840 | Benton County | 26,461 | 06 | | 02252 | Antioch city | 102,372 |
| 05 | | 05290 | Benton city | 30,681 | 06 | 013 | 02252 | Contra Costa County | 102,372 |
| 05 | 125 | 05290 | Saline County | 30,681 | 06 | | 02364 | Apple Valley town | 69,135 |
| 05 | | 05320 | Bentonville city | 35,301 | 06 | 071 | 02364 | San Bernardino County | 69,135 |
| 05 | 007 | 05320 | Benton County | 35,301 | 06 | | 02462 | Arcadia city | 56,364 |
| 05 | | 15190 | Conway city | 58,908 | 06 | 037 | 02462 | Los Angeles County | 56,364 |
| 05 | 045 | 15190 | Faulkner County | 58,908 | 06 | | 03064 | Atascadero city | 28,310 |
| 05 | | 23290 | Fayetteville city | 73,580 | 06 | 079 | 03064 | San Luis Obispo County | 28,310 |
| 05 | 143 | 23290 | Washington County | 73,580 | 06 | | 03162 | Atwater city | 28,168 |
| 05 | | 24550 | Fort Smith city | 86,209 | 06 | 047 | 03162 | Merced County | 28,168 |
| 05 | 131 | 24550 | Sebastian County | 86,209 | 06 | | 03386 | Azusa city | 46,361 |
| 05 | | 33400 | Hot Springs city | 35,193 | 06 | 037 | 03386 | Los Angeles County | 46,361 |
| 05 | 051 | 33400 | Garland County | 35,193 | 06 | | 03526 | Bakersfield city | 347,483 |
| 05 | | 34750 | Jacksonville city | 28,364 | 06 | 029 | 03526 | Kern County | 347,483 |
| 05 | 119 | 34750 | Pulaski County | 28,364 | 06 | | 03666 | Baldwin Park city | 75,390 |
| | | | | | 06 | 037 | 03666 | Los Angeles County | 75,390 |

| State code | Place code | County code | Geographic Area Name | 2010 census population |
|---|---|---|---|---|
| 06 | | 03820 | Banning city | 29,603 |
| 06 | 065 | 03820 | Riverside County | 29,603 |
| 06 | | 04758 | Beaumont city | 36,877 |
| 06 | 065 | 04758 | Riverside County | 36,877 |
| 06 | | 04870 | Bell city | 35,477 |
| 06 | 037 | 04870 | Los Angeles County | 35,477 |
| 06 | | 04982 | Bellflower city | 76,616 |
| 06 | 037 | 04982 | Los Angeles County | 76,616 |
| 06 | | 04996 | Bell Gardens city | 42,072 |
| 06 | 037 | 04996 | Los Angeles County | 42,072 |
| 06 | | 05108 | Belmont city | 25,835 |
| 06 | 081 | 05108 | San Mateo County | 25,835 |
| 06 | | 05290 | Benicia city | 26,997 |
| 06 | 095 | 05290 | Solano County | 26,997 |
| 06 | | 06000 | Berkeley city | 112,580 |
| 06 | 001 | 06000 | Alameda County | 112,580 |
| 06 | | 06308 | Beverly Hills city | 34,109 |
| 06 | 037 | 06308 | Los Angeles County | 34,109 |
| 06 | | 08100 | Brea city | 39,282 |
| 06 | 059 | 08100 | Orange County | 39,282 |
| 06 | | 08142 | Brentwood city | 51,481 |
| 06 | 013 | 08142 | Contra Costa County | 51,481 |
| 06 | | 08786 | Buena Park city | 80,530 |
| 06 | 059 | 08786 | Orange County | 80,530 |
| 06 | | 08954 | Burbank city | 103,340 |
| 06 | 037 | 08954 | Los Angeles County | 103,340 |
| 06 | | 09066 | Burlingame city | 28,806 |
| 06 | 081 | 09066 | San Mateo County | 28,806 |
| 06 | | 09710 | Calexico city | 38,572 |
| 06 | 025 | 09710 | Imperial County | 38,572 |
| 06 | | 10046 | Camarillo city | 65,201 |
| 06 | 111 | 10046 | Ventura County | 65,201 |
| 06 | | 10345 | Campbell city | 39,349 |
| 06 | 085 | 10345 | Santa Clara County | 39,349 |
| 06 | | 11194 | Carlsbad city | 105,328 |
| 06 | 073 | 11194 | San Diego County | 105,328 |
| 06 | | 11530 | Carson city | 91,714 |
| 06 | 037 | 11530 | Los Angeles County | 91,714 |
| 06 | | 12048 | Cathedral City city | 51,200 |
| 06 | 065 | 12048 | Riverside County | 51,200 |
| 06 | | 12524 | Ceres city | 45,417 |
| 06 | 099 | 12524 | Stanislaus County | 45,417 |
| 06 | | 12552 | Cerritos city | 49,041 |
| 06 | 037 | 12552 | Los Angeles County | 49,041 |
| 06 | | 13014 | Chico city | 86,187 |
| 06 | 007 | 13014 | Butte County | 86,187 |
| 06 | | 13210 | Chino city | 77,983 |
| 06 | 071 | 13210 | San Bernardino County | 77,983 |
| 06 | | 13214 | Chino Hills city | 74,799 |
| 06 | 071 | 13214 | San Bernardino County | 74,799 |
| 06 | | 13392 | Chula Vista city | 243,916 |
| 06 | 073 | 13392 | San Diego County | 243,916 |
| 06 | | 13588 | Citrus Heights city | 83,301 |
| 06 | 067 | 13588 | Sacramento County | 83,301 |
| 06 | | 13756 | Claremont city | 34,926 |
| 06 | 037 | 13756 | Los Angeles County | 34,926 |
| 06 | | 14218 | Clovis city | 95,631 |
| 06 | 019 | 14218 | Fresno County | 95,631 |
| 06 | | 14260 | Coachella city | 40,704 |
| 06 | 065 | 14260 | Riverside County | 40,704 |
| 06 | | 14890 | Colton city | 52,154 |
| 06 | 071 | 14890 | San Bernardino County | 52,154 |
| 06 | | 15044 | Compton city | 96,455 |
| 06 | 037 | 15044 | Los Angeles County | 96,455 |
| 06 | | 16000 | Concord city | 122,067 |
| 06 | 013 | 16000 | Contra Costa County | 122,067 |
| 06 | | 16350 | Corona city | 152,374 |
| 06 | 065 | 16350 | Riverside County | 152,374 |
| 06 | | 16532 | Costa Mesa city | 109,960 |
| 06 | 059 | 16532 | Orange County | 109,960 |
| 06 | | 16742 | Covina city | 47,796 |
| 06 | 037 | 16742 | Los Angeles County | 47,796 |
| 06 | | 17568 | Culver City city | 38,883 |
| 06 | 037 | 17568 | Los Angeles County | 38,883 |
| 06 | | 17610 | Cupertino city | 58,302 |
| 06 | 085 | 17610 | Santa Clara County | 58,302 |
| 06 | | 17750 | Cypress city | 47,802 |
| 06 | 059 | 17750 | Orange County | 47,802 |
| 06 | | 17918 | Daly City city | 101,123 |
| 06 | 081 | 17918 | San Mateo County | 101,123 |
| 06 | | 17946 | Dana Point city | 33,351 |
| 06 | 059 | 17946 | Orange County | 33,351 |
| 06 | | 17988 | Danville town | 42,039 |
| 06 | 013 | 17988 | Contra Costa County | 42,039 |
| 06 | | 18100 | Davis city | 65,622 |
| 06 | 113 | 18100 | Yolo County | 65,622 |
| 06 | | 18394 | Delano city | 53,041 |
| 06 | 029 | 18394 | Kern County | 53,041 |
| 06 | | 18996 | Desert Hot Springs city | 25,938 |
| 06 | 065 | 18996 | Riverside County | 25,938 |
| 06 | | 19192 | Diamond Bar city | 55,544 |
| 06 | 037 | 19192 | Los Angeles County | 55,544 |
| 06 | | 19766 | Downey city | 111,772 |
| 06 | 037 | 19766 | Los Angeles County | 111,772 |
| 06 | | 20018 | Dublin city | 46,036 |
| 06 | 001 | 20018 | Alameda County | 46,036 |
| 06 | | 20956 | East Palo Alto city | 28,155 |
| 06 | 081 | 20956 | San Mateo County | 28,155 |

| State code | Place code | County code | Geographic Area Name | 2010 census population | State code | Place code | County code | Geographic Area Name | 2010 census population |
|---|---|---|---|---|---|---|---|---|---|
| 06 | | 21712 | El Cajon city | 99,478 | 06 | | 33182 | Hemet city | 78,657 |
| 06 | 073 | 21712 | San Diego County | 99,478 | 06 | 065 | 33182 | Riverside County | 78,657 |
| 06 | | 21782 | El Centro city | 42,598 | 06 | | 33434 | Hesperia city | 90,173 |
| 06 | 025 | 21782 | Imperial County | 42,598 | 06 | 071 | 33434 | San Bernardino County | 90,173 |
| 06 | | 22020 | Elk Grove city | 153,015 | 06 | | 33588 | Highland city | 53,104 |
| 06 | 067 | 22020 | Sacramento County | 153,015 | 06 | 071 | 33588 | San Bernardino County | 53,104 |
| 06 | | 22230 | El Monte city | 113,475 | 06 | | 34120 | Hollister city | 34,928 |
| 06 | 037 | 22230 | Los Angeles County | 113,475 | 06 | 069 | 34120 | San Benito County | 34,928 |
| 06 | | 22300 | El Paso de Robles (Paso Robles) | 29,793 | 06 | | 36000 | Huntington Beach city | 189,992 |
| 06 | 079 | 22300 | San Luis Obispo County | 29,793 | 06 | 059 | 36000 | Orange County | 189,992 |
| 06 | | 22678 | Encinitas city | 59,518 | 06 | | 36056 | Huntington Park city | 58,114 |
| 06 | 073 | 22678 | San Diego County | 59,518 | 06 | 037 | 36056 | Los Angeles County | 58,114 |
| 06 | | 22804 | Escondido city | 143,911 | 06 | | 36294 | Imperial Beach city | 26,324 |
| 06 | 073 | 22804 | San Diego County | 143,911 | 06 | 073 | 36294 | San Diego County | 26,324 |
| 06 | | 23042 | Eureka city | 27,191 | 06 | | 36448 | Indio city | 76,036 |
| 06 | 023 | 23042 | Humboldt County | 27,191 | 06 | 065 | 36448 | Riverside County | 76,036 |
| 06 | | 23182 | Fairfield city | 105,321 | 06 | | 36546 | Inglewood city | 109,673 |
| 06 | 095 | 23182 | Solano County | 105,321 | 06 | 037 | 36546 | Los Angeles County | 109,673 |
| 06 | | 24638 | Folsom city | 72,203 | 06 | | 36770 | Irvine city | 212,375 |
| 06 | 067 | 24638 | Sacramento County | 72,203 | 06 | 059 | 36770 | Orange County | 212,375 |
| 06 | | 24680 | Fontana city | 196,069 | 06 | | 39220 | Laguna Hills city | 30,344 |
| 06 | 071 | 24680 | San Bernardino County | 196,069 | 06 | 059 | 39220 | Orange County | 30,344 |
| 06 | | 25338 | Foster City city | 30,567 | 06 | | 39248 | Laguna Niguel city | 62,979 |
| 06 | 081 | 25338 | San Mateo County | 30,567 | 06 | 059 | 39248 | Orange County | 62,979 |
| 06 | | 25380 | Fountain Valley city | 55,313 | 06 | | 39290 | La Habra city | 60,239 |
| 06 | 059 | 25380 | Orange County | 55,313 | 06 | 059 | 39290 | Orange County | 60,239 |
| 06 | | 26000 | Fremont city | 214,089 | 06 | | 39486 | Lake Elsinore city | 51,821 |
| 06 | 001 | 26000 | Alameda County | 214,089 | 06 | 065 | 39486 | Riverside County | 51,821 |
| 06 | | 27000 | Fresno city | 494,665 | 06 | | 39496 | Lake Forest city | 77,264 |
| 06 | 019 | 27000 | Fresno County | 494,665 | 06 | 059 | 39496 | Orange County | 77,264 |
| 06 | | 28000 | Fullerton city | 135,161 | 06 | | 39892 | Lakewood city | 80,048 |
| 06 | 059 | 28000 | Orange County | 135,161 | 06 | 037 | 39892 | Los Angeles County | 80,048 |
| 06 | | 28168 | Gardena city | 58,829 | 06 | | 40004 | La Mesa city | 57,065 |
| 06 | 037 | 28168 | Los Angeles County | 58,829 | 06 | 073 | 40004 | San Diego County | 57,065 |
| 06 | | 29000 | Garden Grove city | 170,883 | 06 | | 40032 | La Mirada city | 48,527 |
| 06 | 059 | 29000 | Orange County | 170,883 | 06 | 037 | 40032 | Los Angeles County | 48,527 |
| 06 | | 29504 | Gilroy city | 48,821 | 06 | | 40130 | Lancaster city | 156,633 |
| 06 | 085 | 29504 | Santa Clara County | 48,821 | 06 | 037 | 40130 | Los Angeles County | 156,633 |
| 06 | | 30000 | Glendale city | 191,719 | 06 | | 40340 | La Puente city | 39,816 |
| 06 | 037 | 30000 | Los Angeles County | 191,719 | 06 | 037 | 40340 | Los Angeles County | 39,816 |
| 06 | | 30014 | Glendora city | 50,073 | 06 | | 40354 | La Quinta city | 37,467 |
| 06 | 037 | 30014 | Los Angeles County | 50,073 | 06 | 065 | 40354 | Riverside County | 37,467 |
| 06 | | 30378 | Goleta city | 29,888 | 06 | | 40830 | La Verne city | 31,063 |
| 06 | 083 | 30378 | Santa Barbara County | 29,888 | 06 | 037 | 40830 | Los Angeles County | 31,063 |
| 06 | | 31960 | Hanford city | 53,967 | 06 | | 40886 | Lawndale city | 32,769 |
| 06 | 031 | 31960 | Kings County | 53,967 | 06 | 037 | 40886 | Los Angeles County | 32,769 |
| 06 | | 32548 | Hawthorne city | 84,293 | 06 | | 41124 | Lemon Grove city | 25,320 |
| 06 | 037 | 32548 | Los Angeles County | 84,293 | 06 | 073 | 41124 | San Diego County | 25,320 |
| 06 | | 33000 | Hayward city | 144,186 | 06 | | 41474 | Lincoln city | 42,819 |
| 06 | 001 | 33000 | Alameda County | 144,186 | 06 | 061 | 41474 | Placer County | 42,819 |

| State code | Place code | County code | Geographic Area Name | 2010 census population | State code | Place code | County code | Geographic Area Name | 2010 census population |
|---|---|---|---|---|---|---|---|---|---|
| 06 | | 41992 | Livermore city | 80,968 | 06 | | 48914 | Monterey Park city | 60,269 |
| 06 | 001 | 41992 | Alameda County | 80,968 | 06 | 037 | 48914 | Los Angeles County | 60,269 |
| 06 | | 42202 | Lodi city | 62,134 | 06 | | 49138 | Moorpark city | 34,421 |
| 06 | 077 | 42202 | San Joaquin County | 62,134 | 06 | 111 | 49138 | Ventura County | 34,421 |
| 06 | | 42524 | Lompoc city | 42,434 | 06 | | 49270 | Moreno Valley city | 193,365 |
| 06 | 083 | 42524 | Santa Barbara County | 42,434 | 06 | 065 | 49270 | Riverside County | 193,365 |
| 06 | | 43000 | Long Beach city | 462,257 | 06 | | 49278 | Morgan Hill city | 37,882 |
| 06 | 037 | 43000 | Los Angeles County | 462,257 | 06 | 085 | 49278 | Santa Clara County | 37,882 |
| 06 | | 43280 | Los Altos city | 28,976 | 06 | | 49670 | Mountain View city | 74,066 |
| 06 | 085 | 43280 | Santa Clara County | 28,976 | 06 | 085 | 49670 | Santa Clara County | 74,066 |
| 06 | | 44000 | Los Angeles city | 3,792,621 | 06 | | 50076 | Murrieta city | 103,466 |
| 06 | 037 | 44000 | Los Angeles County | 3,792,621 | 06 | 065 | 50076 | Riverside County | 103,466 |
| 06 | | 44028 | Los Banos city | 35,972 | 06 | | 50258 | Napa city | 76,915 |
| 06 | 047 | 44028 | Merced County | 35,972 | 06 | 055 | 50258 | Napa County | 76,915 |
| 06 | | 44112 | Los Gatos town | 29,413 | 06 | | 50398 | National City city | 58,582 |
| 06 | 085 | 44112 | Santa Clara County | 29,413 | 06 | 073 | 50398 | San Diego County | 58,582 |
| 06 | | 44574 | Lynwood city | 69,772 | 06 | | 50916 | Newark city | 42,573 |
| 06 | 037 | 44574 | Los Angeles County | 69,772 | 06 | 001 | 50916 | Alameda County | 42,573 |
| 06 | | 45022 | Madera city | 61,416 | 06 | | 51182 | Newport Beach city | 85,186 |
| 06 | 039 | 45022 | Madera County | 61,416 | 06 | 059 | 51182 | Orange County | 85,186 |
| 06 | | 45400 | Manhattan Beach city | 35,135 | 06 | | 51560 | Norco city | 27,063 |
| 06 | 037 | 45400 | Los Angeles County | 35,135 | 06 | 065 | 51560 | Riverside County | 27,063 |
| 06 | | 45484 | Manteca city | 67,096 | 06 | | 52526 | Norwalk city | 105,549 |
| 06 | 077 | 45484 | San Joaquin County | 67,096 | 06 | 037 | 52526 | Los Angeles County | 105,549 |
| 06 | | 46114 | Martinez city | 35,824 | 06 | | 52582 | Novato city | 51,904 |
| 06 | 013 | 46114 | Contra Costa County | 35,824 | 06 | 041 | 52582 | Marin County | 51,904 |
| 06 | | 46492 | Maywood city | 27,395 | 06 | | 53000 | Oakland city | 390,724 |
| 06 | 037 | 46492 | Los Angeles County | 27,395 | 06 | 001 | 53000 | Alameda County | 390,724 |
| 06 | | 46842 | Menifee city | 77,519 | 06 | | 53070 | Oakley city | 35,432 |
| 06 | 065 | 46842 | Riverside County | 77,519 | 06 | 013 | 53070 | Contra Costa County | 35,432 |
| 06 | | 46870 | Menlo Park city | 32,026 | 06 | | 53322 | Oceanside city | 167,086 |
| 06 | 081 | 46870 | San Mateo County | 32,026 | 06 | 073 | 53322 | San Diego County | 167,086 |
| 06 | | 46898 | Merced city | 78,958 | 06 | | 53896 | Ontario city | 163,924 |
| 06 | 047 | 46898 | Merced County | 78,958 | 06 | 071 | 53896 | San Bernardino County | 163,924 |
| 06 | | 47766 | Milpitas city | 66,790 | 06 | | 53980 | Orange city | 136,416 |
| 06 | 085 | 47766 | Santa Clara County | 66,790 | 06 | 059 | 53980 | Orange County | 136,416 |
| 06 | | 48256 | Mission Viejo city | 93,305 | 06 | | 54652 | Oxnard city | 197,899 |
| 06 | 059 | 48256 | Orange County | 93,305 | 06 | 111 | 54652 | Ventura County | 197,899 |
| 06 | | 48354 | Modesto city | 201,165 | 06 | | 54806 | Pacifica city | 37,234 |
| 06 | 099 | 48354 | Stanislaus County | 201,165 | 06 | 081 | 54806 | San Mateo County | 37,234 |
| 06 | | 48648 | Monrovia city | 36,590 | 06 | | 55156 | Palmdale city | 152,750 |
| 06 | 037 | 48648 | Los Angeles County | 36,590 | 06 | 037 | 55156 | Los Angeles County | 152,750 |
| 06 | | 48788 | Montclair city | 36,664 | 06 | | 55184 | Palm Desert city | 48,445 |
| 06 | 071 | 48788 | San Bernardino County | 36,664 | 06 | 065 | 55184 | Riverside County | 48,445 |
| 06 | | 48816 | Montebello city | 62,500 | 06 | | 55254 | Palm Springs city | 44,552 |
| 06 | 037 | 48816 | Los Angeles County | 62,500 | 06 | 065 | 55254 | Riverside County | 44,552 |
| 06 | | 48872 | Monterey city | 27,810 | 06 | | 55282 | Palo Alto city | 64,403 |
| 06 | 053 | 48872 | Monterey County | 27,810 | 06 | 085 | 55282 | Santa Clara County | 64,403 |

| State code | Place code | County code | Geographic Area Name | 2010 census population | State code | Place code | County code | Geographic Area Name | 2010 census population |
|---|---|---|---|---|---|---|---|---|---|
| 06 | | 55520 | Paradise town | 26,218 | 06 | | 62000 | Riverside city | 303,871 |
| 06 | 007 | 55520 | Butte County | 26,218 | 06 | 065 | 62000 | Riverside County | 303,871 |
| 06 | | 55618 | Paramount city | 54,098 | 06 | | 62364 | Rocklin city | 56,974 |
| 06 | 037 | 55618 | Los Angeles County | 54,098 | 06 | 061 | 62364 | Placer County | 56,974 |
| 06 | | 56000 | Pasadena city | 137,122 | 06 | | 62546 | Rohnert Park city | 40,971 |
| 06 | 037 | 56000 | Los Angeles County | 137,122 | 06 | 097 | 62546 | Sonoma County | 40,971 |
| 06 | | 56700 | Perris city | 68,386 | 06 | | 62896 | Rosemead city | 53,764 |
| 06 | 065 | 56700 | Riverside County | 68,386 | 06 | 037 | 62896 | Los Angeles County | 53,764 |
| 06 | | 56784 | Petaluma city | 57,941 | 06 | | 62938 | Roseville city | 118,788 |
| 06 | 097 | 56784 | Sonoma County | 57,941 | 06 | 061 | 62938 | Placer County | 118,788 |
| 06 | | 56924 | Pico Rivera city | 62,942 | 06 | | 64000 | Sacramento city | 466,488 |
| 06 | 037 | 56924 | Los Angeles County | 62,942 | 06 | 067 | 64000 | Sacramento County | 466,488 |
| 06 | | 57456 | Pittsburg city | 63,264 | 06 | | 64224 | Salinas city | 150,441 |
| 06 | 013 | 57456 | Contra Costa County | 63,264 | 06 | 053 | 64224 | Monterey County | 150,441 |
| 06 | | 57526 | Placentia city | 50,533 | 06 | | 65000 | San Bernardino city | 209,924 |
| 06 | 059 | 57526 | Orange County | 50,533 | 06 | 071 | 65000 | San Bernardino County | 209,924 |
| 06 | | 57764 | Pleasant Hill city | 33,152 | 06 | | 65028 | San Bruno city | 41,114 |
| 06 | 013 | 57764 | Contra Costa County | 33,152 | 06 | 081 | 65028 | San Mateo County | 41,114 |
| 06 | | 57792 | Pleasanton city | 70,285 | 06 | | 65042 | San Buenaventura (Ventura) | 106,433 |
| 06 | 001 | 57792 | Alameda County | 70,285 | 06 | 111 | 65042 | Ventura County | 106,433 |
| 06 | | 58072 | Pomona city | 149,058 | 06 | | 65070 | San Carlos city | 28,406 |
| 06 | 037 | 58072 | Los Angeles County | 149,058 | 06 | 081 | 65070 | San Mateo County | 28,406 |
| 06 | | 58240 | Porterville city | 54,165 | 06 | | 65084 | San Clemente city | 63,522 |
| 06 | 107 | 58240 | Tulare County | 54,165 | 06 | 059 | 65084 | Orange County | 63,522 |
| 06 | | 58520 | Poway city | 47,811 | 06 | | 66000 | San Diego city | 1,307,402 |
| 06 | 073 | 58520 | San Diego County | 47,811 | 06 | 073 | 66000 | San Diego County | 1,307,402 |
| 06 | | 59444 | Rancho Cordova city | 64,776 | 06 | | 66070 | San Dimas city | 33,371 |
| 06 | 067 | 59444 | Sacramento County | 64,776 | 06 | 037 | 66070 | Los Angeles County | 33,371 |
| 06 | | 59451 | Rancho Cucamonga city | 165,269 | 06 | | 67000 | San Francisco city | 805,235 |
| 06 | 071 | 59451 | San Bernardino County | 165,269 | 06 | 075 | 67000 | San Francisco County | 805,235 |
| 06 | | 59514 | Rancho Palos Verdes city | 41,643 | 06 | | 67042 | San Gabriel city | 39,718 |
| 06 | 037 | 59514 | Los Angeles County | 41,643 | 06 | 037 | 67042 | Los Angeles County | 39,718 |
| 06 | | 59587 | Rancho Santa Margarita city | 47,853 | 06 | | 67112 | San Jacinto city | 44,199 |
| 06 | 059 | 59587 | Orange County | 47,853 | 06 | 065 | 67112 | Riverside County | 44,199 |
| 06 | | 59920 | Redding city | 89,861 | 06 | | 68000 | San Jose city | 945,942 |
| 06 | 089 | 59920 | Shasta County | 89,861 | 06 | 085 | 68000 | Santa Clara County | 945,942 |
| 06 | | 59962 | Redlands city | 68,747 | 06 | | 68028 | San Juan Capistrano city | 34,593 |
| 06 | 071 | 59962 | San Bernardino County | 68,747 | 06 | 059 | 68028 | Orange County | 34,593 |
| 06 | | 60018 | Redondo Beach city | 66,748 | 06 | | 68084 | San Leandro city | 84,950 |
| 06 | 037 | 60018 | Los Angeles County | 66,748 | 06 | 001 | 68084 | Alameda County | 84,950 |
| 06 | | 60102 | Redwood City city | 76,815 | 06 | | 68154 | San Luis Obispo city | 45,119 |
| 06 | 081 | 60102 | San Mateo County | 76,815 | 06 | 079 | 68154 | San Luis Obispo County | 45,119 |
| 06 | | 60466 | Rialto city | 99,171 | 06 | | 68196 | San Marcos city | 83,781 |
| 06 | 071 | 60466 | San Bernardino County | 99,171 | 06 | 073 | 68196 | San Diego County | 83,781 |
| 06 | | 60620 | Richmond city | 103,701 | 06 | | 68252 | San Mateo city | 97,207 |
| 06 | 013 | 60620 | Contra Costa County | 103,701 | 06 | 081 | 68252 | San Mateo County | 97,207 |
| 06 | | 60704 | Ridgecrest city | 27,616 | 06 | | 68294 | San Pablo city | 29,139 |
| 06 | 029 | 60704 | Kern County | 27,616 | 06 | 013 | 68294 | Contra Costa County | 29,139 |

| State code | Place code | County code | Geographic Area Name | 2010 census population | State code | Place code | County code | Geographic Area Name | 2010 census population |
|---|---|---|---|---|---|---|---|---|---|
| 06 | | 68364 | San Rafael city | 57,713 | 06 | | 78582 | Thousand Oaks city | 126,683 |
| 06 | 041 | 68364 | Marin County | 57,713 | 06 | 111 | 78582 | Ventura County | 126,683 |
| 06 | | 68378 | San Ramon city | 72,148 | 06 | | 80000 | Torrance city | 145,438 |
| 06 | 013 | 68378 | Contra Costa County | 72,148 | 06 | 037 | 80000 | Los Angeles County | 145,438 |
| 06 | | 69000 | Santa Ana city | 324,528 | 06 | | 80238 | Tracy city | 82,922 |
| 06 | 059 | 69000 | Orange County | 324,528 | 06 | 077 | 80238 | San Joaquin County | 82,922 |
| 06 | | 69070 | Santa Barbara city | 88,410 | 06 | | 80644 | Tulare city | 59,278 |
| 06 | 083 | 69070 | Santa Barbara County | 88,410 | 06 | 107 | 80644 | Tulare County | 59,278 |
| 06 | | 69084 | Santa Clara city | 116,468 | 06 | | 80812 | Turlock city | 68,549 |
| 06 | 085 | 69084 | Santa Clara County | 116,468 | 06 | 099 | 80812 | Stanislaus County | 68,549 |
| 06 | | 69088 | Santa Clarita city | 176,320 | 06 | | 80854 | Tustin city | 75,540 |
| 06 | 037 | 69088 | Los Angeles County | 176,320 | 06 | 059 | 80854 | Orange County | 75,540 |
| 06 | | 69112 | Santa Cruz city | 59,946 | 06 | | 80994 | Twentynine Palms city | 25,048 |
| 06 | 087 | 69112 | Santa Cruz County | 59,946 | 06 | 071 | 80994 | San Bernardino County | 25,048 |
| 06 | | 69196 | Santa Maria city | 99,553 | 06 | | 81204 | Union City city | 69,516 |
| 06 | 083 | 69196 | Santa Barbara County | 99,553 | 06 | 001 | 81204 | Alameda County | 69,516 |
| 06 | | 70000 | Santa Monica city | 89,736 | 06 | | 81344 | Upland city | 73,732 |
| 06 | 037 | 70000 | Los Angeles County | 89,736 | 06 | 071 | 81344 | San Bernardino County | 73,732 |
| 06 | | 70042 | Santa Paula city | 29,321 | 06 | | 81554 | Vacaville city | 92,428 |
| 06 | 111 | 70042 | Ventura County | 29,321 | 06 | 095 | 81554 | Solano County | 92,428 |
| 06 | | 70098 | Santa Rosa city | 167,815 | 06 | | 81666 | Vallejo city | 115,942 |
| 06 | 097 | 70098 | Sonoma County | 167,815 | 06 | 095 | 81666 | Solano County | 115,942 |
| 06 | | 70224 | Santee city | 53,413 | 06 | | 82590 | Victorville city | 115,903 |
| 06 | 073 | 70224 | San Diego County | 53,413 | 06 | 071 | 82590 | San Bernardino County | 115,903 |
| 06 | | 70280 | Saratoga city | 29,926 | 06 | | 82954 | Visalia city | 124,442 |
| 06 | 085 | 70280 | Santa Clara County | 29,926 | 06 | 107 | 82954 | Tulare County | 124,442 |
| 06 | | 70742 | Seaside city | 33,025 | 06 | | 82996 | Vista city | 93,834 |
| 06 | 053 | 70742 | Monterey County | 33,025 | 06 | 073 | 82996 | San Diego County | 93,834 |
| 06 | | 72016 | Simi Valley city | 124,237 | 06 | | 83332 | Walnut city | 29,172 |
| 06 | 111 | 72016 | Ventura County | 124,237 | 06 | 037 | 83332 | Los Angeles County | 29,172 |
| 06 | | 72520 | Soledad city | 25,738 | 06 | | 83346 | Walnut Creek city | 64,173 |
| 06 | 053 | 72520 | Monterey County | 25,738 | 06 | 013 | 83346 | Contra Costa County | 64,173 |
| 06 | | 73080 | South Gate city | 94,396 | 06 | | 83542 | Wasco city | 25,545 |
| 06 | 037 | 73080 | Los Angeles County | 94,396 | 06 | 029 | 83542 | Kern County | 25,545 |
| 06 | | 73220 | South Pasadena city | 25,619 | 06 | | 83668 | Watsonville city | 51,199 |
| 06 | 037 | 73220 | Los Angeles County | 25,619 | 06 | 087 | 83668 | Santa Cruz County | 51,199 |
| 06 | | 73262 | South San Francisco city | 63,632 | 06 | | 84200 | West Covina city | 106,098 |
| 06 | 081 | 73262 | San Mateo County | 63,632 | 06 | 037 | 84200 | Los Angeles County | 106,098 |
| 06 | | 73962 | Stanton city | 38,186 | 06 | | 84410 | West Hollywood city | 34,399 |
| 06 | 059 | 73962 | Orange County | 38,186 | 06 | 037 | 84410 | Los Angeles County | 34,399 |
| 06 | | 75000 | Stockton city | 291,707 | 06 | | 84550 | Westminster city | 89,701 |
| 06 | 077 | 75000 | San Joaquin County | 291,707 | 06 | 059 | 84550 | Orange County | 89,701 |
| 06 | | 75630 | Suisun City city | 28,111 | 06 | | 84816 | West Sacramento city | 48,744 |
| 06 | 095 | 75630 | Solano County | 28,111 | 06 | 113 | 84816 | Yolo County | 48,744 |
| 06 | | 77000 | Sunnyvale city | 140,081 | 06 | | 85292 | Whittier city | 85,331 |
| 06 | 085 | 77000 | Santa Clara County | 140,081 | 06 | 037 | 85292 | Los Angeles County | 85,331 |
| 06 | | 78120 | Temecula city | 100,097 | 06 | | 85446 | Wildomar city | 32,176 |
| 06 | 065 | 78120 | Riverside County | 100,097 | 06 | 065 | 85446 | Riverside County | 32,176 |
| 06 | | 78148 | Temple City city | 35,558 | 06 | | 85922 | Windsor town | 26,801 |
| 06 | 037 | 78148 | Los Angeles County | 35,558 | 06 | 097 | 85922 | Sonoma County | 26,801 |

# CITES BY COUNTY—*Continued*

| State code | Place code | County code | Geographic Area Name | 2010 census population | State code | Place code | County code | Geographic Area Name | 2010 census population |
|---|---|---|---|---|---|---|---|---|---|
| 06 | | 86328 | Woodland city | 55,468 | 08 | | 46465 | Loveland city | 66,859 |
| 06 | 113 | 86328 | Yolo County | 55,468 | 08 | 069 | 46465 | Larimer County | 66,859 |
| | | | | | | | | | |
| 06 | | 86832 | Yorba Linda city | 64,234 | 08 | | 54330 | Northglenn city | 35,789 |
| 06 | 059 | 86832 | Orange County | 64,234 | 08 | 001 | 54330 | Adams County | 35,777 |
| | | | | | 08 | 123 | 54330 | Weld County | 12 |
| 06 | | 86972 | Yuba City city | 64,925 | | | | | |
| 06 | 101 | 86972 | Sutter County | 64,925 | 08 | | 57630 | Parker town | 45,297 |
| | | | | | 08 | 035 | 57630 | Douglas County | 45,297 |
| 06 | | 87042 | Yucaipa city | 51,367 | | | | | |
| 06 | 071 | 87042 | San Bernardino County | 51,367 | 08 | | 62000 | Pueblo city | 106,595 |
| | | | | | 08 | 101 | 62000 | Pueblo County | 106,595 |
| 08 | | | **COLORADO** | 5,029,196 | | | | | |
| | | | | | 08 | | 77290 | Thornton city | 118,772 |
| 08 | | 03455 | Arvada city | 106,433 | 08 | 001 | 77290 | Adams County | 118,772 |
| 08 | 001 | 03455 | Adams County | 2,849 | 08 | 123 | 77290 | Weld County | 0 |
| 08 | 059 | 03455 | Jefferson County | 103,584 | | | | | |
| | | | | | 08 | | 83835 | Westminster city | 106,114 |
| 08 | | 04000 | Aurora city | 325,078 | 08 | 001 | 83835 | Adams County | 63,696 |
| 08 | 001 | 04000 | Adams County | 39,871 | 08 | 059 | 83835 | Jefferson County | 42,418 |
| 08 | 005 | 04000 | Arapahoe County | 285,090 | | | | | |
| 08 | 035 | 04000 | Douglas County | 117 | 08 | | 84440 | Wheat Ridge city | 30,166 |
| | | | | | 08 | 059 | 84440 | Jefferson County | 30,166 |
| 08 | | 07850 | Boulder city | 97,385 | | | | | |
| 08 | 013 | 07850 | Boulder County | 97,385 | 09 | | | **CONNECTICUT** | 3,574,097 |
| | | | | | | | | | |
| 08 | | 08675 | Brighton city | 33,352 | 09 | | 08000 | Bridgeport city | 144,229 |
| 08 | 001 | 08675 | Adams County | 33,009 | 09 | 001 | 08000 | Fairfield County | 144,229 |
| 08 | 123 | 08675 | Weld County | 343 | | | | | |
| | | | | | 09 | | 08420 | Bristol city | 60,477 |
| 08 | | 09280 | Broomfield city | 55,889 | 09 | 003 | 08420 | Hartford County | 60,477 |
| 08 | 014 | 09280 | Broomfield County | 55,889 | | | | | |
| | | | | | 09 | | 18430 | Danbury city | 80,893 |
| 08 | | 12415 | Castle Rock town | 48,231 | 09 | 001 | 18430 | Fairfield County | 80,893 |
| 08 | 035 | 12415 | Douglas County | 48,231 | | | | | |
| | | | | | 09 | | 37000 | Hartford city | 124,775 |
| 08 | | 12815 | Centennial city | 100,377 | 09 | 003 | 37000 | Hartford County | 124,775 |
| 08 | 005 | 12815 | Arapahoe County | 100,377 | | | | | |
| | | | | | 09 | | 46450 | Meriden city | 60,868 |
| 08 | | 16000 | Colorado Springs city | 416,427 | 09 | 009 | 46450 | New Haven County | 60,868 |
| 08 | 041 | 16000 | El Paso County | 416,427 | | | | | |
| | | | | | 09 | | 47290 | Middletown city | 47,648 |
| 08 | | 16495 | Commerce City city | 45,913 | 09 | 007 | 47290 | Middlesex County | 47,648 |
| 08 | 001 | 16495 | Adams County | 45,913 | | | | | |
| | | | | | 09 | | 49880 | Naugatuck borough | 31,862 |
| 08 | | 20000 | Denver city | 600,158 | 09 | 009 | 49880 | New Haven County | 31,862 |
| 08 | 031 | 20000 | Denver County | 600,158 | | | | | |
| | | | | | 09 | | 50370 | New Britain city | 73,206 |
| 08 | | 24785 | Englewood city | 30,255 | 09 | 003 | 50370 | Hartford County | 73,206 |
| 08 | 005 | 24785 | Arapahoe County | 30,255 | | | | | |
| | | | | | 09 | | 52000 | New Haven city | 129,779 |
| 08 | | 27425 | Fort Collins city | 143,986 | 09 | 009 | 52000 | New Haven County | 129,779 |
| 08 | 069 | 27425 | Larimer County | 143,986 | | | | | |
| | | | | | 09 | | 52280 | New London city | 27,620 |
| 08 | | 27865 | Fountain city | 25,846 | 09 | 011 | 52280 | New London County | 27,620 |
| 08 | 041 | 27865 | El Paso County | 25,846 | | | | | |
| | | | | | 09 | | 55990 | Norwalk city | 85,603 |
| 08 | | 31660 | Grand Junction city | 58,566 | 09 | 001 | 55990 | Fairfield County | 85,603 |
| 08 | 077 | 31660 | Mesa County | 58,566 | | | | | |
| | | | | | 09 | | 56200 | Norwich city | 40,493 |
| 08 | | 32155 | Greeley city | 92,889 | 09 | 011 | 56200 | New London County | 40,493 |
| 08 | 123 | 32155 | Weld County | 92,889 | | | | | |
| | | | | | 09 | | 68100 | Shelton city | 39,559 |
| 08 | | 43000 | Lakewood city | 142,980 | 09 | 001 | 68100 | Fairfield County | 39,559 |
| 08 | 059 | 43000 | Jefferson County | 142,980 | | | | | |
| | | | | | 09 | | 73000 | Stamford city | 122,643 |
| 08 | | 45255 | Littleton city | 41,737 | 09 | 001 | 73000 | Fairfield County | 122,643 |
| 08 | 005 | 45255 | Arapahoe County | 39,328 | | | | | |
| 08 | 035 | 45255 | Douglas County | 28 | 09 | | 76500 | Torrington city | 36,383 |
| 08 | 059 | 45255 | Jefferson County | 2,381 | 09 | 005 | 76500 | Litchfield County | 36,383 |
| | | | | | | | | | |
| 08 | | 45970 | Longmont city | 86,270 | 09 | | 80000 | Waterbury city | 110,366 |
| 08 | 013 | 45970 | Boulder County | 86,240 | 09 | 009 | 80000 | New Haven County | 110,366 |
| 08 | 123 | 45970 | Weld County | 30 | | | | | |

| State code | Place code | County code | Geographic Area Name | 2010 census population | State code | Place code | County code | Geographic Area Name | 2010 census population |
|---|---|---|---|---|---|---|---|---|---|
| 09 | | 82800 | West Haven city | 55,564 | 12 | | 16525 | Daytona Beach city | 61,005 |
| 09 | 009 | 82800 | New Haven County | 55,564 | 12 | 127 | 16525 | Volusia County | 61,005 |
| | | | | | | | | | |
| 10 | | | **DELAWARE** | 897,934 | 12 | | 16725 | Deerfield Beach city | 75,018 |
| | | | | | 12 | 011 | 16725 | Broward County | 75,018 |
| 10 | | 21200 | Dover city | 36,047 | | | | | |
| 10 | 001 | 21200 | Kent County | 36,047 | 12 | | 16875 | DeLand city | 27,031 |
| | | | | | 12 | 127 | 16875 | Volusia County | 27,031 |
| 10 | | 50670 | Newark city | 31,454 | | | | | |
| 10 | 003 | 50670 | New Castle County | 31,454 | 12 | | 17100 | Delray Beach city | 60,522 |
| | | | | | 12 | 099 | 17100 | Palm Beach County | 60,522 |
| 10 | | 77580 | Wilmington city | 70,851 | | | | | |
| 10 | 003 | 77580 | New Castle County | 70,851 | 12 | | 17200 | Deltona city | 85,182 |
| | | | | | 12 | 127 | 17200 | Volusia County | 85,182 |
| 11 | | | **DISTRICT OF COLUMBIA** | 601,723 | | | | | |
| | | | | | 12 | | 17935 | Doral city | 45,704 |
| 11 | | 50000 | Washington city | 601,723 | 12 | 086 | 17935 | Miami-Dade County | 45,704 |
| 11 | 001 | 50000 | District of Columbia | 601,723 | | | | | |
| | | | | | 12 | | 18575 | Dunedin city | 35,321 |
| 12 | | | **FLORIDA** | 18,801,310 | 12 | 103 | 18575 | Pinellas County | 35,321 |
| | | | | | | | | | |
| 12 | | 00950 | Altamonte Springs city | 41,496 | 12 | | 24000 | Fort Lauderdale city | 165,521 |
| 12 | 117 | 00950 | Seminole County | 41,496 | 12 | 011 | 24000 | Broward County | 165,521 |
| | | | | | | | | | |
| 12 | | 01700 | Apopka city | 41,542 | 12 | | 24125 | Fort Myers city | 62,298 |
| 12 | 095 | 01700 | Orange County | 41,542 | 12 | 071 | 24125 | Lee County | 62,298 |
| | | | | | | | | | |
| 12 | | 02681 | Aventura city | 35,762 | 12 | | 24300 | Fort Pierce city | 41,590 |
| 12 | 086 | 02681 | Miami-Dade County | 35,762 | 12 | 111 | 24300 | St. Lucie County | 41,590 |
| | | | | | | | | | |
| 12 | | 07300 | Boca Raton city | 84,392 | 12 | | 25175 | Gainesville city | 124,354 |
| 12 | 099 | 07300 | Palm Beach County | 84,392 | 12 | 001 | 25175 | Alachua County | 124,354 |
| | | | | | | | | | |
| 12 | | 07525 | Bonita Springs city | 43,914 | 12 | | 27322 | Greenacres city | 37,573 |
| 12 | 071 | 07525 | Lee County | 43,914 | 12 | 099 | 27322 | Palm Beach County | 37,573 |
| | | | | | | | | | |
| 12 | | 07875 | Boynton Beach city | 68,217 | 12 | | 28452 | Hallandale Beach city | 37,113 |
| 12 | 099 | 07875 | Palm Beach County | 68,217 | 12 | 011 | 28452 | Broward County | 37,113 |
| | | | | | | | | | |
| 12 | | 07950 | Bradenton city | 49,546 | 12 | | 30000 | Hialeah city | 224,669 |
| 12 | 081 | 07950 | Manatee County | 49,546 | 12 | 086 | 30000 | Miami-Dade County | 224,669 |
| | | | | | | | | | |
| 12 | | 10275 | Cape Coral city | 154,305 | 12 | | 32000 | Hollywood city | 140,768 |
| 12 | 071 | 10275 | Lee County | 154,305 | 12 | 011 | 32000 | Broward County | 140,768 |
| | | | | | | | | | |
| 12 | | 11050 | Casselberry city | 26,241 | 12 | | 32275 | Homestead city | 60,512 |
| 12 | 117 | 11050 | Seminole County | 26,241 | 12 | 086 | 32275 | Miami-Dade County | 60,512 |
| | | | | | | | | | |
| 12 | | 12875 | Clearwater city | 107,685 | 12 | | 35000 | Jacksonville city | 821,784 |
| 12 | 103 | 12875 | Pinellas County | 107,685 | 12 | 031 | 35000 | Duval County | 821,784 |
| | | | | | | | | | |
| 12 | | 12925 | Clermont city | 28,742 | 12 | | 35875 | Jupiter town | 55,156 |
| 12 | 069 | 12925 | Lake County | 28,742 | 12 | 099 | 35875 | Palm Beach County | 55,156 |
| | | | | | | | | | |
| 12 | | 13275 | Coconut Creek city | 52,909 | 12 | | 36950 | Kissimmee city | 59,682 |
| 12 | 011 | 13275 | Broward County | 52,909 | 12 | 097 | 36950 | Osceola County | 59,682 |
| | | | | | | | | | |
| 12 | | 14125 | Cooper City city | 28,547 | 12 | | 38250 | Lakeland city | 97,422 |
| 12 | 011 | 14125 | Broward County | 28,547 | 12 | 105 | 38250 | Polk County | 97,422 |
| | | | | | | | | | |
| 12 | | 14250 | Coral Gables city | 46,780 | 12 | | 39075 | Lake Worth city | 34,910 |
| 12 | 086 | 14250 | Miami-Dade County | 46,780 | 12 | 099 | 39075 | Palm Beach County | 34,910 |
| | | | | | | | | | |
| 12 | | 14400 | Coral Springs city | 121,096 | 12 | | 39425 | Largo city | 77,648 |
| 12 | 011 | 14400 | Broward County | 121,096 | 12 | 103 | 39425 | Pinellas County | 77,648 |
| | | | | | | | | | |
| 12 | | 15968 | Cutler Bay town | 40,286 | 12 | | 39525 | Lauderdale Lakes city | 32,593 |
| 12 | 086 | 15968 | Miami-Dade County | 40,286 | 12 | 011 | 39525 | Broward County | 32,593 |
| | | | | | | | | | |
| 12 | | 16335 | Dania Beach city | 29,639 | 12 | | 39550 | Lauderhill city | 66,887 |
| 12 | 011 | 16335 | Broward County | 29,639 | 12 | 011 | 39550 | Broward County | 66,887 |
| | | | | | | | | | |
| 12 | | 16475 | Davie town | 91,992 | 12 | | 43125 | Margate city | 53,284 |
| 12 | 011 | 16475 | Broward County | 91,992 | 12 | 011 | 43125 | Broward County | 53,284 |

| State code | Place code | County code | Geographic Area Name | 2010 census population | State code | Place code | County code | Geographic Area Name | 2010 census population |
|---|---|---|---|---|---|---|---|---|---|
| 12 | | 43975 | Melbourne city | 76,068 | 12 | | 58050 | Pompano Beach city | 99,845 |
| 12 | 009 | 43975 | Brevard County | 76,068 | 12 | 011 | 58050 | Broward County | 99,845 |
| 12 | | 45000 | Miami city | 399,457 | 12 | | 58575 | Port Orange city | 56,048 |
| 12 | 086 | 45000 | Miami-Dade County | 399,457 | 12 | 127 | 58575 | Volusia County | 56,048 |
| 12 | | 45025 | Miami Beach city | 87,779 | 12 | | 58715 | Port St. Lucie city | 164,603 |
| 12 | 086 | 45025 | Miami-Dade County | 87,779 | 12 | 111 | 58715 | St. Lucie County | 164,603 |
| 12 | | 45060 | Miami Gardens city | 107,167 | 12 | | 60975 | Riviera Beach city | 32,488 |
| 12 | 086 | 45060 | Miami-Dade County | 107,167 | 12 | 099 | 60975 | Palm Beach County | 32,488 |
| 12 | | 45100 | Miami Lakes town | 29,361 | 12 | | 62100 | Royal Palm Beach village | 34,140 |
| 12 | 086 | 45100 | Miami-Dade County | 29,361 | 12 | 099 | 62100 | Palm Beach County | 34,140 |
| 12 | | 45975 | Miramar city | 122,041 | 12 | | 62625 | St. Cloud city | 35,183 |
| 12 | 011 | 45975 | Broward County | 122,041 | 12 | 097 | 62625 | Osceola County | 35,183 |
| 12 | | 49425 | North Lauderdale city | 41,023 | 12 | | 63000 | St. Petersburg city | 244,769 |
| 12 | 011 | 49425 | Broward County | 41,023 | 12 | 103 | 63000 | Pinellas County | 244,769 |
| 12 | | 49450 | North Miami city | 58,786 | 12 | | 63650 | Sanford city | 53,570 |
| 12 | 086 | 49450 | Miami-Dade County | 58,786 | 12 | 117 | 63650 | Seminole County | 53,570 |
| 12 | | 49475 | North Miami Beach city | 41,523 | 12 | | 64175 | Sarasota city | 51,917 |
| 12 | 086 | 49475 | Miami-Dade County | 41,523 | 12 | 115 | 64175 | Sarasota County | 51,917 |
| 12 | | 49675 | North Port city | 57,357 | 12 | | 69700 | Sunrise city | 84,439 |
| 12 | 115 | 49675 | Sarasota County | 57,357 | 12 | 011 | 69700 | Broward County | 84,439 |
| 12 | | 50575 | Oakland Park city | 41,363 | 12 | | 70600 | Tallahassee city | 181,376 |
| 12 | 011 | 50575 | Broward County | 41,363 | 12 | 073 | 70600 | Leon County | 181,376 |
| 12 | | 50750 | Ocala city | 56,315 | 12 | | 70675 | Tamarac city | 60,427 |
| 12 | 083 | 50750 | Marion County | 56,315 | 12 | 011 | 70675 | Broward County | 60,427 |
| 12 | | 51075 | Ocoee city | 35,579 | 12 | | 71000 | Tampa city | 335,709 |
| 12 | 095 | 51075 | Orange County | 35,579 | 12 | 057 | 71000 | Hillsborough County | 335,709 |
| 12 | | 53000 | Orlando city | 238,300 | 12 | | 71900 | Titusville city | 43,761 |
| 12 | 095 | 53000 | Orange County | 238,300 | 12 | 009 | 71900 | Brevard County | 43,761 |
| 12 | | 53150 | Ormond Beach city | 38,137 | 12 | | 75812 | Wellington village | 56,508 |
| 12 | 127 | 53150 | Volusia County | 38,137 | 12 | 099 | 75812 | Palm Beach County | 56,508 |
| 12 | | 53575 | Oviedo city | 33,342 | 12 | | 76582 | Weston city | 65,333 |
| 12 | 117 | 53575 | Seminole County | 33,342 | 12 | 011 | 76582 | Broward County | 65,333 |
| 12 | | 54000 | Palm Bay city | 103,190 | 12 | | 76600 | West Palm Beach city | 99,919 |
| 12 | 009 | 54000 | Brevard County | 103,190 | 12 | 099 | 76600 | Palm Beach County | 99,919 |
| 12 | | 54075 | Palm Beach Gardens city | 48,452 | 12 | | 78250 | Winter Garden city | 34,568 |
| 12 | 099 | 54075 | Palm Beach County | 48,452 | 12 | 095 | 78250 | Orange County | 34,568 |
| 12 | | 54200 | Palm Coast city | 75,180 | 12 | | 78275 | Winter Haven city | 33,874 |
| 12 | 035 | 54200 | Flagler County | 75,180 | 12 | 105 | 78275 | Polk County | 33,874 |
| 12 | | 54700 | Panama City city | 36,484 | 12 | | 78300 | Winter Park city | 27,852 |
| 12 | 005 | 54700 | Bay County | 36,484 | 12 | 095 | 78300 | Orange County | 27,852 |
| 12 | | 55775 | Pembroke Pines city | 154,750 | 12 | | 78325 | Winter Springs city | 33,282 |
| 12 | 011 | 55775 | Broward County | 154,750 | 12 | 117 | 78325 | Seminole County | 33,282 |
| 12 | | 55925 | Pensacola city | 51,923 | 13 | | | **GEORGIA** | 9,687,653 |
| 12 | 033 | 55925 | Escambia County | 51,923 | | | | | |
| 12 | | 56975 | Pinellas Park city | 49,079 | 13 | | 01052 | Albany city | 77,434 |
| 12 | 103 | 56975 | Pinellas County | 49,079 | 13 | 095 | 01052 | Dougherty County | 77,434 |
| 12 | | 57425 | Plantation city | 84,955 | 13 | | 01696 | Alpharetta city | 57,551 |
| 12 | 011 | 57425 | Broward County | 84,955 | 13 | 121 | 01696 | Fulton County | 57,551 |
| 12 | | 57550 | Plant City city | 34,721 | 13 | | 04000 | Atlanta city | 420,003 |
| 12 | 057 | 57550 | Hillsborough County | 34,721 | 13 | 089 | 04000 | DeKalb County | 28,292 |
| | | | | | 13 | 121 | 04000 | Fulton County | 391,711 |

| State code | Place code | County code | Geographic Area Name | 2010 census population | State code | Place code | County code | Geographic Area Name | 2010 census population |
|---|---|---|---|---|---|---|---|---|---|
| 13 | | 19000 | Columbus city | 189,885 | 13 | | 78800 | Valdosta city | 54,518 |
| 13 | 215 | 19000 | Muscogee County | 189,885 | 13 | 185 | 78800 | Lowndes County | 54,518 |
| 13 | | 21380 | Dalton city | 33,128 | 13 | | 80508 | Warner Robins city | 66,588 |
| 13 | 313 | 21380 | Whitfield County | 33,128 | 13 | 153 | 80508 | Houston County | 66,224 |
| | | | | | 13 | 225 | 80508 | Peach County | 364 |
| 13 | | 23900 | Douglasville city | 30,961 | | | | | |
| 13 | 097 | 23900 | Douglas County | 30,961 | 15 | | | **HAWAII** | 1,360,301 |
| 13 | | 24600 | Duluth city | 26,600 | 15 | | 06290 | East Honolulu **CDP** | 49,914 |
| 13 | 135 | 24600 | Gwinnett County | 26,600 | 15 | 003 | 06290 | Honolulu County | 49,914 |
| 13 | | 24768 | Dunwoody city | 46,267 | 15 | | 14650 | Hilo **CDP** | 43,263 |
| 13 | 089 | 24768 | DeKalb County | 46,267 | 15 | 001 | 14650 | Hawaii County | 43,263 |
| 13 | | 25720 | East Point city | 33,712 | 15 | | 22700 | Kahului **CDP** | 26,337 |
| 13 | 121 | 25720 | Fulton County | 33,712 | 15 | 009 | 22700 | Maui County | 26,337 |
| 13 | | 31908 | Gainesville city | 33,804 | 15 | | 23150 | Kailua **CDP** | 38,635 |
| 13 | 139 | 31908 | Hall County | 33,804 | 15 | 003 | 23150 | Honolulu County | 38,635 |
| 13 | | 38964 | Hinesville city | 33,437 | 15 | | 28250 | Kaneohe **CDP** | 34,597 |
| 13 | 179 | 38964 | Liberty County | 33,437 | 15 | 003 | 28250 | Honolulu County | 34,597 |
| 13 | | 42425 | Johns Creek city | 76,728 | 15 | | 51050 | Mililani Town **CDP** | 27,629 |
| 13 | 121 | 42425 | Fulton County | 76,728 | 15 | 003 | 51050 | Honolulu County | 27,629 |
| 13 | | 43192 | Kennesaw city | 29,783 | 15 | | 62600 | Pearl City **CDP** | 47,698 |
| 13 | 067 | 43192 | Cobb County | 29,783 | 15 | 003 | 62600 | Honolulu County | 47,698 |
| 13 | | 44340 | LaGrange city | 29,588 | 15 | | 71550 | Urban Honolulu **CDP** | 337,256 |
| 13 | 285 | 44340 | Troup County | 29,588 | 15 | 003 | 71550 | Honolulu County | 337,256 |
| 13 | | 45488 | Lawrenceville city | 28,546 | 15 | | 79700 | Waipahu **CDP** | 38,216 |
| 13 | 135 | 45488 | Gwinnett County | 28,546 | 15 | 003 | 79700 | Honolulu County | 38,216 |
| 13 | | 49000 | Macon city | 91,351 | 16 | | | **IDAHO** | 1,567,582 |
| 13 | 021 | 49000 | Bibb County | 90,885 | | | | | |
| 13 | 169 | 49000 | Jones County | 466 | 16 | | 08830 | Boise City city | 205,671 |
| | | | | | 16 | 001 | 08830 | Ada County | 205,671 |
| 13 | | 49756 | Marietta city | 56,579 | | | | | |
| 13 | 067 | 49756 | Cobb County | 56,579 | 16 | | 12250 | Caldwell city | 46,237 |
| | | | | | 16 | 027 | 12250 | Canyon County | 46,237 |
| 13 | | 51670 | Milton city | 32,661 | | | | | |
| 13 | 121 | 51670 | Fulton County | 32,661 | 16 | | 16750 | Coeur d'Alene city | 44,137 |
| | | | | | 16 | 055 | 16750 | Kootenai County | 44,137 |
| 13 | | 55020 | Newnan city | 33,039 | | | | | |
| 13 | 077 | 55020 | Coweta County | 33,039 | 16 | | 39700 | Idaho Falls city | 56,813 |
| | | | | | 16 | 019 | 39700 | Bonneville County | 56,813 |
| 13 | | 59724 | Peachtree City city | 34,364 | | | | | |
| 13 | 113 | 59724 | Fayette County | 34,364 | 16 | | 46540 | Lewiston city | 31,894 |
| | | | | | 16 | 069 | 46540 | Nez Perce County | 31,894 |
| 13 | | 66668 | Rome city | 36,303 | | | | | |
| 13 | 115 | 66668 | Floyd County | 36,303 | 16 | | 52120 | Meridian city | 75,092 |
| | | | | | 16 | 001 | 52120 | Ada County | 75,092 |
| 13 | | 67284 | Roswell city | 88,346 | | | | | |
| 13 | 121 | 67284 | Fulton County | 88,346 | 16 | | 56260 | Nampa city | 81,557 |
| | | | | | 16 | 027 | 56260 | Canyon County | 81,557 |
| 13 | | 68516 | Sandy Springs city | 93,853 | | | | | |
| 13 | 121 | 68516 | Fulton County | 93,853 | 16 | | 64090 | Pocatello city | 54,255 |
| | | | | | 16 | 005 | 64090 | Bannock County | 54,239 |
| 13 | | 69000 | Savannah city | 136,286 | 16 | 077 | 64090 | Power County | 16 |
| 13 | 051 | 69000 | Chatham County | 136,286 | | | | | |
| | | | | | 16 | | 64810 | Post Falls city | 27,574 |
| 13 | | 71492 | Smyrna city | 51,271 | 16 | 055 | 64810 | Kootenai County | 27,574 |
| 13 | 067 | 71492 | Cobb County | 51,271 | | | | | |
| | | | | | 16 | | 67420 | Rexburg city | 25,484 |
| 13 | | 73256 | Statesboro city | 28,422 | 16 | 065 | 67420 | Madison County | 25,484 |
| 13 | 031 | 73256 | Bulloch County | 28,422 | | | | | |
| | | | | | 16 | | 82810 | Twin Falls city | 44,125 |
| 13 | | 73704 | Stockbridge city | 25,636 | 16 | 083 | 82810 | Twin Falls County | 44,125 |
| 13 | 151 | 73704 | Henry County | 25,636 | | | | | |

| State code | Place code | County code | Geographic Area Name | 2010 census population | State code | Place code | County code | Geographic Area Name | 2010 census population |
|---|---|---|---|---|---|---|---|---|---|
| 17 | | | **ILLINOIS** | 12,830,632 | 17 | | 14026 | Chicago Heights city | 30,276 |
| | | | | | 17 | 031 | 14026 | Cook County | 30,276 |
| 17 | | 00243 | Addison village | 36,942 | | | | | |
| 17 | 043 | 00243 | DuPage County | 36,942 | 17 | | 14351 | Cicero town | 83,891 |
| | | | | | 17 | 031 | 14351 | Cook County | 83,891 |
| 17 | | 00685 | Algonquin village | 30,046 | | | | | |
| 17 | 089 | 00685 | Kane County | 8,433 | 17 | | 15599 | Collinsville city | 25,579 |
| 17 | 111 | 00685 | McHenry County | 21,613 | 17 | 119 | 15599 | Madison County | 22,573 |
| | | | | | 17 | 163 | 15599 | St. Clair County | 3,006 |
| 17 | | 01114 | Alton city | 27,865 | | | | | |
| 17 | 119 | 01114 | Madison County | 27,865 | 17 | | 17887 | Crystal Lake city | 40,743 |
| | | | | | 17 | 111 | 17887 | McHenry County | 40,743 |
| 17 | | 02154 | Arlington Heights village | 75,101 | | | | | |
| 17 | 031 | 02154 | Cook County | 75,101 | 17 | | 18563 | Danville city | 33,027 |
| 17 | 097 | 02154 | Lake County | 0 | 17 | 183 | 18563 | Vermilion County | 33,027 |
| 17 | | 03012 | Aurora city | 197,899 | 17 | | 18823 | Decatur city | 76,122 |
| 17 | 043 | 03012 | DuPage County | 49,433 | 17 | 115 | 18823 | Macon County | 76,122 |
| 17 | 089 | 03012 | Kane County | 130,976 | | | | | |
| 17 | 093 | 03012 | Kendall County | 6,019 | 17 | | 19161 | DeKalb city | 43,862 |
| 17 | 197 | 03012 | Will County | 11,471 | 17 | 037 | 19161 | DeKalb County | 43,862 |
| 17 | | 04013 | Bartlett village | 41,208 | 17 | | 19642 | Des Plaines city | 58,364 |
| 17 | 031 | 04013 | Cook County | 16,797 | 17 | 031 | 19642 | Cook County | 58,364 |
| 17 | 043 | 04013 | DuPage County | 24,411 | | | | | |
| 17 | 089 | 04013 | Kane County | 0 | 17 | | 20591 | Downers Grove village | 47,833 |
| | | | | | 17 | 043 | 20591 | DuPage County | 47,833 |
| 17 | | 04078 | Batavia city | 26,045 | | | | | |
| 17 | 043 | 04078 | DuPage County | 0 | 17 | | 22255 | East St. Louis city | 27,006 |
| 17 | 089 | 04078 | Kane County | 26,045 | 17 | 163 | 22255 | St. Clair County | 27,006 |
| 17 | | 04845 | Belleville city | 44,478 | 17 | | 23074 | Elgin city | 108,188 |
| 17 | 163 | 04845 | St. Clair County | 44,478 | 17 | 031 | 23074 | Cook County | 24,032 |
| | | | | | 17 | 089 | 23074 | Kane County | 84,156 |
| 17 | | 05092 | Belvidere city | 25,585 | | | | | |
| 17 | 007 | 05092 | Boone County | 25,585 | 17 | | 23256 | Elk Grove Village village | 33,127 |
| | | | | | 17 | 031 | 23256 | Cook County | 33,127 |
| 17 | | 05573 | Berwyn city | 56,657 | 17 | 043 | 23256 | DuPage County | 0 |
| 17 | 031 | 05573 | Cook County | 56,657 | | | | | |
| | | | | | 17 | | 23620 | Elmhurst city | 44,121 |
| 17 | | 06613 | Bloomington city | 76,610 | 17 | 031 | 23620 | Cook County | 0 |
| 17 | 113 | 06613 | McLean County | 76,610 | 17 | 043 | 23620 | DuPage County | 44,121 |
| 17 | | 07133 | Bolingbrook village | 73,366 | 17 | | 24582 | Evanston city | 74,486 |
| 17 | 043 | 07133 | DuPage County | 1,571 | 17 | 031 | 24582 | Cook County | 74,486 |
| 17 | 197 | 07133 | Will County | 71,795 | | | | | |
| | | | | | 17 | | 27884 | Freeport city | 25,638 |
| 17 | | 09447 | Buffalo Grove village | 41,496 | 17 | 177 | 27884 | Stephenson County | 25,638 |
| 17 | 031 | 09447 | Cook County | 13,644 | | | | | |
| 17 | 097 | 09447 | Lake County | 27,852 | 17 | | 28326 | Galesburg city | 32,195 |
| | | | | | 17 | 095 | 28326 | Knox County | 32,195 |
| 17 | | 09642 | Burbank city | 28,925 | | | | | |
| 17 | 031 | 09642 | Cook County | 28,925 | 17 | | 29730 | Glendale Heights village | 34,208 |
| | | | | | 17 | 043 | 29730 | DuPage County | 34,208 |
| 17 | | 10487 | Calumet City city | 37,042 | | | | | |
| 17 | 031 | 10487 | Cook County | 37,042 | 17 | | 29756 | Glen Ellyn village | 27,450 |
| | | | | | 17 | 043 | 29756 | DuPage County | 27,450 |
| 17 | | 11163 | Carbondale city | 25,902 | | | | | |
| 17 | 077 | 11163 | Jackson County | 25,902 | 17 | | 29938 | Glenview village | 44,692 |
| 17 | 199 | 11163 | Williamson County | 0 | 17 | 031 | 29938 | Cook County | 44,692 |
| 17 | | 11332 | Carol Stream village | 39,711 | 17 | | 30926 | Granite City city | 29,849 |
| 17 | 043 | 11332 | DuPage County | 39,711 | 17 | 119 | 30926 | Madison County | 29,849 |
| 17 | | 11358 | Carpentersville village | 37,691 | 17 | | 32018 | Gurnee village | 31,295 |
| 17 | 089 | 11358 | Kane County | 37,691 | 17 | 097 | 32018 | Lake County | 31,295 |
| 17 | | 12385 | Champaign city | 81,055 | 17 | | 32746 | Hanover Park village | 37,973 |
| 17 | 019 | 12385 | Champaign County | 81,055 | 17 | 031 | 32746 | Cook County | 20,636 |
| | | | | | 17 | 043 | 32746 | DuPage County | 17,337 |
| 17 | | 14000 | Chicago city | 2,695,598 | | | | | |
| 17 | 031 | 14000 | Cook County | 2,695,598 | 17 | | 33383 | Harvey city | 25,282 |
| 17 | 043 | 14000 | DuPage County | 0 | 17 | 031 | 33383 | Cook County | 25,282 |

| State code | Place code | County code | Geographic Area Name | 2010 census population | State code | Place code | County code | Geographic Area Name | 2010 census population |
|---|---|---|---|---|---|---|---|---|---|
| 17 | | 34722 | Highland Park city | 29,763 | 17 | | 57225 | Palatine village | 68,557 |
| 17 | 097 | 34722 | Lake County | 29,763 | 17 | 031 | 57225 | Cook County | 68,557 |
| | | | | | 17 | 097 | 57225 | Lake County | 0 |
| 17 | | 35411 | Hoffman Estates village | 51,895 | | | | | |
| 17 | 031 | 35411 | Cook County | 51,895 | 17 | | 57875 | Park Ridge city | 37,480 |
| 17 | 089 | 35411 | Kane County | 0 | 17 | 031 | 57875 | Cook County | 37,480 |
| 17 | | 38570 | Joliet city | 147,433 | 17 | | 58447 | Pekin city | 34,094 |
| 17 | 093 | 38570 | Kendall County | 9,749 | 17 | 143 | 58447 | Peoria County | 0 |
| 17 | 197 | 38570 | Will County | 137,684 | 17 | 179 | 58447 | Tazewell County | 34,094 |
| 17 | | 38934 | Kankakee city | 27,537 | 17 | | 59000 | Peoria city | 115,007 |
| 17 | 091 | 38934 | Kankakee County | 27,537 | 17 | 143 | 59000 | Peoria County | 115,007 |
| 17 | | 41183 | Lake in the Hills village | 28,965 | 17 | | 60287 | Plainfield village | 39,581 |
| 17 | 111 | 41183 | McHenry County | 28,965 | 17 | 093 | 60287 | Kendall County | 2,079 |
| | | | | | 17 | 197 | 60287 | Will County | 37,502 |
| 17 | | 42028 | Lansing village | 28,331 | | | | | |
| 17 | 031 | 42028 | Cook County | 28,331 | 17 | | 62367 | Quincy city | 40,633 |
| | | | | | 17 | 001 | 62367 | Adams County | 40,633 |
| 17 | | 44407 | Lombard village | 43,165 | | | | | |
| 17 | 043 | 44407 | DuPage County | 43,165 | 17 | | 65000 | Rockford city | 152,871 |
| | | | | | 17 | 201 | 65000 | Winnebago County | 152,871 |
| 17 | | 45694 | McHenry city | 26,992 | | | | | |
| 17 | 111 | 45694 | McHenry County | 26,992 | 17 | | 65078 | Rock Island city | 39,018 |
| | | | | | 17 | 161 | 65078 | Rock Island County | 39,018 |
| 17 | | 48242 | Melrose Park village | 25,411 | | | | | |
| 17 | 031 | 48242 | Cook County | 25,411 | 17 | | 65442 | Romeoville village | 39,680 |
| | | | | | 17 | 197 | 65442 | Will County | 39,680 |
| 17 | | 49867 | Moline city | 43,483 | | | | | |
| 17 | 161 | 49867 | Rock Island County | 43,483 | 17 | | 66040 | Round Lake Beach village | 28,175 |
| | | | | | 17 | 097 | 66040 | Lake County | 28,175 |
| 17 | | 51089 | Mount Prospect village | 54,167 | | | | | |
| 17 | 031 | 51089 | Cook County | 54,167 | 17 | | 66703 | St. Charles city | 32,974 |
| | | | | | 17 | 043 | 66703 | DuPage County | 543 |
| 17 | | 51349 | Mundelein village | 31,064 | 17 | 089 | 66703 | Kane County | 32,431 |
| 17 | 097 | 51349 | Lake County | 31,064 | | | | | |
| | | | | | 17 | | 68003 | Schaumburg village | 74,227 |
| 17 | | 51622 | Naperville city | 141,853 | 17 | 031 | 68003 | Cook County | 74,227 |
| 17 | 043 | 51622 | DuPage County | 94,533 | 17 | 043 | 68003 | DuPage County | 0 |
| 17 | 197 | 51622 | Will County | 47,320 | | | | | |
| | | | | | 17 | | 70122 | Skokie village | 64,784 |
| 17 | | 53000 | Niles village | 29,803 | 17 | 031 | 70122 | Cook County | 64,784 |
| 17 | 031 | 53000 | Cook County | 29,803 | | | | | |
| | | | | | 17 | | 72000 | Springfield city | 116,250 |
| 17 | | 53234 | Normal town | 52,497 | 17 | 167 | 72000 | Sangamon County | 116,250 |
| 17 | 113 | 53234 | McLean County | 52,497 | | | | | |
| | | | | | 17 | | 73157 | Streamwood village | 39,858 |
| 17 | | 53481 | Northbrook village | 33,170 | 17 | 031 | 73157 | Cook County | 39,858 |
| 17 | 031 | 53481 | Cook County | 33,170 | | | | | |
| | | | | | 17 | | 75484 | Tinley Park village | 56,703 |
| 17 | | 53559 | North Chicago city | 32,574 | 17 | 031 | 75484 | Cook County | 49,236 |
| 17 | 097 | 53559 | Lake County | 32,574 | 17 | 197 | 75484 | Will County | 7,467 |
| 17 | | 54638 | Oak Forest city | 27,962 | 17 | | 77005 | Urbana city | 41,250 |
| 17 | 031 | 54638 | Cook County | 27,962 | 17 | 019 | 77005 | Champaign County | 41,250 |
| 17 | | 54820 | Oak Lawn village | 56,690 | 17 | | 77694 | Vernon Hills village | 25,113 |
| 17 | 031 | 54820 | Cook County | 56,690 | 17 | 097 | 77694 | Lake County | 25,113 |
| 17 | | 54885 | Oak Park village | 51,878 | 17 | | 79293 | Waukegan city | 89,078 |
| 17 | 031 | 54885 | Cook County | 51,878 | 17 | 097 | 79293 | Lake County | 89,078 |
| 17 | | 55249 | O'Fallon city | 28,281 | 17 | | 80060 | West Chicago city | 27,086 |
| 17 | 163 | 55249 | St. Clair County | 28,281 | 17 | 043 | 80060 | DuPage County | 27,086 |
| 17 | | 56640 | Orland Park village | 56,767 | 17 | | 81048 | Wheaton city | 52,894 |
| 17 | 031 | 56640 | Cook County | 56,583 | 17 | 043 | 81048 | DuPage County | 52,894 |
| 17 | 197 | 56640 | Will County | 184 | | | | | |
| | | | | | 17 | | 81087 | Wheeling village | 37,648 |
| 17 | | 56887 | Oswego village | 30,355 | 17 | 031 | 81087 | Cook County | 37,642 |
| 17 | 093 | 56887 | Kendall County | 30,355 | 17 | 097 | 81087 | Lake County | 6 |

| State code | Place code | County code | Geographic Area Name | 2010 census population | State code | Place code | County code | Geographic Area Name | 2010 census population |
|---|---|---|---|---|---|---|---|---|---|
| 17 | | 82075 | Wilmette village | 27,087 | 18 | | 48798 | Michigan City city | 31,479 |
| 17 | 031 | 82075 | Cook County | 27,087 | 18 | 091 | 48798 | LaPorte County | 31,479 |
| 17 | | 83245 | Woodridge village | 32,971 | 18 | | 49932 | Mishawaka city | 48,252 |
| 17 | 031 | 83245 | Cook County | 0 | 18 | 141 | 49932 | St. Joseph County | 48,252 |
| 17 | 043 | 83245 | DuPage County | 32,949 | 18 | | 51876 | Muncie city | 70,085 |
| 17 | 197 | 83245 | Will County | 22 | 18 | 035 | 51876 | Delaware County | 70,085 |
| 18 | | | **INDIANA** | 6,483,802 | 18 | | 52326 | New Albany city | 36,372 |
| | | | | | 18 | 043 | 52326 | Floyd County | 36,372 |
| 18 | | 01468 | Anderson city | 56,129 | 18 | | 54180 | Noblesville city | 51,969 |
| 18 | 095 | 01468 | Madison County | 56,129 | 18 | 057 | 54180 | Hamilton County | 51,969 |
| 18 | | 05860 | Bloomington city | 80,405 | 18 | | 60246 | Plainfield town | 27,631 |
| 18 | 105 | 05860 | Monroe County | 80,405 | 18 | 063 | 60246 | Hendricks County | 27,631 |
| 18 | | 10342 | Carmel city | 79,191 | 18 | | 61092 | Portage city | 36,828 |
| 18 | 057 | 10342 | Hamilton County | 79,191 | 18 | 127 | 61092 | Porter County | 36,828 |
| 18 | | 14734 | Columbus city | 44,061 | 18 | | 64260 | Richmond city | 36,812 |
| 18 | 005 | 14734 | Bartholomew County | 44,061 | 18 | 177 | 64260 | Wayne County | 36,812 |
| 18 | | 16138 | Crown Point city | 27,317 | 18 | | 68220 | Schererville town | 29,243 |
| 18 | 089 | 16138 | Lake County | 27,317 | 18 | 089 | 68220 | Lake County | 29,243 |
| 18 | | 19486 | East Chicago city | 29,698 | 18 | | 71000 | South Bend city | 101,168 |
| 18 | 089 | 19486 | Lake County | 29,698 | 18 | 141 | 71000 | St. Joseph County | 101,168 |
| 18 | | 20728 | Elkhart city | 50,949 | 18 | | 75428 | Terre Haute city | 60,785 |
| 18 | 039 | 20728 | Elkhart County | 50,949 | 18 | 167 | 75428 | Vigo County | 60,785 |
| 18 | | 22000 | Evansville city | 117,429 | 18 | | 78326 | Valparaiso city | 31,730 |
| 18 | 163 | 22000 | Vanderburgh County | 117,429 | 18 | 127 | 78326 | Porter County | 31,730 |
| 18 | | 23278 | Fishers town | 76,794 | 18 | | 82700 | Westfield town | 30,068 |
| 18 | 057 | 23278 | Hamilton County | 76,794 | 18 | 057 | 82700 | Hamilton County | 30,068 |
| 18 | | 25000 | Fort Wayne city | 253,691 | 18 | | 82862 | West Lafayette city | 29,596 |
| 18 | 003 | 25000 | Allen County | 253,691 | 18 | 157 | 82862 | Tippecanoe County | 29,596 |
| 18 | | 27000 | Gary city | 80,294 | 19 | | | **IOWA** | 3,046,355 |
| 18 | 089 | 27000 | Lake County | 80,294 | | | | | |
| 18 | | 28386 | Goshen city | 31,719 | 19 | | 01855 | Ames city | 58,965 |
| 18 | 039 | 28386 | Elkhart County | 31,719 | 19 | 169 | 01855 | Story County | 58,965 |
| 18 | | 29898 | Greenwood city | 49,791 | 19 | | 02305 | Ankeny city | 45,582 |
| 18 | 081 | 29898 | Johnson County | 49,791 | 19 | 153 | 02305 | Polk County | 45,582 |
| 18 | | 31000 | Hammond city | 80,830 | 19 | | 06355 | Bettendorf city | 33,217 |
| 18 | 089 | 31000 | Lake County | 80,830 | 19 | 163 | 06355 | Scott County | 33,217 |
| 18 | | 34114 | Hobart city | 29,059 | 19 | | 09550 | Burlington city | 25,663 |
| 18 | 089 | 34114 | Lake County | 29,059 | 19 | 057 | 09550 | Des Moines County | 25,663 |
| 18 | | 38358 | Jeffersonville city | 44,953 | 19 | | 11755 | Cedar Falls city | 39,260 |
| 18 | 019 | 38358 | Clark County | 44,953 | 19 | 013 | 11755 | Black Hawk County | 39,260 |
| 18 | | 40392 | Kokomo city | 45,468 | 19 | | 12000 | Cedar Rapids city | 126,326 |
| 18 | 067 | 40392 | Howard County | 45,468 | 19 | 113 | 12000 | Linn County | 126,326 |
| 18 | | 40788 | Lafayette city | 67,140 | 19 | | 14430 | Clinton city | 26,885 |
| 18 | 157 | 40788 | Tippecanoe County | 67,140 | 19 | 045 | 14430 | Clinton County | 26,885 |
| 18 | | 42426 | Lawrence city | 46,001 | 19 | | 16860 | Council Bluffs city | 62,230 |
| 18 | 097 | 42426 | Marion County | 46,001 | 19 | 155 | 16860 | Pottawattamie County | 62,230 |
| 18 | | 46908 | Marion city | 29,948 | 19 | | 19000 | Davenport city | 99,685 |
| 18 | 053 | 46908 | Grant County | 29,948 | 19 | 163 | 19000 | Scott County | 99,685 |
| 18 | | 48528 | Merrillville town | 35,246 | 19 | | 21000 | Des Moines city | 203,433 |
| 18 | 089 | 48528 | Lake County | 35,246 | 19 | 153 | 21000 | Polk County | 203,419 |
| | | | | | 19 | 181 | 21000 | Warren County | 14 |

| State code | Place code | County code | Geographic Area Name | 2010 census population | State code | Place code | County code | Geographic Area Name | 2010 census population |
|---|---|---|---|---|---|---|---|---|---|
| 19 | | 22395 | Dubuque city | 57,637 | 20 | | 62700 | Salina city | 47,707 |
| 19 | 061 | 22395 | Dubuque County | 57,637 | 20 | 169 | 62700 | Saline County | 47,707 |
| 19 | | 28515 | Fort Dodge city | 25,206 | 20 | | 64500 | Shawnee city | 62,209 |
| 19 | 187 | 28515 | Webster County | 25,206 | 20 | 091 | 64500 | Johnson County | 62,209 |
| 19 | | 38595 | Iowa City city | 67,862 | 20 | | 71000 | Topeka city | 127,473 |
| 19 | 103 | 38595 | Johnson County | 67,862 | 20 | 177 | 71000 | Shawnee County | 127,473 |
| 19 | | 49485 | Marion city | 34,768 | 20 | | 79000 | Wichita city | 382,368 |
| 19 | 113 | 49485 | Linn County | 34,768 | 20 | 173 | 79000 | Sedgwick County | 382,368 |
| 19 | | 49755 | Marshalltown city | 27,552 | 21 | | | **KENTUCKY** | 4,339,367 |
| 19 | 127 | 49755 | Marshall County | 27,552 | | | | | |
| 19 | | 50160 | Mason City city | 28,079 | 21 | | 08902 | Bowling Green city | 58,067 |
| 19 | 033 | 50160 | Cerro Gordo County | 28,079 | 21 | 227 | 08902 | Warren County | 58,067 |
| 19 | | 60465 | Ottumwa city | 25,023 | 21 | | 17848 | Covington city | 40,640 |
| 19 | 179 | 60465 | Wapello County | 25,023 | 21 | 117 | 17848 | Kenton County | 40,640 |
| 19 | | 73335 | Sioux City city | 82,684 | 21 | | 24274 | Elizabethtown city | 28,531 |
| 19 | 149 | 73335 | Plymouth County | 6 | 21 | 093 | 24274 | Hardin County | 28,531 |
| 19 | 193 | 73335 | Woodbury County | 82,678 | 21 | | 27982 | Florence city | 29,951 |
| 19 | | 79950 | Urbandale city | 39,463 | 21 | 015 | 27982 | Boone County | 29,951 |
| 19 | 049 | 79950 | Dallas County | 6,337 | 21 | | 28900 | Frankfort city | 25,527 |
| 19 | 153 | 79950 | Polk County | 33,126 | 21 | 073 | 28900 | Franklin County | 25,527 |
| 19 | | 82425 | Waterloo city | 68,406 | 21 | | 30700 | Georgetown city | 29,098 |
| 19 | 013 | 82425 | Black Hawk County | 68,406 | 21 | 209 | 30700 | Scott County | 29,098 |
| 19 | | 83910 | West Des Moines city | 56,609 | 21 | | 35866 | Henderson city | 28,757 |
| 19 | 049 | 83910 | Dallas County | 11,569 | 21 | 101 | 35866 | Henderson County | 28,757 |
| 19 | 153 | 83910 | Polk County | 44,999 | 21 | | 37918 | Hopkinsville city | 31,577 |
| 19 | 181 | 83910 | Warren County | 41 | 21 | 047 | 37918 | Christian County | 31,577 |
| 20 | | | **KANSAS** | 2,853,118 | 21 | | 40222 | Jeffersontown city | 26,595 |
| | | | | | 21 | 111 | 40222 | Jefferson County | 26,595 |
| 20 | | 18250 | Dodge City city | 27,340 | 21 | | 46027 | Lexington-Fayette urban county | 295,803 |
| 20 | 057 | 18250 | Ford County | 27,340 | 21 | 067 | 46027 | Fayette County | 295,803 |
| 20 | | 25325 | Garden City city | 26,658 | 21 | | 56136 | Nicholasville city | 28,015 |
| 20 | 055 | 25325 | Finney County | 26,658 | 21 | 113 | 56136 | Jessamine County | 28,015 |
| 20 | | 33625 | Hutchinson city | 42,080 | 21 | | 58620 | Owensboro city | 57,265 |
| 20 | 155 | 33625 | Reno County | 42,080 | 21 | 059 | 58620 | Daviess County | 57,265 |
| 20 | | 36000 | Kansas City city | 145,786 | 21 | | 58836 | Paducah city | 25,024 |
| 20 | 209 | 36000 | Wyandotte County | 145,786 | 21 | 145 | 58836 | McCracken County | 25,024 |
| 20 | | 38900 | Lawrence city | 87,643 | 21 | | 65226 | Richmond city | 31,364 |
| 20 | 045 | 38900 | Douglas County | 87,643 | 21 | 151 | 65226 | Madison County | 31,364 |
| 20 | | 39000 | Leavenworth city | 35,251 | 22 | | | **LOUISIANA** | 4,533,372 |
| 20 | 103 | 39000 | Leavenworth County | 35,251 | | | | | |
| 20 | | 39075 | Leawood city | 31,867 | 22 | | 00975 | Alexandria city | 47,723 |
| 20 | 091 | 39075 | Johnson County | 31,867 | 22 | 079 | 00975 | Rapides Parish | 47,723 |
| 20 | | 39350 | Lenexa city | 48,190 | 22 | | 05000 | Baton Rouge city | 229,493 |
| 20 | 091 | 39350 | Johnson County | 48,190 | 22 | 033 | 05000 | East Baton Rouge Parish | 229,493 |
| 20 | | 44250 | Manhattan city | 52,281 | 22 | | 08920 | Bossier City city | 61,315 |
| 20 | 149 | 44250 | Pottawatomie County | 146 | 22 | 015 | 08920 | Bossier Parish | 61,315 |
| 20 | 161 | 44250 | Riley County | 52,135 | 22 | | 13960 | Central city | 26,864 |
| 20 | | 52575 | Olathe city | 125,872 | 22 | 033 | 13960 | East Baton Rouge Parish | 26,864 |
| 20 | 091 | 52575 | Johnson County | 125,872 | 22 | | 36255 | Houma city | 33,727 |
| 20 | | 53775 | Overland Park city | 173,372 | 22 | 109 | 36255 | Terrebonne Parish | 33,727 |
| 20 | 091 | 53775 | Johnson County | 173,372 | | | | | |

| State code | Place code | County code | Geographic Area Name | 2010 census population | State code | Place code | County code | Geographic Area Name | 2010 census population |
|---|---|---|---|---|---|---|---|---|---|
| 22 | | 39475 | Kenner city | 66,702 | 25 | | | **MASSACHUSETTS** | 6,547,629 |
| 22 | 051 | 39475 | Jefferson Parish | 66,702 | 25 | | 00840 | Agawam Town city | 28,438 |
| 22 | | 40735 | Lafayette city | 120,623 | 25 | 013 | 00840 | Hampden County | 28,438 |
| 22 | 055 | 40735 | Lafayette Parish | 120,623 | 25 | | 02690 | Attleboro city | 43,593 |
| 22 | | 41155 | Lake Charles city | 71,993 | 25 | 005 | 02690 | Bristol County | 43,593 |
| 22 | 019 | 41155 | Calcasieu Parish | 71,993 | 25 | | 03690 | Barnstable Town city | 45,193 |
| 22 | | 51410 | Monroe city | 48,815 | 25 | 001 | 03690 | Barnstable County | 45,193 |
| 22 | 073 | 51410 | Ouachita Parish | 48,815 | 25 | | 05595 | Beverly city | 39,502 |
| 22 | | 54035 | New Iberia city | 30,617 | 25 | 009 | 05595 | Essex County | 39,502 |
| 22 | 045 | 54035 | Iberia Parish | 30,617 | 25 | | 07000 | Boston city | 617,594 |
| 22 | | 55000 | New Orleans city | 343,829 | 25 | 025 | 07000 | Suffolk County | 617,594 |
| 22 | 071 | 55000 | Orleans Parish | 343,829 | 25 | | 07740 | Braintree Town city | 35,744 |
| 22 | | 70000 | Shreveport city | 199,311 | 25 | 021 | 07740 | Norfolk County | 35,744 |
| 22 | 015 | 70000 | Bossier Parish | 2,702 | 25 | | 09000 | Brockton city | 93,810 |
| 22 | 017 | 70000 | Caddo Parish | 196,609 | 25 | 023 | 09000 | Plymouth County | 93,810 |
| 22 | | 70805 | Slidell city | 27,068 | 25 | | 11000 | Cambridge city | 105,162 |
| 22 | 103 | 70805 | St. Tammany Parish | 27,068 | 25 | 017 | 11000 | Middlesex County | 105,162 |
| 23 | | | **MAINE** | 1,328,361 | 25 | | 13205 | Chelsea city | 35,177 |
| 23 | | 02795 | Bangor city | 33,039 | 25 | 025 | 13205 | Suffolk County | 35,177 |
| 23 | 019 | 02795 | Penobscot County | 33,039 | 25 | | 13660 | Chicopee city | 55,298 |
| 23 | | 38740 | Lewiston city | 36,592 | 25 | 013 | 13660 | Hampden County | 55,298 |
| 23 | 001 | 38740 | Androscoggin County | 36,592 | 25 | | 21990 | Everett city | 41,667 |
| 23 | | 60545 | Portland city | 66,194 | 25 | 017 | 21990 | Middlesex County | 41,667 |
| 23 | 005 | 60545 | Cumberland County | 66,194 | 25 | | 23000 | Fall River city | 88,857 |
| 23 | | 71990 | South Portland city | 25,002 | 25 | 005 | 23000 | Bristol County | 88,857 |
| 23 | 005 | 71990 | Cumberland County | 25,002 | 25 | | 23875 | Fitchburg city | 40,318 |
| 24 | | | **MARYLAND** | 5,773,552 | 25 | 027 | 23875 | Worcester County | 40,318 |
| 24 | | 01600 | Annapolis city | 38,394 | 25 | | 25172 | Franklin Town city | 31,635 |
| 24 | 003 | 01600 | Anne Arundel County | 38,394 | 25 | 021 | 25172 | Norfolk County | 31,635 |
| 24 | | 04000 | Baltimore city | 620,961 | 25 | | 26150 | Gloucester city | 28,789 |
| 24 | 510 | 04000 | Baltimore city | 620,961 | 25 | 009 | 26150 | Essex County | 28,789 |
| 24 | | 08775 | Bowie city | 54,727 | 25 | | 29405 | Haverhill city | 60,879 |
| 24 | 033 | 08775 | Prince George's County | 54,727 | 25 | 009 | 29405 | Essex County | 60,879 |
| 24 | | 18750 | College Park city | 30,413 | 25 | | 30840 | Holyoke city | 39,880 |
| 24 | 033 | 18750 | Prince George's County | 30,413 | 25 | 013 | 30840 | Hampden County | 39,880 |
| 24 | | 30325 | Frederick city | 65,239 | 25 | | 34550 | Lawrence city | 76,377 |
| 24 | 021 | 30325 | Frederick County | 65,239 | 25 | 009 | 34550 | Essex County | 76,377 |
| 24 | | 31175 | Gaithersburg city | 59,933 | 25 | | 35075 | Leominster city | 40,759 |
| 24 | 031 | 31175 | Montgomery County | 59,933 | 25 | 027 | 35075 | Worcester County | 40,759 |
| 24 | | 36075 | Hagerstown city | 39,662 | 25 | | 37000 | Lowell city | 106,519 |
| 24 | 043 | 36075 | Washington County | 39,662 | 25 | 017 | 37000 | Middlesex County | 106,519 |
| 24 | | 45900 | Laurel city | 25,115 | 25 | | 37490 | Lynn city | 90,329 |
| 24 | 033 | 45900 | Prince George's County | 25,115 | 25 | 009 | 37490 | Essex County | 90,329 |
| 24 | | 67675 | Rockville city | 61,209 | 25 | | 37875 | Malden city | 59,450 |
| 24 | 031 | 67675 | Montgomery County | 61,209 | 25 | 017 | 37875 | Middlesex County | 59,450 |
| 24 | | 69925 | Salisbury city | 30,343 | 25 | | 38715 | Marlborough city | 38,499 |
| 24 | 045 | 69925 | Wicomico County | 30,343 | 25 | 017 | 38715 | Middlesex County | 38,499 |
| | | | | | 25 | | 39835 | Medford city | 56,173 |
| | | | | | 25 | 017 | 39835 | Middlesex County | 56,173 |

| State code | Place code | County code | Geographic Area Name | 2010 census population | State code | Place code | County code | Geographic Area Name | 2010 census population |
|---|---|---|---|---|---|---|---|---|---|
| 25 | | 40115 | Melrose city | 26,983 | 26 | | 12060 | Burton city | 29,999 |
| 25 | 017 | 40115 | Middlesex County | 26,983 | 26 | 049 | 12060 | Genesee County | 29,999 |
| 25 | | 40710 | Methuen Town city | 47,255 | 26 | | 21000 | Dearborn city | 98,153 |
| 25 | 009 | 40710 | Essex County | 47,255 | 26 | 163 | 21000 | Wayne County | 98,153 |
| 25 | | 45000 | New Bedford city | 95,072 | 26 | | 21020 | Dearborn Heights city | 57,774 |
| 25 | 005 | 45000 | Bristol County | 95,072 | 26 | 163 | 21020 | Wayne County | 57,774 |
| 25 | | 45560 | Newton city | 85,146 | 26 | | 22000 | Detroit city | 713,777 |
| 25 | 017 | 45560 | Middlesex County | 85,146 | 26 | 163 | 22000 | Wayne County | 713,777 |
| 25 | | 46330 | Northampton city | 28,549 | 26 | | 24120 | East Lansing city | 48,579 |
| 25 | 015 | 46330 | Hampshire County | 28,549 | 26 | 037 | 24120 | Clinton County | 1,969 |
| | | | | | 26 | 065 | 24120 | Ingham County | 46,610 |
| 25 | | 52490 | Peabody city | 51,251 | 26 | | 24290 | Eastpointe city | 32,442 |
| 25 | 009 | 52490 | Essex County | 51,251 | 26 | 099 | 24290 | Macomb County | 32,442 |
| 25 | | 53960 | Pittsfield city | 44,737 | 26 | | 27440 | Farmington Hills city | 79,740 |
| 25 | 003 | 53960 | Berkshire County | 44,737 | 26 | 125 | 27440 | Oakland County | 79,740 |
| 25 | | 55745 | Quincy city | 92,271 | 26 | | 29000 | Flint city | 102,434 |
| 25 | 021 | 55745 | Norfolk County | 92,271 | 26 | 049 | 29000 | Genesee County | 102,434 |
| 25 | | 56585 | Revere city | 51,755 | 26 | | 31420 | Garden City city | 27,692 |
| 25 | 025 | 56585 | Suffolk County | 51,755 | 26 | 163 | 31420 | Wayne County | 27,692 |
| 25 | | 59105 | Salem city | 41,340 | 26 | | 34000 | Grand Rapids city | 188,040 |
| 25 | 009 | 59105 | Essex County | 41,340 | 26 | 081 | 34000 | Kent County | 188,040 |
| 25 | | 62535 | Somerville city | 75,754 | 26 | | 38640 | Holland city | 33,051 |
| 25 | 017 | 62535 | Middlesex County | 75,754 | 26 | 005 | 38640 | Allegan County | 7,016 |
| | | | | | 26 | 139 | 38640 | Ottawa County | 26,035 |
| 25 | | 67000 | Springfield city | 153,060 | 26 | | 40680 | Inkster city | 25,369 |
| 25 | 013 | 67000 | Hampden County | 153,060 | 26 | 163 | 40680 | Wayne County | 25,369 |
| 25 | | 69170 | Taunton city | 55,874 | 26 | | 41420 | Jackson city | 33,534 |
| 25 | 005 | 69170 | Bristol County | 55,874 | 26 | 075 | 41420 | Jackson County | 33,534 |
| 25 | | 72600 | Waltham city | 60,632 | 26 | | 42160 | Kalamazoo city | 74,262 |
| 25 | 017 | 72600 | Middlesex County | 60,632 | 26 | 077 | 42160 | Kalamazoo County | 74,262 |
| 25 | | 73440 | Watertown Town city | 31,915 | 26 | | 42820 | Kentwood city | 48,707 |
| 25 | 017 | 73440 | Middlesex County | 31,915 | 26 | 081 | 42820 | Kent County | 48,707 |
| 25 | | 76030 | Westfield city | 41,094 | 26 | | 46000 | Lansing city | 114,297 |
| 25 | 013 | 76030 | Hampden County | 41,094 | 26 | 045 | 46000 | Eaton County | 4,734 |
| | | | | | 26 | 065 | 46000 | Ingham County | 109,563 |
| 25 | | 77890 | West Springfield Town city | 28,391 | 26 | | 47800 | Lincoln Park city | 38,144 |
| 25 | 013 | 77890 | Hampden County | 28,391 | 26 | 163 | 47800 | Wayne County | 38,144 |
| 25 | | 78972 | Weymouth Town city | 53,743 | 26 | | 49000 | Livonia city | 96,942 |
| 25 | 021 | 78972 | Norfolk County | 53,743 | 26 | 163 | 49000 | Wayne County | 96,942 |
| 25 | | 81035 | Woburn city | 38,120 | 26 | | 50560 | Madison Heights city | 29,694 |
| 25 | 017 | 81035 | Middlesex County | 38,120 | 26 | 125 | 50560 | Oakland County | 29,694 |
| 25 | | 82000 | Worcester city | 181,045 | 26 | | 53780 | Midland city | 41,863 |
| 25 | 027 | 82000 | Worcester County | 181,045 | 26 | 017 | 53780 | Bay County | 157 |
| | | | | | 26 | 111 | 53780 | Midland County | 41,706 |
| 26 | | | **MICHIGAN** | 9,883,640 | 26 | | 56020 | Mount Pleasant city | 26,016 |
| | | | | | 26 | 073 | 56020 | Isabella County | 26,016 |
| 26 | | 01380 | Allen Park city | 28,210 | 26 | | 56320 | Muskegon city | 38,401 |
| 26 | 163 | 01380 | Wayne County | 28,210 | 26 | 121 | 56320 | Muskegon County | 38,401 |
| 26 | | 03000 | Ann Arbor city | 113,934 | 26 | | 59440 | Novi city | 55,224 |
| 26 | 161 | 03000 | Washtenaw County | 113,934 | 26 | 125 | 59440 | Oakland County | 55,224 |
| 26 | | 05920 | Battle Creek city | 52,347 | | | | | |
| 26 | 025 | 05920 | Calhoun County | 52,347 | | | | | |
| 26 | | 06020 | Bay City city | 34,932 | | | | | |
| 26 | 017 | 06020 | Bay County | 34,932 | | | | | |

| State code | Place code | County code | Geographic Area Name | 2010 census population | State code | Place code | County code | Geographic Area Name | 2010 census population |
|---|---|---|---|---|---|---|---|---|---|
| 26 | | 59920 | Oak Park city | 29,319 | 27 | | 08794 | Burnsville city | 60,306 |
| 26 | 125 | 59920 | Oakland County | 29,319 | 27 | 037 | 08794 | Dakota County | 60,306 |
| 26 | | 65440 | Pontiac city | 59,515 | 27 | | 13114 | Coon Rapids city | 61,476 |
| 26 | 125 | 65440 | Oakland County | 59,515 | 27 | 003 | 13114 | Anoka County | 61,476 |
| 26 | | 65560 | Portage city | 46,292 | 27 | | 13456 | Cottage Grove city | 34,589 |
| 26 | 077 | 65560 | Kalamazoo County | 46,292 | 27 | 163 | 13456 | Washington County | 34,589 |
| 26 | | 65820 | Port Huron city | 30,184 | 27 | | 17000 | Duluth city | 86,265 |
| 26 | 147 | 65820 | St. Clair County | 30,184 | 27 | 137 | 17000 | St. Louis County | 86,265 |
| 26 | | 69035 | Rochester Hills city | 70,995 | 27 | | 17288 | Eagan city | 64,206 |
| 26 | 125 | 69035 | Oakland County | 70,995 | 27 | 037 | 17288 | Dakota County | 64,206 |
| 26 | | 69800 | Roseville city | 47,299 | 27 | | 18116 | Eden Prairie city | 60,797 |
| 26 | 099 | 69800 | Macomb County | 47,299 | 27 | 053 | 18116 | Hennepin County | 60,797 |
| 26 | | 70040 | Royal Oak city | 57,236 | 27 | | 18188 | Edina city | 47,941 |
| 26 | 125 | 70040 | Oakland County | 57,236 | 27 | 053 | 18188 | Hennepin County | 47,941 |
| 26 | | 70520 | Saginaw city | 51,508 | 27 | | 22814 | Fridley city | 27,208 |
| 26 | 145 | 70520 | Saginaw County | 51,508 | 27 | 003 | 22814 | Anoka County | 27,208 |
| 26 | | 70760 | St. Clair Shores city | 59,715 | 27 | | 31076 | Inver Grove Heights city | 33,880 |
| 26 | 099 | 70760 | Macomb County | 59,715 | 27 | 037 | 31076 | Dakota County | 33,880 |
| 26 | | 74900 | Southfield city | 71,739 | 27 | | 35180 | Lakeville city | 55,954 |
| 26 | 125 | 74900 | Oakland County | 71,739 | 27 | 037 | 35180 | Dakota County | 55,954 |
| 26 | | 74960 | Southgate city | 30,047 | 27 | | 39878 | Mankato city | 39,309 |
| 26 | 163 | 74960 | Wayne County | 30,047 | 27 | 013 | 39878 | Blue Earth County | 39,305 |
| | | | | | 27 | 079 | 39878 | Le Sueur County | 4 |
| 26 | | 76460 | Sterling Heights city | 129,699 | 27 | 103 | 39878 | Nicollet County | 0 |
| 26 | 099 | 76460 | Macomb County | 129,699 | 27 | | 40166 | Maple Grove city | 61,567 |
| 26 | | 79000 | Taylor city | 63,131 | 27 | 053 | 40166 | Hennepin County | 61,567 |
| 26 | 163 | 79000 | Wayne County | 63,131 | 27 | | 40382 | Maplewood city | 38,018 |
| 26 | | 80700 | Troy city | 80,980 | 27 | 123 | 40382 | Ramsey County | 38,018 |
| 26 | 125 | 80700 | Oakland County | 80,980 | 27 | | 43000 | Minneapolis city | 382,578 |
| 26 | | 84000 | Warren city | 134,056 | 27 | 053 | 43000 | Hennepin County | 382,578 |
| 26 | 099 | 84000 | Macomb County | 134,056 | 27 | | 43252 | Minnetonka city | 49,734 |
| 26 | | 86000 | Westland city | 84,094 | 27 | 053 | 43252 | Hennepin County | 49,734 |
| 26 | 163 | 86000 | Wayne County | 84,094 | 27 | | 43864 | Moorhead city | 38,065 |
| 26 | | 88900 | Wyandotte city | 25,883 | 27 | 027 | 43864 | Clay County | 38,065 |
| 26 | 163 | 88900 | Wayne County | 25,883 | 27 | | 47680 | Oakdale city | 27,378 |
| 26 | | 88940 | Wyoming city | 72,125 | 27 | 163 | 47680 | Washington County | 27,378 |
| 26 | 081 | 88940 | Kent County | 72,125 | 27 | | 49300 | Owatonna city | 25,599 |
| 27 | | | **MINNESOTA** | 5,303,925 | 27 | 147 | 49300 | Steele County | 25,599 |
| 27 | | 01486 | Andover city | 30,598 | 27 | | 51730 | Plymouth city | 70,576 |
| 27 | 003 | 01486 | Anoka County | 30,598 | 27 | 053 | 51730 | Hennepin County | 70,576 |
| 27 | | 01900 | Apple Valley city | 49,084 | 27 | | 54214 | Richfield city | 35,228 |
| 27 | 037 | 01900 | Dakota County | 49,084 | 27 | 053 | 54214 | Hennepin County | 35,228 |
| 27 | | 06382 | Blaine city | 57,186 | 27 | | 54880 | Rochester city | 106,769 |
| 27 | 003 | 06382 | Anoka County | 57,186 | 27 | 109 | 54880 | Olmsted County | 106,769 |
| 27 | 123 | 06382 | Ramsey County | 0 | 27 | | 55852 | Roseville city | 33,660 |
| 27 | | 06616 | Bloomington city | 82,893 | 27 | 123 | 55852 | Ramsey County | 33,660 |
| 27 | 053 | 06616 | Hennepin County | 82,893 | 27 | | 56896 | St. Cloud city | 65,842 |
| 27 | | 07948 | Brooklyn Center city | 30,104 | 27 | 009 | 56896 | Benton County | 6,396 |
| 27 | 053 | 07948 | Hennepin County | 30,104 | 27 | 141 | 56896 | Sherburne County | 6,785 |
| 27 | | 07966 | Brooklyn Park city | 75,781 | 27 | 145 | 56896 | Stearns County | 52,661 |
| 27 | 053 | 07966 | Hennepin County | 75,781 | | | | | |

| State code | Place code | County code | Geographic Area Name | 2010 census population | State code | Place code | County code | Geographic Area Name | 2010 census population |
|---|---|---|---|---|---|---|---|---|---|
| 27 | | 57220 | St. Louis Park city | 45,250 | 29 | | 13600 | Chesterfield city | 47,484 |
| 27 | 053 | 57220 | Hennepin County | 45,250 | 29 | 189 | 13600 | St. Louis County | 47,484 |
| 27 | | 58000 | St. Paul city | 285,068 | 29 | | 15670 | Columbia city | 108,500 |
| 27 | 123 | 58000 | Ramsey County | 285,068 | 29 | 019 | 15670 | Boone County | 108,500 |
| 27 | | 58738 | Savage city | 26,911 | 29 | | 24778 | Florissant city | 52,158 |
| 27 | 139 | 58738 | Scott County | 26,911 | 29 | 189 | 24778 | St. Louis County | 52,158 |
| 27 | | 59350 | Shakopee city | 37,076 | 29 | | 27190 | Gladstone city | 25,410 |
| 27 | 139 | 59350 | Scott County | 37,076 | 29 | 047 | 27190 | Clay County | 25,410 |
| 27 | | 59998 | Shoreview city | 25,043 | 29 | | 31276 | Hazelwood city | 25,703 |
| 27 | 123 | 59998 | Ramsey County | 25,043 | 29 | 189 | 31276 | St. Louis County | 25,703 |
| 27 | | 71032 | Winona city | 27,592 | 29 | | 35000 | Independence city | 116,830 |
| 27 | 169 | 71032 | Winona County | 27,592 | 29 | 047 | 35000 | Clay County | 0 |
| 27 | | 71428 | Woodbury city | 61,961 | 29 | 095 | 35000 | Jackson County | 116,830 |
| 27 | 163 | 71428 | Washington County | 61,961 | 29 | 027 | 37000 | Jefferson City city | 43,079 |
| 28 | | | **MISSISSIPPI** | 2,967,297 | 29 | 027 | 37000 | Callaway County | 22 |
| | | | | | 29 | 051 | 37000 | Cole County | 43,057 |
| 28 | | 06220 | Biloxi city | 44,054 | 29 | | 37592 | Joplin city | 50,150 |
| 28 | 047 | 06220 | Harrison County | 44,054 | 29 | 097 | 37592 | Jasper County | 43,955 |
| 28 | | 14420 | Clinton city | 25,216 | 29 | 145 | 37592 | Newton County | 6,195 |
| 28 | 049 | 14420 | Hinds County | 25,216 | 29 | | 38000 | Kansas City city | 459,787 |
| 28 | | 29180 | Greenville city | 34,400 | 29 | 037 | 38000 | Cass County | 197 |
| 28 | 151 | 29180 | Washington County | 34,400 | 29 | 047 | 38000 | Clay County | 113,415 |
| 28 | | 29700 | Gulfport city | 67,793 | 29 | 095 | 38000 | Jackson County | 302,499 |
| 28 | 047 | 29700 | Harrison County | 67,793 | 29 | 165 | 38000 | Platte County | 43,676 |
| 28 | | 31020 | Hattiesburg city | 45,989 | 29 | | 39044 | Kirkwood city | 27,540 |
| 28 | 035 | 31020 | Forrest County | 41,000 | 29 | 189 | 39044 | St. Louis County | 27,540 |
| 28 | 073 | 31020 | Lamar County | 4,989 | 29 | | 41348 | Lee's Summit city | 91,364 |
| 28 | | 33700 | Horn Lake city | 26,066 | 29 | 037 | 41348 | Cass County | 1,917 |
| 28 | 033 | 33700 | DeSoto County | 26,066 | 29 | 095 | 41348 | Jackson County | 89,447 |
| 28 | | 36000 | Jackson city | 173,514 | 29 | | 42032 | Liberty city | 29,149 |
| 28 | 049 | 36000 | Hinds County | 172,891 | 29 | 047 | 42032 | Clay County | 29,149 |
| 28 | 089 | 36000 | Madison County | 622 | 29 | | 46586 | Maryland Heights city | 27,472 |
| 28 | 121 | 36000 | Rankin County | 1 | 29 | 189 | 46586 | St. Louis County | 27,472 |
| 28 | | 46640 | Meridian city | 41,148 | 29 | | 54074 | O'Fallon city | 79,329 |
| 28 | 075 | 46640 | Lauderdale County | 41,148 | 29 | 183 | 54074 | St. Charles County | 79,329 |
| 28 | | 54040 | Olive Branch city | 33,484 | 29 | | 60788 | Raytown city | 29,526 |
| 28 | 033 | 54040 | DeSoto County | 33,484 | 29 | 095 | 60788 | Jackson County | 29,526 |
| 28 | | 55760 | Pearl city | 25,092 | 29 | | 64082 | St. Charles city | 65,794 |
| 28 | 121 | 55760 | Rankin County | 25,092 | 29 | 183 | 64082 | St. Charles County | 65,794 |
| 28 | | 69280 | Southaven city | 48,982 | 29 | | 64550 | St. Joseph city | 76,780 |
| 28 | 033 | 69280 | DeSoto County | 48,982 | 29 | 021 | 64550 | Buchanan County | 76,780 |
| 28 | | 74840 | Tupelo city | 34,546 | 29 | | 65000 | St. Louis city | 319,294 |
| 28 | 081 | 74840 | Lee County | 34,546 | 29 | 510 | 65000 | St. Louis city | 319,294 |
| 29 | | | **MISSOURI** | 5,988,927 | 29 | | 65126 | St. Peters city | 52,575 |
| | | | | | 29 | 183 | 65126 | St. Charles County | 52,575 |
| 29 | | 03160 | Ballwin city | 30,404 | 29 | | 70000 | Springfield city | 159,498 |
| 29 | 189 | 03160 | St. Louis County | 30,404 | 29 | 043 | 70000 | Christian County | 2 |
| 29 | | 06652 | Blue Springs city | 52,575 | 29 | 077 | 70000 | Greene County | 159,496 |
| 29 | 095 | 06652 | Jackson County | 52,575 | 29 | | 75220 | University City city | 35,371 |
| 29 | | 11242 | Cape Girardeau city | 37,941 | 29 | 189 | 75220 | St. Louis County | 35,371 |
| 29 | 031 | 11242 | Cape Girardeau County | 37,941 | 29 | | 78442 | Wentzville city | 29,070 |
| 29 | 201 | 11242 | Scott County | 0 | 29 | 183 | 78442 | St. Charles County | 29,070 |

| State code | Place code | County code | Geographic Area Name | 2010 census population | State code | Place code | County code | Geographic Area Name | 2010 census population |
|---|---|---|---|---|---|---|---|---|---|
| 29 | | 79820 | Wildwood city | 35,517 | 33 | | 65140 | Rochester city | 29,752 |
| 29 | 189 | 79820 | St. Louis County | 35,517 | 33 | 017 | 65140 | Strafford County | 29,752 |
| 30 | | | **MONTANA** | 989,415 | 34 | | | **NEW JERSEY** | 8,791,894 |
| 30 | | 06550 | Billings city | 104,170 | 34 | | 02080 | Atlantic City city | 39,558 |
| 30 | 111 | 06550 | Yellowstone County | 104,170 | 34 | 001 | 02080 | Atlantic County | 39,558 |
| 30 | | 08950 | Bozeman city | 37,280 | 34 | | 03580 | Bayonne city | 63,024 |
| 30 | 031 | 08950 | Gallatin County | 37,280 | 34 | 017 | 03580 | Hudson County | 63,024 |
| 30 | | 32800 | Great Falls city | 58,505 | 34 | | 05170 | Bergenfield borough | 26,764 |
| 30 | 013 | 32800 | Cascade County | 58,505 | 34 | 003 | 05170 | Bergen County | 26,764 |
| 30 | | 35600 | Helena city | 28,190 | 34 | | 07600 | Bridgeton city | 25,349 |
| 30 | 049 | 35600 | Lewis and Clark County | 28,190 | 34 | 011 | 07600 | Cumberland County | 25,349 |
| 30 | | 50200 | Missoula city | 66,788 | 34 | | 10000 | Camden city | 77,344 |
| 30 | 063 | 50200 | Missoula County | 66,788 | 34 | 007 | 10000 | Camden County | 77,344 |
| 31 | | | **NEBRASKA** | 1,826,341 | 34 | | 13690 | Clifton city | 84,136 |
| | | | | | 34 | 031 | 13690 | Passaic County | 84,136 |
| 31 | | 03950 | Bellevue city | 50,137 | 34 | | 19390 | East Orange city | 64,270 |
| 31 | 153 | 03950 | Sarpy County | 50,137 | 34 | 013 | 19390 | Essex County | 64,270 |
| 31 | | 17670 | Fremont city | 26,397 | 34 | | 21000 | Elizabeth city | 124,969 |
| 31 | 053 | 17670 | Dodge County | 26,397 | 34 | 039 | 21000 | Union County | 124,969 |
| 31 | | 19595 | Grand Island city | 48,520 | 34 | | 21480 | Englewood city | 27,147 |
| 31 | 079 | 19595 | Hall County | 48,520 | 34 | 003 | 21480 | Bergen County | 27,147 |
| 31 | | 25055 | Kearney city | 30,787 | 34 | | 22470 | Fair Lawn borough | 32,457 |
| 31 | 019 | 25055 | Buffalo County | 30,787 | 34 | 003 | 22470 | Bergen County | 32,457 |
| 31 | | 28000 | Lincoln city | 258,379 | 34 | | 24420 | Fort Lee borough | 35,345 |
| 31 | 109 | 28000 | Lancaster County | 258,379 | 34 | 003 | 24420 | Bergen County | 35,345 |
| 31 | | 37000 | Omaha city | 408,958 | 34 | | 25770 | Garfield city | 30,487 |
| 31 | 055 | 37000 | Douglas County | 408,958 | 34 | 003 | 25770 | Bergen County | 30,487 |
| 32 | | | **NEVADA** | 2,700,551 | 34 | | 28680 | Hackensack city | 43,010 |
| | | | | | 34 | 003 | 28680 | Bergen County | 43,010 |
| 32 | | 09700 | Carson City | 55,274 | 34 | | 32250 | Hoboken city | 50,005 |
| 32 | 510 | 09700 | Carson City | 55,274 | 34 | 017 | 32250 | Hudson County | 50,005 |
| 32 | | 31900 | Henderson city | 257,729 | 34 | | 36000 | Jersey City city | 247,597 |
| 32 | 003 | 31900 | Clark County | 257,729 | 34 | 017 | 36000 | Hudson County | 247,597 |
| 32 | | 40000 | Las Vegas city | 583,756 | 34 | | 36510 | Kearny town | 40,684 |
| 32 | 003 | 40000 | Clark County | 583,756 | 34 | 017 | 36510 | Hudson County | 40,684 |
| 32 | | 51800 | North Las Vegas city | 216,961 | 34 | | 40350 | Linden city | 40,499 |
| 32 | 003 | 51800 | Clark County | 216,961 | 34 | 039 | 40350 | Union County | 40,499 |
| 32 | | 60600 | Reno city | 225,221 | 34 | | 41310 | Long Branch city | 30,719 |
| 32 | 031 | 60600 | Washoe County | 225,221 | 34 | 025 | 41310 | Monmouth County | 30,719 |
| 32 | | 68400 | Sparks city | 90,264 | 34 | | 46680 | Millville city | 28,400 |
| 32 | 031 | 68400 | Washoe County | 90,264 | 34 | 011 | 46680 | Cumberland County | 28,400 |
| 33 | | | **NEW HAMPSHIRE** | 1,316,470 | 34 | | 51000 | Newark city | 277,140 |
| | | | | | 34 | 013 | 51000 | Essex County | 277,140 |
| 33 | | 14200 | Concord city | 42,695 | 34 | | 51210 | New Brunswick city | 55,181 |
| 33 | 013 | 14200 | Merrimack County | 42,695 | 34 | 023 | 51210 | Middlesex County | 55,181 |
| 33 | | 18820 | Dover city | 29,987 | 34 | | 55950 | Paramus borough | 26,342 |
| 33 | 017 | 18820 | Strafford County | 29,987 | 34 | 003 | 55950 | Bergen County | 26,342 |
| 33 | | 45140 | Manchester city | 109,565 | 34 | | 56550 | Passaic city | 69,781 |
| 33 | 011 | 45140 | Hillsborough County | 109,565 | 34 | 031 | 56550 | Passaic County | 69,781 |
| 33 | | 50260 | Nashua city | 86,494 | | | | | |
| 33 | 011 | 50260 | Hillsborough County | 86,494 | | | | | |

| State code | Place code | County code | Geographic Area Name | 2010 census population | State code | Place code | County code | Geographic Area Name | 2010 census population |
|---|---|---|---|---|---|---|---|---|---|
| 34 | | 57000 | Paterson city | 146,199 | 36 | | 11000 | Buffalo city | 261,310 |
| 34 | 031 | 57000 | Passaic County | 146,199 | 36 | 029 | 11000 | Erie County | 261,310 |
| 34 | | 58200 | Perth Amboy city | 50,814 | 36 | | 24229 | Elmira city | 29,200 |
| 34 | 023 | 58200 | Middlesex County | 50,814 | 36 | 015 | 24229 | Chemung County | 29,200 |
| 34 | | 59190 | Plainfield city | 49,808 | 36 | | 27485 | Freeport village | 42,860 |
| 34 | 039 | 59190 | Union County | 49,808 | 36 | 059 | 27485 | Nassau County | 42,860 |
| 34 | | 61530 | Rahway city | 27,346 | 36 | | 29113 | Glen Cove city | 26,964 |
| 34 | 039 | 61530 | Union County | 27,346 | 36 | 059 | 29113 | Nassau County | 26,964 |
| 34 | | 65790 | Sayreville borough | 42,704 | 36 | | 32402 | Harrison village | 27,472 |
| 34 | 023 | 65790 | Middlesex County | 42,704 | 36 | 119 | 32402 | Westchester County | 27,472 |
| 34 | | 74000 | Trenton city | 84,913 | 36 | | 33139 | Hempstead village | 53,891 |
| 34 | 021 | 74000 | Mercer County | 84,913 | 36 | 059 | 33139 | Nassau County | 53,891 |
| 34 | | 74630 | Union City city | 66,455 | 36 | | 38077 | Ithaca city | 30,014 |
| 34 | 017 | 74630 | Hudson County | 66,455 | 36 | 109 | 38077 | Tompkins County | 30,014 |
| 34 | | 76070 | Vineland city | 60,724 | 36 | | 38264 | Jamestown city | 31,146 |
| 34 | 011 | 76070 | Cumberland County | 60,724 | 36 | 013 | 38264 | Chautauqua County | 31,146 |
| 34 | | 79040 | Westfield town | 30,316 | 36 | | 42554 | Lindenhurst village | 27,253 |
| 34 | 039 | 79040 | Union County | 30,316 | 36 | 103 | 42554 | Suffolk County | 27,253 |
| 34 | | 79610 | West New York town | 49,708 | 36 | | 43335 | Long Beach city | 33,275 |
| 34 | 017 | 79610 | Hudson County | 49,708 | 36 | 059 | 43335 | Nassau County | 33,275 |
| 35 | | | **NEW MEXICO** | 2,059,179 | 36 | | 47042 | Middletown city | 28,086 |
| | | | | | 36 | 071 | 47042 | Orange County | 28,086 |
| 35 | | 01780 | Alamogordo city | 30,403 | | | | | |
| 35 | 035 | 01780 | Otero County | 30,403 | 36 | | 49121 | Mount Vernon city | 67,292 |
| | | | | | 36 | 119 | 49121 | Westchester County | 67,292 |
| 35 | | 02000 | Albuquerque city | 545,852 | | | | | |
| 35 | 001 | 02000 | Bernalillo County | 545,852 | 36 | | 50034 | Newburgh city | 28,866 |
| | | | | | 36 | 071 | 50034 | Orange County | 28,866 |
| 35 | | 12150 | Carlsbad city | 26,138 | | | | | |
| 35 | 015 | 12150 | Eddy County | 26,138 | 36 | | 50617 | New Rochelle city | 77,062 |
| | | | | | 36 | 119 | 50617 | Westchester County | 77,062 |
| 35 | | 16420 | Clovis city | 37,775 | | | | | |
| 35 | 009 | 16420 | Curry County | 37,775 | 36 | | 51000 | New York city | 8,175,133 |
| | | | | | 36 | 005 | 51000 | Bronx County | 1,385,108 |
| 35 | | 25800 | Farmington city | 45,877 | 36 | 047 | 51000 | Kings County | 2,504,700 |
| 35 | 045 | 25800 | San Juan County | 45,877 | 36 | 061 | 51000 | New York County | 1,585,873 |
| | | | | | 36 | 081 | 51000 | Queens County | 2,230,722 |
| 35 | | 32520 | Hobbs city | 34,122 | 36 | 085 | 51000 | Richmond County | 468,730 |
| 35 | 025 | 32520 | Lea County | 34,122 | | | | | |
| | | | | | 36 | | 51055 | Niagara Falls city | 50,193 |
| 35 | | 39380 | Las Cruces city | 97,618 | 36 | 063 | 51055 | Niagara County | 50,193 |
| 35 | 013 | 39380 | Doña Ana County | 97,618 | | | | | |
| | | | | | 36 | | 53682 | North Tonawanda city | 31,568 |
| 35 | | 63460 | Rio Rancho city | 87,521 | 36 | 063 | 53682 | Niagara County | 31,568 |
| 35 | 001 | 63460 | Bernalillo County | 130 | | | | | |
| 35 | 043 | 63460 | Sandoval County | 87,391 | 36 | | 55530 | Ossining village | 25,060 |
| | | | | | 36 | 119 | 55530 | Westchester County | 25,060 |
| 35 | | 64930 | Roswell city | 48,366 | | | | | |
| 35 | 005 | 64930 | Chaves County | 48,366 | 36 | | 59223 | Port Chester village | 28,967 |
| | | | | | 36 | 119 | 59223 | Westchester County | 28,967 |
| 35 | | 70500 | Santa Fe city | 67,947 | | | | | |
| 35 | 049 | 70500 | Santa Fe County | 67,947 | 36 | | 59641 | Poughkeepsie city | 32,736 |
| | | | | | 36 | 027 | 59641 | Dutchess County | 32,736 |
| 36 | | | **NEW YORK** | 19,378,102 | | | | | |
| | | | | | 36 | | 63000 | Rochester city | 210,565 |
| 36 | | 01000 | Albany city | 97,856 | 36 | 055 | 63000 | Monroe County | 210,565 |
| 36 | 001 | 01000 | Albany County | 97,856 | | | | | |
| | | | | | 36 | | 63418 | Rome city | 33,725 |
| 36 | | 03078 | Auburn city | 27,687 | 36 | 065 | 63418 | Oneida County | 33,725 |
| 36 | 011 | 03078 | Cayuga County | 27,687 | | | | | |
| | | | | | 36 | | 65255 | Saratoga Springs city | 26,586 |
| 36 | | 06607 | Binghamton city | 47,376 | 36 | 091 | 65255 | Saratoga County | 26,586 |
| 36 | 007 | 06607 | Broome County | 47,376 | | | | | |

| State code | Place code | County code | Geographic Area Name | 2010 census population | State code | Place code | County code | Geographic Area Name | 2010 census population |
|---|---|---|---|---|---|---|---|---|---|
| 36 | | 65508 | Schenectady city | 66,135 | 37 | | 28080 | Greenville city | 84,554 |
| 36 | 093 | 65508 | Schenectady County | 66,135 | 37 | 147 | 28080 | Pitt County | 84,554 |
| 36 | | 70420 | Spring Valley village | 31,347 | 37 | | 31060 | Hickory city | 40,010 |
| 36 | 087 | 70420 | Rockland County | 31,347 | 37 | 023 | 31060 | Burke County | 66 |
| | | | | | 37 | 027 | 31060 | Caldwell County | 18 |
| 36 | | 73000 | Syracuse city | 145,170 | 37 | 035 | 31060 | Catawba County | 39,926 |
| 36 | 067 | 73000 | Onondaga County | 145,170 | 37 | | 31400 | High Point city | 104,371 |
| 36 | | 75484 | Troy city | 50,129 | 37 | 057 | 31400 | Davidson County | 5,310 |
| 36 | 083 | 75484 | Rensselaer County | 50,129 | 37 | 067 | 31400 | Forsyth County | 8 |
| 36 | | 76540 | Utica city | 62,235 | 37 | 081 | 31400 | Guilford County | 99,042 |
| 36 | 065 | 76540 | Oneida County | 62,235 | 37 | 151 | 31400 | Randolph County | 11 |
| 36 | | 76705 | Valley Stream village | 37,511 | 37 | | 33120 | Huntersville town | 46,773 |
| 36 | 059 | 76705 | Nassau County | 37,511 | 37 | 119 | 33120 | Mecklenburg County | 46,773 |
| 36 | | 78608 | Watertown city | 27,023 | 37 | | 33560 | Indian Trail town | 33,518 |
| 36 | 045 | 78608 | Jefferson County | 27,023 | 37 | 179 | 33560 | Union County | 33,518 |
| 36 | | 81677 | White Plains city | 56,853 | 37 | | 34200 | Jacksonville city | 70,145 |
| 36 | 119 | 81677 | Westchester County | 56,853 | 37 | 133 | 34200 | Onslow County | 70,145 |
| 37 | | | **NORTH CAROLINA** | 9,535,483 | 37 | | 35200 | Kannapolis city | 42,625 |
| | | | | | 37 | 025 | 35200 | Cabarrus County | 33,194 |
| 37 | | 01520 | Apex town | 37,476 | 37 | 159 | 35200 | Rowan County | 9,431 |
| 37 | 183 | 01520 | Wake County | 37,476 | 37 | | 41960 | Matthews town | 27,198 |
| 37 | | 02080 | Asheboro city | 25,012 | 37 | 119 | 41960 | Mecklenburg County | 27,198 |
| 37 | 151 | 02080 | Randolph County | 25,012 | 37 | | 43920 | Monroe city | 32,797 |
| 37 | | 02140 | Asheville city | 83,393 | 37 | 179 | 43920 | Union County | 32,797 |
| 37 | 021 | 02140 | Buncombe County | 83,393 | 37 | | 44220 | Mooresville town | 32,711 |
| 37 | | 09060 | Burlington city | 49,963 | 37 | 097 | 44220 | Iredell County | 32,711 |
| 37 | 001 | 09060 | Alamance County | 49,308 | 37 | | 46340 | New Bern city | 29,524 |
| 37 | 081 | 09060 | Guilford County | 655 | 37 | 049 | 46340 | Craven County | 29,524 |
| 37 | | 10740 | Cary town | 135,234 | 37 | | 55000 | Raleigh city | 403,892 |
| 37 | 037 | 10740 | Chatham County | 1,422 | 37 | 063 | 55000 | Durham County | 1,067 |
| 37 | 183 | 10740 | Wake County | 133,812 | 37 | 183 | 55000 | Wake County | 402,825 |
| 37 | | 11800 | Chapel Hill town | 57,233 | 37 | | 57500 | Rocky Mount city | 57,477 |
| 37 | 063 | 11800 | Durham County | 2,836 | 37 | 065 | 57500 | Edgecombe County | 17,524 |
| 37 | 135 | 11800 | Orange County | 54,397 | 37 | 127 | 57500 | Nash County | 39,953 |
| 37 | | 12000 | Charlotte city | 731,424 | 37 | | 58860 | Salisbury city | 33,662 |
| 37 | 119 | 12000 | Mecklenburg County | 731,424 | 37 | 159 | 58860 | Rowan County | 33,662 |
| 37 | | 14100 | Concord city | 79,066 | 37 | | 59280 | Sanford city | 28,094 |
| 37 | 025 | 14100 | Cabarrus County | 79,066 | 37 | 105 | 59280 | Lee County | 28,094 |
| 37 | | 19000 | Durham city | 228,330 | 37 | | 67420 | Thomasville city | 26,757 |
| 37 | 063 | 19000 | Durham County | 228,300 | 37 | 057 | 67420 | Davidson County | 26,493 |
| 37 | 135 | 19000 | Orange County | 30 | 37 | 151 | 67420 | Randolph County | 264 |
| 37 | 183 | 19000 | Wake County | 0 | 37 | | 70540 | Wake Forest town | 30,117 |
| 37 | | 22920 | Fayetteville city | 200,564 | 37 | 069 | 70540 | Franklin County | 899 |
| 37 | 051 | 22920 | Cumberland County | 200,564 | 37 | 183 | 70540 | Wake County | 29,218 |
| 37 | | 25480 | Garner town | 25,745 | 37 | | 74440 | Wilmington city | 106,476 |
| 37 | 183 | 25480 | Wake County | 25,745 | 37 | 129 | 74440 | New Hanover County | 106,476 |
| 37 | | 25580 | Gastonia city | 71,741 | 37 | | 74540 | Wilson city | 49,167 |
| 37 | 071 | 25580 | Gaston County | 71,741 | 37 | 195 | 74540 | Wilson County | 49,167 |
| 37 | | 26880 | Goldsboro city | 36,437 | 37 | | 75000 | Winston-Salem city | 229,617 |
| 37 | 191 | 26880 | Wayne County | 36,437 | 37 | 067 | 75000 | Forsyth County | 229,617 |
| 37 | | 28000 | Greensboro city | 269,666 | 38 | | | **NORTH DAKOTA** | 672,591 |
| 37 | 081 | 28000 | Guilford County | 269,666 | 38 | | 07200 | Bismarck city | 61,272 |
| | | | | | 38 | 015 | 07200 | Burleigh County | 61,272 |

| State code | Place code | County code | Geographic Area Name | 2010 census population | State code | Place code | County code | Geographic Area Name | 2010 census population |
|---|---|---|---|---|---|---|---|---|---|
| | | | | | 39 | | 27048 | Findlay city | 41,202 |
| 38 | | 25700 | Fargo city | 105,549 | 39 | 063 | 27048 | Hancock County | 41,202 |
| 38 | 017 | 25700 | Cass County | 105,549 | | | | | |
| | | | | | 39 | | 29106 | Gahanna city | 33,248 |
| 38 | | 32060 | Grand Forks city | 52,838 | 39 | 049 | 29106 | Franklin County | 33,248 |
| 38 | 035 | 32060 | Grand Forks County | 52,838 | | | | | |
| | | | | | 39 | | 29428 | Garfield Heights city | 28,849 |
| 38 | | 53380 | Minot city | 40,888 | 39 | 035 | 29428 | Cuyahoga County | 28,849 |
| 38 | 101 | 53380 | Ward County | 40,888 | | | | | |
| | | | | | 39 | | 31860 | Green city | 25,699 |
| 38 | | 84780 | West Fargo city | 25,830 | 39 | 153 | 31860 | Summit County | 25,699 |
| 38 | 017 | 84780 | Cass County | 25,830 | | | | | |
| | | | | | 39 | | 32592 | Grove City city | 35,575 |
| 39 | | | **OHIO** | 11,536,504 | 39 | 049 | 32592 | Franklin County | 35,575 |
| 39 | | 01000 | Akron city | 199,110 | 39 | | 33012 | Hamilton city | 62,477 |
| 39 | 153 | 01000 | Summit County | 199,110 | 39 | 017 | 33012 | Butler County | 62,477 |
| 39 | | 03828 | Barberton city | 26,550 | 39 | | 35476 | Hilliard city | 28,435 |
| 39 | 153 | 03828 | Summit County | 26,550 | 39 | 049 | 35476 | Franklin County | 28,435 |
| 39 | | 04720 | Beavercreek city | 45,193 | 39 | | 36610 | Huber Heights city | 38,101 |
| 39 | 057 | 04720 | Greene County | 45,193 | 39 | 057 | 36610 | Greene County | 0 |
| | | | | | 39 | 109 | 36610 | Miami County | 959 |
| 39 | | 07972 | Bowling Green city | 30,028 | 39 | 113 | 36610 | Montgomery County | 37,142 |
| 39 | 173 | 07972 | Wood County | 30,028 | | | | | |
| | | | | | 39 | | 39872 | Kent city | 28,904 |
| 39 | | 09680 | Brunswick city | 34,255 | 39 | 133 | 39872 | Portage County | 28,904 |
| 39 | 103 | 09680 | Medina County | 34,255 | | | | | |
| | | | | | 39 | | 40040 | Kettering city | 56,163 |
| 39 | | 12000 | Canton city | 73,007 | 39 | 057 | 40040 | Greene County | 467 |
| 39 | 151 | 12000 | Stark County | 73,007 | 39 | 113 | 40040 | Montgomery County | 55,696 |
| 39 | | 15000 | Cincinnati city | 296,943 | 39 | | 41664 | Lakewood city | 52,131 |
| 39 | 061 | 15000 | Hamilton County | 296,943 | 39 | 035 | 41664 | Cuyahoga County | 52,131 |
| 39 | | 16000 | Cleveland city | 396,815 | 39 | | 41720 | Lancaster city | 38,780 |
| 39 | 035 | 16000 | Cuyahoga County | 396,815 | 39 | 045 | 41720 | Fairfield County | 38,780 |
| 39 | | 16014 | Cleveland Heights city | 46,121 | 39 | | 43554 | Lima city | 38,771 |
| 39 | 035 | 16014 | Cuyahoga County | 46,121 | 39 | 003 | 43554 | Allen County | 38,771 |
| 39 | | 18000 | Columbus city | 787,033 | 39 | | 44856 | Lorain city | 64,097 |
| 39 | 041 | 18000 | Delaware County | 7,245 | 39 | 093 | 44856 | Lorain County | 64,097 |
| 39 | 045 | 18000 | Fairfield County | 9,666 | | | | | |
| 39 | 049 | 18000 | Franklin County | 770,122 | 39 | | 47138 | Mansfield city | 47,821 |
| | | | | | 39 | 139 | 47138 | Richland County | 47,821 |
| 39 | | 19778 | Cuyahoga Falls city | 49,652 | | | | | |
| 39 | 153 | 19778 | Summit County | 49,652 | 39 | | 47754 | Marion city | 36,837 |
| | | | | | 39 | 101 | 47754 | Marion County | 36,837 |
| 39 | | 21000 | Dayton city | 141,527 | | | | | |
| 39 | 113 | 21000 | Montgomery County | 141,527 | 39 | | 48188 | Mason city | 30,712 |
| | | | | | 39 | 165 | 48188 | Warren County | 30,712 |
| 39 | | 21434 | Delaware city | 34,753 | | | | | |
| 39 | 041 | 21434 | Delaware County | 34,753 | 39 | | 48244 | Massillon city | 32,149 |
| | | | | | 39 | 151 | 48244 | Stark County | 32,149 |
| 39 | | 22694 | Dublin city | 41,751 | | | | | |
| 39 | 041 | 22694 | Delaware County | 4,018 | 39 | | 48790 | Medina city | 26,678 |
| 39 | 049 | 22694 | Franklin County | 35,367 | 39 | 103 | 48790 | Medina County | 26,678 |
| 39 | 159 | 22694 | Union County | 2,366 | | | | | |
| | | | | | 39 | | 49056 | Mentor city | 47,159 |
| 39 | | 25256 | Elyria city | 54,533 | 39 | 085 | 49056 | Lake County | 47,159 |
| 39 | 093 | 25256 | Lorain County | 54,533 | | | | | |
| | | | | | 39 | | 49840 | Middletown city | 48,694 |
| 39 | | 25704 | Euclid city | 48,920 | 39 | 017 | 49840 | Butler County | 45,994 |
| 39 | 035 | 25704 | Cuyahoga County | 48,920 | 39 | 165 | 49840 | Warren County | 2,700 |
| 39 | | 25914 | Fairborn city | 32,352 | 39 | | 54040 | Newark city | 47,573 |
| 39 | 057 | 25914 | Greene County | 32,352 | 39 | 089 | 54040 | Licking County | 47,573 |
| 39 | | 25970 | Fairfield city | 42,510 | 39 | | 56882 | North Olmsted city | 32,718 |
| 39 | 017 | 25970 | Butler County | 42,510 | 39 | 035 | 56882 | Cuyahoga County | 32,718 |
| 39 | 061 | 25970 | Hamilton County | 0 | | | | | |

| State code | Place code | County code | Geographic Area Name | 2010 census population | State code | Place code | County code | Geographic Area Name | 2010 census population |
|---|---|---|---|---|---|---|---|---|---|
| 39 | | 56966 | North Ridgeville city | 29,465 | 40 | | 23200 | Edmond city | 81,405 |
| 39 | 093 | 56966 | Lorain County | 29,465 | 40 | 109 | 23200 | Oklahoma County | 81,405 |
| 39 | | 57008 | North Royalton city | 30,444 | 40 | | 23950 | Enid city | 49,379 |
| 39 | 035 | 57008 | Cuyahoga County | 30,444 | 40 | 047 | 23950 | Garfield County | 49,379 |
| 39 | | 61000 | Parma city | 81,601 | 40 | | 41850 | Lawton city | 96,867 |
| 39 | 035 | 61000 | Cuyahoga County | 81,601 | 40 | 031 | 41850 | Comanche County | 96,867 |
| 39 | | 66390 | Reynoldsburg city | 35,893 | 40 | | 48350 | Midwest City city | 54,371 |
| 39 | 045 | 66390 | Fairfield County | 910 | 40 | 109 | 48350 | Oklahoma County | 54,371 |
| 39 | 049 | 66390 | Franklin County | 26,157 | 40 | | 49200 | Moore city | 55,081 |
| 39 | 089 | 66390 | Licking County | 8,826 | 40 | 027 | 49200 | Cleveland County | 55,081 |
| 39 | | 67468 | Riverside city | 25,201 | 40 | | 50050 | Muskogee city | 39,223 |
| 39 | 113 | 67468 | Montgomery County | 25,201 | 40 | 101 | 50050 | Muskogee County | 39,223 |
| 39 | | 70380 | Sandusky city | 25,793 | 40 | | 52500 | Norman city | 110,925 |
| 39 | 043 | 70380 | Erie County | 25,793 | 40 | 027 | 52500 | Cleveland County | 110,925 |
| 39 | | 71682 | Shaker Heights city | 28,448 | 40 | | 55000 | Oklahoma City city | 579,999 |
| 39 | 035 | 71682 | Cuyahoga County | 28,448 | 40 | 017 | 55000 | Canadian County | 44,541 |
| 39 | | 74118 | Springfield city | 60,608 | 40 | 027 | 55000 | Cleveland County | 63,723 |
| 39 | 023 | 74118 | Clark County | 60,608 | 40 | 109 | 55000 | Oklahoma County | 471,671 |
| 39 | | 74944 | Stow city | 34,837 | 40 | 125 | 55000 | Pottawatomie County | 64 |
| 39 | 153 | 74944 | Summit County | 34,837 | 40 | | 56650 | Owasso city | 28,915 |
| 39 | | 75098 | Strongsville city | 44,750 | 40 | 131 | 56650 | Rogers County | 2,614 |
| 39 | 035 | 75098 | Cuyahoga County | 44,750 | 40 | 143 | 56650 | Tulsa County | 26,301 |
| 39 | | 77000 | Toledo city | 287,208 | 40 | | 59850 | Ponca City city | 25,387 |
| 39 | 095 | 77000 | Lucas County | 287,208 | 40 | 071 | 59850 | Kay County | 25,387 |
| 39 | | 77588 | Troy city | 25,058 | 40 | | 66800 | Shawnee city | 29,857 |
| 39 | 109 | 77588 | Miami County | 25,058 | 40 | 125 | 66800 | Pottawatomie County | 29,857 |
| 39 | | 79002 | Upper Arlington city | 33,771 | 40 | | 70300 | Stillwater city | 45,688 |
| 39 | 049 | 79002 | Franklin County | 33,771 | 40 | 119 | 70300 | Payne County | 45,688 |
| 39 | | 80892 | Warren city | 41,557 | 40 | | 75000 | Tulsa city | 391,906 |
| 39 | 155 | 80892 | Trumbull County | 41,557 | 40 | 113 | 75000 | Osage County | 6,136 |
| 39 | | 83342 | Westerville city | 36,120 | 40 | 131 | 75000 | Rogers County | 0 |
| 39 | 041 | 83342 | Delaware County | 7,792 | 40 | 143 | 75000 | Tulsa County | 385,613 |
| 39 | 049 | 83342 | Franklin County | 28,328 | 40 | 145 | 75000 | Wagoner County | 157 |
| 39 | | 83622 | Westlake city | 32,729 | 41 | | | **OREGON** | 3,831,074 |
| 39 | 035 | 83622 | Cuyahoga County | 32,729 | 41 | | 01000 | Albany city | 50,158 |
| 39 | | 86548 | Wooster city | 26,119 | 41 | 003 | 01000 | Benton County | 6,463 |
| 39 | 169 | 86548 | Wayne County | 26,119 | 41 | 043 | 01000 | Linn County | 43,695 |
| 39 | | 86772 | Xenia city | 25,719 | 41 | | 05350 | Beaverton city | 89,803 |
| 39 | 057 | 86772 | Greene County | 25,719 | 41 | 067 | 05350 | Washington County | 89,803 |
| 39 | | 88000 | Youngstown city | 66,982 | 41 | | 05800 | Bend city | 76,639 |
| 39 | 099 | 88000 | Mahoning County | 66,971 | 41 | 017 | 05800 | Deschutes County | 76,639 |
| 39 | 155 | 88000 | Trumbull County | 11 | 41 | | 15800 | Corvallis city | 54,462 |
| 39 | | 88084 | Zanesville city | 25,487 | 41 | 003 | 15800 | Benton County | 54,462 |
| 39 | 119 | 88084 | Muskingum County | 25,487 | 41 | | 23850 | Eugene city | 156,185 |
| 40 | | | **OKLAHOMA** | 3,751,351 | 41 | 039 | 23850 | Lane County | 156,185 |
| 40 | | 04450 | Bartlesville city | 35,750 | 41 | | 30550 | Grants Pass city | 34,533 |
| 40 | 113 | 04450 | Osage County | 3 | 41 | 033 | 30550 | Josephine County | 34,533 |
| 40 | 147 | 04450 | Washington County | 35,747 | 41 | | 31250 | Gresham city | 105,594 |
| 40 | | 09050 | Broken Arrow city | 98,850 | 41 | 051 | 31250 | Multnomah County | 105,594 |
| 40 | 143 | 09050 | Tulsa County | 80,634 | 41 | | 34100 | Hillsboro city | 91,611 |
| 40 | 145 | 09050 | Wagoner County | 18,216 | 41 | 067 | 34100 | Washington County | 91,611 |

| State code | Place code | County code | Geographic Area Name | 2010 census population |
|---|---|---|---|---|
| 41 | | 38500 | Keizer city | 36,478 |
| 41 | 047 | 38500 | Marion County | 36,478 |
| 41 | | 40550 | Lake Oswego city | 36,619 |
| 41 | 005 | 40550 | Clackamas County | 34,066 |
| 41 | 051 | 40550 | Multnomah County | 2,544 |
| 41 | 067 | 40550 | Washington County | 9 |
| 41 | | 45000 | McMinnville city | 32,187 |
| 41 | 071 | 45000 | Yamhill County | 32,187 |
| 41 | | 47000 | Medford city | 74,907 |
| 41 | 029 | 47000 | Jackson County | 74,907 |
| 41 | | 55200 | Oregon City city | 31,859 |
| 41 | 005 | 55200 | Clackamas County | 31,859 |
| 41 | | 59000 | Portland city | 583,776 |
| 41 | 005 | 59000 | Clackamas County | 744 |
| 41 | 051 | 59000 | Multnomah County | 581,485 |
| 41 | 067 | 59000 | Washington County | 1,547 |
| 41 | | 61200 | Redmond city | 26,215 |
| 41 | 017 | 61200 | Deschutes County | 26,215 |
| 41 | | 64900 | Salem city | 154,637 |
| 41 | 047 | 64900 | Marion County | 130,398 |
| 41 | 053 | 64900 | Polk County | 24,239 |
| 41 | | 69600 | Springfield city | 59,403 |
| 41 | 039 | 69600 | Lane County | 59,403 |
| 41 | | 73650 | Tigard city | 48,035 |
| 41 | 067 | 73650 | Washington County | 48,035 |
| 41 | | 74950 | Tualatin city | 26,054 |
| 41 | 005 | 74950 | Clackamas County | 2,862 |
| 41 | 067 | 74950 | Washington County | 23,192 |
| 41 | | 80150 | West Linn city | 25,109 |
| 41 | 005 | 80150 | Clackamas County | 25,109 |
| 42 | | | **PENNSYLVANIA** | 12,702,379 |
| 42 | | 02000 | Allentown city | 118,032 |
| 42 | 077 | 02000 | Lehigh County | 118,032 |
| 42 | | 02184 | Altoona city | 46,320 |
| 42 | 013 | 02184 | Blair County | 46,320 |
| 42 | | 06064 | Bethel Park municipality | 32,313 |
| 42 | 003 | 06064 | Allegheny County | 32,313 |
| 42 | | 06088 | Bethlehem city | 74,982 |
| 42 | 077 | 06088 | Lehigh County | 19,343 |
| 42 | 095 | 06088 | Northampton County | 55,639 |
| 42 | | 13208 | Chester city | 33,972 |
| 42 | 045 | 13208 | Delaware County | 33,972 |
| 42 | | 21648 | Easton city | 26,800 |
| 42 | 095 | 21648 | Northampton County | 26,800 |
| 42 | | 24000 | Erie city | 101,786 |
| 42 | 049 | 24000 | Erie County | 101,786 |
| 42 | | 32800 | Harrisburg city | 49,528 |
| 42 | 043 | 32800 | Dauphin County | 49,528 |
| 42 | | 33408 | Hazleton city | 25,340 |
| 42 | 079 | 33408 | Luzerne County | 25,340 |
| 42 | | 41216 | Lancaster city | 59,322 |
| 42 | 071 | 41216 | Lancaster County | 59,322 |

| State code | Place code | County code | Geographic Area Name | 2010 census population |
|---|---|---|---|---|
| 42 | | 42168 | Lebanon city | 25,477 |
| 42 | 075 | 42168 | Lebanon County | 25,477 |
| 42 | | 50528 | Monroeville municipality | 28,386 |
| 42 | 003 | 50528 | Allegheny County | 28,386 |
| 42 | | 54656 | Norristown borough | 34,324 |
| 42 | 091 | 54656 | Montgomery County | 34,324 |
| 42 | | 60000 | Philadelphia city | 1,526,006 |
| 42 | 101 | 60000 | Philadelphia County | 1,526,006 |
| 42 | | 61000 | Pittsburgh city | 305,704 |
| 42 | 003 | 61000 | Allegheny County | 305,704 |
| 42 | | 61536 | Plum borough | 27,126 |
| 42 | 003 | 61536 | Allegheny County | 27,126 |
| 42 | | 63624 | Reading city | 88,082 |
| 42 | 011 | 63624 | Berks County | 88,082 |
| 42 | | 69000 | Scranton city | 76,089 |
| 42 | 069 | 69000 | Lackawanna County | 76,089 |
| 42 | | 73808 | State College borough | 42,034 |
| 42 | 027 | 73808 | Centre County | 42,034 |
| 42 | | 85152 | Wilkes-Barre city | 41,498 |
| 42 | 079 | 85152 | Luzerne County | 41,498 |
| 42 | | 85312 | Williamsport city | 29,381 |
| 42 | 081 | 85312 | Lycoming County | 29,381 |
| 42 | | 87048 | York city | 43,718 |
| 42 | 133 | 87048 | York County | 43,718 |
| 44 | | | **RHODE ISLAND** | 1,052,567 |
| 44 | | 19180 | Cranston city | 80,387 |
| 44 | 007 | 19180 | Providence County | 80,387 |
| 44 | | 22960 | East Providence city | 47,037 |
| 44 | 007 | 22960 | Providence County | 47,037 |
| 44 | | 54640 | Pawtucket city | 71,148 |
| 44 | 007 | 54640 | Providence County | 71,148 |
| 44 | | 59000 | Providence city | 178,042 |
| 44 | 007 | 59000 | Providence County | 178,042 |
| 44 | | 74300 | Warwick city | 82,672 |
| 44 | 003 | 74300 | Kent County | 82,672 |
| 44 | | 80780 | Woonsocket city | 41,186 |
| 44 | 007 | 80780 | Providence County | 41,186 |
| 45 | | | **SOUTH CAROLINA** | 4,625,364 |
| 45 | | 00550 | Aiken city | 29,524 |
| 45 | 003 | 00550 | Aiken County | 29,524 |
| 45 | | 01360 | Anderson city | 26,686 |
| 45 | 007 | 01360 | Anderson County | 26,686 |
| 45 | | 13330 | Charleston city | 120,083 |
| 45 | 015 | 13330 | Berkeley County | 8,095 |
| 45 | 019 | 13330 | Charleston County | 111,988 |
| 45 | | 16000 | Columbia city | 129,272 |
| 45 | 063 | 16000 | Lexington County | 559 |
| 45 | 079 | 16000 | Richland County | 128,713 |

| State code | Place code | County code | Geographic Area Name | 2010 census population | State code | Place code | County code | Geographic Area Name | 2010 census population |
|---|---|---|---|---|---|---|---|---|---|
| 45 | | 25810 | Florence city | 37,056 | 47 | | 16420 | Collierville town | 43,965 |
| 45 | 041 | 25810 | Florence County | 37,056 | 47 | 047 | 16420 | Fayette County | 0 |
| | | | | | 47 | 157 | 16420 | Shelby County | 43,965 |
| 45 | | 29815 | Goose Creek city | 35,938 | | | | | |
| 45 | 015 | 29815 | Berkeley County | 35,933 | 47 | | 16540 | Columbia city | 34,681 |
| 45 | 019 | 29815 | Charleston County | 5 | 47 | 119 | 16540 | Maury County | 34,681 |
| 45 | | 30850 | Greenville city | 58,409 | 47 | | 16920 | Cookeville city | 30,435 |
| 45 | 045 | 30850 | Greenville County | 58,409 | 47 | 141 | 16920 | Putnam County | 30,435 |
| 45 | | 30985 | Greer city | 25,515 | 47 | | 27740 | Franklin city | 62,487 |
| 45 | 045 | 30985 | Greenville County | 18,635 | 47 | 187 | 27740 | Williamson County | 62,487 |
| 45 | 083 | 30985 | Spartanburg County | 6,880 | | | | | |
| | | | | | 47 | | 28540 | Gallatin city | 30,278 |
| 45 | | 34045 | Hilton Head Island town | 37,099 | 47 | 165 | 28540 | Sumner County | 30,278 |
| 45 | 013 | 34045 | Beaufort County | 37,099 | | | | | |
| | | | | | 47 | | 28960 | Germantown city | 38,844 |
| 45 | | 48535 | Mount Pleasant town | 67,843 | 47 | 157 | 28960 | Shelby County | 38,844 |
| 45 | 019 | 48535 | Charleston County | 67,843 | | | | | |
| | | | | | 47 | | 33280 | Hendersonville city | 51,372 |
| 45 | | 49075 | Myrtle Beach city | 27,109 | 47 | 165 | 33280 | Sumner County | 51,372 |
| 45 | 051 | 49075 | Horry County | 27,109 | | | | | |
| | | | | | 47 | | 37640 | Jackson city | 65,211 |
| 45 | | 50875 | North Charleston city | 97,471 | 47 | 113 | 37640 | Madison County | 65,211 |
| 45 | 015 | 50875 | Berkeley County | 0 | | | | | |
| 45 | 019 | 50875 | Charleston County | 78,393 | 47 | | 38320 | Johnson City city | 63,152 |
| 45 | 035 | 50875 | Dorchester County | 19,078 | 47 | 019 | 38320 | Carter County | 1,252 |
| | | | | | 47 | 163 | 38320 | Sullivan County | 367 |
| 45 | | 61405 | Rock Hill city | 66,154 | 47 | 179 | 38320 | Washington County | 61,533 |
| 45 | 091 | 61405 | York County | 66,154 | | | | | |
| | | | | | 47 | | 39560 | Kingsport city | 48,205 |
| 45 | | 68290 | Spartanburg city | 37,013 | 47 | 073 | 39560 | Hawkins County | 2,854 |
| 45 | 083 | 68290 | Spartanburg County | 37,013 | 47 | 163 | 39560 | Sullivan County | 45,351 |
| 45 | | 70270 | Summerville town | 43,392 | 47 | | 40000 | Knoxville city | 178,874 |
| 45 | 015 | 70270 | Berkeley County | 3,643 | 47 | 093 | 40000 | Knox County | 178,874 |
| 45 | 019 | 70270 | Charleston County | 1,010 | | | | | |
| 45 | 035 | 70270 | Dorchester County | 38,739 | 47 | | 41200 | La Vergne city | 32,588 |
| | | | | | 47 | 149 | 41200 | Rutherford County | 32,588 |
| 45 | | 70405 | Sumter city | 40,524 | | | | | |
| 45 | 085 | 70405 | Sumter County | 40,524 | 47 | | 41520 | Lebanon city | 26,190 |
| | | | | | 47 | 189 | 41520 | Wilson County | 26,190 |
| 46 | | | **SOUTH DAKOTA** | 814,180 | | | | | |
| | | | | | 47 | | 46380 | Maryville city | 27,465 |
| 46 | | 00100 | Aberdeen city | 26,091 | 47 | 009 | 46380 | Blount County | 27,465 |
| 46 | 013 | 00100 | Brown County | 26,091 | | | | | |
| | | | | | 47 | | 48000 | Memphis city | 646,889 |
| 46 | | 52980 | Rapid City city | 67,956 | 47 | 157 | 48000 | Shelby County | 646,889 |
| 46 | 103 | 52980 | Pennington County | 67,956 | | | | | |
| | | | | | 47 | | 50280 | Morristown city | 29,137 |
| 46 | | 59020 | Sioux Falls city | 153,888 | 47 | 063 | 50280 | Hamblen County | 29,131 |
| 46 | 083 | 59020 | Lincoln County | 21,095 | 47 | 089 | 50280 | Jefferson County | 6 |
| 46 | 099 | 59020 | Minnehaha County | 132,793 | | | | | |
| | | | | | 47 | | 51560 | Murfreesboro city | 108,755 |
| 47 | | | **TENNESSEE** | 6,346,105 | 47 | 149 | 51560 | Rutherford County | 108,755 |
| | | | | | 47 | | 55120 | Oak Ridge city | 29,330 |
| 47 | | 03440 | Bartlett city | 54,613 | 47 | 001 | 55120 | Anderson County | 26,271 |
| 47 | 157 | 03440 | Shelby County | 54,613 | 47 | 145 | 55120 | Roane County | 3,059 |
| 47 | | 08280 | Brentwood city | 37,060 | 47 | | 69420 | Smyrna town | 39,974 |
| 47 | 187 | 08280 | Williamson County | 37,060 | 47 | 149 | 69420 | Rutherford County | 39,974 |
| 47 | | 08540 | Bristol city | 26,702 | 47 | | 70580 | Spring Hill city | 29,036 |
| 47 | 163 | 08540 | Sullivan County | 26,702 | 47 | 119 | 70580 | Maury County | 7,023 |
| | | | | | 47 | 187 | 70580 | Williamson County | 22,013 |
| 47 | | 14000 | Chattanooga city | 167,674 | | | | | |
| 47 | 065 | 14000 | Hamilton County | 167,674 | 48 | | | **TEXAS** | 25,145,561 |
| 47 | | 15160 | Clarksville city | 132,929 | 48 | | 01000 | Abilene city | 117,063 |
| 47 | 125 | 15160 | Montgomery County | 132,929 | 48 | 253 | 01000 | Jones County | 5,145 |
| | | | | | 48 | 441 | 01000 | Taylor County | 111,918 |
| 47 | | 15400 | Cleveland city | 41,285 | | | | | |
| 47 | 011 | 15400 | Bradley County | 41,285 | | | | | |

| State code | Place code | County code | Geographic Area Name | 2010 census population | State code | Place code | County code | Geographic Area Name | 2010 census population |
|---|---|---|---|---|---|---|---|---|---|
| 48 | | 01924 | Allen city | 84,246 | 48 | | 19000 | Dallas city | 1,197,816 |
| 48 | 085 | 01924 | Collin County | 84,246 | 48 | 085 | 19000 | Collin County | 46,885 |
| | | | | | 48 | 113 | 19000 | Dallas County | 1,124,296 |
| 48 | | 03000 | Amarillo city | 190,695 | 48 | 121 | 19000 | Denton County | 26,579 |
| 48 | 375 | 03000 | Potter County | 105,486 | 48 | 257 | 19000 | Kaufman County | 0 |
| 48 | 381 | 03000 | Randall County | 85,209 | 48 | 397 | 19000 | Rockwall County | 56 |
| 48 | | 04000 | Arlington city | 365,438 | 48 | | 19624 | Deer Park city | 32,010 |
| 48 | 439 | 04000 | Tarrant County | 365,438 | 48 | 201 | 19624 | Harris County | 32,010 |
| 48 | | 05000 | Austin city | 790,390 | 48 | | 19792 | Del Rio city | 35,591 |
| 48 | 209 | 05000 | Hays County | 2 | 48 | 465 | 19792 | Val Verde County | 35,591 |
| 48 | 453 | 05000 | Travis County | 754,691 | | | | | |
| 48 | 491 | 05000 | Williamson County | 35,697 | 48 | | 19972 | Denton city | 113,383 |
| | | | | | 48 | 121 | 19972 | Denton County | 113,383 |
| 48 | | 06128 | Baytown city | 71,802 | | | | | |
| 48 | 071 | 06128 | Chambers County | 4,116 | 48 | | 20092 | DeSoto city | 49,047 |
| 48 | 201 | 06128 | Harris County | 67,686 | 48 | 113 | 20092 | Dallas County | 49,047 |
| 48 | | 07000 | Beaumont city | 118,296 | 48 | | 21628 | Duncanville city | 38,524 |
| 48 | 245 | 07000 | Jefferson County | 118,296 | 48 | 113 | 21628 | Dallas County | 38,524 |
| 48 | | 07132 | Bedford city | 46,979 | 48 | | 21892 | Eagle Pass city | 26,248 |
| 48 | 439 | 07132 | Tarrant County | 46,979 | 48 | 323 | 21892 | Maverick County | 26,248 |
| 48 | | 08236 | Big Spring city | 27,282 | 48 | | 22660 | Edinburg city | 77,100 |
| 48 | 227 | 08236 | Howard County | 27,282 | 48 | 215 | 22660 | Hidalgo County | 77,100 |
| 48 | | 10768 | Brownsville city | 175,023 | 48 | | 24000 | El Paso city | 649,121 |
| 48 | 061 | 10768 | Cameron County | 175,023 | 48 | 141 | 24000 | El Paso County | 649,121 |
| 48 | | 10912 | Bryan city | 76,201 | 48 | | 24768 | Euless city | 51,277 |
| 48 | 041 | 10912 | Brazos County | 76,201 | 48 | 439 | 24768 | Tarrant County | 51,277 |
| 48 | | 11428 | Burleson city | 36,690 | 48 | | 25452 | Farmers Branch city | 28,616 |
| 48 | 251 | 11428 | Johnson County | 29,111 | 48 | 113 | 25452 | Dallas County | 28,616 |
| 48 | 439 | 11428 | Tarrant County | 7,579 | | | | | |
| | | | | | 48 | | 26232 | Flower Mound town | 64,669 |
| 48 | | 13024 | Carrollton city | 119,097 | 48 | 121 | 26232 | Denton County | 64,457 |
| 48 | 085 | 13024 | Collin County | 2 | 48 | 439 | 26232 | Tarrant County | 212 |
| 48 | 113 | 13024 | Dallas County | 49,352 | | | | | |
| 48 | 121 | 13024 | Denton County | 69,743 | 48 | | 27000 | Fort Worth city | 741,206 |
| | | | | | 48 | 121 | 27000 | Denton County | 7,813 |
| 48 | | 13492 | Cedar Hill city | 45,028 | 48 | 367 | 27000 | Parker County | 7 |
| 48 | 113 | 13492 | Dallas County | 44,477 | 48 | 439 | 27000 | Tarrant County | 733,386 |
| 48 | 139 | 13492 | Ellis County | 551 | 48 | 497 | 27000 | Wise County | 0 |
| 48 | | 13552 | Cedar Park city | 48,937 | 48 | | 27648 | Friendswood city | 35,805 |
| 48 | 453 | 13552 | Travis County | 489 | 48 | 167 | 27648 | Galveston County | 25,510 |
| 48 | 491 | 13552 | Williamson County | 48,448 | 48 | 201 | 27648 | Harris County | 10,295 |
| 48 | | 15364 | Cleburne city | 29,337 | 48 | | 27684 | Frisco city | 116,989 |
| 48 | 251 | 15364 | Johnson County | 29,337 | 48 | 085 | 27684 | Collin County | 72,489 |
| | | | | | 48 | 121 | 27684 | Denton County | 44,500 |
| 48 | | 15976 | College Station city | 93,857 | | | | | |
| 48 | 041 | 15976 | Brazos County | 93,857 | 48 | | 28068 | Galveston city | 47,743 |
| | | | | | 48 | 167 | 28068 | Galveston County | 47,743 |
| 48 | | 16432 | Conroe city | 56,207 | | | | | |
| 48 | 339 | 16432 | Montgomery County | 56,207 | 48 | | 29000 | Garland city | 226,876 |
| | | | | | 48 | 085 | 29000 | Collin County | 266 |
| 48 | | 16612 | Coppell city | 38,659 | 48 | 113 | 29000 | Dallas County | 226,608 |
| 48 | 113 | 16612 | Dallas County | 37,905 | 48 | 397 | 29000 | Rockwall County | 2 |
| 48 | 121 | 16612 | Denton County | 754 | | | | | |
| | | | | | 48 | | 29336 | Georgetown city | 47,400 |
| 48 | | 16624 | Copperas Cove city | 32,032 | 48 | 491 | 29336 | Williamson County | 47,400 |
| 48 | 027 | 16624 | Bell County | 0 | | | | | |
| 48 | 099 | 16624 | Coryell County | 31,457 | 48 | | 30464 | Grand Prairie city | 175,396 |
| 48 | 281 | 16624 | Lampasas County | 575 | 48 | 113 | 30464 | Dallas County | 123,487 |
| | | | | | 48 | 139 | 30464 | Ellis County | 45 |
| 48 | | 17000 | Corpus Christi city | 305,215 | 48 | 439 | 30464 | Tarrant County | 51,864 |
| 48 | 007 | 17000 | Aransas County | 0 | | | | | |
| 48 | 273 | 17000 | Kleberg County | 0 | 48 | | 30644 | Grapevine city | 46,334 |
| 48 | 355 | 17000 | Nueces County | 305,215 | 48 | 113 | 30644 | Dallas County | 0 |
| 48 | 409 | 17000 | San Patricio County | 0 | 48 | 121 | 30644 | Denton County | 0 |
| | | | | | 48 | 439 | 30644 | Tarrant County | 46,334 |

| State code | Place code | County code | Geographic Area Name | 2010 census population | State code | Place code | County code | Geographic Area Name | 2010 census population |
|---|---|---|---|---|---|---|---|---|---|
| 48 | | 30920 | Greenville city | 25,557 | 48 | | 45384 | McAllen city | 129,877 |
| 48 | 231 | 30920 | Hunt County | 25,557 | 48 | 215 | 45384 | Hidalgo County | 129,877 |
| 48 | | 31928 | Haltom City city | 42,409 | 48 | | 45744 | McKinney city | 131,117 |
| 48 | 439 | 31928 | Tarrant County | 42,409 | 48 | 085 | 45744 | Collin County | 131,117 |
| 48 | | 32312 | Harker Heights city | 26,700 | 48 | | 46452 | Mansfield city | 56,368 |
| 48 | 027 | 32312 | Bell County | 26,700 | 48 | 139 | 46452 | Ellis County | 95 |
| | | | | | 48 | 251 | 46452 | Johnson County | 1,652 |
| 48 | | 32372 | Harlingen city | 64,849 | 48 | 439 | 46452 | Tarrant County | 54,621 |
| 48 | 061 | 32372 | Cameron County | 64,849 | 48 | | 47892 | Mesquite city | 139,824 |
| 48 | | 35000 | Houston city | 2,099,451 | 48 | 113 | 47892 | Dallas County | 139,731 |
| 48 | 157 | 35000 | Fort Bend County | 38,124 | 48 | 257 | 47892 | Kaufman County | 93 |
| 48 | 201 | 35000 | Harris County | 2,057,280 | | | | | |
| 48 | 339 | 35000 | Montgomery County | 4,047 | 48 | | 48072 | Midland city | 111,147 |
| | | | | | 48 | 317 | 48072 | Martin County | 0 |
| 48 | | 35528 | Huntsville city | 38,548 | 48 | 329 | 48072 | Midland County | 111,147 |
| 48 | 471 | 35528 | Walker County | 38,548 | | | | | |
| 48 | | 35576 | Hurst city | 37,337 | 48 | | 48768 | Mission city | 77,058 |
| 48 | 439 | 35576 | Tarrant County | 37,337 | 48 | 215 | 48768 | Hidalgo County | 77,058 |
| 48 | | 37000 | Irving city | 216,290 | 48 | | 48804 | Missouri City city | 67,358 |
| 48 | 113 | 37000 | Dallas County | 216,290 | 48 | 157 | 48804 | Fort Bend County | 61,755 |
| | | | | | 48 | 201 | 48804 | Harris County | 5,603 |
| 48 | | 38632 | Keller city | 39,627 | 48 | | 50256 | Nacogdoches city | 32,996 |
| 48 | 439 | 38632 | Tarrant County | 39,627 | 48 | 347 | 50256 | Nacogdoches County | 32,996 |
| 48 | | 39148 | Killeen city | 127,921 | 48 | | 50820 | New Braunfels city | 57,740 |
| 48 | 027 | 39148 | Bell County | 127,921 | 48 | 091 | 50820 | Comal County | 47,586 |
| | | | | | 48 | 187 | 50820 | Guadalupe County | 10,154 |
| 48 | | 39352 | Kingsville city | 26,213 | | | | | |
| 48 | 273 | 39352 | Kleberg County | 26,213 | 48 | | 52356 | North Richland Hills city | 63,343 |
| 48 | | 39952 | Kyle city | 28,016 | 48 | 439 | 52356 | Tarrant County | 63,343 |
| 48 | 209 | 39952 | Hays County | 28,016 | 48 | | 53388 | Odessa city | 99,940 |
| 48 | | 40588 | Lake Jackson city | 26,849 | 48 | 135 | 53388 | Ector County | 98,270 |
| 48 | 039 | 40588 | Brazoria County | 26,849 | 48 | 329 | 53388 | Midland County | 1,670 |
| 48 | | 41212 | Lancaster city | 36,361 | 48 | | 55080 | Paris city | 25,171 |
| 48 | 113 | 41212 | Dallas County | 36,361 | 48 | 277 | 55080 | Lamar County | 25,171 |
| 48 | | 41440 | La Porte city | 33,800 | 48 | | 56000 | Pasadena city | 149,043 |
| 48 | 201 | 41440 | Harris County | 33,800 | 48 | 201 | 56000 | Harris County | 149,043 |
| 48 | | 41464 | Laredo city | 236,091 | 48 | | 56348 | Pearland city | 91,252 |
| 48 | 479 | 41464 | Webb County | 236,091 | 48 | 039 | 56348 | Brazoria County | 86,706 |
| | | | | | 48 | 157 | 56348 | Fort Bend County | 721 |
| 48 | | 41980 | League City city | 83,560 | 48 | 201 | 56348 | Harris County | 3,825 |
| 48 | 167 | 41980 | Galveston County | 81,998 | | | | | |
| 48 | 201 | 41980 | Harris County | 1,562 | 48 | | 57176 | Pflugerville city | 46,936 |
| | | | | | 48 | 453 | 57176 | Travis County | 46,636 |
| 48 | | 42016 | Leander city | 26,521 | 48 | 491 | 57176 | Williamson County | 300 |
| 48 | 453 | 42016 | Travis County | 1,077 | | | | | |
| 48 | 491 | 42016 | Williamson County | 25,444 | 48 | | 57200 | Pharr city | 70,400 |
| 48 | | 42508 | Lewisville city | 95,290 | 48 | 215 | 57200 | Hidalgo County | 70,400 |
| 48 | 113 | 42508 | Dallas County | 841 | 48 | | 58016 | Plano city | 259,841 |
| 48 | 121 | 42508 | Denton County | 94,449 | 48 | 085 | 58016 | Collin County | 254,525 |
| | | | | | 48 | 121 | 58016 | Denton County | 5,316 |
| 48 | | 43012 | Little Elm city | 25,898 | | | | | |
| 48 | 121 | 43012 | Denton County | 25,898 | 48 | | 58820 | Port Arthur city | 53,818 |
| 48 | | 43888 | Longview city | 80,455 | 48 | 245 | 58820 | Jefferson County | 53,814 |
| 48 | 183 | 43888 | Gregg County | 78,585 | 48 | 361 | 58820 | Orange County | 4 |
| 48 | 203 | 43888 | Harrison County | 1,870 | 48 | | 61796 | Richardson city | 99,223 |
| 48 | | 45000 | Lubbock city | 229,573 | 48 | 085 | 61796 | Collin County | 28,569 |
| 48 | 303 | 45000 | Lubbock County | 229,573 | 48 | 113 | 61796 | Dallas County | 70,654 |
| 48 | | 45072 | Lufkin city | 35,067 | 48 | | 62828 | Rockwall city | 37,490 |
| 48 | 005 | 45072 | Angelina County | 35,067 | 48 | 397 | 62828 | Rockwall County | 37,490 |

| State code | Place code | County code | Geographic Area Name | 2010 census population | State code | Place code | County code | Geographic Area Name | 2010 census population |
|---|---|---|---|---|---|---|---|---|---|
| 48 | | 63284 | Rosenberg city | 30,618 | 48 | | 76864 | Weatherford city | 25,250 |
| 48 | 157 | 63284 | Fort Bend County | 30,618 | 48 | 367 | 76864 | Parker County | 25,250 |
| 48 | | 63500 | Round Rock city | 99,887 | 48 | | 77272 | Weslaco city | 35,670 |
| 48 | 453 | 63500 | Travis County | 1,362 | 48 | 215 | 77272 | Hidalgo County | 35,670 |
| 48 | 491 | 63500 | Williamson County | 98,525 | 48 | | 79000 | Wichita Falls city | 104,553 |
| 48 | | 63572 | Rowlett city | 56,199 | 48 | 485 | 79000 | Wichita County | 104,553 |
| 48 | 113 | 63572 | Dallas County | 49,188 | 48 | | 80356 | Wylie city | 41,427 |
| 48 | 397 | 63572 | Rockwall County | 7,011 | 48 | 085 | 80356 | Collin County | 39,957 |
| 48 | | 64472 | San Angelo city | 93,200 | 48 | 113 | 80356 | Dallas County | 415 |
| 48 | 451 | 64472 | Tom Green County | 93,200 | 48 | 397 | 80356 | Rockwall County | 1,055 |
| 48 | | 65000 | San Antonio city | 1,327,407 | 49 | | | **UTAH** | 2,763,885 |
| 48 | 029 | 65000 | Bexar County | 1,327,381 | | | | | |
| 48 | 091 | 65000 | Comal County | 0 | 49 | | 01310 | American Fork city | 26,263 |
| 48 | 325 | 65000 | Medina County | 26 | 49 | 049 | 01310 | Utah County | 26,263 |
| 48 | | 65516 | San Juan city | 33,856 | 49 | | 07690 | Bountiful city | 42,552 |
| 48 | 215 | 65516 | Hidalgo County | 33,856 | 49 | 011 | 07690 | Davis County | 42,552 |
| 48 | | 65600 | San Marcos city | 44,894 | 49 | | 11320 | Cedar City city | 28,857 |
| 48 | 055 | 65600 | Caldwell County | 3 | 49 | 021 | 11320 | Iron County | 28,857 |
| 48 | 187 | 65600 | Guadalupe County | 0 | 49 | | 13850 | Clearfield city | 30,112 |
| 48 | 209 | 65600 | Hays County | 44,891 | 49 | 011 | 13850 | Davis County | 30,112 |
| 48 | | 66128 | Schertz city | 31,465 | 49 | | 16270 | Cottonwood Heights city | 33,433 |
| 48 | 029 | 66128 | Bexar County | 1,157 | 49 | 035 | 16270 | Salt Lake County | 33,433 |
| 48 | 091 | 66128 | Comal County | 845 | 49 | | 20120 | Draper city | 42,274 |
| 48 | 187 | 66128 | Guadalupe County | 29,463 | 49 | 035 | 20120 | Salt Lake County | 40,532 |
| 48 | | 66644 | Seguin city | 25,175 | 49 | 049 | 20120 | Utah County | 1,742 |
| 48 | 187 | 66644 | Guadalupe County | 25,175 | 49 | | 36070 | Holladay city | 26,472 |
| 48 | | 67496 | Sherman city | 38,521 | 49 | 035 | 36070 | Salt Lake County | 26,472 |
| 48 | 181 | 67496 | Grayson County | 38,521 | 49 | | 40360 | Kaysville city | 27,300 |
| 48 | | 68636 | Socorro city | 32,013 | 49 | 011 | 40360 | Davis County | 27,300 |
| 48 | 141 | 68636 | El Paso County | 32,013 | 49 | | 43660 | Layton city | 67,311 |
| 48 | | 69032 | Southlake city | 26,575 | 49 | 011 | 43660 | Davis County | 67,311 |
| 48 | 121 | 69032 | Denton County | 773 | 49 | | 44320 | Lehi city | 47,407 |
| 48 | 439 | 69032 | Tarrant County | 25,802 | 49 | 049 | 44320 | Utah County | 47,407 |
| 48 | | 70808 | Sugar Land city | 78,817 | 49 | | 45860 | Logan city | 48,174 |
| 48 | 157 | 70808 | Fort Bend County | 78,817 | 49 | 005 | 45860 | Cache County | 48,174 |
| 48 | | 72176 | Temple city | 66,102 | 49 | | 49710 | Midvale city | 27,964 |
| 48 | 027 | 72176 | Bell County | 66,102 | 49 | 035 | 49710 | Salt Lake County | 27,964 |
| 48 | | 72368 | Texarkana city | 36,411 | 49 | | 53230 | Murray city | 46,746 |
| 48 | 037 | 72368 | Bowie County | 36,411 | 49 | 035 | 53230 | Salt Lake County | 46,746 |
| 48 | | 72392 | Texas City city | 45,099 | 49 | | 55980 | Ogden city | 82,825 |
| 48 | 071 | 72392 | Chambers County | 0 | 49 | 057 | 55980 | Weber County | 82,825 |
| 48 | 167 | 72392 | Galveston County | 45,099 | 49 | | 57300 | Orem city | 88,328 |
| 48 | | 72530 | The Colony city | 36,328 | 49 | 049 | 57300 | Utah County | 88,328 |
| 48 | 121 | 72530 | Denton County | 36,328 | 49 | | 60930 | Pleasant Grove city | 33,509 |
| 48 | | 74144 | Tyler city | 96,900 | 49 | 049 | 60930 | Utah County | 33,509 |
| 48 | 423 | 74144 | Smith County | 96,900 | 49 | | 62470 | Provo city | 112,488 |
| 48 | | 75428 | Victoria city | 62,592 | 49 | 049 | 62470 | Utah County | 112,488 |
| 48 | 469 | 75428 | Victoria County | 62,592 | 49 | | 64340 | Riverton city | 38,753 |
| 48 | | 76000 | Waco city | 124,805 | 49 | 035 | 64340 | Salt Lake County | 38,753 |
| 48 | 309 | 76000 | McLennan County | 124,805 | 49 | | 65110 | Roy city | 36,884 |
| 48 | | 76816 | Waxahachie city | 29,621 | 49 | 057 | 65110 | Weber County | 36,884 |
| 48 | 139 | 76816 | Ellis County | 29,621 | | | | | |

| State code | Place code | County code | Geographic Area Name | 2010 census population | State code | Place code | County code | Geographic Area Name | 2010 census population |
|---|---|---|---|---|---|---|---|---|---|
| 49 | | 65330 | St. George city | 72,897 | 51 | | 61832 | Petersburg city | 32,420 |
| 49 | 053 | 65330 | Washington County | 72,897 | 51 | 730 | 61832 | Petersburg city | 32,420 |
| 49 | | 67000 | Salt Lake City city | 186,440 | 51 | | 64000 | Portsmouth city | 95,535 |
| 49 | 035 | 67000 | Salt Lake County | 186,440 | 51 | 740 | 64000 | Portsmouth city | 95,535 |
| 49 | | 67440 | Sandy city | 87,461 | 51 | | 67000 | Richmond city | 204,214 |
| 49 | 035 | 67440 | Salt Lake County | 87,461 | 51 | 760 | 67000 | Richmond city | 204,214 |
| 49 | | 70850 | South Jordan city | 50,418 | 51 | | 68000 | Roanoke city | 97,032 |
| 49 | 035 | 70850 | Salt Lake County | 50,418 | 51 | 770 | 68000 | Roanoke city | 97,032 |
| 49 | | 71290 | Spanish Fork city | 34,691 | 51 | | 76432 | Suffolk city | 84,585 |
| 49 | 049 | 71290 | Utah County | 34,691 | 51 | 800 | 76432 | Suffolk city | 84,585 |
| 49 | | 72280 | Springville city | 29,466 | 51 | | 82000 | Virginia Beach city | 437,994 |
| 49 | 049 | 72280 | Utah County | 29,466 | 51 | 810 | 82000 | Virginia Beach city | 437,994 |
| 49 | | 75360 | Taylorsville city | 58,652 | 51 | | 86720 | Winchester city | 26,203 |
| 49 | 035 | 75360 | Salt Lake County | 58,652 | 51 | 840 | 86720 | Winchester city | 26,203 |
| 49 | | 76680 | Tooele city | 31,605 | 53 | | | **WASHINGTON** | 6,724,540 |
| 49 | 045 | 76680 | Tooele County | 31,605 | | | | | |
| 49 | | 82950 | West Jordan city | 103,712 | 53 | | 03180 | Auburn city | 70,180 |
| 49 | 035 | 82950 | Salt Lake County | 103,712 | 53 | 033 | 03180 | King County | 62,761 |
| | | | | | 53 | 053 | 03180 | Pierce County | 7,419 |
| 49 | | 83470 | West Valley City city | 129,480 | 53 | | 05210 | Bellevue city | 122,363 |
| 49 | 035 | 83470 | Salt Lake County | 129,480 | 53 | 033 | 05210 | King County | 122,363 |
| 50 | | | **VERMONT** | 625,741 | 53 | | 05280 | Bellingham city | 80,885 |
| | | | | | 53 | 073 | 05280 | Whatcom County | 80,885 |
| 50 | | 10675 | Burlington city | 42,417 | 53 | | 07380 | Bothell city | 33,505 |
| 50 | 007 | 10675 | Chittenden County | 42,417 | 53 | 033 | 07380 | King County | 17,090 |
| | | | | | 53 | 061 | 07380 | Snohomish County | 16,415 |
| 51 | | | **VIRGINIA** | 8,001,024 | 53 | | 07695 | Bremerton city | 37,729 |
| 51 | | 01000 | Alexandria city | 139,966 | 53 | 035 | 07695 | Kitsap County | 37,729 |
| 51 | 510 | 01000 | Alexandria city | 139,966 | 53 | | 08850 | Burien city | 33,313 |
| 51 | | 07784 | Blacksburg town | 42,620 | 53 | 033 | 08850 | King County | 33,313 |
| 51 | 121 | 07784 | Montgomery County | 42,620 | 53 | | 17635 | Des Moines city | 29,673 |
| 51 | | 14968 | Charlottesville city | 43,475 | 53 | 033 | 17635 | King County | 29,673 |
| 51 | 540 | 14968 | Charlottesville city | 43,475 | 53 | | 20750 | Edmonds city | 39,709 |
| 51 | | 16000 | Chesapeake city | 222,209 | 53 | 061 | 20750 | Snohomish County | 39,709 |
| 51 | 550 | 16000 | Chesapeake city | 222,209 | 53 | | 22640 | Everett city | 103,019 |
| 51 | | 21344 | Danville city | 43,055 | 53 | 061 | 22640 | Snohomish County | 103,019 |
| 51 | 590 | 21344 | Danville city | 43,055 | 53 | | 23515 | Federal Way city | 89,306 |
| 51 | | 35000 | Hampton city | 137,436 | 53 | 033 | 23515 | King County | 89,306 |
| 51 | 650 | 35000 | Hampton city | 137,436 | 53 | | 33805 | Issaquah city | 30,434 |
| 51 | | 35624 | Harrisonburg city | 48,914 | 53 | 033 | 33805 | King County | 30,434 |
| 51 | 660 | 35624 | Harrisonburg city | 48,914 | 53 | | 35275 | Kennewick city | 73,917 |
| 51 | | 44984 | Leesburg town | 42,616 | 53 | 005 | 35275 | Benton County | 73,917 |
| 51 | 107 | 44984 | Loudoun County | 42,616 | 53 | | 35415 | Kent city | 92,411 |
| 51 | | 47672 | Lynchburg city | 75,568 | 53 | 033 | 35415 | King County | 92,411 |
| 51 | 680 | 47672 | Lynchburg city | 75,568 | 53 | | 35940 | Kirkland city | 48,787 |
| 51 | | 48952 | Manassas city | 37,821 | 53 | 033 | 35940 | King County | 48,787 |
| 51 | 683 | 48952 | Manassas city | 37,821 | 53 | | 36745 | Lacey city | 42,393 |
| 51 | | 56000 | Newport News city | 180,719 | 53 | 067 | 36745 | Thurston County | 42,393 |
| 51 | 700 | 56000 | Newport News city | 180,719 | 53 | | 37900 | Lake Stevens city | 28,069 |
| 51 | | 57000 | Norfolk city | 242,803 | 53 | 061 | 37900 | Snohomish County | 28,069 |
| 51 | 710 | 57000 | Norfolk city | 242,803 | | | | | |

| State code | Place code | County code | Geographic Area Name | 2010 census population | State code | Place code | County code | Geographic Area Name | 2010 census population |
|---|---|---|---|---|---|---|---|---|---|
| 53 | | 38038 | Lakewood city | 58,163 | 54 | | | **WEST VIRGINIA** | 1,852,994 |
| 53 | 053 | 38038 | Pierce County | 58,163 | 54 | | 14600 | Charleston city | 51,400 |
| 53 | | 40245 | Longview city | 36,648 | 54 | 039 | 14600 | Kanawha County | 51,400 |
| 53 | 015 | 40245 | Cowlitz County | 36,648 | 54 | | 39460 | Huntington city | 49,138 |
| 53 | | 40840 | Lynnwood city | 35,836 | 54 | 011 | 39460 | Cabell County | 45,214 |
| 53 | 061 | 40840 | Snohomish County | 35,836 | 54 | 099 | 39460 | Wayne County | 3,924 |
| 53 | | 43955 | Marysville city | 60,020 | 54 | | 55756 | Morgantown city | 29,660 |
| 53 | 061 | 43955 | Snohomish County | 60,020 | 54 | 061 | 55756 | Monongalia County | 29,660 |
| 53 | | 47560 | Mount Vernon city | 31,743 | 54 | | 62140 | Parkersburg city | 31,492 |
| 53 | 057 | 47560 | Skagit County | 31,743 | 54 | 107 | 62140 | Wood County | 31,492 |
| 53 | | 51300 | Olympia city | 46,478 | 54 | | 86452 | Wheeling city | 28,486 |
| 53 | 067 | 51300 | Thurston County | 46,478 | 54 | 051 | 86452 | Marshall County | 276 |
| 53 | | 53545 | Pasco city | 59,781 | 54 | 069 | 86452 | Ohio County | 28,210 |
| 53 | 021 | 53545 | Franklin County | 59,781 | 55 | | | **WISCONSIN** | 5,686,986 |
| 53 | | 56625 | Pullman city | 29,799 | 55 | | 02375 | Appleton city | 72,623 |
| 53 | 075 | 56625 | Whitman County | 29,799 | 55 | 015 | 02375 | Calumet County | 11,088 |
| 53 | | 56695 | Puyallup city | 37,022 | 55 | 087 | 02375 | Outagamie County | 60,045 |
| 53 | 053 | 56695 | Pierce County | 37,022 | 55 | 139 | 02375 | Winnebago County | 1,490 |
| 53 | | 57535 | Redmond city | 54,144 | 55 | | 06500 | Beloit city | 36,966 |
| 53 | 033 | 57535 | King County | 54,144 | 55 | 105 | 06500 | Rock County | 36,966 |
| 53 | | 57745 | Renton city | 90,927 | 55 | | 10025 | Brookfield city | 37,920 |
| 53 | 033 | 57745 | King County | 90,927 | 55 | 133 | 10025 | Waukesha County | 37,920 |
| 53 | | 58235 | Richland city | 48,058 | 55 | | 22300 | Eau Claire city | 65,883 |
| 53 | 005 | 58235 | Benton County | 48,058 | 55 | 017 | 22300 | Chippewa County | 1,981 |
| 53 | | 61115 | Sammamish city | 45,780 | 55 | 035 | 22300 | Eau Claire County | 63,902 |
| 53 | 033 | 61115 | King County | 45,780 | 55 | | 25950 | Fitchburg city | 25,260 |
| 53 | | 62288 | SeaTac city | 26,909 | 55 | 025 | 25950 | Dane County | 25,260 |
| 53 | 033 | 62288 | King County | 26,909 | 55 | | 26275 | Fond du Lac city | 43,021 |
| 53 | | 63000 | Seattle city | 608,660 | 55 | 039 | 26275 | Fond du Lac County | 43,021 |
| 53 | 033 | 63000 | King County | 608,660 | 55 | | 27300 | Franklin city | 35,451 |
| 53 | | 63960 | Shoreline city | 53,007 | 55 | 079 | 27300 | Milwaukee County | 35,451 |
| 53 | 033 | 63960 | King County | 53,007 | 55 | | 31000 | Green Bay city | 104,057 |
| 53 | | 67000 | Spokane city | 208,916 | 55 | 009 | 31000 | Brown County | 104,057 |
| 53 | 063 | 67000 | Spokane County | 208,916 | 55 | | 31175 | Greenfield city | 36,720 |
| 53 | | 67167 | Spokane Valley city | 89,755 | 55 | 079 | 31175 | Milwaukee County | 36,720 |
| 53 | 063 | 67167 | Spokane County | 89,755 | 55 | | 37825 | Janesville city | 63,575 |
| 53 | | 70000 | Tacoma city | 198,397 | 55 | 105 | 37825 | Rock County | 63,575 |
| 53 | 053 | 70000 | Pierce County | 198,397 | 55 | | 39225 | Kenosha city | 99,218 |
| 53 | | 73465 | University Place city | 31,144 | 55 | 059 | 39225 | Kenosha County | 99,218 |
| 53 | 053 | 73465 | Pierce County | 31,144 | 55 | | 40775 | La Crosse city | 51,320 |
| 53 | | 74060 | Vancouver city | 161,791 | 55 | 063 | 40775 | La Crosse County | 51,320 |
| 53 | 011 | 74060 | Clark County | 161,791 | 55 | | 48000 | Madison city | 233,209 |
| 53 | | 75775 | Walla Walla city | 31,731 | 55 | 025 | 48000 | Dane County | 233,209 |
| 53 | 071 | 75775 | Walla Walla County | 31,731 | 55 | | 48500 | Manitowoc city | 33,736 |
| 53 | | 77105 | Wenatchee city | 31,925 | 55 | 071 | 48500 | Manitowoc County | 33,736 |
| 53 | 007 | 77105 | Chelan County | 31,925 | 55 | | 51000 | Menomonee Falls village | 35,626 |
| 53 | | 80010 | Yakima city | 91,067 | 55 | 133 | 51000 | Waukesha County | 35,626 |
| 53 | 077 | 80010 | Yakima County | 91,067 | 55 | | 53000 | Milwaukee city | 594,833 |
| | | | | | 55 | 079 | 53000 | Milwaukee County | 594,833 |
| | | | | | 55 | 131 | 53000 | Washington County | 0 |
| | | | | | 55 | 133 | 53000 | Waukesha County | 0 |

# CITIES BY COUNTY—*Continued*

| State code | Place code | County code | Geographic Area Name | 2010 census population | State code | Place code | County code | Geographic Area Name | 2010 census population |
|---|---|---|---|---|---|---|---|---|---|
| 55 | | 54875 | Mount Pleasant village | 26,197 | 13 | | 83728 | Winterville city | 1,122 |
| 55 | 101 | 54875 | Racine County | 26,197 | 13 | | 04200 | Augusta-Richmond county | 200,549 |
| 55 | | 55750 | Neenah city | 25,501 | 13 | 04204 | | Augusta-Richmond county (balance) | 195,844 |
| 55 | 139 | 55750 | Winnebago County | 25,501 | 13 | | 09040 | Blythe city | 694 |
| 55 | | 56375 | New Berlin city | 39,584 | 13 | | 38040 | Hephzibah city | 4,011 |
| 55 | 133 | 56375 | Waukesha County | 39,584 | 18 | | | **INDIANA** | 6,483,802 |
| 55 | | 58800 | Oak Creek city | 34,451 | 18 | | 36000 | Indianapolis city | 829,718 |
| 55 | 079 | 58800 | Milwaukee County | 34,451 | 18 | | 04204 | Beech Grove city | 0 |
| 55 | | 60500 | Oshkosh city | 66,083 | 18 | | 13492 | Clermont town | 1,356 |
| 55 | 139 | 60500 | Winnebago County | 66,083 | 18 | | 16156 | Crows Nest town | 73 |
| 55 | | 66000 | Racine city | 78,860 | 18 | | 16336 | Cumberland town | 2,597 |
| 55 | 101 | 66000 | Racine County | 78,860 | 18 | | 34420 | Homecroft town | 722 |
| 55 | | 72975 | Sheboygan city | 49,288 | 18 | | 36003 | Indianapolis city (balance) | 820,445 |
| 55 | 117 | 72975 | Sheboygan County | 49,288 | 18 | | 42426 | Lawrence city | 42 |
| 55 | | 77200 | Stevens Point city | 26,717 | 18 | | 48456 | Meridian Hills town | 1,616 |
| 55 | 097 | 77200 | Portage County | 26,717 | 18 | | 54612 | North Crows Nest town | 45 |
| 55 | | 78600 | Sun Prairie city | 29,364 | 18 | | 65556 | Rocky Ripple town | 606 |
| 55 | 025 | 78600 | Dane County | 29,364 | 18 | | 72232 | Spring Hill town | 98 |
| 55 | | 78650 | Superior city | 27,244 | 18 | | 80234 | Warren Park town | 1,480 |
| 55 | 031 | 78650 | Douglas County | 27,244 | 18 | | 84374 | Williams Creek town | 407 |
| 55 | | 84250 | Waukesha city | 70,718 | 18 | | 85742 | Wynnedale town | 231 |
| 55 | 133 | 84250 | Waukesha County | 70,718 | 21 | | | **KENTUCKY** | 4,339,367 |
| 55 | | 84475 | Wausau city | 39,106 | 21 | | 46003 | Louisville/Jefferson County | 741,096 |
| 55 | 073 | 84475 | Marathon County | 39,106 | 21 | | 01504 | Anchorage city | 2,348 |
| 55 | | 84675 | Wauwatosa city | 46,396 | 21 | | 02656 | Audubon Park city | 1,473 |
| 55 | 079 | 84675 | Milwaukee County | 46,396 | 21 | | 03376 | Bancroft city | 494 |
| 55 | | 85300 | West Allis city | 60,411 | 21 | | 03556 | Barbourmeade city | 1,218 |
| 55 | 079 | 85300 | Milwaukee County | 60,411 | 21 | | 05068 | Beechwood Village city | 1,324 |
| 55 | | 85350 | West Bend city | 31,078 | 21 | | 05392 | Bellemeade city | 865 |
| 55 | 131 | 85350 | Washington County | 31,078 | 21 | | 05464 | Bellewood city | 321 |
| 56 | | | **WYOMING** | 563,626 | 21 | | 07858 | Blue Ridge Manor city | 767 |
| 56 | | 13150 | Casper city | 55,316 | 21 | | 09532 | Briarwood city | 435 |
| 56 | 025 | 13150 | Natrona County | 55,316 | 21 | | 09847 | Broeck Pointe city | 272 |
| 56 | | 13900 | Cheyenne city | 59,466 | 21 | | 10162 | Brownsboro Farm city | 648 |
| 56 | 021 | 13900 | Laramie County | 59,466 | 21 | | 10198 | Brownsboro Village city | 319 |
| 56 | | 31855 | Gillette city | 29,087 | 21 | | 12066 | Cambridge city | 175 |
| 56 | 005 | 31855 | Campbell County | 29,087 | 21 | | 16395 | Coldstream city | 1,100 |
| 56 | | 45050 | Laramie city | 30,816 | 21 | | 18270 | Creekside city | 305 |
| 56 | 001 | 45050 | Albany County | 30,816 | 21 | | 18766 | Crossgate city | 225 |

The following consolidated cities are included in Table D. They are listed here with their 2010 census populations followed by the separate entities that make up the consolidated city. Data from the American Community Survey include only the "balance", the major city of each consolidated city.

| | | | | | 21 | | 22204 | Douglass Hills city | 5,484 |
|---|---|---|---|---|---|---|---|---|---|
| | | | | | 21 | | 22474 | Druid Hills city | 308 |
| | | | | | 21 | | 27262 | Fincastle city | 817 |
| | | | | | 21 | | 28342 | Forest Hills city | 444 |
| 09 | | | **CONNECTICUT** | 3,574,097 | 21 | | 31348 | Glenview city | 531 |
| | | | | | 21 | | 31402 | Glenview Hills city | 319 |
| 09 | | 47500 | Milford city | 52,759 | 21 | | 31420 | Glenview Manor city | 191 |
| | | 47515 | Milford city (balance) | 51,271 | 21 | | 31870 | Goose Creek city | 294 |
| | | 88050 | Woodmont borough | 1,488 | 21 | | 32523 | Graymoor-Devondale city | 2,870 |
| | | | | | 21 | | 32986 | Green Spring city | 715 |
| 13 | | | **GEORGIA** | 9,687,653 | 21 | | 36102 | Heritage Creek city | 1,076 |
| 13 | | 03436 | Athens-Clark county | 116,714 | 21 | | 36374 | Hickory Hill city | 114 |
| 13 | | 03440 | Athens-Clark county (balance) | 115,452 | 21 | | 36865 | Hills and Dales city | 142 |
| 13 | | 09068 | Bogart town | 140 | 21 | | 37576 | Hollow Creek city | 783 |
| | | | | | 21 | | 37630 | Hollyvilla city | 537 |
| | | | | | 21 | | 38170 | Houston Acres city | 507 |
| | | | | | 21 | | 38814 | Hurstbourne city | 4,216 |
| | | | | | 21 | | 38818 | Hurstbourne Acres city | 1,811 |
| | | | | | 21 | | 39304 | Indian Hills city | 2,868 |
| | | | | | 21 | | 40222 | Jeffersontown city | 26,595 |
| | | | | | 21 | | 42598 | Kingsley city | 381 |
| | | | | | 21 | | 43900 | Langdon Place city | 936 |
| | | | | | 21 | | 46540 | Lincolnshire city | 148 |
| | | | | | 21 | 48006 | | Louisville/Jefferson County (balance) | 597,337 |
| | | | | | 21 | | 48558 | Lyndon city | 11,002 |
| | | | | | 21 | | 48648 | Lynnview city | 914 |
| | | | | | 21 | | 49800 | Manor Creek city | 140 |
| | | | | | 21 | | 50412 | Maryhill Estates city | 179 |
| | | | | | 21 | | 51193 | Meadowbrook Farm city | 136 |

| State code | Place code | County code | Geographic Area Name | 2010 census population | State code | Place code | County code | Geographic Area Name | 2010 census population |
|---|---|---|---|---|---|---|---|---|---|
| 21 | | 51258 | Meadow Vale city | 736 | 21 | | 76380 | Thornhill city | 178 |
| 21 | | 51294 | Meadowview Estates city | 363 | 21 | | 80913 | Watterson Park city | 976 |
| 21 | | 51978 | Middletown city | 7,218 | 21 | | 81372 | Wellington city | 565 |
| 21 | | 52842 | Mockingbird Valley city | 167 | 21 | | 81624 | West Buechel city | 1,230 |
| 21 | | 53328 | Moorland city | 431 | 21 | | 82164 | Westwood city | 634 |
| 21 | | 54660 | Murray Hill city | 582 | 21 | | 83208 | Wildwood city | 261 |
| 21 | | 56550 | Norbourne Estates city | 441 | 21 | | 83784 | Windy Hills city | 2,385 |
| 21 | | 56730 | Northfield city | 1,020 | 21 | | 84486 | Woodland Hills city | 696 |
| 21 | | 56928 | Norwood city | 370 | 21 | | 84576 | Woodlawn Park city | 942 |
| 21 | | 57658 | Old Brownsboro Place city | 353 | 21 | | 84891 | Worthington Hills city | 1,446 |
| 21 | | 59322 | Parkway Village city | 650 | | | | | |
| 21 | | 61554 | Plantation city | 832 | **30** | | | **MONTANA** | 989,415 |
| 21 | | 62370 | Poplar Hills city | 362 | | | | | |
| 21 | | 63264 | Prospect city | 4,636 | 30 | | 11390 | Butte-Silver Bow | 34,200 |
| 21 | | 65208 | Richlawn city | 405 | 30 | | 11397 | Butte-Silver Bow (balance) | 33,525 |
| 21 | | 65766 | Riverwood city | 446 | 30 | | 77650 | Walkerville town | 675 |
| 21 | | 66486 | Rolling Fields city | 646 | | | | | |
| 21 | | 66504 | Rolling Hills city | 959 | **47** | | | **TENNESSEE** | 6,346,105 |
| 21 | | 67944 | St. Matthews city | 17,472 | | | | | |
| 21 | | 67998 | St. Regis Park city | 1,454 | 47 | | 52004 | Nashville-Davidson | 626,681 |
| 21 | | 69384 | Seneca Gardens city | 696 | 47 | | 04620 | Belle Meade city | 2,912 |
| 21 | | 70284 | Shively city | 15,264 | 47 | | 05140 | Berry Hill city | 537 |
| 21 | | 72138 | South Park View city | 7 | 47 | | 27020 | Forest Hills city | 4,812 |
| 21 | | 72770 | Spring Mill city | 287 | 47 | | 29920 | Goodlettsville city | 10,319 |
| 21 | | 72790 | Spring Valley city | 654 | 47 | | 40720 | Lakewood city | 2,302 |
| 21 | | 74064 | Strathmoor Manor city | 337 | 47 | | 52006 | Nashville-Davidson  (balance) | 601,222 |
| 21 | | 74082 | Strathmoor Village city | 648 | 47 | | 54780 | Oak Hill city | 4,529 |
| 21 | | 75190 | Sycamore city | 160 | 47 | | 63140 | Ridgetop city | 48 |
| 21 | | 75963 | Ten Broeck city | 103 | | | | | |

# APPENDIX F
# SOURCE NOTES AND EXPLANATIONS

The following documentation is provided in the order in which items appear in the tables. Internet addresses are provided for the sources of the data. Some of the links refer to the specific data tables. Others provide information about the general data source.

## TABLE A—STATES

Table A presents 355 items for the United States as a whole, for each individual state, and for the District of Columbia. The states are presented in alphabetical order.

### LAND AREA, Items 1 and 4
Source: U.S. Census Bureau—2020 U.S. Gazetteer Files,
https://www.census.gov/geographies/reference-files/time-series/geo/gazetteer-files.html

Land area measurements are shown to the nearest square mile. Land area includes dry land and land temporarily or partially covered by water, such as marshlands, swamps, and river floodplains. The 2020 land areas have been aggregated from the counties as listed in the 2020 Gazetteer files.

### POPULATION AND COMPONENTS OF CHANGE, Items 2–4, 31—41
Source: U.S. Census Bureau—Decennial Censuses and Population Estimates
https://www.census.gov/programs-surveys/popest.html
https://www.census.gov/programs-surveys/decennial-census/data/datasets.2010.html

The population data for 2020 are Census Bureau estimates of the resident population as of July 1, 2020.

The population data for 1990, 2000, and 2010 are from the decennial censuses and represent the resident population as of April 1 of those years.

The change in population between 2010 and 2020 is made up of (a) natural increase—births minus deaths, and (b) net migration—the difference between the number of persons moving into a particular state and the number of persons moving out of the state. Net migration is composed of internal and international migration.

### POPULATION AND POPULATION CHARAC-TERISTICS, Items 5–23 and 45–63
Source: U.S. Census Bureau—Population Estimates and 2019 American Community Survey

https://www.census.gov/programs-surveys/popest.html
http://www.census.gov/acs/www/

Data on age, sex, race, and Hispanic origin are from the Population Estimates program. Median age and data on place of birth are from the 2019 American Community Survey, a nationwide continuous survey designed to replace the long form question-naire used in previous censuses.

The concept "race alone or in combination" includes people who reported a single race alone (e.g., Asian) and people who reported that race in combination with one or more of the other major race groups (e.g., White, Black or African American, American Indian and Alaska Native, Native Hawaiian and Other Pacific Islander, and Some Other Race). The "race alone or in combination" concept, therefore, represents the maximum number of people who reported as that race group, either alone, or in combination with another race(s).

The sum of the four individual race alone or in combina-tion categories in this book may add to more than the total population because people who reported more than one race were tallied in each race category. In this book, the Asian group has been combined with the Native Hawaiian and Other Pacific Islander group, causing double-counting of persons who identify with both groups. This is especially pronounced in Hawaii.

Data on race were derived from answers to the question on race that was asked of all persons. The concept of race, as used by the Census Bureau, reflects self-identification by respondents according to the race or races with which they most closely iden-tify. These categories are sociopolitical constructs and should not be interpreted as being scientific or anthropological in nature. Furthermore, the race categories include both racial and national origin groups.

The **White** population is defined as persons who indicated their race as White, as well as persons who did not classify themselves in one of the specific race categories listed on the questionnaire but entered a nationality such as Irish, German, Italian, Lebanese, Near Easterner, Arab, or Polish.

The **Black** population includes persons who indicated their race as "Black or African American" as well as persons who did not classify themselves in one of the specific race categories but reported entries such as African American, Afro American, Kenyan, Nigerian, or Haitian.

The **American Indian or Alaska Native** population includes persons who indicated their race as American Indian or Alaska Native, as well as persons who did not classify themselves in one of the specific race categories but reported entries such as Cana-dian Indian, French-American Indian, Spanish-American Indian, Eskimo, Aleut, Alaska Indian, or any of the American Indian or Alaska Native tribes.

The **Asian and Pacific Islander** population combines two census groupings: **Asian** and **Native Hawaiian or Other Pacific Islander**. The **Asian** population includes persons who indicated their race as Asian Indian, Chinese, Filipino, Japanese, Korean, Vietnamese, or "Other Asian," as well as persons who provided write-in entries of such groups as Cambodian, Laotian, Hmong, Pakistani, or Taiwanese. The **Native Hawaiian or Other Pacific Islander** population includes persons who indicated their race as "Native Hawaiian," "Guamanian or Chamorro," "Samoan," or "Other Pacific Islander," as well as persons who reported entries such as Part Hawaiian, American Samoan, Fijian, Melanesian, or Tahitian.

The **Hispanic population** is based on a question that asked respondents "Is this person Spanish/Hispanic/Latino?" Persons marking any one of the four Hispanic categories (i.e., Mexican, Puerto Rican, Cuban, or other Spanish) are collectively referred to as Hispanic.

**Age** is defined as age at last birthday (number of completed years since birth), as of April 1 of the census year.

The **female** population is shown as a percentage of total population.

The **median age** is the age that divides the population into two equal-size groups. Half of the population is older than the median age and half is younger. Median age is based on a standard distribution of the population by single years of age and is shown to the nearest tenth of a year.

The **foreign-born** population includes all persons who were not U.S. citizens at birth. Foreign-born persons are those who indicated they were either a U.S. citizen by naturalization or were not a citizen of the United States. Neither the census nor the American Community Survey asked about immigration status. The population surveyed included all persons who indicated that the United States was their usual place of residence. The foreign-born population consists of immigrants (legal permanent residents), temporary migrants (students), humanitarian migrants (refugees), and unauthorized migrants (persons illegally residing in the United States).

**Percent born in state of residence** is shown as a percentage of total population.

## IMMIGRANTS, Item 24

**Source: Department of Homeland Security, U.S. Citizenship and Immigration Services**
**http://www.dhs.gov/yearbook-immigration-statistics**

The number of immigrants by state of intended residence is summarized from the administrative records of the Citizenship and Immigration Services. This information is compiled from immigrant visas and forms granting legal permanent resident status.

An **immigrant** is an alien admitted to the United States as a lawful permanent resident. Immigrants are those persons lawfully accorded the privilege of residing permanently in the United States (**i.e., immigrants who receive a "green card".**) They may be newly arrived individuals who were issued immigrant visas by the Department of State overseas, or they may be U.S. residents who were admitted to permanent resident status in 2019 by the U.S. Citizenship and Immigration Services.

## HOUSEHOLDS, Items 25–30 and 64–68

**Source: U.S. Census Bureau—2019 American Community Survey**
**http://www.census.gov/acs/www/**

A **household** includes all of the persons who occupy a housing unit. Persons not living in households are classified as living in group quarters. A housing unit is a house, an apartment, a mobile home, a group of rooms, or a single room occupied (or, if vacant, intended for occupancy) as separate living quarters. Separate living quarters are those in which the occupants live separately from any other persons in the building and have direct access from the outside of the building or through a common hall. The occupants may be a single family, one person living alone, two or more families living together, or any other group of related or unrelated persons who share living quarters. The number of households is the same as the number of year-round occupied housing units.

The measure of **persons per household** is obtained by dividing the number of persons in households by the number of households or householders. One person in each household is designated as the householder. In most cases, this is the person, (or one of the persons) in whose name the house is owned, being bought, or rented. If there is no such person in the household, any adult household member 15 years old and over can be designated as the householder.

A **family** includes a householder and one or more other persons living in the same household who are related to the householder by birth, marriage, or adoption. All persons in a household who are related to the householder are regarded as members of his or her family. A **family household** may contain persons not related to the householder; thus, family households may include more members than families do. A household can contain only one family for the purposes of census tabulations. Not all households contain families, as a household may comprise a group of unrelated persons or one person living alone. Families are classified by type as either a "married couple family" or "other family" according to the presence of a spouse.

The category **female family householder** includes only female-headed family households with no spouse present.

## POPULATION PROJECTIONS, Items 42–44

**Source: U.S. Census Bureau—Population Projections Branch**
**https://census.gov/programs-surveys/popproj/guidance.html**

Projections are estimates of the population for future dates. They illustrate plausible courses of future population change based on assumptions about future births, deaths, international migration, and domestic migration. Projected numbers are based on an estimated population consistent with the most recent decennial census as enumerated. The Census Bureau does not have a current set of state population projections and currently has no plans to produce them. This volume includes projections released in 2005, based on the 2000 census. The Census Bureau notes that these projections should be used with caution because population trends may have changed substantially since their release.

## HOUSING, Items 69–92

**Source: U.S. Census Bureau—2010 and 2019 American Community Survey**
**http://www.census.gov/acs/www/**

Housing data for 2010 and 2019 are from the American Community Survey, a nationwide continuous survey designed to replace the long form questionnaire used in previous censuses. A sample of households is surveyed to provide estimates.

A **housing unit** is a house, apartment, mobile home or trailer, group of rooms, or single room occupied or, if vacant, intended for occupancy as separate living quarters. Separate living quarters are those in which the occupants do not live and eat with any other person in the structure and which have direct access from the outside of the building or through a common hall. For vacant units, the criteria of separateness and direct access are applied to the intended occupants whenever possible. If that information cannot be obtained, the criteria are applied to the previous occupants.

The occupants of a housing unit may be a single family, one person living alone, two or more families living together, or any other group of related or unrelated persons who share living arrangements. Both occupied and vacant housing units are included in the housing inventory, although recreational vehicles, tents, caves, boats, railroad cars, and the like are included only if they are occupied as a person's usual place of residence.

A housing unit is classified as **occupied** if it is the usual place of residence of the person or group of persons living in it at the time of enumeration, or if the occupants are only temporarily absent (away on vacation). A household consists of all persons who occupy a housing unit as their usual place of residence.

**Housing cost**, as a percentage of income, is shown separately for owners with mortgages, owners without mortgages, and renters. Also shown is the percentage of mortgaged owners and renters who pay 30 percent or more of household income on selected monthly costs. Rent as a percent of income is a computed ratio of gross rent and monthly household income (total household income divided by 12). Selected owner costs include utilities and fuels, mortgage payments, insurance, taxes, etc. In each case, the ratio of housing cost to income is computed separately for each housing unit. The housing cost ratios for half of all units are above the median shown in this book, and half are below the median. Median monthly housing costs divides the monthly housing costs distribution into two equal parts, one-half of the cases falling below the median monthly housing costs and one-half above the median.

**Median value** is the dollar amount that divides the distribution of specified owner-occupied housing units into two equal parts, with half of all units below the median value and half above the median value. Value is defined as the respondent's estimate of what the house would sell for if it were for sale. Data are presented for single-family units on fewer than 10 acres of land that have no business or medical office on the property.

**Median rent** divides the distribution of renter-occupied housing units into two equal parts. The rent concept used in this volume is gross rent, which includes the amount of cash rent a renter pays (contract rent) plus the estimated average cost of utilities and fuels, if these are paid by the renter. The rent is the amount of rent only for living quarters and excludes any business or other space occupied. Single-family houses on lots of 10 or more acres of land are excluded.

**Substandard units** are occupied units that are overcrowded or lack complete plumbing facilities. For the purposes of this item, "overcrowded" is defined as having 1.01 persons or more per room. Complete plumbing facilities include hot and cold piped water, a flush toilet, and a bathtub or shower. These facilities must be located inside the housing unit, but do not have to be in the same room.

**Different house** includes all people 1 year old and over who, a year earlier, lived in a different house or apartment from the one they occupied at the time of interview.

## BUILDING PERMITS, Items 93–95

**Source: U.S. Census Bureau—Building Permits Survey**
**http://www.census.gov/construction/bps/**

These figures represent private residential construction authorized by building permits in approximately 20,000 places in the United States. Valuation represents the expected cost of construction as recorded on the building permit. This figure usually excludes the cost of on-site and off-site development and improvements, as well as the cost of heating, plumbing, electrical, and elevator installations.

National, state, and county totals were obtained by adding the data for permit-issuing places within each jurisdiction. These totals thus are limited to permits issued in the 20,000 place universe covered by the Census Bureau and may not include all permits issued within a state. Current surveys indicate that construction is undertaken for all but a very small percentage of housing units authorized by building permits.

Residential building permits include buildings with any number of housing units. Housing units exclude group quarters (such as dormitories and rooming houses), transient accommodations (such as transient hotels, motels, and tourist courts), "HUD-code" manufactured (mobile) homes, moved or relocated units, and housing units created in an existing residential or nonresidential structure.

## MANUFACTURED HOUSING UNITS, Item 96

**Source: U.S. Census Bureau—Manufactured Housing Survey**
**https://www.census.gov/data/tables/time-series/econ/mhs/shipments.html**

The Manufactured Housing Survey (MHS) is conducted by the U.S. Census Bureau and sponsored by the Department of Housing and Urban Development (HUD). MHS produces monthly regional estimates of the average sales price of new manufactured homes and more detailed annual estimates including selected characteristics of new manufactured homes. In addition, MHS produces monthly estimates of homes shipped to each state.

A manufactured home is defined as a movable dwelling, 8 feet or more wide and 40 feet or more long, designed to be towed on its own chassis, with transportation gear integral to the unit when it leaves the factory, and without need of a permanent foundation. These manufactured homes include multi-wides and expandable manufactured homes. Excluded are travel trailers, motor homes, and modular housing.

## BIRTHS AND DEATHS, Items 97–103

**Source: U.S. Centers for Disease Control and Prevention, National Center for Health Statistics**
**https://www.cdc.gov/nchs/data/nvsr/nvsr70/nvsr70-02-508.pdf**
**https://www.cdc.gov/nchs/data/nvsr/nvsr69/nvsr69-13-508.pdf**

The registration of births, deaths, and other vital events in the United States is primarily a state and local function. The civil laws of every state provide for continuous and permanent birth and death registration systems. Through the National Vital Statistics System, the National Center for Health Statistics (NCHS) obtains data on births and deaths from the registration offices of each state, New York City, and the District of Columbia.

Birth and death statistics are limited to events occurring during the year. The data are by place of residence and exclude events for nonresidents of the United States. Births or deaths occurring outside the United States are excluded.

Birth and death rates represent the number of births and deaths per 1,000 resident population enumerated as of April 1 for decennial census years and estimated as of July 1 for other years.

Figures for infant deaths include deaths of children under 1 year of age but exclude fetal deaths. The infant death rate is per 1,000 live births.

The rates of almost all causes of disease, injury, and death vary by age. Age adjustment is a technique for "removing" the effects of age from crude rates, in order to allow meaningful comparisons across populations with different underlying age structures. For example, comparing the crude death rate in Florida to that of California is misleading, since the relatively older population in Florida will lead to a higher crude death rate. For such a comparison, age-adjusted death rates are preferable.

The population estimates were developed by the Census Bureau's Population Division using a traditional cohort component method. Starting with a basic population from the 2000 census, each component of population change—births, deaths, domestic migration, and international migration—is estimated separately for each birth cohort by sex, race, and Hispanic or Latino origin.

Age-adjusted rates are calculated by applying the age-specific rates of various populations to a single standard population. In this volume, the standard population is 2000. Beginning in 2003, The Centers for Disease Control and Prevention switched to the year 2000, after many years of using the year 1940 as the standard population for age-adjusted death rates.

## PERSONS LACKING HEALTH INSURANCE, Items 104–105

**Source: U.S. Census Bureau—American Community Survey**

**https://www.census.gov/library/publications/2020/demo/p60-271.html**

These estimates are from the American Community Survey, an ongoing nationwide survey that is conducted throughout the year. About 250,000 addresses per month receive the ACS. Respondents are asked whether each household member is currently covered (by specific types of health coverage) at the time of interview. The 2013 estimates were the first to use the ACS. Prior year estimates were based on the Annual Social and Economic Supplement (ASEC) of the Current Population Survey (CPS).

Those lacking coverage are the percentage of the population of each state who were not covered by private health plans purchased directly or provided by an employer, Medicaid, Medicare, or military health care.

## MEDICARE BENEFICIARIES, Item 106

**Source: U.S. Department of Health and Human Services, Centers for Medicare and Medicaid Services**
**https://www.cms.gov/Research-Statistics-Data-and-Systems/Statistics-Trends-and-Reports/CMSProgramStatistics/Dashboard.html**

The Centers for Medicare and Medicaid Services (CMS) administers Medicare, which provides health insurance to persons 65 years old and over, persons with permanent kidney failure, and certain persons with disabilities. Original Medicare has two parts: Hospital Insurance and Supplemental Medical Insurance. In recent years, Medicare has been expanded to include two new programs: Medicare Advantage plans and prescription drug coverage. Medicare Advantage Plans are health plan options that are approved by Medicare but run by private companies. Medicare prescription drug plans can be part of Medicare Advantage plans or stand-alone drug plans.

Persons who are eligible for Medicare can enroll in Part A (Hospital Insurance) at no charge, and can choose to pay a monthly premium to enroll in Part B. Most eligible persons are enrolled in Part A, and most enrollees in Part A are also enrolled in Part B (Supplemental Medical Insurance.) This table includes persons who were enrolled in both Part A and Part B during 2019.

Part B beneficiaries can choose to enroll in **Original Medicare**, a fee-for-service plan administered by the Centers for Medicare and Medicaid Services, or in a **Medicare Advantage** plan. Medicare Advantage plans include private fee-for-service plans, preferred provider organizations, health maintenance organizations, medical savings account plans, demonstration plans, and programs for all-inclusive care for the elderly.

The annual Medicare enrollment counts are determined using a person-year methodology. For each calendar year, total person-year counts are determined by summing the total number of months that each beneficiary is enrolled during the year and dividing by 12. Using this methodology, a beneficiary's partial-year enrollment may be counted in more than one category (i.e., both Original Medicare and Medicare Advantage).

## CRIME, Items 107–110

Source: U.S. Federal Bureau of Investigation—Uniform Crime Reports
https://ucr.fbi.gov/crime-in-the-u.s/2019/crime-in-the-u.s.-2019

Crime data are as reported to the Federal Bureau of Investigation (FBI) by law enforcement agencies and have not been adjusted for underreporting. This may affect comparability between geographic areas or over time.

Through the voluntary contribution of crime statistics by law enforcement agencies across the United States, the Uniform Crime Reporting (UCR) Program provides periodic assessments of crime in the nation as measured by offenses that have come to the attention of the law enforcement community. The Committee on Uniform Crime Records of the International Association of Chiefs of Police initiated this voluntary national data-collection effort in 1930. The UCR Program contributors compile and submit their crime data either directly to the FBI or through state-level UCR Programs.

Seven offenses, because of their severity, frequency of occurrence, and likelihood of being reported to police, were initially selected to serve as an index for evaluating fluctuations in the volume of crime. These serious crimes were murder and nonnegligent manslaughter, forcible rape, robbery, aggravated assault, burglary, larceny-theft, and motor vehicle theft. By congressional mandate, arson was added as the eighth index offense in 1979. The totals shown in this volume do not include arson.

In 2004, the FBI discontinued the use of the Crime Index in the UCR Program and its publications, stating that the Crime Index was driven upward by the offense with the highest number of cases (in this case, larceny-theft) creating a bias against jurisdictions with a high number of larceny-thefts but a low number of other serious crimes, such as murder and forcible rape. The FBI is currently publishing a violent crime total and a property crime total until a more viable index is developed.

In 2013, the FBI adopted a new definition of rape. Rape is now defined as, "Penetration, no matter how slight, of the vagina or anus with any body part or object, or oral penetration by a sex organ of another person, without the consent of the victim." The new definition updated the 80-year-old historical definition of rape which was "carnal knowledge of a female forcibly and against her will." Effectively, the revised definition expands rape to include both male and female victims and offenders, and reflects the various forms of sexual penetration understood to be rape, especially non-consenting acts of sodomy, and sexual assaults with objects. **Violent crimes** include four categories of offenses: (1) Murder and non-negligent manslaughter, as defined in the UCR Program, is the willful (non-negligent) killing of one human being by another. This offense excludes deaths caused by negligence, suicide, or accident; justifiable homicides; and attempts to murder or assaults to murder. (2) Rape is the penetration, no matter how slight, of the vagina or anus with any body part or object, or oral penetration by a sex organ of another person, without the consent of the victim. Assaults or attempts to commit rape by force or threat of force are also included; however, statutory rape (without force) and other sex offenses are excluded. (3) Robbery is the taking or attempting to take anything of value from the care, custody, or control of a person or persons by force or threat of force or violence and/or by putting the victim in fear. (4) Aggravated assault is an unlawful attack by one person upon another for the purpose of inflicting severe or aggravated bodily injury. This type of assault is usually accompanied by the use of a weapon or by other means likely to produce death or great bodily harm. Attempts are included, since injury does not necessarily have to result when a gun, knife, or other weapon is used, as these incidents could and probably would result in a serious personal injury if the crime were successfully completed.

**Property crimes** include three categories: (1) Burglary, or breaking and entering, is the unlawful entry of a structure to commit a felony or theft, even though no force was used to gain entrance. (2) Larceny-theft is the unauthorized taking of the personal property of another, without the use of force. (3) Motor vehicle theft is the unauthorized taking of any motor vehicle.

Rates are based on population estimates provided by the FBI. For some states, reporting is not sufficiently complete to be representative of the state as a whole. The FBI has estimated state totals for those states.

## ELEMENTARY AND SECONDARY SCHOOL ENROLLMENT, Items 111 and 112

Source: U.S. Department of Education, National Center for Education Statistics—Common Core of Data
http://nces.ed.gov/ccd/elsi/

Data on public school enrollment is from the Common Core of Data 2019–2020 survey. Public school enrollment includes pre-kindergarten through grade 12 and ungraded students. The student/teacher ratio is calculated by dividing the number of students in all schools by the number of full-time equivalent teachers employed by all schools and agencies.

## EDUCATIONAL ATTAINMENT, Items 113–116

Source: U.S. Census Bureau—2010 and 2019 American Community Survey
http://www.census.gov/acs/www/

Data on **educational attainment** are tabulated for the population 25 years old and over. The data were derived from a question that asked respondents for the highest level of school completed or the highest degree received. Persons who had passed a high school equivalency examination were considered high school graduates. Schooling received in foreign schools was to be reported as the equivalent grade or years in the regular American school system. Vocational and technical training, such as barber school training; business, trade, technical, and vocational schools; or other training for a specific trade are specifically excluded.

**High school graduate or more.** This category includes persons whose highest degree was a high school diploma or its equivalent, and those who reported any level higher than a high school diploma.

**Bachelor's degree or more**. This category includes persons who have received bachelor's degrees, master's degrees, professional school degrees (such as law school or medical school degrees), and doctoral degrees.

## LOCAL GOVERNMENT EDUCATION EXPENDITURES, Items 117 and 118

**Source: U.S. Department of Education, National Center for Education Statistics—Common Core of Data https://nces.ed.gov/ccd/**

Total expenditure for education includes provision or support of schools and facilities for elementary and secondary education. It encompasses instructional, support, and auxiliary services (school lunch, student activities, and community service) offered by public school systems. Retirement benefits paid to former education employees and interest payments are not included. Current expenditure includes all components of total expenditure except capital outlay. Expenditure data are obtained by the Census Bureau through its annual survey of government finances and are supplied to the National Center for Education Statistics (NCES). Current expenditure per student is current expenditure divided by the number of students enrolled. The number of students enrolled is based on an annual "membership" count of students on or about October 1.

NCES uses the Common Core of Data (CCD) Survey system to acquire and maintain statistical data from each of the 50 states, the District of Columbia, and the outlying areas. State education agencies compile and submit data for approximately 94,000 schools and 17,000 local school districts. Typically, this results in varying interpretation of NCES definitions and different record keeping systems, leading to large amounts of missing data for several states; this absence is reflected in the data in this publication. The numbers in Table A reflect imputations and adjustments as published in *Revenues and Expenditures for Public Elementary and Secondary Education: School Year 2017-2018 (Fiscal Year 2018)*.

## EXPORTS, Items 119–121

**Source: U.S. Department of Commerce, International Trade Administration http://www.census.gov/foreign-trade/statistics/state/ origin_movement/index.html**

The data on exports of goods by state of origin are based on the location of the exporter (the principal party responsible for exportation from the United States). Exporters are often intermediaries, so the data do not necessarily represent the states in which the goods were actually produced. The total includes re-exports of foreign goods.

## INCOME AND POVERTY, Items 122–133

**Source: U.S. Census Bureau—2019 American Community Survey http://www.census.gov/acs/www/**

The data on income were derived from answers to questions which were asked of the population 15 years old and over. **Total income** is the sum of the amounts reported separately for wage or salary income; net self-employment income; interest, dividends, or net rental or royalty income or income from estates and trusts; Social Security or railroad retirement income; Supplemental Security Income (SSI); public assistance or welfare payments; retirement, survivor, or disability pensions; and all other income. Receipts from the following sources are not included as income: capital gains; money received from the sale of property (unless the recipient was engaged in the business of selling such property); the value of income "in kind" from food stamps, public housing subsidies, medical care, employer contributions for individuals, etc.; withdrawal of bank deposits; money borrowed; tax refunds; exchange of money between relatives living in the same household; and gifts and lump-sum inheritances, insurance payments, and other types of lump-sum receipts.

**Per capita income** is the mean income computed for every man, woman, and child in a particular group. It is derived by dividing the aggregate income of a particular group by the total population in that group. Per capita income is rounded to the nearest whole dollar.

**Household income** includes the income of the householder and all other individuals 15 years old and over in the household, whether or not they are related to the householder. Since many households consist of only one person, average household income is usually less than average family income. Although the household income statistics cover the past 12 months, the characteristics of individuals and the composition of households refer to the time of enumeration. Thus, the income of the household does not include amounts received by individuals who were members of the household during all or part of the past 12 months if these individuals no longer resided in the household at the time of interview. Similarly, income amounts reported by individuals who did not reside in the household during the past 12 months but who were members of the household at the time of interview are included. However, the composition of most households was the same during the past 12 months as at the time of interview.

**Median income** divides the income distribution into two equal parts, with half of all cases below the median income level and half of all cases above the median income level. For households and families, the median income is based on the distribution of the total number of households and families, including those with no income. Median income for households is computed on the basis of a standard distribution with a minimum value of less than $2,500 and a maximum value of $200,000 or more and is rounded to the nearest whole dollar.

For **family income**, the incomes of all household members 15 years old and over related to the householder are summed and treated as a single amount. Although the family income statistics cover the past 12 months, the characteristics of individuals and the composition of families refer to the time of interview. Thus, the income of the family does not include amounts received by individuals who were members of the family during all of part of the past 12 months if these individuals no longer resided with the family at the time of interview. Similarly, income amounts reported by individuals who did not reside with the family during the past 12 months but who were members of the family at

**Poverty Thresholds for 2019 by Size of Family and Number of Related Children Under 18 Years**

| Size of family unit | Weighted average thresholds | Related children under 18 years | | | | | | | | |
|---|---|---|---|---|---|---|---|---|---|---|
| | | None | One | Two | Three | Four | Five | Six | Seven | Eight or more |
| One person (unrelated individual): | 13,011 | | | | | | | | | |
| Under age 65 ............................ | 13,300 | 13,300 | | | | | | | | |
| Aged 65 and older.................... | 12,261 | 12,261 | | | | | | | | |
| Two people: ............................... | 16,521 | | | | | | | | | |
| Householder under age 65........ | 17,196 | 17,120 | 17,622 | | | | | | | |
| Householder aged 65 and older ..................................... | 15,468 | 15,453 | 17,555 | | | | | | | |
| Three people............................. | 20,335 | 19,998 | 20,578 | 20,598 | | | | | | |
| Four people............................... | 26,172 | 26,370 | 26,801 | 25,926 | 26,017 | | | | | |
| Five people ............................... | 31,021 | 31,800 | 32,263 | 31,275 | 30,510 | 30,044 | | | | |
| Six people................................. | 35,129 | 36,576 | 36,721 | 35,965 | 35,239 | 34,161 | 33,522 | | | |
| Seven people............................ | 40,016 | 42,085 | 42,348 | 41,442 | 40,811 | 39,635 | 38,262 | 36,757 | | |
| Eight people.............................. | 44,461 | 47,069 | 47,485 | 46,630 | 45,881 | 44,818 | 43,470 | 42,066 | 41,709 | |
| Nine people or more .................. | 52,875 | 56,621 | 56,895 | 56,139 | 55,503 | 54,460 | 53,025 | 51,727 | 51,406 | 49,426 |

**Source: U.S. Census Bureau.**

the time of interview are included. However, the composition of most families was the same during the past 12 months as at the time of interview.

The **poverty status** data were derived from data collected on the number of persons in the household, each person's relationship to the householder, and the income data. The Social Security Administration (SSA) developed the original poverty definition in 1964, which federal interagency committees subsequently revised in 1969 and 1980. The Office of Management and Budget's (OMB) *Directive 14* prescribes the SSA's definition as the official poverty measure for federal agencies to use in their statistical work. Poverty statistics presented in American Community Survey products adhere to the standards defined by OMB in *Directive 14*.

The poverty thresholds vary depending on three criteria: size of family, number of children, and, for one- and two-person families, age of householder. In determining the poverty status of families and unrelated individuals, the Census Bureau uses thresholds (income cutoffs) arranged in a two-dimensional matrix. The matrix consists of family size (from one person to nine or more persons), cross-classified by presence and number of family members under 18 years old (from no children present to eight or more children present). Unrelated individuals and two-person families are further differentiated by age of reference person (under 65 years old and 65 years old and over). To determine a person's poverty status, the person's total family income in the last 12 months is compared to the poverty threshold appropriate for that person's family size and composition. If the total income of that person's family is less than the threshold appropriate for that family, then the person is considered poor or "below the poverty level," together with every member of his or her family. If a person is not living with anyone related by birth, marriage, or adoption, then the person's own income is compared with his or her poverty threshold. The total number of persons below the

poverty level is the sum of persons in families and the number of unrelated individuals with incomes below the poverty level in the last 12 months. The average poverty threshold for a four-person family was $26,172 in 2019.

The data on **poverty status of households** were derived from answers to the income questions. Since poverty is defined at the family level and not the household level, the poverty status of the household is determined by the poverty status of the householder. Households are classified as poor when the total income of the householder's family in the previous 12 months is below the appropriate poverty threshold. (For nonfamily householders, the person's income is compared with the appropriate threshold.) The income of persons living in the household who are unrelated to the householder is not considered when determining the poverty status of a household, nor does their presence affect the family size in determining the appropriate threshold. The poverty thresholds vary depending upon three criteria: size of family, number of children, and, for one- and two-person families, age of the householder.

Poverty status of children by **family type** is the percentage of children living in that particular type of family that has a family income below the poverty threshold based on family size and composition.

## PERSONAL INCOME AND EARNINGS, Items 134–158

**Source: U.S. Bureau of Economic Analysis, Regional Economic Accounts**
**http://www.bea.gov/regional/index.htm#state**

**Total personal income** is the current income received by residents of an area from all sources. It is measured before deductions of income and other personal taxes but after deductions of personal contributions for Social Security, government retirement, and other social insurance programs. It consists of **wage and salary disbursements** (covering all employee earnings, including executive salaries, bonuses, commissions, payments-in-kind, incentive payments, and tips); various types of supplementary earnings, such as employers' contributions to pension funds (termed "other labor income" or "supplements to wages and salaries"); proprietors' income; rental income of persons; dividends; personal interest income; and government and business transfer payments.

**Proprietors' income** is the monetary income and income-in-kind of proprietorships and partnerships (including the independent professions), and the income of tax-exempt cooperatives. **Dividends** are cash payments by corporations to stockholders who are U.S. residents. **Interest** is the monetary and imputed interest income of persons from all sources. **Rent** is the monetary income of persons from the rental of real property, except the income of persons primarily engaged in the real estate business; the imputed net rental income of owner-occupants of nonfarm dwellings; and the royalties received by persons.

**Transfer payments** are income for which services are not currently rendered. They consist of both government and business transfer payments. Government transfer payments include payments under the following programs: Federal Old-Age, Survivors, and Disability Insurance ("Social Security"); Medicare and medical vendor payments; unemployment insurance; railroad and government retirement; federal- and state-government-insured workers' compensation; veterans' benefits, including veterans' life insurance; food stamps; black lung payments; Supplemental Security Income; and Temporary Assistance for Needy Families. Government payments to nonprofit institutions, other than for work under research and development contracts, are also included. Business transfer payments consist primarily of liability payments for personal injury and of corporate gifts to nonprofit institutions.

**Per capita personal income** is based on resident population estimated as of July 1 of the year shown.

**Personal tax payments** include taxes paid by individuals to federal, state, and local governments. Personal taxes include individual income taxes, estate and gift taxes, motor vehicle license taxes, and personal property taxes. Personal contributions to social insurance ("Social Security taxes") are not included, nor are sales taxes.

**Disposable personal income** equals personal income less personal tax payments. It is a measure of the income available to persons for spending or saving.

**Earnings** cover wage and salary disbursements, other labor income, and proprietors' income.

The data for earnings obtained from the Bureau of Economic Analysis (BEA) are based on place of work. In computing personal income, BEA makes an "adjustment for residence" to earnings based on commuting patterns; thus, personal income is presented on a place-of-residence basis.

**Farm earnings** include the income of farm workers (wages and salaries and other labor income) and farm proprietors. Farm proprietors' income includes only the income of sole proprietorships and partnerships.

Farm earnings estimates are benchmarked to data collected in the Census of Agriculture and the revised Department of Agriculture state totals of income and expense items.

**Goods-related** industries include mining, construction, and manufacturing. **Service-related** and other industries includes private-sector earnings in forestry, related activities, and other; utilities; transportation and warehousing; information; wholesale trade; retail trade; finance and insurance; real estate and rental and leasing; and services, which includes professional, scientific, and technical services; management of companies and enterprises; administrative and waste services; educational services; health care and social assistance; arts, entertainment, and recreation; accommodation and food services; and other services, except public administration. Government earnings include all levels of government. Industries are categorized under the North American Industry Classification System (NAICS), and are not directly comparable to years prior to 2002.

## GROSS STATE PRODUCT, Item 159
**Source: U.S. Bureau of Economic Analysis, Regional Economic Accounts**
**http://www.bea.gov/regional/index.htm#state**

Gross state product (GSP) for a state is derived as the sum of gross state product originating in all industries in the state. In concept, an industry's GSP, referred to as its "value added," is equivalent to its gross output (sales or receipts and other operating income, commodity taxes, and inventory changes) minus its intermediate inputs (consumption of goods and services purchased from other industries or imported from other countries). As such, it is often referred to as the state counterpart to the nation's gross domestic product (GDP). In practice, GSP estimates are measured as the sum of distributions by industry of the components of gross domestic income—that is, the sum of the costs incurred (such as compensation of employees, net interest, and indirect business taxes) and the profits earned in production.

## SOCIAL SECURITY AND SUPPLEMENTAL SECURITY INCOME, Items 160–162
**Source: U.S. Social Security Administration**
**http://www.ssa.gov/policy/docs/statcomps/oasdi_sc/**
**http://www.ssa.gov/policy/docs/statcomps/ssi_sc/**

**Social Security beneficiaries** are persons receiving benefits under the Old-Age, Survivors, and Disability Insurance Program. These include retired or disabled workers covered by the program, their spouses and dependent children, and the surviving spouses and dependent children of deceased workers.

**Supplemental Security Income (SSI) recipients** are persons receiving SSI payments. The SSI program is a cash assistance program that provides monthly benefits to low-income aged, blind, or disabled persons.

Data are as of December of the year shown.

## CIVILIAN EMPLOYMENT, Items 163–166

Source: U.S. Census Bureau—2019 American Community Survey
http://www.census.gov/acs/www/
https://www2.census.gov/programs-surveys/acs/tech_docs/code_lists/2018_ACS_Code_Lists.pdf?#

The data on occupation were derived from answers to questions that were asked of all persons 15 years old and over who had worked in the past 5 years. **Occupation** describes the kind of work the person does on the job. For employed persons, the data refer to the person's job during the previous week. For those who worked two or more jobs, the data refer to the job at which the person worked the greatest number of hours. For unemployed persons, the data refer to their last job. The American Community Survey uses the occupational classification system that was developed for the 2000 census and modified in 2002 and again in 2010. This system consists of 539 specific occupational categories for employed persons arranged into 23 major occupational groups. This classification was developed based on the *Standard Occupational Classification (SOC) Manual: 2010*, published by the Executive Office of the President, Office of Management and Budget.

## CIVILIAN LABOR FORCE AND UNEMPLOYMENT, Items 167–171

Source: U.S. Bureau of Labor Statistics—Local Areas Unemployment Statistics
http://www.bls.gov/lau/#tables

Data for the civilian labor force are the product of a federal-state cooperative program in which state employment security agencies prepare labor force and unemployment estimates under concepts, definitions, and technical procedures established by the Bureau of Labor Statistics (BLS). The **civilian labor force** consists of all civilians 16 years old and over who are either employed or unemployed.

**Unemployment** includes all persons who did not work during the survey week, made specific efforts to find a job during the prior four weeks, and were available for work during the survey week (except for temporary illness). Persons waiting to be called back to a job from which they had been laid off and those waiting to report to a new job within the next 30 days are included in unemployment figures.

## PRIVATE NONFARM EMPLOYMENT AND EARNINGS, Items 172–183

Source: U.S. Bureau of Labor Statistics—Current Employment Survey
http://www.bls.gov/ces/#tables

Data for private nonfarm employment and earnings are compiled from payroll information reported monthly on a voluntary basis to the BLS and its cooperating state agencies. More than 350,000 establishments represent all industries except agriculture.

**Employment** is the annual average of monthly totals of persons who received pay for any part of the pay period including the 12th day of the month. Included are all full-time and part-time workers in nonfarm establishments. Not covered are government employees, proprietors, the self-employed, unpaid volunteers or family workers, farm workers, and domestic workers in households. The data by industry conform to the definitions established in the North American Industry Classification System (NAICS).

**Earnings** of **production workers** in **manufacturing** industries are derived from reports of gross payrolls and corresponding paid hours. Payroll is reported before deductions of any kinds. Total hours during the pay period include all hours worked (including overtime hours) and hours paid for holidays, vacations, and sick leave.

## AGRICULTURE, ITEMS 184–202

Source: U.S. Department of Agriculture, National Agricultural Statistics Service—2017 Census of Agriculture
https://www.nass.usda.gov/Publications/AgCensus/2017/index.php

The Census Bureau took a census of agriculture every 10 years from 1840 to 1920; since 1925, this census has been taken roughly once every 5 years. The 1997 Census of Agriculture was the first one conducted by the National Agricultural Statistics Service of the U.S. Department of Agriculture. Over time, the definition of a farm has varied. For recent censuses (including the 2017 census), a farm has been defined as any place from which $1,000 or more of agricultural products were produced and sold or normally would have been sold during the census year. Dollar figures are expressed in current dollars and have not been adjusted for inflation or deflation.

The term **producer** designates a person who is involved in making decisions for the farm operation. Decisions may include decisions about such things as planting, harvesting, livestock management, and marketing. The producer may be the owner, a member of the owner's household, a hired manager, a tenant, a renter, or a sharecropper. If a person rents land to others or has land worked on shares by others, he/she is considered the producer only of the land which is retained for his/her own operation. The census collected information on the total number of male producers, the total number of female producers, and demographic information for up to four producers per farm.

**Government payments** consists of direct payments as defined by the 2002 Farm Bill; payments from Conservation Reserve Program (CRP), Wetlands reserve Program (WRP), Farmable Wetlands Program (FWP), and Conservation Reserve Enhancement Program (CREP); loan deficiency payments; disaster payments; other conservation programs; and all other federal farm programs under which payments were made directly to farm producers, including those specified in the 2014 Agricultural Act (Farm Bill), including Agriculture Risk Coverage (ARC) and Price

Loss Coverage (PLC).. Commodity Credit Corporation (CCC) proceeds, amounts from state and local government agricultural program payments, and federal crop insurance payments were not included in this category.

The acreage designated as **land in farms** consists primarily of agricultural land used for crops, pasture, or grazing. It also includes woodland and wasteland not actually under cultivation or used for pasture or grazing, provided that this land was part of the farm operator's total operation.

Land in farms is an operating-unit concept and includes all land owned and operated, as well as all land rented from others. Land used rent-free is classified as land rented from others. All grazing land, except land used under government permits on a per-head basis, was included as "land in farms" provided it was part of a farm or ranch. Land under the exclusive use of a grazing association was reported by the grazing association and included as land in farms. All land in Indian reservations used for growing crops or grazing livestock is classified as land in farms.

**Irrigated land** includes all land watered by any artificial or controlled means, such as sprinklers, flooding, furrows or ditches, sub-irrigation, and spreader dikes. Included are supplemental, partial, and preplant irrigation. Each acre was counted only once regardless of the number of times it was irrigated or harvested. Livestock lagoon waste water distributed by sprinkler or flood systems was also included.

**Total cropland** includes cropland harvested, cropland used only for pasture or grazing, cropland on which all crops failed or were abandoned, cropland in cultivated summer fallow, and cropland idle or used for cover crops or soil improvement but not harvested and not pastured or grazed.

Respondents were asked to report their estimate of the current market **value of land and buildings** owned, rented, or leased from others and rented and leased to others. Market value refers to the respondent's estimate of what the land and buildings would sell for under current market conditions.

The **value of machinery and equipment** was estimated by the respondent as the current market value of all cars, trucks, tractors, combines, balers, irrigation equipment, etc., used on the farm. This value is an estimate of what the machinery and equipment would sell for in its present condition and not the replacement of depreciated value. Share interests are reported at full value at the farm where the equipment and machinery are usually kept. Only equipment that was physically located at the farm on December 31, 2017, is included.

Market **value of agricultural products sold** by farms represents the gross market value before taxes and the production expenses of all agricultural products sold or removed from the place in 2017, regardless of who received the payment. It is equivalent to total sales and it includes sales by producers as well as the value of any share received by partners, landlords, contractors, and others associated with the operation. It includes value of organic sales, direct sales and the value of commodities placed in the Commodity Credit Corporation (CCC) loan program. Market value of agricultural products sold does not include payments received for participation in other federal farm programs. Also, it does not include income from farm-related sources such as customwork and other agricultural services, or income from nonfarm sources.

The value of crops sold in 2017 does not necessarily represent the sales from crops harvested in 2017. Data may include sales from crops produced in earlier years and may exclude some crops produced in 2017 but held in storage and not sold. For commodities such as sugarbeets and wool sold through a co-op that made payments in several installments, respondents were requested to report the total value received in 2017.

**Organic farms** are those that had organic production according to USDA's National Organic Program (NOP). Respondents reported whether their organic production was certified or exempt from certification and the sales from NOP produced commodities. Not included are farms that had acres transitioning into NOP production.

Farms with **internet access** are those that reported using personal computers, laptops, or mobile devices (e.g., cell phones or tablets) to access the internet. This can be done using services such as dial-up, DSL, cable modem, fiber-optic, mobile internet service for a cell phone or other device (tablet), satellite, or other methods. In 2017 respondents were also able to report connecting with an unknown service type, labeled as "Don't know" in the publication tables.

# LAND USE, ITEMS 203 to 205

**Source: U.S. Department of Agriculture, Natural Resources Conservation Service—2015 National Resources Inventory**
**http://www.nrcs.usda.gov/technical/NRI/**

The National Resources Inventory (NRI) was conducted every five years between 1982 and 1997. Since 2000, NRI data have been gathered annually, using a more complex sampling design of core and rotational subsamples. It provides updated information on the status, condition, and trends of land, soil, water, and related resources on the Nation's non-federal lands. Non-federal lands include privately owned lands, tribal and trust lands, and lands controlled by State and local governments.

The 2015 NRI is based on a sample of about 800,000 locations throughout the United States (excluding Alaska and the District of Columbia). Acreages for federal land and total surface area are established through geospatial processes and administrative records. Total surface area of the contiguous United States is 1,944 million acres.

**Federally-owned lands** include military bases, national forests, wildlife refuges, parks, grassland game preserves, scenic waterways, wilderness areas, monuments, lakeshore, parkways, battlefields, Bureau of Land Management lands, and other federal lands.

The NRI **developed land** category includes (a) large tracts of urban and built-up land; (b) small tracts of built-up land of less than 10 acres; and (c) land outside of these built-up areas that is in a rural transportation corridor (roads, railroads, and associated rights-of-way). Urban and built-up areas consist of residential, industrial, commercial, and institutional land; construction sites; public administrative sites; railroad yards; cemeteries; airports; golf courses; sanitary landfills; sewage treatment plants; water control structures and spillways; other land used for such purposes; small parks (less than 10 acres) within urban and built-up areas; and highways, *railroads*, and other transportation facilities if they are surrounded by urban areas. Also included are tracts of less than 10 acres that do not meet the above definition but are completely surrounded by Urban and built-up land. Two size categories are recognized in the NRI: areas of 0.25 acre to 10 acres, and areas of at least 10 acres.

**Rural land** consists of four primary rural land types: forest, rangeland, cropland, and pasture.

Forest land is at least 10 percent stocked by single-stemmed woody species of any size that will be at least 4 meters (13 feet) tall at maturity. Also included is land bearing evidence of natural regeneration of tree cover (cut over forest or abandoned farmland) and not currently developed for non-forest use. Ten percent stocked, when viewed from a vertical direction, equates to an areal canopy cover of leaves and branches of 25 percent or greater. The minimum area for classification as forest land is 1 acre, and the area must be at least 100 feet wide.

Rangeland is composed principally of native grasses, grass-like plants, forbs or shrubs suitable for grazing and browsing, and introduced forage species that are managed like rangeland. This would include areas where introduced hardy and persistent grasses, such as crested wheatgrass, are planted and such practices as deferred grazing, burning, chaining, and rotational grazing are used, with little or no chemicals or fertilizer being applied. Grasslands, savannas, many wetlands, some deserts, and tundra are considered to be rangeland. Certain communities of low forbs and shrubs, such as mesquite, chaparral, mountain shrub, and pinyon-juniper, are also included as rangeland.

Cropland includes areas used for the production of adapted crops for harvest. Two subcategories of cropland are recognized: cultivated and non-cultivated. Cultivated land comprises land in row crops or close-grown crops, as well as other cultivated cropland; for example, hayland or pastureland that is in a rotation with row or close-grown crops. Non-cultivated cropland includes permanent hayland and horticultural cropland.

Patureland is land managed primarily for the production of introduced forage plants for livestock grazing. Pastureland cover may consist of a single species in a pure stand, a grass mixture, or a grass-legume mixture. Management usually consists of cultural treatments: fertilization, weed control, reseeding, renovation, and control of grazing. For the NRI, this includes land that has a vegetative cover of grasses, legumes, and/or forbs, regardless of whether or not it is being grazed by livestock.

Other rural land includes farmsteads and other farm structures, field windbreaks, *barren land*, and *marshland*.

## WATER CONSUMPTION, Item 206
### Source: U.S. Geological Survey, National Water Use Information Program—2015 Water Use Data
### http://water.usgs.gov/watuse/

Every five years, the U.S. Geological Survey compiles national water-use estimates. This volume includes the total fresh and saline water withdrawals for public water supplies expressed as million gallons per day. Estimate of withdrawals of ground and surface water are given for the following categories of use:

public water supplies, domestic, commercial, irrigation, livestock, industrial, mining, and thermoelectric power. Only public water supply is included in this volume. Public supply refers to water withdrawn from ground and surface sources by public and private water systems for use by cities, towns, rural water districts, mobile-home parks, Native American Indian reservations, and military bases. Public-supply facilities provide water to at least 25 persons or have a minimum of 15 service connections. Water withdrawn by public suppliers may be delivered to users for domestic, commercial, industrial, and thermoelectric-power purposes, as well as to other public-water suppliers. Public-supply water is also used for public services (public uses)—such as pools, parks, and public buildings—and may have unaccounted uses (losses) because of system leaks or such non-metered services as firefighting, flushing of water lines, or backwashing at treatment plants. Some public-supply water may be used in the processes of water and wastewater treatment. Some public suppliers treat saline water before distributing the water. The definition of saline water for public supply refers to water that requires treatment to reduce the concentration of dissolved solids through the process of desalination or dilution.

## MANUFACTURES, Items 207–216
### Source: U.S. Census Bureau— 2017 Economic Census
### https://www.census.gov/programs-surveys/asm.html

Columns 207 through 216 are from the 2017 Economic Census. The Annual Survey of Manufactures (ASM) has been conducted annually every year since 1949, except for years ending in "2" and "7," at which time ASM data are included in the manufacturing sector of the Economic Census. The ASM provides statistics on employment, payroll, worker hours, payroll supplements, cost of materials, value added by manufacturing, capital expenditures, inventories, and energy consumption. It also provides estimates of value of shipments for over 1,400 classes of manufactured products. The Annual Survey of Manufactures includes approximately 50,000 establishments selected from the census universe of 350,000 manufacturing establishments.

The **all employees** number is the average number of production workers for the payroll periods including the 12th of March, May, August, and November plus the number of other employees in mid-March. Included are all persons on paid sick leave, paid holidays, and paid vacations during the pay period. Officers of corporations are included as employees, while proprietors and partners of unincorporated firms are excluded.

**Payroll** figures include the gross annual earnings of all employees on the payroll of operating manufacturing establishments. The definition, which is the same as the one used for calculating the federal withholding tax, includes all forms of compensation, such as salaries, wages, commissions, dismissal pay, bonuses, vacation and sick leave pay, and compensation-in-kind, prior to such deductions as employees' Social Security contributions, withholding taxes, group insurance, union dues, and savings bonds. The total includes salaries of officers of corporations; it excludes payments to proprietors or partners of unincorporated concerns. Also excluded are payments to members of armed forces and to pensioners carried on the active payrolls of manufacturing establishments.

**Production workers** include workers (up through the line-supervisor level) engaged in fabricating, processing, assembling, inspecting, receiving, storing, handling, packing, warehousing, shipping (but not delivering), maintenance, repair, janitorial and guard services, product development, auxiliary production for the plant's own use (for example, power plant), record keeping, and other services closely associated with these production operations at the establishment covered by the report. Employees above the working-supervisor level are excluded.

The number of production workers is the average for the payroll periods including the 12th of March, May, August, and November. Not included in this classification are all other employees, defined as non-production employees, including those engaged in factory supervision above the line-supervisor level.

Production worker hours cover hours worked or paid for at the manufacturing plant, including actual overtime hours (not straight-time equivalent hours). The data exclude hours paid for vacations, holidays, or sick leave when the employee is not at the establishment. Production wages represent all compensation paid to production workers.

**Value added** by manufacture is derived by subtracting the cost of materials, supplies, containers, fuel, purchased electricity, and contract work from the value of shipments (products manufactured plus receipts for services rendered). The result of this calculation is adjusted by the addition of value added by merchandising operations (the difference between the sales value and the cost of merchandise sold without further manufacture, processing, or assembly) plus the net change in finished goods and work-in-process between the beginning- and end-of-year inventories.

**Value of shipments** covers the received or receivable net selling values; free on board plant (excluding of freight and taxes), of all products shipped, both primary and secondary; and all miscellaneous receipts, such as receipts for contract work performed for others, installation and repair, sales of scrap, and sales of products bought and sold without further processing. Included are all items made by or for the establishments from material owned by it, whether sold, transferred to other plants of the same company, or shipped on consignment. The net selling value of products made in one plant on a contract basis from materials owned by another was reported by the plant providing the materials.

In the case of multi-unit companies, the manufacturer was asked to report the value of products transferred to other establishments of the same company at full economic or commercial value, including both the direct cost of production and a reasonable proportion of "all other costs" (including company overhead) and profit (interplant transfers).

The aggregate of the value of shipments figure for industry groups and for all manufacturing industries includes large amounts of duplications, as the products of some industries are used as materials by others. Estimates as to the overall extent of this duplication indicate that the value of manufactured products exclusive of such duplication (the value of finished manufactures) tends to approximate two-thirds of the total value of products reported in the census of manufactures.

**Total cost of materials** refers to direct charges actually paid or payable for items consumed or put into production during the year, including freight charges and other direct charges incurred by the establishment in acquiring these materials. It includes the cost of materials or fuel consumed, whether purchased by the individual establishment from other companies, transferred to it from other establishments of the same company, or withdrawn from inventory during the year. Included in this item are cost of parts, components, containers, etc.; cost of products bought and sold in the same condition; cost of fuels consumed for heat and power; cost of purchased electricity; and cost of contract work. Aggregate of total cost of materials and total value of shipments includes extensive duplication, since products of some industries are used as materials of others.

## 2017 ECONOMIC CENSUS: OVERVIEW, Items 217–308

### Source: U.S. Census Bureau
https://www.census.gov/programs-surveys/economic-census.html

The Economic Census provides a detailed portrait of the nation's economy, from the national to the local level, once every five years. The 2017 Economic Census covers nearly all of the U.S. economy in its basic collection of establishment statistics. The 1997 Economic Census was the first major data source to use the North American Industry Classification System (NAICS); therefore, data are not comparable to economic data from prior years, which were based on the Standard Industrial Classification (SIC) system.

NAICS, developed in cooperation with Canada and Mexico, classifies North America's economic activities at two-, three-, four-, and five-digit levels of detail; the U.S. version of NAICS further defines industries to a sixth digit. The Economic Census takes advantage of this hierarchy to publish data at these successive levels of detail: sector (two-digit); subsector (three-digit); industry group (four-digit); industry (five-digit); and U.S. industry (six-digit). Information in Table A is at the two-digit level, with a few three- and four-digit items.

Several key statistics are tabulated for all industries included in this volume: number of establishments (or companies); number of employees; payroll; and a measure of output (sales, receipts, revenue, value of shipments, or value of construction work done).

**Number of establishments.** An establishment is a single physical location at which business is conducted. It is not necessarily identical with a company or enterprise, which may consist of one establishment or more. Economic Census figures represent a summary of reports for individual establishments rather than companies. For cases in which a census report was received, separate information was obtained for each location where business was conducted. When administrative records of other federal agencies were used instead of a census report, no information was available on the number of locations operated. Each Economic Census establishment was tabulated according to the physical location at which the business was conducted. The count of establishments represents those in business at any time during 2012.

When two activities or more were carried on at a single location under a single ownership, all activities were generally grouped together as a single establishment. The entire establishment was classified on the basis of its major activity and all of its data were included in that classification. However, when distinct and separate economic activities (for which different industry classification codes were appropriate) were conducted at a single location under a single ownership, separate establishment reports for each of the different activities were obtained in the census.

**Number of employees.** Paid employees consist of the full-time and part-time employees, including salaried officers and executives of corporations. Included are employees on paid sick leave, paid holidays, and paid vacations; not included are proprietors and partners of unincorporated businesses. The definition of paid employees is the same as that used by the Internal Revenue Service (IRS) on form 941.

For some industries, the Economic Census gives codes representing the number of employees as a range of numbers (for example, "100 to 249 employees" or "1,000 to 2,499" employees). In this volume, those codes have been replaced by the standard suppression code "D".

**Payroll.** Payroll includes all forms of compensation, such as salaries, wages, commissions, dismissal pay, bonuses, vacation allowances, sick-leave pay, and employee contributions to qualified pension plans paid during the year to all employees. For corporations, payroll includes amounts paid to officers and executives; for unincorporated businesses, it does not include profit or other compensation of proprietors or partners. Payroll is reported before deductions for Social Security, income tax, insurance, union dues, etc. This definition of payroll is the same as that used on IRS form 941.

**Sales, shipments, receipts, revenue, or business done.** This measure includes the total sales, shipments, receipts, revenue, or business done by establishments within the scope of the Economic Census. The definition of each of these items is specific to the economic sector measured.

## CONSTRUCTION, Items 217–221

### Source: U.S. Census Bureau—2017 Economic Census (See Overview of 2017 Economic Census prior to Item 217)

The Construction sector (sector 23) comprises establishments primarily engaged in the construction of buildings and other structures, heavy construction (except buildings), additions, alterations, reconstruction, installation, and maintenance and repairs. Establishments engaged in the demolition or wrecking of buildings and other structures, the clearing of building sites, and the sale of materials from demolished structures are also included. This sector also contains those establishments engaged in blasting, test drilling, landfill, leveling, earthmoving, excavating, land drainage, and other land preparation. The industries within this sector have been defined on the basis of their unique production processes. As with all industries, the production processes are distinguished by their use of specialized human resources and specialized physical capital. Construction activities are generally administered or managed at a relatively fixed place of business,

but the actual construction work can be performed at one or more different project sites. This sector is divided into three subsectors of construction activities: (1) building construction and land sub-division and land development; (2) heavy construction (except buildings), such as highways, power plants, and pipelines; and (3) construction activity by special trade contractors.

## WHOLESALE TRADE, Items 222–226

### Source: U.S. Census Bureau—2017 Economic Census (See Overview of 2017 Economic Census prior to Item 217)

The Wholesale Trade sector (sector 42) comprises establishments engaged in wholesaling merchandise, generally without transformation, and rendering services incidental to the sale of merchandise. The wholesaling process is an intermediate step in the distribution of merchandise. Wholesalers are organized to sell or arrange the purchase or sale of (1) goods for resale (i.e., goods sold to other wholesalers or retailers), (2) capital or durable nonconsumer goods, and (3) raw and intermediate materials and supplies used in production.

Wholesalers sell merchandise to other businesses and normally operate from a warehouse or office. These warehouses and offices are characterized by having little or no display of merchandise. In addition, neither the design nor the location of the premises is intended to solicit walk-in traffic. Wholesalers do not normally use advertising directed to the general public. Customers are generally first reached via telephone, in-person marketing, or by specialized advertising that may include internet and other electronic means. Follow-up orders are either vendor-initiated or client-initiated, are usually based on previous sales, and typically exhibit strong ties between sellers and buyers. In fact, transactions are often conducted between wholesalers and clients that have long-standing business relationships.

This sector is made up of two main types of wholesalers: those that sell goods on their own account and those that arrange sales and purchases for others for a commission or fee.

(1) Establishments that sell goods on their own account are known as wholesale merchants, distributors, jobbers, drop shippers, import/export merchants, and sales branches. These establishments typically maintain their own warehouse, where they receive and handle goods for their customers. Goods are generally sold without transformation, but may include integral functions, such as sorting, packaging, labeling, and other marketing services.

(2) Establishments arranging for the purchase or sale of goods owned by others or purchasing goods on a commission basis are known as agents and brokers, commission merchants, import/export agents and brokers, auction companies, and manufacturers' representatives. These establishments operate from offices and generally do not own or handle the goods they sell.

Some wholesale establishments may be connected with a single manufacturer and/or promote and sell that particular manufacturer's products to a wide range of other wholesalers or retailers. Other wholesalers may be connected to a retail chain or a limited number of retail chains and only provide a variety of products needed by that particular retail operation(s). These wholesalers may obtain the products from a wide range of manufacturers. Still other wholesalers may not take title to the goods but act as agents and brokers for a commission.

Although, in general, wholesaling normally denotes sales in large volumes, durable nonconsumer goods may be sold in single units. Sales of capital or durable nonconsumer goods used in the production of goods and services, such as farm machinery, medium- and heavy-duty trucks, and industrial machinery, are always included in Wholesale Trade.

## RETAIL TRADE, Items 227–235
### Source: U.S. Census Bureau—2017 Economic Census (See Overview of 2017 Economic Census prior to Item 217)

The Retail Trade sector (44–45) is made up of establishments engaged in retailing merchandise, generally without transformation, and rendering services incidental to the sale of merchandise.

The retailing process is the final step in the distribution of merchandise; retailers are, therefore, organized to sell merchandise in small quantities to the general public. This sector comprises two main types of retailers: store and nonstore retailers.

Store retailers operate fixed point-of-sale locations, located and designed to attract a high volume of walk-in customers. In general, retail stores have extensive displays of merchandise and use mass-media advertising to attract customers. They typically sell merchandise to the general public for personal or household consumption; some also serve business and institutional clients. These include establishments, such as office supply stores, computer and software stores, building materials dealers, plumbing supply stores, and electrical supply stores. Catalog showrooms, gasoline service stations, automotive dealers, and mobile home dealers are treated as store retailers.

In addition to retailing merchandise, some types of store retailers are also engaged in the provision of after-sales services, such as repair and installation. For example, new automobile dealers, electronic and appliance stores, and musical instrument and supply stores often provide repair services. As a general rule, establishments engaged in retailing merchandise and providing after-sales services are classified in this sector.

Nonstore retailers, like store retailers, are organized to serve the general public, although their retailing methods differ. The establishments of this subsector reach customers and market merchandise with methods, such as the broadcasting of "infomercials," the broadcasting and publishing of direct-response advertising, the publishing of paper and electronic catalogs, door-to-door solicitation, in-home demonstration, selling from portable stalls (street vendors, except food), and distribution through vending machines. Establishments engaged in the direct sale (nonstore) of products, such as home heating oil dealers and home-delivery newspaper routes are included in this sector.

The buying of goods for resale is a characteristic of retail trade establishments that distinguishes them from establishments in the Agriculture, Manufacturing, and Construction sectors. For example, farms that sell their products at or from the point of production are classified in Agriculture instead of in Retail Trade. Similarly, establishments that both manufacture and sell their products to the general public are classified in Manufacturing

instead of Retail Trade. However, establishments that engage in processing activities incidental to retailing are classified in retail.

Industries in the **Motor Vehicle and Parts Dealers** subsector (441) retail motor vehicle and parts merchandise from fixed point-of-sale locations. Establishments in this subsector typically operate from a showroom and/or an open lot where the vehicles are on display. The display of vehicles and the related parts require little by way of display equipment. Personnel generally include both sales and sales support staff familiar with the requirements for registering and financing a vehicle as well as a staff of parts experts and mechanics trained to provide vehicle repair and maintenance services. Specific industries have been included in this subsector to identify the type of vehicle being retailed. Sales of capital or durable nonconsumer goods, such as medium and heavy-duty trucks, are always included in the Wholesale Trade sector. These goods are virtually never sold through retail methods.

Industries in the **Food and Beverage Stores** subsector (445) usually retail food and beverage merchandise from fixed point-of-sale locations. Establishments in this subsector have special equipment (e.g., freezers, refrigerated display cases, and refrigerators) for displaying food and beverage goods. They have staff trained in the processing of food products to guarantee the proper storage and sanitary conditions, as mandated by regulatory authority.

Industries in the **Clothing and Clothing Accessories Stores** subsector (448) retail new clothing and clothing accessories merchandise from fixed point-of-sale locations. Establishments in this subsector have similar types of display equipment, as well as employees who are knowledgeable regarding fashion trends and who can match styles, colors, and combinations of clothing and accessories to the characteristics and tastes of the customer.

Industries in the **General Merchandise Stores** subsector (452) retail new general merchandise from fixed point-of-sale locations. Establishments in this subsector are unique in that they have the equipment and staff capable of retailing a large variety of goods from a single location. This includes a variety of display equipment and staff trained to provide information on many lines of products.

## INFORMATION, Items 236–246
### Source: U.S. Census Bureau—2017 Economic Census (See Overview of 2017 Economic Census prior to Item 217)

The Information sector (51) comprises establishments engaged in the following processes: (1) producing and distributing information and cultural products, (2) providing the means to transmit or distribute these products as well as data or communications, and (3) processing data.

The main components of this sector are the publishing industries, including software publishing; the motion picture and sound recording industries; the broadcasting and telecommunications industries; and the information services and data processing industries.

For the purpose of NAICS, the transformation of information into a commodity that is produced and distributed by a number of growing industries is at issue. The Information sector groups three types of establishments: (1) those engaged in producing and distributing information and cultural products; (2) those that provide the means to transmit or distribute these products as well as data or communications; and (3) those that process data. Cultural products are those that directly express attitudes, opinions, ideas, values, and artistic creativity; provide entertainment; or offer information and analysis concerning the past and present. Included in this definition are popular, mass-produced products, as well as cultural products that normally have a more limited audience, such as poetry books, literary magazines, or classical records. These activities were formerly classified throughout the existing national classifications. Traditional publishing was in manufacturing; broadcasting in communications; software production in business services; film production in amusement services; and so forth.

Industries in the **Publishing Industries, Except Internet** subsector (511) include establishments engaged in the publishing of newspapers, magazines, other periodicals, and books, as well as database and software publishing. In general, these establishments, which are known as publishers, issue copies of works for which they usually possess copyright. Works may be in one or more formats, including traditional print format, CD-ROM format, or proprietary electronic networks. Publishers may publish works originally created by others for which they have obtained the rights and/or works that they have created in-house. Software publishing is included here because the activity (creation of a copyrighted product and bringing it to market) is equivalent to the creation process for other types of intellectual products.

In NAICS, publishing—the reporting, writing, editing, and other processes that are required to create an edition of a book or a newspaper—is treated as a major economic activity in its own right, rather than as a subsidiary activity to printing, which is a manufacturing activity. Thus, publishing is classified in the Information sector, while printing remains in the NAICS Manufacturing sector. In part, the NAICS classification reflects the fact that publishing increasingly takes place in establishments that are physically separate from the associated printing establishments. More crucially, the NAICS classification of book and newspaper publishing is intended to portray their roles in a modern economy—roles that do not resemble manufacturing activities.

Music publishers are not included in the Publishing Industries subsector, but can be found in the Motion Picture and Sound Recording Industries subsector. Reproduction of prepackaged software is treated in NAICS as a manufacturing activity; online distribution of software products is in the Information sector, and custom design of software to client specifications is included in the Professional, Scientific, and Technical Services sector. These distinctions arise because of the different ways that software is created, reproduced, and distributed.

The Information sector does not include products, such as manifold business forms. Information is not the essential component of these items. Establishments producing these items are included in subsector 323, Printing and Related Support Activities.

Industries in the **Motion Picture and Sound Recording Industries** subsector (512) group establishments involved in the production and distribution of motion pictures and sound recordings. While producers and distributors of motion pictures and sound recordings issue works for sale as traditional publishers do, the processes are different enough to warrant placing the

establishments engaged in these activities in separate subsectors. Production is typically a complex process that involves several distinct types of establishments engaged in activities, such as contracting with performers, creating the film or sound content, and providing technical postproduction services. Film distribution is often to exhibitors, such as theaters and broadcasters, rather than to a wholesale or retail distribution chain. When the product is in a mass-produced form, NAICS treats production and distribution as the major economic activity, rather than as a subsidiary activity to the manufacture of such products.

This subsector does not include establishments primarily engaged in the wholesale distribution of video cassettes and sound recordings, such as compact discs and audio tapes; these establishments are included in the Wholesale Trade sector. Reproduction of video cassettes and sound recordings that is carried out separately from establishments engaged in production and distribution is treated in NAICS as a manufacturing activity.

Industries in the **Broadcasting, except Internet** subsector (515) include establishments that create content or acquire the right to distribute and subsequently broadcast content. The industry groups (Radio and Television Broadcasting and Cable and Other Subscription Programming) are based on differences in the methods of communication and the nature of services provided. The Radio and Television Broadcasting industry group includes establishments that operate broadcasting studios and facilities for over-the-air or satellite delivery of radio and television programs, including entertainment, news, and talk programs. These establishments are often engaged in production and purchase of programs and generating revenues from the sale of air time to advertisers, as well as from donations, subsidies, and/or the sale of programs. The Cable and Other Subscription Programming industry group includes establishments that operate studios and facilities for the broadcasting of limited-format programs (such as news, sports, educational, and youth-oriented programs) that are typically narrowly-focused in nature; these programs are usually available on a subscription or fee basis. The distribution of cable and other subscription programming is included in subsector 517, Telecommunications.

Industries in the **Internet Publishing and Broadcasting and Web search portals** subsector (51913) consist of establishments primarily engaged in 1) publishing and/or broadcasting content on the Internet exclusively or 2) operating Web sites that use a search engine to generate and maintain extensive databases of Internet addresses and content in an easily searchable format (and known as Web search portals). The publishing and broadcasting establishments in this industry do not provide traditional (non-Internet) versions of the content that they publish or broadcast. They provide textual, audio, and/or video content of general or specific interest on the Internet exclusively. Establishments known as Web search portals often provide additional Internet services, such as e-mail, connections to other web sites, auctions, news, and other limited content, and serve as a home base for Internet users.

Establishments that are *not* in this group include those primarily engaged in--

- Providing wired broadband Internet access using own operated telecommunications infrastructure—these are classified in Wired Telecommunications Carriers;

- Providing both Internet publishing and other print or electronic (e.g., CD-ROM, diskette) editions in the same establishment or using proprietary networks to distribute content—these are classified in Publishing Industries (except Internet) based on the materials produced;
- Providing Internet access via client-supplied telecommunications connections—these are classified in All Other Telecommunications;
- Providing streaming services on content owned by others—these are classified in Data Processing, Hosting, and Related Services;
- Wholesaling goods on the Internet—these are classified in Wholesale Trade;
- Retailing goods on the Internet—these are classified in Retail Trade;
- Operating stock brokerages, travel reservation systems, purchasing services, and similar activities using the Internet rather than traditional methods—these are classified with the more traditioal establishments providing these services.

Industries in the **Telecommunications** subsector (517) include establishments that provide telecommunications and services related to that activity (e.g., telephony, including Voice over Internet Protocol (VoIP); cable and satellite television distribution services; Internet access; telecommunications reselling services). The Telecommunications subsector is primarily engaged in operating, maintaining, and/or providing access to facilities for the transmission of voice, data, text, sound, and video. A transmission facility may be based on a single technology or a combination of technologies. Establishments primarily engaged as independent contractors in the maintenance and installation of broadcasting and telecommunications systems are classified in sector 23, Construction.

Industries in the **Data processing, hosting, and related services** subsector (518) are establishments primarily engaged in providing infrastructure for hosting or data processing services. These establishments may provide specialized hosting activities, such as web hosting, streaming services or application hosting; provide application service provisioning; or may provide general time-share mainframe facilities to clients. Data processing establishments provide complete processing and specialized reports from data supplied by clients or provide automated data processing and data entry services.

## UTILITIES, Items 247–252

**Source: U.S. Census Bureau—2017 Economic Census (See Overview of 2017 Economic Census prior to Item 217)**

The Utilities sector (22) comprises establishments engaged in the provision of the following utility services: electric power, natural gas, steam supply, water supply, and sewage removal. Within this sector, the specific activities associated with the utility services provided vary by utility: electric power includes generation, transmission, and distribution; natural gas includes distribution; steam supply includes provision and/or distribution; water supply includes treatment and distribution; and sewage removal includes collection, treatment, and disposal of waste through sewer systems and sewage treatment facilities.

Excluded from this sector are establishments primarily engaged in waste management. These services are classified in subsector 562, Waste Management and Remediation Services, which also collect, treat, and dispose of waste materials; however, establishments in this subsector do not use sewer systems or sewage treatment facilities.

## TRANSPORTATION AND WAREHOUSING, Items 252–256

**Source: U.S. Census Bureau—2017 Economic Census (See Overview of 2017 Economic Census prior to Item 217)**

The Transportation and Warehousing sector (48–49) includes industries that provide transportation of passengers and cargo, warehousing and storage for goods, scenic and sightseeing transportation, and support activities related to modes of transportation. Establishments in these industries use transportation equipment or transportation related facilities as a productive asset. The type of equipment depends on the mode of transportation, which includes air, rail, water, road, and pipeline.

The transportation and warehousing sector distinguishes three basic types of activities: subsectors for each mode of transportation, a subsector for warehousing and storage, and a subsector for establishments providing support activities for transportation. In addition, there are subsectors for establishments that provide passenger transportation for scenic and sightseeing purposes, postal services, and courier services.

## FINANCE AND INSURANCE, Items 257–261

**Source: U.S. Census Bureau—2017 Economic Census (See Overview of 2017 Economic Census prior to Item 217)**

The Finance and Insurance sector (52) comprises establishments primarily engaged in financial transactions (transactions involving the creation, liquidation, or change in ownership of financial assets) and/or in facilitating financial transactions. Three principal types of activities are identified:

(1) Raising funds by taking deposits and/or issuing securities and, in the process, incurring liabilities. Establishments engaged in this activity use raised funds to acquire financial assets by making loans and/or purchasing securities. Putting themselves at risk, they channel funds from lenders to borrowers and transform or repackage the funds with respect to maturity, scale and risk. This activity is known as financial intermediation.

(2) Pooling of risk by underwriting insurance and annuities. Establishments engaged in this activity collect fees, insurance premiums, or annuity considerations; build up reserves; invest those reserves; and make contractual payments. Fees are based on the expected incidence of the insured risk and the expected return on investment.

(3) Providing specialized services facilitating or supporting financial intermediation, insurance, and employee benefit programs.

In addition, monetary authorities charged with monetary control are included in this sector.

# REAL ESTATE AND RENTAL AND LEASING, Items 262–266

## Source: U.S. Census Bureau—2017 Economic Census (See Overview of 2017 Economic Census prior to Item 217)

The Real Estate and Rental and Leasing sector (53) comprises establishments primarily engaged in renting, leasing, or otherwise allowing the use of tangible or intangible assets, and establishments providing related services. The major portion of this sector comprises establishments that rent, lease, or otherwise allow the use of their own assets by others. The assets may be tangible, such as real estate and equipment, or intangible, such as patents and trademarks.

This sector also includes establishments primarily engaged in managing real estate for others, selling, renting, and/or buying real estate for others, and appraising real estate. These activities are closely related to this sector's main activity. In addition, a substantial proportion of property management is self-performed by lessors.

The main components of this sector are the real estate lessors industries; equipment lessors industries (including motor vehicles, computers, and consumer goods); and lessors of nonfinancial intangible assets (except copyrighted works).

# PROFESSIONAL, SCIENTIFIC, AND TECHNICAL SERVICES, Items 267–275

## Source: U.S. Census Bureau—2017 Economic Census (See Overview of 2017 Economic Census prior to Item 217)

The Professional, Scientific, and Technical Services sector (54) is made up of establishments that specialize in performing professional, scientific, and technical activities for others. These activities require a high degree of expertise and training. The establishments in this sector specialize according to expertise and provide services to clients in a variety of industries (and, in some cases, to households). Activities performed include legal advice and representation; accounting, bookkeeping, and payroll services; architectural, engineering, and specialized design services; computer services; consulting services; research services; advertising services; photographic services; translation and interpretation services; veterinary services; and other professional, scientific, and technical services.

This sector excludes establishments primarily engaged in providing a range of day-to-day office administrative services, such as financial planning, billing and record keeping, personnel services, and physical distribution and logistics services. These establishments are classified in sector 56, Administrative and Support and Waste Management and Remediation Services.

**Legal Services** comprises establishments primarily engaged in offering legal services such as offices of lawyers, notaries, title abstract and settlement offices, and all other legal services such as patent agent services, paralegal services, and process serving services.

**Accounting, Tax Preparation, Bookkeeping, and Payroll Services** comprises establishments primarily engaged in providing services, such as auditing of accounting records, designing accounting systems, preparing financial statements, developing budgets, preparing tax returns, processing payrolls, bookkeeping, and billing.

**Architectural, Engineering, and Related Services** comprises establishments primarily engaged in offering (1) architectural services for residential, institutional, leisure, commercial, and industrial buildings and structures as well as for landscape purposes; (2) offering engineering services, including drafting services or building inspection services; (3) offering geophysical surveying and mapping services; (4) surveying and mapping services, except geophysical; and (5) offering testing laboratory services except medical and veterinary (the testing can occur in a laboratory or on-site).

**Computer Systems Design and Related Services** consists of establishments primarily engaged in providing expertise in the field of information technologies through one or more of the following activities: (1) writing, modifying, testing, and supporting software to meet the needs of a particular customer; (2) planning and designing computer systems that integrate computer hardware, software, and communication technologies; (3) on-site management and operation of clients' computer systems and/or data processing facilities; and (4) other professional and technical computer-related advice and services.

# HEALTH CARE AND SOCIAL ASSISTANCE, Items 276–289

## Source: U.S. Census Bureau—2017 Economic Census (See Overview of 2017 Economic Census prior to Item 217)

The Health Care and Social Assistance sector (62) consists of establishments that provide health care and social assistance services to individuals. The sector includes both health care and social assistance, because it is sometimes difficult to distinguish between the boundaries of these two activities. The industries in this sector are arranged on a continuum starting with those that provide medical care exclusively, continuing with those that provide health care and social assistance, and finishing with those that provide only social assistance. The services provided by establishments in this sector are delivered by trained professionals. All industries in the sector share this commonality of process—namely, labor inputs of health practitioners or social workers with the requisite expertise. Many of the industries in the sector are defined based on the educational degree held by the practitioners included in the industry.

In this volume, taxable and tax-exempt establishments are presented separately.

Excluded from this sector are aerobic classes, which can be found in subsector 713, Amusement, Gambling and

Recreation Industries; and nonmedical diet and weight-reducing centers, which can be found in subsector 812, Personal and Laundry Services. Although these can be viewed as health services, they are not typically delivered by health practitioners.

Industries in the **Ambulatory Health Care Services** subsector (621) provide health care services directly or indirectly to ambulatory patients and do not typically provide inpatient services. Health practitioners in this subsector provide outpatient services, and facilities and equipment do not usually play the most significant part in this sector's production process.

Industries in the **Hospitals** subsector (622) provide medical, diagnostic, and treatment services, including physician, nursing, specialized accommodation, and other health services, to inpatients. Hospitals may provide outpatient services as a secondary activity. Many of the services provided by establishments in the Hospitals subsector require the use of specialized facilities and equipment, both of which form a significant and integral part of the production process.

# ARTS, ENTERTAINMENT, AND RECREATION, Items 290–294

**Source: U.S. Census Bureau—2017 Economic Census (See Overview of 2017 Economic Census prior to Item 217)**

The Arts, Entertainment, and Recreation sector (71) includes a wide range of establishments that operate facilities or provide services that meet the diverse cultural, entertainment, and recreational interests of their patrons. This sector is made up of: (1) establishments that are involved in producing, promoting, or participating in live performances, events, or exhibits intended for public viewing; (2) establishments that preserve and exhibit objects and sites of historical, cultural, or educational interest; and (3) establishments that operate facilities or provide services that enable patrons to participate in recreational activities or pursue amusement, hobby, and leisure time interests.

Some establishments that provide cultural, entertainment, or recreational facilities and services are classified in other sectors. Excluded from this sector are: (1) establishments that provide both accommodations and recreational facilities—such as hunting and fishing camps and resort and casino hotels—are classified in subsector 721, Accommodation; (2) restaurants and night clubs that provide live entertainment in addition to the sale of food and beverages are classified in subsector 722, Food Services and Drinking Places; (3) motion picture theaters, libraries and archives, and publishers of newspapers, magazines, books, periodicals, and computer software are classified in sector 51, Information; and (4) establishments that use transportation equipment to provide recreational and entertainment services, such as those operating sightseeing buses, dinner cruises, or helicopter rides, are classified in subsector 487, Scenic and Sightseeing Transportation.

# ACCOMMODATION AND FOOD SERVICES, Items 295–300

**Source: U.S. Census Bureau—2017 Economic Census (See Overview of 2017 Economic Census prior to Item 217)**

The Accommodation and Food Services sector (72) consists of establishments that provide customers with lodging and/or meals, snacks, and beverages for immediate consumption. The sector includes both accommodation and food services establishments because the two activities are often combined at the same establishment. Excluded from this sector are civic and social organizations, amusement and recreation parks, theaters, and other recreation or entertainment facilities providing food and beverage services.

Industries in the **Food Services and Drinking Places** subsector (722) prepare meals, snacks, and beverages to customer order for immediate on-premises and off-premises consumption. There is a wide range of establishments in these industries. Some provide food and drink only; while others provide various combinations of seating space, waiter/waitress services and incidental amenities, such as limited entertainment. The industries in the subsector are grouped based on the type and level of services provided. The industry groups are full-service restaurants; limited-service eating places; special food services, such as food service contractors, caterers, and mobile food services, and drinking places. Food services and drink activities at hotels and motels; amusement parks, theaters, casinos, country clubs, and similar recreational facilities; and civic and social organizations are included in this subsector only if these services are provided by a separate establishment primarily engaged in providing food and beverage services. Excluded from this subsector are establishments operating dinner cruises. These establishments are classified in subsector 487, Scenic and Sightseeing Transportation, because they utilize transportation equipment to provide scenic recreational entertainment.

# OTHER SERVICES, EXCEPT PUBLIC ADMINISTRATION Items 301–308

**Source: U.S. Census Bureau—2017 Economic Census (See Overview of 2017 Economic Census prior to Item 217)**

The Other Services, Except Public Administration sector (81) comprises establishments engaged in providing services not specifically categorized elsewhere in the classification system. Establishments in this sector are primarily engaged in activities such as equipment and machinery repairing, promoting or administering religious activities, grant making, and advocacy; this sector also includes establishments that provide dry-cleaning and laundry services, personal care services, death care services, pet care services, photofinishing services, temporary parking services, and dating services.

Private households that employ workers on or about the premises in activities primarily concerned with the operation of the household are included in this sector.

Excluded from this sector are establishments primarily engaged in retailing new equipment and performing repairs and general maintenance on equipment. These establishments are classified in sector 44–45, Retail Trade.

Industries in the **Repair and Maintenance** subsector (811) restore machinery, equipment, and other products to working order. These establishments also typically provide general or routine maintenance (i.e., servicing) on such products to ensure they work efficiently; this maintenance also helps prevent breakdowns and make certain repairs unnecessary.

The NAICS structure for this subsector brings together most types of repair and maintenance establishments and categorizes them based on production processes (i.e., on the type of repair and maintenance activity performed, and the necessary skills, expertise, and processes required for different repair and maintenance establishments). This NAICS classification does not delineate between repair services provided to businesses versus those provided to households. Although some industries primarily serve either businesses or households, separation by class of customer is limited by the fact that many establishments serve both types. Establishments that repair computers and consumer electronics products are examples of such overlap.

The Repair and Maintenance subsector does not include all establishments engaged in repair and maintenance. For example, a substantial amount of repair is done by establishments that also manufacture machinery, equipment, and other goods. These establishments are included in the Manufacturing sector in NAICS. In addition, the repairing of transportation equipment is often provided by or based at transportation facilities, such as airports and seaports; these activities are included in the Transportation and Warehousing sector.

A particularly unique situation exists with repair of buildings. Plumbing, electrical installation and repair, painting and decorating, and other construction-related establishments are often involved in performing installation or other work on new construction, while also providing repair services on existing structures. Although some establishments do specialize in repair, it is difficult to distinguish between these two types. Thus, all such establishments are included in the Construction sector.

Excluded from this subsector are establishments primarily engaged in rebuilding or remanufacturing machinery and equipment. These are classified in sector 31–33, Manufacturing. Also excluded are retail establishments that provide after-sale services and repair. These are classified in sector 44–45, Retail Trade.

Industries in the **Personal and Laundry Services** subsector (812) include establishments that provide personal and laundry services to individuals, households, and businesses. Services performed include personal care services, death care services, laundry and dry-cleaning services, and a wide range of other personal services, such as pet care (except veterinary) services, photofinishing services, temporary parking services, and dating services.

The Personal and Laundry Services subsector is by no means all-inclusive of the activities that could be termed personal services (i.e., those provided to individuals rather than businesses). There are many other sectors and subsectors that provide services to persons. Establishments providing legal, accounting, tax preparation, architectural, portrait photography, and similar professional services are classified in sector 54, Professional, Scientific,

and Technical Services; those providing job placement, travel arrangement, home security, interior and exterior house cleaning, exterminating, lawn and garden care, and similar support services are classified in sector 56, Administrative and Support and Waste Management and Remediation Services; those providing health and social services are classified in sector 62, Health Care and Social Assistance; those providing amusement and recreation services are classified in sector 71, Arts, Entertainment and Recreation; those providing educational instruction are classified in sector 61, Educational Services; those providing repair services are classified in subsector 811, Repair and Maintenance; and those providing spiritual, civic, and advocacy services are classified in subsector 813, Religious, Grantmaking, Civic, Professional, and Similar Organizations.

Industries in the **Religious, Grantmaking, Civic, Professional, and Similar Organizations** subsector (813) include establishments that organize and promote religious activities, support various causes through grant making, advocate various social and political causes, and promote and defend the interests of their members. This category includes only tax-exempt establishments.

The industry groups within the subsector are defined in terms of their activities, separately grouping establishments that provide funding for specific causes or for a variety of charitable causes, establishments that advocate and actively promote causes and beliefs for the public good, and establishments that have an active membership structure to promote causes and represent the interests of their members. Establishments in this subsector may publish newsletters, books, and periodicals for distribution to their membership.

# GOVERNMENT EMPLOYMENT, Items 309–311
**Source: U.S. Bureau of Economic Analysis—Regional Economic Accounts**
http://www.bea.gov/regional/index.htm#state

Employment is measured as the average annual sum of full-time and part-time jobs. The estimates are on a place-of-work basis. Data for federal civilian employment include civilian employees of the Department of Defense. Military employment includes all persons on active duty status.

# STATE GOVERNMENT EMPLOYMENT AND PAYROLL, Items 312–330
**Source: U.S. Census Bureau—Annual Survey of Public Employment and Payroll**
https://www.census.gov/programs-surveys/apes.html

The annual Survey of Public Employment and Payroll measures the number of federal, state, and local civilian government employees and their gross monthly payroll for March of the survey year for state and local governments and for the Federal Government.

The survey provides state and local government data on full-time and part-time employment, part-time hours worked, full-time equivalent employment, and payroll statistics by governmental function (i.e., elementary and secondary education, higher education, police protection, fire protection, financial administration, central staff services, judicial and legal, highways, public welfare, solid waste management, sewerage, parks and recreation, health, hospitals, water supply, electric power, gas supply, transit, natural resources, correction, libraries, air transportation, water transport and terminals, other education, state liquor stores, social insurance administration, and housing and community development).

Data have been collected annually since 1957. A census is conducted every five years (years ending in '2' and '7'). A sample of state and local governments is used to collect data in the intervening years. A new sample is selected every five years (years ending in '4' and '9').

**State** government employees include all persons paid for personal services performed, including persons paid from federally funded programs, paid elected or appointed officials, persons in a paid leave status, and persons paid on a per meeting, annual, semiannual, or quarterly basis. Unpaid officials, pensioners, persons whose work is performed on a fee basis, and contractors and their employees are excluded from the count of employees. **Full-time employees** are persons employed during the pay period to work the number of hours per week that represents regular full-time employment. Included are full-time temporary or seasonal employees who are working the number of hours that represent full-time employment. **Part-time employees** are persons paid on a part-time basis during the designated pay period. Included are those daily or hourly employees usually engaged for less than the regular full-time workweek, as well as any part-time paid officials. **Full-Time Equivalent employees** is a computed statistic representing the number of full-time employees that could have been employed if the reported number of hours worked by part-time employees had been worked by full-time employees. This statistic is calculated separately for each function of a government by dividing the "part-time hours paid" by the standard number of hours for full-time employees in the particular government and then adding the resulting quotient to the number of full-time employees.

**Full-time payroll** represents gross payroll amounts for the one-month period of March for full-time employees. **Part-time pay** represents gross payroll amounts for the one-month period of March for part-time employees. Gross payroll includes all salaries, wages, fees, commissions, and overtime paid to employees **before** withholdings for taxes, insurance, etc. It also includes incentive payments that are paid at regular pay intervals. It excludes employer share of fringe benefits like retirement, Social Security, health and life insurance, lump sum payments, and so forth.

**Administration** combines **Financial administration** and **Other government administration**. **Financial administration** includes activities concerned with tax assessment and collection, custody and disbursement of funds, debt management, administration of trust funds, budgeting, and other government-wide financial management activities. This function is not applied to school district or special district governments. **Other government administration** applies to the legislative and government-wide administrative agencies of governments. Included here are overall planning and zoning activities, and central personnel and administrative activities. This function is not applied to school district or special district governments.

**Judicial and legal** includes all court and court related activities (except probation and parole activities that are included at the "Correction" function), court activities of sheriff's offices, prosecuting attorneys' and public defenders' offices, legal departments, and attorneys providing government-wide legal service.

**Police includes** all activities concerned, with the enforcement of law and order, including coroner's offices, police training academies, investigation bureaus, and local jails, "lockups", or other detention facilities not intended to serve as correctional facilities.

**Corrections** includes activities pertaining to the confinement and correction of adults and minors convicted of criminal offenses. Pardon, probation, and parole activities are also included here.

**Highways and transportation includes** activities associated with the maintenance and operation of streets, roads, sidewalks, bridges, tunnels, toll roads, and ferries. Snow and ice removal, street lighting, and highway and traffic engineering activities are also included here. Also included are the operation, maintenance, and construction of public mass transit systems, including subways, surface rails, and buses, and the provision, construction, operation, maintenance; support of public waterways, harbors, docks, wharves, and related marine terminal facilities; and activities associated with the operation and support of publicly operated airport facilities.

**Public welfare** includes the administration of various public assistance programs for the needy, veteran services, operation of nursing homes, indigent care institutions, and programs that provide payments for medical care, handicap transportation, and other services for the needy.

**Health** includes administration of public health programs, community and visiting nurse services, immunization programs, drug abuse rehabilitation programs, health and food inspection activities, operation of outpatient clinics, and environmental pollution control activities.

**Hospitals** includes only government operated medical care facilities that provide inpatient care. Employees and payrolls of private corporations that lease and operate government-owned hospital facilities are excluded.

**Social insurance administration** includes the administration of unemployment compensation systems, public employment services, and the Federal Social Security, Medicare, and Railroad Retirement trusts.

**Natural resources and Parks** includes activities primarily concerned with the conservation and development of natural resources (soil, water, energy, minerals, etc.) and the regulation of industries that develop, utilize, or affect natural resources, as well as the operation and maintenance of parks, playgrounds, swimming pools, public beaches, auditoriums, public golf courses, museums, marinas, botanical gardens, and zoological parks.

**Utilities, sewerage, and waste management** includes operation, maintenance, and construction of public water supply systems, including production, acquisition, and distribution of water to general public or to other public or private utilities, for residential, commercial, and industrial use; activities associated with the production or acquisition and distribution of electric power; provision, maintenance, and operation of sanitary and storm sewer systems and sewage disposal and treatment facilities; and refuse collection and disposal, operation of sanitary landfills, and street cleaning activities.

**Elementary and secondary education and libraries includes** activities associated with the operation of public elementary and secondary schools and locally operated vocational-technical schools. Special education programs operated by elementary and secondary school systems are also included as are all ancillary services associated with the operation of schools, such as pupil transportation and food service. **Also included are the** establishment and provision of libraries for use by the general public and the technical support of privately operated libraries. This category includes classroom teachers, principals, supervisors of instruction, librarians, teacher aides, library aides, and guidance and psychological personnel as well as school superintendents and other administrative personnel, clerical and secretarial staffs, plant operation and maintenance personnel, health and recreation employees, transportation and food service personnel, and any student employees.

**Higher education** includes state government degree granting institutions that provide academic training above grade 12. This includes persons engaged in teaching and related academic research as well as administrative, clerical, custodial, cafeteria, health personnel, noninstructional employees engaged in organized research, law enforcement personnel, and paid student employees.

# STATE GOVERNMENT FINANCES, Items 331–350

**Source: U.S. Census Bureau—State Government Finances**
**http://www.census.gov/govs/state/**

Data are from an annual survey conducted by the Census Bureau and pertain to state government fiscal years ending on June 30, except for four states with other ending dates: Alabama and Michigan (September 30), New York (March 31), and Texas (August 31).

The state government finance data presented in this publication may differ from data published by state governments because the Census Bureau may be using a different definition of which organizations are covered under the term, "state government."

For the purpose of Census Bureau statistics, the term "state government" refers not only to the executive, legislative, and judicial branches of a given state, but it also includes agencies, institutions, commissions, and public authorities that operate separately or somewhat autonomously from the central state government but where the state government maintains administrative or fiscal control over their activities as defined by the Census Bureau.

Total **general revenue** includes all revenue except utility, liquor stores, and insurance trust revenue. All tax revenue and intergovernmental revenue, even if designated for employee-retirement or local utility purpose, are classified as general revenue.

**Intergovernmental revenue** covers amounts received from the federal government as fiscal aid, reimbursements for performance of general government functions and specific services

for the paying government, or in lieu of taxes. It excludes any amounts received from other governments from the sale of property, commodities, and utility services.

**Taxes** consist of compulsory contributions exacted by governments for public purposes. However, this category excludes employer and employee payments for retirement and social insurance purposes, which are classified as insurance trust revenue; it also excludes special assessments, which are classified as non-tax general revenue. Sales and gross receipts taxes do not include dealer discounts, or "commissions" allowed to merchants for collection of taxes from consumers. General sales taxes and selected taxes on sales of motor fuels, tobacco products, and other particular commodities and services are included.

**General government expenditure** includes capital outlay, a major portion of which is commonly financed by borrowing. Government revenue does not include receipts from borrowing. Among other things, this distorts the relationship between totals of revenue and expenditure figures that are presented and renders it useless as a direct measure of the degree of budgetary "balance" (as that term is generally applied).

**Direct general expenditure** comprises all expenditures of the state governments, excluding utility, liquor stores, insurance trust expenditures, and any intergovernmental payments.

State government expenditure for **education** is mainly for the provision and general support of schools and other educational facilities and services, including those for educational institutions beyond high school. They cover such related services as student transportation; school lunch and other cafeteria operations; school health, recreation, and library services; and dormitories, dining halls, and bookstores operated by public institutions of higher education.

**Health and hospitals expenditure** includes health research; clinics; nursing; immunization; other categorical, environmental, and general health services provided by health agencies; establishment and operation of hospital facilities; provision of hospital care; and support of other public and private hospitals.

**Highways expenditure** is for the provision and maintenance of highway facilities, including toll turnpikes, bridges, tunnels, and ferries, as well as regular roads, highways, and streets. Also included are expenditures for street lighting and for snow and ice removal. Not included are highway policing and traffic control, which are considered part of police protection

**Public safety expenditure** includes police and correctional institution expenditures.

**Public welfare expenditure** covers support of and assistance to needy persons; this aid is contingent upon the person's needs. Included are cash assistance paid directly to needy persons under categorical (Old Age Assistance, Temporary Assistance for Needy Families, Aid to the Blind, and Aid to the Disabled) and other welfare programs; vendor payments made directly to private purveyors for medical care, burials, and other commodities and services provided under welfare programs; welfare institutions; and any intergovernmental or other direct expenditure for welfare purposes. Pensions to former employees and other benefits not contingent on need are excluded.

**Natural resources, parks, and recreation** includes expenditures for conservation, promotion, and development of natural resources (soil, water, energy, minerals, etc.) and the regulation of industries which develop, utilize, or affect natural resources. It also includes the provision and support of recreational and cultural-scientific facilities, such as golf courses, playgrounds, tennis courts, public beaches, swimming pools, play fields, parks, camping areas, recreational piers and marinas, galleries, museums, zoos, botanical gardens, auditoriums, stadiums, recreational centers, convention centers, exhibition halls, community music, drama, and celebrations.

**Debt outstanding** includes all long-term debt obligations of the government and its agencies (exclusive of utility debt) and all interest-bearing, short-term (repayable within one year) debt obligations remaining unpaid at the close of the fiscal year. It includes judgments, mortgages, and revenue bonds, as well as general obligation bonds, notes, and interest-bearing warrants. This category consists of non-interest-bearing, short-term obligations; inter-fund obligations; amounts owed in a trust or agency capacity; advances and contingent loans from other governments; and rights of individuals to benefit from government-administered employee-retirement funds.

## VOTING AND REGISTRATION, Items 351 and 352

**Source: U.S. Census Bureau—Current Population Survey**
http://www.census.gov/topics/public-sector/voting.html

These estimates are based on the November 2020 Voting and Registration Supplement to the Current Population Survey (CPS).

Voting rates are calculated using the voting-age population, which includes both citizens and noncitizens. The percentages in columns 351 and 352 are based on the citizen population. Statistics from surveys are subject to sampling and nonsampling error. The CPS estimate of overall turnout differs from the "official" turnout reported by the Clerk of the House of Representatives.

## ELECTION STATISTICS, Items 353–355

**Source: U.S. House of Representatives, Statistics of the Presidential and Congressional Election of November 3, 2020**
http://history.house.gov/Institution/Election-Statistics/2020election/

Election results show the percentage of the total vote cast for the Democratic and Republican candidates, as well as the combined percentage for all other candidates in the 2020 presidential election. This information was compiled by the Office of the Clerk, U.S. House of Representatives and published on February 26, 2021.

## TABLE B—STATES AND COUNTIES

Table B presents 199 items for the United States as a whole, each individual state, and the District of Columbia; and every county, county equivalent, and independent city. The counties are presented in alphabetical order within each state, and the states are also presented in alphabetical order. Independent cities, which are found in Maryland, Missouri, Nevada, and Virginia, are placed in alphabetical order at the end of the list of counties for those states. The District of Columbia is included in Table B as both a county and a state. It is also included as a city in Table D.

## LAND AREA, Items 1 and 4

**Source: U.S. Census Bureau—2020 U.S. Gazetteer Files,**
https://www.census.gov/geographies/reference-files/time-series/geo/gazetteer-files.html

Land area measurements are shown to the nearest square mile. Land area is an area measurement providing the size, in square miles, of the land portions of each county.

## POPULATION, Items 2–4

**Source: U.S. Census Bureau—Population Estimates**
https://www.census.gov/programs-surveys/popest.html

The population data are Census Bureau estimates of the resident population as of July 1 of the year shown. The ranks are shown for counties (including independent cities and the District of Columbia).

The Census Bureau released these 2020 estimates for research purposes. The July 1, 2020 estimates are based on the 2010 Census and were created without incorporation or consideration of the 2020 Census results. They are typically used in comparisons with the 2020 Census to make determinations about the accuracy of the estimates. Differences between the estimates and census counts are interpreted as error in the estimates and are used to inform population and housing unit estimates research and methodological improvements over the decade.

## POPULATION AND POPULATION CHARACTERISTICS, Items 5–19

**Source: U.S. Census Bureau—Population Estimates**
https://www.census.gov/programs-surveys/popest.html

The concept of race, as used by the Census Bureau, reflects self-identification by persons according to the race or races with which they most closely identify. These categories are sociopolitical constructs and should not be interpreted as being scientific or anthropological in nature. Furthermore, race categories include both racial and national origin groups.

Beginning with the 2000 census, respondents were offered the option of selecting one or more races. This option was not available in prior censuses; thus, comparisons between censuses should be made with caution. In Table B, Columns 5 through 8 refer to individuals who identified with each racial category, either alone or in combination with other races. The estimates exclude persons of Hispanic or Latino origin from all race groups.

The sum of the four individual race alone or in combination categories in this book will often add to more than the total population because people who reported more than one race were tallied in each race category. In this book, the Asian group has been combined with the Native Hawaiian and Other Pacific Islander group, causing double-counting of persons who identify with both groups. This is especially pronounced in Hawaii.

The **White** population is defined as persons who indicated their race as White, as well as persons who did not classify themselves in one of the specific race categories listed on the questionnaire but entered a nationality such as Irish, German, Italian, Lebanese, Near Easterner, Arab, or Polish.

The **Black** population includes persons who indicated their race as "Black or African American" as well as persons who did not classify themselves in one of the specific race categories but reported entries such as African American, Afro American, Kenyan, Nigerian, or Haitian.

The **American Indian or Alaska Native** population includes persons who indicated their race as American Indian or Alaska Native, as well as persons who did not classify themselves in one of the specific race categories but reported entries such as Canadian Indian, French-American Indian, Spanish-American Indian, Eskimo, Aleut, Alaska Indian, or any of the American Indian or Alaska Native tribes.

The **Asian and Pacific Islander** population combines two census groupings: **Asian** and **Native Hawaiian or Other Pacific Islander**. The **Asian** population includes persons who indicated their race as Asian Indian, Chinese, Filipino, Japanese, Korean, Vietnamese, or "Other Asian," as well as persons who provided write-in entries of such groups as Cambodian, Laotian, Hmong, Pakistani, or Taiwanese. The **Native Hawaiian or Other Pacific Islander** population includes persons who indicated their race as "Native Hawaiian," "Guamanian or Chamorro," "Samoan," or "Other Pacific Islander," as well as persons who reported entries such as Part Hawaiian, American Samoan, Fijian, Melanesian, or Tahitian.

The **Hispanic population** is based on a question that asked respondents "Is this person Spanish/Hispanic/Latino?" Persons marking any one of the four Hispanic categories (i.e., Mexican, Puerto Rican, Cuban, or other Spanish) are collectively referred to as Hispanic.

**Age** is defined as age at last birthday (number of completed years since birth), as of April 1 of the census year.

The **female** population is shown as a percentage of total population.

## POPULATION AND COMPONENTS OF CHANGE, Items 20–26

**Source: U.S. Census Bureau—Decennial Censuses and Population Estimates**
http://www.census.gov/main/www/cen2000.html
https://www.census.gov/programs-surveys/popest.html
https://www.census.gov/programs-surveys/decennial-census/data/datasets.2010.html

The population data for 2000 and 2010 are from the decennial censuses and represent the resident population as of April 1 of those years. The 2010 estimates are based on the 2010 Census and reflect changes to the April 1, 2010 population due to the Count Question Resolution program and geographic program revisions. The components of change are based on Census Bureau estimates of the resident population as of July 1, 2020. The change in population between 2010 and 2020 is made up of (a) natural increase—births minus deaths, and (b) net migration—the difference between the number of persons moving into a particular area and the number of persons moving out of the area. Net migration is composed of internal and international migration.

Because the 2020 population estimates are based on a model that begins with a national population estimate, the county components of change do not always exactly add up to the difference between the 2010 census population and the 2020 estimates.

The Census Bureau released these 2020 estimates for research purposes. The July 1, 2020 estimates are based on the 2010 Census and were created without incorporation or consideration of the 2020 Census results. They are typically used in comparisons with the 2020 Census to make determinations about the accuracy of the estimates. Differences between the estimates and census counts are interpreted as error in the estimates and are used to inform population and housing unit estimates research and methodological improvements over the decade.

## HOUSEHOLDS, Items 27–31

**Source: U.S. Census Bureau—American Community Survey, 2019 5-year Estimates**
http://www.census.gov/acs/www/

A **household** includes all of the persons who occupy a housing unit. (Persons not living in households are classified as living in group quarters.) A housing unit is a house, an apartment, a mobile home, a group of rooms, or a single room occupied (or, if vacant, intended for occupancy) as separate living quarters. Separate living quarters are those in which the occupants live separately from any other persons in the building and have direct access from the outside of the building or through a common hall. The occupants may be a single family, one person living alone, two or more families living together, or any other group of related or unrelated persons who share living quarters. The number of households is the same as the number of year-round occupied housing units.

A **family** includes a householder and one or more other persons living in the same household who are related to the householder by birth, marriage, or adoption. All persons in a household who are related to the householder are regarded as members of his or her family. A **family household** may contain persons not related to the householder; thus, family households may include more members than families do. A household can contain only one family for the purposes of census tabulations. Not all households contain families, as a household may comprise a group of unrelated persons or of one person living alone. Families are classified by type as either a "married-couple family" or "other family," according to the presence or absence of a spouse.

The measure of **persons per household** is obtained by dividing the number of persons in households by the number of households or householders. One person in each household is designated as the householder. In most cases, this is the person (or one of the persons) in whose name the house is owned, being bought, or rented. If there is no such person in the household, any adult household member 15 years old and over can be designated as the householder.

The category **female family householder** includes only female-headed family households with no spouse present.

A nonfamily household consists of a householder living alone or with nonrelatives only. Column 31 shows one-person households as a percentage of all households.

## GROUP QUARTERS, Item 32

**Source:**
**Source: U.S. Census Bureau—Population Estimates**
**https://www.census.gov/programs-surveys/popest.**
**html**

The Census Bureau classifies all persons not living in households as living in group quarters; this category includes both the institutional and noninstitutional populations. The institutionalized population includes persons under formally authorized, supervised care or custody in institutions, such as correctional institutions, nursing homes, mental (psychiatric) hospitals, and juvenile institutions. The noninstitutionalized population includes persons who live in group quarters other than institutions, such as college dormitories, military quarters, and group homes. This volume includes the total number of persons in group quarters.

## DAYTIME POPULATION, Items 33 and 34

**Source: U.S. Census Bureau—American Community**
**Survey, 2019 5-year Estimates**
**http://www.census.gov/acs/www/**

Daytime population refers to the number of persons who are present in an area or place during normal business hours, including workers. This can be contrasted with the "resident" population, which is present during the evening and nighttime hours. The daytime population estimate is calculated by adding the total resident population and the total workers working in the area/place, and then subtracting the total workers living in the area/place from that result. Information on the expansion or contraction experienced by different communities between their nighttime and daytime populations is important for many planning purposes, especially those concerning transportation, disaster, and relief operations.

The employment/residence ratio is a measure of the total number of workers working in an area or place, relative to the total number of workers living in the area or place. It is often used as a rough indication of the jobs-workers balance in an area/place, although it does not take into account whether the resident workers possess the skills needed for the jobs available in their particular area/place. The employment/residence ratio is calculated by dividing the number of total workers working in an area/place by the number of total workers residing in the area/place.

## BIRTHS AND DEATHS, Items 35–38

**Source: U.S. Census Bureau—Population Estimates**
**https://www.census.gov/programs-surveys/popest.**
**html**

The numbers of births and deaths are from the Census Bureau's Population Estimates Program. They represent the total number of live births and deaths occurring to residents of an area as estimated using reports from the National Center for Health Statistics (NCHS) and the Federal-State Cooperative for Population Estimates (FSCPE). The rates measure births and deaths during the specified time period as a proportion of an area's population. Rates are expressed per 1,000 population estimated as of July 1. These numbers and rates do not represent the calendar year, but rather the year-long period ending on July 1.

## PERSONS UNDER 65 WITH NO HEALTH INSURANCE, Items 39 and 40

**Source: U.S. Census Bureau—Small Area Health**
**Insurance Estimates**
**https://www.census.gov/programs-surveys/sahie.html**

The Small Area Health Insurance Estimates (SAHIE) program develops model-based estimates of health insurance coverage for counties and states. This developmental program builds on the work of the Small Area Income and Poverty Estimates (SAIPE) program. The SAHIE program models health insurance coverage by combining survey data with population estimates and administrative records. These estimates combine data from administrative records, postcensal population estimates, and the decennial census with direct estimates from the American Community Survey to provide consistent and reliable single-year estimates. These model-based single-year estimates are more reflective of current conditions than multi-year survey estimates.

## MEDICARE ENROLLMENT, Items 41—43

**Source: U.S. Department of Health and Human Services, Centers for Medicare and Medicaid Services**
**https://www.cms.gov/Research-Statistics-Data-and-Systems/Statistics-Trends-and-Reports/CMSProgramStatistics/Dashboard.html**

The Centers for Medicare and Medicaid Services (CMS) administers Medicare, which provides health insurance to persons 65 years old and over, persons with permanent kidney failure, and certain persons with disabilities. Original Medicare has two parts: Hospital Insurance and Supplemental Medical Insurance. In recent years, Medicare has been expanded to include two new programs: Medicare Advantage plans and prescription drug coverage. Medicare Advantage Plans are health plan options that are approved by Medicare but run by private companies. Medicare prescription drug plans can be part of Medicare Advantage plans or stand-alone drug plans.

Persons who are eligible for Medicare can enroll in Part A (Hospital Insurance) at no charge, and can choose to pay a monthly premium to enroll in Part B. Most eligible persons are enrolled in Part A, and most enrollees in Part A are also enrolled in Part B (Supplemental Medical Insurance.) This table includes persons who were enrolled in both Part A and Part B during 2017.

Part B beneficiaries can choose to enroll in Original Medicare, a fee-for-service plan administered by the Centers for Medicare and Medicaid Services, or in a Medicare Advantage plan. Medicare Advantage plans include private fee-for-service plans, preferred provider organizations, health maintenance organizations, medical savings account plans, demonstration plans, and programs for all-inclusive care for the elderly.

The annual Medicare enrollment counts are determined using a person-year methodology. For each calendar year, total person-year counts are determined by summing the total number of months that each beneficiary is enrolled during the year and dividing by 12. Using this methodology, a beneficiary's partial-year enrollment may be counted in more than one category (i.e., both Original Medicare and Medicare Advantage).

## LAW ENFORCEMENT, Items 44–47

### Source: U.S. Federal Bureau of Investigation—Uniform Crime Reports https://ucr.fbi.gov/crime-in-the-u.s/2019/crime-in-the-u.s.-2019

This table provides the volume of violent crime (murder and nonnegligent manslaughter, rape, robbery, and aggravated assault) and property crime (burglary, larceny-theft, and motor vehicle theft) as reported by law enforcement agencies (such as individual sheriffs' offices and/or county police departments) in counties that contributed data to the UCR Program. Arson is not included in the property crime total in this table. These data do not represent county totals as they exclude crime counts for city or town agencies and other types of agencies that have jurisdiction within each county. This table includes jurisdictions covered by noncity law enforcement agencies located within the county, such as a county police department or a Sherriff's office.

Also included are the total numbers of law enforcement employees—total officers, and total civilians—employed by county agencies.

Crime data are as reported to the Federal Bureau of Investigation (FBI) by law enforcement agencies and have not been adjusted for underreporting or overreporting. This may affect comparability between geographic areas or over time.

Through the voluntary contribution of crime statistics by law enforcement agencies across the United States, the Uniform Crime Reporting (UCR) Program provides periodic assessments of crime in the nation as measured by offenses that have come to the attention of the law enforcement community. The Committee on Uniform Crime Records of the International Association of Chiefs of Police initiated this voluntary national data collection effort in 1930. The UCR Program contributors compile and submit their crime data either directly to the FBI or through state-level UCR Programs.

**Violent crimes** include four categories of offenses: (1) Murder and nonnegligent manslaughter, as defined in the UCR Program, is the willful (nonnegligent) killing of one human being by another. This offense excludes deaths caused by negligence, suicide, or accident; justifiable homicides; and attempts to murder or assaults to murder. (2) Rape is the penetration, no matter how slight, of the vagina or anus with any body part or object, or oral penetration by a sex organ of another person, without the consent of the victim. Assaults or attempts to commit rape by force or threat of force are also included; however, statutory rape (without force) and other sex offenses are excluded. (3) Robbery is the taking or attempting to take anything of value from the care, custody, or control of a person or persons by force or threat of force or violence and/or by putting the victim in fear. (4) Aggravated assault is an unlawful attack by one person upon another for the purpose of inflicting severe or aggravated bodily injury. This type of assault is usually accompanied by the use of a weapon or by other means likely to produce death or great bodily harm. Attempts are included, since injury does not necessarily have to result when a gun, knife, or other weapon is used, as these incidents could and probably would result in a serious personal injury if the crime were successfully completed.

**Property crimes** include three categories: (1) Burglary, or breaking and entering, is the unlawful entry of a structure to commit a felony or theft, even though no force was used to gain entrance. (2) Larceny-theft is the unauthorized taking of the personal property of another, without the use of force. (3) Motor vehicle theft is the unauthorized taking of any motor vehicle.

The information on **law enforcement employees** is derived from law enforcement employee counts (as of October 31, 2018) submitted by participating agencies. The UCR Program defines law enforcement **officers** as individuals who ordinarily carry a firearm and a badge, have full arrest powers, and are paid from governmental funds set aside specifically to pay sworn law enforcement. **Civilian** employees include full-time agency personnel such as clerks, radio dispatchers, meter attendants, stenographers, jailers, correctional officers, and mechanics.

## EDUCATION—SCHOOL ENROLLMENT AND EDUCATIONAL ATTAINMENT, Items 48–51

### Source: U.S. Census Bureau—American Community Survey, 2019 5-year Estimates http://www.census.gov/acs/www/

Data on school enrollment are tabulated for the population 3 years old and over. Persons were classified as enrolled in school if they reported attending a "regular" public or private school (or college) during the three months preceding the interview. The instructions were to include only nursery school, kindergarten, elementary school, and schooling which would lead to a high school diploma or a college degree as regular school. The Census

Bureau defines a public school as "any school or college controlled and supported by a local, county, state, or federal government." Schools primarily supported and controlled by religious organizations or other private groups are defined as private schools.

Data on **educational attainment** are tabulated for the population 25 years old and over. The data were derived from a question that asked respondents for the highest level of school completed or the highest degree received. Persons who had passed a high school equivalency examination were considered high school graduates. Schooling received in foreign schools was to be reported as the equivalent grade or years in the regular American school system.

Vocational and technical training, such as barber school training; business, trade, technical, and vocational schools; or other training for a specific trade are specifically excluded.

**High school graduate or less**. This category includes persons whose highest degree was a high school diploma or its equivalent, and those who reported any level lower than a high school diploma.

Bachelor's degree or more. This category includes persons who have received bachelor's degrees, master's degrees, professional school degrees (such as law school or medical school degrees), and doctoral degrees.

# LOCAL GOVERNMENT EDUCATION EXPENDITURES, Items 52 and 53

### Source: U.S. Department of Education, National Center for Education Statistics, Common Core of Data (CCD), "Local Education Agency (School District) Finance Survey (F33) Data", 2017-2018 v.1a
### https://nces.ed.gov/ccd/

Expenditures are for elementary and secondary education which includes prekindergarten through twelfth grade regular, special, and vocational education, as well as cocurricular, community service, and adult education programs provided by a public school system, including charter schools. The financial activities of these systems for all instruction, support service, and noninstructional activities are included

Current spending comprises current operation expenditure, payments made by the state government on behalf of school systems, and transfers made by school systems into their own retirement funds. Current operation expenditures include direct expenditure for salaries, employee benefits, purchased professional and technical services, purchased property and other services, and supplies. It includes gross school system expenditure for instruction, support services, and noninstructional functions. It excludes expenditure for debt service, capital outlay, and reimbursement to other governments (including other school systems).

Current expenditure per student is current expenditure divided by the number of students enrolled. The number of students enrolled is based on an annual "membership" count of students on or about October 1, collected by the National Center for Education Statistics on the Common Core of Data (CCD) agency

universe file—"Local Education Agency (School District) Universe Survey."

# MONEY INCOME, Items 54–57

### Source: U.S. Census Bureau—American Community Survey, 2019 5-year Estimates
### http://www.census.gov/acs/www/

Total money income is the sum of the amounts reported separately for wage or salary income; net self-employment income; interest, dividends, or net rental or royalty income or income from estates and trusts; Social Security or railroad retirement income; Supplemental Security Income (SSI); public assistance or welfare payments; retirement, survivor, or disability pensions; and all other income. Receipts from the following sources are not included as income: capital gains; money received from the sale of property (unless the recipient was engaged in the business of selling such property); the value of income "in kind" from food stamps, public housing subsidies, medical care, employer contributions for individuals, etc.; withdrawal of bank deposits; money borrowed; tax refunds; exchange of money between relatives living in the same household; and gifts, lump-sum inheritances, insurance payments, and other types of lump-sum receipts.

Money income differs in definition from personal income (item 62). For example, money income does not include the pension rights, employer provided health insurance, food stamps, or Medicare payments that are included in personal income.

Per capita income is the mean income computed for every man, woman, and child in a particular group. It is derived by dividing the aggregate income of a particular group by the resident population in that group as estimated in the American Community Survey. Per capita income is rounded to the nearest whole dollar.

**Household income** includes the income of the householder and all other individuals 15 years old and over in the household, whether or not they are related to the householder. Since many households consist of only one person, median household income is usually less than median family income. Although the household income statistics cover the 12 months preceding the interview, the characteristics of individuals and the composition of households refer to the date of interview. Thus, the income of the household does not include amounts received by individuals who were no longer residing in the household at the time of interview. Similarly, income amounts reported by individuals who did not reside in the household during all of the past 12 months but who were members of the household at the time of interview are included. However, the composition of most households was the same during those 12 months as it was at the time of interview.

**Median income** divides the income distribution into two equal parts, with half of all cases below the median income level and half of all cases above the median income level. For households, the median income is based on the distribution of the total number of households, including those with no income. Median income for households is computed on the basis of a standard distribution with a minimum value of less than $2,500 and a maximum value of $200,000 or more and is rounded to the nearest whole dollar. Median income figures are calculated using linear interpolation if

**Poverty Thresholds for 2019 by Size of Family and Number of Related Children Under 18 Years**

| Size of family unit | Weighted average thresholds | Related children under 18 years | | | | | | | | |
|---|---|---|---|---|---|---|---|---|---|---|
| | | None | One | Two | Three | Four | Five | Six | Seven | Eight or more |
| One person (unrelated individual): | 13,011 | | | | | | | | | |
| Under age 65 ............................. | 13,300 | 13,300 | | | | | | | | |
| Aged 65 and older.................... | 12,261 | 12,261 | | | | | | | | |
| | | | | | | | | | | |
| Two people: ................................. | 16,521 | | | | | | | | | |
| Householder under age 65........ | 17,196 | 17,120 | 17,622 | | | | | | | |
| Householder aged 65 and older ........................................ | 15,468 | 15,453 | 17,555 | | | | | | | |
| | | | | | | | | | | |
| Three people............................. | 20,335 | 19,998 | 20,578 | 20,598 | | | | | | |
| Four people................................ | 26,172 | 26,370 | 26,801 | 25,926 | 26,017 | | | | | |
| Five people ................................ | 31,021 | 31,800 | 32,263 | 31,275 | 30,510 | 30,044 | | | | |
| Six people ................................. | 35,129 | 36,576 | 36,721 | 35,965 | 35,239 | 34,161 | 33,522 | | | |
| Seven people............................. | 40,016 | 42,085 | 42,348 | 41,442 | 40,811 | 39,635 | 38,262 | 36,757 | | |
| Eight people............................... | 44,461 | 47,069 | 47,485 | 46,630 | 45,881 | 44,818 | 43,470 | 42,066 | 41,709 | |
| Nine people or more .................. | 52,875 | 56,621 | 56,895 | 56,139 | 55,503 | 54,460 | 53,025 | 51,727 | 51,406 | 49,426 |

Source: U.S. Census Bureau.

the width of the interval containing the estimate is $2,500 or less. If the width of the interval containing the estimate is greater than $2,500, Pareto interpolation is used.

Income amounts have been adjusted for inflation to represent the final year of multi-year estimates, in this case 2015-2019 estimates. The constant-dollar figures are based on an annual average Consumer Price Index from the Bureau of Labor Statistics. Constant-dollar figures are estimates representing an effort to remove the effects of price changes from statistical series reported in dollar terms. However, the estimates do not reflect the price and cost-of-living differences that may exist between areas.

## INCOME AND POVERTY, Items 58–61
**Source: U.S. Census Bureau—Small Area Income and Poverty Estimates Program**
https://www.census.gov/programs-surveys/saipe.html

The annual income and poverty estimates by county are constructed from statistical models that relate income and poverty to indicators based on summary data from federal income tax returns, data about participation in the Food Stamp program, and the previous census. A regression model predicts the number of people in poverty using county-level observations from the current year's American Community Survey (ACS) and administrative records and census data as the predictors. The 2005 estimates were the first to use the ACS. Prior year models were based on the Annual Social and Economic Supplement (ASEC) of the Current Population Survey (CPS). The ACS is a much larger survey than the ASEC, permitting income and poverty estimates based on a single year for many counties, Because of the differences between the two surveys, caution should be used when comparing these estimates with those from earlier years.

The **poverty status** data were derived from data collected on the number of persons in a household, each person's relationship to the householder, and income data. The Social Security Administration (SSA) developed the original poverty definition in 1964, which federal interagency committees subsequently revised in 1969 and 1980. The Office of Management and Budget's (OMB) *Directive 14* prescribes the SSA's definition as the official poverty measure for federal agencies to use in their statistical work. Poverty statistics presented in American Community Survey products adhere to the standards defined by OMB in *Directive 14*.

Poverty thresholds vary depending on three criteria: size of family, number of children, and, for one- and two-person families, age of householder. In determining the poverty status of families and unrelated individuals, the Census Bureau uses thresholds (income cutoffs) arranged in a two-dimensional matrix. The matrix consists of family size (from one person to nine or more persons), cross-classified by presence and number of family members under 18 years old (from no children present to eight or more children present). Unrelated individuals and two-person families are further differentiated by age of reference person (under 65 years old and 65 years old and over). To determine a person's poverty status, the person's total family income over the previous 12 months is compared with the poverty threshold appropriate for that person's family size and composition. If the total income of that person's family is less than the threshold appropriate for that family, then the person is considered poor or "below the poverty level," together with every member of his or her family. If a person is not living with anyone related by birth, marriage, or adoption, then the person's own income is compared with his or her poverty threshold. The total number of persons below the poverty level is the sum of persons in families and the number of unrelated individuals with incomes below the poverty level over the previous 12 months.

# PERSONAL INCOME AND EARNINGS, Items 62–83

**Source: U.S. Bureau of Economic Analysis, Regional Economic Accounts**

http://www.bea.gov/regional/index.htm#state
http://www.bea.gov/data/income-saving/
personal-income-county-metro-and-other-areas

Total personal income is the current income received by residents of an area from all sources. It is measured before deductions of income and other personal taxes, but after deductions of personal contributions for Social Security, government retirement, and other social insurance programs. It consists of wage and salary disbursements (covering all employee earnings, including executive salaries, bonuses, commissions, payments-in-kind, incentive payments, and tips); various types of supplementary earnings, such as employers' contributions to pension funds (termed "other labor income" or "supplements to wages and salaries"); proprietors' income; rental income of persons; dividends; personal interest income; and government and business transfer payments.

Per capita personal income is based on the resident population estimated as of July 1 of the year shown.

Proprietors' income is the monetary income and income-in-kind of proprietorships and partnerships (including the independent professions) and the income of tax-exempt cooperatives.

Dividends are cash payments by corporations to stockholders who are U.S. residents. Interest is the monetary and imputed interest income of persons from all sources. Rent is the monetary income of persons from the rental of real property, except the income of persons primarily engaged in the real estate business; the imputed net rental income of owner-occupants of nonfarm dwellings; and the royalties received by persons.

Transfer payments are income for which services are not currently rendered. They consist of both government and business transfer payments. Government transfer payments include payments under the following programs: Federal Old-Age, Survivors, and Disability Insurance ("Social Security"); Medicare and medical vendor payments; unemployment insurance; railroad and government retirement; federal- and state-government-insured workers' compensation; veterans' benefits, including veterans' life insurance; SNAP (Supplemental Nutrition Assistance Program, or food stamps); black lung payments; Supplemental Security Income; and Temporary Assistance for Needy Families. Government payments to nonprofit institutions, other than for work under research and development contracts, are also included. Business transfer payments consist primarily of liability payments for personal injury and of corporate gifts to nonprofit institutions.

Personal income differs in definition from money income (items 54–57). For example, personal income includes pension rights, employer-provided health insurance, food stamps, and Medicare. These are not included in the definition of money income.

Earnings cover wage and salary disbursements, other labor income, and proprietors' income.

The data for earnings obtained from the Bureau of Economic Analysis (BEA) are based on place of work. In computing personal income, BEA makes an "adjustment for residence" to earnings, based on commuting patterns; personal income is thus presented on a place-of-residence basis.

Farm earnings include the income of farm workers (wages and salaries and other labor income) and farm proprietors. Farm proprietors' income includes only the income of sole proprietorships and partnerships. Farm earnings estimates are benchmarked to data collected in the Census of Agriculture and the revised Department of Agriculture statistical totals of income and expense items.

Goods-related industries include mining, construction, and manufacturing.

**Mining, Quarrying, and Extracting** comprises establishments that extract naturally occurring mineral solids, such as coal and ores; liquid minerals, such as crude petroleum; and gases, such as natural gas. The term mining is used in the broad sense to include quarrying, well operations, beneficiating (e.g., crushing, screening, washing, and flotation), and other preparation customarily performed at the mine site, or as a part of mining activity.

The **Construction** sector comprises establishments primarily engaged in the construction of buildings or engineering projects (e.g., highways and utility systems). Establishments primarily engaged in the preparation of sites for new construction and establishments primarily engaged in subdividing land for sale as building sites also are included in this sector.

**Manufacturing** comprises establishments engaged in the mechanical, physical, or chemical transformation of materials, substances, or components into new products. The assembling of component parts of manufactured products is considered manufacturing, except in cases where the activity is appropriately classified in **Construction**.

Service-related and other industries include private-sector earnings in agricultural services, forestry, and fisheries; transportation and public utilities; wholesale trade; retail trade; finance, insurance, and real estate; and services. Government earnings include all levels of government.

The Information sector comprises establishments engaged in the following processes: (a) producing and distributing information and cultural products, (b) providing the means to transmit or distribute these products as well as data or communications, and (c) processing data.

Professional, Scientific, and Technical Services comprises establishments that specialize in performing professional, scientific, and technical activities for others. These activities require a high degree of expertise and training. The establishments in this sector specialize according to expertise and provide these services to clients in a variety of industries and, in some cases, to households. Activities performed include: legal advice and representation; accounting, bookkeeping, and payroll services; architectural, engineering, and specialized design services; computer services; consulting services; research services; advertising services; photographic services; translation and interpretation services; veterinary services; and other professional, scientific, and technical services.

The **Retail Trade** sector comprises establishments engaged in retailing merchandise, generally without transformation, and rendering services incidental to the sale of merchandise.

The retailing process is the final step in the distribution of merchandise; retailers are, therefore, organized to sell merchandise in small quantities to the general public. This sector comprises two main types of retailers: store and nonstore retailers.

Store retailers operate fixed point-of-sale locations, located and designed to attract a high volume of walk-in customers. In general, retail stores have extensive displays of merchandise and use mass-media advertising to attract customers. In addition to retailing merchandise, some types of store retailers are also engaged in the provision of after-sales services, such as repair and installation.

Nonstore retailers, like store retailers, are organized to serve the general public, but their retailing methods differ. The establishments of this subsector reach customers and market merchandise with methods, such as the broadcasting of "infomercials," the broadcasting and publishing of direct-response advertising, the publishing of paper and electronic catalogs, door-to-door solicitation, in-home demonstration, selling from portable stalls (street vendors, except food), and distribution through vending machines. Establishments engaged in the direct sale (nonstore) of products, such as home heating oil dealers and home delivery newspaper routes are included here.

Finance and Insurance comprises establishments primarily engaged in financial transactions (transactions involving the creation, liquidation, or change in ownership of financial assets) and/or in facilitating financial transactions. Three principal types of activities are identified: raising funds by taking deposits and/or issuing securities and, in the process, incurring liabilities; pooling of risk by underwriting insurance and annuities; and providing specialized services facilitating or supporting financial intermediation, insurance, and employee benefit programs.

The Real Estate and Rental and Leasing sector comprises establishments primarily engaged in renting, leasing, or otherwise allowing the use of tangible or intangible assets, and establishments providing related services. The major portion of this sector comprises establishments that rent, lease, or otherwise allow the use of their own assets by others. The assets may be tangible, as is the case of real estate and equipment, or intangible, as is the case with patents and trademarks. This sector also includes establishments primarily engaged in managing real estate for others, selling, renting and/or buying real estate for others, and appraising real estate.

Health Care and Social Assistance comprises establishments providing health care and social assistance for individuals. The sector includes both health care and social assistance because it is sometimes difficult to distinguish between the boundaries of these two activities. The industries in this sector are arranged on a continuum starting with those establishments providing medical care exclusively, continuing with those providing health care and social assistance, and finally finishing with those providing only social assistance. The services provided by establishments in this sector are delivered by trained professionals. All industries in the sector share this commonality of process, namely, labor inputs of health practitioners or social workers with the requisite expertise. Many of the industries in the sector are defined based on the educational degree held by the practitioners included in the industry.

Government includes the executive, legislative, judicial, administrative and regulatory activities of Federal, state, local, and international governments. Also included are Government enterprises: government agencies that cover a substantial portion of their operating costs by selling goods and services to the public and that maintain separate accounts.

Industries are categorized under the North American Industry Classification System (NAICS), and are not comparable to years prior to 2002.

## SOCIAL SECURITY AND SUPPLEMENTAL SECURITY INCOME, Items 84–86

**Source: U.S. Social Security Administration**
http://www.ssa.gov/policy/docs/statcomps/oasdi_sc/
http://www.ssa.gov/policy/docs/statcomps/ssi_sc/

Social Security beneficiaries are persons receiving benefits under the Old-Age, Survivors, and Disability Insurance Program. These include retired or disabled workers covered by the program, their spouses and dependent children, and the surviving spouses and dependent children of deceased workers.

Supplemental Security Income (SSI) recipients are persons receiving SSI payments. The SSI program is a cash assistance program that provides monthly benefits to low-income aged, blind, or disabled persons.

Data are as of December of the year shown.

## HOUSING, Items 87–96

**Source: U.S. Census Bureau—Population Estimates Program**
http://www.census.gov/popest/
**Source: U.S. Census Bureau—American Community Survey, 2019 5-year Estimates**
http://www.census.gov/acs/www/

Housing data for 2020 are from the Population Estimates Program. Housing unit characteristics for 2015-2019 are from the American Community Survey.

A housing unit is a house, apartment, mobile home or trailer, group of rooms, or single room occupied or, if vacant, intended for occupancy as separate living quarters. Separate living quarters are those in which the occupants do not live and eat with any other person in the structure and which have direct access from the outside of the building or through a common hall.

The occupants of a housing unit may be a single family, one person living alone, two or more families living together, or any other group of related or unrelated persons who share living quarters. Both occupied and vacant housing units are included in the housing inventory, although recreational vehicles, tents, caves, boats, railroad cars, and the like are included only if they are occupied as a person's usual place of residence.

A housing unit is classified as **occupied** if it is the usual place of residence of the person or group of persons living in it at the time of enumeration, or if the occupants are only temporarily absent (away on vacation). A household consists of all persons who occupy a housing unit as their usual place of residence. Vacant units for sale or rent include units rented or sold but not occupied and any other units held off the market.

**Median value** is the dollar amount that divides the distribution of specified owner-occupied housing units into two equal parts, with half of all units below the median value and half of all units above the median value. Value is defined as the respondent's estimate of what the house would sell for if it were for sale. Data are presented for single-family units on fewer than 10 acres of land that have no business or medical offices on the property.

**Median rent** divides the distribution of renter-occupied housing units into two equal parts. The rent concept used in this volume is gross rent, which includes the amount of cash rent a renter pays (contract rent) plus the estimated average cost of utilities and fuels, if these are paid by the renter. The rent is the amount of rent only for living quarters and excludes amounts paid for any business or other space occupied. Single-family houses on lots of 10 or more acres of land are also excluded.

**Housing cost** as a percentage of income is shown separately for owners with mortgages, owners without mortgages, and renters. Rent as a percentage of income is a computed ratio of gross rent and monthly household income (total household income divided by 12). Selected owner costs include utilities and fuels, mortgage payments, insurance, taxes, etc. In each case, the ratio of housing cost to income is computed separately for each housing unit. The housing cost ratios for half of all units are above the median shown in this book, and half are below the median shown in the book.

**Substandard units** are occupied units that are overcrowded or lack complete plumbing facilities. For the purposes of this item, "overcrowded" is defined as having 1.01 persons or more per room. Complete plumbing facilities include hot and cold piped water, a flush toilet, and a bathtub or shower. These facilities must be located inside the housing unit, but do not have to be in the same room.

## CIVILIAN LABOR FORCE AND UNEMPLOY-MENT, Items 97–100

**Source: U.S. Bureau of Labor Statistics—Local Area Unemployment Statistics**
**http://www.bls.gov/lau/#tables**

Data for the civilian labor force are the product of a federal-state cooperative program in which state employment security agencies prepare labor force and unemployment estimates under concepts, definitions, and technical procedures established by the Bureau of Labor Statistics (BLS). The civilian labor force consists of all civilians 16 years old and over who are either employed or unemployed.

Unemployment includes all persons who did not work during the survey week, made specific efforts to find a job during the previous four weeks, and were available for work during the survey week (except for temporary illness). Persons waiting to be called back to a job from which they had been laid off and those waiting to report to a new job within the next 30 days are included in unemployment figures.

Table B includes annual average data for the year shown. The Local Area Unemployment Statistics data are periodically updated to reflect revised inputs, re-estimation, and controlling to new statewide totals.

## CIVILIAN EMPLOYMENT, Items 101–103

**Source: U.S. Census Bureau—American Community Survey, 2019 5-year Estimates**
**http://www.census.gov/acs/www/**

**Total employment** includes all civilians 16 years old and over who were either (1) "at work"—those who did any work at all during the reference week as paid employees, worked in either their own business or profession, worked on their own farm, or worked 15 hours or more as unpaid workers in a family farm or business; or were (2) "with a job, but not at work" —those who had a job but were not at work that week due to illness, weather, industrial dispute, vacation, or other personal reasons.

The **occupational categories** are based on the occupational classification system that was developed for the 2000 census and revised in 2002 and 2010. This system consists of 539 specific occupational categories for employed persons arranged into 23 major occupational groups. This classification was developed based on the *Standard Occupational Classification (SOC) Manual: 2000*, published by the Executive Office of the President, Office of Management and Budget.

Column 102 includes the Management, business, science, and arts occupations category while Column 103 combines the Natural resources, construction, and maintenance occupations and the Production, transportation, and material moving occupations.

## PRIVATE NONFARM EMPLOYMENT AND EARNINGS, Items 104–112

**Source: U.S. Census Bureau—County Business Patterns**
**https://www.census.gov/programs-surveys/cbp.html**

Data for private nonfarm employment and earnings are compiled from the payroll information reported monthly in the Census Bureau publication *County Business Patterns*. The estimates are based on surveys conducted by the Census Bureau and administrative records from the Internal Revenue Service (IRS).

The following types of employment are excluded from the tables: government employment, self-employed persons, farm workers, and domestic service workers. Railroad employment jointly covered by Social Security and railroad retirement programs, employment on oceanborne vessels, and employment in foreign countries are also excluded.

Annual payroll is the combined amount of wages paid, tips reported, and other compensation (including salaries, vacation allowances, bonuses, commissions, sick-leave pay, and the value of payments-in-kind such as free meals and lodging) paid to employees before deductions for Social Security, income tax, insurance, union dues, etc. All forms of compensation are included, regardless of whether they are subject to income tax or the Federal Insurance Contributions Act tax, with the exception of annuities, third-party sick pay, and supplemental unemployment compensation benefits (even if income tax was withheld). For corporations, total annual payroll includes compensation paid to officers and executives; for unincorporated businesses, it excludes profit or other compensation of proprietors or partners.

# AGRICULTURE, ITEMS 113–132

Source: U.S. Department of Agriculture, National Agricultural Statistics Service—2017 Census of Agriculture
https://www.nass.usda.gov/Publications/AgCensus/2017/index.php

Data for the 2017 Census of Agriculture were collected in 2018, but pertain to the year 2017.

The Census Bureau took a census of agriculture every 10 years from 1840 to 1920; since 1925, this census has been taken roughly once every 5 years. The 1997 Census of Agriculture was the first one conducted by the National Agricultural Statistics Service of the U.S. Department of Agriculture. Over time, the definition of a farm has varied. For recent censuses (including the 2017 census), a farm has been defined as any place from which $1,000 or more of agricultural products were produced and sold or normally would have been sold during the census year. Dollar figures are expressed in current dollars and have not been adjusted for inflation or deflation.

The term **producer** designates a person who is involved in making decisions for the farm operation. Decisions may include decisions about such things as planting, harvesting, livestock management, and marketing. The producer may be the owner, a member of the owner's household, a hired manager, a tenant, a renter, or a sharecropper. If a person rents land to others or has land worked on shares by others, he/she is considered the producer only of the land which is retained for his/her own operation. The census collected information on the total number of male producers, the total number of female producers, and demographic information for up to four producers per farm.

The acreage designated as **land in farms** consists primarily of agricultural land used for crops, pasture, or grazing. It also includes woodland and wasteland not actually under cultivation or used for pasture or grazing, provided that this land was part of the farm operator's total operation. Land in farms is an operating-unit concept and includes all land owned and operated, as well as all land rented from others. Land used rent-free is classified as land rented from others. All grazing land, except land used under government permits on a per-head basis, was included as "land in farms" provided it was part of a farm or ranch. Land under the exclusive use of a grazing association was reported by the grazing association and included as land in farms. All land in Indian reservations used for growing crops or grazing livestock is classified as land in farms.

**Irrigated land** includes all land watered by any artificial or controlled means, such as sprinklers, flooding, furrows or ditches, sub-irrigation, and spreader dikes. Included are supplemental, partial, and preplant irrigation. Each acre was counted only once regardless of the number of times it was irrigated or harvested. Livestock lagoon waste water distributed by sprinkler or flood systems was also included.

**Total cropland** includes cropland harvested, cropland used only for pasture or grazing, cropland on which all crops failed or were abandoned, cropland in cultivated summer fallow, and cropland idle or used for cover crops or soil improvement but not harvested and not pastured or grazed.

Respondents were asked to report their estimate of the current market **value of land and buildings** owned, rented, or leased from others, and rented and leased to others. Market value refers to the respondent's estimate of what the land and buildings would sell for under current market conditions. If the value of land and buildings was not reported, it was estimated during processing by using the average value of land and buildings from similar farms in the same geographic area.

The **value of machinery and equipment** was estimated by the respondent as the current market value of all cars, trucks, tractors, combines, balers, irrigation equipment, etc., used on the farm. This value is an estimate of what the machinery and equipment would sell for in its present condition and not the replacement or depreciated value. Share interests are reported at full value at the farm where the equipment and machinery are usually kept. Only equipment that was physically located at the farm on December 31, 2017, is included.

**Market value of agricultural products sold** by farms represents the gross market value before taxes and the production expenses of all agricultural products sold or removed from the place in 2017, regardless of who received the payment. It is equivalent to total sales and it includes sales by producers as well as the value of any share received by partners, landlords, contractors, and others associated with the operation. It includes value of organic sales, direct sales and the value of commodities placed in the Commodity Credit Corporation (CCC) loan program. Market value of agricultural products sold does not include payments received for participation in other federal farm programs. Also, it does not include income from farm-related sources such as customwork and other agricultural services, or income from nonfarm sources.

**Organic farms** are those that had organic production according to USDA's National Organic Program (NOP). Respondents reported whether their organic production was certified or exempt from certification and the sales from NOP produced commodities. Not included are farms that had acres transitioning into NOP production.

Farms with **internet access** are those that reported using personal computers, laptops, or mobile devices (e.g., cell phones or tablets) to access the internet. This can be done using services such as dial-up, DSL, cable modem, fiber-optic, mobile internet service for a cell phone or other device (tablet), satellite, or other methods. In 2017 respondents were also able to report connecting with an unknown service type, labeled as "Don't know" in the publication tables.

**Government payments** consist of direct payments as defined by the 2002 Farm Bill; payments from Conservation Reserve Program (CRP), Wetlands Reserve Program (WRP), Farmable Wetlands Program (FWP), and Conservation Reserve Enhancement Program (CREP); loan deficiency payments; disaster payments; other conservation programs; and all other federal farm programs under which payments were made directly to farm producers, including those specified in the 2014 Agricultural Act (Farm Bill), including Agriculture Risk Coverage (ARC) and Price Loss Coverage (PLC). Commodity Credit Corporation (CCC) proceeds, amount from State and local government agricultural program payments, and federal crop insurance payments were not included in this category.

## WATER CONSUMPTION, Items 133–134

**Source: U.S. Geological Survey, National Water-Use Information Program, Estimated Use of Water in the United States County-Level Data for 2015, version 1.0**
**http://water.usgs.gov/watuse/**

Every five years, the U.S. Geological Survey compiles county-level water-use estimates. This volume includes the total fresh and saline withdrawals for public water supplies in 2015, expressed as million gallons per day. Estimate of withdrawals of ground and surface water are given for the following categories of use: public water supplies, domestic, commercial, irrigation, livestock, industrial, mining, and thermoelectric power. Only public water supply is included in this volume because the other categories had not been published in time for this book. The number of gallons withdrawn per person is based on the county population in 2015. Public supply refers to water withdrawn from ground and surface sources by public and private water systems for use by cities, towns, rural water districts, mobile-home parks, Native American Indian reservations, and military bases. Public-supply facilities provide water to at least 25 persons or have a minimum of 15 service connections. Water withdrawn by public suppliers may be delivered to users for domestic, commercial, industrial, and thermoelectric-power purposes, as well as to other public-water suppliers. Public-supply water is also used for public services (public uses)—such as pools, parks, and public buildings—and may have unaccounted uses (losses) because of system leaks or such non-metered services as firefighting, flushing of water lines, or backwashing at treatment plants. Some public-supply water may be used in the processes of water and wastewater treatment. Some public suppliers treat saline water before distributing the water. The definition of saline water for public supply refers to water that requires treatment to reduce the concentration of dissolved solids through the process of desalination or dilution.

## 2017 ECONOMIC CENSUS: OVERVIEW, Items 135–166

**Source: U.S. Census Bureau**
**https://www.census.gov/programs-surveys/economic-census.html**

The Economic Census provides a detailed portrait of the nation's economy, from the national to the local level, once every five years. The 2017 Economic Census covers nearly all of the U.S. economy in its basic collection of establishment statistics. The 1997 Economic Census was the first major data source to use the new North American Industry Classification System (NAICS); therefore, data from this census are not comparable to economic data from prior years, which were based on the Standard Industrial Classification (SIC) system.

NAICS, developed in cooperation with Canada and Mexico, classifies North America's economic activities at two-, three-, four-, and five-digit levels of detail; the U.S. version of NAICS further defines industries to a sixth digit. The Economic Census takes advantage of this hierarchy to publish data at these successive levels of detail: sector (two-digit), subsector (three-digit), industry group (four-digit), industry (five-digit), and U.S. industry (six-digit). Information in Table A is at the two-digit level, with a few three- and four-digit items. The data in Table B are at the two-digit level.

Several key statistics are tabulated for all industries in this volume, including number of establishments (or companies), number of employees, payroll, and certain measures of output (sales, receipts, revenue, value of shipments, or value of construction work done).

**Number of establishments.** An establishment is a single physical location at which business is conducted. It is not necessarily identical with a company or enterprise, which may consist of one establishment or more. Economic Census figures represent a summary of reports for individual establishments rather than companies. For cases in which a census report was received, separate information was obtained for each location where business was conducted. When administrative records of other federal agencies were used instead of a census report, no information was available on the number of locations operated. Each Economic Census establishment was tabulated according to the physical location at which the business was conducted. The count of establishments represents those in business at any time during 2017.

When two activities or more were carried on at a single location under a single ownership, all activities were generally grouped together as a single establishment. The entire establishment was classified on the basis of its major activity and all of its data were included in that classification. However, when distinct and separate economic activities (for which different industry classification codes were appropriate) were conducted at a single location under a single ownership, separate establishment reports for each of the different activities were obtained in the census.

**Number of employees.** Paid employees consist of the full-time and part-time employees, including salaried officers and executives of corporations. Included are employees on paid sick leave, paid holidays, and paid vacations; not included are proprietors and partners of unincorporated businesses. The definition of paid employees is the same as that used by the Internal Revenue Service (IRS) on form 941.

For some industries, the Economic Census gives codes representing the number of employees as a range of numbers (for example, "100 to 249 employees" or "1,000 to 2,499" employees). In this volume, those codes have been replaced by the standard suppression code "D".

**Payroll.** Payroll includes all forms of compensation, such as salaries, wages, commissions, dismissal pay, bonuses, vacation allowances, sick-leave pay, and employee contributions to qualified pension plans paid during the year to all employees. For corporations, payroll includes amounts paid to officers and executives; for unincorporated businesses, it does not include profit or other compensation of proprietors or partners. Payroll is reported before deductions for Social Security, income tax, insurance, union dues, etc. This definition of payroll is the same as that used by on IRS form 941.

**Sales, shipments, receipts, revenue, or business done.** This measure includes the total sales, shipments, receipts, revenue, or business done by establishments within the scope of the Economic Census. The definition of each of these items is specific to the economic sector measured.

## WHOLESALE TRADE, Items 135–138
### Source: U.S. Census Bureau—2017 Economic Census (See Overview of 2017 Economic Census prior to Item 135)

The Wholesale Trade sector (sector 42) comprises establishments engaged in wholesaling merchandise, generally without transformation, and rendering services incidental to the sale of merchandise. The wholesaling process is an intermediate step in the distribution of merchandise.

Wholesalers are organized to sell or arrange the purchase or sale of (1) goods for resale (i.e., goods sold to other wholesalers or retailers), (2) capital or durable nonconsumer goods, and (3) raw and intermediate materials and supplies used in production.

Wholesalers sell merchandise to other businesses and normally operate from a warehouse or office. These warehouses and offices are characterized by having little or no display of merchandise. In addition, neither the design nor the location of the premises is intended to solicit walk-in traffic. Wholesalers do not normally use advertising directed to the general public. In general, customers are initially reached via telephone, in-person marketing, or specialized advertising, which may include the internet and other electronic means. Follow-up orders are either vendor-initiated or client-initiated, are usually based on previous sales, and typically exhibit strong ties between sellers and buyers. In fact, transactions are often conducted between wholesalers and clients that have long-standing business relationships.

This sector is made up of two main types of wholesalers: those that sell goods on their own account and those that arrange sales and purchases for others for a commission or fee.

(1) Establishments that sell goods on their own account are known as wholesale merchants, distributors, jobbers, drop shippers, import/export merchants, and sales branches. These establishments typically maintain their own warehouse, where they receive and handle goods for their customers. Goods are generally sold without transformation, but may include integral functions, such as sorting, packaging, labeling, and other marketing services.

(2) Establishments arranging for the purchase or sale of goods owned by others or purchasing goods on a commission basis are known as agents and brokers, commission merchants, import/export agents and brokers, auction companies, and manufacturers' representatives. These establishments operate from offices and generally do not own or handle the goods they sell.

Some wholesale establishments may be connected with a single manufacturer and promote and sell that particular manufacturer's products to a wide range of other wholesalers or retailers. Other wholesalers may be connected to a retail chain or a limited number of retail chains and only provide the products needed by the particular retail operation(s). These wholesalers may obtain the products from a wide range of manufacturers. Still other wholesalers may not take title to the goods, but act instead as agents and brokers for a commission.

Although wholesaling normally denotes sales in large volumes, durable nonconsumer goods may be sold in single units. Sales of capital or durable nonconsumer goods used in the production of goods and services, such as farm machinery, medium- and heavy-duty trucks, and industrial machinery, are always included in Wholesale Trade.

The county table includes only **Merchant wholesalers, except manufacturers' sales branches and offices**, establishments primarily engaged in buying and selling merchandise on their own account. Included here are such types of establishments as wholesale distributors and jobbers, importers, exporters, own-brand importers/marketers, terminal and country grain elevators, and farm products assemblers.

## RETAIL TRADE, Items 139–142
### Source: U.S. Census Bureau—2017 Economic Census (See Overview of 2017 Economic Census prior to Item 135)

The Retail Trade sector (44–45) is made up of establishments engaged in retailing merchandise, generally without transformation, and rendering services incidental to the sale of merchandise.

The retailing process is the final step in the distribution of merchandise; retailers are therefore organized to sell merchandise in small quantities to the general public. This sector comprises two main types of retailers: store and nonstore retailers.

Store retailers operate fixed point-of-sale locations, located and designed to attract a high volume of walk-in customers. In general, retail stores have extensive displays of merchandise and use mass-media advertising to attract customers. They typically sell merchandise to the general public for personal or household consumption; some also serve business and institutional clients. These include establishments such as office supply stores, computer and software stores, building materials dealers, plumbing supply stores, and electrical supply stores. Catalog showrooms, gasoline service stations, automotive dealers, and mobile home dealers are treated as store retailers.

In addition to retailing merchandise, some types of store retailers are also engaged in the provision of after-sales services, such as repair and installation. For example, new automobile dealers, electronic and appliance stores, and musical instrument and supply stores often provide repair services. As a general rule, establishments engaged in retailing merchandise and providing after-sales services are classified in this sector.

Nonstore retailers, like store retailers, are organized to serve the general public, although their retailing methods differ. The establishments of this subsector reach customers and market merchandise with methods including the broadcasting of "infomercials," the broadcasting and publishing of direct-response advertising, the publishing of paper and electronic catalogs, door-to-door solicitation, in-home demonstration, selling from portable stalls (street vendors, except food), and distribution through vending machines. Establishments engaged in the direct sale (nonstore) of products, such as home heating oil dealers and home-delivery newspaper routes are included in this sector.

The buying of goods for resale is a characteristic of retail trade establishments that distinguishes them from establishments in the Agriculture, Manufacturing, and Construction sectors. For example, farms that sell their products at or from the point of production are classified in Agriculture instead of in Retail Trade. Similarly, establishments that both manufacture and sell their products to the general public are classified in Manufacturing instead of Retail Trade. However, establishments that engage in processing activities incidental to retailing are classified in Retail Trade.

# REAL ESTATE AND RENTAL AND LEASING, Items 143–146

## Source: U.S. Census Bureau—2017 Economic Census (See Overview of 2017 Economic Census prior to Item 135)

The Real Estate and Rental and Leasing sector (53) comprises establishments primarily engaged in renting, leasing, or otherwise allowing the use of tangible or intangible assets, and establishments providing related services. The major portion of this sector is made up of establishments that rent, lease, or otherwise allow the use of their own assets by others. The assets may be tangible, such as real estate and equipment, or intangible, such as patents and trademarks.

This sector also includes establishments primarily engaged in managing real estate for others, selling, renting, and/or buying real estate for others, and appraising real estate. These activities are closely related to this sector's main activity. In addition, a substantial proportion of property management is self-performed by lessors.

The main components of this sector are the real estate lessors industries; equipment lessors industries (including motor vehicles, computers, and consumer goods); and lessors of nonfinancial intangible assets (except copyrighted works).

# PROFESSIONAL, SCIENTIFIC, AND TECHNICAL SERVICES, Items 147–150

## Source: U.S. Census Bureau—2017 Economic Census (See Overview of 2017 Economic Census prior to Item 135)

The Professional, Scientific, and Technical Services sector (54) is made up of establishments that specialize in performing professional, scientific, and technical activities for others. These activities require a high degree of expertise and training. The establishments in this sector specialize in one or more areas

and provide services to clients in a variety of industries (and, in some cases, to households). Activities performed include legal advice and representation; accounting, bookkeeping, and payroll services; architectural, engineering, and specialized design services; computer services; consulting services; research services; advertising services; photographic services; translation and interpretation services; veterinary services; and other professional, scientific, and technical services.

This sector excludes establishments primarily engaged in providing a range of day-to-day office administrative services, such as financial planning, billing and record keeping, personnel services, and physical distribution and logistics services. These establishments are classified in sector 56, Administrative and Support and Waste Management and Remediation Services.

# MANUFACTURING, Items 151–154

## Source: U.S. Census Bureau—2017 Economic Census (See Overview of 2017 Economic Census prior to Item 135)

The Manufacturing sector (31–33) is made up of establishments engaged in the mechanical, physical, or chemical transformation of materials, substances, or components into new products. The assembling of component parts of manufactured products is considered manufacturing, except in cases in which the activity is appropriately classified in the Construction sector. Establishments in the Manufacturing sector are often described as plants, factories, or mills, and characteristically use power-driven machines and materials-handling equipment. However, establishments that transform materials or substances into new products by hand or in the worker's home, and establishments engaged in selling to the general public products made on the same premises from which they are sold (such as bakeries, candy stores, and custom tailors) may also be included in this sector. Manufacturing establishments may process materials or contract with other establishments to process their materials for them. Both types of establishments are included in the Manufacturing sector.

The materials, substances, or components transformed by manufacturing establishments are raw materials that are products of agriculture, forestry, fishing, mining, or quarrying, or are products of other manufacturing establishments. The materials used may be purchased directly from producers, obtained through customary trade channels, or secured without recourse to the market by transferring the product from one establishment to another, under the same ownership. The new product of a manufacturing establishment may be finished (in the sense that it is ready for utilization or consumption), or it may be semifinished to become an input for an establishment engaged in further manufacturing. For example, the product of the alumina refinery is the input used in the primary production of aluminum; primary aluminum is the input used in an aluminum wire drawing plant; and aluminum wire is the input used in a fabricated wire product manufacturing establishment.

Data are included for counties with 500 or more employees in the Manufacturing sector.

## ACCOMMODATION AND FOOD SERVICES, Items 155–158

**Source: U.S. Census Bureau—2017 Economic Census (See Overview of 2017 Economic Census prior to Item 135)**

The Accommodation and Food Services sector (72) consists of establishments that provide customers with lodging and/or meals, snacks, and beverages for immediate consumption. This sector includes both accommodation and food services establishments because the two activities are often combined at the same establishment.

Excluded from this sector are civic and social organizations, amusement and recreation parks, theaters, and other recreation or entertainment facilities providing food and beverage services.

## HEALTH CARE AND SOCIAL ASSISTANCE, Items 159–162

**Source: U.S. Census Bureau—2017 Economic Census (See Overview of 2017 Economic Census prior to Item 135)**

The Health Care and Social Assistance sector (62) consists of establishments that provide health care and social assistance services to individuals. The sector includes both health care and social assistance because it is sometimes difficult to distinguish between the boundaries of these two activities. The industries in this sector are arranged on a continuum, starting with establishments that provide medical care exclusively, continuing with those that provide health care and social assistance, and finishing with those that provide only social assistance. The services provided by establishments in this sector are delivered by trained professionals. All industries in the sector share this commonality of process—namely, labor inputs of health practitioners or social workers with the requisite expertise. Many of the industries in the sector are defined based on the educational degree held by the practitioners included in the industry.

Excluded from this sector are aerobic classes, which can be found in subsector 713, Amusement, Gambling, and Recreation Industries; and nonmedical diet and weight-reducing centers, which can be found in subsector 812, Personal and Laundry Services. Although these can be viewed as health services, they are not typically delivered by health practitioners.

## OTHER SERVICES, EXCEPT PUBLIC ADMINISTRATION Items 163–166

**Source: U.S. Census Bureau—2017 Economic Census (See Overview of 2017 Economic Census prior to Item 135)**

The Other Services, Except Public Administration sector (81) comprises establishments engaged in providing services not specifically categorized elsewhere in the classification system. Establishments in this sector are primarily engaged in activities such as equipment and machinery repairing, promoting or administering religious activities, grant making, and advocacy; this sector also includes establishments that provide dry-cleaning and laundry services, personal care services, death care services, pet care services, photofinishing services, temporary parking services, and dating services.

Private households that employ workers on or about the premises in activities primarily concerned with the operation of the household are included in this sector.

Excluded from this sector are establishments primarily engaged in retailing new equipment and performing repairs and general maintenance on equipment. These establishments are classified in sector 44–45, Retail Trade.

## NONEMPLOYER BUSINESSES, Items 167 and 168

**Source: U.S. Census Bureau—Nonemployer Statistics https://www.census.gov/programs-surveys/nonemployer-statistics.html**

Nonemployer Statistics is an annual series that provides subnational economic data for businesses that have no paid employees and are subject to federal income tax. The data consist of the number of businesses and total receipts by industry. Most nonemployers are self-employed individuals operating unincorporated businesses (known as sole proprietorships), which may or may not be the owner's principal source of income.

The majority of all business establishments in the United States are nonemployers, yet these firms average less than 4 percent of all sales and receipts nationally. Due to their small economic impact, these firms are excluded from most other Census Bureau business statistics (the primary exception being the Survey of Business Owners). The Nonemployers Statistics series is the primary resource available to study the scope and activities of nonemployers at a detailed geographic level.

## BUILDING PERMITS, Items 169 and 170

**Source: U.S. Census Bureau—Building Permits Survey http://www.census.gov/construction/bps/**

These figures represent private residential construction authorized by building permits in approximately 20,000 places in the United States. Valuation represents the expected cost of construction as recorded on the building permit. This figure usually excludes the cost of on-site and off-site development and improvements, as well as the cost of heating, plumbing, electrical, and elevator installations.

National, state, and county totals were obtained by adding the data for permit-issuing places within each jurisdiction. Not all areas of the country require a building or zoning permit. The statistics only represent those areas that do require a permit. These totals thus are limited to permits issued in the 20,000 place universe covered by the Census Bureau and may not include all permits issued within a state. Current surveys indicate that construction is undertaken for all but a very small percentage of housing units authorized by building permits.

Residential building permits include buildings with any number of housing units. Housing units exclude group quarters (such as dormitories and rooming houses), transient accommodations (such as transient hotels, motels, and tourist courts), "HUD-code" manufactured (mobile) homes, moved or relocated units, and housing units created in an existing residential or nonresidential structure.

## COUNTY AREA LOCAL GOVERNMENT EMPLOYMENT AND PAYROLL, Items 171-179

### Source: U.S. Census Bureau—2017 Census of Governments
https://www.census.gov/programs-surveys/cog.html

These items include data for all local governments (i.e., counties, municipalities, townships, special districts, and school districts) located within the county. The Census of Governments identifies the scope and nature of the nation's state and local government sector; provides authoritative benchmark figures of public finance and public employment; classifies local government organizations, powers, and activities; and measures federal, state, and local fiscal relationships. The Employment component was mailed **March 2017** to collect information on the number of state and local government civilian employees and their payrolls.

Government employees include all persons paid for personal services performed, including persons paid from federally funded programs, paid elected or appointed officials, persons in a paid leave status, and persons paid on a per meeting, annual, semiannual, or quarterly basis. Unpaid officials, pensioners, persons whose work is performed on a fee basis, and contractors and their employees are excluded from the count of employees. **Full-Time Equivalent employees** is a computed statistic representing the number of full-time employees that could have been employed if the reported number of hours worked by part-time employees had been worked by full-time employees. This statistic is calculated separately for each function of a government by dividing the "part-time hours paid" by the standard number of hours for full-time employees in the particular government and then adding the resulting quotient to the number of full-time employees.

**March payroll** represents gross payroll amounts for the one-month period of March for full-time and part-time employees. Gross payroll includes all salaries, wages, fees, commissions, and overtime paid to employees **before** withholdings for taxes, insurance, etc. It also includes incentive payments that are paid at regular pay intervals. It excludes employer share of fringe benefits like retirement, Social Security, health and life insurance, lump sum payments, and so forth.

**Administration and Judicial and Legal** combines **Financial administration**, **Other government administration, and Judicial and Legal** activities. **Financial administration includes** activities concerned with tax assessment and collection, custody and disbursement of funds, debt management, administration of trust funds, budgeting, and other government-wide financial management activities. This function is not applied to school district or special district governments. **Other government administration** applies to the legislative and government-wide administrative agencies of governments. Included here are overall planning and zoning activities, and central personnel and administrative activities. This function is not applied to school district or special district governments. **Judicial and legal** includes all court and court related activities (except probation and parole activities that are included at the "Correction" function), court activities of sheriff's offices, prosecuting attorneys' and public defenders' offices, legal departments, and attorneys providing government-wide legal service.

**Police and Corrections includes** all activities concerned, with the enforcement of law and order, including coroner's offices, police training academies, investigation bureaus, and local jails, "lockups", or other detention facilities not intended to serve as correctional facilities. **Corrections** includes activities pertaining to the confinement and correction of adults and minors convicted of criminal offenses. Pardon, probation, and parole activities are also included here.

**Fire protection** includes local government fire protection and prevention activities plus any ambulance, rescue, or other auxiliary services provided by a fire protection agency. Volunteer firefighters, if remunerated for their services on a "per fire" or some other basis, are included as part-time employees.

**Highways and transportation includes** activities associated with the maintenance and operation of streets, roads, sidewalks, bridges, tunnels, toll roads, and ferries. Snow and ice removal, street lighting, and highway and traffic engineering activities are also included here. Also included are the operation, maintenance, and construction of public mass transit systems, including subways, surface rails, and buses, and the provision, construction, operation, maintenance; support of public waterways, harbors, docks, wharves, and related marine terminal facilities; and activities associated with the operation and support of publicly operated airport facilities.

**Health and Welfare includes Health, Hospitals, and Public welfare.** **Health** includes administration of public health programs, community and visiting nurse services, immunization programs, drug abuse rehabilitation programs, health and food inspection activities, operation of outpatient clinics, and environmental pollution control activities. **Hospitals** includes only government operated medical care facilities that provide inpatient care. Employees and payrolls of private corporations that lease and operate government-owned hospital facilities are excluded.

**Public Welfare** includes the administration of various public assistance programs for the needy, veteran services, operation of nursing homes, indigent care institutions, and programs that provide payments for medical care, handicap transportation, and other services for the needy.

**Natural resources and Utilities** includes activities primarily concerned with the conservation and development of natural resources (soil, water, energy, minerals, etc.) and the regulation of industries that develop, utilize, or affect natural resources, as well as the operation and maintenance of **parks**, playgrounds, swimming pools, public beaches, auditoriums, public golf courses, museums, marinas, botanical gardens, and zoological parks. **Utilities, sewerage, and waste management** includes operation, maintenance, and construction of public water supply systems, including production, acquisition, and distribution of water to general public or to other public or private utilities, for residential, commercial, and industrial use; activities associated with the production or acquisition and distribution of electric power; provision, maintenance, and operation of sanitary and storm sewer systems and sewage disposal and treatment facilities; and refuse collection and disposal, operation of sanitary landfills, and street cleaning activities.

**Education and libraries includes** activities associated with the operation of public elementary and secondary schools and locally operated vocational-technical schools. Special education programs operated by elementary and secondary school systems are also included as are all ancillary services associated with the operation of schools, such as pupil transportation and food service. **Also included are the e**stablishment and provision of libraries for use by the general public and the technical support of privately operated libraries. This category includes classroom teachers, principals, supervisors of instruction, librarians, teacher aides, library aides, and guidance and psychological personnel as well as school superintendents and other administrative personnel, clerical and secretarial staffs, plant operation and maintenance personnel, health and recreation employees, transportation and food service personnel, and any student employees. Also included are any degree granting institutions that provide academic training above grade 12.

# LOCAL GOVERNMENT FINANCES, Items 180–193

**Source: U.S. Census Bureau—2017 Census of Governments**
**https://www.census.gov/programs-surveys/cog.html**

Data on local government finances are based on result of the 2012 Census of Governments. For each county area, the financial data comprise amounts for all local governments—not only the county government, but also any municipalities, townships, school districts, and special districts within the county. Statistics from governmental units located in two or more county areas are assigned to the county area containing the administrative office.

Revenue and expenditure items include all amounts of money received and paid out, respectively, by a government and its agencies (net of correcting transactions such as recoveries of refunds), with the exception of amounts for debt issuance and retirement and for loan and investment, agency, and private transactions.

Payments among the various funds and agencies of a particular government are excluded from revenue and expenditure items as representing internal transfers. Therefore, a government's contribution to a retirement fund that it administers is not counted as expenditure, nor is the receipt of this contribution by the retirement fund counted as revenue.

Total **general revenue** includes all revenue except utility, liquor stores, and insurance trust revenue. All tax revenue and intergovernmental revenue, even if designated for employee-retirement or local utility purpose, are classified as general revenue.

**Intergovernmental revenue** covers amounts received from the federal government as fiscal aid, reimbursements for performance of general government functions and specific services for the paying government, or in lieu of taxes. It excludes any amounts received from other governments from the sale of property, commodities, and utility services.

**Taxes** consist of compulsory contributions exacted by governments for public purposes. However, this category excludes employer and employee payments for retirement and social insurance purposes, which are classified as insurance trust revenue; it also excludes special assessments, which are classified as non-tax general revenue. Property taxes are taxes conditioned on ownership of property and assessed by its value. Sales and gross receipts taxes do not include dealer discounts, or "commissions" allowed to merchants for collection of taxes from consumers. General sales taxes and selected taxes on sales of motor fuels, tobacco products, and other particular commodities and services are included.

**General government expenditure** includes capital outlay, a major portion of which is commonly financed by borrowing. Government revenue does not include receipts from borrowing. Among other things, this distorts the relationship between totals of revenue and expenditure figures that are presented and renders it useless as a direct measure of the degree of budgetary "balance" (as that term is generally applied).

**Direct general expenditure** comprises all expenditures of the local governments, excluding utility, liquor stores, insurance trust expenditures, and any intergovernmental payments.

Local government expenditure for **education** is mainly for the provision and general support of schools and other educational facilities and services, including those for educational institutions beyond high school. They cover such related services as student transportation; school lunch and other cafeteria operations; school health, recreation, and library services; and dormitories, dining halls, and bookstores operated by public institutions of higher education.

**Health and hospital expenditure** includes health research; clinics; nursing; immunization; other categorical, environmental, and general health services provided by health agencies; establishment and operation of hospital facilities; provision of hospital care; and support of other public and private hospitals.

**Police protection expenditure** includes police activities such as patrols, communications, custody of persons awaiting trial, and vehicular inspection.

**Public welfare expenditure** covers support of and assistance to needy persons; this aid is contingent upon the person's needs. Included are cash assistance paid directly to needy persons under categorical (Old Age Assistance, Temporary Assistance for Needy Families, Aid to the Blind, and Aid to the Disabled) and other welfare programs; vendor payments made directly to private purveyors for medical care, burials, and other commodities and services provided under welfare programs; welfare institutions; and any intergovernmental or other direct expenditure for welfare purposes. Pensions to former employees and other benefits not contingent on need are excluded.

**Highway expenditure** is for the provision and maintenance of highway facilities, including toll turnpikes, bridges, tunnels, and ferries, as well as regular roads, highways, and streets. Also included are expenditures for street lighting and for snow and ice removal. Not included are highway policing and traffic control, which are considered part of police protection

**Debt outstanding** includes all long-term debt obligations of the government and its agencies (exclusive of utility debt) and all interest-bearing, short-term (repayable within one year) debt obligations remaining unpaid at the close of the fiscal year. It includes judgments, mortgages, and revenue bonds, as well as general obligation bonds, notes, and interest-bearing warrants. This category consists of non-interest-bearing, short-term obligations; inter-fund obligations; amounts owed in a trust or agency capacity; advances and contingent loans from other governments; and rights of individuals to benefit from government-administered employee-retirement funds.

## GOVERNMENT EMPLOYMENT, Items 194–196

**Source: U.S. Bureau of Economic Analysis—Regional Economic Accounts**
**http://www.bea.gov/regional/index.htm#state**

Employment is measured as the average annual sum of full-time and part-time jobs. The estimates are on a place-of-work basis. State and local government employment includes person employed in all state and local government agencies and enterprises. Data for federal civilian employment include civilian employees of the federal government, including civilian employees of the Department of Defense. Military employment includes all persons on active duty status.

## INDIVIDUAL INCOME TAX RETURNS, Items 197–199

**Source: U.S. Internal Revenue Servicw, Statistics of Income Program**
**https://www.irs.gov/uac/soi-tax-stats-county-data-2018**

The Revenue Act of 1916 mandated the annual publication of statistics related to "the operations of the internal revenue laws" as they affect individuals, all forms of businesses, estates. non-profit organizations, trusts, and investments abroad and foreign investments in the United States. The Statistics of Income (SOI) division fulfills this function by collecting and processing data so that they become informative and by sharing information about how the tax system works with other government agencies and the general public. Publication types include traditional print sources, Internet files, CD-ROMs, and files sent via e-mail. SOI has an information office, Statistical Information Services, to facilitate the dissemination of SOI data.

SOI bases its county data on administrative records of individual income tax returns (Forms 1040) from the Internal Revenue Service (IRS) Individual Master File (IMF) system. Included in these data are returns filed during the 12-month period, January 1, 2018 to December 31, 2018. While the bulk of returns filed during the 12-month period are primarily for Tax Year 2017, the IRS received a limited number of returns for tax years before 2018 and these have been included within the county data.

Data do not represent the full U.S. population because many individuals are not required to file an individual income tax return. The address shown on the tax return may differ from the taxpayer's actual residence. State and county codes were based on the ZIP code shown on the return. Excluded were tax returns

filed without a ZIP code and returns filed with a ZIP code that did not match the State code shown on the return.

SOI did not attempt to correct any ZIP codes on the returns; however, it did take the following precautions to avoid disclosing information about specific taxpayers: Excluded from the data are items with less than 20 returns within a county. Also excluded are tax returns representing a specified percentage of the total of any particular cell. For example, if one return represented 75 percent of the value of a given cell, the return was suppressed from the county detail. The actual threshold percentage used cannot be released.

Column 197 shows the number of returns. Column 198 shows the mean Adjusted Gross Income for the county and column 199 shows the mean income tax (line 56 on Form 1040) for the county.

## TABLE C—METROPOLITAN AREAS

Table C presents 199 items for the 384 metropolitan statistical areas (MSAs) and 31 metropolitan divisions in the United States. The metropolitan areas are presented in alphabetical order, and the metropolitan divisions are presented in alphabetical order within the appropriate metropolitan area. For some data items, the metropolitan area data have been aggregated from county data sources.

## LAND AREA, Items 1 and 4

**Source: U.S. Census Bureau—2020 U.S. Gazetteer Files,**
**https://www.census.gov/geographies/reference-files/time-series/geo/gazetteer-files.html**

Land area measurements are shown to the nearest square mile. Land area is an area measurement providing the size, in square miles, of the land portions of each county.

## POPULATION, Items 2–4

**Source: U.S. Census Bureau—Population Estimates**
**https://www.census.gov/programs-surveys/popest.html**

The population data are Census Bureau estimates of the resident population as of July 1 of the year shown. The ranks are shown for metropolitan statistical areas, but exclude metropolitan divisions.

The Census Bureau released these 2020 estimates for research purposes. The July 1, 2020 estimates are based on the 2010 Census and were created without incorporation or consideration of the 2020 Census results. They are typically used in comparisons with the 2020 Census to make determinations about the accuracy of the estimates. Differences between the estimates and census counts are interpreted as error in the estimates and are used to inform population and housing unit estimates research and methodological improvements over the decade.

## POPULATION AND POPULATION CHARAC-
## TERISTICS, Items 5–19

**Source: U.S. Census Bureau—Population Estimates**
**https://www.census.gov/programs-surveys/popest.**
**html**

The concept of race, as used by the Census Bureau, reflects self-identification by persons according to the race or races with which they most closely identify. These categories are sociopolitical constructs and should not be interpreted as being scientific or anthropological in nature. Furthermore, race categories include both racial and national origin groups.

Beginning with the 2000 census, respondents were offered the option of selecting one or more races. This option was not available in prior censuses; thus, comparisons between censuses should be made with caution. In Table C, Columns 5 through 8 refer to individuals who identified with each racial category, either alone or in combination with other races. The estimates exclude persons of Hispanic or Latino origin from all race groups. Because respondents could include as many categories as they wished, and because the columns refer to the percentage of the population, the total will often exceed 100 percent.

The **White** population is defined as persons who indicated their race as White, as well as persons who did not classify themselves in one of the specific race categories listed on the questionnaire but entered a nationality such as Irish, German, Italian, Lebanese, Near Easterner, Arab, or Polish.

The **Black** population includes persons who indicated their race as "Black or African American" as well as persons who did not classify themselves in one of the specific race categories but reported entries such as African American, Afro American, Kenyan, Nigerian, or Haitian.

The **American Indian or Alaska Native** population includes persons who indicated their race as American Indian or Alaska Native, as well as persons who did not classify themselves in one of the specific race categories but reported entries such as Canadian Indian, French-American Indian, Spanish-American Indian, Eskimo, Aleut, Alaska Indian, or any of the American Indian or Alaska Native tribes.

The **Asian and Pacific Islander** population combines two census groupings: **Asian** and **Native Hawaiian or Other Pacific Islander**. The **Asian** population includes persons who indicated their race as Asian Indian, Chinese, Filipino, Japanese, Korean, Vietnamese, or "Other Asian," as well as persons who provided write-in entries of such groups as Cambodian, Laotian, Hmong, Pakistani, or Taiwanese. The **Native Hawaiian or Other Pacific Islander** population includes persons who indicated their race as "Native Hawaiian," "Guamanian or Chamorro," "Samoan," or "Other Pacific Islander," as well as persons who reported entries such as Part Hawaiian, American Samoan, Fijian, Melanesian, or Tahitian.

The sum of the four individual race alone or in combination categories in this book will often add to more than the total population because people who reported more than one race were tallied in each race category. In this book, the Asian group has been combined with the Native Hawaiian and Other Pacific Islander

group, causing double-counting of persons who identify with both groups. This is especially pronounced in Hawaii.

The **Hispanic population** is based on a complete-count question that asked respondents "Is this person Spanish/Hispanic/Latino?" Persons marking any one of the four Hispanic categories (i.e., Mexican, Puerto Rican, Cuban, or other Spanish) are collectively referred to as Hispanic.

**Age** is defined as age at last birthday (number of completed years since birth), as of April 1 of the census year.

The **female** population is shown as a percentage of total population.

## POPULATION AND COMPONENTS OF
## CHANGE, Items 20–26

**Source: U.S. Census Bureau—Decennial Censuses**
**and Population Estimates**
**http://www.census.gov/main/www/cen2000.html**
**https://www.census.gov/programs-surveys/popest.**
**html**
**https://www.census.gov/programs-surveys/decennial-**
**census/data/datasets.2010.html**

The population data for 2000 and 2010 are from the decennial censuses and represent the resident population as of April 1 of those years. The components of change are based on Census Bureau estimates of the resident population as of July 1 of 2020. The change in population between 2010 and 2020 is made up of (a) natural increase—births minus deaths, and (b) net migration—the difference between the number of persons moving into a particular area and the number of persons moving out of the area. Net migration is composed of internal and international migration.

Because the 2020 population estimates are based on a model that begins with a national population estimate, the county and metropolitan area components of change do not always exactly add up to the difference between the 2010 census population and the 2020 estimates.

The Census Bureau released these 2020 estimates for research purposes. The July 1, 2020 estimates are based on the 2010 Census and were created without incorporation or consideration of the 2020 Census results. They are typically used in comparisons with the 2020 Census to make determinations about the accuracy of the estimates. Differences between the estimates and census counts are interpreted as error in the estimates and are used to inform population and housing unit estimates research and methodological improvements over the decade.

## HOUSEHOLDS, Items 27–31

**Source: U.S. Census Bureau—American Community**
**Survey, 2019 1-year Estimates**
**http://www.census.gov/acs/www/**

A **household** includes all of the persons who occupy a housing unit. (Persons not living in households are classified as living in group quarters.) A housing unit is a house, an apartment, a mobile home, a group of rooms, or a single room occupied (or, if vacant, intended for occupancy) as separate living quarters. Separate living quarters are those in which the occupants live separately from any other persons in the building and have direct access from the outside of the building or through a common hall. The occupants may be a single family, one person living alone, two or more families living together, or any other group of related or unrelated persons who share living quarters. The number of households is the same as the number of year-round occupied housing units.

A **family** includes a householder and one or more other persons living in the same household who are related to the householder by birth, marriage, or adoption. All persons in a household who are related to the householder are regarded as members of his or her family. A **family household** may contain persons not related to the householder; thus, family households may include more members than families do. A household can contain only one family for the purposes of census tabulations. Not all households contain families, as a household may comprise a group of unrelated persons or of one person living alone. Families are classified by type as either a "husband-wife family" or "other family," according to the presence or absence of a spouse.

The measure of **persons per household** is obtained by dividing the number of persons in households by the number of households or householders. One person in each household is designated as the householder. In most cases, this is the person (or one of the persons) in whose name the house is owned, being bought, or rented. If there is no such person in the household, any adult household member 15 years old and over can be designated as the householder.

The category **single parent households** includes family households with children under 18 with either a male or female householder with no spouse present.

## GROUP QUARTERS, Item 32

**Source: U.S. Census Bureau—Population Estimates**
**https://www.census.gov/programs-surveys/popest.html**

The Census Bureau classifies all persons not living in households as living in group quarters; this category includes both the institutional and noninstitutional populations. The institutionalized population includes persons under formally authorized, supervised care or custody in institutions, such as correctional institutions, nursing homes, mental (psychiatric) hospitals, and juvenile institutions. The noninstitutionalized population includes persons who live in group quarters other than institutions, such as college dormitories, military quarters, and group homes. This volume includes the total number of persons in group quarters.

## DAYTIME POPULATION, Items 33 and 34

**Source: U.S. Census Bureau—American Community Survey, 2019 1-year Estimateshttp://www.census.gov/acs/www/**

Daytime population refers to the number of persons who are present in an area or place during normal business hours, including workers. This can be contrasted with the "resident" population, which is present during the evening and nighttime hours. The daytime population estimate is calculated by adding the total resident population and the total workers working in the area/place, and then subtracting the total workers living in the area/place from that result. Information on the expansion or contraction experienced by different communities between their nighttime and daytime populations is important for many planning purposes, especially those concerning transportation, disaster, and relief operations.

The employment/residence ratio is a measure of the total number of workers working in an area or place, relative to the total number of workers living in the area or place. It is often used as a rough indication of the jobs-workers balance in an area/place, although it does not take into account whether the resident workers possess the skills needed for the jobs available in their particular area/place. The employment/residence ratio is calculated by dividing the number of total workers working in an area/place by the number of total workers residing in the area/place.

## BIRTHS AND DEATHS, Items 35–38

**Source: U.S. Census Bureau—Population Estimates**
**https://www.census.gov/programs-surveys/popest.html**

The numbers of births and deaths are from the Census Bureau's Population Estimates Program. They represent the total number of live births and deaths occurring to residents of an area as estimated using reports from the National Center for Health Statistics (NCHS) and the Federal-State Cooperative for Population Estimates (FSCPE). The rates measure births and deaths during the specified time period as a proportion of an area's population. Rates are expressed per 1,000 population estimated as of July 1. These numbers and rates do not represent the calendar year, but rather the year-long period ending on July 1.

## PERSONS UNDER 65 WITH NO HEALTH INSURANCE, Items 39 and 40

**Source: U.S. Census Bureau—Small Area Health Insurance Estimates**
**https://www.census.gov/programs-surveys/sahie.html**

The Small Area Health Insurance Estimates (SAHIE) program develops model-based estimates of health insurance coverage for counties and states. This developmental program builds on the work of the Small Area Income and Poverty Estimates (SAIPE) program. The SAHIE program models health insurance coverage by combining survey data with population estimates and administrative records. These estimates combine data from administrative records, postcensal population estimates, and the decennial census with direct estimates from the American Community Survey to provide consistent and reliable single-year estimates. These model-based single-year estimates are more reflective of current conditions than multi-year survey estimates. The metropolitan area estimates have been aggregated from the county estimates.

## MEDICARE ENROLLMENT, Items 41–43

**Source: U.S. Department of Health and Human Services, Centers for Medicare and Medicaid Services**
https://www.cms.gov/Research-Statistics-Data-and-Systems/Statistics-Trends-and-Reports/CMSProgramStatistics/Dashboard.html

The Centers for Medicare and Medicaid Services (CMS) administers Medicare, which provides health insurance to persons 65 years old and over, persons with permanent kidney failure, and certain persons with disabilities. Original Medicare has two parts: Hospital Insurance and Supplemental Medical Insurance. In recent years, Medicare has been expanded to include two new programs: Medicare Advantage plans and prescription drug coverage. Medicare Advantage Plans are health plan options that are approved by Medicare but run by private companies. Medicare prescription drug plans can be part of Medicare Advantage plans or stand-alone drug plans.

Persons who are eligible for Medicare can enroll in Part A (Hospital Insurance) at no charge, and can choose to pay a monthly premium to enroll in Part B. Most eligible persons are enrolled in Part A, and most enrollees in Part A are also enrolled in Part B (Supplemental Medical Insurance.) This table includes persons who were enrolled in both Part A and Part B during 2014.

Part B beneficiaries can choose to enroll in Original Medicare, a fee-for-service plan administered by the Centers for Medicare and Medicaid Services, or in a Medicare Advantage plan.

Medicare Advantage plans include private fee-for-service plans, preferred provider organizations, health maintenance organizations, medical savings account plans, demonstration plans, and programs for all-inclusive care for the elderly.

The annual Medicare enrollment counts are determined using a person-year methodology. For each calendar year, total person-year counts are determined by summing the total number of months that each beneficiary is enrolled during the year and dividing by 12. Using this methodology, a beneficiary's partial-year enrollment may be counted in more than one category (i.e., both Original Medicare and Medicare Advantage).

## LAW ENFORCEMENT, Items 44–47

**Source: U.S. Federal Bureau of Investigation—Uniform Crime Reports**
https://ucr.fbi.gov/crime-in-the-u.s/2019/crime-in-the-u.s.-2019

Crime data are as reported to the Federal Bureau of Investigation (FBI) by law enforcement agencies. The Metropolitan Area numbers in this volume are from *Crime in the United States* and have been adjusted by the FBI to include estimates of areas that did not report. Where the FBI analysts determined that the numbers reflected underreporting or overreporting or other factors that made them unreliable, the numbers have not been included.

Through the voluntary contribution of crime statistics by law enforcement agencies across the United States, the Uniform Crime Reporting (UCR) Program provides periodic assessments of crime in the nation as measured by offenses that have come to the attention of the law enforcement community. The Committee on Uniform Crime Records of the International Association of Chiefs of Police initiated this voluntary national data collection effort in 1930. The UCR Program contributors compile and submit their crime data either directly to the FBI or through state-level UCR Programs.

Seven offenses, because of their severity, frequency of occurrence, and likelihood of being reported to police, were initially selected to serve as an index for evaluating fluctuations in the volume of crime. These serious crimes were murder and nonnegligent manslaughter, forcible rape, robbery, aggravated assault, burglary, larceny-theft, and motor vehicle theft. By congressional mandate, arson was added as the eighth index offense in 1979. The totals shown in this volume do not include arson.

In 2004, the FBI discontinued the use of the Crime Index in the UCR Program and its publications, stating that the Crime Index was driven upward by the offense with the highest number of cases (in this case, larceny-theft), creating a bias against jurisdictions with a high number of larceny-thefts but a low number of other serious crimes, such as murder and forcible rape. The FBI is currently publishing a violent crime total and property crime total until a more viable index is developed.

In 2013, the FBI adopted a new definition of rape. Rape is now defined as, "Penetration, no matter how slight, of the vagina or anus with any body part or object, or oral penetration by a sex organ of another person, without the consent of the victim." The new definition updated the 80-year-old historical definition of rape which was "carnal knowledge of a female forcibly and against her will." Effectively, the revised definition expands rape to include both male and female victims and offenders, and reflects the various forms of sexual penetration understood to be rape, especially nonconsenting acts of sodomy, and sexual assaults with objects.

**Violent crimes** include four categories of offenses: (1) Murder and nonnegligent manslaughter, as defined in the UCR Program, is the willful (nonnegligent) killing of one human being by another. This offense excludes deaths caused by negligence, suicide, or accident; justifiable homicides; and attempts to murder or assaults to murder. (2) Rape is the penetration, no matter how slight, of the vagina or anus with any body part or object, or oral penetration by a sex organ of another person, without the consent of the victim. Assaults or attempts to commit rape by force or threat of force are also included; however, statutory rape (without force) and other sex offenses are excluded. (3) Robbery is the taking or attempting to take anything of value from the care, custody, or control of a person or persons by force or threat of force or violence and/or by putting the victim in fear. (4) Aggravated assault is an unlawful attack by one person upon another for the purpose of inflicting severe or aggravated bodily injury. This type of assault is usually accompanied by the use of a weapon or by other means likely to produce death or great bodily harm. Attempts are included, since injury does not necessarily have to result when a gun, knife, or other weapon is used, as these incidents could and probably would result in a serious personal injury if the crime were successfully completed.

**Property crimes** include three categories: (1) Burglary, or breaking and entering, is the unlawful entry of a structure to commit a felony or theft, even though no force was used to gain entrance. (2) Larceny-theft is the unauthorized taking of the personal property of another, without the use of force. (3) Motor vehicle theft is the unauthorized taking of any motor vehicle.

Rates are based on population estimates provided by the FBI. For many of the metropolitan areas, the FBI has estimated violent and property crimes, adjusted to include areas that did not report and to correct for underreporting and overreporting, as published in the FBI's *Crime in the United States*. If a metropolitan area was not included in *Crime in the United States,* it is not included in Table C in this volume. The metropolitan area data includes police departments of counties and cities within the area. For details on police departments included in each metropolitan area and those that did not provide data, see https://ucr.fbi.gov/crime-in-the.u.s/2019/crime-in-the-u.s.-2019/topic-pages/tables/table-6

## EDUCATION—SCHOOL ENROLLMENT AND EDUCATIONAL ATTAINMENT, Items 48–51

**Source: U.S. Census Bureau—American Community Survey, 2019 1-year Estimates**
**http://www.census.gov/acs/www/**

Data on school enrollment and educational attainment were derived from a sample of the population. Persons were classified as enrolled in school if they reported attending a "regular" public or private school (or college) during the three months prior to the survey. The instructions were to "include only nursery school, kindergarten, elementary school, and schooling which would lead to a high school diploma or a college degree" as regular school. The Census Bureau defines a public school as "any school or college controlled and supported by a local, county, state, or federal government." Schools primarily supported and controlled by religious organizations or other private groups are defined as private schools.

Data on **educational attainment** are tabulated for the population 25 years old and over. The data were derived from a question that asked respondents for the highest level of school completed or the highest degree received. Persons who had passed a high school equivalency examination were considered high school graduates. Schooling received in foreign schools was to be reported as the equivalent grade or years in the regular American school system.

Vocational and technical training, such as barber school training; business, trade, technical, and vocational schools; or other training for a specific trade are specifically excluded.

**High school graduate or less**. This category includes persons whose highest degree was a high school diploma or its equivalent, and those who reported any level lower than a high school diploma.

Bachelor's degree or more. This category includes persons who have received bachelor's degrees, master's degrees, professional school degrees (such as law school or medical school degrees), and doctoral degrees.

## LOCAL GOVERNMENT EDUCATION EXPENDITURES, Items 52 and 53

**Source: U.S. Department of Education, National Center for Education Statistics, Common Core of Data (CCD), "Local Education Agency (School District) Finance Survey (F33) Data", 2017-2018 v.1a**
**https://nces.ed.gov/ccd/**

Expenditures are for elementary and secondary Education which includes prekindergarten through twelfth grade regular, special, and vocational education, as well as cocurricular, community service, and adult education programs provided by a public school system. The financial activities of these systems for all instruction, support service, and noninstructional activities are included

Current Spending comprises current operation expenditure, payments made by the state government on behalf of school systems, and transfers made by school systems into their own retirement funds. Current operation expenditures include direct expenditure for salaries, employee benefits, purchased professional and technical services, purchased property and other services, and supplies. It includes gross school system expenditure for instruction, support services, and noninstructional functions. It excludes expenditure for debt service, capital outlay, and reimbursement to other governments (including other school systems).

Current expenditure per student is current expenditure divided by the number of students enrolled. The number of students enrolled is based on an annual "membership" count of students on or about October 1.

## INCOME AND POVERTY Items 54–61

**Source: U.S. Census Bureau—American Community Survey, 2019 1-year Estimates**
**http://www.census.gov/acs/www/**

The data on income were derived from responses of a sample of persons 15 years old and over. Total money income is the sum of the amounts reported separately for wage or salary income; net self-employment income; interest, dividends, or net rental or royalty income or income from estates and trusts; Social Security or railroad retirement income; Supplemental Security Income (SSI); public assistance or welfare payments; retirement, survivor, or disability pensions; and all other income. Receipts from the following sources are not included as income: capital gains; money received from the sale of property (unless the recipient was engaged in the business of selling such property); the value of income "in kind" from food stamps, public housing subsidies, medical care, employer contributions for individuals, etc.; withdrawal of bank deposits; money borrowed; tax refunds; exchange of money between relatives living in the same household; and gifts, lump-sum inheritances, insurance payments, and other types of lump-sum receipts.

Money income differs in definition from personal income (item 62). For example, money income does not include the pension

**Poverty Thresholds for 2019 by Size of Family and Number of Related Children Under 18 Years**

| Size of family unit | Weighted average thresholds | Related children under 18 years | | | | | | | | |
|---|---|---|---|---|---|---|---|---|---|---|
| | | None | One | Two | Three | Four | Five | Six | Seven | Eight or more |
| One person (unrelated individual): | 13,011 | | | | | | | | | |
| Under age 65 ..................... | 13,300 | 13,300 | | | | | | | | |
| Aged 65 and older.................... | 12,261 | 12,261 | | | | | | | | |
| | | | | | | | | | | |
| Two people: .................................. | 16,521 | | | | | | | | | |
| Householder under age 65........ | 17,196 | 17,120 | 17,622 | | | | | | | |
| Householder aged 65 and older ..................................... | 15,468 | 15,453 | 17,555 | | | | | | | |
| | | | | | | | | | | |
| Three people.............................. | 20,335 | 19,998 | 20,578 | 20,598 | | | | | | |
| Four people................................ | 26,172 | 26,370 | 26,801 | 25,926 | 26,017 | | | | | |
| Five people ................................ | 31,021 | 31,800 | 32,263 | 31,275 | 30,510 | 30,044 | | | | |
| Six people.................................. | 35,129 | 36,576 | 36,721 | 35,965 | 35,239 | 34,161 | 33,522 | | | |
| Seven people.............................. | 40,016 | 42,085 | 42,348 | 41,442 | 40,811 | 39,635 | 38,262 | 36,757 | | |
| Eight people............................... | 44,461 | 47,069 | 47,485 | 46,630 | 45,881 | 44,818 | 43,470 | 42,066 | 41,709 | |
| Nine people or more ................... | 52,875 | 56,621 | 56,895 | 56,139 | 55,503 | 54,460 | 53,025 | 51,727 | 51,406 | 49,426 |

Source: U.S. Census Bureau.

rights, employer provided health insurance, food stamps, or Medicare payments that are included in personal income.

Per capita income is the mean income computed for every man, woman, and child in a particular group. It is derived by dividing the aggregate income of a particular group by the resident population in that group in the survey year. Per capita income is rounded to the nearest whole dollar.

**Household income** includes the income of the householder and all other individuals 15 years old and over in the household, whether or not they are related to the householder. Since many households consist of only one person, median household income is usually less than median family income. Although the household income statistics cover the year preceding the survey, the characteristics of individuals and the composition of households refer to the date of the survey. Thus, the income of the household does not include amounts received by individuals who were members of the household during the year if these individuals were no longer residing in the household at the time of the survey. Similarly, income amounts reported by individuals who did not reside in the household during the year but who were members of the household at the time of the survey are included. However, the composition of most households was the same during the year as it was at the time of the survey.

**Mean household income** is the amount obtained by dividing the aggregate income of all households by the total number of households. Mean indivdual income is obtained by dividing the aggregate income of all individuals by the total number of persons age 15 and lder. The mean is based on the distribution of the total number of households or individuals including those with no income. Mean income is rounded to the nearest whole dollar. Care should be exercised in using and interpreting mean income values for small subgroups of the population. Because the mean is influenced strongly by extreme values in the distribution, it is especially susceptible to the effects of sampling variability,

misreporting, and processing errors. The median, which is not affected by extreme values, is, therefore, a better measure than the mean when the population base is small.

Income components were reported for the 12 months preceding the interview month. Monthly Consumer Price Indices (CPI) factors were used to inflation-adjust these components to a reference calendar year (January through December). For example, a household interviewed in March 2012 reports their income for March 2011 through February 2012. Their income is adjusted to the 2012 reference calendar year by multiplying their reported income by 2012 average annual CPI (January-December 2008) and then dividing by the average CPI for March 2011–February 2012. However, the estimates do not reflect the price and cost-of-living differences that may exist between areas.

The **poverty status** data were derived from data collected on the number of persons in a household, each person's relationship to the householder, and income data. The Social Security Administration (SSA) developed the original poverty definition in 1964, which federal interagency committees subsequently revised in 1969 and 1980. The Office of Management and Budget's (OMB) *Directive 14* prescribes the SSA's definition as the official poverty measure for federal agencies to use in their statistical work. Poverty statistics presented in American Community Survey products adhere to the standards defined by OMB in *Directive 14*.

Poverty thresholds vary depending on three criteria: size of family, number of children, and, for one- and two-person families, age of householder. In determining the poverty status of families and unrelated individuals, the Census Bureau uses thresholds (income cutoffs) arranged in a two-dimensional matrix. The matrix consists of family size (from one person to nine or more persons), cross-classified by presence and number of family members under 18 years old (from no children present to eight or more children present). Unrelated individuals and two-person families are further differentiated by age of reference person (under 65 years old and 65 years old and over). To determine a person's poverty status, the person's total family income over the previous 12 months is compared with the poverty threshold appropriate for that person's family size and composition. If the total income of that person's family is less than the threshold appropriate for that family, then the person is considered poor or "below the poverty level," together with every member of his or her family. If a person is not living with anyone related by birth, marriage, or adoption, then the person's own income is compared with his or her poverty threshold. The total number of persons below the poverty level is the sum of persons in families and the number of unrelated individuals with incomes below the poverty level over the previous 12 months.

## PERSONAL INCOME AND EARNINGS, Items 62–83

### Source: U.S. Bureau of Economic Analysis, Regional Economic Accounts
http://www.bea.gov/regional/index.htm#state

Total personal income is the current income received by residents of an area from all sources. It is measured before deductions of income and other personal taxes, but after deductions of personal contributions for Social Security, government retirement, and other social insurance programs. It consists of wage and salary disbursements (covering all employee earnings, including executive salaries, bonuses, commissions, payments-in-kind, incentive payments, and tips); various types of supplementary earnings, such as employers' contributions to pension funds (termed "other labor income" or "supplements to wages and salaries"); proprietors' income; rental income of persons; dividends; personal interest income; and government and business transfer payments.

Per capita personal income is based on the resident population estimated as of July 1 of the year shown.

Proprietors' income is the monetary income and income-in-kind of proprietorships and partnerships (including the independent professions) and the income of tax-exempt cooperatives. Dividends are cash payments by corporations to stockholders who are U.S. residents. Interest is the monetary and imputed interest income of persons from all sources. Rent is the monetary income of persons from the rental of real property, except the income of persons primarily engaged in the real estate business; the imputed net rental income of owner-occupants of nonfarm dwellings; and the royalties received by persons.

Transfer payments are income for which services are not currently rendered. They consist of both government and business transfer payments. Government transfer payments include payments under the following programs: Federal Old-Age, Survivors, and Disability Insurance ("Social Security"); Medicare and medical vendor payments; unemployment insurance; railroad and government retirement; federal- and state-government-insured workers' compensation; veterans' benefits, including veterans' life insurance; food stamps; black lung payments; Supplemental Security Income; and Temporary Assistance for Needy Families. Government payments to nonprofit institutions, other than for

work under research and development contracts, are also included. Business transfer payments consist primarily of liability payments for personal injury and of corporate gifts to nonprofit institutions.

Personal income differs in definition from money income (items 54–57). For example, personal income includes pension rights, employer-provided health insurance, food stamps, and Medicare. These are not included in the definition of money income.

Earnings cover wage and salary disbursements, other labor income, and proprietors' income.

The data for earnings obtained from the Bureau of Economic Analysis (BEA) are based on place of work. In computing personal income, BEA makes an "adjustment for residence" to earnings, based on commuting patterns; personal income is thus presented on a place-of-residence basis.

Farm earnings include the income of farm workers (wages and salaries and other labor income) and farm proprietors. Farm proprietors' income includes only the income of sole proprietorships and partnerships. Farm earnings estimates are benchmarked to data collected in the Census of Agriculture and the revised Department of Agriculture statistical totals of income and expense items.

Goods-related industries include mining, construction, and manufacturing. Service-related and other industries include private-sector earnings in agricultural services, forestry, and fisheries; transportation and public utilities; wholesale trade; retail trade; finance, insurance, and real estate; and services. Government earnings include all levels of government. Industries are categorized under the North American Industry Classification System (NAICS), and are not comparable to years prior to 2002.

## SOCIAL SECURITY AND SUPPLEMENTAL SECURITY INCOME, Items 84–86

### Source: U.S. Social Security Administration
http://www.ssa.gov/policy/docs/statcomps/oasdi_sc/
http://www.ssa.gov/policy/docs/statcomps/ssi_sc/

Social Security beneficiaries are persons receiving benefits under the Old-Age, Survivors, and Disability Insurance Program. These include retired or disabled workers covered by the program, their spouses and dependent children, and the surviving spouses and dependent children of deceased workers.

Supplemental Security Income (SSI) recipients are persons receiving SSI payments. The SSI program is a cash assistance program that provides monthly benefits to low-income aged, blind, or disabled persons.

Data are as of December of the year shown.

## HOUSING, Items 87–96

### Source: U.S. Census Bureau—Population Estimates Program
https://www.census.gov/programs-surveys/popest.html
### Source: U.S. Census Bureau—American Community Survey. 2019 1-year Estimates
http://www.census.gov/acs/www/

Housing data for 2020 are from the Population Estimates Program. Housing unit characteristics for 2019 are from the American Community Survey.

A housing unit is a house, apartment, mobile home or trailer, group of rooms, or single room occupied or, if vacant, intended for occupancy as separate living quarters. Separate living quarters are those in which the occupants do not live and eat with any other person in the structure and which have direct access from the outside of the building through a common hall.

The occupants of a housing unit may be a single family, one person living alone, two or more families living together, or any other group of related or unrelated persons who share living quarters. Both occupied and vacant housing units are included in the housing inventory, although recreational vehicles, tents, caves, boats, railroad cars, and the like are included only if they are occupied as a person's usual place of residence.

A housing unit is classified as occupied if it is the usual place of residence of the person or group of persons living in it at the time of enumeration, or if the occupants are only temporarily absent (away on vacation). A household consists of all persons who occupy a housing unit as their usual place of residence. Vacant units for sale or rent include units rented or sold but not occupied and any other units held off the market.

**Value** is the respondent's estimate of how much the property (house and lot, mobile home and lot (if lot owned), or condominium unit) would sell for if it were for sale. If the house was owned or being bought, but the land on which it sits was not, the respondent was asked to estimate the combined value of the house and the land. Owners of noncondominium multi-unit buildings who live in one of the building's units, like duplexes and small apartment buildings, should report the value of the building, the land, and any additional buildings on the same plot of land. For vacant units, value was the price asked for the property. Value was tabulated separately for all owner-occupied and vacant-for-sale and sold, not occupied housing units, as well as owner-occupied mobile homes.

**Mortgage** refers to all forms of debt where the property is pledged as security for repayment of the debt, including deeds of trust; trust deeds; contracts to purchase; land contracts; junior mortgages; and home equity loans.

**Gross rent** is the contract rent plus the estimated average monthly cost of utilities (electricity, gas, and water and sewer) and fuels (oil, coal, kerosene, wood, etc.) if these are paid by the renter (or paid for the renter by someone else). Gross rent is intended to eliminate differentials that result from varying practices with respect to the inclusion of utilities and fuels as part of the rental payment. The estimated costs of water and sewer, and fuels are reported on a 12-month basis but are converted to monthly figures for the tabulations.

**Housing cost** as a percentage of income is shown separately for owners with mortgages, owners without mortgages, and renters. Rent as a percentage of income is a computed ratio of gross rent and monthly household income (total household income in the past 12 months divided by 12). Selected owner costs include utilities and fuels, mortgage payments, insurance, taxes, etc. In each case, the ratio of housing cost to income is computed separately for each housing unit. The housing cost ratios for half of all units are above the median shown in this book, and half are below the median shown in the book.

## CIVILIAN LABOR FORCE AND UNEMPLOY-MENT, Items 97–100
### Source: U.S. Bureau of Labor Statistics—Local Area Unemployment Statistics
### http://www.bls.gov/lau/#tables

Data for the civilian labor force are the product of a federal-state cooperative program in which state employment security agencies prepare labor force and unemployment estimates under concepts, definitions, and technical procedures established by the Bureau of Labor Statistics (BLS). The civilian labor force consists of all civilians 16 years old and over who are either employed or unemployed.

Unemployment includes all persons who did not work during the survey week, made specific efforts to find a job during the previous four weeks, and were available for work during the survey week (except for temporary illness). Persons waiting to be called back to a job from which they had been laid off and those waiting to report to a new job within the next 30 days are included in unemployment figures.

Table C includes annual average data for the year shown. The Local Area Unemployment Statistics data are periodically updated to reflect revised inputs, reestimation, and controlling to new statewide totals.

## CIVILIAN EMPLOYMENT, Items 101–103
### Source: U.S. Census Bureau—American Community Survey. 2019 5-year Estimates
### http://www.census.gov/acs/www/

**Total employment** includes all civilians 16 years old and over who were either (1) "at work"—those who did any work at all during the reference week as paid employees, worked in either their own business or profession, worked on their own farm, or worked 15 hours or more as unpaid workers in a family farm or business; or were (2) "with a job, but not at work" —those who had a job but were not at work that week due to illness, weather, industrial dispute, vacation, or other personal reasons.

The **occupational categories** are based on the occupational classification system that was developed for the 2000 census. This system consists of 509 specific occupational categories for employed persons arranged into 23 major occupational groups. This classification was developed based on the *Standard Occupational Classification (SOC) Manual: 2000*, published by the Executive Office of the President, Office of Management and Budget.

Column 102 includes the Management, business, science, and arts occupations category while Column 103 combines the Natural resources, construction, and maintenance occupations and the Production, transportation, and material moving occupations.

## PRIVATE NONFARM EMPLOYMENT AND EARNINGS, Items 104–112

**Source: U.S. Census Bureau—County Business Patterns**
**http://www.census.gov/econ/cbp/index.html**

Data for private nonfarm employment and earnings are compiled from the payroll information reported monthly in the Census Bureau publication *County Business Patterns*. The estimates are based on surveys conducted by the Census Bureau and administrative records from the Internal Revenue Service (IRS).

The following types of employment are excluded from the tables: government employment, self-employed persons, farm workers, and domestic service workers. Railroad employment jointly covered by Social Security and railroad retirement programs, employment on oceanborne vessels, and employment in foreign countries are also excluded.

Annual payroll is the combined amount of wages paid, tips reported, and other compensation (including salaries, vacation allowances, bonuses, commissions, sick-leave pay, and the value of payments-in-kind such as free meals and lodging) paid to employees before deductions for Social Security, income tax, insurance, union dues, etc. All forms of compensation are included, regardless of whether they are subject to income tax or the Federal Insurance Contributions Act tax, with the exception of annuities, third-party sick pay, and supplemental unemployment compensation benefits (even if income tax was withheld). For corporations, total annual payroll includes compensation paid to officers and executives; for unincorporated businesses, it excludes profit or other compensation of proprietors or partners.

## AGRICULTURE, ITEMS 113–132

**Source: U.S. Department of Agriculture, National Agricultural Statistics Service—2017 Census of Agriculture**
**https://www.nass.usda.gov/Publications/AgCensus/2017/index.php**

Data for the 2017 Census of Agriculture were collected in 2018, but pertain to the year 2017.

The Census Bureau took a census of agriculture every 10 years from 1840 to 1920; since 1925, this census has been taken roughly once every 5 years. The 1997 Census of Agriculture was the first one conducted by the National Agricultural Statistics Service of the U.S. Department of Agriculture. Over time, the definition of a farm has varied. For recent censuses (including the 2017 census), a farm has been defined as any place from which $1,000 or more of agricultural products were produced and sold or normally would have been sold during the census year. Dollar figures are expressed in current dollars and have not been adjusted for inflation or deflation.

The term **producer** designates a person who is involved in making decisions for the farm operation. Decisions may include decisions about such things as planting, harvesting, livestock management, and marketing. The producer may be the owner, a member of the owner's household, a hired manager, a tenant, a renter, or a sharecropper. If a person rents land to others or has land worked on shares by others, he/she is considered the producer only of the land which is retained for his/her own operation. The census collected information on the total number of male producers, the total number of female producers, and demographic information for up to four producers per farm.

The acreage designated as **land in farms** consists primarily of agricultural land used for crops, pasture, or grazing. It also includes woodland and wasteland not actually under cultivation or used for pasture or grazing, provided that this land was part of the farm operator's total operation. Land in farms is an operating-unit concept and includes all land owned and operated, as well as all land rented from others. Land used rent-free is classified as land rented from others. All grazing land, except land used under government permits on a per-head basis, was included as "land in farms" provided it was part of a farm or ranch. Land under the exclusive use of a grazing association was reported by the grazing association and included as land in farms. All land in Indian reservations used for growing crops or grazing livestock is classified as land in farms.

**Irrigated land** includes all land watered by any artificial or controlled means, such as sprinklers, flooding, furrows or ditches, sub-irrigation, and spreader dikes. Included are supplemental, partial, and preplant irrigation. Each acre was counted only once regardless of the number of times it was irrigated or harvested. Livestock lagoon waste water distributed by sprinkler or flood systems was also included.

**Total cropland** includes cropland harvested, cropland used only for pasture or grazing, cropland on which all crops failed or were abandoned, cropland in cultivated summer fallow, and cropland idle or used for cover crops or soil improvement but not harvested and not pastured or grazed.

Respondents were asked to report their estimate of the current market **value of land and buildings** owned, rented, or leased from others, and rented and leased to others. Market value refers to the respondent's estimate of what the land and buildings would sell for under current market conditions. If the value of land and buildings was not reported, it was estimated during processing by using the average value of land and buildings from similar farms in the same geographic area.

The **value of machinery and equipment** was estimated by the respondent as the current market value of all cars, trucks, tractors, combines, balers, irrigation equipment, etc., used on the farm. This value is an estimate of what the machinery and equipment would sell for in its present condition and not the replacement or depreciated value. Share interests are reported at full value at the farm where the equipment and machinery are usually kept. Only equipment that was physically located at the farm on December 31, 2017, is included.

**Market value of agricultural products sold** by farms represents the gross market value before taxes and the production expenses of all agricultural products sold or removed from the place in 2017, regardless of who received the payment. It is equivalent to total sales and it includes sales by producers as well as the value of any share received by partners, landlords, contractors, and others associated with the operation. It includes value of organic sales, direct sales and the value of commodities placed in the Commodity Credit Corporation (CCC) loan program. Market value of agricultural products sold does not include payments received for participation in other federal farm programs. Also, it does not include income from farm-related sources such as customwork and other agricultural services, or income from nonfarm sources.

**Organic farms** are those that had organic production according to USDA's National Organic Program (NOP). Respondents reported whether their organic production was certified or exempt from certification and the sales from NOP produced commodities. Not included are farms that had acres transitioning into NOP production.

Farms with **internet access** are those that reported using personal computers, laptops, or mobile devices (e.g., cell phones or tablets) to access the internet. This can be done using services such as dial-up, DSL, cable modem, fiber-optic, mobile internet service for a cell phone or other device (tablet), satellite, or other methods. In 2017 respondents were also able to report connecting with an unknown service type, labeled as "Don't know" in the publication tables.

**Government payments** consist of direct payments as defined by the 2002 Farm Bill;

payments from Conservation Reserve Program (CRP), Wetlands Reserve Program (WRP), Farmable Wetlands Program (FWP), and Conservation Reserve Enhancement Program (CREP); loan deficiency payments; disaster payments; other conservation programs; and all other federal farm programs under which payments were made directly to farm producers, including those specified in the 2014 Agricultural Act (Farm Bill), including Agriculture Risk Coverage (ARC) and Price Loss Coverage (PLC). Commodity Credit Corporation (CCC) proceeds, amount from State and local government agricultural program payments, and federal crop insurance payments were not included in this category.

## WATER CONSUMPTION, Items 133–134
**Source: U.S. Geological Survey, National Water-Use Information Program, Estimated Use of Water in the United States County-Level Data for 2015, version 1.0
http://water.usgs.gov/watuse/**

Every five years, the U.S. Geological Survey compiles county-level water-use estimates. This volume includes the total fresh and saline withdrawals for public water supplies in 2015, expressed as million gallons per day. Estimate of withdrawals of ground and surface water are given for the following categories of use: public water supplies, domestic, commercial, irrigation, livestock, industrial, mining, and thermoelectric power. Only public water supply is included in this volume because the other categories had not been published in time for this book. The number of gallons withdrawn per person is based on the metropolitan area population in 2015 but the water is not necessarily used locally, providing an indicator of metropolitan areas that serve as major water sources.

Public supply refers to water withdrawn from ground and surface sources by public and private water systems for use by cities,

towns, rural water districts, mobile-home parks, Native American Indian reservations, and military bases. Public-supply facilities provide water to at least 25 persons or have a minimum of 15 service connections. Water withdrawn by public suppliers may be delivered to users for domestic, commercial, industrial, and thermoelectric-power purposes, as well as to other public-water suppliers. Public-supply water is also used for public services (public uses)—such as pools, parks, and public buildings—and may have unaccounted uses (losses) because of system leaks or such non-metered services as firefighting, flushing of water lines, or backwashing at treatment plants. Some public-supply water may be used in the processes of water and wastewater treatment. Some public suppliers treat saline water before distributing the water. The definition of saline water for public supply refers to water that requires treatment to reduce the concentration of dissolved solids through the process of desalination or dilution.

## 2017 ECONOMIC CENSUS: OVERVIEW, Items 135–166
**Source: U.S. Census Bureau
https://www.census.gov/programs-surveys/economic-census/library/publications.html**

The Economic Census provides a detailed portrait of the nation's economy, from the national to the local level, once every five years. The 2017 Economic Census covers nearly all of the U.S. economy in its basic collection of establishment statistics. Release of the 2017 Economic Census data began in 2019 and was ongoing as this book was developed. Some sections are not available or are incomplete. All tables in this book use the metropolitan area definitions established in 2018, reflecting substantial changes from the previous definitions. Because the Economic Census data refer to 2017, there are no official datasets that use the new metropolitan area definition so the county data were aggregated for this book. Consequently, suppression at the county level resulted in less than complete totals in some areas. Users are encouraged to review the metropolitan area changes and consult the Census Bureau's website for specific metropolitan areas.

The 1997 Economic Census was the first major data source to use the new North American Industry Classification System (NAICS); therefore, data from this census are not comparable to economic data from prior years, which were based on the Standard Industrial Classification (SIC) system.

NAICS, developed in cooperation with Canada and Mexico, classifies North America's economic activities at two-, three-, four-, and five-digit levels of detail; the U.S. version of NAICS further defines industries to a sixth digit. The Economic Census takes advantage of this hierarchy to publish data at these successive levels of detail: sector (two-digit), subsector (three-digit), industry group (four-digit), industry (five-digit), and U.S. industry (six-digit). Information in Table A is at the two-digit level, with a few three- and four-digit items. The data in Tables B and C are at the two-digit level.

Several key statistics are tabulated for all industries in this volume, including number of establishments (or companies), number of employees, payroll, and certain measures of output (sales, receipts, revenue, value of shipments, or value of construction work done).

**Number of establishments.** An establishment is a single physical location at which business is conducted. It is not necessarily identical with a company or enterprise, which may consist of one establishment or more. Economic Census figures represent a summary of reports for individual establishments rather than companies. For cases in which a census report was received, separate information was obtained for each location where business was conducted. When administrative records of other federal agencies were used instead of a census report, no information was available on the number of locations operated. Each Economic Census establishment was tabulated according to the physical location at which the business was conducted. The count of establishments represents those in business at any time during 2002.

When two activities or more were carried on at a single location under a single ownership, all activities were generally grouped together as a single establishment. The entire establishment was classified on the basis of its major activity and all of its data were included in that classification. However, when distinct and separate economic activities (for which different industry classification codes were appropriate) were conducted at a single location under a single ownership, separate establishment reports for each of the different activities were obtained in the census.

**Number of employees.** Paid employees consist of the full-time and part-time employees, including salaried officers and executives of corporations. Included are employees on paid sick leave, paid holidays, and paid vacations; not included are proprietors and partners of unincorporated businesses. The definition of paid employees is the same as that used by the Internal Revenue Service (IRS) on form 941.

For some industries, the Economic Census gives codes representing the number of employees as a range of numbers (for example, "100 to 249 employees" or "1,000 to 2,499" employees). In this volume, those codes have been replaced by the standard suppression code "D".

**Payroll.** Payroll includes all forms of compensation, such as salaries, wages, commissions, dismissal pay, bonuses, vacation allowances, sick-leave pay, and employee contributions to qualified pension plans paid during the year to all employees. For corporations, payroll includes amounts paid to officers and executives; for unincorporated businesses, it does not include profit or other compensation of proprietors or partners. Payroll is reported before deductions for Social Security, income tax, insurance, union dues, etc. This definition of payroll is the same as that used by on IRS form 941.

**Sales, shipments, receipts, revenue, or business done.** This measure includes the total sales, shipments, receipts, revenue, or business done by establishments within the scope of the Economic Census. The definition of each of these items is specific to the economic sector measured.

# WHOLESALE TRADE, Items 135–138

## Source: U.S. Census Bureau—2017 Economic Census (See Overview of 2017 Economic Census prior to Item 135)

The Wholesale Trade sector (sector 42) comprises establishments engaged in wholesaling merchandise, generally without transformation, and rendering services incidental to the sale of merchandise. The wholesaling process is an intermediate step in the distribution of merchandise.

Wholesalers are organized to sell or arrange the purchase or sale of (1) goods for resale (i.e., goods sold to other wholesalers or retailers), (2) capital or durable nonconsumer goods, and (3) raw and intermediate materials and supplies used in production.

Wholesalers sell merchandise to other businesses and normally operate from a warehouse or office. These warehouses and offices are characterized by having little or no display of merchandise. In addition, neither the design nor the location of the premises is intended to solicit walk-in traffic. Wholesalers do not normally use advertising directed to the general public. In general, customers are initially reached via telephone, in-person marketing, or specialized advertising, which may include the internet and other electronic means. Follow-up orders are either vendor-initiated or client-initiated, are usually based on previous sales, and typically exhibit strong ties between sellers and buyers. In fact, transactions are often conducted between wholesalers and clients that have long-standing business relationships.

This sector is made up of two main types of wholesalers: those that sell goods on their own account and those that arrange sales and purchases for others for a commission or fee.

(1) Establishments that sell goods on their own account are known as wholesale merchants, distributors, jobbers, drop shippers, import/export merchants, and sales branches. These establishments typically maintain their own warehouse, where they receive and handle goods for their customers. Goods are generally sold without transformation, but may include integral functions, such as sorting, packaging, labeling, and other marketing services.

(2) Establishments arranging for the purchase or sale of goods owned by others or purchasing goods on a commission basis are known as agents and brokers, commission merchants, import/export agents and brokers, auction companies, and manufacturers' representatives. These establishments operate from offices and generally do not own or handle the goods they sell.

Some wholesale establishments may be connected with a single manufacturer and promote and sell that particular manufacturer's products to a wide range of other wholesalers or retailers. Other wholesalers may be connected to a retail chain or a limited number of retail chains and only provide the products needed by the particular retail operation(s). These wholesalers may obtain the products from a wide range of manufacturers. Still other wholesalers may not take title to the goods, but act instead as agents and brokers for a commission.

Although wholesaling normally denotes sales in large volumes, durable nonconsumer goods may be sold in single units. Sales of capital or durable nonconsumer goods used in the production of goods and services, such as farm machinery, medium- and heavy-duty trucks, and industrial machinery, are always included in Wholesale Trade.

The metropolitan area table includes only **Merchant wholesalers, except manufacturers' sales branches and offices**, establishments primarily engaged in buying and selling merchandise on their own account. Included here are such types of establishments as wholesale distributors and jobbers, importers, exporters, own-brand importers/marketers, terminal and country grain elevators, and farm products assemblers.

## RETAIL TRADE, Items 139–142
**Source: U.S. Census Bureau—2017 Economic Census (See Overview of 2017 Economic Census prior to Item 135)**

The Retail Trade sector (44–45) is made up of establishments engaged in retailing merchandise, generally without transformation, and rendering services incidental to the sale of merchandise.

The retailing process is the final step in the distribution of merchandise; retailers are therefore organized to sell merchandise in small quantities to the general public. This sector comprises two main types of retailers: store and nonstore retailers.

Store retailers operate fixed point-of-sale locations, located and designed to attract a high volume of walk-in customers. In general, retail stores have extensive displays of merchandise and use mass-media advertising to attract customers. They typically sell merchandise to the general public for personal or household consumption; some also serve business and institutional clients. These include establishments such as office supply stores, computer and software stores, building materials dealers, plumbing supply stores, and electrical supply stores. Catalog showrooms, gasoline service stations, automotive dealers, and mobile home dealers are treated as store retailers.

In addition to retailing merchandise, some types of store retailers are also engaged in the provision of after-sales services, such as repair and installation. For example, new automobile dealers, electronic and appliance stores, and musical instrument and supply stores often provide repair services. As a general rule, establishments engaged in retailing merchandise and providing after-sales services are classified in this sector.

Nonstore retailers, like store retailers, are organized to serve the general public, although their retailing methods differ. The establishments of this subsector reach customers and market merchandise with methods including the broadcasting of "infomercials," the broadcasting and publishing of direct-response advertising, the publishing of paper and electronic catalogs, door-to-door solicitation, in-home demonstration, selling from portable stalls (street vendors, except food), and distribution through vending machines. Establishments engaged in the direct sale (nonstore) of products, such as home heating oil dealers and home-delivery newspaper routes are included in this sector.

The buying of goods for resale is a characteristic of retail trade establishments that distinguishes them from establishments in the Agriculture, Manufacturing, and Construction sectors. For example, farms that sell their products at or from the point of production are classified in Agriculture instead of in Retail Trade. Similarly, establishments that both manufacture and sell their products to the general public are classified in Manufacturing instead of Retail Trade. However, establishments that engage in processing activities incidental to retailing are classified in Retail Trade.

## REAL ESTATE AND RENTAL AND LEASING, Items 143–146
**Source: U.S. Census Bureau—2017 Economic Census (See Overview of 2017 Economic Census prior to Item 135)**

The Real Estate and Rental and Leasing sector (53) comprises establishments primarily engaged in renting, leasing, or otherwise allowing the use of tangible or intangible assets, and establishments providing related services. The major portion of this sector is made up of establishments that rent, lease, or otherwise allow the use of their own assets by others. The assets may be tangible, such as real estate and equipment, or intangible, such as patents and trademarks.

This sector also includes establishments primarily engaged in managing real estate for others, selling, renting, and/or buying real estate for others, and appraising real estate. These activities are closely related to this sector's main activity. In addition, a substantial proportion of property management is self-performed by lessors.

The main components of this sector are the real estate lessors industries; equipment lessors industries (including motor vehicles, computers, and consumer goods); and lessors of nonfinancial intangible assets (except copyrighted works).

## PROFESSIONAL, SCIENTIFIC, AND TECHNICAL SERVICES, Items 147–150
**Source: U.S. Census Bureau—2017 Economic Census (See Overview of 2017 Economic Census prior to Item 135)**

The Professional, Scientific, and Technical Services sector (54) is made up of establishments that specialize in performing professional, scientific, and technical activities for others. These activities require a high degree of expertise and training. The establishments in this sector specialize in one or more areas and provide services to clients in a variety of industries (and, in some cases, to households). Activities performed include legal advice and representation; accounting, bookkeeping, and payroll services; architectural, engineering, and specialized design services; computer services; consulting services; research services; advertising services; photographic services; translation and interpretation services; veterinary services; and other professional, scientific, and technical services.

This sector excludes establishments primarily engaged in providing a range of day-to-day office administrative services, such as financial planning, billing and record keeping, personnel services, and physical distribution and logistics services. These establishments are classified in sector 56, Administrative and Support and Waste Management and Remediation Services.

## MANUFACTURING, Items 151–154
**Source: U.S. Census Bureau—2017 Economic Census (See Overview of 2017 Economic Census prior to Item 135)**

The Manufacturing sector (31–33) is made up of establishments engaged in the mechanical, physical, or chemical transformation of materials, substances, or components into new products. The assembling of component parts of manufactured products is considered manufacturing, except in cases in which the activity is appropriately classified in the Construction sector. Establishments in the Manufacturing sector are often described as plants, factories, or mills, and characteristically use power-driven machines and materials-handling equipment. However, establishments that transform materials or substances into new products by hand or in the worker's home, and establishments engaged in selling to the general public products made on the same premises from which they are sold (such as bakeries, candy stores, and custom tailors) may also be included in this sector. Manufacturing establishments may process materials or contract with other establishments to process their materials for them. Both types of establishments are included in the Manufacturing sector.

The materials, substances, or components transformed by manufacturing establishments are raw materials that are products of agriculture, forestry, fishing, mining, or quarrying, or are products of other manufacturing establishments. The materials used may be purchased directly from producers, obtained through customary trade channels, or secured without recourse to the market by transferring the product from one establishment to another, under the same ownership. The new product of a manufacturing establishment may be finished (in the sense that it is ready for utilization or consumption), or it may be semifinished to become an input for an establishment engaged in further manufacturing. For example, the product of the alumina refinery is the input used in the primary production of aluminum; primary aluminum is the input used in an aluminum wire drawing plant; and aluminum wire is the input used in a fabricated wire product manufacturing establishment.

Data are included for counties with 500 or more employees in the Manufacturing sector.

## ACCOMMODATION AND FOOD SERVICES, Items 155–158

**Source: U.S. Census Bureau—2017 Economic Census (See Overview of 2017 Economic Census prior to Item 135)**

The Accommodation and Food Services sector (72) consists of establishments that provide customers with lodging and/or meals, snacks, and beverages for immediate consumption. This sector includes both accommodation and food services establishments because the two activities are often combined at the same establishment.

Excluded from this sector are civic and social organizations, amusement and recreation parks, theaters, and other recreation or entertainment facilities providing food and beverage services.

## HEALTH CARE AND SOCIAL ASSISTANCE, Items 159–162

**Source: U.S. Census Bureau—2017 Economic Census (See Overview of 2017 Economic Census prior to Item 135)**

The Health Care and Social Assistance sector (62) consists of establishments that provide health care and social assistance services to individuals. The sector includes both health care and social assistance because it is sometimes difficult to distinguish between the boundaries of these two activities. The industries in this sector are arranged on a continuum, starting with establishments that provide medical care exclusively, continuing with those that provide health care and social assistance, and finishing with those that provide only social assistance. The services provided by establishments in this sector are delivered by trained professionals. All industries in the sector share this commonality of process—namely, labor inputs of health practitioners or social workers with the requisite expertise. Many of the industries in the sector are defined based on the educational degree held by the practitioners included in the industry.

Excluded from this sector are aerobic classes, which can be found in subsector 713, Amusement, Gambling, and Recreation Industries; and nonmedical diet and weight-reducing centers, which can be found in subsector 812, Personal and Laundry Services. Although these can be viewed as health services, they are not typically delivered by health practitioners.

## OTHER SERVICES, EXCEPT PUBLIC ADMINISTRATION Items 163–166

**Source: U.S. Census Bureau—2017 Economic Census (See Overview of 2017 Economic Census prior to Item 135)**

The Other Services, Except Public Administration sector (81) comprises establishments engaged in providing services not specifically categorized elsewhere in the classification system. Establishments in this sector are primarily engaged in activities such as equipment and machinery repairing, promoting or administering religious activities, grant making, and advocacy; this sector also includes establishments that provide dry-cleaning and laundry services, personal care services, death care services, pet care services, photofinishing services, temporary parking services, and dating services.

Private households that employ workers on or about the premises in activities primarily concerned with the operation of the household are included in this sector.

Excluded from this sector are establishments primarily engaged in retailing new equipment and performing repairs and general maintenance on equipment. These establishments are classified in sector 44–45, Retail Trade.

## NONEMPLOYER BUSINESSES, Items 167 and 168

**Source: U.S. Census Bureau—Nonemployer Statistics https://www.census.gov/programs-surveys/nonemployer-statistics.html**

Nonemployer Statistics is an annual series that provides subnational economic data for businesses that have no paid employees and are subject to federal income tax. The data consist of the number of businesses and total receipts by industry. Most nonemployers are self-employed individuals operating unincorporated businesses (known as sole proprietorships), which may or may not be the owner's principal source of income.

The majority of all business establishments in the United States are nonemployers, yet these firms average less than 4 percent of all sales and receipts nationally. Due to their small economic impact, these firms are excluded from most other Census Bureau business statistics (the primary exception being the Survey of Business Owners). The Nonemployers Statistics series is the primary resource available to study the scope and activities of nonemployers at a detailed geographic level.

## BUILDING PERMITS, Items 169 and 170

Source: U.S. Census Bureau—Building Permits Survey
http://www.census.gov/construction/bps/

These figures represent private residential construction authorized by building permits in approximately 20,000 places in the United States. Valuation represents the expected cost of construction as recorded on the building permit. This figure usually excludes the cost of on-site and off-site development and improvements, as well as the cost of heating, plumbing, electrical, and elevator installations.

National, state, and county totals were obtained by adding the data for permit-issuing places within each jurisdiction. Not all areas of the country require a building or zoning permit. The statistics only represent those areas that do require a permit. These totals thus are limited to permits issued in the 20,000 place universe covered by the Census Bureau and may not include all permits issued within a state. Current surveys indicate that construction is undertaken for all but a very small percentage of housing units authorized by building permits.

Residential building permits include buildings with any number of housing units. Housing units exclude group quarters (such as dormitories and rooming houses), transient accommodations (such as transient hotels, motels, and tourist courts), "HUD-code" manufactured (mobile) homes, moved or relocated units, and housing units created in an existing residential or nonresidential structure.

## METROPOLITAN AREA LOCAL GOVERNMENT EMPLOYMENT AND PAYROLL, Items 171-179

Source: U.S. Census Bureau—2017 Census of Governments
https://www.census.gov/programs-surveys/cog.html

These items include data for all local governments (i.e., counties, municipalities, townships, special districts, and school districts) located within the metropolitan area. The Census of Governments identifies the scope and nature of the nation's state and local government sector; provides authoritative benchmark figures of public finance and public employment; classifies local government organizations, powers, and activities; and measures federal, state, and local fiscal relationships. The Employment component was mailed **March 2012** to collect information on the number of state and local government civilian employees and their payrolls.

Government employees include all persons paid for personal services performed, including persons paid from federally funded programs, paid elected or appointed officials, persons in a paid leave status, and persons paid on a per meeting, annual, semi-annual, or quarterly basis. Unpaid officials, pensioners, persons whose work is performed on a fee basis, and contractors and their employees are excluded from the count of employees. **Full-Time Equivalent employees** is a computed statistic representing the number of full-time employees that could have been employed if the reported number of hours worked by part-time employees had been worked by full-time employees. This statistic is calculated separately for each function of a government by dividing the "part-time hours paid" by the standard number of hours for full-time employees in the particular government and then adding the resulting quotient to the number of full-time employees.

**March payroll** represents gross payroll amounts for the one-month period of March for full-time and part-time employees. Gross payroll includes all salaries, wages, fees, commissions, and overtime paid to employees **before** withholdings for taxes, insurance, etc. It also includes incentive payments that are paid at regular pay intervals. It excludes employer share of fringe benefits like retirement, Social Security, health and life insurance, lump sum payments, and so forth.

**Administration and Judicial and Legal** combines **Financial administration, Other government administration, and Judicial and Legal** activities. **Financial administration includes** activities concerned with tax assessment and collection, custody and disbursement of funds, debt management, administration of trust funds, budgeting, and other government-wide financial management activities. This function is not applied to school district or special district governments. **Other government administration** applies to the legislative and government-wide administrative agencies of governments. Included here are overall planning and zoning activities, and central personnel and administrative activities. This function is not applied to school district or special district governments. **Judicial and legal** includes all court and court related activities (except probation and parole activities that are included at the "Correction" function), court activities of sheriff's offices, prosecuting attorneys' and public defenders' offices, legal departments, and attorneys providing government-wide legal service.

**Police and Corrections includes** all activities concerned, with the enforcement of law and order, including coroner's offices, police training academies, investigation bureaus, and local jails, "lockups", or other detention facilities not intended to serve as correctional facilities. **Corrections** includes activities pertaining to the confinement and correction of adults and minors convicted of criminal offenses. Pardon, probation, and parole activities are also included here.

**Fire protection** includes local government fire protection and prevention activities plus any ambulance, rescue, or other auxiliary services provided by a fire protection agency. Volunteer firefighters, if remunerated for their services on a "per fire" or some other basis, are included as part-time employees.

**Highways and transportation includes** activities associated with the maintenance and operation of streets, roads, sidewalks, bridges, tunnels, toll roads, and ferries. Snow and ice removal, street lighting, and highway and traffic engineering activities are also included here. Also included are the operation, maintenance, and construction of public mass transit systems, including subways, surface rails, and buses, and the provision, construction, operation, maintenance; support of public waterways, harbors, docks, wharves, and related marine terminal facilities; and activities associated with the operation and support of publicly operated airport facilities.

**Health and Welfare includes Health, Hospitals, and Public welfare. Health** includes administration of public health programs, community and visiting nurse services, immunization programs, drug abuse rehabilitation programs, health and food inspection activities, operation of outpatient clinics, and environmental pollution control activities. **Hospitals** includes only government operated medical care facilities that provide inpatient care. Employees and payrolls of private corporations that lease and operate government-owned hospital facilities are excluded.

**Public Welfare** includes the administration of various public assistance programs for the needy, veteran services, operation of nursing homes, indigent care institutions, and programs that provide payments for medical care, handicap transportation, and other services for the needy.

**Natural resources and Utilities** includes activities primarily concerned with the conservation and development of natural resources (soil, water, energy, minerals, etc.) and the regulation of industries that develop, utilize, or affect natural resources, as well as the operation and maintenance of **parks**, playgrounds, swimming pools, public beaches, auditoriums, public golf courses, museums, marinas, botanical gardens, and zoological parks. **Utilities, sewerage, and waste management** includes operation, maintenance, and construction of public water supply systems, including production, acquisition, and distribution of water to general public or to other public or private utilities, for residential, commercial, and industrial use; activities associated with the production or acquisition and distribution of electric power; provision, maintenance, and operation of sanitary and storm sewer systems and sewage disposal and treatment facilities; and refuse collection and disposal, operation of sanitary landfills, and street cleaning activities.

**Education and libraries includes** activities associated with the operation of public elementary and secondary schools and locally operated vocational-technical schools. Special education programs operated by elementary and secondary school systems are also included as are all ancillary services associated with the operation of schools, such as pupil transportation and food service. **Also included are the e**stablishment and provision of libraries for use by the general public and the technical support of privately operated libraries. This category includes classroom teachers, principals, supervisors of instruction, librarians, teacher aides, library aides, and guidance and psychological personnel as well as school superintendents and other administrative personnel, clerical and secretarial staffs, plant operation and maintenance personnel, health and recreation employees, transportation and food service personnel, and any student employees. Also included are any degree granting institutions that provide academic training above grade 12.

## LOCAL GOVERNMENT FINANCES, Items 180–193

**Source: U.S. Census Bureau—2017 Census of Governments**
**https://www.census.gov/programs-surveys/cog.html**

Data on local government finances are based on result of the 2017 Census of Governments. For each metropolitan area, the data are aggregated from its component counties, and the financial data comprise amounts for all local governments—not only the county governments, but also any municipalities, townships, school districts, and special districts within the county. Statistics from governmental units located in two or more county areas are assigned to the county area containing the administrative office.

Revenue and expenditure items include all amounts of money received and paid out, respectively, by a government and its agencies (net of correcting transactions such as recoveries of refunds), with the exception of amounts for debt issuance and retirement and for loan and investment, agency, and private transactions.

Payments among the various funds and agencies of a particular government are excluded from revenue and expenditure items as representing internal transfers. Therefore, a government's contribution to a retirement fund that it administers is not counted as expenditure, nor is the receipt of this contribution by the retirement fund counted as revenue.

Total **general revenue** includes all revenue except utility, liquor stores, and insurance trust revenue. All tax revenue and intergovernmental revenue, even if designated for employee-retirement or local utility purpose, are classified as general revenue.

**Intergovernmental revenue** covers amounts received from the federal government as fiscal aid, reimbursements for performance of general government functions and specific services for the paying government, or in lieu of taxes. It excludes any amounts received from other governments from the sale of property, commodities, and utility services.

**Taxes** consist of compulsory contributions exacted by governments for public purposes. However, this category excludes employer and employee payments for retirement and social insurance purposes, which are classified as insurance trust revenue; it also excludes special assessments, which are classified as non-tax general revenue. Property taxes are taxes conditioned on ownership of property and assessed by its value. Sales and gross receipts taxes do not include dealer discounts, or "commissions" allowed to merchants for collection of taxes from consumers. General sales taxes and selected taxes on sales of motor fuels, tobacco products, and other particular commodities and services are included.

**General government expenditure** includes capital outlay, a major portion of which is commonly financed by borrowing. Government revenue does not include receipts from borrowing. Among other things, this distorts the relationship between totals of revenue and expenditure figures that are presented and renders it useless as a direct measure of the degree of budgetary "balance" (as that term is generally applied).

**Direct general expenditure** comprises all expenditures of the local governments, excluding utility, liquor stores, insurance trust expenditures, and any intergovernmental payments.

Local government expenditure for **education** is mainly for the provision and general support of schools and other educational facilities and services, including those for educational institutions beyond high school. They cover such related services as student transportation; school lunch and other cafeteria operations; school health, recreation, and library services; and dormitories, dining halls, and bookstores operated by public institutions of higher education.

**Health and hospital expenditure** includes health research; clinics; nursing; immunization; other categorical, environmental, and general health services provided by health agencies; establishment and operation of hospital facilities; provision of hospital care; and support of other public and private hospitals.

**Police protection expenditure** includes police activities such as patrols, communications, custody of persons awaiting trial, and vehicular inspection.

**Public welfare expenditure** covers support of and assistance to needy persons; this aid is contingent upon the person's needs. Included are cash assistance paid directly to needy persons under categorical (Old Age Assistance, Temporary Assistance for Needy Families, Aid to the Blind, and Aid to the Disabled) and other welfare programs; vendor payments made directly to private purveyors for medical care, burials, and other commodities and services provided under welfare programs; welfare institutions; and any intergovernmental or other direct expenditure for welfare purposes. Pensions to former employees and other benefits not contingent on need are excluded.

**Highway expenditure** is for the provision and maintenance of highway facilities, including toll turnpikes, bridges, tunnels, and ferries, as well as regular roads, highways, and streets. Also included are expenditures for street lighting and for snow and ice removal. Not included are highway policing and traffic control, which are considered part of police protection

**Debt outstanding** includes all long-term debt obligations of the government and its agencies (exclusive of utility debt) and all interest-bearing, short-term (repayable within one year) debt obligations remaining unpaid at the close of the fiscal year. It includes judgments, mortgages, and revenue bonds, as well as general obligation bonds, notes, and interest-bearing warrants. This category consists of non-interest-bearing, short-term obligations; inter-fund obligations; amounts owed in a trust or agency capacity; advances and contingent loans from other governments; and rights of individuals to benefit from government-administered employee-retirement funds.

# GOVERNMENT EMPLOYMENT, Items 194–196

**Source: U.S. Bureau of Economic Analysis—Regional Economic Accounts**
**http://www.bea.gov/regional/index.htm#state**

Employment is measured as the average annual sum of full-time and part-time jobs. The estimates are on a place-of-work basis. The estimates are on a place-of-work basis. State and local government employment includes person employed in all state and local government agencies and enterprises. Data for federal civilian employment include civilian employees of the federal government, including civilian employees of the Department of Defense. Military employment includes all persons on active duty status.

# INDIVIDUAL INCOME TAX RETURNS, Items 197–199

**Source: U.S. Internal Revenue Service, Statistics of Income Program**
**https://www.irs.gov/uac/soi-tax-stats-county-data-2018**

The Revenue Act of 1916 mandated the annual publication of statistics related to "the operations of the internal revenue laws" as they affect individuals, all forms of businesses, estates. nonprofit organizations, trusts, and investments abroad and foreign investments in the United States. The Statistics of Income (SOI) division fulfills this function by collecting and processing data so that they become informative and by sharing information about how the tax system works with other government agencies and the general public. Publication types include traditional print sources, internet files, CD-ROMs, and files sent via e-mail. SOI has an information office, Statistical Information Services, to facilitate the dissemination of SOI data.

SOI bases its county data on administrative records of individual income tax returns (Forms 1040) from the Internal Revenue Service (IRS) Individual Master File (IMF) system. Included in these data are returns filed during the 12-month period, January 1, 2017 to December 31, 2017. While the bulk of returns filed during the 12-month period are primarily for Tax Year 2016, the IRS received a limited number of returns for tax years before 2016 and these have been included within the county data.

Data do not represent the full U.S. population because many individuals are not required to file an individual income tax return. The address shown on the tax return may differ from the taxpayer's actual residence. State and county codes were based on the ZIP code shown on the return. Excluded were tax returns filed without a ZIP code and returns filed with a ZIP code that did not match the State code shown on the return.

SOI did not attempt to correct any ZIP codes on the returns; however, it did take the following precautions to avoid disclosing information about specific taxpayers: Excluded from the data are items with less than 20 returns within a county. Also excluded are tax returns representing a specified percentage of the total of any particular cell. For example, if one return represented 75 percent of the value of a given cell, the return was suppressed from the county detail. The actual threshold percentage used cannot be released.

Column 197 show the number of returns. Column 198 shows the mean Adjusted Gross Income for the metropolitan area and column 199 shows the mean income tax (line 56 on Form 1040). for the county.

## TABLE D—CITIES

Table D present 146 items of data for cities with populations of 25,000 or more at the time of the 2010 census.

### LAND AREA, Items 1 and 4

**Source: U.S. Census Bureau—2020 U.S. Gazetteer Files,**
**https://www.census.gov/geographies/reference-files/time-series/geo/gazetteer-files.html**

Land area measurements are shown to the nearest square mile. Land area is an area measurement providing the size, in square miles, of the land portions of each county.

### POPULATION, Items 2-4

**Source: U.S. Census Bureau—Population Estimates**
**https://www.census.gov/programs-surveys/popest.html**

The population data are Census Bureau estimates of the resident population as of July 1 of the year shown.

The Census Bureau released these 2020 estimates for research purposes. The July 1, 2020 estimates are based on the 2010 Census and were created without incorporation or consideration of the 2020 Census results. They are typically used in comparisons with the 2020 Census to make determinations about the accuracy of the estimates. Differences between the estimates and census counts are interpreted as error in the estimates and are used to inform population and housing unit estimates research and methodological improvements over the decade.

### POPULATION CHARACTERISTICS, Items 5–22

**Source: U.S. Census Bureau—American Community Survey 2019 1-year Supplemental Estimates**
**http://www.census.gov/acs/www/**

Data on age, sex, race, Hispanic origin, and place of birth are from the 2019 American Community Survey, a nationwide continuous survey designed to replace the long form questionnaire used in previous censuses.

Data on race were derived from answers to the question on race that was asked of all persons. The concept of race, as used by the Census Bureau, reflects self-identification by respondents according to the race or races with which they most closely identify. These categories are sociopolitical constructs and should not be interpreted as being scientific or anthropological in nature. Furthermore, the race categories include both racial and national origin groups.

On the American Community Survey, respondents were offered the option of selecting one or more races. This option was not available prior to the 2000 census; thus, comparisons between censuses should be made with caution. In this table, Columns 5 through 10 refer to individuals who identified with one specific racial category, while Column 11 includes those who selected more than one race.

The **White** population is defined as persons who indicated their race as White, as well as persons who did not classify themselves in one of the specific race categories listed on the questionnaire but entered a nationality such as Irish, German, Italian, Lebanese, Near Easterner, Arab, or Polish.

The **Black** population includes persons who indicated their race as "Black or AfricanAmerican" as well as persons who did not classify themselves in one of the specific race categories but reported entries such as African American, Afro American, Kenyan, Nigerian, or Haitian.

The **American Indian or Alaska Native** population includes persons who indicated their race as American Indian or Alaska Native, as well as persons who did not classify themselves in one of the specific race categories but reported entries such as Canadian Indian, French-American Indian, Spanish-American Indian, Eskimo, Aleut, Alaska Indian, or any of the American Indian or Alaska Native tribes.

The **Asian** population includes persons who indicated their race as Asian Indian, Chinese, Filipino, Japanese, Korean, Vietnamese, or "Other Asian," as well as persons who provided write-in entries of such groups as Cambodian, Laotian, Hmong, Pakistani, or Taiwanese.

The **Native Hawaiian or Other Pacific Islander** population includes persons who indicated their race as "Native Hawaiian," "Guamanian or Chamorro," "Samoan," or "Other Pacific Islander," as well as persons who reported entries such as Part Hawaiian, American Samoan, Fijian, Melanesian, or Tahitian.

**Some Other Race** includes all other responses not included in the "White," "Black or African American," "American Indian or Alaska Native," "Asian," and "Native Hawaiian or Other Pacific Islander" race categories described above. Respondents reporting entries such as multiracial, mixed, interracial, or a Hispanic, Latino, or Spanish group (for example, Mexican, Puerto Rican, Cuban, or Spanish) in response to the race question are included in this category.

**Two or More Races.** People may choose to provide two or more races either by checking two or more race response check boxes, by providing multiple responses, or by some combination of check boxes and other responses. The race response categories shown on the questionnaire are collapsed into the five minimum race groups identified by OMB, and the Census Bureau's "Some Other Race" category.

The **Hispanic population** is based on a separate question that asked respondents "Is this person Spanish/Hispanic/Latino?" Persons marking any one of the four Hispanic categories (i.e., Mexican, Puerto Rican, Cuban, or other Spanish) are collectively referred to as Hispanic.

The Hispanic origin question was placed before the race question and specific instructions indicated that both questions should be answered.

The **foreign-born** population includes all persons who were not U.S. citizens at birth. Foreign-born persons are those who indicated they were either a U.S. citizen by naturalization or were not a citizen of the United States. The foreign-born population consists of immigrants (legal permanent residents), temporary migrants (students), humanitarian migrants (refugees), and unauthorized migrants (persons illegally residing in the United States).

**Age** is defined as age at last birthday (number of completed years since birth), at the time of the interview. The American Community Survey also asked for the specific date of birth of the respondent. Both age and date of birth are used in combination to calculate the most accurate age at the time of the interview.

The **female** population is shown as a percentage of total population.

## POPULATION CHANGE, Items 23–26

**Source: U.S. Census Bureau—Decennial Censuses**

**U.S. Census Bureau—Population Estimates**

http://www.census.gov/main/www/cen2000.html
https://www.census.gov/programs-surveys/decennial-census/data/datasets.2010.html
https://www.census.gov/programs-surveys/popest.html

The population data for 2000 and 2010 are from the decennial censuses and represent the resident population as of April 1 of those years. The data for 2020 are from the Census Bureau's Population Estimates Program and represent the estimated resident population as of July 1.

The change in population from 2000 to 2010 and from 2010 to 2020 is calculated from census data based on city boundaries as they existed in 2000 and 2010, respectively. No attempt was made to adjust the data to reflect boundary changes.

## HOUSEHOLDS, Items 27–34

**Source: U.S. Census Bureau—American Community Survey, 2019 1-year Supplemental Estimates**
http://www.census.gov/acs/www/

A **household** includes all of the persons who occupy a housing unit. (Persons not living in households are classified as living in group quarters.) A housing unit is a house, an apartment, a mobile home, a group of rooms, or a single room occupied (or, if vacant, intended for occupancy) as separate living quarters. Separate living quarters are those in which the occupants live separately from any other persons in the building and have direct access from the outside of the building or through a common hall. The occupants may be a single family, one person living alone, two or more families living together, or any other group of related or unrelated persons who share living quarters. The number of households is the same as the number of year-round occupied housing units.

A **family** includes a householder and one or more other persons living in the same household who are related to the householder by birth, marriage, or adoption. All persons in a household who are related to the householder are regarded as members of his or her family. A **family household** may contain persons not related to the householder; thus, family households may include more members than families do. A household can contain only one family for the purposes of census tabulations. Not all households contain families, as a household may comprise a group of unrelated persons or of one person living alone. Families are classified by type as either a "married couple family" or "other family," according to the presence or absence of a spouse.

A **Married-Couple Family** is a family in which the householder and his or her spouse are listed as members of the same household. The category **female family householder** includes only female-headed family households with no spouse present.

A **Nonfamily Household** includes a householder living alone or with nonrelatives only. Unmarried couples households, whether opposite-sex or same-sex, with no relatives of the householder present are tabulated in nonfamily households.

The measure of **persons per household** is obtained by dividing the number of persons in households by the number of households or householders. One person in each household is designated as the householder. In most cases, this is the person (or one of the persons) in whose name the house is owned, being bought, or rented. If there is no such person in the household, any adult household member 15 years old and over can be designated as the householder.

## CRIME, Items 35–38

**Source: U.S. Federal Bureau of Investigation—Uniform Crime Reports**
https://ucr.fbi.gov/crime-in-the-u.s/2019/crime-in-the-u.s.-2019

Crime data are as reported to the Federal Bureau of Investigation (FBI) by law enforcement agencies and have not been adjusted for underreporting. This may affect comparability between geographic areas or over time.

Through the voluntary contribution of crime statistics by law enforcement agencies across the United States, the Uniform Crime Reporting (UCR) Program provides periodic assessments of crime in the nation as measured by offenses that have come to the attention of the law enforcement community. The Committee on Uniform Crime Records of the International Association of Chiefs of Police initiated this voluntary national data collection effort in 1930. The UCR Program contributors compile and submit their crime data either directly to the FBI or through state-level UCR Programs.

Seven offenses, because of their severity, frequency of occurrence, and likelihood of being reported to police, were initially selected to serve as an index for evaluating fluctuations in the volume of crime. These serious crimes were murder and nonnegligent manslaughter, forcible rape, robbery, aggravated assault, burglary, larceny-theft, and motor vehicle theft. By congressional mandate, arson was added as the eighth index offense in 1979. The totals shown in this volume do not include arson.

In 2004, the FBI discontinued the use of the Crime Index in the UCR Program and its publications, stating that the Crime Index was driven upward by the offense with the highest number of cases (in this case, larceny-theft), creating a bias against jurisdictions with a high number of larceny-thefts but a low number of other serious crimes, such as murder and forcible rape. The FBI is currently publishing a violent crime total and property crime total until a more viable index is developed. This book includes the total Crime Index, as well as violent crime and property crime rates.

In 2013, the FBI adopted a new definition of rape. Rape is now defined as, "Penetration, no matter how slight, of the vagina or anus with any body part or object, or oral penetration by a sex organ of another person, without the consent of the victim." The new definition updated the 80-year-old historical definition of rape which was "carnal knowledge of a female forcibly and against her will." Effectively, the revised definition expands rape to include both male and female victims and offenders, and reflects the various forms of sexual penetration understood to be rape, especially nonconsenting acts of sodomy, and sexual assaults with objects.

**Violent crimes** include four categories of offenses: (1) Murder and nonnegligent manslaughter, as defined in the UCR Program, is the willful (nonnegligent) killing of one human being by another. This offense excludes deaths caused by negligence, suicide, or accident; justifiable homicides; and attempts to murder or assaults to murder. (2) ) Rape is the penetration, no matter how slight, of the vagina or anus with any body part or object, or oral penetration by a sex organ of another person, without the consent of the victim Assaults or attempts to commit rape by force or threat of force are also included; however, statutory rape (without force) and other sex offenses are excluded. (3) Robbery is the taking or attempting to take anything of value from the care, custody, or control of a person or persons by force or threat of force or violence and/or by putting the victim in fear. (4) Aggravated assault is an unlawful attack by one person upon another for the purpose of inflicting severe or aggravated bodily injury. This type of assault is usually accompanied by the use of a weapon or by other means likely to produce death or great bodily harm. Attempts are included, since injury does not necessarily have to result when a gun, knife, or other weapon is used, as these incidents could and probably would result in a serious personal injury if the crime were successfully completed.

**Property crimes** include three categories: (1) Burglary, or breaking and entering, is the unlawful entry of a structure to commit a felony or theft, even though no force was used to gain entrance. (2) Larceny-theft is the unauthorized taking of the personal property of another, without the use of force. (3) Motor vehicle theft is the unauthorized taking of any motor vehicle.

Rates are based on population estimates provided by the FBI. If a city is not in the UCR database, or if the population total for the units aggregated was less than 75 percent of the city's population (as estimated by the Census Bureau), the total was not considered representative of the city as a whole and was not published.

## EDUCATIONAL ATTAINMENT, Items 39–41

**Source: U.S. Census Bureau—American Community Survey, 2019 1-year Supplemental Estimates http://www.census.gov/acs/www/**

Data on **educational attainment** are tabulated for the population 25 years old and over. The data were derived from a question that asked respondents for the highest level of school completed or the highest degree received. Persons who had passed a high school equivalency examination were considered high school graduates. Schooling received in foreign schools was to be reported as the equivalent grade or years in the regular American school system.

Vocational and technical training, such as barber school training; business, trade, technical, and vocational schools; or other training for a specific trade are specifically excluded.

**High school graduate or less.** This category includes persons whose highest degree was a high school diploma or its equivalent, and those who reported any level lower than a high school diploma.

Bachelor's degree or more. This category includes persons who have received bachelor's degrees, master's degrees, professional school degrees (such as law school or medical school degrees), and doctoral degrees.

## INCOME, Items 42–49

**Source: U.S. Census Bureau—American Community Survey, 2019 1-year Supplemental Estimates http://www.census.gov/acs/www/**

Total money income is the sum of the amounts reported separately for wage or salary income; net self-employment income; interest, dividends, or net rental or royalty income or income from estates and trusts; Social Security or railroad retirement income; Supplemental Security Income (SSI); public assistance or welfare payments; retirement, survivor, or disability pensions; and all other income. Receipts from the following sources are not included as income: capital gains; money received from the sale of property (unless the recipient was engaged in the business of selling such property); the value of income "in kind" from food stamps, public housing subsidies, medical care, employer contributions for individuals, etc.; withdrawal of bank deposits; money borrowed; tax refunds; exchange of money between relatives living in the same household; and gifts, lump-sum inheritances, insurance payments, and other types of lump-sum receipts.

**Household income** includes the income of the householder and all other individuals 15 years old and over in the household, whether or not they are related to the householder. Since many households consist of only one person, median household income is usually less than median family income. Although the household income statistics cover the twelve months prior to the survey, the characteristics of individuals and the composition of households refer to the date of the interview. Thus, the income of the household does not include amounts received by individuals who were members of the household during all or part of the year if these individuals were no longer residing in the household at the time of the interview. Similarly, income amounts reported by individuals who did not reside in the household during full year but who were members of the household at the time of the interview are included. However, the composition of most house-holds was the same during the year as it was at the time of the interview.

**Income of Families** – In compiling statistics on family income, the incomes of all members 15 years old and over related to the householder are summed and treated as a single amount.

**Median income** divides the income distribution into two equal parts, with half of all cases below the median income level and half of all cases above the median income level. For households, the median income is based on the distribution of the total number of households, including those with no income. Median income for households and families is computed on the basis of a standard distribution with a minimum value of less than $2,500 and a maximum value of $200,000 or more and is rounded to the nearest whole dollar. Median income figures are calculated using linear interpolation if the width of the interval containing the estimate is $2,500 or less. If the width of the interval containing the estimate is greater than $2,500, Pareto interpolation is used.

Income components were reported for the 12 months preceding the interview month. Monthly Consumer Price Index (CPI) factors were used to inflation-adjust these components to a reference calendar year (January through December). For example, a household interviewed in March 2017 reports their income for March 2016 through February 2017. Their income is adjusted to the 2017 reference calendar year by multiplying their reported income by 2017 average annual CPI (January-December 2017) and then dividing by the average CPI for March 2016-February 2017. In addition, the 3-year estimates are inflation-adjusted to the final year. However, the estimates do not reflect the price and cost-of-living differences that may exist between areas.

**Earnings** – Earnings are defined as the sum of wage or salary income and net income from self-employment. "Earnings" represent the amount of income received regularly for people 16 years old and over before deductions for personal income taxes, Social Security, bond purchases, union dues, Medicare deductions, etc. An individual with earnings is one who has either wage/salary income or self-employment income, or both. Respondents who "break even" in self-employment income and therefore have zero self-employment earnings also are considered "individuals with earnings."

## HOUSING, Items 50–54
### Source: U.S. Census Bureau--American Community Survey, 2019 1-year Supplemental Estimates
http://www.census.gov/acs/www/

The characteristics of occupied housing units are from the 2019 American Community Survey.

A housing unit is a house, apartment, mobile home or trailer, group of rooms, or single room occupied or, if vacant, intended for occupancy as separate living quarters. Separate living quarters are those in which the occupants do not live and eat with any other person in the structure and which have direct access from the outside of the building through a common hall. For vacant units,

the criteria of separateness and direct access are applied to the intended occupants whenever possible. If that information cannot be obtained, the criteria are applied to the previous occupants.

The occupants of a housing unit may be a single family, one person living alone, two or more families living together, or any other group of related or unrelated persons who share living quarters. Both occupied and vacant housing units are included in the housing inventory, although recreational vehicles, tents, caves, boats, railroad cars, and the like are included only if they are occupied as a person's usual place of residence.

A housing unit is classified as occupied if it is the usual place of residence of the person or group of persons living in it at the time of enumeration, or if the occupants are only temporarily absent (away on vacation). A household consists of all persons who occupy a housing unit as their usual place of residence. Vacant units for sale or rent include units rented or sold but not occupied and any other units held off the market.

A housing unit is owner occupied if the owner or co-owner lives in the unit, even if it is mortgaged or not fully paid for. The owner or co-owner must live in the unit and is usually the first person listed on the census or ACS questionnaire.

All occupied housing units that are not owner occupied, whether they are rented for cash rent or occupied without payment of cash rent, are classified as renter occupied.

**Median value** is the dollar amount that divides the distribution of specified owner-occupied housing units into two equal parts, with half of all units below the median value and half of all units above the median value. Value is defined as the respondent's estimate of what the house would sell for if it were for sale. Data are presented for single-family units on fewer than 10 acres of land that have no business or medical offices on the property.

**Median rent** divides the distribution of renter-occupied housing units into two equal parts. The rent concept used in this volume is gross rent, which includes the amount of cash rent a renter pays (contract rent) plus the estimated average cost of utilities and fuels, if these are paid by the renter. The rent is the amount of rent only for living quarters and excludes amounts paid for any business or other space occupied. Single-family houses on lots of 10 or more acres of land are also excluded.

## COMMUTING, Items 55 and 56
### Source: U.S. Census Bureau—American Community Survey, 2019 1-year Supplemental Estimates
http://www.census.gov/acs/www/

**Means of transportation to work** refers to the principal mode of travel or type of conveyance that a worker usually used to get from home to work during the reference week. This question was asked of people who indicated that they worked at some time during the reference week. People who used different means of transportation on different days of the week were asked to specify the one they used most often, that is, the greatest number of days. People who used more than one means of transportation to get to work each day were asked to report the one used for the longest distance during the work trip. The category, "Car, truck, or van," includes workers using a car (including company cars but excluding taxicabs), a truck of one-ton capacity or less, or a van. The category, "**Drove alone**," includes people who usually drove alone to work as well as people who were driven to work by someone who then drove back home or to a non-work destination. Column 55 shows those who drove alone as a percentage of all workers age 16 and older.

The question on **travel time to work** was asked of people who indicated that they worked at some time during the reference week, and who reported that they worked outside their home. Travel time to work refers to the total number of minutes that it usually took the worker to get from home to work during the reference week. The elapsed time includes time spent waiting for public transportation, picking up passengers in carpools, and time spent in other activities related to getting to work. Column 56 shows those whose commute lasted 30 minutes or more as a percentage of all persons who worked outside their homes.

## COMPUTER AND INTERNET USE, Items 57 and 58

**Source: U.S. Census Bureau—American Community Survey, 2019 1-year Supplemental Estimates**
**http://www.census.gov/acs/www/**

The **computer use** question asked if anyone in the household owned or used a computer and included three response categories for a desktop/laptop, a handheld computer, or some other type of computer. Respondents could select all categories that applied.

Another question asked if any member of the household accesses the **internet**. "Access" refers to whether or not someone in the household uses or connects to the Internet, regardless of whether or not they pay for the service. Respondents were to select only ONE of the following choices:

- Yes, with a subscription to an Internet service –This category includes housing units where someone pays to access the Internet through a service such as a data plan for a mobile phone, a cable modem, DSL or other type of service. This will normally refer to a service that someone is billed for directly for Internet alone or sometimes as part of a bundle.
- Yes, without a subscription to an Internet service–Some respondents may live in a city or town that provides free Internet services for their residents. In addition, some colleges or universities provide Internet services. These are examples of cases where respondents may be able to access the Internet without a subscription.
- No Internet access at this house, apartment, or mobile home-This category includes housing units where no one connects to or uses the Internet using a paid service or any free services.

## MIGRATION, Items 59 and 60

**Source: U.S. Census Bureau—American Community Survey, 2019 1-year Supplemental Estimates**
**http://www.census.gov/acs/www/**

**Residence one year ago** is used in conjunction with location of current residence to determine the extent of residential mobility of the population and the resulting redistribution of the population across the various states, metropolitan areas, and regions of the country. **Same house** includes all people 1 year old and over who, a year before the survey date, lived in the same house or apartment that they occupied at the time of interview.

The percent who lived outside current county includes all persons who did not live in their current county 1 year before the interview. However, they may have lived outside their current city but still in the same county.

## CIVILIAN LABOR FORCE AND UNEMPLOY-MENT, Items 61–64

**Source: U.S. Bureau of Labor Statistics—Local Areas Unemployment Statistics**
**http://www.bls.gov/lau/#tables**

Data for the civilian labor force are the product of a federal-state cooperative program in which state employment security agencies prepare labor force and unemployment estimates under concepts, definitions, and technical procedures established by the Bureau of Labor Statistics (BLS). The civilian labor force consists of all civilians 16 years old and over who are either employed or unemployed.

Unemployment includes all persons who did not work during the survey week, made specific efforts to find a job during the previous four weeks, and were available for work during the survey week (except for temporary illness). Persons waiting to be called back to a job from which they had been laid off and those waiting to report to a new job within the next 30 days are included in unemployment figures.

Table D includes annual average data for the year shown. The Local Area Unemployment Statistics data are periodically updated to reflect revised inputs, reestimation, and controlling to new statewide totals.

## EMPLOYMENT, Items 65–68

**Source: U.S. Census Bureau—American Community Survey, 2019 1-year Supplemental Estimates**
**http://www.census.gov/acs/www/**

The **civilian labor force** includes all civilian persons 16 years old and over who were either (1) "at work"—those who did any work at all during the reference week as paid employees, worked in either their own business or profession, worked on their own farm, or worked 15 hours or more as unpaid workers in a family farm or business; or were (2) "with a job, but not at work" — those who had a job but were not at work that week due to illness, weather, industrial dispute, vacation, or other personal reasons.

**Full-year, Full-Time Workers** includes people 16 to 64 years old who usually worked 35 hours or more per week for 50 to 52 weeks in the past 12 months.

## BUILDING PERMITS, Items 69–71

**Source: U.S. Census Bureau—Building Permits Survey**

**http://www.census.gov/construction/bps/**

These figures represent private residential construction authorized by building permits in approximately 20,000 places in the United States. Valuation represents the expected cost of construction as recorded on the building permit. This figure usually excludes the cost of on-site and off-site development and improvements, as well as the cost of heating, plumbing, electrical, and elevator installations.

National, state, and county totals were obtained by adding the data for permit-issuing places within each jurisdiction. These totals thus are limited to permits issued in the 20,000 place universe covered by the Census Bureau and may not include all permits issued within a state. Current surveys indicate that construction is undertaken for all but a very small percentage of housing units authorized by building permits.

Residential building permits include buildings with any number of housing units. Housing units exclude group quarters (such as dormitories and rooming houses), transient accommodations (such as transient hotels, motels, and tourist courts), "HUD-code" manufactured (mobile) homes, moved or relocated units, and housing units created in an existing residential or nonresidential structure.

## 2017 Economic CENSUS: OVERVIEW, Items 72–107

**Source: U.S. Census Bureau**

**https://www.census.gov/programs-surveys/economic-census.html**

The Economic Census provides a detailed portrait of the nation's economy, from the national to the local level, once every five years. The 2017 Economic Census covers nearly all of the U.S. economy in its basic collection of establishment statistics. The 1997 Economic Census was the first major data source to use the new North American Industry Classification System (NAICS); therefore, data from this census are not comparable to economic data from prior years, which were based on the Standard Industrial Classification (SIC) system.

NAICS, developed in cooperation with Canada and Mexico, classifies North America's economic activities at two-, three-, four-, and five-digit levels of detail; the U.S. version of NAICS further defines industries to a sixth digit. The Economic Census takes advantage of this hierarchy to publish data at these successive levels of detail: sector (two-digit), subsector (three-digit), industry group (four-digit), industry (five-digit), and U.S. industry (six-digit). Information in Table A is at the two-digit level, with a few three- and four-digit items. The data in Table D are at the two-digit level.

Several key statistics are tabulated for all industries in this volume, including number of establishments (or companies), number of employees, payroll, and certain measures of output (sales, receipts, revenue, value of shipments, or value of construction work done).

**Number of establishments.** An establishment is a single physical location at which business is conducted. It is not necessarily identical with a company or enterprise, which may consist of one establishment or more. Economic Census figures represent a summary of reports for individual establishments rather than companies. For cases in which a census report was received, separate information was obtained for each location where business was conducted. When administrative records of other federal agencies were used instead of a census report, no information was available on the number of locations operated. Each Economic Census establishment was tabulated according to the physical location at which the business was conducted. The count of establishments represents those in business at any time during 2002.

When two activities or more were carried on at a single location under a single ownership, all activities were generally grouped together as a single establishment. The entire establishment was classified on the basis of its major activity and all of its data were included in that classification. However, when distinct and separate economic activities (for which different industry classification codes were appropriate) were conducted at a single location under a single ownership, separate establishment reports for each of the different activities were obtained in the census.

**Number of employees.** Paid employees consist of the full-time and part-time employees, including salaried officers and executives of corporations. Included are employees on paid sick leave, paid holidays, and paid vacations; not included are proprietors and partners of unincorporated businesses. The definition of paid employees is the same as that used by the Internal Revenue Service (IRS) on form 941. For some industries, the Economic Census gives codes representing the number of employees as a range of numbers (for example, "100 to 249 employees" or "1,000 to 2,499" employees). In this volume, those codes have been replaced by the standard suppression code "D".

**Payroll.** Payroll includes all forms of compensation, such as salaries, wages, commissions, dismissal pay, bonuses, vacation allowances, sick-leave pay, and employee contributions to qualified pension plans paid during the year to all employees. For corporations, payroll includes amounts paid to officers and executives; for unincorporated businesses, it does not include profit or other compensation of proprietors or partners. Payroll is reported before deductions for Social Security, income tax, insurance, union dues, etc. This definition of payroll is the same as that used by on IRS form 941.

**Sales, shipments, receipts, revenue, or business done.** This measure includes the total sales, shipments, receipts, revenue, or business done by establishments within the scope of the Economic Census. The definition of each of these items is specific to the economic sector measured.

## WHOLESALE TRADE, Items 72–75

**Source: U.S. Census Bureau—2017 Economic Census (See Overview of 2017 Economic Census prior to Item 72)**

The Wholesale Trade sector (sector 42) comprises establishments engaged in wholesaling merchandise, generally without transformation, and rendering services incidental to the sale of merchandise. The wholesaling process is an intermediate step in the distribution of merchandise.

Wholesalers are organized to sell or arrange the purchase or sale of (1) goods for resale (i.e., goods sold to other wholesalers or retailers), (2) capital or durable nonconsumer goods, and (3) raw and intermediate materials and supplies used in production.

Wholesalers sell merchandise to other businesses and normally operate from a warehouse or office. These warehouses and offices are characterized by having little or no display of merchandise. In addition, neither the design nor the location of the premises is intended to solicit walk-in traffic. Wholesalers do not normally use advertising directed to the general public. In general, customers are initially reached via telephone, in-person marketing, or specialized advertising, which may include the internet and other electronic means. Follow-up orders are either vendor-initiated or client-initiated, are usually based on previous sales, and typically exhibit strong ties between sellers and buyers. In fact, transactions are often conducted between wholesalers and clients that have long-standing business relationships.

This sector is made up of two main types of wholesalers: those that sell goods on their own account and those that arrange sales and purchases for others for a commission or fee.

(1) Establishments that sell goods on their own account are known as wholesale merchants, distributors, jobbers, drop shippers, import/export merchants, and sales branches. These establishments typically maintain their own warehouse, where they receive and handle goods for their customers. Goods are generally sold without transformation, but may include integral functions, such as sorting, packaging, labeling, and other marketing services.

(2) Establishments arranging for the purchase or sale of goods owned by others or purchasing goods on a commission basis are known as agents and brokers, commission merchants, import/export agents and brokers, auction companies, and manufacturers' representatives. These establishments operate from offices and generally do not own or handle the goods they sell.

Some wholesale establishments may be connected with a single manufacturer and promote and sell that particular manufacturer's products to a wide range of other wholesalers or retailers. Other wholesalers may be connected to a retail chain or a limited number of retail chains and only provide the products needed by the particular retail operation(s). These wholesalers may obtain the products from a wide range of manufacturers. Still other wholesalers may not take title to the goods, but act instead as agents and brokers for a commission.

Although wholesaling normally denotes sales in large volumes, durable nonconsumer goods may be sold in single units. Sales of capital or durable nonconsumer goods used in the production of goods and services, such as farm machinery, medium- and heavy-duty trucks, and industrial machinery, are always included in Wholesale Trade.

The city table includes only **Merchant wholesalers, except manufacturers' sales branches and offices,** establishments primarily engaged in buying and selling merchandise on their own account. Included here are such types of establishments as wholesale distributors and jobbers, importers, exporters, own-brand importers/marketers, terminal and country grain elevators, and farm products assemblers.

## RETAIL TRADE, Items 76–79

### Source: U.S. Census Bureau—2017 Economic Census (See Overview of 2017 Economic Census prior to Item 72)

The Retail Trade sector (44–45) is made up of establishments engaged in retailing merchandise, generally without transformation, and rendering services incidental to the sale of merchandise.

The retailing process is the final step in the distribution of merchandise; retailers are therefore organized to sell merchandise in small quantities to the general public. This sector comprises two main types of retailers: store and nonstore retailers.

Store retailers operate fixed point-of-sale locations, located and designed to attract a high volume of walk-in customers. In general, retail stores have extensive displays of merchandise and use mass-media advertising to attract customers. They typically sell merchandise to the general public for personal or household consumption; some also serve business and institutional clients. These include establishments such as office supply stores, computer and software stores, building materials dealers, plumbing supply stores, and electrical supply stores. Catalog showrooms, gasoline service stations, automotive dealers, and mobile home dealers are treated as store retailers.

In addition to retailing merchandise, some types of store retailers are also engaged in the provision of after-sales services, such as repair and installation. For example, new automobile dealers, electronic and appliance stores, and musical instrument and supply stores often provide repair services. As a general rule, establishments engaged in retailing merchandise and providing after-sales services are classified in this sector.

Nonstore retailers, like store retailers, are organized to serve the general public, although their retailing methods differ. The establishments of this subsector reach customers and market merchandise with methods including the broadcasting of "infomercials," the broadcasting and publishing of direct-response advertising, the publishing of paper and electronic catalogs, door-to-door solicitation, in-home demonstration, selling from portable stalls (street vendors, except food), and distribution through vending machines. Establishments engaged in the direct sale (nonstore) of products, such as home heating oil dealers and home-delivery newspaper routes are included in this sector.

The buying of goods for resale is a characteristic of retail trade establishments that distinguishes them from establishments in the Agriculture, Manufacturing, and Construction sectors. For example, farms that sell their products at or from the point of production are classified in Agriculture instead of in Retail Trade. Similarly, establishments that both manufacture and sell their products to the general public are classified in Manufacturing instead of Retail Trade. However, establishments that engage in processing activities incidental to retailing are classified in Retail Trade.

## REAL ESTATE AND RENTAL AND LEASING, Items 80–83
### Source: U.S. Census Bureau—2017 Economic Census (See Overview of 2017 Economic Census prior to Item 72)

The Real Estate and Rental and Leasing sector (53) comprises establishments primarily engaged in renting, leasing, or otherwise allowing the use of tangible or intangible assets, and establishments providing related services. The major portion of this sector is made up of establishments that rent, lease, or otherwise allow the use of their own assets by others. The assets may be tangible, such as real estate and equipment, or intangible, such as patents and trademarks.

This sector also includes establishments primarily engaged in managing real estate for others, selling, renting, and/or buying real estate for others, and appraising real estate. These activities are closely related to this sector's main activity. In addition, a substantial proportion of property management is self-performed by lessors.

The main components of this sector are the real estate lessors industries; equipment lessors industries (including motor vehicles, computers, and consumer goods); and lessors of nonfinancial intangible assets (except copyrighted works).

## PROFESSIONAL, SCIENTIFIC, AND TECHNICAL SERVICES, Items 84–87
### Source: U.S. Census Bureau—2017 Economic Census (See Overview of 2017 Economic Census prior to Item 72)

The Professional, Scientific, and Technical Services sector (54) is made up of establishments that specialize in performing professional, scientific, and technical activities for others. These activities require a high degree of expertise and training. The establishments in this sector specialize in one or more areas and provide services to clients in a variety of industries (and, in some cases, to households). Activities performed include legal advice and representation; accounting, bookkeeping, and payroll services; architectural, engineering, and specialized design services; computer services; consulting services; research services; advertising services; photographic services; translation and interpretation services; veterinary services; and other professional, scientific, and technical services.

This sector excludes establishments primarily engaged in providing a range of day-to-day office administrative services, such as financial planning, billing and record keeping, personnel services, and physical distribution and logistics services. These establishments are classified in sector 56, Administrative and Support and Waste Management and Remediation Services.

## MANUFACTURING, Items 88–91
### Source: U.S. Census Bureau—2017 Economic Census (See Overview of 2017 Economic Census prior to Item 72)

The Manufacturing sector (31–33) is made up of establishments engaged in the mechanical, physical, or chemical transformation of materials, substances, or components into new products. The assembling of component parts of manufactured products is considered manufacturing, except in cases in which the activity is appropriately classified in the Construction sector. Establishments in the Manufacturing sector are often described as plants, factories, or mills, and characteristically use power-driven machines and materials-handling equipment. However, establishments that transform materials or substances into new products by hand or in the worker's home, and establishments engaged in selling to the general public products made on the same premises from which they are sold (such as bakeries, candy stores, and custom tailors) may also be included in this sector. Manufacturing establishments may process materials or contract with other establishments to process their materials for them. Both types of establishments are included in the Manufacturing sector.

The materials, substances, or components transformed by manufacturing establishments are raw materials that are products of agriculture, forestry, fishing, mining, or quarrying, or are products of other manufacturing establishments. The materials used may be purchased directly from producers, obtained through customary trade channels, or secured without recourse to the market by transferring the product from one establishment to another, under the same ownership. The new product of a manufacturing establishment may be finished (in the sense that it is ready for utilization or consumption), or it may be semifinished to become an input for an establishment engaged in further manufacturing. For example, the product of the alumina refinery is the input used in the primary production of aluminum; primary aluminum is the input used in an aluminum wire drawing plant; and aluminum wire is the input used in a fabricated wire product manufacturing establishment.

Data are included for cities with 500 or more employees in the Manufacturing sector.

## ACCOMMODATION AND FOOD SERVICES, Items 92–95
### Source: U.S. Census Bureau—2017 Economic Census (See Overview of 2017 Economic Census prior to Item 72)

The Accommodation and Food Services sector (72) consists of establishments that provide customers with lodging and/or meals, snacks, and beverages for immediate consumption. This sector includes both accommodation and food services establishments because the two activities are often combined at the same establishment.

Excluded from this sector are civic and social organizations, amusement and recreation parks, theaters, and other recreation or entertainment facilities providing food and beverage services.

## ARTS, ENTERTAINMENT, AND RECREATION, Items 96–99
### Source: U.S. Census Bureau—2017 Economic Census
### (See Overview of 2017 Economic Census prior to Item 72)

The Arts, Entertainment, and Recreation sector (71) includes a wide range of establishments that operate facilities or provide services that meet the diverse cultural, entertainment, and recreational interests of their patrons. This sector is made up of: (1) establishments that are involved in producing, promoting, or participating in live performances, events, or exhibits intended for public viewing; (2) establishments that preserve and exhibit objects and sites of historical, cultural, or educational interest; and (3) establishments that operate facilities or provide services that enable patrons to participate in recreational activities or pursue amusement, hobby, and leisure time interests.

Some establishments that provide cultural, entertainment, or recreational facilities and services are classified in other sectors. Excluded from this sector are: (1) establishments that provide both accommodations and recreational facilities—such as hunting and fishing camps and resort and casino hotels—are classified in subsector 721, Accommodation; (2) restaurants and night clubs that provide live entertainment in addition to the sale of food and beverages are classified in subsector 722, Food Services and Drinking Places; (3) motion picture theaters, libraries and archives, and publishers of newspapers, magazines, books, periodicals, and computer software are classified in sector 51, Information; and (4) establishmentsthat use transportation equipment to provide recreational and entertainment services, such as those operating sightseeing buses, dinner cruises, or helicopter rides, are classified in subsector 487, Scenic and Sightseeing Transportation.

Table D includes only those establishments subject to federal tax.

## HEALTH CARE AND SOCIAL ASSISTANCE, Items 100–103
### Source: U.S. Census Bureau—2017 Economic Census
### (See Overview of 2017 Economic Census prior to Item 72)

The Health Care and Social Assistance sector (62) consists of establishments that provide health care and social assistance services to individuals. The sector includes both health care and social assistance because it is sometimes difficult to distinguish between the boundaries of these two activities. The industries in this sector are arranged on a continuum, starting with establishments that provide medical care exclusively, continuing with those that provide health care and social assistance, and finishing with those that provide only social assistance. The services

provided by establishments in this sector are delivered by trained professionals. All industries in the sector share this commonality of process—namely, labor inputs of health practitioners or social workers with the requisite expertise. Many of the industries in the sector are defined based on the educational degree held by the practitioners included in the industry.

Excluded from this sector are aerobic classes, which can be found in subsector 713, Amusement, Gambling, and Recreation Industries; and nonmedical diet and weight-reducing centers, which can be found in subsector 812, Personal and Laundry Services. Although these can be viewed as health services, they are not typically delivered by health practitioners.

## OTHER SERVICES, EXCEPT PUBLIC ADMINISTRATION Items 104–107
### Source: U.S. Census Bureau—2017 Economic Census
### (See Overview of 2017 Economic Census prior to Item 72)

The Other Services, Except Public Administration sector (81) comprises establishments engaged in providing services not specifically categorized elsewhere in the classification system. Establishments in this sector are primarily engaged in activities such as equipment and machinery repairing, promoting or administering religious activities, grant making, and advocacy; this sector also includes establishments that provide dry-cleaning and laundry services, personal care services, death care services, pet care services, photofinishing services, temporary parking services, and dating services.

Private households that employ workers on or about the premises in activities primarily concerned with the operation of the household are included in this sector.

Excluded from this sector are establishments primarily engaged in retailing new equipment and performing repairs and general maintenance on equipment. These establishments are classified in sector 44–45, Retail Trade.

## CITY GOVERNMENT EMPLOYMENT AND PAYROLL, Items 171-179
### Source: U.S. Census Bureau—2017 Census of Governments
### https://www.census.gov/programs-surveys/cog.html

Data are assembled from the Census Bureau's Individual Units file from the 2017 Census of Governments. Because the Census Bureau has not reviewed the individual unit data as separate time series, caution must be exercised in their use and interpretation. Users are encouraged to review the tables on the Census Bureau's website. The Census Bureau has not sanctioned, conducted, or reviewed any analysis using the Individual Units data files and they may contain high levels of non-sampling error.

These items include data for municipal governments only. They do not include any special district government entities within the city. The Census of Governments identifies the scope and nature of the nation's state and local government sector; provides authoritative benchmark figures of public finance and public employment; classifies local government organizations, powers, and activities; and measures federal, state, and local fiscal relationships. The Employment component was mailed **March 2012** to collect information on the number of state and local government civilian employees and their payrolls.

Government employees include all persons paid for personal services performed, including persons paid from federally funded programs, paid elected or appointed officials, persons in a paid leave status, and persons paid on a per meeting, annual, semiannual, or quarterly basis. Unpaid officials, pensioners, persons whose work is performed on a fee basis, and contractors and their employees are excluded from the count of employees. **Full-Time Equivalent employees** is a computed statistic representing the number of full-time employees that could have been employed if the reported number of hours worked by part-time employees had been worked by full-time employees. This statistic is calculated separately for each function of a government by dividing the "part-time hours paid" by the standard number of hours for full-time employees in the particular government and then adding the resulting quotient to the number of full-time employees.

**March payroll** represents gross payroll amounts for the one-month period of March for full-time and part-time employees. Gross payroll includes all salaries, wages, fees, commissions, and overtime paid to employees **before** withholdings for taxes, insurance, etc. It also includes incentive payments that are paid at regular pay intervals. It excludes employer share of fringe benefits like retirement, Social Security, health and life insurance, lump sum payments, and so forth.

**Administration and Judicial and Legal** combines **Financial administration**, **Other government administration, and Judicial and Legal** activities. **Financial administration includes** activities concerned with tax assessment and collection, custody and disbursement of funds, debt management, administration of trust funds, budgeting, and other government-wide financial management activities. This function is not applied to school district or special district governments. **Other government administration** applies to the legislative and government-wide administrative agencies of governments. Included here are overall planning and zoning activities, and central personnel and administrative activities. This function is not applied to school district or special district governments. **Judicial and legal** includes all court and court related activities (except probation and parole activities that are included at the "Correction" function), court activities of sheriff's offices, prosecuting attorneys' and public defenders' offices, legal departments, and attorneys providing government-wide legal service.

**Police and Corrections includes** all activities concerned, with the enforcement of law and order, including coroner's offices, police training academies, investigation bureaus, and local jails, "lockups", or other detention facilities not intended to serve as correctional facilities. **Corrections** includes activities pertaining to the confinement and correction of adults and minors convicted of criminal offenses. Pardon, probation, and parole activities are also included here.

**Fire protection** includes local government fire protection and prevention activities plus any ambulance, rescue, or other auxiliary services provided by a fire protection agency. Volunteer firefighters, if remunerated for their services on a "per fire" or some other basis, are included as part-time employees.

**Highways and transportation includes** activities associated with the maintenance and operation of streets, roads, sidewalks, bridges, tunnels, toll roads, and ferries. Snow and ice removal, street lighting, and highway and traffic engineering activities are also included here. Also included are the operation, maintenance, and construction of public mass transit systems, including subways, surface rails, and buses, and the provision, construction, operation, maintenance; support of public waterways, harbors, docks, wharves, and related marine terminal facilities; and activities associated with the operation and support of publicly operated airport facilities.

**Health and Welfare includes Health, Hospitals, and Public welfare.** **Health** includes administration of public health programs, community and visiting nurse services, immunization programs, drug abuse rehabilitation programs, health and food inspection activities, operation of outpatient clinics, and environmental pollution control activities. **Hospitals** includes only government operated medical care facilities that provide inpatient care. Employees and payrolls of private corporations that lease and operate government-owned hospital facilities are excluded.

**Public Welfare** includes the administration of various public assistance programs for the needy, veteran services, operation of nursing homes, indigent care institutions, and programs that provide payments for medical care, handicap transportation, and other services for the needy.

**Natural resources and Utilities** includes activities primarily concerned with the conservation and development of natural resources (soil, water, energy, minerals, etc.) and the regulation of industries that develop, utilize, or affect natural resources, as well as the operation and maintenance of **parks**, playgrounds, swimming pools, public beaches, auditoriums, public golf courses, museums, marinas, botanical gardens, and zoological parks. **Utilities, sewerage, and waste management** includes operation, maintenance, and construction of public water supply systems, including production, acquisition, and distribution of water to general public or to other public or private utilities, for residential, commercial, and industrial use; activities associated with the production or acquisition and distribution of electric power; provision, maintenance, and operation of sanitary and storm sewer systems and sewage disposal and treatment facilities; and refuse collection and disposal, operation of sanitary landfills, and street cleaning activities.

**Education and libraries includes** activities associated with the operation of public elementary and secondary schools and locally operated vocational-technical schools. Special education programs operated by elementary and secondary school systems are also included as are all ancillary services associated with the operation of schools, such as pupil transportation and food service. **Also included are the e**stablishment and provision of libraries for use by the general public and the technical support of privately operated libraries. This category includes classroom teachers, principals, supervisors of instruction, librarians, teacher aides, library aides, and guidance and psychological personnel as well as school superintendents and other administrative personnel, clerical and secretarial staffs, plant operation and maintenance personnel, health and recreation employees, transportation and food service personnel, and any student employees. Also included are any degree granting institutions that provide academic training above grade 12.

# CITY GOVERNMENT FINANCES, Items 117–139

## Source: U.S. Census Bureau—2017 Census of Governments
https://www.census.gov/programs-surveys/cog.html

Data are assembled from the Census Bureau's Individual Units file from the 2017 Census of Governments. Because the Census Bureau has not reviewed the individual unit data as separate time series, caution must be exercised in their use and interpretation. Users are encouraged to review the tables on the Census Bureau's website. The Census Bureau has not sanctioned, conducted, or reviewed any analysis using the Individual Units data files and they may contain high levels of non-sampling error.

Revenue and expenditure data are included in Table D for municipal governments only. The data do not include funds of any special district governments located in the city. For example, if a city's school district is a separate governmental unit, it is not included.

Total **general revenue** includes all revenue except utility, liquor stores, and insurance trust revenue. All tax revenue and intergovernmental revenue, even if designated for employee-retirement or local utility purpose, are classified as general revenue.

**Intergovernmental revenue** covers amounts received from other governments as fiscal aid in the form of shared revenues and grants-in-aid, as reimbursements for the performance of general government functions and specific services for the paying government (for example, care of prisoners or contractual research), or in lieu of taxes. It excludes any amounts received from other governments from the sale of property, commodities, and utility services. All intergovernmental revenue is classified as general revenue. Intergovernmental revenue from the state governments includes amounts originally from the federal government but channeled through the state.

**Taxes** consist of compulsory contributions exacted by governments for public purposes. However, this category excludes employer and employee payments for retirement and social insurance purposes, which are classified as insurance trust revenue. All tax revenue is classified as general revenue and comprises amounts received (including interest and penalties, but excluding protested amounts and refunds) from all taxes imposed by a government. Note that local government tax revenue excludes any amounts from shares of state-imposed and collected taxes, which are classified as intergovernmental revenue.

**Property taxes** are based on ownership of property and measured by its value. They include general property taxes related to property as a whole—real and personal, tangible or intangible—whether taxed at a single rate or at classified rates. Also included

are taxes on selected types of property, such as motor vehicles or certain or all intangibles.

**Sales and gross receipts taxes** include "licenses" at more than nominal rates, based on volume or value of transfers of goods or services; taxes upon gross receipts or upon gross income; and related taxes based upon the use, storage, production (other than the severance of natural resources), importation, or consumption of goods. Dealer discounts "commissions," which are allowed to merchants for the collection of taxes from consumers, are excluded.

Total **general expenditure** includes all city expenditure other than specifically enumerated kinds of expenditure, including utility, liquor store, and employee-retirement and other insurance trust expenditure.

**Capital outlays** are direct expenditures for contract of force account construction or buildings, roads, and other improvements, and for purchases of equipment, land, and existing structures. They include amounts for additions, replacements, and major alterations to fixed work and structures. Expenditures for repair to such works and structures, however, is classified as current operation expenditure.

A major portion of capital outlay is commonly financed by borrowing, while governmental revenue does not include receipts from borrowing. Among other things, this distorts the relationship between the totals presented for revenue and expenditure and renders this relationship useless as a direct measure of the degree of budgetary "balance" (as that term is generally applied).

**Public welfare expenditure** covers support of and assistance to needy persons; this aid is contingent upon the person's needs. Included are cash assistance paid directly to needy persons; vendor payments made directly to private purveyors for medical care, burials, and other commodities and services provided under welfare programs; welfare institutions; and any intergovernmental or other direct expenditure for welfare purposes. Pensions to former employees and other benefits not contingent on need are excluded.

**Highway expenditure** is for the provision and maintenance of highway facilities, including toll turnpikes, bridges, tunnels, and ferries, as well as regular roads, highways, and streets. Also included are expenditures for street lighting and for snow and ice removal. Not included are highway policing and traffic control, which are considered part of police protection

**Parking facilities** include the construction, purchase, maintenance, and operation of public-use parking lots, garages, parking meters, and other distinctive parking facilities on a commercial basis.

**Education** is mainly for the provision and general support of schools and other educational facilities and services, including those for educational institutions beyond high school. Elementary and secondary education includes the provision of public kindergarten through high school education by local governments. It encompasses instructional, support, and auxiliary services (school lunch, student activities, and community services) offered by public school systems. Higher education consists of all local institutions of higher education.

**Health expenditures** include outpatient health services other than hospital care, such as public health administration; research and education; categorical health programs; treatment and immunization clinics; nursing; environmental health activities, such as air and water pollution control; ambulance service if provided separately from fire protection services; and other general public health activities, such as mosquito abatement. School health services provided by health agencies (rather than school agencies) are included here. Not included are sewage treatment operations, which are classified as part of sewerage and sanitation. **Hospital expenditures** include financing, construction, acquisition, maintenance and operation of hospital facilities, provision of hospital care, and support of public or private hospitals.

**Police protection** encompasses expenditures for the preservation of law and order, as well as for traffic safety. It includes police patrols and communications, crime prevention activities, detention and custody of persons awaiting trial, traffic safety, and vehicular inspection.

**Sewerage and recreation** include sanitary and storm sewers, sewage disposal facilities and services, and other government activities for such purposes. Street cleaning and the collection and disposal of garbage and other waste are also included.

**Parks and recreation** includes cultural and scientific activities, such as museums and art galleries; organized recreation, including playgrounds and playing fields, swimming pools, and bathing beaches; and municipal parks and special recreation facilities, such as auditoriums, stadiums, auto camps, recreation piers, and boat harbors.

**Housing and community development** includes city housing and redevelopment projects and the regulation, promotion, and support of private housing and redevelopment activities. Data from Arizona, Kentucky, Michigan, New Mexico, New York, and Virginia generally include municipal housing authorities. Housing authorities for other cities are usually classified as independent governments, and data from them are not included.

**Interest** on debt is the amount paid for the use of borrowed money.

Total **debt outstanding** is the total of debt obligations remaining unpaid on the date specified. **Debt issued during the year** is the amount of the outstanding debt that was recently borrowed.

## CLIMATE, Items 140–146

**Source: National Oceanic and Atmospheric Administration**
**https://www.ncdc.noaa.gov/data-access/**
**land-based-station-data/land-based-datasets/**
**climate-normals**

All climate data are average values for the 30-year period from 1971 to 2000.

Mean temperatures for January and July were determined by adding the average daily maximum temperatures and the average daily minimum temperatures and dividing by two.

Temperature limits represent average daily minimum for January and average daily maximum for July.

Annual precipitation values are the average annual water equivalent of all precipitation for the 30-year period.

Heating and cooling degree days are used as relative measures of the energy required for heating and cooling buildings. One heating degree day is accumulated for each whole degree that the mean daily temperature is below 65 degrees Fahrenheit (a mean daily temperature of 62 degrees Fahrenheit will produce three heating degree days). Cooling degree days are accumulated in similar fashion for deviations of the mean daily temperature above 65 degrees Fahrenheit.

## TABLE E—CONGRESSIONAL DISTRICTS OF THE 116TH CONGRESS

**Members of the House of Representatives are for the 117th Congress.**

## LAND AREA, Items 1 and 3

**Source: U.S. Census Bureau—2020 U.S. Gazetteer Files,**
**https://www.census.gov/geographies/reference-files/**
**time-series/geo/gazetteer-files.html**

Land area measurements are shown to the nearest square mile. Land area includes dry land and land temporarily or partially covered by water, such as marshlands, swamps, and river floodplains.

## POPULATION, Items 2–3

**Source: U.S. Census Bureau—American Community Survey**
**http://www.census.gov/acs/www/**

The population data are estimates from the American Community Survey.

## POPULATION AND POPULATION CHARACTERISTICS, Items 4–24

**Source: U.S. Census Bureau—American Community Survey**
**http://www.census.gov/acs/www/**

Data on age, sex, race, Hispanic origin foreign-born residents, and percent born in state of residence are from the 2019 American Community Survey.

Data on race were derived from answers to the question on race that was asked of all respondents. The concept of race, as used by the Census Bureau, reflects self-identification by people according to the race or races with which they most closely identify. These categories are sociopolitical constructs and should not be interpreted as being scientific or anthropological in nature. Furthermore, the race categories include both racial and national origin groups.

In Table E, Columns 4 through 8 refer to individuals who identified with each racial category alone, while column 9 includes persons who identified with two or more races.

The **White** population is defined as persons who indicated their race as White, as well as persons who did not classify themselves in one of the specific race categories listed on the questionnaire but entered a nationality such as Irish, German, Italian, Lebanese, Near Easterner, Arab, or Polish.

The **Black** population includes persons who indicated their race as "Black or African American" as well as persons who did not classify themselves in one of the specific race categories but reported entries such as African American, Afro American, Kenyan, Nigerian, or Haitian.

The **American Indian or Alaska Native** population includes persons who indicated their race as American Indian or Alaska Native, as well as persons who did not classify themselves in one of the specific race categories but reported entries such as Canadian Indian, French-American Indian, Spanish-American Indian, Eskimo, Aleut, Alaska Indian, or any of the American Indian or Alaska Native tribes.

The **Asian and Pacific Islander** population combines two census groupings: **Asian** and **Native Hawaiian or Other Pacific Islander**. The **Asian** population includes persons who indicated their race as Asian Indian, Chinese, Filipino, Japanese, Korean, Vietnamese, or "Other Asian," as well as persons who provided write-in entries of such groups as Cambodian, Laotian, Hmong, Pakistani, or Taiwanese. The **Native Hawaiian or Other Pacific Islander** population includes persons who indicated their race as "Native Hawaiian," "Guamanian or Chamorro," "Samoan," or "Other Pacific Islander," as well as persons who reported entries such as Part Hawaiian, American Samoan, Fijian, Melanesian, or Tahitian.

The **Hispanic population** is based on a question that asked respondents "Is this person Spanish/Hispanic/Latino?" Persons marking any one of the four Hispanic categories (i.e., Mexican, Puerto Rican, Cuban, or other Spanish) are collectively referred to as Hispanic.

The **Non-Hispanic White alone** number in Column 11 includes only those persons who were not Hispanic and whose race was "White only."

The **female** population is shown as a percentage of total population.

The **foreign-born** population includes all persons who were not U.S. citizens at birth. Foreign-born persons are those who indicated they were either a U.S. citizen by naturalization or were not a citizen of the United States. The foreign-born population consists of immigrants (legal permanent residents), temporary migrants (students), humanitarian migrants (refugees), and unauthorized migrants (persons illegally residing in the United States).

**Percent born in state of residence** is shown as a percentage of total population.

**Age** is defined as age at last birthday (number of completed years since birth).

## EDUCATION—SCHOOL ENROLLMENT AND EDUCATIONAL ATTAINMENT, Items 25–27
**Source: U.S. Census Bureau—American Community Survey**
http://www.census.gov/acs/www/

Data on school enrollment and educational attainment were derived from a sample of the population. Persons were classified as enrolled in school if they reported attending a "regular" public or private school (or college) during the year. The instructions were to "include only nursery school, kindergarten, elementary school, and schooling which would lead to a high school diploma or a college degree" as regular school. The Census Bureau defines a public school as "any school or college controlled and supported by a local, county, state, or federal government." Schools primarily supported and controlled by religious organizations or other private groups are defined as private schools.

Data on **educational attainment** are tabulated for the population 25 years old and over. The data were derived from a question that asked respondents for the highest level of school completed or the highest degree received. Persons who had passed a high school equivalency examination were considered high school graduates. Schooling received in foreign schools was to be reported as the equivalent grade or years in the regular American school system.

Vocational and technical training, such as barber school training; business, trade, technical, and vocational schools; or other training for a specific trade are specifically excluded.

**High school graduate or more.** This category includes persons who have received a high school diploma or its equivalent, and those who reported any level higher than a high school diploma.

**Bachelor's degree or more.** This category includes persons who have received bachelor's degrees, master's degrees, professional school degrees (such as law school or medical school degrees), and doctoral degrees.

## HOUSEHOLDS, Items 28–33
**Source: U.S. Census Bureau—American Community Survey**
http://www.census.gov/acs/www/

A **household** includes all persons who occupy a housing unit. (Persons not living in households are classified as living in group quarters.) A housing unit is a house, an apartment, a mobile home, a group of rooms, or a single room occupied (or, if vacant, intended for occupancy) as separate living quarters. Separate living quarters are those in which the occupants live separately from any other persons in the building and have direct access from the outside of the building or through a common hall. The occupants may be a single family, one person living alone, two or more families living together, or any other group of related or unrelated persons who share living quarters. The number of households is the same as the number of year-round occupied housing units.

A **family** includes a householder and one or more other persons living in the same household who are related to the householder by birth, marriage, or adoption. All persons in a household who are related to the householder are regarded as members of his or her family. A **family household** may contain persons not related to the householder; thus, family households may include more members than families do. A household can contain only one family for the purposes of census tabulations. Not all households